American Men & Women of Science

1992-93 • 18th Edition

The 18th edition of *AMERICAN MEN & WOMEN OF SCIENCE* was prepared by the R.R. Bowker Database Publishing Group.

Stephen L. Torpie, Managing Editor
Judy Redel, Managing Editor, Research
Richard D. Lanam, Senior Editor
Tanya Hurst, Research Manager
Karen Hallard, Beth Tanis, Associate Editors

Peter Simon, Vice President, Database Publishing Group
Dean Hollister, Director, Database Planning
Edgar Adcock, Jr., Editorial Director, Directories

American Men & Women of Science

1992-93 • 18th Edition

A Biographical Directory of Today's Leaders in Physical, Biological and Related Sciences.

Volume 1 • A-B

R. R. BOWKER
New Providence, New Jersey

Published by R.R. Bowker, a division of Reed Publishing, (USA) Inc.

International Standard Book Number

Set:	0-8352-3074-0
Volume I:	0-8352-3075-9
Volume II:	0-8352-3076-7
Volume III:	0-8352-3077-5
Volume IV:	0-8352-3078-3
Volume V:	0-8352-3079-1
Volume VI:	0-8352-3080-5
Volume VII:	0-8352-3081-3
Volume VIII:	0-8352-3082-1

International Standard Serial Number: 0192-8570
Library of Congress Catalog Card Number: 6-7326
Printed and bound in the United States of America.

8 Volume Set

ISBN 0-8352-3074-0

9 780835 230742

Contents

Advisory Committee

Dr. Robert F. Barnes
Executive Vice President
American Society of Agronomy

Dr. John Kistler Crum
Executive Director
American Chemical Society

Dr. Charles Henderson Dickens
Section Head, Survey & Analysis Section
Division of Science Resource Studies
National Science Foundation

Mr. Alan Edward Fechter
Executive Director
Office of Scientific & Engineering Personnel
National Academy of Science

Dr. Oscar Nicolas Garcia
Prof Electrical Engineering
Electrical Engineering & Computer Science Department
George Washington University

Dr. Charles George Groat
Executive Director
American Geological Institute

Dr. Richard E. Hallgren
Executive Director
American Meteorological Society

Dr. Michael J. Jackson
Executive Director
Federation of American Societies for Experimental
Biology

Dr. William Howard Jaco
Executive Director
American Mathematical Society

Dr. Shirley Mahaley Malcom
Head, Directorate for Education and Human
Resources Programs
American Association for the Advancement of Science

Mr. Daniel Melnick
Sr Advisor Research Methodologies
Sciences Resources Directorate
National Science Foundation

Ms. Beverly Fearn Porter
Division Manager
Education & Employment Statistics Division
American Institute of Physics

Dr. Terrence R. Russell
Manager
Office of Professional Services
American Chemical Society

Dr. Irwin Walter Sandberg
Holder, Cockrell Family Regent Chair
Department of Electrical & Computer Engineering
University of Texas

Dr. William Eldon Splinter
Interim Vice Chancellor for Research,
Dean, Graduate Studies
University of Nebraska

Ms. Betty M. Vetter
Executive Director, Science Manpower Comission
Commission on Professionals in Science & Technology

Dr. Dael Lee Wolfe
Professor Emeritus
Graduate School of Public Affairs
University of Washington

Preface

American Men and Women Of Science remains without peer as a chronicle of North American scientific endeavor and achievement. The present work is the eighteenth edition since it was first compiled as *American Men of Science* by J. Mckeen Cattell in 1906. In its eighty-six year history *American Men & Women of Science* has profiled the careers of over 300,000 scientists and engineers. Since the first edition, the number of American scientists and the fields they pursue have grown immensely. This edition alone lists full biographies for 122,817 engineers and scientists, 7021 of which are listed for the first time. Although the book has grown, our stated purpose is the same as when Dr. Cattell first undertook the task of producing a biographical directory of active American scientists. It was his intention to record educational, personal and career data which would make "a contribution to the organization of science in America" and "make men [and women] of science acquainted with one another and with one another's work." It is our hope that this edition will fulfill these goals.

The biographies of engineers and scientists constitute seven of the eight volumes and provide birthdates, birthplaces, field of specialty, education, honorary degrees, professional and concurrent experience, awards, memberships, research information and adresses for each entrant when applicable. The eighth volume, the discipline index, organizes biographees by field of activity. This index, adapted from the National Science Foundation's Taxonomy of Degree and Employment Specialties, classifies entrants by 171 subject specialties listed in the table of contents of Volume 8. For the first time, the index classifies scientists and engineers by state within each subject specialty, allowing the user to more easily locate a scientist in a given area. Also new to this edition is the inclusion of statistical information and recipients of theNobel Prizes, the Craaford Prize, the Charles Stark

Draper Prize, and the National Medals of Science and Technology received since the last edition.

While the scientific fields covered by *American Men and Women Of Science* are comprehensive, no attempt has been made to include all American scientists. Entrants are meant to be limited to those who have made significant contributions in their field. The names of new entrants were submitted for consideration at the editors' request by current entrants and by leaders of academic, government and private research programs and associations. Those included met the following criteria:

1. Distinguished achievement, by reason of experience, training or accomplishment, including contributions to the literature, coupled with continuing activity in scientific work;

   or

2. Research activity of high quality in science as evidenced by publication in reputable scientific journals; or for those whose work cannot be published due to governmental or industrial security, research activity of high quality in science as evidenced by the judgement of the individual's peers;

   or

3. Attainment of a position of substantial responsibility requiring scientific training and experience.

This edition profiles living scientists in the physical and biological fields, as well as public health scientists, engineers, mathematicians, statisticians, and computer scientists. The information is collected by means of direct communication whenever possible. All entrants receive forms for corroboration and updating. New entrants receive questionaires and verification proofs before publication. The information submitted by entrants is included as completely as possible within the boundaries of editorial and space restrictions. If an entrant does not return the form and his or her current location can be verified in secondary sources, the full entry is repeated. References to the previous edition are given for those who do not return forms and cannot be located, but who are presumed to be still active in science or engineering. Entrants known to be deceased are noted as such and a reference to the previous edition is given. Scientists and engineers who are not citizens of the United States or Canada are included if a significant portion of their work was performed in North America.

The information in AMWS is also available on CD-ROM as part of *SciTech Reference Plus*. In adition to the convenience of searching scientists and engineers, *SciTech Reference Plus* also includes *The Directory of American Research & Technology*, *Corporate Technology Directory*, sci-tech and medical books and serials from *Books in Print* and *Bowker International Series*. *American Men and Women Of Science* is available for online searching through the subscription services of DIALOG Information Services, Inc. (3460 Hillview Ave, Palo Alto, CA 94304) and ORBIT Search Service (800 Westpark Dr, McLean, VA 22102). Both CD-Rom and the on-line subscription services allow all elements of an entry, including field of interest, experience, and location, to be accessed by key word. Tapes and mailing lists are also available through the Cahners Direct Mail (John Panza, List Manager, Bowker Files 245 W 17th St, New York, NY, 10011, Tel: 800-537-7930).

A project as large as publishing *American Men and Women Of Science* involves the efforts of a great many people. The editors take this opportunity to thank the eighteenth edition advisory committee for their guidance, encouragement and support. Appreciation is also expressed to the many scientific societies who provided their membership lists for the purpose of locating former entrants whose addresses had changed, and to the tens of thousands of scientists across the country who took time to provide us with biographical information. We also wish to thank Bruce Glaunert, Bonnie Walton, Val Lowman, Debbie Wilson, Mervaine Ricks and all those whose care and devotion to accurate research and editing assured successful production of this edition.

Comments, suggestions and nominations for the nineteenth edition are encouraged and should be directed to The Editors, *American Men and Women Of Science*, R.R. Bowker, 121 Chanlon Road, New Providence, New Jersey, 07974.

Edgar H. Adcock, Jr.
Editorial Director

Major Honors & Awards

Nobel Prizes
Nobel Foundation

The Nobel Prizes were established in 1900 (and first awarded in 1901) to recognize those people who "have conferred the greatest benefit on mankind."

1990 Recipients

Chemistry:

Elias James Corey

Awarded for his work in retrosynthetic analysis, the synthesizing of complex substances patterned after the molecular structures of natural compounds.

Physics:

Jerome Isaac Friedman

Henry Way Kendall

Richard Edward Taylor

Awarded for their breakthroughs in the understanding of matter.

Physiology or Medicine:

Joseph E. Murray

Edward Donnall Thomas

Awarded to Murray for his kidney transplantation achievements and to Thomas for bone marrow transplantation advances.

1991 Recipients

Chemistry:

Richard R. Ernst

Awarded for refinements in nuclear magnetic resonance spectroscopy.

Physics:

Pierre-Gilles de Gennes*

Awarded for his research on liquid crystals.

Physiology or Medicine:

Erwin Neher

Bert Sakmann*

Awarded for their discoveries in basic cell function and particularly for the development of the patch clamp technique.

Crafoord Prize
Royal Swedish Academy of Sciences
(Kungl. Vetenskapsakademien)

The Crafoord Prize was introduced in 1982 to award scientists in disciplines not covered by the Nobel Prize, namely mathematics, astronomy, geosciences and biosciences.

1990 Recipients

Paul Ralph Ehrlich

Edward Osborne Wilson

Awarded for their fundamental contributions to population biology and the conservation of biological diversity.

1991 Recipient

Allan Rex Sandage

Awarded for his fundamental contributions to extragalactic astronomy, including observational cosmology.

Charles Stark Draper Prize
National Academy of Engineering

The Draper Prize was introduced in 1989 to recognize engineering achievement. It is awarded biennially.

1991 Recipients

Hans Joachim Von Ohain

Frank Whittle

Awarded for their invention and development of the jet aircraft engine.

National Medal of Science
National Science Foundation

The National Medals of Science have been awarded by the President of the United States since 1962 to leading scientists in all fields.

1990 Recipients:

Baruj Benacerraf
Elkan Rogers Blout
Herbert Wayne Boyer
George Francis Carrier
Allan MacLeod Cormack
Mildred S. Dresselhaus
Karl August Folkers
Nick Holonyak Jr.
Leonid Hurwicz
Stephen Cole Kleene
Daniel Edward Koshland Jr.
Edward B. Lewis
John McCarthy
Edwin Mattison McMillan**
David G. Nathan
Robert Vivian Pound
Roger Randall Dougan Revelle**
John D. Roberts
Patrick Suppes
Edward Donnall Thomas

1991 Recipients

Mary Ellen Avery
Ronald Breslow
Alberto Pedro Calderon
Gertrude Belle Elion
George Harry Heilmeier
Dudley Robert Herschbach
George Evelyn Hutchinson**
Elvin Abraham Kabat
Robert Kates
Luna Bergere Leopold
Salvador Edward Luria**
Paul A. Marks
George Armitage Miller
Arthur Leonard Schawlow
Glenn Theodore Seaborg
Folke Skoog
H. Guyford Stever
Edward Carroll Stone Jr
Steven Weinberg
Paul Charles Zamecnik

National Medal of Technology
U.S. Department of Commerce, Technology Administration

The National Medals of Technology, first awarded in 1985, are bestowed by the President of the United States to recognize individuals and companies for their development or commercialization of technology or for their contributions to the establishment of a technologically-trained workforce.

1990 Recipients

John Vincent Atanasoff
Marvin Camras
The du Pont Company
Donald Nelson Frey
Frederick W. Garry
Wilson Greatbatch
Jack St. Clair Kilby
John S. Mayo
Gordon Earle Moore
David B. Pall
Chauncey Starr

1991 Recipients

Stephen D. Bechtel Jr
C. Gordon Bell
Geoffrey Boothroyd
John Cocke
Peter Dewhurst
Carl Djerassi
James Duderstadt
Antonio L. Elias
Robert W. Galvin
David S. Hollingsworth
Grace Murray Hopper
F. Kenneth Iverson
Frederick M. Jones**
Robert Roland Lovell
Joseph A. Numero**
Charles Eli Reed
John Paul Stapp
David Walker Thompson

*These scientists' biographies do not appear in *American Men & Women of Science* because their work has been conducted exclusively outside the US and Canada.

**Deceased [Note that Frederick Jones died in 1961 and Joseph Numero in May 1991. Neither was ever listed in *American Men and Women of Science*.]

Statistics

Statistical distribution of entrants in *American Men & Women of Science* is illustrated on the following five pages. The regional scheme for geographical analysis is diagrammed in the map below. A table enumerating the geographic distribution can be found on page xvi, following the charts. The statistics are compiled by tallying all occurrences of a major index subject. Each scientist may choose to be indexed under as many as four categories; thus, the total number of subject references is greater than the number of entrants in *AMWS*.

All Disciplines

	Number	Percent
Northeast	58,325	34.99
Southeast	39,769	23.86
North Central	19,846	11.91
South Central	12,156	7.29
Mountain	11,029	6.62
Pacific	25,550	15.33
TOTAL	**166,675**	**100.00**

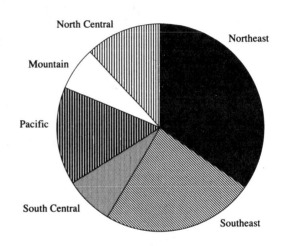

Age Distribution of American Men & Women of Science

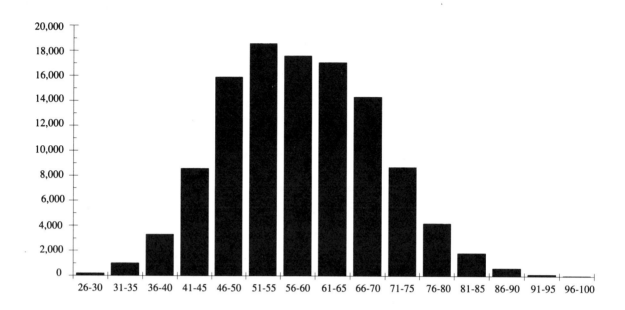

Number of Scientists in Each Discipline of Study

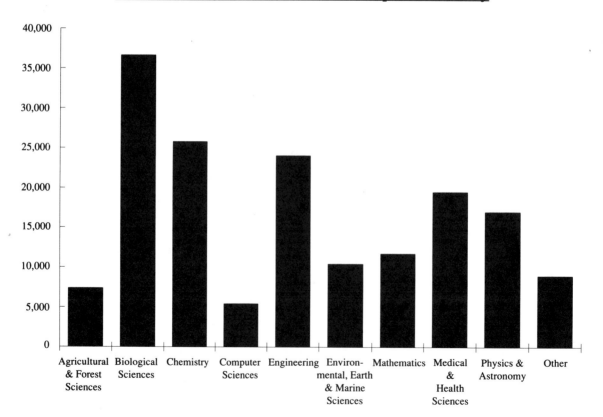

Agricultural & Forest Sciences

	Number	Percent
Northeast	1,574	21.39
Southeast	1,991	27.05
North Central	1,170	15.90
South Central	609	8.27
Mountain	719	9.77
Pacific	1,297	17.62
TOTAL	**7,360**	**100.00**

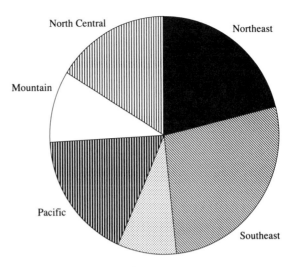

Biological Sciences

	Number	Percent
Northeast	12,162	33.23
Southeast	9,054	24.74
North Central	5,095	13.92
South Central	2,806	7.67
Mountain	2,038	5.57
Pacific	5,449	14.89
TOTAL	**36,604**	**100.00**

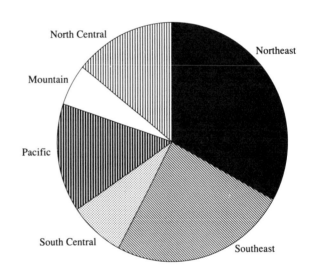

Chemistry

	Number	Percent
Northeast	10,343	40.15
Southeast	6,124	23.77
North Central	3,022	11.73
South Central	1,738	6.75
Mountain	1,300	5.05
Pacific	3,233	12.55
TOTAL	**25,760**	**100.00**

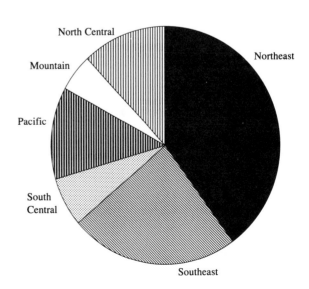

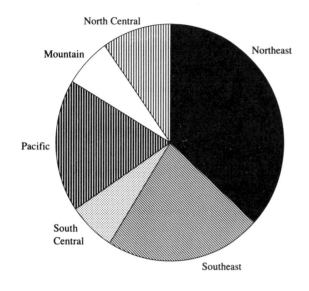

Computer Sciences

	Number	Percent
Northeast	1,987	36.76
Southeast	1,200·	22.20
North Central	511	9.45
South Central	360	6.66
Mountain	372	6.88
Pacific	976	18.05
TOTAL	**5,406**	**100.00**

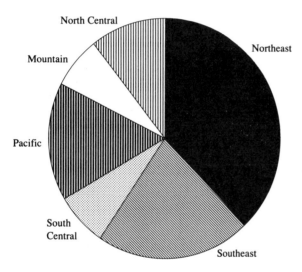

Engineering

	Number	Percent
Northeast	9,122	38.01
Southeast	5,202	21.68
North Central	2,510	10.46
South Central	1,710	7.13
Mountain	1,646	6.86
Pacific	3,807	15.86
TOTAL	**23,997**	**100.00**

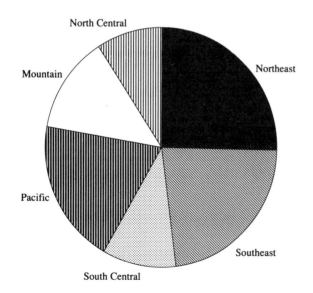

Environmental, Earth & Marine Sciences

	Number	Percent
Northeast	2,657	25.48
Southeast	2,361	22.64
North Central	953	9.14
South Central	1,075	10.31
Mountain	1,359	13.03
Pacific	2,022	19.39
TOTAL	**10,427**	**100.00**

Mathematics

	Number	Percent
Northeast	4,211	35.92
Southeast	2,609	22.26
North Central	1,511	12.89
South Central	884	7.54
Mountain	718	6.13
Pacific	1,789	15.26
TOTAL	**11,722**	**100.00**

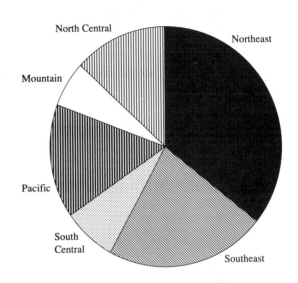

Medical & Health Sciences

	Number	Percent
Northeast	7,115	36.53
Southeast	5,004	25.69
North Central	2,577	13.23
South Central	1,516	7.78
Mountain	755	3.88
Pacific	2,509	12.88
TOTAL	**19,476**	**100.00**

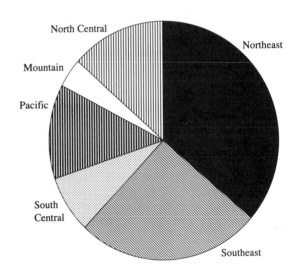

Physics & Astronomy

	Number	Percent
Northeast	5,961	35.12
Southeast	3,670	21.62
North Central	1,579	9.30
South Central	918	5.41
Mountain	1,607	9.47
Pacific	3,238	19.08
TOTAL	**16,973**	**100.00**

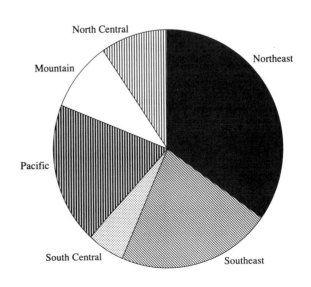

Geographic Distribution of Scientists by Discipline

	Northeast	Southeast	North Central	South Central	Mountain	Pacific	TOTAL
Agricultural & Forest Sciences	1,574	1,991	1,170	609	719	1,297	**7,360**
Biological Sciences	12,162	9,054	5,095	2,806	2,038	5,449	**36,604**
Chemistry	10,343	6,124	3,022	1,738	1,300	3,233	**25,760**
Computer Sciences	1,987	1,200	511	360	372	976	**5,406**
Engineering	9,122	5,202	2,510	1,710	1,646	3,807	**23,997**
Environmental, Earth & Marine Sciences	2,657	2,361	953	1,075	1,359	2,022	**10,427**
Mathematics	4,211	2,609	1,511	884	718	1,789	**11,722**
Medical & Health Sciences	7,115	5,004	2,577	1,516	755	2,509	**19,476**
Physics & Astronomy	5,961	3,670	1,579	918	1,607	3,238	**16,973**
Other Professional Fields	3,193	2,554	918	540	515	1,230	**8,950**
TOTAL	**58,325**	**39,769**	**19,846**	**12,156**	**11,029**	**25,550**	**166,675**

Geographic Definitions

Northeast
Connecticut
Indiana
Maine
Massachusetts
Michigan
New Hampshire
New Jersey
New York
Ohio
Pennsylvania
Rhode Island
Vermont

Southeast
Alabama
Delaware
District of Columbia
Florida
Georgia
Kentucky
Maryland
Mississippi
North Carolina
South Carolina
Tennessee
Virginia
West Virginia

North Central
Illinois
Iowa
Kansas
Minnesota
Missouri
Nebraska
North Dakota
South Dakota
Wisconsin

South Central
Arkansas
Louisiana
Texas
Oklahoma

Mountain
Arizona
Colorado
Idaho
Montana
Nevada
New Mexico
Utah
Wyoming

Pacific
Alaska
California
Hawaii
Oregon
Washington

Sample Entry

American Men & Women of Science (AMWS) is an extremely useful reference tool. The book is most often used in one of two ways: to find more information about a particular scientist or to locate a scientist in a specific field.

To locate information about an individual, the biographical section is most helpful. It encompasses the first seven volumes and lists scientists and engineers alphabetically by last name. The fictitious biographical listing shown below illustrates every type of information an entry may include.

The Discipline Index, volume 8, can be used to easily find a scientist in a specific subject specialty. This index is first classified by area of study, and within each specialty entrants are divided further by state of residence.

Name ——————

Date(s) of marriage ——————

Degrees Earned ——————

Professional Experience ——————

Current Position ——————

———— Birthplace & Date

———— Number of Children

———— Field of Specialty

———— Honorary Degrees

———— Concurrent Positions

———— Membership

———— Areas of research

———— Address

CARLETON, PHYLLIS B(ARBARA), b Glenham, SDak, April 1, 30. m 53, 69; c 2. ORGANIC CHEMISTRY. *Educ:* Univ Notre Dame, BSc, 52, MSc, 54, Vanderbilt Univ, PhD(chem), 57. *Hon Degrees:* DSc, Howard Univ, 79. *Prof Exp:* Res chemist, Acme Chem Corp, 54-59, sr res chemist, 59-60; from asst to assoc prof chem 60-63, prof chem, Kansas State Univ, 63-72; prof chem, Yale Univ, 73-89; CONSULT, CARLETON & ASSOCS, 89-. *Concurrent Pos:* Adj prof, Kansas State Univ 58-60; vis lect, Oxford Univ, 77, consult, Union Carbide, 74-80. *Honors & Awards:* Gold Medal, Am Chem Society, 81; *Mem:* AAAS, fel Am Chem Soc, Sigma Chi. *Res:* Organic synthesis, chemistry of natural products, water treatment and analysis. *Mailing Address:* Carleton & Assocs 21 E 34th St Boston MA 02108

Abbreviations

AAAS—American Association for the Advancement of Science
abnorm—abnormal
abstr—abstract
acad—academic, academy
acct—Account, accountant, accounting
acoust—acoustic(s), acoustical
ACTH—adrenocorticotrophic hormone
actg—acting
activ—activities, activity
addn—addition(s), additional
Add—Address
adj—adjunct, adjutant
adjust—adjustment
Adm—Admiral
admin—administration, administrative
adminr—administrator(s)
admis—admission(s)
adv—adviser(s), advisory
advan—advance(d), advancement
advert—advertisement, advertising
AEC—Atomic Energy Commission
aerodyn—aerodynamic
aeronaut—aeronautic(s), aeronautical
aerophys—aerophsical, aerophysics
aesthet—aesthetic
AFB—Air Force Base
affil—affiliate(s), affiliation
agr—agricultural, agriculture
agron—agronomic, agronomical, agronomy
agrost—agrostologic, agrostological, agrostology
agt—agent
AID—Agency for International Development
Ala—Alabama
allergol—allergological, allergology
alt—alternate
Alta—Alberta
Am—America, American
AMA—American Medical Association
anal—analysis, analytic, analytical
analog—analogue
anat—anatomic, anatomical, anatomy
anesthesiol—anesthesiology
angiol—angiology
Ann—Annal(s)
ann—annual
anthrop—anthropological, anthropology
anthropom—anthropometric, anthropometrical, anthropometry
antiq—antiquary, antiquities, antiquity
antiqn—antiquarian

apicult—apicultural, apiculture
APO—Army Post Office
app—appoint, appointed
appl—applied
appln—application
approx—approximate(ly)
Apr—April
apt—apartment(s)
aquacult—aquaculture
arbit—arbitration
arch—archives
archaeol—archaeological, archaeology
archit—architectural, architecture
Arg—Argentina, Argentine
Ariz—Arizona
Ark—Arkansas
artil—artillery
asn—association
assoc(s)—associate(s), associated
asst(s)—assistant(s), assistantship(s)
assyriol—Assyriology
astrodyn—astrodynamics
astron—astronomical, astronomy
astronaut—astonautical, astronautics
astronr—astronomer
astrophys—astrophysical, astrophysics
attend—attendant, attending
atty—attorney
audiol—audiology
Aug—August
auth—author
AV—audiovisual
Ave—Avenue
avicult—avicultural, aviculture

b—born
bact—bacterial, bacteriologic, bacteriological, bacteriology
BC—British Colombia
bd—board
behav—behavior(al)
Belg—Belgian, Belgium
Bibl—biblical
bibliog—bibliographic, bibliographical, bibliography
bibliogr—bibliographer
biochem—biochemical, biochemistry
biog—biographical, biography
biol—biological, biology
biomed—biomedical, biomedicine
biomet—biometric(s), biometrical, biometry
biophys—biophysical, biophysics

bk(s)—book(s)
bldg-building
Blvd—Boulevard
Bor—Borough
bot—botanical, botany
br—branch(es)
Brig—Brigadier
Brit—Britain, British
Bro(s)—Brother(s)
byrol—byrology
bull—Bulletin
bur—bureau
bus—business
BWI—British West Indies

c—children
Calif—California
Can—Canada, Canadian
cand—candidate
Capt—Captain
cardiol-cardiology
cardiovasc—cardiovascular
cartog—cartographic, cartographical, cartography
cartogr—cartographer
Cath—Catholic
CEngr—Corp of Engineers
cent—central
Cent Am—Central American
cert—certificate(s), certification, certified
chap—chapter
chem—chemical(s), chemistry
chemother—chemotherapy
chg—change
chmn—chairman
citricult—citriculture
class—classical
climat—climatological, climatology
clin(s)—clinic(s), clinical
cmndg—commanding
Co—County
co—Companies, Company
co-auth—coauthor
co-dir—co-director
co-ed—co-editor
co-educ—coeducation, coeducational
col(s)—college(s), collegiate, colonel
collab—collaboration, collaborative
collabr—collaborator
Colo—Colorado
com—commerce, commercial
Comdr—Commander

commun—communicable, communication(s)
comn(s)—commission(s), commissioned
comndg—commanding
comnr—commissioner
comp—comparitive
compos—composition
comput—computation, computer(s),
  computing
comt(s)—committee(s)
conchol—conchology
conf—conference
cong—congress, congressional
Conn—Connecticut
conserv—conservation, conservatory
consol—consolidated, consolidation
const—constitution, constitutional
construct—construction, constructive
consult(s)—consult, consultant(s),
  consultantship(s), consultation, consulting
contemp—contemporary
contrib—contribute, contributing,
  contribution(s)
contribr—contributor
conv—convention
coop—cooperating, cooperation, cooperative
coord—coordinate(d), coordinating,
  coordination
coordr—coordinator
corp—corporate, corporation(s)
corresp—correspondence, correspondent,
  corresponding
coun—council, counsel, counseling
counr—councilor, counselor
criminol—criminological, criminology
cryog—cryogenic(s)
crystallog—crystallographic,
  crystallographical, crystallography
crystallogr—crystallographer
Ct—Court
Ctr—Center
cult—cultural, culture
cur—curator
curric—curriculum
cybernet—cybernetic(s)
cytol—cytological, cytology
Czech—Czechoslovakia

DC—District of Columbia
Dec—December
Del—Delaware
deleg—delegate, delegation
delinq—delinquency, delinquent
dem—democrat(s), democratic
demog—demographic, demography
demogr—demographer
demonstr—demontrator
dendrol—dendrologic, dendrological,
  dendrology
dent—dental, dentistry
dep—deputy
dept—department
dermat—dermatologic, dermatological,
  dermatology
develop—developed, developing,
  development, developmental
diag—diagnosis, diagnostic
dialectol-dialectological, dialectology
dict—dictionaries, dictionary
Dig—Digest

dipl—diploma, diplomate
dir(s)—director(s), directories, directory
dis—disease(s), disorders
Diss Abst—Dissertation Abstracts
dist—district
distrib—distributed, distribution, distributive
distribr—distributor(s)
div—division, divisional, divorced
DNA—deoxyribonucleic acid
doc—document(s), documentary,
  documentation
Dom—Dominion
Dr—Drive
E—east
ecol—ecological, ecology
econ(s)—economic(s), economical, economy
economet—econometric(s)
ECT—electroconvulsive or electroshock
  therapy
ed—edition(s), editor(s), editorial
ed bd—editorial board
educ—education, educational
educr—educator(s)
EEG—electroencephalogram,
  electroencephalographic,
  electroencephalography
Egyptol—Egyptology
EKG—electrocardiogram
elec—elecvtric, electrical, electricity
electrochem-electrochemical, electrochemistry
electroph—electrophysical, electrophysics
elem—elementary
embryol—embryologic, embryological,
  embryology
emer—emeriti, emeritus
employ—employment
encour—encouragement
encycl—encyclopedia
endocrinol—endocrinologic, endocrinology
eng—engineering
Eng—England, English
engr(s)—engineer(s)
enol—enology
Ens—Ensign
entom—entomological, entomology
environ-environment(s), environmental
enzym—enzymology
epidemiol—epideiologic, epidemiological,
  epidemiology
equip—equipment
ERDA—Energy Research & Development
  Administration
ESEA—Elementary & Secondary Education
  Act
espec—especially
estab—established, establishment(s)
ethnog—ethnographic, ethnographical,
  ethnography
ethnogr—ethnographer
ethnol—ethnologic, ethnological, ethnology
Europ—European
eval—evaluation
Evangel—evangelical
eve—evening
exam—examination(s), examining
examr—examiner
except—exceptional
exec(s)—executive(s)

exeg—exegeses, exegesis, exegetic,
  exegetical
exhib(s)—exhibition(s), exhibit(s)
exp—experiment, experimental
exped(s)—expedition(s)
explor—exploration(s), exploratory
expos—exposition
exten—extension

fac—faculty
facil—facilities, facility
Feb—February
fed—federal
fedn—federation
fel(s)—fellow(s), fellowship(s)
fermentol—fermentology
fertil—fertility, fertilization
Fla—Florida
floricult—floricultural, floriculture
found—foundation
FPO—Fleet Post Office
Fr—French
Ft—Fort

Ga—Georgia
gastroenterol—gastroenterological,
  gastroenterology
gen—general
geneal—genealogical, genealogy
geod—geodesy, geodetic
geog—geographic, geographical, geography
geogr—geographer
geol—geologic, geological, geology
geom—geometric, geometrical, geometry
geomorphol—geomorphologic,
  geomorphology
geophys—geophysical, geophysics
Ger—German, Germanic, Germany
geriat—geriatric
geront—gerontological, gerontology
GES—Gesellschaft
glaciol—glaciology
gov—governing, governor(s)
govt—government, governmental
grad—graduate(d)
Gt Brit—Great Britain
guid—guidance
gym—gymnasium
gynec—gynecologic, gynecological,
  gynecology

handbk(s)—handbook(s)
helminth—helminthology
hemat—hematologic, hematological,
  hematology
herpet—herpetologic, herpetological,
  herpetology
HEW—Department of Health, Education &
  Welfare
Hisp—Hispanic, Hispania
hist—historic, historical, history
histol—histological, histology
HM—Her Majesty
hochsch—hochschule
homeop—homeopathic, homeopathy
hon(s)—honor(s), honorable, honorary
hort—horticultural, horticulture
hosp(s)—hospital(s), hospitalization
hq—headquarters

ABBREVIATIONS

HumRRO—Human Resources Research Office
husb—husbandry
Hwy—Highway
hydraul—hydraulic(s)
hydrodyn—hydrodynamic(s)
hydrol—hydrologic, hydrological, hydrologics
hyg—hygiene, hygienic(s)
hypn—hypnosis

ichthyol—ichthyological, ichthyology
Ill—Illinois
illum—illuminating, illumination
illus—illustrate, illustrated, illustration
illusr—illustrator
immunol—immunologic, immunological, immunology
Imp—Imperial
improv—improvement
Inc—Incorporated
in-chg—in charge
incl—include(s), including
Ind—Indiana
indust(s)—industrial, industries, industry
Inf—infantry
info—information
inorg—inorganic
ins—insurance
inst(s)—institute(s), institution(s)
instnl—institutional(ized)
instr(s)—instruct, instruction, instructor(s)
instrnl—instructional
int—international
intel—intellligence
introd—introduction
invert—invertebrate
invest(s)—investigation(s)
investr—investigator
irrig—irrigation
Ital—Italian

J—Journal
Jan—January
Jct—Junction
jour—journal, journalism
jr—junior
jurisp—jurisprudence
juv—juvenile

Kans—Kansas
Ky—Kentucky

La—Louisiana
lab(s)—laboratories, laboratory
lang—language(s)
laryngol—larygological, laryngology
lect—lecture(s)
lectr—lecturer(s)
legis—legislation, legislative, legislature
lett—letter(s)
lib—liberal
libr—libraries, library
librn—librarian
lic—license(d)
limnol—limnological, limnology
ling—linguistic(s), linguistical
lit—literary, literature
lithol—lithologic, lithological, lithology

Lt—Lieutenant
Ltd—Limited

m—married
mach—machine(s), machinery
mag—magazine(s)
maj—major
malacol—malacology
mammal—mammalogy
Man—Manitoba
Mar—March
Mariol—Mariology
Mass—Massachusetts
mat—material(s)
mat med—materia medica
math—mathematic(s), mathematical
Md—Maryland
mech—mechanic(s), mechanical
med—medical, medicinal, medicine
Mediter—Mediterranean
Mem—Memorial
mem—member(s), membership(s)
ment—mental(ly)
metab—metabolic, metabolism
metall—metallurgic, metallurgical, metallurgy
metallog—metallographic, metallography
metallogr—metallographer
metaphys—metaphysical, metaphysics
meteorol—meteorological, meteorology
metrol—metrological, metrology
metrop—metropolitan
Mex—Mexican, Mexico
mfg—manufacturing
mfr—manufacturer
mgr—manager
mgt—management
Mich—Michigan
microbiol—microbiological, microbiology
micros—microscopic, microscopical, microscopy
mid—middle
mil—military
mineral—mineralogical, mineralogy
Minn—Minnesota
Miss—Mississippi
mkt—market, marketing
Mo—Missouri
mod—modern
monogr—monograph
Mont—Montana
morphol—morphological, morphology
Mt—Mount
mult—multiple
munic—municipal, municipalities
mus—museum(s)
musicol—musicological, musicology
mycol—mycologic, mycology

N—north
NASA—National Aeronautics & Space Administration
nat—national, naturalized
NATO—North Atlantic Treaty Organization
navig—navigation(al)
NB—New Brunswick
NC—North Carolina
NDak—North Dakota
NDEA—National Defense Education Act
Nebr—Nebraska

nematol—nematological, nematology
nerv—nervous
Neth—Netherlands
neurol—neurological, neurology
neuropath—neuropathological, neuropathology
neuropsychiat—neuropsychiatric, neuropsychiatry
neurosurg—neurosurgical, neurosurgery
Nev—Nevada
New Eng—New England
New York—New York City
Nfld—Newfoundland
NH—New Hampshire
NIH—National Institute of Health
NIMH—National Institute of Mental Health
NJ—New Jersey
NMex—New Mexico
No—Number
nonres—nonresident
norm—normal
Norweg—Norwegian
Nov—November
NS—Nova Scotia
NSF—National Science Foundation
NSW—New South Wales
numis—numismatic(s)
nutrit—nutrition, nutritional
NY—New York State
NZ—New Zealand

observ—observatories, observatory
obstet—obstetric(s), obstetrical
occas—occasional(ly)
occup—occupation, occupational
oceanog—oceanographic, oceanographical, oceanography
oceanogr—oceanographer
Oct—October
odontol—odontology
OEEC—Organization for European Economic Cooperation
off—office, official
Okla—Oklahoma
olericult—olericulture
oncol—oncologic, oncology
Ont—Ontario
oper(s)—operation(s), operational, operative
ophthal—ophthalmologic, ophthalmological, ophthalmology
optom—optometric, optometrical, optometry
ord—ordnance
Ore—Oregon
org—organic
orgn—organization(s), organizational
orient—oriental
ornith—ornithological, ornithology
orthod—orthodontia, orthodontic(s)
orthop—orthopedic(s)
osteop—osteopathic, osteopathy
otol—otological, otology
otolaryngol—otolaryngological, otolaryngology
otorhinol—otorhinologic, otorhinology

Pa—Pennsylvania
Pac—Pacific
paleobot—paleobotanical, paleontology
paleont—paleontology

Pan-Am—Pan-American
parisitol—parasitology
partic—participant, participating
path—pathologic, pathological, pathology
pedag—pedagogic(s), pedagogical, pedagogy
pediat—pediatric(s)
PEI—Prince Edward Islands
penol—penological, penology
periodont—periodontal, periodontic(s)
petrog—petrographic, petrographical, petrography
petrogr—petrographer
petrol—petroleum, petrologic, petrological, petrology
pharm—pharmacy
pharmaceut—pharmaceutic(s), pharmaceutical(s)
pharmacog—pharmacognosy
pharamacol—pharmacologic, pharmacological, pharmacology
phenomenol—phenomenologic(al), phenomenology
philol—philological, philology
philos—philosophic, philosophical, philosophy
photog—photographic, photography
photogeog—photogeographic, photogeography
photogr—photographer(s)
photogram—photogrammetric, photogrammetry
photom—photometric, photometrical, photometry
phycol—phycology
phys—physical
physiog—physiographic, physiographical, physiography
physiol—physiological, phsysiology
Pkwy—Parkway
Pl—Place
polit—political, politics
polytech—polytechnic(s)
pomol—pomological, pomology
pontif—pontifical
pop—population
Port—Portugal, Portuguese
Pos:—Position
postgrad—postgraduate
PQ—Province of Quebec
PR—Puerto Rico
pract—practice
practr—practitioner
prehist—prehistoric, prehistory
prep—preparation, preparative, preparatory
pres—president
Presby—Presbyterian
preserv—preservation
prev—prevention, preventive
prin—principal
prob(s)—problem(s)
proc—proceedings
proctol—proctologic, proctological, proctology
prod—product(s), production, productive
prof—professional, professor, professorial
Prof Exp—Professional Experience
prog(s)—program(s), programmed, programming
proj—project(s), projection(al), projective

prom—promotion
protozool—protozoology
Prov—Province, Provincial
psychiat—psychiatric, psychiatry
psychoanal—psychoanalysis, psychoanalytic, psychoanalytical
psychol—psychological, psychology
psychomet—psychometric(s)
psychopath—psychopathologic, psycho pathology
psychophys—psychophysical, psychophysics
psychophysiol—psychophysiological, psychophysiology
psychosom—psychosomtic(s)
psychother—psychoterapeutic(s), psycho-therapy
Pt—Point
pub—public
publ—publication(s), publish(ed), publisher, publishing
pvt—private

Qm—Quartermaster
Qm Gen—Quartermaster General
qual—qualitative, quality
quant—quantitative
quart—quarterly
Que—Quebec

radiol—radiological, radiology
RAF—Royal Air Force
RAFVR—Royal Air Force Volunteer Reserve
RAMC—Royal Army Medical Corps
RAMCR—Royal Army Medical Corps Reserve
RAOC—Royal Army Ornance Corps
RASC—Royal Army Service Corps
RASCR—Royal Army Service Corps Reserve
RCAF—Royal Canadian Air Force
RCAFR—Royal Canadian Air Force Reserve
RCAFVR—Royal Canadian Air Force Volunteer Reserve
RCAMC—Royal Canadian Army Medical Corps
RCAMCR—Royal Canadian Army Medical Corps Reserve
RCASC—Royal Canadian Army Service Corps
RCASCR—Royal Canadian Army Service Corps Reserve
RCEME—Royal Canadian Electrical & Mechanical Engineers
RCN—Royal Canadian Navy
RCNR—Royal Canadian Naval Reserve
RCNVR—Royal Canadian Naval Volunteer Reserve
Rd—Road
RD—Rural Delivery
rec—record(s), recording
redevelop—redevelopment
ref—reference(s)
refrig—refrigeration
regist—register(ed), registration
registr—registrar
regt—regiment(al)
rehab—rehabilitation
rel(s)—relation(s), relative
relig—religion, religious
REME—Royal Electrical & Mechanical

Engineers
rep—represent, representative
Repub—Republic
req—requirements
res—research, reserve
rev—review, revised, revision
RFD—Rural Free Delivery
rhet-rhetoric, rhetorical
RI—Rhode Island
Rm—Room
RM—Royal Marines
RN—Royal Navy
RNA—ribonucleic acid
RNR—Royal Naval Reserve
RNVR—Royal Naval Volunteer Reserve
roentgenol—roentgenologic, roentgenologi-cal, roentgenology
RR—Railroad, Rural Route
Rte—Route
Russ—Russian
rwy—railway

S—south
SAfrica—South Africa
SAm—South America, South American
sanit—sanitary, sanitation
Sask—Saskatchewan
SC—South Carolina
Scand—Scandinavia(n)
sch(s)—school(s)
scholar—scholarship
sci—science(s), scientific
SDak—South Dakota
SEATO—Southeast Asia Treaty Organization
sec—secondary
sect—section
secy—secretary
seismog—seismograph, seismographic, seismography
seismogr—seismographer
seismol—seismological, seismology
sem—seminar, seminary
Sen—Senator, Senatorial
Sept—September
ser—serial, series
serol—serologic, serological, serology
serv—service(s), serving
silvicult—silvicultural, silviculture
soc(s)—societies, society
soc sci—social science
sociol—sociologic, sociological, sociology
Span—Spanish
spec—special
specif—specification(s)
spectrog—spectrograph, spectrographic, spectrography
spectrogr—spectrographer
spectrophotom—spectrophotometer, spectrophotometric, spectrophotometry
spectros—spectroscopic, spectroscopy
speleol—speleological, speleology
Sq—Square
sr—senior
St—Saint, Street(s)
sta(s)—station(s)
stand—standard(s), standardization
statist—statistical, statistics
Ste—Sainte
steril—sterility

ABBREVIATIONS

stomatol—stomatology
stratig—stratigraphic, stratigraphy
stratigr—stratigrapher
struct—structural, structure(s)
stud—student(ship)
subcomt—subcommittee
subj—subject
subsid—subsidiary
substa—substation
super—superior
suppl—supplement(s), supplemental,
    supplementary
supt—superintendent
supv—supervising, supervision
supvr—supervisor
supvry—supervisory
surg—surgery, surgical
surv—survey, surveying
survr—surveyor
Swed—Swedish
Switz—Switzerland
symp—symposia, symposium(s)
syphil—syphilology
syst(s)—system(s), systematic(s), systematical

taxon—taxonomic, taxonomy
tech—technical, technique(s)
technol—technologic(al), technology
tel—telegraph(y), telephone
temp—temporary
Tenn—Tennessee
Terr—Terrace
Tex—Texas
textbk(s)—textbook(s)
text ed—text edition
theol—theological, theology
theoret—theoretic(al)
ther—therapy
therapeut—therapeutic(s)
thermodyn—thermodynamic(s)
topog—topographic, topographical,
    topography
topogr—topographer
toxicol—toxicologic, toxicological,

toxicology
trans—transactions
transl—translated, translation(s)
translr—translator(s)
transp—transport, transportation
treas—treasurer, treasury
treat—treatment
trop—tropical
tuberc—tuberculosis
TV—television
Twp—Township

UAR—United Arab Republic
UK—United Kingdom
UN—United Nations
undergrad—undergraduate
unemploy—unemployment
UNESCO—United Nations Educational
    Scientific & Cultural Organization
UNICEF—United Nations International
    Childrens Fund
univ(s)—universities, university
UNRRA—United Nations Relief &
    Rehabilitation Administration
UNRWA—United Nations Relief & Works
  Agency
urol—urologic, urological, urology
US—United States
USAAF—US Army Air Force
USAAFR—US Army Air Force Reserve
USAF—US Air Force
USAFR—US Air Force Reserve
USAID—US Agency for International
    Development
USAR—US Army Reserve
USCG—US Coast Guard
USCGR—US Coast Guard Reserve
USDA—US Department of Agriculture
USMC—US Marine Corps
USMCR—US Marine Corps Reserve
USN—US Navy
USNAF—US Naval Air Force
USNAFR—US Naval Air Force Reserve
USNR—US Naval Reserve

USPHS—US Public Health Service
USPHSR—US Public Health Service Reserve
USSR—Union of Soviet Socialist Republics

Va—Virginia
var—various
veg—vegetable(s), vegetation
vent—ventilating, ventilation
vert—vertebrate
Vet—Veteran(s)
vet—veterinarian, veterinary
VI—Virgin Islands
vinicult—viniculture
virol—virological, virology
vis—visiting
voc—vocational
vocab—vocabulary
vol(s)—voluntary, volunteer(s), volume(s)
vpres—vice president
vs—versus
Vt—Vermont

W—west
Wash—Washington
WHO—World Health Organization
WI—West Indies
wid—widow, widowed, widower
Wis—Wisconsin
WVa—West Virginia
Wyo—Wyoming

Yearbk(s)—Yearbook(s)
YMCA—Young Men's Christian Association
YMHA—Young Men's Hebrew Association
Yr(s)—Year(s)
YT—Yukon Territory
YWCA—Young Women's Christian
    Association
YWHA—Young Women's Hebrew Associa-
    tion

zool—zoological, zoology

American Men & Women of Science

A

AABOE, ASGER (HARTVIG), b Copenhagen, Denmark, Apr 26, 22; m 50; c 4. MATH, HISTORY & PHILOSOPHY OF SCIENCE. *Educ:* Brown Univ, PhD(hist math), 57. *Prof Exp:* Vis lectr math, Washington Univ, 47-48; adj, Birkerod State Sch, Denmark, 48-52; from instr to assoc prof, Tufts Univ, 52-62; assoc prof, 62-67, chmn Dept Hist Sci & Med, 68-71, PROF MATH & HIST SCI & MED, 67-, PROF NEAR EASTERN LANG & LITS, YALE UNIV, 77- *Concurrent Pos:* Vis assoc prof, Yale Univ, 61-62. *Mem:* Royal Danish Soc. *Res:* History of mathematics and mathematical astronomy. *Mailing Add:* Dept Math Yale Univ Box 2155 New Haven CT 06520

AAGAARD, GEORGE NELSON, b Minneapolis, Minn, Aug 16, 13; m 39; c 5. INTERNAL MEDICINE, CLINICAL PHARMACOLOGY. *Educ:* Univ Minn, BS, 34, BMed, 36, MD, 37. *Prof Exp:* Instr internal med, Univ Minn, 42-45, asst prof, 45-48, assoc prof & dir postgrad med educ, 48-51; prof med & dean, Univ Tex Southwestern Med Sch Dallas, 52-54; dean sch med, 54-64, ombudsman, 69-70, prof med & head, Div Clin Pharmacol, 64-79, first distinguished prof, 78-84, EMER PROF MED & PHARMACOL, SCH MED, UNIV WASH, 84- *Concurrent Pos:* Fel internal med, Univ Minn, 41-47; mem health res facil adv coun, NIH, 56-58; mem nat adv coun, Nat Heart Inst, 61-65; mem bd dirs, Am Heart Asn, 58; mem spec med adv group, Vet Admin, 70-74; mem sci adv comt, Pharmaceut Mfg Asn Found, 67-74. *Honors & Awards:* Elliott Award, Am Soc Clin Pharmacol & Therapeut. *Mem:* AMA; Am Heart Asn; Am Pub Health Asn; Am Col Physicians; Asn Am Med Col (pres, 60-61); Am Soc Clin Pharmacol & Therapeut (pres, 77). *Res:* Medical education; hypertension; clinical pharmacology. *Mailing Add:* Sch Med RG-23 Univ Wash Seattle WA 98195

AAGAARD, JAMES S(TUART), b Lake Forest, Ill, July 20, 30; m 63. COMPUTER SCIENCE. *Educ:* Northwestern Univ, BS, 53, MS, 55, PhD(elec eng), 57. *Prof Exp:* From asst prof to assoc prof elec eng, 57-70, assoc prof elec eng & comput sci, 70-76, PROF ELEC ENG & COMPUT SCI, NORTHWESTERN UNIV, EVANSTON, 76-, HEAD INFO SYSTS DEVELOP LIBR, 72- *Concurrent Pos:* Mem bd dirs, Nat Electronics Conf, 58-66. *Honors & Awards:* Lita/Gaylord Award for Achievement in Libr & Info Technol, Am Lib Asn, 85. *Mem:* Fel AAAS; Inst Elec & Electronics Engrs; Asn Comput Mach; Am Soc Eng Educ; Am Soc Info Sci. *Res:* Computer software; electronic circuits; communications theory. *Mailing Add:* 744 Lenox Lane Glenview IL 60025

AAGAARD, KNUT, b Brooklyn, NY, Feb 5, 39; m 63; c 3. PHYSICAL OCEANOGRAPHY. *Educ:* Oberlin Col, AB, 61; Univ Wash, MS, 64, PhD(phys oceanog), 66. *Prof Exp:* Res assoc phys oceanog, Univ Wash, 66; Nat Acad Sci-US Air Force Off Sci Res, res assoc, Geophys Inst, Univ Bergen, 67; res asst prof, Univ Wash, 68-73, res assoc prof, 73-78, res prof phys oceanog, 78-86; OCEANOGRAPHER, NOAA, 86- *Concurrent Pos:* Mem various int comts & working groups. *Mem:* AAAS; fel Arctic Inst NAm; Am Geophys Union. *Res:* Arctic oceanography, current measurements, general physical oceanography, climate. *Mailing Add:* NOAA/PMEL/CARD 7600 Sand Point Way NE Seattle WA 98115-0070

AAGARD, ROGER L, b East Chain, Minn, Aug 2, 31; m 60; c 3. PHYSICS, MATHEMATICS. *Educ:* Univ Minn, BA, 57, MS, 59. *Prof Exp:* Asst atmospheric physics, Univ Minn, 57-59; SR RES SCIENTIST, HONEYWELL RES CTR, 59- *Mem:* Inst Elec & Electronics Engrs; Optical Soc Am. *Res:* Thermal radiation sensors; semiconductor infrared detectors and dielectric lasers; atmospheric infrared radiation and optical properties of lasers; lasers and magneto-optic materials for computer information storage; integrated optics; computer modeling of frost formation on air-to-air heat pumps; thin flim optical components and systems; fiber optic pressure sensor; computer modelling of solid state sensors. *Mailing Add:* Honeywell Inc Corp Phys Sci Ctr 10701 Lyndale Ave S Bloomington MN 55420

AALDERS, LEWIS ELDON, b Kentville, NS, July 19, 33; div. SCIENCE EDUCATION. *Educ:* Acadia Univ, BSc, 52, MSc, 53; Cornell Univ, PhD(cytogenetics), 57. *Prof Exp:* Horticulturist plant breeding, Res Sta, Can Dept Agr, 53-67, res scientist, 67-82; RETIRED. *Mem:* Sigma Xi. *Mailing Add:* 83 Elm Ave Kentville NS B4N 1Z3 Can

AALUND, OLE, b Gentofte, Denmark, April 26, 30; m 55; c 2. LIVESTOCK HEALTH MANAGEMENT. *Educ:* Royal Vet & Agr Univ, DVM, 55, DVSc, 68. *Prof Exp:* Asst specialist immunol, Univ Calif, Davis, 62-64; asst prof immunol, 64-68, from asst prof to assoc prof forensic med, 69-74, PROF PREV MED, ROYAL VET & AGR UNIV, DENMARK, 74- *Mem:* Am Asn Immunologists; Sigma Xi; Scand Soc Allergol & Immunol. *Res:* Livestock health management from a holistic point of view in intensive production units; contributions to the improvement of the situation in developing countries. *Mailing Add:* Royal Vet & Agr Univ Bulowsvej 13 Frederiksberg Denmark

AAMODT, RICHARD E, b St Paul, Minn, Nov 22, 36; c 5. PHYSICS, PLASMA PHYSICS. *Educ:* Univ Mich, BSE, 58, MS, 59, PhD(physics), 63. *Prof Exp:* Staff assoc plasma physics, Gen Atomic Div, Gen Dynamics Corp, 62-64, staff mem, 64-65; from asst prof to assoc prof physics, Univ Tex, Austin, 65-69; vis assoc prof astro-geophys, Univ Colo, 69-70, from assoc prof to prof, 70-75; sr staff scientist, 75-79, dir, Plasma Res Inst, Sci Appln Intl Corp, 79-87; PRES, LODESTAR RES CORP, 87- *Concurrent Pos:* Consult, Lawrence Radiation Lab, Univ Calif, 66-; consult, Boeing Sci Res Lab, 66-70; vis staff mem, Los Alamos Sci Lab, 69-; res physicist, Naval Res Lab, 74. *Mem:* Fel Am Phys Soc. *Res:* Theoretical plasma physics; transport theory. *Mailing Add:* Lodestar Res Corp 2400 Central Ave Boulder CO 80301

AAMODT, ROGER LOUIS, b San Francisco, Calif, Dec 9, 41; m 62; c 2. CANCER DIAGNOSIS, RADIATION BIOLOGY. *Educ:* Univ Utah, BS, 66; Univ Rochester, PhD(radiation biol), 72. *Prof Exp:* Res asst health physics, Dept Radiol Health, Univ Utah, 65-66; sect chief tracer metab, Dept Nuclear Med, Clin Ctr, 71-83, grants assoc, Off of Dir, 83-84, PROG DIR, CANCER DIAG BR, DCBDC, NAT CANCER INST, NIH, ROCKVILLE, MD, 84- *Mem:* Am Asn Pathologists; Sigma Xi; Int Soc Anal Cytol; NY Acad Sci; AAAS. *Res:* Cancer diagnosis in the areas of pathology and cytology: approaches include immunohistochemistry, flow cytometry, automated image analysis, artificial intelligence, fine needle aspiration cytology, etc. *Mailing Add:* Cancer Diag Br DCBDC Nat Cancer Inst NIH EPS Rm 638 6120 Executive Blvd Rockville MD 20892

AARON, CHARLES SIDNEY, b Jonesboro, La, Feb 17, 43; m 87; c 2. MOLECULAR GENETICS, MUTAGENESIS. *Educ:* La Tech Univ, BS, 68; La State Univ, Baton Rouge, PhD(biochem), 72; Am Bd Toxicol, dipl. *Prof Exp:* Teacher chem, Episcopal High Sch, Baton Rouge, 71-72; res assoc chem mutagenesis, La State Univ, Baton Rouge, 72- 76; mgr, analytical & biochem toxicol, Allied Corp, 80-84; sr res scientist, genetic toxicol res, 85-89, DIR, INVESTIGATIVE TOXICOL, UPJOHN CO, KALAMAZOO, MICH, 89- *Concurrent Pos:* Res scientist, Dept Radiation Genetics & Chem Mutagenesis, State Univ, Leiden, Neth, 76-79; genetics & reproduction team leader, Off Toxic Subst, US Environ Protection Agency, Washington, DC, 79-80; mem coun, Environ Mutagen Soc. *Mem:* Am Chem Soc; Sigma Xi; Environ Mutagen Soc; Soc Risk Anal; Int Soc Study Xenobiotics; fel Environ Health Inst. *Res:* Molecular dosimetry of chemical mutagens in in vitro test systems; molecular mechanisms of mutagenesis; quantitive risk assessment; structure activity correlations; non-genotoxic carcinogens. *Mailing Add:* Upjohn Co 301 Henrietta St Kalamazoo MI 49001

AARON, HERBERT SAMUEL, b Minneapolis, Minn, Jan 12, 29; m 59; c 2. ORGANOPHOSPHORUS CHEMISTRY, STEREOCHEMISTRY. *Educ:* Univ Minn, BS, 49; Univ Calif, Los Angeles, PhD(chem), 53. *Prof Exp:* Res chemist, Yerkes Res Lab, E I du Pont de Nemours & Co, 53-54; CHEMIST, CHEM RES LABS, EDGEWOOD ARSENAL-ABERDEEN PROVING GROUND, 54- *Mem:* Am Chem Soc. *Res:* Stereochemistry; conformational analysis; organo-phosphorus chemistry. *Mailing Add:* Chem Res Labs CRDEC Edgewood Area Aberdeen Proving Ground MD 21010-5423

AARON, HOWARD BERTON, b Brooklyn, NY, Dec 2, 39; m 62; c 1. PHYSICAL METALLURGY, MATERIALS SCIENCE. *Educ:* Cornell Univ, BMetE, 62; Univ Ill, Urbana-Champaign, MS, 63, PhD(metall eng), 67; Univ Mich, Ann Arbor, MBA, 73. *Prof Exp:* Res asst metall eng, Univ Ill, Urbana-Champaign, 62-66, res assoc, 66-67; sr staff scientist, Ford Motor Co, 67-78; vpres eng & res, 78-81, VPRES & GEN MGR, Transmission Prod Div, D A B Indust, Inc, 81-85; PRES, H AARON CO, 85- *Concurrent Pos:* Mem exec comt & vchmn, Epilepsy Ctr of Mich; vchmn adv bd, Protection & Advocacy Serv. *Mem:* Metall Soc; Am Soc Metals; Soc Automotive Engrs;

Am Soc Testing & Mat; Am Mgt Asn. *Res:* Physical metallurgy; phase transformations; diffusion; interfaces; precipitate nucleation and growth; dissolution; electron microscopy; Berg-Barrett studies of strain fields; quantitative metallography; low temperature strain aging; zener disaccommodation studies of defects. *Mailing Add:* 4110 Nearbrook Rd Bloomfield Hills MI 48302

AARON, M ROBERT, b Philadelphia, Pa, Aug 21, 22; m 44; c 2. ELECTRICAL ENGINEERING, COMMUNICATIONS. *Educ:* Univ Pa, BS, 49, MS, 51. *Prof Exp:* Res engr, Res Labs, Franklin Inst, 49-51 & Leeds & Northrup Co, 51; mem tech staff, Bell Tel Labs, Inc, 51-54, supvr, 54-69, head, Dept Digital Tech, AT&T Bell Labs, 69-88; CONSULT, 89- *Honors & Awards:* Alexander Graham Bell Medal, Inst Elec & Electronics Engrs, 78, Centennial Medal, 84, Donald W McLellan Award, 85; C & C Prize, C & C Found, Japan, 88. *Mem:* Nat Acad Eng; fel Inst Elec & Electronics Engrs. *Res:* Digital communications systems, especially pulse code modulation; broadband ISDN; digital signal processing; technology forecasting; strategic planning; more than 12 patents. *Mailing Add:* 1700 Presidential Way Apt 203 West Palm Beach FL 33401

AARON, RONALD, b Philadelphia, Pa, June 14, 35; m 62; c 2. THEORETICAL PHYSICS. *Educ:* Temple Univ, AB, 56; Univ Pa, PhD(physics), 61. *Prof Exp:* Res assoc physics, Univ Md, 61-62; res assoc, Ind Univ, 62-63; physicist, Goddard Space Ctr, NASA, 63; asst prof physics, 63-72, PROF PHYSICS, NORTHEASTERN UNIV, 72- *Concurrent Pos:* Consult, Lawrence Radiation Lab, Univ Calif, Livermore, 64-67; vis staff mem, Los Alamos Sci Lab, NMex, 67-74; Med Physics, The Gordon Inst, 86- *Mem:* Am Phys Soc. *Res:* Theoretical high energy physics and scattering theory. *Mailing Add:* Dept Physics Northeastern Univ Boston MA 02115

AARONOFF, BURTON ROBERT, b Brooklyn, NY, July 4, 35; m 66; c 2. ORGANIC CHEMISTRY. *Educ:* Brooklyn Col, BS, 56; Johns Hopkins Univ, MA, 59, PhD(chem), 62. *Prof Exp:* Res assoc chem, Lever Bros Co, 62-63; sr chemist, Merck, Sharp & Dohme Labs, 63-66; asst to vpres res & develop, Block Drug Co, 66-68, assoc res dir, 68-73; dir explor res, Pfizer Consumer Prod Div, 73-84; pharm consult, 85-88; MGR, STATE REGIST, ZENITH LABS, 89- *Mem:* Am Chem Soc; Am Pharmaceut Asn; Regulatory Affairs Prof Soc. *Res:* Organic and medicinal chemistry; proprietary drugs and toiletries; pharmaceutical dosage forms; research administration; regulatory affairs. *Mailing Add:* 20 Otis Pl Verona NJ 07044

AARONS, JULES, b New York, NY, Oct 3, 21; m 43; c 2. IONOSPHERIC PHYSICS. *Educ:* City Col, NY, BS, 42; Boston Univ, MA, 49; Univ Paris, PhD, 54. *Prof Exp:* Physicist, Air Force Cambridge Res Labs, 46-55, chief radio astron br, 55-72, chief trans-ionospheric propagation br & sr scientist, Space Physics Lab, Air Force Geophys Lab, 72-81; RES PROF ASTRON & SPACE SCI, BOSTON UNIV, 81- *Concurrent Pos:* Chmn electromagnetic propagation panel, Adv Group Aerospace Res & Develop, 73-75; int chmn comn G, Int Sci Radio Union. *Honors & Awards:* Harry Diamond Award, Inst Elec & Electronic Engrs, 82. *Mem:* Fel Inst Elec & Electronics Engrs; Am Geophys Union. *Res:* Upper atmosphere physics; ionospheric propagation, solar radio astronomy, solar terrestrial relationships. *Mailing Add:* Ctr Space Physics Boston Univ 725 Commonwealth Ave Boston MA 02215

AARONSON, HUBERT IRVING, b New York, NY, July 10, 24. NUCLEATION KINETICS, GROWTH KINETICS. *Educ:* Carnegie Inst Technol, BS, 48, MS & PhD(metall eng), 54. *Prof Exp:* Res metallurgist, Carnegie Inst Technol, 53-57; prin res scientist, Sci Lab, Ford Motor Co, 58-72; prof metall eng, Mich Technol Univ, 72-79; R F MEHL PROF METALL ENG, CARNEGIE MELLON UNIV, 79- *Concurrent Pos:* Chmn bd, Tech Div, Metall Soc, 70-72; sr res fel, Univ Manchester, UK, 72. *Honors & Awards:* Champion H Mathewson Gold Medal, Metall Soc, 68; Albert Sauveur Achievement Award, Am Soc Metals, 87; R F Mehl Lectr, Metall Soc, 90. *Mem:* Fel Metall Soc; Am Soc Metals; Int Soc Metallog Soc; Inst Metals. *Res:* Morphology, kinetics and mechanism of diffusional nucleation; diffusional growth in metallic alloys. *Mailing Add:* Dept Metall Eng & Mat Sci Carnegie Mellon Univ Pittsburgh PA 15213

AARONSON, SHELDON, b New York, NY, Oct 9, 22; m 48; c 4. MICROBIAL BIOCHEMISTRY. *Educ:* City Col New York, BS, 44; Univ Pa, MA, 48; NY Univ, PhD(protistology), 53. *Prof Exp:* Lectr biol, City Col New York, 48-53; from instr to assoc prof, 53-64, PROF BIOL, QUEENS COL, NY, 65- *Concurrent Pos:* Res assoc, Haskins Labs, Inc, 49-65, sr staff mem, 65-68; pres fel, Am Soc Microbiol, 57; guest res collabr, biol dept, Brookhaven Nat Lab, 57-58; NSF sci fac fel, 59-60; guest collab scientist, Dept Physiol & Anat, Univ Calif, Berkeley, 63; vis scientist, Biochem Dept, Weizmann Inst, 75; vis prof, Biol Dept, Ben Gunion Univ Negev, 75; lectr, UNEP/UNESCO/ICRO, Nairobi, Kenya, 76; Mellon Fel, Queens Col, 82-84; Fulbright sr med lectr, Hacettepe Med Sch, Turkey, 86-87. *Mem:* Fel AAAS; Am Soc Microbiol; Soc Protozool; Soc Gen Microbiol; Phycol Soc Am; Biochem Soc. *Res:* Biochemistry of microorganisms; biochemical phylogeny; cellular action of drugs; microbial ecology; ethnomicrobiology. *Mailing Add:* Dept of Biol Queens Col Flushing NY 11367

AARONSON, STUART ALAN, b Mt Clemens, Mich, Feb 28, 42; m 71; c 3. VIRAL ONCOLOGY, CELL BIOLOGY. *Educ:* Univ Calif, Berkeley, BS, 62, MS, 65, San Francisco, MD, 66. *Prof Exp:* Staff assoc, Viral Carcinogenesis Br, 67-69, sr staff fel, 69-70, head, Molecular Biol Sect, 70-77, CHIEF, LAB CELLULAR & MOLECULAR BIOL, NAT CANCER INST, 77- *Concurrent Pos:* Assoc ed, J Nat Cancer Inst, 77- & Cancer Res, 81-; mem, Sci Adv Comt, Am Cancer Soc, 84-87, Mott Selection Comt, Gen Motors Cancer Res Found, 84-; counr, Soc Exp Biol & Med, 84-89. *Honors & Awards:* Rhoads Mem Award, Am Asn Cancer Res, 82. *Mem:* Am Soc Microbiol; AAAS; Fedn Am Soc Exp Biol; Am Asn Cancer Res; Am Soc Virol. *Res:* Cell biology and molecular genetics of cancer; growth factor signal transduction; retroviruses. *Mailing Add:* Lab Cellular & Molecular Biol Nat Cancer Inst Bldg 37 Rm 1E24 Bethesda MD 20892

AASE, JAN KRISTIAN, b Stavanger, Norway, Nov 21, 35; US citizen; m 65; c 5. SOIL SCIENCE. *Educ:* Brigham Young Univ, BA, 58; Univ Minn, MS, 60; Colo State Univ, PhD(soil physics), 67. *Prof Exp:* RES SOIL SCIENTIST, NORTHERN PLAINS SOIL & WATER RES CTR, AGR RES SERV, USDA, 67-, RES LEADER, 82- *Concurrent Pos:* Sr res assoc & adj prof, Mont State Univ; location coordr, Agr Res Serv, USDA, 83- *Mem:* Am Soc Agron; Soil Sci Soc Am; Int Soil Sci Soc; Soil Conserv Soc Am; Sigma Xi. *Res:* Soil-plant-water relations; plant water use efficiency and microclimatological research on agronomic crops and native rangeland and remote sensing. *Mailing Add:* Agr Res Serv USDA PO Box 1109 Sidney MT 59270

AASLESTAD, HALVOR GUNERIUS, b Birmingham, Ala, Sept 6, 37; m 60; c 4. VIROLOGY. *Educ:* La State Univ, BS, 59, PhD(microbiol), 65; Pa State Univ, 61. *Prof Exp:* Nat Acad Sci-Nat Res Coun fel, US Army Biol Ctr, 65-66, prin investr microbiol, 66-68; asst prof, Univ Ga, 68-70; asst prof, Wistar Inst, 70-73; sr scientist, Cancer Res Ctr, Nat Cancer Inst, Frederick, Md, 73-76; exec secy, 76-81, ASST BR CHIEF, DIV RES GRANTS, NIH, 81- *Mem:* Am Soc Microbiol. *Res:* Scientific review process; virus and phage biochemistry and genetics. *Mailing Add:* 21 Pine Hollow Rd North Branford CT 06471

ABADI, DJAHANGUIR M, b Tehran, Iran, Sept 1, 29; US citizen; m 60; c 2. CLINICAL CHEMISTRY. *Educ:* Doane Col, BA, 52; Univ Nebr, MS, 55; Univ Wash, PhD(biochem), 58; Am Bd Clin Chem, dipl. *Prof Exp:* Biochemist, Swed Hosp, Seattle, Wash, 59-61; clin chemist, Good Samaritan Hosp, Portland, Ore, 61-64; CLIN CHEMIST, VET ADMIN HOSP, IOWA CITY, 64- *Mem:* Fel Am Asn Clin Chemists. *Res:* Proteolytic enzymes; protein chemistry. *Mailing Add:* Vet Admin Hosp Iowa City IA 52246

ABADIE, STANLEY HERBERT, b New Orleans, La, Nov 13, 25; m 62; c 2. MEDICAL PARASITOLOGY, MEDICAL EDUCATION. *Educ:* Loyola Univ, La, BS, 55; La State Univ, New Orleans, MS, 58, PhD(med parasitol), 63. *Prof Exp:* Grad asst med microbiol & parasitol, 56-58, from instr to assoc prof med parasitol, 58-72, MEM GRAD FAC MED PARASITOL, LA STATE UNIV MED CTR, 60-, PROF MED PARASITOL & MICROBIOL, 72-, DEAN, SCH ALLIED HEALTH PROFESSIONS, 75- *Concurrent Pos:* La State Univ Interam Training Prog fel trop med, Costa Rica, 64; Schlieder Educ Found res grant Chagas' dis, 74-77; consult mycol, Touro Infirm, New Orleans, 63-64; scientist, Vis Staff, Charity Hosp, New Orleans, 63-; consult parasitol, Vet Admin Hosp, New Orleans, 70-; lectr parasitic dis, Sch Dent, La State Univ, New Orleans, 71-; contribr, Manual of Microbiol & Manual of Trop Med. *Mem:* AAAS; Am Soc Trop Med & Hyg; Am Soc Parasitol; Am Soc Allied Health. *Res:* Hookworm disease; heartworm; filarial infections. *Mailing Add:* Sch Allied Health La State Univ 1900 Gravier St New Orleans LA 70112-2262

ABAJIAN, PAUL G, b Framingham, Mass, Aug 4, 41; div; c 4. WATER ANALYSIS, PHYSICAL CHEMISTRY. *Educ:* Worcester Polytech Inst, BS, 63; Univ Vt, PhD(chem), 67. *Prof Exp:* Asst anal chem, Univ Vt, 63-67; asst prof chem, Southern Conn State Col, 67-69; asst prof, Johnston State Col, 69-77, assoc dean, 81, assoc prof chem, 77-89, PROF CHEM, JOHNSON STATE COL, 89-, ACTG DEAN, ACAD AFFAIRS, 90- *Res:* Solution calorimetry of transition metal compounds and also polarographic studies of rare earth metals; correlation of metals in water and neurological disorders. *Mailing Add:* Dept Chem Johnson State Col Johnson VT 05656

ABARBANEL, HENRY D I, b Washington, DC, May 31, 43; m 82; c 2. THEORETICAL PHYSICS. *Educ:* Princeton Univ, PhD(physics), 66. *Prof Exp:* Res assoc physics, Princeton Univ, 66, vis fel, 66-67; res assoc, Stanford Linear Accelerator Ctr, 67-68; asst prof, Princeton Univ, 68-72; physicist, Fermi Nat Accelerator Lab, 72-80; physicist, Lawrence Berkeley Lab, Univ Calif, 80-83; PROF PHYSICS, UNIV CALIF, SAN DIEGO, 83-, DIR, INST NONLINEAR SCI, 86-; RES PHYSICIST, SCRIPPS INST OCEANOG, 83- *Concurrent Pos:* NSF fel, Princeton Univ, 67-68, A P Sloan res fel, 69-72; adj prof physics & astron, Northwestern Univ, 77-79; chair, activ group dynamical systs, Soc Indust & Appl Math, 88-90. *Mem:* Am Phys Soc; Am Geophys Union. *Res:* High energy collision phenomena of elementary particles; quantum field theory; classical dynamics of non-linear stochastic systems. *Mailing Add:* Scripps Inst Oceanog Univ Calif San Diego La Jolla CA 92093-0402

ABASHIAN, ALEXANDER, b Binghamton, NY, Mar 11, 30; m 57; c 4. NUCLEAR PHYSICS. *Educ:* Purdue Univ, BS, 52; Johns Hopkins Univ, PhD, 57. *Prof Exp:* Tech engr, IBM, 52; jr instr physics, Johns Hopkins Univ, 52-55; jr res assoc high energy physics, Brookhaven Cosmotron, 55-57; instr physics, Univ Rochester, 57-59; res assoc, Radiation Lab, Univ Calif, 59-61; from asst prof to prof, Univ Ill, Urbana-Champaign, 61-74; prog dir elem particle physics, NSF, 72-80; head, Physics Depts, Va Polytech Inst, 80-82, dir, Inst High Energy Physics, 87-90; CONSULT, 90- *Concurrent Pos:* Vis prof, Univ Hamburg, Ger, 78-79; vis staff, Nat Lab H E Physics (KEK), Japan, 86-87. *Mem:* AAAS; fel Am Phys Soc. *Res:* Pion interactions; nuclear polarization; invariance tests; weak interactions; instrumentation; muon scattering; neutrino interactions; electron-positron interactions. *Mailing Add:* Dept Physics Va Polytech Inst & State Univ Blacksburg VA 24061

ABATE, KENNETH, b Detroit, Mich, Jan 9, 44; m 73; c 2. SILICONE CHEMISTRY. *Educ:* Wayne State Univ, BS, 65; Mich Tech Univ, MS, 67; Univ Fla, PhD(chem), 71. *Prof Exp:* Postdoctoral fel pharm, Col Pharm, Univ Fla, 71-72; group leader, Conchemco, 72-73; res chemist, Celanese, 73-76; vpres chem tech, H H Robertson Co, 76-90; ASSOC DIR SPECIALTY CHEM, PPG INDUSTS, 90- *Mem:* Am Chem Soc; Am Inst Chemists; Nat Asn Corrosion Engrs; Fedn Soc Coatings Technologies. *Res:* Corrosion prevention, coatings, polymers, paint manufacture, surfactants and speciality monomers, metal pretreating and coatings application methods. *Mailing Add:* PPG Industs Inc 440 College Park Dr Monroeville PA 15146

ABBAS, DANIEL CORNELIUS, b Cincinnati, Ohio, Nov 30, 47; m 69; c 2. SOLID STATE PHYSICS. *Educ:* Calvin Col, AB, 69; Univ Ill, MS, 70, PhD(physics), 77. *Prof Exp:* Phys sci asst vehicle signature study, US Army, 71-72, res physicist automotive diag equip, 72-74; RES PHYSICIST SOLID STATE LIGHT EMITTERS, EASTMAN KODAK CO, 77- *Concurrent Pos:* Instr, Wayne County Community Col, 71-72. *Mem:* Am Phys Soc. *Res:* Solid state light emitters. *Mailing Add:* Res Labs Eastman Kodak Co Rochester NY 14650

ABBAS, MIAN MOHAMMAD, b Pakistan, June 22, 33; US citizen; m 60; c 4. ATMOSPHERIC PHYSICS, SPACE PHYSICS. *Educ:* Panjab Univ, BS, 55; Laval Univ, MS, 60; Univ RI, PhD, 67. *Prof Exp:* Res fel, Univ RI, 64-68; asst prof, Ohio Univ, 68-70; assoc prof, Univ Ky, Lexington, 70-74; Nat Acad Sci sr res assoc, Goddard Space Flight Ctr, NASA, 74-80; assoc prof, dept physics, Drexel Univ, 80-87; ASTROPHYSICIST, NASA/MARSHALL SPACE FLIGHT CTR, 87- *Concurrent Pos:* Sr res assoc astron prog, Univ Md. *Mem:* Am Geophys Union. *Res:* Infrared astronomy; planetary atmospheres; stratosphere. *Mailing Add:* ES55 Space Sci Lab NASA/ Marshall Space Flight Ctr Huntsville AL 35812

ABBASABADI, ALIREZA, b Gorgan, Iran, Jan 18, 50; m. FIELD THEORY. *Educ:* Tehran Univ, BS, 73, MS, 75; Mich State Univ, PhD(physics), 85. *Prof Exp:* Instr physics, Aborihan Univ, 75-78; vis asst prof, Mich State Univ, 85-88; ASST PROF PHYSICS, FERRIS STATE UNIV, 88- *Mem:* Am Phys Soc. *Res:* Quantum electrodynamics and study of various implications and phenomenology of the standard model of electroweak and strong interactions. *Mailing Add:* Physics Sci Dept Ferris State Univ Big Rapids MI 49307

ABBASCHIAN, G J, b Zanjan, Iran Jan 23, 44; m 73; c 2. MATERIALS SCIENCE. *Educ:* Univ Teheran, BS, 65; Mich Technol Univ, MS, 68; Univ Calif, Berkeley, PhD(mat sci & eng), 71. *Prof Exp:* Res asst, dept mat sci & eng, Univ Calif, 68-71; from asst prof to assoc prof, dept mat sci & eng, Shiraz Univ, Iran, 72-80, chmn dept, 74-76; assoc prof, 80-83, actg chmn, 86-87, PROF DEPT MAT SCI & ENG, UNIV FLA, 83-, CHMN, 87- *Concurrent Pos:* Res analyst, US Steel Corp, 67; vis assoc prof, dept metall & mining eng, Univ Ill, Urbana, 76-78; vis scientist, Mass Inst Technol, 80, NASA Space Processing Lab, Marshall Space Flight Ctr, 81. *Honors & Awards:* Imp Award, Iranian Govt, 65. *Mem:* Metall Soc, Am Inst Mining, Metall & Petrol Engrs; Am Asn Crystal Growth; Am Asn Metals; Int Metallog Soc; Mat Res Soc. *Res:* Solidification processing, electromagnetic levitation, space processing, electronic materials processing, crystal growth and composites. *Mailing Add:* 1515 NW 51st Terr Gainesville FL 32605

ABBATIELLO, MICHAEL JAMES, b New York, NY, June 19, 21; m 45; c 1. ZOOLOGY, BIOLOGY. *Educ:* Hofstra Col, BA, 54, MA, 57; St John's Univ, NY, PhD(zool), 66. *Prof Exp:* Tech asst hort, 47-53, from jr instr to assoc prof biol, 53-66, PROF BIOL, STATE UNIV NY AGR & TECH COL FARMINGDALE, 66- *Mem:* AAAS; Entom Soc Am; Nat Asn Biol Teachers. *Res:* Biology and ecology of oribatid mites. *Mailing Add:* 29 Elm Dr Farmingdale NY 11735

ABBE, WINFIELD JONATHAN, b Cleveland, Ohio, Feb 27, 39; m 66; c 1. THEORETICAL HIGH ENERGY PHYSICS. *Educ:* Univ Calif, Berkeley, AB, 61; Calif State Col, Los Angeles, MS, 62; Univ Calif, Riverside, PhD(physics), 66. *Prof Exp:* From asst prof to assoc prof physics, Univ Ga, 66-80. *Concurrent Pos:* Res fel, Univ Mich, 66-67. *Res:* Elementary particle theory; S-matrix theory of strong interactions; Regge Pole theory; dispersion relations. *Mailing Add:* 150 Raintree Ct Athens GA 30607

ABBEY, HELEN, b Ann Arbor, Mich, Sept 1, 20. BIOSTATISTICS. *Educ:* Battle Creek Col, BS, 40; Univ Mich, MA, 42; Johns Hopkins Univ, ScD, 51. *Prof Exp:* From instr to assoc prof, 51-69, PROF BIOSTATIST, JOHNS HOPKINS UNIV, 69- *Concurrent Pos:* Mem epidemiol & dis control study sect, Div Res Grants, NIH, 72-76. *Mem:* Fel Am Statist Asn; Pop Asn Am; Am Soc Human Genetics; Am Pub Health Asn; Biomet Soc; Sigma Xi. *Res:* Application of statistics in medicine; biological sciences; epidemiology; human genetics; demography. *Mailing Add:* Sch Hyg & Pub Health Dept Biostat Johns Hopkins Univ 615 N Wolfe St Baltimore MD 21205

ABBEY, KATHLEEN MARY KYBURZ, b Oak Park, Ill, Feb 28, 50. ELECTROCHEMISTRY, LASER LIGHT SCATTERING. *Educ:* Rockford Col, BS, 72; Georgetown Univ, PhD(phys chem), 77. *Prof Exp:* Res fel laser light scattering, State Univ NY, Stony Brook, 76-80; res chemist, Lewis Res Ctr, NASA, Cleveland, 80-84; polymer chemist, Therm-O-Disc Div, Emerson Elec, 84-86, consult, 86-88; CONSULT, 88- *Concurrent Pos:* Co-ed, Membranes & Ionic & Elec Conducting Polymers, Electrochem Soc, 72; comput software consult, Malvern Sci, Inc, 80; tech adv, Shuttle Student Involvment Proj, NASA, 83-84; chmn, Ohio Sect, Electrochem Soc, 84-86. *Mem:* Electrochem Soc; Am Chem Soc; Am Phys Soc. *Res:* Laser light scattering and instrumentation with emphasis on computer interfacing; electrochemical storage systems and fuel cells; polymeric devices for energy storage and electrical switching; colloidal interaction in solution; physical chemistry; aeronautical and astronautical engineering. *Mailing Add:* PO Box 5527 Cary NC 27512-5527

ABBEY, KIRK J, b Monmouth, Ill, Aug, 19, 49. EMULSION POLYMERIZATION, CURE CHEMISTRY. *Educ:* Rockford Col, BA, 71; Johns Hopkins Univ, MS, 76, PhD(org chem), 77. *Prof Exp:* Trainee chem, Northern Regional Res Lab, US Dept Agr, 69-71; chemist, Glidden Coatings & Resins, Div SCM Corp, 76-86, Scientist, 86-91; STAFF SCIENTIST, LORD CORP, 91- *Concurrent Pos:* Chmn, Akron Polymer Lect Group, 83-84, chmn-elect, Polymer Group, NC Sect, Am Chem Soc, 90-91. *Mem:* Am Chem Soc. *Res:* Polymer chemistry and engineering with emphasis on oligomer synthesis and cure chemistry; emulsion polymerization. *Mailing Add:* PO Box 5527 Cary NC 27511

ABBEY, ROBERT FRED, JR, b Klamath Falls, Ore, May 27, 47; m 68; c 5. MARINE METEOROLOGY. *Educ:* Univ Ore, BS & BA, 69, MA, 71; Colo State Univ, MS, 72, PhD(atmospheric sci), 82; Univ Southern Calif, MPA, 83, DPA, 85. *Prof Exp:* Res assoc physics, Southwestern Ore Community Col, 66-67; res asst atmospheric sci, Colo State Univ, 69-72; sr environ scientist meteorol, Atomic Energy Comn, 72-75; dir, Meteorol Res Prog, US Nuclear Regulatory Comn, 75-83; DIR MARINE METEOROL RES PROG, OFF NAVAL RES, US NAVY, 83- *Concurrent Pos:* Mem, Standards Comn Tornadoes & Extreme Winds, Am Nuclear Soc, 76- & Fed Comn Meteorol Serv & Supporting Res, 79-83; Comt Wind Effects on Structures, Am Soc Civil Engrs, 79-85, Turbulence, 81-; vis prof, Univ Md, 81-83; chmn, Severe Local Storms, Am Meteorol Soc, 76-77, Fed Working Group Atmospheric Transport Models, 80-83; consult, Time Life, Nat Geographic, Univ Chicago Press, GEO Magazine; cert consult meteorologist, Am Meteorol Soc, 81-; adj prof, Montgomery Col, 83-; exec secy, Tropical Cyclone Steering Comt, Dept Defense, 85-; adjunct prof, Montgomery Col, Southern Univ, Univ Southern Calif. *Mem:* Fel Am Meteorol Soc; Am Soc Civil Engrs; AAAS. *Res:* Theoretical and empirical studies in atmospheric transport and diffusion of airborne effluents, severe storm (tornado, hurricane and lightning) characterization, probabilistic hazard modeling and climatology; author or coauthor of over 100 papers in technical journals, conference proceedings and chapters in books. *Mailing Add:* 5815 Ogden Ct Bethesda MD 20816

ABBOTT, ANDREW DOYLE, b Waterville, Wash, Nov 20, 13; c 3. PHYSICAL CHEMISTRY. *Educ:* Univ Wash, BS, 37, PhD(phys chem), 42. *Prof Exp:* Res chemist, Procter & Gamble Co, 42-46 & Calif Res Corp, 46-66; res chemist, Chevron Res Co, 66-70, sr res chemist, 70-77; RETIRED. *Mem:* Am Chem Soc. *Res:* Fatty oil chemistry; colloidal chemistry; petroleum chemistry. *Mailing Add:* 1665 NE 78th St Gresham CA 97030

ABBOTT, BERNARD C, b Yeovil, Eng, Oct 13, 20; m 44; c 1. BIOPHYSICS. *Educ:* Univ London, BSc, 41 & 48, PhD(biophys), 50. *Prof Exp:* Lectr biophys, Univ Col, Univ London, 46-53; prin sci officer, Marine Biol Asn, UK, 53-57; assoc res zoologist, Univ Calif, Los Angeles, 57-58, assoc prof zool, 58-62; prof biophys & physiol, Univ Ill, Urbana, 62-68; prof physiol, Sch Med, chmn, Dept Biol Sci, dir, Allan Hancock Found, 68-81, prof, 81-85, EMER PROF BIOL, UNIV SOUTHERN CALIF, 85- *Mem:* Am Physiol Soc; Biophys Soc; Soc Gen Physiol; Am Soc Zoologists; NY Acad Sci; Sigma Xi. *Res:* Muscle physiology; thermodynamics of bioelectric phenomena; toxins present in poisonous red tides. *Mailing Add:* 2025 Fair Oaks Ave South Pasadena CA 91030

ABBOTT, BETTY JANE, b Roanoke, Va, Oct 13, 31. CANCER, AIDS. *Educ:* Radford Col, BS, 52; Va Polytech Inst, MS, 55. *Prof Exp:* Instr high schs, Va, 52-57, chmn dept biol, 53-57; asst biol, Va Polytech Inst, 57-59; biologist, Va Agr Res Sta, Agr Res Serv, USDA, 59-60; biologist, 60-67, HEAD SCREENING SECT, BIOL TESTING BR, DEVELOP THERAPEUT PROG, NAT CANCER INST, 67- *Mem:* AAAS; Am Asn Cancer Res; NY Acad Sci. *Res:* Development and selection of new drugs for trial as potential cancer chemotherapeutic agents or AIDS; ways to increase effectiveness of agents of established clinical usefulness. *Mailing Add:* Biol Testing Br Nat Cancer Inst/FCRF Frederick MD 21702-1201

ABBOTT, DAVID MICHAEL, b Kansas City, Mo, Dec 29, 45; m 71; c 2. DENTISTRY, PERIODONTOLOGY. *Educ:* James Madison Univ, BS, 68; Univ NC, DDS, 72; Univ Minn, MS, 75. *Prof Exp:* Instr dent, Univ NC, 72-73; clin instr periodont, Univ Minn, 74-75; ASSOC PROF PERIODONT & OCCLUSION, MED COL, VA COMMONWEALTH UNIV, 75- *Mem:* Sigma Xi; Am Acad Periodont; Int Asn Dent; Am Dent Asn; Am Asn Dent Schs. *Res:* Microbiology of periodontal disease; biofeedback in dentistry; chronic facial pain. *Mailing Add:* Sch Dent Box 637 Va Commonwealth Univ Richmond VA 23298

ABBOTT, DEAN WILLIAM, b Oak Park, Ill, Nov 28, 63; US citizen; m 88. SIGNAL PROCESSING, CONTROL SYSTEMS. *Educ:* Rensselaer Polytechnic Institute, BS, 85; Univ Va, MAM(applied math), 87. *Prof Exp:* Assoc Engr, Sperry Marine, Sys Div, 86; RES SCIENTIST, BARRON ASSOC, INC, 87- *Res:* Extentions of traditional calculus of variations to optimally guide tactical weapons to fixed or moving targets. *Mailing Add:* 501 B Harris Rd Charlottesville VA 22903

ABBOTT, DONALD CLAYTON, b Lincoln, Kans, Nov 11, 23; m 45; c 3. BIOCHEMISTRY, CEREAL CHEMISTRY. *Educ:* Kans State Univ, BS, 49, MS, 51, PhD, 64. *Prof Exp:* Asst to head exp baking lab, Fleischman Labs, Standard Brands, Inc, NY, 51; cereal technologist, Soft Wheat Qual Lab, Ohio Agr Exp Sta, 51-54; PROF BIOCHEM, OKLA STATE UNIV, 54- *Mem:* Am Asn Cereal Chem; Am Chem Soc. *Res:* Baking quality of wheat varieties; immunological properties of wheat proteins; development of wheat proteins during maturation. *Mailing Add:* Dept Biochem Okla State Univ Stillwater OK 74074

ABBOTT, DOUGLAS E(UGENE), b Glendale, Calif, Apr 20, 34; m 56; c 4. FLUID MECHANICS, APPLIED MATHEMATICS. *Educ:* Stanford Univ, BS, 56, MS, 57, PhD(mech eng), 61. *Prof Exp:* Asst head fluid mech sect, Vidya Div, Itek Corp, 60-64; from assoc to prof fluid mech, Purdue Univ, 64-77, dir, Thermal Sci & Propulsion Ctr, 72-77; prof mech eng & chmn dept, Lehigh Univ, 77-85; ASSOC VCHANCELLOR FOR COMPUT & INFO SYSTS, UNIV MASS, AMHERST, 85- *Concurrent Pos:* Lectr mech eng, Stanford Univ, 63-64; staff consult, Midwest Appl Sci Corp, 64-72, Energy Controls Div, Bendix Corp, 67-75 & Westinghouse Elec Corp, 70-75; hon res fel, Sci Res Coun, UK, 71-72; chmn air breathing propulsion adv comt, Air Force Off Sci Res, 73-83; dir, Comput-Aided Design/Comput Aided Mfg Educ Prog & chmn, Comt Interactive Graphics Educ & Res, Lehigh Univ, 81-83; vice provost, Comput & Info Serv, Lehigh Univ, 83-85. *Mem:* NY Acad Sci; Am Inst Aeronaut & Astronaut; Am Soc Mech Engrs; Am Soc Eng Educ; AAAS; fel Am Asn Adv Sci; fel Am Phys Soc. *Res:* Analytical fluid mechanics; subsonic and supersonic viscous flows; inviscid transonic and hypersonic flows; experimental and theoretical investigations of fluid mechanics of biological systems. *Mailing Add:* Off Exec VChancellor Univ Mass 239 Whitmore Amherst MA 01003

ABBOTT, EDWIN HUNT, b New York, NY, Dec 28, 41; m 64; c 2. INORGANIC CHEMISTRY. *Educ:* Tufts Univ, BS, 64; Tex A&M Univ, PhD(chem), 69. *Prof Exp:* Fel chem, Tex A&M Univ, 69-70; from asst prof to assoc prof, Hunter Col, 74-77; PROF CHEM & HEAD DEPT, MONT STATE UNIV, 77- *Concurrent Pos:* Mem doctoral fac, City Univ New York, 70-77. *Mem:* Am Chem Soc; fel NY Acad Sci; Sigma Xi. *Res:* Coordination chemistry; bioinorganic chemistry; nuclear magnetic resonance. *Mailing Add:* Dept of Chem Mont State Univ Bozeman MT 59717

ABBOTT, FRANK SIDNEY, animal physiology, for more information see previous edition

ABBOTT, HERSCHEL GEORGE, b Woodstock, Maine, May 4, 21; m 43. FORESTRY. *Educ:* Univ Maine, BS, 43; Harvard Univ, MF, 52, MA, 59. *Prof Exp:* Forester, Chadbourne Lumber Co, Maine, 46-50; from instr to assoc prof, 53-70, PROF FORESTRY, UNIV MASS, AMHERST, 70- *Concurrent Pos:* Vpres & treas, Rich Tree Farms & Forestry Corp, 64- *Mem:* Soc Am Foresters. *Res:* Silviculture. *Mailing Add:* 78 or 79 Montague Rd Leverett MA 01054

ABBOTT, ISABELLA AIONA, b Hana, Hawaii, June 20, 19; m 43; c 1. PHYCOLOGY, MARINE BIOLOGY. *Educ:* Univ Hawaii, AB, 41; Univ Mich, MS, 42; Univ Calif, Berkeley, PhD(bot), 50. *Prof Exp:* Lectr bot, Univ Hawaii, 43-46; res assoc & lectr phycol, 60-70, prof biol, Stanford Univ, 71-82; WILDER PROF BOT, UNIV HAWAII, 78- *Honors & Awards:* Darbaker Award, Bot Soc Am, 69. *Mem:* Int Phycol Soc (treas, 67-71); Phycol Soc Am; Bot Soc Am; Brit Phycol Soc; Sigma Xi; fel AAAS. *Res:* Life histories, development and systematics of marine red algae, particularly Pacific marine algae. *Mailing Add:* Dept Bot Univ Hawaii 3190 Malie Way Honolulu HI 96822

ABBOTT, JAMES H, b San Miguel, NMex, Aug 19, 24. MATHEMATICS STATISTICS, INFORMATION THEORY. *Educ:* Univ Colo, BS, 46; SMU, MS, 49, Univ Ill, PhD(math), 59. *Prof Exp:* Instr math, SMU, 47-49; instr math, Tex A&M Univ, 49-51; asst prof math, Purdue Univ, 57-60; assoc prof math, NMex Univ, 60-64; prof math, Univ New Orleans, 65-84, chmn dept, 65-67; RETIRED. *Mem:* Am Math Soc; Am Math Asn; Inst Elec & Electronics Engrs. *Mailing Add:* 2334 New York St New Orleans LA 70122

ABBOTT, JOAN, b Newton, Mass, Aug 5, 32. DEVELOPMENTAL BIOLOGY. *Educ:* Conn Col, BA, 54; Wash Univ, MA, 57; Univ Pa, PhD(develop biol), 65. *Prof Exp:* Asst prof biol, Barnard Col, Columbia Univ, 65-67, asst prof dept biol sci, Columbia Univ, 67-72; assoc, Mem Sloan-Kettering Cancer Ctr, 72-83, asst prof biol, Sloan-Kettering Div, Grad Sch Med Sci, Cornell Univ, 73-84; sr res scientist, 83-84, RES DIR, ELECTRO-BIOL INC, 84- *Mem:* AAAS; Am Soc Cell Biol; Soc Develop Biol; NY Acad Sci. *Res:* Effects of electromagnetic fields on biological systems. *Mailing Add:* Electro Biol Inc Six Upper Pond Rd Parsippany NJ 07054

ABBOTT, JOHN S, b Denver, Colo, April 23, 52. GEOMETRY. *Educ:* Univ Colo, BA, 76; Univ Denver, PhD(math), 79. *Prof Exp:* SR PROJ ENGR, CORNING GLASS WORKS, 78- *Mailing Add:* Corning Glass Works HP ME3 A2 Corning NY 14831

ABBOTT, LAURENCE FREDERICK, b Toronto, Ont, May 15, 50; m 74; c 2. ELEMENTARY PARTICLE PHYSICS. *Educ:* Brandeis Univ, PhD(physics), 77. *Prof Exp:* Res assoc physics, Stanford Linear Accelerator Ctr, 77-79; ASST PROF PHYSICS, BRANDEIS UNIV, 79- *Concurrent Pos:* Elec engr, Honeywell Comput Div, Gilford Instruments, Teredyne Inc; sci assoc, Europ Orgn Nuclear Res, 80-81; Alfred P Sloan Found, fel, 83. *Mem:* Am Phys Soc. *Res:* Elementary particle physics and field theory including weak and strong interaction phenomenology, computational techniques in field theory, supersymmetry, weak interaction models, grand unified models and cosmology. *Mailing Add:* Dept of Physics Brandeis Univ Waltham MA 02154

ABBOTT, LYNN DE FORREST, JR, b Ithaca, NY, Nov 23, 13; m 40; c 3. BIOCHEMISTRY. *Educ:* Wayne Univ, BS, 36, MS, 37; Univ Mich, PhD(biochem), 40. *Prof Exp:* Asst chem, Wayne Univ, 35-36, asst biochem, 36-37; asst, Univ Mich, 37-40; res assoc, 40-43, from asst prof to prof, 43-76, chmn dept, 62-76, EMER PROF BIOCHEM, MED COL VA, 76- *Concurrent Pos:* Ed, Va J Sci, 68-73. *Mem:* AAAS; Am Chem Soc; Am Soc Biol Chemists; Soc Exp Biol & Med; Am Soc Clin Nutrit. *Res:* Blood chemistry and amino acid metabolism; enzyme chemistry; isotopic nitrogen. *Mailing Add:* 607 Horsepen Rd Richmond VA 23229

ABBOTT, MICHAEL MCFALL, b Niagara Falls, NY, Oct 3, 38; m 62; c 1. THERMODYNAMICS. *Educ:* Rensselaer Polytech Inst, BChE, 61, PhD(thermodyn), 65. *Prof Exp:* Chem engr, Esso Res & Eng Co, Standard Oil Co, NJ, 65-67, res engr, 67-69; res assoc fluid, chem & thermal properties, 69-76, ASSOC PROF CHEM ENG, RENSSELAER POLYTECH INST, 76- *Mem:* Am Inst Chem Engrs. *Res:* Thermodynamics and phase equilibria; conductivities of heterogeneous systems; electrochemical engineering. *Mailing Add:* Rensselaer Polytech Inst 204 RI Troy NY 12181

ABBOTT, MITCHEL THEODORE, b Los Angeles, Calif, June 6, 30; m 62; c 3. BIOCHEMISTRY. *Educ:* Univ Calif, Los Angeles, BS, 57, PhD(biochem), 63. *Prof Exp:* Fel biochem, Sch Med, NY Univ, 62-64; asst prof, 64-71, PROF BIOCHEM, SAN DIEGO STATE UNIV, 71- *Concurrent Pos:* Vis scientist, Roche Inst Molecular Biol, 72-73; NSF & NIH basic res grants. *Mem:* Am Soc Biol Chem; AAAS; Protein Soc. *Res:* Enzymology and regulation of pyrimidine metabolism; processing of small nuclear RNA. *Mailing Add:* Dept of Chem San Diego State Univ San Diego CA 92182

ABBOTT, NORMAN JOHN, b Toronto, Ont, Nov 14, 18; m 44; c 3. TEXTILE PHYSICS. *Educ:* Univ Toronto, BA, 40. *Prof Exp:* Elec engr, Amalgamated Elec Corp, 40-41; spec instr armed forces, dept physics, Univ Toronto, 41-43; tech writer, Radar Div, Res Enterprises Ltd, 43-45; fel, Textile Dept, Ont Res Found, 45-47; from asst dir to assoc dir appl physics, Albany Int Res Co, 57-87; CONSULT, 88- *Honors & Awards:* Harold DeWitt Smith Mem Medal, Am Soc Testing & Mat, 87- *Mem:* Fiber Soc; fel Brit Textile Inst; Sigma Xi; Tech Asn Pulp & Paper Indust. *Res:* Mechanical properties of textile structures; behavior of textiles under extreme conditions; coated fabric mechanics; characterization, utilization of high temperature, high performance, nonflammable fibers; design of mechanical structures containing such fibers. *Mailing Add:* 102 Blake St Needham MA 02192

ABBOTT, PATRICK LEON, b San Diego, Calif, Sept 30, 40; div; c 3. SEDIMENTOLOGY, ENVIRONMENTAL GEOLOGY. *Educ:* San Diego State Col, BS, 63; Univ Tex, Austin, MA, 65, PhD(geol), 73. *Prof Exp:* Lectr geol, 68-69, from asst prof to assoc prof, 71-76, PROF GEOL, SAN DIEGO STATE UNIV, 76-, CHMN, 80- *Concurrent Pos:* vis prof, Univ Canterbury, NZ, 84. *Mem:* AAAS; Geol Soc Am; Soc Econ Paleontologists & Mineralogists; Am Asn Petrol Geologists; Int Asn Sedimentologists. *Res:* Interaction of humankind and the earth; sedimentology and tectonics of Mesozoic and Cenozoic rocks, southwestern United States and northwestern Mexico; geologic hazards. *Mailing Add:* Dept Geol Sci San Diego State Univ San Diego CA 92182

ABBOTT, RICHARD NEWTON, JR, b Newton, Mass, Jan 12, 49; m 71; c 3. CRYSTAL-CHEMISTRY. *Educ:* Bowdoin Col, BA, 71; Univ Maine, Orono, MSc, 73; Harvard Univ, PhD(geol), 77. *Prof Exp:* Fel, Dalhousie Univ, 77-79; from asst prof to assoc prof, 79-89, PROF GEOL, APPALACHIAN STATE UNIV, 89- *Concurrent Pos:* Postdoctoral Fel, Harvard Univ, 87. *Honors & Awards:* Hawley Award, Mineral Asn Can, 79. *Mem:* Mineral Asn Can; Mineral Soc Am; Sigma Xi; Am Geophys Union. *Res:* Crystal-chemistry; phase relationships of igneous and metamorphic rocks. *Mailing Add:* Dept Geol Appalachian State Univ Boone NC 28607

ABBOTT, ROBERT CLASSIE, b Syracuse, NY, June 7, 26; m 62; c 2. COMPUTER SCIENCES, STATISTICAL ANALYSIS. *Educ:* Syracuse Univ, BEE, 50, MS, 58, PhD(physics), 59. *Prof Exp:* Elec engr, Wright Air Develop Ctr, Ohio, 50 & Navy Ord Div Eastman Kodak Co, NY, 50-51; asst physics, Syracuse Univ, 51-59; res physicist, Cornell Aeronaut Lab, Inc, 59-61, prin physicist, 61-62, head solid state physics sect, 62-66; assoc prof fac eng & appl sci, State Univ NY Buffalo, 66-73; vpres res, Multitech Res & Invest Corp, 73-78; res asst prof, Biophys Dept, State Univ NY Buffalo, 78-82; TECH STAFF, MITRE CORP, BOSTON, 84- *Mem:* Am Phys Soc; Am Asn Artificial Intel; Soc Comput Simulation. *Res:* Surface physics and biophysics; ion-metal surface interactions; metallic adsorption systems; macromolecular information storage mechanisms; digital signal processing for biological and economic systems; artificial intelligence-expert systems techniques applied to decision aids and simulatios. *Mailing Add:* Mitre Corp Mail Stop L206A, Burlington Rd Bedford MA 01730

ABBOTT, ROBINSON S, JR, b Philadelphia, Pa, June 23, 26; m 56; c 4. ECOLOGY. *Educ:* Bucknell Univ, BS, 49; Cornell Univ, PhD(biol), 56. *Prof Exp:* Teacher pub sch, LI, 50-51; asst bot, Cornell Univ, 53; asst, Wiegand Herbarium, 53-54, econ bot, 55-56; instr bot, Smith Col, 56-61; from asst prof to assoc prof biol, 61-71, chmn div sci & math, 64-68, PROF BIOL, UNIV MINN, MORRIS, 71- *Concurrent Pos:* Vis researcher, Padre Island Nat Seashore, Tex, 81-82. *Mem:* Bot Soc Am; Ecol Soc Am; Phycol Soc Am; Int Phycol Soc. *Res:* General ecology, especially beach and salt marsh ecology; use of filamentous algae as ecological indicators; environmental effects of the growth and development of algae. *Mailing Add:* 205 E Sixth St Morris MN 56267

ABBOTT, ROSE MARIE SAVELKOUL, b Chaska, Minn, Feb 7, 31; m 56; c 4. PLANT MORPHOLOGY. *Educ:* Univ Minn, BA, 53, MS, 55; Cornell Univ, PhD(plant morphol), 59. *Prof Exp:* Instr biol, Smith Col, 57-58; instr plant morphol, 64-66, instr genetics, 66-77, ASST PROF BIOL, UNIV MINN, MORRIS, 77- *Mem:* Bot Soc Am. *Res:* Developmental anatomy of primary xylem elements; development of shoot apices; developmental anatomy of Berberidaceae. *Mailing Add:* Dept Biol Univ Minn Morris MN 56267

ABBOTT, SETH R, b Brooklyn, NY, Mar 14, 44; c 2. CHEMICAL INSTRUMENTATION, SPECTROSCOPY. *Educ:* Brooklyn Col, BS, 64; Mass Inst Technol, PhD(chem), 69. *Prof Exp:* Res chemist anal & polymer chem, Memorex Corp, 69-72; sr res chemist, 72-79, MGR RES, VARIAN ASSOCS, 79- *Concurrent Pos:* Lectr liquid chromatography, Am Chem Soc. *Mem:* Am Chem Soc. *Res:* Liquid chromatography; emission and Raman spectroscopy; mass spectroscopy; electrochemistry; artificial intelligence. *Mailing Add:* 5308 Lightwood Dr Concord CA 94521

ABBOTT, STEPHEN DUDLEY, b Davidson, NC, Oct 8, 64. OPERATIONS RESEARCH. *Educ:* Colgate Univ, BA, 86; Va Univ, MA, 88. *Prof Exp:* PROF MATH, UNIV VA, 87- *Mem:* Am Math Asn. *Mailing Add:* 1517-B Amherst St Charlottesville VA 22901

ABBOTT, THOMAS B, b Pa, June 27, 21; m 45; c 2. SPEECH PATHOLOGY. *Educ:* Muskingum Col, BA, 43; Case-Western Reserve Univ, MA, 48; Univ Fla, PhD, 57. *Prof Exp:* Asst prof speech & hearing, St Cloud State Col, 49-53; instr, Univ Fla, 54-57; dir speech ther prog, Los Angeles County Soc Crippled Children, 57-58; from assoc prof to prof speech path, Baylor Univ, 58-63; dir speech clin, 58-63; asst prof speech path & head speech & hearing clin, 63-66, assoc prof speech path & dir div, 66-78, prof & chmn dept speech, Univ Fla, 78-86; consult, Kuwait Univ, Kuwait, 88-89; RETIRED. *Concurrent Pos:* Lectr, Univ Southern Calif, 57-58. *Mem:* Fel Am Speech-Lang-Hearing Asn; Speech Asn Am. *Res:* Stuttering; language. *Mailing Add:* Dept Speech Univ Fla Gainesville FL 32611-2005

ABBOTT, THOMAS PAUL, b Akron, Ohio, Sept 26, 42; m 61; c 3. POLYMER CHEMISTRY. *Educ:* Univ Akron, BS, 64, MS, 68, PhD(polymer chem), 72. *Prof Exp:* Chemist polymer anal, Cent Res, Gen Tire & Rubber Co, 64-69; res chemist elastomers & plastics, 71-80, res chemist natural polymers, 81-85, RES CHEMIST AGR PROD, USDA, 85- *Concurrent Pos:* Lectr, Ill Cent Col, 87. *Mem:* Am Chem Soc. *Res:* Lignin chemistry and natural products chemistry; detoxification of oilseed meals. *Mailing Add:* 2016 Del Ray Peoria IL 61614

ABBOTT, URSULA K, b Chilliwack, BC, July 30, 27; m 54; c 2. DEVELOPMENTAL GENETICS. *Educ:* Univ BC, BSA, 49, MSA, 50; Univ Calif, PhD(genetics), 55. *Prof Exp:* Asst poultry, Univ BC, 48-50; res asst poultry genetics, Univ Calif, Berkeley, 50-54; prin lab tech, 54-55, lectr & asst specialist, 55-56, instr & jr poultry geneticist, 56-57, from asst prof & asst geneticist to assoc prof & assoc geneticist, 57-67, PROF AVIAN SCI & GENETICIST, UNIV CALIF, DAVIS, 67- *Concurrent Pos:* Guggenheim fel, 60-61; prof dir develop biol, NSF, 68-69; Hon prof, Univ BC, 67. *Mem:* Am Genetic Asn; Poultry Sci Asn; Int Soc Develop Biol; Soc Develop Biol; Am Soc Naturalists. *Res:* Developmental genetics; experimental embryology and teratology. *Mailing Add:* Dept Avian Sci Univ Calif Davis CA 95616

ABBOTT, WILLIAM ASHTON, b Apr 28, 51; m; c 4. CHEMICAL BIOLOGY, BIOCHEMICAL TOXICOLOGY. *Educ:* Univ Rochester, PhD(toxicol), 81. *Prof Exp:* Instr biochem, Med Col, Cornell Univ, 81-85; BIOLOGIST, ABBOTT LABS, 86- *Mem:* Am Soc Biochem & Molecular Biol. *Res:* Biochemistry. *Mailing Add:* Dept 90L Abbott Lab Bldg AP20 Abbott Park IL 60064

ABBOTT, WILLIAM HAROLD, b Atlanta, Ga, Mar 23, 44; m 69; c 1. MICROPALEONTOLOGY, OCEANOGRAPHY. *Educ:* Ga State Univ, Atlanta, BS, 69; Northeast La Univ, MS, 70; Univ SC, Columbia, PhD(geol), 72. *Prof Exp:* Asst prof geol & oceanog, Univ SC, Beaufort, 72-74; staff geologist, SC State Geol Surv, 74-80; geologist, 80-86, PALEONT ADV, MOBIL EXPLOR & PROD SERV, 86- EXPLOR & PROD SERV, 86- *Concurrent Pos:* Geol Soc Am Penrose Bequest grant, 73; adj prof, Univ SC, Columbia & researcher, Belle W Baruch Marine Inst, 73-74. *Mem:* Am Inst Prof Geol; Am Asn Petrol Geologist; Paleont Res Inst. *Res:* Stratigraphic relationships of marine Cenozoic sediments utilizing fossil diatoms and nannoplankton Gulf Coast US. *Mailing Add:* McAlister CV No R Mabank TX 75147

ABBOTT, WILLIAM M, b San Francisco, Calif, Apr 14, 36. VASCULAR SURGERY. *Educ:* Stanford Univ, MD, 61. *Prof Exp:* ASSOC PROF SURG, HARVARD UNIV, 76-; CHIEF VASCULAR SURG, MASS GEN HOSP, 78- *Mem:* Am Col Surgeons; Am Surg Asn; Soc Vascular Surg. *Mailing Add:* Dept Surg Harvard Med Sch 25 Shattuck St Boston MA 02114

ABBOTT, WILTON R(OBERT), b Campbell, Calif, Jan 19, 16; m 38; c 3. ELECTRICAL ENGINEERING, SYSTEMS ENGINEERING. *Educ:* Univ Calif, BS, 37; Iowa State Col, MS, 42, PhD(elec eng), 45. *Prof Exp:* Asst instruction & res, Stanford Univ, 37-38; asst eng, Remler Co, Ltd, Calif, 38-39 & Gen Elec Co, 39-40; from instr to assoc prof elec eng, Iowa State Col, 40-47; asst prof, Univ Calif, 47-52; sr res engr & res specialist, NAm Aviation Co, 52-57; mgr satellite syst integration, Missile & Space Systs Div, 57-63, consult spacecraft scientist, 63-68, spec asst to vpres & gen mgr, Space Systs Div, 68-74, SR CONSULT ENGR, LOCKHEED AIRCRAFT CORP, 74- *Concurrent Pos:* Asst prof continuing educ, San Jose State Univ; trustee, Linfield Col, McMinnville, Ore. *Mem:* Inst Elec & Electronics Engrs. *Res:* Transient and steady state analysis of electrical circuits containing vacuum tubes; numerical machine tool control; spacecraft systems. *Mailing Add:* 27899 Via Ventana Los Altos Hills CA 94022

ABBOUD, FRANCOIS MITRY, b Cairo, Egypt, Jan 5, 31; US citizen; m 55; c 4. INTERNAL MEDICINE. *Educ:* Ain Shams Univ, Cairo, MB, BCh, 55; Am Bd Internal Med, dipl, 64, Am Bd Cardiovascular Dis, dipl, 72. *Prof Exp:* Intern, Demerdash Govt Hosp, Cairo, 55; resident, Milwaukee County Hosp, Wis, 55-58; res assoc & instr med & cardiol, Marquette Univ, 58-60; from asst prof to assoc prof internal med, 61-68, dir cardiovasc div, Dept Internal Med, 70-76, PROF MED, UNIV IOWA, 68-, PROF PHYSIOL & BIOPHYS & DIR CARDIOVASC CTR, 75-, CHMN DEPT INTERNAL MED, 76-, EDITH KING PEARSON PROF CARDIOVASC RES, 88- *Concurrent Pos:* Am Heart Asn res fel, 58-60, adv res fel, 60-62; adv res fel cardiol, Cardiovasc Res Labs, Univ Iowa, 60-61; USPHS res career develop award, 62-71; chmn, Coun on Circulation, Am Heart Asn, 77-80; ed, Circulation Res, 81-86; pres, Cent Soc Clin Res, 85-86; chmn, Syntex Scholars Prog Adv Comt, 84-90; bd mem, Am Bd Internal Med, 87-93. *Honors & Awards:* Award for Exp Therapeut, Am Soc for Pharmacol & Exp Therapeut, 72; Wiggers Award, Cardiovascular Section, Am Physiol Soc. *Mem:* Nat Acad Sci Inst Med; Am Fedn Clin Res (pres, 71-72); Am Heart Asn (pres, 90-91); AMA; Am Physiol Soc; Soc Exp Biol & Med; Am Soc Clin Invest; fel Am Col Physicians; Asn Univ Cardiologists; Asn Am Physicians (treas, 79-84, pres, 90-91). *Res:* Cardiovascular physiology and pharmacology; peripheral circulation; vascular reactivity. *Mailing Add:* Dept Internal Med Univ Iowa Iowa City IA 52242

ABBRECHT, PETER H, b Toledo, Ohio, Nov 27, 30; m 57; c 2. PHYSIOLOGY, PULMONARY MEDICINE. *Educ:* Purdue Univ, BS, 52; Univ Mich, MS, 53, PhD(chem eng), 57, MD, 62. *Prof Exp:* Sr chem engr, Minn Mining & Mfg Co, 56-58; intern med, Univ Calif Hosp, Los Angeles, 62-63; from asst prof to prof physiol, Univ Mich, Ann Arbor, 63-80, resident internal med, 70-72, chmn bioeng prog, 72-78, prof internal med, 75-80; PROF PHYSIOL & MED, UNIFORMED SERV UNIV HEALTH SCI, 80-, CHMN PHYSIOL, 77- *Concurrent Pos:* NIH career develop award, 69-73; vis prof bioeng, Univ Calif, San Diego, 73; fel pulmonary dis, Med Ctr, Univ Mich, Ann Arbor, 74-75; instr, Wayne State Univ, 57-58; actg dir, Physiol & Biomed Eng Prog, Nat Inst Gen Med Sci, NIH, 77-78. *Mem:* Am Physiol Soc; fel Am Col Physicians; fel Am Col Chest Physicians; Biomed Eng Soc; Am Thoracic Soc. *Res:* Cardiopulmonary physiology and modeling; use of systems theory in medical diagnosis and therapy. *Mailing Add:* Dept Physiol & Med Uniformed Serv Univ Health Sci 4301 Jones Bridge Rd Bethesda MD 20814-4799

ABBUNDI, RAYMOND JOSEPH, b New York, NY, Aug 14, 49. LUMINESCENCE, MAGNETO-ELASTIC EFFECTS. *Educ:* Am Univ, BS, 71, MS, 73, PhD(physics), 76. *Prof Exp:* Res asst physics, Am Univ, 73-75; res physicist, Solid State Br, 76-85, RES PHYSICIST, NUCLEAR BR, US NAVAL SURFACE WEAPONS CTR, 85- *Concurrent Pos:* Lectr physics, Am Univ, 73-74; NAS-Nat Res Coun res assoc, US Naval Surface Weapons Ctr, 76-77. *Res:* Magnetic and magnetoelastic properties of rare earth compounds; magnetostriction of rare earth-iron intermetallic compounds; luminescence and optical properties of phosphors for personnel dosimetry. *Mailing Add:* 10103 Ridgeline Dr Gaithersburg MD 20879

ABDALI, SYED KAMAL, b Patna, India, Mar 10, 40; US citizen; m 65; c 2. COMPUTER SCIENCES. *Educ:* Am Univ Beirut, BE, 63; Univ Montreal, MSc, 68; Univ Wis-Madison, PhD(comput sci), 74. *Prof Exp:* Lectr elec eng, Govt Eng Col, Karachi, Pakistan, 63-64; jr engr, Tel Indust Pakistan, Haripur, 64-65; design engr, Int Tel & Tel, Montreal, 65-66; consult, All-Tech Assocs, Montreal, 66-67; teaching asst comput sci, Univ Wis, 68-71; asst prof, NY Univ, 71-73; from asst prof to assoc prof math sci, Rensselaer Polytech Inst, 73-83; prin computer scientist, Tektronic Computer Res Lab, 83-88; PROG DIR, NAT SCI FOUND, 88- *Mem:* Asn Comput Mach; Asn Symbolic Logic. *Res:* Semantics of programming languages; verification methods for programs and machine designs; combinatory logic; computer algebra. *Mailing Add:* Nat Sci Found 1800 G St NW Washington DC 20550

ABDALLAH, ABDULMUNIEM HUSEIN, b Nov 29, 29; US citizen; m 65; c 2. PHARMACOLOGY, BIOCHEMISTRY. *Educ:* Am Univ Cairo, BS, 57; Univ Ky, BS, 62; Ohio State Univ, MS, 64, PhD(pharmacol), 67. *Prof Exp:* Teaching chem, Teacher's Training Col, Libya, 57-59; RES ASSOC PSYCHOPHARMACOL, DOW CHEM CO, 67- *Mem:* Soc Neurosci; Am Soc Exp Biol. *Res:* Psychopharmacology; antidepressants; tranquilizers; appetite suppressants; analgesics; cardiovascular; pharmacology. *Mailing Add:* 10618 E 300 S Zionsville IN 46077

ABDEL-BASET, MAHMOUD B, b 1944. COAL CHEMISTRY, PETROLEUM CHEMISTRY. *Educ:* Cairo Univ, BS, 65; Ain Shams Univ, MS, 69; Univ Tex, PhD(chem), 73. *Prof Exp:* Res asst petrol chem, Nat Res Ctr, Cairo, 66-69; teaching asst, Univ Tex, 69-71; res assoc coal sci, Pa State Univ, 73-77; res chemist, Allied Chem Corp, 77-80; staff chemist coal res, Exxon Res & Eng, 80-; AT KUWAIT INST SCI RES. *Mem:* Am Chem Soc. *Res:* Physical organic chemistry; hydrocarbons and carbonium ions reactions and mechanism; coal chemistry and coal conversion processes. *Mailing Add:* 20641 Camino De Los Cantos Yorba Linda CA 92686

ABDEL-GHAFFAR, AHMED MANSOUR, b Egypt, Apr 30, 47; US citizen; m 75; c 3. EARTHQUAKE ENGINEERING, STRUCTURAL DYNAMICS. *Educ:* Fac Eng, Cairo Univ, BSc, 70; Calif Inst Technol, MSc, 73, PhD(civil eng), 76. *Prof Exp:* Instr structural eng, Fac Eng, Cairo Univ, 70-72; res fel civil eng, Calif Inst Technol, 76-78; asst prof, Univ Ill, Chicago Circle, 78-79; from asst prof to assoc prof struct eng, Princeton Univ, 79-88, prof civil eng, 84-86; PROF & CO-DIR CTR RES EARTHQUAKE & CONSTRUCT ENG, UNIV SOUTHERN CALIF, 88- *Concurrent Pos:* Structural engr, Bakhoum & Moharram Consult, Egypt, 71-72; assoc ed, Intern J Soil Dynamics & Earthquake Eng, 82-86; mem, Wind Eng Res Coun; vis prof struct & earthquake eng, Kyoto Univ, Japan, 86. *Honors & Awards:* Raymond C Reese Res Prize & Thomas A Middlebrooks Award, Am Soc Civil Engrs, 82 & Huber Res Award, 89; Rheinstein Fac Res Award, Princeton Univ, 81, Norman Sollenberger Fac Award, 85. *Mem:* Am Soc Civil Engrs; Am Acad Mech; Earthquake Eng Res Inst; Int Soc Soil Mech & Found Eng; Int Asn Bridge Structure & Structural Eng; Int Asn Safety & Reliability Structs; fel Japan Soc Prom Sci. *Res:* Structural dynamics and earthquake-response analysis and design of large civil engineering structures such as long-span cable supported bridges, buildings, and earth dams and the full scale dynamic testings of these structures. *Mailing Add:* Civil Eng Dept Univ Southern Calif Los Angeles CA 90089-2531

ABDEL-HADY, M(OHAMED) A(HMED), b Cairo, Egypt, Feb 15, 34; m 66; c 3. CIVIL & HIGHWAY ENGINEERING. *Educ:* Ain-Shams Univ, Cairo, BS, 56; Univ Ill, MS, 61, PhD(civil eng), 63. *Prof Exp:* Res engr, Bldg Res Ctr, Cairo, 56-57; instr civil eng, Col Eng, Ain-Shams Univ, Cairo, 57-59; from asst prof to assoc prof, 63-71, PROF CIVIL ENG, OKLA STATE UNIV, 71- *Concurrent Pos:* Spec lectr, US Army Corps Eng, 66-68, consult, 67-; dir, NSF-Okla State Univ Remote Sensing Ctr, Cairo, Egypt, 72-75; bd adv, Pres Sadat, Egypt, 79-80; pres, Acad Sci Res, Cairo, Egypt, 86-88; consult, Wibur Smith & Assoc, US AID, Cairo proj, World Bank & Cairo Govt for Transp Studies; co-chmn, Int Conf Remote Sensing, Cairo, 81; auth & co-auth of four bks in civil eng, remote sensing & outer space. *Mem:* Soc Am Mil Engrs; Am Soc Civil Engrs; Am Soc Photogram. *Res:* Highway and transportation engineering; aerial photographic interpretation and remote sensing; engineering geology and rock engineering. *Mailing Add:* Sch Civil Eng Okla State Univ Stillwater OK 74074

ABDEL-KHALIK, SAID IBRAHIM, b Alexandria, Egypt, Aug 9, 48. NUCLEAR ENGINEERING, MECHANICAL ENGINEERING. *Educ:* Alexandria Univ, Egypt, BS, 67; Univ Wis-Madison, MS, 71, PhD(mech eng), 73. *Prof Exp:* Instr mech eng, Alexandria Univ, Egypt, 67-69; vis asst prof chem eng, Univ Wis-Madison, 74-75; sr engr nuclear eng, Babcock & Wilcox Co, 75; from asst prof to assoc prof, 76-82, prof nuclear eng, Univ Wis-Madison, 82-87; GEORGIA POWER DISTINGUISHED PROF, NUCLEAR ENG, GEORGIA INST TECHNOL, 87-, ASSOC DIR, SCH MECH ENG, 90- *Concurrent Pos:* Fel, Rheol Res Ctr, Univ Wis-Madison, 73-74 & Solar Energy Lab, 74-75; consult, Los Alamos Sci Lab, 77, EG&G Idaho, 77-, Elec Power Res Inst, 81, Dept Energy, Sandia Labs, 81-82, adv comt Reactor Safeguards, 82-83 & Kewaunee Nuclear Power Plant, 83-, United Energy Serv Corp, 90 & Westinghouse Savannah River Co, 90-; vis res scientist, Nuclear Res Ctr, Karlsruhe, WGer, 79; invited prof, Ecole Polytechnique Federale de Lausanne, Switzerland. *Mem:* Am Inst Physics; Am Nuclear Soc; Solar Energy Soc; Combustion Inst; Sigma Xi. *Res:* Fission reactor operations; fission reactor safety; transport phenomena. *Mailing Add:* Nuclear Eng Prog Ga Inst Technol Atlanta GA 30332

ABDELLAH, FAYE G, b New York, NY. MEDICAL ADMINISTRATION. *Educ:* Columbia Univ, BS, 45, MA, 47, EdD, 55. *Hon Degrees:* Var from US univs, 67-90. *Prof Exp:* Nurse consult, Div Nursing Resources, Bur State Serv, USPHS, 49-54, sr consult, 55-58, asst chief, Res Grants Br, 60-61, chief, Bur Health Manpower Educ, NIH, 61-68, dir, Off Res Training, 69-70, asst surgeon gen & chief nurse officer, 70-87, dep surgeon gen, 82-89; RETIRED. *Concurrent Pos:* Consult, WHO, 74, Portuguese Govt, 76, Tel Aviv Univ, 76, Japanese Nursing Assoc; mem, Coun Cardiovasc Nursing, Am Heart Asn & Subcomt Preparing & Estab Criteria Res; bd trustees, Am Nurses' Found; mem bd regents, Nat Libr Med; vis prof, Univ Colo, Univ Wash, Univ Minn; dir, Off Long Term Care, Off Asst Secy Health, USPHS, HEW, 76-77, chief adv, Long Term Care Policy to dep asst secy health policy, res & statist, 77-79, health planning & eval, 79-80. *Honors & Awards:* Nat Defense Serv Medal, 70; Nickolai Ivanovich Pirogov Medal, Acad Sci, USSR, 70; Distinguished Serv Award, Am Pub Health Asn, 78. *Mem:* Inst Med-Nat Acad Sci; AAAS; Am Heart Asn; Am Acad Nursing; NY Acad Sci; Am Pub Health Asn; fel Am Psychol Asn; Am Nurses' Asn; Geront Soc. *Res:* Published numerous articles in various journals. *Mailing Add:* 3713 Chanel Rd Annandale VA 22003

ABDEL-LATIF, ATA A, b Beitunia, Ramallah, Palestine, Jan 22, 33; US citizen; m 57; c 4. BIOCHEMISTRY, NEUROCHEMISTRY. *Educ:* DePaul Univ, BS, 55, MS, 58; Ill Inst Technol, PhD(biochem, physiol), 63. *Prof Exp:* Control chemist, Ninol Chem Labs, 55-56; med res assoc, State of Ill Pediat Inst, 63-67; from assoc prof to prof, 67-87, REGENTS PROF, MED COL GA, 87- *Concurrent Pos:* Res assoc psychiat, Col Med, Univ Ill, 63-67; NIH grants, 65-; vis prof, Univ Nottingham, 75-76. *Honors & Awards:* Merit Award, NIH, 90; Vision Res Award, Alcon Res Inst, 90. *Mem:* AAAS; Am Soc Biol Chem; Am Physiol Soc; Int Soc Eye Res; Am Soc Neurochem; Int Soc Pharmacol Exp Ther; Int Soc Neurochem. *Res:* Experimental therapeutics; role of phosphoinositides in adrenergic and cholinergic muscarinic receptor function in nerve and muscle; glycerolipids and prostaglandin synthesis in ocular tissues. *Mailing Add:* Dept Biochem & Molecular Biol Med Col Ga Augusta GA 30912-2100

ABDELMALEK, NABIN N, b Egypt, 29. MATHEMATICS. *Educ:* Cairo Univ, BSc, 54; Manchester Univ, PhD(theoret physics), 58. *Prof Exp:* Teacher, Cairo Univ, 52-65; researcher, Comput Ctr, Bell Northern Res, 65-67; researcher numerical anal technol, 67-76; mem staff, Nat Res Coun Can, 76-90; RETIRED. *Mem:* Am Math Soc; Can Math Soc; Can Image Processing Soc. *Mailing Add:* 788 E Vale Dr Ottawa ON K1J 7A1 Can

ABDEL-MONEM, MAHMOUD MOHAMED, b Cairo, Egypt, June 7, 38; m 64. MEDICINAL CHEMISTRY. *Educ:* Cairo Univ, BS, 59; Univ Minn, Minneapolis, PhD(med chem), 66. *Prof Exp:* Instr plant chem, Cairo Univ, 60-61; head res, Nile Co Pharmaceut, Cairo, 65-68; res specialist med chem, Univ Minn, Minneapolis, 68-70; asst prof, Univ Ill Med Ctr, Chicago, 70-71; asst prof, 71-75, assoc prof, 75-80, PROF MED CHEM & ASST DEAN, COL PHARM, UNIV MINN, MINNEAPOLIS, 80- *Concurrent Pos:* Assoc ed, J Med Chem, 72-; consult, Zinpro Corp, Excelsior, Minn, 72- *Res:* Study of distribution and metabolism of drugs; structure activity correlation and chemical mechanisms in drug metabolism. *Mailing Add:* Wash State Univ Pullman WA 99164

ABDELNOOR, ALEXANDER MICHAEL, b New York, NY, Jan 18, 41; m 70; c 3. CELLULAR ANTIGENS, TISSUE TYPING. *Educ:* Am Univ Beirut, BS, 64, MS, 66; Univ Mich, PhD(microbiol), 69. *Prof Exp:* Postdoctoral res, Univ Mich, 69-70 & Temple Univ, 70-72; lab dir res, Fanar Res Inst, 73-76; vis researcher, Temple Univ, 76-77; from asst prof to assoc prof, 77-84, PROF, AM UNIV BEIRUT, 84- *Concurrent Pos:* Guest prof, Yale Univ, 83, Univ Pa, 89. *Mem:* Am Asn Immunologists; Am Soc Microbiol; NY Acad Sci; Int Endotoxin Soc; Soc Exp Biol & Med; Am Soc Histocompatibility & Immunogenetics. *Res:* Structure-function relationship of bacterial endotoxins; human leucocyte antigens, transplantation and disease associations; epidemiological study of HIV in Lebanon. *Mailing Add:* Am Univ Beirut 850 Third Ave New York NY 10022

ABDOU, HAMED M, b Cairo, Egypt, Feb 2, 41; US citizen; m 70; c 1. PHARMACEUTICAL MANUFACTURING AND ANALYSIS. *Educ:* Univ Cairo, BSc, 61, MSc, 66; Rutgers Univ, PhD(pharmaceut chem), 77. *Prof Exp:* Head qual control dept, El-Kahira Pharmaceut & Chem, Cairo, 67-69; VPRES PHARMACEUT TECH OPERS, E R SQUIBB & SON, NJ, 69- *Concurrent Pos:* Adj prof, Col Pharm, Rutgers Univ, 78- *Mem:* Am Pharmaceut Asn; Am Chem Soc; Sigma Xi. *Res:* Separation sciences; hydrazine chemistry; automation in pharmaceutical analysis; dissolution kinetics and bioavailability. *Mailing Add:* E R Squibb Sons Inc Georges Rd New Brunswick NJ 08903

ABDOU, NABIH I, b Cairo, Egypt, Oct 11, 34; m 67; c 2. MEDICINE, IMMUNOLOGY. *Educ:* Cairo Univ, MD, 59; Univ Pa, MSc, 67; McGill Univ, PhD(immunol), 69. *Prof Exp:* Asst prof med, Univ Pa, 69-75, assoc prof, 75-78, PROF MED, UNIV KANS, 78- *Mem:* Am Asn Immunol; Am Acad Allergy; Transplantation Am Rheumatism Asn; Cent Soc Clin Res; Am Col Physicians. *Res:* Cellular immunology. *Mailing Add:* Rm 4035B Med Ctr Univ Kans Kansas City KS 66103

ABDULLA, RIAZ FAZAL, b Calcutta, India, Mar 24, 43; US citizen; m 68; c 2. CHEMISTRY, SUPER-COMPUTER. *Educ:* Calcutta Univ, BS, 65; Indian Inst Technol, MS, 66, PhD(chem), 70. *Prof Exp:* Vis Damon Runyon res fel chem, Princeton Univ, 71-73; sr scientist, 73-77, res scientist, 77-84, RES ASSOC, ELI LILLY & CO, 84-, MGR, SUPERCOMPUT APPLN, 88- *Mem:* Fel Am Inst Chemists; Am Chem Soc. *Res:* Synthesis and design of herbicides and plant bioregulators; synthesis of new heterocyclic compounds. *Mailing Add:* MC 797 Lilly Corp Ctr Indianapolis IN 46285

ABDULLAH, MUNIR, CELLULAR & MOLECULAR PHARMACOLOGY. *Educ:* Univ Karachi, Pakistan, MS, 74; Queen's Univ Belfast, PhD(biochem), 84. *Prof Exp:* Postdoctoral fel, Dept Biochem & Oncol, Queen's Univ Belfast, 84 & Dept Cell Biol & Anat, Univ NC, Chapel Hill, 84-87; res assoc, Dept Cell Biol & Anat, Univ NC, 87-90; SR STAFF FEL, LAB CELLULAR & MOLECULAR PHARMACOL, NAT INST ENVIRON HEALTH SCI, 90- *Mem:* Fedn Am Socs Exp Biol. *Res:* Author of 18 technical publications. *Mailing Add:* Lab Cellular & Molecular Pharmacol Nat Inst Environ Health Sci Box 12233 Mail Drop 7-10 Research Triangle Park NC 27709

ABDULRAHMAN, MUSTAFA SALIH, b Sulaimaniah, Iraq, June 15, 30. CIVIL ENGINEERING, THEORETICAL MECHANICS. *Educ:* Univ Baghdad, BS, 52; Rutgers Univ, MS, 58; Iowa State Univ, PhD(civil eng), 64. *Prof Exp:* Chief engr admin, Univ Baghdad, 64-65; asst prof, 65-70, ASSOC PROF CIVIL ENG, UNIV MISS, 70- *Concurrent Pos:* Dir, Abdulrahman & Assoc - Consult Engrs & Land Surveyors. *Mem:* Am Soc Civil Engrs; Am Soc Eng Educ; Am Concrete Inst. *Res:* Effect of ions in solution on the permeability of filter aids; computer applied matrix methods of structural analysis. *Mailing Add:* PO Box 1251 University MS 38677

ABE, RYO, b Tokyo, Japan, Dec, 24, 51; m; c 2. IMMUNOLOGY. *Educ:* Teikyo Univ, MD, 78; Univ Tokyo, DrMedSci, 83. *Prof Exp:* Physician, Kawakita Gen Hosp, 78-79; asst prof, Dept Immunol, Fac Med, Univ Tokyo, 83-85; vis fel, 85-88, vis assoc, 88-89, INVESTR, EXP IMMUNOL BR, DCBD, NAT CANCER INST, NIH, 89- *Mem:* Japanese Soc Immunol; Am Asn Immunologists. *Res:* Author or co-author of over 40 publications. *Mailing Add:* Exp Immunol Br Bldg 10 Rm 4B17 Nat Cancer Inst NIH 9000 Rockville Pike Bethesda MD 20892

ABED, EYAD HUSNI, b Jordan, Jan 24, 59; US citizen; m 82; c 3. NON-LINEAR CONTROL SYSTEMS, AEROSPACE CONTROL SYSTEMS. *Educ:* Mass Inst Technol, SB, 79; Univ Calif, Berkeley, MS, 81, PhD(elec eng), 82. *Prof Exp:* Res asst elec eng, Univ Calif, Berkeley, 80-82; asst prof, 83-87, ASSOC PROF ELEC ENG, UNIV MD, COL PARK, 87- *Concurrent Pos:* Consult, Naval Res Lab, 86-, Gen Elec Co, 87-, BDM Int, 88-; pres young investr award, NSF, 87; assoc ed, Inst Elec & Electronic Engrs Trans Automatic Control, 90- *Mem:* Inst Elec & Electronic Engrs; Soc Indust & Appl Math; Am Inst Aeronaut & Astronaut. *Res:* Theory and application of non-linear control systems; bifurcation analysis, singular perturbations, robust stability, power system dynamics; advanced aircraft control, radar systems, satellite control and jet engine dynamics and control. *Mailing Add:* Elec Eng Dept Univ Md College Park MD 20742

ABEDI, FARROKH, b Tehran, Iran, Aug 20, 49. PROGRAMMING. *Educ:* Okla State Univ, PhD(math), 83. *Prof Exp:* ASSOC PROF MATH, UNIV ARK, MONTICELLO, 82- *Mem:* Am Math Asn. *Mailing Add:* Dept Math & Physics Univ Ark Monticello AR 71655

ABEDIN, ZAIN-UL, cell biology; deceased, see previous edition for last biography

ABEGG, CARL F(RANK), b Peoria, Ill, June 5, 39; m 64; c 2. SCIENCE EDUCATION. *Educ:* Univ Ill, BS, 62; Iowa State Univ, MS, 63, PhD(chem eng), 66. *Prof Exp:* Asst prof chem eng, Univ Cincinnati, 66-73; proj engr, Corning Glass Works, 73-76; mgr eng res, O M Scott & Sons, 76-84; PROF CHEM ENG, ROSE HULMAN INST TECHNOL, 84- *Concurrent Pos:* Consult, Eli Lilly & Co, 85-, Stauffer Chem Co, 85-86, LESCO Inc, 86. *Mem:* Am Inst Chem Engrs; Am Chem Soc. *Res:* Crystallization from solution; controlled release nitrogen fertilizer technology; chemical reaction engineering. *Mailing Add:* Box 50 R Hulman Inst Technol 5500 Wabash Ave Terre Haute IN 47803

ABEGG, VICTOR PAUL, b Torrance, Calif, Feb 14, 45. ORGANIC CHEMISTRY, HISTORY OF SCIENCE. *Educ:* Loyola Univ Chicago, AB, 67; Mass Inst Technol, PhD(chem), 70; Univ Toronto, MDiv, 73. *Prof Exp:* Teaching fel, 74-75, sessional lectr chem, York Univ, 75-76; asst prof, 76-80, ASSOC PROF CHEM, CALIF STATE POLYTECH UNIV, POMONA, 80- *Mem:* AAAS; Am Chem Soc. *Res:* An examination of new synthetic pathways to small-ring carbon compounds is continuing with emphasis on photochemical methods; syntheses leading to steroids and chrysenes are being tested; subsequently subjected to strong-acid-catalyzed rearrangement and product amidine sensitivity to oxidation; ethics. *Mailing Add:* Dept Chem Calif State Polytech Univ 3801 W Temple Ave Pomona CA 91768

ABEL, ALAN WILSON, b Wilkinsburg, Pa, Mar 7, 39. PHYSICAL CHEMISTRY. *Educ:* Univ Pittsburgh, BS, 61, PhD(chem), 67. *Prof Exp:* Asst ed, 67-69, assoc indexer, 69-70, sr assoc ed, 70-87, SR ED P HYS-INORG-ANAL DEPT, CHEM ABSTR SERV, 87- *Mem:* Am Chem Soc. *Res:* Indexing of chemical literature; magnetic properties of metals, alloys and intermetallic compounds; gas-metal reactions. *Mailing Add:* 3015 Stadium Dr Apt 2 Columbus OH 43202

ABEL, CARLOS ALBERTO, b Buenos Aires, Arg, May 7, 30; US citizen; m 59. IMMUNOCHEMISTRY, LECTINS. *Educ:* M Belgrano Col, Arg, BS, 49; Univ Buenos Aires, Arg, MD, 57. *Prof Exp:* Physician, pediat sect, Childrens Hosp, Buenos Aires, 57-59; intern, St Joseph's Hosp, Providence, RI, 59-60; fel pediat, Univ Md Hosp, Baltimore, 60-64; resident, 64-66; res fel immunol, Scripps Clin, La Jolla, Calif, 66-69; vis scientist, Univ Oxford, Eng, 69-70; from assoc prof to assoc prof biophys & genetics, Univ Colo Med Ctr, 70-84, assoc prof med, 79-84; SR SCIENTIST, GBCRI, MED RES INST, SAN FRANCISCO, 84- *Concurrent Pos:* Fel, Arthritis Found, NY & Wellcome Trust, Eng, 69; mem, immunol sect, Nat Jewish Hosp, Denver, Colo, 70-84; vis scholar immunol, dept microbiol, Univ Calif, Berkeley, 82. *Mem:* Am Asn Immunologists; Am Soc Path; Latin Am Soc Immunol; Arg Asn Immunologists; AAAS. *Res:* Isolation, purification and characterization of oligosaccharide moieties from the glycoproteins present on the surface of lymphocytes and relationships of the structures of these components with various immunological functions; biology of HiV infection. *Mailing Add:* 523 Cragmont Ave Berkeley CA 94708

ABEL, ERNEST LAWRENCE, b Toronto, Ont, Feb 10, 43; m 71; c 2. TERATOLOGY, TOXICOLOGY. *Educ:* Univ Toronto, BA, 65, MA, 67, PhD(psychol), 71. *Prof Exp:* Actg dep dir, res inst alcoholism, NY State Dept Ment Hyg, 84-85; prof obstet & dir oper, 85-89, DIR, MOTT CTR HUMAN GROWTH & DEVELOP, WAYNE STATE UNIV, 89- *Concurrent Pos:* Pres, Fetal Alcohol Study Group, Res Soc Alcohol, 85-86; pres, Behav Teratal Soc, 84-85. *Mem:* Behav Teratology Soc (pres, 84-85). *Res:* Consequences of exposure to drugs during pregnancy. *Mailing Add:* Mott Ctr Human Growth & Develop 275 E Hancock Detroit MI 48201

ABEL, FRANCIS LEE, b Iowa City, Iowa, Apr 12, 31; m 74; c 4. CARDIOVASCULAR PHYSIOLOGY. *Educ:* Univ Kans, BA, 52; Harvard Med Sch, MD, 57; Univ Wis, PhD(physiol), 60. *Prof Exp:* Intern pediat, Children's Hosp, Los Angeles, 60-61; res instr, Univ Wash, 61-62; from asst prof to prof physiol, Sch Med, Ind Univ, Indianapolis, 62-75; chmn, Dept Physiol & Pharmacol, 75-78, interim dean, 76, CHMN & PROF PHYSIOL, SCH MED, UNIV SC, 75- *Concurrent Pos:* USPHS trainee physiol, Univ Wis, 59-60; vis prof, Ind Univ-USAID at Jinnah Postgrad Med Ctr, Karachi, 64-65, Univ Calif, 70 & Simon Fraser, Vancouver, BC, 82, Univ Limburg, Meastricht, Neth, 89-90; NIH career develop award, 68-73, Nat Res Serv Award, 89-90. *Mem:* Biomed Eng Soc; Inst Elec & Electronics Engrs; Am Physiol Soc; Am Heart Asn; Shock Soc. *Res:* Cardiovascular physiology; biophysical instrumentation; blood flow studies; reactions to hemorrhage; electro-magnetic flowmeters; shock; venous return; arterial pressure; regulation; cardiac function; coronary blood flow. *Mailing Add:* Dept Physiol Sch Med Univ SC Columbia SC 29208

ABEL, JOHN FREDRICK, b Brooklyn, NY, Oct 24, 40; m 64; c 2. CIVIL & STRUCTURAL ENGINEERING. *Educ:* Cornell Univ, BCE, 63; Stanford Univ, MS, 64; Univ Calif, Berkeley, PhD(civil eng), 68. *Prof Exp:* Res assoc civil eng, Princeton Univ, 70-74, lectr archit, 73-74; from asst prof to assoc prof, 74-82, PROF CIVIL ENG, CORNELL UNIV, 82- *Concurrent Pos:* Prin, 3D-Eye Inc, Ithaca, 81. *Mem:* Am Soc Civil Engrs; Int Asn Bridge & Struct Engrs; Am Concrete Inst; Int Assoc Shell & Spatial Structures. *Res:* Numerical methods in structural engineering, structural mechanics and civil engineering; computer graphics and computer-aided design in civil engineering; behavior and analysis of large frames and shells; structural dynamics and earthquake engineering. *Mailing Add:* Sch of Civil & Environ Eng Hollister Hall Cornell Univ Ithaca NY 14853-3501

ABEL, JOHN H, JR, b Painesville, Ohio, Feb 25,37; m 58; c 3. CELL BIOLOGY. *Educ:* Col Wooster, BA, 59; Brown Univ, MA, 64; PhD(cell biol), 66. *Prof Exp:* Chmn dept high sch, Ohio, 59-61; teacher high sch, Ill, 61-62; instr cell biol, NY Med Col, 66-67; from asst prof to prof cell biol, Colo State Univ, 67-80; PROF & HEAD DEPT ZOOL, UNIV TENN, 80- *Concurrent Pos:* Consult, Ciba Chem & Dye Co, 66-68; ed consult, McGraw-Hill Bk Co, 66-; guest prof, Univ Bonn, Fed Repub Ger, 74-75. *Honors & Awards:* Sr US Scientist Award for Res & Teaching, Alexander von Humboldt Stiftung, Fed Repub Ger, 73. *Mem:* Am Soc Cell Biol; Am Asn Anatomists; NY Acad Sci; Soc Study Reproduction & Fertil; AAAS. *Res:* Endocrine and neuroendocrine control of ovarian and adrenal function; osmoregulation in euryhaline birds. *Mailing Add:* Dept of Biol Lehigh Univ Bethlehem PA 18015

ABEL, LARRY ALLEN, b Pittsburgh, Pa, Dec 21, 49. BIOENGINEERING, ELECTRICAL ENGINEERING. *Educ:* Carnegie-Mellon Univ, BS, 71, MS, 72, PhD(elec eng, bioeng), 76. *Prof Exp:* Res assoc neurol, Sch Med, Univ Pittsburgh, 78-80; asst prof, Univ Hosp, Case-Western Reserve Univ, 80-86; assoc prof biomed eng, Univ Akron, 86-90; ASSOC PROF OPHTHAL, IND UNIV, INDIANAPOLIS, 90- *Concurrent Pos:* Res fel, Bascom Palmer Eye Inst, Sch Med, Univ Miami, 76-78. *Mem:* Asn Res Vision & Ophthal; Inst Elec & Electronics Engrs. *Res:* Ocular motility; visual perception; electrophysiology; neural control systems. *Mailing Add:* Dept Ophthal Sch Med Ind Univ 702 Rotary Circle Indianapolis IN 46202-5175

ABEL, PETER WILLIAM, b Milwaukee, Wis, June 23, 49; m 73; c 2. VASCULAR SMOOTH MUSCLE, ELECTROPHYSIOLOGY. *Educ:* Univ Wis-Madison, BS, 73; WVa Univ, PhD(pharmacol), 78. *Prof Exp:* Fel pharmacol, Univ Iowa, 78-80, res scientist, Div Neurosurg, 81; asst prof pharmacol, Emory Univ, 82-87; ASSOC PROF DEPT PHARMACOL, CREIGHTON UNIV, 88- *Res:* Factors which regulate the sensitivity of cardiac and smooth muscle to drugs; correlation of electrophysiological changes in muscle cells with altered sensitivity; role of altered vascular sensitivity in hypertension and cerebrovasospasm; adrenergic receptors and neuropeptides in smooth muscle. *Mailing Add:* Dept Pharmacol Creighton Univ Omaha NE 68178

ABEL, WILLIAM T, b Marion, Ind, Feb 16, 22; m 49; c 4. FUEL SCIENCE, PHYSICAL CHEMISTRY. *Educ:* Franklin Col, BA, 44; Univ Ill, Urbana, MS, 47. *Prof Exp:* Res asst chem, Ill Geol Surv, 44-47; res asst, Mound Lab, Ohio, 47-53; chemist, Dowell Inc, Okla, 53-56; chemist, Elec Auto-Lite Co, 56-58; supvy chemist, US Bur Mines, 58-60, chem res engr, 60-74; chem engr, Energy Res Develop Admin, Dept Energy, 74-78, chem engr, 78-83; RETIRED. *Concurrent Pos:* Chem engr, Resource Technol Group, Inc, WVa, 86-88. *Mem:* AAAS; Am Chem Soc. *Res:* Corrosion inhibition; gas-solid reaction; dry processes for removal of pyrite from coal; coal liquefaction; fluidized bed combustion. *Mailing Add:* 564 Killarney Dr Morgantown WV 26505

ABELE, LAWRENCE GORDON, b Baltimore, Md, Mar 01, 46; m 66; c 2. ECOLOGY, SYSTEMATICS. *Educ:* Fla State Univ, BS, 68, MS, 70; Univ Miami, PhD(biol), 72. *Prof Exp:* Fel trop biol, Smithsonian Trop Res Inst, 72-73; asst prof, Int Progs, 73-74, from asst prof to assoc prof, Dept Biol Sci, 74-83, PROF BIOL, FLA STATE UNIV, 83- *Mem:* Crustacean Soc; Sigma Xi; Soc Syst Zool; Am Soc Zoologists; fel AAAS. *Res:* Systematics ecology and systematics of decapod crustaceans; community ecology; island biogeography; quantitative morphology of invertebrates. *Mailing Add:* Dept Biol Sci Fla State Univ Tallahassee FL 32306

ABELES, BENJAMIN, b Vienna, Austria, June 23, 25; m 58; c 1. SOLID STATE PHYSICS. *Educ:* Prague Univ, MS, 49; Hebrew Univ, Israel, PhD, 56. *Prof Exp:* Res tech, Meteorol Serv, Israel, 49-51; researcher, Weizmann Inst, Israel, 51-56 & Radio Corp Am, 56-77; MEM STAFF, EXXON RES & ENG CO, 77- *Honors & Awards:* Franklin Inst Medal. *Mem:* Fel Am Phys Soc. *Res:* Solar cells; plasma chemistry. *Mailing Add:* Exxon Res & Eng Co Annandale NJ 08801

ABELES, FRANCINE, b New York, NY, Oct 19, 35; m 57; c 3. COMPUTER INFORMATION & OPERATING SYSTEMS. *Educ:* Barnard Col, Columbia Univ, AB, 57; Columbia Univ, MA, 59, EdD(math), 64; Stevens Inst of Technol, MS(comput sci), 86. *Prof Exp:* Teacher high schs, NY & Ger, 57-63; from asst prof to assoc prof, 64-72, PROF MATH, 72-, HEAD GRAD PROG MATH & COMPUT SCI, KEAN COL NJ, 74- *Concurrent Pos:* Vis mem, Courant Inst Math Sci, NY Univ, 76-77; ed, Math Pamphlets Lewis Carroll, Univ Va Press, 87-; referee, Math Mag; res grant, Am Philos Soc, 91. *Mem:* Am Math Soc; Math Asn Am; Lewis Carroll Soc (treas, 90-); Can Soc Hist & Philos of Math; NY Acad Sci; Asn Comput Mach. *Res:* Geometry; Lewis Carroll; 19th century mathematics; databases; ciphers; linear algebra; logic. *Mailing Add:* Dept Math & Comput Sci Kean Col NJ Union NJ 07083

ABELES, JOSEPH HY, b Brooklyn, NY, Apr 14, 55. OPTICAL SEMICONDUCTOR DEVICES,. *Educ:* Mass Inst Technol, BS, 76; Princeton Univ, PhD(physics), 82. *Prof Exp:* MECH TECH STAFF, BELL COMMUN RES, 82- *Mem:* Am Phys Soc; Inst Elec & Electronic Engrs. *Mailing Add:* Bell Comm Res Inc 331 Newman Springs Rd Red Bank NJ 07701

ABELES, ROBERT HEINZ, b Vienna, Austria, Jan 14, 26; nat US; m 48; c 2. BIOCHEMISTRY. *Educ:* Univ Chicago, MS, 50; Univ Colo, PhD(biochem), 55. *Hon Degrees:* DSc, Univ Chicago. *Prof Exp:* Asst dept pediat, Univ Chicago, 50-51; fel, Nat Found Infantile Paralysis, 55-56; fel chem, Harvard Univ, 56-57; asst prof chem, Ohio State Univ, 57-60; from asst prof to assoc prof biochem, Univ Mich, 60-64; assoc prof, 64-67, chmn grad dept, 73-88, PROF BIOCHEM, BRANDEIS UNIV, 67- *Concurrent Pos:* Ed, J Biol Chem, Am Soc Biol Chem, 67-72; mem coun, Am Cancer Soc, 77; Shen vis prof, Mass Inst Technol, 83; Burroughs Welcome prof, 83. *Honors & Awards:* Steenbock Lectr, 84; Edward E Smissman-Bristol Myers Award Med Chem, 87; Repligen Award Biol Chem, 88; Alfred E Bader Award, 90. *Mem:* Nat Acad Sci; Am Chem Soc; Am Soc Biol Chem; Am Acad Arts & Sci; Fedn Am Soc Exp Biol. *Res:* Mechanism of enzyme action; biological oxidations. *Mailing Add:* Dept Biochem Brandeis Univ Waltham MA 02154

ABELEV, GARRI IZRAILEVICH, b Moscow, USSR, Jan 10, 28; m 49; c 2. IMMUNOCHEMISTRY, CANCER IMMUNOLOGY. *Educ:* Moscow State Univ, dipl biol, 50; Acad Med Sci, PhD(biochem), 55, DSc, 63. *Prof Exp:* Sr technician biochem cancer, NF Gamaleya Inst Epidemiol & Microbiol, USSR Acad Med Sci, 50-55, jr res worker, 55-59, sr res worker, 59-61, head lab cancer immunochem, 61, head lab dept cancer virol & immunol, 66-77; HEAD LAB CANCER IMMUNOCHEM, CANCER RES CTR USSR, ACAD MED SCI, 77- *Concurrent Pos:* Lectr immunochem, Moscow State Univ, 64-; prin investr, Coop Ref Ctr, WHO, 66-85; mem, comt tumor immunodiag, Int Union Against Cancer, adv bd develop biol, USSR Acad Med Sci, 75-, adv bd malignant tumors, 80- *Honors & Awards:* USSR State Award, 78. *Mem:* USSR Asn Immunologist; hon mem Am Asn Immunologists; Int Soc Oncodevelop Biol & Med; corresp mem USSR Acad Sci. *Res:* Cancer immunology and immunodiagnosis; carcino-embryonic proteins, their finding, characterization and use in cancer diagnosis; antigenic structure of virus-induced experimental leukoses; development of new techniques of protein immunochemistry. *Mailing Add:* Cancer Res Ctr Kashirskoye Shosse 24 Moscow 115478 USSR

ABELING, EDWIN JOHN, b Amsterdam, NY, Oct 4, 15; m 45; c 3. FOOD SCIENCE & TECHNOLOGY. *Educ:* Univ Ky, BS, 44. *Prof Exp:* Med technologist, Montgomery County Lab, NY, 39-41; med technologist, St Joseph's Hosp, Lexington, Ky, 41-44; bacteriologist, Beech-Nut Packing Co, NY, 44-47, head, San Jose Lab, Calif, 47-50, food lab, Canajoharie, NY, 50-54, assoc dir res, Beech-Nut Life Savers, 54-57, dir res & develop, Quality Control, 57-61; dir, Peter Paul, Inc, 61-68, vpres, res & develop, 68-81; RETIRED. *Concurrent Pos:* Mem, Exec Prog Bus Admin, Grad Sch Bus, Columbia Univ, 60. *Mem:* Inst Food Technologists. *Res:* Development of new products; food, packaging and market research; quality control; bacteriology. *Mailing Add:* 162 Ansonia Rd Woodbridge CT 06525

ABELL, CREED WILLS, b Charlottesville, Va, July 8, 34; m 56; c 2. BIOCHEMISTRY, ONCOLOGY. *Educ:* Va Mil Inst, BS, 56; Purdue Univ, MS, 58; Univ Wis, PhD(oncol), 62. *Prof Exp:* Chemist, Chem Sect, Carcinogenesis Studies Br, Nat Cancer Inst, 62-67; from assoc prof to prof biochem & molecular biol, Sch Med, Univ Okla, 67-72; PROF HUMAN BIOL CHEM & GENETICS, UNIV TEX MED BR, GALVESTON, 72-, DIR, DIV BIOCHEM, 78- *Mem:* Am Asn Cancer Res; Am Soc Biol Chemists; Sigma Xi; AAAS; NY Acad Sci. *Res:* Regulation of division of normal and leukemic lymphocytes, enzyme therapy of neoplasia biosynthesis and degradation of neurotransmitters. *Mailing Add:* Col of Pharm Div of Med Chem PHR 2222 Univ Tex Austin TX 78712-1074

ABELL, JARED, b Los Angeles, Calif, Sept 5, 28; m 50; c 3. ORGANIC CHEMISTRY. *Educ:* Calif Inst Technol, BS, 50; Univ Calif, Los Angeles, PhD(org chem), 54. *Prof Exp:* Sr res chemist, Chevron Res Co, 54-66; res chemist, Ortho Div, Chevron Chem Co, 66-68, mgr prod res & serv, Div Res & Develop, 68-83, sup, Residue Metab & environ fate of pesticides, 83-86; RETIRED. *Mem:* Am Chem Soc. *Res:* Formulation, residue, metabolism, environmental fate of pesticides and agricultural chemicals in general. *Mailing Add:* 75 Monte Vista Rd Orinda CA 94563

ABELL, LIESE LEWIS, b Frankfurt, Ger, Aug 17, 09; nat US. BIOCHEMISTRY. *Educ:* Univ Frankfurt, PhD, 35. *Prof Exp:* Chemist, Flower & Fifth Ave Hosp, New York Med Col, 40-42; chemist, Wyeth Inst Appl Biochem, Pa, 45-47; asst prof biochem, Columbia Univ, 47-68; sr res scientist, Bur Labs, Pub Health Labs, New York City Dept Health, 68-74; asst to ed, J Lipid Res, 75-78; coordr clin trial, Med Sch, Cornell Univ, 79-81; RETIRED. *Concurrent Pos:* Res fel, Columbia Univ, 38-40 & 42-45. *Mem:* AAAS; Am Soc Biol Chemists. *Res:* Lipid metabolism in connection with arteriosclerosis. *Mailing Add:* 7 Peter Cooper Rd New York NY 10010

ABELL, MURRAY RICHARDSON, b Aylmer, Ont, Oct 14, 20; nat US; m 44; c 4. MEDICINE, PATHOLOGY. *Educ:* Univ Western Ont, MD, 44, PhD(path), 51; Am Bd Path, dipl; FRCPS(C). *Prof Exp:* From instr to assoc prof, Univ Mich, Ann Arbor, 52-59, prof path, 59-; EXEC DIR, AM BD PATH. *Concurrent Pos:* Trustee, Am Bd Path. *Mem:* Am Asn Pathologists; Col Am Pathologists; Int Acad Path; Am Soc Clin Path. *Res:* Hepatic disease; mediastinal tumors; neoplasms of female reproductive tract; antigen-antibody reactions in tissues; testicular neoplasms. *Mailing Add:* Lincoln Ctr 5401 W Kennedy Blvd, PO Box 25915 Tampa FL 33622

ABELL, PAUL IRVING, b Pelham, Mass, July 24, 23; m 51, 80; c 2. ORGANIC CHEMISTRY, GEOCHEMISTRY. *Educ:* Univ NH, BS, 48; Univ Wis, PhD(chem), 51. *Prof Exp:* From instr to assoc prof, 51-64, PROF ORG CHEM, UNIV RI, 64- *Concurrent Pos:* Petrol Res Fund Int Award, Univ Wales, 60-61 & Univ Bristol, 69-70; Fulbright lectr, UAR, 65-66; mem, Omo River Res Exped, 67 & Lake Rudolf Res Expeds, 68-74. *Mem:* Am Chem Soc; Geochem Soc; Brit Chem Soc; Faraday Soc; Sigma Xi. *Res:* Stereochemistry and kinetics of free radical reactions; organic geochemistry; paleontology. *Mailing Add:* Wolf Pack Rd West Kingston RI 02892

ABELLA, ISAAC D, b Toronto, Ont, June 20, 34; m 66; c 2. EXPERIMENTAL PHYSICS, QUANTUM OPTICS. *Educ:* Univ Toronto, BA, 57; Columbia Univ, MA, 59, PhD(physics), 63. *Prof Exp:* Asst physics, Columbia Univ, 57-63, res assoc, 63-65; from asst prof to assoc prof, 65-86, PROF PHYSICS, UNIV CHICAGO, 86-, RESIDENT MASTER, 85- *Concurrent Pos:* Mem laser apparatus comt, Am Inst Physics, 65, vis scientist, 66-71; consult, Mithras, Inc, Mass, 65-70 & Sanders Assoc, NH, 70-; mem laser adv comt, Ill Dept Pub Health, 68-72; vis fel, Joint Inst Lab Astrophys, Univ Colo, Boulder, 72-73; vis scientist, Naval Res Lab, Washington, DC, 81-82; vis fac, physics div, Argonne Nat Lab, 83; Am Soc Eng Educ fac scientist, optical sci, Naval Res Lab, 85-89; consult, Physics Div, Argonne Nat Lab, 82-88; mem, US Nat Comt Int Comn Optics, 82-84; chair, Educ Comt, Laser Sci Tech Group, Am Phys Soc, 88-90; mem, Nat Educ Comt, Am Phys Soc & Isaicson Prize Comt, Advan Optical Sci, 91-93. *Mem:* AAAS; fel Am Phys Soc; fel Optical Soc Am. *Res:* Experimental physics, especially non-linear optics, photon echoes, quantum beats, atomic physics and lasers; energy transfer; rare-earth materials. *Mailing Add:* Dept of Physics Univ Chicago 5734 S Ellis Ave Chicago IL 60637

ABELMANN, WALTER H, b Frankfurt, Ger, May 16, 21; nat US; m 58; c 5. MEDICINE, CARDIOLOGY. *Educ:* Harvard Univ, AB, 43; Univ Rochester, MD, 46. *Prof Exp:* Asst, 51-53, instr, 53-55, assoc, 55-58, asst prof, 58-64, assoc clin prof, 64-69, assoc prof, 69-72, PROF MED, HARVARD MED SCH, 72- *Concurrent Pos:* Res fel, Thorndike Mem Lab, Boston City Hosp, 51- & Am Heart Asn, 53-55; estab investr, Am Heart Asn, 55-60. *Mem:* AAAS; Am Fedn Clin Res; Am Heart Asn; Asn Univ Cardiol; Am Soc Clin Invest; Am Col Cardiol; Asn Am Physicians. *Res:* Cardiovascular pathophysiology. *Mailing Add:* 330 Brookline Ave Boston MA 02215

ABELS, LARRY L, b Freeport, Ill, Feb 16, 37; m 60; c 3. PHYSICS. *Educ:* Knox Col, Ill, BA, 59; Ohio State Univ, MSc, 61, PhD(physics), 65. *Prof Exp:* ASSOC PROF PHYSICS, UNIV ILL, CHICAGO CIRCLE, 65- *Mem:* Optical Soc Am. *Res:* Optical properties of solids. *Mailing Add:* Dept Physics Univ Ill PO Box 4348 Chicago IL 60680

ABELSON, JOHN NORMAN, b Grand Coulee Dam, Wash, Oct 19, 38. MOLECULAR BIOLOGY, MOLECULAR GENETICS. *Educ:* Wash State Univ, BS, 60; Johns Hopkins Univ, PhD(biophys), 65. *Prof Exp:* Fel, Lab Molecular Biol, Cambridge Univ, Eng, 65-68; from asst prof to prof chem, Univ Calif, San Diego, 68-82; PROF BIOL, CALIF INST TECHNOL, 82-, CHMN, DIV BIOL, 89- *Concurrent Pos:* John Simon Guggenheim Mem Found fel, 80-81; mem, Rev Comt Biol, Univ Calif, Santa Cruz, 85 & Adv Comt Molecular Biol, Am Cancer Soc, 85-88; chmn, Sci Adv Bd, Agooron Pharmaceut, Inc, 84-; chmn, comt to visit Dept Biochem & Molecular Biol, Bd Overseers Harvard Col, 87-; lectr, Harvey Soc, 89. *Mem:* Nat Acad Sci; Am Soc Biol Chemists; Am Chem Soc; Am Acad Arts & Sci; Protein Soc. *Res:* Genetic control mechanisms; nucleotide sequences of DNA and RNA; RNA synthesis and post-transcriptional processing; author of numerous technical publications. *Mailing Add:* Div Biol Calif Inst Technol Pasadena CA 91125

ABELSON, PHILIP HAUGE, b Tacoma, Wash, Apr 27, 13; m 36; c 1. ORGANIC GEOCHEMISTRY. *Educ:* Wash State Univ, BS, 33, MS, 35; Univ Calif, PhD(nuclear physics), 39. *Hon Degrees:* DSc, Yale Univ, 64, Southern Methodist Univ, 69, Tufts Univ,76, Duke Univ, 81, Univ Pittsburgh, 82; DHL, Univ Puget Sound, 68. *Prof Exp:* Asst physics, Wash State Univ, 33-35; asst, Univ Calif, 35-38, asst, Radiation Lab, 38-39; asst physicist, Carnegie Inst, 39-41; from assoc physicist to prin physicist, Naval Res Lab, 41-45; chmn biophys sect, Dept Terrestrial Magnetism, Carnegie Inst, 46-53, dir geophys lab, 53-71, pres, 71-78; ed, Sci, AAAS, 61-84, dep ed Sci & Scholar in Residence, Resources for Future Sci, 85-88, SCI ADV, AAAS, 85- *Concurrent Pos:* Mem, Nat Defense Res Comt, 40-42; chmn radiation cataract comt, Nat Res Coun, 49-57; mem biophys & biophys chem study sect, NIH, 56-59, phys biol training grants comt, 58-60; co-ed, J Geophys Res, 58-64; plowshare adv comt, AEC, 59-63, gen adv comt, 60-63; sci counr, Nat Inst Arthritis & Metab Dis, 60-63; mem coun, Nat Acad Sci, 78-81; space studies bd, 87-; bd dirs, Chem Ind Inst Toxicol, 85- *Honors & Awards:* Mod Med Award, 67; Kalinga Prize, UNESCO, 73; Sci Achievement Award, AMA, 74; President's Nat Medal Sci. *Mem:* Nat Acad Sci; Int Union Geol Sci (pres, 72-76); fel Am Geophys Union (pres, 72-74); fel Am Phys Soc; fel Am Acad Arts & Sci; fel Geol Soc Am; fel Mineral Soc Am; Seismol Soc Am; Am Chem Soc; Soc Am Bacteriologists; Am Asn Petrol Geol; Sigma Xi; AAAS. *Res:* Nuclear physics; radioactive tracers; fission products of uranium; characteristic x-rays emitted by radioactive substances; neptunium; separation of uranium isotopes; mechanisms of ion transport into living matter; biosynthesis in microorganisms; plasma volume expanders; petrology; paleobiochemistry; geochemistry; 400 articles and editorials published. *Mailing Add:* AAAS 1333 H St NW Washington DC 20005

ABEND, KENNETH, b New York, NY, Jan 14, 36; m 66; c 4. SIGNAL PROCESSING, INFORMATION SCIENCE. *Educ:* City Col New York, BEE, 58; Univ Pa, MSE, 63, PhD(elec eng), 66. *Prof Exp:* Elec eng trainee, Rome Air Develop Ctr, NY, 56; eng aid, Bell Aircraft Corp, 57; jr engr res lab, Philco-Ford Corp, Ford Motor Co, 58-59, engr, 59-60, sr res engr, 60-63, proj engr, 64-65, eng res specialist, 65-68, sr eng specialist, Commun & Tech Serv Div, 68-76; prin mem eng staff, missile & surface radar systs eng, RCA Govt Systs Div, 76-84; MGR, ADVAN SIGNAL PROCESSING, INTERSPEC INC, 84- *Concurrent Pos:* Inst Elec & Electronics Engrs pattern recognition workshop partic, PR, 66; lectr grad ctr, Pa State Univ, 67-; lectr systs eng dept, Univ Pa, 72- *Honors & Awards:* Spec 50th Anniversary Commemoration Medal Award, Outstanding Young Engr, City Col New York, 69. *Mem:* Sr mem Inst Elec & Electronics Engrs. *Res:* Development and analysis of statistical techniques for communications, radar and pattern classification; adaptive sampled data detector for channels with intersymbol interference; compound decision techniques for pattern recognition using context; modern spectral estimation for radar imaging; adaptive processing for phased array radars; automatic motion compensation for synthetic apertures. *Mailing Add:* 623 Killdeer Lane Huntingdon Valley PA 19006

ABENDROTH, REINHARD P(AUL), b St Louis, Mo, Mar 19, 31; m 55; c 2. GLASS TECHNOLOGY. *Educ:* Mo Sch Mines, BS, 53, MS, 54; Univ Mo, PhD(metall), 57. *Prof Exp:* Res metallurgist, Union Carbide Metals Co, 57-60; sr res scientist, 60-76, Owens-IllInc, 60-76, chief, Eval & Testing Sect, 76-82; MGR, LAB SERV, KIMBLE GLASS INC, 82- *Mem:* Am Inst Mining, Metall & Petrol Engrs; Am Soc Metals; Sigma Xi; Am Ceramic Soc; Soc Glass Technol. *Res:* High temperature physical chemistry of metallurgical reactions; high temperature glass-metal reactions; glass-aqueous solution reactions. *Mailing Add:* Kimble Glass Inc 537 Crystal Ave Vineland NJ 08360

ABENES, FIORELLO BIGORNIA, b Naguilian, La Union, Philippines, Dec 10, 46; Can citizen; m 71; c 2. SWINE PRODUCTION SYSTEMS, COMPUTER APPLICATIONS IN AGRICULTURE. *Educ:* Cent Luzon State Univ, BS, 69; Univ Conn, MS, 71, PhD(physiol), 75. *Prof Exp:* Asst prof animal sci, Univ Philippines, 75-77; swine specialist, Alta Agr, 79-84; prof animal sci, Olds Col, 84-86; PROF ANIMAL SCI, CALIF STATE POLYTECH UNIV, 86- *Concurrent Pos:* Prin investr, Sustainable Agr Prog, Calif State Polytech Univ, Pomona. *Mem:* Am Soc Animal Sci; Sigma Xi. *Res:* Developing ecologically responsible ways of producing foods & fiber. *Mailing Add:* Dept Animal Sci Calif State Polytech Univ Pomona CA 91768

ABER, JAMES SANDUSKY, b Kansas City, Mo, Apr 25, 52; m 77; c 2. GEOMORPHOLOGY, GLACIOLOGY. *Educ:* State Univ NY Binghamton, BS, 74; Univ Kans, MS, 76, PhD(geol), 78. *Prof Exp:* Asst prof geol, Chadron State Col, Nebr, 78-80; asst prof, 80-84, ASSOC PROF GEOL, EMPORIA STATE UNIV, KANS, 84- *Concurrent Pos:* Fulbright-Hays fel, Inst Gen Geol, Univ Copenhagen, 79; Fac Exchange, Univ Regina, Can, 84; Marshall Fund fel, Univ Bergen, Norway, 87. *Mem:* Geol Soc Am; Geol Soc Denmark; Sigma Xi. *Res:* Glaciation, glacial stratigraphy, glaciotectonics and Quaternary environments of central North America and northern Europe. *Mailing Add:* Dept Phys Sci Emporia State Univ 1200 Commercial St Emporia KS 66801

ABERCROMBIE, RONALD FORD, b Greenville, SC, June 10, 46; m 69. BIOPHYSICS, CELL PHYSIOLOGY. *Educ:* Univ NC, Chapel Hill, BS, 68; Univ Md, College Park, MS, 72; Univ Md, Baltimore City, PhD(biophys), 77. *Prof Exp:* asst prof biophys, Sch Med, Univ MD, 76-80; res assoc, dept physiol, Sch Med, Wash Univ, 80-82; asst prof, 82-88, ASSOC PROF PHYSIOL, SCH MED, EMROY UNIV, 88- *Concurrent Pos:* Grass Found fel, 76. *Mem:* Biophys Soc; Am Asn Physics Teachers; NY Acad Sci; Soc Gen Physiol. *Res:* Ion transport mechanisms; active membrane transport; regulation of intracellular ionic environment; the diffusion and buffering of calcium and the interaction between calcium and hydrogen in the intracellular environment. *Mailing Add:* Dept Physiol Emory Univ Sch Med Atlanta GA 30322

ABERE, JOSEPH FRANCIS, b New York, NY, Mar 30, 20; m 44; c 7. POLYMER CHEMISTRY. *Educ:* Queens Col, NY, BS, 41; Polytech Inst Brooklyn, MS, 43, PhD(high polymers), 48. *Prof Exp:* Tutor chem, Queens Col, NY, 43-44, instr, 46-47; res chemist, 48-53, head appl res sect, Cent Res, 53-57, proj mgr chem div, 57-62, assoc mgr tape res, 62-67, mgr pioneering res, Indust Spec Prod Dept, 67-72, tech mgr composites, 70-72, tech mgr overseas opers, 72-78, CORP SCIENTIST, MINN MINING & MFG CO, 78- *Honors & Awards:* Best Symp, Rubber Div, Am Chem Soc, 84. *Mem:* AAAS; Am Chem Soc; Am Inst Chemists; Am Inst Aeronaut & Astronaut. *Res:* High polymers; fluorine containing polymers; vulcanization; adhesives and adhesion; resins; plastics; chemical development; imine chemistry; composites; reinforced plastics; irradiation chemistry. *Mailing Add:* 3M Ctr Bldg 230-3N-02 St Paul MN 55101

ABERHART, DONALD JOHN, b St John, NB, Oct 20, 41. BIOCHEMISTRY. *Educ:* Univ Western Ont, BSc, 63, PhD(org chem), 67. *Prof Exp:* NATO sci fel, Glasgow Univ, 67-68; fel, Worcester Found Exp Biol, 69-70; res assoc, Mass Inst Technol, 70-71; from asst prof to assoc prof chem, Cath Univ Am, 71-77; SR SCIENTIST, WORCESTER FOUND EXP BIOL, 77- *Mem:* AAAS; Am Chem Soc. *Res:* Enzyme chemistry; biosynthetic studies. *Mailing Add:* Worcester Found Exp Biol 222 Maple Ave Shrewsbury MA 01545-2732

ABERLE, ELTON D, b Sabetha, Kans, Aug 30, 40; m 65; c 2. MEAT SCIENCE, FOOD SCIENCE. *Educ:* Kans State Univ, BS, 62; Mich State Univ, MS, 65, PhD(food sci), 67. *Prof Exp:* From asst prof to prof animal sci, Purdue Univ, 67-83; PROF ANIMAL SCI & HEAD DEPT, UNIV NEBR, 83- *Concurrent Pos:* Mem NC-91, Tech Comt Regional Res Proj, Agr Res Serv, USDA, 68-74, NC-131, 75-83, NCA-6, 7, 8, 83-; assoc, Univ Minn, 75. *Honors & Awards:* Meat Res Award, Am Soc Animal Sci, 82; Distinguished Res Award, Am Meat Sci Asn, 86. *Mem:* Am Soc Animal Sci; Inst Food Tech; Am Meat Sci Asn; AAAS. *Res:* Muscle biochemistry and physiology and muscle growth and differentiation in meat animals relating to use of muscle as food; adipose tissue growth in animals; meat processing. *Mailing Add:* Univ Nebr Dept Animal Sci C203 Animal Sci Lincoln NE 68583-0908

ABERLE, SOPHIE D, b Schenectady, NY, July 21, 99; m 40. NUTRITION. *Educ:* Stanford Univ, AB, 23, MA, 25, PhD(genetics), 27; Yale Univ, MD, 30. *Prof Exp:* Asst histol, Stanford Univ, 24-25, asst embryol & neurol, 25-26; instr anthrop, Inst Human Rels, Yale Univ, 27-30, Sterling fel, Sch Med, 30-31, instr, 30-34; assoc res, Carnegie Inst, 34-35; supt, Pueblo Indians, Bur Indian Affairs, US Dept Interior & secy, Southwest Supts Coun, US Indian Serv, 35-44; div med sci, Nat Res Coun, 44-49; spec res dir, Univ NMex, 49-54; chief nutrit, Bernalillo County Indian Hosp, 53-66; mem staff, Dept Psychiat, Med Sch, 66-69, MEM STAFF, LAW SCH, UNIV N MEX, 70- *Concurrent Pos:* Field worker among Pueblo Indians under grant from Comt Res Probs of Sex, 27-35; chief emergency med serv, NMex State Coun Nat Defence, 42-44; mem upper Rio Grande drainage basin comt, Nat Resources Planning Bd, 36-44; NMex Nutrit Comt, 40-; White House Conf Children in Democracy, 40-46; chmn bd dirs, Southwest Field Training Sch for Fed Serv, 37-41; mem, Defence Saving Comt, NMex, 41-44; chmn, Bernalillo County Hosp Comt, 49-50; mem Nat Sci Bd, 50-58, exec comt, 52-54; mus comt, Albuquerque Chamber of Commerce, 50-56; consult, Health Comt, All Pueblo Coun, 53-; mem comt maternal & infant mortality, NMex Med Soc, 54-; actg exec dir, Comn Rights, Liberties & Responsibilities of Am Indian, 57-63; dir surv Indian Educ, Bur Indian Affairs, 63-; bd dir, Bernalillo County Planned Parenthood, 64-; dir, YWCA, 64; mem bd dirs, Planned Parenthood, World Pop, 69-75; consult, Inst Math Studies Soc Sci on Indian Educ, Stanford Univ, 71-; consult, All Indian Pueblo Coun comput assisted instruction prog & soc sci, Sch Law, Univ NMex, 70-; consult bilingual/bicultural proj, Bernalillo Sch Dist Title VII, 75- *Mem:* AAAS; AMA; fel Soc Res Child Develop; fel Am Anthrop Asn; Am Asn Anat; Sigma Xi. *Res:* Anthropology; human nutrition. *Mailing Add:* 1021 Coal SW Apt 10 Albuquerque NM 87102

ABERNATHY, A(TWELL) RAY, b Hiddenite, NC, May 29, 30; m 55; c 2. ENVIRONMENTAL BIOLOGY, ENVIRONMENTAL CHEMISTRY. *Educ:* Lenoir-Rhyne Col, AB, 53; Univ NC, MSPH, 59, PhD(environ sci), 63. *Prof Exp:* From asst prof civil eng to assoc prof environ systs eng, 62-73, PROF ENVIRON SYSTS ENG, CLEMSON UNIV, 73- *Concurrent Pos:* Fel, Univ Wash, 67-68. *Honors & Awards:* Arthur Sidney Bedell Award, Water Pollution Control Fedn, 75; W T Linton Award, Water & Pollution Control Asn SC, 77. *Mem:* Water Pollution Control Fedn; Am Soc Limnol & Oceanog; Sigma Xi. *Res:* Water pollution control; fate of heavy metals in wastewater discharges; nonpoint source pollution control. *Mailing Add:* Dept Environ Systs Eng Clemson Univ 501 Rhodes Ctr Clemson SC 29631

ABERNATHY, BOBBY F, b Athens, Tex, June 25, 33; m 63; c 2. RESERVOIR ENGINEERING, ENHANCED RECOVERY. *Educ:* Univ Tex Austin, BS, 55. *Prof Exp:* Engr, Amoco Prod co, 55-68; chief engr, Amoco Canada Petroleum Ltd, 68-72; vpres Am Quasar Petroleum, 72-76; sr vpres explor & Prod, dir Champlin Petroleum, Co, 76-81; OWNER, ABERNATHY EXPLOR CO, 81- *Concurrent Pos:* Exec vpres, Quasar Petroleum Co, 72-76; pres, Quasar Energy Inc, 74-76; exec comt, Independent Petroleum Asn Am. *Honors & Awards:* Cedrick Ferguson Medal, Soc Petroleum Engr. *Mem:* Am Instit Mining Mettal & Petroleum Engrs. *Res:* Study of reservoir mechanics, fragile environment of arctic areas, ice mechanics. *Mailing Add:* 1204 Commerce Bldg Ft Worth TX 76102

ABERNATHY, CHARLES OWEN, b Brunswick, Ga, Nov 18, 41; m 72. PHARMACOLOGY. *Educ:* Asbury Col, AB, 64; Univ Ky, MS, 67; NC State Univ, PhD(physiol), 70. *Prof Exp:* Res asst entom, Univ Ky, 64-66; USPHS trainee toxicol, NC State Univ, 67-70; res entomologist, Univ Calif, Berkeley, 70-73; PHARMACOLOGIST, LIVER RES UNIT, VET ADMIN HOSP, WASHINGTON, DC, 73- *Res:* Effects of drugs and other compounds; endotoxin on liver function with the goal of establishing in vitro models to study hepatotoxicity. *Mailing Add:* Environ Protection Agency 41 M St SW Washington DC 20460

ABERNATHY, FREDERICK HENRY, b Denver, Colo, June 19, 30; m 61; c 3. FLUID MECHANICS, NUCLEAR ENGINEERING. *Educ:* Newark Col Eng, BS, 51; Harvard Univ, SM, 54, PhD, 59. *Prof Exp:* Develop engr, Oak Ridge Nat Lab, 52-54; asst, Los Alamos Sci Lab, 55; lectr eng, 59, from asst prof to assoc prof, 59-67, assoc dean eng & appl physics, 67-69, GORDON McKAY PROF MECH ENG, HARVARD UNIV, 67- *Concurrent Pos:* Consult, Reactor Proj Div, Oak Ridge Nat Lab, 59, Arthur D Little Co, Mass, 63-, Inst Defense Anal, 66- & NSF, 71-; NSF fel, 62-63; mem, Nat Comt Fluid Mech Films, 65-, chmn, 69-71; master South House, Harvard-Radcliffe Cols, 69-71. *Mem:* Am Soc Mech Engrs; Sigma Xi. *Res:* Incompressible wakes of bluff bodies; forced convection film boiling; nuclear reactor kinetics. *Mailing Add:* 326 Pierce Hall Harvard Univ Cambridge MA 02138

ABERNATHY, GEORGE HENRY, b West Newton, Pa, Nov 9, 29; m 55; c 4. AGRICULTURAL ENGINEERING. *Educ:* NMex State Univ, BS, 52; Univ Calif, ME, 56; Okla State Univ, PhD(soil tillage), 67. *Prof Exp:* Jr res specialist, Univ Calif, 55-57; from asst prof to assoc prof, 57-76, PROF AGR ENG, NMEX STATE UNIV, 76-, HEAD, DEPT AGR ENG, 80- *Mem:* Am Soc Agr Engrs. *Res:* Mechanization of agricultural activities. *Mailing Add:* Dept Agr Eng Star Rte Box 30 Las Cruces NM 88003

ABERNATHY, HENRY HERMAN, b Vidalia, Ga, Oct 7, 13; m 41; c 4. POLYMER CHEMISTRY. *Educ:* Emory Univ, AB, 38, MS, 39. *Prof Exp:* Anal chemist, Jackson Lab, E I Du Pont de Nemours & Co Inc, 39-40, rubber chemist, Rubber Lab, 40-52, tech mgr, Rubber Chem Div, 54-57, mgr mkt develop, Elastomer Chem Dept, 57-64, dir & mkt adv, Du Pont Far East, Inc, Showa Neoprene KK, Tokyo, 64-65, mgr planning & eng, Elastomer Chem Dept, 65-78; consult, 78-81; RETIRED. *Mem:* Am Chem Soc. *Res:* Synthetic rubber and latices. *Mailing Add:* 402 Country Club Dr Wilmington DE 19803-2921

ABERNATHY, JAMES RALPH, b Dadeville, Ala, Jan 8, 26; m 55; c 4. BIOSTATISTICS. *Educ:* Samford Univ, BS, 51; Univ NC, MSPH, 53, PhD(biostatist), 65. *Prof Exp:* Statistician, Jeffer Co Bd Health, Ala, 52-59; biostatistician, NC State Bd Health, 59-61; res assoc, 64-65, from asst prof to assoc prof, 65-75, PROF BIOSTATIST, UNIV NC, CHAPEL HILL, 75- *Mem:* Fel Am Pub Health Asn; fel Am Statist Asn; Pop Asn Am. *Res:* Statistical methodology in perinatal mortality and morbidity; demography and life tables; demographic surveys in developing countries. *Mailing Add:* 110 Virginia Dr Chapel Hill NC 27514

ABERNATHY, RICHARD PAUL, b McCaysville, Ga, Mar 22, 32; m 57; c 3. NUTRITION. *Educ:* Berry Col, BSA, 52; Univ Ga, MS, 57; Cornell Univ, PhD(animal nutrit), 60. *Prof Exp:* Asst prof & nutritionist, Ga Exp Sta, 60-66; from assoc prof to prof nutrit, Va Polytech Inst & State Univ, 66-74; HEAD DEPT FOODS & NUTRIT, PURDUE UNIV, 74- *Mem:* Am Inst Nutrit; Inst Food Technol. *Res:* Metabolic studies and nutrient requirements of preadolescent children. *Mailing Add:* Dept Foods & Nutrit Purdue Univ West Lafayette IN 47907

ABERNATHY, ROBERT O, b Dallas, Tex, Apr 16, 27; m 64. MATHEMATICS. *Educ:* Prairie View Agr & Mech Col, BS, 48; Univ Calif, Berkeley, MA, 54, PhD(math), 62. *Prof Exp:* Asst math, Univ Calif, Berkeley, 53-56; asst prof, Southern Univ, 58-59; asst, Univ Calif, Berkeley, 60-62; prof, Tenn State Univ, 62-69; chmn dept, 69-74, PROF MATH, SC STATE COL, 69- *Mem:* Am Math Soc; Math Asn Am; Soc Indust & Appl Math; Sigma Xi. *Res:* Functional analysis and partial differential equations. *Mailing Add:* SC State Col Box 1675 Orangeburg SC 29117

ABERNATHY, ROBERT SHIELDS, b Gastonia, NC, Nov 18, 23; m 49; c 5. MEDICINE, MICROBIOLOGY. *Educ:* Duke Univ, BS & MD, 49; Univ Minn, PhD, 57. *Prof Exp:* From intern to resident med, Univ Minn, 49-54, instr med & microbiol, 55-57; from asst prof to assoc prof, 57-67, head dept med, 67-76, PROF MED & MICROBIOL, UNIV ARK, LITTLE ROCK, 67-, dir, Div Infectious Diseases, 76-86. *Mem:* Am Fedn Clin Res; Am Col Physicians; Infectious Dis Soc Am. *Res:* Infectious disease. *Mailing Add:* Dept Med Slot 640 Univ Ark Med Sci 230 Kingsrow Dr Little Rock AR 72207

ABERNETHY, JOHN LEO, b San Jose, Calif, Mar 6, 15. BIO-ORGANIC CHEMISTRY. *Educ:* Univ Calif, Los Angeles, BA, 36; Northwestern Univ, MS, 38, PhD(org chem), 40. *Prof Exp:* Instr chem, Univ Tex, El Paso, 40-41; asst prof, Univ Tex, Austin, 41-45 & Washington & Lee Univ, 47-48; assoc prof, Bowling Green State Univ, 48-49 & Univ SC, 49-50; from asst prof to assoc prof, Calif State Col Syst, 51-60, res assoc, Univ Calif, Davis & Univ Calif, Los Angeles, 60-69; prof, 69-80, EMER PROF CHEM, CALIF STATE POLYTECH UNIV, 80- *Concurrent Pos:* NSF grant, 60-61; NIH fel, 60-61; assoc prof, Claremont Men's Col, 61-62; Fulbright fel, San Marcos Univ, Lima, 62-63; Sigma Xi grants, 67-70, 71-73; Res Corp grant, 69-70. *Mem:* Am Chem Soc; Sigma Xi. *Res:* Papain-catalyzed reactions involving resolutions of racemic mixtures, partial asymmetric syntheses and other organic syntheses involving amino acids and amino-containing bases, as well as other papain-catalyzed reactions. *Mailing Add:* 2932 Welcome Dr Durham NC 27705

ABERNETHY, VIRGINIA DEANE, b Havana, Cuba, Oct 4, 34; US citizen; m 80; c 4. MEDICAL ANTHROPOLOGY. *Educ:* Wellesley Col, BA, 55; Harvard Univ, MA, 68, PhD(anthrop), 70. *Prof Exp:* Fel social psychiat, Med Sch, Harvard Univ, 70-72, res assoc anthrop, Dept Psychiat, 72, assoc, 72-75; from asst prof to assoc prof, 75-80, PROF PSYCHIAT, SCH MED, VANDERBILT UNIV, 80- *Concurrent Pos:* Ed, quat jour, Population & Environment. *Mem:* Am Anthrop Asn; AAAS; Pop Asn Am; Soc Med Anthrop. *Res:* Socioeconomic determinants of postneonatal mortality; medical ethics, patient competency; medical and health sciences. *Mailing Add:* Dept Psychiat Sch Med Vanderbilt Univ Nashville TN 37232

ABERS, ERNEST S, b San Mateo, Calif, Dec 31, 36; m 60; c 1. PHYSICS. *Educ:* Harvard Univ, AB, 58; Univ Calif, Berkeley, PhD(physics), 63. *Prof Exp:* Res fel physics, Calif Inst Technol, 63-64; vis scientist, Europ Orgn Nuclear Res, Switz, 64-65; from asst prof to assoc prof, 65-74, PROF PHYSICS, UNIV CALIF, LOS ANGELES, 74- *Concurrent Pos:* Visitor, Ctr Theoret Physics, Mass Inst Technol, 68-69; Alfred P Sloan Found fel, 68-70. *Mem:* Am Phys Soc. *Res:* Theoretical elementary particle physics. *Mailing Add:* Dept Physics Univ Calif Los Angeles CA 90024

ABERTH, OLIVER GEORGE, b Akron, Ohio, July 23, 29; m 58; c 3. MATHEMATICS. *Educ:* City Col New York, BS, 50; Mass Inst Technol, MS, 51; Univ Pa, PhD(math), 62. *Prof Exp:* Engr, Remington Rand Univac Div, Sperry Rand Corp, 51-53 & 56-58; instr math, Swarthmore Col, 60-62; asst prof, Univ Ill, 62-66; assoc prof, Rutgers Univ, 66-69; lectr, City Col New York, 69-70; PROF MATH, TEX A&M UNIV, 70- *Mem:* Am Math Soc; Soc Indust & Appl Math. *Res:* Constructive analysis; geometry; tensor analysis; numerical analysis. *Mailing Add:* Dept of Math Tex A&M Univ College Station TX 77843

ABERTH, WILLIAM H, b Los Angeles, Calif, Jan 4, 33; m 54; c 3. MASS SPECTROMETRY. *Educ:* City Col New York, BS, 54; Columbia Univ, MA, 57; NY Univ, PhD(physics), 63. *Prof Exp:* Lectr physics, City Col New York, 57-60; res fel, 63-65, physicist, Stanford Res Inst, 65-69; assoc prof physics,

Sonoma State Col, 69-70; sr physicist, Stanford Res Inst Int, 70-72, asst mgr, Mass Spectrom Res Ctr, 72-75; res assoc, Space Sci Lab, Univ Calif, Berkeley, 75-76; assoc res prof, Linus Pauling Inst Sci & Med, 76-82; RES PHYSICIST, MASS SPECTRUM FAC, UNIV CALIF, SAN FRANCISCO, 82- *Mem:* Am Phys Soc; Am Soc Mass Spectrom. *Res:* Electron scattering; atomic beams; universal detectors; ionic and atomic collisions; ultra high vacuum techniques; mass spectrometer development; ion source development. *Mailing Add:* 3146 Manchester Ct Palo Alto CA 94303

ABETTI, PIER ANTONIO, b Florence, Italy, Feb 7, 21; US citizen; m 48; c 2. TECHNOLOGICAL INNOVATION, TECHNOLOGY STRATEGY & PLANNING. *Educ:* Univ Pisa, Italy, PhD(indust eng), 46; Ill Inst Technol, MS, 48, PhD(elec eng), 53. *Prof Exp:* Adv develop engr, power transformer div, Gen Elec Co, 48-56, mgr, Proj Extra-High-Voltage, 57-60, Advan Technol Elec & Info Labs, 61-62, Large Comput Systs, 65-66, Data Commun Prods, 67-70, Pvt Tel Systs, 70-73 & Europe Strategic Planning, 74-79, consult res & develop, Tech Systs, 80-81; asst mgr dir, Univac Div, Sperry Rand Int, 63-64; PROF MGT TECHNOL & ENTREPRENEURSHIP, SCH MGT, RENSSELAER POLYTECH INST, 82-, CONSULT, TECHNOL ASSESSMENT GROUP, 82- *Concurrent Pos:* Instr, Mass Inst Technol, 50-57; adj prof, Berkshire Community Col, 58-60; vis prof, Technol Univ Compliegne, France, 88-, Elec Res Inst, Mexico, 89-90; mem, Am Mgt Asn Res & Develop Coun; non-resident prof, Gordon Inst, Wakefield, MA, 87- *Honors & Awards:* Coffin Award, Gen Elec Co, 52. *Mem:* Fel Inst Elec & Electronics Engrs. *Res:* Creation of new high-technology entrepreneurial ventures; innovation and entrepreneurship in large and small companies; technological strategies and diversification; technological innovation and new product development; technological intrepreneurship and intrapreneurship. *Mailing Add:* 1026 Mohegan Rd Schenectady NY 12309

ABEY, ALBERT EDWARD, b Spokane, Wash, Aug 12, 35; m 57; c 2. SOLID MECHANICS. *Educ:* Wash State Univ, BS, 58; Univ Wash, MS, 60; Univ Ariz, PhD(physics), 64. *Prof Exp:* Res assoc physics, Univ Ariz, 64; res assoc, Advan Mat Res & Develop Lab, Pratt & Whitney Aircraft, 64-67; PHYSICIST, LAWRENCE RADIATION LAB, 67- *Mem:* Am Phys Soc. *Res:* High hydrostatic and nearly hydrostatic pressure studies of solid state physics, particularly ionic conductivity; elastic and plastic properties of materials; surface physics in connection with thermionics. *Mailing Add:* Lawrence Livermore Nat Lab L-86 Livermore CA 94551

ABHYANKAR, SHREERAM, b Ujjain, India, July 22, 30; m 58; c 2. MATHEMATICS. *Educ:* Inst Sci, India, BSc, 51; Harvard Univ, AM, 52, PhD(math), 55. *Prof Exp:* Res fel math, Harvard Univ, 54-55; res assoc, Columbia Univ, 55-56, vis asst prof, 56-57; asst prof, Cornell Univ, 57-58; vis asst prof, Princeton Univ, 58-59; assoc prof, Johns Hopkins Univ, 59-63; prof, 63-67, MARSHALL DISTINGUISHED PROF MATH, PURDUE UNIV, 67-, PROF INDUST ENG AND COMP SCI, 87- *Concurrent Pos:* Sloan Found fel, 58- *Honors & Awards:* Chauvenet Award, Math Asn Am, 78. *Mem:* Am Math Soc. *Res:* Algebraic geometry; algebra; several complex variables; circuit theory. *Mailing Add:* Div Math Sci Purdue Univ West Lafayette IN 47907

ABIAN, ALEXANDER, b Tabriz, Iran, Jan 1, 25; US citizen; m 59; c 3. PURE MATHEMATICS. *Educ:* Univ Tehran, BS, 46; Univ Chicago, MS, 54; Univ Cincinnati, PhD(math), 56. *Prof Exp:* Asst prof math, Univ Tenn, 56-57, Queens Col, 57-59 & Univ Pa, 59-62; assoc prof, Ohio State Univ, 62-67; PROF MATH, IOWA STATE UNIV, 67- *Mem:* Am Math Soc. *Res:* Theory of sets, analysis and mathematical logic. *Mailing Add:* Dept of Math Iowa State Univ Ames IA 50011

ABIKOFF, WILLIAM, b New York, NY, Aug 18, 44; div. MATHEMATICS. *Educ:* Polytech Inst Brooklyn, BS, 65, MS, 66, PhD(math), 71. *Prof Exp:* Mem tech staff, Bell Tel Labs, Inc, 65-70; instr math, Columbia Univ, 70-71; res fel, Mittag-Leffler Inst, Swed Royal Acad Sci, 71-72; asst prof, Columbia Univ, 72-75; from asst prof to assoc prof math, Univ Ill, Urbana, 75-81; PROF, DEPT MATH, UNIV CONN, 81- *Concurrent Pos:* Res fel, Inst Higher Sci Studies, 76-77 & Sloan Fel. *Mem:* Am Math Soc. *Res:* Complex analysis; hyperbolic geometry; Riemann surfaces; computation; mathematical physics. *Mailing Add:* Dept Math Box U-9 Univ Conn Storrs CT 06268

ABILDGAARD, CHARLES FREDERICK, b Winfield, Kans, Aug 10, 30; m; c 3. PEDIATRICS. *Educ:* Stanford Univ, AB, 52, MD, 55; Am Bd Pediat, dipl, 59. *Prof Exp:* Intern, Boston City Hosp, 54-55; resident pediat, Stanford Univ Hosps, 55-57; pediatrician, Coco Solo Hosp, Cristobal, CZ, 59-60; from asst prof to assoc prof pediat, Col Med, Univ Ill, 61-68; PROF PEDIAT, SCH MED, UNIV CALIF, DAVIS, 68- *Concurrent Pos:* USPHS trainee hemat, Children's Mem Hosp, Chicago, 60-61. *Mem:* Am Soc Hemat; Am Pediat Soc; Soc Pediat Res; Am Acad Pediat; Am Physiol Soc. *Res:* Pediatric hematology, thrombopoiesis, hemostasis, hemophilia. *Mailing Add:* UC Davis Sacramento Med Ctr 4301 X St Sacramento CA 95817

ABILDSKOV, J(UNIOR) A, b Salem, Utah, Sept 22, 23; m 44; c 4. MEDICINE. *Educ:* Univ Utah, BA, 44, MD, 46. *Prof Exp:* Instr internal med, Tulane Univ, 51-54; cardiologist & chief serv, William Beaumont Army Hosp, 54-56; from asst prof to prof med, State Univ NY Upstate Med Ctr, 56-68; PROF MED, UNIV UTAH, 68- *Concurrent Pos:* Fel internal med, Tulane Univ, 47-51. *Honors & Awards:* Cummings Humanitarian Award, Am Col Cardiol. *Mem:* AAAS; Am Soc Clin Invest; Am Fedn Clin Res; Soc Exp Biol & Med; Asn Am Physicians. *Res:* Improved electrocardiographic examination; recognition of states at high risk of cardiac arrhythmias; physiologic basis of electrocardiographic waveform and disorders of cardiac rhythm. *Mailing Add:* Univ Utah CVRTI Bldg 100 Salt Lake City UT 84112

ABITBOL, CAROLYN LARKINS, b Quincy, Fla, Apr 8, 46; m 74; c 5. PEDIATRIC NEPHROLOGY. *Educ:* James Madison Univ, BS, 67; Univ VA, MD, 71. *Prof Exp:* Intern med & pediat, Emory Univ, 71-72, resident pediat 72-73; fel pediat nephrol, Univ Calif, San Francisco, 73-75; Fulbright Hays scholar, Hospital des Enfants Malades, Paris, France, 75-76; asst prof pediat, Cornell Med Col, 76-78; attending physician, Nassau County Med Ctr, 78-83; asst prof pediat, Stonybrook Univ, 80-83; ASST PROF PEDIAT, UNIV MIAMI, 83- *Honors & Awards:* Ross Pediat Award, Ross Labs & Am Col Nutrit, 87. *Mem:* Am Col Nutrit; Am Acad Pediat; Am Soc Nephrol; Int Soc Nephrol; Int Soc Pediat Nephrol. *Res:* Problems of growth failure related to chronic renal failure in infants and children. *Mailing Add:* Dept Pediat Div Pediat Nephrotoxi Univ Miami Sch Med PO Box 016960 R131 Miami FL 33101

ABKOWITZ, MARTIN A(ARON), b Revere, Mass, Sept 19, 18; m 47; c 3. OCEAN ENGINEERING, NAVAL ARCHITECTURE. *Educ:* Mass Inst Technol, SB, 40; Harvard Univ, AM, 49, PhD, 53. *Prof Exp:* Naval architect, David Taylor Model Basin, 40-42, physicist, 46-47; from asst prof to assoc prof naval archit, 49-59, dir, ship model towing tank, 51-88, PROF OCEAN ENG, MASS INST TECHNOL, 59- *Concurrent Pos:* Consult hydrodynamics, dynamics, control & ocean engineering. *Honors & Awards:* Fulbright Lectr, Tech Univ Denmark, 62-63, France, 71-72 & Israel Technion, 79; Humboldt Award, 90. *Mem:* Soc Naval Architects & Marine Engrs; Int Towing Tank Conf; Am Towing Tank Conf; Sigma Xi. *Res:* Ship dynamics and automatic control; hydrodynamics of naval architecture and underwater bodies; model research. *Mailing Add:* 28 Peacock Farm Rd Lexington MA 02173

ABKOWITZ, MARTIN ARNOLD, b New York, NY, Feb 22, 36; m 57; c 2. EXPERIMENTAL SOLID STATE PHYSICS. *Educ:* City Col New York, BS, 57; Univ Rochester, MA, 59; Syracuse Univ, PhD(physics), 64. *Prof Exp:* Fel physics, Univ Pittsburgh, 64-65; scientist, 65-73, PRIN SCIENTIST, XEROX WEBSTER CORP RES LAB, 73-; ASSOC LECTR PHYSICS, UNIV ROCHESTER, 69- *Concurrent Pos:* Adj prof physics, Univ Rochester, 66-67. *Mem:* Am Phys Soc; Am Chem Soc; fel Am Phys Soc. *Res:* Amorphous semiconductors; magnetic resonance; dielectric spectroscopy; charge generation and transport in disordered molecular solids; polymer physics. *Mailing Add:* 1198 Gatestone Circle Webster NY 14550

ABKOWITZ, STANLEY, b Revere, Mass, Mar 17, 27; m 54; c 3. POWDER METALLURGY, TITANIUM METALLURGY. *Educ:* Mass Inst Technol, SB, 48. *Prof Exp:* Mgr, Dolotiva Metals, Inc, 55-58 & Spec Metals, Inc, 60-62; dir res & develop, Nuclear Metals, Inc, 62-68; PRES, DYNAMET TECHNOL, INC, 69- *Concurrent Pos:* Mem, Comt Export Admin, 88- *Mem:* Fel Am Soc Metals Int; Metall Soc; Metal Powder Industs Fedn; Titanium Develop Asn. *Res:* Titanium technology; development of alloys and processing techniques brought to commercial operation; isostatic pressing. *Mailing Add:* Dynamet Technol Eight A St Burlington MA 01803

ABLE, KENNETH PAUL, b Louisville, Ky, Feb 5, 44; m 67; c 1. ANIMAL ORIENTATION & NAVIGATION. *Educ:* Univ Louisville, BS, 66, MS, 68; Univ Ga, PhD(zool), 71. *Prof Exp:* From asst prof to assoc prof, 71-84, PROF BIOL, STATE UNIV NY ALBANY, 84- *Concurrent Pos:* NSF res grant, 74- *Mem:* Animal Behav Soc; Am Soc Naturalists; Ecol Soc Am; fel Am Ornithologists Union (treas, 81-85); AAAS. *Res:* Migration and orientation of birds; ecology and evolution of animal migration. *Mailing Add:* Dept Biol Sci State Univ NY Albany NY 12222

ABLER, RONALD FRANCIS, b Milwaukee, Wis, May 30, 39; c 2. HISTORY OF GEOGRAPHY. *Educ:* Univ Minn, BA, 63, MA, 66, PhD(geog), 68. *Prof Exp:* Prof geog, Pa State Univ, 67-90; EXEC DIR, ASN AM GEOGR, 90- *Concurrent Pos:* Vis prof, Univ BC, 71, Univ Minn, 72-74 & Stockholm Sch Econ, 82-83; dir geog prog, NSF, 84-87. *Honors & Awards:* Centerary Medal, Royal Scottish Geog Soc, 90. *Mem:* Fel AAAS; Asn Am Geogrs (pres, 85-86); Can Asn Geogrs; Am Geog Soc; Nat Coun Geog Educ. *Res:* Interpersonal communications technologies especially postal and telecommunications networks; urban geography; history of geography and history of science. *Mailing Add:* 1710 16th St NW Washington DC 20009

ABLES, ERNEST D, b Hugo, Okla, Jan 13, 34; m 60. WILDLIFE ECOLOGY, ZOOLOGY. *Educ:* Okla State Univ, BS, 61; Univ Wis, MS, 64, PhD(wildlife ecol), 68. *Prof Exp:* From asst prof to assoc prof ecol, Tex A&M Univ, 67-73; actg dean, 84-85, head, Fish & Wildlife Dept, 82-89, PROF WILDLIFE ECOL, UNIV IDAHO, 73-, ASSOC ACAD DEAN, COL FORESTRY, WILDLIFE & RANGE SCI, 74-82, 90- *Concurrent Pos:* Vis lectr wildlife ecol, Peoples Repub China, 80. *Mem:* Ecol Soc Am; Wildlife Soc; Am Soc Mammal; Am Inst Biol Sci; Wilderness Soc. *Res:* Home range and activity studies of red foxes by radio tracking; radio tracking studies of impala in Africa; ecology of exotic ungulates in Texas; ecological animal behavior. *Mailing Add:* Col Forestry Wildlife & Range Sci Univ of Idaho Moscow ID 83843

ABLES, HAROLD DWAYNE, b Hico, Tex, Feb 18, 38; m 59. ASTRONOMY. *Educ:* Univ Tex, Austin, BA, 61, PhD(astron), 68. *Prof Exp:* ASTRONR, NAVAL OBSERV, FLAGSTAFF STA, 64- *Mem:* Am Astron Soc; Int Astron Union; Sigma Xi. *Res:* Observational studies of the luminosity distributions in Magellanic type irregular galaxies and the low luminosity halos and the globular clusters around early type galaxies. *Mailing Add:* 1512 W University Heights N Flagstaff AZ 86001

ABLIN, RICHARD J, b Chicago, Ill, May 15, 40; m 64; c 1. IMMUNOBIOLOGY, SCIENCE EDUCATION. *Educ:* Lake Forest Col, AB, 62; State Univ NY Buffalo, PhD(microbiol & immunol), 67. *Prof Exp:* Lectr microbiol, Sch Med, State Univ NY Buffalo, 66-68; dir immunol, Millard Fillmore Hosp, 68-70; head immunol, Mem Hosp Springfield, 70-73; dir immunobiol, Cook County Hosp & Hektoen Inst Med Res, 73-75, sci officer immunol, 76, SR SCI OFFICER IMMUNOL, 76-; PROF UROL, STATE UNIV NY HEALTH SCI CTR, STONY BROOK. *Concurrent Pos:* Consult, Millard Fillmore Hosp, 70; consult med staff, Mem Hosp Springfield, 70-73 & St Johns Hosp, 71-73; asst prof, Sch Med, Southern Ill Univ, 71-73; assoc prof, Univ Health Sci/Chicago Med Sch, 73-74. *Honors & Awards:* P D Agarwal Mem Award Lect, Bhoruka Pub Welfare Trust, Calcutta, India, 88. *Mem:* Am Col Cryosurg (vpres, 77-78 & 78-79); Am Asn Cancer Res; Am

Asn Immunologists; Am Soc Microbiol; Int Soc Cryosurg (pres, 77-80, hon lifetime pres, 80); Soc Exp Biol & Med; Sigma Xi. *Res:* Immunobiology of genitourinary tumors, principally of the prostate, and of the knowledge derived thereof for the diagnosis, prognosis and treatment of diseases of the prostate. *Mailing Add:* Dept Urol Immunol Unit State Univ NY Health Sci Ctr Stony Brook NY 11794-8093

ABLOW, CLARENCE MAURICE, b New York, NY, Nov 6, 19; m 45. MATHEMATICS. *Educ:* Univ Calif, Los Angeles, BA, 40, MA, 42; Brown Univ, PhD(appl math), 51. *Prof Exp:* Mathematician, Boeing Airplane Co, 51-55; MATHEMATICIAN, SRI INT, 55- *Mem:* Am Math Soc; Soc Indust & Appl Math; Math Asn Am. *Res:* Solution of differential equations. *Mailing Add:* 193 Portola Rd Portola Valley CA 94025

ABLOWITZ, MARK JAY, b New York, NY, June 5, 45; m; c 3. MATHEMATICAL PHYSICS. *Educ:* Univ Rochester, BS, 67; Mass Inst Technol, PhD(math), 71. *Prof Exp:* From asst prof to prof math, Clarkson Univ, 71-79, prof & chmn, dept math & comput sci, 79-85, dean sci, 85-89; DIR, PROG APPL MATH, UNIV COLO, BOULDER, 89- *Concurrent Pos:* Teaching asst math, Mass Inst Technol, 67-71; vis prof appl math, Princeton Univ, 77-78 & 84; Sloan found fel, 75-77, John Simon Guggenheim Found fel, 84; Nat Acad Sci exchange visit, USSR, 84; assoc ed, J Math Physics, 76-79. *Honors & Awards:* Clarkson Graham Res Award, 76. *Mem:* Sigma Xi; Soc Indust & Appl Math; Math Asn Am; Am Math Soc. *Res:* Author of over 110 technical publications and two books. *Mailing Add:* Prog Appl Math Univ Colo Box 526 Boulder CO 80309-0526

ABNEY, THOMAS SCOTT, b Galatia, Ill, July 25, 38; m 62; c 3. PLANT PATHOLOGY, MYCOLOGY. *Educ:* Southern Ill Univ, BS, 60, MS, 64; Iowa State Univ, PhD(plant path), 67. *Prof Exp:* Res assoc corn dis, Iowa State Univ, 64-66; RES PATHOLOGIST, CROPS RES DIV, AGR RES SERV, USDA, 67-; asst prof, 67-84, ASSOC PROF BOT & PLANT PATH, PURDUE UNIV, WEST LAFAYETTE, 84- *Mem:* Am Phytopath Soc; Sigma Xi. *Res:* Diseases of plants, particularly organisms that attack Midwest field crops; nature of resistance; physiology of parasitism. *Mailing Add:* Dept of Bot & Plant Path Purdue Univ Lafayette IN 47907

ABOLHASSNI, MOHSEN, b Nov 7, 47. MEDICINE. *Educ:* Tehran Univ, BS, 70, MS, 76; Fla State Univ, PhD(immunol), 84. *Prof Exp:* Lab instr, Razapad Hosp, Maragheh, Iran, 70-72; lab technologist, Dept Immunol & Biochem, Pasteur Inst, 73-77; res asst, Dept Biol, Univ SFla, Tampa, 77-78, Dept Biol Sci, Fla State Univ, Tallahassee, 82-83; postdoctoral fel, Dept Large Animal Med, Univ Ga, Athens, 85-86; instr, 87-90, ASST PROF MED, DEPT MED, NY MED COL, VALHALLA, 90- *Mem:* Am Asn Immunologists. *Mailing Add:* Dept Med NY Med Col 207 Vosburgh Valhalla NY 10595

ABOLINS, MARIS ARVIDS, b Liepaja, Latvia, Feb 5, 38; US citizen; m 59; c 2. ELEMENTARY PARTICLE PHYSICS. *Educ:* Univ Wash, BS, 60; Univ Calif, San Diego, MS, 62, PhD(physics), 65. *Prof Exp:* Res asst physics, Univ Calif, San Diego, 60-65; physicist, Lawrence Radiation Lab, Univ Calif, Berkeley, 65- 68; assoc prof, 68-72, PROF PHYSICS, MICH STATE UNIV, 72- *Concurrent Pos:* Vis physicist, Europ Coun Nuclear Res, 76-77, Ctr Atomic Energy, Saclay, France, 77 & Fermi Nat Accelerator Lab, 90-92; chmn, Fermilab Users Orgn, 82-83, Exec Comt, Div Particles & Fields, 84-86, SSC Users Orgn, 89-91; consult, Dept Energy, 83- *Mem:* Fel Am Phys Soc; AAAS. *Res:* High energy physics research with electronic detectors; high energy instrumentation. *Mailing Add:* Dept Physics Mich State Univ East Lansing MI 48823

ABOOD, LEO GEORGE, b Erie, Pa, Jan 15, 22; m 47; c 2. BIOCHEMISTRY. *Educ:* Ohio State Univ, BS, 43; Univ Chicago, PhD(pharmacol), 50. *Prof Exp:* Instr physiol, Univ Chicago, 50-52; from asst prof to prof neurophysiol & biochem, Col Med, Univ Ill, 52-65, dir res labs, Dept Psychiat, 56-65; PROF BIOCHEM, UNIV ROCHESTER, 65-, PROF OF PHARMOCOL, 67- *Concurrent Pos:* Ad hoc mem rev panel, NIH; mem var comts, Nat Res Coun. *Mem:* Am Physiol Soc; Am Chem Soc; Am Col Neuropsychopharmacol; Soc Neurosci; Int Soc Neurochem. *Res:* Chemistry and cellular physiology of nervous system; isolation and characterization of neurotransmitter receptors from mammalian brain, particularly the nicotine, vasopressin and opioid receptors; brain sites and mechanisms of action of nicotine and vasopressin. *Mailing Add:* Dept of Pharmacol Univ of Rochester Med Ctr Rochester NY 14642

ABOU-DONIA, MOHAMED BAHIE, NEUROTOXICOLOGY. *Educ:* Univ Calif, Berkeley, PhD(pharmacol), 67. *Prof Exp:* PROF PHARMACOL, MED CTR, DUKE UNIV, 73- *Mailing Add:* Dept Pharmacol Duke Univ Med Ctr PO Box 3813 Durham NC 27710

ABOU-EL-SEOUD, MOHAMED OSMAN, b Cairo, Egypt, Apr 6, 21; m 49; c 4. MYCOLOGY. *Educ:* Cairo Univ, BSc, 43; Ain Shams Univ, Cairo, Dipl Educ, 59; Ohio State Univ, MSc, 62, PhD(mycol), 64. *Prof Exp:* Sec sch teacher, Egypt, 45-51; tech secy, Sci Div, Egyptian Ministry Educ, 51-54; asst prof bot, Teachers Col, Cairo, 54-60; res assoc, Southern Ill Univ, Carbondale, 64-65; prof, Ain Shams Univ, Cairo, 65-67; lectr, 68-69, asst prof, 69-71, assoc prof, 71-77, PROF NATURAL SCI, MICH STATE UNIV, 77- *Concurrent Pos:* Curric consult, King Abdulaziz Univ, Jedda, Saudi Arabia, 81-82. *Mem:* Am Inst Biol Sci. *Res:* Lipid and protein synthesis by fungi; light effect on sporulation and pigmentation of fungi. *Mailing Add:* Natural/Sci 104n Kedzie Lab Mich State Univ East Lansing MI 48824

ABOU-GHARBIA, MAGID, b Cairo, Egypt, Dec 1, 49; US citizen; m 72; c 3. PHARMACY. *Educ:* Cairo Univ, BSc, 71, MSc, 74; Univ Pa, PhD(org chem), 79. *Prof Exp:* Instr org chem, Cairo Univ, Egypt, 71-74; teaching fel chem, Univ Pa, 74-78; vis res fel med clin chem, Sch Pharm, Temple Univ, 78-79, NIH res fel, Fels Res Inst, 79-81; sr res chemist, 82-84, supvr res, Wyeth Labs, 84-87, sect head, 87-89, assoc dir, 89-91, DIR, WYETH-AYERST RES, 91- *Concurrent Pos:* Lectr, Univ Pa, 79-81. *Mem:* Am Chem Soc; fel Sigma Xi; Royal Inst Chem; Soc Neurosci. *Res:* Synthesis of biologically active compounds for the cardiovascular and central nervous systems; reaction mechanisms. *Mailing Add:* Wyeth-Ayerst Res CN-8000 Princeton NJ 08543

ABOUHALKAH, T A, b Tripoli, Lebanon, May 31, 29; m 54; c 4. PETROLEUM RESERVOIR ENGINEERING. *Educ:* Univ Tex, BA & BS, 54. *Prof Exp:* Petrol Reservoir engr, Panhandle Eastern Pipe Line Co, 56-63, mgr reservoir eng, Underground Storage Div, 63-75; consult petrol engr, Miller & Lents Ltd, 75-86; CONSULT PETROL ENGR, 86- *Concurrent Pos:* Expert witness, fed & state regulatory bodies. *Mem:* Soc Prof Well Log Analysts; Soc Petrol Engrs. *Res:* Electric log interpretation; estimation, evaluation and economic studies of oil and gas reserves; well completion recommendation; direction and supervision of various tests used for injection into and production from storage reservoirs; schedules, graphs and charts for predicting reservoir behavior; fluid flow in porous rocks; well-test analyses. *Mailing Add:* 6522 Apple Valley Lane Houston TX 77069-2407

ABOUKARSH, NAAMA A DK, renal control, blood pressure, for more information see previous edition

ABOU-KHALIL, SAMIR, b Lebanon, July 18, 43; US citizen; m 76; c 2. BIOENERGETICS. *Educ:* Lebanese Univ, Beirut, BS, 70; Claude Bernard Univ, Lyon, France, MS, 73, PhD(biochem), 75. *Prof Exp:* Res assoc plant biochem, Univ Ill, Urbana-Champaign, 76-79; res assoc, Sch Med, Univ Miami, 79-81, res prof med, 81-84, res assoc prof med, 85-88; PRES, EASYCOMP, COMPUTER EDUC & CONSULT, MIAMI, 89- *Concurrent Pos:* Res assoc biochem, Howard Hughes Med Inst, 79-81. *Mem:* Am Soc Biochem & Molecular Biol. *Res:* Exploration of cellular energy in normal and neoplastic tissues by investigating unusual mitochondrial functions in tumors; drug interactions and metabolism in human and animal tissues using state-of-the-art technology. *Mailing Add:* 7121 SW 112 Ave Miami FL 33173

ABOUL-ELA, MOHAMED MOHAMED, b Damanhur, Egypt, Mar 10, 18; US citizen; m 51; c 6. PLANT PHYSIOLOGY. *Educ:* Univ Cairo, BA, 40; Iowa State Univ, MS, 47, PhD(plant physiol), 50. *Prof Exp:* Teacher high sch, Damanhur, Egypt, 40-42; from instr to asst prof crops, Univ Alexandria, 42-55; assoc prof, Univ Bagdad, 55-56; assoc prof, Univ Alexandria, 56-62; fel plant sci, Tex A&M Univ, 62-65; asst prof, 65-67, ASSOC PROF BIOL, TEX WOMAN'S UNIV, 67- *Mem:* Consult, Agrarian Reform Ministry Egypt, 56-62. Mem: AAAS; Am Inst Biol Sci. *Res:* Effect of radiation on growing bean seedlings; cotton defoliation. *Mailing Add:* 1100 Mansell Dr Youngstown OH 44505

ABOUL-ENEIN, HASSAN YOUSSEFF, medicinal chemistry, pharmacology, for more information see previous edition

ABOU-SABE, MORAD A, b Cairo, Egypt, Apr 3, 37; m 60; c 2. MOLECULAR BIOLOGY. *Educ:* Univ Alexandria, BSc, 58; Univ Calif, Berkeley, MSc, 62; Univ Pittsburgh, PhD(bact), 65. *Prof Exp:* Instr agr, Univ Alexandria, 58-59; asst prof genetics, Cairo Univ, 66-68; asst prof, 68-74, ASSOC PROF BACT, RUTGERS UNIV, 74- *Mem:* Am Soc Microbiol; Genetics Soc Am; UAR Soc Microbiol. *Res:* Gene enzyme interactions; specific genetic control mechanism in bacteria; cyclic adenosine monophosphate regulation of cell growth and growth control. *Mailing Add:* Dept Biol Rutgers Univ New Brunswick NJ 08903

ABPLANALP, HANS, b Zurich, Switz, Apr 22, 25; m 54; c 3. GENETICS. *Educ:* Inst Technol, Zurich, Ing Agron, 49; Wash State Col, MS, 51; Univ Calif, PhD(genetics), 55. *Prof Exp:* Res asst genetics, Wash State Col, 49-51 & Univ Calif, 51-53; asst specialist statist & genetics, Div Poultry Husb, Univ Calif, Davis, 53-56, asst prof genetics, 56-59, prof avian sci & univ & poultry geneticist, Exp Sta, 69-91; RETIRED. *Mem:* Genetics Soc Am; Am Genetic Asn; Soc Study Evolution. *Res:* Population genetics; application to breeding; basic research in underlying genetic mechanisms. *Mailing Add:* 24112 County Rd 105 Davis CA 95616

ABRAHAM, BERNARD M, b Kansas City, Mo, Nov 21, 18; m 42; c 3. CHEMISTRY. *Educ:* Univ Chicago, BS, 40, PhD(chem), 46. *Prof Exp:* Jr assoc, Univ Chicago, 41-42 & Northwestern Univ, 42; jr assoc & jr chemist, Univ Chicago, 42-45; sr chemist, Monsanto Chem Co, 45; Universal Oil Prod fel, Northwestern Univ, 46-47; assoc chemist, Argonne Nat Lab, 47-59, sr chemist & group leader, 59-85; RES PROF, DEPT PHYSICS & ASTRON, NORTHWESTERN UNIV, 85- *Concurrent Pos:* Guggenheim Mem fel, 59; Lady Davis fel, Technion, Haifa, Israel, 76; dir, Off Indust Coop, Univ Chicago, 76-78. *Mem:* AAAS; Am Chem Soc; fel Am Phys Soc. *Res:* Thermodynamic and transport properties of liquid helium-3 and liquid helium-4; thermal and magnetic measurements as very low temperatures; suspension of inorganic compounds in liquid metals; thermochemical water splitting cycles; surface chemistry; monomolecular films. *Mailing Add:* 1131 N Oak Park Oak Park IL 60302

ABRAHAM, DONALD JAMES, b Greensburg, Pa, Nov 19, 36; m 62; c 1. ORGANIC CHEMISTRY, MEDICINAL CHEMISTRY. *Educ:* Pa State Univ, BS, 58; Marshall Univ, MS, 59; Purdue Univ, PhD(org chem), 63. *Prof Exp:* Res assoc med chem & fel, Univ Va, 63-64; res assoc, Univ Pittsburgh, 64-67, from assoc prof to prof med chem, 68-88, head dept, 68-88; PROF MED CHEM & DEPT CHMN, MED COL VA, VA COMMONWEALTH UNIV, 88- *Concurrent Pos:* Mem preclin psychopharmacol study sect, NIMH, Alcohol & Drug Abuse Ment Health Admin, 73-77; vis scientist, MRC Lab Molecular Biol, Cambridge, Eng, 80-81. *Honors & Awards:* Sr US Scientist, Alexander von Humboldt Found, 74. *Mem:* Am Chem Soc. *Res:* Antisickling agents and x-ray structures of complexes of antisickling agents and hemoglobin; structure elucidation of biologically important compounds; synthesis of medicinals; x-ray crystallography; molecular action of drug molecules; psychopharmacology research. *Mailing Add:* Dept Med Chem Med Col Va Va Commonwealth Univ Richmond VA 23298-0540

ABRAHAM, EDATHARA CHACKO, b Kottayam, India; m 63; c 3. BIOCHEMISTRY. *Educ:* Univ Kerala, BS, 59; Univ Louisville, PhD(biochem), 71. *Prof Exp:* Clin biochemist, Christian Med Col, Vellore, India, 59-62, res biochemist, 63-67; res biochemist renal biochem, Univ Louisville Sch Med, 67-69, grad res asst biochem, 69-71; from asst prof to assoc prof, 74-86, PROF BIOCHEM, MED COL GA, 86- *Concurrent Pos:* Res assoc, Univ Louisville Sch Med, 72-73 & Med Col Ga, 73-74. *Mem:* Am Soc Hematol; Am Asn Clin Chem; Am Soc Human Genetics; NY Acad Sci; AAAS; Am Diabetes Asn; Geront Soc Am; Asn Res Vision Ophthal; Inst Social Econ Res. *Res:* Posttranslational modifications of proteins, acetylation, glycation, phosphorylation, etc; protein glycation in biological aging and in diabetes; glycation of lens crystallins in diabetic and senile cataractogenesis; protein kinase C mediated membrane phosphoryation; on the mechanism of human fetal hemoglobin acetylation. *Mailing Add:* Dept Biochem & Molecular Biol Med Col Ga Augusta GA 30912

ABRAHAM, FARID FADLOW, b Phoenix, Ariz, May 5, 37; m 59; c 4. PHYSICS. *Educ:* Univ Ariz, BS, 59, PhD(physics), 62. *Prof Exp:* Res assoc high energy physics, Univ Chicago, 62-64; sr physicist, Lawrence Radiation Lab, Univ Calif, 64-66; staff mem, IBM Sci Ctr, 66-71, proj leader air qual dynamics group, 71-72, MEM STAFF, IBM RES LAB, 72- *Mem:* Am Phys Soc. *Res:* Nucleation theory; phase stability, thermodynamics of nonuniform systems, statistical physics. *Mailing Add:* 15433 Kennedy Rd Los Gatos CA 95032

ABRAHAM, GEORGE, b New York, NY, July 15, 18; wid; c 4. SOLID STATE ELECTRONICS. *Educ:* Brown Univ, ScB, 40; Harvard Univ, SM, 42; Univ Md, PhD(physics), 72. *Prof Exp:* Radio engr, RCA, Camden, NJ, 41, electronic scientist, 42-45, head sci educ prog training off, 45-52, head exp devices sect, 52-64, head microelectronics sect & consult, 64-69, RES PHYSICIST & HEAD SYSTS APPLN, OFF RES & TECH APPLN, NAVAL RES LAB, 69- *Concurrent Pos:* Pres, chmn & mem bd dirs, Intercol Broadcasting Syst, Inc, 40-88; lectr sci educ prog, Naval Res Lab, 45-52; res assoc & lectr elec eng & physics, Univ Md, 45-74; lectr elec eng, George Washington Univ, 52-67; indust consult, Nat Cash Regist Co, Borg Warner Co, Maico Electronics, Electro Nuclear Systs & Aero Geo Astro, Inc, 58-66; lectr, Capitol Radio Eng Inst, 62-64; tutor, Naval Postgrad Sch, Monterey, 76-78; res adv, Nat Res Coun, Nat Acad Sci-Nat Acad Eng, 76-78; lectr, Am Univ, 79. *Honors & Awards:* Founders Award, Govt Microcircuit Appln Conf, 86. *Mem:* Fel Inst Elec & Electronics Engrs; Am Phys Soc; fel AAAS; Am Asn Univ Prof; fel NY Acad Sci. *Res:* Microelectronics; multistable semiconductor devices and integrated circuits; nonvolatile electronic devices; analog and digital solid state system applications. *Mailing Add:* 3107 Westover Dr SE Washington DC 20020

ABRAHAM, GEORGE N, BIOCHEMISTRY, MOLECULAR BIOLOGY. *Educ:* Univ Buffalo, MD, 63. *Prof Exp:* ASSOC PROF MED & MICROBIOL, SCH MED, UNIV ROCHESTER, 70- *Res:* Immunology. *Mailing Add:* Immunol Unit Sch Med Univ Rochester Box 695 Rochester NY 14642

ABRAHAM, IRENE, b Chicago, Ill, Dec 12, 46; m 76; c 2. CELL BIOLOGY, SOMATIC CELL GENETICS. *Educ:* Univ Ill, Urbana, BS, 68; Univ NC, Chapel Hill, PhD(zool), 73. *Prof Exp:* Fel biol, Yale Univ, 73-75; asst prof biol, Emory Univ, 75-76; fel biol, Univ Berne, Switz, 76-78; biologist, Nat Cancer Inst, 79-84; BIOLOGIST, CELL BIOL DEPT, UPJOHN CO, 84- *Mem:* Soc Develop Biol; Am Soc Cell Biol; Sigma Xi; Am Asn Cancer Res. *Res:* Gene regulation and developmental genetics of eukaryotes; cyclic adenosine monophosphate dependent protein kinase and the control of cell growth; the role of the cytoskeleton in cell growth and behavior; multidrug resistance. *Mailing Add:* Cell Biol Dept Upjohn Co Kalamazoo MI 49001

ABRAHAM, JACOB A, b Tiruvalla, Kerala State, India, Dec 8, 48; US citizen; m 75; c 2. COMPUTER ARCHITECTURE. *Educ:* Univ Kerala, BSc, 70; Stanford Univ, MS, 71, PhD(elec eng & comput sci), 74. *Prof Exp:* Actg asst prof elec eng, Stanford Univ, 74-75; from asst prof to assoc prof elec eng, 75-83, PROF ELEC ENG, UNIV ILL, URBANA, 83-, PROF COMPUT SCI, 84- *Concurrent Pos:* Co-prin investr, Very High Speed Integrated Circuits Prog, Off Naval Res, 80-84; consult, Gen Elec Co, 83-, Gen Tel & Electronics, 84-, Gen Motors Res Labs, 85- & Gen Res Corp; assoc ed, Trans Comput-Aided Design of Integrated Circuits & Systs, Inst Elec & Electronics Engrs, 84-; sr prin investr, NASA, 85-; dir res prog, Semiconductor Res Corp, 85-; Beckman assoc, Ctr Advan Study, Univ Ill, 85-86. *Mem:* Fel Inst Elec & Electronics Engrs; Asn Comput Mach; Sigma Xi; Int Fedn Info Processing. *Res:* Functional-level testing of very-large-scale integration; design automation; design of testable systems and built-in tests; concurrent error detection; fault-tolerant computer; reconfiguration and recovery in distributed systems; high-performance computer architecture. *Mailing Add:* Comput Eng Res Ctr Univ Tex Austin 2201 Donley Dr Suite 395 Austin TX 78758

ABRAHAM, JERROLD L, b Los Angeles, Calif, Nov 5, 44; m 67; c 3. ENVIRONMENTAL PATHOLOGY, PULMONARY PATHOLOGY. *Educ:* Mass Inst Technol, BS, 66; Univ Calif, San Francisco, MD, 70. *Prof Exp:* Pathologist, Appalachian Ctr Occup Safety & Health, USPHS, 72-74, chief path, 74-75; ASST PROF PATH, SCH MED, UNIV CALIF, SAN DIEGO, 75- *Concurrent Pos:* Instr path, WVa Univ, 72-75. *Mem:* AAAS; Am Thoracic Soc; Int Acad Path; Soc Toxicol Path. *Res:* Diagnostic and investigational studies of environmental occupational disease, especially lung diseases caused by inorganic particulates; development and application of microanalytic techniques, electron probe, ion probe, to pathology of humans and animals. *Mailing Add:* Dept Path SUNY Health Sci Ctr 766 Irving Ave Syracuse NY 13210

ABRAHAM, JOHN, b Kerala, India, Dec 14, 56; m 86. COMBUSTION, NUMERICAL METHODS. *Educ:* Indian Inst Technol, India, BTech, 81; Princeton Univ, MA, 84, PhD(mech & aerospace eng), 86. *Prof Exp:* SR ENGR, JOHN DEERE TECHNOLOGIES INT, DEERE & CO, 86- *Concurrent Pos:* Vis res collabr, Princeton Univ, 89- *Mem:* Am Soc Mech Engrs; Am Inst Aeronaut & Astronaut; Soc Automotive Engrs; Soc Indust & Appl Math; Combustion Inst. *Res:* Combustion; numerical solution of reacting flow problems and fluid mechanics with specific interest in modeling of combustion in reciprocating and rotary engines. *Mailing Add:* Rotary Eng Div John Deere Technologies Int PO Box 128 Wood-Ridge NJ 07075

ABRAHAM, L(EONARD) G(LADSTONE), JR, b Brooklyn, NY, Mar 3, 26; m 50, 78; c 4. RADIO WAVE PROPAGATION,. *Educ:* Cornell Univ, BEE, 49, MEE, 52, PhD(radiowave propagation), 53. *Prof Exp:* Res assoc, Gen Elec Res Lab, 53-60; sr eng specialist, Appl Res Lab, Sylvania Electronic Syst Div, 60-70, sr scientist, Gen Tel & electronics Labs, Gen Tel & Electronics Corp, 70-84; lead engr, 84-89, PRIN SCIENTIST, MITRE CORP, 89- *Mem:* Inst Elec & Electronic Engrs; Sigma Xi. *Res:* Telephone systems, digital processing and transmission; communication systems, interactive terminals; radio wave propagation. *Mailing Add:* 99 Whitman St Stow MA 01775

ABRAHAM, MARVIN MEYER, b New York, NY, Dec 8, 30; m 58; c 3. PHYSICS. *Educ:* City Col New York, BS, 53; Univ Calif, Berkeley, MA, 54, PhD(physics), 58. *Prof Exp:* Asst physics, Univ Calif, 54-58; Fulbright res fel, Clarendon Lab, Oxford Univ, 58-60; res physicist, Lawrence Radiation Lab, Univ Calif, Berkeley, 60-63; RES PHYSICIST, SOLID STATE DIV, OAK RIDGE NAT LAB, 63- *Concurrent Pos:* Pressed steel fel, Oxford Univ, 59-60; Int Atomic Energy Authority vis prof, San Carlos de Bariloche Inst Physics, Arg, 62-63; OAS fel, UNAM, Mexico, 74; vis fel, St Catherine's Col, Oxford Univ, UK, 89. *Honors & Awards:* IR-100 Award, 75 & 82. *Mem:* Fel Am Phys Soc; fel Am Ceramic Soc. *Res:* Experimental solid state physics; electron paramagnetic resonance; nuclear magnetic resonance. *Mailing Add:* Solid State Div Oak Ridge Nat Lab PO Box 2008 Oak Ridge TN 37831-6056

ABRAHAM, NEAL BROADUS, b Hagerstown, MD, Dec 3, 50. QUANTUM OPTICS. *Educ:* Dickinson Col, BS, 72; Bryn Mawr Col, PhD(physics), 77. *Prof Exp:* Instr, Bryn Mawr Col, 76; asst prof physics, Swarthmore Col, 77-80; from asst prof to assoc prof, 80-87, chmn, 84-87, PROF PHYSICS, BRYN MAWR COL, 87-, RACHEL C HALE PROF MATH & SCI, 88- *Concurrent Pos:* Proj dir, Res Corp grant, 77-79; prin investr, NSF grants, 78-87; proj dir, ARO Contract, 86-87; NATO grants, 84-90; Alexander von Humboldt Fel, 87-88; vis prof, Univ Florence, 84, Polytech Torino, 88, Univ Lill, 88, Jilin Univ, 87; Int Conf Organizer, 83-85, 87 & 90; Res Corp grants adv comt, 87-92; ed, Optics Comt, 89-92. *Mem:* Am Phys Soc; Optical Soc Am; Am Asn Physics Teachers; Sigma Xi; fel Opt Soc Am; AAAS. *Res:* High gain gas lasers and laser amplifiers; intensity fluctuations; descriptions of stimulated and spontaneous emission; nonlinear phenomena; optical chaos and dynamics. *Mailing Add:* Dept Physics Bryn Mawr Col Bryn Mawr PA 19010-2899

ABRAHAM, RAJENDER, b Hyderabad, India, May 26, 32; US citizen; m 58; c 1. ENVIRONMENTAL HEALTH, EXPERIMENTAL PATHOLOGY. *Educ:* Osmania Univ, BSc, 52, MSc, 54, PhD(protozool), 57. *Prof Exp:* Prof toxicol, Sch Pub Health, State Univ NY, 84; prof pharmacol, 84-86, prof path & toxicol, Albany Med Col, 86-88; PRES, ABRAHAM ASSOC LTD, 87- *Concurrent Pos:* Dir, Inst Exp Path & Toxicol, Albany Med Col, 81-86; vis prof, Russell Sage Col, Albany, NY, 75-77. *Mem:* Soc Toxicol; Path Soc Gt Brit & Ireland; Fel Royal Micros Soc; AAAS. *Res:* Toxicology; effects of chemicals on molecular mechanisms in carcinogenesis; chemical interfering with reproductive process. *Mailing Add:* 191 Ormond St Albany NY 12208

ABRAHAM, RALPH HERMAN, b Burlington, Vt, July 4, 36; m 58; c 2. MATHEMATICS. *Educ:* Univ Mich, BSE, 56, MS, 58, PhD(math), 60. *Prof Exp:* Lectr math, Univ Calif, Berkeley, 60-62; Off Naval Res res assoc, Columbia Univ, 62-63, asst prof, 63-64; asst prof, Princeton Univ, 64-68; assoc prof, 68-80, PROF NATURAL SCI, UNIV CALIF, SANTA CRUZ, 80- *Mem:* Am Math Soc. *Res:* General relativity, qualitative theory of ordinary differential equations and mechanical systems; analysis on manifolds. *Mailing Add:* Univ Calif Univ of Calif Santa Cruz CA 95064

ABRAHAM, ROBERT THOMAS, b Munhall, Pa, Oct 11, 52. CHEMICAL PHARMACOLOGY, DRUG METABOLISM. *Educ:* Bucknell Univ, BS, 74; Univ Pittsburgh, PhD(pharmacol), 81. *Prof Exp:* Res fel pharmacol, 81-91, ASST PROF & CONSULT, DEPT IMMUNOL & PHARMACOL, MAYO FOUND, 91- *Mem:* Am Soc Mass Spectrometry; AAAS. *Res:* Effects of xenobiotics on the immune system; the interplay between the drug metabolizing system of liver and other organs and the immunomodulatory activity of sulfhydryl containing drugs. *Mailing Add:* Dept Immunol & Pharmacol Mayo Clin 342B Guggenheim Bldg Rochester MN 55905

ABRAHAM, W(AYNE) G(ORDON), b Durban, SAfrica, Mar 23, 21; US citizen; m 50; c 4. ELECTRONICS. *Educ:* Calif Inst Technol, BS, 41; Stanford Univ, MS, 47, PhD(elec eng), 50. *Prof Exp:* Asst proj engr, Sperry Gyroscope Co, 41-44; res physicist, Naval Res Lab, 44-45; sr scientist, Varian Assocs, 48-87; RETIRED. *Mem:* Inst Elec & Electronics Engrs; Sigma Xi. *Res:* Electronics; velocity modulation tubes. *Mailing Add:* 12630 Corte Madera Los Altos Hills CA 94022

ABRAHAM, WILLIAM H, b Englewood, NJ, May 20, 29; m 53; c 5. CHEMICAL ENGINEERING. *Educ:* Cornell Univ, BChE, 52; Purdue Univ, PhD, 57. *Prof Exp:* Res engr, Eng Res Lab, E I du Pont de Nemours & Co, 57-62; from asst prof to assoc prof chem eng, 62-67, PROF CHEM ENG, IOWA STATE UNIV, 67- *Mem:* Am Inst Chem Engrs; Sigma Xi. *Res:* Chemical process dynamics; process control; thermodynamic measurements. *Mailing Add:* Dept Chem Eng Sweeney Hall Iowa State Univ Ames IA 50011

ABRAHAM, WILLIAM MICHAEL, b Clifton Springs, NY, Nov 11, 49; m 72; c 2. ALLERGIC AIRWAY DISEASES, PULMONARY PHYSIOLOGY. *Educ:* Brown Univ, AB, 71; Univ Rochester, MS, 74, PhD(physiol), 76. *Prof Exp:* Sr res assoc, div pulmonary dis, 81-83, ASSOC DIR RES, MT SINAI MED CTR, 83-, ADMIN, DEPT RES, 87- *Mem:* NY Acad Sci; Am Physiol Soc; Am Thoracic Soc. *Res:* Respiratory physiology and pharmacology. *Mailing Add:* Mt Sinai Med Ctr 300 Alton Rd Miami Beach FL 33140

ABRAHAMS, ATHOL DENIS, b New S Wales, Australia, Mar 10, 46; m 76; c 1. FLUVIAL GEOMORPHOLOGY, DESERT GEOMORPHOLOGY. *Educ:* Univ Sydney, BA, 67 & PhD(geomorphol), 71. *Prof Exp:* Post doc fel geog, Univ Alta, 71-73; lectr geog, Univ New S Wales, 73-77; assoc prof, 77-

84, PROF GEOG, STATE UNIV NY BUFFALO, 84-, CHMN DEPT, 88- *Concurrent Pos:* Assoc ed, Annals Asn Am Geogr, 82-87; secy-treas, Geomorphol Specialty Group, Asn Am Geogr, 84-85, chair, 85-86; proj dir, Nat Sci Found Grants, 85, 86, 88-90; NATO grants, 85; distinguished vis prof, Ariz State Univ, 87; vis mem staff, Univ Exeter, Eng, 87. *Honors & Awards:* GK Gilbert Award, Asn Am Geogr, 85; Gladys W Cole Res Award, Geol Soc Am, 85. *Mem:* Asn Am Geogr; Am Geophys Union; Sigma Xi; Geog Soc Am; Brit Geomorphol Res Group; Int Geog Union. *Res:* Hydraulics and erosion mechanics of overland flow on desert hillslopes; morphology and sedimentology of desert hillslopes; morphometry of channel networks; river channel morphology and processes. *Mailing Add:* Dept Geog State Univ NY Buffalo Buffalo NY 14261

ABRAHAMS, ELIHU, b Port Henry, NY, Apr 3, 27; m 53; c 2. SUPERCONDUCTIVITY, QUANTUM STATISTICAL MECHANICS. *Educ:* Univ Calif, PhD(physics), 52. *Prof Exp:* Res assoc physics, Univ Calif, 52-53; res assoc, Univ Ill, 53-55, asst prof, 55-56; from asst prof to assoc prof, 56-64, PROF PHYSICS, RUTGERS UNIV, NEW BRUNSWICK, 64- *Concurrent Pos:* Trustee, Aspen Ctr Physics, 76-82, pres, 79-82. *Mem:* Nat Acad Sci; Am Phys Soc; AAAS. *Res:* Theoretical condensed matter physics; many body problem; critical phenomena; disordered systems; superconductivity. *Mailing Add:* Serin Phys Labs Rutgers Univ Freylinghuysen Rd Piscataway NJ 08855-0849

ABRAHAMS, M(ARVIN) S(IDNEY), b New York, NY, Jan 5, 33; m 58; c 3. MATERIALS SCIENCE. *Educ:* Columbia Univ, BS, 54, MS, 55, ScD(metall eng), 58. *Prof Exp:* Res assoc, Columbia Univ, 56-57; MEM TECH STAFF, MAT RES LAB, DAVID SARNOFF RES CTR, RCA LABS, 57- *Concurrent Pos:* Sr vis fel, Univ Oxford, Eng, 80. *Mem:* Am Phys Soc; Mat Anal Soc; Electron Micros Soc Am; Electrochem Soc. *Res:* Plastic deformation of crystalline solids; brittle fracture of semiconductors; effect of structural imperfections on transport phenomena in semiconductors; defect origin, interactions and control in crystals; crystal growth of display phosphor. *Mailing Add:* eight Evans Dr Cranbury NJ 08512

ABRAHAMS, PAUL W, b Brooklyn, NY, Nov 17, 35; m 80; c 1. PROGRAMMING LANGUAGES, TEXT EDITING. *Educ:* Mass Inst Technol, BS, 56; ScD(math), 63. *Prof Exp:* Sect supvr, ITT Data & Info Systs, 62-64; sr res scientist, Info Inst Inc, 64-64; assoc prof comput sci, NY Univ, 66-80; CONSULT COMPUT SCI, 80- *Concurrent Pos:* Chmn, ACM Spec Interest Group Prog Languages, 78-80; vpres, assoc comput mach, 82-84, chmn, 84-86, pres, 86-88. *Mem:* Asn Comput Mach; Inst Elec & Electronic Engrs. *Res:* Programming language design, definition and implementation; text editing systems, technical writing. *Mailing Add:* 214 River Rd Deerfield MA 01342

ABRAHAMS, SIDNEY CYRIL, b London, Eng, May 28, 24; US citizen; m 50; c 3. CRYSTALLOGRAPHY. *Educ:* Glasgow Univ, BSc, 46, PhD, 49, DSc, 57. *Hon Degrees:* Uppsala Univ, Fil Dr, Master Liberal Arts, 82. *Prof Exp:* Fel, Univ Minn, 49-50; mem staff, Div Indust Coop, Lab Insulation Res, Mass Inst Technol, 50-54; Imp chem indust fel, Univ Glasgow, 54-57; mem tech staff, AT&T Bell Labs, 57-82, distinguished mem tech staff, 82-88; ADJ PROF PHYSICS DEPT, SOUTHERN ORE STATE COL, 90- *Concurrent Pos:* Guest scientist, Brookhaven Nat Lab, 57-, mem vis comt, 80-83, mem High Flux Reactor Adv Comt, 85-89; mem comn crystallog apparatus, Int Union Crystallog, 63-72, chmn, 72-75, chmn comn jour, 78-87, chmn comn crystallog nomenclature, 78-, mem finance comt, 81-87; mem, Nat Comt Crystallog, 66-79, chmn, 70-72; mem comt chem crystallog, Div Chem & Chem Technol, Nat Res Coun, 67-68, chmn, 68-76; mem publ policy comt, Am Inst Physics, 74-81, chmn, 81-; vis prof, Univ Bordeaux, France, 79, 90; mem, Numerical Data Adv Bd, Nat Res Coun, 87-89; Humboldt Found Sr US Scientist Award, Inst Kristallographie, Univ Tübingen, West Germany, 89-90. *Mem:* Am Crystallog Asn (pres, 68); fel Am Phys Soc; Brit Royal Soc Chem; fel AAAS; Sigma Xi. *Res:* Crystal chemistry and solid state physics; dielectric, elastic, optical and magnetic properties in relation to atomic arrangement; crystallographic instrumentation, accuracy and automation. *Mailing Add:* Physics Dept Southern Ore State Col Ashland OR 97520

ABRAHAMS, VIVIAN CECIL, b London, Eng, Oct 19, 27; m 55; c 4. NEUROPHYSIOLOGY. *Educ:* Univ Edinburgh, BSc, 52, PhD(physiol), 55, DSc, 78. *Prof Exp:* Asst lectr physiol, Univ Edinburgh, 52-54, lectr, 54-55; vis instr, Univ Pa, 55-56; Beit fel med res, Nat Inst Med Res, Univ London, 56-59, mem sci staff, Med Res Coun, 59-63; assoc prof, 63-67, head dept, 76-88, PROF PHYSIOL, QUEENS UNIV, ONT, 67-; DIR, MED RES COUN CAN GROUP SENSORY-MOTOR PHYSIOL, 90- *Concurrent Pos:* Med Res Coun vis scientist, Cerebral Functions Group, Dept Anat, Univ Col, Univ London, 70-71; bd mem, Biol Studies, Ont Grad Scholar Prog, chmn, 76-77; mem, Sci Policy Comt, Can Fedn Biol Socs, SCITEC Coun, Rep Can Physiol Soc, 76-, sci comt, Muscular Dystrophy Asn Can, 82-91, med res coun, Comt Neurosci, 82-85, finance comt, Soc Neurosci, 83-86, organizing comt, Int Union Physiol Sci, 84-86, med res coun, 84-89, standing comt on pub affairs, 85-89, med res coun, Working Group on Res Involving Animals, 86-89, med res coun, Standing Comt on Res & Personnel Funding, 87-89, nat organizing comt, Third IBRO/WFNS Cong, 87-91, Soc Neurosci, Govt & Pub Affairs Comt, 87-; chmn, Sci Policy Comt, Can Psychol Soc, Ont Grad Scholar Bd, 77-78; bd dirs, Sci Policy Comt, Can Fedn Biol Socs, 82-85 & vchmn, 83-85; counr, Interciencia Asn Sci Asn of Americas; vpres, Parliamentary & Sci Comt, 79-80; mem & vpres, Sci Side Comt Parliamentarians, Scientists & Engrs, 79-83; mem, Coun Scientists, Human Frontiers Sci Prog Orgn, 90-; prin investr, Network Centres Excellence for Neural Regeneration & Functional Recovery, 90- *Mem:* Am Physiol Soc; Can Physiol Soc (pres, 83-84); Asn Sci, Technol & Eng Commun Can (pres, 78-79); Soc Neurosci; Can Asn Neurosci (pres, 81-82); Can Fedn Biol Socs; Int Asn Study Pain; Physiol Soc. *Res:* Organization of sensory and effector systems in the brain stem. *Mailing Add:* Dept Physiol Queens Univ Kingston ON K7L 3N6 Can

ABRAHAMSON, ADOLF AVRAHAM, b Berlin, Ger, Mar 14, 20; US citizen; m 51; c 1. THEORETICAL PHYSICS. *Educ:* NY Univ, BA, 50, MS, 54, PhD(physics), 60. *Prof Exp:* Tutor physics, 57-60, from instr to assoc prof, 60-74, PROF PHYSICS, CITY COL NEW YORK, 74- *Concurrent Pos:* Consult, Brookhaven Nat Lab, 60-66; NASA res grant, 61-67; nat scientist personnel registr, Am Inst Physics, 63; Israel Atomic Energy Comn fel, 64-65. *Mem:* Am Phys Soc; Am Asn Physics Teachers; NY Acad Sci. *Res:* Interatomic potentials at very small and intermediate separations; applications to radiation damage in solids; gas scattering; astrophysics; statistical model of the atom; three articles on math and eight articles on physics in Encyclopedia Britanica. *Mailing Add:* Dept Physics City Col NY Convent Ave & 139th St New York NY 10031

ABRAHAMSON, DALE RAYMOND, b Washington, DC, June 18, 49; m 71; c 1. ELECTRON MICROSCOPY, IMMUNOCYTOCHEMISTRY. *Educ:* Univ Va, BA, 71, PhD(cell biol), 81; George Mason Univ, BA, 76. *Prof Exp:* Res fel, Med Sch, Harvard Univ, 80-83; asst prof cell biol, 83-87, ASSOC PROF, CELL BIOL, UNIV ALA, BIRMINGHAM, 87- *Concurrent Pos:* Res fel, Brigham & Women's Hosp, 80-83. *Mem:* AAAS; Am Asn Anatomists; Am Soc Cell Biol. *Res:* Use of labeled polyclonal and monoclonal antibodies against laminin, which is the primary noncollagenous glycoprotein of basement membranes, to determine the ultrastructural distribution of laminin with the normal glomerular basement membrane and in several experimental models of kidney disease; investigations of in vivo and in vitro development of nephrons. *Mailing Add:* Dept Cell Biol Univ Ala Univ Sta Birmingham AL 35294

ABRAHAMSON, DEAN EDWIN, b Hasty, Minn, Dec 21, 34; c 2. FORENSIC SCIENCE, ENVIRONMENTAL SCIENCES. *Educ:* Gustavus Adolphus Col, BS, 55; Univ Nebr, MA, 58; Univ Minn, Minneapolis, MD & PhD(anat), 67. *Prof Exp:* First asst physics, Royal Inst Technol, Sweden, 58; reactor physicist, Babcock & Wilcox Co, Va, 58-59; sr scientist, Honeywell Inc, Minn, 59-63; asst prof anat & lab med, 67-70, assoc prof anat, lab med & physics & dir ctr studies phys environ, 70-72, assoc dean, Humphrey Inst, 82-84, PROF PUB AFFAIRS & CHMN ALL-UNIV COUN ENVIRON QUAL, UNIV MINN, 72- *Concurrent Pos:* Mem adv bd, Energy Pol Proj, Ford Found, 72-74; mem bd dirs, Natural Resources Defense Coun, 72-; adv, Swedish Minister Energy, Stockholm, 77-78; vis prof, Inst Theoret Physics, Chalmers Tech Univ & Univ Gothenburg, Sweden, 77-78, Univ Iceland, 84-85; mem, Resource Group Sociopolit Effects, Nat Res Coun Comt Nuclear & Alternative Energy Systs, Nat Acad Sci, 76-79; mem, Int Rev Group, Gorleben Nuclear Waste Storage & Reprocessing Ctr, Minister Social Affairs, Lower Saxony, 78-79; mem spec inquiry, Three Mile Island Nuclear Accident, Peer Rev Group, US Nuclear Regulatory Comn, 79-80; gov, Select Health Adv Comt, Dept Health, Pa, 82-87; vis prof, Univ Iceland, 84-85; vis prof, Lund Univ, Lund, Sweden, 90. *Mem:* Fel, Sci Inst Pub Info; fel AAAS; Am Geophys Union. *Res:* Technology assessment; energy policy with emphasis on nuclear issues; environmental implications of energy policies; policy relating to atmospheric buildup of carbon dioxide and climate change. *Mailing Add:* Humphrey Inst Pub Affairs Univ Minn 301 19th Ave S Minneapolis MN 55455

ABRAHAMSON, EARL ARTHUR, b Aberdeen, SDak, June 5, 24; m 48; c 3. ANALYTICAL CHEMISTRY. *Educ:* Univ NDak, BS, 48; Univ Kans, PhD(chem), 51. *Prof Exp:* Anal res chemist, E I du Pont de Nemours & Co, Inc, 51-58, asst div head, Phys & Anal Div, 58-69, suprv comput serv sect, Cent Res Dept, 69-83, res suprv, 83-85; RETIRED. *Mem:* AAAS; Am Chem Soc; Sigma Xi. *Res:* Analytical and physical chemistry; electrochemistry; on-line computer systems. *Mailing Add:* 85 Superfine Lane Wilmington DE 19802

ABRAHAMSON, EDWIN WILLIAM, physical chemistry, molecular biology; deceased, see previous edition for last biography

ABRAHAMSON, GEORGE R(AYMOND), b Painesville, Ohio, Aug 31, 27; m 48; c 4. ENGINEERING MECHANICS. *Educ:* Stanford Univ, BS, 55, MS, 56, PhD(eng mech), 58. *Prof Exp:* Machinist, 53-54, mech engr, 54-58, head, Explosives Eng Sect, 58-69, dir, Poulter Lab, 69-80, vpres, Phys Sci Div, 81-88, SR VPRES, SCI GROUPS, SR INT, 88- *Concurrent Pos:* Consult, Norweg Defense Estab, 68-69, & US Air Force sci adv bd, 85- *Mem:* Sigma Xi. *Res:* Nuclear weapons effects; dynamics of structures under pulse loads; simulation of pulse loads with explosives; properties of explosives; mechanics of penetration; precision explosive devices. *Mailing Add:* Poulter Lab SRI Int Menlo Park CA 94025

ABRAHAMSON, HARMON BRUCE, b Cokato, Minn, Aug 26, 52; m 82; c 3. PHOTOCHEMISTRY, SPECTROSCOPY. *Educ:* Univ Minn, Minneapolis, BChem, 74; Mass Inst Technol, PhD(chem), 78. *Prof Exp:* Asst prof chem, Univ Okla, 78-84; asst prof chem, 84-88, ASSOC PROF CHEM, UNIV NDAK, 88- *Mem:* Am Chem Soc; Sigma Xi; Inter-Am Photochem Soc. *Res:* Inorganic and organometallic photochemistry; spectroscopy of transition metal coordination complexes; coordination chemistry of small reactive molecules. *Mailing Add:* Dept Chem Univ NDak Univ Sta Box 7185 Grand Forks ND 58202-7185

ABRAHAMSON, LILA, b Paterson, NJ, Dec 26, 27. PLANT BIOCHEMISTRY. *Educ:* Queens Col, NY, BS, 49; Brooklyn Col, AM, 55; Univ Mich, PhD(bot), 60. *Prof Exp:* Instr biol, Brooklyn Col, 53-54; from instr to asst prof, Cornell Col, 59-61; asst prof biol, Univ Chicago, 61-64, asst prof biol & bot, 64-67; assoc prof biol, WVa Univ, 67-78; TECH ED, LAWRENCE LIVERMORE LAB, 78- *Mem:* AAAS; Sigma Xi. *Res:* Growth and development of plant tissue cultures, especially biochemisty of development; purification and characterization of homocysteine methyltransferase from plant seeds; methionone biosynthesis in higher plants. *Mailing Add:* 1633 Bluebell Dr Livermore CA 94550

ABRAHAMSON, SEYMOUR, b New York, NY, Nov 28, 27; m 53; c 1. GENETICS. *Educ:* Rutgers Univ, AB, 51; Ind Univ, PhD(genetics), 56. *Prof Exp:* USPHS fel, Univ Wis, 56-57; from instr to asst prof biol, Rutgers Univ, 57-60; from asst prof to assoc prof zool, 60-69, PROF ZOOL & GENETICS, UNIV WIS-MADISON, 69- *Concurrent Pos:* Mem, UN Sci Comt Effect of Atomic Radiation, 65-66 & 70; USPHS fel, Calif Inst Technol, 69; Ford Found biol adv, Agrarian Univ, Peru; mem radiation protection adv comt, Wis State Bd Health; mem, Nat Comt Radiation Protection & Measurements, adv comt biol effect of ionizing radiation, Nat Acad Sci & comt 1, Int Comn Radiation Protection. *Mem:* Radiation Res Soc; Genetics Soc Am; Environ Mutagen Soc. *Res:* Radiation genetics; induced mutation by radiation and other environmental mutagens. *Mailing Add:* 2012 Wavnona Way Madison WI 53713

ABRAHAMSON, WARREN GENE, II, b Ludington, Mich, Mar 26, 47; m 69; c 1. PLANT ECOLOGY, INSECT HERBIVORY. *Educ:* Univ Mich, Ann Arbor, BS, 69; Harvard Univ, AM, 71, PhD(biol), 73. *Prof Exp:* Res asst plant ecol, Univ Mich, 68-69; teaching asst biol, Harvard Univ, 69-72; tutor, 72-73; from asst prof to assoc prof biol, 73-83, DAVID BURPEE PROF PLANT GENETICS, BUCKNELL UNIV, 83- *Concurrent Pos:* Res assoc & res fel, Archbold Biol Sta, 76-; vis asst prof biol, Mich State Univ, 76; Colo Plateau Scholar-in-residence, Mus Northern Ariz & Northern Ariz Univ, 87-88. *Honors & Awards:* Bradley-Moore Davis Award, Univ Mich, 69; Lindback Award for Distinguished Teaching, Bucknell Univ, 75; William Dutcher Award, Nat Audubon Soc, 84. *Mem:* Ecol Soc Am; Soc Study Evolution; Sigma Xi; Bot Soc Am; Torrey Bot Club; AAAS. *Res:* Life history strategies of plants, vegetative and seed reproduction; environmental influences on life histories; effects of herbivorous insects on life history parameters of goldenrods; fire ecology of south central Florida. *Mailing Add:* Dept Biol Bucknell Univ Lewisburg PA 17837

ABRAIRA, CARLOS, b Buenos Aires, Arg, Mar 25, 36; US citizen; m 63; c 2. ENDOCRINOLOGY, DIABETES & NUTRITION. *Educ:* Univ Buenos Aires Nat Col, Baccalaureate, 53; Univ Buenos Aires Med Sch, MD, 61; Am Bd Internal Med, dipl, 72. *Prof Exp:* Med internship, Inst Med Res, Univ Buenos Aires, 62-63; int med resident, Mt Sinai Hosp, Chicago, 64-66; clin asst med, Chicago Med Sch, 65-66; res fel med, Northwestern Univ-Evanston Hosp, 66-67; fel metab & endocrinol, Michael Reese Hosp, Chicago, 67-69, res assoc, 69-70, attend physician, dept med, 69-70; from asst prof to assoc prof med, Abraham Lincoln Sch, Univ Ill, 70-83,; asst chief, CHIEF ENDOCRINOL & DIABETES, HINES VET ADMIN HOSP, HINES, ILL, 72-; PROF MED, GRAD SCH MED, LOYOLA UNIV, 86- *Concurrent Pos:* Lectr, Cook Ct Grad Sch Med, 75-77; nat chmn, Vet Admin Coop Study Glycemic Control & Complications in Diabetes type II (CSDM), 89- *Mem:* Am Diabetes Asn; fel Am Col Physicians; Am Fedn Clin Res; Am Inst Nutrit; Am Soc Clin Nutrit; Endocrine Soc; fel Am Col Nutrit. *Res:* Carbohydrate metabolism; dietary influences on lipid metabolism in diabetes; drug treatment of diabetes and its complications. *Mailing Add:* Hines Vet Admin Hosp Med Serv 111C 12E Endocrinol-Diabetes Sect Hines IL 60141

ABRAM, JAMES BAKER, JR, b Tulsa, Okla, Dec 5, 37; m 61; c 2. ZOOLOGY, PARASITOLOGY. *Educ:* Langston Univ, BS, 59; Okla State Univ, MS, 63, PhD(zool), 68. *Prof Exp:* Teacher high sch, Okla, 59-62; from instr to assoc prof biol, Md State Col, Princess Anne, 63-70; from assoc prof to prof, Hampton Inst, 73-82; asst dean, Sch Math & Sci, Cent State Univ, Edmond, Okla, 82-83; prof & chmn, Biol Dept, Norfolk State Univ, 83-88; PROF, DEPT BIOL, DILLARD UNIV, 88- *Mem:* Am Soc Parasitol; Am Soc Mammal. *Res:* Helminths of muskrats; helminths of small mammals; ecological influences upon parasites. *Mailing Add:* Div Natural Sci Dillard Univ 2601 Gentilly Blvd New Orleans LA 70122

ABRAMOFF, PETER, b Montreal, Que, Dec 28, 27; nat US; m 52, 88; c 6. ZOOLOGY, IMMUNOLOGY. *Educ:* Univ Western Ont, BA, 50; Univ Detroit, MSc, 52; Univ Wis, PhD(zool), 55. *Prof Exp:* Asst, Univ Detroit, 50-52; from instr to prof biol, Marquette Univ, 55-78, chmn dept, 65-89, WEHR DISTINGUISHED PROF BIOL, MARQUETTE UNIV, 78-; CLIN PROF IMMUNOL, DEPT SURG, MED COL WIS, 72- *Concurrent Pos:* Consult, Faust Sci Supply Co, 66-70; US Comn Undergrad Educ Biol Sci, 66-70, Educ Prods Div, 68-74, US Off Biol Educ, 70-78, NSF, 70-78, NIH Pulmonary Spec Res, 73-86 & Vet Admin Hosp, Wood, Wis, 74-86; sect ed, J Reticuloendothelial Soc, 77-87; pres, Int Union Reticuloendothelial Soc, 84-87; sr educ consult, Fotodyne, Inc, 89-; secy, Am Soc Zoologists, Div Comparative Immunol, 79-81. *Mem:* Am Asn Immunologists; Nat Asn Biol Teachers; Nat Sci Teachers Asn; Reticuloendothelial Soc (pres, 80); Int Soc Exp Hemat. *Res:* Immunobiology of the lung; hypersensitivity pneumonitis. *Mailing Add:* Dept Biol Marquette Univ Milwaukee WI 53233

ABRAMOVICI, MIRON, b Bacau, Romania, May 19, 34; m 67; c 2. ORGANIC CHEMISTRY, PHYSICAL ORGANIC CHEMISTRY. *Educ:* Univ Sao Paulo, BS, 60; Northwestern Univ, MS, 68; Loyola Univ Chicago, PhD(chem), 70. *Prof Exp:* Res chemist, Uniao Oil Refinery Co, Brazil, 60-63; res chemist, Cent Res Lab, GAF Corp, Pa, 69-70; res chemist, Catalytic & Coord Chem Sect, Res & Develop, 70-78, tech mgr, 78-81, TECH DIR, CHEMCATALYSTS DEPT, ENGELHARD INDUSTS DIV, ENGELHARD MINERALS & CHEM CORP, 81- *Mem:* Am Chem Soc; Catalysis Soc. *Res:* Preparative and synthetic organic chemistry; analytical instrumentation for organic compounds; homogeneous and heterogeneous catalysis in organic chemistry; catalysis; chemical and physical characterization of heterogeneous catalysts. *Mailing Add:* 209 Berger St Somerset NJ 08873

ABRAMOVITCH, RUDOLPH ABRAHAM HAIM, b Alexandria, Egypt, July 19, 30; m 52, 77. ORGANIC CHEMISTRY. *Educ:* Univ London, BSc, 50, PhD(chem), 53, DSc, 64. *Prof Exp:* Med Res Coun fel, Univ Exeter, 54; res chemist, Weizmann Inst Sci, Israel, 54-55; Imp Chem Indust res fel, King's Col, Univ London, 55-57; from asst prof to prof chem, Univ Sask, 57-67; prof, Univ Ala, 67-70, res prof chem, 70-77; prof chem & geol & head dept, 77-82, PROF CHEM, CLEMSON UNIV, 82- *Concurrent Pos:* Consult, Mead Johnson Res Ctr, 73-78; chmn, Org Div, Chem Inst, Can, 66-67; assoc prof, Marseille, France, 83, Fulbright grant, 83; Emilio Noelting chair, Mulhouse, France, 83-84. *Honors & Awards:* Outstanding Res Scientist, Sigma Xi, 81. *Mem:* Chem Soc; Am Chem Soc; Chem Inst Can; Int Soc Heterocyclic Chem. *Res:* Reactive intermediates; heterocyclic chemistry; medicinal and agricultural chemistry; polymer chemistry. *Mailing Add:* Dept Chem Clemson Univ Clemson SC 29634-1905

ABRAMOWICZ, DANIEL ALBERT, b Levittown, Pa, July 16, 57; m 83; c 2. BIODEGRADATION & CHEMICAL BIOSYNTHESIS, ENZYMOLOGY. *Educ:* St Francis Col, BS(chem), 79 & BA(math), 79; Princeton Univ, MA, 81, PhD(phys chem), 84. *Prof Exp:* Comput programmer, Rohm & Haas, 79; instr chem, Princeton Univ, 81-84; res chemist, Corp Res & Develop, 84-88, MGR, ENVIRON TECHNOL, GEN ELEC, 88- *Mem:* Am Chem Soc; AAAS. *Res:* Use of microorganisms to degrade hazardous chemicals; application of enzymes to chemical synthesis; utilizing whole cells or immobilized enzymes. *Mailing Add:* Corp Res & Develop Gen Elec PO Box 8 Schenectady NY 12301

ABRAMOWITZ, STANLEY, b Brooklyn, NY, Sept 18, 36; m 57; c 3. CHEMICAL PHYSICS. *Educ:* Brooklyn Col, BS, 57; Polytech Inst Brooklyn, PhD(phys chem), 63. *Prof Exp:* Asst chem, Polytech Inst Brooklyn, 57-59; Nat Acad Sci-Nat Res Coun fel, 62-64, PHYS CHEMIST, MOLECULAR SPECTROS, NAT BUR STANDARDS, 64- *Res:* Molecular spectroscopy and structure. *Mailing Add:* Nat Bur of Standards Gaithersburg MD 20899

ABRAMS, ADOLPH, b Springfield, Mass, Apr 30, 19; m 51; c 3. BIOCHEMISTRY. *Educ:* Univ Wis, PhD(biochem), 49. *Prof Exp:* actg chmn dept biochem, 73-78, from asst to prof, 54-90, EMER PROF BIOCHEM, SCH MED, UNIV COLO, DENVER, 90- *Concurrent Pos:* Nat Res Coun fel, Cancer Soc, Carlsberg Lab & Copenhagen Univ, 49-50; Lilly fel biochem, Calif Inst Technol, 50-54. *Mem:* Am Chem Soc; Am Soc Biol Chemists. *Res:* Physical chemistry and biosynthesis of proteins; intermediary metabolism of amino and nucleic acids; chemistry and enzymology of bacterial cell membranes; relations between membranes, nucleic acids and protein synthesis; membrane permeability and active transport; bacterial membrane ATPase. *Mailing Add:* 13691 E Marina Dr No 402 Aurora CO 80014-3724

ABRAMS, ALBERT, b St Joseph, La, Sept 2, 21; m 54; c 2. PHYSICAL CHEMISTRY, PETROLEUM ENGINEERING. *Educ:* City Col NY, BS, 48; NY Univ, PhD(chem), 52. *Prof Exp:* Control chemist, Stauffer Chem Co, 48; chemist, 52-63, res chemist, 63-71, RES ASSOC, EXPLOR & PROD RES DIV, SHELL DEVELOP CO, 71- *Mem:* Am Chem Soc; Am Inst Mining, Metall & Petrol Eng. *Res:* Fluid flow; transport phenomena; surface chemistry; flow of fluids through porous media; oil formation impairment; oil production problems. *Mailing Add:* 4630 O'Meara Houston TX 77035

ABRAMS, ALBERT MAURICE, b Los Angeles, Calif, June 7, 31; m 56; c 2. PATHOLOGY, ORAL PATHOLOGY. *Educ:* Univ Calif, DDS, 57; Univ Southern Calif, MS, 61. *Prof Exp:* Instr periodont & oral path, 58-59, PROF PATH & CHMN ORAL PATH DEPT, SCH DENT, UNIV SOUTHERN CALIF, 65- *Concurrent Pos:* Fel oral path, Armed Forces Inst Path, 62-64; mem attend staff, Dept Path, Los Angeles County Univ Southern Calif Med Ctr, 65-; consult, Long Beach Vet Hosp, 65-; Wadsworth Va Hosp, 79- *Mem:* Am Acad Oral Path (pres, 83); fel Am Col Dent; Am Soc Clin Path. *Res:* Clinical-pathological studies of odontogenic tumors, salivary gland diseases and fibro-osseous lesions. *Mailing Add:* Sch Dent Univ Southern Calif 925 W 34th St Los Angeles CA 90089-0641

ABRAMS, CHARLIE FRANK, JR, b Edgecombe Co, NC, June 3, 44; m 69; c 2. AGRICULTURAL ENGINEERING, BIOENGINEERING. *Educ:* NC State Univ, BS, 66, MS, 69, PhD(bioeng & agr eng), 71. *Prof Exp:* Asst exten prof agr eng, Univ Ky, 71-72; PROF BIOL & AGR ENG, NC STATE UNIV, 72- *Mem:* Am Soc Agr Engrs; Am Soc Eng Educ; Sigma Xi; Eng Med Biol Soc of Inst Elec & Electronic Engrs; Biomed Eng Soc. *Res:* Mechanics of biological tissues as related to human and animal medicine; occupational biomechanics; agriculture. *Mailing Add:* NC State Univ PO Box 7625 Raleigh NC 27695-7625

ABRAMS, ELLIS, b Pittsburgh, Pa, Jan 24, 17; m 46; c 2. INDUSTRIAL CHEMISTRY. *Educ:* Univ Pittsburgh, BS, 38, PhD(chem), 42. *Prof Exp:* Asst anal chem, Univ Pittsburgh, 38-42; res chemist, Quaker Chem Prod Corp, 42-49; dir textile res, E F Houghton & Co, 49-54, supvr org textile & paper res, 54-58, asst mgr res, 59-69, mgr res, 70-72, dir res, 72-79, mgr technol, 80-82; RETIRED. *Concurrent Pos:* Tech consult, 89- *Mem:* Am Chem Soc; Am Soc Lubrication Engrs. *Res:* Surface active agents; textile finishes; derivatives of m-diphenyl benzene; chemical defoamers: cellulose reactants; cutting and rolling oils; pollution control. *Mailing Add:* 601 Hanover Ct Wayne PA 19087

ABRAMS, GERALD DAVID, b Detroit, Mich, Apr 27, 32; m 54; c 2. PATHOLOGY. *Educ:* Wayne State Univ, AB, 51; Univ Mich, MD, 55. *Prof Exp:* Instr path, Med Sch, Univ Mich, 59-60; asst chief dept exp path, Walter Reed Army Inst Res, 61-62; from asst prof to assoc prof, 63-69, dir anat path, 85-89, PROF PATH, MED SCH, UNIV MICH, ANN ARBOR, 69- *Concurrent Pos:* Markle scholar acad med, 63-68; attend consult, Vet Admin Hosp, Ann Arbor, Mich, 63-; dep med exam, Washtenaw Co, Mich, 63- *Honors & Awards:* Kaiser Permanente Award, 78. *Mem:* AAAS; Am Soc Exp Path; Am Asn Pathologists; Int Acad Path. *Res:* Pathologic anatomy; experimental pathology; gnotobiotics; effects of normal flora on host resistance; comparative pathology. *Mailing Add:* Dept Path Rm 2G 332 Univ Hosp Univ Mich Med Sch Ann Arbor MI 48109-0054

ABRAMS, GERALD STANLEY, b New York, NY, May 30, 41; m 63, 80; c 2. HIGH ENERGY PHYSICS. *Educ:* Cornell Univ, BA, 62; Univ Md, PhD(physics), 67. *Prof Exp:* Res assoc physics, Univ Ill, Urbana, 67-69; PHYSICIST, LAWRENCE BERKELEY LAB, 69- *Concurrent Pos:* Guest physicist, Superconducting Super Collider Lab. *Mem:* Am Phys Soc. *Res:* Experimental high energy physics. *Mailing Add:* Lawrence Berkeley Lab MS 50A/2160 One Cyclotron Rd Berkeley CA 94720

ABRAMS, HERBERT L, b New York, NY, Aug 16, 20; m 43; c 2. DIAGNOSTIC RADIOLOGY. *Educ:* Cornell Univ, BA, 41; State Univ NY Downstate Med Ctr, MD, 46; Am Bd Radiol, cert, 51. *Hon Degrees:* MA, Howard Univ, 67. *Prof Exp:* Intern, Long Island Col Hosp, 46-47; resident, Montefiore Hosp, 47-48; resident radiol, Stanford Univ Hosp, 48-51; from instr to assoc prof radiol, Sch Med, 51-61, actg exec dir, Dept Radiol, 61-62, dir, Div Diag Roentgenol, 61-67, prof radiol, Stanford Univ, 62-67; Philip H Cook prof radiol & chmn dept, Harvard Med Sch, 67-85; PROF RADIOL, SCH MED, STANFORD UNIV, 85- *Concurrent Pos:* Nat Cancer res fel, 50; Nat Heart Inst spec res fel, Univ Lund, 60; mem radiation study sect, NIH, 63-67, exec comt, Coop Study Renovascular Hypertension; mem med adv bd, Coun High Blood Pressure Res, AHC, 66, chmn prog comt, Coun Cardiovasc Radiol, 72-75, mem exec comt, 74-; radiologist-in-chief, Peter Bent Brigham Hosp, 67-85; mem exec & adv comts, Inter-Soc Comn Heart Dis Resources, 70; mem, Joint Comn Accreditation Hosps, 72; NIH res fel, Cardiovasc Res Inst, Univ Calif, San Francisco, 73-74; radiologist-in-chief, Sidney Farber Cancer Inst, 74-; consult surv renovascular hypertension, Nat Heart Inst; consult US Naval Hosp & Vet Admin, Calif & coun pharmacol & chem, AMA; mem overseas adv comt, Tel Aviv Univ Med Sch. *Honors & Awards:* Caldwell Lectr, Am Roentgen Ray Soc, 82; Dotter Lectr, Soc Cardiovascular & Interventional Radiol, 88; Hodes Lectr, Thomas Jefferson Univ, 88. *Mem:* Inst Med-Nat Acad Sci; Asn Univ Radiol; AAAS; fel Am Col Radiol; Am Asn Cancer Res; hon fel Royal Soc Radiol Eng, 80; hon fel Royal Soc Surg, Ireland, 85. *Res:* Cardiovascular radiology and physiology. *Mailing Add:* Dept Radiol Stanford Univ Sch Med Stanford CA 94305

ABRAMS, IRVING M, b St Paul, Minn, Feb 25, 17; m 52, 78; c 3. INDUSTRIAL CHEMISTRY. *Educ:* Univ Minn, BCh, 38, PhD(biochem), 42. *Prof Exp:* Res chemist, Econ Lab, Inc, St Paul, 38-39; Bristol-Myers fel, Stanford Univ, 42-44, & Nat Adv Comt Aeronaut, 42-44; sr res chemist, Int Minerals & Chem Corp, Chicago, 44-47; lab dir, Chem Process Co, 47-55, assoc tech dir, 55-62; tech mgr, dept ion exchange, Duolite Int Inc, Subsid Diamond Shamrock Corp, 62-75; mgr tech develop, 75-82; CONSULT & LECTR ION EXCHANGE, 82- *Concurrent Pos:* Instr human rels & effective speech, Simmons Inst, San Francisco & Los Angeles, 48-68; consult, Stanford Res Inst, 48 & Aptech Eng Serv, Inc, 79-81; mem adv comt, Int Water Conf, 62-83; chmn-elect, Gordon Res Conf ion exchange, 85. *Mem:* Emer mem Am Chem Soc; Am Soc Enologists; Am Soc Mech Engrs. *Res:* Anomalous osmosis and collodion membranes; foaming phenomena; chemistry of sugar beets; ion exchange resins; adsorption phenomena; water treatment and wastewater renovation; counter-current ion exchange; condensate polishing; production of ultrahigh purity water for semiconductor industry. *Mailing Add:* 511 Castano Corte Los Altos CA 94022

ABRAMS, ISRAEL JACOB, b Syracuse, NY, June 11, 26. MATHEMATICAL STATISTICS. *Educ:* Univ Calif, BA, 50, MA, 52, PhD(math statist), 57. *Prof Exp:* Asst, Univ Calif, 50-57; math statistician, Space Tech Labs, 57-61; vpres technol, Mkt Res Corp Am, 61-69; pres, Datanamics Corp Am, 69-71; VPRES, MRCA INFO SERV, 71- *Concurrent Pos:* Lectr, Exten, Univ Calif, Los Angeles, 59-60; ed comput sect, J Mkt Res, 67-72. *Mem:* Am Mkt Asn; Inst Math Statist; Am Statist Asn; Inst Food Technologists. *Res:* Marketing research; applied statistics; computer systems design; operations research. *Mailing Add:* MRCA Info Serv 2215 Sanders Rd Northbrook IL 60062

ABRAMS, JOEL IVAN, b Baltimore, Md, Sept 7, 28; m 53; c 3. CIVIL ENGINEERING. *Educ:* Johns Hopkins Univ, BS, 49, MS, 50, Dr Eng, 56. *Prof Exp:* Instr civil eng, Johns Hopkins Univ, 50-53, asst, 53-55, asst prof, 55-56; from asst prof to assoc prof, Yale Univ, 56-65; PROF CIVIL ENG & CHMN DEPT, UNIV PITTSBURGH, 65- *Concurrent Pos:* Consult, Am Mach & Foundry Co, 61- & Skorsky Aircraft Co, 62- *Mem:* Am Soc Civil Engrs; Am Soc Eng Educ; Am Concrete Inst; Int Asn Shell Struct. *Res:* Dynamics of structures; plate and shell structures; numerical methods. *Mailing Add:* Dept Civil Eng 949 Benedum Univ Pittsburgh Main Campus Pittsburgh PA 15260

ABRAMS, JONATHAN, b New York, NY, Feb 15, 39; m 62; c 2. CARDIOLOGY. *Educ:* Univ Calif, Berkeley, BA, 60; Univ Calif, San Francisco, MD, 64. *Prof Exp:* Physician internal med & cardiol, Sandia Base Army Hosp, Albuquerque, 68-70; from asst prof to assoc prof, 70-80, PROF MED, SCH MED, UNIV NMEX, 81-, CHIEF, DIV CARDIOL, 73- *Concurrent Pos:* Staff cardiologist, Albuquerque Vet Admin Hosp, 70-71, actg chief cardiol, 71-72, chief cardiol sect, 72-74; trustee, Univ NMex Hosp, 79- *Mem:* Fel Am Col Physicians; fel Am Col Cardiol. *Mailing Add:* Dept Med Univ NMex Sch Med Albuquerque NM 87131

ABRAMS, LLOYD, b New York, NY, June 9, 39; m 71; c 2. THERMODYNAMICS & MATERIAL PROPERTIES. *Educ:* City Col NY, BChE, 61; Rutgers Univ, PhD(phys chem), 67. *Prof Exp:* Anal engr, Conn Aircraft Nuclear Engine Lab, Pratt & Whitney Aircraft, 61, exp engr, Adv Powers Systs, Conn, 61-62, Apollo Proj, 62-63; teaching asst chem, Rutgers Univ, 63-64; res assoc, Brookhaven Nat Lab, 66-68; res chemist, Cent Res Dept, Exp Sta, E I du Pont de Nemours & Co Inc, 68-72, sr res chemist, Pigments Div, 72-75, res assoc, 75-79, STAFF SCIENTIST, CENT RES DEPT, EXP STA, E I DU PONT DE NEMOURS & CO, INC, 79- *Concurrent Pos:* Fel radiation chem, Brookhaven Nat Lab, 66-68; chmn, Gordon Res Conf, 87. *Mem:* AAAS; Am Chem Soc; Sigma Xi. *Res:* Gas-solid reactions; adsorption studies; phenomena in micropores; zeolites; solid state chemistry; oxides. *Mailing Add:* 555 Holly Knoll Rd Hockessin DE 19707

ABRAMS, MARSHALL D, b Jersey City, NJ, Nov 7, 40; m 62; c 2. INFORMATION SECURITY. *Educ:* Carnegie Inst Technol, BSEE, 62; Univ Pittsburgh, MSEE, 63, PhD(elec eng), 66. *Prof Exp:* From asst prof to assoc prof elec eng, Univ Md, 66-76; mem prof staff, Nat Bur Standards, Inst Comput Sci & Technol, Washington, DC, 76-81; PRIN SCIENTIST, MITRE CORP, MCLEAN, VA, 81- *Concurrent Pos:* Vis adj prof, George Mason Univ, 89-90. *Honors & Awards:* Appl res award, Nat Bur Standards, 79. *Mem:* Inst Elec & Electronic Engrs; Asn Comput Mach. *Res:* Computer and network security; generalized framework for access control; security policy models. *Mailing Add:* 7525 Colshire Dr McLean VA 22102-3184

ABRAMS, MARVIN COLIN, b Reno, Nev, Sept 6, 30; m 56; c 2. PHYSICAL CHEMISTRY. *Educ:* Univ Nev, BS, 52, MS, 54; Wash State Univ, PhD(chem), 58. *Prof Exp:* Res chemist, AEC, Univ Nev, 52-54; asst chem, Wash State Univ, 54-58; staff scientist chem & physics, 59-67; sr staff scientist, 67-70, chief mfg technol, 70-79, mgr, mfg technol, Gen Dynamics, Pomona, 79-85. *Mem:* Am Chem Soc; Combustion Inst; Sigma Xi; Spectros Soc. *Res:* Chemical mechanism and kinetics of combustion processes; spectroscopic studies of solids and reacting gaseous media. *Mailing Add:* 1989 Yorba Dr Pomona CA 91767

ABRAMS, MICHAEL J, b Cleveland, Ohio, Mar 3, 48; m 73. GEOLOGY, REMOTE SENSING. *Educ:* Calif Inst Technol, BSc, 70, MSc, 73. *Prof Exp:* Jr chemist, Calif Inst Technol, 72; SR SCIENTIST GEOL, JET PROPULSION LAB, NASA, 73- *Concurrent Pos:* Vpres consult, GeoImages, Inc, 74- *Mem:* Soc Econ Geol; Int Asn on Genesis Ore Deposits. *Res:* Mineral exploration. *Mailing Add:* NASA Jet Propulsion Lab 4800 Oak Grove Pasadena CA 91109

ABRAMS, RICHARD, b Chicago, Ill, Sept 19, 17; m 47; c 4. BIOCHEMISTRY. *Educ:* Univ Chicago, SB, 38, PhD(chem), 41. *Prof Exp:* Res instr chem, Univ Chicago, 41-42, group leader, Manhattan Proj, 42-46; Donner Found fel, Karolinska Inst, Sweden, 46; asst prof biochem, Inst Radiobiol & Biophys, Univ Chicago, 47-51; assoc dir inst res, Montefiore Hosp, 51-58, dir res, 58-65; prof biochem & nutrit & head dept, Grad Sch Pub Health, 65-70, PROF BIOCHEM & CHMN DEPT, FAC ARTS & SCI, UNIV PITTSBURGH, 70- *Mem:* AAAS; Am Chem Soc; Am Soc Biol Chemists; Sigma Xi. *Res:* Isolation and properties of respiratory enzymes; metabolism of fission products and transuranic elements; tracer studies of nucleic acid metabolism. *Mailing Add:* 5315 Westminster Pl Pittsburgh PA 15232

ABRAMS, RICHARD LEE, b Cleveland, Ohio, Apr 20, 41; m 62; c 2. APPLIED PHYSICS. *Educ:* Cornell Univ, BEP, 64, PhD(appl physics), 68. *Prof Exp:* Mem tech staff laser physics, Bell Tel Labs, 68-71; sr staff physicist, Hughes Aircraft Co, 71-72, head Electro-Optics Sect, 72-75, mgr, Optical Physics Dept, 75-83, chief scientist, Govt Systs Div, 83- 88, CHIEF SCIENTIST, RES LAB, HUGHES AIRCRAFT CO, 88- *Honors & Awards:* Centennial Medal, Inst Elec & Electronics Engrs. *Mem:* Fel Optical Soc Am; fel Inst Elec & Electronics Eng. *Res:* Quantum electronics; atomic physics; infrared physics; laser communications; nonlinear optics; electro-optics. *Mailing Add:* Hughes Res Lab 3011 Malibu Lyn Rd Malibu CA 90265

ABRAMS, ROBERT A, b 1939. OPERATIONS RESEARCH, RESEARCH ADMINISTRATION. *Educ:* Mass Univ, BA, 61, Northwestern Univ, MA, 65, PhD(indust eng), 69. *Prof Exp:* Assoc prof, oper res & mgt, Univ Chicago, 66-69 & Northwestern Univ, 70-75; assoc prof, Univ Ill, Urbana, 69-70, PROF & HEAD INFO & DECISION SCI, UNIV ILL, CHICAGO, 79- *Mem:* Inst mgt scientist; Oper Res Soc; Soc Indust & Appl Math. *Mailing Add:* Dept Infor & Decision Sci Univ Ill Chicago MIC 294 Box 4348 Chicago IL 60680

ABRAMS, ROBERT JAY, b Chicago, Ill, June 16, 38; m 66; c 3. EXPERIMENTAL HIGH ENERGY PHYSICS. *Educ:* Ill Inst Technol, BS, 59; Univ Ill, Urbana, MS, 61, PhD(physics), 66. *Prof Exp:* Res assoc physics, Brookhaven Nat Lab, 66-68, asst physicist, 68-70; asst prof, 70-74, ASSOC PROF PHYSICS, UNIV ILL, CHICAGO CIRCLE, 74- *Mem:* Am Phys Soc. *Res:* High energy multiparticle reactions; K meson decays. *Mailing Add:* 6716 N Kedvale Ave Chicago IL 60646

ABRAMS, ROBERT MARLOW, b Rome, NY, June 8, 31; m 57; c 4. PHYSIOLOGY. *Educ:* Hamilton Col, AB, 53; Univ Pa, MS & DDS, 60, PhD(physiol), 63. *Prof Exp:* Asst prof oral health & epidemiol, Yale Univ, 68-69; from asst prof to assoc prof, 69-86, PROF OBSTET & GYNEC, 86-, PROF PEDIAT, COL MED, UNIV FLA, 87- *Concurrent Pos:* USPHS career develop award, 63-68; asst fel, John B Pierce Found Lab, Med Sch, Yale Univ, 63-69; mem study sect, Human Embryol & Develop, NIH, 75-78; Fogarty sr int fel, Clin Res Ctr, Harrow, UK, 78-79. *Mem:* Am Physiol Soc; Soc Gynecol Invest. *Res:* Reproductive physiology; reproduction. *Mailing Add:* Dept Obstet & Gynec Univ Fla Col Med Gainesville FL 32601

ABRAMS, STEVEN ALLEN, b Columbus, Ohio, Dec 27, 57; m 87; c 2. CALCIUM METABOLISM, STABLE ISOTOPES. *Educ:* Mass Inst Technol, BS, 79; Ohio State, MD, 82. *Prof Exp:* Resident pediat, Children's Hosp Akron, 82-85; fel neonatology, Baylor Col Med, 85-88; res fel nutrit, Nat Inst Child Health & Human Develop, 88-91; fel nutrit, 85-88, ASST PROF NUTRIT & NEONATOLOGY, CHILDREN'S NUTRIT RES CTR, AGR RES SERV, USDA, 91- *Concurrent Pos:* Asst prof neonatology, George Washington Univ, 88-91. *Mem:* Am Inst Nutrit; Am Soc Clin Nutrit; Am Soc Bone & Mineral Res. *Res:* Assessment of nutritional needs especially calcium needs of infants and children using balance techniques; photon densitometry and stable isotopes; development of mathematical models of calcium disteasation in children and adults. *Mailing Add:* One Baylor Plaza Houston TX 77030

ABRAMS, SUZANNE ROBERTA, b Goderich, Ont, Apr 8, 47; m 81; c 2. SYNTHETIC ORGANIC CHEMISTRY. *Educ:* Carleton Univ, BSc, 72; Dalhousie Univ, PhD(org chem), 77. *Prof Exp:* Fel, Univ Alberta, 77-78; res assoc, 78-80, asst res officer, 80-84, ASSOC RES OFFICER, PRAIRIE REGIONAL LAB, NAT RES COUN CAN, 84- *Mem:* Chem Inst Can. *Res:* Acetylene chemistry, specific deuteration of aromatic, long chain compounds and plant hormones; Abscisic acid analogs, metabolism of abscisic acid in plants. *Mailing Add:* Plant Biotechnol Inst Nat Res Coun 110 Gymnasium Pl Saskatoon SK S7N 0W9 Can

ABRAMS, WILLIAM R, b Philadelphia, Pa, May 7, 45. PROTEASE & PROTEASE INHIBITOR INTERACTIONS. *Educ:* Univ Pa, PhD(biochem), 71. *Prof Exp:* Mem staff, Albert Einstein Med Ctr, 75-84; instr, Temple Univ, 81-84; res asst prof, Dept Med, Grad Hosp, 84-87, ASST PROF, UNIV PA SCH DENT MED, 87- *Mem:* AAAS; Am Asn Pathologists; Am Chem Soc; Protein Soc. *Res:* Connective tissue/elastin. *Mailing Add:* Dept Anat/Histology Univ Pa Sch Dent Med, 4001 Spruce St Philadelphia PA 19104

ABRAMSKY, TESSA, biochemistry, for more information see previous edition

ABRAMSON, ALLAN LEWIS, b New York, NY, Nov 8, 40; m 64; c 2. OTOLARYNGOLOGY. *Educ:* Colgate Univ, BA, 62; State Univ NY Downstate Med Ctr, MD, 67. *Prof Exp:* Resident otolaryngol, Mt Sinai Sch Med, 69-72; dir div otolaryngol, Naval Hosp, Camp LeJeune, NC, 72-74; assoc prof otolaryngol, State Univ NY Stony Brook, 74-; dir otolaryngol, Long Island Jewish-Hillside Med Ctr, 74-; mem staff, Dept Otolaryngol, State Univ NY Health Sci Ctr, Stony Brook; PROF & CHMN, DEPT OTOLARYNGOL, LONG ISLAND JEWISH MED CTR, 82- *Mem:* Fel Am Col Surgeons; Am Broncho-Esophagol Soc; Am Acad Ophthal & Otolaryngol; Am Acad Facial Plastic & Reconstruct Surg. *Res:* Molecular and antibiotic transfer across normal and infected sinus membranes; bone grafting into the frontal sinus cavity. *Mailing Add:* Dept of Otolaryngol Long Island Jewish Med Ctr New Hyde Park NY 11042

ABRAMSON, DAVID C, b Philadelphia, Pa, Mar 30, 38; m 60; c 3. PEDIATRICS, NEONATOLOGY. *Educ:* Muhlenberg Col, BS, 60; Georgetown Univ, MD(med, physiol), 66. *Prof Exp:* DIR NURSERIES, GEORGETOWN UNIV HOSP, 69- *Concurrent Pos:* Consult, Columbia Hosp Women, Providence Hosp & Arlington Hosp, 68-; prin investr, NIH, 68-; dir nurseries, Fairfax County Hosp, 70-; consult med law, 76- *Mem:* Soc Obstet Anesthesia & Perinatology; Sigma Xi. *Res:* Placental transfer of iron; disseminated intravascular coagulation in the newborn; carbohydrate metabolism in the newborn; congenital heart disease; effect of microtransmitters on cerebral palsy. *Mailing Add:* 4200 42nd St NW Washington DC 20016

ABRAMSON, DAVID IRVIN, b New York, NY, Oct 14, 05; m 40; c 2. CARDIOVASCULAR DISEASES. *Educ:* Long Island Col Med, MD, 29; Am Bd Int Med Cardiovasc Dis, dipl. *Prof Exp:* Asst physiol, Long Island Col Med, 28-29; intern, Bushwick Hosp, Brooklyn, 29-30; instr physiol & pharm, Long Island Col Med, 30-36; dir cardiovasc res, May Inst Med Res, Jewish Hosp, Cincinnati, 37-42; from asst prof to prof med, 46-72, prof phys med & rehab & head dept, 55-72, EMER PROF MED, COL MED, UNIV ILL, 72- *Concurrent Pos:* Fels, Michael Reese Hosp, Chicago, 36 & Mt Sinai Hosp, N Y, 36-37; dir, Dept Electrocardiography, Jewish Hosp, Cincinnati, 38-42; instr, Sch Med, Univ Cincinnati, 38-42; attend physician, Hines Vet Hosp, Ill, 46-53, consult, 63-; attend physician, W Side Vet Hosp, Chicago, 46-53, consult, 63-; attend physician, Michael Reese Hosp, Chicago, 47- & Mt Sinai Hosp, 49- *Honors & Awards:* Conrad Jobst Award Vascular Res, Am Cong Phys Med & Rehab, 70. *Mem:* Fel Am Col Physicians; Am Soc Clin Invest; Soc Exp Biol & Med; Am Physiol Soc. *Res:* Physiology of peripheral circulation; peripheral vascular disorders; convalescence; rate of peripheral blood flow in man under normal and abnormal conditions and in various pathologic states, using the venous occlusion plethysmographic method. *Mailing Add:* 916 N Oak Park Ave Oak Park IL 60302

ABRAMSON, EDWARD, b Philadelphia, Pa, Mar 2, 33; m 54; c 3. HIGH SPEED SIGNAL PROPAGATION, OPTICAL FIBER SYSTEMS. *Educ:* Ursinus Col, BS, 54; Univ Pa, MS, 57, PhD(physics), 62. *Prof Exp:* Physicist, Cent Res Dept, 62-67, res physicist, Org Chem Dept, 67-71, SR RES PHYSICIST, ELECTRONICS DEPT, E I DU PONT DE NEMOURS & CO, 71- *Mem:* Optical Soc Am. *Res:* Ultraviolet sources and optics; photochemistry; optical imaging systems; applications of lasers; electrooptics and fiber optics; high speed electronic signal propagation. *Mailing Add:* 5 Woodcrest Way Landenberg PA 19350

ABRAMSON, EDWARD E, b Brooklyn, NY, July 7, 44; div; c 2. CLINICAL PSYCHOLOGY, HEALTH PSYCHOLOGY. *Educ:* State Univ NY, Stony Brook, BA, 65; Catholic Univ, PhD(clin psychol), 71. *Prof Exp:* From asst prof to assoc prof, 70-78, PROF PSYCHOL, CALIF STATE UNIV, CHICO, 78- *Concurrent Pos:* Clin psychologist, Sutter-Yuba Mental Health Serv, 73-75, York Clinic, Guys Hosp Med Sch, London, 78, Alcohol Unit, Glen Gen Hosp, 81-83 & pvt pract, 85-; dir, Eating Disorders Ctr, 83-85, clin psychologist, N State Ctr Psychiat, Chico Community Hosp, 88-; consult, Cypress Ctr, 85-88. *Mem:* Am Psychol Asn; Asn Adv Behav Therap. *Res:* Psychological variable affecting weight regulation, obesity and eating disorders; anxiety disorders; sexual dysfunction. *Mailing Add:* Dept Psychol Calif State Univ Chico CA 95929

ABRAMSON, FRED PAUL, b Columbus, Ohio, 41; m 63; c 2. PHARMACOKINETICS, MASS SPECTROMETRY. *Educ:* Ohio State Univ, PhD(chem), 65. *Prof Exp:* PROF PHARMACOL, GEORGE WASHINGTON UNIV, 81- *Mem:* Am Soc Pharmacol & Exp Therapeut; Am Soc Mass Spectrometry; AAAS. *Res:* Pharmacology. *Mailing Add:* Dept Pharmacol George Washington Univ 2300 Eye St NW Washington DC 20037

ABRAMSON, FREDRIC DAVID, b Philadelphia, Pa, Nov 9, 41; m 64, 79; c 5. COMMUNICATION SYSTEMS, MANAGEMENT. *Educ:* Univ Pa, AB, 63; Univ Rochester, MS, 66; Univ Mich, PhD(genetics), 71; Mass Inst Technol, MSM, 77; American Univ, JD, 77. *Prof Exp:* Genetic counr birth defects, Wayne State Univ, 71-72; asst prof community med, Col Med, Univ Ky, 72-75; PRES, HEALTH MGT SYSTS, INC, 75- *Concurrent Pos:* Consult, Gen Elec Co, 78-80 & Nuclear Regulatory Comn, 80-; adj prof mgt, Loyola Col, Baltimore, 80-81; bd trustees, N Charles Hosp, 82-; chmn, gov comn on workmen's comp laws, Md, 83-; adj prof mgt, Georgetown Univ, 83-84; assoc prof mgt, Col Bus, Am Univ, 83-85. *Res:* Development of decision support systems; recruitment and staffing; program planning and evaluation; basis of public policy formulation; strategic and long range planning systems; productivity management; performance evaluation; organization management; expert software systems; law & technology. *Mailing Add:* Health Mgt Systs Inc 21155 Burnham Rd Gaithersburg MD 20879

ABRAMSON, H(YMAN) NORMAN, b San Antonio, Tex, Mar 4, 26; m 47; c 2. ENGINEERING, MATHEMATICS. *Educ:* Stanford Univ, BS, 50, MS, 52; Univ Tex, PhD(mech eng, math), 56. *Prof Exp:* Proj anal engr, Chance-Vought Aircraft, 51-52; assoc prof aeronaut eng, Agr & Mech Col Tex, 52-56; dir, Dept Mech Sci, Southwest Res Inst, 56-72, vpres, Eng Sci Div, 72-85, exec vpres, 85-91; RETIRED. *Concurrent Pos:* Indust consult, 52-56; consult, govt agencies; assoc ed, Appl Mech Rev, 54-; vis prof, Incarnate Word Col, 58-61; adj prof mech eng, Univ Tex-Austin, 83- *Mem:* Nat Acad Eng; Am Soc Mech Engrs; Soc Exp Mech; Am Inst Aeronaut & Astronaut; Soc Naval Architects & Marine Engrs. *Res:* Applied mathematics and engineering mechanics; aerodynamics; fluid flow; shock and vibration; aeroelasticity; dynamics; nonlinear mechanics. *Mailing Add:* Southwest Res Inst PO Drawer 28510 San Antonio TX 78228-0510

ABRAMSON, HANLEY N, b Detroit, Mich, June 10, 40; m 67; c 3. MEDICINAL CHEMISTRY. *Educ:* Wayne State Univ, BS, 62; Univ Mich, MS, 63, PhD(pharmaceut chem), 66. *Prof Exp:* NSF fel, Hebrew Univ Jerusalem, 66-67; from asst prof to assoc prof, 67-78, PROF PHARMACEUT CHEM, COL PHARM & ALLIED HEALTH PROFESSIONS, WAYNE STATE UNIV, 78-, CHAIR, DEPT PHARMACEUT SCI, 86- *Concurrent Pos:* Interim dir, Inst Chem Toxicol, 86-88; interim dean, Col Pharm & Allied Health Professions, 87-88. *Mem:* Am Pharmaceut Asn; Am Chem Soc; Acad Pharmaceut Sci; Am Asn Col Pharm; Am Asn Pharmaceut Scientists. *Res:* Synthesis and biochemical pharmacology of natural products and their analogs. *Mailing Add:* Dept Pharmaceut Sci Wayne State Univ Shapiro Hall Detroit MI 48202

ABRAMSON, I JEROME, b Philadelphia, Pa, Aug 3, 35; m 59; c 2. MICROBIOLOGY, BACTERIOLOGY. *Educ:* Pa State Univ, BS, 57; Va Polytech Inst & State Univ, PhD(microbiol), 71. *Prof Exp:* Chief microbiol, Philadelphia Labs, Inc, 63-67; staff fel & res microbiologist, Baltimore Cancer Res Ctr, Nat Cancer Inst, 71-73; sr microbiologist, BBL Div, Becton Dickinson Corp, 73-77; consult pharmaceut, 72-74, 77-78; sci reviewer, Div Clin Lab Device, 78-88, EVAL ANTIMICROBIAL AGENTS, CTR DRUG EVAL & RES, FOOD & DRUG ADMIN, 88- *Concurrent Pos:* Consult anaerobe lab apparatus, Kontes Glass Co, 72-74; mid-level assoc, Mgt & Sci Training, Food & Drug Admin, 81-82; subcomt anaerobic bacteria, Nat Comt Clin Lab Standards, 79-85; mem, Task Force Vet Educ, Md, 78-81; mem comt, Am Acad Microbiol, 88-; rep public policy, Am Soc Microbiol, 88- *Honors & Awards:* US Pub Serv Award, 73. *Mem:* Am Soc Microbiol; fel Am Acad Microbiol; Sigma Xi. *Res:* Development, design and testing of media and devices for collection, transport, and isolation of anaerobic and aerobic microorganisms from a clinical specimen; investigation of methods to improve growth of anaerobes on commercial media; effect of antimicrobial agents on treponemes; veterinary research; inventions; patent. *Mailing Add:* 12405 Knollcrest Rd Reisterstown MD 21136-5609

ABRAMSON, JON STUART, b NY, NY Oct 31, 50; m 85; c 2. PEDIATRICS INFECTIOUS DISEASES. *Educ:* Boston Univ, BA, 72; Bowman Med Ctr, MD, 76. *Prof Exp:* Resident pediat, NC Baptist Hosp, 79; from asst prof to assoc prof, 81-90, PROF PEDIAT INFECTIOUS DIS & CROSS APPOINTMENT DEPT MICROBIOL & IMMUNOL, BOWMAN GRAY SCH MED, 90- *Concurrent Pos:* Fel infectious dis, Univ Minn, 81-; prin investr, res career develop award, Nat Inst Allergy Infectious Dis, 85- & effect on influenza virus on neotrophils, NIH, 84-; secy, Soc Pediat Res. *Honors & Awards:* Res Career Develop Award, CNIAID. *Mem:* Am Soc Microbiol; Am Asn Immunologist; Fel Infectious Dis Soc Am; Soc Pediat Res; Am Fedn Clin Res; Fel Acad Pediat. *Res:* Determine the mechanism by which viruses depress neutrophil function predisposing people to bacterial and fungal superinfections; pathogenesis of serious bacterial infections. *Mailing Add:* Dept Pediat Bowman Gray Sch Med 300 S Hawthorne Winston-Salem NC 27103

ABRAMSON, LEE RICHARD, b New York, NY, May 19, 33; m 54; c 2. MATHEMATICAL STATISTICS. *Educ:* Columbia Univ, AB, 54, PhD(math statist), 63. *Prof Exp:* Sr res mathematician, Electronics Res Lab, Columbia Univ, 57-63, prin res mathematician, 64-67; mem res staff, Riverside Res Inst, 67-71; opers res analyst, Off Emergency Preparedness, 71-73; sr opers res analyst, AEC, 73-75; analyst, Energy Res & Develop Admin, 75-77; STATIST ADV, NUCLEAR REGULATORY COMN, 77- *Concurrent Pos:* Lectr, Sci Hons Prog, Columbia Univ, 64-69 & Math Prog High Sch Students, City Univ New York, 70-71; consult, Sandoz Pharmaceut, 69-71 & Schering Corp, 72-; prof lectr, Am Univ, 79- *Mem:* AAAS; Inst Math Statist; Am Math Soc; Opers Res Soc Am; Am Statist Asn; Sigma Xi. *Res:* General statistics; reliability. *Mailing Add:* Off Res Nuclear Regulatory Comn Washington DC 20555

ABRAMSON, MORRIS BARNET, b New York, NY, July 8, 10; m 38; c 2. PHYSICAL BIOCHEMISTRY, NEUROCHEMISTRY. *Educ:* NY Univ, BS, 31, MS, 34, PhD(chem), 39. *Prof Exp:* Res assoc, 62-73, ASSOC PROF PHYS CHEM, DEPT NEUROL, ALBERT EINSTEIN COL MED, 73-

Concurrent Pos: Adj assoc prof, NY Univ, 42-; Sir Ernest Oppenheimer fel, Cambridge Univ, 61 & Polytech Inst Brooklyn, 63; NIH fel, 63; vis prof, Univ Fla, 82-84. *Mem:* Am Chem Soc. *Res:* Physical chemistry of lipids; surface chemistry of particles in aqueous media; model membrane systems; physical chemistry of neurotransmitters and enzymes acting on these. *Mailing Add:* 53-35 Hollis Ct Blvd Flushing NY 11365

ABRAMSON, N(ORMAN), b Boston, Mass, Apr 1, 32; m 54; c 2. COMMUNICATION THEORY. *Educ:* Harvard Univ, AB, 53; Univ Calif, Los Angeles, MA, 55; Stanford Univ, PhD(elec eng), 58. *Prof Exp:* Systs engr, Hughes Aircraft Co, 53-56; asst prof, Stanford Univ, 59-62, assoc prof elec eng,62-65; vis lectr, Harvard Univ, 65-66; chmn info sci prog, 68-70, PROF ELEC ENG, UNIV HAWAII, 66-, PROF INFO SCI & DIR, ALOHA SYST, 68- *Concurrent Pos:* Consult, indust & govt labs, 57, UN Develop Prog, 75- & IBM World Trade Corp, 76-; vis prof, Univ Calif, Berkeley, 64-65; pres, Systs Res Corp, Honolulu, 68-77; vis prof, Mass Inst Technol, 81; chmn bd, Technol Educ Assocs, Sydney, Australia, 80- *Mem:* AAAS; Inst Elec & Electronics Engrs. *Res:* Satellite communications; communication and detection theory; pattern recognition; information processing; computer nets. *Mailing Add:* Dept Info Sci Univ Hawaii Manoa Honolulu HI 96822

ABRAMSON, NORMAN JAY, b Los Angeles, Calif, May 4, 27; m 51; c 1. FISHERIES, BIOMETRICS. *Educ:* Univ Wash, BS, 55. *Prof Exp:* Marine biologist, Calif State Fisheries Lab, 55-57, head biomet anal sect, 57-70; surv biometrician, Nat Marine Fisheries Serv, 71-72; fishery biologist, 72-74, dir, Tiburon Lab, Southwest Fisheries Ctr, 74-87; FISHERY BIOLOGIST, CALIF DEPT FISH & GAME, INTERGOVERNMENTAL PERSONNEL ACT, 88- *Concurrent Pos:* Consult, Fisheries Res Inst, Univ Wash, 69; consult, Food & Agr Orgn, UN, 70, 86; mem, bd of dir, Inst for Biol Explor, 82- *Mem:* Biomet Soc; Am Statist Asn; Am Inst Fishery Res Biologists. *Res:* Population dynamics of marine fishes and related statistical problems. *Mailing Add:* Tiburon Lab 3150 Paradise Dr Tiburon CA 94920

ABRAMSON, STANLEY L, b Akron, Ohio, Mar 29, 25; m 49; c 3. PHYSICS, OPERATIONS RESEARCH. *Educ:* Univ Akron, BS, 47, MS, 48. *Prof Exp:* Instr physics, Univ Akron, 46-49; asst prof, Nebr Wesleyan Univ, 49-50; chief ballistics sect, US Army Chem Ctr, 50-53, mem opers res group, 53-55; sr engr, Air Arm Div, Westinghouse Elec Corp, 55-58; sr engr, Arma Div, Am Bosch Arma Corp, 58-60; adv engr, Res Lab, Westinghouse Elec Corp, 60-63, mgr tech utilization prog res & develop, Astronuclear Lab, 63-67, planning consult, Westinghouse Power Systs Planning, 67-69, consult, Westinghouse Nuclear Fuel Div, 69-78, sr consult technologist to corp staff, Corp Planning, Westinghouse Elec Corp, Pittsburgh, 78-85; RETIRED. *Concurrent Pos:* Adj prof, Carnegie-Mellon Univ, Grad Sch Indust Admin, 89-, Univ Pittsburgh, Katz Grad Sch Bus, 89. *Mem:* AAAS; Am Phys Soc; Inst Elec & Electronic Engrs. *Res:* Ballistics and aerodynamics of various blunt configurations; operations research applied to chemical, biological and radiological warfare and to selection of new products and new business ventures. *Mailing Add:* 1190 Colgate Dr Monroeville PA 15146

ABRAMSON, STEPHAN B, b Newark, NJ, Aug 17, 45; m; c 2. CLINICAL RESEARCH. *Educ:* Calif Inst Technol, BS, 67; Harvard Univ, PhD(biochem), 75. *Prof Exp:* Teaching fel biol chem, Div Med Sci, Harvard Univ, 68-74; postdoctoral fel, Div Hemat-Oncol, Ctr Health Sci, Univ Calif, 75-77; postdoctoral res assoc biochem, Sect Gastroenterol, Sch Med, Univ Southern Calif, 77-81; investr biochem & cell biol, Huntington Med Res Insts, 81-86; clin res assoc, 86, clin proj mgr, 86-89, ASSOC DIR, CLIN RES, ALPHA THERAPEUT CORP, 89- *Concurrent Pos:* Postdoctoral fel, Nat Heart, Lung & Blood Inst, USPHS, 75-77. *Mem:* Sigma Xi; AAAS; Am Chem Soc; Am Soc Cell Biol; Am Soc Microbiol; NY Acad Sci. *Mailing Add:* Dept Clin Res Alpha Therapeut Corp 5555 Valley Blvd Los Angeles CA 90032

ABRASH, HENRY I, b Paterson, NJ, Sept 27, 35; m 62; c 1. ORGANIC CHEMISTRY, BIOCHEMISTRY. *Educ:* Harvard Univ, BA, 56; Calif Inst Technol, PhD(chem), 61. *Prof Exp:* Proj asst chem, Univ Wis, 60-61; from asst prof to assoc prof, 61-70, PROF CHEM, CALIF STATE UNIV, NORTHRIDGE, 70- *Concurrent Pos:* USPHS grant, 62-68; Hays-Fulbright travel grant, Carlsberg Labs, Copenhagen Univ, 68-69. *Mem:* Am Chem Soc; The Chem Soc; Sigma Xi. *Res:* Mechanism of enzyme action, particularly on hydrolytic enzymes; mechanism of hydrolytic reactions; proteins in solution; oxidation of polyphenols. *Mailing Add:* Dept of Chem Calif State Univ Northridge CA 91324

ABRELL, JOHN WILLIAM, b Martinsburg, WVa, Aug 30, 36; m 67. BIOORGANIC CHEMISTRY. *Educ:* Duke Univ, BS, 58; Dartmouth Col, MA, 60; Wash Univ, PhD(chem), 65. *Prof Exp:* NIH fel, 65-68; res chemist, Nat Inst Arthritis & Metab Dis, NIH, 68-71; biochemist, Litton-Bionetics, Inc, 71-74; sci adminr, Div Lung Dis, Nat Heart & Lung Inst, 74-76, Ctr Rev, Nat Inst Arthritis, Metab & Digestive Dis, 76-80, HEALTH SCIENTIST ADMINR, GRANTS REV BR, NAT CANCER INST, NIH, 80- *Mem:* Sigma Xi; Am Asn Cancer Res; AAAS; NY Acad Sci. *Res:* Enzymology; ribonuclease I from E coli; mechanism of action; nucleic acid metabolism, in particular DNA synthesis by the DNA polymerases found in RNA viruses; isolation and characterization of mammalian cellular DNA polymerases; biomedical research administration. *Mailing Add:* Div Extramural Activ Nat Cancer Inst Bethesda MD 20892

ABREMSKI, KENNETH EDWARD, b Baltimore, MD, Jan 3, 48. SEQUENCE ANALYSIS, MOLECULAR BIOLOGY. *Educ:* Loyola Col, MD, BS, 69; Univ MD Med Sch, PhD(biochem), 77. *Prof Exp:* Fel NIH, Lab Molecular Biol, 77-80; scientist, Frederick Cancer Res Facil, 80-84; prin investr, E I DuPont & Co, 84-90; PRIN INVESTR, DUPONT-MERCK PHARMACEUT, 91- *Concurrent Pos:* vis lectr, Hood Col, Frederick, MD, 83- *Mem:* Int Neural Network Soc; Asn Comput Mach. *Res:* Protein DNA interactions, computer analysis of DNA and protein sequences; neural networks. *Mailing Add:* DuPont Merck Pharmaceut E328/150 Box 80328 Wilmington DE 19880-0328

ABREU, SERGIO LUIS, b Habana, Cuba, Sept 9, 44. BIOMATHEMATICS, INTELLIGENT SYSTEMS. *Educ:* Long Island Univ, BS, 70, MS, 72; Univ Conn, PhD (biochem), 76. *Prof Exp:* Chmn dept, 86-90, ASSOC PROF BIOL SCI, FORDHAM UNIV, 79-, DIR GRAD PROGS, DEPT BIOL SCI, 90- *Mem:* AAAS; Am Soc Microbiol; Am Soc Cell Biol; NY Acad Sci; Int Soc Differentiation. *Res:* Biology and biochemistry of differentiating systems; computer modeling of biological phenomena. *Mailing Add:* Dept Biol Sci Fordham Univ Bronx NY 10458

ABRHAM, JAROMIR VACLAV, b Prague, Czech, Apr 9, 31; Can citizen; m 57; c 1. MATHEMATICS. *Educ:* Charles Univ, Prague, BSc, 54, PhD(math), 58. *Prof Exp:* From asst prof to assoc prof, Tech Univ Prague, 57-68; assoc prof, 69-81, PROF INDUST ENG, UNIV TORONTO, 81- *Concurrent Pos:* Invited prof, Ctr Operations Res & Ecouometrits, Univ Catholique de Louvain, Belgium, 68-69 & Ctr de Recherche de Mathematiques Appliquees, Univ de Montreal, 79. *Mem:* Math Prog Soc; Oper Res Soc Am; Can Oper Res Soc; Am Math Soc. *Res:* Perfect systems of difference sets: inequalities for such systems, enumeration of systems with few components; additive permutations: bases with the minimum number of negative elements, symmetric bases and related permutations; continuous programming. *Mailing Add:* Dept Indust Eng Univ Toronto Toronto ON M5S 1A4 Can

ABRIOLA, LINDA MARIE, b Philadelphia, Pa, Nov 17, 54. SANITARY & ENVIRONMENTAL ENG, HYDROLOGY. *Educ:* Drexel Univ, BS, 76; Princeton Univ, MS, 79, MA, 80, PhD(civil eng), 83. *Prof Exp:* Proj engr, Procter & Gamble Mfg Co, 76-78; postdoctoral researcher, Princeton Univ, 83-84; asst prof, 84-90, ASSOC PROF CIVIL ENG, UNIV MICH, 90- *Concurrent Pos:* Presidential young investr award, NSF, 85; mem pub info comt, Am Geophys Union, 90-; core consult, environ eng comt, US Environ Protection Agency sci adv bd, 90-; mem bd dirs, Asn Environ Eng Professors, 90-; vis assoc prof, Dept Petrol Eng, Univ Tex, Austin, 91. *Mem:* Asn Environ Eng Professors; Am Geophys Union; Int Asn Hydrol Sci; Asn Ground Water Scientists & Engrs; Int Asn Hydraul Res; Asn Women Geoscientists. *Res:* Mathematical modeling of flow and transport in porous media, particularly the use of models to elucidate processes involved in the multiphase migration and transformation of organic chemical pollutants and in aquifer remediation. *Mailing Add:* Dept Civil & Environ Eng Univ Mich Ann Arbor MI 48109-2125

ABROMSON-LEEMAN, SARA R, PATHOLOGY, IMMUNOLOGY. *Educ:* Simmons Col, Boston, BS, 75; Harvard Univ, PhD(immunol), 82. *Prof Exp:* Postdoctoral fel immunol, Dana-Farber Cancer Inst, 82-84; postdoctoral fel immunol, 85-88, INSTR, DEPT PATH, MED SCH, HARVARD UNIV, 88- *Concurrent Pos:* Cancer Res Inst fel, 84-87; ad hoc reviewer, J Immunol, 88- *Mem:* Am Asn Immunologists. *Res:* Immunology; pathology. *Mailing Add:* Dept Path Harvard Med Sch 25 Shattuck St Boston MA 02115

ABRUNA, HECTOR D, b Santurce, PR, Nov 8, 53; US citizen. ELECTROCHEMISTRY. *Educ:* Rensselaer Polytech Inst, BS, 75, MS, 76; Univ NC, Chapel Hill, PhD(chem), 80. *Prof Exp:* Fel, Univ Tex, Austin, 80-81; asst prof, Univ PR, Rio Piedras, 82-83; ASST PROF CHEM, CORNELL UNIV, 83- *Concurrent Pos:* Sloan Fel, 87. *Honors & Awards:* Presidential Young Investr Award, NSF, 84; Tajima Prize, 87. *Mem:* Am Chem Soc; Soc Electroanal Chem; Electrochem Soc. *Res:* Chemically modified electrodes; electrocatalysis; sensors; electrochemistry of transition metal complexes; electrochemistry in novel solvent media; extended x-ray absorption of fine structure at the electrode solution interface. *Mailing Add:* Dept Chem Cornell Univ Baker Lab Ithaca NY 14853-1301

ABRUTYN, DONALD, b New York, NY, Nov 13, 34; m 64; c 4. VETERINARY MEDICINE, TOXICOLOGY. *Educ:* Univ Pa, VMD, 61. *Prof Exp:* Scientist toxicol, E R Squibb Corp, 65-67; res scientist, Bristol-Myers Corp, 67-68; sr res assoc, Warner-Lambert Corp, 68-77 & Pennwalt Corp, 77; group leader, 78-81, sect head, 81-85, RES MGR, ORTHO PHARMACEUT CORP, 85- *Mem:* Am Vet Med Asn; Soc Comp Ophthal (secy-treas, 70-); Indust Vet Asn Genetic Toxicol; Assoc Soc Toxicol. *Res:* Regulatory toxicology; ophthalmologic toxicology. *Mailing Add:* Petticoat Lane Annandale NJ 08801

ABRUZZO, JOHN L, b West New York, NJ, Apr 27, 31; m; c 7. MEDICINE. *Educ:* St Peter's Col, NJ, BS, 53; Georgetown Univ, MD, 57. *Prof Exp:* Jr intern, Childrens Convalescent Hosp, DC, 56-57; med intern, Jersey City Med Ctr, NJ, 57-58; asst in med, Seton Hall Col Med, 58-60, instr, 63-64; asst prof, NJ Col Med, 64-67; from asst prof to assoc prof, 67-74, PROF MED, JEFFERSON MED COL, 74- *Concurrent Pos:* USPHS fel, Columbia-Presby Med Ctr, 60-61, vis fel, Col Physicians & Surgeons, 63-64; Arthritis Found fel, 63-66; NIH res grants, 64-70; from jr asst med resident to sr asst med resident, Jersey City Med Ctr, 58-60; spec investr, Arthritis Found, 66-67; asst ed, Annals Internal Med, 71-78, assoc ed, 78-; attend physician, Thomas Jefferson Univ Hosp. *Mem:* Am Fedn Clin Res; AMA; fel Am Col Physicians; Am Rheumatism Asn; Sigma Xi. *Mailing Add:* Dept Med Jefferson Med Col 1025 Walnut St Philadelphia PA 19107

ABSE, DAVID WILFRED, b Cardiff, Wales, Mar 15, 15; US citizen; c 2. PSYCHIATRY, PSYCHOANALYSIS. *Educ:* Univ Wales, BSc, 35, MD, 48; Welsh Nat Sch Med, MB, BCh, 38; Univ London, DPM, 40. *Prof Exp:* Dep med supt, Monmouthshire Ment Hosp, Abergavenny, UK, 47-48; chief asst psychiatrist, Med Sch, Charing Cross Hosp, London, 49-51; clin dir, Dorothea Dix State Hosp, Raleigh, NC, 52-53; from assoc prof to prof psychiat, Univ NC, Chapel Hill, 53-62; prof psychiat, Sch Med, 62-87, EMER PROF, UNIV VA, 87- *Concurrent Pos:* Lectr psychol, City Col, London, 50-51; lectr psychol & clin assoc prof psychiat, Univ NC, Chapel Hill, 52-53; teaching analyst, Washington Psychoanal Inst, 62- & Washington Sch Psychiat, 71-73. *Mem:* Fel Royal Soc Med; fel Am Psychiat Asn; Brit Med Asn; Am Psychoanal Asn; Int Psychoanal Asn. *Res:* Clinical evaluation of psychotropic drugs; psychiatric aspects of cancer; psychotherapy of schizophrenics. *Mailing Add:* 1852 Winston Rd Charlottesville VA 22903

ABSHIER, CURTIS BRENT, b Tallapoosa, Mo, June 11, 45; m 78. DIGITAL SIGNAL PROCESSING, COMPUTER SYSTEM INTEGRATION. *Educ:* Univ Mo, Rolla, BSEE, 70, MSEE, 71, PhD(elec eng), 76. *Prof Exp:* Geophysicist, Gulf Res & Develop, 76-80; sr res geophysicist, Cities Serv, 80-85; vpres, Mountain Systs Serv, 85-90; PRES, SOFTWARE SCI, INC, 90- *Mem:* Inst Elec & Electronic Engrs; Soc Explor Geophysicists; Europ Asn Explor Geophysicists. *Res:* Adaptive filtering for the oil exploration industry. *Mailing Add:* Software Sci Inc 11172 Huron St Suite 20B Northglenn CO 80234

ABSHIRE, CLAUDE JAMES, b Kaplan, La, Nov 19, 33; Can citizen. BIOCHEMISTRY, MICROBIOLOGY. *Educ:* Univ Southwestern La, BSc, 55; Univ Tex, Austin, MA, 57, PhD(org chem), 60. *Prof Exp:* Res assoc chem, Univ Ill, Urbana, 59-60; res chemist, fabrics and finishes dept, Du Pont Chem Co, Pa, 60-61; res assoc chem, Mass Inst Technol, 61-62; fel biochem, Laval Univ, 62-64, from asst prof to assoc prof, 64-72; PROF CHEM, UNIV QUE, MONTREAL, 72- *Concurrent Pos:* Scholar Med Res Coun Can, 65-68. *Res:* Synthesis of new unnatural amino acids and the study of their metabolic action in microbial systems; food science research. *Mailing Add:* Dept Chem Univ Que Montreal CP8888 Succ A Montreal PQ H3C 3P8 Can

ABSOLON, KAREL B, b Brno, Czech, Mar 21, 26; US citizen; m 54; c 4. THORACIC & CARDIOVASCULAR SURGERY, WRITER & PUBLISHER. *Educ:* Masaryk Univ, BM, 48; Yale Univ, MD, 52; Univ Minn, MS & PhD(physiol & surg), 63; Am Bd Surg, dipl; Am Bd Thoracic Surg, dipl; Am Bd Cardiovasc Dis, dipl. *Prof Exp:* Asst prof surg, Med Col, Univ Minn, 63-69; asst prof, Univ Tex Health Sci Ctr Dallas, 69-70; assoc prof surg, Med Sch, George Washington Univ, 70-75 & chmn dept surg, Washington Hosp Ctr, 70-75; prof surg & acad chmn, Sch Med, Univ Ill, 75-80; expert appointment, Artificial Heart Prog, Nat Heart, Lung & Blood Inst, NIH, 80-84; PUBL & AUTHOR, KABEL PUBL, ROCKVILLE, MD, 85-; CONSULT SURGEON, 85- *Concurrent Pos:* Am Cancer Soc fel, 54-56; USPHS trainee, 58-61, spec fel, 61-63; chg surg, Cardiopulmonary Inst, Methodist Hosp, Dallas, 69-70. *Mem:* Am Col Angiol; NY Acad Sci; Am Heart Asn; Am Col Surg. *Res:* Clinical and experimental surgery; experimental physiology and surgery; cardiovascular and pulmonary and hepatic physiology and surgery; history of surgical development. *Mailing Add:* 11225 Huntover Dr Rockville MD 20852

ABT, HELMUT ARTHUR, b Helmstedt, Ger, May 26, 25; US citizen; c 1. ASTRONOMY. *Educ:* Northwestern Univ, BS, 46, MS, 48; Calif Inst Technol, PhD(astron), 52. *Prof Exp:* Jr res astronomer, Lick Observ, Univ Calif, 52-53; res assoc, Yerkes Observ, Univ Chicago, 53-55, asst prof, 55-59; assoc astronomer, 59-63, ASTRONOMER, KITT PEAK NAT OBSERV, 63- *Concurrent Pos:* Assoc ed, Astron J, 67-69; consult, NASA, 67-; managing ed, Astrophys J, 71- *Mem:* Fel AAAS; Am Astron Soc; Astron Soc Pac (pres, 66-68). *Res:* Stellar spectroscopy and rotation; spectroscopic binaries; special classification; astrosociology. *Mailing Add:* Kitt Peak Nat Observ Box 26732 Tucson AZ 85726-6732

ABU-ISA, ISMAT ALI, b Tarshiha, Palestine, Dec 12, 38; m 77; c 4. PHYSICAL CHEMISTRY, POLYMER CHEMISTRY. *Educ:* Am Univ Beirut, BSc, 59, MS, 61; Northwestern Univ, PhD(phys chem), 65. *Prof Exp:* Asst prof chem, P E I Univ, 64-67; assoc sr res chemist, 67-69, sr res chemist, 69-77, staff res scientist, 77-80, sr staff res scientist, 80-87, PRIN RES SCIENTIST, RES LABS, GEN MOTORS CORP, 87- *Concurrent Pos:* Nat Res Coun Can res awards, 65-67. *Honors & Awards:* John M Campbell Award. *Mem:* Sigma Xi; Am Chem Soc. *Res:* Thermodynamic properties of polymers, oxidation and surface characteristics of polymers; processing and properties of polymers; permeation characteristics of polymer; flammability of polymers; polymer aging; polymer blends. *Mailing Add:* Polymers Dept GM Res Labs GM Tech Ctr 30500 Mound Rd Warren MI 48090-9055

ABUL-HAJJ, JUSUF J, b Jerusalem, Palestine, Apr 3, 40; m 65; c 2. BIOCHEMISTRY, ORGANIC CHEMISTRY. *Educ:* Am Univ, Beirut, BSc, 62, MSc, 64; Univ Wis-Madison, PhD(pharmaceut biochem), 68. *Prof Exp:* Asst chem, Am Univ, Beirut, 62-64; asst pharmaceut biochem, Univ Wis-Madison, 64-68; assoc prof, 68-80, actg chmn dept med chem & pharmacog, 84-87, PROF PHARMACOG, UNIV MINN, MINNEAPOLIS, 80-,. *Concurrent Pos:* Gustavus A Pfeiffer Mem Res fel, 84. *Mem:* AAAS; Am Asn Cancer Res; Am Soc Pharmacog; Sigma Xi. *Res:* Mechanisms of enzymic reactions; mechanisms in steroid metabolism; biotransformation of steroids and alkaloids; estrogen carcinogenesis, steroid metabolism in mammary tumors. *Mailing Add:* Col Pharm Med Chem Univ Minn Health Sci Unit Minneapolis MN 55455

ABURDENE, MAURICE FELIX, b Bethlehem, Palestine, May 22, 46; US citizen; m; c 2. ELECTRICAL ENGINEERING. *Educ:* Univ Conn, BSEE, 68, MSEE, 70, PhD(elec eng), 74. *Prof Exp:* Proj mgr, Bristol Div, Am Chain & Cable, 68-70; asst prof comput sci, State Univ NY, Oswego, 74-75; asst prof comput eng, Swarthmore Col, 75-81; prof, 81-90, T JEFFERSON MIERS PROF ELEC ENG, BUCKNELL UNIV, 90-, CHMN DEPT, 81- *Concurrent Pos:* Electron engr, Rome Air Develop Ctr, 74 & 75; NASA/Am Soc Eng Educ fel, 77; vis scholar, Stanford Univ, 78; vis res assoc, Mass Inst Technol, 78-79; vis Dept Energy & Am Soc Eng Educ fel, Jet Propulsion Lab, Calif Inst Technol, 81; acad vis, Imperial Col, 87-88. *Mem:* Inst Elec & Electron Engrs; Asn Comput Mach; Am Soc Eng Educ; Sigma Xi. *Res:* Design and application of distributed algorithms for computer networks and graphics simulation for performance analyses; real time data acquistion; control of laboratory and industrial processes; parallel algorithms for signal processing. *Mailing Add:* Dept Elec Eng Bucknell Univ Lewisburg PA 17837

ABUSHANAB, ELIE, b Damascus, Syria, Nov 6, 36; US citizen; m 65; c 3. SYNTHETIC ORGANIC CHEMISTRY. *Educ:* Am Univ Beirut, BSc, 60; Univ Wis, MSc, 62, PhD(pharmaceut chem), 65. *Prof Exp:* Asst prof pharmaceut chem, Univ Md, Baltimore, 65-67; staff chemist, Chas Pfizer & Co, Inc, Conn, 67-70; asst prof, to assoc prof,70-79, PROF MED CHEM, UNIV RI, 79- *Mem:* Am Chem Soc; Sigma Xi; Cong Heterocycles Chem. *Res:* The use of carbohydrates as chiral templates for the synthesis of 1',2'-seco-nucleosites as inhibitors of target enzymes in cancer and viral chemotherapy; of interest are adenosien deaminase and purine and pyrimidine phosphorylase inhibitors. *Mailing Add:* 230 Fogarty Hall Col Pharm RI Univ Kingston RI 02881-0809

ABU-SHUMAYS, IBRAHIM KHALIL, b Jaffa, Palestine, Feb 11, 37; m 68; c 2. NUMERICAL ANALYSIS, APPLIED MATHEMATICS. *Educ:* Am Univ Beirut, BS, 60; Harvard Univ, PhD(physics), 66. *Prof Exp:* Res asst, Harvard Univ, 60-61, teaching fel, 61-64, Off Naval Res asst, 64-66; res assoc appl math, Argonne Nat Lab, 66-69, asst mathematician, Appl Math Div, 69-72; vis assoc prof eng sci dept, Northwestern Univ, 72-73; SR FEL & ADV MATHEMATICIAN NUMERICAL ANALYST, BETTIS ATOMIC POWER LAB, 73- *Concurrent Pos:* Res fel, Harvard Univ, 67; vchmn & chmn, Math & Comput Div, Am Nuclear Soc, 85-87, chmn, Pittsburgh Sect, 86-87. *Mem:* Soc Indust & Appl Math; Am Nuclear Soc. *Res:* Reactor mathematics; mathematical analysis; differential equations; approximation theory; reactor science and engineering; numerical analysis; computational methods for supercomputers. *Mailing Add:* Bettis Atomic Power Lab PO Box 79 West Mifflin PA 15122-0079

ABUZZAHAB, FARUK S, SR, b Beirut, Lebanon, Oct 12, 32; m 62; c 4. PSYCHIATRY, PHARMACOLOGY. *Educ:* Am Univ Beirut, BS, 55, MD, 59; Univ Minn, Minneapolis, PhD, 68. *Prof Exp:* Pharmaceut Mfr Asn Found fac career develop award & asst prof clin pharmacol, 67-69, asst clin prof psychiat & pharmacol, 69-73, ASSOC CLIN PROF PSYCHIAT, PHARMACOL & FAMILY PRACT, UNIV MINN, 73-; PRES, CLIN PSYCHOPHARMACOL CONSULTS, P A, 73- *Concurrent Pos:* Consult, Northern Pines Ment Health Clin, Turtle Lake, Wis, 63-70; chmn pharm subcomt, Hastings State Hosp, Minn, 64-78, chief psychiat, 70-78; clin dir, Clin Psychopharmacol Serv, Univ Minn Hosp, Minneapolis, 67-69; mem subcomt on alcoholism & drug abuse, Minn State Med Soc, 71-; consult polydrug abuse proj, Multi-Resource Ctr, 74- *Mem:* AMA; Am Psychiat Asn; Am Soc Clin Pharmacol & Therapeut; Int Col Neuropsychopharmacol; Sigma Xi. *Res:* Clinical psychopharmacology research in the area of psychoactive drugs mainly in depression, schizophrenia and memory; neurochemical correlates of behavior and drug action; toxicity of lithium carbonate in animals and man. *Mailing Add:* Univ Minn Box 393 Mayo Minneapolis MN 55455

ABZUG, M(ALCOLM) J, b New York, NY, Apr 13, 20; m 46; c 2. FLIGHT MECHANICS. *Educ:* Mass Inst Technol, BS, 41; Univ Calif, Los Angeles, MS, 59, PhD(eng), 62. *Prof Exp:* Asst aeronaut engr, Aircraft Lab, Aerodyn Br, US Air Corps, 41-43; proj engr, Flight Control Dept, Aeronaut Div, Sperry Gyroscope Co, 49-52; aeroodyn engr, Stability & Control Group, Aerodyn Sect, Douglas Co, 46-48, design specialist, 52-55, supvr, 55-62, chief, Astrodyn Br, Missile & Space Systs Div, Santa Monica, 63-64, chief engr, Flight Mech Dept, Res & Develop Directorate, 64-67; sr staff engr, TRW Systs, Inc, 67-72; CONSULT, 76 - *Concurrent Pos:* Consult, Cambridge Univ Press, 51-57; guest lectr, Univ Calif, Los Angeles, 58, lectr, 62-70; adj prof, Univ Southern Calif, 80-86. *Honors & Awards:* Douglas Aircraft Co Award, 57. *Mem:* Fel Am Inst Aeronaut & Astronaut. *Res:* Flight mechanics of airplanes and other vehicles; fluid mechanics contributions; effects of automatic control and stabilization. *Mailing Add:* 14951 Camarosa Dr Pacific Palisades CA 90272

ACARA, MARGARET A, b Buffalo, NY, Jan 11, 34; m 57; c 4. PHARMACOLOGY, PHYSIOLOGY. *Educ:* D'Youville Col, BA, 55; State Univ NY, Buffalo, PhD(pharmacol), 71. *Prof Exp:* Prof & tech asst clin chem, Roswell Park Mem Inst, 55-57; res asst med chem, 61-64, res asst biochem pharmacol, 64-67, fel & res assoc pharmacol, 73-74, from asst prof to assoc prof pharmacol, 74-90, PROF PHARMACOL, STATE UNIV NY, BUFFALO, 90- *Concurrent Pos:* United Health Found fel, 71-73; Nat Women Sci grant-in-aid, 72; Pharmaceut Mfrs Asn Found award, 74. *Mem:* Am Soc Pharmacol & Exp Therapeut; Am Soc Nephrol; Am Chem Soc; Sigma Xi. *Res:* Renal transport and metabolism of organic cations; the role of choline and betaine in the kidney; effect of drugs on renal function. *Mailing Add:* Dept of Pharmacol & Therapeut State Univ of NY Sch Med Buffalo NY 14214

ACCIARRI, JERRY A(NTHONY), b Filbert, Pa, Jan 6, 32; m 56; c 3. CHEMICAL ENGINEERING. *Educ:* Pa State Univ, BS, 53, MS, 55, PhD(chem eng), 57. *Prof Exp:* From res chem engr to res group leader, 57-66, supvr petrochem process develop, 66-70, supvr fuels technol div, 70-73, COKE COORDR, WORLDWIDE REFINING, CONTINENTAL OIL CO, 73- *Concurrent Pos:* Fulbright lectr, Naples, 61-62. *Mem:* Am Chem Soc; Am Inst Chem Engrs. *Res:* Thermodynamics of vapor-liquid equilibria. *Mailing Add:* PO Box 6166 Kingwood TX 77339

ACEDO, GREGORIA N, b Albay, Philippines, Apr 24, 36; m 65; c 1. PLANT GENETICS, PLANT PATHOLOGY. *Educ:* Univ Philippines, Los Banos, BSA, 59, MS, 65; Univ Mass, Amherst, PhD(plant path), 69. *Hon Degrees:* BSA Honor's Curriculum, 59. *Prof Exp:* Researcher plant path, Bur Plant Indust, Philippines, 59-62; res instr, Univ Philippines, 65-66; fel plant path, Univ Mo, Columbia, 70-71, res assoc, 73-85; SCIENTIST, ENVIRON PROTECTION AGENCY, DURHAM, NC, 85- *Concurrent Pos:* Nat Sci Develop Bd fel plant path, 62-64. *Mem:* Philippine Phytopath Soc; Environ Mutagenesis Soc. *Res:* Genetic aspect of disease resistance in plants; genetic control of flower differentiation; search for auxotrophs; tissue culture; mutagenicity testing; plant metabolic activation. *Mailing Add:* 22 Westridge Rd Durham NC 27713

ACERBO, SAMUEL NICHOLAS, b Port Chester, NY, Jan 19, 29. ORGANIC CHEMISTRY. *Educ:* Iona Col, BS, 55; Fordham Univ, MS, 57, PhD(org chem), 60. *Prof Exp:* From instr to assoc prof, 58-70, PROF CHEM, IONA COL, 70- *Concurrent Pos:* NIH res grant, 61-62. *Mem:* Am Chem Soc. *Res:* Trace quantities of boron produce profound effects on plant growth; the use of C14 on cell wall formation in plants grown in media of varying boron content. *Mailing Add:* One Mitchell Pl Port Chester NY 10573

ACETO, MARIO DOMENICO, b Providence, RI, Oct 26, 30; m 59, 82; c 2. CENTRAL NEROUS SYSTEM PHARMACOLOGY, DRUG DEPENDENCE. *Educ:* Univ RI, BS, 53; Univ Md, MS, 55; Univ Conn, PhD(pharmacol), 59. *Prof Exp:* From instr to asst prof pharmacol, Sch Pharm, Univ Pittsburgh, 58-62; assoc res biologist, Sterling Winthrop Res Inst, 62, res biologist & group leader, 63-66, sr res biologist & head sect pharmacol, 66-72, proj dir, 71-72; assoc prof, 72-86, PROF PHARMACOL, MED COL VA, VA COMMONWEALTH UNIV, 86- *Concurrent Pos:* Mem, Ad Hoc Comt Narcotic Interaction, NIMH, 73, Drug Eval Comt, CPDD, 88- & Tech Rev Comt Preclin Testing, Nat Inst Drug Abuse, 90; consult, Lederle Labs, 78, Key Pharmaceut, 84-86 & Biomet Res Inst, 90. *Honors & Awards:* Honor Achievement Award, Am Col Angiol. *Mem:* Am Soc Pharmacol & Exp Therapeut; fel Am Col Angiol; Neuropharmacol Soc. *Res:* Central nervous system pharmacology and drug dependence. *Mailing Add:* Dept Pharmacol MCV Box 613 Med Col Va Commonwealth Univ Richmond VA 23298-0613

ACHAR, B N NARAHARI, b India, Jan 5, 39. SOLID STATE PHYSICS. *Educ:* Mysore Univ, BS, 60; Pa State Univ, MS, 65, PhD(solid state sci), 68. *Prof Exp:* Res assoc, Argonne Nat Lab, 69-72; res assoc, Pa State Univ, 73-81; vis asst prof physics, Bucknell Univ, 81-84; PROF, DEPT PHYSICS, MEMPHIS STATE UNIV, 84- *Mem:* Am Phys Soc. *Mailing Add:* Dept Physics Memphis State Univ Memphis TN 38152

ACHARI, RAJA GOPAL, b West Bengal, India, Apr 22, 43; US citizen; m 71; c 1. ORGANIC CHEMISTRY, ANALYTICAL CHEMISTRY. *Educ:* Ranchi Univ, India, BSc, 61, MSc, 63; Bradford Univ, Eng, PhD(bioorg chem), 70. *Prof Exp:* Res asst org chem, Nottingham Univ, Eng, 66-67; res fel anal & org chem, Bradford Univ, Eng, 70-72; res assoc anal chem, SC State Col, 73-74; group leader anal chem, Cooper Labs, Inc, 74-79; SECT HEAD ANAL CHEM, BRISTOL-MYERS PROD, 79- *Mem:* Am Chem Soc; Royal Inst Chem London; Am Asn Pharm Scientists; Am Inst Chem. *Res:* Pharmaceutically active drugs in dosage forms and biological fluids; biosynthetic route of naturally occuring drugs and metabolism of various drugs. *Mailing Add:* Bristol-Myers Products 1350 Liberty Ave Hillside NJ 07207

ACHARYA, SEETHARAMA A, b Ubipi, India, Aug 23, 41. PROTEIN CHEMISTRY. *Educ:* India Inst Sci, Bangalore, PhD(biochem), 70. *Prof Exp:* ASST PROF BIOCHEM, ROCKEFELLER UNIV, 78- *Mailing Add:* Dept Biochem Rockefeller Univ 1230 York Ave New York NY 10021

ACHE, HANS JOACHIM, b Steinau, Ger, Jan 18, 31; m 63; c 1. RADIOCHEMISTRY, RADIATION CHEMISTRY. *Educ:* Univ Cologne, BS, 55, MS, 57, PhD(chem), 59. *Prof Exp:* Res assoc nuclear chem & radiochem, Inst Nuclear Chem, Univ Cologne, 59-62; res assoc radiochem, Brookhaven Nat Lab, 62-64, vis assoc chemist, 64-65; from asst prof to assoc prof, 65-71, PROF RADIOCHEM, VA POLYTECH INST & STATE UNIV, 71- *Mem:* Am Chem Soc; Am Nuclear Soc; Soc Ger Chem. *Res:* Hot atom chemistry; positronium chemistry; production of radioisotopes; radioactive labeling of organic compounds. *Mailing Add:* Inst fur Radiochem Postfa 3640 Kernforschungszentrum Karlsruhe 1 7500 Germany

ACHENBACH, JAN DREWES, b Leeuwarden, Neth, Aug 20, 35; m 61. APPLIED MECHANICS & MATHEMATICS, ULTRASONICS. *Educ:* Tech Univ Delft, Neth, MS, 59; Stanford Univ, PhD(aeronaut, astronaut), 62. *Prof Exp:* Preceptor civil eng, Columbia Univ, 62-63; from asst prof to assoc prof civil eng, 63-69, prof civil eng & appl math, 69-81, WALTER P MURPHY PROF CIVIL ENG, MECH ENG & APPL MATH, NORTHWESTERN UNIV, EVANSTON, 81- *Concurrent Pos:* Vis assoc prof, Univ Calif, San Diego, 69; vis prof, Tech Univ Delft, 70-71; mem at large, US Nat Comt Theoret & Appl Mech, 72-78, 87-; consult prof, Huazhong Inst Sci & Technol, China, 81- *Honors & Awards:* C Gelderman Found Award, 70; C W McGraw Res Award, Am Soc Eng Educ, 75. *Mem:* Nat Acad Eng; fel Am Acad Mech (pres, 78-79); fel Am Soc Mech Engrs; Am Geophys Union; Acoust Soc Am; Japanese Soc Adv Sci; Am Soc Nondestructive Testing. *Res:* Theoretical and applied mechanics and mathematics, particularly propagation of mechanical disturbances in solids, mechanics of fracture of solids, and quantitative non-destructive evaluation by ultrasonic wave-scattering techniques. *Mailing Add:* Ctr for Qual Eng & Failure Prev Northwestern Univ Evanston IL 60208

ACHESON, DONALD THEODORE, b Waltham, Mass, June 18, 35; m 61; c 2. METEOROLOGY, INSTRUMENTATION. *Educ:* Harvard Univ, BA, 57; Univ Md, MS, 65, PhD(meteorol), 74. *Prof Exp:* Gen engr physics, Engr Res & Develop Labs, Dept Army, 57-61; gen engr physics, Nat Weather Serv, Nat Oceanic & Atmospheric Admin, 61-72, gen engr physics & meteorol, Ctr Exp Design & Data Anal, 72-77, supvry electronics engr, 77-90; PROJ MGR, NS&R CO, 91- *Mem:* Am Meteorol Soc. *Res:* Origin and propagation of instrumental errors and optimal processing of observations in meteorology. *Mailing Add:* 3512 Beret Lane Silver Spring MD 20906

ACHESON, LOUIS KRUZAN, JR, b Brazil, Ind, Apr 2, 26; m 58; c 2. SPACE SCIENCES, REMOTE SENSING. *Educ:* Case Inst Technol, BS, 46; Mass Inst Technol, PhD(physics), 50. *Prof Exp:* Mem tech staff systs anal, Hughes Aircraft Co, 50-58; mem tech staff ballistic missile defense, Inst Defense Anal, 58-59; sr scientist space systs, Hughes Aircraft Co, 60-85, chief scientist Systs Design Lab, Space & Commun Group, 85-89; RETIRED. *Mem:* Am Phys Soc; Am Inst Aeronaut & Astronaut; Am Geophys Union; Brit Interplanetary Soc; AAAS. *Res:* Systems for defense against ballistic missiles, sensors and systems for meteorological, earth resources, solar terrestrial and planetary exploration spacecraft; air traffic control systems. *Mailing Add:* 17721 Marcello Pl Encino CA 91316-4328

ACHESON, WILLARD PHILLIPS, b Cairo, Egypt, Nov 24, 27; US citizen; m 49; c 3. PHYSICS, APPLIED MATHEMATICS. *Educ:* Westminster Col, Pa, BS, 48; Pa State Univ, MS, 50; Univ Pittsburgh, PhD(appl math), 61. *Prof Exp:* Teacher sci & Head dept, Assiut Col, Egypt, 50-53; fel struct clay prod, Mellon Inst, 53-60; asst prof physics, Muskingum Col, 60-62; sr res eng

petrophys, Gulf Res & Develop Co, 62-75, res assoc energy resources, 75-81, dir, Alt Resource Recovery Sect, 81-83, sr res assoc, 83-85; CONSULT, 85- *Mem:* Am Phys Soc; fuel technology and petroleum engineering; Soc Petrol Engrs; Soc Indust & Appl Math. *Res:* Plasticity; rheology; fluid flow; acoustics; petrophysics; enhanced recovery technology of petroleum and alternative fuels. *Mailing Add:* 128 N Craig St Pittsburgh PA 15206

ACHEY, FREDERICK AUGUSTUS, b Lancaster, Pa, Sept 10, 29; m 58; c 3. ANALYTICAL CHEMISTRY. *Educ:* Franklin & Marshall Col, BS, 51; Lehigh Univ, MS, 53, PhD, 56. *Prof Exp:* Asst, Lehigh Univ, 51-56; vpres, Serfass Corp, Pa, 56-60; supvr processes & temperature, Bethlehem Steel Corp, 60-73, asst sect mgr, 73-82, tech consult, 82-85. *Mem:* Am Chem Soc (treas, 63); Instrument Soc Am. *Res:* Analytical instrumentation; x-ray spectroscopy. *Mailing Add:* 3345 Nazareth Pike Bethlehem PA 18017

ACHEY, PHILLIP M, b Lancaster, Pa, June 25, 39; c 4. RADIATION BIOLOGY. *Educ:* Franklin & Marshall Col, AB, 61; Pa State Univ, MS, 64, PhD(biophys), 66. *Prof Exp:* Fel, M D Anderson Hosp & Tumor Inst Houston, 66-67; asst prof biophys, 67-75, assoc prof, 75-80, PROF MICROBIOL, UNIV FLA, 80- *Mem:* Biophys Soc; Am Soc Microbiol; Radiation Res Soc; Sigma Xi. *Res:* Biological and physical action of radiation on cells and DNA; repair of radiation damage. *Mailing Add:* Dept Microbiol & Cell Sci Univ Fla 403 Nuclear Sci Bldg Gainesville FL 32611

ACHILLES, ROBERT F, b Inman, Kans, July 5, 24; m 46; c 2. SPEECH PATHOLOGY, AUDIOLOGY. *Educ:* McPherson Col, BS, 49; Wichita State Univ, MA, 55, PhD(logopedics), 68. *Prof Exp:* Speech pathologist, Inst Logopedics, 49-52; dir speech & hearing, Children's Rehab Ctr, Md, 52-53; clin supvr, Inst Logopedics, 53-58, dir clin serv, 58-60; assoc prof, 63-68, PROF SPEECH, LAMAR UNIV, 68-, DIR SPEECH PATH, 63- *Concurrent Pos:* Lectr logopedics, Wichita State Univ, 55-57, instr, 57-60. *Mem:* Am Speech & Hearing Asn. *Res:* Cerebral palsy and speech; voice science. *Mailing Add:* Dept Commun AAS & Sci Lamar Univ Beaumont TX 77710

ACHOR, WILLIAM THOMAS, b Birmingham, Ala, June 18, 29; m 58; c 2. PHYSICS EDUCATION, NOISE. *Educ:* Ala Polytech Inst, BS, 52; Vanderbilt Univ, MS, 54, PhD, 58. *Prof Exp:* Instr physics, Western Reserve Univ, 57-59; mem tech staff, Radio Corp Am Labs, 59-62; asst prof physics, Earlham Col, 63-65; dept chmn, 65-87, PROF PHYSICS, WESTERN MD COL, 65- *Concurrent Pos:* Consult, Carroll County Bd Educ, 77-78; vis prof acoust res, Pa State Univ, 78-79; lectr, Hood Col, 82-83 & Loyola Col, 83-85; mem, Fundamental Particles & Interactions Chart Comt, 87- *Mem:* Am Asn Physics Teachers. *Res:* Nuclear spectrometry; Auger effect; physical electronics; science education; noise. *Mailing Add:* Dept Physics Western Md Col Westminster MD 21157

ACHORN, FRANK P, b Biloxi, Miss, Mar 23, 23; m 45; c 3. CHEMICAL ENGINEERING. *Educ:* Univ Louisville, BChE, 47. *Prof Exp:* Process engr, Tenn Valley Authority, Wilson Dam, 47-50, process & prod engr, 52-57; asst plant mgr, Coastal Chem Corp, Miss, 57-60; field engr, 60-63, engr-in-chg eng field servs, 63-66, HEAD PROCESS & PROD IMPROV SECT, NAT FERTILIZER DEVELOP CTR, TENN VALLEY AUTHORITY, 66-, SR SCIENTIST, 76- *Concurrent Pos:* Consult, var US & foreign firms, 60- & C F Braun Co, Calif, 63-64. *Honors & Awards:* Authority, 76.. *Mem:* Hon mem Nat Fertilizer Solutions Asn; Am Chem Soc. *Res:* Fertilizer technology, especially distribution systems and technical service. *Mailing Add:* Box 176 Rte 6 Killen AL 35645

ACKELL, EDMUND FERRIS, b Danbury, Conn, Nov 29, 25; m 53; c 4. MEDICAL & EDUCATIONAL ADMINSTRATION, MAXILLOFACIAL SURGERY. *Educ:* Col Holy Cross, BS, 49; Tufts Univ, DMD, 53; Western Reserve Univ, MD, 62; Am Bd Oral Surg, dipl. *Prof Exp:* From asst prof to assoc prof oral surg & chmn dept, Western Reserve Univ, 57-66, asst prof oral path, 62-66; dean col dent, Univ Fla, 66-69, vpres health affairs, J Hillis Miller Health Ctr, 69-74; vpres health affairs, Univ Southern Calif, 74-78; pres, 78-90, EMER PRES, VA COMMONWEALTH UNIV, 90- *Concurrent Pos:* Am Cancer Soc clin res fel, 55-56; mem rev comt, Health Prof Educ Act, 64-; HEW Comt Health Educ, 68; Health Planning, govt Yugoslavia, 73; mem, Nat Avd Coun, Educ Health Prof, 69. *Mem:* Am Dent Asn; AMA. *Res:* Dental education; education in the health professions. *Mailing Add:* Va Commonwealth Univ 600 W Franklin Richmond VA 23284-3025

ACKENHUSEN, JOHN GOODYEAR, b South Charleston, WVa. ELECTRICAL ENGINEERING. *Educ:* Univ Mich, Ann Arbor, BS & BSE, 75, MS & MSE, 76, PhD(nuclear eng & exp plasma physics), 77. *Prof Exp:* Res assoc plasma physics, Laser Plasma Interaction Lab, Univ Mich, 76-77, interim lab dir, 77-78; MEM TECH STAFF ELEC ENG, BELL LABS, 78- *Concurrent Pos:* Scholar, Laser Plasma Interaction Lab, Univ Mich, 77-78. *Mem:* Am Phys Soc; Optical Soc Am; Sigma Xi; sr mem Inst Elec & Electronics Engrs. *Res:* Implementation of man-machine communications, including information display and pattern recognition; optics and optical communications; experimental laser-plasma interaction. *Mailing Add:* Bell Labs 78 New England Ave 32 Summit NJ 07901

ACKER, ANDREW FRENCH, III, b New London, Conn, May 9, 43. PARTIAL DIFFERENTIAL EQUATIONS, FREE BOUNDARY PROBLEMS. *Educ:* Union Col, Schenectady, NY, BS, 65; Boston Univ, PhD(math), 72; Univ Karlsruhe, Ger, Prof Dr, 81. *Prof Exp:* Asst prof math, Univ New Orleans, 72-73; asst, Univ Karlsruhe, 73-83; assoc prof, Iowa State Univ, 84-87; assoc prof, 87-91, PROF MATH, WICHITA STATE UNIV, 91- *Concurrent Pos:* Spec lectr, Boston Univ, 69-72; vis assoc prof, Iowa State Univ, 83-84. *Mem:* Am Math Soc. *Res:* Free-boundary problems in elliptic partial differential equations, with emphasis on uniqueness, the geometric form of the free boundary, especially convexity, and the successive approximation of the free boundary; multilayer fluid problem; plasma confinement in a magnetic field. *Mailing Add:* Dept Math & Statist Wichita State Univ Wichita KS 67208

ACKER, DUANE CALVIN, b Atlantic, Iowa, Mar 13, 31; m 52; c 2. ANIMAL NUTRITION. *Educ:* Iowa State Univ, BS, 52, MS, 53; Okla State Univ, PhD(animal nutrit), 57. *Prof Exp:* Asst animal nutrit, Iowa State Univ, 52-53; instr animal husb, Okla State Univ, 53-55; from instr to asst prof animal sci, Iowa State Univ, 55-58, assoc prof animal sci in-chg farm oper curriculum, 58-62; assoc dean agr & dir resident instr, Kans State Univ, 62-66, asst dir Agr Exp Sta, 65-66; dean agr & biol sci & dir agr exp sta, SDak State Univ, 66-74, dir coop exten serv, 71-74; vchancellor agr & natural resources, Univ Nebr-Lincoln, 74-75; pres, Kans State Univ, 75-86; dir & asst to adminr, US Agency Int Develop, 86-90; adminr, Int Coop & Develop, 90-91, ADMINR, FOREIGN AGR SERV, USDA, 91- *Concurrent Pos:* exec comt, Great Plains Agr Coun, 66-67; US team for review of Marshall Plan Aid to WGer, 67; co-chmn, USDA, Exp Sta Task Force Qual of Environ, 67-68; dir, Northwest Nat Bank, Souix Falls, 70-74, Omaha, 74-75, Fed Reserve Bank Kans City, 82-86, KPL Gas Serv Co, 82-86. *Mem:* AAAS; Am Soc Animal Sci. *Res:* Meat quality as influenced by nutrients, hormones, drugs; protein and amino acid requirements and metabolism. *Mailing Add:* 1801 Crystal Dr No 608 Arlington VA 22202

ACKER, FRANK EARL, b Pittsburgh, Pa, Mar 11, 37; m. ENGINEERING, ELECTRICAL ENGINEERING. *Educ:* Carnegie-Mellon Univ, BS, 59, MS, 61, PhD(elec eng), 67. *Prof Exp:* Electronics sect head, Naval Res Lab, 67-69; asst prof elec eng, Univ Pittsburgh, 69-74; sr res engr, US Steel Res Labs, 74-80; engr, res & develop ctr, Westinghouse Corp, 80-85; ASSOC PROF, ROSE MULMAN INST TECH, 85- *Honors & Awards:* Award of Merit for Group Achievement, Dept Navy, 68. *Mem:* Sr mem Inst Elec & Electronics Engrs. *Res:* Industrial control systems; electrical engineering and robotics applied to industrial and mining equipment; mine power systems. *Mailing Add:* Rose-Hulman Inst Tech 5500 Wabash Ave Terre Haute IN 47803

ACKERBERG, ROBERT C(YRIL), b Minneapolis, Minn, Sept 14, 34; c 2. APPLIED MATHEMATICS, FLUID DYNAMICS. *Educ:* Mass Inst Technol, BS, 56; Univ Mich, MSE, 57; Harvard Univ, MA, 60, PhD(appl math), 63. *Prof Exp:* Sr res scientist physics, Raytheon Corp, 62-63; res assoc aerospace eng, 63, from asst prof to assoc prof, 63-72, assoc prof chem eng, 72-73, PROF CHEM ENG, POLYTECH INST NEW YORK, 73- *Concurrent Pos:* Res fel, Sci Res Coun of Gt Brit, 71-72; prin investr, NSF, 74- *Mem:* Am Inst Chem Engrs. *Res:* Boundary-layer theory; theory of boundary-layer separation; nonlinear theory of thin jets; potential flow; singular perturbations; matched asymptotic expansions; thermodynamics; heat transfer. *Mailing Add:* Dept Chem Eng Polytech Univ Long Island Ctr Rte 110 Farmingdale NY 11735

ACKERHALT, JAY RICHARD, b Passaic, NJ, Nov 22, 47; m 69; c 1. QUANTUM OPTICS, PHYSICS. *Educ:* Hobart Col, BS, 69; Univ Rochester, MA, 72, PhD(physics), 74. *Prof Exp:* Res worker physics, Inst Theoret Physics, Univ Warsaw, 74; res assoc, Dept Physics, Univ Rochester, 75-76 & Johns Hopkins Univ, 76; STAFF MEM, LOS ALAMOS NAT LAB, 77- *Concurrent Pos:* Consult, Lawrence Livermore Nat Lab, 76, Allied-Signal Corp, 87- *Mem:* Am Phys Soc. *Res:* Quantum optics, multiple photon excitation and dissociation of polyatomic molecules; theoretical optics modelling; chaos in optical systems. *Mailing Add:* Univ Calif LASL MS J569 Los Alamos NM 87545

ACKERLIND, E(RIK), b New York, NY, July 9, 10; m 44; c 1. ELECTRICAL ENGINEERING. *Educ:* Polytech Inst, Brooklyn, BEE, 32, DEE, 37; Columbia Univ, MS, 34. *Prof Exp:* Electronic engr, Hazeltine Serv Corp, NY, 39-41; subsect head, Naval Res Lab, Washington, DC, 41-46; sect supvr, Northrop Aircraft, Inc, 46-49; group supvr, Jet Propulsion Lab, Calif Inst Technol, 49-51; tech dir, Ackerlind Corp, 51-53; mgr systs eng, Radio Corp Am, 53-57; electronics consult, 57-65; mgr systs perform anal, ITT Fed Elec Co, 65-69; consult, 69-76; WRITER, 76- *Concurrent Pos:* Instr, Univ Exten, Univ Calif, Los Angeles, 48-49. *Mem:* Inst Elec & Electronics Engrs. *Mailing Add:* PO Box 1587 Bellingham WA 98227-1587

ACKERMAN, BERNICE, b Chicago, Ill, July 24, 24. CLOUD PHYSICS, CLOUD DYNAMICS. *Educ:* Univ Chicago, BS, 48, MS, 55, PhD(geophys sci), 65. *Prof Exp:* Meteorologist & hydrologist, US Weather Bur, 48-53; res assoc cloud physics, Univ Chicago, 53-65, asst prof meteorol, 65-67; assoc prof, Tex A&M Univ, 67-70; assoc meteorologist, Argonne Nat Lab, 70-72; sr meteorologist cloud physics, meso-meteorol & boundry layer, 72-78, PRIN SCIENTIST, ILL STATE WATER SURV, 78-, HEAD, METEOROL SECT, 80- *Mem:* Fel AAAS; fel Am Meteorol Soc; Am Geophys Union; Royal Meteorol Soc. *Res:* Dynamics of clouds; meso-meteorology; weather modification; boundary layer physics; urban meteorology. *Mailing Add:* Ill State Water 2204 Griffith Dr Champaign IL 61820-7407

ACKERMAN, BRUCE DAVID, b New York, NY, Mar 12, 34; m 57; c 2. PEDIATRICS. *Educ:* NY Univ, AB, 54; Univ Chicago, MD, 58. *Prof Exp:* Fel neonatology, Magee-Womens Hosp, Univ Pittsburgh, 63-65; from instr to assoc prof pediat, Univ Calif, Irvine, 65-72; assoc prof, State Univ NY Stony Brook, 72-74; assoc prof, Mt Sinai Sch Med, 74-75; ASSOC PROF PEDIAT, STATE UNIV NY AT BROOKLYN, 75- *Mem:* Am Thoracic Soc; Soc Pediat Res; Am Acad Pediat; NY Pediat Soc. *Res:* Perinatal physiology; fetal, placental and neonatal blood flow; effect of narcotic drugs on the fetus and newborn; use of expressed breast milk for nutritional support of premature infants in less advaced societies. *Mailing Add:* Maimonides Med Ctr Brooklyn NY 11219

ACKERMAN, C(ARL) D(AVID), b Edwardsville, Ill, Nov 17, 18; m 42; c 1. PROCESS DESIGN, SYNTHETIC FUELS. *Educ:* Univ Ill, BS, 42. *Prof Exp:* Chem engr, Gulf Res & Develop Co, 42-46 & 48-53, group leader, 53-61, sr proj engr, 61-66, tech assoc, 66-72; chem engr, Standard Oil Co, Ind, 46-47; chem engr res & develop, Pittsburgh & Midway Coal & Mining Co, 73-75, dir process develop, 76-81; RETIRED. *Mem:* Am Inst Chem Engrs; Instrument Soc Am. *Res:* Pilot plant design; instrumentation; automation. *Mailing Add:* 118 19th Ave Olympia WA 98501

ACKERMAN, DONALD GODFREY, JR, analytical chemistry, for more information see previous edition

ACKERMAN, ERIC J, b Coral Gables, Fla, Feb 6, 54. DNA REPAIR, IMMUNOTOXINS. *Educ:* Univ Fla, BSc, 75; Univ Chicago, PhD(biophys), 80. *Prof Exp:* Postdoctoral fel, MRC Lab Molecular Biol, 81-85; SR STAFF FEL MOLECULAR BIOL, NAT INST DIABETES & DIGESITIVE & KIDNEY DIS, NIH, 85- *Concurrent Pos:* Fel, Anna Fuller Found, 81, Helen Hay Whitney Found, 82-84 & Deutsche Akademischeaustaush Dienst, 85. *Res:* Elucidating the molecular mechanisms used by toxins to kill cells; molecular biology of eukaryotic DNA repair. *Mailing Add:* NIH Bldg 10 Rm 9D15 Bethesda MD 20892

ACKERMAN, EUGENE, b Brooklyn, NY, July 8, 20; m 43; c 3. HEALTH & COMPUTER SCIENCES. *Educ:* Swarthmore Col, BA, 41; Brown Univ, ScM, 43; Univ Wis, PhD(biophys), 49. *Prof Exp:* Asst, Brown Univ, 41-43; asst, Univ Wis, 46-47; res asst, Wis Alumni Res Found, 47-49; assoc biophys, Johnson Res Found, Univ Pa, 49-51; from asst prof to assoc prof physics, Pa State Univ, 51-58, prof biophys, 58-60; assoc prof biophys, Mayo Grad Sch Med, 60-65, prof, 65-67, prof biomed comput & biomet, Col Med Sci, 67-69, dir, Div Health Comput Sci, 69-78, PROF LAB MED & PATH, MED SCH, UNIV MINN, MINNEAPOLIS, 69-, dir grad prog biophys sci, 79-88. *Concurrent Pos:* Johnson Res Found fel, Univ Pa, 57-58; consult, Mayo Found, 60-67, dir comput facil, 63-65; mem var comts, NIH, 64-74; ed, Biophys J, 83-87. *Mem:* Asn Comput Mach; NY Acad Sci; Inst Elec & Electronics Engrs; Sigma Xi; Am Physiol Soc; Biophys Soc; fel Am Inst Med Info. *Res:* Health computer sciences; biomathematics; Monte Carlo models; biophysics; biomedical engineering. *Mailing Add:* Box 511 UMHC Univ Minn 420 Delaware St SE Minneapolis MN 55455

ACKERMAN, GUSTAVE ADOLPH, JR, b Columbus, Ohio; m 56; c 2. ANATOMY, MEDICINE. *Educ:* Ohio State Univ, BA, 48, MSc, 49, MD & PhD(anat), 54. *Prof Exp:* Intern med, Salt Lake County Hosp, 54-55; from asst prof to assoc prof, 57-64, PROF ANAT, OHIO STATE UNIV, 64- *Concurrent Pos:* Lederle Med Fac award, 61-63. *Honors & Awards:* Borden Award, 54. *Mem:* Am Asn Anat; Histochem Soc; Am Soc Cell Biol. *Res:* Morphology and histochemistry of the cells of the blood and hemopoietic system; cellular differentiation and origin. *Mailing Add:* Dept Anat Ohio State Univ Col Med 370 W Ninth Ave Columbus OH 43210

ACKERMAN, HERVEY WINFIELD, JR, b Easton, Pa, Feb 28, 29; m 47, 64; c 6. ORGANIC CHEMISTRY. *Educ:* Lafayette Col, BS, 49; Yale Univ, PhD(org chem), 53. *Prof Exp:* Res chemist, Film Dept, E I du Pont de Nemours & Co, 52-54; res chemist, Nat Aniline Div, Allied Chem Corp, 54-62; res chemist, Air Prod & Chem, Inc, 62-64; res chemist, M & T Chem, Inc, 64-68; opers res, Am Can Co, 68-70, adminr mgt sci, 70-72; assoc, ROI Controls Corp, 72-75; consult, 75-77, CONTROL SYSTS SPECIALIST, CONSOLIDATED EDISON CO NY, INC, 77- *Mem:* Am Chem Soc. *Res:* Management sciences. *Mailing Add:* 350 65th St Brooklyn NY 11220

ACKERMAN, JAMES HOWARD, b Detroit, Mich, Oct 4, 28; m 54, 76; c 2. MEDICINAL CHEMISTRY. *Educ:* Univ Mich, BS, 50; Univ Wis, PhD(org chem), 54. *Prof Exp:* Chemist, Mallinckrodt Chem Works, 54-58; RES INVESTR, STERLING RES GROUP, 58- *Mem:* Am Chem Soc. *Res:* Radiopaques; steroids. *Mailing Add:* Sterling Res Group Rensselaer NY 12144-3493

ACKERMAN, JAMES L, b Elizabeth, NJ, Feb 2, 38; m 70. ORTHODONTICS. *Educ:* Univ Pa, DDS, 60; Harvard Univ, cert orthod, 62; Am Bd Orthod, dipl, 70. *Prof Exp:* Res fel orthod, Harvard Univ, 60-62; res assoc orthod, Nat Inst Dent Res, 62-64; dir res, Fairleigh Dickinson Univ, 64-66; assoc prof, 68-71, chmn dept, 68-73, assoc dean student affairs, 70-72, PROF ORTHOD, UNIV PA, 71-, PROF PEDIAT DENT & CHMN DEPT, 73- *Concurrent Pos:* Pvt practr, Westfield, NJ, 64-72; consult, Children's Specialized Hosp, Mountainside, NJ, 65-72 & Univ Ky, 66-68; sr dentist, Children's Hosp, Philadelphia, 73-; vis prof pediat dent, Harvard Univ, 74-75. *Mem:* AAAS; Am Dent Asn; Am Asn Orthod; Int Asn Dent Res; Am Soc Dent Children; Sigma Xi. *Res:* Skeletal morphogenesis at the cellular, tissue and gross levels; relationship of form and function in bone; growth and development. *Mailing Add:* 102 W Montgomery Ave, Unit B Ardmore PA 19003-1422

ACKERMAN, JEROME LEONARD, b New York, NY, Jan 4, 50; m 73, 86; c 2. NUCLEAR MAGNETIC RESONANCE, MAGNETIC RESONANCE IMAGING. *Educ:* State Univ NY Stony Brook, BS, 71; Mass Inst Techol, PhD(phys chem), 76. *Prof Exp:* Res chemist, dept chem, Univ Calif, Berkeley, 76-77; assoc prof, chem dept, Univ Cincinnati, 77-85; ASST PHYSICS & DIR SPECTROS NUCLEAR MAGNETIC RESONANCE, DEPT RADIOL, MASS GEN HOSP, 85-; ASST PROF RADIOL, HARVARD MED SCH, 85- *Concurrent Pos:* Scientist appointee, Argonne Nat Lab, 86; res affil, Mass Inst Technol, 90. *Mem:* Am Chem Soc; AAAS; Soc Magnetic Resonance Med; Am Ceramic Soc; Mat Res Soc. *Res:* Nuclear magnetic resonance imaging techniques for biological and nonbiolgical systems; development of magnetic resonance imaging of solid materials; selected problems in physical chemistry, through the application of high resolution solid state nuclear magnetic resonance. *Mailing Add:* NMR Ctr Mass Gen Hosp 149 13th St Charleston MA 02129

ACKERMAN, JOSEPH JOHN HENRY, b Tulsa, Okla, July 17, 49; m 77; c 2. NUCLEAR MAGNETIC RESONANCE, SUPERCOOLED SOLUTIONS. *Educ:* Boston Univ, BA, 72; Colo State Univ, PhD(chem), 77. *Prof Exp:* Res asst, Dept Chem, Colo State Univ, 77-78; NIH fel, Dept Biochem, Oxford Univ, 78-79; from asst prof to assoc prof chem, 79-88, res asst prof, 80-86, RES ASSOC PROF, SCH MED, WASHINGTON UNIV, 86-, PROF CHEM, 88- *Concurrent Pos:* Vpres, Dan Broida-Sigma-Aldrich Corp Scholarship Fund, Inc, 84-; mem bd trustees, Soc Magnetic Resonance in Med, 86-; mem exec comt, Exp Nuclear Magnetic Resonance Conf, Inc, 88- *Res:* Nuclear magnetic resonanace studies of intact biological systems including in vivo and perfused in vitro tissues, chemical exchange in supercooled solutions, and metalloenzymes. *Mailing Add:* Dept Chem Campus Box 1134 Washington Univ St Louis MO 63130

ACKERMAN, LARRY JOSEPH, b Hanover, Pa, Jan 8, 39; m 63; c 3. VETERINARY PATHOLOGY, ELECTRON MICROSCOPY. *Educ:* Univ Pa, VMD, 63; Univ Mo, Columbia, 68, 70; Am Col Vet Pathologists, dipl. *Prof Exp:* Asst chief, Path Div & prin investr, US Army Med Res & Nutrit Lab, Denver, 64-66; teaching & res asst vet path & prin investr, Sch Vet Med, Univ Mo, Columbia, 66-68; asst chief, Path Div, US Army Med Res & Nutrit Lab, Denver, 68-71; chief, Path Div & prin investr, US Army Med Res Unit, Panama, 71-74; chief vet path, Biomed Lab Edgewood, Army Proving Ground, US Army, Md, 74-76; vet pathologist, 76-78, dir path, 78-87, DIR CORP RES, EXP PATH LAB, INC, 87- *Concurrent Pos:* Consult, Consumer Prod Safety Comn Studies, 74-76; lectr path, Exp Path Labs, Inc, 76- *Mem:* Soc Toxicol Pathologists; Int Acad Pathologists; Soc Comp Ophthal; Am Vet Med Asn. *Res:* Experimental pathology: viral, parasitic and nutritional diseases, safety and carcinogenic potential of various environmental, industrial and pharmaceutical compounds. *Mailing Add:* Exp Path Labs Inc PO Box 474 Herndon VA 22070

ACKERMAN, NORMAN BERNARD, b New York, NY, Nov 27, 30; m 53; c 4. SURGERY, CANCER. *Educ:* Harvard Univ, AB, 52; Univ Pa, MD, 56; Univ Minn, PhD(surg), 64. *Prof Exp:* From asst prof to assoc prof surg, Sch Med, Boston Univ, 65-69; assoc clin prof surg, Univ Kans, 69-74; prof surg, State Univ NY Upstate Med Ctr, 74-; PROF & CHIEF SURG, METROP HOSP CTR, NEW YORK. *Concurrent Pos:* Nat Cancer Inst fel, Univ Minn, 60-63; Am Cancer Soc fac res assoc award, 66-69; attend surgeon, Univ Hosp, Boston & Boston Vet Admin Hosp, 65-69; consult, Providence Vet Admin Hosp, RI, 65-69; chmn dept surg, Menorah Med Ctr, Kansas City, 69-74; prof surg, Univ Mo-Kansas City, 72-74; consult, Syracuse Vet Hosp, 74- *Mem:* Am Col Surg; Soc Univ Surg; Am Surg Asn; Int Soc Surg. *Res:* Cancer vascularity; gastrointestinal physiology; carcinogenesis. *Mailing Add:* Dept Surg Metrop Hosp Ctr 1901 First Ave New York NY 10029

ACKERMAN, RALPH AUSTIN, b Mar 4, 45; m 73, 87; c 2. ENVIRONMENTAL PHYSIOLOGY, REGULATORY PHYSIOLOGY. *Educ:* Univ Fla, PhD(zool), 75. *Prof Exp:* Asst res physiologist, Scripps Inst Oceanog, Univ Calif, 77-81; asst prof, 81-85, ASSOC PROF ZOOL, IOWA STATE UNIV, 85- *Mem:* Am Physiol Soc; AAAS; Am Soc Zoologists. *Res:* Environmental physiology of embryonic and developing organisms. *Mailing Add:* Dept Zool Iowa State Univ Ames IA 50011

ACKERMAN, ROY ALAN, b Brooklyn, NY, Sept 9, 51; m; c 2. BIOENGINEERING. *Educ:* Polytechnic Inst Brooklyn, BS, 72; Mass Inst Technol, SM, 75; Univ Va, Birmingham, PhD, 86. *Prof Exp:* Asst mgr chem eng, Arlee, Inc, 67-72; process engr bioeng, Corning Glass Works, 73; chief ed med sci, Linguistics Systems, Inc, 74-75; chem engr, Tri-Flo Res Labs, Ltd, New York, 72-74; sr proj engr, Thetford Corp, 75; TECH DIR, ASTRE CORP GROUP, CHARLOTTESVILLE, VA, 76- *Concurrent Pos:* Educ assoc, George Washington Univ, 76; consult, Indust Microgenics, Ltd, 81- & Bio Filtration Technologies, Inc, 81-84; consult, Bicarbolyte Corp, 84-; Samuel Ruben Scholar in Chem Eng. *Mem:* Water Pollution Control Fedn; Am Inst Chem Engrs; Am Chem Soc; Am Soc Artificial Internal Organs; Am Soc Microbiol; Sigma Xi. *Res:* Development of biosupport apparatus drawing upon chemical engineering principles, hemodialysis, aerosol therapy, and membrane oxygenators; development of recycling systems for water and waste utilizing microbial and physicochemical techniques. *Mailing Add:* Astre 809 Princes St Alexandria VA 22314-2223

ACKERMAN, STEVEN J, b New York, NY, Jan 13, 51. IMMUNOLOGY, BIOLOGY. *Educ:* McGill Univ, PhD(path), 77. *Prof Exp:* Asst prof immunol, Mayo Clin, 78-83; PROF MED, HARVARD MED SCH-BETH ISRAEL HOSP, 84- *Mailing Add:* Harvard Med Sch-Beth Israel Hosp 330 Brookline Ave Boston MA 02215

ACKERMANN, GUENTER ROLF, b Fellbach, Ger, Mar 20, 24; US citizen; m 51; c 3. ORGANIC CHEMISTRY, PHARMACEUTICAL CHEMISTRY. *Educ:* Ursinus Col, BS, 51; Univ Md, MS, 54; Univ Mich, PhD(pharmaceut chem), 56. *Prof Exp:* Chemist, Synthetic Org Chem, Atlantic Refining Co, Pa, 56-62; sr res chemist, Rohm & Haas Co, Philadelphia, 62-78, mgr, Govt Regulatory Affairs, 78-89; RETIRED. *Mem:* AAAS; Am Chem Soc. *Res:* Polymer synthesis; synthesis, spinning and testing of synthetic fibers; synthesis of coagulants and flocculents; development of ion-exchange resins; development of desalination processes; pollution control research; removal of pollutants from waste streams. *Mailing Add:* 616 Andover Rd Newtown Square PA 19073

ACKERMANN, HANS WOLFGANG, b Berlin, Ger, June 16, 36. MEDICAL MICROBIOLOGY. *Educ:* Free Univ Berlin, MD, 62. *Prof Exp:* Asst gynec, Rudolf Virchow Hosp, Berlin, Ger, 61, asst surg & internal med, 62-63; sci asst, Inst Hyg & Med Microbiol, Free Univ Berlin, 63-67; from asst prof to assoc prof, 67-76, PROF, FAC MED, LAVAL UNIV, 76- *Concurrent Pos:* Airlift Mem Found res fel, Pasteur Inst, Paris, France, 61-62; French Govt spec training fel med microbiol, 63-64; chmn bacterial virus subcomt, Int Comt Taxon Viruses, 75-81 & 87-90, vpres, 84-90, vchmn, 90- *Mem:* Can Soc Microbiol; French Soc Microbiol; Am Soc Microbiol; Can Soc Micros; Am Soc Virol. *Res:* Bacteriophages, also their implications with bacterial and viral taxonomy; airborne fungi; human hepatitis viruses; baculoviruses. *Mailing Add:* Dept Microbiol Laval Univ Fac Med Quebec PQ G1K 7P4 Can

ACKERMANN, MARTIN NICHOLAS, b Philadelphia, Pa, Feb 19, 41; m 63; c 2. INORGANIC CHEMISTRY. *Educ:* Carnegie-Mellon Univ, BS, 63; Univ Calif, Berkeley, PhD(chem), 66. *Prof Exp:* From asst prof to assoc prof, 66-78, PROF CHEM, OBERLIN COL, 78- *Concurrent Pos:* Res assoc, Univ Ga, 72-73; vis fel, Gen Elec Res & Develop Ctr, 79-80. *Mem:* Am Chem Soc. *Res:* Transition metal organometallic synthesis; complexes of N-coordinating ligands; chemistry of diazenes. *Mailing Add:* Dept of Chem Oberlin Col Oberlin OH 44074

ACKERMANN, NORBERT JOSEPH, JR, b Chattanooga, Tenn, July 3, 42; div; c 4. NUCLEAR ENGINEERING, INSTRUMENTATION. *Educ:* Univ Tenn, Knoxville, BS, 65, MS, 67, PhD(nuclear eng), 71. *Prof Exp:* Res instr nuclear eng, Univ Tenn, Knoxville, 65-68; develop sect head, I&C Div, Oak Ridge Nat Lab, Union Carbide Corp, 68-76; pres, Technol Energy Corp, 75-88; SPEC INSTRUMENTS LAB INC, 88- *Concurrent Pos:* Fel, Atomic Energy Comn, 65-68; consult, Atomic Energy Comn-Nuclear Regulatory Comn, 68-76, Oak Ridge Nat Lab, 76-77 & Dept Energy, 76-78; adj prof nuclear eng, Univ Tenn, Knoxville, 70- *Mem:* Am Nuclear Soc; Inst Elec & Electronics Engrs; AAAS. *Res:* Measurement and control techniques and analysis methods for surveillance and diagnosis of power plant system's performance and operation. *Mailing Add:* Qualte Q 9041 9041 Executive Park Dr Suite 500 Knoxville TN 37923

ACKERMANN, PHILIP GULICK, b Waterville, Minn, May 3, 09; m 45. BIOCHEMISTRY. *Educ:* Ore State Col, BS, 31; Johns Hopkins Univ, PhD(phys chem), 36. *Prof Exp:* Chief chemist, Cole Chem Co, 39-45; asst biochem, Sch Med, Washington Univ, 45-54, res assoc, 54-58, res asst prof, 58-64; biochemist, De Paul Hosp, St Louis, Mo, 64-79; RETIRED. *Concurrent Pos:* Biochemist, St Louis Chronic Hosp, 47-64; biochem consult, Faith Hosp & Homer G Phillips Hosp, 65-79; consult clin chem, Alexian Bros Hosp. *Mem:* AAAS; Am Chem Soc; Am Asn Clin Chemists; NY Acad Sci. *Res:* Determination of molecular structure by electron diffraction; statistical mechanics; analysis of pharmaceutical preparations; biochemistry of aging in man; nutrition and hormonal changes in the aged; blood lipoproteins and atherosclerosis. *Mailing Add:* 12501 Village Circle Dr-338 Sunset Hills MO 63127

ACKERMANN, UWE, b Greifswald, Ger, Dec 24, 39. MAMMALIAN PHYSIOLOGY. *Educ:* Univ Toronto, BASc, 66, MASc, 68, PhD(physiol), 73. *Prof Exp:* Asst prof physiol, Fac Med, Mem Univ, Nfld, 73-76; from asst prof to assoc prof, 76-87, PROF PHYSIOL, FAC MED, UNIV TORONTO, 87- *Concurrent Pos:* Vis lectr biomed instrumentation, Univ Lagos, Nigeria, 79. *Mem:* Can Hypertension Soc. *Res:* Mechanisms of extracellular fluid volume regulation. *Mailing Add:* Dept of Physiol Fac of Med Univ of Toronto Toronto ON M5S 1A8 Can

ACKERS, GARY K, b Dodge City, Kans, Oct 25, 39; m 59; c 3. PHYSICAL BIOCHEMISTRY. *Educ:* Harding Col, BS, 61; Johns Hopkins Univ, PhD(biochem), 64. *Prof Exp:* NIH fel physial chem, Sch Med, Johns Hopkins Univ, 64-65, instr, 65-66; from asst prof to prof biochem, Univ Va, 66-77; PROF BIOL, MCCOLLUM PRATT INST, 77- & PROF BIOPHYSICS, JOHNS HOPKINS UNIV, 85- *Concurrent Pos:* Guggenheim fel, 72-73. *Mem:* AAAS; Biophys Soc. *Res:* Interacting systems of macromolecules; molecular sieve methods. *Mailing Add:* Dept Biochem & Molecular Biophys Wash Univ Sch Med 660 S Euclid St Louis MO 63112

ACKERSON, BRUCE J, b Omaha, Nebr, Feb 13, 48; m 69; c 2. CHEMICAL PHYSICS. *Educ:* Univ Nebr, BS, 70; Univ Colo, MS, 72, PhD(physics), 76. *Prof Exp:* From asst prof to assoc prof, 77-86, PROF PHYSICS, OKLA STATE UNIV, 86- *Concurrent Pos:* Nat Bur Standards-Nat Res Coun grant, Cryog Div, Inst Basics Standards, Nat Bur Standards, Dept Com, 75-77. *Mem:* Am Phys Soc. *Res:* Complex fluids; scattering studies of structure and dynamics. *Mailing Add:* Dept Physics Okla State Univ Stillwater OK 74074

ACKLES, KENNETH NORMAN, b Hamilton, Ont, June 29, 35. PHYSIOLOGY. *Educ:* Queen's Univ, Ont, BSc, 59, MSc, 61; Univ Alta, PhD(physiol), 66. *Prof Exp:* Sci officer hyperbaric physiol, 67-70, head pressure physiol group, Defence Res Estab, Toronto, 70-79; DEFENCE RES & DEVELOP LIAISON OFFICER, CAN DEFENCE LIAISON STAFF, 79- *Concurrent Pos:* Defence Res Bd Can fel aviation med, Royal Air Force, Eng, 66-67. *Mem:* Aerospace Med Asn; Undersea Med Soc. *Res:* Physiological measurement of inert gas narcosis in diving; bubble detection and hematological aspects of decompression sickness; physiological effects of exposure to hyperbaric environment; assessment of stress in operational environments; life support systems in high performance aircraft. *Mailing Add:* Area Space Psychol Sect Defense & Civil Inst Environ Med PO Box 2000 Downsview ON M3M 3B9 Can

ACKLEY, STEPHEN FRED, b Nashua, NH, Jan 22, 44; m 65; c 3. GEOPHYSICS, GLACIOLOGY. *Educ:* Cornell Univ, BS, 65. *Prof Exp:* Teacher physics & math, Manning's Sch, Jamaica, 65-67; physicist ocean acoustics, Sanders Assocs, NH, 67-68; sci & eng asst ice physics, US Army Cold Regions Res & Eng Lab, 68-70, res physicist geophysics, 70-81, chief, snow & ice br, 81-90, RES GEOPHYSICIST, US ARMY COLD REGIONS RES & ENG LAB, 91- *Concurrent Pos:* Mem, Weather & Climate Working Group, NASA, 76-77; co-chmn glaciol, NSF Workshop on the Weddell Gyre, 77; mem comn, Polar Meteorol & Oceanog, Am Meteorol Soc, 81-85, Snow, Ice, Permafrost and Polar Science, Am Geophys Union, 85-; adj asst prof, 81-85, adj assoc prof eng, Dartmouth Col, 85-; vis prof, Arctic Marine Sci Chair, Naval Postgrad Sch, 85-86. *Honors & Awards:* Antarctic Serv Medal, Nat Sci Found, 79. *Mem:* Am Geophys Union; Int Glaciol Soc. *Res:* Geophysics of sea ice; sea ice climate and sea ice ocean interactions; ice physics; ice accretion and adhesion. *Mailing Add:* US Army Cold Regions Res & Eng Lab 72 Lyme Rd Hanover NH 03755

ACKLEY, WILLIAM BENTON, b Portis, Kans, June 5, 18; m 41; c 2. HORTICULTURE. *Educ:* Kans State Col, BS, 40, MS, 47; State Col Wash, PhD(hort), 53. *Prof Exp:* Tech asst, US Bur Census, Washington, DC, 40-43; asst, Kans State Col, 46-47; res asst, Wash State Univ, 47-48, horticulturist, 48-64, prof hort, 48-83, chmn dept, 64-74; RETIRED. *Concurrent Pos:* Consult, 84- *Mem:* Fel Am Soc Hort Sci; Int Soc Hort Sci. *Res:* Water relations of plants; physiological disorders of tree fruits; orchard pest control. *Mailing Add:* SE 740 Derby Rd Pullman WA 99163

ACKMAN, DOUGLAS FREDERICK, b Sept 15, 35; Can citizen; m 58; c 5. UROLOGY. *Educ:* Mt Allison Univ, BSc, 56; McGill Univ, MDCM, 60; FRCS(C), 67; FACS, 69. *Prof Exp:* ASST PROF SURG, MCGILL UNIV, 73- *Concurrent Pos:* Surgeon, Montreal Gen Hosp, 67-; dir, Can Ski Patrol Syst, 68-; consult fertil, Fed Royal Comn, 77-78 & KLI Corp, Penn, 78- *Mem:* Royal Col Surgeons; Am Col Surgeons; Can Urol Asn; Am Urol Asn. *Res:* Computer systems for medical data management; artificial insemination techniques; fertility control techniques, especially male. *Mailing Add:* 4690 Westmont Ave Montreal PQ H3Y 1X1 Can

ACKMAN, ROBERT GEORGE, b Dorchester, NB, Sept 27, 27; m 57; c 2. MARINE OILS & LIPIDS, ANALYSES OF FATTY ACIDS. *Educ:* Univ Toronto, BA, 50; Dalhousie Univ, MS, 52; Univ London, PhD(org chem), 56; Imp Col, dipl(org chem), 56. *Prof Exp:* Asst scientist, Atlantic Fisheries Exp Sta, Halifax, NS, 52-53; assoc scientist, Halifax Lab, Fisheries Res Bd Can, 56-62, group leader marine lipids, 62-70, proj mgr, 70-72, head, Marine Oils Sect, Technol Br, 73-79; PROF, TECH UNIV NOVA SCOTIA, HALIFAX, 79- *Concurrent Pos:* Mem comt fats & oils, Nat Res Coun Can, 62-66, chmn, 67-84; mem, expert comt plant prod & expert comt fats & other lipids, Can Comt Food; mem tech comt ISO/TC 34/SC II, subcomt animal & veg fats & oils, Can Adv Comt, Food & Agr Orgn/UN Int Orgn Standardization; assoc referee fats & oils, Asn Off Anal Chemists; assoc ed, Lipids. *Honors & Awards:* H P Kaufman Mem Medal, Int Soc Fat Res, 80. *Mem:* Fel Chem Inst Can; Am Oil Chemists Soc; Asn Off Anal Chemists. *Res:* Analysis and utilization of marine lipids and oils; gas-liquid chromatography of fatty acids and derivatives. *Mailing Add:* Dept Food Sci Tech Univ Nova Scotia PO Box 1000 Halifax NS B3J 2X4 Can

ACKRELL, BRIAN A C, RESPIRATORY FLAVO-ENZYMES, BIOENERGETICS. *Educ:* Univ Bristol, UK, PhD(biochem), 61. *Prof Exp:* RES BIOCHEMIST, UNIV CALIF, SAN FRANCISCO, 72- *Mailing Add:* Dept Biochem & Biophys Vet Admin Med Ctr Univ Calif 3rd & Parnassus Ave San Francisco CA 94121

ACOSTA, ALLAN JAMES, b Anaheim, Calif, Aug 10, 24; m 53; c 2. MECHANICAL ENGINEERING. *Educ:* Calif Inst Technol, BS, 45, MS, 49, PhD(mech eng), 52. *Prof Exp:* Res engr, 46-48, sect chief, Hydrodyn Lab, 52-54, from asst prof to assoc prof mech eng, 54-66, PROF MECH ENG, CALIF INST TECHNOL, 66-, EXEC OFFICER MECH ENG, 88- *Concurrent Pos:* NSF fel, 61-62; consult, Able Corp & Byron Jackson Pump Co. *Honors & Awards:* Fluids Eng Award, Am Soc Mech Engrs, 88. *Mem:* AAAS; Int Asn Hydraul Res; Am Soc Mech Engrs. *Res:* Fluid mechanics; hydrodynamics; turbomachines. *Mailing Add:* PO Box 578 Seal Beach CA 90740

ACOSTA, DANIEL, JR, b El Paso, Tex, Mar 25, 45; m 73; c 3. PHARMACOLOGY. *Educ:* Univ Tex, Austin, BS, 68; Univ Kans, PhD(pharmacol), 74. *Prof Exp:* Res asst pharmacol, Sch Pharm, Univ Kans, 70-74; from asst prof to assoc prof, 74-83, PROF PHARMACOL, UNIV TEX, AUSTIN, 83- *Concurrent Pos:* NSF trainee, 70-72; Ford Found fel, Nat Chicano Coun Higher Educ, 78-79; Eli Lilly-C R Sublet+Centennial fel, Univ Tex. *Mem:* Am Heart Asn, Basic Sci; AAAS; Tissue Cult Asn; Soc Toxicol; Am Soc Pharmacol & Exp Therapeut. *Res:* Cellular toxicology, particularly use of cell culture techniques to evaluate the effects of drugs and toxicants at the cellular and subcellular level; study of ischemic myocardial cell injury with cultured heart cells; in vitro approach to target organ toxicology. *Mailing Add:* Col of Pharm Univ of Tex Austin TX 78712

ACOSTA, PHYLLIS BROWN, b Postell, NC, Dec 27, 33; wid; c 3. GENETICS. *Educ:* Andrews Univ, BA, 55; Univ Iowa, MS, 57; Univ Calif, Los Angeles, MPH, 65, DPH, 69. *Prof Exp:* Teaching dietitian & instr, White Mem Hosp, Loma Linda Univ, 57-59, asst prof & asst dir, Sch Dietetics, 59-62, assoc prof, Sch & dir dietary serv, White Mem Med Ctr, 62-64; prof nutrit, Sch Home Econ & Sch Med Sci, Univ Nev, Reno, 70-74; assoc prof foods & nutrit, Univ Ga, 74-75; assoc prof pediat, Sch Med, Univ NMex, 75-77; from assoc prof to prof allied health, Sch Med, Emory Univ, 77-81; prof & head, Dept Nutrit & Food Sci, Fla State Univ, 81-87; DIR, METAB DIS, ROSS LABS, 87- *Concurrent Pos:* Co-investr, NIH Grant Study Phenylketonuria, 61-64; nutrition consult, Child Develop Clin, Children's Hosp Soc, Los Angeles, 62-70, nutritionist, 65-83; consult nat collab study phenylketonuria, Maternal & Child Health Sect, Health Serv & Ment Health Admin, US Dept Health, Educ & Welfare; consult pediat, Div Med Genetics, Sch Med, Emory Univ, 75-81; mem adv comt, Nat Inst Child Health & Human Develop, Maternal PKU Study, 84- *Mem:* Am Inst Nutrit; Am Dietetic Asn; Soc Inherited Metab Dis; Soc Study Inborn Errors Metab. *Res:* Relationship of diet to mental retardation in phenyketonuria and galactosemia; nutrition and growth of infants, maternal nutrition. *Mailing Add:* Med Dept Ross Labs Columbus OH 43215

ACREE, TERRY EDWARD, b West Hamlin, WVa, Aug 6, 40; m 66. FOOD CHEMISTRY. *Educ:* Univ Calif, Berkeley, BA, 63; Cornell Univ, MS, 66, PhD(biochem), 68. *Prof Exp:* from asst prof to assoc prof, 68-81, PROF BIOCHEM, NY STATE AGR EXP STA, CORNELL UNIV, 81- *Concurrent Pos:* Fulbright-Hays res scholar, Portugal, 72-73. *Mem:* Am Chem Soc; Inst Food Technol; AAAS; Sigma Xi. *Res:* Chemistry and biochemistry of the flavor components of fruit and vegetable foods; chemistry of sensory perception. *Mailing Add:* NY State Agr Exp Sta Cornell Univ Geneva NY 14456

ACRES, ROBERT BRUCE, b Kelowna, BC, June 14, 51; m 84; c 2. CYTOKINE BIOLOGY, CANCER BIOLOGY. *Educ:* Univ BC, BSc, 74, MSc, 77; Univ Toronto, PhD(immunol), 82. *Prof Exp:* Fel immunol, Imp Cancer Res Fund, London, 82-84; asst staff scientist immunol, Immunex Corp, Seattle, 84-86, assoc staff scientist, 86-88; STAFF SCIENTIST IMMUNOL, TRANSGENE SA, STRASBOURG, FRANCE, 88- *Res:* Immunobiology and tumour biology, particularly the regulation of the immune system by soluble mediators and the impact of this regulation on tumour development, rejection, therapy and possibly prevention by vaccination; animal models for vaccine research. *Mailing Add:* 11 rue de Molsheim Strasbourg 67082 France

ACRIVOS, ANDREAS, b Athens, Greece, June 13, 28; US citizen; m 56. SUSPENSION RHEOLOGY, TWO-PHASE FLOW. *Educ:* Syracuse Univ, BChE, 50; Univ Minn, MS, 51, PhD(chem eng), 54. *Prof Exp:* Asst, Univ Minn, 50-54; from instr to assoc prof chem eng, Univ Calif, Berkeley, 54-62; prof chem eng, Stanford Univ, 62-88; ALBERT EINSTEIN PROF CHEM ENG, CITY COL NEW YORK, 88- *Concurrent Pos:* Guggenheim Found fel, 59 & 76; mem, US Nat Comt Theoret & Appl Mech, 80-; ed, Am Inst Physics: Physics of Fluids A, 82-; S Fairchild distinguished scholar, Calif Inst Technol, 83-84; dir, Levich Inst PCH, City Col New York, 88- *Honors & Awards:* Fluid Dynamics Prize, Am Phys Soc, 91. *Mem:* Nat Acad Sci; Nat Acad Eng; fel Am Inst Chem Engrs; Am Chem Soc; fel Am Phys Soc; Soc Rheology. *Res:* Fundamental research, experimental and theoretical, on various aspects of viscous flow, including sedimentation, the rheology of suspensions and effective medium theories; fluid mechanics; heat and mass transfer; surface phenomena. *Mailing Add:* Levich Inst City Col New York Steinman Hall No 202 New York NY 10031

ACRIVOS, JUANA LUISA VIVO, b June 24, 28; US citizen; m 56. SOLID STATE PHYSICS, QUANTUM CHEMISTRY. *Educ:* Univ Minn, PhD(phys chem, math), 56; Univ Havana, DSc(phys chem), 56. *Prof Exp:* Res assoc biophys, Hanasen Labs, Stanford Univ, 57-59; fel phys chem, Univ Calif, 59-62; fac res grant, 62, from asst prof to assoc prof, 62-72, PROF PHYS CHEM, SAN JOSE STATE UNIV, 72- *Concurrent Pos:* Res Corp grants, 64, 66 & 71; NSF res grants, 66-91; fel, San Jose State Univ, 74-75; vis fel, Trinity Col, Cambridge, 83 & Lucy Cavendish Col, Cambridge, 90; NSF vis prof, Univ Calif, Berkeley, 87-88. *Mem:* Am Chem Soc; Am Phys Soc. *Res:* Molecular structure determinations of semiquinones, organic conductors, copper oxide superconductors and biologically active free radicals by synchrotron x-ray and electron spin resonance spectroscopy; alkali metal and alkali halide solutions in ammonia, amines and ethers; metal-insulator transitions in metal ammonia systems and in layer transition element dichalcogenides. *Mailing Add:* Dept Chem San Jose State Univ San Jose CA 95192

ACS, GEORGE, b Dunaszerdahely, Hungary, Aug 14, 23; m 48; c 1. BIOCHEMISTRY. *Educ:* Orvosegyetem, Budapest, MD, 50, PhD, 53. *Prof Exp:* Demonstr, Orvosivegytan, Szeged, 47-48; demonstr med chem, Orvosivegytan, Budapest, 48-50, from asst prof to assoc prof, 50-56; res assoc, Rockefeller Inst, 57-60; sr res chemist, NY Psychiat Inst, 60-61; asst prof, Columbia Univ, 61; mem, Inst Muscle Disease, 61-74; prof pediat, 74-86, PROF BIOCHEM, MT SINAI SCH MED, 75-, PROF NEOPLASTICS, 86- *Concurrent Pos:* Res fel, Mass Gen Hosp, 56-57. *Mem:* Am Soc Cell Biol; Fedn Am Soc Exp Biol; Am Cancer Soc. *Res:* Viral Replication. *Mailing Add:* Dept Biochem & Neoplastic Dis Mt Sinai Sch Med New York NY 10029

ACTON, DONALD FINDLAY, b Lemberg, Sask, June 16, 34; m 60; c 2. PEDOLOGY, GEOMORPHOLOGY. *Educ:* Univ Sask, BSA, 57, MSc, 61; Univ Ill, PhD(soil sci), 71. *Prof Exp:* Res officer, 57-68, head, Soil Surv, Agr Can, 68-; AT DEPT SOIL SCI, UNIV SASK. *Concurrent Pos:* Assoc prof, Dept Soil Sci, Univ Sask, 68- *Mem:* Am Soc Agron; Agr Inst Can; Can Soc Soil Sci. *Res:* Soil survey; soil genesis; soil classification; mineralogy. *Mailing Add:* Soil Surv Univ Sask Rm 5C 26 Agr Bldg Saskatoon SK S7N 0W0 Can

ACTON, EDWARD MCINTOSH, b Morgan Hill, Calif, May 30, 30. BIO-ORGANIC CHEMISTRY. *Educ:* Stanford Univ, BS, 51; Mass Inst Technol, PhD(org chem), 57. *Prof Exp:* Chemist, Merck & Co, NJ, 51-53; from org chemist to sr org chemist, 57-75, prog mgr, SRI Int, 75-85; prof med chem, M D Anderson Hosp & Cancer Inst, Univ Tex, 86-87; expert, Nat Cancer Inst, 87-91; CONSULT, 91- *Concurrent Pos:* Vis res sci, Dept Pharmacol, Yale Univ, 81-; adj prof chem, Univ Houston, 86-87. *Mem:* Am Chem Soc; The Chem Soc; AAAS; Am Asn Cancer Res. *Res:* Synthetic organic chemistry for anticancer drug development; drug design based on biochemical mechanisms; anthracyclines, nucleosides, heterocycles, sugars; new artificial sweeteners. *Mailing Add:* 281 Arlington Way Menlo Park CA 94025

ACTON, LOREN WILBER, b Lewistown, Mont, Mar 7, 36; m 57; c 2. SOLAR PHYSICS. *Educ:* Mont State Col, BS, 59; Univ Colo, PhD(astro-geophys), 65. *Hon Degrees:* DSc, Mont State Univ, 88. *Prof Exp:* Jr scientist, Hanford Atomic Prod Oper, Gen Elec Co, 59; physicist, Cent Radio Propagation Lab, Nat Bur Standards, Colo, 60-61; physicist, US Naval Res Lab, DC, 62; res asst, High Altitude Observ, Colo, 62-64; assoc res scientist x-ray astron, 64-66, res scientist, 66-70, dir, Lockheed Solar Observ, 68-71, staff scientist & leader space astron group, 70-85, CONSULT SCIENTIST, SPACE SCI LAB, LOCKHEED MISSILES & SPACE CO, 85- *Concurrent Pos:* Assoc astronomer, Inst Astron, Univ Hawaii, 68-69; vis scientist, Group Phys Space Res, WGer, 71-72; payload specialist, Spacelab 2, 78-86; chmn, solar physics spacelab team, NASA, 75-79, mem spacelab ad hoc working group, 76 & Space & Earth Sci Adv Comt, 87-; chmn, Sci Prog Eval Comt, High Altitude Observ, 84. *Honors & Awards:* Spaceflight Achievement Award, Am Astronaut Soc, 86. *Mem:* AAAS; Am Astron Soc; Int Astron Union; Sigma Xi. *Res:* X-ray astronomy; solar physics. *Mailing Add:* 2913 Ramona St Palo Alto CA 94306

ACTON, RONALD TERRY, b Birmingham, Ala, Oct 19, 41; m 62; c 2. MOLECULAR GENETICS, MEDICAL LABORATORY IMMUNOLOGY. *Educ:* Birmingham Southern Col, Ala, AB, 65; Univ Ala, Birmingham, PhD(microbiol), 70; Am Bd Med Lab Immunol, cert, 85. *Prof Exp:* Res fel microbiol, Div Clin Immunol-Rheumatology, Univ Ala, Birmingham, 69-70; res fel biol, Calif Inst Technol, Pasadena, 70-72; spec fel biochem, Univ Oxford, Eng, 72-73; asst prof, 72-75, assoc prof microbiol, 75-78, co-dir, Diabetes OP Clin Lab Med, 79-85, dep dir, DRTC, 79-84, PROF MICROBIOL & EPIDEMIOL, UNIV ALA, BIRMINGHAM, 78-, PROF MED & DIR, IMMUNOGENETICS PROG, 82- *Concurrent Pos:* Mem bd dirs, Creative Schs, Inc, 84; mem, Cancer Ctr Support Grant Rev Comt, NIH, 78-82, mem Ad Hoc Cancer Construct Rev Comt, 84-85; consult, Separation Processes Br, George C Marshall Space Flight Ctr, NASA, 77-79; ed, Pediat Oncol Vols 1,2 & 3, 81-83 & Sports Med, 85- *Mem:* Sigma Xi; Am Asn Immunologists; Am Soc Biol Chemists; Am Soc Histocompatibility &

Immunogenetics; Am Soc Human Genetics; Am Asn Clin Res; Am Soc Microbiol. *Res:* The role of genes within the major histocompatibility complex in predisposing individuals to autoimmune diseases; identifying and cloning genes. *Mailing Add:* Diabetes Bldg Rm 802 Univ Ala Birmingham Sta Birmingham AL 35294

ACTOR, PAUL, b New York, NY, 1933; m; c 3. ANTIMICROBIAL CHEMOTHERAPY. *Educ:* Hunter Col, AB, 55; Rutgers Univ, MA, 58, PhD(zool), 59. *Prof Exp:* Asst zool, Rutgers Univ, 55-59; sr res scientist, Squibb Inst Med Res, 59-63; sr microbiologist, Smith Kline & French Labs, 63-66, group leader microbiol, 66-68, from asst dir to dir microbiol, 68-81, dir Dept Natural Prod Pharmacol, 81-84, dir strategic planning & technol eval, 84-86, dir compound & technol acquisition, Smith Kline & French Labs, 86-90; PRES, PAUL ACTON ASSOC, 90- *Concurrent Pos:* Res prof microbiol & immunol, Med Sch, Temple Univ, Philadelphia, 81-; chmn Antimicrobiol Chemother Sect, Am Soc Microbiol, 84-85; pres, Am Soc Microbiol, E Pa Br, 88-91. *Mem:* Fel Am Acad Microbiol; Am Soc Microbiol; Am Soc Parasitol; Am Soc Trop Med & Hyg; fel Infectious Dis Soc Am. *Res:* Chemotherapy and immunology of bacterial, fungal and parasitic infection. *Mailing Add:* 632 Pickering Lane Phoenixville PA 19460

ACUNA, MARIO HUMBERTO, b Cordoba, Arg, Mar 21, 40; m 66; c 4. SPACE SCIENCE, PHYSICS. *Educ:* Nat Univ Tucuman, MS, 67; Cath Univ Am, PhD(space sci), 74. *Prof Exp:* Res asst ionospheric res, Nat Univ Tucuman, 63-67; sr engr electronics, Fairchild Hiller Corp, 67-69; design engr res & develop, 69-70, RES SCIENTIST, GODDARD SPACE FLIGHT CTR, NASA, 70-, PROJ SCIENTIST, PRIN INVESTR, NASA MISSIONS, 79- *Concurrent Pos:* Distinguished lectr, Magnetics Soc, Inst Elec & Electronics Engrs, 85; pres, Nanotesla, Inc, 82- *Honors & Awards:* Sci Award, NASA, 69 & Group Achievement Award, 70, 74 & 75; Except Performance Award, Goddard Space Flight Ctr, NASA, 75; Moe Schneebaum Mem Award, 80; Except Sci Achievement Medal, NASA, 81, Except Serv Medal, 86. *Mem:* Am Geophys Union; Inst Elec & Electronics Engrs; Sigma Xi. *Res:* Magnetic fields and plasmas in space; planetary magnetospheres; geophysical and space research instrumentation. *Mailing Add:* Planetary Magnetospheres Br Code 695 Greenbelt MD 20771

ACZÉL, JÁNOS D, b Budapest, Hungary, Dec 26, 24; m 46; c 2. APPLIED MATHEMATICS. *Educ:* Eotvos Lorand Univ, Budapest, PhD(math), 47; Hungarian Acad Sci, Habil, 52, DSc(functional equations), 57. *Hon Degrees:* Dr, Univ Karlsruhe, 90. *Prof Exp:* Asst prof math, Univ Szeged, 48-50; assoc prof & dept head, Miskolc Inst Technol, 50-52; from assoc prof to prof & dept head, Debrecen Univ, 52-65; prof, 65-69, DISTINGUISHED PROF MATH, UNIV WATERLOO, 69- *Concurrent Pos:* Vis prof, Univ Fla, 63-64, 80, Stanford Univ, 64, Univ Cologne, 65, Univ Giessen, 66, 70, Univ Bohum, 68, Nat Inst Math, Rome, 71, Monash Univ, 72, Ahmadu Bello Univ, Nigeria, 75-76, Univ Lecce, 76, Calif Inst Technol & Univ Ariz, 78, Graz Univ, 79, 86,91, Okayama Univ, Japan, 84, Univ Milan, 85, 91 Univ Hamburg, 85, Tech Univ Barcelona, 86, 92, Univ Karlsruhe, 92 & Univ Bern, 86; NSF grant, 63-64; Nat Res Coun Can grant, 66-; Naval Ocean Systs Ctr, San Diego, 79-81. *Honors & Awards:* M Beke Award, 61; foreign fel Hungarian Acad Sci; Santiago Ramon Cajal Medal, Nat Coun Sci Res, Spain, 88. *Mem:* Fel Royal Soc Can; Can Math Soc; Am Math Soc; Austrian Math Soc; NY Acad Sci. *Res:* Functional equations; inequalities; mean values; geometric objects; generalized groups; webs; rational decision making; allocation, taxation theory; production functions; analytic hierarchy processes; probability theory; statistical distributions; theory of measurement; theory of information; economic index numbers. *Mailing Add:* Fac Math Univ Waterloo Waterloo ON N2L 3G1 Can

ACZEL, THOMAS, b Nagykanizsa, Hungary, Dec 18, 30; US citizen; m 62; c 4. ANALYTICAL CHEMISTRY. *Educ:* Univ Trieste, DSc(phys chem), 54. *Prof Exp:* Chemist, Aquila Co, Italy, 54-55; tech adv, Petroli Aquila Co, 55-58; from res chemist to sr res chemist, Humble Oil & Ref Co, 59-66; res specialist, 66-69, res assoc, 69-76, sr res assoc, 76-83, SCI ADV, EXXON RES & ENG CO, 83- *Concurrent Pos:* Chmn, comt D-2, Res & Develop Div IV, sect M, Am Soc Testing & Mat, 71-89, chmn & officer, comt E-14, 77-83, mem comt, Div Petrol Chem, Am Chem Soc & chmn, SE Tex Sect, 84, chmn elect, Div Petrol Chem Am Chem Soc, 90- *Honors & Awards:* Bituminous Coal Res Award, Am Chem Soc, Nat Award Petrol Chem, 89. *Mem:* Am Chem Soc (coun, 86-87); Am Soc Testing & Mat; Am Soc Mass Spectrometry. *Res:* Mass spectrometry; gas chromatography; high resolution mass spectrometry; application to characterization of complex petroleum mixtures; automatic data acquisition and handling systems required by such analyses; application to characterization of complex petroleum and coal liquids. *Mailing Add:* Exxon Res Lab Rte 22E Annandale NJ 08801

ADACHI, KAZUHIKO, ABNORMAL HEMAGLOBIN, SICKLE CELL DISEASE. *Educ:* Tokyo Med Univ, PhD(biochem), 72. *Prof Exp:* ASSOC RES PROF, DEPT PEDIAT, UNIV PA, 84- *Mailing Add:* Pa Children's Hosp 34th St & Civic Ctr Blvd Philadelphia PA 19104

ADAIR, DENNIS WILTON, b San Jose, Calif, Dec 7, 39; m 66; c 1. BIOPHARMACEUTICS. *Educ:* Univ Calif, San Francisco, PharmD, 64, PhD(pharm chem), 76. *Prof Exp:* Sr scientist pharmaceut, Ciba-Geigy Pharmaceut Div, 74-79, asst dir pharmaceut, 79-81; res group chief pharmaceut res & develop, Hoffman-La Roche Pharmaceut, Nutley, NJ, 81-83; MGR VET FORMULATION RES, SCHERING PLOUGH CORP, KENILWORTH, NJ, 83- *Mem:* Sigma Xi; Am Pharmaceut Asn; Acad Pharmaceut Sci. *Res:* Gastrointestinal absorption of drugs; drug delivery system with in vivo-in vitro evaluation. *Mailing Add:* 4371 Edinburg Ct Suisan CA 94585

ADAIR, ELEANOR R, b Arlington, Mass, Nov 28, 26. EXPERIMENTAL BIOLOGY. *Educ:* Mt Holyoke Col, BA, 48; Univ Wis, MA, 51, PhD(psychol & physics), 55. *Prof Exp:* Res assoc psychol, Mt Holyoke Col, 48-50; res asst, Univ Wis, 50-53; res asst, Yale Univ, 60-62, res assoc, 63-77; asst fel, 66-70, assoc fel, 70-86, FEL, JOHN B PIERCE FOUND, 86- *Concurrent Pos:* Sr res assoc psychol, Yale Univ, 77-89, sr res scientist, 89-, lectr, 85-; consult, Bioelectromagnetics Res Lab, Dept Rehab Med, Sch Med, Univ Wash, Seattle, 83-85; Sigma Xi nat lectr, 85-88; secy, Comt Man & Radiation, Inst Elec & Electronics Engrs, 88-89. *Mem:* Fel AAAS; fel Am Psychol Asn; Sigma Xi; Am Physiol Soc; Am Soc Primatologists; Bioelectromagnetics Soc (secy-treas, 83-86); fel NY Acad Sci; sr mem Inst Elec & Electronics Engrs; Am Asn Lab Animal Sci; fel Am Psychol Soc. *Res:* Author or co-author of over 60 publications and 1 book. *Mailing Add:* John B Pierce Found Lab 290 Congress Ave New Haven CT 06519

ADAIR, GERALD MICHAEL, b St Louis, Mo, June 8, 49; m 72; c 1. SOMATIC CELL GENETICS, MOLECULAR GENETICS. *Educ:* Washington Univ, St Louis, BA, 72, PhD(biol), 75. *Prof Exp:* Res assoc, Lawrence Livermore Nat Lab, 75-78; asst biologist, 78-82, asst prof, 82-86, ASSOC PROF BIOL, SCI PARK RES DIV, UNIV TEX M D ANDERSON CANCER CTR, 86-, ASSOC BIOLOGIST, 82- *Concurrent Pos:* Mem grad fac, Univ Tex Grad Sch Biomed Sci, 83- *Mem:* Genetics Soc Am; Environ Mutagen Soc; Am Soc Microbiol; Am Soc Human Genetics. *Res:* DNA repair and mutagenesis in mammalian cells; molecular analysis of mutation in DNA repair-proficient versus deficient genetic backgrounds; targeted homologous recombination in mammalian cells; DNA-mediated gene transfer. *Mailing Add:* Univ Tex M D Anderson Cancer Ctr Sci Park Res Div Smithville TX 78957

ADAIR, JAMES EDWIN, b Farmington, Ky, Jan 18, 35; m 66; c 1. ELECTRICAL ENGINEERING. *Educ:* Univ Mich, BSE, 61, MSE, 63, PhD(elec eng), 68. *Prof Exp:* Asst, Univ Mich, 60-63, res assoc, 63-67; asst prof elec eng, Univ Mo-Rolla, 67-73; elec engr, Air Force Avionics Lab, Wright-Patterson AFB, 73-80; MEM STAFF, SPERRY GYROSCOPE, 80- *Mem:* Inst Elec & Electronics Engrs; Sigma Xi. *Res:* Physical electronics-electron beams, traveling wave tubes, millimeter waves; guides waves-coupled transmission lines, microwave circuits and solid state devices. *Mailing Add:* 2254 Willowbrook Clearwater FL 33546

ADAIR, KENT THOMAS, b Allegan, Mich, Jan 21, 33; m 55; c 3. FORESTRY, MANAGEMENT SCIENCES. *Educ:* Colo State Univ, BS, 58, PhD(forest econ), 68; Ore State Univ, MF, 61. *Prof Exp:* Forester, US Forest Serv, 58-59; forester & consult, Bigley & Feiss, Foresters, Inc, 60-61; res forester, Colo State Univ, 61-63; asst prof forest econ, Univ Mont, 63-66; prof forestry, Univ Mo-Columbia, 67-77; DEAN, SCH FORESTRY, STEPHEN F AUSTIN STATE UNIV, 77- *Concurrent Pos:* Dir, Ctr Appl Studies, 77- *Mem:* Soc Am Foresters; Forest Prod Res Soc. *Mailing Add:* Forestry Box 6109 Stephen F Austin State Univ Nacogdoches TX 75962

ADAIR, ROBERT KEMP, b Ft Wayne, Ind, Aug 14, 24; m 52; c 2. ELEMENTARY PARTICLE PHYSICS, NUCLEAR PHYSICS. *Educ:* Univ Wis, PhD(physics), 51. *Prof Exp:* Instr physics, Univ Wis, 51-53; from assoc physicist to physicist, Brookhaven Nat Lab, 53-58; from assoc prof to prof physics, Yale Univ, 59-72, chmn dept, 67-70, dir, Div Phys Sci, 77-81, Eugene Higgins prof, 72-88, STERLING PROF PHYSICS, YALE UNIV, 88- *Concurrent Pos:* Guggenheim fel, 53; Ford Found fel, Europ Orgn Nuclear Res, 62; assoc ed, Phys Rev, 63-66 & Phys Rev Lett, 78-84; assoc dir high energy & nuclear physics, Brookhaven Nat Lab, 87-88. *Mem:* Nat Acad Sci; fel Am Phys Soc. *Res:* Nuclear and particle physics. *Mailing Add:* Gibbs Lab Yale Univ 219 Prospect St Box 666 New Haven CT 06511

ADAIR, SUZANNE FRANK, b Milwaukee, Wis, Aug 17, 41; m 66; c 1. BIOPHARMACEUTICS. *Educ:* Univ Wis, BS, 63; Univ Calif, San Francisco, PhD(pharmaceut chem), 76. *Prof Exp:* Actg asst prof pharm, Univ Calif, San Francisco, 74; asst prof pharm, Rutgers Univ, Busch Campus, 74-77; sr assoc, 77-78, asst dir clin pharmacol, 79-81, assoc dir human pharmacol, 81-83, ASSOC DIR CLIN RES, CIBA-GEIGY PHARMACEUT CORP, 83- *Mem:* Sigma Xi; NY Acad Sci; Acad Pharmaceut Sci; AAAS. *Res:* Clinical pharmacology; bioavailability and pharmacokinetics; clinical trials of anti-inflammatory and central nervous system agents. *Mailing Add:* 4560 Horton St Emeryville CA 94608-2916

ADAIR, THOMAS WEYMON, III, b Houston, Tex, Jan 23, 35; m 62. SOLID STATE PHYSICS, LOW TEMPERATURE PHYSICS. *Educ:* Tex A&M Univ, BS, 57, PhD(physics), 65; Rice Univ, MA, 60. *Prof Exp:* Res physicist, Prod Res Lab, Humble Oil & Refining Co, 59-62; asst prof, 66-72, asst to pres, 72-75, assoc prof, 72-78, PROF PHYSICS, TEX A&M UNIV, 78- *Concurrent Pos:* NSF fel, Kamerlingh Onnes Lab, Univ Leiden, 65-66. *Mem:* Am Phys Soc; Sigma Xi. *Res:* Magnetic properties of single crystals; magnetic properties amorphous metal films. *Mailing Add:* 1018 Rose Circle College Station TX 77840

ADAIR, WINSTON LEE, JR, b Chicago, Ill, May 27, 44; m 72; c 1. BIOCHEMISTRY. *Educ:* Brown Univ, ScB, 66; Georgetown Univ, PhD(biochem), 72. *Prof Exp:* Asst prof, 75-80, ASSOC PROF BIOCHEM, COL MED, UNIV SOUTH FLA, 80- *Concurrent Pos:* NIH fel, Dept Hematol-Oncol, Sch Med, Wash Univ, 73-75. *Mem:* Sigma Xi; Am Chem Soc; Soc Complex Carbohydrates. *Res:* Cell surface glycoproteins; membrane biochemistry; biosynthesis of long chain polyisoprenoid alcohols; interactions of plant lectins with cell membranes. *Mailing Add:* Dept Biochem Univ SFla Col Med Bruce B Downs Blvd Tampa FL 33612

ADAM, DAVID PETER, b Berkeley, Calif, May 18, 41; m 67; c 3. PALYNOLOGY, CLIMATOLOGY. *Educ:* Harvard Univ, BA, 62; Univ Ariz, MS, 65, PhD(geochronology), 70. *Prof Exp:* Res assoc palynology, Univ Ariz, 66-69; teaching asst, 70, res assoc dendrochronology, Lab Tree-ring Res, 70-71; res assoc palynology, 71-72, GEOLOGIST, US GEOL SURV, 71- *Concurrent Pos:* Pres, Long Terrestrial Records, Int Quaternary Asn, 87- *Mem:* Am Quaternary Asn; Ecol Soc Am; Brit Freshwater Biol Asn; Am Quaternary Asn; Sigma Xi. *Res:* Theoretical palynology; causes of climatic change; computer applications in paleoecology; climatic history of California; Chrysophyte cysts. *Mailing Add:* 750 Cedar St San Carlos CA 94070

ADAM, JOHN ANTHONY, b Salisbury, Eng, Dec 18, 49; m 71; c 3. THEORY OF WAVE MOTION, MATHEMATICAL MODELS OF TUMOR GROWTH. *Educ:* Univ London, BSc, 71, PhD(appl math & astrophysics), 75. *Prof Exp:* Res fel, Univ Sussex, Eng, 74-76; res asst, Univ St Andrews, Scotland, 76-78; lectr, math, New Univ Ulster, Ireland, 78-83, sr lectr, 83-84; PROF MATH & STATS, OLD DOMINION UNIV, 84- *Concurrent Pos:* Res fel, theoret physics, Dublin Inst Advan Studies, 77-80; vis asst, prof & Fulbright Scholar, Univ Rochester, 83. *Mem:* Soc Math Biol; Inst Math & Applications UK. *Res:* Mathematical wave theory; astrophysical fluid dynamics; mathematical biology; models of tumor growth; differential operators arising in wave problems. *Mailing Add:* Dept Math & Stats Old Dominion Univ Norfolk VA 23529-0077

ADAM, KLAUS, b Merzig, WGer, Jan 8, 43; US citizen; m 72; c 2. ORGANIC CHEMISTRY. *Educ:* St Edward's Univ, BS, 64; Tex A&M Univ, MS, 66, PhD(chem), 71. *Prof Exp:* Fel chem, Univ Saarland, 71-72 & Calif Inst Technol, 72-74; sr res chemist, Monsanto Co, 74-80; staff chemist, Exxon Chem Co, 80-83; ASST PROF, GALVESTON COL, 88- *Mem:* Am Chem Soc. *Res:* Organic and biochemical synthesis; reaction mechanism; process improvement products development for agricultural application; water treatment application; lubrication and fire resistance application; refinery process aids. *Mailing Add:* 1706 Bowline Houston TX 77062

ADAM, RANDALL EDWARD, b Cheyenne, Wyo, Sept 3, 43; m 70; c 2. PHYSICAL CHEMISTRY, POLYMER PHYSICS. *Educ:* Univ Minn, BChem, 70; Univ Calif, San Diego, PhD(phys chem), 76. *Prof Exp:* Fel polymer physics, IBM Res Lab, 76-78; AT 3M CO. *Mem:* Am Chem Soc; Soc Rheology. *Res:* Stress-strain and fracture of polymers; time-dependent polymer properties and their molecular origins; rheology of dispersions; diffusion of gases; vapors in polymers. *Mailing Add:* 10672 Hallwood Dr Temple City CA 91780

ADAMANTIADES, ACHILLES G, b Volos, Greece, Apr 22, 34; m 63; c 3. REACTOR SAFETY. *Educ:* Athens Tech Univ, dipl, 57; Mass Inst Technol, PhD(nuclear eng), 66. *Prof Exp:* Opers engr, Pub Power Corp Greece, 60; asst biol effect nuclear radiation, Mass Inst Technol, 61-62, asst reactor & nuclear physics, 62-63, asst neutron slowing down, 63-66; asst prof nuclear eng, Iowa State Univ, 66-71; vis prof, Univ Patras, 71-73; mem staff, Fed Power Comn, 73; mem tech staff, 74-81, PROJ MGR, ELEC POWER RES INST, 81- *Concurrent Pos:* Consult, World Bank; assoc prof lectr, George Washington Univ, 74-79; consult prof, Stanford Univ, 82. *Mem:* Am Nuclear Soc; AAAS; Sigma Xi; Am Asn Univ Professors; Tech Chamber of Greece. *Res:* Pulsed neutron research; neutronics; nuclear reactor safety; power systems planning; power economics. *Mailing Add:* 5305 Moorland Lane Bethesda MD 20814

ADAMCIK, JOE ALFRED, b Taylor, Tex, June 28, 30. ORGANIC CHEMISTRY. *Educ:* Univ Tex, BS, 51, MA, 54, Univ Ill, PhD(chem), 58. *Prof Exp:* From asst prof to assoc prof chem, Tex Tech Univ, 57-88; RETIRED. *Mem:* AAAS; Am Chem Soc; Am Geophys Union; Royal Soc Chem. *Mailing Add:* 5223 42nd St Lubbock TX 79414

ADAMCZAK, ROBERT L, b Buffalo, NY, Aug 16, 27; m 58; c 3. PHYSICAL CHEMISTRY. *Educ:* Univ Buffalo, BA, 51, MA, 55, PhD(chem), 56. *Prof Exp:* Res analyst propellant chem, Olin Mathieson Chem Corp, 51-53; res chemist fuel chem, Esso Res & Eng Co, Stand Oil Co NJ, 56-58; res phys chemist, Wright Air Develop Ctr, US Air Force Wright Aeronaut Labs, 59-60, res phys chemist, Br Fluids & Lubricants, Res & Technol Div, US Air Force Mat Lab, 60-68, sci adminr, 68-85, Chief Plans, 75-86; RETIRED. *Concurrent Pos:* Mem, Lubricants Comt, Res & Eng Div, US Dept Defense, 61- *Mem:* Am Chem Soc; fel Am Soc Mech Engrs; Am Soc Lubrication Engrs; fel Am Inst Chemists. *Res:* Energy transfer fluids; lubricants; conventional and propellant lubrication fuels; inorganic polymer chemistry; theory of friction and wear; properties of liquid metals and inorganics. *Mailing Add:* 7556 Beldale Ave Dayton OH 45424

ADAMEK, EDUARD GEORG, b Mistelbach, Austria, Apr 19, 25; Can citizen; m 54. ENVIRONMENTAL MANAGEMENT, AIR POLLUTION. *Educ:* Univ Vienna, BSc, 50; Innsbruck Univ, Drs, 52; McGill Univ, PhD(org chem), 59. *Prof Exp:* Res chemist, Starch & Chem Div, Ogilvie Flour Mills Co, Ltd, 52-58; sr develop chemist, Textile Fibres Div, Du Pont of Can, 59-60, from res chemist to sr res chemist, Cent Res Labs, 60-70; scientist, Air Mgt Br, Ont Dept Energy & Resources, 70-71; ORG LAB SUPVR, ONT MINISTRY ENVIRON, 71- *Mem:* AAAS; Am Chem Soc; Chem Inst Can; NY Acad Sci; Sigma Xi. *Res:* New organic products and processes; polymers; organic air and water pollutants research; waste product utilization; metallo-organic compounds; analytical methods development; carbohydrates, protein and amino acids. *Mailing Add:* 50 Quebec Ave Suite 1902 Toronto ON M6P 4B4 Can

ADAMES, ABDIEL JOSE, b Panama, Repub Panama, Mar 18, 38. ENTOMOLOGY, ECOLOGY. *Educ:* Univ Panama, BS, 62; Univ Calif, Los Angeles, PhD(zool), 70. *Prof Exp:* Entomologist arbovirol, 70-72, asst head, 72-74, HEAD ECOL, GORGAS MEM LAB, UNIV PANAMA, 74-, DIR POSTGRAD STUDIES & RES DIV, 77- *Concurrent Pos:* Mem adv bd, Cent Am Prog Develop Sci, Confederation Cent Am Univ, 78-; mem, Joint Environ Comn, Panama Canal Treaties, 80; govt expert, Environ Prog, Great Caribbean, 80-81; mem, Nat Comn Energy, 80- *Mem:* Am Soc Trop Med & Hyg; Panamanian Asn Microbiologists. *Res:* Biomedical studies in man made lakes; environmental assessments for hydroelectric projects; arbovirus ecology; culicidology, vector biology; bionomics and control of blackflies. *Mailing Add:* Apartado 36 Estafeta de Balboa Ancon Panama

ADAMKIEWICZ, VINCENT WITOLD, b Poland, Nov 27, 24; m 54; c 5. PHYSIOLOGY, IMMUNOLOGY. *Educ:* Bristol Univ, BSc, 48 & 49, MSc, 50; Univ Montreal, PhD(endocrinol), 53. *Prof Exp:* Agr engr, Poland, 44; res assoc, Sch Vet Med, Bristol Univ, 50; assoc prof physiol, 56-67, PROF IMMUNOPHYSIOL, FAC MED, UNIV MONTREAL, 67- *Mem:* Am Physiol Soc; Can Physiol Soc; Asn French Speaking Physiologists; Can Soc Immunol. *Res:* MagnetobiologyBacterial adsorbtions; allergies of the dental pulp; pulpal hypersensitivity reactions, including Arthus reaction and delayed cellular hypersensitivity of dental pulp; modulation of the sorbtion of bacterial exopolysaccharides to surfaces by weak, static magnetic and geomagnetic fields and their effects on macrophage migration; pharmacology. *Mailing Add:* Dept Microbiol & Immunol Univ Montreal Fac of Med Montreal PQ H3C 3J7 Can

ADAMO, JOSEPH ALBERT, b Jersey City, NJ, Oct 22, 38; m 64, 78; c 3. NEMATOLOGY. *Educ:* Jersey City State Col, BA, 64; Fairleigh Dickinson Univ, MS, 67; Rutgers Univ, PhD(nematol), 75. *Prof Exp:* Instr bot, Fairleigh Dickinson Univ, 65-66; asst prof bot & zool, Jersey City State Col, 66-67; from assoc prof to prof microbiol, 67-85, CHMN, SCI DEPT, OCEAN COUNTY COL, 81-, coordr environ ctr, 75-81. *Concurrent Pos:* Vis prof biol res, Rutgers Univ, 85-87. *Mem:* Soc Nematol; Am Phytopath Soc; Am Microbiol Soc; AAAS; Nat Sci Teachers Asn. *Res:* Nematode behavior and physiology specifically related to the vertical migration, quiescence, and senescence of above-ground phytoparasitic forms; nematode parasites of beach grass. *Mailing Add:* Dept of Biol Ocean County Col Toms River NJ 08753

ADAMS, ALAYNE A(MERCIER), b Niagara Falls, NY, Jan 31, 38; m 61; c 2. PHYSICAL CHEMISTRY, ELECTROCHEMISTRY. *Educ:* Niagara Univ, BS, 59, MS, 61; Am Univ, PhD(chem), 71. *Prof Exp:* Chem abstractor phys chem, Chem Abstracts Serv, 62-70; res asst electrochem, Am Univ, 68-71, asst res scientist phys chem, 72-74, assoc res scientist, 74-76; RES CHEMIST ELECTROCHEM, US ARMY MOBILITY EQUIP RES & DEVELOP COMMAND, 76- *Concurrent Pos:* Fel, US Army, 72-74; adj prof, Am Univ, 78-79. *Honors & Awards:* Melvin Romanoff Award, North East Sect, Nat Asn Corrosion Engrs, 77. *Mem:* Electrochem Soc; Nat Asn Corrosion Engrs; Am Chem Soc; Sigma Xi. *Res:* Electrochemistry of fuel cell processes; use of fuel cells as energy conversion devices; corrosion processes occuring at metallic interfaces. *Mailing Add:* 8436 Rushing Creek Ct Springfield VA 22153

ADAMS, ALBERT WHITTEN, b Ft Scott, Kans, Oct 9, 27; m 54; c 4. ANIMAL NUTRITION. *Educ:* Kans State Col, BS, 51, MS, 55; SDak State Univ, PhD(animal sci), 65. *Prof Exp:* Asst hatchery mgr, Swift & Co, 51-52, hatchery mgr, 52-54; asst poultry sci, Kans State Col, 54-55; from instr to asst prof, SDak State Univ, 55-60; mgr, Stant's Turkey Farms, 60-62; asst prof dairy & poultry sci, 62-66, assoc scientist, Agr Exp Sta, 66-72, assoc prof, 66-75, PROF ANIMAL SCI & INDUST, KANS STATE UNIV, 75-, RES POULTRY SCIENTIST, AGR EXP STA, 73- *Mem:* Poultry Sci Asn; World Poultry Sci Asn. *Res:* Relationship between the effects of various nutritional and environmental factors on performance of cage layers. *Mailing Add:* Dept Animal Sci Kans State Univ Call Hall Manhattan KS 66506

ADAMS, ARLON TAYLOR, b Bottineau, NDak, Apr 26, 31; m 58; c 4. ANTENNAS, MICROWAVES. *Educ:* Harvard Univ, BA, 53; Univ Mich, Ann Arbor, MS, 61, PhD(elec eng), 64. *Prof Exp:* Engr, Sperry Gyroscope Co, 57-59; res assoc, Univ Mich, 59-63; from asst prof to assoc prof, 63-74, PROF ELEC ENG, SYRACUSE UNIV, 74- *Concurrent Pos:* Consult, Hughes Aircraft Corp, Int Bus Mach, USN, NCR & Syracuse Res Corp, 63-; assoc ed, Inst Elec & Electronic Engrs Electromagnet Compatibility Trans & Antennas Propagation Soc Newsletter, 75-88; vis scholar, Univ Calif, Berkeley, 76-77; Fulbright researcher, Univ Banja Luka, Yugoslavia, 84. *Honors & Awards:* Achievement Award, Inst Elec & Electronic Engrs, 73. *Mem:* Inst Elec & Electronic Engrs; Inst Elec & Electronic Engrs Antennas & Propagation Soc. *Res:* Numerical methods for electromagnetics, especially with application to microstrip transmission lines, antennas, and waveguides. *Mailing Add:* 102 Kensington Rd Syracuse NY 13210

ADAMS, ARNOLD LUCIAN, b Aurora, Ill, Dec 24, 52; m 78; c 1. SURFACE SCIENCE. *Educ:* Univ Ill, Urbana, BS, 75; Univ Calif, Santa Barbara, PhD(physics), 80. *Prof Exp:* res fel, Univ Calif, Santa Barbara, 80-82; staff, Santa Barbara Res, 82-85; PRES, ADAMS RES CORP, 85- *Mem:* Am Phys Soc. *Res:* Electromagnetic interactions of molecules and atoms with various surfaces; surface properties of semiconductors. *Mailing Add:* Adams Res PO Box 783 Goleta CA 93116

ADAMS, BRENT LARSEN, b Provo, Utah, Sept 26, 49; m 71; c 3. MECHANICAL BEHAVIOR OF SOLIDS. *Educ:* Univ Utah, BS, 74; Ohio State Univ, MS, 76, PhD(metall eng), 79. *Prof Exp:* Sr res engr, Lynchburg Res Ctr, Babcock & Wilcox Co, 76-79; advan scientist, Hanford Eng & Develop Lab, US Dept Energy, 79-80; asst prof mat sci & eng, Univ Fla, 80-83; assoc prof mech eng, Brigham Young Univ, 83-88; ASSOC PROF, MECH ENGR, YALE UNIV, 88- *Concurrent Pos:* Consult, Aluminum Co Am, 82- *Honors & Awards:* Henry Marion Howe Medal, Am Soc Metals Int, 87. *Mem:* Metall Soc, Am Inst Mining, Metall & Petrol Engrs; Am Soc Metals. *Res:* Mechanics and physics of flow and fracture in polycrystallic metals; texture and microstructural effects upon mechanical properties. *Mailing Add:* Yale Univ 84 Norwood Ave PO Box 1968 Yale Sta Hamden CT 06518

ADAMS, C(HARLES) HOWARD, b Bergen, NY, May 20, 17; m 41; c 3. CHEMICAL ENGINEERING. *Educ:* Univ Ill, BS, 39, MS, 45. *Prof Exp:* Mem staff prod control & res, Am Plastics Corp, 40-41; asst chem eng res, Univ Ill, 41-45; sr group leader plastics res, Plastics Div, 45-56, mgr plastics prod develop, Res & Eng Div, 56-60, mgr prod eng plastics div, 60-62, mgr mat eng cent res dept, 62-68, mgr systs eng dir, New Enterprise Div, 68-70, mgr prod eng, 70-72, sr res specialist rubber chem div, 72-73, com develop mgr, Plastics & Resins Co, 75-78, PRIN ENG SPECIALIST, CORP ENG DEPT, MONSANTO CO, 78-; res assoc, Ctr Fire Res, Nat Bur Standards, 73-75. *Concurrent Pos:* Chmn US Plastics Comt, Int Orgn Standardization, 56-58. *Mem:* Am Inst Chem Eng; hon mem Am Soc Test & Mat. *Res:* Theoretical and applied mechanics; physics and engineering mechanics of high polymers; polymeric materials engineering; plastics standards; plastics design processing and fabrication; acoustic emission engineering. *Mailing Add:* 5376 Royal Hills Dr St Louis MO 63129

ADAMS, CHARLES HENRY, b Burdick, Kans, Nov 7, 18; m 43. ANIMAL SCIENCE, MEAT SCIENCE. *Educ:* Kans State Univ, BS, 41, MS, 42; Mich State Univ, PhD, 64. *Prof Exp:* Asst instr animal husb, Kans State Univ, 46-47; from asst prof to prof, 47-84, mem grad fac, 70-84, asst dean, Col Agr, 73-84, EMER PROF ANIMAL SCI, UNIV NEBR, LINCOLN, 84- *Mem:* AAAS; fel Am Soc Animal Sci; Am Meat Sci Asn; Inst Food Technol; Am Inst Biol Sci; Sigma Xi. *Res:* Meat and meat products; beef and pork carcass evaluation; packaging of frozen meat and meat products; tenderness of meat. *Mailing Add:* 7101 Colby St Lincoln NE 68505

ADAMS, CHARLES REX, b Mt Vernon, Tex, Aug 19, 30; m 51, 67; c 4. PETROLEUM CHEMISTRY. *Educ:* ETex State Teachers Col, BA, 50; Rice Inst, MA, 52, PhD(chem), 54. *Prof Exp:* Res chemist, 54-64, on assignment to Shell Grundlagenforschung Gmb H, Siegburg bei, Bonn, WGer, 64-66, res supvr, Petrol Chem Dept, 66-77, MGR, CHEM RES & APPLNS, SHELL DEVELOP CO, 77- *Honors & Awards:* Ipatieff Prize, Am Chem Soc, 68. *Mem:* Am Chem Soc; fel Am Inst Chem; Catalysis Soc. *Res:* Mechanisms of heterogeneous catalysis; structure of solid catalysts; structure of the solid state. *Mailing Add:* 11918 Knippwood Lane Houston TX 77024

ADAMS, CLARK EDWARD, b Algona, Iowa, Mar 18, 42; m 62; c 3. MAMMALIAN ECOLOGY. *Educ:* Concordia Teachers Col, BS, 64; Univ Ore, MS, 66; Univ Nebr-Lincoln, PhD(zool), 73. *Prof Exp:* Instr biol, St John's Acad & Jr Col, 64-65; assoc prof ecol, Concordia Teachers Col, 66-77, assoc prof biol, 77-; AT WILDLIFE-FISHERIES SCI, TEX A&M UNIV. *Mem:* Am Soc Mammalogists; Wildlife Soc; Sigma Xi; Nat Asn Biol Teachers. *Res:* Measuring territoriality in fox squirrels; Sciurus niger rufiventer pre and post weaning and pre and post mating. *Mailing Add:* Wildlife-Fisheries Sci Tex A&M Univ College Station TX 77843

ADAMS, CLIFFORD LOWELL, b Knox Co, Ind, Jan 28, 15; m 43; c 2. PHYSICS. *Educ:* Ind State Teachers Col, BS, 40; Mo Sch Mines & Metals, MS, 50. *Prof Exp:* Teacher high sch, Ind, 44-46; instr physics, Morehead State Col, 42-44; asst prof, Tri-State Col, 46-48; instr, Mo Sch Mines & Metal, 48-50; assoc prof, Union Univ, 50-51; asst prof & assoc prof physics & elec eng, US Naval Acad, 51-56; assoc prof physics, Northeastern La State Col, 56-58; head, Dept Physics, 58-68, assoc provost res, 68-74, prof physics, 58-79, EMER PROF PHYSICS, OLD DOMINION UNIV, 79- *Concurrent Pos:* Mem, Nat Coun Univ Res Adminr; Nat Conf on Admin of Res. *Mem:* AAAS; Am Phys Soc; Am Asn Physics Teachers; Sigma Xi. *Res:* Energy levels of potassium chloride; optics. *Mailing Add:* 1325 Monterey Ave Norfolk VA 23508

ADAMS, CURTIS H, b DeKalb, Miss, Sept 12, 17; m 42; c 1. ENTOMOLOGY, ZOOLOGY. *Educ:* Miss State Univ, BS, 41, PhD(entom), 65; Henderson State Teachers Col, MSEd, 63. *Prof Exp:* Assoc prof mil sci, The Citadel, 51-54; prof, Henderson State Teachers Col 61-63; head life sci sect & chmn biol fac, 68-73, from assoc prof to prof, 65-82, EMER PROF, BIOL & CHMN ENVIRON SCI, UNIV ALA, HUNTSVILLE, 82- *Concurrent Pos:* Fel systs ecol, Univ Ga, 73-74. *Mem:* Wildlife Soc; AAAS; Am Inst Biol Sci; Entom Soc Am. *Res:* Insecticide resistance in beneficial insect parasites. *Mailing Add:* PO Box 1121 Huntsville AL 35807

ADAMS, DALE W, b Pleasant Grove, Utah, June 21, 34; m 60; c 3. AGRICULTURAL ECONOMICS. *Educ:* Utah State Univ, BS, 56; Mich State Univ, MS, 61, PhD(agr econ), 64. *Prof Exp:* Asst prof agr econ, Univ Wis, 64-66; PROF AGR ECON, OHIO STATE UNIV, 66- *Concurrent Pos:* Staff economist, AID, 68-69; vis prof, Stanford Univ, 74-75. *Mem:* Am Agr Econ Asn. *Res:* Rural savings and credit activities in low income countries. *Mailing Add:* Agr Econ 103 Agr Admin Bldg Ohio State Univ Main Campus Columbus OH 43210

ADAMS, DANIEL OTIS, b Portland, Maine, Mar 14, 18; m 45; c 4. CHEMISTRY, PAPER TECHNOLOGY. *Educ:* Oberlin Col, AB, 39; Lawrence Col, MS, 41, PhD(pulp & paper), 43. *Prof Exp:* Res chemist, WVa Pulp & Paper Co, 43-44; res proj leader, 44-48; res chemist, Bird & Son, Inc, 48-50, chief chemist, Res Div, 50-55; res dir, WVa Pulp & Paper Co, 55-59, supt tech serv, 59-64, mgr tech serv, 64-79, TECH MGR, KRAFT DIV, WESTVACO CORP, 79- *Mem:* Fel Tech Asn Pulp & Paper Indust. *Res:* Morphology of paper making fibers as related to their papermaking characteristics; air and water pollution abatement; resin solution saturation of paper; pulp and paper manufacture; manufacture of paperboard having structural strength. *Mailing Add:* 1446 Burningtree Rd Charleston SC 29412

ADAMS, DARIUS MAINARD, b Los Angeles, Calif, July 5, 44; m 66. FOREST ECONOMICS. *Educ:* Humboldt State Univ, BS, 66; Yale Univ, MFS, 68; Univ Calif, Berkeley, PhD(wildland resource sci), 73. *Prof Exp:* Asst prof forestry, Univ Wis-Madison, 71-74; asst prof, 75-77, ASSOC PROF FOREST MGT, ORE STATE UNIV, 77- *Mem:* Soc Am Foresters; Am Econ Asn. *Res:* Forest products, especially markets for public and private timber and logs; optimal management programs for uneven-aged stands and forests. *Mailing Add:* Dept Forest Univ Wash Seattle WA 98195

ADAMS, DAVID B, b Cincinnati, Ohio, June 1, 45; m 84; c 2. PSYCHOSOMATIC DISEASE. *Educ:* Univ Cincinnati, BA, 67; Xavier Univ, MA, 69; Univ Ala, FdS, 71 & PhD(psychol), 73. *Prof Exp:* Assoc prof behav med, Univ SC Sch Med, 75-84; PVT PRACT CLIN PSYCHOL, ATLANTA MED & NEUROPSYCHOL, 85- *Concurrent Pos:* Postdoctoral fel psychol, Devereux Found, 75; mem bd dir, Bridge Counseling Ctr, 87-; vis prof behav sci, Oglethorpe Univ, 87-; consult, Ciba Vision Corp & Upjohn Labs, 88. *Mem:* Fel Acad Psychosomat Med; fel Am Orthopsychiat Asn; Am Psychol Asn. *Res:* Adult children of mentally ill mothers; impact of psychopathology on work related injuries. *Mailing Add:* Suite 251 Med Quarters 5555 Peachtree-Dunwoody Rd NE Atlanta GA 30342-1710

ADAMS, DAVID GEORGE, b Mansfield, Ohio, June 17, 36; m 63; c 2. POLYMER CHEMISTRY, PHYSICAL CHEMISTRY. *Educ:* Univ Chicago, MS, 59; Ohio State Univ, PhD(chem), 64. *Prof Exp:* Sr scientist adhesion, Uniroyal Inc, 64-67, group leader polyester fiber, 67-69; res dir, St Clair Rubber Co, 69-72; res mgr synthetic fibers, Norton Co, 72-85; CONSULT, 85- *Mem:* Am Chem Soc; AAAS; Soc Mfg Engrs. *Res:* Adhesion to textiles, rubber and abrasives; technology of non-woven textiles. *Mailing Add:* 1112 Fifth Ave S Apt 7 Devils Lake ND 58301

ADAMS, DAVID LAWRENCE, b Brockton, Mass, Oct 10, 45; m 66; c 2. SCIENCE EDUCATION. *Educ:* Univ Mass, BS, 67; Univ Conn, MS, 69, PhD(org chem), 71. *Prof Exp:* Instr org chem, Pa State Univ, 71-72; instr chem & chmn, Div Sci & Allied Health, 72-77, asst dean, 77-83, assoc dean fac, North Shore Community Col, 83-86; ASSOC PROF CHEM, BRADFORD COL, 86- *Concurrent Pos:* Textbook consult, Wadsworth Publ Co, 73- & Prentice Hall Publ Co, 75-; consult ed, McGraw-Hill Publ Co, 76-80; vis lectr, Salem State Col, 85-88. *Mem:* Am Chem Soc; Sigma Xi; Nat Sci Teachers Asn. *Res:* Determination and quantitative evaluation of environmental pollutants as influenced by various other environmental parameters; synthesis of bicyclic terpines and acid catalyzed rearrangements of same; science and chemical education for non-specialists and history of chemistry. *Mailing Add:* Chem Dept Babson Col Babson Park MA 02157-0901

ADAMS, DAVID S, BIOLOGY. *Educ:* Okla State Univ, BS, 74; Univ Houston, MS, 76; Univ Tex, Austin, PhD(molecular biol), 79. *Prof Exp:* Teaching asst, Dept Biophys Sci, Univ Houston, 75, res asst, 76-77; res asst, Dept Zool, Univ Tex, Austin, 77-79, instr, 79; fel, Lab Cell Biol, Rockefeller Univ, 80-84; asst prof, 84-90, ASSOC PROF, DEPT BIOL-BIOTECHNOL, WORCESTER POLYTECH INST, 90-; SCI ADV, NAR LABS, WESTBORO, 91- *Concurrent Pos:* Am Cancer Soc individual res fel, 81-82; NIH individual res fel, 82-84; consult, Transgenic Sci Inc, 88- *Mem:* Am Soc Cell Biol; AAAS; Sigma Xi. *Res:* Numerous publications; biology. *Mailing Add:* Dept Biol & Biotechnol Worcester Polytech Inst 100 Institute Rd Worcester MA 01609

ADAMS, DENNIS RAY, b Celina, Ohio, Sept 13, 49; m 71; c 2. EDUCATION ADMINISTRATION. *Educ:* Bowling Green State Univ, BS, 71; Univ Ala, Tuscaloosa, MA, 75, PhD(math educ), 77. *Prof Exp:* Teacher math, New Bremen Sch, Ohio, 71-72; Colegio Karl C Parrish, Barranquilla, Colombia, 72-73; Colegio Nueva Granada, Bogota, 73-75; asst, Univ Ala, 75-77; instr, State Tech Inst, Knoxville, 77-81, head dept math, 81-86, asst dean humanities & sci, 86-88; dir, Div St Progs, 88-90, ASST DEAN INSTR, PELLISSIPPI STATE TECH COMMUNITY COL, KNOXVILLE, TENN, 90- *Mem:* Nat Coun Teachers Math; Am Math Asn. *Mailing Add:* PO Box 22990 Knoxville TN 37933-0990

ADAMS, DOLPH OLIVER, b Montezuma, Ga, Apr 12, 39; m 69. PATHOLOGY, IMMUNOLOGY. *Educ:* Duke Univ, AB, 60; Med Col Ga, MS & MD, 65; Univ NC, PhD(exp path), 72. *Prof Exp:* Chief path, US Army Joint Laser Safety Team, Frankford Arsenal, 70-72; from asst prof to assoc prof, & dir autopsy serv, 72-81, PROF PATH, MED CTR, DUKE UNIV, 81-, CHIEF, DIV AUTOPSY PATH, 81- *Honors & Awards:* Res Recognition Award, Noble Found, 85. *Mem:* Am Asn Pathologists & Bacteriologists; Am Soc Exp Path; Reticuloendothelial Soc; Am Soc Cell Biol; Am Asn Immunologists; Am Asn Cancer Res. *Res:* Role of macrophages in cellular immunology and tumor biology. *Mailing Add:* Dept of Path Duke Univ Med Ctr Durham NC 27710

ADAMS, DON, b Oklahoma City, Okla, Apr 2, 40; m 59. CHEMICAL ENGINEERING. *Educ:* Okla State Univ, BS, 62, MS, 64, PhD(chem eng), 68. *Prof Exp:* Chem engr, Phillips Petrol Co, 67-68; assoc prof chem eng, McNeese State Univ, 68-78; prof, Dept Mech Tech, Okla State Univ, 78-80, AT DEPT PETROL. *Mem:* AAAS; Am Inst Chem Engrs; Am Chem Soc. *Res:* Heat transfer and pressure drop for flow of non-Newtonian fluids across ideal tube banks; heat transfer, fluid flow and analog/digital hybrid computer application. *Mailing Add:* Div Eng Technol Okla State Univ 294 Cordell St Stillwater OK 74078-0233

ADAMS, DONALD E, b Burlington, Vt; m 65; c 3. ENERGY ENGINEERING, AEROSPACE GROUND TESTING. *Educ:* Univ Vt, BS, 60; Cornell Univ, MS, 63. *Prof Exp:* Jr engr, Texaco Res Ctr, 60; NSF res asst, Cornell Univ, 60-62; jr engr, Naval Ordnance Lab, 61; asst engr, Cornell Aeronaut Lab, 62-65, assoc engr, 65-69; sr engr, 69-78, PRIN ENGR, CALSPAN CORP, 78- *Mem:* Am Soc Mech Engrs; Sigma Xi. *Res:* Heat transfer; high temperature instrumentation; energy saving devices (safety and energy conservation potential); gas-fired appliances; railroad tank car safety; rocket stage separation testing; hypersonic wind tunnel testing; heating and erosion of ordnance. *Mailing Add:* PO Box 400 Buffalo NY 14225

ADAMS, DONALD F, b Streator, Ill, Sept 25, 35; m 57; c 4. COMPOSITE MATERIALS, APPLIED MECHANICS. *Educ:* Univ Ill, BS, 57, PhD(appl mech), 63; Univ Southern Calif, MS, 60. *Prof Exp:* Engr, Northrop Corp, 57-60; instr appl mech, Univ Ill, 60-63; supvr, Ford Aerospace & Commun, 63-67; mem staff, Rand Corp, 67-72; PROF MECH ENG, UNIV WYO, 72-; PRES, WYO TEST FIXTURES INC, 88- *Concurrent Pos:* Consult & expert witness, indust groups, 72-; vis fac, NSF, 74; mem, Nat Res Coun, 79-81 & UN Independent Develop Orgn, 79- *Honors & Awards:* Ralph Teetor Award, Soc Automotive Engrs, 78. *Mem:* Am Soc Mech Engrs; Am Soc Composites; Am Soc Testing & Mats; Soc Aerospace Mat Engrs; Am Soc Eng Educ. *Res:* Composite materials research, both analytical and experimental for government and industry. *Mailing Add:* 421 S 19th St Laramie WY 82070

ADAMS, DONALD ROBERT, b Red Bluff, Calif, Aug 3, 37; m 74; c 3. RESPIRATORY MORPHOLOGY. *Educ:* Univ Calif, Davis, AB, 60, PhD(anat), 70; Chico State Col, Calif, MS, 67. *Prof Exp:* Govt secy educ officer, Tanganyika, 61-63; teacher biol, physiol & health sci, C K McClatchy Sr High Sch, Calif, 64-65; asst prof anat, Col Vet Med, Mich State Univ,

70-74; from asst prof to assoc prof, 74-85, PROF ANAT, COL VET MED, IOWA STATE UNIV, 85- *Mem:* Am Asn Anatomists; Am Asn Vet Anatomists; World Asn Vet Anatomists. *Res:* Structure and function of the upper respiratory system with emphasis on secretory cells and blood vascular system of the mammalian nasal cavity. *Mailing Add:* 1034 Vet Anat Col Vet Med Iowa State Univ Ames IA 50011

ADAMS, EARNEST DWIGHT, b Carrollton, Ga, Feb 16, 33; m 53; c 3. LOW TEMPERATURE PHYSICS. *Educ:* Berry Col, AB, 53; Emory Univ, MS, 54; Duke Univ, PhD, 60. *Prof Exp:* Res assoc physics, Stanford Univ, 60-62; from asst prof to assoc prof, 62-70, PROF PHYSICS, UNIV FLA, 70- *Concurrent Pos:* Vis prof, Helsinki Univ Technol, 71. *Honors & Awards:* Jesse W Beams Award, Am Phys Soc, 78. *Mem:* Fel Am Phys Soc; Am Asn Physics Teachers. *Res:* Low temperature physics, especially properties of helium-3 and helium-4. *Mailing Add:* Dept Physics Univ Fla Gainesville FL 32611

ADAMS, EDWARD FRANKLIN, b Washington, DC, Oct 15, 36; m 64. CERAMICS, CHEMICAL ENGINEERING. *Educ:* Univ Md, BS, 58; Rensselaer Polytech Inst, PhD(chem eng), 63. *Prof Exp:* Sr chem engr, Ceramic Res, Corning Glass Works, 62-65, res chem engr, 65-68, proj supvr, 68-71, plant eng mgr, 71-73, mgr, Corstar Prod, 73-77; dir res & develop, Anchor Hocking Corp, 77-79, vpres & engr, 77-83; FINANCIAL REP, FIRST FINANCIAL GROUP, 83- *Mem:* Am Chem Soc; Am Ceramic Soc; Am Inst Chem Engrs. *Res:* Glass contact refractories; ceramic processing, including slip casting, isostatic pressing, ceramic coatings on metals; cellular structures for heat exchangers; crystal growth from melt; glass manufacturing technology; glass melting. *Mailing Add:* First Financial Group 703 Giddings Ave Annapolis MD 21401

ADAMS, EMORY TEMPLE, JR, b Tampico, Mex, Sept 5, 28; US citizen. BIOPHYSICAL CHEMISTRY. *Educ:* Rice Univ, BA, 49; Baylor Col Med, MS, 52; Univ Wis-Madison, PhD(biochem), 62. *Prof Exp:* Res assoc chem, Univ Wis-Madison, 62-63; res chemist, Hercules Res Ctr, Del, 63-64; res assoc biol, Brookhaven Nat Lab, 64-65; staff fel, Nat Inst Arthritis & Metab Dis, 66; asst prof chem, Ill Inst Technol, 66-70; assoc prof, 70-76, PROF CHEM, TEX A&M UNIV, 76- *Concurrent Pos:* Vis prof, Dept Med, Univ Chicago, 83-84. *Mem:* Am Chem Soc; Biophys Soc; Am Soc Biol Chemists. *Res:* Analysis of self-associating proteins and mixed associations of biopolymers by sedimentation equilibrium and osmometry; molecular weights and molecular-weight distributions from sedimentation equilibrium experiments; sedimentation equilibrium theory. *Mailing Add:* Dept Chem Tex A&M Univ College Station TX 77843

ADAMS, ERNEST CLARENCE, b Meridian, Miss, Mar 27, 25. MEDICAL DIAGNOSTICS, IMMUNOCHEMISTRY. *Educ:* Miss State Col, BS, 48; Northwestern Univ, PhD(biochem), 50. *Prof Exp:* Asst prof chem, Miss State Col, 50-51; res biochemist & group leader, Miles-Ames Res Lab, Miles Labs, 51-59, sr res biochemist & group leader, Ames Res Lab, 59-61, sect head, 61-70, adminr immunochem opers, 70-72, staff scientist, 72-76, prin investr, Ames Res Lab, 76-78 & Res Prod Div, Miles Lab, 78-82; dir res & develop, Chem-Elec, Inc, 83-86; SR RES SCIENTIST, GENESIS LABS, 86- *Mem:* AAAS; Am Chem Soc; NY Acad Sci; Asn Clin Scientists; Sigma Xi; Am Inst Chemists. *Res:* Immunochemistry; antisera production; protein isolation; immunogen preparation; medical diagnostics; clinical chemistry; solid state chemistry. *Mailing Add:* 5182 W 76th St Minneapolis MN 55439-2904

ADAMS, EVELYN ELEANOR, b Georgetown, Ill, Feb 17, 25; m 51; c 2. OCCUPATIONAL MEDICINE. *Educ:* Univ Chicago, BS, 46, MD, 49. *Prof Exp:* Intern med, Univ Chicago Hosp, 49-50; resident gen surg, Univ Iowa Hosp, 50-51; dir employee health, Univ Chicago Clin, 51-55; sr physician chest med, Suburban Cook Co Sanitarium, 60-68; assoc med dir employee health, Abbott Lab, 68-74; med dir, Johnson & Johnson, Chicago, 74-78; MED DIR, CTR HUMAN RADIO BIOL, ARGONNE NAT LAB, 78- *Mem:* Am Occup Med Asn. *Res:* Effects of internally deposited radionuclides in humans. *Mailing Add:* 7255 Caldwell Ave Niles IL 60648

ADAMS, FORREST HOOD, b Minneapolis, Minn, Sept 20, 19; m 43, 69; c 10. PEDIATRIC CARDIOLOGY. *Educ:* Univ Minn, BA, 41, MB, 43, MD, 44, MS, 46; Am Bd Pediat, dipl, 48, cert cardiol, 61. *Prof Exp:* Intern, Univ Minn Hosp, 43-44, from instr to asst prof pediat, 48-52; from assoc prof to prof, Sch Med, Univ Calif, Los Angeles, 52-78, actg chmn dept, 58-59 & 64-65, vchmn dept, 62-64; clin prof pediat, Sch Med, Univ Hawaii, 81-82; RETIRED. *Concurrent Pos:* Nat Res fel pediat, Univ Minn Hosp, 48-49; asst dir heart sect, Crippled Children's Prog, St Paul, 49-50; physician chg pediat, Sister Elizabeth Kenny Inst, 49-50; assoc physician chg pediat, Minneapolis Gen Hosp, 49-50, chief, 50-52; dir pediat heart clins, Univ Minn Hosp & Variety Club Heart Hosp, 51-52; med dir, Marion Davies Children's Clin, 52-55; mem staff, St John's Hosp & Harbor Gen Hosp, 52-78; consult, State Bd Pub Health, Calif, 63-78; pres, Sub-Specialty Bd Pediat Cardiol, 67, emer mem, 69-; mem adv bd, Inter-Soc Comn Heart Dis Resources, 68-, mem comt congenital heart dis, 68-71, mem exec comt, 68-71; consult, Off Surgeon Gen; Career Res Award, US Pub Health Serv, 62-67; chmn, Health Benefits Adv Coun, Calif Pub Employee's Retirement Syst, 89- *Honors & Awards:* Theodore & Susan Cummings Humanitarian Award, 64, 65, 66, 67, 71 & 72; Distinguished Fel Award, Am Col Cardiol, 74. *Mem:* Soc Pediat Res; Am Acad Pediat; Am Col Cardiol (vpres, 68-69, pres-elect, 70, pres, 71-72); Am Pediat Soc; Am Heart Asn. *Res:* Toxoplasmosis; congenital heart disease; rheumatic fever; hyaline membrane disease; pediatric cardiology; hypertension; exercixe physiology. *Mailing Add:* 16767 Via Lago Azul PO Box 8066 Rancho Santa Fe CA 92067

ADAMS, FRANCIS L(EE), b Talladega, Ala, June 28, 06; m 38; c 2. ENGINEERING. *Educ:* Univ NC, BS, 29. *Prof Exp:* Struct designer, Stone & Webster Eng Corp, 29-31; asst engr, Off Munic Archit, DC, 31-32; engr, Pub Utilities Comn, 32-33; assoc engr, Nat Power Surv, Fed Power Comn, 34-36, hydraul engr, Denver Regional Off, 36-38, sr engr & asst regional dir, 38-41, regional adminr, Ft Worth Regional Off, 41-45, asst chief, Bur Power, DC, 45-51, chief, 51-58, chief engr, 58-62; asst vpres, Pac Power & Light Co, 63-71; CONSULT ENGR, 71-; assoc, Overseas Adv Assoc, 75-80; RETIRED. *Concurrent Pos:* Mem US deleg, World Power Conf, India, 51; mem, Fourth Int Cong Large Dams & Int Conf Large Elec High-Tension Systs, France, 52; US deleg to Orgn for Europ Econ Coop Mission on Elec Energy, 56; US deleg chmn, World Power Conf, Belgrade, Yugoslavia, 57, Montreal, Can, 58 & Madrid, Spain, 60; comnr, Int Joint Comn, US & Can, 58-62; chmn & mem exec bd, US Nat Comt, World Power Conf, 60-61. *Mem:* Fel Am Soc Civil Engrs. *Res:* Power supply and requirements; power system planning; river basin development. *Mailing Add:* 3900 Watson Pl NW Washington DC 20016

ADAMS, FRANK WILLIAM, b Billings, Mont, Feb 17, 25; m 47; c 7. ANIMAL NUTRITION. *Educ:* Mont State Col, BS, 48; Ore State Univ, MS, 50, PhD(zool, biochem), 65. *Prof Exp:* Instr, 53-65, ASST PROF AGR CHEM, ORE STATE UNIV, 65- *Mem:* AAAS; Am Chem Soc; Nutrit Today Soc; Sigma Xi. *Res:* Mammalian embryology; normal and nutritionally caused abnormalities in placentae; trace element relationships in animal nutrition; animal fluorosis. *Mailing Add:* 1153 NW 16th St Corvallis OR 97330

ADAMS, FRED, b Marion, La, Mar 1, 21; m 43; c 2. SOIL SCIENCE. *Educ:* La State Univ, BS, 43, MS, 48; Univ Calif, PhD(soil sci), 51. *Prof Exp:* Prof soil sci, Am Univ, Beirut, 52-55; from assoc prof to prof soils, 55-77, prof agron & soils, 77-84, EMER PROF AGRON & SOILS, AUBURN UNIV, 84- *Mem:* Fel Am Soc Agron; fel Soil Sci Soc Am. *Res:* Chemistry and fertility of soils. *Mailing Add:* Dept Agron & Soils Auburn Univ Auburn AL 36849

ADAMS, GABRIELLE H M, b Mateszalka, Hungary, Oct 29, 39; Can citizen; m 64; c 2. MOLECULAR BIOLOGY. *Educ:* McMaster Univ, BSc, 63; Carleton Univ, PhD(cell biol), 68. *Prof Exp:* Fel biochem, Nat Res Coun Can, 67-69; publ asst, Can J Bot, Nat Res Coun Can, 69-70; res assoc biochem, Nat Res Coun Can & Dept Biol, Carleton Univ, 70-75; publ supvr res journals, Div Biol Sci, Nat Res Coun Can, 75-78, mgr res jour, 79-82, admin, 83-85, indust liaison, 85-86, head, Med Biosci Lab, 86-89, DIR INDUST AFFAIRS, INST BIOL SCI, NAT RES COUN CAN, 89- *Mem:* Can Soc Cell Biol. *Res:* Chemistry and function of nuclear chromatin proteins. *Mailing Add:* Inst Biol Sci Nat Res Coun Can Montreal Rd Ottawa ON K1A 0R6 Can

ADAMS, GAIL DAYTON, b Cleveland, Ohio, Jan 27, 18; m 42, 66, 76; c 3. RADIOLOGICAL PHYSICS. *Educ:* Case Inst Technol, BS, 40; Univ Ill, MS, 42, PhD(physics), 43; Am Bd Radiol, dipl; Am Bd Health Physics, dipl. *Prof Exp:* Asst physics, Univ Ill, 40-43, physicist, 43-45, asst prof physics, 45-51; res physicist & assoc dir radiol lab, Med Sch, Univ Calif, San Francisco, 51-65, lectr radiol, 51-53, clin prof physics, 53-65; prof radiol & radiation physics, Univ Okla, 65-85, vcdmn dept radiol sci, 70-85, campus radiation safety officer, 71-84, EMER PROF RADIOL & RADIATION PHYSICS, HEALTH SCI CTR, UNIV OKLA, 85- *Concurrent Pos:* Consult, Vet Admin Hosp; mem staff, Univ Okla Hosp & Okla Childrens Mem Hosp; AEC & Off Sci Res & Develop; ed, Am Asn Physicist in Med, 73-78. *Honors & Awards:* Williams Award, Am Col Med Physics, 89. *Mem:* Fel Am Phys Soc; Radiation Res Soc (secy-treas, 63-69); fel Am Asn Physicists Med (pres, 58-60); fel Am Col Radiol; NY Acad Sci; fel Am Col Med Physics. *Res:* Absorption of high energy quanta; magnet design and testing; radiography at 20-mev; effect of magnetic fields on biological systems; application of x-rays, particularly 70 mev x-rays to radiology and to radiation therapy; general medical radiation physics. *Mailing Add:* Box 4370 Br 1 Pagosa Springs CO 81157-4370

ADAMS, GAYLE E(LDREDGE), b Cherry Valley, Ill, Nov 24, 27; m 51; c 2. COMPUTER ENGINEERING. *Educ:* Univ Wis, BS, 49, MS, 50, PhD(elec eng), 52. *Prof Exp:* Eng analyst, Gen Elec Co, NY, 52-59, opers analyst, Ariz, 59, consult comput engr, 60-63; prof eng, Ariz State Univ, 63-66; dir, Comput Eng & Develop Ctr, 66-73, PROF ELEC ENG, UNIV MO-COLUMBIA, 66-, PROF COMP ENG, 73-, DIR GRAD STUDIES, 88- *Concurrent Pos:* Consult, Control Systs Assocs, 63-66, Integrated Circuits Eng Corp & Salt River Proj, 66, Gen Learning Corp, 67 & Health Serv & Ment Health Admin, 70; dir data eval & comput simulation proj, Mo Regional Med Prog, 67-73, Mede Quip Corp, 83. *Mem:* Inst Elec & Electronics Engrs. *Res:* Digital systems and computers; health care systems engineering and design; systems engineering; electrical power engineering; FEM analysis of electrical machinery. *Mailing Add:* Dept Elec Eng-139 Ee Bldg Univ Mo Columbia MO 65211

ADAMS, GEORGE BAKER, JR, b Kentfield, Calif, Feb 23, 19. PHYSICAL CHEMISTRY. *Educ:* Univ Calif, BS, 41, MS, 47; Ohio State Univ, PhD(phys chem), 51. *Prof Exp:* Asst chem, Univ Calif, 46-47; res assoc, Univ Ore, 53-55, from instr to asst prof, 55-58; sr scientist, Lockheed Missiles & Space Co, Inc, 58-59, res scientist, 59-62, sr staff scientist, 62-69, sr mem res lab, 62- 88, consult scientist, 70-88; RETIRED. *Res:* Electrochemical energy conversion; chemical thermodynamics; environmental chemistry; hydrometallurgical chemistry; anodic oxide film formation kinetics. *Mailing Add:* 22069 McClellan Rd Cupertino CA 95014

ADAMS, GEORGE G, b Brooklyn, NY, Sept 12, 48; m 84; c 2. TRIBOLOGY, MOVING LOADS. *Educ:* Cooper Union Col, BS, 69; Univ Calif, Berkeley, MS, 72, PhD(mech eng), 75. *Prof Exp:* Assoc engr vibrations, Curtiss-Wright Co, 69-70; asst prof solid mech & comput lang, Clarkson Univ, 75-79; from asst prof to assoc prof, Solid Mech & Design, 79-86, mech div head, 82-85, PROF SOLID MECH & DESIGN, NORTHEASTERN UNIV, 86- *Concurrent Pos:* NSF fel, 70-73; prin investr, NSF, 76-78, 78-79, CDC, 82-83, IBM, 83-84; vis scientist, res lab, IBM, Calif, 78-79; vis scholar, Univ Calif, Berkeley, 79; mem exec comt, Boston Sect, Am Soc Mech Engrs. *Mem:* Am Acad Mech; Am Soc Mech Engrs; Am Soc Eng Educ; Soc Eng Sci. *Res:* Contact problems in elasticity and structural mechanics; deformation of thin-walled flexible bodies, and the response of systems to moving loads; tribology of magnetic storage systems. *Mailing Add:* Dept Mech Eng Northeastern Univ Boston MA 02115

ADAMS, HAROLD ELWOOD, b Altoona, Pa, Dec 10, 16; m 40; c 5. POLYMER CHEMISTRY, PHYSICAL CHEMISTRY. *Educ:* Dickinson Col, BS, 38; Lafayette Col, MS, 39; Mass Inst Technol, PhD(phys chem), 42. *Prof Exp:* Res chemist, Armstrong Cork Co, 42-48; res chemist, Firestone Tire & Rubber Co, 48-70, div mgr chem, 70-73, sr res assoc, 73-79; RETIRED. *Mem:* Am Chem Soc. *Res:* Oxidation of drying oils; polymer characterization; polymerization; kinetics of polymerization. *Mailing Add:* 1930 12th St Cuyahoga Falls OH 44223

ADAMS, HARRY, b Ft Riley, Kans, June 13, 24; m 53; c 4. NUCLEAR PHYSICS. *Educ:* Kans State Univ, BS, 45, MS, 47; Univ Minn, PhD(physics), 62. *Prof Exp:* Instr physics, Pac Lutheran Univ, 47-52; assoc scientist optical eng, Aeronaut Labs, Univ Minn, 53-55; res assoc & vis prof physics, Fla State Univ, 60-62; PROF PHYSICS, PAC LUTHERAN UNIV, 62- *Mem:* Am Phys Soc; Sigma Xi. *Res:* Nuclear reactions and structure. *Mailing Add:* Dept of Physics Pac Lutheran Univ Tacoma WA 98447

ADAMS, HELEN ELIZABETH, b Cowra, Australia, Feb 21, 45; m 67. MATHEMATICS. *Educ:* Univ New Eng, Australia, BS, 66; Monash Univ, Australia, PhD(math), 71. *Prof Exp:* Sr tutor math, Univ Melbourne, 66-68; fel, Mt Holyoke Col, 72-73; ASST PROF MATH, SMITH COL, 73- *Mem:* Am Math Soc; Math Asn Am. *Res:* Commutative algebra. *Mailing Add:* 3 Fisher Ave Wahroonga 2076 Australia

ADAMS, HENRY RICHARD, b Dallas, Tex, Apr 12, 42; m 65; c 2. VETERINARY PHARMACOLOGY. *Educ:* Tex A&M Univ, BS, 65, DVM, 66; Univ Pittsburgh, PhD(pharmacol), 72. *Prof Exp:* Res vet, US Army, Ft Detrick, Md, 66-68; res assoc pharmacol, Univ Pittsburgh, 68-72; asst prof, 72-75, ASSOC PROF PHARMACOL, SOUTHWESTERN MED SCH, UNIV TEX HEALTH SCI CTR DALLAS, 75-, PROF & CHMN, DEPT VET BIOMED SCI, UNIV MO. *Mem:* Am Vet Med Asn; Am Soc Pharmacol & Exp Therapeut; Am Soc Vet Physiol & Pharmacol. *Res:* Comparative cardiovascular pharmacology. *Mailing Add:* Dept Vet Biomed Sci Univ Mo Col Vet W116 Vet Med Bldg Columbia MO 65211

ADAMS, HERMAN RAY, b Abilene, Tex, Sept 15, 39; m 63; c 2. CLINICAL CHEMISTRY. *Educ:* Tex A&M Univ, BS, 62, MS, 65; Univ Tex Southwestern Med Sch Dallas, PhD(pharmacol), 72; Am Bd Clin Chem, cert, 76. *Prof Exp:* DIR, SPEC CHEM & TOXICOL LAB, SCOTT & WHITE CLIN, 72- *Mem:* Am Acad Forensic Sci; Am Asn Clin Chemists; Am Chem Soc; Sigma Xi. *Mailing Add:* Scott & White Clin Temple TX 76508

ADAMS, J MACK, b Marfa, Tex, Aug 14, 33; m 52; c 2. COMPUTER SCIENCE. *Educ:* Tex Western Col, BS, 54; NMex State Univ, MS, 60, PhD(math), 63. *Prof Exp:* Assoc scientist, Bettis Atomic Power Div, Westinghouse Elec Corp, 54-56; mathematician, Flight Simulation Lab, White Sands Missile Range, NMex, 56-58; supvr mathematician, 58-60, mathematician, 61-62, res mathematician, Electronics Res & Develop Activity, 63; assoc prof math, Tex Western Col, 64-65; prog dir, NSF, 87-90; head dept, 65-73 & 75-80, PROF COMPUTER SCI, NMEX STATE UNIV, 65-, ASSOC DEAN, COL ARTS & SCI, 90- *Concurrent Pos:* Vis fel, Wolfson Col, Oxford Univ, 78. *Honors & Awards:* US Govt Award, 64; Fulbright-Hays sr lectr, Cath Univ, Chile, 72. *Mem:* Math Asn Am; Asn Comput Mach; Sigma Xi; Inst Elec & Electronic Engrs. *Res:* Programming languages; theory of algorithms; computer science education. *Mailing Add:* 905 Conway No Seven Las Cruces NM 88005

ADAMS, JACK DONALD, b Gary, Ind, June 3, 19; m 47; c 2. GENETICS. *Educ:* Purdue Univ, BS, 50, MS, 52; Wash State Univ, PhD(radiation genetics), 56. *Prof Exp:* Asst bot, Purdue Univ, 50-52; exp aide radiation genetics, Wash State Univ, 52-53, asst, 53-56; asst prof agron, NMex State Univ, 56-59; from assoc prof to prof biol, Cent Mich Univ, 59-80; RETIRED. *Res:* Plant genetics; cytology; cytogenetics; radiation genetics. *Mailing Add:* 2665 S Vandecar Rd Mt Pleasant MI 48858

ADAMS, JACK H(ERBERT), b Hamburg, NY, July 11, 24; m 45; c 2. PHARMACOLOGY, PHYSIOLOGY. *Educ:* Univ Buffalo, BA, 58, PhD(pharmacol), 61. *Prof Exp:* Asst pharmacol, Sch Med, Univ Buffalo, 52-57; sr pharmacologist, SKF Labs, 61-67; sect head gen pharmacol, 67-78, corp compliance, 78-80, dir drug regulatory affairs, 80-82, ASSOC DIR CLIN RES, ASTRA PHARMACEUT PROD, INC, 82- *Mem:* Am Soc Regional Anesthesia; Am Pharmaceut Asn; Assocs Clin Pharmacol; Acad Pharmaceut Sci; Regulatory Affairs Prof Soc. *Res:* Absorption, distribution and excretion of drugs; neuromuscular blocking agents; antiparkinson agents; antiarrhythmic agents; local anesthetic agents; tocolytic agents. *Mailing Add:* Med Dept Astra Pharmaceut Prod Inc 50 Otis St Westborough MA 01581

ADAMS, JAMES ALAN, b Fayetteville, Ark, Feb 11, 36; m 58; c 3. FLUID MECHANICS, COMPUTER GRAPHICS. *Educ:* Univ Ark, BSME, 58; Purdue Univ, MSME, 63, PhD(mech eng), 65. *Prof Exp:* Instr mech eng, Purdue Univ, 63-64; assoc prof, 64-80, PROF ENG, US NAVAL ACAD, 80- *Mem:* Am Soc Eng Educ; Am Soc Mech Engrs. *Res:* Combined heat and mass transfer in boundary layer flow; density variations at a liquid-air interface; mathematics of computer graphics; computer aided design. *Mailing Add:* Dept of Mech Eng US Naval Acad Annapolis MD 21402

ADAMS, JAMES HALL, JR, b Statesville, NC, Aug 7, 43; m 68; c 2. COSMIC RAY PHYSICS. *Educ:* NC State Univ, BS, 66, MS, 68, PhD(physics), 72. *Prof Exp:* Res asst cosmic ray physics, Johnson Space Ctr, 69-72; res assoc, Nat Res Coun, 72-74; RES PHYSICIST COSMIC RAY PHYSICS, NAVAL RES LAB, 74- *Mem:* Am Phys Soc; AAAS; Am Geophys Union; Sigma Xi. *Res:* Measurements of cosmic ray, elemental composition and charge states; energy spectra; studies of heavy ion nuclear interactions; cosmic ray effects on microelectronics. *Mailing Add:* 7733 Wellington Rd Alexandria VA 22306

ADAMS, JAMES MILLER, b Cleveland, Ohio, Sept 9, 24; m 60; c 7. ANALYTICAL CHEMISTRY, METHOD DEVELOPMENT. *Educ:* Case Inst Technol, BS, 45; Va Polytech Inst, BS, 48, MS, 49, PhD(biochem), 54. *Prof Exp:* Instr, Va Polytech Inst, 48-51; res chemist, Visking Corp, Ill, 54-55; chief chemist, Lab Vitamin Technol, 55-58; res chemist, Chemagro Corp, Mo, 58-63; res assoc drug residues & atherosclerosis, Merck Inst Therapeut Res, NJ, 63-67; sect leader phys & anal res dept, Merck, Sharp & Dohme Res Labs, 67-69; regional dir lab div, US Customs Serv, 69-82, chief, Res Br, US Customs Serv Lab, NY, 82-86; CONSULT, 86- *Concurrent Pos:* Instr, Univ Kansas City, 58-63 & Middlesex County Col, 67-69; safety officer, US Customs, Region III, 69-75; charter mem, Mid Atlantic Asn Forensic Scientists, mem bd dirs, 73-74. *Honors & Awards:* Intergovt Rels Award, 72; Comnr Citation, US Customs Serv, 82. *Mem:* Am Chem Soc; Am Soc Microbiol; Soc Appl Spectros; Asn Off Anal Chemists. *Res:* Saran coatings; large cellulose casings; vitamin assays; radiation sterilization of meat; organophosphate insecticide assay; drug residue assay; bacterial metabolism; atherosclerosis; narcotics detection; forensic analysis; adulteration of fruit juices, vegetable oils and lime oil; aromatic components of petroleum products. *Mailing Add:* 307 Sheffield Ct Joppa MD 21085-4747

ADAMS, JAMES MILLS, b Sioux Falls, SDak, Aug 4, 36; m 58; c 2. CHEMICAL ENGINEERING. *Educ:* SDak Sch Mines & Technol, BS, 58; Univ Wash, MS, 60, PhD(eng), 62. *Prof Exp:* Sr res engr spectros, Aerojet-Gen Corp, 62-68; sr scientist, Hoffmann La Roche Inc, 68-72, mgr eng, 72-75; plant mgr, Aroma Chem Mfg, 75-77, vpres, Aroma Chem Div, 77-78, exec vpres, 78-79, PRES, HAARMANN & REIMER CORP, 80- *Concurrent Pos:* Chmn of the bd, Res Inst Fragrance Mat; mem, adv bd, Cook Col, Rutgers Univ; mem, Bd Fragrance Mat Asn; bd, Flavor/Extract Mfr Asn. *Mem:* Am Mgt Asn; Am Phys Soc; AAAS; Sigma Xi; NY Acad Sci. *Res:* Optics; spectroscopy; heat transfer; sensor-machine systems; non-destructive testing; medical and dental instrumentation. *Mailing Add:* 17 Hwy 34 Matawan NJ 07747

ADAMS, JAMES MILTON, b Roanoke, Va, Sept 22, 49; m 71; c 2. BIOENGINEERING, BIOMEDICAL ENGINEERING. *Educ:* Va Polytech Inst & State Univ, BS, 71; Univ Va, PhD(biomed eng), 76. *Prof Exp:* Res assoc & instr physiol, Albany Med Col, 76-78; asst prof, Univ Va, 78-84, ASSOC PROF BIOMED ENG, UNIV VA, 84- *Concurrent Pos:* Vis scientist, Dartmouth Med Sch, 84-85. *Mem:* Biomed Eng Soc; Am Physiol Soc. *Res:* Pulmonary physiology. *Mailing Add:* Med Ctr Univ Va Box 377 Charlottesville VA 22908

ADAMS, JAMES RUSSELL, zoology; deceased, see previous edition for last biography

ADAMS, JAMES RUSSELL, b North Platte, Nebr, Apr 10, 45; m 67; c 1. MICROELECTRONICS, RADIATION EFFECTS. *Educ:* Univ Nebr, Lincoln, BS, 67, MS, 69, PhD(elec eng), 74. *Prof Exp:* Mem tech staff, Sandia Nat Labs, 69-79; mgr, Prod Tech Develop, Inmos Corp, 79-83, vpres, 83-85; vpres, Triad Semiconductors, 85-86; dir, Monolithic Memories Inc, 86-87; sr prin prog mgr, 87-88, DIR, STRATEGIC PLANNING, UNITED TECHNOLOGIES MICROELECTRONICS, 88- *Mem:* Sr mem Inst Elec & Electronic Engrs; Electrochem Soc; Sigma Xi. *Res:* Optical properties of materials; radiation effects in microelectronics; mos vlsi wafer fabrication technology; strategic technology planning. *Mailing Add:* 1038 Oak Hills Dr Colorado Springs CO 80919

ADAMS, JAMES WILLIAM, b Conover, Wis, Oct 29, 21; c 2. INDUSTRIAL CHEMISTRY. *Educ:* Univ Wis, BS, 43. *Prof Exp:* Control chemist, US Rubber Co, WVa, 43, res group leader, 43-47, chief chemist, Conn, 48-51; sr scientist, Marathon Div, Am Can Co, 51-57, res assoc, Marathon Div, 57-63, sr res assoc, 63-71, res supvr II, 71-80; mgr process develop, Reed Lignih, 80-88; CHEM SPECIALIST, J & B CONSULTS, 88- *Mem:* NY Acad Sci; Am Chem Soc. *Res:* Lignin chemicals; grafting vinyl polymers on cellulose. *Mailing Add:* 2008 Clarberth St Schofield WI 54476

ADAMS, JANE N, b Marion, Ind, Aug 28, 29; m 54. MICROBIOLOGY, GENETICS. *Educ:* Earlham Col, AB, 51; Purdue Univ, MS, 54; Univ Ill, PhD(bact), 59. *Prof Exp:* Instr bact, Univ Ill, 59-61; asst prof, 61-69, ASSOC PROF MICROBIOL, SAN FRANCISCO STATE UNIV, 69- *Res:* Bacterial and viral genetics. *Mailing Add:* Dept of Biol San Francisco State Univ San Francisco CA 94132

ADAMS, JASPER EMMETT, JR, b Smith Co, Tex, Aug 27, 42; m 65; c 2. MATHEMATICAL STATISTICS, MATHEMATICS. *Educ:* Stephen F Austin State Univ, BS, 64, MS, 65; Tex Tech Univ, PhD(math statist), 71. *Prof Exp:* Instr math, Stephen F Austin State Univ, 65-68 & Tex Tech Univ, 68-69; from asst prof to assoc prof, 71-78, PROF MATH STATIST, STEPHEN F AUSTIN STATE UNIV, 78- *Concurrent Pos:* NSF sci scholar fac fel, 70-71. *Mem:* Am Statist Asn; Am Soc Eng Educ; Nat Coun Teachers Math; Sigma Xi. *Res:* Bias reduction techniques and application of statistics. *Mailing Add:* Dept of Math & Statist Stephen F Austin State Univ Nacogdoches TX 75962

ADAMS, JEAN RUTH, b Edgewater Park, NJ, Aug 17, 28. INSECT PATHOLOGY, ELECTRON MICROSCOPY. *Educ:* Rutgers Univ, BS, 50, PhD(entom), 62. *Prof Exp:* Technician, Rohm and Haas Co, 51-57; res fel zool, Univ Pa, 61-62; RES ENTOMOLOGIST, AGR RES CTR, USDA, 62- *Honors & Awards:* Ed Bd, J Invert Path, 86-89. *Mem:* Electron Micros Soc Am; Entom Soc Am; Soc Invert Path (secy, 82-84); Am Soc Cell Biol; NY Acad Sci; Am Soc Microbiol. *Res:* Electron microscopic investigations on histopathology of pathogens of insects; virus invasion and replication of insect pathogens in insect tissues and in insect tissue culture cells. *Mailing Add:* USDA Insect Biocontrol Lab Agr Res Ctr-W Blag 011A Rm 214 Beltsville MD 20705

ADAMS, JOHN ALLAN STEWART, SR, b Independence, Mo, Nov 1, 26; div; c 4. GEOCHEMISTRY. *Educ:* Univ Chicago, PhB, 46, BS, 48, MS, 49, PhD(geol), 51. *Prof Exp:* Proj assoc geochem, dept chem, Univ Wis, 51-54; from asst prof to prof geol,Rice Univ,54-88, chmn dept, 65-71. *Concurrent Pos:* Lectr, Univ Wis, 53 & Am Asn Petrol Geologists, 55; consult, Shell Develop Co, 54-58 & Humble Oil & Ref Co, 58-; exec ed, Geochimica et Cosmochimica Acta, 60-66; adj prof, Sch Pub Health, Univ Tex, 73- *Mem:* Am Chem Soc; fel Geol Soc Am; Geochem Soc; Am Asn Petrol Geologists. *Res:* Geochemistry of thorium and uranium; physical geology; mineralogy; petrology; analytical chemistry of rocks and minerals; geochronology; remote sensing and environmental management studies; lunar samples; atmospheric carbon dioxide; use of videotapes in instruction. *Mailing Add:* Dept Geol 2368 Dunstan Houston TX 77005

ADAMS, JOHN CLYDE, b Angie, La, Dec 11, 47. FORESTRY, SILVICULTURE. *Educ:* La State Univ, BSF, 69, MS, 73, PhD(forestry), 76. *Prof Exp:* PROF FORESTRY, LA TECH UNIV, 76- *Mem:* Soc Am Foresters; Hardwood Res Coun; Int Union Forestry Res Orgn. *Res:* Forestry tree improvement, especially genetic improvement of water and willow oak for timber production and study of species diversity throughout the natural range; regeneration of hardwood stands to provide more usable high quality material. *Mailing Add:* Sch Forestry La Tech Univ PO Box 3168 Ruston LA 71272

ADAMS, JOHN COLLINS, b New Albany, Ind, Sept 18, 38; m 60; c 2. MICROBIAL ECOLOGY, WATER POLLUTION. *Educ:* Purdue Univ, BS, 62; Iowa State Univ, MS, 64; Wash State Univ, PhD(bact), 69. *Prof Exp:* From asst prof to assoc prof, 69-86, PROF MICROBIOL, UNIV WYO, 86- *Mem:* Am Soc Microbiol; Water Pollution Control Fedn; AAAS; Am Water Works Asn; Int Ozone Asn. *Res:* Microbiology of process waters associated with oil shale; microbiology of coal strip mines in the western United States; microbiology of fresh waters and sediments; microbiology of game meat; fluorescent and denitrifying bacteria; causes of tastes and odors in drinking water; drinking water bacteriology; biofouled wells; direct viable count methodology. *Mailing Add:* Div Molecular Biol Univ Wyo Box 3944 Univ Sta Laramie WY 82071-3944

ADAMS, JOHN EDGAR, b Curtis, Nebr, Jan 18, 18; m 42; c 3. SOIL PHYSICS. *Educ:* Univ Nebr, BSc, 42, MSc, 49; Iowa State Univ, PhD(soil physics), 56. *Prof Exp:* Soil scientist sedimentation & hydrol, Qual Water Br, US Geol Surv, 48-50; SOIL SCIENTIST SOUTHERN REGION, OKLA-TEX AREA, SCI EDUC ADMIN, FED SERV, USDA, 55- *Mem:* Am Soc Agron; Soil Sci Soc Am; Soil Conserv Soc Am; Sigma Xi. *Res:* Water intake and erodibility of heavy clay soils; management factors affecting evaporation and water use efficiency of row crops in dryland areas. *Mailing Add:* 2213 49th St Temple TX 76501

ADAMS, JOHN EDWIN, b Berkeley, Calif, Apr 18, 14; m 35; c 3. NEUROSURGERY. *Educ:* Univ Calif, AB, 35; Harvard Univ, MD, 39; Am Bd Neurol Surg, dipl. *Prof Exp:* Asst neuroanat, 47, from instr to assoc prof neurosurg, 48-65, res assoc med physics, 54-55, chmn div neurosurg, 57-65, prof neurol surg & chmn div, 65-68, GUGGENHEIM PROF NEUROL SURG, MED CTR, UNIV CALIF, SAN FRANCISCO, 68- *Concurrent Pos:* Attend neurosurgeon, Hosps, 50-; consult, Naval Hosp, 52- *Mem:* AAAS; Soc Univ Surg; Am Asn Neurol Surg; Am Col Surg; AMA. *Res:* Stereotactic surgery; cerebral metabolism and circulation; hypothermia. *Mailing Add:* Div Neurol Surg Univ Calif Med Ctr 786 Moffitt Hosp M787 San Francisco CA 94143

ADAMS, JOHN EVI, b Durham, NC, May 23, 37; m 65; c 2. PSYCHIATRY. *Educ:* Swarthmore Col, BA, 59; Univ NC, 59-60; Cornell Univ, MD, 64. *Prof Exp:* Intern med, Vanderbilt Univ Hosp, Tenn, 64-65; resident psychiat, Stanford Univ Hosp, Calif, 65-68; spec asst to the dir, NIMH, 68-69, assoc dir, Div Manpower & Training, 69-70; asst prof psychiat, Sch Med, Stanford Univ, 70-74; PROF PSYCHIAT & CHMN DEPT, COL MED, UNIV FLA, 74- *Concurrent Pos:* USPHS res training fel, Dept Psychiat, Stanford Univ, 67-68; consult, Vet Admin Hosps, Palo Alto, Calif, 70-74 & Gainesville, Fla, 74-; mem training rev comt, Nat Inst Alcoholism & Alcohol Abuse, 71-, chmn, 74-75; chmn nat adv comt, Nat Ctr Alcohol Educ, 74-; mem, Fla Adv Coun Ment Health, 75- *Mem:* AAAS; Am Psychiat Asn; Am Psychosom Soc. *Res:* Coping behavior in acute and transitional crisis; aggressive behavior and conflict resolution; ego psychology; alcoholism and aggression. *Mailing Add:* 2834 or 2934 NW 31st Terr Gainesville FL 32605

ADAMS, JOHN GEORGE, b Pittsburgh, Pa, Jan 31, 21; m 53; c 2. PHARMACOLOGY. *Educ:* Duquesne Univ, BS, 47; Univ Ill, MS, 52, PhD, 55. *Prof Exp:* Instr pharmacol & pharmacog, Sch Pharm, Duquesne Univ, 47-49, asst prof pharmacol, 52-55, prof & dean, 55-61; prof pharmacol, Univ Conn, 61-65; dir off sci activ, Pharmaceut Mfrs Asn, 65-68, vpres res, Off Sci & prof rels, 68-83; RETIRED. *Mem:* AAAS; Am Pharmaceut Asn; NY Acad Sci. *Res:* Relationship of molecular architecture to biological activity; biogenesis of neurohumoral agents in nerve impulse transmission and muscular contraction; drug enzymology; biochemical pharmacology. *Mailing Add:* 8107 Whites Ford Way Rockville MD 20854

ADAMS, JOHN HOWARD, b Portland, Maine, May 7, 39; div; c 3. ORGANIC CHEMISTRY. *Educ:* Union Col, BS, 61; Univ Colo, PhD(org chem), 65. *Prof Exp:* Asst chem, Univ Colo, 61-64; SR RES CHEMIST, CHEVRON RES CO, 65- *Res:* Fluorocarbons; reaction of fluorohalocyclobutenes with nucleophiles; alkyl flavanoid synthesis; polymer antioxidant and stabilizers; thermal and photo oxidation of polyolefins; lubricating oil additives; borate dispersions; solar radiation; tribology; mechanisms of wear. *Mailing Add:* 595 Market San Francisco CA 94104

ADAMS, JOHN KENDAL, b Willimantic, Conn, June 14, 33; m 67. GEOLOGY. *Educ:* Univ Conn, BA, 55; Rutgers Univ, MS, 57, PhD(geol), 59. *Prof Exp:* Asst prof geol, Univ Del, 58-63; geologist, US Peace Corps, 63-65; ASSOC PROF GEOL, TEMPLE UNIV, 65- *Mem:* Geol Soc Am; Soc Econ Paleont & Mineral; Sigma Xi. *Res:* Environmental stratigraphy; engineering soils mapping; diagenesis. *Mailing Add:* Dept of Geol Temple Univ Philadelphia PA 19122

ADAMS, JOHN PLETCH, b Ashburn, Mo, Feb 2, 22; m 48; c 1. ORTHOPEDIC SURGERY. *Educ:* Univ Mo, BS, 43; Wash Univ, MD, 45; Harvard Univ, MPH, 78; Am Bd Orthop Surg, dipl, 51. *Prof Exp:* PROF ORTHOP SURG, SCH MED, GEORGE WASHINGTON UNIV, 53-, CHMN DEPT, 69-, CHIEF ORTHOP SERV, UNIV HOSP, 53-; DIR, CLIN LABS, MED COL GA, 79-, PROF PATH, 84- *Concurrent Pos:* Consult orthop surg, Vet Hosp, Washington, DC, 53; dir clin labs, Franklin Sq Hosp, Baltimore, Md, 72-76; consult, Vet Admin Hosp, Augusta, Ga, 79-; consult, Vet Admin Hosp, WVa, Col Med, Howard Univ, Dept Child & Maternal Welfare, DC, Walter Reed Army Med Ctr, Hand Surg & Div Indian Health, USPHS & US Navy Med Ctr. *Mem:* Orthop Res Soc; AMA; Am Soc Surg of Hand; fel Am Col Surg; Am Acad Cerebral Palsy. *Res:* Children's orthopedics; office practice laboratories; vascular problems in the extremities. *Mailing Add:* Dept Orthop Surg George Washington Univ 2150 Pennsylvania Ave NW Washington DC 20037

ADAMS, JOHN QUINCY, physical chemistry, for more information see previous edition

ADAMS, JOHN R, b Eureka, Calif, Jan 6, 18; m 42; c 4. MEDICINE, PSYCHIATRY. *Educ:* Northwestern Univ, 35-39; McGill Univ, MD & CM, 43. *Prof Exp:* Staff psychiatrist, Menninger Found, Kans, 49-54, dir admis, 52-54; chief psychiat, Passavant Mem Hosp, 54-72; PROF PSYCHIAT, MED SCH, NORTHWESTERN UNIV, CHICAGO, 54-, SR PSYCHIAT, INST PSYCHIAT, 75- *Concurrent Pos:* Consult, US Army Hosp, Ft Riley, Kans, 53-54, Vet Admin Res Hosp, Chicago, 54-61 & Ill State Psychiat Inst, 60- *Mem:* AMA; Am Psychiat Asn. *Res:* Psychiatry teaching; curriculum design; general hospital psychiatry. *Mailing Add:* Dept Psychol Behav Sci 707 N Fairbanks Ct Chicago IL 60611

ADAMS, JOHN RODGER, b Milwaukee, Wis, Apr 15, 37; m 62; c 2. CIVIL ENGINEERING, FLUID MECHANICS. *Educ:* Marquette Univ, BCE, 59; Mich State Univ, MS, 61, PhD(porous media), 66. *Prof Exp:* From instr to asst prof civil eng, Lehigh Univ, 65-70; assoc prof scientist, 70-84, PROF SCIENTIST, ILL STATE WATER SURV, 84- *Mem:* Am Soc Civil Engrs; Am Geophys Union; Int Asn Hydraul Res; Int Water Resources Asn. *Res:* Open channel flow; sediment transportation; flow in porous media; hydraulic models; groundwater; regional water supply systems; cost of water supply; navigation on inland waterways. *Mailing Add:* 2204 Griffith Dr Champaign IL 61820-7495

ADAMS, JOHN WAGSTAFF, b New York, NY, Sept 17, 15; m 42; c 2. ECONOMIC GEOLOGY. *Educ:* Colo Sch Mines, GeolE, 41. *Prof Exp:* Student engr, Empire Zinc Co, 42-43; asst res geologist, Pine Creek Unit, US Vanadium Corp, 46-47; geologist & engr, Am Zinc Lead & Smelting Co, 47; jr geologist, US Geol Surv, 43-45 & 47-48, from assoc geologist to geologist, 48-75; RETIRED. *Concurrent Pos:* Geologist-mineralogist, Colo Sch Mines Res Found, 58; consult, Derry, Michener & Booth, Inc, Golden, Colo, 79-80. *Mem:* Fel Mineral Soc Am; Soc Econ Geologists; Mineral Asn Can. *Res:* Geology and mineralogy of the rare-earth elements. *Mailing Add:* 2950 Crabapple Rd Golden CO 80401-1537

ADAMS, JOHN WILLIAM, b Hot Springs, SDak, Sept 24, 28. MATHEMATICAL STATISTICS. *Educ:* Univ Nebr, BS, 52; Univ NC, PhD(statist), 62. *Prof Exp:* Eng trainee, Dow Chem Co, 53-55; statist asst, Westinghouse Elec Co, 55-57; res assoc, Univ NC, Chapel Hill, 62-65; ASSOC PROF INDUST ENG, LEHIGH UNIV, 65- *Mem:* Inst Math Statist; Am Statist Asn. *Res:* Operations research; industrial applications; inventory control programming; queueing theory. *Mailing Add:* Dept Indust Eng Lehigh Univ Mehler Lab 200 Bethlehem PA 18015

ADAMS, JULIAN PHILIP, b Windsor, Eng, Oct 30, 44. POPULATION GENETICS. *Educ:* Univ Wales, BSc, 65; Univ Calif, Davis, PhD(genetics), 69. *Prof Exp:* Fel genetics, Dept Pomol, Univ Calif, Davis, 69-70; asst prof bot, 70-75, assoc prof biol, 75-84, PROF BIOL, UNIV MICH, ANN ARBOR, 84- *Concurrent Pos:* Prof genetics, Fed Univ Rio de Janeiro, Brazil, 78-79; Alexander von Humboldt Fel, WGer, 84. *Mem:* Genetics Soc Am; Am Soc Naturalists. *Res:* Population genetics of microorganisms; human population genetics and demography. *Mailing Add:* 4042a Nat Sci Biol Univ Mich Main Campus Ann Arbor MI 48109-1048

ADAMS, JUNIUS GREENE, III, b Chicago, Ill, May 30, 43; m 76; c 2. HEMOGLOBINOPATHIES, THALASSEMIA. *Educ:* Univ NC, Chapel Hill, AB, 66; Univ Mich, Ann Arbor, MS, 70, PhD(human genetics), 71. *Prof Exp:* Asst prof, Ctr Genetics, Col Med, Univ Ill, Chicago, 72-76, asst prof, Dept Med, Univ, 73-76; asst prof, 76-83, ASSOC PROF GENETICS, DEPT PREV MED, SCH MED, UNIV MISS, 83-, ASSOC PROF, DEPT MED, 76-, ASST PROF BIOCHEM, 80- *Concurrent Pos:* Chief, Hemoglobin Res Lab, Vet Admin Westside Hosp, Chicago, 72-76 & Vet Admin Ctr, Jackson, Miss, 76-; mem, Japan-USA Coop Hemogloblin Study Group, 74-76, CDC Task Force Abnormal Hemoglobins & Thalassemia, 76-78; chmn, Bd Dir, Davis Planetarium Found, 81- *Mem:* Am Soc Human Genetics; Am Soc Hemat; Sigma Xi; Am Fedn Clin Res; Undersea Med Soc; Fedn Am Socs Exp Biol; Am Soc Biol Chem. *Res:* Structure, function, synthesis and molecular biology of variant hemoglobins; hemoglobin variants which mimic the phenotype of thalassemia. *Mailing Add:* 426 Canterbury Ct Jackson MS 39211

ADAMS, KENNETH ALLEN HARRY, b Melville, Sask, Nov 16, 34; m 60; c 2. ORGANIC CHEMISTRY. *Educ:* Univ Man, BSc, 55; McMaster Univ, MSc, 58; Univ Sask, PhD(chem), 61. *Prof Exp:* Res assoc chem, Ind Univ, 61-62; res & teaching fel, McMaster Univ, 62-63; asst prof, 63-68, ASSOC PROF CHEM, MT ALLISON UNIV, 68- *Concurrent Pos:* Nat Res Coun Can grants, 63-65. *Mem:* Chem Inst Can; Royal Soc Chem. *Res:* Synthesis and chemistry of nitrogenous heterocyclic compounds, including natural products; chemistry and synthetic application of nitrene intermediates. *Mailing Add:* Dept Chem Mt Allison Univ Sackville NB E0A 3C0 Can

ADAMS, KENNETH H, b San Francisco, Calif, Dec 4, 37; m 59; c 1. MECHANICAL ENGINEERING, MATERIALS SCIENCE. *Educ:* Calif Inst Technol, BSME, 59, MSME, 60, PhD(mat sci), 65. *Prof Exp:* Asst prof eng, Ariz State Univ, 64-66; asst prof, 66-68, ASSOC PROF MECH ENG, TULANE UNIV, 68- *Concurrent Pos:* Consult, Tex Instruments, 65-66. *Mem:* AAAS; Am Soc Metals; Am Inst Mining Metall & Petrol Engrs. *Res:* Dynamic mechanical behavior of metals and alloys. *Mailing Add:* 6804 Memphis St New Orleans LA 70124

ADAMS, KENNETH HOWARD, b Elgin, Ill, July 7, 06; m 34, 67; c 1. ORGANIC CHEMISTRY. *Educ:* Univ Chicago, BS, 28, PhD(chem), 32. *Prof Exp:* Instr & cur chem, Univ Chicago, 33-40; assoc prof, Harris Teachers Col, Mo, 40-43; from asst prof to prof, 43-75, EMER PROF CHEM, ST LOUIS UNIV, 75- *Mem:* Am Chem Soc. *Res:* Mechanism of carbinol dehydrations and rearrangements; acid catalysis in nonaqueous solvents; pinacol rearrangements; aromatic nucleophilic substitution. *Mailing Add:* 723 Laclede Station Rd No 147 St Louis MO 63119-4911

ADAMS, LAURENCE J, ENGINEERING. *Prof Exp:* INDEPENDENT CONSULT. *Mem:* Nat Acad Eng. *Mailing Add:* 13401 Beall Creek Ct Potomac MD 20854

ADAMS, LEON MILTON, b Waco, Tex, Mar 11, 13; wid; c 2. ORGANIC CHEMISTRY. *Educ:* Agr & Mech Col Tex, BS, 33, MS, 34; Univ Nebr, PhD(phys chem), 37. *Prof Exp:* Res fel, Mellon Inst, 37-38; res chemist, Pittsburgh Plate Glass Co, 38-43; asst chief chemist, Taylor Ref Corp, Tex, 43-46; group leader, Am Oil Co, 46-60; mgr, Org & Polymers Sect, 60-78, inst scientist, Southwest Res Inst, 78-83; RETIRED. *Concurrent Pos:* Consult, 83- *Mem:* Am Chem Soc. *Res:* Encapsulation; polymers; organic synthesis; gelation; radiation grafting; reverse osmosis. *Mailing Add:* 2707 Gainesborough San Antonio TX 78230-4528

ADAMS, LEONARD C(ALDWELL), b Saluda, SC, Nov 7, 21; m 45; c 3. ELECTRICAL ENGINEERING. *Educ:* Clemson Univ, BSEE, 43; Okla State Univ, MS, 49; Univ Fla, PhD(elec eng, physics), 56. *Prof Exp:* From instr to prof elec eng, Clemson Univ, 46-58; nuclear reactor physicist, Savannah River Lab, E I du Pont de Nemours & Co, 58-59; dir eng exp sta, Clemson Univ, 59-61; prof elec eng & head dept, La State Univ, Baton Rouge, 61-77. *Concurrent Pos:* Asst, Okla State Univ, 49-50 & Univ Fla, 51-53. *Mem:* Am Soc Eng Educ; Inst Elec & Electronics Engrs. *Res:* Electron ballistics; travelling wave tubes; large heavy-water moderated nuclear reactor neutron economy research. *Mailing Add:* Dept Elec Eng La State Univ Baton Rouge LA 70803

ADAMS, LESLIE GARRY, b Brownwood, Tex, Aug 17, 41; m 65; c 2. VETERINARY PATHOLOGY. *Educ:* Tex A&M Univ, BS, 63, DVM, 64, PhD(vet path), 68; Am Col Vet Pathologists, dipl, 70. *Prof Exp:* From asst prof to assoc prof vet path & trop vet med, 68-78, PROF VET PATH, TEX A&M UNIV, 78- *Mem:* Am Col Vet Pathologists; Int Acad Path; AAAS; Am Vet Med Asn. *Res:* Veterinary pathology, immunopathology, intracellular parasitism, hepatic diseases and toxicology. *Mailing Add:* Dept of Vet Path Tex A&M Univ Col Vet Med College Station TX 77843

ADAMS, LOUIS W, JR, b Youngstown, Ohio, May 1, 51. APPLIED MATHEMATICS. *Educ:* Youngstown State Univ, BS, 73 NMex State Univ, MS, 75; La State, PhD(physics), 80. *Prof Exp:* RES PHYSICIST, MILLIKEN RES CORP, 80- *Mem:* Am Phys Soc. *Res:* Computer aided design; low temperature plasmas; color science. *Mailing Add:* Milliken Res Corp PO Box 1927 M-420 Spartanburg SC 29304

ADAMS, LOWELL WILLIAM, b Harrisonburg, Va, Aug 8, 46; m 70; c 1. URBAN WILDLIFE ECOLOGY & MANAGEMENT. *Educ:* Va Polytech Inst & State Univ, BSc, 68; Ohio State Univ, MSc, 73, PhD(zool), 76. *Prof Exp:* Teaching assoc biol, zool & mammal, Ohio State Univ, 72-73; res assoc, Ohio Coop Wildlife Res Unit, 73-76; biologist, Urban Wildlife Res Ctr, 76-80, res dir, 80-83, VPRES RES, NAT INST URBAN WILDLIFE, 83- *Concurrent Pos:* Chmn, Urban Wildlife Comt, Wildlife Soc, 84-86; Instr, Univ Md, 87- *Honors & Awards:* Chevron Coserv Award, 87. *Mem:* Sigma Xi; Wildlife Soc; Am Soc Mammalogists; Wilson Ornith Soc; Am Ornithologists Union. *Res:* Urban wildlife ecology and management; human-wildlife relationships in the urban environment; wildlife-habitat associations in disturbed environments; tritium kinetics in freshwater wetlands. *Mailing Add:* Nat Inst Urban Wildlife 10921 Trotting Ridge Way Columbia MD 21044

ADAMS, MARTHA LOVELL, b Springfield, Ill, May 8, 25. ANALYTICAL CHEMISTRY. *Educ:* Col William & Mary, BS, 46; Univ Md, MS, 52. *Prof Exp:* Instr inorg chem, St Helena Exten, Col William & Mary, 46-48; res chemist coating & chem lab, Aberdeen Proving Ground, 51-74; res chemist, US Army Mobility Equip Res & Develop Ccommand, Ft Belvoir, Va, 74-85; RETIRED. *Mem:* Am Chem Soc; Soc Appl Spectros. *Res:* Analytical methods for coating materials, especially the use of ultraviolet and infrared spectrophotometry. *Mailing Add:* 1860 White Oak Dr Unit 323 Houston TX 77009

ADAMS, MAURICE WAYNE, b Rosedale, Ind, June 23, 18; m 45; c 5. PLANT BREEDING. *Educ:* Purdue Univ, BSA, 41; Univ Wis, PhD, 49. *Prof Exp:* Prof agron, SDak State Col, 47-58, agronomist, Exp Sta, 47-58; assoc prof farm crops, 58-64, prof crop sci, 64-88, EMER PROF CROP SCI, MICH STATE UNIV, 88- *Res:* Breeding behavior and genetics of legumes; genetic bases of yield structures and processes in edible dry beans. *Mailing Add:* Dept Crop Soil Sci Mich State Univ East Lansing MI 48824

ADAMS, MAX DAVID, b St Marys, WVa, July 25, 41; m 63; c 2. PHARMACOLOGY, TOXICOLOGY. *Educ:* WVa Univ, BS, 65; Purdue Univ, MS, 68, PhD(pharmacol), 71; Am Bd Toxicol, dipl. *Prof Exp:* From instr to asst prof pharmacol, Med Col Va, 71-77; sr res pharmacologist, 77-81, res mgr, pharmacol & toxicol, 81-84, assoc dir, pharmaceut & device res & develop, 84-85, ASSOC DIR, PROD DEVELOP, MALLINCKRODT, INC,

85- *Mem:* Am Soc Pharmacol & Exp Therapeut. *Res:* Cardiovascular pharmacology; hypertension and antihypertensive drugs; evaluation of x-ray contrast media and magnetic resonance contrast media; in vivo radiopharmaceuticals. *Mailing Add:* 3069 Winding River Dr St Charles MO 63303

ADAMS, MAX DWAIN, b Red Oak, Iowa, May 23, 26; m 47; c 3. INORGANIC CHEMISTRY. *Educ:* Tarkio Col, BA, 48; Okla State Univ, MS, 50; St Louis Univ, PhD(chem), 55. *Prof Exp:* Anal chemist, Mallinckrodt Chem Works, 50-52; res assoc, St Louis Univ, 52-55; chemist, Argonne Nat Lab, 55-73; proprietor, M D Adams Assocs, 73-90; RETIRED. *Mem:* Am Chem Soc; Sigma Xi; Am Inst Chemists; Am Nuclear Soc. *Res:* High temperature chemistry; nuclear materials and fission product elements; molten salt chemistry; microanalytical chemistry; particle identification; trace contamination detection methods. *Mailing Add:* 313 Tamarack Ave Naperville IL 60540

ADAMS, MICHAEL STUDEBAKER, b Houston, Tex, Aug 26, 38; m 64. PLANT ECOLOGY. *Educ:* Univ Calif, Davis, BS, 62, MS, 64; Univ Calif, Riverside, PhD(biol), 68. *Prof Exp:* NASA trainee, Univ Calif, Riverside, 65-68; asst prof, 68-72, assoc prof, 72-78, PROF BOT, UNIV WIS-MADISON, 78- *Mem:* AAAS; Ecol Soc Am; Am Inst Biol Sci; Am Soc Limnol & Oceanog; Int Asn Theoret & Appl Limnol. *Res:* Physiological ecology of aquatic macrophytes; analysis of littoral ecosystems. *Mailing Add:* Dept Bot 132 Birge Hall Univ Wis 430 Lincoln Dr Madison WI 53706

ADAMS, OTTO EUGENE, JR, b Baltimore, Md, Jan 2, 27; m 51; c 4. MECHANICAL ENGINEERING, APPLIED MECHANICS. *Educ:* Cornell Univ, BME, 49; Univ Rochester, MS, 52; Lehigh Univ, PhD(mech eng), 57. *Prof Exp:* Engr, Delco Appliance Div, Gen Motors Corp, NY, 49-51; instr mech eng, Lehigh Univ, 53-57; res engr, Hanford Atomic Prod Oper, Gen Elec Co, Wash, 57-61; mem tech staff, Bell Tel Labs, Pa, 61-63; assoc prof mech eng, 63-69, chmn dept, 67-74, PROF MECH ENG, OHIO UNIV, 69- *Mem:* Am Soc Mech Engrs; Soc Exp Stress Anal; Am Soc Eng Educ. *Res:* Mechanical vibrations; stress analysis; machine design; nonlinear mechanics; dynamics. *Mailing Add:* Dept of Mech Eng Eng Bldg Ohio Univ Athens OH 45701

ADAMS, P B, b Elmira, NY, Sept 4, 29; m 61; c 2. CERAMICS ENGINEERING. *Educ:* Hobart Col, BS, 51. *Prof Exp:* From jr chemist to sr chemist, Corning Inc, 51-62, res chemist, 62-75, res supvr, Tech Staffs Div, 75-88; OWNER, PRECISION ANAL, 88- *Mem:* AAAS; Am Chem Soc; fel Am Ceramic Soc; Am Soc Testing & Mat; Risk Anal Soc; Sigma Xi; Human Factors Soc. *Res:* Silicate chemistry; degradation processes; product safety; surface chemistry and fractography. *Mailing Add:* Precision Anal 300 S Madison Ave Watkins Glen NY 14891

ADAMS, PAUL ALLISON, b Davenport, Iowa, May 30, 40; m 65; c 2. PLANT PHYSIOLOGY. *Educ:* Calvin Col, BA, 62; Univ Mich, MA, 64, PhD(bot), 69. *Prof Exp:* Res assoc plant physiol, Mich State Univ/AEC Plant Res Lab, 68-70; asst prof, 70-74, ASSOC PROF, UNIV MICH, FLINT, 74- dept chmn, biol, 85-88. *Mem:* AAAS; Am Soc Plant Physiologists; Sigma Xi; Am Sci Affil. *Res:* Physiology of plant development; mechanism of gibberellic acid-promoted growth; creation-evolution controversy. *Mailing Add:* Dept Biol Univ Mich Flint MI 48502

ADAMS, PAUL LIEBER, b Broken Bow, Okla, Jan 22, 24; m 46; c 3. PSYCHIATRY, SOCIOLOGY. *Educ:* Centre Col, AB, 43; Columbia Univ, MA, 48, MD, 55; Am Bd Psychiat & Neurol, dipl, 63, cert child psychiat, 64. *Prof Exp:* Instr soc sci, Bennett Col, 47-51; from asst to instr psychiat, Duke Univ, 56-60; from asst prof to prof psychiat & pediat, Col Med, Univ Fla, 60-74, dir, Children's Ment Health Unit, 66-72; PROF PSYCHIAT & V CHMN GRAD EDUC, SCH MED, UNIV LOUISVILLE, 74- *Concurrent Pos:* Fel child psychiat, Duke Univ Hosp & Durham Child Guid Clin, 58-60; vis prof, NC Col, 57-60; dir, Cumberland County Guid Ctr, 57-60; consult, Jacksonville Hosp Educ Prog, 60-; dir educ, Sch Med, Univ Miami, 72-74. *Mem:* Fel Am Psychiat Asn; fel Am Acad Child Psychiat; fel Am Col Psychiat; fel Am Orthopsychiat Asn; Acad Psychosom Med. *Res:* Child psychiatry. *Mailing Add:* Dept Psychiat Univ Tex Med Sch 301 University Blvd Galveston TX 77550

ADAMS, PAUL LOUIS, b Memphis, Tenn, May 15, 48. MEDICAL PHYSICS. *Educ:* Rice Univ, BA, 70; Stanford Univ, MS, 72, PhD(appl physics), 75. *Prof Exp:* Instr med physics, Dept Diag Radiol, 75-80, ASST PROF RADIOL, UNIV TENN, 80- *Res:* Applications of computed tomography and radiological imaging in general. *Mailing Add:* Dept Radiol Univ Tenn Col Med 800 Madison Ave Memphis TN 38163

ADAMS, PETER D, b Cardiff, Wales, Nov 18, 37; m 60; c 2. CONDENSED MATTER PHYSICS, SCIENTIFIC EDITOR. *Educ:* Univ Wales, BS, 59; Univ London, PhD(physics), 64; Imperial Col, DIC, 64. *Prof Exp:* Asst physicist, Brookhaven Nat Lab, 64-69; DEP ED CHIEF, AM PHYSICS SOC, 80- *Concurrent Pos:* Ed, Phys Rew B, 69- *Mem:* AAAS; fel Am Phys Soc. *Mailing Add:* One Research Rd PO Box 1000 Ridge NY 11961

ADAMS, PETER FREDERICK GORDON, b Halifax, NS, Feb 4, 36; m 57; c 3. CIVIL ENGINEERING. *Educ:* NS Tech Col, BE, 58, ME, 61; Lehigh Univ, PhD(civil eng), 66. *Prof Exp:* Design engr, Int Nickle Co Can, Ltd, 58-59; asst prof civil eng, Univ Alta, 60-63; res asst, Lehigh Univ, 63-66; assoc prof, 66-71, PROF CIVIL ENG, UNIV ALTA, 71-, DEAN ENG, 76- *Concurrent Pos:* Design engr, Dominion Bridge Co, 73-74. *Mem:* Am Soc Civil Engrs. *Res:* Behavior of structural members and frames in the elastic and inelastic range; elastic and inelastic instability; behavior of structures subjected to blast and earthquake loadings. *Mailing Add:* Tech Univ Nova Scotia PO Box 1000 Halifax NS B3J 2X4 Can

ADAMS, PHILLIP, b Brooklyn, NY, June 2, 25; m 55; c 3. ORGANIC CHEMISTRY. *Educ:* Univ Md, BS, 45; Cornell Univ, PhD(chem), 50. *Prof Exp:* Lab asst org chem, Cornell Univ, 45-50; res chemist, Am Cyanamid Co, NJ, 50-51; asst dir res & develop, Berkeley Chem Co, 51-68, dir res & develop, Berkeley Chem Div & asst dir corp res & develop, Corp, 68-69, mem exec comt, 69-75, mem operating comt, 75-87, DIR ORG CHEM ONYX DIV, MILLMASTER ONYX CORP, 87- *Concurrent Pos:* Lectr, applied hydrogenation technol; pres, Applied Org Chem & consult practical org custom res, develop & mfg, 87- *Mem:* Am Chem Soc; Asn Consult Chemists & Engrs. *Res:* Custom synthesis in organic chemistry. *Mailing Add:* 27 Burlington Rd Murray Hill NJ 07974

ADAMS, PHILLIP A, b Los Angeles, Calif, Jan 13, 29. ENTOMOLOGY. *Educ:* Univ Calif, BS, 51; Harvard Univ, AM & PhD(biol), 58. *Prof Exp:* From instr to asst prof zool, Univ Calif, Santa Barbara, 58-63; from asst prof to assoc prof, 63-71, PROF ZOOL, CALIF STATE UNIV, FULLERTON, 71- *Concurrent Pos:* Vis assoc prof biol & vis lectr cur insects, Yale Univ, 68-69. *Mem:* AAAS; Am Entom Soc; World Asn Neropterists (pres, 88-). *Res:* Insect evolution; phylogeny and taxonomy of Neuroptera; Myrmeleontidae and Chrysopidae; insect thermo-regulation. *Mailing Add:* Dept Biol Calif State Univ Fullerton CA 92634

ADAMS, PRESTON, b Madison, Fla, Jan 12, 30; m 63. BOTANY. *Educ:* Univ Ga, BS, 54, MS, 56; Harvard Univ, PhD(biol), 59. *Prof Exp:* Asst bot, Univ Ga, 54-56 & Gray Herbarium, Harvard Univ, 56-57; res assoc bot, Fla State Univ, 59-61; from asst prof to assoc prof, 61-73, PROF BOT, DePAUW UNIV, 74- *Mem:* Bot Soc Am; Am Soc Plant Taxon. *Res:* Evolution of vascular plants; monographic studies of flowering plants, especially the family Guttiferae. *Mailing Add:* Rte 2 Box 348 Greencastle IN 46135

ADAMS, RALPH M, b New York, NY, Mar 31, 33; c 2. BIOLOGY. *Educ:* Univ Miami, BS, 55, MS, 59, PhD, 68. *Prof Exp:* Res instr marine physiol, Inst Marine Sci, Univ Miami, 59-60; res biologist microbiol, Seafood Res Lab, Univ Calif, San Francisco, 60-65; from asst prof to assoc prof, 68-80, PROF ECOL, FLA ATLANTIC UNIV, 80- *Mem:* Ecol Soc Am; Soc Study Evolution. *Res:* Plant-animal interactions, particularly insect-orchid pollination and insect chemoreception also Caribbean orchid taxonomy, ecology and biogeography. *Mailing Add:* Dept Biol Sci Fla Atlantic Univ Boca Raton FL 33431

ADAMS, RALPH MELVIN, b College Place, Wash, Apr 16, 21; m 43; c 2. NUCLEAR MEDICINE INSTRUMENTATION. *Educ:* La Sierra Col, BA, 47; Am Bd Radiol, dipl, 51. *Prof Exp:* Physicist, White Mem Med Ctr, La, 51-62 & 68-70, Cedars-Sinai Med Ctr, 62-68; PHYSICIST & ASSOC RES PROF RADIOL, LOMA LINDA UNIV, 70- *Concurrent Pos:* Int Atomic Energy Agency tech assistance expert to Iraq, 61, Israel, 71, Pakistan, 74-75, Thailand, 81, 82 & 85, Ecuador, 88. *Mem:* Radiol Soc NAm; Am Col Radiol; Soc Nuclear Med; Am Asn Physicists in Med. *Res:* Development of scanning and imaging equipment for nuclear medicine; color-coded imaging display methods; development of computer applications for data acquisition processing, and display with the scintillation camera; development of methods to evaluate count rate performance of scintillation cameras. *Mailing Add:* Nuclear Radiol Loma Linda Univ Loma Linda CA 92350

ADAMS, RALPH NORMAN, b Atlantic City, NJ, Aug 26, 24; m 53; c 3. ANALYTICAL CHEMISTRY, NEUROCHEMISTRY. *Educ:* Rutgers Univ, BS, 50; Princeton Univ, PhD(anal chem), 53. *Prof Exp:* Instr anal chem, Princeton Univ, 53-55; from asst prof to assoc prof, 55-63, PROF CHEM, UNIV KANS, 63- *Honors & Awards:* Fisher Award, Am Chem Soc, 82. *Mem:* Am Chem Soc; Am Soc Neurosci. *Res:* Electroanalytical methods applied to neurochemistry, chemistry of the central nervous system. *Mailing Add:* 2125 Terrace Rd Lawrence KS 60615

ADAMS, RANDALL HENRY, b Milburn, Okla, Aug 24, 39; m 62; c 2. ENTOMOLOGY, PLANT SCIENCE. *Educ:* Okla State Univ, BS, 61, MS, 66, PhD(entom ecol), 72. *Prof Exp:* Res asst entom, Okla State Univ, 66-68; asst prof, NMex State Univ, 68-70, agron, 70-74; ASST PROF ENTOM & PLANT SCI, SOUTHERN ARK UNIV, 74- *Concurrent Pos:* Res consult entom & agron, Shell Chem Co, Dow Chem USA, PPG Chem Co, Thuron Industs & Cooper USA, 70- *Mem:* Sigma Xi; Entom Soc Am. *Res:* Evaluation of pesticides on animal and plant pests. *Mailing Add:* Dept of Agr Southern Ark Univ Magnolia AR 71753

ADAMS, RAYMOND F, b Pittsburgh, Pa, Apr 15, 24; m 50; c 1. ELECTRICAL ENGINEERING, ELECTRONICS. *Educ:* Villanova Univ, BEE, 45; Univ Del, MEE, 55. *Prof Exp:* From instr to asst prof, 47-55, ASSOC PROF ELEC ENG, VILLANOVA UNIV, 55- *Concurrent Pos:* Consult, Polyphase Instrument Co, Pa. *Mem:* Sr mem Inst Elec & Electronics Engrs. *Res:* Electronics; electrical circuit analysis; computer aided design. *Mailing Add:* Dept of Elec Eng Villanova Univ Villanova PA 19085

ADAMS, RAYMOND KENNETH, b Denver, Colo, Apr 23, 28; m 50; c 3. ELECTRICAL ENGINEERING, MEASUREMENTS & CONTROL ENGINEERING. *Educ:* Univ Colo, BS, 51; Univ Tenn, MS, 60. *Prof Exp:* Design engr instrumentation, Oak Ridge Gaseous Diffusion Plant, Union Carbide Corp, 51-53, development engr instrumentation & controls, Oak Ridge Nat Lab, 54-61, group leader, Instrumentation & Controls Div, 62-76, staff consult, Instrumentation & Controls Div, 77-84; sr staff engr & group leader, Instrumentation & Controls Div, Oak Ridge Nat Lab, Martin Marietta Energy Systs Corp, 84-86; FAC ASSOC, ELEC & COMPUTER ENG DEPT, UNIV TENN, KNOXVILLE, 86- *Concurrent Pos:* Gen chmn, Joint Automatic Control Conf, 60-61; mem steering comt, Am Automatic Control Coun, 60-62; expert witness & spec master, Embedded Microprocessor Intellectual Property Litigation. *Mem:* Fel Instrument Soc Am; sr mem Inst Elec & Electronic Engrs. *Res:* Measurement and control devices and theory, development and application for energy RanD; pioneered mini-microcomputer acquisition and control systems application for scientific/engineering RanD; electric utility load control communications systems; modernizing the sophomore EE laboratory curriculum; student teams coached have won national alternate fuels automobile design and implementation contests since 1986- electronic/computer automotive engine controls. *Mailing Add:* 102 Wendover Circle Oak Ridge TN 37830

ADAMS, RICHARD A, b New York, NY. IMMUNOLOGY & PATHOLOGY, GENETICS. *Educ:* City Col New York, BA, 49; Boston Univ, AM, 51, PhD(biol), 56; Am Bd Toxicology, dipl, 82. *Prof Exp:* Res assoc, Children's Cancer Res Found, 56-74; res assoc, 74-84, sci dir toxicol, 84-86, CONSULT, BIO-RES CONSULT, INC, 86- *Concurrent Pos:* Res assoc, Children's Hosp, Boston, 59-74, & Med Sch, Harvard Univ, 60-74; ed, Hamster Info Serv, 79- *Mem:* Am Asn Cancer Res; Am Soc Exp Path; Am Col Toxicol; Sigma Xi; Hamster Soc; Soc of Toxicol. *Res:* Cancer research with principal interest in human tumor and leukemic hetero (xeno) transplantation and immunologic aspects; carcinogenesis, particularly in Syrian hamsters with emphasis on the use of this application for safety testing. *Mailing Add:* 10 Mount Pleasant St Rockport MA 01966

ADAMS, RICHARD DARWIN, b Reading, Pa, July 26, 47; m 72; c 2. MOLECULAR STRUCTURE, CATALYSIS. *Educ:* Pa State Univ, BS, 69; Mass Inst Technol, PhD(inorg chem), 73. *Prof Exp:* Asst prof chem, State Univ NY, Buffalo, 73-75; from asst prof to assoc prof, Yale Univ, 75-84; PROF CHEM, UNIV SC, 84- *Honors & Awards:* A P Sloan Fel, 79-83. *Mem:* Am Chem Soc; Am Crystallog Asn; AAAS. *Res:* Chemistry of transition metal cluster compounds and x-ray crystallography. *Mailing Add:* Dept Chem Univ SC Columbia SC 29208

ADAMS, RICHARD E, b Springfield, Ohio, Aug 14, 21; c 3. AEROSPACE ENGINEERING. *Educ:* Purdue Univ, BS, 42. *Prof Exp:* Vpres eng, Convair Div, San Diego, Gen Dynamics Corp, 70-71, vpres & gen mgr, Fort Worth Div, 71-74, corp vpres, 74-81, exec vpres aerospace, Gen Dynamics, 81-87; MGT CONSULT, 87- *Mem:* Nat Acad Eng. *Mailing Add:* 4200 S Hullen St Suite 319 Ft Worth TX 76109

ADAMS, RICHARD JAMES, physical chemistry, physics, for more information see previous edition

ADAMS, RICHARD LINWOOD, b Dixfield, Maine, Sept 17, 28; m 53; c 2. POULTRY NUTRITION. *Educ:* Univ Maine, BS, 52; Purdue Univ, MS, 59, PhD(poultry nutrit), 61. *Prof Exp:* Fieldman, Swift & Co, Ore, 52-53 & C M T Co, Maine, 53-55; 4-H Club agent, York County Exten Serv, 55-57; POULTRY EXTEN SPECIALIST, PURDUE UNIV, WEST LAFAYETTE, 62-, ASSOC PROF ANIMAL SCI, 77- *Mem:* Poultry Sci Asn. *Res:* Poultry nutrition. *Mailing Add:* 2825 Henderson St West Lafayette IN 47906

ADAMS, RICHARD MELVERNE, b Gary, Ind, Nov 1, 16; m 44; c 2. INORGANIC CHEMISTRY, REACTOR SAFETY. *Educ:* Univ Chicago, BS, 39, MS, 48; Ill Inst Technol, PhD(chem), 53. *Prof Exp:* Chemist, Portland Cement Asn, 40-42; res assoc, Off Sci Res & Develop Proj, Univ Chicago, 42-43, jr chemist, Metall Lab, Manhattan Dist, 43-46; asst group leader, Los Alamos Sci Lab, 46; lectr chem, Ind Univ, 46-49; assoc chemist, 49-59, sci asst to dir, 59-65, asst lab dir, 65-80, dep dir, Off Int Energy Develop Prog, Argonne Nat Lab, 80-83,; RETIRED. *Concurrent Pos:* Mem, Joint US-Korea Standing Comt on Energy Technol, 76-83. *Mem:* AAAS; Am Chem Soc; Sigma Xi; Am Nuclear Soc; Korean Nuclear Soc. *Res:* Etherates of boron trifluoride; radiochemistry; fission products; kinetics of exchange reactions; fluorine chemistry; reactor safety. *Mailing Add:* 4744 Kimbark Ave Chicago IL 60615

ADAMS, RICHARD OWEN, b Garden Home, Ore, Mar 5, 33; m 64. SOLID STATE PHYSICS. *Educ:* Willamette Univ, BA, 55; Wash State Univ, PhD(physics), 64. *Prof Exp:* Sr develop specialist, 63-67, sr res physicist, 67-68, res mgr, 68-75, sr res specialist, 75-87, ASSOC SCIENTIST, ROCKY FLATS PLANT, ROCKWELL INT, 87- *Mem:* Am Vacuum Soc. *Res:* Surface science, particularly the structure and composition of surfaces, the interactions of gases with metal surfaces and the growth of films on metal substrates; acoustic emission, its sources and propagation through metals. *Mailing Add:* 2790 Regis Dr Boulder CO 80303

ADAMS, RICHARD SANFORD, b Lewiston, Maine, May 5, 28; m 51; c 4. ANIMAL NUTRITION, DAIRY HUSBANDRY. *Educ:* Univ Maine, BS, 50; Univ Minn, PhD(dairy husb), 55. *Prof Exp:* Asst prof dairy sci, Pa State Univ, 54-57; res assoc ruminant nutrit, Gen Mills, Inc, 57-58; from asst prof to assoc prof, 58-65, PROF DAIRY SCI, PA STATE UNIV, 65- *Concurrent Pos:* Consult, US Feed Grains Coun, SEurope, 69 & Univ Hawaii, 71. *Mem:* Am Dairy Sci Asn; Am Soc Animal Sci. *Res:* Dairy cattle nutrition; forage evaluation; extension teaching methods. *Mailing Add:* Dept Dairy/Animal Sci Penn State Univ University Park PA 16802

ADAMS, ROBERT D, b Grand Rapids, Mich, Dec 10, 26; m 56; c 2. MATHEMATICS. *Educ:* Univ Minn, PhD(math), 60. *Prof Exp:* ASSOC PROF MATH, UNIV KANS, 60- *Mem:* Am Math Soc; Math Asn Am. *Res:* Application of functional analysis to partial differential equations. *Mailing Add:* Univ Kans Lawrence KS 66045

ADAMS, ROBERT EDWARD, b Memphis, Tenn, Nov 4, 29; m 49; c 1. HAZARDOUS WASTE TECHNOLOGY. *Educ:* Memphis State Univ, BS, 54; Univ Miss, MS, 56. *Prof Exp:* Res staff mem gas adsorption, Oak Ridge Nat Lab, 56-66, res group leader aerosol sci, 66-70; consult, Air-Tec, Inc, 70-75; proj leader aerosol sci, Oak Ridge Nat Lab, 75-86; PROJ MGR, DEMONSTRATION HAZARDOUS WASTE REMEDIATION TECHNOL, MARTIN-MARIETTA ENERGY SYST CTR STAFF, OAK RIDGE, TENN, 87- *Mem:* Am Chem Soc; Am Nuclear Soc. *Res:* Management of technical projects to demonstrate remediation of hazardous wastes and/or waste sites. *Mailing Add:* 112 Amanda Dr Oak Ridge TN 37830

ADAMS, ROBERT JOHN, b Solon, Iowa, May 24, 15; m 41; c 2. RADAR SYSTEMS & RADAR ANTENNAES. *Educ:* Univ Iowa, BA, 36; Univ Wis, PhD(physics), 41. *Prof Exp:* Physicist, Corning Glass Works, NY, 40-42; head antenna sect, Naval Res Lab, 42-54, head search radar br, 54-76, consult, 76-90; RETIRED. *Concurrent Pos:* Assoc ed, Trans Antennas & Propagation, Inst Elec & Electronic Engrs. *Mem:* AAAS; Am Geophys Union; Am Phys Soc; Inst Elec & Electronic Engrs; Sigma Xi. *Res:* Radar antennas and systems; radio physics. *Mailing Add:* 10114 Livingston Rd SE Ft Washington MD 20744-4930

ADAMS, ROBERT JOHNSON, b Charleston, SC, April 27, 50; m 74; c 2. CARDIOVASCULAR PHARMACOLOGY, MEMBRANE BIOCHEMISTRY. *Educ:* Col Charleston, BS, 72; Univ Ark Med Sci, PhD(pharmacol), 78. *Prof Exp:* Lab technician res & develop, Med Univ SC, 72-73; grad asst pharmacol, Univ Ark Med Sci, 73-77; fel, Univ Cincinnati Med Ctr, 77-80, res asst prof, 80-81; asst prof anesthesiol, Med Col Ga, Augusta, 81-; AT DIV CARDIOVASC RES, MEDICAL COL GA. *Concurrent Pos:* Asst dir, Div Cardiovasc Res, Anesthesiol Med Col Ga, 82- *Mem:* Sigma Xi; Int Soc Oxygen Transp Tissue; Int Anesthesia Res Soc. *Res:* Cardiovascular physiology and biochemistry, mechanisms of cellular regulation of calcium transport, calcium utilization and muscle contractin; drug-receptor interactions and cellular mechanisms of drug action; pharmacology and membrane pathophysiology of the myocardium and vascular smooth muscle in myocardial ischemia, heart failure, hypertension. *Mailing Add:* Div Cardiovasc Res Sch Med Med Col Ga Box 1151 Augusta GA 30912-0352

ADAMS, ROBERT MCLEAN, chemical engineering, for more information see previous edition

ADAMS, ROBERT R(OYSTON), b McConnelsville, Ohio, May 26, 17; m 41; c 2. METALLURGICAL ENGINEERING. *Educ:* Antioch Col, BS, 39; Ohio State Univ, MS, 41. *Prof Exp:* Asst foundry metal, Battelle Mem Inst, 36-39, res engr, 39-42, admin asst, 42-81; RETIRED. *Concurrent Pos:* With AEC, Off Sci Res & Develop & Off Prod Res & Develop, 42-46; dir, Battelle Mem Inst Ger, 52-53; dir, Indust Inst Beirut, Lebanon, 56-58; with Advan Res Proj Agency, Field Off Beirut, Lebanon, 64-70. *Res:* Research administration. *Mailing Add:* 1812 Sandalwood Pl Columbus OH 43229

ADAMS, ROBERT W, b New York, NY, July 8, 34; m 57; c 3. GEOLOGY. *Educ:* Univ Rochester, BA, 56; Johns Hopkins Univ, PhD(geol), 64. *Prof Exp:* Explor geologist, Shell Oil Co, 64-67; from asst prof to assoc prof geol, 67-74, PROF GEOL, STATE UNIV NY COL BROCKPORT, 74-, CHMN, 80- *Mem:* Geol Soc Am; Soc Econ Paleontologists & Mineralogists; Am Asn Petrol Geologists; Sigma Xi. *Res:* Sedimentology and sedimentary petrology. *Mailing Add:* Dept of Earth Sci State Univ of NY Col Brockport NY 14420

ADAMS, ROBERT WALKER, JR, b Ashburn, Ga, Nov 9, 20; m 45; c 4. MEDICINE. *Educ:* Vanderbilt Univ, MD, 48. *Prof Exp:* Asst prof, 55-70, ASSOC PROF PSYCHIAT, MED SCH, VANDERBILT UNIV, 71- *Concurrent Pos:* Clin asst prof, Meharry Med Sch, 55-68. *Mem:* AMA; Am Psychiat Asn. *Res:* Clinical psychiatry; teaching. *Mailing Add:* 109 Groome Dr Nashville TN 37205

ADAMS, ROGER JAMES, b Rio, Wis, Aug 27, 36; m 58; c 3. PHYSICAL CHEMISTRY. *Educ:* Univ Wis, BS, 58; Iowa State Univ, PhD(phys chem), 61. *Prof Exp:* Sr chemist, Minn Mining & Mfg Co, 61-67, res supvr, 67-68, tech mgr res & develop, 68-74, lab mgr med prod, 74-78, tech dir, Med Prod Div, 78-86, tech dir biosci lab, 86-90, TECH DIR LIFE SCI SECTOR LAB, MINN MINING & MFG CO, 90- *Mem:* Am Chem Soc; Health Indust Mfrs Asn. *Res:* Chemical kinetics; surface chemistry of metals; electrochemistry; displacement plating, electroplating, vapor plating; applications of tagged tracers in surface chemistry; materials science. *Mailing Add:* Minn Mining & Mfg Co 3M Ctr Bldg 1270-2A-08 St Paul MN 55144

ADAMS, ROGER OMAR, b Willows, Calif, Apr 17, 33; c 3. IMMUNOGENETICS. *Educ:* Fla State Univ, Tallahassee, BS, 66, MS, 68, PhD(biol), 80. *Prof Exp:* RES ASSOC, FLA STATE UNIV, 81- *Concurrent Pos:* Pres, Marisyst Res, Inc, Fla, 77- *Mem:* AAAS; Am Soc Zoologists. *Res:* Development, functional morphology, and genetics of Gorgonian Octocorals (anthozoa), with special interest in their self/not-self histocompatibility; symbiotic relationships of various animals with octocorals. *Mailing Add:* Sci Bldg D 100 College Blvd Niceville FL 32578

ADAMS, ROY MELVILLE, b Cheung Chau, Hong Kong, Oct 19, 19; m 46; c 4. CHEMISTRY. *Educ:* Sterling Col, AB, 40; Univ Kans, AM, 42, PhD(chem), 49. *Prof Exp:* From instr to assoc prof, 46-58, chmn dept, 58-85, PROF CHEM, GENEVA COL, 58- *Concurrent Pos:* Chief dept chem, Callery Chem Co, 52-53, res coordr, 53-58; mem inorg nomenclature comn, Int Union Pure & Appl Chem, 68-78. *Mem:* Am Chem Soc; Am Sci Affil; The Chem Soc; Creation Res Soc; Sigma Xi. *Res:* Boron; inorganic nomenclature. *Mailing Add:* Dept Chem Geneva Col Beaver Falls PA 15010

ADAMS, RUSSELL S, JR, b Kincaid, Kans, Mar 9, 26; m 58; c 2. SOIL CHEMISTRY. *Educ:* Kans State Univ, BS, 58, MS, 59; Univ Ill, PhD(soil biochem), 62. *Prof Exp:* Asst agron, Univ Ill, 61-62; res assoc, 62-63, from asst prof to assoc prof, 63-71, PROF SOILS, UNIV MINN, ST PAUL, 71- *Concurrent Pos:* Dist supvr, Ramsey Soil & Water Conserv, 71-85. *Honors & Awards:* Hon res grant, 3M Environ Sci, 85; travel grant, Int Soil Cong, Nat Sci Found & Am Soc Agron, Adelaide, Australia, 68. *Mem:* Am Soc Agron; Am Chem Soc; Soil Sci Soc Am; Weed Sci Soc Am; Soc Environ Geochem & Health; Int Soil & Water Conserv, Soc Am, 90. *Res:* Contamination of soils by petroleum hydrocarbons; fixation of ammonium ions by soil-forming minerals; the fate of pesticide residues in soils; soybean nutrition; disposal of solid wastes in soils. *Mailing Add:* 3264 New Brighton Rd St Paul MN 55112

ADAMS, S KEITH, b Salisbury, Md, Mar 1, 38; m 66; c 2. INDUSTRIAL & HUMAN FACTORS ENGINEERING. *Educ:* Rensselaer Polytech Inst, BMgte, 60; Ariz State Univ, MSE, 62, PhD(indust eng), 66. *Prof Exp:* Res asst, Ariz State Univ, 60-62; from asst prof to assoc prof indust eng, Okla State Univ, 65-76; ASSOC PROF INDUST ENG, IOWA STATE UNIV, 76- *Concurrent Pos:* NSF res grant, 66-67. *Mem:* Am Inst Indust Engrs; Human Factors Soc; Am Soc Eng Educ; Sigma Xi. *Res:* Ergonomics in work system design and safety engineering; human factors in nuclear power plant operations and maintenance; human/computer interaction; human factors in maintainability; resource and energy recovery from municipal and industrial solid waste. *Mailing Add:* Dept Indust Eng Iowa State Univ Ames IA 50011

ADAMS, SAM, b Walthall, Miss, Mar 14, 16; m 44; c 1. SCIENCE EDUCATION. *Educ:* Delta State Teachers Col, BS, 36; La State Univ, MA, 40, PhD(sci educ), 51. *Prof Exp:* Teacher high sch, Miss, 36-38; instr, Univ Ala, 40-42, Univ exten supvr, 42; civilian instr radio, US Air Force, SDak, 42-43; elec supply foreman, Tenn Eastman Corp, Oak Ridge, 43-44; physicist and spectroscopist, Carbide & Carbon Chem Co, 46-49; assoc prof physics, McNeese State Col, 51-54; assoc prof sci educ, La State Univ, Baton Rouge, 54-62, prof sci educ, 62-81, assoc dean acad affairs, 62-65; RETIRED. *Concurrent Pos:* Teacher high sch, Ala, 40-42. *Mem:* Nat Sci Teachers Asn. *Res:* Emission spectra of uranium isotopes; science education. *Mailing Add:* 2035 W Magna Carta Place Baton Rouge LA 70815-5523

ADAMS, SAMUEL S, b Lincoln, NH, July 26, 37. MINERAL DEPOSIT. *Educ:* Harvard Univ, PhD(geol), 76. *Prof Exp:* Vpres, Anaconda Co, 67-77; consult, 77-86; PROF & HEAD, GEOL & GEOL ENG, COLO SCH MINES, 86- *Mem:* Fel Geol Soc Am; Am Asn Petrol Geologists; Am Inst Petrol Geologists. *Mailing Add:* Dept Geol Colo Sch Mines Golden CO 80401

ADAMS, STEVEN PAUL, b Oceanside, Calif, Dec 31, 52; m 74; c 5. PEPTIDE CHEMISTRY, NUCLEOTIDE CHEMISTRY. *Educ:* Univ Utah, BA, 76; Univ Wis, PhD(chem), 80. *Prof Exp:* Res asst chem, Univ Utah, 75-76, Univ Wis, 76-80; sr res chemist, Monsanto Co, 80-83, res specialist, 83-84, res group leader, 84-85, assoc fel, 85-88, FEL, CORP RES LAB, BIOL SCI, MONSANTO CO, 88-, DEPT HEAD, BIOL CHEM, 90- *Mem:* Am Chem Soc; Sigma Xi. *Res:* Chemical applications to biotechnology; peptide, nucleotide and oligosaccharide chemistry; medicinal chemistry, structural biology. *Mailing Add:* Monsanto Co 700 Chesterfield Village Pkwy BB3E Chesterfield MO 63198

ADAMS, TERRANCE STURGIS, b Los Angeles, Calif, Dec 14, 38; div; c 2. INSECT PHYSIOLOGY. *Educ:* Calif State Col Los Angeles, BA, 61; Univ Calif, Riverside, PhD(entom), 66. *Prof Exp:* RES ENTOMOLOGIST, METAB LAB, AGR RES SERV, USDA, 66-, MEM STAFF, BIOSCI RES LAB. *Mem:* Entom Soc Am; Int Soc Insect Reproduction; Sigma Xi. *Res:* Factors affecting insect reproduction; endocrine control of reproduction in dipterous insects. *Mailing Add:* Biosci Res Lab Univ Sta Fargo ND 58105-5674

ADAMS, THOMAS, b Buffalo, NY, Oct 3, 30; m 53; c 2. PHYSIOLOGY. *Educ:* Univ Md, BS, 52; Purdue Univ, MS, 55; Univ Wash, PhD(physiol), 63. *Prof Exp:* Chief thermal sect, Civil Aeromed Inst, Fed Aviation Agency, Okla, 63-66; PROF PHYSIOL, MICH STATE UNIV, 66- *Mem:* AAAS; Am Physiol Soc; Soc Exp Biol & Med; Am Asn Univ Prof; Sigma Xi. *Res:* Body temperature regulation; neurophysiology of temperature regulation; mechanisms of thermal acclimatization; sweat gland physiology. *Mailing Add:* Dept Physiol Mich State Univ East Lansing MI 48823

ADAMS, THOMAS C, b San Francisco, Calif, Nov 4, 18; m 57; c 2. FOREST ECONOMICS, INTERNATIONAL ECONOMICS. *Educ:* Univ Calif, Berkeley, BS, 40, AB, 41; Univ Mich, MA, 51, PhD(forestry), 52. *Prof Exp:* Forest economist, Pac Northwest Forest & Range Exp Sta, US Forest Serv, 52-55; asst prof forestry, Ore State Univ, 55-57; prin economist, Pac Northwest Forest & Range Exp Sta, US Forest Serv, 57-84; RETIRED. *Concurrent Pos:* Consult, SRI Int, Ore, 55. *Mem:* Soc Am Foresters; Forest Prod Res Soc. *Res:* Marketing, market development and international trade of forest products; economic analysis of new logging systems, thinning operations and forest residue reduction; wood energy research and development. *Mailing Add:* 7640 SE 28 Ave Portland OR 97202

ADAMS, THOMAS EDWARDS, b Neenah, Wis, July 10, 47; m 75; c 2. BIOCHEMICAL ENDOCRINOLOGY, REPRODUCTIVE NEUROENDOCRINOLOGY. *Educ:* Wash State Univ, BS, 72, MS, 75; Colo State Univ, PhD(endocrinol), 79. *Prof Exp:* Res asst physiol, Wash State Univ, 72-74; endocrinol, Colo State Univ, 74-79; NIH fel reprod endocrinol, Ore Regional Primate Res Ctr, 79-80; asst prof, 81-86, ASSOC PROF, ANIMAL SCI, UNIV CALIF, DAVIS, 86- *Mem:* AAAS; Soc Study Reprod; Am Soc Animal Sci. *Res:* Temporal changes in gonadotropin releasing hormone receptor concentration in anterior pituitary tissue through the reproductive cycle of the normal female and in response to estradiol; the role of gonadotropin releasing hormone receptor dynamics in modulation of gonadotroph responsiveness and anterior pituitary function. *Mailing Add:* Dept Animal Sci Univ Calif Davis CA 95616

ADAMS, WADE J, b Calhoun Co, Mich, Jan 9, 40; m 65; c 1. STRUCTURAL CHEMISTRY, CHEMICAL PHYSICS. *Educ:* Western Mich Univ, BS, 65; Univ Mich, MS, 68, PhD(phys chem), 69. *Prof Exp:* Fel, Univ Mich, 71-72; asst prof chem, Macalester Col, 72-73; asst prof chem, Western Mich Univ, 73-77; MEM STAFF DRUG METAB RES, THE UPJOHN CO, 77- *Mem:* Am Chem Soc. *Res:* Experimental studies of the structure and force field of molecules by gaseous electron diffraction; molecular mechanics calculations of physical and chemical properties. *Mailing Add:* Dept of Drug Metab Res The Upjohn Co Kalamazoo MI 49001

ADAMS, WALTER C, b Newtown, Pa, Aug 22, 36; m 62; c 2. ENDOCRINOLOGY, PHYSIOLOGY. *Educ:* Drew Univ, AB, 58; Rutgers Univ, MS, 62, PhD(endocrinol), 63. *Prof Exp:* Res scientist, NJ Bur Res Neurol & Psychiat, 63-67; from asst prof to assoc prof, 67-82, PROF BIOL SCI, KENT STATE UNIV, 82-, DIR SCH BIOMED SCI, 79-, ASSOC DEAN, GRAD COL, 87- *Mem:* AAAS; Am Physiol Soc; Soc Study Reproduction; Soc Exp Biol Med. *Res:* Control of gonadotrophin secretion and ovarian physiology with respect to abnormal ovarian responses; physiological influences on parturition; physiology of relaxin. *Mailing Add:* Dept of Biol Sci Kent State Univ Kent OH 44242

ADAMS, WILLIAM ALFRED, b Toronto, Ont, June 15, 41; m 64; c 2. GLACIOLOGY. *Educ:* McMaster Univ, BSc, 63; Univ Ottawa, PhD(chem), 68. *Prof Exp:* Fel, Div Appl Chem, Nat Res Coun Can, 67-69; res scientist, Hydrol Sci Div, Inland Waters Br, Environ Can, 69-71, res scientist, Water Qual Br, 71-75, res scientist, Glaciol Div, Inland Waters Directorate, 75-77; DEFENCE SCIENTIST, ENERGY CONVERSION DIV, DEFENCE RES ESTAB OTTAWA, DEPT NAT DEFENCE, 78- *Concurrent Pos:* Ed, Electrochem Soc, 75-78. *Mem:* Am Chem Soc; Chem Inst Can; Spectros Soc Can (pres, 75-76); Electrochem Soc. *Res:* Effects of pressure and temperature on the interactions of ions and polar molecules in solution using spectroscopic and electrochemical methods; high energy density battery; molten salt electrochemistry; impact of oil spills on Arctic aquatic ecosystems; deep well disposal. *Mailing Add:* 26 Foxmeadow Lane Nepean ON K2G 3W2 Can

ADAMS, WILLIAM EUGENE, b Mt Vernon, Ohio, Oct 18, 30; m 53; c 2. MECHANICAL ENGINEERING. *Educ:* Ohio State Univ, BME, 55. *Prof Exp:* Res engr, Ethyl Corp, 55-70, res adv, 70-74, dir auto res, 74-76, mgr Detroit Res Labs, 76-84, asst dir, Air Conserv Corp Staff, 84-86; RETIRED. *Honors & Awards:* Horning Mem Award, Soc Automotive Engrs, 64. *Mem:* Soc Automotive Engrs. *Res:* Fuel and engine research; fuel and lubricant additives. *Mailing Add:* Rte 3 Box 2403 McMillan MI 49853

ADAMS, WILLIAM HENRY, b Baltimore, Md, Dec 21, 33; m 85; c 2. THEORETICAL CHEMISTRY, QUANTUM CHEMISTRY. *Educ:* Johns Hopkins Univ, AB, 55; Univ Chicago, SM, 56, PhD(chem phys), 60. *Prof Exp:* NSF fel, Quantum Chem Group, Univ Uppsala, 60-62; asst prof chem, Pa State Univ, 62-66; from asst prof to assoc prof, 66-75, PROF CHEM, RUTGERS UNIV, 75- *Concurrent Pos:* Vis prof, Tech Univ, Munich, 70-71. *Mem:* Am Phys Soc; Am Chem Soc. *Res:* Quantum theory, particularly interactions between many electron systems. *Mailing Add:* Dept Chem Rutgers State Univ New Brunswick NJ 08903

ADAMS, WILLIAM LAWRESON, b Gay, Mich, July 9, 32; m 56; c 11. ELECTRICAL ENGINEERING. *Educ:* Mich Technol Univ, BS, 54; Mass Inst Technol, SM, 56; Purdue Univ, PhD, 62. *Prof Exp:* Instr elec eng, Mass Inst Technol, 54-56; proj engr, Gen Electronic Labs, Inc, Mass, 56-58; from instr to asst prof elec eng, Purdue Univ, 58-63; sr engr, 63-65, mgr electronic systs sect, New Prod Develop Dept, 65-66, mgr, Indust Systs Res & Develop Dept, 66-77, vpres eng, 77-80, sr vpres mfg & eng, 80-87, SR VPRES MEASUREMENT & ACTUATOR SYSTS, COMBUSTION ENG, INDUST NUCLEONICS CORP, 87- *Concurrent Pos:* Consult, Addressograph Multigraph, Ohio, 55-56 & Gen Electronic Labs, Inc, Mass, 58-63. *Mem:* Inst Elec & Electronics Engrs; sr mem Instrument Soc Am; Sigma Xi. *Res:* Electrical properties of materials; determination of moisture content in sheet materials; computerized process control systems; modeling and control of industrial processes. *Mailing Add:* 9600 Olentangy River Rd Powell OH 43065

ADAMS, WILLIAM MANSFIELD, b Kissimmee, Fla, Feb 19, 32; m 55, 76; c 5. GEOPHYSICS. *Educ:* Univ Chicago, BA, 51; Univ Calif, BA, 53; St Louis Univ, MS, 55, PhD, 57; Univ Santa Clara, MBA, 64. *Prof Exp:* Geophys trainee, Stanolind Oil & Gas Co, 53; computer, Western Geophys Co, 54; seismologist, Geotech Corp, 57-58; physicist, Lawrence Radiation Lab, Univ Calif, 59-62; pres, Planetary Sci, Inc, 62-64; prof geophysics, Univ Hawaii, 65-90; RES ASSOC, WESTERN WASH UNIV, 90- *Concurrent Pos:* Fulbright scholar, Italy, 56-57; dir tsunami res, Univ Hawaii, 67-72; vis fel, Coop Inst Trs Environ Sci, Univ Colo, 70-71; NATO grant, Geothermal Inst, Pisa, Italy, 73; vis prof gophys, Ind Univ, 75-76; sismol expert, Int Inst Seismol & Earthquake Eng, UNESCO, Tokyo, Japan, 71-72; oceanographer, Inst Physics Atmosphere, Nat Oceanic & Atmospheric Admin, Miami, Fla, 79-80; vis geophysicist, Desert Res Inst, Univ Nev, Las Vegas, 83; partic, Regional Crustal Stability & Natural Hazards, UNESCO, Beijing, China, 85 & Strong-motion Seismol, NATO, Ankara, Turkey, 85. *Mem:* Seismol Soc Am; Geol Soc Am; Soc Explor Geophys; Am Geophys Union; Tsunami Soc. *Res:* Direct exploration of the interior of the earth; indexing and retrieval of scientific literature; seismic holography; tsunamis; systems for alerting society to natural hazards; hydrogeophysics. *Mailing Add:* 1333 Lincoln St No 177 Bellingham WA 98226-6243

ADAMS, WILLIAM S, b Sodus, NY, May 28, 19; m 47; c 4. METABOLISM. *Educ:* Univ Rochester, MD, 43. *Prof Exp:* Intern, Strong Mem Hosp, Rochester, NY, 44; asst resident med, Univ Rochester, 44-45, instr, 47-48; resident, Vet Admin Ctr Hosp, Los Angeles, 48-49, admin sect chief, asst resident & physician full grade, 49-50; asst clin prof, 48-49, from asst prof to assoc prof, 49-59, vchmn dept, 67, PROF MED, SCH MED, UNIV CALIF, LOS ANGELES, 59- *Concurrent Pos:* Res fel, Univ Rochester, 46-47; consult, Atomic Bomb Casualty Comn. *Mem:* Asn Am Physicians; Am Col Physicians; Sigma Xi. *Res:* Metabolic aspects of malignant disease. *Mailing Add:* Dept Med Univ Calif Ctr Health Sci Los Angeles CA 90024-1736

ADAMS, WILLIAM SANDERS, b Williamsport, Pa, Jan 16, 34; m 60; c 3. ELECTRICAL ENGINEERING, MATHEMATICS. *Educ:* Pa State Univ, BS, 56, MS, 58, PhD(elec eng), 63. *Prof Exp:* From instr to assoc prof, 58-73, PROF ELEC ENG, PA STATE UNIV, 73-, DIR ENG COMPUT LAB, 70- *Mem:* Inst Elec & Electronics Engrs; Am Soc Eng Educ. *Res:* Computation, especially applied to biomedical simulation; non-linear analysis. *Mailing Add:* Rm 11 Elec Eng Bldg Pa State Univ University Park PA 16801

ADAMS, WILLIAM WELLS, b Redlands, Calif, July 23, 37; m 64; c 2. DIOPHANTINE APPROXIMATIONS, COMPUTATIONAL NUMBER THEORY. *Educ:* Univ Calif, Los Angeles, AB, 59; Columbia Univ, PhD(math), 64. *Prof Exp:* From instr to asst prof math, Univ Calif, Berkeley, 64-69; assoc prof, 69-71, PROF MATH, UNIV MD, COLLEGE PARK, 71- *Concurrent Pos:* NSF fel, Inst Advan Study, Princeton, NJ, 66-67; vis prof, Technion, Israel, 83-84, Imperial Col, London, 75; prog dir, Nat Sci Found, 86-88; ed, Proc Am Math Soc, 88- *Mem:* Am Math Soc; Math Asn Am. *Res:* Asympotitic theory of diophantine approximations; primality testing; Groebner bases. *Mailing Add:* Math Dept Univ MD College Park MD 20742-4015

ADAMSKI, ROBERT J, b Newark, NJ, Mar 26, 35; m 56; c 3. MEDICINAL CHEMISTRY. *Educ:* Univ Wis-Madison, BS, 60; Univ Iowa, PhD(pharmaceut chem), 64. *Prof Exp:* Fel, Ciba Pharmaceut Co, NJ, 64-66; res assoc, Inst Microbiol, Rutgers Univ, 66-67; sr chemist, Carter-Wallace, Inc, 67-68; MGR OPHTHAL NEW PROD RES, ALCON LABS, INC, 68-, DIR RES, 80-, VPRES RES, 85- *Mem:* Am Chem Soc; Am Pharmaceut Asn; NY Acad Sci; Am Inst Chemists; Am Mgt Asn; Licensing Execs Soc. *Res:* Chemical modification of known and structural elucidation of unknown antimicrobial materials; design and synthesis of agents having potential medicinal value as antiglaucomics, anti-inflammatories, antibacterial, anticataract agents. *Mailing Add:* Alcon Labs Inc 6201 S Freeway Ft Worth TX 76134-2099

ADAMSON, ALBERT S, JR, b Cincinnati, Ohio, Sept 30, 47; m 69. PERFUMERY, ESSENTIAL OILS. *Prof Exp:* Apprentice perfumer, Procter & Gamble Co, 67-75; perfumer, 75-77, res mgr, 77-86, RES DIR PERFUME, DIAL CORP, 86- *Concurrent Pos:* Lectr, Int Perfumery Congress, 84, Am Chem Soc, Am Inst Chem Engrs, & dept chem & bioeng, Ariz State Univ, 86, Am Oil Chem Soc & Am Soc Perfumers Symp, 89. *Mem:* Am Soc Perfumers. *Res:* Creation of fragrances, perfume quality control, odor quality assurance and perfume cost control. *Mailing Add:* Dial Corp 15101 N Scottsdale Rd Scottsdale AZ 85254-2199

ADAMSON, ARTHUR WILSON, b Shanghai, China, Aug 15, 19; m 42; c 3. PHYSICAL CHEMISTRY. *Educ:* Univ Calif, BS, 40; Univ Chicago, PhD(phys chem), 44. *Prof Exp:* Res asst, Metall Lab, Univ Chicago, 42-44; res assoc, Plutonium Proj, Tenn, 44-46; from asst prof to prof chem, 46-89, chmn dept, 72-75, EMER PROF CHEM, UNIV SOUTHERN CALIF, 89- *Concurrent Pos:* NSF sr fel, 62-63; Unilever prof, Bristol Univ, 65-66; Australian Acad Sci sr fel, 69; Foster lectr, State Univ NY Buffalo, 70; Venable lectr, Univ NC, 75; Vis prof, Univ Queensland, Australia, 80; Bikerman lectr, Case Western Univ, 82; Reilly lectr, Notre Dame Univ, 84. *Honors & Awards:* Tolman Award, Am Chem Soc, 67, Kendall Award in Surface Chem, 79; Res Assocs Award, Univ Southern Calif, 71; Honor Scroll Award, Am Inst Chemists, 76; Alexander von Humboldt Sr Scientist Award, 77; Distinguished Serv Award inorg chem, Am Chem Soc, 82; Chem Educ Award, Am Chem Soc, 84. *Mem:* AAAS; fel Am Inst Chemists; Am Chem Soc; Sigma Xi. *Res:* Coordination chemistry; photochemistry of coordination compounds; surface chemistry. *Mailing Add:* Dept of Chem Univ of Southern Calif Los Angeles CA 90089

ADAMSON, IAN YOUNG RADCLIFFE, b Wishaw, Scotland, Mar 16, 41; m 67; c 2. EXPERIMENTAL PATHOLOGY. *Educ:* Univ Glasgow, BSc, 63, PhD(chem), 66. *Prof Exp:* From lectr to assoc prof path, 67-79, PROF PATH, UNIV MANITOBA, 79-, MEM SCI STAFF, HEALTH SCI CTR, 78- *Honors & Awards:* Wild Leitz Jr Pathologist Award, 76. *Mem:* Am Asn Pathologists; Am Thoracic Soc. *Res:* Experimentally induced lung injury and repair, with emphasis on changes in ultrastructure, cell kinetics and the factors involved in producing pulmonary fibrosis; lung development. *Mailing Add:* Dept Path Univ Man 770 Bannatyne Ave Winnipeg MB R3E 0W3 Can

ADAMSON, JEROME EUGENE, b Columbus, Ohio, Oct 2, 27; m 59; c 3. PLASTIC & RECONSTRUCTIVE SURGERY. *Educ:* WVa Univ, BA, 50; Duke Univ, MD, 54. *Prof Exp:* Resident gen surg, Duke Univ Med Ctr, 54-57, plastic surg, 57-60; chief, Dept Plastic Surg, Med Ctr Hosps, 73-75; PROF PLASTIC SURG, HAND SURG, EASTERN VA MED SCH, 74- *Concurrent Pos:* Consult, Med Ctr Hosps, Naval Regional Med Ctr & Vet Admin Hosp; assoc ed, J Plastic & Reconstruct Surg, 80-85; pvt practice, Plastic Surg Specialists, Inc, 60- *Mem:* Am Soc Plastic & Reconstruct Surgeons (pres, 80-81); Am Col Surgeons; Am Asn Plastic Surgeons; Am Soc Surg Hand; Am Soc Maxillofacial Surgeons. *Res:* Peripheral nerve healing and clinical improvement of upper extremity rehabilitative methods. *Mailing Add:* Dept Plastic Surg 900 Wainwright Bldg 229 W Bute St Norfolk VA 23501

ADAMSON, JOHN DOUGLAS, b St Boniface, Man, Mar 1, 32; m 57; c 6. PSYCHIATRY, PSYCHOPHYSIOLOGY. *Educ:* Univ Man, MD, 56; Royal Col Physicians & Surgeons Can, cert psychiat, 61. *Prof Exp:* Asst psychiat, Sch Med & Dent, Univ Rochester, 59-60, instr, 60-61; lectr, 61-63, asst prof, 63-65, assoc prof, 65-86, PROF PSYCHIAT, FAC MED, UNIV MAN, 86- *Concurrent Pos:* Mem subcomt rehab, Dept Nat Health & Welfare Can, 63-65 & mem subcomt ment health res, 65-68; coordr psychiat res, Dept Health & Social Develop, Govt Prov Man, 65-77; dir psychiat clin invest, Health Sci Gen Ctr, 65-77; chmn prof & tech adv comt, Alcoholism Found Man, 65-75; chmn, Man Ment Health Res Found, 74-84, vchmn & secy, 84-86, mem exec comt, 86-88. *Mem:* Can Med Asn; Can Psychiat Asn. *Res:* Biology and psychopathology of schizophrenia; clinical and biological studies on alcoholism; psychophysiology of emotion; sleep. *Mailing Add:* Health Sci Ctr 820 Sherbrook St Winnipeg MB R3A 1R9 Can

ADAMSON, LUCILE FRANCES, b Chetopa, Kans, Nov 10, 26. NUTRITIONAL BIOCHEMISTRY, ENVIRONMENTAL SCIENCES. *Educ:* Kans State Col, BS, 48; Univ Iowa, MS, 50; Univ Calif, PhD(biochem), 56. *Prof Exp:* Asst prof nutritionist, Univ Hawaii, 56-60; asst prof pediat & biochem, Univ Mo, 60-64; chief biochemist, Thorndike Mem Lab, Harvard Med Sch, 64-70; res fel biochem, Monash Univ, Australia, 70-72; staff scientist, Environ Defense Fund, Washington, DC, 72-74; prof & chmn prog macroenviron studies, Howard Univ, 74-80; consult, Biomed Sci, 81-87; COMP PROGRAMMER, ERL SOFTWARE, INC, 87- *Mem:* Am Chem Soc; Am Inst Nutrit. *Res:* Health and nutritional aspects of environmental pollution. *Mailing Add:* PO Box 698 Sequim WA 98382

ADAMSON, RICHARD H, b Council Bluffs, Iowa, Aug 9, 37; m 63; c 2. ONCOLOGY, PHARMACOLOGY. *Educ:* Drake Univ, BA, 57; Univ Iowa, MS, 59, PhD(pharmacol), 61; George Washington Univ, MA, 68. *Prof Exp:* Asst, Col Med, Univ Iowa, 57-58; sr investr, Lab Chem Pharmacol, 63-69, head, Pharmacol & Exp Therapeut Sect, 69-73, chief, Lab Chem Pharmacol, 73-81, DIR, DIV CANCER CAUSE & PREV, NAT CANCER INST, 80- *Concurrent Pos:* Lectr, Col Med, George Washington Univ, 63-70; Fulbright vis scientist, Dept Biochem, St Mary's Hosp Med Sch, Univ London, 65-66; sr policy analyst, Off Sci & Technol Policy, Exec Off Pres, 79-80. *Honors & Awards:* Superior Serv Award, USPHS, 76. *Mem:* AAAS; Am Asn Cancer Res; Am Soc Pharmacol & Exp Therapeut; NY Acad Sci; Soc Toxicol. *Res:* Carcinogenesis, drug metabolism; cancer chemotherapy; antibiotics; toxicology; marine biology; comparative pharmacology; science and government. *Mailing Add:* Div Cancer Prev & Control Nat Cancer Inst Bldg 31 Rm 11A03 Bethesda MD 20892

ADAMSON, S LEE, b Ajax, Ont, Mar 23, 54. EXPERIMENTAL BIOLOGY. *Educ:* Univ Guelph, BS, 77; Univ Western Ont, MS, 80, PhD(biophys), 84. *Prof Exp:* Postdoctoral fel, Med Res Coun, Univ Western Ont, London, 84-85; SCIENTIST, SAMUEL LUNENFELD RES INST, MT SINAI HOSP, TORONTO, 85-; ASST PROF OBSTET & GYNEC, PEDIAT & PHYSIOL, UNIV TORONTO, 85- *Concurrent Pos:* Lab instr biophys, Univ Western Ont, 77-79, physics, 79-82; lectr physiol, Univ Toronto, 89-, path, 89-; res scholar, Heart & Stroke Found, Ont, 90- *Mem:* Soc Gynec Invest; Am Physiol Soc; Biophys Soc Can. *Mailing Add:* Res Inst Mt Sinai Hosp 600 University Ave Rm 138-P Toronto ON M5G 1X5

ADAMSON, THOMAS C(HARLES), JR, b Cicero, Ill, Mar 24, 24; m 49; c 3. AEROSPACE ENGINEERING. *Educ:* Purdue Univ, BS, 49; Calif Inst Technol, MS, 50, PhD(aeronaut), 54. *Prof Exp:* Res engr, Guggenheim Aeronaut Lab, Calif Inst Technol, 49-50 & Jet Propulsion Lab, 52-54; assoc res engr & lectr aeronaut eng, Res Inst, 54-56, from asst prof to assoc prof, 56-61, PROF AEROSPACE ENG, UNIV MICH, ANN ARBOR, 61-, CHMN, 83- *Mem:* Fel Am Inst Aeronaut & Astronaut; Combustion Inst; Sigma Xi. *Res:* Fluid Mechanics; combustion; gas dynamics. *Mailing Add:* Dept Aerospace Eng Univ Mich Ann Arbor MI 48109-2140

ADAMSONS, KARLIS, JR, b Riga, Latvia, Oct 30, 26; US citizen; m 68; c 2. OBSTETRICS & GYNECOLOGY. *Educ:* Univ Gottingen, MD, 52; Columbia Univ, PhD(pharmacol), 56. *Prof Exp:* Intern, St Vincents Hosp, NY, 52-53; asst resident obstet & gynec, Columbia-Presby Med Ctr, 56-58, from resident to chief resident, 59-61, asst prof, 61-65, assoc obstetrician & gynecologist, 65-69; from assoc prof to prof, Col Physicians & Surgeons, Columbia Univ, 65-70; prof obstet, gynec & pharmacol, Mt Sinai Sch Med, City Univ New York, 70-75; Obstetrician & gynecologist-in-chief, Women & Infants Hosp RI, 75-79; PROF OBSTET & GYNEC & CHMN DEPT, BROWN UNIV, 75-; PROF & CHMN, DEPT OB-GYN & DIR, UNIV HOSP, UNIV PR, 79- *Concurrent Pos:* Macy fel, Nuffield Inst Med Res, Oxford Univ, Eng, 58-59; Wellcome sr res fel, 61-; vis scientist, NIH, 62, consult, Nat Inst Child Health & Human Develop, 66-71; vis prof obstet & gynec, Univ PR, 63; consult, Coun Drugs, AMA, 63; mem adv coun, Food & Drug Admin, 65-70; Fulbright prof, Univ Uruguay, 69; consult Ob-Gyn, Beth Israel Hosp, NY, 80- *Honors & Awards:* Outstanding Residency Prog Dir, Am Col Obstetricians & Gynecologists, 84. *Mem:* Soc Pediat Res; Am Col Obstetricians & Gynecologists; Soc Gynec Invest. *Res:* Perinatal physiology; biochemistry and physiology of neurohypophysical hormones. *Mailing Add:* Dept Obstet/Gynec Univ PR Med Sci Box 5067 San Juan PR 00936

ADASKAVEG, JAMES ELLIOTT, b Waterbury, Conn, Feb 27, 60; m 90. MYCOLOGY, TAXONOMY OF FUNGI. *Educ:* Univ Conn, BS, 82; Univ Ariz, MS, 84, PhD(plant path), 86. *Prof Exp:* Postdoctoral researcher, Dept Plant Path, 87-90, lectr mycol, Dept Bot, 90-91, RES PLANT PATHOLOGIST, UNIV CALIF, DAVIS, 91- *Concurrent Pos:* Mem, Mycol Comt, Am Phytopath Soc, 89-92. *Mem:* Am Phytopath Soc; Brit Mycol Soc; Mycol Soc Am; Sigma Xi. *Res:* Biology, ecology, and taxonomy of plant pathogenic fungi of forest, fruit, and nut trees; identification, ultrastructure, and classical genetics of fungi; epidemiology of foliar diseases; degradation processes of wood. *Mailing Add:* Dept Plant Path Univ Calif Davis CA 95616

ADASKIN, ELEANOR JEAN, b Manitoba, Can, July 27, 37; m 60; c 2. NURSING RESEARCH DEVELOPMENT, SOCIAL PSYCHOLOGY OF HEALTH. *Educ:* Univ Saska, BScN, 60, BA, 70; Univ Wash, MA, 76; Univ Tex, Austin, PhD(nursing), 87. *Prof Exp:* Teacher psychiat nursing, 71-73, counr, Misericordia Sch Nursing, Can, 73-74; teaching asst, psycho-physiol nursing, Univ Wash, 74-75; clin nursing head, 76-78, clin nurse specialist, 78-88, DIR NURSING RES, ST BONIFACE GEN HOSP, CAN, 88- *Concurrent Pos:* Mem, Task Force Educ Requirements, Can Nurses Asn, 78-79; asst prof psychiat, nursing, Univ Man, 78-79 & clin demonstrator family therap, 80-88; pvt pract counr, 79-; teaching asst, family therap, Univ Tex, Austin, 84-85; prin investr, Clin Nursing Res Prog, 86-; lectr, Univ Winnipeg, 88- *Honors & Awards:* Award Prof Nursing Res, Manitoba Asn Registered Nurses, 90. *Mem:* Sigma Xi; Am Asn Marriage & Family Therap; Can Nurses Found; Can Nursing Res Group. *Res:* Family health; family strengths which may assist members to remain well under stress; psychosocial and physical aspects of stress-resistance. *Mailing Add:* St Boniface Gen Hosp Res Ctr 351 Tache Ave Winnipeg MB R2H 2A6 Can

ADAWI, IBRAHIM (HASAN), b Palestine, Apr 18, 30; US citizen; m 56; c 4. THEORETICAL SOLID STATE PHYSICS. *Educ:* Wash Univ, BS, 53; Cornell Univ, PhD(eng physics), 57. *Prof Exp:* Teacher math & physics, Terra Santa Sch, Syria, 49-51; physicist, Radio Corp Am Labs, 56-60; res consult,

Battelle Mem Inst, 60-68; PROF PHYSICS, UNIV MO-ROLLA, 68- *Concurrent Pos:* Adj prof elec eng, Ohio State Univ, 65-68; vis prof, Univ Hamburg, 77-78; Fulbright Lectr, Rabat, 82; vis scientist & res leader, Int Ctr Theoret Physics, Trieste, Italy, 82, 83 & 85; vis prof, Sch Math & Physics, Univ East Anglia, Norwich, UK, 82. *Mem:* Am Phys Soc. *Res:* Penetration of charged particles and radiation in matter; transport theory; magnetic monopoles; solid state theory. *Mailing Add:* Dept of Physics Univ of Mo Rolla MO 65401

ADCOCK, JAMES LUTHER, b Crane, Tex, June 26, 43; m 67; c 3. SYNTHETIC FLUORINE CHEMISTRY, AEROSOL DIRECT FLUORINATION. *Educ:* Univ Tex, Austin, BS, 66, PhD(chem), 71. *Prof Exp:* Res assoc fluorine chem, Mass Inst Technol, 71-74; ASSOC PROF CHEM, UNIV TENN, KNOXVILLE, 74- *Concurrent Pos:* Chmn, div fluorine chem, Am Chem Soc, 89. *Mem:* Am Chem Soc; Sigma Xi. *Res:* Synthetic organofluorine and organometallic chemistry; elemental fluorine as a synthetic reagent; aerosol direct fluorination methodology; preparation of fluorinated monomers and polymers; physical-organic chemistry of elemental fluorine. *Mailing Add:* Dept Chem Univ Tenn Knoxville TN 37996-1600

ADCOCK, LOUIS HENRY, b Durham, NC, Mar 28, 29; m 60; c 3. ANALYTICAL CHEMISTRY. *Educ:* Duke Univ, BS, 51, MA, 53; La State Univ, PhD, 70. *Prof Exp:* From asst prof to assoc prof phys sci & chem, 56-77, PROF CHEM, UNIV NC, WILMINGTON, 77- *Concurrent Pos:* Pres, NC Inst Chemists. *Mem:* AAAS; Am Chem Soc; fel Am Inst Chemists. *Res:* Analysis of trace elements; pesticides in marine environment; lecture demonstration. *Mailing Add:* Dept Chem Univ NC Wilmington NC 28401

ADCOCK, WILLIS ALFRED, b St Johns, Que, Nov 25, 22; nat US; m 43. SEMICONDUCTORS, ELECTRONICS. *Educ:* Hobart Col, BS, 43; Brown Univ, PhD(chem), 48. *Hon Degrees:* PhD, Hobart Col, 89. *Prof Exp:* Chemist, Stanolind Oil & Gas Co, 48-53; res group supvr, Tex Instruments, 53-55, dir device res, 55-58, mgr res & eng dept, 58-61, mgr qual assurance dept, 60-61, mgr integrated circuit dept, 61-64; tech dir, Sperry Semiconductor Div, Sperry Rand Corp, 64-65; mgr, Advan Planning Dept, Semiconductor Div, Texas Instruments Inc, 65-67, vpres tech develop, Components Group, 67-69, vpres strategic planning, Corp Develop, 69-72, asst vpres & tech dir consumer prod, 72-78, asst vpres corp res develop & eng, 78-82, vpres corp staff, 82-86, prin fel, 78-86; PROF, UNIV TEX, AUSTIN, 87- *Mem:* Nat Acad Eng; Am Chem Soc; fel Inst Elec & Electronic Engrs; fel AAAS. *Res:* Semiconductors. *Mailing Add:* 3414 Mt Bonnell Dr Austin TX 78731

ADDA, LIONEL PAUL, b Allentown, Pa, Jan 17, 22. SOLID STATE SCIENCE. *Educ:* Lehigh Univ, BS, 49, MS, 50, PhD(physics), 62. *Prof Exp:* Engr fed tel telecommun labs, Int Tel & Tel Co, 50-54; distinguished mem tech staff, Bell Labs, 54-87; RETIRED. *Mem:* Electrochem Soc. *Res:* Measurement of electrical and infrared properties; crystalline perfection and impurities in semiconductors; role of imperfections on device behavior; insulator properties and metal-insulator-semiconductor systems; process development of Very Large Scale Integration circuits. *Mailing Add:* 804 Woodbury Lane Whitehall PA 18052-7852

ADDAMIANO, ARRIGO, b Molfetta, Italy, Feb 16, 23; US citizen; m 52; c 4. INORGANIC CHEMISTRY, PHYSICAL CHEMISTRY. *Educ:* Univ Rome, PhD(chem), 44, PhD(physics), 46. *Prof Exp:* Asst prof gen & inorg chem, Univ Rome, 44-48; Ital Bd Educ scholar res x-ray crystallog, Res Assoc Dept X-ray, Oxford Univ, 48-50; asst prof gen & inorg chem, Univ Rome, 51-53; Fulbright fel & res assoc x-ray & crystal structure lab, Pa State Univ, 53-54; phys chemist, Gen Elec Co, 54-69; res chemist, US Naval Res Lab, 69-85; RETIRED. *Concurrent Pos:* Nat Res Coun fel, Rome, 51-53. *Mem:* Am Chem Soc; Am Crystallog Asn; Electrochem Soc. *Res:* Crystal structure determination; crystal physics; single crystal growth; inorganic syntheses; phosphors; semiconductors. *Mailing Add:* 4222 Robertson Blvd Alexandria VA 22309

ADDANKI, SOMASUNDARAM, b Lakkavaram, India, Mar 7, 32; m 54; c 3. CLINICAL CHEMISTRY, BIOCHEMISTRY. *Educ:* Madras Univ, BVS, 57; Ohio State Univ, MSc, 62, PhD(nutrit biochem), 64; Am Bd Clin Chem, cert, 70. *Prof Exp:* From instr to asst prof, 64-70, ASSOC PROF PEDIAT & PHYSIOL CHEM, OHIO STATE UNIV, 70-; DIR LABS, CLIN STUDY CTR, COLUMBUS CHILDREN'S HOSP, 66- *Concurrent Pos:* Consult, Pathologists, Inc, Licking Mem Hosp, Newark, Ohio, 73- *Mem:* Fel Am Asn Clin Chem; Am Fedn Clin Res; Brit Biochem Soc. *Res:* Metabolic pediatric chemistry; radio-immuno assays-ketotic hypoglycemia; catecholamine chemistry and metabolism; endocrinology. *Mailing Add:* PhD Consult 15109 Nashua Lane Bowie MD 20716

ADDICOTT, FREDRICK TAYLOR, b Oakland, Calif, Nov 16, 12; m 35; c 4. ABSCISSION, ABSCISIC ACID. *Educ:* Stanford Univ, AB, 34; Calif Inst Technol, PhD(plant physiol), 39. *Prof Exp:* Asst, Stanford Univ, 34-37; teaching fel, Calif Inst Technol, 37-39; from instr to asst prof bot, Santa Barbara State Col, 39-46; assoc physiologist, Spec Guayule Res Proj, USDA, Calif, 43-44; from asst prof to prof bot, Univ Calif, Los Angeles, 46-61; prof agron, 61-72, prof bot, 72-77, EMER PROF BOT, UNIV CALIF, DAVIS, 77- *Concurrent Pos:* Sigma Xi grant-in-aid, 40; Fulbright res scholar, Victoria Univ, Wellington, NZ, 57 & Royal Botanic Gardens, Kew, UK, 76; vis prof, Univ Adelaide, 66 & Univ Natal, 70. *Honors & Awards:* Charles Reid Barnes Life Mem Award, Am Soc Plant Physiol, 90. *Mem:* Australian Soc Plant Physiol; Am Soc Plant Physiol; Bot Soc Am; Int Plant Growth Substances Asn; SAfr Asn Bot; fel AAAS. *Res:* Physiology, ecology, cytology and morphology of abscission in plants; physiological and morphological responses to abscisic acid. *Mailing Add:* Dept Bot Univ Calif Davis CA 95616-8537

ADDICOTT, JOHN FREDRICK, b Santa Barbara, Calif, Dec 28, 44. ECOLOGY. *Educ:* Univ Calif, Davis, AB, 67; Univ Mich, MSc, 68, PhD(zool), 72. *Prof Exp:* Res asst prof biol, Univ Utah, 72-73; from asst prof to assoc prof, 73-86, PROF ZOOL, UNIV ALTA, 86- *Mem:* Ecol Soc Am; Soc Study Evolution. *Res:* Competition in patchy environments; effects of predation on prey community structure; mutualism; pollination-seed predation mutualism; applied ant mutualism. *Mailing Add:* Dept Zool Univ Alberta Edmonton AB T6G 2E9 Can

ADDICOTT, WARREN O, b Fresno, Calif, Feb 17, 30; m 55, 74; c 2. STRATIGRAPHY. *Educ:* Pomona Col, AB, 51; Stanford Univ, MA, 52; Univ Calif, PhD(paleont), 56. *Prof Exp:* Asst, Univ Calif, 52-54; geologist, Gen Petrol Corp, 54-62, explor coordr, 58-60, dist geologist, 60-62; geologist, 62-70, res geologist, 70-82, supvry geologist, 82-86, RES GEOLOGIST, US GEOL SURV, 86- *Concurrent Pos:* Consult prof, Stanford Univ, 71-82; dep chmn, Circum-Pacific Map Proj, 79-82, gen chmn, 82-86, proj adv, 86-; dir, Cicum -Pacific Coun Energy & Mineral Resources, 82-86; adj prof, S Ore State Col, 90- *Mem:* AAAS; Paleont Soc (secy, 71-77, pres, 80-81); Geol Soc Am; Am Asn Petrol Geol; Paleont Res Inst. *Res:* Cenozoic marine stratigraphy; molluscan zoogeography; paleoclimatology; molluscan taxonomy. *Mailing Add:* MS 952 Off Int Geol US Geol Surv 345 Middlefield Rd Menlo Park CA 94025

ADDINK, SYLVAN, b Sheldon, Iowa, Dec 23, 41; m 62; c 2. WEED SCIENCE. *Educ:* Ore State Univ, BS, 69; NDak State Univ, MS, 72, PhD(agron), 73. *Prof Exp:* Plant sci rep, Eli Lilly & Co, 73-74, regional res rep, 74-78; OWNER, ASC SPRAYING & CONSULT, 78- *Mem:* Coun Agr Sci & Technol. *Res:* Control of pests that are harmful to crops, animals, mankind and his environment. *Mailing Add:* PO Box 1457 Iowa City IA 52244

ADDIS, PAUL BRADLEY, b Honolulu, Hawaii, Feb 13, 41; m 67; c 2. NUTRITION. *Educ:* Wash State Univ, BS, 62; Purdue Univ, PhD(food sci), 67. *Prof Exp:* Fel, Max Planck Soc, Ger, 67; from asst prof to assoc prof food sci, 67-75, PROF FOOD SCI & NUTRIT, UNIV MINN, ST PAUL, 75- *Concurrent Pos:* Fulbright travel grant, 67; consult, Totino's Finer Foods, Inc & Brother Redi-Roast Prod, Inc, Minn, 70-72; researcher, Muscle Biol Lab, Univ Calif, Davis, 71; researcher chem & med depts, Univ Calif, San Diego, 76; res, Biochem Dept, Univ Wash, Seattle, 79-80. *Mem:* AAAS; Am Chem Soc; Inst Food Technologists; Sigma Xi; Am Oil Chem Soc; Int Asn Milk, Food & Environ Sanitarians. *Res:* Lipid oxidation in foods; atherogenicity of cholesterol oxides and fatty acid hydroperoxides; human serum lipoproteins; omega 3 fatty acids in Lake Superior fishes; incorporation and stabilization of omega 3 fatty acids in foods. *Mailing Add:* Dept Food Sci & Nutrit Univ Minn St Paul MN 55108

ADDISON, ANTHONY WILLIAM, b Sydney, NSW, June 24, 46; m 72; c 2. BIOINORGANIC CHEMISTRY. *Educ:* Univ New South Wales, BSc, 68; Univ Kent, PhD(chem), 71. *Prof Exp:* Res asst chem, Univ Kent, 68-70; res fel, Northwestern Univ, 70-72; asst prof chem, Univ BC, 72-78; ASSOC PROF CHEM, DREXEL UNIV, 78- *Concurrent Pos:* Chmn, Philadelphia Sect, Am Chem Soc, 89. *Honors & Awards:* J von Geuns Fonds Lectr, Univ Amsterdam, 85; C & M Lindback Award, 87. *Mem:* Royal Soc Chem; Am Chem Soc; Chem Inst Can; Sigma Xi. *Res:* Chemistry of metalloproteins and of synthetic models for their properties particularly oxygen transport and redox chemistry. *Mailing Add:* Dept of Chem Drexel Univ Philadelphia PA 19104

ADDISON, JOHN WEST, JR, b Washington, DC, Apr 2, 30; m 55; c 4. MATHEMATICAL LOGIC. *Educ:* Princeton Univ, AB, 51; Univ Wis, MS, 53, PhD(math), 55. *Prof Exp:* Instr math, Univ Mich, 54-56; asst prof math, Univ Mich, 57-62; assoc prof, 62-68, chmn group in logic & methodology of sci, 63-65 & 81-85, chmn dept, 68-77, PROF MATH, UNIV CALIF, BERKELEY, 68-, CHMN DEPT, 85- *Concurrent Pos:* NSF fel, 56-57; mem, Inst Advan Study, Princeton, 56-57 & Math Inst, Polish Acad Sci, 57; vis lectr, Univ Calif, Berkeley, 59-60; vis scholar, Math Inst, Univ Oxford, 72-73 & vis fel, Wolfson Col, 79. *Mem:* Fel AAAS; Am Math Soc; Math Asn Am; Asn Symbolic Logic. *Res:* Theory of definability, including recursive function theory and descriptive set theory; foundations of mathematics; theory of models; axiomatic set theory. *Mailing Add:* Dept Math Univ Calif 2120 Oxford St Berkeley CA 94720

ADDISON, RICHARD FREDERICK, b Belfast, Northern Ireland, May 7, 41; Can citizen; m 65; c 2. BIOCHEMISTRY, ANALYTICAL CHEMISTRY. *Educ:* Queens Univ, Belfast, BSc, 63, PhD(agr), 66. *Prof Exp:* Res scientist lipids, Fisheries Res Bd Can, 66-71; res scientist pollution, 71-76, HEAD ENVIRON DIV, MARINE ECOL LAB, CAN DEPT ENVIRON, 71- *Concurrent Pos:* Chmn environ div, Chem Inst Can, 83-84. *Mem:* Royal Inst Chem, UK; Chem Inst Can. *Res:* Distribution, metabolism and effects of marine environmental contaminants. *Mailing Add:* 6253 Oakland Rd Dartmouth NS B2Y 3Y1 Can

ADDISS, RICHARD ROBERT, JR, b Jersey City, NJ, May 13, 29; m 52; c 1. SOLID STATE PHYSICS, SOLAR PHOTOVOLTAIC ENERGY SYSTEMS. *Educ:* Rensselaer Polytech Inst, 51; Cornell Univ, MS, 56, PhD(physics), 58. *Prof Exp:* Asst mech, heat, sound, elec & magnetism, Cornell Univ, 51-56, asst oxidation metals, 56-58; mem tech staff evaporated films, Radio Corp Am Labs, 58-65; mem sci staff photoelectronic mat, Itek Corp, 65-75; mgr res & develop, Solar Power Corp, 75-77, tech dir, 78-84; consult, 84-89; RETIRED. *Concurrent Pos:* Mem steering comt, Int Conf Thin Films, Mass, 69; tech adv, US Nat Comt, Int Electrotech Comn, 84- *Mem:* Am Phys Soc; Inst Elec & Electronics Engrs; Sigma Xi. *Res:* Photoelectronic processes in crystals; properties of solids relating to defect structure; structure and properties of surfaces and thin film; photovoltaic devices. *Mailing Add:* 6 Dewey Rd Bedford MA 01730-1212

ADDOR, ROGER WILLIAMS, b New Rochelle, NY, Apr 30, 26; m 50; c 2. HETEROCYCLIC CHEMISTRY, ORGANO-SULFUR CHEMISTRY. *Educ:* Univ Maine, BS, 49, MS, 50; Ohio State Univ, PhD(chem), 54. *Prof Exp:* Res chemist, 54-61, sr res chemist, 61-68, GROUP LEADER, AM CYANAMID CO, 68- *Mem:* Sigma Xi; Am Chem Soc; AAAS; Soc Heterocyclic Chem. *Res:* Novel and better pesticides. *Mailing Add:* 97 Woosamonsa Rd Pennington NJ 08534

ADDUCI, JERRY M, b Rochester, NY, June 5, 34; m 58; c 2. ORGANIC CHEMISTRY, POLYMER CHEMISTRY. *Educ:* Univ Rochester, BS, 57; Univ Pa, PhD(org chem), 63. *Prof Exp:* Lab asst chem, Eastman Kodak Co, 52-53; instr math, Rochester Inst Technol, 58; lab technician, Strong Mem Hosp, Rochester, 58-59; res chemist, Fabrics & Finishes Dept, E I du Pont de Nemours & Co, Del, 63-65; res assoc & fel, Univ Md, 65-66, instr chem, 66; from asst prof to assoc prof, 66-84, PROF CHEM, ROCHESTER INST TECHNOL, 84- *Concurrent Pos:* Guest prof, Inst Kemiindustri, Tech Univ Denmark, Lyngby, 79-80. *Mem:* Am Chem Soc. *Res:* Reactions mechanisms; polymer synthesis. *Mailing Add:* Dept Chem Rochester Inst Tech Rochester NY 14623

ADDY, ALVA LEROY, b Dallas, SDak, Mar 29, 36; m 58. MECHANICAL ENGINEERING, FLUID DYNAMICS. *Educ:* SDak Sch Mines & Technol, BS, 58; Univ Cincinnati, MS, 60; Univ Ill, PhD(mech eng), 63. *Prof Exp:* Engr, Gen Elec Co, 58-60; from asst prof to assoc prof, 63-73, assoc head res & grad study, 80-86, PROF MECH ENG, UNIV ILL, URBANA-CHAMPAIGN, 73-, HEAD DEPT, MECH & INDUST ENG, 87- *Concurrent Pos:* Consult, Aerodyn Br, US Army Missile Command, 64- & various aerospace firms; vis res prof, US Army, 73. *Honors & Awards:* Educ Ralph Coates Roe Award, Am Soc Eng, 90. *Mem:* Assoc fel Am Inst Aeronaut & Astronaut; fel Am Soc Mech Engrs; Am Soc Eng Educ; Sigma Xi. *Res:* Aerodynamics and high velocity compressible flows, including base drag, turbulent mixing, ejector and propulsive nozzle design and performance problems; high energy lasers. *Mailing Add:* 1706 Golfview Dr Urbana IL 61801

ADDY, JOHN KEITH, b Sheffield, Eng, June 30, 37; m 63; c 2. PHYSICAL ORGANIC CHEMISTRY. *Educ:* Univ London, BSc, 58; Univ Southampton, PhD(phys org chem), 62. *Prof Exp:* Res assoc phys chem, Univ Ore, 61-62; vis lectr chem, Northeast Essex Tech Col, Eng, 62-63; lectr phys chem, John Dalton Col, 63-66; from asst prof to assoc prof, 66-74, chmn, Chem Dept, 85-89, PROF CHEM, WAGNER COL, 74-, CHMN, PHYS SCI, 90- *Mem:* The Chem Soc; fel Royal Soc Chem; Am Asn Univ Professors; Sigma Xi; Am Chem Soc. *Res:* Reaction kinetics, especially mechanisms of epoxide ring fission and allylic rearrangements; plasticization and structure of polymers, especially nylon. *Mailing Add:* Dept Chem Wagner Col Staten Island NY 10301

ADDY, TRALANCE OBUAMA, ü Kumasi, Ghana, Aug 24, 44; US citizen; m 79; c 3. INFECTION CONTROL TECHNOLOGY, MEDICAL INSTRUMENT STERILILZATION. *Educ:* Swarthmore Col, BA(Chem) & BS(mech eng) 69; Univ Mass, Amherst, MSME, 73, PhD(biomechanical eng), 74. *Prof Exp:* Teaching asst mech, Univ Mass, Amherst, 69-71, res asst food eng, 71-73; sr res engineer, Scott Paper Co, 73-76, res scientist, 76-79, prog leader, 79-80; dir appl res, Surgikos Inc, 80-85, dir technol ventures, 85-88; vpres res & gen mgr, Surgikos Inc, 88-89; vpres, 89-90, VPRES & GEN MGR, ASP DIV, JOHNSON & JOHNSON MED, 90- *Mem:* ASME; AAAS; Sigma Xi. *Res:* Activities in the area of low temperature gas plasma and applications to surface modification and sterilization; also fibrous structures for use as microbial barriers; antimicrobial technology. *Mailing Add:* 1904 Rockcliff Ct Arlington TX 76012

ADEL, ARTHUR, b Brooklyn, NY, Nov 22, 08; m 35. ASTROPHYSICS. *Educ:* Univ Mich, AB, 31, PhD(physics), 33. *Prof Exp:* Res assoc, Lowell Observ, 33-35; fel, Johns Hopkins Univ, 35-36; mem staff, Lowell Observ, 36-42; asst prof physics, Univ Mich, 42-46, asst prof astron, McMath Hulbert Observ, 46-48; prof physics & dir res observ, 48-76, EMER PROF PHYSICS, NORTHERN ARIZ UNIV, 76- *Mem:* Fel AAAS; fel Am Phys Soc; Am Astron Soc; Sigma Xi. *Res:* Atmospheres of the solar system; infrared spectra and molecular structure; far infrared spectroscopy of the solar and terrestrial atmospheres; composition and temperatures of upper atmosphere; infrared spectroscopy of atmospheric minor constituents; discovered atmospheric nitrous oxide, atmospheric heavy water and the 20-micron window; prepared first maps of solar-telluric spectrum, 7 to 14 microns. *Mailing Add:* PO Box 942 Flagstaff AZ 86002

ADELBERG, ARNOLD M, b Brooklyn, NY, Mar 17, 36; m 62; c 2. GEOMETRY, ALGEBRA. *Educ:* Columbia Univ, BA, 56; Princeton Univ, MA, 62. *Prof Exp:* Instr math, Columbia Univ, 59-62; from instr to assoc prof, 62-72, chmn dept math, 69-71 & 77-79 & 85-87, chmn sci div, 71-73, chmn fac, 74-76, PROF MATH GRINNELL COL, 72- *Concurrent Pos:* Hon res fel, Harvard Univ, 68-69; vis scholar, Univ Chicago, 83-84; vis prof, Univ Chicago, 89-90. *Mem:* Am Math Soc; Math Asn Am; Am Asn Univ Prof. *Res:* Algebraic geometry, especially rationality questions of algebraic groups; Koszul resolutions; Bezout's theorem; elementary metric geometry; combinatoric identities. *Mailing Add:* Dept of Math Grinnell Col Grinnell IA 50112

ADELBERG, EDWARD ALLEN, b Cedarhurst, NY, Dec 6, 20; m 42; c 3. MEMBRANE TRANSPORT SYSTEMS. *Educ:* Yale Univ, BS, 42, MS, 47, PhD(microbiol), 49. *Prof Exp:* From instr to prof bact, Univ Calif, Berkeley, 49-60, chmn dept, 57-61; prof microbiol, 61-74, chmn dept, 61-64 & 70-72, dir biol sci, 64-69, PROF HUMAN GENETICS, YALE UNIV, 74-, DEP PROVOST BIOMED SCI, 83- *Concurrent Pos:* Guggenheim fels, Pasteur Inst, Paris, 56-57 & Nat Ctr Sci Res, 65-66; consult, Chem Res Div, Eli Lilly & Co, 59-67; ed, J Bact, 64-67; ed-in-chief, Bact Rev, 67-70; mem genetics training comt, Nat Inst Gen Med Sci, 70-73; consult, Genetics Br, NSF, 71-74; bd chmn, Med Sci Res Ctr, Brandeis Univ, 76-79. *Mem:* Nat Acad Sci; fel Am Acad Arts & Sci. *Res:* Genetics of cultured mammalian cells; membrane transport. *Mailing Add:* Provosts Off Yale Univ 320 York St New Haven CT 06520

ADELBERGER, ERIC GEORGE, b Bryn Mawr, Pa, June 26, 38; m 61; c 2. EXPERIMENTAL NUCLEAR PHYSICS. *Educ:* Calif Inst Technol, BS, 60, PhD(physics), 67. *Prof Exp:* Res fel physics, Calif Inst Technol, 67-68; res assoc, Stanford Univ, 68-69; asst prof, Princeton Univ, 69-71; from asst prof to assoc prof, 71-75, PROF PHYSICS, UNIV WASH, 75- *Concurrent Pos:* Mem adv panel physics, NSF, 73-; fac adv nuclear physics, Max Planck Soc, Ger, 75-79; assoc ed, Phys Rev Lett, 78-81; mem adv panel physics, Argonne Univ Assoc, 79-; chmn, Div Nuclear Phys, Am Phys Soc, 87. *Honors & Awards:* von Humboldt Sr Sci Award, 82; Tom W Bonner Prize, 85. *Mem:* Am Phys Soc. *Res:* Experimental studies of fundamental symmetries in nuclei and atoms; nuclear structure; experimental gravitation. *Mailing Add:* Neuro Physics Lab G1-10 Univ Wash Seattle WA 98195

ADELBERGER, REXFORD E, b Cleveland, Ohio, Mar 30, 40; m 66; c 2. NUCLEAR PHYSICS. *Educ:* Col William & Mary, BSc, 61; Univ Rochester, PhD(physics), 67. *Prof Exp:* Asst prof physics, State Univ NY Col Geneseo, 67-73; ASSOC PROF PHYSICS, GUILFORD COL, 73- *Concurrent Pos:* Ed, J Undergraduate Research in Physics, 80- *Mem:* Am Asn Physics Teachers; Am Phys Soc. *Res:* Experimental study of the few nucleon problem at moderate energies; computer controlled machinery. *Mailing Add:* Dept of Physics Guilford Col Greensboro NC 27410

ADELI, HOJJAT, b Langrood, Iran, June 3, 50; m 79; c 3. STRUCTURAL ENGINEERING, EARTHQUAKE ENGINEERING. *Educ:* Univ Tehran, BS, 72, MS, 73; Stanford Univ, PhD(structural eng), 76. *Prof Exp:* Res assoc appl mechs, Stanford Univ, 76; asst prof civil eng, Northwestern Univ, 77; from asst prof to assoc prof, Univ Tehran, 78-82; res assoc prof, Univ Utah, 82-83; assoc prof, 83-88, PROF CIVIL ENG & CHMN STRUCT FAC, OHIO STATE UNIV, 88- *Concurrent Pos:* Ed in chief, Int J Microcomput Civil Eng, 86-; vis scientist, Air Force Wright Aeronaut Labs, 85; mem, Tech Coun Comput Pract Educ Comt, Am Soc Civil Engrs, 85-, Comt Optimal Design, 86-, Shock & Vibratory Effects Comt, 86-90, Comt Comput Graphics, 86-, Comt Personal Computs & Worksta, 86-, Comt Methods Anal, 86-, Comt Automated Anal & Design, 86-, Aerospace Structures & Mats, 86-, Comt Inelastic Behav, 87-, Spec Metal Structures Comt, 87-89, Composite & Metallic Mats Comt, 87-, Mats Appln Tech Comt, 87-, Task Comt Parallel Processing & Supercomput, & chmn, 87-90; ed-in-chief & founder, Microcomputers Civil Eng, Int J, 86; hon adv, Res Bd Adv, Am Biog Inst, 87-; mem, Tech Comt Comput Med, Inst Elec & Electronic Engrs Comput Soc, 88-; res award, Col Eng, Ohio State Univ, 90. *Honors & Awards:* Distinguished Leadership Award, Am Biog Inst, 86 & Commemorative Medal of Honor, 87. *Mem:* Am Asn Artificial Intelligence; Asn Comput Mach; Comput Soc Inst Elec & Electronic Engrs; Earthquake Eng Res Inst; Am Soc Civil Engrs. *Res:* Has authored or edited over 220 research and technical publications in the fields of computer-aided engineering, artificial intelligence, structural engineering, applied mechanics, mathematical optimization, earthquake engineering and numerical analysis. *Mailing Add:* Dept Civil Eng Ohio State Univ 470 Hitchcock Hall 2070 Neil Ave Columbus OH 43210

ADELMAN, ALBERT H, b New York, NY, Dec 17, 30; m 52; c 2. PHYSICAL CHEMISTRY. *Educ:* Brooklyn Col, BS, 51; Polytech Inst Brooklyn, PhD(chem), 56. *Prof Exp:* Assoc res scientist, NY Univ, 56-60; sr scientist, 60-70, assoc chief, Chem Dept, 70-72, sect mgr org chem, 72-75, mgr, Chem Dept, 75-79, assoc dir, Columbus Div, Battelle Mem Inst, 79-85; pres, Oread Labs Inc, 86-87; dir, Nat Labs, Ensr, 88-89; CONSULT, BURT ASSOCS, 89- *Concurrent Pos:* NSF fel, 56-58; instr, City Col New York, 59- *Mem:* AAAS; Am Phys Soc; Am Chem Soc. *Res:* Charge-transfer photochemistry; stepwise excitation of luminescence; solar energy utilization employing physical and biological systems. *Mailing Add:* Burt Assocs PO Box 719 Westford MA 01886

ADELMAN, BARNET REUBEN, b Helena, Ark, Dec 17, 25; m 48; c 3. JET PROPULSION & ORDINANCE, ROCKETS, MISSILES & EXPLOSIVES. *Educ:* Columbia Univ, BS, 47, MS, 48. *Prof Exp:* Res engr, Picatinny Arsenal, NJ, 48-49; sr res engr, Jet Propulsion Lab, Calif Inst Technol, 49-51; tech dir, Rocket Fuels Div, Phillips Petrol Co, Tex, 51-55; dir, Vehicle Eng Lab, Ramo Wooldridge Corp, Calif, 55-59; exec vpres, United Technol Ctr, United Aircraft Corp, 59-62, div pres, 62-80, vpres, Power Group, United Technol Corp, 80-84, pres, Biother Systs, 84-86; PRES, ADELMAN ASSOC, 84- *Concurrent Pos:* Adv & US deleg, Mutual Weapons Group, NATO, 55-56; pres, United Space Boosters, Inc, 74-80. *Honors & Awards:* C N Hickman Award, Am Rocket Soc, 58; Propulsion Award, Am Inst Aeronaut & Astronaut. *Mem:* NY Acad Sci; fel Am Inst Aeronaut & Astronaut; Am Defense Preparedness Asn; Am Inst Chem Engrs. *Mailing Add:* 88 Stern Ln Atherton CA 94027

ADELMAN, IRA ROBERT, b New York, NY, Nov 14, 41; m 66; c 1. FISHERIES BIOLOGY. *Educ:* Univ Vt, BA, 63; Univ Minn, PhD(fisheries), 69. *Prof Exp:* Res fel, 69-70, res assoc, 70-74, from asst prof to assoc prof, 74-82, PROF FISHERIES & DEPT HEAD, UNIV MINN, ST PAUL, 83- *Concurrent Pos:* Spec asst dir Fish & Wildlife, Minn Dept Natural Resources, 90. *Mem:* Am Fisheries Soc; AAAS; Sigma Xi; Am Inst Fisheries Res Biologists. *Res:* Fisheries ecology effects of environmental variables and water quality on fish physiology and in particular fish growth. *Mailing Add:* Dept Fisheries & Wildlife Univ Minn St Paul MN 55108

ADELMAN, MARK ROBERT, b Philadelphia, Pa, 1942; m 71; c 1. MOLECULAR BIOLOGY, BIOPHYSICS. *Educ:* Princeton Univ, AB, 63; Univ Chicago, PhD(biophys), 69. *Prof Exp:* asst prof anat, Duke Univ, 71-79; ASSOC PROF ANAT, MICROANAT, UNIFORMED SERV UNIV HEALTH SCI, 79- *Mem:* Am Soc Cell Biol; Biophys Soc; Am Asn Anat; AAAS; NY Acad Sci. *Res:* Biochemistry of cytoskeletal proteins; dynamics of cytoskeletal changes during amoebaflagellate transformation in Physarum. *Mailing Add:* Dept Anat Uniformed Serv Univ Health Sci 4301 Jones Bridge Rd Bethesda MD 20814-4799

ADELMAN, RICHARD CHARLES, b Newark, NJ, Mar 10, 40; m 63; c 2. GERONTOLOGY, BIOCHEMISTRY. *Educ:* Kenyon Col, AB, 62; Temple Univ, MA, 65, PhD(biochem), 67. *Prof Exp:* From asst prof to prof biochem Fels Res Inst, Med Sch, Temple Univ, 69-82, dir, Inst on Aging, 78-82; dir, Div Biomed Res, Philadelphia Geriat Ctr, 78-82; PROF BIOL CHEM & DIR, INST GERONT, UNIV MICH, 82- *Concurrent Pos:* Am Cancer Soc fel, Albert Einstein Col Med, 67-69; Am Can Soc res grant aging, Fels Res Inst, Med Sch, Temple Univ, 70-73; NIH res grants, 70-; mem Study Sect Pathobiol Chem, NIH, 75-78; chmn, Gordon Res Conf Biol of Aging, 76; chmn, Vet Admin Nat Adv Comt, Geront & Geriat, 87- *Mem:* AAAS; fel Geront Soc (secy-treas 72-75, vpres, 75-76, pres, 87-88); Am Soc Biol Chemists; Am Chem Soc. *Res:* Biology of aging; hormonal regulation of enzyme activity. *Mailing Add:* Inst Geront Univ Mich 300 N Ingalls Ann Arbor MI 48109

ADELMAN, ROBERT LEONARD, b Chicago, Ill, May 20, 19. ORGANIC CHEMISTRY. *Educ:* Univ Chicago, BS, 41, PhD(org chem), 45. *Prof Exp:* Res chemist, Nat Defense Res Comt & Off Sci Res & Develop, Univ Chicago, 42-45; from res chemist to sr chemist, 45-61, staff scientist, 61-66, res assoc, Indust Chem Dept, 72-74, res assoc, Plastics Prod & Resins Dept, 74-79, res assoc, Polymer Prod Dept, E I Du Pont De Nemours & Co, Inc, 79-84; RETIRED. *Mem:* Am Chem Soc; Sigma Xi. *Res:* Reaction mechanism; synthesis and characterization of monomers, polymers, adhesives, protective coatings, structural plastics, plated plastics, polymers in electroplating and electronics, synthetic elastomers and paper. *Mailing Add:* 2422 Riddle Ave Wilmington DE 19806

ADELMAN, SAUL JOSEPH, b Atlantic City, NJ, Nov 18, 44; m 70; c 3. ASTRONOMY, ASTROPHYSICS. *Educ:* Univ Md, BS, 66; Calif Inst Technol, PhD(astron), 72. *Prof Exp:* Resident res assoc astron, Nat Acad Sci-Nat Res Coun, NASA Goddard Space Flight Ctr, 72-74; asst prof, Boston Univ, 74-78; from asst prof to assoc prof, 78-89, PROF PHYSICS, THE CITADEL, 89- *Concurrent Pos:* Guest investr, Int Ultraviolet Satellite, 80, 82-87 & 89-91; guest observer, Dominion Astrophys Observ, 84-91; res assoc, Goddard Space Flight Ctr, Nat Res Coun, NASA, 84-86. *Mem:* Int Astron Union; Am Astron Soc; Astron Soc Pac; Sigma Xi; Royal Astron Soc. *Res:* Magnetic peculiar A stars; coude spectroscopy; photoelectric spectrophotometry; abundance analyses; line identification techniques; atomic spectroscopy; HgMn stars; main sequence B, A and F stars; space astronomy; horizontal-branch stars. *Mailing Add:* 1434 Fairfield Ave Charleston SC 29407

ADELMAN, WILLIAM JOSEPH, JR, b Mt Vernon, NY, Jan 29, 28; m 51; c 3. PHYSIOLOGY, BIOPHYSICS. *Educ:* Fordham Univ, BS, 50; Univ Vt, MS, 52; Univ Rochester, PhD(physiol), 55. *Prof Exp:* Asst physiol & biophys, Univ Vt, 50-52; aviation physiologist, US Air Force Sch Aviation Med, 55-56; from instr to asst prof physiol, Sch Med, Univ Buffalo, 56-59; physiologist, Biophys Lab, Nat Inst Neurol Dis & Blindness, 59-62; from assoc prof to prof physiol, Sch Med, Univ Md, Baltimore, 62-71; CHIEF LAB BIOPHYS, NAT INST NEUROL COMMUN DIS & STROKE, 71- *Concurrent Pos:* Nat Inst Neurol Dis & Stroke spec fel, Marine Biol Lab, Woods Hole, 69-70. *Mem:* AAAS; Am Physiol Soc; Biophys Soc; Soc Gen Physiol; Soc Neurosci. *Res:* Neurophysiology and electrobiology; role of ions in membrane phenomena in nerve; modeling of neural behavior; ultra structure of neurons. *Mailing Add:* Lab Biophys NINCDS NIH Bldg Nine Rm 1E124 Bethesda MD 20092

ADELSON, BERNARD HENRY, b Tampa, Fla, Mar 16, 20; m 50; c 3. MEDICINE, CHEMISTRY. *Educ:* Northwestern Univ, BS, 41, PhD(chem), 46, BM, 50, MD, 51. *Prof Exp:* Asst chem, 42-44, res assoc, 44-46, instr, 46-47, lectr, 47, clin asst med, 54-57, from instr to asst prof med, 58-73, assoc prof, 73-81, PROF CLIN MED, NORTHWESTERN UNIV, CHICAGO, 81- *Concurrent Pos:* Resident, Evanston Hosp, Ill, 51-53, dir, Artificial Kidney Unit, chief nephrology & assoc chmn, Dept Med; resident, Cook County Hosp, 53-54; dir, Geriat Serv, Evanston Hosp, prog med ethics; clinician laureate, Ill Chap, Am Col Physicians. *Mem:* AAAS; AMA; Royal Soc Med; Am Soc Nephrology. *Res:* Organic chemistry; synthesis of 5-substituted quinolines; cardiac outputs in myocardial infarcts and medical shock; extracorporeal hemodialysis. *Mailing Add:* 595 Lincoln Ave Glencoe IL 60022

ADELSON, HAROLD ELY, b New York, NY, May 5, 31. TECHNICAL MANAGEMENT. *Educ:* City Col New York, BS, 53; Univ Calif, MA, 54, PhD(physics), 59. *Prof Exp:* Asst physics, Univ Calif, 53-55, asst nuclear physics, Radiation Lab, 55-59; sr res physicist, Convair Astronaut, 59-61, sr staff scientist, Gen Dynamics/Astronaut, 61-65; mgr environ res satellite dept, TRW Systs Group, 65-67, from asst dir to dir res appln lab, 67-71, mgr Viking Lander biol & meteorol instruments prog, 71-75, mgr appl Technol Div Design Rev Off, 75-80, MGR APPL TECHNOL DIV SR TECH STAFF, TRW SYSTS, INC, 80- *Mem:* Am Phys Soc; Am Geophys Union; Am Inst Aeronaut & Astronaut; Sigma Xi. *Res:* Spacecraft technology; space instrumentation; computer aided design. *Mailing Add:* Apt 1902 2170 Century Park E Los Angeles CA 90067

ADELSON, LESTER, b Chelsea, Mass, Aug 20, 14; m 42; c 2. FORENSIC PATHOLOGY. *Educ:* Harvard Univ, AB, 35; Tufts Univ, MD, 39. *Prof Exp:* From asst prof legal med to assoc prof, 53-69, EMER PROF FORENSIC PATH, SCH MED, CASE WESTERN RESERVE UNIV, 69-; pathologist & chief dep coroner, Cuyahoga County, 50-87. *Concurrent Pos:* Res fel path & legal med, Harvard Med Sch, 49-50. *Mem:* AMA; Am Soc Clin Path; Sigma Xi. *Mailing Add:* 23005 Beachwood Blvd Beachwood OH 44122-1401

ADELSON, LIONEL MORTON, marine ecology, histology, for more information see previous edition

ADELSTEIN, PETER Z, b Montreal, Que, Sept 1, 24; nat US; m 47; c 3. PHYSICAL CHEMISTRY. *Educ:* McGill Univ, BE, 46, PhD(chem), 49. *Prof Exp:* Chemist, 49-63, unit dir, Eastman Kodak Co, 58-86; CONSULT, IMAGE PERMANENCE INST, 86- *Concurrent Pos:* Chmn, Am Nat

Standards Inst, 67-; chmn, WG-5 Int Standards Orgn, TC/42, 73-; chmn, rec preserv comt, NAS, 84-86. *Mem:* Am Chem Soc; Am Soc Testing & Mat; Nat Fire Protection Asn; Soc Photog Sci & Eng. *Res:* Behavior of high polymers in solution; physical properties of high polymer materials in solid states; physical behavior of photographic film; permanence of photographic materials. *Mailing Add:* 1629 Clover St Rochester NY 14618

ADELSTEIN, ROBERT SIMON, b New York, NY, Jan 16, 34; m 61; c 3. MEDICAL SCIENCE, MOLECULAR BIOLOGY. *Educ:* Princeton Univ, AB, 55, Harvard Med Sch, MD, 59. *Prof Exp:* Intern & resident med, Duke Med Ctr & Bellevue Hosp, 59-61, sr med researcher, 64-65; res assoc, Nat Heart Inst, 61-64, sr investr biochem, Nat Heart Lung & Blood Inst, 66-72, sect head, 72-81, LAB CHIEF BIOCHEM, NAT HEART LUNG & BLOOD INST, NIH, 81- *Concurrent Pos:* Fel biochem, Univ Wash 65-66; chmn, Comt Concern Scientist, 77-79, Gordon Conf Muscle Protein, 81; mem molecular cytol, NIH, 81-85; sabbatical, Hebrew Univ Sch Med, Jerusalem, 83-84. *Mem:* Am Soc Biochem & Molecular Biol; Am Soc Cell Biol; AAAS; Biophys Soc; Am Soc Clin Invest. *Res:* Regulation of contractile activity in vertebrate muscle and non muscle cell; regulation of the expression of contractile protein in those cells. *Mailing Add:* NHBLI Bldg 10 Rm 8N-202 Bethesda MD 20892

ADELSTEIN, STANELY JAMES, b New York, NY, Jan 24, 28; m 57; c 2. RADIATION BIOPHYSICS, NUCLEAR MEDICINE. *Educ:* Mass Inst Technol, BS & MS, 49, PhD(biophys), 57; Harvard Univ, MD, 53. *Prof Exp:* House officer, Peter Bent Brigham Hosp, Boston, 53-54, sr asst resident, 57-58, chief resident physician, 59-60; assoc anat, 63-65, from asst prof to prof, 65-89, DEAN ACAD PROG, HARVARD MED SCH, 78-, PAUL C CABOT PROF MED BIOPHYSICS, 89- *Concurrent Pos:* Moseley traveling fel, Harvard Med Sch, 58-59, HA&C Christian fel, 59-60, Med Found fel, 60-63, P H Cook fel radiol, 60-68; USPHS career develop award, 65-68; assoc med & radiol, Peter Bent Brigham Hosp, 63-68, dir nuclear med, 68-; vis fel, Johns Hopkins Med Inst, 68; chief nuclear med servs, Children's Hosp Med Ctr, Boston, 70-78; mem, Am Bd Nuclear Med, 72-78. *Honors & Awards:* Aebersold Award, Soc Nuclear Med, Blumgart Award. *Mem:* Biophys Soc; Radiation Res Soc; Am Soc Cell Biol; Soc Nuclear Med; Inst Med; fel AAAS. *Res:* Cellular and molecular radiation biology. *Mailing Add:* Off Dean Acad Prog Harvard Med Sch 25 Shattuck St Boston MA 02115

ADEM, JULIAN, meteorology, applied mathematics, for more information see previous edition

ADEN, DAVID PAUL, b Quincy, Ill, Feb 23, 46; m 71. IMMUNOBIOLOGY. *Educ:* Quincy Col, BS, 68; Mont State Univ, MS, 70, PhD(microbiol), 73. *Prof Exp:* Res investr, 73-76; asst prof, Wistar Inst Anat & Biol, 76-90; DENTIST, MARILYN & ADEN, 90- *Concurrent Pos:* Nat Cancer Inst fel, 74-76. *Mem:* Am Asn Immunologists. *Res:* Cell surface antigens coded for by specific human chromosomes; genetic control of human serum proteins; chromosomes involved in tumorgenicity. *Mailing Add:* 802 Schemmer Dr Prescott AZ 86301

ADENT, WILLIAM A, b Chicago, Ill, May 27, 23. OIL & GAS. *Educ:* Stanford Univ, BA, 49. *Prof Exp:* CHIEF, DEVELOP UNIT, PAC OFFSHORE REGION, US GOVT, 78- *Mem:* Am Asn Petrol Geologist; fel Geol Soc Am. *Mailing Add:* 5813 Washington Ave Whittier CA 90601

ADER, ROBERT, b New York, NY, Feb 20, 32; m 57; c 4. PSYCHONEUROIMMUNOLOGY, PSYCHOSOMATICS. *Educ:* Tulane Univ, BS, 53; Cornell Univ, PhD(exp psychol), 57. *Prof Exp:* From instr to assoc prof, 57-68, PROF PSYCHIAT & PSYCHOL, DEPT PSYCHIAT, SCH MED & DENT, UNIV ROCHESTER, 68-, PROF MED & GEORGE L ENGEL PROF PSYCHOSOCIAL MED, 83-, DIR, DIV BEHAV & PSYCHOSOCIAL MED, 82- *Concurrent Pos:* Vis prof, Rudolf Magnus Inst Pharmacol, Univ Utrecht, Neth, 70-71; ed-in-chief, Brain, Behav & Immunity, 86- *Mem:* Am Psychosom Soc (pres, 79-80); Int Soc Develop Psychobiol (pres, 81-82); Acad Behav Med Res (pres, 84-85); Soc Behav Med; Am Asn Univ Prof. *Res:* Brain, behavior and immune system interactions in health and disease. *Mailing Add:* Dept Psychiat Med Ctr Univ Rochester Rochester NY 14642

ADES, EDWIN W, b Kew Gardens, NY, Mar 30, 49; m. BIOLOGY PRODUCTION. *Educ:* Emory Univ, BS, 71, PhD(immunol-path), 77; Ga State Univ, MS, 73. *Prof Exp:* Res microbiologist, Ctr Dis Control, Atlanta, Ga, 74-76; postdoctoral res fel, Cellular Immunobiol Unit, Comprehensive Cancer Ctr, Univ Ala, Birmingham, 77-79; asst prof, Dept Basic & Clin Immunol & Microbiol, Med Univ SC, 79-81; sr scientist immunol, Eli Lilly & Co, 81-85; assoc prof & dir, Barton Immunomorphol Lab, Dept Path, Med Col Ga, 85-87; CHIEF, BIOL PROD BR, SRP, CID, CTR DIS CONTROL, ATLANTA, GA, 87- *Concurrent Pos:* Res assoc, Comprehensive Cancer Ctr, Univ Ala, Birmingham, 77-79; spec fel, Leukemia Soc Am, 78-80; chmn, Grad Fac Comt, Dept Basic & Clin Immunol & Microbiol, Med Univ SC, 79-81; adj asst prof, Med Univ SC, 81-86, Pasteur Inst, France, 81-86, Sch Med, Ind Univ, 82-85, Med Col Ga, 87-; mem ad hoc spec study sect, Nat Cancer Inst, 86; adj assoc prof, Sch Med, Emory Univ, 87-; mem adv comt, Clin invest, Immunol & Immunother, Am Cancer Soc, 89-90; lectr, Am Soc Microbiol, 90-91. *Mem:* AAAS; Sigma Xi; NY Acad Sci; Am Soc Microbiol; Am Asn Cancer Res; Am Asn Immunologists; Soc Exp Biol & Med; Am Asn Pathologists; Am Soc Clin Oncologists; Soc Biol Ther. *Res:* Cellular immunobiology; 3 patents. *Mailing Add:* Biol Prod Br OSS CID 3202 Ctr Dis Control 1600 Clifton Rd NE Atlanta GA 30333

ADES, IBRAHIM Z, b Cairo, Egypt, May 24, 46; m; c 1. CELL BIOLOGY, BIOLOGY. *Educ:* Univ Calif, Los Angeles, PhD(biol), 76. *Prof Exp:* Post doctoral fel, Dept Biochem, Univ Tex Health Sci Ctr, Dallas, 76-79; asst prof, dept Biochem, Bowman Gray Sch Med, 79-82; asst prof, 82-87, ASSOC PROF ZOOL, UNIV MD, COLLEGE PARK, 87-, ASSOC PROF, MOLECULAR & CELL BIOL PROG, 89- *Concurrent Pos:* Am Cancer Soc post doctoral fel, 77-78; cell biologist, Nat Cancer Inst, Bethesda, Md, 89-90. *Mem:* Am Soc Biol Chem & Molecular Biol; Am Soc Cell Biol. *Res:* Biosynthesis and processing of proteins and mechanisms of assembly of cell organelles; pathways of porphyrin production and regulation of biogenesis of their enzymes. *Mailing Add:* Dept Zool Univ Md College Park MD 20742

ADESNIK, MILTON, b May 28, 43. CELL BIOLOGY. *Educ:* City Col NY, BS, 64; Mass Inst Technol, PhD(biophys), 69. *Prof Exp:* Predoctoral fel biophys, Mass Inst Technol, 65-68; instr biol sci, Columbia Col, 68-70, res assoc, Dept Biol Sci, 70-72; from asst prof to assoc prof, 72-86, PROF CELL BIOL, SCH MED, NY UNIV, 86- *Concurrent Pos:* Postdoctoral fel, Damon Runyon Fund Cancer Res, 70-72. *Res:* Regulation of expression of genes encoding liver microsomal cytochromes P450; mechanisms of insertion of proteins into membranes and for targetting these proteins to the specific membranes in the cell where they function; co-author of over 50 publications. *Mailing Add:* Dept Cell Biol Sch Med NY Univ 550 First Ave New York NY 10016

ADEY, W ROSS, b Adelaide, Australia, Jan 31, 22; nat US. RESEARCH SERVICE. *Educ:* Univ Adelaide, BS & MB, 43, MD, 49. *Prof Exp:* Lectr, sr lect & reader anat, Univ Adelaide, Australia, 46-53, sr lectr, 55-56; asst prof anat, Univ Calif, Los Angeles, 54, prof anat & physiol, 57-77; DISTINGUISHED PROF NEUROL, SCH MED, LOMA LINDA UNIV, CALIF, 88-, ASST DEAN RES, 87-; ASSOC CHIEF STAFF, RES & DEVELOP, PETTIS MEM VET HOSP, LOMA LINDA, CALIF. *Concurrent Pos:* Consult, NIH, NASA, Dept Energy, Vet Admin, WHO; Nuffield Found Dominion traveling fel med, Oxford, 50; fel, Royal Soc London & Nuffield Found, 56; dir, Space Biol Lab, Brain Res Inst, Univ Calif, Los Angeles, 61-74; prin investr, Long Duration Biosatellite Exp, NASA, 63-70; assoc, Neurosci Res Prog, Mass Inst Technol, 64-; mem, Space Sci Panel, Pres Sci Adv Comt, 69-71; mem, Telecommun Panel, Nat Acad Eng, 72-74; proj leader, Med Hazards Microwave Exposure, US/USSR Exchange Prog, 76-; mem, Assembly Life Sci, Nat Acad Sci, 76-, panel biosphere effects extremely low frequency radiation, 76; distinguished prof physiol, Sch Med, Loma Linda Univ, 87-, surg, 88- *Honors & Awards:* D'Arsonval Medal, Bioelectromagnetics Soc, 89. *Mem:* Fel Am Acad Arts & Sci; Am Physiol Soc; fel Am Asn Anatomists; fel Inst Elec & Electronics Engrs; Am EEG Soc; AAAS; Am Asn Neurol Surgeons; Biomed Eng Soc. *Res:* Behavioral neurophysiology; organization of cerebral systems and cerebral cellular mechanisms; computer applications in physiological data analysis and in models of brain systems; bioinstrumentation and bioengineering; aerospace medicine and physiology. *Mailing Add:* Res Serv Pettis Mem Vet Hosp 11201 Benton St Loma Linda CA 92357

ADEY, WALTER HAMILTON, b Stoneham, Mass, Apr 9, 34; m 85; c 2. SYNTHETIC ECOLOGY, MARICULTURE. *Educ:* Mass Inst Technol, BS, 55; Univ Mich, PhD(marine bot & geol), 63. *Prof Exp:* RES SCIENTIST & CUR MARINE BIOL & GEOL, SMITHSONIAN INST, 64-, DIR, MARINE SYSTS LAB, 75- *Concurrent Pos:* Res assoc, Colo Sch Mines, 63-64 & Hawaiian Inst Geophys, 71-75; adj prof, George Washington Univ, Georgetown Univ & Univ Md, 75-; prof, Univ Maine, 76- *Mem:* AAAS; Fedn Am Scientists. *Res:* Synthetic ecology; coral reef geology and ecology; theoretical marine biogeography; biology of coralline algae; simulation and modeling of marine ecosystems; mariculture; bioengineering. *Mailing Add:* 600 Water St Washington DC 20024

ADEY, WILLIAM ROSS, b Adelaide, Australia, Jan 31, 22; US citizen; m 70; c 3. NEUROPHYSIOLOGY. *Educ:* Univ Adelaide, MB & BS, 43, MD, 49. *Prof Exp:* Lectr, sr lectr & reader anat, Univ Adelaide, 46-53; asst prof, Univ Calif, Los Angeles, 53-54; sr lectr, Univ Melbourne, 54-57; prof anat & physiol, Univ Calif, Los Angeles, 57-77, dir environ lab, Brain Res Inst, 75-77; DIR RES SERV, LOMA LINDA VET ADMIN MED CTR, 77-, PROF PHYSIOL & SURG, LOMA LINDA UNIV, 77- *Concurrent Pos:* Nuffield Found Dom traveling fel, 50; Rockefeller Found grants, 51, 55; Royal Soc London & Nuffield Found traveling bursary, 56-57; mem space sci panel, President's Sci Adv Comt, 66-, biol & med sci panel, 69-72; mem electromagnetic radiation coun, Exec Off of President, 69-78; chmn, Cow Freq Radiation Comt, Nat Coun Radiation Protection, 85- *Honors & Awards:* Herrick Award, Am Asn Anat, 63. *Mem:* AAAS; Am Physiol Soc; Am Asn Anat; fel Am Acad Arts & Sci; fel Inst Elec & Electronics Eng. *Res:* Neuroanatomy and neurophysiology of brain functions in behavioral mechanisms; bioeffects non-ionizing radiation; molecular biology of cell membranes in cancer. *Mailing Add:* Res Serv 151 Vet Admin Hosp Loma Linda CA 92357

ADHAV, RATNAKAR SHANKAR, b Sakri Boombay, India, Oct 30, 27; US citizen; m 53; c 3. CRYSTAL GROWTH FOR LASERS. *Educ:* Univ Poona, India, BSc, 49, MSc, 52; Gujarat Univ, Ahmedabad, PhD(physics), 58. *Prof Exp:* Lectr physics, Gujarat Col, 52-58; fel, Nat Res Coun, Can, 58-60; res physicist, Can govt, 60-63; dir res, EDO (Can) Ltd, 64-68; FOUNDER & PRES, QUANTUM TECHNOL, INC, 69- *Mem:* Optical Soc Am; Inst Elec Electronics Engrs. *Res:* Electro-optics; electronics engineering; engineering physics; electro-optics. *Mailing Add:* Quantum Technol Inc 108 Commerce St Lake Mary FL 32746

ADHIKARI, P K, b Nagpur, India, May 18, 28; m 66. INTERNAL MEDICINE. *Educ:* Univ Calcutta, MB, BS, 52; FRCP(C), 65. *Prof Exp:* Resident med, Univ Vt, 55-57; res assoc, Pulmonary Function Lab, Wayne State Univ, 60-62; res assoc, 62-64, ASST PROF PEDIAT, UNIV MAN, 65-; INTERNAL MED, PEIKOFF CLIN. *Concurrent Pos:* Fel cardio-respiration, Univ Vt, 57-59. *Res:* Respiratory physiology and diseases. *Mailing Add:* 400309 Hargrove St Winnipeg MB R3C 0N8 Can

ADHOUT, SHAHLA MARVIZI, b Tehran, Iran, Apr 30, 54; m 82; c 1. MATHEMATICS. *Educ:* Arya-Mehr Univ, BS, 79; MIT, PhD(math), 81. *Prof Exp:* Post doc, Inst Adv Study, Princeton, 81-82; post doc, Math Res Inst, Berkeley, 82-83; prof, Univ Calif, Berkeley, 83-85; PROF, LONG ISLAND UNIV, CW POST COL, 85- *Mem:* Am Math Soc. *Res:* Geometry and analysis; symplictic geometry and inverse spectral problems. *Mailing Add:* Dept Math C W Post Ctr, Long Island Univ Greenvale NY 11548

ADHYA, SANKAR L, b Calcutta, India, Oct 4, 37. CONTROL OF GENE EXPRESSION. *Educ:* Univ Calcutta, PhD(biochem), 63; Univ Wis, PhD(biochem), 66. *Prof Exp:* CHIEF, DEVELOP GENETICS SECT, NAT CANCER INST, NIH, 80- *Mem:* Am Soc Biol Chemists; Am Soc Microbiol; Am Soc Virol. *Mailing Add:* Sec Lab Molecular Biol Nat Cancer Inst NIH Bldg 37 Rm 4B04 Bethesda MD 20892

ADIARTE, ARTHUR LARDIZABAL, b San Nicolas, Philippines, Oct 27, 43; m 72; c 2. BIOPHYSICAL CHEMISTRY, ENERGY. *Educ:* Univ Philippines, BS, 63; Univ Pittsburgh, PhD(biophys), 72. *Prof Exp:* Instr physics math, Univ Philippines, 63-66; teaching asst physics, Univ Pittsburgh, 66-67, res asst biophys, 67-72; fel, Univ Regensburg, 72-74 & Lab Biophys Chem, Dept Chem, Univ Minn, Minneapolis, 75-77; res scientist, Minn Energy Agency, 77-90; INDUST ECONOMIST, MINN OFF TOURISM, MINN DEPT TRADE & ECON DEVELOP, 90- *Mem:* Biophys Soc; Sigma Xi; NY Acad Sci; AAAS. *Res:* Protein structure and function, thermodynamics and kinetics of biological processes; the role of water in protein systems and enthalpy-entropy compensation phenomena; energy and environment. *Mailing Add:* 767 Fairmont Ave St Paul MN 55105

ADIBI, SIAMAK A, b Tehran, Iran, Mar 17, 32; US citizen; m 63; c 3. CLINICAL NUTRITION, GASTROENTEROLOGY. *Educ:* Johns Hopkins Univ, BA, 55; Jefferson Med Col, MD, 59; Mass Inst Technol, PhD(nutrit biochem), 67. *Prof Exp:* PROF MED, UNIV PITTSBURGH SCH MED, 74-, PROF CLIN NUTRIT, 80-, HEAD, CLIN NUTRIT UNIT, MONTEFIORE UNIV HOSP, 80- *Concurrent Pos:* Mem various study sections, NIH, 75-; chmn, Gastroenterol Sect, Am Physiol Soc, 80-81. *Mem:* Asn Am Physicians; Am Inst Nutrit; Am Fedn Clin Res; Am Soc Clin Invest; Am Physiol Soc; Am Gastroenterol Asn. *Res:* Transport and metabolism of oligo peptides; metabolic and nutritional regulation of branched-chain amino acid metabolism; treatment of obesity. *Mailing Add:* Montefiore Univ Hosp Univ Pittsburgh 3459 Fifth Ave Pittsburgh PA 15213

ADICKES, H WAYNE, b Cuero, Tex, Sept 6, 40. SYNTHETIC ORGANIC CHEMISTRY, PACKAGING TECHNOLOGY. *Educ:* Stephen F Austin State Univ, BS, 62; Tex Christian Univ, PhD(chem), 68. *Prof Exp:* Instr chem, Univ Tex , Arlington, 62-65; fel, La State Univ, 68-69 & Yale Univ, 69-70; res chemist, St Regis Corp, 70-71 & AM Int, 71-74; dir, St Regis Corp, 74-83; VPRES ENG & DEVELOP, PACKAGING CORP AM, 83- *Mem:* Tech Asn Pulp & Paper Indust; NY Acad Sci; Soc Plastics Engrs; Am Chem Soc; Sigma Xi. *Res:* Reactions of halothiophenes with metal amides; use of cyclopropanones in the synthesis of betta-lactoms; new synthetic routes to oldehydes; development of new organic photoconductors. *Mailing Add:* 3310 E Third St Tucson AZ 85716

ADICOFF, ARNOLD, b New York, NY, Oct 13, 23; m 50; c 2. SOLID ROCKET PROPELLANTS, SURFACE CHEMISTRY. *Educ:* Univ Calif, Los Angeles, BS, 49; Polytechnic Inst NY, Brooklyn, PhD(phys chem), 55. *Prof Exp:* Teaching fel chem, Polytechnic Inst NY, 49-50, res fel, 50-54; proj leader polymers & testing, Air Reduction Co, 54-56; res engr chem, Sci Lab, Ford Motor Co, Dearborn, Mich, 56-57; chemist polymers, Naval Weapons Ctr, China Lake, Calif, 57-68, head, polymer sci br, 68-79, assoc head, chem div, 73-79, scientist chem, 82-83; assoc dir res, defense sci, Off Dep Under Secy Defense, 79-81; tech adv propulsion, Naval Sea Systs Command, Washington, DC, 81-82; SCIENTIST PROPULSION, TRW INC, 83- *Concurrent Pos:* Chmn, structures & mech behav, Joint Army-Navy-Air Force-NASA Subcomts, 68, 73, 75-78. *Mem:* Res Soc Am; AAAS; Am Chem Soc; Rheology Soc. *Res:* Research and development of solid rocket motor propellants; new polymeric materials; specialties, publications and patents include polymers, composite materials, adhesion, surface chemical effects, combustion, electrostatic effects and material mechanical properties. *Mailing Add:* PO Box 4691 Mission Viejo CA 92690

ADIN, ANTHONY, b Sheffield, Eng, Nov 11, 42; m 69; c 2. PHYSICAL INORGANIC CHEMISTRY. *Educ:* Leeds Univ, BSc, 64, PhD(chem), 67. *Prof Exp:* Fel, Brookhaven Nat Lab, 67-69 & Ames Lab, AEC, 69-71; sr res chemist, 71-77, RES ASSOC, EASTMAN KODAK CO, 78- *Res:* Application of inorganic and organometallic chemistry to unconventional photography; new technologies for silver-halide based photography. *Mailing Add:* 164 Belvista Dr Rochester NY 14625

ADINOFF, BERNARD, b Toronto, Ont, Mar 9, 19; US citizen; m 48; c 3. MATERIAL ENGINEERING, GENERAL COMPUTER SCIENCES. *Educ:* Univ Chicago, BS, 39, PhD(phys chem), 43. *Prof Exp:* Res asst, Univ Chicago, 43-44; chemist, Dayton Rubber Co, 44-59; chief chem engr, Fruehauf Trailer Co, 59-60; chief chem engr automotive opers, Rockwell Int, 60-82; lectr computer sci, Calif Lutheran Univ, 85-89; RETIRED. *Concurrent Pos:* Consult, 85- *Mem:* Am Chem Soc. *Res:* Lubrication; corrosion; plastics; adhesives; rubber; paints; latex foam; urethane foam; industrial fabrics. *Mailing Add:* 895 Tupelo Wood Ct Thousand Oaks CA 91320-3649

ADINOLFI, ANTHONY M, b New Haven, Conn, May 5, 39; m 63; c 2. NEUROCYTOLOGY. *Educ:* Yale Univ, BA, 60; Columbia Univ, MA, 65, PhD(anat), 67. *Prof Exp:* Asst prof anat, 69-74, asst prof anat & psychiat, 74-77, ASSOC PROF ANAT & PSYCHIAT, SCH MED, UNIV CALIF, LOS ANGELES, 77- *Concurrent Pos:* Fel anat, Med Ctr, Univ Calif, Los Angeles, 67-69; chief neurocytol res lab, Vet Admin Hosp, 69-74. *Mem:* Am Asn Anat; Electron Micros Soc Am; Soc Neurosci. *Res:* Computer morphometric and electron microscopic analysis of neuronal organization in the central nervous system. *Mailing Add:* Dept Anat Univ Calif Sch Med 405 Hilgard Ave Los Angeles CA 90024

ADISESH, SETTY RAVANAPPA, b Y N Hosakote, Mysore, India, June 17, 26; m 52; c 1. PHYSICAL CHEMISTRY, INORGANIC CHEMISTRY. *Educ:* Univ Mysore, BSc, 52, MSc, 56; Kent State Univ, PhD(chem), 65. *Prof Exp:* Lectr chem, Univ Mysore, 52-60; from asst prof to assoc prof, 65-72, PROF CHEM, ST LEO COL, 72- *Concurrent Pos:* Res assoc, Kent State

Univ, 66; res appointment, Tex A&M Univ, 68 & Ga Inst Technol, 69; NSF res grant, 69. *Mem:* Am Chem Soc. *Res:* X-ray diffraction; voltometry; kinetic and reaction mechanisms. *Mailing Add:* Dept of Chem St Leo Col St Leo FL 33574

ADISMAN, I KENNETH, b New York, NY, Aug 3, 19; m 57; c 2. PROSTHODONTICS. *Educ:* Univ Buffalo, DDS, 40; NY Univ, MS, 60. *Prof Exp:* Prof maxillofacial prosthodont, NY Univ Dent Ctr, 71-78, PROF DEPT PROSTHODONT, COL DENT, NY UNIV, 78- *Concurrent Pos:* USPHS fel cancer control, NY Univ, 66-70; dir training prog, Nat Cancer Inst, 74-76; consult, Vet Admin Hosp, 69-; sect ed, J Prosthetic Dent, 69-; consult, Clemson Univ, 72-; pres, Gtr NY Acad Prosthodont Res Found, 78-83; pres & dir, Int Circuit Courses, Am Prosthodontic Soc; vchmn, Ed Coun J Prosthetic Dent, 87- *Honors & Awards:* Ackerman Award, Am Acad Maxillofacial Prosthetics, 73. *Mem:* Fel Am Col Dent; fel Am Col Prosthodont; Int Asn Dent Res; fel Acad Dent Prosthetics (pres, 88-); Am Cleft Palate Asn; fel Int Col Dentists; fel Acad Maxillofacial Prosthetics. *Res:* Cleft palate prosthetics; maxillofacial prosthetics; bio-materials. *Mailing Add:* 40 E 66th St New York NY 10021

ADJEMIAN, HAROUTIOON, b Sept 20, 48; US citizen; c 3. ELECTRICAL ENGINEERING. *Educ:* Worcester Polytechnic Inst, BSEE, 70; Univ Mich, MS, 72, MBA, 74. *Prof Exp:* Engr, Omnispectra, 75-80; dir mkt opers, Nat Micronetics, 80-86; opers mgr, NH Col, 88; DIR, ADVAN DEVELOP GROUP & SALES & MKT, M/A-COM ADAMS RUSSELL, 86- *Mem:* Inst Elec & Electronic Engrs; Int Soc Hybrid Microelectronics; Am Optical Co; Armed Forces Commun & Electronics Asn. *Mailing Add:* PO Box 6536 Portsmouth NH 03802

ADKINS, DEAN AARON, b Wayne, WVa, April 3, 46; m 69; c 1. AQUATIC ENTOMOLOGY. *Educ:* Marshall Univ, AB, 68, MS, 71; Univ Louisville, PhD(aquatic biol), 81. *Prof Exp:* Biol teacher, Wayne High Sch, 68-72; ASST PROF BIOL, MARSHALL UNIV, 72- *Concurrent Pos:* Consult, Dept Health, Cabell-Huntington, 78-80. *Mem:* NAm Benthological Soc; Entom Soc Am; Nat Sci Teachers Asn. *Res:* Aquatic entomology with emphasis upon community interaction and life history data; effects of surface mining on aquatic communities. *Mailing Add:* 4383 Fifth St Rd Huntington WV 25701

ADKINS, JAMES SCOTT, PROTEIN QUALITY RESEARCH, NUTRIENT INTERACTION. *Educ:* Univ Wis, PhD(biochem & nutrit), 61. *Prof Exp:* PROF & CHMN DEPT HUMAN NUTRIT & FOOD, HOWARD UNIV, 76- *Mailing Add:* Dept Human Nutrit & Foods Howard Univ 2400 6th St NW Washington DC 20059

ADKINS, JOHN EARL, JR, b Lynchburg, Va, June 20, 37; m 64; c 2. INDUSTRIAL HYGIENE MONITORING & ANALYSIS. *Educ:* Davidson Col, BS, 58; Univ Tenn, PhD(anal chem), 63; Am Bd Indust Hyg, cert, 79. *Prof Exp:* Res chemist, Plastics Dept, 63-68, div chemist, Tech Dept, 68-72, SR CHEMIST, PETROCHEM DEPT, E I DU PONT DE NEMOURS & CO, INC, 72- *Concurrent Pos:* Chmn, anal chem comt, Am Indust Hyg Asn 87-88. *Mem:* Am Indust Hyg Asn; Am Chem Soc; fel Am Inst Chem. *Res:* Atomic absorption and infrared spectrophotometry; gas chromatography, ambient air quality monitoring and analysis; industrial hygiene monitoring and analytical method development and analysis. *Mailing Add:* E I du Pont de Nemours & Co Box 1089 Orange TX 77631-1089

ADKINS, RONALD JAMES, b Bremerton, Wash, June 28, 32; m 55; c 2. NEUROPHYSIOLOGY. *Educ:* Univ Wash, BS, 54, MS, 57, PhD(physiol, psychol), 65. *Prof Exp:* Res asst neurophysiol, Univ Wash, 57-59; instr psychol & sociol, Lower Columbia Col, 59-61; behavioral trainee physiol & psychol, Univ Wash, 61-65; instr physiol, NY Med Col, 65-66; asst prof zool & physiol, 66-71, ASSOC PROF ZOOL & PHYSIOL, WASH STATE UNIV, 71- *Mem:* AAAS; assoc Am Physiol Soc. *Res:* Physiology of behavior; interaction among descending and ascending systems within the sensory apparatus, currently emphasizing the pyramidal tract and the dorsal column-medial lemniscal system in cats. *Mailing Add:* 605 Fisk Rd Pullman WA 99163

ADKINS, THEODORE ROOSEVELT, JR, b San Antonio, Tex, Dec 26, 30; m 49; c 3. FORENSIC ENTOMOLOGY. *Educ:* Auburn Univ, BS, 52, MS, 54, PhD(entom), 58. *Prof Exp:* Asst entom, Auburn Univ, 56-57; asst prof entom & zool & asst entomologist, Clemson Univ, 57-61, assoc prof entom & zool, 61-68, prof entom, 68-85; LITIGATION CONSULT, 85- *Concurrent Pos:* Entom Soc Am travel grant, 13th Int Cong Entom, Moscow, 68; ed, J Agr Entom, 83- *Mem:* Entom Soc Am; Am Mosquito Control Asn; Sigma Xi. *Res:* Medical-veterinary entomology. *Mailing Add:* One Poplar Dr Clemson SC 29631

ADKINSON, NEWTON FRANKLIN, JR, b NC, May 18, 43; m 66; c 2. CLINICAL IMMUNOLOGY, ALLERGY. *Educ:* Univ NC, Chapel Hill, BA, 65; Johns Hopkins Univ, MD, 69, MA, 84. *Prof Exp:* Intern & resident med, Johns Hopkins Hosp & Univ, 69-71; clin assoc immunol, Lab Immunol, Nat Cancer Inst, Bethesda, Md, 71-73; asst prof, 73-80, ASSOC PROF, MED ALLERGY & CLIN IMMUNOL, 80-, PROF MED, SCH MED, JOHNS HOPKINS UNIV, 87- *Concurrent Pos:* Mem, Immunol Sci Study Sect, NIH, 79-83; physician-in-chg, Asthma & Allergy Clins, Johns Hopkins Hosp, 84-; mem, Allergy & Immunol Rev Comt, NIH, 87-91. *Mem:* Am Asn Immunologists; Am Acad Allergy; Am Thoracic Soc; Am Soc Clin Investr; Am Acad Allergy Immunol; Am Soc Clin Res. *Res:* Drug hypersensitivity states, especially penicillin allergy; immediate hypersensitivity reactions; IgE antibody; prostaglandins and inflammatory reactions, especially in the lung; design and execution of clinical trials. *Mailing Add:* Prof Dept Med John's Hopkins Asthma & Allergy Ctr 301 Bayview Blvd Baltimore MD 21224

ADKINS-REGAN, ELIZABETH KOCHER, b Washington, DC, July 12, 45. BEHAVIORAL BIOLOGY. *Educ:* Univ Md, BS, 67; Univ Pa, PhD(psychol), 71. *Prof Exp:* Asst prof psychol, State Univ NY Col Cortland, 74-75; asst prof, 75-80, ASSOC PROF PSYCHOL, NEUROBIOL & BEHAV, CORNELL

UNIV, 80- *Honors & Awards:* Fulbright Res Scholar Award, 86. *Mem:* Fel AAAS; Animal Behav Soc; Am Ornithologists Union; Inst Soc Psychoneuroendocrinol; Soc Neuroscience; Sigma Xi. *Res:* Role of sex hormones in the development and maintenance of social behavior in vertebrates, especially reproductive behavior. *Mailing Add:* 1878 Ellis Hollow Rd Ithaca NY 14850

ADKISON, CLAUDIA R, b Montgomery, Ala, Dec 9, 41. ANATOMY. *Educ:* Huntingdon Col, BA, 64; Tulane Univ, PhD(anat), 69. *Prof Exp:* Instr myocardial biol, Baylor Col Med & anat, Sch Med, Tulane Univ, 70-71; asst prof, 71-80, ASSOC PROF ANAT, SCH MED, EMORY UNIV, 80- *Concurrent Pos:* NIH training grant pharmacol, Baylor Col Med, 69-70. *Mem:* Am Soc Cell Biol; Electron Micros Soc Am; Am Heart Asn. *Res:* Pepsinogen secretion by chief cells; heart ultrastructure; heart organelle function; parietal cell function. *Mailing Add:* Dept Anat Emory Univ Atlanta GA 30322

ADKISON, DANIEL LEE, b Pulaski, Tenn, Dec 27, 50; m 78; c 2. CRUSTACEAN SYSTEMATICS. *Educ:* Univ W Fla, BS, 74, MS, 81, PhD, 90. *Prof Exp:* Fisheries aid, Nat Marine Fisheries Serv, 72-73; res asst, Univ W Fla, 74-76; res assoc, Dauphin Island Sea Lab, 77-80; grad asst invert zool, Tulane Univ, 80-82; field engr, GEO Viking, 83-85; technician, Tex Agr Anal Serv, 86; res assoc, geochem & environ res group, Tex A&M Univ, 86-88; res assoc, Jackson Lab, 88-89; RES ASSOC, MERCER UNIV SCH MED, 89- *Mem:* Am Asn Zool Nomenclature; Crustacean Soc. *Res:* Systematics of stomatopods, caridean shrimp and epicaridean isopods; crustaceans of deep sea oil and natural gas seep communities. *Mailing Add:* 1699 Wesleyan Bowman Rd Macon GA 31210

ADKISSON, CURTIS SAMUEL, b Little Rock, Ark, Feb 25, 42; m 69; c 1. ORNITHOLOGY. *Educ:* Oberlin Col, BA, 65; Miami Univ, MA, 67; Univ Mich, PhD(zool), 72. *Prof Exp:* Teaching fel zool, Univ Mich, 68-72; asst prof, 72-78, ASSOC PROF ZOOL, VA POLYTECH INST & STATE UNIV, 78- *Concurrent Pos:* Res fel Alexander von Humboldt Found, Bonn, WGer, 79-80. *Mem:* Am Ornith Union; AAAS. *Res:* Dialects in bird vocalizations; behavioral ecology of seed dispersal by birds; systematics and behavior of cardueline finches. *Mailing Add:* Dept Biol Va Polytech Inst & State Univ Blacksburg VA 24061

ADKISSON, PERRY LEE, b Blytheville, Ark, Mar 11, 29; m 56; c 1. ECONOMIC ENTOMOLOGY. *Educ:* Univ Ark, BS, 50, MS, 54; Kans State Col, PhD(entom), 56. *Prof Exp:* Asst prof entom, Univ Mo, 56-57; from assoc prof to prof, Tex A&M Univ, 58-79, head dept, 68-78, vpres agr & renewable resources, 78-80, dep chancellor agr, 80-83, dep chancellor, 83-86, DISTINGUISHED PROF ENTOM, TEX A&M UNIV, 79-, CHANCELLOR, 86- *Concurrent Pos:* Fel, Harvard Univ, 63-64; consult, Int Atomic Energy Agency, Vienna; chmn sci adv panel agr chem, Gov of Tex, 70-71; mem panel of experts on integrated pest control, Food & Agr Orgn, Rome, Italy, 70-74; consult, Hazardous Mat Adv Comt, US Environ Protection Agency, 71 & UC-USAID Pest Mgt Proj, 72-; mem exec comt study prob pest mgt, Environ Sci Bd, Nat Acad Sci- Nat Acad 72-76; 72-; mem, Struct Pest Control Bd Tex, 72-78; chmn comt biol pest species, Nat Res Coun, 73-; mem ed bd, Ann Rev Entom, 74-; mem, US Insect Control Deleg to People's Repub of China, 75; mem UN-Food & Agr Orgn consult on pesticides in agr & pub health, Rome, Italy, 75; mem, US Directorate to UNESCO Man & Biosphere Prog, 75; mem, US Plant Protection Team to USSR, 76; mem, Nat Res Coun Study Comt world food & nutrition, 76; vchmn, Nat Acad Sci Comt use of sci & tech info in regulatory decision-making, 76-77; mem, Nat Sci Bd, 85- *Honors & Awards:* J Everett Bussart Award, Entom Soc Am, 67; Alexander von Humboldt Award, 80. *Mem:* Nat Acad Sci; Entom Soc Am (pres, 74); Am Inst Biol Sci; Sigma Xi; Am Registry Prof Entomologists (pres, 77); Am Acad Arts & Sci. *Res:* Basic and applied research on cotton insects; insect photoperiodism. *Mailing Add:* Chancellor Tex A&M Univ Syst College Station TX 77843-1122

ADLAKHA, RAMESH CHANDER, polyamines, protein phosphorylation & dephosphorylation, for more information see previous edition

ADLDINGER, HANS KARL, b Munich, WGer. VIROLOGY, ONCOGENESIS. *Educ:* Univ Munich, dipl vet med, 61, Dr med vet, 62; Cornell Univ, PhD(vet microbiol), 71. *Prof Exp:* Res assoc, Bavarian State Vaccination Inst, Munich, 61-63; sr res assoc virol, Inst Comp Trop Med, Univ Munich, 64-65; sr res assoc, NY State Vet Col, Cornell Univ, 66-68; res assoc virol, Albert Einstein Med Ctr, Philadelphia, 71; assoc prof, 72-77, PROF MICROBIOL, COL VET MED, UNIV MO, 78-, DIR GRAD STUDIES VET MICROBIOL, 74- *Concurrent Pos:* NATO vis scientist, Plum Island Animal Dis Ctr, USDA, 65-66; vis prof virol, Albert-Ludwig Univ Freiburg, WGer, 82-83. *Mem:* Am Soc Microbiol. *Res:* Virology; immunology; pathogenesis of animal virus diseases; mechanisms of oncogenesis and Herpes virus latency; Epstein-Barr virus transforming genes; gene expression in baculovirus-insect cell systems. *Mailing Add:* 2317 Shepard Blvd Columbia MO 65011

ADLER, ALAN DAVID, b Nyack, NY, Oct 5, 31; m 54; c 3. BIOPHYSICAL CHEMISTRY, MOLECULAR BIOLOGY. *Educ:* Univ Rochester, BS, 53; Univ Pa, PhD(chem), 60. *Prof Exp:* Asst prof molecular biol, Univ Pa, 61-67; staff scientist & assoc prof chem sci & chmn div, New Eng Inst, 67-74; assoc prof chem, 74-77, PROF CHEM, WESTERN CONN STATE COL, 77- *Concurrent Pos:* Mem, Shroud of Turin Res Proj; consult. *Mem:* AAAS; Am Chem Soc; Am Phys Soc; Am Asn Clin Chemists; Sigma Xi; NY Acad Sci. *Res:* Physical, chemical and biological aspects of porphyrin materials; thermodynamics of electrolytic solutions; physical chemistry; history of science. *Mailing Add:* Dept Chem Western Conn State Univ Danbury CT 06810

ADLER, ALEXANDRA, b Vienna, Austria, Sept 24, 01; nat US; m 59. NEUROLOGY, PSYCHIATRY. *Educ:* Univ Vienna, MD, 26. *Prof Exp:* Intern, resident & vis physician, Univ Vienna Hosp, 26-34; res fel, asst & instr neurol, Harvard Univ, 35-44; clin prof psychiat, Duke Univ, 44-46; from asst to assoc clin prof neurol, 46-69, CLIN PROF PSYCHIAT, SCH MED, NY UNIV, 69- *Concurrent Pos:* Vis physician, Goldwater Mem Hosp, 46-56 & Bellevue & Univ Hosps, 46-; psychiatrist, Dept Correction, New York City, 48-72; med dir, Alfred Adler Ment Hyg Clin, 54- *Honors & Awards:* Goldenes Ehrenzeichen der Stadt Wien, Austria, 77. *Mem:* AMA; Am Psychiat Asn; Asn Res Nerv & Ment Dis; Am Acad Neurol; Int Asn Individual Psychol (past pres). *Res:* Psychotherapy; general psychiatry; organic mental syndromes; psycho-pharmacol-therapy. *Mailing Add:* 30 Park Ave Apt 12M New York NY 10016

ADLER, ALFRED, b Ger, Feb 21, 30; US citizen; m 54; c 3. MATHEMATICS. *Educ:* Mass Inst Technol, BS, 52; Univ Calif, Los Angeles, PhD(math), 56. *Prof Exp:* Instr math, Princeton Univ, 56-58; lectr, Mass Inst Technol, 58-60; asst prof, Rutgers Univ, 60-61; vis prof, Univ Bonn, 61-63; from assoc prof to prof, Purdue Univ, 63-67; PROF MATH, STATE UNIV NY STONY BROOK, 67- *Res:* Geometry of complex manifolds. *Mailing Add:* Dept Math State Univ Ny Stony Brook Main Stony Brook NY 11794

ADLER, ALICE JOAN, b Jersey City, NJ, Dec 8, 35; wid; c 3. BIOCHEMISTRY. *Educ:* Barnard Col, AB, 56; Harvard Univ, PhD(phys chem), 61. *Prof Exp:* Fel biochem, Children's Cancer Res Found, Boston, 61-62; fel biol, Mass Inst Technol, 62-64; fel biochem, Oxford Univ, 64-65; sr res assoc biochem, Brandeis Univ, 65-76; assoc scientist, 76-84, SR SCIENTIST, EYE RES INST, BOSTON, 85- *Mem:* AAAS; Am Chem Soc; Asn Res Vision & Ophthal; Int Soc Eye Res; Am Soc Biol Chem. *Res:* Biochemistry of retina and bordering tissues: retinoids and their binding proteins, biochemical studies of retinal adhesion and macular degeneration; physical studies of biological macromolecules and model compounds. *Mailing Add:* Eye Research Inst 20 Staniford St Boston MA 02114

ADLER, BEATRIZ C, b Havana, Cuba, Oct 29, 29; US citizen; m 51. BIOLOGY. *Educ:* Univ Havana, Dr nat sci, 53. *Prof Exp:* Parochial sch teacher, Cuba, 48-61 & Pa, 61-64; from instr to assoc prof biol & phys sci, 64-73, PROF BIOL, CALDWELL COL, 73- *Res:* Geographical survey and methods of fishery of early Cuban Indians. *Mailing Add:* Dept Nat Sci Caldwell Col Caldwell NJ 07006

ADLER, CARL GEORGE, b Buffalo, NY, Oct 3, 39; m 63; c 2. EDUCATIONAL ADMINISTRATION. *Educ:* Univ Notre Dame, BS, 61, PhD(physics), 66. *Prof Exp:* Assoc prof, 65-72, PROF PHYSICS, ECAROLINA UNIV, 72-, CHAIR, 85- *Mem:* Am Asn Physics Teachers; Int Solar Energy Soc. *Res:* Solar energy; quantum mechanics; physics history; relativity. *Mailing Add:* Dept Physics ECarolina Univ Greenville NC 27834

ADLER, ERIC, b Vienna, Austria, Oct 20, 37; US citizen; m 58; c 1. SOLID STATE PHYSICS. *Educ:* City Col New York, BS, 59; Columbia Univ, PhD(physics), 64. *Prof Exp:* Res asst physics, Watson Lab, Columbia Univ, 64-65; asst prof, City Col New York, 65-68; SR ENGR, IBM CORP, 68- *Res:* Theoretical solid state physics; electron-phonon interaction; optical properties of semiconductors; semiconductor device and circuit design and reliability; FET integrated circuits. *Mailing Add:* Dept M65-9721 IBM Corp 1000 River ST Essex Junction VT 05452

ADLER, ERIC L, b Alexandria, Egypt, Dec 10, 30; Can citizen; m 61. ELECTRICAL ENGINEERING, WAVE SCIENCES. *Educ:* Univ London, BSc, 55; Univ Toronto, MASc, 59; McGill Univ, PhD(elec eng), 66. *Prof Exp:* Demonstr elec eng, Univ Toronto, 57-58; instr elec eng, 59-60, lectr, 60-61, from asst prof to assoc prof, 60-73, assoc dean, 77-85, PROF ELEC ENG, MCGILL UNIV, 73- *Mem:* Fel Inst Elec & Electronic Engrs. *Res:* Acoustoelectric effects; thin film transducers; acoustic surface waves and waveguiding; surface wave devices for signal processing applications. *Mailing Add:* Dept Elec Eng McGill Univ 3480 University St Montreal PQ H3A 2A7 Can

ADLER, FRANK LEO, b Graz, Austria, Aug 17, 22; nat US; m 49; c 2. BACTERIOLOGY, IMMUNOLOGY. *Educ:* City Col New York, BS, 47; Univ Ky, MS, 49; Wash Univ, PhD(bact, immunol), 52. *Prof Exp:* Res assoc, Depts Med, Bact & Immunol, Peter Bent Brigham Hosp & Harvard Med Sch, 52-54; mem, Dept Immunol, Pub Health Res Inst City New York, 54-75; chmn, St Jude Children's Res Hosp, 75-84, mem div immunol, 75-87; RES PROF, DARTMOUTH MED SCH, 88- *Concurrent Pos:* Res prof path, Med Sch, NY Univ, 56-75; adj prof microbiol, Univ Tenn, 75-85. *Mem:* Am Asn Immunologists. *Res:* Cellular immunology; immunochemistry; immunoregulation. *Mailing Add:* Dept Microbiol Dartmouth Med Sch Hanover NH 03756

ADLER, GEORGE, b Germany, Nov 25, 20; US citizen; m 49; c 2. SOLID STATE & PHYSICAL ORGANIC CHEMISTRY. *Educ:* City Col New York, BS, 48; Brooklyn Col, MA, 52. *Prof Exp:* Asst lectr, Brooklyn Col, 49-51; rubber technologist, Mat Lab, NY Naval Shipyard, 50-56; chemist, Brookhaven Nat Lab, 56-78; adj res prof, State Univ New York, Stony Brook, 78-83; RETIRED. *Mem:* AAAS; Am Chem Soc; Am Crystallog Asn. *Res:* Solid state polymerization; electron spin resonance; organic solid state chemistry; colloids and monolayers; gas-solid reactions; photo and radiation chemistry of organic solids; polymers. *Mailing Add:* 21 Harvard Rd Shoreham NY 11786

ADLER, HOWARD IRVING, b New York, NY, July 1, 31; m 53; c 4. BACTERIOLOGY. *Educ:* Cornell Univ, BS, 53, MS, 55, PhD, 56. *Prof Exp:* Res assoc, Oak Ridge Nat Lab, 56-57; biologist, 57-69, dir, Biol Div, 69-75, sr staff mem, 75-88; DIR, MICROBIOL, MED DIV, ORAU, 88- *Concurrent Pos:* Adj prof, Univ Tenn, 69-; sr staff mem, Inst Energy Anal, Oak Ridge Assoc Univs, 75-76. *Honors & Awards:* Indust Res-100 Award, 84. *Mem:* Am

Soc Microbiol; Radiation Res Soc; Sigma Xi. *Res:* Mechanisms of radiation damage in microorganisms and control of cell division; anaerobic microbiology. *Mailing Add:* ORAU Med Div PO Box 117 Oak Ridge TN 37831-0117

ADLER, IRVING LARRY, b Philadelphia, Pa, Sept 12, 43; m 66; c 2. ORGANIC CHEMISTRY, ANALYTICAL CHEMISTRY. *Educ:* Temple Univ, BA, 64; Pa State Univ, PhD(org chem), 68. *Prof Exp:* Sr chemist, 68-75, res proj leader, 75-, RES ADM GEN, ROHM & HAAS CO, SPRING HOUSE. *Mem:* Am Chem Soc. *Res:* Organolithium chemistry; pesticide residue analysis; isolation and identification of metabolites; gas chromatography; radiotracers; formulation. *Mailing Add:* 10823 Hawley Rd Philadelphia PA 19154

ADLER, IRWIN L, b New York, NY, July 30, 28; m 54; c 3. FOOD BEVERAGE STABILITY, NEW PRODUCT DEVELOPMENT. *Educ:* Newark Col Eng, BS, 50; NY Univ, MS, 53, ScD, 61. *Prof Exp:* Res mgr, Gen Foods, 54-68; mgr technol ctrs, BeechNut, 68-72; PRES, FFI CORP CONSULTS, 72- *Mem:* Am Chem Soc; Am Inst Chem Engrs; Inst Food Technologists. *Res:* Coffee extraction, roasting and decaffeination; rheology of chocolate coating, puffed food products, stable beverage alcohol emulsion, shelf stable wine and juice beverages and plant by-products utilization. *Mailing Add:* 643 Primrose Lane River Vale NJ 07675

ADLER, JOHN G, b Budapest, Hungary, Sept 29, 35; Can citizen; m 54; c 3. SOLID STATE PHYSICS. *Educ:* Univ BC, BSc, 59; Univ Alta, MSc, 61, PhD(physics), 63. *Prof Exp:* Geophysicist, Imp Oil Ltd, 57-59; asst prof physics, Dalhousie Univ, 63-64; res fel, Case Western Reserve Univ, 64-65, from asst prof to assoc prof, 65-72, PROF PHYSICS, UNIV ALTA, 72- *Concurrent Pos:* NSERC sr indust res fel, 85-86. *Mem:* Am Phys Soc. *Res:* Electron tunneling spectroscopy; superconductivity; low temperature physics; ion mobility in solids; scanning and tunneling microscopy. *Mailing Add:* Dept Physics Univ Alta Edmonton AB T6G 2E1 Can

ADLER, JOHN HENRY, b Brooklyn, NY, 48; m 88. PLANT STEROID CHEMISTRY. *Educ:* Univ Md, College Park, BS, 70, MS, 73, PhD(plant physiol), 75. *Prof Exp:* Fel, 75-76, res assoc biochem, 76-77, asst prof biol sci, Drexel Univ, 77-84; ASSOC PROF BIOL SCI, MICH TECH UNIV, 84-, ASSOC PROF CHEM & HEAD DEPT, 90- *Mem:* Am Chem Soc; Am Soc Plant Physiologists; Am Oil Chemists Soc; Sigma Xi; Am Soc Microbiologists; Mycol Soc Am. *Res:* Examination of the role of sterols and steroids in the form and function of biological membranes; the ontogenetic and phylogenetic implications of different lipid structures; ecdysteroids in plants. *Mailing Add:* Dept Chem Michigan Tech Univ Houghton MI 49931

ADLER, JULIUS, b Edelfingen, Ger, Apr 30, 30; nat US; m 63; c 2. BIOCHEMISTRY, GENETICS. *Educ:* Harvard Univ, AB, 52; Univ Wis, MS, 54, PhD(biochem), 57. *Hon Degrees:* DSc, Univ Tübingen, 87. *Prof Exp:* Fel microbiol, Wash Univ, 57-59; fel biochem, Stanford Univ, 59-60; from asst prof to prof, 60-72, EDWIN BRET HART PROF BIOCHEM & GENETICS, UNIV WIS, MADISON, 72-, STEENBOCK PROF MICROBIOL SCI, 82- *Honors & Awards:* Selman A Waksman Microbiol Award, Nat Acad Sci, 80; Otto-Warburg Medal, Ger Soc Biol Chem, 86; R H Wright Award, 88. *Mem:* Nat Acad Sci; Am Acad Arts & Sci; Am Chem Soc; Am Soc Biol Chemists; fel AAAS; Am Philos Soc. *Res:* Biochemistry and genetics of behavior, especially in microorganisms. *Mailing Add:* Dept Biochem Univ Wis Madison WI 53706

ADLER, KENNETH B, b New York, NY, Nov 21, 45; m 72; c 2. CELL BIOLOGY. *Educ:* Queens Col, BA, 69; Adelphi Univ, MS, 75; Univ Vt, PhD(cell biol), 78. *Prof Exp:* Biophysicist, Brooklyn Vet Admin Med Ctr, NY, 69-75; asst prof cell biol, Univ Vt, Burlington, 80-86; assoc prof, 87-90, PROF CELL BIOL, COL VET MED, NC STATE UNIV, 90- *Concurrent Pos:* Estab investr, Am Heart Asn, 87. *Honors & Awards:* Award for Res Excellence, Smith Kline Beecham Corp, 91. *Mem:* Am Thoracic Soc; Am Soc Cell Biol; Am Asn Pathologists. *Res:* Study of the structure and function of airway epithelial cells in health and disease. *Mailing Add:* Dept Anat Col Vet Med NC State Univ 4700 Hillsborough St Raleigh NC 27606

ADLER, KRAIG (KERR), b Lima Ohio, Dec 6, 40; m 67; c 1. ANIMAL BEHAVIOR, EVOLUTION. *Educ:* Ohio Wesleyan Univ, 62; Univ Mich, MS, 65, PhD(zool), 68. *Prof Exp:* Asst prof, Univ Notre Dame, 68-72; assoc prof, 72-80, chmn neurobiol & behav, Div Biol Sci, 76-79, PROF BIOL, CORNELL UNIV, 80- *Concurrent Pos:* Ed, Jour, Soc Study Amphibians & Reptiles, 58-63, miscellaneous publ, 61-; temporary cur, Mus Zool, Univ Mich, 65; co-ed, Int J Interdisciplinary Cycle Res, Neth, 70-78; assoc ed, Am Midland Nat, 77-82; visiting prof zool, Ariz State Univ, 80; secy-gen World Cong Herpet, 82-89; distinguished scholar, China Prog, US Nat Acad, 84-85; vis fel, Pembroke Col, Univ Cambridge UK, 85; acad lectr, USSR Acad Sci, 86. *Honors & Awards:* Baer mem lectr, 77; Hefner lectr, 80; Anderson mem lectr, 82. *Mem:* Fel AAAS; Am Soc Ichthyologists & Herpetologists; Soc Study Evolution; Animal Behav Soc; Soc Study Amphibians & Reptiles (pres, 82). *Res:* Photoreception, orientation, navigation and circadian rhythms of vertebrates; evolution, systematics and zoogeography of amphibians and reptiles; paleo- and archeozoology. *Mailing Add:* Neurobiol & Behav Seeley G Mudd Hall Cornell Univ Ithaca NY 14853-2702

ADLER, KURT ALFRED, b Vienna, Austria, Feb 25, 05; US citizen; m; c 1. PSYCHIATRY. *Educ:* Univ Vienna, PhD(physics), 35; Long Island Col Med, MD, 41. *Prof Exp:* Chief psychiat serv, US Army, 42-46; DEAN & DIR, ALFRED ADLER INST, NEW YORK, 52-; PSYCHIAT CONSULT, ADVAN CTR PSYCHOTHER, 58- *Concurrent Pos:* Consult psychiatrist, Youth-House, New York, 47-48; emer psychiatrist, Lenox Hill Hosp, New York, 47-; consult, Alfred Adler Inst Minn, 70-; dir, Bowie State Col, 75 & Alfred Adler Inst, Ottawa, Ont; dean, Avan Inst Anal Psychother, 69- *Honors & Awards:* Physicians Recognition Award, AMA, 70 & 78. *Mem:* AMA; fel Am Psychiat Asn; Am Soc Adlerian Psychol; Int Asn Individual Psychol (pres, 63-70); fel Asn Advan Psychother. *Mailing Add:* 30 E 60th St New York NY 10022

ADLER, LASZLO, b Debrecen, Hungary, Dec 3, 32; US citizen; m 63; c 2. PHYSICAL ACOUSTICS. *Educ:* Mich State Univ, MS, 61; Univ Tenn, Knoxville, PhD(physics), 69. *Prof Exp:* Assoc prof physics, Gen Motors Inst, Univ Mich, 60-64; instr physics, Univ Tenn, Knoxville, 66-69, res assoc asst prof, 71-76, res asst prof, 76-80, assoc prof, 80; PROF WELDING ENG & ENG MECH, OHIO STATE UNIV, 80 - *Concurrent Pos:* Consult, Metal & Ceramic Div, Oak Ridge Nat Lab, Tenn, 69 -; vis prof, Univ Paris, 80. *Mem:* Fel Acoust Soc Am; Am Soc Nondestructive Testing. *Res:* Non-linear mechanisms of ultrasonic waves in liquids, especially harmonic and fractional harmonics generation and correlation to liquid structure; defects in metals by interference of a multifrequency pulse echo; physics of non-destructive testing. *Mailing Add:* Ohio State Univ 190 W 19th Ave Columbus OH 43210

ADLER, LAWRENCE, b New York, NY, June 6, 23; m 57; c 3. MINING ENGINEERING. *Educ:* NY Univ, AB, 46; Columbia Univ, BS, 49; Univ Utah, MS, 53; Univ Ill, PhD, 64. *Prof Exp:* Asst prof mining eng, Univ Mo, 55-56, Lehigh Univ, 56-58 & Mich Col Mining & Technol, 58-61; assoc prof mining eng, Va Polytech Inst & State Univ, 63-69, prof, 69-; AT DEPT MINING, WVA UNIV. *Concurrent Pos:* Gen consult. *Mem:* Am Inst Mining, Metall & Petrol Engrs; Am Soc Civil Engrs. *Res:* Design of underground openings; excavating and materials handling. *Mailing Add:* Dept Mining Eng WVa Univ Box 6070 Morgantown WV 26506

ADLER, LOUISE TALE, b Brooklyn, NY, Nov 6, 25; m 49; c 2. IMMUNOREGULATION, TRANSPLANTATION. *Educ:* Univ Miami, BS, 47; Univ Ky, Lexington, MS, 49; Wash Univ, St Louis, Mo, PhD(bact immunol), 53. *Prof Exp:* Teaching fel bact immunol, Harvard Med Sch, 52-54; adj asst prof microbiol, Adelphi Univ, 67-69; assoc immunol, Pub Health Res Inst, New York, 69-75; asst mem, 75-77, assoc mem immunol, St Jude Children's Res Hosp, Memphis, 77-88; RES PROF, DEPT MICROBIOL, DARTMOUTH MED SCH, HANOVER, NH, 88- *Concurrent Pos:* Adj asst prof, Ctr Health Sci Microbiol, Univ Tenn, 77-78, assoc prof, 78-85; consult, ad hoc biomed sci study sect, NIH, 80-85, mem, 85- *Mem:* Am Asn Immunol. *Res:* Regulation of the immune system, using in vitro and in vivo tolerance, allotype suppression systems and genetically defined rabbits; immunocompetence in bone marrow transplantation using the rabbit model. *Mailing Add:* Dept of Microbiol Dartmouth Med School Hanover NH 03756

ADLER, MARTIN E, b Philadelphia, Pa, Oct 30, 29; m 53; c 2. NEUROPHARMACOLOGY, DRUGS OF ABUSE. *Educ:* NY Univ, BA, 49; Brooklyn Col Pharm, BS, 53; Columbia Univ, MS, 57; Albert Einstein Col Med, PhD(pharmacol), 60. *Prof Exp:* From instr to assoc prof, 60-73, PROF PHARMACOL, SCH MED, TEMPLE UNIV, 73- *Concurrent Pos:* NIDA grants, 70-, consult, 75-; chmn, Drug Abuse Biomed Res Review Comt, Nat Inst Drug Abuse, 80- & Comt on Substance Abuse, Am Soc Pharmacol & Exp Therapeut, 84-88; exec secy, Comt on Problems of Drug Dependence, Inc, 86- *Mem:* Fel AAAS; fel Am Col Neuropsychopharmacology; Am Soc Pharmacol & Exp Therapeut; Am Pain Soc; Sigma Xi. *Res:* Neuropharmacology; psychopharmacology; endogenous and exogenous opioid substances; narcotic receptors; brain lesions and narcotic dependence; interaction of drugs of abuse; thermo-regulation. *Mailing Add:* 3247 W Bruce Dr Dresher PA 19025

ADLER, MICHAEL, b Mezoladany, Hungary, Mar 3, 48; nat US; m; c 2. PHARMACOLOGY, NEUROTOXICOLOGY. *Educ:* City Col NY, BS, 71; State Univ NY, PhD(pharmacol), 76. *Prof Exp:* Res assoc, Dept Pharmacol & Exp Therapeut, Sch Med, Univ Md, 76-77; staff fel, Lab Biochem Genetics, Nat Heart, Lung & Blood Inst, NIH, 78-80; sr staff fel, Lab Preclin Studies, Nat Inst Alcohol Abuse & Alcoholism, Alcohol, Drug Abuse & Ment Health Admin, 81-82; PHARMACOLOGIST, NEUROTOXICOL BR, US ARMY MED RES INST CHEM DEFENSE, 82-; ADJ PROF, DEPT BIOL SCI, UNIV MD, BALTIMORE COUNTY, 85- *Concurrent Pos:* Mem, Grant Rev Comt, Nat Inst Drug Abuse, Chem Accident Response & Rescue Team, US Army Med Res Inst Chem Defense & Atropine Equivalence Study Comt. *Mem:* Inst Elec & Electronics Engrs, Eng Med & Biol Soc; Am Soc Pharmacol & Exp Therapeut; NY Acad Sci. *Res:* Synaptic transmission in smooth and skeletal muscle; excitation-secretion coupling in anterior pituitary cells; action of metabolic inhibitors on nerve and muscle fibers; actions of neurotoxins on electrically excitable tissues. *Mailing Add:* Neurotoxicol Br US Army Med Res Inst Chem Defense Aberdeen Proving Ground MD 21010

ADLER, MICHAEL STUART, b Detroit, Mich, Sept 13, 43; m 65; c 2. SEMICONDUCTORS, COMMUNICATION. *Educ:* Mass Inst Technol, BS, 65, MS, 67, PhD(elec eng), 71. *Prof Exp:* Physicist, 71-77, mgr, Advan Archival Memory, 77-79, mgr, Device Physics Unit, 79-81, mgr, Power Semiconductor Br, 81-84, MGR, POWER ELEC LAB, GEN ELECT CO, 84-; MGR POWER ELECTRONICS LAB, 88- *Honors & Awards:* Region I Award, Inst Elec & Electronics Engrs, 80. *Mem:* Fel Inst Elec & Electronic Engrs. *Res:* Semiconductor device physics and design; high doping and high current effects; advanced device concepts; materials and process development; semiconductor packaging; computer aided design. *Mailing Add:* Rm KWC-328 Gen Elec Co PO Box Eight Schenectady NY 12301

ADLER, NORMAN, b Brooklyn, NY, June 19, 28. ANALYTICAL CHEMISTRY, PHYSICAL CHEMISTRY. *Educ:* Brooklyn Col, BS, 49; Polytech Inst Brooklyn, PhD(anal chem), 54. *Prof Exp:* Res analyst, Control Div, Merck & Co, Inc, 53 & 55, chief instrumental anal, 56-64; consult, Arthur D Little, Inc, 64-69; dir radiopharmaceut res & develop, New Eng Nuclear Corp, 70-71, div mgr, Radiopharm Div, 72-75; tech dir & vpres, Clin Assays, Inc, 75-76; vpres, RIA Projs Inc, 77-78. *Concurrent Pos:* Consult, 79- *Mem:* Am Chem Soc; Soc Nuclear Med. *Res:* Pharmaceutical and trace analysis; gas and liquid chromatography; isotope generators; diagnostic parenteral radiopharmaceuticals; in vitro radioactive clinical test kits. *Mailing Add:* 48 Bigelow Ave Watertown MA 02172

ADLER, PHILIP N(ATHAN), b New York, NY, Feb 20, 35; m 56; c 3. METALLURGY, MATERIALS SCIENCE. *Educ:* City Col New York, BChE, 56; Stevens Inst Technol, MS, 59; NY Univ, DEngSc(phys metall), 64. *Prof Exp:* Asst res metallurgist, Gen Cable Corp, 56-59; res engr, Radio Corp Am, 59-61; res asst phys metall, NY Univ, 61-64; eng specialist, Gen Tel & Electronics Labs, Inc, 64-68; LAB HEAD, GRUMMAN AEROSPACE CORP, BETHPAGE, 69- *Concurrent Pos:* Adj lectr, Polytech Inst Brooklyn, 66-67; adj assoc prof, NY Univ, 69-73. *Mem:* Am Inst Mining, Metall & Petrol Engrs; Am Soc Metals. *Res:* Physical metallurgy of nonferrous metals; high-pressure research; thermodynamics and kinetics of phase transformations; electronic materials; stress corrosion; strengthening mechanisms; hydrogen embrittlement. *Mailing Add:* 40 Flower Lane Roslyn Heights NY 11577

ADLER, RALPH PETER ISAAC, b Bombay, India, Mar 10, 37; US citizen; m 60; c 2. RAPID SOLIDIFICATION TECHNOLOGY, MANUFACTURING SCIENCES. *Educ:* Stanford Univ, BS, 58, MS, 59; Yale Univ, DEng, 65. *Prof Exp:* Assoc scientist, Mat Dept, Lockheed Aircraft Corp, 58-59; res scientist, Mat Res Lab, Martin Marietta Corp, 64-68; group leader, Mat Res & Develop Sect, Brunswick Corp, 68-75; mem tech staff, Ceramics & Metall Tech Ctr, GTE Labs Inc, 75-85; SUPVRY METALLURGIST CHIEF, METALS RES BR, ARMY MAT TECHNOL LAB, 85- *Concurrent Pos:* Lectr, Fac Col Eng & Grad Sch Eng, Northeastern Univ, 81- *Mem:* Am Soc Metals; Metall Soc Am; Inst Mining & Metall Eng. *Res:* Research and development of advanced materials and processes such as by rapid solidification technology for defense and civilian applications; cost effective product improvements by concurrent engineering and computer aided intelligent manufacturing science; interaction of federal science policy; education and technical resource allocation on national techno-economic interests. *Mailing Add:* US Army Mat Technol Lab SLCMT-EMM Watertown MA 02172

ADLER, RICHARD, b Lebanon, Pa, Jan 7, 48; m 70; c 4. VIROLOGY. *Educ:* Pa State Univ, BS, 69, PhD(microbiol), 73. *Prof Exp:* Res fel, Roche Inst Molecular Biol, 73-75; res assoc II, Ann Arbor, 75-77, asst prof biol, 77-83, ASSOC PROF BIOL & MICROBIOL, UNIV MICH, DEARBORN, 83- *Concurrent Pos:* Vis assoc prof microbiol, Univ Western Ont, London, Ont, 84. *Mem:* Am Soc Microbiol. *Res:* Regulation of coronavirus replication. *Mailing Add:* Dept Natural Sci Univ Mich Dearborn MI 48128

ADLER, RICHARD B(ROOKS), electrical engineering; deceased, see previous edition for last biography

ADLER, RICHARD H, b Buffalo, NY, July 2, 22; m 70; c 5. SURGERY. *Educ:* Univ Buffalo, MD, 45; Univ Colo, MS, 55; Am Bd Surg, cert, 51; Am Bd Thoracic Surg, cert, 54. *Prof Exp:* Asst physicist, Sch Med, Univ Buffalo, 49; instr surg, Med Sch, Univ Mich, 50-52; hon registr, Brompton Hosp, London, Eng, 52-53; from asst prof to assoc prof, 55-67, dir thoracic surg, 75-88, PROF SURG, SCH MED, STATE UNIV NY, BUFFALO, 67- NY BUFFALO, 67-; DIR THORACIC SURG & ATTEND SURGEON, BUFFALO GEN HOSP, 57- *Concurrent Pos:* Markle scholar, 57-; attend thoracic surgeon, Vet Hosp, Buffalo, 55-; asst attend surgeon, Buffalo Children's Hosp, 56-; consult, Roswell Park Mem Inst, 57-; vis prof surg, Nat Defense Med Ctr, Taipei, Taiwan, 68-69; assoc consult, Erie County Med Ctr; consult thoracic surg, Millard Fillmore, Brooks Mem & Tri-County Hosps. *Mem:* AAAS; Am Col Surg; Am Asn Thoracic Surg; Am Heart Asn; Am Col Chest Physicians; Soc Univ Surgeons. *Res:* General and thoracic surgery. *Mailing Add:* Buffalo Gen Hosp 100 High St Buffalo NY 14203

ADLER, RICHARD JOHN, b Port Alberno, BC, Feb 1, 55; m 73; c 2. INTENSE CHARGED PARTICLE BEAMS. *Educ:* Univ Alta, BSc, 76; Cornell Univ, PhD(elec eng), 80. *Prof Exp:* Res asst, Defence Res Estab Pac, 76, Dept Elec Eng, Cornell Univ, 77-80; tech staff mem, Mission Res Corp, 80-85; staff scientist, Pulse Sci, Inc, 86-87; PRES, NORTH STAR RES CORP, 88- *Mem:* Am Physiol Soc; Mat Res Soc. *Res:* Experimental physics of intense beams in vacuum, including magnetized beams, induction accelerators, electron and ion sources; collective acceleration theories of macroscopic beam behavior; plasma physics; high power electronics. *Mailing Add:* North Star Res Corp 5555 Zuni SE Suite 345 Albuquerque NM 87108

ADLER, RICHARD R, MAMMALIAN EMBRYOLOGY, CELL SURFACE INTERACTIONS. *Educ:* Univ Chicago, PhD(develop biol), 84. *Prof Exp:* RES ASSOC, WORCESTER FOUND EXP BIOL, 84- *Mailing Add:* Dept Cell Biol 222 Maple Ave Worcester Found Exp Biol Shrewsbury MA 01604

ADLER, ROBERT, b Vienna, Austria, Dec 4, 13; nat US; m 46. ACOUSTICS, OPTICS & ELECTRONOPTICS. *Educ:* Univ Vienna, PhD(physics), 37. *Prof Exp:* Asst to patent attorney, Vienna, Austria, 37-38; in chg lab, Sci Acoust, Ltd, London, 39-40 & Assoc Res, Inc, Chicago, 40-41; res engr, Zenith Radio Corp, 41-52, assoc dir res, 52-63, vpres res, 63-77; dir res, Extel Corp, 78-82; TECH CONSULT, ZENITH ELECTRONICS CORP, 82- *Concurrent Pos:* Vis prof, Univ Ill, Urbana, 78; regent lectr, Univ Calif, Santa Barbara, 85. *Honors & Awards:* Edison Medal, Inst Elec & Electronics Engrs, 80. *Mem:* Nat Acad Eng; AAAS; fel Inst Elec & Electronics Engrs. *Res:* Electron beam parametric amplifiers; opto-acoustic interaction devices; surface wave amplifiers; video disc recording and playback; image display devices. *Mailing Add:* Zenith Electronics Corp Glenview IL 60025

ADLER, ROBERT ALAN, b Somerville, NJ, Jan 11, 45; m 72; c 2. INTERNAL MEDICINE, ENDOCRINOLOGY. *Educ:* Johns Hopkins Univ, BA, 67, MD, 70. *Prof Exp:* Clin investr & asst chief endocrinol, Fitzsimons Army Med Ctr, 74-76; asst prof med, endocrinol, Dartmouth Med Sch, 72-84; ASSOC PROF MED ENDOCRINOL, MED COL, VA, 84- *Concurrent Pos:* Clin & res fel, Mass Gen Hosp, 72-73 & Walter Reed Army Med Ctr, 73-74; clin instr med, Univ Colo Med Sch, 75-76; staff physician, Mary Hitchcock Mem Hosp, 76-84; consult endocrinol, Vet Admin Hosp, White River Jct, Vt, 78-84; chief, endocrinol & metabol, McGuire Vet Admin Med Ctr, Richmond, Va, 84- *Mem:* Endocrine Soc; Am Soc Bone & Mineral Res; Am Fedn Clin Res; AAAS; fel Am Col Physicians; Southern Soc Clin Investigation. *Res:* Pituitary physiology, pathology and biochemistry. *Mailing Add:* McGuire Vet Admin Med Ctr 1201 Broad Rock Blvd Richmond VA 23249

ADLER, ROBERT FREDERICK, b West Reading, Pa, Jan 19, 44; m 72; c 2. METEOROLOGY. *Educ:* Pa State Univ, BS, 65, MS, 67; Colo State Univ, PhD(atmospheric sci), 74. *Prof Exp:* Res meteorologist, Navy Weather Res Fac, Dept of Defense, 67-71; METEOROLOGIST, GODDARD SPACE FLIGHT CTR, NASA, 74- *Concurrent Pos:* Res assoc meteor, Colo State Univ, 74. *Mem:* Am Meteorol Soc. *Res:* Dynamics and physics of intense atmospheric convection using satellite and aircraft remote sensing; application to detection of severe thuderstorms and to precipitation estimation from satellite observations. *Mailing Add:* 12532 Woodridge Lane Highland MD 20777

ADLER, ROBERT GARBER, b Upland, Calif, July 11, 29; m 63; c 3. ANALYTICAL CHEMISTRY. *Educ:* Calif Inst Technol, BSc, 51; Univ Southern Calif, MSc, 55; Univ Calif, Riverside, PhD(inorg chem), 68; Am Bd Inst Hyg, Cert Chem Aspects of Indust Hyg. *Prof Exp:* Sr chemist, Whittier Res Lab, Am Potash & Chem Corp, 57-63; chemist, Sci Ctr, NAm Aviation, Inc, Calif, 63-64; asst prof chem, Bethel Col, Kans, 68-71; anal chemist, Occup Safety & Health Admin, US Dept Labor, Utah, 72-74, dir, Div Qual Control, 74-86, METHODS DEVELOP CHEMIST, OCCUP SAFETY & HEALTH ADMIN, US DEPT LABOR, UTAH, 86- *Mem:* Am Chem Soc; Am Indust Hyg Asn; Am Conf Govt Indust Hygienists. *Res:* Boron hydrides, especially preparations and properties; improvement of analytical chemical techniques for industrial hygiene analyses. *Mailing Add:* 1781 S 300 W Salt Lake City UT 84165-0200

ADLER, ROBERT J, b Staten Island, NY, June 22, 31; m 61; c 2. CHEMICAL ENGINEERING. *Educ:* Lehigh Univ, BS, 54, MS, 57, PhD(chem eng), 59. *Prof Exp:* Mem staff, Alloy Tile Corp, 49-54; polymer chemist, Shell Chem Corp, 54-55; asst prof chem eng, 59-64, assoc prof-in-chg, 64-69, head div, 69-74, PROF CHEM ENG, CASE WESTERN RESERVE UNIV, 69- *Concurrent Pos:* NSF res grants, 60-63 & 64-; consult var indust; mem bd, Helipump Corp, 74- *Mem:* fel Am Inst Chem Engrs; Am Chem Soc; Am Soc Eng Educ; Sigma Xi. *Res:* Separation, mixing, helical flow devices and processes. *Mailing Add:* 3068 Van Aken Blvd Shaker Heights OH 44120

ADLER, RONALD JOHN, b Pittsburgh, Pa, Apr 17, 37. THEORETICAL PHYSICS, GENERAL PHYSICS. *Educ:* Carnegie Mellon Univ, BS, 59; Stanford Univ, PhD(physics), 65. *Prof Exp:* Res asst prof physics, Univ Wash, 64-66; res assoc, Univ Colo, 66-67; asst prof, Va Polytech Inst, 67-70; assoc prof, Am Univ, 70-73; prof physics, Fed Univ Pernambuco, Brazil, 74-77; SR SCIENTIST, LOCKHEED PALO ALTO LAB, 79- *Concurrent Pos:* mem, Inst Adv Study, 73-74. *Mem:* Am Phys Soc; fel Brit Interplanetary Soc. *Res:* High energy theory; two nucleon problem; general relativity; mathematical physics. *Mailing Add:* Lockheed Res Lab 97-40 Bldg 202 3251 Hanover St Palo Alto CA 94304

ADLER, ROY LEE, b Newark, NJ, Feb 22, 31; m 53; c 2. MATHEMATICS. *Educ:* Yale Univ, BS, 52; Columbia Univ, AM, 54; Yale Univ, PhD(math), 61. *Prof Exp:* Jr engr, Nat Union Radio, 52 & Bendix Aviation Corp, 53-54; RES STAFF MEM MATH, INT BUS MACH CORP, 60- *Concurrent Pos:* Adj prof, Columbia Univ, 63-65 & Yeshiva Univ, 65-66; vis prof, Stanford Univ, 70- *Mem:* Am Math Soc; Math Asn Am; Sigma Xi. *Res:* Ergodic theory; classification and structure problems of dynamical systems. *Mailing Add:* Watson Res Ctr IBM Box 218 Yorktown Heights NY 10598

ADLER, RUBEN, b Los Toldos, Argentina, Nov 10, 40. SURVIVAL & DIFFERENTIATION OF PHOTORECEPTORS. *Educ:* Univ Buenos Aires, MD, 63. *Prof Exp:* Asst res biologist, Univ Calif, San Diego, 78-83; ASSOC PROF NEUROSCI & OPHTHAL, SCH MED, JOHNS HOPKINS UNIV, 83- *Mem:* Neurosci Soc; Am Soc Cell Biol; Asn Res Vision & Opthal. *Mailing Add:* Wilmer Inst Johns Hopkins Univ 601 N Broadway Baltimore MD 21205

ADLER, SANFORD CHARLES, b Yonkers, NY, Jan 19, 40; m 64; c 1. OPERATIONS RESEARCH, SAFETY ENGINEERING. *Educ:* NY Univ, BS, 63, MS, 65, MS, 71. *Prof Exp:* Sci programmer comput appln physics, Nevis Cyclotron Labs, 63-64; res asst & asst res scientist ergonomics, Sch Eng & Sci, NY Univ, 64-66; systs engr comput systs, Int Bus Mach Corp, 66-68; prof mgt sci, Indust Col Armed Forces, 68-70; opers res analyst safety & data processing, Nat Bur Standards, 70-84; OPERS RES ANALYST, DEPT JUSTICE, 84- *Mem:* AAAS. *Res:* Health and safety; accessibility of buildings to the disabled; ergonomics; risk analysis. *Mailing Add:* 14238 Briarwood Terr Rockville MD 20853

ADLER, SEYMOUR JACOB, b New York, NY, May 12, 18; m 47; c 3. ANALYTICAL CHEMISTRY. *Educ:* Cooper Union, BChE, 41; Columbia Univ, MA, 50, PhD(chem), 54. *Prof Exp:* Anal chemist, Columbia Mineral Beneficiation Lab, 53-56, Brookhaven Nat Lab, 56-57 & Radio Corp Am, NJ, 57-69; asst prof chem, Trenton State Col, 70-88; RETIRED. *Concurrent Pos:* Adj prof, Trenton State Col, 88- *Mem:* Am Chem Soc. *Res:* Trace analysis; spectrophotometry; radio-chemistry; emission spectrography; solvent extraction; ion exchange. *Mailing Add:* Dept of Chem Trenton State Col Trenton NJ 08625

ADLER, SOLOMON STANLEY, b New York, NY, May 26, 45; m 67; c 5. HEMATOLOGY, ONCOLOGY. *Educ:* City Col New York, BS, 66; Albert Einstein Col Med, MD, 70. *Prof Exp:* Instr med, 73-75, chief, Spec Hemat Lab, 74-77, asst prof, 75-78, ASSOC PROF MED, RUSH MED COL, 78-; CHIEF, SPEC MORPHOL LAB, SECT HEMAT, PRESBY-ST LUKE'S MED CTR, 76- *Concurrent Pos:* Adj attend physician, Presbyterian-St Luke's Hosp, 74-75, asst attend physician, 75-78, assoc attend physician, 78-82; NIH res fel, Nat Cancer Inst, 74-77, sr attend physician, 82- *Mem:* Fel Am Col Physicians; Am Soc Clin Oncol; Am Fedn Clin Res; Am Soc Hematol. *Res:* Hemopoiesis in vitro and hemopoietic microenvironment; bone marrow transplantation. *Mailing Add:* 1753 W Congress Pkwy Chicago IL 60612

ADLER, STEPHEN FRED, b Berlin, Ger, Sept 27, 30; nat US; m 51; c 2. INDUSTRIAL CHEMISTRY. *Educ:* Roosevelt Univ, BS, 51; Northwestern Univ, MS, 53, PhD(inorg chem), 54. *Prof Exp:* Res chemist, Am Cyanamid Co, 54-60, group leader, 60-69; sect mgr, Stauffer Chem Co, 68-70, mgr chem res dept, 70-76, asst dir, 76-79, dir, Eastern Res Ctr, Stauffer Chem Co, 79-87; dir, 87-90, SITE DIR, DUBBS FERRY RES CTR, AKZO CHEMICALS INC, 90- *Mem:* Am Chem Soc; NY Acad Sci. *Res:* Product development of inorganic chemicals; catalysts. *Mailing Add:* Akzo Chem Dubbs Ferry Res Ctr Livingston Ave Dobbs Ferry NY 10522

ADLER, STEPHEN L, b New York, NY, Nov 30, 39; div; c 3. THEORETICAL HIGH ENERGY PHYSICS. *Educ:* Harvard Univ, AB, 61; Princeton Univ, PhD(physics), 64. *Prof Exp:* Jr fel, Soc Fel, Harvard Univ, 64-66; mem, 66-69, prof theoret physics, 69-79, ALBERT EINSTEIN PROF, INST ADVAN STUDY, PRINCETON, NJ, 79- *Honors & Awards:* JJ Sakurai Prize, Am Phys Soc, 88. *Mem:* Nat Acad Sci; fel Am Phys Soc; fel Am Acad Arts & Sci; fel AAAS. *Res:* Theoretical research on problems in elementary particle physics and quantum field theory. *Mailing Add:* 287A Nassau St Princeton NJ 08540

ADLER, VICTOR EUGENE, b New York, NY, Jan 18, 24; m 65; c 3. ENTOMOLOGY. *Educ:* Memphis State Univ, BS, 50; Kans State Univ, MS, 56. *Prof Exp:* Biol aide plant pest control, Agr Environ Qual Inst, Insect Chem Ecol Lab, Agr Res Ctr, USDA, 56-57, entomologist, 57-61, res entomologist, Grain & Forage Insects Res Br, 61-65, Pesticide Chem Res Br, 65-66, Pesticide Chem Res Div, 66-72 & Agr Environ Qual Inst, Insect Chem Ecol Lab, 72-86; RETIRED. *Mem:* Entom Soc Am. *Res:* Safer chemical and practical applications of insecticide for controlling insect pests; study of insect attractant and repellents electrophysiologically and behaviorally; cause of homosexuality in all animals where two sexes are to be found, mainly humans and insects. *Mailing Add:* 8540 Pineway Ct Laurel MD 20707

ADLER, WILLIAM, b Buffalo, NY, Sept 3, 39. CLINICAL IMMUNOLOGY. *Educ:* State Univ NY, Buffalo, MD, 65. *Prof Exp:* CHIEF CLIN IMMUNOL, NIA, NIH, BALTIMORE, MD, 74- *Concurrent Pos:* Assoc prof pediat, Sch Med, Johns Hopkins Univ. *Mem:* Am Asn Immunol; Soc Pediat Res; Geront Soc Am; Clin Immunol Soc; Soc Anal Cytometry. *Res:* Investigation of the immunodeficiencies associated with ageing and secondary to the HIV infection and the use of abused drugs. *Mailing Add:* Francis Scott Key Med Ctr Nat Inst Ageing NIH Baltimore MD 21224

ADLER, WILLIAM FRED, b Chicago, Ill, Aug 19, 37; m 58; c 1. PARTICULATE EROSION OF MATERIALS, ENGINEERING MECHANICS. *Educ:* Ill Inst Technol, BS, 58, MS, 61; Columbia Univ, PhD(eng mech), 65. *Prof Exp:* Teaching asst, Ill Inst Technol, 58-61; sr res engr, Denver Div, Martin-Marietta Corp, 61-62; sr res scientist, Columbus Labs, Battelle Mem Inst, 65-71; prin res scientist, Bell Aerospace Co, 71-76; dir, Mat Sci Sect, Effects Technol, Inc, 76-79, assoc mgr, 79-81, mgr mat group, 81-82; MGR, MAT SCI SECT, GEN RES CORP, 83- *Concurrent Pos:* Chmn, Nat Mat Adv Bd Comt, Nat Acad Sci, 76-77; Int adv, Fifth Int Conf, 79, Sixth Int Conf, 83 & Seventh Int Conf, Erosion by Liquid & Solid Impact, Cambridge, UK, 87. *Mem:* Am Ceramic Soc; Am Soc Testing & Mat; Soc Eng Sci; Sigma Xi. *Res:* Application of continuum mechanical to problems in technology; theoretical and experimental studies in erosion of material surfaces by solid and fluid particles; fracture mechanics; development of testing procedures for evaluating the dynamic response of materials; non-destructive evaluation of materials and structures; structural analysis; microstructural analysis of materials. *Mailing Add:* General Research Corp 5383 Hollister Ave Santa Barbara CA 93111

ADLERSTEIN, MICHAEL GENE, b New York, NY. ELECTRONIC DEVICES. *Educ:* Polytech Inst, Brooklyn, BS, 66, MS, 66; Harvard Univ, SM, 71, PhD(appl physics), 71. *Prof Exp:* Sr scientist, 71-80, prin scientist, 80-87, CONSULT SCIENTIST, RES DIV, RAYTHEON CO, 87- *Concurrent Pos:* Assoc ed, Inst Elec & Electronics Engrs Electron Divices, 81-; Conf tech Comt, Cornell Univ. *Mem:* Am Phys Soc; sr mem Inst Elec & Electronics Engrs; Sigma Xi. *Res:* Microwave semiconductor devices; impact avalanche and transit time diodes and components for frequencies up to millimeter waves; gallium arsenide field effect transistors; millimeter wave monolithic integrated circuits; heterojunction bipolar transistors. *Mailing Add:* Raytheon Res Div 131 Spring St Lexington MA 02173

ADLERZ, WARREN CLIFFORD, entomology; deceased, see previous edition for last biography

ADLOF, RICHARD OTTO, b Vienna, Austria, Dec 14, 47; US citizen; m 76; c 2. FATS & FATTY ACIDS. *Educ:* Wabash Col, BS, 70; La State Univ, MS, 75. *Prof Exp:* RES CHEMIST, NAT CTR AGR UTIL RES, AGR RES SERV, USDA, 69- *Mem:* Am Chem Soc; Am Oil Chemists Soc. *Res:* Synthesis of deuterium-labelled fats; utilization of deuterium-labelled fats to study the metabolism of fatty acids and fatty acid isomers in humans; develop new methods of lipid and lipoprotein fractionation and analysis. *Mailing Add:* USDA Nat Ctr Agr Util Res 1815 N University St Peoria IL 61604

ADMAN, ELINOR THOMSON, b New York, NY, Jan 3, 41; c 2. BIOLOGICAL STRUCTURE. *Educ:* Col Wooster, BA, 62; Brandeis Univ, MA, 64, PhD(phys chem), 67. *Prof Exp:* Sr fel crystallog, Univ Wash, 67-71, res assoc, 71-77, from res asst prof to res assoc prof, 77-90, RES PROF BIOL STRUCT, UNIV WASH, 90- *Concurrent Pos:* Mem, Metallobiochem Study Sect, NIH, 87-91 & US Nat Comt Crystallog, 90; adj res prof biochem, Univ Wash, 90- *Mem:* AAAS; NY Acad Sci; Am Crystallog Asn. *Res:* Determination and refinement of protein structures, iron-sulfur proteins, blue copper proteins; nitrite reductase. *Mailing Add:* Dept Biol Struct SM-20 Univ Wash Seattle WA 98195

ADMAN, RAYMOND LANCE, biochemistry, for more information see previous edition

ADNEY, JOSEPH ELLIOTT, JR, b DeLand, Fla, Aug 20, 23; m 52; c 2. MATHEMATICS. *Educ:* Stetson Univ, BS, 44; Ohio State Univ, MA, 49, PhD(math), 54. *Prof Exp:* Assoc, Res Found, Ohio State Univ, 52-54, instr math, 54-55; asst prof, Purdue Univ, 55-64; assoc prof, 64-69, PROF MATH, MICH STATE UNIV, 69-, CHMN DEPT, 74- *Concurrent Pos:* Consult, US Air Force, 57, 59. *Mem:* Am Math Soc. *Res:* Abstract algebra; ground theory. *Mailing Add:* Dept of Math Mich State Univ East Lansing MI 48824

ADOLPH, ALAN ROBERT, b New York, NY, Feb 5, 32; m 57; c 3. NEUROPHYSIOLOGY. *Educ:* Rensselaer Polytech Inst, BEE, 53; Mass Inst Technol, SM, 57; Rockefeller Inst, PhD(biol sci), 63. *Prof Exp:* Sr scientist, Bolt, Beranek & Newman, Inc, 63-64; HEAD NEUROSCI LAB, EYE RES INST OF RETINA FOUND, 64- *Concurrent Pos:* Consult, Lockheed Missile & Space Co Labs, 58-59 & Stanford Res Inst, 62. *Mem:* Am Physiol Soc; Soc Neurosci; Asn Res Vision & Ophthal. *Res:* Neurophysiology and pharmacology of the retina; electrophysiology of sensory neurons; marine neurobiology; mathematical biophysics; visual behavior; psychophysiology of vision. *Mailing Add:* Eye Res Inst Retina Found 20 Staniford St Boston MA 02114

ADOLPH, EDWARD FREDERICK, b Philadelphia, Pa, July 5, 95; m 21; c 3. PHYSIOLOGY, GENERAL. *Educ:* Harvard Col, Ab, 16; Harvard Univ, PhD(biol), 20. *Prof Exp:* Fel physiol, Oxford Univ, Eng, 20-21; instr zool, Marine Biol Lab, Univ Pittsburgh, 21-24; fel biol, Johns Hopkins Univ, 25-25; from asst prof to prof physiol, Univ Rochester, 25-60, emer prof, 60-80; RETIRED. *Concurrent Pos:* Res assoc, Harvard Univ, 36-37; investr, Off Sci Res & Develop, 42-45; mem, Panel Physiol, Res & Develop Bd, 48-53. *Honors & Awards:* Presidential Award of Merit, 48. *Mem:* Am Physiol Soc (pres, 53); Am Acad Arts & Sci; Am Soc Zoologists; Soc Gen Physiologists; Soc Develop Biol; Am Soc Naturalists. *Res:* Self-regulation of bodily processes; changes in body water; responses to heat, cold and low oxygen; physiological adaptations and developments from exposures and withdrawals. *Mailing Add:* 120 Renn Wood Dr Rochester NY 14625

ADOLPH, HORST GUENTER, b Pforzheim, Ger, Nov 27, 32; m 60; c 2. ORGANIC CHEMISTRY. *Educ:* Univ Tuebingen, Dr rer nat, 59. *Prof Exp:* Res asst, Inst Appl Chem & asst to ed, Houben-Weyl, Univ Tubingen, 61; RES CHEMIST, US NAVAL SURFACE WARFARE CTR, 61- *Mem:* Am Chem Soc. *Res:* Synthetic organic chemistry; reaction mechanisms; spectroscopy; chemistry of explosives; polymer chemistry. *Mailing Add:* 23 Scarlet Sage Court Burtonsville MD 20866

ADOLPH, KENNETH WILLIAM, b Oldham, Eng, Dec 11, 44; US citizen. ELECTRON MICROSCOPY, PROTEIN CHEMISTRY. *Educ:* Univ Wis, Milwaukee, BS, 66, MS, 67; Univ Chicago, PhD(biophysics), 72. *Prof Exp:* Fel, Lab Molecular Biol, Med Res Coun, Eng, 73-74, Rosenstiel Basic Med Sci Res Ctr, Brandeis Univ, 75, & Dept Biochem Sci, Princeton Univ, 76-78; asst prof, 78-84, ASSOC PROF, DEPT BIOCHEM, MED SCH, UNIV MINN, MINNEAPOLIS, 84- *Mem:* Am Soc Biochem & Molecular Biol; Am Soc Cell Biol; Am Soc Virol. *Res:* Electron microscopy and image analysis of chromosome organization; role of non-histone proteins; virus assembly. *Mailing Add:* Dept Biochem Med Sch 4-225 Millard Hall Univ Minn 435 Delaware St SE Minneapolis MN 55455

ADOLPH, ROBERT J, b Chicago, Ill, May 12, 27; m 58; c 3. INTERNAL MEDICINE. *Educ:* Univ Ill, MD, 52; Am Bd Internal Med, cert, 63; Am Bd Cardiovasc Dis, dipl. *Prof Exp:* Asst prof med, Univ Ill, 59-60; from asst prof to assoc prof, 62-70, PROF MED, COL MED, UNIV CINCINNATI, 70- *Concurrent Pos:* Am Heart Asn res fel, Col Med, Univ Ill, 56-58; NIH res fel physiol & biophys, Sch Med, Univ Wash, 60-62; mem staff, Dept Internal Med, Cardiac Res Lab, Cincinnati Gen Hosp, 62-, dir lab, 70-; mem med adv bd, Sect on Circulation & Coun Clin Cardiol, Am Heart Asn. *Mem:* Am Fedn Clin Res; Am Col Cardiol. *Res:* Cardio-vascular physiology and investigation; myocardial contractility; effect of digitalis on myocardial electrolytes; high out-put states; cardiac pacing; biomedical engineering developments. *Mailing Add:* Cardiac Res Lab Rm 3407 Med Sci Bldg Univ Cincinatti Col Med Cincinnati OH 45267

ADOMAITIS, VYTAUTAS ALBIN, ecological chemistry, geochemistry, for more information see previous edition

ADOMIAN, GEORGE, b Buffalo, NY; m 56; c 4. NONLINEAR STOCHASIC DYNAMICAL SYSTEMS. *Educ:* Univ Mich, BS, MS; Univ Calif, Los Angeles, PhD(theoret physics), 63. *Prof Exp:* Res engr, Gen Elec Co, 48-49; instr math, Wayne State Univ, 51-52; res assoc, Univ Mich, 52-53; sr scientist, Hughes Aircraft Co, 53-64; prof math & res prof eng, Pa State Univ, 64-66; distinguished prof math & dir, Ctr Appl Math, Univ Ga, 66-90; CHIEF SCIENTIST, GENERAL ANALYTICS CORP, 90- *Concurrent Pos:* Consult govt & indust; assoc ed, J Math Anal & Appln & J Nonlinear Anal-Theory & Appln, 86-; consult ed, Soviet J Contemp Anal, 81-, Soviet J Contemp Phys, 81-; guest ed, Pergamon Press, 86- *Honors & Awards:* Richard E Bellman Prize, 89. *Mem:* Fel AAAS; Am Phys Soc; Am Math Soc; sr mem Inst Elec & Electronics Engrs; Soc Indust & Appl Math. *Res:* Nonlinear stochastic partial differential equations and stochastic operator theory; theoretical physics and engineering applications. *Mailing Add:* 155 Clyde Rd Athens GA 30605

ADOMIAN, GERALD E, CARDIAC TISSUE, NERVE TISSUE. *Educ:* Univ Calif, Los Angeles, PhD(biol), 80. *Prof Exp:* ASSOC PROF MED & DIR FUNCTIONAL MORPHOLOGY, RES & EDUC INST, HARBOR-UCLA MED CTR, 75- *Mailing Add:* Dir Funct Morphol Lab Res & Educ Inst A E Mann Found Sci Res 12744 San Fernando Rd Sylmar CA 91342

ADORNO, DAVID SAMUEL, mathematics, statistics, for more information see previous edition

ADRAGNA, NORMA C, b Cordoba, Arg. ION TRANSPORT. *Educ:* Nat Univ Cordoba, BCh, 69, PhD(biophys), 73. *Prof Exp:* Teaching asst, physico-chem, Nat Univ Cordoba, 66-69; res fel, Nat Res Coun Arg, 70-74; tech asst, biomed phys, Nat Univ Buenos Aires, Arg, 74-76; int res fel, NIH Fogarty Int Ctr, 77-79; res assoc biomed phys, Harvard Univ, 79-82; med res asst prof, Duke Univ, 82-85; ASSOC PROF PHARMACOL, WRIGHT STATE UNIV, 85- *Concurrent Pos:* Lab asst, Nat Univ Cordoba, Arg, 69; res assoc, Nat Univ Buenos Aires, Arg, 74-76; vis prof, biomed phys, Ctr Nuclear Studies, Saclay, France, 76-77. *Honors & Awards:* John Lawrence Mem Award, Am Heart Asn, 86. *Mem:* Biophys Soc; Am Physiol Soc; Soc Gen Physiologists. *Res:* Studies of ion and water transport across artificial and biological membranes; red cells, cultured endothelial cells and frog bladder; the relationship between ion transport and exercise training and/or essential hypertension. *Mailing Add:* Dept Pharmacol/Toxicol Wright State Univ Dayton OH 45435

ADRIAN, ALAN PATRICK, b Kaukauna, Wis, Feb 13, 16; m 42; c 2. CHEMISTRY. *Educ:* Lawrence Univ, AB, 38, MS, 40, PhD(paper chem & technol), 42. *Prof Exp:* Develop chemist, Detroit Sulphite Pulp & Paper Co, 42-45, tech dir, 45-50, oper mgr, 50-58; chief bond papers & allied prods, Res & Develop, Kimberly Clark Corp, 58-60, dir bus paper mfg, 60-62, prod mgr, Neenah Paper Div, 62-70, opers & develop mgr, 70-81; RETIRED. *Mem:* Am Chem Soc; Am Tech Asn Pulp & Paper Indust; Paper Indust Mgt Asn. *Res:* Chemistry and technology of pulp and paper; evaluating the optical constants of pigments in paper; resin application; specialty papers. *Mailing Add:* 107 Courtney G Neenah WI 54956

ADRIAN, ERLE KEYS, JR, b Temple, Tex, Apr 18, 36; m 62; c 3. ANATOMY. *Educ:* Rice Inst, BA, 58; Univ Tex, Galveston, MA, 61, PhD(anat), 67; Harvard Med Sch, MD, 63. *Prof Exp:* Res asst anat, 59-60, from instr to assoc prof, Med Br, Univ Tex, Galveston, 63-69; assoc prof, 69-74, actg chmn, Dept Anat, 80-81, PROF ANAT, UNIV TEX HEALTH SCI CTR, SAN ANTONIO, 74-, DEPT CHMN, 81- *Mem:* Am Asn Anat. *Res:* Neuropathology; gerontology. *Mailing Add:* Dept Cellular & Structural Biol Univ Tex Health Sci Ctr San Antonio TX 78284

ADRIAN, FRANK JOHN, b Brooklyn, NY, Oct 7, 29; m 69; c 2. PHYSICAL CHEMISTRY. *Educ:* Cath Univ Am, AB, 51; Cornell Univ, PhD(phys chem), 55. *Prof Exp:* PRIN PROF STAFF CHEMIST, APPL PHYSICS LAB, JOHNS HOPKINS UNIV, 55- *Concurrent Pos:* Assoc ed, J Chem Physics, 75-78. *Mem:* Fel Am Phys Soc; Am Chem Soc. *Res:* Magnetic and nuclear quadrupole resonance; structure of molecules, free radicals and paramagnetic imperfections in solids; photochemistry; chemically induced magnetic polarization; surface enhanced Raman scattering; high-temperature superconductivity. *Mailing Add:* 17716 Queens Elizabeth Dr Olney MD 20832

ADRIAN, RONALD JOHN, b Minneapolis, Minn, June 16, 45; m 69; c 3. TURBULENT FLOW, LASER VELOCIMETRY. *Educ:* Univ Minn, BME, 67, MS, 69; Cambridge Univ, PhD(physics), 72. *Prof Exp:* Engr, Boeing Co, Seattle, 67; from asst prof to assoc prof, 72-81, PROF THEORET & APPL MECH, UNIV ILL, URBANA-CHAMPAIGN, 81- *Concurrent Pos:* Churchill Scholar, 69-72; consult, TSI, INC, St Paul, Minn, 75-; vis prof, Stanford Univ, 82, Nalt Coun Supercomp Appl, 89. *Honors & Awards:* A T Colwell Award, 90; D C Drucker Award, 89. *Mem:* Am Soc Mech Engrs; Am Phys Soc; Am Optical Soc; Am Inst Aeronaut & Astronaut; Laser Inst Am. *Res:* Experimental study of turbulent shear flows and turbulent thermal convection; scientific instruments for measurement of fluid velocity fields. *Mailing Add:* Dept Theoret & Appl Mech Univ Ill 216 Talbot Lab 104 S Wright St Urbana IL 61801

ADRIAN, THOMAS E, b London, UK, Apr 26, 50; m; c 2. PHYSIOLOGY, BIOCHEMISTRY. *Educ:* Inst Biol, MIBiol, 74; Brunel Univ, MSc, 76; Univ London, PhD(biochem), 80; MRCPath, 84. *Prof Exp:* Sr technician, Dept Chem Path, Devonport Lab, London, 71-73, chief technician, 73-74; res biochemist, Gastrointestinal Unit, Greenwich Dist Hosp, Eng, 74-75; res officer, Dept Med, Royal Postgrad Med Sch, Hammersmith Hosp, London, 75-80, sr res officer & lectr, 80-85; assoc res scientist surg, Sch Med, Yale Univ, 85-87, res scientist & assoc prof, 87-88; PROF PHYSIOL & BIOCHEM, MED SCH, CREIGHTON UNIV, OMAHA, 88-, DIV HEAD PHYSIOL, DEPT BIOMED SCI, 91- *Concurrent Pos:* Dir, Gastrointestinal Surg Res Lab, Vet Admin Med Ctr, West Haven, Conn, 85-88; prin investr, Am Inst Cancer Res, 89-91, NIH, 90-93 & 91-92, Nat Dairy Coun, 91-93 & Nat Livestock & Meat Bd, 91-92; mem, Rev Comt Gastrointestinal Core Ctr Grants, Nat Inst Diabetes & Digestive Kidney Dis, NIH, 90. *Mem:* Royal Col Pathologists; Asn Clin Chem; Endocrine Soc; Am Physiol Soc; Am Gastroenterol Soc; Brit Diabetic Asn; Am Pancreatic Asn; Int Asn Pancreatology. *Res:* Cholecystokinin effect on pancreatic growth and tumors; hormonal mechanism of potentiation of pancreatic cancer by polyunsaturated rather than saturated dietary fat; early diagnosis of pancreatic cancer; author of more than 450 technical publications. *Mailing Add:* Dept Physiol Med Sch Creighton Univ 2500 California St Omaha NE 68178

ADRION, WILLIAM RICHARDS, b Alexandria, La, Nov 2, 43; m 71; c 2. PROGRAMMING SYSTEMS, SOFTWARE ENGINEERING. *Educ:* Cornell Univ, BS, 66, MEE, 67; Univ Tex, Austin, PhD(elec eng & comput sci), 71. *Prof Exp:* Res engr, Electron Data Processing, Honeywell, Inc, 69-79; sr res engr, 67-71, asst prof elec eng, Electronic Res Ctr, Univ Tex, Austin, 71-72; asst prof & group chmn, elec & comput engr, Ore State Univ, 72-78; prog dir theoret comput sci, NSF, 76-78; mgr, Software Eng Group, Nat Bur Standards, 78-80; prog dir, spec proj comput sci, 80-82, coord exp res, 82- 85, dep div dir, comput res, 85-86; chief scientist, Comp Int Sci & Engr, Nat Sci Found, 86; PROF & CHMN, COMP & INFO SCI, UNIV MASS, AMHERST, 86- *Concurrent Pos:* Consult, Electron Data Processing Technol Ctr, Honeywell, Inc, 69-70, Tektronix, Inc, 73-74, Appl Theory Assocs, 73-78, Radio Free Europe & Radio Liberty, 81, Lawrence Livermore Labs, 85-87, Superconductivity Super Collider Lab, 90-; chmn bd, ACSIOM, Inc ACSIOM, Lab, 89-; dir, Computing Res Assoc, 87-; chair NSF adv panel for CISE/CDA, 89-; chmn, Asn Comput Mach/SIGSoft, 85-; dir, Comput Res Bd, 88, Texas Higher Educ Coun, 88- *Mem:* AAAS; Asn Comput Mach; Inst Elec & Electron Engrs; NY Acad Sci; Soc Indust & Appl Math; Sigma Xi. *Res:* Programming systems and software engineering with particular emphasis on programming environments, program verification and testing and user interfaces to programming systems. *Mailing Add:* COINS-221LGRC Univ Mass Amherst MA 01003

ADROUNIE, V HARRY, b Battle Creek, Mich, Apr 29, 15; m 43, 81; c 2. PUBLIC & ENVIRONMENTAL HEALTH, INDUSTRIAL HYGIENE. *Educ:* St Ambrose Col, BS, 40, BA, 59; Am Bd Indust Hyg, dipl, 62; Am Acad Sanitarians, cert sanit, 67; Western States Univ Prof Studies, MS & PhD(environ health pub health), 84. *Prof Exp:* Chief labs & chief prev med, Med Serv Corps, US Air Force, 50-51, chief, Planning & Test Br, Field Test & Meteorol Div, 51-52, mem staff, Biol & Chem Spec Weapons Br & bioenviron specialist, Off Inspector Gen, 53-56, environ med officer, Off Surgeon Gen, 57-61, comdr, Detachment 10, 1st Aeromed Transport Group, 61-63; vis lectr environ health, Sch Pub Health, Am Univ Beirut, 63-64, vis assoc prof & actg chmn dept, 64-66; dep comdr, 1st Aeromed Evacuation Group, Pope AFB, 66-68; tech dir, ARA Environ Serv, 68-70; dir, Div Environ Protection, Chester County Health Dept, Pa, 70-75 & Div Environ Health, Berrien County Health Dept, Mich, 75-78; prof environ health, Sch Pub Health, Univ Hawaii, 78-80; RETIRED. *Concurrent Pos:* US Air Force environ health rep, Nat Acad Sci, 57-61; mem, Nat Cong Environ Health, 59-60; US Air Force rep, US Interdept Comt Nutrit for Nat Defense, 59-61; mem, President's Coun Youth Fitness, 60; Nat Environ Health Asn del, Nat Health Coun, 61; consult, Health Mobilization Prog, USPHS, 61-62 & UN Relief, Works Agency Educ Div, 64-66 & Dist Health Dept, 81-; chmn, Bd Registr Sanitarians Pa, 70-77; mem bd dirs, Chester Co Pa Water Resources Auth, 70-76, chmn, 74-76; mem, Chester Co Bd Health, 75-78; mem policy & sci adv bds, SWest Mich Ground Water Surv & Monitoring Prog, 81-, chmn policy bd, 88-90; assoc ed, J Environ Health, 58-70; adj prof environ health, Ferris State Col, 76-79; bd dirs, World Safety Orgn, 84- & mem cert bd, 85- *Honors & Awards:* Mangold Award, 63. *Mem:* Fel & emer mem Am Pub Health Asn; fel Royal Soc Health; Asn Mil Surgeons US; Am Indust Hyg Asn; emer mem Int Health Soc US; Nat Asn Sanitarians (pres, 61-62); Nat Environ Health Asn. *Res:* Environmental and public health through the media of research, teaching and direction; beneficial control of the environment for benefit of humanity; industrial hygiene; author of over 50 publications and journal articles in the preventive, occupational medicine and public health area. *Mailing Add:* 1905 N Broadway Hastings MI 49058

ADROUNY, GEORGE ADOUR (KUYUMJIAN), b Kilis, Turkey, Apr 1, 16; US citizen; m 45; c 3. BIOCHEMISTRY. *Educ:* Am Univ Beirut, BA, 34, PhC, 40; Emory Univ, PhD(biochem), 54. *Prof Exp:* Instr & res assoc biochem, Emory Univ, 54-57; from asst prof to assoc prof, Sch Med, Tulane Univ, 57-72, prof biochem, 72-81, emer prof, 81-; RETIRED. *Mem:* AAAS; Am Chem Soc. *Res:* Isolation and characterization of fire ant venom; effects of nutritional and hormonal factors on the carbohydrate metabolism of cardiac and skeletal muscles; the glycogen storing effect of somatotrophin. *Mailing Add:* 103 Amberfield Lane Gaithersburg MD 20878

ADT, ROBERT (ROY), b Brooklyn, NY, June 21, 40; m 63; c 2. MECHANICAL ENGINEERING. *Educ:* Univ Miami, BSME, 62; Mass Inst Technol, SM, 65, ScD(mech eng), 67. *Prof Exp:* Res engr, Northern Res & Eng Corp, 62-65; from asst prof to assoc prof, 67-77, PROF MECH ENG, UNIV MIAMI, 77- *Concurrent Pos:* Consult & legal expert. *Honors & Awards:* Ralph R Teetor Award, Soc Automotive Engrs. *Mem:* Soc Automotive Engrs; Am Soc Mech Engrs; Sigma Xi. *Res:* Alternate fuels for automotive engines; automotive engineering; engineering analysis and forensics of accidents; heat transfer thermodynamics. *Mailing Add:* 6391 Snapper Creek Dr Miami FL 33143-8046

ADUSS, HOWARD, b Brooklyn, NY, Mar 31, 32; m 53; c 4. ORTHODONTICS, CRANIOFACIAL ANOMALIES. *Educ:* Purdue Univ, BS, 54; Northwestern Univ, DDS, 57; Univ Rochester, MS, 62, Eastman Dent Ctr, cert orthod, 62. *Prof Exp:* PROF ORTHOD, DEPT PEDIAT, CTR CRANIOFACIAL ANOMALIES, UNIV ILL, 66-, DIR CTR, 84- *Concurrent Pos:* Prof orthod, dept surg, Div Plastic Surg, Med Sch, Rush Univ. *Mem:* Am Dent Asn; Am Asn Orthod; Am Cleft Palate Asn (pres, 75-76); fel Am Col Dent; Int Asn Dent Res. *Res:* Craniofacial malformations, particularly cleft lip and palate; craniofacial growth and development in syndromes. *Mailing Add:* 237 Lakeside Pl Highland Park IL 60035

ADVANI, SURESH GOPALDAS, b Pune, India, Aug 19, 59; m 87; c 1. POLYMER & COMPOSITES PROCESSING, PROCESS MODELS FOR VARIOUS FLUIDS. *Educ:* Indian Inst Technol, Bombay, BTech, 82; Univ Ill, PhD(mech eng), 87. *Prof Exp:* Res asst turbo-expander, Indian Inst Technol, Bombay, 81-82; res asst polymer proc, Univ Ill, Urbana-Champaign, 83-87, teaching fel mech mach, 86-87; ASST PROF THERMO-FLUIDS, UNIV DEL, 87- *Concurrent Pos:* Consult, Tex Instruments, Dallas, Tex, 88-90; ed, Am Soc Mech Engrs-Heat & Mass Transfer in Solidification Processing, 91; co-chmn, Int Conf Composite Mat, 92; chmn, Rheology in Composite Processing, 91 & Polymer Mat Div, Am Soc Mech Engrs, 91-93. *Mem:* Am Soc Mech Engrs; Polymer Processing Soc; Soc Rheology; Soc Plastic Engrs. *Res:* Rheology of short and continuous fiber reinforced polymer composites and incorporations of these concepts into process models to gain insight about the physical system; polymer and composites processing; transport phenomena in manufacturing processes; non-Newtonian fluid mechanics, rheology and heat transfer. *Mailing Add:* Mech Eng Dept Univ Del Newark DE 19716-3140

AEBERSOLD, RUEDI H, b Switz, Sept 12, 54; m 81; c 3. PROTEIN BIOCHEMISTRY. *Educ:* Univ Basel, Switz, dipl, 79, PhD(cell biol), 83. *Prof Exp:* Postdoctoral protein chem, Calif Inst Technol, 84-86, sr res fel, 86-88; SR SCIENTIST BIOCHEM, BIOMED RES CTR, UNIV BC, 88-, ASST PROF, 89- *Mem:* Protein Soc; Am Soc Biochem & Molecular Biol. *Res:* Development of improved technologies for protein primary structure analysis; investigation of signal transduction process induced by the activation of lymphocytes. *Mailing Add:* Biomed Res Ctr Univ BC Vancouver BC V6R 1W5

AEIN, JOSEPH MORRIS, b Washington, DC, Jan 26, 36; div; c 2. COMMUNICATIONS. *Educ:* Mass Inst Technol, SB & SM, 58; Purdue Univ, PhD(elec eng), 62. *Prof Exp:* Instr elec eng, Purdue Univ, 58-61; MEM TECH STAFF, SCI & TECHNOL DIV, INST DEFENSE ANAL, 62- *Concurrent Pos:* Vis lectr, Univ Calif, Berkeley, 65-66; prof lectr, George Washington Univ, 79- *Mem:* Fel Inst Elec & Electronics Engrs. *Res:* Communications satellites; signal processing; navigation sensors; communications theory. *Mailing Add:* MTS Rand Corp 2100 M St NW Washington DC 20037

AEPPLI, ALFRED, b Zurich, Switz, Nov 8, 28; nat US; m 55; c 2. MATHEMATICS. *Educ:* Swiss Fed Inst Technol, dipl, 51, Dr math, 56. *Prof Exp:* Asst math, Swiss Fed Inst Technol, 52-57; from instr to asst prof, Cornell Univ, 57-62; assoc prof, 62-66, PROF MATH, UNIV MINN, MINNEAPOLIS, 66- *Concurrent Pos:* Guest prof, Univ Heidelberg, 74-75; pres, Univ Minn Educ Asn, 76- *Mem:* Math Asn Am; Asn Women Math; Am Math Soc; Swiss Math Soc; Soc Indust & Appl Math. *Res:* Topology; differential geometry; complex variables. *Mailing Add:* Sch Math Univ Minn Minneapolis MN 55455

AFANADOR, ARTHUR JOSEPH, b Tampa, Fla, Nov 2, 42; m 65; c 2. OPTOMETRY, PHYSIOLOGICAL OPTICS. *Educ:* Southern Col Optom, BS & OD, 65; Univ Calif, Berkeley, PhD(physiol optics), 72. *Prof Exp:* From asst prof to assoc prof optom, Ind Univ, Bloomington, 72-81; DEAN, SCH OPTOM, INTER-AM UNIV, PR, 81- *Mem:* Fel Am Acad Optom. *Res:* Neurophysiology of the retina; single cell recording of goldfish ganglion cells. *Mailing Add:* Inter-Am Univ PR PO Box 1293 Hato Rey PR 00919

AFFELDT, JOHN E, b Lansing, Mich. MEDICAL ADMINISTRATION. *Prof Exp:* Resident internal med, White Mem Hosp, 46-49; chief resident, Commun Dis Serv, Los Angeles County Hosp, 49-50; res assoc physiol, Sch Pub Health, Harvard Univ, 50-52; chief, Phys Respiratory Ctr Poliomyletis, Rancho Los Amigos Hosp, 52-56, med dir, 56-64; med dir, Los Angeles County Hosp, 64-72 & Dept Health Serv, Los Angeles County, 72-77; pres, Joint Comn Accreditation Hosp, 77-86; RETIRED. *Concurrent Pos:* Assoc prof med, Loma Linda Univ Sch Med, 62-66. *Mem:* Inst Med-Nat Acad Sci; AMA; fel Am Col Physicians; Am Cong Rehab Med. *Mailing Add:* 5140 Bareback Sq Box 8432 Rancho Santa Fe CA 92067

AFFENS, WILBUR ALLEN, b New York, NY, Jan 9, 18; m 45; c 2. CHEMISTRY. *Educ:* City Col New York, BS, 37; Polytech Inst Brooklyn, MS, 46; Georgetown Univ, PhD(phys chem), 60. *Prof Exp:* Chemist, Rubatex Prod Inc, NY, 37-38; chemist, Maltbie Chem Co, NJ, 38-42; asst chemist, Corps Engrs, US War Dept, 42-45; assoc chemist, USDA, 45-53, phys chemist, Entom Res Br, Agr Res Serv, 53-56; res chemist, US Naval Res Lab, 56-83; Geo-Centers Inc, Washington DC, 84-87; RETIRED. *Mem:* Am Chem Soc; Combustion Inst; Am Soc Testing & Mat. *Res:* Hydrocarbon chemistry; combustion; ignition; flammability; petroleum; fuels. *Mailing Add:* 8302 26th Pl Adelphi MD 20783

AFFRONTI, LEWIS FRANCIS, b Rochester, NY, Aug 12, 28; m 56; c 4. MICROBIOLOGY, IMMUNOLOGY. *Educ:* Univ Buffalo, BA, 50, MA, 51; Duke Univ, PhD(microbiol), 58. *Prof Exp:* Asst, Univ Buffalo, 50-51; res assoc, Vet Admin Hosp, 51-52; med entomologist, US Air Force, 52-54; res assoc, Roswell Park Mem Inst, 54; dir vaccine lab, Duke Univ Hosp, 56; res assoc tuberc, Henry Phipps Inst, Univ Pa, 57-58; from sr asst scientist to scientist, USPHS, 58-62; from asst prof to assoc prof, 62-72, PROF MICROBIOL, SCH MED, GEORGE WASHINGTON UNIV, 72-, CHMN DEPT, 73- *Concurrent Pos:* Spec consult, Vet Admin Hosp, Wilmington, Del, 66-; consult, Vet Admin Ctr, Martinsburg, WVa; US rep, Int Meeting Stand & Uses of Tuberculins, WHO, Geneva, Switz, 66; USPHS spec res fel, Inst Superiore Sanita, Italy, 69-70; US-Japanese res in tuberc grant, 69-; WHO exchange of res workers fel, Univ Goteborg, 70; Nat Tuberc Asn travel fel, 71; mem med adv bd, Wilmington Vet Admin Hosp Ctr, 72-; Nat Acad Sci interacad exchange fel, Prague, 80. *Honors & Awards:* Unione Cavalleresca Europa Award, Contribusion to Sci, 80. *Mem:* Fel, Am Acad Microbiol; Asn Med Sch Microbiol (secy-treas, 76-87); Am Soc Microbiol; Am Asn Immunol; Protein Soc. *Res:* Antigens of mycobacteria; immunochemistry; bacterial physiology; isolation and characterization of bacterial tumor isolates; cellular mechanisms in delayed allergy. *Mailing Add:* 4316 John Silver Rd Virginia Beach VA 21455

AFGHAN, BADERUDDIN KHAN, b Shikarpur, Pakistan, Dec 12, 40; Can citizen; m 69; c 3. ANALYTICAL CHEMISTRY, ENVIRONMENTAL CHEMISTRY. *Educ:* Sind Univ, Pakistan, BSc, 62, Hons, 63; Univ London, DIC, 64, PhD(anal chem), 66. *Prof Exp:* Fel org reagents for fluorometric analysis, Dalhousie Univ, 66-68; res assoc solution chem, Univ Montreal, 68-69; res scientist anal methods develop, Dept Energy Mines & Resources, Fed Govt, Ottawa, Ont, 69-72; res scientist environ anal chem, Can Centre for Inland Waters, Dept Environ, 72-75, res mgr anal methods res, 75-77, head anal methods res sect, 77-83; head, anal chem res sect, Nat Water Res Inst, 83-85; CHIEF, NAT WATER RES LAB, CLIW, BURLINGTON, 85- *Mem:* Fel Chem Inst Can; Spectros Soc Can; Am Soc Testing & Mat; Int Standards Orgn. *Res:* Modern polarographic and related electroanalytical techniques; high speed liquid chromatography and high performance gas chromatography; atomic and molecular absorption as well as fluorescence spectroscopy; trace analysis; environmental analytical chemistry; pesticides and industrial chemicals analysis; trace and ultra trace organic analysis using high resolution gas chromatography. *Mailing Add:* Nat Water Res Inst PO Box 5050 867 Lakeshore Rd Burlington ON L7R 4A6 Can

AFNAN, IRAJ RUHI, b Beirut, Lebanon, May 27, 39; m 66; c 2. NUCLEAR THEORY, SCATTERING THEORY. *Educ:* Am Univ, Beirut, BSc, 61; Mass Inst Technol, PhD(physics), 66. *Prof Exp:* Res assoc physics, Univ Minn, 66-68 & Univ Calif, Davis, 68-70; vis prof physics, Univ BC, 74-75; lectr physics, 70-73, sr lectr physics, 73-77, READER PHYSICS, FLINDERS UNIV, 77- *Concurrent Pos:* Consult & collabr, Los Alamos Nat Lab, 79- *Mem:* Am Phys Soc. *Res:* Quantum theory of scattering with application to nuclear and particle physics. *Mailing Add:* Sch Phys Sci Flinders Univ Bedford Park SA 5042 Australia

AFONSO, ADRIANO, b Goa, India, Mar 5, 35; m 66; c 2. ORGANIC CHEMISTRY. *Educ:* Univ Bombay, MS, 57; Univ Wis, PhD(pharmaceut chem), 61. *Prof Exp:* Res assoc, dept chem, Ind Univ, Bloomington, 61-63; RES FEL, CHEM RES DIV, SCHERING CORP, 63- *Honors & Awards:* Lunsford Richardson Award, 60. *Mem:* Am Chem Soc. *Res:* Synthetic organic chemistry; structural studies on veratrum alkaloids; syntheses of diterpene resin acids; chemical modifications of steroids for biological activity; peptide structure elucidations; peptide syntheses; chemical modifications of aminoglycoside antibiotics; syntheses of beta lactam antibacterials. *Mailing Add:* 10 Woodmere Rd West Caldwell NJ 07006

AFREMOW, LEONARD CALVIN, b Chicago, Ill, July 11, 33; m 60; c 1. ANALYTICAL CHEMISTRY, RESEARCH MANAGEMENT. *Educ:* Univ Ill, BS, 59; Univ Wis, MS, 61; Univ Chicago, MBA, 72. *Prof Exp:* Anal chemist, Universal Oil Prod, 61; anal chemist, DeSoto Inc, 61-63, group leader, 63-65, sect leader, 65-66, mgr res serv, 66-69, tech dir, 69-70, mgr indust, 70-71, dir, 71-74, dir res, 74-80; VPRES RES & ADMIN, MIDLAND DIV, DEXTER CORP. *Concurrent Pos:* Instr chem, Roosevelt Univ, 68-71. *Mem:* Nat Paint & Coatings Asn; Fedn of Soc for Coatings Technol; Am Chem Soc; Paint Res Inst (vpres, 79). *Res:* Analytical instrumentation; chemical coatings; research management. *Mailing Add:* Midland Div Dexter Corp E Water St Waukegan IL 60085

AFSAR, MOHAMMED NURUL, PHYSICS. *Educ:* Univ Dacca, MS, 67; Univ London, MS, 72, PhD(physics), 78. *Prof Exp:* Lectr Physics, Nizampur Col, Chittagong, 68-70; higher sci officer, Nat Phys Lab, Eng, 75-78; STAFF SCIENTIST & PHYSICIST, MASS INST TECHNOL, 78- *Concurrent Pos:* Guest scientist, Nat Phys Lab, UK, 72-75; res fel, Univ London, 78. *Honors & Awards:* Duddell Premium, Inst Elec Engrs, UK, 77. *Mem:* Inst Elec & Electronics Engrs; Inst Elec Engrs UK; Inst Physics. *Res:* Millimeter and submillimeter wave techniques; high precision measurement techniques and measurements of millimeter, submillimeter and infrared materials; characterization of impurities in high purity gallium arsemide and related compounds; far infrared ziman spectroscopy. *Mailing Add:* Halligan Hall Tufts Univ Medford MA 02155

AFSHAR, SIROOS K, b Ghom, Iran, July 3, 49; US citizen; m 75; c 2. SOFTWARE ENGINEERING MANAGEMENT, COMPLEXITY THEORY OF ALGORITHMS. *Educ:* Sharif Univ Technol, Tehran, BSEE, 72; Univ Mich, Ann Arbor, MSE, 76, PhD(elec & computer eng), 79. *Prof Exp:* Design engr, Nat Iranian Oil Co, 72-75; res asst computer eng, Univ Mich, 76-79; asst prof elec & computer eng, La State Univ, 79-82; mem tech staff, 82-87, SUPVR, AT&T BELL LABS, 87- *Concurrent Pos:* Vis prof, Monmouth Col, 85. *Mem:* Inst Elec & Electronic Engrs; Asn Comput Mach. *Res:* Leading a research and development group to understand the needs, to document the necessary features, to define standards, and to define and develop products for the management of voice and data networks. *Mailing Add:* Ten Constitution Ct Englishtown NJ 07726

AFT, HARVEY, b Chicago, Ill, June 17, 29; m 53; c 4. ORGANIC CHEMISTRY. *Educ:* Univ Southern Calif, AB, 50; Univ Puget Sound, MS, 52; Ore State Univ, PhD(chem), 62. *Prof Exp:* Res chemist, Forest Prod Res Lab, 57-61; asst prof forest prod chem, Forest Res Lab, Ore State Univ, 61-65; asst prof chem, Mich Technol Univ, 65-66; assoc prof, Ore State Univ, 66-69; assoc prof, 69-72, chmn dept, 72-80, PROF CHEM, UNIV MAINE, FARMINGTON, 80- *Mem:* AAAS; Am Chem Soc; Brit Chem Soc. *Res:* Chemistry of phenolic and oxygen containing heterocyclic compounds and natural products chemistry. *Mailing Add:* Dept of Chem Univ Maine Farmington ME 04938

AFTERGOOD, LILLA, b Krakow, Poland, Jan 10, 25; nat US; m 49; c 3. BIOCHEMISTRY, NUTRITION. *Educ:* Univ Paris, Sorbonne, Lic en Sc, 48; Univ Southern Calif, MS, 51, PhD(biochem), 56. *Prof Exp:* Asst biochem, Univ Southern Calif, 50-56, res assoc, 56-62; res biochemist, Sch Pub Health, Univ Calif, Los Angeles, 62-80; RETIRED. *Concurrent Pos:* Fel Coun Arteriosclerosis, Am Heart Asn. *Honors & Awards:* Bond Award, Am Oil Chemists Soc, 74. *Mem:* AAAS; Am Inst Nutrit; Am Oil Chemists Soc; Sigma Xi. *Res:* Essential fatty acids; cholesterol, their metabolic interrelationship; effect of sex hormones on lipid metabolism; metabolic effects of oral contraceptives; vitamin E; nutrition and cancer, trans fatty acids. *Mailing Add:* 7050 Arizona Ave Los Angeles CA 90045

AFTERGUT, SIEGFRIED, b Frankfurt am Main, Ger, Jan 7, 27; nat US; m 51; c 2. ORGANIC CHEMISTRY, POLYMER CHEMISTRY. *Educ:* Syracuse Univ, BA, 50, MS, 54, PhD(org chem), 56. *Prof Exp:* Asst instr chem, Syracuse Univ, 51-56; proj engr, Advan Tech Labs, 56-66, mgr photorecording mat, 66-73, mgr display mat, 73-76, MEM STAFF, RES & DEVELOP CTR, GEN ELEC CO, 76- *Mem:* Am Chem Soc; Soc Info Display. *Res:* Mannich reaction; aromatic silanes and ethers; electrical properties of organic compounds; photoplastic recording; organic photoconductors; liquid crystals; dichroic dyes; color filters; electronic displays. *Mailing Add:* 1063 Nott St Schenectady NY 12308

AGABIAN, NINA, PHARMACOLOGY. *Educ:* Adelphi Univ, BA, 66, MS, 68; Albert Einstein Col Med, PhD(molecular biol), 71. *Prof Exp:* Teaching asst, Dept Biol, Adelphi Univ, 66-68; res fel, Dept Molecular Biol, Albert Einstein Col Med, 68-71; from asst prof to prof biochem, Sch Med, Univ Wash, Seattle, 73-84; actg sci dir, Naval Biosci Lab, 84-87, PROF BIOMED & ENVIRON HEALTH SCI, SCH PUB HEALTH, UNIV CALIF, BERKELEY, 84-, PROF PHARMACEUT CHEM, SAN FRANCISCO, 86-, DIR, MOLECULAR PARASITOL, INTERCAMPUS PROG, 87- *Concurrent Pos:* Res fel, NIH, 68-71 & 71-73; Guggenheim fel, 79-80; Burroughs Wellcome award molecular parasitol, 84-89; mem bd dirs, Drug & Vaccine Develop Corp, 86- *Mem:* Fel AAAS; Am Soc Biol Chemists. *Res:* Author or co-author of over 90 publications. *Mailing Add:* Dept Pharmaceut Chem Sch Pharm Univ Calif Laurel Heights Campus Box 1204 Suite 150 3333 California St San Francisco CA 94143-1204

AGALLOCO, JAMES PAUL, b Queens, NY, Apr 19, 48; m 74; c 3. PROCESS VALIDATION, SYSTEMS RELIABILITY. *Educ:* Pratt Inst, Brooklyn, NY, BE, 68; Polytechnic Inst NY, Brooklyn, MS, 79; Fairleigh-Dickinson Univ, Teaneck, NJ, MBA, 83. *Prof Exp:* Chem eng, Merck & Co; mgr, Pfizer, Inc, 73-80; DIR, BRISTOL-MYERS SQUIBB, 80- *Concurrent Pos:* Co-founder, NJ Validation Discussion Group, 80; comt chmn, Int Soc Pharmaceut Eng, Tampa, Fla, 82. *Honors & Awards:* b Queens, Ny, April 19, 48. *Mem:* Parenteral Drug Asn, Philadelphia, Pa (pres, 88-89); Am Soc Qual Control; Parenteral Soc Great Brit. *Res:* Process vilidation; process automation; pharmaceutical manufacturing; aseptic processing; sterilization processes. *Mailing Add:* Bristol-Myers Squibb PO Box 191 New Brunswick NJ 08903

AGAMANOLIS, DIMITRIS, b Mytilene, Greece, Nov 3, 38. NEUROPATHOLOGY. *Educ:* Univ Thesaloniki, Greece, MD, 62. *Prof Exp:* DIR NEUROPATH, CHILDREN'S HOSP MED CTR. *Mailing Add:* Children's Hosp Med Ctr 281 Locust St Akron OH 44308

AGAN, RAYMOND JOHN, b Knoxville, Iowa, July 8, 19; m 39; c 2. AGRICULTURE. *Educ:* Iowa State Univ, BS, 40, MS, 50; Univ Mo, EdD(agr), 55. *Prof Exp:* Asst agr educ, Univ Mo, 54-55; from instr to asst prof, Ore State Univ, 55-58; from assoc prof to prof, Kans State Univ, 58-71; PROF VOC EDUC & DIR INT EDUC PROGS, SAM HOUSTON STATE UNIV, 71- *Concurrent Pos:* Consult, Univ Costa Rica, 63-64, Indonesian & Malaysian ministries of agr; UNESCO consult, Colombian projs & ODECA/OCE PLAN, Cent Am, 65-70; pres, Agan Consultancy Int Agency. *Mem:* Asn Teachers Educators Agr (vpres, 63-64). *Res:* Agricultural occupations and training needs of Oregon Indian tribes; agricultural needs in Central and South America. *Mailing Add:* Dept Voc Educ Sam Houston State Univ Huntsville TX 77341

AGAR, G(ORDON) E(DWARD), b Kindersley, Sask, Sept 17, 31; Can citizen; m 56; c 3. MINERAL PROCESSING. *Educ:* Univ BC, BASc, 55; Mass Inst Technol, ScD(mineral eng), 61. *Prof Exp:* Metallurgist, Britannia Mining & Smelting, 55; asst, Mass Inst Technol, 55-56, from instr to asst prof, 57-62; metallurgist, Union Carbide Nuclear Corp, 56; mgr mineral res, Int Minerals & Chem Corp, 62-67; res engr, Duval Corp, 67-72; SECT HEAD MINERAL PROCESSING RES, INCO LTD, 72- *Concurrent Pos:* Adj prof, Univ Toronto, 77-, Univ BC, 90-; distinguished lectr, Can Inst Mining & Metall. *Mem:* Am Inst Mining, Metall & Petrol Engrs; fel Can Inst Mining & Metall. *Res:* Mineral processing and extractive metallurgy, especially processes such as flotation, electrostatic separation and leaching; surface and colloid chemistry related to separation processes. *Mailing Add:* Inco Ltd Sheridan Park Mississauga ON L5K 1Z9 Can

AGARD, EUGENE THEODORE, b Christ Church, WI, Aug 15, 32; m 60; c 4. MEDICAL PHYSICS, RADIOLOGICAL HEALTH. *Educ:* Univ WI, BSc, 56; Univ London, MSc, 58; Univ Toronto, PhD(physics), 70; Am Bd Radiol, cert radiol physics, 82. *Prof Exp:* Asst Physicist, Mass Gen Hosp, 59-63; lectr & asst prof physics, Univ WI, Trinidad, 63-72; dir, Med Physics, PR Nuclear Ctr, 72-76; radiation physicist, Kettering Med Ctr, 76-84; consult radiol physics, Radgard Inc, 84-90; ASSOC PROF MED PHYSICS, MED COL OHIO, 90- *Concurrent Pos:* Consult, Ministry Health, Trinidad & Tobago, 66-67, Vet Admin Med Ctr, PR, 73-76 & Dayton, Ohio, 78-90; rep, Comt n-44, Am Nat Standards Inst, 72-76; mem, Int Rels Comt, Am Asn Physicists Med, 74-77, Pub Educ Comt, 80-86; prof, Kettering Col Med Arts, 78-82; assoc clin prof, Dept Radiol Sci, Wright State Univ, 87-90. *Mem:* Am Asn Physicists Med; Health Physics Soc; Am Col RAdiol; Radiol Soc NAm. *Res:* Medical applications of radiation, especially in the practice of the radiological sciences; measurements of radiation doses and quality control of equipment used in radiation oncology, diagnostic radiology and nuclear medicine. *Mailing Add:* Dept Radiation Oncol Med Col Ohio 3000 Arlington Ave Dayton OH 45429

AGARDY, FRANKLIN J, b New York, NY, Mar 23, 33; m 55; c 2. SANITARY ENGINEERING, ENVIRONMENTAL ENGINEERING. *Educ:* City Col New York, BCE, 55; Univ Calif, Berkeley, MS, 57, PhD(sanit eng), 63. *Prof Exp:* Prof civil eng, San Jose State Univ, 62-69; vpres, URS Res Co, 69-73, pres, 73-74, vpres, URS Corp, 74-80, sr vpres, 84-85, pres, 80-88, exec vpres & chief operating officer, 85-88; PRES, FORENSIC MGT ASSOCS INC, 88- *Concurrent Pos:* Consult radiation shielding, Off Civil Defense & US Govt, 61-69; forensic engr, 64-; consult indust pollution, US Environ Protection Agency, 72-73; hazardous & toxic waste investr, 80-; pres & chief exec officer, In-Process Technol Inc, 88-90, chmn bd, 90- *Honors & Awards:* George A Elliot Mem Award, Am Water Works Asn, 73, Water Utility Educ Award, 74. *Mem:* Am Soc Civil Engrs; Water Pollution Control Fedn; Am Water Works Asn; Nat Fire Protection Asn; Am Soc Testing & Mat. *Res:* Pollution control, including biological treatment of industrial wastes; engineering principles of gamma radiation shielding; continency planning for water and waste water utilities. *Mailing Add:* Forensic Mgt Assocs Inc 400 S El Camino Real Suite 1050 San Manteo CA 94402

AGARWAL, ARUN KUMAR, b Lucknow, India, Mar 17, 44; m 67; c 2. PURE MATHEMATICS. *Educ:* Lucknow Univ, BSc, 60, MSc 63, PhD(math), 67. *Prof Exp:* Coun Sci Indust Res sr res fel, New Delhi, India, 67-68; fel, WVa Univ, 68-69; assoc prof, 69-80, PROF MATH & COMPUTER SCI, GRAMBLING STATE UNIV, 80- *Concurrent Pos:* Reviewer, Math Rev, 70- & Zentrablatt für Math, 70- *Mem:* Am Math Soc; Math Asn Am; Asn Comput Mach. *Res:* Entire functions of a single and several complex variables; special functions; computer sciences, general. *Mailing Add:* Box 679 Dept of Math & Computer Sci Grambling State Univ Grambling LA 71245

AGARWAL, ASHOK KUMAR, b Firdzabad, India, Aug 14, 50; US citizen. PROCESS DEVELOPMENT OF DRUGS, MANAGEMENT. *Educ:* Indian Inst Technol, BTech, 71, MTech, 74; WVa Univ, PhD(chem eng), 78. *Prof Exp:* Res asst, Indian Inst Technol, 72-74; proj eng, Indian Consult Bur, New Delhi, 74; res asst, WVa Univ, 75-78, res assoc, 78; sr res chem engr, Monsanto Res Corp, Miamisburg, Ohio, 78-84, contract mgr, Monsanto Co, 84-85, PROJ SUPVR, MONSANTO CO, DAYTON, OHIO, 85- *Honors & Awards:* Shalimar Gold Medal, Indian Pulp & Paper Tech Asn, 74. *Mem:* Am Inst Chem Eng. *Res:* Process development; reaction kinetics; catalysis; reaction engineering; synthetic fuels development; in situ coal gasification; state-of-the-art instrumentation development; hazardous wastes and environmental engineering; process instrumention and data acquisition; computer simulation and process analysis; drug manufacturing; agricultural chemicals manufacturing. *Mailing Add:* Maulana Acad Med Col Delhi Univ 54 Meadowood Carbondale IL 62901

AGARWAL, GYAN C, b Bhagwanpur, India, Apr 22, 40; m 65; c 2. BIOENGINEERING, ELECTRICAL ENGINEERING. *Educ:* Agra Univ, BSc, 57; Univ Roorkee, BE, 60; Purdue Univ, MSEE, 62, PhD(elec), 65. *Prof Exp:* Instr elec eng, Purdue Univ, 62-65; from asst prof to assoc prof systs eng, 65-73, from asst prof to assoc prof physiol, Med Ctr, 69-73, PROF ELEC ENG & COMPUT SCI & BIOENG, UNIV ILL CHICAGO & PROF PHYSIOL, MED CTR, 73- *Concurrent Pos:* From asst to assoc biomed eng, Presby St Luke Hosp, Chicago, 65-73; vis prof physiol, Rush Med Col, Rush Univ, 76-; panel mem, Neurol Sect, Bur Med Devices, Food & Drug Admin, 81-; consult ed, J Motor Behav, 81-; assoc ed, Trans on Biomed Eng, Inst Elec & Electronic Engrs, 88- *Mem:* Fel AAAS; fel Inst Elec & Electronic Engrs; Soc Neurosci. *Res:* Application of control theory in physiological control systems; human motor control system; modeling and simulation; bio-instrumentation; control theory; optimization; man-machine systems. *Mailing Add:* Dept Elec Eng & Comput Sci MC-154 Univ Chicago Ill PO Box 4348 Chicago IL 60680-4348

AGARWAL, JAGDISH CHANDRA, b Karachi, India, Sept 8, 26; US citizen; m 47; c 2. CERAMICS ENGINEERING, METALLURGICAL ENGINEERING. *Educ:* Benares Hindu Univ, BS, 45; Polytech Inst New York, MChE, 47, DChE, 51. *Prof Exp:* Res engr, Vulcan Copper & Supply Co, 48-49; process engr, Fleischmann Labs, Standard Brands, Inc, 49-51; sr process engr, Blaw-Knox Co, 51-54; res assoc, Appl Res Lab, US Steel Corp, 54-64, div chief process analysis, 64-69; asst dir develop, Ledgemont Lab, Kennecott Copper Corp, 69-73, dir develop, 73-79; vpres, 79-82; vpres tech, Amax Inc, 82-85; VPRES TECH ASSOCS, CHARLES RIVER ASSOC, 85- *Honors & Awards:* fel Am Inst Chem Engrs; Henry Krumb Lectr, 86. *Mem:* Am Inst Mining, Metall & Petrol Engrs; Am Inst Chem Engrs; Asn Iron & Steel Engrs; Am Chem Soc; Metal Soc. *Res:* Process plant design; process economic evaluation; methods of production of iron and steel; fluidization; heat transfer; blast furnaces; magnesium and other light metals; adv ceramics. *Mailing Add:* 125 Ford Rd Sudbury MA 01776

AGARWAL, JAI BHAGWAH, b Sonepat, India, Oct 15, 38. CARDIOLOGY-INTERVENTIONAL, CARDIAC-PHYSIOLOGY. *Educ:* All India Inst Med Sci, BS & MB, 60; FRCP(C); Am Bd Internal Med, dipl, 69, dipl, 79. *Prof Exp:* Lectr cardiol, All India Inst Med Sci, 71-74, asst prof, 74-77; asst instr med, 77-78, asst prof, 78-82, asst prof clin med, 82-85, ASSOC PROF CLIN MED, UNIV PA SCH MED, 85- *Concurrent Pos:* Mem Basic Sci Coun, Am Heart Asn. *Mem:* Am Col Cardiol; Am Fedn Clin Res. *Res:* Myocardial blood flow and contraction relationship; digital subtraction angiography; use of laser and angioscopy in coronary artery disease. *Mailing Add:* Philadelphia Heart Inst Presby Univ Penn Med Ctr 51 N 39th St Philadelphia PA 19104

AGARWAL, KAILASH C, b Uttarpradesh, India; m 63; c 2. BIOCHEMISTRY. *Educ:* Allahabad Univ, India, MSc, 63; Agra Univ, India, PhD(sci), 67. *Prof Exp:* Lectr biochem, Kanpur Med Col, 66-68; res assoc biomed, 68-71, instr, 71-74, ASSOC PROF BIO-MED, BROWN UNIV, 79- *Mem:* Am Heart Asn; Am Soc Exp Pharmacol & Therapeut. *Res:* Blood platelet physiology and pharmacology. *Mailing Add:* Dept Molecular & Biochem Pharmacol Brown Univ Brown Station Providence RI 02912

AGARWAL, KAN L, STRUCTURE OF NUCLEIC ACIDS, GENE EXPRESSION. *Educ:* Rajasthan Univ, India, PhD(org chem), 67. *Prof Exp:* PROF BIOCHEM & CHEM, UNIV CHICAGO, 75- *Res:* Chemistry and molecular biology of nucleic acids. *Mailing Add:* Dept Biochem 920 E 58th St Chicago IL 60637

AGARWAL, PAUL D(HARAM), electrical engineering; deceased, see previous edition for last biography

AGARWAL, RAMESH CHANDRA, b Gwalior, India, May 8, 46; m 81; c 3. MATHEMATICAL SOFTWARE, VECTOR & PARALLEL ALGORITHMS. *Educ:* Indian Inst Technol, B Tech, 68; Rice Univ, MS, 70, PhD(elec eng), 74. *Prof Exp:* Fel, IBM, T J Watson Res Ctr, 74-76, res staff mem, 76-77; prin sci officer, Indian Inst Technol, 77-81; RES STAFF MEM, IBM, T J WATSON RES CTR, YORKTOWN HEIGHTS, NY, 82- *Honors & Awards:* Sr Award, Inst Elec & Electronic Engrs, 74. *Mem:* Fel Inst Elec & Electronic Engrs; Soc Indust & Appl Math. *Res:* Developing vector and parallel algorithms and software for supercomputers, that are used in most engineering and scientific computations. *Mailing Add:* T J Watson Res Ctr IBM PO Box 218 Yorktown Heights NY 10598

AGARWAL, RAMESH K, b India, Jan 4, 47; US citizen; m 76; c 2. COMPUTATIONAL FLUID DYNAMICS, APPLIED MATHEMATICS. *Educ:* Indian Inst Technol, BTech, 68; Univ Minn, MS, 69; Stanford Univ, PhD(aeronaut sci), 75. *Prof Exp:* Sr engr, Rao & Assocs, Calif, 75-76; Nat Res Coun assoc, Ames Res Ctr, NASA, Moffett Field, 76-78; PROF DIR, MCDONNELL DOUGLAS RES LABS, MO, 78- *Concurrent Pos:* Consult, Joint Inst Aeronaut & Acoust, Stanford Univ, 75-76; affil prof mech eng, Wash Univ, St Lous, 86-; adj prof aerospace eng, Parks Col, St Lous Univ, 86-; mem, Tech Comt Fluid Dynamics, Am Inst Aeronaut & Astronaut, 86-89, Multidisciplinary Optimization Comt, 89- & Comput Fluid Dynamics, Papers Standards Task Force, 87-88; assoc ed, J Aircraft & J Fluids Eng, Am Soc Mech Engrs, 91- *Mem:* Sigma Xi; Am Inst Aeronaut & Astronaut; Am Soc Mech Engrs; Am Helicopter Soc. *Res:* Computational fluid dynamics, super computing, parallel computing, fixed and rotary-wing aircraft aerodynamics, asymptotic methods in applied mathematics. *Mailing Add:* 2358 Sterling Pointe Dr Chesterfield MO 63005-4510

AGARWAL, SHYAM S, b Bareilly, India, July 5, 41; m 71; c 2. MEDICAL GENETICS & CLINICAL IMMUNOLOGY, HAEMATOLOGICAL DISORDERS & BONE MARROW TRANSPLANTATION. *Educ:* Lucknow Univ, BSc, 58, MBBS(Hons), 63, MD(Hons), 66; FRCP(C), 76. *Hon Degrees:* FASc, Acad Sci, Bangalore; FAMS, Acad Med Sci, New Delhi. *Prof Exp:* Res physician genetics & oncol, Inst Cancer Res & Univ Pa, Philadelphia, 69-70; lectr med, K G Med Col, Lucknow, 70-73, reader, 73-86; PROF & HEAD, DEPT GENETICS & IMMUNOL, SANJAY GANDHI POSTGRAD INST MED SCI, LUCKNOW, 86- *Concurrent Pos:* Vis scientist & res physician, Inst Cancer Res, Philadelphia, 74-77; vis res assoc, Mem Sloan Kettering Cancer Ctr, NY, 79-81. *Honors & Awards:* Shanti Swarup Bhatnagar Prize, 86. *Mem:* Fel Indian Col Physicians; Am Soc Human Genetics; Am Asn Immunologists. *Res:* Lymphocyte biology; DNA replication and repair; effect of elevated temperatures on lymphocyte function. *Mailing Add:* Dept Genetics & Immunol Sanjay Gandhi Postgrad Inst Med Sci PO Box 375 Lucknow 226001 India

AGARWAL, SOM PRAKASH, b Aligarh, India, Apr 6, 30; US citizen; m 55; c 3. COSMIC RAY PHYSICS, NUCLEAR PHYSICS. *Educ:* Agra Univ, BSc, 49; Aligarh Muslim Univ, MSc, 51; Temple Univ, PhD(physics), 62. *Prof Exp:* Lectr physics, MMH Col, India, 51-52; res scholar, Aligarh Muslim Univ, India, 52-56; res fel, Bartol Res Found, Franklin Inst, 56-62; assoc prof, Univ Tex, El Paso, 62-64; physicist, Braddock, Dunn & McDonald, Inc, 64-65; asst prof, 65-66, ASSOC PROF PHYSICS, UNIV MINN, MORRIS, 66- *Concurrent Pos:* Consult, Braddock, Dunn & McDonald, Inc, 63-64; adj assoc prof, Solid State Physics Lab, Univ Tex, El Paso, 78-79. *Mem:* Am Phys Soc; Am Geophys Union; Am Asn Physics Teachers; Nat Educ Asn. *Res:* Cosmic rays; interplanetary particles and fields; nuclear reactions; experimental solid state physics; thin film transport properties; elementary particles; radiation effects and electronic physics. *Mailing Add:* Eight N Court St Morris MN 56267

AGARWAL, SURESH KUMAR, b Pilibhit, India, June 22, 32; m 50; c 4. RADIOLOGICAL PHYSICS. *Educ:* Agra Univ, BS, 50; Lucknow Univ, MS, 52, PhD(physics), 65; Am Bd Radiol, cert radiol physics, 72; Am Bd Med Psychotherapists, cert radiol therapy Physics, 89. *Prof Exp:* Physicist, GM & Assoc Hosps, India, 52-58; lectr radiol physics, Lucknow Univ, 58-69; asst prof, 69-74, assoc prof, 74-80, PROF RADIOL PHYSICS, UNIV VA, 80- *Concurrent Pos:* Clin fel, Mem Cancer Hosp, New York, 57-58. *Mem:* Am Asn Physicists Med; fel Am Col Radiol; Health Physics Soc; Sigma Xi. *Res:* Radiation treatment planning of irregular fields and its computerization; radiation dosimetry; quality control in diagnostic radiology; role of hyperthermia in radiation oncology; radiation risk estimates; use of linear accelerators in treating small brain lesions (radio-surgery). *Mailing Add:* Radiol Physics Div Univ Va Sch Med Box 375 Charlottesville VA 22908

AGARWAL, VIJENDRA KUMAR, b Kotah, India, Dec 15, 48; m 79; c 2. SOLAR ENERGY RESEARCH, APPLICATION THIN FILMS. *Educ:* Agra Univ, BS, 66; Meerut Univ, MS, 68; Univ Roorkee, PhD(physics), 72. *Prof Exp:* Fel physics, Univ Parma, Italy, 73-74; lectr, Kashmir Univ, India, 74-75; fel, Shizuoka Univ, Japan, 75-76; res assoc, Ecole Superieure d Electricite, France, 76-78; vis asst prof, Drexel Univ, 78-80, Stockton State Col, 81; from asst prof to assoc prof physics, Tex Tech Univ, 81-88; assoc prof, 88-90, PROF, DEPT PHYSICS, MOORHEAD STATE UNIV, 90- *Concurrent Pos:* Vis scientist, Int Ctr Theoret Physics, Italy, 74 & 77. *Mem:* Am Phys Soc; Inst Elec & Electronics Engrs; Indian Physics Asn; Am Asn Phys Teachers. *Res:* Experimental solid state physics; characterization and applied research on organic thin films; photothermal solar-energy conversion; solar cell encapsulation; solar energy; charge accumulation, electrical breakdown and aging of insulators under high voltage in space environment. *Mailing Add:* Dept Physics Moorhead State Moorhead MN 56563

AGARWAL, VINOD KUMAR, b India, Apr 30, 52; m 78; c 2. FAULT TOLERANT COMPUTING. *Educ:* Birla Inst Tech & Sci, BEng, 73; Univ Pittsburgh, MEng, 74; Johns Hopkins Univ, PhD(elec eng), 77. *Prof Exp:* Asst prof, Wayne State Univ, 77-78; res asst elec eng, Johns Hopkins Univ, 78; from asst prof to assoc prof, 78-87, RES CHAIR, MCGILL UNIV, 87-, PROF ELECT ENG, 88- *Concurrent Pos:* Prin investr, strategic res grant, Nat Sci & Eng Res Coun, 85-88; vis prof, Indian Inst Tech, Delhi, 86; Indust Res Chair Prof. *Mem:* Inst Elec Electronic Engr; Sigma Xi; Asn Comput Mach. *Res:* Design and test integrated circuits and printed circuits boards; design for testability, built-in self test, and fault tolerant schemes. *Mailing Add:* VLSI Design Lab Dept Elec Eng McGill Univ 3480 Univ St Montreal PQ H3A 2A7 Can

AGARWAL, VIPIN K, b Badaun, India, Jan 22, 55; US citizen; m 82; c 2. HIGH PERFORMANCE LIQUID, ANALYTICAL METHOD DEVELOPMENT CHROMATOGRAPHY. *Educ:* Lucknow Univ, BSc, 73, MSc, 75, PhD(org chem), 80; Univ New Brunswick, PhD(org chem), 82; Rensselaer Polytech Int, MBA, 85. *Prof Exp:* Postdoctoral assoc anal chem, Rensselaer Polytech Inst, 83-85; ANAL CHEMIST, CONN AGR EXP STA, 85- *Mem:* Am Chem Soc. *Res:* Development of analytical methods for monitoring trace levels of residues in biological and nonbioloical matrices. *Mailing Add:* Conn Agr Exp Sta 123 Huntington St New Haven CT 06510

AGARWALA, VINOD SHANKER, b Sitapur, India, Aug 9, 39; US citizen; m 67; c 2. CORROSION, ELECTROCHEMISTRY. *Educ:* Banaras Hindu Univ, India, MS, 61, PhD(corrosion), 68; Mass Inst Technol, MS, 66. *Prof Exp:* Can Dept Energy Mines & Res fel corrosion, 67-69; metallurgist prod develop, Superior Steel Ball Co, 70-71; res assoc corrosion, Univ Conn, 72-76; sr res assoc, Air Force Mat Lab, Wright Patterson AFB, 76-77; RES METALLURGIST, NAVAL AIR DEVELOP CTR, 77- *Concurrent Pos:* Tech consult, Corning Glass Works, 74, Crucible Steel Co, 74-75, Bethlehem Steel Corp, 74-75 & Allied Chem, 75; res grant atmospheric corrosion, Am Iron & Steel Inst, Washington, DC, 74-75; adj asst prof, Wright State Univ, Ohio, 76-77; Nat Res Coun fel, Washington, DC, 76-77; chmn, Nat Asn Corrosion Engrs, Tech Pract Comt, T-9B; sect chmn, Am Standard Testing Mat, sub comt, G-105.02; mem, Develop Oper Div, Tri Serv Planning Comt Corrosion. *Honors & Awards:* Corrosion Specialist Award, Nat Asn Corrosion Engrs, 74. *Mem:* Nat Asn Corrosion Engrs; Am Standard Testing Mat; Am Chem Soc; Electrochem Soc; Am Soc Metals. *Res:* Corrosion of metals; electrochemistry; metallurgy; mechanism of corrosion and material failure; stress corrosion cracking; hydrogen embrittlement, corrosion fatigue in aircraft alloys. *Mailing Add:* 1006 Marian Rd Warminster PA 18974-5000

AGATHOS, SPIROS NICHOLAS, b Zante, Greece, July 31, 50; Greek & Can citizen; m 85; c 2. BIOENGINEERING & BIOMEDICAL ENGINEERING. *Educ:* Nat Tech Univ Athens, dipl eng, 73; McGill Univ, MEng, 76; Mass Inst Technol, PhD(biochem eng), 83. *Prof Exp:* Res & teaching asst chem eng, Dept Chem Eng, McGill Univ, Montreal, Que, Can, 73-75; res asst appl biol, Naval Res Lab, Dept Hemat, Boston City Hosp, Mass, 75-77; teaching asst biochem eng, Dept Appl Biol Sci, Mass Inst Technol, Cambridge, 78-82; lectr chem & biochem eng, Univ Western Ont, London, Can, 82-83, asst prof, 83-85; ASST PROF CHEM & BIOCHEM ENG, RUTGERS STATE UNIV NJ, NEW BRUNSWICK, 85- *Concurrent Pos:* Prin investr, Natural Sci & Eng Res Coun, Ottawa, Can, 83-85 & NSF, Washington, DC, 87-; external reviewer, Natural Sci & Eng Res Coun, Ottawa, Can, 83-90, Acad Press, Hanser Publ, 88-; organizer & chmn sci sessions ann meetings, Am Inst Chem Engrs, 84-; consult, var biotechnol co, 84-; mem, Waksman Inst Microbiol, Rutgers Univ, 85- & Int Sci Bd Int Union Microbiol Socs 1994 World Cong Microbiol 89-; dir, Grad Cert Prog Biotechnol, Rutgers Univ, 86-; vis prof, Inst Molecular Biol & Biotechnol, Univ Crete, Iraklion, Greece, 87; panel mem & reviewer, NSF, Washington, DC, 87. *Mem:* Am Inst Chem Engrs; Am Chem Soc; Soc Indust Microbiol; AAAS; NY Acad Sci; Sigma Xi. *Res:* Biochemical engineering; applied microbiology; molecular cell biology; fungal immunosuppressants; insect cell cultivation scaleup; recombinant microbe dynamics; enzyme stability. *Mailing Add:* Dept Chem & Biochem Eng Busch Campus Rutgers Univ Piscataway NJ 08855-0909

AGATSTON, ROBERT STEPHEN, b New York, NY, Apr 20, 23; m 57. GEOLOGY. *Educ:* Ohio State Univ, BS, 43; Columbia Univ, MA, 47, PhD(geol), 52. *Prof Exp:* Div res geologist, Atlantic Refining Co, 48-67, DIR GEOL SCI, ATLANTIC RICHFIELD CO, 67- *Mem:* Am Asn Petrol Geologists. *Res:* Pennsylvanian of Wyoming. *Mailing Add:* 11421 Cromwell Ct Dallas TX 75229

AGBABIAN, MIHRAN S, b Larnacca, Cyprus, Dec 9, 23; nat US; m 53; c 3. STRUCTURAL & EARTHQUAKE ENGINEERING. *Educ:* Am Univ, Beirut, Lebanon, BA, 44, BS, 47; Calif Inst Technol, MS, 48; Univ Calif, Berkeley, PhD(struct & fluid mech), 51. *Prof Exp:* Proj engr & tech dir, Ralph M Parsons Co, Los Angeles, Calif, 55-62; FRED CHAMPION PROF CIVIL ENG & CHMN DEPT, UNIV SOUTHERN CALIF, 84-; pres, 63-91, MEM BD DIRS, AGBABIAN & ASSOCS, 91-; FRED CHAMPION PROF CIVIL ENG & CHMN DEPT, UNIV SOUTHERN CALIF, 84- *Mem:* Nat Acad Eng; Earthquake Eng Res Inst (pres, 83-85); fel Am Soc Civil Engrs; AAAS; fel Inst Advan Eng; NY Acad Sci; Am Soc Testing & Mat; Am Concrete Inst; Am Inst Aeronaut & Astronaut; Am Defense Preparedness Asn. *Mailing Add:* 1111 S Arroyo Pkwy No 405 Pasadena CA 91105-3254

AGEE, ERNEST M(ASON), b Richmond, Ky, Oct 2, 42; m 63; c 2. DYNAMIC METEOROLOGY, FLUID MECHANICS. *Educ:* Eastern Ky Univ, BS, 64; Univ Mo-Columbia, MS, 66, PhD(atmospheric sci), 68. *Prof Exp:* From asst prof to assoc prof, 68-78, PROF ATMOSPHERIC SCI, PURDUE UNIV, 78-, HEAD EARTH & ATMOSPHERIC SCI, 90- *Concurrent Pos:* NSF res grant, 87-; contract, Dept Defense-Oceanic Nat Res, 86-, IBM, 89-; US rep, Int Asn Meteorol & Atmospheric Physics, 80-88. *Mem:* Fel Am Meteorol Soc; Meteorol Soc Japan; Sigma Xi. *Res:* Thunderstorms and tornadoes; mesoscale cellular convection; theoretical and laboratory study of convective patterns and vortex features; field study of convective clouds in the atmosphere; climate. *Mailing Add:* Dept Earth & Atmospheric Sci Purdue Univ West Lafayette IN 47907

AGEE, HERNDON ROYCE, b Cottonburg, Ky, Dec 21, 33; m 53; c 2. INSECT PHYSIOLOGY. *Educ:* Berea Col, BA, 58; Univ Minn, MS, 60; Tufts Univ, PhD, 68. *Prof Exp:* Asst entom, Univ Minn, 58-60; entomologist, Cotton Insect Br, 60-75, PROF ENTOM, UNIV FLA, 74-, RES ENTOMOLOGIST INSECT ATTRACTANTS, BEHAV & BASIC BIOL RES LAB, AGR RES SERV, USDA, 75- *Mem:* Entom Soc Am; Am Soc Zoologists; Am Soc Neurosci. *Res:* Electrophysiological and behavioral studies of the sensory systems of insects; visual and acoustic responses; quality control; neurobiology of central nervous system. *Mailing Add:* 401 NW 91st St Gainesville FL 32607

AGEE, JAMES KENT, b Alameda Calif, July 7, 45; m 69; c 2. FOREST ECOLOGY. *Educ:* Univ Calif, Berkeley, BS, 67, MS, 68, PhD(wildland resource sci), 73. *Prof Exp:* Lectr, Univ Calif, Berkeley, 73, forest ecologist, Nat Park Serv, San Francisco, 74-78; RES BIOLOGIST FOREST FIRE MGT, COOP PARK STUDIES UNIT, NAT PARK SERV, UNIV WASH, 78- *Mem:* Soc Am Foresters; Forest Hist Soc. *Res:* Role of fire in natural ecosystems; management of natural disturbances in park ecosystems. *Mailing Add:* Dept Forestry Univ Wash Seattle WA 98195

AGEE, MARVIN H, b Floyd, Va, Nov 7, 31; c 2. INDUSTRIAL ENGINEERING, MATERIALS HANDLING. *Educ:* Va Polytech Inst, BS, 52, MS, 60; Ohio State Univ, PhD(indust eng), 69. *Prof Exp:* Engr, Eastman Kodak Co, NY, 52, 55; owner retail bus, 55-57; instr indust eng, Va Polytech Inst, 57-58, from asst prof to assoc prof, 59-64; instr, Ohio State Univ, 64-68; assoc prof, 68-81, head, Dept Indust Eng & Opers Res, 70-73, PROF INDUST ENG & OPERS RES, VA POLYTECH INST & STATE UNIV, 81- *Concurrent Pos:* Consult in materials handling, facil design and economic analysis, strategic planning. *Honors & Awards:* Region III Award, Am Inst Indust Engrs, 79; Western Elec Fund Award, Am Soc Engrs Educ, 81. *Mem:* Soc Mfg Eng; fel Inst Indust Eng. *Res:* Material handling systems and logistics systems. *Mailing Add:* Dept ISE Va Polytech Inst & State Univ Blacksburg VA 24061

AGENBROAD, LARRY DELMAR, b Nampa, Idaho, Apr 3, 33; m 55; c 2. GEOLOGY, HYDROGEOLOGY. *Educ:* Univ Ariz, BS, 59, MS, 62, PhD(geol), 67, MA, 70. *Prof Exp:* Explor geophysicist, Pan Am Petrol Corp, 60-62; geologist AEC shoal event, Nev Bur Mines, 62-63; teaching asst geol, Univ Ariz, 63-65, teaching asst geol & anthrop, 65-67; prof earth sci, Chadron State Col, 67-78; PROF GEOL, NORTHERN ARIZ UNIV, 78-, DIR, QUATERNARY STUDIES PROG. *Concurrent Pos:* Dir, Hudson-Meng Paleo-Indian Bison Kill, 70-75; asst dir, Lehner Ranch, 74-75; dir, Hot Springs Mammoth Site, 74-; consult, environ impact statement, cultural resource, hydrogeol, quaternary deposits, fauna & geochronology. *Mem:* Soc Am Archaeol; Am Quaternary Asn; Geol Soc Am; Am Geol Inst. *Res:* Quaternary geology and its relation to early man in the New World; present paleohydrology and ground water resources; archaeology; Pleistocene megafauna, particularly mammoth and paleoenvironment. *Mailing Add:* Dept Geol Box 6030 Flagstaff AZ 86011

AGER, DAVID JOHN, b Northampton, Eng, Apr 10, 53; m 71; c 4. ORGANOSILICON CHEMISTRY. *Educ:* Univ London, BSc, 74; Imp Col, London, ARCS, 74; Univ Cambridge, PhD(org chem), 77. *Prof Exp:* Res fel org chem, Sci Res Coun, Univ Southampton, 77-79; sr demonstr org chem, Univ Liverpool, 79-82; ASST PROF ORG CHEM, UNIV TOLEDO, 82- *Concurrent Pos:* Career develop award, Royal Soc Chem, 82-83. *Mem:* Am Chem Soc; Royal Soc Chem; Sigma Xi. *Res:* Synthesis of natural products; new synthetic methods, particularly those using organosilicon, organosulfur and organotin chemistry. *Mailing Add:* Dept Chem Univ Toledo 2801 W Bancroft St Toledo OH 43606

AGER, JOHN WINFRID, b Birmingham, Ala, Nov 2, 26; m 56; c 2. CHEMISTRY. *Educ:* Harvard Univ, BA, 49; Univ NC, MA, 51; Oxford Univ, PhD(org chem), 55. *Prof Exp:* Res chemist, Olin Mathieson Chem Corp, NY, 56-60; RES CHEMIST, FMC CORP, 60- *Mem:* Am Chem Soc. *Res:* Boron based high energy fuels; synthesis in the field of natural products; process development of commercial chemicals; agricultural chemicals. *Mailing Add:* FMC Corp Box 8 Princeton NJ 08540

AGER, THOMAS ALAN, b Detroit, Mich, July 31, 46. GEOLOGY, PALEONTOLOGY. *Educ:* Wayne State Univ, BS, 68; Univ Alaska, MS, 72; Ohio State Univ, PhD(geol), 75. *Prof Exp:* Res assoc geol, Inst Polar Studies, Ohio State Univ, 75-76; GEOLOGIST PALYNOLOGY, US GEOL SURV, 76- *Mem:* Am Quaternary Asn; Am Asn Stratig Palynologists; Ecol Soc Am. *Res:* Quaternary and Neogene palynology and paleo-ecology of Alaska and Eastern USA; paleoclimatology; biostratigraphy. *Mailing Add:* PO Box 3175 Evergreen CO 80439

AGERSBORG, HELMER PARELI KJERSCHOW, JR, b Decatur, Ill, Dec 2, 28; m 52; c 3. PHYSIOLOGY, TOXICOLOGY. *Educ:* Southern Ill Univ, AB, 54; Univ Tenn, PhD, 57. *Prof Exp:* Asst physiol, Univ Tenn, 54-57, instr clin physiol & obstet & gynec & dir obstet & gynec, 57-58; clin physiologist, Med Div, Wyeth Labs, Inc, 58-61; mgr toxicol & comp pharmacol, Res Div, 61-69, assoc dir, 69-76, vpres res & develop, 76-85, PRES, WYETH AYERST RES, 85- *Mem:* Am Soc Pharmacol & Exp Therap; Am Physiol Soc; Am Soc Zool; Soc Toxicol. *Res:* Metabolic disease; cardiovascular physiology; ion and fluid dynamics; drug safety evaluation. *Mailing Add:* Wyeth Labs Inc Philadelphia PA 19101

AGGARWAL, BHARAT BHUSHAN, b Batala, India; m 82; c 2. RECOMBINANT DNA TECHNOLOGY. *Educ:* Univ Delhi, BS, 70; Benaras Hindu Univ, India, MS, 72; Univ Calif, Berkeley, PhD(biochem), 77. *Prof Exp:* Fel hormone res, Univ Calif, San Francisco, 77-80; sr scientist protein biochem, Genentech Inc, S San Francisco, 80-89; PROF MED, BIOCHEM, M D ANDERSON CANCER CTR, UNIV TEX, 89- *Mem:* Am Endocrine Soc; AAAS; Soc Biochem & Molecular Biol; Am Asn Immunologist. *Res:* Purification and physiochemical characterization of proteins derived from natural and recombinant DNA technology; relationship of the structure of a protein and carbohydrates to its function. *Mailing Add:* Dept Clinical Immunol & Biol Therapy Univ Texas M D Anderson Cancer Ctr 1515 Holcomb Blvd Houston TX 77030

AGGARWAL, H(ANS) R(AJ), b Jullundur, India, Dec 12, 25; m 51; c 4. ENGINEERING MECHANICS, APPLIED MATHEMATICS. *Educ:* Panjab Univ, BA, 48, MA, 50; Cornell Univ, PhD(eng mech), 62. *Prof Exp:* Lectr math, Govt Col, Hoshiarpur, India, 50-57; teaching asst, Cornell Univ, 57-60; mem tech staff, Nat Eng Sci Co, Calif, 61-63; res mathematician, Stanford Res Inst, 63-70; sr res assoc appl mech, Ames Res Ctr, NASA, 71-73; sr res assoc, Univ Santa Clara, Calif, 73-80; SR RES ASSOC, AMES RES CTR, NASA, 80- *Mem:* Sigma Xi; Soc Indust & Appl Math. *Res:* Transonic flow; helicopter aerodynamics; aeroacoustics and helicopter noise; computational analysis; numerical solution of partial differential equations; planetology; numerical solution of partial differential equations. *Mailing Add:* 3374 Tryna Dr Mountain View CA 94040

AGGARWAL, ISHWAR D, b Delhi, India, May 18, 45; m 69; c 2. CHEMICAL ENGINEERING, OPTICAL ENGINEERING. *Educ:* Indian Inst Technol, BS, 69; Catholic Univ Am, MS, 71, PhD(mat sci), 74. *Prof Exp:* Sr mat scientist, Corning Glass Works, 74-75; mgr mat res, Galileo Electro Optic Corp, Sturbridge, Mass, 75-78; vpres res, Optical Commun, Valtec, 78-84; vpres eng, Lasertron Inc, Burlington, Mass, 84-86; RESEARCHER, NAVAL RES LAB, 86- *Mem:* Optical Soc Am; Am Ceramic Soc; Nat Inst Ceramic Engrs. *Res:* Optical fiber communication; glass science; engineering management; optical engineering research; polymeric coating; laser and detector materials; quality engineering management. *Mailing Add:* 6505-A Naval Res Lab Washington DC 20375-5000

AGGARWAL, JAGDISHKUMAR KESHORAM, b Amritsar, India, Nov 19, 36; m 65; c 2. ELECTRICAL ENGINEERING, COMPUTER SCIENCE. *Educ:* Univ Bombay, BSc, 56; Univ Liverpool, BEng, 60; Univ Ill, MSc, 61, PhD(elec eng), 64. *Prof Exp:* Asst elec eng, Coord Sci Lab, Univ Ill, 61-64; from asst prof to assoc prof, 64-72, PROF ELEC ENG, UNIV TEX,

AUSTIN, 72- *Concurrent Pos:* Vis assoc prof, Univ Calif, Berkeley, 69-70. *Mem:* Fel Inst Elec & Electronics Engrs; Pattern Recognition Soc. *Res:* Digital filtering; image processing; computer vision. *Mailing Add:* Dept Elec & Comput Eng Univ Tex Austin TX 78712

AGGARWAL, ROSHAN LAL, b Salala, India, Feb 15, 37; m 58; c 2. QUANTUM OPTICS, PHYSICS OF SEMICONDUCTORS. *Educ:* Punjab Univ, India, BSc, 57, MSc, 58; Purdue Univ, PhD(physics), 65. *Prof Exp:* Lectr, Punjab Univ, 58-60; res assoc physics, Purdue Univ, 65; staff mem, Mass Inst Technol, 65-71, proj leader quantum optics, 71-74, group leader quantum optics & plasma physics, 74-76, group leader quantum optics, 76-84, assoc dir, 77-84, sr res scientist, Physics Dept, Francis Bitter Nat Magnet Lab, 75-90, group leader, Semiconductor Group, 84-86, STAFF MEM, MASS INST TECHNOL, LINCOLN LAB, 86-, SR LECTR, PHYSICS DEPT, 90- *Mem:* Fel Am Phys Soc; Sigma Xi; Optical Soc Am. *Res:* Near and far infrared spectroscopy; modulation magnetospectroscopy; lasers and nonlinear optics; light scattering in solids; physics of semiconductors. *Mailing Add:* Mass Inst Technol Lincoln Lab PO Box 73 244 Wood St Lexington MA 02173

AGGARWAL, SHANTI J, b Balrampur, India, July 1, 33; m 65; c 2. BIOMEDICAL ENGINEERING. *Educ:* Banaras Hindu Univ, India, BSc & MSc, 52; Univ Mich, Ann Arbor, MS, 58, PhD(microbiol), 62. *Prof Exp:* Res fel, Nat Res Coun Can, Ottawa, 62-63; res assoc microbiol, Univ Ill, Urbana, 63-64; res assoc zool, 65-70, microbiol, 70-78, lectr, 74-78, SR RES SCIENTIST BIOMED ENG, UNIV TEX, AUSTIN, 83- *Concurrent Pos:* Res assoc, Space Sci Res Labs, Univ Calif, Berkeley, 69-70. *Mem:* Soc Cryobiol; AAAS; Microcirculatory Soc. *Res:* Cryopreservation of tissue for transplantation and tissue banking; cryopreservation of human pancreatic islets and human skin for tissue banking and subsequent transplantation in diabetic patients and grafting on burn sites, respectively; microcirculatory response after burn injury; computer vision analysis of biomedical scenes; 3-D reconstruction of biomedical images using light microscopy and stereo-microscopy; laser tissue interactions. *Mailing Add:* Biomed Eng Prog ENS 612 Univ Tex Austin TX 78712-1084

AGGARWAL, SUNDAR LAL, b Jullundur, India, Oct 15, 22; nat US; m 48; c 3. POLYMER CHEMISTRY. *Educ:* Panjab Univ, India, BSc, 42, MSc, 43; Cornell Univ, PhD(phys chem), 49. *Prof Exp:* Lectr anal chem, Govt Col, India, 43-45; res assoc polymer chem, Cornell Univ, 49-50; sci officer, Nat Chem Lab, India, 50-52; sr scientist, Film Res Sect, Olin Industs, 52-54, group leader, 54-55, sect chief, Phys Properties & Polymer Sect, 55-57; head chem physics res, Gen Tire & Rubber Co, 57-62, mgr basic res, 62-68, mat res & tech serv, 68-75, vpres & dir, Res Div, 75-87; RETIRED. *Concurrent Pos:* Chmn, Gordon Res Conf Elastomers, 69; assoc ed, Int J Polymer Mat; chmn, Symp Block Polymers. *Mem:* Fel AAAS; Am Chem Soc; Am Phys Soc; Soc Rheol; NY Acad Sci. *Res:* Structure and properties of block polymers and elastomers; dynamic mechanical and strength properties of polymers; electron microscopy and molecular spectroscopy methods for polymer structure; polymer physics of composite materials. *Mailing Add:* Global Polymer Technol Assocs Inc 1947 Burlington Rd Akron OH 44313-5345

AGGARWAL, SURINDER K, b Panjab, India, June 3, 38; m 67; c 3. TOXICOLOGY, TERATOLOGY. *Educ:* Panjab Univ, BSc, 57, MSc, 58, PhD(zool), 64. *Prof Exp:* Res assoc zool, Dept Biol Sci, Northwestern Univ, 64-67; PROF ZOOL, MICH STATE UNIV, 67- *Concurrent Pos:* Fulbright exchange fel, 87-88. *Mem:* Histochem Soc; Electron Micros Soc Am; Cell Biol Soc Am; Cancer Res Asn Am. *Res:* Mechanism of action of various heavy metal coordination complexes as antitumor agents, using hitochemical and electron microscopical approach. *Mailing Add:* Dept Zool Mich State Univ East Lansing MI 48824

AGGARWALA, BHAGWAN D, b India, Mar 9, 31. APPLIED MATHEMATICS. *Educ:* Punjab Univ, India, BA, 50, MA, 52, PhD(math), 59. *Prof Exp:* Instr math, Indian Inst Technol, Kharagpur, 53-55; instr, Carnegie Inst Technol, 55-57; asst prof mech, Rensselaer Polytech Inst, 57-60; asst prof math, McGill Univ, 60-66; assoc prof, 66-71, PROF MATH, STATIST & COMPUT SCI, UNIV CALGARY, 71- *Mem:* Can Math Cong; Soc Indust & Appl Math; Am Acad Mech. *Res:* Stress analysis of elastic and viscoelastic materials; heat transfer. *Mailing Add:* Dept Math Statist Univ Calgary 2500 University Dr Calgary AB T2N 1N4 Can

AGGELER, JUDITH, b San Francisco, Calif, July 18, 44; c 2. ELECTRON MICROSCOPY, CELL BIOLOGY. *Educ:* Univ Calif, San Francisco, PhD(anat & cell biol), 82. *Prof Exp:* Res biologist, Lab Radiobiol, Univ Calif, Davis, 72-84, asst prof, Dept Human Anat, 84-91; ASST RES CELL BIOLOGIST, DEPT MED, UNIV CALIF, SAN FRANCISCO, 91- *Mem:* Am Soc Cell Biol; AAAS; Asn Women Sci; Soc Develop Biol; Tissue Cult Asn; Electron Micros Soc Am. *Mailing Add:* Vet Admin Med Ctr 4150 Clement St Box 151-E San Francisco CA 94121

AGGEN, GEORGE, b Rotterdam, NY, Sept 28, 24; m 52; c 5. METALLURGICAL ENGINEERING. *Educ:* Rensselaer Polytech Inst, BME, 51, DSc(metall eng), 63. *Prof Exp:* Jr metallurgist Res & Develop, 51-54, res metallurgist, 54-56, sr res metallurgist, 56-57, supv metallurgist, 57-62, proj supvr, 62-63, res assoc, 63-65, sr res specialist, 65-68, sr supv metallurgist, 68-70, sr supv metallurgist, Allegheny Ludlum Industs, Inc, 70-75, MGR RES, ALLEGHENY LUDLUM STEEL CORP, 75-, MGR STAINLESS RES, 75- *Mem:* Am Soc Metals; Am Inst Mining, Metall & Petrol Engrs; Am Soc Testing & Mat. *Res:* High temperature alloys; stainless steels, particularly precipitation-hardening types. *Mailing Add:* 245 Edgewood Dr Sarver PA 16055

AGGOT, J DESMOND, drug disposition, clinical pharmacology, for more information see previous edition

AGHAJANIAN, GEORGE KEVORK, b Beirut, Lebanon, Apr 14, 32; m 59; c 4. NEUROPHARMACOLOGY. *Educ:* Cornell Univ, AB, 54; Yale Univ, MD, 58. *Prof Exp:* Intern, Jackson Mem Hosp, 58-59; resident, 59-63, from asst prof psychiat to assoc prof psychiat & pharmacol, 65-74, PROF PSYCHIAT & PHARMACOL, SCH MED, YALE UNIV, 74-, FOUND FUND PROF RES PSYCHIAT, 85- *Honors & Awards:* Efron Award, 75; Scheele Medal, 81; Res Prize, Found Fund, 81; Merit Award, NIMH, 90. *Mem:* Am Soc Pharmacol & Exp Therapeut; Psychiat Res Soc; fel Am Col Neuropsychopharmacol; Soc Neurosci; Int Brain Res Orgn. *Res:* Neurotransmitters and the mechanism of action of psychotropic drugs. *Mailing Add:* Dept of Psychiat Yale Univ Sch of Med 34 Park St New Haven CT 06508

AGIN, DANIEL PIERRE, b New York, NY, May 19, 30; m 56; c 2. NEUROPHYSIOLOGY, BIOPHYSICS. *Educ:* City Col New York, BA, 53; Univ Rochester, PhD(psychol), 61. *Prof Exp:* NIH fel, 61-62; from instr to asst prof, 62-67, actg chmn dept, 68-69, ASSOC PROF PHYSIOL, UNIV CHICAGO, 67- *Concurrent Pos:* Vis prof, Max Planck Inst Biophys, 69-70. *Res:* Membrane biophysics; biophysical pharmacology. *Mailing Add:* Dept Cell Biol & Genetics Univ Chicago Pritzker Sch Med 920 E 58th St Chicago IL 60637

AGIN, GARY PAUL, b Kansas City, Mo, Dec 22, 40. NUCLEAR PHYSICS. *Educ:* Univ Kans, BS, 63; Kans State Univ, MS, 67, PhD(physics), 68. *Prof Exp:* Asst prof, 68-87, ASSOC PROF PHYSICS, MICH TECHNOL UNIV, 87- *Concurrent Pos:* Zone 8 counr, Soc Physics Students, 81-88, coun pres, 89- *Mem:* AAAS; Am Asn Physics Teachers; Am Phys Soc. *Res:* Low energy nuclear physics; beta and gamma ray spectroscopy; computer applications. *Mailing Add:* Dept Physics Mich Technol Univ Houghton MI 49931-1295

AGINS, BARNETT ROBERT, b New York, NY, May 19, 22; m 45; c 2. MATHEMATICS. *Educ:* NY Univ, BEE, 52, MEE, 56; Stanford Univ, MSc, 61. *Prof Exp:* US Air Force, 50-, commun off Air Defense Command, 50-54, proj engr electronic syst, 56-59, chief appl math div, Off Sci Res, 61-67, asst to dir, Courant Inst Math Sci, 67-69, prog dir appl math & statist, NSF, 69-77; asst to dir, Courant Inst Math Sci, 67-69; prog dir appl math & statist, NSF, 69-77; ADJ PROF, FLA ATLANTIC UNIV, 77- *Concurrent Pos:* Prof lectr, Am Univ, 62-; assoc ed, J Optimization Theory & Applns, 67-; assoc ed, Comput & Math Applications, 74-87. *Mem:* Am Math Soc; Inst Elec & Electronic Engrs; Soc Indust & Appl Math. *Res:* Sophisticated electronic warfare techniques; mathematics with respect to ordinary differential equations, especially those of celestial mechanics. *Mailing Add:* Dept Math Florida Atlantic Univ Boca Raton FL 33431-7831

AGNELLO, ARTHUR MICHAEL, b Ithaca, NY, Sept 25, 52; m 81; c 1. ECONOMIC ENTOMOLOGY, EXTENSION. *Educ:* Cornell Univ, BS, 74; Univ Fla, Gainesville, MS, 79; NC State Univ, PhD(entom), 85. *Prof Exp:* Aquatic weeds res officer, min agr, US Peace Corp, Repub Botswana, Africa, 74-76; surv entomologist, Ill Coop Ext Serv, Univ Ill, 80-81; res assoc, Dept Entom, NC State Univ, 85-86; ASST PROF ENTOM, CORNELL UNIV, 86- *Mem:* Entom Soc Am; Sigma Xi. *Res:* Control of arthropod pests of fruit; monitoring and sampling strategies for fruit mites and insects; development of pest management recommendations and educational materials; pesticide application technology. *Mailing Add:* Dept Entom Barton Lab NY State Agr Exp Sta Geneva NY 14456

AGNELLO, EUGENE JOSEPH, b Rochester, NY, Oct 3, 19; m 59; c 5. ORGANIC CHEMISTRY. *Educ:* State Univ NY Albany, AB, 41; Univ Rochester, PhD(chem), 50. *Prof Exp:* Teacher pub sch, NY, 41-46; asst, Univ Rochester, 46-50; fel, Univ Ill, 50-51; res chemist, Chas Pfizer & Co, Inc, 51-63; asst prof chem, Waynesburg Col, 63-64; from asst prof to prof chem, 64-90, chmn chem dept, 79-90, EMER PROF CHEM, HOFSTRA UNIV, 90- *Mem:* Am Chem Soc. *Res:* Steroids; medicinal chemistry; natural products. *Mailing Add:* Ten Stirrup Run Stirrup Farms Newark NJ 19711

AGNELLO, VINCENT, b Brooklyn, NY, Aug 1, 38. MEDICINE. *Educ:* Rennsselaer Polytech Inst, BS, 60; Univ Rochester, MD, 64; Am Bd Internal Med, dipl, 74. *Prof Exp:* Intern, Vanderbilt Hosp, 64-65; from asst prof to assoc prof, Hosp Rockefeller Univ, 67-73; chief, Div Rheumat & Clin Immunol, New Eng Med Ctr, 73-81, dir, Clin Immunol Lab, 81-84; PHYSICIAN, ARTHRITIS GERONT CTR, EDITH NOURSE ROGERS VET HOSP, 82-; DIR, CLIN IMMUNOL LAB, LAHEY CLIN MED CTR, 84- *Concurrent Pos:* Postdoctoral fel, Rockefeller Univ, 67-71, asst prof, 71-73; resident, NY Hosp, 72-73; assoc prof, Sch Med, Tufts Univ, 73-85; career develop award, NIH, 74-79; mem adv panel, Am Bd Med Lab Immunol, 78- *Mem:* Fel Am Col Physicians; fel Am Acad Microbiol; Sigma Xi; Am Asn Immunologists. *Res:* Author or co-author of over 60 publications. *Mailing Add:* Dept Lab Med Lahey Clin Med Ctr 41 Mall Rd PO Box 541 Burlington MA 01805

AGNEW, ALLEN FRANCIS, b Ogden, Ill, Aug 24, 18; m 46; c 4. PUBLIC POLICY. *Educ:* Univ Ill, AB, 40, MS, 42; Stanford Univ, PhD(geol), 49. *Prof Exp:* Asst geologist, Ill State Geol Surv, 39-42; geologist, US Geol Surv, Wis, 42-47 & 49-55; asst prof geol, Univ Ala, 48-49; from assoc prof to prof, Univ SDak, 55-63, geologist, SDak Geol Surv, 55-57, dir & state geologist, 57-63; prof geol & dir water resources res ctr, Ind Univ, Bloomington, 63-69; prof geol & dir water res ctr, Wash State Univ, 69-73; sr specialist environ policy/mining, Cong Res Serv, Libr Cong, 74-81; COURTESY PROF, ORE STATE UNIV, 82- *Honors & Awards:* Robert Peele Award, Am Inst Mining Engrs; Pub Serv Award, Am Inst Prof Geologists, Cert Merit. *Mem:* AAAS; Asn Eng Geol; Geol Soc Am; Soc Econ Geologists; Soc Environ Geochem Health; Soc Mining Engrs; Am Geophys Union; Am Inst Prof Geologists; Nat Water Well Asn; hon mem Asn Am State Geologists. *Res:* Hydrogeology; engineering geology. *Mailing Add:* 33125 White Oak Rd Apt 24 Corvallis OR 97333

AGNEW, HAROLD MELVIN, b Denver, Colo, Mar 28, 21; m 42; c 2. PHYSICS. *Educ:* Univ Denver, AB, 42; Univ Chicago, MS, 48, PhD(physics), 49; Col Santa Fe, 81. *Prof Exp:* Physicist, Manhattan Dist, Univ Calif, Los Alamos, NMex, 42-46; Nat Res fel, Physics Div, Los Alamos Sci Lab, 49-50, asst to tech assoc dir, 51-53 & Theoret Div, 54-56, leader, Alt Weapons Div, 56-61; sci adv to Supreme Allied Comdr, Europe, 61-64; leader, Weapons Div, Los Alamos Nat Lab, 64-70, dir, 70-79; pres, GA Technol, Inc, San Diego, 79-85; RETIRED. *Concurrent Pos:* Mem, US Air Force Sci Adv Bd, 57-68; chmn, US Army Combat Develop Command Sci Adv Group, 65-66; mem, President's Sci Adv Comt, 65-73; mem, Defense Sci Bd, 66-70; chmn, US Army Sci Adv Panel, 66-70, mem, 70-74; mem, Aerospace Safety Adv Panel, NASA, 68-74, 85-87; mem, gen adv comt, US Arms Control & Disarmament Agency, 74-82, chmn, 74-78; mem, Coun Foreign Rels, 75-; mem bd dirs, Fedn Rocky Mtn States, Inc, 75-; mem, Army Sci Bd, 78-84; White House Sci Coun, 81-88; adj prof, Univ Calif, San Diego, 88- *Honors & Awards:* Ernest Orlando Lawrence Award, Am Eng Coun, 66; Enrico Fermi Award, Dept Energy, 78. *Mem:* Nat Acad Sci; Nat Acad Eng; fel AAAS; fel Am Phys Soc; Sigma Xi. *Res:* Neutron physics; light particle reactions; particle accelerators. *Mailing Add:* 322 Punta Baja Dr Solana Beach CA 92075

AGNEW, JEANNE LE CAINE, b Port Arthur, Ont, May 3, 17; US citizen; m 42; c 5. MATHEMATICS. *Educ:* Queen's Univ, Ont, BA, 37, MA, 38; Harvard Univ, PhD(math), 41. *Hon Degrees:* LLD Queen's Univ, Can, 88, Lakehead Univ, Thunder Bay, Ont, 90. *Prof Exp:* Instr math, Smith Col, 41-42; res physicist, Nat Res Coun Can, 42-45; instr math, Cambridge Jr Col, 46-47; from asst prof to prof math, Okla State Univs, 56-84; RETIRED. *Concurrent Pos:* Vis assoc prof, Ga State Col, 66-67. *Mem:* Nat Coun Teachers & Maths; Math Asn Am. *Res:* Applications of undergraduate level mathematics to industrial problems; preparation of written material based on this research for use in high school and college. *Mailing Add:* 1216 N Lincoln St Stillwater OK 74075

AGNEW, LESLIE ROBERT CORBET, b Newcastle-on-Tyne, Eng, Nov 18, 23. HISTORY OF MEDICINE, EXPERIMENTAL PATHOLOGY. *Educ:* Glasgow Univ, MB, ChB, 46, MD, 50; Harvard Univ, AM, 57. *Prof Exp:* Mem staff path, Rowett Res Inst, Scotland, 47-49; res assoc prof, Univ Fla, 53-55; assoc nutrit, Sch Pub Health, Harvard Univ, 55-56, resident tutor, Harvard Col, 55-57; from assoc prof to prof hist med & chmn dept, Med Ctr, Univ Kans, 57-65; sr lectr, 65-66, assoc prof, 66-84, EMER PROF MED HIST, SCH MED, UNIV CALIF, LOS ANGELES, 85- *Concurrent Pos:* Trent Mem lectr, Duke Univ, 62; Shuman Mem lectr, Univ Calif, Los Angeles, 64; Davis lectr, Univ Ill, 65; mem fac med hist, Worshipful Co of Apothecaries; mem study sect hist life sci, NIH, 62-67. *Mem:* Am Asn Hist Med; Hist Sci Soc; NY Acad Sci; corresp mem Int Acad Hist Med; Path Soc Gt Brit & Ireland. *Res:* Eighteenth and nineteenth century British and American medicine; vitamin deficiency states; hormones and cancer. *Mailing Add:* Div Med Hist Univ Calif 405 Hilgard Ave Los Angeles CA 90024

AGNEW, LEWIS EDGAR, JR, b Glendale, Mo, June 20, 26; m 52; c 4. PHYSICS. *Educ:* Univ Mo-Rolla, BS, 50; Univ Calif, Berkeley, MA, 56, PhD(physics), 60. *Prof Exp:* Res asst, Los Alamos Sci Lab, 50-53, staff mem, 53-54; res asst, Lawrence Radiation Lab, Univ Calif, 56-59; staff mem, Lab, 59-69, GROUP LEADER, LAMPF ACCELERATOR PROJ, LOS ALAMOS NAT LAB, 70- *Concurrent Pos:* Head physics sect, Div Res & Labs, Int Atomic Energy Agency, Vienna, Austria, 66-68. *Mem:* Fel AAAS; Am Phys Soc; Sigma Xi. *Res:* Medium energy physics; plasma spectroscopy. *Mailing Add:* 292 Cascabel Los Alamos NM 87544

AGNEW, ROBERT A, b Ga, Jan 9, 42. CORPORATE BUSINESS ANALYSIS. *Educ:* North Western Univ, PhD(math), 68; Univ Chicago, MBA, 85. *Prof Exp:* MGR, FMC BUSINESS ANALYSIS CORP, 76- *Mailing Add:* 2626 N Lakeview No 1503 Chicago IL 60614

AGNEW, ROBERT MORSON, b Cardigan, PEI, Nov 20, 32. IMMUNOBIOLOGY OF TUMORS. *Educ:* Dalhousie Univ, BSc, 53, MSc, 55; Cambridge Univ, PhD(bact), 59. *Prof Exp:* Lectr bact, Univ Regina 58-64, asst prof biol, 64-69, head, Dept Biol, 81-87, assoc prof biol, 69-88; RETIRED. *Concurrent Pos:* Vis mem staff, Rheumatic Dis Unit, Univ Toronto, 70-71 & Div Biochem, Univ Tex Med Br, Galveston, 77-78. *Mem:* AAAS; Can Col Microbiologists. *Res:* Immunology; medical bacteriology; tumor immunology. *Mailing Add:* 216 Roland Rd Fulford Harbour BC V0S 1C0 Can

AGNEW, WILLIAM FINLEY, b Greenville, SC, Aug 28, 25; m 58; c 2. PHYSIOLOGY. *Educ:* Wheaton Col, AB, 49; Univ Ill, Urbana, MS, 54; Univ Southern Calif, PhD(physiol), 64. *Prof Exp:* Res asst physiol, Baxter Labs, 52-53; res asst biol, Calif Inst Technol, 55; SR INVESTR, INST MED RES, HUNTINGTON MEM HOSP, 63-, DIR NEUROL RES, HUNTINGTON MED RES INST, 76- *Concurrent Pos:* Laband neurosurg res fel physiol, Inst Med Res, Huntington Mem Hosp, 55-63; res asst, Univ Southern Calif, 57-60; NIH res grant, 61-63; Nat Inst Neurol Dis & Stroke fel, Univ Copenhagen, 65-66, res grant, 69-71; prin investr, NIH contract, 76-79, 80-82, 83-85, 86-88 & 87-90; NIH res grants, 77-80 & 81-84; ed, Neural Prostheses: Fundamental Studies, 90. *Mem:* Teratology Soc; Am Physiol Soc; Soc Exp Biol & Med; Soc Neurosci. *Res:* Cerebral circulation; blood-brain and cerebrospinal fluid barriers and development and testing of neuroprosthetic devices. *Mailing Add:* Huntington Med Res Insts 234 Fairmount Ave Pasadena CA 91105

AGNEW, WILLIAM G(EORGE), b Oak Park, Ill, Jan 12, 26; m 57; c 3. MECHANICAL ENGINEERING. *Educ:* Purdue Univ, BSME, 48, MSME, 50, PhD(mech eng), 52. *Prof Exp:* Mem staff, Eng Proj Squid, US Navy, Purdue Univ, 48-50; sr res engr, Fuels & Lubricants Dept, Gen Motors Corp, 52-56, res assoc, 56-63, asst dept head, 63-67, head, 67-70, head, Emissions Res Dept, 70-71, tech dir, 71-87, dir prog & plans, Gen Motors Res Labs, 87-89; CONSULT, 89- *Honors & Awards:* Horning Mem Award, Soc Automotive Engrs, 60. *Mem:* Nat Acad Eng; fel AAAS; fel Soc Automotive Engrs; Sigma Xi; Am Soc Mech Engrs; Combustion Inst (dir, 60-76); Am Soc

Eng Educ. *Res:* The effects of atmospheric conditions of engine performance; radiant energy and radiation temperature measurement in gas turbine combustions; end-gas temperature measurement in spark-ignition engines; cool flames and pre-flame reactions; engine knock and antiknock mechanisms. *Mailing Add:* 3450 31 Mile Rd Romeo MI 48065

AGNEW-MARCELLI, G(LADYS) MARIE, b Albany, NY, Mar 20, 27. BIOMEDICAL SCIENCES, EPIDEMIOLOGY. *Educ:* Rensselaer Polytech Inst, BSCh, 48, MS, 52; Univ Sussex, PhD(biol sci & pub admin), 74. *Prof Exp:* Tech asst, Pharm Div, Sterling-Winthrop Res Inst, 48-51, res asst, Chem-Biol Coord Sect, 51-53, res assoc new drug develop, Off Dir Res, 54-57, sci liaison officer, Europ-Australian Tech Opers, Sterling Drug, Inc, 58-67, assoc mem new drug develop, Sterling-Winthrop Res Inst, 61-65, res chemist, 65-71, mem new prod develop comt, 70-71, dir spec projs, Prod Develop & Regulatory Affairs, 71-72; consult sci & med info, 72-83; RES FEL & COORDR, SCI ANAL SECT, JACOB, MEDINGER & FINNEGAN, 84- *Concurrent Pos:* Curric adv comt, State Univ NY, Hudson Valley Community Col, 56-71; spec proj asst mild analgesics, Sterling Drug Res Bd, 67-70; sci consult, Ritual Object Sources & Authentication, Yeshiva Univ, Mus, 80 & mem, Mus Adv comt, 81-86; consult, Essex County, NY, 90- *Mem:* Am Chem Soc; Am Soc Metals; Sigma Xi; Am Pub Health Asn; Biomet Soc; Soc Epidemiol Res. *Res:* Mild analgesics; barbiturates; toxicity of chemicals, occupational and environmental pollutants; epidemiology of chronic diseases; development of pharmaceuticals, toiletries and household products; scientific, medical and regulatory aspects of consumer product liability. *Mailing Add:* 61 Maple Ave Hastings-on-Hudson NY 10706

AGNIHOTRI, RAM K, b Kanpur, India, Oct 15, 33; c 2. ORGANIC CHEMISTRY. *Educ:* Agra Univ, BS, 54, MS, 56; Purdue Univ, PhD, 63. *Prof Exp:* Lectr chem, India, 57-58; res fel & asst, Purdue Univ, 58-63; res chemist, E I du Pont de Nemours & Co, 63-67; staff chemist, IBM Corp, 67-69, proj mgr, 69-70, develop chemist & mgr photolithographic mat, 70-71, adv engr, 71-80, proj mgr heads, 81-84, proj mgr media, 84-87, sr engr, 84-88, mgr chem process, 88-90. *Mem:* Am Chem Soc. *Res:* Photostabilizers; polymers; resins; organometallic chemistry; optical rotatory power; polyipides in semiconductors; coatings; photolithography; semiconductor personalization; semiconductor packaging; manufacturing tape heads and thin film. *Mailing Add:* 5149 Calle Bosque Tucson AZ 85718

AGNISH, NARSINGH DEV, b Jullundur, India, Jan 3, 40; m 69; c 4. TOXICOLOGY. *Educ:* Panjab Univ, BSc, 59, MSc, 60; Univ Sask, PhD(anat & cancer res), 69; Am Bd Toxicol, dipl, 82. *Prof Exp:* Sr scientist, 77-78, asst res group chief, 79-83, RES GROUP CHIEF, HOFFMAN-LA ROCHE, INC, 83- *Concurrent Pos:* Adj asst prof, Jefferson Med Col, 79- *Mem:* Teratol Soc; Am Asn Anatomists; Soc of Toxicol. *Res:* Preclinical testing of candidate drugs for effects on reproduction and developing fetus; normal and abnormal development of mammalian limbs. *Mailing Add:* 43 Morris Ave West Milford NJ 07480

AGOSIN, MOISES, b Marseille, France, Dec 1, 22; m 48; c 3. BIOCHEMISTRY, PARASITOLOGY. *Educ:* Univ Chile, MD, 48. *Hon Degrees:* Dr, Peruvian Univ, Cayetano Heredia. *Prof Exp:* From asst prof to assoc prof parasitol, Sch Med, Univ Chile, 48-57, prof biochem, 57-61 & chem, 61-68; vis prof, 68-69, RES PROF ZOOL, UNIV GA, 69- *Concurrent Pos:* Rockefeller Found fel, NIH, 52-54, res assoc, 54-55, res grants, 58-90; Rockefeller Found res grant, 56; vis prof, Univ Calif, Berkeley, 61 & Univ London, 65. *Honors & Awards:* Fel Chilean Acad Sci, 78; Bueding-von Brand Award, Am Soc Parisitol, 90. *Mem:* Fel Am Acad Microbiol; Am Chem Soc; Biochem Soc; Chilean Acad Sci. *Res:* Biochemistry of parasitic organisms; biochemistry of insecticide resistance. *Mailing Add:* Dept of Zool Univ of Ga Athens GA 30602

AGOSTA, VITO, b New York, NY, July 26, 23; m 52; c 3. COMBUSTION OF ALTERNATIVE FUELS, PROPULSION. *Educ:* Polytech Inst Brooklyn, BS, 46; Univ Mich, MS, 49; Columbia Univ, PhD(mech eng), 59. *Prof Exp:* Thermodynamacist, DeLaval Steam Turbine Co, 46-47; prof mech eng, 50-62, prof, 62-86, EMER PROF AEROSPACE ENG, POLYTECH INST NEW YORK, 86- *Concurrent Pos:* Consult, Gen Appl Sci Labs, Inc, 56-57, Appl Phys Lab, Johns Hopkins Univ, 56-79, Curtiss-Wright Corp, 62-63, Marquardt Corp, 64-67, Aerojet-Gen Corp, 64-67 & Jet Propulsion Labs, Calif Inst Technol, 64-68; Fulbright scholar, Univ London, 66-67; pres, Fuels Systs Design Corp, 76- *Mem:* Am Inst Aeronaut & Astronaut; Am Soc Mech Engrs; Combustion Inst. *Res:* Combustion; design of fuel services for the combustion of toxic and waste fuels in boilers and alternate fuels in engines; aerothermochemistry of subsonic and supersonic combustion systems in propulsion devices; air pollution. *Mailing Add:* 42 Cherry Ln Huntington NY 11743

AGOSTA, WILLIAM CARLETON, b Dallas, Tex, Jan 1, 33; m 58; c 2. ORGANIC CHEMISTRY, PHOTOCHEMISTRY. *Educ:* Rice Inst, BA, 54; Harvard Univ, AM, 55, PhD(chem), 57. *Prof Exp:* Nat Res Coun fel org chem, Dyson Perrins Lab, Oxford Univ, 57-58; Pfizer fel, Univ Ill, 58-59; asst prof chem, Univ Calif, 59-61; sci liaison officer, US Naval Forces Europe, 61-63; from asst prof to assoc prof, 63-74, PROF CHEM, ROCKEFELLER UNIV, 74- *Concurrent Pos:* Alfred P Sloan Found fel, 69-71; John Angus Erskine fel, Univ Canterbury, NZ, 81. *Mem:* Inter-Am Photochem Soc; Am Chem Soc; The Chem Soc; Am Soc Photobiol; Europ Photochem Asn; AAAS. *Res:* Photochemistry; physical organic chemistry; pheromone chemistry. *Mailing Add:* Rockefeller Univ New York NY 10021-6399

AGOSTINI, ROMAIN CAMILLE, b Luxembourg, Mar 4, 59. USER INTERFACES & HI-SPEED NETS, ENGINEERING MANAGEMENT. *Educ:* Univ Karlsruve, Ger, BS, 82, MS, 87. *Prof Exp:* COMPUTER SCIENTIST RES & DEVELOP, STANFORD LINEAR ACCELERATOR CTR, 87- *Res:* Local area networks; data communications; software engineering; artificial intelligence; engineering management. *Mailing Add:* 1476 Arbor Ave Los Altos CA 94024

AGOSTON, MAX KARL, b Stockerau, Austria, Mar 25, 41; US citizen. MATHEMATICS. *Educ:* Reed Col, BA, 62; Yale Univ, MA, 64, PhD(math), 67; Stanford Univ, MS, 78. *Prof Exp:* Lectr math, Wesleyan Univ, 66-67, asst prof, 67-75; asst prof dept math & comp sci, 76-80, ASSOC PROF, SAN JOSE STATE UNIV, 80- *Concurrent Pos:* Vis prof math, Univ Heidelberg, 70-71; vis fel, Univ Auckland, 73-74. *Mem:* Am Math Soc; Asn Comp Mach; Inst Elect & Electronics Engr. *Res:* Geometric modeling; general software development; artificial intelligence. *Mailing Add:* Dept Math & Comp Sci San Jose State Univ San Jose CA 95192

AGRANOFF, BERNARD WILLIAM, b Detroit, Mich, June 26, 26; m 57; c 2. BIOCHEMISTRY. *Educ:* Wayne State Univ, MD, 50; Univ Mich, BS, 54. *Prof Exp:* Intern, Robert Packer Hosp, 50-51; res fel, Mass Inst Technol, 51-52; from asst officer chg to officer chg, Dept Chem, US Naval Med Sch, 52-54; biochemist, Nat Inst Neurol Dis & Blindness, NIH, 54-60; assoc prof biochem, res biochemist & chief sect biochem, 61-65, assoc dir, 77-83, PROF BIOCHEM, MENT HEALTH RES INST, UNIV MICH, ANN ARBOR, 65-, DIR, 83- *Concurrent Pos:* Mem staff, Max Planck Inst Cell Chem, 58-59; adv ed, Advan in Lipid Res, 62-; fel comt biochem & nutrit, NIH, 64-67, mem study sect neurol A, 67-71 & neurol B, 75-79; mem adv comn fundamental res, Nat Multiple Sclerosis Soc, 70-73; chmn panel biochem, Nat Comn Multiple Sclerosis, 73; vis scientist, Med Res Coun, Mill Hill, 74-75; ed, J Biol Chem, J Neurochem & Brain Res; chmn, Int Soc Neurochem, 89-91. *Honors & Awards:* Fogarty Scholar-in-Residence Award, Nat Insts of Health, 1988. *Mem:* Inst Med-Nat Acad Sci; fel Am Psychol Soc; Soc Neurosci; Am Soc Neurochem (pres, 73-75); fel Am Col Neuropsychopharmacol; Int Soc Neurochem (treas, 85-89); fel NY Acad Sci. *Res:* Biochemistry of lipids; nerve regeneration; neurochemistry; biochemical correlates of behavior; non-invasive biochemical techniques (PET). *Mailing Add:* Univ Mich Neurosci Lab Bldg 1103 E Huron Ann Arbor MI 48104-1687

AGRAS, WILLIAM STEWART, b London, Eng, May 17, 29; Can citizen; m 55; c 2. MEDICINE, EXPERIMENTAL PSYCHIATRY. *Educ:* Univ London, MB, BS, 55. *Prof Exp:* Demonstr psychiat, McGill Univ, 60-61; instr, Col Med, Univ Vt, 61-62, from asst prof to assoc prof, 62-69; prof & chmn dept, Med Ctr, Univ Miss, 69-73; PROF PSYCHIAT, SCH MED, STANFORD UNIV, 73- *Concurrent Pos:* Ed, J Appl Behav Anal, 74-77; fel, Ctr Advan Study Behav Sci, 76-77, 90-91. *Mem:* Soc Behav Med (pres, 78); Psychiat Res Soc; Soc Exp Analysis Behav; Asn Advan Behav Therapy (pres, 85). *Res:* Behavioral medicine, particularly applications of behavior change procedures to disease prevention; eating disorders, essential hypertension, stress prevention. *Mailing Add:* 515 Gerona Rd Stanford CA 94305

AGRAWAL, ARUN KUMAR, b Agra, India, Nov 24, 45; m 74; c 1. MANUFACTURING AUTOMATION, MAN-MACHINE INTERFACE. *Educ:* Indian Inst Technol, Kanpur, BS, 67; Case Western Reserve Univ, MS, 68, PhD(comput sci), 74. *Prof Exp:* Electronic engr, Gould Inc, 69-71, sr engr, 74-76, group leader, 76-79, prog mgr, Gould Labs, 79-; ENG MGR, GEA CORP-INDUST SYSTS GROUP. *Mem:* Inst Elec & Electronics Engrs Comput Soc; Robotics Int Soc Mech Engrs. *Res:* Robotic technology; speech recognition/synthesis; digital signal processing. *Mailing Add:* Motorola Inc 1295 E Algonquin Rd Schaumburg IL 60196

AGRAWAL, DHARMA PRAKASH, b Balod, India, April 12, 45; m 71; c 2. PARALLEL PROCESSING, COMPUTER ARCHITECTURE. *Educ:* Ravishankar Univ, India, BE, 66; Roorkee Univ, ME, 68; Fed Inst Technol, Switz, DScTech, 75. *Prof Exp:* Lectr elec eng, MNR Eng Col, India, 68-72 & Roorkee Univ, 72-73; asst mini & micro comput, Fed Inst Tech, Switz, 73-75; lectr elec eng, Univ Technol, Iraq, 76; instr comput sci, Southern Methodist Univ, Tex, 76-77; from asst prof to assoc prof comput eng, Wayne State Univ, 77-80; PROF COMPUT ENG, DEPT ELEC ENG, NC STATE UNIV. *Concurrent Pos:* Reviewer, Nat Sci Found, 80-, Army Res Off, 82-; adv, Grad Reliability Eng. *Honors & Awards:* Cert Appreciation, Inst Elec & Electronic Engrs Computer Soc. *Mem:* Asn Comput Mach; fel Inst Elec & Electronic Engrs. *Res:* VLSI; fault-tolerant computing. *Mailing Add:* Elec & Comp Eng Box 7911 NC State Univ Raleigh NC 27695

AGRAWAL, GOVIND P(RASAD), b Kashipur, India, July 24, 51; c 3. SEMICONDUCTOR LASERS, OPTICAL COMMUNICATIONS. *Educ:* Univ Lucknow, BS, 69; Indian Inst Technol, MS, 71, PhD(physics), 74. *Prof Exp:* Full researcher polymers, Polytech Sch, France, 74-76; res assoc laser physics, City Univ New York, 77-80; staff scientist quantum electronics, Quantel, France, 80-81; mem tech staff semiconductor lasers, AT&T Bell Labs, 82-88; ASSOC PROF, UNIV ROCHESTER, 89- *Concurrent Pos:* Consult, AT&T Bell Labs, 81-82 & Quantel, France , 82. *Honors & Awards:* Gold Medal, Univ Lucknow, 70. *Mem:* Am Phys Soc; Inst Elec & Electronics Engrs; Optical Soc Am; fel Optical Soc Am. *Res:* Optical communications; quantum electronics; nonlinear optics; long wavelength semiconductor lasers; nonlinear fiber optics. *Mailing Add:* Inst Optics Univ Rochester Rochester NY 14627

AGRAWAL, HARISH C, b Allahabad, India, June 14, 39; US citizen; m 60; c 2. MYELIN, OLIGODENDROGLIA. *Educ:* Allahabad Univ, India, BS, 57, MS, 59, PhD(biochem), 64. *Prof Exp:* Lectr res, Charing Cross Hosp, London, UK, 68-70; from res asst prof to assoc prof, 70-79, PROF PEDIAT & NEUROL, DEPT PEDIAT, SCH MED, WASH UNIV, ST LOUIS, MO, 79- *Concurrent Pos:* RCDA, Nat Inst Neurol Dis & Stroke, NIH, 74-79; mem neurol B study sect, NIH, 79-82; vis prof, dept pediat, Univ Lussane, Switz, 82. *Mem:* Int Brain Res Orgn; Int Soc Neurochem; Am Soc Neurochem; Am Soc Biochem & Molecular Biol; Am Physiol Soc. *Res:* Isolation, characterization, localization, turnover and posttranslational modification(s) of myelin-glial cell specific proteins. *Mailing Add:* Dept Pediat Sch Med Washington Univ 400 S Kingshighway St Louis MO 63110

AGRAWAL, JAGDISH CHANDRA, applied mathematics, for more information see previous edition

AGRAWAL, KRISHNA CHANDRA, b Calcutta, India, Mar 15, 37; m 60; c 3. MEDICINAL CHEMISTRY, MOLECULAR BIOLOGY. *Educ:* Andhra Univ, India, BS, 59, MS, 60; Univ Fla, PhD(pharmaceut chem), 65. *Prof Exp:* Res assoc, Yale Univ, 66-69, from instr to asst prof pharmacol, 69-76; assoc prof, 76-81, PROF PHARMACOL, SCH MED, TULANE UNIV, 81- *Concurrent Pos:* NIH fel grant, Univ Fla, 65-66; mem sci adv comt, Instnl Res Grants, Am Cancer Soc, 80-84 & 86-88; sci adv, Phase II Comt, SE Cancer Study Group, 83-86; mem, Study Sect AIDS Res, NIH, 88-94. *Mem:* Am Soc Pharmacol & Exp Therapeut; Am Asn Cancer Res; Am Chem Soc; Radiation Res Soc; Am Cancer Soc. *Res:* Design and synthesis of chemical agents for chemotherapy of cancer and AIDS; development of radiosensitizing agents for use in combination with radiotherapy; studies of structure-activity relationship and biochemical mechanisms involved in cell death; study of hyperthermia and drug interactions. *Mailing Add:* Dept Pharmacol Tulane Univ Sch Med New Orleans LA 70112

AGRAWAL, PRADEEP KUMAR, b Meerut, India, Oct 26, 46; m 74; c 1. ELECTROMAGNETISM. *Educ:* Agra Univ, India, BSc, 63; Indian Inst Technol, Kanpur, BTech, 68; Ohio State Univ, MSc, 69, PhD(elec eng), 72. *Prof Exp:* Grad res assoc, Electro Sci Lab, Ohio State Univ, 68-72; asst prof, Ohio Inst Technol, 72-74; asst prof, Cleveland State Univ, 74-75; res scientist, Joint Inst Advan Flight Sci, NASA, 75-80; ASST PROF CHEM ENG, SCH CHEM ENGRS, GA INST TECHNOL, 80- *Honors & Awards:* NSG-1110, NASA Langley Res Ctr, 75. *Res:* Wire antennas; microstrip antennas; reflector antennas; computer simulation of antennas. *Mailing Add:* Sch Chem Engrs Ga Inst Technol Atlanta GA 30332

AGRAWAL, SUPHAL PRAKASH, b Bareilly, India, Nov 10, 46; US citizen; m 76; c 3. TECHNICAL MANAGEMENT, SUPERPLASTIC FORMING. *Educ:* Agra Univ, India, BSc, 63; Indian Inst Technol, Kanpur, BTech, 68; Univ Ky, Lexington, MS, 70, PhD(metal eng & mat sci), 73. *Prof Exp:* Scholar, Mat & Metall Eng, Univ Mich, 73-76; mem tech staff & prog mgr, N Am Aircraft Opers, Rockwell Int, 76-80; sr tech specialist & prog mgr, Aircraft Div, 80-89, DIR ADV TECH BUS DEVELOP, NORTHROP CORP AIRCRAFT DIV, 89- *Concurrent Pos:* Scholar, Metall eng & mat sci, Univ Ky, 76; spec course lectr, dept mat sci, Univ Southern Calif, Los Angeles, 80; spec course lectr, Am Soc Metals Int, 90. *Mem:* Fel Am Soc Metals; Sigma Xi; Indian Inst Metals; Metall Soc. *Res:* Advanced materials and processes research and development; innovations fundamental to the development and implementation of superplastic forming process for fabricating titanium and aluminum structures; diffusion bonding; advanced materials and processes technologies; long-range planning for advanced technology thrusts. *Mailing Add:* Dept 2380/Zone MF Advan Tech Bus Develop Northrop Corp-Aircraft Div One Northrop Ave Hawthorne CA 90250

AGRAWALA, ASHOK KUMAR, b Meerut, India, June 28, 43; m; c 2. COMPUTER SCIENCE. *Educ:* Indian Inst Sci, Bangalore, BE, 63, ME, 65; Harvard Univ, AM, 70, PhD(appl math), 70. *Prof Exp:* Prin engr, Data Systs Div, Honeywell Inc, 70-71; ASST PROF COMPUT SCI, UNIV MD, COLLEGE PARK, 71- *Concurrent Pos:* Chmn bd, AT&T Systs, 86- *Mem:* Fel Inst Elec & Electronic Engrs; Asn Comput Mach; AAAS; Sigma Xi. *Res:* Analysis, modelling, measurement and evaluation of computer systems, their design and architecture; distributed computing; real time systems; network protocols. *Mailing Add:* Dept Comput Sci Univ Md College Park MD 20742

AGRE, COURTLAND LEVERNE, b Boyd, Minn, Sept 11, 13; m 46; c 6. CHEMISTRY. *Educ:* Univ Minn, BChE, 34, PhD(org chem), 37. *Prof Exp:* Res chemist, E I du Pont de Nemours & Co, Inc, 37-40 & Minn Mining & Mfg Co, St Paul, 41-46; prof chem, St Olaf Col, 46-58; prof chem, Augsburg Col, 59-78; vis prof chem, Fla Southern Col, 79-82; RETIRED. *Concurrent Pos:* NSF fac fel, Univ Calif, 58-59; consult, Minn Mining & Mfg Co, 46-78. *Honors & Awards:* Minnesota Award, Am Chem Soc. *Mem:* AAAS; Am Chem Soc. *Res:* Synthetic resins; organic chemicals; organo-silicon compounds; use of benzoin and diacetyl as catalysts for light activated polymerization of unsaturated compounds; preparation of polyvinyl ketals; preparation of nitroesters. *Mailing Add:* 2297 Mailand Rd Maplewood St Paul MN 55119

AGRE, KARL, b Feb 24, 32; US citizen; m 55; c 3. CLINICAL PHARMACOLOGY, PEDIATRICS. *Educ:* Villanova Col, BS, 52; Hahnemann Med Col, MS, 54, PhD(pharmacol), 56; Duke Univ, MD, 59. *Prof Exp:* Instr pharmacol, Sch Med, Duke Univ, 56-59; intern-resident pediat, Bronx Munic Hosp Ctr, 59-62, clin instr, 64-66; dir clin pharmacol, Bristol Labs, Inc, 66-75; VPRES MED RES, PHARMACEUT DIV, SEARLE LABS, 75- *Concurrent Pos:* Clin instr, State Univ NY Upstate Med Ctr, 66-75; pvt med pract, NY, 64-66. *Mem:* Fel Am Acad Pediat. *Mailing Add:* 1794 Stanford Ave Menlo Park CA 94025

AGRESS, CLARENCE M, b Knoxville, Tenn, Mar 10, 12; m 72; c 2. CARDIOLOGY. *Educ:* Harvard Univ, AB, 33; Univ Tex, MD, 37. *Prof Exp:* Chief, Dept Cardiol, Cedars Lebanon Hosp, Los Angeles, 60-67; emer prof cardiol, Sch Med, Univ Calif, Los Angeles; mem staff, Century City Hosp; MEM STAFF, CEDARS-SINAI MED CTR. *Concurrent Pos:* Cardiologist. *Mem:* Fel Am Col Angiol; fel Am Col Cardiol; fel Am Col Chest Physicians; fel Am Col Physicians; fel AMA; Am Physiol Soc. *Mailing Add:* 150 N Robertson Blvd Beverly Hills CA 90210

AGRESTA, JOSEPH, b Long Island City, NY, June 13, 29; m 53; c 7. PHYSICS, APPLIED MATHEMATICS. *Educ:* Cooper Union, BEE, 50; NY Univ, MS, 52, PhD(physics), 58. *Prof Exp:* Asst physics, NY Univ, 50-53; physicist, Curtiss-Wright Corp, 53-56; res assoc physics, NY Univ, 56-58; physicist, United Nuclear Corp, 58-62; physicist, Union Carbide Res Inst, 62-68, consult, 69-81, MGR OPERS RES, UNION CARBIDE CORP, 82- *Concurrent Pos:* Lectr, City Col New York, 58-62; adj asst prof, NY Univ, 59-62. *Mem:* Sigma Xi. *Res:* Radiation transport; nuclear reactor theory; computing machine methods; hydrodynamics; operations research; systems analysis; simulation; computer science; computer graphics. *Mailing Add:* 23 Tanglewood Dr Danbury CT 06811

AGRESTI, DAVID GEORGE, b Washington, DC, Aug 8, 38. NUCLEAR SPECTROSCOPY, MATERIALS SCIENCE. *Educ:* Ohio State Univ, BSc, 59; Calif Inst Technol, MS, 62, PhD(physics), 67. *Prof Exp:* Asst prof physics, Calif State Col, Los Angeles, 67-69; asst prof, 69-74, chmn physics dept, 74-78, ASSOC PROF PHYSICS, UNIV ALA, BIRMINGHAM, 74- *Mem:* Am Phys Soc. *Res:* Mossbauer spectroscopy; computer modeling of biomolecules. *Mailing Add:* Dept Physics Univ Ala Birmingham AL 35294

AGRESTI, WILLIAM W, b Erie, Pa, Oct 19, 46. COMPUTER SCIENCES. *Educ:* Case Inst Technol, BS, 68; NY Univ, MS, 71, PhD(comput sci), 73. *Prof Exp:* Systs analyst oper res, Lord Corp, 68-69; asst dir res, Traffic Safety Systs, Hwy Safety Comn, PR, 70; from asst prof to assoc prof indust & systs eng, Univ Mich, Dearborn, 73-83; sr computer scientist, Computer Sci Corp, 83-88; CONSULT SCIENTIST, MITRE CORP, 88- *Concurrent Pos:* Dir comput & info sci, Univ Mich-Dearborn, 75-78. *Honors & Awards:* Group Achievement Award, NASA, 89. *Mem:* Asn Comput Mach; Inst Elec & Electronic Engrs Computer Soc. *Res:* Software engineering; simulation and dynamic programming. *Mailing Add:* 4016 Shallow Brook Lane Olney MD 20832

AGRIOS, GEORGE NICHOLAS, b Galarinos, Greece, Jan 16, 36; US citizen; m 62; c 3. PLANT PATHOLOGY, VIROLOGY. *Educ:* Univ Thessaloniki, BS, 57; Iowa State Univ, PhD(plant path & genetics), 60. *Prof Exp:* From asst prof to prof plant path, Univ Mass, Amherst, 63-88; PROF & CHMN DEPT PLANT PATHOL, UNIV FLA, GAINESVILLE, 88- *Concurrent Pos:* Ed-in-chief, Am Phytopath Soc Press. *Honors & Awards:* Fel, Am Phytopath Soc, 83. *Mem:* Fel Am Phytopath Soc (pres, 90-91); Can Phytopath Soc; AAAS. *Res:* Plant viruses, transmission and identification; physiological effects of viruses on host plants; methods for virus detection; behavior of viruses in tissue culture. *Mailing Add:* Dept of Plant Path Univ of Fla Gainesville FL 32611

AGRIS, PAUL F, b July 8, 44. BIOCHEMISTRY, MOLECULAR BIOPHYSICS. *Educ:* Bucknell Univ, BS, 66; Mass Inst Technol, PhD(biochem), 71. *Prof Exp:* Postdoctoral fel, Dept Molecular Biophys & Biochem, Yale Univ, 71-73; from asst prof to prof biol sci, Univ Mo, Columbia, 73-87, chmn, Nuclear Magnetic Resonance Facil, 81-86, prof, Dept Med, 83-87, dir, Div Biol Sci Protein Sequencing Lab, 85-87; PROF & HEAD, DEPT BIOCHEM & SUPVR, BIOMACROMOLECULER NUCLEAR MAGNETIC RESONANCE FACIL, NC STATE UNIV, 88- *Concurrent Pos:* Consult, New Eng Nuclear Corp, 71; assoc scientist & consult biochem, Cancer Res Ctr & Ellis Fischel State Cancer Hosp, 73-87; pres & bd mem, Am Found Aging Res, 79-; mem bd, Am Fedn Aging Res, 80-; vis scientist, Biochem Dept, Univ Oxford, Eng, 80-81 & Okla Med Res Found, 81-82; Fogarty sr int fel, Univ Oxford, 80-81; sr fel, Exeter Col, Oxford, 80-; Nat Res Serv spec fel, 81-82; mem, US-Japan Conf Werner's Syndrome, NSF, Japan, 82; res grants, NIH, 83-89 & 86-91, NSF, 88-91; exchange lectr & scientist, Nat Acad Sci & Polish Acad Sci, 84, spec vis scientist, 86 & 91. *Mem:* AAAS; Am Chem Soc; Am Soc Biochem & Molecular Biol; Am Soc Cell Biol; Tissue Cult Asn; Int Soc Magnetic Resonance. *Res:* Structure-function relationships for nucleic acids and proteins in biology and in medical problems. *Mailing Add:* Dept Biochem NC State Univ 128 Polk Hall PO Box 7622 Raleigh NC 27695-7622

AGRON, SAM LAZRUS, b Russia, Nov 27, 20; nat US; m 44; c 2. GEOLOGY. *Educ:* Northwestern Univ, BS, 41; Johns Hopkins Univ, PhD(geol), 49. *Prof Exp:* Instr geol, Brown Univ, 49-51; from asst prof to prof geol, 51-84, dir, NSF Earth Sci Inst, 64-67, EMER PROF GEOL, RUTGERS UNIV, NEWARK, 84- *Mem:* Geol Soc Am; Nat Asn Geol Teachers; Sigma Xi. *Res:* Structural, economic and environmental geology. *Mailing Add:* 386 Irving Ave South Orange NJ 07079

AGUAYO, ALBERT J, b Buenos Aires, Arg. MEDICINE. *Educ:* Univ Cordoba, MD, 59. *Prof Exp:* DIR, CTR RES NEUROSCI, MONTREAL GEN HOSP, 85- *Mem:* Inst Med-Nat Acad Sci; fel Royal Soc Can; Soc Neurosci; AAAS. *Res:* Medicine. *Mailing Add:* Ctr Res Neurosci Montreal Gen Hosp 1650 Cedar Ave Montreal PQ H3G 1A4 Can

AGUIAR, ADAM MARTIN, b Newark, NJ, Aug 11, 29; m 80; c 5. ORGANIC CHEMISTRY. *Educ:* Fairleigh Dickinson Univ, BS, 55; Columbia Univ, MA, 57, PhD(chem), 60. *Prof Exp:* Org chemist, Otto B May, 47-55; asst, Columbia Univ, 55-59; asst prof chem, Fairleigh Dickinson Univ, 59-63; from asst prof to prof, Tulane Univ, 63-72, chmn, chem dept, Newcomb Col, 70-72; dean grad progs & res, William Paterson Col, 72-73; chmn dept, 84-89, PROF CHEM, FAIRLEIGH DICKINSON UNIV, 73- *Concurrent Pos:* Fel, NIH, 59; sabbatical, Europe, 69-70 & Roche Inst Molecular Biol, 82; hon res prof, Birkbeck Col, Univ London, 70; res specialist, Rutgers Univ, Newark, 73-75; consult, 74- *Mem:* AAAS; Am Chem Soc; NY Acad Sci; Sigma Xi. *Res:* Organo-phosphorus chemistry; medicinal chemistry; biochemistry and pharmacology. *Mailing Add:* Chem Dept Fairleigh Dickinson Univ Madison NJ 07974

AGUILAR, RODOLFO J, b San Jose, Costa Rica, Sept 28, 36; US citizen; m 56; c 4. CIVIL ENGINEERING. *Educ:* La State Univ, BS, 58, MS, 60, BArch, 61; NC State Univ, PhD(civil eng), 64; Tulane Univ, MBA, 89. *Prof Exp:* From asst prof to prof, 61-75, consult prof civil eng, 75-77, prof, 77-85, EMER PROF ARCHIT & FINANCE, 85- *Concurrent Pos:* Consult, Iberia Parish Sch Bd, Perry Brown, Inc & C E Newman, La, 64, Gulf South Res Inst, La & Smith, Hinchman & Grylls, Mich, 67; struct examr, La State Bd Archit Exam, 65-66; consult systs eng & opers res; pres, A D H Systs, Inc, Planners, Archit, Engrs, 70-, Corp Develop Group Inc, 76-77 & Aguilar & Assocs, 77-; adj prof, Int & Free Enterprise Studies, 85-88; adj prof real estate, A B Freeman Sch Bus, Tulane Univ, 89- *Mem:* Am Soc Civil Engrs; Am Inst Architects; Am Planning Asn; Am Soc Appraisers. *Res:* Dynamic response of steel plate girder bridges with partially cracked lightweight concrete decks; stability of framed domes systems of concentrated loads; structural dynamics; real estate planning, finance and development; author of over 40 publications in architectural design. *Mailing Add:* 100 France St Baton Rouge LA 70802

AGUILERA, JOSE MIGUEL, food technology, for more information see previous edition

AGUILO, ADOLFO, b Buenos Aires, Arg, Sept 6, 28; m; c 4. INDUSTRIAL CHEMISTRY. *Educ:* Univ Buenos Aires, MS, 53, PhD(chem), 55. *Prof Exp:* Asst res chemist, Steel Factory, Arg, 50-51; res chemist, Arg Air Force, 51-53; group leader, Arg AEC, 53-60; from res chemist to sr res chemist, Celanese Chem Co, 60-66, res assoc, 66-70, sect leader, 70-90, SR RES ASSOC, CELANESE CHEM CO, 90- *Concurrent Pos:* Lab supvr, Univ Buenos Aires, 55-58, assoc prof, 58-60. *Mem:* Am Chem Soc; AAAS. *Res:* Organometallics; catalysts; coordination compounds in catalysis; liquid and vapor phase oxidation of parafins and olefins; petrochemicals from synthesis gas; hydroformylation; carbonylation. *Mailing Add:* Hoechst Celanese PO Box 9077 Corpus Christi TX 78469

AGULIAN, SAMUEL KEVORK, b Beirut, Lebanon, Apr 9, 43; US citizen; m 76; c 3. ION-SENSITIVE ELECTRODES, BIOSENSORS. *Educ:* Am Univ Beirut, BSc, 65, MSc, 73, PhD(med sci), 80. *Prof Exp:* Res asst physiol, Free Univ Berlin, 65-66; res asst, Am Univ Beirut, 66-76, res assoc, 77-80, asst prof, 80-85; proj dir, World Precision Instruments, 85-90; res assoc physiol, 76-77, RES SCIENTIST, YALE UNIV, 90- *Concurrent Pos:* Consult, Yale Univ, 85-90. *Res:* Developed first intracellular double-barelled electrodes to measure electro-chemical potentials of major intracellular ions; micro sensors, such as for carbon dioxide, oxygen and pH. *Mailing Add:* 47 Park Ave Hamden CT 06517

AGUS, ZALMAN S, b Chicago, Ill, Apr 3, 41; m 63; c 3. PHYSIOLOGY, NEPHROLOGY. *Educ:* Johns Hopkins Univ, BA, 61; Univ Md, MD, 65. *Hon Degrees:* MA, Univ Pa. *Prof Exp:* Intern, Sch Med, Univ Md, 65-66, resident, 66-68; res fel, Sch Med, Univ Penn, 68-71; attend physician, nephrology, US Air Force, Lackland AFB, Tex, 71-73; asst prof med nephrology, 73-78, prof med, 79-88, ASSOC PROF MED NEPHROL, SCH MED, UNIV PA, 78-, CHIEF, RENAL SECT, 88- *Concurrent Pos:* NIH fel, 69-71; clin investigatorship, Vet Admin, 73-76; NIH res career develop award, 77; fel, Am Col Physicians, 77. *Mem:* Am Fedn Clin Res; AAAS; NY Acad Sci; Am Soc Nephrology; Int Soc Nephrology; Am Soc Clin Invest. *Res:* Renal and electrolyte physiology; calcium and phosphate metabolism. *Mailing Add:* A-401 Richards Bldg Univ Pa Philadelphia PA 19104

AGYILIRAH, GEORGE AUGUSTUS, b Ghana, Sept 15, 47; m 80; c 3. PHARMACEUTICS, PHARMACEUTICAL TECHNOLOGY & DRUG DELIVERY. *Educ:* Univ Sci & Technol, China, BPharm, 75; Univ London, MSc, 78, PhD(pharmaceut), 81. *Prof Exp:* Postdoctoral fel, Purdue Univ, 82-85; res assoc indust pharm, 85-89, ASST PROF INDUST PHARM & PHARMACEUT CALCULATIONS, COL PHARM, UNIV MINN, 90- *Concurrent Pos:* Consult, Biocontrol Inc, 90- *Mem:* Am Asn Pharmaceut Scientists; Am Asn Cols Pharm. *Res:* Design and delivery of novel pharmaceutical drug delivery systems; microcapsules, tablets and tablet coatings, bioadhesive dosage forms and topical-transdermal products. *Mailing Add:* Col Pharm Univ Minn 308 Harvard St SE Minneapolis MN 55455

AH, HYONG-SUN, b Suwon, Korea, May 27, 31; US citizen; m 58; c 3. VETERINARY PARASITOLOGY, ACAROLOGY. *Educ:* Seoul Nat Univ, DVM, 55; Univ Ga, PhD(med entom), 68. *Prof Exp:* Vet res assoc, Diag & Res Lab, Tifton, 69-71, ASST PROF PARASITOL, COL VET MED, UNIV GA, 71- *Mem:* Am Soc Parasitologists; Acarological Soc Am; Int Filariasis Asn; World Fedn Parasitologists. *Res:* Experimental filariasis in the areas of pathology, chemotherapy, immunology and host-parasite relationships. *Mailing Add:* 1475 Whit Davis Rd Athens GA 30605

AHARONI, SHAUL MOSHE, b Tel Aviv, Israel, Dec 3, 33; US citizen; m 56; c 2. POLYMER CHEMISTRY. *Educ:* Univ Wis-Eau Claire, BS, 67; Case Western Reserve Univ, MS, 69, PhD(polymer sci), 72. *Prof Exp:* Asst chief chemist, Nat Presto Industs, Wis, 66-67, consult, 68-69; sr chemist, Gould Labs, Gould Inc, Ohio, 71-73; res assoc, Allied Chem Corp, 73-85, SR RES ASSOC, ALLIED-SIGNAL CORP, 85- *Mem:* Am Chem Soc; Am Phys Soc. *Res:* Structure property relationships in polymers; rigid backbone polymers; organization of the amorphous state, packing density and free volume; condensation polymers; liquid crystalline polymers; polymer networks. *Mailing Add:* Allied-Signal Corp 45 Northview Dr PO Box 1087R Morris Plains NJ 07950

AHARONY, DAVID, m; c 2. PHARMACOLOGY. *Educ:* Bar-Ilan Univ, Israel, BS, 75; Tel-Aviv Univ, MS, 77; Thomas Jefferson Univ, PhD(pharmacol), 81. *Prof Exp:* Teaching asst, Dept Biochem, Tel-Aviv Univ, 75-77; res technician, Div Miles Labs, Miles-Yeda, Rehovot, Israel, 77-78; res asst, Cardeza Found Hemat Res, Thomas Jefferson Univ, 78-81; res pharmacologist, Dept Pharmacol, Pulmonary Sect, ICI Pharmaceut Group, 82-86, sr res pharmacologist, 87-89, prin pharmacologist, 90, GROUP LEADER, DEPT PHARMACOL, PULMONARY SECT, ICI PHARMACEUT GROUP, 90- *Mem:* Am Soc Pharmacol & Exp Therapeut; AAAS; Sigma Xi; Int Soc Immunopharmacol; NY Acad Sci. *Res:* Author or co-author of over 40 publications. *Mailing Add:* Dept Pharmacol ICI Pharmaceuticals Group Concord Pike & Murphy Rd Wilmington DE 19897

AHEARN, DONALD G, b Grove City, Pa, Feb 1, 34; m 59; c 3. MYCOLOGY, MARINE MICROBIOLOGY. *Educ:* Mt Union Col, BS, 57; Univ Miami, MS, 59, PhD(microbiol), 64. *Prof Exp:* Instr microbiol & marine biol, Univ Miami, 63-64, asst prof, Sch Med & Inst Marine Sci, 64-66; from asst prof to assoc prof microbiol, 67-72, dean, Grad Div, Sch Arts & Sci, 70-72, PROF MICROBIOL, GA STATE UNIV, 72-, ASST VPRES RES, 80- *Concurrent Pos:* Mem, Int Oceanog Found. *Mem:* Am Soc Microbiol; Soc Indust Microbiol (pres-elect, 81); Mycol Soc Am; Int Soc Human & Animal Mycol; Am Acad Microbiol. *Res:* Ecology, physiology and systematics of fungi, chiefly yeasts, in aquatic habitats; contamination of industrial products, mascaras, SCP; epidemology of yeast-like fungi pathogenic to man. *Mailing Add:* Dept Biol Ga State Univ University Plaza Atlanta GA 30303

AHEARN, GREGORY ALLEN, b Cambridge, Mass, Nov 28, 43; m 67. COMPARATIVE PHYSIOLOGY, CELL PHYSIOLOGY. *Educ:* Univ Calif, Los Angeles, BA, 65; Univ Hawaii, MS, 67; Ariz State Univ, PhD(zool), 70. *Prof Exp:* Teaching asst gen zool, Ariz State Univ, 67-68, teaching assoc, 68-69; res fel, Zoophysiol Lab A, August Krogh Inst, Univ Copenhagen, 70-72; asst marine biologist, Hawaii Inst Marine Biol, 72-76; asst prof, 76-79, ASSOC PROF ZOOL, UNIV HAWAII, 79- *Concurrent Pos:* Fel, Danish Res Coun, 70; NSF instnl grants, 75, 77 & 79; Nat Oceanic & Atmospheric Admin instnl sea grant, 76. *Mem:* Am Soc Zoologists; NY Acad Sci; Am Physiol Soc. *Res:* Membrane transport, epithelial biology, osmotic and ionic regulation, water balance, nutritional physiology, thermoregulation, environmental physiology. *Mailing Add:* Dept Zool Univ Hawaii Manoa 2500 Campus Rd 2538 The Mall Honolulu HI 96822

AHEARN, JAMES JOSEPH, JR, b Beverly, Mass, May 21, 43. ANALYTICAL CHEMISTRY, PROCESS RESEARCH & DEVELOPMENT. *Educ:* Boston Col, BS, 65, PhD(chem), 69. *Prof Exp:* Res chemist, Org Chem Dept, Res Div, E I du Pont de Nemours & Co, 69-70; res group leader anal chem, 70-82, SR MGR CHEM DEVELOP LAB, POLAROID CORP, 82- *Mem:* Am Chem Soc. *Res:* Chromatographic procedures for dyes, dye intermediates and photographic chemicals; chemical process research and development. *Mailing Add:* Seven Ellsworth Ave Beverly MA 01915

AHEARN, JAYNE NEWTON, developmental genetics, for more information see previous edition

AHEARN, JOHN STEPHEN, b New York, NY, June 14, 44; m 70; c 3. METRIALS SCIENCE. *Educ:* St Lawrence Univ, BS, 66; Univ Va, MS, 68, PhD(physics), 72. *Prof Exp:* Fel physics, Inst Metall, Univ Gottingen, 71-73; fel mat, Dept Metall & Mat Sci, Univ Pa, 73-75; scientist mat, 75-78, sr scientist, 78-83, mgr mat sci, 83-87, sr mgr res & develop, 87-88, ASSOC DIR, PHOTONICS & SURF SCI, MARTIN MARIETTA LABS, 88- *Concurrent Pos:* Vis asst prof, Swarthmore Col, 75; fel, Alexander von Humboldt Found, 71-73; fel, Univ Va, 66-69; lectr, Johns Hopkins Univ, 85- *Mem:* Am Phys Soc; Metall Soc. *Res:* Mechanical properties of materials; dislocations in metals, ceramics and semiconductors; transmission electron microscopy of material defects; surface and interface analysis, corrosion and adhesive bonding; molecular beam epitaxy; electronic materials. *Mailing Add:* 1450 S Rolling Rd Baltimore MD 21227

AHEARN, MICHAEL JOHN, b Jacksonville, Tex, June 22, 36; m 64. CELL BIOLOGY. *Educ:* Univ Tex, BA, 58, MA, 62, PhD(zool), 65. *Prof Exp:* Lectr zool, Univ Tex, 64-65; assoc biologist, 65-80, ASSOC PROF LAB MED, UNIV TEX M D ANDERSON HOSP & TUMOR INST, TEX MED CTR, 80-, LAB MED, SYST CANCER CTR. *Mem:* Int Soc Exp Hemat; Am Asn Cancer Res; Am Soc Cell Biol; Electron Micros Soc Am; Sigma Xi. *Res:* Hematology, especially cellular alterations induced by chemotherapeutic agents; electron microscopy; flow cytometry; cytological techniques. *Mailing Add:* 2200 Willowick 6B Houston TX 77027

AHEARNE, JOHN FRANCIS, b New Britain, Conn, June 14, 34; m 56; c 5. RESOURCE MANAGEMENT. *Educ:* Cornell Univ, BEngPhys, 57, MS, 58; Princeton Univ, MA, 63, PhD(plasma phys), 66. *Prof Exp:* Analyst nuclear weapons effects, Weapons Lab, US Air Force, Kirtland AFB, NMex, 59-61; from instr to assoc prof physics, US Air Force Acad, 64-69; syst analyst, Off Asst Secy, Defense for Systs Anal, 69-70, dir tactical air prog, 70-72, dep asst secy defense, Off Asst Secy Defense for Prog Anal & Eval, 72-74, prin dept asst secy defense, Off Asst Secy Defense for Manpower & Reserve Affairs, 75-76; syst analyst, White House Energy Off, 77; dep asst secy, Energy for Resource Applications, 78; comn, 78-83, chmn, US Nuclear Regulatory Comn, 79-81; vpres & sr fel, Resources For Future, 84-89; EXEC DIR, SIGMA XI, SCI RES SOC, 89- *Concurrent Pos:* Consult, Comptroller Gen US, 83-84; chmn, Comt Risk Perception & Commun, Nat Res Coun, 87-90, Adv Comt Nuclear Facil Safety, Dept Energy, 88-; chmn comt future nuclear power, Nat Res Coun, 90- *Mem:* Am Phys Soc; AAAS; Sigma Xi; Soc Risk Anal; Am Nuclear Soc. *Res:* Resource allocation and public policy. *Mailing Add:* PO Box 13975 Research Triangle Park NC 27709

AHERN, DAVID GEORGE, b Winchester, Mass, May 21, 36; m 65. SYNTHESIS OF LABELLED COMPOUNDS. *Educ:* Merrimack Col, BS, 58; Northeastern Univ, MS, 65; Boston Univ, PhD(biochem), 72. *Prof Exp:* Jr chemist, TracerLab Inc, 58-60; jr chemist, New Eng Nuclear Corp, 60-62, chemist, 62-66, sr chemist, 72-74, group leader, 74-78, sect leader, 78-86; area supvr, Dupont, New Eng Nuclear Prod, 87-91; CONSULT, 91- *Mem:* Am Oil Chemists Soc; Int Isotope Soc. *Res:* Chemistry and biochemistry of lipid metabolism synthesis of carbon 14 and tritium labelled lipids; synthesis of 3H labelled ligands for receptor site analysis. *Mailing Add:* 26 Dewey Rd Lexington MA 02173

AHERN, FRANCIS JOSEPH, b New York, NY, June 21, 44; m 70; c 2. APPLIED PHYSICS. *Educ:* Cornell Univ, AB, 66; Univ Md, PhD(astron), 72. *Prof Exp:* Fel astrophys, Univ Toronto, 72-74; RES SCIENTIST APPL PHYSICS, CAN CTR REMOTE SENSING, 75- *Res:* Research into methods of remote sensing of the earth for environmental and resource management with emphasis on spectrometry in visual and near infrared regions; radiometric and atmospheric correction and calibration of multipartral scanner data; research into forestry applications of remotely sensed data; microwave backscattering research. *Mailing Add:* Can Ctr for Remote Sensing 2464 Sheffield Rd Ottawa ON K1A 0Y7 Can

AHERN, JUDSON LEWIS, b Canton, Ohio, Apr 11, 51; m 76; c 2. SOLID-EARTH GEOPHYSICS, GEOMECHANICS. *Educ:* Ohio Wesleyan Univ, BA, 73; Ohio State Univ, MSc, 75; Cornell Univ, PhD(geophysics), 80. *Prof Exp:* Geophysicist, Chevron Oil Co, New Orleans, La, 74; asst prof, 79-85, ASSOC PROF GEOPHYS, SCH GEOL & GEOPHYS, UNIV OKLA, 85- *Mem:* Sigma Xi; Am Geophys Union; Soc Explor Geophysics. *Res:* Mechanical and thermal aspects of the solid earth, including problems such as convection in magma bodies, thermal and mechanical subsidance of sedimentary basins, and frictional heating along faults. *Mailing Add:* Sch Geol & Geophysics Univ Okla 660 Parrington Oval Norman OK 73019

AHL, ALWYNELLE S, b Leesville, La, Mar 18, 41; m 63; c 2. ZOOLOGY, MAMMALOGY. *Educ:* Centenary Col, BS, 61; Univ Wyo, MS, 63, PhD(zool), 67; Mich State Univ, DVM, 87. *Prof Exp:* Res asst biochem, Univ Wyo, 65-67; from asst prof to prof natural sci, Mich State Univ, 67-87; vet med officer, Animal & Plant Health Inspection Serv, 87-88, DEP DIR SCI & PROF EDUC & DEVELOP, ANIMAL & PLANT HEALTH INSPECTION SERV, USDA, 88- *Concurrent Pos:* Supply asst prof biochem & human med, Univ Wyo, 76-77. *Mem:* Am Vet Med Asn; Am Soc Mammalogists; Nat Asn Fed Vet; Sigma Xi. *Res:* Physiological studies of mammals in their natural environment; environmental epidemiology, particularly as concerns drug and pesticide residues in food animals; veterinary medical continuing education. *Mailing Add:* Fed Bldg Rm 265 USDA 6505 Belcrest Hyattsville MD 20782

AHLBERG, DAN LEANDER, b Georgetown, Tex, Jan 6, 26; m 47; c 2. PHYSICAL CHEMISTRY, PETROLEUM CHEMISTRY. *Educ:* Southwestern Univ, BS(physics) & BS(chem), 51. *Prof Exp:* Chemist, E I du Pont de Nemours & Co, Inc, 51-54 & Eastern States Petrol Co, 54-60; anal mgr res, Signal Oil & Gas Co, 60-68; chief chemist labs, Charter Int Oil Co, 68-84, supt labs, 73-84, supt refinery environ serv, 81-84; CONSULT, 84- *Concurrent Pos:* Legal consult, injury & prod. *Mem:* Am Soc Testing & Mat; Am Chem Soc. *Res:* Gas chromotography; mass spectroscopy; infrared spectroscopy, atomic absorption spectroscopy, general analytical chemistry and petroleum process chemistry. *Mailing Add:* 1126 Marshall Deer Park TX 77536

AHLBERG, HENRY DAVID, b Boston, Mass, Oct 9, 39; m 62; c 2. BIOLOGY, MOLECULAR BIOLOGY. *Educ:* NPark Col, AB, 61; Boston Univ, PhD(biol), 69. *Prof Exp:* Asst prof biol, Northeastern Univ, 67-76; assoc prof, 76-85, PROF BIOL, AM INT COL, 85-, CHMN, DEPT BIOL, 89- *Mem:* Wildlife Soc; Am Soc Zoologists; Am Soc Mammalogists. *Res:* Geographic variation of North American porcupine; seasonal variation of genital systems in rodents; aging in natural populations of rodents; oxygen toxicity. *Mailing Add:* Dept of Biol Am Int Col Springfield MA 01109

AHLBERG, JOHN HAROLD, b Middletown, Conn, Dec 10, 27. APPLIED MATHEMATICS, NUMERICAL ANALYSIS. *Educ:* Yale Univ, BA, 50, MA, 54, PhD(math), 56; Wesleyan Univ, MA, 52. *Prof Exp:* Chief, Math Anal, United Aircraft Res Labs, 56-68; PROF APPL MATH, BROWN UNIV, 68- *Mem:* Am Math Soc; Math Asn Am; Soc Indust & Appl Math. *Res:* Application of splines to the numerical solution of boundary and initial value problems and their application to the representation of curves and surfaces. *Mailing Add:* PO Box 65 Amston CT 06231

AHLBORN, BOYE, b Kampen, Ger, July 16, 33; m 61; c 3. PLASMA PHYSICS. *Educ:* Univ Kiel, dipl physics, 60; Munich Tech Univ, Dr rer nat, 64. *Prof Exp:* Sci asst, Inst Plasma Physics, Garching, Ger, 62-64; instr, 64-65, from asst prof to assoc prof, 65-72, PROF PLASMA PHYSICS, UNIV BC, 72- *Mem:* Can Asn Physicists; Ger Phys Soc. *Res:* Plasma flow with heat sources--arcs, detonations, radiation gas dynamics; shock waves; gas dynamical and chemical lasers. *Mailing Add:* Dept Physics Univ BC Vancouver BC V6T 1W5 Can

AHLBRANDT, CALVIN DALE, b Scottsbluff, Nebr, Aug 13, 40; m 61; c 3. DIFFERENTIAL & DIFFERENCE EQUATIONS, NUMERICAL LINEAR ALGEBRA. *Educ:* Univ Wyo, BS, 62; Univ Okla, MA, 65, PhD(math), 68. *Prof Exp:* From asst prof to assoc prof, 68-82, PROF MATH, UNIV MO-COLUMBIA, 82- *Mem:* Am Math Soc; Soc Indust & Appl Math. *Res:* Behavior of solutions of linear difference equations and associated nonlinear (Riccati) equations, these results are applied to matrix continued fractions. *Mailing Add:* Dept of Math Univ of Mo Columbia MO 65211

AHLBRANDT, THOMAS STUART, b Torrington, Wyo, May 31, 48; m 69; c 2. GEOLOGY, SEDIMENTOLOGY. *Educ:* Univ Wyo, BA, 69, PhD(geol), 73. *Prof Exp:* Sr res geologist, Esso Prod Res Co, Tex, 73-74; RES GEOLOGIST, OIL & GAS RESOURCES BR, US GEOL SURV, 74- *Concurrent Pos:* NSF fel, 70, 71 & 72. *Mem:* Soc Econ Paleontologists & Mineralogists. *Res:* Modern and ancient Eolian deposits; development of equipment and techniques to recognize and describe dune and interdune deposits and practical applications of such work. *Mailing Add:* US Geol Surv 915 Nat Ctr 12201 Sunrise Valley Dr Reston VA 22092

AHLEN, STEVEN PAUL, b Oak Park, Ill, Dec 24, 49; m 78. ASTROPHYSICS, NUCLEAR PHYSICS. *Educ:* Univ Ill, Chicago, BS, 70, MS, 71; Univ Calif, Berkeley, PhD, 76. *Prof Exp:* Res physicist, Space Sci Lab, Univ Calif, Berkeley, 76-77, physicist, Lawrence Berkeley Lab, 77-78, asst res physicist, Space Sci Lab, 78-83; prof, Dept Physics, Ind Univ, Bloomington, 83-85, PROF DEPT PHYSICS, BOSTON UNIV, 85- *Honors & Awards:* Sloan Fel, 84. *Mem:* Am Phys Soc; Radiation Res Soc. *Res:* Experimental cosmic ray astrophysics; experimental and theoretical work related to charged particle penetration in matter; experimental high energy nuclear physics. *Mailing Add:* Dept Physics Boston Univ Boston MA 02215

AHLERS, GUENTER, b Bremen, Ger, Mar 28, 34; US citizen; m. PHYSICAL CHEMISTRY, SOLID STATE PHYSICS. *Educ:* Univ Calif, Riverside, BA, 59; Univ Calif, Berkeley, PhD(phys chem), 63. *Prof Exp:* Chemist silicate chem, Riverside Cement Co, 56-58; mem tech staff solid state physics, Bell Tel Labs, 63-80; PROF, DEPT PHYSICS, UNIV CALIF, SANTA BARBARA, 79-, DIR, CTR NONLINEAR SCI, QUANTUM INST, 89- *Concurrent Pos:* Mem, Condensed Matter Sci Oversight Rev Comt, NSF, 82, Mat Res Adv Comt, NSF, 85-88, External Adv Comt Ctr Nonlinear Studies, Los Alamos Nat Lab, 88-, Steering Comt, Soviet-US Conf Chaos, 90 & Sci Adv Group, Jet Propulsion Lab, NASA, 90-91; Erma & Jakob Michael vis prof, Rehovot, Israel, 85; presidential chair physics, Univ Calif, Santa Barbara, 89-, fac res lectr, 89-90; Alexander von Humboldt sr US scientist award, 89-90. *Honors & Awards:* Tenth Fritz London Mem Award in Low Temperature Physics, Am Phys Soc, 78; Morris Loeb Lectr, Harvard Univ, 79. *Mem:* Nat Acad Sci; fel AAAS; fel Am Phys Soc. *Res:* Thermodynamic properties of solidified gases; hydrodynamic instabilities; heat capacities; critical phenomena; liquid helium; transport properties. *Mailing Add:* Dept Physics Univ Calif Santa Barbara CA 93106

AHLERT, ROBERT CHRISTIAN, b New York, NY, Jan 22, 32; m 54; c 3. CHEMICAL ENGINEERING. *Educ:* Polytech Inst Brooklyn, BChE, 52; Univ Calif, Los Angeles, MSEng, 58; Lehigh Univ, PhD(chem eng), 64. *Prof Exp:* Res engr, Rocketdyne Div NAm Aviation, Inc, 54-56, sr res engr, 56-58, eng supvr, 58-62, propulsion specialist, 62-64, res group leader, 64; prof chem & biochem eng & exec dir eng res, 64-80, DIST PROF CHEM & BIOCHEM ENG, RUTGERS UNIV, NEW BRUNSWICK, 80- *Concurrent Pos:* Mem, Joint Army-Navy-Air Force Panel Liquid Propellant Test Methods, 58-64; consult, ad hoc panel liquid propellant test methods, US Dept Defense, 66-70; consult water qual, 71-; mem, Task Force Source Red, NJ Hazardous Waste Fac Siting Com, 84- *Honors & Awards:* Cecil Award, Am Inst Chem Engrs, 88. *Mem:* Am Inst Chem Engrs; Nat Soc Prof Engrs; Am Soc Eng Educ; Am Inst Chemists; Sigma Xi; Am Acad Environ Eng. *Res:* Ignition and combustion phenomena; chemical and biochemical processes; thermodynamics; biochemical reactor analysis; water resources analysis and simulation; hazardous waste management. *Mailing Add:* PO Box 27 Three Creek Rd Buttzville NJ 07829

AHLFELD, CHARLES EDWARD, b Aug 9, 40; m 62; c 2. REACTOR PHYSICS. *Educ:* Univ Fla, BS, 62; Fla State Univ, MS, 64, PhD(physics), 68. *Prof Exp:* Res physicist, Savannah River Plant, 67-76, staff physicist, 76-78, asst chief supvr, reactor physics, 78-82, chief supvr, 82-84, asst dep supt, 84-85, res mgr, 85-86, prog mgr reactors, 87-89,NPR DEPT MGR, SAVANNAH RIVER SITE, WESTINGHOUSE SAVANNAH RIVER CO, 89- *Mem:* Am Phys Soc; Am Nuclear Soc. *Res:* Experimental and theoretical studies of static and kinetic nuclear reactor behavior; cross sections and nuclear data pertinent to reactor design; application of computers to reactor experimentation; reactor core design; reactor operations; management of nuclear reactor operations. *Mailing Add:* Westinghouse Savannah River Co 37 Varden Dr Aiken SC 29803

AHLFORS, LARS VALERIAN, b Helsinki, Finland, Apr 18, 07; nat US; m 33; c 3. COMPLEX ANALYSIS, CONFORMAL MAPPING. *Educ:* Univ Helsinki, PhD(math), 30. *Hon Degrees:* AM, Harvard Univ, 38 & ScD, 89; LLD, Boston Col, 51; DrPhil, Univ Zurich, 78; ScD, London Univ, 78, Univ Md, 86. *Prof Exp:* Adj math, Univ Helsinki, 33-36; asst prof, Harvard Univ, 36-38; prof, Univ Helsinki, 38-44 & Univ Zurich, 45-46; prof, 46-77, EMER PROF MATH, HARVARD UNIV, 77- *Concurrent Pos:* Rockefeller fel, Paris, 32. *Honors & Awards:* Field's Medal, 36; Int Vihuri Prize, 68; Wolf Prize, 81. *Mem:* Nat Acad Sci; Am Math Soc; Finnish Acad Sci; Danish Royal Soc; Swedish Royal Soc; Soviet Acad Sci. *Res:* Theory of functions of a complex variable; conformal mapping; Riemann surfaces. *Mailing Add:* Dept Math Harvard Univ Cambridge MA 02138

AHLGREN, CLIFFORD ELMER, b Toimi, Minn, Apr 22, 22; m 54; c 2. FORESTRY. *Educ:* Univ Minn, BS, 48, MS, 53. *Hon Degrees:* DSc, Cornell Col, 76. *Prof Exp:* Forester, Iron Range Resources & Rehab, 48; res forester, Wilderness Res Found, 48-52, dir, Quetico-Super Wilderness Res Ctr, 52-63, dir res, Wilderness Res Found, 63-91; RETIRED. *Concurrent Pos:* Res assoc, Col Forestry, Univ Minn, St Paul, 63-91. *Mem:* Am Phytopath Asn; fel Soc Am Foresters; Am Forestry Asn; Ecol Soc Am; corresp mem Forestry Soc Finland. *Res:* Vegetational succession following natural disturbances in wilderness areas; field grafting and breeding of northern coniferous species; ecological effect of fire on northern coniferous forests. *Mailing Add:* 12734 Copperstone Dr Sun City W AZ 85375

AHLGREN, GEORGE E, b Cloquet, Minn, Dec 20, 31; m 61; c 2. PLANT PHYSIOLOGY. *Educ:* Univ Minn, BS, 59, MS, 62, PhD(agr, plant physiol), 66. *Prof Exp:* Asst prof, 66-71, ASSOC PROF BIOL, UNIV MINN, DULUTH, 71- *Mem:* Am Soc Plant Physiol. *Res:* Absorption and translocation of mineral ions and other substances by the plant. *Mailing Add:* Dept of Biol Univ of Minn Duluth MN 55812

AHLGREN, HENRY LAWRENCE, b Wyoming, Minn, Oct 3, 08; m 36; c 2. AGRONOMY. *Educ:* Univ Wis, BS, 31, MS, 33, PhD(agron), 35. *Prof Exp:* Asst agron & soils, 29-35, from instr to prof agron, 35-74, chmn dept, 49-52, assoc dir agr exten, 52-67, asst chancellor, Univ Exten, 66-67, vchancellor, 67-69, chancellor, 69-74, dir coop exten serv, 69-70, EMER PROF AGRON & EMER CHANCELLOR UNIV EXTEN, UNIV WIS-MADISON, 74- *Concurrent Pos:* Traveling fel, Europe, 36; dep undersecy rural develop, USDA, 70. *Mem:* Fel Am Soc Agron; fel Royal Swed Acad Agr & Forestry. *Res:* Pasture improvement, particularly the fertilization, management and ecological aspects; effect of various fertilizers, cutting treatments and irrigation on yield of forage and chemical composition of rhizomes of Kentucky bluegrass. *Mailing Add:* 10402 Sierra Dawn Dr Sun City AZ 85351

AHLGREN, ISABEL FULTON, b Viroqua, Wis, Feb 11, 24; m 54; c 2. BOTANY. *Educ:* DePauw Univ, AB, 46; Ind Univ, PhD(bot), 50. *Prof Exp:* Asst bot, Ind Univ, 46-49, instr, 50; instr, Wheaton Col, Mass, 50-52 & Wellesley Col, 52-54; res assoc, Quetico-Superior Wilderness Res Ctr, Wilderness Res Found, 55-63, res assoc, 63-78; RETIRED. *Concurrent Pos:* Lectr, Univ Minn, Duluth, 55, 62-75; res specialist, Col Forestry, Univ Minn, St Paul, 63-79, res assoc, 80- *Res:* Boreal forest ecology; plant migration into wilderness areas; curation of herbarium. *Mailing Add:* 12734 Copperstone Dr Sun City West AZ 85375

AHLGREN, MOLLY O, b Duluth, Minn, Nov 27, 57. DETRITIVORY, FEEDING ECOLOGY. *Educ:* Univ Idaho, BS, 80; Mich Tech Univ, MS, 84, PhD(aquatic ecol), 89. *Prof Exp:* ASST PROF AQUATIC RESOURCES, SHELDON JACKSON COL, 90- *Mem:* NAm Benthological Soc; Ecol Soc Am; Am Fisheries Soc; Sigma Xi. *Res:* Significance of detritus in the diet and nutrition of freshwater fish and role of marine invertebrates in processing organic matter in nearshore sediments. *Mailing Add:* Sci Div Sheldon Jackson Col 801 Lincoln St Sitka AK 99835

AHLRICHS, JAMES LLOYD, b Palmer, Iowa, Sept 13, 28; m 52; c 4. SOIL CHEMISTRY & MINERALOGY. *Educ:* Iowa State Univ, BS, 50, MS, 55; Purdue Univ, PhD(soil chem), 61. *Prof Exp:* Teacher, Pierson Pub Sch, 50-51; from instr to assoc prof, 57-68, PROF SOIL CHEM, PURDUE UNIV, WEST LAFAYETTE, 68- *Concurrent Pos:* Vis scientist, Macaulay Inst Soil Sci, Scotland, 66-67; sr Fulbright lectr, Coun Sci Invest & Autonomous Univ Madrid, 73-74; Agency Int Develop tech adv, Univ Tras-os-Montes e Alto Douro, Port, 81-83. *Mem:* Am Chem Soc; Soil Sci Soc Am; Soil Conserv Soc Am; Am Soc Agron; Clay Minerals Soc; Sigma Xi. *Res:* Aluminum toxicity in soils and cultivator tolerance to aluminum. *Mailing Add:* Dept Agron Purdue Univ West Lafayette IN 47906

AHLSCHWEDE, WILLIAM T, b Lincoln, Nebr, Jan 31, 42; m 64; c 2. ANIMAL BREEDING, SWINE PRODUCTION. *Educ:* Univ Nebr, BS, 64; NC State Univ, MS, 67, PhD(animal sci), 70. *Prof Exp:* Instr animal sci, NC state Univ, 67-69; res assoc med genetics, Univ Wis, 69-70; asst prof, 70-75, ASSOC PROF ANIMAL SCI, UNIV NEBR, 75- *Mem:* Am Soc Animal Sci. *Res:* Evaluation of crossbreeding systems in swine; development of specialized breeding systems. *Mailing Add:* Dept of Animal Sci Univ Nebr-Lincoln Lincoln NE 68583-0908

AHLSTROM, HARLOW G(ARTH), b Yakima, Wash, Aug 20, 35; m 56; c 2. AERONAUTICS, FLUID DYNAMICS. *Educ:* Univ Wash, BS, 57, MS, 59; Calif Inst Technol, PhD(aeronaut), 63. *Prof Exp:* Assoc res engr, Boeing Airplane Co, 57; aerodynamicist, Douglas Aircraft Co, 59; res fel magneto fluid dynamics, Calif Inst Technol, 62 & 63; from asst prof aeronaut & astronaut to prof, Univ Wash, 63-74; group leader, laser plasma interaction group, Lawrence Livermore Lab, Unif Calif, 74-75; assoc prog leader fusion exp, 75-81, asst dep assoc dir inertial fusion, 81-85; ASST DIR ENG TECHNOL, BOEING AEROSPACE CO, 85- *Concurrent Pos:* Consult, Douglas Aircraft Co, 59-64, Boeing Co, 63-69; Math Sci Northwest, 69- & Nat Acad Sci, 69-; spec consult, NASA Hq, Washington, DC, 67; vis prof, Univ D'Aix-Marseille, France, 71. *Mem:* Am Inst Aeronaut & Astronaut; Am Phys Soc. *Res:* Magneto fluid dynamics; fluid dynamics; lasers; fusion. *Mailing Add:* 21814 234 Ave SE Maple Valley WA 98038

AHLUWALIA, BALWANT SINGH, b India, Mar 12, 32; m 62; c 2. ENDOCRINOLOGY. *Educ:* Univ Minn, MS, 59, PhD(reproduction), 62. *Prof Exp:* Scientist, Worcester Found Exp Biol, 67-68; scientist, Nat Inst Arthritis & Metab Dis, 68-70; asst prof, 70-72, ASSOC PROF OBSTET & GYNEC, COL MED, HOWARD UNIV, 72- *Concurrent Pos:* Fel physiol, Hormel Inst, Univ Minn, 63-67. *Mem:* Am Vet Med Asn; Endocrine Soc; Am Inst Nutrit; Am Oil Chemists Soc. *Res:* Lipid metabolism in the reproductive organs with respect to fertility and sterility in animals. *Mailing Add:* Dept Obstet & Gynec Howard Univ Col Med 520 W St NW Washington DC 20059

AHLUWALIA, DALJIT SINGH, b Sialkot, India, Sept 5, 32; m 60; c 4. APPLIED MATHEMATICS. *Educ:* Punjab Univ, India, BA, 52, MA, 55; Ind Univ, Bloomington, MS, 65, PhD(appl math), 65. *Prof Exp:* Lectr math, R G Col, Phagwara, Punjab, 55-57, Khalsa Col, Bombay, 57-62 & Univ Bombay, 59-62; part time teaching assoc, Ind Univ, Bloomington, 62-65, asst prof, 65-66; vis mem & adj assoc prof math, Courant Inst Math Sci, NY Univ, 66-68, from asst prof to prof, 68-87; PROF MATH & DIR, CTR APPL MATH & STATIST, NJ INST TECHNOL, 86- *Concurrent Pos:* Prof math, Univ SFla, Tampa, 72-74; consult, Lamont Doherty Geol Observ, Columbia Univ, 77-, Inst Comput Appln Sci & Eng, 78-79 & Univ Miami, 79; vis prof, Northwestern Univ, 84 & Stanford Univ, 85; chmn, Ctr Appl Math & Statist, NJ Inst Technol, 86-89. *Mem:* Am Math Soc; Soc Indust & Appl Math; Tensor Soc; Acoustic Soc Am; Int Sci Radio Union. *Res:* Plastic flow and fracture in solids; uniform theories of diffraction for the edges and convex bodies; study of wave propagation in elastic media and acoustics; asymptotic methods in underwater acoustics. *Mailing Add:* Dept Math NJ Inst Technol Newark NJ 07102

AHLUWALIA, GURPREET S, PHARMACOLOGY. *Educ:* Delhi Univ, India, BS, 76; Howard Univ, MS, 78, PhD(biochem), 81. *Prof Exp:* Postdoctoral res assoc, Dept Biochem, Col Med, Howard Univ, 81-82; NIH fel, Lab Med Chem & Pharmacol, Nat Cancer Inst, NIH, 82-84; res scientist, 84-87, res assoc, 87-89, SR RES ASSOC, GILLETTE RES INST, 89- *Mem:* Am Soc Pharmacol & Exp Therapeut; NY Acad Sci. *Res:* Clinical pharmacology and drug development; correlating drug disposition to its pharmacological effects; mechanism of action of compounds; formulations for effective drug delivery; author of 30 technical publications. *Mailing Add:* Gillette Res Inst 401 Professional Dr Gaithersburg MD 20879

AHLUWALIA, HARJIT SINGH, b Bombay, India, May 13, 34; m 64; c 2. COSMIC RAY PHYSICS, SOLAR PHYSICS. *Educ:* Panjab Univ, India, BSc Hons, 53, MSc, 54; Gujarat Univ, India, PhD(physics), 60. *Prof Exp:* Sr res asst cosmic rays, Phys Res Lab, Ahmedabad, Gujarat, India, 54-62; tech asst expert, UNESCO, France, 62; res assoc cosmic rays & geomagnetism, Southwest Ctr Advan Studies, Tex, 63-64; vis prof physics, Int Atomic Energy Agency, Austria & sci dir, Lab Cosmic Physics, Univ La Paz, 65-67; vis prof physics, Pan Am Union, Washington, DC, 67; assoc prof, 68-73, PROF PHYSICS, UNIV NMEX, 73- *Concurrent Pos:* NASA fel, 63-64; chmn, Bolivian Space Res Comt, 65-67; prin investr, US Air Force res proj, 65-68, NSF res proj, 68-81 & Sandia Corp res proj, Albuquerque, 69-71, NASA res proj, 88-; Bolivian nat rep, Comt Space Res & Int Union Pure & Appl Physics, 66-69; mem, Cosmic Ray Comn, Int Union Pure & Appl Physics, 66-69; gen secy high energy cosmic ray group, Int Coun Sci Unions, 76-85; Vis sr scientist, NASA head quaters, Wash,DC, 87-88. *Honors & Awards:* Distinguished Serv Award, Am Geophys Union, 88. *Mem:* AAAS; Am Geophys Union; Inst Elec & Electronics Engrs; Am Meteorol Soc; Am Phys Soc; Int Astron Union; Am Astron Soc; Sigma Xi. *Res:* Cosmic rays; geomagnetism; nuclear electronics; space physics; plasma physics; astrophysics. *Mailing Add:* 13000 Cedar Brook NE Albuquerque NM 87111-3018

AHMAD, FAZAL, LIPID SYNTHESIS, ENZYME REGULATION. *Educ:* Univ Edinburgh, Scotland, PhD(chem), 61. *Prof Exp:* PROF BIOCHEM, SCH MED, UNIV MIAMI, 72- *Mailing Add:* Dept Biochem & Molecular Biol 1600 NW Tenth Ave PO Box 016129 Miami FL 33101

AHMAD, IRSHAD, b Azamgarh, India, Nov 1, 39; m 69; c 3. NUCLEAR CHEMISTRY. *Educ:* Univ Punjab, Pakistan, BSc, 58; Univ Peshawar, MSc, 62; Univ Pac, MS, 65; Univ Calif, Berkeley, PhD(chem), 66. *Prof Exp:* Demonstr chem, Edwardes Col, Peshawar, 58-60; res fel, Lawrence Radiation Lab, Univ Calif, Berkeley, 66; res assoc, 66-68, asst chemist, 68-70, CHEMIST, ARGONNE NAT LAB, 70- *Mem:* Am Phys Soc; Am Chem Soc. *Res:* Synthesis of new actinide isotopes; nuclear structure studies of transuranium nuclei by high-resolution alpha, beta and gamma spectroscopy and nuclear reactions; investigation of high spin states by heavy-ion reactions. *Mailing Add:* D-203-F-157 Argonne Nat Lab Argonne IL 60439

AHMAD, JAMEEL, b Lahore, Pakistan, May 22, 41; m 65; c 1. CIVIL ENGINEERING, SOLID MECHANICS. *Educ:* Univ Punjab, WPakistan, BS, 61; Univ Hawaii, MS, 64; Univ Pa, PhD(civil eng), 67. *Prof Exp:* Asst engr, WAPDA, Pakistan, 61-62; asst prof eng, PMC Cols, 67-68; assoc prof, 68-80, PROF & CHMN, DEPT CIVIL ENG, COOPER UNION, 80- *Concurrent Pos:* Dir, PMA Corp, 78-; proj dir, Cooper Union Res Found, 80- *Mem:* Am Soc Civil Engrs; Am Concrete Inst; Am Soc Eng Educ. *Res:* Mechanics of materials and structural mechanics; environmental engineering; energy engineering. *Mailing Add:* Dept Civil Eng The Cooper Union New York NY 10003

AHMAD, MOID UDDIN, b Agra, India, Aug 18, 27; m; c 2. GROUNDWATER HYDROLOGY, GEOPHYSICS. *Educ:* Agra Univ, BSc, 44, LLB, 47; NMex Inst Mining & Technol, MS, 61; Univ London, PhD(geol, seismol), 66. *Prof Exp:* Asst seismologist, Geophys Inst Quetta, Pakistan, 49-58; head dept physics, Battersea Grammar Sch, London, Eng, 61-67; geophysicist, Food & Agr Orgn, UN & consult to Govt of Kuwait, 67-69; assoc prof hydrol & geophys, 69-74, PROF HYDROL & GEOPHYS, OHIO UNIV, 74- *Concurrent Pos:* Spec consult, Libyan Govt Reclamation & Resettlement Dept. *Mem:* Am Geophys Union; Soc Explor Geophys; Europ Asn Explor Geophys; Am Water Resources Asn. *Res:* Exploration of groundwater in arid regions; acid mine drainage control, development of a hydrological approach to control acid discharge from strip and drift mines and techniques to map acid producing areas by thermal mapping; well field design; designed Sarir Well Field, consisting of 250 wells, one of the largest desert farms in Sahara. *Mailing Add:* Dept Geol Ohio Univ Main Campus Athens OH 45701

AHMAD, NAZIR, b Pakistan, Feb 22, 36; US citizen; m 55; c 2. HISTOLOGY, REPRODUCTIVE ENDOCRINOLOGY. *Educ:* San Francisco State Col, BA, 55; Univ Calif, Berkeley, MA, 59; Univ Calif, San Francisco, PhD(anat), 68. *Prof Exp:* Asst prof histol, Sch Med, Georgetown Univ, 68-69; ASSOC PROF HISTOL, SCH MED, UNIV SOUTHERN CALIF, 75- *Mem:* AAAS; Soc Study Reproduction; Am Asn Anat; Endocrine Soc. *Res:* Reproductive endocrinology, utilizing pituitary as well as ovarian hormones essential for the maintenance of gestation and lactation; maintenance of spermatogenesis; effects of toxic substances on male reproduction including developmental studies in human histology. *Mailing Add:* Dept Anat Univ Southern Calif 2025 Zonal Ave Los Angeles CA 90033

AHMAD, SHAIR, b Kabul, Afghanistan, June 19, 34; US citizen; m 58, 74; c 3. MATHEMATICS. *Educ:* Univ Utah, BS, 60, MS, 62; Case Western Reserve Univ, PhD(math), 68. *Prof Exp:* Asst math, Univ Utah, 60-62; instr, SDak State Univ, 62-64; asst, Case Western Reserve Univ, 64-65, instr, 66-68; asst prof, Univ NDak, 65-66; from asst prof to prof math, Okla State Univ, 68-78, chmn dept, 78-80; CHMN, DEPT MATH & COMPUT SCI, UNIV MIAMI, 80- *Mem:* Am Math Soc; Math Asn Am. *Res:* Dynamical systems; differential equations. *Mailing Add:* Dept Math Univ Miami PO Box 249085 Coral Gables FL 33124

AHMAD, SHAMIM, b Pakistan, June 8, 44; US citizen; m 71; c 3. ELASTOMER TECHNOLOGY, TIRE ENGINEERING. *Educ:* Tri State Univ, BS, 68; Univ Akron, MS, 73; Kent State Univ, MBA, 82. *Prof Exp:* Prod engr mat develop, B F Goodrich Co, 68-73; sr prod engr, 73-78, tire scientist, 78-79, mgr passenger compound develop, 79-82, group mgr textile develop, 82-83, group mgr compound develop, 83-85, dir, 85-87, SR DIR, PROD DEVELOP, UNIROYAL GOODRICH TIRE CO, 87- *Mem:* Soc Automotive Engrs. *Res:* Mixing, vulcanization, calendering and extrusion of elastomeric compound; rubber and tire technology; awarded six US patents. *Mailing Add:* 2211 Donner St NW North Canton OH 44720

AHMADJIAN, VERNON, b Whitinsville, Mass, May 19, 30; m 56; c 3. BOTANY. *Educ:* Clark Univ, AB, 52, MA, 56; Harvard Univ, PhD, 60. *Prof Exp:* From asst prof to prof bot, Clark Univ, 59-68; prof, Univ Mass, 68-69; assoc dean grad sch, 69-71, coordr res, 69-75, dean, Grad Sch, 71-75, PROF BOT, CLARK UNIV, 69- *Concurrent Pos:* Vis prof, Univ Calif, Berkeley, 65-66, Wellesley Col, 85-86. *Honors & Awards:* New York Bot Garden Award, 68; Antarctic Medal. *Mem:* Fel AAAS; Bot Soc Am; Phycol Soc Am; Mycol Soc Am; Brit Lichen Soc. *Res:* Symbiosis; lichenology; molecular biology of lichens. *Mailing Add:* Dept of Biol Clark Univ Worcester MA 01610-1477

AHMANN, DONALD H(ENRY), b Struble, Iowa, Jan 9, 20; m 45; c 6. MATERIALS SCIENCE, INORGANIC CHEMISTRY. *Educ:* Iowa State Univ, BS, 41, PhD(phys & inorg chem), 48. *Prof Exp:* Jr chemist, Manhattan Proj, Iowa State Univ, 42-48; res chemist, Knolls Atomic Power Lab, Gen Elec Co, NY, 48-55, mgr chem & chem eng, 55-57, mgr chem, Vallecitos Nuclear Ctr, Calif, 57-60, mgr chem & chem eng, 60-66, mgr chem & metall, 66-67, mgr, Nuclear Mat & Propulsion Opers, Ohio, 67-68, mgr mat sci & technol, Nuclear Systs Progs, 68-69, mgr eng, neutron devices dept, 69-85; CONSULT, 88- *Mem:* Am Chem Soc; Am Nuclear Soc; Am Soc Metals; AAAS; Am Vacuum Soc. *Res:* Material science studies related to nuclear applications and vacuum components; chemistry of nuclear reactors and fuels; nuclear fuels reprocessing. *Mailing Add:* 660 Bluffview Dr Belleair Bluffs FL 34640

AHMED, A RAZZAQUE, b Hyderabad, India, Oct 27, 48; US citizen. SKIN DISEASES. *Educ:* Nagpur Univ, India, BS, 66; All India Inst Med Sci, New Delhi, 72; Harvard Univ, DSc(oral biol), 88. *Prof Exp:* Asst prof, dermat, Univ Calif Sch Med, Los Angeles, 78-85; RES ASSOC, ORAL PATH, HARVARD SCH DENT MED, BOSTON, 85-, CONSULT, HARVARD HEALTH, 85- *Honors & Awards:* Gold Award, Am Acad Dermat, 82. *Mem:* Soc Invest Dermat; Am Asn Immunologists; Am Fedn Clin Res; Am Soc Clin Invest; Royal Soc Med. *Res:* Pathogenesis of actummune blistering skin disease. *Mailing Add:* Ctr Blood Res Harvard Univ 800 Huntington Ave Boston MA 02115

AHMED, ASAD, b Saharanpur, India, Nov 7, 39; Can citizen; m 78. MOLECULAR GENETICS. *Educ:* Aligarh Muslim Univ, India, BSc, 56, MSc, 58, PhD(plant path), 61; Yale Univ, PhD(biochem genetics), 64. *Prof Exp:* Fulbright scholar, Yale Univ, 60-64, univ fel, 61-62, Wadsworth fel, 62-63, Sterling fel, 63-64; res assoc, Inst Molecular Biol, Univ Ore, 64-65; res scientist biochem, Dept Nat Health & Welfare, Can, 66-67; from asst prof to assoc prof, 67-77, PROF GENETICS, UNIV ALTA, 77- *Concurrent Pos:* Sect ed virol, genetics & molecular biol, Can J Microbiol, 80. *Mem:* AAAS. *Res:* Transposable genetic elements in Escherichia coli; Recombinant DNA technology; cloning and sequencing vectors. *Mailing Add:* Dept Genetics Univ Alta Edmonton AB T6G 2E9 Can

AHMED, ESAM MAHMOUD, b Cairo, UAR, Oct 7, 25; US citizen; m 56, 67; c 4. FOOD SCIENCE. *Educ:* Univ Cairo, BS, 45; Univ Alexandria, MS, 53; Univ Md, PhD(hort), 57. *Prof Exp:* From instr to asst prof veg crops physiol, Univ Alexandria, 45-59; res assoc hort physiol, Univ Md, 59-64; from asst prof to assoc prof, 64-75, PROF FOOD SCI, UNIV FLA, 75- *Concurrent Pos:* Joseph H Gourley res award, 64. *Mem:* Am Soc Host Sci; Inst Food Technologists. *Res:* Food quality; psychophysical aspects of foods; food color and texture measurements; food irradiation; post-harvest physiology. *Mailing Add:* Dept of Food Sci Univ of Fla Gainesville FL 32611

AHMED, ISMAIL YOUSEF, b Silwad, Jordan, Sept 27, 39. INORGANIC CHEMISTRY. *Educ:* Cairo Univ, BS, 60; Pa State Univ, MS, 65; Southern Ill Univ, Carbondale, PhD(chem), 68. *Prof Exp:* High sch teacher, Jordan, 60-64; lectr chem, Southern Ill Univ Carbondale, 68-69; asst prof, 69-73, ASSOC PROF CHEM, UNIV MISS, 73- *Mem:* Am Chem Soc. *Res:* Lewis acid-Lewis base interactions; interactions of metal halides with compounds containing N-S bond; characterization of solute species in nonaqueous solvents; solvation and rates of solvent exchange by nuclear magnetic resonance. *Mailing Add:* Dept of Chem Univ of Miss University MS 38677

AHMED, KHALIL, b Lahore, Pakistan, Nov 30, 34; US citizen; m 69; c 2. BIOCHEMISTRY. *Educ:* Univ Panjab, Pakistan, BSc, 54, MSc, 55; McGill Univ, PhD(biochem), 60. *Prof Exp:* Res chemist, WRegional Labs, Pakistan, 55-57; res asst biochem, Montreal Gen Hosp & Res Inst, McGill Univ, 57-60; res fel, Wistar Inst, Univ Pa, 60-61, res assoc, 61-63; asst prof metab res, Chicago Med Sch, 63-67; sr staff mem & res biochemist, Lab Pharmacol, Baltimore Cancer Res Ctr, Nat Cancer Inst, 67-72; assoc prof, 73-77, PROF LAB MED & PATH, UNIV MINN, MINNEAPOLIS, 77-; RES BIOCHEMIST & CHIEF, CELLULAR & MOLECULAR BIOCHEM LAB, DEPT VET AFFAIRS MED CTR, MINNEAPOLIS, 70-, RES CAREER SCIENTIST, 77- *Concurrent Pos:* Vis lectr, Chicago Med Sch, 68-69; mem, Path B Study Sect, NIH, 78-81; mem, Pharmacol Study Sect, 88- *Mem:* AAAS; Am Soc Biochem & Molecular Biol; Am Soc Pharmacol & Exp Therapeut; Am Soc Cell Biol; Sigma Xi; Soc Gen Physiol. *Res:* Enzymic mechanism of drug action; biochemistry of ion transport; biochemistry of androgen action in male sex glands; neurochemistry; phosphoproteins; biochemistry of gene action; protein kinases and phosphatases. *Mailing Add:* Cellular & Molecular Biochem Lab 151 VA Med Ctr One Vet Dr Minneapolis MN 55417

AHMED, MAHMOUD S, b Cairo, Egypt, Mar 17, 42; m 68; c 2. DRUG ABUSE DURING PREGNANCY, OPIOIDS & COCAINE. *Educ:* Cairo Univ, BSc, 64, MSc, 69; Univ Tenn, PhD(biochem), 74. *Prof Exp:* Fel biochem, Johns Hopkins Univ, 75-77; asst prof biochem, Univ Tenn, Sch Med, 77-86; ASSOC PROF OBSTET/GYNEC, SCH MED, UNIV MO, KANSAS CITY, 86-, ASSOC PROF SCH NURSING & ASSOC PROF BIOCHEM, SCH BASIC LIFE SCI, 86- *Concurrent Pos:* Vis scientist, Werner Klee Lab, NIH, 80. *Mem:* Sigma Xi; Am Soc Biol Chemists; Soc Perinatal Obstetricians. *Res:* Membrane structure function relationships; role of opiate peptides in analgesia during labor; mechanism of drug dependence and addiction. *Mailing Add:* Univ Mo Sch Med 2411 Holmes Kansas City MO 64108-2792

AHMED, MOGHISUDDIN, b Dhanbad, India, July 1, 50; m 85; c 1. LIPID CHEMISTRY, ORGANIC-ANALYTICAL CHEMISTRY. *Educ:* Aligarh Muslim Univ, India, BSc, 71, MSc, 73, MPhil, 75, PhD(org chem), 78. *Prof Exp:* Fel lipid chem, Aligarh Muslim Univ, India, 78-79; res assoc fatty acid chem, Tex A&M Univ, 79-81; res assoc food chem & toxicol, Ore State Univ, 81-88; CHEMIST LIPIDS DEPT, SIGMA CHEM CO, 88- *Mem:* Am Oil Chemists Soc; Am Chem Soc; Sigma Xi. *Res:* Biological and organic chemistry of lipids; chromatograpic and spectroscopic analysis; organic synthesis; chemistry of vegetable oils; chemistry of nitrosamines; new product development. *Mailing Add:* Lipids Dept Sigma Chem Co PO Box 14508 St Louis MO 63178

AHMED, NAHED K, cancer research, experimental therapeutics, for more information see previous edition

AHMED, NASIR, b Bangalore, India, Aug 11, 40; m 65; c 1. SYSTEMS THEORY, INFORMATION SCIENCE. *Educ:* Col Eng, Bangalore, India, BS, 61; Univ NMex, MS, 63, PhD(elec eng), 66. *Prof Exp:* Prin res engr, Systs & Res Ctr, Honeywell, Inc, Minn, 66-68; PROF ELEC ENG, DEPT ELEC & COMPUT ENG, UNIV NMEX. *Mem:* Inst Elec & Electronics Engrs. *Res:* System and communication theory; information processing. *Mailing Add:* Dept Elec & Comput Eng Univ NMex Albuquerque NM 87131

AHMED, NASIR UDDIN, b Jan 1, 34; Can citizen; m 58; c 5. SYSTEMS & CONTROL THEORY. *Educ:* Univ Dhaka, BSc, 56; Univ Ottawa, MASc, 63, PhD(elec eng), 65. *Prof Exp:* Asst engr, Power Syst Control Water & Power Develop Authority, Pakistan, 56-57; sr sci officer, Reactor Physics & nucleonics, Pakistan Atomic Energy Comn, 58-67; fel syst & control, Univ Alta, 67-68; from asst prof to assoc prof, 68-75, PROF SYST OPTIMAL CONTROL, UNIV OTTAWA, 76- *Concurrent Pos:* Vis prof, Univ Calif, Berkeley, 77-78, Univ New South Wales, Australia, 80; consult, Dept Nat Defence, Can, 84-87. *Mem:* Inst Elec & Electronics Eng; Soc Indust & Appl Math; Am Math Soc; Asn Prof Engrs Ont; Can Asn Univ Teachers. *Res:* Systems and control theory for abstract evolution equations on Banach space with applications to ordinary, delay-differential, partial differential and stochastic differential equations and nonlinear filtering; mathematical modeling of spacecraft and space stations with flexible appendages. *Mailing Add:* Dept Elec Eng Univ Ottawa Ottawa ON K1N 6N5 Can

AHMED, S SULTAN, b Delhi, India, Sept 13, 37; US citizen; m 67; c 2. ADULT CARDIOLOGY, CARDIOVASCULAR RESEARCH. *Educ:* D J Science Col, Pakistan, BSc, 58; Dow Med Col, Pakistan, MB BS, 63. *Hon Degrees:* FRSM, Royal Soc Med, UK, 74, FACP, 73, FACP, 74, FACC, FACCP, FACA, 76, FAMS, Pakistan, 88, FECE, 90. *Prof Exp:* Asst in med & cardiol, 68-69, from instr to assoc prof med, 70-80, fel, 71-72, PROF MED & CARDIOL, UNIV MED & DENT NJ, NJ MED SCH, 80- *Concurrent Pos:* Attend physician, Univ Med & Dent NJ, NJ Med Sch, 72-; educ coordr cardiol, St Joseph's Hosp, 74-76 & consult physician, 74-; consult physician, St Michael's Med Ctr, 75-; vis prof, Dow Med Col, 84-, Post-Grad Inst Med & Air Force Inst Cardiol, 88- & Punjab Med Col, 90; mem affil res, Peer Rev Comt, Am Heart Asn, 85-86. *Mem:* Am Fedn Clin Res; Am Heart Asn; Am Col Cardiol; AAAS. *Res:* Cardiovascular function and diseases, both in the basic research laboratory as well as part of patient care activities; diagnostic tests in the cardiac catherization lab and management as in a medical ward or intensive care unit; cardiac effects of cigarette smoking; alcohol and the heart; LV functional indices; systolic time intervals; diabetes and the heart; coronary artery disease and blood flow; hypertension; cardiac catherization and angiography. *Mailing Add:* UMDNJ-NJ Med Sch 100 Bergen St Rm I-534 Newark NJ 07103

AHMED, SAAD ATTIA, b Giza, Egypt, July 28, 50; m 81; c 2. GAS TURBINE, BIOENGINEERING. *Educ:* Cairo Univ, BSc, 73; Carleton Univ, Can, MEng, 78; Ga Inst Technol, PhD(eng), 81. *Prof Exp:* Instr mech eng, Cairo Univ, Egypt, 73-76; demostrator, Carleton Univ, Can, 76-78; res asst, Ga Inst Technol, 78-81, teaching fel, 82-83; res assoc, Ariz State Univ, 83-84, res analyst, 84-86; SR RES ENGR, GARRETT TURBINE ENGINE CO, 86- *Mem:* Am Inst Aeronaut & Astronaut; Sigma Xi; Am Soc Eng Educ; Am Phys Soc; Am Soc Mech Eng. *Res:* Coronary fluid dynamics; coherent structures; experimental fluid mechanics; two dimensioal and vaneless diffusers; aerodynamics of combustion processes; gas turbine and ram-jet combustors; turbulent and jet flows; experimental methods. *Mailing Add:* 1111 E University Dr No 236 Tempe AZ 85281

AHMED, SAIYED I, b Desna, Pakistan, Jan 5, 41; m 64; c 2. MICROBIOLOGY, BIOCHEMICAL GENETICS. *Educ:* Univ Karachi, BSc, 60; Univ Frankfurt, PhD(microbiol, biochem, bot), 63. *Prof Exp:* Sr res off microbiol & fermentation res, Pakistan Coun Sci & Indust Res, 64; Nat Res Coun Can fel biochem genetics, Univ Man, 64-66; NIH sr fel, Yale Univ, 66-69, res assoc, 69-70; vis scientist, Sch Med, 70-73, res assoc, 73-76, sr res assoc, 76-78, res assoc prof, 75-85, RES PROF, SCH OCEANOG, UNIV WASH, 85- *Concurrent Pos:* Assoc prog dir biotechnol,nat sea grant, Nat Oceanic & Atmospheric Admin, 86-88; fel fac Univ Wash,85. *Mem:* Am Soc Microbiol; Am Soc Plant Physiologists; Am Soc Limnol & Oceanog; NY Acad; AAAS. *Res:* Mutation genetics in serratia and E coli; control mechanisms and regulations; enzymology; mutagenicity; pollution and environmental studies; marine bacteria and marine phytoplankton metabolism; metabolic studies in anoxic marine environments; application of biotechnology in marine sciences. *Mailing Add:* Dept Oceanog Univ Wash Seattle WA 98195

AHMED, SHAFFIQ UDDIN, b Howrah, India, Mar 1, 34; m 63; c 1. METALLURGICAL ENGINEERING, MATERIALS SCIENCE. *Educ:* Univ Calcutta, BE, 54; Univ Ill, MS, 58; Case Inst Technol, PhD(metall eng), 65. *Prof Exp:* Res officer metall eng, Univ Ill, 55-60; assoc prof, 60-69, chmn dept, 68-72, PROF METALL ENG, YOUNGSTOWN STATE UNIV, 69- *Concurrent Pos:* Pres, Dr S Ahmed & Assocs, Youngstown, 60- *Honors & Awards:* Distinguished Serv Award, Am Soc Metals, 68, Prof Man of Year, 71; Western Electric Fund Award, Am Soc Eng Educ. *Mem:* Fel AAAS; fel Am Inst Chemists; Am Soc Metals; Am Inst Mining, Metall & Petrol Engrs; Am Soc Eng Educ. *Res:* Alloys; x-rays; electron microscopy; phase transformation. *Mailing Add:* 269 Redondo Rd Youngstown OH 44504

AHMED, SUSAN WOLOFSKI, b Detroit, Mich, Apr 25, 46; m 72; c 2. MULTIVARIATE ANALYSIS. *Educ:* Kalamazoo Col, BS, 68; Univ Mich, Ann Arbor, MPH, 70; Univ NC, Chapel Hill, PhD(biostatist), 75. *Prof Exp:* Res asst prof, Univ NC, Chapel Hill, 76-77; ASST PROF BIOSTATIST, GEORGETOWN UNIV, 77- *Mem:* Am Statist Asn; Biometric Soc; Sigma Xi. *Mailing Add:* Community Family Med Georgetown Univ Sch Med 3900 Reservoir NW Washington DC 20007

AHMED, SYED MAHMOOD, b Hyderabad, India, Mar 30, 49; m 78; c 3. SURFACE CHEMISTRY. *Educ:* Osmania Univ, Hyderabad, India, BTech, 72; Lehigh Univ, MS, 76, PhD(polymer sci & eng), 79. *Prof Exp:* Res asst, Emulsion Polymers Inst, Lehigh Univ, 74-79; SR RES ENGR, RES CTR, HERCULES, INC, 79- *Mem:* Am Chem Soc; Am Inst Chem Engrs. *Res:* Synthesis and applications of natural and synthetic water soluble polymers; emulsion polymers and emulsion polymerization; suspension polymerization; electrotheology. *Mailing Add:* Hercules Inc Res Ctr 1313 N Market St Wilmington DE 19894

AHMED, WASE U, b Moradabad, India, Sept 27, 31; m 62; c 4. MICROSCOPY, MATERIAL SCIENCE. *Educ:* Univ Sind, BS, 57; Univ Ill, 61. *Prof Exp:* MINERALOGIST, BUEHLER LTD, 62- *Mem:* Int Metallog Soc; Geol Soc Am; Int Cement Micros Asn. *Res:* New product development to study structure of materials; new methods to prepare specimens for microstructure studies. *Mailing Add:* 720 Newport North Roselle IL 60172

AHMED-ANSARI, AFTAB, b Karachi, Pakistan, Feb 18, 44; US citizen; m; c 2. PATHOLOGY. *Educ:* Univ Karachi, Pakistan, MS, 63; Univ Calif, Los Angeles, MS, 67; Univ Ariz, PhD, 70. *Prof Exp:* Res assoc, NIH fel, Univ Ariz, 70-71; res assoc, Nat Res Coun, Naval Med Res Inst, Nat Naval Med Ctr, 71-73, head, Cellular Immunol Div, 73-74, chmn, res comt, 76-78, sci dir, Dept Immunol, 78-79; dir, Dept Immunol, Merck Inst Therapeut Res, 79-81; sci dir, Naval Med Res Unit No Three, Cairo, 81-84; sr res scientist, Rocky Mountain Lab, Nat Inst Allergy & Infectious Dis, NIH, 84-85; PROF, DEPT PATH & LAB MED, WINSHIP CANCER CTR, SCH MED, EMORY UNIV, 85-, EXEC MEM, 85- *Mem:* Am Asn Immunologists; Am Soc Microbiol; AAAS; Sigma Xi; Transplantation Soc; Int Soc Exp Hemat; Asn Clin Scientist. *Res:* Numerous publications; pathology. *Mailing Add:* Dept Path Emory Univ Sch Med Rm 5008 Winship Cancer Ctr Atlanta GA 30322

AHMED-ZAID, SAID, b Algiers, Algeria, May 6, 56; m 83. ELECTRIC POWER ENGINEERING, POWER ELECTRONICS. *Educ:* Univ Ill, Urbana-Champaign, BS, 79, MS, 81, PhD(elec eng), 84. *Prof Exp:* Prin res engr, Nat Elec & Gas Co, Algeria, 85-88; ASSOC PROF ELEC ENG, CLARKSON UNIV, POTSDAM, 89- *Concurrent Pos:* Vis res assoc, Univ Ill, Urbana-Champaign, 84, vis asst prof, 89; consult, Gen Elec Co, Schenectady, 84; asst prof, Electronics Inst, Univ Blida, Algeria, 85-87; NSF presidential young investr award, 90. *Mem:* Inst Elec & Electronic Engrs; Inst Elec & Electronic Engrs Power Eng Soc. *Res:* Power system modeling and simulation; power system stability; computational methods; optimization; advanced control applications in power systems. *Mailing Add:* Dept Elec & Computer Eng Clarkson Univ Potsdam NY 13699

AHN, HO-SAM, b Seoul, South Korea, Feb 25, 40; US citizen; m 68; c 2. CARDIOVASCULAR PHARMACOLOGY, BIOCHEMICAL PHARMACOLOGY. *Educ:* Seoul Nat Univ, BS, 63; Princeton Univ, MA, 67; Rutgers Univ, PhD(zool), 72. *Prof Exp:* Res assoc & fel, Einstein Col Med, 73-77, res asst prof, 77-78; adj assoc prof biochem, Fairley Dickinson Univ, 86-90; sr scientist, 78-84, PRIN SCIENTIST, SCHERING-PLOUGH RES, 84- *Concurrent Pos:* Ed vpres, WeSearch Toastmaster Int, 87-88. *Mem:* Am Soc Pharmacol & Exp Therapeut; AAAS; NY Acad Sci; Sigma Xi. *Res:* Elucidation of the active site of calcium/calmodulin-dependent cyclic nucleotide phosphodiesterase by determining activity and inhibitor binding after selective modification of amino acid residues; calcium/calmodulin-dependent cyclic nucleotides phosphodiesterase inhibitors as potential antihypertensive agents. *Mailing Add:* Dept Pharmacol Schering-Plough Res Bloomfield NJ 07003

AHN, K(IE) Y(EUNG), b Pyongnam, Korea, Apr 6, 30; m 61; c 3. ELECTRICAL ENGINEERING. *Educ:* Ore State Univ, BS, 58; Rensselaer Polytech Inst, MEE, 59, PhD(elec eng), 63. *Prof Exp:* MEM RES STAFF, THOMAS J WATSON RES CTR, IBM CORP, 63- *Mem:* Inst Elec & Electronics Engrs. *Res:* Solid state electronics; applied research in magnetic-films applications and technology; VLSI technology. *Mailing Add:* Res Div Thomas J Watson Res Ctr IBM Corp PO Box 218 Yorktown Heights NY 10598

AHN, TAE IN, b Kyungbuk, Korea, Jan 15, 47; m 77; c 2. MEMBRANE BIOLOGY. *Educ:* Seoul Nat Univ, BS, 74; Univ Tenn, MS, 77, PhD(zool), 81. *Prof Exp:* Teacher biol, Ohryu Girl's Jr High Sch, 74-75; res & teaching asst cell biol, Univ Tenn, 75-81; asst prof, 81-87, ASSOC PROF CELL BIOL, SEOUL NAT UNIV, 87-, ASSOC DEAN ACAD AFFAIRS, COL EDUC, 90- *Concurrent Pos:* Gen secy, Korean Zool Soc, 83-87; ed, Korean J Genetics, 85-87; vis res prof cell biol, Univ Tenn, 87-88. *Mem:* Am Soc Cell Biol; Korean Soc Zool; Korean Soc Genetics. *Res:* Cellular and molecular biological studies on symbiotic amoeba; protein chemistry; actin cDNA cloning; XGroE cloning and characterization. *Mailing Add:* Dept Biol Educ Seoul Nat Univ Seoul 151-742 Republic of Korea

AHO, PAUL E, b Worcester, Mass, June 29, 34; m 57; c 3. PLANT PATHOLOGY. *Educ:* Univ Mass, BS, 56; Yale Univ, MF, 57; Ore State Univ, PhD(plant pathology), 76. *Prof Exp:* Plant pathologist, Forestry Sci Lab, Forest Serv, USDA, 57-; AT DEPT FOREST SCI, ORE STATE UNIV. *Mem:* Soc Am Foresters; Am Phytopath Soc. *Res:* Heart rots of western conifers; dwarf mistletoes; development of methods for estimating or predicting the extent of damage. *Mailing Add:* 223 NW 30th St Corvallis OR 97330

AHOKAS, ROBERT A, CARDIOVASCULAR PHYSIOLOGY, NUTRITION. *Educ:* Univ NDak, PhD(biol), 73. *Prof Exp:* ASSOC PROF PHYSIOL, UNIV TENN, 75- *Res:* Perinatal physiology. *Mailing Add:* Dept Obstet & Gynec Univ Tenn Univ Physicians 853 Jefferson Memphis TN 38103

AHR, WAYNE MERRILL, b San Antonio, Tex, Dec 11, 38; m 61; c 2. GEOCHEMISTRY OF CARBONATE ROCKS. *Educ:* Univ Tex, El Paso, BS, 60; Tex A&M Univ, MS, 65; Rice Univ, PhD(geol), 67. *Prof Exp:* Res geologist, Shell Develop Co Lab, 63-65; NASA fel geol, Rice Univ, 66-67; explor geologist, Shell Oil Co, 67-70; asst prof, 70-83, PROF GEOL, TEX A&M UNIV, 83- *Concurrent Pos:* Vis prof, La State Univ, 78; vis scholar, Leicester Univ, London, 84; consult, domestic & foreign petrol indust, 70-; Fulbright scholar, Cath Univ Louvain, Belg, 88. *Mem:* Am Asn Petrol Geologists; Soc Econ Paleontologists & Mineralogists; Int Asn Sedimentologists; Fulbright Asn. *Res:* Geology of lower carboniferous reefs of Europe and North America; diagenesis of carbonate rocks and carbonate sedimentology. *Mailing Add:* Dept Geol Tex A&M Univ College Station TX 77843

AHRENHOLZ, H(ERMAN) WILLIAM, b New York, NY, Nov 1, 16; m 39; c 1. MINING ENGINEERING. *Educ:* Lehigh Univ, BA & BS, 38, EM, 49. *Prof Exp:* Teaching fel, Mass Inst Technol, 38-39; mining engr, NJ Zinc Co, 39-49; assoc prof mining eng, WVa Univ, 49-54; mine supt, NJ Zinc Co, 54-57; prof, Univ Ala, 57-73, head, Sch Mines, 61-69, dir engr exten, 70-73; resident chief engr, Paul Weir Co, Zonguldak, Turkey, 73-75, vpres, Chicago, 75-82; RETIRED. *Concurrent Pos:* Consult, Ala Power Co, 57-65; consult & actg mine supt, Appalachian Sulphides, Inc, 59; consult, Int Paper Co, 64, Ga Marble Co, 64-67, Mining Res Div, US Bur Mines, 65, Mineral Resources Div, 65- & NJ Zinc Co, 67-70. *Mem:* Am Soc Eng Educ; Am Inst Mining, Metall & Petrol Engrs; Am Mining Cong. *Res:* Blasting, ground control and mining methods. *Mailing Add:* 4313 Stonehill Lane Tuscaloosa AL 35405

AHRENKIEL, RICHARD K, b Springfield, Ill, Jan 7, 36; m 56; c 1. SOLID STATE PHYSICS. *Educ:* Univ Ill, BS, 59, PhD(physics), 64. *Prof Exp:* Res physicist, Eastman Kodak Co, 64-70, res assoc, Res Labs, 70-77; mem staff, Los Alamos Nat Lab, 77-; AT SOLAR ENERGY RES INT. *Mem:* Am Phys Soc. *Res:* Transport phenomena in high impedance semiconductors, particularly the silver and alkali halides and studies of defect centers; optical properties of magnetic semiconductors. *Mailing Add:* Solar Energy Res Int 1617 Cole Blvd 16/3 Rm 305 Golden CO 80401

AHRENS, EDWARD HAMBLIN, JR, b Chicago, Ill, May 21, 15; m 40; c 3. METABOLISM. *Educ:* Harvard Univ, BS, 37, MD, 41. *Hon Degrees:* MD, Lund, Sweden, Univ Edinbourgh, Scotland. *Prof Exp:* Intern, Babies Hosp, 42-43, chief resident, 51-52; asst, Rockefeller Inst, 46-49; assoc, 52-58, from assoc prof to prof, 58-85, EMER PROF BIOCHEM & MED, ROCKEFELLER UNIV, 85- *Concurrent Pos:* Nat Res Coun sr fel, 49-50; Nat Found Infantile Paralysis sr fel, 50-52; mem,Metab Study Sect, USPHS, 56-61, chmn, 59-61; NSF sr fel, 58-59; founder, J Lipid Res, 58, mem adv bd, 58-74; mem bd sci counsr, Nat Heart Inst, 63-67, chmn diet-heart rev panel, 67-68; mem sci adv comt, New Eng Regional Primate Ctr, 63-69; mem tech adv comt, Inst Human Nutrit, Columbia Univ, 66-71; mem,Gen Clin Res Ctrs Comt, NIH, 70-74; master prof, Cornell Univ Med Col, 70-74; mem sci adv comt, Hirschl Trust, 72-74; mem bd dirs & exec comt, Regional Plan Asn, New York, 73-77 & 79-81; mem adv comt, Ernst Klenk Found, 75-85; mem, Assembly Life Sci, Nat Res Coun, 78-84; pres, Onteora Arboretum, 79-; mem bd dirs, NY Bot Garden, 81- *Honors & Awards:* Heart & Vascular Res Award, Mitchell Found Int, 68; McCollum Award, Am Soc Clin Nutrit, 69; 25th Ann Res Award, Am Heart Asn, 78. *Mem:* Nat Acad Sci; Am Soc Clin Invest; Asn Am Physicians; Am Soc Biol Chem. *Res:* Biochemistry of lipids and clinical investigation in field of lipid metabolism. *Mailing Add:* Rockefeller Univ York Ave & 66th St New York NY 10021

AHRENS, FRANKLIN ALFRED, b Leigh, Nebr, Apr 27, 36; m 60; c 4. PHARMACOLOGY, TOXICOLOGY. *Educ:* Kans State Univ, BS & DVM, 59; Cornell Univ, MS, 65, PhD(pharmacol), 68. *Prof Exp:* Chmn dept, 82-90, from asst prof to assoc prof, 68-75, PROF PHARMACOL, IOWA STATE UNIV, 75- *Mem:* NY Acad Sci; Sigma Xi. *Res:* Toxicity of chelating agents and lead; enteric disease and drug absorption. *Mailing Add:* Dept of Vet Physiol & Pharmacol Iowa State Univ Ames IA 50011

AHRENS, JOHN FREDERICK, b Bellmore, NY, Nov 21, 29; m 52; c 4. WEED SCIENCE, WOODY PLANTS. *Educ:* Univ Ga, BS, 54; Iowa State Univ, MS, 55, PhD(plant physiol), 57. *Prof Exp:* From asst plant physiologist to assoc plant physiologist, 57-70, PLANT PHYSIOLOGIST, VALLEY LAB, CONN AGR EXP STA, 70- *Concurrent Pos:* Veg mgt consult, pvt, 70-, L I Railroad, 76-89. *Mem:* Fel Weed Sci Soc Am; Int Plant Propagators Soc; Coun Agr Sci & Technol; Plant Growth Regulator Soc Am; Northeastern Weed Sci Soc (pres, 70); Weed Sci Soc Am (pres 88). *Res:* Weed control; residual effects of herbicides in soils; plant growth; inactivation of herbicide residues in soil. *Mailing Add:* Valley Lab Conn Agr Exp Sta PO Box 248 Windsor CT 06095

AHRENS, M(ILES) CONNER, b Everson, Wash, June 17, 22; m 48; c 3. AGRICULTURE, ENGINEERING. *Educ:* Wash State Univ, BS, 47, 48 & 54, MS, 52. *Prof Exp:* Test engr, Gen Elec Co, NY, 48-49; proj engr, Agr Eng Res Div, Wash, 49-59, asst br chief farm electrification br, 59-66, asst dir prog, planning & budgeting, 66-72, PROG ANAL, NE REGION, AGR RES SERV, USDA, 72- *Concurrent Pos:* Mem bd exam, US Civil Serv Comn, 61- *Mem:* Am Soc Agr Engrs; Inst Elec & Electronics Engrs. *Res:* Farm electrification; refrigeration and heat pumps; light and radiation applied to agriculture. *Mailing Add:* 1936 Kimberly Rd Silver Spring MD 20903

AHRENS, RICHARD AUGUST, b Manitowoc, Wis, Sept 18, 36; m 61; c 3. NUTRITION, PHYSIOLOGY. *Educ:* Univ Wis, BS, 58; Univ Calif, Davis, PhD(nutrit), 63. *Prof Exp:* Res physiologist human nutrit, Agr Res Serv, USDA, 63-66; assoc prof, 66-75, PROF NUTRIT, UNIV MD, COLLEGE PARK, 75-, COORDR, GRAD PROG NUTRIT SCI, 89- *Concurrent Pos:* Vis lectr nutrit, Univ London, U K, 73- *Honors & Awards:* Fulbright lectr, Egerton Univ, Kenya, 87-88. *Mem:* Am Inst Nutrit; Am Home Econ Asn; NY Acad Sci; Soc Nutrit Educ; Am Dietetic Asn. *Res:* Diet and atherosclerosis; enzymes and energy metabolism; dietary carbohydrate and the effects of physical activity on health; sucrose and high blood pressure; fermentation of weaning diets. *Mailing Add:* Dept Human Nutrit & Food Systs Univ Md College Park MD 20742-7521

AHRENS, ROLLAND WILLIAM, b Clarkson, Nebr, Sept 28, 33; m 63; c 2. PHYSICAL CHEMISTRY. *Educ:* Univ Nebr, BSc, 54, MS, 55; Univ Wis, PhD(phys chem), 59. *Prof Exp:* Chemist, Separations Div, Savannah River Lab, Dacron Qual Assurance, 59-64, chemist, Seaford Textile Fibers Plant, 65-67, anal res supvr, 67-77, sr res chemist, Spunbonded Res & Develop, 77-79, SR CHEMIST, DACRON QUAL ASSURANCE, E I DU PONT DE NEMOURS & CO, 79, FIBER SURFACES RES, KINGSTON, NC, 87- *Concurrent Pos:* Traveling lectr, Oak Ridge Inst Nuclear Studies, 63-64. *Mem:* Am Chem Soc. *Res:* Radiation chemistry of aqueous solutions and liquid ammonia; recovery of transplutonium elements by solvent extraction; nylon carpet new products development; polyester spunbonded product development; ion exchange. *Mailing Add:* 501 Bremerton Dr Greenville NC 27858

AHRENS, RUDOLF MARTIN (TINO), b St Louis, Mo, Feb 6, 28; m; c 3. THEORETICAL PHYSICS. *Educ:* Wash Univ, PhD, 52. *Prof Exp:* Sr nuclear engr, Convair, 52-54; staff specialist, Lockheed Aircraft Corp, 54-56, dept mgr nuclear anal, 56-57; assoc prof, 57-66, PROF THEORET PHYSICS, GA INST TECHNOL, 66- *Concurrent Pos:* Vis prof, Max-Planck Inst Physics & Astrophys, 59-60 & Univ SC, 67-68; consult & vpres, Advan Res Corp Ind, 62-71. *Res:* elementary particles; symmetries; field theory. *Mailing Add:* Dept Physics Ga Inst Technol Atlanta GA 30332

AHRENS, THOMAS J, b Wichita Falls, Tex, Apr 25, 36; m 58; c 3. GEOPHYSICS, HIGH-PRESSURE PHYSICS. *Educ:* Mass Inst Technol, BS, 57; Calif Inst Technol, MS, 58; Rensselaer Polytech Inst, 62. *Prof Exp:* Intermediate explor geophysicist, Pan Am Petrol Corp, 58-59; asst geophys, Rensselaer Polytech Inst, 62; geophysicist, Poulter Res Labs, Stanford Res Inst, 62-66, head geophys group, Shock Wave Physics Div, 66-67; assoc prof, 67-76, PROF GEOPHYS, CALIF INST TECHNOL, 76- *Concurrent Pos:* Assoc ed, Rev Sci Instruments & J Geophys Res, 71-74; mem earth sci adv panel, NSF, 72-75; ed, J Geophys Res, 79- *Mem:* AAAS; Am Geophys Union; Soc Explor Geophys; Am Phys Soc; fel Royal Astron Soc. *Res:* Shock and ultrasonic wave propagation in solids; exploration seismology; high-pressure physics of the earth's interior; impact effects on planetary surfaces; measurement of stress and tilt in the earth's crust. *Mailing Add:* Calif Inst Technol 252-21 Pasadena CA 91125

AHRING, ROBERT M, b Lincoln, Kans, Oct 4, 28; m 51; c 4. AGRONOMY, SOILS. *Educ:* Okla State Univ, BS, 52, MS, 58; Univ Nebr, PhD, 72. *Prof Exp:* Instr, 52-56, res asst, 56-57, asst prof, 57-75, ASSOC PROF AGRON, OKLA STATE UNIV, 75-; RES AGRONOMIST, FORAGE & RANGE, SOUTHERN REGION, SCI EDUC ADMIN, USDA, 57- *Mem:* Crop Sci Soc Am; Am Soc Agron; Am Soc Range Mgt; Int Herbage Seed Asn. *Res:* Seed production and technology; native and introduced grasses and legumes; crop physiology. *Mailing Add:* Dept Agronomy Okla State Univ Stillwater OK 74078

AHSANULLAH, MOHAMMAD, b Tangra, India; Can citizen; c 3. STATISTICAL INFERENCE, DISTRIBUTION THEORY. *Educ:* Calcutta Univ, BSc, 57, MSc, 59; NC State Univ, PhD(statist), 69. *Prof Exp:* Asst prof statist, NDak State Univ, 68-69; statistician, Nat Health & Welfare, 69-74 & 76-79; assoc prof, Univ Brasilia, 74-76 & 79-81; assoc prof, Temple Univ, 82-84; assoc prof statist, 84-87, PROF, RIDER COL, 87- *Concurrent Pos:* Syst analyst, InterAm Inst Coop Agr, Orgn Am States, 74-76; vis prof statist, Univ Nat Autonoma, Mex, 81-82. *Mem:* Int Statist Inst; fel Royal Statist Soc; Inst Math Statist; Am Statist Asn. *Res:* Theory of inferences; record value theory; order statistics; characterizations of statistical distributions; time series and reliability theory. *Mailing Add:* Dept Mgt Sci Rider Col 2083 Lawrenceville Rd Lawrenceville NJ 08648-3099

AHSHAPANEK, DON COLESTO, b Milton, Del, Apr 29, 32; m 62; c 3. PLANT ECOLOGY, MICROBIOLOGY. *Educ:* Cent State Univ, Okla, BS, 56; Univ Okla, MS, 59, PhD(bot), 62. *Prof Exp:* Asst prof biol, Kans State Teachers Col, 62-67, assoc prof, 67-71; instr biol, 71-73, chmn, Div Native Am Cult, 73-76, INSTR BIOL, HASKELL AM INDIAN JR COL, 76- *Concurrent Pos:* Mem comt equal opportunities sci & technol, NSF. *Mem:* Am Indian Sci & Eng Soc; Nat Indian Educ Asn; Nat Cong Am Indians. *Res:* Ecological research in causes of plant succession in grasslands with studies in phenology, plant inhibition and the inter-relations between micro-organisms and plant roots in the rhizosphere and rhizoplane; studies on halophilic bacteria. *Mailing Add:* Dept Biol Haskell Am Indian Jr Col PO Box H1305 Lawrence KS 66046

AHUJA, JAGAN N, b Rawalpindi, WPakistan, Nov 12, 35; m 60; c 3. BIOCHEMISTRY, CLINICAL CHEMISTRY. *Educ:* Agra Univ, BSc, 54, MSc, 56; Mich State Univ, PhD(biochem), 61. *Prof Exp:* NIH res asst, Mich State Univ, 58-61; instr biochem, Sch Med, Univ Miami, 61-64; reader, Maulana Azad Med Col, India, 66; NIH sr fel clin chem, Univ Hosp, Seattle, Wash, 66-68; CHIEF CLIN CHEM, KAISER FOUND HOSP, 68- *Mem:* AAAS; Am Asn Clin Chemists; Am Chem Soc. *Res:* Effects of cytotoxic alkylating agents on the metabolism of pyridine nucleotides in ascites tumor cells; techniques for the isolation and measurement of nucleotides; measurements of protein bound iodine and serum thyroxine. *Mailing Add:* 76 Moore Ct San Ramon CA 94583

AHUJA, JAGDISH C, b Rawalpindi, India; m 55; c 2. STATISTICS, MATHEMATICS. *Educ:* Banaras Hindu Univ, BA, 53, MA, 55; Univ BC, PhD(math), 63. *Prof Exp:* Teacher high sch, Nairobi, Kenya, 55-56; teacher math, dept educ, Tanganyika, 56-58; lab instr statist, dept econ, Univ BC, 59-61, lectr, 61-63; asst prof math, Univ Alta, 63-66; assoc prof, 66-69, PROF MATH, PORTLAND STATE UNIV, 69- *Concurrent Pos:* Teaching fel, Univ BC, 61-63; Can Math Cong fel, Res Inst, 64-66. *Mem:* Inst Math Statist. *Res:* Distribution theory; regression analysis; estimation and statistical inference. *Mailing Add:* Dept Math Portland State Univ Portland OR 97207

AHUJA, NARENDRA, b Achnera, India, Dec 8, 50; m 83; c 1. COMPUTER VISION, ROBOTICS. *Educ:* Birla Inst Technol & Sci, BE, 72; Indian Inst Sci, ME, 74; Univ Md, PhD(comput sci), 79. *Prof Exp:* Sci officer software develop, Dept Elec, Govt India, New Delhi, 74-75; from asst prof to assoc prof elec & comput eng, 79-88, PROF ELEC & COMPUT ENG, UNIV ILL, URBANA-CHAMPAIGN, 88- *Concurrent Pos:* Prin investr, Nat Sci Found, 81-, Air Force Off Sci Res, 82-; assoc ed, Comput Vision, Graphics & Image Processing, 87-; prog chair, Inst Elec & Electronics Engrs, Workshop Comput Vision, 87; prog comt Conf Comput Vision & Pattern Recognition, 88, mem, Tech Comt Pattern Anal & Mach Intel, 84-, Tech Comt Robotics, 84-; pres young investr award, Nat Sci Found, 84. *Mem:* Sr mem Inst Elec & Electronics Engrs; Asn Comput Mach; Am Asn Artificial Intel. *Res:* Computational realization of visual perception: three dimensional sensing, object recognition, automatic inspection, visual monitoring and autonomous navigation; visual pattern recognition, immage processing, robotic manipulation and computer architectures for vision. *Mailing Add:* Coord Sci Lab Univ Ill 1101 W Springfield Ave Urbana IL 61801

AHUJA, SATINDER, b Jehlum, Pakistan, Sept 11, 33; m 62; c 2. CHROMATOGRAPHY. *Educ:* Banaras Univ, BPharm, 55, MPharm, 56; Philadelphia Col Pharm, PhD(pharmaceut anal chem), 64. *Prof Exp:* Asst prof pharmaceut chem & pharm, Univ Nagpur, 57-58; teaching asst pharmaceut anal chem, Philadelphia Col Pharm, 58-64; assay develop chemist, Lederle Labs, NY, 64-66; res assoc pharmaceut anal chem, Geigy Chem Corp, Ardsley, 66-69, group leader, 69-73, sr staff scientist, 73-90, SR RES FEL, CIBA-GEIGY CORP, 91- *Mem:* Am Chem Soc. *Res:* Various modes of chromatography; kinetic studies; ultratrace methods; dosage variation and availability of drug products; automation and computerization; discovery of new compounds by chromatography; good manufacturing practice; good lab practice. *Mailing Add:* Ciba-Geigy Corp Old Mill Rd Suffern NY 10901

AHUMADA, ALBERT JIL, JR, b Los Angeles, Calif, Jan 12, 40; m 60; c 3. VISUAL PSYCHOPHYSICS. *Educ:* Stanford Univ, BS, 61; Univ Calif, Los Angeles, PhD(psychol), 67. *Prof Exp:* Asst prof psychol, Sch Soc Sci, Univ Calif, Irvine, 67-75; sr res assoc psycho-acoustics, dept aeronaut & astronaut, Stanford Univ, 75-80; RES PSYCHOLOGIST, AERO-SPACE HUMAN FACTORS, NASA AMES RES CTR, 80- *Concurrent Pos:* Instr, San Jose State Univ, 80-82. *Mem:* Optical Soc Am; Asn Res Vision & Ophthal; Soc Info Display; Int Soc Optical Eng; Soc Math Psychol. *Res:* Mathematical and computer modelling of perceptual processes. *Mailing Add:* NASA Ames Res Ctr Mail Stop 262-2 Moffett Field CA 94035

AICHELE, DOUGLAS B, b Flushing, NY, Nov 2, 42. GEOMETRY. *Educ:* Columbia Col, BA, 64, MA, 66, EdD(math), 69. *Prof Exp:* PROF MATH, OKLA STATE UNIV, 69- *Mem:* Math Asn Am. *Mailing Add:* 302 Gundersen Hall Okla State Univ Stillwater OK 74078

AICHELE, MURIT DEAN, b Freewater, Ore, July 17, 28; m 52, 88; c 1. DIAGNOSTICS, VARIETY EVALUATION. *Educ:* Wash State Col, BS, 52. *Prof Exp:* Plant pathologist, Wash State Dept Agr, 55-87; exec dir, Cent Wash Nursery Improvement Inst, 87-90; FRUIT TREE CONSULT, AICHELE CONSULTS, INC, 90- *Concurrent Pos:* Fruit tree consult, plant patents, Adjustment Serv. *Honors & Awards:* Nat Serv Award, Nat Asn State Depts Agr, 87. *Mem:* Am Phytopath Soc; Am Soc Hort Sci. *Res:* Horticulture; pomology; ornamental horticulture; fruit tree variety evaluation; fruit tree diseases; stone and pome fruit viruses. *Mailing Add:* Aichele Consults Inc 1420 S 25th Ave Yakima WA 98902

AICKIN, MIKEL G, b Seattle, Wash, Jan 17, 44; m 67; c 2. BIOMETRICS, BIOSTATISTICS. *Educ:* Univ Washington, BSc, 66, PhD(biomath), 76. *Prof Exp:* Asst prof statist, Ariz State Univ, 76-78, Univ Kans, 78-79; ASST PROF STATIST, ARIZ STATE UNIV, 79-, DEPT MATH. *Concurrent Pos:* Dir, Statist Consult Serv, 81- *Mem:* Inst Math Statist; Biomet Soc; Soc Clin Trials. *Res:* Statistical analysis, particularly for contingency tables, and simultaneous inference; adaptive design of clinical trials. *Mailing Add:* 333 E Glenhurst Dr Tucson AZ 85704-6660

AIELLO, ANNETTE, b Manhattan, NY, May 1, 41; m 81. TAXONOMIC BOTANY, BEHAVIORAL ENTOMOLOGY. *Educ:* Brooklyn Col, BA, 72; Harvard Univ, MA, 75, PhD(biol), 78. *Prof Exp:* Res assoc entom, Smithsonian Trop Res Inst, 78-80; res assoc entom, mus comparative zool, Harvard Univ, 80-82; AT SMITHSONIAN TROP RES INST. *Mem:* Asn Trop Biol; AAAS; Lepidopterists' Soc; Sigma Xi. *Res:* Plant taxonomy; arthropod behavior, especially insect behavior; insect life history. *Mailing Add:* Smithsonian Trop Res Inst APO Miami FL 34002-0011

AIELLO, EDWARD LAWRENCE, b Flushing, NY, June 12, 28; m 53; c 3. PHARMACOLOGY. *Educ:* St Peters Col, BS, 53; Columbia Univ, MA, 54, PhD(zool), 60. *Prof Exp:* Instr anat & physiol, Flint Jr Col, 57-58; instr pharmacol, NY Med Col, 58-60, from asst prof to assoc prof, 60-65; assoc prof, 65-73, chmn 78-81, PROF BIOL SCI, FORDHAM UNIV, 73- *Mem:* AAAS; Am Soc Zoologists; Am Soc Pharmacol & Exp Therapeut. *Res:* Invertebrate neuropharmacology; ciliary movement; mucociliary transport. *Mailing Add:* Dept of Biol Sci Fordham Univ Bronx NY 10458

AIKAWA, JERRY KAZUO, b Stockton, Calif, Aug 24, 21; m 44; c 1. CLINICAL MEDICINE. *Educ:* Univ Calif, AB, 42; Wake Forest Col, MD, 45. *Prof Exp:* From intern to asst resident, NC Baptist Hosp, 45-57; instr med, Wake Forest Col, 51-53; Am Heart Asn estab investr, 53-58; from asst prof to assoc prof med, Sch Med, 53-67, dir technol, 67-69, dir lab serv, 60-83, PROF MED, SCH MED, UNIV COLO MED CTR, DENVER, 67-, DIR ALLIED HEALTH PROG, 69-, ASSOC DEAN, 82- *Concurrent Pos:* Nat Res Coun fel, Med Sch, Univ Calif, 48-49; Nat Res Coun & AEC fel, Duke Univ & Bowman Gray Sch Med, Wake Forest Col, 49-51, Am Heart Asn res fel, Wake Forest Col, 51-53. *Res:* Cell physiology; electrolyte metabolism; computer applications in medicine; allied health training programs. *Mailing Add:* 6170 Montgomery Place San Jose CA 95135-1428

AIKAWA, MASAMICHI, b Japan, Sept 24, 31; m 68; c 2. PATHOLOGY, ELECTRON MICROSCOPY. *Educ:* Kyoto Univ, MD, 58, DSc, 65; Georgetown Univ, MS, 63. *Prof Exp:* Assoc pathologist, Walter Reed Army Inst Res, 64-68; asst prof, 68-71, assoc prof, 71-74, PROF PATH, CASE WESTERN RESERVE UNIV, 74- *Concurrent Pos:* Consult, Mt Sinai Hosp & Vet Admin Hosp, Cleveland, 77-; vis prof, Kyoto Univ, Japan, 80. *Mem:* Am Asn Pathologists; Am Soc Cell Biol; Am Soc Parasitol; Am Soc Trop Med Hyg; Royal Soc Med Hyg. *Res:* Interaction between host cells and parasites. *Mailing Add:* Inst Path Case Western Reserve Univ Cleveland OH 44106

AIKEN, JAMES G, b Midway, Pa, Jan 11, 40; m 65; c 2. ELECTRICAL ENGINEERING. *Educ:* Univ Cincinnati, BS, 63; Carnegie-Mellon Univ, MS, 64, PhD(elec eng), 67. *Prof Exp:* Mem tech staff semiconductors, Tex Instruments Inc, 66-84; PRES, IRVING CIRCUITBOARD CO, 84- *Res:* Effects of radiation on semiconductor devices and materials; electrical transport properties of semiconductors; ion implanted semiconductor devices; correlation between semiconductor processing and device parameters. *Mailing Add:* 5915 Bentwood Dr Richardson TX 75080

AIKEN, JAMES WAVELL, b Burlington, Vt, Jan 10, 43; m 65; c 2. PHARMACOLOGY, PHYSIOLOGY. *Educ:* Dartmouth Col, BA, 65; Univ Vt, PhD (pharmacol), 69. *Prof Exp:* Fel Am Thoracic Soc Pharmacol, Royal Col Surgeons Eng, 69-71; res scientist pharmacol, Eli Lilly & Co, 71-77; sr res scientist exp biol, 77-83, ASSOC DIR THERAPEUT, UPJOHN CO, 83- *Concurrent Pos:* Consult, Nat Heart, Lung & Blood Inst, 73-; adj asst prof med, Sch Med, Ind Univ, 75-77; adj prof pharmacol, Sch Med, Mich State Univ, 85-; bd trustees, Am Heart Asn, Mich. *Mem:* Am Soc Pharmacol & Exp Therapeut; Brit Pharmacol Soc; Am Heart Asn. *Res:* Physiological importance of neurohumoral substances in control of blood pressure, thrombosis, atherosclerosis, and neuro-endocrine function; manipulation of these systems with chemicals for therapeutic benefit. *Mailing Add:* Metabolic Dis Res Upjohn Co Kalamazoo MI 49001

AIKEN, LINDA H, b July 29, 43. MEDICINE. *Educ:* Univ Fla, BSN, 64, MN, 66; Univ Tex, Austin, PhD (sociol & demog), 74. *Prof Exp:* Nurse, J Hillis Miller Health Ctr, 64-65, instr, Col Nursing, Univ Gainesville, 66-67; instr, Sch Nursing, Univ Mo, Columbia, 68-70; clin nurse spec, Cardiac Surg, Univ Mo Med Ctr, 67-70; lectr, Univ Wis, Madison, 73-74; prog officer, The Robert Wood Johnson Found, 74-76, dir res, 76-79, asst vpres res, 79-81, vpres, The Robert Wood Johnson Foundation, 81-87; TRUSTEE PROF NURSING & SOCIOL & ASSOC DIR NURSING AFFAIRS, LEONARD DAVIS INST HEALTH ECON, UNIV PA, 88- *Concurrent Pos:* Fel, NIH, Univ Fla, 65-66; fel, NIH Nurse Scientist, Univ Tex, 70-73; vis lectr, Sci Human Affairs Prog, Princeton Univ, 75-87; Acad Sci Social Med, Hanover, Ger, 80; assoc ed, Jour Health & Social Behav, 78-81, Image, Jour Nursing Scholar, 80-; Transaction/Soc, 85-; mem, coun on nursing, Am Hosp Asn, 83-85; mem, Comt Sci, Eng & Pub Policy, Nat Acad Sci, 83-84; mem, Task Force on Long-Term Health Care Policies, US Dept Health & Human Servs, 86-87, Secy Comm on Nursing, 88; mem panel, Nat Health Care Surv, Comt Nat Statist, Nat Res Coun, Nat Acad Sci, 90- *Honors & Awards:* Jessie M Scott Award, Am Nurses' Asn, 84; Beverly Lectr, Asn Geront Higher Educ, 90. *Mem:* Inst Med-Nat Acad Sci; fel Am Acad Nursing (pres, 79-80); Am Sociol Asn; Am Nurses Asn; AAAS; Am Pub Health Asn. *Res:* Health policy and evaluation research; health services for persons with AIDS; author of over 100 articles, journals & books. *Mailing Add:* Univ Pa 420 Guardian Dr Philadelphia PA 19104-6096

AIKEN, ROBERT MCLEAN, b Springfield, Ill, May 24, 41; m 73; c 1. COMPUTER SCIENCE. *Educ:* Northwestern Univ, BS, 63, MS, 65, PhD (indust eng), 68. *Prof Exp:* from asst prof to accoc prof comput sci, Univ Tenn, Knoxville, 68-84; assoc prof, Dept Comput Sci, 84-87, PROF COMPUT SCI, TEMPLE UNIV, 87- *Concurrent Pos:* Vis prof, Univ Info Ctr, Univ Geneva, Switz, 75-76; Fulbright prof, Nat Inst Statist & Appl Econ, Rabat, Morocco, 82-83; US delegate, Int Fedn Info Processing, Tech Comt Educ, 87-, vchmn, 91-; vis prof, Inst Informatics, Univ Zurich, Switz, 90-91. *Mem:* Inst Elec & Electronic Engrs Comput Soc; Asn Comput Mach; Asn Develop Comput-Based Instrnl Systs. *Res:* Intelligent tutoring systems; effective uses of computers as teaching tools. *Mailing Add:* Dept Comput Sci 038-24 Temple Univ Philadelphia PA 19122

AIKENS, DAVID ANDREW, b Boston, Mass, Apr 27, 32; m 58; c 6. ELECTROANALYTICAL CHEMISTRY. *Educ:* Northeastern Univ, BS, 54; Mass Inst Technol, PhD, 60. *Prof Exp:* Fel, Univ NC, 60-62; from asst prof to assoc prof, 62-69, PROF CHEM, RENSSELAER POLYTECH INST, 69- *Mem:* Am Chem Soc. *Res:* Electrode reaction mechanisms; ion-selective electrodes; coordination chemistry. *Mailing Add:* Dept Chem Rensselaer Polytech Inst Troy NY 12180-3590

AIKIN, ARTHUR COLDREN, b Gettysburg, Pa, Dec 21, 32; m 64; c 2. ATMOSPHERIC PHYSICS, ASTRONOMY. *Educ:* Gettysburg Col, AB, 54; Pa State Univ, MS, 56, PhD (physics), 60. *Prof Exp:* Physicist, Serv d'Aeronomie du Ctr Nat Res Sci, Meudon, France, 60-61; ionospheric physicist, 61-70, head chemosphere br, 70-74, assigned US Senate staff, 75-76, SR SCIENTIST, GODDARD SPACE FLIGHT CTR, NASA, 76- *Mem:* Sigma Xi; AAAS; Am Geophys Union. *Res:* Physics of the ionosphere, planetary atmospheres cometary structure, atmospheric chemistry. *Mailing Add:* Goddard Space Flight Ctr Greenbelt MD 20741

AIKMAN, GEORGE CHRISTOPHER LAWRENCE, b Ottawa, Ont, Nov 11, 43; m 83, 90; c 3. ASTROPHYSICS. *Educ:* Bishop's Univ, BSc, 65; Univ Toronto, MSc, 68. *Prof Exp:* Res asst, 68-80, RES ASSOC ASTROPHYS, NAT RES COUN CAN, 80- *Mem:* Can Astron Soc (secy, 83-89); Am Astron Soc; Astron Soc Pac; Royal Astron Soc Can. *Res:* Origin of abundance anomalies in chemically peculiar stars; galactic structure; spectroscopic binary stars; comets. *Mailing Add:* Herzberg Inst Nat Res Coun Can 5071 W Saanich Rd Victoria BC V8X 4M6 Can

AILION, DAVID CHARLES, b London, Eng, Mar 21, 37; US citizen; m 64. SOLID STATE PHYSICS, NUCLEAR MAGNETIC RESONANCE. *Educ:* Oberlin Col, AB, 56; Univ Ill, MS, 58, PhD (physics), 64. *Prof Exp:* From asst prof to assoc prof, 64-78, PROF PHYSICS, UNIV UTAH, 78- *Concurrent Pos:* Prin investr, NSF grants & NIH grants; vis assoc, Calif Inst Technol, 71-72; NIH spec fel, 71-72; David P Garder fel, 80. *Mem:* Fel Am Phys Soc; Sigma Xi. *Res:* Biophysics; nuclear magnetic resonance imaging of lungs. *Mailing Add:* Dept Physics Univ Utah Salt Lake City UT 84112

AINBINDER, ZARAH, b Brooklyn, NY, July 12, 37; m 62; c 2. ORGANIC CHEMISTRY. *Educ:* Brooklyn Col, BS, 59; Univ Chicago, PhD (chem), 63. *Prof Exp:* Res chemist, 63-69, sr res chemist, 69-80, RES ASSOC, E I DU PONT DE NEMOURS & CO, INC 81- *Mem:* Am Chem Soc. *Res:* Polymer intermediates; catalysis; petro chemicals. *Mailing Add:* 1300 Grinnell Green Acres Wilmington DE 19803

AINES, PHILIP DEANE, b Lancaster, Pa, Mar 29, 25; m 51; c 2. NUTRITION. *Educ:* Rutgers Univ, BS, 49; Cornell Univ, MNS, 51, PhD (nutrit), 54. *Prof Exp:* Asst biochem, Cornell Univ, 53-54; res chemist, Procter & Gamble Co, 54-55; head prod res dept, Buckeye Cellulose Corp, 56-58, res chemist, Procter & Gamble Co, 59, Head basic develop, Food Prod Div, 60-65, from assoc dir to dir Food Prod Develop Div, 65-73; vpres res & eng, Pillsbury Co, 73-80. *Mem:* AAAS; Am Chem Soc; Sigma Xi. *Res:* Protein nutrition and metabolism; food research and development; animal nutrition. *Mailing Add:* 63905 Quail Haven Dr West Bend OR 97701

AINLEY, DAVID GEORGE, b Bridgeport, Conn, Apr 3, 46. MARINE ECOLOGY, ORNITHOLOGY. *Educ:* Dickinson Col, Pa, BS, 68; Johns Hopkins Univ, PhD (ecol), 71. *Prof Exp:* Res mem antarctic ecol, Johns Hopkins Univ, 68-70; BIOLOGIST MARINE ORNITH, POINT REYES BIRD OBSERV, BOLINAS, 71- *Concurrent Pos:* Polar Res Medal, US Antarctic Res Prog, 74-76. *Mem:* Am Ornithologists' Union; Cooper Ornith Soc; Wilson Ornith Soc. *Res:* Marine ecology of seabirds in California, the Gulf of California and Antarctica; trophic relationships of seabirds in marine communities. *Mailing Add:* 144 Overlook Dr Bolinas CA 94924

AINSLIE, HARRY ROBERT, b Hartwick, NY, Dec 2, 23; m 47; c 5. ANIMAL NUTRITION. *Educ:* Kans State Univ, BS, 49, MS, 50, PhD (animal nutrit), 65. *Prof Exp:* From asst prof to prof dairy husb, Cornell Univ, 50-83, exten leader, dept animal sci, 69-83; RETIRED. *Concurrent Pos:* Supt off testing, NY, 54-66; vis prof exten educ, Univ Philippines, 66-67; mem northeast subcomt, Nat Dairy Herd Improv Coord Group, 70. *Honors & Awards:* Delaval Exten Award, Am Dairy Sci Assoc, 79. *Mem:* Am Diary Sci Asn. *Res:* Dairy husbandry extension; educational programs in dairy cattle nutrition; breeding, management and production testing; organization of inservice training programs for extension personnel. *Mailing Add:* 30 Horizon Dr Ithaca NY 14850

AINSWORTH, CAMERON, b Alta, Can, Nov 13, 20; US citizen; M 48; c 1. ORGANIC CHEMISTRY, MEDICINE. *Educ:* Univ Alta, BS, 45; Univ Rochester, PhD (chem), 49; Univ Cd Juarez, MD, 81; Univ Hawaii, MPH, 83. *Prof Exp:* Instr chem, Univ Colo, 49-51; sr org chemist, Eli Lilly & Co, 51-66; assoc prof chem, Colo State Univ, 66-71; prof chem, San Francisco State Univ, 79-83. *Mem:* AAAS; Am Chem Soc; Sigma Xi; AMA. *Res:* Synthetic organic compounds of nitrogen related to the natural products histamine and serotonin; primary health care in developing countries. *Mailing Add:* 3879 Vineyard Dr Redwood City CA 94061

AINSWORTH, EARL JOHN, b Indianapolis, Ind, May 18, 33; m 60; c 3. TOXICOLOGY, RADIOBIOLOGY. *Educ:* Butler Univ, AB, 55; Brown Univ, ScM, 57, PhD (biol), 59. *Prof Exp:* Trainee, Nat Cancer Inst, 57-59; resident res assoc, Argonne Nat Lab, 59-61; sr investr cellular radiobiol br, US Naval Radiol Defense Lab, 61-64, actg head, 64-65, head mammalian radiobiol sect, 65-69, prin investr, 69; biologist, Div Biol & Med Res, 69-73, group leader, Neutron & Gamma Radiation Toxicity Prog, Argonne Nat Lab, 73-77, actg group leader radiol sci, 77; mgr Bevalac biomed progs, 77-86, actg dept dir, Biol & Med Div, 85-86, SR BIOPHYSICIST, RES MED & RADIATION BIOPHYS DIV, LAWRENCE BERKELEY LAB, 80- *Concurrent Pos:* Mem comt genetic stand, Nat Res Coun, 75; mem radiation study sect, NIH, 77-81, FREIR comt, 79-80 & comt on Space Res, Int Coun Sci Unions, 85; consult, Nat Coun Radiation Protection & Measurements, 83; guest scientist & Fogarty sr int fel, Soc Heavy Ion Res, WGermany, 84-85. *Honors & Awards:* Gold Medal Sci Achievement, US Naval Radiol Defense Lab, 69. *Mem:* Radiation Res Soc (secy-treas, 87); Sigma Xi; Int Soc Exp Hemat; Cell Kinetics Soc. *Res:* Radiation toxicology; life shortening neoplasia and late functional injury to the hematopoietic system. *Mailing Add:* Biol & Med Div Lawrence Berkeley Lab 74-303 One Cyclotron Rd Berkeley CA 94720

AINSWORTH, LOUIS, b Preston, Eng, June 4, 37; m 59; c 2. REPRODUCTIVE ENDOCRINOLOGY. *Educ:* Univ Leeds, BSc, 59; McGill Univ, MSc, 61, PhD (steroid biochem), 64. *Prof Exp:* Res assoc obstet & gynec, Case Western Reserve Univ, 64-65, res fel, 65-67, instr reproductive biol, 67-69; res scientist, 69-75, SR RES SCIENTIST, ANIMAL RES CTR, AGR CAN RES BR, CENT EXP FARM, 75- *Concurrent Pos:* Sr lectr, dept biochem, Univ Ottawa, 70- *Mem:* Am Soc Animal Sci; Soc Study Reproduction. *Res:* Mammalian reproduction; hormonal control of reproductive cycles; mechanism of ovulation; development of procedures for increasing reproductive efficiency in domestic animals. *Mailing Add:* Animal Res Centre Cent Exp Farm Ottawa ON K1A 0C6 Can

AINSWORTH, OSCAR RICHARD, b Vicksburg, Miss, July, 28, 22; m 47. MATHEMATICS. *Educ:* Univ Miss, BA & MA, 46; Univ Calif, PhD (math), 51. *Prof Exp:* From asst prof to assoc prof, 50-60, PROF MATH, UNIV ALA, 60-; MATHEMATICIAN, REDSTONE ARSENAL, NASA, 53- *Mem:* Am Math Soc; Sigma Xi. *Res:* Special functions; elasticity; missile guidance systems. *Mailing Add:* Box 870350 Tuscaloosa AL 35487-0350

AINSWORTH, STERLING K, b Meridian, Miss, Oct 23, 39. MEDICAL SCIENCE. *Educ:* Univ Miss, PhD, 69. *Prof Exp:* ASSOC PROF PATH, MED UNIV SC, 72- *Mailing Add:* Dept Path 1066 Ft Sumter Dr Charleston SC 29425

AIRD, STEVEN DOUGLAS, b Ann Arbor, Mich, June 12, 52; m 78; c 3. MOLECULAR BIOLOGY, BIOCHEMISTRY. *Educ:* Mont State Univ, BS, 74; Northern Ariz Univ, MS, 77; La State Univ, PhD (zool), 84. *Prof Exp:* Teaching asst plant taxon, Northern Ariz Univ, 74-75 & cell physiol, 75-76; res asst & teaching asst biochem, Med Ctr, La State Univ, 76-77; teaching asst herpet & cell biol, Colo State Univ, 78-79; res assoc, Dept Biochem, Univ Wyo, 83-84; postdoctoral res assoc, Dept Molecular Biol, 85-88; RES SCIENTIST, NATURAL PROD SCI, INC, 88- *Concurrent Pos:* Mem, Grants Herpet Comt, Soc Study Amphibians & Reptiles, 86- & Provisional Comt Toxin Nomenclature, Int Soc Toxinology, 90; lectr, Am Vet Med Asn, 87; tech consult, Colo Div Wildlife, 87; assoc researcher, Univ Católica de Goi0ls, Brazil. *Mem:* Int Soc Toxinol; Am Soc Biochem & Molecular Biol; Herpetologists' League; Soc Study Amphibians & Reptiles. *Res:* Isolation of toxins from spider, scorpion and snake venoms for bioassay and structural-functional studies; comparative biochemical studies of snake venoms; author of more than 50 technical publications. *Mailing Add:* Natural Prod Sci Inc 420 Chipeta Way Suite 240 Salt Lake City UT 84108

AIREE, SHAKTI KUMAR, b Hoshiarpur, India, Nov 12, 34; m 63; c 3. PHYSICAL CHEMISTRY, BIOCHEMISTRY. *Educ:* Panjab Univ, BSc, 56, MSc, 58; Okla State Univ, PhD(chem), 67. *Prof Exp:* From asst prof to assoc prof, 66-76, PROF CHEM & DIR, MUS/ARCHIVES, UNIV TENN, MARTIN, 76- *Concurrent Pos:* Co-ed, KIMAT-Chemex. *Mem:* Sigma Xi; Am Chem Soc; Nat Sci Teachers Asn. *Res:* Glucose determinations; effects of vitamins C & E. *Mailing Add:* Dept Chem Univ Tenn Martin TN 38238

AISEN, PHILIP, b New York, NY, Mar 28, 29; m 51; c 3. BIOCHEMISTRY, MEDICINE. *Educ:* Columbia Univ, AB, 49, MD, 53. *Prof Exp:* Assoc med, 60-63, from asst prof to assoc prof, 63-73, actg chmn dept biophys, 70-78, prof biophys & med, 73-81, PROF BIOPHYS PHYSIOL & MED, ALBERT EINSTEIN COL MED, 81- *Concurrent Pos:* Am Cancer Soc fel, 54-55; Nat Res Coun fel, 57-58; Guggenheim fel, Univ Gothenburg, 67-68; mem staff, Watson Lab, IBM Corp, 64-70; bioanalytical & metallobiochemistry study sect, NIH, 78-83, chmn, 82-83; Donders professorship, Univ Utrecht, Netherlands, 88. *Mem:* Am Soc Clin Invest; Asn Am Physicians; Am Soc Biol Chemists; NY Acad Sci; Am Fedn Clin Res. *Res:* Protein chemistry and function; physics of metal proteins; biophysics. *Mailing Add:* Dept of Biophys Albert Einstein Col of Med Bronx NY 10461

AISENBERG, ALAN CLIFFORD, b New York, NY, Dec 7, 26; m 52; c 2. ONCOLOGY. *Educ:* Harvard Univ, SB, 45, MD, 50; Univ Wis, PhD, 56. *Prof Exp:* Instr, 57-59, assoc, 59-62, asst prof, 62-69, assoc prof med, 69-84, PROF MED, HARVARD MED SCH, 84- *Concurrent Pos:* Clin asst, Mass Gen Hosp, 57-61, asst physician, 61-69, assoc physician, 69-81, physician, 81- *Mem:* Am Col Physicians; Am Asn Immunol; Am Soc Clin Oncol. *Res:* Malignant lymphoma; cellular immunology. *Mailing Add:* Dept Med Mass General Hosp, Fruit St Boston MA 02114

AISENBERG, SOL, b New York, NY, Aug 26, 28; m 56; c 3. PHYSICS. *Educ:* Brooklyn Col, BS, 51; Mass Inst Technol, PhD(physics), 57. *Prof Exp:* Res staff mem, Res Div, Raytheon Co, 57-61, sr res scientist, 61-64; sr scientist, Space Sci, Inc, 64-65, actg head physics dept, 65-69, vpres, 68-70, gen mgr, 69-70, GEN MGR, SPACE SCI DIV, WHITTAKER CORP, 70-, CHIEF RES ENGR, 74- *Mem:* AAAS; Am Phys Soc; Am Vacuum Soc; Am Inst Aeronaut & Astronaut; Asn Adv Med Instrumentation; Sigma Xi. *Res:* Plasma diagnostics; gaseous discharges; thin films; lasers; energy conversion; electron emission; plasma accelerators; ultra high vacuum; physical electronics; atmospheric physics; electro-optics; solid state; medical instrumentation; diagnostic instrumentation. *Mailing Add:* 36 Bradford Rd Natick MA 01760

AISSEN, MICHAEL ISRAEL, b Istanbul, Turkey, Jan 16, 21; nat US; m 44; c 3. MATHEMATICS. *Educ:* City Col New York, BS, 47; Stanford Univ, PhD(math), 51. *Prof Exp:* Instr & res assoc math, Univ Pa, 49-51; res scientist, Radiation Lab, Johns Hopkins Univ, 51-61; from assoc prof to prof math, Fordham Univ, 61-70; PROF MATH & CHMN DEPT, NEWARK CAMPUS, RUTGERS UNIV, 70- *Mem:* Am Math Soc; Math Asn Am; Soc Indust & Appl Math; NY Acad Sci; Sigma Xi; Math Assoc Am. *Res:* Zeroes of polynomials; completion processes; combinatorics. *Mailing Add:* Dept Math Naval Postgrad Sch Monterey CA 93943

AIST, JAMES ROBERT, b Cheverly, Md, Feb 20, 45; m 67; c 3. PLANT PATHOLOGY, PLANT CYTOLOGY. *Educ:* Univ Ark, BS, 66, MS, 68; Univ Wis, PhD(plant path), 71. *Prof Exp:* NATO fel, Swiss Fed Inst Technol, 71-72; asst prof, 72-78, ASSOC PROF PLANT PATH, CORNELL UNIV, 78- *Mem:* AAAS; Am Phytopath Soc; Am Soc Cell Biol. *Res:* Time course, experimental and cytochemical studies of a plant disease resistance response, wall apposition formation, during fungal attack; enzymatic degradation of plant cell walls; nuclear studies of plant pathogenic fungi. *Mailing Add:* Dept Plant Path Cornell Univ 334 Plant Sci Bldg Ithaca NY 14853

AITA, CAROLYN RUBIN, b Brooklyn, NY, Sept 7, 43; m 65; c 1. THIN FILM GROWTH, SURFACE SCIENCE. *Educ:* Brooklyn Col, City Univ NY, BA, 66; Utica Col, Syracuse Univ, BS, 70; Queens Col, City Univ NY, MA, 74; Northwestern Univ, PhD(mat sci), 77. *Prof Exp:* Sr staff mem, US Gypsum Co, 77-78; sr scientist, Elec & Electronic Res, Gould Labs, Gould, Inc, 78-81; from asst prof to assoc prof mat sci, 81-87, SCIENTIST, LAB SURFACE STUDIES, 81-, PROF MAT SCI, UNIV WIS-MIWAUKEE, 88- *Concurrent Pos:* Lectr, Queens Col, City Univ New York, 70-74. *Mem:* Am Vacuum Soc; Am Ceramic Soc; Materials Res Soc; Sigma Xi. *Res:* Growth of compound semiconductor and ceramic thin films near room temperature by reactive sputter deposition; in situ discharge diagnostics by mass and optical spectrometry to obtain fundamental plasma-film growth relationships and for process monitoring and control; understanding film behavior in terms of microstructure, atomic order, chemistry and interfacial effects; film characterization by x-ray diffraction, electron spectroscopy, electron microscopy, spectrophotometry, and electrical measurements; optics; ceramic engineering. *Mailing Add:* Mat Dept Col Eng & Appl Sci Univ Wis Milwaukee WI 53201

AITA, JOHN ANDREW, b Council Bluffs, Iowa, June 20, 14; m 37; c 2. NEUROLOGY, PSYCHIATRY. *Educ:* Univ Iowa, MD, 37; Univ Minn, PhD(psychiat), 44. *Prof Exp:* Intern, US Marine Hosp, Calif, 37-38; asst instr, Sch Med, Yale Univ, 38-40; instr, Mayo Found, Univ Minn, 40-44; from assoc prof to prof neurol & psychiat, Col Med, Univ Nebr Med Ctr, Omaha, 46-89; RETIRED. *Concurrent Pos:* Res physician & ment hygienist, Yale Univ, 38-40; first asst, Mayo Clin, 43-44. *Mem:* Am Psychiat Asn; Am Acad Neurol. *Res:* Brain injury. *Mailing Add:* 2302 N 55th St Omaha NE 68104

AITKEN, ALFRED H, b New York, NY, June 22, 25; m 50; c 1. ELECTROMAGNETISM. *Educ:* Lehigh Univ, BS, 49; Ind Univ, MS, 50, PhD(physics), 55. *Prof Exp:* Physicist, US Naval Res Lab, 54-85; PHYSICIST, CONSOLIDATED CONTROLS CORP, 85- *Concurrent Pos:* Lectr, Univ Col, Univ Md, 55-67. *Mem:* Am Phys Soc. *Mailing Add:* 7604 Elba Rd Alexandria VA 22306

AITKEN, DONALD W, JR, b Hilo, Hawaii, Mar 14, 36; m 58; c 2. PHYSICS, ASTROPHYSICS. *Educ:* Dartmouth Col, AB, 58; Stanford Univ, MS, 61, PhD(physics), 63. *Prof Exp:* Res assoc physics, Stanford Univ, 63-70; PROF ENVIRON STUDIES, SAN JOSE STATE UNIV. *Mem:* AAAS. *Res:* X-ray astronomy; interaction mechanisms of x-rays in solids; nucleon structure from high energy electron scattering experiments. *Mailing Add:* Dept of Environ Sci San Jose State Univ San Jose CA 95192

AITKEN, G(EORGE) J(OHN) M(URRAY), b Orangeville, Ont, Sept 26, 36; m 61; c 3. IMAGE & SIGNAL PROCESSING. *Educ:* Queen's Univ, Ont, BSc, 59, MSc, 61, PhD(elec eng), 65. *Prof Exp:* Lectr elec eng, Queen's Univ, Ont, 63-65; from asst prof to assoc prof elec eng, 66-76, PROF ELEC ENG, QUEEN'S UNIV, ONT, 76-, HEAD DEPT, 86- *Concurrent Pos:* NATO fel, 65-66; foreign investr, radio astron, Observ Paris, France, 65-66 & 72-73; infrared interferometry, CERGA, Grasse, 79-80; France-Canada exchange fel, 79-80. *Mem:* Inst Elec & Electronics Engrs; Can Asn Physicists; Optical Soc Am; Soc Photo-Optical Instrumentation Engrs; Can Asn Univ Teachers; Can Astron Soc. *Res:* high angular resolution imaging in astronomy; interferometry; optical phase measurement and retrieval; 3-D optical imaging; image restoration. *Mailing Add:* Dept Elec Eng Queen's Univ Kingston ON K7L 3N6 Can

AITKEN, JAMES HENRY, b Glasgow, Scotland, Mar, 25, 27; Can citizen; m 55; c 3. NUCLEAR PHYSICS. *Educ:* Univ Edinburgh, BSc, 52; Univ Toronto, PhD, 69. *Prof Exp:* Asst lectr physics, Univ Edinburgh, 53-56; physicist, Comput Devices Can, Ltd, 56-57; physicist res off, Nat Res Coun Can, 57-65; res assoc, Univ Toronto, 65-70; analyst, Ont Hydro, Ont Govt, 70-72, chief health physics, 72-88; ADJ PROF, UNIV TORONTO, 75- *Mem:* Can Asn Physicists. *Res:* Methods of nuclear radiation detection and measurement; reactor physics; reactor safety; health physics. *Mailing Add:* 137 Santa Barbara Rd Willowdale ON M2N 2C6 Can

AITKEN, JANET M, b Millers Hall, Mass, Dec 23, 16. GEOPHYSICS. *Educ:* Johns Hopkins Univ, PhD(geol), 48. *Prof Exp:* Prof geol, Univ Conn, 41-71; RETIRED. *Mailing Add:* PO Box 17 Storrs CT 06268

AITKEN, JOHN MALCOLM, b Staten Island, NY, Jan 1, 45; m 68; c 3. SOLID STATE PHYSICS. *Educ:* Fordham Univ, BS, 66; Rensselaer Polytech Inst, MS & PhD(physics), 73. *Prof Exp:* Assoc solid state physics, Rensselaer Polytech Inst, 73-74; RES STAFF MEM, T J WATSON RES CTR, IBM CORP, 74- *Mem:* Electrochemical Soc; Inst Elec & Electronics Engrs. *Res:* Metal-oxide-semiconductor device physics; electron and hole trapping in silicon dioxide; radiation damage in metal-oxide-semiconductor circuits and devices; hot-election effects in metal-oxide-semiconductor field-effect transistors; device measurements and characterization. *Mailing Add:* 63 Boulder Trail Bronxville NY 10708

AITKEN, THOMAS HENRY GARDINER, b Porterville, Calif, Aug 31, 12; m 48; c 2. MEDICAL ENTOMOLOGY, PARASITOLOGY. *Educ:* Univ Calif, BS, 35, PhD(med entom), 40. *Prof Exp:* Asst entom & parasitol, Univ Calif, 36-40; staff mem, Rockefeller Found, 46-74; sr res scientist, 71-83, EMER SR RES SCIENTIST EPIDEMIOL, YALE ARBOVIRUS RES UNIT, SCH MED, YALE UNIV, 83- *Concurrent Pos:* Consult, US Pub Health Serv, Ctr Dis Control, Arbovirus Ecol Br, Ft Collins Co, 74, Gorgas Mem Lab, Panama, 76, Int Develop Res Ctr, Can, yellow fever Trinidad, 81-82. *Honors & Awards:* Medal, USA Typhus Comn, 44; Gold Medal, Govt Sardinia, 51; Gilt Medal, Hort Soc Trinidad & Tobago, 74; Richard M Taylor Medal, Am Comt Arthropod-borne Viruses, 84. *Mem:* Am Soc Trop Med & Hyg; Am Mosquito Control Asn; Entom Soc Am. *Res:* Bloodsucking diptera, particularly mosquitoes; malaria and arthropod-borne viruses; experimental infection transmission. *Mailing Add:* Yale Arbovirus Res Unit Med Sch Yale Univ New Haven CT 06510

AIVAZIAN, GARABED HAGPOP, b Turkey, Dec 11, 12; US citizen; m 39; c 2. PSYCHIATRY. *Educ:* Am Univ Beirut, MD, 35; Am Bd Psychiat & Neurol, cert psychiat, 59. *Prof Exp:* Intern internal med, Am Univ Beirut Hosp, 35-37; resident psychiat, Am Univ Beirut & Lebanon Hosp Ment & Nerv Dis, 38-41; clin asst, Am Univ Beirut, 41-52, clin asst prof, 52-54; from asst prof to prof psychiat, 54-79, actg chmn to chmn dept, 63-74, dir residency training prog, 61-79, EMER PROF PSYCHIAT, COL MED, UNIV TENN, MEMPHIS, 79- *Concurrent Pos:* Consult, Am Univ Beirut Hosp, 41-54, John Gaston Hosp, Memphis, Tenn, 55-74, Gailor Ment Health Ctr, 56-75, Memphis Ment Health Inst, 62-75 & Methodist Hosp, 74-; Rockefeller Found fel psychiat, Col Med, Univ Tenn, Payne Whitney Clin & Med Col, Cornell Univ, 48; dir continuing psychiat educ, Univ Tenn Ctr Health Sci, 78-79. *Mem:* AMA; fel Am Psychiat Asn; Sigma Xi. *Res:* Medical education and psychopharmaco-therapy; clinical evaluation of psychotropic drugs. *Mailing Add:* 92 E Charlotte Circle Memphis TN 38117

AIZLEY, PAUL, b Boston, Mass, Feb 16, 36; m 69; c 3. MATHEMATICS. *Educ:* Harvard Univ, AB, 57; Univ Ariz, MS, 59; Ariz State Univ, PhD(math), 69. *Prof Exp:* Instr math, Tufts Univ, 59-61; vis asst prof, Ariz State Univ, 67-68; from asst prof to assoc prof 68-80, admin asst to pres, 74-79, dir summer session & mini term, 76-79, PROF MATH, UNIV NEV, LAS VEGAS, 80- *Mem:* Am Math Soc; Math Asn Am. *Res:* Convolution algebras, number theory. *Mailing Add:* Extended Educ Univ Nev Las Vegas NV 89154

AJAMI, ALFRED MICHEL, b Caracas, Venezuela, May 18, 48; US citizen; m 79. STABLE ISOTOPE CHEMISTRY. *Educ:* Harvard Univ, AB, 70, AM, 72, PhD(biol), 73. *Prof Exp:* Vpres, KOR, Inc, 71-; AT TRACER TECHNOLOGIES INC. *Concurrent Pos:* Dir res, KOR Isotopes; consult, Environ Protection Agency, 74-75 & NSF, 76. *Mem:* NY Acad Sci; Am Chem Soc; Am Asn Clin Chemists; AAAS; Sigma Xi. *Res:* Chemical and biochemical studies on pharmaceuticals and environmental chemical labeled with stable isotopes; non-radioactive, non-destructive, non-invasive isotopic tracer systems for the health care, chemical process and performance materials industries. *Mailing Add:* Tracer Technologies Inc 20 Assembly Square Dr Somerville MA 02145

AJAX, ERNEST THEODORE, b Salt Lake City, Utah, Oct 11, 26; m 50; c 4. CLINICAL NEUROLOGY. *Educ:* Univ Utah, BS, 49, MD, 51. *Prof Exp:* Instr psychiat, 58-65, from instr to assoc prof neurol, 59-71, instr med, 61-69, asst prof, 65-72, ASSOC PROF PSYCHIAT, COL MED, UNIV UTAH, 72-PROF NEUROL, 71-, ASST NEUROLOGIST, UNIV HOSP, 65- *Concurrent Pos:* Asst chief neurol serv, Vet Admin Hosp, Salt Lake City, 57-62, chief, 62-; assoc neurologist, Salt Lake County Gen Hosp, 59-65; neurologist, Univ Utah Med Ctr. *Mem:* Am Acad Neurol; Am Electroencephalog Soc. *Res:* Electroencephalography; disorders of language. *Mailing Add:* Dept Neurol Univ of Utah Salt Lake City UT 84112

AJELLO, LIBERO, b New York, NY, Jan 19, 16; m 42; c 1. MEDICAL MYCOLOGY. *Educ:* Columbia Univ, AB, 38, MA, 40, PhD(mycol), 48. *Prof Exp:* Med mycologist, Off Sci Res & Develop, Ft Benning, Ga, 42-43 & Johns Hopkins Hosp, 43-46; scientist dir chg Med Mycol Div, Ctr Infectious Dis, Ctrs Dis Control, USPHS, 47-67, dir, Div Mycotic Dis, 67-90; PRIN INVEST HEAD, WHO COLL CTR MYC DIS, 79- *Concurrent Pos:* Burroughs-Wellcome vis prof microbiol, 85. *Mem:* AAAS; Am Soc Microbiol; Mycol Soc Am; Brit Mycol Soc; Int Soc Human & Animal Mycol. *Res:* Fungi pathogenic to man. *Mailing Add:* Emory Univ Eye Ctr Ophthalmic Res 3704 S 1327 Clifton Rd NE Atlanta GA 30322

AJEMIAN, MARTIN, b Harpoot, Armenia, May 15, 07; nat US; m 46; c 1. ANATOMY. *Educ:* Boston Univ, BS, 37; Univ Ark, MA, 51; Georgetown Univ, PhD(anat), 53. *Prof Exp:* Teacher, Pub Sch, Mass, 41-42; instr biol, Ark Polytech Col, 48-50; instr anat, Sch Med & Dent Sch, Georgetown Univ, 50-53; prof & head dept, South Col Optom, 53-54; prof biol, Ark State Univ, 54-55; assoc prof anat, Univ Miss, 55-74; RETIRED. *Res:* Degeneration and regeneration of peripheral nerves. *Mailing Add:* Rte 3 Box 199B Carthage MS 39051

AJERSCH, FRANK, b Poland, Feb 7, 41; Can citizen; m 69; c 2. METALLURGICAL ENGINEERING. *Educ:* McGill Univ, BE, 63; Univ Toronto, MASc, 68, PhD(metall eng), 71. *Prof Exp:* Res engr process metall, Noranda Res Ctr, 63-67; fel thermodynamics, Univ Grenoble, France, 71-72; res assoc chem metall, 72-73, from assoc prof to prof metall kinetics, 73-89, CHMN, DEPT METALL ENG, ECOLE POLYTECHNIQUE, MONTREAL, CAN, 89- *Mem:* Can Soc Chem Eng; Metall Soc; Can Inst Mining & Metall; Am Inst Mining & Metall; Sigma Xi. *Res:* Kinetics of metallurgical processes; transport phenomena; gaseous reduction of porous oxides; gas-solid and gas-liquid reactions applied to metallurgy. *Mailing Add:* Ecole Polytech CP 6079 Sta A Montreal PQ H3C 3A7 Can

AJL, SAMUEL JACOB, b Poland, Nov 15, 23; nat US; m 46; c 3. MICROBIOLOGY. *Educ:* Brooklyn Col, BA, 45; Iowa State Col, PhD(physiol bact), 49; Dropsie Univ, DHL, 68. *Prof Exp:* Asst prof bact & immunol, Sch Med, Wash Univ, 49-52; chief microbiol chem sect, dept bact, Army Med Serv Grad Sch, Walter Reed Army Med Ctr, 52-54, asst chief, 54-58; prog dir metab biol, NSF, 59-60; dir res, Albert Einstein Med Ctr, 60-73; VPRES RES, NAT FOUND MARCH DIMES, 73- *Concurrent Pos:* NSF spec fel, Hebrew Univ, Jerusalem & Oxford Univ, Eng, 68; consult, NIH & NSF; prof biol & microbiol, Sch Med, Temple Univ, 60-73. *Mem:* Fel AAAS; Am Soc Biol Chemists; Am Soc Microbiol; Am Acad Microbiol; fel NY Acad Sci. *Res:* Respiratory mechanisms of microorganisms; energy transfer reactions; microbial toxins; purification, mechanism of action and immunology of microbial toxins. *Mailing Add:* March of Dimes Birth Defects Found 1275 Mamaroneck Ave White Plains NY 10605

AJMONE-MARSAN, COSIMO, b Cossato, Italy, Jan 2, 18; nat US; m 43; c 2. NEUROPHYSIOLOGY, NEUROSCIENCES. *Educ:* Univ Turin, MD, 42. *Prof Exp:* Asst neurol, Clin Nerv Dis, Italy, 41-47, chief EEG & neurophysiol, 48-49; Rockefeller fel, Montreal Neurol Inst, McGill Univ, 50-51, lectr, Inst, 52-53; chief neurosci br, Nat Inst Neurol Communicative Dis & Stroke, 54-; AT DEPT NEUROL, JACKSON MEM MED CTR, MIAMI, FLA. *Concurrent Pos:* Clin instr, George Washington Univ, 55-58, assoc, 58-64, asst prof neurol, 65-73, assoc prof, 74-; chief ed, EEG J, 59-69; chief ed, J Clin Neurophysiol, 83-89. *Honors & Awards:* Lennox Award; H Jasper Award. *Mem:* Am Epilepsy Soc (pres, 73); Am Acad Neurol; Am Physiol Soc; Int Brain Res Orgn; Am EEG Soc (pres, 62-63); Int Fedn Soc EEG Clin Neurophysiol (pres, 69-72). *Res:* Electroencephalography; physiology of cerebral cortex and thalamus; clinical electroencephalography; epilepsy; clinical neurophysiology. *Mailing Add:* Chief EEG Lab Univ Miami Sch Med 1611 NW 12th Ave Miami FL 33136

AJZENBERG-SELOVE, FAY, b Berlin, Ger, Feb 13, 26; m 55. NUCLEAR PHYSICS. *Educ:* Univ Mich, BSE, 46; Univ Wis, MS, 49, PhD(physics), 52. *Prof Exp:* From asst prof to assoc prof physics, Boston Univ, 52-57; from assoc prof to prof, Haverford Col, 57-70, actg chmn dept, 60-61 & 67-69, comnr, Comn Col Physics, 68-70; res prof, 70-73, PROF PHYSICS, UNIV PA, 73-, ASSOC CHAIR, 89- *Concurrent Pos:* Smith-Mundt fel, US Dept State, 55; vis asst prof, Columbia Univ, 55; vis prof, Nat Univ Mex, 55; vis assoc physicist, Brookhaven Nat Lab, 56; lectr, Univ Pa, 57; Guggenheim fel, Lawrence Radiation Lab, Univ Calif, 65-66; consult, Calif Inst Technol, 70-72, exec secy comt physics in cols, Am Inst Physics, 62-65, mem adv comt vis scientist prog, 63-65 & adv comt manpower, 64-68; mem exec comt, Div Nuclear Physics, Am Phys Soc, 70-75, chmn, 73-74; exec secy ad hoc panel nuclear data compilation, Nat Acad Sci-Nat Res Coun, 71-75; mem comn nuclear physics, Int Union Pure & Appl Physics, 72-81, chmn, 78-81; Sigma Xi Nat lectr, 74-75; mem, Nuclear Sci Adv Comt, Dept Energy, NSF, 77-80; mem, Comt Pub Educ & Info, Am Inst Physics, 80-83. *Mem:* Fel AAAS; fel Am Phys Soc. *Res:* Neutron spectra; nuclear structure. *Mailing Add:* Dept Physics Univ Pa Philadelphia PA 19104-6396

AKAGI, JAMES MASUJI, b Seattle, Wash, Dec 23, 27; wid; c 2. MICROBIAL PHYSIOLOGY. *Educ:* Univ Ill, BS, 51; Univ Kans, MA, 55, PhD, 59. *Prof Exp:* Instr bact, Univ Kans, 58-59; fel microbiol, Sch Med, Western Reserve Univ, 59-61; from asst prof to assoc prof, 61-64, chmn dept, 76-85, PROF MICROBIOL, UNIV KANS, 67- *Concurrent Pos:* Res career develop award, 67-76. *Mem:* Am Soc Microbiol; Am Soc Biol Chemists; Am Acad Microbiol. *Res:* Biochemistry. *Mailing Add:* Dept Microbiol 7042 Haworth Hall Univ Kans Lawrence KS 66045-2103

AKAMINE, ERNEST KISEI, b Hilo, Hawaii, May 10, 12; m 38; c 3. POST HARVEST PHYSIOLOGY. *Educ:* Univ Hawaii, BS, 35, MS, 41. *Prof Exp:* Asst agr, 35-36, agron, 36-38, plant physiol, 38-42, assoc plant physiol, 42-44, jr physiologist, 44-52, asst plant physiologist, 52-62, assoc plant physiologist & assoc prof plant physiol, 62-68, plant physiologist & prof plant physiol, 68-77, EMER PLANT PHYSIOLOGIST & EMER PROF PLANT PHYSIOL, UNIV HAWAII, 77- *Concurrent Pos:* Mem sci invests, Ryukyu Islands Prog, Nat Res Coun, 52; cousult, Nat Lexicog Bd, Ltd, 53-; tech consult, New Wonder World Encycl, 57; vis consult, US Civil Admin Ryukyu Islands, US Dept Army, 62; vis scientist, Commonwealth Sci & Indust Res Orgn, Australia, 74-75. *Mem:* Am Soc Plant Physiol; Am Soc Hort Sci. *Res:* Commodity shipment studies; postharvest physiology. *Mailing Add:* 2255 Hulali Pl Honolulu HI 96819

AKASAKI, TAKEO, b Long Beach, Calif, Dec 19, 36; m 73; c 2. ALGEBRAIC TOPOLOGY. *Educ:* Univ Calif, Los Angeles, BA, 58, MA, 61, PhD(math), 64. *Prof Exp:* Actg asst prof math & asst res mathematician, Air Force Off Sci Res grant, Univ Calif, Los Angeles, 64; asst prof math, Rutgers Univ, 64-66; asst prof, 66-72, ASSOC PROF MATH, UNIV CALIF, IRVINE, 72- *Mem:* Am Math Soc. *Res:* Integral group rings; projective modules/semi local rings. *Mailing Add:* Dept of Math Univ of Calif Irvine CA 92664

AKASOFU, SYUN-ICHI, b Nagano-ken, Japan, Dec 4, 30; m 61; c 2. AERONOMY. *Educ:* Tohoku Univ, BS, 53, MS, 57; Univ Alaska, PhD(geophys), 61. *Prof Exp:* Asst geophys, Nagasaki Univ, 53-55; asst geophys, 58-61, res geophysicist, 61-62, from assoc prof to prof, 62-86, DIR, GEOPHYS INST, UNIV ALASKA, FAIRBANKS, 86- *Concurrent Pos:* Assoc ed, J Geomagnetism & Geoelec, 72 & J Geophys Res, 72-74. *Honors & Awards:* Chapman Medal, Royal Astron Soc, Eng, 76; Japan Acad Award, 77; John Adam Fleming Medal, Am Geophys Union, 79; Spec lectr for Emperor Hirohito of Japan, 85. *Mem:* Fel Am Geophys Union; Soc Terrestrial Magnetism & Elec Japan; Int Acad Aeronaut; fel Arctic Inst N Am. *Res:* Geomagnetic storms; aurora polaris; physics of the magnetosphere; solar-terrestrial relationships. *Mailing Add:* Geophys Inst Univ Alaska Fairbanks AK 99775-0800

AKAWIE, RICHARD ISIDORE, b New York, NY, June 11, 23; m 45; c 3. ORGANIC CHEMISTRY, POLYMER CHEMISTRY. *Educ:* Univ Calif, Los Angeles, AB, 42, AM, 43, PhD(chem), 47. *Prof Exp:* Asst chem, Univ Calif, Los Angeles, 42-44, 46-47; org res chemist, Gasparcolor, Inc, 47-49; org chemist, US Vet Admin, 49-56; sr chemist, Atomics Int, 56-61; mem tech staff, 61-77, tech group leader, 77-79, sr staff chemist, Hughes Aircraft Co, 79-87; RETIRED. *Mem:* Am Chem Soc; Sigma Xi. *Res:* Synthesis of labeled compounds; pharmaceuticals; organic dyes; Grignard reagents; antimalarials; radiation chemistry; silicones; fluorinated compounds; heat-resistant polymers; thermal control coatings. *Mailing Add:* 12301 Deerbrook Lane Los Angeles CA 90049

AKAY, ADNAN, b Turkey. NOISE CONTROL. *Educ:* NC State Univ, BS, 71, MME, 72, PhD(mech eng), 76. *Prof Exp:* Res asst mech eng, NC State Univ, 72-75; res fel, Nat Inst Environ Sci, 76-78; asst prof to assoc prof, 78-86, PROF MECH ENG, WAYNE STATE UNIV, 86- *Concurrent Pos:* Prin investr, NSF grants, 80-; consult, Bendix Corp, 81- *Mem:* Am Soc Mech Engrs; Acoustical Soc Am; Soc Mfg Engrs. *Res:* Analytical and experimental investigation of sound radiation from mechanical systems; friction induced vibration and noise; impact mechanics and impact induced sound and vibration; dynamics of high-speed machinery. *Mailing Add:* Mech Eng Dept Wayne State Univ Detroit MI 48202

AKBAR, HUZOOR, HEMATOLOGY, BIOCHEMISTRY. *Educ:* Australian Nat Univ, PhD(clin sci), 78. *Prof Exp:* ASST PROF PHARMACOL, OHIO UNIV, 81- *Mailing Add:* Dept Zool & Biomed Sci Ohio Univ Col Osteop Med Irvine Hall Athens OH 45701

AKCASU, AHMET ZIYAEDDIN, b Aydin, Turkey, Aug 26, 24; m 54; c 3. NUCLEAR ENGINEERING, STATISTICAL MECHANICS. *Educ:* Tech Univ Istanbul, MS, 48; Int Sch Nuclear Sci & Eng, Argonne Nat Lab, dipl, 57; Univ Mich, PhD(nuclear eng), 63. *Prof Exp:* Asst prof electronics, Tech Univ Istanbul, 48-53, assoc prof, 54-59; resident res assoc nuclear eng, Argonne Nat Lab, 59-61; from asst prof to assoc prof, 63-68, PROF NUCLEAR ENG, UNIV MICH, ANN ARBOR, 68- *Concurrent Pos:* Vis res scientist, Brookhaven Nat Lab, Upton, NY, 65-69; vis prof, Nagoya Univ, Japan, 74, Argonne Nat Lab, Ill, 75, Inst Phys Sci & Technol, Univ Md, 77-78 & Univ Konstanz, Ger, 86; physicist, Ctr Mat Res Polymer Sci & Standards Div, Nat Bur Standards, Wash, DC, 78-84, Nat Inst Standards & Technol, Gaithersburg, Md, 89; vis scientist, Univ Freiburg, Germany, 84, Univ Strasbourg, France & Nat Tsing-Hua Un iv, Taiwan, 85; res scientist, Alcoa Labs, Pa, 87; guest prof, Physics Dept, Univ Konstanz, Ger, 90; Humboldt res award, sr US scientist, 91. *Honors & Awards:* Class of 38E Distinguished Serv Award, Univ Mich, 65. *Mem:* Am Phys Soc; fel Am Nuclear Soc; Turkish Phys Soc. *Res:* Nonlinear reactor dynamics; reactor noise analysis; correlation analysis in liquids and calculation of transport coefficients; kinetic theory of liquids and plasmas; polymer solution dynamics; interpretation of dynamic scattering from dense fluids and polymer solutions; stochastic differential equations. *Mailing Add:* 2820 Pebble Creek Univ Mich North Campus Ann Arbor MI 48108

AKE, THOMAS BELLIS, III, b Woodbury, NJ, Mar 15, 49; m 69; c 3. ASTRONOMY. *Educ:* Case Western Reserve Univ, BS, 71, MS, 74, PhD(astron), 77. *Prof Exp:* Res fel astron, Calif Inst Technol, 78-80; mem staff, Goddard Space Flight Ctr, NASA, 80-; AT SPACE TELESCOPE SCI INST. *Concurrent Pos:* Staff mem, Hale Observ, 78- *Mem:* Am Astron Soc. *Res:* Stellar spectroscopy. *Mailing Add:* 13311 Yarland Lane Bowie MD 21218

AKELEY, DAVID FRANCIS, b Presque Isle, Maine, Apr 13, 28; m 56; c 2. POLYMER CHEMISTRY. *Educ:* Univ Maine, BS, 49; Univ Wis, PhD(chem), 53. *Prof Exp:* Res chemist, Textile Fibers Dept, E I du Pont de Nemours & Co, Inc, 54-63, patent chemist, 63-89; RETIRED. *Res:* Patent law; textile fibers. *Mailing Add:* 1607 Shadybrook Rd Wilmington DE 19803

AKELL, ROBERT B(ERRY), b Boston, Mass, May 14, 21; m 88; c 3. CHEMICAL ENGINEERING. *Educ:* Northeastern Univ, BS, 42; Univ Louisville, MChE, 46. *Prof Exp:* Asst, Polaroid Corp, 39-42; prod supvr, Joseph E Seagrams & Sons, Inc, 42-44, asst maintenance supt, 44-45, dir educ, 45-46; asst dir personnel, Calvert Distilling Co, 46-48, res & develop supvr, 48-50, chief chemist, 50-56; sr engr, E I Du Pont De Nemours, 56-62, consult, 62-68, sr consult, Dept Eng, 68-78, prin consult, 78-82; CONSULT, 82- *Concurrent Pos:* Adj prof, Univ Del, 79-82. *Mem:* Fel Am Inst Chem Engrs. *Res:* Mass transfer; transport properties and phenomena; phase equilibria; liquid-liquid extraction; adsorption. *Mailing Add:* 1506 Fresno Rd Wilmington DE 19803-5124

AKELLA, JAGANNADHAM, b Kakinada, Andhra, India, Sept 13, 37; US citizen; m 67; c 3. GEOCHEMISTRY, MATERIALS SCIENCES. *Educ:* Andhra Univ, Waltair-India, BScHons, 57, MSc, 58; Indian Inst Technol, Kharagpur, PhD(geochem), 63. *Prof Exp:* Sr lectr petrol, Univ Rajastan, Udiapur, India, 62-63; res geophysicist high pressure geophys, Univ Calif, Los Angeles, 65-71; Carnegie res assoc exp petrol, Geophys Lab, Carnegie Inst, Washington, DC, 71-74; sr resident res assoc, Nat Res Coun & Nat Acad Sci, NASA Johnson Space Ctr, Houston, 74-77; SR SCIENTIST EXP PETROL & HIGH PRESSURE PHYSICS, UNIV CALIF, LIVERMORE, 77- *Concurrent Pos:* Fel, Ger acad exchange scholar, Univ Gottingen, WGermany 63-65; Adv UN Develop Proj, Geophys Res Inst , India. *Mem:* Am Geophys Union; fel Mineral Soc Am; fel Indian Mineral Soc; Asn Geoscientists for Int Develop. *Res:* High pressure and temperature phase equilibria properties of materials at elevated pressures and temperatures; equation of state studies; crystal growth, lunar and terrestrial mantle rock studies; high pressure and temperature work on trans-uranics. *Mailing Add:* Lawrence Livermore Nat Labs Univ Calif L-201 Livermore CA 94550

AKER, FRANKLIN DAVID, b Norristown, Pa, Apr 26, 43; m 68; c 2. NEUROANATOMY, ANATOMY. *Educ:* Gettysburg Col, BA, 65; Temple Univ, PhD(anat), 70. *Prof Exp:* ASST PROF ANAT SCI, SCH DENT, TEMPLE UNIV, 70- *Mem:* Am Asn Anat. *Res:* Peripheral and central nervous system in conjunction with fluorescence microscopy and ultra-freezing and drying techniques; development of hard tissues of head region. *Mailing Add:* Dept Anat Temple Univ Sch Dent/Health Sci Campus Broad & Ont Philadelphia PA 19140

AKERA, TAI, b Wakayama, Japan, July 13, 32; m 62; c 3. PHARMACOLOGY. *Educ:* Keio Univ, Japan, MD, 58, PhD(pharmacol), 65. *Prof Exp:* Intern med, Keio Univ Hosp, 58-59, asst pharmacol, Sch Med, 59-62, instr, 64-66, asst prof, 66-71; assoc prof, 71-74, PROF PHARMACOL, MICH STATE UNIV, 74- *Concurrent Pos:* Vis asst prof, Mich State Univ, 67-70; vis prof, Sch Med, Tokai Univ, 77. *Mem:* Am Soc Pharmacol & Exp Therapeut; Japanese Med Asn; NY Acad Sci; Int Soc Heart Res; Nat Inst Drug Abuse. *Res:* Biochemical and biophysical aspects of the action of narcotics, tranquilizers and cardiac glycosides. *Mailing Add:* Natl Children's Hosp Med Res Ctr 3-35-31 Taishido Setagaya-ku Tokyo Japan

AKERBOOM, JACK, b Bridgeton, NJ, Apr 8, 28; m 56; c 2. FOOD TECHNOLOGY. *Educ:* Lehigh Univ, BS, 49. *Prof Exp:* Chemist qual control, Best Foods Inc, 49-53, food technologist prod develop, 53-60, dir qual assurance, Knorr Prod, Best Foods Div, CPC Int, 60-62, dir res & develop, Corn Prod, Food Technol Inst, 62-70, asst dir food technol, 70-76, V PRES RES & DEVELOP, BEST FOODS DIV, CPC INT, 76- *Res:* New food product development; vegetables, pasta, shelf stable foods, frozen chicken products, beverages, and egg products. *Mailing Add:* Best Foods Div CPC Int 1120 Commerce Ave Union NJ 07083

AKERLOF, CARL W, b New Haven, Conn, Mar 5, 38; m 65; c 2. ELEMENTARY PARTICLE PHYSICS, ASTROPHYSICS. *Educ:* Yale Univ, BA, 60; Cornell Univ, PhD(physics), 67. *Prof Exp:* Res assoc, 66-68, asst prof, 68-72, assoc prof, 72-78, PROF PHYSICS, UNIV MICH, ANN ARBOR, 78- *Mem:* Am Phys Soc; Am Astron Soc. *Res:* Strong and electromagnetic interactions of elementary particles; gamma-rays from compact stellar objects. *Mailing Add:* Dept of Physics Univ of Mich Ann Arbor MI 48109

AKERS, ARTHUR, b Smethwick, UK, Mar 24, 27; m 78; c 2. TRIBOLOGICAL ASPECTS OF FLUID POWER. *Educ:* Univ London, UK, BSc, 53, PhD(mech eng), 69; Univ Cranfield, MSc, 55. *Prof Exp:* Sr lectr aeronaut & mech eng, Univ Bath, UK, 60-64; prin lectr mech engr, Lords of the Admiralty Royal Naval Col, 65-68 & Her Britannic Majesty's Govt Royal Mil Col Sci, 68-73 & 75-76; from asst prof mech engr, 75-80, assoc prof eng mech & sci dept eng sci & mech, 80-87, PROF AEROSPACE ENG & ENG MECH, IOWA STATE UNIV, 90- *Concurrent Pos:* Examr gen cert educ in eng sci, Univ London, UK, 71-75; vis prof, dept mech eng, Univ Va, 73-74; consult, US Navy, Dahlgren, Va, 75-78, Fisher Controls, Marshalltown, Iowa, 78-82, Marshalltown Instruments, 78-85, Vector Corp, Marion, 82-83, 3M Corp, 82-83 & Sundstrand, Ames, 82-85; consult, Pack Corp USA, Tampa, Ia, 83; consult, Midwest Carbide, Keokuk, Ia, 84; consult, Shivvers Inc, Corydon, Ia, 86; consult, Servus Rubber Louisville, Ky, 87; consult, Eastern Iowa Hog Producers, Earlville, Ia, 87; tech adv, Ang Eng, Sunstrand Adv, Tech group, 87-88. *Honors & Awards:* Bliss Medal, Am Soc Mil Engrs, 83. *Mem:* Am Soc Mech Engrs; Soc Exp Mech; Soc Am Mil Engrs; fel Royal Aeronaut Soc UK. *Res:* Fluid power technology and control of fluid power systems; experimental fluid mechanics; tribological conditions in complex mechanisms. *Mailing Add:* 1519 Stone Brooke Rd Ames IA 50010-4100

AKERS, CHARLES KENTON, b Eugene, Ore, May 11, 42; m 71; c 4. SURFACE CHEMISTRY, ENVIRONMENTAL CHEMISTRY. *Educ:* Willamette Univ, BA, 65; State Univ NY, Buffalo, PhD(biophys), 72. *Prof Exp:* Cancer scientist, Roswell Park Mem Inst, 67-72; engr, Town Amherst, 73-74; scientist, Calspan Corp, 74-85, prog dir, 85-90; PRES, AKERS ASSOCS, 90- *Concurrent Pos:* Chmn, Chem Opers Sect, Am Defense Preparedness Asn, 84- *Mem:* Fel Am Inst Chemists; Am Chem Soc; Am Defense Preparedness Asn. *Res:* Survivability in hazardous environment, chemical warfare, hazardous waste; surface chemistry for biomaterial development, biological- material interface; chemical detection technology; test methodology development for toxic materials. *Mailing Add:* 73 Oakgrove Dr Williamsville NY 14221

AKERS, LAWRENCE KEITH, b Ashburn, Ga, Apr 16, 19; m 46; c 4. PHYSICS. *Educ:* Univ Ga, BS, 49, MS, 50; Vanderbilt Univ, PhD, 55. *Prof Exp:* Scientist, Oak Ridge Inst Nuclear Studies, 54-55, sr scientist, 55-58, res scientist & actg chmn univ rels div, 58-59; head training unit, Int Atomic Energy Agency, 59-60; prin scientist, Spec Training Div, Oak Ridge Inst Nuclear Studies, 61-64, asst chmn, 64-65, chmn spec training div, Oak Ridge Assoc Univs, 65-75; prog mgr, Dept Energy, 76-86; actg head physics dept, Univ Tenn, Chattanooga, 86-87, distinguished vis prof, 87, prof physics & astron, 88-90; RETIRED. *Concurrent Pos:* Consult, Cordoba Nat Univ; mem steering comt tech-physics proj, Am Inst Physics; dir res grad prog, Univ Tenn, 73-76. *Mem:* Am Phys Soc; Sigma Xi. *Res:* Infrared and Raman spectra of corrosive compounds; nuclear spectroscopy. *Mailing Add:* 401 Crewdson Ave Chattanooga TN 37405-4111

AKERS, LEX ALAN, b Washington, DC, May 7, 50; m 69; c 1. ELECTRICAL ENGINEERING, SOLID STATE ELECTRONICS. *Educ:* Tex Tech Univ, BSEE, 71, MSEE, 73, PhD(elec eng), 75. *Prof Exp:* Res asst integrated circuits, Dept of Elec Eng, Tex Tech Univ, 71-74, instr elec eng, 74-75; res engr digital design, Southwest Res Inst, 75-76; asst prof elec eng, Univ Nebr, 76-; ASSOC PROF, DEPT ELEC ENG, ARIZ STATE UNIV. *Concurrent Pos:* Assoc ed elec devices, Soc Comput Simulation, 77- *Mem:* Inst Elec & Electronics Engrs; Soc Comput Simulation; Sigma Xi. *Res:* Small geometry and nonisothermal MOSFET models; simulation of semiconductor devices; microprocessor design and applications. *Mailing Add:* Dept Elec Eng Ariz State Univ Tempe AZ 85287

AKERS, MICHAEL JAMES, b Beech Grove, Ind, Aug 24, 46; m 68; c 3. PHYSICAL PHARMACY, PARENTERAL TECHNOLOGY. *Educ:* Wabash Col, BA, 68; Univ Iowa, PhD(pharm), 72. *Prof Exp:* Res investr pharm, Searle Labs, Inc, 72-74; sr scientist, Alcon Labs Inc, 74-77; assoc prof pharm, Univ Tenn, Memphis, 77-81; res scientist, 81-84, head, Dry Prod Develop, Lilly Res Lab, head, Parenteral & Liquid Ointment Prod Dev, 87-89, mgr, Parenteral Qual Control, 89-90, MGR, CORP QUAL ASSURANCE, ELI LILLY & CO, 91- *Concurrent Pos:* Adj prof, Sch Pharm, Purdue Univ, 83- & Col Pharm, Univ Ill, 84- *Honors & Awards:* Res Award, Parenteral Drug Asn, 80. *Mem:* Am Asn Pharm Sci; Parenteral Drug Asn; Sigma Xi. *Res:* Parenteral delivery systems; sterilizer validation; kinetics of drug degradation; parenteral quality control; formulation of polypeptides. *Mailing Add:* Corp Qual Assurance Eli Lilly & Co Indianapolis IN 46285

AKERS, SHELDON BUCKINGHAM, JR, b Washington, DC, Oct 22, 26; m 53; c 4. COMPUTER SCIENCE. *Educ:* Univ Md, BS, 48, MA, 52. *Prof Exp:* Electronics scientist, Nat Bur Stand, Washington, DC, 48-50; radio engr, US Coast Guard Hq, 50-53; comput engr, Nat Bur Stand, 53-54; mathematician, Avion Div, ACF Industs, Inc, Va, 54-56; mathematician & comput scientist, Electronics Lab, Gen Elec Co, 56-85; PROF ELEC & COMPUT ENG, UNIV MASS, 85- *Concurrent Pos:* Adj prof, Syracuse Univ, 75-85, vis prof, 80. *Mem:* Fel Inst Elec & Electronics Engrs; Sigma Xi; Math Asn Am. *Res:* Switching circuit theory; design and application of digital computers; operations research; design automation; combinatorial analysis; graph theory. *Mailing Add:* 78 Larkspur Dr Amherst MA 01002

AKERS, THOMAS GILBERT, b Oakland, Calif, Jan 5, 28; m 54; c 3. VIROLOGY, MICROBIOLOGY. *Educ:* Univ Calif, Berkeley, BS, 50, MPH, 56; Mich State Univ, PhD(virol), 63; Southern Ill Univ, Edwardsville, MBA, 73. *Prof Exp:* US Navy, 50-, pub health officer, Formosa, 51-54, res virologist, Naval Biomed Res Lab, Univ Calif, Berkeley, 54-56, res virologist, Cairo, Egypt, 56-60, res virologist, Naval Biomed Res Lab, Univ Calif, Berkeley, 63-76; assoc dean, Sch Pub Health & Trop Med, 76-81, PROF ENVIRON HEALTH SCI, TULANE UNIV, 81- *Mem:* Am Pub Health Asn; Soc Exp Biol & Med; Am Asn Immunol; Am Soc Trop Med & Hyg; NY Acad Sci; Sigma Xi. *Res:* Public health microbiology; biophysical properties of virus nucleic acids; epidemiology of respiratory and enteric virus; cytochemical techniques. *Mailing Add:* Sch of Pub Health & Trop Med Tulane Univ New Orleans LA 70112

AKERS, THOMAS KENNY, b Brooklyn, NY, Jan 16, 31; m 56; c 3. COMPARATIVE PHYSIOLOGY, ENVIRONMENTAL PHYSIOLOGY. *Educ:* DePaul Univ, BS, 56; Loyola Univ Chicago, MS, 59, PhD(physiol), 61. *Prof Exp:* Instr physiol, Norweg Am Sch Nursing, 58-59; instr, St Anthony of Padua, 59-60; from instr to asst prof pharmacol, Stritch Sch Med, Loyola Univ Chicago, 61-65; from asst prof to assoc prof , 69-77, PROF PHYSIOL, SCH MED, UNIV NDAK, 77-, INTERIM CHAIR, 90- *Concurrent Pos:* Res assoc, Inst Study Mind, Drug & Behav, 63-65, AMTE, Portsmouth, Eng, 80; dir, High Pressure Life Lab, 74-79. *Honors & Awards:* Peck Fund Res Award, 62; B C Gamble Award, 87; Sigma Xi Medal, 90. *Mem:* Am Physiol Soc; AAAS; Undersea Med Soc; Sigma Xi. *Res:* Hyperbaric studies; oxygen toxicity; adrenergic function; bioengineering. *Mailing Add:* Dept Physiol Univ NDak Sch Med Grand Forks ND 58202

AKESON, RICHARD ALLAN, b Primghar, Iowa, Nov 14, 47; m 69; c 2. NEUROSCIENCE, DEVELOPMENTAL BIOLOGY. *Educ:* Iowa State Univ, BS, 69; Univ Calif, Los Angeles, PhD(neurosci), 74. *Prof Exp:* Fel immunol, Univ Calif, Los Angeles, 74-76; ASST PROF RES PEDIAT & BIOL CHEM, CHILDRENS HOSP & UNIV CINCINNATI, 76- *Mem:* AAAS; Soc Neurosci. *Res:* Cellular and molecular specificity in the development of the nervous system; role of cell membrane surface in cell-cell interaction. *Mailing Add:* Childrens Hosp Res Found/Dept Cell Biol Elland & Bethesda Aves Cincinnati OH 45229

AKESON, WALTER ROY, b Chappell, Nebr, Nov 2, 37; m 59; c 3. BIOCHEMISTRY, AGRONOMY. *Educ:* Univ Nebr, BS, 59, MS, 61; Univ Wis, PhD(biochem), 66. *Prof Exp:* Asst prof agron, Univ Nebr, Lincoln, 65-69; sr plant physiologist, Agr Res Ctr, Great Western Sugar Co, 69-85; SR PLANT PHYSIOLOGIST, MONO HY SUGAR BEET SEED, INC, 85- *Mem:* AAAS; Am Soc Agron; Am Soc Plant Physiol; Am Soc Sugar Beet Technol; Sigma Xi. *Res:* Metabolism of aromatic compounds; extraction and evaluation of leaf protein concentrates; nature of biochemical resistance of plants to disease and insects; sugar beet storage losses; seed physiology; tissue culture. *Mailing Add:* 1815 Duchess Dr Longmont CO 80501

AKESON, WAYNE HENRY, b Sioux City, Iowa, May 5, 28; m 51, 69; c 4. ORTHOPEDIC SURGERY. *Educ:* Univ Chicago, MD, 53; Am Bd Orthop Surg, dipl. *Prof Exp:* Instr orthop surg, Univ Chicago, 57-58; asst prof, Sch Med, Univ Wash, 61-70; actg chmn, Dept Surg, 80-81, PROF ORTHOP & HEAD DIV, UNIV CALIF, SAN DIEGO, 70- *Concurrent Pos:* Consult, Vet Admin. *Honors & Awards:* Nicholas Andry Award, 66; Sports Med Award, Am Ortho Soc, 83. *Mem:* Am Orthop Asn; Orthop Res Soc; Am Acad Orthop Surg; Orthop Res & Educ Found. *Res:* Collagen and mucopolysaccharides and their interaction; biochemical and biophysical basis of joint stiffness; metabolism of intercellular substances of articular cartilage and of arthroplastic surfaces. *Mailing Add:* 225 W Dickinson St San Diego CA 92103

AKESSON, NORMAN B(ERNDT), b Grandin, NDak, June 12, 14; m 46; c 2. AGRICULTURAL ENGINEERING, PESTICIDE CHEMICAL SAFETY ENGINEERING. *Educ:* NDak Agr Col, BS, 40; Univ Idaho, MS, 42. *Prof Exp:* Asst res, Univ Idaho, 40-42; physicist bur ord, US Dept Navy, 42-47; from instr to prof, 47-84, EMER PROF AGR ENG, AGR EXP STA, UNIV CALIF, DAVIS, 84- *Concurrent Pos:* Fulbright fel, Eng & E Africa, 57-58; consult, WHO & FAO, Israel, 68, Japan, 72, China, 85, Can, 87, assoc ed, Am Soc Agr Eng, 83- *Mem:* Fel Am Soc Agr Engrs; Am Entom Soc; Brit Soc Res Agr Eng; Am Soc Testing & Mat; Weed Sci Soc Am; Am Mosquito Control Asn. *Res:* Machines for aerial and ground application of plant protection materials; spray atomization and distribution from air carrier and aircraft equipment; small particle analysis and behavior and air pollution meteorology as applied to particle drift and deposit; evaluation of pesticide application and reduction of potential damages to workers, crops and wildlife through safe application procedures. *Mailing Add:* Dept Agr Eng Univ Calif Davis CA 95616

AKGERMAN, AYDIN, b Izmir, Turkey, July 12, 45; US citizen; m 69; c 2. REACTION ENGINEERING, SUPERCRITICAL TECHNOLOGY. *Educ:* Robert Col, Istanbul, Turkey, BS, 68; Univ Va, MS, 69, PhD(chem eng), 71. *Prof Exp:* Asst prof chem eng, Bogazici Univ, Turkey, 71-72; from asst prof to assoc prof, Ege Univ, Turkey, 72-79; res & develop mgr, Cimentas Izmir Cement Plant, 76-80; assoc prof, 80-86, PROF CHEM ENG, TEX A&M UNIV, 86- *Concurrent Pos:* Fulbright scholar, Fulbright Comt France, 89; Halliburton prof, 89; consult various cos, 89- *Mem:* Am Inst Chem Engrs; Am Chem Soc. *Res:* Reaction engineering in relation to multiphase reactors, catalysis, and fuel upgrading; supercritical extraction from aqueous solutions and solid matrices; environmental remediation techniques such as vacuum stripping, biodegradation, and supercritical extraction. *Mailing Add:* Chem Eng Dept Texas A&M Univ College Station TX 77843

AKHTAK, RASHID AHMED, LIPID METABOLISM, MEMBRANE RECEPTORS. *Educ:* Univ London, PhD(biochem), 75. *Prof Exp:* ASSOC PROF CELLULAR & MOLECULAR BIOL & BIOCHEM, MED COL GA, 76- *Mailing Add:* Dept Cellular & Molecular Biol Med Col Ga 1120 15th St Augusta GA 30912

AKHTAR, MOHAMMAD HUMAYOUN, b Patna, India, Jan 31, 43; Can citizen; m 74; c 2. ORGANIC CHEMISTRY, ANALYTICAL CHEMISTRY. *Educ:* Univ Karachi, Pakistan, BSc Hons, 62; Univ Moncton, MSc, 66; Simon Fraser Univ, PhD(org chem), 70. *Prof Exp:* Demonstr, lectr chem, Haroon Col, Karachi, 62-64; res asst, Univ Moncton, 64-66 & Simon Fraser Univ, 66-70; fel org chem, Univ Alta, 70-75; SR RES SCIENTIST, FOOD & PROD SAFETY ANIMAL RES CTR, AGR CAN, 75- *Mem:* Am Chem Soc; Chem Inst Can. *Res:* Assess the nature, extent and safety of agrichemicals in animal feedstuffs; determine the biochemical fate of these compounds in poultry and livestock; transmission of residues to animal derived products. *Mailing Add:* Animal Res Ctr Neatby Bldg Ottawa ON K1A 0C6 Can

AKHTAR, SALIM, mineral engineering, for more information see previous edition

AKI, KEIITI, b Yokohama, Japan, Mar 3, 30; m 56; c 2. SEISMOLOGY. *Educ:* Univ Tokyo, BS, 52, PhD(geophys), 58. *Prof Exp:* Fulbright res fel geophys, Calif Inst Technol, 58-60; res fel seismol, Earthquake Res Inst, Univ Tokyo, 60-62; vis assoc prof geophys, Calif Inst Technol, 62-63; assoc prof seismol, Earthquake Res Inst, Univ Tokyo, 63-65; prof geophys, Mass Inst Technol, 66-84; PROF GEOL SCI, WM KECK, 84- *Concurrent Pos:* Lectr, Int Inst Seismol & Earthquake Eng, UNESCO, 60-65; geophysicist, Nat Ctr Earthquake Res, US Geol Surv, 67-; vis prof, Univ Chile, 70 & 72, Univ Paris, 83; vis scientist, Royal Norweg Res Coun; consult, Sandia Labs & Los Alamos Sci Labs, 76-79; vis scholar, Japan Soc Promotion Sci, 78; chmn comt seismol, Nat Res Coun, 78-82; distinguished vis prof, Univ Alaska, 81; lectr, Int Asn Seismol & Earth's Interior, 81; consult, NSF, 81; ed-in-chief, Pure & Appl Geog, 83-87. *Honors & Awards:* Medal, Seismol Soc Am, 87. *Mem:* Nat Acad Sci; Am Geophys Union; Seismol Soc Am (pres, 79-80); Soc Explor Geophys; fel Am Acad Arts & Sci; hon mem Europ Union Geosci. *Res:* Geophysical research directed toward predicting, preventing and controlling earthquake hazard; seismic wave propagation; earthquake statistcs; structure of earth's crust and upper mantle; thermal processes in the earth; geothermal energy source exploration. *Mailing Add:* Dept Geol Sci Univ Southern Calif Los Angeles CA 90089

AKIL, HUDA, b May 19, 45; US citizen; m 72; c 2. NEUROCHEMISTRY, NEUROSCIENCE. *Educ:* Am Univ Beirut, BA, 68; Univ Calif, Los Angeles, PhD(neurosci), 72. *Prof Exp:* Fel dept of psychiat, Stanford Univ, 74-77, res assoc, 77-78; asst prof, 78-81, ASSOC PROF NEUROCHEM, MENT HEALTH RES INST, UNIV MICH, 81- *Concurrent Pos:* Fels, NIH, 74-77 & Alfred P Sloan 74-78; mem, Study Sect Preclin Psychopharmacol, NIH & adv bd, Ment Health Inst Med; chmn Mode Action Opiates, Gordon conf, 83; organizer, Decade of Brain. *Mem:* Neurosci Soc; Int Asn Study Pain; Int Narcotic Res Conf; fel Am Col Neuropsychopharmacol. *Res:* Endogenous opiate peptides, endorphins; pain control; neuroregulation and behavior; steroid receptors; stress; dopamine receptors; molecular neurobiology. *Mailing Add:* Dept Psych Ment Health Res Inst Univ Mich Ann Arbor MI 48109

AKIN, CAVIT, b Nigde, Turkey, Feb 28, 31; m 78; c 4. BIOTECHNOLOGY, YEAST TECHNOLOGY. *Educ:* Univ Ankara, MS, 54; Univ Ill, MS, 59, PhD(food technol), 61. *Prof Exp:* Asst chem technol, Univ Ankara, 54-55; res engr, Technol Lab, Sugar Corp, Turkey, 56; NIH res grant, Univ Mass, 61-62; sr bioengr, Dept Res & Develop, Falstaff Brewing Corp, 62-67; sr res engr, Am Oil Co, Whiting, Ind, 67-70; sr res engr, Corp Res, Standard Oil Co Ind, Naperville, 70-74, res supvr, Food & Yeast Sci, 74-77; mem staff, Amoco Foods Co, Naperville, Ill, 77-79; res assoc, Corp Res, Standard Oil Co, Inc, Ind, 79-85; res assoc, corp res, Amoco Corp, 85-87; ASSOC DIR BIOTECHNOL RES, INST GAS TECHNOL, 87- *Mem:* AAAS; Am Chem Soc; Am Soc Microbiol; Soc Indust Microbiol; Inst Food Technol. *Res:* Polymerization kinetics of acrylonitrile; batch and continuous propagation of yeasts; microbial food and fodder production; vapor phase sterilization; fermentation kinetics of beer; fermenter design; bioengineering and biotechnology; food from petroleum; torula yeast from ethanol; food applications of yeast; SCP texturization; functional proteins; textured meat alternates; genetic engineering plant tissue culture; biocatalysis, immobilized cells; environmental bioremediation; energy biotechnology; bio-activation of methane; coal and oil bio-desulfurization; biofuel cells. *Mailing Add:* 1462 Inverrary Dr Naperville IL 60563

AKIN, FRANK JERREL, b Carroll Co, Ga, Oct 8, 41; m 65; c 3. PHARMACOLOGY, TOXICOLOGY. *Educ:* Univ Ga, BS, 64, MS, 69, PhD(pharmacol), 71; Am Bd Toxicol, cert. *Prof Exp:* Instr pharmacol, Univ Ga, 69-71; res pharmacologist, Richard B Russell Res Ctr, USDA, 71-77; dir pharmacol & toxicol, Plough Div, 77-85, SR DIR, CLIN & TOXICOL RES, CONSUMER OPERS, SCHERING-PLOUGH CORP. *Concurrent Pos:* Adj asst prof Sch Pharm, Univ Ga, 76-77; adj prof, Univ Tenn Ctr Health Sci, 82-85. *Mem:* Am Soc Pharmacol & Exp Therapeut; Fedn Am Soc Exp Biol; AAAS; Proprietary Asn; Cosmetic Toiletry & Fragrance Asn. *Res:* Pharmacology of analgesics, antifungals and topical steroids; sunscreen agents; toxicology of chemicals causing acute and subacute phototoxic and photoallergic reactions; toxicology of proprietary medications; natural toxicants. *Mailing Add:* Schering-Plough Corp 3030 Jackson Ave Memphis TN 38112

AKIN, GWYNN COLLINS, b Century, Fla, Mar 24, 39; m 63, 85; c 2. PHARMACEUTICAL INDUSTRY POLICY, HEALTH SCIENCE POLICY. *Educ:* Fla State Univ, BS, 61; Tulane Univ, PhD(human anat), 65. *Prof Exp:* Res fel exp embryol, Marine Biol Lab, Woods Hole, 63; instr gross anat & physiol, Loyola Univ Sch Dent, 65-66; asst to dean, La State Univ Sch Med, 66-71, asst prof anat, 68-71; head spec studies, Off Chancellor, Univ Calif, San Francisco, 71-72, dir spec proj, 72, acad asst to vpres health affairs, Univ Calif Systemwide Admin, 73-75; spec asst to dir, Neuropsychiat Inst, Univ Calif, Los Angeles, 75-76; policy & planning consult, Univ Wash, 76-79; staff dir, Nat Sci Bd, 78-81; sr prof assoc, Asn Acad Health Ctr, 79-81; asst to chmn, pres & chief exec off, 81-85, dir health policy, 85-89, DIR, SYNTEX SCHOLARS PROG, SYNTEX CORP, PALO ALTO, CALIF, 85-, VPRES PUB POLICY, 89- *Concurrent Pos:* Adj prof human anat & physiol, Univ West Fla, 68-69, asst prof biol & marine sci, 69-70; assoc adj prof, Sch Med, Univ Calif, San Francisco, 72-; mem gov bd, Univ Calif Retirement Syst, 73-76; consult, Fla Bd Regents, 74-75; mem, Sch Nursing Vis Comt, Univ Wash, 76-, Gen Res Support Prog Adv Comt, NIH, 73-, biomed res develop grant prog, NIH, 77- & policy analyst group, Pharmaceut Mfr Asn, 81-; mem bd dirs, Nat Asn Biomed Res, 87-90, Nat Leadership Coalition AIDS, 88-; mem bd overseers, Univ Calif, San Francisco, 89-; mem overseer's comt to vis Harvard Sch Pub Health, 90- *Honors & Awards:* Res Award, Student Res Forum, 67. *Mem:* Asn Am Med Col; Am Asn Anatomists; AAAS. *Res:* Health science policy; experimental embryology; growth inhibition; regeneration. *Mailing Add:* Syntex Corp 3401 Hillview Ave Palo Alto CA 94304

AKIN, JIM HOWARD, b McAllen, Tex, Mar 8, 37; m 59; c 2. MECHANICAL ENGINEERING. *Educ:* Univ Tex, Austin, BSME, 60, MSME, 61, PhD(mech eng), 67. *Prof Exp:* From asst prof to assoc prof mech eng, 63-77, PROF ENG SCI, UNIV ARK, FAYETTEVILLE, 77- *Mem:* Am Soc Mech Engrs; Am Soc Eng Educ; Nat Soc Prof Engrs; Sigma Xi. *Res:* Vibrational and systems analysis; machine design and analysis; structural dynamics. *Mailing Add:* Dept Mech Eng Univ Ark Fayetteville AR 72701

AKIN, LEE STANLEY, b Portland, Ore, Aug 5, 27; m 49; c 2. GEAR DESIGN ANALYSIS & RESEARCH, TRIBOLOGY. *Educ:* WCoast Univ, BS, 59, MS, 60, PhD(mech eng), 65; Univ Calif, Los Angeles, cert bus mgt, 68. *Prof Exp:* Gear design engr, Vard, Inc, 54-56; sr proj engr, Gear Prod Div, Waste King Corp, 56-59; sr engr, Librascope Group, Gen Precision, Inc, 59-60; sr syst engr, Gen Dynamics/Pomona, 60-64; staff scientist, Missile & Space Systs Div, Douglas Aircraft Co, Inc, 64-67 & McDonnell Douglas Corp, 67-69; consult engr advan gear technol, marine turbine & gear dept, Gen Elec Co, 69-76; dir res & develop, Appl Technol Div, Western Gear Corp, 76-81; mgr gear & bearing design, Aerojet Electro Systems Co, 81-88; OWNER, GEARSEARCH ASSOC, 79- *Concurrent Pos:* Instr, Pasadena City Col, 55-56, Univ Southern Calif, 68-69 & Northeastern Univ, 71-76; Douglas Prof Achievement awards, 66-68; adj prof & dir advan mach design & gear, Res Lab, Calif State Univ, Long Beach, 79- *Honors & Awards:*

Douglas Prof Achievement Awards, 66-68. *Mem:* Fel Am Soc Mech Engrs. *Res:* EHD lubrication and wear; design and analysis of geared mechanisms; gear scuffing; machinery noise and vibration; gear cooling; general rotating machine design. *Mailing Add:* 750 Indian Wells Rd Banning CA 92220

AKIN, WALLACE ELMUS, b Murphysboro, Ill, May 18, 23; m 48; c 2. PHYSICAL GEOGRAPHY, HYDROLOGY RESOURCES. *Educ:* Southern Ill Univ, BA, 48; Ind Univ, MA, 49; Northwestern Univ, PhD(geog), 52. *Prof Exp:* Field team chief, Rural Land Classification Prog PR, Commonwealth Dept Agr & Com, 50-51; instr geog, Univ Ill, Chicago, 52-53; chmn dept, 53-83, PROF GEOG & GEOL, DRAKE UNIV, 53- *Concurrent Pos:* Fulbright res grant, Univ Copenhagen, 61-62; consult water resources, Iowa Natural Resources Coun; ed, Iowa State Water Plan; fac res grant for study of subsistence agr, Solomon Islands, 80. *Mem:* Asn Am Geogr; Royal Danish Geog Soc; Arctic Inst NAm; Soil Conserv Soc Am; Asn Ground Water Scientists & Engrs. *Res:* The role of physical geography in resource inventory and management; water resources. *Mailing Add:* Dept of Geog & Geol Drake Univ Des Moines IA 50311

AKINS, RICHARD G(LENN), b Harlan, Ky, Dec 14, 34; m 58; c 2. CHEMICAL ENGINEERING. *Educ:* Univ Louisville, BS, 57, MS, 58; Northwestern Univ, PhD(chem eng), 62. *Prof Exp:* Res assoc, Chem Eng Div, Argonne Nat Lab, 62-63; asst prof, 63-77, assoc prof, 77-80, PROF CHEM ENG, KANS STATE UNIV, 80- *Concurrent Pos:* NSF res grant, 64-66; Off Saline Water res grant, 65-67. *Mem:* Am Inst Chem Engrs; Sigma Xi. *Res:* Laminar flow heat transfer in circular tubes; natural convection in liquid metals; Non-Newtonian flow. *Mailing Add:* 1832 Virginia Dr Manhattan KS 66502

AKINS, VIRGINIA, b Lafayette Co, Wis. BIOLOGY. *Educ:* Univ Wis, BA, 37, PhD(bot), 40. *Prof Exp:* Asst bot, Univ Wis, 37-40; agent forage crops & dis, USDA, Wis, 42-43; asst forest prod technologist, Forest Prod Lab, US Forest Serv, 43-47; from teacher to assoc prof, 47-71, PROF BIOL, UNIV WIS-RIVER FALLS, 71- *Mem:* AAAS. *Res:* Cytology of flagellates and grasses; detection and effect of gelatinous fibers in wood; cytological study of Carteria crucifera. *Mailing Add:* 550 N Main St River Falls WI 54022

AKISKAL, HAGOP S, b Beirut, Lebanon, Jan 16, 44; m. PSYCHIATRY, PSYCHOPHARMACOLOGY. *Educ:* Am Univ Beirut, BSc, 65, MD, 69. *Prof Exp:* Res psychiatrist, Alcohol & Drug Dependence Clin, Memphis Ment Health Inst, 72-84; dir med student educ psychiat, 74-78, assoc prof, 77-80, dir, Mood Clin & Affective Dis Prog, 75, DIR, CONTINUING MED EDUC PSYCHIAT, UNIV TENN, 79-, PROF PSYCHIAT & PHARMACOL, 80- *Concurrent Pos:* Co-dir, Sleep Disorder Ctr, Baptist Hosp, 77-; mem Royal Col Physicians Can Speakers, Univ Toronto, 84; mem French Ministry External Rels Prog Invitation Am Scientists, Univ Paris, 84. *Honors & Awards:* Eli Roberts Lectr, Wash Univ, St Louis, 80; Clin Res Award, Am Acad Clin Psychiatrists, 81. *Mem:* AMA; Am Psychiat Asn; Soc Biol Psychiat; Sigma Xi; Am Psychopath Asn; Psychiat Res Soc; Am Col Psychiat. *Res:* Diagnosis, etiology and psychopharmacology of affective disorders; sleep disorders. *Mailing Add:* Univ Tenn Col Med 66 N Pauline St Memphis TN 38163

AKIYAMA, STEVEN KEN, m. CELL BIOLOGY, CELL ATTACHMENT. *Educ:* Cornell Univ, PhD(phys chem), 81. *Prof Exp:* ASST PROF ONCOLOGY, 85-, ASST PROF BIOCHEM, CANCER CTR, HOWARD UNIV, 87- *Mem:* AAAS; Am Cancer Soc; Am Soc Cell Biol. *Res:* Fibronectin. *Mailing Add:* 9207 Bulls Run Pkwy Bethesda MD 20817

AKKAPEDDI, MURALI KRISHNA, b Vijayawada, India, Aug 31, 42; m 70; c 2. POLYMER CHEMISTRY. *Educ:* Osmania Univ, India, BSc, 60, MSc, 62; Univ Pa, PhD(org chem), 71. *Prof Exp:* Lectr chem, Osmania Univ, India, 62-67; sr chemist, Polysci Inc, Warrington, Pa, 71-74; res chemist, 74-78, sr res chemist, 78-82, res assoc, corp res & develop, 82-85, MGR POLYMER RES, ENG PLASTICS RES & DEVELOP, ALLIED SIGNAL CORP, 85- *Mem:* Am Chem Soc. *Res:* Polymer synthesis and structure-property correlation; biomedical application of polymers; nylon reaction injection molding; barrier resins; high performance engineering plastics and polymer blends. *Mailing Add:* Seven Manor Dr Morristown NJ 07960

AKKAPEDDI, PRASAD RAO, b Andhra Prudesh, India, Dec 23, 43; US citizen. HIGH ENERGY LASERS, SPECTROSCOPY. *Educ:* Andhra Univ, Waltair, India, BSc, 61, MSc,63; Univ Southern Calif, MS, 70, PhD(elec eng), 73. *Prof Exp:* Res asst physics & elec eng, Univ Southern Calif, 66-73, asst elec eng, 73; tech staff physics, Comput Serv Corp, NASA, 73-76; mgr, adv tech high energy lasers, 76-81, group leader, lasers & devices, 81-83, ASSOC DIR RES, PERKIN-ELMER CORP, DANBURY, CONN, 83- *Concurrent Pos:* Adj prof elec eng, Univ Bridgeport, Cpnn, 78-80. *Mem:* Optical Soc Am; Laser Inst Am; Am Inst Physics. *Res:* Diode lasers; integrated and non-linear optics; thin films; surface sciences. *Mailing Add:* MS 953 Perkin-Elmer Corp 100 Wooster Heights Rd Danbury CT 06810

AKLONIS, JOHN JOSEPH, b Elizabeth, NJ, Sept 28, 40; m 66, 85. PHYSICAL CHEMISTRY, POLYMER CHEMISTRY. *Educ:* Rutgers Univ, BS, 62; Princeton Univ, MA, 64, PhD(phys chem), 65. *Prof Exp:* Fel polymer physics, Princeton Univ, 65-66; from asst prof to assoc prof, 66-81, PROF CHEM, UNIV SOUTHERN CALIF, 81- *Concurrent Pos:* Fulbright fel polymer chem, Cath Univ Louvain, 66-67; res assoc, Macromolecule Res Ctr, Strasbourg, France, 74-75; vis prof, Danish Tech Univ, 81-82. *Honors & Awards:* Fel, Am Phys Soc. *Mem:* Am Chem Soc; Am Phys Soc. *Res:* Physics and physical chemistry of high polymeric systems. *Mailing Add:* Dept of Chem Univ of Southern Calif Los Angeles CA 90007

AKO, HARRY MU KWONG CHING, b Honolulu, Hawaii, Mar 16, 45; div; c 2. AQUACULTURE NUTRITION. *Educ:* Univ Calif, Berkeley, AB, 67; Wash State Univ, PhD(biochem), 73. *Prof Exp:* Asst prof, 75-82, ASSOC PROF BIOCHEM, UNIV HAWAII, 82- *Concurrent Pos:* Sr fel biochem, Univ Wash, 75. *Mem:* World Aquacult Soc. *Res:* Aquaculture nutrition. *Mailing Add:* Dept Agr Biochem Univ of Hawaii Honolulu HI 96822

AKONTEH, BENNY AMBROSE, b Bafut, Cameroon, Dec 7, 47; Can citizen; m 70; c 4. INTELLIGENT CONTROL DECISIONS EMULATING COMPUTER ARCHITECTURE. *Educ:* Univ NH, Durham, BS & BA & MA, 72; Stanford Univ, MS, 76, PhD(comput eng), 78. *Prof Exp:* Syst consult, World Bank, Washington, DC, 73-79; vis prof comput archit, Cath Univ Rio de Janeiro, 78-80; prof digital syst & digital control, Mil Inst Eng, Rio de Janeiro, 79-80; systs consult, Secretariat Presidency, Brazil, 79-81; assoc prof math prog & comput archit, Univ Brazil, 80-84; CHIEF SCIENTIST EXPERT SYST, AKONTEH INST TECHNOL, BRAZIL, 83- *Concurrent Pos:* Tech consult, Ministry Telecommun, Brazil, 80-83; assoc prof comput sci, Northrop Univ, Los Angeles, 84-88; sr res scientist, C E Harris & Co, Los Angeles, 84-88; assoc prof med informatics, Drew Med Sch, Los Angeles, 88-; assoc prof comput sci, Calif State Univ, Long Beach, 88-; res asst, Mass Inst Technol, 72-73; teaching asst - Stanford Univ, 74-75; res assoc, ENWATS, Cambridge, Mass, 72-73; pres, HiTech Assocs, Inc, 87. *Mem:* Asn Comput Mach; Math Soc Am. *Res:* Heuristic programming, emulation and control decisions; expert systems applications in new computer architecture, manufacturing automation and control decisions (management science). *Mailing Add:* PO Box 91150 Los Angeles CA 90009

AKOVALI, YURDANUR A, b Ankara, Turkey; div; c 2. NUCLEAR PHYSICS. *Educ:* Univ Ankara, BS, 58; Univ Md, PhD(nuclear physics), 67. *Prof Exp:* Lab asst elec, Dept Physics, Sci Fac, Univ Ankara, 58-59; teaching asst physics, Univ Md, 59-61; MEM RES STAFF, PHYSICS DIV, OAK RIDGE NAT LAB, 67- *Concurrent Pos:* Consult & lectr, 78. *Mem:* Am Phys Soc; Sigma Xi. *Res:* Investigation of nuclear structure through alpha, beta and delayed-proton decays, in-beam reaction gammas and particle transfers. *Mailing Add:* Physics Div Oak Ridge Nat Lab PO Box 2008 Oak Ridge TN 37831-6371

AKRABAWI, SALIM S, b Amman, Jordan, Jan 20, 41; US citizen; m 67; c 3. MEDICINE. *Educ:* Vanderbilt Univ, MD, 77; Univ Calif, Davis, PhD(nutrit), 68. *Prof Exp:* Asst prof nutrit, Am Univ, Lebanon, 71-74; ASSOC DIR MED AFFAIRS, MEAD JOHNSON & CO, 80- *Mem:* Am Soc Parenteral & Enteral Nutrit; Sigma Xi. *Res:* Adult nutritional needs in trauma and illness. *Mailing Add:* AMA-Am Col Nutrit Mead Johnson Nutrit Div 867 College Hwy Evansville IN 47721

AKRE, ROGER DAVID, b Grand Rapids, Minn, Mar 27, 37; m 56; c 2. ENTOMOLOGY. *Educ:* Univ Minn, Duluth, BS, 60; Kans State Univ, MS, 62, PhD(entom), 64. *Prof Exp:* From asst prof to assoc prof, 64-74, PROF ENTOM, WASH STATE UNIV, 74- *Mem:* Entom Soc Am; Entom Soc Can; Sigma Xi; Int Union Study Social Insects (pres, 88). *Res:* Biology and behavior of arthropods associated with ants; behavior of vespine wasps; carpenter ant biology and control; aggressive house spider, a new venomous spider in the Pacific Northwest. *Mailing Add:* Dept Entom Wash State Univ Pullman WA 99164-6432

AKRUK, SAMIR RIZK, b Beirut, Lebanon, Feb 2, 43; m 77. REPRODUCTIVE BIOLOGY, ELECTRON MICROSCOPY. *Educ:* Am Univ Beirut, BS, 65, MS, 67; Univ Ga, PhD(zool), 77. *Prof Exp:* Instr & coordr biol, Beirut Univ Col, 66-71; ASST PROF BIOL, MARIETTA COL, 78- *Concurrent Pos:* Res assoc biochem, Univ Ga, 77-78. *Mem:* Electron Micros Soc Am; Soc Study Reprod; Sigma Xi. *Res:* In vitro capacitation and fertilization in mammals; mammalian sperm acrosome reaction; scanning and transmission electron microscopy of gametes; ultrastructural localization of acrosomal enzymes. *Mailing Add:* 561 Heritage Oak Dr Yardley PA 19067-5622

AKSAMIT, ROBERT ROSOOE, CHEMOTAXIS OF MAMMALIAN CELLS. *Educ:* Okla State Univ, PhD(chem), 71. *Prof Exp:* RES CHEMIST, NIMH, 80- *Mailing Add:* PhD Res Chem Lab Gen & Comp Biochem NIMH 9000 Rockville Pike Bldg 36 Rm 3DO6 Bethesda MD 20892

AKSELRAD, ALINE, b Wilno, Poland, Apr 14, 35; US citizen; m 62; c 1. EXPERIMENTAL SOLID STATE PHYSICS, MAGNETISM. *Educ:* McGill Univ, BSc, 57; Univ Toronto, MSc, 59, PhD(solid state spectros), 61. *Prof Exp:* Asst physics, Univ Toronto, 59-61; grant & mem tech staff, Weizmann Inst Sci, 61-62; mem tech staff, RCA Labs, 62-77; MEM RES STAFF, AT&T ENG RES CTR, 77- *Mem:* Am Phys Soc; AAAS; Sigma Xi. *Res:* Electronic structure and properties of solids; magnetic insulators and their properties; propagation and applications of acoustic waves; solar energy conversion; plasma deposition; semiconductor lasers; luminescence; semiconductors, electronic and optical properties. *Mailing Add:* 960 Lawrenceville Rd Princeton NJ 08540

AKST, GEOFFREY R, b New York, NY, July 15, 43. MATHEMATICAL STATISTICS. *Educ:* Columbia Univ, EdD, 76. *Prof Exp:* PROF MATH, BOROUGH MANHATTAN COMMUNITY COL, 66- *Mem:* Am Math Asn Two Yr Col. *Mailing Add:* 833 Lexington Ave New York NY 10021

AKTIK, CETIN, b Erzincan, Turkey, Nov 14, 53; Can citizen; m 84. III-V COMPOUND SEMICONDUCTORS FABRICATION, VLSI CIRCUIT FABRICATION. *Educ:* Inst Nat Sci Appl, Toulouse, France, BS, 77; Univ Paul Sabatier, Toulouse France, DEA, 78, DIng, 80. *Prof Exp:* Res assoc solid state physics, Ecole Polytechnique Montreal, 80-84; prof engr, OMVPE Inc, Montreal, 85-88; ATTACHÉ RECHERCHE MICROELECTRONICS, UNIV SHERBROOKE, 89- *Concurrent Pos:* Vpres, Digiser Ltee, 88- *Mem:* Am Phys Soc. *Res:* Compound semiconductor technology and fabrication; solid state device physics; electronic transport properties of semiconductors; microelectronics; crystal growth by metal-organic chemical vapor deposition; deep level transient spectroscopy; infrared spectroscopy; very-large-scale integration circuit fabrication; ion implantation and vacuum techniques. *Mailing Add:* Elec Eng Dept Univ Sherbrooke Sherbrooke PQ J1K 2R1 Can

AKTIPIS, STELIOS, b Athens, Greece, July 14, 35; US citizen; m 65; c 2. BIOCHEMISTRY, BIOPHYSICS. *Educ:* Nat Univ Athens, dipl, 59; Brown Univ, ScM, 62; Brandeis Univ, PhD(phys org chem), 65. *Prof Exp:* Res fel biochem, Harvard Med Sch, 65-66; from asst prof to assoc prof, 66-74, PROF BIOCHEM, STRITCH SCH MED, LOYOLA UNIV CHICAGO, 74-, ASST CHMN DEPT, 70- *Mem:* AAAS; Am Soc Biol Chemists; Am Chem Soc. *Res:* Circular dichroism of macromolecules; interaction of nucleic acids with intercalating molecules; inhibition of RNA polymerase by template inactivators. *Mailing Add:* Stritch Sch Med Loyola Univ Chicago Med Ctr Maywood IL 60153

ALADJEM, FREDERICK, b Vienna, Austria, Feb 8, 21; nat US; m 57; c 3. IMMUNOCHEMISTRY, BIOPHYSICS. *Educ:* Univ Calif, AB, 44, PhD(biophys), 54. *Prof Exp:* Physicist, Radiation Lab, Univ Calif, 50-54; physicist, Nat Microbiol Inst, USPHS fel & res fel div chem, Calif Inst Technol, 54-57; from instr to assoc prof, 56-65, PROF MED MICROBIOL, SCH MED, UNIV SOUTHERN CALIF, 65- *Concurrent Pos:* Consult path, Los Angeles County Gen Hosp, 57- *Mem:* AAAS; Am Asn Immunol; Am Soc Microbiol; Biophys Soc. *Res:* Antigen-antibody reaction; immunochemistry of lipoproteins; chemical aspects of allergic reactions. *Mailing Add:* Dept Microbiol Univ Southern Calif Sch Med 2025 Zonal Ave Los Angeles CA 90033

ALAGAR, VANGALUR S, b Srivilliputtur, Tamil Nadu, India, Mar 15, 40; Can citizen; m 62; c 2. THEORY, COMPUTER SCIENCE. *Educ:* Madras Univ, BA, 59, MSc, 61; State Univ NY, Stony Brook, MA, 71; McGill Univ, PhD(comput sci), 75. *Prof Exp:* Lectr comput sci, McGill Univ, 74-75; from asst prof to assoc prof, 75-87, PROF COMPUT SCI, CONCORDIA UNIV, MONTREAL, 87- *Concurrent Pos:* Lectr math, Vivekananda Col, Madras Univ, 62-70 & Annamalai Univ, India, 61-62; vis scientist, Vikram Sarabhai Space Ctr, Indian Space Res Orgn, Trivandrum, India, 86. *Mem:* Asn Comput Mach; Inst Elec & Electronic Engrs. *Res:* Investigation of formal models for the specification and proof of correctness for the functional behavior of real time distributed systems; formal specification methods for capturing geometric, physical and functional properties of objects to be manipulated in robotic and CAD systems; study on pure applicative real time language also useful for distributed computing. *Mailing Add:* Concordia Univ Montreal PQ H3G 1M8 Can

ALAIMO, ROBERT J, b Rochester, NY, Mar 2, 40; m 63; c 3. ORGANIC CHEMISTRY. *Educ:* Univ Miami, BS, 61; Cornell Univ, PhD(org chem), 65. *Prof Exp:* Sr res chemist, Norwich Pharmacal co, 65-79, unit leader, 79-83, MGR, TECH OPERS NORWICH-EATON PHARMACEUT, 83- *Mem:* Am Chem Soc; The Chem Soc; Nat Safety Coun. *Res:* Synthetic organic chemistry involving the synthesis of medicinal agents. *Mailing Add:* 29 Hillview Dr Norwich NY 13815

AL-AISH, MATTI, DIAGNOSTIC IMAGING RESEARCH. *Prof Exp:* PROG DIR, DIAG IMAGING RES BR, NAT CANCER INST, NIH, 89- *Mailing Add:* NIH Nat Cancer Inst Diag Imaging Res Br Exec Plaza N Rm 800 6130 Executive Blvd Rockville MD 20852

ALAM, ABU SHAFIUL, b June 1, 45; US citizen. PHARMACEUTICS. *Educ:* Philadelphia Col Pharm & Sci, BS, 67; Univ Iowa, MS, 69, PhD(pharm), 71. *Prof Exp:* Scientist pharm res develop, Rohm & Haas, 71-77; mgr pharm res & develop, Adria Labs Inc, 77-81; SECT HEAD PROD DEVELOP, AM CRIT CARE, 81- *Mem:* Acad Pharmaceut Sci; Am Pharmaceut Asn. *Res:* Develop pharmaceutical products which exhibit optimum stability and formula; process conditions and bioavailability; anticancer research, psychotropic; cardiovascular; analgesic and antiinflammatory areas. *Mailing Add:* 719 Mullady Pkwy Libertyville IL 60048

ALAM, BASSIMA SALEH, b Alexandria, Egypt, Aug 16, 35; US citizen. NUTRITIONAL BIOCHEMISTRY. *Educ:* Alexandria Univ, Egypt, BSc, 58; Mass Inst Technol, MS, 64; Fla State Univ, PhD(nutrit), 67. *Prof Exp:* Instr, Alexandria Univ, Egypt, 58-60; asst prof, Cairo Univ, Egypt, 67-69; res assoc lab carcinogenesis & toxicol, Children's Cancer Res Found, 69-71; res assoc surg, 71-73, instr biochem, Tulane Med Sch, 73-75; RES ASST PROF BIOCHEM, MED CTR, LA STATE UNIV, 75- *Mem:* Am Inst Nutrit; Am Soc Biochem & Molecular Biol. *Res:* Effects of nutrition on membrane structures and function; nutrition and cancer. *Mailing Add:* Dept Biochem Med Ctr La State Univ 1100 Florida Ave New Orleans LA 70119

ALAM, JAWED, b Patna, India, Oct 16, 55; m 87. TRANSCRIPTIONAL REGULATION, INFLAMMATION & ACUTE PHASE RESPONSE. *Educ:* Clemson Univ, BS, 78; Purdue Univ, PhD(biochem), 83. *Prof Exp:* Postdoctoral fel genetics, NC State Univ, 84-86 & biochem & molecular biol, La State Univ Med Ctr, 87-89; STAFF SCIENTIST MOLECULAR GENETICS, ALTON OCHSNER MED FOUND, 89- *Concurrent Pos:* Adj asst prof, La State Univ Med Ctr, 90- *Mem:* Am Soc Biochem & Molecular Biol; Am Soc Microbiol; AAAS. *Res:* Role of hemopexin in heme transport to liver cells and iron conservation; heme-dependent regulation of gene transcription; regulation of the expression of acute phase proteins by inflammatory cytokines. *Mailing Add:* Dept Molecular Genetics Alton Ochsner Med Found 1516 Jefferson Hwy New Orleans LA 70121

ALAM, M KHAIRUL, US citizen. MATERIAL SYNTHESIS, MATERIAL PROCESSING & ENGINEERING. *Educ:* Indian Inst Technol, Kanpur, BTech, 78; Calif Inst Technol, MS, 79, PhD(mech eng), 84. *Prof Exp:* Asst prof, 83-88, ASSOC PROF MECH ENG, OHIO UNIV, 88- *Honors & Awards:* Cert of Recog, NASA, 84. *Mem:* Am Soc Mech Engrs; Am Asn Aerosol Res; Am Asn Combustion Synthesis; Metall Soc. *Res:* Transport processes; materials processing; solidification processing; CVD and modified CVD processes; aerosol science and technology; combustion and air pollution; synthesis of powders and ceramics. *Mailing Add:* Ohio Univ 255 Stocker Ctr Athens OH 45701-2979

ALAM, MAKTOOB, b Allahabad, India, Nov 15, 42; m 73. MARINE NATURAL PRODUCTS, MARINE BIOLOGY. *Educ:* Govt Col, Pakistan, BSc, 62; Univ Karachi, Pakistan, MSc, 64; Univ NH, PhD(biochem), 72. *Prof Exp:* Asst Biochemist, Nat Health Labs, Islamabad, Pakistan, 64-66; lectr chem, Islamia Col, Karachi, Pakistan, 66-68; asst prof med chem, 77-82, ASSOC PROF CHEM, UNIV HOUSTON, 82- *Mem:* Am Soc Pharmacol; Am Asn Col Pharm. *Res:* Marine natural products with pharmacological activities and marine biotoxins. *Mailing Add:* Col Pharmacy Univ Houston Houston TX 77204-5515

ALAM, MOHAMMED ASHRAFUL, b Rajshahi, Bangladesh, Jan 1, 32; m 65; c 2. ANALYTICAL CHEMISTRY, PHYSICAL CHEMISTRY. *Educ:* Rajshahi Govt Col, BSc, 51; Univ Dacca, MSc, 53; La State Univ, PhD(chem), 62. *Prof Exp:* Asst chemist, Jute Res Inst, Pakistan, 54-58; res assoc, Tulane Univ, 62; sr lectr phys chem, Univ Dacca, 64-65; assoc prof chem, Elizabeth City State Col, 65-68; reader phys chem, Islamabad Univ, 68-69; assoc prof chem, 69-76, PROF CHEM, ELIZABETH CITY STATE UNIV, 76- *Concurrent Pos:* NSF res fel, Univ Fla, 66 & 67. *Mem:* Am Chem Soc; Sigma Xi. *Res:* Physicochemical studies of proteins and enzymes; hydrogen bonding studies; metal ion interaction with nucleic acid and polynucleotides. *Mailing Add:* 104 Quail Run Elizabeth City NC 27909

ALAM, SYED QAMAR, b Jagrawan, India, June 22, 32; US citizen. NUTRITIONAL BIOCHEMISTRY, DENTAL RESEARCH. *Educ:* Univ Punjab, WPakistan, BSc, 52; Univ Calif, Berkeley, MS, 61; Mass Inst Technol, PhD(nutrit biochem), 65. *Prof Exp:* Assoc prof nutrit, Fla A&M Univ, 65-67; res assoc nutrit & oral sci, Mass Inst Technol, 67-71; from asst prof to assoc prof, 71-84, PROF BIOCHEM, LA STATE UNIV MED CTR, NEW ORLEANS, 84- *Concurrent Pos:* Consult UN Develop Prog, Pakistan, 81; dir salivary res group, Int Asn Dent Res & Am Asn Dent Res, 90-92. *Mem:* Am Inst Nutrit; Int Asn Dent Res; Am Asn Dent Res; Am Oil Chemists Soc; Am Soc Biochem & Molecular Biol. *Res:* Lipid biochemistry; nutrional effects on structure and function of oral tissues and fluids; membrane lipids and enzymes as influenced by nutritional modifications; carotenoids, retinoids and carcinogenesis. *Mailing Add:* Dept Biochem 1100 Florida Ave New Orleans LA 70119

ALARIE, YVES, b Montreal, Que, Mar 18, 39; m 60; c 4. PHYSIOLOGY. *Educ:* Univ Montreal, BS, 60, MS, 61, PhD(physiol), 63. *Prof Exp:* Physiologist, Hazleton Labs, Inc, 63-70; PROF TOXICOL, UNIV PITTSBURGH, 70- *Concurrent Pos:* Nat Inst Environ Health Sci spec fel, 70. *Honors & Awards:* F R Blood Award, 74 & 81. *Mem:* AAAS; Soc Toxicol; Am Bd Toxicol; Am Col Toxicol. *Res:* Respiratory physiology, effects of air pollutants on respiratory system; sensory irritation; chemoreceptors. *Mailing Add:* Grad Sch of Pub Health Univ of Pittsburgh Pittsburgh PA 15261

AL-ASKARI, SALAH, b Baghdad, Iraq, Nov 16, 27; m 56; c 2. IMMUNOLOGY, UROLOGY. *Educ:* Royal Col Med, Baghdad, MB, ChB, 51; NY Univ, MSc, 59. *Prof Exp:* Instr, 58, univ fel transplantation immunol, 60, USPHS fel, 61, instr urol, 61-63, from asst prof to assoc prof, 63-70, PROF UROL, NY UNIV, 70- *Concurrent Pos:* Co-investr, res contract transplantation immunol, NY Health Res Coun, 61-64, career scientist award, 64- *Mem:* Harvey Soc; Soc Exp Biol & Med; Am Asn Immunol; Transplantation Soc; Am Urol Asn. *Res:* Ileal segments in genitourinary surgery; homograft rejection mechanism and transplantation antigens. *Mailing Add:* Dept of Urol NY Univ Med Ctr 550 1st Ave New York NY 10016

ALAUPOVIC, PETAR, b Prague, Czech, Aug 3, 23; m 47; c 1. BIOCHEMISTRY, ORGANIC CHEMISTRY. *Educ:* Univ Zagreb, ChemE, 48, PhD(chem), 56. *Hon Degrees:* DHC, Univ Lille, France, 87. *Prof Exp:* Researcher, Pharmaceut Res Lab, Chem Corp, Prague, 48-49; researcher, Org Lab, Inst Indust Res, Yugoslavia, 49-50; asst, Agr Fac, Univ Zagreb, 51-54 & Chem Inst Med Fac, 54-56; res biochemist, Univ Ill, 57-60; MEM CARDIOVASC SECT, OKLA MED RES FOUND, 60-, HEAD, LIPOPROTEIN LAB, 72-, HEAD, LIPOPROTEIN & ATHEROSCLEROSIS RES PROG; PROF RES BIOCHEM, SCH MED, UNIV OKLA, 60- *Concurrent Pos:* NIH grants, 61-91; assoc ed, Lipids, 74-78. *Mem:* AAAS; Am Soc Biol Chemists; Am Chem Soc; Am Heart Asn; Am Oil Chem Soc. *Res:* Chemistry of naturally occurring macromolecular lipid compounds such as serum and tissue lipoproteins and bacterial endotoxins; biochemistry of red cell membranes; isolation and characterization of tissue lipases. *Mailing Add:* Lipoprotein & Atherosclerosis Res Prog Okla Med Res Found Oklahoma City OK 73104

ALAVANJA, MICHAEL CHARLES ROBERT, b Brooklyn, NY, Sept 15, 48; m 70; c 2. EPIDEMIOLOGY, ENVIRONMENTAL HEALTH. *Educ:* Brooklyn Col, BS, 70; Hunter Col, MS, 72; Columbia Univ, PhD(epidemiol), 77. *Prof Exp:* From instr to assoc prof indust hyg & epidemiol, City Univ NY, 75-78; sr epidemiol, Nat Inst Occup Safety, 78-81; health sci admn, Nat Inst Health, 81-83; SPEC ASST, NAT CANCER INST, 83- *Mem:* AAAS; Sigma Xi; Soc Epidemiol Res; Am Col Epidemiol. *Res:* Principal investigator environmental and occupational epidemiologic research projects concerning exposures to pesticides, industrial chemical and domestic radon. *Mailing Add:* Epidemiol & Biostatist Prog, Nat Cancer Inst Rm 543 6130 Exec Blvd Exec Plaza N Bethesda MD 20892

ALAVI, MISBAHUDDIN ZAFAR, b Hyderabad State, India, Feb 22, 46; Can citizen; m 77; c 3. EXTRA CELLULAR MATRIX BIOLOGY. *Educ:* Univ Karachi, Pakistan, BSc, 67, BSc(hons), 72, MSc, 73; Univ Heidelberg, Ger, dipl biol, 76 & Univ Ulm, ScD(Human biol), 79. *Prof Exp:* Sci asst, Dept Internal Med, Div Nutrit & Metab, Univ Med Clin, Ulm, Donau, Ger, 79-80; res fel, Ont Heart & Stroke Found, dept path, McMaster Univ, 80-82, asst prof arterial wall metab, 82-85; asst prof pathobiol atherosclerosis, Dept Path, 85-91, ASSOC PROF PATHOBIOL OF ARTERIAL WALL & ATHEROSCLEROSIS, MCGILL UNIV, 91- *Concurrent Pos:* Prin investr, Ont Heart & Stroke Found, 82-; med res scholar, Can Heart Found, 85-90. *Mem:* Am Heart Asn; Am Asn Pathologists; Can Heart Found; Can

Atherosclerosis Soc; Europ Lipoprotein Club; NY Acad Sci. *Res:* Exploration of factors like injury and dietary cholesterol which may act to bring about the changes in arterial wall leading to the clinical syndromes associated with atherosclerosis; interactions of matrix proteoglycan with plasma lipoproteins in health and disease; pathogenesis of atherosclerosis. *Mailing Add:* Dept Path McGill Univ Lyman Duff Med Sci Bldg 3775 University St Montreal PQ H3A 2B4 Can

ALAVI, YOUSEF, b 1929. MATHEMATICS. *Educ:* Mich State Univ, BS, 53, MS, 55, PhD(math), 58. *Prof Exp:* From asst prof to assoc prof, 58-68, PROF MATH, WESTERN MICH UNIV, 68- *Mem:* Am Math Soc; Sigma Xi; Math Asn Am. *Res:* Graph theory. *Mailing Add:* Dept Math & Statist Western Mich Univ Kalamazoo MI 49008

AL-AWQATI, QAIS, b Baghdad, Iraq, Aug 18, 39; m. PHYSIOLOGY, BIOPHYSICS. *Educ:* Univ Baghdad, Iraq, MB, ChB, 62; Am Bd Internal Med, cert, 72; Am Bd Nephrology, cert, 72. *Prof Exp:* Fel med, Johns Hopkins Med Sch, 67-70; fel med, Harvard Med Sch, Mass Gen Hosp, 70-71, instr, 71-73; asst prof med, Col Med, Univ Iowa, 73-77; assoc prof, 77-84, PROF MED & PHYSIOL, COLUMBIA UNIV COL PHYSICIANS & SURGEONS, 84-, R F LOEB PROF, 88- *Concurrent Pos:* NIH res award, 76- *Mem:* Soc Gen Physiologists; Am Soc Clin Invest; Am Physiol Soc; Am Fedn Clin Res; Am Soc Nephrology; Am Asn Physicians. *Res:* Mechanisms of water and ion transport across biological membranes; particularly the energetics of active transport. *Mailing Add:* Dept of Med Columbia Univ 630 W 168th St New York NY 10032

ALBACH, RICHARD ALLEN, b Chicago, Ill, Mar 3, 31; m 53; c 7. CELL PHYSIOLOGY, MICROBIOLOGY. *Educ:* Univ Ill, BS, 56, MS, 58; Northwestern Univ, PhD(microbiol), 63. *Prof Exp:* Res assoc microbiol, Lutheran Gen Hosp, 63-67; assoc, Inst Med Res, 67-68, from asst prof to assoc prof, 68-73, PROF MICROBIOL, CHICAGO MED SCH, 73-, V CHMN DEPT, 74- *Concurrent Pos:* Co-prin investr, Nat Inst Allergy & Infectious Dis grant, 65-68 & 68-73, prin investr, 74-78. *Honors & Awards:* Bd Trustees Res Award, Chicago Med Sch, 68. *Mem:* AAAS; Am Soc Microbiol; Soc Protozool; NY Acad Sci; Soc Parasitologists. *Res:* Aspects of DNA and RNA metabolism; nutritional requirements; metabolism of entamoeba histolytica. *Mailing Add:* Dept Microbiol/ Immun Univ Health Sci Chicago Med 3333 Green Bay Rd N Chicago IL 60064

ALBACH, ROGER FRED, b Chicago, Ill, Mar 25, 32; m 57; c 3. ENVIRONMENTAL & FOOD CHEMISTRY. *Educ:* Fresno State Col, BS, 57; Univ Calif, Davis, MS, 60, PhD(chem), 64. *Prof Exp:* Range aide, Agr Res Serv, USDA, 54-55; field inspector, Calif Spray Chem Corp, 56; lab technician chem, Fresno State Col, 57; RES CHEMIST, SUBTROP AGR RES LAB, AGR RES SERV, USDA, 63- *Mem:* Am Chem Soc; AAAS; Sigma Xi; Am Soc Hort Sci; Phytochem Soc NAm; Int Soc Chem Ecol. *Res:* Phytochemistry; natural products; food chemistry; groundwater pollution. *Mailing Add:* 2413 E Highway 83 Weslaco TX 78596-8344

ALBANESE, ANTHONY AUGUST, b New York, NY, Feb 12, 08; m 33. BIOCHEMISTRY. *Educ:* NY Univ, BS, 30; Columbia Univ, PhD(biochem), 40; Am Bd Nutrit, dipl, 51. *Prof Exp:* Res asst path, Col Physicians & Surgeons, Columbia Univ, 33-40; res assoc, Div Chem, NIH, 40-41; assoc pediat & res, Johns Hopkins Univ, 41-45; from asst prof to assoc prof pediat biochem, Col Med, NY Univ, 45-49; dir Res & Clin Lab, St Luke's Convalescent Hosp, 49-60; DIR GERIAT NUTRIT & CLIN LAB, MIRIAM OSBORN MEM HOME, 50-; DIR NUTRIT & METAB RES DIV & CLIN LAB, BURKE REHAB CTR, 59-; LAB DIR, NY STATE DEPT HEALTH, 66-, ED-IN-CHIEF, NUTRIT REPORTS INT. *Concurrent Pos:* Corn Indust fel, NIH, 40-41; responsible investr, Off Sci Res & Develop, 41-45, Off Naval Res, 45-72 & Air Res & Develop Command, 60-65; guest lectr, Univ Brazil, 59, Univ Tokyo & Univ Osaka, 60 & Univ Istanbul, 70; assoc ed, NY State J Med, 59-; lab dir, NY State Dept Health, 66-; ed-in-chief, Nutrit Reports Int, 70-; mem, President's Sci Adv Comt Toxicol Info Prog. *Mem:* Fel Am Inst Chemists; Am Chem Soc; Fedn Am Socs Exp Biol; Am Soc Biol Chemists; Soc Exp Biol & Med; Sigma Xi; Am Inst Nutrit; Am Soc Clin Nutrit; AMA; Am Therapeut Soc. *Res:* Nutritional effects of steroids; fat metabolism; metabolic effects of enzymes; amino acid chemistry; biological value of various proteins in the human; electrolytic method for determination of basic amino acids in proteins; carbohydrate and cholesterol metabolisms; nutritional needs, problems and requirements of infants, pre-school children, elderly people and stroke patients; food habits of teenagers, college students and crippled children; metabolic management of wounded military personnel and post-surgical patients; effects of fats, cholesterol and other factors in heart disease; nutritional problems of hypokinesia of prolonged submarine and space travel; radiographic and therapeutic aspects of bone loss. *Mailing Add:* Ed-in-Chief Nutrit Reports Int PO Box 788 Harrison NY 10528

ALBANO, ALFONSO M, b Laoag, Ilocos Norte, Philippines, Aug 2, 39; US citizen; m 67; c 2. NONLINEAR DYNAMICS CHAOS, STATISTICAL PHYSICS. *Educ:* Univ Philippines, BS, 59; Univ Iowa, MS, 64; State Univ NY, Stony Brook, PhD(physics), 69. *Prof Exp:* Asst instr physics, Univ Philippines, 60-62; instr physics, State Univ NY, Stony Brook, 69-70; lectr physics, Bryn Mawr Col, 70-71, from asst prof to prof, 71-85, MARION REILLY PROF PHYSICS, BRYN MAWR COL, 85- *Concurrent Pos:* Vis physicist, Lorentz Inst, Univ Leiden, Neth, 74-75 & 78-79; Sloan vis fel, Dept Civil Eng, Princeton Univ, 85, 87, 88 & 89; consult, UN Develop Prog, 88; fac, Naval Air Develop Ctr, Navy-Am Soc Elec Eng prog, 89 & 90. *Mem:* Am Phys Soc; Am Asn Physics Teachers; Sigma Xi; Philippine-Am Acad Sci & Eng (pres, 88). *Res:* Nonlinear dynamics and chaos; nonlinear dynamical characterization of electrophysiological signals. *Mailing Add:* Dept Physics Bryn Mawr Col Bryn Mawr PA 19010

ALBANO, MARIANITA MADAMBA, b Dingras, Ilocos Norte, Philippines, Nov 15, 41. MICROBIOLOGY. *Educ:* Univ Philippines, BSc, 64; Ohio State Univ, MSc, 69, PhD(med mycol), 77. *Prof Exp:* Cur asst, algal herbarium, Univ Philippines, 64-65, instr bot, 65-66; MEM FAC BIOL & MICROBIOL, UNIV MD, EASTERN SHORE, 73- *Mem:* Sigma Xi; Am Soc Microbiologists; Med Mycol Soc. *Res:* Evaluation of antimicrobial properties of extracts from Ganyaulax monilata; effects of microwaves on aflatoxin production and growth of Aspergillus flavus on corn. *Mailing Add:* Dept Natural Science Univ Md Eastern Shore Princess Anne MD 21853

ALBATS, PAUL, b Latvia, Dec 31, 41; US citizen; m 69; c 2. COSMIC RAY PHYSICS. *Educ:* Univ Chicago, BS, 64; Cornell Univ, PhD(physics), 71. *Prof Exp:* From res assoc to sr res assoc physics, 69-73, asst prof physics, Case Western Reserve Univ, 74-76; MEM STAFF, SCHLUMBERGER-DOLL RES CTR, 77- *Mem:* Am Phys Soc; Sigma Xi. *Res:* Experimental high energy astrophysics gamma ray astronomy; experiments to measure atmospheric high energy neutron flux; experimental search for solar neutrons. *Mailing Add:* Schlumberger-Doll Res Old Quarry Rd Ridgefield CT 06877

ALBAUGH, A HENRY, b Chicago, Ill, Aug 21, 22; m 50; c 4. MATHEMATICS, STATISTICS. *Educ:* Mich State Univ, BS, 48; Univ Mich, MA, 50. *Prof Exp:* Teacher high schs, Mich, 52-57; asst prof, 57-60, head dept, 57-75, assoc prof math, Hillsdale Col, 60-84; CONSULT, 84- *Mem:* Nat Coun Teachers Math. *Res:* Educational games; thermocycling. *Mailing Add:* 26 N Norwood Hillsdale MI 49242-1511

ALBAUGH, FRED WILLIAM, b Albia, Iowa, Apr 13, 17; m 44; c 3. NUCLEAR ENGINEERING, TECHNICAL MANAGEMENT. *Educ:* Univ Calif, Los Angeles, BA, 35; Univ Mich, MA, 38, PhD(chem), 42. *Prof Exp:* Res engr, Union Oil Co, Calif, 41-43 & 45-47, Manhattan Proj, Univ Chicago, 43-45; mgr res & develop, Gen Elec Co, 47-65; mgr res & develop & lab dir, Northwest Lab, Battelle Mem Inst, 65-71, corp tech dir, 71-76; chmn, Utility Adv Bd, Richland, Wash, 77-87; consult, 76-88; RETIRED. *Mem:* Nat Acad Eng; Am Chem Soc; AAAS; fel Am Nuclear Soc; Sigma Xi. *Res:* Nuclear energy; nuclear chemistry; petroleum production research. *Mailing Add:* 2534 Harris Ave Richland WA 99352

AL-BAZZAZ, FAIQ J, b Baghdad, Iraq, July 1, 39; c 3. MEDICINE, PHYSIOLOGY. *Educ:* Univ Baghdad, MBChB, 62; FRCP(C). *Prof Exp:* Clin & res fel, Mass Gen Hosp, Harvard Univ, 69-71; DIR RESPIRATORY PHYSIOL LAB, VET ADMIN W SIDE MED CTR, 72-, CHIEF RESPIRATORY & CRITICAL CARE SECT, 77-; PROF MED, COL MED, UNIV ILL, 86- *Mem:* Fel Am Col Chest Physicians; fel Am Col Physicians; Am Thoracic Soc; Am Physiol Soc; AAAS. *Res:* Ion and water transport by respiratory mucosa; roles of arachidonic acid metabolites, cyclic nucleotides and calcium in regulation of ion transport. *Mailing Add:* Med Serv MP 111 Vet Admin W Side Med Ctr 820 S Damen Ave Chicago IL 60612

ALBEE, ARDEN LEROY, b Port Huron, Mich, May 28, 28; m 53, 78; c 8. GEOLOGY, PLANETARY SCIENCE. *Educ:* Harvard Univ, BA, 50, MA, 51, PhD(geol), 57. *Prof Exp:* Geologist, US Geol Surv, 50-59; chief scientist, Jet Propulsion Lab, NASA, 78-84; from vis asst prof to assoc prof, 59-66, PROF GEOL, CALIF INST TECHNOL, 66-, DEAN GRAD STUDIES, 84- *Concurrent Pos:* Consult, Los Angeles Dept Water & Power, 64-70, Corral Canyon nuclear plant site & Pac DC Intertie, Converse, Davis & Assocs, 64-73 & Fugro, 74-78; assoc ed, Bull Geol Soc Am, 70-, Am Mineralogist, 72-76, Proc 5th Lunar Sci Conf, J Geophys Res, 77-82, Ann Rev Earth & Space Sciences, 78-; NASA serv: chmn, Lunar Sci Rev Panel, 72-77, Mars Sci Working Group, 77-78 & Mars Sci Steering Group, 77-, mem, Lunar Planning Comt, 73-75, Terrestrial Bodies Sci Working Group, 76-77, Phys Sci Comt, 76-77, Mars '84 Sci Working Group, 77 & Space Sci Adv Comt, 78-81, leader, Mars Prog, 78-79, Solar System Explor Comt, 80-83 & Mgt Coun, 83-88, proj sci, Mars Observation Mission, 84-; mem, Task Force Planetary & Lunar Explor, Nat Acad Sci, 84-86; mem, Mars Rover Sample Return Sci Working Group, 87-; mem, US-USSR Joint Working Group Solar Syst Explor, 87- *Honors & Awards:* NASA Medal for Except Sci Achievement, 76. *Mem:* Geol Soc Am; Microbeam Anal Soc; Mineral Soc Am; Geochem Soc; Am Geophys Union. *Res:* Planetary science; metamorphic petrology; electron microprobe; lunar rock investigations. *Mailing Add:* Div Geol & Planetary Sci Calif Inst Technol Pasadena CA 91125

ALBEE, HOWARD F, b Fruitland, Idaho, Feb 1, 15. ECONOMICS, COAL. *Educ:* Idaho State Univ, BA, 50. *Prof Exp:* Dist geologist, Coal Explor, 50-84; RETIRED. *Mem:* Sr fel Geol Soc Am. *Mailing Add:* 5051 Niagara Circle Salt Lake City UT 84118

ALBEN, JAMES O, b Seattle, Wash, July 13, 30; m 54; c 3. BIOCHEMISTRY. *Educ:* Reed Col, BA, 51; Univ Ore, MS, 57, PhD(biochem), 59. *Prof Exp:* USPHS res fel physiol chem, Sch Med, Johns Hopkins Univ, 59-62; from asst prof to assoc prof, 62-73, PROF PHYSIOL CHEM, OHIO STATE UNIV, 73- *Concurrent Pos:* NIH grant, 64- *Mem:* Am Chem Soc; NY Acad Sci; Coblentz Soc; Brit Chem Soc; Am Soc Biol Chemists. *Res:* Molecular spectroscopy and structure of heme proteins; Fourier transform infrared interferometry; hemoglobin structure and function; oxygen binding and transport; physical inorganic biochemistry of metal proteins; mechanisms of metal-enzyme catalysis. *Mailing Add:* Dept of Physiol Chem Ohio State Univ 1645 Neil Ave Columbus OH 43210

ALBEN, RICHARD SAMUEL, b Brooklyn, NY, July 12, 44; m 71; c 3. SOLID STATE PHYSICS, THERMODYNAMICS. *Educ:* Harvard Col, AB, 64; Harvard Univ, AM, 65, PhD(physics), 67. *Prof Exp:* Res assoc physics, NSF fel, Univ Osaka, 67-68; from asst prof to assoc prof eng, Yale Univ, 68-77; tech staff, Gen Elec Res Ctr, 77-79, mgr technol eval, 79-82, consult technol eval, 83-89, MGR ELECTRONICS PLANNING, GEN ELECTRIC RES CTR, 89- *Concurrent Pos:* Consult, IBM Watson Res Ctr, 73; vis scholar, US-France Exchange Scientists Prog-Ctr Nat Res Sci, Grenoble, 74; chmn prog comt, Conf Magnetism & Magnetic Mat, 76; mem, Gas Res Inst Res Coord Coun, 83-86, Indust Planning Coun, Carnegie Mellon Univ Eng Res Ctr, 88- *Mem:* Am Phys Soc; AAAS. *Res:* Magnetism in insulators, amorphous semiconductors, disordered alloys, optical properties, thermodynamic cycles, solar energy systems, energy technology evaluation, utility economics, computer-aided design. *Mailing Add:* Gen Elec Res Ctr Schenectady NY 12301

ALBERCH, PERE, b Badalona, Spain, Nov 2, 54; m 75. EVOLUTION, MORPHOGENESIS. *Educ:* Univ Kans, BA, 76; Univ Calif Berkeley, PhD(zool), 80. *Prof Exp:* ASST PROF BIOL, HARVARD UNIV, 80-, ASST CUR, MUS COMPARATIVE ZOOL, 80- *Mem:* AAAS; Soc Study Evolution; Am Soc Ichthyologists & Herpetologists; Soc Systs Zool; Soc Study Amphibians & Reptiles. *Res:* Relationships between development and evolution; comparative and experimental vertebrate embryology. *Mailing Add:* Mus Vert Zool Univ Calif 2593 Life Sci Bldg Berkeley CA 94720

ALBERDA, WILLIS JOHN, b Bozeman, Mont, Feb 7, 36; m; c 2. MATHEMATICAL STATISTICS. *Educ:* Calvin Col, AB, 59; Mont State Univ, MS, 63, Nat Defense Educ Act fel & PhD(math statist), 64. *Prof Exp:* PROF MATH, DORDT COL, 64- *Mem:* Am Math Soc; Am Statist Asn; Math Asn Am; Am Sci Affil. *Res:* Central limit theorems in probabilistic models; design and analysis of sampling procedures and population models. *Mailing Add:* Dordt Col Fourth Ave NE Dordt Col Fourth Ave NE Sioux Center IA 51250

ALBERGOTTI, JESSE CLIFTON, b Columbia, SC, Jan 4, 37. EXPERIMENTAL NUCLEAR PHYSICS. *Educ:* Wheaton Col, BS, 58; Univ NC, PhD(physics), 63. *Prof Exp:* Asst prof physics, Davidson Col, 62-64; from asst prof to assoc prof, 64-80, PROF PHYSICS, UNIV SAN FRANCISCO, 80-, CHMN DEPT, 68- *Mem:* Am Phys Soc; Am Asn Physics Teachers. *Mailing Add:* Dept Physics Univ San Francisco San Francisco CA 94117

ALBERNAZ, JOSE GERALDO, b Januaria, Brazil, Dec 3, 23; m 50; c 5. NEUROLOGY, NEUROSURGERY. *Educ:* Univ Minas Gerais, BS, 40, MD, 46, MS, 55; Univ Guanabara, PhD(neurosurg), 58. *Prof Exp:* Prof neurol & actg head dept, Univ Minas Gerais, 56-58, asst prof neuropath, 58-59; Rockefeller Found res fel & travel grant, Sch Med, Univ Wash, 59-60; prof neurol & actg head dept, Univ Minas Gerais, 61-62, prof neurol & chmn dept, 62-66, prof neurol & neurol surg & head dept, 66-68; prof neuroanat, Med Col Ohio, 68-72; mem staff, Dept Neurol & Neurol Surg, Marion Gen Hosp, Marion, Ohio, 72-79; MEM STAFF, DEPT NEUROL & NEUROL SURG, MERCY HOSP, TIFFIN, OHIO, 79- *Concurrent Pos:* Exec comt mem, Int Cong Neurol Surg, Denmark, 66; mem courtesy staff, Marion Gen Hosp & Bellevue Hosp. *Mem:* Am Asn Anat; cor mem Am Asn Neurol Surg; fel Am Col Surg; Pan Am Asn Anat (secy, 70-); Soc Neuroscience. *Res:* Significance of supraspinal control of the gamma system; structure and function of cerebral arteries; causes of reactions to spinal puncture and pneumoencephalography. *Mailing Add:* Mercy Hosp 485 W Market St PO Box 727 Tiffin OH 44883-0727

ALBERS, EDWIN WOLF, b Schenectady, NY, July 29, 30; m 57; c 3. PHYSICAL CHEMISTRY. *Educ:* Clarkson Univ, BS, 52; Rensselaer Polytech Inst, PhD(phys chem), 62. *Prof Exp:* Chemist, Houston Refinery, Shell Oil Co, 52-55; asst chem, Rensselaer Polytech Inst, 55-57 & 58-60, instr, 60-61, from res asst to res assoc, 61-65; phys chemist, Inst Gas Technol, Ill Inst Technol Res Inst Ctr, 65-67; sr res scientist, W R Grace & Co, 67-69, supvr catalytic prep, Davidson Chem Div, 69-82; CONSULT, ALBERS ASSOC, MD, 85-; PRES, CMP INC, 91- *Concurrent Pos:* Adj prof chem, Anne Arundel Community Col, 78- *Mem:* Am Chem Soc; Am Phys Soc. *Res:* Adsorption; gas solid systems; low temperature physics and chemistry; chemical kinetics related to the upper atmosphere; interstellar space research; zeolite and zeolitic promoted catalysts for petroleum and petrochemical applications. *Mailing Add:* CMP Inc 1922 Benhill Rd Baltimore MD 21226

ALBERS, FRANCIS C, b Mar 3, 16. MATERIALS SCIENCE ENGINEERING, METALLURGY & PHYSICAL METALLURGICAL ENGINEERING. *Educ:* Univ Wis, BS, 40. *Prof Exp:* Chief metall, Chicago Pneumatic Tool Co, 49-82; RETIRED. *Mem:* Fel Am Soc Metals. *Mailing Add:* 20 Hills Dr Utica NY 13501

ALBERS, HENRY, b Andover, Mass, Nov 17, 25; m 50; c 3. ASTRONOMY. *Educ:* Harvard Univ, AB, 50; Univ Minn, MA, 52; Case Inst Technol, PhD(astron), 56. *Prof Exp:* Instr astron, Univ Minn, 53-55 & Case Inst Technol, 55-56; asst prof math & astron, Butler Univ, 56-58; from asst prof to assoc prof, Vassar Col, 58-77, prof & chmn dept, 77-91; RETIRED. *Concurrent Pos:* Vis lectr, Macalester Col, 54-55; NSF sci fac fel, 66. *Mem:* Am Astron Soc; Sigma Xi; AAAS. *Res:* Objective prism spectroscopy; galactic structure; Magellanic Clouds. *Mailing Add:* 62 Prospect St Falmouth MA 02540

ALBERS, JAMES RAY, b Tacoma, Wash, Sept 4, 34; m 61; c 2. THEORETICAL PHYSICS. *Educ:* Wash State Univ, BS, 56; George Washington Univ, MS, 58; Univ Wash, PhD(physics), 62. *Prof Exp:* Physicist, Nat Bur Standards, 57-58; lectr physics, Seattle Univ, 63-66; from asst prof to assoc prof, 66-71; assoc prof, 71-74; vprovost instruct & planning, 74-83, PROF, HUXLEY COL ENVIRON STUDIES, WESTERN WASH UNIV, 74-, ASSOC VPRES ACAD AFFAIRS, 83- *Mem:* AAAS; Am Phys Soc; Soc Comput Simulation; Sigma Xi. *Res:* Field theory as it pertains to the elementary particle mass spectrum; neutrinos and weak interactions; alternate energy resources; computer modeling. *Mailing Add:* 120 Sea Pines Rd Bellingham WA 98226

ALBERS, JOHN J, b Alton, Ill, Oct 5, 41. MICROBIOLOGY, IMMUNOLOGY. *Educ:* Univ Ill, PhD(microbiol & immunol), 69. *Prof Exp:* Lab Dir, NW Lipid Res Clin, Div Metab, Endocrin & Nutrit, Seattle, Wash, 71-81; from res asst prof to res assoc prof med, Univ Wash, 71-82, assoc dir, NW Lipid Res Ctr, Div Metab, Endocrin & Nutrit, 82-89, RES PROF MED, UNIV WASH, DIV METAB, ENDOCRIN & NUTRIT, 82- *Concurrent Pos:* Estab investr, Am Heart Asn, 79-84; adj res prof path, Univ Wash, 85-; dir NW Lipid Res Labs, Div Metab & Endocrin & Nutrit, Univ Wash, Seattle, 90- *Honors & Awards:* Irvine H Page Res Award Young Invest, Council Ateriosclerosis, Am Heart Asn, 80. *Mem:* Am Asn Immunologists; Am Chem Soc; AAAS; Sigma Xi; Am Oil Chemist Soc; Am Asn Clin Chem. *Res:* Lipoprotein structure; metabolism; pathophysiology. *Mailing Add:* Northwest Lipid Res Lab 2121 N 35th St Seattle WA 98103

ALBERS, JOHN P, geology, for more information see previous edition

ALBERS, PETER HEINZ, b Staten Island, NY, May 4, 43; m 72; c 2. ENVIRONMENTAL CONTAMINANTS, WILDLIFE BIOLOGY. *Educ:* Univ Mont, BS, 65; Univ Guelph, MS, 66; Univ Mich, Ann Arbor, PhD(wildlife mgt), 75. *Prof Exp:* Instr wildlife mgt, Sch Forestry, Univ Maine, Orono, 75; WILDLIFE RES BIOLOGIST ENVIRON CONTAMINANTS, PATUXENT WILDLIFE RES CTR, US FISH & WILDLIFE SERV, LAUREL, MD, 76- *Mem:* Wildlife Soc; Ecol Soc Am; Sigma Xi. *Res:* Effects of environmental contaminants on animal populations and individual animal function,; petroleum; metals; acid deposition; conservation tillage. *Mailing Add:* US Fish & Wildlife Serv Patuxent Wildlife Res Ctr Laurel MD 20708

ALBERS, ROBERT CHARLES, b San Jose, Calif, Feb 11, 49; m 78; c 2. SOLID STATE PHYSICS. *Educ:* Univ Santa Clara, BS, 71; Cornell Univ, MA, 74, PhD(physics), 77. *Prof Exp:* Fel, 77, MEM STAFF, THEORET PHYSICS DIV, LOS ALAMOS NAT LAB, 77- *Mem:* Am Phys Soc; AAAS. *Res:* Anharmonic phonon theories; theory of metals; electronic band structure of solids; x-ray absorption in solids; multiple-scattering theory. *Mailing Add:* Los Alamos Nat Lab Group T-11 M/S B262 PO Box 1663 Los Alamos NM 87545

ALBERS, ROBERT EDWARD, b Orange, NJ, Apr 19, 31; m 56; c 4. CHEMICAL ENGINEERING. *Educ:* Lehigh Univ, BS, 53. *Prof Exp:* Engr nuclear reactor testing, Savannah River Plant, 53-54, shift supvr, 54-55, hydraul supvr, 55, control room supvr, 55-57, res engr, pigments dept, Res Div, 57-58, develop engr, 59-60, res engr, 60-65, from res engr to sr res engr, Textile Fibers Dept, 65-80, sr engr, eng dept, 81-84, CONSULT, E I DU PONT DE NEMOURS & CO, INC, 84- *Res:* Nuclear reactor engineering and operation; refractory metal extractive and process metallurgy; titanium and columbium metals compounds and minerals; powder metallurgy; process planning and economic analyses; synthetic fiber processing, nonwovens; melt fiber spinning; ceramic fiber spinning; melt crystalization; gas permeation. *Mailing Add:* 2840 Kennedy Rd Wilmington DE 19810

ALBERS, ROBERT WAYNE, b Hebron, Nebr, Aug 5, 28. NEUROCHEMISTRY. *Educ:* Univ Nebr, BS, 50; Wash Univ, PhD(pharmacol), 54. *Prof Exp:* Biochemist, Lab Neuroanat Sci, 54-61, head sect enzymes, 61-77, CHIEF SECT ENZYMES, LAB NEUROCHEM, NIH, 77- *Concurrent Pos:* Prof lectr, George Washington Univ, 64-65; lectr neurochem, Found Adv Educ Sci, 70-80. *Mem:* Am Soc Biol Chem; Am Soc Neurochem; Int Soc Neurochem; Fedn Am Socs Exp Biol. *Res:* Biochemistry of the nervous system; molecular biology of cell membranes and membrane transport systems. *Mailing Add:* Lab Neurochem Nat Inst Neurol & Commun Disorders & Stroke Bldg 36 Rm 4D 20 Bethesda MD 20892

ALBERS, WALTER ANTHONY, JR, b McKeesport, Pa, July 19, 30; m 52; c 5. RESEARCH ADMINISTRATION, SOLID STATE PHYSICS. *Educ:* Wayne State Univ, BS, 52, MS, 54, PhD(physics), 59. *Prof Exp:* Mem tech staff, Bell Tel Labs, 54-55; physicist, Res Labs, Bendix Corp, 55-57; res assoc, Wayne State Univ, 57-59; sr physicist, Res Labs, Bendix Corp, 59-62; supvry physicist, 62-73, head societal anal dept, 73-87, HEAD OPERATING SCI DEPT, RES LABS, GEN MOTORS CORP, 87- *Mem:* AAAS; Am Phys Soc; Sigma Xi. *Res:* Surface and chemical physics; optics; quantitative social sciences; marketing sciences; manufacturing sciences. *Mailing Add:* 2068 S Hammond Lake Dr West Bloomfield MI 48033

ALBERSHEIM, PETER, b New York, NY, Mar 30, 34; m 58; c 3. PLANT BIOCHEMISTRY, COMPLEX CARBOHYDRATES SCIENCE. *Educ:* Cornell Univ, BS, 56; Calif Inst Technol, PhD(biochem), 59. *Prof Exp:* Fel biochem, Calif Inst Technol, 59; NSF fel, Swiss Fed Inst Technol, 59-60; from instr to asst prof biol, Harvard Univ, 60-64; from assoc prof to prof biochem, Univ Colo, 64-85, prof molecular biol, 70-85; RES PROF BIOCHEM & DIR COMPLEX CARBOHYDRATE RES CTR, UNIV GA, 85- *Concurrent Pos:* Fac fel, Univ Colo, 70-71 & 75-76, fac res lectr, 80. *Honors & Awards:* Charles A Shull Award, Am Soc Plant Physiologists, 73; Ruth & Tracey Storer life sci lectr, Univ Calif, Davis, 77; Robert L Stearns Award, 79; Kenneth A Spencer Award, Am Chem Soc, 84. *Mem:* AAAS; Am Chem Soc; Am Soc Plant Physiol; Am Soc Biol Chem; Am Phytopath Soc; Soc Complex Carbohydrates; Int Soc Plant Molecular Biol. *Res:* Structure and function of biologically active complex carbohydrates; function and nature of plant cell walls; mechanisms underlying disease resistance in plants and virulence in pathogens; development of new methods for the structural analysis of complex carbohydrates. *Mailing Add:* Complex Carbohydrates Res Ctr Univ Ga 220 Riverbend Rd Athens GA 30602

ALBERT, ANTHONY HAROLD, b Los Angeles, Calif, Nov 24, 40; m 69. PHARMACEUTICAL CHEMISTRY. *Educ:* Occidental Col, BA, 63; San Diego State Univ, MS, 65; Ariz State Univ, PhD(org chem), 71. *Prof Exp:* Res assoc med chem, Nucleic Acid Res Inst, ICN Pharmaceut, 69-71, res chemist, 71-73, head process res, Prod Develop, 73-76; instr, Saddleback Col, 76-77; prin res chem RIA, Diag Oper, Beckman Instrument, 77-79; chemist, Fluorochem Inc, 79-80; dir org synthesis & scale up, Morton Thiokol/Dynachem, 80-83; dir process develop, Newport Pharmaceut, 83-84; group leader chem process develop, 84-89, MGR PROCESS ENG & DEVELOP, MORTON INT/DYNACHEM, 89- *Mem:* Am Chem Soc. *Res:* Direct development of processes for photoresist intermediate preparation; design equipment and processes for photoresist productions. *Mailing Add:* 18241 Montana Circle Villa Park CA 92667

ALBERT, ARTHUR EDWARD, b New York, NY, Nov 6, 35; m 59; c 3. MATHEMATICAL STATISTICS, APPLIED MATHEMATICS. *Educ:* Mass Inst Technol, BS, 56; Stanford Univ, MS, 57, PhD(math statist), 59. *Prof Exp:* NSF fel math & statist, Inst Math Sci, Stockholm, 59-60; asst prof, Columbia Univ, 60-61; res assoc elec eng, Mass Inst Technol, 61-62; sr scientist, Arcon Corp, 62-70; prof math, Boston Univ, 70-84; RETIRED. *Mem:* Inst Math Statist; Soc Indust & Appl Math. *Res:* Statistical decision theory; stochastic processes; time series analysis; pattern recognition; statistical estimation procedures; design of experiments. *Mailing Add:* RFD 1 Box 100 Plymouth NH 03264

ALBERT, ERNEST NARINDER, b Gujarat, WPakistan, July 21, 37; m 60; c 5. ANATOMY, CELL BIOLOGY. *Educ:* High Point Col, BS, 58; Univ Pittsburgh, MS, 63; Georgetown Univ, PhD(anat), 65. *Prof Exp:* Instr gen & oral histol, Sch Med & Dent, Georgetown Univ, 65-66; res assoc, Univ Calif, Los Angeles, 66-67; asst prof anat, Jefferson Med Col, 67-68; from asst prof to assoc prof anat,68-76, PROF ANAT, SCH MED, GEORGE WASHINGTON UNIV, 76- *Mem:* Am Asn Anat; Electron Micros Soc Am; Am Heart Asn; Bioelectromagnetic Soc (pres, 79-80). *Res:* Initiating factors in atherogenesis; electromagnetic interactions with biological systems. *Mailing Add:* Dept Anat George Washington Univ Med Sch 2300 Eye St NW Washington DC 20037

ALBERT, EUGENE, b New York, NY, Jan 9, 30; m 60; c 1. MATHEMATICS. *Educ:* Brooklyn Col, BA, 50, MA, 51; Univ Va, PhD(math), 61. *Prof Exp:* Instr math, Brooklyn Col, 53-56; engr, Gen Elec Corp, NY, 56-57; asst prof math, Union Col, NY, 57-58; instr, Univ Va, 58-61; John Wesley Young res instr, Dartmouth Col, 61-63; asst prof, Univ Calif, Davis, 63-67; ASSOC PROF MATH, CALIF STATE UNIV, LONG BEACH, 67- *Concurrent Pos:* Ed, "Ideal-Mate Rev", 83- *Mem:* Math Asn Am. *Res:* Probability theory; Markov chains; chess problems. *Mailing Add:* Calif State Univ Calif State Univ 1250 Bell Flower Blvd Long Beach CA 90840

ALBERT, HAROLD MARCUS, b Russellville, Ala, Mar 5, 19; m 47; c 2. CARDIOVASCULAR SURGERY. *Educ:* Univ Chicago & Tulane Univ, BS, 40; Tulane Univ, MD, 44. *Prof Exp:* Intern, Touro Infirmary, 44-45, resident surg, 45-47; sr fel, Ochsner Found, 47-48; sr surgeon, Huey P Long Mem Hosp, 48-49; clin instr, 49-56, from asst prof to assoc prof, 56-73, prof, 73-80, EMER PROF SURG, SCH MED, LA STATE UNIV, NEW ORLEANS, 80- *Concurrent Pos:* Preceptorship, Huey P Long Mem Hosp, 49-51; La Heart Asn fel, Northwestern Univ, 51-52; mem staff, Charleston Naval Hosp & Portsmouth Naval Hosp, 54-56; mem cardiovasc surg coun, Am Heart Asn; sr vis surgeon, Charity Hosp La, New Orleans; sr assoc, Touro Infirmary; active staff mem, Hotel Dieu, New Orleans, Methodist Hosp & West Jefferson Hosp, Marrero. *Mem:* Am Col Surg; Soc Vascular Surg; Am Col Chest Physicians; Am Soc Artificial Internal Organs; Am Col Cardiol. *Res:* Congenital and acquired heart disease; shock. *Mailing Add:* 1111 Ave D Suite 501 Marrero LA 70072

ALBERT, HARRISON BERNARD, b Oakland, Calif, Dec 12, 36; m 59; c 4. CHEMICAL INSTRUMENTATION, COMPUTER SCIENCES. *Educ:* Univ Calif, Berkeley, BS, 59; Univ Colo, Boulder, PhD(phys chem), 69. *Prof Exp:* Mem res staff, Thomas J Watson Res Ctr, Int Bus Mach Corp, NY, 69-70; res assoc, Univ Colo, Boulder, 74-77, asst prof chem, 70-80, sr res assoc, 77-80; MEM STAFF, BALL AEROSPACE SYST, 80- *Concurrent Pos:* NSF fel, 63-66. *Mem:* AAAS; Am Chem Soc; Sigma Xi. *Res:* Analytical chemistry instrumentation and computing science; gas and high pressure liquid chromatography and disc gel chromatography data acquisition hardware; algorithms for storage, compression, deconvolutions, displays and removing noise and drift; innovative microcalorimeter configurations and electronics. *Mailing Add:* Ball Aerospace Syst PO Box 1062 Boulder CO 80306

ALBERT, JERRY DAVID, b Milwaukee, Wis, June 6, 37; m 61; c 2. BIOCHEMISTRY, HYDROLOGY & WATER RESOURCES. *Educ:* Occidental Col, BA, 59; Iowa State Univ, PhD(biochem), 64. *Prof Exp:* Sr scientist appl res labs, Aeronutronic Div, Philco-Ford Corp, 64-67; fel prebiotic chem, Salk Inst Biol Studies, 67-68; staff res assoc IV, Univ Hosp, San Diego County, 68-73; res biochemist, Mercy Hosp & Med Ctr, 73-88; CHEMIST, CITY SAN DIEGO, 88- *Concurrent Pos:* Instr Chem, San Diego Mesa Col. *Mem:* Am Chem Soc. *Res:* Analysis of water. *Mailing Add:* 11223 Cascada Wy San Diego CA 92124-2878

ALBERT, LUKE SAMUEL, b Palmyra, Pa, Feb 19, 27; m 50; c 2. PLANT PHYSIOLOGY. *Educ:* Lebanon Valley Col, BS, 50; Rutgers Univ, MS, 52, PhD(plant physiol), 58. *Prof Exp:* Res plant physiologist, Biol Res Lab, Ft Detrick, Md, 53-55; plant physiologist, Am Cyanamid Co, 58-60; from asst prof to prof, 60-90, EMER PROF BOT, UNIV RI, 90- *Mem:* AAAS; Am Soc Plant Physiol; Bot Soc Am. *Res:* Physiological ecology; boron physiology; growth and development; leaf morphogenesis. *Mailing Add:* Dept Bot Univ RI Kingston RI 02881

ALBERT, MARY ROBERTS FORBES (DAY), b Manchester, NH, Mar 2, 26; m 55; c 2. CYTOLOGY, BIOCHEMISTRY. *Educ:* Univ NH, BS, 48; Bryn Mawr Col, MA, 50; Brown Univ, PhD, 55. *Prof Exp:* Asst cytol, Brown Univ, 51-55; res fel med, Mass Gen Hosp, 55-61; instr, Northeastern Univ, 61-62 & Wellesley Col, 63-64; asst prof biol sci, Newton Col Sacred Heart, 64-75; DIR, BIOL LABS, BOSTON COL, 75- *Res:* Irradiation effects on mitosis in the rat and mouse; effects of tetrahydrocannabinol on the reproductive organs of male and female rats; cytological studies of mouse brain tumors and hippocampi. *Mailing Add:* Dept Biol Boston Col Chestnut Hill MA 02167

ALBERT, PAUL A(NDRE), b Van Buren, Maine, Apr 14, 26; m 55; c 8. PHYSICAL METALLURGY. *Educ:* Univ Maine, BS, 49; NY Univ, MS, 54, ScD(metall), 56. *Prof Exp:* Res asst, Res Div, NY Univ, 50-55; from engr to supvr engrs, Res & Develop Ctr, Westinghouse Elec Corp, 55-64; mgr electrochem, Thomas J Watson Res Ctr, Int Bus Mach Corp, 64-66, mgr plated films, 66-67, mgr advan device processing, Components Div, 67-71, res staff mem, Inorg Mat Dept, Res Div, 71-81; consult, Magnetic Recording & Mat, 81; PRES, ALBERT CONSULT INC. *Mem:* Inst Elec & Electronics Engrs. *Res:* Magnetism; metallurgy of magnetic materials; high density magnetic recording systems with film media and transducers. *Mailing Add:* Albert Consult Inc 10473 Anderson Rd San Jose CA 95127

ALBERT, PAUL JOSEPH, b Edmundston, Can, Nov 11, 46. SENSORY PHYSIOLOGY. *Educ:* Univ NB, BSc, 68, PhD(biol), 72. *Prof Exp:* Fel, Univ Sask, 72-73; ASSOC PROF BIOL, CONCORDIA UNIV, 73- *Mem:* Can Soc Zool; Entom Soc Can. *Res:* Structure and physiology of insect sense organs. *Mailing Add:* Dept Biol Concordia Univ 1455 DeMaisonneuve Blvd W Montreal PQ H3G 1M8 Can

ALBERT, R(OBERT) E(YER), b Barberton, Ohio, May 21, 21; m 43; c 1. CHEMICAL ENGINEERING. *Educ:* Ohio State Univ, BChE, 43, PhD(chem eng), 50. *Prof Exp:* Chem engr, Phillips Petrol Co, 43-46; instr chem eng, Ohio State Univ, 46-50; res engr, E I Du Pont De Nemours & Co, Inc, 50-56, res supvr, 56-64, develop mgr, 64-69, eng assoc, 69-82; RETIRED. *Res:* Liquid-vapor equilibrium, binary systems; synthetic fibers; polymers; structural property relationships; ceramics; refractories; inorganics; new products; agricultural chemicals and formulations and applications. *Mailing Add:* 12442 Mcgregor Woods Circle Ft Myers FL 33408-2442

ALBERT, RICHARD DAVID, b Elmira, NY, Aug 9, 22; m 46; c 4. NUCLEAR PHYSICS. *Educ:* Univ Mich, BA, 43; Columbia Univ, AM, 46, PhD(physics), 51. *Prof Exp:* Asst, Columbia Univ, 46-49; sr scientist, Westinghouse Elec Corp, 49-51; res assoc, Knolls Atomic Power Lab, 51-55; sr physicist, Lawrence Radiation Lab, Univ Calif, 55-65, res physicist, Space Sci Lab, Univ Calif, 66-70; physicist, Terradynamics, Inc, 70-71; PRES, X-RAY SYSTS, INC, 71-; PRES, DIGIRAY CORP, 81-; ASSOC PROF PHYSICS, UNIV PAC, 81- *Concurrent Pos:* Consult, Space Sci Lab, Univ Calif, 70-; NASA grant; consult x-ray field instrumentation, govt & indust, 73- *Mem:* AAAS; Am Phys Soc. *Res:* Low energy nuclear physics research and instrumentation; x-ray spectroscopy intrumentation; inventor of reverse geometry x-ray scanning system for low dosage digital radiography and industrial inspection; twenty-one US Patents. *Mailing Add:* 317 Hartford Rd Danville CA 94526

ALBERT, RICHARD K, b Los Angeles, Calif, Sept 30, 45; m 67; c 2. MEDICINE. *Prof Exp:* From instr to assoc prof, 76-88, PROF, DEPT MED, SCH MED, UNIV WASH, 88-, SECT HEAD, PULMONARY & CRITICAL CARE MED, MED CTR, 90- *Concurrent Pos:* Asst chief, Pulmonary & Critical Care Sect, Vet Admin Med Ctr, Seattle, Wash, 76-90, dir, Med Intensive Care Unit, 76-90, actg chief, 85-86; vis res prof, Unit Res Physio-Path Respiratoire, Nat Inst Health & Med Res, France, 84-85; prin investr grant, Vet Admin Merit Rev, 88-, Nat Heart, Lung & Blood Inst, 89-; ad hoc mem, Respiratory & Appl Physiol Study Sect, NIH, 90. *Mem:* Fel Am Col Physicians; fel Am Col Chest Physicians; Am Thoracic Soc; Am Fedn Clin Res; Int Union Against Tuberculosis; Am Physiol Soc; Am Heart Asn; Am Soc Clin Invest. *Mailing Add:* Dept Med Univ Wash Med Ctr Rm-12 Seattle WA 98195

ALBERT, ROBERT LEE, b Pittsburgh, Pa, June 4, 46; m 74; c 2. METALLURGY, MATERIAL SCIENCE. *Educ:* Carnegie-Mellon Univ, BS, 68, MS, 71. *Prof Exp:* Researcher metall, NSF, 65; PRIN ENGR PROCESS DEVELOP, BETTIS ATOMIC POWER LAB, WESTINGHOUSE ELEC CORP, 68- *Concurrent Pos:* Abstractor, Int Am Soc Metals. *Mem:* Am Soc Metals; Am Powder Metall Inst. *Res:* Process development in metal and ceramic systems; physical metallurgy, powder metallurgy. *Mailing Add:* Bettis Atomic Power Lab PO Box 79 West Mifflin PA 15122

ALBERT, ROY ERNEST, b New York, NY, Jan 11, 24; m 45; c 4. ENVIRONMENTAL MEDICINE. *Educ:* Columbia Univ, AB, 43; NY Univ, MD, 46. *Prof Exp:* Med officer, NY Opers Off, AEC, 52-54, chief med br, Div Biol & Med, 54-56; asst prof med, Sch Med, George Washington Univ, 56-59; from assoc prof to prof environ med, Med Ctr, NY Univ, 66-85; PROF & CHMN, DEPT ENVIRON HEALTH, UNIV CINCINNATI MED CTR, 85- *Concurrent Pos:* USPHS fel, 49-51. *Mem:* Radiation Res Soc; Soc Epidemiol Res; Am Asn Cancer Res. *Res:* Radiation and chemical carcinogenesis; aerosol deposition and clearance; cancer and environmental toxicants epidemiology. *Mailing Add:* Dept Environ Health Univ Cincinnati Col Med 231 Bethesda Ave Cincinnati OH 45267-0056

ALBERTE, RANDALL SHELDON, b Newark, NJ, June 7, 47. PLANT PHYSIOLOGY, BIOCHEMISTRY. *Educ:* Gettysburg Col, BA, 69; Duke Univ, PhD(bot), 74. *Prof Exp:* Res assoc, 73-75, NSF energy-related fel biochem, Univ Calif, Los Angeles, 75-76; NIH fel biochem, 76-77; asst prof, 77-80, ASSOC PROF BIOL, UNIV CHICAGO, 81- *Concurrent Pos:* Consult, NSF grant oceanog, Univ Wash, 77-81 & Encycl Britannica, 80-81; Andrew Mellon Found fel, 79-80. *Mem:* Am Soc Plant Physiologists; Soc Exp Biol; Bot Soc Am; AAAS; Am Inst Biol Sci. *Res:* Photobiology and adaptive physiology of photosynthesis; chloroplast development; organization of chlorophyll in vivo; membrane biochemistry; biochemistry and physiology of water stress and of the control of development and differentiation. *Mailing Add:* 5630 S Ingleside Ave Chicago IL 60637

ALBERTI, PETER W R M, b Aug 23, 34; m 61; c 3. OTOLARYNGOLOGY. *Educ:* Univ Durham, MB, BS, 57; Washington Univ, PhD(anat), 63; FRCS, 65; FRCPS(C), 68. *Prof Exp:* Demonstr anat, Med Sch, Univ Durham, 58-59, first asst otolaryngol, 64-67; instr anat, Emory Univ, 60-61; clin teacher, 67-68, from asst prof to assoc prof, 68-77, assoc prof prev med & biostatist, 77-80, PROF OTOLARYNGOL, UNIV TORONTO, 77-, PROF OCCUP & ENVIRON MED, 78-, CHMN OTOLARYNGOL, 82- *Honors & Awards:* Wilde Medalist, Irish Otolaryngol Soc, 86; Jozhi Medalist, Asn Otolaryngologists, India, 88. *Mem:* AAAS; Am Asn Anat; Can Otolaryngol Soc; Royal Soc Med; Am Acad Otolaryngol; hon mem Brazilian Otolaryngol Soc; hon mem Irish Otolaryngol Soc; hon mem Indian Otolaryngol Soc. *Res:* Anatomy of middle ear; teaching techniques; hearing testing; industrial deafness. *Mailing Add:* 600 Univ Ave Toronto ON M5G 1X5 Can

ALBERTINE, KURT H, b Newark, NJ, Nov 29, 52. ANATOMY, PHYSIOLOGY. *Educ:* Lawrence Univ, BA, 75; Loyola Univ, Chicago, PhD(anat), 79. *Prof Exp:* Adj asst prof anat, Univ Calif, San Francisco, 82-83; asst prof, Univ SFla, 84-86; res asst prof path, Univ Pa, 86-87; asst prof, 87-90, ASSOC PROF MED & PHYSIOL, JEFFERSON MED COL, 90- *Concurrent Pos:* Vis scientist, Univ Calif, San Francisco, 86 & 90. *Mem:* AAAS; Am Asn Anatomists; Am Physiol Soc; NY Acad Sci; Electron Micros Soc Am; Sigma Xi; Am Thoracic Soc. *Res:* Structure and function of the pulmonary circulation and lymphatic system; pathophysiology of pulmonary edema; neutrophil activation. *Mailing Add:* Dept Med 804 Col Bldg Jefferson Med Col 1025 Walnut St Philadelphia PA 19107-5587

ALBERTINI, DAVID FRED, b Hudson, Mass, Mar 19, 49; m 72; c 2. CELL BIOLOGY, REPRODUCTIVE BIOLOGY. *Educ:* Marquette Univ, BS, 70; Univ Mass, Ms, 72; Harvard Univ, PhD(anat & cell biol), 75. *Prof Exp:* Res assoc biophysics, Univ Conn Health Ctr, 75-77; asst prof anat, Harvard Med Sch, 77-83; ASSOC PROF ANAT, TUFTS MED SCH, 84- *Concurrent Pos:* Ctr Environ Mgt, Tufts Univ. *Mem:* Am Soc Cell Biol; Am Asn Anatomists; Soc Study Reproduction; NY Acad Sci. *Res:* Female fertility/infertility; evaluation of gamete/embryo viability; role of hormones in reproduction of mammals; assisted reproductive technologies; environmental effects on reproduction. *Mailing Add:* Dept Anat & Cell Biol Tufts Med Sch 136 Harrison Ave Boston MA 02111

ALBERTINI, RICHARD JOSEPH, b Racine, Wis, Mar 15, 35; m 57; c 4. MUTAGENESIS, IMMUNOGENETICS. *Educ:* Univ Wis-Madison, BS, 60, MD, 63, PhD(med genetics), 72. *Prof Exp:* Instr med, Univ Wis-Madison, 70-72; from asst prof to assoc prof, 72-79, PROF MED, COL MED, UNIV VT, 79-, ADJ PROF MICROBIOL & MOLECULAR GENETICS, 80- *Concurrent Pos:* Sci dir, Vt-NH Red Cross Blood Ctr, 83-; dir, Vt Regional Cancer Ctr, Genetics Lab, Univ Vt, 84- & Genetics Unit, Dept Med, 88-; pres, Vt Div, Am Cancer Soc, 88-90; mem, Environ Health Sci Rev Comt, NIH, 88-; ed-in-chief, Environ & Molecular Mutagenesis, 88-; vis prof, Oncol & Pediat, Univ Wis-Madison, 90-91. *Honors & Awards:* Alexander Hollaender Award, Environ Mutagen Soc, 90; St George Medal, Am Cancer Soc, 90. *Mem:* Environ Mutagen Soc (pres, 84-85); Am Asn Cancer Res; Am Asn Clin Oncol; Am Soc Human Genetics. *Res:* Nature and consequences of somatic cell gene mutations that arise spontaneously or as a result of mutagen/carcinogen exposure in humans; mutations that involve immunocompetent lymphocytes and their effects in vivo. *Mailing Add:* Vt Regional Cancer Ctr Genetics Lab Univ Vt 32 N Prospect St Burlington VT 05401

ALBERTS, ALFRED W, b New York, NY, May 16, 31. BIOCHEMISTRY OF LIPIDS. *Prof Exp:* SR SCIENTIST BIOCHEM, MERCK SHARP & DOHME RES LABS, 75- *Mailing Add:* Merck Sharp & Dohme Res Labs P O Box 2000 Rahway NJ 07065

ALBERTS, ARNOLD A, b Davis, SDak, May 28, 06; m 31; c 2. ORGANIC CHEMISTRY. *Educ:* Univ SDak, AB, 28; Univ Okla, MS, 30; Ohio State Univ, PhD(org chem), 34. *Prof Exp:* Asst gen & org chem, Univ Okla, 28-30; asst Ohio State Univ, 30-31, asst gen chem, 31-34, instr chem, 35-36, res asst metall eng exp sta, 35-37, investr indust probs, 36-37; asst prof org & phys chem, Otterbein Col, 37; from instr to asst prof chem, Washington & Jefferson Col, 37-42; from asst prof to assoc prof, Purdue Univ, 43-52; res chemist, Western Co, 52-56; prof chem & head dept, Hastings Col, 56-76; vis prof chem, Sterling Col, Kans, 78-79; RETIRED. *Concurrent Pos:* Admin asst, Res Found, Purdue Univ, 46-49, supvr chem, Exten Ctr, 47-52. *Mem:* Am Chem Soc. *Res:* Corrosion and corrosion inhibitors. *Mailing Add:* 1629 George Washington Way Apt 319 Richland WA 99352-5716

ALBERTS, BRUCE MICHAEL, b Chicago, Ill, Apr 14, 38; m 60; c 3. MOLECULAR BIOLOGY. *Educ:* Harvard Univ, AB, 60, PhD(biophys), 65. *Prof Exp:* NSF res fel, 65-66; asst prof chem, Princeton Univ, 66- 71, from assoc prof to prof biochem sci, 73-76; prof biochem & chmn dept, 76-81, Am Cancer Soc res prof, 81-85, PROF & CHMN, DEPT BIOCHEM & BIOPHYS, UNIV CALIF, SAN FRANCISCO, 85- *Concurrent Pos:* Mem, adv panel human cell biol, NSF, 74-76, bd sci counr, Div Arthritis & Metab Dis, NIH, 74-78, bd sci adv, Jane Coffin Childs Mem Fund Med Res, 78-85, Adv Coun Dept Biochem Sci & Molecular Biol, Princeton Univ, 79085, Molecular Cytol Study Sect, NIH, 82-86 & chmn, 84-86; chmn, comt mapping & sequencing human genome, Nat Res Coun, 86-88; mem, Corp vis comt, dept biol, Mass Inst Technol, 78-, vis comt, dept embryol, Carnegie Inst Wash, 83, Sci adv bd, Markey Found, 84, sci adv comt, Marine Biol Lab, Woodshole, 88-, chmn, vis comt, dept biochem & molecular biol, Harvard Col, 83-; counr, Am Soc Biol Chemists, 84-; assoc ed, Ann Reviews Cell Biol, 84- *Honors & Awards:* Eli Lilly Award, Am Chem Soc, 72; US Steel Award, Nat Acad Sci, 75; Lifetime Res Prof Award, Am Cancer Soc, 80. *Mem:* Nat Acad Sci; Am Chem Soc; Am Soc Biol Chem; fel Am Acad Arts & Sci; Am Soc Microbiol; Am Soc Cell Biol; Genetics Soc Am. *Res:* Molecular genetics; DNA replication; eukaryotic gene control; author of approximately 100 publications on biochemistry. *Mailing Add:* Dept Biochem & Biophys Univ Calif Box 0448-HSE 920 San Francisco CA 94143-0448

ALBERTS, GENE S, b Brookings, SDak, Sept 20, 37; m 59. ANALYTICAL CHEMISTRY. *Educ:* Wash State Univ, BS, 59; Univ Wis, PhD(chem), 63. *Prof Exp:* Staff chemist, Systs Develop Div, Int Bus Mach Corp, 63-66, res staff mem, T J Watson Res Lab, 66-67, develop chemist, Component Div, IBM Corp, 67-69, sr eng, 69-72, sr engr, Systs Prod Div, 72-77, sr engr-mgr, Advan Monolithic Technol, 77-81, mgr lab staff, Burlington Gen Technol Div, 81-86, SR TECH STAFF MEM, IBM CORP, 87- *Mem:* Sigma Xi. *Res:* Electroanalytical and electrokinetic chemistry; magnetic film deposition; semi conductor process engineering. *Mailing Add:* 9499 NE Shore Dr PO Box 41 Indianola WA 98342

ALBERTS, JAMES JOSEPH, b Chicago, Ill, May 23, 43. GEOCHEMISTRY. *Educ:* Cornell Col, Iowa, AB, 65; Dartmouth Col, NH, MS, 67; Fla State Univ, PhD(chem oceanog), 70. *Prof Exp:* Res assoc, Kans State Geol Surv, 70; res assoc, Univ Ga, 70-72, asst res prof, 72-74; asst chemist, Argonne Nat Lab, 74-77; assoc res ecologist, Savannah River Ecol Lab, 77-84; DIR & SR RES SCIENTIST, UNIV GA MARINE INST, 84- *Honors & Awards:* Sr US Scientist Award, Alexander Von Humboldt Found, 89. *Mem:* AAAS; Am Soc Limnol & Oceanog; Estuarine Res Fedn. *Res:* Mechanisms and processes by which inorganic elements, organic compounds, and radioisotopes interact and cycle in aquatic environments. *Mailing Add:* Marine Inst Univ Ga Sapelo Island GA 31327

ALBERTS, WALTER WATSON, b Los Angeles, Calif, Dec 31, 29; m 59; c 2. NEUROSCIENCES, RESEARCH ADMINISTRATION. *Educ:* Univ Calif, AB, 51, PhD(biophys), 56. *Prof Exp:* Res physiologist, Univ Calif Med Ctr, San Francisco, 55-56; biophysicist, Mt Zion Hosp & Med Ctr, 56-72; spec asst to assoc dir, Nat Inst Neurol & Commun Dis & Stroke, NIH, 73-74, head res contracts sect, 74-75, asst dir contract res progs, 75-77; admin dir, Smith-Kettlewell Inst & Dept Visual Sci, Univ Pac, 77-78; dep dir, Div Fundamental Neurosci, Nat Inst Neurol & Commun Dis & Stroke, 79-89, actg dep dir, Nat Inst Neurol Dis & Stroke, 89-90, DEP DIR, DIV FUNDAMENTAL NEUROSCI, NAT INST NEUROL DIS & STROKE, NIH, 90- *Concurrent Pos:* Consult, Donner Lab, Univ Calif, Berkeley, 59-64; res career prog award, Nat Inst Neurol Dis & Blindness, NIH, 63-68; lectr, Dept Physiol, Univ Calif Med Ctr, San Francisco, 69; grant assoc, NIH, 72-73; sr scientist, inst Med Sci, San Francisco, 77-78. *Mem:* Am Physiol Soc; fel AAAS; Biophys Soc; Soc Neurosci. *Res:* Neurophysiology; biological and medical physics, particularly the central nervous system of man. *Mailing Add:* Nat Inst of Neurolog Nat Inst Health Bethesda MD 20892

ALBERTSON, CLARENCE E, b Rockwell, Tenn, Oct 7, 22; m 47; c 3. FRICTION MATERIALS, WEAR-RESISTANT MATERIALS. *Educ:* Miami Univ, Ohio, AB, 43; Univ Cincinnati, MS, 48, PhD(chem), 49. *Prof Exp:* Teaching fel chem, Purdue Univ, 49-50; phys chemist, Robertshaw Fulton Controls Co, Pa, 50-51; phys chemist, Borg Warner Corp, Ill, 51-66, mgr phys chem, 66-78, sr scientist, 78-86; RETIRED. *Concurrent Pos:* Indust res assoc hard coatings, Argonne Nat Lab, Ill, 86-87. *Mem:* Am Soc Lubrication Engrs; Am Electroplaters Soc; Am Chem Soc; Soc Automotive Engrs; Soc Advan Mat & Process Eng; Am Ceramic Soc. *Res:* Development of wet and dry clutch linings for automobiles and light trucks and electroplated, porous metal surfaces to augment boiling heat transfer; measurement of friction and wear of materials for use in industrial equipment. *Mailing Add:* 240 S Monterey Ave Villa Park IL 60181

ALBERTSON, HAROLD D, b Parsons, Kans, Dec 28, 31; m 53; c 1. COMPUTER SCIENCE, ENGINEERING. *Educ:* Univ Houston, BS, 53; Southern Methodist Univ, MS, 60; Univ Tex, Austin, PhD(elec eng), 68. *Prof Exp:* Design engr, Chance Vought Aircraft, Inc, 53-58; design engr, Tex Instruments, Inc, 58-60, prog mgr, 60-63, syst analyst, 63-65 & 68-73; instr, 73-77, chmn div, 77-78, dean instrnl serv, 78-83, PROF, TECHNOL DEPT, RICHLAND COL, 83- *Concurrent Pos:* Mem, Simulation Coun. *Mem:* Inst Elec & Electronics Eng; Fluid Power Soc; Robotics Int; Soc Mfg Engrs. *Res:* Study of the modal properties of linear dynamical systems and the construction of irreducible (minimum state) realizations of fixed and time varying systems; application to the practical simulation of transfer function matrices; robotics and automated systems. *Mailing Add:* 7660 Chalkstone Dallas TX 75240

ALBERTSON, JOHN NEWMAN, JR, b New Haven, Conn, Jan 18, 33; m 55; c 7. LABORATORY MANAGEMENT, MICROBIOLOGY. *Educ:* Univ Conn, AB, 54; Hahnemann Med Col, MS, 64; Am Bd Microbiol, dipl; Indust Col Armed Forces, Dipl, 75. *Prof Exp:* Comndg officer med dispensary, US Army, Munich, Ger, 56-57, asst chief dept immunol & serol, Med Lab, Landstuhl, Ger, 57-59, chief clin path & bacteriologist, Valley Forge Gen Hosp, Phoenixville, Pa, 59-62, res assoc, Hahnemann Med Col, 62-64, chief bact & virol div, First Med Lab, Ft George C Meade, Md, 64-68, chief med & biol sci br, Off Chief Res & Develop, Hq Dept Army, Washington, DC, 68-70, exec for res, 70-71, comndg officer, 9th Med Lab, Repub Vietnam, 71-72, exec officer, Walter Reed Army Inst Res, Washington, DC, 72-74, exec officer, Armed Forces Inst Path, 75-76, chief of staff, US Army Res & Develop Command, 76- 79, dir, US Army Med Bioeng Res & Develop Lab, Fort Detrick, Md, 79-84; RETIRED. *Concurrent Pos:* Consult, Europ Command & partic coop study on Treponemal pallidum immobilization & other tests, WHO, Ger, 57-59; mem, Nat Adv Res Resources Coun, NIH, Washington, DC, 80-84. *Mem:* Fel AAAS; Am Soc Microbiol; fel Am Acad Microbiol; Sigma Xi. *Res:* Bioengineering (teleradiography/telencephalography); chemotherapy and bacteriology of tuberculosis; research management; biomechanics; microbial mining and metal recovery. *Mailing Add:* PO Box 100 Harpers Ferry WV 25425-0100

ALBERTSON, MICHAEL OWEN, b Philadelphia, Pa, June 24, 46; m 69; c 3. COMBINATORIAL MATHEMATICS. *Educ:* Mich State Univ, BS, 66; Univ Pa, PhD(math), 71. *Prof Exp:* Asst prof math, Swarthmore Col, 72-73; from asst prof to assoc prof math, Smith Col, 73-84, chmn dept, 85-88, chmn computer sci dept, 87-88, PROF MATH, SMITH COL, 84- *Concurrent Pos:* Prin investr, Cottrell grant, Res Corp, 74; co-prin investr, NSF grant, 77-79, 80, prin investr, 84-88; vis scholar, Doshisha Univ, Kyoto, 89. *Mem:* Am Math Soc; Soc Indust & Appl Math. *Res:* Independent sets in graphs; chromatic graph theory; topological graph theory; perfect graphs; duality theory; graph homomorphisms. *Mailing Add:* Dept Math Smith Col Northampton MA 01063

ALBERTSON, NOEL FREDERICK, b New Haven, Conn, Oct 29, 15; m 42; c 3. MEDICINAL CHEMISTRY. *Educ:* Polytech Inst Brooklyn, BS 36, MS, 38; Ohio State Univ, PhD, 41. *Prof Exp:* Jr chemist, Merck & Co, 38-39; asst, Ohio State Univ, 39-41, res assoc & fel, Univ Res Found, 41-43; group leader, Winthrop Chem Co, 43-46; group leader, Sterling-Winthrop Res Inst, 46-64, sect head, 64-66, asst dir, Chem Div, 66-77, res scientist, 77-80; RETIRED. *Mem:* Am Chem Soc. *Res:* Molecular addition compounds of sulphur dioxide; production of sulphur from natural gas; synthesis of amino acids, peptides, piperidines and benzomorphans. *Mailing Add:* 155 Ida Red Ave Sparta MI 49345-1715

ALBERTSON, VERNON D(UANE), b Syre, Minn, Sept 28, 28; m 51; c 5. ELECTRICAL ENGINEERING. *Educ:* NDak State Univ, BSEE, 50; Univ Minn, Minneapolis, MSEE, 56; Univ Wis, PhD(elec eng), 62. *Prof Exp:* Test engr, Gen Elec Co, NY, 50-51, field engr, Minn, 51-52; from instr to asst prof elec eng, NDak State Univ, 56-58; syst planning engr, Otter Tail Power Co, Minn, 58-59; instr elec eng, Univ Wis, 59-60, asst prof, 62-63; from asst prof to assoc prof, 63-70, PROF ELEC ENG, UNIV MINN, MINNEAPOLIS, 70- *Mem:* Am Soc Eng Educ; Inst Elec & Electronics Engrs. *Res:* Electric power system analysis and transients; power system planning; environmental effects of electric and magnetic fields; effects of geomagnetically-induced currents in power systems. *Mailing Add:* Dept Elec Eng Univ Minn 200 Union St SE Minneapolis MN 55455

ALBERTY, ROBERT ARNOLD, b Winfield, Kans, June 21, 21; m 44; c 3. CHEMISTRY. *Educ:* Univ Nebr, BS, 43, MS, 44; Univ Wis, PhD(phys chem), 47. *Hon Degrees:* DSc, Lawrence Col, Univ Nebr. *Prof Exp:* Asst, Univ Wis-Madison, 44-46, from instr to prof phys chem, 47-62, assoc dean letters & sci, 62-63, dean grad sch, 63-67; dean sch sci, 67-82, PROF CHEM, MASS INST TECHNOL, 82- *Concurrent Pos:* Guggenheim fel, Calif Inst Technol, 50-51; chmn, Comn Human Resources, Nat Res Coun, 74-77. *Honors & Awards:* Eli Lilly Award, 56. *Mem:* Inst Med-Nat Acad Sci; Am Acad Arts & Sci; AAAS; Am Chem Soc. *Res:* Thermodynamics and kinetics of complex systems. *Mailing Add:* Rm 6-215 Sch of Sci Mass Inst Technol Cambridge MA 02139

ALBICKI, ALEXANDER, b Warsaw, Poland, May 24, 41; m 70; c 2. LARGE SCALE INTEGRATION DESIGNS, DESIGN FOR TESTABILITY. *Educ:* Tech Univ Warsaw, BS/MS, 64, PhD(telecommunications), 73, Habilitation(elec), 80. *Prof Exp:* Design engr, Tech Univ Warsaw, 64-73, asst prof telecommun, 73-80; asst prof, 83-88, ASSOC PROF ELEC ENG, UNIV ROCHESTER, 88- *Concurrent Pos:* Prin Investr, Ministry Telecommun, Poland, 73-80, Semiconductor Res Corp, 85-88 & NY State Sci & Technol Found, 86-87; lectr, Nat Higher Sch Telecommun, Paris, 78; vis prof, Univ Rochester, 81-83. *Mem:* Inst Elec & Electronic Engrs. *Res:* Design for testability with emphasis on built-in-self-test; VLSI designs; developing protocols for computers communication. *Mailing Add:* Dept Elec Eng Univ Rochester Rochester NY 14627

ALBIN, ROBERT CUSTER, b Beaver City, Okla, May 19, 39; m 60; c 2. ANIMAL NUTRITION. *Educ:* Tex Technol Col, BS, 61, MS, 62; Univ Nebr, PhD(animal nutrit), 65. *Prof Exp:* From asst prof animal husb to prof animal sci, 64-80, PROF ANIMAL SCI & DEAN, COL AGR SCI, TEX TECH UNIV, 80- *Mem:* Am Soc Animal Sci; Sigma Xi. *Res:* Ruminant nutrition; beef cattle nutrition and management. *Mailing Add:* 1803 Bangor Lubbock TX 79416

ALBIN, SUSAN LEE, b New York, NY, Sept 25, 50; m 79; c 1. QUALITY ENGINEERING, QUEUEING. *Educ:* NY Univ, BS, 71, MS, 73; Columbia Univ, DSc(opers res), 81. *Prof Exp:* Res asst community health, Albert Einstein Col Med, 71-74, res assoc neurosci, 74-77; mem tech staff, Bell Labs, 77-81; asst prof, 81-87, ASSOC PROF INDUST ENG, RUTGERS NJ STATE UNIV, 87- *Concurrent Pos:* Prin investr, NSF, 82-84 & 91-92, Coun Solid Waste Solutions, 90-91; chair, Int Conf Queueing Networks, 87; assoc ed, Mgt Sci, 87-; panel mem, NSF, 89, 90 & 91; consult, Gen Cable Co, 90-91; vis prof, Tel Aviv Univ, 90. *Mem:* Opers Res Soc Am; Inst Mgt Sci; Inst Indust Engrs; Am Soc Qual Control. *Res:* Quality engineering; stochastic modeling and queueing theory; semiconductor manufacturing; plastics recycling; aviation; telecommunications systems; public health and epidemiology. *Mailing Add:* Dept Indust Eng Rutgers Univ PO Box 909 Piscataway NJ 08855-0909

ALBINAK, MARVIN JOSEPH, b Detroit, Mich, June 21, 28; m 61; c 3. ENVIRONMENTAL SCIENCES. *Educ:* Univ Detroit, AB, 49, MS, 52; Wayne State Univ, PhD(inorg chem), 59. *Prof Exp:* Chemist explor res, Ethyl Corp, 52-54; from instr to asst prof chem, Univ Detroit, 54-61; sr res scientist, Elec Autolite, 61-62; res chemist, Owens-Ill Glass Co, 62-65; from asst prof to assoc prof chem, Wheeling Col, 65-68; prof & chmn div sci & math, 68-77, dir spec projs, 77-79, PROF CHEM, ESSEX COMMUNITY COL, 79-, HEAD DEPT, 85- *Mem:* Sigma Xi; Am Chem Soc. *Res:* Inorganic chemistry of glass; synthesis and resolution of coordination compounds; absorption and fluorescence of inorganics; science for the layman; science and public policy. *Mailing Add:* 819 Providence Rd Towson MD 21204

ALBINI, FRANK ADDISON, b Madera, Calif, May 15, 36; m 58; c 3. MECHANICAL ENGINEERING, APPLIED PHYSICS. *Educ:* Calif Inst Technol, BS, 58, MS, 59, PhD(mech eng philos), 62. *Prof Exp:* Mem tech staff aerodyn, Hughes Aircraft Co, 60-63; mem tech staff, Defense Res Corp, 66-70, dir defensive systs dept, Gen Res Corp, 70-71; mech engr, Intermountain Forest & Range Exp Sta, US Forest Serv, USDA, 73-85; sr mem tech staff systs anal, Res & Eng Support Div, Inst Defense Anal, 63-66, mem res staff, 71-73 & 85-86; consult, 86-91; SR ANALYST, SCI APPLN INT CORP, 91- *Concurrent Pos:* Consult, Cosmodyne Corp, Calif, 59-60; lectr, Univ Southern Calif, 60-62. *Mem:* NY Acad Sci; Combustion Inst; Sigma Xi; Soc Am Foresters; Am Soc Mech Engrs. *Res:* Electromagnetic theory; statistical and fluid mechanics; combustion; plasma physics; applied mechanics; weapon systems analysis; wildland fire behavior and effects. *Mailing Add:* 9485 Old Mill Trail Missoula MT 59802

ALBISSER, ANTHONY MICHAEL, b Johannesburg, Africa, Sept 5, 41; m 64; c 4. BIOMEDICAL ENGINEERING. *Educ:* McGill Univ, BEng, 64; Univ Toronto, MASc, 66, PhD(biomed eng), 68. *Prof Exp:* Assoc dir med eng, 68-71, dir, 71-75, investr biomed res, 75-78, sr scientist, Dept Surg & dir, Hosp For Sick Children, 78-83; DEPT ELEC ENG, UNIV TORONTO, 83- *Concurrent Pos:* Spec lectr elec eng, Univ Toronto, 68-71, asst prof, 72-75, adj prof, 75-77, assoc prof, Dept Med, 77-87, prof, 87-; co-ed-in-chief, Med Progress Technol, 75-80; mem biomed eng subcomt, Med Res Coun, 73-78 & Inst Med Sci, Univ Toronto, 78-; consult, Life Sci Div, Miles Labs, 74-80. *Honors & Awards:* David Rumbough Awards, Juv Diabetes Found, 81. *Mem:* Can Med Biol Eng Soc; Am Diabetes Asn; Asn Advan Med Instrumentation; Europ Asn Study Diabetes. *Res:* Study of the pathophysiology of diabetes and how this relates to the development of an artificial endocrine pancreas; development of portable artificial insulin delivery systems. *Mailing Add:* Dir Loyal True Blue & Orange Res Inst PO Box 209 Richmond Hill ON L4C 4Y2 Can

ALBO, DOMINIC, JR, b Helper, Utah, Dec 26, 34; m 60; c 7. SURGERY. *Educ:* Carbon Col, AS, 55; Univ Utah, BS, 65; St Louis Univ, MD, 60, Am Bd Surg, dipl, 69. *Hon Degrees:* DHH, Col Eastern Utah. *Prof Exp:* Intern, NC Mem Hosp, Chapel Hill, 61; resident, Univ Utah Affil Hosps, Salt Lake City, 61-62, 64-68; US Army post physician, Ft Douglas, 62-64; from instr to asst prof, 68-76, ASSOC PROF SURG, COL MED, UNIV UTAH, 76- *Concurrent Pos:* Grants, Resnick Award, prin investr, 68-69, NSF, co-prin investr, 71- & Nat Inst Gen Med Sci, co-investr, Univ Utah, 78-81; coordr gen surg, Holy Cross Hosp, Salt Lake City, 68-,; consult staff, Vet Admin Hosp, Salt Lake City, 68-; instr postgrad med, Col Med, Univ Utah, 69-73; chief peripheral vascular serv, Col Med, Univ Utah & Vet Admin Hosp, 72-78; dir med educ, Holy Cross Hosp. *Mem:* Am Cancer Soc; AMA; Asn Academic Surg; Am Col Surgeons; Int Cardiovasc Soc. *Res:* Polymer implant materials. *Mailing Add:* Dept Surg Univ Utah Salt Lake City UT 84112

ALBOHN, ARTHUR R(AYMOND), b New York, NY, Dec 27, 21; m 44; c 3. CHEMICAL ENGINEERING. *Educ:* Columbia Univ, AB, 42, BS, 43. *Prof Exp:* Chem engr, Res & Develop Depts, Goodyear Tire & Rubber Co, 43-50; sr chem engr, Summit Res Labs, Celanese Corp Am, 50-56; supvr film sect, Eastern Res Div, Rayonier, Inc, 56-59, supvr fiber sect, 59-60, asst res mgr, 60-66; group leader spinning res, Celanese Res Co, NJ, 66-69, staff assoc mat sci res, 69-70; mgr tech serv, Komline-Sanderson Eng Corp, 71-90; ASSEMBLYMAN, STATE NJ, 80- *Concurrent Pos:* Mem, NJ Comn Sci & Technol, 85- *Mem:* Am Chem Soc; Am Inst Chem Eng. *Res:* Cellulose and synthetic polymer processing, especially in the fields of films and fibres; pollution control; solids/liquid separations. *Mailing Add:* 29 Hilltop Circle Whippany NJ 07981

ALBRECHT, ALBERTA MARIE, b Reading, Pa, June 30, 30. MICROBIAL PHYSIOLOGY, CANCER. *Educ:* Seton Hill Col, BA, 51; Fordham Univ, MS, 52; Rutgers Univ, PhD(microbiol), 61. *Prof Exp:* Biologist, Res Div, Am Cyanamid, 52-58; res fel, Sloan-Kettering Inst Cancer Res, 61-63, res assoc, 63-65, assoc, 65-75, assoc mem, 75-82; ASSOC PROF, BIOL DEPT, MANHATTANVILLE COL, 82- *Concurrent Pos:* From instr to prof microbiol, Sloan-Kettering Div, Grad Sch Med Sci, Cornell Univ, 64-82; vis scholar biochem dept, Vanderbilt Univ, 90-91. *Mem:* Am Soc Biol Chemists; AAAS; Sigma Xi; NY Acad Sci. *Res:* Interests in folic acid and vitamin A metabolism, nutrition of malignant cells. *Mailing Add:* Biol Dept Manhattanville Col Purchase NY 10577

ALBRECHT, ANDREAS CHRISTOPHER, b Berkeley, Calif, June 3, 27; m 51; c 4. QUANTUM CHEMISTRY, SOLID STATE CHEMISTRY. *Educ:* Univ Calif, BS, 50; Univ Wash, PhD(chem), 54. *Prof Exp:* Res assoc, Mass Inst Technol, 54-56; from instr to assoc prof, 56-65, PROF CHEM, CORNELL UNIV, 65- *Concurrent Pos:* Consult, Eastman Kodak, 66-; partic, US-USSR Cult Exchange, 63-64 & US-USSR Acad Sci Exchange Prog, 74 & 83; NSF sci fac fel, 70-71; mem adv comt on USSR & Eastern Europe, Nat Acad Sci, 70-73; vis prof, Univ Calif, Santa Cruz, 71, SUNY Col at Purchase, 79-80, Rockefeller Univ, 79-80; mem rev comt, Dept Chem, State Univ NY Stony Brook, 85, Bd Chem, Univ Calif, Santa Cruz, 87, Rev Comt, Chem Div, Argonne Nat Lab, Univ Chicago, 86-89. *Honors & Awards:* Photochemistry Award, Polychrom Corp, 86; Ellis R Lippincott Medal, 88; Earle K Plyler Prize, Am Physical Soc. *Mem:* Fedn Am Scientists; Am Physical Soc; AAAS. *Res:* Theoretical and experimental molecular spectroscopy-laser based nonlinear spectroscopies in particular; ultrashort time photophysics, ionization and transient spectroscopies; author of over 160 papers. *Mailing Add:* Baker Lab of Chem Cornell Univ Ithaca NY 14853-1301

ALBRECHT, BOHUMIL, b Teplicka, Czech, May 6, 21; nat US; m 51. ENGINEERING MECHANICS. *Educ:* Slovak Inst Technol, CE, 44; Columbia Univ, MS, 49, PhD(eng), 56. *Prof Exp:* Instr bridge eng, Slovak Inst Technol, 44-47; asst eng mat, Columbia Univ, 48-51, res assoc, 51-56, from asst prof to assoc prof civil eng, 56-65; PROF MECH ENG, UNIV NMEX, 65- *Mem:* Soc Rheol; Soc Exp Stress Anal. *Res:* Mechanical properties of engineering materials. *Mailing Add:* Dept Mech Eng 7016 Veranda Rd NE Albuquerque NM 87110

ALBRECHT, BRUCE ALLEN, b Aug 14, 48; m 70; c 2. ATMOSPHERIC SCIENCE. *Educ:* Ill State Univ, BS, 72; Colo State Univ, MS, 74, PhD(atmospheric sci), 77. *Prof Exp:* ASSOC PROF METEOROL, PA STATE UNIV, 77- *Concurrent Pos:* Vis scientist, Nat Ctr Atmospheric Res, 80 & 82; mem, Am Meteorol Soc Comt, atmospheric radiation, 84-90; ed, Monthly Weather Rev, 85-90; lead scientist, First Int Satellite Cloud Climat Proj Regional Exp Sci Exp, 86-90, Atlantic Stratocumulus Exp, Intensive Observations, 89- *Mem:* Am Meteorol Soc; Sigma Xi; Am Geophys Union. *Res:* Theoretical and observational studies of clouds; cloud-climate interactions. *Mailing Add:* Dept Meteorol Pa State Univ University Park PA 16802

ALBRECHT, EDWARD DANIEL, b Kewanee, Ill, Feb 11, 37; m 88; c 1. MICROSTRUCTURAL ANALYSIS TO ENTREPRENEURS, CORPORATE MANAGEMENT CONSULTANT. *Educ:* Univ Ariz, Tucson, BSMetE, 59, MS, 61, PhD(metall & solid state physics), 64. *Prof Exp:* Sr physicist, Lawrence Livermore Labs, 64-71; pres, Metall Innovations Inc, Pleasanton, Calif, 69-72; vpres & gen mgr, 71-76, pres & chief exec officer, 76-90, CHMN, BUEHLER INT INC, LAKE BLUFF, ILL, 76- *Concurrent Pos:* Mem, Nat Adv Bd, Heard Mus Anthrop, Phoenix, Ariz, 80-, chmn, 90-; mem bd trustees, Millicent Rogers Mus, Taos, NMex, 82- *Honors & Awards:* Pres Award, Int Metallog Soc, 81, Henry Clifton Sorby Award, 90. *Mem:* Int Metallog Soc (pres & vpres, 70-81); fel Soc Mat Int; Metals Soc UK; fel Royal Micros Soc UK. *Res:* Design and development of equipment, supplies and technological procedures for preparation, examination and interpretation of the microstructure of solid materials including metals, minerals, coal, ceramics, refractories, glasses, composites and non-metallics; author of various publications. *Mailing Add:* 485 N Oakwood Ave Apt 2B Lake Forest IL 60045

ALBRECHT, FELIX ROBERT, b Cernauti, Bucovina, Romania, Apr 19, 26; US citizen; m 47. MATHEMATICS. *Educ:* Dipl math, Univ Bucharest, Romania, 51. *Prof Exp:* From res fel to sr res fel, res math, Inst Math of Romanian Acad, 51-63; from assoc prof math to prof, Wesleyan Univ, Middletown, Conn, 64-68; PROF MATH, UNIV ILL, URBANA, 68- *Mem:* Am Math Soc; Math Asn Am. *Res:* Geometric control theory; dynamical systems; singularity theory. *Mailing Add:* Dept Math Univ Ill 1409 W Green St Urbana IL 61801

ALBRECHT, FREDERICK XAVIER, b New York, NY, Oct 25, 43. PHYSICAL ORGANIC CHEMISTRY. *Educ:* State Univ NY Albany, BS, 67; State Univ NY Buffalo, PhD(chem), 72. *Prof Exp:* Asst prof chem, State Univ NY Col Canton, 69-70; fel chem, Univ Wis-Madison, 72-74; RES CHEMIST, EASTMAN KODAK CO, 74- *Mem:* Am Chem Soc; Sigma Xi. *Res:* Mechanistic study of organic photoconductive materials for use in electrophotography. *Mailing Add:* 50 Staub Rd Rochester NY 14626

ALBRECHT, G(EORGE) H(ENRY), b Rochester, NY, Feb 14, 32. CHEMICAL ENGINEERING. *Educ:* Univ Cincinnati, ChE, 54; Rensselaer Polytech Inst, MS, 55. *Prof Exp:* Tech writer, Sperry Gyroscope Co, Inc, NY, 55-56 & Burson-Marsteller Assocs, Inc, 56; chem engr, Appl Physics Lab, 59-65, Airtronics, Inc, 65-67, Booz-Allen Appl Res, Inc, 67-70 & Va Res, Inc, Arlington, 70-78. CONSULT, 78-; GEN ENGR, ELECTRONICS RES & DEVELOP COMMAND, US ARMY, ADELPHI, MD, 78- *Mem:* Am Chem Soc; Am Inst Aeronaut & Astronaut; Am Defense Prep Asn. *Res:* Cost schedule control systems criteria; contractor performance measurement; value engineering; engineering analysis; technical writing; design to cost. *Mailing Add:* 4701 Willard Ave Apt 1711 Chevy Chase MD 20815

ALBRECHT, HERBERT RICHARD, b Kenosha, Wis, Nov 14, 09; m 36; c 2. AGRONOMY. *Educ:* Univ Wis, BS, 32, MS, 33, PhD(plant genetics), 38. *Hon Degrees:* DAgr, Purdue Univ, 62; DSc, NDak State Univ, 72. *Prof Exp:* Asst genetics, Univ Wis, 31-36; from asst prof to assoc prof agron, Auburn Univ, 36-44, from asst agronomist to assoc agronomist, Exp Sta, 36-44; assoc prof agron, Purdue Univ, 44-45, asst chief agron, 45-47; prof & head dept, Pa State Univ, 47-52, dir, Coop Agr & Home Econ Exten Serv, 53-61; pres, NDak State Univ, 61-68; dir, Int Inst Trop Agr, Ibadan, Nigeria, 68-75; RETIRED. *Mem:* Fel AAAS; fel Crop Sci Soc Am (pres, 53); fel Am Soc Agron; Sigma Xi. *Res:* Legume and turf grass breeding; inoculation and fertilizer studies with legumes; crop rotation; insect resistance in legumes; inheritance of resistance to bacterial wilt in alfalfa. *Mailing Add:* 68 Lake Point Dr Merifield Acres Clarksville VA 23927

ALBRECHT, KENNETH ADRIAN, b Sheboygan, Wis, Feb 23, 53. TROPICAL GRASSES & LEGUMES. *Educ:* Univ Wis-Oshkosh, BS, 76; Univ Minn, MS, 78; Iowa State Univ, PhD(agron), 83. *Prof Exp:* ASST PROF AGRON, UNIV FLA, 85- *Mem:* Am Soc Agron; Crop Sci Soc Am; Am Forage & Grassland Coun. *Res:* Identify and investigate specific physiological processes that limit the productivity and nutritive value of tropical and temperate forage crops in the subtropics; water relations; photosynthesis. *Mailing Add:* Dept Agron 1575 Linden Dr Madison WI 53706

ALBRECHT, MARY LEWNES, b New York, NY, 1953; m; c 1. FLORICULTURE, GREENHOUSE MANAGEMENT. *Educ:* Rutgers Univ, BS, 75; Ohio State Univ, MS, 77, PhD(hort), 80. *Prof Exp:* Asst prof, 80-86, ASSOC PROF HORT, KANS STATE UNIV, 86- *Concurrent Pos:* Teacher fel, Nat Asn Cols & Teachers Agr, 88. *Mem:* Sigma Xi; Am Soc Hort Sci; Asn Women Sci. *Res:* Floral induction and development in herbaceous perennials; adaptation of native plant material for use in commercial horticulture. *Mailing Add:* Dept Hort Waters Hall Kans State Univ Manhattan KS 66506-4002

ALBRECHT, PAUL, b Jan 27, 25; US Citizen; m 53; c 2. VIROLOGY, TISSUE CULTURE. *Educ:* Comenius Univ Bratislava, MD, 49; Czech Acad Sci, PhD, 60. *Prof Exp:* Asst prof path, Med Fac, Comenius Univ Bratislava, 50-54; from res assoc to chief lab, Inst Virol, Czech Acad Sci, 55-65; res assoc microbiol, Nat Inst Neurol Dis & Blindness, 65-68; res virologist, Div Biol Standards, NIH, 68-72; RES VIROLOGIST, BUR BIOLOGICS, FOOD & DRUG ADMIN, 72- *Concurrent Pos:* Czech Acad Sci Award, 58-63. *Honors & Awards:* Czech Med Asn Prize, 64; Award of Merit, Dept Health & Human Servs, 86. *Mem:* Am Soc Microbiol; Soc Exp Biol & Med; NY Acad Sci. *Res:* Immunogenesis of viral infections; slow virus infections; development and control of vaccines. *Mailing Add:* 1201 Fallsmead Way Rockville MD 20854

ALBRECHT, ROBERT WILLIAM, b Cleveland, Ohio, Mar 31, 35; m 57; c 2. NUCLEAR ENGINEERING. *Educ:* Purdue Univ, BSEE, 57; Univ Mich, MS, 58, PhD(nuclear eng), 61. *Prof Exp:* PROF NUCLEAR ENG, UNIV WASH, 61- *Concurrent Pos:* Researcher, Inst Für Neutronenphysik und Reaktortechnik, Kernforschungs Zentrum, Karlsruhe, Ger, 67-68. *Honors & Awards:* Mark Mills Award, Am Nuclear Soc, 61. *Mem:* Am Nuclear Soc; Am Soc Eng Educ. *Res:* Nuclear reactor dynamics; stochastic processes in nuclear reactors. *Mailing Add:* Dept Elec Eng Univ Wash Ft-10 Seattle WA 98195

ALBRECHT, RONALD FRANK, b Chicago, Ill, Apr 17, 37; m 62; c 3. ANESTHESIOLOGY. *Educ:* Univ Ill, Urbana, BA, 58, BS, 59, MD, 61. *Prof Exp:* Intern, Univ Cincinnati Hosp, 61-62; resident anesthesiol, Univ Ill Res & Educ Hosp, 62-64; clin assoc, NIH, Bethesda, 64-66; asst prof anesthesiol, Univ Ill Col Med, 66-69, clin assoc prof, 69-73; prof, 73-76, CLIN PROF ANESTHESIOL, PRITZKER SCH MED, UNIV CHICAGO, 76-; CHMN & ATTEND PHYSICIAN, DEPT ANESTHESIOL, MICHAEL REESE MED CTR, 71- *Mem:* Am Soc Anesthesiologists; Am Physiol Soc; Int Anesthesia Res Soc. *Res:* Effect of anesthetics upon the brain, especifically blood flow, intracranial pressure and brain metabolism; effect of anesthetics on cardiovascular system. *Mailing Add:* Dept Anesthesiol Michael Reese Med Ctr 31st St Lake Shore Dr Chicago IL 60616

ALBRECHT, STEPHAN LAROWE, b April 14, 43. NITROGEN FIXATION, MICROBIAL ECOLOGY. *Educ:* Univ NMex, BS & BA, 68, MS, 71; Kings Col, Univ London, PhD(plant physiol), 74. *Prof Exp:* Res asst, Biochem Dept, Univ Wis, 74-77 & Lab Nitrogen Fixation Res, Ore State Univ, 77-78; plant physiologist, USDA, 78-90; ASSOC RES SCIENTIST, UNIV FLA, 90- *Concurrent Pos:* Adj asst prof, Agron Dept, Univ Fla, 78- *Mem:* Am Soc Microbiol; Am Soc Plant Physiologists; Crop Sci Soc Am; Soil Sci Soc Am; Am Soc Agron; Sigma Xi. *Res:* Biological nitrogen fixation in symbiotic systems and free-living soil bacteria associated with the roots of crop plants; microbial ecology; physiology of roots; bioenergetics of the rhizosphere. *Mailing Add:* 6020 NW 32nd St Gainesville FL 32611-0311

ALBRECHT, THOMAS BLAIR, b Philadelphia, Pa, July 31, 43; m 67; c 3. VIROLOGY, CELL BIOLOGY. *Educ:* Brigham Young Univ, BS, 67, MS, 69; Pa State Univ, PhD(microbiol & biochem), 73. *Prof Exp:* Res fel virol, Harvard Univ, 74-75; from asst prof to assoc prof, 76-86, PROF MICROBIOL, UNIV TEX MED BR, GALVESTON, 86- *Concurrent Pos:* Prin investr, NIH, Nat Inst Allergy & Infectious Dis, & US Environ Protection Agency, 81; mem sci rev panel, US Environ Protection Agency, 81 - *Mem:* Am Soc Microbiol; Am Soc Cell Biol; Soc Exp Biol & Med; Sigma Xi; NY Acad Sci; Soc Gen Microbiol; Environ Mutagen Soc. *Res:* Cellular responses to human herpesvirus infections; regulation of cytomegalovirus replication and expression; antiviral drugs; interaction of herpesvirus and genotoxic chemicals in initiation and progression of cancers. *Mailing Add:* Dept Microbiol Univ Tex Med Br Galveston TX 77550

ALBRECHT, WILLIAM LLOYD, b Aurora, Ill, May 9, 33; m 56; c 4. PHYSICAL CHEMISTRY. *Educ:* Oberlin Col, AB, 55; Univ Wis, PhD(phys chem), 60. *Prof Exp:* Sr res chemist, 60-66, group leader, 66-69, mgr water treat chem res, 69-70, tech dir, 70-71, mgr technol & mkt, 71-75, mgr res, Pulp & Paper Chem, 75-76, corp qual assurance mgr, 76-82, MGR PURCHASES, NALCO CHEM CO, 82- *Mem:* Am Chem Soc. *Res:* Radiochemistry and tracer techniques; colloid chemistry; paper process chemicals. *Mailing Add:* Nalco Chem Co One Nalco Ctr Naperville IL 60566-1024

ALBRECHT, WILLIAM MELVIN, b Hungerford, Pa, Feb 18, 26; m 48; c 8. PHYSICAL CHEMISTRY, MATERIALS ENGINEERING. *Educ:* Lebanon Valley Col, BS, 48; Univ Cincinnati, MS, 50. *Prof Exp:* Prin chemist physics div, Battelle Mem Inst, 50-57, asst div chief ferrous metall div, 57-60; staff chemist, IBM Corp, 60-62, adv chemist, 60-66, develop engr & mgr advan technol, 66-73, sr engr & mgr, 73-86; CONSULT, 86- *Mem:* Am Chem Soc; Electrochem Soc; Am Soc Metals; fel Am Inst Chem; AAAS; Sigma Xi. *Res:* Analysis, diffusion, kinetics, thermodynamics and sorption in gas metal systems; electrochemistry; surface chemistry; research and development of new materials for electronic packaging. *Mailing Add:* 414 Corey Ave Endicott NY 13760

ALBRECHT-BUEHLER, GUENTER WILHELM, b Berlin, WGer, Mar 8, 42; m 67; c 3. CELL BIOLOGY. *Educ:* Univ Munich, BSc, 63, dipl physics, 67, PhD(physics), 71. *Prof Exp:* Investr elec physiol, Ger Radiol Soc, 67-70; fel cell biol, Friedrich Miescher Inst, Basel, Switz, 70-73; fel, Univ Fla, 73-74; staff investr cell biol, 74-82, HEAD CELL BIOL DEPT, COLD SPRING HARBOR LAB, NORTHWESTERN UNIV MED SCH, 82- *Concurrent Pos:* Training fel, Int Agency for Res of Cancer, Lyon, France, 73-74. *Res:* Motility phenomena in the surface of animal cells and their relation to cell communication and transformation. *Mailing Add:* Dept Cell Bio/Anat Northwestern Univ Med Sch 303 E Chicago Ave Chicago IL 60611

ALBRECHTSEN, RULON S, b Emery, Utah, Mar 12, 33; m 59; c 6. GENETICS, PLANT BREEDING. *Educ:* Utah State Univ, BS, 56, MS, 57; Purdue Univ, PhD(genetics), 65. *Prof Exp:* Res agronomist, Agr Res Serv, USDA, 57-59; res asst agron, Purdue Univ, 59-63; from asst prof to assoc prof, SDak State Univ, 63-69; PROF AGRON, UTAH STATE UNIV, 69- *Mem:* Am Soc Agron. *Res:* Genetics and breeding of barley, safflower, birdsfoot trefoil, oats, flax, rye, and wheat. *Mailing Add:* Dept Plant Sci Utah State Univ Logan UT 84322-4820

ALBREGTS, EARL EUGENE, b Earl Park, Ind, May 30, 29; m 49; c 3. SOIL CHEMISTRY. *Educ:* Purdue Univ, BS, 64, PhD(soils), 68. *Prof Exp:* Asst prof, 67-73, assoc prof, 73-80, PROF SOIL CHEM, UNIV FLA, 80- *Mem:* Am Soc Agron; Am Soc Hort Sci. *Res:* Soil fertility; plant nutrition. *Mailing Add:* 603 Valle Vista Dr Brandon FL 33511

ALBRETHSEN, A(DRIAN) E(DYSEL), b Carey, Idaho, June 20, 29; m 61; c 3. METALLURGY, MINERAL ENGINEERING. *Educ:* Univ Idaho, BS, 52, MS, 58; Mass Inst Technol, PhD(mineral eng), 63. *Prof Exp:* Chemist, Bunker Hill Co, Idaho, 54-55 & 57; asst mining engr, Anaconda Co, Mont, 55-57; sr engr res & develop, Hanford Labs, Gen Elec Co, 63-65; sr res engr, Pac Northwest Labs, Battelle Mem Inst, 65-66; sr res engr, Centr Res Dept, Asarco, Inc, 66-86; consult, 86-89; PLANT METALLURGIST, NORD ILMENITE CORP, JACKSON, NJ, 89- *Mem:* Am Inst Mining, Metall & Petrol Eng; Am Soc Metals; Electrochem Soc. *Res:* Surface chemistry; extractive metallurgy; hydrometallurgy; electrometallurgy; mineral processing. *Mailing Add:* 485 Vicki Dr Bridgewater NJ 08807-1941

ALBRIDGE, ROYAL, b Lima, Ohio, Jan 20, 33; m 57; c 2. NUCLEAR PHYSICS, ATOMIC PHYSICS. *Educ:* Ohio State Univ, BS, 55; Univ Calif, Berkeley, PhD(nuclear physics), 60. *Prof Exp:* From asst prof to assoc prof, 61-73, PROF PHYSICS, VANDERBILT UNIV, 73- *Concurrent Pos:* Res officer, US Air Force, 58-60; sabbatical leave, Univ Uppsala, Sweden, 66-67 & 72-73. *Mem:* AAAS; Am Asn Physics Teachers; Am Phys Soc; NY Acad Sci. *Res:* Photoelectron, beta and gamma ray spectroscopy. *Mailing Add:* Dept Physics Box 1815 Sta B Vanderbilt Univ Nashville TN 37235

ALBRIGHT, BRUCE CALVIN, b Kansas City, Mo, July 30, 46; m 69; c 2. NEUROSCIENCES. *Educ:* Univ Md, BS, 69; Med Col Va, Va Commonwealth Univ, MS, 72, PhD(anat), 74. *Prof Exp:* Asst prof, 74-80, ASSOC PROF ANAT, SCH MED, UNIV NDAK, 80- *Mem:* Am Asn Anatomists; Sigma Xi. *Res:* Comparative prosimian neuroanatomy. *Mailing Add:* Dept Anat Sch Med Univ NDak Box 8135 Univ Station Grand Forks ND 58202

ALBRIGHT, CARL HOWARD, b Allentown, Pa, June 1, 33; m 70. PARTICLE PHYSICS. *Educ:* Lehigh Univ, BS, 55; Princeton Univ, PhD(physics), 60. *Prof Exp:* Instr physics, Princeton Univ, 59-61; res assoc, Northwestern Univ, 61-62, asst prof, 62-68; assoc prof, 68-71, PROF PHYSICS, NORTHERN ILL UNIV, 71- *Concurrent Pos:* NSF grants, 65-68 & 74- *Mem:* Am Phys Soc. *Res:* Theoretical particle physics; weak and electromagnetic interactions. *Mailing Add:* Dept Physics Northern Ill Univ DeKalb IL 60115

ALBRIGHT, DARRYL LOUIS, b Chicago, Ill, Oct 8, 37; m 60; c 2. METALLURGICAL ENGINEERING. *Educ:* Univ Ill, Urbana, BS, 59; Rensselaer Polytech Inst, MS, 62; Lehigh Univ, PhD(metall eng), 65. *Prof Exp:* Res engr, Mat Sect, Res Labs, United Aircraft Corp, 59-62; instr & res asst metall eng, Lehigh Univ, 62-65; res assoc eng res, Int Harvester Co, 65-68; ASSOC PROF METALL ENG, ILL INST TECHNOL, 68- *Concurrent Pos:* Gen indust consult, 68- *Mem:* Am Soc Metals; Am Inst Mining, Metall & Petrol Engrs; Am Soc Eng Educ; Sigma Xi. *Res:* Structural evaluation of materials by x-ray diffraction; controlled solidification processes. *Mailing Add:* Climax Molybdenum 1600 Heron Pk PO Box 1568 Ann Arbor MI 48105

ALBRIGHT, EDWIN C, medicine; deceased, see previous edition for last biography

ALBRIGHT, FRED RONALD, b Talmage, Pa, Feb 16, 44; m 66; c 1. MICROBIOLOGY, NUTRITION. *Educ:* Muhlenberg Col, BS, 66; Univ Ill, PhD(biochem), 70. *Prof Exp:* Fel, Johns Hopkins Univ, 70-72; asst tech dir, 72-76, vpres & tech dir, 76-84, vpres & lab dir, 84-87, SR VPRES & DIR, HEALTH SCI DIV, LANCASTER LABS, INC, 87- *Concurrent Pos:* Damon Runyon Mem Fund Cancer Res fel, 70-71; Am Cancer Soc fel, 71-72. *Mem:* Am Asn Cereal Chemists; Am Oil Chemists Soc; Inst Food Technologists. *Res:* Analytical method development for detection of trace elements, pesticides, nutrients, and toxins in agricultural products and our environment; effect of these materials on plant and animal life. *Mailing Add:* Lancaster Labs Inc 2425 New Holland Pike Lancaster PA 17601

ALBRIGHT, JACK LAWRENCE, b San Francisco, Calif, Mar 14, 30; m 57; c 2. ANIMAL PHYSIOLOGY. *Educ:* Calif Polytech State Univ, BS, 52; Wash State Univ, MS, 54, PhD(animal sci), 57; Mich State Univ, cert(animal behav), 64. *Prof Exp:* Actg instr dairy sci, Wash State Univ, 54; from instr to asst prof dairy husb, Calif Polytech State Univ, 55-59; asst prof dairy sci, Univ Ill, 59-63; assoc prof, 63-66, PROF ANIMAL SCI, PURDUE UNIV, WEST LAFAYETTE, 66- *Concurrent Pos:* Partic, Int Dairy Cong, Copenhagen, 62, Munich, 66, Sydney, 70 & Paris, 78; Fulbright sr res fel, Ruakura Animal Res Sta, Hamilton, NZ, 70-71; vis prof zool, Univ Reading, UK, 77-78; secy-treas, Comn Farm Animal Care, 81-; vis prof animal behav, Univ Ill, 88-89. *Honors & Awards:* Dairy Mgt Res Award, 86; Animal Mgt Res Award, 88. *Mem:* Fel AAAS; Animal Behav Soc; Soc Vet Ethology; Am Dairy Sci Asn; Am Soc Animal Sci. *Res:* Analysis and measurement of management; life cycle management, housing and behavior; improving large dairy herd management practices; animal behavior; bovine physiology; dairy herd health; animal welfare. *Mailing Add:* Dept of Animal Sci Purdue Univ West Lafayette IN 47906

ALBRIGHT, JAMES ANDREW, b Amsterdam, NY, Jan 31, 45; m 64; c 5. ORGANIC CHEMISTRY. *Educ:* State Univ NY, BS, 66; Clark Univ, PhD(chem), 69. *Prof Exp:* Res chemist pesticides, Air Prod & Chem Inc, 69-70; res chemist, Glidden-Durkee Div SCM Corp, 70-72; group leader flame retardants, Mich Chem Corp, 72-77, dir synthesis, Velsicol Chem Corp, 77-80; mem staff, Rexham Corp, 80-83; DIR TECHNOL, JAMES RIVER CORP, 83- *Mem:* Am Chem Soc. *Res:* Synthesis of novel heterocycles and the stereo chemical studies of cyclic organophosphorous compounds. *Mailing Add:* James River Corp 8044 Montgomery Rd Suite 650 Cincinnati OH 45236

ALBRIGHT, JAMES CURTICE, b Madison, Wis, Sept 8, 29; m 51; c 4. OIL WELL LOGGING. *Educ:* Univ Wichita, BA, 50; Univ Okla, MS, 52, PhD(physics), 56. *Prof Exp:* Res engr, Conoco, Inc, 55-57, sr res engr, 57-61, res group leader, 61-64, res assoc, 64-70; tech data div coordr, Oasis Oil Co, 70-74; res assoc, 74-82, SR RES ASSOC, CONOCO, INC, 82- *Mem:* Soc Petrol Eng; Soc Prof Well Log Anal. *Res:* Oil well logging; computer systems; physics. *Mailing Add:* Conoco Inc PO Box 1267 Ponca City OK 74603

ALBRIGHT, JAY DONALD, b Lancaster Co, Pa, Apr 28, 33; m 54; c 3. ORGANIC CHEMISTRY. *Educ:* Elizabethtown Col, BS, 55; Univ Ill, PhD(chem), 58. *Prof Exp:* Asst org chem, Univ Ill, 55-57; from res chemist to sr res chemist, 58-74, GROUP LEADER, LEDERLE LABS, AM CYANAMID CO, 74- *Mem:* Am Chem Soc; Int Soc Heterocyclic Chem. *Res:* Synthetic organic and medicinal chemistry; heterocycles; alkaloids; hypotensive and anxiolytic agents. *Mailing Add:* Five Clifford Ct Nanuet NY 10954

ALBRIGHT, JOHN GROVER, b Winfield, Kans, June 29, 34; m 60; c 2. LIQUID STATE DIFFUSION, OPTICAL INTERFEROMETERS. *Educ:* Wichita Univ, BA, 56; Univ Wis, PhD(chem), 62. *Prof Exp:* PROF CHEM, TEX CHRISTIAN UNIV, 66- *Concurrent Pos:* Vis prof, Lawrence Livermore Nat Lab, 68, 71, 74, 77-81, 85-88, Australian Nat Univ, 81, Univ Naples, 88. *Mem:* Am Chem Soc; Sigma Xi. *Res:* Transport processes in liquid systems. *Mailing Add:* Dept Chem Tex Christian Univ Ft Worth TX 76129

ALBRIGHT, JOHN RUPP, b Wilkes-Barre, Pa, June 10, 37; m 60, 84; c 5. PHYSICS, PHYSICAL MATHEMATICS. *Educ:* Susquehanna Univ, AB, 59; Univ Wis, MS, 61, PhD, 64. *Prof Exp:* From asst prof to assoc prof, 63-78, assoc chmn, 80-85, PROF PHYSICS, FLA STATE UNIV, 78- *Concurrent Pos:* Consult, Fermi Nat Accelerator Lab, 75-; sr res fel, Cavendish Lab, Cambridge, Eng, 76; vis scholar, Chicago Ctr Relig & Sci, 90. *Mem:* Am Phys Soc; Am Asn Physics Teachers. *Res:* Elementary particles; digital computer programming; theoretical and elementary particle physics. *Mailing Add:* Dept Physics Fla State Univ Tallahassee FL 32306

ALBRIGHT, JOSEPH FINLEY, b New Tazewell, Tenn, Mar 9, 27; m 51, 73; c 2. ZOOLOGY. *Educ:* Southwestern Univ, Tex, BS, 49; Ind Univ, PhD(zool), 56. *Prof Exp:* Res assoc biol, Oak Ridge Nat Lab, 56-57; NIH fel, Nat Cancer Inst, 57-58; asst prof surg, Med Col Va, 58-61; biologist, Oak Ridge Nat Lab, 61-70; sr immunologist, Smith, Kline & French Labs, 70-71; prof life sci, Ind State Univ, Terre Haute, 71-84, prof microbiol, Terre Haute Ctr Med Educ, 73-84; CHIEF, BASIC IMMUNOL BR, NIAID, NIH, 84- *Concurrent Pos:* sr fel, Carnegie Inst Wash, 75-76; consult, Div Biol Med, Argonne Nat Lab, 77-79; adj prof, Dept Microbiol and Immunol, George Washington Univ Med Sch, 84- *Mem:* AAAS; Am Asn Immunologists. *Res:* Immunology; immunoparasitology; interactions of parasites and the immune system of their hosts (mechanisms of immune reactions, immunodepression, lymphocyte activation); age-associated decline of immunological competence and aberrations of the immune system. *Mailing Add:* Div Allergy Immunol & Transplantation NIAID NIH 757 Westwood Bldg Bethesda MD 20892

ALBRIGHT, JULIA W, m. MICROBIOLOGY. *Educ:* E Tenn State Univ, BS; Univ Akron, MS; Ind State Univ, PhD(microbiol-immunol), 78. *Prof Exp:* Res asst, Dept Biochem, Mem Res Ctr, Univ Tenn, 63-65; res biologist, Biol Div, Radiation Immunol, Oak Ridge Nat Lab, 65-68; res asst, Infectious Dis-Immunol Unit, Sch Med, Univ Rochester, 68-70; microbiologist, Lab Cellular & Comp Biol, Nat Inst Aging, NIH, 72-76; teaching fel, Dept Life Sci, Ind State Univ, 76-78, asst prof, 78-80, assoc prof, 81-84, dir, Immunol & Cellular Studies Aging, 81-85, prof, 84-85; assoc prof, 80-81, PROF, DEPT MICROBIOL, MED CTR, GEORGE WASHINGTON UNIV, 85- *Concurrent Pos:* Adj prof, Dept Life Sci, Ind State Univ, 85-; grants, NSF, 88-91 & 91, NIH, 90- *Honors & Awards:* Outstanding Res Award, Geront Soc Am, 76. *Mem:* Sigma Xi; Tissue Cult Asn; Geront Soc Am; Am Asn Immunologists; Am Soc Microbiol; AAAS. *Res:* Age-related changes in competence for immune responses; all aspects of the interactions between parasites and the immune system of their hosts; biotechnology, hybridoma production and molecular genetics. *Mailing Add:* Dept Microbiol George Washington Univ Med Ctr 2300 Eye St NW Washington DC 20037

ALBRIGHT, LAWRENCE JOHN, b Owen Sound, Ont, July 22, 41; c 3. AQUACULTURE. *Educ:* McGill Univ, BSc, 63; Ore State Univ, MS, 65, PhD(marine microbiol), 67. *Prof Exp:* From asst prof to assoc prof, 67-81, PROF SALMONID CULT & DIS, SIMON FRASER UNIV, 82- *Concurrent Pos:* Grants, Nat Sci & Eng Res Coun Can, 67- & Sci Coun BC; acting dir, Master Aquacult Prog, Simon Fraser Univ. *Mem:* Can Soc Microbiol; Am Soc Microbiol; Aquaculture Asn Can; Am Fisheries Soc. *Res:* Ecology of aquatic microorganisms, particularly diseases of farmed and wild salmonids. *Mailing Add:* 00001031x Sci Simon Fraser Univ Burnaby BC V5A 1S6 Can

ALBRIGHT, LYLE F(REDERICK), b Bay City, Mich, May 3, 21; m 50; c 2. CHEMICAL ENGINEERING. *Educ:* Univ Mich, BS, 43, MS, 44, PhD(chem eng), 50. *Prof Exp:* Lab technician, Dow Chem Co, 39-41; chem engr, Manhattan Proj, E I du Pont de Nemours & Co, Inc, 44-46; res assoc, Univ Mich, 48-49; res chem engr, Colgate-Palmolive Co, 50-51; from asst prof to assoc prof chem eng, Univ Okla, 51-55; assoc prof, 55-58, PROF CHEM ENG, PURDUE UNIV, 58- *Concurrent Pos:* Vis prof, Univ Tex, 52 & Tex A&M Univ, 85; indust consult, 56-; ed chem processing & eng ser, Marcel Dekker, Inc, 72-80. *Honors & Awards:* Potter Award, 88; Shreve Prize, 60, 70 & 88. *Mem:* Fel Am Inst Chem Eng; Am Chem Soc. *Res:* Chemical processes and kinetics of alkylation, pyrolysis, partial oxidation, hydrogenation and chlorination; thermodynamics; fuel cells. *Mailing Add:* Sch Chem Eng Purdue Univ Lafayette IN 47907

ALBRIGHT, MELVIN A, b Oklahoma City, Okla, Feb 13, 29; m 49; c 2. CHEMICAL ENGINEERING. *Educ:* Okla State Univ, BS, 50, MS, 56. *Prof Exp:* Jr res engr, Continental Oil Co, 52-54; math engr, Phillips Petrol Co, 57-58, develop engr, 58-68, mgr, Thermodyn Sect, Res & Develop Dept, 68-77, mem res dir's staff, 77-85; INDEPENDENT CONSULT, 85- *Concurrent Pos:* Mem, subcomt tech data, Am Petrol Inst; chmn, tech comt, Fractionation Res, Inc & physical property develop comt, Gas Processors Asn; mem, Thermodyn Adv Bd, Thermodyn Res Lab, Wash Univ; internal consult fractionation & thermodyn & phys property data; mem, Chem Eng Assessment Bd, Nat Res Coun. *Honors & Awards:* Hanlon Award, Gas Processors Asn. *Mem:* Fel Am Inst Chem Engrs; AAAS; Gas Processors Asn. *Res:* Application of mathematics to chemical engineering process design; analysis optimization and development; fractionation; thermodynamics; physical properties. *Mailing Add:* 1853 S Madison Blvd Bartlesville OK 74006

ALBRIGHT, RAYMOND GERARD, anatomy, zoology, for more information see previous edition

ALBRIGHT, ROBERT LEE, b Leola, Pa, Jan 28, 32; m 59; c 2. ORGANIC & POLYMER CHEMISTRY. *Educ:* Elizabethtown Col, BS, 54; Univ Ill, PhD(org chem), 58. *Prof Exp:* Asst, Univ Ill, 54-56; res chemist, 58-70, sr chemist & consult, 70-78, sr scientist & group leader, 78-82, RES FEL, ROHM & HAAS RES LABS, 82- *Concurrent Pos:* Instr, Elizabethtown Col, 59; lectr, Holy Family Col, 63. *Honors & Awards:* John C Vaaler Award. *Mem:* Am Chem Soc. *Res:* Reaction mechanisms; anchimerically assisted reactions; reactions of iodonium salts and organophosphorus compounds; polymerization, especially mechanism of formation of macrostructure and microstructure; polymer morphology; mechanism of reactions of polymers; synthesis and properties of ion exchange polymers; catalysis by porous polymers; polymeric adsorbents; bio-catalysts. *Mailing Add:* 36 Autumn Rd Churchville PA 18966

ALBRIGO, LEO GENE, b Palmdale, Calif, Aug 24, 40; m 59; c 3. HORTICULTURE, PLANT PHYSIOLOGY. *Educ:* Univ Calif, Davis, BS, 62, MS, 64; Rutgers Univ, PhD(hort), 68. *Prof Exp:* From res asst to res assoc pomol, Rutgers Univ, 64-68; asst horticulturist, 68-74, assoc horticulturist, 74-79, HORTICULTURIST, CITRUS RES & EDUC CTR, UNIV FLA, 79- *Concurrent Pos:* Consult, Food & Agr Orgn. *Mem:* Am Hort Soc; Int Soc Citricult; Am Soc Plant Physiol; Int Soc Hort Sci. *Res:* Fresh fruit quality; environmental influences on fruit development; stress physiology. *Mailing Add:* Citrus Res & Educ Ctr 700 Exp Sta Rd Lake Alfred FL 33850

ALBRINK, MARGARET JORALEMON, b Bisbee, Ariz, Jan 6, 20; m 44; c 3. INTERNAL MEDICINE, ENDOCRINOLOGY. *Educ:* Radcliffe Col, BA, 41; Yale Univ, MD, 46, MPH, 51. *Prof Exp:* Asst med, Yale Univ, 46-47, from instr internal med to asst prof med, 51-61; from assoc prof to prof med, 61-90, EMER PROF MED, SCH MED, WVA UNIV, 90- *Concurrent Pos:* Intern, New Haven Hosp, 46-47; estab investr, Am Heart Asn, 58-63; Res Career Award, NIH, 63-90. *Honors & Awards:* McCollum Award, Am Soc Clin Nutrit, 78. *Mem:* Am Soc Clin Invest; Am Fedn Clin Res; Am Soc Clin Nutrit; fel Am Col Physicians; fel Am Col Nutrit. *Res:* Serum lipids in metabolic diseases; lipid metabolism in atherosclerosis, diabetes, mellitus, and obesity; nutrition and serum lipids. *Mailing Add:* Dept Med Health Sci Ctr WVa Univ Morgantown WV 26506

ALBRINK, WILHELM STOCKMAN, b Napoleon, Ohio, Aug 22, 15; m 44; c 3. PATHOLOGY. *Educ:* Oberlin Col, AB, 37; Yale Univ, PhD(zool), 41, MD, 47. *Prof Exp:* Asst zool, Yale Univ, 37-40, asst path, Off Sci Res & Develop Proj, Sch Med, 42-45, asst, Univ, 47-49, from instr to assoc prof, 51-61; chmn dept, 61-69, prof, 61-81, EMER PROF PATH, MED CTR, WVA UNIV, 81- *Concurrent Pos:* Intern, New Haven Hosp, Conn, 47-48, asst resident, 48-49; Am Cancer Soc fel, 49-51. *Mem:* AAAS; emer mem Am Asn Path & Bact; emer mem Am Asn Cancer Res; NY Acad Sci; emer mem Am Soc Exp Path. *Res:* Electromotive force of living tissues; pathological physiology of war gases; synthetic membrane behavior; biology of neoplasia; pathology of anthrax; localization of particulate matter in the lung. *Mailing Add:* Dept Path WVa Univ Med Ctr Morgantown WV 26506

ALBRITTON, CLAUDE CARROLL, JR, archaeological geology; deceased, see previous edition for last biography

ALBRITTON, WILLIAM LEONARD, b Andalusia, Ala, Dec 1, 41; m 63; c 2. MICROBIOLOGY & PEDIATRICS. *Educ:* Univ Ala, BS, 64; Univ Tenn, PhD(biochem), 68; Med Col Ala, MD, 70. *Prof Exp:* Med officer, Ctr Dis Control, USPHS, 73-75; from asst prof to assoc prof infectious dis, Univ Manitoba, 76-82; dir, Sexually Transmitted Dis Lab Prog, USPHS, 82-83; prof microbiol & head dept, Univ Sask, 84-90; DIR PROV LAB PUB HEALTH, UNIV ALTA, 90- *Mem:* Can Infectious Dic Soc (pres, 82); Infectious Dis Soc Am; Am Soc Microbiol; Am Acad Pediat; Royal Col Physicians & Surgeons Can. *Res:* Pediatric infectious diseases; molecular epidemiology of sexually transmitted diseases; microbiology of Haemophilus. *Mailing Add:* Prov Lab Public Health Univ Alta Edmonton AB T6G 2J2 Can

ALBRO, PHILLIP WILLIAM, b Geneva, NY, Aug 24, 39; c 4. BIOCHEMISTRY. *Educ:* Univ Rochester, BA, 61; St Louis Univ, PhD(biochem), 68. *Prof Exp:* Chemist, US Army Biol Labs, Md, 64-65; BIOCHEMIST, NAT INST ENVIRON HEALTH SCI, 68- *Concurrent Pos:* Adj asst prof, Duke Univ, 75-; adj assoc prof, Univ NC, 80- *Mem:* AAAS; Am Chem Soc; NY Acad Sci. *Res:* Analytical biochemistry; intermediary metabolism of xenobiotics, especially environmental pollutants. *Mailing Add:* 808 Griffis St Cary NC 27511

ALBU, EVELYN D, b Mt Vernon, NY, Sept 25, 38. MEDICAL EDUCATION. *Educ:* Ohio Univ, BS, 60; Georgetown Univ Med & Dent Sch, PhD(microbiol), 73. *Prof Exp:* Microbiol biochemist res, Squibb Inst Med Res, 60-64, med writer, 64-65; clin res assoc int clin res, Schering Corp, 65-69; teaching asst microbiol, Georgetown Univ, 69-73; dir prof commun, 73-79, assoc dir prof serv, 79-82, DIR PROF SERV, SCHERING CORP, 82- *Mem:* Am Soc Microbiol. *Res:* Clinical use of antibiotics, resistance development, epidemiology. *Mailing Add:* 160 Canor Brook Pkwy Summit NJ 07901

ALBUQUERQUE, EDSON XAVIER, b Recife, Brazil, Jan 22, 36; US citizen; wid; c 2. MOLECULAR PHARMACOLOGY, NEUROBIOLOGY. *Educ:* Salesiano Col, Brazil, BS, 53; Univ Recife, MD, 59; Univ Sao Paulo, PhD(physiol & pharmacol), 62. *Prof Exp:* Asst prof anat & physiol, Fac Med Sci, Recife, 57-59; instr pharmacol, Univ Ill Col Med, 64-65; asst prof Int Brain Res Orgn, UNESCO fel, Univ Lund, 65-67; from asst prof to prof, State Univ NY Buffalo, 68-73, actg chmn dept, 73-74; PROF PHARM & CHMN DEPT, UNIV MD SCH MED, BALTIMORE, 74- *Concurrent Pos:* Trainee fel, Rockefeller Found, Univ Ill, 63-65; asst prof, UNESCO fel, Royal Vet Col, Stockholm, 65-67; res asst prof, State Univ NY Buffalo, 67-68; fel, Paulista Sch of Sao Paulo, 69-72. *Mem:* AAAS; Am Physiol Soc; Am Soc Pharmacol & Exp Therapeut; Latin Am Soc Physiol Sci. *Res:* Neurophysiology; electrophysiology; biophysics of excitable membranes. *Mailing Add:* Dept Pharmacol & Exp Therapeut Univ Md 655 W Redwood St Baltimore MD 21201

ALBURGER, DAVID ELMER, b Philadelphia, Pa, Oct 6, 20; m 45; c 4. BETA RAY & GAMMA RAY SPECTROSCOPY. *Educ:* Swarthmore Col, BA, 42; Yale Univ, MS, 46, PhD(physics), 48. *Prof Exp:* Proj engr, Naval Res Lab, 42-45; lab asst physics, Yale Univ, 45-47; sr physicist, Brookhaven Nat Lab, 48-90; RETIRED. *Concurrent Pos:* NSF fel, Nobel Inst Physics, Sweden, 52-53. *Mem:* Fel Am Phys Soc. *Res:* Radioactivity and Van de Graaff accelerator research on nuclear energy levels; gamma-ray pair spectrometer design; new isotopes from heavy-ion reactions. *Mailing Add:* Brookhaven Nat Lab Upton NY 11973

ALBUS, JAMES S, b Louisville, Ky, May 4, 35. INTELLIGENT SYSTEMS, INDUSTRIAL & MANUFACTURING ENGINEERING. *Educ:* Wheaton Col, BS, 57; Ohio State Univ, MS, 58; Univ Md, PhD(elec eng), 72. *Prof Exp:* Sect chief, cybernet & subsysts develop, Goddard Space Flight Ctr, NASA, 58-73; prof mgr, sensors & comput control technol, 73-79, group leader, programmable automation, 79-80, CHIEF, ROBOT SYSTS DIV, NAT BUR STANDARDS, 80- *Honors & Awards:* IR-100 Award, 76. *Mailing Add:* 9520 W Stanhope Rd Kensington MD 20895

ALCALA, JOSE RAMON, b Ponce, PR, May 1, 40; m 64; c 1. ANATOMY, BIOCHEMISTRY. *Educ:* Univ Mo-Columbia, BA, 64, MA, 66; Univ Ill Med Ctr, PhD(anat), 72. *Prof Exp:* PROF ANAT & CELL BIOL & OPHTHAL, SCH MED, WAYNE STATE UNIV, 72- *Concurrent Pos:* Consult, subcomt cataract res, Nat Adv Eye Council, Dept Health & Human Serv, NIH, 80- *Mem:* NY Acad Sci; Am Asn Clin Anatomists; Asn Res Vision & Ophthal; Int Soc Eye Res; Am Asn Anatomists. *Res:* Biochemistry and immunochemistry of lens plasma membranes. *Mailing Add:* Dept of Anat Wayne State Univ Sch of Med Detroit MI 48201

ALCAMO, I EDWARD, b New York, NY, Oct 16, 41; m 64; c 3. MICROBIOLOGY, BIOLOGY. *Educ:* Iona Col, BS, 63; St John's Univ, NY, MS, 65, PhD(microbiol), 71. *Prof Exp:* Teaching fel, St John's Univ, NY, 63-65; PROF BIOL, STATE UNIV NY AGR & TECH COL, FARMINGDALE, 65- *Concurrent Pos:* NSF res grant, 66; assoc prof biol, C W Post Col, Long Island Univ, 71-75 & City Univ New York, 73-75; prof microbiol, NY Chiropractic Col, 74-; assoc prof microbiol, Adelphi Univ, 75; Whitmire Res Corp res grant, 75-78. *Mem:* Sigma Xi; Am Soc Microbiol; Nat Asn Biol Teachers; Asn Col & Univ Biologists. *Res:* Arthropod vectors of disease; effects of ultrasonic vibrations on bacterial spores; lysosomes and hyaluronidase of trichomonads. *Mailing Add:* Dept Biol State Univ NY Tech Col Farmingdale Melville Rd Farmingdale NY 11735

ALCANTARA, EMERITA N, b Philippines, Jan 16, 43. NUTRITIONAL SCIENCES. *Educ:* Univ Philippines, BS, 61; Univ Wis-Madison, MS, 67, PhD(nutrit sci), 70. *Prof Exp:* Intern dietetics, Yale New Haven Hosp, Conn, 62-63; therapeut dietitian, Univ Wis Hosp, Madison, 63-64; res asst human metab studies, Univ Wis, Madison, 64-69; res asst, Nat Dairy Coun, 70-74, asst dir, Nutrit Res Div, 75-85, dir, Nutrit Policy Issues, 85, vpres, Nutrit Res, 85-90; DIR MFR RELS, DAIRY COUN WIS, 90- *Mem:* Am Dietetic Asn; Inst Food Technologists; Sigma Xi; Am Inst Nutrit; AAAS; Coun Agr Sci & Technol. *Res:* Protein and mineral metabolism; amino acid requirements; nutrition research and public policy development. *Mailing Add:* Dairy Coun Wis 999 Oakmount Plaza Dr Westmont IL 60559

ALCANTARA, VICTOR FRANCO, b Davao City, Philippines, Nov 4, 25; US citizen; m 62; c 2. SANITARY ENGINEERING, ENGINEERING SURVEYS. *Educ:* Mapua Inst Technol, Phillipines, BS, 51; Miss State Univ, MS, 69. *Prof Exp:* Civil engr & survr, Alcantara & Assocs, Philippines, 51-57; dean, Col Eng, Cent Mindanao Univ, Philippines, 57-70; assoc prof civil eng technol, 70-75, chmn, 75-85, ASSOC PROF CIVIL ENG TECHNOL, HARTFORD STATE TECH COL, 85- *Concurrent Pos:* Mem vis fac, Yale Univ, 79-80; civil engr, Arlombardi Assocs, 85-86, Hayden-Wegman, 87-89; WMC Consult Eng, 89- *Mem:* Nat Soc Prof Engrs; Am Soc Civil Engrs; Am Soc Eng Educ; Am Inst Steel Construct; Soc Hist Technol. *Mailing Add:* Six Cadwell St West Hartford CT 06107

ALCARAZ, ERNEST CHARLES, b Coronado, Calif, Dec 4, 35; m 60; c 1. LASER, ATMOSPHERIC CHEMISTRY & PHYSICS. *Educ:* San Diego State Col, BA, 58, MS, 60; Wayne State Univ, PhD(physics), 68. *Prof Exp:* Instr physics, San Diego State Col, 60-62; from teaching asst to instr, Wayne State Univ, 62-68; res physicist, Signature & Propagation Lab, US Army Ballistic Res Labs, 68-75; scientist, Sci Appln, Inc, 75-78; div mgr, Jaycor, 78-82; sr prog consult, Berkeley Res Assoc, 82-85; SR PROG CONSULT, WJ SCHAFER ASSOC, 85- *Mem:* Optical Soc Am; Sigma Xi; Am Inst Aeronaut & Astronaut; Inst Elec & Electronic Engrs. *Res:* Properties of liquid helium; coherent light propagation through the near earth atmosphere; micrometeorological structure of the near earth atmosphere; field test evaluation of laser/eo systems; survivability of space and ground systems. *Mailing Add:* 10209 Westford Dr Vienna VA 22182

ALCORN, STANLEY MARCUS, b Modesto, Calif, June 18, 26; m 49; c 4. PHYTOPATHOLOGY. *Educ:* Univ Calif, BSc, 48, PhD(plant path), 54. *Prof Exp:* Res asst plant path, Univ Calif, 49-53, Merck & Co fel, 54-55; plant pathologist, USDA, 55-63; from assoc prof to prof, 63-89, EMER PROF PLANT PATH, UNIV ARIZ, 89- *Mem:* Fel AAAS; Am Soc Microbiol; Am Phytopath Soc; Am Soc Hort Sci; Guayule Rubber Soc (past pres); Crop Sci Soc; Sigma Xi. *Res:* Verticillium dahliae; cactus-rotting Erwinia; Macrophomina phaseolina; cacti; guayule; jojoba; diseases of new crops and native plants. *Mailing Add:* Dept of Plant Path Univ of Ariz Tucson AZ 85721

ALCORN, WILLIAM R(OBERT), b Chicago, Ill, Nov 11, 35; m 88; c 2. CATALYSIS. *Educ:* Mass Inst Technol, SB, 57, SM, 60, ScD(chem eng), 65. *Prof Exp:* Indust engr, Tousey Varnish Co, Ill, 57-58; asst prof chem eng, Mass Inst Technol, 61-62; proj chem engr, Northern Res & Eng Corp, Mass, 64-66; res group leader, Leesona Moos Labs, NY, 66-69; res assoc catalyst div, Harshaw Chem Co, 71-73; prog mgr synthetic fuels, 74-76, mgr catalyst res, 77-81, tech dir catalyst dept, 82-83; Consult, 84-85; MGR, RES & DEVELOP, CAMET CO, 86- *Mem:* Am Chem Soc; Am Inst Chem Engrs; Soc Automotive Engrs. *Res:* Applied chemical kinetics and heterogeneous catalysis; heat and mass transport; fuel cells and batteries; research and development management. *Mailing Add:* 289 Orchard St Chagrin Falls OH 44202

ALDAG, ARTHUR WILLIAM, JR, b Chicago, Ill, Oct 13, 41; m 61; c 3. CHEMICAL ENGINEERING, PHYSICAL CHEMISTRY. *Educ:* Univ Ill, Urbana, BS, 63; Stanford Univ, MS, 65, PhD(chem eng), 68. *Prof Exp:* Res scientist, Universal Oil Prod Co, Ill, 67; actg asst prof & fel chem eng, Univ Minn, 67-69; asst prof chem eng & mat sci, Univ Okla, 69-76, assoc prof, 76-80; ENGR ASSOC, PHILLIPS PETROL, 80- *Concurrent Pos:* Adj prof chem eng & mat sci, Univ Okla, 80- *Mem:* AAAS; Am Inst Chem Engrs; Am Chem Soc; Am Soc Eng Educ; Sigma Xi. *Res:* Heterogeneous catalysis; coupled heat and mass transfer in reacting systems; clean surface chemistry; enzyme catalysis; enzyme reactor technology. *Mailing Add:* 6636 E 88th Pl Tulsa OK 74133

ALDEN, CARL L, b Centerville, Iowa, Oct 14, 44; c 3. NEPHROTOXICOLOGY, INDIRECT CARCINOGENESIS. *Educ:* Ohio State Univ, DVM, 68, MS, 76; Am Col Vet Pathologists, dipl, 78. *Prof Exp:* Vet pathologist, 77-79, group leader, 80-81, sect head path, 81-84, sect head pathobiol, 84-87, SECT HEAD IN-VITRO TOXICOL, MIAMI VALLEY LABS, HUMAN & ENVIRON SAFETY DIV, PROCTER & GAMBLE CO, 87- *Concurrent Pos:* Vet path officer, US Army Med Res & Nutrit Lab, Fitzsimmons Gen Hosp, Denver, Colo, 68-71; vet pathologist, Vet Diag Lab, Ohio Dept Agr, Reynoldsburg, 72-76; chmn div comparative path, Col Med, WVA Univ, Morgantown, 76- 77. *Mem:* Am Col Vet Pathologists; Soc Toxicol Pathologists; Am Asn Vet Lab Diagnosticians; Am Asn Lab Animal Sci; Am Vet Med Asn; Am Col Toxicol; AAAS; Soc Environ Health Prof. *Res:* Nephrotoxicology, mechanisms of toxicity linked with non- genotoxic tumorigenicity; extrapolation of laboratory animal response in human risk assessment; animal model development; standard settling in toxicologic pathology. *Mailing Add:* Miami Valley Labs Proctor & Gamble Co PO Box 398707 Cincinnati OH 45239-8707

ALDEN, RAYMOND W, III, b Daytona, Fla, Dec 29, 49; m 71; c 3. MARINE POLLUTION ECOLOGY. *Educ:* Stetson Univ, BS, 71; Univ Fla, PhD(marine biol), 76. *Prof Exp:* Grad res asst, Univ Fla, 71-73, marine biologist II, 74-75; from asst prof to assoc prof biol, 76-88, PROF BIOL, OLD DOMINION UNIV, 88- *Concurrent Pos:* Dir appl marine res lab, Old Dominion Univ, 82- *Mem:* Estuarine Res Fed; Am Chem Soc; Soc Environ Toxicol & Analytical Chem; Limnology & Oceanog; Sigma Xi. *Res:* Marine pollution ecology; effect of contaminants on marines and estuarine communities; marine toxicology. *Mailing Add:* Appl Marine Res Lab Old Dominion Univ Norfolk VA 23529

ALDEN, ROLAND HERRICK, b Champaign, Ill, Feb 4, 14; m 37; c 3. ANATOMY. *Educ:* Stanford Univ, AB, 36; Yale Univ, PhD(zool), 41. *Prof Exp:* Instr zool, Yale Univ, 41-42; mem fac, Div Anat, 49-51, chief, 51-61, assoc dean grad sch med sci, 60-68, chancellor pro tem, 70, dean col basic med sci, 61-79, dean grad sch med sci, 68-79, EMER DEAN & EMER PROF, CTR HEALTH SCI, UNIV TENN, MEMPHIS, 79- *Concurrent Pos:* Mem anat comt, Nat Bd Med Examrs, 59-62; spec consult, Anat Sci Training Comt, USPHS, 60-64; secy-treas, Tenn Bd Basic Sci Examrs, 63-75; mem comt grad educ, Nat Asn State Univ & Land-Grant Cols, 68-72. *Mem:* Am Asn Anat (pres, 69-70); Am Physiol Soc; Soc Exp Biol & Med; Am Soc Zoologists. *Res:* Mammalian reproduction; experimental embryology; implantation. *Mailing Add:* 1718 Airport Ct Placerville CA 95667

ALDEN, THOMAS H(YDE), b Philadelphia, Pa, Oct 20, 33; m 57; c 4. PHYSICAL METALLURGY. *Educ:* Amherst Col, AB, 55; Mass Inst Technol, MS, 57, PhD(phys metall), 60. *Prof Exp:* Mem res staff metall, Gen Elec Res Lab, 60-67; from assoc prof to prof Metall, Univ BC, 67-91; CONSULT, 91- *Honors & Awards:* Grossman Award, Am Soc Metals, 69; Hofmann Prize, 74; Mathewson Gold Medal, TMS-AIME, 90. *Res:* Fatigue fracture; strain hardening; theory of creep; superplasticity. *Mailing Add:* Dept Metall Univ BC Vancouver BC V6T 1W5 Can

ALDER, BERNI JULIAN, b Duisburg, Ger, Sept 9, 25; nat US; m 56; c 3. CHEMICAL PHYSICS, STATISTICAL MECHANICS. *Educ:* Univ Calif, BS, 47, MS, 48; Calif Inst Technol, PhD(chem), 51. *Prof Exp:* Instr chem, Univ Calif, Berkeley, 51-54, THEORET PHYSICIST, LAWRENCE LIVERMORE LAB, UNIV CALIF, LIVERMORE, 55-, PROF, DEPT APPL SCI, UNIV CALIF, DAVIS, 87- *Concurrent Pos:* Guggenheim fel, 54-55; NSF fel, 63-64; ed, J Computational Physics, 66-; Van der Waals prof, Univ Amsterdam, 70-71, Hinshelwood prof, Oxford, 86, Sorentz prof, Leiden, 90; lectr, Japanese Promotion Sci, 89. *Honors & Awards:* Hildebrand Award, Am Chem Soc, 85; G N Lewis lectr, Univ Calif, Berkeley, 85; G Kistiakowski Lectr, Harvard Univ, 90; Royal Soc Lectr, 91. *Mem:* Nat Acad Sci; Am Chem Soc; Am Phys Soc. *Res:* Computational classical and quantum mechanical statistical mechanics. *Mailing Add:* 1245 Contra Costa Dr El Cerrito CA 94530

ALDER, EDWIN FRANCIS, b Hugo, Okla, Sept 1, 27; m 51; c 4. PESTICIDES. *Educ:* Univ Okla, BS, 51, PhD(plant sci), 56; Univ Chicago, MS, 52. *Prof Exp:* Instr sci, Ark State Col, 53; from asst to instr plant sci, Univ Okla, 54-55; instr bot, Univ Ark, 55-57; sr plant physiologist, Eli Lilly & Co, 57-59, asst head agr res, 59-61, head plant sci res, 61-65, dir, plant sci res, 65-66 & agr res, 66-69, vpres agr res & develop, 69-73, vpres, Lilly Res Lab Div, 73-86; RETIRED. *Concurrent Pos:* Mem bd dirs, Eli Lilly & Co, 78-86. *Res:* Pesticide research and development. *Mailing Add:* 10140 E Troy Ave Indianapolis IN 46239

ALDER, HENRY LUDWIG, b Duisburg, Ger, Mar 26, 22; nat US; m 63; c 1. MATHEMATICS, STATISTICS. *Educ:* Univ Calif, Berkeley, AB, 42, PhD(math), 47. *Prof Exp:* Asst math, Univ Calif, Berkeley, 42-43, assoc math & jr instr meteorol, 43-44. instr math, 47-48; from instr to assoc prof, 48-65, PROF MATH, UNIV CALIF, DAVIS, 65- *Concurrent Pos:* Chmn, Coun Sci Soc Presidents, 80; mem, Calif State Bd Educ, 82-85. *Honors & Awards:* Lester R Ford Award, Math Asn Am, 70, Award for Distinguished Serv to Math, 80; Cert of Merit, Nat Coun Teachers Math, 75. *Mem:* Am Math Soc; Math Asn Am (secy, 60-74, pres-elect, 76, pres, 77-78). *Res:* Number theory; existence and nonexistence of certain identities in the theory of partitions and compositions. *Mailing Add:* Dept of Math Univ of Calif Davis CA 95616

ALDERDICE, MARC TAYLOR, b Crane, Tex, May 26, 48; m 75; c 2. NEUROPHARMACOLOGY, NEUROPHYSIOLOGY. *Educ:* Univ Tex, Arlington, BS, 70; Univ Tex Health Sci Ctr Dallas, PhD(pharmacol), 75. *Prof Exp:* Asst prof pharmacol, La State Univ Med Ctr, New Orleans, 77-82; asst dir clin studies, Bristol-Meyers US Pharmaceut Group, 82-89; DIR CLIN RES, SIGMA-TAU PHARMACEUT INC, 89- *Concurrent Pos:* NIH fel, Nat Inst Neurol & Commun Dis & Stroke, Univ Conn Health Ctr Farmington, 75-77. *Mem:* Am Soc Pharmacol & Exp Therapeut; Soc Neurosci; AAAS. *Res:* Clinical studies involving buspirone (anxiolytic) and trazodone (antidepressant); investigation of synaptic mechanisms electrophysiological and pharmacokinetictechniques. *Mailing Add:* Sigma-Tau Pharmaceut 200 Orchard Ridge Suite 300 Gaithersburg MD 20878

ALDERFER, JAMES LANDES, NUCLEAR MAGNETIC RESONANCE SPECTROSCOPY. *Educ:* Univ Ky, PhD(org chem), 70. *Prof Exp:* ASSOC CANCER RES SCIENTIST, ROSWELL PARK MEM INST, 77- *Res:* Structure and confirmation of biological molecules; nucleic acids. *Mailing Add:* Dept Biophys Roswell Park Mem Inst 666 Elm St Buffalo NY 14263

ALDERFER, RONALD GODSHALL, b Harleysville, Pa, July 14, 43; m 65; c 2. ENVIRONMENTAL SCIENCES. *Educ:* Wash Univ, AB, 65, PhD(biol), 69. *Prof Exp:* Asst prof biol, Univ Chicago, 69-75; chief ecologist, Harland Bartholomew & Assocs, 75-81; midwest regional mgr, ESE Inc, 81-87; PDC TECHNICAL SERV INC, 88- *Mem:* Ecol Soc Am; Nat Asn Environ Professionals; Soc Environ Toxicol & Chem. *Res:* Environmental biology; applied ecology; chemical stabilization of hazardous waste; environmental assessment. *Mailing Add:* 15574 Highcroft St Louis MO 63017

ALDERFER, RUSSELL BRUNNER, b Lansdale, Pa, Sept 27, 13; m 41; c 2. SOIL & PLANT WATER RELATIONS. *Educ:* Pa State Col, BS, 36, MS, 40, PhD(soil technol), 47. *Prof Exp:* Soil technologist, Soil Conserv Serv, USDA, 36-43; instr soil technol & coop agent, Pa State Univ, 43-47, from asst prof to prof soil technol, 47-54; prof & chmn dept, 54-62, res prof soils, 62-75, EMER PROF SOILS, RUTGERS UNIV, NEW BRUNSWICK, 75- *Mem:* Fel AAAS; Soil Sci Soc Am; Am Soc Agron; Soil Conserv Soc Am; Int Soc Soil Sci. *Res:* Soil physics; soil, water and waste management. *Mailing Add:* 390 Independence Blvd North Brunswick NJ 08902

ALDERMAN, MICHAEL HARRIS, b New Haven, Conn, Mar 26, 36; m 68; c 4. PUBLIC HEALTH, INTERNAL MEDICINE. *Educ:* Harvard Col, BA, 58; Yale Univ, MD, 62. *Prof Exp:* Fel human genetics, New York Hosp, 67-68; chief renal serv, Lincoln Hosp, 68-70; from asst prof to assoc prof med & pub health, 70-79, PROF PUB HEALTH, MED SCH, CORNELL UNIV, 79- *Concurrent Pos:* Coordr pub health, Rural Health Proj, Jamaica, WI, 71-; mem comt mother & child health, Jamaican Govt, 71-; mem NY comt, US Civil Rights Comn, 72-76; WHO traveling fel, 73 & 76; assoc ed, Milbank Mem Quart Health & Soc, 73-77; contrib ed, Sci Yearbk, 73-; mem bd dirs, Int Med & Res Found, 74-, chmn bd dirs, 81-; chmn commun progs, NY Heart Asn, 75-79; physician to outpatients, New York Hosp, 75-; chmn adv comt hypertension, New York City Health Dept, 75-77; sr asst ed, Cardiovasc Rev & Reports, 80- *Mem:* Fel Am Col Physicians; Am Heart Asn; fel Am Soc Clin Nutrit; Am Soc Nephrol; Am Soc Trop Med & Hyg. *Res:* Developing and establishing methods of health care delivery; malnutrition in Jamaica and detection and treatment of hypertension in the United States. *Mailing Add:* Dept Epidemiol Pub Health Albert Einstein Col Med 1300 Morris Park New York NY 10461

ALDERS, C DEAN, b Nebr, 24. STATISTICS. *Educ:* Peru State Univ, BA, 49; Iowa State Univ, MA, 50 & 74, Phd (statist), 76. *Prof Exp:* Instr statist & undergrad math, Mankato State Univ, 56-; RETIRED. *Mem:* Math Asn Am; Am Statist Asn. *Mailing Add:* Dept Math Mankato State Univ Box 41 Mankato MN 56001

ALDERSON, NORRIS EUGENE, b Columbia, Tenn, Aug 13, 43; m 68; c 2. ANIMAL NUTRITION, RESEARCH ADMINISTRATION. *Educ:* Univ Tenn, BS, 65; Univ Ky, MS, 68, PhD(animal nutrit), 70. *Prof Exp:* Res assoc animal sci, Univ Ky, 70-71; animal husbandman, Bur Vet Med, 71-80, dep dir, 80-83, dir, Div Vet Med Res, 83-88, DIR, OFF SCI, FOOD & DRUG ADMIN, 88- *Mem:* Am Soc Animal Sci; Am Dairy Sci Asn. *Mailing Add:* Ctr for Vet Med HFV-500 5600 Fishers Lane Rockville MD 20857

ALDERSON, THOMAS, b New York, NY, Aug 18, 17; m 42; c 3. ORGANIC CHEMISTRY,TEXTILE TECHNOLOGY. *Educ:* Ripon Col, BA, 39; Ohio State Univ, MS, 41, PhD(org chem), 47. *Prof Exp:* Asst chem, Ohio State Univ, 39-40 & 46-47; res chemist, Exp Sta, 47-61, res assoc, textile fibers dept, E I Du Pont de Nemours & Co, Inc, 61-82; RETIRED. *Concurrent Pos:* Consult, textile technol, 82- *Mem:* Am Chem Soc. *Res:* Fine chemicals; petroleum; natural gas; organic fluorine compounds; organic synthesis; preparation of trifluoracetic acid and a study of ethyltrifluoro acetate; preparation and evaluation of addition and condensation polymers; catalytic chemistry; synthetic fibers; textile technology. *Mailing Add:* 522 Brighton Rd Wilmington DE 19809

ALDO-BENSON, MARLENE ANN, b Bridgeport, Conn, July 31, 39; m 65; c 4. IMMUNOLOGY TOLERANCE SYSTEMIC, LUPUS FRYTHEMATOSUS. *Educ:* Univ Vt, MD, 65. *Prof Exp:* PROF MED, SCH MED, IND UNIV, 84-; PROF, MICROBIOL & IMMUNOL, 86- *Mem:* Am Rheumatism Asn (pres cent region, 83-84); Am Asn of Immunologists; Cent Soc Clin Res; Fel Am Col Soc; Sigma Xi. *Res:* Research interest is regulation of antibody production by B lymphocytes, in health and disease; this includes mechanism of self tolerance and autoimmunity and suppression of B cells by toxins such as alcohol. *Mailing Add:* Dept Med Long Clin 492 Ind Univ Sch Med 541 Clinical Dr Indianapolis IN 46223

ALDON, EARL F, b Chicago, Ill, Mar 5, 30; wid; c 2. FORESTRY, RESEARCH ADMINISTRATION. *Educ:* Univ Mich, BSF, 52, MS, 53. *Prof Exp:* Forester, 53-56, RES FORESTER, ROCKY MOUNTAIN FOREST & RANGE EXP STA, US FOREST SERV, 56- *Mem:* AAAS; Soc Am Foresters; Am Geophys Union; Am Soc Surface Mining & Reclamation; AAAS. *Res:* Watershed rehabiliation; stabilizing mine spoils in the Southwest. *Mailing Add:* 2937 Santa Cruz SE Albuquerque NM 87106

AL-DOORY, YOUSEF, b Baghdad, Iraq, 1924. CLINICAL MYCOLOGY, MOULD ALLERGY. *Educ:* Univ Baghdad, BS, 45; Univ Tex, MA, 51; La State Univ, PhD(bot, plant path), 54; Duke Univ, dipl, 60. *Prof Exp:* Teacher high sch, Baghdad, 45-49; teaching & res, Univ Baghdad, 54-58; res assoc med mycol, Univ Okla, 58-59, res assoc microbiol, Sch Med, 59-60, fel med mycol, 60-61; fel, NY State Dept Health, 61-62; sr res scientist in-chg diag med mycol lab, New York City Dept Health, 62-64; chmn dept mycol, Southwest Found Res & Educ, 64-70; asst prof epidemiol & environ health, George

Washington Univ, 70-75, assoc prof path, Sch Med, 75-; RETIRED. *Concurrent Pos:* Consult, Santa Rosa Med Ctr, San Antonio, Tex, 66-; chief microbiol lab med & mycol-serol labs, George Washington Univ Hosp, 75- *Mem:* Mycol Soc Am; Am Soc Microbiol; Int Soc Human & Animal Mycol; Med Mycol Soc of the Americas; fel Am Acad Microbiol; Am Col Allergists; Am Acad Allergy & Immunol. *Res:* Pathogenics dematiaceous fungi; application of fluorescent antibody procedures; lipid of H capsulatum; mycoflora of subhuman primates; soil and air; electron microscopy studies of host-parasite in mycotic diseases. *Mailing Add:* 4001 N Ninth St No 1225 Arlington VA 22203

ALDOUS, DUANE LEO, b Albuquerque, NMex, Nov 2, 30; m 55; c 5. ORGANIC CHEMISTRY, PHARMACY. *Educ:* Univ NMex, BS, 53, PhD(chem), 61. *Prof Exp:* res chemist, E I du Pont de Nemours & Co, NC, 62-68; res chemist, E I du Pont de Nemours & Co, NC, 62-68; from asst prof to assoc prof, 68-80, dir, Res Pharm & Dean, Col Pharm, 74-79, Res Assoc, Nat Ctr for Health Serv Res, 79-80; prof pharmaceut chem, Xavier Univ La, 80-90; DIR, NUSKIN LABS, 90- *Concurrent Pos:* Fel, Univ NMex, 61-62; Consult, 80-; Res assoc, Nat Ctr for Health Serv Res, 79-80. *Mem:* Am Chem Soc; Sigma Xi. *Res:* Synthetic organic and heterocyclic chemistry; synthesis of potential antitumor compounds; formulation of cosmetics. *Mailing Add:* 145 E Center St Provo UT 84606

ALDRED, ANTHONY T, b Rainhill, Eng, Feb 5, 35; m 64; c 2. MATERIALS SCIENCE, SOLID STATE PHYSICS. *Educ:* Univ Birmingham, BSc, 56, PhD(phys metall), 59. *Prof Exp:* Fel phys metall, Univ Birmingham, 59-60; resident res assoc, 60-62, asst metallurgist, 62-67, assoc metallurgist, 68-72, METALLURGIST, ARGONNE NAT LAB, 72- *Concurrent Pos:* Res assoc, Atomic Energy Res Estab, Harwell, Eng, 68-69. *Mem:* Am Inst Mining, Metall & Petrol Engrs. *Res:* Electronic structure and magnetic properties of actinide compounds and transition metal alloys; conduction processes in high-temperature oxides; structures and physical properties of complex oxides. *Mailing Add:* 1615 Briarcliffe Blvd Wheaton IL 60187

ALDRED, J PHILLIP, physiology, pharmacology, for more information see previous edition

ALDRETE, JORGE ANTONIO, b Mexico City, Mex, Feb 28, 37; c 3. ANESTHESIOLOGY. *Educ:* Nat Univ Mex, BS, 53, MD, 60; Univ Colo, MS, 67. *Prof Exp:* Med dir surg care, Jackson Mem Hosp, Miami, Fla, 71; prof anesthesiol & chmn dept, Sch Med, Univ Louisville, 71-75; prof anesthesia & chmn dept, Univ Colo Med Ctr, 75-80; prof anesthesia, Univ Ala, Birmingham, 80-86; chmn anesthesia, Cook County Hosp, Chicago, 86-89; MED DIR, PAIN MGT CTR, HUMANA HOSP-DESTIN, FLA, 90- *Concurrent Pos:* Chmn drug & therapeut comt, Vet Admin Hosp, Denver, Colo, 69-70 & Jackson Mem Hosp, 70-71; attend anesthesiol, Jackson Mem Hosp, 71; chief anesthesiol, Louisville Gen Hosp, 71-75, Children's Hosp, 72-75, Vet Admin Hosp, 72-, Norton Mem Infirmary, 72-75, Kosair Crippled Children Hosp, 72-75 & Community Hosp, 74-75; chmn intensive care comt, Kosair Crippled Children Hosp, 74-75. *Mem:* Am Soc Anesthesiologists; Int Anesthesia Res Soc; Am Soc Magnesium Res; Soc Educ Anesthesia; Am Soc Critical Care Anesthesiologists. *Res:* Exploring new therapeutic forms, such as new routes of administration of anesthetic drugs, exploring hemodynamic and oxygenation changes that occur as hemodilution takes place to lower levels of hematocrit and application of the pharmacokinetic concepts to the uptake of inhalation anesthetics as produced by closed circuit anesthesia; new approaches to pain management. *Mailing Add:* Rte 2 Box 7510 Santa Rosa Beach FL 32459

ALDRICH, CLARENCE KNIGHT, b Chicago, Ill, Apr 12, 14; m 42; c 4. PSYCHOTHERAPY. *Educ:* Wesleyan Univ, BA, 35; Northwestern Univ, MD, 40. *Prof Exp:* Staff mem, USPHS, 40-46; asst prof neuropsychiat, Univ Wis, 46-47; from asst prof to assoc prof psychiat, Sch Med, Univ Minn, 47-55; prof, Sch Med, Univ Chicago, 55-69, chmn dept, 55-64; prof & chmn dept, NJ Col Med & Dent, 69-73; dir, Blue Ridge Community Ment Health Ctr, 73-76, mem, Ctr Advan Studies, 81-84; prof psychiat & family pract, Sch Med, Univ Va, 73-84; RETIRED. *Concurrent Pos:* Vis prof, Univ Edinburgh, 63-64; Erskine fel, Univ Canterbury, 71; Mayne fel, Univ Queensland, 86. *Mem:* Am Psychiat Asn; Am Orthopsychiat Asn; Am Col Psychiatrists. *Res:* Psychiatric and family practice education; community mental health. *Mailing Add:* 905 Cottage Lane Charlottesville VA 22903

ALDRICH, DAVID VIRGIL, b Jamestown, NY, Oct 22, 28. MARINE ECOLOGY, AQUACULTURE. *Educ:* Kenyon Col, AB, 50; Rice Inst, MA, 52, PhD(parasitol), 54. *Prof Exp:* Physiologist, US Fish & Wildlife Serv, 56-66; head, Dept Marine Biol, 78-79, ASSOC PROF BIOL, TEX A&M UNIV, GALVESTON, 66-, ASSOC PROF WILDLIFE & FISHERY SCI, 70-, PROF MARINE BIOL, 78- *Honors & Awards:* Piper Found Achievement Award, 77. *Mem:* AAAS; Am Soc Limnol & Oceanog; Ecol Soc Am; Animal Behav Soc; Marine Biol Asn UK; World Mariculture Soc. *Res:* Ecology and behavior of penaeid shrimp, dinoflagellates and parasites; mariculture. *Mailing Add:* Dept Marine Biol Tex A&M Univ Bldg 311 Ft Crockett Galveston TX 77550

ALDRICH, FRANKLIN DALTON, b Detroit, Mich, Jan 25, 29; m 52, 84; c 3. TOXICOLOGY, INTERNAL MEDICINE. *Educ:* Mich State Univ, BS, 50; Ore State Univ, MA, 53, PhD(plant physiol), 54; Western Reserve Univ, MD, 62. *Prof Exp:* Plant physiologist, US Dept Agr, Denver, Colo, 56-58; intern, State Univ Iowa Hosps, 62-63; resident, Vet Admin Hosp, Denver, 63-64; staff physician, Upjohn Co, Mich, 64-65; prin investr, Colo Community Study Pesticides, Colo Dept Health, Greeley, 66-69; from resident to chief resident med, Lemuel Shattuck Hosp, Boston, 69-71; asst med dir, med dept & physician-in-chg, Environ Med Serv, Mass Inst Technol, 71-76; corp med toxicologist, 76-81, MGR HEALTH EFFECTS RES, IBM CORP, 81- *Concurrent Pos:* Res fel, Sch Med, Western Reserve Univ, 58-62; fel med, Univ Colo, Denver, 65-66; Mead Johnson scholar med, 70-71. *Mem:* Fel Am Acad Clin Toxicol (pres, 80-82); fel Am Col Physicians; Am Col Occup Med. *Res:* Environmental and clinical toxicology; chemical human ecology. *Mailing Add:* IBM Corp 415/021C PO Box 1900 Boulder CO 80301-9191

ALDRICH, FREDERICK ALLEN, b Butler, NJ, May 1, 27; m 52. INVERTEBRATE ZOOLOGY. *Educ:* Drew Univ, BA, 49; Rutgers Univ, MS, 53, PhD(zool), 54. *Prof Exp:* Asst zool, Drew Univ, 47-49; teaching asst, Rutgers Univ, 49-54; asst cur limnol, Acad Nat Sci, Philadelphia, 54-57, assoc cur, Div Estuarine Sci & in chg marine invert, 58-61; assoc prof biol, Mem Univ Nfld, 61-63, head dept, 63-65, dir marine sci res lab, 65-70, dean grad studies, 70-87, PROF BIOL, MEM UNIV NFLD, 65-, CHMN, PRESIDENTIAL TASK FORCE OCEAN STUDIES, 87-, MOSES HARVEY PROF MARINE BIOL, 90- *Concurrent Pos:* Mem bd dirs, Huntsman Marine Lab, St Andrews, NB; pres, Can Asn Grad Schs, 84-85. *Mem:* Fel AAAS; Am Soc Limnol & Oceanog; Nat Shellfisheries Asn; Can Soc Zool; fel Linnean Soc. *Res:* Marine invertebrate zoology; functional morphology of marine invertebrates, especially the Asteroidea and decapod Cephalopoda, paticularly the Architeuthids; invertebrate ecology; giant squid of the family Architeuthidae, particularly in the western North Atlantic; sensory behavior of ommastrephid squid; marine education. *Mailing Add:* Ocean Studies Task Force Mem Univ Nfld St Johns NF A1C 5S7 Can

ALDRICH, HARL P, JR, b 1923. ENGINEERING. *Educ:* Mass Inst Technol, BS, 47, DSc, 51. *Prof Exp:* Teaching asst, Mass Inst Technol, 47-49, instr soil mech, 49-51, asst prof, 51-577; chmn, 57-, EMER CHMN, HALEY & ALDRICH INC. *Concurrent Pos:* Vis lectr, Harvard Univ, 55-56, Mass Inst Technol, 72. *Mem:* Am Soc Civil Engrs; Asn Soil & Found Engr; Am Soc Testing & Mat; Sigma Xi. *Res:* Published numerous articles in various journals. *Mailing Add:* Haley & Aldrich Inc 58 Charles St Cambridge MA 02141

ALDRICH, HAVEN SCOTT, b Middletown, Conn, Dec 4, 43; m 69; c 3. PHYSICAL CHEMISTRY, THEORETICAL CHEMISTRY. *Educ:* Millsaps Col, BS, 67; Tulane Univ, PhD(phys chem), 72. *Prof Exp:* Instr chem, Mass Col Pharm, 72-73; res assoc, Northeastern Univ, 72-73; asst prof chem, St Mary's Dominican Col, 73-77, assoc prof, 77-80; staff chemist, 80-83, SR STAFF CHEMIST, PROD RES DIV, EXXON RES & ENG CO, 83- *Concurrent Pos:* NIH Nat Inst Neurol Dis & Stroke res grant, 75-77; researcher, Am Heart Asn La Inc, 77-78; acad consult, Exxon Prod Res Co, Houston, 77. *Mem:* Sigma Xi; Am Chem Soc; Soc Automotive Eng; Soc Tribologists & Lubricant Engrs. *Res:* Semiempirical molecular orbital studies of small and large molecules including heavier elements, and compounds of biological or pharmacological interest; research and development of aviation lubricants; antioxidants; thermal and thermal oxidation chemistry. *Mailing Add:* 961 Rahway Ave Westfield NJ 07090-2026

ALDRICH, HENRY CARL, b Beaumont, Tex, Feb 17, 41; m 62; c 2. BOTANY, MYCOLOGY. *Educ:* Univ Tex, BA, 63, PhD(bot), 66. *Prof Exp:* Asst prof, 66-71, assoc prof bot, 71-77, PROF MICROBIOL, UNIV FLA, 77- *Mem:* Mycol Soc Am (pres, 85); Bot Soc Am; Am Soc Cell Biol. *Res:* Ultrastructure of myxomycetes, algae, fungi; membranes, freeze-etching; bacteria. *Mailing Add:* Microbiol & Cell Sci Univ of Fla G041 McCarty Hall Gainesville FL 32611

ALDRICH, JEFFREY RICHARD, b Columbus, Ohio, June 14, 49; m 80; c 1. PHEROMONAL HUSBANDRY, CHEMOTAXONOMY. *Educ:* Univ Mo, BS, 71, MS, 74; Univ Ga, PhD(entom), 77. *Prof Exp:* Post doctoral fel entom, Univ Ga, 77-78; entom fel, NY State Agr Exp Sta, Cornell Univ, 78-80; RES ASSOC ENTOM, AGR RES STA, INSECT & NEMATODE HORMONE LAB, USDA, 80- *Concurrent Pos:* Prog chmn, Entom Soc Wash, 82-85. *Mem:* Entom Soc Am; Sigma Xi; Int Soc Chem Ecol; Am Soc Zoology. *Res:* Chemical communication of hemiptera and other insects; hormonal regulation of pheromone production and the mode of action of hormone inhibitors; chemotaxonomy of beneficial insects. *Mailing Add:* 4713 Muskogee St College Park MD 20740

ALDRICH, JOHN WARREN, ornithology, ecology, for more information see previous edition

ALDRICH, LYMAN THOMAS, b Hopkins, Minn, June 28. 17; m 41; c 2. GEOPHYSICS. *Educ:* Univ Minn, PhD(physics), 48. *Prof Exp:* Asst physicist, Naval Ord Lab, 40-45, assoc physicist, 45; asst & res assoc, Univ Minn, 45-48; asst prof physics, Univ Mo, 48-50; from asst dir to assoc dir, 65-74, actg dir, 74-75, mem staff, 50-84, EMER MEM, CARNEGIE INST DEPT TERRESTRIAL MAGNETISM, 84- *Concurrent Pos:* Vis prof, Kyoto Univ, 62. *Mem:* AAAS; Seismol Soc Am; Am Phys Soc; Am Geophys Union (gen secy, 74-80). *Res:* Properties of earth's crust and upper mantle from electrical conductivity measurements and use of controlled and natural seismic sources in South American Andes. *Mailing Add:* Carnegie Inst 5241 Broad Branch Rd NW Washington DC 20015

ALDRICH, MICHELE L, b Seattle, Wash, Oct 6, 42; m 65. HISTORY OF SCIENCE. *Educ:* Univ Calif, Berkeley, AB, 64; Univ Tex, Austin, PhD(hist sci), 74. *Prof Exp:* Lectr hist of sci, Smith Col, 69-70; staff women's studies, Valley Women's Ctr, 70-73; asst ed hist of sci, Joseph Henry Papers, Smithsonian Inst, 74-75; fieldworker women's studies, Women's Hist Sources Surv, Univ Minn, 76-77; proj dir, Women in Sci, 77-84, mgr Comput Serv & Archivist, 85-90, DIR INFO SERVS, 91- *Concurrent Pos:* Consult, Aaron Burr Papers, NY Hist Soc, 75-76; Biomed Res Opportunities for Women, NIH, 78-79; deleg, Mass Govs Conf on Libr, 78-; chmn, hist geol div, Geol Soc Am, 79-80, secy-treas, 85-; res assoc, herpet dept, Calif Acad Sci, 80-; vchmn, Women's Comt, Hist Sci Soc, 84-86; US Nat Comt His Geol, Nat Acad Sci, 85-90. *Mem:* Hist Sci Soc; Geol Soc Am; fel AAAS. *Res:* Nineteenth century American geologists and geological surveys; women in American science. *Mailing Add:* 24 Elm St Hatfield MA 01038

ALDRICH, PAUL E, b Springfield, Ohio, June 29, 28; m 63; c 2. ORGANIC CHEMISTRY. *Educ:* Mass Inst Technol, BS, 52; Univ Wis, PhD(chem), 58. *Prof Exp:* Res org chemist, Merck & Co, 52-54; res org chemist, 58-78, RES ASSOC, E I DU PONT DE NEMOURS & CO, 78- *Mem:* Am Chem Soc. *Res:* Fluorocarbons. *Mailing Add:* 306 Spalding Rd Sharpley Wilmington DE 19803

ALDRICH, RALPH EDWARD, b Worcester, Mass, Jan 29, 40; m 68. ACTIVE & ADAPTIVE OPTICS. *Educ:* Amherst Col, AB, 61; Univ Nebr, MS, 63; Tufts Univ, PhD(physics), 67. *Prof Exp:* Sr mem tech staff, Cent Res Labs, Itek Corp, 67-73, mgr mat sci, 73-75, mgr device technol, Itek Optical Systs, 75-83; mgr device technol, 83-89, asst dir, 89-90, DIR RES, LITTON-ITEK OPTICAL SYSTS, 90- *Concurrent Pos:* Vis prof, Electro-Optics Technol Ctr, Tufts Univ, 88- *Mem:* Am Phys Soc. *Res:* Active and adaptive optics. *Mailing Add:* 23 Birch Ridge Rd Acton MA 01720

ALDRICH, RICHARD JOHN, b Fairgrove, Mich, Apr 16, 25; m 43; c 3. AGRONOMY, WEED SCIENCE. *Educ:* Mich State Col, BS, 48; Ohio State Univ, PhD(agron), 50. *Prof Exp:* Agronomist & coordr weed control res, Northeastern Region, USDA, 50-57, asst dir, Mich Agr Exp Sta, 57-64; assoc dir agr exp sta, Univ Mo-Columbia, 64-67, assoc dean col agr, 67-76; adminr, Coop State Res Serv, USDA, Washington, DC, 76-77, actg dep dir, Coop Res, Sci & Educ Admin, USDA, Washington, DC, 78; prof agron, 79-87, res agronomist, Agr Res Serv, 81-87, EMER PROF AGRON, UNIV MO-COLUMBIA, 88- *Concurrent Pos:* Mem gov bd, Agr Res Inst, Nat Res Coun-Nat Acad Sci, 64-67 & 67-70, pres elect, 74, pres, 74-75; chmn, North Cent Agr Exp State Dir Asn, 70-71, Nat Soybean Res Coord Comt, 72-76 & Div Agr, Nat Asn State Univs & Land Grant Cols, 74-75; mem, Int Sci & Educ Coun, 76-78, US/USSR Joint Comt, Coop Field Agr, 77-78 & Joint Coun Food & Agr Sci, 78; ed, Weed Sci, 89- *Mem:* Fel AAAS; Weed Sci Soc Am. *Res:* Weed crop ecology: influence of changes in crop production practices on composition of weed community; relationship between production practices and weed emergence and competitiveness; reducing the number of viable weed propagules in soil. *Mailing Add:* 701 Wildwood Dr Columbia MO 65203

ALDRICH, ROBERT A(DAMS), b Horseheads, NY, Apr 25, 24; m 46; c 4. AGRICULTURAL ENGINEERING. *Educ:* Wash State Univ, BS, 50, MS, 52; Mich State Univ, PhD(eng), 58. *Prof Exp:* Instr & jr agr engr, Wash State Univ, 51-54, asst prof & asst agr engr, 54-56; asst, Mich State Univ, 56-58; assoc prof agr eng, Univ Ky, 58-59 & Mich State Univ, 59-62; assoc prof agr eng, PA State Univ, 62-79; head, 79-88, prof, 88-90, PROF EMER AGR ENG, UNIV CONN, 90- *Concurrent Pos:* Consult, agr bldg indust. *Mem:* Am Soc Agr Engrs; Nat Soc Prof Engrs; Am Soc Heating, Refrigerating & Air Conditioning Engrs. *Res:* Structural research; teaching in agricultural engineering; agricultural structures and environment. *Mailing Add:* 295 Wormwood Hill Rd Mansfield Center CT 06250

ALDRICH, ROBERT ANDERSON, b Evanston, Ill, Dec 13, 17; m 40; c 3. PEDIATRICS. *Educ:* Amherst Col, BA, 39; Northwestern Univ, MD, 44. *Prof Exp:* Instr pediat, Grad Sch, Univ Minn, 50; asst prof, Sch Med, Univ Ore, 51-53, assoc prof pediat & res assoc biochem, 53-56; prof & exec officer pediat, Sch Med, Univ Wash, 56-62; dir, Nat Inst Child Health & Human Develop, 62-64; prof pediat, Sch Med, Univ Wash, 64-70; vpres health affairs, 70-74, prof pediat & prev med, Univ Colo, Denver, 70-81; CLIN PROF PEDIAT, UNIV WASH, 80- *Concurrent Pos:* Mem training comt, NIH, 59; vchmn, President's Comt Ment Retardation, 66-71; chief pediat, King County Hosp, Seattle; pediatrician-in-chief, Univ Wash Hosp; assoc chief med serv, Children's Orthop Hosp, 57-62. *Mem:* AAAS; fel Am Psychiat Asn; Soc Pediat Res; Am Pediat Soc. *Res:* Biochemistry of porphyrins; chemistry of bilirubin; mechanisms of heme synthesis; inborn errors of metabolism; health services research. *Mailing Add:* Dept Pediat RD-20 Univ Wash Seattle WA 98195

ALDRICH, ROBERT GEORGE, b Watertown, NY, July 2, 40; m 64; c 4. PHYSICAL METALLURGY. *Educ:* Rensselaer Polytech Inst, BMetE, 62; Syracuse Univ, PhD(solid state sci), 66. *Prof Exp:* Scientist, Res Corp, Syracuse, 66-67, dir mat sci, 67-80; PRES, HALOMET INC, EAST SYRACUSE, 80- *Mem:* AAAS; Am Soc Metals; Am Inst Mining, Metall & Petrol Engrs; NY Acad Sci. *Res:* Surface phenomena of metals and alloys; strength properties and interfacial behavior of composite alloys; corrosion and embrittlement phenomena of alloys. *Mailing Add:* Owahgena Rd Manlius NY 13104

ALDRICH, SAMUEL ROY, b Fairgrove, Mich, June 12, 17; m 40; c 1. AGRONOMY. *Educ:* Mich State Col, BS, 38; Ohio State Univ, PhD(agron), 42. *Prof Exp:* Asst agron, Ohio State Univ, 38-42, instr, 42; from asst prof to assoc prof, Exten, Cornell Univ, 42-52, prof field crops, 52-57; prof soils exten agron, 57-73, asst dir, Ill Agr Exp Sta, 73-80, EMER PROF AGRON, UNIV ILL, URBANA-CHAMPAIGN, 80- *Concurrent Pos:* Mem, Ill Pollution Control Bd, 70-71. *Honors & Awards:* Agron Educ Award, Am Soc Agron, 65. *Mem:* AAAS; fel Am Soc Agron; Soil Sci Soc Am. *Res:* Corn maturity; culture of corn and small grains, land use, rotations; soil fertility; environmental problems. *Mailing Add:* V3 Country Club Village Lake Wales FL 33853-5235

ALDRICH, THOMAS K, b Minneapolis, Minn, Sept 11, 50; m 72; c 2. RESPIRATORY PHYSIOLOGY, MUSCLE PHYSIOLOGY. *Educ:* Swarthmore Col, BA, 72, Univ Minn, MD, 75. *Prof Exp:* Intern & resident intern med, Univ Calif, Irvine, 75-78; fel pulmonary & allergy, Univ Va, 78-80; post doctoral respiratory physiol, Univ Pa, 80-82; asst prof med, 82-87, ASSOC PROF MED, ALBERT EINSTEIN COL MED, 87- *Concurrent Pos:* Asst attend pulmonary med, 82-87, adj attend pulmonary med, Montefiore Med Ctr, 87- *Honors & Awards:* J Burns Amberson Award, NY Lung Asn, 83; Edward Livingston Trudeau Award, Am Lung Asn, 86. *Mem:* Am Thoracic Soc; Am Col Chest Physicians; Am Physiol Soc; Am Fedn Clin Res. *Res:* Studies of respiratory muscle function and fatigue. *Mailing Add:* Pulmonary Med Div Montefiore Med Ctr 111 E 210th St Bronx NY 10467

ALDRIDGE, DAVID WILLIAM, b Honolulu, Hawaii, July 26, 52; m 82; c 3. PHYSIOLOGICAL ECOLOGY. *Educ:* Univ Tex, Arlington, BS, 74, MA, 76; Syracuse Univ, PhD(biol), 80. *Prof Exp:* Asst prof, 82-87, ASSOC PROF BIOL, NC A&T STATE UNIV, 87- *Concurrent Pos:* Prin investr, NSF, 83-86 & 89-92; consult, Army Corp Engrs, 83. *Mem:* Am Soc Zoologists; Am Malacol Union; Malacological Soc London. *Res:* Physiological ecology of freshwater molluscs. *Mailing Add:* Dept Biol NC A&T State Univ Greensboro NC 27411

ALDRIDGE, JACK PAXTON, III, b Greenwood, Miss, Sept 6, 38; m 65; c 2. EXPERT SYSTEM DEVELOPMENT. *Educ:* Rice Univ, BA, 60, MA, 62, PhD(physics), 65. *Prof Exp:* Res assoc nuclear physics, Fla State Univ, 65-68, asst prof physics, 69-72; staff-group leader, Los Alamos Nat Lab, 72-86; TECH MGR, A I MCDONNELL ASTRONAUT CO, 86- *Concurrent Pos:* Consult, Advan Technol Assocs, 82-86. *Mem:* Am Asn Artificial Intel; Laser Inst Am (pres, 82); Am Phys Soc; AAAS; Optical Soc Am; Am Defense Preparedness Asn. *Res:* Design of expert systems; rule-systems; natural language interfaces; handling constraints. *Mailing Add:* 810 Baronridge Seabrook TX 77586

ALDRIDGE, MARY HENNEN, b Ark, Jan 11, 19; m 41; c 1. ORGANIC CHEMISTRY, BIOCHEMISTRY. *Educ:* Univ Ga, BS, 39; Duke Univ, MA, 41; Georgetown Univ, PhD(chem), 54. *Prof Exp:* Chemist, E I du Pont de Nemours & Co, NY, 41-47; asst prof chem, Univ Md, 47-55; from assoc prof to prof chem, Am Univ, 55-86, chmn dept, 79-83; RETIRED. *Concurrent Pos:* Res grants, Eve Star, 60, US Army Med Res & Develop Command, 61-64, 66-69, 76-79, NIH, 67-68, Water Resources Res Ctr, 75-76 & Naval Res Lab, 76-77. *Mem:* Am Chem Soc; AAAS; Am Asn Univ Professors. *Res:* Structure-activity relationships; trace organics in drinking water; synthesis of novel electrolytes; synthesis of biologically active compounds. *Mailing Add:* 7904 Hackamore Dr Potomac MD 20854

ALDRIDGE, MELVIN D(AYNE), b Crab Orchard, WVa, July 20, 41; m 63; c 2. ELECTRICAL ENGINEERING. *Educ:* WVa Univ, BS, 63; Univ Va, MEE, 65, DSc(elec eng), 68. *Prof Exp:* Aerospace technologist, Langley Res Ctr, NASA, 63-68; from asst prof to assoc prof elec eng, WVa Univ, 68-76, actg dir, 78-79, prof, 76-84, dir, Energy Res Ctr, 79-84; assoc dean, Res & Dir Eng, 84-89, ASSOC DEAN CD PROG & DIR TW CTR TECHNOL MGT, EXP STA, COL ENG, AUBURN UNIV, ALA, 89- *Concurrent Pos:* Mem, Secy of Interior's Adv Comt Coal Mine Safety Res, 76-77. *Mem:* Indust Appln Soc; fel Inst Elec & Electronic Engrs; Am Soc Eng Educ. *Res:* Quantum effects in communication systems; photon statistics for signal plus noise; phase noise in traveling wave tube power amplifiers; detection of neural signals; electronic monitoring and communications in coal mines; technology management. *Mailing Add:* Tiger Dr Rm 104 Auburn Univ Auburn AL 36849

ALDRIDGE, ROBERT DAVID, b St Louis, Mo, July 15, 44. REPTILIAN REPRODUCTION. *Educ:* Univ Mo, BS, 66, MA, 69; Univ N Mex, PhD(biol), 73. *Prof Exp:* From asst prof to assoc prof biol, 73-85, chmn dept, 81-89, PROF BIOL, ST LOUIS UNIV, 85- *Concurrent Pos:* Publ secy, Soc Amphibians & Reptiles, 88- *Mem:* Am Soc Ichthyologists & Herpetologists; Herpetologists League; Soc Study of Amphibians & Reptiles. *Res:* Environmental and hormonal control of reptilian reproduction. *Mailing Add:* Dept Biol St Louis Univ St Louis MO 63103

ALDRIDGE, WILLIAM GORDON, b Gladstone, NJ, Apr 28, 34; m 55; c 5. ANATOMY. *Educ:* Rutgers Univ, AB, 60; Univ Rochester, NIH fel, 60-62, PhD(anat), 62. *Prof Exp:* From instr to asst prof, 63-71, ASSOC PROF ANAT & RADIATION BIOL & DIR MULTIDISCIPLINE LABS, UNIV ROCHESTER, 71-, ASSOC PROF BIOPHYSICS, MED EDUC & COMMUN & ASST DIR AUXILIARY SERV, ENERGY CONSERV OFF, MED CTR, 76- *Mem:* Sigma Xi; Asn Multidiscipline Educ Health Sci; Histochem Soc; Am Soc Cell Biologists; Am Asn Anatomists. *Res:* Human anatomy; biology of nucleic acids and correlation of biochemical and electron morphological observations; evaluation of medical education programs; educational facility design and operation. *Mailing Add:* Sch Med Dent Univ Rochester Box 708 Rochester NY 14642

ALEEM, M I HUSSAIN, b Lyallpur, WPakistan, Jan 2, 24; m 64; c 1. MICROBIAL BIOCHEMISTRY. *Educ:* Univ Panjab, WPakistan, BSc, 45, MSc, 50; State Univ Groningen, dipl, 53; Cornell Univ, PhD(microbiol), 59. *Prof Exp:* Asst bacteriologist, Punjab Govt, Pakistan, 51-52; res assoc biochem, Johns Hopkins Univ, 58-61; asst prof microbiol, Univ Man, 61-64; biochemist, Res Inst Advan Studies, 64-66; assoc prof microbiol, 66-69, PROF MICROBIOL, UNIV KY, 69- *Concurrent Pos:* Brit Coun vis prof, Oxford & Bristol Univs, 63; chief sci officer & guest prof, Lab Microbiol, Free Univ Amsterdam, 75-76; Fulbright prof agr biochem, Univ Adelaide, 76; consult, UN Develop Prog, Pakistan, 81; Brit Coun Award, 63, Fulbright Award, 54, UNESCO-Neth Govt Award, 52, Fulbright Sr Award, 76. *Honors & Awards:* Distinguished Res Award, Univ Ky Res Found, 74. *Mem:* Am Soc Microbiol; Am Soc Biochem & Molecular Biol. *Res:* Energy conversions in autotrophic bacteria. *Mailing Add:* Sch Biol Sci Univ Ky Lexington KY 40506

ALEGNANI, WILLIAM CHARLES, bacteriology, for more information see previous edition

ALEINIKOFF, JOHN NICHOLAS, b Washington, DC, Apr 11, 50; m 80; c 2. GEOLOGY. *Educ:* Beloit Col, BA, 72; Dartmouth Col, AM, 75, PhD(geol), 78. *Prof Exp:* Geol field asst econ geol, Bear Creek Mining Co, 69-71, 75; GEOLOGIST GEOCHRONOLOGY, US GEOL SURV, 78- *Mem:* Geol Soc Am. *Res:* U-Th-Pb geochronology; Pb/Pb isotope tracer geochemistry; tectonics. *Mailing Add:* 1760 Locust St Denver CO 80220

ALEKMAN, STANLEY L, b New York, NY, Mar 21, 38; m 61; c 3. PHYSICAL ORGANIC CHEMISTRY. *Educ:* City Col New York, BA, 62; Univ Del, PhD(phys org chem), 68. *Prof Exp:* Res chemist, US Army Ballistics Res Lab, 63-64 & Atlas Chem Co, Inc, 65; res chemist, E I du Pont de Nemours & Co, Inc, 68-80; mem staff, Air Products Chem Inc, 80-; AT NUODEX. *Mem:* AAAS; Sigma Xi; Am Chem Soc; Royal Soc Chem; NY Acad Sci. *Res:* Kinetics of chromic acid oxidation; carbonium ion structure; kinetics of polycondensation reactions; kinetics of fast reactions. *Mailing Add:* Huls Am Inc Turner Pl PO Box 365 Piscataway NJ 08855-0365

ALEKSANDROV, GEORGIJ NIKOLAEVICH, b Leningrad, USSR, Jan 7, 30; m 54; c 3. HIGH VOLTAGE TECHNIQUES. *Educ:* Leningrad Polytech Inst, Cand Dr Tech Sci, 57; All Union Electrotech Inst, Moscow, Dr Tech Sci, 67. *Prof Exp:* Asst prof high voltage tech, Leningrad Polytech Inst, 53-58, sci worker, 58-62, lectr, 62-70, prof, 71-74, prorector sci activ, 77-82, SCI DIR LAB, LENINGRAD POLYTECH INST, 60-, DIR, ELEC APPARATUS DEPT, 74- *Concurrent Pos:* Pres, Energetic State Comt Sci & High Educ Russian Repub, 69-; mem sci coun, All Union Electrotech Inst, 80-; vis prof, Chong Qing Tech Univ, China & Wuhan Inst Hydraul & Elec Eng, 88-, Xian Tech Univ & Tsinghua Univ, Beijing, 89- *Mem:* Fel Inst Elec & Electronics Engrs. *Res:* Dielectric behavior of large air insulating configurations; ultra high voltage transmission lines design and compact lines with increased transmission capacity. *Mailing Add:* Leningrad Tech Univ Polytechnicheskaya St 29 St Petersburg USSR

ALEKSOFF, CARL CHRIS, b Flint, Mich, Apr 3, 40; m 71; c 2. OPTICS, ELECTRICAL ENGINEERING. *Educ:* Univ Mich, BSE, 62, MSE, 63, PhD(lasers & holography), 69. *Prof Exp:* Res asst mod optics, Radar & Optics Lab, Univ Mich, 64-67, res assoc coherent optics, 67-69, Inst Sci & Technol fel, 69-70, res engr lasers & holography, 69-71; RES ENGR, ENVIRON RES INST MICH, 72- *Mem:* Inst Elec & Electronics Engrs; Optical Soc Am. *Res:* Modern coherent optics; holography; lasers; coherence; electro-optics; optical processing. *Mailing Add:* Environ Res Inst Mich Box 8618 Ann Arbor MI 48107

ALEO, JOSEPH JOHN, b Wilkes-Barre, Pa, Oct 8, 25; m 49; c 2. PATHOLOGY. *Educ:* Bucknell Univ, BS, 48; Temple Univ, DDS, 53; Univ Rochester, PhD(path), 65. *Prof Exp:* Asst biochem, Hahnemann Med Sch, 48-49; asst dent surgeon, USPHS, 53-54; pvt pract, 54-60; USPHS fel path, Univ Rochester, 60-65; chmn dept, 65-70, asst dean, 70-78, assoc dean, 78-86, PVT CONSULT, ADVAN EDUC & RES, SCH DENT, 86-, EMER PROF PATH, SCH MED, TEMPLE UNIV, 86- *Concurrent Pos:* Consult, NIH, 69- & Food & Drug Admin, 72-79; vis scholar Univ Cambridge, 71, 72 & 82. *Mem:* Fel AAAS; Am Dent Asn; Am Soc Exp Path; Tissue Cult Asn; Int Asn Dent Res; fel NY Acad Sci. *Res:* Connective tissue diseases; experimental carcinogenesis; tissue culture; periodontal diseases. *Mailing Add:* 6610 Seawind Dr Ft Myer FL 33908

ALERS, GEORGE A, b Bisbee, Ariz, Nov 22, 28; m 56; c 3. MATERIALS SCIENCE ENGINEERING, ACOUSTICS. *Educ:* Rice Univ, BA, 50; State Univ Iowa, MS, 52, PhD(physics), 54. *Prof Exp:* Res engr metallurgy, Westinghouse Res Labs, 54-56; res engr physics, Ford Motor Co Sci Labs, 56-68; mem tech staff, mat sci, Rockwell Int Sci Ctr, 68-79; res prof physics, Univ NMex, 79-81; PRES, MAGNASONICS INC, NMex, 81- *Concurrent Pos:* Nat lectr, Sonics & Ultrasonics Group, Inst Elec & Electronics Engrs, 83-84, tech chmn, Sonics & Ultrasonics Group Symposium, 76 & 79, pres, 80-82; chmn, Physics Metals Comt, Am Inst Mech Engrs, 66-68. *Mem:* Am Soc Nondestructive Testing; Inst Elec & Electronics Engrs; Am Physical Soc; Am Assn Adv Sci. *Res:* Commercial application of non-contact, electromagnetic transducers; development of noval ultrasonic nondestructive testing techniques; effects of temperature and pressure on elastic properties of materials; origins of mechanical strength of metals. *Mailing Add:* 13108 Sandstone Plane Albuquerque NM 87111

ALERS, PERRY BALDWIN, b Bisbee, Ariz, Mar 24, 26; m 55; c 3. PHYSICS, GEOPHYSICS. *Educ:* Rice Univ, BA, 48, MA, 50; Univ Md, PhD(physics), 55. *Prof Exp:* Res physicist low temperature physics, Marine Systs, Naval Res Lab, 50-59, high pressure physics, 59-66, head Crystal Physics Br Admin, 66-71, physics consult geophysics, 71-75, physics systs analyst, 75-83; MGR ADV PROGS, EASTPORT INT, INC, 83- *Concurrent Pos:* Sabbatical, Brigham Young Univ, 64-65 & Univ Calif, San Diego, 71-72; res & develop plans officer, Off of Chief Naval Opers, 78-79. *Mem:* Fel Am Phys Soc; Am Geophys Union; Sigma Xi. *Res:* Superconductivity; magnetic properties of metals and insulators at low temperatures; luminescence at high pressures; geomagnetism of ocean floor; geomagnetic noise; ocean instrumentation; ocean vehicles. *Mailing Add:* 9405 Caldran Dr Clinton MD 20735

ALEVIZON, WILLIAM, b Brooklyn, NY, Sept 30, 43. ICHTHYOLOGY, MARINE BIOLOGY. *Educ:* Calif State Univ, Fullerton, BA, 69; Univ Calif, Santa Barbara, MA, 71, PhD(biol), 73. *Prof Exp:* Fel, Harbor Br Found, Inc, 73-74; asst prof, 75-80, assoc prof, 81-86, PROF ICHTHYOL & MARINE ECOL, FLA INST TECHNOL, 87- *Mem:* Am Soc Ichthyol & Herpet. *Res:* Population and community ecology of reef fishes; artificial reefs. *Mailing Add:* Dept Biol Sci Fla Inst Technol 150 W University Blvd Melbourne FL 32901

ALEX, JACK FRANKLIN, b Rutland, Sask, Aug 20, 28; m 57; c 3. PLANT TAXONOMY, ECOLOGY. *Educ:* Univ Sask, BSA, 50, MSc, 52; Wash State Univ, PhD(bot), 59. *Prof Exp:* Asst plant ecol, Univ Sask, 50-52; field asst, Div Bot, Can Dept Agr, 52; asst bot, State Col Wash, 52-54; lectr agr bot, Univ Ceylon, 54-56; res assoc, Univ Sask, 57-58; weed ecologist, Plant Res Inst, Can Dept Agr, 58-62, ecologist, Regina Res Sta, 62-68; assoc prof, 68-79, actg chair dept, 89-90, PROF TAXON, UNIV GUELPH, 79- *Concurrent Pos:* Weed biol, Sri Lanka, Indonesia & Thailand. *Mem:* Ont Inst Agrol; Weed Sci Soc Am; Agr Inst Can; Can Bot Asn. *Res:* Ecological investigations on weeds and native vegetation; taxonomy and ecology of weedy species; biological control of weeds. *Mailing Add:* Dept Environ Biol Univ Guelph Guelph ON N1G 2W1 Can

ALEX, LEO JAMES, b Rochester, Minn, Dec 3, 42; m 67; c 3. EXPONENTIAL DIOPHANTINE EQUATIONS. *Educ:* Univ Minn, BA, 64, PhD(math), 70. *Prof Exp:* From asst prof to assoc prof, 70-78, PROF MATH, STATE UNIV COL, ONEONTA, 78-, DEPT CHMN, 88- *Concurrent Pos:* Vis asst prof, Univ Va, 74. *Mem:* Am Math Soc; Math Asn Am. *Res:* Finite group representation theory; solving exponential diophantine equations. *Mailing Add:* State Univ NY Oneonta NY 13820

ALEXANDER, A ALLAN, b Hudson, Mass, July 19, 28; m 52; c 1. VERTEBRATE ANATOMY, TAXONOMY. *Educ:* Univ Mass, BS, 50; Springfield Col, MS, 47; State Univ NY, Buffalo, PhD(biol), 66. *Prof Exp:* Asst biol, Springfield Col, 54-55, asst physiol, 55-57; asst biol, State Univ NY, Buffalo, 58-60; from instr to asst prof, 61-68, assoc prof, 68-76, chmn dept, 71-78, PROF BIOL, CANISIUS COL, 76- *Mem:* Am Soc Zoologists; Soc Study Evolution; Am Soc Ichthyologists & Herpetologists; Asn Study Animal Behav. *Res:* Herpetological taxonomy and comparative herpetological morphology and development. *Mailing Add:* Dept of Biol Canisius Col Buffalo NY 14208

ALEXANDER, AARON D, b New York, NY, Jan 14, 17; m 41; c 3. MEDICAL MICROBIOLOGY. *Educ:* City Col New York, BS, 38; George Washington Univ, MS, 53, PhD(microbiol), 61; Am Bd Microbiol, dipl, 65. *Prof Exp:* Bacteriologist antibiotics, Food & Drug Admin, 46-49; res bacteriologist div vet med, Walter Reed Army Inst Res, 49-53, chief res sect dept vet bact, 53-60, from asst chief to chief dept vet microbiol, 60-74; prof, 74-86, EMER PROF MICROBIOL, CHICAGO COL OSTEOP MED, 86-; ADJ PROF, PHILADELPHIA COL MED, PA, 87- *Concurrent Pos:* Mem expert comt leptospirosis, WHO-Food Agr Argn, Orgn, 57-73; leptospira subcomt, Int Comt Bact Nomenclature & Taxon, 58-; chief, Leptospirosis Ref Lab, 60-74. *Mem:* Am Soc Microbiol; Am Soc Exp Biol; Am Asn Immunol. *Res:* Microbial zoonoses; leptospirosis, melioidosis. *Mailing Add:* 771 Old Eagle Sch Rd Stafford PA 19087

ALEXANDER, ALLEN LEANDER, b Statesville, NC, Feb 12, 10; m 41; c 2. CHEMISTRY. *Educ:* Univ NC, BS, 31, MS, 32, PhD(chem), 36. *Prof Exp:* Teaching fel chem, Univ NC, 31-32 & 33-36; res chemist, Sherwin-Williams Co, Ohio, 36-38; chemist, US Naval Res Lab, 38-40, head org chem br, 40-72; RETIRED. *Mem:* Am Chem Soc; Nat Asn Corrosion Engrs; Fedn Socs Paint Technol. *Res:* Paints, varnishes, lacquers; fungicides; antifouling and temperature indicating paints; chemical compounds obtained from destructive distillation of tobacco; wood preservation; corrosion and biodegradation in tropical environments. *Mailing Add:* 4216 Sleepy Hollow Rd Annandale VA 22003

ALEXANDER, ARCHIBALD FERGUSON, b Minneapolis, Minn, Oct 13, 28; m 53; c 4. VETERINARY PATHOLOGY. *Educ:* Univ Minn, BS & DVM, 51; Colo State Univ, MS, 58, PhD(animal path), 62. *Prof Exp:* Chief animal supply, Dugway Proving Ground, Utah, 51-56; from instr to assoc prof, 56-65, head, Dept Path, 66-81, PROF VET PATH, COLO STATE UNIV, 65-, DIR, DIAG LABS, 81- *Concurrent Pos:* Nat Heart Inst spec fel, Glasgow, 63-64. *Mem:* AAAS; Am Vet Med Asn; Am Col Vet Path; Int Acad Path; Sigma Xi. *Res:* Experimental veterinary medicine; cardiovascular-pulmonary pathology in relation to high altitude acclimatization. *Mailing Add:* Pathol Dept Col Vet Med Colo State Univ Ft Collins CO 80523

ALEXANDER, BENJAMIN H, b Roberta, Ga, Oct 18, 21; m 48; c 2. BIOCHEMISTRY, ENVIRONMENTAL CHEMISTRY. *Educ:* Univ Cincinnati, BA, 43; Bradley Univ, MS, 50; Georgetown Univ, PhD(chem), 57. *Hon Degrees:* LLD, Bradley Univ, 79. *Prof Exp:* Technician, Cincinnati Chem Works, 44-45; chemist, Agr Res Serv, USDA, Ill, 45-54, res chem, Md, 54-62, Walter Reed Army Inst Res, 62-67; health scientist admin, NIH, 67-68, spec asst to dir for disadvantaged, Nat Ctr Health Serv Res & Develop, Health Serv & Ment Health Admin, USPHS, 68-69, adminr, New Health Career Projs & Dep Equal Employ Off, 69-70, prog officer, Health Care Orgn & Resources Div, 70-74; pres, Chicago State Univ, 74-82, Univ DC, 82-83; dep asst secy, Dept Educ, Interim, 84; PRES, DREW DAWN ENTERPRISES, INC, 83- *Concurrent Pos:* Adj prof, Am Univ, 58-74; lectr, Grad Sch, USDA, 60-68; mem, comt women higher educ, Am Coun Educ, 75-77; comnr, NCent Comn Inst Higher Educ, 75-78; consult, Nat Ctr Health Serv Res & NSF, 77-81; distinguished vis prof, Nat Grad Univ, Arlington, Va, 83-; res prof, Am Univ, Washington, DC; vpres, Wash Acad Sci, 85-86; chmn, Joint Bd on Sci & Eng Educ, 85-86; mem, Acad Joint Bd Sci & Eng Educ. *Honors & Awards:* Cert Achievement, Am Chem Soc. *Mem:* Am Chem Soc; fel Acad Scis. *Res:* Syntheses of organophosphorus compounds for medicinal purposes; syntheses of pesticide chemicals including insect attractants and repellents, dextrans and glucuronic acids; preparation of useful compounds from agricultural wastes. *Mailing Add:* Drew Dawn Enterprises, Inc PO Box 41126 Washington DC 20018-0526

ALEXANDER, CHARLES EDWARD, JR, b Port Washington, NY, Nov 25, 30; m 60; c 3. MEDICINE, PUBLIC HEALTH. *Educ:* Yale Univ, BA, 51; Univ Pa, MD, 55; Johns Hopkins Univ, MPH, 59; DrPH(chronic dis), 64; Am Bd Prev Med, dipl, 64. *Prof Exp:* Intern med, Geisinger Hosp, Danville, Pa, 55-56; Med Corps, US Navy, 56-, med officer, 56-57, clinician, Naval Air Sta, Alameda, Calif, 57-58, resident pub health, Johns Hopkins Univ, 58-60, asst officer in chg, Prev Med Unit 6, 60-62, head venereal dis & tuberc control sects, Bur Med & Surg, 62-64, head commun dis br, 63-65, officer in chg, Prev Med Univ 7, 65-67, officer in chg, Navy Prev Med Unit, Danang, Vietnam, 67-68, prev med officer, US Mil Assistance Command, Thailand, 68-71, dep surgeon, 68-71, dir prev med div, Bur Med & Surg, 71-75, dir occup & prev med div, Bur Med & Surg, Navy Dept, 75-76; chief bur tuberc serv, 76-81, chief, bur commun dis serv, 81-84, chief, bur epidemiol, 84-87, chief, bur AIDS/STD Control, Tex Dept Health, 87-88; DEP DIR HEALTH SERVS, TEX DEPT CRIMINAL JUSTICE-INSTNL DIV, 88- *Mem:* Am Asn Pub Health Physicians; fel Am Col Prev Med; Soc Med Consult Armed Forces. *Res:* Communicable diseases; venereal disease; tuberculosis; military preventive medicine; public health administration; epidemiology. *Mailing Add:* Tex Dept Criminal Justice Box 99 Huntsville TX 77342

ALEXANDER, CHARLES WILLIAM, b Olathe, Kans, June 2, 31; m 53; c 2. AGRONOMY. *Educ:* Kans State Col, BS, 53, MS, 54; NC State Col, PhD(agron), 57. *Prof Exp:* Res agronomist, Humid Pasture Mgt, 57-62, agr res admin, 62-75, AREA DIR, AGR RES SERV, USDA, COLUMBIA, MO, 75- *Mem:* Crop Sci Soc Am; Am Soc Agron; Am Soc Pub Admin; Sigma Xi. *Res:* Plant physiology; forage crop ecology; agricultural research administration; public administration. *Mailing Add:* 1610 Stoney Brook Pl Columbia MO 65201

ALEXANDER, CHESTER, JR, b Tarboro, NC, Nov 6, 37; m 61; c 2. PHYSICS. *Educ:* Davidson Col, BS, 60; Emory Univ, MS, 62; Duke Univ, PhD(physics), 68. *Prof Exp:* Instr physics, Emory Univ, 61-62; physicist, Feltman Labs, Picatinny Arsenal, 62-64; from asst prof to assoc prof, 68-78, PROF PHYSICS, UNIV ALA, 79- *Concurrent Pos:* Vis researcher, Univ Oslo, 75-76. *Res:* Electron spin resonance; microwave spectroscopy. *Mailing Add:* Box 870324 Tuscaloosa AL 35487-0324

ALEXANDER, CLAUDE GORDON, b San Diego, Calif, Sept 15, 24; m 54; c 3. ZOOLOGY. *Educ:* Ore State Col, BS, 48, MS, 50; Univ Calif, Los Angeles, PhD, 55. *Prof Exp:* From asst prof to assoc prof, 55-69, PROF BIOL, SAN FRANCISCO STATE UNIV, 69- *Mem:* Am Soc Parasitol; Am Soc Zoologists; Sigma Xi. *Res:* Helminth parasites comparative physiology of Elasmobranch fishes. *Mailing Add:* 935 Canada Rd Woodside CA 94062

ALEXANDER, DAVID MICHAEL, b Cleveland, Ohio, Nov 4, 41; m 66. OTHER MEDICAL & HEALTH SCIENCES. *Educ:* Case Inst Technol, BS, 63, MS, 65, PhD(biomed eng), 68. *Prof Exp:* Asst prof elec eng, Christian Bros Col, Tenn, 68-71; res engr, Pulmonary Lab, Baptist Mem Hosp, 71-83; Comput Specialist, Vet Admin Med Ctr, Memphis, Tenn, 83-86; COMPUT SPECIALIST, VET ADMIN MED CTR, BIRMINGHAM, ALA, 86- *Concurrent Pos:* Res assoc clin physiol, Med Units, Tenn, 68-69. *Mem:* AAAS; Inst Elec & Electronics Engrs; Am Soc Mech Eng Educ; Sigma Xi. *Res:* Mathematical modeling of physiological systems; statistical and time series analysis of biological variables. *Mailing Add:* Vet Admin Med Ctr 700 S 19th St 10ba3/Adp Birmingham AL 35205

ALEXANDER, DENTON EUGENE, b Potomac, Ill, Dec 18, 17; m 43. PLANT BREEDING, AGRONOMY. *Educ:* Univ Ill, BS, 41, PhD(agron), 50. *Prof Exp:* Instr aircraft engine mechs, Army Air Force, 41-44; prod supvr Manhattan proj, Oak Ridge Nat Lab, 44-47; fel bot, 50-51, from instr & asst prof to assoc prof, 51-63, PROF PLANT BREEDING, UNIV ILL, URBANA-CHAMPAIGN, 63- *Concurrent Pos:* Hybrid maize expert, Food & Agr Orgn, Yugoslavia, 57; Ford Found consult, Latin Am, 65-67; mem bd dir, Funk Bros Seed Co, 69-74; consult, Dept Agr, Repub SAfrica, 76. *Honors & Awards:* Crop Sci Award, Am Soc Agron, 70. *Mem:* Am Soc Agron; Soviet Acad Agr Sci; hon mem Asn Genetic Soc Yugoslavia. *Res:* Breeding and genetics of oil and protein in maize. *Mailing Add:* 701 W Pennsylvania Ave Urbana IL 61801

ALEXANDER, DREW W, b Peoria, Ill, Dec 21, 48. MEDICINE. *Educ:* Earlham Col, BA, 70; Med Col Ohio, MD, 73; Albert Einstein Col Med, cert pediat, 76. *Prof Exp:* Asst prof, Univ Tex Health Sci Ctr, Dallas, 77-, health team physician, West Dallas Youth Clinic, Children & Youth Proj, 77-; partner, Adolescent Health Assocs, 81-; AT DEPT PEDIAT, UNIV TEX HEALTH SCI CTR. *Concurrent Pos:* Fel adolescent med, Univ Tex Health Sci Ctr Dallas, 76-77; consult, multidisciplinary adolescent health training proj, 77- *Mem:* Soc Adolescent Med; Am Acad Pediat; AMA. *Res:* Adolescent growth and development; adolescent health care delivery; adolescents as parents; adolescent medicine. *Mailing Add:* Dept Pediat Univ Tex Southwestern Med Sch 5323 H Hines Blvd Dallas TX 75235

ALEXANDER, DUANE FREDERICK, b Baltimore, MD, Aug 11, 40; m 63; c 2. PEDIATRICS, DEVELOPMENT DISABILITIES. *Educ:* Pa State Univ, BS, 62; Johns Hopkins Univ Sch Med, 66. *Prof Exp:* Med officer, US Dept Health Educ & Welfare, 74-78; asst sci, deputy dir, 82-86, DIR, NAT INST CHILD HEALTH & HUMAN DEVELOP, 86- *Honors & Awards:* Commendation Medal, Pub Health Serv, 70, Meritorious Serv Medal, 85, Spec Recognition Award, 85; Surgeon General's Exemplary Serv Medal, 90. *Mem:* Am Acad Pediat; Soc Develop Pediat; Asn Retarded Citizens; Am Pediat Soc. *Mailing Add:* Nat Inst Child Health & Human Develop NIH Bldg 31 Rm 2A03 Bethesda MD 20892

ALEXANDER, EARL L(OGAN), JR, b Flint, Mich, May 22, 20; wid; c 3. CHEMICAL ENGINEERING. *Educ:* Purdue Univ, BS, 42, MS, 48, PhD(chem eng), 52. *Prof Exp:* Plant operator, Joseph E Seagram & Sons, 42-43; chem engr, Streptomycin Pilot Plant, Parke Davis, 46-47; asst chem engr, Purdue Univ, 49-51; chem engr, Distillation Design & Textile Polymers, Exp Sta, E I du Pont de Nemours & Co, 52-59; prin scientist, Advan Concepts Composite Mat, Res Div, Rocketdyne, 59-71; mem tech staff composites, mat & processes, Los Angeles Div, 71-81, SYSTS ANALYST, SOLAR POWER, SHALE OIL PROCESSING, ENERGY TECHNOL ENG CTR, NORTH AM ROCKWELL, 81- *Mem:* Am Chem Soc; Am Inst Chem Engrs. *Res:* Advanced solid propellants; filament reinforced composite structures anisotropic solid propellant combustion; high temperature polymers; heat and mass transfer; chemical process computer simulating. *Mailing Add:* 19380 Halsted St Northridge CA 91324

ALEXANDER, EDWARD CLEVE, b Knoxville, Tenn, Nov 20, 43; c 3. ORGANIC CHEMISTRY. *Educ:* City Col New York, BS, 65; State Univ NY, Buffalo, PhD(org chem), 69. *Prof Exp:* Fel, Iowa State Univ, 69-70; asst prof chem, Univ Calif, San Diego, 70-78; lectr chem, Calif State Univ, Los Angeles, 78-82; SUPVR DEPT MATH & SCI, SAN DIEGO COMMUNITY COL DIST, BOOST PROG, NAVAL TRAINING CTR, 82- *Mem:* Am Chem Soc; NY Acad Sci. *Res:* Organic photochemistry; highly strained ring systems; small ring heterocyclic chemistry; organic synthesis. *Mailing Add:* 9777 Genesee Ave San Diego CA 92121

ALEXANDER, EMMIT CALVIN, JR, b Lawton, Okla, July 4, 43; m 66; c 3. GEOCHRONOLOGY. *Educ:* Okla State Univ, BS, 66; Univ Mo-Rolla, PhD(chem), 70. *Prof Exp:* Asst res chemist, Univ Calif, Berkeley, 70-73; from asst prof to assoc prof, 73-87, PROF UNIV MINN, MINNEAPOLIS, 87- *Concurrent Pos:* Prin investr & co-investr, Lunar Sample Anal Prog, NASA, 70-80. *Mem:* Am Geophys Union; AAAS; Geochem Soc; Sigma Xi; Geol Soc Am; Meteoritical Soc; Nat Speleological Soc; Nat Water Well Asn. *Res:* Geohydrology; isotope geochemistry; ground water pollution and environmental geology; karst hydrology and geomorphology; geochronology; cosmochronology; rare-gas isotope studies. *Mailing Add:* Dept of Geol & Geophys Univ of Minn Minneapolis MN 55455-0219

ALEXANDER, FORREST DOYLE, b Trenton, Tenn, Oct 27, 27; m 51; c 3. MATHEMATICS. *Educ:* Union Univ, Tenn, BS, 50; George Peabody Col, MA, 55, PhD(math), 61. *Prof Exp:* Teacher high schs, Tenn, 51-56; instr, 56-59, from asst prof to assoc prof math, 61-67, admin asst, 68-79, PROF MATH, STEPHEN F AUSTIN STATE UNIV, 67-, ASST CHMN, 79- *Concurrent Pos:* NSF grant, 61-62, lectr, NSF Coop Col-Sch Sci Prog, Stephen F Austin State Univ, 70-71; consult math staff, ETex Baptist Col & Nacogdoches County Pub Schs, 63-64. *Mem:* Math Asn Am. *Res:* Elementary modern mathematics; abstract algebra; geometry; analysis. *Mailing Add:* Dept Math Box 13040 Stephen F Austin State Univ Nacogdoches TX 75962-3040

ALEXANDER, FRANK CREIGHTON, JR, b Aspinwall, Pa, Nov 30, 18; m 42; c 3. PHYSICS, ELECTRONICS ENGINEERING. *Educ:* Carnegie Inst Technol, 42. *Prof Exp:* Physicist, Gulf Res & Develop Co, 47-56; head, Electronic Lab, Res & Develop Div, Am Viscose Corp, 56-58, instrumentation group, FMC Corp, 58-62, head appl physics & instrumentation, Res & Develop Lab, 62-78; CONSULT, TECH PHYSICS CO, 78- *Concurrent Pos:* Instr, Carnegie Inst Technol, 50-51; instr physics & math, Friends Cent Sch, 78-79; instr antenna instrumentation, Rois Mfg Co, Inc, 79-81. *Mem:* Am Phys Soc; Inst Elec & Electronics Engrs. *Res:* Design and development of electronic instrumentation; electrical properties of organic materials; digital computer application. *Mailing Add:* 570 Juniata Ave Swarthmore PA 19081-2413

ALEXANDER, FRED, b NJ, Sept 17, 17; m 50; c 4. INTERNAL MEDICINE. *Educ:* St John's Col, Md, AB, 37; Univ Md, MD, 41. *Prof Exp:* Intern med, Univ Pittsburgh Hosps, 42-43; asst resident pediat, Children's Hosp, Philadelphia, 43-45; resident med, Jefferson Hosp, 45-46; resident cardiol, Mass Gen Hosp, 46-50; sr scientist med, Los Alamos Sci Lab, 50-52; ASST PROF MED, UNIV PA, 52-; DIR CLIN LAB, SMITH KLINE & FRENCH LABS, 62- *Concurrent Pos:* Consult, Pa Mutual Ins Co, 50-51. *Mem:* Am Col Physicians; Sigma Xi; AMA; Am Col Cardiol. *Res:* Initial human pharmacological trials of potential therapeutic agents; cardiovascular diseases, hypertension and renal pathology; chemotherapy; metabolism; psychopharmacological entities and diseases of nervous system. *Mailing Add:* 1400 Youngsford Rd Gladwyne PA 19035

ALEXANDER, GEORGE JAY, b Paris, France, June 27, 25; nat US; m 58. NEUROTOXICOLOGY. *Educ:* Hobart Col, BS, 49; Rutgers Univ, PhD(microbiol & chem), 53. *Prof Exp:* Res assoc, Worcester Found Exp Biol, 53-58; assoc, 58-60, asst prof biochem, Col Physicians & Surgeons, 60-84, ASSOC PROF CLIN BIOCHEM PSYCHIAT, COLUMBIA UNIV, 84- *Concurrent Pos:* Assoc res scientist, NY State Psychiat Inst, 58-60, res scientist V, 60- *Mem:* NY Acad Sci; Fedn Am Socs Exp Biol; Am Soc Biol Chem; Soc Exp Biol & Med. *Res:* Cholesterol synthesis; transmethylation; brain metabolism; neurotoxicology; chemistry of epilepsy. *Mailing Add:* Columbia Univ 722 W 168th St New York NY 10032

ALEXANDER, GERALD CORWIN, b Corvallis, Ore, Mar 8, 30; m 53. ELECTRICAL & ELECTRONICS ENGINEERING. *Educ:* Ore State Col, BS, 51; Mass Inst Technol, ScM, 60; Univ Calif, Berkeley, PhD, 73. *Prof Exp:* Eng trainee, Allis-Chalmers Mfg Co, Inc, 54-55; asst prof elec eng, 55-63, ASSOC PROF ELEC ENG, ORE STATE UNIV, 64- *Concurrent Pos:* Power systs anal, Bohn Power Admin; econ generator selection CHZM. *Mem:* Inst Elec & Electronics Engrs; Nat Soc Prof Engrs; Am Soc Eng Educ. *Res:* Power systems; electric machines; power electronics; on-line data acquisition and processing. *Mailing Add:* Dept Elec & Comput Eng Ore State Univ Corvallis OR 97331

ALEXANDER, GUY B, b Ogden, Utah, May 31, 18; m 41; c 2. CHEMISTRY, METALS. *Educ:* Univ Utah, BS, 41, MS, 42; Univ Wis, PhD(inorg chem), 47. *Prof Exp:* Res chemist, Manhattan Proj, Monsanto Chem Co, Ohio, 44-45; instr chem, Univ Utah, 46-47; res chemist, E I du Pont de Nemours & Co, 47-50, res supvr, 50-68; mgr chem res, Fansteel Inc, Md, 68-69, dir res, Vr-Wesson Div, Ill, 69-70 & San Fernando Labs, 70-73, dir res & lab mgr, Fansteel Res Ctr, 73-79; dir energy dept, Univ Utah Res Inst, 80-83; pres, Tech Res Assoc, Inc, 83-90; RETIRED. *Concurrent Pos:* Mem adv bd, Chem Mag, 72-75; adj prof, Univ Utah, 73- *Mem:* Am Chem Soc. *Res:* Colloid chemistry; sintered carbides; arc reactions; hot pressing; carbide powder processes; cermets; metals for high temperatures. *Mailing Add:* 4381 S Fortuna Way Salt Lake City UT 84124

ALEXANDER, HENRY R(ICHARD), b Scranton, Pa, June 13, 25; m 52; c 5. ELECTRICAL ENGINEERING. *Educ:* Rutgers Univ, BSEE, 45, MSEE, 47; Princeton Univ, PhD(elec eng), 52. *Prof Exp:* Instr elec eng, Rutgers Univ, 45; asst, Princeton Univ, 48-50, res assoc, 50-51; physicist, Off Naval Res, 51-54, sci staff, Naval Attache, London, 55-56; mem tech staff, Inst Defense Analysis, DC, 56-59; mgr systs design, Gen Dynamics Electronics, NY, 59-61; systs prog mgr, 61-80, TECH STAFF, EQUIP DIV, RAYTHEON CO, 80- *Concurrent Pos:* Consult, Mitre Corp, Mass, 60. *Mem:* Inst Elec & Electronics Engrs. *Res:* Circuit analysis; wave propagation; information processing; underwater sound systems; reconnaissance systems; radar. *Mailing Add:* Equip Div Raytheon Co Boston Post Rd Wayland MA 01778

ALEXANDER, HERMAN DAVIS, b Sweetwater, Tex, Dec 19, 19; m 46; c 2. NUTRITION, BIOCHEMISTRY. *Educ:* Auburn Univ, BS, 50, MS, 52, PhD(nutrit, biochem), 55. *Prof Exp:* Asst nutrit & biochem, 50-55, asst prof nutrit, 55-60, from asst prof to assoc prof physiol, 60-66, assoc prof pharmacol, 66-76, ASSOC PROF ZOOL & ENTOM, AUBURN UNIV, 76- *Mem:* Am Chem Soc. *Res:* Nutrition, effects of various nutritional deficiencies. *Mailing Add:* 498 Cary Dr Auburn AL 36830

ALEXANDER, IRA H(ENRIS), b Ft Leavenworth, Kans, July 9, 20; m 49; c 2. GEODESY, CIVIL ENGINEERING. *Educ:* Univ Calif, BS, 42. *Prof Exp:* Hwy engr, US Pub Roads Admin, 42-43 & 46; from civil engr asst to sr civil engr asst, Dept County Engr, County of Los Angeles, 47-53, civil engr assoc & asst geod sect chief, 53-56, assoc civil engr & geod sect chief, 56-60, civil engr, 60-63, asst div engr, Surv Div, 63-65, div engr, Mapping Div, 65-71, asst chief dep county engr, 71-78; CONSULT GEOD, 78- *Concurrent Pos:* Vpres, Metrex Systs Corp, 84-89. *Honors & Awards:* Surv & Mapping Award, Am Soc Civil Engrs, 80. *Mem:* Fel Am Soc Civil Engrs; Am Cong Surv & Mapping (pres-elect, 81, pres, 82); Am Geophys Union. *Res:* Application of geodetic techniques to subsidence problems; analysis of triangulation networks with time and movement parameters; criteria for intensification of mapping control. *Mailing Add:* 3812 Shannon Rd Los Angeles CA 90027

ALEXANDER, JAMES CRAIG, b Ont, Feb 12, 26; m 53; c 3. BIOCHEMISTRY, TOXICOLOGY. *Educ:* Univ Toronto, BSA, 49, MSA, 51; Univ Wis, PhD(biochem), 54. *Prof Exp:* Res biochemist, Procter & Gamble Co, 54-66; assoc prof biochem, 66-69, actg chmn, dept nutrit, 71-72 & 80-81, PROF BIOCHEM, UNIV GUELPH, 69- *Concurrent Pos:* Indust consult, 68-91. *Mem:* Fel AAAS; Am Inst Nutrit; Am Oil Chemists' Soc; Sigma Xi; Can Inst Food Sci & Technol; Can Soc Nutrit Sci. *Res:* Nutritional, biochemical and physiological studies on the effect of lipids on biological systems; nutritional value of germinated cereals. *Mailing Add:* Dept Nutrit Sci Univ Guelph Guelph ON N1G 2W1 Can

ALEXANDER, JAMES CREW, b Zanesville, Ohio, Mar 22, 42; m; c 2. MATHEMATICS. *Educ:* Johns Hopkins Univ, BA, 64, PhD(math), 68. *Prof Exp:* Instr math, Johns Hopkins Univ, 67-69; res assoc, 69-70, asst prof, 70-73, assoc prof, 73-79, PROF MATH, UNIV MD, COLLEGE PARK, 79- *Mem:* Am Math Soc; Soc Indust Appl Math. *Res:* Topological and functional methods applied to differential equations and dynamic systems of physical interest; mathematical modelling and simulation of systems in biology and engineering. *Mailing Add:* Dept Math Univ Md College Park MD 20742

ALEXANDER, JAMES KERMOTT, b Evanston, Ill, Dec 25, 20; m 45, 77; c 3. INTERNAL MEDICINE. *Educ:* Amherst Col, AB, 42; Harvard Univ, MD, 46; Am Bd Internal Med, dipl, 55. *Prof Exp:* Intern, First Med Div, Bellevue Hosp, NY, 46-47; chief, Radiol Sect, US Army Hosp, Guam, 47-49; resident internal med, Bellevue Hosp, 50-51; asst med, Columbia Univ, 53-54; from asst prof to assoc prof med, Baylor Col Med, Houston, 54-60, chief, Cardiac Sect & dir, Cardiopulmonary Lab, 60-70, head, Cardiac Sect, Dept Med, 63-72, PROF MED, BAYLOR COL MED, HOUSTON, TEX, 65-, DIR CARDIOVASC TEACHING, 82-; CHIEF CARDIOL, BEN TAUB GEN HOSP, 70- *Concurrent Pos:* Res fel, Cardiopulmonary Lab, Bellevue Hosp, NY, 49-50; Am Heart Asn res fels, Dept Physiol, Harvard Med Sch, 51-52 & Cardiopulmonary Lab, Presby Hosp, 52-53; dir cardiac lab, Ben Taub Gen Hosp, 54-; asst attend physician, Presby Hosp, 53-54; assoc internal med, Methodist Hosp, 55; actg chief cardiol, Vet Admin Hosp, Houston, Tex, 84- *Mem:* AMA; Am Heart Asn; Am Fedn Clin Res; Am Physiol Soc; Am Soc Clin Invest. *Res:* Cardiopulmonary function in a variety of chronic lung and circulatory diseases, including obesity; cardiopulmonary function at high altitude. *Mailing Add:* Dept Internal Med Baylor Col Med 6550 Fannin MS SM-1246 Houston TX 77030

ALEXANDER, JAMES KING, b Hysham, Mont, Jan 9, 28; m 55; c 3. MICROBIOLOGY, BIOCHEMISTRY. *Educ:* Univ Mont, BA, 50; Mont State Univ, MS, 54, PhD(bact), 59. *Prof Exp:* Instr bact, NDak State Univ, 54-55; res assoc microbiol, 58-60, from instr to assoc prof, 60-72, PROF BIOL CHEM, HAHNEMANN MED COL, 72-, PROF MICROBIOL, 73- *Concurrent Pos:* Vis prof biol, Mass Inst Technol, 73. *Mem:* AAAS; Am Soc Biol Chemists; Am Soc Microbiol. *Res:* Carbohydrate metabolism of microorganisms; regulation of enzyme synthesis; cellulose utilization; energy from biomass. *Mailing Add:* Dept Biochem Hahnemann Med Col MS 411 Broad & Vine Philadelphia PA 19102

ALEXANDER, JAMES WESLEY, b El Dorado, Kans, May 23, 34; m 84; c 7. SURGERY, IMMUNOLOGY. *Educ:* Univ Tex, MD, 57; Univ Cincinnati, ScD(surg), 64; Am Bd Surg & Bd Thoracic Surg, cert, 65. *Prof Exp:* Chief trauma study br, Surg Res Unit, Ft Sam Houston, Tex, 65-66; dir res, Shriners Burn Inst, Cincinnati Unit, 79-90; from asst prof to assoc prof surg, 66-75, actg dir, Blood Ctr, 72-79, PROF SURG, MED CTR, UNIV CINCINNATI, 75-, DIR, TRANSPLANTATION DIV, 67- *Concurrent Pos:* Ad hoc consult, NIH, 67-; NIH study sect, Scholastic Aptitude Tests, 83-, chmn, 90- *Honors & Awards:* Cuthbertson Lectr, Europ Soc Parenteral & Enteral Nutrit, 89; Strauss lectr, Univ Wash, 90. *Mem:* Fel Am Col Surg; Am Asn Surg Trauma; Am Soc Transplant Surgeons (pres, 87-88); Int Soc Surg; Transplantation Soc; Am Burn Asn (pres, 84-85); Surg Infection Soc (pres, 86-87); Am Asn Immunol; Asn Acad Surg; Int Soc Burn Injuries; Am Surg Asn. *Res:* Infections; host defense mechanisms; transplantation; burn injury; nutrition. *Mailing Add:* Dept Surg Univ Cincinnati Med Ctr Cincinnati OH 45267-0558

ALEXANDER, JOHN J, b Indianapolis, Ind, Apr 13, 40. INORGANIC CHEMISTRY, ORGANOMETALLIC CHEMISTRY. *Educ:* Columbia Col, AB, 62; Columbia Univ, MA, 63, NSF fel & PhD(chem), 67. *Prof Exp:* Res assoc chem, Ohio State Univ, 67-69, fel, 67-68; from asst prof to assoc prof, 70-79, PROF CHEM, UNIV CINCINNATI, 79- *Concurrent Pos:* Vis prof, Ohio State Univ, 85-86. *Mem:* Am Chem Soc; Sigma Xi. *Res:* Electronic structures of transition metal complexes; synthetic organometallic chemistry. *Mailing Add:* Dept Chem Univ Cincinnati Cincinnati OH 45221-0172

ALEXANDER, JOHN MACMILLAN, JR, b Columbia, Mo, Aug 17, 31; m 53; c 4. PHYSICAL CHEMISTRY, NUCLEAR SCIENCE & CHEMISTRY. *Educ:* Davidson Col, BS, 53; Mass Inst Technol, PhD(phys chem), 56. *Prof Exp:* Res assoc chem, Mass Inst Technol, 56-57; chemist, Lawrence Radiation Lab, Univ Calif, Berkeley, 57-63; assoc prof, 63-68, chmn dept, 70-72, PROF CHEM, STATE UNIV NY STONY BROOK, 68- *Concurrent Pos:* Researcher, AEC-ERDA, Dept Energy, res collabr, Brookhaven Nat Lab, 64-; Sloan fel, 64-66; chmn, Gordon Res Conf Nuclear Chem, 66; assoc ed, Am Chem Soc Monographs, 68-69; Guggenheim fel, 69-70; vis scientist, Nuclear Study Ctr, Bordeaux, France, 69-70; mem exec comt, Berkeley Superhilac Accelerator, 75-78; prog adv comt, Van de Graaff Accelerator, Brookhaven Nat Lab, 77-81; Holifield Heavy Ion Res Facil, Oak Ridge Nat Lab, 86-89, Système Accélérateur Rhône-Alpes, Inst des Sci

Nucldrires de Grenoble, Grenoble, France, 88-89; vis prof, Nuclear Study Ctr, Bordeaux-Gradignan & Inst Nuclear Physics, Orsay, France, 78-79. *Honors & Awards:* Nuclear Chem Award, Am Chem Soc, 91. *Mem:* Am Chem Soc; Am Phys Soc. *Res:* Nuclear chemistry and reactions; stopping of heavy atoms; research on radioactivity; high-energy nuclear reactions, especially fission, spallation, and fragmentation; heavy ion reactions; elastic scattering, complete fusion and reactions cross sections, energy and spin dissipation, evaporative deexcitation. *Mailing Add:* Dept of Chem State Univ of NY Stony Brook NY 11794-3400

ALEXANDER, JOSEPH KUNKLE, b Staunton, Va, Jan 9, 40; m 62; c 3. SPACE PHYSICS, RADIO ASTRONOMY. *Educ:* Col William & Mary, BS, 60, MA, 62. *Prof Exp:* Physicist, 62-70, sect head, 70-76, br head, Goddard Space Flight Ctr, 76-83, sr policy analyst, White House Sci Off, 84-85, assoc lab chief, Goddard Space Flight Ctr, 85, dep chief scientist, 85-87, ASST ASSOC ADMIN, SCI & APPLN, NASA, HQ, 87- *Concurrent Pos:* Vis scientist, Univ Colo, 73-74; mem, space sci adv comt, NASA, 80-82, space & earth sci adv comt, 82-84, proj scientist, Origins Plasmas Earth's Neighborhood Proj, 81-83 & Int Solar-Terrestrial Physics Proj, 83-85. *Honors & Awards:* Sci Achievement Medal, NASA, 81. *Mem:* Am Geophys Union; Int Astron Union. *Res:* Planetary magnetospheres and solar-planetary relationships, especially in the area of radio-wave and plasma-wave phenomena and their relation to the interactions of charged particles and fields in space. *Mailing Add:* Code S NASA Hq Washington DC 20546

ALEXANDER, JOSEPH WALKER, b Washington, DC, Jan 22, 47; m 70; c 2. VETERINARY MEDICINE & ADMINISTRATION, MARINE MAMMAL MEDICINE. *Educ:* Univ Ariz, BS, 69; Colo State Univ, DVM, 73; Univ Tenn, MS, 81; Am Col Vet Surgeons, dipl, 79. *Prof Exp:* Staff vet, Cheshire Vet Clin, 76-77; from asst prof to assoc prof vet surg, Univ Tenn, 77-81; prof vet surg, Va Tech, 81-85, chmn div agr & urban pract, 82-85, dir Vet Med Teaching Hosp, 83-85; PROF VET SURG & DEAN COL VET MED, OKLA STATE UNIV, 85- *Concurrent Pos:* Consult, Marine Animal Prod, 86- *Mem:* Asn Am Vet Med Col (pres, 90-91); Comn Vet Med; Asn Aquatic Animal Med (pres-elect, 91-92). *Res:* Veterinary orthopedics and veterinary orthopedic diseases; author of three books and over 60 publications. *Mailing Add:* 6310 Coventry Stillwater OK 74074

ALEXANDER, JUSTIN, b Ulm, Ger, Oct 1, 21; m 43; c 3. PHYSICAL THERAPY, PROSTHETICS. *Educ:* NY Univ, BS, 50, MA, 59, PhD(admin higher educ), 68. *Prof Exp:* Phys therapist, Vet Admin, NY, 50-54 & Jamaica Med Group, Health Ins Plan, 54-55; assoc dir phys ther, 58-70, PROF & DIR PHYS THER, ITHACA COL, 70-; ASSOC PROF & DIR PHYS THER, ALBERT EINSTEIN COL MED, 55- *Concurrent Pos:* Co-investr, training grant, Bur State Serv, 67-72, Strength Norms for Children, 72-77 & Reliability & Validity, 79-80; chmn, NY State Bd Phys Ther. *Mem:* Am Phys Ther Asn; Am Cong Rehab Med. *Res:* Improvement of delivery of patient services, and investigations of variety of problems. *Mailing Add:* 75 Hawthorne St Brooklyn NY 11225

ALEXANDER, KENNETH ROSS, b Seattle, Wash, Feb 23, 45; m 81. VISION. *Educ:* Univ Wash, BA & BS, 67, MS, 68, PhD(psychol), 72. *Prof Exp:* Fel vision, Univ Rochester, 72-73; asst assoc prof physiol optics, Ill Col Optom, 73-81; RES ASSOC PHYSIOL OPTICS, ILL EYE & EAR INFIRMARY, 81- *Concurrent Pos:* Instr, Sch Art Inst Chicago, 80-81. *Mem:* AAAS; Asn Res Vision & Ophthal; Optical Soc Am. *Res:* Psychophysical investigation of visual characteristics and underlying mechanisms of various types of inherited retinal disorders, such as retinitis pigmentosa and cone dystrophy, with an emphasis on rod and cone mechanisms. *Mailing Add:* Eye & Ear Infirmary Univ Ill 1855 W Taylor St Chicago IL 60680

ALEXANDER, LESLIE LUTHER, b Kingston, Jamaica, Oct 10, 17; US citizen; m 51; c 5. MEDICINE, RADIOLOGY. *Educ:* NY Univ, AB, 47, AM, 48; Howard Univ, MD, 52. *Prof Exp:* From instr to prof radiol, Col Med, State Univ NY Downstate Med Ctr, 56-77; PROF RADIOL, HEALTH SCI CTR, STATE UNIV NY STONY BROOK & DIR RADIOL, LONG ISLAND JEWISH-HILLSIDE MED CTR AFFIL, QUEENS HOSP CTR, 77- *Concurrent Pos:* Nat Med Fels, Inc fel, 54-56; consult, Brooklyn Vet Hosp, 62-71; Cath Med Ctr Brooklyn & Queens, 67- & Bur Health Prof Educ & Manpower Training, NIH, 70-74; asst ed, J Nat Med Asn, 64-; dir radiation ther, North Shore Hosp, Manhasset, NY, 70-77; consult, Bur Med Devices, Food & Drug Admin, HEW, 75- & med adv comt, Bur Radiol Health, State NY, 77-81; mem, Bd Chancellors, Am Col Radiol, Chicago, 80- *Mem:* AMA; Asn Univ Radiologists; fel Am Col Radiol; Am Soc Therapeut Radiologist; fel NY Acad Med. *Res:* Radiology, including radiation therapy, nuclear medicine, radiobiology and cancer research. *Mailing Add:* Queens Hosp Ctr 82-68 164th St Jamaica NY 11432

ALEXANDER, LLOYD EPHRAIM, b Salem, Va, Aug 17, 02; m 34; c 2. EMBRYOLOGY. *Educ:* Univ Mich, AB, 27, AM, 28; Univ Rochester, PhD(embryol), 36. *Prof Exp:* From instr to assoc prof biol, Fisk Univ, 30-48; prof & head dept, 49-72, EMER PROF BIOL & HEAD DEPT, KY STATE COL, 73- *Res:* Experimental embryology; grafting and transplanting tissues in vertebrates; production of lenses in chick embryos; capacities of optic tissues of Gallus domesticus for induction and regeneration. *Mailing Add:* 1400 Willow Ave Apt 903 Louisville KY 40204-1463

ALEXANDER, MARTIN, b Newark, NJ, Feb 4, 30; m 51; c 2. SOIL MICROBIOLOGY, MICROBIAL ECOLOGY. *Educ:* Rutgers Univ, BS, 51; Univ Wis, MS, 53, PhD(bact), 55. *Prof Exp:* From asst prof to assoc prof, 55-64, PROF SOIL MICROBIOL, CORNELL UNIV, 64-, LIBERTY HYDE BAILEY PROF, 77- *Concurrent Pos:* Consult to var pvt industs, int & nat agencies. *Honors & Awards:* Fisher Award. *Mem:* Fel AAAS; fel Am Soc Agron; fel Am Acad Microbiol; Am Soc Microbiol. *Res:* Biodegradation; bioremediation; environmental fate; pesticide decomposition; biochemical ecology; environmental pollution. *Mailing Add:* 708 Bradfield Hall Cornell Univ Ithaca NY 14853

ALEXANDER, MARTIN DALE, b Grants, NMex, Nov 27, 38; m 68; c 1. INORGANIC CHEMISTRY. *Educ:* NMex State Univ, BSc, 60; Ohio State Univ, PhD(inorg chem), 64. *Prof Exp:* From asst prof to assoc prof, 64-74, PROF CHEM, NMEX STATE UNIV, 74- *Mem:* Am Chem Soc; fel Brit Chem Soc. *Res:* Mechanisms of reactions of coordination compounds of transition metals; synthesis of transition metal coordination compounds. *Mailing Add:* Dept Chem NMex State Univ Box 30001 Dept 3C Las Cruces NM 88003

ALEXANDER, MARY LOUISE, b Ennis, Tex, Jan 15, 26. GENETICS. *Educ:* Univ Tex, BA, 47, MA, 49, PhD(zool, genetics), 51. *Prof Exp:* Fel, Oak Ridge Nat Lab, AEC, 51-52; res assoc zool, Genetics Found, Univ Tex, 52-55, instr, 54, asst prof, Med Sch & assoc biologist, M D Anderson Hosp & Tumor Inst, 56-62, res scientist, Genetics Found, 62-67; assoc prof, 67-70, PROF BIOL, SOUTHWEST TEX STATE COL, 70- *Concurrent Pos:* Consult, Brookhaven Nat Lab, 55; res partic, Oak Ridge Inst Nuclear Studies, 56-77; NIH fel, Univ Scotland, 60-62. *Mem:* Genetics Soc Am; Soc Study Evolution; Am Soc Naturalists; Am Soc Human Genetics; Radiation Res Soc. *Res:* Genetics of Drosophila; evolution and population genetics; radiation and chemical genetics. *Mailing Add:* Dept Biol Southwest Tex State Univ San Marcos TX 78666

ALEXANDER, MAURICE MYRON, b South Onondaga, NY, Dec 18, 17; m 43; c 3. VERTEBRATE ECOLOGY. *Educ:* NY State Col Forestry, BS, 40, PhD(wildlife mgt), 50; Univ Conn, MS, 42. *Prof Exp:* Res asst, State Bd Fisheries & Game, Conn, 45-46; instr forestry & wildlife mgt, Univ Conn, 46-47; teaching fel, 47-49, from instr to prof, 49-83, chmn dept, 65-77, EMER PROF FOREST ZOOL, STATE UNIV NY COL ENVIRON SCI & FORESTRY, 83- *Concurrent Pos:* Dir, Roosevelt Wildlife Forest Exp Sta, 65-83. *Mem:* Wildlife Soc; Am Soc Mammalogists; Ecol Soc Am; Soc Study Amphibians & Reptiles. *Res:* Animal ecology; aging techniques; furbearer management; wetland ecology. *Mailing Add:* 4039 Tanner Rd Rd 2 Syracuse NY 13215-9728

ALEXANDER, MICHAEL NORMAN, b Washington, DC, Mar 27, 41; m 64; c 2. SOLID STATE PHYSICS, MATERIALS SCIENCE. *Educ:* Harvard Univ, AB, 62; Cornell Univ, PhD(physics), 67. *Prof Exp:* Res asst physics, Cornell Univ, 64-67; res physicist, Mat Sci Div, Army Mat & Mech Res Ctr, Watertown, Mass, 67-79; sr mem of tech staff, GTE Labs, Inc, 79-86; prin scientist, Res & Develop & New Bus Ctr, Thermo Electron Corp, Waltham, Mass, 86-88; branch chief, 88-89, DIV CHIEF, ROME AIR DEVELOP CTR, HANSCOM, MASS, 89-, ROME LAB, 91- *Concurrent Pos:* Mem adv comt on wetlands, Planning Bd, Lexington, Mass, 75-76; exec comt, New Eng Sect Am Phys Soc, 84-86; co-chmn, High Sch Guidance Adv Comt, 84-87. *Mem:* AAAS; Sigma Xi; Am Phys Soc; Inst Elec & Electronic Engrs; Mat Res Soc. *Res:* Electronic structure of metals, alloys and semiconductors; optical properties of crystalline insulators, glasses and semiconductors; high temperature superconductors; III-V semiconductor materials fabrication and properties. *Mailing Add:* 66 Baskin Rd Lexington MA 02173

ALEXANDER, MILLARD HENRY, b Boston, Mass, Feb 17, 43; div; c 1. THEORETICAL CHEMISTRY. *Educ:* Harvard Col, BA, 64; Univ Paris, PhD(chem), 67. *Prof Exp:* Res fel chem, Harvard Univ, 67-71; asst prof molecular physics, 71-73, from asst prof to assoc prof chem, 73-79, PROF CHEM, UNIV MD, COLLEGE PARK, 79- *Mem:* Am Chem Soc; fel Am Phys Soc. *Res:* Theoretical study of rotationally and electronically inelastic collisions between atoms and molecules; chemical reactions of electronically excited atoms and molecules. *Mailing Add:* Dept of Chem Univ of Md College Park MD 20742

ALEXANDER, NANCY J, b Cleveland, Ohio, Dec 1, 39; m 70; c 2. REPRODUCTIVE PHYSIOLOGY, IMMUNOLOGY. *Educ:* Miami Univ, BS, 60, MA, 61; Univ Wis, PhD(entom), 65. *Prof Exp:* Asst prof zool, Miami Univ, 66-67; fel electron micros, Ore Health Sci Univ, 67-69, from asst scientist to assoc scientist, 69-78, scientist reproductive biol & behav, Ore Regional Primate Ctr, 78-86, dir andrology infertility serv, 75-86, prof anat & cell biol & obstet & gynec & urol, 80-86, prof microbiol & immunol, 84-86; prof obstet & gynec, Eastern Va Med Sch, 86-90, dir appl & basic res, Contraceptive Res & Develop Prog, 86-90; SPEC ASST, CONTRACEPTIVE DEVELOP PROG, NAT INST CHILD HEALTH & HUMAN DEVELOP, 90- *Concurrent Pos:* Mem animal resources adv comt, NIH, 73-77, sci adv comt, Prog Applied Res Fertility Regulation, 76-83, Nat Adv Res Rescources Coun, NIH, 80-84. *Mem:* Am Soc Cell Biol; Am Fertility Soc; Am Soc Andrology (treas, 75-77, vpres, 78, pres, 79); Soc Study Reproduction; AAAS. *Res:* Semen banking, infertility and immunoreproduction; immunological effects of vasectomy. *Mailing Add:* Dept CDP-NICHHD 6120 Executive Blvd Rm 420 Bethesda MD 20892

ALEXANDER, NANCY J, b Ithaca, NY, Jan 14, 47. GENETICS. *Educ:* State Univ NY, Oswego, BS, 68; Duke Univ, MA, 71, PhD(genetics), 77. *Prof Exp:* Res & teaching asst, Dept Bot, Duke Univ, 68-70, Dept Bot & Genetics, 70-73 & Dept Zool, 73-77; res assoc, Dept Genetics, Ohio State Univ, 77-80; MICROBIOLOGIST, AGR RES SERV, USDA, PEORIA, 80- *Mem:* Sigma Xi; Genetics Soc Am; Soc Indust Microbiol; Am Soc Microbiol; Am Phytopath Soc. *Res:* Molecular and physiological studies of yeast; fermentations; biomass conversion; xylose conversion to ethanol; biocontrol of weeds using fungi. *Mailing Add:* Nat Ctr Agr Utilization Res USDA 1815 N University St Peoria IL 61604

ALEXANDER, NATALIE, b Los Angeles, Calif, Aug 31, 26; m 50; c 2. NEURAL CONTROL CIRCULATION, HYPERTENSION. *Educ:* Univ Southern Calif, MS, 48, PhD(med physiol), 52. *Prof Exp:* From instr to asst prof physiol, 50-58, res assoc med physiol, 58-70, from asst prof med to assoc prof med, 70-82, PROF MED, SCH MED, UNIV SOUTHERN CALIF, 82- *Concurrent Pos:* USPHS career develop grant, 60-; mem staff, Los Angeles County Hosp, 55; mem nat adv comt, Nat Heart & Lung Inst, 73-75, 79, 81-84, CV & Renal Study Sect, 76-80. *Mem:* AAAS; Am Physiol Soc; Am Heart Asn; Soc Exp Biol Med; Sigma Xi. *Res:* Circulatory physiology and hypertension. *Mailing Add:* Clin Pharmacol Sect Sch Med HRB 801 Univ Southern Calif 2011 Zonal Ave Los Angeles CA 90033

ALEXANDER, NICHOLAS MICHAEL, b Boise, Idaho, June 30, 25; m 53; c 3. BIOCHEMISTRY. *Educ:* Univ Calif, AB, 50, PhD(biochem), 55. *Prof Exp:* Lectr biochem, Yale Univ, 56-64, from asst prof to assoc prof, Sch Med, 64-70; assoc adj prof, 70-74, PROF PATH, UNIV CALIF, SAN DIEGO, 74-, DIR CLIN CHEM, UNIV HOSP, 85- *Concurrent Pos:* Prin scientist biochem & assoc dir radioisotope serv, Vet Admin Hosp, 55-63. *Honors & Awards:* Van Meter Award, Am Thyroid Asn, 60. *Mem:* Am Chem Soc; Am Thyroid Asn; Am Soc Biol Chem; Endocrine Soc; Acad Clin Lab Physicians & Scientists; Am Asn Clin Chem. *Res:* Iodine and amino acid metabolism; thyroid hormone biosynthesis; fatty acid and protein metabolism; clinical chemistry, thyroid hormones and liver regeneration. *Mailing Add:* Univ Calif Med Ctr 225 Dickinson St San Diego CA 92103

ALEXANDER, PAUL MARION, b Akron, Ohio, Aug 21, 27; m 55; c 3. PLANT PATHOLOGY, HORTICULTURE. *Educ:* Calif State Polytech Col, BS, 53; Ohio State Univ, MSc, 55, PhD(bot, plant path), 58. *Prof Exp:* Asst plant pathologist, Clemson Univ, 58-66, from asst prof to assoc prof hort, 66-69; agronomist, Green Sect, US Golf Asn, 69-70; dir educ, Golf Course Supt Asn Am, 70-73; staff vpres agron, Sea Pines Co, 73-74; chief agronomist & turf mgr, Goltra, Inc, Winston-Salem, NC, 75-77; staff agronomist, Porter Bros, Inc, 77-79; vpres, Golf Operations, Sea Pines Co, 79-81; PRES, GRASS ROOTS, INC, 81-; INSTR, HORRY—GEORGETOWN TECH COL, 87- *Concurrent Pos:* Proj leader, SC Turfgrass Res Proj, 58-69; pres, PM's Corp, Clemson, 81-83; nat training dir & agronomist, Chemlawn Serv Corp, 82-87. *Mem:* Am Phytopath Soc; Soc Nematologists; Golf Course Supt Asn Am. *Res:* Diseases of turf grasses and ornamental plants; phytonematology; turf fungicides; weed control; soil amendments; irrigation practices; turf insects. *Mailing Add:* Grass Roots Inc 102 Timberline Dr Conway SC 29526

ALEXANDER, RALPH WILLIAM, b Schley, Ohio, Mar 14, 11; m 37; c 5. MEDICINE. *Educ:* Marietta Col, AB, 32; Univ Rochester, MD, 36. *Prof Exp:* Rotating intern, Jefferson Hosp, Philadelphia, Pa, 36-38; resident, William Pepper Lab Clin Med, Hosp Univ Pa, 38-39, student health serv, 39-40, staff physician, 40-46; from asst prof to assoc prof clin & prev med, Cornell Univ, 46-61, attend physician clin & infirmary, 53-61, dep dir & attend physician, Gannett Clin & Sage Infirm, 61-69, actg dir, Dept Univ Health Serv, 69-71, prof clin med, Dept Univ Health Serv, 61-77, dep dir dept, 71-77; RETIRED. *Concurrent Pos:* Ed, Student Med, 52-62 & J Am Col Health Asn, 62-73; mem, Am Bd Internal Med. *Honors & Awards:* Ruth Boynton Award, Am Col Health Asn, 70, Edward Hitchcock Award, 73. *Mem:* AMA; Am Col Physicians; fel Am Col Health Asn. *Res:* Internal medicine; university students health. *Mailing Add:* Bldg 5 Apt 7 Quail's Run Blvd Englewood FL 34223

ALEXANDER, RALPH WILLIAM, JR, b Philadelphia, Pa, May 17, 41; m 65; c 2. SOLID STATE PHYSICS, SPECTROSCOPY. *Educ:* Wesleyan Univ, BA, 63; Cornell Univ, PhD(physics), 68. *Prof Exp:* Ger Res Asn fel, Univ Freiburg, 68-69; fel, 70, from asst prof to assoc prof, 70-80, PROF PHYSICS, UNIV MO-ROLLA, 80-, CHMN, 84- *Mem:* Am Phys Soc; Am Asn Physics Teachers. *Res:* Spectroscopy of solid surfaces; millimeter and submillimeter spectroscopy. *Mailing Add:* Physics Dept Univ Missouri-Rolla Rolla MO 65401

ALEXANDER, RENEE R, b Leipzig, Ger, Jan 23, 32; US citizen; m 51; c 2. MOLECULAR BIOLOGY, BIOCHEMISTRY. *Educ:* Univ Wis, BS, 54, MS, 55; Cornell Univ, PhD(microbiol), 58. *Prof Exp:* Asst bact, Univ Wis, 54-55; asst bact, Cornell Univ, 55-58, res assoc microbiol, 58, phys biol, 62-65 & biochem & molecular biol, 65-70; asst prof genetics, State Univ NY Col Cortland, 70-71; lectr, 71-81, SR LECTR BIOCHEM, CORNELL UNIV, 81- *Mem:* Am Soc Microbiol; Sigma Xi. *Res:* Genetic studies of Escherichia coli; Enzyme deficiencies related to genetic disorders; chromosome mapping and study of regulation of leucine biosynthesis in Salmonella typhimurium; regulation and catalytic properties of yeast invertase. *Mailing Add:* 301 Winthrop Dr Ithaca NY 14850

ALEXANDER, RICHARD DALE, b White Heath, Ill, Nov 18, 29; m 50; c 2. ZOOLOGY. *Educ:* Ill State Univ, BSc, 50; Ohio State Univ, MSc, 51, PhD(entom), 56. *Hon Degrees:* LHD, Ill State Univ. *Prof Exp:* Res assoc, Rockefeller Found, 56-57; from instr to assoc prof, 57-69, PROF ZOOL, UNIV MICH, ANN ARBOR, 69-, CUR INSECTS, MUS ZOOL, 57-, HUBBELL DISTINGUISHED UNIV PROF EVOLUTIONARY BIOL, 90- *Concurrent Pos:* J S Guggeheim, 68-69. *Honors & Awards:* Daniel Giraud Elliot Medal, 71; Newcomb Cleveland Prize, AAAS, 61. *Mem:* Nat Acad Sci; Am Soc Naturalists; fel AAAS; Animal Behav Soc. *Res:* Acoustical communication in insects; systematics of Orthoptera and Cicadidae; evolution and human behavior; evolution of insect and social behavior. *Mailing Add:* Mus Zool Univ Mich Ann Arbor MI 48109

ALEXANDER, RICHARD RAYMOND, b Covington, Ky, Feb 2, 46; m 72. PALEOECOLOGY. *Educ:* Univ Cincinnati, BS, 68; Ind Univ, MA, 70, PhD(geol), 72. *Prof Exp:* From asst prof to assoc prof geol, Utah State Univ, 72-81; assoc prof, 81-86, PROF GEOL, RIDER COL, 86-, DEPT CHMN, 83- *Concurrent Pos:* Chmn, northeast sect, Paleont Soc, 88. *Mem:* Soc Econ Paleont & Mineral; Paleont Soc; Paleont Asn; Sigma Xi; Int Paleont Union. *Res:* Paleoautecology of brachiopods and bivalves, particularly demographic and morphologic adaptations to sedimentologic influences and intraspecific-interspecific competition, and secondly, the relationship of functional morphology to generic longevity. *Mailing Add:* Dept Geosci Rider Col PO Box 6400 Lawrenceville NJ 08648

ALEXANDER, ROBERT ALLEN, animal husbandry; deceased, see previous edition for last biography

ALEXANDER, ROBERT BENJAMIN, physical chemistry; deceased, see previous edition for last biography

ALEXANDER, ROBERT L, b London, Eng, Feb 9, 23; US citizen; m 57; c 2. CIVIL ENGINEERING. *Educ:* Rensselaer Polytech Inst, BArch, 49; Harvard Univ, MSc, 56; Univ Calif, DEng(transp), 64. *Prof Exp:* Architect, Stephenson & Turner, Australia, 49-51; archit engr, Western Elec Co, NY, 51-53; struct engr, Skidmore, Owings & Merrill, Okinawa, 53-54; found engr, Koppers Co, Pa, 56-57; asst prof civil eng, Univ Conn, 57-60; assoc prof, 64-67, PROF CIVIL ENG, CALIF STATE UNIV, LONG BEACH, 67- *Concurrent Pos:* Assoc mem hwy res bd, Nat Acad Sci-Nat Res Coun, 64- *Mem:* Am Soc Civil Engrs. *Res:* Architectural sciences; transportation engineering; phenomenological viscoelasticity. *Mailing Add:* Dept Civil Eng Calif State Univ 1250 Bellflower Blvd Long Beach CA 90840

ALEXANDER, ROBERT SPENCE, b Melrose, Mass, June 14, 17; m 42; c 5. PHYSIOLOGY. *Educ:* Amherst Col, AB, 38, MA, 40; Princeton Univ, PhD(gen physiol), 42. *Hon Degrees:* DSc, Albany Med Col, 82. *Prof Exp:* Asst physiol, Princeton Univ, 40-42; instr, Med Col, Cornell Univ, 42-45; sr instr, Sch Med, Western Reserve Univ, 45-47, from asst prof to assoc prof, 47-53; assoc prof, Med Col, Univ Ga, 53-55; chmn dept, 55-73, prof physiol, 55-79, EMER PROF PHYSIOL, ALBANY MED COL, 79-, EMER PROF PHYSIOL. *Concurrent Pos:* Res consult, Vet Admin Hosp, 48-52 & Off Naval Res, 65-81. *Mem:* AAAS; Am Physiol Soc; Harvey Soc. *Res:* Dynamics of arterial pulses; regulation of venous system; smooth muscle tone. *Mailing Add:* 69 Murray Ave Delmar NY 12054

ALEXANDER, ROGER KEITH, b Wichita, Kans, Dec 17, 46; m 68; c 2. NUMERICAL ANALYSIS, NONLINEAR ANALYSIS. *Educ:* Univ Kans, BA, 68; Univ Calif, Berkeley, MA, 74, PhD(math), 75. *Prof Exp:* Asst prof math, Univ Colo, Boulder, 75-77; asst prof math sci, Rennselaer Polytech Inst, 77-; ASSOC PROF MATH SCI, IOWA STATE UNIV. *Mem:* Math Asn Am; Am Math Soc; Soc Indust & Appl Math. *Res:* Analysis of methods and development of numerical software for stiff differential equations; analysis of nonlinear differential equations, especially nonlinear eigenvalve problems and free boundary problems. *Mailing Add:* Iowa State Univ Ames IA 50011

ALEXANDER, SAMUEL CRAIGHEAD, JR, b Upper Darby, Pa, May 3, 30; m 51; c 3. ANESTHESIOLOGY, PHARMACOLOGY. *Educ:* Davidson Col, BS, 51; Univ Pa, MD, 55. *Prof Exp:* Intern, Philadelphia Gen Hosp, 55-56; sr asst surgeon, USPHS, 56-58; instr pharmacol, Univ Pa, 58-60, instr anesthesiol, 60-63, assoc, 63-65, asst prof, 65-70; prof & head dept, Univ Conn, 70-72; chmn dept, 72-77, PROF ANESTHESIOL, MED SCH, UNIV WIS-MADISON, 72- *Concurrent Pos:* Pharmaceut Mfrs fel clin pharmacol, 57-58; USPHS career develop award, 65-70; consult, Philadelphia Vet Admin Hosp, 64-68 & Madison Vet Admin Hosp, 72-; vis scientist clin physiol, Bispebjerg Hosp, Copenhagen, Denmark, 68-69. *Mem:* Am Soc Pharmacol & Exp Therapeut; Am Soc Anesthesiol; Asn Univ Anesthetists; Sigma Xi. *Res:* Respiratory control; effects of anesthetics on brain metabolism. *Mailing Add:* B6-387 Clin Sci Ctr 600 Highland Ave Madison WI 53792

ALEXANDER, STEPHEN, b Chicago, Ill, Sept 28, 48; m 76; c 3. INTERCELLULAR INTERACTIONS, REGULATION OF DEVELOPMENTAL TIMING. *Educ:* Univ Ill, Urbana, BS, 71; Brandeis Univ, PhD(biol), 76. *Prof Exp:* Res fel biochem, Dana-Farber Cancer Inst, Harvard Med Sch, 76-80; res assoc develop biol, dept immunol, 80-83, asst mem develop biol, Dept Molecular Biol, Res Inst Scripps Clin, La Jolla Calif, 83-87; ASSOC PROF, DIV BIOL SCIS, UNIV MO, COLUMBIA, 88- *Concurrent Pos:* NIH Cell Biol Study Sect, 88-92. *Mem:* Am Soc Microbiol; Soc Develop Biol. *Res:* Molecular and genetic basis of developmentally regulated gene expression; intercellular interactions during multicellular morphogenesis; control of developmental timing. *Mailing Add:* Div Biol Scis Tucker Hall Univ Mo Columbia MO 65211

ALEXANDER, STUART DAVID, b Brooklyn, NY, Nov 18, 38; m 62; c 2. PULP & PAPER TECHNOLOGY, PROCESS CONTROL. *Educ:* Cornell Univ, BChE, 60; State Univ NY Col Environ Sci & Forestry, PhD(pulp & paper technol), 66. *Prof Exp:* Tech serv engr, Papermill, Westvaco Corp, NY, 60-62; sr develop engr, St Regis Paper Co, 66-74, staff specialist-papermaking field serv, Tech Ctr, 74-77; res sect head, 77-83, RES & DEVELOP MGR, SCOTT PAPER CO, 83- *Honors & Awards:* Res & Develop Div Leadership & Serv, Tech Asn Pulp & Paper Indust, 86. *Mem:* Tech Asn Pulp & Paper Indust; Am Chem Soc. *Res:* Paper grade development based on paper and fiber rheology and chemistry; paper making process control; papermaking chemistry. *Mailing Add:* Scott Paper Co Scott Plaza Philadelphia PA 19113

ALEXANDER, TAYLOR RICHARD, b Hope, Ark, May 27, 15; m 40; c 2. BOTANY. *Educ:* Ouachita Col, AB, 36; Univ Chicago, MS, 38, PhD(physiol), 41. *Prof Exp:* From instr to assoc prof, Univ Miami, 40-46, chmn dept, 47-65, prof bot, 47-77; RETIRED. *Mem:* Am Inst Biol Sci. *Res:* Plant nutrition in subtropical conditions; minor element deficiencies and subtropical ecology. *Mailing Add:* 6900 SW 73rd Ct Miami FL 33143

ALEXANDER, THOMAS GOODWIN, b Washington, DC, Sept 6, 28; m 51; c 3. PHARMACEUTICAL CHEMISTRY. *Educ:* Univ Md, BS, 50; George Washington Univ, MS, 57. *Prof Exp:* Chem analyst, NY State Dept Agr, Cornell Univ, 50; anal chemist dairy prod lab, USDA, DC, 52-56; res analysis chemist, Div Pharmaceut Chem, Food & Drug Admin, 56-67, supvry chemist, Div Pharmaceut Sci, 67-70, supvry chemist, Nat Ctr Antibiotic Analysis, 70-82, supvry chemist, Div Drug Biol, 82-89; RETIRED. *Concurrent Pos:* Res chemist, Food & Drug Admin, 90-91. *Mem:* Am Chem Soc; fel Asn Off Analytical Chem (secy-treas, 86-88). *Res:* Elemental analysis milk and milk products; analysis of ergot alkaloids; chromatographic methods; development of nuclear magnetic resonance methods for pharmaceuticals; analysis of particulate matter in parenterals; interfacing computers and analytical instruments. *Mailing Add:* 16716 Huron St Accokeek MD 20607-9710

ALEXANDER, THOMAS KENNEDY, b Vancouver, BC, Mar 6, 31; m 55; c 3. NUCLEAR PHYSICS. *Educ:* Univ BC, BA, 53, MSc, 55; Univ Alta, PhD(physics), 64. *Prof Exp:* Res officer reactor physics, Atomic Energy Can Ltd, 55-56, electronics, 56-61, res officer nuclear physics, 64-91; RETIRED. *Mem:* Can Asn Physicists; fel Am Phys Soc. *Res:* Nuclear spectroscopy; measurements of lifetimes, moments and reactions with heavy ions. *Mailing Add:* PO Box 1634 Deep River ON K0J 1P0 Can

ALEXANDER, VERA, b Budapest, Hungary, Oct 26, 32; m 53, 67; c 2. AQUATIC ECOLOGY. *Educ:* Univ Wis, BA, 55, MS, 62; Univ Alaska, PhD(marine sci), 65. *Prof Exp:* Sr res asst, 63-65, from asst prof to assoc prof, 64-74, actg dean, Col Environ Sci, 77-79, dean, 82-84, PROF MARINE SCI, UNIV ALASKA, FAIRBANKS, 74-, DIR, INST MARINE SCI, 80-, DEAN, SCH FISHERIES & OCEAN SCI, 79- *Concurrent Pos:* Mem oceanog panel, NSF, 75-76; chmn, Alpha Helix Rev Comt, Univ Nat Oceanog Lab Syst, 75-78; mem, Ocean Sci Bd, Nat Res Coun, 76-79 & Polar Res Bd; adv comt, Ocean Sci, NSF, 80-84, chair, 83-84, vchmn, Arctic Ocean Sci Bd, 87-88; actg dean, Sch Fisheries & Ocean Sci, Univ Alaska, Fairbanks, 87-89. *Mem:* AAAS; Sigma Xi; Am Soc Limnol & Oceanog; Explorer's Club. *Res:* Primary production and nitrogen cycle processes in marine and freshwater systems. *Mailing Add:* Inst of Marine Sci Univ of Alaska Fairbanks AK 99775-1080

ALEXANDER, WILLIAM CARTER, b Wichita Falls, Tex, June 19, 37; m 76; c 3. CARDIOVASCULAR CONTROL SYSTEMS, RESEARCH & DEVELOPMENT MANAGEMENT. *Educ:* NTex State Col, Denton, BA, 60; NTex State Univ, Denton, MA, 62; Bowman Gray Sch Med, Winston-Salem, NC, 69, PhD(physiol), 69. *Prof Exp:* Aerospace technologist physiol, NASA Johnson Space Ctr, Tex, 62-79, Kennedy Space Ctr, Fla, 79-82; dir crew technol physiol, USAF Sch Aerospace Med, 82-91; DIR PLANS & PROG PHYSIOL, H G ARMSTRONG LAB, 91- *Concurrent Pos:* Adj prof, Univ Tex San Antonio, 83-; consult, Fulbright-Jaworski Atty at Law, 85-; exec coun, Aerospace Med Asn, 87-90; distinguished vis prof, USAF Acad, Colo, 90-91. *Mem:* Fel Aerospace Med Asn (vpres 90-91); Am Physiol Soc. *Res:* Integration of man into weapon system or civilian spacecraft; cardiovascular control of fluid and electrolyte balance in hypo and hypergravic environments; development of manual and/or physiologic countermeasures for crew protection. *Mailing Add:* Hq H G Armstrong Lab XP Brooks AFB TX 78235

ALEXANDER, WILLIAM DAVIDSON, III, b Charlotte, NC, June 20, 11; m 36; c 1. CIVIL ENGINEERING. *Educ:* Va Mil Inst, BS, 34; NC State Univ, CE, 53. *Prof Exp:* Engr, High Point, NC, 34-40; partner, Eng Serv, Idaho, 45-46; proj mgr, Urbahn-Roberts-Seeyle-Moran, New York, 62-63; from vpres to vchmn bd dir, Seelye Stevenson Value & Knecht, New York, 63-75; asst gen mgr transit syst develop, Metrop Atlanta Regional Transit Authority, 75-79; CONSULT, 79- *Concurrent Pos:* Mem, Eng Found, New York, 63-75, chmn, 66-75. *Mem:* Nat Acad Eng; fel Am Soc Civil Engrs. *Mailing Add:* PO Box 770 Pawleys Island SC 29585

ALEXANDER, WILLIAM NEBEL, b Pittsburgh, Pa, May 22, 29; m 58; c 5. ORAL MEDICINE, ORAL PATHOLOGY. *Educ:* Univ Pittsburgh, DDS, 53; Jackson State Univ, MS, 82. *Prof Exp:* Assoc prof, 78-81, actg chmn, 81-82, prof community & oral health & dir patient admis, 82-86, vchmn, Dept Diag Sci, 86-87, actg chmn, 87-88, CHMN, DEPT DIAG SCI, UNIV MISS MED CTR, 88- *Mem:* Sigma Xi; AAAS; Am Dent Asn. *Res:* Oral medicine; oral pathology; geriatric dentistry. *Mailing Add:* Dept Diagnostic Sci Univ Miss 2250 N State St Jackson MS 39216

ALEXANDER-JACKSON, ELEANOR GERTRUDE, bacteriology; deceased, see previous edition for last biography

ALEXANDERSON, GERALD LEE, b Caldwell, Idaho, Nov 13, 33. MATHEMATICS. *Educ:* Univ Ore, BA, 55; Stanford Univ, MS, 58. *Prof Exp:* From instr to prof math, 58-79, mem, bd trustees, 79-86, dir, div math & natural sci, 81-90, vchmn, 84-86, CHAIR, MATH, SANTA CLARA UNIV, 67-, MICHAEL & ELIZABETH VALERIOTE PROF MATH, 79- *Concurrent Pos:* Assoc dir, William Lowell Putman Math Competition, 75-; assoc ed, Col Math J, 79-84 & Am Math Monthly, 83-86; ed, Math Mag, 86-90. *Mem:* Am Math Soc; Math Asn Am (vpres, 84-86, secy, 90-). *Res:* Combinatorial geometry; theory of partitions. *Mailing Add:* Dept Math Santa Clara Univ Santa Clara CA 95053

ALEXANDRATOS, SPIRO, b New York, NY, Dec 11, 51. ORGANIC CHEMISTRY, POLYMER CHEMISTRY. *Educ:* Manhattan Col, BS, 73; Univ Calif, Berkeley, PhD(chem), 77. *Prof Exp:* Sr chemist, Rohm and Haas Co, 77-81; asst prof, 81-87, ASSOC PROF CHEM, UNIV TENN, KNOXVILLE, 87- *Concurrent Pos:* Consult, Oak Ridge Nat Lab, 82-, Argonne Nat Lab, 84-; chmn, Separations Sci Subdiv, Am Chem Soc, 91. *Mem:* Sigma Xi; Am Chem Soc. *Res:* Polymers; ion-exchange resins; polymeric catalysts; novel uses of membranes, physical organic chemistry; quantum chemistry. *Mailing Add:* Chem Dept Univ Tenn Knoxville TN 37996-1600

ALEXANDRIDIS, ALEXANDER A, b Patras, Greece, Jan 17, 49; m 77; c 2. VEHICLE DYNAMICS & CONTROL THEORY. *Educ:* Princeton Univ, BSE, 71, MA, 74, PhD(aerospace & mech eng), 77. *Prof Exp:* Tech consult, Aereon Corp, 70-76; res & teaching asst math & internal combustion eng, Princeton Univ, 72-76; SR STAFF RES ENGR, GEN MOTORS RES LABS, 76- *Concurrent Pos:* Tech consult, Dynalysis Princeton, 74-76. *Mem:* Soc Automotive Engrs; Inst Elec & Electronics Engrs. *Res:* Vehicle dynamics, stability and control; powertrain dynamics and control; control systems design and optimization; active vehicle vibration isolation (active suspensions design); computer-aided design of optimal control systems; human factors. *Mailing Add:* Power Syst Res Dept Gen Motors Res Labs Warren MI 48090-9055

ALEXANDRU, LUPU, b Vaslui, Romania, Apr 6, 23; Can citizen; m 49; c 1. POLYMER SCIENCE & TECHNOLOGY. *Educ:* Bucharest Polytech Inst, Engr, 49; Moscow Univ, PhD(polymer chem), 53. *Prof Exp:* Assoc prof synthetic fibers, Polytech Inst Jassy-Romania, 53-56; prin scientist & res dept mgr polymer sci, Chem Res Inst, Bucharest, 56-72; res assoc, Univ Toronto, 73-75; PRIN POLYMER SCIENTIST, XEROX RES CTR CAN, 75-, MGR POLYMER SYNTHESIS LAB, 83- *Mem:* Am Chem Soc; Chem Inst Can; Can Soc Chem Eng; Asn Prof Engrs. *Res:* Polymer synthesis, characterization and processing, free radical and ionic polymerization; polycondensation;

graft, block, charge transfer polymers; molecular, thermal, electrical properties of polymers; films, fibers, coatings, information storage devices, photoreceptors, xerographic toners. *Mailing Add:* Xerox Res Ctr Can LTD 2660 Speakman Dr Mississauga ON L5K 2L1 Can

ALEXANIAN, VAZKEN ARSEN, b Aleppo, Syria, Nov 6, 43; US citizen. ORGANIC CHEMISTRY, PATENT LAW. *Educ:* Am Univ Beirut, BS, 68, MS, 70; Univ Ala, PhD(chem), 74. *Prof Exp:* Res assoc organic chem, Univ Colo, 74-75 & State Univ NY, Binghamton, 75-78; res chemist synthesis & process develop, Am Cyanamid Co, 78-80, sr res chemist, 78-88; CONSULT, PATENT LAW, 88- *Mem:* Am Chem Soc. *Res:* Intermolecular and intramolecular reactions of sulfonyl carbenes, pyridinium ylides, azirines, heterocyclic chemistry of sulfur, nitrogen and phosphorus, plastics additives, urethane chemistry and coatings; polymer chemistry; chemistry of crosslinking; process chemistry. *Mailing Add:* Am Cyanamid Co 1937 W Main St Stamford CT 06904-0060

ALEXEFF, IGOR, b Pittsburgh, Pa, Jan 5, 31; m 54; c 2. NUCLEAR PHYSICS. *Educ:* Harvard Univ, BA, 52; Univ Wis, MS, 55, PhD(physics), 59. *Prof Exp:* Res engr nuclear physics, Westinghouse Res Labs, 52-53; NSF fel, Univ Zurich, 59-60; group leader turbulent heating, basic physics & levitated multipole, Oak Ridge Nat Lab, 60-71; PROF ELEC ENG, UNIV TENN, KNOXVILLE, 67- *Concurrent Pos:* Consult, Oak Ridge Nat Lab & Space Inst, Tenn; chmn, Gordon Res Conf, 74; organizer, First Int Inst Elec & Electronic Engrs Conf Plasma Sci, Tenn, 74; vis prof, Physics Dept, Univ Fluminense, Rio de Janeiro, Phys Res Lab, Ahmedabad, India, Inst Plasma Physics, Nagoya Univ, Japan & Univ Natal, Durban, SAfrica; chmn, Plasma Div, Nuclear & Plasma Sci; Chancellor's Res Scholar, 84- *Honors & Awards:* Centennial Medal, Inst Elec & Electronic Engrs, 84; R&D 100 Award, 89. *Mem:* Fel Inst Elec & Electronic Engrs; fel Am Phys Soc. *Res:* Experimental plasma physics; controlled thermonuclear fusion; low energy nuclear physics; ultra-high vacuum techniques; ultra-high frequency microwave production; author of over 100 publications and one book. *Mailing Add:* Dept Elec Eng Univ Tenn Knoxville TN 37996-2100

ALEXIOU, ARTHUR GEORGE, b Manchester, NH, Feb 26, 30. OCEAN CLIMATE, MARINE SCIENCE. *Educ:* Univ NH, BS, 51, MS, 64. *Prof Exp:* Meteorologist, USAF, 51-53; dep dir instrumentation, US Naval Oceanog Off, 53-65, proj mgr spacecraft oceanog, Joint USN-Nat Aeronaut & Space Admin Proj, 65-67; dir inst progs, NSF, 67-70; assoc dir progs, Off Sea Grant, Nat Oceanic & Atmospheric Admin, 70-85; ASST SECY, COMT CLIMATE CHANGES IN THE OCEAN, IOC, UNESCO, 85- *Concurrent Pos:* Res assoc, Scripps Inst Oceanog, Univ Calif, 77-78. *Honors & Awards:* Silver Medal, Dept Com, 73; Meritorious Serv Award, Marine Technol Soc, 78. *Mem:* Am Geophys Union. *Res:* Ocean climate research , ocean engineering, including instrumentation, remote sensing, and coastal processes. *Mailing Add:* Amer Embassy Paris AD2 APO New York NY 09777

ALFANO, MICHAEL CHARLES, b Newark, NJ, Aug 8, 47; m 69; c 2. PERIODONTICS, DENTAL PRODUCT DEVELOPMENT. *Educ:* Univ Med & Dent NJ, DMD, 71; Mass Inst Technol, PhD(nutrit), 75. *Prof Exp:* From asst prof to prof periodont, Sch Dent, Fairleigh Dickinson Univ, 74-82, dir, Oral Health Res Ctr, 77-82, asst dean res & grad affairs, 81-82; vpres dent res, 82-85, pres prof dent prod, 85-88, SR VPRES RES TECHNOL & CORP OFF, BLOCK DRUG CO, INC, 87- *Concurrent Pos:* Consult dent, Inst Nutrit Cent Am & Panama, 72, Nat Inst Dent Res, NIH, 76-82, Am Dent Asn, 79-82, Int Life Sci Inst, 80-82 & numerous food & pharmaceut co; instr, Sch Dent Med, Tufts Univ, 73-74; Nat Inst Dent Res Indust & Found grants, Fairleigh Dickinson Univ, 74-; prin investr numerous grants, NIH & indust, 74-82; vis prof, Nat Dairy Coun, 81; mem, Sci Adv Coun, State of NJ, 81-84; chmn, prog comt, Found Univ Med & Dent NJ, 90. *Mem:* Am Dent Asn; Int Asn Dent Res; Am Acad Periodont; Am Inst Nutrit; fel Am Col Dentists; Am Asn Dent Res; Am Col Prosthodontics; Am Asn Oral Biologists. *Res:* Nutritional and chemotherapeutic aspects of periodontal disease; pathogenesis of dental caries and periodontal disease; nutritional and pharmacologic modulation of epithelial permeability; development of new dental restorative materials and devices. *Mailing Add:* Block Drug Co 257 Cornelison Ave Jersey City NJ 07302

ALFANO, ROBERT R, b New York, NY, May 5, 41. BIOMEDICAL OPTICS, MEDI PHOTONICS. *Educ:* Fairleigh Dickinson Univ, BS, 63, MS, 64; NY Univ, PhD(physics), 72. *Prof Exp:* Physicist lasers, Gen Tel & Electronics, 64-72; DISTINGUISHED PROF SCI & ENG, CITY COL NY, 72- *Concurrent Pos:* Sloan fel, 75. *Mem:* Fel Am Phys Soc. *Res:* Picosecond laser spectroscopy, in solids, liquids and biological materials, medical materials. *Mailing Add:* 3777 Independence Ave Bronx NY 10463

ALFELD, PETER, b Feb 20, 50; m; c 3. MATHEMATICS. *Educ:* Univ Hamburg, WGer, Vordiplom, 74; Univ Dundee, Scotland, MS, 75, PhD, 77. *Prof Exp:* Instr Math, 77-79, from asst prof assoc prof 79-87, PROF MATH, UNIV UTAH, SALT LAKE CITY, 87- *Concurrent Pos:* Vis assoc prof, math, Math Res Ctr, Univ Wis-Madison, 83-84. *Mem:* Am Math Soc; Asn Comput Mach; AAAS. *Mailing Add:* Univ Utah Salt Lake City UT 84112

ALFERT, MAX, b Vienna, Austria, Apr 23, 21; nat US; m 45; c 1. CYTOCHEMISTRY. *Educ:* Wagner Col, BS, 47; Columbia Univ, MA, 48, PhD(zool), 51. *Prof Exp:* From instr to prof, 50-90, EMER PROF ZOOL, UNIV CALIF, BERKELEY, 90- *Concurrent Pos:* Guggenheim fel, 56. *Mem:* Int Soc Cell Biol; Am Soc Cell Biol; Europ Acad Sci, Art & Letters. *Res:* Quantitative cytochemical studies of nuclear composition; development of cytochemical techniques. *Mailing Add:* Dept Integrative Biol Univ Calif Berkeley CA 94720

ALFIERI, CHARLES C, b Groton, Conn, July 19, 22; m 53; c 2. ORGANIC CHEMISTRY, INORGANIC CHEMISTRY. *Educ:* Brown Univ, ScB, 44; Purdue Univ, MS, 48, PhD(org chem), 50. *Prof Exp:* Jr chemist, Shell Oil Co, 43-44; asst chemist, Ansco Div, Gen Aniline & Film Corp, 44, 46-47; sr res chemist, Cent Res Labs, US Rubber Co, 50-52, group leader explosives & agr

chem res, Naugatuck Chem Div, 52-58; head chem res sect, Elkton Div, Thiokol Chem Corp, 58-71; counr, Tech Serv Div, Univ Del, 71-80; RETIRED. *Concurrent Pos:* Consult chem, 81- *Mem:* Am Chem Soc. *Res:* Use of high energy chemicals for solid rocket propellants and high explosives; synthesis and application of new liquid polymers. *Mailing Add:* 24 Chestnut Hill Sq Groton CT 06340

ALFIERI, GAETANO T, b Brooklyn, NY, Feb 7, 26; c 1. PHYSICAL CHEMISTRY. *Educ:* Fordham Univ, BS, 50; NY Univ, MS, 55; Polytech Inst Brooklyn, PhD(chem), 69. *Prof Exp:* From instr to assoc prof, 53-69, chmn dept, 70-79, PROF CHEM, NEW YORK CITY COMMUNITY COL, 69- *Mem:* Am Chem Soc. *Res:* High pressure chemistry; pressure dependence of magnetic phase transitions. *Mailing Add:* Five Putnam Ave Jerico NY 11753

ALFIN-SLATER, ROSLYN BERNIECE, b New York, NY, July 28, 16; m 48. BIOCHEMISTRY. *Educ:* Brooklyn Col, AB, 36; Columbia Univ, AM, 42, PhD(biochem), 46. *Prof Exp:* Lectr asst, Chem Dept, Brooklyn Col, 38-43, tutor, 43; asst instr, Columbia Univ, 43-45; res chemist enzymes, Takamine Labs, 46-47; res fel cancer, Sloan-Kettering Inst, 47-48; res assoc, Dept Biochem & Nutrit, Med Sch, Univ Southern Calif, 48-52, vis asst prof, 52-56, vis assoc prof, 56-59; assoc prof nutrit, Univ Calif, Los Angeles, 59-65, head, Div Environ & Nutrit Sci, 69-77, prof nutrit, Sch Pub Health, prof biol chem, sch med, 71-87, asst dean, acad affairs, 83-87, EMER PROF, UNIV CALIF LOS ANGELES, 87- *Concurrent Pos:* Mem US nat comt, Int Union Nutrit Sci, Nat Res Coun, 74-80, vchmn, 75-78, chmn comt biol eval of fat, 78-; mem expert panel food safety & nutrit, Inst Food Technol, 80-83; mem, Coun Arteriosclerosis, Am Heart Asn. *Honors & Awards:* Osborne Mendel Award, Am Inst Nutrit, 70, Borden Award, 81; Bond Award, Am Oil Chemists' Soc, 74. *Mem:* Fel AAAS; Am Pub Health Asn; Am Soc Biol Chemists; fel Am Inst Nutrit (treas, 77-80). *Res:* Biochemical and nutritional aspects of lipid metabolism; cholesterol and essential fatty acids; saturated and unsaturated fats; carbohydrate-lipid interrelationships; essential fatty acids and vitamin E; experimental atherosclerosis; studies with aflatoxin; nutrition and aging. *Mailing Add:* 986 Somera Rd Los Angeles CA 90077-2624

ALFORD, BETTY BOHON, b St Louis, Mo, June 9, 32; m 58; c 2. NUTRITION. *Educ:* Tex Woman's Univ, BS, 54, MA, 56, PhD(nutrit), 65. *Prof Exp:* Instr nutrit & home econ, Baylor Univ, 56-57; instr foods & nutrit, Tex Woman's Univ, 58-63, res assoc nutrit, 63-66, from asst prof to assoc prof nutrit, 66-78, chmn nutrit & food sci, 75-77, dean, Col Nutrit, Textile & Human Develop, 78-88, PROF NUTRIT, TEX WOMAN'S UNIV, 78- *Concurrent Pos:* Consumer adv, Cotton Bd, 85-; extramural assoc, Nat Inst Health, 88; vis prof, Div Nutrit Res Coord, NIH, 89. *Mem:* Am Dietetic Asn; Am Inst Nutrit; Am Soc Clin Nutrit. *Res:* Nutrition of elderly, assessment and requirement; cottonseed and plant protein for human needs; effectiveness of nutrition education for low income families; nutrition during the life cycle; obesity. *Mailing Add:* Box 23564 Tex Woman's Univ Denton TX 76204

ALFORD, BOBBY R, b Dallas, Tex, May 30, 32; m 53; c 3. OTOLARYNGOLOGY. *Educ:* Baylor Col Med, MD, 56. *Prof Exp:* From asst prof to assoc prof, Baylor Col Med, 62-66, PROF OTOLARYNGOL, 66-, CHMN DEPT, 67-, DISTINGUISHED SERV PROF, BAYLOR COL MED, 85- *Concurrent Pos:* NIH fel, Sch Med, Johns Hopkins Univ, 61-62; consult to Surgeon Gen, US Army, 64-70 & USPHS, 65-68 & 70-; chief ed, Arch Otolaryngol, AMA, 70-80; mem, Nat Adv Neurol & Commun Disorders & Stroke Coun, NIH, 77-80. *Honors & Awards:* Herman Johnson Award, Baylor Col Med, 56. *Mem:* Am Otol Soc; Am Laryngol, Rhinol & Otol Soc; Soc Univ Otolaryngol (secy, 65-69); Am Col Surg; Am Acad Head & Neck Surg & Otolaryngol. *Res:* Otology; otophysiology; otopathology. *Mailing Add:* Exec Vpres & Dean Med Baylor Col of Med One Baylor Plaza Houston TX 77030

ALFORD, CECIL ORIE, b Gay, Ga, Sept 28, 33; m 59; c 3. ELECTRICAL ENGINEERING, COMPUTER SCIENCE. *Educ:* Ga Inst Technol, BEE, 56, MSEE, 57; Miss State Univ, PhD(elec eng), 66. *Prof Exp:* Tech asst, Western Elec Co, 52-56; engr, Radiation, Inc, 57-59, res engr, Res Div, 60-61; sr engr, Martin Co, 61-63; asst prof elec eng, Miss State Univ, 63-66; assoc prof, Tenn Technol Univ, 66-68; assoc prof, 68-77, PROF ELEC ENG, GA INST TECHNOL, 77- *Concurrent Pos:* Res engr, Res Sta, Ga Inst Technol, 56, Aerospace Corp, Patrick Air Force Base, Fla, 64, Boeing Co, Ala, 65, ARO, Inc, Tenn, 66, Oak Ridge Assoc Univs, 67 & Western Elec, Atlanta, 69. *Mem:* Inst Elec & Electronics Engrs. *Res:* Computer architecture and organization; microprogramming; real-time computer control; robotics. *Mailing Add:* Sch Elec Eng Ga Inst Technol Atlanta GA 30332

ALFORD, CHARLES AARON, JR, b Birmingham, Ala, Dec 8, 28; m 62; c 2. VIROLOGY. *Educ:* Univ Ala, BS, 51; Med Col Ala, MD, 55. *Prof Exp:* From intern to chief resident pediat, Med Col Ala, 55-58; pediatrician in chief, Sta Naval Hosp, Sasebo, Japan, 58-60; instr pediat, Med Col Ala, 60-62; Nat Inst Allergy & Infectious Dis spec fel virol, Dept Trop Pub Health, Sch Pub Health, Harvard Univ & res fel med, Children's Hosp Med Ctr, Boston, 62-65; from asst prof to assoc prof pediat, 65-67, MEYER PROF PEDIAT, SCH MED, UNIV ALA, BIRMINGHAM, 67-, ASSOC PROF CLIN PATH, 69-, PROF MICROBIOL & SR SCIENTIST CANCER RES TRAINING PROG, 73- *Concurrent Pos:* Mem basic res adv comt, March of Dimes Birth Found, 76- & Nat Adv Child Health & Human Develop Coun, NIH, 79-83. *Honors & Awards:* P R Edwards Award, Am Soc Microbiol, 77. *Mem:* Soc Pediat Res; Am Fedn Clin Res; Infectious Dis Soc Am; Am Pediat Soc; Am Asn Immunologist; Am Acad Microbiol. *Res:* Pediatric virology and immunology in the study of congenital infections. *Mailing Add:* Dept Pediat Suite 752 Univ Ala Sch Med Childrens Hosp Tower Birmingham AL 35294

ALFORD, DONALD KAY, b Long Beach, Calif, Dec 30, 36; m 67; c 2. PLANT PHYSIOLOGY, HORTICULTURE. *Educ:* Whittier Col, BA, 58, MS, 61; Univ Wis-Madison, PhD(bot & hort), 70. *Prof Exp:* Instr biol defense, US Army Chem Corps, 61-63; res scientist microbiol, Space Gen Corp, 64-65; res engr space res, NAm Rockwell Corp, 65-67; mem staff circadian rhythms, Univ Calif, Los Angeles, 70; PROF BIOL, METROP STATE COL, 70-

Concurrent Pos: Consult, Oshkosh Paving Co, 75 & Controlled Environ Life Support Syst Prog, NASA, 80; dir, Summer Sci Inst, 85. *Honors & Awards:* Fulbright-Hays lectr, Philippines, 80. *Mem:* Am Soc Plant Physiologists; Am Inst Biol Sci; AAAS. *Res:* Circadian rhythms of bean plants under controlled environmental conditions; tissue culture studies. *Mailing Add:* Dept Biol Metropolitan State Col 1006 11th St Denver CO 80204

ALFORD, GEARY SIMMONS, b McComb, Miss, April 11, 45; m 76; c 2. EXPERIMENTAL CLINICAL PSYCHOLOGY. *Educ:* Millsaps Col, BA, 68; Univ Ariz, MA, 71, PhD (clin psych), 72. *Prof Exp:* Psych resident, Dept Psych, Univ Miss Med Ctr, 71-72, asst prof psych & psychiat, 73-75; fel, Nat Inst Drug Abuse & Nat Inst Alcoholism & Alcohol Abuse, 75-77; assoc prof psych & psychiat, 80-87, PROF PSYCH & PSYCHIAT, UNIV MISS MED CTR, 87- *Concurrent Pos:* Assoc prof, Dept Pharmacol & Toxicol, Univ Miss Med Ctr, 77-; Nat Adv Bd, Am Coun Alcoholism, 86-; psychol consult, Miss Bur Narcotics, 72-75 & 78- *Mem:* Fel Am Psychol Asn; Asn Adv Behavior Therapy; Behavior Therapy & Res Soc. *Res:* Behavior therapy, application of experimental psychology to clinical psychopathology; psychopharmacology and chemical substance misuse and addiction. *Mailing Add:* Dept Psychiat & Human Behavior Univ Medical Ctr 2500 N State St Jackson MS 39216

ALFORD, HARVEY EDWIN, b Ashtabula, Ohio, Aug 30, 24; m 47; c 2. ORGANIC CHEMISTRY. *Educ:* Hiram Col, BA, 47; Western Reserve Univ, MS, 52. *Prof Exp:* Res assoc, Standard Oil Co Ohio, 47-75, sr res assoc, 75-83; RETIRED. *Mem:* Am Chem Soc. *Res:* Petroleum processing; coke production; activated carbon production and use; waste treatment; microballoons; resid upgrading. *Mailing Add:* 177 Orchard Hill Amherst OH 44001

ALFORD, WILLIAM LUMPKIN, b Albertville, Ala, Oct 6, 24; m 48; c 5. NUCLEAR PHYSICS. *Educ:* Vanderbilt Univ, AB, 48; Calif Inst Technol, MS, 49, PhD(physics), 53. *Prof Exp:* From asst prof to assoc prof physics, Auburn Univ, 53-58; nuclear physicist, US Army Missile Command, 58-64; assoc dean, Sch Arts & Sci, 80-86, PROF PHYSICS, AUBURN UNIV, 86-, ASSOC DEAN, COL SCI & MATH & DIR, NUCLEAR SCI CTR, 86- *Honors & Awards:* Pegran Award, Am Phys Soc, 83; Algernon Sydney Sullivan Award, 85. *Mem:* Fel Am Phys Soc. *Res:* Nuclear reactions induced by neutrons in the zero to twenty MeV energy range; measurement of reaction cross sections and investigation of the gamma activities of reaction product nuclei. *Mailing Add:* Dept of Physics Auburn Univ Auburn AL 36849-5311

ALFORD, WILLIAM PARKER, b London, Ont, Mar 22, 27; m 49; c 3. NUCLEAR PHYSICS. *Educ:* Univ Western Ont, BSc, 49; Princeton Univ, PhD(physics), 54. *Prof Exp:* Instr physics, Princeton Univ, 54-55; from instr to prof, Univ Rochester, 55-73; chmn dept, 73-84, PROF PHYSICS, UNIV WESTERN ONT, 73- *Concurrent Pos:* Vis prof, Univ Munich, 70-71, Univ Colo, 78-79 & Univ BC, 84-85. *Honors & Awards:* Lyle Fel, Univ Melbourne, 88. *Mem:* Fel Am Phys Soc; Am Asn Physics Teachers; Can Asn Physicists. *Res:* Nuclear reactions and spectroscopy. *Mailing Add:* Dept of Physics Univ of Western Ont London ON N6A 3K7 Can

ALFORS, JOHN THEODORE, b Reedley, Calif, Nov 24, 30; m 69; c 2. GEOLOGY. *Educ:* Univ Calif, Berkeley, BA, 52, MA, 56, PhD(geol), 59. *Prof Exp:* GEOLOGIST, CALIF DEPT MINES-GEOL, 60- *Mem:* Geol Soc Am; Am Asn Petrol Geol; Mineral Soc Am. *Res:* Mineralogy and petrology of glaucophane schists; geology and mineralogy of the barium silicate deposits of Fresno County, California; geochemical study of the Rocky Hill granodiorite, Tulare County, California; urban geology master plan for California; geology of Tiburon Peninsula, Marin County, California. *Mailing Add:* 4337 Vista Way Davis CA 95616

ALFRED, LOUIS CHARLES ROLAND, b Mauritius, June 2, 29; m 64; c 3. SOLID STATE PHYSICS, PHYSICAL METALLURGY. *Educ:* Univ Bombay, BSc, 52, MSc, 54; Univ Sheffield, PhD(theoret physics), 57. *Prof Exp:* Res assoc phys chem, Brandeis Univ, 57-59; res fel physics, Brookhaven Nat Lab, 59-61 & Atomic Energy Res Estab, Eng, 61-65; assoc physicist, Argonne Nat Lab, 65-70; PROF PHYSICS, TRENT UNIV, 70- *Res:* Defect properties and structure of solids; electron states in solids; magnetic resonance; atomic structure and surface physics; mathematical methods. *Mailing Add:* Dept Physics Trent Univ Peterborough ON K9J 7B8 Can

ALFREY, CLARENCE P, JR, b Brownwood, Tex, May 25, 30; m 55; c 4. HEMATOLOGY, BIOENGINEERING. *Educ:* Baylor Univ, MD, 55; Univ Minn, PhD(med), 66. *Prof Exp:* From instr to assoc prof, 61-71, prof med, Baylor Col Med, 71-, co-dir hemat res, 66-; AT HEMATOL SOC, METHODIST HOSP, TEX MED CTR. *Concurrent Pos:* Chief radioisotope & hemat serv, Vet Admin Hosp, Houston, 61-66, assoc chief of staff res & educ, 62-66; adj assoc prof bioeng, Rice Univ, 67-71, adj prof, 71- *Mem:* Am Soc Hemat; Am Fedn Clin Res; AMA; Sigma Xi. *Res:* Medical uses of radioisotopes; iron metabolism in man; effects of physical forces on red blood cells; ultrasound in medical diagnosis. *Mailing Add:* Hematol Soc Methodist Hosp Texas Med Ctr Houston TX 77030

ALFVEN, HANNES OLOF GOSTA, b 1908. GEOPHYSICS, PLASMA PHYSICS. *Educ:* Univ Uppsala, PhD(physics), 34. *Prof Exp:* Prof theory elec, Royal Inst Technol, Stockholm, 40-45, prof electronics, 45-63, prof plasma physics, 63-73; res physicist, 67-73, prof, 73-75, EMER PROF ELEC ENG & COMPUTER SCI, UNIV CALIF, SAN DIEGO, 75- *Honors & Awards:* Nobel Prize in Physics, 70; Gold Medal, Royal Astron Soc, UK, 67; Lomonsov Gold Medal, USSR Acad Sci, 71. *Mem:* Foreign Assoc Nat Acad Sci; Swed Acad Sci. *Mailing Add:* c/o Debbie Ezell Univ Calif San Diego ECE Dept EBUI-0407 La Jolla CA 92093-0407

ALGARD, FRANKLIN THOMAS, embryology; deceased, see previous edition for last biography

ALGER, ELIZABETH A, b Plainfield, NJ, Apr 20, 39; m 69. INTERNAL MEDICINE, ANATOMY. *Educ:* Seton Hall Univ, MD, 64. *Prof Exp:* ASSOC PROF ANAT & ASST PROF MED, UNIV MED & DENT NJ, 66-, ASSOC DEAN EDUC, 81- *Mem:* Am Asn Anatomists; Am Fedn Clin Res; Am Med Women's Asn; Am Soc Andrology. *Res:* Endocrine regulation of spermatogenesis; computers in medicine and medical education; clinical andrology. *Mailing Add:* UMDNJ Univ Hosp NJ 532 Dudley Ct Westfield NJ 07090

ALGER, NELDA ELIZABETH, b Ithaca, NY, Dec 14, 23. ZOOLOGY. *Educ:* Univ Mich, BS, 45, MS, 47; NY Univ, PhD(zool), 62. *Prof Exp:* Instr zool, Highland Park Jr Col, 47-48; instr zool, Hunter Col, 49-50; lab instr parasitol, Columbia Univ, 50-51; from instr to asst prof prev med, NY Univ, 52-64; asst prof, 64-70, ASSOC PROF ZOOL, UNIV ILL, URBANA-CHAMPAIGN, 70- *Mem:* Am Soc Parasitol; Am Soc Trop Med & Hyg; Am Soc Invert Path; Soc Protozool; Tissue Cult Asn; Sigma Xi. *Res:* Malaria; trypanosomiasis; parasitic immunology. *Mailing Add:* 549 Morrill-Zool Univ Ill Urbana IL 61801

ALGERMISSEN, SYLVESTER THEODORE, b St Louis, Mo, May 9, 32; m 68; c 1. PROBABILISTIC EARTHQUAKE HAZARD & RICK STUDIES. *Educ:* Mo Sch Mines, BS, 53; Wash Univ, AM, 55, PhD(geophys), 57. *Prof Exp:* Proj engr, Sinclair Res Labs, Inc, 57-59; asst prof geophys, Univ Utah, 59-63; chief data anal & res br, Coast & Geod Surv, Environ Sci Serv Admin, 63-65, chief geophys res group, 65-71; dir seismol res group, Environ Res Labs, Nat Oceanic & Atmospheric Admin, 71-73; chief seismicity & risk anal br, 73-75, res geophysicist, 75-79, SUPVRY GEOPHYSICIST, US GEOL SURV, 80- *Concurrent Pos:* Mem sci & eng task force, Fed Reconstruct & Develop Comn for Alaska, 64; mem comt Alaska earthquake, Nat Acad Sci-Nat Res Coun, 64-; vis prof, Univ Chile, 70; mem US deleg, UNESCO Intergovt Conf Mitigation of Seismic Risk, 76; dir, Seismol Soc Am, 73-75 & Earthquake Eng Res Inst, 77-79. *Honors & Awards:* Meritorious Serv Award, US Dept Interior, 85. *Mem:* Fel AAAS; Soc Explor Geophys; Seismol Soc Am; Am Geophys Union. *Res:* Earthquake and risk (loss) assessment; seismicity; engineering seismology; interpretation of geophysical data; structure of the earth. *Mailing Add:* US Geol Surv Stop 966 Box 25046 DFC Denver CO 80225

ALHADEFF, JACK ABRAHAM, b Vallejo, Calif, May 9, 43; m 81. GLYCOCONJUGATES, ENZYMOLOGY. *Educ:* Univ Chicago, BA, 65; Univ Ore Med Sch, PhD(biochem), 72. *Prof Exp:* Fel, Univ Calif, San Diego, 72-74, asst res neuroscientist, 74-75, from asst prof to assoc prof, 75-82; PROF BIOCHEM, LEHIGH UNIV, 82- *Concurrent Pos:* Res Career Develop Award, NIH, 78. *Mem:* Sigma Xi; AAAS; Am Chem Soc; Biochem Soc. *Res:* Biochemical studies on glycoconjugate metabolism in normal and pathological (cancer, diabetes, cystic fibrosis) human tissues. *Mailing Add:* Dept Chem Lehigh Univ Bethlehem PA 18015

ALHOSSAINY, EFFAT M, b Mosul, Iraq, Sept 15, 22; m. BIOCHEMISTRY. *Educ:* Univ Tex, Austin, MED, 63, MA, 70. *Prof Exp:* Res asst microbiol, Univ Tex, 78-81; CHEMIST, TEX DEPT HEALTH, 82- *Mem:* Sigma Xi; Am Asn Chem; Am Soc Biol. *Mailing Add:* 7302 Sevilla Dr Austin TX 78752

ALI, KERAMAT, b Benipur, Bangladesh; Can citizen; c 5. ATOMIC PHYSICS, MOLECULAR PHYSICS. *Educ:* Univ Peshawar, BSc Hons, 64, MSc, 65; Mem Univ Nfld, MSc, 69; Univ Western Ont, PhD(physics), 75. *Prof Exp:* Res physicist spectros, Pakistan Coun Sci & Indust Res, 65-67; fel intermolecular forces, Univ Western Ont, 75-77; res assoc solitón solutions, Nat Res Coun Can, 77-81; PROF PHYSICS, UNIV LETHBRIDGE, ALTA, 81- *Mem:* Can Asn Physicists; assoc mem Chem Inst Can; Am Phys Soc; NY Acad Sci; Am Asn Physicists. *Res:* Studies of order and chaos in dynamical systems (of special interest is chaos in microscopic systems); studies of quasicrystals (dynamical trace maps of generalized Fibonacci lattices); scattering of atoms and molecules from corrugated surfaces. *Mailing Add:* Dept Physics Univ Lethbridge Lethbridge AB T1K 3M4 Can

ALI, MAHAMED ASGAR, b Burdwan, India, Nov 1, 34; m 68; c 2. PHYSICAL CHEMISTRY, QUANTUM CHEMISTRY. *Educ:* Presidency Col, Calcutta, India, BSc, 54; Univ Col Sci, Calcutta, MSc, 56; Oxford Univ, DPhil(theoret chem), 60. *Prof Exp:* Fel, Quantum Chem Group, Royal Univ Uppsala, 60-61; Imp Chem Indust res fel, Univ Keele, 61-62; prof chem, Presidency Col, Calcutta, 62-65; lectr, Univ Sheffield, 65-66 & Univ York, 66-68; res fel, Battelle Mem Inst, 68-69; vis assoc prof, Vanderbilt Univ, 69; res fel metals & ceramics, Oak Ridge Nat Lab, 69; PROF CHEM, HOWARD UNIV, 69- *Mem:* Am Phys Soc; fel Brit Inst Physics & Phys Soc; Am Chem Soc; fel Inst Physics. *Res:* Study of atomic structure and spectra by quantum mechanical methods; theoretical study of molecular electronic structure; potential scattering involving atoms and molecules; theoretical molecular spectroscopy. *Mailing Add:* 2523 13th St Washington DC 20009

ALI, MIR MASOOM, b Patuakhali, Bangladesh, Mar 1, 29; Can citizen; m 62; c 5. MATHEMATICAL STATISTICS. *Educ:* Univ Dacca, Pakistan, BSc, 48, MSc, 50; Univ Mich, MS, 58; Univ Toronto, PhD, 61. *Prof Exp:* Lectr statist, Univ Dacca, Pakistan, 51-52; mem actuarial dept, Norwich Union Ins Soc, Eng, 52-56; group actuarial dept, Can Life Assurance Co, 56-57; from asst prof to assoc prof math, 61-66, PROF MATH, UNIV WESTERN ONT, 66- *Mem:* Inst Math Statist; Can Math Cong; fel Royal Statist Soc; Am Statist Asn; Statist Sci Asn Can. *Res:* Theory of estimation; statistical inference; foundation of probability theory; multivariate statistics. *Mailing Add:* Dept Math Sci Ball State Univ Muncie IN 47306

ALI, MOHAMED ATHER, b Bangalore, India, July 16, 32; nat Can. ZOOLOGY. *Educ:* Presidency Col, Madras, India, BSc, 52; Univ Madras, MSc, 54, DSc(physiol), 69; Univ BC, PhD(zool), 58; Dr es Sc Nat, Sorbonne, 71. *Prof Exp:* demonstr invert embryol, Madras, 52-54; demonstr comp anat, Univ BC, 54-56, demonstr physiol, 56-58; asst prof physiol & histol, McGill Univ, 58-59; asst prof biol, Mem Univ, 59-61; from asst prof to assoc prof, 61-64, PROF BIOL, UNIV MONTREAL, 68- *Concurrent Pos:* Res fel, Queen's

Univ, Ont, 59; Alexander von Humboldt sr fel, Max-Planck Inst Physiol Behav, Ger, 64-65; NATO investr, Mus Nat, Paris, 65; Brit Coun vis prof, UK, 65; chief scientist, Te Vega Exped 15, Stanford Univ, 67; vis investr, Harvard Univ, 68, 78 & Carnegie-Mellon Univ, 69; vis prof, Univ Sao Paulo, Univ Rio de Janeiro & Rio Grande do Sul, Brazil, 71-74, Harvard Univ, 78, Iceland, 78-, Japan, 78, 85, & 90, Norway, 80, 82, Austria, 87-88 & Republic of China, 90; dir, NATO-Advan Study Inst, 74, 77, 79, 82, 86 & 91; ed, Revue Can de Biologie, 78-83 & Mikroopie, 82 & 86; ed-in-chief, Exp Biol, 84-89. *Honors & Awards:* Mem Int Comt Photobiol Res Award, Arctic Inst NAm, 58, 59; Albert Monaco Medal, 86. *Mem:* Fel AAAS; Am Soc Ichthyologists & Herpetologists; Sigma Xi. *Res:* Sensory physiology; behavior; neurophysiology; histology; embryology; experimental biology and ecology. *Mailing Add:* Dept Biol Stn A Univ Montreal CP 6128 Montreal PQ H3C 3J7 Can

ALI, MONICA MCCARTHY, b Boston, Mass, Nov 22, 41; c 2. CHEMISTRY. *Educ:* Emmanuel Col, AB, 63; Georgetown Univ, MS, 66, PhD(org anal chem), 71. *Prof Exp:* Teacher chem, biol & gen sci, Gate of Heaven High Sch, South Boston, Mass, 63-64; org chemist drug synthesis, Arthur D Little, Inc, Mass, 66-67; teacher chem, Georgetown Visitation Prep Sch, Washington, DC, 71-72; lectr gen & org chem, George Mason Univ, 72-75; ASST PROF GEN & ORG CHEM, OXFORD COL, EMORY UNIV, 75- *Concurrent Pos:* Lectr org chem, Northern Va Community Col, 74-75. *Mem:* Am Chem Soc. *Res:* Formation of aziridinium ion intermediates in the reactions of beta chloro amines. *Mailing Add:* PO Box 1576 Snellville GA 30278

ALI, MOONIS, b Gunnaur, India, Sept 21, 44; US citizen; m 75; c 1. FAULT MONITORING, ROCKET & JET ENGINE DIAGNOSITICS. *Educ:* Aligarh Univ, India, BSc, 64, MSc, 66, PhD, 69. *Prof Exp:* Sr programmer & lectr, Aligarh Univ, 68-74; Fulbright postdoctoral fel, Univ Tex, Austin, 74-75, fac mem & res engr, 75-77; sr researcher, Al-Hazen Res Inst, 77-78; prof computer sci, Mosul Univ, 79-80; assoc prof, Dept Computer Sci, Old Dom Univ, 80-83; prof computer sci & chmn dept, Univ Tenn Space Inst, 83-91; PROF & CHMN COMPUTER SCI, SOUTHWEST TEX STATE UNIV, 91- *Concurrent Pos:* NASA-Am Soc Eng Educ fel, Langley Res Ctr, 83; numerous res grants, NASA, Intel Corp & Arnold Eng Develop Ctr, 85-; ed-in-chief, Int J Appl Intel; prog chmn, Int Conf Indust & Eng Applications Artificial Intel & Expert Systs, 88, gen chmn, 89, 90, 91 & 92. *Mem:* Sr mem Inst Elec & Electronic Engrs; Asn Comput Mach; Inst Elec & Electronic Engrs Computer Soc; Am Asn Artificial Intel; Sigma Xi. *Res:* Development of expert systems to perform automatic data analysis to identify anomalous events in complex physical mechanisms where faults develop rapidly and early detection and diagnosis is critically important; architecture of these expert systems integrates design-based reasoning with mechanism behavior analysis and automatic learning approaches; intelligent robotics; diagnosing systems; author of numerous publications. *Mailing Add:* Dept Computer Sci Southwest Tex State Univ San Marcos TX 78666

ALI, NAUSHAD, b Dubra, India, Mar 10, 53; m 83; c 2. HIGH-TECHNETIUM SUPERCONDUCTIVITY, MAGNETISM. *Educ:* Aligarh Univ, India, BSc Hons, 73, MSc, 75; Mem Univ, Can, MS, 78; Univ Alta, Can, PhD(physics), 84. *Prof Exp:* Postdoctoral fel physics, McMaster Univ, Can, 85-86; asst prof, 86-90, ASSOC PROF PHYSICS, SOUTHERN ILL UNIV, 90- *Concurrent Pos:* Prin investr, Ill Dept Energy & Natural Resources, 88- *Mem:* Am Phys Soc. *Res:* High temperature superconductivity; magnetism of rare earths and rare earth alloys to investigate the magnetic structures. *Mailing Add:* Physics Dept Southern Illinois Univ Carbondale IL 62901

ALI, RIDA A, b Cairo, Egypt, Dec 22, 38; US citizen; m 61; c 3. CLINICAL NUTRITION, FOOD SCIENCE AND TECHNOLOGY. *Educ:* Cairo Univ, BSc, 59; Univ Calif, Davis, MSc, 63; Rutgers Univ, PhD(nutrit), 66. *Prof Exp:* From asst prof to assoc prof nutrit, Cairo Univ, 67-71; res assoc clin res, Pharmacia Inc, US, 71-72, dept head, 72-73, res mgr clin & biol res, US & Sweden, 73-76; corp res mgr nutrit & health sci, Tech Ctr, Gen Foods Corp, 76-81; dir, Bristol-Myers Int, 81-83, vpres nutrit & res develop, 83-89; VPRES BUS & TECH DEV, MEAD JOHNSON INT, 89- *Concurrent Pos:* Fullbright fel, 62-64; fel & res assoc, Rutgers Univ, 66-67 & 70-71; consult clin nutrit, Pharmacia Can Ltd, 73-76; vchmn, Food & Nutrit Coun, Am Health Found, 82- *Mem:* Am Inst Nutrit; Inst Food Technologists; Am Soc Clin Path; Am Mgt Asn; Am Chem Soc; fel Am Col Nutrit; Am Soc Parenteral & Enteral Nutrit. *Res:* Clinical and experimental nutrition in infants and adults; biochemical and nutritional factors affecting body composition, energy metabolism, aging and development; physiology of the gastrointestinal tract. *Mailing Add:* Mead Johnson Int 2400 W Lloyd Expressway Evansville IN 47721

ALI, SHAHIDA PARVIN, b Peshawar, Pakistan, Oct 10, 42; Can citizen; m 67; c 5. QUANTUM CHEMISTRY. *Educ:* Univ Peshawar, Pakistan, BSc, 65, MSc, 65; Mem Univ Newfoundland, MSc, 71; Univ Western Ont, PhD(physics), 76. *Prof Exp:* Fel physics, Univ Ottawa, 78-80, res assoc, 80-82; FEL CHEM, UNIV LETHBRIDGE, 82- *Res:* Charge exchange total cross-section measurements of singly and doubly charged ions with inert gas targets at intermediate energy range. *Mailing Add:* 330 Laval Blvd Lethbridge AB T1K 3W5 Can

ALI, YUSUF, b Dohad, Gujarat, India, Dec 9, 53; US citizen; m 85. OPHTHALMIC DRUG DEVELOPMENT, OPHTHALMIC DRUG DELIVERY. *Educ:* MS Univ, India, BS, 76; Polytech Inst NY, MS, 78; Univ Fla, PhD(chem eng), 84. *Prof Exp:* Postdoctoral assoc drug delivery, 84-85, sr scientist prod develop, 85-89, MGR PROD DEVELOP, ALCON LABS, INC, 90- *Mem:* Am Asn Pharmaceut Scientists; Am Chem Soc; Am Inst Chem Engrs. *Res:* Pharmaceutical product development of various ophthalmic drugs, identification of ophthalmic drug delivery systems, enhancement of stability of drugs in aqueous media and improvement of bioavailability of ophthalmic drugs using appropriate excipients; 1 United States patent. *Mailing Add:* 6904 Wicks Trail Ft Worth TX 76133

ALIC, JOHN A, b Oak Park, Ill, Nov 24, 41. INTERNATIONAL COMPETITIVENESS. *Educ:* Cornell Univ, BME, 64; Stanford Univ, MS, 65; Univ Md, PhD(mat eng), 72. *Prof Exp:* Case writer, Stanford Univ, 65-66; instr mech eng, Univ Md, 66-72; from asst prof to assoc prof, Wichita State Univ, 72-79; PROJ DIR, OFF TECHNOL ASSESSMENT, 79- *Honors & Awards:* Wright Bros Medal, Soc Automotive Engrs, 75. *Mem:* AAAS; Am Soc Metals; Soc Automotive Engrs; Am Soc Eng Educ. *Res:* Technology assessment, particularly international competitiveness of United States industries, services as well as manufacturing. *Mailing Add:* Off Technol Assessment US Congress Washington DC 20510

ALICH, AGNES AMELIA, b International Falls, Minn, June 10, 32. ORGANOMETALLIC CHEMISTRY. *Educ:* Marquette Univ, BS, 60, MS, 61; Northwestern Univ, Evanston, PhD(inorg chem), 72. *Prof Exp:* Instr chem, Col St Scholastica, 61-64; instr, Gerard High Sch, Phoenix, 64-67; PROF CHEM & CHAIRPERSON DEPT, COL ST SCHOLASTICA, 67- *Mem:* Am Chem Soc; Sigma Xi; The Chem Soc; Int Union Pure & Appl Chem. *Res:* Synthesis of low- and zero-valent compounds of transition metals, mainly organometallic carbonyls; trace metal drug interactions. *Mailing Add:* Dept of Chem Col of St Scholastica Duluth MN 55811

ALICINO, NICHOLAS J, b New York, NY, Apr 29, 21; m 44; c 8. ANIMAL SCIENCE. *Educ:* Fordham Univ, BS, 42. *Prof Exp:* Anal chemist, Calco Div, Am Cyanamid Co, 42-44; anal chem, Nopco Chem Co, 46-50, chief analyst, 50-58, dir anal chem, 58-70; mgr res fire chem div, Diamond Shamrock Chem Co, NJ, 70, mgr res & develop, Nutrit & Animal Health Div, 70-82, CONSULT, TECH & REGULATORY, DIAMOND SHAMROCK CHEM CO, OHIO, 82- *Concurrent Pos:* President's pvt sector survey, Grace Comn, 82-83. *Mem:* Am Chem Soc; Am Microchem Soc. *Res:* Analytical research; chemical method by means of functional groups; new drug research for animal diseases and growth promotion. *Mailing Add:* 79 Carriage Dr Chargin Falls OH 44022

ALIG, ROGER CASANOVA, b Indianapolis, Ind, Nov 7, 41; m 63; c 3. SOLID STATE PHYSICS, ELECTRON OPTICS. *Educ:* Wabash Col, BA, 63; Purdue Univ, MS, 65, PhD(physics), 67. *Prof Exp:* MEM TECH STAFF PHYSICS, DAVID SARNOFF RES CTR, SARNOFF RES INT, 67- *Concurrent Pos:* Vis prof, Sao Carlos Sch Eng, Brazil, 70-71; adj prof, Rider Col, 90- *Honors & Awards:* Sarnoff Medal, RCA, 83. *Mem:* Am Phys Soc; Soc Info Display. *Res:* Electron optics in television picture tubes; electron scattering in semiconductors. *Mailing Add:* 17 Landing Ln Princeton Junction NJ 08550

ALI-KHAN, AUSAT, b India, Mar 3, 39; m 65; c 2. POLYMER CHEMISTRY, ORGANIC CHEMISTRY. *Educ:* Aligarh Muslim Univ India, BSc, 57, MSc, 59; Nat Univ of the South, Arg, PhD(org chem), 62. *Prof Exp:* Asst prof org chem, Nat Univ Cordoba, 62-64; fel, Univ Del, 64-66; SR RES ASSOC, E I DU PONT DE NEMOURS & CO, INC, 66- *Mem:* NY Acad Sci; Am Chem Soc. *Res:* Fluorocarbon polymers and monomers; hydrocarbon rubbers; emulsion polymerization and colloidal stability of lattices. *Mailing Add:* E I Dupont ESL 323-129 Wilmington DE 19880-0323

ALIKHAN, MUHAMMAD AKHTAR, b Lyallpur, Pakistan, Apr 1, 32; Eng citizen; m 61; c 3. ANIMAL PHYSIOLOGY, BIOCHEMISTRY. *Educ:* Univ Panjab, Pakistan, BSc, 52; Univ Leeds, Eng, MSC, 58; Marie Curie-Sklodowska Univ, Poland, PhD(physiol), 61. *Prof Exp:* Lectr zool, Agr Col, Lyallpur, 52-57; assoc prof, WPakistan Agr Univ, 61-64; asst prof biol, Univ Calgary, 64-65 & Lethbridge Univ, 65-67; from asst prof to assoc prof, 67-73, PROF BIOL, LAURENTIAN UNIV, 73- *Concurrent Pos:* Polish Acad Sci res fel, 64, 68 & 69; Nat Res Coun fel physiol & travel grant, 66 & res grant, 66-; vis scientist, Czechoslovak Acad Sci, Prague, 72, 73, 80, Univ Lodz, Poland, 76, & UN Develop Prog, Pakistan, 81; vis prof, State Univ Gent, Belg, 73, Univ Agr, Lyallpur, Pakistan Univ Mosul, Iraq, 74, Univ Tech Sci, Montpellier, France, 78, Univ Garyounis, Beida, Libya, 81, Warsaw Agr Univ, Poland, 83, Univ Agr, Faisalabad, Pakistan, 88 & German Sci Soc, 88. *Mem:* Fel Royal Entom Soc; Asn Sci Eng & Technol Community Can; Int Asn Ecologists; Asn Can Fr Advan Sci; Can Biochem Soc; Zool Soc Can. *Res:* Biochemical adaptations in crustacea to their environments; environmental toxicology. *Mailing Add:* Dept Biol Laurentian Univ Ramsey Lake Rd Sudbury ON P3E 2C6 Can

ALIN, JOHN SUEMPER, b LaMoure, NDak, Aug 14, 40; m 77; c 5. ALGEBRA. *Educ:* Concordia Col, BA, 63; Univ Nebr, MA, 65, PhD(math), 67. *Prof Exp:* Vis asst prof math, Univ Southern Calif, 67-68; asst prof, Univ Utah, 68-72; assoc prof, 72-81, PROF MATH, LINFIELD COL, 81- *Concurrent Pos:* Consult, India Prog, NSF, 70; adj prof math, Portland State Univ, 72-; adj prof, C S Ore Grad Ctr, 86- *Mem:* Am Math Soc; Math Asn Am; Am Comput Mach. *Mailing Add:* Dept Math Linfield Col McMinnville OR 97128

ALINCE, BOHUMIL, b Bratislava, Czech; Can citizen; m 65; c 2. COLLOID CHEMISTRY IN PAPERMAKING PROCESS. *Educ:* Tech Univ, Bratislava, Dipl Ing, 54; Acad Sci, Bratislava, PhD(chem), 65. *Prof Exp:* Scientist, Water Res Inst, Bratislava, 54-59, Acad Sci, Bratislava, 59-69; SR SCIENTIST PAPER CHEM, PULP & PAPER RES INST CAN, 69- *Concurrent Pos:* Adj prof chem eng, McGill Univ, Montreal, Can. *Mem:* Tech Asn Pulp & Paper Indust; Can Pulp & Paper Asn; Am Chem Soc; Can Chem Soc. *Res:* Colloid and surface phenomena in papermaking; applications and performance of pigments, their optical and rheological properties; interaction of pigments and cellulosic fibers; modification of cellulosic fibers by polyelectrolytes. *Mailing Add:* 3420 Univ St Montreal PQ H3A 2A7 Can

ALIPRANTIS, CHARALAMBOS DIONISIOS, b Cephalonia, Greece, May 12, 46; m 74; c 2. MATHEMATICAL ANALYSIS. *Educ:* Univ Athens, dipl math, 68; Calif Inst Technol, MS, 71, PhD(math), 73. *Prof Exp:* Lectr math, Occidental Col, 73-74, asst prof, 74-75; from asst to assoc prof, 75-81, PROF MATH, IND UNIV-PURDUE UNIV, INDIANAPOLIS, 81- *Concurrent Pos:* Res scientist, STD Res Corp, 73-74. *Mem:* Am Math Soc; Math Asn Am. *Res:* Functional analysis; operator theory; mathematical economics. *Mailing Add:* Ind Univ-Purdue Univ 1125 E 28th St Indianapolis IN 46205

ALIRE, RICHARD MARVIN, b Mogote, Colo, July 5, 32; m 57; c 2. CHEMISTRY, MANAGEMENT OF RESEARCH IN CHEMISTRY. *Educ:* York Col, Nebr, BA, 54; Univ Nebr, Lincoln, MS, 56; Univ NMex, PhD(chem), 62. *Prof Exp:* Staff mem chem res, Los Alamos Sci Lab, 61-69, sect leader, 69-76; div leader, 76-83, dept head, 83-86, assoc div leader, 86-88, PROG MGR, LAWRENCE LIVERMORE LAB, 88- *Mem:* Am Chem Soc; fel Am Inst Chemists; AAAS. *Res:* Molecular spectroscopy of rare earth chelates; kinetics of gas-solid reactions; application of microbalance techniques to studying gas-solid interactions; materials for controlled thermonuclear reactors; supervising research on tritium technology; plutonium processing; energetic materials. *Mailing Add:* 1462 Groth Circle Pleasanton CA 94566

ALIVISATOS, A PAUL, b Chicago, Ill, Nov 12, 59. MATERIALS SCIENCE. *Educ:* Univ Chicago, BA, 81; Univ Calif, Berkeley, PhD(chem), 86. *Prof Exp:* Mem tech staff, AT&T Bell Labs, 86-88; ASST PROF CHEM, UNIV CALIF, BERKELEY, 88- *Concurrent Pos:* NSF presidential young investr, 90; Alfred P Sloan fel, 91. *Honors & Awards:* Exxon Solid State Chem Award, Am Chem Soc, 91. *Mem:* Mat Res Soc; Am Chem Soc. *Res:* Preparation and properties of very small crystals of semiconductors, comprised of a few hundred to tens of thousands of atoms. *Mailing Add:* Dept Chem Univ Calif Berkeley CA 94720

ALIVISATOS, SPYRIDON GERASIMOS ANASTASIOS, biochemistry, physiology, for more information see previous edition

AL-JURF, ADEL SAADA, US citizen. SURGICAL ONCOLOGY. *Educ:* Cairo Univ Col Med, MB, ChB(med), 66. *Prof Exp:* Intern, Cairo Univ Col Med, 66-67, St Vincent Charity Hosp, 72-73; resident fel Cleveland Clin Found, 73-77; AT UNIV IOWA HOSP & CLINS, DEPT SURG, 77- *Mem:* Fel Am Col Surgeons; Asn Acad Surg; Soc Surg Alimentary Tract; Am Soc Parenteral & Enteral Nutrit; Am Soc Clin Oncol. *Mailing Add:* Dept Surg Univ Iowa Hosp & Clins 1521 JCP Iowa City IA 52242

ALKADHI, KARIM A, b Baghdad, Iraq, July 29, 38; US citizen; m 67; c 3. GANGLIONIC TRANSMISSION. *Educ:* Univ Baghdad, BSc, 60; Univ Conn, MSc, 67; State Univ NY Buffalo, PhD(pharmacol), 72. *Prof Exp:* Pharmacist, Repub Hosp, Baghdad, Iraq, 60-65; res asst pharmacol, Sch Med, Yale Univ, 67-68; teaching fel pharmacol, Sch Med, State Univ NY Buffalo, 72-73; asst prof pharmacol, Sch Med, Univ Benghazi, 73-75; from asst prof to assoc prof pharmacol, Sch Med, Mustansiria Univ, 75-80; asst prof, 81-87, ASSOC PROF PHARMACOL, COL PHARM, UNIV HOUSTON, 87- *Concurrent Pos:* Vis prof, dept pharmacol, Sch Med, Univ Conn, Farmington, 76-81. *Mem:* Brit Pharmacol Soc; Am Soc Pharmacol Exp Therapeut; Soc Exp Biol & Med; Soc Neurosci; Int Brain Res Orgn. *Res:* Pharmacology of synaptic transmission at autonomic ganglia and neuromuscular junction; pharmacological characterization of dopaminergic and muscarinic cholinergic receptors in ganglia; impact of chemical neurotoxicants, such as lead, on transmission; study of mechanisms of epilepsy in brain slices. *Mailing Add:* Dept Pharmacol Univ Houston Houston TX 77204-5515

ALKEZWEENY, ABDUL JABBAR, b Amarah, Iraq, Sept 1, 35. CLOUD PHYSICS, AIR POLLUTION. *Educ:* Univ Baghdad, BS, 58; Univ Calif, Santa Barbara, MA, 63; Univ Wash, PhD(atmospheric sci), 68. *Prof Exp:* Scientist, Meteorol Res Inc, 68-71; sr res scientist, Battelle-Pac Northwest Lab, 71-86; CERT, 86- *Mem:* Am Meteorol Soc; Sigma Xi. *Res:* Physical and chemical transformations of pollutants in clear and cloudy atmosphere using airborne and ground instrumentation. *Mailing Add:* 112 S Kellogg Kennewick WA 99336

AL-KHAFAJI, AMIR WADI NASIF, b Baghdad, Iraq, Aug 14, 49; m 81; c 3. ORGANIC SOILS. *Educ:* Wayne State Univ, BSCE, 73, MSCE, 74; Mich State Univ, MS, 76, PhD(civil eng), 79. *Prof Exp:* Sr engr, Mich Testing, 79-81; asst prof civil eng, Univ Detroit, 80-82; assoc prof & chmn dept, Univ Evansville, 82-86; PROF CIVIL ENG & CHMN DEPT, BRADLEY UNIV, 86- *Concurrent Pos:* Vis prof, Mich State Univ, 79-80; vpres, Corbin Drilling & Testing, 81-82; consult, Gerling Law Off, 82-84, Morley & Assocs, 82- & Nat Labs, 82- *Mem:* Am Soc Civil Engrs. *Res:* Numerical method; computer programming; geotechnical engineering programs; organic soils. *Mailing Add:* Dept Civil Eng Bradley Univ Peoria IL 61625

ALKIRE, RICHARD COLLIN, b Easton, Pa, Apr 19, 41; wid; c 2. CHEMICAL ENGINEERING. *Educ:* Lafayette Col, BS, 63; Univ Calif, Berkeley, MS, 65, PhD(chem eng), 68. *Prof Exp:* Fel, Max Planck Inst Phys Chem, 68-69; from asst prof to assoc prof, 69-77, PROF CHEM ENG, UNIV ILL, URBANA, 77-, HEAD, DEPT CHEM ENG, 86- *Concurrent Pos:* Chmn, Indust Electrolytic Div, Electrochem Soc, 78-80, long range planning comt, 86-88, ed, J Electrochem Soc, 73-; chmn, Heat Transfer & Energy Conversion Div, Am Inst Chem Engr, 82-83, tech prog, 85, Nat Heat Transfer Conf, Pittsburgh, 87, coord conf, San Francisco, 86, electrochem fundamentals, Nat Prog Comt Area, 74, consult ed, Am Inst Chem Engr J, awards comt, 86-, dir coun, 88-; mem, comt surv chem eng, Panel Surface & Interfacial Engr, Nat Res Coun, 85, Nat Mats Adv Bd, 85-88, chmn, comt fuel cell mats technol vehicular propulsion, Nat Mats Adv Bd, 83 & comt new horizons electrochem sci & technol, 86; keynote lectr, Int Soc Electrochem, 84; plenary lect, Conf Corrosion Chem, 84 & Int Soc Electrochem, 89. *Honors & Awards:* Prof Progress Award, Am Inst Chem Engrs, 85; Res Award, Electrochem Soc, 83, Carl Wagner Mem Award, 85; McGabe Lectr, NC State Univ, 85; Lindsey Lectr, Tex A&M, 86; Berkeley Lect, Univ Calif, 87; Eltech Lect, Case-Western Reserve Univ, 87; Tech Achievement Award, Nat Asn Corrosion Engrs, 90; Lacey Lectr in Chem Eng, Calif Inst Technol, 90; E V Murphree Award in Indust & Eng Chem, Am Chem Soc, 91. *Mem:* Nat Acad Eng; hon mem Electrochem Soc (pres, 85-86); Am Inst Chem Engrs. *Res:* Electrochemical reaction engineering; localized corrosion and etching; plasma reactor design; author over 105 publications. *Mailing Add:* Dept Chem Eng Univ Ill 1209 W California St Urbana IL 61801

ALKJAERSIG, NORMA KIRSTINE (MRS A P FLETCHER), b Ikast, Denmark, Dec 25, 21; nat US; m 61. BIOCHEMISTRY. *Educ:* Tech Univ Denmark, 49; Univ Copenhagen, PhD, 65. *Prof Exp:* Res assoc, Biol Inst, Carlsberg Found, 49-51 & Col Med, Wayne State Univ, 51-55; res asst, Jewish Hosp, 55-58; res asst prof, 58-69, RES ASSOC PROF MED, SCH MED, WASH UNIV, 69- *Concurrent Pos:* Mem, Thrombosis Coun, Am Heart Asn. *Mem:* Am Physiol Soc; Cent Soc Clin Res; Am Fedn Clin Res; Int Soc Hematol. *Res:* Fibrinolytic enzymes; blood clotting. *Mailing Add:* Wash Univ Sch Med 30 Oak Bend Ct St Louis MO 63124

ALKS, VITAUTS, b Bauska, Latvia, Sept 28, 38; US citizen; m 67. MEDICINAL CHEMISTRY. *Educ:* Univ Buffalo, BS, 61; State Univ NY, Buffalo, MS, 64, PhD(med chem), 74. *Prof Exp:* Res chemist, Starks Assoc Inc, 63-65; fel, 73-74, cancer res scientist I, 74-79, CANCER RES SCIENTIST II, ROSWELL PARK MEM INST, 79- *Mem:* Am Chem Soc; AAAS; Royal Soc Chem; NY Acad Sci. *Res:* Chemical problems of biological and medicinal interest; design and synthesis of enzyme inhibitors; synthesis of steroid analogs; design of active site alkylating agents; studies of metabolic pathways; synthesis of polynuclear aromatic hydrocarbon derivatives; chemotherapeutic agents; amino acid analogues; especially methionine analogues; photosensitizers. *Mailing Add:* Roswell Park Cancer Inst Dept Dermat Buffalo NY 14263

ALLABEN, WILLIAM THOMAS, b Chicago, Ill, Nov 23, 39; m 69; c 2. BIOCHEMISTRY, NUTRITION. *Educ:* Southern Ill Univ, BA, 66, MS, 68, PhD(physiol), 75. *Prof Exp:* Actg dir res, Endocrinol-Pharmacol Res Lab, 67-68; scientist, Mead Johnson & Co Pharmaceut, 68-72; res asst, Southern Ill Univ, 72-74, instr, Dept Physiol, 75; fel, 76-77, res toxicologist, 77-84, dep dir comp toxicol, 84-87, ASSOC DIR SCI COORD, NAT CTR TOXICOL RES, FOOD & DRUG ADMIN, 87- *Concurrent Pos:* Actg dir, Endocrinol-Pharmacol Res Lab, Southern Ill Univ, 75; chem mgr, Nat Toxicol Prog, 80-; dir, Nat Toxicol Prog Study, Nat Ctr Toxicol Res, 84- *Mem:* AAAS; Am Col Toxicol; Sigma Xi; Am Asn Cancer Res; Soc Toxicol; Am Chem Soc. *Res:* Efffect of caloric restriction on the expression of non-neoplastic and neoplastic disease and studies to determine the mechanism(s) by which caloric restrictions prevents the expression of toxic endpoints; determining specific genetic or epigenetic events critical for the expression of neoplastic disease; whole body toxicology studies to determine risks associated with exposure to chemicals, food addititives, drugs, and pesticides; other chemicals of interest to government regulatory agencies; risk assessment. *Mailing Add:* Sci Coordr HFT-30 Nat Ctr Toxicol Res Jefferson AR 72079

ALLAIRE, FRANCIS RAYMOND, b Fall River, Mass, Nov 18, 37; m 61; c 4. DAIRY SCIENCE. *Educ:* Univ Mass, BVA, 59, MS, 63; Cornell Univ, PhD(animal breeding), 65. *Prof Exp:* From fel to assoc prof, 65-82, PROF DAIRY SCI, OHIO STATE UNIV, 82- *Concurrent Pos:* Fulbright res scholar, Ireland, 78. *Mem:* Am Soc Animal Sci; Am Dairy Sci Asn; Biomet Soc; AAAS. *Res:* Selection and breeding methods for the economic improvement of dairy cattle populations; studies include theoretical and empirical analyses of alternative multiple trait selection procedures among cows and bulls; developing breeding goals within an economic framework. *Mailing Add:* Dept Dairy Sci Ohio State Univ 2027 Coffey Rd Columbus OH 43210

ALLAIRE, PAUL EUGENE, b Bristol, Conn, Sept 23, 41; m 71; c 1. MECHANICAL ENGINEERING, BIOMEDICAL ENGINEERING. *Educ:* Yale Univ, BE, 63, ME, 64; Northwestern Univ, PhD(mech eng), 72. *Prof Exp:* Asst prof mech eng, Mem Univ Nfld, 71-72; asst prof, 72-77, assoc prof mech eng, 77, DIR ROTATING MACH & CONTROLS INDUST RES PROG, UNIV VA, 81- *Concurrent Pos:* Chmn bearing educ, Am Soc Lubrication Engrs, 78. *Mem:* Am Soc Mech Engrs; Sigma Xi; Am Soc Eng Educ; Am Soc Lubrication Engrs. *Res:* Turbomachinery flows; dynamics of rotating machinery; finite element analysis; lubrication. *Mailing Add:* Dept Mech Eng Univ Va Charlottesville VA 22093

ALLAMANDOLA, LOUIS JOHN, b New York, NY, Aug 28, 46; m 68; c 4. INTERSTELLAR SPECTROSCOPY, PHYSICAL CHEMISTRY. *Educ:* St Peter's Col, BS, 68; Univ Calif, Berkeley, PhD(phys chem), 74. *Prof Exp:* Fel energy transfer low temperature solids, Ore State Univ, 74-75, NSF fel, 75-76, instr molecular spectros, 76; from asst to assoc prof low temperature spectros, Lab Astrophysics Group, Univ Leiden, 76-83, instr molecular physics, optics, 78-82; LAB ASTROPHYSICS GROUP LEADER, SPACE SCI, NASA-AMES RES CTR, 83- *Honors & Awards:* H Julian Allen Award, NASA, 86. *Mem:* Am Chem Soc; Sigma Xi; Am Astron Soc. *Res:* Spectroscopy of low temperature (IOK) solid analogs of celestial materials and the chemical reactions possible at this temperature in vacuum; ultra-violet irradiated samples; infra-red astronomy; interstellar molecule formation; presence and properties of large aromatic molecules in space; astrophysics. *Mailing Add:* Ames Res Ctr NASA M/S 245-6 Moffett Field CA 94035-1000

AL-LAMI, FADHIL, b Baghdad, Iraq, Sept 6, 32; m 75; c 2. ANATOMY, PHYSIOLOGY. *Educ:* Univ Baghdad, BSc, 55; Ind Univ, MSc, 59, PhD(anat), 64. *Prof Exp:* Res assoc electron micros, Ind Univ, 66-67; researcher, Putnam Hosp, 67-71; asst prof, Ind Univ, 74-77 & Baghdad Univ, 78-79; PROF, MONTEVALLO UNIV, 80- *Concurrent Pos:* Sabbatical leave, Case Western Reserve Univ, 79. *Mem:* Am Asn Anatomists; Sigma Xi. *Res:* Ultrastructure of the carotid body of normal anoxic mammals as well as the study of the effects of some cholenergic drugs on this organ. *Mailing Add:* Dept Biol Montevallo Univ Montevallo AL 35115

ALLAN, A J GORDON, b London, Eng, Mar 3, 26; US citizen; m 50; c 3. CHEMICAL ENGINEERING, SURFACE CHEMISTRY. *Educ:* Univ London, BSc, 47; Cambridge Univ, PhD(surface chem), 51. *Prof Exp:* Chemist, Imp Chem Industs Ltd, 50-53; chemist, E I Du Pont de Nemours & Co, 53-57, res assoc plastics, 57-58, Europ res assoc int dept, 58-62, mgr licensing & spec studies, 62-64, patent negotiator, 64-76, patent consult, 76-90; RETIRED. *Concurrent Pos:* Imperial Col Scholar, 44-47. *Res:* Surface properties of polymers; adhesion, friction and wear; monomolecular films. *Mailing Add:* 703 Westover Rd Wilmington DE 19807

ALLAN, BARRY DAVID, b Steubenville, Ohio, Jan 20, 35; m 61; c 2. PHYSICAL CHEMISTRY, LASERS. *Educ:* Ariz State Univ, BS, 56; Univ Ala, MS, 64, PhD(chem), 68. *Prof Exp:* Proj officer propellants, Army Rocket & Guided Missile Agency, 56-58, from chemist to res chemist, 58-70, res staff, Chem Laser Br, 70-75, chief high energy laser directorate, Device-Components Br, 75-78, mem staff high energy laser syst proj off, 78-80, MEM STAFF ADVAN TECHNOL FUNCTION, PROPULSION DIRECTORATE, ARMY MISSILE COMMAND, 80- *Concurrent Pos:* Mem, Army Explosives & Propellant Res & Develop Liaison Group, 58-62; mem panel on liquid test methods, Chem Propellant Info Agency, 62-; mem Army Ad Hoc Comt Sensitivity of New Mat, 62-; reviewer, NSF, 74- *Honors & Awards:* Army Res & Develop Award, Dept Army,62. *Mem:* Am Chem Soc. *Res:* Combustion and oscillatory combustion; physico-chemical properties of high energy compounds; sensitivity and desensitization of high energy compounds; theoretical propellant performance analysis; physico-chemical and rheological analysis of thixotropic gels. *Mailing Add:* 7803 Michael Circle Huntsville AL 35802

ALLAN, DAVID WAYNE, b Mapleton, Utah, Sept 25, 36. PHYSICS, MATHEMATICS. *Educ:* Brigham Young Univ, BS, 60; Univ Colo, MS, 65. *Prof Exp:* Physicist, 60-68, asst sect chief, Time & Frequency Div, 68-77, chief, Time & Frequency Coord Group, 79-88, SR SCIENTIST, TIME & FREQUENCY DIV, NAT INST STANDARDS & TECHNOL, DEPT COM, 88-; CONSULT, 67- *Concurrent Pos:* Ital Govt vis prof grant, 69; consult, UN Develop Prog, New Delhi, India, 81; guest scientist, Peoples Repub of China, 82; vis scientist, Int Astron Union Symp, Leningrad, USSR, 85; assoc ed, Frequency & Time, Inst Elec & Electronic Engrs, 85; lectr & consult to dir, Nat Phys Lab, Israel, 87; group leader, Nat Inst Standards & Technol. *Honors & Awards:* Silver Medal Award & Sustained Super Performance Award, Dept Commerce, 68; IR-100 Award, 76. *Mem:* Sci Res Soc Am; Inst Elec & Electronic Engrs; Sigma Xi; Int Radio Consultative Comt; Int Astron Union. *Res:* Atomic time scale; classification of the statistical characteristics of atomic and molecular frequency standards; international time and frequency metrology; Air Force invention award and patent. *Mailing Add:* Time & Frequency Div Nat Inst Standards & Technol Boulder CO 80303-3328

ALLAN, FRANK DUANE, b Salt Lake City, Utah, May 19, 25; m 46; c 7. ANATOMY, EMBRYOLOGY. *Educ:* Univ Utah, BS, 47, MS, 49; La State Univ, PhD(anat), 54. *Prof Exp:* Asst anat, Univ Utah, 48-49; from asst prof to assoc prof, 54-68, PROF ANAT, SCH MED, GEORGE WASHINGTON UNIV, 68-, DIR AV SERV, 62-, DIR AV EDUC, 73- *Concurrent Pos:* Vis scientist, Carnegie Inst Washington, 62-64, mem, Int Anatomical Nomenclature Subcomt Embryology. *Mem:* Am Asn Anatomists; Asn Biomed Commun Dirs (pres, 81-82); Health Sci Commun Asn. *Res:* Early human embryology; cardiovascular physiology; electrophysiological studies on the development of tolerance to alcohol; anomalous arteries; innervation of the ducts arteriosus; anatomy of the fetus and newborn; audiovisual technology in medical education; gross and microscopic anatomy; neuroanatomy. *Mailing Add:* Dept Anat George Washington Univ Sch Med 2300 Eye St NW Washington DC 20037

ALLAN, GEORGE B, b London, Ont, May 8, 35; m 62; c 2. COMPUTER SCIENCE. *Educ:* McGill Univ, BS, 59; Bowdoin Col, MA, 67. *Prof Exp:* Acad vpres, 75-81, pres, 81-84, PROF MATH, LAMBTON COL, 67- *Mem:* Math Asn Am. *Mailing Add:* Technol Div Lambton Col PO Box 969 Sarnia ON N7T 7K4 Can

ALLAN, GEORGE GRAHAM, b Glasgow, Scotland, Nov 21, 30; m 55; c 6. ORGANIC CHEMISTRY. *Educ:* Royal Col Sci & Technol, Scotland, dipl, 51, assoc, 52; Glasgow Univ, BSc, 53, PhD(org chem), 55; Univ Strathclyde, DSc(fiber & polymer sci), 70; FRSC. *Prof Exp:* Asst to Prof F S Spring, Royal Col Sci & Technol, Scotland, 55-56; res chemist electrochem, E I du Pont de Nemours, 56-62; sr scientist, pioneering res dept, Weyerhaeuser Co, 62-66; PROF FIBER SCI & POLYMER CHEM, DEPT CHEM ENG & COL FOREST RESOURCES, UNIV WASH, 66- *Concurrent Pos:* Head sci leather technologists, David Dale Tech Col, Glasgow & lectr, Paisley Tech Col, Scotland, 52-56. *Mem:* Am Chem Soc; Am Inst Chem Engrs; Tech Asn Pulp & Paper Indust. *Res:* Lignin and forest products chemistry; polymers; biologically active organic heterocycles; epoxidation; nonwovens; fiber surface modification; adhesion; controlled release pesticides; pollution control; marine polymers. *Mailing Add:* Univ Wash AR-10 Seattle WA 98195

ALLAN, J DAVID, b London, Ont, Feb 5, 45; m 68. ECOLOGY. *Educ:* Univ BC, BSc, 66; Univ Mich, MS, 68, PhD(zool), 71. *Prof Exp:* Fel biol, Univ Chicago, 71-72; asst prof, 72-76, ASSOC PROF ZOOL, UNIV MD, COLLEGE PARK, 76- *Mem:* AAAS; Ecol Soc Am; Int Soc Theoret & Appl Limnol; Am Soc Naturalists; Soc Study Evolution. *Res:* Ecology and aquatic biology; population dynamics and life histories; competition and predation; distributional patterns. *Mailing Add:* Dept Zool Univ Md College Park MD 20742-4415

ALLAN, JOHN R, b Detroit, Mich, Feb 6, 37; m 63; c 4. PLANT PHYSIOLOGY. *Educ:* Univ Mich, BS, 59; Ind Univ, MA, 62; Univ Sask, PhD(biol, plant physiol), 67. *Prof Exp:* Teaching asst aquatic plants, Mich Biol Sta, 59; res asst plant physiol, Argonne Nat Lab, 61; teaching asst, Ind Univ, 61-62; res asst, Univ Sask, 62-64; RES SCIENTIST PLANT PHYSIOL, AGR CAN RES STA, 66- *Mem:* Phytochem Soc NAm; Weed Sci Soc Am; Aquatic Plant Mgt Soc NAm. *Res:* Aquatic plant physiology; limnology; aquatic plant taxonomy and ecology; management of aquatic vegetation to provide maximum water efficiency with minimum ecological damage; integrated Aquatic Vegetation Management Program using chemical, biological, mechanical and nutrient limitation techniques to maintain aquatic plant populations at non-nuisance levels. *Mailing Add:* Can Agr Res Sta PO Box 3000 Main Lethbridge AB T1J 4B1 Can

ALLAN, JOHN RIDGWAY, b New Jersey, Apr 1, 47; m 70; c 2. CARBONATE GEOLOGY, INORGANIC GEOCHEMISTRY. *Educ:* Lehigh Univ, BA, 69, MS, 72; Brown Univ, PhD(geol), 78. *Prof Exp:* Production geologist, Gulf Oil Co, 69-70; lectr geol, RI Jr Col, 77; from res geologist to sr res geologist, 77-86, SR RES ASSOC, CHEVRON OIL FIELD RES CO, 86- *Concurrent Pos:* proj leader, Petrol Res Proj, 80-83; Group leader, Petrol Res Group, Chevron Oil Field Res Co, 86- *Mem:* Am Asn Petrol Geologists; Soc Econ Paleontologists & Mineralogists; Int Asn Geochem & Cosmochem. *Res:* Sedimentology, diagenesis and geochemistry of carbonate petroleum reservoir rocks. *Mailing Add:* Chevron Oil Field Res Co Box 446 LaHabra CA 90631

ALLAN, ROBERT EMERSON, b Morris, Ill, Jan 12, 31; c 4. AGRONOMY. *Educ:* Iowa State Univ, BS, 52; Kans State Univ, MS, 56, PhD(agron), 58. *Prof Exp:* Res asst, Kans State Univ, 54-57; agronomist, 57-59, GENETICIST & RES LEADER WHEAT BREEDING & PROD, AGR RES SERV, USDA, 59-; PROF AGRON, WASH STATE UNIV, 59- *Concurrent Pos:* Adj prof agron, Univ Idaho, 79- *Mem:* Genetics Soc Am; fel Am Soc Agron; Am Genetic Asn; Am Phytopathological. *Res:* Genetics and cytogenetics of wheat and its relatives; genetics of host, parasite relationships; nuclear and cytoplasmic interactions in Triticum species. *Mailing Add:* 209 Johnson Hall Dept Agron Wash State Univ Pullman WA 99163

ALLAN, ROBERT K, b Hamilton Ont, Jan 25, 40; m 62; c 3. BIOCHEMISTRY. *Educ:* McMaster Univ, BSc, 62, PhD(biochem), 67. *Prof Exp:* Asst prof, 66-70, ASSOC PROF CHEM, YORK UNIV, 70-, ASSOC DEAN, FAC SCI, 72- *Mem:* Inst Can. *Res:* Biochemistry of microorganisms; chloroplast mutagenesis; mechanism of action of chemical mutagens. *Mailing Add:* Dept Chem York Univ 4700 Keele St Downsview ON M3J 1P6 Can

ALLANSMITH, MATHEA R, IMMUNOLOGY, ALLERGIES. *Educ:* Univ Calif, San Francisco, MD, 55. *Prof Exp:* ASSOC PROF OPHTHAL, MED SCH, HARVARD UNIV, 77-; SR SCIENTIST, EYE RES INST, 81- *Mailing Add:* Dept Ophthal Med Sch Harvard Univ 20 Staniford St Boston MA 02114

ALLARA, DAVID LAWRENCE, b Vallejo, Calif, Nov 3, 37; m 68; c 2. SURFACE CHEMISTRY. *Educ:* Univ Calif, Berkeley, BS, 59; Univ Calif, Los Angeles, PhD(chem), 64. *Prof Exp:* NSF fel Oxford Univ, 64-65; fel, Stanford Res Inst, 65-66, org chemist, 66-67; assoc prof chem. San Francisco State Col, 67-69; mem tech staff, Bell Tel Labs, 69-84, res mgr, Bell Commun Res, Inc, 84-87; PROF CHEM & MAT SCI, PA STATE UNIV, 87- *Mem:* AAAS; Am Chem Soc; Am Phys Soc; Mat Res Soc. *Res:* Interfacial chemistry; surface spectroscopy. *Mailing Add:* Mat Sci Dept 309 Steidle Penn State Univ University Park PA 16802

ALLARD, DOUGLAS LEE, wood science and technology, for more information see previous edition

ALLARD, GILLES OLIVIER, b Rougemont, Que, Dec 12, 27; m 52; c 3. GEOLOGY, ECONOMIC GEOLOGY. *Educ:* Univ Montreal, BA, 48, BS, 51; Queen's Univ, Ont, MA, 53; Johns Hopkins Univ, PhD(geol), 56. *Prof Exp:* Consult, Que Dept Natural Resources, 52-55; field mgr, Chibougamau Mining & Smelting, Que, 55-58; asst prof geol, Univ Va, 58-59; prof, Ctr Res & Develop Petrol-Petrobras, Salvador, Brazil, 59-64; vis lectr, Univ Calif, Riverside, 64-65; assoc prof , 65-69, actg head dept, 69-70, PROF GEOL, UNIV GA, 70- *Concurrent Pos:* Consult, Que Dept Energy & Resources, 66-; Sandy Beaver Professorship, 78-82; Can Inst Mining Distinguished lectr, 84-85; pres, Southeastern Sect Geol Soc Am, 85-86. *Mem:* Fel Geol Soc Am; fel Mineral Soc Am; Can Inst; Soc Econ Geologists. *Res:* Structural and economic geology of Chibougamau, Quebec; discovery of Henderson Copper Mine, Chibougamau; Dore Lake layered complex; Propria geosyncline, Brazil; geologic link Brazil-Gabon and continental drift; volcanic ore deposits; metamorphism and ores. *Mailing Add:* Dept of Geol Univ of Ga Athens GA 30602

ALLARD, JOHANE, b Montreal, Sept 29, 56; m; c 2. MEDICINE. *Educ:* Ahuntsic Col, DEC, 75; Univ Montreal, MD, 81; Am Bd Internal Med, cert, 85; FRCP(C). *Prof Exp:* Resident I & I internal med, Maisonneuve-Rosemont Hosp, 81-83; resident III, Hotel-Dieu de Montreal Hosp, 83-84; resident IV, McGill Univ, 84-86; res fel nutrit, 87-89, ASST PROF, DEPT MED, UNIV TORONTO, 89-, STAFF PHYSICIAN, DIV GASTROENTEROL, TORONTO GEN HOSP, 89- *Concurrent Pos:* Chief resident, Maisonneuve-Rosemont Hosp, 82-83; mem training comt, McGill Univ, 85-86; mem nutrit comt, Toronto Gen Hosp, 89- *Mem:* Am Gastroenterol Asn; Can Asn Gastroenterol; Am Soc Enteral & Parenteral Nutrit; Int Soc Burn Injuries; Am Soc Clin Nutrit. *Mailing Add:* Dept Med Toronto Gen Hosp 200 Elizabeth St 7 Bell Wing Rm 637 Toronto ON M5G 2C4

ALLARD, NONA MARY, b Minneapolis, Minn, Dec 23, 28. MATHEMATICS. *Educ:* Col St Catherine, BA, 50; Cath Univ Am, MA. 60, PhD(math), 67. *Prof Exp:* Teacher, Cromwell High Sch, Minn, 50-52, Alexander Ramsey High Sch, 53-54 & Visitation High Sch, Ill, 56-58; instr, 60-61, from instr to asst prof, 65-73, ASSOC PROF MATH, ROSARY COL, 74- *Mem:* Math Asn Am. *Res:* Functional analysis; convex topological algebras and other generalizations of Banach algebras. *Mailing Add:* 7900 W Division St River Forest IL 60305

ALLARD, ROBERT WAYNE, b Los Angeles, Calif, Sept 3, 19; m 44; c 5. GENETICS & PLANT BREEDING, AGRONOMY. *Educ:* Univ Calif, BS, 41; Univ Wis, PhD, 46. *Prof Exp:* From asst prof to prof genetics, 46-86, chmn dept genetics, 67-75, chmn, Acad Senate, 74-76, EMER PROF GENETICS, UNIV CALIF, DAVIS, 86- *Concurrent Pos:* Guggenheim fels, 54-55 & 60-61; Fulbright sr res fel, 55; centennial prof, Am Univ, Beirut, 67; Nat Res Coun comt Managing Global Plant Genetics Resources, chair, 86-89; fac res lectr, Univ Calif, 74. *Honors & Awards:* Crop Sci Award, Am Soc Agron, 64; Nilsson Ehle Lectr, Mendelian Soc, Sweden, 80; De Kalb-Phzer Distinguished Career Award, Crop Sci Soc, 83; Wilhelmine Key Lectr, Am

Genetics Asn, 87. *Mem:* Nat Acad Sci; Genetics Soc Am (pres, 84); Am Genetic Asn (pres, 88); Am Soc Naturalists (pres, 75); Am Soc Agron; Am Acad Arts & Sci. *Res:* Population, ecological, and evolutionary genetics; genetic resource conservation; genetics of host-pathogen interactions; plant breeding. *Mailing Add:* Dept Agron & Range Sci Univ of Calif Davis CA 95616

ALLARD, WILLIAM KENNETH, b Lowell, Mass, Oct 29, 41; m 68; c 2. PURE MATHEMATICS. *Educ:* Villanova Univ, ScB, 63; Brown Univ, PhD(math), 68. *Prof Exp:* Asst prof to math, Princeton Univ, 68-78; chmn dept, 85-86, PROF MATH, DUKE UNIV, 76- *Concurrent Pos:* Sloan Found fel, 70-72; man ed, Duke Math J, 83-85. *Mem:* Am Math Soc. *Res:* Application of geometric measure theoretic techniques to the study of elliptic variational problems. *Mailing Add:* Duke Univ Durham NC 27707

ALLAUDEEN, HAMEEDSULTHAN SHEIK, b Madras, India, Apr 12, 43; m 69; c 2. CANCER. *Educ:* Madras Univ, BSc, 63, MSc, 65; Indian Inst Sci, PhD(biochem), 71. *Prof Exp:* Res staff scientist, nucleic acid biochem, Molecular Biophysics & Biochem Dept, 71-73, Leukemia Soc Am fel viral etiology human cancer, 73-75, lectr, 75-77, ASST PROF, DEPT PHARMACOL, YALE UNIV, 77- *Concurrent Pos:* Guest researcher tumor virol, Nat Cancer Inst, 73-74; co-prin investr, Am Cancer Soc, 75-78, prin investr, 78-80. *Mem:* Soc Pharmacol & Exp Med (treas, 66-67); Biochem Soc (secy, 66-67); Indian Soc Biol Chemists; Am Asn Cancer Res; Am Soc Microbiol. *Res:* Regulation of DNA replication in neoplastic cells; DNA polymerases of human cancer cells, herpesviruses and retroviruses; viral etiology of human cancer; biochemical mechanism of action of certain antitumor and antiherpesviral compounds. *Mailing Add:* 10 Sheffield Rd North Haven CT 06473

ALLAWAY, NORMAN C, biostatistics; deceased, see previous edition for last biography

ALLCOCK, HARRY R(EX), b Loughborough, Eng, Apr 8, 32; m 59. INORGANIC CHEMISTRY, POLYMER CHEMISTRY. *Educ:* Univ London, BSc, 53, PhD(chem), 56. *Prof Exp:* Fel org chem, Purdue Univ, 56-57; res fel phys-org chem, Nat Res Coun Can, 58-60; res chemist, Stamford Res Labs, Am Cyanamid Co, 61-65; sr res chemist, 65-66; from assoc prof to prof, 66-85, EVAN PUGH PROF CHEM, PA STATE UNIV, 85- *Concurrent Pos:* Guggenheim fel, 86-87. *Honors & Awards:* Nat Award in Polymer Chem, Am Chem Soc, 84; Pioneer Award, Am Inst Chemists, 89. *Mem:* Am Chem Soc; Chem Soc; Royal Soc Chem. *Res:* Inorganic-organic polymer chemistry; solid state chemistry; biomedical polymers; synthesis; reaction mechanisms; molecular structure studies; phosphorus-nitrogen, organosilicon and transition metal organometallic systems. *Mailing Add:* Dept Chem Pa State Univ University Park PA 16802

ALLDAY, CHRISTOPHER J, b Sutton Coldfield, Eng, Sept 5, 43. ALGEBRAIC TOPOLOGY. *Educ:* Cambridge Univ, Eng, BA, 65, MA, 69; Univ Calif, PhD(math), 70. *Prof Exp:* PROF MATH, UNIV HAWAII, 70- *Mem:* Am Math Soc. *Mailing Add:* Dept Math Univ Hawaii Honolulu HI 96822

ALLDREDGE, ALICE LOUISE, b Denver, Colo, Feb 1, 49; m 78. BIOLOGICAL OCEANOGRAPHY, MARINE ECOLOGY. *Educ:* Carleton Col, BA, 71; Univ Calif, Davis, PhD(ecol), 75. *Prof Exp:* NATO fel marine ecol, Australian Inst Marine Sci, 75-76; ASST PROF BIOL, UNIV CALIF, SANTA BARBARA, 76- *Concurrent Pos:* Prin investr, NSF grant biol oceanog, 77- *Mem:* AAAS; Am Asn Limnol & Oceanog; Sigma Xi. *Res:* Ecology of gelatinous marine zooplankton; macroscopic aggregates and the particle size distribution of particulate organic matter in the sea; behavior and ecology of demersal reef zooplankton. *Mailing Add:* Dept Biol Sci Univ Calif 7295 Almeda Ave Gol Santa Barbara CA 93106

ALLDREDGE, GERALD PALMER, b Hereford, Tex, Oct 20, 35. SURFACE PHYSICS, SURFACE CHEMISTRY. *Educ:* Tex Technol Col, BA, 58, MS, 60; Mich State Univ, PhD(physics), 66. *Prof Exp:* From instr to asst prof physics, Southern Ill Univ, 64-68; res scientist & lectr physics, Univ Tex, Austin, 68-75; sr res investr, Grad Ctr Mat Res, Univ Mo, Rolla, 75-77, assoc prof physics, Rolla, 76-79 & Columbia, 79-85; prof physics, Northeast Mo State Univ, 85-86; PRIN STAFF MEM, BDM INT, 86- *Concurrent Pos:* Partic, Prof Activ Continuing Educ Prog, Argonne Nat Lab, resident res assoc, 65 & 66; consult, Columbia Sci Res Inst, Tex, 69-75; vis scholar, Univ Tex, Austin, 75-83; guest prof, Tech Univ of Denmark, 83-84; vis res scientist, UK AERE-Harwell, 84; vis prof, Univ Calif, Irvine, 84; vis scholar, Univ Mo, Columbia, 85. *Mem:* AAAS; Am Phys Soc; Am Asn Physics Teachers; Am Chem Soc; Sigma Xi; Mat Res Soc; Am Vacuum Soc; Asn Comput Mach; Inst Elec & Electronics Engrs; Comput Soc; Soc Indust & Appl Math. *Res:* Theory of condensed matter; physics and chemistry of surfaces and interfaces; chemical physics; statistical mechanics; computational physics; mathematical physics; materials science, quantum chemistry. *Mailing Add:* MS-C-23 BDM Int Inc 1801 Randolph Rd SE Albuquerque NM 87106

ALLDREDGE, LEROY ROMNEY, b Mesa, Ariz, Feb 6, 17; m 40; c 7. GEOPHYSICS. *Educ:* Univ Ariz, MS, 40; Harvard Univ, ME, 52; Univ Md, PhD, 55. *Prof Exp:* Instr physics, Univ Ariz, 40-41; radio inspector, Fed Commun Comn, 41-44; radio engr, Dept Terrestrial Magnetism, Carnegie Inst Washington, 44-45; physicist & div chief, Res Dept, Naval Ord Lab, 45-55; analyst, Opers Res Off, Johns Hopkins Univ, 55-59; geophysicist, Coast & Geodetic Surv, US Dept Com, 59-65; from actg dir to dir, Inst Earth Sci, Environ Sci Serv Admin, 65-70, dir, Earth Sci Labs, Nat Oceanic & Atmospheric Admin Environ Res Labs, 70-73; res geophysicist, US Geol Surv, Dept Interior, Denver, 73-88; RETIRED. *Mem:* Am Geophys Union; Int Asn Geomag & Aeronomy. *Res:* Electricity and magnetism. *Mailing Add:* 4475 Chippewa Dr Boulder CO 80303

ALLDRIDGE, NORMAN ALFRED, b Eugene, Ore, July 28, 24; m 48; c 4. BOTANY. *Educ:* Utah State Agr Col, BS, 50; Univ Tex, PhD(bot), 56. *Prof Exp:* Instr bot, Univ Tex, 56-57; asst prof biol, 57-69, dir, Biol Field Sta, 61-69, exec officer dept biol, 64-69, ASSOC PROF BIOL, CASE WESTERN RESERVE UNIV, 69- *Mem:* AAAS; Am Soc Plant Physiol; Am Soc Cell Biologists. *Res:* Plant growth, development and metabolism; physiology of dwarfism; nature, distribution and function of plant peroxidases; metabolic pathways involved in germination. *Mailing Add:* Dept Biol Case Western Reserve Univ University Circle Cleveland OH 44106

ALLEE, MARSHALL CRAIG, b Rome, Ga, July 18, 41; m 60; c 3. ENTOMOLOGY, ZOOLOGY. *Educ:* Shorter Col, Ga, BA, 63; Clemson Univ, MS, 65, PhD(entom), 68. *Prof Exp:* Instr biol, Clemson Univ, 66-67; assot prof biol, Shorter Col, Ga, 68-, dean, Div Student Serv, 72-; AT DEPT BIOL, FLOYD JR COL. *Mem:* Entom Soc Am. *Res:* Response of the face fly, Musca autumnalis DeGeer, to various concentrations of aphotate and tepa; potential of Aleochara tristis Gravenhorst as a means of controlling the face fly. *Mailing Add:* 135 Harpers Way Carrolltown GA 30117

ALLEGRE, CHARLES FREDERICK, b Osage City, Kans, Oct 6, 11; m 38. BIOLOGY. *Educ:* Kans State Teachers Col, BS, 36; Univ Iowa, MS, 41, PhD(zool), 47. *Prof Exp:* Assoc prof biol & actg head dept, Gustavus Adolphus Col, 47-50; from asst prof to prof, 50-80, EMER PROF BIOL, UNIV NORTHERN IOWA, 80- *Mem:* AAAS; assoc Am Soc Zool; Am Micros Soc. *Res:* Protozoology; fresh water and parasitic taxonomy; microbiology; life history of a gregarine parasitic in grasshoppers. *Mailing Add:* 1403 W Seventh St Cedar Falls IA 50613

ALLEGRETTI, JOHN E, b Weehawken, NJ, Aug 23, 26; m 51; c 2. CHEMICAL ENGINEERING. *Educ:* NY Univ, BChE, 51. *Prof Exp:* Supvr pilot plant, Merck Sharp & Dohme Res Labs, 51-53, assoc chemist process develop, 53-55, res assoc eng develop, 55-57, group leader, 57-59, sr res engr, 59-61, res assoc, Electronic Chem Div, Merck & Co, Inc, 61-63, mgr process res, 63-64, sect mgr, Merck Sharp & Dohme Res Labs, Rahway, NJ, 64-66, mgr & chem engr res & develop, 66-69, dir pharmaceut develop, West Point, Pa, 69-75, SR DIR PHARMACEUT RES & DEVELOP, MERCK SHARP & DOHME RES LABS, WEST POINT, PA, 75- *Mem:* Am Inst Chem Engrs; NY Acad Sci; AAAS. *Res:* Materials science; epitaxial growth of single crystal silicon layers on silicon substrates for producing unique electronic device structures; optical isomer separation; automation of pharmaceutical processes. *Mailing Add:* 31 Hamlin Rd East Brunswick NJ 08816

ALLEMAND, CHARLY D, b Soengeiliat, Banka, Sept 2, 24; m; c 1. APPLIED PHYSICS. *Educ:* Swiss Fed Inst Technol, dipl, 49; Univ Neuchatel, PhD(semiconductors), 54. *Prof Exp:* Asst solid state physics, Univ Neuchatel, 51-52, chief adv, 52-54; scientist, Fabrique Suisse de Ressorts d'Horlogerie, SA, 54-58, tech mgr metall, 58-60, gen mgr method, time & motion orgn prog, 60-61; mgr eng res, Jarrell-Ash Co, Switz, 62-66; staff scientist, Jarrell-Ash Div, 66-69, Fisher Res Labs, 69-73, SR STAFF SCIENTIST, JARRELL-ASH DIV, FISHER SCI CO, 73- *Honors & Awards:* Meggers Award, Soc Appl Spectros, 71. *Mem:* Optical Soc Am; NY Acad Sci; Soc Appl Spectros. *Res:* Underwater telephony; electronic conduction in dielectric crystals; high tensile strength alloys; automation; controlled electrolytic polishing; excitation sources for spectroscopy; geometrical and physical optics; micro-analysis systems; automated computer controlled spectrometric analyser. *Mailing Add:* 11 Oakwood Rd Newtonville MA 02160

ALLEN, ALEXANDER CHARLES, b Morristown, NJ, Dec 27, 33; m 60; c 3. PEDIATRICS, NEONATOLOGY. *Educ:* Haverford Col, BS, 55; McGill Univ, MD, CM, 59; FRCP(C); Am Pediat, dipl, 70. *Prof Exp:* Intern med, Royal Victoria Hosp, Montreal, Que, 59-60; jr asst resident pediat, Montreal Children's Hosp, 60-61; jr asst resident internal med, Royal Victoria Hosp, 61-62; sr asst resident pediat, Montreal Children's Hosp, 62-63; res assoc neonatology, McGill Univ & Royal Victoria Hosp, Montreal, 63-66; asst prof pediat, obstet & gynec, Sch Med, Univ Pittsburgh, 69-75; ASSOC PROF PEDIAT & ASST PROF OBSTET & GYNEC, FAC MED, DALHOUSIE UNIV, 75-; HEAD NEONATAL PEDIAT DEPT, GRACE MATERNITY HOSP, 76- *Concurrent Pos:* Dir, Infant Referral Ctr, Magee-Women's Hosp, 73-75. *Mem:* Fel Am Acad Pediat; Can Pediat Soc; Soc Pediat Res. *Res:* Idiopathic respiratory distress syndrome of prematurity; asphyxia neonatorum; pulmonary blood flow and pulmonary water changes during neonatal adaptation to extra-uterine life; intraventricular hemorrage in neonates. *Mailing Add:* Dept Neonatal Pediat Grace Maternity Hosp 5821 University Ave Halifax NS B3H 1W3 Can

ALLEN, ANNEKE S, b Dronrijp, Netherlands, Sept 14, 30; US citizen; m 57; c 3. PHYSICAL CHEMISTRY. *Educ:* Tulane Univ, PhD(phys chem), 55. *Prof Exp:* Instr chem, Tulane Univ, 54-55; chemist, Cent Labs, Dow Chem Co, Tex, 55-58; pres, Alchem, Fla, 59-63; asst prof, 64-71, assoc dean lab, 77-81, ASSOC PROF CHEM, WICHITA STATE UNIV, 71- *Concurrent Pos:* Asst prof, Orlando Jr Col, 62-63. *Mem:* AAAS; Am Chem Soc; Sigma Xi. *Res:* Inorganic reactions in solution; electrochemistry of organometallic compounds. *Mailing Add:* Dept of Chem Wichita State Univ Wichita KS 67208

ALLEN, ANTON MARKERT, b Augusta, Ga, Feb 9, 31; m 51; c 4. VETERINARY PATHOLOGY. *Educ:* Univ Ga, DVM, 55; Univ Wis, PhD(path), 61. *Prof Exp:* Asst vet officer, Comp Path Sect, Nat Cancer Inst, 55-57; vet officer, Comp Path & Cytol, Sch Med, Univ Wis, 57-59; asst chief, 59-61, CHIEF COMP PATH SECT, VET RESOURCES BR, NIH, 61- *Mem:* AAAS; Int Primatol Soc; Am Vet Med Asn; Am Asn Path & Bact. *Res:* Pathology of primate diseases; mitosis and binucleation of mast cells; pathogenesis of naturally occurring diseases of laboratory animals. *Mailing Add:* Dir Vet Serv Microbiol Assocs Inc 5221 River Rd Bethesda MD 20816

ALLEN, ARCHIE C, b Ash, NC, Dec 23, 29; m 59; c 2. ZOOLOGY, GENETICS. *Educ:* Univ NC, BS, 55, MA, 58; Univ Pittsburgh, PhD(zool), 61. *Prof Exp:* NIH fel genetics, Univ Tex, 61-63; asst prof biol, 63-67, assoc prof biol, Tex Tech Univ 67-; RETIRED. *Mem:* AAAS; Genetics Soc Am. *Res:* Recombination, population and lethal studies of Drosophila melanogaster. *Mailing Add:* 3413 23rd St Lubbock TX 79401

ALLEN, ARNOLD ORAL, b Malcolm, Nebr, Aug 2, 29; m 61; c 1. MATHEMATICS. *Educ:* Univ Nebr, BA, 50; Univ Calif, Los Angeles, MA, 59, PhD(math), 62. *Prof Exp:* Asst physics, Univ Nebr, 50; comput programmer, IBM Corp, 56-58; asst math, Univ Calif, Los Angeles, 58-59; systs engr, IBM Corp, 62-64, mathematician, 64-68, PhD col rels rep, 68-71, sr staff mem, Systs Sci Inst, 71-84; WITH HEWLETT-PACKARD, 86- *Concurrent Pos:* Lectr, Univ Calif, Los Angeles, 65-71. *Mem:* Math Asn Am; Sigma Xi. *Res:* Applications of probability, statistics and queueing theory to analysis of computer systems. *Mailing Add:* 1420 Wabash Way Roseville CA 95678

ALLEN, ARTHUR, b Philadelphia, Pa, Nov 20, 28; m 56; c 2. BIOCHEMISTRY. *Educ:* Temple Univ, BA, 50, MA, 53, PhD(chem), 56. *Prof Exp:* Chief, div chem & toxicol, First US Army Med Lab, NY, 56-58; from asst prof to assoc prof biochem, 58-83, PROF BIOCHEM & MOLECULAR BIOL, JEFFERSON MED COL, 83- *Concurrent Pos:* Lectr chem, Temple Univ, 60-67, adj assoc prof, 67-69. *Mem:* AAAS; Am Chem Soc; Sigma Xi. *Res:* Intermediary metabolism of carbohydrates and lipids. *Mailing Add:* Dept Biochem & Molecular Biol Jefferson Med Col Philadelphia PA 19107

ALLEN, ARTHUR (SILSBY), b Brewer, Maine, Mar 5, 34; m 58; c 2. PLANT PATHOLOGY, MYCOLOGY. *Educ:* Univ Maine, BS, 56, MS, 60; Mich State Univ, PhD(phytopath), 68. *Prof Exp:* Asst prof bot, Univ Maine, 58-60; asst phytopath, Mich State Univ, 60-63, res assoc & asst instr, 65-68; asst prof bot, Humboldt State Col, 63-65; assoc prof phytopath, 68-74, PROF BOT & FORESTRY, NORTHWESTERN STATE UNIV, 74- *Concurrent Pos:* NSF fel. *Mem:* Am Phytopath Soc; Sigma Xi; Am Soc Soil Conserv; Am Forestry Asn. *Res:* Host-pathogen relations; epidemiology and control of pecan diseases; biocontrol aquatic plants; forest restoration; post lignite mining operations. *Mailing Add:* Dept of Biol Sci Northwestern State Univ Natchitoches LA 71497

ALLEN, ARTHUR CHARLES, b Trenton, NJ, Dec 16, 10; wid. PATHOLOGY. *Educ:* Univ Calif, MD, 36. *Prof Exp:* Asst med examr, New York, 46-48; assoc pathologist, Mem Ctr Cancer, 48-57; prof path, Sch Med, Univ Miami, 57-61; clin prof path, 61-77, prof path, State Univ NY Downstate Med Ctr, 77-; dir labs, Jewish Hosp Brooklyn, 61-88; RETIRED. *Concurrent Pos:* Consult, Armed Forces Inst Path, 46-49, Bronx Vet Admin Hosp, 46-57, Coral Gables Vet Admin Hosp, 57-61, Hunterdon Med Ctr, 56-, Brooklyn Vet Admin Hosp, Phelps Mem Hosp, Nassau Hosp & Huntington Hosp, Long Island, 69-; assoc prof, Sch Med, Cornell Univ, 53-56; asst med examiner, New York City, 46-48. *Mem:* AAAS; Am Soc Clin Path; Asn Am Med Cols; Am Asn Path & Bact; Col Am Path. *Res:* Diseases of kidney, skin and gastrointestinal tract. *Mailing Add:* Dept Pathol State Univ NY Downstate Med Ctr 555 Prospect Pl Brooklyn NY 11238

ALLEN, ARTHUR LEE, b Los Angeles, Calif, June 7, 44; c 2. SOIL CHEMISTRY, PLANT PHYSIOLOGY. *Educ:* Univ Ark, Pine Bluff, BS, 66; Okla State Univ, MS, 68; Univ Ill, PhD(soil chem), 71. *Prof Exp:* Lab technician soil & water testing, Okla State Univ, 64-67, asst clay minerol, 66-71; assoc prof chem, Langston Univ, Okla, 71-75; res agronomist & assoc prof, 76-78, ASST DEAN DIV AGR & TECHNOL, 78-, DIR COOP RES PROG & COOP EXTEN SERV, UNIV ARK, PINE BLUFF, 78- *Concurrent Pos:* Reviewer proposals minority div, Environ Protection Agency, 75-76; asst dir Sci & Educ Admin-Coop Res Prog & asst state coordr 1890 inst, Coop Exten Serv, Univ Ark, Pine Bluff, 78- *Mem:* Am Soc Agron; Soil Sci Soc Am; Crop Sci Soc Am. *Res:* Nitrate in soil and water; aluminum-hydroxide polymers in soil clays; chemical distribution and residual fertilizer derived nitrogen in soil; nitrogen forms under feedlots; slow release fertilizers; irrigation studies with soybeans; foliar fertilization; nitrogen uptake in nutrients solution cultures. *Mailing Add:* 7929 Claremont Dr North Little Rock AR 72116

ALLEN, ARTHUR T, JR, b Darlington, SC, Sept 8, 17; m 40; c 1. GEOLOGY. *Educ:* Emory Univ, AB, 39; Univ Tenn, MS, 47; Univ Colo, PhD(geol), 50. *Prof Exp:* Mining geologist, Am Zinc Co, 40-46; from asst prof to assoc prof geol, 46-58, PROF GEOL, EMORY UNIV, 58-, CHMN DEPT, 64- *Mem:* Am Asn Petrol Geologists; Geol Soc Am. *Res:* Sedimentation; stratigraphy; paleontology. *Mailing Add:* Dept of Geol Emory Univ 1364 Clifton Rd NE Atlanta GA 30322

ALLEN, AUGUSTINE OLIVER, b San Rafael, Calif, July 16, 10; m 38; c 5. PHYSICAL CHEMISTRY, RADIATION CHEMISTRY. *Educ:* Univ Calif, BS, 30; Harvard Univ, PhD(phys chem), 38. *Prof Exp:* Res chemist, Bell Tel Labs, NY, 30-31; fel, Harvard Univ, 31-35; res chemist, Ethyl Corp, Detroit, 37-43; group leader & assoc sect chief, metall lab, Univ Chicago, 43-46; prin chemist, Oak Ridge Nat Lab, 46-48; sr scientist, Brookhaven Nat Lab, 48-76; RETIRED. *Concurrent Pos:* Vis scientist, Int Atomic Energy Agency, Greece, 65-66, Danish Atomic Energy Comn Lab, 70-71, Interuniv Reactor Inst, Delft, Netherlands, 75 & Hahn-Meitner Inst, Berlin, Ger, 77-78; consult, US Nuclear Regulatory Comn, 80-82, Am Phys Soc Comt Source Terms, 83-84. *Honors & Awards:* Humboldt Award, 77. *Mem:* AAAS; Am Chem Soc; Radiation Res Soc (pres, 63-64). *Res:* Reaction kinetics; electronic properties of liquids; radiation chemistry. *Mailing Add:* Box 173 Shoreham NY 11786

ALLEN, BONNIE L, b Hillsboro, Tex, Aug 11, 24. SOIL GENESIS, SOIL MINERALOGY. *Educ:* Tex Tech Univ, BS, 48; Mich State Univ, MS, 51, PhD(soil sci), 60. *Prof Exp:* Asst prof agron, Eastern NMex Univ, 52-57, assoc prof soils & geol, 57-59; assoc prof, 59-72, PROF SOILS, TEX TECH UNIV, 72- *Mem:* Fel Am Soc Agron; fel Soil Sci Soc Am; Am Chem Soc; Clay Minerals Soc; Int Soc Soil Sci; Int Soc Study Clays. *Res:* Mineralogical transformations and micromorphological changes with soil development; soil-native vegetation relationships. *Mailing Add:* Plant/Soil Sci Dept Tex Tech Univ Lubbock TX 79406

ALLEN, C EUGENE, b Burley, Idaho, Jan 25, 39. AGRICULTURAL RESEARCH. *Educ:* Univ Idaho, Moscow, BS, 61; Univ Wis, MS, 63, PhD(meat & animal sci), 65. *Prof Exp:* Res asst, Univ Wis, 61-65; NSF fel, Div Food Res, Commonwealth Sci & Indust Res Orgn, Sidney, Australia, 66-67; from asst prof to prof animal sci, Food Sci & Nutrit, Univ Minn, 67-84, dean, Col Agr & assoc dir, Minn Agr Exp Sta, 84-88, actg vpres, Inst Agr, Forestry & Home Econ & actg dir, Minn Agr Exp Sta, 88-90, VPRES, INST AGR, FORESTRY & HOME ECON & DIR, MINN AGR EXP STA, 90- *Concurrent Pos:* Res Award, Except Achievement, Midwestern Sect Am Soc Animal Sci, 72, Meat Sci, Am Soc Animal Sci, 77, Distinguished Meats, Am Meat Sci Asn, 80; vis prof animal sci, Pa State Univ, 78. *Honors & Awards:* C Glen King Lectr, Nutrit Grad Prog, Wash State Univ, 81; Signal Serv Award, Am Meat Sci Asn, 85. *Mem:* Fel Inst Food Technol. *Res:* Animal growth biology and the functional and nutritional characteristics of animal food products. *Mailing Add:* Univ Minn Coffey Hall 1420 Eckles Ave St Paul MN 55108

ALLEN, CHARLES A, b Bairdford, Pa, Apr 25, 33; m 54; c 7. ELECTRONICS ENGINEERING. *Educ:* Univ Pittsburgh, BS, 56; Calif Inst Technol, MS, 61; Stanford Univ, PhD(elec eng), 66. *Prof Exp:* Sr engr, Int Bus Mach Corp, 56-69; dir prod develop, Cogar Corp, 69-72, Amdahl Corp, 72-73; pres, Falcon Electronics, 73-88; dir IC Design, Seagate Corp, 88; PROG MGR, MEGATEST CORP, 89- *Mem:* Inst Elec & Electronic Engrs. *Res:* Application-specific integrated circuit design, particularly complementary metal-oxide semiconductor transistor. *Mailing Add:* 220 Whippet Run Corralitos CA 95076

ALLEN, CHARLES C(ORLETTA), b New York, NY, Mar 1, 21; m 44, 73; c 4. MICROWAVE ANTENNA SYSTEMS, COMMUNICATIONS. *Educ:* Worcester Polytech Inst, BS, 49, MS, 50. *Prof Exp:* Develop engr, Gen Res & Develop Ctr, Gen Elec Co, Schenectady, NY, 50-55, microwave engr, 55-59, mgr microwave eng, 59-61, microwave engr, 61-75, tech staff engr, Space Div, Valley Forge Space Ctr, Pa, 75-87; CONSULT ENGR, ALLEN CONSULT, 87- *Concurrent Pos:* Asst prof elec eng, Union Col, NY, 67-68, vis assoc prof, 68-69, adj assoc prof, 69-70; mem antenna standards comt, Inst Elec & Electronics Engrs, 59- *Mem:* Inst Elec & Electronics Engrs; Sigma Xi. *Res:* Ultra high frequency current measurement; microwave systems and antennas analysis and development; computer-aided antenna pattern and noise temperature calculations; microwave breakdown; microwave microelectronic circuits; satellite antenna systems; space based radar. *Mailing Add:* 604 Hickory Lane Berwyn PA 19312

ALLEN, CHARLES EUGENE, b Burley, Idaho, Jan 25, 39; m 60; c 2. ANIMAL SCIENCE, FOOD SCIENCE. *Educ:* Univ Idaho, BS, 61; Univ Wis, MS, 63, PhD(animal sci), 66. *Prof Exp:* NSF fel, Div Food Res, Commonwealth Sci & Indust Res Orgn, Australia, 66-67; from asst prof to assoc prof, 67-72, prof animal sci, food sci & nutrit, 72-84, DEAN COL AGR, UNIV MINN, ST PAUL, 84- *Concurrent Pos:* Consult, Am Cyanamid Co, 74-85; vis prof animal sci, Pa State Univ, 78; fel, Inst Food Technol, 87. *Honors & Awards:* Meats Res Award, Am Soc Animal Sci, 77 & Am Meat Sci Asn, 80. *Mem:* Am Soc Animal Sci; Am Meat Sci Asn; Inst Food Technol; Am Inst Nutrit; Sigma Xi. *Res:* Lipid deposition and composition; mechanisms related to growth and development of muscle and adipose tissue. *Mailing Add:* 3319 Katie Lane St Paul MN 55112

ALLEN, CHARLES FREEMAN, b Berkeley, Calif, Feb 16, 28; m 50; c 3. ORGANIC CHEMISTRY, BIOCHEMISTRY. *Educ:* Univ Calif, BS, 48; Univ Wis, PhD(org chem), 52. *Prof Exp:* Res chemist, Univ Calif, 52-54, instr, 53-54; from asst prof to assoc prof chem, 54-66, PROF CHEM, POMONA COL, 66- *Concurrent Pos:* NSF fac fel, Univ Cologne, 61-62, Kettering Res Lab, Ohio & Scripps Inst Oceanog, 68-69 & Nat Bur Standards, 81-82. *Mem:* Am Chem Soc; AAAS. *Res:* Air pollution chemistry; lipid biochemistry. *Mailing Add:* Dept Chem Pomona Col Claremont CA 91711

ALLEN, CHARLES MARSHALL, JR, b Cortland, NY, Sept 13, 38; m 63; c 2. BIOCHEMISTRY. *Educ:* Syracuse Univ, BS, 60; Brandeis Univ, PhD(biochem), 64. *Prof Exp:* USPHS fel chem, Harvard Univ, 64-67; from asst prof to assoc prof, 67-79, PROF BIOCHEM, UNIV FLA, 79- *Mem:* Am Chem Soc; Am Soc Biol Chemists. *Res:* Biosynthesis of long chain polyprenyl phosphates which function in glycosyl transfer reactions in bacterial and mammalian cells; enzyme action of isoprenoid metabolism. *Mailing Add:* Dept of Biochem Univ of Fla Gainesville FL 32601

ALLEN, CHARLES MARVIN, b Mt Gilead, NC, July 31, 18; m 43. CYTOLOGY, INVERTEBRATE ZOOLOGY. *Educ:* Wake Forest Col, BS, 39, MA, 41; Duke Univ, PhD, 55. *Prof Exp:* From instr to assoc prof, 41-62, PROF BIOL, WAKE FOREST UNIV, 62-, DIR CONCERTS & LECTS, 74- *Mem:* AAAS; Sigma Xi. *Res:* Chromosome morphology and germ cell cycles; morphology and cytology of marine annelids. *Mailing Add:* Dept Biol Wake Forest Univ Reynolds Sta PO Box 7325 Winston-Salem NC 27109

ALLEN, CHARLES MICHAEL, b New Castle, Pa, Sept 1, 42; c 4. MEDICAL IMAGING, PROCESS CONTROL. *Educ:* Carnegie Mellon Univ, 64, MS, 65; State Univ NY, Buffalo, PhD(systs eng), 68. *Prof Exp:* Asst prof elec eng, State Univ NY, Buffalo, 68-74; dir res & develop, DataSpan, Inc, 82-85, pres, 85-87; PROF COMPUTER ENG, UNIV NC, CHARLOTTE, 74- *Concurrent Pos:* Consult eng design, Int Bus Mach, Kodak, AT&T, Sylvania & numerous eng firms, 65-; sect chmn, Inst Elec & Electronic Engrs, 81. *Mem:* Inst Elec & Electronic Engrs; Asn Comput Mach. *Res:* Embedded micro-computer control of processors; medical imaging; process control. *Mailing Add:* 3816 Smokerise Hill Dr Charlotte NC 28277

ALLEN, CHARLES W(ILLARD), b Akron, Ind, Jan 7, 32; m 59; c 4. PHYSICAL METALLURGY. *Educ:* Univ Notre Dame, BS, 54, MS, 56, PhD(metall), 58. *Prof Exp:* Air Force Off Sci Res fel phys metall, 58, Off Naval Res fel, 58-59, from asst prof to assoc prof, 59-68, prof metall eng, Univ Notre Dame, 68-86; MAT SCIENTIST, ARGONNE NAT LAB, 86- *Concurrent Pos:* Fulbright res fel, SCK-Ctr Study Nuclear Energy, Mol, Belg, 70-71; fac leave for res at Argonne, 82-83. *Mem:* ASM Int; Am Phys Soc; Electron Miscros Soc Am; Mat Res Soc. *Res:* Dislocations and ferromagnetism; phase transformations, radiation effects; control of dislocation substructure; defects in intermediate alloy phases. *Mailing Add:* MSD 212 Argonne Nat Lab Argonne IL 60439

ALLEN, CHARLES WILLIAM, b Newbury, Eng, July 24, 32; US citizen; m 57; c 2. MECHANICAL ENGINEERING, TRIBOLOGY. *Educ:* Univ London, BS, 57; Case Inst Technol, MS, 62; Univ Calif, Davis, PhD(mech eng), 66. *Prof Exp:* Design engr, Lear Siegler Inc, Ohio, 57-62; group leader bearing develop, Aerojet Gen Corp, 62-63; assoc eng, Univ Calif, Davis, 63-66; assoc prof, 66-71, head, Dept Mech Eng, 76-79 & 82-84, PROF MECH ENG, CALIF STATE UNIV, CHICO, 71- *Concurrent Pos:* Fac fel, Stanford Univ, 67; res fel, Lewis Res Lab, NASA, 68 & 69; vis fel, Univ Leicester, 74. *Mem:* Am Soc Mech Engrs. *Res:* Traction and film thickness in elastohydrodynamic lubrication, squeeze film lubrication, melt lubrication, extending life of mechanical devices. *Mailing Add:* Mech Eng Dept Calif State Univ Chico CA 95929

ALLEN, CHERYL, b Detroit, Mich, Jan 30, 39. BIOCHEMISTRY, MEDICAL ETHICS. *Educ:* Nazareth Col, BS, 69; Univ Ill Med Ctr, PhD(biochem), 74. *Prof Exp:* Sec teacher math & sci, Detroit Parochial Sch Syst, 59-68; teaching asst, Univ Ill Med Ctr, 69-73; asst prof biochem, Nazareth Col, 74-84; bioethics consult, 84-89; CONGREGATIONAL ADMINR, SISTERS ST JOSEPHS HEALTH SYST, 84-89. *Mem:* Inst Soc, Ethics & Life Sci; Am Soc Law & Med. *Res:* The role of serum proteins, especially the alpha-2 macro-globulins, in immunity and host defense; ethical aspects of issues of life sciences. *Mailing Add:* Sisters St Joseph Health Syst 3427 Gull Rd Nazareth MI 49074

ALLEN, CHRISTOPHER WHITNEY, b Waterbury, Conn, Oct 19, 42; m 65; c 2. INORGANIC CHEMISTRY, POLYMER CHEMISTRY. *Educ:* Univ Conn, BA, 64; Univ Ill, MS, 66, PhD(inorg chem), 67. *Prof Exp:* From asst prof to assoc prof, 67-76, PROF CHEM, UNIV VT, 76- *Mem:* The Chem Soc; Am Chem Soc. *Res:* Synthesis, structure and spectroscopic properties of non-metal and organometallic compounds; inorganic polymers. *Mailing Add:* 20 Grandview Ave Essex Junction VT 05452

ALLEN, CLARENCE RODERIC, b Palo Alto, Calif, Feb 15, 25. GEOLOGY, SEISMOLOGY. *Educ:* Reed Col, BA, 49; Calif Inst Technol, MS, 51, PhD(geol), 54. *Prof Exp:* Asst prof geol, Univ Minn, 54-55; from asst prof to assoc prof geol, Calif Inst Technol, 55-64, interim dir, Seismol Lab, 65-67, actg chmn, Div Geol Sci, 67-68, chmn fac, 70-71, prof, 64-90, EMER PROF GEOL & GEOPHYS, CALIF INST TECHNOL, 90- *Concurrent Pos:* Consult, Calif Dept Water Resources, US Geol Surv & UNESCO; mem adv panel earth sci, NSF, 65-68, chmn, 67-68, mem adv panel environ sci, 70-; mem, Calif State Mining & Geol Bd, 65-75, chmn, 75; mem task force earthquake hazard reduction, Off Sci & Technol, Exec Off President, 70; chmn, Panel Earthquake Prediction, Nat Acad Sci, 73-76, Nat Earthquake Prediction Eval Coun, 79-84 & Am Plate Tectonics Deleg People's Repub China, 79; vchmn, Am Seismol Deleg People's Repub China, 74; consult, Bur Reclamation. *Honors & Awards:* G K Gilbert Award in Seismic Geol, 60. *Mem:* Nat Acad Sci; Nat Acad Eng; fel Geol Soc Am (pres, 73-74); Seismol Soc Am (pres, 75-76); fel Am Acad Arts & Sci; fel Am Geophys Union. *Res:* Mechanics of faulting; San Andreas fault system; relation of seismicity to geologic structure; geophysical exploration of glaciers; tectonics of Southern California and Baja California; Circum-Pacific earthquakes and faulting; micro-earthquakes; geologic hazards; nuclear waste management. *Mailing Add:* Seismol Lab 252-21 Calif Inst Technol Pasadena CA 91125

ALLEN, CLIFFORD MARSDEN, b Winnipeg, Man, Dec 14, 27; m 60; c 3. VOLCANOLOGY, METAMORPHISM. *Educ:* Univ Man, BSc, 49, MSc, 51. *Prof Exp:* From asst prof to assoc prof, 54-75, PROF GEOL, MT ALLISON UNIV, 75-, HEAD DEPT, 87- *Mem:* Geol Soc Am; Nat Asn Geol Teachers; fel Geol Asn Can. *Res:* Igneous and metamorphic petrology; volcanic geology of the Maritime Provinces of Canada. *Mailing Add:* Dept of Geol Mt Allison Univ Sackville NB E0A 3C0 Can

ALLEN, DAVID MITCHELL, b Sebree, Ky, July 15, 38; m 64; c 3. STATISTICS. *Educ:* Univ Ky, BS, 61; NC State Univ, MES, 66, PhD(statist), 68. *Prof Exp:* Asst prof, 67-72, ASSOC PROF STATIST, UNIV KY, 72- *Concurrent Pos:* Consult, Mead Johnson & Co, 68-; vis assoc prof, Cornell Univ, 74-75; assoc ed, Biometrics, 75- & Commun Statist, Algorithms Sect, 78- *Mem:* Biomet Soc; Inst Math Statist; Am Statist Asn. *Res:* Statistical computation; linear models; prediction. *Mailing Add:* Dept Statist Univ Ky Lexington KY 40506

ALLEN, DAVID THOMAS, b Pittsburgh, Pa, Jan 23, 58; m 86; c 2. ENERGY & ENVIRONMENT. *Educ:* Cornell Univ, BS, 79; Calif Inst Technol, MS, 81, PhD(chem eng), 83. *Prof Exp:* ASSOC PROF CHEM ENG, UNIV CALIF, LOS ANGELES, 83- *Concurrent Pos:* NSF presidential young investr, 87; Mem, Am Inst Pollution Prev, Environ Protection Agency, 90-92 & Nat Adv Comt Environ Policy & Technol, 90-93. *Mem:* Am Inst Chem Engrs; Am Asn Aerosol Res; Am Chem Soc. *Res:* Energy and the environment; characterization of hazardous chemical mixtures; thermodynamic and kinetic modeling of heavy fuels; development of a catalytic hydrodechlorination process for recycling chlorinated organics; establishing new methods for characterizing atmospheric aerosols. *Mailing Add:* 5531 Boelter Hall Univ Calif Los Angeles CA 90024

ALLEN, DAVID WEST, RED CELL MEMBRANES, RED CELL ENZYMES. *Educ:* Harvard Univ, MD, 54. *Prof Exp:* PROF MED, UNIV MINN, 74- *Mailing Add:* Hemacol-Oncol Dept Vet Admin Affairs Med Ctr One Veterians Dr Minneapolis MN 55417

ALLEN, DELL K, b Cove, Utah, Dec 1, 31; m; c 8. COMPUTER INTEGRATED MANUFACTURING. *Educ:* Utah State Univ, BS, 54; Brigham Young Univ, MS, 60; Utah State Univ, EdD(industr educ), 73. *Prof Exp:* Process engr, EIMCO Corp, 59-60; PROF, MFR ENG TECHNOL, BRIGHAM YOUNG UNIV, 60- *Concurrent Pos:* Mech engr, Lawrence Livermore Lab, 75; res engr, Boeing Com Airplane Co, 76. *Honors & Awards:* Ray J Spies Award, CAM-I Inc, Cannes, 80. *Mem:* Fel Nat Acad Eng. *Res:* Integrated manufacturing; published numerous articles in various journals. *Mailing Add:* Dept Mfg Eng & Technol Brigham Young Univ Provo UT 84602

ALLEN, DELMAS JAMES, b Hartsville, SC, Aug 13, 37; m 58; c 3. HUMAN ANATOMY, NEUROANATOMY. *Educ:* Am Univ Beirut, BSc, 65, MSc, 67; Univ NDak, PhD(anat), 74. *Prof Exp:* Asst prof biol & chmn dept, Clarke Col, Iowa, 68-72; asst prof anat, Univ SAla, 74-75; prof anat, Med Col Ohio, 75-86; prof & assoc dean acad affairs & res, Ga State Univ, Atlanta, 86-88; VPRES ACAD AFFAIRS, NGA COL, 88- *Concurrent Pos:* Dir, NSF Res Grant, 70-72; grant-in-aid, Ala Heart Asn, 74-75 & Am Heart Asn, 76-78; Am Cancer Soc res grant, 77-79; NSF fel; NSF grant, 81; vis prof, Col Med, Riyad Univ, Kingdom of Saudi Arabia, 81. *Honors & Awards:* Acad Med Award, Brazilian Acad Med, 80; Res Award, Electron Micros Soc, 80; Res Award, Sigma Xi, 74. *Mem:* AAAS; Sigma Xi; Electron Micros Soc Am; Am Asn Anatomists; British Brain Res Asn. *Res:* Cranial meninges; ventricular system of the brain, choroid plexus; electron microscopy of dura mater cardiac valves recovered at surgery; pineal gland; smooth muscle in the ovary; scanning electron microscopy; transmission electron microscopy; x-ray microanalysis; cat scan and mri interpretation. *Mailing Add:* Off Dean NGa Col Dahlonega GA 30597

ALLEN, DELORAN MATTHEW, b Cherryvale, Kans, Feb 28, 39; m 65. ANIMAL SCIENCE, MEAT SCIENCE. *Educ:* Kans State Univ, BSAgr, 61; Univ Idaho, MSAgr, 63; Mich State Univ, PhD(meat sci), 66. *Prof Exp:* Asst prof animal sci, Kans State Univ, 66-70, assoc prof animal sci & indust & assoc animal scientist, 70-80, prof animal sci & indust, 80-88; VPRES QUAL & TRAINING, EXCEL CORP, 88- *Concurrent Pos:* Consult, ITAL, Campinas, Sao Paulo, Brazil, 78 & Chicago Mercantile Exchange, 80-81. *Mem:* Am Soc Animal Sci; Am Meat Sci Asn. *Res:* Beef carcass composition and cutability. *Mailing Add:* Excel Corp PO Box 2159 Wichita KS 67201-2519

ALLEN, DON LEE, b Burlington, NC, Mar 13, 34; m 58; c 3. PERIODONTICS. *Educ:* Univ NC, DDS, 59; Univ Mich, MS, 64. *Prof Exp:* From instr to prof periodont, Univ NC, 59-70, assoc dean admin, 69-70; assoc dean, Col Dent, Univ Fla, 70-73, dean, 73-82; DEAN, DENT BR, UNIV TEX, 82- *Concurrent Pos:* Consult, USPHS, 60-62, Philip Morris Co, 68-69 & Dent Educ, Am Dent Asn, 70-78; mem nat adv comt, Dept Health & Human Servs, 78-82; mem coun dent educ, Am Dent Asn, 78-86, chmn, 84-86; chmn, Comn Dent Accredidation, 78-86; chair, Fla Sect, Am Col Dentists, 80-81; pres, Southern Conf Dent Deans & Examiners, 83-84; chair, Int Dent Fedn, 90-91. *Mem:* Int Col Dentists (pres, 81-82); Am Asn Dent Sch (pres, 82-83); Am Dent Asn; Am Col Dentists; Int Dent Fedn. *Res:* Bone development, peridontal surgery, preventive dentistry and dental education. *Mailing Add:* Dean Dent Sch Univ Tex Health Sci Ctr PO Box 20036 Houston TX 77225

ALLEN, DONALD ORRIE, b Belding, Mich, Jan 11, 39; m 61; c 4. PHARMACOLOGY, LIPOLYSIS CONTROL. *Educ:* Ferris State Col, BS, 62; Marquette Univ, PhD(pharmacol), 67. *Prof Exp:* Res assoc pharmacol, Sch Med, Ind Univ, Indianapolis, 67-68, from asst prof to assoc prof, 68-75; PROF & CHMN PHARMACOL, SCH MED, UNIV SC, 75- *Mem:* Am Soc Pharmacol & Exp Therapeut. *Res:* Autonomic pharmacology and drug interaction; metabolic pharmacology; cyclic nucleotides. *Mailing Add:* Dept of Pharmacol Univ of SC Sch of Med Columbia SC 29208

ALLEN, DONALD STEWART, b Saugus, Mass, Sept 9, 11; m 42; c 2. CHEMISTRY. *Educ:* Dartmouth Col, AB, 32, AM, 34; Yale Univ, PhD(chem), 43. *Prof Exp:* Instr chem, Dartmouth Col, 32-34; master sci & math, Tex Country Day Sch, 34-40; from asst instr to instr chem, Yale Univ, 41-43; asst prof chem, Bates Col, 43-45; prof chem & chmn div natural sci, State Teachers Col, New Paltz, 45-59; prof chem, State Univ NY Albany, 59-68, chmn dept, 59-63; prof chem & dir div sci & math, 68-77, EMER PROF CHEM, EISENHOWER COL, 78- *Concurrent Pos:* Res scientist, Oak Ridge Inst Nuclear Studies, 59; consult, State Univ NY Ford Found Proj, Indonesia, 63-65; vis scholar hist & philos sci, Cambridge Univ, 74-75; vis fel, SEAP, Cornell Univ, 79-84; Paul Harris fel, Rotary, 85; Fulbright fel, Indonesia, 81. *Mem:* Am Chem Soc; fel Am Inst Chemists. *Res:* Friedel-Crafts reaction; latex research; standard electrode potentials. *Mailing Add:* 2976 Cayuga Lake Blvd Seneca Falls NY 13148

ALLEN, DOUGLAS CHARLES, b Brattleboro, Vt, Mar 8, 40; m 65; c 2. INSECT ECOLOGY. *Educ:* Univ Maine, BS, 62, MS, 65; Univ Mich, PhD(forest entom), 68. *Prof Exp:* ASSOC PROF FOREST INSECT ECOL, STATE UNIV NY COL ENVIRON SCI & FORESTRY, 68- *Mem:* Entom Soc Am; Soc Am Foresters; Entom Soc Can; Sigma Xi. *Res:* Population dynamics and bionomics of forest insects. *Mailing Add:* State Univ of NY Col of Environ Sci & Forestry Syracuse NY 13210

ALLEN, DUFF SHEDERIC, JR, b St Louis, Mo, Dec 8, 28; m 60; c 3. ORGANIC CHEMISTRY. *Educ:* Princeton Univ, BA, 49; Univ Wis, PhD(org chem), 60. *Prof Exp:* Res chemist, Monsanto Chem Co, 52-53; asst org chem, Univ Wis, 54-58, res assoc, 58-60; res chemist, Org Chem Div, 60-62, group leader, Lederle Labs, 63-65, HEAD DEPT ORG CHEM PROCESS RES, PROCESS & PREP RES SECT, LEDERLE LABS, AM CYANAMID CO, 65- *Mem:* Am Chem Soc; AAAS; NY Acad Sci. *Res:* Organic synthesis of steroids; heterocycles; natural products; medicinal chemicals. *Mailing Add:* Lederle Labs Div Am Cyanamid Co Pearl River NY 10965

ALLEN, DURWARD LEON, b Uniondale, Ind, Oct 11, 10; m 35; c 3. VERTEBRATE ECOLOGY. *Educ:* Univ Mich, AB, 32; Mich State Univ, PhD(zool), 37. *Hon Degrees:* LHD, Northern Mich Univ, 71; DAgr, Purdue Univ, 85. *Prof Exp:* Res biologist, Game Div, Mich State Dept Conserv, 35-46; biologist in charge wildlife invest on agr lands, US Fish & Wildlife Serv, Washington, DC, 46-50, asst chief, Br Wildlife Res, 51-53, actg chief, Br Wildlife, 53-54; assoc prof wildlife mgt, 54-57, prof wildlife ecol, 57-76, EMER PROF WILDLIFE ECOL, PURDUE UNIV, 76- *Concurrent Pos:* Asst secy gen, Inter-Am Conf Renewable Natural Resources, 48; mem, comt conserv, Nat Coun, Boy Scouts Am, 48-72; counr wildlife mgr, Nat Merit Badge, 52-64; mem, bd trustees, Nat Parks Asn, 64-75; mem & ed comt, wildlife & land-use relationships, Nat Acad Sci, 65-76; mem, Nat Park Syst Adv Bd, 66-72, chmn, 72-, coun of the bd, 72-84; mem bd dir, Nat Audubon Soc, 83-89. *Honors & Awards:* Leopold Mem Medal, 69; Audubon Medal, 90. *Mem:* Wildlife Soc (pres, 56-57); Ecol Soc Am; Am Soc Mammalogists; Sigma Xi; fel AAAS; Am Inst Biol Sci; Am Forestry Asn; George Wright Soc. *Res:* Ecological and management studies on fox squirrel, skunk, rabbit, pheasant, moose and wolf; management of renewable natural resources; ecology of the wolf and its prey in Isle Royale Nat Park. *Mailing Add:* Dept Forestry & Natural Resources Purdue Univ West Lafayette IN 47907

ALLEN, EDWARD DAVID, b Madison, Wis, Apr 3, 42; m 67; c 2. CRYOBIOLOGY, DEVELOPMENTAL BIOLOGY. *Educ:* Univ Wis, BS, 64, MS, 67; Univ Mich, PhD(bot), 72. *Prof Exp:* Consult electron micros, 70-72, RES SCIENTIST DEVELOP BIOL, UNIV MICH, 74-; RES BIOLOGIST CRYOBIOL, VET ADMIN HOSP, 72- *Mem:* Am Soc Cell Biologists; Soc Cryobiol. *Res:* Freeze preservation of blood components; developmental systems in fungi. *Mailing Add:* Vet Admin Hosp 2215 Fuller Rd Ann Arbor MI 48105

ALLEN, EDWARD FRANKLIN, b Denver, Colo, Aug 28, 07; m 30; c 2. PHYSICS. *Educ:* Acad New Church, BS, 28; Univ Pa, MA, 32. *Prof Exp:* Instr math & physics, 28-45, head dept math & phys sci, 52-69, prof math, 45-78, prof philos & chmn dept, 69-78, EMER PROF, ACAD OF NEW CHURCH, 78- *Concurrent Pos:* Res physicist, Franklin Inst, 42-45; ed, New Philos, 61-70. *Honors & Awards:* Naval Ord Develop Award, 45; Glencairn Award, 58. *Mem:* Am Asn Physics Teachers. *Res:* History and philosophy of science; instrumentation; specialization in the philosophy of Swedenborg. *Mailing Add:* PO Box 550 Bryn Athyn PA 19009-0550

ALLEN, EDWARD PATRICK, b Dallas, Tex, Sept 20, 43; m 65; c 5. PERIODONTICS. *Educ:* Baylor Univ, DDS, 69, PhD(physiol), 72. *Prof Exp:* Res fel physiol, Grad Sch, Baylor Univ, 69-72; asst prof periodont, Med Col Va, 72-76; assoc prof, 76-88, PROF PERIODONT, COL DENT, BAYLOR UNIV, 88- *Concurrent Pos:* Consult, Div Oral Surg, Southwestern Med Sch, 78- & J Periodont, 80- *Mem:* Am Acad Periodont; Int Asn Dent Res; Am Dent Asn; Sigma Xi; Fel, Int Col Dentists; Am Acad Restor Dent. *Res:* Evaluation of diagnostic methods including immunofluorescent analysis of biopsies for differentiation of severe and unusual inflammatory disorders affecting the oral mucosa so that effective therapy may be developed; development of periodontal plastic surgery procedures. *Mailing Add:* 3302 Gaston Ave Dallas TX 75246

ALLEN, ELIZABETH MOREI, GENETICS, IMMUNOLOGY. *Educ:* Univ Tenn, MA, 72. *Prof Exp:* Res assoc, Med Col Wis & Vet Admin Med Ctr, 75-87, RES SCIENTIST INSTR, DEPT DERMATOL, MED COL WIS, 90- *Mailing Add:* Dept Dermatol & Med Col Wis 8701 Watertown Plank Rd Milwaukee WI 53226

ALLEN, EMORY RAWORTH, b Augusta, Ga, Jan 21, 35; m 59; c 4. ANATOMY, CELL BIOLOGY. *Educ:* Univ Md, BS; Univ Pa, PhD(develop anat), 64. *Prof Exp:* Asst instr anat, Univ Pa, 62-65; from instr to asst prof anat & cell biol, Sch Med, Univ Pittsburgh, 65-71; adj assoc prof anat, Dept Occup Ther, 86, ASSOC PROF ANAT, LA STATE UNIV MED CTR, NEW ORLEANS, 71- *Concurrent Pos:* Fel anat, Univ Pa, 64-65. *Mem:* Sigma Xi; Int Acad Path; Am Asn Anatomists; Electron Micros Soc Am. *Res:* Ultrastructure of developing muscle; myosin synthesis. *Mailing Add:* Dept Anat La State Univ Med Ctr New Orleans LA 70112

ALLEN, ERIC RAYMOND, b Teneriffe, Canary Isles, Apr 12, 32; m 61; c 2. ATMOSPHERIC & AIR POLLUTION CHEMISTRY. *Educ:* Univ Liverpool, BSc, 56, PhD(phys chem), 59. *Prof Exp:* Fel photochem, Nat Res Coun, Can, 59-61; fel gas kinetics, Univ Calif, Riverside, 61-63; sr staff scientist, Nat Ctr Atmospheric Res, 63-68, prog scientist, 68-74; sr res assoc, Atmospheric Sci Res Ctr, State Univ NY Albany, 74-81; PROF ENVIRON ENG SCI, UNIV FLA, 81- *Concurrent Pos:* Vis assoc prof dept civil eng, Univ Tex, Austin, 71; mem, panel on carbon monoxide, Div Med Sci, Nat Acad Sci-Nat Res Coun, lectr, Environ Studies Prog, State Univ NY Albany, 74-77 & assoc prof innovative & interdisciplinary studies, 74-75; adj prof chem, State Univ NY Albany, 77-81. *Mem:* Air Pollution Control Asn; Am Chem Soc; Soc Appl Spectros; Sigma Xi; fel Am Inst Chemists. *Res:* Gas phase kinetics; atomic and free radical reactions; photochemistry; atmospheric and air pollution chemistry; water chemistry; solar energy storage systems; combustion and oxidations of hydrocarbons; source and ambient-air sampling and analysis, air toxics and odors, fluoride pollution, indoor air pollution; biofiltration, acid rain, "greenhouse" gas emissions and climate change. *Mailing Add:* 9108 SW First Place Gainesville FL 32607

ALLEN, ERNEST E, b Detroit, Mich, Sept 7, 33; m 57; c 3. MATHEMATICS EDUCATION. *Educ:* Wayne State Univ, BA, 55; Mich State Univ, BA, 57, MA, 59; Univ Detroit, MATM, 63; Univ Northern Colo, EdD(math educ), 70. *Prof Exp:* Pub sch teacher, 58-60; asst prof math educ, Eastern Mich Univ, 60-63; assoc prof, 63-74, PROF MATH, UNIV SOUTHERN COLO, 74-, DEAN, SCH OF SCI & MATH, 79- *Concurrent Pos:* Dir, Colo Coun Teachers Math, 68-70, ed, 70-, pres, 72. *Res:* Selected characteristics of junior high level mathematics teachers; methods of teaching mathematics. *Mailing Add:* Dept Math Univ Southern Colo 2200 N Bonforte Blvd Pueblo CO 81001

ALLEN, ERNEST MASON, b Terrell, Tex, Dec 1, 04; m 28; c 3. PUBLIC HEALTH ADMINISTRATION. *Educ:* Emory Univ, PhB, 26, MA, 39, DSc, 56. *Hon Degrees:* LLD, Clemson Univ, 68. *Prof Exp:* Instr French, Jr Col Augusta, 26-41; proj mgr, Nat Youth Admin, 41-43; sr pub health rep, Div Venereal Dis, USPHS, 43-46; asst chief div res grants, NIH, 46-51, chief div res grant, 51-60, assoc dir, Insts, 60-63; grants policy officer, USPHS, 63-68, dir off extramural prog, Off Asst Secy Health & Sci Affairs, 68-69, dir div grants admin policy, 69-70, dep asst secy grant admin policy, 70-73, assoc dir nat libr med, 73-82; RETIRED. *Honors & Awards:* Dirs Award, NIH, 77. *Mem:* AAAS; Biol Sci Info Exchange; Nat Health Forum. *Mailing Add:* 1010 Hickman Rd No B6 Augusta GA 30904-6312

ALLEN, ESTHER C(AMPBELL), b Tecumseh, Okla, Mar 28, 05. CHEMISTRY, BACTERIOLOGY. *Educ:* Pomona Col, AB, 26; Univ Calif, MS, 32. *Prof Exp:* Asst zool, Northwest Univ, 28-30; res asst, Scripps Inst, Univ Calif, 32-37; clin lab technician, Tulare-King Counties Tuberc Sanitarium, 37-38; supvry laboratorian, Camarillo State Hosp, 38-73; RETIRED. *Mem:* Am Soc Microbiol; Am Chem Soc; Am Asn Bioanalysts; fel Am Inst Chem. *Res:* Medical laboratory technology. *Mailing Add:* 3228 Marguerite Ave Corning CA 96021-9652

ALLEN, EUGENE (MURRAY), b Newark, NJ, Nov 7, 16; m 37; c 2. COLOR SCIENCE. *Educ:* Columbia Univ, BA, 38; Stevens Inst Technol, MS, 44; Rutgers Univ, PhD(chem), 52. *Prof Exp:* Res chemist, Utility Color Co, 38-39, United Color & Pigment Co, 39-41 & E R Squibb & Sons, 41-42; analytical res chemist, Picatinny Arsenal, 42-45; anal res chemist, Am Cyanamid Co, NJ, 45-56, group leader, 56-61, sr res scientist, 61-63, res assoc, 63-66, res fel, 66-67,; prof chem, 67-82, EMER PROF CHEM, LEHIGH UNIV, 82- *Concurrent Pos:* Consult, Int Comn Illum, 67-; mem US nat comt, Int Comn on Illumination, 67- *Honors & Awards:* Armin J Bruning Award, Fedn Socs Coatings Technol, 82; Godlove Award, Intersoc Color Coun, 83. *Mem:* Am Chem Soc; fel Optical Soc Am; Am Asn Textile Chem & Colorists; Sigma Xi. *Res:* Measurement and specification of color; colorimetry of fluorescent substances; color matching by digital computer; radiative transfer theory. *Mailing Add:* 2100 Main St Bethlehem PA 18017-3752

ALLEN, FRANCES ELIZABETH, b Peru, NY, Aug 4, 32; m 72. COMPUTER SCIENCE. *Educ:* State Univ NY, Albany, BA, 54; Univ Mich, MA, 57. *Hon Degrees:* DSc, Univ Alta, 91. *Prof Exp:* MEM RES STAFF, IBM, YORKTOWN HEIGHTS, 57- *Concurrent Pos:* Indust sabbatical fel, Courant Inst Math Sci, 69-70; adj assoc prof computer sci, NY Univ, 70-71 & 72-73; nat lectr, Asn Comput Mach, 72-73, chmn, Software Syst Award Comt, 83; gen chmn, Conf Principles Prog Languages, 90, mem, Comt Women & Minorities, 90-; mem computer sci adv bd, NSF, 72-75, consult, 75; instr, Peking Inst Computer Sci, 77; consult prof, Stanford Univ, 77-78; assoc ed, Trans Prog Languages & Systs J, 79-90, J Computer Languages, 81-; IBM fel, 89; distinguished lectr, Computer Sci Dept, Mass Inst Technol, 91. *Mem:* Nat Acad Eng; Asn Comput Mach; Inst Elec & Electronics Engrs. *Res:* Optimizing compilers; software for parallel systems; numerous publications. *Mailing Add:* TJ Watson Res Ctr IBM PO Box 704 Yorktown Heights NY 10598

ALLEN, FRANK B, b Mt Vernon, Ill, Oct 19, 09; m 43; c 3. MATHEMATICS, PHYSICS. *Educ:* Southern Ill Univ, BEd, 29; Univ Iowa, MS, 34. *Prof Exp:* High sch teacher, Ill, 29-41; teacher math, Lyons Twp High Sch & Jr Col, 41-42, 46-56, chmn dept, 56-68; from assoc prof to prof math, Elmhurst Col, 68-76, chmn dept, 70-74, adj prof, 76-79; RETIRED. *Mem:* Math Asn Am. *Res:* Application of elementary logic to the improvement of exposition in school mathematics. *Mailing Add:* 567 Berkley Ave Elmhurst IL 60126

ALLEN, FRANK LLUBERAS, information science, for more information see previous edition

ALLEN, FRED ERNEST, b Everett, Mass, Dec 27, 10; m 33; c 8. VETERINARY MEDICINE. *Educ:* Univ NH, BS, 32; Ohio State Univ, DVM, 36. *Prof Exp:* Jr veterinarian, US Bur Animal Indust, 36-37; munic veterinarian, Columbus, Ohio, 37-40; prof animal sci & veterinarian, Agr Exp Sta, 40-76, EMER PROF ANIMAL SCI, UNIV NH, 76- *Mem:* NY Acad Sci. *Res:* Bovine mastitis; immunization studies of staphylococcal type. *Mailing Add:* Packers Fall Rd Newmarket NH 03857

ALLEN, FREDDIE LEWIS, b Jellico, Tenn, Dec 6, 47; m 69; c 2. AGRONOMY. *Educ:* Tenn Technol Univ, BS, 70; Va Polytech Inst & State Univ, MS, 72; Univ Minn, PhD(plant breeding), 75. *Prof Exp:* From asst prof to assoc prof, 75-90, PROF PLANT & SOIL SCI, UNIV TENN, 90- *Concurrent Pos:* Consult to EPA on acid precipitation on soybeans, 85; consult, Nat Biol Impact Assessment Prog-Soybeans, 89; vis prof, Henan & Shandong Acad Agr Scis, Narying Agr Univ, People's Repub China, 90. *Mem:* Am Soc Agron; Crop Sci Soc Am; Am Soybean Asn; Sigma Xi. *Res:* Soybean breeding and genetics with emphasis on improved efficiency of breeding and selection methodology; effects of biotic and abiotic stresses on yield. *Mailing Add:* Dept Plant & Soil Sci Univ Tenn Knoxville TN 37916

ALLEN, FREDERICK GRAHAM, b Boston, Mass, Feb 2, 23; m 49; c 3. ELECTRICAL ENGINEERING. *Educ:* Cornell Univ, BME, 44; Harvard Univ, PhD(appl physics), 56. *Prof Exp:* Mem tech staff, Bell Tel Res Labs, NJ, 55-65; dept head manned space flight exp prog, Bellcomm, Inc, Washington, DC, 67-69; chmn dept, 69-76, 85-87, PROF ELEC SCI & ENG, UNIV CALIF, LOS ANGELES, 69- *Mem:* Fel Am Phys Soc; Inst Elec & Electronics Engrs. *Res:* Physics of semiconductor and metal surfaces; photoemission; molecular beam epitaxy of silicon. *Mailing Add:* 823 Westholme Ave Los Angeles CA 90024

ALLEN, GARLAND EDWARD, III, b Louisville, Ky, Feb 13, 36; m 66. HISTORY OF SCIENCE, HISTORY OF BIOLOGY. *Educ:* Univ Louisville, AB, 57; Harvard Univ, AMT, 58, AM, 64, PhD(hist sci), 66. *Prof Exp:* Allston-Burr sr tutor & instr hist sci, Harvard Univ, 65-67; from asst prof to assoc prof, 67-79, PROF BIOL, WASH UNIV, 79- *Concurrent Pos:* Comnr, Comn Undergrad Educ Biol Sci, 65-68; consult, Educ Res Coun, Ohio, 67; mem div soc sci, NSF Panel, 68-70; Charles Warren fel, Harvard Univ, 81-82; trustee, Marine Biol Lab, Woods Hole, Mass, 85-; vis prof, hist sci, Harvard Univ, 89-91. *Mem:* AAAS; Hist Sci Soc; Am Asn Hist Med; Brit Soc Hist Sci. *Res:* History of late 19th and early 20th century biology, especially genetics and evolution; history of eugenics in the 20th century. *Mailing Add:* Dept Biol Wash Univ St Louis MO 63130

ALLEN, GARY CURTISS, b Stockton, Calif, July 18, 39; m 65; c 2. GEOCHEMISTRY, PETROLOGY. *Educ:* Stanford Univ, BS, 61; Rice Univ, MA, 63; Univ NC, PhD(geochem), 68. *Prof Exp:* Geologist, Va Div Mineral Resources, 66-68; asst prof earth sci, La State Univ, New Orleans, 68-72; assoc prof, 72-78, PROF GEOL, UNIV NEW ORLEANS, 78- *Concurrent Pos:* Consult, indust firms & sch systs, 71 & consult mineralogy, Med Sch, Tulane Univ, 72-80; pres, Sunbelt Assocs, Inc, 78-; dir, Amguy Mining Ltd, 83-85, Placer Explor Ltd, 85-, Holocene Res Inst. *Mem:* Nat Asbestos Coun; Geol Soc Am; Sigma Xi; Mineral Soc Am. *Res:* Petrogenic geochemistry; mineralogy; analytical chemistry of major and trace elements in rocks and minerals; medical mineralogy of silica and asbestos; archeological geochemistry. *Mailing Add:* Dept Geol & Geophys Univ New Orleans New Orleans LA 70148

ALLEN, GARY WILLIAM, b Washington, DC, Mar 3, 44; m 73; c 1. PHYSICAL ORGANIC & PHOTOGRAPHIC CHEMISTRY. *Educ:* Univ Wash, BA, 67; Wesleyan Univ, PhD(chem), 71. *Prof Exp:* NIH res fel chem, Harvard Univ, 71-73; asst prof, Ariz State Univ, 73-75; SR RES CHEMIST, EASTMAN KODAK CO, 75- *Mem:* AAAS; Am Chem Soc; The Chem Soc; Sigma Xi. *Res:* Physical organic chemistry of the photographic process. *Mailing Add:* Eastman Kodak EPD Admin/RL Bldg 65 FL 1 Rochester NY 14650-1823

ALLEN, GEORGE E, entomology, invertebrate pathology, for more information see previous edition

ALLEN, GEORGE HERBERT, b Zurich, Switz, Aug 16, 23; m 55; c 3. WASTEWATER AQUACULTURE. *Educ:* Univ Wyo, BS, 50; Univ Wash, PhD(fisheries), 56. *Prof Exp:* From asst prof to prof fisheries, Humboldt State Univ, 56-82, prog leader, 67-75, chmn, 78-82; aquacult prog dir, 77-87, CONSULT, CITY ARCATA, CALIF, 87-; EMER PROF FISHERIES, HUMBOLDT STATE UNIV, 82- *Mem:* AAAS; Am Inst Biol Sci; Am Inst Fish Res Biol. *Res:* Development of public salmon culture system using treated wastewater. *Mailing Add:* Col Nat Resources Humboldt State Univ Arcata CA 95521

ALLEN, GEORGE PERRY, b Frankfort, Ky, Dec 16, 41; m 72; c 2. VIROLOGY. *Educ:* Georgetown Col, BS, 63; Univ Ky, PhD(microbiol), 75. *Prof Exp:* Lab technician microbiol, Ky State Dept Health, 64-66; lab technician, Univ Cincinnati, 66-68 & Univ Ky, 68-70; Am Cancer Soc fel, Univ Miss Med Ctr, 75-77, res asst microbiol, 75-78; from asst prof to assoc prof, 78-85, PROF VET SCI, UNIV KY, 86- *Concurrent Pos:* Mem, Conf Res Workers Animal Dis; postdoctoral training, Med Ctr, Univ Miss, 75-78. *Mem:* Am Soc Microbiol; Sigma Xi; Am Soc Virol. *Res:* Pathogenesis, epidemiology, and control of equine herpes virus disease; molecular characterization of antigens of equine herpes virus. *Mailing Add:* 5702 Applegate Lane Louisville KY 40219

ALLEN, GEORGE RODGER, JR, b Port Arthur, Tex, Nov 8, 29; m 55; c 2. MEDICINAL CHEMISTRY. *Educ:* Univ Tex, BS, 49, AM, 51, PhD(chem), 53. *Prof Exp:* Instr, Lamar Col, 51-52; assoc, Univ Tex, 52-53; res chemist, Calco Div, Am Cyanamid Co, 53-54, Lederle Labs, 55-58; res chemist, Mead Johnson & Co, 58-60; res chemist, Lederle Labs, 60-65, group leader, 65-74, dept head, 74-75, assoc dir new prod acquisitions, 75-76, dir pharmaceut & mech res & develop, Lederle Labs, 76-82; mgr, Info Systs, Admin/Planning, 82-90, DIR PLANNING, MED RES DIV, AM CYANAMID CO, 90- *Mem:* Am Chem Soc; Am Asn Pharmaceut Sci. *Res:* Inositols; steroids; antibiotics and related compounds; synthetic antibacterials and heterocyclic compounds. *Mailing Add:* Lederle Labs Pearl River NY 10965

ALLEN, GORDON AINSLIE, b London, Ont, May 24, 22; m 54; c 3. ENVIRONMENTAL MANAGEMENT. *Educ:* Univ Western Ont, BA, 43; Univ Rochester, PhD(chem), 49. *Prof Exp:* Demonstr chem, Univ Toronto, 43-44; develop chemist, L-Air Liquide Soc, Montreal, 44-46; res asst chem, Univ Rochester, 46-49; fel, Nat Res Coun Can, 49-50; chemist, Pulp & Paper Res Inst Can, 50-53; asst dir res, Fraser Co, Ltd, 53-58, spec asst admin dept, 58-62, coordr com develop, 62-65; dir res & develop, Great Lakes Paper Co, 65-74, dir environ serv, Great Lakes Forest Prod, Ltd, 74-87; CONSULT, 87- *Mem:* Am Chem Soc; Am Tech Asn Pulp & Paper Industs; Can Pulp & Paper Asn; Chem Inst Can. *Res:* Pulp and paper. *Mailing Add:* 622 Rosewood Crescent Thunder Bay ON P7E 2R7 Can

ALLEN, HAROLD DON, b Montreal, Que, July 2, 31; m 55; c 4. CURRICULUM DEVELOPMENT, TEACHER DEVELOPMENT. *Educ:* McGill Univ, BSc, 52; Univ Santa Clara, MSTM, 67; Rutgers Univ, EdM, 68, EdD(math educ), 77. *Prof Exp:* Teacher math, Protestant Sch Bd Greater Montreal, 53-63; teaching prin, Chibougamau Protestant Sch Bd, 63-65; supv prin, Saguenay Valley Sch Bd, 65-68; teaching asst math, Rutgers Univ, 68-69; from assoc prof to prof math & sci, NS Teachers Col, 69-87; TEACHER & COORDR MATH, ST GEORGE'S SCH MONTREAL, 87- *Concurrent Pos:* Lectr, Prince Edward Island Teacher Educ, 60, Protestant Sch Bd Teacher Educ, 60-63, Govt Que Prog Teachers, 68 & Col Cape Breton Educ, 77-79; comt mem, Task Force Metric Usage Textbooks, Govt Can, 74-77; counr, Can Col Teachers, 79-81; consult math curric, NS Adult Educ Prog, 80-81; demonstration teacher, McGill Univ Gifted-Talent Summer Schs, 83-, adj prof, Fac Educ, 91- *Honors & Awards:* Metric Serv Award, Govt Can, 83; Damon Award, Am Cryptogram Asn, 90. *Mem:* Math Asn Am; Nat Coun Teachers Math. *Res:* Heuristics; problem solving; mathematical recreations; recreational cryptography; mathematics and its applications for non-technical audiences. *Mailing Add:* 6150 Ave Bienville Brossard PQ J4Z 1W8 Can

ALLEN, HARRY CLAY, JR, b Saugus, Mass, Nov 26, 20; m 48; c 2. CHEMICAL PHYSICS. *Educ:* Northeastern Univ, BS, 48; Brown Univ, MS, 49; Univ Wash, PhD(phys chem), 51. *Prof Exp:* Atomic Energy Comn fel, Harvard Univ, 51-53; asst prof physics, Mich State Col, 53-54; physicist, Radiometry Sect, Atomic & Radiation Physics Div, Nat Bur Standards, 54-61, chief, Analytical & Inorg Chem Div, 61-63, chief, Inorg Mat Div, 63-65, dep dir, Inst Mat Res, 65-66; asst dir minerals res, Bur Mines, US Dept Interior, 66-69; chmn, chem dept, 69-78 & 84-85, dean, Grad Sch & coodr res, 78-80, actg provost, 80-81, assoc provost & dean res, 81-83, prof chem, 69-86, EMER PROF, CLARK UNIV, 86- *Concurrent Pos:* Vis prof, Univ Wash, 58 & Univ NC, Chapel Hill, 78; res fel theoret chem, Cambridge Univ, 59-60; vis lectr, Univ Md, 64. *Honors & Awards:* Stratton Award, 65. *Mem:* Am Chem Soc; fel Am Phys Soc; fel Am Inst Chemists. *Res:* Solid state; molecular spectroscopy. *Mailing Add:* Box 1092 Seven Lakes West End NC 27376

ALLEN, HARRY PRINCE, b New York, NY, July 31, 38; m 62; c 2. MATHEMATICS. *Educ:* Brooklyn Col, BS, 60; Yale Univ, MA, 62, PhD(math), 65. *Prof Exp:* Actg instr math, Yale Univ, 64-65; instr, Mass Inst Technol, 65-67; NATO res fel, Math Inst, State Univ Utrecht, 67-68; asst prof, Rutgers Univ, 68-70; ASSOC PROF MATH, OHIO STATE UNIV, 70- *Mem:* Am Math Soc. *Res:* Nonassociative and lie algebras. *Mailing Add:* Ohio State Univ 231 W 18th Ave 416 Columbus OH 43210

ALLEN, HENRY L, b Philadelphia, Pa, Apr 26, 45; div; c 1. VETERINARY PATHOLOGY. *Educ:* Pa State Univ, BA, 67; Univ Pa, VMD, 71. *Prof Exp:* From resident to instr vet path, Sch Vet Med, Univ Pa, 71-75; res fel path, 75-77, sr res fel path, 77-79, dir, 79-85, SR INVESTR, TOXICOL & PATH, MERCK SHARP & DOHME RES LABS, 85- *Concurrent Pos:* Consult & vet surg pathologist. *Mem:* Am Vet Med Asn. *Res:* Naturally occurring and drug-induced diseases in laboratory and domestic animals; comparative aspects. *Mailing Add:* Merck Sharp & Dohme Res Labs 45-207 West Point PA 19486

ALLEN, HERBERT, mechanical engineering; deceased, see previous edition for last biography

ALLEN, HERBERT CLIFTON, JR, b Richmond, Va, Jan 7, 17; m 49; c 6. INTERNAL MEDICINE, NUCLEAR MEDICINE. *Educ:* Univ Richmond, BSc, 37; Med Col Va, MD, 41. *Prof Exp:* Intern, Philadelphia Gen Hosp, 41-42; resident path, Pa Hosp, 46-47; resident med, Med Col Va, 47-48; asst chief, Radioisotope Sect, Dept Med & Surg, Vet Admin, Washington, DC, 48-49; chief metab serv & asst dir, Birmingham Vet Admin Hosp, Van Nuys, Calif, 49-50; asst dir, Radioisotope Univ, Wadsworth Vet Admin Hosp, Los Angeles, 50-51; asst prof med, Baylor Col Med, 51-77; PRES, NUCLEAR MED LABS TEX, 77- *Concurrent Pos:* Dir, Radioisotope Univ, Vet Admin Hosp, Houston, 51-55, actg dir, 55-57; dir, Dept Nuclear Med, Methodist Hosp, 52-, Hermann Hosp, 56- & Mem Baptist Hosp, 65-; mem, Radiation Adv Bd; secy, Tex State Dept Health, 61-65; pres, Atomic Energy Indust Labs Southwest, Inc & Atomic Food Processing Corp Am. *Mem:* AMA; Am Thyroid Asn; NY Acad Sci. *Res:* Diagnostic and therapeutic atomic medicine; diseases of thyroid; development of diagnostic instrumentation for nuclear medicine, particularly thyroid, brain and renal scanning. *Mailing Add:* 100 Hermann Prof Bldg Houston TX 77030

ALLEN, HERBERT E, b Sharon, Pa, July 19, 39; m 63, 84; c 4. AQUATIC RESOURCES, AQUATIC CHEMISTRY. *Educ:* Univ Mich, BS, 62, PhD(environ chem), 74; Wayne State Univ, MS, 67. *Prof Exp:* Chemist, US Bur Com Fisheries, 62-70; lectr environ chem, Sch Pub Health, Univ Mich, 70-74; from asst prof to prof, dept environ eng, Ill Inst Technol, 74-84; PROF CHEM & DIR ENVIRON STUDIES INST, DREXEL UNIV, 84. *Concurrent Pos:* Fac assoc, Argonne Nat Lab, 78-79; vis Prof, Water Res Ctr, Eng, 80-81. *Mem:* Am Chem Soc; AAAS; Int Asn Water Pollution Res & Control; Water Pollution Control Fedn; Am Soc Limnol & Oceanog; Am Pub Health Asn. *Res:* Trace metal fate and effects in aquatic systems; speciation of metals. *Mailing Add:* Environ Studies Inst Drexel Univ Philadelphia PA 19104

ALLEN, HOWARD JOSEPH, b Gloversville, NY, Mar 11, 41; m 59; c 5. GLYCOCONJUGATES, CARBOHYDRATE-BINDING PROTEINS. *Educ:* State Univ NY, Albany, BS, 65; State Univ NY, Buffalo, PhD(biol), 70. *Prof Exp:* Teaching fel, Fla State Univ, 70-72; CANCER RES SCIENTIST, ROSWELL PARK, BUFFALO, 72-; ASST PROF, STATE UNIV NY, BUFFALO, 74-; RES PROF, NIAGARA UNIV, 76- *Mem:* Am Chem Soc; Am Asn Biol Chemists; Am Asn Cancer Res; Soc Complex Carbohydrates; Am Soc Cell Biol. *Res:* Chemistry, function and genetics of animal lectins; role of lectenis in neoplasia. *Mailing Add:* Roswell Park Mem Inst Buffalo NY 14263

ALLEN, J FRANCES, b Arkville, NY, Apr 14, 16. ECOLOGY. *Educ:* Radford Col, BS, 38; Univ Md, MS, 48, PhD(zool), 52. *Prof Exp:* Teacher various high schs, 38-47; asst zool, Univ Md, 47-48, instr, 48-55, asst prof, 55-58; prof asst, Syst Biol Prog, NSF, 58-61, assoc prog dir, 61-67; chief br water qual requirements, Fed Water Pollution Control Admin, US Dept Interior, 67-68, asst dir biol sci, Div Water Qual Res, 68-70; chief biol sci br, Div Water Qual Res, Environ Protection Agency, 70-71, chief ecol effects, Processes & Effects Div, 71-73, staff scientist-ecologist, sci adv bd, 73-82; RETIRED. *Concurrent Pos:* Biol Exam, NY State Dept Educ, 47. *Mem:* Fel AAAS; Am Soc Zool; Am Fisheries Soc; Am Inst Biol Sci. *Res:* Fishery biology; shellfish; gastropod ecology and distribution; taxonomy of the gastropods; aquatic biology; pollution; biology; ecological effects. *Mailing Add:* Meeker Hollow Rd Roxbury NY 12474-9749

ALLEN, JACK C, JR, b Dallas, Tex, Sept 17, 35; m 61; c 2. GEOLOGY. *Educ:* Southern Methodist Univ, BS, 58; Princeton Univ, MA, 60, PhD(geol), 61. *Prof Exp:* Asst prof geol, Wesleyan Univ, 61-63; from asst prof to assoc prof, 63-74, chmn coun univ courses, 76-80, PROF GEOL, BUCKNELL UNIV, 74-, CHMN DEPT, 82- *Mem:* AAAS; Geol Soc Am; fel Mineral Soc Am. *Res:* Structure and petrology of granitic rocks; stability of amphibole in andesite and basalt. *Mailing Add:* PO box 207 Lewisburg PA 17837

ALLEN, JAMES DURWOOD, b Commerce, Tex, Nov 22, 35; m 58; c 3. ORGANIC POLYMER CHEMISTRY. *Educ:* ETex State Col, BA, 58; Rice Univ, PhD(chem), 62. *Prof Exp:* Res chemist, Shell Oil Co, 62-63; NIH fel chem, Northwestern Univ, 63-64; res chemist, Phillips Petrol Co, 64-69; mem tech staff, NAm Rockwell Corp, 69-75; mgr res & develop, 75-81, TECH DIR, ADVAN COMPOSITE DIV, FIBERITE CORP, 81- *Mem:* Am Chem Soc; Soc Aerospace Mat & Process Engrs; Sigma Xi. *Res:* Polymer chemistry as applied to graphite fiber reinforced advanced composites. *Mailing Add:* 3926 E Lavender Ln Phoenix AZ 85044

ALLEN, JAMES FREDERICK, b London, Eng, Mar 25, 50; Can citizen. ARTIFICIAL INTELLIGENCE, COMPUTATIONAL LINGUISTICS. *Educ:* Univ Toronto, BSc, 73, MSc, 75, PhD(comput sci), 79. *Prof Exp:* From asst prof to assoc prof, 79-87, chmn, 87-90, PROF COMPUT SCI, UNIV ROCHESTER, 87- *Concurrent Pos:* Prin investr, NSF, 80-, Off Naval Res, 82- & Rome Air Develop Ctr, 84-; ed-in-chief, Computational Ling, 83-; auth, Natural Language Understanding, Benjamin-Cummings Pub Co, 88. *Honors & Awards:* Presidential Young Investr Award, NSF, 84- *Mem:* Asn Computational Ling; fel Am Asn Artificial Intel; Asn Comput Mach. *Res:* Artificial intelligence, especially in the areas of natural language understanding (dialogue understanding) and the presentations of knowledge (time, action and planning). *Mailing Add:* Dept Comput Sci Univ Rochester Rochester NY 14627

ALLEN, JAMES LAMAR, b Graceville, Fla, Sept 25, 36; m 59; c 2. MICROWAVE TECHNIQUES, COMPUTER APPLICATIONS. *Educ:* Ga Inst Technol, BEE, 59, MSEE, 61, PhD(elec eng), 66. *Prof Exp:* Sr engr, Sperry Microwave Electronics, 61-63; instr elec eng, Ga Inst Technol, 63-66; res staff consult, Sperry Microwave Electronics, 66-68; assoc prof elec eng, Univ SFla, 68-70, Colo State Univ, 70-72; assoc prof, Univ SFla, 72-74, prof, 74-81; pres, Lamar Allen Enterprises, Inc, 81-83; CHIEF ENGR & VPRES, KAMAN SCI CORP, 83- *Concurrent Pos:* Chief ed, Transactions Microwave Theory & Techniques, Inst Elec & Electronics Engrs, 77-79. *Mem:* Fel Inst Elec & Electronics Engrs. *Res:* Electromagnetic interference; microprocessor applications; microwave theory and techniques; high power microwave weapon systems. *Mailing Add:* PO Box 112 Manitou Springs CO 80829-0112

ALLEN, JAMES R, JR, b Mars Hill, NC, Dec 13, 27; m 48; c 4. PATHOLOGY. *Educ:* Univ Tenn, BS, 50, MS, 51; Univ Ga, DVM, 55; Univ Wis, PhD(path), 61. *Prof Exp:* Asst nutrit, Univ Tenn, 50-51; dir field serv, Res Div, Cent Soya Co, Inc, 55-59; asst prof path, Sch Vet Med, Univ Ga, 61-62; asst prof, 62-70, PROF PATH, SCH MED, UNIV WIS-MADISON, 70- *Concurrent Pos:* NIH res grant, 63-; sr scientist & assoc, Wis Regional Primate Res Ctr, 62- *Mem:* AAAS; Soc Toxicol; Am Vet Med Asn; Am Soc Cell Biol; Int Acad Path. *Res:* Biochemical pathologic and ultramicroscopic changes in the liver and cardiovascular system from consumption of certain fats and alkaloids. *Mailing Add:* 226 Gabbrieles Creek Rd Mars Hill NC 28754

ALLEN, JAMES RALSTON, b Jamaica, WI, Oct 4, 38; US citizen; m 63; c 2. CROP PHYSIOLOGY, PLANT PHYSIOLOGY. *Educ:* Tuskegee Inst, BS, 68; Pa State Univ, MS, 70, PhD(crop physiol), 74. *Prof Exp:* Head plant & soil sci dept, Tenn State Univ, 73-76; asst prof plant & soil sci, Tuskegee Inst, 76-89; DIR AGR EXP STA, COL LIFE SCI, UNIV DC, 89- *Mem:* Am Soc Agron; Sigma Xi. *Res:* Nitrogen fixation, micronutrients in agronomic crop plant physiology. *Mailing Add:* Col Life Sci Agr Exp Sta Univ DC 4200 Connecticut Ave NW MB 4404 Washington DC 20008

ALLEN, JAMES ROY, b Kingston, Ont, Aug 27, 26; m 58; c 2. THEORETICAL PHYSICS. *Educ:* Queen's Univ, Ont, BA, 47, MA, 49; Univ Manchester, PhD(theoret physics), 63. *Prof Exp:* Exhib of 1851 res scholar, 49-52; lectr physics, Queen's Univ, Ont, 53-55, asst prof, 56-61; res assoc theoret physics, Univ Manchester, 62-63; assoc prof physics, 64-70, PROF PHYSICS, QUEEN'S UNIV, ONT, 71- *Mem:* Am Phys Soc; Can Asn Physicists; fel Brit Inst Physics & Phys Soc. *Res:* Nuclear emulsions; cosmic rays; passage of fast charged particles through solids; photonuclear reactions; hydrodynamic and magnetohydrodynamic turbulence; quasi-stationary states and resonant reactions. *Mailing Add:* Dept of Physics Queen's Univ Kingston ON K7L 3N6 Can

ALLEN, JAMES WARD, b Livingston, Mont, May 4, 41; m 62; c 2. SOLID STATE PHYSICS. *Educ:* Stanford Univ, BS, 63, MS, 65, PhD(elec eng), 68. *Prof Exp:* Staff mem, Lincoln Lab, Mass Inst Technol, 68-73; staff mem, 73-85, sr staff mem, Xerox Palo Alto Res Ctr, 85-87; PROF PHYSICS, UNIV MICH, 87-, GRAD CHAIR PHYSICS DEPT, 91- *Concurrent Pos:* Acad guest, Swiss Fed Inst Technol, Zurich, 78, Max Planck Inst Solid State Physics, Stuttgart, 89-90. *Honors & Awards:* Sr US Scientist Award, Alexander von Homboldt Found, Ger, 89. *Mem:* Fel Am Phys Soc. *Res:* Optical properties of transition metal and rare earth compounds; magnetic anisotropy and magnetoelastic effects; insulator-to-metal transition materials; synchrotron radiation and electron spectroscopy of narrow band and other materials. *Mailing Add:* Randall Lab Univ Mich Ann Arbor MI 48109-1120

ALLEN, JOE FRANK, b Hogansville, Ga, June 3, 34; m 55; c 2. INORGANIC CHEMISTRY, SCIENCE EDUCATION. *Educ:* Berry Col, AB, 55; Univ Miss, MS, 59; Ga Inst Technol, PhD(inorg chem), 63. *Prof Exp:* Instr chem, Ga Inst Technol, 57-61; radiochemist, Oak Ridge Nat Lab, 62-64; from asst prof to assoc prof inorg chem, 64-78, PROF CHEM, CLEMSON UNIV, 78-, ASSOC HEAD, 89- *Concurrent Pos:* Consult, mobile lab prog, Oak Ridge Assoc Univs, 66-72; Am Soc Eng Educ fac fel, NASA, Houston, 69-70; consult develop learning mat, Humbolt State Univ, Utah State Univ, Charles County Community Col, Ulster County Community Col & Environ Protection Agency, 72-82; consult, nuclear chem & lab skills workshops, Westinghouse SRS, 80-, Duke Power Co, 81-83 & Ga Power Co, 82- *Mem:* Am Chem Soc. *Res:* Chemistry of complex compounds; application of radioisotopes; chemical separations using ion exchange and solvent extraction; development of individualized instructional systems for science and technical training. *Mailing Add:* Dept Chem Clemson Univ Clemson SC 29633-1905

ALLEN, JOE HASKELL, b Oklahoma City, Okla, June 6, 39; m 61; c 3. GEOMAGNETISM, INDICES. *Educ:* Univ Okla, BS, 61; Univ Calif, Berkeley, MS, 66. *Prof Exp:* Teacher math, Casady Sch, 61-63; geophysicist geomagnetism, 63-71, chief data studies div, 73-81, CHIEF, SOLAR-TERRESTRIAL PHYSICS DIV & DIR WORLD DATA CTR-A FOR SOLAR TERRESTRIAL PHYSICS, US DEPT COM, 81- *Concurrent Pos:* Head, Int Magnetospheric Study Cent Info Exchange Off, 76-80. *Mem:* Am Geophys Union; Soc Terrestrial Magnetism & Elec Japan; Int Asn Geomagnetism & Aeronomy; Int Coun Sci Unions. *Res:* Measurement and classification of geomagnetic variations associated with natural and man-caused phenomena; promotion of international cooperation in science; direct national and international data exchange for solar, interplanetary, ionospheric and geomagnetic data. *Mailing Add:* NOAA/NGDC E/GC2 325 Broadway Boulder CO 80303

ALLEN, JOHN BURTON, b Detroit, Mich, Mar 21, 40; m 66; c 1. ELECTRICAL ENGINEERING. *Educ:* Univ Mich, BSE & MS, 63; Ga Inst Technol, MS, 69, PhD(elec eng), 73. *Prof Exp:* Res eng, NAm Aviation Co, 64-65; assoc scientist, Lockheed-Ga Co, 65-71; SR MEM TECH STAFF ELEC ENG, TEX INSTRUMENTS INC, 71- *Mem:* Sigma Xi. *Res:* Coherent optics; holography; laser communication systems; image processing; optical missile guidance systems; thermal imaging systems; statistical estimation and decision theory. *Mailing Add:* Texas Instruments Inc PO Box 869305 MS 8507 Plano TX 75086

ALLEN, JOHN CHRISTOPHER, b Dallas, Tex, Sept 17, 35; m 61; c 2. VOLCANOLOGY, EXPERIMENTAL PETROLOGY. *Educ:* Southern Methodist Univ, BS, 58; Princeton Univ, MA, 60, PhD(geol), 61. *Prof Exp:* Instr geol, Wesleyan Univ, 61, asst prof, 61-63; from asst prof to assoc prof, 63-74, PROF GEOL, BUCKNELL UNIV, 74-, CHMN DEPT, 82- *Concurrent Pos:* Vis prof, Univ, Tex, 82-84. *Mem:* Fel Mineral Soc Am; Geol Soc Am; AAAS; Sigma Xi. *Res:* Stability of amphiboles in andesite and basalt at high pressures and temperatures. *Mailing Add:* Dept Geol Bucknell Univ Lewisburg PA 17837

ALLEN, JOHN ED, b Danville, La, Nov 18, 37; m 60; c 3. MATHEMATICS. *Educ:* La Polytech Inst, BS, 58; Okla State Univ, MS, 60, PhD(math), 63. *Prof Exp:* Asst prof, 63-72, ASSOC PROF MATH, NORTH TEX STATE UNIV, 72-, CHMN DEPT, 76- *Concurrent Pos:* NSF sci fac fel, Purdue Univ, 71-72. *Mem:* Soc Indust & Appl Math; Math Asn Am; Am Math Soc. *Res:* Convexity; numerical analysis. *Mailing Add:* Dept Math Univ NTex Denton TX 76203

ALLEN, JOHN EDWARD, JR, b Lakeland, Fla, Apr 12, 45; m 62, 90; c 2. PHOTODISSOCIATION DYNAMICS, MOLECULAR DYNAMICS. *Educ:* Univ Fla, BS, 67, MS, 70, PhD(appl physics), 76. *Prof Exp:* Res asst aeronomy, Ctr Aeronomy & Atmospheric Sci, dept physics, Univ Fla, 69-70, res asst appl physics, dept eng sci & appl math, 70-73, res assoc, 73-76; res assoc, Nat Acad Sci-Nat Res Coun, Chem Physics Br, US Army Ballistic Res Lab, Aberdeen Proving Ground, 76-78; space scientist, Atmospheric Physics Br, 78-84, ASTROPHYSICIST, ASTROCHEM BR, GODDARD SPACE FLIGHT CTR, NASA, 84- *Mem:* Am Phys Soc; Optical Soc Am; Am Geophys Union; Am Astron Soc. *Res:* Application of lasers and nonlinear optical techniques to chemical physics; chemical kinetics, laser-matter interactions, light scattering, isotope separation; application of chemical physics to astrophysical and atmospheric systems; comets, upper atmosphere, circumstellar shells, planetary atmospheres, interstellar matter. *Mailing Add:* Code 691 NASA Goddard Space Flight Ctr Greenbelt MD 20771

ALLEN, JOHN ELIOT, b Seattle, Wash, Aug 12, 08; m 33; c 1. GEOLOGY. *Educ:* Univ Ore, BA, 31, MA, 32; Univ Calif, Berkeley, PhD(geol), 44. *Prof Exp:* Field geologist, Rustless Iron & Steel Co, Md, 35-38; field geologist, Ore Dept Geol & Mineral Indust, 38-39, chief geologist-in-chg field invests & assoc ed publ, 39-47; head dept geol, NMex Inst Mining & Technol, 49-51; chief, Navaho Mineral Surv, 51-54; econ geologist, Bur Mines, NMex, 54-56; prof & head, dept earth sci, 56-74, EMER PROF GEOL, PORTLAND STATE UNIV, 74- *Concurrent Pos:* SEATO prof & chmn dept, Univ Peshawar, 63-64; prof geol, Whitman Col, 75; res geologist, Nev Bur Mines & Geol, 75-76, NMex Bur Mines, 76-77; columnist, Oregonian, 83-87. *Honors & Awards:* Niel Miner Award, Nat Asn Geol Teachers, 75. *Mem:* Fel Geol Soc Am; Nat Asn Geol Teachers; Am Inst Prof Geologists; Sigma Xi. *Res:* Structural relations of chromite deposits; Oregon economic geology and areal geology of the Coos Bay and Wallowa Mountains; geology of San Juan Bautista Quadrangle, California and Northern Sacramento Mountains, New Mexico; glaciation in Northwest; volcanoes of Cascade Range; geology of Columbia River gorge. *Mailing Add:* Dept Geol Portland State Univ Portland OR 97207-0751

ALLEN, JOHN KAY, b Rochdale, Eng, Mar 14, 36; m 67; c 2. POLYMER CHEMISTRY. *Educ:* Univ Birmingham, BS, 57, PhD(chem), 60. *Prof Exp:* Res asst chem, Univ Louisville, 60-62; res chemist, Shell Chem Co, Eng, 62-65; proj chemist, 65-68, sr proj chemist, 68-70, res chemist, 70-78, staff res chemist, 78-88, SR RES SCIENTIST, AMOCO CHEM CORP, 88- *Mem:* Am Chem Soc. *Res:* Use of radioactive tracers in polymer chemistry; properties of ion-exchange resins; chemistry of polyether polyols and polyurethanes; preparation of fire resistant cellular plastics; plasticizer alcohols synthesis and aromatic acids synthesis; chemical waste disposal; pyrolysis of aromatic acids; waste water treatment; unsaturated polyesters and polyurethanes used in reinforced plastics. *Mailing Add:* Amoco Chem Corp Res & Develop PO Box 3011 Naperville IL 60566-7011

ALLEN, JOHN L(OYD), b Estherville, Iowa, June 13, 31; m 52; c 4. ELECTRONICS. *Educ:* Pa State Univ, BS, 58; Mass Inst Technol, SM, 62, PhD(commun biophys), 68. *Prof Exp:* Jr engr, H R B Singer, Inc, 54-58; staff mem, Lincoln Lab, 58-60, assoc group leader spec radars, 60-62, group leaders array radars, 62-65, group leader, Radar Div, 65-68 & tactical radars, 68-69, assoc head, Radar Maesurements Div, Mass Inst Technol, 69-71; assoc dir res electronics, Naval Res Lab, 71-74; dep dir, Defense Res & Eng, 74-77; vpres, Gen Res Corp, 77-78, pres, 78-81; PVT CONSULT, 81- *Concurrent Pos:* VPres, Flow Gen, Inc, 77-81; mem, Air Force Studies Bd, Nat Res Coun, 83-85 & sci adv bd, US Air Force, 85- *Mem:* Fel Inst Elec & Electronics Engrs. *Res:* Antennas; radar system design; optics of invertebrates; aerospace systems. *Mailing Add:* 1708 Besley Rd Vienna VA 22180

ALLEN, JOHN RYBOLT, b Indianapolis, Ind, Sept 14, 26; m 53. CHEMISTRY, BIOCHEMISTRY. *Educ:* Ball State Teachers Col, AB, 49; Univ Ill, Urbana, PhD(biochem), 54. *Prof Exp:* Res assoc biochem, Northwestern Univ, 53-56; asst prof, Col Med, Baylor Univ, 56-59; sr scientist, Warner-Lambert Pharm Co, NJ, 59-60; res assoc dent sch, Wash Univ, 60-62; prof chem & head dept, Union Col, Ky, 62-64; clin assoc clin chem, Univ Hosp, Case Western Reserve Univ, 64-65; asst prof path & radiol, Col Med, Ohio State Univ, 65-68; clin chemist, St John's Mercy Hosp, St Louis, Mo, 68-69, Decatur Mem Hosp, Ill, 69-70, San Diego Inst Path, 70 & San Bernardino County Hosp, 70-75; instr chem, Phoenix Col, 75-80. *Mem:* Fel AAAS; fel Am Asn Clin Chem; Am Chem Soc; Acad Clin Lab Physicians & Scientists; fel Am Inst Chem. *Res:* Quality control and methods; creatine phosphokinase; vitamin E deficiency, lipid metabolism and structure. *Mailing Add:* 4645 N 22nd St Apt 220W Phoenix AZ 85016

ALLEN, JON CHARLES, b Ventura, Calif, Nov 24, 40; m 64; c 2. ENTOMOLOGY, MATHEMATICAL MODELING. *Educ:* Univ Calif, Santa Barbara, BA, 66, MA, 67; Univ Calif, Riverside, PhD(pop biol), 73. *Prof Exp:* asst prof, 73-80, ASSOC PROF ENTOM, UNIV FLA, 80- *Mem:* Entom Soc Am; Int Soc Citricult. *Res:* Mathematical modeling of plant-pest interactions; simulation of plant growth and development; modeling micrometeorological dynamics and population effects; biological system dynamics, chaos. *Mailing Add:* Dept Entomol Univ Fla Gainesville FL 32611

ALLEN, JONATHAN, b Hanover, NH, June 4, 34; m 60; c 2. ELECTRICAL ENGINEERING. *Educ:* Dartmouth Col, AB, 56, MS, 57; Mass Inst Technol, PhD(elec eng), 68. *Prof Exp:* Supvr human factors eng, Bell Tel Labs, 61-67; asst prof elec eng, 68-76, ASSOC PROF ELEC ENG & ASSOC DIR RES LAB ELECTRONICS, MASS INST TECHNOL, 76- *Concurrent Pos:* Consult, Lincoln Lab & Sperry Rand Res Corp, 68- *Mem:* Inst Elec & Electronics Engrs; Sigma Xi. *Res:* Computation; linguistic analysis; speech analysis and synthesis. *Mailing Add:* 3 Creek Rim Dr Titusville NJ 08560

ALLEN, JONATHAN, b Fall River, Mass, Dec 29, 42; m 71; c 1. PHYSICS, ENVIRONMENTAL SCIENCE. *Educ:* Colby Col, AB, 64; Southeastern Mass Univ, MS, 71; Washington Univ, PhD(physics), 77. *Prof Exp:* Engr res & develop, Trans-Sonics, Inc, 64-65, advan develop, Hamilton Standard Div, United Aircraft, 65-68; physicist consult, High Energy Processing Corp, 68-71; res scientist physics, Aerochem Res Labs, Inc. 78-83; sr scientist process develop, Chronar Corp, 83-90; DIR RF & PROCESS ELECTRONICS, ADVAN PHOTOVOLTAIC SYSTS, INC, 90- *Concurrent Pos:* Jr fel, Ctr Biol Natural Systs, 72-77; lectr physics, Harris-Stowe Col, 73-74; physics dept fel, Washington Univ, 73-76; contrib ed, NJ Hazardous Waste News, 81- *Mem:* Am Phys Soc; Fedn Am Scientists; Optical Soc Am; NY Acad Sci; Sigma Xi; Inst Elec & Electronic Engrs. *Res:* Process research and development for manufacture of amorphous photovotaic panels including radio frequency systems, controls, instrumentation and electrical measurements. *Mailing Add:* Advan Photovoltaic Systs Inc PO Box 7093 Princeton NJ 08543-7093

ALLEN, JOSEPH GARROTT, b Elkins, WVa, June 5, 12; m; c 5. MEDICINE. *Educ:* Washington Univ, AB, 34; Harvard Univ, MD, 38. *Prof Exp:* Smith fel, Univ Chicago, 40-43, instr surg, 43-44, res assoc, Manhattan Proj, 44-46, chief surg res, 46-47, from instr to prof surg, 46-59; PROF SURG, SCH MED, STANFORD UNIV, 59- *Concurrent Pos:* Mem panel blood coagulation, Nat Res Coun; mem surg study sect, NIH, 54-59, med sci training comt, 63-; chmn study sect cancer chemother, Am Cancer Soc, 60-63. *Honors & Awards:* Abel Prize, 48; Silver Medal, Am Roentgen Ray Soc, 48; Gold Medal, AMA, 48; Gross Surg Prize, Philadelphia Acad Surg, 55; Elliot Award, Am Asn Blood Banks, 56. *Mem:* Fel AAAS; fel AMA; fel Am Col Surg; fel Am Physiol Soc; fel Soc Exp Biol & Med. *Res:* Surgery; ionizing irradiation injury; parenteral protein nutrition; coagulation of blood; posttransfusion hepatitis in relation to paid donors and the need for an all-volunteer national blood program; hazards of nuclear warfare to plant and animal life; advocated non-icterogenic pooled plasma as an inexpensive source of fibrinolysin for treatment of acute intravascular occlusion in place of synthetic products. *Mailing Add:* 583 Salvatierra St Stanford CA 94305

ALLEN, JOSEPH HUNTER, b St Joseph, Mo, Aug 31, 25; m 56; c 1. RADIOLOGY. *Educ:* Westminster Col, 42-43; Washington Univ, MD, 48. *Prof Exp:* From asst prof to assoc prof, 58-69, PROF RADIOL, VANDERBILT UNIV, 69- *Mem:* AMA; Am Soc Neuroradiol; fel Am Col Radiol; Asn Univ Radiol. *Res:* Neuroradiology. *Mailing Add:* Dept Radiol Vanderbilt Univ Hosp 21st Ave Nashville TN 37232

ALLEN, JULIA NATALIA, b Seattle, Wash. VETERINARY MEDICINE & PARASITOLOGY, TERRESTRIAL ECOLOGY. *Educ:* Mich State Univ, BS, 70; Wash State Univ, PhD(zool), 78, DVM, 86. *Prof Exp:* Tech specialist plant tissue cult, Wash State Univ, 78-80; tech specialist virol, USDA, 80; field biologist, US Army Corps Engrs, 80-81; tech specialist, Fruit & Trees, USDA, 81-82; TECH SPECIALIST, WADDL PARASITOL DATA COMPUTARIZATION, 85- *Mem:* Sigma Xi; Am Vet Med Asn. *Res:* Ecology and behavior of long billed curlews; plant tissue culture; prey base study and habitat analysis in Columbia Basin; frost tolerance of fruit tree buds; pea seed borne mosaic virus; veterinary parasitology. *Mailing Add:* Interbay Animal Hosp 3040 16th Ave West Seattle WA 98119

ALLEN, JULIUS CADDEN, b New London, Conn, Oct 18, 38; m 61; c 2. CELL PHYSIOLOGY, VASCULAR CELL BIOLOGY. *Educ:* Amherst Col, BA, 60; Univ Mass, MA, 62; Univ Alta, PhD(pharmacol), 67. *Prof Exp:* Asst prof pharmacol, 70-72, asst prof cell biophys, 72-73, ASSOC PROF CELL BIOPHYS, BAYLOR COL MED, 73-, ASSOC PROF & ASSOC HEAD, SECT CARDIOVASC SCI, DEPT MED, 76-, PROF MED & PHYSIOL, 83. *Concurrent Pos:* Fel, Baylor Col Med, 67-69, Nat Heart & Lung Inst fel pharmacol, 69-70, Nat Heart & Lung Inst grants, 74-77 & 80-; Am Heart Asn grant, 70-73; mem, Int Study Group Res Cardiac Metab, 72-; estab investr, Am Heart Asn, 74-79; prin investr, Nat Heart Lung Inst Training grant, 77 -; ed, J Molecular & Cellular Cardiol, 77-80; assoc prof physiol, Baylor Col Med, 79- *Mem:* Am Heart Asn; Am Soc Pharmacol & Exp Therapeut; Biophys Soc; Am Physiol Soc. *Res:* Biology of smooth muscle membrane systems; molecular basis of smooth muscle contractile heterogeneity; biochemistry and biophysics of ion transport; Napump function in smooth muscle; molecular control of contraction & cell growth. *Mailing Add:* Dept Med Baylor Col Med Houston TX 77030

ALLEN, KENNETH G D, b Colwyn Bay, Wales, UK, July 30, 43; m 84; c 3. TRACE METAL BIOCHEMISTRY. *Educ:* Univ London, BS, 65; Mont State Univ, PhD(biochem), 73. *Prof Exp:* From asst prof to assoc prof, 78-87, PROF, NUTRIT, COLO STATE UNIV, 87- *Concurrent Pos:* Res biochem, USDA Human Nutrit Lab, NDak, 75-78; sabbatical leave, Rowett Res Inst, Aberdeenn, Scotland, 85-86. *Mem:* Am Inst Nutrit; Nutrit Soc (UK); Sigma Xi. *Res:* Trace metal biochemistry and nutrition with particular emphasis on copper. *Mailing Add:* Dept Food Sci & Human Nutrit Colo State Univ Ft Collins CO 80523

ALLEN, KENNETH WILLIAM, b St Stephens, NB, June 20, 30; nat US; m 52; c 4. ZOOLOGY. *Educ:* Wheaton Col, BS, 52; Univ Maine MS, 56; Rice Univ, PhD, 59. *Prof Exp:* Fel biol, Rice Univ, 59-60; asst prof zool, Univ Calif, Los Angeles, 60-63; prof zool & oceanog & head dept, Univ Maine, Orono, 63-77, actg dean col arts & sci, 74-76, actg pres, Univ Maine, Augusta, 76-77 & Univ Southern Maine, 77-78, prof zool, Univ Maine, Orono, 77-78, actg pres, 79-80. *Mem:* Am Soc Zoologists; Am Soc Parasitol. *Res:* Comparative biochemistry and parasitology. *Mailing Add:* Dept Zool Univ Maine Murray Hall Orono ME 04469

ALLEN, LARRY MILTON, b Los Angeles, Calif, Nov 22, 42; m 66; c 1. ONCOLOGY, DERMATOLOGY. *Educ:* Univ Pac, BS, 66, MS, 67; Univ Ore, PhD(chem), 71; Univ Colo, MBA, 87. *Prof Exp:* Res pharmacologist, NCI/Vet Admin Med Oncol, 70-75; sr scientist cancer pharmacol, Roswell Park Mem Cancer Inst, 75-77; assoc prof cancer pharmacol, Sch Med, Univ Miami, 77-82; from assoc dir res to vpres corp develop & sci affil, Chemex Pharmaceut, 82-89; PRES MED PROD, ELANTEC INC, 89- *Concurrent Pos:* Travel grant, Fedn Am Soc Exp Biol & Am Soc Pharmacol & Exp Therapeut, 73; mem comt, Phase II Clin Res Grants-NIH, 84 & SBIR-NIH, 87. *Mem:* Am Col Clin Pharmacol; Am Asn Cancer Res; Am Soc Pharmacol & Exp Therapeut; Soc Exp Dermat. *Mailing Add:* Elantec Inc 85 S Union Blvd Suite G-160 Lakewood CO 80228-2207

ALLEN, LAWRENCE HARVEY, b Ottawa, Ont, Mar 27, 43; Can citizen; m 68; c 4. PHYSICAL CHEMISTRY, COLLOID CHEMISTRY. *Educ:* Carleton Univ, BSc, 65; Clarkson Univ, PhD(phys chem), 70. *Prof Exp:* Res scientist colloid chem, 72-80, DIR, PROCESS CHEM DIV & DIR RES CHEM SCI, PULP & PAPER INST CAN, 90- *Concurrent Pos:* Nat Res Coun Can fel, Inland Waters Directorate, Environ Can, 71-72; postdoctoral assoc, Clarkson Univ, 69-71. *Mem:* Am Chem Soc; fel Chem Inst Can; Can Pulp & Paper Asn; fel Tech Asn Pulp & Paper Indust. *Res:* Chemistry of pulping and papermaking processes; colloid and surface chemistry; chemistry of wood resins; pitch control; foam control; retention and drainage on paper machines; research management. *Mailing Add:* Pulp & Paper Res Inst Can 570 St John's Rd Pointe Claire PQ H9R 3J9 Can

ALLEN, LELAND CULLEN, b Cincinnati, Ohio, Dec 3, 26; m 60; c 3. CHEMISTRY. *Educ:* Univ Cincinnati, BS, 49, EE, 55; Mass Inst Technol, PhD(theoret physics), 57. *Prof Exp:* Fel, Mass Inst Technol, 57-59; NSF fel physics, Univ Calif, Berkeley, 59-60; from asst prof to assoc prof, 60-65, PROF CHEM, PRINCETON UNIV, 65- *Concurrent Pos:* NSF sr fel, Centre Mecanique Ondulatoire Appliquee, Paris, 67; Guggenheim fel, Math Inst, Oxford Univ, 67; NIH spec fel biochem sci, Princeton Univ, 72. *Mem:* AAAS; Am Phys Soc; Am Chem Soc. *Res:* Electronic structure theory, applications to inorganic, organic and biochemistry. *Mailing Add:* Dept Chem Princeton Univ Princeton NJ 08544

ALLEN, LEROY RICHARD, b Garden Grove, Calif, Sept 14, 13; m 38; c 3. RESEARCH ADMINISTRATION, PUBLIC HEALTH. *Educ:* Univ Redlands, AB, 36; Univ Southern Calif, MD, 41; Univ Mich, MPH, 48; Am Bd Prev Med & Pub Health, dipl, 49. *Prof Exp:* Tuberc control officer, State Dept Pub Health, Md, 45-48; tuberc consult, Regional Off San Francisco, USPHS, 49-50; med dir, US Tech Coop Mission to Burma, 50-53, prof prev & social med, Vellore Med Col, South India, 54-56; med educ & pub health rep, Rockefeller Found, New Delhi, India & Quezon City, Philippines, 57-72; from dep dir to dir, Atomic Bomb Casualty Comn, Nat Acad Sci, 72-75, vchmn, Radiation Effects Res Found, Japan, 75-78; RETIRED. *Concurrent Pos:* Consult, Indian Pub Health Asn, USPHS, 44-53; vis prof community med, Med Col, Univ Philippines, 68-71; on leave, consult, Atomic Bomb Casualty Comn, US Nat Acad Sci-Japanese Nat Inst Health, 71-72; mem bd dirs, Hospice Monterey Peninsula, 79-, vpres, 81-82 & 85-89; consult health sci, Rockefeller Found, 79. *Mem:* Am Thoracic Soc; Am Pub Health Asn; hon mem Indian Pub Health Asn. *Res:* Medical education and care; problems related to population control. *Mailing Add:* 3765 Raymond Way Carmel CA 93923

ALLEN, LEW, JR, b Miami, Fla, Sept 30, 25; m 49; c 5. PHYSICS. *Educ:* US Mil Acad, BS, 46; Univ Ill, MS, 52, PhD(physics), 54. *Prof Exp:* Mem staff, Los Alamos Sci Lab, 54-57; sci adv, Air Force Spec Weapons Ctr, 57-61, mem staff defense dir res & eng, 61-65, dep dir spec projs, Off of Secy of Air Force,

65-68, dir space systs, 68-70, dir spec projs, 71-73; dir, Nat Security Agency, 73-77; comdr, Air Force Systs Command, USAF, 77-78, chief staff, 78-82; dir, Jet Propulsion Lab & vpres, Calif Inst Technol, 82-91; CHMN, CHARLES STARK DRAPER LAB, 91- *Honors & Awards:* Thomas White Award, Nat Geog Soc; George Goddard Award, Int Soc Opt Eng; Von Karman Lect, Am Inst Astronaut & Aeronaut; Distinguished Serv Award, NASA. *Mem:* Nat Acad Eng; fel Am Phys Soc; Am Geophys Union; Am Phys Soc. *Mailing Add:* Charles Stark Draper Lab 1040 S Arroyo Blvd Pasadena CA 91105

ALLEN, LEWIS EDWIN, b Monroe, La, Aug 9, 37; m 64; c 3. ORGANIC CHEMISTRY, ENVIRONMENTAL CHEMISTRY. *Educ:* Queens Col, BS, 58; Syracuse Univ, PhD(org chem), 70. *Prof Exp:* Assoc prof chem, Fla A&M Univ, 63-70; sr chemist, 70-74, coordr environ serv, 74-82, UNIT DIR, ENVIRON ANALYSIS SERV, EASTMAN KODAK CO, 82- *Mem:* Am Chem Soc; Soc Photog Sci & Eng; Water Pollution Control Fedn. *Res:* Physical organic chemistry; stereochemistry; organometallic compounds; photographic science; disposal of photographic waste solutions. *Mailing Add:* 34 Circle Wood Dr Rochester NY 14625

ALLEN, LINDSAY HELEN, b Chippenham, Eng, July 24, 46; m 89; c 1. NUTRITION. *Educ:* Univ Nottingham, BSc, 67; Univ Calif, PhD(nutrit), 73. *Prof Exp:* Res asst nutrit, Med Res Coun, Cambridge Univ, 67-69, Univ Calif Davis, 69-73; fel nutrit, Univ Calif, Berkeley, 73-74; from asst prof to assoc prof, 74-84, PROF NUTRIT, UNIV CONN, 84- *Concurrent Pos:* Vis prof, Nat Inst Nutrit, Mexico City. *Mem:* Brit Nutrit Soc; Am Inst Nutrit; AAAS; Am Dietetic Asn; Am Soc Clin Nutrit; Am Anthrop Asn. *Res:* Calcium and trace mineral metabolism; assessment of nutritional status; nutrition and development; nutrition and function; international nutrition. *Mailing Add:* Dept Nutrit Sci Univ Conn Storrs CT 06269-4017

ALLEN, LINUS SCOTT, b Dallas, Tex, Dec 9, 35; m 54; c 2. APPLIED PHYSICS. *Educ:* Southern Methodist Univ, BS, 59, MS, 61, PhD(physics), 71. *Prof Exp:* Scientist reactor physics, Jet Propulsion Lab, Calif Inst Technol, 61-62; SCIENTIST, MOBIL RES & DEVELOP CORP, 62- *Mem:* Am Nuclear Soc; Soc Prof Well Log Analysts; Sigma Xi. *Res:* Development of new or improved methods for locating and assessing petroleum and mineral occurances. *Mailing Add:* Mobil Res & Develop Corp 13777 Midway Rd Dallas TX 75244

ALLEN, LOIS BRENDA, b Martin, Ky, July 4, 39. VIROLOGY, CHEMOTHERAPY. *Educ:* Georgetown Col, Ky, BS, 61; Mich State Univ, MS, 63; Univ Mich, Ann Arbor, PhD(epidemiol), 71. *Prof Exp:* From virologist to sr virologist, ICN Nucleic Acid Res Inst, 71-75, head dept virol, 75-76; ASST PROF MICROBIOL & DIR VIRUS LAB, N TEX STATE UNIV HEALTH SCI CTR, 76- *Mem:* AAAS; Am Asn Immunol; NY Acad Sci; Am Soc Microbiol. *Res:* Susceptibility and resistance factors of disease with emphasis on development of new antiviral drugs; development of Virazole and ara-HxMP. *Mailing Add:* Dept Microbiol & Immunol 605 Fairlane Ct Hurst TX 76053

ALLEN, MALWINA I, US citizen; c 2. NUCLEAR MAGNETIC RESONANCE SPECTROSCOPY. *Educ:* Mass Inst Technol, SB, 52, SM, 53, PhD(chem), 60. *Prof Exp:* Res chemist, Moleculon Res Corp, 61-63; consult, Riverside Res Lab, 66-68; PROF CHEM, FRAMINGHAM STATE COL, 68- *Concurrent Pos:* Vis assoc, Calif Inst Technol, 76-77; vis scientist, Weizman Inst Sci, Israel, 86. *Mem:* Am Chem Soc. *Res:* Application of nuclear magnetic resonance investigation to the investigation of dynamic processes such as hydrogen bonding, segmental motion and host guest complexation; guest dynamics in solid inclusion compounds. *Mailing Add:* Framingham State Col HA 325 Framingham Center MA 01701

ALLEN, MARCIA KATZMAN, b New York, NY, Jan 10, 35; m 57; c 3. MICROBIAL GENETICS. *Educ:* Bryn Mawr Col, BA, 56; Stanford Univ, PhD(biol), 62. *Prof Exp:* NIH res fel, Mass Inst Technol, 63-65; instr molecular biol, Syntex Corp, 65-67; sr lectr biol sci, Stanford Univ, 67-82; clin studies coordr, Syva Co, 82-84; tech publ coordr, Nellcor Co, 84-85; bionet sci consult, Intelligenetics Inc, 85-86; CONSULT, 86- *Res:* Biochemical genetics; biochemistry; gene action. *Mailing Add:* 325 Chatham Way Mountain View CA 94040

ALLEN, MARSHALL B, JR, b Long Beach, Miss, Oct 19, 27; m 57; c 3. NEUROSURGERY. *Educ:* Univ Miss, BA, 49; Harvard Med Sch, MD, 53. *Prof Exp:* Intern, Jefferson Med Col, 53-54; asst thoracic surg, Miss State Sanitorium, 54-55; gen pract med, Houston Hosp Inc, Miss, 55-56; resident gen surg, Sch Med, Univ Miss, 56-57, resident neurosurg, 57-61; chief serv neurosurg, Vet Admin Hosp, Jackson, Miss, 62-64; PROF NEUROSURG & CHIEF DIV, MED COL GA, 65- *Concurrent Pos:* Nat Inst Neurol Dis & Blindness fel neurophysiol, Hosp Henri-Rouselle, Paris, France, 61-62; consult, Vet Admin Hosp, Augusta & Ga State Penitentiary, 65- & Milledgeville State Hosp, 66- *Mem:* Am Med Asn; Cong Neurol Surgeons; Am Asn Neurol Surgeons; Soc Neurol Surgeons; Am Col Surgeons. *Res:* Neurophysiology; clinical neurosurgery. *Mailing Add:* Dept Surg Med Col Ga Augusta GA 30912

ALLEN, MARTIN, b New York, NY, Mar 26, 18; m 42; c 4. PHYSICAL CHEMISTRY. *Educ:* Brooklyn Col, AB, 38; Univ Minn, MS, 41, PhD(phys chem), 44. *Prof Exp:* Asst, Univ Minn, 40-43, instr phys chem, 43-45; res assoc, Allegany Ballistics Lab, Md, 45; sr technician, B F Goodrich Chem Co, Ohio, 45-47; assoc prof chem, Butler Univ, 47-56; from assoc prof to prof, Univ St Thomas, 56-84, chmn dept, 75-84, dir sci & math div, 77-82, EMER PROF CHEM, UNIV ST THOMAS, 84- *Concurrent Pos:* Abstractor, Chem Abstr Serv, 57-77; vis prof, Univ Minn, 81-82, adj prof chem eng, 82-83, sr res assoc, 84-86; assoc dir interfacial eng, Univ Minn, 89-90. *Mem:* AAAS; Am Chem Soc; Fedn Am Scientists. *Res:* Thermodynamics; solutions of electrolytes; properties of high polymers; solubility of silver acetate in mixed solvents containing other electrolytes; surfactants. *Mailing Add:* Dept Chem Rm 139 Smith Hall 207 Pleasant St SE Minneapolis MN 55455

ALLEN, MARVIN CARROL, b Searcy, Ark, Nov 30, 39; m 65; c 2. ANALYTICAL CHEMISTRY. *Educ:* Ark State Univ, BS, 61; Univ Ark, MS, 64, PhD(phys chem), 67. *Prof Exp:* RES SCIENTIST CHEM, CONTINENTAL OIL CO, 69- *Mem:* Am Chem Soc; Sigma Xi. *Res:* Separation and quantitation of organic compounds by thin layer chromatography and high performance liquid chromatography; characterization of polynuclear aromatic compounds in petroleum fractions by fluorescence spectrophotometry and/or chromatography. *Mailing Add:* 2308 El Camino Ponca City OK 74604

ALLEN, MARY A MENNES, b Owatonna, Minn, Apr 26, 38; m 61. MICROBIOLOGY. *Educ:* Univ Wis-Madison, BS, 60, MS, 61; Univ Calif, Berkeley, PhD(microbiol), 66. *Prof Exp:* USPHS trainee, Univ Calif, Berkeley, 62-63; from instr to assoc prof, 68-80, William R Kenan Jr prof, 82-90, PROF BIOL SCI, WELLESLEY COL, 80-, JEAN GLASSCOCK PROF, 89- *Concurrent Pos:* Res assoc dept biol sci, Tufts Univ, 68; Brown-Hazen res grant, Res Corp, 69; NSF res grants, 71-; Cottrel Col sci grants, Res Corp, 73 & 80; NIH res grant, 90-93; chair, Biol Coun, Coun Undergrad Res, 89-91. *Mem:* Fel AAAS; Phycol Soc Am; Am Soc Cell Biol; Am Soc Microbiol; Coun Undergrad Res. *Res:* Biochemistry and physiology of cyanobacteria; growth and cell division of unicellular blue-green algae; nitrogen chlorosis and heterotrophy in cyanobacteria; effect of changes in environment on cyanobacteria. *Mailing Add:* Dept Biol Sci Wellesley Col Wellesley MA 02181

ALLEN, MATTHEW ARNOLD, b Edinburgh, Scotland, Apr 27, 30; US citizen; m 57; c 3. PHYSICS. *Educ:* Univ Edinburgh, BSc, 51; Stanford Univ, PhD(physics), 59. *Prof Exp:* Res assoc microwave physics, Hansen Labs, Stanford Univ, 55-61; mgr res tube div, Microwave Assocs, Inc, Mass, 61-65; mem tech staff & head dept, 65-90, ASSOC DIR, KLYSTRON/MICROWAVE, STANFORD LINEAR ACCELERATOR CTR, STANFORD UNIV, 90- *Concurrent Pos:* Consult, Microwave Assocs, Inc, 65-70, Bechtel Corp, 70-72 & SAIC, 90- *Mem:* Am Phys Soc; fel Inst Elec & Electronics Engrs. *Res:* Microwave electronics; accelerator and plasma physics; high energy electron-positron storage ring research. *Mailing Add:* Stanford Linear Accelerator Ctr Stanford Univ Stanford CA 94309

ALLEN, MAX SCOTT, b Chanute, Kans, Jan 13, 11; m 37; c 3. INTERNAL MEDICINE. *Educ:* Univ Wichita, AB, 33, Univ Kans, MD, 37. *Prof Exp:* Assoc prof, 50-59, prof, 59-81, EMER PROF MED, UNIV KANS MED CTR, KANSAS CITY, 81- *Mem:* AMA; fel Am Col Physicians; Am Asn Hist Med. *Res:* Clinical research in internal medicine. *Mailing Add:* 5103 W 96th Terr Shawnee Mission KS 66207

ALLEN, MERRILL JAMES, b San Antonio, Tex, Aug 2, 18; m 42; c 2. VISION, TRANSPORTATION RESEARCH. *Educ:* Ohio State Univ, BSc, 41, MSc, 42, PhD(physiol optics), 49. *Prof Exp:* Grad asst, Ohio State Univ, 41-43, asst prof, 49-53; physicist, Frankfort Arsenal, Pa, 43-44; from asst prof to prof, 53-88, EMER PROF OPTOM, IND UNIV, BLOOMINGTON, 88- *Concurrent Pos:* Visual consult to attys, 52-; mem adv panel automotive safety, US Gen Serv Admin, 66-69; consult, Potters Indust, Inc, 79-87 & Gen Motors Tech Ctr, 85-87; guest speaker, Kikuchi Optical Co, Nagoya, Japan, 80; transp res ctr assoc, Ind Univ, 90. *Honors & Awards:* Res Medal, Brit Optical Asn, 70; Apollo Award, Am Optom Asn, 71, Cert of Commendation, 73; Mike Drain Mem Award, Mid Am Vision Conf, 89. *Mem:* Am Optom Asn; AAAS; Sigma Xi; Physicians Automotive Safety; Am Asn Automotive Med; Am Asn Univ Profs; Am Acad Optom. *Res:* Vision related problems in drivers and in the design of vehicles, highways and railroads; transportation, nutrition and therapy of vision problems. *Mailing Add:* 1311 Valley Forge Rd Bloomington IN 47401

ALLEN, MICHAEL THOMAS, b Raleigh, NC, Nov 9, 49; m 71; c 3. PSYCHOPHYSIOLOGY. *Educ:* Univ NC, Chapel Hill, AB, 71; Appalachian State Univ, MA, 74; Univ Tenn, Knoxville, PhD(exp psychol), 82. *Prof Exp:* Psychologist, Blue Ridge Community Mental Health Ctr, Asheville, NC, 74-78; trainee epidemiol, Sch Pub Health, Univ Pittsburgh, 82-83; fel psychophysiol, dept psychiat, Univ NC, 83-85; ASST PROF PSYCHOL, UNIV SOUTHERN MISS, 85- *Concurrent Pos:* Prin investr, NIH, 85-88. *Mem:* Soc Psychophysiol Res; Sigma Xi. *Res:* Cardiovascular and respiratory adjustments to both physical and psychological stressors and how these responses may be related to hypertension and asthma. *Mailing Add:* Dept Psychiat Univ Pittsburgh 3811 O'Hara St Pittsburgh PA 15213

ALLEN, NEIL KEITH, b Elburn, Ill, July 24, 44; m 79. POULTRY NUTRITION & ANIMAL NUTRITION, ANIMAL PHYSIOLOGY. *Educ:* Univ Ill, Urbana, BS, 67, MS, 68, PhD(animal sci), 71. *Prof Exp:* Res asst animal nutrit, Univ Ill, Urbana, 67-71; res assoc poultry nutrit, Cornell Univ, 71-73; formulation mgr, Dawe's Lab, Inc, 73-75; asst prof animal sci, Univ Minn, St Paul, 75-82; NUTRITIONIST, GOLDSBORO MILLING CO, NC, 85- *Mem:* Poultry Sci Asn. *Res:* Protein and amino acid requirements of chickens as a function of age and production; effects of fusariummycotoxins on poultry; nutrition of geese; nutrition of turkeys. *Mailing Add:* Goldsboro Milling Co PO Box 10009 Goldsboro NC 27532

ALLEN, NINA STRMGREN, b Copenhagen, Denmark, Sept 17, 35; m 58, 70; c 5. LIGHT MICROSCOPY & VIDEO IMAGES, PLANT CELL BIOLOGY. *Educ:* Univ Wis, BS, 57; Univ Md, Col Park, MS, 70, PhD(plant physiol), 73. *Prof Exp:* Teaching asst bot, Univ Wis-Madison, 57-58; teaching asst, Univ Md, College Park, 64-67, res asst, 67-70; res asst biol, State Univ NY, Albany, 70-73, res assoc, 73-75; vis asst prof biol, Dartmouth Col, 75-76, asst prof, 76-83; ASSOC PROF BIOL, WAKE FOREST UNIV, 84- *Concurrent Pos:* NIH fel, 67-70; Trustee exec comt, Corp Marine Biol Lab, Woods Hole, Mass, 79-82; chmn, Women in Cell Biol, Am Soc Cell Biol, 84-85; series ed, Plant Biol, Alan R Liss, Inc, 85-; mem Biomed Res Technol Rev Comt, NIH, 88, 89; NSF vis prof women, Stamford Univ, 90-91. *Mem:* Am Soc Plant Physiologists; fel AAAS; Sigma Xi; Am Soc Cell Biol; Phycological Soc Am; Am Soc Gravitational & Space Biol; Electron Micros Soc Am. *Res:* Cell biology and cell motility; study of movements in and of plant (especially algae) and animal cells with a particular interest in the cytoskeleton and the endoplasmic reticulum and its relation to cytoplasmic streaming and intracellular particle motions, signal transduction; high resolution video-enhanced light microscopy, electron microscopy and vibrating probe analysis. *Mailing Add:* Dept Biol Box 7325 Wake Forest Univ Winston-Salem NC 27109

ALLEN, NORRIS ELLIOTT, b Baltimore, Md, Sept 15, 39; m 64; c 1. MICROBIOLOGY. *Educ:* Univ Md, BS, 64, MS, 66, PhD(microbiol), 69. *Prof Exp:* Res assoc biol, Univ Pa, 69-71; sr microbiologist, 71-78, res scientist, 79-87, SR RES SCIENTIST, LILLY RES LABS, ELI LILLY & CO, 87- *Mem:* Am Soc Microbiol. *Res:* Mechanisms of antibiotic action and resistance; mechanisms of patheogenesis. *Mailing Add:* 1481 E 77th Indianapolis IN 46240

ALLEN, PETER MARTIN, b Chicago, Ill, Sept 18, 47; m 79; c 1. HYDROLOGY, GEOLOGY. *Educ:* Denison Univ, BA, 70; Baylor Univ, MS, 72; Southern Methodist Univ, PhD(geol), 77. *Prof Exp:* Geologist & planner, Dept Planning, City Dallas, 72-78; lectr, 78-79, ASST PROF GEOL, BAYLOR UNIV, 79- *Concurrent Pos:* Consult, Dept Urban Planning, Dallas, 78-82. *Mem:* Asn Eng Geologists; Asn Prof Geol Scientists; Am Planning Asn. *Res:* Analysis of geological information for land use planning, urban geology and geomorphology: stream chanel enlargement due to urbanization, hillslope ordinances, shrink-swell soils, archaeology and flood plain use, bed load transport and gravel mining. *Mailing Add:* Dept Geol Baylor Univ PO Box 97354 Waco TX 76706

ALLEN, PETER ZACHARY, b New York, NY, July 21, 25. IMMUNOLOGY, IMMUNOCHEMISTRY. *Educ:* Columbia Col, AB, 49; Columbia Univ, PhD(microbiol), 58. *Prof Exp:* Nat Res Coun fel, Lister Inst, Eng, 58-59; NIH fel, 59-60; from asst prof to assoc prof, 60-71, PROF MICROBIOL, SCH MED, UNIV ROCHESTER, 71- *Concurrent Pos:* Adv ed, Immunochem, 71-75. *Mem:* Am Asn Immunologists; Biochem Soc; Sigma Xi. *Res:* Immunochemistry of polysaccharides and proteins. *Mailing Add:* Dept Microbiol Univ Rochester Sch Med Rochester NY 14627

ALLEN, PHILIP B, b Boston, Mass, Sept 22, 42; m 64; c 4. ELECTRONIC STRUCTURE THEORY, ELECTRON-PHONON INTERACTIONS. *Educ:* Amherst Col, BA, 64; Univ Calif, Berkeley, PhD(physics), 69. *Prof Exp:* Mem tech staff res, Bell Tel Labs, 70-71; from asst prof to assoc prof, 71-81, PROF PHYSICS, STATE UNIV NY, STONY BROOK, 81- *Concurrent Pos:* Vis researcher, Cavendish Lab, Cambridge, Eng, 73-74; vis assoc prof, Univ Calif, Berkeley, 76-77; vis scientist, Max-Planck Inst Solid State Res, 79-80 & 84, Naval Res Lab, 87; B Mathias vis scholar, Los Alamos Nat Lab, 90. *Honors & Awards:* US Sr Scientist Award, Alexander von Humboldt Found, 84. *Mem:* Fel Am Phys Soc; Am Asn Physics Teachers; AAAS; Mat Res Soc. *Res:* Theory of solids; electronic structure and electron-phonon interactions; transport properties; superconducting transition temperature; thermal shifts of energy bands. *Mailing Add:* Dept Physics State Univ NY Stony Brook NY 11794-3800

ALLEN, PHILIP G, ANATOMIC BIOLOGY. *Educ:* Colby Col, AB, 83; Harvard Univ, PhD(cell & develop biol), 89. *Prof Exp:* Postdoctoral res fel, Dept Anat & Cellular Biol, LHRRB, Harvard Med Sch, 90-91; POSTDOCTORAL RES FEL, DEPT MOLECULAR MED, BRIGHAM & WOMEN'S HOSP, 91- *Mem:* Am Soc Cell Biol; AAAS. *Mailing Add:* Molecular Med LMRC Brigham & Women's Hosp 221 Longwood Ave Boston MA 02115

ALLEN, RALPH ORVILLE, JR, b Nashville, Tenn, Aug 3, 43; m 65; c 3. NEUTRON ACTIVATION ANALYSIS, ARCHAEOLOGICAL CHEMISTRY. *Educ:* Cornell Col, BA, 65; Univ Wis-Madison, PhD(chem), 70. *Prof Exp:* From asst prof to assoc prof, 70-85, PROF CHEM, UNIV VA, 85-, DIR ENVIRON HEALTH & SAFETY, 81-, PROF ENVIRON SCI, 90- *Concurrent Pos:* Vis scientist, Chem Div, Argonne Nat Lab, 70-76 & Norweg Inst Atomic Energy, 77-78; res fel, Norweg Marshall Fund, Univ Trondheim, 85. *Honors & Awards:* Erickson Award, Nat Acad-Fed Bur Invest, 84. *Mem:* Am Chem Soc; Sigma Xi. *Res:* Development of analytical techniques for monitoring air pollutants; application of nuclear techniques to study trace element geochemistry and archaeological materials. *Mailing Add:* 2415 Hillwood Pl Charlottesville VA 22901

ALLEN, RAYMOND A, b Lyman, Utah, Nov 7, 21; m 60; c 2. PATHOLOGY. *Educ:* Univ Utah, BS, 46; Univ Louisville, MD; Univ Minn, MS, 54. *Prof Exp:* Asst path & microbiol, Rockefeller Inst, 54-56; from asst prof to prof path, Univ Calif, Los Angeles, 56-68, assoc dir, Jules Stein Eye Inst, 66-68; pathologist, Good Samaritan Hosp, Phoenix, Ariz, 68-70; chief path & dir labs, St Dominic-Jackson Mem Hosp, 70-75; pathologist, Centinela Hosp, Inglewood, 75-76; PROF PATH, UNIV CALIF, IRVINE, 76-, PROF OPHTHAL, 77- *Concurrent Pos:* Attend physician, Los Angeles County Harbor Gen Hosp, 55-67; consult, Vet Admin Ctr, Los Angeles, Calif, 54-67; staff physician, Long Beach Veterans Admin Hosp, 76- *Mem:* AAAS; Am Soc Clin Path; Col Am Path; Int Acad Path. *Res:* Clinical pathology; pathology of eye disease. *Mailing Add:* Dept Path Univ Calif Med Col Irving CA 92717

ALLEN, RICHARD CHARLES, b Nov 22, 42; m 66; c 5. MEDICINAL CHEMISTRY. *Educ:* Med Col Va, BS, 65, PhD(med chem), 68. *Prof Exp:* Res fel, Univ NC, Chapel Hill, 68-69, instr med chem, Sch Pharm, 69-70; res assoc, chem res dept, Hoechst-Roussel Pharmaceut, Inc, NJ, 70-72, res group supvr, 73-74, dir chem res, 74-89, GROUP DIR, NEUROSCI, HOECHST-ROUSSEL PHARMACEUT, INC, NJ, 89- *Mem:* Sigma Xi; Am Chem Soc; Int Soc Heterocyclic Chem; Asn Res Vision & Ophthal; NY Acad Sci; Am Chem Soc. *Mailing Add:* Hoechst-Roussel Pharmaceut Inc Rte 202-206 N Somerville NJ 08876

ALLEN, RICHARD CRENSHAW, JR, b Detroit, Mich, July 17, 33; m 56; c 2. APPLIED MATHEMATICS. *Educ:* Murray State Univ, BS, 59; Univ Mo-Columbia, MA, 60; Univ Colo, 64-66; Univ NMex, PhD(math), 68. *Prof Exp:* Mathematician, Martin-Marietta Corp, Colo, 60-64; ASST PROF MATH, UNIV NMEX, 68- *Mem:* Am Math Soc; Soc Indust & Appl Math. *Res:* Transport theory; analytical and numerical studies of partial differential integral equations arising from transport theory; numerical solution of two-point boundary value problems, linear and nonlinear; integral equations and ordinary differential equations. *Mailing Add:* Numerical Math Div Sandia Nat Labs Div 1422 Albuquerque NM 87185

ALLEN, RICHARD DEAN, b Dallas Center, Iowa, Sept 20, 35; m 58; c 3. CELL BIOLOGY, PROTOZOOLOGY. *Educ:* Greenville Col, BA, 57; Univ Ill, MS, 60; Iowa State Univ, PhD(cell biol), 64. *Prof Exp:* Teacher sci & math, Niagara Christian Col, Ont, 57-59; instr biol & bot, Greenville Col, 60-61; asst prof biol, Messiah Col, 65-68; lectr biol & res assoc electron micros, Biol Labs, Harvard Univ, 68-69; assoc prof, 69-75, PROF MICROBIOL, PAC BIOMED RES CTR, UNIV HAWAII, MANOA, 75-, DIR, BIOL EM FACIL, 85- *Concurrent Pos:* NIH res fel cell biol, Biol Labs, Harvard Univ, 64-65; vis prof, Univ Colo, Boulder, 75-76; prin investr, NIH grants, 72-79 & 80-82; Am Heart Asn grant, 76-78; mem grad fac, trop med, Univ Hawaii, 77-, cellular, molecular & neurosci, 88-; NSF grants, 79-82, 81-84, 82-85, 83-84, 84-87, 84-85, 85-88, 88-90, 88-91, 89-, 89-91 & 91-; bd reviewers, J Protozool, 75-79 & 81-86, J Histochem & Cytochem, 75-81; co-ed, Europ J Cell Biol, 79-84; ed, Protoplasma, 90- *Honors & Awards:* Seymour H Hutner Prize, Soc Protozoologists, 78. *Mem:* Fel AAAS; Soc Protozool (pres, 81-82); Am Soc Cell Biologists; Sigma Xi. *Res:* Fine structure, function and morphogenesis of cell organelles and organelle systems with emphasis on ciliated protozoans; studies on intracellular digestion and membrane reutilization. *Mailing Add:* Dept Microbiol 2538 The Mall Univ Hawaii Honolulu HI 96822

ALLEN, RICHARD GLEN, b Sioux City, Iowa, Mar 8, 52; m 79; c 3. EVAPOTRANSPIRATION, IRRIGATION MANAGEMENT. *Educ:* Iowa State Univ, Ames, BS, 74; Univ Idaho, Moscow, MS, 77, PhD(civil eng), 84. *Prof Exp:* Res assoc water resources, Univ Idaho, 77-83; asst prof civil eng, Iowa State Univ, 84-85; asst prof, 86-90, ASSOC PROF IRRIG ENG, UTAH STATE UNIV, 91- *Concurrent Pos:* Consult, US Agency Int Develop, 88-90, UN Food & Agr Org, 88-91, UN Develop Prog, 90-91; chmn, Irrig Water Req Comm, Am Soc Civil Eng, 88-91, Int Symp Lysimeters Eng, Honoluluu, HI, 88-91; expert panel, Crop Water Req, UN Food & Agr Org, Rome, Italy, 90-91; co-ed, Am Soc Civil Eng Manual Evapotranspiration, 90. *Mem:* Am Soc Civil Eng; Am Soc Agr Eng; Am Geophys Union; Am Water Resources Asn; Int Comn Irrig & Drainage. *Res:* Evapotranspiration measurement and equation development; irrigation water requirements; water management; computer modeling; soil moisture movement; finite element; sprinkler design; climate change. *Mailing Add:* Agr & Irrig Eng Dept Utah State Univ Logan UT 84322-4105

ALLEN, ROBERT B, b Pittsburgh, Pa, Sept 16, 51. COMPUTER SCIENCE. *Educ:* Reed Col, BA, 73; Univ Calif, San Diego, PhD(exp psychol), 79. *Prof Exp:* Mem tech staff, Bell Labs, 78-85; MEM TECH STAFF, BELL COMMUN RES, 85- *Concurrent Pos:* Ed-in-chief, Asn Comput Mach Trans & Info Systs, 84- *Mem:* Asn Comput Mach. *Mailing Add:* Bell Commun Res 2A-367 Morristown NJ 07962-1910

ALLEN, ROBERT C(HADBOURNE), marine engineering, for more information see previous edition

ALLEN, ROBERT CARTER, b Natick, Mass, Feb 5, 30; m 56; c 2. PATHOLOGY, ANIMAL MEDICINE. *Educ:* Univ Vt, BS, 52, MS, 55; Va Polytech Inst, PhD(bact), 59. *Prof Exp:* Asst prof vet path, Va Polytech Inst, 55-60; NIH training grant, Jackson Mem Lab, 60-62; sr scientist radiation path, Oak Ridge Nat Lab, 62-68; sr scientist life sci, Ortec Inc, 68-73; assoc prof path, 73-76, prof & chmn, 76-85, PROF PATH, DEPT LAB ANIMAL MED, MED UNIV SC, 81- *Concurrent Pos:* Consult biol div, Oak Ridge Nat Lab, 68-69; guest prof, Univ Heidelberg, 71-72, 74 & 82; consult, Vet Admin Hosp, Charleston, 76-85. *Mem:* Am Asn Pathologists; NY Acad Sci; Am Asn Lab Animal Sci; Histochem Soc; Am Soc Clin Path; Electrophoresis Soc. *Res:* Chronic obstructive lung disease; isozymes, proteinases and antiproteinases; molecular pathology. *Mailing Add:* Dept Path Med Univ SC Charleston SC 29425-2645

ALLEN, ROBERT CHARLES, b Pueblo, Colo, Aug 12, 45; m 76; c 3. INFLAMMATION, CHEMILUMINESCENCE. *Educ:* Southeastern La Univ, BS, 67; Tulane Univ, PhD(biochem), 73, MD, 77; Am Bd Pathol, dipl, 85. *Prof Exp:* Res assoc biochem, Tulane Univ, New Orleans, La, 73-76; intern med, Brooke Army Med Ctr, Ft Sam Houston, Tex, 77-78, resident clin path 83-85; med officer infectious dis, US Army Inst Surg Res, Ft Sam Houston, 78-83, med officer infectious dis & path, 85-87; SCI DIR, EXOXEMIS, INC, SAN ANTONIO, TEX, 87- *Concurrent Pos:* Consult ed, J Leukocyte Biol, 84-85; ed, Wiley-Intersci Press, 86- & J Bioluminescence Chemiluminescence. *Mem:* Am Asn Immunologists; Am Soc Biol Chemists; Am Soc Clin Pathologists; Am Soc Microbiol; Biophys Soc; Am Asn Path; fel Col Am Pathologists. *Res:* Chemistry of oxygen and biological redox reactions; oxygenation reactions yielding excited products and chemiluminescence; granulocytic leukocyte and macrophage microbicidial action; humoral immune mechanisms; role of oxygen in inflammation and immune defense; reactions; macrophage microbicidial immune role of oxygen. *Mailing Add:* Exoxemis Inc 18585 Sigma Rd San Antonio TX 78258

ALLEN, ROBERT ERWIN, b Lufkin, Tex, Oct 9, 41; m 69; c 2. MEDICAL PHYSIOLOGY, RESEARCH ADMINISTRATION. *Educ:* Stephen F Austin State Col, BA, 63; Vanderbilt Univ, PhD(med physiol), 68. *Prof Exp:* Mem staff human factors eng, NASA, 68-72, sect chief biomet sect, 72-74, br chief biotechnol br, 74-76; spec asst to vpres, Univ Ala Med Sch, Birmingham, 76; fel NIH, Bethesda, Md, 76-78; spec asst to dir, Med Res Serv, 78-87, DEP DIR, AIDS PROG OFF, VET ADMIN, WASHINGTON, DC, 88- *Res:* Exercise and aerospace physiology, transfer RNA and protein synthesis. *Mailing Add:* Dept Vet Affairs 146B 810 Vermont Ave NW Washington DC 20420-0001

ALLEN, ROBERT H, b Detroit, Mich, Apr 15, 38; m; c 2. HEMATOLOGY. *Educ:* Amherst Col, BA, 60; Wash Univ, St Louis, MD, 66; Am Bd Internal Med, cert, 74, cert hemat, 76. *Prof Exp:* Intern med, Parkland Mem Hosp, Dallas, 66-67; staff assoc, Nat Inst Arthritis, Metab & Digestive Dis, 67-69; sr resident med, Barnes Hosp, St Louis, 69-70; fel hemat-oncol, Sch Med, Wash Univ, St Louis, 70-72, from asst prof to assoc prof med, 72-77; PROF MED & DIR, DIV HEMAT, HEALTH SCI CTR, UNIV COLO, DENVER, 77-, PROF BIOCHEM, BIOPHYS & GENETICS, 80- *Concurrent Pos:* Am Cancer Soc fac res award, 74-78; mem, Vet Admin Hemat Study Sect, 75-77, chmn, 77; mem, Task Force Vitamin B12 Assay, Nat Comt Clin Lab Standards, 79-82. *Mem:* Sigma Xi; Asn Am Physicians; Am Soc Clin Invest; Am Soc Hemat; Am Soc Biol Chemists; Am Fedn Clin Res. *Res:* Cobalamin metabolism and the cellular uptake of macromolecules; author of more than 80 technical publications. *Mailing Add:* Div Hemat Campus Box B170 Univ Colo Health Sci Ctr 4200 E Ninth Ave Denver CO 80262

ALLEN, ROBERT HARRY, b New York, NY, Aug 25, 56. EXPERT SYSTEMS, TACTILE SENSING. *Educ:* State Univ NY, Stony Brook, BE, 77; Univ Calif, Berkeley, MS, 78; Carnegie-Mellon Univ, PhD(civil eng), 84. *Prof Exp:* Staff engr, Simpson Gumperta & Neger Inc, 78-80; asst prof mech eng, Univ Houston, 84-88; ASST PROF MECH ENG, UNIV DEL, 88- *Concurrent Pos:* Forensic engr, Forensic Consults & Eng Inc, 82; fac fel, NASA, 85-88; evaluator, Spantran Educ Serv, 85-; prin investr, Whitaker Found, 86-89; vis asst prof, Mech Eng Dept, Rice Univ, 88. *Honors & Awards:* Charles Martin Hall Award, Aluminum Asn, 85; Herbert Allen Award, Am Soc Mech Engrs, 86; Ralph R Teetor Award Soc Automotive Engrs, 86; Interacad Exchange Award, Nat Acad Sci, 86. *Mem:* Am Soc Mech Engrs; Am Soc Civil Engrs; Nat Soc Prof Engrs. *Res:* Investigation of birth injuries; design and design methods; developing aids for the handicapped; vibrations and dynamics; engineering with computers; forensics. *Mailing Add:* 126 Spencer Lab Newark DE 19716

ALLEN, ROBERT MAX, b Ill, June 5, 21; m 46; c 3. FORESTRY. *Educ:* Iowa State Univ, BS, 47, MS, 51; Duke Univ, PhD(tree physiol), 58. *Prof Exp:* Forester, Southern Forest Exp Sta, US Forest Serv, 48-57, plant physiologist, 57-66; Belle W Baruch prof forestry, 66-70, head dept, 70-82, PROF FORESTRY, CLEMSON UNIV, 70- *Mem:* Fel Soc Am Foresters. *Res:* Tree physiology and genetics; growth of southern pine. *Mailing Add:* Dept Forestry Clemson Univ Clemson SC 29634-1003

ALLEN, ROBERT PAUL, b Milwaukee, Wis, Jan 13, 41; m 63; c 3. ANALYTICAL CHEMISTRY. *Educ:* Univ Wis-Milwaukee, 65; Univ NDak, PhD(org chem), 70. *Prof Exp:* Chemist, 70-76, sr chemist, 76-90, DEVELOP ASSOC, TEX EASTMAN CO, 90- *Res:* Resin chemistry; vapor-phase processes; heterogeneous catalysis; mass spectrometry. *Mailing Add:* Tex Eastman Co PO Box 7444 Longview TX 75607-7444

ALLEN, ROBERT RAY, b Potwin, Kans, Sept 2, 20; m 44; c 5. ORGANIC CHEMISTRY. *Educ:* Kans State Col, BS, 47, MS, 48, PhD(Chem), 50. *Prof Exp:* Chemist, Armour & Co, 50-56; head chem res dept, Foods Div, Anderson Clayton Foods, Anderson, Clayton & Co, 56-68, dir explor res, 68-78, prin scientist, 78-83, consult, 81-83; RETIRED. *Concurrent Pos:* Vis prof, Nat Taiwan Univ, 84-85. *Honors & Awards:* Bailey Award, Am Oil Chemists Soc, Supelco Award. *Mem:* Am Chem Soc; Am Oil Chem Soc (pres, 72-73). *Res:* Fats and oils. *Mailing Add:* 1314 Timber Creek Allen TX 75002

ALLEN, ROBERT THOMAS, b Farmerville, La, Dec 14, 39; m 62. SYSTEMATIC ENTOMOLOGY. *Educ:* La State Univ, BS, 62, MS, 64; Univ Ill, PhD(entom), 68. *Prof Exp:* From asst prof to assoc prof, 67-76, PROF ENTOM, UNIV ARK, FAYETTEVILLE, 76- *Mem:* Entom Soc Am; Soc Syst Zool; Soc Study Evolution. *Res:* Systematics and ecology of Coleoptera, especially the family Carabidae. *Mailing Add:* Dept Entom Univ Ark Fayetteville AR 72701

ALLEN, ROBERT WADE, b 1950; c 2. MOLECULAR IMMUNOLOGY, IMMUNOGENETICS. *Educ:* Univ Tulsa, BS, 72; Purdue Univ, PhD(cell biol), 77. *Prof Exp:* Fel molecular immunol, Scripps Clin & Res Found, 77-80; asst sci dir, Am Red Cross, 81-85; ASST PROF, SCH MED, ST LOUIS UNIV, 81-; SCI DIR, AM RED CROSS, 85- *Concurrent Pos:* Am Red Cross res grant awardee. *Mem:* AAAS; NY Acad Sci; Am Soc Hemat; Am Soc Cell & Molecular Biol. *Res:* Isolation/characterization of DNA probes detecting polymorphisms in the human genome and their use in paternity testing & forensics eg: "DNA Fingerprinting". *Mailing Add:* Dept Res Am Red Cross 4050 Lindell Blvd St Louis MO 63108

ALLEN, ROBIN LESLIE, b Tauranga, NZ, June 4, 43; m 68; c 2. FISHERIES, STATISTICS. *Educ:* Victoria Univ Wellington, BSc, 65, BSc, Hons, 66; Univ Otago, NZ, dipl sci statist, 68; Univ BC, PhD(zool), 72. *Prof Exp:* Statistician, Fisheries Div, Marine Dept, NZ, 65-72; scientist, Fisheries Res Div, Ministry Agr & Fisheries, NZ, 72-76; scientist, Intert-Am Trop Tuna Comn, 76-81; asst dir, 81-83, dir, Fisheries Res Div, 83-86, DEP GROUP DIR, MINISTRY AGR & FISHERIES, NZ, 86- *Mem:* Marine Sci Soc, NZ. *Res:* Assessment and management of fish and marine mammal populations. *Mailing Add:* Ministry Agr & Fisheries PO Box 2526 Wellington New Zealand

ALLEN, ROGER BAKER, b Portland, Maine, Mar 26, 29; m 54; c 3. PHYSICAL CHEMISTRY, OPERATION RESEARCH. *Educ:* Univ Idaho, BS, 51; Univ NH, MS, 57, PhD(chem), 59. *Prof Exp:* Res chemist, Electrochem Syst, US Naval Ord Lab, Md, 59-61; RES CHEMIST, S D WARREN CO DIV, SCOTT PAPER CO, 61- *Mem:* Am Chem Soc; Tech Asn Pulp & Paper Indust. *Res:* Kinetic studies of electrochemical systems for ordnance hardware; specialty coatings for paper, pressure sensitive adhesive systems, grease resistance, lithographic and electrophotographic. *Mailing Add:* 23 Mitchell Rd South Portland ME 04106

ALLEN, ROGER W, JR, b Montgomery, Ala, Sept 25, 41; m; c 3. MATHEMATICS. *Educ:* Auburn Univ, BS, 63; Univ Va, MA, 66, PhD(math), 68. *Prof Exp:* PROF & CHMN MATH, FRANCIS MARION COL, 71- *Mem:* Am Math Soc; Math Asn Am; Nat Coun Teachers Math. *Mailing Add:* Dept Math Francis Marion Col Florence SC 29501-0547

ALLEN, ROLAND E, b Houston, Tex, Sept 3, 41; m; c 1. THEORETICAL SOLID STATE PHYSICS. *Educ:* Rice Univ, BA, 63; Univ Tex, Austin, PhD(physics), 68. *Prof Exp:* Res assoc physics, Univ Tex, Austin, 69-70; from asst prof to assoc prof 70-83, PROF PHYSICS, TEX A&M UNIV, 83- *Concurrent Pos:* Scientist, Solar Energy Res Inst, Golden, Colo, 79-80, vis assoc prof physics, Univ Ill-Urbana, 80-81. *Mem:* AAAS; Am Phys Soc. *Res:* Surface physics; computer simulation; superconductivity; fundamental physics. *Mailing Add:* Dept Physics Tex A&M Univ College Station TX 77843

ALLEN, ROSS LORRAINE, public health, for more information see previous edition

ALLEN, SALLY LYMAN, b New York, NY, Aug 3, 26; c 1. GENETICS. *Educ:* Vassar Col, AB, 46; Univ Chicago, PhD(zool), 54. *Prof Exp:* Asst, R B Jackson Mem Lab, Bar Harbor, 46-48; res asst, 53-54, res assoc zool, 55-73, from assoc prof to prof bot, 67-75, prof zool, 73-75, chairperson cell & molecular biol dept, Div Biol Sci,75-76, PROF BIOL SCI & BIOL, UNIV MICH, 75-, PROF BIOL, 75-, ASSOC DEAN, 88- *Concurrent Pos:* Assoc prof bot, Univ Mich, 67-70; vis prof genetics, Ind Univ, 67; assoc ed, Genetics, 73-; mem, Coun Am Soc Cell Biol, 73-75; consult, Am Type Cult Collection, 76-; assoc ed, Developmental Genetics, 90- *Mem:* Genetics Soc Am; Am Soc Cell Biol; Am Soc Zool; Soc Protozool; Sigma Xi; fel AAAS. *Res:* Genetics of tumor transplantation; bacteriophage and protozoan genetics; isozymes of protozoa; phenotypic drift; genomic exclusion; DNA hybridization between species of Tetrahymena; comparison of DNA sequences of micronucleus and macronucleus of Tetrahymena thermophila. *Mailing Add:* Biol Dept Univ Mich Ann Arbor MI 48109

ALLEN, SAMUEL MILLER, b Washington, DC, July 1, 48; m 73; c 2. PHASE TRANSFORMATIONS. *Educ:* Stevens Inst Technol, BE, 70; Mass Inst Technol, SM, 71, PhD(metall), 75. *Prof Exp:* Instr metall, Mass Inst Technol, 74-75, res assoc, 75-76; res engr, Univ Calif, Berkeley, 76; res assoc, 77-79, asst prof, 79-83, ASSOC PROF METALL, MASS INST TECHNOL, 83- *Concurrent Pos:* Vis lectr, Univ Calif, Berkeley, 76. *Mem:* Am Soc Metals; Am Inst Mech Engrs Metall Soc; Electron Micros Soc Am; Sigma Xi; Mat Res Soc. *Res:* Physical metallurgy; the theory of phase transformations, the structure and behavior of interfaces in materials and the experimental study of these phenomena using electron microscopy. *Mailing Add:* 52 Forest St Stoneham MA 02180

ALLEN, SANDRA LEE, b Kingston, Ont, Jan 18, 48; m 68; c 4. PHYSICAL CHEMISTRY, COLLOID CHEMISTRY. *Educ:* Univ Western Ont, BSc, 68; Clarkson Col Technol, MS, 71, PhD(phys chem), 72. *Prof Exp:* Res assoc phys chem, Pulp & Paper Res Inst Can, McGill Univ, 72-74; proj leader nonwoven fabrics & air filters, Johnson & Johnson, Can, Ltd, 74-77; EDUC PRESENTATIONS, 87- *Res:* Colloidal stability; defoamers; adsorption from solution; fibrous structures. *Mailing Add:* 107 Seigniory Ave Pointe Claire PQ H9R 1J6 Can

ALLEN, SEWARD ELLERY, b Macedon, NY, Mar 21, 20; m 49; c 2. PLANT PHYSIOLOGY, SOIL CHEMISTRY. *Educ:* Cornell Univ, BS, 42; Univ Wis, MS, 47, PhD(bot), 50. *Prof Exp:* Res asst, Univ Wis, 47-50; plant physiologist, Tex Res Found, 50-52; agr chemist, Midwest Res Inst, 52-57; res chemist, US Rubber Co, 57-62; plant physiologist, Tenn Valley Auth, 62-85; RETIRED. *Mem:* Am Soc Agron; Soil Sci Soc Am; Am Soc Plant Physiol. *Res:* Plant growth regulators; soil fertility; utilization of fertilizers by isotope technique; nutrition and physiology of Hevea rubber; agronomic evaluation and development of experimental plant nutrients. *Mailing Add:* 2934 Alexander St Florence AL 35632

ALLEN, STEPHEN GREGORY, b Portland, Ore, Nov 22, 52. AGRONOMY, PLANT PHYSIOLOGY. *Educ:* Mont State Univ, BS, 79, MS, 82; Univ Ariz, PhD(agron & plant genetics), 84. *Prof Exp:* RES PHYSIOLOGIST, AGR RES SERV, USDA, 84- *Mem:* Am Soc Agron; Crop Sci Soc Am. *Res:* Physiological responses of crop species to arid environments. *Mailing Add:* US Water Conserv Lab USDA Agr Res Serv 4331 E Broadway Phoenix AZ 85040

ALLEN, STEPHEN IVES, b Holyoke, Mass, Dec 13, 14; m 43; c 2. MATHEMATICS. *Educ:* Amherst Col, BA, 37; Harvard Univ, MA, 46; Univ Pittsburgh, PhD(math, anal), 63. *Prof Exp:* Instr math & physics, Ruston Acad, Havana, Cuba, 38-39 & Wilbraham Acad, 39-40; instr & prin, Grahmam-Eckes Sch, Fla, 40-41 & 46-48; from instr to asst prof, Univ Mass, Amherst, 48-67, assoc dean, Col Arts & Sci, 73-76, assoc prof math, 67-; RETIRED. *Mem:* Math Asn Am; Am Math Soc. *Res:* Algebraic functions; algebra. *Mailing Add:* 17 Fairfield St Amherst MA 01002

ALLEN, STEVEN, b Vienna, Austria, Apr 28, 27; US citizen; m 51; c 2. UNDERSEA APPLICATIONS, MAN IN THE SEA. *Educ:* NY Univ, BME, 50; Mass Inst Technol, SM, 51, ScD, 59. *Prof Exp:* Res asst metall, Mass Inst Technol, 51-59; sr scientist staff mat res, ManLabs, Inc, 59-62; prin eng staff, C S Draper Lab, Inc, 62-75, mem staff, Ocean & Sci Syst, 75-89; CONSULT, 89- *Concurrent Pos:* Lectr, Mass Inst Technol, 71-90. *Mem:* Am Soc Metals; Marine Technol Soc; Soc Advan Mat & Process Eng; Am Soc Composites. *Res:* Underwater vehicles, underwater research and diving; autonomous underwater vehicles; materials engineering for state of the art high reliability, mechanical applications; advanced composites fabrication. *Mailing Add:* Seven Country Corners Rd Wayland MA 01778-3605

ALLEN, SUSAN DAVIS, b Jacksonville, Fla, Sept 13, 43; m 62; c 2. CHEMICAL PHYSICS, OPTICS. *Educ:* Colo Col, BS, 66; Univ Southern Calif, PhD(chem physics), 71. *Prof Exp:* Res assoc chem physics, Univ Southern Calif, 71-73; mem tech staff chem physics, Hughes Res Lab, 73-77; res scientist, 77-80, SR RES SCIENTIST CHEM PHYSICS, CTR LASER STUDIES, UNIV SOUTHERN CALIF, 80- *Concurrent Pos:* Consult, Hughes Res Labs, 77-78, Xerox Corp, 79- & TRW, 79- *Mem:* Am Phys Soc; Am Chem Soc; Am Vacuum Soc; AAAS; Optical Soc Am; Sigma Xi. *Res:*

Laser/interface interactions including laser chemical vapor deposition, laser induced desorption and laser induced interfacial mixing in semiconductors; surface and bulk optical properties of solids. *Mailing Add:* Ctr Laser Sci & Eng Univ Iowa Iowa City IA 52242

ALLEN, SYDNEY HENRY GEORGE, b Cambridge, Mass, Aug 26, 29; m 52; c 2. BIOCHEMISTRY. *Educ:* Tufts Univ, BS, 51; Univ Mass, MS, 53; Purdue Univ, PhD, 57. *Prof Exp:* Instr bact, Univ Conn, 57-60; from asst prof to assoc prof biochem, 63-75, PROF BIOCHEM, ALBANY MED COL, 75- *Concurrent Pos:* Fel biochem, Sch Med, Western Reserve Univ, 60-63; Lederle med fac award, 65-68. *Mem:* AAAS; Am Soc Microbiol; Am Soc Biol Chemists. *Res:* Subunit structure and function of biotin containing enzymes especially propionyl coenzyme A carboxylase; metabolic pathways involving biotin and vitamin B12; lactate and malate dehydrogenases as well as other enzymes in clinical chemistry; relationship between vitamin and enzyme activity; fermentation pathways; membrane-bound enzymes and intestinal lipid absorption. *Mailing Add:* 385 Morris St Albany NY 12208

ALLEN, TED TIPTON, b McKenzie, Tenn, Mar 22, 32; m 58; c 2. VERTEBRATE ZOOLOGY. *Educ:* Murray State Col, BA, 54; Univ Wis, MS, 58; Univ Fla, PhD(zool), 62. *Prof Exp:* From instr to assoc prof, 61-74, PROF BIOL, JACKSONVILLE UNIV, 74-, HEAD DEPT, 63- *Concurrent Pos:* Vert biol surv, Ariz, 78-80. *Honors & Awards:* Adams Environ Award, 79. *Res:* Ornithology; ecology and myology of birds. *Mailing Add:* Dept Biol Jacksonville Univ 2800 Univ Blvd N Box 3 Jacksonville FL 32211

ALLEN, THERESA O, b Torrington, Conn, Apr 27, 48; m 72; c 2. NEUROENDOCRINOLOGY. *Educ:* Univ Conn, BA, 70; Villanna Univ, MS, 75; Duke Univ, PhD(psychol), 78. *Prof Exp:* Fel, Inst Neurol Sci, Univ Pa, 78-80, Dept Psychol, 80-81, res assoc, 81-84; sci dir image processing, Drexel Univ, 83-84; CONSULT, 84- *Concurrent Pos:* Lectr, Dept Psychol, Univ Pa, 81. *Mem:* AAAS; Animal Behav Soc; Soc Neurosci. *Res:* Neurobiology of reproduction; neural processing of sensory input necessary for reproductive success. *Mailing Add:* 728 Dodds Lane Gladwyne PA 19035

ALLEN, THOMAS CHARLES, b Sept 6, 48; m; c 3. HUMAN REPRODUCTION, BIOCHEMICAL ENDOCRINOLOGY. *Educ:* Miami Univ, Ohio, AB, 70; Univ Wyo, MS, 73; Univ Cincinnati, PhD(physiol), 77. *Prof Exp:* Asst prof biochem, Health Sci Ctr, Univ Louisville, 80-85, assoc ophthal, 84-85; sci dir, Androl Lab & co-dir, In Vitro Fertilization Prog, Southeastern Fertil Ctr, 85-88; VPRES, NOVO VIVO CORP, SPRING VALLEY, WIS, 88-; PRES, INT FERTIL LAB, GREENVILLE, SC, 88- *Concurrent Pos:* Res assoc/asst prof, Ben May Lab Cancer Res, Univ Chicago, 80-81; assoc, James Graham Brown Cancer Ctr, Louisville, Ky, 80-85; instr, Jefferson Community Col, 84; dir, Carolina Fertil Lab, Charleston, SC, 86-88; gen partner & lab dir, Reproductive Lab Assoc, 89- & Qual Assurance Lab, 91- *Mem:* Endocrine Soc; Am Asn Cancer Res; Am Soc Cell Biol; Am Soc Andrology; Soc Study Reproduction; Am Fertil Soc; Sigma Xi. *Res:* Human reproduction and endocrinology. *Mailing Add:* Southeastern Fertil Ctr 900 Bowman Rd Mt Pleasant SC 29464

ALLEN, THOMAS CORT, JR, b Madison, Wis, Oct 28, 31; m 53; c 1. PLANT PATHOLOGY. *Educ:* Univ Wis, BS, 53; Univ Calif, PhD(plant path), 56. *Prof Exp:* Plant pathologist, Biol Warfare Labs, Ft Detrick, Md, 56-58; head plant path sect, Agr Res & Develop Labs, Stauffer Chem Co, Calif, 58-62; from asst prof to assoc prof, 62-73, PROF BOT & PLANT PATH, ORE STATE UNIV, 73- *Concurrent Pos:* NSF fel, 64; NATO sr fel sci, 70. *Mem:* Am Phytopath Soc; Sigma Xi; Am Inst Biol Sci; Electron Micros Soc Am; fel Royal Hort Soc. *Res:* Ultrastructural events in virus-infected plant cells; virus diagnosis; tissue culture manipulations to control viruses of vegetatively propagated plants. *Mailing Add:* Dept Bot & Plant Path Ore State Univ Corvallis OR 97331-2902

ALLEN, THOMAS HUNTER, b Sept 24, 14. AEROSPACE PHYSIOLOGY, BLOOD VOLUME OXIDATED ENZYMES. *Educ:* Univ Iowa, PhD(zool), 41. *Prof Exp:* Chief, Physiol Div, Sch Aerospace Med, Brooks AFB, Tex, 61-71, sci adv, 71-74; RETIRED. *Mailing Add:* 633 Balfour Dr San Antonio TX 78239

ALLEN, THOMAS JOHN, b Newark, NJ, Aug 20, 31; m 61; c 3. DESIGN ENGINEERING, MANAGEMENT. *Educ:* Upsala Col, BS, 54; Mass Inst Technol, SM, 63, PhD(mgt), 66. *Hon Degrees:* Dr Mgt honoris causa, Rijksuniversiteit, Gent, Belgium, 90. *Prof Exp:* Design engr, Tung-Sol Elec Co, Bloomfield, 56-57; res engr, Boeing Co, Seattle, Wash & Lexington, Mass, 57-65; res assoc, 63-66, from asst prof to assoc prof, 66-79, assco fac chmn, 83-85, PROF ORGN PSYCHOL & MGT, SLOAN SCH MGT, MASS INST TECHNOL, 74- *Concurrent Pos:* USMC, 54-56; vis sr lectr, Manchester Bus Sch, Univ Manchester, Eng, 69, hon sr res fel, 70- *Mem:* AAAS; Inst Elec & Electronics Engr; Am Psychol Asn; Sigma Xi. *Res:* Communication pattern among engineers and scientist; organizational structure for research and development organizations; architecture of research and development laboratories and its impact on communication; nature of engineering design; author of numerous books, articles and reviews in scientific publications. *Mailing Add:* Sloan Sch Mgt Mass Inst Technol Cambridge MA 02139

ALLEN, THOMAS LOFTON, b San Jose, Calif, Jan 20, 24; m 44; c 4. THEORETICAL CHEMISTRY. *Educ:* Univ Calif, BS, 44; Calif Inst Technol, PhD(chem), 49. *Prof Exp:* Asst chem, Calif Inst Technol, 46-47; asst prof chem, Univ Idaho, 48-49; lectr chem, Univ Calif, Davis, 49-51; res chemist, Calif Res Corp, 51-52; from asst prof to assoc prof, 52-63, PROF CHEM, UNIV CALIF, DAVIS, 63- *Concurrent Pos:* Vis res scholar, Ind Univ, 59-60, vis prof, 70-71; vis prof, Univ Nottingham, 71; vis scholar, Univ Calif, Berkeley, 79 & 85-86. *Mem:* Am Chem Soc; Sigma Xi; Am Phys Soc; AAAS. *Res:* Chemical bonding; molecular quantum mechanics. *Mailing Add:* Dept Chem Univ Calif Davis CA 95616

ALLEN, THOMAS OSCAR, b Weimar, Tex, Dec 8, 14; m 38; c 2. PETROLEUM ENGINEERING. *Educ:* Tex A&M Univ, BS, 36. *Prof Exp:* Petrol engr, Exxon Co USA, 37-42, sr petrol engr, 46-49, supv petrol engr, 49-53, sr supv petrol engr, 53-58; res mgr petrol prod res, Jersey Prod Res Co, Exxon Co, 58-63, sr eng assoc, 64-65; pres, 65-84, CHIEF EXEC OFFICER, OIL & GAS CONSULT INT, INC, 84- *Concurrent Pos:* Chmn prod comt, Southern Dist, Am Petrol Inst, 55-56, nat chmn, petrol subcomt, Smithsonian Hall Petrol, 61-67; vpres, Int Petrol Inst Ltd, 69-83. *Mem:* Soc Petrol Engrs; Am Inst Mining, Metall & Petrol Engrs. *Res:* Petroleum production, with emphasis on drilling, well completion and well stimulation. *Mailing Add:* 4632 N Cincinnati Pl Tulsa OK 74126

ALLEN, VERNON R, b Nashville, Tenn, Mar 25, 28; m 51; c 4. PHYSICAL CHEMISTRY, POLYMER CHEMISTRY. *Educ:* Tenn Polytech Inst, BS, 52; Univ Akron, MS, 57, PhD(polymer chem), 60. *Prof Exp:* Analyst, Carbide & Carbon Chem Co, 53-54; instr gen chem, Univ Wis, Milwaukee Exten, 54-55; res fel polymer physics, Mellon Inst, 60-63; asst prof phys chem, 63-69, assoc prof chem, 69-74, PROF CHEM, TENN TECHNOL UNIV, 74- *Mem:* Am Chem Soc. *Res:* Rheology of polymeric system; physical properties of rubbery materials. *Mailing Add:* Dept Chem Tenn Technol Univ Box 5055 Cookeville TN 38505

ALLEN, VIVIEN GORE, b Nashville, Tenn, Mar 31, 40; m 62; c 3. MINERAL INTERRELATIONSHIPS, GRAZING MANAGEMENT. *Educ:* Univ Tenn, BS, 62; La State Univ, MS, 73, PhD(agron/animal sci), 79. *Prof Exp:* Postdoctoral res, Agron Dept, La State Univ, 79-80; asst prof, 80-86, ASSOC PROF, AGRON DEPT, VA POLYTECH INST & STATE UNIV, 86- *Concurrent Pos:* Consult, Gabon, Africa & Sri Lanka. *Honors & Awards:* Merit Award, Am Forage & Grassland Coun, 89. *Mem:* Sigma Xi; Am Soc Agron; Am Soc Animal Sci; Soc Range Mgt; Crop Sci Soc Am; AAAS; Am Forage & Grassland Coun. *Res:* Forage management and forage systems, soil plant-animal interrelationships, utilization and system for high forage diets; mineral interrelationships in forage-livestock grazing systems, related to magnesium, sulphur and aluminum. *Mailing Add:* Dept Crop & Soil Environ Sci Va Polytech Inst & State Univ Blacksburg VA 24061

ALLEN, WAYNE ROBERT, b New Castle, Pa, Feb 28, 34; m 63; c 3. PLANT PATHOLOGY. *Educ:* Hiram Col, BA, 56; Cornell Univ, PhD(plant path), 62. *Prof Exp:* RES SCIENTIST, AGR CAN, 61-, HEAD PLANT PATH, 66-, ORNAMENTALS PROG, 89- *Mem:* Am Phytopath Soc; Can Phytopath Soc. *Res:* Chemical and physical properties of plant viruses; epidemiology of nematode (and insect) transmitted plant viruses; serological and biochemical assessment of insect predator-prey interactions; virus-free plant production. *Mailing Add:* Res Sta Agr Can PO Box 6000 Vineland Station ON L0R 2E0 Can

ALLEN, WENDALL E, b Elizabethtown, Ky, Nov 16, 36; div; c 1. MICROBIOLOGY, MICROBIAL GENETICS. *Educ:* Vanderbilt Univ, BA, 58, MA, 61; Univ Ky, PhD(microbiol), 68. *Prof Exp:* Instr microbiol, Ala Col, 61-63; res assoc bact genetics, Univ Ky, 65-67; PROF BIOL & UNIV COORDR MOLECULAR BIOL & BIOTECHNOL, E CAROLINA UNIV. *Concurrent Pos:* Assoc dean, Gen Col, 70-81. *Mem:* Am Soc Microbiol; AAAS; Sigma Xi. *Res:* Genetics of staphylococci and their bacteriophage; effects of lysogeny and conversion on extracellular products of staphylococci; phage and plasmid relationships among the species of staphylococci. *Mailing Add:* Dept Biol E Carolina Univ Greenville NC 27858-4353

ALLEN, WILLARD M, b Macedon, NY, Nov 5, 04; m 27, 46; c 1. OBSTETRICS & GYNECOLOGY, ENDOCRINOLOGY. *Educ:* Hobart Col, BS, 26; Univ Rochester, MS, 29, MD, 32. *Hon Degrees:* ScD, Hobart Col, 40 & Univ Rochester, 57. *Prof Exp:* From instr to asst prof obstet & gynec, Sch Med & Dent, Univ Rochester, 36-40; prof & chmn dept, 40-71, EMER PROF OBSTET & GYNEC, SCH MED, WASH UNIV, 71-; PROF OBSTET & GYNEC, SCH MED, UNIV MD, BALTIMORE CITY, 71- *Honors & Awards:* Eli Lilly Award in Biochem, 35. *Mem:* AMA; Am Gynec Soc; Am Asn Obstet & Gynec; Endocrine Soc; Am Physiol Soc. *Res:* Female sex hormones and endocrinology. *Mailing Add:* 14180 Burntwoods Rd Glenwoode MD 21738

ALLEN, WILLIAM CORWIN, b Geneva, Ohio, Sept 30, 34; m 57; c 3. ORTHOPEDIC SURGERY, BIOMEDICAL ENGINEERING. *Educ:* Hiram Col, BA, 56; Univ Chicago, MD, 60. *Prof Exp:* From asst prof to prof orthop, Univ Fla, 65-77; PROF & CHIEF ORTHOP, UNIV MO, 77- *Concurrent Pos:* Spec fel, Case Western Reserve Univ, 67-68. *Mem:* Am Acad Orthop Surgeons; Am Orthop Asn; Int Soc Orthop & Traumatic Surg; AMA; Int Soc Knee. *Res:* Biomechanics and sports medicine. *Mailing Add:* Div Orthop Surg Univ Mo Med Ctr Columbia MO 65212

ALLEN, WILLIAM F, JR, b North Kingstown, RI, June 22, 19. SCIENCE ADMINISTRATION. *Educ:* Brown Univ, ScB, 41; Harvard Univ, MS, 47. *Hon Degrees:* DTech, Wentworth Inst Technol, 83, DEng, Northeastern Univ, 90. *Prof Exp:* Instr, Brown Univ, 41-44; teaching fel, mech eng, Harvard Univ, 48; vpres & dir eng, 68-71, sr vpres eng, construct & qual assurance, 71, pres, 72 & 86-87, chief exec off, 73-86, CHIEF EXEC OFFICER, STONE & WEBSTER, INC 86-, CHMN, 88- *Concurrent Pos:* Trustee, Northwestern Univ, Thayer Acad, 75-84, chmn, 80-82; mem adv comt, Jr Achievement Eastern Mass, dir nat bd; dir, US Coun Energy Awareness; dir & chair, Blue Cross & Blue Shield, Mass; mem, Res Comt on Properties Steam, Am Soc Mech Engrs, 54-62, finance comt, 72-77, chmn, 76-77. *Mem:* Nat Acad Eng; AAAS; fel Am Soc Mech Engrs; Nat Soc Prof Engrs; Sigma Xi. *Res:* Author and co-author of twelve articles. *Mailing Add:* Stone & Webster Inc One Penn Plaza 250 W 34th St New York NY 10119

ALLEN, WILLIAM MERLE, b San Luis Obispo, Calif, Oct 9, 39; m 63; c 2. ORGANIC REACTION KINETICS & MECHANISMS. *Educ:* Loma Linda Univ, BA, 61; Univ Md, College Park, PhD(org chem), 67. *Prof Exp:* Asst prof chem, Andrews Univ, 66-68; from asst prof to prof chem, Loma Linda Univ, La Sierra Campus, 68-84, chmn dept, 71-77, dir, Div Natural Sci &

Math, 78-81, dean grad sch, Dir Sponsored Res, 87-88; vpres acad admin, Southern Col Seventh-Day Adventists, Tenn, 84-87; DIR, CTR LIFELONG LEARNING, LOMA LINDA UNIV, RIVERSIDE, 88-, WORLD MUS NATURAL HIST, 88- Mem: Am Asn Higher Educ. Res: Synthetic and mechanistic studies of organic compounds; synthesis of organo-cadmium reagents as intermediates in organic synthesis. Mailing Add: 11760 Valiant St Riverside CA 92505

ALLEN, WILLIAM PETER, b Buffalo, NY, Jan 15, 27; m 51; c 3. MEDICAL MICROBIOLOGY. Educ: Univ Buffalo, BA, 49, MA, 51; Univ Mich, PhD(med bact), 56. Prof Exp: Res asst immunol, Univ Mich, 54-55; microbiologist, US Army Biol Labs, 55-68; supvry microbiologist, US Dept Army, 68-71; res scientist, Delta Primate Ctr, Tulane Univ, 71-75; health scientist, adminr, Nat Inst Allergy & Infectious Dis, NIH, 75-89; CONSULT, 90- Concurrent Pos: Prog officer, Viral Dis Panel, US-Japan Coop Med Sci Prog; exec secy, Nat Inst AID Task Force on Virol; adj assoc prof microbiol & immunol, Tulane Univ, 73-75; consult, viral res admin, 89-91. Honors & Awards: Res Award, Sci Res Soc Am, 62. Mem: AAAS; Am Soc Microbiol; Sigma Xi; Am Soc Virol. Res: Natural and acquired immunity to microbial diseases; administrator of grants on viral diseases. Mailing Add: 7101 Ridge Crest Dr Frederick MD 21702

ALLEN, WILLIAM WESTHEAD, b Santa Cruz, Calif, Oct 13, 21; m 54; c 2. ECONOMIC ENTOMOLOGY, ACAROLOGY. Educ: Univ Calif, BS, 43, PhD(entom), 52. Prof Exp: From asst entomologist to entomologist, 52-85, ASSOC DEAN RES, COL NATURAL RESOURCES, DEPT ENTOM SCI, UNIV CALIF, BERKELEY, 85- Mem: Entom Soc Am; Sigma Xi; AAAS. Res: Insect pest management of agricultural crops with emphasis on economic injury levels and the use of selective pesticides. Mailing Add: 11 Edgewood Rd Orinda CA 94563

ALLENBACH, CHARLES ROBERT, b Buffalo, NY, Aug 3, 28; m 55; c 2. INORGANIC CHEMISTRY, PHYSICAL CHEMISTRY. Educ: Univ Buffalo, BA, 49, PhD(chem), 52. Prof Exp: Instr chem, Univ Buffalo, 51-52; sect mgr, chem eng process develop div, Tech Dept, Metals Div, Union Carbide Corp, 52-63, sr develop chemist, Develop Dept, Consumer Prod Div, 63-66, sect mgr, res & develop, Mining & Metals Div, 66-70, staff engr environ control, Eng Dept, Ferroalloys Div, 70-75, mgr, environ health & prod safety affairs, Metals Div, 75-81, mgr, environ health & prod safety affairs, Elkem Metals Co, 81-87, DIR, ENVIRON, HEALTH & PROD SAFETY AFFAIRS, ELKEM METALS CO, 87- Concurrent Pos: Corp specialist, Elkem Metals, Norway. Mem: Am Chem Soc; Am Inst Mining, Metall & Petrol Eng; Am Soc Test Mat; Chem Mfg Asn. Res: Environmental control activities; new product development; powder technology; aerosol technology; metal chemicals and compounds and reactive metals. Mailing Add: 88 Cherrywood Dr Williamsville NY 14221

ALLENBY, RICHARD JOHN, JR, b Chicago, Ill, July 4, 23; m 46; c 3. GEOPHYSICS, TECTONOPHYSICS. Educ: Dartmouth Col, AB, 43, MA, 48; Univ Toronto, PhD(geophys), 52. Prof Exp: Instr physics, Dartmouth Col, 46-48; demonstr, Univ Toronto, 49-52; res seismologist, Calif Res Corp, Standard Oil Co, Calif, 52-55, lead geophysicist, 59-62; chief geophysicist, Richmond Petrol Co, 55-59; staff scientist, Off Space Sci, NASA, 62-63, chief planetology, 63-64, dep dir, Manned Space Sci Div, 64-68, asst dir lunar sci, 68-75, RES GEOPHYSICIST, GODDARD SPACE FLIGHT CTR, 75- Honors & Awards: Apollo 8 Letter of Commendation; Apollo Achievement Award; NASA Medal Exceptional Sci Achievement, 71. Mem: AAAS; Am Geophys Union; Seismol Soc Am. Res: Geophysics and geology of earth moon and planets; spacecraft scientific instrumentation; explourational petroleum geophysics and geology. Mailing Add: Goddard Greenbelt Rd Greenbelt MD 20771

ALLENDER, DAVID WILLIAM, b Mt Pleasant, Iowa, Nov 20, 47; m 70; c 2. THEORETICAL SOLID STATE PHYSICS. Educ: Univ Iowa, BA, 70; Univ Ill, MS, 71, PhD(physics), 75. Prof Exp: Res assoc physics, Brown Univ, 74-75; ASST PROF PHYSICS, KENT STATE UNIV, 75-, MEM STAFF DIV PHYSICS & ASTRON, LIQUID CRYSTAL INST, 77- Mem: Am Phys Soc. Res: Liquid crystal; phase transitions; superconductivity. Mailing Add: Dept Physics Kent State Univ Kent OH 44242

ALLENDER, ERIC WARREN, b Mt Pleasant, Iowa, Nov 19, 56; m 85. COMPLEXITY THEORY. Educ: Univ Iowa, BA, 79; Ga Inst Technol, PhD (comput sci), 85. Prof Exp: ASST PROF COMPUT SCI, RUTGERS UNIV DEPT COMPUT SCI, 85- Concurrent Pos: Prin investr, NSF Grant, Appl Ko Imegorov Complexity, 88-90, Comput Complexity Theory & Circuit Complexity, 90-; vis prof, Univ Würzburg, 89. Mem: Asn Comput Mach; Soc Indust & Appl Math; Asn Symbolic Logic; AAAS. Res: Computational problems and resources required to solve problems. Mailing Add: Dept Comput Sci Rutgers Univ New Brunswick NJ 08903

ALLENDORF, FREDERICK WILLIAM, b Philadelphia, Pa, Apr 29, 47; m; c 2. POPULATION GENETICS. Educ: Pa State Univ, BS, 71; Univ Wash, MS, 73, PhD(fisheries, genetics), 75. Prof Exp: Lectr, Genetics Inst, Aarhus Univ, Denmark, 75-76; from asst prof to assoc prof, 76-84, PROF ZOOL, UNIV MONT, 84- Concurrent Pos: NATO fel, Genetics Dept, Univ Nottingham, Eng, 78-79; vis scientist, Univ Calif, Davis, 83-84; prog dir, NSF, 89-90. Mem: Genetics Soc Am; Soc Study Evolution; Am Genetic Asn; Am Soc Naturalists; fel AAAS; Soc Conserv Biol. Res: Evolutionary genetics; application of population genetics to the management of fish species; conservation of genetic resources. Mailing Add: Div Biol Sci Univ Mont Missoula MT 59812-1002

ALLENSON, DOUGLAS ROGERS, b Oak Park, Ill, Sept 23, 26; m 48; c 5. ANALYTICAL CHEMISTRY, PHYSICAL CHEMISTRY. Educ: Oberlin Col, AB, 48; Ore State Univ, MS, 51; Duke Univ, PhD(phys chem), 63. Prof Exp: Asst prof chem, La State Univ, 54-55; res chemist electrochem, Union Carbide Corp, 55-68; assoc prof chem, Heidelberg Col, 68-69; chief labs, Dept Utilities, Cleveland, 69-71; CHIEF CHEMIST, HERROH TESTING LABS,

71- Concurrent Pos: Mem Cuyohoga River Basin Water Facil Comt, 75- Mem: Am Chem Soc; Am Indust Hyg Asn; Sigma Xi. Res: Instrumental analytical chemistry; electrochemistry of metal oxide cathodes and of alkaline and solid electrolytes; water pollution analysis; methods of industrial hygiene. Mailing Add: 4160 E Sprague Rd Cleveland OH 44147

ALLENSPACH, ALLAN LEROY, b Campbell, Minn, Sept 23, 35; m 57; c 3. DEVELOPMENTAL BIOLOGY. Educ: Sioux Falls Col, BS, 57; Iowa State Univ, MS, 59, PhD(embryol, bot), 61; Marine Biol Lab, cert, 61. Prof Exp: Asst prof biol, Albright Col, 61-66; ASSOC PROF ZOOL & PHYSIOL, MIAMI UNIV, 66- Concurrent Pos: Res consult, Bernville Biol Labs, 64; USPHS fel, 65. Mem: AAAS; Am Soc Zoologists; Am Soc Cell Biol; Soc Develop Biol; Am Inst Biol Sci. Res: Normal and experimental studies on organogenesis in the chick embryo. Mailing Add: Dept Zool Miami Univ Oxford OH 45056

ALLENSTEIN, RICHARD VAN, b Lamont, Iowa, Aug 28, 30; m 55; c 3. ANALYTICAL CHEMISTRY. Educ: Univ Minn, BS, 52, MA, 57; Wash Univ, EdD, 61. Prof Exp: Dir NSF acad yr inst, Wash Univ, 62-66; asst prof chem, Univ Wis, Waukesha, 66-69; from asst prof to assoc prof, 69-80, PROF CHEM, NORTHERN MICH UNIV, 80- Mem: Fel AAAS; Am Chem Soc; Nat Sci Teachers Asn. Res: Measuring the effectiveness of science teaching and determining its correlates; energy education; aerospace education. Mailing Add: 61 W Elder Dr Marquette MI 49855

ALLENTOFF, NORMAN, b New York, NY, Nov 16, 23; m 56; c 4. PHOTOGRAPHIC CHEMISTRY. Educ: Univ Toronto, BA, 50, MA, 51, PhD(org chem), 56. Prof Exp: Res chemist, Chem Div, Can Dept Agr, 51-53, 55-58; sr chemist, Photog Chem Res Labs, Eastman Kodak Co, 58-63, res assoc, 64-65, sr tech assoc, Film Emulsion Div, 65-80, sr res assoc, Kodak Res Labs, 80-86, PATENT CONSULT, 86- Mem: Am Chem Soc. Res: Plant biochemistry; stereochemistry of the Grignard reaction; chemistry; photographic emulsion chemistry-patent liaison. Mailing Add: 17 Stuyvesant Rd Pittsford NY 14534

ALLENTUCH, ARNOLD, b Worcester, Mass, Dec 4, 30; m 56. SOLID MECHANICS. Educ: Worcester Polytech Inst, 53; Cornell Univ, MS, 59; Polytech Inst Brooklyn, PhD(appl mech), 62. Prof Exp: Instr appl mech, Cornell Univ, 54-55; res asst, Polytech Inst Brooklyn, 55-58, res assoc, 58-62, res group leader, 62-63; fel & preceptor, Columbia Univ, 63-64; asst prof, Cooper Union, 64-66; assoc prof mech eng, Newark Col Eng, 66-70, PROF & ASSOC VPRES GRAD STUDIES & RES, NJ INST TECHNOL, 70- Concurrent Pos: Consult, Lincoln Labs, Mass Inst Technol, 57-58 & Polytech Inst Brooklyn, 63- Mem: Am Soc Mech Engrs; Sigma Xi. Res: Stress analysis of reinforced shells and equilibrium crack; random vibrations; stability of discontinuous plates and rings. Mailing Add: 49 Fairview St Huntington NY 11743

ALLER, HAROLD ERNEST, b New York, NY, May 13, 34; m 57; c 3. ENTOMOLOGY. Educ: City Col New York, BS, 55; Cornell Univ, PhD(entom), 62. Prof Exp: Res entomologist, Niagara Chem Div, FMC Corp, 62-67; SR ENTOMOLOGIST, ROHM & HAAS CO, 67- Mem: Entom Soc Am; Am Mosquito Control Asn. Res: Chemical control; pesticides; toxicology; acaricides. Mailing Add: C/O Rohn/Hass Res Lab Spring House PA 19477

ALLER, JAMES CURWOOD, b Yakima, Wash, Aug 19, 21; m 43; c 5. ELECTRICAL ENGINEERING, COMPUTER SCIENCE. Educ: US Naval Acad, BS, 42; Harvard Univ, MA, 49, MES, 54; George Washington Univ, DSc(biomed), 68. Prof Exp: Mem prof staff, Ctr Naval Anal, 63-66, sect head command & control, 66-67; prof phys sci, Naval War Col, RI, 68-70; assoc prof clin eng, Sch Med, George Washington Univ, 70-73; sr assoc, Ketron Inc, 73-75; var prog off assignments, NSF, 75-85; prog dir, Eng Res Ctr, 85-86; CONSULT, 86- Concurrent Pos: Res assoc prof clin eng, Sch Med, George Washington Univ, 73-74; adj prof, Sch Eng & Appl Sci, 74-75; mem comt, Environ Quality, Inst Elec & Electronics Eng, 74-75 & Career & Prof Develop, Am Soc Eng Educ. Mem: Fel AAAS; sr mem Inst Elec & Electronics Engrs; NY Acad Sci. Res: Computer-based systems particularly for use in measurement control; instrumentation particularly as applied in clinical medicine; technology for aiding the physically disabled; biosaline research. Mailing Add: 9111 Deer Park Lane Great Falls VA 22066-4010

ALLER, LAWRENCE HUGH, b Tacoma, Wash, Sept 24, 13; m 41; c 3. SPECTROSCOPY OF STARS & NEBULAE. Educ: Univ Calif, AB, 36; Harvard Univ, AM, 38, PhD(astron), 43. Prof Exp: Lectr astron, Tufts Col, 40; instr physics, Harvard Univ, 42-43, res assoc, Observ, 42-43, 47; physicist, Univ Calif, 43-45; asst prof astron, Ind Univ & res assoc, W J McDonald Observ, 45-48; from assoc prof to prof, Univ Mich, 48-62; prof, 62-84, EMER PROF ASTRON & RES ASTRONR, UNIV CALIF, LOS ANGELES, 84- Concurrent Pos: Pres stellar spectros comn, Int Astron Union, 58-64; vis prof, Australian Nat Univ, 60-61 & Univ Toronto, 61-62; NSF sr fel, Australia, 60-61 & 68-69; vis prof, Univ Sydney & Univ Tasmania, 61-68; res assoc, Commonwealth Sci & Indust Res Orgn, Australia, 68, 69 & 71; vis prof, Univ Queensland, 77-78 & Raman Res Inst, Bangalore, India, 78; vis res prof, Observ Astrophys Arcetri, Florence Italy & Univ Florence, 81. Mem: Fel Nat Acad Sci; fel AAAS; fel Am Acad Arts & Sci; Am Astron Soc; Royal Astron Soc Gt Brit; Int Astron Union. Res: Spectroscopic and theoretical studies of the gaseous nebulae and stellar atmospheres; transition probabilities for spectral lines; cosmic abundances of elements. Mailing Add: 18118 W Kingsport Malibu CA 90265

ALLER, MARGO FRIEDEL, b Springfield, Ill, Aug 27, 38; m 64; c 1. ASTRONOMY. Educ: Vassar Col, BA, 60; Univ Mich, MS, 64, PhD(astron), 69. Prof Exp: Mathematician programmer astron, Smithsonian Astrophys Observ, 60-62; teaching fel, 62-66, res asst, 66-68, teaching fel, 69, res assoc astron, 70-76, assoc res scientist, 76-85, RES SCIENTIST, UNIV MICH, 85- Mem: Sigma Xi; Am Astron Soc; Int Astron Union. Res: Microwave extragalactic astronomy to study active extragalactic sources. Mailing Add: Dept Astron David Dennison Bldg Univ Mich Ann Arbor MI 48109-1090

ALLER, ROBERT CURWOOD, b Chelsea, Mass, May 17, 50; m; c 2. LOW TEMPERATURE GEOCHEMISTRY. *Educ:* Univ Rochester, BS, 72, BA, 72; Yale Univ, MS, 74, PhD(geol), 77. *Prof Exp:* From asst prof to assoc prof, Univ Chicago, 77-85, prof geophys sci & geochem, 85-86; PROF GEOCHEM, SUNY STONY BROOK, 86- *Concurrent Pos:* Alfred P Sloan Found fel, 78-80; assoc ed, Geochimica et Cosmochimica Acta; teacher scholar grant chem, Dreyfus Found, 82-87. *Mem:* Am Soc Limnol & Oceanog; Geochem Soc; AAAS; Sigma Xi; Am Geophys Union; Estuarine Res Fedn. *Res:* Early chemical diagenesis of marine sediments; animal-sediment interactions and controls on sediment water exchange of dissolved material. *Mailing Add:* Marine Sci Res Ctr State Univ NY Stony Brook NY 11794-5000

ALLERHAND, ADAM, b Krakow, Poland, May 23, 37; c 2. PHYSICAL CHEMISTRY, BIOPHYSICAL CHEMISTRY. *Educ:* State Tech Univ, Chile, BS, 58; Princeton Univ, PhD(chem), 62. *Prof Exp:* Res assoc chem, Univ Ill, 62-63, instr phys chem, 63-64, res assoc, 64-65; asst prof chem, Johns Hopkins Univ, 65-67; from asst prof to assoc prof, 67-72, chmn dept, 78-81, PROF CHEM, IND UNIV, BLOOMINGTON, 72- *Mem:* Am Chem Soc; Am Soc Biol Chemists. *Res:* Nuclear magnetic resonance. *Mailing Add:* Dept Chem Ind Univ Bloomington IN 47406

ALLERTON, JOSEPH, b New York, NY, Mar 17, 19; m 44; c 4. FOOD ENGINEERING, RESEARCH ADMINISTRATION. *Educ:* NY Univ, BChE, 41, MChE, 44; Univ Mich, PhD(chem eng), 48. *Prof Exp:* Asst, Off Sci Res & Develop, Columbia Univ, 41-42; instr chem eng, Cooper Union, 42-44; chem engr, Manhattan Proj, Kellex Corp, NY, 44-45; chem engr, Nat Dairy Res Labs, Inc, 48-51; sect head, Res Ctr, Gen Foods Corp, 52-63; mgr tech servs-Fulton, Nestle Co, Inc, 63-72, mgr technol develop, 72-84; RETIRED. *Concurrent Pos:* Consult, 84- *Mem:* Am Chem Soc; Am Inst Chem Engrs; Inst Food Technol; Sigma Xi. *Res:* Engineering aspects of food processing; research administration; chocolate research and development; environmental sciences. *Mailing Add:* 827 Forest Ave Fulton NY 13069-3337

ALLERTON, SAMUEL E, b Three Rivers, Mich, Aug 21, 33; m 66; c 2. BIOPHYSICAL CHEMISTRY, MEDICAL SCIENCE. *Educ:* Kalamazoo Col, BA, 55; Harvard Univ, PhD(biochem), 62. *Prof Exp:* Res assoc biochem, Rockefeller Inst, 61-64; asst prof biophys, Sch Dent, Univ Southern Calif, 65-67, asst prof biochem, 77-84, ASSOC PROF BIOCHEM, SCH DENT, UNIV SOUTHERN CALIF, 69- *Concurrent Pos:* Consult med devices & nutrit. *Mem:* AAAS; Sigma Xi; NY Acad Sci. *Res:* Isolation and characterization of proteins; ultracentrifugation and electrophoresis; biochemical pathology; acidic glycosaminoglycans in tumors; pediatric oncology; nutrition. *Mailing Add:* Sch Dent MC-0641 Univ Southern Calif University Park Los Angeles CA 90089-0641

ALLESSIE, MAURITS, b Gemert, Neth, Aug 14, 45; m 68; c 3. REENTRANT CARDIAC ARRHYTHMIAS, FLUTTER FIBRILLATION. *Educ:* Univ Amsterdam, MD, 71; Univ Limburg, PhD(cardiac arrhythmias), 77. *Prof Exp:* From asst prof to assoc prof, 70-84, PROF PHYSIOL, DEPT PHYSIOL, UNIV LIMBURG, 84- *Concurrent Pos:* Vis prof, Columbia Univ, NY, 80; mem bd, Working Group Arrhythmias & Intracardiac Electrophysiol, Europ Soc Cardiol, 87-; Dutch Sci Res Orgn, NWO, 88-; div head, Cardiovasc Res Inst, Maastricht, 88-; sci coun, ICIN, 88-; consult, Medtronic, 89-; chmn, Working Group Heart Function, NWO, 89- *Mem:* Am Physiol Soc. *Res:* Electrophysiological mechanisms of cardiac arrhythmias; development of animal models of reentrant mechanisms underlying atrial flutter-fibrillation and ventricular tachycardia-fibrillation; high resolution mapping, 256 points, of reentrant impulse propagation in the heart. *Mailing Add:* Dept Physiol Univ Limburg PO Box 616 Maastricht 6200 MD Netherlands

ALLEVA, FREDERIC REMO, b Norristown, Pa, Oct 16, 33; m 62. IMMUNOTOXICOLOGY, PERINATAL TOXICOLOGY. *Educ:* Gettysburg Col, AB, 56; George Washington Univ, MPhil, 70, PhD(zool), 71. *Prof Exp:* Neuropharmacologist & toxicologist, Merck Sharp & Dohme, 59-66; PHARMACOLOGIST, FOOD & DRUG ADMIN, 66- *Mem:* Sigma Xi; Endocrine Soc; Soc Toxicol. *Res:* Determine the effects of perinatally administered pediatric drugs on various parameters of development in animals. *Mailing Add:* 321 Locust Thorn Ct Millersville MD 21108

ALLEVA, JOHN J, b Norristown, Pa, Apr 17, 28; m 60; c 4. REPRODUCTIVE ENDOCRINOLOGY, BIOLOGICAL RHYTHMS. *Educ:* Univ Pa, AB, 50; Univ Mo, MS, 52; Harvard Univ, PhD(biol), 59. *Prof Exp:* Sr scientist biochem, Smith Kline & French Labs, 59-62; instr physiol, Albany Med Col, 62-63; RES BIOLOGIST & PHARMACOLOGIST, DIV DRUG BIOL, FOOD & DRUG ADMIN, 63- *Mem:* AAAS; Endocrine Soc; Soc Study Reproduction; Soc Exp Biol Med; Soc Res Biol Rhythms; Int Soc Chronobiol. *Res:* Reproductive endocrinology: the biological clock, pineal gland, and other central mechanisms; controlling fertility and biorhythms in hamsters. *Mailing Add:* Div of Res & Testing HFD-472 Food & Drug Admin Washington DC 20204

ALLEWELL, NORMA MARY, b Hamilton, Ont; m 70; c 1. PHYSICAL BIOCHEMISTRY, PROTEIN CHEMISTRY. *Educ:* McMaster Univ, BSc, 65; Yale Univ, PhD, 69. *Prof Exp:* Fel, Nat Inst Arthritis & Metab Dis, 69-70; asst prof biochem, Polytech Inst Brooklyn, 70-73; from asst prof to prof biol, 73-84, chmn, molecular biol & biochem, 84-87, PROF MOLECULAR BIOL & BIOCHEM, WESLEYAN UNIV, 84- *Concurrent Pos:* Instr, Marine Biol Labs, Woods Hole, Mass, 81-83. *Mem:* Am Chem Soc; Biophys Soc; fel AAAS; Am Soc Cell Biol; Am Soc Biol Chemists; Sigma Xi. *Res:* Energetics of macromolecular interactions; biomolecular regulatory mechanisms. *Mailing Add:* Dept Molecular Biol & Biochem Wesleyan Univ Middletown CT 06457

ALLEY, CURTIS J, viticulture, enology; deceased, see previous edition for last biography

ALLEY, EARL GIFFORD, b Corsica, SDak, Dec 10, 35; m 62; c 2. PESTICIDE CHEMISTRY, PHOTOCHEMISTRY. *Educ:* Miss State Univ, BS, 59, MS, 61; Univ Ill, Urbana, PhD(org chem), 68. *Prof Exp:* Chemist, Dow Chem Co, 67-69; DIR, RES DIV, MISS STATE CHEM LAB, 70- *Concurrent Pos:* Prof chem, Chem Dept, Miss State Univ; consult. *Mem:* Am Chem Soc; AAAS; Sigma Xi. *Res:* Photochemistry and degradation of agricultural chemicals; involvement of charge transfer and adsorption in these processes; analyses of trace metals and organic materials. *Mailing Add:* Rte 5 Box 49 Southgate Starkville MS 39759

ALLEY, FORREST C, b Tuskegee, Ala, Aug 14, 29; m 52; c 3. ENVIRONMENTAL ENGINEERING. *Educ:* Auburn Univ, BS, 51, MS, 55; Univ NC, PhD(environ eng), 62. *Prof Exp:* Eng technologist, Shell Oil Co, La, 55-58; from asst prof to assoc prof, 58-69, PROF CHEM ENG, CLEMSON UNIV, 69- *Mem:* Am Inst Chem Engrs; Air Pollution Control Asn. *Res:* Process development; industrial pollution control; atmospheric chemistry. *Mailing Add:* Continuing Eng Educ Clemson Univ Clemson SC 29631

ALLEY, HAROLD PUGMIRE, b Cokeville, Wyo, Mar 26, 24; m 46; c 2. WEED SCIENCE. *Educ:* Univ Wyo, BS, 49, MS, 56; Colo State Univ, PhD(bot sci), 65. *Prof Exp:* Voc agr instr, La Grange Sch Dist, Wyo, 49-55; instr plant sci, 56-59, from asst prof to assoc prof, 59-68, PROF WEED CONTROL & EXTEN WEED SPECIALIST, UNIV WYO, 68- *Mem:* Am Soc Agron; Weed Sci Soc Am. *Res:* Weed physiology; morphological and physiological effects of herbicides; chemistry and mode of action of herbicides; control of undesirable plants. *Mailing Add:* 1121 E Reynolds St Laramie WY 82070

ALLEY, KEITH EDWARD, b Palm Springs, Calif, June 27, 43; m 66; c 1. NEUROBIOLOGY. *Educ:* Univ Ill, DDS, 68, PhD(anat), 72. *Prof Exp:* Fel anat, Univ Ill, 68-72; asst prof oral biol, Univ Iowa, 72-74; asst prof anat, Case Western Reserve Univ, 74-80, assoc prof oral biol, 80-85, asst dean res, 84-85; PROF ORAL BIOL & CHMN DEPT, OHIO STATE UNIV, 85- *Concurrent Pos:* Fel neurobiol, Univ Iowa, 72-74; mem, Oral Biol & Med Study Sect, NIH, 88-90. *Mem:* Soc Neurosci; Am Asn Anatomists; Int Asn Dent Res; AAAS. *Res:* Molecular and physiologic aspects of neuromuscular development; metamorphic transformation of cranial systems; developmental mechanisms and evolutionary biology. *Mailing Add:* Ohio State Univ 305 W 12th Ave Columbus OH 43210-1239

ALLEY, PHILLIP WAYNE, b Chicago, Ill, Mar 11, 32; m 58; c 4. SOLID STATE PHYSICS. *Educ:* Lawrence Col, BS, 53; Rutgers Univ, PhD(physics), 58. *Prof Exp:* Asst physics, Rutgers Univ, 53-56, instr, 56-58; asst prof, Franklin & Marshall Col, 58-64; PROF PHYSICS, STATE UNIV NY COL GENESEO, 64- *Mem:* Am Phys Soc; Am Asn Physics Teachers; Sigma Xi. *Res:* Electrical conductivity in dilute metal alloys; teaching physics at undergraduate and graduate level. *Mailing Add:* Three Main St Geneseo NY 14454

ALLEY, REUBEN EDWARD, JR, b Petersburg, Va, July 16, 18; m 49; c 1. PHYSICS. *Educ:* Univ Richmond, BA, 38; Princeton Univ, EE, 40, PhD(elec eng), 49. *Prof Exp:* Instr physics, Univ Richmond, 40-42; staff mem, radiation lab, Mass Inst Technol, 42-43; instr elec eng, Princeton Univ, 43-44; assoc prof physics & chmn dept, Univ Richmond, 49-51; mem tech staff, Bell Tel Labs, 51-53; assoc prof physics & chmn dept, Univ Richmond, 53-55; assoc prof physics, Washington & Lee Univ, 55-57; mem tech staff, Bell Tel Labs, 57-59; sr proj engr, semiconductor-components div, Tex Instruments, Inc, 59-60; prof physics, Vassar Col, 60-62, chmn dept, 61-62; prof elec eng, Univ SC, 62-65; PROF ELEC ENG, US NAVAL ACAD, 65- *Concurrent Pos:* Bd of Gov, Am Inst Physics, 75-78. *Mem:* Am Phys Soc; Am Asn Physics Teachers; Sigma Xi. *Res:* Direct current controlled reactors; electrical analogy networks for solution of nonlinear differential equations; magnetic materials and high frequency magnetic properties; nuclear radiation effects. *Mailing Add:* Dept Elec Eng US Naval Acad Annapolis MD 21402

ALLEY, RICHARD BLAINE, b Columbus, Ohio, Aug 18, 57; m 80; c 2. GLACIER DYNAMICS, ICE-CORE PALEOCLIMATES. *Educ:* Ohio State Univ, BSc, 80, MSc, 83; Univ Wis-Madison, PhD(geol), 87. *Prof Exp:* Asst scientist, Geophys & Polar Res Ctr, Univ Wis-Madison, 87-88; ASST PROF GEOSCI, DEPT GEOSCI & EARTH SYST SCI CTR, PA STATE UNIV, 88- *Concurrent Pos:* NSF presidential young investr, 90. *Mem:* Int Glaciol Soc; Am Geophys Union. *Res:* Flow, stability and deposits of ice sheets and glaciers; paleoclimatic records in ice cores. *Mailing Add:* Dept Geosci 306 Deike Bldg Pa State Univ University Park PA 16802

ALLEY, STARLING KESSLER, JR, b Crab Orchard, Tenn, Sept 20, 30; m 52; c 2. PETROLEUM CHEMISTRY. *Educ:* Berea Col, AB, 52; Univ Calif, Los Angeles, PhD(phys chem), 61. *Prof Exp:* Res chemist, Olin Mathieson Chem Corp, NY, 52-55, res chemist, high energy fuels div, Calif, 55-57; asst chem, Univ Calif, Los Angeles, 57-61; sr res chemist, 61-69, res assoc, 70-73, supvr catalyst res, 73-78, mgr refining res, 78-81, VPRES, REFINING & PROD RES, UNOCAL CORP, 81- *Mem:* Am Chem Soc; Am Petrol Inst; Soc Automotive Engrs; Indust Res Inst. *Res:* High energy fuels; boron chemistry; thermodynamics of nonelectrolytes; nuclear magnetic resonance; hydrogen bonding; heterogeneous catalysis. *Mailing Add:* Sci & Technol Div Unocal Corp 376 S Valencia Ave Brea CA 92621

ALLFREY, VINCENT GEORGE, b June 28, 21; m 43; c 2. CELL BIOLOGY, GENETICS. *Educ:* Col City New York, BS, 43; Columbia Univ, MS, 48, PhD(chem), 49. *Prof Exp:* Res asst, Rockefeller Inst Med Res, 49-52, res assoc, 52-57; assoc prof, 57-63, PROF, ROCKEFELLER UNIV, 63- *Concurrent Pos:* Vis prof cell biol, Yale Univ, 64-65; mem, Adv Comts, Am Cancer Soc, 66-88 & Nat Cancer Inst, NIH, 71-89, coun, Am Cancer Soc, 69-72 & Molecular Cytol Study Sect, NIH, 82; chmn, Coun Res, Am Cancer Soc, 71-74 & Gordon Res Conf Nuclear Proteins, Chromatin Struct & Gene Regulation, 74; adj prof cell biol & genetics, Grad Sch Med Sci, Cornell Univ, 73- *Mem:* Am Soc Biochem & Molecular Biol; Am Soc Cell Biol; Am Soc

Microbiol; Am Asn Cancer Res; AAAS; Biochem Soc. *Res:* DNA-binding proteins and their post-synthetic modifications in controlling chromosome structure and RNA synthesis; separation of transcriptionally active and inactive DNA sequences of oncogenes and anti-oncogenes; differences in carcinogen-induced DNA damage and repair in active and inactive genes. *Mailing Add:* Rockefeller Univ 1230 York Ave New York NY 10021

ALLGAIER, ROBERT STEPHEN, b Union City, NJ, Nov 29, 25; m 54; c 2. GALVANOMAGNETIC PROPERTIES, IV-VI SEMICONDUCTORS. *Educ:* Columbia Univ, AB, 50, AM, 52; Univ Md, PhD(physics), 58. *Prof Exp:* Asst physics, Columbia Univ, 50-51; physicist, Naval Ord Lab, 51-73, chief, solid state div, 73-74; adminr mat technol prog, Off Chief of Naval Develop, 74-75 & 76; sr scientist, Solid State Br, Naval Surface Weapons Ctr, Silver Spring, 75-81; CONSULT, 81- *Concurrent Pos:* Vis scientist, Cavendish Lab, Cambridge Univ, 65-66; mem tech rev comt, Joint Serv Electronics Prog, Dept of Defense, 72-81; lectr physics, Univ Md, 75-79; consult, 84-; vis prof, Linz Univ, Austria, 87; tech ed, Nat Acad Sci, 89; consult, Technol Base Off, Naval Surface Weapons Ctr, 90- *Mem:* Am Phys Soc. *Res:* Transport properties of semiconductors and metals in liquid, solid, amorphous and crystalline forms; electronic properties of IV-VI semiconductors. *Mailing Add:* 11034 Seven Hill Lane Potomac MD 20854

ALLGOOD, JOSEPH PATRICK, b Calvert, Ala, June 27, 27. PHYSIOLOGY. *Educ:* Univ Ala, BS, 49; Northwestern Univ, DDS, 53, MS, 60, PhD(physiol), 63. *Prof Exp:* Instr, 62-63, ASST PROF PHYSIOL, DENT SCH, NORTHWESTERN UNIV, CHICAGO, 63- *Mem:* AAAS; Int Asn Dent Res; Sigma Xi. *Res:* Masticatory efficiency; electromyography; temporomandibular joint disturbances; intro-oral force measurements. *Mailing Add:* 311 E Chicago Ave Chicago IL 60611

ALLGOWER, EUGENE L, b Chicago, Ill, Aug 11, 35; m 58; c 1. MATHEMATICS. *Educ:* Ill Inst Technol, BS, 57, MS, 59, PhD(math), 64. *Prof Exp:* Instr math, Ill Inst Technol, 60-62 & Univ Ariz, 62-64; asst prof, Sacramento State Col, 64-65; assoc prof, Univ Tex, El Paso, 65-67; assoc prof, 67-74, PROF MATH, COLO STATE UNIV, 67- *Concurrent Pos:* Marathon Oil Co Indust fel, 65; vis prof, Swiss Fed Inst Technol, Zurich, 72 & Univ Bonn, 77-78 & 81; Richard Merton prof, Univ Stuttgart, 78; vis prof, Univ Hamburg, 83, Ore State Univ, 87; Sr Hubbolt Fel Award, WGer, 88-89. *Mem:* Am Math Soc; Soc Indust & Appl Math. *Res:* Analysis; numerical analysis; approximation of fixed points; differential equations; continuation methods. *Mailing Add:* Dept Math Colo State Univ Ft Collins CO 80521

ALLIEGRO, RICHARD ALAN, b New York, NY, Mar 20, 30; m 52; c 6. CERAMICS ENGINEERING. *Educ:* Alfred Univ, BS, 51, MS, 52. *Prof Exp:* Res assoc ceramics, Alfred Univ, 52-54; develop engr, 57-59, supvr process eng, 59-61, tech develop, 61-63, mgt eng, 63-66, chief armor res & develop, 66-67, asst dir res & develop, 67-71, mgr res & develop, Indust Ceramics Div, 71-73, dir res & new bus develop, 73-78, dir mkt & sales, 78-82, vpres & gen mgr, Indust Ceramics Div, 82-84, VPRES, HIGH PERFORMANCE CERAMICS, NORTON CO, 84- *Concurrent Pos:* Trustee, Alfred Univ. *Honors & Awards:* Ceramist of the Year, Amer Ceramic Soc, 82. *Mem:* Fel Am Ceramic Soc; Nat Inst Ceramic Engrs (pres, 75-76); US Advan Ceramics Asn (pres, 87-88). *Res:* Development of high temperature materials, usually fabrication by way of hot pressing; development of monolithic boron carbide ballistic armor for use on helicopter crewmen in Vietnam; patented furnace components for silicon diffusion processing. *Mailing Add:* 26 Woodstone Rd Northborough MA 01532

ALLIET, DAVID F, b Rochester, NY, May 9, 38; m 60; c 3. LASER DIFFRACTION. *Educ:* Rochester Inst Technol, BS, 67. *Prof Exp:* Chemist, 65-70, sr chemist, 70-73, PROJ LEADER, XEROX CORP, 73- *Concurrent Pos:* Adj assoc prof chem, Monroe Community Col, 81- *Mem:* Am Chem Soc; fel Am Inst Chemists. *Res:* Investigating methods for measuring very fine particles; laser diffraction; laser velocimetry; quasi elastic light scattering methods. *Mailing Add:* Six Cypress Circle Fairport NY 14450

ALLIN, EDGAR FRANCIS, b Edmonton, Alta, Jan 31, 39; m 67. ANATOMY. *Educ:* Univ Alta, MD, 63, BSc, 66. *Prof Exp:* Instr anat, Univ Alta, 61-62; intern, Vancouver Gen Hosp, 63-64; jr asst resident surg, Univ Hosp, Edmonton, Alta, 64-65; from proj specialist & teaching asst to asst prof anat, Sch Med, Univ Wis-Madison, 67-75; asst prof, Univ Ill Med Ctr, Chicago, 75-78; assoc prof anat, 78-84, PROF, CHICAGO COL OSTEOPATH MED, 84- *Concurrent Pos:* Hon res assoc geol, Field Mus Natural Hist, 79- *Mem:* AAAS; Am Asn Anat; Soc Vert Paleontol. *Res:* Evolution of the middle ear and feeding apparatus of mammal-like reptiles and early mammals; skeletal muscle fiber type histochemistry. *Mailing Add:* Dept Anat Chicago Col Osteopath Med 5200 S Ellis Ave Chicago IL 60615

ALLIN, ELIZABETH JOSEPHINE, b Blackwater, Ont, July 8, 05. PHYSICS. *Educ:* Univ Toronto, BA, 26, MA, 27, PhD(spectros), 31. *Prof Exp:* Asst demonstr physics, Univ Toronto, 26-27, demonstr & asst, 30-33; Royal Soc Can fel, Cambridge Univ, 33-34; lectr, Univ Toronto, 34-41, from asst prof to prof, 41-72, emer prof physics, 72-; RETIRED. *Mem:* Can Asn Physicists (secy, 47 & 48). *Res:* Underwater spark spectra; hyperfine structure of spectral lines; x-ray; structure of alloys; spectroscopy; low temperature spectroscopy. *Mailing Add:* 36 Willowbank Blvd Toronto ON M4R 1B6 Can

ALLING, DAVID WHEELOCK, b Rochester, NY, July 5, 18; m 48; c 1. MATHEMATICAL STATISTICS, MEDICINE. *Educ:* Univ Rochester, BA, 40, MD, 48; Cornell Univ, PhD(statist), 59. *Prof Exp:* Intern med, Arnot-Ogden Hosp, 48-49; resident, Hermann M Biggs Mem Hosp, 49-56; med officer statist, Nat Cancer Inst, 59-60, med officer statist, 60-64, RES MATH STATISTICIAN, NAT INST ALLERGY & INFECTIOUS DIS, NIH, 64- *Mem:* Inst Math Statist; Am Statist Asn; Biomet Soc. *Res:* Partition of chi-squared variables; sequential tests of hypotheses; stochastic models of chronic diseases. *Mailing Add:* Bldg 10 Rm 11N238 NIH Bethesda MD 20205

ALLING, NORMAN LARRABEE, b Rochester, NY, Feb 8, 30; m 57; c 2. MATHEMATICS. *Educ:* Bard Col, BA, 52; Columbia Univ, MA, 54, PhD(math), 58. *Prof Exp:* Lectr math, Columbia Univ, 55-57; from asst prof to assoc prof, Purdue Univ, 57-65; assoc prof, 65-70, PROF MATH, UNIV ROCHESTER, 70- *Concurrent Pos:* NSF fel, Harvard Univ, 61-62; lectr, Mass Inst Technol, 62-64, NSF fel, 64-65; vis prof, math inst, Univ Würzburg, 71. *Mem:* Am Math Soc. *Res:* Analysis and algebra of real algebraic curves; surreal numbers. *Mailing Add:* 215 Saudringham Rd Rochester NY 14610

ALLINGER, NORMAN LOUIS, b Alameda, Calif, Apr 6, 28; m 52; c 3. ORGANIC CHEMISTRY. *Educ:* Univ Calif, BS, 51; Univ Calif, Los Angeles, PhD, 54. *Prof Exp:* Fel, Univ Calif, Los Angeles, 54-55, NSF fel, 55-56; from asst prof to prof chem, Wayne State Univ, 56-69; PROF CHEM, UNIV GA, 69- *Honors & Awards:* Herty Medal, Am Chem Soc, Atlanta, 82, Arthur C Cope, 88, James Flack Norris, 89. *Mem:* Nat Acad Sci; Am Chem Soc. *Res:* Conformational analysis; organic quantum chemistry; physical properties of organic compounds; macro-rings; computational chemistry. *Mailing Add:* Dept Chem Univ Ga Athens GA 30602

ALLINGTON, ROBERT W, b Madison, Wis, Sept 18, 35; m 76; c 5. ELECTRICAL ENGINEERING, CHEMICAL INSTRUMENTATION. *Educ:* Univ Nebr, BS, 59, MS, 61. *Hon Degrees:* DSc, Univ Nebr, 85. *Prof Exp:* PRES, ISCO INC, 61- *Honors & Awards:* Indust Res 100 Award, Res & Develop Mag, 78 & 85. *Mem:* Inst Elec & Electronics Engrs; Instrument Soc Am; Am Chem Soc; fel Am Inst Chem; Nat Soc Prof Engrs. *Res:* Instrumentation and measurement methods in the fields of liquid chromatography, electrophoresis, supercritical fluids and spectrophotometry; data processing and handling; laboratory automation. *Mailing Add:* Isco Inc 4700 Superior Lincoln NE 68504-1398

ALLINSON, MORRIS JONATHAN CARL, b New Haven, Conn, Feb 20, 12; m 48; c 4. RADIOLOGY. *Educ:* Yale Univ, BS, 32; Boston Univ, PhD(biochem), 38; Univ Ark, MD, 45; Am Bd Radiol, dipl. *Prof Exp:* Res biochemist, Res Lab, Sharp & Dohme, 34-35; instr biochem, Sch Med, La State Univ, 39-40; instr physiol & pharmacol, Sch Med, Univ Ark, 40-45; intern, Grace-New Haven Hosp, 45-46; resident radiol, Hosp St Raphael's New Haven, 50-52 & Bridgeport Hosp, 52; resident radiation ther, Hosp Joint Dis, 53, staff therapist, 53-57; practicing radiologist, 53-57; radiologist, Franklin Hosp, 57-76; RETIRED. *Res:* Metabolism and colorimetric methods of analysis of creatine and creatinine; action of bacterial enzymes on creatine and nicotinic acid analysis of inositol and penicillin; relation of cyanamid to cancer therapy; use of zymomonas bacteria in producing alcohol from sugar and oranges. *Mailing Add:* 550 S Ocean Blvd Apt 306 Boca Raton FL 33432

ALLIS, JOHN W, b Buffalo, NY, Apr 24, 39; m 65; c 2. MEMBRANE BIOCHEMISTRY, MACROMOLECULAR PHYSICAL CHEMISTRY. *Educ:* Syracuse Univ, BS, 60; Univ Wis, PhD(phys chem), 65. *Prof Exp:* Res assoc protein chem, Georgetown Univ, 67-69; res chemist, USPHS, 70-71; res chemist, 71-76, SUPVRY RES CHEMIST, ENVIRON PROTECTION AGENCY, 76- *Concurrent Pos:* In US Army, 65-67. *Honors & Awards:* Bronze Medal, Environ Protection Agency, 85. *Mem:* Am Chem Soc; Bioelectromagnet Soc; NY Acad Sci; Biophys Soc. *Res:* Physical chemistry of macromolecules and biological membranes; investigation of the conformational properties of proteins and structural properties of biological membranes and their relation to chemical reactivity and biological function; effects of nonionizing radiation. *Mailing Add:* Health Effects Res Lab MD-74 Environ Protection Agency Research Triangle Park NC 27711

ALLIS, WILLAM PHELPS, b Menton, France, Nov 15, 01; US citizen; m 35; c 3. PLASMA PHYSICS, ELECTRON PHYSICS. *Educ:* Mass Inst Technol, BS, 23, MS, 24; Univ Nancy, DSc(high frequency resonance), 25. *Hon Degrees:* MA, Oxford Univ, 68; Dr, Univ Paris, 79. *Prof Exp:* Res assoc plasma physics, 25-29, from instr to prof, 31-67, sr lectr, 67-77, EMER PROF PLASMA PHYSICS, MASS INST TECHNOL, 67- *Concurrent Pos:* Vis prof, Harvard Univ, 58, Univ Tex, 60, St Catherine's Col, Oxford Univ, 68, Univ Paris, Sud, 69-83, Mid East Tech Univ, Ankara, 70 & Univ S Fla, 71; chmn, Gaseous Electronics Conf, 49-62, hon chmn, 66-; consult, Los Alamos Sci Lab, 52-; asst secy gen sci affairs, NATO, France, 62-64; Fulbright sr lectr, Univ Innsbruck, 74-75; consult, Lawrence Livermore Lab, Univ Calif, 78; vis fel, Joint Inst Lab Astrophys, Boulder, Colo, 79-80. *Honors & Awards:* Legion of Honor, France, 68. *Mem:* Fel Am Phys Soc; Am Acad Arts & Sci (vpres, 61-62); fel Brit Inst Physics; Royal Soc Arts; fel AAAS. *Res:* Free electrons in gases; ionized gases; electron distributions and processes in lasers. *Mailing Add:* 33 Reservoir St Cambridge MA 02138-3335

ALLISON, ANTHONY CLIFFORD, b East London, SAfrica; UK citizen. IMMUNOLOGY, PATHOLOGY. *Educ:* Oxford Univ, DPhil(biol), 50, BM & BCh, 52; FRCPath, 70. *Prof Exp:* Mem sci staff virol, Nat Inst Med Res, Mill Hill, London, 57-67; head div path, Clin Res Ctr, Harrow, Eng, 67-77; dir parasitol, Int Lab Res Animal Dis, Nairobi, Kenya, 77-80; VPRES RES, SYNTEX RES, PALO ALTO, CALIF, 81- *Honors & Awards:* Murgatroyd Prize, Royal Col Physicians, 70; Heberden Medal, Heberden Soc, 73. *Mem:* Europ Molecular Biol Orgn; Am Asn Immunologists. *Res:* Development of an adjuvant formulation to elicit cell-mediated immunity against antigens produced by recombinant DNA technology and peptide synthesis; development of novel immunosuppressive drugs for prevention of allograft rejection; treatment of patients with autoimmune disease. *Mailing Add:* Syntex Res S3-5 3401 Hillview Ave Palo Alto CA 94304

ALLISON, CAROL WAGNER, b Fairbanks, Alaska, Jan 15, 32; m 67; c 2. MICROPALEONTOLOGY, STRATIGRAPHY. *Educ:* Univ Calif, Berkeley, BA, 53, MA, 63, PhD(paleont), 70. *Prof Exp:* Micropaleontologist, Shell Oil Co, 53-54; res asst, Univ Calif, 63-68; cur mus & assoc prof paleont, Univ Alaska, Fairbanks, 70-86; RETIRED. *Concurrent Pos:* Res assoc, Los Angeles County Mus Nat History, 71-; prin investr NSF grants, 74-88; geologist, US Geol Surv, 78-86. *Res:* Paleontology, stratigraphy and paleoenvironments of Late Proterozoic; tectonic history of northwestern North America; evolution and distribution of North Pacific echinoids; metal-biologic interactions. *Mailing Add:* 5334 Berkeley Rd Santa Barbara CA 93111

ALLISON, DAVID C, b Monmouth, Ill, Feb 25, 31; m 54; c 3. PLANT GENETICS, CYTOLOGY. *Educ:* Univ Ill, BS, 56, MS, 57; Pa State Univ, PhD(genetics, breeding), 60. *Prof Exp:* Asst plant breeder, Univ Ariz, 61-62; from asst prof to assoc prof, 62-73, PROF BIOL, MONMOUTH COL, ILL, 73- *Mem:* Am Soc Agron; Crop Sci Soc Am; Am Inst Biol Sci. *Res:* Taxonomy and cytotaxonomy of the Gramineae; plant breeding. *Mailing Add:* Dept Biol Monmouth Col Monmouth IL 61462

ALLISON, DAVID COULTER, b Detroit, Mich, June 2, 42; c 3. MICROBIOLOGY, MEDICINE. *Educ:* Univ Mich, MD, 67; Univ Chicago, PhD(microbiol), 76. *Prof Exp:* Intern, Univ Chicago, 67-68; med officer, US Army, 68-70; resident, Univ Chicago, 74-77, assoc instr & chief resident, 77-78; ASST CHIEF, VET ADMIN MED CTR, 78-; ASST PROF SURG, UNIV NMEX, 78- *Mem:* Cell Kinetics Soc; Histochem Soc; Soc Anal Cytol; Asn Vet Admin Surgeons. *Res:* Tumor cell biology and cytometry. *Mailing Add:* Dept Surg Univ NMex Albuquerque NM 87131

ALLISON, FRED, JR, b Abingdon, Va, Sept 8, 22; m 49; c 4. MEDICINE. *Educ:* Ala Polytech Inst, BS, 44; Vanderbilt Univ, MD, 46. *Prof Exp:* Instr microbiol, Sch Med, La State Univ, 50-51; instr med, 51-52; fel, Div Infectious Dis, Univ Wash, 52-53; instr med, 53-54, from instr to asst prof prev med, 54-55; from asst prof to prof med, Univ Miss, 55-68, chief div infectious dis, 55-68, assoc prof microbiol, 64-68; PROF MED & HEAD DEPT, MED CTR, LA STATE UNIV, NEW ORLEANS, 68- *Concurrent Pos:* Consult infectious dis, Jackson Vet Hosp, 55-67; vis investr, Rockefeller Univ, 66-67; consult med, New Orleans Vet Hosp, 77- *Mem:* AAAS; Asn Am Physicians; Am Soc Clin Invest; Am Col Physicians; Am Clin & Climatol Asn. *Res:* Mechanisms relating to acute inflammatory reaction; clinical aspects of infectious diseases. *Mailing Add:* 418 Fairfax Ave Nashville TN 37212

ALLISON, IRA SHIMMIN, geomorphology, glaciology; deceased, see previous edition for last biography

ALLISON, JEAN BATCHELOR, b Teague, Tex, Dec 16, 31; m 53; c 2. RESEARCH ADMINISTRATION. *Educ:* Rice Univ, BA, 53; Univ Houston, MS, 58, PhD(molecular spectros), 62. *Prof Exp:* Chemist, med res, Vet Admin Hosp, Houston, Tex, 53-57; mem staff, 62-78, sr res chemist, 78-81, supvr oil recovery res, 81-87, RES CONSULT, TEXACO EXPLOR & PROD TECH DIV, 87- *Mem:* Am Chem Soc; Sigma Xi. *Res:* Electronic and vibrational molecular spectroscopy of metalloporphyrins; development of inorganic and organic analytical chemistry techniques; enhanced oil recovery research. *Mailing Add:* 107 Manor Pl Lake Jackson TX 77566

ALLISON, JERRY DAVID, b Brevard, NC, June 18, 48; m 69; c 2. RADIATION PHYSICS, HEALTH PHYSICS. *Educ:* NC State Univ, BS, 70; Old Dominion Univ, MS, 74; Univ Fla, PhD, 78. *Prof Exp:* Radiol control engr, health physics, Newport News Shipbuilding, 70-75; grad asst radiol, Univ Fla, 75-78; ASST PROF RADIOL, MED COL GA, 78- *Mem:* Health Physics Soc; Am Asn Physicists Med; Sigma Xi. *Res:* Nuclear medicine instrumentation. *Mailing Add:* 2262 Overton Rd Augusta GA 30904

ALLISON, JOHN, b Philadelphia, Pa, Nov 14, 51; m 74. ANALYTICAL & PHYSICAL CHEMISTRY. *Educ:* Widener Col, BS, 73; Univ Del, PhD(chem), 77. *Prof Exp:* Res fel chem, Univ Del, 73-77; NSF fel chem, Stanford Univ, 77-79; asst prof, 79-84, assoc prof, 84-89, PROF ANALYTICAL CHEM, MICH STATE UNIV, 89- *Mem:* Am Chem Soc; Am Soc Mass Spectrometry. *Res:* Ion cyclotron resonance spectroscopy; organometallic chemistry in the gas phase; ion-molecule reactions; laser-induced fluorescence as a probe of chemical processes; electron impact dissociative excitation; mass spectrometry. *Mailing Add:* Dept Chem Mich State Univ East Lansing MI 48824

ALLISON, JOHN EVERETT, b Mont, Aug 16, 17; m 43; c 1. ANATOMY. *Educ:* Concordia Col, BA, 40; Univ Minn, MA, 47; Univ Iowa, PhD(zool), 52. *Prof Exp:* Instr biol, Drake Univ, 47-49; asst zool, Univ Iowa, 49-52; from instr to asst prof anat, Med Sch, St Louis Univ, 52-57; assoc prof, 57-69, PROF ANAT, COL MED, UNIV OKLA, 69- *Mem:* Am Asn Anatomists; Endocrine Soc. *Res:* Reproductive endocrinology. *Mailing Add:* 9640 Sandy Lane Oklahoma City OK 73131

ALLISON, JOHN P, b Beckenham, Kent, Eng, Feb 17, 36; US citizen; m 66; c 2. ORGANIC CHEMISTRY, POLYMER CHEMISTRY. *Educ:* Univ Birmingham, BSc, 58, PhD(chem), 61. *Prof Exp:* Fel chem, Univ Ariz, 61-63; sr res chemist, polymer dept, res labs, Gen Motors Tech Ctr, 63-66; staff mem, polymer chem, res div, Raychem Corp, 66-70; RES CHEMIST, KIMBERLY-CLARK CORP, 70- *Mem:* Am Chem Soc; Royal Soc Chem. *Res:* Synthetic polymer chemistry; fiber bonding; chemistry of natural polymers and polymer reactions; flexible foams. *Mailing Add:* 2584 N Arbor Trail Marietta GA 30066

ALLISON, MILTON JAMES, b South Shore, SDak, May 10, 31; m 53; c 3. MICROBIOLOGY. *Educ:* SDak State Col, BS, 53, MS, 54; Univ Md, PhD, 61. *Prof Exp:* Instr bact, SDak State Col, 54-55; bacteriologist, Dairy Cattle Res Br, Agr Res Serv, 57-62, RES MICROBIOLOGIST, NAT ANIMAL DIS CTR, USDA, 62 - *Mem:* Am Soc Microbiol; Sigma Xi; Am Soc Animal Sci; AAAS; Am Acad Microbiol. *Res:* Ecology of microorganisms; bacterial physiology and nutrition; biosynthesis of microbial amino acids and lipids; anaerobic bacteria of the mammalian digestive system and their interactions with toxic substances. *Mailing Add:* Nat Animal Dis Ctr USDA Box 70 Ames IA 50010

ALLISON, RICHARD C, b Seattle, Wash, Oct 19, 35; m 67; c 2. INVERTEBRATE PALEONTOLOGY. *Educ:* Univ Wash, BS, 57, MS, 59; Univ Calif, Berkeley, PhD(paleont), 67. *Prof Exp:* Instr geol, Col San Mateo, 65-68; assoc prof, 68-75, chmn, dept geol geopys, 81-84, prof geol, Univ Alaska, 75-88; RETIRED. *Concurrent Pos:* Consult to petrol co Alaskan Cenozoic & Mesozoic Molluscan fossils, 69- *Mem:* Paleont Soc; Brit Palaeont Asn; Sigma Xi; Paleont Res Inst; Soc Econ Paleont & Mineral. *Res:* Mesozoic molluscan biostratigraphy of Alaska; Cenozoic stratigraphy of the Gulf of Alaska perimeter; Mesozoic and Cenozoic invertebrate paleontology; systematic paleontology of turritellid gastropods; Mesozoic and Cenozoic invertebrate paleontology. *Mailing Add:* 5334 Berkeley Rd Santa Barbara CA 93111

ALLISON, RICHARD GALL, b Hanover, Pa, Jan 28, 43; m 66; c 1. NUTRITIONAL BIOCHEMISTRY. *Educ:* Pa State Univ, BS, 64; Univ Calif, Davis, PhD(nutrit), 68. *Prof Exp:* Res biochemist, Dept Food Sci & Technol, Univ Calif, Davis, 69; biochemist, Walter Reed Army Inst Res, Walter Reed Army Med Ctr, 69-73; res assoc, Life Sci Res Off, Fedn Am Socs Exp Biol, 74-78, sr staff scientist, 78-84; EXEC OFFICER & SCI OFFICER, AM INST NUTRIT, 84- *Concurrent Pos:* Asst ed, Am J Clin Nutrit, 79-81. *Mem:* AAAS; Am Inst Nutrit; Am Soc Clin Nutrit. *Res:* Biomedical research evaluation; nutrition; food ingredients; water. *Mailing Add:* Am Inst Nutrit 9650 Rockville Pike Bethesda MD 20814

ALLISON, ROBERT DEAN, b Farmersville Station, NY, Feb 29, 32; m 81; c 3. CLINICAL PHARMACOLOGY, BIOENGINEERING. *Educ:* Hartwick Col, BA, 54; Wayne State Univ, MS, 60, Col Med, PhD(physiol & pharmacol), 62. *Prof Exp:* Instr, Anesthesiol Dept, Grace Hosp, Mich, 61-62; staff physiologist & assoc, Dept Physiol, Lovelace Found Med Educ & Res, NMex, 62-66; asst prof, Dept Biol, Univ NMex, 65-66; ASSOC PROF & SPEC LECTR, LA STATE UNIV MED SCH, 70-; PRES, DYNAMIC RES ENTERPRISES INC, 73- *Concurrent Pos:* Chief, Sect Cardiovascular Physiol & dir, Cardiovasc Labs, Scott & White Clin, Tex, 66-73; consult clin res, Santa Fe Hosp, Tex & Vet Admin Med Ctr, Tex, 66-73; res physiologist, Vet Med Ctr, Tex, 73-75; consult, Clin Path Lab, Grand Prairie Community Hosp, Tex, 78-80; adj assoc prof physiol, Tex Health Sci Ctr, Southwestern Med Sch, 76-; adj assoc prof bioeng, Univ Tex, Arlington, 76-; assoc prof biol, Tex Wesleyan Col, 76- & Brookhaven Col, 82-; assoc prof & div chmn basic sci & res, Parker Col, 83- *Mem:* Fel Am Geriat Soc; Am Physiol Soc; fel Int Col Angiol; Sigma Xi; Int Soc Nephrology. *Res:* Development of non-invasive (no needle or catheter) techniques for measuring cardiovascular and pulmonary function in human subjects, the systems utilize ultrasound, plethysmographs and other parameters that do not require skin puncture. *Mailing Add:* PO Box 538 Groesbeck TX 76642

ALLISON, RONALD C, b San Angelo, Tex, Jan 11, 45. PULMONARY MEDICINE. *Educ:* Univ Tex, San Antonio, MD, 71. *Prof Exp:* Asst Prof, 80-85, ASSOC PROF MED & PHYSIOL, COL MED, UNIV S ALA, 85- *Mem:* Am Col Physicians; Am Col Chest Physicians; Am Thoracic Soc; Am Physiol Soc. *Mailing Add:* Col Med & Physiol Univ S Ala 2451 Fillingine St Mobile AL 36617

ALLISON, STEPHEN WILLIAM, b Pocahontas, Ark, Nov 3, 50; m 75; c 2. FIBEROPTIC SENSORS, THERMOMETRY. *Educ:* Harding Univ, BS, 72; Memphis State Univ, MS, 74; Univ Va, PhD(eng physics), 79. *Prof Exp:* Develop staff physicist, Union Carbide Nuclear Div, 78-84; DEVELOP STAFF PHYSICIST, ENRICHMENT TECHNOL APPLN CTR, OAK RIDGE GAS DIFFUSION PLANT, MARTIN MARIETTA ENERGY SYSTS, 84- *Concurrent Pos:* Vis assoc res prof, Dept Nuclear Eng & Eng Physics, Univ Va, 89-90; adj prof elec eng, Univ Tenn, 87- *Mem:* Optical Soc Am; Laser Inst Am; Soc Photo-Optical Instrumentation Engrs. *Res:* Diagnostic and remote measurement techniques with lasers, fiberoptics, and electro-optics; spectroscopy; rotational systems; nonlinear optics; thermometry; received three patent awards related to thermometry and fiber sensors. *Mailing Add:* Appln Technol Div Martin Marietta Energy Systs PO Box 7280 Oak Ridge TN 37831-7280

ALLISON, TRENTON B, b Sacramento, Calif, Mar 17, 42; m 64; c 5. PHARMACOKINETICS, THERAPEUTIC DRUG MONITORING. *Educ:* Univ Utah, BS, 66, PhD(biochem), 72. *Prof Exp:* Res assoc med, Med Ctr, Univ Nebr, 72-76; asst prof pharmacol, Health Ctr, Univ Conn, 76-78; dir, Radioimmunoassay Lab, Vet Admin Ctr, Augusta, Ga, 78-82; asst prof pharmacol, Med Col Ga, 78-82; ASSOC PROF, DEPT PATH, MED COL VA, 82- *Mem:* Am Physiol Soc; Int Soc Heart Res; Am Soc Pharmacol & Exp Therapeut; Sigma Xi; Am Asn Clin Chemists. *Res:* Cardiovascular pharmacology relating to ischemic heart disease; pharmacokinetics and therapeutic drug monitoring. *Mailing Add:* Vet Admin Med Ctr Lab 113 1201 Broad Rock Rd Richmond VA 23249

ALLISON, WILLIAM EARL, b Claremore, Okla, Oct 2, 32; m 56; c 2. ENTOMOLOGY. *Educ:* Okla A&M Col, BS, 57; Okla State Univ, MS, 58; Tex A&M Univ, PhD(biochem), 63. *Prof Exp:* Res entomologist, Dow Chem Co, Seal Beach, Calif, 63-67, group leader entom, Midland, Mich, 67-71, group leader entom & nematol, Walnut Creek, 71-73, sr res entomologist, Dow Chem Co Japan Ltd, 74 & Dow Chem Co Pac Ltd, Malaysia, 75, RES MGR, INSECTICIDES DISCOVERY GROUP, DOW CHEM CO, WALNUT CREEK, 83-, MGR, WESTERN TECH SERV & DEVELOP, DOW ELANCO, 90- *Concurrent Pos:* Sr develop specialist entom, Plant Path & Plant Nutrit, Dow Chem Co, Walnut Creek, 75-, regional mgr tech serv & develop, Agr Prod Dept, 78- *Mem:* Entom Soc Am; AAAS. *Res:* Structure versus activity to search for new insecticides. *Mailing Add:* 2719 W San Bruno Ave Fresno CA 93711

ALLISON, WILLIAM HUGH, b Harrison Twp, Pa, Nov 25, 34; m 58; c 3. MYCOLOGY. *Educ:* Pa State Univ, BS, 56, MS, 57, PhD(bot), 63. *Prof Exp:* Plant pathologist crops div, US Army Biol Labs, Md, 60-63; dir res clean mushroom cultivation, Brandywine Mushroom Co Div, Borden Co, 63-66; mgr, Great Lakes Spawn Co, 66-68; PROF BIOL, DEL VALLEY COL SCI & AGR, 68- *Mem:* AAAS; Mycol Soc Am; Bot Soc Am; Am Phytopath Soc; Sigma Xi; Am Soc Plant Taxonomists. *Res:* General biology; commercial mushroom production; epidemiology of rice blast disease. *Mailing Add:* 4581 Landisville Rd Doylestown PA 18901

ALLISON, WILLIAM S, b North Adams, Mass, June 16, 35; m 64; c 2. PROTEIN CHEMISTRY. *Educ:* Dartmouth Col, AB, 57, MA, 59; Brandeis Univ, PhD(biochem), 63. *Prof Exp:* USPHS fel, Lab Molecular Biol, Cambridge Univ, 64-65; res assoc, Brandeis Univ, 65-66, asst prof biochem, 66-69; from asst to assoc prof, 73-80, PROF CHEM, UNIV CALIF, SAN DIEGO, 80- *Honors & Awards:* Career Develop Award, NIH, 67-69. *Mem:* Am Soc Biochem & Molecular Biol; Biophys Soc; Am Chem Soc. *Res:* Enzymology; comparative biochemistry; functional groups of enzymes; mechanism of enzyme action. *Mailing Add:* Dept Chem Univ Calif San Diego La Jolla CA 92093-0601

ALLISTON, CHARLES WALTER, b Florence, Miss, May 27, 30; m 53; c 2. ENVIRONMENTAL PHYSIOLOGY, REPRODUCTIVE PHYSIOLOGY. *Educ:* Miss State Col, BS, 51, MS, 57; NC State Col, PhD(physiol), 60. *Prof Exp:* Res asst, Miss State Col, 55-56, res technician, 56-57; res asst, NC State Col, 57-60, from asst prof to assoc prof zool, 60-67; assoc prof, 67-75, PROF ANIMAL SCI, PURDUE UNIV, 75- *Concurrent Pos:* NIH res grant, 62-; David Ross res grant, 67-69 & 77-79; NSF grant, 74- *Honors & Awards:* Sigma Xi Res Award, 66. *Res:* Environmental and reproductive physiology, particularly influence of the physical environment upon reproductive efficiency of the mammalian female. *Mailing Add:* Dept Animal Sci Lilly Hall Purdue Univ West Lafayette IN 47907

ALLMAN, JOHN MORGAN, b Columbus, Ohio, May 17, 43. NEUROPHYSIOLOGY, PRIMATOLOGY. *Educ:* Univ Va, BA, 65; Univ Chicago, PhD(anthrop), 71. *Prof Exp:* Fel neurophysiol, Univ Wis, 70-73; res asst prof psychol, Vanderbilt Univ, 73-74; assoc prof, 74-84, PROF BIOL, CALIF INST TECHNOL, 84-, HIXON PROF PSYCHOBIOL, 89- *Concurrent Pos:* Sloan fel, 74; US Pub Health Serv career develop, 76-81. *Honors & Awards:* Golden Brain Award, Minerva Found. *Mem:* Soc Neurosci; Asn Res Vision & Ophthal; Int Brain Res Orgn; Am Soc Primatologists. *Res:* Functional organization of the visual system in primates; evolution of brain and behavior in primates. *Mailing Add:* Div Biol 216-76 Calif Inst Technol Beckman Lab Pasadena CA 91125

ALLMANN, DAVID WILLIAM, b Peru, Ind, May 20, 35; m 56; c 2. BIOCHEMISTRY. *Educ:* Ind Univ, BS, 58, PhD(biochem), 64. *Prof Exp:* Res asst bact, Ind Univ, 57, res asst biochem, 58-60; asst prof, Univ Wis-Madison, 66-70; assoc prof biochem, 70-80, assoc prof, Dent Sch, 73-80, PROF BIOCHEM, SCH MED & PROF DENT SCH, IND UNIV-PURDUE UNIV, INDIANAPOLIS, 80- *Concurrent Pos:* NIH fel biochem, Univ Wis-Madison, 64-66; res biochemist, Vet Admin Hosp, 70-72; consult, Am Dent Asn Accreditation Comt & mem, Study Sect, Nat Inst Dent Res, 80; rev, J Dent Res & Arch Oral Biol, 80-; pres, Pharmacol, Toxicol Therapeut Sect, Int Asn Dent Res, 85-86. *Mem:* Am Inst Nutrit; Int Asn Dent Res; Am Soc Biol Chemists; fel Am Inst Chemists; Am Soc Cell Biologists; Europ Orgn Caries Res. *Res:* Effect of F-ions on adenylate cyclase and glucose metabolism in vivo. *Mailing Add:* Dept Biochem - MS B56 635 Barnhill Dr Ind Univ Sch Med Indianapolis IN 46223

ALLMARAS, RAYMOND RICHARD, b New Rockford, NDak, Sept 11, 26; m 52; c 6. SOIL SCIENCE. *Educ:* NDak State Univ, BS, 52; Univ Nebr, MS, 56; Iowa State Univ, PhD(soil sci), 60. *Prof Exp:* Soil scientist, Agr Res Serv, USDA, Nebr, 52-56; asst, Iowa State Univ, 57-60; soil scientist, Columbia Plateau Conserv Res Ctr, Agr Res Serv, USDA, 60-72, res leader & tech adv, 72-80, res & location leader, 80-; AT DEPT SOIL SCI, ORE STATE UNIV. *Mem:* Am Soc Agron; Soil Sci Soc Am; Biomet Soc; Am Statist Asn; AAAS. *Res:* Tillage, soil structure, plant rooting, soil and plant environment and plant-water relations. *Mailing Add:* Dept Soil Sci 439 Borlaug Hall Ore State Univ 1991 Upper Buford Circle Corvallis OR 97331

ALLMENDINGER, E EUGENE, b Dearborn, Mich, July 4, 17; m 49; c 3. NAVAL ARCHITECTURE, MECHANICAL ENGINEERING. *Educ:* Univ Mich, BSE, 41; Univ NH, MS, 50. *Prof Exp:* Asst prof naval eng, US Naval Acad, 50-53; asst prof naval archit, Mass Inst Technol, 53-58; assoc prof mech eng, Univ NH, 58-73, assoc prof naval archit, 73-83; RETIRED. *Concurrent Pos:* Prof, Univ Sao Paulo, 59-61; Gulf Oil Corp grant, 63; vis assoc prof, Univ Hawaii, 68-69 & Univ Trondheim, Norway, 76; chmn, NH Oceanog Found; vis prof, Chalmers Inst Technol, Gothenburg, Sweden, 80, 81. *Mem:* Soc Naval Archit & Marine Eng: Am Soc Naval Engrs; Marine Technol Soc; Am Soc Eng Educ; Sigma Xi. *Res:* Submarine and submersible vehicle systems design--both manned annd unmanned systems. *Mailing Add:* 46 Oyster River Rd Durham NH 03824

ALLNATT, ALAN RICHARD, b Portsmouth, Eng, July 18, 33; m 73. STATISTICAL MECHANICS, SOLID STATE. *Educ:* Univ London, BSc, 56, PhD(phys chem), 59. *Prof Exp:* NATO fel chem, Univ Chicago, 59-61; lectr, Univ Manchester, 61-69; PROF CHEM, UNIV WESTERN ONT, 69- *Concurrent Pos:* Fel, Chem Inst Can, 74. *Mem:* Royal Soc Chem; fel Can Inst Chem. *Res:* Statistical mechanics of matter transport and thermodynamic properties of imperfect solids and simple liquids. *Mailing Add:* Dept Chem Univ Western Ont London ON N6A 5B7 Can

ALLRED, ALBERT LOUIS, b Mt Airy, NC, Sept 19, 31; m 58; c 3. INORGANIC CHEMISTRY. *Educ:* Univ NC, BS, 53; Harvard Univ, AM, 55, PhD(chem), 56. *Prof Exp:* Instr chem, Northwestern Univ, 56-58, from asst prof to assoc prof, 58-69, assoc dean arts & sci, 70-74, chmn dept, 80-86, interim dean, Col Arts & Sci, 87-88, PROF CHEM, NORTH WESTERN UNIV, 69- *Concurrent Pos:* Alfred P Sloan res fel, 63-65; consult organometallic chem, UNESCO, Colombia, 79; vis scholar, Cambridge Univ, 87. *Mem:* Am Chem Soc; Royal Soc Chem; Sigma Xi; fel AAAS. *Res:* Synthetic inorganic and organometallic chemistry; electrochemistry. *Mailing Add:* Dept Chem Northwestern Univ Evanston IL 60208-3113

ALLRED, DAVID R, PARASITOLOGY. *Educ:* Univ Calif, Riverside, PhD(cell biol), 82. *Prof Exp:* Res assoc, Univ Colo, 83-86; asst res sci, 86-90, ASST PROF, UNIV FLA, 90- *Mem:* Am Soc Cell Biol; Am Soc Parasitol; AAAS. *Res:* Molecular biology of host-parasite interactions involving blood-borne rickettsial and protozoan parasites. *Mailing Add:* Dept Infectious Dis Univ Fla Bldg 471 Mowry Rd Gainesville FL 32611-0633

ALLRED, DORALD MERVIN, b Lehi, Utah, July 11, 23; m 52; c 5. ENTOMOLOGY, ECOLOGY. *Educ:* Brigham Young Univ, AB, 50, MA, 51; Univ Utah, PhD(entom), 54. *Prof Exp:* Field entomologist, State Exten Serv, Utah, 48-50; ranger-naturalist, US Nat Park Serv, 50-51; res fel entom, Brigham Young Univ, 51-53; instr biol, St Mary-of-the-Wasatch Acad, 53; instr biol, Univ Utah, 53-54, assoc ecologist & chief entom & arachnid sect, ecol res, 54-56; from asst prof to assoc prof, 56-66, PROF ZOOL, BRIGHAM YOUNG UNIV, 66- *Res:* Parasitic acarology; parasitology; medical entomology and ecology. *Mailing Add:* 2797 N 700 East Provo UT 84604

ALLRED, E(VAN) R(ICH), b St Charles, Idaho, May 6, 16; m 39; c 3. CIVIL & AGRICULTURAL ENGINEERING. *Educ:* Utah State Univ, BS, 39; Univ Minn, MS, 41. *Prof Exp:* Civil engr, US Bur Reclamation, 41-43 & US Engrs Off, 43-45; from asst prof to assoc prof, 45-57, PROF AGR ENG, UNIV MINN, ST PAUL, 57- *Mem:* Am Soc Civil Engrs; Am Soc Agr Engrs. *Res:* Water supply; irrigation, agricultural drainage and hydrology. *Mailing Add:* Dept Agr Eng Univ Minn St Paul MN 55108

ALLRED, EVAN LEIGH, b Deseret, Utah, May 22, 29; m 55; c 4. ORGANIC CHEMISTRY, STRUCTURAL CHEMISTRY. *Educ:* Brigham Young Univ, BS, 51, MS, 56; Univ Calif, Los Angeles, PhD(chem), 59. *Prof Exp:* Res chemist, Phillips Petrol Co, 51-54; asst org chem, Brigham Young Univ, 54-55 & Univ Calif, Los Angeles, 55-59; instr chem, Univ Wash, 60-61; sr res chemist, Rohm and Haas Co, 61-63; from asst prof to assoc prof, 63-70, PROF CHEM, UNIV UTAH, 70- *Concurrent Pos:* NSF fel, Univ Colo, 59-60; David P Gardner fac fel, Univ Utah, 76. *Mem:* Am Chem Soc. *Res:* Physical-organic chemistry; factors affecting the reactivity of molecules; synthesis and reactivity of structures of theoretical interest; reaction mechanisms. *Mailing Add:* Dept Chem Univ Utah Salt Lake City UT 84112

ALLRED, JOHN B, b Oklahoma City, Okla, June 17, 34; m 79; c 4. BIOCHEMISTRY, NUTRITION. *Educ:* Okla State Univ, BS, 55; Wash State Univ, MS, 57; Univ Calif, Davis, PhD(biochem), 62. *Prof Exp:* Asst nutrit, Wash State Univ, 55-57; asst nutrit & biochem, Univ Calif, Davis, 58-60; from asst prof to assoc prof chem, Okla City Univ, 61-68, chmn dept, 62-63; dir, Inst Nutrit, 68-71, assoc prof food sci & nutrit, 71-74, PROF FOOD SCI & TECHNOL, OHIO STATE UNIV, 74- *Mem:* Am Soc Biochem & Molecular Biol; Am Inst Nutrit; Brit Biochem Soc. *Res:* Intermediary metabolism; metabolic control mechanisms; enzymology; nutritional interest in lipid metabolism. *Mailing Add:* Dept Food Sci & Technol 2121 Fyffe Rd Columbus OH 43210

ALLRED, JOHN C(ALDWELL), b Breckenridge, Tex, Apr 24, 26; m 50; c 3. ACOUSTICS. *Educ:* Tex Christian Univ, BA, 44; Univ Tex, MA, 48, PhD(physics), 50. *Prof Exp:* Asst, Los Alamos Sci Lab, 48-49, mem staff, 49-55; res scientist, Convair, 55-56; from assoc prof to prof physics & biophys sci, Univ Houston, 56-79, assoc dean, Col Arts & Sci, 59-61, asst to pres, 61-62, vpres & dean faculties, 62-68; CONSULT, LOS ALAMOS NAT LAB, 79- *Mem:* Fel Am Phys Soc; Am Nuclear Soc; Acoust Soc Am; Sigma Xi. *Res:* Light particle scattering and interactions; neutron scattering and spectra; fluid dynamics; reactor design; architectural acoustics. *Mailing Add:* 798 47th St Los Alamos NM 87544

ALLRED, KEITH REID, b Spring City, Utah, Feb 19, 25; m 45; c 6. CROP PRODUCTION, PLANT BIOCHEMISTRY. *Educ:* Brigham Young Univ, BS, 51; Cornell Univ, PhD(crop prod), 55. *Prof Exp:* Asst, Cornell Univ, 51-54; res assoc, Co-op Grange League Fedn Exchange, Inc, 54-57; from asst prof to assoc prof agron, 57-65, from assoc prof to prof plant sci, 65-70, PROF AGRON, UTAH STATE UNIV, 70-, ASSOC DEAN, 75-, HEAD DEPT PLANT SCI, 77- *Concurrent Pos:* Utah State Univ SAm assignments, Bolivia, 67-71 & 72-75; Ford Found travel grant, Brazil; consult, SAm & Africa. *Mem:* Am Soc Agron. *Res:* Forage crop physiology and production; study of relationship between dodder, parasitic plant, and alfalfa as host. *Mailing Add:* Dept Plant Sci Utah State Univ Logan UT 84322-4820

ALLRED, KELLY WAYNE, b Sacramento, Calif, April 23, 49; m 72; c 3. PLANT TAXONOMY. *Educ:* Brigham Young Univ, BS, 74, MS, 75; Tex A&M Univ, PhD(range sci), 79. *Prof Exp:* Instr bot, State Univ NY, Geneseo, 78-79; ASST PROF RANGE SCI, NMEX STATE UNIV, 79- *Mem:* Am Soc Plant Taxonomists; Int Asn Plant Taxonomists; Soc Range Mgt; Sigma Xi. *Res:* Plant taxonomy, particularly of new world grasses with emphasis in the tribe Andropogoneae and genus Bothriochloa; floristics of the southwestern United States with emphasis on New Mexico. *Mailing Add:* 2015 Jordan Rd Las Cruces NM 88001

ALLRED, LAWRENCE ERVIN (JOE), b Matewan, WVa, June 9, 46; m 70; c 2. CELL BIOLOGY, DERMATOLOGY. *Educ:* Univ Houston, BS, 68; Univ Tex, MS, 70, PhD(cellular biophys), 73. *Prof Exp:* Fel, dept molecular cell develop biol, Univ Colo, 73-77; asst prof, dept pharmacol, Ohio State Univ, 78-80, clin asst prof, 80; SR RES ASSOC, EXPLOR RES & NEW TECHNOL DEPT, S C JOHNSON & SON, INC, 80- *Concurrent Pos:* Adj assoc prof, Univ Wis-Parkside, 81; assoc clin prof, dept med, Med Col Wis, 84. *Mem:* Am Soc Cell Biol; Soc Toxicol; Skin Pharmacol Soc. *Res:* Cell biology and biochemistry of human skin tissues; experimental and clinical pharmacology of dermatological drugs; artificial organ generation. *Mailing Add:* S C Johnson & Son Inc 1525 Howe St Racine WI 53403

ALLRED, V(ICTOR) DEAN, b Delta, Utah, Sept 13, 22; m 44; c 6. FUEL ENGINEERING. *Educ:* Univ Utah, BChE, 48, PhD(fuel technol), 51. *Prof Exp:* Res engr, Catalytic Construct Co, 53-54; res chemist, Union Carbide Nuclear Co, Oak Ridge Nat Lab, 54-56; sr res engr, Denver Res Ctr, Ohio Oil Co, 56-62, mgr, Anal Dept, Marathon Oil Co, 62-66, res assoc, Denver Res Ctr, Marathon Oil Co, 66-86; RETIRED. *Concurrent Pos:* Adj prof fuels eng, Univ Utah, 85- *Mem:* Am Inst Chem Engrs; Am Chem Soc; Soc Petrol Engrs. *Res:* Catalytic chemistry; synthetic fuels and lubricants; petroleum production; oil shale and bituminous sand processing; petroleum coke production and processing. *Mailing Add:* 1925 E Otero Lane Littleton CO 80122

ALLUM, FRANK RAYMOND, b Melbourne, Australia, Jan 31, 36; m 68. COSMIC RAY PHYSICS, SPACE PHYSICS. *Educ:* Univ Melbourne, BS, 55, MS, 58, PhD(nuclear physics), 63. *Prof Exp:* Sr demonstr, Dept Physics, Univ Melbourne, 61-63; res assoc auroral physics, Space Sci Dept, Rice Univ, 64-66; vis res assoc cosmic rays, Univ Tex, Dallas, 66, res assoc, 66-69, asst prof, 69-71, res scientist, 71-76, mem cosmic ray group, 76-80. *Mem:* Am Geophys Union; Am Phys Soc. *Res:* Propagation characteristics of low energy solar cosmic rays in the interplanetary medium using temporal, spectral and anisotropic data from the satellites, Explorers 34 and 41; solar-terrestrial relationships. *Mailing Add:* 633 Timberlake Circle Richmond TX 75080

ALM, ALVIN ARTHUR, b Albert Lea, Minn, July 30, 35; m 60; c 2. FORESTRY. *Educ:* Univ Minn, BS, 61, MS, 65, PhD(forestry), 71. *Prof Exp:* Assoc forester, Forestry Consult Serv, Inc, 61-62; res asst forestry, Univ Minn, 63-65; appraiser real estate, Bur Pub Rds, 65-66; res fel forestry, 66-71, from asst prof to assoc prof, 71-80, PROF FORESTRY, UNIV MINN, 80- *Mem:* Soc Am Foresters; Sigma Xi. *Res:* Forest regeneration; vegetative successional changes; silvicultural practices such as site preparation, thinnings and harvesting; Christmas tree management; moisture and tree growth relationships. *Mailing Add:* Cloquet Forestry Ctr Univ Minn Cloquet MN 55720

ALM, ROBERT M, b Princeton, Ill, Sept 19, 21; m 43; c 3. ORGANIC CHEMISTRY. *Educ:* Monmouth Col, BS, 43; Ohio State Univ, PhD(chem), 48. *Prof Exp:* Chemist, Standard Oil Co (Ind), 48-53, group leader, 53-63, res assoc, 63-69, dir anal res, Naperville, 70-74, sr consult chemist, 74-76, sr res assoc, 76-79; RETIRED. *Mem:* Am Chem Soc. *Res:* Shale oil; synthetic fuels. *Mailing Add:* 927 Stoddard Ave Wheaton IL 60187

ALMASI, GEORGE STANLEY, b Budapest, Hungary, Oct 8, 38; US citizen; m 65; c 2. ELECTRICAL ENGINEERING. *Educ:* Syracuse Univ, BSc, 61; Mass Inst Technol, MSc, 62, PhD(elec eng), 66. *Prof Exp:* MEM RES STAFF, EXPLOR MAGNETICS, THOMAS J WATSON RES CTR, IBM CORP, 66-, MGR, ADVAN VERY LARGE SCALE INTEGRATION DESIGN, 80- *Mem:* Sr mem Inst Elec & Electronics Engrs. *Res:* Photovoltaic effects in semiconductors; mercury telluride-cadmium telluride heterostructures; memory design and sensing; low-temperature behavior of magnetic films; magnetic bubble-domain devices; very large scale integration design. *Mailing Add:* Thomas J Watson Res Ctr IBM Corp PO Box 218 Yorktown Heights NY 10598

ALMASON, CARMEN CRISTINA, b Cluj-Napoca, Romania, Dec 9, 54; US citizen; m 79. SUPERCONDUCTIVITY, MAGNETISM. *Educ:* Univ Bucharest, Romania, BS, 77, MS, 79; Univ SC, PhD(physics), 89. *Prof Exp:* Res assoc, Inst Physics & Technol of Mat, 79-83; teaching asst res, Univ SC, Columbia, 84-88, Amelia Earhart fel res, 87-89, Link fel, 88-89, postdoctoral res, 89; PHYSICIST, UNIV CALIF, SAN DIEGO, 90- *Concurrent Pos:* Vis res scientist, Matsushita Int Res & Develop Ctr, Japan, 89. *Mem:* Am Phys Soc; AAAS. *Res:* Normal and superconducting properties of the high transition temperature cuprates; dynamic and steady-state magnetic behaviors; electric and magnetic critical fluctuations; coexistence of superconductivity and magnetism; high pressure physics. *Mailing Add:* Inst Pure & Appl Phys Sci 0075 Univ Calif La Jolla CA 92093

ALMEIDA, SILVERIO P, b Hudson, Mass, July 27, 33; m 71; c 2. ELECTROOPTICS, BIOPHYSICS. *Educ:* Clark Univ, BA, 57; Mass Inst Technol, MS, 59; Cambridge Univ, PhD(physics), 64. *Prof Exp:* Res asst physicist, Lawrence Radiation Lab, Univ Calif, Berkeley, 60-62; res asst physicist, Europ Orgn Nuclear Res, Geneva, Switz, 62-63, fel, 63-64; sr res assoc elem particle physics, Cavendish Lab, Cambridge Univ, 64-66; sr scientist, Aeronutronic-Philco Ford, Calif, 66-67; from asst prof to prof physics, Va Polytech Inst, 68-89; PROF PHYSICS & CHMN, UNIV NC, CHARLOTTE, 89- *Mem:* Am Optical Soc. *Res:* Coherent optics; holography; pattern recognition; image analysis; light scattering; biophysics. *Mailing Add:* Dept Physics Univ NC Charlotte Charlotte NC 28223

ALMENAS, KAZYS KESTUTIS, b Lithuania, Apr 11, 35; US citizen; m 68; c 3. NUCLEAR ENGINEERING. *Educ:* Univ Nebr, Lincoln, BS, 57; Univ Warsaw, PhD(nuclear eng), 68. *Prof Exp:* Res engr, Argonne Nat Lab, 59-65; from asst prof to assoc prof, 68-89, PROF NUCLEAR ENG, UNIV MD, 90- *Concurrent Pos:* Consult, Argonne Nat Lab, 72 & Bechtel Power Corp, 76, Nuclear Regulatory Comn, 77-85, Elec Power Res Indust, Brookhaven Nat Lab, Los Alamos Lab, 88, Gen Physics, 90. *Mem:* Am Nuclear Soc; Am Soc Mech Engrs. *Res:* Boiling water and fast reactor development; light water reactor safety; containment analysis; evaluation of fluid mass and energy transfer between large, complex geometry volumes; analysis of condensing and convective heat transfer. *Mailing Add:* Dept Chem & Nuclear Eng Univ Md College Park MD 20742

ALMERS, WOLFHARD, b Helmstedt, WGer, May 29, 43. BIOPHYSICS, PHYSIOLOGY. *Educ:* Univ Rochester, PhD(physiol), 71. *Prof Exp:* Res assoc physiol, Duke Univ, 67-69; res worker, Cambridge Univ, 71-74; from asst prof to assoc prof, 74-82, PROF PHYSIOL, UNIV WASH, 82- *Concurrent Pos:* Fel, Muscular Dystrophy Asn, 71-74; tutor, Churchill Col, Cambridge Univ, 72-73; Guggenheim fel, 84; Alexander von Humbolt Sr Scientist Award, 84. *Mem:* Biophys Soc; Soc Neurosci; Physiol Soc Gt Brit; Soc Gen Physiologists. *Res:* Electrical properties of and ion transport by biological membranes; mechanism of secretion. *Mailing Add:* Dept Physiol & Biophys Univ Wash Sch Med Seattle WA 98195

ALMETER, FRANK M(URRAY), b Rochester, Minn, Mar 23, 29; m 61; c 1. PHYSICAL METALLURGY, MATERIALS ENGINEERING. *Educ:* Mo Sch Mines, BS, 53; Univ London, DIC, 56, MS, 58, PhD(phys metall), 59. *Prof Exp:* Consult metallurgist, London, Eng, 55-59; process engr, Gen Dynamics/Convair, 59-60; assoc res scientist, Lockheed Missiles & Space Co, 60-61; res metallurgist, Res Lab, Int Nickel Co, 61-63; staff scientist, Aerospace Res Ctr, Gen Precision, Inc, 63-65; group leader mat res & res specialist, Airplane Div, Boeing Co, 65-68; chief metallurgist, Burndy Corp, 68-71; metallurgist, Off Saline Water, US Dept Interior, 71-74; mat engr, US Nuclear Regulatory Comn, 74-80; GEN ENGR, US DEPT ENERGY, 80- *Concurrent Pos:* Mem corrosion adv comt, Elec Power Res Inst, 77- *Mem:* Am Inst Mining, Metall & Petrol Engrs; Am Soc Metals; Am Mgt Asn; Brit Inst Metals; Nat Asn Corrosion Engrs. *Res:* Corrosion of metals and alloys; surface science of electrical contact surfaces; structure and properties of magnetic oxides; microplasticity and piezoresistivity of materials; fracture and surface damage; fatigue and stress corrosion cracking. *Mailing Add:* 12004 Settle Ct Fairfax VA 22033

ALMGREN, FREDERICK JUSTIN, JR, b Birmingham, Ala, July 3, 33; m 73; c 3. MATHEMATICS. *Educ:* Princeton Univ, BSE, 55; Brown Univ, PhD(math), 62. *Prof Exp:* Instr math, Princeton Univ, 62-63; mem staff, Inst Advan Study, 63-65; from asst prof to assoc prof, 65-74, PROF MATH, PRINCETON UNIV, 74- *Concurrent Pos:* Mem staff, Inst Advan Study, 69, 74-75, 78, 81-82 & 85; Nat Acad Sci exchange visitor, Steklov Math Inst, Leningrad, USSR, 70; Alfred P Sloan fel, 68-70; John Simon Guggenheim Mem fel, 74-75. *Mem:* Am Math Soc; Math Asn Am; Soc Indust Appl Math; AAAS. *Res:* Geometric measure theory; calculus of variations. *Mailing Add:* Dept Math Princeton Univ Princeton NJ 08544

ALMOG, RAMI, b Israel; Israeli & US citizen; m; c 1. BIOPHYSICS. *Educ:* Univ Calif, Los Angeles, BS, 59; Calif State Univ, MS, 74; State Univ NY, PhD, 76. *Prof Exp:* Chemist, Res Org & Inorg Chem Corp, 70-72; postdoctoral res assoc, State Univ NY, Binghamton, 78-79; ASST PROF, SCH PUB HEALTH, STATE UNIV NY, ALBANY, 89-; RES SCIENTIST, WADSWORTH CTR LABS & RES, NY STATE DEPT HEALTH, ALBANY, 79- *Mem:* Am Chem Soc. *Res:* Structure and function of cell membranes and their role in cellular action and pathological conditions. *Mailing Add:* Dept Biophys Wadsworth Ctr Labs & Res PO Box 509 Albany NY 12201

ALMON, RICHARD REILING, b San Bernardino, Calif, Aug 24, 46; m 67; c 3. BIOCHEMISTRY, PHARMACOLOGY. *Educ:* Univ Ill, Urbana-Champaign, BS, 68, MS, 70, PhD(physiol & biochem), 72. *Prof Exp:* Teaching asst, dept physiol, Univ Ill, Urbana-Champaign, 68-70; fel med biochem, Duke Univ, 71-75; asst res prof biol, Univ Calif, San Diego, 75-77; ASSOC PROF CELL & MOLECULAR BIOL, STATE UNIV NY, BUFFALO, 77- *Concurrent Pos:* Prin investr, grant, 77- & Muscular Dystrophy grant, 78-, NIH & NASA grants, 77- *Mem:* Soc Neurosci; Am Physiol Soc; Am Soc Pharmacol & Exp Therapeut. *Res:* Neurochemistry; membrane biochemistry; muscle biochemistry; endocrinology; biological information processes. *Mailing Add:* Biol Sci State Univ NY N Campus 220 Hochstetter Buffalo NY 14260

ALMOND, HAROLD RUSSELL, JR, b Oakland, Calif, Dec 21, 34; m 56; c 2. COMPUTATIONAL CHEMISTRY. *Educ:* Calif Inst Technol, BS, 56, PhD(org chem), 61. *Prof Exp:* Sr res chemist, Janssen Res Found, Johnson & Johnson, 60-63, group leader anal chem, 64-76, res fel, McNeil Pharmaceut Div, 76-86, res fel, 87-89, RES FEL, R W JOHNSON PHARMACEUT RES INST, 90- *Mem:* AAAS; assoc mem Sigma Xi; Am Chem Soc; NY Acad Sci. *Res:* Computer aided drug design using molecular modeling with computers; quantitative structure-activity relationship analysis. *Mailing Add:* R W Johnson Pharmaceut Res Inst McKean & Welsh Roads Spring House PA 19477-0776

ALMOND, HY, b Chattanooga, Tenn, June 11, 14; m 49; c 2. INORGANIC CHEMISTRY. *Educ:* Univ Chicago, BS, 38. *Prof Exp:* Explosives supvr, Army Ord, 42-43; chemist, US Geol Surv, 46-57; process control supvr, Autonetics Group, Rockwell Int Corp, 57-81; RETIRED. *Mem:* Am Chem Soc; Geochem Soc. *Res:* Printed circuitry; geochemistry; analytical chemistry; general electroplating; thin films; thick films. *Mailing Add:* 5021 Hamer Lane Placentia CA 92670

ALMOND, PETER R, b Downton Wiltshire, Eng, Sept 10, 37; m 60; c 2. MEDICAL PHYSICS. *Educ:* Nottingham Univ, BSc, 58; Rice Univ, MA, 62, PhD(med physics), 65. *Prof Exp:* Dep head physics, M D Anderson Hosp & Tumor Inst, Univ Tex, 72-74, physicist, 72-85, prof biophys, 72-85, head radiation physics, 72-85, dir physics, 80-85; VCHMN & PROF RES, DEPT RADIATION ONCOL, UNIV LOUISVILLE, 85- *Concurrent Pos:* Mem, Nat Coun Radiation Protection & Measurements, 84-; comt mem, Radiation Study Sect, NIH, 86-, chmn, 88-; study sect mem, Construct Grant Study Sect, Nat Cancer Inst, 86-; consult & task group mem, Int Comn Radiation Units & Measurements, 67-84; vchmn, Int Soc Coun Radiation Oncol, 82-; consult, Nuclear Regulatory Comn, 78-; chmn, Am Col Med Physics, 86. *Mem:* Am Soc Therapeut Radiol & Oncol; Am Phys Soc; Am Asn Physicists Med (pres, 70-71); Sigma Xi; Am Col Med Physics; Am Bd Med Physics; Am Bd Radiol. *Res:* Application of imaging techniques to radiation oncology treatment planning using computers; 2D treatment planning is done using a CT scanner as input; plan to use magnetic resonance imaging input for 3D planning; basic measurements of radiation; applications to radiation therapy and radiobiology. *Mailing Add:* Brown Cancer Ctr 529 S Jackson St Louisville KY 40202

ALMS, GREGORY RUSSELL, b Sycamore, Ill, July 8, 47; m 69; c 4. POLYMER CHEMISTRY, CHEMICAL PHYSICS. *Educ:* Monmouth Col, BA, 69; Stanford Univ, PhD(chem), 73. *Prof Exp:* Fel chem, Univ Ill, 73-75; asst prof chem, Fordham Univ, 75-78; SR RES SUPVR, POLYMER PROD DEPT, WASHINGTON WORKS, E I DU PONT DE NEMOURS & CO, INC, 78- *Mem:* Am Phys Soc. *Res:* Laser light scattering spectroscopy studies on polymer systems; emulsion polymerization studies; fluoropolymers; engineering plastics. *Mailing Add:* One Ashleaf Ct Hockessin DE 19707

ALMY, CHARLES C, JR, b 1935. GEOPHYSICS, HYDROGEOLOGY OF GROUNDWATER. *Educ:* Rice Univ, MA, 60, PhD(geol), 65. *Prof Exp:* PROF GEOL, GUILFORD COL, 72- *Mailing Add:* Dept Geol Guilford Col Greensboro NC 27410

ALMY, THOMAS PATTISON, b New York, NY, Jan 10, 15; m 43; c 3. INTERNAL MEDICINE. *Educ:* Cornell Univ, AB, 35, MD, 39. *Hon Degrees:* MA, Dartmouth Col, 70. *Prof Exp:* Asst med, Med Col, Cornell Univ, 40-42, from instr to asst prof, 42-48, assoc prof neoplastic dis, 48-54, from assoc prof to prof med, 54-68; Nathan Smith prof & chmn dept, Dartmouth Med Sch, 68-74, third century prof med, 74-86, prof community & family med, 81-86, EMER PROF MED & COMMUNITY & FAMILY MED, DARTMOUTH MED SCH, 86- *Concurrent Pos:* asst attend physician, Mem Hosp, 48-68 & James Ewing Hosp, 50-68; vis physician & dir 2nd med div, Bellevue Hosp, 54-68; dir med, Dartmouth-Hitchcock Affiliated Hosps, 68-74; consult med, Vet Admin Hosp, NY, 54-68 & Vet Admin Ctr, White River Junction, Vt, 68-; Vet Admin distinguished physician, 82-85. *Honors & Awards:* Friedenwald Medal, Am Gastroenterol Asn, 76. *Mem:* Inst Med-Nat Acad Sci; Asn Am Physicians; Am Gastroenterol Asn (pres, 64); Am Physiol Soc; Master Am Col Physicians. *Res:* Psychosomatic medicine; physiology of human gastrointestinal tract; methods of medical teaching. *Mailing Add:* Dept Med-Community & Family Med Dartmouth Med Sch Hanover NH 03756

ALO, RICHARD ANTHONY, b Erie, Pa; m 60; c 4. COMPUTER SCIENCES. *Educ:* Gannon Col, BA, 60; Pa State Univ, MA, 65, PhD(math), 65. *Prof Exp:* Systs test retrofit engr, Defense Projs Div, Western Elec Co, 60-61; teaching asst math, Pa State Univ, 61-63, res asst, 63-64, instr, 65-66; from asst prof to assoc prof, Carnegie-Mellon Univ, 66-76; prof math & head dept, Lamar Univ, 76-82; PROF, DEPT APPL MATH SCI, UNIV HOUSTON-DOWNTOWN, 82- *Concurrent Pos:* Kanpur Indo-Am Prog math sci adv, Indian Inst Technol, 69-70; Nat Res Coun Italy vis fel, 74; vis prof math, Univ Parma, 75; Nat Acad Sci vis fel, 75; consult appl math, Math Asn Am, 79- *Mem:* Am Math Soc; Math Asn Am; Ital Math Union; Soc Indust & Appl Math; Sigma Xi. *Res:* General topology; analysis and functional analysis; biomathematics; non-linear analysis; software systems; operations research; computer science theory. *Mailing Add:* Dept Appl Math Sci Univ Houston-Downtown One Main St Houston TX 77002

ALOIA, ROLAND C, b Newark, NJ, Dec 21, 43; m 74. CHEMISTRY. *Educ:* St Mary's Col, BS, 65; Univ Calif, Riverside, PhD(cell biol), 70. *Prof Exp:* Res biologist, Univ Calif, Riverside, 75-76; from asst prof to assoc prof, 76-89, PROF, DEPT ANESTHESIOL, SCH MED, LOMA LINDA UNIV, 89-; RES CHEMIST, ANESTHESIA SERV, PETTIS MEM VET HOSP, 79- *Concurrent Pos:* Chemist, Vet Admin, Loma Linda Univ, 79-; fel, Calif Heart Asn, 71-73, bd dirs, 73-80, pres, Riverside Chap, 79-80, 84-86. *Mem:* Am Chem Soc; NY Acad Sci; Sigma Xi. *Mailing Add:* Dept Anesthesiol Sch Med Loma Linda Univ Loma Linda CA 92350

ALONSO, CAROL TRAVIS, b Montreal, Que, Dec 5, 41; US citizen; m 69; c 2. ASTROPHYSICS, FLUIDS. *Educ:* Allegheny Col, BS, 63; Bryn Mawr Col, MS, 65; Mass Inst Technol, PhD(nuclear physics), 70. *Prof Exp:* Mem res staff heavy ion physics, Yale Univ, 70-72; staff physicist heavy ion physics, Lawrence Berkeley Lab, 72-75, PROG MGR RES, LAWRENCE LIVERMORE NAT LAB, 75- *Concurrent Pos:* Vis comt physics, Mass Inst Technol, 81- *Mem:* Am Phys Soc. *Res:* Hydrodynamics; thermonuclear fusion; particle transport; plasma physics; computer modelling of liquid drop and heavy nucleus collisions; astrophysical applications. *Mailing Add:* Lawrence Livermore Nat Lab PO Box 808 Livermore CA 94550

ALONSO, JOSE RAMON, b San Francisco, Calif, July 26, 41; m 69; c 2. ACCELERATOR PHYSICS, HEAVY ION PHYSICS. *Educ:* Mass Inst Technol, SB, 62, PhD(physics), 67. *Prof Exp:* Instr physics, Mass Inst Technol, 67-68; res physicist, Yale Univ, 68-72; res physicist physics, 72-82, Bevalac Opers mgr, 82-89, ASST DIV DIR, ACCEL & FUSION RES DIV, LAWRENCE BERKELEY LAB, 89- *Honors & Awards:* R & D 100 Award, 87. *Mem:* Am Phys Soc. *Res:* Accelerator physics; nuclear, biomedical and atomic physics aspects of heavy ion physics; medical physics. *Mailing Add:* Lawrence Berkeley Lab Bldg 50-149 Berkeley CA 94720

ALONSO, KENNETH B, b Tampa, Fla, Nov 26, 42; m 67. PATHOLOGY, NUCLEAR MEDICINE. *Educ:* Princeton Univ, AB, 64; Univ Fla, MD, 68. *Prof Exp:* Intern surg, Riverside Hosp, Newport News, Va, 68-69; resident, Univ Fla, 69, resident path, 69-70 & Fitzsimons Gen Hosp, Denver, 71-73, chief resident path, 73, actg comdr, US Army Regional Med Lab, Atlanta, 74-75; pathologist, Clayton Gen Hosp, Riverdale, Ga, 75-76; dir, Lab Procedures South, Upjohn Co, Atlanta, 76-79; chief staff, Physicians & Surgeons Commun Hosp, 78-79; chief staff, Henry Gen Hosp, Stockbridge, Ga, 81-82, pathologist, 79-84; DIR, LAB ATLANTA, RIVERDALE, GA, 84- *Concurrent Pos:* Clin assoc prof chem, Auburn Univ, 77-79; Clin assoc prof path, Morehouse Col, 77-83, prof, 84-; chief med examr, DOFS, GBI, Decatur, Ga, 85-88. *Mem:* Am Col Physicians; Am Soc Clin Path; Col Am Pathologists; Soc Nuclear Med; Am Col Nuclear Physicians; Nat Asn Med Examrs; Am Soc Clin Oncol. *Res:* immunologic disorders, especially receptors, cytokines, membrane transport, human tumor stem cell assay & monoclonal antibodies; oncology; hyperthermia. *Mailing Add:* 2921 Margaret Mitchell Ct NW Altanta GA 30327

ALONSO, MARCELO, b Havana, Cuba, Feb 6, 21; US citizen; m 43; c 4. NUCLEAR PHYSICS, QUANTUM THEORY. *Educ:* Univ Havana, PhD(physics), 43. *Prof Exp:* Fel physics, Yale Univ, 43-44; prof physics, Univ Havana, 44-60, chmn dept, 56-60; lectr physics, Georgetown Univ, 61-71; dep dir, Orgn Am States, 60-73, dir sci & technol, 73-80; exec dir res & eng, 80-88, PRIN RES SCIENTIST, FLA INST TECHNOL, 88- *Concurrent Pos:* Dir, Nuclear Energy Comn, Cuba, 55-60; tech dir, Nat Bank Econ Develop, Cuba, 58-60; hon prof, Univ Guadalajara, Mex, 71. *Mem:* Hon mem, Guatemala Acad Sci; Am Phys Soc; Am Asn Physics Teachers; AAAS; Am Nuclear Soc. *Res:* Energy policy & planning; science education. *Mailing Add:* 509 Third Ave Melbourne Beach FL 32951

ALONSO, RICARDO N, b Salta, Arg, Sept 22, 54; m 87. BORATE DEPOSITS, CENTRAL ANDES GEOLOGY. *Educ:* Nat Univ Salta, Arg, Geologist, 78, Dr Geol Sc, 86. *Prof Exp:* Geologist mineral explor, Boroquimica Samicaf, RTZ Co Group, 80-84; SCI INVESTR BORATE DEPOSITS, CONSEJO NAT INVEST SCI & TECHNOL, ARG, 84-; PROF ORE DEPOSITS, NAT UNIV SALTA, ARG, 88- *Concurrent Pos:* Res asst, Nat Univ Salta, Arg, 74-78; field asst, Inst Antarctica Arg, 76, sci investr, 87; vis scientist, Cornell Univ, NY, 86 & 89, Consejo Super Sci Invest, Spain, 87 & Dokuz Eylul Univ, Turkey, 88. *Honors & Awards:* Bernard Houssay, Nat Prize Sci, Arg, 87. *Mem:* Fel Geol Soc Am; Soc Mining, Metall & Explor, Inc; Asn Geol Arg; Asn Paleont Arg; Asn Arg Mineral, Petrol & Sediment. *Res:* Borate deposits; continental evaporites; general geology of Puna and central Andes; vertebrate paleoichnology. *Mailing Add:* Casilla de Correo 362 Salta 4400 Argentina

ALONSO-CAPLEN, FIRELLI V, b Cebu City, Philippines, Oct 19, 56; US citizen; m 82; c 2. INFLUENZA VIRUS, MRNA SPLICING. *Educ:* Univ Ala, Birmingham, PhD(microbiol), 84. *Prof Exp:* Res fel, Mem Sloan Kettering Cancer Inst, 85-90; ASST RES PROF, RUTGERS UNIV, 90- *Mem:* Am Soc Virol; Am Soc Microbiol; Am Soc Cell Biol. *Res:* Regulation of influenza viral NSI MRNA splicing in vivo. *Mailing Add:* Dept Molecular Biol & Biochem Rutgers Univ Ctr Advan Biotechnol & Med 302 679 Hoes Lane Piscataway NJ 08854

ALOUSI, ADAWIA A, b Baghdad, Iraq; US citizen. MOLECULAR CARDIOLOGY. *Educ:* Univ Baghdad, BS, 55; Univ Mich, MS, 59; State Univ NY, PhD(pharmacol), 64. *Prof Exp:* Fel pharmacol, Med Sch, Harvard Univ, 64-66; asst prof, Univ Baghdad, 66-67 & State Univ NY Upstate Med Ctr, 67-68; res scientist, Geigy Pharmaceut, 68-70; res biologist, 71-74, sr res biologist, 74-78, sect head, 79-84, INST FEL PHARMACOL, STERLING-WINTHROP RES INST, 84- *Concurrent Pos:* vis scholar, Univ Calif, San Diego, 89- *Mem:* Am Soc Pharmacol & Exp Therapeut; Am Soc Cell Biol; NY Acad Sci; Am Heart Asn; Int Soc Heart Res. *Res:* Cardiovascular pharmacology and biochemistry, with special emphasis on congestive heart failure; new cardiotonic agents; etiology of congestive heart failure and new approaches to its management; signal transduction; G-proteins. *Mailing Add:* Dept Pharmacol Sch Med M-036 Univ Calif San Diego La Jolla CA 92093

ALPEN, EDWARD LEWIS, b San Francisco, Calif, May 14, 22; m 45; c 2. PHYSIOLOGY, RADIOBIOLOGY. *Educ:* Univ Calif, BS, 46, PhD(pharmaceut chem), 50. *Prof Exp:* Asst prof pharmacol, Sch Med, George Washington Univ, 50-51; investr & head thermal injury br, Biol & Med Sci Div, US Naval Radiol Defense Lab, San Francisco, 53-55, head biophys br, 55-58, head div, 58-69; assoc lab dir, 69-73, lab dir, Pac Northwest Labs, Battelle Mem Inst, 73-75; Donner Lab, 75-85, PROF BIOPHYS, UNIV CALIF, BERKELEY, 75- *Concurrent Pos:* NSF fel, Oxford Univ, 58-59; Guggenheim fel, Univ Paris, 65-66; dir, Univ Calif Study Ctr, London, 88-90. *Honors & Awards:* Mem Award, Asn Mil Surg US, 61; Sci Medal, US Secy Navy, 62; Distinguished Civilian Serv Medal, US Dept Defense, 63. *Mem:* Am Physiol Soc; Radiation Res Soc; Biophys Soc; Bioelectromagnetics Soc (pres, 78-); Soc Exp Biol & Med; Royal Soc Med. *Res:* Radiation biology and biophysics; cellular kinetics; erythropoetic mechanisms; regulation of erythropoesis; environmental science. *Mailing Add:* Donner Lab Univ Calif Berkeley CA 94720

ALPER, ALLEN MYRON, b New York, NY, Oct 23, 32; m 59; c 2. MATERIALS SCIENCE ENGINEERING. *Educ:* Brooklyn Col, BS, 54; Columbia Univ, PhD(petrol, mineral), 57. *Prof Exp:* Instr phys geol, Brooklyn Col, 56-57; sr mineralogist, Ceramic Res Lab, Corning Glass Works, 57-59, res mineralogist, 59-62, res mgr, 62-69; mgr chem & electronic mat, 69-70, mgr res & develop, 70-71, chief engr, 71-72, dir res & eng, Chem & Metall Div, GTE Sylvania Inc, 72-80, PRES, GTE WALMET, 80-, VPRES & GEN MGR, GTE PROD CORP. *Concurrent Pos:* Kemp Mem res grant, 54; lectr, St John's Seminary, NY, 58 & Elmira Col, 59-60; mem adv bd, Mat Res Lab, Pa State Univ, 78- *Mem:* Explorers Club; fel Am Ceramic Soc; Am Chem Soc; Am Inst Mining, Metall & Petrol Eng; fel Geol Soc Am. *Res:* Superalloys; rhenium; tungsten; molybdenum; powder metallurgy; rare-earth oxides and metals; iodides; metallurgical extraction; photochemical machining technology; phosphors; crystal growth; ceramics; phase studies; crystal chemistry; silicon nitride; metallurgy and physical metallurgical engineering. *Mailing Add:* GTE Sylvania Inc Hawes St Tawanda PA 18848

ALPER, CARL, b Hoboken, NJ, May 28, 20; m 49; c 4. CLINICAL BIOCHEMISTRY. *Educ:* Drew Univ, BA, 41; Tulane Univ, MS, 43, PhD(chem), 47. *Prof Exp:* Instr chem, Tulane Univ, 43-46; res assoc nutrit, E R Squibb & Sons, 47-48, res assoc immunochem develop, 48-49; from asst prof to assoc prof biochem, Hahnemann Med Col, 49-66, biochemist, Hahnemann Hosp, 62-66; assoc prof biochem & dir dept clin biochem, Temple Univ, 66-70; dir, Philadelphia Br, Bio-Sci Labs, 70-85; CONSULT, 85- *Concurrent Pos:* Adj assoc prof, Temple Univ, 70-73. *Mem:* Am Chem Soc; Am Soc Clin Path; Am Asn Clin Chem; Am Inst Nutrit; Soc Acad Clin Lab Physicians & Scientists. *Res:* Biochemistry of disease; protein nutrition and metabolism; body composition; enzymology; immunochemistry. *Mailing Add:* 1126 S Park Ave Haddon Heights NJ 08035-1494

ALPER, CHESTER ALLAN, b New York, NY, May 21, 31; m 61; c 1. SERUM PROTEIN GENETICS. *Educ:* Harvard Col, AB, 52; Harvard Med Sch, MD, 56. *Prof Exp:* Res fel med, 62-64, res assoc, 64-66, assoc, 66-69, assoc prof, 69-75, PROF PEDIAT, HARVARD MED SCH, 75-; SCI DIR, CTR BLOOD RES, 72- *Concurrent Pos:* Sr assoc med, Children's Hosp Med Ctr, 74-; consult med, Brigham & Women's Hosp, 75-; mem Microbiol & Immunol Study Comt, Am Heart Asn, 79-82; consult & panel mem vitro diagnosis, Food & Drug Admin, Health & Human Serv, 74-80; assoc ed, J Immunol, 77-81. *Mem:* Am Soc Clin Invest; Asn Am Physicians; Am Asn Immunologists; Am Asn Clin Chemists; Am Fedn Clin Res. *Res:* Control of metabolism of complement proteins. *Mailing Add:* Dept Pediat Sci Dir Blood Res Harvard Med Sch 800 Huntington Ave Boston MA 02115

ALPER, HOWARD, b Montreal, Que, Oct 17, 41; m 66; c 2. ORGANIC CHEMISTRY, ORGANOMETALLIC CHEMISTRY. *Educ:* Sir George Williams Univ, BSc, 63; McGill Univ, PhD(chem), 67. *Prof Exp:* NATO fel, Princeton Univ, 67-68; from asst prof to assoc prof chem, State Univ NY Binghamton, 68-74; assoc prof, 75-78, chmn dept, 82-85, PROF CHEM, UNIV OTTAWA 78-, CHMN DEPT 88- *Concurrent Pos:* J S Guggenheim fel. *Honors & Awards:* E W R Steacie Award, Natural Sci & Eng Res Coun, Can; Alcan Award, Chem Inst Can, Catalysis Award; Killam Res Fel; Fel, Royal Soc Can; Alfred Bader Award Org Chem. *Mem:* Am Chem Soc; Chem Inst Can; Royal Soc Chem. *Res:* Metal complexes as catalysts and reagents in organic chemistry; mechanistic organometallic chemistry; phase transfer catalysis. *Mailing Add:* Dept Chem Univ Ottawa Ottawa ON K1N 9B4 Can

ALPER, JOSEPH SETH, b Brooklyn, NY, Aug 2, 42; m 68. THEORETICAL CHEMISTRY. *Educ:* Harvard Univ, AB, 63; Yale Univ, PhD(chem), 68. *Prof Exp:* Fel chem, Mass Inst Technol, 68-69; ASSOC PROF CHEM, UNIV MASS, DORCHESTER, 69- *Concurrent Pos:* Petrol Res Fund-Am Chem Soc type G grant, 70-73. *Mem:* Am Phys Soc. *Res:* Applications of group theory to problems in atomic and molecular structure and spectra. *Mailing Add:* Dept Chem Harbor Campus Univ Mass Boston MA 02146

ALPER, MARSHALL EDWARD, b New York, NY, Feb 9, 30; m 56; c 4. SYSTEMS DESIGN. *Educ:* Mass Inst Technol, SB, 51, ScD(civil eng), 62. *Prof Exp:* Res asst civil eng, Mass Inst Technol, 51-55; develop engr, 55-60, group supvr structures & dynamics res, 60-64, chief, Appl Mech Sect, 64-71, mgr, Civil Systs Proj Off, 71-72, environ syst prog, 72-74, energy systs prog develop, 74-75, mgr, Solar Energy Prog, 75-85, Energy Environ & Health Prog, 85-88, MGR, INSTNL COMPUT & INFO SERV OFF, JET PROPULSION LAB, CALIF INST TECHNOL, 88- *Concurrent Pos:* Mem res adv comt missile & spacecraft structures, NASA, 59-62. *Honors & Awards:* Except Serv Medal, NASA. *Mem:* Am Inst Aeronaut & Astronaut; AAAS. *Res:* Systems engineering; mission studies leading to definition of future programs and required supporting research and advanced development; development and applications of solar energy. *Mailing Add:* 2286 Lambert Dr Pasadena CA 91107

ALPER, MILTON H, b Lynn, Mass, July 26, 30; m 54; c 3. ANESTHESIOLOGY, PHARMACOLOGY. *Educ:* Harvard Univ, AB, 50, MD, 54. *Prof Exp:* NIH training grant & res fel pharmacol, Harvard Med Sch, 61-62, instr anesthesia, 62-64, clin assoc, 64-69, assoc prof anesthesia, 69-80; anesthesiologist-in-chief, Boston Hosp Women, 69-80; PROF & ANESTHESIOLOGIST-IN-CHIEF, CHILDRENS HOSP, BOSTON, 80- *Concurrent Pos:* From jr assoc to assoc, Peter Bent Brigham Hosp, 62-69; attend, West Roxbury Vet Admin Hosp, 63- *Honors & Awards:* Mead Johnson Training Award Anesthesiol, 59. *Mem:* Am Soc Anesthesiol; Am Soc Pharmacol & Exp Therapeut; Asn Univ Anesthetists. *Res:* Pharmacology of anesthetic drugs; obstetrical and pediatric anesthesia. *Mailing Add:* Anesthesia Dept Harvard Med Sch Childrens Hosp 300 Longwood Ave Boston MA 02115

ALPER, RICHARD H, b Suffern, NY, Sept 27, 53. CARDIOVASCULAR PHYSIOLOGY, NEUROENDOCRINOLOGY. *Educ:* State Univ NY, Stony Brook, BS, 75; Mich State Univ, PhD(pharmacol & toxicol), 81. *Prof Exp:* Postdoctoral fel physiol, Univ Calif, San Francisco, 81-84; postdoctoral res assoc pharmacol, Univ Iowa, 84-86; ASST PROF PHARMACOL, TOXICOL & THERAPEUT, UNIV KANS MED CTR, 86- *Concurrent Pos:* Mem, High Blood Pressure Coun, Am Heart Asn. *Mem:* Soc Neurosci; Am Soc Pharmacol & Exp Therapeut; Am Heart Asn; NY Acad Sci; AAAS. *Res:* Central neurotransmitters in the integrated cardiovascular and endocrine responses to pharmacologic and physiologic stimuli. *Mailing Add:* Dept Pharmacol Univ Kans Med Ctr 39th & Rainbow Blvd Kansas City KS 66103

ALPER, ROBERT, b New York, NY, Sept 8, 33; m 70; c 4. CONNECTIVE TISSUE BIOCHEMISTRY. *Educ:* Utica Col, BA, 59; State Univ NY Upstate Med Ctr, PhD(biochem), 69. *Prof Exp:* Res asst biochem, Vet Admin Hosp, Syracuse, NY, 60-64; res assoc, State Univ NY, Buffalo, 68-69 & Fla State Univ, 69-71; res assoc med, Univ Pa Div, Philadelphia Gen Hosp, 71-72, ASST PROF BIOCHEM, SCH DENT MED, UNIV PA, 72-; ASST DIR, CONNECTIVE TISSUE RES INST, UNIV CITY SCI CTR, 77- *Concurrent Pos:* Consult, Advan Technol Ctr, Southeastern Pa, 83-89. *Mem:* AAAS; Am Chem Soc; Asn Res Vision & Ophthal. *Res:* Involvement of glycoproteins and mucopolysaccharides in atherosclerosis; corneal structural glycoproteins; identification and characterization of cell surface glycoproteins in neoplastic cells; biochemistry of basement membrane; collagen biochemistry; connective tissue and skin diseases; cell biology of the cardiovascular system. *Mailing Add:* 27 Morningside Dr Yardley PA 19067

ALPERIN, HARVEY ALBERT, b New York, NY, Mar 13, 29; m 56; c 2. MATERIALS SCIENCE, NEUTRON SCATTERING & PHYSICS EDUCATION. *Educ:* Rensselaer Polytech Inst, BS, 49; Univ Mich, MS, 50; Univ Conn, PhD(physics), 60. *Prof Exp:* Res assoc physics, Willow Run Res Ctr, Univ Mich, 50-53; physicist, Underwater Sound Lab, USN, 53-57; US Naval Surface Weapons Ctr, 57-88; physicist, Vela Assocs, 89-90; PHYSICS TEACHER, BLAIR HIGH SCH, 90- *Concurrent Pos:* Guest scientist, Brookhaven Nat Lab, 60-68; vis scientist, Weizmann Inst Sci, 70-71; guest scientist, Nat Bur Standards, 68-84. *Mem:* AAAS; Am Phys Soc; Am Asn Physics Teachers; Sigma Xi. *Res:* Neutron scattering; magnetic materials; computer applications; hydrogen in metals; amorphous metals; very-large-scale integration; parallel processing. *Mailing Add:* 317 Westlawn Dr Ashton MD 20861

ALPERIN, JACK BERNARD, b Memphis, Tenn, June 19, 32; m 60; c 1. VITANIN K, ANTICOAGULATION. *Educ:* Univ Tenn, MD, 57. *Prof Exp:* Intern med, Michael Reese Hosp, 57-58, resident, 58-61, fel hematol, 61-62; instr fac med, Chicago Med Sch, 62-63; fel hematol, 63-65; FAC HEMAT & MED, UNIV TEX MED BR, 65- *Concurrent Pos:* Prof internal med, human biochem & genetics, Univ Tex Med Br, Galveston, Tex, 65- *Mem:* Am Soc Hemat; Am Soc Clin Nutrit; Nutrit Soc Am; Am Med Asn; Am Asn Blood Banks; Southern Soc Clin Invest. *Res:* Traditional hematology including nutritional anemias, coagulathids and inherited blood diseases; hemoglobin. *Mailing Add:* 4164 E63 Univ Tex Med Br Galveston TX 77550

ALPERIN, JONATHAN L, b Boston, Mass, June 2, 37. MATHEMATICS. *Educ:* Harvard Univ, AB, 59; Princeton Univ, MA, 60, PhD(math), 61. *Prof Exp:* Inst math, Mass Inst Technol, 62-63; asst prof, 63-73, PROF MATH, UNIV CHICAGO, 73- *Mem:* Am Math Soc; Math Asn Am. *Mailing Add:* Dept Math Eckhart Hall Univ Chicago 1118 E 58th St Chicago IL 60637

ALPERIN, MORTON, aeronautics, mathematics; deceased, see previous edition for last biography

ALPERIN, RICHARD JUNIUS, b Philadelphia, Pa, Dec 16, 41. CYTOCHEMISTRY, DEVELOPMENTAL BIOLOGY. *Educ:* Univ Pa, AB, 59, PhD(biol), 69. *Prof Exp:* From asst prof to assoc prof, 75-81, PROF BIOL, COMMUN COL PHILADELPHIA, 81- *Concurrent Pos:* Fel, Prof I Gersh's Lab, Dept Animal Biol, Univ Pa, 69-74; consult, Bahamas mariculture activ, 90, Maritek Corp; guest lectr electron micros, Sch Vet Med, Univ Pa, 71-74; chmn fac sen, Community Col Philadelphia, 74 & 75; mem health delivery planning bd, AFL-CIO Cent Comt, Philadelphia, 76-; lectr, McCrone Res Inst, 80. *Mem:* Am Micros Soc; Am Soc Zool; Pattern Recognition Soc; Am Soc Testing & Mat; Am Fish Genetics Soc. *Res:* Assay of biomaterials as mutagens; materials science; methods of in vitro testing of biomaterials in blood and blood simulating fluids during preclinical trials; role of hypoblast in avian embryos; author of one book. *Mailing Add:* 842 Lombard St Philadelphia PA 19147-1317

ALPERN, HERBERT P, b New York, NY, Oct 3, 40. PSYCHOPHARMACOLOGY, NEUROSCIENCES. *Educ:* City Col New York, BS, 63; Univ Ore, MA, 65; Univ Calif, Irvine, PhD(psychobiol), 68. *Prof Exp:* Nat Inst Gen Med Sci grant, 69-71, from asst to assoc prof, 68-79, PROF PSYCHOL, UNIV COLO, BOULDER, 79-, FAC FEL, INST BEHAV GENETICS, 69- *Concurrent Pos:* NIMH grant, 71-74; mem, Coun Tobacco res grant, 76-79. *Mem:* AAAS; Soc Neurosci; Int Brain Res Orgn. *Res:* Behavioral pharmacogenetics, emphasizing drugs that facilitate and disrupt memory processes; animal models of behavior pathology, especially convulsive disorders, hyperkinesis and drug tolerance and dependence. *Mailing Add:* Dept Psychol Univ Colo Boulder Box 345 Boulder CO 80309-0345

ALPERN, MATHEW, b Akron, Ohio, Sept 22, 20; m 51; c 4. VISUAL PHYSIOLOGY. *Educ:* Univ Fla, BME, 46; Ohio State Univ, PhD(physics), 50. *Hon Degrees:* DSc, State Univ New York, 88. *Prof Exp:* Res assoc physiol optics, 49-51; asst prof optom, Pac Univ, 51-55; from instr to asst prof, 55-58, assoc prof ophthal & psychol, 58-63, PROF PHYSIOL OPTICS & PSYCHOL, VISION LAB, UNIV HOSP, UNIV MICH, 63- *Concurrent Pos:* Mem vis sci study sect, NIH, 70-74. *Honors & Awards:* Jonas S Friedenwald Award, Asn Res Vision & Ophthal, 74; Edgar D Tillyer Med Optical Soc Am, 84; Sr Sci Investr Award for Res to Prevent Blindness, 88; Charles F Prentice Medal, Am Acad Optom, 88. *Mem:* Nat Acad Sci; Am Physiol Soc; Biophys Soc; Asn Res Vision Ophthal; Am Psychol Asn; Optical Soc Am. *Res:* Electrophysiology of the retina; intensity time relations of the visual stimulus; critical flicker frequency; factors influencing size of the pupil; accomodation-convergence relations; human visual pigments in normal and abnormal eyes; color vision, color matching and color blindness. *Mailing Add:* W K Kellogg Eye Ctr Rm 347 Univ Mich Med Ctr 1000 Wall St Ann Arbor MI 48105-1994

ALPERN, MILTON, b US citizen, June 25, 25; m 46; c 2. CIVIL & STRUCTURAL ENGINEERING. *Educ:* Cooper Union, BCE, 45; Columbia Univ, MSCE, 50. *Prof Exp:* From instr to asst prof civil eng, Cooper Union, 48-59; consult engr, 59-70; PRIN, ALPERN & SOIFER, CONSULT ENGRS, 70- *Concurrent Pos:* Consult, Alexander Crosett, 50-52, Ammann & Whitney, 52-53, Edwards & Hjorth, 53, NY State Bldg Code Comn, 54 & Norden-Ketay Corp, 54-56; adj prof struct, NY Inst Technol, 77-79; mem, NY State Bd Engrs & Land Surv, 78-87, chmn, 84-86; vis prof, Pratt Inst, 86- *Honors & Awards:* Lincoln Arc Welding Award, 61. *Mem:* Fel Am Soc Civil Engrs; Am Concrete Inst; Nat Soc Prof Engrs; Am Soc Testing & Mat; Am Consult Engrs Coun. *Res:* Collate information on structural parameters of burst protection; structural engineering; engineering education. *Mailing Add:* Alpern & Soifer Consult Engrs 2635 Pettit Ave Bellmore NY 11710

ALPERS, DAVID HERSHEL, b Philadelphia, Pa, May 9, 35; m 77; c 3. CELL BIOLOGY, NUTRITION. *Educ:* Harvard Univ, BA, 56, MD, 60. *Prof Exp:* From instr to asst prof med, Harvard Med Sch, 64-69; from asst to assoc prof, 69-73, PROF MED, WASH UNIV MED CTR, 73- *Concurrent Pos:* Assoc ed, J Clin Invest, 77-82; mem, Nat Inst Diabetes & Digestive & Kidney Dis, 75-81, prin investr grants, 69-; ed, Am J Physiol, 91- *Mem:* Asn Am Physicians; Am Gastroenterol Asn, (pres, 90-91); Am Soc Biochem & Molecular Biol; Am Soc Cell Biol; Am Soc Clin Nutrit. *Res:* Physiology of cobalamin absorption, structure and cell biology of cobalamin binding proteins, secretion of intestinal proteins. *Mailing Add:* Dept Med Box 8124 Wash Univ Med Sch, 660 S Euclid Ave St Louis MO 63110

ALPERS, JOSEPH BENJAMIN, b Salem, Mass, Aug 24, 25; m 56; c 2. BIOLOGICAL CHEMISTRY, METABOLISM. *Educ:* Yale Univ, AB, 49; Columbia Univ, MD, 53; NY Univ, PhD(biochem), 62. *Prof Exp:* Res fel med, Sloan-Kettering Inst, NY, 53-55; from intern to asst resident med, NY Hosp, 55-57; res fel biochem, New York City Pub Health Res Inst, 57-60, asst biochem, 60-61; assoc, 61-65, ASSOC PROF BIOCHEM, HARVARD MED SCH, 66-; DIR CLIN LABS, CHILDREN'S HOSP, 71- *Mem:* Am Soc Biol Chemists. *Res:* Intermediary metabolism; enzymology of carbohydrates. *Mailing Add:* Children's Hosp 300 Longwood Ave Boston MA 02115

ALPERT, DANIEL, b Hartford, Conn, Apr 10, 17; m 42; c 2. COMPUTER-BASED SYSTEMS, SCIENCE POLICY. *Educ:* Trinity Col, Conn, BS, 37; Stanford Univ, PhD(physics), 42. *Hon Degrees:* DSc, Trinity Col, Conn, 57. *Prof Exp:* Res physicist, Westinghouse Res Labs, 42-50, mgr, Physics Dept, 50-55, assoc dir, 55-57; dir,Co-ordinated Sci Lab, 59-65, dean, Grad Col, 65-72, dir, Ctr Advan Study & assoc dir, Comput-Based Educ Res Lab, 72-87, PROF PHYSICS, UNIV ILL, URBANA-CHAMPAIGN, 57-, DIR PROG SCI, TECHNOL & SOC, 86- *Concurrent Pos:* Civilian with AEC, 44; mem, Defense Sci Bd, 63-72; ed, J Vacuum Sci & Technol, 64-66; charter trustee, Trinity Col, Conn, 64- *Honors & Awards:* Newcomb Cleveland Award, AAAS, 54; Gaede-Longmuir Award, Am Vacuum Soc, 80. *Mem:* Fel AAAS; fel Am Phys Soc; Am Vacuum Soc; Sigma Xi. *Res:* Ultrahigh vacuum technology; surface physics; computer-based education; study of higher education. *Mailing Add:* Dir Prog Sci Technol & Soc Univ Ill Urbana-Champaign 912 W Ill St Urbana IL 61801

ALPERT, ELLIOT, b New York, NY, July 8, 36. MEDICINE. *Educ:* City Univ NY, BS, 57; State Univ NY, MD, 61. *Prof Exp:* Intern, II & IV, Harvard Med Serv, Boston City Hosp, 61-62, asst resident, 62-63; preceptor med, Dept Med, Albany Med Col, 63-65; teaching fel, Harvard Med Sch, 65-66, instr, 69-71; asst med, Mass Gen Hosp, 70-74, asst physician, 75-81; attend physician, Methodist Hosp, Houston, Tex, 81-90, head, Gastroenterol Serv, 81-90; PHYSICIAN-IN-CHIEF, SIR MORTIMER B DAVIS JEWISH GEN HOSP, MONTREAL, 90- *Concurrent Pos:* Epidemiologist, WHO Med Exped Easter Island, 64-65; res fel med, Harvard Med Sch, 66-67, 67-69 & 69-71, assoc prof, 76-81; lectr physiol chem, Mass Inst Technol, 73-81; mem, Comt Cancer, Am Gastroenterol Asn, 74-78; mem, Res Comt, Am Asn Study Liver Dis, 75-78; prof med & chief, Gastroenterol Sect, Baylor Col Med, 81-89, dir, NIH, Inst Res Training Grant, 83-89, prof cell biol, 89; consult staff, St Luke's Episcopal Hosp, Houston, Tex, 82-90; assoc ed, Tumor Biol, 84-; mem, Res Comt, Am Col Gastroenterol, 87-90; consult staff, M D Anderson Hosp & Tumor Inst, Houston, Tex, 87-90. *Mem:* Am Fedn Clin Res; AAAS; Am Asn Study Liver Dis; Am Gastroenterol Asn; NY Acad Sci; Am Asn Immunol; Am Asn Cancer Res; Am Soc Clin Invest; Int Asn Study Liver; Int Soc Oncodevelop Biol (vpres, 83-). *Res:* Author or co-author of over 140 publications. *Mailing Add:* Dept Med M B Davis Jewish Gen Hosp 3755 St Catherine Rd Montreal PQ H3T 1E2 Can

ALPERT, JOEL JACOBS, b New Haven, Conn, May 9, 30; m; c 3. PEDIATRICS. *Educ:* Yale Univ, AB, 52; Harvard Med Sch, MD, 56; Am Bd Pediat, dipl, 61. *Prof Exp:* Intern in med, Children's Hosp Med Ctr, Boston, Mass, 56-57, jr asst resident, 57-58, chief resident ambulatory serv & fel, 61-62, asst dir, Family Health Care Prog, 62-64, asst med, 63-64, assoc, 64-66, chief, Family & Child Health Div, 64-72, sr assoc med, 66-73; res fel pediat, Sch Pub Health, Harvard Med Sch, 62-63, asst dir, Family Health Care Prog, 62-64, instr pediat, 63-66, assoc, 64-68, med dir, Family Health Care Prog, 64-72, from asst prof to assoc prof, 68-72, LECTR PEDIAT, HARVARD MED SCH, 72-; PROF & CHMN, DEPT PEDIAT, BOSTON UNIV SCH MED, 72- *Concurrent Pos:* Exchange registrar, St Mary's Hosp Med Sch, London, Eng, 58-59, vis consult pediatrician, 71-72; lectr med, Simmons Col, Boston, 64-72; assoc med, Beth Israel Hosp, Boston, 64-72; exec secy, Boston Poison Control Info Ctr, 64-71, chmn exec comt, 76-78; assoc pediatrician, Boston Hosp for Women, 66-76; mem, Sect Child Develop, Am Acad Pediat, 66-, consult oper head start, 66-72, mem, Community Pediat, 69-, Adolescent Med, 79-, chmn, Subcomt Accidental Poisoning, 68-71, mem, Comt Community Health Serv, 86-; mem, Health Serv Res Study Sect, HEW, 68-73; hon res fel, Univ London, Bedford Col, 71-72; dir, Pediat Serv, Boston City Hosp, 72-; consult child & family med, Children's Hosp Med Ctr, 72-, consult pediat, Franciscan Children's Hosp & Rehab Ctr, 72-, Carney Hosp, 72-80, Ctr Health Serv Res & Develop, Dept Health & Human Serv, 72-, Pub Health Serv, Health Resources Admin, Div Med, Bur Health Professions, 76-; pediatrician-in-chief, Univ Hosp, 72-; vis res assoc, Dept Econ, Northeastern Univ, 75-80; prof, Dept Socio-Med Sci, Boston Univ Sch Med, 76-85, Dept Community Med, 76-80, adj prof pub health law, Sch Pub Health, 80-85, prof socio-med sci & community med, 85-; vis lectr health serv, Harvard Sch Pub Health, 76-79; Dozer vis prof, Ben Gurion Sch Med, Beer Sheva, Israel, 79; Raine Found vis scholar, Dept Child Health, Univ Western Australia & Princess Margaret Hosp, 83; Leonard Erlich vis prof, North Shore Children's Hosp, Manhasset, NY, 84; assoc ed, Pediat Rev & Commun, 85-; Stacey White vis prof, Emory Med Sch, Atlanta, Ga, 86; vis prof pediat, Columbia Col Physicians & Surgeons, New York, 89, NY Univ Sch Med, 89. *Honors & Awards:* Colin Stewart Lectr, Dartmouth Med Sch, 86; George Armstrong Award, Ambulatory Pediat Asn, 89. *Mem:* Inst Med-Nat Acad Sci; Ambulatory Pediat Asn (pres, 69); Am Acad Pediat; Am Pub Health Asn; AAAS; emer mem Soc Pediat Res; Am Pediat Soc; AMA. *Mailing Add:* Dept Pediat Taldot 134 Boston Univ Sch Med 818 Harrison Ave Boston MA 02118

ALPERT, LEO, b Boston, Mass, Oct 31, 15; m 43; c 2. ENVIRONMENTAL SCIENCES, METEOROLOGY. *Educ:* Mass State Col, BS, 37; Clark Univ, MA, 39, PhD(climat), 46. *Prof Exp:* Staff weather officer, USAF, Panama, 41-43; res weather officer, Hqs Air Weather Serv, DC, 44-46; environ specialist, Engr Intel Div, Corps Engrs, 50-51; chief climat lab, Geophys Directorate, Air Force Cambridge Res Labs, 52; geogr, Engr Strategic Intel Div, Corps Engrs, US Army, 53-60, trop environ scientist, Environ Sci Div, Army Res Off, DC, 61-62, chief scientist, Tropic Test Ctr, Panama, CZ, 63-65, meteorologist, Environ Sci Div, Army Res Off, Washington, DC, 65-71, chief atmospheric sci bd, Army Res Off, Res Triangle Park, 71-78, dir, Geosci Div, 78-81; PRES, WEATHER-ENVIRON SURVEYS, INC, 81- *Concurrent Pos:* Consult climatologist, US, SAm & Cent Am, 52-; deleg, Pan Am Inst Geog & Hist, DC, 52, Mex, 55; Nat Res Coun deleg, Int Geog Cong, Brazil, 56; chief engr, Field Party, Mex, 57; deleg, Comn Climat, DC, 57; Geophys Union deleg, Int Union Geod & Geophys, Finland, 60; deleg, Int Geog Cong, Sweden, 60; Nat Res Coun deleg, Pac Sci Cong, Hawaii, 61; Int Geog Union deleg, Comn Agr Climat, Can, 62; deleg, Int Union Conserv Nature & Natural Resources, Galapagos Island Symp, Ecuador, 64; cert consult meteorologist, 81- *Mem:* Am Meteorol Soc. *Res:* Environmental research and testing of material; tropical environment forest; applied meteorology; climate of the Galapagos Islands and eastern Tropical Pacific Ocean Area; weather related accidents and storm damage. *Mailing Add:* 4122 Cobblestone Pl Durham NC 27707

ALPERT, LOUIS KATZ, b New York, NY, Sept 8, 07; m 32; c 1. MEDICINE. *Educ:* Yale Univ, BS, 28, MD, 32. *Prof Exp:* Asst path, Yale Univ, 32-33, asst med, 33-35; from asst to instr, Univ Chicago, 35-38; Nat Res Coun fel, Rockefeller Inst, NY, 38-39; instr, Johns Hopkins Univ, 39-43; from adj clin prof to clin prof, 48-54, prof, 54-74, EMER PROF MED, SCH MED, GEORGE WASHINGTON UNIV, 74- *Mem:* Am Fedn Clin; Am Cancer Res; Endocrine Soc; Am Diabetes Asn; Am Col Physicians. *Res:* Cancer chemotherapy; endocrinology; endocrinology. *Mailing Add:* 13338 Arroya Vista Rd Poway CA 92064

ALPERT, NELSON LEIGH, b New Haven, Conn, June 14, 25; m 50; c 2. MEDICAL TECHNOLOGY, SPECTROSCOPY INSTRUMENTATION. *Educ:* Yale Univ, BS, 45; Mass Inst Technol, PhD(physics), 48. *Prof Exp:* Asst prof physics, Rutgers Univ, 48-52; physicist, White Develop Corp, 52-58; group leader, Perkin-Elmer Corp. 58-61, chief engr, Spectros Prod Develop, 61-67; dir instrument develop, Becton Dickinson & Co, 67-68; INDEPENDENT CONSULT, 68- *Concurrent Pos:* Consult, Dept Path, Hartford Hosp, 69-; ed, Clinical Instrument Systems, 80- *Honors & Awards:* Seligson-Golden Award, Am Asn Clin Chem, 88. *Mem:* AAAS; Soc Appl Spectros; Am Asn Clin Chem. *Res:* Analytical and clinical instruments, particularly spectroscopic, ultraviolet, visible, infrared atomic absorbtion and emission, flame, fluoresence. *Mailing Add:* 447 Glenbrook Rd Stamford CT 06906

ALPERT, NORMAN ROLAND, b Stamford, Conn, July 28, 22; m 52. PHYSIOLOGY. *Educ:* Wesleyan Univ, AB, 43; Columbia Univ, PhD(physiol), 51. *Prof Exp:* From asst to instr physiol, Columbia Univ, 38-53; from asst prof to prof, Col Med, Univ Ill, 53-66; PROF PHYSIOL & CHMN DEPT, COL MED, UNIV VT, 66- *Mem:* Am Physiol Soc; Biophys Soc; Harvey Soc; Am Soc Gen Physiol; Am Soc Exp Biol & Med. *Res:* Anaerobiosis in vivo; metabolism, respiration and circulation; thermodynamics, chemistry and mechanics of skeletal and cardiac muscle contraction. *Mailing Add:* Harbor Rd Shelburne VT 05482

ALPERT, RONALD L, b Rochester, NY, Apr 27, 41; m 69; c 2. HEAT TRANSFER, FIRE DYNAMICS. *Educ:* Mass Inst Technol, SB & SM, 65, ScD(mech engr), 70. *Prof Exp:* Res asst mech eng, Mass Inst Technol, 64-69; sr res scientist basic fire res, Factory Mutual Res Corp, 69-74, asst mgr basic res, 74-86, mgr, basic res, 86-90, MGR, FIRE & EXPLOSION RES, FACTORY MUTUAL RES CORP, 91- *Concurrent Pos:* Mem, Comt Heat Transfer in Fires & Combustion Systs, Heat Transfer Div, Am Soc Mech Engrs; mem tech comt, Halon Alternatives Res Corp, 77- *Mem:* Combustion Inst; Am Soc Mech Engrs. *Res:* Fire detection, growth, suppression, modeling; heat transfer in two phase and buoyant flows; spray dynamics. *Mailing Add:* Factory Mutual Res Corp PO Box 9102 Norwood MA 02062

ALPERT, SEYMOUR, b New York, NY, Apr 20, 18; m 41. MEDICINE, ANESTHESIOLOGY. *Educ:* Columbia Univ, AB, 39; Long Island Col Med, MD, 43; Am Bd Anesthesiol, dipl, 49. *Prof Exp:* Intern, Beth Israel Hosp, 43-44; chief med officer anesthesiol, Gallinger Munic Hosp, 46-47; from instr to assoc prof, 48-61, from asst dir to assoc dir, 61-69, PROF ANESTHESIOL, SCH MED, GEORGE WASHINGTON UNIV, 61-, VPRES DEVELOP, 69- *Concurrent Pos:* Univ fel anesthesiol, Sch Med, George Washington Univ, 48; consult, Walter Reed Army Hosp, 48-, DC Gen Hosp, 48-69 & Mt Alto Vet Hosp, 59-69. *Mem:* Am Soc Anesthesiologists; AMA; fel Am Col Anesthesiologists; Asn Am Med Cols; Pan-Am Med Asn. *Mailing Add:* 9 E 63rd St New York NY 10021

ALPERT, SEYMOUR SAMUEL, b Amsterdam, NY, Dec 21, 30; div; c 3. PHYSICS. *Educ:* Univ Calif, Berkeley, AB, 53, PhD(physics), 62. *Prof Exp:* Physicist, Lawrence Radiation Lab, 61-62; mem tech staff, Bell Tel Labs, Inc, 62-64; res assoc laser scattering, Columbia Univ, 64-66; from asst prof to assoc prof, 66-74, PROF PHYSICS, UNIV NMEX, 74- *Mem:* Am Phys Soc; Am Asn Physics Teachers. *Res:* Laser light scattering; energetic modelling of human body; optical spectroscopy of liquids; critical opalescence. *Mailing Add:* Dept Physics Univ NMex Main Campus Albuquerque NM 87131

ALPHA, ANDREW G, b Leth Bridge Alberta, Can, May 1, 12. TECTONICS, GROUND WATER. *Educ:* Univ NDak, MA, 35. *Prof Exp:* Mgr exploration, Mobil Oil Corp, 43-77; INDEPENDENT CONSULT, 77- *Mem:* Fel Geol Soc Am; Geol Soc Am; Am Asn Petrol Geologists. *Mailing Add:* 1101 Monaco Pkwy Denver CO 80220

ALPHER, RALPH ASHER, b Washington, DC, Feb 3, 21; m 42; c 2. PHYSICS. *Educ:* George Washington Univ, BS, 43, PhD(physics), 48. *Prof Exp:* Physicist, US Naval Ord Lab & USN Bur Ord, 40-44; physicist, Appl Physics Lab, Johns Hopkins Univ, 44-55; physicist, Gen Elec Co, 55-86; DISTINGUISHED RES PROF PHYSICS, UNION COL, 87-; ADMINR & DISTINGUISHED SR SCIENTIST, DUDLEY OBSERV, 87- *Concurrent Pos:* Adj prof, Rensselaer Polytech Inst, 60-64 & 87-; assoc ed, Physics of Fluids, 60-62; mem, steering comt, sect B (Physics), AAAS, 82-86; Am Phys Soc Panel Pub Affairs, 78-82, Comt Opportunities Physics, 78-82; Am Inst Physics Comt Manpower, 77-82. *Honors & Awards:* Magellanic Premium, Am Philos Soc, 75; Prix Georges Vanderlinden, Royal Acad Sci, Lett & Fine Arts, Belg, 75; Physics & Math Sci Award, NY Acad Sci, 81; John Price Wetherill Medal, Franklin Inst, 80. *Mem:* Fel Am Phys Soc; fel AAAS; Sigma Xi; fel Franklin Inst; Fedn Am Scientist; fel Am Acad Arts & Sci. *Res:* Cosmology; astrophysics; physics of fluids. *Mailing Add:* Dept Physics Union Col Schenectady NY 12308

ALPHIN, REEVIS STANCIL, b Mt Olive, NC, Apr 21, 29; m 55; c 2. PHARMACOLOGY, BIOCHEMISTRY. *Educ:* Univ NC, BA, 51; Duke Univ, MA, 56; Med Col Va, PhD(pharmacol), 66. *Prof Exp:* Res asst physiol, Duke Univ, 53-56; pharmacologist, Lilly Res Labs, Eli Lilly & Co, 56-60; pharmacologist, 60-64, head sect gastroenterol, 64-73, ASSOC DIR PHARMACOL, A H ROBINS CO, INC, 73- *Mem:* AAAS; assoc Am Gastroenterol Asn; Am Physiol Soc; NY Acad Sci; Am Soc Pharmacol & Exp Therapeut. *Res:* Gastrointestinal tract; peptic ulcer; anorectic agents. *Mailing Add:* Dept Pharmacol A H Robins Co Inc Res Labs 1211 Sherwood Ave Richmond VA 23220

ALQUIRE, MARY SLANINA (HIRSCH), b Garrett, Ind, Feb 14, 43; m 65, 88; c 2. STATISTICAL-COMPUTER DATA ANALYSIS. *Educ:* Marquette Univ, BS, 65; Univ Ark, MS, 77. *Prof Exp:* Res assoc statist, Dept Math Sci, 77-79, Agr Statist Lab, 79-82, STATISTICIAN & SYSTS ANALYST, BUR BUS & ECON RES, UNIV ARK, 82- *Concurrent Pos:* Lectr statist, Dept Agron, Univ Ark, 79-82, instr statist & programming, Computer Info Systs/ Quant Anal Dept, 82- *Mem:* Sigma Xi; Am Statist Asn. *Res:* Analytical techniques in international marketing; spectral analysis of public utility rates; time series analysis of water quality and filter field system; computer access to commercial and government databases. *Mailing Add:* Bur Bus & Econ Res Univ Ark Fayetteville AR 72701

ALRUTZ, ROBERT WILLARD, b Pittsburgh, Pa, Aug 20, 21; m 46; c 1. ECOLOGY. *Educ:* Univ Pittsburgh, BS, 43; Univ Ill, MS, 47, PhD, 51. *Prof Exp:* Instr biol, Univ Minn, 51-52; from asst prof to assoc prof, 52-64, PROF BIOL, DENISON UNIV, 64- DIR BIOL RESERVE, 69- *Concurrent Pos:* Chmn dept biol, Denison Univ, 64-65, chmn biol reserve, 65-89. *Mem:* AAAS; Ecol Soc Am; Am Inst Biol Sci; Nat Asn Biol Teachers; Conserv Educ Asn. *Res:* Biology of wild pepulations of peromyscus. *Mailing Add:* Dept Biol Denison Univ Granville OH 43023

AL SAADI, A AMIR, b Baqubah, Iraq, Oct 20, 35; US citizen; m 61; c 4. GENETICS, CYTOLOGY. *Educ:* Univ Baghdad, BA, 55; Univ Kans, MA, 59; Univ Mich, PhD(biol), 63. *Prof Exp:* Res assoc radiation genetics, Univ Mich Sch Med, 62-65, asst prof cellular biol, 65-69; CHIEF CLIN GENETICS, WILLIAM BEAUMONT HOSP, 70-, ASSOC DIR RES CANCER CYTOGENETICS, 71- *Concurrent Pos:* Dir, Sch Histotechnol, William Beaumont Hosp, 77- *Mem:* Am Soc Human Genetics; Am Asn Cancer Res; Am Thyroid Asn; Am Soc Cell Biol; Tissue Cult Asn; AAAS; Am Fedn Clin Res. *Res:* Chromosomal role in the development of human malignancies; the role of chromosome change in human disease; familial conditions including cancer families and prevalence of diseases associated with it; chromosome abnormalities and birth defects; prenatal diagnosis. *Mailing Add:* Dept Anat Path William Beaumont Hosp Royal Oak MI 48072

AL-SAADI, ABDUL A, b Iraq, Oct 20, 35; US citizen; m 61; c 4. CELL BIOLOGY, MICROSCOPIC ANATOMY. *Educ:* Univ Baghdad, BA, 55; Univ Kans, MA, 59; Univ Mich, PhD(zool), 63; Am Bd Med Genetics, cert. *Prof Exp:* Res assoc radiation effects, Sch Med, Univ Mich, 62-66, asst prof cellular biol, 66-70; CHIEF CYTOGENETICS & ASSOC DIR RES, DEPT ANAT PATH, WILLIAM BEAUMONT HOSP, 70-, DIR, SCH HISTOTECHNOL, 77- *Concurrent Pos:* NIH grant, 62-67; Univ Mich Cancer Inst grants, 67, 68-69; Am Cancer Soc grant; asst clin prof, Health Ctr, Oakland Univ, 77-; mem, Genetics Adv Bd, State Mich, 80- *Mem:* AAAS; Am Soc Human Genetics; NY Acad Sci; Am Asn Cancer Res; Am Fedn Clin Res; Am Soc Cell Biol; Tissue Cult Asn. *Res:* Cytogenetics in cancer research; clinical cytogenetics; ultrastructure changes in carcinogenesis; thyroid physiology in health and disease; radiation biology; relationship between chromosomal abnormalities and development of cancer; cytogenetic studies of human brain tumors; clinical genetic studies on the role of chromosomal abnormalities in fetal malformation and birth defects. *Mailing Add:* Dept Anat Path William Beaumont Hosp 3601 W 13 Mile Rd Royal Oak MI 48072

AL-SAADOON, FALEH T, b Baghdad, Iraq, Jan 11, 43; US citizen; m 72; c 3. RESERVOIR ENGINEERING, PETROLEUM PRODUCTION ENGINEERING. *Educ:* WVa Univ, BS, 65, MS, 66; Univ Pittsburgh, PhD(chem eng), 70. *Prof Exp:* Head, dept petrol eng, Baghdad Univ, 70-75; asst prof petrol eng, Al-Fateh Univ, Libya, 75-78; assoc prof, WVa Univ, 78-82; prof chem & natural gas eng & chmn dept, 82-87, interim dir, Off Sponsored Res, 87-89, PROF NATURAL GAS ENG, TEX A&I UNIV, 89- *Mem:* Soc Petrol Engrs; Am Inst Chem Engrs; Soc Prof Well Log Analysts. *Res:* Unconventional gas resources; thermal recovery of heavy crude oils and tar sands; enhanced oil recovery. *Mailing Add:* 1124 W Yoakum Ave Kingsville TX 78363

AL-SARRAF, MUHYI, b Baghdad, Iraq, Sept 15, 38; US citizen. ONCOLOGY. *Educ:* Univ Baghdad, MD, 61; Royal Col Physicians Can, cert internal med, 69. *Prof Exp:* PROF MED, SCH MED, WAYNE STATE UNIV, 81-; CHIEF HEAD & NECK CANCER, HARPER-GRACE HOSP, 83- *Concurrent Pos:* Vis prof, Miss Med Ctr, Univ Miss, 72-, Div Oncol, Albany Med Col, NY, 73, dept med, Univ Iraq, Baghdad & Cancer Control Agency, BC, 75. *Mem:* Am Col Radiol. *Res:* Chemotherapy; clinical trial research; cancer immunology; immunotherapy and tumor markers. *Mailing Add:* Dept Internal Med Oncol Div Wayne State Univ 3990 John R Detroit MI 48201

ALSBERG, HENRY, b Ger, Oct 6, 21; US citizen; m 49; c 2. POLYMER CHEMISTRY, INDUSTRIAL CHEMISTRY. *Educ:* Univ Toronto, BASc, 47; Purdue Univ, MS, 49. *Prof Exp:* Group leader polymers, Koppers Co, 51-64; MGR RES & DEVELOP, RICHARDSON CO, 64- *Mem:* Am Chem Soc; Soc Plastics Engrs; Am Soc Testing & Mat. *Res:* Development of specialty suspension and emulsion polymers; pilot plant operations; plant process development. *Mailing Add:* 1847 Tanglewood Dr Apt 2A Glenview IL 60025

ALSCHER, RUTH, b Chicago, Ill, July 17, 43; c 2. PLANT PHYSIOLOGY. *Educ:* Trinity Col, BA, 65; Washington Univ, MA, 68; Univ Calif, Davis, PhD(plant physiol), 72. *Prof Exp:* Fel, Dept Biochem, Univ Calif, Davis, 73; Nat Res Serv Awards, Lab Chem Biodynamics, Sect Genetics Develop & Physiol, Cornell Univ, 74-75 &75-77; res assoc, NY State Agr Exp Sta, 77-79; res assoc, Boyce Thompson Inst, Cornell Univ, 79-88; ASSOC PROF, DEPT PLANT PATH PHYSIOL & WEED SCI, VA POLYTECH & STATE UNIV, 88- *Mem:* Am Soc Plant Physiologists; Am Chem Soc; Sigma Xi. *Res:* Chloroplast physiology; light regulation of carbon metabolism; effects of air pollutants on photosynthesis; mechanisms of plant resistance to air pollutants; effects of herbicides on photosynthesis. *Mailing Add:* Dept Plant Path Physiol & Weed Sci Va Polytech & State Univ Blacksburg VA 24061

AL-SHAIEB, ZUHAIR FOUAD, b Damascus, Syria, Sept 23, 40; US citizen; m 70; c 1. DIAGENESIS OF CLASTIC ROCKS. *Educ:* Univ Damascus, BS, 63; Univ Mo-Rolla, MS, 67, PhD(geol), 72. *Prof Exp:* Instr phys sci, Shaiwk Col, Kuwait, 65-66; asst prof, 72-76, assoc prof, 76-81, PROF GEOL, OKLA STATE UNIV, 81- *Concurrent Pos:* Consult, Erico, Inc, Tulsa, 70-, Mid-Am Pipeline Co, 77-80, Shell Develop Co, Houston & Amereda Hess Corp, Tulsa, 80-81; prin investr, AEC, Dept Engery, 75-80, Environ Protection Agency, 81- *Mem:* Soc Econ Paleontologists & Mineralogists; Am Inst Mineral Petrol & Metall Engrs; Sigma Xi. *Res:* Energy (hydrocarbon and uranium); relationship between hydrocarbon migration and generation of secondary porosity in sandstone reservoir which has application in hydrocarbon exploration. *Mailing Add:* Sch Geol Okla State Univ Stillwater OK 74074

ALSMEYER, RICHARD HARVEY, b Sebring, Fla, Feb 4, 29; m 53; c 2. MEAT SCIENCE, BIOCHEMISTRY. *Educ:* Univ Fla, BSA, 52, MSA, 56, PhD(animal sci & nutrit), 60. *Prof Exp:* Res technician meat sci, Agr Exp Sta, Univ Fla, 56-57; res animal husbandman, Meat Qual Lab, Agr Res Serv, USDA, 60-67, head standards group, Tech Serv Div, 67-71, sr staff officer, Prod Standards, Sci & Tech Serv, 71-75, dir prog eval & spec reports, Coop Res Sci & Educ Admin, 75-79, nat res prog leader, Animal Prod Processing & Mkt, 79-83; pres, Alsmeyer Consult, 84-88, PRES, ALSMEYER FOOD CONSULT, 89- *Concurrent Pos:* Res fel, Brit Meat Res Inst, Cambridge Univ, Eng, 66; mem adv bd marine safety & ecol, Int Mariners, Inc; Am Meat Sci Asn rep, Nat Res Coun-Nat Acad Sci, 72-75; food sci consult, 83 -84; consult, Webb Tech Group, Webb Foodlab Inc. *Mem:* Am Meat Sci Asn; Inst Food Sci & Technol; Am Soc Animal Sci; Am Soc Agr Consults; assoc mem Am Meat Inst. *Res:* Meat and poultry products, seafood, and other protein foods of plant and animal origin; specialties are: meat and poultry products, seafood, and other protein foods of plant and animal origin. *Mailing Add:* 8905 Maxwell Dr Potomac MD 20854-3125

ALSMILLER, RUFARD G, JR, b Louisville, Ky, Nov 16, 27; m 52. PHYSICS, MATHEMATICS. *Educ:* Univ Louisville, BS, 49; Purdue Univ, MS, 52; Univ Kans, PhD(physics), 57. *Prof Exp:* Fire control systs analyst, Naval Ord Plant, Ind, 52-53; physicist, Allison Div, Gen Motors Corp, 56; assoc dir div, 69-73, GROUP LEADER HIGH & MEDIUM ENERGY SHIELDING, ENG PHYSICS DIV, OAK RIDGE NAT LAB, 57- *Mem:* Am Phys Soc; fel Am Nuclear Soc. *Mailing Add:* Eng Physics Div Oak Ridge Nat Lab PO Box X Oak Ridge TN 37830

ALSON, ELI, b New York, NY, Aug 1, 29; m 56; c 2. BEHAVIORAL MEDICINE, APPLIED PSYCHOPHYSIOLOGY. *Educ:* Brooklyn Col, BA, 52; Univ Buffalo, PhD(psychol), 59. *Prof Exp:* Clin psychologist, Vet Admin Med Ctr, Lyons, NJ, 69-75, dir psychol training, 75-84, dir behav med clin, 82-86; DIR STRESS & PAIN MGT CTR, ST CLAIRES-RIVERSIDE MED CTR, NJ, 87- *Concurrent Pos:* Dir, Biofeedback Clin, Bernardsville, 80-87; consult, NJ Dept Vet Servs, 85-86; med advisor, Off Hearings & Appeals, Social Security Admin, 86- *Mem:* Am Psychol Asn; Biofeedback Soc Am. *Res:* Biofeedback; psychophysiology; interpersonal perception; post-traumatic stress disorder. *Mailing Add:* 25 Kitchell Rd Denville NJ 07834

ALSOP, DAVID W, b Lindsay, Ont, Nov 15, 39. CYTOLOGY, MORPHOLOGY. *Educ:* Cornell Univ, BS, 64, PhD(biol), 70. *Prof Exp:* ASSOC PROF BIOL, QUEENS COL, NY, 70- *Mem:* Entom Soc Am; Sigma Xi. *Res:* Comparative studies on cockroaches. *Mailing Add:* Dept Biol Queens Col Flushing NY 11367

ALSOP, FREDERICK JOSEPH, III, b Owensboro, Ky, Jan 9, 42; div. ORNITHOLOGY, VERTEBRATE ECOLOGY. *Educ:* Austin Peay State Univ, BS, 64; Univ Tenn, Knoxville, MS, 68, PhD(zool), 72. *Prof Exp:* Sec sch teacher sci & fine arts, Ft Campbell Dependent Sch Syst, 64-66; asst zool, Univ Tenn, 66-72; from asst prof to assoc prof, 72-83, PROF BIOL, ETENN STATE UNIV, 83-, CHMN BIOL SCI, 83- *Concurrent Pos:* Primary investr, environ impact statement-vertebrates, Tenn Valley Authority, 75-76 & grant, Tenn endangered & threatened bird species, Tenn Wildlife Resources Agency, 78-79; consult environ res & technol, 81- & US Army Corps Engrs, 81. *Honors & Awards:* Nat Lectr, Sigma Xi, 84. *Mem:* Am Ornithologists Union; Wilson Soc. *Res:* Life history, distribution population dynamics of birds, primarily of Tennessee; avian photography; tropical ornithology; author of book. *Mailing Add:* Dept Biol Sci E Tenn State Univ PO Box 23590 A Johnson City TN 37614

ALSOP, JOHN HENRY, III, b Okmulgee, Okla, Oct 4, 24; m 51; c 2. ANALYTICAL CHEMISTRY. *Educ:* Okla State Univ, BS, 50; Univ Tex, PhD(anal chem), 57. *Prof Exp:* Chemist, Eagle-Picher Co, Okla, 50-51, Am Window Glass Co, 51-52 & Stanolind Oil & Gas Co, 56-58; sr chemist, Pan Am Petrol Corp, 58-59, Chemstrand Corp, Ala, 59-60 & Chemstrand Res Ctr, NC, 60-62; res chemist, Richmond Res Labs, Texaco Inc, 62-76; PROJ CHEMIST, PORT ARTHUR RES LABS, TEXACO INC, 76- *Mem:* Am Chem Soc; Nat Asn Corrosion Engrs; Soc Appl Spectros; Sigma Xi. *Res:* Analytical methods development in air and water conservation. *Mailing Add:* 8150 Kahilan PO Box 1139 Boerne TX 78006

ALSOP, LEONARD E, b Oroville, Calif, Apr 10, 30; m 55; c 3. GEOPHYSICS. *Educ:* Columbia Univ, AB, 51, AM, 56, PhD(physics), 61. *Prof Exp:* Res scientist, Lamont Geol Observ, 59-65, adj assoc prof geol, 65-73, mem acad staff, Lamont-Doherty Geol Observ, Columbia Univ, 73- *Concurrent Pos:* Mem staff, IBM Corp, 65-73. *Mem:* Am Geophys Union; Am Seismol Soc; Royal Astron Soc. *Res:* Free oscillations of the earth; use of a laser as a transducer for seismograph; tides of the solid earth. *Mailing Add:* Watson Res Ctr Div IBM Corp PO Box 218 Yorktown Heights NY 10598

ALSPACH, BRIAN ROGER, b Minot, NDak, May 29, 38; m 61, 80; c 2. GRAPH THEORY. *Educ:* Univ Wash, BA, 61; Univ Calif, Santa Barbara, MA, 64, PhD(math), 66. *Prof Exp:* Teacher math, Cordell Hull Jr High Sch, 61-62; from asst prof to assoc prof, 66-80, PROF MATH, SIMON FRASER UNIV, 80- *Concurrent Pos:* Ed, Discrete Math, 81-; Ed, Ars Combin, 84-

Mem: Am Math Soc; Math Asn Am; Can Math Soc; Combinatorial Math Soc Australia; Can Appl Math Soc. *Res:* Graph theory with emphasis on automorphism groups of graphs, graph decompsitions and tournaments. *Mailing Add:* Dept Math Simon Fraser Univ Burnaby BC V5A 1S6 Can

ALSPACH, DANIEL LEE, b Wenatchee, Wash, July 4, 40; m 65; c 2. CONTROL THEORY, STOCHASTIC SYSTEMS. *Educ:* Univ Wash, BS, 62, MS, 66; Univ Calif, San Diego, PhD(eng sci), 70. *Prof Exp:* Design engr, Honeywell, Inc, 63-66; staff assoc, Gen Atomic Div Gen Dynamics Corp, 66-67; assoc prof elec eng, Colo State Univ, 70-73; vis fac mem, Dept Appl Mech & Eng Sci, Univ Calif, San Diego, 73-74; vpres & tech dir, 73-80, PRES, ORINCON CORP, 80- *Mem:* Inst Elec & Electronics Engrs; Sigma Xi. *Res:* Signal processing; surveillance systems; stochastic control; estimation theory; digital systems; system modeling and simulation. *Mailing Add:* 9363 Town Ctr Dr San Diego CA 92121

ALSPAUGH, DALE W(ILLIAM), b Dayton, Ohio, May 25, 32; m 55; c 4. AERONAUTICAL ENGINEERING. *Educ:* Univ Cincinnati, ME, 55; Purdue Univ, MS, 58, PhD(eng sci), 65. *Prof Exp:* Jr proj engr, Frigidaire Div, Gen Motors Corp, 55-56; instr, Purdue Univ, 57-58 & 59-64, from asst prof to assoc prof eng sci, 64-81, prof aeronaut & astronaut, 81, CHANCELLOR, PURDUE UNIV N CENT, 82- *Concurrent Pos:* Consult, Midwest Appl Sci Corp, 64-70, Los Alamos Lab, 77- & US Army Missile Res & Develop Command, 78- *Mem:* Am Soc Eng Educ; Am Inst Aeronaut & Astronaut. *Res:* Dynamics; optimization of engineering systems; shell theory; applied mathematics. *Mailing Add:* Chancellor Purdue Univ N Cent Westville IN 46391

ALSTADT, DON MARTIN, b Erie, Pa, July 29, 21; m; c 1. PHYSICS. *Educ:* Univ Pittsburgh, BS, 47. *Prof Exp:* Develop engr, 47-50, chief physicist, 50-52, head basic res, 52-56, mgr cent res, 56-61, vpres & gen mgr, 64-66, exec vpres & mem bd dirs, 66-68, VPRES, LORD MFG CO, 61-, GEN MGR, HUGHSON CHEM CO DIV, 58-, PRES, LORD CORP, 68-, VCHMN, LORD CORP, 75- *Concurrent Pos:* Consult, Carborundum Co, 47-48, Sci Specialties Corp, 51-52 & Transistor Prod Corp, 52-56; metrop chmn, Nat Alliance Businessmen, 69-; mem bd dirs, Lord Corp, Kiethley Instruments, Inc,; Schs Mgt, Univ Pittsburgh, Case Western Reserve Univ & Univ Denver, 69-; mem bd adv, Case Inst Technol, Sch Eng, Tulane Univ & Mellon Inst, 70-; mem bd trustees, Polytech Univ NY, Mercyhurst Col; mem corp, Conf Bd, NY; vchmn, Res & Develop Div, Nat Asn Mfrs; mem adv coun, Kolff Found; guest lectr, Int Inst Mgt Sci Ctr, Berlin, 79 & Swed Royal Acad Eng Sci, 84; mem sci & technol adv bd & hardwoods working comt, Pa Gov Casey's Economic Develop Partnership. *Honors & Awards:* Behrend Medallion, Pa State Univ; Francis Wright Davis Distinguished Lectr, Acad Appl Sci, 83. *Mem:* Am Chem Soc; Am Phys Soc; Electrochem Soc; fel Am Inst Chemists; Inst Mgt Sci. *Res:* Adhesives and adhesion; dielectrics and rheology of materials; physics and chemistry of surfaces; frictional phenomena; bioengineering. *Mailing Add:* 228 Rosemont Ave Erie PA 16505

ALSTON, JIMMY ALBERT, b Temple, Tex, Mar 23, 42; m 67; c 2. PLANT BREEDING. *Educ:* Tex A&M Univ, BS, 64, MS, 67, PhD(plant breeding), 68. *Prof Exp:* Res asst, Tex A&M Univ, 64-68; supvr plant breeding dept, Seed Prod Div, Baker Castor Oil Co, Tex, 68-70; plant breeder, 70-79, DIR RES, GEORGE W PARK SEED CO, INC, 79- *Mem:* Am Soc Agron; Crop Sci Soc Am. *Res:* Cyclic gametic selection as a breeding method in corn; inheritance of female characters in Ricinus communis; development of new floricultural varieties by selective breeding methods; devising more efficient methods of seed production. *Mailing Add:* Rte 8 Belle Meade 206 Yosemite Greenwood SC 29649

ALSTON, PETER VAN, b Windsor, NC, Apr 27, 46; m 68; c 3. PHYSICAL ORGANIC CHEMISTRY, ANALYTICAL CHEMISTRY. *Educ:* Univ NC, BS, 68; Va Commonwealth Univ, PhD(chem), 74. *Prof Exp:* Res chemist, 74-80, sr res chemist, 81-90, RES ASSOC, E I DU PONT DE NEMOURS & CO, INC, 91- *Concurrent Pos:* Adj fac, J Sargeant Reynolds Community Col, 74-77. *Mem:* Am Chem Soc. *Res:* Mechanism of cycloaddition reactions; perturbation molecular orbital theory in organic chemistry; open-end-spinning of polyester fibers. *Mailing Add:* 305 Plantation Circle Kinston NC 28501

ALSTON-GARNJOST, MARGARET, b Ashtead, Eng, Jan 23, 29; m 66. HIGH ENERGY PHYSICS. *Educ:* Univ Liverpool, BSc, 51, Hons, 52, PhD(physics), 55. *Prof Exp:* Demonstr physics, Univ Liverpool, 54-56, Imp Chem Industs res fel, 56-58; physicist, 59-72, SR STAFF SCIENTIST, LAWRENCE BERKELEY LAB, UNIV CALIF, 72- *Mem:* Am Phys Soc. *Res:* High energy particle physics; experimental high energy physics. *Mailing Add:* 41505 Little Netucca River Rd Cloverdale OR 97112

ALSUM, DONALD JAMES, b Randolph, Wis, June 27, 37; m 61; c 2. REPRODUCTIVE PHYSIOLOGY, NEURO-RECEPTOR PHYSIOLOGY. *Educ:* Calvin Col, BA, 61; Purdue Univ, MS, 66; Univ Minn, PhD(animal physiol), 74. *Prof Exp:* Jr high sch teacher sci, Western Suburbs Sch, Ill, 61-63; sec sci teacher, Timothy Christian High Sch, Ill, 63-70; res asst physiol, Univ Minn, Minneapolis, 70-74; asst prof, 74-76, assoc prof biol, 76-85, chmn dept, 80-85, PROF BIOL, ST MARY'S COL, 83- *Concurrent Pos:* Physiol consult, Mcmillan Publ Co & Charles E Merrill; vis scientist neuro physiol, Mayo Found, 84-85. *Mem:* Am Soc Allied Health Prof; Am Inst Biol Sci. *Res:* Testicular functions with particular interests in spermatozial maturation and levels of acid and alkaline phosphatases in certain regions of the male reproductive tracts of specific mammals. *Mailing Add:* Dept Biol St Mary's Col Winona MN 55987

ALT, DAVID D, b St Louis, Mo, Sept 17, 33. GEOLOGY, GEOCHEMISTRY. *Educ:* Wash Univ, St Louis, AB, 55; Univ Minn, MS, 58; Univ Tex, PhD(geol), 61. *Prof Exp:* Sr res assoc chem, Univ Leeds, 61-62; asst prof geol, Univ Fla, 62-65; from asst prof to assoc prof, 65-73, PROF GEOL, UNIV MONT, 73- *Mem:* Geol Soc Am; Soc Econ Paleont & Mineral; Nat Asn Geol Teachers. *Res:* Emergent shorelines and coastal plain geomorphology; karst landscapes. *Mailing Add:* Dept Geol Univ Mont Missoula MT 59812

ALT, FRANZ L, b Vienna, Austria, Nov 30, 10. APPLIED MATHEMATICS, COMPUTER SCIENCE. *Educ:* Univ Vienna, Austria, PhD(math), 32. *Prof Exp:* Dep dir, Div Appl Math, Nat Bur Standards, 48-67 & Info Div, Am Inst Physics, 67-73; RETIRED. *Mem:* Asn Comput Mach. *Mailing Add:* 350 Cabrini Apt 9H New York NY 10033

ALT, GERHARD HORST, b Berlin, Ger, Jan 24, 25; m 59; c 2. ORGANIC CHEMISTRY. *Educ:* Univ London, BSc, 49, PhD(org chem), 54. *Prof Exp:* Res fel, Wayne State Univ, 54-56; res chemist, Monsanto Co, 56-64, res specialist, Agr Div, 64-67; sr res specialist, Monsanto Agr Prod Co, 67-; AT AGR RES DEPT, MONSANTO CHEM CO. *Mem:* Am Chem Soc; The Chem Soc. *Res:* Enamine chemistry; steroids; conformational analysis; reaction mechanisms; heterocycles. *Mailing Add:* Agr Res Dept Monsanto Chem Co 700 Chesterfield Villa Pkwy St Louis MO 63198

ALTARES, TIMOTHY, JR, physical chemistry, polymer chemistry; deceased, see previous edition for last biography

ALTEMEIER, WILLIAM ARTHUR, III, b Detroit, Mich, Oct 3, 36; m 61; c 3. PEDIATRICS, NEONATOLOGY. *Educ:* Univ Cincinnati, BA, 58; Vanderbilt Univ, MD, 62. *Prof Exp:* Fel infectious dis, Univ Fla, 64-66, asst prof pediat, 69-72; res assoc immunol, Walter Reed Res Inst, 66-68; PROF PEDIAT, VANDERBILT UNIV, 72-, VCHMN DEPT, 83- *Mem:* Soc Pediat Res; Am Asn Immunologists. *Res:* Causes and prevention of child abuse. *Mailing Add:* Dept Pediat/Nursing Vanderbilt Univ Nashville TN 37240

ALTEMOSE, VINCENT O, b Nazareth, Pa, Aug 8, 28; m 50; c 4. PHYSICS, MATERIALS SCIENCE. *Educ:* Lafayette Col, BS, 52; Lehigh Univ, MS, 55. *Prof Exp:* Instr, Lafayette Col, 53-56; RES SCIENTIST PHYSICS, CORNING GLASS WORKS, 56- *Concurrent Pos:* Mem comt, Int Comn Glass, 77-83. *Mem:* Am Ceramic Soc; Am Vacuum Soc; Am Soc Mass Spectrometry; Sigma Xi. *Res:* Gases in glass and ceramics; gas diffusion and out gassing of glassy material by various forms of energy; high temperature vaporization studies. *Mailing Add:* Res & Develop Div Sullivan Park-FR-1-8 Corning Glass Works Corning NY 14831

ALTENAU, ALAN GILES, b Cincinnati, Ohio, May 16, 38; m 74; c 2. ANALYTICAL CHEMISTRY. *Educ:* Univ Cincinnati, BS, 59; Purdue Univ, MS, 61, PhD(anal chem), 64. *Prof Exp:* Group leader, Anal Div, Firestone Tire & Rubber Co, 66-69, mgr, 69-74, asst dir, Cent Res, 74-79, dir, 79-90; DIR, INT TECH CTR, ROME, ITALY, 90- *Mem:* Am Chem Soc. *Res:* Research and development of tires. *Mailing Add:* 6763 St James Circle Hudson OH 44236

ALTENBURG, LEWIS CONRAD, b Houston, Tex, May 12, 42; m 65; c 1. CYTOGENETICS. *Educ:* Rice Univ, BA, 65; Univ Tex Grad Sch Biomed Sci, MS, 68, PhD(biochem), 70. *Prof Exp:* ASST PROF MED GENETICS, UNIV TEX GRAD SCH BIOMED SCI, HOUSTON, 73- *Concurrent Pos:* Fel, Max Planck Inst Exp Med, Gottingen, WGer, 70-71; fel, Univ Tex Grad Sch Biomed Sci, 71-72; NIH fel, Dept Biol, Univ Tex M D Anderson Hosp & Tumor Inst, 71-73; res assoc med genetics, Univ Tex Grad Sch Biomed Sci, Houston, 73. *Mem:* Am Chem Soc; Am Soc Microbiol; Am Soc Cell Biol; Am Soc Human Genetics; Genetics Soc Am. *Res:* Molecular cytogenetics of human chromosomes. *Mailing Add:* 5800 Lost Forest Dr Houston TX 77092

ALTENKIRCH, ROBERT AMES, b St Louis, Mo, May 13, 48; m 70; c 2. COMBUSTION, THERMAL SCIENCES. *Educ:* Purdue Univ, BSME, 70, PhD(mech eng), 75; Univ Calif, Berkeley, MS, 71. *Prof Exp:* From asst prof to assoc prof, 75-85, PROF MECH ENG, UNIV KY, 85-, CHMN DEPT, 84- *Concurrent Pos:* Prin investr, NSF, 76-78, NASA, 82-; co-prin investr, Nat Aeronaut & Space Admin, 76-81 & NSF, 77-79 & 80-82; prin investr, NASA, 82- *Honors & Awards:* Teetor Award, Soc Advan Educ, 79; Gustus L Larson Award, Am Soc Mech Engrs, 84. *Mem:* Combustion Inst; Am Soc Mech Engrs. *Res:* Pollutant formation during coal combustion; buoyancy effects on flames; flame spreading; two phase flow; continuous flow combustion. *Mailing Add:* Dept Mech Eng Miss State Univ Mississippi State MS 39762

ALTER, AMOS JOSEPH, b Rensselaer, Ind, Aug 4, 16; m 40; c 4. SANITARY ENGINEERING, PUBLIC HEALTH. *Educ:* Purdue Univ, BS, 38, CE, 49; MPH, Univ Mich, 48. *Prof Exp:* Sanit engr, Ind State Bd Health, 38-42; comn officer sanit engr, USPHS, 42-44; chief engr, Alaska Dept Health, 44-67; environ res engr, State of Alaska, 67-75; environ eng consult, 75-85; RETIRED. *Concurrent Pos:* Fel, WHO, Scandinavia, 50-51; mem sanit eng & environ comt, Nat Res Coun, 50-66; mem, Conf State Sanit Engrs; vis lectr, Univ Alaska, 70-75. *Honors & Awards:* Arthur Sidney Bedell Award, Water Pollution Control Fedn, 70; Pub Works, Man-of-the-year-Am Pub Works Asn, 67; Hon mem, Am Soc Civil Engrs, 90. *Mem:* Arctic Inst NAm; Am Soc Civil Engrs; Am Pub Health Asn; Am Water Works Asn; Am Acad Environ Engrs; Nat Soc Prof Engrs. *Res:* Civil engineering; cold weather design and construction problems. *Mailing Add:* PO Box 20304 Juneau AK 99802-0304

ALTER, GERALD M, PHYSICAL BIOCHEMISTRY, ENZYMOLOGY. *Educ:* Wash State Univ, PhD(biol chem), 75. *Prof Exp:* ASSOC PROF BIOL CHEM, WRIGHT STATE UNIV, 77- *Res:* Confirmational properties of proteins and enzymes; in vivo enzymology. *Mailing Add:* Dept Biol Chem Wright State Univ Dayton OH 45435

ALTER, HARVEY, b New York, NY, Sept 4, 32; m 57; c 2. PHYSICAL CHEMISTRY, ENVIRONMENTAL CHEMISTRY. *Educ:* City Univ New York, BS, 52; Univ Cincinnati, MS, 54, PhD(phys chem), 57. *Prof Exp:* Physicist plastics dept, Union Carbide Corp, 57-59; sr chemist, Harris Res Labs, Gillette Co Res Inst, 59-63, group leader, 64-65, assoc dir res, Toni Div, 65-66, tech dir res, 66-68, mgr,68-72, vpres, 72; dir res progs, Nat Ctr for Resource Recovery, Inc, 72-79; MGR, RESOURCES POLICY DEPT, US CHAMBER COMMERCE, 79- *Concurrent Pos:* Lectr, City Col New York,

58-59; prof lectr, Am Univ, 70-74, adj prof, 74-87; mem US Air Force PBI Technol Rev Comt, 70, ad hoc adhesion Adv Comt, 71, Environ Res Guid Comt Md, 77-79; chmn, Gordon Res Conf Sci Adhesion, 72, Am Soc Test & Mat Comt E-38 Resource Recovery, 73-78, Outside Evaluation Panel Recycled Mat, Nat Bureau Stand, 79-82; Gov's Task Force Solid Waste Mgt, Md, 87-, bd dir, WTE, Bedford, Mass, 82-, Nat Inst Urban Wildlife, Columbia, Md, 84-; ed, Resources Conserv, 72-87, Resources Conserv & Recycling, 88-; mem Gov adv coun recycling, 89, adv panel Assessment of Mat Technol, 90. *Honors & Awards:* Award of Merit, Am Soc Testing & Mat, 80. *Mem:* Fel Am Soc Testing & Mat; Am Chem Soc. *Res:* Materials and energy recovery from wastes; science and public policy; chemistry of non-metallic materials; recovery of materials and energy from municipal solid waste, environmental policy. *Mailing Add:* US Chamber Commerce 1615 H St NW Washington DC 20062

ALTER, HARVEY JAMES, b New York, NY, Sept 12, 35; m 65; c 2. INTERNAL MEDICINE, HEMATOLOGY. *Educ:* Univ Rochester, BA, 56, MD, 60; Am Bd Path, cert, 78. *Prof Exp:* From intern to resident med, Strong Mem Hosp, Univ Rochester, 60-61; clin assoc blood bank, NIH, 61-64; resident med, Seattle Univ Hosp, 64-65; from instr to asst prof med, Georgetown Univ Hosp, 66-69; SR INVESTR, NIH, 69-, CHIEF IMMUNOL SECT, BLOOD BANK, 78-, FAC, FOUND ADVAN EDUC SCI, 76- *Concurrent Pos:* Univ fel hemat, Georgetown Univ Hosp, 65-66, clin assoc prof, 69-86, clin prof, 87- *Honors & Awards:* Superior Serv Award, Dept Health, Educ & Welfare, 74 & Distinguished Serv Medal, 77. *Mem:* Fedn Clin Res; Am Soc Hemat; Am Asn Blood Banks; Int Soc Hemat; Am Asn Study Liver Dis. *Res:* Studies relating to nature and significance of the hepatitis B virus and its associated antigens and to the non-A/non-B hepatitus virus. *Mailing Add:* Dept Transfusion Med NIH Bldg 10 Rm 1C-711 Bethesda MD 20892

ALTER, HENRY WARD, b Taxila, Punjab, India, Dec 26, 23; US citizen; m 44, 79; c 3. PHYSICAL CHEMISTRY. *Educ:* Univ Calif, AB, 43, PhD(chem), 48. *Prof Exp:* Asst chem, Univ Calif, 42-44, 46-47, chemist, Manhattan Proj, 44-48; mgr separations chem & res assoc, Knolls Atomic Power Lab, Gen Elec Co, 48-57; tech specialist & mgr nuclear & inorg chem, Vallecitos Atomic Lab, 57-68; mgr nucleonics lab, Gen Elec Co, 68-71, mgr nuclear technol & appln oper, 71-74; pres, Terradex Corp, 74-86; consult, Tech-Ops Laudauer, 86-88; RETIRED. *Mem:* Am Chem Soc; fel Am Nuclear Soc; Sigma Xi. *Res:* Nuclear energy; nuclear fuels; plutonium technology; nuclear chemistry; fuel reprocessing; uranium exploration; environmental monitoring; radon. *Mailing Add:* 110 Haslemere Ct Lafayette CA 94549

ALTER, JOHN EMANUEL, b Davenport, Iowa, June 25, 45. ANALYTICAL RESEARCH, PROTEIN SEPARATIONS. *Educ:* Univ Iowa, BS, 68; Cornell Univ, MS, 70, PhD(phys chem), 74. *Prof Exp:* Res scientist phys chem, 74-80, supvr separations res, 80- 82, dir, Separations/Anal Res, 82-88, DIR O A DEVELOP, MILES LABS, INC, 88- *Mem:* Am Chem Soc; AAAS. *Res:* Properties of immobilized enzymes; texturization of vegetable protein; analytical chemistry; protein separations; analytical chemistry; protein separations; process development; process instrumentation. *Mailing Add:* Miles Labs Inc 1127 Myrtle St Elkhart IN 46514

ALTER, JOSEPH DINSMORE, b Lawrence, Kans, Apr 19, 23; m 46, 81; c 3. GERIATRICS, COMMUNITY MEDICINE. *Educ:* Hahnemann Med Col, MD, 50; Univ Calif, Berkeley, MPH, 61; Am Bd Prev Med, cert, 68. *Prof Exp:* Med staff, Group Health Coop, Seattle, 51-60; lectr, Sch Pub Health, Univ Calif, Berkeley, 61-62; from lectr to asst prof, Sch Hygiene & Pub Health, Johns Hopkins Univ, 62-68; assoc & field prof community med, Col Med, Univ Ky Med Ctr, 68-70; med dir, Pilot City Health Serv, Cincinnati, Ohio, 70-73; med dir, Health CareLouisville Inc, Ky, 73-75; chief domiciliary med serv, Vet Admin Ctr, Dayton, Ohio, 75-77; clin prof & actg chmn dept, 67-77, PROF COMMUNITY MED & CHMN DEPT, SCH MED, WRIGHT STATE UNIV, 77-, CHMN, CTR AGING RES EDUC & SERV, 80- *Concurrent Pos:* Clin assoc prof community med, Col Med, Univ Ky Med Ctr, 70-73; field dir, rural health res projs, Narangwal, India, Johns Hopkins Univ, 62-67; clin prof family pract, Sch Med, Univ Louisville, 73-75. *Mem:* Am Acad Family Physicians; Am Col Prev Med; fel Am Pub Health Asn; Am Geriat Soc; Geront Soc. *Res:* Consumer health education; geriatrics; self-care and health promotion for older persons. *Mailing Add:* Dept Community Med Wright State Univ Sch Med Box 927 Dayton OH 45401

ALTER, MILTON, b Buffalo, NY, Nov 11, 29; m 52; c 5. NEUROLOGY, EPIDEMIOLOGY. *Educ:* State Univ NY, Buffalo, BA, 51, MD, 55; Univ Minn, PhD, 64. *Prof Exp:* Mem staff epidemiol br, Nat Inst Neurol Dis & Blindness, 56-62; consult neurol, Vet Hosp, St Cloud & Ancker Hosp, St Paul, 64-65; consult neurologist, Vet Admin Hosp, Fargo, NDak & Faribault State Hosp, Minn, 65-67; chief neurol serv, Minneapolis Vet Admin Hosp, Minneapolis, 67-76; from assoc prof to prof neurol, Med Sch, Univ Minn, Minneapolis, 71-76; prof neurol, Temple Univ Hosp, 76-89, chmn dept, 76-87; PROF NEUROL, MED COL PA, 89- *Concurrent Pos:* USPHS fel, 62-64; Minn Asn Retarded Children grant, 63-64; Off Int Res grant, NIH, 63-66; consult, Honeywell, Inc, 63-72; USPHS grant, 65-68; Multiple Sclerosis Soc grants, 68, 71, 73 & 78; mem task force epidemiol of epilepsy, NIH, 70-73; ed, Neurol, 73-78 & J Neurol, 73-; mem adj comt, Nat Mult Sclerosis Soc; NIH grant neuroepidemiol, Safety Monitoring Comt, 85-; assoc ed, Neuroepidemiol, 83-89, ed-in-chief, 89- *Mem:* Am Acad Neurol; AMA; Soc Epidemiol Res; fel Am Neurol Asn; World Fedn Neurol. *Res:* epidemiology and genetics of neurological disorders, especially multiple sclerosis, myasthenia gravis and congenital malformations of the nervous system; dermatoglyphics of mental retardation; dementia, especially transmissable types, and stroke. *Mailing Add:* Med Col Pa 3300 Henry Ave Philadelphia PA 19129

ALTER, RALPH, b Toronto, Ont, July 26, 38; US citizen; div; c 2. COMPUTER COMMUNICATIONS, DISTRIBUTED COMPUTER SYSTEMS. *Educ:* Mass Inst Technol, SB & SM, 61, ScD(elec eng), 65. *Prof Exp:* Teaching asst elec eng, Mass Inst Technol, 59-61, instr, 61-65, mem staff comput systs, Lincoln Lab, 65-68; tech dir comput systs, Telcomp Div, Bolt Beranek & Newman Inc, 68-70 & Telcomp Corp Am, 70, sr comput scientist, Comput Systs Div, 70-72; dir & vpres opers, Packet Commun Inc, 72-75; network specialist, Boeing Comput Serv, Inc, 75-77; eng mgr, 77-81, SR COMPUT SCIENTIST, COMPUTER SYSTEM DIV, BOLT BERANEK & NEWMAN INC, 81- *Concurrent Pos:* Consult, H-W Electronics Inc, Mass, 61-64, Missile Systs Div, Raytheon Co, 64-65, Gen Tel Elec, Info Syst, 74, Mitre Corp, 74-75 & Bell-Norten Res, 74-75. *Mem:* Asn Comput Mach. *Res:* Computer and data communications networks; packet-switched network systems; distributed computer systems; multiprocessor computer systems. *Mailing Add:* 121 Worcester Lane Waltham MA 02154

ALTER, RONALD, b New York, NY, Mar 27, 39; m 63; c 3. MATHEMATICS, COMPUTER SCIENCE. *Educ:* City Col New York, BS, 60; Univ Pa, MA, 62, PhD(math), 65. *Prof Exp:* Asst prof math, Univ Calif, Los Angeles, 64-67; res fel, Syst Develop Corp, 67-68, assoc res scientist, Res & Technol Div, 68-69; asst prof math & computer sci, Univ Ky, 69-70, assoc prof computer sci, 70-83; RETIRED. *Concurrent Pos:* Math analyst, Litton Systs, Inc, 66; fac res fel, Oak Ridge Nat Lab, 75. *Mem:* Am Math Soc; Asn Comput Mach; Math Asn Am. *Res:* Analytic, algebraic and elementary number theory; combinatorial mathematics; graph theory; algebraic coding theory; computer applications to mathematics. *Mailing Add:* 521 N Broadway Lexington KY 40508

ALTER, THEODORE ROBERTS, b Toledo, Ohio, Sept 16, 46; m 68; c 2. PUBLIC FINANCE, RURAL DEVELOPMENT. *Educ:* Univ Rochester, BA, 68; Mich State Univ, MS, 73, PhD(resource econ), 76. *Prof Exp:* Asst prof, 76-82, assoc prof, 82-89, PROF, DEPT AGR ECON & RURAL SOCIOL, PA STATE UNIV, 89-, REGIONAL DIR SOUTHEAST, PA STATE COOP EXT, 89- *Concurrent Pos:* Chmn, Exten Comt Orgn & Policy Task Force on Local Govt Educ, 82-85; consult, Orgn Econ Coop & Develop, 87-89. *Honors & Awards:* Distinguished Exten Prog Award, Am Agr Econ Asn, 85. *Mem:* Am Agr Econ Asn; Asn Evolutionary Econ; Community Develop Soc. *Res:* The economics of community issues in agricultural and rural areas; comparative analysis of rural development policy. *Mailing Add:* Regional Dir Southeast Pa State Univ Coop Ext 401 Agr Admin Bldg University Park PA 16802

ALTER, WILLIAM A, III, b Flushing, NY, Sept 24, 42. CARDIOVASCULAR PHYSIOLOGY. *Educ:* Univ NMex, Alberquerque, PhD(physiol), 73. *Prof Exp:* Chief, Aerospace Med Div, 85-88, ASST PROF MIL MED, CHEM WARFARE DEFENSE DIV, RES & TECHNOL DIRECTORATE, BROOKS AFB, 78- *Concurrent Pos:* Adj assoc prof, Dept Physiol & res coordr, Sch Sci & Eng, Univ Tex, 88- *Mem:* Radiation Res Soc; Aerospace Med Asn; Fedn Am Socs Exp Biol. *Mailing Add:* Div Life Sci Univ Tex Sci Bldg Rm 10358B San Antonio TX 78249

ALTERA, KENNETH P, b Chicago, Ill, June 30, 36; m 63; c 2. VETERINARY PATHOLOGY. *Educ:* Univ Ill, BS, 58, DVM, 60; Univ Calif, PhD(comp path), 65. *Prof Exp:* Asst pathologist, Univ Calif, 64-65; asst prof vet path, 65-71, assoc prof vet path & vet pathologist, Colo State Univ, 71-72; pathologist, United Med Labs, 72-73; PATHOLOGIST, SYNTEX CORP, 73- *Mem:* Am Col Vet Path; Sigma Xi. *Res:* Comparative and experimental oncology; drug safety evaluation. *Mailing Add:* Syntex Corp 3401 Hillview Ave Palo Alto CA 94304

ALTES, RICHARD ALAN, b New York, NY, Aug 23, 41; m 69; c 2. MATHEMATICAL MODELS OF ANIMAL SENSORY SYSTEMS, NEURAL NETWORKS. *Educ:* Cornell Univ, BEE, 64, MEE, 65; Univ Rochester, MS, 66, PhD (elec eng), 71. *Prof Exp:* Res engr, Raytheon Res Div, 66-67; eng specialist, Electromagnetic Systs Lab, Inc, 70-76; sr scientist, Sci Applications Int Corp, 76-78; sr prin engr, Orincon Corp, 78-86; PRES, CHIRP CORP, 86- *Concurrent Pos:* Res assoc neurobiol, Scripps Inst Oceanog, Univ Calif, San Diego, 78-86, res engr, Ctr Res Lang, 86-88; res assoc, Hubbs-Sea World Res Inst, 80-86; adj prof, San Diego State Univ, 87- *Mem:* Inst Elec & Electronics Engrs; Acoust Soc Am; Int Neural Network Soc. *Res:* Radar/sonar signal processing and signal design; acoustic analysis; sensor fusion; pattern recognition; multidimensional signal processing; speech recognition; biomedical data processing (brain waves, medical ultrasound, tomography); models of animal sensory systems from a statistical viewpoint; applications and innovations with respect to neural networks. *Mailing Add:* 8248 Sugarman Dr La Jolla CA 92037

ALTEVEER, ROBERT JAN GEORGE, b Rheden, Neth, June 5, 35; US citizen; m 74; c 4. PHYSIOLOGY, BIOPHYSICS. *Educ:* Acad Phys Educ, Amsterdam, BPE, 57; Springfield Col, MS, 58; Univ Minn, PhD(educ, physiol), 65. *Prof Exp:* Instr physiol hyg, Univ Minn, 62-65; lectr physiol, Ind Univ, Bloomington, 65, asst prof physiol & med, 65-67; asst prof, 67-72, ASSOC PROF PHYSIOL, HAHNEMANN UNIV SCH MED, 72- *Concurrent Pos:* Adj asst prof, Univ Nev, Las Vegas, 66-67; vis prof bioeng, Rose Polytech Inst. *Honors & Awards:* Lindback Award. *Mem:* AAAS; Am Asn Univ Professors; Am Col Sports Med; Sigma Xi; Shock Soc. *Res:* Physiology of circulatory shock; environmental stress; biophysics of respiration and circulation; exercise physiology. *Mailing Add:* Dept Physiol & Biophys Hahnemann Univ Sch Med Philadelphia PA 19102

ALTEVOGT, RAYMOND FRED, b St Louis, Mo, Oct, 9, 40; m 69; c 2. ECONOMIC BOTANY, GENETICS. *Educ:* Wash Univ, AB, 63, PhD(bot), 70. *Prof Exp:* Asst prof biol, bot & math, Southern Ill Univ, Edwardsville, 70-71 & Lewis & Clark Community Col, 71-73; consult agr & germ plasm, Iranian Dept Environ, 74-75; lead analyst agr, World Food & Nutrit Study, Nat Acad Sci, 76; mem staff bot, Off Endangered Species, US Dept Interior, 76-77; asst dir & head, Agr Sect, Nat Food Processors Asn, 77-81; CONSULT, 81- *Mem:* Soc Econ Bot. *Res:* Germ plasm conservation; crop evolution; agricultural research policy and pesticide regulation. *Mailing Add:* 7049 Eastern Ave Silver Spring MD 20912

ALTGELT, KLAUS H, b Ger, Jan 30, 27; m 51; c 2. PHYSICAL CHEMISTRY, PETROLEUM CHEMISTRY. *Educ:* Univ Mainz, BS, 52, MS, 55, PhD(phys chem), 58. *Prof Exp:* Instr phys chem, Inst Phys Chem, Univ Mainz, 54-59; res assoc phys chem, Mass Inst Technol, 59-61; res chemist, Chevron Res Co, 61-65, sr res chemist, 65-68, sr res assoc, 68-86; CONSULT, 86- *Concurrent Pos:* Fel, Univ Mainz, 58-59; pres, Ger Sch San Francisco, 68-71. *Mem:* Am Chem Soc. *Res:* Polymeric fuel and oil additives in petroleum chemistry; chemical composition of asphalt, heavy petroleum fractions, and coal; compatibility in and between asphalts; thermodynamics of wax in fuels; mechanism and suppression of oil misting; mechanism of detergent action in combustion engines; stability of coal slurries; oxidation stability of lubricants in engines; deoiling industrial wastes. *Mailing Add:* Chevron Res Co 555 Appleberry Dr San Rafael CA 94903

ALTHAUS, RALPH ELWOOD, b Bluffton, Ohio, May 27, 25; m 49; c 3. PLANT PATHOLOGY. *Educ:* Bluffton Col, BS, 49; Ohio State Univ, MS, 51. *Prof Exp:* Asst, Boyce-Thompson Inst, 51-53; prod develop rep, B F Goodrich Chem Co, 53-57 & Merck & Co, Inc, 57-60; prod develop rep, 60-68, mgr prod develop dept, Agr Prod, 68-80, tech dir, Peoples Repub China, Monsanto Co, 80-85; AGR BUS CONSULT, 85- *Mem:* Weed Sci Soc Am. *Res:* Herbicides for agricultural and industrial vegetation control. *Mailing Add:* 11107 Oak Lake Ct Creve Coeur MO 63146

ALTHOEN, STEVEN CLARK, b Dayton, Ohio, Mar 3, 46; m 66; c 2. FINITE-DIMENSIONAL REAL DIVISION ALGEBRAS. *Educ:* Kenyon Col, BA, 69; City Univ NY, PhD(math), 73. *Prof Exp:* Prof math, Hofstra Univ, 73-75; PROF MATH, UNIV MICH, FLINT, 75- *Mem:* Am Math Soc; Math Asn Am. *Res:* Classification of non-associative, finite-dimensional real division algebras. *Mailing Add:* Dept Math Univ Mich Flint MI 48502-2186

ALTHUIS, THOMAS HENRY, b Kalamazoo, Mich, June 21, 41; m 67; c 3. ORGANIC & GENERAL CHEMISTRY, MEDICAL SCIENCE. *Educ:* Western Mich Univ, BS, 63, MA, 65; Mich State Univ, PhD(org chem), 68. *Prof Exp:* From res scientist to sr res scientist, Med Chem Res Dept, 68-78, MGR SCI AFFAIRS & DIR SCI POLICY AFFAIRS, PUB AFFAIRS DIV, PFIZER, INC, 80- *Concurrent Pos:* Am Chem Soc Cong Fel, Sci & Technol Comt, US House Rep, 78-79; mem, Technol Task Force, Coun Competitiveness, 87-88, Comt Chem & Pub Affairs, Am Chem Soc & bd dirs, Am Inst Chemists. *Mem:* Am Chem Soc; AAAS; NY Acad Sci; fel Am Inst Chemists; Soc Indust Chem. *Res:* Synthesis of nitrogen heterocycles; natural product synthesis; asymmetric synthesis; drugs for the treatment of allergies and pain; science and technology policy issues affecting research, development, approval and marketing of drugs; drugs for rare diseases; science education; US industrial competitiveness. *Mailing Add:* Pfizer Inc 235 E 42nd St New York NY 10017

ALTICK, PHILIP LEWIS, b Los Angeles, Calif, June 27, 33; m 55; c 4. ATOMIC PHYSICS. *Educ:* Stanford Univ, BS, 55; Univ Calif, Berkeley, MA, 60, PhD(atomic physics), 63. *Prof Exp:* Res physicist, Lawrence Radiation Lab, Univ Calif, Berkeley, 63; from asst prof to assoc prof, 63-75, chmn dept, 76-79 & 85-86, PROF PHYSICS, UNIV NEV, RENO, 75- *Concurrent Pos:* Consult, Lockheed Missile & Space Co, 63-; res assoc, Univ Chicago, 67-68; vis prof, Univ Trier-Kaiserslautern, 72-73; sr vis fel, Daresbury Lab & Univ Col, London, 81-82; prin investr, NASA G471 & NSF, 84- *Mem:* Fel Am Phys Soc. *Res:* Theoretical study of atomic processes, especially electron impact ionization. *Mailing Add:* Dept Physics Univ Nev Reno NV 89507

ALTIERI, A(NGELO) M(ICHAEL), b Watertown, Mass, Oct 11, 07; m 34; c 2. CHEMICAL ENGINEERING. *Educ:* Mass Inst Technol, SB, 29. *Prof Exp:* Res chemist, Dennison Mfg Co, 29-31; foreman, US Rubber Co, 31-32; from chem engr to asst res dir, Fitchburg Paper Co, 33-46; chief chemist & tech dir, Tileston & Hollingsworth Co, Diamond-Int, 46-70; engr, Fay, Spofford & Thorndike, 70-73; RETIRED. *Mem:* Tech Asn Pulp & Paper Indust; Am Inst Chemists. *Res:* Papermaking; de-inking and waterproofing processes; paper coating; printing; laminating; resins and adhesives; gelatin foils; latex thread; emulsions; design and operation of chemical processing equipment; water purification; waste water treatment. *Mailing Add:* 67 Copeland St Watertown MA 02172

ALTIERI, DARIO CARLO, b Milan, Italy, July 20, 58. CLINICAL HEMATOLOGY. *Educ:* Univ Milan, MD, 82. *Prof Exp:* Resident clin hemat, ABB Hemophilia & Thrombosis Ctr, Milan, 83-86; res fel, 87-89, ASST MEM, SCRIPPS CLIN, LA JOLLA, CALIF, 89. *Concurrent Pos:* Adj prof hemat, Univ Milan Sch Med, 90- *Mem:* Am Asn Pathologists; Am Asn Immunologists; AAAS. *Res:* Adhesion receptors in the immune system; role of proteases and protease receptors in the modulation of the immune response. *Mailing Add:* Dept Immunol Res Inst Scripps Clin 10666 N Torrey Pines Rd La Jolla CA 92037

ALTIERO, NICHOLAS JAMES, JR, b Youngstown, Ohio, Sept 22, 47; m 78; c 1. THEORETICAL MECHANICS, APPLIED MECHANICS. *Educ:* Univ Notre Dame, BS, 69; Univ Mich, MSE, 70, AM, 71, PhD(aerospace eng), 74. *Prof Exp:* Scholar aerospace eng, Univ Mich, 74-75; from asst prof to assoc prof, 75-86, PROF MECH, MICH STATE UNIV, 86-, ASSOC DEAN ENG, 90- *Concurrent Pos:* Consult mech engr, US Bur Mines; Fulbright scholar; Alexander von Humboldt fellow. *Mem:* Am Soc Mech Engrs; Soc Eng Sci; Int Soc Comput Methods Eng; Am Acad Mech; Am Soc Eng Educ; Int Soc Boundary Elements. *Res:* Numerical methods in mechanics; fracture mechanics; composite materials. *Mailing Add:* Col Eng Mich State Univ East Lansing MI 48824

ALTLAND, HENRY WOLF, b Takoma Park, Md, Feb 26, 45; m 68; c 1. ORGANIC CHEMISTRY, EXPERIMENTAL DESIGN. *Educ:* Gettysburg Col, AB, 67; Princeton Univ, PhD(org chem), 71, Rochester Inst Tech, MS, 87. *Prof Exp:* NIH fel, Univ Va, 71-73; sr res chemist, 73-81, RES ASSOC EASTMAN KODAK RES LABS, 81- *Concurrent Pos:* Asst instr eve exten, Univ Rochester, 77-80. *Mem:* Am Chem Soc; Am Soc Qual Control. *Res:* Synthesis and reactions of heterocyclic compounds, heterocyclic rearrangements; application of organometallic reactions to organic synthesis; application of experimental designs, (mixture taguchi central composite, etc) to chemical problems; emulsion addenda. *Mailing Add:* 547 Pinegrove Ave Rochester NY 14617

ALTLAND, PAUL DANIEL, b York, Pa, Apr 1, 13; m 44; c 3. ZOOLOGY, PHYSIOLOGY. *Educ:* Gettysburg Col, BS, 34; Duke Univ, AM, 36, PhD(cytol), 37. *Prof Exp:* Asst zool, Duke Univ, 34-36; from instr to asst prof biol, Gettysburg Col, 37-44; from asst physiologist to physiologist, 44-50, sr physiologist, 50-57, physiologist, 57-62, chief, Physiol Sect, Lab Phys Biol & Lab Chem Physics, 62-81, EMER SCIENTIST, NIH, USPHS, 81- *Mem:* Am Soc Zool; Soc Exp Biol & Med; Am Physiol Soc; Am Inst Biol Sci; Int Soc Biometeorol; Sigma Xi. *Res:* Physiology, pathology, and tolerance of animals exposed to simulated altitude, exercise and cold; reptile physiology; experimental endocarditis in animals; cigarette smoke, carbon monoxide, nicotine and restraint effects in rats. *Mailing Add:* 9319 W Parkhill Dr Bethesda MD 20814

ALTMAN, ALBERT, b New York, NY, Dec 11, 32; c 2. THEORETICAL NUCLEAR PHYSICS, SOLID STATE PHYSICS. *Educ:* Brooklyn Col, BS, 54; Univ Md, MS, 58, PhD(nuclear physics), 62. *Prof Exp:* Sr physicist, Gen Dynamics Convair, 60-61; asst prof nuclear theory, Univ Va, 62 & Univ Md, 63-64; physicist, Naval Ord Lab, 64-66; prof physics Lowell Technol Inst, 66-; AT DEPT PHYSICS, UNIV LOWELL. *Mem:* Sigma Xi. *Res:* Nuclear reactions and nuclear structure; quantum mechanics and scattering theory; molecular physics. *Mailing Add:* 124 Bellingham Rd Brookline MA 02167

ALTMAN, ALLEN BURCHARD, b Berkeley, Calif, Nov 16, 42; m 64. GEOMETRY, ALGEBRA. *Educ:* Columbia Univ, PhD(math), 68. *Prof Exp:* Actg asst prof math, Univ Calif, San Diego, 68-69, asst prof, 69-75; prof math, Univ Simon Bolivar, 75-86; PROF MATH, SIMON'S ROCK COL, 86- *Concurrent Pos:* Fulbright fel, 71-72; res assoc, Mass Inst Technol, 81-82; vis prof, Univ Nacional de Pernambuco, 85. *Mem:* AAAS; Am Math Soc; Math Asn Am. *Res:* Algebraic geometry. *Mailing Add:* Dept Math Simon's Rock Col Great Barrington MA 01230

ALTMAN, DAVID, b Paterson, NJ, Feb 13, 20; m 47; c 3. CHEMISTRY. *Educ:* Cornell Univ, AB, 40; Univ Calif, PhD(chem), 43. *Prof Exp:* Teaching asst chem, Univ Calif, 40-43; res assoc, Radiation Lab, Manhattan Proj, 43-45; chief chemist, Jet Propulsion Lab, 45-56; mgr vehicle technol, Aeronutronic Systs, Inc, 56-59; sr vpres chem systs div, 59-81, AEROSPACE CONSULT, UNITED TECHNOL CORP, 81- *Concurrent Pos:* Consult, Dept Defense, USAF & NASA. *Honors & Awards:* Propulsion Award, Am Inst Aeronaut & Astronaut, 64. *Mem:* Am Chem Soc; fel Am Inst Aeronaut & Astronaut; Aerospace Indust Asn Am; Sigma Xi. *Res:* High temperature; equilibrium and kinetics; thermodynamics; aerochemistry; surface chemistry; propulsion; missile systems. *Mailing Add:* 1670 Oak Ave Menlo Park CA 94025

ALTMAN, JACK, b Antwerp, Belg, Feb 17, 23; nat US; m 47; c 4. PLANT PATHOLOGY. *Educ:* Rutgers Univ, BSc, 54, PhD(plant path), 57. *Prof Exp:* From asst prof & asst plant pathologist to assoc prof & assoc plant pathologist, 57-70, PROF BOT & PLANT PATH & PLANT PATHOLOGIST, EXP STA, COLO STATE UNIV, 70- *Concurrent Pos:* Tech rep, Regional Res Proj W-56, 58-; mem Colo Gov Comt Air Pollution, 63-; Colo State Rep, Fed Mold Metabolites, 64-; NSF awards, 63-65; Indust Res grants, 64-68; Dept Health, Educ & Welfare Air Pollution Proj res award, 67-69; pest control adv & plant pathologist, Khuzestan Water & Power Authority, Iran, 70-72; Sr Scientist Award, Alexander von Humboldt Found, 77-78; partic, Distinguished Scholar Exchange Prog, People's Repub China, Nat Acad Sci, 81. *Mem:* Am Phytopath Soc; Soc Nematologists; Sigma Xi; Int Soc Plant Pathologists. *Res:* Associated nematode complex of root rot diseases of plants; use of antibiotics to control plant diseases; side effects of chlorinated hydrocarbons on nitrogenous transformations in soil; use of chemicals for Rhizoctonia control; turf diseases; changes in amino acids in soil following fumigation. *Mailing Add:* 1321 Springfield Dr Ft Collins CO 80521

ALTMAN, JOSEPH, b Budapest, Hungary, Oct 7, 25; m 50; c 1. NEUROPSYCHOLOGY. *Educ:* NY Univ, PhD(physiol psychol), 59. *Prof Exp:* Asst neurophysiol, Mt Sinai Hosp, NY, 57-59; res fel neuroanat, Col Physicians & Surg, Columbia Univ, 59-60; res assoc physiol psychol, Sch Med, NY Univ, 60-61; res assoc psychophysiol lab, Mass Inst Technol, 61-62, assoc prof psychol, 62-68; prof psychol sci, 74-80, PROF, DEPT BIOL SCI, PURDUE UNIV, 68- *Mem:* Soc Neurosci; Int Soc Develop Psychobiol. *Res:* Development of the nervous system and behavior under normal and experimental conditions. *Mailing Add:* Dept Biol Sci Purdue Univ Lafayette IN 47907

ALTMAN, JOSEPH HENRY, b Cambridge, Mass, Oct 11, 21; m 48; c 2. PHYSICS. *Educ:* Mass Inst Technol, BS, 42. *Prof Exp:* Physicist, Res Labs, Eastman Kodak Co, 46-59, res assoc, 59-70, lab head, 70-75, sr lab head, 75-78, sr res assoc, 78-85; RETIRED. *Concurrent Pos:* Part-time lectr, Univ Rochester, 68-; adj fac, Rochester Inst Technol, 85- *Honors & Awards:* Jour Award, Soc Motion Picture & TV Eng, 68. *Mem:* Fel Optical Soc Am; fel Soc Photog Sci & Eng. *Res:* Structures of optical and photographic images; microphotography; microminiaturization; properties of photographic materials; photographic sensitometry. *Mailing Add:* 2856 Elmwood Ave Rochester NY 14618

ALTMAN, KURT ISON, b Breslau, Ger, Oct 13, 19; US citizen; m 45; c 3. BIOCHEMISTRY, RADIOBIOLOGY. *Educ:* Univ Chicago, BS, 41, MS, 42; Univ Rochester, PhD(biochem), 63. *Prof Exp:* From res assoc to instr biochem, Univ Chicago, 44-57; res assoc radiation biol, 47-65, head div radiation chem, Atomic Energy Proj, 61-65, from asst prof to assoc prof biochem, 65-82, ASSOC PROF EXP RADIOL, SCH MED & DENT, UNIV ROCHESTER, 65- *Concurrent Pos:* Vis sr scientist, Ctr Study Nuclear Energy, Europ AEC, Mol & Geel, Belg, 64-65; vis prof physiol chem, Fac Med, Univ Düsseldorf, 73; mem adv bd, Advan in Radiation Biol; consult-in-residence, Sandoz Res Inst, Vienna, Austria, 79, consult, Univ Waterloo, Prog Aging, Waterloo, Can, 83-85; assoc ed, Radiation Res, 79-82. *Mem:* AAAS; Radiation Res Soc Am; Am Soc Biol Chemists; NY Acad Sci. *Res:* Enzymology; intermediary metabolism; isotope technology; biosynthesis of porphyrins and bile pigments; biochemical effects of ionizing radiations; physiological and radiological aging; connective tissue metabolism; bio-organic aspects of organic chemistry; erythrocyte metabolism; splenic function in hemolysis; antimycotic agents. *Mailing Add:* Dept Radiation Biol & Biophys Univ Rochester Sch Med & Dent 601 Elmwood Ave Rochester NY 14642

ALTMAN, LANDY B, JR, b Georgetown, SC, Mar 26, 19; m 49; c 1. AGRICULTURAL ENGINEERING. *Educ:* Univ NC, BS, 40; Univ Ga, MS, 47; Iowa State Univ, PhD(eng), 60. *Prof Exp:* Engr, Rural Electrification Admin, USDA, 40-48, res engr curing tobacco, Agr Res Serv, 48-50, res engr elec demand, 50-67; asst chief, Farm Electrification Res Br, 66-68, chief, 68-77; ENERGY RES COORDR, AGR RES SERV CTR, USDA, 77- *Mem:* Am Soc Agr Engrs; Inst Elec & Electronics Engrs. *Res:* Prediction of electrical demands of farms for sizing distribution transformers, services, and farm wiring circuits. *Mailing Add:* 11904 Renick Lane Silver Spring MD 20904

ALTMAN, LAWRENCE JAY, b Chicago, Ill, Nov 12, 41; m 65; c 2. ORGANIC CHEMISTRY. *Educ:* Calif Inst Technol, BS, 62; Columbia Univ, PhD(org chem), 65. *Prof Exp:* Air Force Off Sci Res fel photochem, Calif Inst Technol, 65-66; asst prof org chem, Stanford Univ, 66-72; from assoc prof to prof chem, State Univ NY, Stony Brook, 72-80; GROUP LEADER, MOBIL RES & DEVELOP CORP, 80- *Concurrent Pos:* Grants, Res Corp, 66-67 & 70-71, Am Chem Soc, 66-71, Lilly Res Found, 67-69 & NIH, 72-80; A P Sloan Found fel, 71-75; NSF fel, 75-80. *Mem:* NY Acad Sci; Am Chem Soc. *Res:* Process analysis; near infrared spectroscopy; petroleum process and chemistry; catalysis. *Mailing Add:* Mobil Res & Develop Corp PO Box 480 Paulsboro NJ 08066-0480

ALTMAN, LAWRENCE KIMBALL, b Quincy, Mass, June 19, 37. MEDICAL JOURNALISM. *Educ:* Harvard Univ, BA, 58; Tufts Univ, MD, 62; Am Vet Epidemiol Soc, dipl. *Prof Exp:* Intern, Mt Zion Hosp, San Francisco, 62-63, epidemic intelligence serv off, USPHS, 63-66, consult US aid, dept state, Pub Health Serv, measles immunization program, WAfrica & resident internal med, Univ Washington Affil Hosps, 66-68, sr fel, dept med & med genetics, 68-69; MED CORRESP, NY TIMES, 69- *Concurrent Pos:* Sr asst med, Univ Washington, Seattle, 66-69; attending physician & clin assoc prof med, NY Univ, Bellevue Med Ctr, 70-; vis scientist, Univ Wash, 71; vis physician, Serafimer Hosp, Karolinska Inst, 73; bd dir, Josiah Macy Jr Found, NY, 84. *Honors & Awards:* George Polk Award, 86; Claude Bernard Award, Nat Soc Med Res, 71, 74; Howard Blekeslee Award, Am Heart Asn, 82-83. *Mem:* Inst Med-Nat Acad Sci; Fel NY Acad Sci; fel Am Col Epidemiol; fel Am Col Physicians; Am Soc Trop Med & Hyg. *Mailing Add:* 140 W End Ave Apt 18-G New York NY 10023

ALTMAN, LEONARD CHARLES, b Fresno, Calif, Sept 1, 44; m 70; c 3. IMMUNOLOGY, ALLERGY. *Educ:* Univ Pa, BA, 65; Harvard Med Sch, MD, 69. *Prof Exp:* From intern to resident med, Univ Wash Affil Hosps, 69-71, chief resident, Harborview Med Ctr, 74-75; asst prof, 75-78, assoc prof med, Dept Med, Div Allergy & Infectious Dis, 78-88, affil assoc prof oral biol, 85-88, CLIN PROF MED & AFFIL PROF ORAL BIOL & ENVIRON HEALTH, UNIV WASH, 88- *Concurrent Pos:* From res assoc to sr res assoc, Lab Microbiol & Immunol, Nat Inst Dent Health, NIH, 71-74. *Mem:* Am Asn Immunologists; fel Am Acad Allergy; fel Am Bd Internal Med; Am Fedn Clin Res. *Res:* Host defense mechanisms with emphasis on white cell function, leukocyte chomotaxis; leukocyte function in allergic and infectious diseases; inflammatory aspects of allergic diseases. *Mailing Add:* Div Allergy & Infectious Dis Dept Med Univ Wash RM-13 Seattle WA 98195

ALTMAN, PHILIP LAWRENCE, b Kansas City, Mo, Jan 6, 24; m 46; c 2. SCIENCE COMMMUNICATIONS, ZOOLOGY. *Educ:* Univ Southern Calif, AB, 48; Western Reserve Univ, MS, 49. *Prof Exp:* Technician histopath, Univ Kans Med Ctr, 51; biologist mycol, Kansas City Field Sta, USPHS, 52-54; biologist virol, NIH, 55-56; res analyst, Nat Acad Sci, 57-59; dir off biol handbks, 59-79, sr staff scientist, Life Sci Res Off, Fedn Am Socs Exp Biol, 79-86, exec ed, Biol Data Book Ser, 83-86, exec dir, Coun Biol Ed, 81-89; RETIRED. *Concurrent Pos:* Int Union Biol Sci del, Comt on Data for Sci & Technol, 70-78, chmn adv panel biosci, 73-78, chmn task group presentation biol data primary lit, 75-78. *Honors & Awards:* Cert of Appreciation, Coun Biol Ed, 84, Meritorious Award, 91. *Mem:* Coun Biol Ed. *Res:* Comparative vertebrate anatomy; histoplasmosis; virus crystallization; analysis, editing and compilation of data in biological and medical sciences; information retrieval in life sciences. *Mailing Add:* 9206 Ewing Dr Bethesda MD 20817-3314

ALTMAN, ROBERT LEON, b Brooklyn, NY, Apr 27, 31; m 58; c 2. MATERIALS SCIENCE ENGINEERING. *Educ:* NY Univ, BA, 52; Univ Southern Calif, PhD(physical chemistry), 59. *Prof Exp:* Sr res engr, Rocketdyne Div, NAm Aviation, Inc, 56-59; fel phys chem, Univ Calif, Berkeley, 59-60, chemist, Lawrence Radiation Lab, 60-62; asst prof, Calif State Univ, Hayward, 62-66; RES SCIENTIST, NASA AMES RES CTR, 66- *Concurrent Pos:* Vis asst prof, Univ Calif, Berkeley, 64. *Mem:* Am Chem Soc; Combustion Inst. *Res:* Protein chemistry; calculation of thermodynamic properties; high temperature chemistry; improvement of fire extinguishment by use of dry chemical powders; polymer research for development of anti-misting kerosene to reduce incidence of post-crash fires; fuel technology and petroleum engineering; development of emissive coatings stable in high temperature airflow. *Mailing Add:* NASA Ames Res Ctr Moffett Field CA 94035

ALTMAN, SAMUEL PINOVER, b Atlantic City, NJ, Apr 15, 21; m 43; c 2. SPACE SCIENCE, CONTROL SYSTEMS. *Educ:* City Col New York, BME, 42. *Prof Exp:* Mech engr ord prod, Rochester Ord Dist, NY, 42; ord shop officer ord maintenance, 206th Ord Co, Ft Leonard Wood, Mo, 42-43; design & develop officer aircraft avionics, Eng Div, Army Air Corps, Wright Field, Dayton, Ohio, 43-46; develop engr tractors, Wilkes-Barre Carriage Co, Philadelphia, Pa, 46; mech engr aircraft flight control systs, Eng Div, Wright Air Develop Ctr, Wright-Patterson AFB, Ohio, 46-50, electronic scientist mil aircraft testing & eval, All-Weather Flying Div, 50-51, design & develop officer, 51-53; sr proj engr aircraft flight control systs, Eng Div, Lear Inc, Grand Rapids, Mich, 53-54; res specialist missile systs, Res & Develop Div, Lockheed Missile Systs, Van Nuys, Calif, 54; sr engr Ramjet engines, Tech Div, Marquardt Aircraft Co, Van Nuys, Calif, 54-55; sr systs anal missile systs, Missile Sect, Bendix Prod, Mishawaka, Ind, 55-56; sr scientist aerospace systs, Elec Eng Res Div, Armour Res Found, Chicago, Ill, 56-57; sr systs engr ICBM defense systs, Tech Develop Sect, Bendix Prod, South Bend, Ind, 57-58; prin engr missile systs, Martin Co, Waterton, Colo, 58-61; supvr space systs & sci, United Aircraft Corp Systs Ctr, Farmington, Conn, 61-63; consult engr spacecraft systs, Missile & Space Div, Gen Elec, King of Prussia, Pa, 63-69; chief systs scientist aerospace systs seismol, State-Space Systs Co, Norristown, Pa, 69; sr staff engr mil aerospace systs, Syst Develop Corp, Santa Monica, Calif, 69-72; dir space mech, Commun Res Ctr, Can Fed Govt, Ottawa, Ont, 72-85; vpres eng, Adv Beam Control Inc, Woodbridge, Ont & Dir Govt Rels, Proj Res Group Ltd, Woodbridge, Ont, 85-86; PRES OPTILOGIC INC, OTTAWA, ONT, 86-; MKT DIR, DYNACON ENTERPRISES LTD, DOWNSVIEW, ONT, 86- *Mem:* Am Astronaut Soc; Am Inst Aeronaut & Astronaut; Can Aeronaut & Space Inst; Inst Elec & Electronics Engrs. *Res:* Aircraft, missile and space research and development with specializations in astrodynamics; spacecraft control and stabilization; missile guidance and control; aircraft avionics; aircraft flight dynamics; missile and aircraft intercept dynamics. *Mailing Add:* 465 Richmond Rd No 2107 Ottawa ON K2A 1Z1 Can

ALTMAN, SIDNEY, b Montreal, Que, May 7, 39. MOLECULAR BIOLOGY. *Educ:* Mass Inst Technol, BS, 60; Univ Colo, PhD(biophys), 67. *Prof Exp:* Teaching asst, Columbia Univ, 60-62; Damon Runyon Mem Fund Cancer Res fel molecular biol, Harvard Univ, 67-69; Anna Fuller Fund fel, Med Res Coun Lab Molecular Biol, 69-70, med res coun fel, 69-71; from asst prof to assoc prof, 71-80, chmn dept, 83-85, dean, 85-89, PROF BIOL, YALE UNIV, 80- *Concurrent Pos:* Tutor biol, Radcliffe Col, 68-69. *Honors & Awards:* Nobel Prize in Chem, 89; Rosentiel Award for Basic Biomed Res, 89. *Mem:* Nat Acad Sci; Am Soc Biol Chemists; Genetics Soc Am; fel AAAS. *Res:* Effects of acridines on T4 DNA replication; mutants; precursors of tRNA; RNA processing by catalytic RNA and ribonuclease function. *Mailing Add:* Dept Biol Yale Univ New Haven CT 06520

ALTMANN, STUART ALLEN, b St Louis, Mo, June 8, 30; m 59; c 2. ANIMAL BEHAVIOR. *Educ:* Univ Calif, Los Angeles, BA, 53, MA, 54; Harvard Univ, PhD(biol), 60. *Prof Exp:* Biologist, NIH, 56-58; from asst prof to assoc prof zool, Univ Alta, 60-65; sociobiologist, Yerkes Regional Primate Res Ctr, 65-70; prof anat, 70-80, PROF BIOL, DEPT ECOL & EVOLUTION, UNIV CHICAGO, 70- *Concurrent Pos:* Res grants, Nat Res Coun Can, 61-63, NIMH, 62-64 & 70-82 & NSF, 63-65 & 69-; mem, Comt Evolutionary Biol, Univ Chicago, 70- *Mem:* Fel AAAS; Am Soc Primatology; fel Animal Behav Soc; Int Primatology Soc; Am Soc Naturalists. *Res:* Behavioral ecology of animals; field studies of primate behavior; foraging behavior and nutrition. *Mailing Add:* 940 E 57th St Univ Chicago Chicago IL 60637

ALTMILLER, DALE HENRY, b Shattuck, Okla, Sept 11, 40; m 64; c 2. CLINICAL CHEMISTRY, MEDICAL GENETICS. *Educ:* WTex State Univ, BS, 63; Univ Tex, PhD(genetics), 69; Am Bd Clin Chem, cert, 79. *Prof Exp:* Asst prof human genetics, Univ Tex Med Br, 69-75; asst prof path, 75-79, ASSOC PROF PATH, COL MED, UNIV OKLA, 79-; dir, clin endocrinol & toxicol serv lab, 75-82,; DIR, GENETICS & METABOLIC CHEM SERV LAB, CHILDREN'S HOSP OKLA, 82- *Concurrent Pos:* Res grants, NSF, 70-72 & Robert A Welch Found, 71-74. *Mem:* Am Asn Clin Chem; Am Soc Human Genetics; Sigma Xi; Am Chem Soc; Clin Ligand Assay Soc. *Res:* Development of new molecular diagnostic; molecular basis of inherited disorders; human gene characterization and mapping; genetic factors in cancer. *Mailing Add:* Childrens Hosp PO Box 26307 Oklahoma City OK 73126

ALTMILLER, HENRY, b Ft Smith, Ark, July 13, 41; c 2. PHYSICAL CHEMISTRY, RADIATION CHEMISTRY. *Educ:* Univ Notre Dame, BS, 64, PhD(chem), 69. *Prof Exp:* Teacher, Holy Cross High Sch, Tex, 64-65; from asst prof to assoc prof, 69-82, exec vpres, 78-82, acad dean, 74-82, PROF CHEM, ST EDWARD'S UNIV, 82- *Mem:* Am Chem Soc; Sigma Xi. *Res:* Energy transfer; production of excited species in photo and radiation chemistry. *Mailing Add:* Dept Chem St Edward's Univ Austin TX 78704

ALTNER, PETER CHRISTIAN, b Starnberg, Ger, April 19, 32; m 59; c 3. ORTHOPAEDIC SURGERY. *Educ:* Univ Wurzburg, Ger, Physicum, 54; Univ Kiel, Ger, MD, 57. *Prof Exp:* Intern, Univ Kiel, Ger, 57-58 & Muhlenberg Hosp, NJ, 58-59; resident, Evangelic Luthern Diakonissenanstalt, 60-61, Borgess Hosp, Mich, 61-62 & Univ Chicago Hosps & Clins, 62-66; PROF & CHIEF, DIV ORTHOP SURG, CHICAGO MED SCH, UNIV HEALTH SCI, 73-, VET ADMIN MED CTR, NORTH CHICAGO, 74- *Concurrent Pos:* Prof orthop surg, Cook County Grad Sch Med, Chicago, 67-; lectr, Dept Orthop Surg, Northwestern Univ, 68-; clin prof, Ill Col Podiatric Med, 79- *Mem:* Am Acad Orthopaed Surgeons; Am Col Surgeons; Int Col Surgeons; Am Med Asn. *Res:* Clinical appraisal of amputees, prosthetics and rehabilitation; author or coauthor of numerous publications. *Mailing Add:* Dept Orhtopaedic Surg Vet Admin Med Ctr Middleville Rd Northport NY 11768

ALTON, ALVIN JOHN, b Linden, Wis, Mar 31, 13; m 40; c 1. FOOD SCIENCE. *Educ:* Univ Wis, BS, 36. *Prof Exp:* Asst dairy indust, Univ Wis, 36-37; food chemist, Beatrice Creamery Co, Chicago, 36-48; dir prod res, Taylor Freezer, Wis, 48-49; dir prod res, Beatrice Foods Co, 49-65, gen mgr, 65-76, group mgr, 76-78; RETIRED. *Mem:* Am Dairy Sci Asn; Inst Food Technol. *Res:* Dairy and allied food products. *Mailing Add:* 18219 Gardenview Dr Sun City West AZ 85375

ALTON, DONALD ALVIN, b Chicago, Ill, Sept 28, 43; div; c 4. PARALLEL GRAPH PROCESSING, DESIGN ANALYSIS ALGORITHMS. *Educ:* Rice Univ, BA, 65; Cornell Univ, PhD(math), 70. *Prof Exp:* Asst prof math, 70-71, from asst prof to assoc prof, 71-81, PROF COMPUT SCI, UNIV IOWA, 81- *Concurrent Pos:* NSF grants, 72-; Fulbright-Hays lectureship, Novosibirsk State Univ, USSR, 79; fac scholar, Univ Iowa, 80- *Mem:* Asn Comput Mach; Am Math Soc; Asn Symbolic Logic; Inst Elec & Electronics Engrs. *Res:* Design and analysis of efficient algorithms; parallel graph processing; computational complexity; recursion-theoretic computational complexity in the spirit of work of Manuyel Blum and others; recursion theory. *Mailing Add:* Dept Comput Sci Univ Iowa Iowa City IA 52242

ALTON, EARL ROBERT, b Oelwein, Iowa, Oct 28, 33; m 78; c 3. INORGANIC CHEMISTRY, GENERAL CHEMISTRY. *Educ:* St Olaf Col, BA, 55; Univ Mich, MS, 58, PhD(chem), 61. *Prof Exp:* From asst prof to assoc prof, Augsburg Col, 60-69, chmn, Nat Sci Div, 72-83, actg dean, 81 & 85, PROF CHEM, AUGSBURG COL, 69-, DIR GEN EDUC, 89- *Concurrent Pos:* NSF sci fac fel, 67-68; vis instr, 3M Co, 69-71, 77-78, 84-85 & 87- *Mem:* AAAS; Am Chem Soc; Sigma Xi. *Res:* Chemistry of phosphorus trifluoride, especially reactions with group II and III halides and with transition metal halides; inorganic complexes; kinetics and bonding; chemistry of coordination compounds and of hydrides. *Mailing Add:* Dept Chem Augsburg Col Minneapolis MN 55454

ALTON, ELAINE VIVIAN, b Watertown, NY, Aug 30, 25. MATHEMATICS EDUCATION. *Educ:* State Univ NY Albany, BA, 42; St Lawrence Univ, MEd, 51; Univ Mich, MA, 58; Mich State Univ, PhD(math educ), 65. *Prof Exp:* Teacher math, Cobleskill High Sch & Fultonville High Sch, 46-48; assoc prof math, Ferris State Col, 48-62; asst instr, Mich State Univ, 62-64; asst prof, 64-71, assoc prof, 71-77, PROF MATH EDUC, IND UNIV-PURDUE UNIV, INDIANAPOLIS, 77- *Concurrent Pos:* Mem, Adv Comt Math, State Ind, 68-; NSF proposal panelist, Coop Col-Sch Proj, 71. *Mem:* Nat Coun Teachers Math; Math Asn Am; Asn Teacher Educrs; Am Math Asn Two Yr Cols. *Res:* Methods and materials for teaching the metric system; commercial and teacher made materials for teaching concepts and computational skills to a wide range of abilities; changing teacher attitudes towards mathematics; remedial algebra; reasons for failure in college calculus. *Mailing Add:* Dept Math Sci Ind Univ-Purdue Univ Indianapolis IN 46223

ALTON, EVERETT DONALD, b Frederika, Iowa, Jan 8, 15; m 41; c 2. ELECTRICAL ENGINEERING. *Educ:* Iowa State Teachers Col, BA, 39; Univ Iowa, MSEE, 55. *Prof Exp:* Teacher pub sch, Iowa, 47-54; from instr to assoc prof elec eng, Univ Iowa, 55-68, assoc chmn dept, 68-81, supvr, Hybrid Comput Lab, 71-81; RETIRED. *Concurrent Pos:* Mem, Simulations Coun & rev comt, Electronics & Commun in US, Inst Elec & Electronics Eng & NSF, 64-; Univ Iowa rep, Am Power Conf, annually. *Mem:* Inst Elec & Electronics Engrs; Am Soc Eng Educ. *Res:* Nuclear reactor control; phototransistors at the temperature of liquid nitrogen; analog computer simulation and computation. *Mailing Add:* 4410 Eng Bldg Dept Elec Eng Univ Iowa Iowa City IA 52242

ALTON, GERALD DODD, b Murray, Ky. EXPERIMENTAL ATOMIC PHYSICS, THEORETICAL ATOMIC PHYSICS. *Educ:* Murray State Univ, Ky, BS, 60; Univ Tenn, Knoxville, MS, 67, PhD(physics), 72. *Prof Exp:* Develop physicist atomic physics, Isotopes Div, 60-72, RES STAFF PHYSICIST EXP & THEORET ATOMIC PHYSICS, PHYSICS DIV, OAK RIDGE NAT LAB, 73- *Concurrent Pos:* Consult, GCA Corp, Sunnyvale, Calif, 74- *Res:* Negative and multiply charged positive ion source research and development; accelerating systems and theoretical and experimental atomic physics. *Mailing Add:* Rte 2 Box 233E Kingston TN 37767

ALTOSE, MURRAY D, b Winnipeg, Can, Oct 1, 41; US citizen; m 73; c 3. PULMONARY DISEASES. *Educ:* Univ Man, BSc & MD, 65; FRCP(C), 72; Am Bd Internal Med, cert internal med, 74, cert pulmonary dis, 76. *Prof Exp:* Fel pulmonary, Univ Pa, 71-73, asst prof med, 73-77; assoc prof, 77-84, PROF MED, CASE WESTERN RESERVE UNIV, 84-; CHIEF STAFF, CLEVELAND VET ADMIN MED CTR, 88- *Concurrent Pos:* Chief, Cleveland Metro Gen Hosp, 77-88; assoc dean, Case Western Reserve Univ, 88-; chmn, Sect Respiratory Neurobiol & Sleep, Am Thoracic Soc, 89. *Mem:* Am Fedn Clin Res; Am Physiol Soc; Cent Soc Clin Res; Am Thoracic Soc; Am Col Chest Physicians. *Res:* Effects of aging, lung disease and neuromuscular diseases on the reflex and behavioral control of breathing. *Mailing Add:* Vet Admin Med Ctr 10701 East Blvd Cleveland OH 44106

ALTPETER, LAWRENCE L, JR, b New York, NY, Nov 4, 32. TECHNICAL MANAGEMENT, CHEMISTRY. *Educ:* Manhattan Univ, BS, 54; Iowa State Univ, MS, 59; Tex A&M Univ, PhD(phys chem), 67. *Prof Exp:* Sr cehmist, Honeywell Corp Res Ctr, 67-71 & N Star Res & Develop Inst, 71-76; prin chemist, Midwest Res Inst, 76-78; anal lab mgr, Environ Sci & Eng Co, 78-81; proj mgr, 81-84, sr proj mgr, 84-86, PROG MGR SAFETY INSTRUMENTATION, GAS RES INST, 86- *Concurrent Pos:* Welch fel, 63-65. *Mem:* Am Chem Soc. *Res:* Development of trace analytical instrumentation. *Mailing Add:* 503 W Brittany Dr Arlington Heights IL 60004

ALTROCK, RICHARD CHARLES, b Omaha, Nebr, Dec 20, 40; m 79; c 4. SOLAR PHYSICS. *Educ:* Univ Nebr, BSc, 62; Univ Colo, PhD(astro-geophys), 68. *Prof Exp:* ASTROPHYSICIST, AIR FORCE SYSTS COMMAND GEOPHYS LAB, NAT SOLAR OBSERV, 67- *Concurrent Pos:* Vis res fel, Dept Appl Math, Univ Sydney, 71-72; guest investr, NASA Solar Maximum Mission Cornagraph/Polarimeter, 84-86; res adv, Res Assoc Prog, Nat Res Coun, 80-; co-investr, Orbiting Solar Lab, 81-; proj mgr, Solar Mass Ejection Imager, 86- *Mem:* Fel AAAS; Am Astron Soc; Sigma Xi; Int Astron Union; Am Geophys Union. *Res:* Global variations in solar physical conditions; solar effects on the geomagnetic field; solar corona. *Mailing Add:* Sacramento Peak Observ Sunspot NM 88349

ALTSCHER, SIEGFRIED, b Berlin, Germany, July 18, 22; US citizen; m 46; c 3. POLYMER CHEMISTRY, ORGANIC CHEMISTRY. *Educ:* Brooklyn Col, BA, 42; Polytech Inst Brooklyn, MS, 50, PhD(chem), 65. *Prof Exp:* Prod supvr fine chem, Wallace & Tiernan Prod Inc, 46-54; group leader res & develop, Nopco Chem Co, 54-66; mgr, Polymer & Anal Dept, 66-79, asst dir, Eastern Res Ctr, 79-84, DIR, RES ADMIN, STAUFFER CHEM CO, 84- *Mem:* Am Chem Soc; Fedn Socs Paint Technol; Soc Plastics Engrs. *Res:* Development of commercial specialty and surfactant chemicals; basic studies of polymer and plastic properties to develop new compositions which satisfy newer commercial requirements of improved performance. *Mailing Add:* 81 Ludvigh Rd Bardonia NY 10954

ALTSCHUL, AARON MAYER, b Chicago, Ill, Mar 13, 14; m 37; c 2. NUTRITION. *Educ:* Univ Chicago, BS, 34, PhD(chem), 37. *Hon Degrees:* DSc, Tulane Univ, 68. *Prof Exp:* Instr chem, Univ Chicago, 37-41; biochemist, Southern Regional Res Lab, USDA, 41-52, head oil seed sect, Agr Res Serv, 52-58, chief chemist, Seed Protein Pioneering Res Lab, 58-67, spec asst int nutrit improv, 67-69, Off Secy, 69-71; prof community & family med, 71-83, head div nutrit, 75-83, dir, Diet Mgt & Eating Disorders Prog, 78-87, EMER PROF, DEPT MED & COMMUNITY & FAMILY MED, GEORGETOWN UNIV, 83-, EMER DIR, 87- *Concurrent Pos:* Res consult, Tulane Univ, 43-58, consult prof, 58-64, prof, 64-66; consult food & nutrit policy, Nat Coun Res & Develop, Israel, 71-76; mem, Select Comt Substances Gen Recognized As Safe, 72-73 & Fedn Am Socs Exp Biol Select Comt Flavor Eval Criteria, 75-; consult, UN Agencies, 51-72. *Honors & Awards:* Charles Spencer Award, 65; Int Award, Inst Food Technologists, 70; Distinguished Serv Award, USDA, 70. *Mem:* Am Chem Soc; Am Soc Biochem & Molecular Biol; Inst Food Technologists; Am Inst Nutrit; Am Soc Clin Nutrit; Am Col Nutrit. *Res:* Cytochrome c peroxidase; seed enzymes and proteins; food utilization of plant proteins; fortification of foods; community nutrition; obesity. *Mailing Add:* 700 New Hampshire Ave NW Washington DC 20037

ALTSCHUL, ROLF, b Duesseldorf, Ger, Jan 24, 18; nat US; m 55; c 2. PHYSICAL ORGANIC CHEMISTRY. *Educ:* Harvard Univ, AM, 39, PhD(chem), 41. *Prof Exp:* Fel, Harvard Univ, 41, Pittsburgh Plate Glass fel, 41-44; lectr, Bryn Mawr Col, 44-45; CHMN DEPT CHEM, SARAH LAWRENCE COL, 45- *Concurrent Pos:* Res assoc, Brandeis Univ, 54- *Mem:* Am Chem Soc. *Res:* Kinetics and mechanisms of liquid phase reactions. *Mailing Add:* 405 E 56th St Apt 6D New York NY 10022

ALTSCHULE, MARK DAVID, b New York, NY, July 16, 06; m 34. MEDICINE. *Educ:* City Col New York, BS, 27; Harvard Univ, MD, 33; Am Bd Internal Med, dipl, 42. *Prof Exp:* Technician, Montefiore Hosp, NY, 26-28; intern, Peter Bent Brigham Hosp, Boston, 32; intern med, Beth Israel Hosp, 32-34; res fel, Harvard Med Sch, 34-35, from asst to asst prof med, 35-52, from asst prof to prof clin med, 52-73; dir internal med & res clin physiol, McLean Hosp. Belmont, Mass, 47-68; consult clin physiologist, McLean Hosp, Belmont, Mass, 68-78; vis prof med, 73-78, Harvard Med Sch, hon cur & photog collections, Francis A Countway Libr Med, 70-78; RETIRED. *Concurrent Pos:* Resident med, Beth Israel Hosp, 34-36, assoc, 37-43 & 46-66, actg dir med res, 43-46, assoc vis physician, 38-46, vis physician, 46-66; consult internist, McLean Hosp, 45-47; attend physician, Vet Admin Hosp, Boston, 53-; assoc, Thorndike Lab, staff consult & vis physician, Boston City Hosp 55-74; ed-in-chief, Lippincott's Med Sci, 59-68 & Med Counterpoint, 69-73; consult, Childrens Hosp Med Ctr, Boston, 64-67, Naval Blood Res Lab, Chelsea, Mass, 68-75 & Yale-New Haven Hosp, Conn, 70-76; lectr med, Sch Med, Yale Univ, 66-76; biomed consult, Off Naval Res, 76-82; pres, Tott's Gae Inst, 80- Int Catacomb Soc, 83- *Mem:* Emer mem Asn Am Physicians; Am Heart Asn; emer mem NY Acad Sci; Am Col Clin Pharmacol & Chemother; emer mem Am Soc Clin Invest. *Res:* Physiology of mental disease; clinical significance of cardiac and respiratory adjustments in chronic anemia. *Mailing Add:* 43 Brown Rd Harvard MA 01451

ALTSCHULER, HELMUT MARTIN, b Mannheim, Ger, Feb 13, 22; US citizen; m 42; c 3. ELECTROPHYSICS. *Educ:* Polytech Inst Brooklyn, BEE, 47, MEE, 49, PhD(electrophys), 63. *Prof Exp:* Res asst, Microwave Res Inst, Polytech Inst Brooklyn, 48-51, jr res assoc, 51-52, sr res assoc, Electrophys Dept, 52-63, res assoc prof, 63-64; chief radio standards eng div, Nat Bur Standards, 64-69, sr res scientist, Electromagnetics Div, 69-77, sr standards specialist, Off Domestic & Int Measurement Standards, 77-; CONSULT. *Concurrent Pos:* Mem US nat comt, Int Union Radio Sci, chmn comn A. *Mem:* Fel Inst Elec & Electronics Engrs. *Res:* Impedance measurement techniques; equivalent network representations; strip transmission lines; electromagnetic and transmission line and modal theory; dielectric constant measurement techniques; microwave bridge measurement methods; nonreciprocal and active two-ports; electromagnetic standards; measurements, standards and laboratories. *Mailing Add:* 2250 Bluebell Ave Boulder CO 80302

ALTSCHULER, MARTIN DAVID, b Brooklyn, NY, Feb 25, 40; m 62; c 3. RADIOLOGY, RADIATION THERAPY. *Educ:* Polytech Inst Brooklyn, BS, 60; Yale Univ, MS, 61, PhD(astron), 64. *Prof Exp:* Instr astron, Yale Univ, 64-65; solar physicist, High Altitude Observ, Nat Ctr Atmospheric Res, 65-76; res assoc prof, Dept Comput Sci, State Univ NY Buffalo, 76-80; RES ASSOC PROF, DEPT RADIATION THERAPY, HOSP UNIV PENN, 80- *Concurrent Pos:* Adj assoc prof astrogeophys, Univ Colo, 74-; mem comts 10 & 12, Int Astron Union. *Mem:* Am Phys Soc; NY Acad Sci; Inst Elec & Electronics Engrs. Am Astron Soc; Am Asn Physicists in Med; Sigma Xi; Soc Med Decision Making. *Res:* Solar magnetic fields; structure and dynamics of solar corona; medical image reconstruction; computer tomography; surface imaging; 3-D radiation therapy treatment planning, medical decisions; anthropometry. *Mailing Add:* 3401 Walnut St Rm 440A Dept Radiation Therapy HUP Philadelphia PA 19104-6228

ALTSHULER, BERNARD, b Newark, NJ, June 22, 19; m 44; c 2. BIOMATHEMATICS, ENVIRONMENTAL HEALTH. *Educ:* Lehigh Univ, BS, 40; NY Univ, PhD(math), 53. *Prof Exp:* Asst math, NY Univ Med Ctr, 49-51, from instr to assoc prof, 51-69, assoc dir, Inst Environ Med, 75-86, PROF ENVIRON MED, NY UNIV MED CTR, 69- *Mem:* Am Math Soc; Am Phys Soc; Soc Risk Anal; Soc Indust & Appl Math. *Res:* Environmental health; pulmonary handling of aerosols; quantitative aspects of carcinogenesis. *Mailing Add:* Dept Environ Med NY Univ Med Ctr New York NY 10016

ALTSHULER, CHARLES HASKELL, b Detroit, Mich, Apr 14, 19; m 61; c 2. PATHOLOGY. *Educ:* Univ Mich, BS, 39, MD, 43. *Prof Exp:* Intern, Mt Sinai Hosp, New York, NY, 43-44; instr bact, Univ Mich, 46-47, res assoc, 50-52; resident path, Univ Wis, 47-50; PATHOLOGIST, ST JOSEPH'S HOSP, MILWAUKEE, 53- *Concurrent Pos:* Clin asst prof, Col Med, Wayne State Univ, 50-52; assoc pathologist, Wayne County Gen Hosp, 50-52; clin prof, Med Col Wis. *Mem:* Fel Col Am Pathologists; AMA; Am Soc Clin Pathologists; Am Soc Exp Path; Am Fedn Clin Res. *Mailing Add:* Dept Path St Joseph's Hosp 5000 W Chambers St Milwaukee WI 53210

ALTSHULER, EDWARD E, b Boston, Mass, Jan 10, 31; m 58; c 4. PHYSICS, ELECTRONICS. *Educ:* Northeastern Univ, BS, 53; Tufts Univ, MS, 54; Harvard Univ, PhD(appl physics), 60. *Prof Exp:* Antenna engr, Sylvania Elec Prod, Inc, 54-57; electronic scientist, Air Force Cambridge Res Labs, 60-61; dir eng antennas, Gabriel Electronics Div, Gabriel Co, 61-63; chief, Transmission Br, Air Force Cambridge Res Labs, 63-69, chief, Millimeter Wave Br, 69-76; CHIEF, MICROWAVE PROPAGATION SECT, ROME AIR DEVELOP CTR, 76- *Concurrent Pos:* Lectr, Grad Sch, Northeastern, 64- *Mem:* Fel Inst Elec & Electronics Engrs; Sigma Xi; Union Radio Sci Int. *Res:* Applied physics; antennas; propagation; electromagnetics; microwaves. *Mailing Add:* 55 Montrose Newton MA 02158

ALTSHULER, HAROLD LEON, b New York, NY, June 8, 41; div; c 2. PHARMACOLOGY. *Educ:* Cornell Univ, BS, 64; Univ Calif, Davis, PhD(pharmacol), 72. *Prof Exp:* Res asst pharmacol, Sandoz Pharmaceut, 63-64; res asst pharmacol & toxicol, Parke, Davis & Co, 64-65; lab supvr clin biochem, Nat Ctr Primate Biol, 67-71; lab supvr psychopharmacol, Univ Calif, Davis, 71-72; sect head neuropsychopharmacol, Tex Res Inst Ment Sci, 72-85; asst prof, 72-78, ASSOC PROF PHARMACOL, BAYLOR COL MED, 78-, HEAD NEUROPSYCHOPHARMACOL RES LAB, 85-, DIR GRAD STUDIES PHARMACOL, DEPT PHARMACOL, 85- *Concurrent Pos:* Spec assoc, Univ Tex Grad Sch Biomed Sci Houston, 73-; clin prof, Univ Houston, 74- *Mem:* AAAS; Am Col Clin Pharmacol; Am Asn Pharmacol & Exp Ther; Soc Neurosci; Res Soc Alcoholism. *Res:* Neuropharmacology; psychopharmacology; drug abuse; alcoholism; electroencephalography; drug self-administration; clinical psychopharmacology; epilepsy; laboratory primate husbandry and behavior. *Mailing Add:* Dept Pharmacol Baylor Col Med One Baylor Plaza Houston TX 77030

ALTSTETTER, CARL J, b Lima, Ohio, Oct 26, 30; m 54; c 2. PHYSICAL METALLURGY. *Educ:* Univ Cincinnati, MetE, 53; Ill Inst Technol, MS, 54; Mass Inst Technol, ScD(metall), 58. *Prof Exp:* From asst prof to assoc prof, 59-70, PROF PHYS METALL, UNIV ILL, URBANA, 70-, ASSOC DEAN ACAD AFFAIRS, 90- *Mem:* Am Inst Mining, Metall & Petrol Engrs; Am Soc Metals; Mat Res Soc. *Res:* Phase transformations; thermodynamics of gas-metal systems; fatigue and hydrogen embrittlement of steel; hydrogen in metals. *Mailing Add:* Dept Mat Sci & Eng Univ Ill 1304 W Green St Urbana IL 61801

ALTSZULER, NORMAN, b Suwalki, Poland, Nov 20, 24; nat US; m 56; c 2. PHARMACOLOGY, ENDOCRINOLOGY. *Educ:* George Washington Univ, BS, 50, MS, 51, PhD(physiol), 54. *Prof Exp:* From instr to assoc prof, 55-69, PROF PHARMACOL, SCH MED, NY UNIV, 69- *Mem:* Am Physiol Soc; Am Soc Pharm & Exp Therapeut; Endocrine Soc; NY Acad Sci; Am Diabetes Asn. *Res:* Endocrine regulation of metabolism. *Mailing Add:* Sch Med NY Univ New York NY 10016

ALTURA, BELLA T, b Solingen, Ger, US citizen; m 61; c 1. PHYSIOLOGY, PHARMACOLOGY. *Educ:* Hunter Col, BA, 53, MA, 64; City Univ NY, PhD(physiol), 68. *Prof Exp:* Clin biochemist, Mem Hosp-Sloan Kettering Cancer Inst, 57-64; NASA fel, 64-67; from res assoc to instr anesthesiol, Albert Einstein Col Med, 69-74; asst prof, 74-81, ASSOC PROF PHYSIOL, STATE UNIV NY MED CTR, 81- *Mem:* Am Physiol Soc; Am Soc Magnesium Res; Am Soc Pharmacol Exp Therapeut. *Res:* Physiology, pharmacology and biochemistry of blood vessels; excitation-contraction coupling mechanisms of vascular smooth muscles; drug abuse and the cardiovascular system. *Mailing Add:* Dept Physiol State Univ NY Health Sci Ctr Brooklyn NY 11203

ALTURA, BURTON MYRON, b New York, NY, Apr 9, 36; m 61; c 1. PHYSIOLOGY, PHARMACOLOGY. *Educ:* Hofstra Univ, BA, 57; NY Univ, MS, 61, PhD(physiol), 64. *Prof Exp:* Res asst physiol, Sch Med, NY Univ, 61-62, instr exp anesthesiol, 64-65, asst prof anesthesiol, 65-66; from asst prof to assoc prof anesthesiol & physiol, Albert Einstein Col Med, 70-74; PROF PHYSIOL, STATE UNIV NY DOWNSTATE MED CTR, 74- *Concurrent Pos:* Res fel, Bronx Munic Hosp Ctr, 67-; NIH travel fel, Gothenburg, Sweden, 68; USPHS res career develop award, 68-72; prin investr, NIH grant, 68-, Nat Inst Alcohol Abuse & Alcoholism grant, 90-; mem, Coun Stroke, Circulation & Basic Sci, Am Heart Asn, 73-; vis prof anesthesiol & physiol, Albert Einstein Col Med, 74- 78; mem spec study sect, Nat Inst Environ Health Sci, 77-; consult, NSF, NIH, NIMH, Alcohol Drug Abuse & Ment Health Admin, Nat Inst Alcohol Abuse & Alcoholims, City Univ NY & Upjohn Co; ed, var med jour; ed-in-chief & founder, Microcirculation & Magnesium, 81- *Honors & Awards:* Medaille Vermeil (Gold-Silver), French Nat Acad Med, 84. *Mem:* Fel Am Physiol Soc; Am Soc Pharmacol & Exp Therapeut; Am Asn Pathologists; Am Inst Nutrit; fel Am Heart Asn; fel Am Col Nutrit; Soc Exp Biol & Med; Microcirc Soc. *Res:* Local regulation of blood flow; physiology and pharmacology of vascular smooth muscle; excitation, contraction coupling mechanisms in smooth muscle; pathology physiology of reticuloendothelial system; physiology and therapeutics in experimental shock; cardiovascular actions of drugs of abuse. *Mailing Add:* 162-01 Powells Cove Blvd Apt 1R Beechhearst NY 11357

ALTWICKER, ELMAR ROBERT, b Wolfen-Bitterfeld, Ger, Apr 4, 30; US citizen; m 58; c 3. AIR POLLUTION RESEARCH. *Educ:* Univ Dayton, BS, 52; Ohio State Univ, PhD(org chem), 57. *Prof Exp:* Asst, Ohio State Univ, 52-57; res chemist, Am Cyanamid Co, NJ, 57-62; res assoc, Univ Vt, 62-63; res chemist, Princeton Chem Res, Inc, 63-65, group leader, UOP Chem Div, 65-68; assoc prof chem, Bio-Environ Div, 68-75, prof chem & environ eng, 75-86, PROF, DEPT CHEM ENG, RENSSELAER POLYTECH INST, 86- *Concurrent Pos:* Guest scientist, Dechema Inst, Frankfurt, 74-75; vis scholar, LSGC-ENSIC, Nancy, France, 89-90; Fulbright travel grant, 89-90. *Mem:* Am Chem Soc; Air Pollution Control Asn; Sigma Xi. *Res:* Air pollution; atmospheric chemistry; combustion/incineration; sulfur dioxide mass transfer and oxidation kinetics; acid rain. *Mailing Add:* Dept Chem Eng Rensselaer Polytech Inst Troy NY 12180

ALVAGER, TORSTEN KARL ERIK, b Halsberg, Sweden, Aug 22, 31; US citizen; m 58; c 2. BIOPHYSICS, NUCLEAR PHYSICS. *Educ:* Univ Stockholm, PhD(physics), 60. *Prof Exp:* Asst prof physics, Univ Stockholm, 60-66; vis prof, Princeton Univ, 66-68; PROF PHYSICS, IND STATE UNIV, TERRE HAUTE, 68- *Mem:* Am Phys Soc; AAAS; Sigma Xi. *Res:* Relativity. *Mailing Add:* Dept Physics Ind State Univ Terre Haute IN 47809

ALVARADO, RONALD HERBERT, b San Bernardino, Calif, Dec 26, 33; m 55; c 3. ZOOLOGY. *Educ:* Univ Calif, Riverside, BA, 56; Wash State Univ, MS, 59, PhD(zool physiol), 62. *Prof Exp:* Res fel physiol, Wash State Univ, 62; from asst prof to prof zool, Ore State Univ, 62-74; chmn dept, 74-81, PROF ZOOL, ARIZ STATE UNIV, 74- *Mem:* AAAS; Am Physiol Soc; Soc Gen Physiologists; Am Soc Zoologists. *Res:* Osmotic and ionic regulation in amphibians; mechanism of ion transport. *Mailing Add:* Dept Zool Ariz State Univ Tempe AZ 85287

ALVARES, ALVITO PETER, b Bombay, India, Dec 25, 35; US citizen; m 69; c 2. PHARMACOLOGY, TOXICOLOGY. *Educ:* Univ Bombay, BSc, 55, BSc(tech), 57; Univ Detroit, MS, 61; Univ Chicago, PhD(pharmacol), 66. *Prof Exp:* USPHS grant, Univ Minn, 65-67; sr res biochemist, Burroughs Wellcome & Co, 67-70; res assoc biochem pharmacol, 70-71, asst prof, 71-74; assoc prof, Rockefeller Univ, 74-77; adj assoc prof pharmacol, Med Col, Cornell Univ, 75-77; assoc prof, 77-78, PROF, DEPT PHARMACOL, UNIFORMED SERV UNIV, 78- *Concurrent Pos:* Hirschl career scientist award, 74-78; NIH career develop award, 75-78. *Mem:* Am Soc Biochem & Molecular Biol; Am Soc Clin Pharmacol & Therapeut; NY Acad Sci; Soc Toxicol; Am Soc Pharmacol & Exp Therapeut. *Res:* Effects of drugs, carcinogens and environmental chemicals on hepatic microsomal enzymes and heme biosynthetic pathway; cytochrome P-450 and mechanisms of enzyme induction; clinical pharmacology. *Mailing Add:* Dept Pharmacol Uniformed Serv Univ 4301 Jones Bridge Rd Bethesda MD 20814-4799

ALVARES, NORMAN J, b Oakland, Calif, Mar 13, 33; m 60; c 3. MECHANICAL ENGINEERING. *Educ:* San Francisco State Univ, AB, 56; Univ Minn, MS, 69. *Prof Exp:* Thermal physicist, US Naval Radiol Defense Lab, 56-69; mech engr, Stanford Res Inst, 69-75; sr mech engr, 75-78, fire sci group leader, Lawrence Livermore Lab, 78-88; INDEPENDENT CONSULT, 88- *Concurrent Pos:* US Navy fel, Univ Minn, 67-69; consult, Fire Protection Eng Group, 75-76; Failier Analysis Asn, 78; Nuclear Regulatory Comn, 78-, Univ San Francisco, 78- & Fed Emergency Mgt Agency, 78-; adv, Fusion-Laser Safety Coord Comt, Dept of Energy, 76-88; N Am ed, Jour Fire & Mat. *Mem:* Sigma Xi; Combustion Inst; Soc Fire Protection Engrs; Am Soc Testing & Mat; Int Asn Fire Safety Sci. *Res:* Ignition and thermal balance in fires; experimental modeling of fires; parametric analysis of fire hazards in industrial size enclosures; physics and chemistry of extinguishants; fire accident analysis. *Mailing Add:* 3628 Oak Knoll Dr Redwood City CA 94062

ALVAREZ, ANNE MAINO, b Rochester, Minn, Apr 14, 41; m. PLANT PATHOLOGY. *Educ:* Stanford Univ, BA, 63; Univ Calif, Berkeley, MS, 66, PhD(plant path), 72. *Prof Exp:* Instr plant path, Univ Neuquen, Arg, 69-70; jr specialist, 73-75, RESEARCHER & EDUCATOR PLANT PATH, UNIV HAWAII, 75- *Concurrent Pos:* Field res, Argentina, Costa Rica, Mexico. *Mem:* Am Phytopath Soc; Am Soc Microbiol. *Res:* Bacterial diseases of vegetable and fruit crops, epidemiology and disease control; survival of Xanthomonas in tropical soils; detection and serological identification of bacterial pathogens; orchard and post-harvest diseases of papaya. *Mailing Add:* Univ Hawaii Dept Plant Path 3190 Maile Way Honolulu HI 96822

ALVAREZ, LAURENCE RICHARDS, b Jacksonville, Fla, Sept 27, 37; m 60; c 3. MATHEMATICS. *Educ:* Univ of the South, BS, 59; Yale Univ, MA, 62, PhD(graph theory), 64. *Prof Exp:* Instr math, Trinity Col, Conn, 63-64; coordr prog planning & budgeting, 72-88, FROM INSTR TO PROF MATH, UNIV OF THE SOUTH, 64-, ASSOC PROVOST, 88- *Concurrent Pos:* Am Coun Educ acad admin intern, Pomona Col. *Mem:* Math Asn Am; Sigma Xi. *Res:* Graph theory. *Mailing Add:* RFD 1 Box 216 Sewanee TN 37375

ALVAREZ, RAYMOND ANGELO, JR, b Mobile, Ala, May 15, 34; m 61; c 2. PHYSICS, PARTICLE ACCELERATORS. *Educ:* La State Univ, BS, 55; Stanford Univ, PhD(physics), 64. *Prof Exp:* Res assoc physics, Stanford Univ, 60-61; from instr to asst prof, Mass Inst Technol, 61-68; PHYSICIST, LAWRENCE LIVERMORE LAB, 68- *Mem:* Am Phys Soc. *Res:* Microwave propagation; high-frequency electric discharges. *Mailing Add:* Lawrence Livermore Nat Lab L-288 Livermore CA 94550

ALVAREZ, ROBERT, b New York, NY, Feb 6, 21; m 45; c 2. ANALYTICAL CHEMISTRY, PHYSICAL CHEMISTRY. *Educ:* City Col New York, BS, 42. *Prof Exp:* Chemist, Tenn Valley Authority, 42-43 & Manhattan Proj, 44-46; sr chemist, Carbide & Carbon Chem Corp & Union Carbide & Carbon

Corp, Tenn, 46-49; phys chemist radiol div, Army Chem Corps, Md, 49-51; phys chemist, 51-55, ANAL CHEMIST, SPECTROCHEM ANALYSIS SECT, NAT BUR STANDARDS, 55- *Mem:* Am Chem Soc; Soc Appl Spectros; Am Soc Testing & Mat; Sigma Xi. *Res:* Development of stable isotope dilution techniques for determinations by spark source mass spectrometry; chemical preconcentration techniques applied to spectrochemical analysis; developing optical emission and x-ray fluorescent methods of analysis. *Mailing Add:* 1505 N 60th St Milwaukee WI 53208

ALVAREZ, RONALD JULIAN, b Brooklyn, NY, Feb 17, 35; m 56; c 3. CIVIL & STRUCTURAL ENGINEERING. *Educ:* Manhattan Col, BCE, 57; Univ Wash, MS, 60; NY Univ, PhD(eng), 67. *Prof Exp:* Jr engr, Boeing Co, 57-58, assoc engr, 58-60, eng designer, 60-61; sr res engr, Repub Aviation Corp, 61-62; from asst prof to assoc prof eng sci, 62-74, dir eng sci & mech progs, 68-78, PROF ENG, HOFSTRA UNIV, 74-, DIR CONTINUING ENG EDUC, 70- *Concurrent Pos:* Consult engr, 67-; dir, Appl Systs, Inc, 68-71; consult, City of Glen Cove; consult solid waste mgt, NY State Dept Environ Conserv, 71-74. *Mem:* Am Soc Mech Engrs; Am Soc Civil Engrs; Nat Soc Prof Engrs. *Res:* Behavior of multistory, multibay plane rigid frames subjected to static in-plane loadings; solid waste disposal; solid waste management. *Mailing Add:* Nine Duke Pl Glen Cove NY 11542

ALVAREZ, VERNON LEON, b Dec 19, 46. BIOCHEMISTRY. *Educ:* Univ Colo, BA, 69; Univ Utah, PhD(phys chem), 72. *Prof Exp:* Prof chem, Northwest Nazarene Col, 73-75; from instr to asst res prof, Dept Path, Col Med, Univ Utah, 75-77; mem, Basel Inst Immunol, Switz, 79-81; expert, Protein Chem Sect, Lab Biochem, Nat Cancer Inst, NIH, 81-83; prin res scientist, 83-85, group leader biochem, 85-88, DIR CHEM RES, CYTOGEN CORP, 88- *Concurrent Pos:* Vis lectr chem, Col Idaho, 74; NIH postdoctoral fel, 75-77; vis scientist, Lab Cell Biol, Nat Cancer Inst, NIH, 78-79. *Mem:* Soc Nuclear Med; AAAS; Am Asn Immunologists. *Res:* Biotechnology; conjugates of monoclonal antibodies for in vivo human clinical applications; membrane receptors; immunology; protein chemistry; molecular recognition units; peptides; author of 24 technical publications. *Mailing Add:* Dept Chem Res Cytogen Corp 600 College Rd E Princeton NJ 08540

ALVAREZ, VINCENT EDWARD, b Shanghai, China, Aug 15, 46; US citizen; m 72. INORGANIC CHEMISTRY. *Educ:* Calif State Univ, Hayward, BS, 69; Univ Calif, Santa Barbara, MA, 72, PhD(chem), 74. *Prof Exp:* Res chemist, E I du Pont de Nemours & Co, Inc, 74-76; PROJ LEADER, CLOROX CO, 76- *Concurrent Pos:* Asst prof chem, Del Tech & Community Col, 74-, Chabot Community Col, 77. *Mem:* Am Chem Soc. *Res:* Reactions at metal oxide surfaces; reactions of singlet oxygen; oxidation chemistry; surfactants. *Mailing Add:* Clorox Tech Ctr 7200 Johnson Dr Pleasanton CA 94566

ALVAREZ, WALTER, b Berkeley, Calif, Oct 3, 40; m 65. STRATIGRAPHIC & STRUCTURAL GEOLOGY. *Educ:* Carleton Col, BA, 62; Princeton Univ, PhD(geol), 67. *Prof Exp:* Geologist, Am Overseas Petrol, Ltd, Netherlands, 67-68, sr geologist, Libya, 68-70; NATO fel, Brit Sch Archaeol, Rome, 70-71; res scientist,Lamont-Doherty Geol Observ, 71-73, Res assoc, 73-77; asst prof, 77-79, assoc prof, 79-81, PROF, DEPT GEOL & GEOPHYS, UNIV CALIF, BERKELEY, 81- *Concurrent Pos:* Lectr geol, Columbia Univ, 73-74; vis sr res assoc, Lamont-Doherty Geol Observ, 77- *Mem:* Nat Acad Sci; Geol Soc Am; Am Geophys Union; AAAS; Ital Geol Soc. *Res:* Plate tectonics and mountain belt structure; causes of mass extinctions; paleomagnetism; Alpine-Mediterranean region. *Mailing Add:* Dept Geol & Geophys Univ Calif Berkeley CA 94720

ALVAREZ-BUYLLA, RAMON, b Oviedo, Spain, June 22, 19; Mex citizen; m 56; c 4. NEUROENDOCRINOLOGY. *Educ:* Ashkhabad State Univ, USSR, MD, 43; Med Acad Sci, Moscow, USSR, PhD(physiol), 46. *Prof Exp:* Prof & investr, Nat Polytech Inst, Mex, 47-54; investr, Nat Cardiol Inst Mex, 50-56 & Nat Inst Neumol Mex, 58-61; prof physiol, Ctr Invest & Advan Study, Nat Polytech Inst, 61-80; HEAD, BASIC RES DIV, NAT INST RESPIRATORY DIS, MEX, 80-; PROF NAT UNIV MEX, 84- *Concurrent Pos:* Vis prof human anat, Oxford Univ, 69-70; vis prof, Dept Anat, Kingston Univ, Can. *Honors & Awards:* Guggenheim Award, 51, Rockefeller, 53; Alfonso Rivera Prize, Mex Nutrit & Endocrinol; Elias Sourasky Prize, Nat Invest 3, 84. *Mem:* Arg Med Asn; Physiol Soc Mex; Latin Am Physiol Soc; Mex Acad Sci; Mex Nutrit & Endocrinol; Spain Royal Med Acad; USSR Acad Sci. *Res:* Neuroendocrinic integration; neuroendocrinic mechanism of the homeostasis; conditioned reflexes. *Mailing Add:* Unidad de Invest INEIR Calz Tlalpan 4502 Tlalpan 14000 DF Mexico

ALVAREZ-GONZALEZ, RAFAEL, b Michoacan, Mex, Nov 25, 58; m 77; c 2. ENZYMOLOGY, PROTEIN CHEMISTRY. *Educ:* Michoacan Univ, BS, 79; N Tex State Univ, MS, 82, PhD(biochem), 85. *Prof Exp:* Post-doctoral fel biochem & pharmacol, Inst Biochem & Pharmacol, Univ Zurich, Switz, 85-86; post-doctoral fel biochem, Biomed Div, Samuel Roberts Nobel Found, Inc, 87-88; ASST PROF, DEPT MICROBIOL & IMMUNOL, TEXAS COL OSTEOP MED, 89- *Mem:* Am Soc Biochem & Molecular Biol; Am Asn Cancer Res; NY Acad Sci; AAAS; Am Chem Soc; Am Soc Microbiol. *Res:* Biological role of enzyme catalyzed post-translational covalent modification of eukaryotic proteins with residues of ADP-ribose in modulating chromatin structure and function; membrane-receptor mediated signal transduction mechanisms. *Mailing Add:* Dept Microbiol & Immunol Texas Col Osteop Med Ft Worth TX 76107

ALVARINO DE LEIRA, ANGELES, b El Ferrol, Spain, Oct 3, 16; US citizen; m 40; c 1. BIOLOGICAL OCEANOGRAPHY, MARINE BIOLOGY. *Educ:* Univ Santiago, BSLett, 33; Univ Complutense, MS Hons, 41, Cert Dr, 51, DSc, 67. *Hon Degrees:* DSc, Univ Madrid, 67. *Prof Exp:* Prof biol, El Ferrol Col, 41-47; fishery biologist, Dept Sea Fisheries, Ministry Commerce, Madrid, 48-52; histologist, Super Coun Sci Res, 49-51; biol oceanogr, Span Inst Oceanog, Madrid, 50-57; biologist, Scripps Inst Oceanog, Univ Calif, La Jolla, 58-69; fishery res biologist, 70-87, EMER SCI, NAT MARINE FISHERIES SERV, FISHERY-OCEANOG CTR, LA JOLLA, 87- *Concurrent Pos:* Brit Coun grant, Marine Biol Lab, Plymouth, Eng, 53-54; Fulbright grant, Woods Hole Oceanog Inst, 56-57; NSF grants, 61-69; Calif Coop Ocean & US Naval Off grants, NSF; dir doctoral thesis cand, Cent Univ Venezuela, 73- & Univ Nat Mex, 77-; assoc prof, Nat Univ Mex, 76-; coordr oceanic res, Hispano-Am Countries, 77-79; assoc prof, San Diego State Univ, 79-82, Antarctic res grants, 79-82, FAO grant, 79; assoc res, Univ San Diego, 82-85; vis prof, Univ Fed de Parana, Brazil, 82, Nat Polytech Inst of Mexico, 82- *Mem:* fel Am Inst Fishery Res Biologists. *Res:* Zooplankton; Chaetognatha; Siphonophorae; Medusae; Ichthyoplankton; taxonomy; zoogeography; indicator organisms in water dynamics; predation in the plankton realm; sea fisheries; biology and ecology of Thunnidae; discoverer of twelve new species of Chaetognatha nine new species of Siphonophora and one new Medusa; ecology of Plankton; biotic environment of fish spawning; historical research on faunistic and geographic accounts of early navigators and expeditions; more than 100 published papers on original scientific research. *Mailing Add:* 7535 Cabrillo Ave La Jolla CA 92037

ALVERSON, DAVID ROY, b Spartanburg, SC, Dec 25, 46; m 67; c 1. INSECT PEST MANAGEMENT. *Educ:* Clemson Univ, BS, 68, MS, 76; Univ Ga, PhD(entom), 79. *Prof Exp:* ASST PROF ENTOM, CLEMSON UNIV, 78- *Mem:* Sigma Xi; Entom Soc Am. *Res:* Economic entomology: insect pests of pecans, forage and field crops, small grains, ornamental plants; apiculture; insect vectors of plant and animal diseases. *Mailing Add:* Dept Entom Fisheries & Wildlife Clemson Univ Clemson SC 29631

ALVERSON, DAYTON L, b San Diego, Calif, Oct 7, 24; m 51; c 2. MARINE BIOLOGY. *Educ:* Univ Wash, BS, 50, PhD, 67. *Prof Exp:* Specialist method & equip, US Fish & Wildlife Serv, 50-53; biologist, Wash State Dept Fisheries, 53-58; chief, Explor Fishing & Gear Res Unit, US Fish & Wildlife Serv, 58-60, dir res base, Seattle, 60-67, Bur Com Fisheries, 67-69, assoc dir fisheries, Washington, DC, 69-70; assoc regional dir resource progs, Nat Oceanic & Atmospheric Admin, 70, dir Northwest Fisheries Ctr, Seattle, 71-80; AT DEPT MARINE STUDIES, UNIV WASH; PRES, NATURAL RESOURCES CONSULT, 81- *Concurrent Pos:* Affil prof, Inst Marine Studies, Univ Wash; mem, Marine Bd, Nat Res Coun; chmn, Adv Comt Marine Resources Res, Food & Agr Orgn, UN; chief of staff, US/Can Salmon Interceptions Negotiations, US State Dept; comnr, US Sect Int N Pacific Fisheries Comn; deleg, US Law of the Sea Conf. *Honors & Awards:* Distinguished Serv Award, Dept Interior, 66; Nat Oceanic & Atmospheric Admin Award; Gold Medal Award Distinguished Serv, Dept of Com, 76. *Mem:* Am Inst Fisheries Res Biologists; Sigma Xi. *Res:* Author of 130 papers, books and articles covering fisheries, management, population diagnosis, resource distribution and behavior. *Mailing Add:* Dept Marine Studies Univ Wash Seattle WA 98195

ALVES, LEO MANUEL, b Philadelphia, Pa, May 21, 45. PLANT PHYSIOLOGY, PLANT PATHOLOGY. *Educ:* St Norbert Col, BS, 68; Univ Chicago, PhD(biol), 75. *Prof Exp:* Asst prof biol, Lincoln Univ, Pa, 75-76; Nat Res Coun res assoc & res plant pathologist, ERegional Res Ctr, Agr Res Serv, USDA, Pa, 76-78; asst prof biol, Lab Plant Morphogenesis, Manhattan Col, 78-84, assoc prof & chmn biol dept, Col Mt St Vincent, 84-88, ASSOC PROF, BIOL DEPT, LAB PLANT MORPHOGENESIS, MANHATTAN COL, 88- *Concurrent Pos:* Sigma Xi res grant-in-aid, 74-75, 80-81; USDA Specific Coop Agreement, 80-83; Scopas Technol Sponsored Proj, 84-89; vis assoc prof, Rockefeller Univ, New York, 88-89. *Mem:* Am Phytopath Soc; Am Soc Plant Physiologists; Japanese Soc Plant Physiologists; Sigma Xi. *Res:* Investigations of physiology and biochemistry of phytoalexin biosynthesis in white potato tuber; role phytoalexins play in expression of resistance to various pathogens; role played by cyclic nucleotides in regulating cell division in oncogenetically-transformed tissue. *Mailing Add:* Lab Plant Morphogenesis Manhattan Col Bronx NY 10471-4099

ALVES, RONALD V, b Oakland, Calif, Apr 11, 35; m 80; c 4. QUANTUM ELECTRONICS, LASERS IN MEDICINE. *Educ:* Univ San Francisco, BS, 58; Univ Calif, Berkeley, PhD(physics), 68; Stanford Univ, MSEE, 73. *Prof Exp:* Physicist, US Navy Radiol Defense Lab, 57-60; res scientist, Lockheed Missiles & Space Co, Lockheed Aircraft Corp, 68-74; mem staff, Coherent Radiation, Palo Alto, Calif, 74-76; develop mgr & head, Lab Res, Develop & Applications, Uthe Technol, Inc, Sunnyvale, Calif, 76-77; chief scientist, Quantex Corp, Sunnyvale, Calif, 77; cofounder & vpres Technol, Luxtron Corp, Mountain View, Calif, 78-83; mgr, Advanced Develop Coherent Med Group, Palo Alto, Calif, 83-88; MGR, LASERS RES & DEVELOP, HEWLETT-PACKARD, SANTA CLARA DIV, SANTA CLARA, CALIF, 89- *Concurrent Pos:* Consult in lasers, optics, physics. *Mem:* Am Phys Soc; Optical Soc Am; Soc Photo-Optical Instrumentation Eng; Am Soc Lasers Med & Surg. *Res:* Solid state laser materials and rare earth activated luminescent materials; lasers in medicine. *Mailing Add:* 475 Marion Ave Palo Alto CA 94301

ALVEY, DAVID DALE, b Harrisburg, Ill, Jan 2, 32; m 55; c 3. AGRONOMY. *Educ:* Univ Ill, BS, 57, MS, 58; Purdue Univ, PhD(genetics), 62. *Prof Exp:* Asst agron, Purdue Univ, 58-62, instr, 62; res agronomist, DeKalb Agr Asn, Inc, 62-70; CORN BREEDER, FARMERS FORAGE RES, 70- *Mem:* Am Soc Agron; Am Genetic Asn; Sigma Xi. *Res:* Genetics and plant breeding; coordinating corn breeding programs; developing commercial corn varieties. *Mailing Add:* 237 Myrtle Dr West Lafayette IN 47906

ALVI, ZAHOOR M, b Sept 11, 32; US citizen; div; c 2. RADIATION PHYSICS, LASER TECHNOLOGY. *Educ:* Carnegie-Mellon Univ, BS, 56; Univ Calif, Los Angeles, MS, 66, PhD(med physics, nuclear eng), 68; Am Bd Health Physics, dipl, 65, recert, 82; Am Bd Radiol, dipl & cert radiol physics, 71. *Prof Exp:* Radiological physicist, Radiation Ther & Nuclear Med, Cedars of Lebanon Hosp, Los Angeles, 58-64 & Dept Radiother, Kaiser-Permanente Med Ctr, Hollywood, 64-72; DIR MED PHYSICS, RADIOLOGICAL SYSTS, INC, 72- *Concurrent Pos:* Mem eng exten fac, Univ Calif, Los Angeles, 71-80; laser physicist, Northrop Corp, Hawthorne, Calif; attending med physicist, Martin Luther King, Jr, Gen Hosp, 74-; res affil, Jet Propulsion Lab, Calif Inst Technol, 74- *Mem:* Am Asn Physicists in Med; Health Physics

Soc; Int Radiation Protection Asn; Biomed Eng Soc. *Res:* Mathematical modeling of human physiological systems; mathematical model of non-equilibrium kinetics of four major electrolytes in the human body; catheter probe system for gastrointestinal motility automatic monitoring; in-vivo neutron activation analysis using microgram Californium-252 sealed ug sources; enhanced visualization through digital radiography; neutron dosimetry techniques; optimization of hyperthermia in conjunction with radiation therapy; designing compact magnetic resonance systems; Excimer laser systems. *Mailing Add:* RadioLogical Systs Inc 13224 G Admiral Ave Marina Del Rey CA 90292

ALVING, CARL RICHARD, b Chicago, Ill, July 10, 39; m 70; c 1. PHOSPHOLIPIDS, GLYCOLIPIDS. *Educ:* Haverford Col, BS, 61; Univ Miami, MD, 66. *Prof Exp:* Intern med, Barnes Hosp, Wash Univ Sch Med, 66-67, resident, 67-68 & fel pharmacol, 68-70; res investr immunol, 70-73, CHIEF DEPT MEMBRANE BIOCHEM, WALTER REED ARMY INST RES, 78- *Concurrent Pos:* Adj asst prof microbiol, Georgetown Univ Sch Med, 78- *Mem:* Am Soc Biol Chemists; Am Asn Immunologists; Soc Complex Carbohydrates; NY Acad Sci; AAAS; Int Soc Pharmacol. *Res:* Antigen, antibody, and complement interactions at the surface of liposomes; receptor properties of lipids; use of liposomes as carriers of drugs or antigens. *Mailing Add:* Dept Membrane Biochem Walter Reed Army Inst Res Washington DC 20307

ALVINO, WILLIAM MICHAEL, b New York, NY, Jan 12, 39; m 60; c 2. POLYMER CHEMISTRY. *Educ:* Iona Col, BS, 60; Seton Hall Univ, MS, 66. *Prof Exp:* Sr tech aide polymers, Bell Tel Labs, 60-66, assoc mem tech staff, 66; from res chemist to sr res chemist, 66-82, FEL SCIENTIST, WESTINGHOUSE ELEC CORP, 82- *Mem:* Am Chem Soc. *Res:* Thermal and ultraviolet stability of polymers; development and formulation of polymers as coating, films, adhesives and laminating resins for electrical insulation; synthesis and modification of polymers for high temperature applications; electrophoretic deposition of polymers from non-aqueous systems; composites; printed wiring boards; radiation curing of polymers. *Mailing Add:* Westinghouse Res Labs Beulah Rd Pittsburgh PA 15235

ALVORD, DONALD C, b Newton, Mass, Nov 12, 22; m 50; c 2. ECONOMIC GEOLOGY, CHEMISTRY. *Educ:* Univ Idaho, BS, 51. *Prof Exp:* Geologist, US Geol Surv, 51-83, US Bur Land Mgt, 83-86; gen mgr Ord Prototype Mfg, 86-88; KANAB ORD WORKS, 89- *Mem:* Am Chem Soc. *Res:* Geology and chemical characterization of coal; chemistry of organic mineral deposits; physical organic chemistry; research and development of ultra yield artillery. *Mailing Add:* Kanab Ord Works PO Box 46 1142 Grand Canyon Dr Kanab UT 84741-0046

ALVORD, ELLSWORTH CHAPMAN, JR, b Washington, DC, May 9, 23; m 43; c 4. NEUROPATHOLOGY. *Educ:* Haverford Col, BS, 44; Cornell Univ, MD, 46. *Prof Exp:* Asst path, Med Col, Cornell Univ, 47-48; instr neurol, Sch Med, Georgetown Univ, 50-55; assoc prof neurol & path, Col Med, Baylor Univ, 55-60; assoc prof, 60-62, PROF PATH, SCH MED, UNIV WASH, 62- *Concurrent Pos:* Prof lectr, Sch Med, George Washington Univ, 50-51; neurologist, Neurophysiol Sect, Army Med Serv Grad Sch, 51-53; instr, Wash Sch Psychiat, 52-55; chief clin neuropath sect, Nat Inst Neurol Dis & Blindness, NIH, 53-55; consult, Seattle hosps. *Mem:* AAAS; Am Asn Neuropath (pres, 64-65); Am Soc Exp Biol & Med; Asn Res Nervous & Ment Dis; assoc Am Neurol Asn. *Res:* Allergic encephalomyelitis. *Mailing Add:* 5601 NE Ambleside Rd Seattle WA 98105

ALWAN, ABDUL-MEHSIN, b Baghdad, Iraq, Nov 22, 27; m 59; c 2. ENGINEERING MECHANICS. *Educ:* Bournemouth Munic Col, Eng, BS, 51; Univ Wis, MS, 56, PhD(eng mech), 63. *Prof Exp:* Supvr, Port Directorate of Basra, Iraq, 52-53; engr, Kellogg Int Corp, 53-54; struct designer, Erik Floor & Assoc, Ill, 57-58; instr eng mech, Univ Wis, 58-63; assoc prof, 63-77, PROF ENG MECH, US NAVAL ACAD, 77- *Mem:* Am Soc Eng Educ. *Res:* Elasticity; theory of plates and shells; rigid body mechanics. *Mailing Add:* 223 Cardamon Dr Edgewater MD 21037

ALWARD, RON E, b St Thomas, Can, July 9, 41; m 69; c 2. SOLAR THERMAL SYSTEMS, BIOMASS ENERGY SYSTEMS. *Educ:* Univ Western Ont, BEngSc, 63, MEngSc, 68. *Prof Exp:* Res assoc solar wind energy, Brace Res Inst, 68-76; appropriate technol specialist, Bu-Ali Sina Univ, Iran, 76-77; solar engr III, Nat Ctr Appropriate Technol, Mont, 78-79; ASST DIR, BRACE RES INST, 79- *Concurrent Pos:* Consult, various nat & int govt agencies, 69-, Lab Solar Energy, Univ Algiers, 81; mem, Memphremagog Community Technol Group, 79-; mem, UN Univ, 84- *Mem:* Solar Energy Soc Can; Int Solar Energy Soc. *Res:* Solar distillation; solar agricultural dryers; solar and fuel conserving cookers; air to air heat exchangers, passive solar houses, greenhouses. *Mailing Add:* RR 4 Mansonville PQ J0E 1X0 Can

ALWITT, ROBERT S(AMUEL), b New York, NY, Nov 4, 33; m 60; c 2. ELECTROCHEMISTRY, OXIDE FILMS. *Educ:* Cooper Union, BChE, 54; Rensselaer Polytech Inst, MChE, 58, PhD(chem eng), 63. *Prof Exp:* Jr engr electrochem processing, Eng Labs, Sprague Elec Co, 54-58, dept head, 58-61, mgr res electrochem res & develop, 63-73; mgr res & develop, United Chemi-Con, Inc, Northbrook, Ill, 74-84; PRES, BOUNDARY TECHNOL INC, BUFFALO GROVE, ILL, 84- *Concurrent Pos:* Sr vis fel, Sci Res Coun, UK, 72-73; div ed, J Electrochem Soc, 72-90; mem, Honors & Awards Comt, Electrochem Soc, 85-89 & Finance Comt, 90-94. *Mem:* AAAS; Am Chem Soc; Electrochem Soc; Am Electroplaters & Surf Finish Soc; Mat Res Soc; Metall Soc. *Res:* Properties of metal and oxide surfaces; growth and properties of anodic oxide films; electrochemical processes; aluminum surface treatment; electrolytic capacitor technology. *Mailing Add:* 1136 Jeffrey Ct Northbrook IL 60062

ALWORTH, WILLIAM LEE, b Twin Falls, Idaho, Jan 3, 39; m 59; c 2. BIO-ORGANIC CHEMISTRY. *Educ:* Harvard Univ, AB, 60; Univ Calif, Berkeley, PhD(org chem), 64. *Prof Exp:* NIH fel, Harvard Univ, 64-65; asst prof org chem, 65-71, assoc prof chem, 71-78, PROF CHEM, TULANE UNIV, 78- *Concurrent Pos:* NIH career develop award, 72; adj assoc prof biochem, Tulane Univ, 75- *Mem:* Am Chem Soc; Am Soc Biol Chemists; Sigma Xi. *Res:* Biosynthesis of vitamins; chemical carcinogenesis; enzyme mechanisms. *Mailing Add:* Dept Chem Tulane Univ New Orleans LA 70118

ALY, ABDEL FATTAH, b Mallawi, Egypt, Nov 28, 42; US citizen. CHEMICAL ENGINEERING. *Educ:* Cairo Univ, BS, 64, MS, 67; Columbia Univ, ScD(chem eng), 70. *Prof Exp:* Res engr, 69-72, sr res engr, 72-76, assoc tech serv, 76-78, supv engr tech serv, 78-80, sr refinery tech adv, 80-84, ADV TECH SERV ENGR, MOBIL OIL CORP, 84- *Mem:* Am Inst Chem Engrs. *Res:* Kinetics; fluidization; dynamics; catalysis; scale-up; simulation and modeling; petroleum refinery operations. *Mailing Add:* Mobil Oil Corp 3225 Gallows Rd Fairfax VA 22037-0001

ALY, ADEL AHMED, b Fayoum, Egypt, Oct 30, 44; US citizen. INDUSTRIAL ENGINEERING, OPERATIONS RESEARCH. *Educ:* Univ Cairo, BSc, 66; NC State Univ, MS, 72; Va Polytech Inst & State Univ, PhD(indust eng), 74. *Prof Exp:* Jr consult mgt sci, Mgt Consult Ctr, 66-67; instr prod eng, Univ Cairo, 67-69; From asst prof to assoc prof, 75-81, PROF INDUST ENG, UNIV OKLA, 81- *Honors & Awards:* Halliburton lectr, Halliburton Found Inc, 82. *Mem:* Am Inst Indust Eng; Soc Mfg Engrs; Nat Soc Prof Engrs; Inst Mgt Sci; Am Soc Opers Res. *Res:* Mathematical programming; facility design and location analysis; distribution systems; application of operations research in the public sector. *Mailing Add:* Dept Indust Eng Univ Okla Main Campus Norman OK 73019

ALY, HADI H, b Baghdad, Iraq, Aug 8, 30; US citizen; m 67; c 2. HIGH ENERGY PHYSICS. *Educ:* Univ Calif, Berkeley, BS, 52, MS, 53; Univ Bristol, Eng, PhD(physics), 60. *Prof Exp:* Instr, Univ Baghdad, Iraq, 53-56; res fel, Nat Res Coun, Can, 61-62; res assoc, Univ Rochester, NY, 62-64; res fel, Max Planck Inst, Munich, 64-65; assoc prof, Am Univ Beirut, Lebanon, 65-69; PROF PHYSICS, SOUTHERN ILL UNIV, EDWARDSVILLE, 69- *Res:* Quantum theory of scattering. *Mailing Add:* 805 Linden Ave Boulder CO 80304

ALY, RAZA, b Quetta, Pakistan, June 3, 35; US citizen; c 2. SKIN MICROBIOLOGY, DERMATOMYCOLOGY. *Educ:* Punjab Univ, BSc, 58; Univ Mich, MS, 62, MPH, 65; Univ Okla, PhD(microbiol), 69. *Prof Exp:* Asst res microbiologist, 69-72, from asst prof to assoc prof, 73-83, PROF MICROBIOL & DERMTOMYCOL, UNIV CALIF, SAN FRANCISCO, 83- *Concurrent Pos:* Vis prof, Harvard Univ, 87- *Mem:* Fel Pakistan Acad Med Sci; Am Soc Microbiol; Soc Invest Dermat; Int Soc Human & Animal Mycol; Am Acad Microbiol; Am Fedn Clin Res. *Res:* Pathogenesis of skin infections with emphasis on ecological mechanisms of microbial colonization; bacterial interference, microbial adherence and antimicrobial properties of skin lipids; product testing of topical antibiotics, surgical soaps and acne preparation. *Mailing Add:* Dept Dermat Univ Calif San Francisco Med Sch 513 Parnassus Ave San Francisco CA 94143-0536

ALYEA, ETHAN DAVIDSON, JR, b Orange, NJ, Mar 7, 31; m 57; c 4. PHYSICS. *Educ:* Princeton Univ, AB, 53; Calif Inst Technol, PhD(physics), 62. *Prof Exp:* From asst prof to assoc prof, 62-71, PROF PHYSICS, IND UNIV, BLOOMINGTON, 71- *Mem:* Am Asn Physics Teachers; Am Phys Soc; Sigma Xi. *Res:* Experimental high energy physics; hybrid bubble chamber and proportional wire chamber experiments. *Mailing Add:* Dept Physics Ind Univ Bloomington IN 47401

ALYEA, FRED NELSON, b Harlingen, Tex, Oct 26, 38; m 62; c 2. DYNAMIC METEOROLOGY. *Educ:* Univ Wis-Madison, BS, 64; Colo State Univ, PhD(atmospheric sci), 72. *Prof Exp:* Res assoc meteor, Mass Inst Technol, 72-79; PRIN RES SCIENTIST, GA INST TECHNOL, 79- *Concurrent Pos:* Consult, Mfg Chemists Asn, 75- *Mem:* Am Meteorol Soc; Am Geophys Union; Sigma Xi. *Res:* Stratospheric modeling and climatology; large-scale dynamic processes in the atmosphere; numerical weather prediction methods; atmospheric predictability; paleoclimatology; climatological effects of anthropogenic pollution sources. *Mailing Add:* 4170 Chestnut Ridge Dr Dunwoody GA 30338

ALYEA, HUBERT NEWCOMBE, b Clifton, NJ, Oct 10, 03; m 29; c 1. PHYSICAL CHEMISTRY. *Educ:* Princeton Univ, AB, 25, AM, 26, PhD(chem), 28. *Hon Degrees:* DSc, Beaver Col, 70. *Prof Exp:* Nat Res Coun fel, Univ Minn, 29; int res fel, Kaiser Wilhelm Inst, Berlin, 30; from instr to prof EMER PROF CHEM, PRINCETON UNIV, 72-; AT FRICK CHEM LAB. *Concurrent Pos:* Vis prof, Univ Hawaii, 48-49; lectr, Int Expos, Brussels, 58, Seattle, 62 & Montreal, 67; Fulbright, NSF, AEC, Dept Com, Dept State, US AID, Asia Found & UNESCO lectr, Europe, Africa, MidE, Orient, Mex & Cent Am, (80 countries), 49-91. *Honors & Awards:* Sci Apparatus Makers Asn Award in Chem Educ, 70; James Flack Norris Award, 70; Priestley Award, 84. *Mem:* Am Chem Soc; Sigma Xi. *Res:* Chain reactions; inhibition and catalysis; radiochemistry; polymerization; armchair chemistry; TOPS method of projecting chemical experiments. *Mailing Add:* 337 Harrison St Princeton NJ 08540

ALZERRECA, ARNALDO, b La Paz, Bolivia, Mar 5, 43; m 69; c 1. ORGANIC CHEMISTRY. *Educ:* Univ Karlsruhe, WGer, Dipl Chem, 70; Univ Marburg, WGer, Dr rer nat, 74. *Prof Exp:* Teaching asst, Univ Marburg, WGer, 71-74; PROF ORG CHEM, INTERAM UNIV PR, 76- *Concurrent Pos:* Res assoc, Univ PR, Rio Piedras, 81-; consult, Exp Rum Plant, 81-; vis prof, Univ Giessen, WGer, 83-84. *Mem:* Am Chem Soc; AAAS; Sigma Xi. *Res:* Natural products; steroid alkaloids; synthesis of small ring compounds; synthesis of spiroannelated oxa-aza ring systems; SAR-studies of biologically active glycoalkaloids; development of molluscicides. *Mailing Add:* Div Sci & Technol InterAm Univ Puerto Rico Rd 1 Km 16-3 Rio Piedras PR 00926

AL-ZUBAIDY, SARIM NAJI, b Baghdad, Iraq, July 1, 53; Irish citizen; m 86; c 2. MECHANICAL ENGINEERING. *Educ:* Baghdad Univ, BSc, 76; Liverpool Polytech Inst, MSc, 77; Hatfield Polytech Inst, PhD(mech eng), 82. *Prof Exp:* Asst lectr, 82-83, asst prof, 83-89, ASSOC PROF MECH ENG, UNITED ARAB EMIRATES UNIV, 89- *Concurrent Pos:* Res fel, Hatfield Polytech Inst, 82-86. *Mem:* Fel Inst Diag Engrs; Soc Automotive Engrs; Am Soc Mech Engrs; assoc mem Royal Aeronaut Soc. *Res:* Aero-thermodynamic design of radial turbomachinery; computer aided design of high pressure ratio centrifugal flow compressors. *Mailing Add:* Mech Eng Dept United Arab Emirates Univ PO Box 17555 Al-Ain United Arab Emirates

AMACHER, DAVID E, CELLULAR TOXICOLOGY, BIOCHEMICAL TOXICOLOGY. *Educ:* Kent State Univ, MS, 71, BA, 87, PhD(mammalian physiol), 73; Univ N Haven, MBA, 81, Am Bd Toxicol, dipl. *Prof Exp:* Grad asst biol, Kent State Univ, 67-72; res assoc, Dept Pharmacol, Univ Sch Med, 73-74; NCI postdoctoral fel, Nat Inst Environ Health Sci, 74-76; genetic toxicologist, 76-83, CELLULAR BIOCHEM TOXICOLOGIST DRUG SAFETY EVAL, PFIZER CENT RES, 84- *Concurrent Pos:* Consult, NCI carcinogenesis prog, Nat Coun prog, Nat Inst Health; adj assoc prof dept zool, 82-88, asst res prof, sch pharm, Univ Conn, 88. *Mem:* Am Asn Cancer Res; Soc Toxicol; Soc Biochem & Molecular Biol; Sigma Xi. *Mailing Add:* Drug Safety Dept Pfizer Cent Res Ctr Groton CT 06340

AMACHER, PETER, b Portland, Ore, Jan 29, 32; m 73; c 2. INFORMATION SCIENCE. *Educ:* Amherst Col, BA, 54; Univ Wash, PhD(hist sci), 62. *Prof Exp:* Fel hist neurol, Brain Res Inst, Univ Calif, Los Angeles Ctr Health Sci, 62-64, asst & assoc prof hist med, Ctr Health Sci, 64-71, asst to dir, Brain Info Serv, 64-69, dir, 69-71; prin, Amacher & Assocs Info Serv, 71-72; dir, Conf Prog, Kroc Found Advan Med Sci, 72-81; CONSULT, MED COMMUN, 82- *Mailing Add:* 118 Rim Rd Santa Fe NM 87501

AMADO, RALPH, b Los Angeles, Calif, Nov 23, 32; m; c 2. THEORETICAL PHYSICS. *Educ:* Stanford Univ, BS, 54; Oxford Univ, PhD(physics), 57. *Prof Exp:* Res assoc physics, 57-59, from asst prof to assoc prof, 59-65, PROF PHYSICS, UNIV PA, 65- *Concurrent Pos:* Consult, Arms Control & Disarmament Agency. *Mem:* Fel Am Phys Soc; fel AAAS. *Res:* Theoretical nuclear physics; many-body problem; particle physics; scattering theory; three-body problem. *Mailing Add:* Dept Physics Univ Pa Philadelphia PA 19104

AMADOR, ELIAS, b Mexico City, Mex, June 8, 32; m 61; c 3. MEDICINE, PATHOLOGY. *Educ:* Cent Univ Mex, BS, 59; Nat Univ Mex, MD, 56; Am Bd Path, dipl, 63. *Prof Exp:* Resident path, Peter Brent Brigham Hosp, Boston, 56-58; intern med, Pa Hosp, 58-59; resident, Univ Hosps, Boston Univ, 59-60; assoc path, Peter Bent Brigham Hosp, 64-66; assoc prof, Inst Path, Case Western Reserve Univ, 66-72; PROF PATH & CHMN DEPT, CHARLES R DREW POSTGRAD MED SCH, 72-; PROF UNIV SOUTHERN CALIF, 72-, CHIEF PATH, MARTIN LUTHER KING JR GEN HOSP, 72- *Concurrent Pos:* Teaching fel, Harvard Med Sch, 57-58 & Sch Med, Boston Univ, 59-60; Dazian Found fel, 60-61; fel med, Peter Bent Brigham Hosp & Biophysics Res Lab, 60-64; Med Found, Inc fel, 61-64. *Mem:* Am Chem Soc; Am Asn Path & Bact; Am Soc Exp Path; Am Asn Clin Chem; Sigma Xi. *Res:* Development of accurate and sensitive methods for diagnosis and detection of disease. *Mailing Add:* Dept of Path Martin Luther King Jr Gen Hosp 12021 S Wilmington Ave Los Angeles CA 90059

AMAI, ROBERT LIN SUNG, b Hilo, Hawaii, Oct 19, 32; m 59; c 1. ORGANIC CHEMISTRY. *Educ:* Univ Hawaii, BA, 54, MS, 56; Iowa State Univ, PhD(org chem), 62. *Prof Exp:* Res assoc, 60-62, from asst prof to assoc prof, 62-74, chmn dept, 70-74, PROF CHEM, NMEX HIGHLANDS UNIV, 74- *Mem:* AAAS; Sigma Xi; Am Chem Soc. *Res:* Natural products; medicinal chemistry; correlation of structure with activity; psychopharmacology. *Mailing Add:* PO Box 859 Las Vegas NM 87701

AMANN, CHARLES A(LBERT), b Thief River Falls, Minn, Apr 21, 26; m 50; c 4. MECHANICAL & AERONAUTICAL ENGINEERING. *Educ:* Univ Minn, BS, 46, MSME, 48. *Prof Exp:* Instr, Univ Minn, 46-49; res engr, Gen Motors Res Lab, 49-54, supvry res engr, 54-71, asst head, Gas Turbine Res Dept, 71-73, head, Engine Res Dept, 73-89, res fel, 89-91; RETIRED. *Concurrent Pos:* Spec instr, Wayne State Univ, 52-55; vis prof, Univ Ariz, 83; guest lectr, Mich State Univ, 83- *Honors & Awards:* Colwell Merit Award, Soc Automotive Engrs, 72 & 84; James Clayton Fund Prize, Brit Inst Mech Engrs, 75; Richard T Woodbury Award, Am Soc Mech Engrs, 89. *Mem:* Nat Acad Eng; Am Soc Mech Engrs; Sigma Xi; Combustion Inst; fel Soc Automotive Engrs. *Res:* Engines; automotive emissions; thermodynamics; fluid mechanics; turbomachinery. *Mailing Add:* 984 Satterlee Rd Bloomfield Hills MI 48304

AMANN, JAMES FRANCIS, b Buffalo, NY, Jan 8, 45. INTERMEDIATE ENERGY, NUCLEAR PHYSICS. *Educ:* Fordham Univ, BS, 66; State Univ NY Stony Brook, MS, 68, PhD(physics), 72. *Prof Exp:* Res assoc, Carnegie-Mellon Univ, 72-77; MEM STAFF, LOS ALAMOS NAT LAB, 77- *Res:* Medium energy physics; nuclear structure using proton and pions as probes. *Mailing Add:* 1171 San Idlefonso Rd Los Alamos NM 37544

AMANN, RUPERT PREYNOESSL, b Boston, Mass, Dec 27, 31; m 63; c 2. REPRODUCTIVE PHYSIOLOGY. *Educ:* Univ Maine, BS, 53; Pa State Univ, MS, 57, PhD(dairy sci), 61. *Prof Exp:* Asst reprod physiol, Pa State Univ, 55-61, res assoc, 61-62; NIH res fel, Royal Vet & Agr Col, Denmark, 62-63; from asst prof to prof dairy physiol, Pa State Univ, 63-79; PROF PHYSIOL, COLO STATE UNIV, 79-, HEAD DEPT, 89- *Concurrent Pos:* Vis prof, Dept Physiol Biophys, Colo State Univ, 75-76; Vis prof, Dept Biochem, Washington State Univ, 90. *Honors & Awards:* Physiol & Endocrinol Award, Am Soc Animal Sci, 79; Res Award, Nat Asn Animal Breeders, 82; Physiol Award, Am Dairy Sci Asn, 84. *Mem:* AAAS; Am Dairy Sci Asn; Am Soc Animal Sci; Soc Study Reprod (dir, 67-69); Brit Soc Study Fertil; Am Soc Andrology (dir, 77-80, secy, 83-86, pres, 89-90). *Res:* Male reproductive physiology; testicular and epididymal physiology; spermatogenesis; spermatozoan maturation; semen physiology; reproductive capacity and sexual behavior of the male. *Mailing Add:* Animal Reproduction Lab Colo State Univ Ft Collins CO 80523

AMARA, ROY C, b Boston, Mass, Apr 7, 25; m 49; c 3. SYSTEMS SCIENCE, POLICY RESEARCH. *Educ:* Mass Inst Technol, SB, 48; Harvard Univ, AM, 49; Stanford Univ, PhD(elec eng), 58. *Prof Exp:* Jr res engr control systs, Stanford Res Inst, 52-53, res engr digital comput, 53-55, sr res engr systs eng, 55-61, mgr systs eng, 61-64, asst to pres, 63, exec dir systs sci, 64-68, vpres inst progs, 68-69; exec vpres, 70-71, pres, RES FEL, INST FOR FUTURE, 90- *Concurrent Pos:* US deleg, First Int Fedn Automatic Control, Moscow. *Mem:* AAAS; Inst Elec & Electronics Engrs; Inst Mgt Sci; Opers Res Soc Am. *Res:* Information systems; communication networks; technology assessment; methodologies for long range planning; decision analysis; private-public sector interfaces; future of the city; corporate planning. *Mailing Add:* 491 La Mesa Dr Menlo Park CA 94025

AMARANATH, L, b Mysore, India, July 27, 39; US citizen; m; c 2. ANESTHESIOLOGY. *Educ:* Mysore Univ, BS & MB, 63; Christian Med Col, India, DA, 67; Am Bd Anesthesiol, dipl, 72. *Prof Exp:* Intern, K R Hosp, Mysore, India, 64-65; resident, CMC Hosp, Vellore, India, 66-67, Parkland Hosp, Dallas, Tex, 67-69, Univ Hosp, Cleveland, Ohio, 69-70, fel, 70-71; researcher, Sch Med, Case Western Reserve Univ, Cleveland, Ohio, 71-72, from instr to asst prof, 72-78; STAFF, DEPT GEN ANESTHESIOL, CLEVELAND CLIN FOUND, OHIO, 78- , DEPT ANESTHESIOL, CLEVELAND CLIN FLA, FT LAUDERDALE, 88- *Concurrent Pos:* Lectr, Bellary Med Col, Karnataka St, India, 65-66; prin investr or co-investr grants, NIMH, 74-77, Nat Inst Gen Med Sci, 75-78, Hoffman-LaRoche Co, 80-81, Cleveland Clin Found, 84-86 & 88-89; sect head, pain ther, Cleveland Clin Found, 81-82, orthop anesthesia, 85-88. *Mem:* Fel Am Col Anesthesiologists; Am Acad Pain Mgt; AMA; Am Soc Anesthesiologists; Am Soc Clin Pharmacol & Therapeut; Am Soc Pharmacol & Exp Therapeut; Int Asn Study Pain. *Res:* Non-surgical control of musculoskeletal cancer pain; anesthesia for patients with multiple neurofibromatosis; postoperative pain management with epidural morphine in pediatric patients; atracurium and phenochromocytoma. *Mailing Add:* Dept Anesthesiol Cleveland Clin Fla 3000 W Cypress Creek Rd Ft Lauderdale FL 33309

AMAREL, S(AUL), b Saloniki, Greece, Feb 16, 28; nat US; m 53; c 2. COMPUTER SCIENCE. *Educ:* Israel Inst Technol, BS, 48, IngEE, 49; Columbia Univ, MS, 53, DEngSc, 55. *Prof Exp:* Res engr, Sci Dept, Israeli Ministry of Defense, 48-52 & 55-57; res engr, Electronics Res Labs, Columbia Univ, 53-55; head comput theory res, RCA Labs, 57-69; dir, Rutgers Res Resource Comput Biomed, 71-83, info sci & technol off, Defense Advan Res Proj Agency, 85-87, ALAN M TURING PROF COMPUT SCI, RUTGERS UNIV, 87- *Concurrent Pos:* Comt mem, Int Joint Artificial Intel Conf, 81-89, gen chmn, 83; trustee, Ramapo Col NJ, 69-73; mem chem/bio info handling rev comt, Nat Inst Health, 71-75, mem exec & adv comt, SUMEX-AIM, 74-; dir, Rutgers Res Resource Comput in Biomed, 71-83; vis scholar, Stanford Univ, 79, vis sr res scientist, Carnegie Mellon Univ, 85, vis res fel, SRI Intern, 83. *Mem:* Fel AAAS; Asn Comput Mach; Soc Indust & Appl Math; Sigma Xi; fel Inst Elec & Electronics Engrs; founding fel Am Asn Artificial Intel. *Res:* Artificial intelligence; computer linguistics; theory of algorithms; applications of artificial intelligence to medicine, science and engineering. *Mailing Add:* Dept Comput Sci Busch Campus Rutgers Univ New Brunswick NJ 08903

AMAROSE, ANTHONY PHILIP, b Oneonta, NY, Mar 17, 32. GENETICS, BIOETHICS. *Educ:* Fordham Univ, BS, 53, MS, 57, PhD(cytol), 59. *Prof Exp:* Asst biol, Fordham Univ, 56-57; instr, Marymount Col, NY, 57-59; resident res assoc, Argonne Nat Lab, 59-61; res assoc radiobiol, Cancer Res Inst, New Eng Deaconess Hosp, Boston, 61-62; asst prof cytogenetics & lectr path, Albany Med Col, 63-66, res assoc prof, 66-67; asst prof, Univ Chicago, 67-71, dir, Cytogenetics Lab, 77-90, ASSOC PROF OBSTET & GYNEC, CHICAGO LYING-IN HOSP, 71-, ASSOC PROF, BIOL SCI DIV & BIOL SCI COL DIV, UNIV CHICAGO, 90- *Concurrent Pos:* Fel gen path, Harvard Med Sch, 61; personal consult to Shields Warren, MD, Cancer Res Inst, Mass, 62-67; abstractor, Excerpta Medica Found, 63-88, mem int ed bd, 73-88; consult, Albany Med Ctr Hosp, 63-67 & Inst Defense Anal, Arlington, Va, 65-67; hon trustee Charles A Berger Scholar Fund, Fordham Univ, 66-; mem int ed bd, Excerpta Medica Found, 73-; pres, Genetics Task Force, Ill, 89-90. *Mem:* AAAS; Am Soc Cell Biol; Am Inst Biol Sci; Environ Mutagen Soc; Am Soc Human Genetics; Sigma Xi. *Res:* Pharmacogenetics, radiation cytology; prediction of fetal sex; chromosomology; human bone marrow; contemporary issues in biomedical ethics. *Mailing Add:* Biol Sci Div & Biol Sci Col Div HM-501 Univ Chicago 1116 E 59th St Chicago IL 60637

AMASINO, RICHARD M, US citizen. PLANT HORMONES, FLOWERING. *Educ:* Pa State Univ, BS, 77; Ind Univ, PhD(biol & biochem), 82. *Prof Exp:* Postdoctoral, Univ Wash, 82-85; asst prof, 85-90, ASSOC PROF BIOCHEM, UNIV WIS-MADISON, 90- *Concurrent Pos:* NSF presidential young investr award, 88. *Res:* Plant development; regulation of cell division and differentiation by cytokinins and the control of flowering. *Mailing Add:* Dept Biochem Univ Wis Madison WI 53706

AMASSIAN, VAHE EUGENE, b Paris, France, Nov 11, 24; nat US; m 56; c 1. NEUROPHYSIOLOGY. *Educ:* Cambridge Univ, BA, 46, MB, 48. *Prof Exp:* House physician, Middlesex Hosp, London, 48-49; from instr to assoc prof physiol, Sch Med, Univ Wash, 49-55; prof, Albert Einstein Col Med, 57-72; PROF PHYSIOL & CHMN DEPT, STATE UNIV NY DOWNSTATE MED CTR, 72- *Concurrent Pos:* Markle Found scholar, 52-59; mem postgrad training comt, USPHS, 59-63. *Mem:* Am Physiol Soc. *Res:* Neurophysiology of cerebral cortex; brainstem and sensory systems. *Mailing Add:* Dept Physiol SUNY Health Sci Ctr Box 31 Brooklyn NY 11203

AMATA, CHARLES DAVID, b Agata, Italy, Feb 11, 41; US citizen; m 61; c 3. PHYSICAL CHEMISTRY. *Educ:* John Carroll Univ, BS, 64; Univ Notre Dame, PhD(chem), 68. *Prof Exp:* Res asst chem, Parma Tech Ctr, Union Carbide Corp, 60-64, res scientist, Carbon Prod Div, 67-71; sr res scientist, Addressograph-Multigraph Corp, 71-74; mgr corp res, 74-80, tech dir, Plastics Div, 80-83, VPRES, CONWED PLASTICS CORP, 83- *Mem:* Am Chem Soc; Soc Plastics Engrs. *Res:* Molecular luminescence and energy transfer; thermal degradation of polymeric systems; organic photoconductors; high temperature inorganic fibers; composites; specialty plastics. *Mailing Add:* 770 29th Ave S E Minneapolis MN 55414

AMATO, JOSEPH CONO, b Hartford, Conn, Jan 21, 45; m 66; c 2. SUPERCONDUCTIVITY, LOW TEMPERATURE PHYSICS. *Educ:* Univ Conn, BA, 66; Stevens Inst Technol, MS, 68; Rutgers State Univ NJ, PhD(physics), 75. *Prof Exp:* Mem tech staff grad study, Bell Labs, 66-70; res assoc physics, McGill Univ, 75-77; res assoc physics, Cornell Univ, 81-84; vis asst prof, 77-81, asst prof, 84-86, ASSOC PROF PHYSICS, COLGATE UNIV, 86- *Concurrent Pos:* Vis asst prof, Cornell Univ, 85, vis assoc prof, 86-, vis scientist, 87-88; mem nat adv bd, Continuous Electron Beam Accelerator Facil, Newport News, Va, 86-88, consult, 87-89. *Mem:* Am Phys Soc. *Res:* Basic studies and applications of superconductivity; microwave response of thin films; applications to particle accelerators; low temperature condensed matter physics. *Mailing Add:* Dept Physics & Astron Colgate Univ Hamilton NY 13346

AMATO, R STEPHEN S, b Brooklyn, NY, July 11, 36; m 83; c 3. HUMAN GENETICS, PEDIATRICS. *Educ:* Manhattan Col, BS, 58; Columbia Univ, MA, 59; NY Univ, PhD(genetics), 68; Univ Nebr, MD, 73. *Prof Exp:* Instr basic sci, St Johns Riverside Hosp, Yonkers, NY, 60-63; asst prof biol, State Univ NY Westchester, 63-65; fel genetics, NY Univ & Beth Israel Med Ctr, 65-67; cytogeneticist, Div Labs, Beth Israel Med Ctr, 68; asst prof human genetics, Univ Nebr Med Ctr, Omaha, 68-72, instr pediat & anat, 72-76, dir lab med & molecular genetics, 70-76, med geneticist, Muscular Dystrophy Clin, 74-76; from assoc prof to prof pediat, Med Ctr, WVA Univ, 76-86, dir med genetics, 76-86; CHMN PEDIAT, GREATER BALTIMORE MED CTR, 86-, CLIN PROF PEDIAT, UNIV MD, 86- *Concurrent Pos:* Lectr pediat, Johns Hopkins Univ, 86- *Mem:* Am Soc Human Genetics; Am Soc Cell Biol; Am Genetics Soc; Pub Health Asn. *Res:* Chromosome structure and function; cell cycle and nucleic acid synthesis; slow virus infections and the central nervous system; genetic counseling and delivery of genetics and pediatrics services. *Mailing Add:* Greater Baltimore Medical Ctr 6701 N Charles St Baltimore MD 21204

AMATO, VINCENT ALFRED, b New York, NY, July 20, 15; m 43. PLANT PATHOLOGY, HORTICULTURE. *Educ:* Mich State Univ, BS, 54, MS, 55; Tex A&M Univ, PhD(plant path), 64. *Prof Exp:* Res instr hort, Mich State Univ, 54-56; asst exten horticulturist, WVa Univ, 56-57; instr floricult, Tex A&M Univ, 59-60 & plant path, 60-64, res asst plant physiol & microtech, 64; from asst prof to assoc prof, 64-73, PROF HORT, SAM HOUSTON STATE UNIV, 73- *Mem:* Am Phytopath Soc; Am Soc Hort Sci. *Res:* Horticultural crops; microtechniques; pathological and physiological plant parasitism. *Mailing Add:* 3353 Winter Way Huntsville TX 77340

AMAZEEN, PAUL GERARD, b Portsmouth, NH, Mar 24, 39; m 61; c 1. ELECTRICAL ENGINEERING. *Educ:* Univ NH, BS, 61; Worcester Polytech Inst, MS, 64, PhD(elec eng), 71. *Prof Exp:* Res assoc physics, Univ NH, 61-62; instr elec eng, Worcester Polytech Inst, 64-70; chief engr, Vitacomp, Inc, 70-72; prog mgr, Raytheon Med Electronics, 72-74; opers mgr, Rohe Sci Corp, 74-76; ENG MGR, FOLSOM OPER, GEN ELEC CO, 76- *Concurrent Pos:* Partner, United Eng Consult, 67- *Mem:* Inst Elec & Electronics Engrs; Am Inst Ultrasound Med; Sigma Xi; Am Soc Echocardiography. *Res:* Communication-information theory; signal processing, especially in biomedical situations such as processing of ultrasound echo signals. *Mailing Add:* 27195 Via Aurora Mission Viejo CA 92691-2161

AMBEGAOKAR, VINAY, b Nagpur, India, Jan, 16, 34; m 56; c 2. SOLID STATE PHYSICS, THEORETICAL PHYSICS. *Educ:* Mass Inst Technol, MS & SB, 56; Carnegie Inst Technol, PhD(physics), 60. *Prof Exp:* Res assoc physics, Niels Bohr Inst, Copenhagen, 60-62; from asst prof to assoc prof, 62-68, PROF PHYSICS, CORNELL UNIV, 68- *Concurrent Pos:* Dir, Inst Theoret Physics, Univ Helsinki, Finland, 69-71; vis scientist, IBM Watson Res Lab, 76-77; Guggenheim fel, Univ Calif, Santa Barbara, 83-84. *Mem:* Fel Am Phys Soc. *Res:* Theory of condensed matter physics; superconducting tunnel junctions; equilibrium and transport properties of superconductors; superfluid helium 3 and superfluid helium 4 films; properties of semiconductors and disordered systems. *Mailing Add:* Clark Hall Cornell Univ Ithaca NY 14853

AMBELANG, JOSEPH CARLYLE, b Bellevue, Ohio, Nov 7, 14; m 48; c 2. ORGANIC CHEMISTRY, RUBBER CHEMISTRY. *Educ:* Univ Akron, BS, 35; Yale Univ, PhD(chem), 38. *Prof Exp:* Asst chem, Yale Univ, 38-40; instr, D'Youville Col, 40-42; res chemist, Firestone Tire & Rubber Co, 42-56; sr compounder, Goodyear Tire & Rubber Co, 56-61, prin compounder, 61-83, res & develop assoc, 83-84; CONSULT, 84- *Honors & Awards:* Dinsmore Award, 61. *Mem:* Am Chem Soc; Am Soc Testing Mat. *Res:* Pyrimidines; synthesis and catalytic hydrogenation; rubber chemistry; softeners; tackifiers; age-resistors; textile treating; antiozonants; curing agents; vulcanization; heat flow; analysis of test data. *Mailing Add:* 366 Dorchester Rd Akron OH 44320

AMBERG, HERMAN R(OBERT), b Germ, Feb 10, 20; nat US; m 47; c 3. SANITARY & CHEMICAL ENGINEERING. *Educ:* Syracuse Univ, BS, 43; Rutgers Univ, MS, 49, PhD(sanit eng), 51. *Prof Exp:* Res engr, Johns-Manville Res Ctr, 46-48 & Eng Exp Sta, Ore State Col, 51-56; proj leader cent res dept, Crown Zellerbach Corp, 56-61, mgr chem & biol res, 61-71, dir environ serv, Cent Res Div, 71-85; RETIRED. *Concurrent Pos:* Consult environ eng, 85- *Honors & Awards:* Individual Achievement Award, Pollution Control Asn, 71; Environ Award, Tech Asn Pulp & Paper Indust, 74. *Mem:* AAAS; Am Chem Soc; Am Tech Asn Pulp & Paper Indust; Water Pollution Control Fedn; fel Tech Asn Pulp Paper Indust. *Res:* Chemical and sanitary engineering research on industrial wastes from pulp and paper industry. *Mailing Add:* 2713 Balboa Dr Vancouver WA 98684-9182

AMBLER, ERNEST, b Bradford, Eng, Nov 20, 23; nat US; m 55; c 2. CRYOGENICS, SCIENCE ADMINISTRATION. *Educ:* Oxford Univ, BA, 44, MA & PhD(physics), 53. *Prof Exp:* Physicist, Metall Lab, Armstrong Siddeley Motors, Ltd, Eng, 44-48; physicist, Cryogenic Physics Sect, 53-65, head div inorg mat, 65-68, dir, Inst Basic Stand, 68-73, dep dir, 73-75, actg dir, 75-78, DIR, NAT BUR STANDARDS, 78- *Concurrent Pos:* Guggenheim fel, 63; US Rep to Int Comt Weights & Measures, 72; chmn, Consult Comt Ionizing Radiation, 74; exofficio pres, Nat Conf Weights & Measures, 78; Dept Com mem bd gov, Israel/US Bi-Nat Indust Res & Develop Found, 81. *Honors & Awards:* Arthur S Flemming Award, 60; John Price Wetherill Medal, Franklin Inst, 62; Samuel Wesley Stratton Award, 64; William A Wildhack Award, 76; President's Award for Distinguished Fed Civilian Serv, 77. *Mem:* Am Phys Soc; fel AAAS. *Mailing Add:* Nat Inst Standards & Technol Washington DC 20234

AMBLER, JOHN EDWARD, b Bidwell, Ohio, Apr 5, 17; m 62; c 1. PLANT NUTRITION. *Educ:* Marshall Univ, BS, 52, MS, 53; Univ Md, PhD(plant physiol), 69. *Prof Exp:* Plant physiologist, Stress Lab, USDA, 56-78; RETIRED. *Mem:* Soc Am Plant Physiologists; Am Soc Agron. *Res:* Effect of stress-nutrition, temperature, water, ultraviolet B light, on growth and development of food and fiber crops. *Mailing Add:* 28 Manteo Ave Hampton VA 23661-3443

AMBLER, MICHAEL RAY, b Wichita, Kans, Feb 20, 47; m 68; c 1. POLYMER CHEMISTRY. *Educ:* Wichita State Univ, BS, 68; Akron Univ, MS, 71, PhD(polymer sci), 75. *Prof Exp:* Develop engr, 68-75, proj leader, 75-77, sect mgr, 77-87, DEPT MGR, GOODYEAR TIRE & RUBBER CO, 87- *Mem:* Am Chem Soc. *Res:* Polymer characterization; polymer rheology; physical chemistry of colloidal and macromolecular species; analytical chemistry; spectroscopy; emulsion polymerization, structure property relationships; process engineering. *Mailing Add:* Corp Res Goodyear Tire & Rubber Co Akron OH 44305

AMBORSKI, LEONARD EDWARD, b Buffalo, NY, Aug 23, 21; m 44; c 2. PHYSICAL CHEMISTRY, POLYMER CHEMISTRY. *Educ:* Canisius Col, BS, 43; State Univ NY Buffalo, AM, 49, PhD(chem), 52. *Prof Exp:* Civilian instr physics, US Army Air Force, Canisius Col & civilian res physicist magnetism, Carnegie Inst, 44-45; res chemist polymer chem, 45-59, staff scientist, 59-62, group mgr, bldg mat div, 62-68, STAFF SCIENTIST, SPECIALTY MKT DIV, FILM DEPT, E I DU PONT DE NEMOURS & CO, INC, 68-, ENVIRON COORDR, 72-, OCCUP HEALTH COORDR, 77- *Concurrent Pos:* Lectr occup health & environ, Canisius Col & State Univ NY, Buffalo; consult environ health, Toxicol Res Ctr, State Univ NY, Buffalo. *Mem:* Am Chem Soc; Am Indust Hyg Asn; Am Soc Testing & Mat; Air Pollution Control Asn; Water Pollution Control Fedn; Am Acad Indust Hyg; Am Lung Asn; Sigma Xi; Am Soc Safety Engrs. *Res:* Chemistry of high polymers; physical and chemical properties of synthetic fibers; properties of polymeric films; electrical properties of polymers; structure-property relationship of polymers; rubber chemistry; mechanism of reinforcement plastic building materials; environmental health. *Mailing Add:* 62 Wedgewood Dr Buffalo NY 14221

AMBRE, JOHN JOSEPH, b Aurora, Ill, Sept 14, 37; m 62; c 4. CLINICAL PHARMACOLOGY. *Educ:* Notre Dame Univ, BS, 59; Loyola Univ, Chicago, MD, 63; Univ Iowa, PhD(pharmacol), 72. *Prof Exp:* Clin fel internal med, Mayo Clin, 66-68; fel clin pharmacol, Univ Iowa, 68-72, from asst prof to assoc prof med & pharmacol, 72-78; dir, Clin Bio-Tox Labs, Inc, 78-80; assoc prof med, Med Sch, Northwestern Univ, 80-90; DIR, DEPT TOXICOL, AMA, 90- *Concurrent Pos:* Clin investigatorship, Vet Admin, 73; consult, Met Path Inc, 80- *Mem:* Am Soc Clin Pharmacol & Therapeut; Am Fedn Clin Res; Am Soc Pharmacol & Exp Therapeut; Am Acad Clin Toxicol; AAAS. *Res:* Drug metabolism in man; toxicology. *Mailing Add:* AMA 515 N State St Chicago IL 60610

AMBROMOVAGE, ANNE MARIE, b Gilberton, Pa, Aug 27, 36. PHYSIOLOGY. *Educ:* Susquehanna Univ, BA, 58; Jefferson Med Col, MS, 61, PhD(physiol), 68. *Prof Exp:* Res instr, 64-66, 68-69, res asst prof, 69-70, ASST PROF HUMAN PHYSIOL, HAHNEMANN MED COL, 70-, RES ASSOC, 63- *Concurrent Pos:* Merck Found fac develop grant, 70. *Mem:* AAAS; NY Acad Sci. *Res:* Gastrointestinal physiology; water of electrolyte absorption; pancreatic enzymes; mesenteric blood flow; physiology of shock and vasoactive substances. *Mailing Add:* Dept Physiol-Biophysics 100 Lombard St Philadelphia PA 19147

AMBRON, RICHARD THOMAS, b New York, NY, June 15, 43; m 66; c 2. NEUROBIOLOGY, NEUROCHEMISTRY. *Educ:* Villanova Univ, BS, 65; Med Sch, Temple Univ, PhD(biochem), 71. *Prof Exp:* Fel neurobiol, Sch Med, NY Univ, 71-74; instr, 74-76, ASST PROF ANAT & CELL BIOL, COL PHYSICIANS & SURGEONS, COLUMBIA UNIV, 76- *Concurrent Pos:* Mem, Div Neurobiol & Behav, Columbia Univ, 74-; career develop award, NIH, 79. *Mem:* Soc Neurosci; NY Acad Sci; Sigma Xi. *Res:* Synthesis, assembly and distribution of membrane components in single identified neurons of the marine mollusc aphysia californica, to elucidate the mechanisms underlying the sorting and movement of organelles in neurons, to maritan neurons in the differentiated state. *Mailing Add:* Anat Dept Col Physicians & Surgeons 630 W 168 St New York NY 10032

AMBROSE, CHARLES T, b Indianapolis, Ind, Nov 29, 29. IMMUNOLOGY. *Educ:* Ind Univ, AB, 51; Johns Hopkins Univ, MD, 55. *Prof Exp:* Intern med, New Eng Med Ctr, Boston, 55-56; asst resident infectious dis, Mass Mem Hosp, 56-57; resident, New Eng Med Ctr, 57-59; from instr to assoc prof bacteriol & immunol, Harvard Med Sch, 62-72; assoc prof, Col de France, Paris, 72-73; PROF CELL BIOL, SCH MED, UNIV KY, 73- *Concurrent Pos:* NSF fel, 59-60; res fel bacteriol & immunol, Harvard Med Sch, 59-62; dir res, INSERM, Paris, 72-73. *Mem:* Am Asn Immunologists; Brit Soc Immunol; Soc Fr Immunol; Reticuloendothelial Soc; Soc Exp Biol & Med; Sigma Xi. *Res:* Regulation of antibody synthesis in vitro; organ cultures; antigenic competition; antimetabolites; salicylates; insecticides; corticosteroids. *Mailing Add:* 163 N Arcadia Park Lexington KY 40503

AMBROSE, ERNEST R, b Montreal, Que, May 12, 26; m 49; c 7. DENTISTRY. *Educ:* McGill Univ, DDS, 50; FRCD(C). *Prof Exp:* From instr to assoc prof oper dent, McGill Univ, 50-77, chmn dept, 58-70, dean fac dent, 70-77; dean, 77-85, PROF RESTORATIVE DENT, COL DENT, UNIV SASK, 77- *Concurrent Pos:* Consult, Montreal Gen Hosp, 58-77. *Mem:* Fel Am Col Dent; Can Acad Restorative Dent; fel Int Col Dent. *Res:* Restorative dentistry; clinical use of dental materials; assessing and maintaining the pulp potential of human teeth. *Mailing Add:* Col Dent Univ Sask Saskatoon SK S7N 0W0 Can

AMBROSE, HARRISON WILLIAM, III, b Winter Haven, Fla, Feb 26, 38; m; c 2. ECOLOGY, ANIMAL BEHAVIOR. *Educ:* Univ Fla, BS, 60; Univ Ky, MS, 62; Cornell Univ, PhD(ecol), 67. *Prof Exp:* From instr to asst prof biol, Cornell Univ, 66-73; asst prof biol, Univ Ill, Urbana, 73-76; assoc prof biol, Univ Tenn, 76-80; PART-TIME TEACHING. *Mem:* AAAS; Am Soc Mammal; Ecol Soc Am; Animal Behav Soc. *Res:* Behavioral animal ecology; social and orientation behavior; population regulation; predator-prey interactions. *Mailing Add:* Zool Dept Univ Tenn Knoxville TN 37916

AMBROSE, JOHN AUGUSTINE, nutritional biochemistry, biochemical genetics; deceased, see previous edition for last biography

AMBROSE, JOHN DANIEL, b Detroit, Mich, June 20, 43; m 71; c 1. CONSERVATION BIOLOGY, RESTORATION ECOLOGY. *Educ:* Univ Mich, BS, 65, MS, 66; Cornell Univ, PhD(bot), 75. *Prof Exp:* Asst hydrographic officer, US Navy, 67-68; CUR ARBORETUM, UNIV GUELPH, 74-, GRAD FAC, 89- *Concurrent Pos:* Mem, Comt Status Endangered Wildlife Can, 84- & Grand River Conserv Authority, 86- *Mem:* AAAS; Am Asn Bot Gardens & Arboreta; Soc Conserv Biol; Can Bot Asn; Sigma Xi; Soc Restoration Ecol. *Res:* Conservation biology of the rare woody plants of Ontario; conservation roles of botanical gardens; forest restoration ecology. *Mailing Add:* Univ Guelph The Arboretum Guelph ON N1G 2W1 Can

AMBROSE, JOHN E, chemical engineering, for more information see previous edition

AMBROSE, JOHN RUSSELL, b Orange, NJ, Feb 25, 40; m 62; c 2. PHYSICAL CHEMISTRY, MATERIALS SCIENCE. *Educ:* Washington & Lee Univ, BSc, 61; Univ Md, College Park, PhD(phys chem), 72. *Prof Exp:* High sch teacher chem, Va Beach City Sch Bd, 62-64; res chemist, Newport News Shipbuilding & Dry Dock Co, 64-66; res chemist, Nat Bur Standards, 66-78; ASSOC PROF, UNIV FLA, 78- *Honors & Awards:* Romanoff Award, Nat Asn Corrosion Engrs, 74. *Mem:* Electrochem Soc; Nat Asn Corrosion Engr; Am Soc Testing & Mat. *Res:* Relationships between susceptibility to localized corrosion attack and interfacial properties; repassivation kinetics and electrochemistry of solutions. *Mailing Add:* Dept Mat Sci Univ Fla Gainesville FL 32611

AMBROSE, RICHARD JOSEPH, b Youngstown, Ohio, July 4, 42; m 65; c 3. POLYMER SCIENCE. *Educ:* Bowling Green State Univ, BS, 64; Univ Akron, PhD(polymer sci), 68. *Prof Exp:* Res scientist, Plastics Div, Firestone Tire & Rubber Co, 68-73, group leader, 73-79; mgr mat & process res & develop, 79-85, ASSOC DIR RES, LORD CORP, 85- *Mem:* Am Chem Soc. *Res:* Chemical reactions of polymers; hydrolysis of acrylate and methacrylate polymers; structure physical property relationships of block copolymers; thermosetting resins; polar-nonpolar block polymers; cationic and anionic polymerization; polymer processing; composites; morphology of multiphase systems; surface chemistry of organic-metal systems. *Mailing Add:* 405-407 Gregson Dr Thomas Lord Res Ctr PO Box 8225 Cary NC 27512-8225

AMBROSE, ROBERT T, b Palmdale, Calif, Sept 21, 33. ANALYTICAL CHEMISTRY, CLINICAL CHEMISTRY. *Educ:* Univ Calif, Riverside, BA, 60; Northwestern Univ, PhD(anal chem), 65. *Prof Exp:* RES ASSOC, EASTMAN KODAK CO RES LABS, 64- *Concurrent Pos:* Instr, Rochester Inst Technol, 72-74. *Mem:* Am Chem Soc; Sigma Xi. *Res:* Analytical approaches to reference methods in clinical chemistry; application of chromatographic procedures to the analysis of diverse analytes in human serum. *Mailing Add:* 35 Pine Knoll Dr Rochester NY 14624

AMBROSIANI, VINCENT F, b Oneida, NY, Aug 4, 29; m 52; c 8. PHYSICAL CHEMISTRY. *Educ:* Le Moyne Col, NY, BS, 52; Syracuse Univ, MS, 57, PhD(phys chem), 64. *Prof Exp:* Res chemist, Exp Sta, E I du Pont de Nemours & Co, Inc, 59-63 & Textile Res Lab, 63-66; tech dir fiber & fabric develop, Blue Ridge-Winkler Textiles, Pa, 66-68; dir res & develop, Vanity Fair Mills, Inc, 68-70, dir res & develop, V F Corp, Reading, 70-82; sr scientist, Burlington Indust, 84, res mgr res & develop, 85-87; OWNER, AMBROSIANI CONSULTS, TEXTILE PROD DEVELOP, 87- *Mem:* Am Chem Soc; Am Asn Textile Chem & Colorists. *Res:* Fibers, yarns and fabrics in new and improved fabrics for apparel and industrial applications. *Mailing Add:* 47 Brighton Pl Greensboro NC 27410-9332

AMBRUS, CLARA MARIA, b Rome, Italy, Dec 28, 24; nat US; m 45; c 7. HEMATOLOGY, PEDIATRICS. *Educ:* Univ Zurich, MD, 49; Jefferson Med Col, PhD, 55; Am Bd Clin Chem, dipl. *Prof Exp:* Asst histol, Budapest Med Sch, 43-46, demonstr pharmacol, 46-47; asst, Med Sch, Univ Zurich, 47-49; asst therapeut chem & virol, Pasteur Inst, Univ Paris, 49; asst pharmacol, Philadelphia Col Pharm, 50-51, from asst prof to assoc prof, 51-55; asst prof pharmacol & assoc med, Sch Med, State Univ NY Buffalo, 55-65, asst res prof pediat, 65-69, assoc res prof pediat, 69-77; assoc cancer res scientist, Roswell Park Mem Inst, 55-69, assoc res prof pharmacol, 66-69, prin cancer res scientist, 69-83, PROF PHARMACOL, ROSWELL PARK MEM INST, 70-; RES PROF PEDIAT, SCH MED, STATE UNIV NY BUFFALO, 69-, OBSTET/GYNEC, 83- *Mem:* Am Asn Cancer Res; Am Physiol Soc; Am Soc Pharmacol & Exp Therapeut; Am Soc Exp Biol; fel Am Col Physicians. *Res:* Pediatric hematology and oncology; hyaline membrane disease of infants; biochemistry and pathology of the blood coagulation and fibrinolysin systems; immobilized enzymes; transplantation and regeneration of hemic tissue; environmental health; experimental and clinical pharmacology; clinical detoxification. *Mailing Add:* 143 Windsor Ave Buffalo NY 14209

AMBRUS, JULIAN LAWRENCE, b Budapest, Hungary, Nov 29, 24; nat US; m 45; c 7. HEMATOLOGY, ONCOLOGY. *Educ:* Univ Zurich, MD, 49; Jefferson Med Col, PhD(med sci), 54; Am Bd Clin Chem, dipl, 54. *Hon Degrees:* ScD, Niagara Univ, 84. *Prof Exp:* Asst histol, Med Sch, Univ Budapest, 43-46, demonstr pharmacol, 46-47; asst, Med Sch, Univ Zurich, 47-49; asst therapeut chem & tropical med virol, Pasteur Inst, Paris, 49; from asst prof to prof pharmacol, Philadelphia Col Pharm, 50-55; from prin cancer res scientist to dir, Cancer Res (Pathoaphysiol), State Dept Health, NY, 55-89; assoc prof pharmacol, Grad Sch & from asst prof to prof internal med, Sch Med, 55-65,; PROF & CHMN, EXP PATH & INTERNAL MED, SCH MED & GRAD SCH, STATE UNIV NY, BUFFALO, 65- *Concurrent Pos:* Consult adv comt coagulation components, Comn Plasma Fractionation & Related Processes; consult, Bur Drugs, Food & Drug Admin; mem coun drugs, AMA; ed-in-chief, J Med & Hematol Rev; mem comt thrombolytic agents, Nat Heart Inst; fel, Am Col Physicians & coun clin cardiol, Am Heart Asn; mem, Dept Med, Roswell Park Mem Inst, 55-, head, Dept Pathophysiol, 75-89. *Honors & Awards:* Nelson Hacket Gold Medal, 72; Cheplow Award Med, Milard Filmore Hosp, 84. *Mem:* Fel AAAS; Am Soc Pharmacol & Exp Therapeut; Am Fedn Clin Res; Asn Am Med Cols; fel NY Acad Sci; fel Am Col Physicans. *Res:* Hemorrhagic and thromboembolic diseases; blood coagulation, platelet and fibrinolysin systems; physiology of the leukocytes; leukemias; radiation sickness; biochemistry and chemotherapy of neoplastic diseases; experimental and clinical pharmacology; resistance to drugs; pathophysiology of AIDS; immunotherapy. *Mailing Add:* 143 Windsor Ave Buffalo NY 14209

AMBS, LAWRENCE LACY, b St Paul, Minn, Aug 14, 37; m 57; c 3. MECHANICAL ENGINEERING. *Educ:* Univ Minn, BSME, 60, MSME, 64, PhD(mech eng), 68. *Prof Exp:* Teaching assoc mech eng, Univ Minn, 66-67; asst prof, 68-72, ASSOC PROF MECH ENG, UNIV MASS, AMHERST, 72- *Honors & Awards:* Ralph Teetor Award, Soc Automotive Engrs, 67. *Mem:* Am Soc Mech Engrs; Combustion Inst: Am Soc Eng Educ. *Res:* Combustion research involving gases and solids; combustion in propulsion applications; reduction of exhaust emissions in combustion devices; solar energy; ocean thermal energy conversion; energy conservation. *Mailing Add:* Dept Mech Eng Univ Mass Amherst Campus Amherst MA 01003

AMBS, WILLIAM JOSEPH, b Philadelphia, Pa, Dec 3, 29; m 57; c 1. SURFACE CHEMISTRY, PHYSICAL MATHEMATICS. *Educ:* Villanova Univ, BS, 52; Stevens Inst Technol, MS, 54; Cath Univ Am, PhD(phys chem), 61. *Prof Exp:* Asst chem, Stevens Inst Technol, 52-54 & Cath Univ Am, 54-56; phys chemist, Nat Bur Standards, 56-63; phys chemist, Air Prod & Chem, Inc, 63-71, sr prin res chemist, 71-84; sci programmer analyst, 85-88, SCI SYSTS ENGR, GEN ELEC, INC, 88- *Mem:* Am Chem Soc; Catalysis Soc; Sigma Xi. *Res:* Catalysis; chemical kinetics; surface chemistry of solids; field emission microscopy; thin film optics; x-ray diffraction; scientific software design and programming. *Mailing Add:* 505 E Lancaster Ave Apt 419 St Davids PA 19087-5136

AMBUDKAR, INDU S, b Madras, India, Feb 17, 54. PATIENT CARE. *Educ:* Isabella Thoburn Col, India, BS, 73; Lucknow Univ, MS, 75; Madurai-Kamaraj Univ, PhD(biochem), 79. *Prof Exp:* Jr res asst, Res Doc, Indust Toxicol Res Ctr, Lucknow, India, 75; jr res fel, Dept Biochem, Madurai-Kamaraj Univ, Madurai, India, 76-77, sr res fel, 77-79; res assoc, Dept Biol Chem, Sch Med, Univ Md, 80-83, res asst prof, Dept Path, 84-85; vis assoc, 85-88, SR STAFF FEL, CIPCB, NAT INST DENT RES, NIH, 88- *Concurrent Pos:* Res supvr, Madurai-Kamaraj Univ, Madurai, India, 77-79; teaching asst, grad level biochem, Univ Madurai, India, 77-79, biochem tutorials & conf, Univ Md, 81-83 & biochem course grad students, Dept Path, 84-85; postdoctoral fel award, Am Heart Asn, 81-82. *Mem:* Am Soc Biochem & Molecular Biol; Biophys Soc; Soc Gen Physiologists; AAAS. *Res:* Author or co-author of over 40 publications. *Mailing Add:* Clin Invest & Patient Care Br Nat Inst Dent Res NIH Bldg 10 Rm 1A05 Bethesda MD 20892

AMBUEL, JOHN PHILIP, b Broadus, Mont, Mar 23, 18; m 46; c 2. PEDIATRICS. *Educ:* Luther Col, BA, 41; Univ Chicago, MD, 46. *Prof Exp:* Intern, Doctors Hosp, Seattle, 46-47; resident pediat, Children's Hosp, Detroit, 49-51; fel, Univ Chicago, 51-53; from asst prof to prof pediat, Ohio State Univ, 53-74; prof pediat, Sch Med, Northwestern Univ, Chicago, 74-78; PROF FAMILY PRACT, DEPT PEDIAT, MED COL WIS, MILWAUKEE, 78- *Mem:* Am Acad Pediat; Ambulatory Pediat Asn; Am Pediat Soc. *Res:* Erythroblastosis; behavior development; medical care programs. *Mailing Add:* 2490 Anita Dr Brookfield WI 53005

AMBURGEY, TERRY L, b Trenton, NJ, Dec 11, 40; m 61; c 2. WOOD BIODETERIORATION, WOOD PRESERVATION. *Educ:* State Univ Col Forestry, Syracuse Univ, BS, 63, MS, 65; NC State Univ, PhD(plant path), 69. *Prof Exp:* Res plant pathologist, Southern Forest Exp Sta, 69-79; ASSOC PROF, MISS STATE UNIV, 79- *Concurrent Pos:* Mem, Int Res Group Wood Preserv. *Mem:* Asn Preserv Technol; Am Wood Preservers Asn; Forest Prod Res Soc. *Res:* Prevention and control of wood decay in structures; interactions between wood-inhabiting fungi and termites. *Mailing Add:* Drawer Fr-Forestry Miss State Univ Mississippi Station MS 39762

AMDAHL, GENE M, b Flandreau, SDak, Nov 16, 22; m 46; c 3. COMPUTER SCIENCE. *Educ:* SDak State Univ, BSEP, 48; Univ Wis, MS & PhD(theoret physics), 52. *Hon Degrees:* DEng, SDak State Univ, 74; DSc, Univ Wis-Madison, 79, Luther Col, Iowa, 80, Augustana Col, SDak, 84. *Prof Exp:* Wis Alumni Res Found res assoc comput develop, Univ Wis, 51-52; sr engr, Int Bus Mach Corp, NY, 52-55; mem tech staff comput res & develop, Ramo-Wooldridge, Calif, 56; lab mgr, Aeronutronic, Inc, 56-60; dir comput res & develop lab, IBM Corp, 60-71; pres, Amdahl Corp, 70-74, chmn bd, 70-79; CHMN BD, TRILOGY/ELXSI CORP, 80-; CHMN, PRES & CHIEF EXEC OFFICER, ANDOR INT INC, 87- *Concurrent Pos:* Vis asst prof, Stanford Univ, 65-67; IBM fel, 65-71; lectr, NATO Sch, 69 & 76. *Honors & Awards:* W Wallace McDowell Award, Inst Elec & Electronics Engrs Computer Soc, 76; Comput Sci "Man of the Year" Award, Data

Processing Mgt Asn, 76; Michelson-Morley Award, Western Reserve Univ, 77; Harry Goode Mem Award, Am Fedn Info Processing Soc, 83. *Mem:* Nat Acad Eng; Am Phys Soc; fel Inst Elec & Electronics Engrs; fel British Comput Soc; Marconi Soc, Italy. *Res:* Internal machine and system organization of high performance computing systems suited to office-like environments. *Mailing Add:* Andor Int Inc 10131 Bubb Rd Cupertino CA 95014

AMDUR, MARY OCHSENHIRT, b Pittsburgh, Pa, Feb 18, 21; m 44. TOXICOLOGY. *Educ:* Univ Pittsburgh, BS, 43; Cornell Univ, PhD(biochem), 46. *Prof Exp:* Asst ophthal res, Howe Lab, Med Sch, Harvard, 47-48; res biochemist, Vet Admin Hosp, 48-49; res assoc, 49-57, asst prof physiol, 57-63, ASSOC PROF TOXICOL, SCH PUB HEALTH, HARVARD UNIV, 63-; LECTR NUTRIT & FOOD SCI, MASS INST TECHNOL, 76-, SR RES SCIENTIST, ENERGY LAB, 80- *Honors & Awards:* Donald E Cummings Mem Award, Am Indust Hyg Asn, 74. *Mem:* AAAS; Am Indust Hyg Asn; NY Acad Sci; Soc Toxicol; Sigma Xi. *Res:* Bone formation in the rat; effect of manganese and choline on liver fat in the rat; effect of inhalation of sulfur dioxide and sulfuric acid mist on guinea pigs and humans; interaction of irritant gases and aerosols; analyses for lead in air and urine; physiologic response to respiratory irritants. *Mailing Add:* 89 Rock Meadow Rd Westwood MA 02090

AMDUR, MILLARD JASON, b Pittsburgh, Pa, Aug 23, 37; m 63; c 2. PSYCHIATRY. *Educ:* Univ Pittsburgh, BA, 59; Yale Univ, MD, 64. *Prof Exp:* Fel psychiat, Sch Med, Yale Univ, 65-68; from instr to asst prof, 68-70; dir student ment health serv & asst prof psychiat, 70-72, asst clin prof 72-81, ASSOC CLIN PROF PSYCHIAT, SCH MED, UNIV CONN, 81-; CHMN DEPT PSYCHIAT, WINDHAM COMMUNITY MEM HOSP, 75-; ASSOC DIR & CHIEF MED OFFICER, UNITED SOCIAL & MENT HEALTH SOC, INC, 78- *Concurrent Pos:* Dir, Psychiat Outpatient Div & Dana Clin, Yale-New Haven Hosp, 68-70; consult, Undercliff Ment Health Ctr, 68-70, New Fairview Hall Convalescent Hosp, 68-70 & Student Ment Health Serv, Univ Conn, 68-70 & 72-73; dir ment health clin, Windham Community Natchaug Mem Hosp, 72-77; psychiat staff, Natchaug Hosp, 73-; pres, Eastern Conn Parent-Child Resource Syst, Inc, 74-76 & med staff, Natchaug Hosp, 76-78; mem psychiat staff, Catchment Area Coun mental health, 75-81 & Eastern Regional Ment Health Bd, 75-77. *Mem:* Fel Am Psychiat Asn; Am Group Psychother Asn. *Res:* Adolescent psychiatry; guilt and conscience; community and rural psychiatry, third party reimbursement. *Mailing Add:* 28 Bush Hill Rd Willmantic CT 06226

AMEER, GEORGE ALBERT, b Norwalk, Conn, July 26, 31; m 72; c 6. OPTICAL PHYSICS. *Educ:* Univ Maine, BS, 53; Univ Pittsburgh, PhD(physics), 61. *Prof Exp:* Sr physicist, J W Fecker Div, Am Optical Co, Pa, 61-64; asst prof physics, Am Univ Beirut, 64-67; tech specialist optics, Autonetics Div, NAm Rockwell Corp, Anaheim, 67-73, supvr advan sensor technol autonetics group, 73-78; VPRES, OPTICAL SCI CO, 78- *Mem:* Am Phys Soc; Optical Soc Am. *Res:* Molecular spectroscopy; optical instrumentation; physical optics. *Mailing Add:* 12872 Bubbling Well Rd Santa Ana CA 92705

AMEIN, MICHAEL, b Kabul, Afghanistan, Jan 15, 26; US citizen; m 55; c 3. CIVIL ENGINEERING, HYDRODYNAMICS. *Educ:* Stanford Univ, BS, 52; Cornell Univ, MS, 54, PhD(civil eng), 55. *Prof Exp:* Design engr, Corbett-Tinghir & Co, NY, 55-56 & F L Ehasz, NY 56-57; hydraul engr, Anderson-Nichols & Co, Mass, 57-60; from asst prof to assoc prof, 60-70, PROF CIVIL ENG, NC STATE UNIV, 70- *Concurrent Pos:* Prin res investr, US Bur Yards & Docks, 63-64, Off Water Res, 65-69, US Weather Bur, 70-73, Sea Grant, 70-75, Environ Protection Agency, 76-78 & State NC, 79-80; consult, Parana River develop, Arg, 80-81. *Mem:* Am Soc Civil Engrs; Am Geophys Union; Int Asn Hydraulic Res. *Res:* Flow in open channels; numerical analysis; computer applications to hydrodynamical problems; long period waves; water quality modeling; flood movements; ocean outfall dispersion. *Mailing Add:* Dept Civil Eng NC State Univ Raleigh Main Campus Box 7908 Raleigh NC 27695-7908

AMELIN, CHARLES FRANCIS, b Worcester, Mass. MATHEMATICAL ANALYSIS. *Educ:* Col of the Holy Cross, AB, 64; Univ Calif, Berkeley, PhD(math), 72. *Prof Exp:* From asst prof to assoc prof, 69-80, PROF MATH, CALIF STATE POLYTECH UNIV, POMONA, 80- *Concurrent Pos:* Vis asst prof math, Univ Md, College Park, 72-73 & Ariz State Univ, 73-74. *Mem:* Am Math Soc. *Res:* Operator theory; functional analysis. *Mailing Add:* Dept Math Calif State Polytech Univ 3801 W Temple Ave Pomona CA 91768

AMELIO, GILBERT FRANK, b New York, NY, Mar 1, 43; m 63; c 3. SOLID STATE ELECTRONICS. *Educ:* Ga Inst Technol, BS, 65, MS, 67, PhD(physics), 69. *Prof Exp:* Mem tech staff device physics, Bell Tel Labs, Inc, 68-71; mgr charge coupled devices, Faidchild Camera & Instrument Corp, 71-; PRES, SEMICONDUCTOR PROD DIV, ROCKWELL INT. *Mem:* Am Phys Soc; Inst Elec & Electronics Engr; Sigma Xi. *Res:* Semiconductor surfaces with Auger spectroscopy; hot electron transport theory; silicon diode array camera tube; theory and experiment of charge coupled devices; management of design and manufacture of charge coupled devices. *Mailing Add:* Semiconductor Prod Div Rockwell Int PO Box C Newport Beach CA 92660

AMELL, ALEXANDER RENTON, b North Adams, Mass, Mar 3, 23; m 45; c 4. PHYSICAL CHEMISTRY. *Educ:* Univ Mass, BS, 47; Univ Wis, PhD(chem), 50. *Prof Exp:* Instr phys chem, Hunter Col, 50-52; asst prof chem, Lebanon Valley Col, 52-55; asst prof, 55-60, chmn dept, 60-77, PROF CHEM, UNIV NH, 60-, INTERIM DEAN, COL ENG & PHYS SCI, 80- *Concurrent Pos:* Fulbright lectr, San Marcos Univ, Lima, 63. *Mem:* Am Chem Soc. *Res:* Radiation chemistry; kinetics. *Mailing Add:* Dept Chem Univ NH Parsons Hall Durham NH 03824

AMELUNXEN, REMI EDWARD, b Kansas City, Mo, Jan 27, 28. MICROBIOLOGY. *Educ:* Rockhurst Col, BS, 49; Univ Kans, MA, 57, PhD(microbiol), 59. *Prof Exp:* USPHS fel biochem, 59-62, instr, 62-64, asst prof microbiol & biochem, 64-66, assoc prof microbiol, 66-71, PROF MICROBIOL, SCH MED, UNIV KANS, 71- *Concurrent Pos:* Res grant, USPHS; NIH grant, 64-69. *Mem:* Am Soc Microbiol; Sigma Xi. *Res:* Physico-chemical characterization of thermophilic enzymes and studies on the genetic bases of thermophily. *Mailing Add:* Dept Microbiol Univ Kans Med Ctr 39 Rainbow Blvd Kansas City KS 66103

AMEMIYA, FRANCES (LOUISE) CAMPBELL, b Riverside, Calif, June 16, 15; m 52; c 2. MATHEMATICS. *Educ:* Univ Calif, Los Angeles, AB, 35, AM, 36; Univ Mich, PhD(math), 45. *Prof Exp:* Teaching fel, Univ Calif, Los Angeles, 35-37; prof, George Pepperdine Col, 37-58; assoc prof, Calif Western Univ, 58-62 & Parsons Col, 62-64; from assoc prof to prof, 64-80, EMER PROF MATH, CALIF STATE UNIV, HAYWARD, 80- *Concurrent Pos:* Prof, chmn, dept math & physics & chmn, Div Sci & Math, George Pepperdine Col, 37-58; vis prof, Ibaraki Christian College, Japan, 49-50. *Mem:* Nat Coun Teachers Math; Math Asn Am; Sigma Xi. *Res:* Mathematical statistics; truncated bivariate normal distributions; elementary geometry for liberal studies. *Mailing Add:* 4428 Tivoli St San Diego CA 92107-3830

AMEMIYA, KEI, GENE REGULATION, RNA POLYMERASE. *Educ:* Rutgers Univ, PhD(microbiol), 73. *Prof Exp:* INSTR MOLECULAR BIOL, ALBERT EINSTEIN COL MED, YESHIVA UNIV, 84- *Res:* Transcription. *Mailing Add:* Dept Molecular Biol Albert Einstein Col Med Yeshiva Univ 1300 Morris Park Ave Bronx NY 10461

AMEMIYA, KENJIE, REPRODUCTIVE TOXICOLOGY. *Educ:* Univ Calif, Davis, BS, 82, PhD(nutrit pharmacol & physiol chem), 87. *Prof Exp:* Res asst, Calif Primate Res Ctr, Univ Calif, Davis, 83-85, teaching asst, 83-87; sr scientist, 87-89, RES SCIENTIST III, DIV TOXICOL & PATH, REPRODUCTIVE TOXICOL DEPT, CIBA-GEIGY CORP, 89- *Mem:* Teratology Soc; Behav Teratology Soc; Am Inst Nutrit. *Mailing Add:* Reproductive Toxicol Dept Ciba-Geigy Pharmaceut Div 556 Morris Ave Summit NJ 07901

AMEMIYA, MINORU, b San Francisco, Calif, Mar 17, 22; m 47; c 2. SOIL CONSERVATION, AGRONOMY. *Educ:* Univ Calif, Berkeley, BS, 42; Ohio State Univ, MS, 48, PhD(agron), 50. *Prof Exp:* Lab technician, Carpenter Bros, Inc, Wis, 43-44; res asst, Ohio Agr Exp Sta, 48-50; soil scientist, Agr Res Serv, USDA, Colo, 50-58 & Tex, 58-60, res soil scientist, Iowa State Univ, 60-68, from assoc prof agron to prof agron, 68-88, exten agronomist, 68-88, PROF EMER AGRON, IOWA STATE UNIV, 88- *Concurrent Pos:* Assoc agronomist, Colo Agr Exp Sta, 50-58; mem, Coun Agr Sci & Tech. *Mem:* Fel Soil Sci Soc Am; fel Am Soc Agron; fel Soil Conserv Soc Am; fel AAAS; Int Soc Soil Sci. *Res:* Reclamation of saline and alkali soils; soil, water and crop management; soil and water conservation; soil-water-plant relationships; soil tilth and tillage. *Mailing Add:* 304 Hilltop Rd Ames IA 50010

AMEN, RALPH DUWAYNE, b Cheyenne, Wyo, Feb 26, 28; m 52; c 4. PHILOSOPHY OF BIOLOGY & MEDICINE, BIOMEDICAL ETHICS. *Educ:* Univ Northern Colo, AB, 52, AM, 54; Univ Colo, MBS, 59, PhD(bot), 62. *Prof Exp:* Instr high sch, Colo, 55-58; instr biol, Univ Colo, 59-60, teaching assoc, 61-62; asst prof plant physiol, 62-67, chmn dept, 67-72, assoc prof, 67-80, PROF BIOL, WAKE FOREST UNIV, 80- *Concurrent Pos:* Sigma Xi res grant, 61-62; NC Bd Sci & Technol grant, 68-69; assoc ed, Ecology, 71-72. *Honors & Awards:* ISI citation classic, current contents Sept 15, 80. *Mem:* Ecol Soc; Sigma Xi. *Res:* Seed germination and dormancy; biomedical philosophy. *Mailing Add:* Dept Biol Wake Forest Univ Box 7325 Winston-Salem NC 27109

AMEN, RONALD JOSEPH, b Brooklyn, NY, Mar 3, 43; m 67; c 2. FOOD SCIENCE, NUTRITION. *Educ:* Bethany Col, BS, 63; Rutgers Univ, MS, 69, PhD(food sci), 71. *Prof Exp:* Food technologist, Dietetic Food Co Inc, Brooklyn, 64-67; res asst food sci, Rutgers Univ, 67-71; food scientist protein res, Thomas J Lipton Inc, NJ, 71-72; dept head nutrit, Syntex Res Inc, Palo Alto, 72-76; dir pharmaceut prod develop, McGaw Labs, Irvine, 76-80; VPRES & GEN MGR, ABIC INT CONSULTS, INC, ORANGE. *Mem:* AAAS; Am Pub Health Asn; Am Inst Nutrit; Sigma Xi; Inst Food Technol. *Res:* Effect of special dietary foods in clinical situations where specific nutritional adjuncts are indicated; including the effects and mechanism of fiber and peptide absorption and transport in gastrointestinal tract; effect of intravenously administered specific amino acid patterns on the exacerbation of certain disease states and on the metabolic changes in the traumatized patient. *Mailing Add:* 18101 Catherine Circle Villa Park CA 92667

AMEND, DONALD FORD, b Portland, Ore, Jan 8, 39; m 62; c 3. FISH PATHOLOGY. *Educ:* Ore State Univ, BS, 60, MS, 65; Univ Wash, PhD(fish path), 73. *Prof Exp:* Aquatic biologist, Ore Fish Comn, 60-63; res asst food toxicol, Ore State Univ, 63-65; proj leader virol, US Fish & Wildlife Serv, 66-73, sect leader microbiol, 74-76; dir res, Tavolek, Inc, Redmond, 76-81; assoc prof, Univ Calif, Davis, 81-83; GEN MGR, SOUTHERN SOUTHEAST REGIONAL AQUACULT ASN, 83- *Concurrent Pos:* Affil asst prof, Univ Wash, 74-78. *Mem:* Am Fisheries Soc; Sigma Xi; Am Soc Microbiol; AAAS. *Res:* Controlling infectious diseases of fish including chemotherapy, immunization, environmental and biological control. *Mailing Add:* 1621 Tongass Suite 103 Ketchikan AK 99901

AMEND, JAMES FREDERICK, b Mount Vernon, Wash, Jan 6, 43; m 64; c 2. CARDIOVASCULAR PHYSIOLOGY, AUTONOMIC PHARMACOLOGY. *Educ:* Pac Lutheran Univ, BS, 65; Baylor Col Med, PhD(physiol), 69; Wash State Univ, DVM, 77. *Prof Exp:* Asst prof physiol, Baylor Col Med, 70-71; asst prof zool & physiol, Univ Idaho, 72-74; res assoc physiol, Wash State Univ, 74-76; asst prof vet med, Univ Mo-Columbia, 76-78; ASSOC PROF VET SCI, UNIV NEBR, LINCOLN, 78- *Mem:* Am

Physiol Soc; Sigma Xi; NY Acad Sci; Am Acad Vet Pharmacol & Therapeut; Am Soc Vet Physiologists & Pharmacologists. *Res:* Role of pulmonary fluid transport disturbances in the susceptibility of bovine lung to respiratory disease. *Mailing Add:* Dept Anat & Physiol Atlantic Col Univ Prince Edward Island PO Box 2000 Charlottetown PE C1A 7N8 Can

AMEND, JOHN ROBERT, b Yakima, Wash, July 20, 38; m 66; c 2. ANALYTICAL CHEMISTRY. *Educ:* Pac Lutheran Univ, BA, 60; Mont State Univ, MS, 64; Univ Tex, Austin, PhD(sci educ), 67. *Prof Exp:* Dept head physics & chem, Curtis Sr High, Tacoma, Wash, 60- 65; res assoc chem, Univ Tex Austin, 66-67; PROF CHEM, MONT STATE UNIV, 67- *Concurrent Pos:* Vis scientist, Oak Ridge Assoc Univs, 70 & 73; assoc dept head, Mont State Univ, 88- *Mem:* Am Chem Soc. *Res:* Instrument design for analytical chemistry; author of various publications. *Mailing Add:* Dept Chem Mont State Univ Bozeman MT 59717

AMENDE, LYNN MERIDITH, June 24, 50; m. LIPID MOBILIZATION, MAST CELL SECRETION. *Educ:* Univ Md, PhD(zool), 79. *Prof Exp:* Sr staff fel, 83-87, Prog Officer, Fogarty Int, 87-89, EXEC SECY, NAT HEART LUNG BLOOD INST, NIH, 89- *Mem:* AAAS; Am Soc Cell Biol; Am Soc Zool; Biophys Soc; Sigma Xi. *Mailing Add:* NIH 9000 Rockville Pike Bethesda MD 20892

AMENOMIYA, YOSHIMITSU, b Tokyo, Japan, May 13, 24; Can citizen; m 48. CATALYSIS. *Educ:* Tokyo Inst Technol, BEng, 47; Hokkaido Univ, DSc, 59. *Prof Exp:* Asst, Tokyo Inst Technol, 51-61; assoc res officer, 61-67, sr res officer, Nat Res Coun Can, 68-87; HON VIS PROF, UNIV OTTAWA, 87- *Concurrent Pos:* Nat Res Coun Can fel, 59-61; chmn, Catalysis Div, Chem Inst Can, 78-79; counr, Int Cong Catalysis, 80- *Honors & Awards:* Catalysis Award, Chem Inst Can, 77. *Mem:* Chem Inst Can; Am Chem Soc; Chem Soc Japan; Catalysis Soc Am; Catalysis Soc Japan. *Res:* Heterogeneous catalysis; chemisorption. *Mailing Add:* Dept Chem Univ Ottawa Ottawa ON K1N 6N5 Can

AMENT, ALISON STONE, b Danbury, Conn, Dec 7, 48; m; c 2. INVERTEBRATE ZOOLOGY. *Educ:* Conn Col, AB, 70; Univ Pa, PhD(biol), 78. *Prof Exp:* Res asst virol, Ctr Oral Health Res, Univ Pa, 70-71; investr biol, Woods Hole Oceanog Inst, 78-80; educ intern, Green Briar Sci Ctr, Sandwich, Mass, 86; BIOL TEACHER, FALMOUTH ACAD, FALMOUTH, MASS, 87- *Concurrent Pos:* Sci writer, 80-81; guest investr, Woods Hole Oceanog Inst, 84-85. *Res:* Forest ecology; population genetics; evolution; physiology of adaptation; larval biology and ecology of marine invertebrates. *Mailing Add:* 23 Two Ponds Rd Falmouth MA 02540

AMENT, MARVIN EARL, b   Dec   5,   38.   PEDIATRICS, GASTROENTEROLOGY. *Educ:* Ill Inst Technol, BS, 59; Univ Minn, Minneapolis, MD, 63. *Prof Exp:* Resident pediat, Hosps, Univ Wash, 64-65; sr resident, Sch Med, Univ Calif, Los Angeles, 65-66; instr gastroenterol, Dept Med, Univ Wash, 70-71, actg asst prof med, 71-73; asst prof pediat & med, 73-80, PROF PEDIAT & GASTROENTEROL, CTR HEALTH SCI, UNIV CALIF, LOS ANGELES, 80- *Concurrent Pos:* NIH fel, Univ Wash, 68-70; mem attend staff, Children's Orthop Hosp, Seattle, 69-70, head gastroenterol clin, 71; mem attend staff gastroenterol, Olive View Med Ctr, 73- *Mem:* Am Fedn Clin Res; Am Gastroenterol Asn; Soc Pediat Res. *Res:* Intractable diarrhea in neonates; development of gastric secretion in infancy; peptic ulcer disease in childhood; inflammatory bowel disease in childhood; endoscopy in children; parenteral nutrition. *Mailing Add:* Gastroenterol & Nutrit UCLA Med Ctr MDCC 22-340 Los Angeles CA 90024

AMENTA, PETER SEBASTIAN, b Cromwell, Conn, Mar 26, 27; m 53; c 2. ANATOMY. *Educ:* Fairfield Univ, BS, 52; Marquette Univ, MS, 54; Univ Chicago, PhD(anat), 58. *Prof Exp:* Asst biol, Fairfield Univ, 49-52; asst zool, Marquette Univ, 52-54; asst anat, Univ Chicago, 55-58; from instr to assoc prof, Hahnemann Med Col, 58-71, actg chmn dept, 73-75, dir electron micros lab, 70-80, PROF ANAT, HAHNEMANN MED COL, 71-, CHMN DEPT, 75- *Concurrent Pos:* Instr, Marine Biol Lab, Woods Hole Oceanog Inst, 56; consult, Franklin Inst, NSF; dir, Raymond C Truex Mus, 81-; assoc ed, J Histol & Histopathology. *Honors & Awards:* Cert Distinction, Am Cancer Soc. *Mem:* Fel AAAS; Tissue Cult Asn; NY Acad Sci; Am Soc Photobiol; Am Asn Anat; Sigma Xi; Asn Anal Chemists. *Res:* Culture; histology; cytology; hematology; nucleocytoplasmic relationships; effects of microbeams of ultraviolet and infrared light on parts of cells; nucleolar activity; ed and publ several books on histology. *Mailing Add:* Dept Anat Hahnemann Univ Broad & Vine MS 408 Philadelphia PA 19102-1192

AMER, MOHAMED SAMIR, b Tanta, Egypt, Sept 2, 35; m 58, 87; c 4. PHARMACOLOGY, BIOCHEMISTRY. *Educ:* Univ Alexandria, BS, 56, dipl hosp pharm, 57; Univ Ill, MS, 60, PhD(pharmacol), 62; Columbia Univ, MS, 80. *Prof Exp:* Assoc prof pharmacol, Univ Alexandria, 62-66; sr res pharmacologist & proj coordr, Wilson Labs Div, Wilson Pharmaceut & Chem Corp, Ling-Temco-Vought, Inc, 66-69; prin investr pharmacol, Mead Johnson Res Ctr, Bristol-Myers Labs, Inc, 69, dir, Biol Res Labs, 75-80; dir biol res, Int Div, Bristol Myers Co, 80-83; PRES, AMER & CO, 83- *Concurrent Pos:* Chmn & CEO, Amstell Inc, 85-89. *Mem:* Am Fedn Clin Res; Acad Pharmaceut Sci; Soc Exp Biol & Med; Am Soc Pharmacol & Exp Therapeut; Am Soc Biol Chemists. *Res:* Biochemical pharmacology of the gastrointestinal tract; mechanisms of hormonal regulation; schistosomiasis. *Mailing Add:* Amer & Co PO Box 5685 Montecito CA 93150

AMER, NABIL MAHMOUD, b Alexandria, Egypt, May 1, 42; m 67; c 2. SOLID STATE PHYSICS, CHEMICAL PHYSICS. *Educ:* Univ Alexandria, BS, 57; Univ Calif, Berkeley, PhD(biophys), 65. *Prof Exp:* Res assoc biophys, Lawrence Radiation Lab, 66-68, PHYSICIST, LAWRENCE BERKELEY LAB, UNIV CALIF, BERKELEY, 68- *Concurrent Pos:* Lectr physics, Univ Calif, Berkeley, 70- *Mem:* AAAS; Am Phys Soc. *Res:* Light scattering from liquid crystals and solids; laser optoacoustic spectroscopy and fluctuation spectroscopy of condensed matter. *Mailing Add:* IBM Watson Res Ctr PO Box 218 Yorktown Heights NY 10598

AMER, PAUL DAVID, b New York, NY, Dec 25, 53; m; c 1. LOCAL COMPUTER NETWORKS, FORMAL DESCRIPTION TECHNIQUES. *Educ:* State Univ NY, Albany, BS, 74; Ohio State Univ, MS, 76, PhD(comput sci), 79. *Prof Exp:* Data Processor statist, Comput Ctr, State Univ NY, 71-74; teaching/res assoc, Ohio State Univ, 78-79; asst prof, 79-83, ASSOC PROF COMPUT SCI, UNIV DEL, 83- *Concurrent Pos:* Computer scientist, Nat Bur Standards, 78-85; assoc ed, Simulation, 81-90; vis scientist, Agence de l'Informatique, Paris, France, 85-86. *Mem:* Asn Comput Mach; Inst Elec & Electronics Engrs. *Res:* protocol specification, testing and verification, random number generation. *Mailing Add:* Dept Comput & Info Sci Univ Del Newark DE 19716

AMERINE, MAYNARD ANDREW, b San Jose, Calif, Oct 30, 11. PSYCHO-PHYSICS OF SENSORY RESPONSE, ENOLOGY. *Educ:* Univ Calif, BS, 32, PhD(plant physiol), 36. *Prof Exp:* From jr enologist to prof, 36-73, EMER PROF ENOL, UNIV CALIF, DAVIS, 73- *Concurrent Pos:* Guggenheim fel, 54; consult, Wine Inst, San Francisco, 73-85. *Mem:* Am Chem Soc; Sigma Xi; Inst Food Technol; Am Soc Enol & Viticulture; AAAS. *Res:* Wine making, including grape varieties, fermentation, finishing and sensory evaluation of foods and wines. *Mailing Add:* PO Box 208 St Helena CA 94574

AMERMAN, CARROLL R(ICHARD), b Danville, Ill, July 25, 30; m 56; c 5. AGRICULTURAL ENGINEERING. *Educ:* Purdue Univ, BS, 57, MS, 58, PhD, 69. *Prof Exp:* Res hydraul engr, Agr Res Serv, USDA, 58-77, res leader, 77-85, nat prog leader, Nat Prog Staff, 85-87, assoc dir, Midwest Region, 87-89, EXEC SECY, WATER GROUP, USDA, 89- *Mem:* Am Soc Agr Engrs; Am Geophys Union; Soil Conserv Soc Am; Sigma Xi. *Res:* Hydrology of agricultural lands, both surface and subsurface. *Mailing Add:* 1348 Windmill Lane Silver Spring MD 20905

AMERO, BERNARD ALAN, b Malden, Mass, Feb 29, 48. PHYSICAL PROCESS CONTROL. *Educ:* Providence Col, BS, 70; Ohio State Univ, PhD(chem), 76. *Prof Exp:* SR PROCESS CONTROL ENGR, ST REGIS PAPER CO, 76- *Mem:* Am Chem Soc; Tech Asn Pulp & Paper Indust. *Mailing Add:* Normandy Village Bldg 15 Apt 2 Nanuet NY 10954

AMERO, R(OBERT) C, b Rochester, NY, Oct 25, 17; m 42; c 2. FUEL TECHNOLOGY, IN-TRANSIT CRUDE OIL LOSS. *Educ:* Univ Rochester, BS, 39. *Prof Exp:* Asst, Floridin Co, 39-41, asst tech dir, 41-46, tech dir, vpres & mem bd dirs, 46-48; sr res engr, res ctr, Johns-Manville, 48-51; group leader, Gulf Res & Develop Co, 51-57, sec head, 57-60, fuels engr, 60-66; staff engr, 66-82; CONSULT 82- *Concurrent Pos:* Chmn, Fuels Div, Am Soc Mech Engrs, 74-75. *Mem:* Am Inst Chem Engrs; Am Soc Mech Engrs. *Res:* Adsorbents; granular desiccants; diatomaceous earth; synthetic silicates; colloidal clays; properties and combustion of petroleum fuels and coal-derived liquid fuels; in-transit crude oil loss. *Mailing Add:* 212 Lucille St Glenshaw PA 15116

AMERO, SALLY ANN, b Pittsburgh, Pa, Feb 24, 52. CHROMATIN STRUCTURE, GENE REGULATION. *Educ:* Ind Univ Pa, BS, 74; WVa Univ, PhD(molecular genetics), 79. *Prof Exp:* Res assoc, Dept Chem, 79-80, RES ASSOC, DEPT BIOL, UNIV VA, 80- *Mem:* AAAS; Am Chem Soc; NY Acad Sci. *Res:* Elucidating the primary structures of transcribed and of inactive chromatin, and the relationship of nucleosome structure to gene regulation. *Mailing Add:* Microbiol Dept Box 441 Univ Va Sch Med Charlottesville VA 22908

AMERSON, GRADY MALCOLM, b Gordon, Ga, Feb 20, 39; m 62; c 2. ENTOMOLOGY, INVERTEBRATE ZOOLOGY. *Educ:* Berry Col, BS; Clemson Univ, MS, 66, PhD(entom), 68. *Prof Exp:* Asst prof biol, 68-69, asst dean of col, 69-70, DEAN OF COL, OGLETHORPE UNIV, 70- *Mem:* Entom Soc Am. *Res:* Chemical sterilants. *Mailing Add:* Dept Biol Oglethorpe Univ 4484 Peachtree Rd NE Atlanta GA 30319

AMES, ADELBERT, III, b Boston, Mass, Feb 25, 21; m 48; c 3. NEUROPHYSIOLOGY, NEUROCHEMISTRY. *Educ:* Harvard Univ, MD, 45. *Prof Exp:* Intern & sr resident internal med, Presby Hosp, 45-52; res assoc, 55-69, PROF PHYSIOL, DEPT SURG, HARVARD MED SCH, 69-; NEUROPHYSIOLOGIST IN NEUROSURG, MASS GEN HOSP, 69- *Concurrent Pos:* Res fel biochem, Harvard Med Sch, 52-55; NIMH res scientist award, 68- *Mem:* Am Physiol Soc; Am Soc Neurochem; Soc Neurosci; Int Soc Neurochem. *Res:* Brain function; correlation of electrophysiology with metabolism of electrolytes, proteins and neurotransmitters in an in vitro preparation of retina; formation of cerebrospinal fluid; cerebral ischemia. *Mailing Add:* Dept Surg 467 Warren Bldg Harvard Univ Med Ctr Mass Gen Hosp Boston MA 02114

AMES, BRUCE NATHAN, b New York, NY, Dec 16, 28; m 60; c 2. CANCER RESEARCH, MOLECULAR BIOLOGY. *Educ:* Cornell Univ, BA, 50; Calif Inst Technol, PhD(biochem), 53. *Prof Exp:* USPHS fel, NIH, 53-54, biochemist, Nat Inst Arthritis & Metab Dis, 54-67; chmn dept, 84-89, PROF BIOCHEM & MOLECULAR BIOL, UNIV CALIF, BERKELEY, 68-, DIR, NAT INST ENVIRON HEALTH SCI, 79. *Honors & Awards:* Eli Lilly Award, 64; Arthur Flemming Award, 66; Lewis Rosenstiel Award, 76; Fedn Am Soc Exp Biol/3M Award, 76; ERDA Distinguished Assoc Award, 76; Environ Mutagen Soc Award, 77; Simon Shubitz Cancer Prize, 78; Felix Wankel Award, 78; John Scott Medal, 79; Bolton L Corson Medal, 80; Wadsworth Award, 81; Mott Prize, Gen Motors Cancer Res Found, 83; Tyler Prize, 85; Gairdner Award, Can, 83. *Mem:* Nat Acad Sci; Am Soc Biol Chemists; Am Acad Arts & Sci; Am Chem Soc; Genetics Soc Am; Soc Toxicol; Am Soc Microbiol; Am Asn Cancer Res; Environ Mutagen Soc. *Res:* Identifying agents damaging human DNA and the consequences for aging and cancer; endogenous oxidants and defenses against them; mutagenesis and carcinogenesis. *Mailing Add:* 1324 Spruce St Berkeley CA 94709

AMES, DENNIS BURLEY, mathematics, for more information see previous edition

AMES, DONALD PAUL, b Brandon, Man, Sept 13, 22; m 49; c 2. SPECTROSCOPY & SPECTROMETRY, ATOMIC & MOLECULAR PHYSICS. *Educ:* Univ Wis, BS, 44, PhD, 49. *Hon Degrees:* DL, Univ Mo-St Louis, 78. *Prof Exp:* Staff chemist, Los Alamos Sci Lab, 50-52; asst prof, Univ Ky, 52-54; res chemist, E I du Pont de Nemours & Co, 54-56; res chemist, Monsanto Chem Co, 56-59, scientist, 59-61; sr scientist, McDonnell Aircraft Corp, 61-68, dep dir res, McDonnell Res Labs, 68-70, dir res, 70-76, staff vpres, 76-86, gen mgr, MDC distinguished fel, McDonnell Douglas Res Labs, 86-89; RETIRED. *Concurrent Pos:* Adv, Air Force Off Sci Res, 71-76; mem, Am Inst Physics Comt on Corp Assocs, 77-80, 82-85. *Mem:* Am Chem Soc; Am Phys Soc; Sigma Xi; Soc Eng Sci; Combustion Inst. *Res:* Radiochemistry; exchange kinetics; diffusion; complex ions; microwave masers; electron and nuclear resonance spectroscopy; nuclear chemistry; actinide elements chemistry; beta and gamma ray spectroscopy; chemical and molecular lasers; polymers; fluorescence; positron annhilation in polymers; technical management. *Mailing Add:* 914 Black Twig Lane Kirkwood MO 63122-2212

AMES, EDWARD R, b Denver, Colo, Apr 10, 35; m 57; c 3. MEDICAL EDUCATION. *Educ:* Colo State Univ, BS, 57, DVM, 59, PhD(vet parasitol), 68. *Prof Exp:* Field sta mgr, eval vet therapeut, Merck Sharp & Dohme Res Labs, 60-68; from asst prof to assoc prof vet parasitol, Col Vet Med, Univ Mo-Columbia, 68-73; dir continuing educ, 73-80, STAFF CONSULT CONTINUING EDUC, AM VET MED ASN, 80- *Mem:* Am Vet Med Asn; Am Soc Vet Parasitol; Asn Am Vet Med Cols; Asn Teachers Vet Pub Health & Prev Med. *Res:* Human motivity and applied principles of adult learning; self-assessment evaluation; definition and measurement of competencies for veterinarians. *Mailing Add:* 600 N Taylor Oak Park IL 60302

AMES, GIOVANNA FERRO-LUZZI, b Rome, Italy, Jan 20, 36; US citizen; m 60; c 2. BACTERIAL GENETICS, MEMBRANE FUNCTION. *Educ:* Univ Rome, Italy, DrBiol, 58. *Prof Exp:* Res biochemist, NIH, 62-68; res biochemist, 68-78, adj prof, 78-80, PROF BIOCHEM, UNIV CALIF, BERKELEY, 80- *Concurrent Pos:* Prin investr, NIH grant, 68- & NSF grant, 83-88; ed, Microbiological Reviews, 79-81, J Sci Molecular Structure, 79-88. *Honors & Awards:* Agnes Fay Morgan Award, 75. *Mem:* Am Soc Biol Chemists; Am Soc Microbiol; Genetics Soc Am. *Res:* Studies on active transport of amino acids in bacteria; studies on the structure and function of bacterial membrane; studies on cellular regulation; studies on bacterial chromosome structure. *Mailing Add:* Biochem Dept Univ Calif Berkeley CA 94720

AMES, IRA HAROLD, b Brooklyn, NY, Apr 27, 37; m 58; c 2. CELL BIOLOGY. *Educ:* Brooklyn Col, AB, 59; NY Univ, MS, 62, PhD(biol), 66. *Prof Exp:* Instr biol, Brooklyn Col, 60-63, lectr, 63-64; from asst prof to assoc prof anat, 68-89, PROF ANAT, HEALTH SCI CTR, STATE UNIV NY, SYRACUSE, 89- *Concurrent Pos:* Fel, Brookhaven Nat Lab, 66-68; NSF grant, 72-76. *Mem:* AAAS; Am Soc Cell Biol; Am Asn Anat; Sigma Xi. *Res:* Role of natural killer and lymphokine-activated killer cells in tumor surveillance; experimental oncology. *Mailing Add:* Dept Anat & Cell Biol Health Sci Ctr State Univ NY Syracuse NY 13210

AMES, IRVING, b New York, NY, June 15, 29; m 52; c 2. PHYSICS. *Educ:* Syracuse Univ, BS, 51; Cornell Univ, PhD(physics), 55. *Prof Exp:* Teaching & res asst physcis, Brown Univ, 51 & Cornell Univ, 51-55; res staff mem, IBM res div, T J Watson Res Ctr, 55-83; ASSOC ED, IBM JOUR RES DEVELOP, 83- *Honors & Awards:* IBM Awards, 88; Cledo Brunetti Award. *Mem:* Am Phys Soc; Inst Elec & Electronics Engrs. *Res:* Surface phenomena; superconducting thin film devices and circuits; electromigration in aluminium thin films. *Mailing Add:* 500 High Pt Dr Apt 401 Hartsdale NY 10530

AMES, JOHN WENDELL, b Lincoln, Nebr, Aug 13, 36; m 57; c 2. COMMUNICATION, RADAR. *Educ:* Stanford Univ, BS, 59, MS, 60, PhD(elec eng), 64. *Prof Exp:* Res engr, Granger Assocs, 64-70; res engr, 70-82, STAFF SCIENTIST, STANFORD RES INST, 82- *Mem:* Inst Elec & Electronics Engrs; Am Geophys Union; Armed Forces Commun & Electronics Asn. *Res:* High frequency ionospheric radio propagation; radio fading and multipath studies; ionospheric radar; radio communication system design. *Mailing Add:* 333 Ravenswood Ave SRI Int Menlo Park CA 94025

AMES, LAWRENCE LOWELL, b Washington, DC, Sept 10, 51; m 77; c 1. X-RAY FLUORESCENCE, INSTRUMENTATION. *Educ:* Univ Ariz, BS, 72; Univ Wis, MS, 74, PhD(physics), 79. *Prof Exp:* Sr physicist, Measurex, 79-80 & Envirotech Measurement Systs, 80-81; SR PHYSICIST, TRACOR X-RAY, 81- *Mem:* Am Phys Soc; Sigma Xi. *Res:* X-ray fluorescence elemental analysis instruments. *Mailing Add:* Org 62-46 Box 3504 Bldg 151 Lockheed Missles & Space Co Sunnyvale CA 94088

AMES, LLOYD LEROY, JR, b Norwich, Conn, Aug 23, 27; m 53; c 4. MINERALOGY. *Educ:* Univ NMex, BA, 52; Univ Utah, MS, 55, PhD(mineral), 56. *Prof Exp:* Res fel, Univ Utah, 56-57; sr scientist, Gen Elec Co, 57-65; staff scientist, Water & Waste Mgt Dept, 65-81, SR RES SCIENTIST, ECOL SCI DEPT, PAC NORTHWEST LABS, BATTELLE MEM INST, 81- *Mem:* Mineral Soc Am. *Res:* Mineralogy and geochemistry applied to removal of radioisotopes from radioactive wastes; inorganic ion exchangers; migration of radionuclides in the natural environment. *Mailing Add:* Rte 1 Box 5310 Harrington Rd Richland WA 99352

AMES, LYNFORD LENHART, b Fresno, Ohio, May 20, 38; m 63; c 2. PHYSICAL CHEMISTRY. *Educ:* Muskingum Col, BS, 60; Ohio State Univ, PhD(phys chem), 65. *Prof Exp:* NSF fel, 65-66; from asst prof to assoc prof, 66-78, from asst head to head dept, 76-87, PROF CHEM, NMEX STATE UNIV, 78-, ASSOC DEAN, 88- *Mem:* AAAS; Am Chem Soc; Sigma Xi. *Res:* High temperature chemistry; matrix isolation; infrared spectroscopy; mass spectroscopy; electronic band spectroscopy. *Mailing Add:* Dept Chem NMex State Univ Las Cruces NM 88001

AMES, MATTHEW MARTIN, b Richland, Wash, Dec 17, 47; m 77. DRUG METABOLISM, CANCER CHEMOTHERAPY. *Educ:* Whitman Col, BA, 70; Univ Calif, San Francisco, PhD(pharmaceut chem), 76. *Prof Exp:* Res asst, Pomona Col, 70-71; res fel, NIH, 75-77; instr, 78-79, asst prof, 79-81, ASSOC PROF PHARMACOL, MAYO MED & GRADSCH, 81-, CHMN, DIV DEVELOP ONCOL RES, DEPT ONCOL, 87- *Concurrent Pos:* Res assoc pharmacol, Div Develop Oncol Res, Dept Oncol, Mayo Clin, 77-78, assoc consult, 78-80, consult, 80- *Mem:* Am Chem Soc; Am Asn Cancer Res; Am Soc Pharmacol & Exp Therapeut; Am Soc Mass Spectrometry; Am Soc Clin Pharmacol Therapeut. *Res:* Disposition and metabolism of anticancer agents, particularly the role of metabolism in the antitumor activity and/or toxicity; clinical pharmacokinetics of anticancer agents. *Mailing Add:* Dept Pharmacol Mayo Med & Grad Schs 200 First St SW Rochester MN 55905

AMES, PETER L, b St Paul, Minn, May 2, 31; m 58; c 2. ENVIRONMENTAL SCIENCES. *Educ:* Harvard Col, BA, 58; Yale Univ, MS, 62, PhD(biol), 65. *Prof Exp:* Pharmacologist, Smith, Kline & French Labs, 58-59; asst prof zool & asst cur birds, Mus Vert Zool, Univ Calif, Berkeley, 65-68; assoc ed life sci, Encycl Britannica, 68-72; head, Terrestrial Biol Sect, 87-90, HEAD, ENVIRON SCI SECT, HARZA ENG CO, 73-, ASST HEAD, ENVIRON SCI DEPT, 91- *Concurrent Pos:* Res assoc, Field Mus of Natural Hist, Chicago, 87- *Mem:* AAAS; Am Ornith Union; Brit Ornith Union; Cooper Ornith Soc; Wilson Ornith Soc. *Res:* Avian anatomy; ecology of raptorial birds; environmental management. *Mailing Add:* Harza Eng Co 150 S Wacker Dr Chicago IL 60606-4288

AMES, ROBERT A, psychophysical nutrition, for more information see previous edition

AMES, ROGER LYMAN, b Northampton, Mass, Nov 6, 33; m 57; c 4. GEOCHEMISTRY. *Educ:* Williams Col, BA, 55; Yale Univ, MS, 59, PhD(econ geol), 63. *Prof Exp:* SPEC RES ASSOC, AMOCO PROD CO RES CTR, 62- *Mem:* AAAS; Geol Soc Am; Am Geochem Soc. *Res:* Application of stable isotope studies to exploration for petroleum and metallic deposits; geochemistry in petroleum exploration and exploitation. *Mailing Add:* Amoco Prod Co Res Ctr PO Box 3385 Tulsa OK 74102

AMES, STANLEY RICHARD, b Madison, Wis, Dec 18, 18; m 43; c 4. COMPUTER SCIENCE, NUTRITIONAL BIOCHEMISTRY. *Educ:* Univ Mont, BA, 40; Columbia Univ, AM, 42, PhD(chem), 44. *Prof Exp:* Asst chem, Univ Mont, 37-40; asst, Columbia Univ, 40-41, statutory asst, 41-42, asst, 42-43; Rockefeller fel biochem, Univ Wis, 43-46; sr res chemist, Distillation Prod Industs, Eastman Kodak Co, 46-53, res assoc, 53-63, head biochem res dept, 63-65, dir, Biochem Res Labs, 65-69, sr res assoc, Tenn Eastman Res Labs, 69- 80, head, Biochem Res Labs, Health & Nutrit Res Div, Eastman Chem Div, 69-82, res fel, 80-82, consult, Kodak Res Labs, 82-83; NUTRIT CONSULT, 83- *Concurrent Pos:* Mem adv bd, Off Biochem Nomenclauture, Nat Acad Sci-Nat Res Coun, 65-68; mem comt nomenclature, Int Union Nutrit Sci, 67-86, actg chmn, 69, 75, chmn, 76-86, mem comn I, nomenclature, procedures & stand, 76-86; actg chmn adv comt, NIH Guide to Nutrit Terminol, 68-69; mem, Interdiv Comt on Nomenclature & Symbols, Int Union Pure & Appl Chem, 76-86. *Mem:* Fel AAAS; Am Chem Soc; Am Soc Biol Chem; fel Am Inst Nutrit; Nutrit Soc. *Res:* Vitamins A and E; biochemistry of monoglycerides; acetylated monoglycerides; bioassay; nutrition; lipid metabolism; enzymes; organic and biological oxidations; feed preservatives; ruminant nutrition. *Mailing Add:* 61 Biltmore Dr Rochester NY 14617

AMEY, RALPH LEONARD, b Huntington Park, Calif, June 5, 37; m 64. PHYSICAL CHEMISTRY. *Educ:* Pomona Col, AB, 59; Brown Univ, PhD(phys chem), 64. *Prof Exp:* Instr chem, Barrington Col, 62-63; mat res & develop specialist missile & space systs div, Douglas Aircraft Co, Inc, 63-65; asst prof, 65-74, ASSOC PROF CHEM, OCCIDENTAL COL, 74- *Mem:* Am Chem Soc. *Res:* Properties of the condensed state; intermolecular forces and local liquid structure; dielectric behavior of materials; thermal analysis of solids and solutions. *Mailing Add:* Dept of Chem Occidental Col 1600 Campus Rd Los Angeles CA 90041

AMEY, WILLIAM G(REENVILLE), b Baltimore, Md, Feb 24, 18; m 41, 65; c 3. ELECTRICAL ENGINEERING. *Educ:* Johns Hopkins Univ, BEE, 38, DEE, 47. *Prof Exp:* Instr math, Johns Hopkins Univ, 39-41, res assoc elec eng, 46-47; instr elec eng, US Naval Acad, 41-43; asst sect head measure & components, Naval Res Lab, Washington, DC, 43-46; engr, Res Div, Leeds & Northrup Co, 47-49, supvry res engr, 49-54, asst sect head elec res, 54-55, nucleonics develop, 55-56, sect head elec res, 56-57, mgr res div, 57-68, assoc dir corp res, Res & Develop Ctr, 68, assoc dir res planning, Corp Res Dept, 68-72; asst chief, 72-74, chief control systs engr, Bechtel Assoc Prof Corp, Control Systs Engr Sect, Ann Arbor, Mich, 74-83; RETIRED. *Concurrent Pos:* Consultant. *Mem:* Fel Instrument Soc Am; fel Inst Elec & Electronics Engrs; Sigma Xi. *Res:* Electrical and electronic measurements associated with precision potentiometers and bridges; spectroscopy; electro-chemistry and control of complex systems. *Mailing Add:* 2036 SW Stratford Way Palm City FL 34990

AMICK, CHARLES JAMES, b Cleveland, Ohio, Mar 3, 52; m; c 2. MATHEMATICS. *Educ:* Carnegie-Mellon Univ, BS(math), 74, BS(physics), 74; Univ Cambridge, PhD, 77. *Prof Exp:* Instr math, 79, asst prof math, 81, ASSOC PROF MATH, UNIV CHICAGO, 84- *Concurrent Pos:* vis prof, Math Res Ctr, Madison, Wis, 82, Inst Math, Minneapolis, 84, Univ Stuttgart, 85, Sci res coun, Univ Bath, 85. *Res:* Published numerous articles in various journals. *Mailing Add:* Dept Math Univ Chicago 1118-32 E 58th St Chicago IL 60637

AMICK, JAMES ALBERT, b Lawrence, Mass, Feb 18, 28; m 61; c 1. SOLID STATE ELECTRONICS, SEMICONDUCTORS. *Educ:* Princeton Univ, AB, 49, AM, 51, PhD(phys chem), 52. *Prof Exp:* Res assoc electron micros, Princeton Univ, 52-53; res engr, Radio Corp Am Labs, 53-56, res engr, Zurich Labs, 56-57, res phys chemist, 57-68, head process res, RCA Labs, 68-71, mgr mat & processes, Solid State Div, RCA, 71-76; res assoc, Exxon Res & Eng, 76-79, sr res assoc, 79-85, group head, coatings & composites, 85-87; SECT HEAD, CELL PROCESSING RES & DEVELOP, MOBIL SOLAR ENERGY CORP, 87- *Concurrent Pos:* Ethyl Corp Fel, 51-52; secy, Dielectrics & Insulation Div, Am Chem Soc, 80-81, vchmn, 81-82, chmn, 82-83, 83-84. *Honors & Awards:* David Sornoff Outstanding Achievement Award, 65. *Mem:* Electrochem Soc (secy, 85-); Am Chem Soc (actg secy-treas, 53); fel Am Inst Chemists; Sigma Xi. *Res:* Infrared absorption intensities; electron microscopy of aerosols; mass spectrometry of hydrocarbons; x-ray diffraction; electrophotographic processes; surface chemistry of semiconductors; epitaxial growth of semiconductors; processes for electronic components; processing and modeling silicon for solar cells; thin film strain gauges. *Mailing Add:* 76 Leabrook Lane Princeton NJ 08540

AMICK, JAMES L(EWIS), b Detroit, Mich, May 18, 25; m 47; c 5. AERODYNAMICS, VEHICLE DESIGN. *Educ:* Univ Mich, BSE, 45; Univ Va, MAeron, 52. *Prof Exp:* Aeronaut res scientist, Langley Aeronaut Labs, Nat Adv Comt Aeronaut, 46-51; res assoc, 51-57, assoc res engr, 57-59, res engr, 59-70, CONSULT AERODYN, UNIV MICH, 70-; prod develop engr, Photon Sources, Inc, 82-86; CONSULT, LUMONICS, 88- *Concurrent Pos:* Mem, Bumblebee Aerodyn Panel, 60-63 & rocket nozzles & jet effects panel, Adv Comt Aeroballistics, Bur Weapons, 60-69; consult, Ammann & Whitney, Inc, 66-67, US Army Missile Command, Redstone Arsenal, 66-69, US Naval Weapons Ctr, 69, Camcheck, Inc, 65- & KMS Fusion, Inc, 78. *Mem:* Int Human-Powered Vehicle Asn. *Res:* Interaction of side jets with supersonic streams; supersonic flow past wings and bodies; flow thru nozzles; sailboat performance; wind-tunnel testing; windmobiles; fast-flow lasers; automotive cam design & evaluation. *Mailing Add:* 1464 Cedar Bend Dr Ann Arbor MI 48105-2305

AMIDON, ROGER WELTON, b Rockford, Ill, Nov 29, 14; m 38; c 3. CHEMISTRY. *Educ:* Univ Minn, BChem, 36, MS, 40, PhD(org chem), 49. *Prof Exp:* Petrol chemist, Midcontinent Petrol corp, 36-39; org res chemist, Dow Chem Co, 40-41; chemist, Gen Labs, US Rubber Co, 49-54 & Naugatuck Chem Div, 54-68; sr info scientist, Chem Div, Uniroyal Inc, 68-79; RETIRED. *Mem:* AAAS; Am Chem Soc; The Chem Soc; Sigma Xi. *Res:* New rubbers and elastomers; plastics and resins; rubber chemicals; polyester resins. *Mailing Add:* 227 Ponce de Leon Spartanburg SC 29302

AMIDON, THOMAS EDWARD, b Seneca Falls, NY, Sept 30, 46; m 68; c 2. SILVICULTURE, QUANTITATIVE GENETICS. *Educ:* State Univ NY, BS, 68, MS, 72, PhD(silvicult), 75. *Prof Exp:* Res asst paper sci, Empire State Paper Res Inst, 75-77; res assoc, 77-79, sr res assoc, 79-80, RES ANALYST FOREST SCI, INT PAPER CO, 80- *Mem:* Tech Asn Pulp & Paper Indust; Soc Am Foresters. *Res:* Silviculture; genetics; asexual propagation; plant growth regulation; physiology; regeneration; computer modeling of biological processes; research evaluation. *Mailing Add:* CRC Long Meadow Rd Tuxedo NY 10987

AMIN, OMAR M, b Minya El Kamh, Egypt, Jan 23, 39; nat US; c 2. PARASITOLOGY, MEDICAL ENTOMOLOGY. *Prof Exp:* Res asst med zool, US Naval Med Res Unit, Cairo, 60-64; assoc agr & zool, Ariz State Univ, 66-67; res assoc & instr biol, Old Dominion Univ, 67-69; vis fel virol, Ctr Dis Contr, Ga, 69-70; from asst prof to assoc prof, 71-83, PROF BIOL, UNIV WIS-PARKSIDE, 83- *Concurrent Pos:* US Army res grant, 68; Sigma Xi res grant, 69; Univ-Wis WARF grant, 72; Am Motors Environ Quality Prog grant, 76; NIH Inst grant, 76, Sea Grant, 76,77 & 84. *Mem:* Entom Soc Am; Am Soc Parasitologists; Acarol Soc Am. *Res:* Helminth parasites of fresh water fishes, particularly Acanthocephalans; bio ecology of ectoparasites, of vertebrates, vectors of diseases, particularly ticks and fleas; host-parasite interrelationships; taxonomy, ecology and host-parasite relationships of helminth parasites of fishes particularly acanthocephalans, trematodes, cestodes, nematodes, leeches and crustacea; ecology and host relationships of arthopod ecto-parasites of mammals particularly ticks and fleas. *Mailing Add:* Dept Biol Sci Univ Wis-Parkside PO Box 2000 Kenosha WI 53141

AMIN, SANJAY INDUBHAI, b Baroda, India, Mar 4, 48; m 72; c 2. CHEMICAL ENGINEERING, INDUSTRIAL CHEMISTRY. *Educ:* Univ Baroda, BE, 70; Mass Inst Technol, SM, 71, ScD(chem eng), 75. *Prof Exp:* Asst dir chem eng, Mass Inst Technol, Prac Sch, Oak Ridge Nat Lab, 71-72, teaching asst, Mass Inst Technol, 72-73, res asst, 73-75; from res scientist to sr res scientist, Chem Process Res Develop, 75-83, GROUP MGR, FINE CHEM PROD, 75-, DIR, SPECIALTY CHEM PROD, UPJOHN CO, 75- *Honors & Awards:* Cert of Merit, All India Stud Asn, 70. *Mem:* Am Inst Chem Engrs; Sigma Xi; Am Chem Soc. *Res:* Applied chemical engineering; industrial production of fine and speciality chemicals; industrial production of bulk drugs and related chemicals. *Mailing Add:* Dir Specialty Chem Prod Upjohn Co 1120-66-1 Kalamazoo MI 49008

AMINOFF, DAVID, b Kokand, Russia, Jan 16, 26; US citizen; m 56; c 3. BIOCHEMISTRY, IMMUNOGENETICS. *Educ:* Univ London, BSc, 45, PhD(biochem), 49, DSc(biochem), 74. *Prof Exp:* Res asst biochem, Lister Inst Prev Med, 49-50; res assoc, Israeli Inst Biol Res, 50-55, 56-57; exchange scientist, Col Physicians & Surgeons, Columbia Univ, 55-56; res assoc, Pub Health Res Inst, New York, 57-60; res assoc, Rackham Arthritis Res Unit, 60-62, res assoc, Simpson Mem Inst, 62-66, asst prof, 66-70, assoc prof, 70-80, PROF BIOL CHEM, DEPT INTERNAL MED, SIMPSON MEM INST, UNIV MICH, ANN ARBOR, 80- *Concurrent Pos:* Consult, Marcus Inst, Israeli Red Cross, 53-57. *Mem:* Am Fedn Clin Res; Am Soc Hemat; Am Chem Soc; Am Soc Biol Chem; Brit Biochem Soc. *Res:* Immunochemistry and biosynthesis of the blood group substances; glycoprotein chemistry and metabolism; erythrocyte viability in circulation and role of glycoconjugates in health and disease. *Mailing Add:* Biochem Inst Gerontol Univ Mich 300 N Ingalls Ann Arbor MI 48109

AMIRAIAN, KENNETH, b Richmond Hill, NY, Sept 29, 26; m 51; c 2. BIOCHEMISTRY, IMMUNOCHEMISTRY. *Educ:* Brooklyn Col, BS, 49; Columbia Univ, MA, 50, PhD, 59. *Prof Exp:* Asst biochem, Sloan-Kettering Inst, 50-53; asst, Columbia Univ, 53-55; asst immunochem, Rutgers Univ, 55-57; from res scientist to assoc res scientist biochem, NY State Dept Health, 57-85; from lectr to asst prof, 60-84, ADJ PROF MICROBIOL, ALBANY MED COL, 84-; ASST DIR LAB DIAG IMMUNOL, NY STATE DEPT HEALTH, 85- *Mem:* AAAS; Am Asn Immunol; Am Soc Microbiol. *Res:* Immunochemistry research on immune hemolytic reaction and gamma globulins; human immunodeficiency virus; correlation of western blot densitometric analysis with clinical status. *Mailing Add:* Div Labs & Res NY State Dept Health Albany NY 12201

AMIRIKIAN, ARSHAM, precast & prestressed concrete, welded structures; deceased, see previous edition for last biography

AMIRKHANIAN, JOHN DAVID, b Julfa-Esfahan, Iran, Nov 10, 27; US citizen; m 57; c 3. NEONATAL LUNG RESEARCH, LUNG SURFACTANT DEVELOPMENT. *Educ:* Univ Tehran, BSc, 73; Univ London, PhD(genetics), 77. *Prof Exp:* WHO sr res fel, Mainz Univ, WGer, 76; asst prof, Univ Tehran, Iran, 77-79; SR RES ASSOC FAC, DEPT PEDIAT, KING-DREW MED CTR, SCH MED, UNIV CALIF, LOS ANGELES, 81- *Concurrent Pos:* Vis prof molecular biol & classic genetics, Univ Southern Calif, 79-82; res fel, Mus Natural Hist, Calif, 79- *Mem:* Genetic Asn Am; AAAS; NY Acad Sci; fel Royal Micros Soc. *Res:* Neonatal lung research; synthetic lung surfactant replacement models; fungal genetics: production of diploid strains and genetic characterization of radiation-sensitive mutants of Coprinus cinereus; mosquito genetics: genetic manipulation of insect pests for eradication purposes; oxidative damage to biological macromolecules; DNA repair and anti-oxidant defense mechanisms; mechanisms of surfactant inactivation; adult respiratory distress syndrome. *Mailing Add:* 3310 Sparr Blvd Glendale CA 91208

AMIR-MOEZ, ALI R, b Tehran, Iran, Apr 7, 19; US citizen. MATHEMATICS. *Educ:* Univ Tehran, BA, 43; Univ Calif, Los Angeles, MA, 51, PhD(math), 55. *Prof Exp:* Instr math, Tehran Technol Col, 43-47; asst prof, Univ Idaho, 55-56, Queens Col, NY, 56-60, Purdue Univ, 60-61 & Univ Fla, 61-63; res & writing, 63-64; prof math, Clarkson Col Technol, 64-65; prof math, Tex Tech Univ, 65-88; RETIRED. *Mem:* Am Math Soc; Math Asn Am; Sigma Xi; NY Acad Sci. *Res:* Singular values of linear transformations and matrices. *Mailing Add:* Dept Math Tex Tech Univ Lubbock TX 79409

AMIRTHARAJAH, APPIAH, b Colombo, Sri Lanka, Apr 4, 40; US citizen; m 68; c 2. WATER TREATMENT PROCESSES, HYDRAULICS & DESIGN. *Educ:* Univ Ceylon, Colombo, BSc(Eng), 63; Iowa State Univ, Ames, MS, 70; PhD(environ eng), 71. *Prof Exp:* Engr, Dept Water Supply & Drainage, Sri Lanka, 63-67, sr engr, 67-68, chief training officer, 73-74; chief engr, Nat Water Supply & Drainage Bd, Sri Lanka, 75-76; from assoc prof to prof environ engr, Mont State Univ, 77-86, coordr, Environ Eng Progs, 79-86; PROF ENVIRON ENG, GA INST TECHNOL, 86- *Concurrent Pos:* Vis prof, Univs Sri Lanka, Peradeniya & Katubedde, 72-76, Johns Hopkins Univ, 84-85 & Univ Sao Paulo, Brazil, 89; prin investr, US Environ Protection Agency, NSF, Am Water Works Res Found, 80-; mem, Coagulation Comn & Student Act Comn, Am Water Works Asn, 80- & Res Adv Coun, 88-91; environ sci & eng fel, AAAS, Environ Protection Agency, 85; ed, J Environ Eng, 86-88; res fel, Univ Col, London, 89. *Honors & Awards:* Rudolph Hering Medal, Am Soc Civil Engrs, 85; G W Fuller Award, Am Water Works Asn, 86. *Mem:* Am Soc Civil Engrs; Am Water Works Asn; Asn Environ Eng Professors; Int Asn Water Pollution Res & Control; Am Filtration Soc. *Res:* Environmental engineering with special reference to the physiochemical processes of coagulation, mixing, flocculation, filtration and backwash of filters; drinking water regulations; design of water treatment plants. *Mailing Add:* Sch Civil Eng Ga Inst Technol Atlanta GA 30332

AMIS, ERIC J, b Topeka, Kans, March 2, 54; m 76; c 1. LIGHT SCATTERING, POLYMER VISCOELASTICITY. *Educ:* Willamette Univ, Salem Ore, BS, 76; Univ Wis-Madison, PhD(chem), 81. *Prof Exp:* Res assoc, Dept Chem, Univ Wis, 82-84; Nat Res Coun res assoc, Nat Bur Standards, 81-82; asst prof, 84-90, ASSOC PROF, DEPT CHEM, UNIV SOUTHERN CALIF, 90- *Mem:* Am Chem Soc; Am Phys Soc; AAAS; Sigma Xi; Soc Rheology; Mat Res Soc. *Res:* Physical chemistry of polymer solutions by quasielastic and elastic light scattering and neutron scattering; synthetic as well as biological polymers; light scattering characterization of biomembranes; dilute solution dynamic viscoelasticity; forced Rayleigh scattering. *Mailing Add:* Dept Chem Univ Southern Calif Los Angeles CA 90089-0482

AMITAY, NOACH, b Tel-Aviv, Israel, Apr 30, 30; US citizen; m 60; c 4. PERSONAL COMMUNICATIONS SYSTEMS, ELECTROMAGNETICS. *Educ:* Technion, Israel Inst Technol, BSc, 53, dipl Ing, 54; Carnegie Inst Technol, MSc, 58, PhD(elec eng), 60. *Prof Exp:* Head, Elec Measurement Dept, Signal Corps, Defense Army Israel, 54-56; asst prof elec eng, Carnegie Inst Technol, 60-62; mem tech staff, 62-84, DISTINGUISHED MEM TECH STAFF, AT&T BELL LABS, 84- *Concurrent Pos:* Consult, Electronic Inst Magnetics Inc, 57-58 & 60-61 & New Prod Lab, Westinghouse Elec Corp, 60-62. *Mem:* Fel Inst Elec & Electronics Engrs; Int Union Radio Sci. *Res:* Microcellular mobile and personal radio and microwave communications; urban propagation measurements and modeling; system concepts and architecture; single mode optical fiber up-tapers and hardware; fiber based local communications systems and networks; electromagnetic theory; numerical methods; antenna and phased arrays design; computer-aided design; author of 50 publications and co-author of one book; holder of ten patents. *Mailing Add:* 57 Willshire Dr Tinton Falls NJ 07724

AMITH, AVRAHAM, b Przemysl, Poland, Mar 16, 29; US citizen; m 52; c 2. SOLID STATE PHYSICS. *Educ:* Univ Mich, BS, 51; Harvard Univ, MA, 53, PhD(chem phys), 56. *Prof Exp:* Mem tech staff, RCA Labs, 56-70; prof physics, City Univ New York, 71-72; sr engr & leader, RCA Thermoelec Opers, 72-74; sr scientist & prog mgr, Inst Energy Conversion, Univ Del, 74-76; vis lectr & sr res scientist, Princeton Univ, 76-78; mgr mat & devices lab III-V, ITT-Electro-Optical Div, Va, 78-82, mgr opto-electronics dept, 82-87, fel, 87-89, SR SCI FEL, ITT-ELECTRO-OPTICAL DIV, VA, 89-*Concurrent Pos:* Adj prof, City Univ New York, 72-74, Va Inst Tech & State Univ, 91-; mem, Indust Adv Bd, Lab Nanostructured Mat Res, Rutgers Univ, 91- *Honors & Awards:* RCA Outstand Achievement Award. *Mem:* Am Phys Soc; Inst Elec & Electronics Engrs. *Res:* Molecular quantum mechanics; photoeffects, lifetimes and transport in semiconductors; electrical properties of thin films and surface layers; thermoelectric properties of semiconductors; galvanomagnetic and thermomagnetic properties of semiconductors and semimetals; magnetism and thin films; surface properties of gallium arsenide; LED's and junction lasers; photodetectors; photoemission; image intensifiers; optical properties of galium arsenide. *Mailing Add:* 5049 Falcon Ridge Rd SW Roanoke VA 24014

AMJAD, ZAHID, b Lahore, Pakistan, Aug 10, 46; nat US; m 78; c 1. KINETICS OF CRYSTAL GROWTH, CHEMICAL TREATMENT WATER PURIFICATION. *Educ:* Panjab Univ, Lahore, Pakistan, BSc, 66, MS, 67; Western Mich Univ, MA, 72; Glasgow Univ, PhD (physical chem), 76. *Prof Exp:* Lectr chem, Panjab Univ, 67-69; teaching asst chem, Western Mich Univ, 70-72; demonstrator chem, Glascow Univ, Scotland, 73-75; asst res prof chem, State Univ NY, Buffalo, 76-79; gr leader, prod develop, Calgon Corp, Pittsburgh, 79-82; SR, RES & DEVELOP, THE B F GOODRICH CO, 82- *Concurrent Pos:* Res & develop fel & supvr, Polymers & Chemicals Div, Brecksville Res Ctr, 82- *Mem:* Nat Assoc Corrosion Engr; Cooling Tower Inst; NY Acad Sci; Soc Cosmetic Scientists. *Res:* Surface and colloid chemistry of hydrophilic polymers-sparingly soluble salt systems; performance evaluation and definition of the role of hydrophilic polymers in water treatment cosmetics and personal care; new raw materials development; product formulation. *Mailing Add:* 32611 Redwood Blvd Avon Lake OH 44012

AMKRAUT, ALFRED A, b Saarbruecken, Ger, Sept 21, 26; US citizen; m 55; c 2. IMMUNOLOGY, IMMUNOCHEMISTRY. *Educ:* Univ Calif, Berkeley, BA, 49; NY Univ, MS, 55; Stanford Univ, PhD(med microbiol), 63. *Prof Exp:* Res asst immunohemat, Med Sch, Stanford Univ, 55-62; vis scientist, Ore Regional Primate Res Ctr, 64-67; asst prof immunol, Med Sch, Univ Ore, 65-67; asst prof med microbiol, Sch Med, Stanford Univ, 67-74; HEAD MICROBIOL, ALZA RES, 74-, PRIN SCIENTIST, 87-, DIR BIOL SCI, 87- *Concurrent Pos:* Res fel immunochem, Calif Inst Technol, 62-64; Nat Acad Sci-Nat Res Coun fel, 63-64; NIH fel, 64- *Mem:* Fel AAAS; Am Asn Immunol; Am Chem Soc; Am Soc Microbiol; Sigma Xi; Soc Investigative Dermat; Soc Clin Immunol. *Res:* Antigen-antibody interactions and their biological manifestations; mechanism of antibody formation; central nervous system and the immune response; contact sensitivity. *Mailing Add:* Alza Res 950 Page Mill Rd Palo Alto CA 94303

AMLANER, CHARLES JOSEPH, JR, b Philadelphia, Pa, Nov 26, 51; m 73; c 2. SLEEP PHYSIOLOGY, BIOTELEMETRY & RADIO TRACKING. *Educ:* Andrews Univ, Mich, BS, 74, MA, 76; Oxford Univ, UK, DPhil(zool), 82. *Prof Exp:* Asst prof behav physiol, Dept Biol Sci, Walla Walla Col, 80-85; chmn, Dept Zool, 86-90, ASSOC PROF BEHAV SCI, DEPT ZOOL, UNIV ARK, FAYETTEVILLE, 86-, PROG CHMN, DEPT BIOL SCI, 90-; DIR, ANIMAL SLEEP RES GROUP, 81- *Concurrent Pos:* Adj assoc prof biomed eng, Dept Eng, Walla Walla Col, 81-85 & Univ Ark, Fayetteville, 86-; prin investr, NIMH, 82-89 & US Dept Interior, 87-; comnr, Ark Game & Fish Comn, 86-91; mem, Fish & Wildlife Res Comt, Nat Asn State Univs & Land Grant Cols, 87- *Mem:* Int Soc Biotelemetry (pres, 72-); Animal Behav Soc; Asn Prof Sleep Socs; Sigma Xi; Sleep Res Soc; Int Asn Fish & Wildlife Agencies. *Res:* Biotelemetry and use in evaluating temporal and spatial patterns of behavior and physiology in unrestrained animals; sleep physiology of birds and mammals; simulation modeling of behavior; animal population dynamics. *Mailing Add:* Dept Biol Sci Rm SE 632 Univ Ark Fayetteville AR 72701

AMLING, HARRY JAMES, b Baltimore, Md, Jan 22, 31; 53, 78; c 5. HORTICULTURE. *Educ:* Rutgers Univ, BS, 52; Univ Del, MS, 54; Mich State Univ, PhD(hort), 58. *Prof Exp:* Asst, Univ Del, 52-54; res asst, Mich State Univ, 56-58; from asst prof to assoc prof, 58-68, PROF HORT, AUBURN UNIV, 68- *Concurrent Pos:* Consult, pecan grower. *Honors & Awards:* Gold Pecan Award, Nat Pecan Shellers & Processors Asn, 64. *Mem:* Am Soc Hort Sci. *Res:* Practical and fundamental aspects of use of herbicides and growth regulators on pecans; growth and development; mineral nutrition of pecans; high density pecan culture; tissue culture. *Mailing Add:* 753 Brenda Ave Auburn AL 36849

AMMA, ELMER LOUIS, b Cleveland, Ohio, Feb 13, 29. BIOINORGANIC CHEMISTRY, BIOPHYSICAL CHEMISTRY. *Educ:* Case Inst, BS, 52; Iowa State Univ, PhD(chem), 57. *Prof Exp:* Asst chem, Iowa State Univ, 52-54; asst, Ames Lab, AEC, 54-57; res assoc, Radiation Lab, Univ Pittsburgh, 57-58, lectr & res assoc, Radiation Lab & Physics Dept, 58-59, asst prof chem, 59-65; assoc prof, 65-70, PROF CHEM, UNIV SC, 70- *Concurrent Pos:* Consult, Exxon Res, 65-70. *Honors & Awards:* Russell Award in Res, 70; Stone Award, Am Chem Soc, 80. *Mem:* AAAS; Am Inst Chem; Am Chem Soc; Am Crystallog Asn; Am Phys Soc. *Res:* Protein crystallography, structure of model enzymes and proteins; structure of carriers of molecular oxygen and nitrogen; structure of molecules of importance to homogeneous catalysis, structure and metal nuclear magnetic resonance. *Mailing Add:* Dept Chem Univ SC Columbia SC 29208

AMMAN, GENE DOYLE, b Greeley, Colo, Feb 26, 31; m 54; c 5. FOREST ENTOMOLOGY. *Educ:* Colo State Univ, BS, 56, MS, 58; Univ Mich, PhD(forest insect ecol), 66. *Prof Exp:* Assoc entomologist, Southeastern Forest Exp Sta, 58-66, prin entomologist, 66-83, PROJ LEADER, INTERMOUNTAIN FOREST & RANGE EXP STA, US FOREST SERV, 84- *Mem:* AAAS; Int Union Forestry Res Orgn; Entom Soc Am; Entom Soc Can; Sigma Xi; Am Registry Prof Entomologists. *Res:* Population ecology; natural and cultural control of forest insects. *Mailing Add:* Intermt Forest & Range Exp Sta US Forest Serv 507 25th St Ogden UT 84401

AMMANN, E(UGENE) O(TTO), b Portland, Ore, June 26, 35; m 70. LASERS, NONLINEAR OPTICS. *Educ:* Univ Portland, BS, 57; Stanford Univ, MS, 59, PhD(elec eng), 63. *Prof Exp:* Res assoc elec eng, Stanford Univ, 63; adv develop engr, 63-64, eng specialist, 64-75, SR MEM TECH STAFF, SYLVANIA ELECTRONIC SYSTS, GEN TEL & ELECTRONICS CORP, 75- *Mem:* Inst Elec & Electronics Engrs; Am Phys Soc; Optical Soc Am. *Res:* Microwave masers; optical birefringent filters; modulation and demodulation of light; lasers; nonlinear optics; optical parametric oscillators; stimulated Raman scattering. *Mailing Add:* Lockheed PARL, Dept 97-01 3251 Hanover St Bldg 201 Palo Alto CA 94304

AMMANN, HARRIET MARIA, b Munich, Germany, Feb 2, 39; US citizen. ZOOLOGY, ENVIRONMENTAL HEALTH ASSESSMENT. *Educ:* Univ Dayton, BS, 61; NMex Highlands Univ, MS, 65; NC State Univ, PhD(zool), 78. *Prof Exp:* Asst prof gen med sci, Bowman Gray Sch Med, 70-73; assoc prof biol & physiol, NC Cent Univ, 76-82; vis prof physiol, NC State Univ, 82-84; biologist, Environ Criteria & Assessment Off, US Environ Protection Agency, 84-90; LEAD TOXICOLOGIST, OFF TOXIC SUBSTANCES, WASH STATE DEPT HEALTH, 90- *Mem:* AAAS; Sigma Xi. *Res:* Write health assessments for hazardous air pollutants, hydrogen sulfide, styrene, woodsmoke and indoor air pollution. *Mailing Add:* Off Toxic Substances Wash State Dept Health MS-LD11 Olympia WA 98504

AMMAR, RAYMOND GEORGE, b Kingston, Jamaica, July 15, 32; US citizen; m 61; c 3. EXPERIMENTAL ELEMENTARY PARTICLE PHYSICS. *Educ:* Harvard Univ, AB, 53; Univ Chicago, SM, 55, PhD(physics), 59. *Prof Exp:* Res assoc physics, Enrico Fermi Inst, Univ Chicago, 59-60; from asst prof to assoc prof, Northwestern Univ, 60-69; PROF PHYSICS, UNIV KANS, 69-, CHMN, DEPT PHYSICS & ASTRON, 89- *Concurrent Pos:* Consult, High Energy Physics Div, Argonne Nat Lab, 65-69, vis scientist, 71-72; vis scientist, Fermi Lab, Batavia, Ill, 76-81, 84-85; Ger Electron Synchrotron, Hamburg, Ger, 82-88 & Lab Nuclear Studies, Cornell Univ, 89-90; prin investr & NSF grant, Res High Energy Physics, 82- *Mem:* Fel Am Phys Soc. *Res:* High energy physics. *Mailing Add:* Dept Physics & Astron Univ Kans Lawrence KS 66045-2151

AMME, ROBERT CLYDE, b Ames, Iowa, May 15, 30; m 51; c 2. EXPERIMENTAL ATOMIC PHYSICS, ATMOSPHERIC PHYSICS. *Educ:* Iowa State Univ, BS, 53, MS, 55, PhD(physics), 58. *Prof Exp:* Res physicist, Humble Oil & Refining Co, 58-59; res physicist, Denver Res Inst, 59-68, from asst prof to assoc prof, 61-68, actg dean arts & sci, 75-76, dir acad res, 77-81, dean, Grad Sch, 78-81, chmn dept, 81-85, PROF PHYSICS, UNIV DENVER, 68- *Concurrent Pos:* Sr res physicist, Denver Res Inst, 68-77. *Mem:* Fel Am Phys Soc; Sigma Xi; Mat Res Soc; Am Asn Physics Teachers. *Res:* Atomic and molecular physics; collision phenomena; energy and charge transfer; ionization; gaseous transport properties; acoustics; stratospheric composition. *Mailing Add:* Dept Physics Univ Denver Denver CO 80208

AMMERMAN, CHARLES R(OYDEN), electrical engineering, for more information see previous edition

AMMERMAN, CLARENCE BAILEY, b Cynthiana, Ky, May 21, 29; m 50; c 3. ANIMAL NUTRITION. *Educ:* Univ Ky, BS, 51, MS, 52; Univ Ill, PhD(animal sci), 56. *Prof Exp:* Asst animal nutrit, Univ Ky, 51-52; asst animal sci, Univ Ill, 54-56; Med Serv Corps, US Army Med Res & Nutrit Lab, Fitzsimons Gen Hosp, Denver, 56-58; from asst prof to assoc prof, 58-70, PROF ANIMAL SCI & ANIMAL NUTRITIONIST, UNIV FLA, 70- *Concurrent Pos:* Moorman travel fel, animal nutrit res, Nat Feed Ingredients Asn, 69. *Honors & Awards:* Gustav Bohstedt Award, Am Soc Animal Sci, 73; Am Feed Mfrs Asn Nutrit Res Award, 77; Morrison Award, Am Soc Animal Sci, 89. *Mem:* AAAS; Am Soc Animal Sci; Am Inst Nutrit; Am Dairy Sci Asn; Soc Exp Biol & Med; Sigma Xi. *Res:* Animal nutrition mineral elements; forage utilization by ruminants. *Mailing Add:* 3756 SW Fifth Pl Gainesville FL 32607

AMMERMAN, GALE RICHARD, b Sullivan, Ind, Mar 6, 23; m 43; c 5. NEW FOOD PRODUCT DEVELOPMENT, FOOD CANNING & FREEZING. *Educ:* Purdue Univ, BS, 50, MS, 53, PhD(food sci), 57. *Prof Exp:* Prof food sci, Purdue Univ, 54-57, actg head dept, 57-58; assoc dir res, Libby McNeill & Libby Co, 58-67; prof, Miss State Univ, 67-85, head, Dept Food Sci, 85-88; RETIRED. *Concurrent Pos:* Pres, Coun Agr Sci & Tech, 86-87. *Honors & Awards:* Res Award, Inst Food Technologists, 77. *Mem:* Inst Food Technologists. *Res:* Packaging, processing and preservation of the nutritional quality of mainly channel catfish but also southern fruits and vegetables. *Mailing Add:* RR 1 Box 41 Eutaw AL 35462

AMMIRATI, JOSEPH FRANK, JR, b Dunsmuir, Calif, Jan 10, 42; m 86; c 4. MYCOLOGY, FUNGAL SYSTEMATICS & ECOLOGY. *Educ:* San Francisco State Univ, BA, 65, MA, 67; Univ Mich, PhD(bot), 72. *Prof Exp:* Res mycologist, US Dept Agr, 73-74; asst prof biol & bot, Univ Toronto, 74-79; from asst prof to assoc prof, 79-86, PROF BOT, UNIV WASH, 86- *Concurrent Pos:* Adj prof bot, Univ Toronto, 79-80; assoc prof forest resources, Univ Wash, 81-86, prof, 86- *Mem:* Mycological Soc Am; NAm Mycological Asn; Am Inst Biol Sci. *Res:* Classification, distribution and ecology of higher fungi, especially agaricales; montane, boreal, arctic and alpine fungi; poisionous mushrooms; ectomycorrhizal fungi. *Mailing Add:* Dept Bot KB-15 Univ Wash Seattle WA 98195

AMMIRATO, PHILIP VINCENT, b New York, NY, Nov 22, 43. PLANT PHYSIOLOGY. *Educ:* City Col New York, BS, 64; Cornell Univ, PhD(plant physiol), 69. *Prof Exp:* NY State Col teaching fel, Cornell Univ, 64-65; asst prof bot, Rutgers Univ, 69-74; ASST PROF BIOL SCI, BARNARD COL, COLUMBIA UNIV, 74- *Concurrent Pos:* Assoc ed, Torreya, 78- *Mem:* Tissue Cult Asn; Torrey Bot Club; Bot Soc Am; Am Soc Plant Physiol; Sigma Xi. *Res:* Plant growth and development; tissue culture; economic botany; morphogenesis; embryogenesis. *Mailing Add:* Biol Dept Barnard Col New York NY 10027

AMMLUNG, RICHARD LEE, b Havre de Grace, Md, May 7, 52; m 77. INORGANIC CHEMISTRY. *Educ:* Univ Del, BS, 74; Northwestern Univ, MS, 76, PhD(chem), 79. *Prof Exp:* Sr res chemist, 78-81, res specialist, 81-85, SR RES SPECIALIST CHEM, MONSANTO CO, 85- *Mem:* Am Chem Soc. *Res:* Solid fast ion conductors; new processes for the production of electronic grade silicon. *Mailing Add:* 161 Weatherstone Dr Worcester MA 01604

AMMON, HERMAN L, b Passaic, NJ, Nov 24, 36; m 58; c 2. ORGANIC CHEMISTRY. *Educ:* Brown Univ, ScB, 58; Univ Wash, PhD(chem), 63. *Prof Exp:* NIH fel, 63-64; res instr x-ray crystallog, Univ Wash, 64-65; asst prof chem, Univ Calif, Santa Cruz, 65-69; from asst prof to assoc prof, 69-76, PROF CHEM, UNIV MD, COLLEGE PARK, 76- *Mem:* Am Chem Soc; Am Crystallog Asn. *Res:* Organic compound and macromolecular structure; computer programming; x-ray crystallography; conformational analysis. *Mailing Add:* Dept Chem Univ Md College Park MD 20742

AMMON, VERNON DALE, b Mendon, Mo, May 15, 41; m 62; c 1. PLANT PATHOLOGY. *Educ:* Univ Mo, BS, 64, MS, 65, PhD(plant path), 72. *Prof Exp:* Res assoc plant path, Univ Ky, 72-73; asst prof, 73-78, ASSOC PROF PLANT PATH, MISS STATE UNIV, 78- *Mem:* Am Phytopath Soc. *Res:* Disease physiology, ultrastructure of diseased plants. *Mailing Add:* Dept Plant Pathol Miss State Univ PO Drawer PG Mississippi State MS 39762

AMMONDSON, CLAYTON JOHN, physical chemistry; deceased, see previous edition for last biography

AMOLS, HOWARD IRA, b New York, NY, Feb 11, 49; m 70; c 2. RADIOLOGICAL PHYSICS, MEDICAL PHYSICS. *Educ:* Cooper Union, BS, 70; Brown Univ, MS, 73, PhD(physics), 74. *Prof Exp:* Mem staff radiol physics & radiobiol, Los Alamos Sci Lab, Univ Calif, 74-79; asst prof radiology, Univ N Mex, 79-81; asst prof, 81-84, ASSOC PROF RADIATION MED, BROWN UNIV, 84- *Concurrent Pos:* Nat Cancer Inst fel, Los Alamos Sci Lab, 74-76; guest scientist, Karlsruhe Nuclear Res Ctr, WGer, 77-78; assoc physicist, 81-84, chief physicist radiation ther, RI Hosp, 84- *Mem:* Am Asn Physicists Med; Radiation Res Soc. *Res:* Physical and biological studies related to the use of x-rays, electrons, and heavy particles in radiotherapy microdosimetry; dose modeling, treatment planning, effects of dose fractionation and cell heterogeneity; modeling radiation effects in tissue. *Mailing Add:* Dept Radiation Med Rhode Island Hosp 593 Eddy St Providence RI 02902

AMON, MAX, b New York, NY, Feb 28, 39; c 3. OPTICS, ENGINEERING. *Educ:* Hofstra Univ, BS, 63. *Prof Exp:* Optical designer, Goerz Am Optical Co, 63-65, Kollsman Instrument Corp, 65-72, Photronics Corp, 72-73 & Opcom Assocs, 73; OPTICAL DESIGNER, MARTIN MARIETTA CORP, 73- *Concurrent Pos:* Cosult various industs. *Honors & Awards:* Improved Ritchey Chretien Telescope Award, NASA, 72. *Mem:* Optical Soc Am; Soc Photo-Optical Instrumentation Engrs; Inst Elec & Electronics Engrs. *Res:* Color corrected and coma corrected Mangin Mirror; large objective for night observation; extending the field of the Ritchey Chretien Telescope; guidance beam assembly, command optics, zoom projection optics, TIR window; Tactical Air Defense Systems/Pilot Night Vision System boresight module, telescope stabilization, laser television, infrared optics; compact optical ware length discriminator. *Mailing Add:* 1727 Lake Waumpi Dr Maitland FL 32751

AMON PARISI, CRISTINA HORTENSIA, b Montevideo, Uruguay, Oct 12, 56; m 80; c 2. COMPUTATION FLUID MECHANICS, THERMAL-FLUID SCIENCES. *Educ:* Univ Simon Bolivar, Caracas, Venezuela, Mech Eng, 81; Mass Inst Technol, MSME, 85. *Hon Degrees:* DSc, Mass Inst Technol, 88. *Prof Exp:* Instr, Univ Simon Bolivar, 81-83; researcher & teaching asst, Mass Inst Technol, 84-88; ASST PROF MECH ENG, CARNEGIE-MELLON UNIV, 88- *Mem:* Am Soc Mech Engrs; Am Inst Aeronaut & Astronaut; Sigma Xi; Am Soc Eng Educ; Soc Women Engrs. *Res:* Computational fluid dynamics and heat transfer; stability and turbulence; numerical algorithms and experimental techniques and cooling electronic equipment. *Mailing Add:* Dept Mech Eng Carnegie-Mellon Univ Pittsburgh PA 15213

AMOORE, JOHN ERNEST, b Derbyshire, Eng, Apr 28, 30; US citizen; m 59; c 2. OLFACTORY PHYSIOLOGY, ENVIRONMENTAL CHEMISTRY. *Educ:* Oxford Univ, BA, 52, MA & PhD(biochem), 58. *Prof Exp:* Sci officer microanal, John Innes Hort Inst, Eng, 58-59; Nuffield Found res fel plant biochem, Univ Edinburgh, 59-62; Jane Coffin Childs Mem Fund res fel zool, Univ Calif, Berkeley, 62-63; res chemist, Western Regional Res Lab, USDA, Berkeley, Calif, 63-78; PARTNER, OLFACTO-LABS, 79- *Concurrent Pos:* Consult chemist, Hekimian & Assoc, Huntington Beach, Calif, 83-; mem, sci adv comt acid deposition, State Calif, 83- *Mem:* Asn Chemoreception Sci; Air Pollution Control Asn. *Res:* Stereochemical theory of olfaction; smell-blindness; primary odors; on-site chemical analysis for hazardous waste. *Mailing Add:* Olfacto Labs 7701 Potrero Ave El Cerrito CA 94530

AMORY, DAVID WILLIAM, b Newark, NJ, Jan 27, 28; m 60; c 2. ANESTHESIOLOGY, PHARMACOLOGY. *Educ:* St John's Univ, NY, BS, 52, MS, 55; Univ Wash, PhD(pharmacol), 61; Univ BC, MD, 67. *Prof Exp:* Pharmacologist toxicol, Ciba Pharmaceut Co, Inc, 55-57; res asst pharmacol, Univ Wash, 57-61; USPHS fel neuropharmacol, Univ Calif, San Francisco,

61-62; intern med, Virginia Mason Hosp, Seattle, 67-68; USPHS scholar, Cardiovasc Res Inst, Univ Calif, San Francisco, 68-69; resident anesthesiol, Sch Med, Univ Wash, 69-71, assoc prof anesthesiol & pharmacol, 75-82, prof, 82-88; PROF, CHIEF CARDIAC ANESTHESIA & DIR RES, DEPT ANESTHESIOL, ROBERT WOOD JOHNSON MED SCH, 88- *Concurrent Pos:* Nat Heart & Lung Inst res grants, 72-75 & 79-86; Cardiovasc Inst res grant, 88-91. *Honors & Awards:* Walter Stewart Baird Mem Prize, Sch Med, Univ BC, 67; Residents' Res Prize, Am Soc Anesthesiologists, 70; Res Prize in Anesthesiol, Mt Sinai Sch Med, 70. *Mem:* Am Soc Anesthesiologists; Sigma Xi; Soc Cardiovasc Anesthegiologists; Int Anethesia Res Soc. *Res:* Near infrared spectroscopy for assessing cerebral oxygen metabolism; excitatory amino acids and spinal cord ischemia; cardioplegia and right ventricular function; iontophoretic transdermal delivery of drugs. *Mailing Add:* Dept Anesthesiol Robert Wood Johnson Med Sch New Brunswick NJ 08903-0019

AMOS, DENNIS BERNARD, b Bromley, Kent, Eng, Apr 16, 23; m 49; c 5. IMMUNOGENETICS. *Educ:* Univ London, MB, BS, 51, MD, 63. *Hon Degrees:* Dr, Univ Curitiba, Brazil, 88. *Prof Exp:* Intern, Guy's Hosp, London, 51; from assoc cancer res scientist to prin cancer res scientist, Roswell Park Mem Inst, 56-62; chief div immunol, 62-88, JAMES B DUKE PROF IMMUNOL & EXP SURG, MED CTR, DUKE UNIV, 61- *Concurrent Pos:* Res fel, Guy's Hosp, London, 52-55; sr res fel, Roswell Park Mem Inst, 55-56; chmn, Human Lymphocyte-Antigen Standards Comt, AACHT & NIH Nat Inst Allergy & Infectious Dis & nomenclature comt on Leukocyte Antigens, WHO, Int Union Immunol Socs. *Honors & Awards:* Rose Payne Award, 88. *Mem:* Nat Acad Sci; Am Asn Immunologists; Am Asn Cancer Res; Transplantation Soc; Am Asn Clin Histocompatability Test; AAAS; Inst Med. *Res:* Tissue and tumor transplantation genetics; immunology; mouse genetics; antigen differentiation; human lymphocyte-antigen and disease. *Mailing Add:* Med Ctr Duke Univ Box 3010 Durham NC 27710

AMOS, DEWEY HAROLD, b Harrisville, WVa, Feb 27, 25; m 48; c 3. PETROLOGY. *Educ:* Marietta Col, BS, 49; Univ Ill, MA, 50, PhD(geol), 58. *Prof Exp:* Asst geol, Univ Ill, 49-51; geologist, US Geol Surv, 52-55; asst prof geol, Southern Ill Univ, 55-65; PROF GEOL, EASTERN ILL UNIV, 65- *Concurrent Pos:* Consult, US Geol Surv & Defense Minerals Explor Admin, 55-64; consult geologist, US Geol Surv, 64- *Mem:* Fel Geol Soc Am; Soc Econ Geologists. *Res:* Structural geology; geology of base metals and nonmetallic deposits. *Mailing Add:* Dept Geol & Geog Eastern Ill Univ Charleston IL 61920

AMOS, DONALD E, b Lafayette, Ind, Feb 28, 29; m 52; c 2. MATHEMATICS, CHEMICAL ENGINEERING. *Educ:* Purdue Univ, BS, 51; Ore State Univ, MS, 56, PhD(math), 60. *Prof Exp:* Chem engr, Gen Elec Co, 51-53; staff mem appl math, Sandia Corp, 60-64; assoc prof math, Univ Mo-Columbia, 64-66; mathematician, 66, STAFF MEM, SANDIA LABS, 66- *Concurrent Pos:* Ed, Communications in Statist, 72-81. *Mem:* Asn Comput Mach; Am Math Asn; Soc Indust & Appl Math. *Res:* Special functions; numerical analysis and solution of ordinary and partial differential equations. *Mailing Add:* 7704 Pickard Ave NE Albuquerque NM 87110

AMOS, HAROLD, b Pennsauken, NJ, Sept 7, 19. BACTERIOLOGY. *Educ:* Springfield Col, BS, 41; Harvard Univ, PhD(bact), 52. *Prof Exp:* Instr bact, Springfield Col, 48-49; assoc, 54-59, from asst prof to assoc prof, 59-70, PROF MICROBIOL & MOLECULAR GENETICS, HARVARD MED SCH, 70-, CHMN, DEPT MICROBIOL, 79- *Concurrent Pos:* Fulbright res fel, Pasteur Inst, France, 51-52; res fel, Harvard Med Sch, 52-54; USPHS fel, 52-54, sr res fel, 58; mem, Nat Cancer Adv Bd, 72-; trustee, Josiah Macy Found, 73- *Mem:* Inst Med-Nat Acad Sci; Am Soc Biol Chemists; Tissue Cult Asn; Am Soc Microbiol. *Res:* Hexose metabolism in mammalian cells; surface changes and hormonal influences. *Mailing Add:* Dept Microbiol & Molecular Genetics Harvard Med Sch 25 Shattuck St Boston MA 02115

AMOS, HENRY ESTILL, b Bowling Green, Ky, Aug 20, 41; m 69; c 2. RUMINANT NUTRITION, AGRICULTURE. *Educ:* Western Ky Univ, BS, 62; Univ Ky, MS, 64, PhD(ruminant nutrit), 69. *Prof Exp:* Instr animal nutrit, Western Ky Univ, 64-67; res assoc, Univ Ky, 69-71; RES PHYSIOLOGIST ANIMAL NUTRIT, AGR RES, USDA, 71- *Concurrent Pos:* Res fel, Univ Ky, 69-71. *Mem:* Am Soc Animal Sci; Am Dairy Sci Asn; Am Registry Cert Animal Scientists. *Res:* Primary emphasis on intraruminal protein transformations; microbial protein syntesis, digestion and utilization and rumen protein by-pass. *Mailing Add:* Dept Animal Sci Univ Ga Athens GA 30602

AMOSS, MAX ST CLAIR, b Baltimore, Md, May 9, 37; m 87; c 3. ENDOCRINOLOGY. *Educ:* Pa State Univ, BS, 62; Tex A&M Univ, MS, 65; Baylor Col Med, PhD(physiol), 69. *Prof Exp:* Fel, Baylor Col Med, 69, asst prof endocrinol, 69-70; asst res prof neuroendocrinol, Salk Inst, 70-75; from asst prof to assoc prof, 75-85, PROF VET PHYSIOL, TEX A&M UNIV, 85- *Concurrent Pos:* Instr, San Diego Eve Col, 72-75. *Mem:* Endocrine Soc; Am Physiol Soc; Soc Study Reproduction; Int Soc Neuroendocrinol; Soc Neurosci. *Res:* Hypothalamic control of hypophyseal function with special emphasis on reproduction; development of radioimmunoassay procedures for use in veterinary medicine; melanoma in swine as model of human disease. *Mailing Add:* Dept Vet Physiol & Pharmacol Tex A&M Univ Col Vet Med College Station TX 77843-4466

AMPLATZ, KURT, b Weistrach, Austria, Feb 25, 24; US citizen; m 56; c 3. RADIOLOGY. *Educ:* Univ Innsbruck, MD, 50. *Prof Exp:* From instr to assoc prof, 58-70, PROF RADIOL, SCH MED, UNIV MINN, MINNEAPOLIS, 70- *Res:* Detection of cardiac shunts with radioactive and nonradioactive gases; blood flow measurements by radiographic technics; radiographic evaluation of renovascular hypertension; pneumotomography; development of see-through film changers; magnification coronary angiography; development of angiographic injection; nonthrombogenic catheter; development of angiographic equipment and endovrologic technics. *Mailing Add:* Dept Radiol Univ Minn Hosp & Clin Harvard St & E River Rd Minneapolis MN 55455

AMPULSKI, ROBERT STANLEY, b Grand Rapids, Mich, Sept 22, 42; m 65; c 3. CELLULOSE CHEMISTRY, MASS SPECTROMETRY. *Educ:* Aquinas Col, BS, 64; Marquette Univ, MS, 69; Fla State Univ, PhD(phys chem), 72. *Prof Exp:* Res chemist, Milwaukee Blood Ctr Inc, 67-68; group leader mass spectrometry, Miami Valley Labs, 72-78, GROUP LEADER CATALYSIS PULP PROCESS DEVELOP, PAPER DIV, PROCTER & GAMBLE CO, 80- *Mem:* AAAS; Am Soc Mass Spectrometry; Sigma Xi. *Res:* Stable isotopes; gas phase kinetics; ion molecule reactions; catalysis, cellulose chemistry and wood pulping; chemical bonding; surface science; molecular design. *Mailing Add:* 5674 Red Oak Dr Fairfield OH 45014

AMPY, FRANKLIN, b Dinwiddie, Va, June 22, 36. MUTAGENESIS, BIOSTATISTICS. *Educ:* Va State Col, BS, 58; Ore State Univ, MS, 60, PhD(genetics), 62. *Prof Exp:* Asst prof genetics, Am Univ Beirut, Lebanon, 62-68; fel, Univ Calif, Davis, 68-70; actg chmn, 73-75, & 84-86, ASSOC PROF ZOOL, HOWARD UNIV, 71-, VCHMN, 81- *Concurrent Pos:* Assoc dean educ opportunity prog, Univ Calif, Davis, 70-71; NASA-AMES fac res fel, 76; prin investr, Minority Biomed Prog, Howard Univ, NIH, 78-; NIH consult, Site Visit Fla Int Univ, 81 & 83; reviewer, Va Sci Talent Search, Va Jr Acad Sci, 81- *Mem:* Environ Mutagenesis Soc; Genetics Soc Am; Am Genetics Asn; Am Soc Cell Biol. *Res:* Environmetnal mutagenesis, specifically in the hormonal regulation of environmental mutagens. *Mailing Add:* Dept Zool Howard Univ 415 Col St NW Washington DC 20059

AMR, SANIA, HORMONE ACTION, PROTEIN CHEMISTRY. *Educ:* Univ Toulouse, France, MD, 77. *Prof Exp:* VIS ASSOC, DEPT HEALTH & HUMAN SERVS, NIH, 83- *Mailing Add:* 4033 The Alamda Baltimore MD 21218

AMREIN, YOST URSUS LUCIUS, b Arosa, Switz, Jan 3, 18; nat US; m 48; c 1. ANIMAL PARASITOLOGY. *Educ:* Univ Calif, Los Angeles, BA, 47, MA, 48, PhD(parasitol), 51. *Prof Exp:* Teaching asst zool, Univ Calif, Los Angeles, 47-51; from instr to assoc prof, 51-59, chmn dept, 59-68, prof, 59-81, EMER PROF ZOOL, POMONA COL, 81- *Concurrent Pos:* USPHS fel, Nat Inst Med Res, London, 57-58; NIH res grants, 59-64 & 66-74; res fel, Swiss Trop Inst, 64-65; lectr, Calif State Polytech Univ, Pomona, 81-84. *Mem:* AAAS; Am Soc Parasitol; fel Royal Soc Trop Med & Hyg. *Res:* Physiology of hemoflagellates; ecology of parasites in arid areas. *Mailing Add:* Dept Biol Seaver Lab Pomona Col Claremont CA 91711-6339

AMRINE, HAROLD THOMAS, b Moline, Ill, June 9, 16; m 42; c 1. INDUSTRIAL ENGINEERING. *Educ:* Univ Iowa, BS, 38, MS, 39. *Hon Degrees:* Dr Aeronaut Eng, Embry-Riddle Aeronaut Univ, 72. *Prof Exp:* Jr engr, Swift & Co, Iowa, 39-40; planning engr, W A Sheaffer Pen Co, 40; instr eng drawing, Ohio State Univ, 40-41 & 45-46; from asst prof to assoc prof indust eng, 46-54, chmn sect 52-55, prof, 54-81, head, Sch Indust Eng, 55-69, assoc dir div sponsored prog, Res Found, 70-73 head, Dept Freshman Eng, 73-81, EMER PROF INDUST ENG, PURDUE UNIV, WEST LAFAYETTE, 81- *Concurrent Pos:* Consult engr, 47-81; chmn & mem, Col Indust Comt Mat Handling Educ, 52-59; Fulbright lectr, Univ Melbourne, 59; vis prof, Ariz State Univ, 69-70; vpres, Pritsker & Assoc, 81-82. *Mem:* Inst Indust Engrs (vpres, 57-58); Am Soc Eng Educ; Am Soc Mech Engrs; fel Inst Indust Engrs; Sigma Xi. *Res:* Methods engineering; human factors. *Mailing Add:* 2601 Trace 26 West Lafayette IN 47906

AMROMIN, GEORGE DAVID, b Gomel, Russia, Feb 27, 19; US citizen; m 42; c 5. NEUROPATHOLOGY. *Educ:* Northwestern Univ, BS, 40, MD, 43. *Prof Exp:* Asst pathologist, Michael Reese Hosp, Chicago, 49; pathologist, Mem Hosp, Exeter, Calif, 50-53; asst dir & assoc pathologist, Michael Reese Hosp, Chicago, 54-56; chmn div path, City of Hope Med Ctr, 56-71, chmn, Dept Neuropath & Res Path, 71-77; prof, 77-84, EMER PROF PATH, UNIV MO MED CTR, COLUMBIA, 84- *Concurrent Pos:* Nat Inst Neurol Dis & Blindness fel, Bellevue Med Ctr, NY Univ, 68-69; attend pathologist, Tulare Kings Joint Hosp, Springville, Calif, 51-54; pathologist, Tulare County Hosp & Visalia Munic Hosp, 51-54; consult staff path, Tulare Dist Hosp, 51-54; clin prof, Loma Linda Univ Med Sch, 70. *Mem:* AMA; fel Am Soc Clin Path; fel Am Col Path; fel Am Col Physicians; Am Asn Neuropathologists; Sigma Xi. *Res:* Neuropathology. *Mailing Add:* 139 N Wetherly Dr Beverly Hills CA 90211

AMRON, IRVING, b Bayonne, NJ, Jan 18, 21; m 42. BATTERY DESIGN, COMPUTER MODELING & COST ANALYSIS. *Educ:* NY Univ, BA, 41, MS, 44, PhD(phys chem), 49. *Prof Exp:* Sr chemist, SAM Labs, Manhattan Dist, NY, 44-46; instr qual anal chem, NY Univ, 46-47; res assoc textile chem, Textile Found, 47-49; res assoc combustion, NY Univ, 49-50 & Gen Elec Co Inc, 50-52; sr engr cathode coatings, Sylvania Elec Prod Inc, 52-54; mem tech staff electronic mat, processes, devices, batteries & systs eng, AT&T BELL LABS, 54-87; CONSULT, 87- *Concurrent Pos:* Consult, batteries & process economet. *Mem:* Sigma Xi; Am Chem Soc; Electrochem Soc; Asn Comput Mach; fel Am Inst Chemists; AAAS; Asn Consult Chemists & Chem Engrs. *Res:* Gas phase kinetics; heterogeneous equilibria; econometrics; electrochemical energy sources. *Mailing Add:* 19 Lexington Dr Livingston NJ 07039-4303

AMSBURY, DAVID LEONARD, b Topeka, Kans, Dec 30, 32; m 54. GEOLOGY. *Educ:* Sul Ross State Col, BS, 52; Univ Tex, Austin, PhD(geol), 57. *Prof Exp:* Geologist, Shell Develop Co, 56-64, res geologist, 64-66; GEOLOGIST, JOHNSON SPACE CTR, NASA, 67- *Concurrent Pos:* Lectr, Univ Houston, 70- *Mem:* Soc Econ Paleont & Mineral; Geol Soc Am; Sigma Xi; Am Inst Prof Geologists; AAAS. *Res:* Physical stratigraphy; remote sensing in environmental geology. *Mailing Add:* 1422 Brookwood Ct Seabrook TX 77586

AMSDEN, THOMAS WILLIAM, b Wichita, Kans, Jan 31, 15; m 40; c 2. INVERTEBRATE PALEONTOLOGY. *Educ:* Univ Wichita, AB, 39; Univ Iowa, MA, 41; Yale Univ, PhD(geol), 47. *Prof Exp:* Geologist, US Geol Surv, 43-45; asst geol, Yale Univ, 42-43 & 45-46, Nat Res fel, 46-47; geologist, Md State Geol Surv, 47; from instr to assoc prof geol, Johns Hopkins Univ, 47-55; geologist, 55-85, EMER GEOLOGIST, OKLA GEOL SURV, 85- *Concurrent Pos:* NSF grants, Wales, Sweden & Czech, 64-66, Russia, 68, Poland, 71. *Honors & Awards:* Levorsen Award, Am Asn Petrol Geologists, 73. *Mem:* Paleont Soc; Geol Soc Am; Am Asn Petrol Geologists; Soc Econ Paleontologists & Mineralogists; Brit Palaeont Asn. *Res:* Late Ordovician, Silurian and early Devonian stratigraphy and paleontology; brachiopods; middle Paleozoic analysis of petroliferous deep sedimentary basins. *Mailing Add:* Okla Geol Surv Univ Okla 100 E Boyd Rm N-131 Norman OK 73019

AMSEL, LEWIS PAUL, b New York, NY, Mar 23, 42; m 64; c 2. PHARMACEUTICAL CHEMISTRY. *Educ:* Columbia Univ, BSc, 63; State Univ NY Buffalo, PhD(pharm), 69. *Prof Exp:* Assoc res pharmacist, Sterling-Winthrop Res Inst, 69-74; mgr, 74-81, DIR PHARMACEUT DEVELOP, PHARMACEUT DIV, PENNWALT CORP, 81- *Mem:* AAAS; Am Pharmaceut Asn; Acad Pharmaceut Sci. *Res:* Biopharmaceutics and pharmacokinetics; drug dosage design and evaluation. *Mailing Add:* Searle 700 E Business Center Dr Mt Prospect IL 60056

AMSTER, ADOLPH BERNARD, b New York, NY, Nov 22, 24; m 53; c 1. PHYSICAL CHEMISTRY. *Educ:* City Col New York, BS, 43; Columbia Univ, AM, 47; Ohio State Univ, PhD(phys chem), 51. *Prof Exp:* Asst chem, Columbia Univ, 47 & Ohio State Univ, 49-51; chemist, US Bur Standards, 51-54; res assoc, Naval Ord Lab, 54-57, supv chemist, 57-60; mgr phys chem, Stanford Res Inst, 61-65, chmn phys & inorg chem div, 65-68; chief explosives & pyrotech br, Naval Sea Systs Command, 68-78; head chem div, 78-85, SR SCIENTIST, NAVAL WEAPONS CTR, CHINA LAKE, CALIF, 85- *Mem:* AAAS; Sigma Xi; Am Phys Soc; Combustion Inst. *Res:* Experimental physical chemistry; thermochemistry; explosives and propellant sensitivity; detonations; pollution abatement; hazard analysis; chemiluminescence; accident investigation. *Mailing Add:* PO Box 1106 Ridgecrest CA 93555

AMSTER, HARVEY JEROME, b Cleveland, Ohio, Sept 30, 28; c 2. THEORETICAL PHYSICS, APPLIED MATHEMATICS. *Educ:* Calif Inst Technol, BS, 50; Mass Inst Technol, PhD, 54. *Prof Exp:* Adv scientist, Westinghouse Bettis Plant, 54-61; from assoc prof to prof engr, Univ Calif, Berkeley, 61-79; RES, WESTERN RESERVE UNION & UNIV WIS, 79- *Concurrent Pos:* Res, Lawrence Livermore Lab, Argonne Nat Lab, Inst Defense Anal, Univ London & Northrop Space Labs. *Mem:* AAAS; Am Nuclear Soc. *Res:* Nuclear physics; reactor theory; environmental effects, especially of nuclear technology. *Mailing Add:* 4208 Ridgemont Ct Oakland CA 94619-3727

AMSTERDAM, ABRAHAM, b Haifa, Israel, Apr 28, 39; m 69; c 4. CANCER RESEARCH, MOLECULAR ENDOCRINOLOGY. *Educ:* Hebrew Univ, BSc, 64, MSc, 66, PhD(biochem), 71. *Prof Exp:* Res asst biochem, Hebrew Univ, Jerusalem, 65-68, instr, 69- 70; res assoc cell biol, Rockefeller Univ, 71-72; sr scientist, 75-78, ASSOC PROF HORMONE RES, WEIZMANN INST SCI, 79- *Concurrent Pos:* Vis scientist, NIH, Bethesda, 79-81; vis prof, Johns Hopkins Univ, Baltimore, 86-87, Univ Basel, 89-90, Max-Planck Inst, Munich, 91- *Honors & Awards:* Pfizer Lectr, Clin Res Inst, Montreal, 83. *Mem:* Am Soc Cell Biol; Am Soc Endocrinol; Israel Soc Electron Micros; Israel Biochem Soc; Israel Endocrine Soc; Fedn Am Socs Exp Biol. *Res:* Structure and function of exocrine and endocrine cells; role of peptide and steroid hormones and growth factors in ovarian physiology; effect of oncogene expression on ovarian cell differentiation; ovarian cancer; induction of steroidogenesis. *Mailing Add:* Dept Hormone Res Weizmann Inst Sci PO Box 26 Rehovot Israel

AMSTERDAM, DANIEL, MICROBIOLOGY. *Educ:* City Col NY, BS, 55; Brooklyn Col, MA, 57; NY Univ, PhD, 65; Am Bd Microbiol, dipl, 81. *Prof Exp:* Bacteriologist, Mt Sinai Hosp, NY, 55-57, res asst, Dept Microbiol, 58-60; sr microbiologist, Kingsbrook Jewish Med Ctr, Brooklyn, NY, 60-61, chief microbiol, Isaac Albert Res Inst, 61-80; DIR CLIN MICROBIOL & IMMUNOL, ERIE COUNTY LAB, ERIE COUNTY MED CTR, BUFFALO, NY, 80-; PROF MICROBIOL & ASSOC PROF MED, SCHS MED & DENT, STATE UNIV NY, BUFFALO, 80- *Concurrent Pos:* Assoc, Dept Microbiol, Mt Sinai Sch Med, City Univ NY, 69-80; head, Human Cell Biol Prog, Neurosci Ctr, Kingsbrook Jewish Med Ctr & Downstate Med Ctr, 71-80; adj prof, Dept Biol, Long Island Univ, Brooklyn, NY, 72-80; consult, microbiol, Prof Exam Serv, New York, NY, 77-80; dir, postdoctoral training prog, Med & Pub Health Lab Microbiol, Buffalo, NY, 81- *Honors & Awards:* NSF-NATO Award Cell Biol, Stresa, Italy, 70. *Mem:* Fel Am Acad Microbiol; AAAS; Am Asn Pathologists; Am Soc Cell Biol; Am Soc Microbiol; fel Infectious Dis Soc Am; Tissue Cult Asn. *Res:* Author or co-author of over 90 publications. *Mailing Add:* Dept Microbiol & Immunol Erie County Med Ctr & State Univ NY 462 Grider St Buffalo NY 14215

AMSTUTZ, HARLAN CABOT, b Santa Monica, Calif, July 17, 31; m; c 3. ORTHOPEDICS, BIOENGINEERING. *Educ:* Univ Calif, Los Angeles, BA, 53, MD, 56. *Prof Exp:* Intern, Los Angeles County Gen Hosp, 56-57; resident surg, Med Ctr Hosp, Univ Calif, Los Angeles, 57-58; resident orthop, Hosp Spec Surg, 58-61; res asst, Inst Orthop, London, 63-64; PROF ORTHOP & CHIEF, DIV ORTHOP, SCH MED, UNIV CALIF, LOS ANGELES, 70- *Concurrent Pos:* Am-Brit-Can traveling fel, 70; lectr, Polytech Inst Brooklyn; chief prosthetics & orthotics, dir bioeng, assoc scientist & asst attend, Hosp Spec Surg; asst prof, Cornell Med Col; hon registr, Royal Nat Orthop Hosp, London, 63-64; mem comt skeletal systs & mem subcomt bioeng, Nat Acad Sci-Nat Res Coun; mem subcomt prosthetic & orthotics Educ, Nat Acad Sci; mem comt interplay eng biol & med, Nat Mat Adv Bd; chief orthop surg, Vet Admin Ctr, Los Angeles. *Mem:* Am Acad Orthop Surgeons; Orthop Res Soc; NY Acad Sci; Am Soc Testing & Mat; Am Col Surgeons. *Res:* Total joint replacement; congenital anomalies. *Mailing Add:* Div Orthop Surg Univ Calif Sch Med Acad Med & Hosp Clin 10833 Le Conte Ave Los Angeles CA 90024

AMSTUTZ, HAROLD EMERSON, b Barrs Mill, Ohio, June 21, 19; m 49; c 4. VETERINARY MEDICINE. *Educ:* Ohio State Univ, BS, 42, DVM, 45. *Prof Exp:* From instr to prof vet med, Ohio State Univ, 47-61, actg chmn dept, 55-56, chmn, 56-61; prof vet sci & head dept large animal clins, 61-89, EMER PROF, PURDUE UNIV, WEST LAFAYETTE, 89. *Concurrent Pos:* Organizer, VI Int Conf Cattle Dis, US, 70. *Honors & Awards:* Borden Award, Am Vet Med Asn, 78; Amstutz Williams Award, Am Asn Bovine Pratitioners, 86. *Mem:* Am Asn Vet Clinicians (pres, 62-64); World Asn Buiatrics (pres, 72-82); Asn Bovine Practitioners; Am Col Internal Med (pres, 73). *Res:* Cattle respiratory diseases; cattle lameness. *Mailing Add:* Lynn Hall Purdue Univ West Lafayette IN 47907

AMSTUTZ, LARRY IHRIG, b Wooster, Ohio, July 11, 41. SOLID STATE PHYSICS. *Educ:* Col Wooster, BA, 63; Duke Univ, PhD(physics), 69. *Prof Exp:* Physicist, Electrotech Lab, 69-76, PHYSICIST, ELEC EQUIP DIV, US ARMY MOBILITY EQUIP RES & DEVELOP COMMAND, 76- *Mem:* Am Phys Soc; Sigma Xi. *Res:* Magnetic resonance; solid hydrogen; thermal and electrical conductivity in alloys; magnetic behavior of rare-earth alloys; applied superconductivity; numerical methods; high energy laser power supplies. *Mailing Add:* 24 W Rosemont Ave Alexandria VA 22301

AMUNDRUD, DONALD LORNE, b Calgary, Alta, May 22, 43; m 69; c 2. SAFEGUARDS, REACTOR PHYSICS. *Educ:* Univ Sask, Regina, BA Hons, 66; Purdue Univ, MS, 69. *Prof Exp:* Nuclear design engr, Ont Hydro, 69-72; reactor physicist, Atomic Energy Can Ltd Whiteshell Nuclear Res Estab, 72-77, safeguards analyst reactor physics, 77-79; reactor & start up & licensing engr, New Brunswick Elec Power Comn, 79-85; SAFEGUARDS OFFICER & INSPECTOR, IAEA, VIENNA, AUSTRIA, 85- *Mem:* Can Asn Physicists; Chem Inst Can. *Res:* Safeguards methods and techniques on all aspects of the nuclear fuel cycle; start up planning, licensing engineering and emergency planning at point lepreau; tests of point lepreau; 600 mw candu; safeguards equipment, performance monitoring analysis of safeguards equipment. *Mailing Add:* c/o Mrs H Amundrud 14 Dunning Cres Regina SK S4S 3W1 Can

AMUNDSEN, CLIFFORD C, b Norwalk, Conn, Dec 21, 33; m 56; c 2. PHYSIOLOGICAL ECOLOGY, MICROCLIMATOLOGY. *Educ:* Idaho State Col, BS, 61; Univ Colo, Boulder, PhD(bot, physiol ecol), 67. *Prof Exp:* Res ecologist, Inst Arctic & Alpine Res, Univ Colo, Boulder, 65-67; res assoc, 67-69, asst prof, 69-74, ASSOC PROF BOT, UNIV TENN, KNOXVILLE, 74- *Concurrent Pos:* Consult, Battelle Mem Inst, US Forest Serv; prin investr Aleutian res, US Dept of Energy, 68-; consult, Nat Park Serv, 75- *Mem:* Ecol Soc Am; Am Polar Soc; Am Quaternary Asn. *Res:* Arctic-alpine and subarctic-subalpine physiological plant ecology and remote sensing; reservoir ecology. *Mailing Add:* Dept Bot Univ Tenn Knoxville TN 37996-1610

AMUNDSEN, LAWRENCE HARDIN, b Pine River, Wis, Sept 23, 09; m 34; c 4. ORGANIC CHEMISTRY. *Educ:* Col Ozarks, BS, 31; Univ Fla, PhD(org chem), 35. *Prof Exp:* Asst org chem, Univ Fla, 31-35; instr chem, Adelphi Col, 35-36; from instr to prof, 36-73, actg head dept, 45-46, EMER PROF CHEM, UNIV CONN, 73- *Concurrent Pos:* Nat Defense Res Comt fel, Purdue Univ, 42-43; chemist, Venereal Dis Res Lab, USPHS, Staten Island, NY, 43-44. *Mem:* AAAS; Am Chem Soc. *Res:* Preparation and properties of aliphatic diamines; preparation of sulfonamides; allylic rearrangements in amination reactions. *Mailing Add:* 128 S Eagleville Rd Storrs CT 06268

AMUNDSON, CLYDE HOWARD, b Nekoosa, Wis, Aug 15, 27; m 52; c 2. FOOD SCIENCE, AQUACULTURE. *Educ:* Univ Wis-Madison, BS, 55, MS, 56, PhD(food sci, biochem), 60. *Prof Exp:* Chemist, Sherwin-Williams Co, 50-52; from instr to assoc prof, 56-70, from assoc chmn to chmn dept, 64-85, DIR FOOD ENG, UNIV WIS-MADISON, 70-, DIR AQUACULT RES CTR, 81- *Concurrent Pos:* Consult, Bur Com Fisheries, Wash; prof food sci, 70-, prof agr eng, 82-, prof oceanog & limnol, Univ Wis-Madison, 80- *Honors & Awards:* Am Dairy Sci Asn Res Award, 71; Pfizer Res Award, 86. *Mem:* Fel AAAS; fel NY Acad Sci; fel Am Inst Chem; Inst Food Technol; Am Chem Soc; World Aquacult Soc. *Res:* Flavor chemistry of food and food products; dehydration of foods and biological materials; fermentation chemistry of foods and food products; fat and oil chemistry and food enzymes. *Mailing Add:* Dept Food Sci Univ Wis 123 Babcock Hall 1605 Linden Dr Madison WI 53706

AMUNDSON, KARL RAYMOND, b Minneapolis, Minn, Mar 11, 61; m 83. BLOCK COPOLYMERS, LIQUID CRYSTAL POLYMERS. *Educ:* Univ Minn, BS(chem eng) & BS(physics), 83; Univ Calif, Berkeley, PhD(chem eng), 89. *Prof Exp:* RES SCIENTIST, AT&T BELL LABS, 89- *Mem:* Am Phys Soc; AAAS. *Res:* Morphologies in thermotropic liquid crystalline polymers by means of nuclear magnetic resonance spectroscopy; effect of shear flow on polymer morphology by means of nuclear magnetic resonance spectroscopy; effect of electric fields on block copolymer morphology. *Mailing Add:* 50 Crescent Dr New Providence NJ 07974-1718

AMUNDSON, MARY JANE, b Keokuk, Iowa, Jan 23, 36; m 58; c 1. PSYCHIATRIC NURSING. *Educ:* Univ Iowa, BS, 57; Univ Calif, Los Angeles, MS, 66, PhD(spec educ), 74. *Prof Exp:* Head nurse, Mendocino State Hosp, Talmage, Calif, 58-60; staff nursing, Langley Porter Inst, San Francisco, 60-61; clinical specialist & lectr nursing, Univ Calif, Los Angeles, Neuropsychiat Inst, 61-74; asst prof, Univ Ore, 74-75; ASSOC PROF NURSING, UNIV HAWAII, 75- *Mem:* Asn Women Sci; Am Asn Univ Prof. *Res:* Network for Pacific Island Mental Health Workers. *Mailing Add:* Dept Nursing Univ Hawaii Manoa Honolulu HI 96822

AMUNDSON, MERLE E, b Sioux Falls, SDak, Aug 21, 36; m 58; c 3. ANALYTICAL & PHARMACEUTICAL CHEMISTRY, CLINICAL SCIENCE. *Educ:* SDak State Univ, BS, 58; Mass Col Pharm, MS, 59, PhD(pharm), 61. *Prof Exp:* Sr anal chemist, Eli Lilly & Co, 61-67, res scientist, 67-68, head agr prod develop dept, 68-71, head agr biochem, 71-75, dir agr anal chem & agr biochem, 75-78, dir toxicol div, 78-83, EXEC DIR, LILLY RES LABS, DIV ELI LILLY & CO, 84- *Mem:* Am Chem Soc; Soc Environ Toxicol & Chem; NY Acad Sci. *Res:* Degradation of pesticides in soil and plants; analytical methodology for determination of residues in soil, plant tissue, and animal tissue; toxicology of chemicals; clinical trials. *Mailing Add:* Lilly Res Labs Lilly Corp Ctr Indianapolis IN 46285

AMUNDSON, NEAL R, b St Paul, Minn, Jan 10, 16. CHEMICAL ENGINEERING. *Educ:* Univ Minn, BCE, 37, MS, 41, PhD(math), 45. *Hon Degrees:* ScD, Univ Minn, 85, EngD, Univ Notre Dame, 86. *Prof Exp:* Process engr, Standard Oil Co, NJ, 37-39; from teaching asst to asst prof, Dept Math, Univ Minn, 39-47, from assoc prof to prof, Dept Chem Eng, 47-67, Regents' prof, 67-77, head, 49-77; vpres, 87-89, CULLEN PROF CHEM ENG, UNIV HOUSTON, 77-, PROF MATH, 82- *Concurrent Pos:* Fulbright scholar, Cambridge Univ, Eng, 54-55, Guggenheim fel, 55; Guggenheim fel, NATO sr fel, 75. *Honors & Awards:* Indust & Eng Chem Award, Am Chem Soc, 60; William H Walker Award, Am Inst Chem Engrs, 61, Warren K Lewis Award, 71, Richard H Wilhelm Award, 73; Vincent Bendix Award, Am Soc Eng Educ, 70. *Mem:* Nat Acad Eng; Am Chem Soc; Am Inst Chem; Am Math Soc; fel Am Inst Chem Engrs. *Res:* Author 5 books and 300 journal publications. *Mailing Add:* 5327 Cherokee St Houston TX 77005

AMUNDSON, ROBERT GALE, b Sunnyside, Wash, Oct 20, 46; m 80. PLANT GAS EXCHANGE, AIR POLLUTION. *Educ:* Whitman Col, BA, 69; Univ Wash, MS, 73, PhD(bot), 78. *Prof Exp:* Fel, Boyce Thompson Inst Plant Res, Cornell Univ, 78-80, res assoc, 80-83, asst scientist, 83-87, ASSOC SCIENTIST, BOYCE THOMPSON INST PLANT RES, CORNELL UNIV, 87- *Mem:* Am Soc Plant Physiologists; Sigma Xi; Air Pollution Control Asn; AAAS; Am Forestry Asn. *Res:* Effects of air pollution on the net photosynthesis, transpiration and pollutant uptake of plants and how these effects are manifested in plant growth and yield. *Mailing Add:* Boyce Thompson Inst Plant Res Cornell Univ Ithaca NY 14853

AMY, JONATHAN WEEKES, b Delaware, Ohio, Mar 3, 23; m 47; c 3. CHEMISTRY, CHEMICAL INSTRUMENTATION. *Educ:* Ohio Wesleyan Univ, BA, 48; Purdue Univ, MS, 50, PhD(molecular spectros), 55. *Prof Exp:* Res assoc chem, 54-60, from assoc prof to prof chem, 63-88, dir instrumentation 75-88, ASSOC DIR LABS, PURDUE UNIV, 60-, EMER PROF CHEM, 88- *Concurrent Pos:* Consult chem instrumentation, Asn Am Univs Adv Panel, 80 & AAAS Adv Panel, 81; dir res, Finnigan MAT, 83-84. *Honors & Awards:* Instrumentation Award, Am Chem Soc, 81. *Mem:* Am Chem Soc; Sigma Xi; AAAS. *Res:* Molecular and mass spectroscopy; gas chromatography; chemical instrumentation; surface chemistry. *Mailing Add:* Dept Chem Purdue Univ 356 Overlook Dr West Lafayette IN 47906

AMY, NANCY KLEIN, b Philadelphia, Pa, June 23, 47; m 71. BIOCHEMISTRY, ENZYMOLOGY. *Educ:* Gwynedd Mercy Col, AB, 69; Univ Va, PhD(enzym), 73. *Prof Exp:* Fel membrane biochem, Univ Va Med Ctr, 73-74; res assoc enzym, Dept Biol, Univ Va, 74-77; res assoc biochem, Duke Univ Med Ctr, 77-81; ASST PROF BIOCHEM, UNIV CALIF, BERKELEY, 81- *Mem:* Am Soc Microbiol; Am Soc Biochem & Molecular Biol. *Res:* Role of metals in enzyme function; molybdenum cofactor and incorporation into nitrate reductase; molybdenum carrier protein. *Mailing Add:* Dept Nutrit Sci Univ Calif Berkeley CA 94720

AMY, ROBERT LEWIS, b Jamestown, Pa, July 18, 19; m 44; c 3. ZOOLOGY. *Educ:* Thiel Col, BS, 41; Univ Pittsburgh, MS, 49; Univ Va, PhD(biol), 55. *Prof Exp:* From asst prof to assoc prof, Susquehanna Univ, 49-58; assoc prof, 58-61, PROF BIOL, SOUTHWESTERN UNIV, MEMPHIS, 61-, CHMN DEPT, 68- *Concurrent Pos:* Res partic, Oak Ridge Nat Lab, 55 & 56; USPHS fel, Nat Blood Transfusion Ctr, France, 64-65; coinvestr, Habrobracon proj, NASA Biosatellite Prog, 66-69; USPHS fel, Inst Cellular Path, France, 71-72. *Mem:* AAAS; Radiation Res Soc; Am Soc Zool; Soc Develop Biol. *Res:* Effects of radiation on development in insects. *Mailing Add:* Dept Biol Southwestern Univ Memphis Memphis TN 38112

AMZEL, L MARIO, b Buenos Aires, Arg, Oct 25, 42; US citizen; c 2. X-RAY DIFFRACTION ANALYSIS, PROTEIN CONFORMATION. *Educ:* Univ Buenos Aires, lic, 65, PhD(phys chem), 68. *Prof Exp:* Grad teaching asst, Univ Buenos Aires, 65-66; instr phys chem, Univ Cent Venezuela, 67-69; teaching fel, 69-70, instr, 70-73, from asst prof to assoc prof, 73-83, PROF BIOPHYS, MED SCH, JOHNS HOPKINS UNIV, 84- *Mem:* Am Crystallog Asn; Biophys Soc; AAAS; Soc Latin Am Biophysicists. *Res:* Structure and conformation of proteins, immunoglobulins and proteins involved in ATP synthesis; modeling of the conformation of the combining site of immunoglobulins. *Mailing Add:* Dept Biophys Johns Hopkins Med Sch 720 Rutland Ave Baltimore MD 21205

AN, LINDA HUANG, information science & systems, for more information see previous edition

ANACKER, ROBERT LEROY, RICKETTSIAE ANTIGENS. *Educ:* Univ Wash, Seattle, PhD(microbiol), 56. *Prof Exp:* RES MICROBIOLOGIST, NAT INST ALLERGY & INFECTIOUS DIS, NIH, ROCKY MOUNTAIN LAB, HAMILTON, MT, 60- *Mailing Add:* Rocky Mountain Lab NW 220 Hill Top Dr Hamilton MT 59840

ANAGNOSTAKIS, SANDRA LEE, b Coffeyville, Kans, May 14, 39; m 69; c 1. MYCOLOGY, GENETICS. *Educ:* Univ Calif, Riverside, BA, 61; Univ Tex, Austin, MA, 66; Justus Liebig Univ, Giessen, PhD, 85. *Prof Exp:* PLANT PATHOLOGIST & GENETICIST FUNGAL PLANT PATHOGENS, CONN AGR EXP STA, 66- *Mem:* Mycol Soc Am; Am Phytopath Soc; AAAS. *Res:* Genetic studies of fungal plant pathogens, including maize smut, chestnut blight and Dutch elm disease; techniques for production of haploids of higher plants. *Mailing Add:* Conn Agr Exp Sta PO Box 1106 New Haven CT 06504

ANAGNOSTOPOULOS, CONSTANTINE E, b Greece, Nov 1, 22; nat US; m 49; c 1. ORGANIC CHEMISTRY, POLYMER CHEMISTRY. *Educ:* Brown Univ, ScB, 49; Harvard Univ, MA, 50, PhD(chem), 52. *Prof Exp:* Res chemist, 52-57, scientist, 57-61, from asst dir to dir res, 61-68, dir res, Org Chem Div, 68-69, Rubber Chem, 69-70 & Tire Textiles Div, 70-71, gen mgr, New Enterprise Div, 71-75, gen mgr Rubber Chem Div, 75-80, vpres & managing dir, Europe-Africa & chmn, Monsanto Europe, 81-82, CORP VPRES, CORP DEVELOP, MONSANTO CO, ST LOUIS, 82- *Concurrent Pos:* Mem nat inventors coun, Brit Intel Serv, 41-43; mem, Presidential Panel Prizes for Innovation; US-USSR Trade & Econ Coun, 80-82; chmn bd, St Louis Technol Ctr, Mo, 85- *Mem:* Am Chem Soc; Sigma Xi (vpres, 59); Soc Chem Indust; Com Develop Asn. *Res:* Synthesis of steroids, derivatives, amino acids and analogs; plasticization of vinyl polymers; polymer-diluent interactions; photocatalyzed oxidation of polyalkylenes. *Mailing Add:* 8000 Maryland Suite 1190 St Louis MO 63105

ANAGNOSTOPOULOS, CONSTANTINE N, b Patras, Greece, July 4, 44; US citizen; m 69; c 2. MICROELECTRONICS. *Educ:* Merrimack Col, BS, 67; Brown Univ, ScM, 71; Univ RI, PhD(elec eng), 75. *Prof Exp:* Res physicist, Eastman Kodak Co, 75-79, sr res physicist, 79-82, group leader, CMOS/BICMOS Tech Group, 85-89, RES ASSOC & MEM SR STAFF, EASTMAN KODAK CO, 82-, GROUP LEADER ASIC DESIGN GROUP, 89- *Concurrent Pos:* Founder, Custom Integrated Circuits Conf, 79, tech prog chmn, 79, 80, conf chmn, 81, gen chmn, 82, mem tech prog & steering comts, 79-91; guest ed J Solid State Circuits Spec issues Technol for custom Integrated Circuits, 84, 86 & 87, assoc ed, 90. *Honors & Awards:* Centennial Medal, Inst Elec & Electronic Engrs. *Mem:* Inst Elec & Electronics Engrs. *Res:* Solid state imaging devices, custom integrated circuits, ASIC's and imaging systems. *Mailing Add:* Eastman Kodak Co Image Aquisition Prod Div Rochester NY 14650

ANAGNOU, NICHOLAS P, b Dire-Dawa, Ethiopia, Nov 15, 47. TRANSACTING FACTORS, GENE EXPRESSION REGULATION. *Educ:* Univ Athens, MD, 72, PhD(biochem), 77. *Prof Exp:* VIS SCIENTIST, NAT HEART, BLOOD & LUNG INST, NIH, 80- *Mem:* Am Soc Cell Biol; Am Soc Hemat; Am Fedn Clin Res. *Mailing Add:* Molecular Biol Dept Basic Sci Univ Crete Sch Med Heraklion Crete 711 10 Greece

ANAND, AMARJIT SINGH, b New Delhi, India, June 1, 40; US citizen; m; c 2. REPRODUCTIVE PHYSIOLOGY, BIOCHEMISTRY. *Educ:* Univ Panjab, India, DVM, 62; Univ Wis-Madison, MS, 65, PhD(physiol), 67. *Prof Exp:* Vet, Vet Med Sch, Univ Panjab, India, 58-62; res asst physiol & biochem, Univ Wis-Madison, 62-67; from asst prof to assoc prof physiol, Univ Wis-Oshkosh, 67-73, assoc prof biol, 73-78, univ vet, 77-85, PROF BIOL, UNIV WIS-OSHKOSH, 79- *Concurrent Pos:* Co-recipient, Wis State Univ Bd Regents res grant, 68-69; NSF instnl res grant & Ford Found fel, 70-71. *Mem:* Am Vet Med Asn. *Res:* Veterinary medicine. *Mailing Add:* Dept Biol Univ Wis 800 Algoma Blvd Oshkosh WI 54901

ANAND, RAJEN S(INGH), b Kohat, India, June 8, 37; US citizen; m 69; c 2. FETAL PHYSIOLOGY. *Educ:* Agra Univ, BSc, 56; Marathwada Agr Univ, Parbhani, DVM, 60; Univ Calif, Davis, PhD(physiol), 69. *Prof Exp:* Instr biol chem, Col Vet Sci & Res Inst, Mhow, India, 60-63; res asst physiol chem, 63-69, res physiologist, Univ Calif, Davis, 69-70; prof physiol & biol, 70-85, chmn, Dept Anat & Physiol, 85-89, CHAIR, COMMUN DIS DEPT, CALIF STATE UNIV, LONG BEACH, 90- *Concurrent Pos:* New fac res grant, Calif State Univ, Long Beach, 70-71; fel endocrinol, Sch Med, Univ Calif, Los Angeles, 77-78. *Honors & Awards:* Hertzendorf Mem Physiol Award, 69. *Mem:* Am Physiol Soc; AAAS; Sigma Xi; Am Heart Asn; Am Inst Biol Sci. *Res:* Energy metabolism in the fetus in utero; placental transfer of glucose; glucose metabolism in the fetal sheep. *Mailing Add:* Dept Anat & Physiol Calif State Univ Long Beach CA 90840

ANAND, SATISH CHANDRA, b India, Aug 9, 30; US citizen; m 50; c 3. PLANT BREEDING. *Educ:* Univ Delhi, BS, 51; NC State Univ, MAgr, 59; Univ Wis, PhD(agron), 62. *Prof Exp:* Res assoc, Univ Ill, Urbana, 62-63; wheat breeder, Punjab Agr Univ, India, 64-71, sr geneticist, 72-73; soybean breeder, McNair Seed Co, 73-79; assoc prof, 80-84, PROF, UNIV MO-COLUMBIA, 85- *Concurrent Pos:* Mem cereals & pulses subcomt, Indian Stand Inst, 64-72; zonal coordr wheat, Indian Coun Agr Res, 68-72; mem, Nat Soybean Variety Rev Bd, Am Soc Agron. *Mem:* Am Soc Agron; fel Indian Soc Genetics & Plant Breeding; Crop Sci Soc Am; Soc Nematologists. *Res:* Breeding and development of high yielding soybean varieties resistant to diseases and nematodes. *Mailing Add:* Univ Mo Columbia PO Box 160 Portageville MO 63873

ANAND, SUBHASH CHANDRA, b Lyallpur, India, July 27, 33; m 65; c 2. CIVIL ENGINEERING, ENGINEERING MECHANICS. *Educ:* Banaras Hindu Univ, BS, 55; Northwestern Univ, MS, 65, PhD(struct eng), 68. *Prof Exp:* Asst engr, Pub Works Dept, India, 56 & Indian Iron & Steel Co, 56-58; engr-in-training, Hein, Lehmann & Co, WGer, 58-59; struct designer, Kloeckner-Humboldt-Deutz Co, 60-64; teaching asst, Northwestern Univ, 65-68; asst prof, Calif State Univ, Sacramento, 68-70; assoc prof, Ill Inst Technol, 70-72; assoc prof, 72-76, PROF CIVIL ENG, CLEMSON UNIV, 76- *Concurrent Pos:* Educ grant, NSF, 72, res grants, 73, 78, 81, 85 & 88; Off Water Resources & Technol grant, 80; Fulbright Lectr, Uruguay, 80. *Mem:* Am Acad Mech; fel Am Soc Civil Engrs; Am Soc Eng Educ; Int Asn Comp Mech. *Res:* Elastic-plastic analysis of plane stress, plane strain, axisymmetric and plate bending problems; shakedown loads on rolling disks, cylinders and spheres; biomechanics; seismic loads on buried pipes; mathematical and finite element modelling of composite masonry; numerical techniques of groundwater flow and mass transport problems. *Mailing Add:* Dept Civil Eng Clemson Univ Clemson SC 29634-0911

ANANDALINGAM, G, b Cambridge, Eng, Oct 27, 53; US citizen; m 87. SYSTEMS SCIENCE. *Educ:* Univ Cambridge, Eng, BA & MA, 75; Harvard Univ, SM, 77, PhD(opers res & econ), 81. *Prof Exp:* Engr & economist, Brookhaven Nat Lab, 81-84; asst prof systs eng, Univ Va, 84-87; asst prof, 87-90, ASSOC PROF SYSTS ENG, UNIV PA, 90- *Concurrent Pos:* Consult, Int Develop & Energy Assocs, 83-; assoc ed, Int Abstracts Opers Res, 86-; Environ Protection Agency fel, 87; mem grad fac, Dept Energy Policy & Mgt, SAsia Regional Studies, Univ Pa, 89-, dir, Exec Eng Prog, 90- *Mem:* Fel AAAS; sr mem Inst Elec & Electronics Engrs; Opers Res Soc; Inst Mgt Sci. *Res:* Optimization and design of networks with application to telecommunications, computer, and electric power networks; multi-agent multi-attribute decision problems; environmental systems analysis; games on networks; technology policy. *Mailing Add:* Dept Systs Univ Pa Philadelphia PA 19104-6315

ANANDAN, MUNISAMY, b India, July 1, 39; m 68; c 2. PLASMA DISPLAYS, LIQUID CRYSTAL DISPLAYS. *Educ:* Voorhees Col, Vellore, India, BS, 60; Annamali Univ, India, MS, 64; Indian Inst Technol, Bombay, MTech, 67; Indian Inst Sci, Bangalore, PhD(plasma display), 80. *Prof Exp:* Sci officer electronics, Atomic Energy Estab, Bombay, 63-64; develop engr electron devices, Bharat Electronics Ltd Bangalore, 67-72, dept mgr display devices, 72-82, mgr res & develop, liquid crystal display, 82-87; vis scientist plasma displays, Bell Commun Res, Red Bank, NJ, 87-88; RES SCIENTIST PLASMA, BACKLIGHT DEVICE FOR FULL COLOR, LIQUID CRYSTAL DISPLAY, THOMAS ELECTRONICS INC, WAYNE, NJ, 88- *Concurrent Pos:* Mat adv, Randl Tube Inc & Dept Health & Human Serv, Fla Int Univ, 90- *Honors & Awards:* First Prize in Res & Develop Excellence, Electronics Industs Asn India, 85. *Mem:* Am Phys Soc; Inst Elec & Electronics Engrs; Soc Info Display; Illumination Eng Soc. *Res:* Plasma display; electric field pattern study of coplanar DC plasma display; new self-aligned process for color plasma display; liquid crystal display; novel three terminal true analogue liquid crystal display with new electrode structure-detailed investigation; backlighting devices; plasma row backlight device development for ferro electric column shutter liquid crystal display; novel flat fluorescent backlight device development for full color LCD; author of many publications; six patents. *Mailing Add:* Thomas Electronics Inc 100 Riverview Dr Wayne NJ 07470

ANAND-SRIVASTAVA, MADHU BALA, BIOCHEMICAL PHARMACOLOGY. *Educ:* Delhi Univ, India, MSc, 72; Univ Man, Can, PhD(physiol), 78. *Prof Exp:* Res assoc, Vanderbilt Univ, 78-80; ASST PROF, UNIV MONTREAL, 81- *Concurrent Pos:* Sr investr, Clin Res Inst, Montreal, 81-90; scholar, Can Heart Found, 82-88; scientist, Med Res Coun, 90- *Mem:* Int Soc Heart Res; Can Biochem Soc; Am Soc Pharmacol & Exp Therapeut; Am Soc Biochem & Molecular Biol. *Res:* Hormonal regulation of adenylate cyclase in cardiovascular system in health and disease; G-binding proteins and gene expression; atrial natriuretic factor and second messengers. *Mailing Add:* Dept Physiol Fac Med Univ Montreal CP 6128 Succursale A Montreal PQ H3C 3J7 Can

ANANTHA, NARASIPUR GUNDAPPA, b Hassan, Mysore, India, Feb 4, 34; US citizen; m 64; c 2. SOLID STATE PHYSICS. *Educ:* Mysore Univ, India, BScHons, 54; Indian Inst Sci, DIISc, 58; Mass Inst Technol, SM, 60; Harvard Univ, PhD(appl physics), 65. *Prof Exp:* Design engr semiconductor devices, Gen Elec Co, 65-67; staff engr, Components Div, 67-69, develop eng, Systs Prods Div, 69-74, sr engr mgr devices, Gen Tech Div, 74-82, consult, semiconductor technol, CHQ, 82-86, SR TECH STAFF, GTD, IBM, 86- *Mem:* Sr mem Inst Elec & Electronics Engrs. *Res:* Study of semiconductor device physics and operation with emphasis on silicon high speed switching devices for very large scale integrated circuits applications; study of physical limits, process limits of microelectronics. *Mailing Add:* Gen Tech Div IBM East Fishkill Facil Rte 52 Hopewell Junction NY 12533

ANANTHA NARAYANAN, VENKATARAMAN, b Madras City, India, Oct 22, 36; m 67; c 1. PHYSICS, SPECTROSCOPY. *Educ:* Annamalai Univ, Madras, MA, 57, MSc, 58; Indian Inst Sci, Bangalore, PhD(physics, spectros), 62. *Prof Exp:* Vis fel, Mellon Inst, 63-64; Robert A Welch Found fel chem, Tex A&M Univ, 64-65; PROF PHYSICS & THEORET VIBRATIONAL SPECTRAL STUDIES, SAVANNAH STATE COL, 65- *Concurrent Pos:* Coop res partic, Oak Ridge Nat Lab, 70, 88; summer res, Bruceton Lab, 85, 86; Oak Ridge Nat Lab, 87. *Honors & Awards:* Hon Chem Abstractor, Am Chem Soc, 69- *Mem:* Optical Soc Am; Am Asn Physics Teachers; fel Brit Inst Physics; Am Asn Univ Prof; Indian Inst Sci Alumni Asn. *Res:* Vibrational spectra of crystals and hydrogen bonded systems; far infrared and Raman techniques; use of spectral data for evaluation of force constants and other chemical bond properties; vibrational spectra; molecular constants; crystal fields; computer oriented physics teaching; innovative general physics curriculum materials; laser spectroscopic diagnostics. *Mailing Add:* Dept Math & Physics PO Box 20473 Savannah State Col Savannah GA 31404

ANANTHANARAYANAN, VETTAIKKORU S, b Madras, India, Nov 29, 38; m 64; c 2. BIOPHYSICAL CHEMISTRY, PROTEIN PHYSICAL CHEMISTRY. *Educ:* Madras Univ, BSc, 58, MSc, 60, PhD(phys chem), 65. *Prof Exp:* Teaching fel biochem, Univ Western Ont, 66-68; teaching fel chem, Cornell Univ, 68-71; asst prof molecular biophys, Indian Inst Sci, Bangalore, 72-78; assoc prof, 79-82, PROF BIOCHEM, MEM UNIV NFLD, 82-; PROF BIOCHEM, MCMASTER UNIV, ONT. *Concurrent Pos:* Vis asst prof, Cornell Univ, 72; Can commonwealth res fel & vis asst prof, Mem Univ Nfld, 76-77; prin investr grants, Med Res Coun Can, 79-, NIH, 80-82 & Can Heart Found, 80-; vis scientist, Med Res Coun Can, 85-86; mem, Biochem Grants Comt, Med Res Coun Can, 87-90. *Mem:* Can Biochem Soc; Am Soc Biol Chemists; NY Acad Sci; AAAS; Can Biophys Soc. *Res:* Biomolecular spectroscopy; conformation of peptides, polypeptides and proteins; conformational transitions; structure-function relationship in proteins; structural aspects of collagen biosynthesis; structure-function studies on antifreeze proteins; structure of hormones and drugs. *Mailing Add:* Dept Biochem McMaster Univ 1200 Main St W Hamilton ON L8N 3Z5 Can

ANAST, CONSTANTINE SPIRO, pediatrics; deceased, see previous edition for last biography

ANASTASIO, SALVATORE, b Brooklyn, NY, Mar 12, 32; m 57; c 2. MATHEMATICS, COMPUTER SCIENCE. *Educ:* Cathedral Col, AB, 54; NY Univ, MS, 61, PhD(math), 64. *Prof Exp:* Instr math, Iona Col, 59-61; asst prof, Fordham Univ, 64-70; assoc prof, 70-72, PROF MATH, STATE UNIV NY COL NEW PALTZ, 72- *Mem:* Am Math Soc; Math Asn Am; Hist Sci Soc. *Res:* History of mathematics; von Neumann algebras; history and philosophy of science. *Mailing Add:* Dept Math State Univ NY Col New Paltz NY 12561

ANASTASSIOU, DIMITRIS, b Athens, Greece, Mar 31, 52; m 78. SIGNAL PROCESSING, COMMUNICATIONS. *Educ:* Nat Tech Univ Athens, Grad dipl, 74; Univ Calif, Berkeley, MS, 75, PhD(elec eng), 79. *Prof Exp:* Res asst, Electron Res Lab, Univ Calif, Berkeley, 74-75 & 76-78; mem res staff, TJ WATSON RES CTR, IBM CORP, 78-; L K Comput Systs Consults Ltd, Nicosia, Cyprus. *Concurrent Pos:* Teaching asst, Dept Elec Eng & Comput Sci, Univ Calif, Berkeley, 76; mem staff, LK comput systs consults Ltd, Nicosia,cyprus. *Mem:* Sigma Xi; Inst Elec & Electronics Engrs. *Res:* Signal processing and communications; development and implementation of algorithms for image processing and data compression for teleconferencing systems. *Mailing Add:* Dept Elec Eng Columbia Univ Main Div New York NY 10027

ANBAR, MICHAEL, b Danzig, June 29, 27; m 53; c 2. MASS SPECTROMETRY. *Educ:* Hebrew Univ, Jerusalem, MSc, 50, PhD(phys org chem), 53. *Prof Exp:* Instr chem, Univ Chicago, 53-55; sr scientist, Weizmann Inst Sci, 55-60, prof, Frienberg Grad Sch, 60-67; sr res assoc exobiol, Ames Res Ctr, 67-68; dir phys sci, SRI Int, 68-72, dir mass spectrometry, Res Ctr, 72-77; prof biophys sci & chmn dept, 77-90, FAC PROF, SCH MED, STATE UNIV NY, BUFFALO, 90- *Concurrent Pos:* Dir, Radioisotope Training Ctr, Weizmann Inst Sci, 56-59 & Radiation Res Dept, AEC Soreq Nuclear Res Ctr, Israel, 59-66; head chem div, Israel AEC Res Lab, 62-66; sr res assoc, Argonne Nat Lab, 63-64; prof inorg chem, Univ Tel Aviv, 66-67; assoc dean appl res & dir, Health Care Instrument & Device Inst, 83-85; dir, Interdept Clin Biophys Group, 90- *Honors & Awards:* Zondek Award, Israel Med Asn, 62. *Mem:* Am Chem Soc; Royal Chem Soc; Am Soc Mass Spectrom; NY Acad Sci; Biophys Soc; Inst Elec & Electronics Engrs; Clin Chem Soc; Am Acad Thermology; Am Asn Clin Chem; Am Phys Soc; Soc Photo-Optical Instrumentation Engrs. *Res:* Mechanisms of organic and inorganic reactions; isotope methodology in biology and medicine; radiation chemistry and molecular radiobiology; hydrated electron chemistry; mass spectrometry, field ionization mass spectrometry in particular; molten salt electrochemistry; ultrasonics and sonochemistry; hard tissue biochemistry; evolvement of life; computer assisted learning; clinical thermography. *Mailing Add:* Sch Med 120 Cary Hall State Univ NY Buffalo NY 14214

ANCEL, FREDRIC D, b Chicago, Ill, Aug 2, 43. TOPOLOGY. *Educ:* Amherst Col, BA, 65; Univ Wis-Madison, PhD(math), 76. *Prof Exp:* ASSOC PROF MATH, UNIV WIS, MILWAUKEE, 86- *Mem:* Am Math Soc. *Mailing Add:* Dept Math Univ Wis Milwaukee WI 53201

ANCHEL, MARJORIE WOLFF, b New York, NY, May 6, 10; m 42. BIOCHEMISTRY. *Educ:* Columbia Univ, BA, 31, MA, 33, PhD(biol chem), 39. *Prof Exp:* Fel, Col Physicians & Surg, Columbia Univ, 37; asst chem, Queens Col, NY, 39-41 & Columbia Univ, 41-43; res assoc, Squibb Inst Med Res, 43-46; sr chemist & adminr lab, 72-77, EMER SR SCIENTIST, NY BOT GARDEN, 77- *Mem:* Fel AAAS; Am Chem Soc; Mycol Soc Am; fel NY Acad Sci; Am Soc Biol Chemists. *Res:* Antibiotics; natural polyacetylenes; chemistry and biogenesis of fungal products. *Mailing Add:* 147-01 Third Ave Whitestone NY 11357

ANCKER, CLINTON J(AMES), JR, b Cedar Falls, Iowa, June 21, 19; m 47; c 4. APPLIED MATHEMATICS, MILITARY OPERATIONS RESEARCH. *Educ:* Purdue Univ, BSME, 40; Univ Calif, Berkeley, MS, 49, ME, 50; Stanford Univ, PhD(eng mech), 55. *Prof Exp:* Cadet engr, Detroit Edison Co, 40-41, jr engr, 46; instr eng mech, Purdue Univ, 46-47; asst prof eng design, Univ Calif, Berkeley, 47-55; opers analyst, Opers Res Off, Md, 55-56; sr engr, Booz-Allen Appl Res, Inc, Ill, 56-58; mgr, Analco Serv Co, Ill, 58-59; head math & opers res staff, Syst Develop Corp, 59-67; dir, Nat Hwy Safety Inst, US Dept of Transp, 67-68; prof indust & systs eng & chmn dept, 66-84, EMER PROF, UNIV SOUTHERN CALIF, 84- *Concurrent Pos:* Consult, Opers Res Off, 56-57; mem exec comt, Mil Opers Res Symp, 61-65; mem nat coun, Opers Res Soc, 70-73. *Mem:* Opers Res Soc Am. *Res:* Theory of elasticity; theory of queues; stochastic models of combat; traffic flow theory. *Mailing Add:* 23908 Malibu Knolls Rd Malibu CA 90265-4823

ANCKER-JOHNSON, BETSY, b St Louis, Mo, Apr 27, 29; m 58; c 4. ENVIRONMENTAL SCIENCE POLICY. *Educ:* Wellesley Col, BA, 49; Tubingen Univ, PhD(physics), 53. *Hon Degrees:* DSc, NY Polytech Inst, 79, Univ Southern Calif, 84; LLD, Bates Col, 80. *Prof Exp:* Jr res physicist & lectr physics, Univ Calif, Berkeley, 53-54; mem staff, Inter-Varsity Christian Fel, 54-56; sr res physicist, Microwave Physics Lab, Sylvania Elec Prods, Inc, 56-58; mem tech staff, David Sarnoff Res Ctr, Radio Corp Am, 58-61; res specialist, Electronic Sci Lab, Boeing Sci Res Labs, 61-70, supvr solid state & plasma electronics, 70-71, mgr advan energy syst, Boeing Aerospace Co, 71-73; asst secy com for sci & technol, US Dept Com, 73-77; assoc lab dir phys res, Argonne Nat Lab, 77-79; VPRES, ENVIRON ACTIV STAFF, GM TECH CTR, GEN MOTORS CORP, 79- *Concurrent Pos:* Mem, US-USSR Comn Sci Technol, 73-77, Comn Energy, 75-77; mem bd dirs, Soc Automotive Engrs, 79-82, Motor Vehicle Mfrs Asn, 82-; chair, Eng Policy Comt, Motor Vehicle Mfrs Asn, 81-84; mem, US Safety Rev Panel Antarctica, NSF, 87-88; regents lectr, dept elec eng & computer sci, Univ Calif, Berkeley, 88-89; chair bd dirs, World Environ Ctr, 88-; mem Relative Risk Reduction Strategies Comt, US Environ Protection Agency, 89-, Sci Adv Bd & Int Environ Technol Transfer Adv Bd, 89- *Mem:* Nat Acad Eng; fel AAAS; fel Am Phys Soc; fel Inst Elec & Electronics Engrs; Soc Automobile Engrs; Sigma Xi. *Res:* Solid state physics; plasma in solids; microwaves and molecular electronics; ferromagnetism and nonreciprocal effects; x-ray studies of imperfections in nearly perfect crystals; author of more than 70 scientific papers and holds several patents in the field of solid state physics, microwave and semiconductor electronics. *Mailing Add:* GM Tech Ctr Gen Motors Corp 30400 Mound Rd Warren MI 48090-9015

ANCSIN, JOHN, b Hungary, May 1, 33; Can citizen. PHYSICS. *Educ:* Univ Ottawa, BSc, 59, MSc, 60, PhD(physics), 65. *Prof Exp:* RES OFFICER PHYSICS, NAT RES COUN CAN, 66- *Res:* Low temperature fixed points of the temperature scale. *Mailing Add:* Navan Rd Ottawa ON K4B 1H8 Can

ANCTIL, MICHEL, b Warwick, Que, June 18, 45; m 74; c 2. NEUROTRANSMITTERS, BIOLUMINESCENCE. *Educ:* Univ Montreal, BSc, 67, MSc, 69; Univ Calif, Santa Barbara, PhD(neurobiol), 76. *Prof Exp:* Asst prof physiol, 76-82, assoc prof comp physiol, 82-88, PROF COMP PHYSIOL, UNIV MONTREAL, 88- *Mem:* AAAS; Asn Physiologists; Can Soc Zoologists. *Res:* Comparative neurobiology of bioluminescent animals; evolution of nervous systems; neurotransmitters of primative nervous systems. *Mailing Add:* Dept Biol Sci Univ Montreal CP 6128 Montreal PQ H3C 3J7 Can

ANDALAFTE, EDWARD ZIEGLER, b Springfield, Mo, Aug 7, 35. ORTHOGONALITY. *Educ:* Southwest Mo State Col, BS, 56; Univ Mo-Columbia, AM, 59, PhD(math), 61. *Prof Exp:* Assoc prof math, Southwest Mo State Col, 61-64; assoc prof, 64-85, chmn dept, 65-67 & 68-69, PROF MATH, UNIV MO, ST LOUIS, 85-, CHMN DEPT, 90- *Concurrent Pos:* NSF fac fel, 70-71; vis assoc prof, Math Res Ctr, Univ Wis-Madison, 70-71 & Univ Tex San Antonio, 78-79. *Mem:* Am Math Soc; Math Asn Am; Am Asn Univ Professors; Sigma Xi. *Res:* Geometry of metric spaces and their generalizations; metric characterizations of Banach and Euclidean spaces; orthogonality in normed linear spaces. *Mailing Add:* Dept Math Univ Mo St Louis MO 63121-4499

ANDEEN, RICHARD E, b Chicago, Ill, Oct 14, 27; m 62; c 3. TECHNICAL MANAGEMENT. *Educ:* Northwestern Univ, BS, 49, MS, 51, PhD(elec eng), 58. *Prof Exp:* Engr, Sperry Gyroscope Co, Sperry Corp, 51-56 & 58-80, eng dir, Sperry Flight Systs Div & gen mgr, Sperry Space Systs, 81-87; VPRES, CENT TECH OPERS, HONEYWELL INC, 88- *Mem:* Am Inst Aeronaut & Astronaut. *Res:* Automatic and flight control systems; computers. *Mailing Add:* 5936 N Hummingbird Lane Paradise Valley AZ 85253

ANDELMAN, JULIAN BARRY, b Boston, Mass, Sept 23, 31; m 53; c 3. WATER CHEMISTRY, ENVIRONMENTAL CHEMISTRY. *Educ:* Harvard Univ, AB, 52; Polytech Inst Brooklyn, PhD(chem), 60. *Prof Exp:* Res fel chem, NY Univ, 59-61; mem tech staff, Bell Tel Labs, Inc, 61-63; from asst prof to assoc prof water chem, 63-73, PROF WATER CHEM, GRAD SCH PUB HEALTH, UNIV PITTSBURGH, 73- *Concurrent Pos:* Consult, WHO, 69, 71 & 74 & Nat Res Coun-Nat Acad Sci, 71-85; vis lectr, Univ Col, London, 70-71. *Mem:* AAAS; Am Water Works Asn; Am Chem Soc; Int Asn Water Pollution Res; Air & Waste Mgt Asn; Soc Risk Analysis. *Res:* Chemistry of trace constituents in natural and treated waters. *Mailing Add:* Grad Sch Pub Health Univ Pittsburgh Pittsburgh PA 15261

ANDELSON, JONATHAN GARY, b Chicago, Ill, Feb 10, 49. PHYSICAL ANTHROPOLOGY, ETHNOLOGY. *Educ:* Grinnell Col, BA, 70; Univ Mich, Ann Arbor, MA, 73, PhD(anthrop), 74. *Prof Exp:* Asst prof, 74-80, ASSOC PROF ANTHROP, GRINNELL COL, 80- *Mem:* Am Anthrop Asn; Soc Med Anthrop. *Res:* Small scale communal societies of the nineteenth and twentieth centuries. *Mailing Add:* Dept Anthrop Grinnell Col Box 805 Grinnell IA 50112

ANDER, PAUL, b Brooklyn, NY, Apr 19, 31; m 56; c 2. PHYSICAL CHEMISTRY. *Educ:* City Col New York, BS, 53; Polytech Inst Brooklyn, MS, 54; Rutgers Univ, PhD(chem), 61. *Prof Exp:* Instr chem, Rutgers Univ, 60-61; from asst prof to assoc prof, 61-70, PROF CHEM, SETON HALL UNIV, 70- *Mem:* Am Chem Soc. *Res:* Physical chemistry of macromolecules; solution properties of synthetic and biological polyelectrolytes. *Mailing Add:* 853 Mitchell Dr Union NJ 07085

ANDEREGG, DOYLE EDWARD, b Uhrichsville, Ohio, Jan 19, 30; m 57; c 2. LICHENOLOGY. *Educ:* Ohio State Univ, BSc, 52, MSc, 57, PhD(bot, plant path), 59. *Prof Exp:* Asst bot, Ohio State Univ, 55-59; from instr bot & microbiol to assoc prof, Univ Okla, 59-67, chmn dept, 63-67; head dept biol sci, 67-75, asst grad dean, 69-70, fiscal & personnel asst to dean col lett & sci, 78-82, PROF BIOL SCI, UNIV IDAHO, 67-, ASST DEAN COL LETT & SCI, 82- *Mem:* AAAS; Mycol Soc Am; Int Soc Plant Taxonomists; Am Bryol & Lichenological Soc. *Res:* Biogeography, taxonomy and ecology of lichens. *Mailing Add:* 825 West C St Moscow ID 83843

ANDEREGG, JOHN WILLIAM, b White Lake, Wis, Nov 12, 23. BIOPHYSICS. *Educ:* Univ Wis, BS, 47, MS, 49, PhD(physics), 52. *Prof Exp:* Asst physics, 46-52, fel & instr, 53-56, proj assoc oncol, 56-57, asst prof zool & physics, 57-62, assoc prof physics, 62-69, PROF PHYSICS & BIOPHYS, UNIV WIS-MADISON, 69- *Concurrent Pos:* Fulbright scholar, Royal Inst, London, 52-53. *Mem:* Am Phys Soc; Biophys Soc; Am Crystallog Asn. *Res:* Virus structure; neutralization of viruses; x-ray scattering. *Mailing Add:* 5706 Lake Mendota Dr Madison WI 53705

ANDEREGG, ROBERT JAMES, b Antigo, Wis, May 25, 51. MASS SPECTROMETRY, CHROMATOGRAPHY. *Educ:* Univ Wis-Madison, BS, 73; Mass Inst Technol, PhD(anal chem), 77. *Prof Exp:* Res assoc, Mass Inst Technol, 77-79; from asst prof to prof chem, Univ Maine, Orono, 79-89; SR RES INVESTR, GLAXO, INC, RESEARCH TRIANGLE PARK, 90- *Concurrent Pos:* Coop asst prof, 80-83, coop assoc prof, Univ Maine, Orono, 83-; vis assoc prof chem, Addis Abada Univ, Ethiopia, 84-85; vis consult, Smith Kline & French Res Labs, Philadelphia, 86-87. *Mem:* Am Chem Soc; Am Soc Mass Spectrometry; Sigma Xi. *Res:* Biomedical and biochemical applications of chromatography and mass spectrometry; mass spectrometric peptide and protein sequencing; computer techniques for the selective reduction of mass spectrometric data. *Mailing Add:* Glaxo Res Labs Five Moore Dr Research Triangle Park NC 27709

ANDERLE, RICHARD, b New York, NY, Oct 8, 26; wid. MATHEMATICS, PHYSICS. *Educ:* Brooklyn Col, BA, 48. *Hon Degrees:* DSc, Ohio State Univ, 81. *Prof Exp:* Mathematician, Exterior Ballistics Br, Naval Surface Weapons Ctr, 48-60, head, Astronaut & Geod Div, 60-81, res assoc, Strategic Systs Dept, 81-85; SR CONSULT, GEN ELEC CO, 85- *Concurrent Pos:* Lectr, Am Univ, 64-65; pres, Geod Sect, Am Geophys Union, 80-82; pres, Advan Space Technol Sect, Int Asn Geod, 83-87; corresp astronr, Royal Belg Observ, 85. *Honors & Awards:* Superior Civilian Award, US Dept Navy, 60; John A Dahlgren Award, Naval Surface Weapons Ctr, 77. *Mem:* AAAS; Inst Elec & Electronics Engrs; Am Inst Navig; Am Soc Photogram; NY Acad Sci; fel Am Geophys Union; Int Asn Geod. *Res:* Geodesy; celestial mechanics. *Mailing Add:* Space Systs Div General Electric Co Box 8555-43E20 Philadelphia PA 19101

ANDERMANN, GEORGE, b Szarvas, Hungary, Oct 14, 24; US citizen; m 50; c 1. SPECTROSCOPY. *Educ:* Univ Calif, Los Angeles, BS, 49; Univ Southern Calif, MS, 61, PhD(chem), 65. *Prof Exp:* Chemist, Southern Pac Co, 49-51; spectroscopist, Appl Res Labs, 51-53, lab supvr, 53-60, sr physicist, 60-61; sr chemist, Austin Robinson Lab, 61-62; instrument consult mat testing div, Magnaflux Corp, 62-63; res asst chem, Univ Southern Calif, 63-65, res assoc mat sci, 65; from asst prof to assoc prof, 65-74, PROF CHEM, UNIV HAWAII, MANOA, 74- *Concurrent Pos:* Assoc chemist, Hawaii Inst Geophys; NSF grant, 66-71. *Mem:* Am Chem Soc; Soc Appl Spectros. *Res:* Optical properties of solids; x-ray emission spectroscopy, fundamental and analytical; infrared spectroscopy; lattice dynamics, particularly electrical anharmonicity. *Mailing Add:* Dept Chem Univ Hawaii Manoa Honolulu HI 96822

ANDERS, EDWARD, b Latvia, June 21, 26; nat US; m 55; c 2. COSMOCHEMISTRY, METEORITES. *Educ:* Columbia Univ, AM, 51, PhD(chem), 54. *Prof Exp:* Instr chem, Univ Ill, 54-55; from asst prof to prof, 55-73, HORACE B HORTON PROF CHEM, UNIV CHICAGO, 73- *Concurrent Pos:* Vis prof, Calif Inst Technol, 60; Nat Acad Sci resident res assoc, Goddard Space Flight Ctr, 61; consult, NASA, 61-69 & 73; NSF sr fel & vis prof, Univ Berne, 63-64, vis prof, 70, 78, 80, 83, 87 & 89; res assoc, Field Mus Natural Hist, 69-; chmn comt meteorites, Int Astron Union, 70-73, pres, Comn on Moon, 76-79; Guggenheim fel, 73-74; distinguished vis lectr, Univ Rochester, 76; chmn div planetary sci, Am Astron Soc, 71-72; assoc ed, Icarus, 70-, Earth, Moon & Planets, 74-; chmn, Div Planetary Sci, Am Astron Soc, 71-72. *Honors & Awards:* Newcomb Cleveland Prize, AAAS, 59; Univ Medal for Excellence, Columbia Univ, 66; J Lawrence Smith Medal, Nat Acad Sci, 71; Medal Exceptional Sci Achievement, NASA, 73; Leonard Medal, Meteoritical Soc, 74; USCD Found lectr, Univ Calif, San Diego, 75; Priestley Lectr, Pa State Univ, 79; Philips Fund Lectr, Haverford Col, 79; Marple-Schweitzer Lectr, Northwestern Univ, 81; Goldschmidt Medal, Geochem Soc, 90. *Mem:* Nat Acad Sci; fel Am Acad Arts & Sci; fel Am Geophys Union; fel AAAS; fel Meteoritical Soc (vpres, 68-72, 89-90, pres, 91-92); Am Chem Soc; Am Astron Soc; assoc Royal Astron Soc; hon mem, Geochem Soc (vpres, 88-89). *Res:* Origin and age of meteorites; interstellar matter, mass extinctions caused by meteorite impact. *Mailing Add:* Enrico Fermi Inst Univ Chicago 5640 S Ellis Ave Chicago IL 60637-1433

ANDERS, EDWARD B, b Chatham, La, Jan 7, 30; m 63; c 2. MATHEMATICAL ANALYSIS. *Educ:* La Polytech Univ, BS, 50; Pa State Univ, BS, 54; Northwestern State Univ, MS & ME, 58; Auburn Univ, PhD(math), 65. *Prof Exp:* High sch teacher, La, 50; instr, Northeastern La State Col, 58-60, from asst prof to assoc prof, 63-66; assoc prof & chmn div math & physics, Ark State Univ, 66-67; from assoc prof to prof math, Northwestern State Univ, 67-86; RETIRED. *Mem:* Nat Acad Sci; Am Math Soc; Math Asn Am; Asn Comput Mach. *Res:* Numerical and error analysis; numerical filter theory; least squares curve fitting and numerical integration theory. *Mailing Add:* 300 E Seventh St Natchitoches LA 71457

ANDERS, MARION WALTER, b Alden, Iowa, May 10, 36; m 58; c 3. TOXICOLOGY. *Educ:* Iowa State Univ, DVM, 60; Univ Minn, PhD(pharmacol), 64. *Prof Exp:* Instr pharmacol, Univ Minn, 64-65; from asst prof to assoc prof, Cornell Univ, 65-69; assoc prof, Univ Minn, Minneapolis, 69-75, prof pharmacol, 75-; AT DEPT PHARMACOL & TOXICOL, UNIV ROCHESTER. *Concurrent Pos:* Mem ad hoc sci adv comt, Food & Drug Admin, 70-71; mem toxicol study sect, NIH, 73-77; USP comt of revision, 75-; mem, Environ Health Sci Rev Comt, 81- *Mem:* AAAS; Soc Toxicol; Am Soc Biol Chem; Am Soc Pharmacol & Exp Therapeut. *Res:* Relationship of metabolism to toxicity; analytical chemistry. *Mailing Add:* Pharmacol Dept Sch Med & Dent-Univ Rochester 601 Elmwood Ave Rochester NY 14642

ANDERS, OSWALD ULRICH, b Karlsruhe, Ger, Nov 10, 28; nat US; m 53; c 3. COLLOID CHEMISTRY, RADIOCHEMISTRY. *Educ:* Georgetown Univ, BS, 52; Univ Mich, MS, 54, PhD(chem), 57. *Prof Exp:* Asst, Univ Mich, 54-57; radiochemist, 57-63, assoc scientist, radiochem res lab, 63-78, res scientist, anal lab & reactor supvr, Triga Reactor, 67-80, RES SCIENTIST, DESIGNED LATEXES & RESINS LAB, MICH DIV APPL SCI & TECHNOL DEPT, DOW CHEM CO, 80- *Concurrent Pos:* Mem, adv comt anal chem, Nat Res Coun, 65-67, exec comt int conf activation anal, 72-75 & atomic scattering factor subcomt nuclear & radiochem, 78-81; prin investr, develop technol complete decontamination Dreden-1 Nuclear Plant, 72-84. *Mem:* Am Inst Chem; Sigma Xi; fel Am Nuclear Soc; Am Chem Soc. *Res:* Activation analysis; particle accelerators; nuclear reactor; nuclear instrumentation; nuclear reactions; ion exchange separations; tracer chemistry, radioactive waste solidification; decontamination of nuclear power plants. *Mailing Add:* 801 Linwood Dr Midland MI 48640

ANDERS, WILLIAM A, b Hong Kong, Oct 17, 33; m; c 6. ASTRONAUTICS. *Educ:* US Naval Acad, BS, 55; US Air Force Inst Technol, MS, 62. *Prof Exp:* Astronaut, Apollo Space Prog, NASA, 63-69; exec secy, Nat Aeronaut & Space Coun, 69-73; comnr, AEC, 73-75, chmn, Nuclear Regulatory Comm, 75-76; vpres & gen mgr, Nuclear Energy Prods Div, Gen Elec Co, 77-80, vpres & gen mgr, Aircraft Equip Div, 80-84; exec vpres aerospace, Textron Inc, 84-86, sr exec vpres opers, 86-89; vchmn, 90-91, CHMN & CHIEF EXEC OFFICER, GEN DYNAMICS CORP, 91- *Concurrent Pos:* US Ambassador to Norway, 76-77. *Honors & Awards:* Distinguished Serv Medals, USAF, NASA & Nuclear Regulatory Comn; Hubbard Medal for Explor, Nat Geog Soc; Flight Achievement Award, Am Astronaut Soc; Collier, Harmon, Goddard & White Flight Trophies. *Mem:* Nat Acad Eng; Soc Exp Test Pilots. *Res:* Crew member astronaut on Apollo 8 and backup pilot for Apollo 11. *Mailing Add:* Gen Dynamics Corp Pierre LaClede Ctr St Louis MO 63105

ANDERSEN, ARNOLD E, b Brooklyn, NY, June 12, 43; c 3. PSYCHIATRY, DEVELOPMENTAL NEUROBIOLOGY. *Educ:* Cornell Univ, BA, 64, MD, 68. *Prof Exp:* Res assoc develop neurobiol, Lab Biomed Sci, Nat Inst Child Health & Develop, 70-74; clin assoc drug abuse & anorexia nervosa, Adult Psychiat Br, NIH, 74-75, staff psychiatrist, 75-76; fel psychiat, 75-76, from instr to asst prof, 76-84-, ASSOC PROF PSYCHIAT, JOHN HOPKINS SCH MED, 84- *Concurrent Pos:* Attend physician, Henry Phipps Psychiat Clin, Johns Hopkins Hosp, 77-, dir, Eating & Weight Disorder Clin & Clin Clerkship Psychait. *Mem:* AAAS; Am Psychiat Asn; Am Psychopath Asn. *Res:* Anorexia nervosa; Bulimia nervosa; psychopharmacology; psychological tests. *Mailing Add:* Dept Psychiat Meyer 3-181 Johns Hopkins Hosp 600 N Wolfe St Baltimore MD 21205

ANDERSEN, AXEL LANGVAD, b Askov, Minn, Sept 24, 14; m 42; c 2. PLANT PATHOLOGY. *Educ:* Univ Minn, BS, 37; Mich State Univ, MS, 41, PhD(plant path), 47. *Prof Exp:* Forest guard, Stanislaus Nat Forest, Calif, 37; asst plant pathologist, Exp Sta, Mich State Univ, 39-42; plant pathologist, Camp Detrick, Md, 43-48; res plant pathologist, Crops Res Div, Agr Res Serv, USDA, 49-65, res coordr res prog develop & eval staff, Off of Dir Sci & Educ, 65-68; assoc prof bot & plant path, 51-64, head exten plant path, 68-74, prof bot & plant path, 64-77, asst to dir, Agr Exp Sta, 71-75, EMER PROF BOT & PLANT PATH, MICH STATE UNIV, 77- *Concurrent Pos:* Mem, Nat Steering Comt Plant Dis Mgt, 74-75; consult plant health probs, 77-86; collabr, Plant Health Inspection Serv, USDA, 78-80. *Mem:* Am Phytopath Soc; Sigma Xi. *Res:* Plant diseases; bean disease and breeding investigations; turfgrass diseases; disease management, detection, development, loss appraisal and control strategies; epidemiology, air pollution injury, remote sensing. *Mailing Add:* 902 Whitman Dr East Lansing MI 48823-2448

ANDERSEN, BLAINE WRIGHT, b Salt Lake City, Utah, Dec 29, 25; m 52; c 5. MECHANICAL ENGINEERING. *Educ:* Univ Utah, BS, 49; Univ Ill, Urbana, MS, 51, PhD(eng mech), 53. *Prof Exp:* Instr eng mech, Univ Ill, Urbana, 52-53; sr res engr, Stress & Vibrations Anal, NAm Aviation, Calif, 53-55; asst prof mech eng, Brigham Young Univ, 55-57; sr eng specialist pneumatic systs anal, AiResearch Mfg Co, Ariz, 57-66, supvr dynamic systs anal, 66-70; PROF MECH ENG, BRIGHAM YOUNG UNIV, 70- *Concurrent Pos:* Consult, AiResearch Mfg Co, Ariz, 56-57 & 70-; fac assoc, Ariz State Univ, 60-66. *Mem:* Fluid Power Soc. *Res:* Pneumatic systems dynamics; analytical techniques for predicting the performance of pneumatic systems. *Mailing Add:* 242 CB Mech Eng Dept Brigham Young Univ Provo UT 84602

ANDERSEN, BURTON, b Chicago, Ill, Aug 27, 32; m 86; c 3. INTERNAL MEDICINE, IMMUNOLOGY. *Educ:* Univ Ill, BS, 55, MD & MS, 57. *Prof Exp:* Rotating intern, Minneapolis Gen Hosp, 57-58; resident internal med, Univ Hosps, Univ Ill, 58-59, assoc prof med & microbiol, Col Med, Northwestern Univ, 67-70; assoc prof, 70-72, PROF MED & MICROBIOL, COL MED, UNIV ILL, 72-; CHIEF SECT INFECTIOUS DIS, UNIV ILL, 86- *Concurrent Pos:* Fel infectious dis, Univ Ill, 59-61; Nat Inst Allergy & Infectious Dis fel, 61-64; fel, Med Ctr, Univ Rochester, 64-67. *Mem:* Am Asn Immunol; Am Fedn Clin Res; Am Soc Microbiol; Infectious Dis Soc Am. *Res:* Infectious diseases. *Mailing Add:* Sect Infectious Dis Univ Ill Col Med 840 S Wood St Chicago IL 60612

ANDERSEN, CARL MARIUS, b Detroit, Mich, Mar 10, 36; m 59; c 3. THEORETICAL PHYSICS. *Educ:* Univ Mich, BS, 58, MS, 59; Univ Pa, PhD(physics), 64. *Prof Exp:* Res assoc theoret physics, Purdue Univ, 64-65, asst prof, 65-67; asst prof theoret physics, 67-72, SR RES ASSOC MATH & COMPUT SCI, COL WILLIAM & MARY, 72- *Mem:* Am Phys Soc; Asn Comput Mach; Soc Indust & Appl Math. *Res:* Applied mathematics; algebraic computation by computer; structural mechanics computations; nonlinear mechanics; theory of group representations. *Mailing Add:* 3210 Dye Rd Falls Church VA 22042

ANDERSEN, DEAN MARTIN, b Monroe, Utah, Dec 14, 31; m 61; c 5. MEDICAL ENTOMOLOGY, ENVIRONMENTAL BIOLOGY. *Educ:* Univ Utah, BS, 60, MS, 62, PhD(entom), 66. *Prof Exp:* Inspector entomologist, Salt Lake City Mosquito Abatement, 58-62; dir field res, Inst Environ Res, Univ Utah, 63-66; assoc prof, 66-74, PROF BIOL, BRIGHAM YOUNG UNIV, HAWAII CAMPUS, 74- *Concurrent Pos:* Vis prof, Brigham Young Univ, 73-74. *Mem:* Am Mosquito Control Asn. *Res:* Ecological control of mosquitoes through water management; host preferences of mosquitoes; audio-tutorial systems of teaching biological sciences and individualized approaches to learning for both science and nonscience majors. *Mailing Add:* 55697 Wahinepee St Laie HI 96762

ANDERSEN, DEWEY RICHARD, b Harlan, Iowa, Oct 15, 27; m 54; c 4. SANITARY ENGINEERING, NUCLEAR SCIENCE. *Educ:* Univ Iowa, BS, 58, MS, 59; Iowa State Univ, PhD(civil eng), 67. *Prof Exp:* Pub health engr, Iowa State Dept Health, 59-68; from asst prof to assoc prof civil eng, 68-76, PROF CIVIL ENG, UNIV NEBR, LINCOLN, 76- *Honors & Awards:* Fuller Award, Am Water Works Asn. *Mem:* Am Water Works Asn; Water Pollution Control Fedn; Am Soc Civil Eng. *Res:* Industrial and municipal wastewater treatment; water treatment; radiological health. *Mailing Add:* W348 Nebraska Hall Univ Nebr Lincoln NE 68588

ANDERSEN, DONALD EDWARD, b New Haven, Conn, June 3, 23; m 47; c 1. PHYSICAL CHEMISTRY, RESEARCH ADMINISTRATION. *Educ:* Brown Univ, ScB, 48, PhD(phys chem), 52; Stanford Univ, MS, 49. *Prof Exp:* Res chemist, E I Du Pont de Nemours & Co, Inc, 52-56, res supvr, 56-60, supt develop, 60-61, gen supt res & develop, 61-63, asst plant mgr, 63-65, plant mgr, 65-70, lab dir & mgr elastomers res & develop, 70-81, consult mgt resources, 82-85; PRES, MYTHOS CONSULTS, 85- *Mem:* AAAS; Am Chem Soc. *Mailing Add:* 1608 N Rodney St Wilmington DE 19806

ANDERSEN, EMIL THORVALD, b Alta, Can, Jan 25, 17; m 43; c 3. HORTICULTURE, POMOLOGY. *Educ:* Univ Alta, BSc, 41, MSc, 43; Univ Minn, PhD(hort), 62. *Prof Exp:* Res off hort, Can Exp Farm, Alta, 43-44; from asst prof to assoc prof hort, Univ Man, 44-56; from instr to assoc prof pomol, Univ Minn, St Paul, 57-67; chief prod res, 67-81, chief, hort prod lab, Hort Res Inst, Ont, 81-83; HORT CONSULT, 83- *Mem:* Am Soc Hort Sci; Am Pomol Soc; Can Soc Hort Sci; Int Soc Hort Sci; Am Soc Enol. *Res:* Management systems studies with apples and grapes; rootstock problems of apples and grapes; iron uptake problems of fruits and plants. *Mailing Add:* RR 1 Sann Rd Beamsville ON L0R 1B0 Can

ANDERSEN, FERRON LEE, b Howell, Utah, July 10, 31; m 58; c 6. PARASITOLOGY. *Educ:* Utah State Univ, BS, 57, MS, 60, PhD(zool), 63; Univ Ill, MS, 62. *Prof Exp:* Asst prof vet med sci & assoc staff mem, Zoonoses Res Ctr, Univ Ill, 63-67; assoc prof, 67-72, PROF ZOOL, BRIGHAM YOUNG UNIV, 72- *Mem:* Am Soc Parasitol. *Res:* Coccidiosis; micrometeorology and ecology of parasites; hydatid disease. *Mailing Add:* Dept Zool Brigham Young Univ Provo UT 84602

ANDERSEN, FRANK ALAN, b Lewiston, Maine, May 20, 44; m 75; c 5. RADIATION BIOPHYSICS, REGULATION. *Educ:* Muhlenberg Col, BS, 66; Pa State Univ, MS, 69, PhD(biophys), 72. *Prof Exp:* Chief, Genetics Studies Sect, Exp Studies Br, Food & Drug Admin, 71-76, chief, Standards Support Staff, 76-79, assoc dir biol effects, Bur Radiol Health, 79-81, dir standards, Bur Med Devices, 81-83, dep dir, Standards & Regulations, 83-85, dir off sci & technol, ctr devices & radiol health, 85-90, DIR OFF SCI & TECHNOL, FOOD & DRUG ADMIN, 90- *Concurrent Pos:* Chmn, Standards Mgt Comt, Nat Comt Clin Lab Standards, 84-88, chmn, Med Device Standards Bd, Am Nat Standards Inst, 85-86; mem bd dir, Nat Comt Clin Lab Standards, 87-, secy, 90- *Honors & Awards:* Super Serv Award, USPHS, 89. *Mem:* Am Soc Photobiol; Sigma Xi. *Res:* Biological effects of radiation, including acoustic energy with special reference to emissions from electronic products; development of criteria in support of performance standards to assure product safety and effectiveness. *Mailing Add:* Food & Drug Admin HFZ-100 5600 Fishers Lane Rockville MD 20857

ANDERSEN, HANS CHRISTIAN, b Brooklyn, NY, Sept 25, 41; m 67; c 2. PHYSICAL CHEMISTRY, STATISTICAL MECHANICS. *Educ:* Mass Inst Technol, BS, 62, PhD(phys chem), 66. *Prof Exp:* Jr fel, soc fels, Harvard Univ, 65-68; from asst to assoc prof chem, 68-80, PROF CHEM, STANFORD UNIV, 80- *Concurrent Pos:* Vis prof chem, Columbia Univ, 81-82; co-dir, Ctr Mat Res, Stanford Univ, 88-89, dir, 89- *Honors & Awards:* Joel Henry Hildebrand Award, Am Chem Soc, 88. *Mem:* Am Phys Soc; AAAS; Am Chem Soc. *Res:* Statistical mechanics; structure and properties of amorphous materials; structure of fluids; computer simulation methods; energy transfer in condensed phases. *Mailing Add:* Dept Chem Stanford Univ Stanford CA 94305

ANDERSEN, HOWARD ARNE, b Minneapolis, Minn, June 10, 16; m 42; c 3. MEDICINE. *Educ:* Univ Minn, BS, 40, BM, 42, MD, 43, MSc, 50. *Prof Exp:* From assoc prof to prof med, Mayo Grad Sch Med, 69-73, prof, 73-81, EMER PROF MED, MAYO MED SCH, UNIV MINN, 81- *Concurrent Pos:* Consult internal med, Thoracic Dis & Bronco-Esophagology, Mayo Clin, 50-85; pres, World Cong Bronchology, 82. *Honors & Awards:* Distinguished Fel, Am Col Chest Physicians, 82. *Mem:* Am Thoracic Soc; AMA; fel Am Col Physicians; fel Am Col Chest Physicians; Sigma Xi; Am Broncho-Esophagical Asn (pres, 74-75). *Res:* Pulmonary and esophageal diseases. *Mailing Add:* Mayo Clin Rochester MN 55905

ANDERSEN, JOHN R(OBERT), civil & sanitary engineering; deceased, see previous edition for last biography

ANDERSEN, JON ALAN, b Oelwein, Iowa, Mar 13, 38. MICROBIOLOGY, BIOCHEMISTRY. *Educ:* Cornell Col, BA, 61; Univ Colo, PhD(physiol), 64. *Prof Exp:* Sr res scientist microbiol & biochemist, 64-73, sect chief microbiol, 73-81, RES AREA TEAM CHMN & DEVELOP TEAM PROJ MGR, NORWICH EATON, 81- *Mem:* Sigma Xi; Int Soc Quantum Biol. *Res:* Antimicrobials, antivirals, immunomodulators and research on kidney stones. *Mailing Add:* RD 2 Norwich NY 13815

ANDERSEN, JONNY, b Oslo, Norway, Aug 13, 35; m 58; c 3. ELECTRICAL ENGINEERING, BIOENGINEERING. *Educ:* Univ Colo, BS, 60; Mass Inst Technol, SM, 62, PhD(elec eng), 65. *Prof Exp:* Asst prof elec eng, Mass Inst Technol, 64-67; from asst prof to assoc prof, 67-78, PROF ELEC ENG, UNIV WASH, 78- *Concurrent Pos:* Ford fel, 65-67; consult, Sylvania Appl Res Lab, 66, Boeing Aerospace Co, 69-76 & Boeing Com Airplane Co, 78- *Honors & Awards:* TV Shares Mgt, Mass Inst Technol, 63; Baker Award, Inst Elec & Electronics Eng, 67. *Mem:* AAAS; Inst Elec & Electronics Eng. *Res:* CAD/CAM; nonlinear device modeling; circuits and systems design; large scale systems simulation; broadband matching. *Mailing Add:* Dept Elec Eng FT-10 Univ Wash Seattle WA 98195

ANDERSEN, KENNETH F, b Innisfail, Alta Apr 5, 45. FOURIER ANALYSIS, MATHEMATICS. *Educ:* Univ BC, BS, 67; Univ Toronto, MS, 68, PhD(math), 70. *Prof Exp:* PROF MATH, UNIV ALBERTA, 71- *Mem:* Am Math Soc; Can Math Soc; Math Asn Am. *Mailing Add:* Dept Math Univ Alta Edmonton AB T6G 2G1 Can

ANDERSEN, KENNETH J, b Brooklyn, NY, Dec 24, 36; m 62; c 2. RESEARCH MANAGEMENT. *Educ:* Davis & Elkins Col, BS, 59; Syracuse Univ, MS, 62, PhD(microbiol), 65. *Prof Exp:* Bacteriologist, Syracuse Univ Res Corp, 60-62, asst microbiol, Syracuse Univ, 62-65; res microbiologist, Battelle Mem Inst, 65-67, sr microbiologist, 67-70, assoc chief microbiol & environ biol div, 70-72; assoc dir diag serv, Johnson & Johnson Int, 72-73; dir chemother, 73-75, dir biol res, Norwich-Eaton Pharmaceut, Inc, 75-82; PRES, ANDERSEN LABORATORIES INC, 83- *Honors & Awards:* Outstanding Scientist, Life Scis, Battelle Mem Inst, 66. *Mem:* Fel AAAS; Am Soc Microbiol; Am Inst Biol Sci. *Res:* Microbiology; virology; bacteriology; immunology; anti-infectious and cardiovascular drugs; microbial genetics and physiology; new drugs for human diseases; drug metabolism, pathology and toxicology; pharmaceutical and health care products. *Mailing Add:* RD 2 Box 165A South New Berlin NY 13843

ANDERSEN, KENNETH K, b Perth Amboy, NJ, May 13, 34; m 57; c 3. ORGANIC CHEMISTRY. *Educ:* Rutgers Univ, BS, 55; Univ Minn, PhD(chem), 59. *Prof Exp:* USPHS fel, 59-60; from asst prof to assoc prof, 60-70, PROF CHEM, UNIV NH, 70- *Concurrent Pos:* NSF sci fac fel, Univ EAnglia, 66-67; Fulbright lectr, Tech Univ Denmark, 71; vis lectr, Polish Acad Sci, 73; vis prof, Colo State Univ, 79; guest prof, Tech Univ, Denmark, 88. *Mem:* Am Chem Soc; fel AAAS. *Res:* Chemistry of organosulfur compounds with emphasis on stereochemistry. *Mailing Add:* Dept Chem Univ NH Durham NH 03824

ANDERSEN, L(AIRD) BRYCE, b Madison, SDak, Sept 16, 28; m 61; c 2. CHEMICAL ENGINEERING. *Educ:* Univ Minn, BS, 50, MS, 51; Univ Ill, PhD(chem eng), 54. *Prof Exp:* Asst prof chem eng, Lehigh Univ, 54-59; assoc prof, Rice Univ, 59-60 & Univ Nebr, 61-63; prof chem eng, NJ Inst Technol, 63-80, from assoc dean to dean eng, Newark Col Eng, 63-75; DEAN, COL ENG, SOUTHEASTERN MASS UNIV, 80- *Mem:* Am Inst Chem Engrs; Am Soc Eng Educ; Sigma Xi. *Res:* Transport phenomena; applied statistics. *Mailing Add:* Col Eng Southeastern Mass Univ North Dartmouth MA 02747

ANDERSEN, NEIL RICHARD, b Lynn, Mass, Sept 22, 35; m 56; c 2. CHEMICAL OCEANOGRAPHY. *Educ:* Clark Univ, AB, 60; Mass Inst Technol, PhD(anal chem), 65. *Prof Exp:* Res asst oceanog, Woods Hole Oceanog Inst, 60-65; prin investr, US Naval Oceanog Off, 65-67, br head, 67-70; chief appl chem oceanog br, Off Res & Develop, US Coast Guard, 70-72; chem oceanogr, Off Naval Res, 72-73, prog dir, Chem Oceanog Prog, 73-75; head marine pollution, Intergovt Oceano Comn, UNESCO, Paris, 81-83; prog dir, 75-79, 80-81, PROG DIR, CHEM OCEANOG PROG, NSF, 83- *Concurrent Pos:* Fel, Woods Hole Oceanog Inst, 62-65; assoc prof, George Washington Univ, 68-78, prof, 78-80; dep US deleg, Global Invest Pollution in Marine Environ Comt of Intergovt Oceanog Comn, UNESCO, 74-76, head US deleg, 76-81, chmn, Group Experts on Methods, Standards & Intercalibration, 76-81 & US Nat coordr, 76-81; vis scholar, Scripps Inst Oceanog, 79-81; chmn sci comt, Global Invest Pollution in Marine Environ, 84-; vis scholar, Scripps Inst Oceanog, La Jolla, 80-81. *Mem:* AAAS; Oceanog Soc; Am Geophys Unio. *Res:* Trace metal and radioisotope distributions in the oceans; chemical, geological and biological mechanisms causing variations such that measured distributions can aid in solving oceanographic problems and in identifying and regulating marine pollution. *Mailing Add:* Ocean Sci Div NSF 1800 G St NW Washington DC 20550

ANDERSEN, NIELS HJORTH, b Copenhagen, Denmark, Oct 9, 43; US citizen; m 63; c 2. ORGANIC CHEMISTRY, BIOCHEMISTRY. *Educ:* Univ Minn, Minneapolis, BA, 63; Northwestern Univ, PhD(org chem), 67. *Prof Exp:* NIH res fel chem, Harvard Univ, 67-68; asst prof, Univ Wash, 68-70; prin scientist, Alza Corp, Calif, 70; from asst prof to assoc prof, 70-76, PROF CHEM, UNIV WASH, 76- *Concurrent Pos:* Consult, Worchester Found Exp Biol, 68-69 & Alza Corp, 70-75; Alfred P Sloan res fel, 72-74; Dreyfus Found scholar, 74-79; NIH career develop award, 75-80. *Mem:* Am Chem Soc; The Chem Soc; Phytochem Soc; Intra-Sci Res Found; NY Acad Sci. *Res:* New synthetic methods; synthesis, structure elucidation and biogenesis of sesquiterpenes; circular dichroism of olefins; chemical models for physiological interactions; prostaglandin analogs, particularly synthesis and structure-activity correlations; molecular biochemistry. *Mailing Add:* Dept Chem Univ Wash Seattle WA 98195

ANDERSEN, OLAF SPARRE, b Hellerup, Denmark, Sept 10, 45; m 73; c 2. ION PERMEABLE CHANNELS. *Educ:* Univ Copenhagen, MD, 71. *Prof Exp:* Fel biophys, Univ Copenhagen, 71-72; fel biophys, Rockefeller Univ, 72-73; from asst prof to assoc prof, 73-82, PROF PHYSIOL, CORNELL UNIV MED COL, 82- *Concurrent Pos:* Assoc ed, J Gen Physiol, 84-; mem NIH Physiol Study Sect, 82-86; mem res coun, NY Heart Asn, 80-84, chmn, 84-, dir, 85- *Mem:* Royal Soc Chem; Biophys Soc; NY Acad Sci; Soc Gen Physiologists; Am Physiol Soc. *Res:* Structure function studies in ion selective channels using single amino acid replacements; physical chemistry of ion movement through channels. *Mailing Add:* Dept Physiol & Biophys Cornell Univ Med Col 1300 York Ave New York NY 10021

ANDERSEN, RICHARD NICOLAJ, b Oakland, Calif, Mar 11, 30; m 58; c 3. REPRODUCTIVE ENDOCRINOLOGY. *Educ:* Abilene Christian Col, BS, 56; Baylor Univ, PhD(biochem), 60. *Prof Exp:* Asst res prof surg, Med Col Va, 63; res assco biochem, Mayo Found, 63-66; asst prof pharmacol, Univ NC, Chapel Hill, 66, asst prof obstet & gynec, 67-69; assoc prof, 70-77, PROF BIOCHEM, OBSTET & GYNEC, UNIV TENN, MEMPHIS, 77-, DIR REPRODUCTIVE ENDOCRINOL LAB, 76- *Concurrent Pos:* USPHS fels, Univ Calif, Berkeley, 60-61 & Med Col Va, 61-63. *Mem:* AAAS; Soc Study Reproduction; Endocrine Soc; Clin Ligand Assay Soc; Am Soc Clin Chem. *Res:* Control of gonadal function; steroid radioimmunoassay; control of adrenal androgen secretion; role of C-19 steroids in insulin resistance. *Mailing Add:* Div Reproductive Endocrinol Univ Tenn 956 Court Rm D-322 Memphis TN 38163

ANDERSEN, ROBERT NEILS, b Steele City, Nebr, June 8, 28; m 52; c 2. PLANT PHYSIOLOGY. *Educ:* Univ Nebr, BS, 51, MS, 53; Univ Minn, PhD, 60. *Prof Exp:* Asst agronomist, Weed Invest Sect, Field Crops Res Br, Agr Res Serv, USDA, NDak, 53-57; res agronomist, USDA & Univ Minn, 57-61, plant physiologist, Agr Res Serv, 61-; RETIRED. *Mem:* Weed Sci Soc Am. *Res:* Weed control. *Mailing Add:* Dept Agron & Plant Genetics Inst Agr Univ Minn St Paul MN 55108

ANDERSEN, ROGER ALLEN, b Milwaukee, Wis, June 24, 30; m 63; c 3. BIOCHEMISTRY, ONCOLOGY. *Educ:* Univ Wis, BS, 53, PhD(oncol), 63; Marquette Univ, MS, 58. *Prof Exp:* Anal chemist, Allis-Chalmers Mfg Co, Wis, 53-58; res asst oncol, McArdle Mem Labs, Univ Wis, 58-63; USPHS fel biochem, Roswell Park Mem Inst, NY, 63-65; RES CHEMIST, AGR RES SERV, USDA, 65-; ASSOC PROF AGRON, UNIV KY, 71- *Concurrent Pos:* Asst prof agron, Univ Ky, 65-71. *Mem:* Am Chem Soc; Am Asn Cancer Res. *Res:* Biochemistry of chemical carcinogens, plant volatiles, plant alkaloids, plant secondary metabolites. *Mailing Add:* US Dept Agr-ARS Dept Agron Univ Ky Lexington KY 40506

ANDERSEN, RONALD, b Omaha, Nebr, June 15, 39; c 1. SCIENCE ADMINISTRATION. *Educ:* Univ Santa Clara, BS, 60; Purdue Univ, MS, 62, PhD(sociol), 68. *Prof Exp:* Res assoc, farm cardiac proj, Purdue Univ, 62-63; res assoc, Ctr Health Admin Studies, Univ Chicago, 63-77, assoc dir, 77-80, assoc prof, 74-77, prof sociol, Grad Sch Bus, 77-91, dir Grad Prog Health Admin, Ctr Health Admin Studies, 80-90; PROF HEALTH SERV, UNIV CALIF, LOS ANGELES, 90- *Mem:* Inst Med-Nat Acad Sci; Am Sociol Asn; Am Pub Health Soc; Asn Health Serv Res; Asn Progs Health Admin. *Mailing Add:* Sch Pub Health Univ Calif Los Angeles 10833 LeCote Ave Los Angeles CA 90024-1772

ANDERSEN, ROY STUART, b Springfield, Mass, Oct 16, 21; m 44; c 3. ELECTRON SPIN RESONANCE, BIOPHYSICS. *Educ:* Clark Univ, BA, 43; Dartmouth Col, AM, 48; Duke Univ, PhD(physics), 51. *Prof Exp:* Res engr, Stanford Res Inst, Calif, 51-52; from asst prof to assoc prof physics, Univ Md, 52-60; chmn dept, 60-70, 71-73 & 77-78, dean, Grad Sch, 70-71, PROF PHYSICS, CLARK UNIV, 60- *Concurrent Pos:* Res assoc, Duke Univ, 51, 53 & 54, Univ Calif, Berkeley, 58-59, Woods Hole Oceanog Inst, 60, Biophysics Inst, Univ Oslo, Norway, 73; consult, United Aircraft Corp. *Mem:* Fel Am Phys Soc; Am Asn Physics Teachers; Biophys Soc; Hist Sci Soc. *Res:* Microwave and radio spectroscopy; electron spin resonance; radiation damage in single crystals. *Mailing Add:* Dept Physics Clark Univ Worcester MA 01610

ANDERSEN, TERRELL NEILS, b Central, Idaho, Apr 22, 37; m 57; c 3. PHYSICAL CHEMISTRY. *Educ:* Univ Utah, BS, 58, PhD(phys chem), 62. *Prof Exp:* Fel & res assoc electrochem, Univ Utah, 62 & Univ Pa, 62-63; res assoc, Univ Utah, 63-64, asst res prof chem & metall, 64-69; sr scientist, process technol, Kennecott Copper Co, 69-85; PRIN CHEMIST, KERR-MCGEE TECH CTR, 85- *Concurrent Pos:* NSF fel, 58-60; adj prof metall, Univ Utah. *Honors & Awards:* Sigma Xi Award, Univ Utah, 62. *Mem:* Am Chem Soc; Sigma Xi; Int Battery Mat Asn. *Res:* Electrochemical kinetics and double layer; electrowinning; electrorefining; corrosion; batteries; industrial electrochemical processes; photocatalysis. *Mailing Add:* 500 Concord Lane Edmond OK 73034

ANDERSEN, THORKILD WAINO, b Copenhagen, Denmark, Nov 13, 20; m 40; c 4. ANESTHESIOLOGY. *Educ:* Univ Copenhagen, MD, 47. *Prof Exp:* Intern, Bispebjerg Hosp, Copenhagen, 48, resident, 49; PROF ANESTHESIOL, COL MED, UNIV FLA, 69- *Concurrent Pos:* Clin fel, Mass Gen Hosp, 51-52, sr fel, 52-56; NIH res grant, 68-72. *Mem:* AMA; Am Soc Anesthesiologists. *Res:* Drug interactions in anesthesia in man. *Mailing Add:* Dept Anesthesiol Cert Anesthesiol 62 Univ Kobenhavn JHM Health Ctr Box J254 Gainesville FL 32610

ANDERSEN, WILFORD HOYT, b Central, Idaho, Dec 15, 24; m 56; c 2. PHYSICAL CHEMISTRY, FLUID DYNAMICS. *Educ:* Univ Utah, BS, 49, PhD(chem), 52. *Prof Exp:* Res assoc chem, Univ Utah, 52-53 & Brown Univ, 53-55; tech specialist ordnance, Aerojet-Gen Corp, 55-64; prin res scientist, Heliodyne Corp, 64-65; prin scientist, Shock Hydrodynamics Div, Whittaker Corp, 65-77, chief scientist, 77-82; SR PROJ ENGR, DEFENSE SYSTS GROUP, TRW, 82- *Mem:* NY Acad Sci; AAAS; Combustion Inst. *Res:* Fundamental and applied research in shock, detonation, combustion and related ordnance phenomena; physical chemistry; fluid dynamics; molecular theory of biological processes; nuclear detonation and related phenomena; ballistic missiles. *Mailing Add:* 1165 East Comstock Ave Glendora CA 91740

ANDERSEN, WILLIAM RALPH, b Rexburg, Idaho, Dec 11, 30; m 54; c 5. BIOCHEMICAL GENETICS. *Educ:* Utah State Univ, BS, 56, MS, 58; Univ Calif, Davis, PhD(genetics), 63. *Prof Exp:* Asst prof hort & genetics, Univ Minn, 63-66; PROF BOT, BOT & RANGE SCI DEPT, BRIGHAM YOUNG UNIV, 66- *Concurrent Pos:* Fel & grants, Univ Calif, 67-69. *Mem:* Am Soc Plant Physiol; Genetics Soc Am; Sigma Xi. *Res:* Genetics of photosynthesis. *Mailing Add:* Bot Range Sci Dept Brigham Young Univ 297 WIDB Provo UT 84602

ANDERSLAND, MARK STEVEN, b Lansing, Mich, June 9, 61. DECENTRALIZED CONTROL, COORDINATION THEORY. *Educ:* Univ Mich, BSE, 83, MSE, 84, PhD(elec eng systs), 89. *Prof Exp:* ASST PROF ELEC & COMPUTER ENG, UNIV IOWA, 89- *Mem:* Inst Elec & Electronics Engrs; Soc Indust & Appl Math. *Res:* Decentralized control, coordination theory and distributed information processing; estimation, communication and information theory; event trace recovery in intrusively monitored parallel and distributed computing systems. *Mailing Add:* Dept Elec & Computer Eng 4408 Eng Bldg Univ Iowa Iowa City IA 52242

ANDERSLAND, ORLANDO BALDWIN, b Albert Lea, Minn, Aug 15, 29; m 58; c 3. CIVIL ENGINEERING, GEOTECHNICAL ENGINEERING. *Educ:* Univ Minn, BCE, 52; Purdue Univ, MSCE, 56, PhD(civil eng), 60. *Prof Exp:* Officer, US Army Corps Engrs, US & Korea, 52-55; staff engr, Nat Acad Sci-Am Asn State Hwy Off Rd Test, Ottawa, Ill, 56-57; res engr, Purdue Univ, 57-59; from asst prof to assoc prof civil eng, 60-68, PROF CIVIL ENG, MICH STATE UNIV, 68- *Concurrent Pos:* NSF res grants, 63-66, 67-70, 76-78, 80-82 & 84-86; Royal Norweg Coun Sci & Indust Res fel, Norweg Geotech Inst, Oslo, 66; Nat Coun Paper Indust Air & Stream Improv res grant, 69-71; Environ Protection Agency res grant, 71-73; ed geotech eng, Cold Regions, 78. *Mem:* Am Soc Civil Engrs; Am Soc Test & Mat; Am Soc Eng Educ; Int Soc Soil Mech & Found Engr. *Res:* Soil mechanics; mechanical properties of frozen ground and organic soils. *Mailing Add:* Dept Civil & Environ Eng Mich State Univ East Lansing MI 48824-1212

ANDERSON, A EUGENE, b Lima, Ohio, Apr 22, 16; m 42; c 5. ELECTRICAL ENGINEERING. *Educ:* Ohio State Univ, BS & MS, 39. *Prof Exp:* Engr, Bell Tel Labs, NJ, 39-49, supvr electronic apparatus develop, 49-53, head subdept, 53-55, head dept & dir develop, Pa, 55-57; supt mfg eng, Western Elec Co, 57-59, works engr mgr, 59-68, dir eng, 68-75; RETIRED. *Concurrent Pos:* Mem, Signal Corps Radar Lab, US Dept Defense, 42-47, adv panel electronics, 54-60. *Mem:* Fel AAAS; fel Inst Elec & Electronics Engrs. *Res:* Electron tube and semiconductor development and manufacture. *Mailing Add:* 1621 E Sawmill Rd Quakertown PA 18951-5703

ANDERSON, A KEITH, b Monticello, Calif, Mar 14, 24; m 52; c 1. MATHEMATICS. *Educ:* Pac Union Col, BA, 49; Loma Linda Univ, MD, 53; Colo State Univ, MA, 64, PhD(statist, math), 67. *Prof Exp:* Pvt pract, 55-60, 63-64; asst prof, 65-67, assoc prof, 67-78, PROF MATH, PAC UNION COL, 78- *Mem:* Math Asn Am. *Res:* Non-parametric statistics. *Mailing Add:* 2828 Loganrita Ave Arcadia CA 91006

ANDERSON, ALBERT DOUGLAS, b New York, NY, Jan 11, 28; m 73; c 2. PHYSICAL & REHABILITATION MEDICINE. *Educ:* Columbia Univ, BA, 48; Harvard Med Sch, MD, 52; Am Bd Phys Med & Rehab, dipl, 61. *Prof Exp:* Asst attend, Med Div, Montefiore Hosp, 57-59, adj attend rehab med, 60-65, coordr amputee serv, 63-65; from asst prof to prof clin rehab med, Col Physicians & Surgeons, Columbia Univ, 66-79, David Gurewitsh prof, 79-89, emer prof, 89; RETIRED. *Concurrent Pos:* Nat Med fel, Med Div, Montefiore Hosp, 55-56, Off Vocab Rehab fel phys med & rehab, 57-59; United Cerebral Palsy consult, Suffolk Rehab Ctr, 62-65; consult phys med & rehab, Hebrew Home Aged, 62-70; consult, Loeb Ctr Nursing & Rehab, 63-65 & Health Ins Plan Greater New York, 63-; clin consult prog, NY State Dept Health, 64-; asst attend, Presby Hosp-Columbia Univ, 67-; asst prof clin rehab med, Albert Einstein Col Med, Yeshiva Univ, 63-65; instr lower extremity orthotics, Post-Grad Med Sch, NY Univ, 65-69; chmn, New Construct Comt, Harlem Hosp Ctr, 67-70, vpres med bd, 70-73, pres, 74-75; secy, adv coun med bds, NY City Health & Hosps Corp, 71-72; prof adv coun, Easter Seal Soc, 75-78. *Mem:* Fel Am Col Physicians; Am Heart Asn; Am Rheumatism Asn; Am Acad Phys Med & Rehab; fel NY Acad Sci. *Res:* Work physiology in the elderly, the disabled and the chronically ill; disability in the poor, the black, and the ghetto dweller. *Mailing Add:* Dept of Rehab Med Harlem Hosp Ctr New York NY 10037

ANDERSON, ALBERT EDWARD, b Jamestown, NY, July 4, 28; m 50; c 1. COMPUTER & MEDICAL SCIENCES. *Educ:* Allegheny Col, BS, 50; Univ Rochester, PhD(org chem), 53. *Prof Exp:* Res chemist, 53-59, sr res chemist, 59-65, res assoc, 65-77, TECH ASSOC, INFO SERV, EASTMAN KODAK CO, 77- *Mem:* Am Chem Soc; Sigma Xi. *Res:* Synthetic organic chemistry; photographic chemistry; film emulsion making. *Mailing Add:* 63 Canyon Trail Rochester NY 14625

ANDERSON, ALBERT GORDON, b Evanston, Ill, Nov 21, 45; m 73; c 2. ORGANIC CHEMISTRY. *Educ:* Calif State Univ, Sacramento, BA, 67; Univ Utah, PhD(chem), 77. *Prof Exp:* RES CHEMIST PHOTOCHEM, CATALYSIS, INTERGRATED OPTICS, ANTIBIOTICS, E I DU PONT DE NEMOURS & CO INC, 77- *Mem:* Am Chem Soc; Optical Soc Am. *Res:* Physical organic chemistry, organic synthesis; elucidation of physical relations between chemical structure and physical properties such as heat of fusion, melting point, stability and refractive index; catalysis; antibiotics. *Mailing Add:* Ctr Res & Develop Dept E I du Pont de Nemours & Co Inc Wilmington DE 19880-0309

ANDERSON, ALFRED TITUS, JR, b Port Jefferson, NY, Aug 3, 37; m; c 2. PETROLOGY, VOLCANOLOGY. *Educ:* Northwestern Univ, BA, 59; Princeton Univ, PhD(geol), 63. *Prof Exp:* NSF res fel, Univ Chicago, 63-66; geologist, US Geol Surv, 66-68; from asst prof to assoc prof, 68-82, PROF GEOL, UNIV CHICAGO, 82- *Mem:* Am Geophys Union. *Res:* Physical-chemical conditions of crystallization of igneous rocks as revealed by field and mineralogical studies; role of gases in volcanism; igneous sources of excess volatiles. *Mailing Add:* Dept Geophys Sci Univ Chicago 5734 Ellis Ave Chicago IL 60637

ANDERSON, ALLAN GEORGE, b New York, NY, Oct 30, 23; m 49, 55; c 4. MATHEMATICS, GENETICS. *Educ:* Univ Fla, BS, 46, MA, 47; Univ Mich, PhD(math), 51. *Prof Exp:* From instr to asst prof math, Oberlin Col, 50-53; from asst prof to assoc prof & chmn dept, Duquesne Univ, 53-55; mathematician, Jones & Laughlin Steel Corp, 55-56 & Pittsburgh Plate Glass Co, 56-57; chief statistician, Gen Tire & Rubber Co, 57-58; prof math & head dept, Western Ky State Col, 58-65; prof & chmn dept, Upsala Col, 65-66; prof, Parsons Col, 66-68; chmn dept, 72-74, prof, 68-86, EMER PROF MATH, QUEENSBOROUGH COMMUNITY COL, CITY UNIV NEW YORK, 86- *Concurrent Pos:* Vis lectr, Math Asn Am, 61-63 & Ky Acad Sci, 64-65. *Mem:* Am Math Soc; Math Asn Am; Inst Math Stat; Soc Indust & Appl Math; Am Statist Asn; Nat Coun Teachers Math. *Res:* Mathematical genetics and statistics. *Mailing Add:* Dept Math Queensborough Community Col Bayside NY 11364

ANDERSON, AMOS ROBERT, b Delavan, Wis, Feb 11, 20; m 45; c 2. ORGANIC CHEMISTRY. *Educ:* Adrian Col, BS, 42, PhD, 65. *Hon Degrees:* DSc, Adrian Col, 65. *Prof Exp:* Anal chemist, Parker Rust-Proof Co, 41, res chemist, 46; res chemist gas process div, Girdler Corp, 43; chief chemist, Houdaille Hershey Corp, Manhattan Proj, 44-45; pres, Anderson Chem Co, 46-58, vpres & gen mgr, Anderson Chem Div, Stauffer Chem Co, 58-65, vpres & gen mgr, Silicone Div, 64-66; pres, Anderson Develop Co, 66-90; PRES, OCEANA MINERALS LTD, 90- *Concurrent Pos:* Pres & gen mgr, Tex Alkyls, Inc, 60-65; pres, Indust Accessories, Inc, 64-66; pres, Valjo Corp, 66-; vpres, AvTec Co, 67-69; vpres, Aquaphase Labs, 67-; dir, Am Dynamics Int, 68-72, pres, 72; vpres, Fabridyne Corp & Kemerica Inc, 73-75; pres, Interamerican Zinc, Inc, 74- *Mem:* NY Acad Sci; Am Chem Soc; Armed Forces Chem Asn. *Res:* Organometallics; silicones; organic chemicals; pharmaceutical intermediates; activated carbon. *Mailing Add:* 11 Lakeridge Rt 6 Adrian MI 49221

ANDERSON, ANNE JOYCE, b Eng, Aug 3, 46. MOLECULAR BIOLOGY, MICROBIOLOGY. *Educ:* Univ London, BSc, 67, PhD(biochem), 70. *Prof Exp:* From asst to assoc prof biol, 79-88, PROF BIOL, UTAH STATE UNIV, 88- *Mem:* Am Plant Physiol Soc; Am Phytopath Soc. *Res:* Plant-microbe interaction; the molecular basis of colonization is being studied to promote plant interaction with beneficial organisms and deter plant pathogen growth. *Mailing Add:* Dept Biol Utah State Univ Logan UT 84322-5305

ANDERSON, ANSEL COCHRAN, b Warren, Pa, Sept 17, 33; m 55; c 2. LOW TEMPERATURE PHYSICS. *Educ:* Allegheny Col, BS, 55; Wesleyan Univ, MA, 57; Univ Ill, PhD(physics), 61. *Prof Exp:* Res assoc, 61-62, from asst prof to assoc prof physics, 62-66, dept head, 86-91, PROF PHYSICS, UNIV ILL, URBANA, 69- *Concurrent Pos:* Guggenheim fel, 66; Fulbright-Hays res grant, 66; assoc, Ctr Advan Studies, 71-72. *Mem:* Fel Am Phys Soc. *Res:* Acoustics of quartz plates; properties of materials at temperatures below 1 Kelvin; ultralow temperature refrigeration and instrumentation. *Mailing Add:* Dept Physics Univ Ill 1110 W Green St Urbana IL 61801

ANDERSON, ANTHONY, b London, Eng, June 8, 35; Can citizen; m 61; c 2. MOLECULAR SPECTROSCOPY, SOLID STATE PHYSICS. *Educ:* Oxford Univ, BA, 56, MA & PhD(low temperature physics), 60. *Prof Exp:* Sci officer far infrared spectros, Nat Phys Lab, Eng, 60-63; res assoc infrared & Raman spectros, Princeton Univ, 63-65; vis asst prof, Kans State Univ, 65-66; from asst prof to assoc prof, 66-77, PROF INFRARED & RAMAN SPECTROS, UNIV WATERLOO, 77- *Concurrent Pos:* Consult, Parsons & Co, Ltd, Eng, 63-65; vis prof, Univ Grenoble, 72-73, Univ Utah, 79-80, Ariz State Univ, 86, Univ Canterbury, New Zealand, 87. *Mem:* Can Asn Physicists; Spectros Soc Can. *Res:* Infrared and Raman spectroscopic studies of molecular crystals, especially lattice vibrations and crystal field splittings; Brillouin scattering; lattice dynamics; collision-induced absorption. *Mailing Add:* Dept Physics Univ Waterloo Waterloo ON N2L 3G1 Can

ANDERSON, ARNOLD LYNN, b Ft Dodge, Iowa, Aug 26, 40; m 68; c 2. FIRE RETARDANT POLYMERS, ORGANIC FLUORINE CHEMISTRY. *Educ:* Luther Col, Decorah, Iowa, BA, 62; Univ Iowa, MS, 67, PhD(chem), 71. *Prof Exp:* Chemist, Sinclair Res, Inc, 62-64; sr res chemist, Mich Chem Corp, 70-75; group leader, Stepan Chem Co, 75-78; CONSULT & OWNER, A LYNN ANDERSON & ASSOCS, 78- *Concurrent Pos:* Cong Sci Coun, Am Chem Soc, 79-81; consult, David W Young & Assocs, Inc, 81-82. *Mem:* Am Chem Soc; Am Inst Chemists; AAAS. *Res:* New products and processes; organic halogen chemistry; fire retardant polymers; organic synthesis. *Mailing Add:* 501 Longview Antioch IL 60002

ANDERSON, ARTHUR G, JR, b Sioux City, Iowa, July 1, 18; m 44; c 3. ORGANIC CHEMISTRY. *Educ:* Univ Ill, AB, 40; Univ Mich, MS, 42, PhD(chem), 44. *Prof Exp:* From jr to sr chemist, Tenn Eastman Corp, Oak Ridge, 44-45; spec asst, Univ Ill, 46; from instr to prof chem, 46-88, assoc chmn dept, 78-88, EMER PROF CHEM, UNIV WASH, 88- *Concurrent Pos:* NSF sr fel, 60-61; vis prof, Heidelberg Univ, 60-61 & Australian Nat Univ, 66. *Mem:* Am Chem Soc; The Chem Soc; fel NY Acad Sci; fel Am Inst Chem. *Res:* Nonclassical aromatic compounds; novel heterocyclic systems; polynuclear alicyclic compounds; new synthetic reactions. *Mailing Add:* Dept Chem BG-10 Univ Wash Seattle WA 98195

ANDERSON, ARTHUR GEORGE, b Evanston, Ill, Nov 22, 26; m 47; c 6. PHYSICS. *Educ:* Univ San Francisco, BS, 49; Northwestern Univ, MS, 51; NY Univ, PhD(physics), 58. *Prof Exp:* Tech engr, IBM Corp, 51-53, comput & physics res, Watson Lab, 53-58, physicist, 58-59, res physicist 59-61, dir, San Jose Res Lab, 61-63, asst dir res, 63-65, staff dir corp tech comt, 65-67, dir res, 67-70, vpres, 69-72, vpres & pres, Gen Prod Div, 72-79, vpres & group exec, DP Prod Group, 79-81, vpres & group exec, Info Syst & Technol Group, IBM Corp,51-84; RETIRED. *Concurrent Pos:* Vis fel, Ctr Study Democratic Insts, 70-71; consult, R & D Mgt, 84-; dir, Compression Labs Inc, 84- *Honors & Awards:* Pake Award, Am Phys Soc. *Mem:* Nat Acad Eng; fel Am Phys Soc; fel Inst Elec & Electronics Engrs; AAAS. *Res:* Nuclear magnetic resonance in solids; computer engineering. *Mailing Add:* Box 8747 Incline Village NV 89450

ANDERSON, ARTHUR O, b Staten Island, NY, Mar 12, 45; m 69; c 1. RESPIRATORY MUCOSAL IMMUNITY, IMMUNOPATHOLOGY. *Educ:* Wagner Col, BS, 66; Univ Md, MD, 70. *Prof Exp:* Intern path, Johns Hopkins Hosp, 70-71, resident, 71-73, fel, Johns Hopkins Univ, 70-74, instr arts appl med, 73-74; asst prof path & biol, Sch Med, Univ Pa, 80-83; prin investr immunol, 74-80; CHIEF DEPT RESPIRATORY & MUCOSAL IMMUNOL, US ARMY MED RES INST INFECTIOUS DIS, FT DETRICK, 83- *Concurrent Pos:* Lectr, Johns Hopkins Immunol Coun, 70-83 & Found Advan Educ Sci, NIH, 79-; vis fel, Geront Res Ctr, NIH, 72-74. *Mem:* Am Asn Immunologists; Am Asn Pathologists; Int Acad Path; NY Acad Sci; AAAS; Soc Mucosal Immunol. *Res:* Physiological mechanisms that regulate invivo immune responses; lymph node microvasculature; angiogenesis; lymphocyte-endothelial interactions; lymphocyte recirculation, motility and chemotaxis; cellular and molecular mechanisms that regulate mucosal immune responses, especially with regard to the respiratory tract. *Mailing Add:* 122 W Third St Frederick MD 21701

ANDERSON, ARTHUR R(OLAND), b Tacoma, Wash, Mar 11, 10; m 38; c 5. ENGINEERING. *Educ:* Univ Wash, BS, 34; Mass Inst Technol, MS, 35, ScD(civil eng), 38. *Hon Degrees:* LLD, Gonzaga Univ, 83. *Prof Exp:* Asst, Mass Inst Technol, 36-38, res assoc, 39-41; res engr, Bauer & Schaurte, Neuss Rhein, Ger, 38; designing engr, Aug Klonne Steel Works, 38-39; hull engr, Cramp Shipbuilding Co, 41-43, head tech div, 43-46, consult engr, 46-51; co-founder & partner, Concrete Eng Co, 51-56; pres, Anderson Enterprises, 57-84; co-founder & chmn bd, Abam Engrs, 51-88; SR VPRES, CONCRETE TECHNOL CORP, 56-, DIR, 73- *Concurrent Pos:* Lectr, Univ Wash, 54-55; mem vis comt, Mass Inst Technol, 63; dir, Tacoma Pub Utility Bd, 64-68, chmn, 68-69; consult, Struct Design Adv Bd, Off Chief Engrs, Dept Army, 69-71; mem, Eng Col Accreditation Comt, Engrs Joint Coun. *Honors & Awards:* Am Concrete Inst Award, 63, Lindau Medal, 68, Corbetta Award, 73, Whitney Award, 75 & Henry Turner Award, 78; T Y Lin Award, Am Soc Civil Engrs, 71; Medal of Honor, Prestressed Concrete Inst, 75. *Mem:* Nat Acad Eng; fel & hon mem Am Soc Civil Engrs; Int Asn Bridge & Struct Engrs; hon mem Am Concrete Inst (vpres, 64-66, pres, 66); Prestressed Concrete Inst (pres, 70); Soc Exp Stress Anal; hon mem Japan Concrete Inst. *Res:* Subaqueous concrete; photoelasticity; welding; experimental stress analysis; secondary factors contributing to stiffness of truss structures; prestressed concrete. *Mailing Add:* 502 N Tacoma Ave Tacoma WA 98403-2741

ANDERSON, ARTHUR W, b Lisbon, NDak, Dec 2, 14; m 48; c 3. BACTERIOLOGY. *Educ:* NDak State Col, BS, 43; Univ Wis, MS, 47; Ore State Col, PhD, 52. *Prof Exp:* Asst prof bact, Univ Calif, Berkeley, 52-53; assoc prof food microbiol, 53-77, prof, 77-80, EMER PROF MICROBIOL, ORE STATE UNIV, 80- *Mem:* Inst Food Technol; Am Soc Microbiol. *Res:* Food borne pathogens and irradiation microbiology. *Mailing Add:* 982 Normandy Ave S Salem OR 97302

ANDERSON, AUBREY LEE, b Ft Worth, Tex, Apr 24, 40; m 60; c 2. ACOUSTICS, OCEANOGRAPHY. *Educ:* Baylor Univ, BS, 62; Univ Tex, Austin, PhD(eng acoust), 74. *Prof Exp:* Physicist, Naval Ord Lab, Corona, Calif, 62-63 & Navy Electronics Lab, San Diego, 63-64; res scientist underwater sound, Appl Res Labs, Univ Tex, Austin, 65-77; physicist, Off Naval Res, Naval Ocean Res & Develop Activ, 77-81, dep dir, Environ Sci Directorate, 81-83; PROF, DEPT OCEANOG, TEX A&M UNIV, COL STATION, 84- *Mem:* Acoust Soc Am; Sigma Xi; Am Inst Physics; Marine Technol Soc; Am Geographical Union; Oceanog Soc. *Res:* Underwater acoustics; mathematical modeling of ocean sound and noise fields and of oceanographic variables; use of acoustics as a tool in geophysics and oceanography; marine sediment studies; physical and geological oceanography. *Mailing Add:* Dept Oceanog Tex A&M Univ College Station TX 77843-3146

ANDERSON, BERNARD A, b Thornton, Wash, Aug 23, 09; m 42; c 8. SPEECH PATHOLOGY, AUDIOLOGY. *Educ:* Univ Wash, BA, 33, dipl, 37; Univ Wis, PhD(speech path), 50. *Prof Exp:* Actg dir speech & hearing, Univ Okla, 39-40; instr speech path, Ind Univ, 46-48; assoc prof, Univ Utah, 49-57; prof, 58-74, PROF EMER SPEECH PATH & DIR SPEECH & HEARING CLIN, UNIV NEV, RENO, 74- *Concurrent Pos:* Consult, Easter Seal Soc, 51-71, Shriner's Hosp Crippled Children, 55-57. *Mem:* Fel Am Speech & Hearing Asn. *Res:* Photography of the human larynx; disorders of speech and hearing. *Mailing Add:* 220 Brownstone Reno NV 89512

ANDERSON, BERNARD JEFFREY, b Waco, Tex, Feb 23, 44; m 71; c 8. SPACE VEHICLE, ENVIRONMENT INTERACTIVE. *Educ:* Univ Nev, Reno, BS, 66, MS, 68, PhD(physics), 74. *Prof Exp:* Res assoc cloud physics, Desert Res Inst, Lab Atmospheric Physics, Univ Nev, 74-75; AEROSPACE ENGR, GEORGE C MARSHALL SPACE FLIGHT CTR, NASA, 75- *Concurrent Pos:* Interagency Nuclear Safety Rev Panel. *Mem:* Am Meteorol Soc. *Res:* Space environment definition for space station; hydrogen ignition hazards in space vehicle launch and test environment; environment effects from solid rocket launch and test; meteorological impacts on space vehicle operations. *Mailing Add:* ES-44 G C Marshall Space Flight Ctr Marshall Space Flight Ctr AL 35812

ANDERSON, BERTIN W, b Luverne, Minn, Dec 15, 39; m 69; c 1. ZOOLOGY. *Educ:* Mankato State Col, BS, 62; Univ Minn, MS, 66; Univ SDak, PhD(zool), 69. *Prof Exp:* Asst prof biol, Northwestern State Col, Okla, 69-72; vis prof zool, Eastern Mich Univ, 72-73; ASSOC PROF ZOOL, ARIZ STATE UNIV, 73- *Mem:* Am Ornith Union; Cooper Ornith Soc; Ecol Soc; Wildlife Soc; Soc Study Evolution. *Res:* Ornithology; evolutionary and systematic biology; community ecology. *Mailing Add:* 201 S Palm Dr Blythe CA 92225

ANDERSON, BEVERLY, b Milwaukee, Wis, Mar 6, 33. BIOMEDICAL RESEARCH. *Educ:* Univ Ore, PhD(biol), 71. *Prof Exp:* SR BIOLOGIST, BIOSCI RES LAB, 3M CO, 81- *Mem:* Am Soc Cell Biol. *Mailing Add:* 4816 Aldrich Ave S Minneapolis MN 55409

ANDERSON, BRADLEY DALE, b Columbus, Nebr, Nov 23, 48; m 72. PHARMACEUTICAL CHEMISTRY, PHYSICAL CHEMISTRY. *Educ:* Univ Kans, BA, 71, MS, 77, PhD(pharmaceut chem), 78. *Prof Exp:* Qual control chemist pharmaceut, Haver-Lockhart Labs, 71-74; RES SCIENTIST PHARM, UPJOHN CO, 77- *Mem:* Am Pharmaceut Asn. *Res:* Design of physicochemical properties of drugs through molecular modification; solution thermodynamics. *Mailing Add:* Dept Pharmaceutics Univ Utah 421 Wakara Rm 305 Salt Lake City UT 84112

ANDERSON, BRUCE HOLMES, b Raymond, Alta, Aug 30, 17; US citizen; m 41; c 6. AGRICULTURAL & IRRIGATION ENGINEERING. *Educ:* Utah State Univ, BS, 50, MS, 54; Univ Calif, Davis, DEng, 63. *Prof Exp:* Agr engr, Utah State Univ-Iran Contract, Shiraz, Iran, 51-52, chief agr sect, Fars Ostan, 52-53, regional dir, Prov Fars, 53-54; head agr eng dept, Int Coop Admin, Tehran, 54-56; sr adv to Ministry Agr, Utah State Univ-Iran Contract, 57-60; dir, Inter-Am Ctr Land & Water Resource Develop, Orgn Am States-Utah State Univ, Venezuela, 64-69; coordr, Inter-Am Ctr Integral

Develop Water & Land Resources & field dir water mgt res prog, 69-70; exec dir, Consortium Int Develop, 73-80; irrig exten specialist, Utah State Univ, 56-57; water use specialist, 63-64; prog irrig eng & dir int progs & studies, 70-77; RETIRED. Concurrent Pos: Coordr joint agr projs, Ministry Agr, Iran, 52-56; consult, US Off Sci & Technol, Latin Am Bur & US Agency Int Develop; team leader, On-Farm Water Mgt Proj, AID-USU-Dominican Repub, 85-87. Mem: Am Soc Agr Engrs; Soil Conserv Soc Am. Res: Consumptive use of water by crops under irrigation; water resource planning and management. Mailing Add: 1486 Maple Dr Logan UT 84321

ANDERSON, BRUCE MARTIN, b Charleston, WVa, Aug 11, 42; m 69; c 2. TITANIUM DIOXIDE PIGMENT TECHNOLOGY. Educ: Morris Harvey Col, Charleston, BS, 64; WVa Univ, PhD(inorg chem), 69. Prof Exp: Process chemist, Am Cyanamid Co, 69-81, process develop mgr, 81-85; qual control mgr, 85-88, asst tech dir, 88-90, TECH DIR, KEMIRA, INC, 90- Concurrent Pos: Chmn, Coastal Empire Subcomt, Am Chem Soc, 74. Mem: Am Chem Soc. Res: Titanium dioxide pigment technology; chemistry of various titanium compounds; environmental control process relating to manufacture of titanium dioxide. Mailing Add: 202 Salisbury Rd Savannah GA 31410

ANDERSON, BRUCE MURRAY, b Detroit, Mich, July 14, 29; m 50; c 3. BIOCHEMISTRY. Educ: Ursinus Col, BS, 53; Purdue Univ, MS, 54; Johns Hopkins Univ, PhD(biochem), 58. Prof Exp: NIH fel biochem, Brandeis Univ, 58-60; asst prof, Univ Louisville, 60-63; from assoc prof to prof, Univ Tenn, 63-70; head dept, 70-82, PROF BIOCHEM, VA POLYTECH INST & STATE UNIV, 70- Concurrent Pos: NSF res grants, 60-; NIH res grants, 64-69. Mem: AAAS; Am Chem Soc; Am Soc Biol Chem. Res: Mechanism of enzyme action; function of pyridine nucleotide coenzymes. Mailing Add: Dept Biochem & Nutrit Va Polytech Inst & State Univ Blacksburg VA 24061

ANDERSON, BRYON DON, b Ft Riley, Kans, Dec 23, 44; m 69; c 2. NUCLEAR PHYSICS, NUCLEAR ASTROPHYSICS. Educ: Univ Idaho, BS, 66; Case Western Reserve Univ, PhD(physics), 72. Prof Exp: Res asst nuclear physics, Idaho Nuclear Corp, 66-67; res assoc, Calif Inst Technol, 71-73; sr res assoc, Case Western Reserve Univ, Los Alamos Nat Lab, 73-75; from asst prof to assoc prof, 75-86, PROF PHYSICS, KENT STATE UNIV, 86- Mem: Am Phys Soc. Res: Neutron and charged-particle spectrometry at intermediate energies, nuclear structure and reaction mechanisms; neutrons from relativistic heavy-ion reactions. Mailing Add: Dept Physics Kent State Univ Kent OH 44242

ANDERSON, BURTON CARL, b Kewanee, Ill, Oct 8, 30; m 52; c 3. ORGANIC POLYMER CHEMISTRY. Educ: Univ Ill, AB, 52; Mass Inst Technol, PhD(org chem), 55. Prof Exp: Res chemist, Cent Res Dept, Exp Sta, 55-66, res chemist, Elastomers Dept, 66-69, div head res, 69-72, develop supt, Chambers Works, 72-74, assoc dir, Cent Res & Develop Dept, 74-79, dir, Polymer Sci Cent Res & Develop Dept, 79-81, ASST DIR, RES FABRICS & FINISHES DEPT, E I DU PONT DE NEMOURS & CO, INC, 81- Concurrent Pos: Assoc dir cent res & develop, E I du Pont de Nemours & Co, Inc. Mem: Am Chem Soc. Res: Organic fluorine chemistry; polymer physical chemistry; polymer synthesis. Mailing Add: 114 Woodhill Rd Wilmington DE 19809

ANDERSON, BYRON, b Hammond, Ind, Dec 30, 41; m 64. BIOCHEMISTRY, IMMUNOLOGY. Educ: Kalamazoo Col, BA, 63; Univ Mich, PhD(biochem), 68. Prof Exp: From asst prof to assoc prof, 71-84, PROF BIOCHEM, MED & DENT SCHS, NORTHWESTERN UNIV, 84- Concurrent Pos: Helen Hay Whitney & Nat Cystic Fibrosis Res Found fel, Col Physicians & Surgeons, Columbia Univ, 68-71; NIH career develop award, 74-79; sr investr, Arthritis Found, 73-74; consult, Abbott Labs, 74- Mem: AAAS; Am Rheumatism Asn; Am Chem Soc; Am Soc Biol Chem; Am Asn Immunologists. Res: Autoimmune diseases, especially rheumatoid arthritis; cancer research and membrane glycoprotein antigens. Mailing Add: Dept Molecular Biol Northwestern Univ Med & Dent Schs 303 E Chicago Chicago IL 60611

ANDERSON, CARL DAVID, cosmic rays, fundamental particles; deceased, see previous edition for last biography

ANDERSON, CARL EINAR, b Brooklyn, NY, Sept 15, 23; m 57. PHYSICS. Educ: Rensselaer Polytech Inst, BS, 49; Yale Univ, MS, 50, PhD(physics), 53. Prof Exp: Res assoc electron physics res lab, Gen Elec Co, 52-54; res assoc nuclear physics, Yale Univ, 54-61; sr scientist appl physics lab, Johns Hopkins Univ, 61; mgr radiation & space physics re-entry systs dept, 61-65, mgr appl physics, Space Sci Lab, 65-81, mgr Electro-Optics & sensors, 81-83, CONSULT PHYSICIST, SPACE DIV, GEN ELEC CO, 83- Concurrent Pos: Consult, Wallingford Steel Co, 56-59. Mem: Am Phys Soc; Soc Photo-Optical Instrumentation Engrs; Am Inst Aeronaut & Astronaut; Sigma Xi. Res: Nuclear, reentry and space physics; lasers. Mailing Add: Rm 45T48 Bldg 20 Astrospace Gen Elec Co PO Box 8555 Philadelphia PA 19101

ANDERSON, CARL FREDERICK, NUTRITIONAL ASSESSMENT, NUTRITION & RENAL DISEASE. Educ: State Univ Iowa, MD, 59. Prof Exp: PROF MED, MAYO MED SCH, 83- Res: Chronic renal failure. Mailing Add: Mayo Med Sch Mayo Clin Rochester MN 55901

ANDERSON, CARL LEONARD, b Ironwood, Mich, Feb 28, 01; m 30; c 3. HYGIENE, PUBLIC HEALTH. Educ: Univ Mich, BS, 28, MS, 32, DrPH, 34. Prof Exp: Dir health educ, Mich Bd Educ, 28-31; pub health adminr, Mich State Dept Health, 31-32; rural health adminr, Mich State Dept Health & USPHS, 33-35; from asst prof to prof physiol & pub health, Utah State Univ, 35-45; prof biol sci, Mich State Univ, 45-49; prof, 49-77, EMER PROF HYG & HEALTH EDUC, ORE STATE UNIV, 77- Mem: AAAS; fel Am Pub Health Asn; fel Soc Res Child Develop; Am Genetic Asn; Am Asn Health, Phys Educ & Recreation. Res: Experimental production of convulsive seizures; inheritance of cataract; metabolism of the human skin; physiological indices of the human; neurone impulse; physiology. Mailing Add: 3907 N Gantenbein Ave Portland OR 97227

ANDERSON, CARL WILLIAM, b Washington, DC, May 19, 44; m 68; c 2. MOLECULAR BIOLOGY, ANIMAL VIROLOGY. Educ: Harvard Col, BA, 66; Wash Univ, PhD(microbiol), 70. Prof Exp: Fel, DNA cancer viruses, Cold Spring Harbor Lab, 70-73, mem staff, 73-75; asst geneticist, 75-76, assoc geneticist, 76-79, GENETICIST, BROOKHAVEN NAT LAB, 79- Concurrent Pos: NIH fel, 70-72; consult, NSF, 75-79; adj asst prof, Dept Microbiol, State Univ NY, Stony Brook, 76-79, adj assoc prof, 79-90, adj prof, 90- Mem: Am Soc Microbiol; AAAS; Am Soc Virol; Protein Soc; Am Soc Biochem & Molecular Biol. Res: Control of gene expression in mammalian cells; protein kinases. Mailing Add: Dept Biol Brookhaven Nat Lab Upton NY 11973

ANDERSON, CARL WILLIAM (BILL), b Schenectady, NY, Apr 7, 50. ANALYTICAL CHEMISTRY, BIOCHEMISTRY. Educ: Univ Mass, BS, 72; Univ Cincinnati, MS, 75, PhD(anal chem) & PhD(biochem), 78. Prof Exp: Res assoc phys chem, Amherst Col, 72-73; ASST PROF ANALYTICAL CHEM, DUKE UNIV, 78- Mem: Am Chem Soc. Res: Applications of electrochemistry to biochemical problems with respect to quantitative, mechanistic and thermodynamic behavior of biochemically significant compounds. Mailing Add: 1002 Atkinson Ave Hampden Sydney VA 23943

ANDERSON, CAROL PATRICIA, b Bluefield, WVa, May 19, 46; m 78. ENVIRONMENTAL ANALYSIS, PATTERN RECOGNITION TECHNIQUES. Educ: Concord Col, BS, 68; Univ Tenn, PhD(chem), 73. Prof Exp: Fel, 73-75, asst prof, 75-80, ASSOC PROF CHEM, UNIV CONN, AVERY POINT, 80- Mem: Am Chem Soc; Nat Sci Teachers Asn. Res: Detecting and quantifying spilled hazardous chemicals; pattern recognition techniques for spilled oils; permeation of protective clothing material; more effective methods of presenting chemistry to the general public. Mailing Add: Univ Conn Avery Point Groton CT 06340

ANDERSON, CHARLES ALFRED, geology; deceased, see previous edition for last biography

ANDERSON, CHARLES DEAN, b Redwood Falls, Minn, Mar 8, 30; m 53; c 3. BIO-ORGANIC CHEMISTRY. Educ: St Olaf Col, BA, 52; Harvard Univ, MA, 54, PhD(org chem), 59. Prof Exp: Org chemist, Stanford Res Inst, Calif, 56-59; assoc prof org chem, Pac Lutheran Univ, 59-62, chmn, Dept Chem, 61-66 & 78-80, dean, Col Arts & Sci, 66-70, Regency professorship, 74-75, PROF CHEM, PAC LUTHERAN UNIV, 62- Concurrent Pos: NSF fac fel, Univ Minn, 64-65; Northwest Col & Univ Asn Sci vis fac appointment, Pac Northwest Labs, Battelle Mem Inst, 70-71; vis scholar, Col Forest Resources, Univ Wash, 81 & 88-89; co-dir, Chengdu Exchange Prog, 86-89; vis fac appointment, Chengdu Univ of Sci & Tech, China, 87. Mem: Am Chem Soc; Royal Soc Chem; Am Asn Univ Professors. Res: Synthetic methods; natural products and analogs; lignin chemistry. Mailing Add: Dept Chem Pac Lutheran Univ Tacoma WA 98447

ANDERSON, CHARLES E, JR, b Baltimore, Md, Mar 7, 46; m 74; c 2. COMPUTATIONAL MODELING, COMPUTATIONAL MECHANICS. Educ: Va Polytech Inst, BS, 68; Rensselaer Polytech Inst, MS & PhD(physics), 72. Prof Exp: Res physicist, US Army Ballistic Res Lab, 72-80; sr res physicist, 80-84, MGR, SOUTHWEST RES INST, 84- Concurrent Pos: Lectr, Univ Tex, San Antonio, 83-89. Mem: Hypervelocity Impact Soc (pres, 89-); AAAS; Sigma Xi; Am Phys Soc. Res: Physical sciences: dynamic modeling of fire spread, intumescent paint, warhead mechanics, penetration mechanics, shock mechanics; dynamic material response, hypervelocity impact. Mailing Add: Southwest Res Inst 6220 Culebra Rd PO Drawer 28510 San Antonio TX 78228-0510

ANDERSON, CHARLES EDWARD, b St Louis, Mo, Aug 13, 19; m 43; c 3. PHYSICAL METEOROLOGY. Educ: Lincoln Univ, Mo, BS, 41; Univ Chicago, cert meteorol, 43; Polytech Inst Brooklyn, MS, 48; Mass Inst Technol, PhD(meteorol), 60. Prof Exp: Chief, Cloud Physics Br, Air Force Cambridge Res Ctr, Mass, 48-61 & Atmospheric Sci Br, Douglas Aircraft Co, Calif, 61-65; dir, Off Fed Coord Meteorol, Environ Sic Serv Admin, US Dept Com, Washington, DC, 65-66; prof space sci & eng, 67-69, PROF METEOROL & CHMN CONTEMP TRENDS COURSE, UNIV WIS-MADISON, 66-, PROF AFRO-AM STUDIES & CHMN DEPT, 70-, ASSOC DEAN, 78- Concurrent Pos: Prin investr, NSF grant, 69-; consult, Sci Ctr, NAm Rockwell Corp, Calif, 67- & Rand Corp, 68-; chmn, Aviation Adv Panel, Nat Ctr Atmospheric Res, Colo, 63- Mem: AAAS; Sigma Xi. Res: Cloud and aerosol physics; high polymer chemistry; space and planetary science; meteorology of other planets; science and society; science, technology and race. Mailing Add: 6301 W Gate Rd Madison WI 53706

ANDERSON, CHARLES EUGENE, b Winfield, Kans, Dec 3, 34; m 54; c 3. MORPHOLOGY. Educ: Purdue Univ, BSA, 56, MS, 61, PhD(biol sci), 63. Prof Exp: Asst prof bot, Univ Okla, 63-66; from asst prof to assoc prof, 69-78, PROF BOT, NC STATE UNIV, 78- Mem: AAAS; Bot Soc Am. Res: Developmental morphology of salt marsh and dune plants; environmental factors influencing cell and tissue differentiation. Mailing Add: 3969 Wendy Lane Raleigh NC 27606

ANDERSON, CHARLES HAMMOND, b Mineola, NY, May 31, 35; m 60; c 3. NEUROSCIENCE. Educ: Calif Inst Technol, BS, 57; Harvard Univ, PhD(molecular physics), 62. Prof Exp: Res assoc, Harvard Univ, 61-62, instr physics, 62-63; mem tech staff, RCA Lab9, 63-77, fel staff, 77-78, group head appl math & phys sci, 78-82, fel staff, 82-87; MEM TECH STAFF, JET PROPULSION LAB, 87- Concurrent Pos: NSF fel, 62-63; RCA fel, Clarendon Lab, Oxford, Eng, 71-72; res assoc, Calif Inst Tech, 84, vis assoc, Biol Div, 88- Mem: Fel Am Phys Soc; AAAS; Am Res Vision & Ophthal. Res: Ion physics in solids; liquid helium; acoustics; low temperature physics; flat television displays; electron optics; mathematical models of human vision-pattern recognition; general applied math. Mailing Add: Jet Propulsion Lab 4800 Oak Grove Dr, MS-23 Pasadena CA 91109

ANDERSON, CHARLES THOMAS, b Fairmont, WVa, Feb 26, 21; m 50; c 2. PHYSICAL INORGANIC CHEMISTRY. *Educ:* Fairmont State Col, AB, 42; Ohio State Univ, PhD(chem), 55. *Prof Exp:* Teaching asst chem, Ohio State Univ, 43-46; instr, Ohio Univ, 46-51; from asst prof to prof, 55-86, EMER PROF CHEM, EASTERN MICH UNIV, 86- *Mem:* Am Chem Soc; Sigma Xi. *Res:* Coordination compounds; equilibrium constants. *Mailing Add:* 720 Kewanee Ave Ypsilanti MI 48197-1741

ANDERSON, CHARLES V, b Little Sioux, Iowa, Aug 18, 33; m 66; c 1. AUDIOLOGY. *Educ:* Univ Nebr, BS, 55, MA, 57; Univ Pittsburgh, PhD(psychoacoust), 62. *Prof Exp:* Instr hearing ther, Univ Nebr, 56-58; asst prof audiol, Purdue Univ, 62-66; asst prof, 66-68, ASSOC PROF AUDIOL, UNIV IOWA, 68- *Mem:* AAAS; fel Am Speech-Lang-Hearing Asn; Acoust Soc Am; Sigma Xi. *Res:* Hearing loss and aging; auditory phenomena and noise. *Mailing Add:* Dept Speech Path & Audiol 16 Woodland Heights Iowa City IA 52242

ANDERSON, CHRISTIAN DONALD, b Ft Dodge, Iowa, Apr 10, 31; m 57; c 6. EXPLORATION GEOPHYSICS. *Educ:* Univ Minn, Duluth, BA, 57; Univ NMex, MS, 61; Univ Utah, PhD(geophys), 66. *Prof Exp:* Physicist, Los Alamos Sci Lab, 58-61; geophysicist, Newmont Mining Corp, 66-67; ASSOC PROF GEOPHYS, UNIV MAN, 67- *Concurrent Pos:* Consult explor geophys, 67- *Mem:* Can Explor Geophysicists Soc; Sigma Xi; Soc Explor Geophysicists; Am Inst Mining Eng. *Res:* Development of geophysical methods and exploration strategies to improve discovery probabilities in mineral exploration. *Mailing Add:* Univ NC Box 5135 Florence AL 35632

ANDERSON, CHRISTOPHER MARLOWE, b Las Cruces, NMex, Feb 21, 41; m 70. ASTRONOMY. *Educ:* Univ Ariz, BS, 63; Calif Inst Technol, PhD(astron), 68. *Prof Exp:* Asst prof, 68-74, assoc prof, 74-80, PROF ASTRON, UNIV WIS-MADISON, 80- *Mem:* Am Astron Soc; Int Astron Union. *Res:* Photoelectric spectrophotometry; interstellar extinction; stellar rotation and chromospheric activity; spectrum variability. *Mailing Add:* Dept Astron Univ Wis 475 N Charter St 5534 Sterling Hall Madison WI 53706

ANDERSON, CLIFFORD HAROLD, b Peoria, Ill, Jan 11, 39; m 62; c 2. MATHEMATICAL ANALYSIS. *Educ:* Ill Inst Technol, BS, 61; Purdue Univ, MS, 63; Univ Mo, PhD(math), 68. *Prof Exp:* Asst prof math, Ohio Univ, 67-74; from asst prof to assoc prof math, Calif State Univ, Los Angeles, 74-81; RETIRED. *Mem:* Am Math Soc; Soc Indust & Appl Math. *Res:* Functional differential equations. *Mailing Add:* 3915 Park Vista Dr Pasadena CA 91107

ANDERSON, CLYDE LEE, b El Paso, Tex, Sept 17, 26; m 50; c 3. CHEMISTRY. *Educ:* Univ Utah, BS, 52, MS, 53; PhD(org chem), 67. *Prof Exp:* Head chem dept, Dixie Jr Col, 53-57; assoc prof, 57-81, PROF CHEM, SAN BERNARDINO VALLEY COL, 81-, CHMN DEPT, 68- *Concurrent Pos:* Lectr, Westminster Col, 65-66; NSF sci fac fel, 65-66. *Mem:* Soc Appl Spectros; Am Chem Soc; Sigma Xi. *Res:* Organic thesis; natural products. *Mailing Add:* Dept Chem San Barnardino Valley Col 701 S Mount Vernon Ave San Bernardino CA 92410

ANDERSON, CURTIS BENJAMIN, b Fargo, NDak, Feb 19, 32. ORGANIC CHEMISTRY. *Educ:* Univ Minn, BS, 56; Univ Calif, Los Angeles, PhD(org chem), 63. *Prof Exp:* Lectr, Univ Calif, Santa Barbara, 62, assoc, 62-63, lectr, 63-64, actg asst prof, 64-65, asst prof, 65-69, ASSOC PROF CHEM, UNIV CALIF, SANTA BARBARA, 69- *Mem:* Am Chem Soc. *Res:* Physical organic chemistry; conformational analysis of heterocyclic compounds; reactions of olefins complexes of metal salts; neighboring group participation. *Mailing Add:* Dept Chem Univ Calif Santa Barbara CA 93106

ANDERSON, DALE ARDEN, b Alta, Iowa, Aug 11, 36; m 58; c 2. AEROSPACE & ELECTRICAL ENGINEERING. *Educ:* St Louis Univ, BS, 57; Iowa State Univ, MS, 59, PhD(aerospace & elec eng), 64. *Prof Exp:* From instr to asst prof aerospace eng, Iowa State Univ, 57-64; mem tech staff, Aerospace Corp, Calif, 64-65; from assoc prof to prof aerospace eng, Iowa State Univ, 65-84, dir, CompFluid Dynamics Inst, 80-84,; PROF AEROSPACE ENG, UNIV TEX, ARLINGTON, 84- *Mem:* Am Inst Aeronaut & Astronaut. *Res:* Unsteady aerodynamics; nonequilibrium gas dynamics; finite difference methods in fluid mechanics. *Mailing Add:* Dept Aerospace Eng Univ Tex Arlington TX 76019-0018

ANDERSON, DANIEL D, b Ft Dodge, Iowa, Dec 20, 48; m 70; c 1. COMMUTATIVE ALGEBRA, LATTICE THEORY. *Educ:* Univ Iowa, BA, 71; Univ Chicago, MS, 71, PhD(math), 74. *Prof Exp:* From asst prof to assoc prof, 74-83, PROF MATH, UNIV IOWA, 83- *Concurrent Pos:* Asst prof math, Va Polytech Inst, 75-76 & Univ Mo, Columbia, 76-77. *Mem:* Am Math Soc; Math Asn Am. *Res:* Divisibility theory (Euclidean, factorial and Krull domains, class groups); multiplicative ideal theory; rings with zerodivisors; lattice-theoretic questions. *Mailing Add:* Dept Math Univ Iowa Univ Iowa Iowa City IA 52242

ANDERSON, DANIEL WILLIAM, b Underwood, NDak, Feb 5, 39; m 71; c 3. WILDLIFE ECOLOGY, POLLUTION BIOLOGY. *Educ:* NDak State Univ, BS, 61; Univ Wis-Madison, MS, 67, PhD(wildlife ecol, zool), 70. *Prof Exp:* Res biologist, US Fish & Wildlife Serv, 70-75; AVIAN ECOLOGIST & PROF, UNIV CALIF, DAVIS, 76- *Concurrent Pos:* Regional rep/exec coun, Pac Seabird Group, 73-75, 85-86, chmn, 78, 85; mem bd dirs, Bodega Bay Inst Pollution Ecol, 71-; bd mem, Cooper Ornith Soc, 85-87. *Mem:* Wildlife Soc; AAAS; fel Am Ornithologists Union; Soc Environ Toxicol & Chem; Ecol Soc Am; Soc Conserv Biol. *Res:* Effects of environmental pollution on wildlife populations and individuals; ecological and physiological effects of pollutants; study of avian ecology, with emphasis on marine birds. *Mailing Add:* 501 Isla Pl Davis CA 95616

ANDERSON, DARWIN WAYNE, b Winnipeg, Man, Dec 12, 40; m 67; c 3. SOIL SCIENCE. *Educ:* Univ Man, BSA, 63, MSc, 66; Univ Sask, PhD(soil sci), 72. *Prof Exp:* Soil survr, Univ Man, 63-66; soil scientist, Ore State Univ, 66-68; vis scientist, Colo State Univ, 79-80; res scientist soil surv, 72-79, 81-90, PROF SOIL, DEPT SOIL SCI, UNIV SASK, 90- *Concurrent Pos:* Tech adv, Comt on Reclamation, Sask Power Corp, 76-77; assoc ed, Can J Soil Sci, 80- *Mem:* Agr Inst Can; Can Soc Soil Sci; Can Land Reclamation Asn; Soil Conserv Soc Am. *Res:* Soil organic matter; humus transformations and clay-humus interactions; nutrient cycling in forest ecosystems. *Mailing Add:* 2237 Lansdowne Saskatoon SK S7J 1G7 Can

ANDERSON, DAVID BENNETT, b Hutchinson, Minn, Nov, 21, 44. ANIMAL SCIENCE & NUTRITION, FOOD SCIENCE & TECHNOLOGY. *Educ:* SDak State Univ, BS, 66; Univ Wis, MS, 68, PhD(animal sci), 71. *Prof Exp:* Asst prof animal sci, Univ Ill, 75-79; sr scientist, 79-83, res scientist, 83-88, SR RES SCIENTIST, LILLY RES LABS, ELI LILLY & CO, 88- *Concurrent Pos:* Adj prof animal sci, Purdue Univ, W Lafayette, 88- *Mem:* Am Meat Sci Asn; Am Soc Animal Sci; Am Inst Nutrit; AAAS; Sigma Xi. *Res:* Growth and development of domestic animials for growth, efficiency and leanness enhancers through the study of muscle, adipose tissue metabolism and development. *Mailing Add:* Lilly Res Lab PO Box 708 Greenfield IN 46140

ANDERSON, DAVID EUGENE, b Ashby, Mass, May 16, 26; m 46; c 2. CANCER GENETICS. *Educ:* Univ Mass, BS, 50; Univ Conn, MS, 52; Iowa State Univ, PhD(animal breeding & genetics), 54. *Prof Exp:* Assoc biologist & assoc prof biol, 58-65, actg head, Dept Biol, 75-80, GENETICIST & ASHBEL SMITH PROF GENETICS, UNIV TEX M D ANDERSON CANCER CTR, HOUSTON, 65- *Concurrent Pos:* Nat Cancer Inst fel, Okla State Univ, 54-57 & spec fel human genetics, Univ Mich, 63-65. *Mem:* AAAS; Genetics Soc Am; Am Soc Human Genetics. *Res:* Human genetics as it relates to cancer and allied diseases. *Mailing Add:* Dept Molecular Genetics M D Anderson Cancer Ctr Houston TX 77030

ANDERSON, DAVID EVERETT, b Bloomington, Ill, Oct 17, 39; m 64; c 3. BEHAVIORAL MEDICINE. *Educ:* Columbia Univ, BA, 61; Univ Ore, MA, 64, PhD(psychol), 66. *Prof Exp:* Asst prof psychiat, Johns Hopkins Univ Sch Med, 70-81; assoc prof psychiat, Univ S Fla Col Med, 81-86, prof psychiat & behavioral med, 86-88; RES PSYCHOLOGIST, NAT INST AGING, 87- *Concurrent Pos:* Prin invest, Nat Heart, Lung & Blood Inst, 75-87; mem res comt, Am Heart Assoc, 83-86; chmn, Task force Restrained animals, NIH, 84. *Honors & Awards:* Pavlovian Award, Pavlovian Soc N Am. *Mem:* Fel Soc Behavioral Med; fel Acad Behavioral Med Res; Pavlovian Soc NAM; Am Psychosomatic Soc. *Res:* Behavioral influences in cardiovascular regulation; developed a model of experimental hypertension in genetically normotensive animals resulting from interaction of task performances and high sodium intake. *Mailing Add:* Nat Inst Aging 4940 Eastern Ave Baltimore MD 21224

ANDERSON, DAVID G, b Iron Mountain, Mich, Sept 30, 28; m 50; c 4. OBSTETRICS & GYNECOLOGY. *Educ:* Univ Mich, AB, 50, MD, 53. *Prof Exp:* From instr to assoc prof, 60-80, ASSOC CLIN PROF OBSTET & GYNEC, MED CTR, UNIV MICH, ANN ARBOR, 80-; DIR GYNEC, ST JOSEPH MERCY HOSP, ANN ARBOR, MICH, 85- *Mem:* AMA; Am Col Obstet & Gynec; Am Col Surg; Am Fertility Soc. *Res:* Gynecologic cancer chemotherapy; gynecologic cancer therapy; sterility; cancer therapy; operative gynecology. *Mailing Add:* Barron Prof Bldg 4870 W Clark Rd Ypsilanti MI 48197

ANDERSON, DAVID H, b Louisville, Ky, Dec 15, 39; m 70. MATHEMATICS. *Educ:* Univ Miss, BS, 61; Univ Calif, Berkeley, MA, 66; Duke Univ, PhD(math), 69. *Prof Exp:* Asst prof math, Millsaps Col, 66-67; asst prof, 69-74, ASSOC PROF MATH, SOUTHERN METHODIST UNIV, 74- *Mem:* Math Asn Am; Soc Indust & Appl Math. *Res:* Compartmental analysis in biology and medicine; population mathematics; numerical problems in mathematical modeling. *Mailing Add:* Dept Math Southern Methodist Univ Dallas TX 75275

ANDERSON, DAVID J, b Rochester, NY, Apr 16, 39; m 67; c 2. NEUROPHYSIOLOGY, BIOENGINEERING. *Educ:* Rensselaer Polytech Inst, BSEE, 61; Univ Wis, MS, 63, PhD(elec eng), 67. *Prof Exp:* Trainee, Lab Neurophysiol, Univ Wis-Madison, 67-69; PROF ELEC & COMPUT ENG, UNIV MICH, ANN ARBOR, 81-, CHMN, GRAD PROG BIOENG, 79-; PROF OTORHINOLARYNGOLOGY, KRESGE HEARING RES INST, 81- *Concurrent Pos:* Fel life sci, NASA-Johnson Space Ctr, 75-76. *Mem:* Soc Neurosci; Acoust Soc Am; Inst Elec & Electronics Engr; Asn Res Otolaryngol. *Res:* Neurophysiology of auditory and vestibular systems; application of digital computers and digital signal processing to medicine and biomedical research; implantable devices for stimulation and recording in the nervous system. *Mailing Add:* Kresge Hearing Res Inst Univ Mich Ann Arbor MI 48109

ANDERSON, DAVID JOHN, b Baraboo, Wis, Jan 7, 42; m 66; c 2. DRUG METABOLISM, TECHNICAL MANAGEMENT. *Educ:* Univ Wis, Superior, BS, 65; Univ Wis, Madison, MS, 67, PhD(pharmacol), 71; Lake Forest Sch Mgt, MBA, 86. *Prof Exp:* Pharmacologist, 69-71, group leader, 71-73, SECT HEAD DRUG METAB, ABBOTT LABS, 73- *Mem:* Sigma Xi. *Res:* Metabolism methods development; identification of metabolic profiles and metabolities; pharmacokinetic analysis. *Mailing Add:* D-463 AP9 Abbott Labs North Chicago IL 60064

ANDERSON, DAVID LEONARD, b Portland, Ore, Dec 19, 19; m 47; c 4. PHYSICS. *Educ:* Harvard Univ, SB, 41, AM, 47, PhD, 50. *Prof Exp:* Scientist, Manhattan Proj, Los Alamos Lab, NMex, 43-46; from asst prof to assoc prof physics, 48-62, chmn dept, 62-68, PROF PHYSICS, OBERLIN COL, 62- *Concurrent Pos:* Res fels, Univ Birmingham, 54-55, Harvard Univ, 61-62 & 75-76 & Univ Edinburgh, 68-69. *Mem:* AAAS; Am Phys Soc; Am Asn Physics Teachers; Hist Sci Soc; Sigma Xi. *Res:* Nuclear physics; history of modern physics. *Mailing Add:* Dept of Physics Oberlin Col Oberlin OH 44074

ANDERSON, DAVID MARTIN, b Boston, Mass, July 19, 30; m 58; c 4. ENVIRONMENTAL HEALTH. *Educ:* Northeastern Univ, BS, 53; Harvard Univ, SM, 55, PhD(indust hyg), 58. *Prof Exp:* Pub health engr, USPHS, 58-60; indust health engr, 60-67, asst mgr, 67-71, MGR ENVIRON QUAL CONTROL, BETHLEHEM STEEL CORP, 71- *Concurrent Pos:* Chmn coun tech adv, Pa Dept Health, 64- *Mem:* AAAS; Am Chem Soc; Am Inst Chem Eng; Indust Hyg Asn; Air Pollution Control Asn; Sigma Xi. *Res:* Air cleaning methods and devices; air pollution studies; control of industrial airborne contaminants. *Mailing Add:* Bethlehem Steel Corp Bethlehem PA 18016

ANDERSON, DAVID PREWITT, b Twin Falls, Idaho, Sept 14, 34; m 62; c 2. VETERINARY MICROBIOLOGY. *Educ:* Wash State Univ, BS, 59, DVM, 61; Univ Wis, MS, 64, PhD(vet sci, med microbiol), 65. *Prof Exp:* Asst prof vet sci & asst dir biotron, Univ Wis, 65-69; prof med microbiol & dir poultry dis res ctr, 69-70, chmn dept avian med, 70-71, assoc dean res & grad affairs, 71-73, prof avian med, 73-75, DEAN COL VET MED, UNIV GA, 75- *Concurrent Pos:* Ed, J Avian Dis, 73- *Mem:* Vet Med Asn; Am Asn Avian Path; Am Col Vet Microbiol. *Res:* Avian diseases, particularly those caused by mycoplasmas. *Mailing Add:* Col Vet Med Univ Ga Athens GA 30602

ANDERSON, DAVID R, b Norton, Kans, Dec 13, 42; m 88; c 2. WILDLIFE SCIENCE, QUANTITATIVE ECOLOGY. *Educ:* Colo State Univ, BS, 65, MS, 67; Univ Md, PhD(theoret ecol), 74. *Prof Exp:* Res biologist migratory birds, US Fish & Wildlife Serv, 67-75; res biologist, Utah State Univ, 75-80, prof wildlife sci & leader, Coop Wildlife Res Unit, 80-85; UNIT LEADER & PROF, COOP FISH & WILDLIFE RES UNIT, COLO STATE UNIV, 85- *Concurrent Pos:* Consult, Nat Marine Fisheries Serv, 76-78. *Honors & Awards:* Outstanding Performance Award, Bur Sport Fisheries & Wildlife, 68 & US Fish & Wildlife Serv, 72. *Mem:* Wildlife Soc; Ecol Soc Am. *Res:* Theoretical ecology; biometrics; estimation theory; optimization theory; migratory bird research; population ecology; sampling theory; statistics. *Mailing Add:* Colo Coop Fish Wildlife Res Unit Colo State Univ Ft Collins CO 80523

ANDERSON, DAVID ROBERT, b Cuba, NY, Feb 10, 40; m 63. ZOOLOGY, PARASITOLOGY. *Educ:* Lycoming Col, BA, 62; Pa State Univ, MS, 64; Colo State Univ, PhD(parasitol, zool), 67. *Prof Exp:* From asst prof to assoc prof zool, Pa State Univ, Fayette, 67-80; MEM FAC, UTAH STATE UNIV, 80- *Concurrent Pos:* NSF inst grant, 68-69. *Mem:* Am Soc Parasitol; Soc Protozool; Wildlife Dis Asn; Wildlife Soc; Am Inst Biol Sci. *Res:* Ecology of helminths in relation to host's environment; all phases of research on coccidia, especially ecology, immunity and life cycle. *Mailing Add:* Dept Zool Pa State Univ Box 519 Uniontown PA 15401

ANDERSON, DAVID VINCENT, b Chicago, Ill, Aug 12, 41; div; c 2. PLASMA PHYSICS. *Educ:* Johns Hopkins Univ, BA, 63; Univ Nev, MS, 68; Univ Calif, Davis, PhD(plasma physics), 71. *Prof Exp:* Res physicist, Naval Res Lab, 71-74; PLASMA PHYSICIST, LAWRENCE LIVERMORE LAB, 74- *Mem:* Am Phys Soc. *Res:* Computational physics applications to plasma physics, particularly multidimensional fluid models for the study of stability and transport phenomena. *Mailing Add:* Lawrence Livermore Lab L-561 PO Box 5509 Livermore CA 94550

ANDERSON, DAVID W, JR, food chemistry, for more information see previous edition

ANDERSON, DAVID WALTER, b Heron Lake, Minn, June 18, 37; m 60; c 2. NUCLEAR PHYSICS, HEALTH PHYSICS. *Educ:* Hamline Univ, BA, 59; Iowa State Univ, PhD(physics), 65; Am Bd Radiol, cert radiol physics, 75. *Prof Exp:* Prof radiol sci, Health Sci Ctr, Univ Okla, 76-82; prof physics, Univ Okal, 78-82; prof & dir radiol physics, Med Sch, Oral Roberts Univ, 82-88; PROF & DIR RADIOL PHYSICS, TULSA REGIONAL MED CTR, COL MED, OKLA STATE UNIV, 88- *Concurrent Pos:* AEC fel physics, Iowa State Univ, 65-66. *Mem:* Am Asn Physicists Med; Am Phys Soc; fel Am Col Radiol; Am Sci Affil; Soc Nuclear Med; Health Phys Soc; Sigma Xi. *Res:* Photonuclear reactions; radiation dosimetry; radiation spectroscopy. *Mailing Add:* Tulsa Regional Med Ctr Radiol Dept Tulsa OK 74127

ANDERSON, DAVID WESLEY, b Independence, Mo, Apr 2, 52; m 75; c 2. IMMUNOPHARMACOLOGY, MOLECULAR BIOLOGY. *Educ:* Graceland Col, BA, 74; Univ Mo-Columbia, PhD(immunol), 78. *Prof Exp:* res assoc immunol, Webb-Waring Lung Inst, Univ Colo, 78-81; res fel immunol, Eleanor Roosevelt Inst Cancer Res, 81-83; res specialist immunopharmacol, Monsanto, 83-85; group leader immunopharmacol, Monsanto/Searle, 85-88; res mgr immunopharmacol, R W Johnson Pharmacol Res Inst, J&J, 88-91; ASST DIR EXP THERAPEUT, JOHNSON & JOHNSON, 91- *Concurrent Pos:* Leukemia res fel, Lady Tata Found, 81; consult, Wilkerson Group, 91- *Mem:* Am Asn Immunologists; NY Acad Sci; Drug Info Serv; AAAS. *Res:* Research and publications have focused on the mechanisms of cellular immunity and tumor immunology; research programs for developing novel antirheumatics and immunomodulatory drugs for autoimmune diseases and cancer; granted 3 patents and 4 development awards. *Mailing Add:* Dept Immunopharmacol R W Johnson Pharmaceut Res Inst Rte 202 Box 300 Raritan NJ 08869

ANDERSON, DEAN MAURITZ, b Pueblo, Colo, Feb 14, 47; m 88. ETHOLOGY. *Educ:* Univ S Colo, BS, 70; Colo State Univ, MS, 72; Tex A&M Univ, PhD(range sci), 77. *Prof Exp:* RESEARCHER RANGE ANIMAL NUTRIT & ETHOLOGY, AGR RES SERV, USDA, 77- *Mem:* Soc Range Mgt; Sigma Xi. *Res:* Range animal nutrition and ethology research directed toward developing multispecies grazing management systems for arid rangeland with special emphasis on animal performance and grazing behavior. *Mailing Add:* US Dept Agr-Agr Res Serv Jornada Exp Range Box 3003 NMSU Dept 3 JER Las Cruces NM 88003-0003

ANDERSON, DEBORAH JEAN, b Ft Belvoir, Va, Dec 28, 49; m 90; c 1. REPRODUCTIVE IMMUNOLOGY. *Educ:* Univ Tex, Houston, PhD(immunol), 76. *Prof Exp:* From instr to assoc prof, 81-86, ASSOC PROF, HARVARD MED SCH, 86- *Mem:* Am Asn Immunologists; Am Asn Pathologists. *Res:* Interactions between the immune system, reproductive processes and sexually transmitted diseases. *Mailing Add:* Harvard Med Sch 250 Longwood Ave Rm 205 Boston MA 02115

ANDERSON, DEBRA F, FETAL PHYSIOLOGY. *Educ:* Univ Wis, PhD(physiol), 79. *Prof Exp:* ASST PROF ANESTHESIOL, ORE HEALTH SCI UNIV, 81- *Mailing Add:* Dept Physiol Ore Health Sci-Univ 3181 SW Sam Jackson Park Rd Portland OR 97201

ANDERSON, DENNIS ELMO, b Dunnell, Minn, Sept 2, 34; m 54; c 3. PLANT TAXONOMY. *Educ:* Univ Northern Iowa, BA, 55; Iowa State Univ, MS, 58, PhD(plant taxon), 60. *Prof Exp:* Instr bot, Fla Presby Col, 60-61; from asst prof to assoc prof, 61-70, PROF BOT, HUMBOLDT STATE UNIV, 70- *Mem:* AAAS; Int Asn Plant Taxon. *Res:* Plant anatomy and morphology; cytology; biosystematics; computer sciences and software; computer aided biological research; biometrics and biostatistics. *Mailing Add:* Dept Biol Humboldt State Univ Arcata CA 95521

ANDERSON, DON LYNN, b Frederick, Md, Mar 5, 33; m 56; c 2. GEOPHYSICS, SEISMOLOGY. *Educ:* Rensselaer Polytech Inst, BS, 55; Calif Inst Technol, MS, 59, PhD(geophys, math), 62. *Prof Exp:* Geophysicist, Air Force Cambridge Res Ctr, Bedford, Mass, 56-58 & Arctic Inst NAm, Boston, Mass, 58-62; res fel, 62-63, asst prof to assoc prof geo phys, 63-68, dir seismol lab, 67-90, PROF GEOPHYS, CALIF INST TECHNOL, 68- *Concurrent Pos:* Consult, var NASA, Nat Acad Sci & NSF comts; consult, Chevron Oil Co, 71-78; Fulbright-Hayes Found sr scholar, 75. *Honors & Awards:* Arthur L Day Medal, Am Geol Soc, 87; Gold Medal, Royal Astron Soc, 88. *Mem:* Nat Acad Sci; Am Geophys Union (pres, 88-90); fel Am Acad Arts & Sci; fel Sigma Xi; AAAS; Geol Soc Am; fel Royal Astron Soc; Seismol Soc Am; Am Philos Soc. *Res:* Structure and composition of the Earth and planets. *Mailing Add:* Seismol Lab 252-21 Calif Inst Technol Pasadena CA 91125

ANDERSON, DONALD ARTHUR, electronic physics; deceased, see previous edition for last biography

ANDERSON, DONALD E, b Fullerton, NDak, Sept 22, 31; m 55; c 5. AGRICULTURAL POLICY, AGRICULTURAL MARKETING. *Educ:* NDak State Univ, BS, 53, MS, 57; Univ Minn, PhD(agr econ), 68. *Prof Exp:* Grad asst, 55-57, from instr to prof agr econ, 57-79, DIR RES, NDAK STATE UNIV, 79- *Res:* Agricultural and related disciplines. *Mailing Add:* Agr Exp Sta NDak State Univ Box 5435 Fargo ND 58105

ANDERSON, DONALD GORDON MARCUS, b Sarnia, Ont, Jan 4, 37. APPLIED MATHEMATICS. *Educ:* Univ Western Ont, BSc, 59; Harvard Univ, AM, 60, PhD(appl math), 63. *Prof Exp:* Lectr & res fel, 63-65, asst prof, 65-69, PROF APPL MATH, HARVARD UNIV, 69- *Mem:* AAAS; Soc Indust & Appl Math. *Res:* Numerical mathematics and scientific computation; applications of mathematics and computing in the physical sciences and engineering. *Mailing Add:* Aiken Comput Lab Harvard Univ Cambridge MA 02138

ANDERSON, DONALD HERVIN, b Seattle, Wash, July 5, 16; m 40; c 2. ANALYTICAL CHEMISTRY. *Educ:* Univ Wash, BS, 38, PhD(chem), 43. *Prof Exp:* Instr physics, Univ Wash, 43; from instr to asst prof chem, Univ Idaho, 44-47; mem staff, Eastman Kodak Co, 47-56, dir indust lab, 56-77; RETIRED. *Mem:* Am Chem Soc. *Res:* Absorption and emission spectroscopy; instrumental methods of analysis; analytical chemical research; corrosion and surface treatment of metals; oceanographic chemistry; absorption phenomena. *Mailing Add:* 719 Marine Dr Sequim WA 98382

ANDERSON, DONALD K(EITH), b Iron Mountain, Mich, July 15, 31; m 57; c 2. CHEMICAL ENGINEERING. *Educ:* Univ Ill, BS, 56; Univ Wash, MS, 58, PhD(chem eng), 60. *Prof Exp:* From asst prof to assoc prof chem eng, 60-69, PROF CHEM ENG & PHYSIOL, MICH STATE UNIV, 69-, CHMN CHEM ENG DEPT, 78- *Mem:* Am Inst Chem Engrs; Am Chem Soc; Am Soc Eng Educ; Sigma Xi. *Res:* Mass transfer in chemical processes and in biological systems. *Mailing Add:* Dept Chem Eng A202 Eng Bldg East Lansing MI 48824-1226

ANDERSON, DONALD LINDSAY, b Cambridge, Mass, July 2, 25; m 48; c 3. ANIMAL NUTRITION. *Educ:* Univ Mass, BS, 50; Univ Conn, MS, 52; Cornell Univ, PhD, 55. *Prof Exp:* Asst nutrit, Univ Conn, 50-52; asst, Cornell Univ, 52-54, res assoc, 54-55; from asst prof to prof nutrit, Univ Mass, 55-87; RETIRED. *Mem:* AAAS; Am Inst Nutrit; Poultry Sci Asn; Nutrit Today Soc; Sigma Xi. *Res:* Poultry nutrition; physiology; animal biochemistry; relationships between environmental stress and nutritional physiology. *Mailing Add:* Laurel Lane PO Box 175 Elkins NH 03233

ANDERSON, DONALD MORGAN, b Washington, DC, Dec 27, 30. SYSTEMATIC ENTOMOLOGY. *Educ:* Miami Univ, BA, 53; Cornell Univ, PhD(entom), 58. *Prof Exp:* Res asst entom, Ohio Agr Exp Sta, 53; asst prof biol, State Univ NY Col Buffalo, 59-60; RES ENTOMOLOGIST, AGR RES SERV, USDA, 60- *Concurrent Pos:* Res assoc, Buffalo Mus Sci, 72-, Dept Entom, Smithsonian Inst, 78- *Mem:* Coleopterists Soc; Entom Soc Am; Am Inst Biol Sci; Soc Syst Zool; Sigma Xi. *Res:* Taxonomy of immature Coleoptera in general, and of adult and immature Curculionidae. *Mailing Add:* Syst Entom Lab USDA c/o Nat Mus Natural Hist Washington DC 20560

ANDERSON, DONALD OLIVER, b Vancouver, BC, May 22, 30; m 56; c 4. EPIDEMIOLOGY, MEDICAL CARE ADMINISTRATION. *Educ:* Univ BC, BA, 50, MD, 54; FRCP(C), 59; Harvard Univ, SM, 61. *Prof Exp:* Asst prof prev med & med, 61-64, assoc prof, 64-68, head dept, 68-70, PROF

HEALTH CARE & EPIDEMIOL, FAC MED, UNIV BC, 68-, DIR DIV HEALTH SERV RES & DEVELOP, HEALTH SCI CTR, 72- *Concurrent Pos:* Mead Johnson residency scholar med, 57-58; McLaughlin traveling fel, 60-61; Can Coun fel, 64; mem task force on cost of health serv, Dept Nat Health & Welfare, Can, 69; mem health serv res study sect, Nat Ctr Health Serv Res & Develop, Dept Health, Educ & Welfare, 67-71; chmn specialty comt pub health, Royal Col Physicians & Surgeons Can, 70-; fac mem, London Sch Hyg & Trop Med, 70-71; chmn health serv demonstration comt, Nat Health Grant, Can, 71-74; BC rep, Fed-Provincial Health Manpower Comt, 72-; consult epidemiol & statist, BC Med Ctr, 73-; mem exec comt, Int Epidemiol Asn; consult & secy health manpower planning, Province of BC; mem, Expert Adv Panel on Air Pollution, WHO, 68-, temp adv, Div Strengthening Health Servs, Columbia, 71-, Geneva, 75; mem study of int migration of health workers, Geneva, 75; mem steering comt children's dent health res study for minister of health, BC, 74-76. *Mem:* Am Col Physicians; Am Pub Health Asn; Asn Teachers Prev Med; Can Med Asn. *Res:* Epidemiology of non-infectious disease, especially non-tuberculous respiratory diseases; health manpower studies; health sciences education; medical school selection practices; hospital utilization; health resources planning; international studies of medical care utilization; internal medicine. *Mailing Add:* 8739 Crest Burnaby BC V6B 3P7 Can

ANDERSON, DONALD T, b Lampman, Sask, Oct 2, 25; m 59; c 5. ECONOMIC GEOLOGY. *Educ:* Queen's Univ, Ont, BSc, 51; Univ Man, MSc, 59, PhD(geol), 63. *Prof Exp:* Field geologist, Falconbridge Nickel Mines, Ltd, 50-62; photogeologist, Geol Surv Can, 62-65; asst head dept, Univ Man, 68, assoc prof photogeol, 65-88; RETIRED. *Mem:* Fel Geol Asn Can; Can Inst Min & Metall. *Res:* Application of photo interpretation to solution of regional geological structure in areas of economic interest. *Mailing Add:* 190 Centennial St Winnipeg MB R3N 1P3 Can

ANDERSON, DONALD WERNER, b Atlantic, Iowa, Apr 16, 38; m 86; c 2. MATHEMATICS. *Educ:* Calif Inst Technol, BS, 60; Univ Calif, Berkeley, PhD(math), 64. *Prof Exp:* Nat Acad Sci-Nat Res Coun fel, Oxford Univ, 64-65; from asst prof to assoc prof math, Mass Inst Technol, 65-71; chmn dept, 73-77, PROF MATH, UNIV CALIF, SAN DIEGO, 71-, DIR ACAD COMPUT, 85- *Concurrent Pos:* Sloan fel, 67-68. *Mem:* Am Math Soc; Asn Comput Mach. *Res:* Computer graphics. *Mailing Add:* Dept Math Univ Calif San Diego La Jolla CA 92037

ANDERSON, DOUGLAS I, b Salt Lake City, Utah, Sept 22, 47; m 75; c 3. CIVIL ENGINEERING, MATERIALS SCIENCE ENGINEERING. *Educ:* Univ Utah, BS, 71, MS, 73. *Prof Exp:* Res engr, 71-78, res coordr, 78-83, RES & DEVELOP ENGR, UTAH DEPT TRANSP, 83- *Concurrent Pos:* Comt mem, Pavement Mgt Task Force, Utah Dept Transp, 82-, chmn, Res Adv Comt, 83-, adv, Motor Vehicle Carrier Adv Comt, 83-; Utah state rep, Transp Res Bd, 83-, Comt Pavement Design, 85- *Res:* Material science, pavement management, transportation economics, viscoelastic materials. *Mailing Add:* 5401 S 2700 W Salt Lake City UT 84119-5998

ANDERSON, DOUGLAS K, CENTRAL NERVOUS SYSTEM TRAUMA, PATHOPHYSIOLOGY. *Educ:* Mich State Univ, PhD(physiol), 72. *Prof Exp:* ASSOC PROF NEUROL & PHYSIOL, COL MED, UNIV CINCINNATI, 81-; RES PHYSIOLOGIST, VET ADMIN MED CTR, CINCINNATI, 78- *Res:* Spinal cord injuries. *Mailing Add:* 7594 Glenover Dr Cincinnati OH 45236

ANDERSON, DOUGLAS POOLE, b Portland, Ore, Jul 11, 39. IMMUNOLOGY, MARINE MICROBIOLOGY. *Educ:* Willamette Univ, BS, 61; Indiana Univ, MA, 63; Univ Maryland, PhD(immunol), 79. *Prof Exp:* Biologist, Ore State Game Comn, 64; IMMUNOLOGIST, US FISH & WILDLIFE SERV, 65- *Concurrent Pos:* Auxillary fac, Univ Wash, 68-72; guest scientist, Univ Miyazaki, Japan, 86; consult, Kuwait Inst Sci Res, 88. *Honors & Awards:* Fisheries Scientist, Am Fisheries Soc, 79; F Snieszko Distinguished Scientist Award, 89. *Mem:* Am Fisheries Soc, Fish Health Sect (pres 88); Am Soc Zool; Am Asn Immunologist; NY Acad Sci; fel Am Inst Fishery Res Biologist. *Res:* The immune response and protection against disease in salmonids and other fish; fish immunology; serodiagnostic identification of fish disease agents. *Mailing Add:* Nat Fish Health Res Lab Box 700 Kearneysville WV 25430

ANDERSON, DOUGLAS RICHARD, b Memphis, Tenn, Apr 7, 38; m 64; c 3. OPHTHALMOLOGY. *Educ:* Univ Miami, AB, 58; Wash Univ, MD, 62. *Prof Exp:* Rotating intern, Univ Hosps, Western Reserve Univ, 62-63; staff assoc, Nat Cancer Inst, 63-65; resident ophthal, Med Ctr, Univ Calif, San Francisco, 65-68; from asst prof to assoc prof, 69-82, PROF OPHTHAL, SCH MED, UNIV MIAMI, 82- *Concurrent Pos:* Res fel, Howe Lab Ophthal, Mass Eye & Ear Infirmary, Boston, 68-69; Nat Eye Inst res grant, 69-91; Res to Prevent Blindness, Inc prof, 69-74, William & Mary Greve int scholar, 78; assoc ed, Am J Ophthal, 73-90; glaucoma sect ed, Investigative Ophthal, 78-82; mem, Int Glaucoma Comt. *Mem:* Am Acad Ophthal; Int Perimetric Soc; Pan-Am Glaucoma Soc (secy-treas, 79-85); Pan-Am Ophthal Soc; Asn Res Vision & Ophthal (pres, 87); Am Ophthal Soc; Am Glaucoma Soc (pres, 90-92). *Res:* Electron microscopy of cartilage, viruses, mycoplasma, leukemia, eye and pathophysiology of optic nerve; eye diseases; especially papilledema, optic atrophy and glaucoma, using electron microscopy, physiologic and pharmacologic techniques. *Mailing Add:* Bascom Palmer Eye Inst Univ Miami Sch Med PO Box 016880 Miami FL 33101-6880

ANDERSON, DUWAYNE MARLO, b Lehi, Utah, Sept 9, 27; m 53; c 3. EARTH SCIENCES, SOIL PHYSICS. *Educ:* Brigham Young Univ, BS, 54; Purdue Univ, PhD(soil chem), 58. *Prof Exp:* From asst prof to prof soil physics, Univ Ariz, 58-63; geologist, US Army Cold Regions Res & Eng Lab, 63-67, chief earth sci br, 67-76; chief scientist, Div Polar Progs, NSF, 76-79; DEAN, FAC NATURAL SCI & MATH, STATE UNIV NY BUFFALO, 79- *Concurrent Pos:* Soil physicist, Ariz Agr Exp Sta, 61-63; vis prof, Dartmouth Col, 65, adj prof earth sci, 75-76; guest researcher, Royal Inst Technol, Sweden & vis prof, Univ Stockholm, 67; vis scientist, Soil Sci Soc Am, 68-71; mem molecular anal team, Viking Sci Team, 69-77; distinguished vis prof, Univ Wash, 70, vis prof, 71, Sloan Found distinguished vis lectr, 75; mem, US Deleg 2nd Int Conf on Permafrost, Yakutsk, USSR, 73; mem comt on permafrost, Nat Acad Sci, 73; NSF consult, 79-80. *Honors & Awards:* Dept of Army Commendation for Res Achievement, US Army Cold Regions Res & Eng Lab, 64, 65 & 68, Sci Achievement Award, 68. *Mem:* AAAS; fel Soil Sci Soc Am; Am Geophys Union; Int Clay Minerals Soc; Antarctican Soc. *Res:* Physical chemistry of soils; thermodynamics of soil processes; surface phenomena in soils; phase relationships in permafrost and frozen ground; physicochemical processes in cold environments; physics and chemistry of planetary surface materials. *Mailing Add:* Dept Geol Tex A&M Univ College Station TX 77843

ANDERSON, DWIGHT LYMAN, b Cokato, Minn, Oct 28, 35; m 60; c 3. MICROBIOLOGY. *Educ:* Univ Minn, BA, 57, MS, 59, PhD(microbiol), 61. *Prof Exp:* Asst bact, 57-60, from instr microbiol to asst prof microbiol & dent, 61-66, assoc prof, 66-74, PROF MICROBIOL & DENT, UNIV MINN, MINNEAPOLIS, 74- *Concurrent Pos:* USPHS career develop awardee, 66-73; ed, J Virol, 74-83. *Mem:* Am Soc Microbiol; Electron Micros Soc Am; Am Soc Virol; AAAS. *Res:* Viral morphogenesis. *Mailing Add:* Moos Tower Rm 18-256 Univ Minn Minneapolis MN 55455

ANDERSON, EDMUND GEORGE, b Seattle, Wash, Sept 1, 28; m 60; c 4. PHARMACOLOGY. *Educ:* Univ Wash, BS, 50, MS, 55, PhD(pharmacol), 57. *Prof Exp:* Asst pharmacol, Univ Wash, 53-54; instr, Seton Hall Col Med, 57-60; from asst prof to assoc prof, 60-70, PROF PHARMACOL, UNIV ILL COL MED, 70-, HEAD DEPT, 76- *Mem:* AAAS; Am Soc Pharmacol & Exp Therapeut; Am Soc Neurosci. *Res:* Neuropharmacology; central synaptic transmission and biogenic amines. *Mailing Add:* Dept Pharmacol Col Med Univ Ill PO Box 6998 Chicago IL 60680

ANDERSON, EDMUND HUGHES, b Camden, Ark, Jan 8, 24; m 75; c 6. FERMATS LAST THEOREM. *Educ:* Tex A&M Univ, BS, 48; La State Univ, MS, 64, PhD(math), 67. *Prof Exp:* Geophysicist, Shell Oil Co, 48-62; assoc prof math, Miss State Univ, 68-81; sr geophysicist, El Paso Explor Co, 81-85; CONSULT, 85- *Res:* Mathematics; topology. *Mailing Add:* 2075 S Wright St Lakewood CO 80228

ANDERSON, EDWARD EVERETT, b Peabody, Mass, Dec 20, 19; m 48; c 5. FOOD SCIENCE, NUTRITION. *Educ:* Mass State Col, BS, 41, MS, 42; Univ Mass, PhD, 49. *Prof Exp:* Fed food inspector, War Food Admin, USDA, 42-44; from asst res prof to assoc res prof food technol, Univ Mass, 48-56; food technologist, A D Little, Inc, 56-63; chief plant prod br, Food Div, US Army Natick Labs, 63-69, spec asst, Dept of Defense Food Prog, US Army Natick Res & Develop Command, 69-80; DIR, ANDERSON ASSOC, 80- *Mem:* Fel AAAS; fel Inst Food Technol; fel Pub Health Asn; Soc Nutrit Educ; Sigma Xi. *Res:* Administration and evaluation of food research and development programs; design and development of new food and ration components and processing methods; sensory and objective methods of quality evaluation and improvement of existing canned or frozen foods; technological factors associated with flavor and packaging problems in foods. *Mailing Add:* Anderson Assocs 47 Winthrop Rd Lexington MA 02173

ANDERSON, EDWARD EVERETT, b Algona, Iowa, Jan 9, 41; m; c 3. ENERGY. *Educ:* Iowa State Univ, BS, 64, MS, 66; Purdue Univ, PhD(mech eng), 72. *Prof Exp:* Asst prof mech eng, Univ Southwestern La, 72-74 & SDak Sch Mines & Technol, 74-76; from assoc prof to prof mech eng, Univ Nebr, 76-86; PROF MECH ENG, TEX TECH UNIV, 86- *Concurrent Pos:* Asst dean, Univ Nebr, 80-86; chmn, Tex Tech Univ, 86- & K-6 Tech Comt, Am Soc Mech Engrs. *Mem:* Am Soc Mech Engrs; Am Inst Astronaut & Aeronaut; Am Soc Eng Educators. *Res:* Radioative heat transfer; energy conversion; solar energy. *Mailing Add:* Dept Mech Eng Tex Tech Univ Lubbock TX 79409

ANDERSON, EDWARD FREDERICK, b Covina, Calif, June 17, 32; m 56; c 7. BOTANY, PLANT TAXONOMY. *Educ:* Pomona Col, BA, 54; Claremont Grad Sch, MA, 59, PhD(bot), 61. *Prof Exp:* Instr bot, Pomona Col, 61-62; from asst prof to assoc prof, 62-76, PROF BIOL, WHITMAN COL, 76- *Concurrent Pos:* Fulbright lectr, Ecuador, 65-67 & Malaysia, 69-70. *Mem:* Soc Econ Bot; Int Orgn Succulent Plant Study. *Res:* Taxonomy of Cactaceae; ethnobotany of hill tribes of northern Thailand. *Mailing Add:* Dept Biol Whitman Col Walla Walla WA 99362

ANDERSON, EDWARD P(ARLEY), b Ogden, Utah, Jan 29, 22; m 45; c 8. ELECTRICAL ENGINEERING. *Educ:* Univ Utah, BS, 48, MS, 51; Stanford Univ, EE, 58. *Prof Exp:* Jr design engr, Westinghouse Elec Corp, 48-49; instr elec eng, Univ Utah, 49-52; asst prof, Brigham Young Univ, 52-53; from asst prof to assoc prof, 55-61, chmn dept, 69-77, PROF ELEC ENG, SAN JOSE STATE UNIV, 61- *Concurrent Pos:* Asst, Stanford Electronics Lab, 53-57; assoc res scientist, Lockheed Missiles & Space Div, 57-59, consult, 59-60 & 63-67; consult, Lawrence Radiation Lab, 62-63, Varian Assocs, 69-71, Hewlett-Packard, 77-78, Fairchild Test Systs, 79-80 & Anderson-Fox, 83-85. *Mem:* Am Soc Eng Educ; sr mem Inst Elec & Electronics Engrs. *Res:* Radio propagation in the ionosphere; radio reflectivity; sea backscatter; solid state electronics; digital logic; computers; automatic testing. *Mailing Add:* Dept Elec Eng San Jose State Univ San Jose CA 95192

ANDERSON, EDWIN J, b Concord, Mass, Mar 16, 39; m 79; c 3. STRATIGRAPHY, PALEOECOLOGY & SEDIMENTOLOGY. *Educ:* Cornell Univ, AB, 61; Brown Univ, ScM, 64, PhD(geol), 67. *Prof Exp:* Instr geol, Univ NH, 65-66; asst prof, 67-75, assoc prof, 75-84, PROF GEOL, TEMPLE UNIV, 84- *Concurrent Pos:* Res Corp Grant, 67; Leverhulm fel, 74-75; NSF Grants, 81-83, 84-87, NSF US-UK Grant, 84-85. *Mem:* Int Asn Sedimentologists; Soc Econ Paleontologists & Mineralogists; Int Paleont Union. *Res:* Paleoecology and paleoenvironments of Paleozoic skeletal limestones for the purpose of providing a base for community analysis and evaluation of speciation within selected invertebrate groups; punctuated aggradational cycles; a technique for environmental and stratigraphic analysis of sedimentary sequences; coauthor of the hypothesis of punctuated aggradational cycles a general model of strategraphic accumulation; a new method for stratigraphic correlation and paleoenvironmental and paleogeographic analysis. *Mailing Add:* Dept Geol Temple Univ Philadelphia PA 19122

ANDERSON, EDWIN MYRON, b St Paul, Minn, Nov 14, 20; m 42; c 4. ELECTRICAL ENGINEERING. *Educ:* Univ Denver, BS, 48, MS, 49. *Hon Degrees:* DSc, NDak State Univ, 85. *Prof Exp:* From instr to prof elec eng, NDak State Univ, 49-82, actg chmn dept, 54-55 & 57-58, chmn, 58-78, actg chmn, Physics Dept, 62-63, emer prof elec eng, 82-86; RETIRED. *Concurrent Pos:* NSF fel, Purdue Univ, 59-60. *Mem:* Inst Elec & Electronics Engrs; Am Soc Eng Educ. *Res:* Advanced mathematical concepts in electrical engineering. *Mailing Add:* Dept Elec Eng NDak State Univ Fargo ND 58102

ANDERSON, ELMER EBERT, b Ottawa, Ill, June 28, 22; m 43, 76; c 7. MAGNETISM, OPTICS. *Educ:* Occidental Col, AB, 50; Univ Ill, MS, 56; Univ Md, PhD(physics), 64. *Prof Exp:* Chief electromagnets div, US Naval Ordnance Lab, 57-65; prof & chmn dept physics, dean arts & sci, Clarkson Univ, 65-77; dean arts & sci & prof physics, Purdue Univ, 77-79; prof physics & vpres acad affairs, 79-85, UNIV PROF, UNIV ALA, HUNTSVILLE, 85- *Concurrent Pos:* Sr Fulbright-Hays fel, 72-73; provost, Univ Ala, Huntsville, 89-90. *Mem:* Fel Am Phys Soc; Sigma Xi. *Res:* Solid state physics; area of magnetism; crystal growth; superconductivity. *Mailing Add:* Dept Physics Univ Alabama Huntsville AL 35899

ANDERSON, EMORY DEAN, b Kenmare, NDak, Sept 11, 39; m 61; c 3. FISH BIOLOGY. *Educ:* Dana Col, BS, 61; Univ Minn, PhD(fishery biol), 69. *Prof Exp:* Res assoc, Inst Water Res, Mich State Univ, 69-70; FISHERY RES BIOLOGIST, NAT MARINE FISHERIES SERV, NORTHEAST FISHERIES CTR, 70- *Concurrent Pos:* Mem, Standing Comt Res & Statist, Int Comn Northwest Atlantic Fisheries, 72-76, Mackerel Working Group, Int Coun Explor Sea, 74- & Sci & Statist Comt, Mid-Atlantic Fishery Mgt Coun, 77- *Honors & Awards:* Spec Achievement Award, US Dept Com, 72. *Mem:* Am Fisheries Soc; Am Inst Biol Sci; Am Inst Fishery Res Biologists. *Res:* Fish population dynamics; Northwest Atlantic fish stock assessments. *Mailing Add:* Int Coun Explor Sea Palaegade 2-4 Copenhagen K OK-1261 Denmark

ANDERSON, ERNEST R, b Mt Kisco, NY, Dec 25, 32. GEOLOGY. *Educ:* Wash Univ, PhD(geol), 61. *Prof Exp:* SUPVRY GEOLOGIST, US GEOL SURV, 61- *Mem:* Fel Geol Soc Am. *Res:* Geology. *Mailing Add:* US Geol Survey Mail Stop 966 Box 25046 Fed Ctr Denver CO 80225

ANDERSON, ETHEL IRENE, b Brooklyn, NY, June 20, 24. BIOCHEMISTRY. *Educ:* Ursinus Col, BS, 45; Univ Del, MS, 47, PhD(bio-org chem), 50. *Prof Exp:* Res fel biochem, Inst Cancer Res, Philadelphia, 50-52, res assoc, 52-58; sr res biochemist, Colgate Palmolive Co, 58-62; RES ASSOC, DEPT OPHTHAL, COLUMBIA UNIV, 62- *Concurrent Pos:* NIH res grant, Nat Eye Inst, 71-76. *Mem:* Am Soc Biol Chemists; AAAS; Am Chem Soc; Asn Res Vision & Ophthal; Sigma Xi. *Res:* Effects of the limiting cell layers on corneal function; the metabolic basis of transendothelial fluid transport and the nature and influence of epithelial proteolytic enzymes in pathophysiologies. *Mailing Add:* 16 Harrison Ave Red Bank NJ 07701

ANDERSON, FLETCHER N, b Kansas City, Mo, Nov 5, 30; m 46; c 3. CHEMICAL ENGINEERING. *Educ:* Univ Mo, BS, 51; Wash Univ, MS, 56. *Prof Exp:* Proj engr, Uranium Div, Mallinckrodt Inc, 51-53, sect supvr, 53-58, proj eng supvr, 58-60, maintenance supt, 60-62, maintenance mgr, 62-63, mgr eng & maintenance, 63-64, opers, 64-66, opers, 64-66, plant serv, 66-68, paper & plastics, 68-70, gen mgr, Calsicat Div, 70-72, Catalysts & Spec Projs, 72-73, vpres & gen mgr, Drug & Cosmetic Chem Div, 73-75, group vpres, food, drug & chem group, 75-76, chem group, 76-78, sr vpres chem group, 78-81, dir, 79-81; pres & dir, Chomerics Inc, 82-85; pres, chief exec officer & dir, Chem Tech Industs Inc, 86-89 & Brulin Corp, 90; CONSULT, 91- *Mem:* Am Inst Chem Engrs. *Res:* Uranium refining and specialty chemical business management. *Mailing Add:* 752 Fairfield Lake Dr Chesterfield MO 63017

ANDERSON, FLOYD EDMOND, b Racine, Wis, Oct 16, 15; m 47; c 1. ORGANIC CHEMISTRY. *Educ:* Univ Wis, BS, 39; Univ Mich, MS, 46, PhD(pharmaceut chem), 49. *Prof Exp:* Analyst, Ditzler Color Co, Mich, 39-40; analyst, Lakeside Labs, Inc, Wis, 40-41, org res chemist, 41-44; res engr, Off Sci Res & Develop, Battelle Mem Inst, 44-45; asst prof chem, WVa Univ, 48-49; sr org res chemist, Nepera Chem Co, 49-50, dir org chem res, 51-57; sr res assoc, Warner Lambert Res Inst, 57-60; head org res & develop, Pharmaceut Div, Wallace & Tiernan, Inc, 60-63; assoc prof med chem, Northeastern Univ, 64-66; chemist, Bur Med, Div New Drugs, Food & Drug Admin, Washington, DC, 66-70, drug control specialist, Bur Narcotics & Dangerous Drugs, 70-73; chief chem-biol coordr, Spec Progs Div, Drug Enforcement Admin, Dept Justice, Washington, DC, 73-78; MEM STAFF, ANDERSON & ASSOCS, CONSULT CHEMISTS, 78- *Concurrent Pos:* Prof chem & chmn dept, Luther Rice Col, 67-77. *Mem:* Am Chem Soc. *Res:* Development of analytical methods; synthesis of plastics and polymers; structure activity relationships in biological activities; antituberculous agents; local anesthetics; insecticides; hypnotics; diuretics; antimetabolites; narcotics; enzyme inhibition; synthesis of organic heterocyclics. *Mailing Add:* 1205 N Warren Ave Tucson AZ 85719

ANDERSON, FRANCES JEAN, b Watertown, SDak, Apr 23, 37; m 59. PHYSICS. *Educ:* SDak State Univ, BS, 59; Univ Minn, MS, 63, PhD(physics), 68. *Prof Exp:* Res assoc, Univ Minn, 68-77; INSTR PHYSICS & CHEM, ABBOTT-NORTHWESTERN HOSP, MINN, 74-; SR SCIENTIST, LIGHTNING & TRANSIENTS RES INST, ST PAUL, MINN, 85- *Mem:* Am Geophys Union; Am Asn Physics Teachers. *Res:* Anesthesia into quantitative physical terms; cloud physics; atmospheric-electricity; initiation of lightning discharges. *Mailing Add:* 1830 W Larpenteur Ave Apt 201 Falcon Heights MN 55113

ANDERSON, FRANCIS DAVID, b Cabano, Que, May 12, 25; m 48; c 3. GEOLOGY. *Educ:* Univ NB, BSc, 48; McGill Univ, MSc, 51, PhD(geol), 56. *Prof Exp:* Geologist, Geol Surv Can, 52-86, asst dir, Econ Geol Div, 72-82; CONSULT, 86- *Concurrent Pos:* Coordr, Fed-Prov Mineral Agreement, New Brunswick, 82-86; prog coordr, Fed Geosci Asbestos Proj, Que, 82-86. *Mem:* fel Geol Soc Am; fel Geol Asn Can. *Res:* General problems of field geology, particularly structural and economic geology. *Mailing Add:* Rural Rte 2 Clayton ON K0A 1P0 Can

ANDERSON, FRANK A(BEL), b Bridgeport, Conn, June 22, 14; m 42; c 2. CHEMICAL ENGINEERING. *Educ:* Univ Southern Calif, BS, 36; Univ Maine, MS, 40; La State Univ, PhD(chem eng), 47. *Prof Exp:* Jr technologist, Shell Oil Co, Calif, 36-38; asst prof chem, Univ Miss, 40-45; res assoc, Gaylord Container Corp, La, 45-47; from actg chmn to chmn dept chem and dept chem eng & prof, 47-55, chmn dept chem eng, 55-63 & 65-78, prof chem eng, 55-79, assoc dean, Sch Eng, 63-79, EMER ASSOC DEAN, UNIV MISS, 79- *Concurrent Pos:* Res assoc, La State Univ, 45-47. *Mem:* Am Chem Soc; Am Soc Eng Educ; Am Inst Chem Engrs. *Res:* Tall oil; fluid mechanics; heat transfer; solar energy. *Mailing Add:* 410 S 11th St Oxford MS 38655

ANDERSON, FRANK DAVID, b Duncan, Okla, May 24, 27; m 54; c 3. ANATOMY. *Educ:* Westminster Col, Mo, BA, 48; Cornell Univ, MS, 50, PhD, 52. *Prof Exp:* Instr anat, Med Col, Cornell Univ, 52-56; from asst prof to assoc prof, Seton Hall Col Med & Dent, 56-67; chmn dept, 67-83, prof, 67-90, EMER PROF ANAT, MED COL WIS, 90- *Concurrent Pos:* Consult, Med Ctr, Univ Ala, 59 & Inst Crippled & Disabled, Med Ctr, NY Univ, 63-64. *Mem:* Am Acad Neurol; Am Asn Anat; Soc Neurosci. *Res:* Neuroanatomy; anatomy of ductus arteriosus in newborn; central pathways of vagus nerve and of trigeminal nerve; experimental degeneration studies of spinal cord and brain stem; segmental anatomy of the spinal cord; ascending fiber systems in the spinal cord and brain stem. *Mailing Add:* Dept Anat Med Col Wis PO Box 26509 Milwaukee WI 53226

ANDERSON, FRANK WALLACE, b Milwaukee, Wis, Feb 12, 21; m 49; c 2. PHYSICAL CHEMISTRY, ELECTRON MICROSCOPY. *Educ:* Univ Ariz, BS, 51, MS, 52. *Prof Exp:* Res chemist, Martinez Res Lab, Shell Oil Co, 52-61, Shell Develop Co, Calif, 61-62; sr res chemist, Sprague Elec Co, Mass, 62-68; SR CHEMIST, IBM CORP, E FISHKILL, 68- *Mem:* Am Chem Soc; Sigma Xi. *Res:* Secondary ion mass spectrometry; ion scattering spectrometry; gas chromatography-mass spectrometry; transmission and scanning electron microscopy; analytical methods development; organic separations; electrophoresis in non-aqueous solvents, colloids and surfaces; ion cyclotron resonance mass spectrometry; gas chromatography fourier transform infrared spectroscopy. *Mailing Add:* PO Box 133 Hopewell Junction NY 12533

ANDERSON, FRANK WYLIE, b Omaha, Nebr, Feb 5, 28. MATHEMATICS. *Educ:* Univ Iowa, BA, 51, MS, 52, PhD, 54. *Prof Exp:* From instr to asst prof math, Univ Nebr, 54-57; from asst prof to assoc prof, 57-69, PROF MATH, UNIV ORE, 69- *Mem:* Am Math Soc; Math Asn Am. *Res:* Algebra; topological algebra. *Mailing Add:* Univ Ore Eugene OR 97403

ANDERSON, FRANZ ELMER, b Cleveland, Ohio, July 23, 38; m 63; c 3. OCEANOGRAPHY, ESTUARINE SEDIMENTATION. *Educ:* Ohio Wesleyan Univ, BA, 60; Northwestern Univ, MS, 62; Univ Wash, PhD(oceanog), 67. *Prof Exp:* Res asst oceanog, Univ Wash, 62-67, lectr, 66; interim dir, Jackson Estuarine Lab, 79-80, dir, 87-89; dir, Romberg-Tiburon Ctrs, 89-90; assoc prof oceanog, 67-83, actg dir, Seagrant, 79, PROF OCEANOG, UNIV NH, 83- *Concurrent Pos:* NSF res grants, 66-68, 78-79 & 84-85; Cottrell res grant, 70-72; Off Naval Res grant, 72-74; Fulbright grant, Turkey, 73-74; Nat Ocean & Atmospheric Admin grant, 79-82, 84-85 & 86-89; fel, Nat Environ Res, Stirling, Scotland, 80-81. *Mem:* Fel Geol Soc Am; Soc Econ Paleontologists & Mineralogists; Estuarine Res Fedn; Int Asn Sedimentologist; Marine Educ Asn. *Res:* Distribution of particulate matter in estuarine and marine environments; effects of small amplitude waves on estuarine erosion; effects of Mya arenaria and Arenicola marina digging on estuarine sediment transportation; effects of rainfall on estuarine sediment transport; measurement of estuarine currents-a low cost solution. *Mailing Add:* Dept Earth Sci Univ of NH Durham NH 03824

ANDERSON, G HARVEY, b Provost, Alta, 1941; m 64. NUTRITION, NUTRIENT REQUIREMENTS. *Educ:* Univ Ill, PhD(nutrit sci), 69. *Prof Exp:* PROF NUTRIT SCI & CHMN DEPT, UNIV TORONTO, 81-, PROF, PHYSIOL DEPT & INST MED SCI, 82- *Concurrent Pos:* Chmn bd, Nat Inst Nutrit, 84-89; vis prof, Dept Clin Nutrit, Sun Yat Sen Univ Med Sci, Guanzhou, People's Repub China. *Honors & Awards:* Borden Award, Can Soc Nutrit Sci, McHenry Award. *Mem:* Soc Neurosci; Am Inst Nutrit; Am Soc Clin Nutrit; Can Soc Nutrit Sci. *Res:* Amino acid nutrition; diet and brain function; infant nutrition; control of feeding behavior. *Mailing Add:* Dept Nutrit Sci Fac Med Univ Toronto Fitzgerald Bldg Toronto ON M5S 1A8 Can

ANDERSON, G(EORGE) M(ULLEN), b Rochester, Pa, May 7, 21; m 47; c 3. ELECTRICAL ENGINEERING. *Educ:* Carnegie Inst Technol, BS, 42, DSc(elec eng), 48. *Prof Exp:* Asst prof elec eng, Carnegie Inst Technol, 48-51; sr scientist, Atomic Power Div, Westinghouse Elec Corp, 51-53; head, Eng Dept, Res Lab, McGraw-Edison Co, 53-60, pres, 60-64; Dept Head Systs Anal, Bellcomm, Inc, 64-76; mem staff, Bell Tel Labs, Holmdel, NJ, 76-; DEPT HEAD, AT&T INFO SYSTS. *Mem:* Sr mem Inst Elec & Electronics Engrs. *Res:* Circuit theory; control systems; electricity and magnetism. *Mailing Add:* 17 Oakes Rd Rumson NJ 07760

ANDERSON, GARY, b Norristown, Pa, Oct 5, 48; m 70; c 2. PHYSIOLOGICAL ECOLOGY, AQUACULTURE. *Educ:* Univ RI, BS, 70; Univ SC, MS, 72, PhD(marine sci), 74. *Prof Exp:* Asst prof zool, Univ Maine, Orono, 74-75; from asst prof to assoc prof, 75-84, PROF BIOL SCI, UNIV SOUTHERN MISS, 84- *Concurrent Pos:* US Army Engr, Waterways Exp Sta. *Mem:* Am Inst Biol Sci; Crustacean Soc; Am Soc Zoologists; Sigma Xi; World Aquaculture Soc. *Res:* Ecology of host-symbiont interrelationships, host finding behavior, energy flow in host-parasite systems, bopyrid isopods and anostracans shrimp aquaculture. *Mailing Add:* Dept Biol Sci Univ Southern Miss Box 5018 Southern Sta Hattiesburg MS 39406-5018

ANDERSON, GARY BRUCE, b Kent City, Mich, Apr 21, 47; m 77; c 1. ANIMAL SCIENCE. *Educ:* Mich State Univ, BS, 69; Cornell Univ, PhD(vert physiol), 73. *Prof Exp:* Asst prof, 74-78, ASSOC PROF ANIMAL SCI & PHYSIOL, UNIV CALIF, DAVIS, 78- *Mem:* Soc Study Reprod; Am Soc Animal Sci; Int Embryo Transfer Soc. *Res:* Increase of reproductive performance and efficiency in farm animals; development, culture and transfer of preimplantatis-mammalian embryos. *Mailing Add:* Dept Animal Sci Univ Calif Davis CA 95616

ANDERSON, GARY DON, b Purcell, Okla, Sept 23, 43. ORGANIC CHEMISTRY, X-RAY CRYSTALLOGRAPHY. *Educ:* Univ Okla, BS, 64, MS, 65; Fla State Univ, PhD(org chem), 72. *Prof Exp:* Res assoc org chem, Fla State Univ, 72-73; res assoc, Stanford Univ, 73-74, NIH fel, 74-75; asst prof org chem, Univ Mo-Kansas City, 75-; AT DEPT CHEM, MARSHALL UNIV. *Mem:* Am Chem Soc; Sigma Xi. *Res:* Chemistry of plant natural products; terpene synthesis; x-ray crystal structures. *Mailing Add:* Dept Chem Marshall Univ Huntington WV 25701

ANDERSON, GARY LEE, MICRO-CIRCULATION, HYPERTENSION. *Educ:* Univ Ariz, PhD(physiol), 80. *Prof Exp:* Instr, 82-86, ASST PROF PHYSIOL & BIOPHYS, UNIV LOUISVILLE, 86- *Mailing Add:* Dept Physiol Univ Louisville Health Sci Ctr Louisville KY 40292

ANDERSON, GEORGE ALBERT, b New Britain, Conn, May 3, 37. MATHEMATICS. *Educ:* Trinity Col, BS, 59; Yale Univ, MA, 61, PhD(math), 64. *Prof Exp:* Actg instr, Yale Univ, 63-64; from instr to asst prof math, Trinity Col, Conn, 64-71; assoc prof, 71-80, PROF MATH, RI COL, 80- *Concurrent Pos:* Consult res div, United Aircraft Corp, 61-64. *Mem:* Math Asn Am; Inst Math Statist. *Res:* Mathematics of sound theory in an inhomogeneous media; mathematical statistics; multivariate analysis; asymptotic distributions for latent roots of Wishart matrices. *Mailing Add:* Dept Math RI Col 600 Mt Pleasant Ave Providence RI 02908

ANDERSON, GEORGE BOINE, marine engineering, military systems, for more information see previous edition

ANDERSON, GEORGE CAMERON, b Vancouver, BC, Oct 8, 26; m 57. BIOLOGY. *Educ:* Univ BC, BA, 47, MA, 49; Univ Wash, PhD(zool, bot), 54. *Prof Exp:* Teaching asst zool, Univ BC, 47-49; teaching fel, res asst & actg instr, Univ Wash, 50-54, res assoc, 54-58, from res asst prof to res prof, 58-70, actg asst chmn res, 70-71; marine biologist, AEC, 71-72; res prof zool, 72-77, assoc chmn res, 77-81, prof, 77-83, dir, Sch Oceanog, 81-83, EMER PROF OCEANOG, UNIV WASH, 83- *Concurrent Pos:* Consult, US Bur Reclamation, 55, Seattle, Wash, 59 & State Fisheries Dept, Wash, 63-83; chief scientist, Deep Ocean Mining Environ Study, Nat Oceanic & Atmospheric Admin, 75-78; mem ship opers panel, NSF, 70-72, ocean sci panel, 73; mem & admin judge atomic safety & licensing bd panel, US Nuclear Regulatory Comn, 73-; mem, Univ Nat Oceanog Lab Syst Adv Coun, 78-81, chmn, 79-81. *Mem:* Am Soc Limnol & Oceanog. *Res:* Biological oceanography and limnology; phytoplankton ecology; marine primary productivity. *Mailing Add:* Sch Oceanog Univ Wash Seattle WA 98195

ANDERSON, GEORGE ROBERT, b Burlington, Iowa, Jan 10, 34; m 58; c 2. PHYSICAL CHEMISTRY, MICROWAVE HEATING TECHNOLOGY. *Educ:* Augustana Col, Ill, BA, 56; Univ Iowa, PhD(phys chem), 61. *Prof Exp:* Res chemist, E I du Pont de Nemours & Co, until 68; collabr, Univ Groningen, Neth, 68-69; res fel chem, Wesleyan Univ, 69-70; instr, Bowdoin Col, 70-71, asst prof, 71-76; SR RES SCIENTIST, PILLSBURY RES & DEVELOP LAB, 76- *Concurrent Pos:* Vis lectr, Swarthmore Col, 65-66; vis prof, Univ Minn, 74-75. *Mem:* Am Chem Soc; Int Microwave Power Inst. *Res:* Molecular structure and spectroscopy; electronic, infrared and Raman spectroscopy; hydrogen bonding, studies of vibrational states; physical chemistry of food products; microwave heating technology; microwave dielectrics. *Mailing Add:* 1100 W 53 St Minneapolis MN 55419

ANDERSON, GEORGE WILLIAM, b Chicago, Ill, Jan 17, 24; m 47; c 1. ELECTRICAL ENGINEERING. *Educ:* Northwestern Univ, BSEE, 48. *Prof Exp:* Sect chief, Aerial Measurements Lab, Northwestern Univ, 47-55; proj mgr, J B Rea Co, 55-56; dept mgr, Aeronutronic Div, Ford Motor Co, 56-62; mgr electronic systs, Mobile Mid-Range Ballistic Missile, 62, dir, Guid & Command Systs, 62-64, dir, Guid & Command Systs, Minuteman, 64-65, assoc group dir, Advan Weapon Systs, 65-66, group dir, WS 120A Integration & Airborne Systs, 66-68, assoc gen mgr, Missile Systs Div, 68-70 & Defense Systs Div, 70-71, assoc gen mgr, Advan Systs Div, 71-72 & Develop Planning Div, 72-76, prin dir develop, Advan Prog Div, 76-79, PRIN DIR, LONGBOW OFF, AEROSPACE CORP, LOS ANGELES, 79- *Mem:* Inst Elec & Electronics Engrs; Sigma Xi. *Res:* Engineering management; automatic control systems. *Mailing Add:* 18792 Silver Maple Way Santa Ana CA 92705

ANDERSON, GERALD CLIFTON, b Barre, Vt, Dec 13, 20; m 47; c 4. ANIMAL NUTRITION. *Educ:* Univ Mass, BS, 43; Univ Mo, MS, 48, PhD(animal nutrit), 50. *Prof Exp:* Asst instr, Univ Mo, 47-50; assoc prof, 50-58, PROF ANIMAL SCI, WVA UNIV, 58-, ANIMAL HUSBANDMAN, AGR EXP STA, 58- *Concurrent Pos:* Assoc animal husbandman, Agr Exp Sta, WVa Univ, 50-58, head dept animal husb, 58-63. *Mem:* Sigma Xi. *Res:* Nutritional problems pertinent to swine, poultry and beef cattle. *Mailing Add:* Dept Animal Sci WVa Univ Evansdale Campus Rte 1 Box 185 Morgantown WV 26506

ANDERSON, GERALD M, b Shanghai, China, Oct 14, 35; US citizen; m 59; c 2. SYSTEMS THEORY, OPERATIONS RESEARCH. *Educ:* US Naval Acad, BS, 57; Mass Inst Technol, MS, 61; Univ Mich, PhD(aerospace eng), 66. *Prof Exp:* US Air Force, 57-78, planning study officer command & control, Electronic Systs Div, Air Force Systs Command, 61-64, from asst prof to prof mech, US Air Force Inst Technol, 66-77; sr prin engr, 77-78, prog mgr tracking, navig, guid & control, 78-87, TECH DIR, HAWAII OPERS, ORINCON CO, 87- *Mem:* Am Inst Aeronaut & Astronaut; Inst Elec & Electronics Engrs. *Res:* Optimal control; differential games; estimation theory; guidance and control; personnel systems and economics systems. *Mailing Add:* Orincon Corp 970 N Kalaheo Ave Suite C-215 Kailua HI 96734

ANDERSON, GERALD S, b Montevideo, Minn, July 28, 30; m 57; c 2. SOLID STATE PHYSICS. *Educ:* Luther Col, Iowa, BA, 52; Iowa State Univ, PhD(physics), 57. *Prof Exp:* Res assoc, Iowa State Univ, 57-58; prin scientist, Electronics Div, Gen Mills, Inc, 58-63; tech specialist, Appl Sci Div, Litton Industs, Minn, 63-69; RES MGR, MINN MINING & MFG CO, 69- *Mem:* Am Phys Soc; Am Vacuum Soc. *Res:* Thin film physics; sputtering. *Mailing Add:* 2453 Cohansey St St Paul MN 55113

ANDERSON, GLEN DOUGLAS, b Madison, SDak, Oct 18, 30; m 53; c 5. ANALYSIS & FUNCTIONAL ANALYSIS. *Educ:* Drury Col, AB, 58; Univ Mich, AM, 59, PhD(math), 65. *Prof Exp:* From asst prof to assoc prof math, Eastern Mich Univ, 60-65; asst prof, 65-70, assoc prof, 70-78, PROF MATH, MICH STATE UNIV, 78- *Concurrent Pos:* Fulbright grant, Univ Helsinki, 72-73; vis prof, Univ Mich, 79-80; vis prof, Stanford, 80, Univ Auckland, 86-87, ITM/MUCIA Coop Program Malaysia, Indiana Univ, 88-90. *Honors & Awards:* Fulbright grant, Univ Helsinki, 72-73. *Mem:* Am Math Soc; Math Asn Am; New Zealand Math Soc; Finnish Math Soc. *Res:* Complex variables; quasiconformal mappings. *Mailing Add:* Dept Math Mich State Univ East Lansing MI 48824-1027

ANDERSON, GLENN ARTHUR, b Mt Vernon, Wash, Nov 30, 24; m 49. ZOOLOGY. *Educ:* Wash State Univ, BS, 53, MA, 58; Ore State Univ, PhD(parasitol), 64. *Prof Exp:* Instr zool, Ore State Univ, 63-64; asst prof biol, Ariz State Col, 64-67; from assoc prof to prof, 67-87, EMER PROF BIOL, NORTHERN ARIZ UNIV, 81- *Mem:* Am Soc Parasitol; Am Soc Zool; Sigma Xi. *Res:* Life cycles of parasitic helminths. *Mailing Add:* Dept Biol Sci Northern Ariz Univ Flagstaff AZ 86011

ANDERSON, GLORIA LONG, b Altheimer, Ark, Nov 5, 38; m 60; c 1. ORGANIC CHEMISTRY. *Educ:* Agr, Mech & Norm Col, Ark, BS, 58; Atlanta Univ, MS, 61; Univ Chicago, PhD(org chem), 68. *Prof Exp:* Instr chem, SC State Col, 61-62 & Morehouse Col, 62-64; assoc prof, 68-73, Calloway prof chem, 73-84, CHMN NATURAL SCI DIV & DEAN ACAD AFFAIRS, MORRIS BROWN COL, 84- *Mem:* AAAS; Am Chem Soc; Nat Sci Teachers Asn; Nat Inst Sci. *Res:* Synthetic organic fluorine chemistry; fluorine-19 nuclear magnetic resonance spectroscopy; mechanism of transmission of substituent effects, synthetic organic chemistry. *Mailing Add:* 560 Lynn Valley Rd SW Atlanta GA 30311

ANDERSON, GORDON FREDERICK, b Grand Rapids, Mich, Oct 16, 34; m 57; c 2. AUTONOMIC NERVOUS SYSTEM, SMOOTH MUSCLE. *Educ:* Ferris State Col, BS, 56; Wayne State Univ, MS, 58, PhD(physiol & pharmacol), 61. *Prof Exp:* Instr physiol & pharmacol, 61-64, from asst prof to assoc prof, 64-75, PROF PHARMACOL, WAYNE STATE UNIV, 75- *Concurrent Pos:* Prin investr, NIH grants, 78- *Mem:* Pharmacol Soc Can; Am Physiol Soc. *Res:* Physiology and pharmacology of the urinary bladder and its autonomic control. *Mailing Add:* Dept Pharmacol Sch Med Wayne State Univ 540 E Canfield Detroit MI 48201

ANDERSON, GORDON WOOD, b Evanston, Ill, Mar 8, 36; m 69. SOLID STATE DEVICES, SIGNAL PROCESSING & PHOTO DETECTORS. *Educ:* Cornell Univ, BEE, 59; Univ Ill, Urbana-Champaign, MS, 61, PhD(physics), 69. *Prof Exp:* Res asst & teaching asst physics, Univ Ill, Urbana-Champaign, 59-69; instr physics, Tougaloo Col, Miss, 65; res assoc solid state physics, Nat Res Coun, 69-70; RES PHYSICIST ELECTRONICS ENG, NAVAL RES LAB, 71- *Concurrent Pos:* Ford Found fel, Univ Ill, Urbana-Champaign, 59-61; NSF grant, 63; consult, Planning Human Systs, Inc, 78-79; Naval Res Lab awards, 80, 83, 85, 86, 88 & 89. *Mem:* Inst Elec & Electronics Engrs; Am Phys Soc; AAAS; Found Sci & Handicapped; Soc Photo Optical Instrumentation Engrs; Sigma Xi. *Res:* Solid state optical detectors for optical signal processing and communications; acousto-optical and opto-electronic devices; metal-insulator-semiconductor device design-fabrication for high-speed, infrared radiation and interface problems; signal processing; metallic, semiconducting, insulating thin-films, materials; fabrication and optical, electrical, magnetic, and strength properties, thin films, materials, devices; technical research program management and supervision; contract management; safety management. *Mailing Add:* Naval Res Lab Code 6813 Washington DC 20375-5000

ANDERSON, GREGOR MUNRO, b Montreal, Que, Aug 18, 32; m 57; c 2. GEOCHEMISTRY. *Educ:* McGill Univ, BEng, 54; Univ Toronto, MASc, 56, PhD(geol), 61. *Prof Exp:* Geologist, Ventures, Ltd, 56-57; res assoc, Pa State Univ, 61-64, asst prof geol, 64-65; from asst prof to assoc prof, 65-72, PROF GEOL, UNIV TORONTO, 72- *Concurrent Pos:* Nuffield fel, 73; assoc ed, Can Mineralogist. *Mem:* Mineral Asn Can; Int Asn Biochem & Cosmochem (secy); Geochem Soc; Soc Econ Geologists; Geol Asn Can. *Res:* Chemistry of solutions and melts at high temperatures and pressures. *Mailing Add:* Dept Geol Univ Toronto Toronto ON M5S 1A1 Can

ANDERSON, GREGORY JOSEPH, b Chicago, Ill, Nov 26, 44; m 66; c 2. BIOSYSTEMATICS, POLLINATION BIOLOGY. *Educ:* St Cloud State Univ, BS, 66; Ind Univ, Bloomington, MA, 68, PhD(bot), 71. *Prof Exp:* Asst prof bot, Univ Nebr, Lincoln, 71-73; from asst to assoc prof, 73-82, PROF BIOL SCI, UNIV CONN, 82-, DEPT HEAD, 90- *Concurrent Pos:* Secy, Orgn Trop Studies, NSF, 83-86 & Bot Soc Am, 89-91. *Honors & Awards:* George R Cooley Award, Am Soc Plant Taxonomists, 81. *Mem:* Ecol Soc Am; Soc Econ Bot (secy, 81-84, vpres, 85-86, pres, 85-86); Soc Study Evolution; Bot Soc Am; Am Soc Plant Taxon. *Res:* Systematics and evolution of vascular plants; reproductive biology. *Mailing Add:* Ecol & Evolutionary Biol Univ Conn Storrs CT 06268

ANDERSON, HAMILTON HOLLAND, EXPERIMENTAL THERAPEUTICS. *Educ:* Univ Calif, Berkeley, MD, 30. *Prof Exp:* Dir pharmacol, Univ Calif, San Francisco, 43-59; RETIRED. *Mailing Add:* 8545 Carmel Valley Rd Carmel CA 93923

ANDERSON, HARLAN JOHN, b Pierre, SDak, Mar 15, 26; m 51; c 2. NUCLEAR MATERIALS. *Educ:* Univ SDak, Vermillion, AB, 51. *Prof Exp:* Tech Sgt commun, US Army Signal Corps, 44-46; nuclear chemist, Gen Elec, Hanford, 51-56, mgr, Anal Labs, 56-60, specialist educ training, 60-61, engr nuclear fuels, 61-65; consult nuclear fuels, Battelle Mem Inst, Richland, 65-70; MGR RES MAT, WESTINGHOUSE HANFORD CO, 70- *Concurrent Pos:* Partic, USA Int Symp Anal Chem Nuclear Fuels, Austria, 70-74; consult, Environ Protection Agency, 76-78. *Honors & Awards:* Harlan J Anderson Award, Am Soc Testing & Mat. *Mem:* Fel Am Soc Testing & Mat; Am Inst Chemists; Am Nat Standards Inst. *Res:* Ceramic nuclear fuels and technical specifications of nuclear materials; nuclear standards for national and international standards organization. *Mailing Add:* 2212 S Vancouver St Kinnewick WA 99337

ANDERSON, HARLAN U(RIE), b Cedar City, Utah, July 15, 35; m 57; c 2. CERAMICS ENGINEERING. *Educ:* Univ Utah, BS, 57; Univ Calif, Berkeley, PhD(eng sci), 62. *Prof Exp:* Engr ceramics, Lawrence Radiation Lab, Univ Calif, Berkeley, 61-62; sr chemist, Sprague Elec Co, 62-69; assoc prof ceramics, Ore Grad Ctr, 69-70; assoc prof, 70-76, PROF CERAMIC ENG, UNIV MO-ROLLA, 76- *Mem:* Fel Am Ceramic Soc. *Res:* Kinetics of gas-solid reactions; sintering of oxides; thermodynamics of high temperature reactions; high melting inorganic solids; ferroelectric oxides; vapor pressure determination; electronic materials. *Mailing Add:* 45 Johnson Rolla MO 65401

ANDERSON, HAROLD D, b Holdrege, Nebr, May 23, 40; m 62; c 2. INORGANIC CHEMISTRY, PHYSICS. *Educ:* Augustana Col, Ill, BA, 62; Univ Northern Iowa, MA, 66; Univ Iowa, PhD(sci ed), 70. *Prof Exp:* Sci instr chem & physics, Cent Col Iowa, 66-68; PROF CHEM & PHYSICS, STEPHENS COL, 70- *Mem:* Am Chem Soc; Am Asn Physics Teachers. *Res:* Neutron activation analysis applied to medically related problems; x-ray crystallography of clays. *Mailing Add:* Dept Sci Box 2072 Columbia MO 65215

ANDERSON, HAROLD J, b Green Bay, Wis, Sept 4, 28; m 50; c 6. PHYSICAL CHEMISTRY. *Educ:* St Norbert Col, BS, 52. *Prof Exp:* Scientist, Marathon Div, James River Corp, 52-58, sr scientist, 58-62, group leader, 62-66, supvr, 66-72, mgr prod develop & improv, 72-75, mgr maj proj consumer prod, 75-79, assoc dir, paperboard process develop, Am Can Co, 79-83, dir paperboard develop, 83-91; CONSULT, 91- *Mem:* Tech Asn Pulp & Paper Indust. *Res:* Decorative and functional coatings for paper and paperboard; air laid, nonwoven technology; paperboard process technology. *Mailing Add:* 1334 Westwood Ct Neenah WI 54956

ANDERSON, HARRISON CLARKE, b Louisville, Ky, Sept 2, 32; m 61; c 4. PATHOLOGY. *Educ:* Univ Louisville, BA, 54, MD, 58. *Prof Exp:* Intern path, Mass Gen Hosp, Boston, 58-59; Nat Cancer Inst path trainee, Univ Louisville, 59-60; fel Mem Hosp, New York, 60-61, chief resident, 61-62; res fel, Sloan-Kettering Inst, 62-63; from asst prof to prof path, State Univ NY Downtown Med Ctr, 63-78; prof path & oncol & chmn dept, 78-90, HARRINGTON PROF ORTHOP RES, UNIV KANS MED CTR, KANSAS CITY, 90- *Concurrent Pos:* Grants, Nat Inst Cancer Res, Nat Inst Arthritis, Metabolism & Digestive Dis, Nat Inst Dent Res, NIH, Air Force Off Sci Res & Kyocera Corp, 67-68; fel, Nat Inst Arthritis, Metab & Digestive Dis, NIH, 71-72; mem study sect, Nat Inst Dent Res, NIH, 78-82; jour ed, J Bone & Joint Surg, 81-, J Exp Pathol, 82-, Am J Pathol, 82- & J Orthop Res, 82- *Honors & Awards:* Elizabeth Winston Lanier Award, Kappa Delta, 82; Biol Mineralization Res Award, Int Asn Dent Res, 85. *Mem:* Am Soc Cell Biol; Am Asn Pathologists; Royal Soc Med; Am Soc Bone & Mineral Res; Int Acad Pathol; Orthop Res Soc. *Res:* Experimental bone induction; calcium metabolism; fine structure of cartilage and bone; mechanism of biological and pathological calcification; development and differentiation of skeletal tissues. *Mailing Add:* Dept Path Univ Kans Med Ctr Rainbow Blvd at 39th St Kansas City KS 66103

ANDERSON, HARVEY L(AWRENCE), b Wahoo, Nebr, Dec 26, 27; m 48; c 4. CHEMICAL ENGINEERING. *Educ:* Univ Nebr, BSc, 48. *Prof Exp:* Res engr, 48-52, group supvr, 52-57, head appl res, 57-62, tech dir, New Prod Div, 62-67, tech dir, Med Prod Div, 67-73, MGR, DENT PROD LAB, MINN MINING & MFG CO, 73- *Mem:* Am Inst Chem Engrs; Am Chem Soc. *Res:* Chemical engineering kinetics and application to process design; synthesis, process and product development in high polymers; dental materials. *Mailing Add:* 2071 Spruce Pl St Paul MN 55110

ANDERSON, HENRY WALTER, b Des Moines, Iowa, Dec 27, 11; m 40; c 3. HYDROLOGY, WATERSHED MANAGEMENT. *Educ:* Univ Calif, BS, 43, MS, 47. *Prof Exp:* Res forester, Univ Calif, 43-46; hydrologist, Soil Conserv, US Forest Serv, 46-51, soil scientist, 51-55, snow res leader, 55-62, water source hydrol proj leader, 63-71, chief res hydrologist, 71-76; RETIRED. *Concurrent Pos:* Mem nat comt, Int Asn Sci Hydrol, 72-76; consult, Calif State Bd Forest. *Honors & Awards:* Hydrol Award, Am Geophys Union, 46; US Forest Serv Award, 58. *Mem:* Fel Am Geophys Union; Soc Am Foresters; Int Water Resources Asn; Sigma Xi. *Res:* Forest effects on floods; sedimentation and water yield; snow accumulation; snow melt. *Mailing Add:* 3168 Tice Creek Dr Apt 3 Walnut Creek CA 94595

ANDERSON, HERBERT GODWIN, JR, b Roanoke, Ala, Dec 29, 31. BIOLOGICAL SCIENCES. *Educ:* Auburn Univ, BS, 58, MS, 60; Univ Miami, PhD(marine sci), 64. *Prof Exp:* Fishery biologist, Sandy Hook Marine Lab, US Bur Sport Fisheries, 62-64; assoc prof, 64-74, PROF BIOL SCI, CENT CONN STATE COL, 74- *Mem:* Am Soc Parasitol; Sigma Xi. *Res:* Animal parasites; invertebrate taxonomy. *Mailing Add:* Dept Biol Cent Conn State Col 1615 Stanley St New Britain CT 06050

ANDERSON, HERBERT HALE, b Dayton, Ohio, Nov 6, 13; m 71; c 1. INORGANIC CHEMISTRY. *Educ:* Harvard Univ, AB, 34; Mass Inst Technol, PhD(inorg chem), 37. *Prof Exp:* Res assoc, Harvard Univ, 38-42, Mass Inst Technol, 42-43, Manhattan Dist, Univ Chicago, 43-46 & Harvard Univ, 46-47, 50-51; from asst prof to prof 51-79, EMER PROF CHEM, DREXEL UNIV, 79- *Mem:* Am Chem Soc. *Res:* Organometallic derivatives of silicon, germanium, and tin; halides of silicon, phosphorus and plutonium. *Mailing Add:* PO Box 271 Austin TX 78767-0271

ANDERSON, HERBERT L, physics, elementary particles; deceased, see previous edition for last biography

ANDERSON, HERBERT RUDOLPH, JR, b Athol, Mass, May 24, 20; m 45; c 5. PHYSICAL CHEMISTRY. *Educ:* Univ NH, BS, 48; Cornell Univ, PhD(phys chem), 52. *Prof Exp:* Asst chem, Los Alamos Sci Lab, 44-46; asst, Cornell Univ, 48-52; res assoc, Knolls Atomic Power Lab, Gen Elec Co, 52-53; res chemist, Phillips Petrol Co, 54-61; adhesion sci res collabr, Lehigh Univ, 74-75; res chemist res div, IBM Corp, Yorktown Heights, NY, 61-65, mgr printing & polymer chem, Systs Develop Div, Boulder, Colo, 65-70, mgr nat sci dept, Gen Prod Div, San Jose, Calif, 70-74, mgr polymer sci & technol, Gen Technol Div, East Fishkill, NY, 75-88, SR ENGR, GEN TECHNOL DIV, IBM CORP, EAST FISHKILL, NY, 88- *Concurrent Pos:* Fel, Cornell Univ, 53-54; cong sci counr, Am Chem Soc, 73- *Honors & Awards:* Texaco Res Award, Am Chem Soc. *Mem:* Am Chem Soc; Sigma Xi; Adhesion Soc (vpres, pres); Soc Plastics Engrs. *Res:* Photochemistry; nuclear fuel elements; low-angle x-ray scattering; rubber and polymer chemistry; radiation chemistry; colloids; adhesion science. *Mailing Add:* RR No 4 Box 44 Patterson NY 12563

ANDERSON, HOWARD BENJAMIN, b Escanaba, Mich, Dec 13, 14; m 46; c 2. APPLIED MATHEMATICS. *Educ:* Northern Mich Univ, AB, 38; Univ Mich, MA, 39; Mich Technol Univ, BS, 54. *Prof Exp:* Teacher pub sch, Mich, 39-42 & 45-46; from assoc prof to prof, 46-81, EMER PROF MATH, MICH TECHNOL UNIV, 81- *Concurrent Pos:* Chmn calculus distinguished teacher, Mich Technol Univ. *Mem:* Math Asn Am; Am Astron Soc. *Res:* Mathematics education; astronomy. *Mailing Add:* 1034 Second St Hancock MI 49930

ANDERSON, HOWARD W(AYNE), b Cincinnati, Ohio, Feb 18, 34; m 58; c 2. CHEMICAL ENGINEERING. *Educ:* Univ Del, BChE, 56; Univ Calif, Berkeley, MS, 58. *Prof Exp:* Chem engr, Del, 58-67, res mgr, 80-87, SR ENGR, PETROCHEM DEPT, E I DU PONT DE NEMOURS & CO, 67-, DIV SUPT, 75-, PRIN CONSULT PROCESS ENG, 87- *Mem:* Am Inst Chem Engrs; Sigma Xi. *Res:* Research and development of processes for manufacture and/or use of olefin high polymers and nylon intermediates. *Mailing Add:* 809 Grande Lane Hockessin DE 19707

ANDERSON, HUGH JOHN, b Winnepeg, Man, Mar 17, 26. SYNTHETIC ORGANIC CHEMISTRY. *Educ:* Univ Man, BSc, 47, MSc, 49; Northwestern Univ, PhD(chem), 52. *Prof Exp:* Lectr chem, Univ Man, 47-49; Nat Res Coun Can fel & Corday-Morgan Commonwealth fel, Oxford Univ, 52-53; from assoc prof to prof, 53-91, EMER PROF CHEM, MEM UNIV NFLD, 91- *Honors & Awards:* Union Carbide Award for Chem Educ, Chem Inst Can, 81. *Mem:* Chem Inst Can; Chem Soc; Am Chem Soc. *Res:* Heterocyclic compounds; history of chemistry. *Mailing Add:* Dept Chem Mem Univ Nfld St John's NF A1B 3X7 Can

ANDERSON, HUGH RIDDELL, b Iowa City, Iowa, June 16, 32; m 56. SPACE PHYSICS. *Educ:* Univ Iowa, BA, 54, MS, 58; Calif Inst Technol, PhD(physics), 61. *Prof Exp:* Sr scientist, Jet Propulsion Lab, Calif Inst Technol, 62-65; from asst prof to prof space sci, Rice Univ, 65-81; RETIRED. *Mem:* Am Phys Soc; Am Geophys Union. *Res:* Primary cosmic radiation variations in space and time caused by solar modulation; ionizing radiation in space; auroral and ionospheric physics. *Mailing Add:* Sci Appln Inc 13400 Northup St Bellevue WA 98005

ANDERSON, HUGH VERITY, b Chicago, Ill, Aug 25, 21; m 46; c 1. ORGANIC CHEMISTRY. *Educ:* Kalamazoo Col, AB, 46; Univ Ill, PhD(org chem), 49. *Prof Exp:* Fel, Ohio State Univ, 49-50; res chemist, Mich, 50-58, sect head, 58-65, mgr chem process res & develop, 65-69, gen mgr, Lab Procedures Div, King of Prussia, 69-73, dir corp purchasing, 73-84, EXEC DIR CORP PURCHASING, UPJOHN CO, 84- *Concurrent Pos:* VChmn, Kalamozoo Col Bd Trustees. *Mem:* Am Chem Soc; Am Asn Clin Chem. *Res:* Steroids; pharmaceutical chemistry; clinical chemistry. *Mailing Add:* 9820 West H Ave Kalamazoo MI 49009

ANDERSON, INGRID, b Sioux Falls, SDak, Oct 7, 30. NUTRITION. *Educ:* Col St Benedict, Minn, BS, 53; Univ Minn, MS, 55, PhD(nutrit), 67. *Prof Exp:* Instr, St Benedicts High Sch, Minn, 53-54; instr home econ, Col St Benedict, Minn, 53-54, instr home econ & biol, 55-60; res asst nutrit, Univ Minn, 63-65; from asst prof to assoc prof nutrit & physiol, 65-77, PROF NUTRIT, COL ST BENEDICT, MINN, 77- *Concurrent Pos:* Vis assoc prof, Univ Minn, 74-75. *Mem:* AAAS; Am Pub Health Asn; Am Home Econ Asn; Am Dietetic Asn; Soc Nutrit Educ; Sigma Xi. *Res:* Congenital abnormalities in folic acid deficiency; growth factors; nutrition education; competency-based dietetic education. *Mailing Add:* Dept Nutrit Col St Benedict St Joseph MN 56374

ANDERSON, IRVIN CHARLES, b Iowa, Apr 4, 28; m 56; c 2. PLANT PHYSIOLOGY. *Educ:* Iowa State Univ, BS, 51; NC State Col Agr & Eng, MS, 54, PhD, 56. *Prof Exp:* Fel biol, Brookhaven Nat Lab, 56-58; from asst prof to assoc prof agron, 58-72, PROF AGRON & BOT, IOWA STATE UNIV, 72- *Mem:* Am Soc Plant Physiol; Am Soc Agron. *Res:* Biochemical aspects of genetic mutants; plant growth regulatory chemicals and nodulation. *Mailing Add:* Dept Agron Iowa State Univ Ames IA 50011

ANDERSON, J(OHN E(RLING), b Quincy, Mass, Mar 12, 29; m 51; c 4. CHEMICAL ENGINEERING. *Educ:* Mass Inst Technol, BS, 50, DSc(chem eng), 55; Ill Inst Technol, MS, 51. *Prof Exp:* Res engr, 54-57, res group leader, 57-59, res supvr, 59-63, asst mgr res, 63-64, sr res assoc, 64-65, sect mgr develop, 65-69, develop fel, 69-82, SR DEVELOP FEL, LINDE DIV, UNION CARBIDE CORP, 82- *Mem:* Nat Acad Eng; Am Inst Chem Engrs; Am Chem Soc; Combustion Inst. *Res:* Arc technology; high temperature chemistry; welding; combustion; solid waste disposal. *Mailing Add:* Linde Div Union Carbide Corp Tarrytown Tech Ctr Tarrytown NY 10591

ANDERSON, J(OHN) EDWARD, b Chicago, Ill; m 75; c 3. MICROPROCESSOR CONTROL SYSTEMS, VEHICLE DYNAMICS. *Educ:* Iowa State Univ, BS, 49; Univ Minn, MS, 55; Mass Inst Technol, PhD(astronaut), 62. *Prof Exp:* Aero res scientist, Nat Adv Comt Aeronaut, 49-51; sr design engr, Aeronaut Div, Honeywell, 51-54, sr res engr, 54-55, res projs engr, 55-57, prin res engr, 57-59, staff engr, 62-63; from assoc prof to prof mech eng, Univ Minn, 63-86; PROF MECH ENG, BOSTON UNIV, 86- *Concurrent Pos:* Consult, Honeywell Mil Prod Group, 63-71, 3M Co, 66-67, Regional Transp Dist, Colo, 74-75, Raytheon Missile Systs Div, 75-76, MBB plus Demag, Fed Repug Ger, 77-80 & State Ind, 80-81; exchange prof, Soviet Acad Sci, 67-68; pres, Taxi 2000 Corp, 86- *Honors & Awards:* Outstanding Inventor, Intellectual Properties Asn, 88. *Mem:* AAAS; Am Soc Mech Engrs; Fedn Am Scientists; Union Concerned Scientists. *Res:* Development and implementation of automated transit systems of type called personal rapid transit. *Mailing Add:* 474 Revere Beach Blvd No 802 Revere MA 02151

ANDERSON, J ROBERT, b Ames, Iowa, Sept 29, 34. PHYSICS. *Educ:* Iowa State Univ, BS, 55, PhD(physics), 62. *Prof Exp:* Asst prof, 64-72, assoc prof physics, 72-80, PROF PHYSICS & ASTRON, UNIV MD, COLLEGE PARK, 80- *Concurrent Pos:* NSF fel, Cambridge Univ, 63-64; AEC fel, Iowa State Univ, 64. *Mem:* Am Phys Soc. *Res:* Fermi surface studies; de Haas-van Alphen effect. *Mailing Add:* Dept Physics & Astron Univ Md College Park MD 20742

ANDERSON, JACK R(OGER), physical chemistry, for more information see previous edition

ANDERSON, JAMES B, b Cleveland, Ohio, Nov 26, 35; m 58; c 3. PHYSICAL CHEMISTRY, CHEMICAL ENGINEERING. *Educ:* Pa State Univ, BS, 57; Univ Ill, MS, 58; Princeton Univ, MA, 62, PhD(chem eng), 63. *Prof Exp:* Process engr, Shell Chem Co, 58-60; res assoc chem eng, Princeton Univ, 62-64, asst prof, 64-68; assoc prof eng & appl sci, Yale Univ, 68-74; PROF CHEM, PA STATE UNIV, 74- *Concurrent Pos:* Vis lectr, Rutgers Univ, 64; vis prof, Cambridge Univ, 86; grad fel, NSF. *Honors & Awards:* Evan Pugh Medal. *Mem:* AAAS; Am Chem Soc; Am Inst Chem Engrs; Am Phys Soc; Sigma Xi. *Res:* Chemical kinetics; molecular dynamics; molecular beams; rarefied gas dynamics; quantum chemistry; theoretical physics. *Mailing Add:* 152 Davey Lab Pa State Univ University Park PA 16802

ANDERSON, JAMES E, b Hartford, Conn, July 31, 38; m 62; c 4. PHYSICAL CHEMISTRY. *Educ:* Union Col, NY, BS, 60; Princeton Univ, PhD(chem), 63. *Prof Exp:* Appointee, Bell Tel Labs, Inc, 63-65; STAFF SCIENTIST, SCI LAB, FORD MOTOR CO, 65- *Concurrent Pos:* Vis scientist biophys, Max-Planck Inst, 75-76; vis scientist phys chem, Univ Mainz, Ger, 75-76; adj res assoc nuclear eng, Univ Mich, 77-, adj res prof chem, 79- *Mem:* Am Chem Soc; AAAS; Am Phys Soc. *Res:* Dielectric and nuclear relaxation in solids and liquids; high polymer physics; stochastic processes; reverse osmosis; membrane transport. *Mailing Add:* 2872 Glacier Way Ann Arbor MI 48105

ANDERSON, JAMES EDWARD, b Perth, Ont, Feb 23, 26; m 57. PSYCHIATRY, PHYSICAL ANTHROPOLOGY. *Educ:* Univ Toronto, MD, 53. *Prof Exp:* Lectr anat, Univ Toronto, 56-58, asst prof, 58-61, assoc prof anat & anthrop, 61-63; assoc prof phys anthrop, State Univ NY, Buffalo, 63-65, prof dept anthrop, 65-67; chmn dept, 67-77, PROF ANAT, MCMASTER UNIV, 67-, ASSOC PROF PSYCHIAT, 77- *Concurrent Pos:* Consult, Nat Mus Can, 60-63, Burlington Orthod Res Centre, 62-; mem coun res, Can Dent Asn, 60-63. *Honors & Awards:* Starr Medal Res Basic Sci, 61. *Mem:* Can Asn Anat. *Res:* Use of hereditary variations in the human skeleton to illustrate microevolution; paleopathology of New World Indians; growth at adolescence; adolescent psychiatry; specific learning disabilities. *Mailing Add:* 200 Bay St S Apt 2409 Hamilton ON L8R 2P9 Can

ANDERSON, JAMES GERARD, b Ceylon, Minn, July 27, 24; m 70; c 1. SOLID STATE PHYSICS, ELECTRONICS. *Educ:* Univ Minn, BS, 45; St Mary's Col, Minn, BA, 51; Univ Colo, MS, 59, PhD(physics), 63. *Prof Exp:* Res elec engr, Univ Minn, 45-46; aircraft electronic engr, Northwest Orient Airlines, Minn, 46-48; instr physics, St John's Univ, Minn, 53-58; asst prof, Wis State Univ, Oshkosh, 63-64; asst prof, Colo State Univ, 64-66; assoc prof, 66-74, PROF PHYSICS, UNIV WIS-EAU CLAIRE, 74- *Mem:* Sigma Xi; Am Asn Physics Teachers. *Res:* De-Haas-Van Alphen effect in metals measured in high pulsed magnetic fields. *Mailing Add:* Dept Physics & Astron Univ Wis Eau Claire WI 54702-4004

ANDERSON, JAMES HOWARD, b Joliet, Ill, Nov 17, 44; m 70. PHYSIOLOGY, RADIOLOGY. *Educ:* Ill Wesleyan Univ, BA, 66; Univ Ill, MS, 68, PhD(physiol), 71. *Prof Exp:* Asst prof radiol, M D Anderson Hosp & Tumor Inst, Univ Tex Syst Cancer Ctr, 71-, head sect exp diag radiol, biol sci div, 74-, asst prof physiol, Univ Tex Med Sch Houston, 72-, asst prof radiol, 74-; ASSOC PROF, DEPT RADIOL & RADIOL SCI, JOHNS HOPKINS MED SCH, JOHNS HOPKINS HOSP. *Mem:* Am Physiol Soc. *Res:* Experimental diagnostic radiology; gastrointestinal physiology; scanning electron microscopy. *Mailing Add:* Dept Radiol Johns Hopkins Univ Med Sch 720 Rutland Ave Baltimore MD 21205

ANDERSON, JAMES JAY, b Seattle, Wash, July 29, 46. FISH DIVERSION, RISK ANALYSIS. *Educ:* Univ Wash, BS, 69, cert, 75, PhD(oceanog), 77. *Prof Exp:* Oceanogr, Univ Wash, 69-81, res assoc, 81-82, from res asst prof to res assoc prof, 82-91, ASSOC PROF, UNIV WASH, 91- *Concurrent Pos:* Vis prof, Nat Oceanog Lab, Indonesia, 80-81; consult, Exxon, Nat Oceanic & Atmospheric Admin, Tech Arts, Chelan Pub Utility, Bonneville Power Admin, Mont Dept Fish Wildife & Parks & Army Corps Engrs. *Mem:* Sigma Xi; Am Soc Limnol & Oceanog; Animal Behav Soc; Am Fisheries Soc. *Res:* Stochastic models in fisheries and ecology; risk analysis; animal and human decision processes; fish toxicology; decision support software for fisheries management. *Mailing Add:* Fisheries Res Inst WH-10 Univ Wash Seattle WA 98195

ANDERSON, JAMES LEROY, b Chicago, Ill, Apr 16, 26; m 50; c 2. THEORETICAL PHYSICS. *Educ:* Univ Chicago, BS, 46, MS, 49; Syracuse Univ, PhD(physics), 52. *Prof Exp:* Asst instr physics, Ill Inst Technol, 47-49; instr, Rutgers Univ, 52-53; res assoc, Univ Md, 53-56; assoc prof, 56-66, PROF PHYSICS, STEVENS INST TECHNOL, 66- *Concurrent Pos:* Guggenheim fel, 60-61; res fel, Woods Hole Oceanog Inst, 66 & 69; vis res scientist, Cambridge Univ, 68-69; consult, Ramo-Wooldridge, 57, Combustion Engr, 57-58 & Sci Teaching Ctr, Mass Inst Technol, 66 & 67; Fulbright fel, 76; res scientist, Inst of Theoret Physics, Univ Amsterdam. *Mem:* NY Acad Sci. *Res:* General relativity; relativistic astrophysics. *Mailing Add:* Dept Physics Stevens Inst Technol Hoboken NJ 07030

ANDERSON, JAMES LEROY, b Detroit, Mich, Jan 21, 46; m 68; c 1. ELECTROANALYTICAL CHEMISTRY, SPECTROELECTROCHEMISTRY. *Educ:* Kalamazoo Col, BA, 67; Univ Wis, Madison, PhD(analytical chem), 74. *Prof Exp:* Asst prof chem, NDak State Univ, 75-79, Univ Ga, 79-83; LECTR CHEM, OHIO STATE UNIV, 74-; ASSOC PROF CHEM, UNIV GA, 83- *Concurrent Pos:* Postdoctoral assoc chem, Ohio State Univ, 74-75; panelist US Environ Protection Agency, Chem, Physics Water Res, 81-83, consult, Analytical Chem text, Am Chem Soc Analytical Div, 81-; comt mem, Am Chem Soc Exam Comt, Grad Analytical Chem, 82-; div ed, J Electrochem Soc, 86-, juror Am Chem Soc Analytical Div Awards Comt, 87- *Mem:* Am Chem Soc; ElectroChem Soc; Sigma Xi; Soc Electroanalytical Chemists; Int Soc Electrochem. *Res:* Development and fundamental characterization and modeling of sensitive and selective electrochemical methods including microelectrode array flow sensors and spectroelectrochemistry; electrochemistry and analytical chemistry; electrodeposition and characterization of novel materials. *Mailing Add:* Dept Chem Univ Ga Sch Chem Sci Athens GA 30602

ANDERSON, JAMES T(REAT), b Washington, DC, June 18, 21; m 45; c 4. MECHANICAL ENGINEERING. *Educ:* Mich State Univ, BSME, 43, MSME, 48; Univ London, PhD & Imp Col, London, dipl, 52. *Prof Exp:* From instr to prof mech eng, Mich State Univ, 46-60; chmn dept, WVa Univ, 60-63; dean col eng, 63-72, actg vpres acad affairs, 69-70, vpres acad affairs, 71-76, actg pres, 73-74, prof mech eng, 76-83,chmn dept, 78-83, EMER VPRES, UNIV NEV, RENO, 83 - *Concurrent Pos:* Consult, US Army Detroit Arsenal, 53-56, jet engine dept, Gen Elec Co, 57-59, aircraft nuclear propulsion dept, 59-61 & Allegany Ballistic Labs, 61-63; vis prof, Tulane Univ, 60; partic, Lab Human Develop Study of Prenatal Dis & Conditions, 65-66; prin investr res, Naval Ord Test Sta, 67-68; consult, Nev Dept Rehab, 84- *Mem:* Am Soc Mech Engrs; Nat Soc Prof Engrs. *Res:* Spatial and time truncation errors in heat transfer; transient numerical solutions; analytical and numerical calculation of transient temperatures at ultra-short time periods. *Mailing Add:* 2170 Royal Dr Reno NV 89503

ANDERSON, JAMES WINGO, b Hinton, WVa, Aug 6, 36; m 57; c 2. ENDOCRINOLOGY, NUTRITION. *Educ:* WVa Univ, BA, 58; Northwestern Univ, MD, 61; Mayo Grad Sch, MS, 65. *Prof Exp:* Asst prof med, Univ Calif Med Sch, San Francisco, 68-73; assoc prof med, Univ Ky, 73-78; chief med serv, Vet Admin Med Ctr, Lexington, Ky, 80-83; vis scientist, Mass Inst Technol, 83-84; CHIEF, METAB ENDOCRINE SECT, VET ADMIN MED CTR, UNIV KY, 73-; PROF MED, 78- *Concurrent Pos:* Prin investr, Diabetes Res Fund, 74-, Nat Inst Arthritis, Metab & Digestive Dis, 77-80, VA Med Ctr, 85-88, NIH HL 37902 Prog Proj, 86-89. *Mem:* Fel Am Col Nutrit; fel Am Col Physicians; Am Diabetes Asn; Am Fedn Clin Res; Am Inst Nutrit; Am Soc Clin Nutrit. *Res:* Applied clinical research applying high-fiber diets to individuals with diabetes, obesity and high blood fat levels. *Mailing Add:* Dept Vet Affairs Med Ctr Lexington KY 40511

ANDERSON, JAY ENNIS, b Logan, Utah, Oct 18, 37; m 59; c 1. PLANT ECOLOGY. *Educ:* Mont State Univ, BS, 59; Syracuse Univ, MS, 67, PhD(bot), 71. *Prof Exp:* Teacher biol, Powell County High Sch, Deer Lodge, Mont, 63-68; consult ecol, Biol Sci Curric Study, 71-74; res assoc, dept chem eng, Univ Colo, 74-75; from asst prof to assoc prof, 75-85, PROF ECOL, DEPT BIOL SCI, IDAHO STATE UNIV, 85- *Concurrent Pos:* Asst attend prof, Dept EPO Biol, Univ Colo, 71-75, res assoc, Inst Arctic & Alpine Res, 74-75. *Mem:* AAAS; Ecol Soc Am; Sigma Xi; Soc Range Mgt; Am Inst Biol Sci; Australian Soc Plant Physiol. *Res:* Factors affecting transpiration, photosynthesis, and water status of terrestrial plants; vegetation dynamics in cold-desert ecosystems; water use and water use efficiency of cold-desert species. *Mailing Add:* 3566 Stockman Rd Pocatello ID 83201

ANDERSON, JAY LAMAR, b Madison, Wis, Apr 22, 31; m 55; c 4. POMOLOGY, WEED SCIENCE. *Educ:* Utah State Univ, BS, 55; Univ Wis, PhD(plant path), 61. *Prof Exp:* From asst prof to assoc prof, 61-75, PROF HORT, UTAH STATE UNIV, 75- *Mem:* Weed Sci Soc Am; fel Am Soc Hort Sci. *Res:* Pomology; fruit growth and development; physiology of herbicides and plant growth regulators; weed control in fruit crops; orchard floor management; cherry rootstocks; modeling of fruit trees and weed species. *Mailing Add:* 856 Juniper Dr Logan UT 84322

ANDERSON, JAY MARTIN, b Paterson, NJ, Oct 16, 39; m 63; c 1. COMPUTER SCIENCE, PHYSICAL CHEMISTRY. *Educ:* Swarthmore Col, BA, 60; Harvard Univ, AM, 61, PhD(chem), 64. *Prof Exp:* From asst prof to assoc prof, 63-75, PROF CHEM & DIR COMPUT SCI, BRYN MAWR COL, 75- *Concurrent Pos:* Kettering vis lectr, Univ Ill, 68-69; Alfred P Sloan res fel, 67-70; secy-treas, Exp Nuclear Magnetic Resonance Conf, 67-71; vis assoc prof, Sloan Sch Mgt, Mass Inst Technol, 71-72. *Mem:* Asn Comput Mach. *Res:* Computer science; nuclear magnetic resonance; computing machinery; environmental systems simulation. *Mailing Add:* 20 N Plum St Lancaster PA 17602

ANDERSON, JAY OSCAR, b Brigham, Utah, Dec 5, 21; m 44; c 6. POULTRY NUTRITION. *Educ:* Utah State Univ, BS, 43; Univ Md, MS, 48, PhD(poultry nutrit), 50. *Prof Exp:* Chemist, Merck & Co, 50-51; prof animal sci, Utah State Univ, 51-84; PROF, ANDERSON ANIMAL CLIN, 84- *Concurrent Pos:* Secy-treas, Utah Feed Mfrs & Dealers Asn, 66-; assoc ed, Poultry Sci. *Mem:* Poultry Sci Asn. *Res:* Amino acid and protein requirements of chickens and turkeys; amino acid content of feedstuffs. *Mailing Add:* Anderson Animal Clin 1550 N Stapley 132 Mesa AZ 85203

ANDERSON, JEFFREY JOHN, b Chicago, Ill, Dec 30, 53; m 77; c 1. AGRICULTURAL & FOOD CHEMISTRY. *Educ:* Ill State Univ, BS, 75; Univ Minn, PhD(biochem), 80. *Prof Exp:* RES CHEMIST, BIOCHEM DEPT, E I DU PONT DE NEMOURS & CO INC, 80- *Mem:* Am Chem Soc. *Res:* Elucidation of metabolic pathways for aromatic compounds in truchosporm cutoneum; determining the environmental fate of experimental pesticides. *Mailing Add:* 1300 Tulane Rd Wilmington DE 19803

ANDERSON, JOHN ANSEL, b Sidcup, Kent, Eng, Aug 16, 03; Can citizen; m 31; c 2. CEREAL CHEMISTRY. *Educ:* Univ Alta, BSc & MSc, 26; Leeds Univ, Eng, PhD(org chem), 30. *Hon Degrees:* DSc, Univ Sask, 64 & Univ Man, 78; LLD, Univ Alta, 65. *Prof Exp:* Biologist, Nat Res Coun, Ottawa, 30-39; dir, Grain Res Lab, Can Grain Comn, 39-62, Winnipeg Res Sta, Can Dept Agr, 62-63, dir-gen, Res Br, Ottawa, 63-68; res prof, Dept Plant Sci, Univ Man, 68-71; RETIRED. *Honors & Awards:* Osborne Medal, Am Asn Cereal Chemists, 57. *Mem:* Fel Royal Soc Can; fel Chem Inst Can; fel Agr Inst Can; Am Asn Cereal Chemists (pres, 52-53); Int Asn Cereal Chemists (pres, 59-60). *Res:* Biochemistry and technology of wheat, durum and barley. *Mailing Add:* 1801-2008 Fullerton Ave North Vancouver BC V7P 3G7 Can

ANDERSON, JOHN ARTHUR, b Chicago, Ill, Mar 6, 32; m 56; c 2. BIOCHEMISTRY. *Educ:* Colo State Univ, BS, 52, MS, 54; Ore State Univ, PhD(biochem), 62. *Prof Exp:* Asst prof chem, Univ Tex, Arlington, 54-55; from asst prof to assoc prof, 66-81, PROF CHEM, TEX TECH UNIV, 81- *Mem:* AAAS; Am Chem Soc; Am Soc Biochem & Molecular Biol. *Res:* Enzyme mechanisms; biosynthesis of mycotoxins. *Mailing Add:* Dept Chem & Biochem Tex Tech Univ Lubbock TX 79409

ANDERSON, JOHN B, b Mobile, Ala, Sept 5, 44; m 65; c 2. MARINE GEOLOGY. *Educ:* Univ SAla, BS, 68; Univ NMex, MS, 70; Fla State Univ, PhD(geol), 72. *Prof Exp:* Asst prof geol, Hope Col, 72-75; asst prof, 75-80, ASSOC PROF GEOL, RICE UNIV, 80- *Mem:* Sigma Xi; Geol Soc Am. *Res:* Glacio-marine geology of antarctic regions; environmental geology of coastal waters. *Mailing Add:* Dept Geol Rice Univ Box 1892 Houston TX 77251

ANDERSON, JOHN D, b Jamestown, NY, Dec 10, 30; m 55; c 3. NUCLEAR PHYSICS. *Educ:* San Diego State Col, BA, 51; Univ Calif, Berkeley, MA, 53, PhD(physics), 56. *Prof Exp:* Sr staff physicist, 56-70, dep div leader, 70-73, div leader, 73-76, dept head, 76-78, ASSOC DIR PHYSICS, LAWRENCE LIVERMORE LAB, UNIV CALIF, 78- *Honors & Awards:* Tom W Bonner Prize, 72. *Mem:* Fel Am Phys Soc; Sigma Xi. *Res:* Fast neutron physics; polarization phenomena; nuclear reactions. *Mailing Add:* 572 Tyler Ave Livermore CA 94550

ANDERSON, JOHN DENTON, b July 15, 1912; m 43; c 4. PRIMITIVE MOBILE SYSTEMS. *Educ:* Stanford Univ, PhD(biol), 49. *Prof Exp:* Assoc dean, Med Sch, Basic Sci, Univ Ill, Urbana, 72-77; RETIRED. *Mailing Add:* 1305 W Clark St Champaign IL 61821-3155

ANDERSON, JOHN DONALD, b Swansea, Wales, UK, Mar 2, 35; m 60; c 3. MEDICAL MICROBIOLOGY. *Educ:* Univ Bristol, UK, BSc, 56, PhD(biochem), 59, MBChB, 70, MD, 74. *Prof Exp:* Fel biochem, Univ Calif, Berkeley, 59-60; sr sci officer med microbiol, Microbiol Res Estab, 60-66; intern, Univ Bristol, 66-71, lectr & sr registrar, Bristol Royal Infirmary, Eng, 71-74; consult microbiologist, York Dist Hosp, Eng, 74-77; PROF MED MICROBIOL, UNIV BC & HEAD, DIV MED MICROBIOL, BC'S CHILDREN'S HOSP, 78- *Concurrent Pos:* consult microbiologist, Health Sci Hosp, 77- *Mem:* Royal Col Pathologists; fel Royal Col Physicians. *Res:* Pathogenesis and treatment of human urinary tract infections; effect of antibiotics on the epidemiology of resistant organisms in the human gut. *Mailing Add:* Dept Pathol BC's Children's Hosp 4480 Oak St Vancouver BC V6M 3V4 Can

ANDERSON, JOHN EDWARD, b Chicago, Ill, May 15, 27; m 75; c 3. URBAN TRANSPORTATION, MECHANICAL ENGINEERING. *Educ:* Iowa State Univ, BSME, 49; Univ Minn, MSME, 55; Mass Inst Technol, PhD(aeronaut & astronaut), 62. *Prof Exp:* Aeronaut res scientist, Nat Adv Comt Aeronaut, 49-51; develop engr, Aero Div, Honeywell, Inc, 51-53, proj engr, 53-54, sr res engr aero res, 54-55, res proj engr, 55-56, prin res engr, 56-57, staff engr, 57-58, adv to dir of mil prod res, 62-63, proj mgr, systs & res div, 63; lectr, Univ Minn, Minneapolis, 62-63, assoc prof thermodyn, 63-71, prof mech eng, 71-86, dir, Indust Eng Div, 78-86; PROF AEROSPACE & MECH ENG, BOSTON UNIV, 86- *Concurrent Pos:* Consult, Systs & Res Div, Honeywell, Inc, 63-67; Space Sci Ctr grant, 65-67; Nat Acad Sci exchange prof, Soviet Union, 67-68; coordr, Task Force New Concepts in Urban Transp, 68-78, Colo Regional Transp Dist, 74-75, The Raytheon Co, 75-76, Mannesmann Demag, 77-78, Indianapolis Transit Comn, 78-81 & Arthur D Little, Inc, 81-82; founder, Anderson MacDonald, Inc, 80- & Davy McKee Corp, 84-85; pres, Taxi 2000 Corp. *Mem:* AAAS; Am Soc Mech Engrs; Sierra Club. *Res:* New concepts in urban transportation; transportation systems analysis; innovative interdisciplinary education on environmental problems; science and engineering policy in government; transit system theory; wind energy; research on determination of optimum characteristics of public transportation through study of economic, social, environmental and technical characteristics. *Mailing Add:* Dept Aerospace & Mech Eng 110 Cummington St Boston MA 02215

ANDERSON, JOHN FRANCIS, b Hartford, Conn, July 25, 36; m 63. PHYSIOLOGY. *Educ:* Cent Conn State Col, BS, 62; Univ Fla, MS, 65, PhD(zool), 68. *Prof Exp:* Asst prof, 69-74, ASSOC PROF ZOOL, UNIV FLA, 74- *Mem:* Am Soc Zool; Sigma Xi. *Res:* Arthropod physiology; energetics in arachnids. *Mailing Add:* Dept Zool Univ Fla 609 Barai Bldg Gainesville FL 32611

ANDERSON, JOHN FREDRIC, b Fargo, NDak, Feb 25, 36; m 58; c 3. ENTOMOLOGY. *Educ:* NDak State Univ, BS, 57, MS, 59; Univ Ill, PhD(entom), 63. *Prof Exp:* NSF fel, 63-64; from asst entomologist to assoc entomologist, 64-69, head dept entom, 69-87, DIR, CONN AGR EXP STA, 87- *Honors & Awards:* Author's Citation Award, Int Soc Arboriculture. *Mem:* Entom Soc Am; Soc Invert Path; Am Mosquito Control Asn; Am Soc Parasitol; Am Soc Microbiol; Am Soc Trop Med Hyg. *Res:* Medical entomology; insect pathology; forest entomology. *Mailing Add:* Conn Agr Exp Sta Box 1106 New Haven CT 06504

ANDERSON, JOHN G(ASTON), b Dante, Va, Aug 21, 22; m 48; c 1. HIGH VOLTAGE ENGINEERING. *Educ:* Va Polytech Inst, BS, 43. *Prof Exp:* Engr, High Voltage Lab, Gen Elec Co, Pittsfield, 47-53, develop engr, 53-57, res engr systs insulation, 57-63, tech dir, Proj Extra-High Voltage, 64-67, mgr, 67-70, mgr, Proj Ultra-High Voltage, Pittsfield, 70-72, mgr, AC Transmission Studies, Schenectady, 72-75, mgr, High Voltage Lab, 75-80, consult engr, elec power transmission systs, Schenectady, 80-84; SR CONSULT, POWER TECHNOLOGIES INC, SCHENECTADY, 85- *Concurrent Pos:* Dir elec & mech res; consult eng, 84-88. *Honors & Awards:* Centennial Medal, Inst Elec & Electronics Engrs; Herman Halperin Award Inst Elec & Electronics Engrs, 91. *Mem:* Nat Acad Eng; fel Inst Elec & Electronics Engrs; Int Conf Large Elec High Tension Systs. *Res:* Dielectric strength of liquid and solid insulations; lightning; extra-high voltage transmission; ultra-high voltage transmission; electric power transmission; saftwave systems. *Mailing Add:* 1101 West St Pittsfield MA 01201

ANDERSON, JOHN GREGG, b Niagara Falls, NY, Aug 31, 48; m 74; c 2. STRONG MOTION SEISMOLOGY, GEOPHYSICS. *Educ:* Mich State Univ, BS, 70; Columbia Univ, PhD(geophysics), 76. *Prof Exp:* Res fel, Calif Inst Technol, 76-77; res assoc, Univ Southern Calif, 77-80; asst res seismologist, Univ Calif, San Diego, 80-84, assoc res seismologist, 84-88; ASSOC PROF GEOPHYS, UNIV NEV, RENO, 88- *Concurrent Pos:* Mem, Comt Probabilistic Seismic Hazard Assessment, Nat Acad Sci, 85-88. *Mem:* Seismol Soc Am; Royal Astron Soc; Am Geophys Union; Earthquake Eng Res Inst; AAAS. *Res:* Strong ground shaking caused by earthquakes; recording systems; physical understanding; numerical modeling; average properties; probability of occurence; relationship of earthquakes and earthquake occurence rates to geological processes. *Mailing Add:* Seismol Lab Univ Nevada Reno Reno NV 89557

ANDERSON, JOHN HOWARD, b Northfork, WVa, Jan 26, 24; m 46; c 3. CHEMICAL PHYSICS. *Educ:* Univ NC, BS, 44, MS, 47; Univ Chicago, PhD(chem), 55. *Prof Exp:* Asst prof physics & chem, Talladega Col, 46-49; fel glass sci, Mellon Inst, 54-58; assoc prof, 58-66, PROF PHYSICS, UNIV PITTSBURGH, 66- *Concurrent Pos:* Fulbright-Hays fel, Coun Int Exchange Scholars, 70. *Mem:* Am Phys Soc; Am Asn Physics Teachers; AAAS. *Res:* Electron spin resonance; color centers; energy conversion; energy utilization. *Mailing Add:* 231 Woodside Rd Pittsburgh PA 15221

ANDERSON, JOHN JEROME, b Port Arthur, Tex, Oct 10, 30; m 64; c 2. GEOLOGY. *Educ:* Carleton Col, BA, 52; Univ Minn, MS, 62; Univ Tex, Austin, PhD(geol), 65. *Prof Exp:* Assoc prof, 65-72, PROF GEOL, KENT STATE UNIV, 72- *Concurrent Pos:* Prin investr, NSF grants, Kent State Univ, 67-71; vis scientist, Am Geol Inst, 70; consult, US Environ Protection Agency, 74, Am Petroleum Inst, 75 & US Geol Survey, 78-81. *Mem:* Geol Soc Am; Sigma Xi. *Res:* Geology of Antartica and Ellsworth Mountains, Antarctica; Cenozoic geologic evolution of southern high plateaus, Utah. *Mailing Add:* Dept Geol Kent State Univ Main Campus Kent OH 44242

ANDERSON, JOHN JOSEPH BAXTER, b Cleveland, Ohio, June 12, 34; m 57; c 3. PHYSIOLOGY, NUTRITION. *Educ:* Williams Col, BA, 56; Harvard Univ, MAT, 58; Boston Univ, MA, 62; Cornell Univ, PhD(phys biol), 66. *Prof Exp:* Teacher high sch, Maine, 58-59; instr biol, Bradford Jr Col, 59-62; asst prof vet physiol & pharmacol, Col Vet Med, Univ Ill, Urbana, 66-71; assoc prof, 72-77, dept chmn, 84-87, PROF NUTRIT, SCH PUB HEALTH, UNIV NC, CHAPEL HILL 77- *Concurrent Pos:* NIH fel phys biol, Cornell Univ, 66; vis scientist, USDA, Beltsville, 80, Univ Copenhagan, Glostrup Hosp, 85 & Karolinska Hosp & Inst, Stockholm, 90. *Mem:* AAAS: Am Physiol Soc; Am Inst Nutrit; Am Pub Health Asn; Brit Nutrit Soc; Am Soc Clin Nutrit; Am Col Nutrit. *Res:* Metabolism of calcium and phosphorus, especially bone, intestine, kidneys and endocrines; nutrition and body composition; diet and bone health; nutrition and cancer; determinants of bone health in pre and post menopausal women, especially dietary, lifestyle and activity variables. *Mailing Add:* 15 Rogerson Dr Chapel Hill NC 27514

ANDERSON, JOHN LEONARD, b Wilmington, Del, Sept 29, 45; m 68; c 2. BIOENGINEERING, BIOPHYSICS. *Educ:* Univ Del, BChE, 67; Univ Ill, Urbana, MS, 69, PhD(chem eng), 71. *Prof Exp:* Asst prof, Cornell Univ, 71-76; assoc prof chem eng, 76-79, dir biomed eng, 80-85, PROF CHEM ENG, CARNEGIE-MELLON UNIV, 79-, DEPT HEAD, 83- *Concurrent Pos:* Vis prof chem eng, Mass Inst Technol, 82-83; Guggenheim fel, 82-83; Fifth Berkeley lectr, Univ Calif, 89; NSF commemorative lectr, 89. *Honors & Awards:* Prof Progress Award, Am Inst Chem Engrs, 89; Holtz lectr, Johns Hopkins Univ, 90. *Mem:* AAAS; Am Inst Chem Engrs; Am Chem Soc; Sigma Xi; Coun Chem Res. *Res:* Theory of transport through porous membranes, emphasizing the role of solute-pore interactions; transport phenomena in colloidal suspensions; transport of polymers in micropores and near surfaces. *Mailing Add:* Dept Chem Eng Carnegie-Mellon Univ Pittsburgh PA 15213-3890

ANDERSON, JOHN M(ELVIN), b Kansas City, Mo, Oct 9, 24; m 50; c 4. ELECTRICAL ENGINEERING. *Educ:* Univ Ill, BSEE, 47, MSEE, 48, PhD(elec eng), 55. *Prof Exp:* Physicist gaseous electronics, Res Lab, Gen Elec Co, 55-87; RETIRED. *Mem:* Fel Am Phys Soc; fel Inst Elec & Electronics Engrs. *Res:* Gaseous electronics; microwave electronics; electro-magnetic and plasma interaction; lamp research and development. *Mailing Add:* 17 Cedar Lane Scotia NY 12302

ANDERSON, JOHN MURRAY, b Toronto, Ont, Sept 3, 26; m 51, 85; c 4. AQUACULTURE. *Educ:* Univ Toronto, BScF, 51, PhD(physiol), 58. *Hon Degrees:* LLD, St Thomas Univ, 74; Univ Maine, 76; Dalhouse Univ, 79. *Prof Exp:* Res assoc, Univ Toronto, 58; asst prof biol, Univ NB, 58-63; assoc prof animal physiol, Carleton Univ, 63-67; dir biol sta, Fisheries Res Bd Can, 67-72; dir gen, Fisheries Res & Develop, Can Dept Environ, 72-73; pres, Univ NB, Fredericton, 73-79; PRES, ANDERSON CONSULTS, INC, 79-; VPRES OPERS, ATLANTIC SALMON FEDN, 84- *Concurrent Pos:* Treas & mem bd dirs, Huntsman Marine Lab, NB, 69-72, pres & chmn bd dir, 74-77; mem comt on Canada's Energy Opportunities, Sci Coun Can, 73-74; mem bd gov, Inst Can Bankers, 74-79; mem exec comt, St John Mus, 74-; vpres, Biol Coun Can, 77-79; pres, Aquaculture Assn Can, 83-85; mem bd dirs, Bradfield Educ Fund, Noranda Mines, 78-; trustee, MacKenzie King Scholar Trust, 77-, chmn bd trustees, 86-; mem bd dirs, New Brunswick Nature Trust, 87-; mem, Sci Coun Can, 88- *Mem:* AAAS; Am Fisheries Soc; Can Soc Zool (vpres, 71, pres, 73-74); Aquacult Asn Can (pres, 83-85). *Res:* Marine biology; physiological and behavioural effects of sub-lethal pollutants on aquatic organisms; salmonid aquaculture. *Mailing Add:* PO Box 547 St Andrews NB E0G 2X0 Can

ANDERSON, JOHN NICHOLAS, b Pittsburgh, Pa, Sept 25, 46; m 69; c 1. MOLECULAR BIOLOGY, ENDOCRINOLOGY. *Educ:* Slippery Rock State Col, BA, 69; Purdue Univ, PhD(molecular biol), 73. *Prof Exp:* Res assoc, Stanford Univ, 73-76; ASST PROF MOLECULAR BIOL, PURDUE UNIV, 76- *Concurrent Pos:* Am Cancer Soc fel, 73-76 & grant, 76-78. *Mem:* Endocrine Soc. *Res:* Mechanisms of gene expression in higher organisms. *Mailing Add:* Dept Biol Sci Purdue Univ West Lafayette IN 47907

ANDERSON, JOHN NORTON, b Mannington, WVa, Aug 28, 37; m 62; c 2. POLYMER CHEMISTRY. *Educ:* Fairmont State Col, BS, 59; Univ Pittsburgh, PhD(acid-base equilibria), 64; Univ New Haven, MBA, 82. *Prof Exp:* Res chemist, Firestone Tire & Rubber Co, 65-69, sr res chemist, 69-71, group leader, 71-76; tech dir, Duracote Corp, 76-78; mgr, Corp Res Lab, Raybestos-Manhattan, 78-84; MGR, PROD DEVELOP, RAYTECH, 84- *Mem:* Am Chem Soc; Soc Plastics Engrs; Soc Advan Mat & Process Eng; Am Indust Hygiene Asn. *Res:* Polymerization of alpha olefins; structural characterization of elastomers; thermal analysis; anionic polymerization of dienes; solution properties of polymers; polymer physics, laminates, coated fabrics, plastisols and organosols; gel permeation chromatography, composites; friction materials. *Mailing Add:* Valeo Inc 37564 Amrhein Livonia MI 48150

ANDERSON, JOHN P, b New Orleans, La, Mar 27, 39; m 62; c 5. ENGINEERING MECHANICS. *Educ:* Ga Inst Technol, BS, 61, MS, 63 & 64, PhD(eng mech), 66; Univ Ala, Birmingham, MBA, 75. *Prof Exp:* Asst prof eng mech, Ga Inst Technol, 65-68; asst to vpres for Univ Col, Univ Ala Birmingham, 71-73, asst to vpres fiscal affairs, 73-74, dir financial planning, 73-75, asst to vpres for finance, 74-75, dir, Div Spec Studies, 74-78, assoc prof math, 69-, prof eng, 73-, asst to pres & dir, long-range planning, 77-, dean, spec studies, 78-84; VPRES, ADMIN & PLANNING, WAKE FOREST UNIV, 84- *Concurrent Pos:* Instr, USAF Acad, 66-68; consult, Kaman Nuclear Co, 66- & Rust Eng Co, 68-71. *Res:* Elastic stability theory; shell theory; applied mathematics. *Mailing Add:* Wake Forest Univ PO Box 7249 Reynolds Sta Winston Salem NC 27109

ANDERSON, JOHN R, b Stromsburg, Nebr, Aug 1, 28; m 50; c 2. MATHEMATICS EDUCATION. *Educ:* Univ Nebr, BA & MA, 51; Purdue Univ, PhD(math), 70. *Prof Exp:* High sch math teacher, 54-60; prog engr, Allison Div GM, 54-60; PROF MATH & CHMN DEPT, DE PAUW UNIV, 60- *Concurrent Pos:* NSF Grant, 65; Danforth teacher fel, 64. *Mem:* Math Asn Am; Nat Coun Teachers Math. *Mailing Add:* Dept Math & Comput Sci DePauw Univ Greencastle IN 46135

ANDERSON, JOHN RICHARD, b Fargo, NDak, May 5, 31; m 55; c 3. MEDICAL ENTOMOLOGY, PARASITOLOGY. *Educ:* Utah State Univ, BS, 57; Univ Wis, MS, 58, PhD(med entom, vet sci), 60. *Prof Exp:* Biol aide, USPHS, 55; proj assoc arbo viruses, Dept Vet Sci, Univ Wis, 60-61; asst prof, 61-67, assoc prof, 67-71, chmn div, 70-71, PROF ENTOM, UNIV CALIF, BERKELEY, 71- *Concurrent Pos:* Trustee & past chmn, Mosquito Abatement Dist, Alameda Co, Calif, 61-73 & 79-; assoc dean res, Col Nat Resources, 78-85; prin investr, NIH & other grants. *Honors & Awards:* Woodworth Award. *Mem:* AAAS; Entom Soc Am; Can Entom Soc; Am Mosquito Control Asn; NY Acad Sci; Soc Vector Ecol. *Res:* Ecology and behavior of Diptera of medical and veterinary importance; host-vector-parasite interrelationships of Hematozoa; anthropod parasites of deer. *Mailing Add:* Dept Entom Univ Calif Berkeley CA 94720

ANDERSON, JOHN SEYMOUR, b Kearney, Nebr, Oct 27, 36; m 76; c 1. BIOCHEMISTRY. *Educ:* Nebr State Teachers Col, AB, 58; Univ Nebr, Lincoln, MS, 60, PhD(chem), 63. *Prof Exp:* NSF res fels, Sch Med, Washington Univ, 63-64 & Sch Med, Univ Wis-Madison, 64-65; USPHS fel, Med Res Coun Lab Molecular Biol, Cambridge, Eng, 65-67; asst prof biochem, 67-73, assoc prof 73-86, PROF BIOCHEM, UNIV MINN-ST PAUL, 86- *Concurrent Pos:* Hon vis assoc prof, Dept Chem, Univ BC, 83. *Mem:* Am Chem Soc; Am Soc Microbiol; Am Soc Bichem & Molecular Biol. *Res:* Biosynthesis and structure of bacterial cell walls and membranes; biosynthesis of polysaccharides; intermediary metabolism of monosaccharides. *Mailing Add:* Dept Biochem Univ Minn St Paul MN 55108

ANDERSON, JOHN THOMAS, b Paterson, NJ, Apr 7, 45; m 68. LOW TEMPERATURE PHYSICS. *Educ:* Case Inst Technol, BS, 67; Univ Minn, PhD(physics), 71. *Prof Exp:* From teaching asst physics to res asst, Dept Physics, Univ Minn, 67-71; res physicist, Hansen Lab Physics, Stanford Univ, 71-74, res physicist, 74-; AT HEWLETT PACKARD LABS, PALO ALTO, CALIF. *Mem:* Am Phys Soc; Sigma Xi. *Res:* Application of superconducting quantum interference device magnetometry to measurement of very small angles, specifically in a gyroscope with a London moment readout, or to precise angle measurement; noise characteristics of superconducting quantum interference devices. *Mailing Add:* Hewlett Packard Labs PO Box 10350 Palo Alto CA 94303-0867

ANDERSON, JOHN WALBERG, b New York, NY, Dec 3, 27; m 64; c 3. ANATOMY, HISTOLOGY. *Educ:* Swarthmore Col, BA, 50; Univ NH, MS, 52; Cornell Univ, PhD(histol), 56. *Prof Exp:* From instr to assoc prof, 56-69, PROF ANAT, UNIV WIS-MADISON, 69- *Mem:* Am Soc Cell Biol; Am Asn Anat; Soc Study Reproduction. *Res:* Histophysiology, especially of reproduction; materno-fetal transfer of immunity. *Mailing Add:* Dept Anat Univ Wis Madison WI 53706

ANDERSON, JULIUS HORNE, JR, b Harrisburg, Pa, June 20, 39; m 63; c 3. PHARMACOLOGY, BIOCHEMISTRY. *Educ:* Princeton Univ, AB, 61; Yale Univ, PhD(pharmacol), 67, MD, 68. *Prof Exp:* Asst prof pharmacol, Sch Med, Univ Pittsburgh, 70-77; RESIDENT INTERNAL MED, ALLEGHENY GEN HOSP, 77- *Concurrent Pos:* Johnson Res Found fel, Univ Pa, 68-70; internship in med, Univ Health Ctr, Univ Pittsburgh, 73-74. *Mem:* Fedn Am Socs Exp Biol. *Res:* Biochemical pharmacology. *Mailing Add:* 16 Fairmont Ave West Newton MA 02158

ANDERSON, KARL E, b Buffalo, NY, May 16, 40; m 83; c 3. GASTROENTEROLOGY. *Educ:* John Hopkins Univ, MD, 65. *Prof Exp:* Assoc prof & physician, Rockefeller Univ, 79-85; prof, Dept Med & Pharmacol & dir, Clin Res Ctr, NY Med Col, Valhalla, 85-87; assoc dir, div Human Nutrit, 87-90, dir, div Human Nutrit, 90-91, PROF, DEPTS PREV MED & COMMUNITY HEALTH, INTERNAL MED & PHARMACOL, UNIV TEX MED BR, GALVESTON, TEX, 87- *Concurrent Pos:* Mem, Gastroenterol & Clin Nutrit Rev Group, NIH, 80-82, Scientific Adv Bd, Am Porphyria Found, Montgomery, Ala, 82-, chmn, 82-87, Gastroenterol Adv Panel, US Pharmacopeial Convention, Inc, 91- *Mem:* Fel Am Col Physicians; Am Asn Study Liver Dis; Am Gastroenterol Asn; Am Soc Clin Pharmacol & Therapeut; Am Soc Pharmacol & Exp Therapeut; Am Soc Clin Nutrit. *Res:* Nutritional pharmacology; drug-nutrient intereactions; effects of diet on cytochrome P450 and drug metabolism; porphyrias and other disorders of heme metabolism. *Mailing Add:* Univ Texas Med Br 700 The Strand Ewing Hall Bldg Rm 3-102 Rte J09 Galveston TX 77550

ANDERSON, KEITH PHILLIPS, b Morgan, Utah, Oct 6, 19; m 42; c 3. PHYSICAL CHEMISTRY. *Educ:* Brigham Young Univ, AB, 46; Cornell Univ, PhD, 50. *Prof Exp:* Mem staff, Univ Calif at Los Alamos, NMex, 50-51; instr, exten div, Univ NMex, 51; chief radiol warfare div, Dugway Proving Ground, Utah, 51-52; assoc res prof physics & develop dir radiol res, Univ Utah, 52-53; chmn dept chem, 58-60, PROF CHEM, BRIGHAM YOUNG UNIV, 53- *Concurrent Pos:* Fulbright lectr, Valencia & Barcelona, Spain, 60-61; Colombia, SAm, 66. *Mem:* AAAS; Am Chem Soc; Sigma Xi; AAAS. *Res:* Tracer chemistry; calorimetry; amino acid chelates; equilibria in slightly soluble salt solutions in aqueous-nonaqueous media. *Mailing Add:* 3758 Foothill Dr Provo UT 84604

ANDERSON, KELBY JOHN, b Ogden, Utah, Feb 15, 44. ELEMENTARY PARTICLE PHYSICS. *Educ:* Idaho State Univ, BS, 66; Univ Colo, PhD(physics), 71. *Prof Exp:* SR RES ASSOC, ENRICO FERMI INST, UNIV CHICAGO, 72- *Res:* Experimental elementary particle physics, counter experiments. *Mailing Add:* Enrico Fermi Inst Univ Chicago 5630 S Ellis Chicago IL 60637

ANDERSON, KENNETH ELLSWORTH, b Ithaca, NY, Dec 21, 14; m 42; c 11. BACTERIOLOGY. *Educ:* Cornell Univ, BS, 37, PhD(bact), 43; Univ NH, MS, 40. *Prof Exp:* Asst chemist & bacteriologist, NY Water Serv Corp, 37-38; asst bot, Univ NH, 38-40; asst bact, Cornell Univ, 41-42; bacteriologist, Genesee Brewing Co, 46; asst prof, 46-48, prof & chmn dept, 48-67, dean, Col Arts & Sci, 66-69, dean, Grad Sch, 69-74, prof, 74-80, EMER PROF BIOL, ST BONAVENTURE UNIV, 80- *Concurrent Pos:* Dir, Med Technol Prog. *Mem:* AAAS; Am Soc Microbiol; Am Pub Health Asn; NY Acad Sci. *Res:* Physiology of bacteria; amino acid metabolism by the genus Proteus; bacterial genus Desulphovibrio; cysteine transaminase; bacteriophage for Clostridia. *Mailing Add:* 96 Rock City Rd Olean NY 14760

ANDERSON, KENNETH VERLE, b La Crosse, Wis, Sept 14, 38; m 67. NEUROPSYCHOLOGY, NEUROANATOMY. *Educ:* Carleton Col, AB, 60; Brown Univ, MSc, 62, PhD(exp psychol), 64. *Prof Exp:* Res psychologist, US Naval Electronics Lab, Calif, 62; lab instr neuroanat, Yale Univ, 65-66; from instr to assoc prof neurol, Emory Univ, 66-77, prof anat, 77-80; mem fac, dept anat, Univ Miss Med Ctr, 80-84; dir res & business mgr, Riverview Phys Therapy & Rehab Ctr, 84-88; DIR BIOL & COMP SCIENCE, BRANDON ACAD, 88- *Concurrent Pos:* US Pub Health Serv fel neurophysiol & neuroanat, Yale Univ, 64-66; NIMH res scientist develop award, 68-72, Nat Inst Neurol Dis & Blindness res grant, 69-71; grants, NIMH, 73-74, Nat Inst Dent Res, 75-78 & 79-82 & Nat Eye Inst, 77-79. *Mem:* AAAS; Am Psychol Asn; Am Asn Anat; Soc Neurosci; Int Asn Study Pain. *Res:* Analysis of neuroanatomical correlates of perception and learning, including microelectrode assessment of sensory codes and rigorous analysis of behavior concomitant with changes in sensory transmission; visual system and the somatosensory system. *Mailing Add:* 103 Hillview Dr Brandon MS 30492

ANDERSON, KENNING M, b Oak Park, Ill, July 17, 33; m 64; c 3. ENDOCRINOLOGY, ONCOLOGY. *Educ:* Northwestern Univ, Evanston, BS, 54; Northwestern Univ, Chicago, MD, 58, MSc, 65; Univ Chicago, PhD(biochem), 69; FRCP(C), 76. *Prof Exp:* Intern med, Northwestern Univ, 58-59, resident, 59-62; capt, US Army, 62-64; fel, Univ Chicago, 64-68; from asst to assoc prof clin biochem, Univ Toronto, Ont, 69-76; ASSOC PROF BIOCHEM, RUSH MED COL, CHICAGO & RUSH-PRESBY ST LUKE'S MED CTR, 76- *Mem:* Am Soc Biol Chem; Am Chem Soc; Am Soc Cell Biol; Biophys Soc. *Res:* Endocrinology; medical oncology; mechanisms of hormone action. *Mailing Add:* Sect Med Oncol Rush Med Col 1753 W Congress Pkwy Chicago IL 60612

ANDERSON, KINSEY AMOR, b Preston, Minn, Sept 18, 26; m 54; c 2. SOLAR PHYSICS, SPACE PHYSICS. *Educ:* Carleton Col, BA, 49; Univ Minn, Minneapolis, PhD(physics), 55. *Hon Degrees:* Dr, Univ Paul Sabatier, 89. *Prof Exp:* Res assoc, Univ Minn, Minneapolis, 55; res assoc, Univ Iowa, 55-58, asst prof physics, 58-60; from asst prof to assoc prof, 60-66, assoc dir, Space Sci Lab, 68-70, dir, 70-79, PROF PHYSICS, UNIV CALIF, BERKELEY, 66- *Concurrent Pos:* Guggenheim fel, Royal Inst Technol, Stockholm, Sweden & Nat Observ, Athens, Greece, 59-60; Fulbright lectr & res award, France, 81; mem, Lunar Sci Rev Panel, NASA, 76, Solar Syst Explor Comt, Off Space Sci, 80-84, Class Mem Comt, Nat Acad Sci, 85 & NASA Ctr Sci Assessment Team, 87; chmn, Res Assoc Prog, Nat Res Coun, 84 & Ad Hoc Comt Rev Inst Geophys & Planetary Physics, 89. *Honors & Awards:* Space Sci Award, Am Inst Aeronaut & Astronaut, 68; Except Sci Achievement Medal, NASA, 72; Alexander von Humboldt Award, 78. *Mem:* Nat Acad Sci; fel Am Phys Soc; fel Am Geophys Union; Am Astron Soc; Int Astron Union; fel AAAS; Int Union Radio Sci; Am Asn Physics Teachers; Sigma Xi. *Res:* Solar cosmic rays; auroral zone phenomena; radiation zone studies using high altitude balloons and satellites; tensiometer for measurements on biaxially stressed films; published numerous articles in various journals; awarded one patent; numerous technical publications. *Mailing Add:* Space Sci Lab Univ Calif Berkeley CA 94720

ANDERSON, L(AWRENCE) K(EITH), b Toronto, Ont, Oct 2, 35; m 63; c 2. OPTICAL ENGINEERING. *Educ:* McGill Univ, BEng, 57; Stanford Univ, PhD(elec eng), 62. *Prof Exp:* Mem tech staff, Solid State Device Develop Div, Bell Tel Labs, Inc, 61-63, supvr, 63-68, head, Optical Control Device Dept, 68-77, head photomask, Pattern Gen Dept, 77-80, HEAD, LIGHTWAVE SUBSYSTEM DEPT, BELL LABS, ALLENTOWN, PA, 80- *Mem:* Inst Elec & Electronics Engrs; Optical Soc Am. *Res:* Optical modulation and detection at microwave frequencies; optical properties of semiconductors; optical memories and displays. *Mailing Add:* Sandia Nat Lab PO Box 5800 Orgn 7481 Albuquerque NM 87185

ANDERSON, LARRY A, b Hindes, Iowa, Nov 2, 45. NUMERICAL ANALYSIS. *Educ:* Wheaton Col, BS, 67; Purdue Univ, MS, 69, PhD(math), 74. *Prof Exp:* PROF MATH & COMPUT SCI, LA TOURNEAU COL, 74- *Mem:* Math Asn Am; Asn Comput Mach. *Mailing Add:* Le Tourneau Univ Longview TX 75607

ANDERSON, LARRY BERNARD, b Cottonwood, Idaho, July 18, 37; m 62; c 3. ANALYTICAL CHEMISTRY. *Educ:* Univ Wash, Seattle, BSc, 59; Syracuse Univ, PhD(chem), 64. *Prof Exp:* Res assoc chem, Univ NC, 64-66; asst prof, 66-72, acad vchmn dept, 68-72, 78-80, ASSOC PROF CHEM, OHIO STATE UNIV, 72- *Concurrent Pos:* Consult, Allyn Corp, Gen Elec, DuPont & Yellow Springs Inst. *Mem:* Am Chem Soc; Electrochem Soc. *Res:* Electroanalytical chemistry; electrode kinetics; electrochemistry of organic and inorganic materials in non-aqueous system; photovoltaic solar energy conversion; semiconductor electrolyte interfaces; filar microlelectrodes for analysis. *Mailing Add:* Dept Chem Ohio State Univ 120 W 18th Columbus OH 43210

ANDERSON, LARRY DOUGLAS, b Traverse City, Mich, Sept 15, 47; m 71; c 2. ENDOCRINOLOGY, MOLECULAR BIOLOGY. *Educ:* Oakland Univ, BS, 70; Wayne State Univ, MS, 73, PhD(physiol), 76. *Prof Exp:* NIH fel res-reproduction endocrinol, Dept Physiol, Univ Md, 76-78, asst prof, 78-83, dir, In Vitro Fertilization Unit, 87, ASSOC PROF, DEPT ANAT, SCH MED, UNIV MD, 83- *Concurrent Pos:* Vis prof, Dept Physiol, Univ Goteborg, Sweden, 81 & Mich State Univ, 81. *Mem:* Endocrine Soc; Soc Study Reproduction; AAAS. *Res:* Basic mechanisms by which cells differentiate and the regulation by hormones and their control of steroid and protein synthesis and secretion in ovary, pituitary gland and mammary gland. *Mailing Add:* Dept Anat Sch Med Univ Md 655 W Baltimore Baltimore MD 21201

ANDERSON, LARRY ERNEST, b Corvallis, Ore, Jan 30, 43; m 67; c 9. NEUROCHEMISTRY, RESEARCH ADMINISTRATION. *Educ:* Brigham Young Univ, BSc, 68; Univ Ill, PhD(biochem), 73. *Prof Exp:* NIH sr fel biochem, Univ Wash, 73-77; res assoc, 77-78, sr scientist, 78-85, STAFF SCIENTIST, BATTELLE NORTHWEST, 85-, PROG DIR, BIOELECTROMAGNETICS, 85. *Concurrent Pos:* mem, Comt on Non-Ionizing Radiation, WHO, 85; deleg, US/USSR Coop Prog on Biol Effects of Phys Factors in Environ, 85-; bd dirs, Bioelectromagnetics Soc, 86-88. *Mem:* Bioelectromagnetics Soc (vpres, 88-89, pres, 89-90); Soc Neurosci. *Res:* Axoplasmic transport; proteases-isolation and mechanism of action; growth regulation-neural cells; neurotoxicity; biological effects of electric and magnetic fields; biological rhythms and pineal gland function. *Mailing Add:* Dept Biol & Chem MSK4-28 Battelle Northwest Richland WA 99352

ANDERSON, LARRY GENE, b Terre Haute, Ind, Nov 23, 46; div; c 2. PHYSICAL CHEMISTRY, ATMOSPHERIC CHEMISTRY. *Educ:* Rose Polytech Inst, BS, 68; Ind Univ, Bloomington, PhD(phys chem), 72. *Prof Exp:* Fel, Advan Study Prog, Nat Ctr Atmospheric Res, 72-73; assoc sr res scientist, Gen Motors Res Lab, 73-82, staff res scientist, environ sci dept & group leader, atmospheric chen, 82; ASSOC PROF, DEPT CHEM, UNIV COLO, DENVER, 82- *Concurrent Pos:* Vis scientist, Jet Propulsion Labs, 90-91. *Mem:* Am Chem Soc; Inter-Am Photochem Soc; Air & Waste Mgt Asn; Sigma Xi. *Res:* Investigation of the kinetics and mechanisms of chemical reactions of atmospheric importance; measurements of atmospheric pollutants to develop a more complete understanding of the chemistry of the atmosphere. *Mailing Add:* Dept Chem Univ Colo PO Box 173364 Denver CO 80217-3364

ANDERSON, LARS WILLIAM JAMES, b Pasadena, Calif, Apr 22, 45. PLANT PHYSIOLOGY, WEED SCIENCE. *Educ:* Univ Calif, Irvine, BA, 67; Calif State Univ, San Diego, MA, 70; Univ Calif, Santa Barbara, PhD(biol), 74. *Prof Exp:* Consult-marine biologist, Oceanog Serv Inc, 72-73; aquatic plant physiologist herbicides, Environ Protection Agency, 74-76; RES LEADER, AQUATIC WEED RES LAB, SCI & EDUC ADMIN-AGR

RES SERV, UNIV CALIF, USDA, 76- *Mem:* AAAS; Phycol Soc Am; Weed Sci Soc Am; Am Soc Plant Physiologists; Sigma Xi. *Res:* Phytoplankton physiology pertaining to vertical distribution; intracellular ion regulation in relation to cell buoyancy; control of aquatic weeds; function of plant growth regulators in aquatic macrophytes; uptake and translocation of herbicides in aquatic macrophytes; methods for assessing toxicity of pesticides to aquatic microorganisms. *Mailing Add:* Aquatic Weed Res Lab Agr Res Serv Bot Dept Univ Calif USDA Davis CA 95616

ANDERSON, LAURENS, b Belle Fourche, SDak, May 19, 20; m 45; c 3. CARBOHYDRATE CHEMISTRY. *Educ:* Univ Wyo, BS, 42; Univ Wis, MS, 48, PhD(biochem), 50. *Prof Exp:* Merck fel natural sci, Swiss Fed Inst Technol, 50-51; from asst prof to assoc prof, Univ Wis-Madison, 51-61, prof biochem, 61-86, Steenbock prof biomolecular structure, 81-86, EMER PROF BIOCHEM, UNIV WIS- MADISON, 86- *Concurrent Pos:* NIH sr fel, Ind Univ, 71-72; assoc ed, Carbohydrate Res, 84- *Honors & Awards:* Claude S Hudson Award, Carbohydrate Chem, Am Chem Soc, 84. *Mem:* Am Chem Soc; Am Soc Biochem & Molecular Biol; Int Endotoxin Soc; Int Carbohydrate Orgn; Sigma Xi. *Res:* Organic chemistry of carbohydrates; chemical synthesis of oligosaccharides; mutarotation of sugars. *Mailing Add:* Dept Biochem Univ Wis 420 Henry Mall Madison WI 53706-1569

ANDERSON, LAWRENCE SVEN, b Boston, Mass, Mar 4, 44; m 72; c 1. ASTRONOMY. *Educ:* Calif Inst Technol, BS, 66; Univ Calif, Berkeley, MA, 74, PhD(astron), 77. *Prof Exp:* Instr astron, Chabot Col, Hayward, Calif, 74-75; res astron, Univ Calif, Berkeley, 77-78; ASST PROF ASTRON & PLANETARIUM DIR, UNIV TOLEDO, 78-, AT RITTER OBSERV. *Concurrent Pos:* Ritter Observ. *Mem:* Am Astron Soc; AAAS. *Res:* The theory of the structure and appearance of contact binary stars; the reflection effect in close binary stars; convection in stellar photospheres. *Mailing Add:* Dept Physics/Astron Univ Toledo 2801 W Toledo OH 43606

ANDERSON, LEIGH, b Minneapolis, Minn, Aug 23, 49. ORAL BIOLOGY. *Educ:* Univ Minn, PhD(oral biol), 79. *Prof Exp:* ASST PROF ORAL BIOL, EMORY UNIV, ATLANTA, GA. *Mailing Add:* Dept Oral Biol Univ Wash Seattle WA 98195

ANDERSON, LEONARD MAHLON, b Palo Alto, July 11, 44; m 83. ELEMENTARY PHYSICS, APPLIED PHYSICS. *Educ:* Univ Calif, Berkeley, BS, 66, PhD(physics), 78. *Prof Exp:* Postdoctoral res, Ecole Polytech, 78-81; PRIN PHYSICIST, MEASUREX CORP, 81- *Mem:* Am Phys Soc. *Res:* On-line measurement of paper moisture, coating weight and strength. *Mailing Add:* 1355 Stockbridge Dr San Jose CA 95130

ANDERSON, LEWIS L, b Deadwood, SDak, Jan 22, 35; m 56; c 2. PHYSICAL BIOCHEMISTRY. *Educ:* SDak Sch Mines & Technol, BS, 56; Univ NMex, PhD(chem), 61. *Prof Exp:* RES CHEMIST, E I DU PONT DE NEMOURS & CO, 61- *Mem:* Sigma Xi. *Res:* Isotopic exchange; absorption spectroscopy; ultracentrifugation applied to biochemical systems; isolation of proteins; physical characterization and study of biological systems. *Mailing Add:* 223 S Summit Ave Quarryville PA 17566

ANDERSON, LLOYD JAMES, b Salt Lake City, Utah, Dec 18, 17; m 41; c 3. RADIOPHYSICS. *Educ:* Univ Calif, Los Angeles, AB, 39; Univ Calif, MA, 42. *Prof Exp:* Asst physicist, USN Electronics Lab, 42-44, head radio-meteorol sect, 50-53, head environ studies br, 53-55; supvry physicist, Smyth Res Assocs, 55-61; prin physicist, Calspan Corp, 61-82; RETIRED. *Mem:* Sigma Xi; AAAS. *Res:* Effects of weather and terrain on tropospheric radio propagation; atmospheric effects on refraction of tracking radar beams; radar/optical analysis of satellites. *Mailing Add:* 7948 Represa Circle Carlsbad CA 92009

ANDERSON, LLOYD L, b Nevada, Iowa, Nov 18, 33; m 70; c 2. ANIMAL SCIENCE, PHYSIOLOGY. *Educ:* Iowa State Univ, BS, 57, PhD(animal sci), 61. *Prof Exp:* Asst animal reprod, 57-58, res assoc, 58-61, from asst prof to assoc prof, 61-65, PROF ANIMAL SCI, IOWA STATE UNIV, 71- *Concurrent Pos:* NIH res fel, 61-62; Lalor Found res fel, Nat Inst Agron Res, France, 63-64; counr, Soc Exp Biol & Med, 80-83; sect ed endocrinol & physiol, J Animal Sci, 83-86; mem, Reproductive Biol Study Sect, NIH, 84-88; mem peer review panel, animal health spec res grants beef & dairy cattle reproductive dis, US Dept Agr, 86-88; mem NIH Reviewers, 88-92. *Honors & Awards:* Animal Physiology & Endocrinol Award, Am Soc Animal Sci, 88. *Mem:* Hon fel Am Soc Animal Sci; Am Asn Anat; Am Physiol Soc; Endocrine Soc; Soc Exp Biol & Med; Soc Study Reproduction; Soc Study Fertil; Sigma Xi; AAAS. *Res:* Central nervous system, pituitary, ovarian, uterine relationships primarily in farm animals, especially endocrine and neural influences in rhythmic regulation of reproduction cycle, hypothalamic regulation of prolactin, luteinizing hormone and growth hormone secretion in beef cattle and pigs, cytoplasmic localization and physiology of porcine relaxin, physiology of relaxin on parturition in cattle, maternal nutrition on fetal development in farm animals, neuroendocrine regulation of puberty in beef calves and pigs. *Mailing Add:* 11 Kildee Hall Dept Animal Sci Iowa State Univ Ames IA 50011-3150

ANDERSON, LORAN C, b Idaho Falls, Idaho, Feb 7, 36; m 57; c 2. BOTANY. *Educ:* Utah State Univ, BS, 58, MS, 59; Claremont Grad Sch, PhD(bot), 62. *Prof Exp:* Asst prof bot, Mich State Univ, 62-63 & Kans State Univ, 63-74; assoc prof, 74-85, PROF BIOL SCI & CUR HERBARIUM, FLA STATE UNIV, 85- *Concurrent Pos:* Assoc prog dir, NSF, 70-71. *Mem:* Am Soc Plant Taxonomists (treas, 74-); Bot Soc Am; Am Inst Biol Sci. *Res:* Comparative anatomy of Chrysothamnus and related genera of tribe Astereae of Compositae; forensic work with Cannabis. *Mailing Add:* Dept Biol Sci Fla State Univ Tallahassee FL 32306

ANDERSON, LOUIS WILMER, b Houston, Tex, Dec 24, 33; m 62; c 3. ATOMIC PHYSICS. *Educ:* Rice Inst Technol, BA, 56; Harvard Univ, AM, 57, PhD(physics), 60. *Prof Exp:* From asst prof to assoc prof, 60-68, PROF PHYSICS, UNIV WIS-MADISON, 68- *Mem:* Fel Am Phys Soc. *Res:* Radio frequency spectroscopy; atomic hyperfine structure; atomic structure; nuclear magnetic resonance in solids; atomic collisions; atom ion charge changing collisions; production of negative ions; magnetism; lasers; optical pumping; polarized ion sources; electron excitation collisions. *Mailing Add:* Dept Physics Univ Wis Madison WI 53706

ANDERSON, LOUISE ELEANOR, b Cleveland, Ohio, May 18, 34. BIOCHEMISTRY. *Educ:* Augustana Col, AB, 56; Cornell Univ, PhD(biochem), 61. *Prof Exp:* Res assoc bot, Wash Univ, 60-62; res assoc microbiol, Dartmouth Med Sch, 62-64, 66-67; Kettering Int fel, Food Preservation Div, Plant Physiol Unit, Commonwealth Sci Indust & Res Orgn, Australia, 64-65; res asst prof, Oak Ridge Biomed Grad Sch, Univ Tenn, 67-68; from asst prof to assoc prof, 68-75, PROF BIOL SCI, UNIV ILL, CHICAGO, 75- *Concurrent Pos:* Katzir-Katchalsky fel, Weizmann Inst Sci, Israel, 74-75; Fogarty Sensor Int fel, Univ Lund, Sweden, 81-82. *Mem:* AAAS; Am Soc Biol Chemists & Molecular Biol; Am Soc Plant Physiol. *Res:* Carbohydrate metabolism; metabolic control in plants and photosynthetic bacteria; enzymology. *Mailing Add:* Dept Biol Sci Univ of Ill Chicago Chicago IL 60680

ANDERSON, LOWELL LEONARD, b Spokane, Wash, Sept 3, 30; m 55; c 2. BIOPHYSICS. *Educ:* Whitworth Col, Spokane, Wash, BS, 53; Univ Rochester, PhD(biophys), 58. *Prof Exp:* From asst biophysicist to assoc biophysicist, Argonne Nat Lab, 58-69; assoc attend physicist, Mem Sloan-Kettering Cancer Ctr, Med Col, Cornell Univ, 69-77, asst prof physics in radiol, 70-77; asst prof psychiat, Sch Med, NY Univ, 77-; AT MED PHYSICS DEPT, MEM HOSP, NEW YORK. *Mem:* AAAS; Health Physics Soc; Am Phys Soc; Sigma Xi; Am Asn Physicists in Med. *Res:* Neutron measurement methods for health physics; nuclear accident dosimetry; intermediate-energy neutron calibration source; medical radiation dosimetry. *Mailing Add:* Med Physics Dept Mem Hosp 1275 York Ave New York NY 10021

ANDERSON, LOWELL RAY, b Burnt Prairie, Ill, Nov 4, 34; m 60. ORGANIC CHEMISTRY. *Educ:* Southern Ill Univ, BA, 56; Ohio State Univ, PhD(chem), 64. *Prof Exp:* Res chemist, Allied Chem Corp, 64-66, sr res chemist, 66-68, tech supvr, 68-69, res group leader, 69-79; TECH ASSOC, GAF CORP, 79- *Mem:* Am Chem Soc. *Res:* Synthesis of fluorinated compounds, particularly fluorinated hypohalites, peroxides and ethers; catalysis; acetylenic compounds and derivatives; pyrrolidones, vinyl ethers and other vinyl compounds. *Mailing Add:* 27 Sunderland Dr Morristown NJ 07960

ANDERSON, LUCIA LEWIS, b Pittsburgh, Pa, Aug 9, 22; m 43, 55; c 3. MICROBIOLOGY. *Educ:* Univ Pittsburgh, BS, 43, MS, 44, PhD(microbiol), 46. *Prof Exp:* Asst prof, Univ Pittsburgh, 49-53 & Duquesne Univ, 54-55; res assoc, Univ Pittsburgh, 55-56; from assoc prof to prof biol, Western Ky State Col, 58-65; sr info res scientist, Squibb Inst Med Res, NJ, 65-66; assoc prof biol, Parsons Col, 66-68; assoc prof, 68-70, PROF BIOL, QUEENSBORO COL, CITY UNIV NEW YORK, 70- *Concurrent Pos:* NSF guest lectr for high schs. *Mem:* Am Soc Microbiol; Am Acad Microbiol; AAAS; Nat Sci Teachers Asn. *Res:* Antibiotics; microbiology education. *Mailing Add:* 26 Dunemere Lane East Hampton NY 11937

ANDERSON, LUCY MACDONALD, b Huntington, WVa, Oct 10, 42; m 64; c 2. CANCER, ETIOLOGY. *Educ:* Bryn Mawr Col, AB, 64; Univ Pa, PhD(biol), 68. *Prof Exp:* Res assoc biol, Univ Pa, 68-70; asst prof biol, Carleton Col, 71; asst prof biol, Macalester Col, 73; asst mem carcinogenesis, Sloan-Kettering Inst Cancer Res, 73-82; spec expert, 83-89, res biologist, 89-90, SECT CHIEF, NAT CANCER INST, 90- *Concurrent Pos:* NIH training grant biochem, Univ Minn, Minneapolis, 70-72; lectr, Bryn Mawr Col, 68-69. *Mem:* Am Asn Cancer Res; Soc Toxicol. *Res:* Insect egg follicle development; metabolism of environmental carcinogens; mechanistic factors in transplacental carcinogenesis; promotion of lung tumors. *Mailing Add:* Natl Cancer Inst Bldg 538 Ft Detrick Frederick MD 21701

ANDERSON, MARILYN P, BONE PATHOLOGY, REPTILIAN & AVIAN PATHOLOGY. *Educ:* Ohio State Univ, PhD(comp biol), 76. *Prof Exp:* PATHOLOGIST, ZOOL SOC SAN DIEGO, 77- *Mailing Add:* Dept Path Zool Soc San Diego P O Box 551 San Diego CA 92112-0551

ANDERSON, MARION C, b Concordia, Kans, Oct 9, 26; m 49; c 4. SURGERY. *Educ:* Northwestern Univ, BM, 50, MD, 53, MS, 62. *Prof Exp:* Resident surg, Passavant Mem Hosp, Chicago, 54-55; resident, Vet Admin Res Hosp, 55-56; resident, Passavant Mem Hosp, 56-57; assoc, Cook County Hosp, 57-58; clin asst, Sch Med, Northwestern Univ, 58-59, instr, 59-60, assoc, 60-61, from asst prof to assoc prof, 61-69; chmn, Dept Med, 69-72, prof surg, Med Col Ohio, 69-77, pres, Col, 72-77; prof & chmn dept surg, 77-88, EXEC ASSOC DEAN, COL MED, MED UNIV SC, 88- *Concurrent Pos:* Kemper res scholar, Am Col Surgeons, 61-64; assoc dean, grad med educ & continuing educ, Med Univ SC. *Mem:* AMA: Am Asn Surg of Trauma; Soc Univ Surg; Am Surg Asn; Soc Surg Alimentary Tract. *Res:* Biliary and pancreatic diseases. *Mailing Add:* Dean's Off Med Univ SC Charleston SC 29425

ANDERSON, MARJORIE ELIZABETH, NEUROPHYSIOLOGY, MOTOR CONTROL. *Educ:* Univ Wash, PhD(physiol), 69. *Prof Exp:* PROF PHYSIOL & BIOPHYSIOL, UNIV WASH, 71- *Res:* Basalganglia function. *Mailing Add:* Dept Physiol Biophys & Rehab Univ Wash Mail Stop RJ-30 Seattle WA 98195

ANDERSON, MARLIN DEAN, b Stanton, Iowa, Nov 30, 34; m 68; c 1. ANIMAL NUTRITION, ANIMAL PHYSIOLOGY. *Educ:* Iowa State Univ, BS, 59, MS, 62, PhD(animal nutrit), 64. *Prof Exp:* Animal nutritionist, Hess & Clark Div, Rohdia, Inc, 65-76; mgr antibiotic prod develop, SDS Biotech Corp, 77-83; DIR, FEED ADDITIVE DEVELOP, FERMENTA ANIMAL HEALTH CO, 84- *Mem:* AAAS; Am Soc Animal Sci; Am Soc Prof Animal Scientists; Am Asn Swine Practr. *Res:* Animal health and nutrition; drug efficacy; safety. *Mailing Add:* 1575 Rte 511 RD 7 Ashland OH 44805

ANDERSON, MARSHALL W, b Lynchburg, Va. MOLECULAR BIOLOGY. *Educ:* Emory & Henry Col, BS, 61; Univ Tenn, PhD(math), 66. *Prof Exp:* Asst prof, Dept Math, Univ Tenn, 66-67; tech staff, Bell Telephone Labs, 67-69; postdoctoral fel, Biomath Dept, NC State Univ, 69-71; sr staff, Biometry Br, 75-84, head, Molecular Toxicol Sect, Lab Biochem Risk Anal, 84-88, CHIEF, LAB MOLECULAR TOXICOL, DEPT BIOCHEM RISK ANALYSIS, NAT INST ENVIRON HEALTH SCI, RESEARCH TRIANGLE PARK, NC, 89- *Concurrent Pos:* Adj prof, Toxicol & Biomath Dept, NC State Univ & Path Dept, Sch Med, Univ NC, Chapel Hill; thesis adv, NC State Univ & Med Col, Toledo, Ohio; mem, Comt Eval Toxicity Testing; mem, Grant Rev Panel, Health Effects Lab, State Calif & Am Cancer Soc. *Honors & Awards:* Outstanding Performance Award, NIH, 87-91, Award of Merit, 88. *Mem:* Am Asn Cancer Res; AAAS; Int Soc Quartum Biol. *Res:* Chemical carcinogenesis; environmental toxicology; detection and identification of activated proto-oncogenes in human and rodent tumors, with particular emphasis on rodent models for liver and lung neoplasia; genetic factors related to differential susceptibility of rodent tumors between mice strains. *Mailing Add:* 111 Alexander Dr MD A3-02 Research Triangle Park NC 27709

ANDERSON, MARVIN DAVID, b Detroit, Mich, Apr 30, 38; m 79; c 2. INTERNAL MEDICINE. *Educ:* Univ Notre Dame, BS, 60; Georgetown Univ, MD, 64. *Prof Exp:* Intern rotating, Harper Hosp, Detroit, 64-65, resident med, 65-66; res assoc cancer, NIH, Bethesda, Md, 66-68; sr resident med, Harlem Hosp Ctr, Col Physicians & Surgeons, Columbia Univ, 68-70, res assoc microbiol, 70-71; fel med, 71-72, SR STAFF PHYSICIAN, HENRY FORD HOSP, 73- *Concurrent Pos:* Asst clin prof med, Sch Med, Univ Mich, Ann Arbor, 74- *Mem:* Am Col Physicians. *Res:* Nutrition. *Mailing Add:* 2799 W Grand Blvd Detroit MI 48202

ANDERSON, MARY ELIZABETH, b San Antonio, Tex, Jan 2, 56. BIOCHEMISTRY. *Educ:* Hollins Col, BA, 77; Cornell Univ, PhD(biochem), 83. *Prof Exp:* Postdoctoral fel, 83-87, instr, 87-89, ASST PROF, DEPT BIOCHEM, MED COL, CORNELL UNIV, 89- *Concurrent Pos:* Postdoctoral fel training grant, NIH, 83-86. *Mem:* Am Chem Soc; Sigma Xi; Am Soc Biochem & Molecular Biol. *Mailing Add:* Dept Biochem Med Col Cornell Univ 1300 York Ave New York NY 10021

ANDERSON, MARY LOUCILE, b Kansas City, Kans, Feb 16, 40; c 1. REPRODUCTIVE ENDOCRINOLOGY. *Educ:* Univ Kans, AB, 62; Univ Calif, San Francisco, MA, 68, PhD(endocrinol), 73. *Prof Exp:* Res fel antifertil agents, Pop Res Inst, Oak Ridge Assoc Univs, 72-75; res assoc reproduction, Vanderbilt Univ, 75-77, instr, 77-78; asst prof, 78-80, ASSOC ANAT, TULANE UNIV, 80- *Mem:* Soc Study Reproduction. *Res:* In vitro regulation of relaxin synthesis in corpora lutea of pregnancy; investigation and characterization of hormone binding sites for relaxin in its target tissues. *Mailing Add:* Dept Anat Tulane Univ Med Sch 1430 Tulane Ave New Orleans LA 70112

ANDERSON, MARY PIKUL, b Buffalo, NY, Sept 30, 48; m 73. GROUNDWATER HYDROLOGY. *Educ:* State Univ NY Buffalo, BA, 70; Stanford Univ, MS, 71, PhD(hydrogeol), 73. *Prof Exp:* Adj asst prof stratig & hydrogeol, Southampton Col, Long Island Univ, 73-75; from asst prof to assoc prof, 75-85, PROF HYDROGEOL, UNIV WIS-MADISON, 85- *Concurrent Pos:* Assoc ed, EOS, Am Geophys Union, 82-84; mem, Water Sci & Technol Bd, Nat Acad Sci, 85-87. *Mem:* Am Geophys Union; Nat Water Well Asn. *Res:* Analysis of mathematical models of subsurface flow systems; groundwater-surface water interaction; contaminant transport. *Mailing Add:* 2114 Bascom St Madison WI 53706

ANDERSON, MARY RUTH, b Bellingham, Wash, Feb 3, 39; m 66; c 2. INDUSTRIAL ENGINEERING, STATISTICS. *Educ:* Hope Col, BA, 61; Univ Iowa, MS, 63, PhD(math & statist), 66. *Prof Exp:* Teaching asst math, Univ Iowa, 63-66; lectr math, 66-72, asst to dean, Col Lib Arts, 72-79, asst prof eng, 74-78, ASSOC PROF ENG, FAC INDUST ENG, ARIZ STATE UNIV, 79-, DIR, GRAD REENTRY PROG, 81-, ASSOC CHAIRPERSON & GRAD STUDENT COORDR, 83- *Concurrent Pos:* Consult, US Indian Health Serv, 73-74, Ariz Solar Energy Comn, 80-, Rogers Corp, 81-, NIH, 82, Digital Equip Corp, 82, Sperry Flight Systs, 83, Motorola, 84 & Dept Transp, 85-; chairperson, Women Academia Comt, Soc Women Engrs, 84-86. *Mem:* Am Inst Indust Engrs; Soc Women Engrs; Asn Women Math; Am Soc Eng Educ. *Res:* Exponential class characterizations; applied statistics in engineering, nursing and transportation; economic analysis of solar water heating; reentry programs for women and men in engineering; women in engineering. *Mailing Add:* Dept Eng Ariz State Univ Tempe AZ 85287

ANDERSON, MAURITZ GUNNAR, b Chicago, Ill, Aug 11, 18; m 42; c 2. FOOD SCIENCE & TECHNOLOGY, INDUSTRIAL MICROSCOPY. *Educ:* Univ Mich, AB, 42; Ind Univ, MS, 62; Va Polytech Inst & State Univ, PhD, 76. *Prof Exp:* Supvr histol & micros div, Swift & Co, 49-57; chief histologist, Norwich Pharmacal Co, 57; res histologist & microscopist, Swift & Co, 58-63; from instr to prof, 63-88, EMER PROF BIOL, TOWSON STATE UNIV, 88- *Concurrent Pos:* Mem, Deleg Appl Microbiologists to China, 84. *Mem:* Bot Soc Am; Mycol Soc Am; Am Asn Feed Micros; Int Asn Aerobiol; Inst Food Technol; Sigma Xi. *Res:* Application of microscopy and microchemistry to biology, pollution and food technology; mycology. *Mailing Add:* Dept Biol Sci Towson State Univ Baltimore MD 21204

ANDERSON, MELVIN JOSEPH, b Heber City, Utah, Feb 14, 27; m 55; c 5. ANIMAL NUTRITION. *Educ:* Utah State Univ, BS, 50; Cornell Univ, MS, 57, PhD(animal nutrit), 59. *Prof Exp:* Asst animal nutrit, Cornell Univ, 55-58; asst prof animal sci, Univ Maine, 58-61; res animal sci, Off Int Coop & Develop, Dept Agr, Riyadh, Saudi Arabi, 82-85; dairy husbandman, Agr Res Serv, 61-82, RES ASSOC PROF, UTAH STATE UNIV, 86- *Concurrent Pos:* Assoc prof, Utah State Univ, 71. *Mem:* Am Dairy Sci Asn; Am Soc Animal Sci. *Res:* Dairy nutrition; nutritional evaluation of forages, hay cubes and level of concentrates for livestock; utilization of liquid whey for feeding dairy animals; effect of exercising pregnant dairy cows; feeding whole cottonseed to dairy animals; testing infrared spectroscopy to evaluate forages and feeds; effect of bovine somatotropin on lactating cows. *Mailing Add:* ADVS UMC 4815 Utah State Univ Logan UT 84322

ANDERSON, MELVIN LEE, b Perkins, Mich, Jan 19, 28; m 55; c 2. INORGANIC CHEMISTRY. *Educ:* Mich Technol Univ, BS, 53, MS, 55; Mich State Univ, PhD(inorg chem), 65. *Prof Exp:* Resident assoc actinide ion solution chem, Argonne Nat Lab, 53-54, jr chemist, 54-55; chemist, Dow Chem Co, Midland, 56-61, res chemist, 63-65, Rocky Flats Div, Colo, 65-69; from asst prof to assoc prof, 69-85, PROF CHEM, LAKE SUPERIOR STATE UNIV, 85- *Concurrent Pos:* vis assoc prof, Gustavus Adolphus Col, St Peter, Minn, 80-81. *Mem:* Am Chem Soc; Sigma Xi. *Res:* Actinides and lanthanides; coordination complexes; separation and purification; nonaqueous solvents; chemical education; organometallics; nitrogen-fluorine oxidizers. *Mailing Add:* Dept Biol & Chem Lake Superior State Univ Sault Ste Marie MI 49783

ANDERSON, MELVIN W(ILLIAM), b Baltimore, Md, Jan 9, 37; m 60; c 2. WATER RESOURCES. *Educ:* Va Mil Inst, BS, 59; Carnegie-Mellon Univ, MS, 63, PhD(water resources), 67. *Prof Exp:* From asst prof to assoc prof civil eng, La State Univ, 63-69; assoc prof water resources, 69-72, chmn, dept civil eng & mech, 79-89, PROF WATER RESOURCES ENG, UNIV SFLA, 72-, ASSOC DEAN ACAD AFFAIRS, 89- *Mem:* Am Soc Civil Engrs; Am Soc Eng Educ; Am Water Resources Asn; Nat Soc Prof Engrs. *Res:* Multidisciplinary studies involved in sustaining and managing the quality and quantity of water available for human activities. *Mailing Add:* Col Engr 118 Univ SFla Tampa FL 33620

ANDERSON, MICHAEL PETER, algebra, for more information see previous edition

ANDERSON, MILES EDWARD, b Ft Worth, Tex, Dec 13, 26; m 48; c 3. ACOUSTICS, ELECTROMAGNETISM. *Educ:* NTex State Univ, BS, 49, MS, 50; Stanford Univ, PhD(physics), 63. *Prof Exp:* Instr physics, NTex State Univ, 50-52; asst, Willow Run Res Ctr, Univ Mich, 52-54; asst prof, 54-56, assoc vpres acad affairs, 69-75, vpres acad affairs, 75-79, PROF PHYSICS, UNIV NTEX, 60- VPRES ACAD AFFAIRS, 75- *Concurrent Pos:* Lectr, Washington Univ, 62-63. *Mem:* Am Phys Soc; Am Asn Physics Teachers; Acoust Soc Am. *Mailing Add:* Dept Physics Univ NTex Box 5368 Denton TX 76203

ANDERSON, MILO VERNETTE, b Wolsey, SDak, Dec 24, 24; m 43; c 4. MICROWAVE PHYSICS, MEDICAL PHYSICS. *Educ:* Union Col, Nebr, BA, 49; Univ Nebr, MA, 55; Univ Colo, Boulder, PhD(elec eng), 71. *Prof Exp:* From instr to asst prof physics, Union Col, Nebr, 49-57, chmn dept, 57-59; physicist, Nat Bur Stand, 59-64; from assoc prof to prof physics, Pac Union Col, 64-86, chmn dept, 73-81; INDEPENDENT CONSULT, MED PHYSICS, 87- *Concurrent Pos:* High sch teacher sci & math, Platte Valley Acad, Shelton, Nebr, 49-52; res physicist, IBM, San Jose, Calif, 81; vis prof, Andrews Univ, Berrien Springs, Mich, 89. *Mem:* Am Asn Physics Teachers; Am Asn Physicists Med; Sigma Xi. *Res:* Applications of microwave resonant structures. *Mailing Add:* Dept Physics Pac Union Col Angwin CA 94508

ANDERSON, MILTON MERRILL, b Litchfield, Minn, Dec, 5, 43; m 69; c 3. ELECTRONICS ENGINEERING. *Educ:* Univ Minn, BS, 65, MS, 67. *Prof Exp:* Mem tech staff, Bell Tel Labs, 67-73, supvr, 73-82; supvr, Am Bell, 82-84; DIST MGR, BELL COMMUN RES, 84- *Mem:* Inst Elec & Electronics Engrs. *Res:* Compressed digital video and image coding; audio coding; speech processing; telecommunications network applications of digital signal coding and processing. *Mailing Add:* Bellcore MRE 2A257 445 South St Morristown NJ 07962-1910

ANDERSON, NEIL ALBERT, b Minneapolis, Minn, Oct 21, 28; m 60; c 2. PLANT PATHOLOGY, MYCOLOGY. *Educ:* Univ Minn, BS, 51, MS, 57, PhD(plant path), 60. *Prof Exp:* From instr to assoc prof, 59-70, PROF MYCOL & PLANT PATHOGEN GENTICS, UNIV MINN, ST PAUL, 70- *Concurrent Pos:* Vis prof, Waite Inst, Univ Adelaide, 70-71. *Mem:* Am Phytopath Soc; Mycol Soc Am; Sigma Xi; Potato Asn Am; AAAS. *Res:* Genetic studies on the plant pathogen, Rhizoctonia solani; plant pathology studies on Hypoxylon mammosum and Verticillium on potato. *Mailing Add:* Dept Plant Path Univ Minn 206 Plant Path St Paul MN 55108

ANDERSON, NEIL OWEN, b Chicago, Ill, Oct 10, 35; m 57; c 2. CELL PHYSIOLOGY. *Educ:* Ripon Col, BA, 57; Drake Univ, MS, 64; Univ Ill, PhD(physiol), 70. *Prof Exp:* PROF BIOL, EAST STROUDSBURG STATE COL, 70- *Mem:* AAAS; Sigma Xi (secy). *Res:* An examination of the nuclear proteins and their role in the cell. *Mailing Add:* 14 Spangenburg Ave East Stroudsburg PA 18301

ANDERSON, NEIL VINCENT, b Monterey, Minn, Apr 14, 33; m 53; c 6. VETERINARY MEDICINE. *Educ:* Mankato State Col, BS, 53; Univ Minn, St Paul, BS, 59, DVM, 61, PhD(vet path), 68. *Prof Exp:* Res asst vet med, Univ Minn, St Paul, 61; NIH res fels, 62-63, res asst vet path, 63-65; asst vet med, Auburn Univ, 65-67; from asst prof to assoc prof, 67-75, PROF VET MED, KANS STATE UNIV, 75- *Mem:* Am Vet Med Asn; Com Gastroenterol Soc (pres, 70-71); Am Col Vet Internal Med. *Res:* Clinical investigation and experimental study of canine pancreas and small intestine. *Mailing Add:* Dept Vet Med Kansas State Univ Manhattan KS 66506

ANDERSON, NELS CARL, JR, b Detroit Lakes, Minn, June 15, 36; m 63; c 2. ENDOCRINOLOGY, BIOPHYSICS. *Educ:* Concordia Col, Moorhead, Minn, BA, 58; Kans State Univ, MS, 60; Purdue Univ, PhD(endocrinol), 64. *Prof Exp:* NIH fel, 64-66; ASST PROF OBSTET, GYNEC, PHYSIOL & PHARMACOL, MED CTR, DUKE UNIV, 66- *Mem:* Soc Gen Physiol; Am Physiol Soc; Sigma Xi. *Res:* Hormonal control of myometrial structure and function, particularly the hormonal and ionic basis of excitation in uterine smooth muscle. *Mailing Add:* Phys/Obstet-Gynec Duke Univ Med Ctr Durham NC 27710

ANDERSON, NORMAN GULACK, b Davenport, Wash, Apr 21, 19; m 43; c 2. PHYSIOLOGY. *Educ:* Duke Univ, BA, 47, MA, 49, PhD(physiol), 52. *Prof Exp:* Biologist, Oak Ridge Nat Lab, 52-66, coordr joint NIH-AEC Zonal Centrifuge Develop Prog, Biol Div, 66-69, dir joint NIH-AEC Molecular Anat Prog, 69-74; mem fac, Dept Basic & Clin Immunol & Microbiol, Med Univ SC, 74-77; head, Molecular Anat Prog, Argonne Nat Lab, 77-84; CHMN & CHIEF SCIENTIST, LARGE SCALE BIOL CORP, 84- *Mem:* Am Physiol Soc; Am Soc Zool; Biophys Soc. *Res:* Cellular physiology; biochemistry; isolation of cell components; effects of ionizing radiations; embryonic antigens in cancer; development of zonal centrifuges; rotating fast chemical analyzers; isopyenometric serology; early cancer detection. *Mailing Add:* 9620 Medical Ctr Dr Suite 201 Rockville MD 20850

ANDERSON, NORMAN HERBERT, b Edam, Sask, Mar 17, 33; m 56; c 4. AQUATIC ECOLOGY, ENTOMOLOGY. *Educ:* Univ BC, BSA, 55; Ore State Col, MS, 58; Imp Col, Univ London, dipl, 61; Univ London, PhD(entom), 61. *Prof Exp:* Res officer entom, Can Dept Agr, 55-62; from asst prof to assoc prof, 62-75, PROF ENTOM, ORE STATE UNIV, 75- *Concurrent Pos:* NSF, US Dept Interior & Agr Res Serv, US Dept Agr grants. *Mem:* Ecol Soc Am; Entom Soc Am; Entom Soc Can; Brit Freshwater Biol Asn; NAm Benthological Soc. *Res:* Aquatic entomology; insect and animal ecology; ecology and taxonomy of aquatic insects. *Mailing Add:* Dept Entom Cordley Hall Ore State Univ Corvallis OR 97331-2907

ANDERSON, NORMAN LEIGH, b Durham, NC, Aug 3, 49; m 84. MOLECULAR BIOLOGY. *Educ:* Yale Univ, BA, 71; Cambridge Univ, PhD(molecular biol), 75. *Prof Exp:* Fel, Med Univ SC, 75-76; fel, Argonne Nat Lab, 76-78, asst biophysicist molecular anat, 78-85; vpres res, Proteus Technologies, 85-88; VPRES RES LARGE SCALE BIOL CORP, 85- *Honors & Awards:* Young Investr Award, Am Asn Clin Chem. *Mem:* Am Soc Cell Biol; Electrophoresis Soc; Am Asn Clin Chem; Soc Toxicol. *Res:* Molecular anatomy; gene product mapping; cataloging all human protein gene products. *Mailing Add:* Large Scale Biol Corp 9620 Medical Center Dr Rockville MD 20850

ANDERSON, NORMAN RODERICK, b Tacoma, Wash, Oct 12, 21; m 49; c 2. GEOLOGY & GEOMORPHOLOGY, ELEMENTARY SCIENCE EDUCATION. *Educ:* Univ Puget Sound, BS, 46; Univ Wash, MS, 54; Univ Utah, PhD(geol), 65. *Prof Exp:* Geologist, US Army Engrs, 47-48; from instr to prof, 46-84, chmn dept, 57-84, EMER PROF GEOL, UNIV PUGET SOUND, 84- *Concurrent Pos:* Dir, NSF Prog, Elementary Sci Teachers, 85-, NSF grant, TPE 8855548. *Mem:* Fel Geol Soc Am; Soc Econ Paleont & Mineral; Nat Asn Geol Teachers; Sigma Xi. *Res:* Pleistocene geology; glacial geology of Western Washington; geology of the Snake River Plain, Southwestern Idaho. *Mailing Add:* Dept Geol Univ Puget Sound Tacoma WA 98416

ANDERSON, O(RVIL) ROGER, b East St Louis, Ill, Aug 4, 37. CELL BIOLOGY, MARINE BIOLOGY. *Educ:* Washington Univ, AB, 59, MAEd, 61, EdD(bot), 64. *Prof Exp:* From asst prof to assoc prof, 64-70, PROF NATURAL SCI, TEACHERS COL, COLUMBIA UNIV, 70- *Concurrent Pos:* From res scientist to sr res scientist, Lamont-Doherty Geol Observ, 66-; biol consult, Bd Higher Educ, State NJ, 68; ed, J Res in Sci Teaching; vis prof cellular micropaleont, Tubingen Univ, 79, 84 & 87; assoc ed, J Protozool, 88- *Honors & Awards:* Diatom species named in honor, Cocconeis andersonii. *Mem:* AAAS; Sigma Xi; NY Acad Sci; Am Inst Biol Sci; Nat Asn Res Sci Teaching (pres, 76); Soc Protozoologists. *Res:* Ultrastructure and physiological ecology of marine microplankton; author of textbooks on protozoology. *Mailing Add:* 525 W 120th St New York NY 10027

ANDERSON, OLIVO MARGARET, b Omaha, Nebr, June 17, 41; m 71. DEVELOPMENTAL BIOLOGY. *Educ:* Augustana Col, BA, 63; Stanford Univ, PhD(biol), 67. *Prof Exp:* Postdoctoral fel biol, Harvard Univ, Cambridge, Mass, 68-70; res assoc neurobiol, Lab Neurobiol, Univ PR, 70-71; affil asst prof biol, Clark Univ, Worcester, Mass, 72-75; instr biol, Bennington Col, Bennington, Vt, 73; from asst prof to assoc prof, 73-85, PROF BIOL, SMITH COL, NORTHAMPTON, MASS, 85- *Concurrent Pos:* Prin investr, NIH grants, 75-85; mem, Bingham Selection Comt, Excellence Teaching, Transylvania Univ, Lexington, Ky, 87-; assoc ed, Advances Physiol Educ, Am Physiol Soc, 89-; co-prin investr, NSF grant, 89-90. *Mem:* Am Physiol Soc; Soc Gen Physiologists; Biophys Soc; Soc Neurosci. *Res:* Polarity in the unfertilized cockroach egg using electrophysiological techniques and scanning and transmission electron microscopy. *Mailing Add:* Dept Biol Sci Smith Col Northampton MA 01063-0010

ANDERSON, OREN P, b Iowa City, Iowa, June 19, 42; m 67; c 2. INORGANIC CHEMISTRY. *Educ:* Carleton Col, Northfield, Minn, BA, 64; Northwestern Univ, Evanston, Ill, PhD(chem), 68. *Prof Exp:* Res fel chem, Univ Bergen, Norway, 68-70; Univ res fel chem, Univ Leicester, England, 70-72; asst prof chem, Univ Mich, Ann Arbor, 72-74; from asst prof to assoc prof, 74-86, PROF CHEM, COLO STATE UNIV, 86- *Concurrent Pos:* Vis prof, Univ Minn, 89; chmn, Dept Chem, Colo State Univ, 89- *Mem:* Am Chem Soc; Royal Soc Chem; Am Crystallog Asn. *Res:* Structural and synthetic studies on transition metal complexes, especially those of biological interest. *Mailing Add:* Dept Chem Colo State Univ Ft Collins CO 80523

ANDERSON, ORSON LAMAR, b Price, Utah, Dec 23, 24; m 76; c 2. GEOPHYSICS, MINERAL PHYSICS. *Educ:* Univ Utah, BS, 48, MS, 49, PhD(physics), 51. *Prof Exp:* Instr mech eng, Univ Utah, 50-51, asst res prof physics, 51; mem tech staff, Bell Tel Labs, 51-60; mgr mat dept, Res Div, Am Standard Co, 60-63; res assoc geophys, Lamont Geol Observ, Columbia Univ, 63-64; adj prof, 64-65; prof mineral, Dept Geol & Lamont Geol Observ, 66-71; prof geophys, Inst Geophys & Planetary Physics & Dept Geophys & Space Physics, 66-74; DIR, SYSTEMWIDE INST GEOPHYS & PLANETARY PHYSICS, UNIV CALIF, LOS ANGELES, 78- *Concurrent Pos:* Ed-in-chief, J Geophys Res, 67-73. *Honors & Awards:* Meyer Award, Am Ceramic Soc, 54. *Mem:* Am Ceramic Soc; Am Phys Soc; Am Geophys Union. *Res:* Mechanical properties of solids; equations of state; high pressure physics; acoustics; physics of the earth's interior. *Mailing Add:* Dept Earth Sci Univ Calif 405 Hilgard Ave Los Angeles CA 90024

ANDERSON, OSCAR EMMETT, b Englewood, Fla, Nov 4, 16; m 46; c 2. AGRONOMY, SOIL FERTILITY. *Educ:* Univ Fla, BS, 41, Rutgers Univ, PhD(soils), 55. *Prof Exp:* Asst soil scientist, Ga Exp Sta & Ga Br Exp Sta, 55-58, assoc soil scientist, 58-65; head soil sect, 55-67, head dept, 71-81, PROF & SOIL SCIENTIST, DEPT AGRON, GA EXP STA, UNIV GA, 65- *Concurrent Pos:* Consult soil problems, Nigeria, 76, 80, 81 & 83. *Mem:* Sigma Xi; Am Soc Agron; Am Soc Crop Sci; Am Soc Soil Sci; Int Soc Soil Sci. *Res:* Soil-plant relationships, particularly trace elements in soils and their availability to plants; microelements. *Mailing Add:* 627 Windy Hill Rd Griffin GA 30223

ANDERSON, OWEN THOMAS, b Adams, Minn, May 11, 31; m 56; c 5. PHYSICS. *Educ:* St Olaf Col, BA, 53; Univ Wis, MS, 55, PhD(physics), 60. *Prof Exp:* Assoc engr, Adv Systs Develop Div, Int Bus Mach Corp, 58-60, adv engr, 60-61; from asst prof to assoc prof physics, 61-75, PROF PHYSICS, BUCKNELL UNIV, 75- *Mem:* Am Phys Soc; Am Asn Physics Teachers. *Res:* Low temperature physics; properties of the liquid helium II film; atmospheric optical effects. *Mailing Add:* Dept Physics Bucknell Univ Lewisburg PA 17837

ANDERSON, PAUL DEAN, b Grand Junction, Colo, Oct 28, 40; m 62; c 2. ANALYTICAL CHEMISTRY, COMPUTER APPLICATIONS. *Educ:* Univ Calif, Berkeley, BS, 62; Mass Inst Technol, PhD(anal chem), 66. *Prof Exp:* Res chemist, 66-68, sr res chemist, 68-78, supvr, 78-82, SR RES ASSOC, PHILLIPS PETROL CO, 82- *Mem:* Am Chem Soc; Sigma Xi. *Res:* Computer applications to analytical chemistry; laboratory automation; data acquistion and process control; computer applications to research. *Mailing Add:* 2301 Avalon Rd Bartlesville OK 74006

ANDERSON, PAUL J, b Akron, Ohio, Oct 11, 25. NEUROPATHOLOGY, NEUROLOGY. *Educ:* Ohio Univ, BS, 49; Univ Chicago, MD, 53; Am Bd Psychiat & Neurol, dipl, 59; Am Bd Path, dipl, 60. *Prof Exp:* Clin instr neurol, NY Univ, 60-61; asst prof neuropath, Col Physicians & Surgeons, Columbia Univ, 61-67; assoc prof path, 65-66, assoc prof neurol, 66-68, PROF NEUROPATH, 67-, PROF NEUROL, MT SINAI SCH MED, 87- *Concurrent Pos:* Nat Inst Neurol Dis & Blindness spec fel, 57-59; asst attend pathologist, Mt Sinai Hosp, New York, 58-61, from asst attend neurologist to assoc attend neurologist, 59-68, assoc attend pathologist in charge, Div Neuropath, 61-74, attend pathologist in charge, Div Neuropath, 74-; ed, J Neuropath & Exp Neurol, 61-81; consult neuropathologist, City Hosp Elmhurst, 64-; ed-in-chief, J Histochem & Cytochem, 73- *Mem:* Fel Am Acad Neurol; Am Asn Neuropath; Am Asn Path & Bact; Asn Res Nerv & Ment Dis; Am Neurol Asn. *Res:* Electron microscopy and electron histochemistry of neuromuscular diseases. *Mailing Add:* Mt Sinai Sch Med Fifth Ave & 100th St New York NY 10029

ANDERSON, PAUL KNIGHT, b Ludlow, Mass, Aug 6, 27; m 57, 68; c 2. VERTEBRATE ECOLOGY. *Educ:* Cornell Univ, AB, 49; Tulane Univ, MS, 51; Univ Calif, PhD(zool), 58. *Prof Exp:* Assoc zool, Univ Calif, 58, instr, 58-59; res assoc, Columbia Univ, 59-61; asst prof, Univ Alta, 61-65, assoc prof zool & asst dean arts & sci, 65-67; sr res fel, Nat Res Coun Can, 67-68; vis assoc prof zool, Stanford Univ, 67-68; from assoc prof to prof, 68-84, EMER PROF ZOOL, UNIV CALGARY, 84- *Concurrent Pos:* Exchange scientist, USSR-Can, 67; hon res fel, James Cook Univ, 74-75 & Univ Western Australia, 81-82 & 87-89. *Mem:* Am Soc Mammalogists; Am Soc Ichthyologists & Herpetologists. *Res:* Population structure, dynamics and evolution in vertebrate populations, particularly rodents; behavior and ecology of sirenians. *Mailing Add:* Dept Biol Sci Univ Calgary Calgary AB T2N 1N4 Can

ANDERSON, PAUL LEROY, b Magnolia, Ill, Sept 20, 35; m 66. ORGANIC CHEMISTRY. *Educ:* Univ Ill, BS, 62; Univ Mich, PhD(org chem), 66. *Prof Exp:* Fel, Syntex Steroid Inst, Mex, 66-67; RES CHEMIST, SANDOZ PHARMACEUT CORP, HANOVER, 67- *Mem:* Am Chem Soc; NY Acad Sci. *Res:* Studies of carbene reactions, dipolar cycloaddition reactions; synthesis of small ring compounds, steroids, terpenes and heterocyclic compounds of medicinal interest. *Mailing Add:* Sandoz Res Inst Hanover NJ 07936

ANDERSON, PAUL M, b Jackson, Minn, May 1, 38; m 59; c 3. BIOCHEMISTRY. *Educ:* Univ Minn, BS, 59, PhD(biochem), 64. *Prof Exp:* NSF res fel biochem, Sch Med, Tufts Univ, 64-65, NIH res fel, 65-66; from asst prof to assoc prof, Southern Ill Univ, 66-70; res biochemist, Molecular Biol Dept, Miles Labs, Inc, 70-71; assoc prof, 71-76, head dept, 71-85, PROF BIOCHEM, UNIV MINN-DULUTH, 76- *Concurrent Pos:* Vis assoc prof, Univ Wash, 77-78; vis prof, Univ Wash, 87-88. *Mem:* AAAS; Am Chem Soc; Am Soc Biol Chemists; Sigma Xi; Protein Soc. *Res:* Purification, mechanisms of catalytic action and regulation of enzymes of the cyanase operon and of enzymes related to nucleotide and urea biosynthesis; nitrogen and energy metabolism in fish (the urea cycle and associated enzymes). *Mailing Add:* Biochem Dept Univ Minn Duluth MN 55812-2487

ANDERSON, PAUL MAURICE, b Des Moines, Iowa, Jan 22, 26; m 50; c 4. ELECTRIC POWER SYSTEM DYNAMICS & RELIABILITY. *Educ:* Iowa State Univ, BS, 49, MS, 58, PhD (elec eng), 61. *Prof Exp:* Elec engr, Iowa Pub Serv Co, 49-55; from instr to prof elec eng, Iowa State Univ, 55-75; prog mgr, Elec Power Res Inst, 75-78; prof elec eng, Ariz State Univ, 80-84; PRES, POWER MATH ASSOC, 78- *Concurrent Pos:* Consult, NSF Summer Inst, Kharagpur, India & Ford Found, Univ Philippines, 69 & Pac Gas & Elec Co, 71-72; chmn, Syst Dynamic Performance Subcomt, Inst Elec & Electronics Engrs, 73-; mem, Task Force, Dept Energy Nat Power Grid Study, 78-79; consult ed, Encycl Sci & Technol, 79-; mem, Int Conf des Grands Réseaux Electriques. *Mem:* Fel Inst Elec & Electronics Engrs; Am Soc Mech Engrs. *Res:* Analytical methods for large power systems; author of techncal books in power systems. *Mailing Add:* Power Math Assocs Inc 12625 High Bluff Dr Suite 103 San Diego CA 92130

ANDERSON, PAUL MILTON, b Penna, Ind, June 27, 27; m 51; c 3. AGRICULTURAL ENGINEERING. *Educ:* Pa State Univ, BS, 50, MS, 57. *Prof Exp:* Farm mgr, Woodville State Hosp, Pa, 50-51; from instr to asst prof, 51-64, ASSOC PROF AGR ENG, PA STATE UNIV, 64- *Mem:* Am Soc Agr Engrs; Am Soc Eng Educ. *Res:* Electric power for production and processing of agricultural products; equipment for harvesting forage crops and for the application of feed additives, including anhydrous ammonia, to improve feed quality and reduce storage losses. *Mailing Add:* Rm 204 Agr Eng Bldg Pa State Univ University Park PA 16802

ANDERSON, PAUL NATHANIEL, b Omaha, Nebr, May 30, 37; m 65; c 2. INTERNAL MEDICINE, ONCOLOGY. *Educ:* Univ Colo, BA, 59, MD, 63; Am Bd Internal Med, dipl, 73, dipl(oncol), 75. *Prof Exp:* Intern internal med, Johns Hopkins Hosp, 63-64, resident, 64; res assoc embryol, Sect Exp Embryol, Lab Neuroanat Sci, Nat Inst Neurol Dis & Blindness, 65-67, staff assoc carcinogenesis sect, Lab Biol, Nat Cancer Inst, 67-70; staff assoc, Div Oncol, Dept Med, Johns Hopkins Hosp, 70-76; asst prof med & oncol, Johns Hopkins Univ, 72-76; dir med oncol, Penrose Cancer Hosp, Colorado Springs, 76-86, dir, 79-86; DIR, CANCER CTR COLORADO SPRINGS, 86- *Concurrent Pos:* Fel internal med, Johns Hopkins Hosp, 63-64; asst physician, Johns Hopkins Hosp; attend physician, Baltimore City Hosps; consult, Mem Hosp & St Francis Hosp, Colorado Springs, 76-; lectr cancer treat & rehab, Belo Horizonte, Brazil, sponsored by Partners Am, 78; med dir, Southern Colo Cancer Prog, 80-; mem, Colo Cancer Control & Res Panel, 80-, mem, Nat Cancer Inst Ad Hoc Comt Eval Community Hosp Oncol Progs, 82-83, mem bd dirs, Med Mgt Inc, 84-86, Health Care Standards Comt, Colo Found Med Care, 85-, Steering Comt Mamograph Screening Proj, Colo Dept Health, 87-; consult oncol, US Air Force Acad Hosp, 80-, Ft Carson Army Hosp, 84-, Ont Cancer Found, 87-; prin investr, Colo Cancer Communications Network & Cancer Info Serv, 81-87, Colo Community Clin Oncol Prog, 83-84; mem, Colo Govs Panel, Rocky Flats Employee Health Assessment Group, 83-84, bd adv, Colo Tumor Registry, Colo State Dept Health, 83-; bd mem, Share Health Plan Colo, Inc, 85-; founding mem, bd trustees, Southern Colo AIDS Proj, 86-, founding dir, Timberline Med Asssoc, 86-87; pres Preferred Physicians Incorp, 86-, asst med dir, Pikes Peak Hospice, 86-; ed bd, J Cancer Prog Mgt, 87-, ed bd, Health Care Mgt Rev, 88-; mem, Nat bd med dir, Fox Chase Ltd Fox Chase Comprehensive Cancer Ctr, Philadelphia, Pa, 87- *Mem:* AAAS; AMA; Am Soc Clin Oncol; Am Asn Cancer Res; NY Acad Sci; Am Asn Prof Consult; Am Asn Med Dirs; Am Mgt Asn; Am Hospice Asn; Int Platform Asn; Am Soc Int Med; Coalition Cancer Res. *Res:* Cancer chemotherapy; clinical investigation in cancer treatment; immunotherapy of cancer; combined modality cancer therapies; tumor immunology; anti-leukemia cellular immunity; efficacy of granulocyte transfusions during therapy-induced aplasia; cancer control res methodology; gall bladder infections in convalescent cholera patients; optimal antibiotic therapy in cholera; plasma cell tumor induction in BALB/c mice; cancer patient symptoms and disease mgt. *Mailing Add:* Cancer Ctr Colo Springs 320 E Fontanero No 100 Colorado Springs CO 80907

ANDERSON, PETER ALEXANDER VALLANCE, b Labuan, North Borneo, July 28, 1951; Brit citizen; m 76; c 1. CNIDARIAN NEUROBIOLOGY, PHYSIOLOGY OF CNIDOCYTES. *Educ:* Univ St Andrews, Scotland, BSc, 73; Univ Calif, Santa Barbara, MA & PhD(biol), 76. *Prof Exp:* Lectr zool, St Andrews Univ, 77-79; asst specialist neurobiology, Univ Calif, Irvine, 79-81; asst prof, 81-85, assoc prof physiol,; assoc prof, 85-90, PROF PHYSIOL, UNIV FLA, 90- *Concurrent Pos:* Vis prof, Univ Sao Paolo, 85; summer investr, Marine Biol Lab, 81-84. *Mem:* Soc Neurosciences; Am Soc Zool; Soc Exp Biol. *Res:* The physiology, pharmacology and structure of simple nervous systems with particular reference to evolution of higher nervous systems. *Mailing Add:* Whitney Lab 9505 Ocean Shore Blvd St Augustine FL 32086-8623

ANDERSON, PETER GLENNIE, b Oxford, NY, Dec 28, 54; m 88. CARDIAC HYPERTROPHY, ISCHEMIC INJURY. *Educ:* Univ Wash, BA, 77; Wash State Univ, DVM, 81; Univ Ala-Birmingham, PhD(path), 86. *Prof Exp:* Fel comp path, 81-84, fel hypertension, 84-86, ASST PROF DEPT PATH, UNIV ALA, BIRMINGHAM, 86-, ASST PROF DEPT COMP MED, 86- *Concurrent Pos:* Vis scientist, Univ Health Ctr, Univ Ariz, 88-; mem Basic Sci Coun, Am Heart Asn. *Mem:* Am Vet Med Asn; Soc Cardiovasc Path; Sigma Xi. *Res:* The changes that occur during cardiac hypertrophy and how these changes make the hypertrophied heart more susceptable to ischemic injury; use morphologic techniques to correlate structural changes with physiologic changes in normal and hypertrophied hearts. *Mailing Add:* Dept Path Univ Ala Birmingham G023 VH Birmingham AL 35294

ANDERSON, PETER GORDON, b Bay City, Mich, Jan 3, 40; m 62; c 2. NEURAL NETWORKS. *Educ:* Mass Inst Technol, SB, 62, PhD(math), 64. *Prof Exp:* Assoc prof computer sci, NJ Inst Technol, 71-78; prof, Seton Hall Univ, 78-80; PROF COMPUTER SCI, ROCHESTER INST TECHNOL, 80- *Concurrent Pos:* Prof, Shanghai Univ Technol & Zhejiang Univ, 86. *Mem:* Inst Elec & Electronics Engrs; Asn Comput Mach; Mat Asn Am; Int Neural Network Soc. *Res:* Neural networks and application to image processing and image understanding; developing reliable Monte Carlo methods. *Mailing Add:* Computer Sci Dept Rochester Inst Technol Rochester NY 14623-0887

ANDERSON, PHILIP CARR, b Grand Rapids, Mich, Dec 25, 30; m 62; c 3. DERMATOLOGY. *Educ:* Univ Mich, AB, 51, MD, 55; Am Bd Dermat, dipl, 62. *Prof Exp:* From jr clin instr to sr clin instr dermat, Med Ctr, Univ Mich, 60-63; asst prof, 63-68, dir res training, 65-68, ASSOC PROF DERMAT & ASST DEAN FOR PLANNING, SCH MED, UNIV MO-COLUMBIA, 68-, CHMN DEPT, 70-, MEM BD DIRS, 75- *Concurrent Pos:* NIH res fel, 61-63; Markle scholar, 66-; spec asst to dir, NIH, 67-68; mem bd dirs, Nat Prog Dermat, 70- *Mem:* Fel Am Acad Dermat; Am Dermat Asn; Soc Invest Dermat; Tissue Cult Asn; Am Asn Prof Dermat. *Res:* Tissue culture; education; management; cholesterol in cell walls. *Mailing Add:* Dept Dermat Univ Mo Sch Med Columbia MO 65211

ANDERSON, PHILIP WARREN, b Indianapolis, Ind, Dec 13, 23; m 47; c 1. THEORETICAL PHYSICS. *Educ:* Harvard Univ, BS, 43, MA, 47, PhD, 49; Cambridge Univ, MA, 78. *Hon Degrees:* ScD, Univ Ill. *Prof Exp:* Mem tech staff, Bell Tel Labs, 49-84; prof theoret physics, Cambridge Univ, 67-75; PROF PHYSICS, PRINCETON UNIV, 75- *Honors & Awards:* Nobel Prize in Physics, 77; Dannie Heinemann Prize, German Acad Sci, 75; Guthrie Medal, Inst Physics, 78; Nat Medal Sci, 82. *Mem:* Nat Acad Sci; fel Am Phys Soc; AAAS; Royal Soc London. *Res:* Quantum theory; theoretical physics of solids; spectral line broadening; magnetism; superconductivity. *Mailing Add:* Dept Physics Princeton Univ Princeton NJ 08544

ANDERSON, R F V, b Montreal, Que, Dec 23, 43; m 71. MATHEMATICS, DIFFERENTIAL EQUATIONS & TIME DELAY PROBLEMS. *Educ:* McGill Univ, BSc, 64; Princeton Univ, MA, 66, PhD(math), 67. *Prof Exp:* Instr math, Mass Inst Technol, 67-69; asst prof, 69-74, ASSOC PROF MATH, UNIV BC, 74- *Mem:* Am Math Soc; Can Math Soc. *Res:* Spectral theory of non-commuting operators; stability of systems with time delays. *Mailing Add:* Dept Math Univ BC Vancouver BC V6T 1Y4 Can

ANDERSON, R(ICHARD) L(OUIS), b Minneapolis, Minn, Feb 4, 27; m 50; c 3. SEMICONDUCTOR DEVICES, SOLID STATE ELECTRONICS. *Educ:* Univ Minn, BEE, 50, MS, 52; Syracuse Univ, PhD, 60. *Hon Degrees:* Dr, Univ Sao Paulo, 70. *Prof Exp:* Asst, Univ Minn, 51-52; assoc physicist, res ctr, Int Bus Mach Corp, 52-60; from assoc prof to prof elec eng, Syracuse Univ, 61-78; consult, Int Telecommun Union, 75-80; PROF ELEC ENG & MAT SCI, UNIV VT, 78-, DIR, PROG MAT SCI, 81- *Concurrent Pos:* Adj prof, Syracuse Univ, 57-60; Fulbright vis prof, Univ Madrid, 60-61; vis prof, Univ Sao Paulo, 67-69 & Univ Vt, 78-79; consult, UN Develop Prog, 90. *Honors & Awards:* Brazilian Prize for Microelectronics, 80. *Mem:* Fel Inst Elec & Electronics Engrs; Am Phys Soc; Electrochem Soc; Brazilian Phys Soc. *Res:* Investigation of physical electronics of semiconductor devices; field effect transistors at low cryogenic temperatures with applications to integrated circuits; heterojunction bipolar transistors. *Mailing Add:* Dept Elec Eng Univ Vt Burlington VT 05405

ANDERSON, R(OBERT) M(ORRIS), JR, b Crookston, Minn, Feb 15, 39; m 60; c 2. ELECTRICAL ENGINEERING, SOLID STATE PHYSICS. *Educ:* Univ Mich, BSEng, 61, MSEng, 63, MS, 65, PhD(elec eng), 67. *Prof Exp:* Asst res engr, Dept Elec Eng, Univ Mich, 64-67; lectr integrated circuits, Purdue Univ, 68, from asst prof to assoc prof elec eng, 70-76, eng coordr continuing educ, 73-79, Ball Bros prof eng, 76-79; MGR TECH EDUC OPER, GEN ELEC CO, 79- *Concurrent Pos:* Lectr, Western Elec Co, Ind, 69-71. *Mem:* Inst Elec & Electronics Engrs; Am Soc Eng Educ; Am Vacuum Soc; AAAS; Am Soc Training & Develop. *Res:* Electrical and optical properties of amorphous, binary, non-stoichiometric, semiconducting thin films; professionalism in engineering; engineering ethics; continuing engineering education. *Mailing Add:* Tech Educ Oper-Gen Elec Co 1285 Boston Ave Bridgeport CT 06602

ANDERSON, RALPH ROBERT, b Fords, NJ, Nov 1, 32; m 61; c 2. DAIRY SCIENCE. *Educ:* Rutgers Univ, BS, 53, MS, 58; Univ Mo, PhD(endocrinol), 61. *Prof Exp:* Instr endocrinol, Univ Mo, 61-62; asst prof reproductive physiol, Iowa State Univ, 62-64; trainee endocrinol, Univ Wis, Madison, 64-65; from asst prof to assoc prof, 65-76, PROF DAIRY SCI, UNIV MO, COLUMBIA, 76- *Concurrent Pos:* Sr Fulbright scholar, Ruakura, NZ, 73-74. *Mem:* Am Dairy Sci Asn; Am Soc Animal Sci; Soc Exp Biol & Med; Endocrine Soc; Soc Study Reproduction; Am Physiol Soc. *Res:* Endocrine control of mammary gland growth and lactation. *Mailing Add:* Dept Dairy Sci S139 ASRC Univ Mo Columbia MO 65211

ANDERSON, RAY HAROLD, b Minneapolis, Minn, Mar 23, 15; m 46. FOOD SCIENCE. *Educ:* St Thomas Col, BS, 36; Univ Minn, PhD(org chem), 50. *Prof Exp:* Chemist, Gen Mills Inc, 37-46; asst instr, Univ Minn, 46-48; PRIN SCIENTIST & HEAD CHEM SERV & NUTRIT RES, GEN MILLS, INC, 50- *Mem:* Am Chem Soc; Am Asn Cereal Chemists; Am Oil Chem Soc; Inst Food Technologists. *Res:* Nutritional enrichment of food products; food preservation; shelf-stable product development; food additives; chemistry of food ingredients and processes. *Mailing Add:* 10930 W River Rd N Champlin MN 55316

ANDERSON, RAYMOND KENNETH, b White Lake, Wis, May 2, 28; m 53; c 4. WILDLIFE ECOLOGY. *Educ:* Wis State Univ, Stevens Point, BS, 54; Univ Mich, MA, 58; Univ Wis, PhD(wildlife ecol), 69. *Prof Exp:* High sch instr, Wis, 54-58; conserv, 58-61, PROF WILDLIFE, UNIV WIS-STEVENS POINT, 66- *Mem:* Am Inst Biol Sci; Wilson Ornith Soc; Wildlife Soc. *Res:* Behavior and management of wild animals associated with grasslands; anuran inventory methods; habitat use, movements via radio-telemetry and behavior of wild mammals, birds, and amphibians; hunter behavior. *Mailing Add:* Dept Natural Resources Univ Wis Stevens Point WI 54481

ANDERSON, REBECCA JANE, b Ft Madison, Iowa, Aug 21, 49. NEUROPHARMACOLOGY, NEUROTOXICOLOGY. *Educ:* Coe Col, BA, 71; Georgetown Univ, PhD(pharmacol), 75. *Prof Exp:* Consult chem, Argonne Nat Lab, 70; fel physiol, Med Res Coun Can, 75-77; asst prof, 77-80, assoc prof pharmacol, George Washington Univ, 80-83; pharmacol dept, Parke-Davis Div, Warner-Lambert Co, 83-88; BOEHRINGER INGELHEIM PHARMACEUT, INC, 88- *Concurrent Pos:* Res grants, Pharmaceut Mgrs Asn Found, 77-78, Dreyfus Med Found, 77-78 & NIH, 78-81; consult, Bur Med Devices, Food & Drug Admin, 80-83 & Life Systems, Inc, 80-88; adj assoc prof toxicol, Univ Mich, 83-88. *Mem:* Sigma Xi; Soc Neuroscience; Am Epilepsy Soc; Am Chem Soc; Am Soc Pharmacol Exp Ther; Soc Toxicol; AAAS. *Res:* Pharmacology of sensorimotor systems; neurotoxicology of environmental agents; mechanisms of action of anticonvulsant agents. *Mailing Add:* Boehringer Ingelheim Pharmceut 90 E Ridge PO Box 368 Ridgefield CT 06877

ANDERSON, RICHARD ALAN, b Rock Rapids, Iowa, Aug 9, 30; m 57; c 1. ATOMIC & MOLECULAR PHYSICS. *Educ:* Augustana Col, BA, 52; Kans State Univ, MS, 54, PhD(physics), 59. *Prof Exp:* From asst prof to assoc prof, 58-69, PROF PHYSICS, UNIV MO, ROLLA, 69- *Mem:* Am Phys Soc; Am Asn Physics Teachers; Optical Soc Am; Soc Photo-Optical Instrumentation Engrs; Am Asn Univ Professors. *Res:* Lasers and optoelectronics; optical propagation of coherent and incoherent radiation; radiative transfer problems. *Mailing Add:* 403 S Adrian Rolla MO 65401

ANDERSON, RICHARD ALLEN, b Middle River, Minn, May 28, 46; m 66; c 2. NUTRITION, BIOCHEMISTRY. *Educ:* Concordia Col, BA, 68; Iowa State Univ, PhD(biochem), 73. *Prof Exp:* Res assoc biochem, Harvard Med Sch, 73-75; RES CHEMIST, HUMAN NUTRIT INST, USDA, 75- *Concurrent Pos:* Adj vis prof, Naval Med Res Inst, Bethesda, Md. *Honors & Awards:* Merit Awards, USDA, 83 & 85; Labcatral Award, 91. *Mem:* Am Inst Nutrit; NY Acad Sci; Int Soc Trace Element Res in Humans. *Res:* Nutritional role of chromium in man; diabetes and cardiovascular diseases; exercise and trace elements; chromium, copper and zinc. *Mailing Add:* Bldg 307 Rm 224 Human Nutrit Inst Beltsville MD 20705

ANDERSON, RICHARD CHARLES, b Moline, Ill, Apr 22, 30; m 53; c 3. GLACIAL GEOLOGY. *Educ:* Augustana Col, AB, 52; Univ Chicago, MS, 53, PhD(geol), 55. *Prof Exp:* Geologist, Geophoto Servs, Colo, 55-57; from asst prof to assoc prof geol & geog, 57-65; chmn dept, 69-87, PROF GEOL, AUGUSTANA COL, ILL, 65- *Concurrent Pos:* Res affil, Ill Geol Surv, 59-; asst dir, Earth Sci Summer Inst, Iowa State Univ, 66-71. *Mem:* Geol Soc Am; Am Geophys Union; Int Union Quaternary Res; Am Quaternary Asn; Am Meteorol Soc. *Res:* Glacial geology; geomorphology; Tertiary and Quaternary geomorphic history. *Mailing Add:* Dept Geol Augustana Col Rock Island IL 61201

ANDERSON, RICHARD DAVIS, b Hamden, Conn, Feb 17, 22; m 43; c 5. MATHEMATICS. *Educ:* Univ Minn, BA, 41; Univ Tex, PhD(math), 48. *Prof Exp:* Instr math, Univ Tex, 41-42 & 45-48; from instr to assoc prof, Univ Pa, 48-56; prof, 56-59, BOYD PROF MATH, LA STATE UNIV, BATON ROUGE, 59- *Concurrent Pos:* Inst Advan Study, 51-52 & 55-56; Alfred P Sloan res fel, 60-63; colleague, Mathematisch Centrum, Amsterdam, 62-63; panel mem, Sch Math Study Group & mem comt undergrad prog in math, chmn, 66-68. *Mem:* AAAS; Am Math Soc; Math Asn Am. *Res:* Topology; point set theory; collections of continua; dimension theory; transformation groups. *Mailing Add:* 2954 Fritchie Baton Rouge LA 70809

ANDERSON, RICHARD ERNEST, b Princeton, Ind, Mar 7, 45; m 67; c 3. INTEGRATED CIRCUIT FAILURE ANALYSIS & RELIABILITY. *Educ:* Univ Ill, Urbana, BS, 67, MS, 70, PhD(elec eng), 74. *Prof Exp:* Mem tech staff, 74-81, SUPVR, SANDIA NAT LABS, 81- *Mem:* Electrochem Soc. *Res:* Development of advanced failure analysis techniques for integrated circuits and other semiconductor components. *Mailing Add:* Div 2142 Sandia Nat Labs PO Box 5800 Albuquerque NM 87185-5800

ANDERSON, RICHARD GILPIN WOOD, b Bryn Mawr, Pa, Mar 25, 40; m 64; c 2. CELL BIOLGOY. *Educ:* Ore State Univ, BS, 65; Univ Ore Sch Med, PhD(anat), 70. *Prof Exp:* Res assoc reprod physiol, Ore Regional Primate Res Ctr, 70-73, asst scientist, 73; from asst prof to assoc prof, 73-81, PROF CELL BIOL, UNIV TEX HEALTH SCI CTR, DALLAS, 81- *Mem:* Am Asn Cell Biol; Am Asn Anat. *Res:* Receptor mediated endocytosis; structure and function of clathrin coated membrane; intracellular membrane movement; receptor coupled transmembrane transport. *Mailing Add:* 6446 Orchid Lane Dallas TX 75230

ANDERSON, RICHARD GREGORY, b Philadelphia, Pa, Mar 11, 46; m 82. PRESTRESSED CONCRETE, REINFORCED CONCRETE. *Educ:* Univ Ill, BS, 68; Mass Inst Technol, MS, 72. *Prof Exp:* Develop engr, 72-82, MANAGING DIR RES & DEVELOP, CONCRETE TECHNOL CORP, 82- *Concurrent Pos:* Res engr, Norges Tekniske Hogskole, 80-81. *Mem:* Am Concrete Inst. *Res:* Investigation of concrete for structural applications, with particular emphasis on the load-resisting capabilities of plant-cast, prestressed concrete elements. *Mailing Add:* 7010 S Alder St Tacoma WA 98409

ANDERSON, RICHARD JOHN, b Chicago, Ill, June 11, 38; m 60; c 2. ATOMIC PHYSICS. *Educ:* Coe Col, BA, 59; DePaul Univ, MS, 62; Univ Okla, PhD(physics), 66. *Prof Exp:* Appl physicist, Eastman Kodak Co, 61-62; from asst prof to prof physics, Univ Ark, Fayetteville, 66-89; PROG MGR, EXP PROG STIMULATE COMPETITIVE RES, NSF, 89- *Concurrent Pos:* Dir hons studies, Univ Ark, Fayetteville, 82- *Mem:* Am Phys Soc; Am Asn Physics Teachers; Sigma Xi. *Res:* Spectroscopic investigation of electronic and atomic collision processes. *Mailing Add:* 1060 N Royal St Alexandria VA 22314-1530

ANDERSON, RICHARD L, b North Liberty, Ind, Apr 20, 15; m 46; c 2. EXPERIMENTAL STATISTICS. *Educ:* DePauw Univ, AB, 36; Iowa State Col, MS, 38, PhD(math statist), 41. *Prof Exp:* Asst math, Iowa State Col, 36-41; from instr to prof statist, Inst Statist, NC Univ, 41-66; res mathematician, Princeton Univ, 44-45; vis res prof statist, Univ Ga, 66-67; prof statist, Univ Ky, 67-80, chmn dept, 67-79, asst to dean, Col Agr, 80-85; VPRES, STATIST CONSULT INC, 85- *Concurrent Pos:* Consult, Off Sci Res & Develop, US Army & US Navy, 44-45; prof, Purdue Univ, 50-51 & London Sch Econ, 58; consult, Pan-Am World Airways Guided Missile Range Div, 63-67; partic, NSF-Japan Soc Promotion Sci Seminar Sampling of Bulk Mat, Tokyo, 65; Inaugural Conf for Sci Comput Ctr, Univ Cairo, 69; mem adv bd, Int Math & Statist Libr, 71-; mem census adv comt, Am Statist Asn, 72-77; consult, Merrill Labs, 73-81, Health Sci Comput Ctr, Univ Calif, Los Angelos, 73-81; chmn, Sect U, AAAS, 79. *Mem:* AAAS; fel Inst Math Statist; fel Am Statist Asn (pres-elect, 82, pres, 83); Int Statist Inst; Biomet Soc (pres, NE Am Region, 66); Int Asn Statist Comput. *Res:* Experimental designs and estimating procedures for variance components; time series analysis, including serial correlation; econometrics; regression analysis, selection of predictors, prior information, intersecting straight lines. *Mailing Add:* Statist Consults Inc 300 E Main St Suite 400 Lexington KY 40507

ANDERSON, RICHARD LEE, b Grinnell, Iowa, Feb 24, 45; m 65; c 2. OPHTHALMOLOGICAL SURGERY. *Educ:* Grinnell Col, BA, 67; Univ Iowa, MD, 71, cert ophthal specialist, 75. *Prof Exp:* Resident ophthal, Univ Iowa, 72-75; fel plastic surg, Albany Med Ctr, 75 & Univ Calif Med Ctr, San Francisco, 76; Head fel, Head Ophthalmic Found, 76; asst prof, 76-81, ASSOC PROF OPHTHAL PLASTIC SURG, DEPT OPHTHAL, UNIV IOWA, 81- *Concurrent Pos:* NSF grants, 65 & 66. *Mem:* Am Med Asn; fel Am Col Surgeons; fel Int Col Surgeons. *Res:* Acquired Ptosis, levator aponeurosis defects and their repair; nasolacrimal duct obstructions; anatomy of the levator muscle; blepharospasm eyelid and orbital tumors. *Mailing Add:* Univ Utah Hosp - Ophthal 50 N Medical Dr Salt Lake City UT 84132

ANDERSON, RICHARD LEE, b Astoria, Ore, June 14, 33. CARBOHYDRATE METABOLISM, ENZYMOLOGY. *Educ:* Univ Wash, BS, 54, PhD(microbiol), 59. *Prof Exp:* Asst, Univ Wash, 54-59; NIH fel, 59-61, from asst prof to assoc prof, 61-70, PROF BIOCHEM, MICH STATE UNIV, 70- *Concurrent Pos:* Ed, J Bact, 70-75. *Mem:* AAAS; Am Chem Soc; Am Soc Microbiol; Am Soc Biol Chem; Soc Complex Carbohydrates. *Res:* Enzymology and metabolism of plants, animals and microorganisms; pathways, mechanisms and control of carbohydrate metabolism. *Mailing Add:* Dept Biochem Mich State Univ East Lansing MI 48824

ANDERSON, RICHARD LOUIS, b Hale Center, Tex, Nov 27, 35; m 60; c 2. PHYSICAL CHEMISTRY. *Educ:* Mass Inst Technol, BS, 58; Rice Univ, PhD(chem), 66. *Prof Exp:* Res chemist, Heat Div, Nat Bur Standards, 63-71; scientist, Nat Phys Inst, Bur Standards, Brunswick, Germany, 71-74; HEAD METEOROL, RES & DEVELOP LAB, OAK RIDGE NAT LAB, 74-, GROUP LEADER MEASUREMENT RES, INSTRUMENTATION & CONTROLS DIV, 79- *Concurrent Pos:* Res prof, Univ Tenn. *Honors & Awards:* IR-100 Award, 82, 87. *Mem:* Am Vacuum Soc; Sigma Xi; Inst Elec & Electronics Engrs; Am Physical Soc; Soc Photo Instrument Engrs; Optical Soc Am. *Res:* Low temperature calorimetry and thermometry; gas thermometry; high temperature resistance thermometry; vacuum devices; pressure measurement; thermocouple thermometry; automated, precision measurements; microcomputers; computer graphics, ultrasonics. *Mailing Add:* 2002 Key West Cove Austin TX 78746

ANDERSON, RICHARD ORR, b Evanston, Ill, Oct 23, 29; m 51; c 2. AQUATIC ECOLOGY. *Educ:* Univ Wis, BS, 51; Univ Mich, MS, 53, PhD(fisheries), 59. *Prof Exp:* Fishery biologist, Mich Inst Fisheries Res, 52-58; instr dept fisheries, Univ Mich, 58-59; biologist in charge, Wolf Lake Fish Hatchery, 59-63; leader, Mo Coop Fishery Res Unit, 63-85; dir, 85-89, SR SCIENTIST, NAT FISH HATCHERY & TECHNOL CTR, SAN MARCOS, TEX, 89- *Concurrent Pos:* Prof, Sch Forestry, Fisheries & Wildlife, Univ Mo, 63-85; vis grad fac, Tex A&M Univ, 87- *Honors & Awards:* Gulf Conservation Award, 84. *Mem:* Am Fisheries Soc; Am Soc Limnol & Oceanog; Int Asn Theoret & Appl Limnol; NAm Benthological Soc; Int Asn Fish & Wildlife Agencies. *Res:* Aquatic and fishery biology and ecology; fish production and growth; aquaculture. *Mailing Add:* Nat Fishery Ctr Rte 1 Box 159-D San Marcos TX 78666

ANDERSON, RICHMOND K, b Bawgacong, India, Dec 6, 07. BIOCHEMISTRY. *Educ:* Northwestern Univ, MD & PhD(biochem); Johns Hopkins Univ, MPH. *Prof Exp:* prof, Sch Pub health, Univ NC, 71-; RETIRED. *Mailing Add:* 190 Carol Woods Chapel Hill NC 27514

ANDERSON, ROBBIN COLYER, b DeRidder, La, June 8, 14; m 46; c 3. PHYSICAL CHEMISTRY. *Educ:* La State Univ, BS, 34, MS, 36; Univ Wis, PhD(phys chem), 39. *Prof Exp:* From instr to prof chem, Univ Tex, Austin, 39-67, assoc dean grad sch, 66-67; dean, Col Arts & Sci, 67-79; prof, 67-81, DEAN & EMER PROF CHEM, UNIV ARK, FAYETTEVILLE, 81- *Concurrent Pos:* Mem adv coun col chem, AAAS, 63-66. *Mem:* AAAS; Am Chem Soc; Nat Sci Teachers Asn. *Res:* Complex ions; mechanism of flame propagation and carbon formation. *Mailing Add:* 1599 Halseu Fayetteville AR 72701

ANDERSON, ROBERT ALAN, b Princeton, Ind, Sept 14, 42; m 65; c 2. DIELECTRIC BREAKDOWN, HYDROGEN IN SILICON. *Educ:* Univ Ill, Urbana, BS, 65, MS, 66, PhD(physics), 71. *Prof Exp:* Res asst & fel low-temperature physics, Univ Ill, Urbana, 67-71; SR MEM TECH STAFF, SANDIA NAT LABS, 71- *Mem:* Am Phys Soc; Mat Res Soc. *Res:* Electrical breakdown phenomena, including surface flashover of insulators in vacuum, breakdown of solid dielectrics, and high-field conduction mechanisms; piezoelectricity and pyroelectricity in polymers; kinetics of hydrogen in silicon. *Mailing Add:* Div 1815 Sandia Nat Labs PO Box 5800 Albuquerque NM 87185

ANDERSON, ROBERT AUTHUR, b Philadelphia, Penn, Apr 3, 32; m 57; c 3. PREVENTIVE MEDICINE, HEALTH PROMOTION. *Educ:* Willamette Univ, BA, 54; Univ Wash, MD, 57; Am Bd Fam Practice, dipl, 73. *Prof Exp:* Founding chief staff, med admin, Stevens Mem Hosp, 63-64, chief psychiat med staff admin, 76-78; CLIN ASST PROF, DEPT FAMILY MED, UNIV WASH, 76- *Mem:* Fel Am Acad Family Physicians; Am Med Asn; Am Holistic Med Asn (pres 86-88); Am Col Prev Med. *Res:* Stress management; wellness medicine; preventive medicine; psychosynthesis. *Mailing Add:* 7935 216th SW Suite E Edmonds WA 98026

ANDERSON, ROBERT CHRISTIAN, b Perth Amboy, NJ, Sept 26, 18; m 44; c 2. ORGANIC CHEMISTRY. *Educ:* Middlebury Col, AB, 40; Princeton Univ, PhD(chem), 48. *Prof Exp:* Asst chemist, Merck & Co, Inc, NJ, 42-45; assoc chemist, 48-57, ASST DIR, BROOKHAVEN NAT LAB, 57- *Concurrent Pos:* Fel, Woodrow Wilson Int Ctr Scholars, 75-; chmn coun, State Univ NY Stony Brook, 75-; mem sci adv coun, NY Legis; mem Bd Atomic Energy Comn Nuclear Sci & Eng Fels. *Mem:* AAAS; Am Chem Soc; Am Soc Eng Educ; Am Nuclear Soc. *Res:* Synthesis and structure determination of physiologically active organic compounds; stereo-chemistry of organic compounds; nuclear transformations in organic systems. *Mailing Add:* 277 Friar Lane Mountainside NJ 07092

ANDERSON, ROBERT CLARK, b Galesburg, Ill, July 18, 26; m; c 2. METALLURGICAL ENGINEERING. *Educ:* Univ Ill, BS, 51. *Prof Exp:* Trainee, Sheffield Steel Div, Armco Steel Corp, 51-52; chief metallurgist, WKM Co Div, ACF Industs, 52-57; PRES, ANDERSON & ASSOCS, INC, 57- *Concurrent Pos:* Consult, NASA Manned Spacecraft Ctr, 68- *Mem:* Nat Soc Prof Engrs; fel Am Soc Metals; Am Inst Mining, Metall & Petrol Engrs; Nat Asn Corrosion Engrs; Int Metallog Soc. *Res:* Analyzing metal failures; application of scientific analysis to resolution of legal problems. *Mailing Add:* 8231 Renmark Lane Houston TX 77070

ANDERSON, ROBERT CLARKE, b Salem, Ind, Feb 19, 11; m 34; c 3. TOXICOLOGY. *Educ:* Purdue Univ, BS, 31, DSc, 53. *Prof Exp:* Lab asst, Eli Lilly & Co, 31-34, pharmacologist, 34-36, head dept toxicol, 61-66, dir toxicol div, 61-66, dir res biol, 66-70, dir res toxicol, 70-72; RETIRED. *Mem:* AAAS; Am Pharmaceut Asn (secy-treas, 58-64); Am Soc Toxicol; Am Soc Pharmacol & Therapeut. *Res:* Toxicology of antibiotics, analgesics, steroids and pesticides. *Mailing Add:* 318 N Franklin Rd Indianapolis IN 46219

ANDERSON, ROBERT CURTIS, b Columbus, Ohio, May 10, 41; m 67; c 2. ENTOMOLOGY. *Educ:* Calif State Col, Long Beach, BS, 63; Purdue Univ, MS, 66, PhD(entom), 68. *Prof Exp:* Instr entom, Purdue Univ, 68-69; from asst prof to assoc prof, 69-82, chmn dept, 74-82, PROF BIOL, IDAHO STATE UNIV, 82 - *Concurrent Pos:* Sr Fulbright Scholar, Egypt, 83-84; adv coun, Nat Inst Health, Environ Health Sci, 85. *Mem:* AAAS; Sigma Xi; Am Soc Zoologists; Entom Soc Am. *Res:* Biology and ecology of insect parasites, Scorpionida, Acarina; biological control; medical entomology; applied ecology; invertebrate zoology; parasitology. *Mailing Add:* Dept Biol Idaho State Univ Pocatello ID 83209

ANDERSON, ROBERT E(DWIN), b Wharton, NJ, Feb 22, 19; m 42; c 2. ELECTRICAL ENGINEERING, EDUCATION. *Educ:* Newark Col Eng, BSEE, 39; Univ NH, MSEE, 48. *Prof Exp:* Elec engr, Fed Shipbldg & Dry Dock Co, 38-43; instr high sch, 46-47; asst prof elec eng, Univ NH, 47-49; from asst prof to assoc prof, Newark Col Eng, 49-57, PROF ELEC ENG & CHMN DEPT, NJ INST TECHNOL, 57-, ACTG ASSOC DEAN, 80-; DIR MICROLABS, FAX INC, 69- *Concurrent Pos:* Consult, Design Serv Inc, NY Tel Co, NJ Bell Tel Co & Am Tel & Tel Co; spec lectr, Int Bus Mach Co; eng educ consult, NY Tel Co, 78- *Mem:* Inst Elec & Electronics Engrs; Am Soc Eng Educ; Nat Soc Prof Engrs; Computer Soc. *Res:* Digital and logic circuits; engineering education; minimization techniques for digital circuits. *Mailing Add:* 21 Theodore St Wharton NJ 07885

ANDERSON, ROBERT E, b Carrizo Springs, Tex, Aug 26, 26; m 47; c 3. SOLID STATE PHYSICS. *Educ:* Tex Col Arts & Indust, BS, 47, MS, 49; Univ Tex, PhD(physics), 55. *Prof Exp:* Teacher math & physics, San Antonio Jr Col, 49-52; asst physics, Univ Tex, 52-55; res engr, Tex Instruments, Inc, 55-57; co res fel low temperature physics, Univ Tex, 57-58; prof physics, Tex Col Arts & Indust, 58-66; chmn dept, 66-77, PROF PHYSICS, SOUTHWEST TEX STATE UNIV, 66- *Concurrent Pos:* Res asst, Defense Res Lab, Tex, 54-55; consult, Tex Instruments, Inc, 60- *Mem:* Am Asn Physics Teachers. *Res:* Semiconductor physics; electronics; vacuum techniques; acoustics; transistor design. *Mailing Add:* Dept Physics & Astron Southwest Tex State Univ San Marcos TX 78666

ANDERSON, ROBERT E, b Livingston, Tex, Oct 5, 40; m 62; c 3. BIOCHEMISTRY, OPHTHALMOLOGY. *Educ:* Tex A&M Univ, BA, 63, MS, 65, PhD(biochem), 68; Baylor Col Med, MD, 75. *Prof Exp:* From asst prof to assoc prof, 69-81, PROF BIOCHEM, BAYLOR COL MED, 81- *Concurrent Pos:* Fel biochem, Oak Ridge Assoc Univs, 68-69; sr investr award, res to prevent blindness, 90. *Honors & Awards:* MacGee Award, Am Oil Chem Soc, 66; Dolly Green Spec Scholar Award for Res to Prevent Blindness, 82; Alcon Res Inst Award, 85. *Mem:* AAAS; Am Chem Soc; Am Soc Biol Chemists; Biophys Soc; Int Soc Neurochem; Am Soc Biochem & Molecular Biol; Asn Res Vision & Ophthal. *Res:* Biochemistry of the visual process; chemistry of photoreceptor membranes; relationships between lipid structure and membrane function; biochemical mechanisms of retinal degeneration. *Mailing Add:* Dept Ophthal Baylor Col Med Houston TX 77030

ANDERSON, ROBERT EDWIN, b Los Angeles, Calif, Aug 20, 31; m 53; c 3. PATHOLOGY. *Educ:* Col Wooster, BA, 53; Western Reserve Univ, MD, 57. *Prof Exp:* Intern med, Strong-Mem-Rochester Munic Hosps, NY, 57-58; resident path, Univ Hosps, Cleveland, Ohio, 58-59; resident, Med Ctr, Univ Calif, Los Angeles, 59-62, from instr to asst prof, 60-64, consult, Clin Labs, 60-62; asst prof, 64-69, PROF PATH, SCH MED, UNIV NMEX, 69-, CHMN DEPT, 68- *Concurrent Pos:* Consult, Community Blood & Plasma Inc, 61-62 & Bernalillo County Indian Hosp, NMex, 64- *Mem:* AAAS; Am Asn Pathologists; Am Soc Exp Path; Am Asn Path & Bact (secy-treas, 74-); Asn Path Chmn (pres, 75). *Res:* Hematologic pathology; radiation in the immune response. *Mailing Add:* Dept Path Univ NMex Sch Med Albuquerque NM 87131

ANDERSON, ROBERT EMRA, b Mentone, Ind, Jan 6, 24; m 45; c 2. WATER CHEMISTRY, ION EXCHANGE. *Educ:* Ind Univ, BS, 49, MA, 51. *Prof Exp:* Chemist, Dow Chem Co, 50-54, proj leader ion exchange, 54-56, group leader, 56-68; group leader ion exchange, 68-78, SR RES SCIENTIST, DIAMOND SHAMROCK CHEM CO, 78- *Concurrent Pos:* Chmn, Gordon Res Conf Ion Exchange, AAAS, 71. *Mem:* Am Chem Soc; Sigma Xi. *Res:* Study of ion exchange and ion-exchange resins, particularly as applied to water treatment and industrial processing. *Mailing Add:* 1617 Wright Ave Sunnyvale CA 94087

ANDERSON, ROBERT GLENN, b Stratton, Ont, Jan 2, 24; m 46; c 4. PLANT GENETICS, PLANT BREEDING. *Educ:* Univ Man, BSA, 50, MSc, 52; Univ Sask, PhD, 55. *Prof Exp:* Asst prof plant breeding & statist, Univ Sask, 54-56; agr res officer, Res Sta, Can Dept Agr, Winnipeg, 56-64; wheat breeder, Int Maize & Wheat Improv Ctr, Rockefeller Found, 64, joint coordr, All India Wheat Improv Prog, 64-71, assoc dir, 71-80; PROF

BOTANY, DEPT BIOL, UNIV MO, KANSAS CITY, 80- *Concurrent Pos:* Mem genetic stock cmt, Int Wheat Genetics Symp, 58-; sci ed, Agr Inst Can, 63-64. *Honors & Awards:* Gold Medal, Univ Man, 50; Recognition Award; Agr Inst Can, 74. *Mem:* Agr Inst Can; Can Soc Agron; Genetics Soc Can; fel Indian Soc Genetics & Plant Breeding; AAAS. *Res:* Wheat breeding; international agriculture in developing countries. *Mailing Add:* Dept Biol Univ Missouri Kansas City MO 64110

ANDERSON, ROBERT GORDON, b Elwood, Ind, Sept 5, 26; m 56; c 5. BOTANY. *Educ:* Youngstown State Univ, BA, 53; Univ Nebr, MS, 55, PhD(bot, bact), 58. *Prof Exp:* Instr bot, biol & bact, Univ Nebr, Omaha, 58-59; assoc prof biol, Davis & Elkins Col, 59-60; PROF BOT, UNIV MO, KANSAS CITY, 60-, CHMN DEPT BIOL, 80- *Concurrent Pos:* Vis prof, Dakota State Col, 79-83. *Mem:* Phycological Soc Am; Sigma Xi. *Res:* Physiology of algae; Chara. *Mailing Add:* Dept Biol Univ Mo Kansas City MO 64110

ANDERSON, ROBERT HUNT, b Kansas City, Kans, Oct 20, 24; m 50; c 6. COMPUTER SCIENCE, CHEMISTRY. *Educ:* Baker Univ, AB, 46; Columbia Univ, MA, 48, PhD(chem), 54. *Prof Exp:* Instr chem, Bloomfield Col, 51-56; asst res specialist, Rutgers Univ, 56-57; from asst prof to prof chem, Western Mich Univ, 57-89; RETIRED. *Mem:* AAAS; Am Chem Soc; NY Acad Sci. *Res:* Neutron spectroscopy in chemical analysis; phase rule; corrosion inhibitors; computers in chemistry. *Mailing Add:* Dept Chem Western Mich Univ Kalamazoo MI 49008-3929

ANDERSON, ROBERT L, b Chicago, Ill, Feb 22, 32; m 58; c 2. BIOCHEMISTRY, NUTRITION. *Educ:* Colo State Univ, BS, 57; NC State Col, MS, 59, PhD(nutrit), 62. *Prof Exp:* RES BIOCHEMIST, PROCTER & GAMBLE CO, 62- *Mem:* Am Chem Soc; Am Inst Nutrit; AAAS. *Res:* Toxicology and metabolism. *Mailing Add:* 454 Whitestone Ct PO Box 39175 Cincinnati OH 45231

ANDERSON, ROBERT LESTER, b Parkersburg, WVa, Dec 30, 33; m 56; c 4. PHYSICS INSTRUMENTATION. *Educ:* Marietta Col, AB, 55; Pa State Univ, MS, 59, PhD(physics), 61. *Prof Exp:* From asst prof to assoc prof physics, Marietta Col, 61-71; head dept, 66-77, prof, 71-; AT DEPT PHYSICS, UNIV GA. *Mem:* AAAS; Am Phys Soc; Am Asn Physics Teachers. *Res:* Viscosity and density of pure hydrocarbon liquids at moderately elevated pressures; electrical and thermal transport properties of metals and semiconductors. *Mailing Add:* Dept Physics Univ Ga Physics Bldg FPO518-381186 Athens GA 30602

ANDERSON, ROBERT LEWIS, b Springfield, Ill, Sept 30, 33; m 55; c 3. BIOCHEMISTRY, PROTEIN CHEMISTRY. *Educ:* Bradley Univ, BS, 55, MS, 67; Univ Nebr, PhD, 86. *Prof Exp:* BIOCHEMIST, NAT CTR AGR UTILIZATION RES, USDA, 57- *Concurrent Pos:* Res asst, Univ Nebr, 67-70. *Mem:* Am Chem Soc; Am Asn Cereal Chem; Inst Food Technol; Sigma Xi. *Res:* Soybean proteins; electrophoresis. *Mailing Add:* 400 Mississippi Ave Morton IL 61550

ANDERSON, ROBERT NEIL, b San Jose, Calif, Nov 8, 33; m 65; c 2. CHEMICAL ENGINEERING, METALLURGICAL ENGINEERING. *Educ:* Univ San Francisco, BS, 56; Univ Calif, Berkeley, BS, 58, MS, 59; Stanford Univ, PhD(metall), 69. *Prof Exp:* Res engr, US Naval Radiol Defense Lab, 59-66; opers res analyst, 66-69; res metallurgist, Stanford Univ, 69-73; assoc prof metall, 75-78, chmn dept mat eng, 84-86, PROF METALL, SAN JOSE STATE UNIV, 78- *Concurrent Pos:* Consult, 69-; actg assoc prof, Stanford Univ, 73-76. *Mem:* Am Chem Soc; Am Inst Chem Engrs; Am Soc Metals; Am Inst Mining, Metall & Petrol Engrs; Am Nuclear Soc. *Res:* Physical metallurgy; chemical metallurgy; high temperature chemistry; chemical thermodynamics and kinetics; forensic engineering. *Mailing Add:* 27820 Saddle Ct Los Altos Hills CA 94022-1810

ANDERSON, ROBERT PHILIP, b Norwood, Mass, Aug 28, 50; m 72; c 2. GENETICS. *Educ:* Johns Hopkins Univ, BA, 72; Univ Calif, Berkeley, PhD(molecular biol), 77. *Prof Exp:* Fel, Med Res Coun Lab Molecular Biol, Cambridge, Eng, 78-81; asst prof, 81-, ASSOC PROF GENETICS, UNIV WIS-MADISON. *Mailing Add:* Dept Genetics Univ Wis 445 Henry Mall Madison WI 53706

ANDERSON, ROBERT SIMPERS, b Bryn Mawr, Pa, Jan 4, 39; m 64; c 2. IMMUNOTOXICOLOGY, COMPARATIVE IMMUNOLOGY. *Educ:* Drexel Univ, Philadelphia, Pa, BS, 61; Hahnemann Med Univ, MS, 68; Univ Del, Newark, Del, PhD(immunol), 71. *Prof Exp:* Fel comp immunol, dept path, Univ Minn, 70-73; asst mem, Sloan-Kettering Inst Cancer Res, 73-82; asst prof, Cornell Univ Grad Sch Med Sci, 75-82; immunologist immunotoxicol, US Army Chem Res & Develop Ctr, Aberdeen Proving Ground, 82-86; PROF IMMUNOTOXICOL, CHESAPEAKE BIOL LAB, UNIV MD, 86- *Concurrent Pos:* Mem, Comt Animal Models & Genetic Stocks, Nat Res Coun, Nat Acad Sci, 77-79; actg ed-in-chief, J Invert Path, 86-87 & Critical Reviews in Aquatic Toxicol, 86- *Honors & Awards:* Except Performance Award, US Army, 84, 86 Outstanding In-House Lab Independent Res Award, 86. *Mem:* Am Asn Immunologists; Am Soc Zoologists; Int Soc Develop & Comp Immunol; Soc Invert Path; Sigma Xi; Soc Toxicol. *Res:* Immunotoxicology; effects of xenobiotics on the immune response of aquatic invertebrates; comparative carcinogenesis; biotransformation of environmental oncogens by acquatic invertebrates. *Mailing Add:* Chesapeake Biol Lab Univ Md Box 38 Solomons MD 20688-0038

ANDERSON, ROBERT SPENCER, internal medicine; deceased, see previous edition for last biography

ANDERSON, ROGER ARTHUR, b Madison, SDak, Nov 7, 35; m 57; c 4. BOTANY, BIOLOGY. *Educ:* Augustana Col, SDak, AB, 58; Univ Colo, MA, 61, PhD(bot), 64. *Prof Exp:* Asst prof bot, Univ Mont, 64-65; asst prof bot & biol, 65-74, ASSOC PROF BIOL SCI, UNIV DENVER, 74- *Mem:* AAAS;

Am Bryol Soc; Am Soc Plant Taxon; Bot Soc Am; Sigma Xi. *Res:* Taxonomy and world distribution of the lichenized fungi of North America, particularly western North America. *Mailing Add:* Dept Biol Sci Univ Denver Denver CO 80208

ANDERSON, ROGER CLARK, b Wausau, Wis, Oct 30, 41; m 67; c 2. BOTANY, ECOLOGY. *Educ:* Wis State Univ, La Crosse, BS, 63; Univ Wis-Madison, MS, 65, PhD(bot). 68. *Prof Exp:* Asst prof bot, Southern Ill Univ, Carbondale, 68-70; asst prof bot & managing dir arboretum, Univ Wis-Madison, 70-73; assoc prof biol, Cent State Univ, Okla, 73-76; assoc prof, 76-80, PROF BIOL, ILL STATE UNIV, 80- *Concurrent Pos:* Vis assoc prof, Univ Okla Biol Sta, 74, 75 & 79, Univ Mich Biol Sta, 80; consult, Metrop Sanitary Dist Great Chicago, 77-81, Kerr McGee Corp, 85; mem, Environ Adv Comt, Fermi Nat Accelerator Lab. *Mem:* Ecol Soc Am; Sigma Xi; Bot Soc Am. *Res:* Plant community ecology including studies of prairie plants and their associated mycorrhizal fungi. *Mailing Add:* Dept Biol Ill State Univ Normal IL 61761

ANDERSON, ROGER E, b Donovan, Ill, May 14, 30; m 59; c 2. PHYSICAL CHEMISTRY, COMPUTER SCIENCES. *Educ:* Univ Ill, BS, 51; Wash State Univ, PhD(phys chem), 61. *Prof Exp:* CHEMIST, LAWRENCE RADIATION LAB, UNIV CALIF, LIVERMORE, 59-, AT LAWRENCE LIVERMORE NAT LAB. *Mem:* Inst Elec & Electronics Eng; Asn Comput Mach; Sigma Xi. *Res:* Instrumentation; laboratory automation; on-line computer utilization; data processing techniques; minicomputer systems analysis. *Mailing Add:* 4354 Guilford Ave Livermore CA 94550

ANDERSON, ROGER FABIAN, b St Paul, Minn, Sept 14, 14; m 50; c 2. FOREST ENTOMOLOGY. *Educ:* Univ Minn, PhD(entom), 45. *Prof Exp:* Entomologist, USDA, 40-48; asst prof entom, Univ Ga, 49-50; assoc prof, 50-59, prof, 59-80, EMER PROF FOREST ENTOM, DUKE UNIV, 80- *Honors & Awards:* A D Hopkins Award, 83. *Mem:* Entom Soc Am. *Res:* Ecology of forest insects; biology, behavior, and physiological relation between boring insects and host trees; pollution damage to trees by acid rain and other toxins. *Mailing Add:* 2528 Perkins Rd Durham NC 27706

ANDERSON, ROGER HARRIS, b Seattle, Wash, Feb 3, 30; m 81; c 2. LOW TEMPERATURE PHYSICS. *Educ:* Univ Wash, BSc, 51, PhD(physics), 61. *Prof Exp:* Asst, Univ Wash, 52-58; res scientist, Boeing Co, 58-61; prof physics, Seattle Pac Col, 61-80; DEPT PHYSICS, SEATTLE PAC COL, 80- *Concurrent Pos:* NSF sci fac fel, Univ Ill, 68-69. *Mem:* Am Phys Soc; Am Asn Physics Teachers. *Res:* Cloud chamber study of scattering of cosmic-ray muons; calculation of thermal properties of monomolecular layers of liquid helium-3; transport phenomena in Landau theory of Fermi liquids; spin-polarized Fermi systems; 2-Dim Helium-3 systems. *Mailing Add:* Dept Physics Seattle Pac Col Seattle WA 98119

ANDERSON, ROGER W(HITING), b Minneapolis, Minn, July 2, 24; m 50; c 1. PHYSICAL CHEMISTRY, CHEMICAL ENGINEERING. *Educ:* Univ Minn, BChEng, 48; Univ Tenn, PhD(phys chem), 63. *Prof Exp:* Develop staff mem II, 48-50, res & develop group leader enrichment technol, 50-69, CONSULT TO PLANT DIV, MAT & SYSTS DEVELOP DEPT, GASEOUS DIFFUSION PLANT, NUCLEAR DIV, UNION CARBIDE CORP, 69- *Mem:* Am Chem Soc; Sigma Xi; AAAS. *Res:* Colloid, surface, and electrochemistry; isotope exchange; gas-solid reactions; ion exchange; water treatment; adsorption. *Mailing Add:* 304 Boundary Lane Knoxville TN 37922

ANDERSON, ROGER W, b Fargo, NDak, Jan 9, 43. PHYSICAL CHEMISTRY. *Educ:* Carleton Col, BA, 64; Harvard Univ, MA, 65, PhD(chem), 68. *Prof Exp:* Asst prof, 69-75, ASSOC PROF CHEM, UNIV CALIF, SANTA CRUZ, 75- *Mem:* AAAS; Am Phys Soc; Sigma Xi. *Res:* Molecular beam studies of nonadiabatic atomic and molecular collisions; collisional excitation; electronic quenching, reaction of excited atoms; theory of nonreactive scattering in reactive systems and scattering of excited atoms. *Mailing Add:* Dept Chem Univ Calif Santa Cruz CA 95064

ANDERSON, ROGER YATES, b Oak Park, Ill, Oct 4, 27; m 48; c 4. GEOLOGY. *Educ:* Univ Ariz, BS, 54, MS, 55; Stanford Univ, PhD, 60. *Prof Exp:* Asst biochem, Univ Ariz, 55-56; from instr to assoc prof, 56-74, PROF GEOL, UNIV NMEX, 74- *Concurrent Pos:* NSF prin investr varve res. *Res:* Paleoecology; palynology; varves. *Mailing Add:* Dept Geol Univ NMex Main Campus Albuquerque NM 87131

ANDERSON, ROLAND CARL, b Trenton, NJ, Nov 27, 34; m 53; c 2. OPTICAL SYSTEMS, IMAGE PROCESSING. *Educ:* Univ Fla, BS, 60, MS, 61, PhD(turbulence), 65. *Prof Exp:* Assoc engr, Martin Co, Fla, 60; instr aerodyn, Univ Fla, 61-62, teaching assoc, 63-65; instr, WVa Univ, 62-63; Ford Found fel astrophys, Johns Hopkins Univ, 65-66; from asst prof to assoc prof, 66-73, asst dean res, Col Eng, 74-75, PROF ENG SCI, UNIV FLA, 73- *Mem:* Optical Soc Am. *Res:* Experimental and theoretical research is done in interferometric flow visualization using holography and other methods; large scale field tests for IR equipment are conducted. *Mailing Add:* Univ Florida Grad Ctr Box 1918 Eglin AFB FL 32542-0918

ANDERSON, RONALD E, b Sikeston, Mo, June 14, 41. COMPUTER SCIENCE, COMPUTER APPLICATIONS. *Educ:* Stanford Univ, PhD(sociol), 70. *Prof Exp:* PROF SOCIOL, UNIV MINN, 68- *Concurrent Pos:* Mem coun & chair, Spec Interest Group Computers & Soc, Asn Comput Mach, 87- *Mem:* Asn Comput Mach. *Mailing Add:* Dept Sociol 909 Sociol Sci Bldg Univ Minn Minneapolis MN 55455

ANDERSON, RONALD EUGENE, b Sioux City, Iowa, Sept 15, 20; m 48; c 4. GENETICS, PLANT BREEDING. *Educ:* Univ Nebr, BA, 48; Univ Wis, MS, 49, PhD(genetics), 52. *Prof Exp:* Asst prof agron, Univ Ky, 52-54; asst prof plant breeding, 54-69, ASSOC PROF PLANT BREEDING & BIOMET, CORNELL UNIV, 69- *Res:* Cytology; breeding of forage crops species; effects of radiation on forage species. *Mailing Add:* 252 Emerson Hall Plant Breed Cornell Univ Main Campus Ithaca NY 14853

ANDERSON, RONALD KEITH, b Duluth, Minn, Sept 12, 41; m 66; c 2. NUCLEAR PHYSICS, ACOUSTICS. *Educ:* Augustana Col, SDak, BA, 63; SDak Sch Mines & Technol, MS, 65; Univ Kans, PhD(nuclear physics), 71. *Prof Exp:* Inst physics, SDak Sch Mines & Technol, 63-65; instr, 65-68, asst prof, 71-73, ASSOC PROF PHYSICS & CHMN DEPT, BEMIDJI STATE UNIV, 73- *Mem:* Am Phys Soc; Audio Eng Soc; Am Asn Physics Teachers; Sigma Xi. *Res:* Theoretical low-energy nuclear structure; isobaric spin; acoustics-vibrations and its engineering applications. *Mailing Add:* Dept Physics Bemidji State Univ Bemidji MN 56601

ANDERSON, RONALD M, b Mower County, Minn, Nov 14, 35. HYDROLOGY & WATER RESOURCES. *Educ:* Luther Col, BA, 57; Iowa State Univ, MS, 59, PhD(math), 62. *Prof Exp:* Collins Radio Co, 62-65; PROF & CHMN MATH DEPT, TEX TECH UNIV, 65- *Mem:* Am Math Soc; Soc Indust & Appl Math; Math Asn Am. *Res:* Porous media, with application to water uptake in plants; solar power. *Mailing Add:* Dept Math Tex Tech Univ Lubbock TX 79409-1042

ANDERSON, ROSALIND COOGAN, b Cleveland, Ohio; m 63; c 3. COMBUSTION TOXICOLOGY, INHALATION TOXICOLOGY. *Educ:* Mt Holyoke Col, BA, 59; Yale Univ, MA, 64, PhD(physiol), 65. *Prof Exp:* Res assoc endocrinol, Sch Med, Yale Univ, 65-68; assoc prof physiol, Duquesne Univ, 70-75; res assoc toxicol, Grad Sch Pub Health, Univ Pittsburgh, 76-79; head toxicol unit, Arthur D Little, Inc, 79-87; PRES, ANDERSON LABS INC, 88- *Concurrent Pos:* Nat Acad Sci Comt Fire Toxicol. *Honors & Awards:* Nininger Metorite Award, 70. *Mem:* Am Soc Testing & Mat; Nat Fire Protection Asn; Soc Toxicol. *Res:* Developing methods for the evaluation of the effects of inhalation exposure to products of thermal decomposition; application of combustion toxicology to building codes. *Mailing Add:* Anderson Labs Inc 30 River St Dedham MA 02026

ANDERSON, ROY CLAYTON, b Camrose, Alta, Apr 26, 26; m 48; c 2. PARASITOLOGY. *Educ:* Univ Alta, BSc, 50; Univ Toronto, MA, 52, PhD(parasitol), 56; Sch Advan Studies, Paris, dipl helminth, 58. *Prof Exp:* Sr res scientist parasitol, Ont Res Found, 54-65; chmn dept, 79-89, PROF ZOOL, UNIV GUELPH, 65- *Concurrent Pos:* Fel Rothmansted Exp Sta, London Sch Hyg & Trop Med, 56-57; med fac, Univ Paris, 57-58. *Honors & Awards:* Henry Baldwin Ward Medal, Am Soc Parasitol, 68; Distinguished Serv Award, Wildlife Dis Asn, 78; Wardle Medal, Can Soc Zoologists, 88. *Mem:* Am Soc Parasitol; Wildlife Dis Asn; Sigma Xi; Wildlife Soc; Royal Soc Trop Med & Hyg; Can Soc Zoologist. *Mailing Add:* Dept Zool Univ Guelph Guelph ON N1G 2W1 Can

ANDERSON, ROY E, b Batavia, Ill, Oct 30, 18; m 43; c 4. EXPERIMENTAL DEMONSTRATION OF CONCEPTS FEASIBILITY, TECHNICAL & REGULATORY STANDARDS. *Educ:* Augustana Col, Ill, BA, 43; Union Col, MSEE, 52. *Prof Exp:* Physics instr, Augustana Col, 43-44 & 46-47; consult engr, Corp Res & Develop, Gen Elec Co, 47-83; co-founder & vpres, Mobil Satellite Corp, 83-88; OWNER & PRES, ANDERSON & ASSOCS, 88- *Concurrent Pos:* Coolidge fel, Gen Elec Co, 70. *Mem:* Fel Inst Elec & Electronics Engrs; fel AAAS; Am Inst Aeronaut & Astronaut; Soc Satellite Prof Int; Inst Navig. *Res:* Development of concepts and systems engineering of mobile satellite communications and position fixing systems for land, aeronautical and maritime users. *Mailing Add:* PO Box 2531 Glenville NY 12325-2531

ANDERSON, RUSSELL D, b Salt Lake City, Utah, Dec 27, 27; m 58; c 3. ENTOMOLOGY, INVERTEBRATE ZOOLOGY. *Educ:* Univ Utah, BS, 54, MS, 56, PhD(entom), 61. *Prof Exp:* Entomologist, South Salt Lake County Mosquito Abatement Dist, 55-58; from instr to assoc prof zool, Church Col, Hawaii, 58-64, head dept biol sci, 59-64; from asst prof to assoc prof zool, 64-71, PROF ZOOL, SOUTHERN UTAH STATE COL, 71-, HEAD, DEPT LIFE SCI, 81- *Concurrent Pos:* Consult environ impact statements. *Mem:* AAAS; Entom Soc Am. *Res:* Taxonomy of the Dytiscidae; revisionary study of the genus Hygrotus. *Mailing Add:* Dept Biol Southern Utah State Col Cedar City UT 84720

ANDERSON, RUSSELL K, b LeRoy, Minn, Mar 13, 24; m 48; c 2. ANIMAL NUTRITION, PHYSIOLOGY. *Educ:* Univ Minn, BS, 48; Iowa State Univ, MS, 50, PhD(animal nutrit), 56. *Prof Exp:* Instr animal sci, Iowa State Univ, 48-55; from asst prof to assoc prof, 55-64, PROF ANIMAL SCI, CALIF POLYTECH STATE UNIV, SAN LUIS OBISPO, 64- *Mem:* Am Soc Animal Sci. *Res:* Ruminant nutrition; mineral requirement for rumen microorganisms and sources of phosphorus. *Mailing Add:* Dept Animal Sci Calif Polytech State Univ San Luis Obispo CA 93407

ANDERSON, RUTH MAPES, b Cleveland, Ohio. NURSING. *Educ:* Mount Holyoke Col, AB, 42; Western Reserve Univ, MN, 45, MSN, 54; Cornell Univ, PhD(organ theory & behav), 66. *Prof Exp:* Supvr & asst dir nursing, Cleveland Clin Hosp, 46-50; with US Navy Nurse Corps, 50-52; from asst prof to prof nursing, Case Western Reserve Univ, 54-85, from asst dean to assoc dean, 73-85. *Concurrent Pos:* Assoc nursing, Univ Hosps, Cleveland, 66-85; prin investr, Planned Change Nursing, 68-73; contrib ed, J Nursing Admin, 70-78. *Mem:* Am Nurses Asn; Nat League Nursing; Am Sociol Asn. *Res:* Planned change in the delivery of nursing care to clients; educational preparation of nurses in leader roles and their subsequent contributions in health care institutions. *Mailing Add:* 2381 Overlook Rd Cleveland OH 44106-2445

ANDERSON, SABRA SULLIVAN, b Philadephia, Pa, June 10, 39; m 65; c 2. MATHEMATICS, EDUCATION ADMINISTRATION. *Educ:* Smith Col, AB, 61; Univ Mich, MA, 63. *Prof Exp:* Fulbright fel, Univ Manchester, Eng, 61-62; res fel math, Inst Social Res, Univ Mich, 65-66; asst prof, Eastern Mich Univ, 66-69; PROF MATH, UNIV MICH, 82-, ASSOC DEAN, COL SCI & ENG, 87- *Concurrent Pos:* Vis scholar math, Western Mich Univ, 79 & Stanford Univ, 80; assoc ed, Math Asn Am, 83-86. *Mem:* Am Math Soc; Math Asn Am; Asn Women Am; Asn Women Sci; Sigma Xi. *Res:* Discrete math specifically finite graph theory; combinatorial and algebraic problems. *Mailing Add:* Math Sci Dept Univ Minn Duluth MN 55812

ANDERSON, SAMUEL, b Clarksdale, Miss, Aug 26, 34; m 65; c 2. ANALYTICAL CHEMISTRY, INORGANIC CHEMISTRY. *Educ:* Univ Mo, Kansas City, BS, 57; Iowa State Univ, MS, 60, PhD(anal chem), 62. *Prof Exp:* Asst prof chem, Tenn State Univ, 62-63; PROF CHEM, NORFOLK STATE COL, 63- *Concurrent Pos:* Res partic, NSF, 64-65. *Mem:* Am Chem Soc; Sigma Xi. *Res:* Preparation, physical and chemical properties of vanadium B-diketone complexes; evaluation of an immobilized chelate for extraction of metal ions from solutions. *Mailing Add:* Dept Chem Norfolk State Col Norfolk VA 23504

ANDERSON, SCOTT, b Johnsonville, Ill, June 26, 13; m 41; c 5. ELECTROLUMINESCENCE. *Educ:* Ill Wesleyan Univ, BS, 35; Univ Ill, MS, 36, PhD(physics), 40. *Hon Degrees:* Ill Wesleyan Univ, DSc,60. *Prof Exp:* Asst prof physics, Carleton Col, 43-44; physicist, Aluminum Res Labs, 40-43; sub-consult, Univ Chicago Metall Lab, 44-45; owner & dir, Anderson Physics Labs, 44-83; PHYSICS CONSULT, APL, INC, 83- *Mem:* Fel Am Phys Soc; Am Ceramic Soc; Optical Soc Am; AAAS. *Res:* Purification of salts; other problems relating to metal halide lamps; author of journal publications. *Mailing Add:* 1116 W Church St Champaign IL 61821

ANDERSON, SONIA R, b Omaha, Nebr, Mar 14, 39. BIOCHEMISTRY, BIOPHYSICS. *Educ:* Univ Nebr, BS, 61; Univ Ill, PhD(biochem), 64. *Prof Exp:* Postdoctoral, Univ Ill, 64-67, Univ Rome, Italy, 67-68; from asst prof to assoc prof, 68-81, PROF, ORE STATE UNIV, 81- *Concurrent Pos:* Vis asst prof, Univ Ill, 68; res grant, USPHS, 70-, res career develop award, 70-75; mem, bd sci counselors, Nat Heart, Lung & Blood Inst, NIH, 91- *Mem:* Am Soc Biol Chemists; Protein Soc; AAAS; Sigma Xi. *Res:* Biophysical characterization of physiologically relevant protein-protein interactions; fluorescence spectroscopy; associations of glycolytic enzymes with contractile proteins; calcium-dependent protein kinases; calmodulin; polypeptide hormones and neurotransmitters; venom peptides; author or co-author of over 50 publications. *Mailing Add:* Ore State Univ Corvallis OR 97331

ANDERSON, STANLEY H, b San Francisco, Calif, Aug 6, 39; m 65; c 2. ECOLOGY, ENVIRONMENTAL SCIENCES. *Educ:* Univ Redlands, BSc, 61; Ore State Univ, MA, 68, DrPhil(ecol), 70. *Prof Exp:* Asst prof biol, Kenyon Col, 70-75; ecologist, Oak Ridge Assoc Univs, 75-76; MAT SCIENTIST & PROG MGR, SRI INT, 78-; LEADER, WYO COOP, FISH & WILDLIFE UNIT, LARAMIE, 80- *Concurrent Pos:* Regist prof metall engr, Calif. *Mem:* AAAS; Am Inst Biol Sci; Am Ornith Union; Am Defense Preparedness Asn; Ecol Soc Am. *Res:* Community interaction of avifauna; determination of effects of habitat alteration on vertebrate populations. *Mailing Add:* 1062 Arapaho Dr Laramie WY 82070

ANDERSON, STANLEY ROBERT, b Rudyard, Mich, Mar 11, 20; m 46; c 3. AGRONOMY. *Educ:* Mich State Col, BS, 46, MS, 49; Iowa State Col, PhD, 54. *Prof Exp:* Field rep, Mich Crop Improv Asn, Mich State Col, 46-49; instr pub schs, Mich, 49-50; instr agron, Iowa State Col, 50-54; from asst prof to prof, Ohio State Univ, 54-67; dean agr, Tex A&I Univ, 67-75; pres, Abraham Baldwin Agr Col, 75-85; RETIRED. *Concurrent Pos:* Chmn, Am Asn Univ Agr Adminr, 75. *Mem:* Am Soc Agron; Crop Sci Soc Am. *Res:* Crop production; management and utilization of pasture and forage establishment. *Mailing Add:* 1815 Columbine Pl Sun City Center FL 33573

ANDERSON, STANLEY WILLIAM, b Stromsburg, Nebr, Aug 2, 21; m 47; c 1. ELECTRICAL ENGINEERING, MATHEMATICS. *Educ:* Ill Inst Technol, BS, 49, MS, 50, PhD(elec eng), 57. *Prof Exp:* Engr, Commonwealth Edison Co, 48-49 & Gen Elec Co, 50-52; consult elec power systs, Mid West Serv Co, 52-68; staff asst, Syst Planning Dept, Commonwealth Edison Co, 68-84; RETIRED. *Concurrent Pos:* Prof, Ill Inst Technol, 57-58 & 70- *Mem:* Fel Inst Elec & Electronics Engrs; Opers Res Soc Am; Sigma Xi; Am Soc Eng Educ. *Res:* Electric power systems; analogue and digital computers; operations research; probability and statistics; use of operations research methods in solving interconnection capability problems of electric utility systems; reliability research activities on electric power systems. *Mailing Add:* 630 Shelton Rd Ridgewood NJ 07450

ANDERSON, STEPHEN WILLIAM, b Evanston, Ill, Apr 3, 56. PHOTOCHEMISTRY. *Educ:* Carthage Col, BA, 78; Marquette Univ, MS, 81; Northern Ill Univ, PhD(chem), 85. *Prof Exp:* Lab technician, Universal Oil Prod, 77, res chemist, 78; postdoctoral res fel gen & org chem, Univ Toronto, 84-87, lectr, 86-87; ASST PROF CHEM, UNIV WIS-WHITEWATER, 87- *Mem:* Sigma Xi; Am Chem Soc; Inter-Am Photochem Soc. *Res:* Thermal solvolytic reactions; cationic and anionic photoinitiators for polymerizations; solvent nucleophilicity in the excited state. *Mailing Add:* Dept Chem Univ Wis-Whitewater Whitewater WI 53190

ANDERSON, STEVEN CLEMENT, b Grand Canyon Nat Park, Ariz, Sept 7, 36; m 60; c 1. SYSTEMATICS, ECOLOGY. *Educ:* Univ Calif, Riverside, BA, 57; San Francisco State Col, MA, 62; Stanford Univ, PhD(biol), 66. *Prof Exp:* Lab technician entom, Univ Calif, Riverside, 60; res asst herpet, Calif Acad Sci, 63-64, from asst cur to assoc cur, 64-70; from asst prof to assoc prof, 70-80, PROF BIOL & ENVIRON SCI, UNIV PACIFIC, 80- *Concurrent Pos:* Vis res assoc, Iran Dept Environ, 75-79; res assoc, Calif Acad Sci, 77-; consult ed, Encycl Iranica, 77- *Mem:* Am Soc Ichthyol & Herpetol; Soc Study Amphibians & Reptiles; fel Herpetologists League; AAAS; Sigma Xi. *Res:* Systematics, ecology, zoogeography of amphibians and reptiles of Southwest Asia. *Mailing Add:* Dept Biol Sci Univ Pacific Stockton CA 95211

ANDERSON, SUE ANN DEBES, b Harrisburg, Ill, Jan 8, 47; m 78; c 2. NUTRITION, FOOD SAFETY. *Educ:* Univ Ill, BS, 69; Auburn Univ, MS, 72; Purdue Univ, PhD(nutrit), 74. *Prof Exp:* Res home economist sensory eval, Armour Food Res Labs, 69-70; asst prof nutrit, Sch Home Econ, Auburn Univ, 75-78; SR STAFF SCIENTIST, LIFE SCI RES OFF, FEDN AM SOCS EXP BIOL, 78- *Concurrent Pos:* Adj prof, Va Polytech Inst & State Univ, 87. *Mem:* Am Inst Nutrit; Am Dietetic Asn. *Res:* Nutrition and public health; food safety issues. *Mailing Add:* Life Sci Res Off 9650 Rockville Pike Bethesda MD 20014

ANDERSON, SYDNEY, b Topeka, Kans, Jan 11, 27; m 51; c 3. VERTEBRATE ZOOLOGY. *Educ:* Univ Kans, AB, 50, MA, 52, PhD, 59. *Prof Exp:* Asst instr, Univ Kans, 50-52; NSF fel, 52-54; asst cur mammals, Mus Natural Hist, Kans, 54-59; from asst cur to assoc cur, 60-69, CUR, AM MUS NATURAL HIST, 69- *Concurrent Pos:* Instr, Univ Kans, 54-59; consult, Bk Div, Time, Inc & C S Hammond Co. *Honors & Awards:* Jackson Award, Am Soc Mammal, 85. *Mem:* Fel AAAS; Am Soc Mammal; Ecol Soc Am; Soc Syst Zool. *Res:* Distribution, variation and relationships of mammals; instrumentation and methodology; bibliography and information retrieval. *Mailing Add:* Am Mus Natural Hist Central Park W at 79th St New York NY 10024

ANDERSON, TED L, CELL BIOLOGY, GENE EXPRESSION. *Educ:* Univ Southern Miss, BS, 76, MS, 78; Vanderbilt Med Sch, PhD, 85. *Prof Exp:* Res assoc, Baylor Col Med, 85-86; instr, Eastern Va Med Sch, 86-88, asst prof reproductive biol, 88-89. *Mem:* Am Soc Cell Biol. *Res:* Hormone action in the uterus. *Mailing Add:* Dept Anat Vanderbilt Univ Sch Med Nashville TN 37232

ANDERSON, TERA LOUGENIA, b Dothan, Ala, Nov 26, 49; m 79; c 1. DATABASE MANAGEMENT SYSTEMS. *Educ:* Univ Wash, BS (physics) & BS(math), 73, MS, 77, PhD(comput sci), 81. *Prof Exp:* Mgr info mgt res, Austin Res Ctr, Burroughs Corp, 80-85; prin scientist, Tektronix, Inc, 85-91; PROG MGR OBJECT TECH, SEQUENT COMPUTER SYSTS INC, 91- *Concurrent Pos:* Adj fac mem, Ore Grad Ctr. *Mem:* Asn Comput Mach. *Res:* Data models and data semantics for database management systems; object-oriented database management; representation of time in information systems; description of user interfaces for database management systems; CAO/CAM databases. *Mailing Add:* Sequent Computer Systs Inc 15450 SW Koll Pkwy Beaverton OR 97006

ANDERSON, TERRY LEE, b Lincoln, Nebr, Sept 11, 47; m 71. THEORETICAL PHYSICS, COMPUTER SCIENCE. *Educ:* Pac Union Col, BS, 69, MA, 69; Univ Neb, Lincoln, MS, 72, PhD(physics), 75. *Prof Exp:* Asst instr physics, Univ Nebr, Lincoln, 69-72; from instr to assoc prof, Walla Walla Col, 72-81, prof physics & comput sci, 81-86, chmn, Dept Comput Sci, 79-86; MEM TECH STAFF, AT&T BELL LABS, 86- *Mem:* Am Asn Physics Teachers; Am Phys Soc; Asn Comput Mach; Soc Computer Simulation. *Res:* Computer languages; instructional computer software; theoretical relativistic astrophysics; properties of ice nucleation; learning theory applied to science instruction and application of computers to science instruction. *Mailing Add:* 24 Hill St Bernardsville NJ 07924

ANDERSON, TERRY ROSS, b Newmarket, Ont, April 8, 44; m 68; c 2. SOIL BORNE DISEASE, BIOLOGICAL CONTROL. *Educ:* Univ Guelph, BSA, 69; Univ Hawaii, MSc, 75; Univ Toronto, PhD(plant path), 79. *Prof Exp:* RES SCIENTIST, AGR CAN, 78- *Mem:* Am Phytopath Soc; Can Phytopath Soc. *Res:* Mycophagous amoeboid organisms in soil; control of phytophtora rot of soybeans. *Mailing Add:* Agr Can Res Br Harrow ON N0R 1G0 Can

ANDERSON, THEODORE GUSTAVE, b New Haven, Conn, Nov 6, 02; m 36. BACTERIOLOGY. *Educ:* Brown Univ, PhB, 31; Yale Univ, PhD(bact), 35; dipl, Am Bd Med Microbiol. *Prof Exp:* Assoc dairy husb, Univ Calif, 36-37; from instr to asst prof bact, Pa State Univ, 37-43, 46-47; bacteriologist, Vet Admin Hosp, New York, 47; from assoc prof bact & immunol to prof microbiol & bacteriologist, Sch Med, Temple Univ, 47-70, emer prof microbiol, 70-; dir microbiol lab, Episcopal Hosp, 70-82; RETIRED. *Mem:* Am Soc Microbiol; fel Am Acad Microbiol; Sigma Xi. *Res:* Medical bacteriology; antibiotic susceptibility testing; urinary tract infection; dental caries; dairy microbiology. *Mailing Add:* 71 Bethlehem Pike Philadelphia PA 19118

ANDERSON, THEODORE WILBUR, b Minneapolis, Minn, June 5, 18; m 50; c 3. MATHEMATICAL STATISTICS, ECONOMETRICS. *Educ:* Northwestern Univ, BS, 39; Princeton Univ, MA, 42, PhD(math), 45. *Hon Degrees:* DL, N Park Col & Theol Sem, 88; DSc, Northwestern Univ, 89. *Prof Exp:* Asst math, Northwestern Univ, 39-40; instr, Princeton Univ, 41-43; res assoc, Appl Math Panel Contract, Nat Defense Res Comt, 43-45 & Cowles Comn, Chicago, Ill, 45-46; from instr to prof math statist, Columbia Univ, 46-67; prof, 67-88, EMER PROF STATIST & ECON, STANFORD UNIV, 88- *Concurrent Pos:* Consult, Cowles Found, 46-60, Bur Appl Social Res, New York, 47 & Rand Corp, 49-66; Guggenheim fel, Univ Stockholm & Cambridge Univ, 47-48; ed, Ann Math Statist, 50-52; vis assoc prof, Stanford Univ, 54; mem, Nat Acad Sci-Nat Res Coun Comt Basic Res, adv to Off Ord Res, 55-58; fel, Ctr Advan Study Behav Sci, 57-58; actg exec officer, Columbia Univ, 50-51 & 63, chmn, dept math statist, 56-60 & 64-65; mem, comt statist, Nat Res Coun, 60-63, chmn, 61-63, comt pres statist soc, 62-64; mem, Nat Acad Sci-Nat Res Coun Panel Appl Math, adv to Nat Bur Standards, 64-65; mem, comt support res math sci, Nat Acad Sci, 65-68; acad vis, Imp Col, Univ London, & vis prof, Univs Moscow & Paris, 67-68; vis scholar, Ctr Advan Study Behav Sci, 72-73 & 80; sci dir, Advan Study Inst Discriminant Anal & Its Appln, NATO, Kifissia, Greece, 72; acad vis, London Sch Econ & Polit Sci, 74-75; Wesley C Mitchell vis prof econ, Columbia Univ & NY Univ, 83-84; IBM Systems Res Inst, 84; Sherman Fairchild distinguished scholar, Calif Inst Technol, 80; res assoc, Naval Postgrad Sch, 86-87. *Honors & Awards:* R A Fisher Award, 85; S S Wilkes Medal, 88. *Mem:* Nat Acad Sci; Inst Math Statist (pres, 63); Am Math Soc; fel Am Statist Asn (vpres, 71-73); fel Economet Soc; Am Acad Arts Sci; Psycmet Soc; Int Statist Inst. *Res:* Multivariate statistical analysis; time series analysis; theory of statistical inference; econometric methodology. *Mailing Add:* Dept Statist Stanford Univ Stanford CA 94305

ANDERSON, THOMAS ALEXANDER, b Elkhart, Ind, Feb 23, 28; m 52; c 2. BIOCHEMISTRY, NUTRITION. *Educ:* Univ Wash, BA, 50; Calif State Polytech Col, BS, 56; Univ Ariz, MS, 61, PhD(agr biochem & nutrit), 62. *Prof Exp:* Sr scientist, Sci Info Dept, Smith, Kline & French Labs, Pa, 62-63; asst prof biochem, Univ SDak, 63-65; chief nutrit res lab, H J Heinz Co, 65-70; from assoc prof to prof pediat, Col Med, Univ Iowa, 70-85; RETIRED. *Concurrent Pos:* Partic, White House Conf Food, Nutrit & Health, 69. *Mem:* AAAS; Am Chem Soc; Am Soc Clin Nutrit; Brit Nutrit Soc; Am Soc Animal Sci. *Res:* Pediatric nutrition,especially influence of the pattern of food intake on body composition and digestibility of complex carbohydrates by the human infant. *Mailing Add:* 17719 Elk Trail Rd Weed CA 96044-9420

ANDERSON, THOMAS EDWARD, b Evergreen Park, Ill, Dec 9, 47; m 71; c 3. NEUROSCIENCE, NEUROPHYSIOLOGY. *Educ:* Hamline Univ, BA, 68; Univ Tex, Austin, MA, 71; Univ Mich, Ann Arbor, PhD(neurosci), 77. *Prof Exp:* Fel neurobiol, Dept Zool, Univ Tex, Austin, 77-79; from asst sr res scientist to staff res scientist, 79-86, SR STAFF RES SCIENTIST & SECT MGR NEUROL SYST, BIOMED SCI DEPT, GEN MOTORS RES LABS, WARREN, 86- *Concurrent Pos:* Vis asst prof, Col Eng, Wayne State Univ, 80-84, adj prof bioeng, 84- *Mem:* Soc Neurosci; Sigma Xi; AAAS; Am Asn Automotive Med; Bioelectromagnetic Soc. *Res:* Study of mechanisms for progressive pathology and dysfunction following brain or spinal cord traumatic injury; neurotoxic effects and mechanisms; biomedical effects of non-ionizing electromagnetic radiation. *Mailing Add:* Biomed Sci Dept Gen Motors Res Labs Warren MI 48090-9058

ANDERSON, THOMAS ERNEST, b Chicago, Ill, Jan 28, 52; m 77; c 2. PESTICIDE DEVELOPMENT, PEST MANAGEMENT. *Educ:* Univ Ill, Urbana, BS, 74, MS, 77; NC State Univ, PhD(entom), 81. *Prof Exp:* Teaching asst, Dept Entom, Univ Ill, 74-77; res asst, NC Agr Exp Sta, 77-81; asst entomologist, Boyce Thompson Inst Plant Res, 81-85; INSECTICIDE GROUP LEADER, BASF CORP, 85- *Mem:* Entom Soc Am; Sigma Xi; AAAS; Soc Invert Path. *Res:* Discovery and development of new insecticides and nematicides; bioassays of microbial insecticides; investigations of entomopathogen-pesticide interactions in pest management systems. *Mailing Add:* BASF Corp PO Box 13528 Res Triangle Park NC 27709-3528

ANDERSON, THOMAS FOXEN, biophysics; deceased, see previous edition for last biography

ANDERSON, THOMAS FRANK, b Chicago, Ill, Dec 7, 39; m 63. GEOCHEMISTRY. *Educ:* DePauw Univ, BA, 61; Columbia Univ, PhD(geochem), 67. *Prof Exp:* Res assoc geochem, Enrico Fermi Inst, Univ Chicago, 67-68; asst prof, 67-73, assoc prof, Geol, 73-82, assoc dean, Col Lib Arts & Sci, 87-90, PROF GEOL, UNIV ILL, URBANA, 82- *Concurrent Pos:* Exchange scientist, NSF and Nat Ctr Sci Res, France, 75-76; vis res lectr, Univ Leicester (UK), 84; vis prof, Chengdu Col Geol, 86. *Mem:* AAAS; Am Geophys Union; Geochem Soc; Geol Soc Am; Sigma Xi. *Res:* Stable isotope geochemistry; geochemistry of sediments and natural waters. *Mailing Add:* 2116 Burlison Dr Urbana IL 61801

ANDERSON, THOMAS L(EONARD), b Everett, Wash, Feb 2, 36; m 58; c 3. STRUCTURAL DYNAMICS, EARTHQUAKE ENGINEERING. *Educ:* Univ Idaho, BS, 58, MS, 61; Univ Colo, PhD(civil eng), 67. *Prof Exp:* Instr civil eng, Univ Idaho, 58-61; assoc res engr, Com Div, Boeing Airplane Co, 61-62; from asst prof to assoc prof civil eng, Univ Idaho, 62-70; sr res engr & mgr struct & mech, Pac Northwest Labs, Battelle Mem Inst, 70-73; supv struct engr, Fluor Engrs & Constructors, Irvine, Calif, 73-89; ENG, TRS GREENVILLE, 89- *Mem:* Seismol Soc Am; Am Soc Civil Engrs; Earthquake Eng Res Inst. *Res:* Engineering mechanics, including fracture mechanics; computer code development; aseismic structures design and innovative materials uses, including composites; stress analysis of nuclear fuel elements and related hardware; earthquake engineering and structural dynamics activities on all petrochemical, pipeline, liquefied natural gas & nuclear projects; computer program development and advanced-level training. *Mailing Add:* TRS 330-A Pelham Rd Greenville SC 29615

ANDERSON, THOMAS PAGE, b Anadarko, Okla, Aug 3, 18; m 48; c 3. PHYSICAL MEDICINE & REHABILITATION. *Educ:* Univ Okla, BS, 40, MD, 43; Univ Minn, MS, 51; Am Bd Phys Med & Rehab, dipl, 53. *Prof Exp:* Intern, St Paul's Hosp, Dallas, Tex, 44; from clin instr to asst clin prof phys med & rehab, Dartmouth Med Sch, 51-64; from assoc clin prof to clin prof, Med Sch, Univ Minn, Minneapolis, 64-70, from assoc prof to prof med & rehab, 70-85; SR PSYCHIAT, SPAULDING REHAB HOSP, 88- *Concurrent Pos:* Consult, Hitchcock Clin, Hanover, NH, 51-64 & Vet Admin Hosp, 52-64; staff physiatrist, Kenny Rehab Inst, 64-70; dir dept med educ, Am Rehab Found, 65-70, med dir, New Medico Head Injury Syst, 85-88. *Honors & Awards:* Coulter lectr, Am Cong Rehab Med, 87; Licht lectr, Minn Psychiatric Soc, 88. *Mem:* Fel Am Acad Phys Med & Rehab; fel Am Col Physicians; Am Cong Rehab Med. *Res:* Stroke; degenerative joint disease; group communications and leadership skills; disorders of the back; spinal cord injury. *Mailing Add:* Spaulding Rehab Hosp 125 Nashua St Boston MA 02114

ANDERSON, THOMAS PATRICK, b Chicago, Ill, Oct 22, 34; m 60; c 2. MECHANICAL ENGINEERING. *Educ:* Northwestern Univ, BSME, 56, MS, 58, PhD(mech eng), 61. *Prof Exp:* Assoc prof mech eng & astron sci, Northwestern Univ, 58-66; prof mech eng, Univ Iowa, 66-75, chmn dept, 66-70; dir indust prog, NSF, 74-78; PROF ENG & DEAN, SCH SCI & ENG, SOUTHERN ILL UNIV, EDWARDSVILLE, 78- *Concurrent Pos:* Assoc dir interdept energy study, US Off Sci & Technol, 63-65. *Mem:* AAAS; Am Inst Aeronaut & Astronaut; Am Phys Soc; Am Soc Mech Engrs; Sigma Xi. *Res:* Gas dynamics; thermodynamics; magnetohydrodynamics; energy. *Mailing Add:* Sch Sci & Eng Southern Ill Univ Edwardsville IL 62026

ANDERSON, VICTOR CHARLES, b Shanghai, China, Mar 31, 22; m 43; c 3. PHYSICS. *Educ:* Univ Redlands, AB, 43; Univ Calif, Los Angeles, MA, 50, PhD(physics), 53. *Prof Exp:* Tech asst, Radiation Lab, Univ Calif, 43-45; asst, Los Alamos, NMex, 45-46; asst, Univ Calif, Los Angeles, 46-47; from asst res physicist to res physicist, 47-68, actg dir, Marine Physical Lab, 64, 68 & 74, PROF APPL PHYSICS & ASSOC DIR MARINE PHYSICAL LAB, SCRIPPS INST, UNIV CALIF, SAN DIEGO, 68- *Concurrent Pos:* Res fel, Harvard Univ, 54-55. *Mem:* Fel Acoust Soc Am; Inst Elec &

Electronics Engrs. *Res:* Acoustics; experimental and theoretical work in field of underwater sound and electronics; methods and instrumentation for digital signal processing; remotely controlled sea floor work systems. *Mailing Add:* Dept Appl Physics Marine Phys Lab San Deigo CA 92132

ANDERSON, VICTOR ELVING, b Stromsburg, Nebr, Sept 6, 21; m 46; c 4. HUMAN GENETICS, BEHAVIORAL GENETICS. *Educ:* Univ Minn, BA, 45, MS, 49, PhD(zool), 53. *Prof Exp:* From instr to prof biol, Bethel Col, Minn, 46-60, chmn dept, 52-60; vis scientist, NIH, 60-61; asst scientist, 49-54, asst dir, Dight Inst Human Genetics, 54-78, actg dir, 78-84, assoc prof zool, 61-65, assoc prof genetics, 65-66, PROF GENETICS, UNIV MINN, MINNEAPOLIS, 66- *Concurrent Pos:* Regent, Bethel Col & Seminary, St Paul, 69-74 & 82-87; mem, Develop Behav Sci Study Sect, NIH, 72-75, chmn, 74-75. *Mem:* Fel AAAS; Am Sci Affil (pres, 63-65); Am Soc Human Genetics; Behav Genetics Asn (secy, 72-74, pres 79-80); Sigma Xi (pres, 82-83). *Res:* Genetic factors in epilepsy, mental retardation, psychotic disorders, and other human behavioral problems. *Mailing Add:* 1775 N Fairview Ave St Paul MN 55113

ANDERSON, VIRGIL LEE, b North Liberty, Ind, May 2, 22; m 43; c 3. MATHEMATICS, STATISTICS. *Educ:* Iowa State Univ, BS, 47, PhD(statist), 53. *Prof Exp:* From asst prof to assoc prof statist, 51-60, dir statist lab, 56-66, PROF STATIST, PURDUE UNIV, WEST LAFAYETTE, 60- *Concurrent Pos:* Statist consult to var govt, mil & pvt orgn, 58- *Mem:* Fel Am Statist Asn; Biomet Soc. *Res:* Quantitative genetics; design of experiments; transportation; environmental effects of cadmium. *Mailing Add:* Dept Statist Purdue Univ West Lafayette IN 47907

ANDERSON, W FRENCH, b Tulsa, Okla, Dec 31, 36; m 61. BIOCHEMISTRY & MEDICINE, GENETICS. *Educ:* Harvard Univ, AB, 58, MD, 63; Cambridge Univ, MA, 60. *Prof Exp:* Intern pediat med, Children's Hosp Med Ctr, Boston, 63-64; res assoc, Lab Biochem Genetics, Nat Heart, Lung & Blood Inst, 65-67, res med officer, 67-68, head, Sect Human Biochem, 68-71 & Sect Molecular Hemat, 71-73, CHIEF, MOLECULAR HEMAT BR, NAT HEART, LUNG & BLOOD INST, 73-, CHIEF, LAB MOLECULAR HEMAT, 77- *Concurrent Pos:* Am Cancer Soc res fel bact & immunol, Harvard Med Sch, 64-65; vol asst, Children's Hosp Med Ctr, 64-65; prof lectr, Sch Med, George Washington Univ, 67-75; mem fac, Dept Genetics, NIH Grad Prog, 67-, fac, Dept Med & Physiol, 81- & chmn, 84-; mem heart fel bd, Nat Heart & Lung Inst, 68-70; mem task force hemoglobinopathies, Nat Heart, Lung & Blood Inst, NIH, 72, Nat Task Group on Cooley's Anemia, 77-78, pres, Assembly Scientists, 82; chmn, Inter-Inst Coord Comt on Cooley's Anemia, NIH, 72-77; mem hemoglobin subcomt, Am Soc Hemat, 72-74; mem Med Resources Coun, Cooley's Anemia Blood & Res Found for Children, 74-77; chmn, Inter-Agency Coord Comt on Cooley's Anemia, Dept Health, Educ & Welfare, 75-77; med adv bd, Cooley's Anemia Found, Inc, 77-; mem, Coun Acad Socs, Asn Am Med Col, 77-80, Ad Hoc Adv Group, Inst Med, Nat Acad Sci, 84, Exec Comt Bd Dirs, Found Adv Ed in Sci, Inc, NIH, 84-, Working Group Human Gene Ther, Recombinant DNA Adv Comt, 84-87 & Working Group on Viruses, 85, Sci Adv Comt, Children's Hosp Res Found, Cincinnati, Ohio, 85-89; sr exec sci serv, Dept Health & Human Serv, 80-; consult, President's Comn Study Ethical Probs Med & Biomed Behav Res, 81-82, Human Gene Ther Ctr for Bioethics, Kennedy Inst Ethics, Washington, DC, 82-; hemat prog dir, Lab Molecular Hemat, Nat Heart, Lung & Blood Inst, NIH-George Washington Univ-Vet Med Ctr, 85; chmn, Sci Adv Bd, Genetic Ther Inc, Gaithersburg, Md, 86-87; sci adv bd, S/L Health Care Ventures, NY, 86-88; mem, NIH Coord Comt Human Genome, 88-; consult Human Gene Therapy: St Jude Childrens Res Hosp, Memphis Tenn, 90- & Univ Pittsburgh, Pittsburgh, 90-; chmn, Res Adv Coun, Children's Nat Med Ctr, Washington, DC, 90-; Outstanding Performance Award, Huguenot Historical Soc, 82, 84-90. *Honors & Awards:* Sci Achievement Biol Sci Award, Wash Acad Sci, 71; Thomas B Cooley Award Sci Achievement, Cooley's Anemia Blood & Res Found Children, 77; Mary Ann Liebert Biotherapeut Award, 91; President's Award lectr, Am Thoracic Soc, 91; Maude L Menten, Univ Pittsburgh. *Mem:* Asn Am Physicians; Am Soc Clin Invest; Am Soc Hemat; Am Soc Human Genetics; Am Soc Biol Chem; Am Fedn Clin Res. *Res:* Human biochemical genetics; hematology; regulation of RNA and protein synthesis; hemoglobin biosynthesis; thalassemia and hemoglobinopathies; gene expression in mammalian cells; genetic engineering of mammalian cells; human gene therapy; first HGT protocol for ADA. *Mailing Add:* Bldg 10 7D-18 Nat Heart Lung & Blood Inst Bethesda MD 20892

ANDERSON, W(ENDELL) L, b Altoona, Pa, Sept 8, 22; m 45; c 2. CHEMICAL ENGINEERING. *Educ:* Pa State Univ, BS, 44. *Prof Exp:* Sect head aerosols, Naval Res Lab, 44-71; dept head & assoc tech dir, Naval Surface Weapons Ctr, 71-83; RETIRED. *Concurrent Pos:* Indust consult; vis lectr. *Honors & Awards:* Kratel Award, Germany, 70. *Mem:* Am Chem Soc; Am Asn Contamination Control; Am Acad Arts & Sci; Sci Res Soc Am; Am Ord Asn; Am Filtration Soc. *Res:* Aerosols, filters, microfibers, chemical biological warfare, light scattering; air pollution contamination control; ion effects; naval weapons; technical management. *Mailing Add:* RR 4 Box 4172 LaPlata MD 20646

ANDERSON, W ROBERT, c 6. PULMONARY PATHOLOGY, RENAL PATHOLOGY. *Educ:* Univ Pa, MD, 58. *Prof Exp:* PROF PATH, DEPT PATH, UNIV MINN, 67-; CHMN, DEPT PATH, HENNIPEN COUNTY MED CTR, 84- *Mem:* Col Am Pathologists; Int Acad Path; Am Asn Path; Sigma Xi. *Res:* Electron-microscopy; surgical pathology. *Mailing Add:* Dept Path Hennipen County Med Ctr 701 Park Ave S Minneapolis MN 55415

ANDERSON, WALLACE ERVIN, b Florence County, SC, Oct 28, 13; m 39; c 2. PHYSICS. *Educ:* The Citadel, BS, 34; Univ Ky, MS, 36; Univ Mich, PhD(physics), 49. *Hon Degrees:* DSc, The Citadel, 81. *Prof Exp:* From asst prof to prof physics, The Citadel, 36-66, head dept, 53-66, acad dean, 66-70, vpres acad affairs, 70-79, interim pres, 78-79; RETIRED. *Honors & Awards:* Legion of Merit, 46. *Mem:* Am Phys Soc; Am Asn Physics Teachers; Am Soc Eng Educ. *Res:* Infrared and microwave spectroscopy; determination of molecular structure; optics. *Mailing Add:* 11 Country Club Dr Charleston SC 29412

ANDERSON, WALLACE L(EE), b Adams, NDak, Sept 2, 22; m 56; c 2. ELECTRICAL ENGINEERING, PHYSICS. *Educ:* Univ NDak, BS, 48; Rice Univ, MA, 57; Univ NMex, ScD(elec eng), 61. *Prof Exp:* Seismog operator, McCollum Explor Co, Tex, 48-50, seismog party mgr, 50-51, res engr, 54-55; res assoc elec eng, Univ NMex, 57-60, lectr, 60-61; assoc prof, NY Univ, 61-64; sr res engr, Southwest Res Inst, 64-68, staff scientist, 68-69; chmn dept, 72-77, PROF ELEC ENG, UNIV HOUSTON, 69- *Mem:* AAAS; Am Phys Soc; Inst Elec & Electronics Engrs; Optical Soc Am. *Res:* Coherent optics; optical information processing; biomedical engineering; electromagnetic and acoustic wave propagation; non-destructive evaluation of metals. *Mailing Add:* Dept Elec Eng Univ Houston Houston TX 77004

ANDERSON, WALTER L(EONARD), b St Paul, Minn, Aug 6, 22; ї 46; c 4. ELECTRONICS, TECHNICAL MANAGEMENT. *Educ:* Univ Minn, BEE, 44, MS, 48. *Prof Exp:* Asst, Univ Minn, 43-44; staff elec engr, Eng Res Assoc Div, Remington Rand, Inc, 46-55; vpres opers, Gen Kinetics, Inc, 55-63, exec vpres & chmn bd dirs, 63-65, pres, 65-74; from assoc dir to sr assoc dir, US Gen Acct Off, 74-85, sr adv, 85-90; CONSULT, 90- *Concurrent Pos:* Dir, Comput Test Corp, 62-66, Tape Serv Co, 63-67, Ocean Space Systs, 64-68, Allo Precision Metals Eng, Inc, 65-67, Food Technol Corp, 66-74 & Nat Instrument Labs, Inc, 68-72; data processing consult, Fairfax Hosp, 64-65; dir, Am Fedn Info Processing Socs, 62-66, mem exec comt, 64-66, chmn admissions comt, 66-68, mem financial comt, 68-71, vpres, 71-72, pres, 73-74; mem, Info Systs Coun, Am Mgt Asn, 75-87; Europ community fel, German Marshall Fund, 79. *Honors & Awards:* Chester Morrill Mem Award, Am Systs Mgt, 80; Supreme Achievement Award, Interagency Comt ADP, 82; Centennial Award, Inst Elect & Electronics Engrs, 84. *Mem:* Inst Elec & Electronics Engrs; Asn Comput Mach. *Res:* Electronic digital computers and communication systems; digital test systems. *Mailing Add:* 2427 Lexington Rd Falls Church VA 22043

ANDERSON, WALTER T(HEODORE), b Iron River, Mich, Mar 30, 23; m 45; c 8. ELECTRICAL ENGINEERING. *Educ:* Mich Technol Univ, BS, 43, MS, 50. *Prof Exp:* Test engr, Gen Elec Co, 43-44; process engr, Tenn Eastman Co, 44-45; instr physics, Mich Technol Univ, 45-46, from instr to asst prof elec eng, 50-51, proj engr, Union Carbide Nuclear Co, Tenn, 52-54; from asst prof to assoc prof elec eng, 54-64, admin asst dept, 70-81, PROF ELEC ENG, MICH TECHNOL UNIV, 64-, ASST DEPT HEAD, 81- *Concurrent Pos:* Dir, Appelt, Criner & Wedeven, Consult Engrs, Grand Rapids, 75-; consult engr; mem, Mich Bd Registr Prof Engrs, Mich Bd Registr Land Surveyors & Nat Coun Eng Examrs, 65-81. *Honors & Awards:* Distinguished Serv Award, Nat Coun Eng Examrs, 78. *Mem:* Sr mem Inst Elec & Electronics Engrs; Am Soc Eng Educ; Nat Soc Prof Engrs; Illum Eng Soc NAm. *Res:* Electrical power and machinery; electrical power distribution systems and illumination engineering. *Mailing Add:* Sch Technol Mich Technol Univ Houghton MI 49931

ANDERSON, WARREN BOYD, b Oak City, Utah, Nov 1, 29; m 53; c 6. SOIL CHEMISTRY & FERTILITY. *Educ:* Brigham Young Univ, BS, 58; Colo State Univ, MS, 62, PhD(soil sci), 64. *Prof Exp:* Phys sci technician, Agr Res Serv, USDA, 60; res assoc soil fertil, Colo Agr Exp Sta, 60-64; asst prof, 64-71, ASSOC PROF SOIL & CROP SCI, TEXAS A&M UNIV, 71- *Mem:* Am Soc Agron; Int Soc Soil Sci; Soil Sci Soc Am; Sigma Xi. *Res:* Soil fertility and plant nutrition; micronutrient nutrition of crops. *Mailing Add:* Soil & Crop Sci Tex A&M Univ College Station TX 77843

ANDERSON, WARREN R(ONALD), b Houston, Minn, July 31, 14; m 45; c 3. ELECTRICAL ENGINEERING. *Educ:* Univ Minn, BS, 39; La State Univ, BS, 44. *Prof Exp:* Prof elec eng, 46-79, head dept electronic & elec eng, 76-79, EMER PROF ELEC ENG, CALIF POLYTECH STATE UNIV, SAN LUIS OBISPO, 79- *Concurrent Pos:* Design engr, Automatic Elec Co, 46 & Gen Elec Co, 51; res analyst, Northrup Aircraft Co, 52; consult, Western Gear Corp, 55 & Gen Elec Co, 56. *Mem:* Inst Elec & Electronics Engrs; Am Soc Eng Educ. *Res:* Engineering problem analysis; creative development part of engineering. *Mailing Add:* Dept Elec & Electronic Eng Calif Polytech State Univ San Luis Obispo CA 93407

ANDERSON, WAYNE ARTHUR, b Jamestown, NY, May 20, 38; m 62; c 2. SEMICONDUCTORS, THIN-FILMS. *Educ:* Univ Buffalo, BS, 61; State Univ NY, MS, 65, PhD(elec eng), 70. *Prof Exp:* Proj supvr, Great Lakes Carbon Corp, 61-65; instr electronics, Univ Buffalo, 65-70; asst prof electronics, Rutgers Univ, 70-78; ASSOC & PROF ELECTRONICS, STATE UNIV NY, BUFFALO, 78- *Concurrent Pos:* Dir, Ctr Electronic & Electro-optic Mat, 86- *Mem:* Inst Elec & Electronics Engrs; Am Inst Physics; Mat Res Soc. *Res:* III-V semiconductor surfaces, Schottky barriers and defects; silicon photovoltaic devices and MOS interfaces; thin-film superconductors, resistors and capacitors. *Mailing Add:* 39 Sleepy Hollow Lane Orchard Park NY 14127

ANDERSON, WAYNE I, b Montrose, Iowa, Sept 14, 35; m 58; c 4. GEOLOGY. *Educ:* Univ Iowa, BA, 58, MS, 61, PhD(geol), 64. *Prof Exp:* Assoc prof, 63-73, PROF EARTH SCI, UNIV NORTHERN IOWA, 73-, HEAD DEPT, 63- *Mem:* Geol Soc Am; Soc Econ Paleontologists & Mineralogists; AAAS; Nat Asn Geol Teachers; Sigma Xi. *Res:* Paleozoic stratigraphy and paleontology; geology of Iowa; history of geology. *Mailing Add:* Dept Earth Sci Univ Northern Iowa Cedar Falls IA 50614

ANDERSON, WAYNE KEITH, b Pine Falls, Man, Apr 4, 41; m 62; c 3. MEDICINAL CHEMISTRY, ORGANIC CHEMISTRY. *Educ:* Univ Man, BSc, 62, MSc, 65; Univ Wis-Madison, PhD(med chem), 68. *Prof Exp:* Instr med chem, Univ Wis, 67 & 68; from asst prof to assoc prof, 68-81, PROF MED CHEM, STATE UNIV NY BUFFALO, 81- *Mem:* Am Chem Soc; Chem Soc, London; Am Pharmaceut Asn; Am Asn Cancer Res. *Res:* Synthetic approaches to and structural modifications of naturally occurring antitumor compounds; design and synthesis of new classes of antitumor agents; synthetic heterocyclic chemistry; synthesis and structure of tumor inhibitory metal coordination compounds; the design of new anti-cancer drugs using computer graphics. *Mailing Add:* Dept Med Chem Rm 429A Cooke Hall SUNY Buffalo-North Campus Buffalo NY 14260

ANDERSON, WAYNE PHILPOTT, b Jamestown, NY, Apr 1, 42; m 72. COMPUTATIONAL CHEMISTRY. *Educ:* Harpur Col, BA, 64; Univ Ill, MS, 66, PhD(chem), 68. *Prof Exp:* Asst prof chem, Univ Del, 68-75; from asst prof to assoc prof, Bloomsburg State Col, 75-84, PROF CHEM, BLOOMSBURG UNIV, PA, 84- *Concurrent Pos:* Chmn, chem dept, Bloomsburg Univ, 87- *Mem:* Am Chem Soc. *Res:* Synthesis and determination of the electronic structure of transition metal complexes; applications of molecular mechanics to chemical systems. *Mailing Add:* Dept Chem Bloomsburg Univ Bloomsburg PA 17815

ANDERSON, WESTON ARTHUR, b Kingsburg, Calif, Mar 28, 28; m 52; c 2. PHYSICS. *Educ:* Stanford Univ, BS, 50, MS, 53, PhD(physics), 55. *Prof Exp:* Res physicist, Europ Coun Nuclear Res, 54-55; res physicist, Varian Assocs, 55-63, dir res, Anal Instrument Div, 63-72, dir systs lab, 72-87, PRIN SCIENTIST, VARIAN ASSOCS, 87- *Mem:* Inst Elec & Electronics Engrs; Am Asn Physics Teachers; Am Phys Soc; Soc Photo-optical Instrumentation Engrs. *Res:* Magnetic resonance; biomedical engineering; vacuum microelectronics. *Mailing Add:* 763 La Para Ave Palo Alto CA 94306

ANDERSON, WILLARD EUGENE, b Vilas, SDak, Aug 27, 33; m 59. APPLIED PHYSICS. *Educ:* Huron Col, SDak, BS, 55. *Prof Exp:* Res asst & tech asst, Corp Res Ctr, Honeywell Inc, 55-56; teaching asst math, SDak State Univ, 56-59; res scientist, Aeronaut Div, Res Dept, 59-64, sr res scientist, Mil Prod Group, Res Dept, 64-65, prin res scientist, 65-74, prin systs engr, Aerospace & Defense Group, Systs & Res Ctr, 74-76, prin develop engr, Energy Resources Ctr, Advan Power Systs, 74-79, PRIN DEVELOP ENGR, TECHNOL STRATEGY CTR, BLDG & INDUST SYST, HONEYWELL INC, 79- *Concurrent Pos:* Consult gen physics, 59-; consult, Manned Orbiting Lab, Controls Subcontract, Aerospace Div, Honeywell Inc, 66-69, prin investr, Technol Strategy Ctr, 79-80. *Mem:* AAAS. *Res:* Heat transfer and radiant thermometry; gyroscopes, precision spheres, rotor spin-up, pickoffs, voltage breakdown; density, atmosphere, ocean; trace gas sensing, chemistry, ionic mobility; optics, laser Doppler, opdar, correlators, spatial filtering, scattering; space physics; atmospheric physics; guidance and control, missile simulations; solar energy systems, photovoltaic solar cell modules; combined cycle and cogeneration. *Mailing Add:* 2418 Ione St Lauderdale MN 55113

ANDERSON, WILLARD WOODBURY, b Boston, Mass, July 22, 39; m 57; c 5. MECHANICAL ENGINEERING, AEROSPACE SYSTEMS. *Educ:* Northeastern Univ, BS, 62; Mass Inst Technol, MS, 64, ScD(mech eng), 67. *Prof Exp:* Engr, Dynatech Corp, 60-62; res asst fiber & polymer res, Mass Inst Technol, 62-65; sr engr & partner, Cohoon & Heasley, Inc, 64-66; asst head, 66-72, head, Stability & Control Br, 72-76, asst chief, 76-79, CHIEF, FLIGHT DYNAMICS & CONTROL DIV, LANGLEY RES CTR, NASA, 79- *Res:* Attitude control of large manned spacecraft utilizing momentum storage systems; automated air traffic control systems research; flight dynamics and control of aircraft and spacecraft. *Mailing Add:* Langley Res Ctr NASA Mail Stop 479 Hampton VA 23665-5225

ANDERSON, WILLIAM ALAN, b St Peter's Bay, PEI, Can, Jan 5, 41; m 63; c 3. MOLECULAR BIOLOGY. *Educ:* Queen's Univ, Ont, BSc, 62; Mass Inst Technol, PhD(microbiol), 69. *Prof Exp:* Fel molecular biol, Univ Geneva, 69-70; from asst prof to assoc prof, 71-83, PROF BIOL, LAVAL UNIV, 83- *Concurrent Pos:* Sabbatical fel, CNRS, Gif-sur-Yvette, 77-78; mem, Laval Univ Cancer Res Ctr, Hotel-Dieu Quebec, 80. *Mem:* Can Biochem Soc; Genetics Soc Can; Am Soc Microbiol. *Res:* Diversity, function and expression of mammalian cytochrome P450 genes; promutagen activation by P450 synthesized in mammalian cells transfected by P450 expression nectors. *Mailing Add:* Dept Biol Laval Univ Quebec PQ G1K 7P4 Can

ANDERSON, WILLIAM B, b Montevideo, Minn, Nov 8, 23; m; m 59; c 3. PSYCHIATRY. *Educ:* Univ Minn, BS, 45, MD, 48. *Prof Exp:* Instr pediat & asst dir pediat clin, Univ Pa, 54-56; asst prof pediat, asst dir state serv crippled children & dir pediat clin, State Univ Iowa, 56-61; ASST PROF PSYCHIAT, MED CTR, DUKE UNIV, 65- *Concurrent Pos:* Fel pediat, Tulane Univ, 49-50; fel child psychiat, Med Ctr, Duke Univ, 63-65. *Res:* Adult and child psychiatry. *Mailing Add:* Dept Psychiat Duke Univ Durham NC 27710

ANDERSON, WILLIAM DEWEY, JR, b Columbia, SC, June 4, 33; m 63; c 2. ICHTHYOLOGY. *Educ:* Univ SC, BS, 53, MS, 55, PhD(biol), 60. *Prof Exp:* Asst prof biol, Susquehanna Univ, 60-61; fishery biologist, Biol Lab, Bur Commercial Fisheries, US Fish & Wildlife Serv, Ga, 61-65; interim asst prof biol sci, Univ Fla, 65-66; assoc prof, Univ Chattanooga, 66-69; assoc prof, 69-79, PROF BIOL, COL CHARLESTON, 79- *Concurrent Pos:* Gen ichthyol ed, Copeia, 76-79. *Mem:* Am Soc Ichthyol & Herpet; Am Fisheries Soc; Soc Syst Zool; Am Soc Zool; Soc Study Evolution. *Res:* Systematics of the fishes of the families Lutjanidae, Serranidae, and Callanthiidae. *Mailing Add:* Grice Marine Biol Lab 205 Ft Johnson Charleston SC 29412

ANDERSON, WILLIAM J, b Brooklyn, NY, Oct 7, 22; m 51; c 2. MECHANICAL ENGINEERING, MATHEMATICS. *Educ:* Mass Inst Technol, SB, 50; Case Inst Technol, MS, 56. *Prof Exp:* Res engr lubrication, Nat Adv Comt Aeronaut, 50-53, head bearings sect, 53-57; chief mech, Technologies Br, 57-81, CONSULT ENGR, LEWIS RES CTR, NASA, 81- *Concurrent Pos:* Lectr, Univ Calif, Los Angeles, 63-80, Case Inst Technol, 66, Univ Tenn, Va Polytech Inst & State Univ, Carnegie-Mellon Univ & Univ Pittsburg; vis prof, Univ Akron, 81. *Honors & Awards:* Hunt Mem Award, Am Soc Lubrication Engrs, 62. *Mem:* Fel Am Soc Mech Engrs; fel Am Soc Lubrication Engrs. *Res:* Lubrication; friction and wear; bearing dynamics. *Mailing Add:* 5031 Devon Dr North Olmsted OH 44070

ANDERSON, WILLIAM JOHN, b Boston, Mass, Dec 9, 38; m 68. NEUROANATOMY. *Educ:* Univ Miami, BA, 62; Purdue Univ, MS, 73, PhD(anat), 75. *Prof Exp:* Res assoc psychol, Mass Inst Technol, 62-68; fac instr neurobiol, Purdue Univ, 68-71, grad instr anat, Sch Vet Med, 71-75; asst prof neuroanat, 75-82, assoc prof anat, Ind Univ Sch Med, Terre Haute Ctr Med Educ, 82-84, PROF ANAT, IND STATE UNIV, 84- *Mem:* Int Soc Develop Psychobiol; Am Asn Anatomists; Soc Neurosci; Am Asn Vet Anatomists; Sigma Xi; World Asn Vet Anatomists. *Res:* Morphological analysis of brain development and brain abnormalities due to toxic elements, with special reference to postnatal development. *Mailing Add:* Ind State Univ 135 Homestead Hall Terre Haute IN 47809

ANDERSON, WILLIAM LENO, b Carney, Okla, Mar 13, 35; m 62; c 2. WILDLIFE BIOLOGY. *Educ:* Okla State Univ, BS, 58; Southern Ill Univ, MA, 64. *Prof Exp:* Field asst pop ecol, 58-62, res asst, 62-63, res assoc ecol & physiol, 63-68, assoc wildlife specialist, Ill Natural Hist Surv, 68-77; CONSERV RESOURCE PROG MGR, ILL DEPT CONSERV, 77- *Mem:* Wildlife Soc; Wilson Ornith Soc; Cooper Ornith Soc. *Res:* Population dynamics; behavioral and physiological responses of birds and mammals to pesticides, lead, trace elements, heated water, power plants and power lines; hunter activity and opinion surveys. *Mailing Add:* Ill Dept Conserv 607 E Peabody Champaign IL 61820

ANDERSON, WILLIAM LOYD, b Minneapolis, Minn, Aug 8, 47; m 70; c 2. IMMUNOLOGY. *Educ:* Univ Minn, BA, 69, PhD(biochem), 74. *Prof Exp:* Teaching asst biochem, Univ Minn, 69-74, res assoc, 74-75; fel, Dept Immunol, Mayo Clinic, 75-78, assoc consult, 78-81; ASST PROF CELL BIOL, UNIV NMEX, 81- *Mem:* Am Chem Soc; Sigma Xi; AAAS. *Res:* Role of soluble factors in the regulation of immunological responses; effect of aging on the immune system. *Mailing Add:* 5609 Cambria NW Albuquerque NM 87120

ANDERSON, WILLIAM MCDOWELL, b Richmond, Va, Sept 13, 51; m 81; c 2. PULMONARY MEDICINE, CRITICAL CARE MEDICINE. *Educ:* Univ Va, BA, 73; Old Dominion Univ, MS, 75; Eastern Va Med Sch, MD, 78. *Prof Exp:* Med intern & resident, Univ Hosp, Jacksonville, Fla, 78-81, chief resident internal med, 81-82, fel critical care med, 82-83; fel pulmonary med, Univ Fla, 83-85; DIR MED INTENSIVE CARE UNIT, VET ADMIN MED CTR, 85-, SECT CHIEF PULMONARY & CRITICAL CARE MED, 88- *Concurrent Pos:* Asst prof med, La State Univ Sch Med, Shreveport, 85-, asst prof physiol & biophys, 88- *Res:* Investigation of the mechanisms involved in respiratory muscle fatigue; investigation of the roles of sleep fragmentation versus hypoxemia in symptoms related to obstructive sleep apnea. *Mailing Add:* Dept Med Physiol & Biophys La State Univ Med Sch Shreveport LA 71104

ANDERSON, WILLIAM NILES, JR, b Pittsburgh, Pa, Nov 27, 39; m 69; c 2. APPLIED MATHEMATICS. *Educ:* Carnegie-Mellon Univ, BS, 60, MS, 67, PhD(math), 68. *Prof Exp:* Res assoc math, Rockefeller Univ, 68-70; asst prof, Univ Md, 70-75; assoc prof math & comput sci, WVa Univ, 75-79; prof & chmn, East Tenn State Univ, 79-84; PROF, FAIRLEIGH DICKINSON UNIV, 84- *Mem:* Soc Indust & Appl Math; Asn Comput Mach; Inst Elec & Electronics Engrs; Sigma Xi. *Res:* Linear algebra. *Mailing Add:* Dept Math & Comput Sci Fairleigh Dickinson Univ Teaneck NJ 07666

ANDERSON, WILLIAM RAYMOND, b St Charles, Ill, Mar 12, 11; wid. PHYSICS. *Educ:* Univ Ill, BS, 33; DePaul Univ, MS, 47. *Prof Exp:* Pvt & pub acct, 34-42; pub sch teacher, Ill, 43-47; from instr to asst prof physics, 47-59, ASSOC PROF PHYSICS, UNIV ILL, CHICAGO CIRCLE, 59- *Mem:* Am Asn Physics Teachers. *Res:* Spectroscopy; molecular physics. *Mailing Add:* Dept Physics Univ Ill Box 4348 Chicago IL 60680

ANDERSON, WILLIAM RUSSELL, b Tucson, Ariz, Sept 25, 42; m 67; c 2. SYSTEMATIC BOTANY. *Educ:* Duke Univ, BS, 64; Univ Mich, MS, 66, PhD(bot), 71. *Prof Exp:* Assoc cur, New York Bot Garden, 71-74; assoc cur & asst prof, 74-80, assoc prof bot, Univ Mich, 80-86; PROF, CUR & DIR, UNIV MICH HERBARIUM, 86- *Concurrent Pos:* Gen ed, Flora Novo-Galiciana. *Honors & Awards:* Greenman Award, Mo Bot Garden, 73; Cooley Award, Am Soc Plant Taxonomists, 75. *Mem:* Am Soc Plant Taxonomists; Int Asn Plant Taxon; Asn Trop Biol; Sigma Xi. *Res:* Systematics and evolution of flowering plants, especially the family Malpighiaceae; Mexican floristics. *Mailing Add:* Univ Mich Herbarium North Univ Bldg Ann Arbor MI 48109-1057

ANDERSON, WILLIAM W, b Tacoma, Wash, Feb 26, 33; m 54; c 3. ELECTRICAL ENGINEERING, SOLID STATE ELECTRONICS. *Educ:* Mass Inst Technol, BS & MS, 56; Stanford Univ, PhD(elec eng), 59. *Prof Exp:* Instr elec eng, Stanford Univ, 57-58; mem tech staff solid state electronics, Bell Tel Labs, 58-60; asst prof elec eng, Stanford Univ, 60-65; prof, Ohio State Univ, 65- 75; CONSULT SCIENTIST ELECTRO-OPTICS, LOCKEED PALO ALTO RES LAB, 75- *Concurrent Pos:* Consult to Watkins-Johnson, Battelle Mem Inst, Corning, Varian Assoc. *Mem:* Inst Elec & Electronics Engrs; Mat Res Soc. *Res:* Solid state electronic devices. *Mailing Add:* Electro-optics Lockheed Palo Alto Res Lab 3251 Hanover St Palo Alto CA 94304

ANDERSON, WILMER CLAYTON, b Waco, Tex, Nov 24, 09; m 33; c 3. PHYSICS. *Educ:* Baylor Univ, AB, 29, AM, 30; Harvard Univ, PhD(physics), 36. *Prof Exp:* Instr physics, chem & math, Lamar Col, 31-33; instr physics, Harvard Univ, 35-38, fel, 38-40; physicist, Tex Co, 41-42 & Div Res, Columbia Univ, 42-44; sr res & develop engr, Arma Corp, NY, 44-45; chief engr, Aireon Mfg Corp, Conn, 45-46; head phys res group, Deering Milliken Res Trust, 46-49; head physics group & mem staff, New Prod Res Lab, Remington Rand, Inc, 49-52; dir electronic res, Talon, Inc, 52-54; dir res, Liquidometer Corp, 54-55; dir res & develop, Gen Time Corp, 55-69; vpres & tech dir, Quipu Corp, Conn, 69-71; CONSULT, 71- *Concurrent Pos:* Bayard Cutting fel, Harvard Univ, 37; with Off Sci Res & Develop, 44. *Mem:* AAAS; Am Phys Soc; Optical Soc Am; Inst Elec & Electronics Engrs. *Res:* Industrial applications of physics and electronics; magnetometers for magnetic surveys; sonic compensators for land-mine detection; voltage regulators for A-C; various textile applications; new high frequency method for measuring velocity of light; aerospace and oceanographic instrumentation. *Mailing Add:* 56 Sound View Dr Greenwich CT 06830

ANDERSON, WYATT W, b New Orleans, La, Mar 27, 39; m 62; c 3. POPULATION GENETICS. *Educ:* Univ Ga, BS, 60, MS, 62; Rockefeller Univ, PhD(life sci), 67. *Prof Exp:* From lectr to assoc prof biol, Yale Univ, 66-72; assoc prof, 72-75, prof zool, 75-80, PROF & HEAD GENETICS, UNIV GA, 80- *Mem:* Nat Acad Sci; Genetics Soc Am; Am Soc Naturalists; Ecol Soc Am; Am Soc Human Genetics; Soc Study Evolution. *Res:* Genetic mechanisms in evolution; selection in natural and experimental populations; relationships between genetical and ecological aspects of populations; molecular evolution. *Mailing Add:* Dept Genetics Univ Ga Biol Sci Bldg Athens GA 30602

ANDERSON-MAUSER, LINDA MARIE, b South Bend, Ind, Apr 27, 54; m. MOLECULAR BIOLOGY. *Educ:* Ind Univ, BA, 77; Univ NC, PhD(genetics), 84. *Prof Exp:* SR RES SCIENTIST, MILES LAB, 87- *Mem:* Am Soc Microbiol. *Res:* Molecular biology and genetics of procaryotes, with special emphasis on bacteriophage; fermentation products research immuno assay development. *Mailing Add:* PO Box 70 Elkhart IN 46515

ANDERSON OLIVO, MARGARET, b Omaha, Nebr, June 17, 41; m 71. PHYSIOLOGY. *Educ:* Augustana Col, SDak, BA, 63; Stanford Univ, PhD(biol), 67. *Prof Exp:* Fel neurobiol, Harvard Univ, 68-70; res assoc, Univ PR, 70-71; vis asst prof, Clark Univ, 72; asst prof, Bennington Col, 73; from asst prof to assoc prof, 73-85, PROF BIOL, SMITH COL, 85- *Concurrent Pos:* Prin investr res grants, NIH, 75-85. *Mem:* Biophys Soc; Soc Gen Physiol; Soc Neurosci. *Res:* Electrophysiology of membrane channels. *Mailing Add:* Dept Biol Sci Smith Col Northampton MA 01063

ANDERSON-SHAW, HELEN LESTER, b Lexington, Ky, Oct 18, 36; m 88. AMINO ACIDS, ZINC. *Educ:* Univ Ky, BS, 58; Univ Wis, MS, 65, PhD(nutrit sci), 69; Am Bd Nutrit dipl. *Prof Exp:* Dietitian, Roanoke Mem Hosp, Va, 59-60; Santa Barbara Cottage Hosp, 60-61; unit mgr, De La Guerra Commons, Univ Calif, Santa Barbara, 61-63; from asst prof to prof nutrit, Univ Mo, Columbia, 69-88, assoc dean res & grad studies home econ, 77-83; PROF & CHAIR, DEPT FOOD NUTRIT & FOOD SERV MGT, UNIV NC, GREENSBORO, 89- *Concurrent Pos:* Dietetic internship, Univ Mich Hosps, 58-89. *Mem:* Am Inst Nutrit; Am Soc Clin Nutrit; Am Dietetic Asn; Am Home Econ Asn; Soc Nutrit Educ. *Res:* Amino acid nutrition in humans and small animals; zinc bioavailability in humans. *Mailing Add:* A4 Park Bldg Univ NC Greensboro NC 27412-5001

ANDERSON-STOUT, ZOE ESTELLE, b Glenview, Ill, Jan 20, 13; m 76. NUTRITION. *Educ:* Ill Inst Technol, BSAS, 39; Univ Ill, MS, 47, PhD(animal nutrit), 50. *Prof Exp:* Mgr plant cafeteria, Hydrox Ice Cream Co, 35-39; fac mem cafeteria mgt & home econ, Frankfort Community High Sch, 39-42; dietician sch food serv, J Sterling Morton High Sch, 42-43; asst dept nutrit res, Nat Dairy Coun, 49-51, dir, 51-60; asst prof home econ & chmn dept, Wayne State Univ, 60-62; assoc prof nutrit, Dept Internal Med, Col Med, Iowa State Univ, 62-64; assoc prof, 65-78, EMER PROF FOODS & NUTRIT, SAN DIEGO STATE UNIV, 78- *Mem:* Am Dietetic Asn; Am Inst Nutrit; Am Home Econ Asn; Soc Nutrit Educ. *Res:* Nutrition in health and disease; effects of diet on voluntary activity; bone ash; growth; tissues; blood and urinary values. *Mailing Add:* 5183 Roxbury Rd San Diego CA 92116

ANDERTON, LAURA GADDES, b Providence, RI, Sept 6, 18. EXPERIMENTAL EMBRYOLOGY, HUMAN CYTOGENETICS. *Educ:* Wellesley Col, BA, 40; Brown Univ, MS, 48; Univ NC, PhD, 59. *Prof Exp:* Instr biol & chem, Howard Sem, 41-43; instr biol & counsr, 48-55, instr biol, 56-58, from asst prof biol & embryol to assoc prof biol, 58-68, DIR CYTOGENETICS LAB, UNIV NC,GREENSBORO,65-, PROF BIOL, 68- *Concurrent Pos:* NIH grant, 68-71; consult, Moses H Cone Mem Hosp, 69- *Mem:* AAAS; Am Inst Biol Sci; Am Soc Human Genetics; Am Soc Zoologists; Int Union Against Cancer; Sigma Xi. *Res:* Cell renewal system in the colon mucosa; familial polyposis; tissue culture of adenomas of the colon. *Mailing Add:* Dept Biol 310 Eberhart Bldg Univ NC Greensboro NC 27412

ANDES, W(ILLARD) ABE, b Miami, Fla, Feb 19, 42; m 72; c 2. HEMATOLOGY. *Educ:* Univ NC, BA, 64; Tulane Univ Sch Med, MD, 68. *Prof Exp:* From instr to assoc prof, 75-85, PROF MED, TULANE UNIV SCH MED, 85- *Mem:* NY Acad Sci; Am Fedn Clin Res; Am Soc Clin Oncol; Am Soc Hemat. *Mailing Add:* Dept Med Tulane Univ Sch Med 1430 Tulane Ave New Orleans LA 70112

ANDIA-WALTENBAUGH, ANA MARIA, endocrinology, for more information see previous edition

ANDONIAN, ARSAVIR TAKFOR, b Istanbul, Turkey, July 26, 50; m 76; c 2. ENGINEERING MECHANICS. *Educ:* Bogazici Univ, BS; Va Polytech Inst & State Univ, MS, 75, PhD(eng mech), 78. *Prof Exp:* Instr mech mat, Dept Eng Sci & Mech, Va Polytech Inst & State Univ, 77-78; asst prof exp solid mech, Dept Math Eng, Univ Ill, Chicago Circle, 79-84; SR RES SCIENTIST, GOODYEAR, AKRON, OH, 84- *Mem:* Soc Exp Stress Analysis. *Res:* Experimental applications of moire, three dimensional photoelasticity, birefrigent coating, brittle coating and holography in fracture mechanics and biomechanics; predicting cracking patterns in reinforced concrete; tire mechanics; cord/rubber composites and material characterization. *Mailing Add:* Dept 410-F Goodyear Res 142 Goodyear Blvd Akron OH 44305

ANDOSE, JOSEPH D, b Philadelphia, Pa, July 26, 44. SCIENTIFIC PROGRAMMING. *Educ:* Temple Univ, BA, 66; Princeton Univ, MA, 68, PhD(org chem), 71. *Prof Exp:* Res assoc phys org chem, Princeton Univ, 71-74; systs analyst comput-assisted molecular modeling, 74-75, systs assoc comput-assisted org synthesis, 75-77, proj mgr sci systs, 77-80, MGR SCI PROG, MERCK & CO, INC, 80- *Mem:* Am Chem Soc; AAAS; Asn Comput Mach. *Res:* Computer-assisted molecular modeling; computer-assisted organic synthesis. *Mailing Add:* Merck & Co Inc PO Box 2000 Rahway NJ 07065

ANDOW, DAVID A, b Cleveland, Ohio. INSECT ECOLOGY, AGRICULTURAL ECOLOGY. *Educ:* Brown Univ, ScB, 77; Cornell Univ, PhD(ecol), 83. *Prof Exp:* Assoc, Ecosysts Res Ctr, Cornell Univ, 82-84; ASST PROF INSECT ECOL, UNIV MINN, 84- *Concurrent Pos:* Vis scientist, Nat Inst Agr Sci, Japan, 82-83 & Nat Inst Agroenviron Sci, Japan, 83-86. *Mem:* Ecol Soc Am; Entom Soc Am; AAAS; Sigma Xi. *Res:* Ecological interactions agricultural systems; development of resource-conserving, sustainable, thought-intensive food production technologies. *Mailing Add:* Dept Entom 219 Hodson Hall Univ Minn 1980 Folwell Ave St Paul MN 55108

ANDRADE, JOHN ROBERT, b Providence, RI, Aug 23, 44; m 69; c 6. PHARMACEUTICS. *Educ:* Col Holy Cross, BS, 66, MS, 67; Brown Univ, PhD(org chem), 79. *Prof Exp:* Instr chem, US Naval Acad, 71-72 & Brown Univ, 77; res chemist, Med Res Div, 77-83, group leader, 83-90, MGR PROCESS DEVELOP & ANAL SERV, AM CYANAMID CO, 90-91. *Mem:* Am Chem Soc; Sigma Xi. *Res:* Synthesis of bulk chemicals for preclinical and clinical evaluation; development processes for the scale-up and manufacture of new drugs; improvement of processes for the manufacture of existing drugs; pharmaceutical process development management. *Mailing Add:* 322 Church St Bound Brook NJ 08805

ANDRADE, JOSEPH D, b Hayward, Calif, July 13, 41; m 66; c 2. BIOENGINEERING, MATERIALS SCIENCE. *Educ:* San Jose State Col, BS, 65; Univ Denver, PhD(metall, mat), 69. *Prof Exp:* Dean Col Eng, 83-87, PROF BIOENG, MAT SCI, ENG & PHARM, UNIV UTAH, 69-, CHMN DEPT, 78-80 & 88- *Honors & Awards:* Ebert Prize, Acad Pharmaceut Soc, 78; Clemson Award, Soc Biomat, 85. *Mem:* AAAS; NY Acad Sci; Am Chem Soc; Am Soc Higher Educ; Biomed Eng Soc. *Res:* Interface between non-living materials and living systems; bloodmaterials interface; biosensors; polymer surface chemistry; adsorption from solution; biomaterials; science education. *Mailing Add:* Dept Bioeng Z480 MEB Col Eng Univ Utah Salt Lake City UT 84112

ANDRADE, MANUEL, b Fall River, Mass, Mar 14, 39; m 69; c 4. METAL CORROSION. *Educ:* RI Sch Design, BS, 60; Southern Conn State Univ, MS, 85. *Prof Exp:* Exp engr, United Technol Corp, 62-69; RES ASSOC, WARNER-LAMBERT CO, 70- *Mem:* Am Chem Soc; fel Am Inst Chemists; Nat Asn Corrosion Engrs; Am Soc Testing & Mat; NY Acad Sci. *Res:* Electrochemical aspects of metal corrosion; infrared spectroscopy; quantitative determination of surface cleanliness; metal cleaning; metal surface treatment; polymeric and metallic thin film adhesion and deposition. *Mailing Add:* Warner-Lambert Co Ten Webster Rd Milford CT 06460

ANDRADY, ANTHONY LAKSHMAN, b Sri Lanka; m 73; c 2. CROSSLINKED ELASTOMERS, PERMEABILITY OF POLYMERS. *Educ:* Univ Ceylon, BSc, 72; Univ Akron, Ohio, MS, 75; Polytech NLondon, Eng, PhD(polymer sci), 78. *Prof Exp:* Res assoc, Univ Cincinnati, 78-80; sr res chemist, Monsanto Co, Mo, 80-82; res scientist, contract consult, Kuwait Inst Sci Res, 82-84; DIR DEPT POLYMER SCI, RES TRIANGLE INST, NC, 85- *Concurrent Pos:* Consult, numerous co. *Mem:* Fel Nat Col Rubber Technol Eng; Royal Inst Chem. *Res:* Investigation of structure-property relationships in crosslinked polymers; transport phenomena in polymeric material; weathering and stabilization of polymers and plastics; plastics and marine environment; plastic solid waste management. *Mailing Add:* PO Box 12194 Research Triangle Park NC 27709

ANDRAKO, JOHN, b Perth Amboy, NJ, Jan 19, 24; m 78; c 3. PHARMACEUTICAL CHEMISTRY. *Educ:* Rutgers Univ, BS, 47, MS, 49; Univ NC, PhD(pharmaceut chem), 53. *Prof Exp:* Instr pharmaceut chem, Univ NC, 49-50, instr pharm, 50-53, from asst prof to assoc prof pharmaceut chem, 53-56; assoc prof, 56-62, asst dean sch pharm, 65-73, asst vpres health sci, 73-75, asst provost, 75-78, asst vpres health sci, 78-82, PROF PHARMACEUT CHEM, MED COL VA, VA COMMONWEALTH UNIV, 62-, ASSOC VPRES HEALTH SCI, 82- *Mem:* Am Chem Soc; Am Pharmaceut Asn. *Res:* Synthesis of derivatives of various heterocyclic systems as antihypertensives and psychotropic agents. *Mailing Add:* Off VPres Health Sci Med Col Va VCU Box 549 Richmond VA 23298

ANDRE, MICHAEL PAUL, b Des Moines, Iowa, April 25, 51; m 88. ULTRASOUND, MEDICAL IMAGING. *Educ:* Cent Univ Iowa, BA, 72; Univ Calif, Los Angeles, MS, 75, PhD(med physics), 80. *Prof Exp:* Prog consult comput sci, univ comput ctr, Cent Univ Iowa, 69-72; tech staff aerospace eng, N Island Naval Air Sta, San Diego, 70-71; res assoc atmospheric physics, Inst Atmospheric Physics, Univ Ariz, 72-73; tech staff systs eng, Hughes Aircraft Co, 73-74; teaching researcher physics, Univ Calif, Los Angeles, 74-77; physicist radiol safety officer radiol physics, Los Angeles County Med Ctr, 77-81; sr physicist radiol, Cedars-Sinai Med Ctr, 79-85; PHYSICIST RADIOL, VET ADMIN MED CTR, 81-; ASSOC PROF RADIOL, UNIV CALIF, SAN DIEGO, 81- *Concurrent Pos:* Physicist, Wadsworth Vet Admin Med Ctr, Los Angeles, 77-81; lectr, Sch Pharm, Univ Southern Calif, 79-85; consult, Centinela Hosp Med Ctr, Inglewood, 79-85, Univ Calif, Los Angeles-Los Angeles County Med Ctr, 81- & Children's Hosp & Health Ctr, San Diego, 84-; prin investr, Vet Admin Res Serv, 89-, Univ Calif, San Diego Found, 88-; ed, J Investigative Radiol. *Honors & Awards:* Silverman Award, Health Physics Soc, 79. *Mem:* Am Asn Physicists Med; Soc Photo-Optical Instrumentation Engrs; Inst Elec & Electronics Engrs; Am Inst Physics; Am Inst Ultrasound Med; Am Coll Radiol. *Res:* Medical imaging systems including x-ray, computed tomography, ultrasound and nuclear magnetic resonance; biological effects and dosimetry of ultrasound; assessment of skeletal strength by non-invasive methods; observer performance analysis. *Mailing Add:* Dept Radiol V-114 Univ Calif La Jolla CA 92093

ANDRE, PETER P, b Glenridge, NJ May 8, 41. ALGEBRA. *Educ:* Develis Col, BA, 63; Yale Col, MS, 65, PhD(math), 70. *Prof Exp:* PROF MATH, US NAVAL ACAD, 74- *Mem:* Am Math Soc; Am Med Asn. *Res:* Algebra. *Mailing Add:* Dept Math US Naval Academy Annapolis MD 21402

ANDREA, STEPHEN ALFRED, b Cuba, NY, July 10, 38. MATHEMATICS. *Educ:* Oberlin Col, BA, 60; Calif Inst Technol, PhD(math), 64. *Prof Exp:* Instr math, Harvard Univ, 64-67; asst prof, Univ Calif, San Diego, 67-71; PROF, SIMON BOLIVAR UNIV, VENEZUELA, 71- *Mem:* Am Math Soc. *Res:* Differential topology; topological dynamics. *Mailing Add:* Apartado Postal 80-184 Zona Postal 1080-A Caracas Venezuela

ANDREADIS, THEODORE GEORGE, b Chelsea, Mass, Mar 22, 50; m 76; c 2. ENTOMOLOGY, INSECT PATHOLOGY. *Educ:* Univ Mass, BS, 72, MS, 75; Univ Fla, PhD(entom), 78. *Prof Exp:* Res asst entom, Univ Mass, 73-74; instr biol & field naturalist, Cape Cod Mus Natural Hist, 75; res & teaching asst entom, Univ Fla, 75-78; asst res entomologist, 78-82, assoc res entomologist, 82-86, RES ENTOMOLOGIST, CONN AGR EXP STA, 86- *Concurrent Pos:* Lectr epidemiol & pub health, Yale Univ Sch Med, 87-; mem, Conn State Salt Marsh Mgt Comt, 85-, Conn State Mosquito Control Adv Bd, 86-; chmn, Div Microsporida, Soc Invertebrate Path, 86-88; bd reviewers, Jour Protozool, 90- *Mem:* Entom Soc Am; Sigma Xi; Soc Invertebrate Path; Am Mosquito Control Asn. *Res:* Development of microbial agents for control of insects of public health importance; mosquitoes; biology of Microsporida; electron microscopy. *Mailing Add:* Dept Entom Conn Agr Exp Sta 123 Huntington St Box 1106 New Haven CT 06504

ANDREADIS, TIM DIMITRI, b Suez, Egypt, Oct 3, 51; US citizen. SURFACE SCIENCE. *Educ:* Univ Md, College Park, BS, 74, MS, 77, PhD(nuclear eng), 81. *Prof Exp:* Spacecraft analyst, Comput Sci Technicolor Assocs, 74-77; res asst, Dept Nuclear Eng, Univ Md, 77-81; nuclear engr, Nat Bur Standards, 81-82; RES ASSOC & LECTR DEPT CHEM & NUCLEAR ENG, UNIV MD, COLLEGE PARK, 82- *Mem:* Am Nuclear Soc; Am Vacuum Soc. *Res:* Near surface phenomena brought about by ion bombardment (implantation, diffusion, sputtering, and composition changes); investigations using Monte Carlo computer simulation of near surface interactions and with surface science techniques such as x-ray and Auger spectroscopy and quartz resonator microbalance. *Mailing Add:* Dept Chem & Nuclear Eng Univ Md College Park MD 20742

ANDREAE, MEINRAT OTTO, b Augsburg, Ger. MARINE CHEMISTRY, BIOGEOCHEMISTRY. *Educ:* Univ Karlsruhe, BSc, 70; Univ Gottingen, MSc, 74; Univ Calif, San Diego, PhD(oceanog), 78. *Prof Exp:* Res asst geochem, Univ Gottingen, 71-74 & marine chem, Scripps Inst Oceanog, Univ Calif, San Diego, 74-78; from asst prof to assoc prof oceanog, Fla State Univ, 82; CHEMIST, MAX PLANCK INST, WGER. *Mem:* Sigma Xi; AAAS; Am Chem Soc; Am Geophys Union. *Res:* Chemistry of trace elements in the marine environment and the atmosphere; biosynthesis of trace compounds by marine plankton; atmospheric sulfur and nitrogen cycles. *Mailing Add:* c/o Max Plank Inst Fur Chemie Post Fach 3060 D-6500 Mainz Germany

ANDREAS, BARBARA KLOHA, b Dundee, Ohio, Sept 21, 46; m 66; c 2. BIOSYSTEMATICS. *Educ:* Kent State Univ, BA, 68, MA, 70, PhD(plant taxon), 80. *Prof Exp:* Instr bot, Kent State Univ, 70-71; ASSOC PROF BIOL, CUYAHOGA COMMUNITY COL, 74- *Concurrent Pos:* Field botanist, Div Natural Areas & Preserves, ODNR, 80-; NSF grant award bot, 81-; botanist, US Fish & Wildlife, 85. *Mem:* Am Inst Biol Sci; Am Soc Plant Taxonomists; Sigma Xi. *Res:* Biosystematics of angiosperms, especially Asclepiadaceae and Gentianaceae; Midwest floristics; bog and fen floristics and ecology. *Mailing Add:* Dept Bus & Technol Cuyahoga CC Eastern 4250 Richmond Rd Warrenville TWSP OH 44122

ANDREAS, JOHN M(OORE), b New York, NY, June 17, 12; m; c 2. PHOTOGRAPHIC CHEMISTRY. *Educ:* Princeton Univ, BS, 35; Mass Inst Technol, SM, 37, ScD(chem eng), 38. *Prof Exp:* Sr res engr, res & develop div, Technicolor Corp Am, 38-69, chem engr, systs res div, 69-70; CONSULT CHEM ENG, 70- *Mem:* AAAS; Am Chem Soc; Soc Motion Picture & TV Eng; Am Inst Chem Engrs; Am Defense Preparedness Asn; Am Inst Chem; Am Asn Textile Chem & Colorists; Sigma Xi. *Res:* Friction for air flow; boundary tension by pendant drops; color motion picture photography; color measurements; dyestuffs; photographic processes and processing; interfacial tension by pendant drops; inks; dye transfer; azo dyes. *Mailing Add:* 890 Hillcrest Pl Pasadena CA 91106

ANDREASEN, ARTHUR ALBINUS, b Springfield, Ill, Dec 18, 17; m 44; c 2. INDUSTRIAL MICROBIOLOGY. *Educ:* Univ Ill, BS, 40, MS, 42; Ind Univ, PhD(cytophysiol), 53. *Prof Exp:* From bacteriologist to chief chemist & head fermentation res, Joseph E Seagram & Sons, Inc, 41-61, dir res, 61-81; RETIRED. *Concurrent Pos:* In chg fermentation, Exp Wood Hydrolysis Distillery, Vulcan Copper & Supply Co, Ore, 46-47. *Mem:* Am Chem Soc; Am Soc Brewing Chem. *Res:* Anaerobic growth and nutrition of yeast; yeast and mold fermentations; enzyme production. *Mailing Add:* 508 Hill Ridge Rd Louisville KY 40214

ANDREASEN, NANCY COOVER, b Lincoln, Nebr, Nov 11, 38; c 2. PSYCHIATRY. *Educ:* Univ Nebr, BA, 58 &PhD(English), 63; Radcliffe Col, MA, 59; Univ Iowa, MD, 70. *Prof Exp:* Instr English, Nebr Wesleyan Univ, 60-61 & Univ Nebr, 62-63; asst prof English, 63-66, from asst prof to assoc prof, 77-81, PROF PSYCHIAT, UNIV IOWA, 81- *Concurrent Pos:* Prin investr, NIMH Collab Study Depression, 73-; mem, Am Psychiat Asn Task Force Diag & Statist Manual, 74-; mem, Nat Merit Rev Bd Behav Sci, 78-; consult, training grant study sect, NIMH, 78-79. *Honors & Awards:* Menninger Award Psychiat Res, Menninger Found, 73; Hibbs Award, Am Psychiat Asn; Nelson Urban Res Award. *Mem:* Am Psychopath Asn (vpres); Am Col Psychiatrists; Psychiat Res Soc; Am Psychiat Asn. *Res:* Brain imaging in the major psychoses and schizophrenia; study of depression; speech, language, thought and communication in mania, depression and schizophrenia; family history and psychiatric symptoms in creative individuals; psychiatric aspects of facial deformity; study of depression. *Mailing Add:* Dept Psychiat 500 Newton Rd Iowa City IA 52242

ANDREATCH, ANTHONY J, b Newark, NJ, June 3, 24; m 49; c 1. PHYSICAL CHEMISTRY, CHEMICAL ENGINEERING. *Educ:* SDak Sch Mines & Technol, BS, 49; Univ SDak, MS, 50. *Prof Exp:* Res chemist, Naval Res Lab, DC, 51-56; head appln eng, Liston Becker Plant, Beckman Instruments Inc, 56-59; res instrumentation scientist, Am Cyanimid Co, 59-67; mgr instrumentation, Scott Res Lab, Inc, 67-69; MGR AIR POLLUTION CONTROL LAB, NJ DEPT ENVIRON PROTECTION, 69-; MGR MOBILE SOURCE LAB, 69-, MGR ENG & INSTRUMENTATION. *Concurrent Pos:* Mem, Coord Res Coun Inc, 64. *Mem:* Am Chem Soc; Instrument Soc Am; Air Pollution Control Asn. *Res:* Development of atmospheric gas analyzer for atomic submarines and flame ionization detector for chromatographic and exhaust analysis; instrumentation for catalytic muffler design; exhaust emission testing of vehicles at New Jersey inspection stations. *Mailing Add:* 15 Cold Spring Rd Trenton NJ 08619

ANDREE, RICHARD VERNON, mathematics; deceased, see previous edition for last biography

ANDREEN, BRIAN H, b Superior, Wis, Aug 15, 34; m 56; c 4. ANALYTICAL CHEMISTRY. *Educ:* Wis State Univ, Superior, BS, 56; Fla State Univ, MS, 59. *Prof Exp:* Assoc chemist, Inst Gas Technol, 59-60, res chemist, 60-62, supvr chem res, 62-64; midwest rep, Res Corp, Midwest, 64-69, regional dir, 69-87, grants prog coordr, 87-89, DIR SCI ADVAN PROGS, RES CORP, MIDWEST, 90- *Mem:* AAAS; Am Chem Soc; Coun Undergrad Res. *Res:* Gas chromatographic analysis; detection of air pollutants; nurture of undergraduate research. *Mailing Add:* Res Corp 6840 E Broadway Blvd Tucson AZ 85710

ANDREGG, CHARLES HAROLD, b Mansfield, Ohio, July 13, 17; m 40. SURVEYING & MAPPING, PHYSICAL SCIENCE ADMINISTRATION. *Educ:* Kent State Univ, BS, 39; Indust Col Armed Forces, MS, 67. *Prof Exp:* Chief, Oper & Planning, Army Map Serv, US Govt, 42-62; technical dir, Defense Intelligence Agency, 62-72; deputy dir, Defense Mapping Agency, 72-79; CONSULT, 79- *Concurrent Pos:* Dep, US mem, Pan Am Inst Geogr Hist. *Honors & Awards:* Luis Struck Award, Am Soc Photogrammetry & Remote Sensing, 74. *Mem:* Hon mem, Am Soc Photogrammetry & Remote Sensing; Am Congress Surveying & Mapping (dir, 54-70, pres 69); Accreditation Bd Engr & Technol; fel Soc Am Military Engrs. *Mailing Add:* 18304 Gulf Blvd PH 4 Redington Shores FL 33708

ANDREIS, HENRY JEROME, b Milwaukee, Wis, Sept 1, 31; m 54; c 5. SOIL FERTILITY. *Educ:* Univ Wis, BS, 54, MS, 55. *Prof Exp:* Soil technologist res, 55-69, FIRST ASST TO DIR & VPRES RES DEPT, US SUGAR CORP, 69- *Mem:* Am Soc Agron; Int Soc Sugarcane Technologists; Am Soc Sugarcane Technologists; Am Soc Agron. *Res:* Sugarcane and pasture fertility studies, silage research, seed cane germination studies; water table and irrigation studies and citrus nursery. *Mailing Add:* 1012 Palmetto St Clewiston FL 33440

ANDREKSON, PETER A(VO), b Gothenburg, Sweden, May 31, 60. OPTICAL COMMUNICATIONS, NONLINEAR FIBER OPTICS. *Educ:* Chalmers Univ Technol, MSc, 84, PhD(optoelectron), 88. *Prof Exp:* Asst prof, Chalmers Univ Technol, 88-89; POSTDOCTORAL MEM STAFF, AT&T BELL LABS, 89- *Concurrent Pos:* Secy, Europ Conf Optical Commun, 89. *Mem:* Optical Soc Am. *Res:* Fiber-optic communications, specifically using solitons for ultra-long distances transmission and ultra-high bit-rates; all-optical swithing and logic based on fiber nonlinearities; semiconductor laser characterization. *Mailing Add:* AT&T Bell Labs Rm 7C-203 600 Mountain Ave Murray Hill NJ 07974-2070

ANDREOLI, ANTHONY JOSEPH, b New York, NY, Sept 13, 26; m 52. BACTERIOLOGY, BIOCHEMISTRY. *Educ:* Univ Southern Calif, AB, 50, PhD(bact), 55. *Prof Exp:* Asst bact, Univ Southern Calif, 50-52, clin lab technician, 52-53, lectr, 54-55; from asst prof to assoc prof chem & microbiol, 55-64, PROF CHEM, CALIF STATE UNIV, LOS ANGELES, 64- *Mem:* Am Soc Microbiol; Am Soc Biol Chemists. *Res:* Bacterial metabolism and physiology; NAD metabolism; bacterial sporulation. *Mailing Add:* Dept Chem Calif State Univ 5151 State Univ Dr Los Angeles CA 90032

ANDREOLI, KATHLEEN GAINOR, b Albany, NY, Sept 22, 35; div; c 3. NURSING ADMINISTRATION. *Educ:* Georgetown Univ, BSN, 57; Vanderbilt Univ, MSN, 59; Univ Ala, DSN, 79. *Prof Exp:* Instr, St Thomas Hosp Sch Nursing, Nashville, 58-59, Georgetown Univ Sch Nursing, 59-60, Duke Univ Sch Nursing, 60-61 & Bon Secours Hosp Sch Nursing, Baltimore, 62-64; clin assoc prof cardiovascular nursing, Sch Nursing, Univ Ala Med Ctr, Birmingham, 70, educ dir, Physician Asst Prof, Dept Med, 70-75, asst prof nursing, 71-72, assoc prof, Sch Pub & Allied Health, 73, assoc dir, Family Nurse Practitioner Prog, Sch Nursing, 76-77, assoc prof, Community Health Nursing Grad Prog, 77, assoc prof, Dept Pub Health, 78, prof, Sch Nursing, 79; spec asst to pres educ affairs, Univ Tex Health Sci Serv Ctr, Houston, 79, actg dean, Sch Allied Health Sci, 80-81, spec asst to pres & dir acad affairs, 81, exec dir acad affairs, 81-84, vpres educ servs, Interprof Educ & Int Progs, assoc dir, Ctr Health Prom Res & Develop, 84-87; VPRES NURSING AFFAIRS & DEAN, COL NURSING, RUSH-PRESBY ST LUKE'S MED CTR, CHICAGO, 87- *Concurrent Pos:* Consult, 66-; ed, Heart & Lung, J Total Care, St Louis, 71; mem comt educ health prof, Int Med-Nat Acad Sci, 73, mem study comt effectiveness of Community Ment Health Centers, 74-75, mem membership comt, 74-77; mem clin assoc prog, Univ Ky, 74; mem physician's asst prog, Western Mich Univ, 75 & Long Island Univ & Brooklyn-Cumberland Med Ctr, 76; mem cancer prevention & control adv comt, Univ Tex Syst Cancer Ctr, 81; mem, Coun Cardiovascular Nursing, Am Heart Asn; bd mem, Houston Acad Med, Tex Med Ctr, Inc, Libr, 82-; mem bd trustees, Found Mus Med Sci, Houston, 82-87; mem, Nat Adv Comt, Robert Wood Johnson Clin Scholars Prog, 88- *Mem:* Inst Med-Nat Acad Sci; fel Am Acad Nursing; Am Nurses Asn; Nat League Nursing; Am Asn Critical Care Nurses; Am Heart Asn; Am Orgn Nurse Execs. *Res:* Coauthor or author of numerous publications. *Mailing Add:* Rush Presby St Luke's Med Ctr 1653 W Congress Pkwy Chicago IL 60612

ANDREOLI, THOMAS E, b Bronx, NY, Jan 9, 35. INTERNAL MEDICINE. *Educ:* St Vincent Col, BA, 56; Georgetown Univ, MD, 60; Am Bd Internal Med, dipl, nephrology, dipl. *Hon Degrees:* ScD, St Vincent Col, 87. *Prof Exp:* From intern to jr asst resident med, Duke Hosp, 60-61; mem, Lab Intermediary Metab, Geront Sect, NIH, Baltimore City Hosps, 61-64; assoc med & physiol, Duke Univ, 65-66, asst prof med & assoc physiol, 66-67, asst prof med & physiol, 67-69, assoc prof med & asst prof physiol, 69-70; dir, Nephrology Res & Training Ctr, prof med & dir, Div Nephrology, Sch Med, Univ Ala, 70-79; prof & chmn, Dept Internal Med, Med Sch, Univ Tex, Houston, 79-86, Edward Randall, III prof, 86-88, chmn, 86-87; PROF & CHMN, DEPT INTERNAL MED & PROF PHYSIOL & BIOPHYS, COL MED, UNIV ARK, LITTLE ROCK, 88-, NOLAN CHAIR INTERNAL MED, 91- *Concurrent Pos:* Jr asst resident med, Duke Hosp, 64; chief resident & instr med, Duke Univ & Durham Vet Admin Hosp, 65; res career develop award, Nat Inst Gen Med Sci, NIH, 67 & 70; from assoc prof to prof physiol & biophys, Sch Med, Univ Ala, 70-79; ed, Am J Physiol, 76-83; assoc ed, Am J Med, 79-86; prof physiol & cell biol, Med Sch, Univ Tex, Houston, 79-88; co-chmn, External Monitoring Comt, Modification of Diet Renal Dis Study, Nat Inst Diabetes, Digestive & Kidney Dis, NIH, 89-, mem, Gen Clin Res Ctrs Comt, Div Res Resources, 89- *Mem:* Fel Am Col Physicians; Am Fedn Clin Res; Am Physiol Soc; Am Soc Clin Invest (secy-treas, 77-80); Am Soc Nephrology; Asn Am Physicians; Asn Prof Med; Biophys Soc; Int Soc Nephrology; Soc Gen Physiologists. *Mailing Add:* Dept Internal Med Col Med Univ Ark 4301 W Markham Slot 640 Little Rock AR 72205

ANDRES, GIUSEPPE A, b Torino, Italy, Mar 11, 24; m 64; c 2. CLINICAL IMMUNOLOGY. *Educ:* Univ Pisa, Italy, MD, 49; Univ Roma, PhD(med), 65. *Prof Exp:* Asst prof med, Univ Pisa, 54-55; fel, Dept Med, Col Physicians & Surgeons, Columbia Univ, 55-56, res fel, Univ, 58-59, res fel microbiol, Col Physicians & Surgeons, 60-61, vis prof, 64-70; prof microbiol, 70-74, PROF MED, STATE UNIV NY, BUFFALO, 74- *Concurrent Pos:* Asst prof med, Univ Roma, 56-70; Fogarty Sr Int fel, NIH, 77-78; mem, Path Study Sect, NIH, 77-81. *Mem:* Harvey Soc; Am Soc Cell Biol; Am Asn Immunologists; Transplantation Soc. *Res:* Human and experimental immunopathology with special interest in diseases of the kidney and the lung; biology and immunopathology of organ transplantation, especially kidney and liver. *Mailing Add:* Dept Path E Pathol Res 7th Floor Harvard Med Sch Mass Gen Hosp 149 13th St Charlestown MA 02129

ANDRES, JOHN MILTON, b Santa Ana, Calif, Feb 4, 27; m 56; c 4. MICROWAVE ELECTRONICS. *Educ:* Calif Inst Technol, BS, 49, MS, 50; Mass Inst Technol, PhD(physics), 53. *Prof Exp:* Mem tech staff, Calif Res Corp, 53-54; mem tech staff, Space Technol Labs, Thompson-Ramo-Wooldridge, Inc, 54-60, sect head, TRW Systs Group, 60-63, mgr quantum electronics dept, 62-70, mem sr staff, Electronic Systs Div, 70-87; RETIRED. *Mem:* Am Phys Soc; Sigma Xi; Inst Elec & Electronics Engrs. *Res:* Field effect transistor amplifiers; oscillators; laser communications. *Mailing Add:* 1340 via Margarita Palos Verdes Estates CA 90278

ANDRES, KLAUS, b Zurich, Switz, Mar 1, 34; c 3. PHYSICS. *Educ:* Swiss Fed Inst Technol, PhD(physics), 63. *Prof Exp:* Mem tech staff physics, Bell Tel Labs, 63-80; PROF TECH UNIV MUNICH, GER, 80- *Concurrent Pos:* Acad guest, Swiss Fed Inst Technol, 68-69; guest prof, Tech Univ Munich, 73. *Mem:* Fel Am Phys Soc. *Res:* Superconductivity; magnetism; rare earth compounds; nuclear magnetism at very low temperatures. *Mailing Add:* Walther Meissner Inst Fur Tieftemperatureforschung Garching Germany

ANDRES, LLOYD A, b Santa Ana, Calif, May 17, 28; m 56; c 3. ENTOMOLOGY, WEED SCIENCE. *Educ:* Univ Calif, Berkeley, PhD(entom), 57. *Prof Exp:* Res entomologist, Univ Calif, Riverside, 56-58; res entomologist, 58-64, leader biol control weeds invests, 64-72, res leader & nat tech adv biol control weeds, Sci & Educ Admin-Agr Res, 72-82, PROJ LEADER BIOL CONTROL RES UNIT, USDA, 82- *Mem:* AAAS; Entom Soc Am; Am Inst Biol Sci; Weed Sci Soc Am. *Res:* Biological control of weeds. *Mailing Add:* 1324 Arch Berkeley CA 94708

ANDRES, REUBIN, b Dallas, Tex, June 13, 23; m 48; c 4. MEDICINE, GERONTOLOGY. *Educ:* Southwestern Med Col, MD, 44; Am Bd Internal Med, dipl. *Prof Exp:* Intern, Gallinger Munic Hosp, Washington, DC, 45; resident med, Vet Admin Hosp, McKinney, Tex, 47-50; from instr to asst prof, 55-63, assoc prof, 63-76, PROF MED, JOHNS HOPKINS UNIV, 76-; CHIEF CLIN PHYSIOL BR, NAT INST ON AGING, 62- CLIN DIR, 75- *Concurrent Pos:* Fel med, Johns Hopkins Univ, 50-55; asst physician, Outpatient Dept, Johns Hopkins Hosp, 50-58; instr, Sch Hyg & Pub Health, Johns Hopkins Univ, 53-55, asst prof, 55-60, lectr, 60-61; vis physician, Baltimore City Hosps, 55-57, 62-, asst chief med, 58-62; asst prof, Univ Md, 58-62. *Honors & Awards:* Kleemeire Award, Geront Soc Am, 74; Edward Henderson Award, Am Geriat Soc, 85. *Mem:* Am Physiol Soc; Am Soc Clin Invest; Geront Soc; Am Diabetes Asn; Endocrine Soc; Asn Am Physicians. *Res:* Carbohydrate and lipid metabolism; physiology of aging. *Mailing Add:* Clin Dir Gerontol Res Ctr NIA NIH Francis Scott Key Med Ctr 4940 Eastern Ave Baltimore MD 21224

ANDRES, RONALD PAUL, b Chicago, Ill, Jan 9, 38; m 61; c 3. CHEMICAL ENGINEERING. *Educ:* Northwestern Univ, BS, 59; Princeton Univ, PhD(chem eng), 63. *Prof Exp:* From asst prof to prof chem, Princeton Univ, 62-81; prof & head chem eng, 81-88, PROF ENG RES, PURDUE UNIV, 88- *Concurrent Pos:* Vis prof, Indian Inst Technol, Kanpur, 70; distinguished prof, Univ Eindhoven, 74. *Mem:* Am Inst Chem Engrs; Am Chem Soc; AAAS; Sigma Xi; Mat Res Soc; Am Asn Aerosol Res. *Res:* Aerosol physics; physics and chemistry of molecular clusters; nucleation and rapid phase change; chemical kinetics and catalysis. *Mailing Add:* Sch Chem Eng Purdue Univ W Lafayette IN 47907-1283

ANDRES, SCOTT FITZGERALD, b Ann Arbor, Mich, July 23, 45; div; c 2. GRANTS & CONTRACTS REVIEW, PROTEIN CHEMISTRY. *Educ:* Mich State Univ, BS, 67; Wayne State Univ, PhD(biochem), 72. *Prof Exp:* Res assoc, Wayne Co Gen Hosp, 72-77; chemist, US Food & Drug Admin, 77-78, consumer safety officer, 78-82, chemist, 82-86; HEALTH SCIENTIST ADMINR, NIH, 86- *Concurrent Pos:* Prin investr, Wayne Co Gen Hosp, 76-77. *Res:* Investigation of the roles of specific amino acid residues in the conformation and immunochemistry of sperm whale myoglobin; active site characterization of parathyroid hormone; radioimmunoassay. *Mailing Add:* Off Rev NIH-NCRR Westwood Bldg 5333 Westbard Ave Bethesda MD 20892

ANDRE-SCHWARTZ, JANINE, ELECTRON MICROSCOPY, IMMUNOLOGY. *Educ:* Toulouse Univ, France, MD, 59. *Prof Exp:* ASSOC PROF MED, DIV HEMAT-ONCOL, NEW ENG MED CTR HOSP INC & TUFTS UNIV, 69- *Res:* Blood and lymphatic system. *Mailing Add:* Dept Med Div Hematol & Oncol Box 245 Tufts New England Med Ctr, 750 Washington St Boston MA 02111

ANDRESEN, BRIAN DEAN, b Reed City, Mich, Jan 20, 47. MASS SPECTROMETRY, ORGANIC CHEMISTRY. *Educ:* Fla State Univ, BS, 69; Mass Inst Technol, SM, 71, PhD(chem), 74; Woods Hole Oceanog Inst, SM, 72. *Prof Exp:* Asst prof pharmaceut chem, Col Pharm, Univ Fla, 74-80; MEM FAC PHARMACOL, OHIO STATE UNIV, 80- *Mem:* Am Chem Soc; Int Oceanog Found. *Res:* Application of gas chromatography, mass spectrometry and computer analysis for the identification of biologically active compounds; application of synthetic organic chemistry for the preparation of useful pharmaceuticals; clinical chemistry and toxicology. *Mailing Add:* 7423 Golden Springs Dr Columbus OH 43235

ANDRESEN, MICHAEL CHRISTIAN, b Lynwood, Calif, Dec 1, 49; m 74; c 2. NEURAL CONTROL OF CIRCULATION, HYPERTENSION. *Educ:* Univ Calif, Irvine, BS, 71; San Diego State Univ, MS, 73; Univ Tex Med Br, PhD(physiol), 78. *Prof Exp:* Teaching asst biol, San Diego Univ, 72-73; teaching & res asst physiol, Med Br, Univ Tex, 75-77, res assoc, 77-78, NIH fel, 78-79, res scientist, 79-80; vis scientist, Baker Med Res Inst, 80-81; asst prof, 81-89, ASSOC PROF PHYSIOL, MED BR, UNIV TEX, 90- *Concurrent Pos:* Prin investr, NIH res grants, 82-; Am Heart Asn res grants, 81-; ad hoc mem Heart Lung & Blood Comt A, NIH, 89, Heart Lung & Blood Cardiovasc & Plumonary Study Sect 89, Vet Admin Merit Review, 90. *Mem:* Biophys Soc; AAAS; Am Physiol Soc; Soc Neurosci. *Res:* Nature of mechanotransduction by arterial baroreceptors and cardiopulmonary mechanoreceptors; the role of receptor properties in the reflex regulation of cardiovascular system; synaptic transmission and medullary neurons in the solitary tract nucleus and cardiovascular integration. *Mailing Add:* Dept Physiol & Biophysics Univ Texas Med Br Galveston TX 77550

ANDRESEN, NORMAN A, b Chicago, Ill, Aug 12, 43; m 66; c 1. ALGAL TAXONOMY, ALGAL ECOLOGY. *Educ:* Taylor Univ, AB, 65; Mich State Univ, MS, 68; Univ Mich, PhD(bot), 76. *Prof Exp:* Asst prof biol, Ball State Univ, 76-77; vis lectr, bot, Eastern Mich Univ, 77-79, 90; chief aquatic biologist, Bionetics Corp, 81-88; RES ASSOC, UNIV MICH, 89- *Mem:* AAAS; Phycol Soc Am; Brit Phycol Soc; Int Soc Diatom Res; Int Phycol Soc. *Res:* Taxonomy and ecology of diatoms paleoecological reconstructions of environments from sediment cores and biological fossils; population dynamics in Great Lakes phytoplankton assemblages. *Mailing Add:* 5742 Princeton Pl Ypsilanti MI 48197

ANDREW, BRYAN HAYDN, b Glasgow, Scotland, Feb 26, 39; m 62; c 2. RADIO ASTRONOMY. *Educ:* Glasgow Univ, BSc, 61; Cambridge Univ, PhD(radio astron), 66. *Prof Exp:* Asst res officer, Nat Res Coun Can, 65-72, assoc res officer, 72-79, sr res officer, 79-84, chief, Prog Serv Br, 84-86, dir, Mgt Serv Br, 86-87, dir, Off Natural Facil Sci & asst dir, 87-90, DIR, RADIO ASTRON, HERZBERG INST ASTROPHYS, NAT RES COUN CAN, 90- *Concurrent Pos:* Vis lectr, Univ Toronto, 74-76. *Mem:* Can Astron Soc. *Res:* Molecules; extragalactic variables; planets; comets. *Mailing Add:* Six Florette Ottawa ON K1J 7L4 Can

ANDREW, DAVID ROBERT, b Wink, Tex, Nov 10, 35; m 58; c 3. MATHEMATICS. *Educ:* Univ Southwestern La, BS, 58; Iowa State Univ, MS, 59; Univ Pittsburgh, PhD(math), 61. *Prof Exp:* From asst prof to assoc prof, Univ Southwestern La, 61-66, head dept, 69-75, prof math, 66-87, dean Col Sci, 75-87; RETIRED. *Concurrent Pos:* Consult, Minn Sch Math & Sci Teaching Proj, 64-66; mem, Coun Cols Arts & Sci, 79; prof math, 89- *Mem:* Math Asn Am; AAAS; Am Statist Asn. *Res:* Point set topology. *Mailing Add:* 412 Kim Dr Lafayette LA 70503

ANDREW, GEORGE MCCOUBREY, b New Glasgow, PEI, Sept 8, 29; m 53; c 3. PHYSIOLOGY, PHYSICAL EDUCATION. *Educ:* McGill Univ, BSc, 52, MSc, 63, PhD(physiol), 67. *Prof Exp:* Athletic dir, YMCA, Charlottetown, PEI, 52-53 & Prince of Wales Col, 53-57; lectr educ, McGill Univ, 58-67; from asst prof to assoc prof phys educ, 67-74, asst prof physiol, 67-79, PROF PHYS EDUC, QUEEN'S UNIV, ONT, 74-, ASSOC PROF, DEPT PHYSIOL, 79-, DIR, SCH PHYS EDUC, 84- *Honors & Awards:* Queen's Jubilee Medal, 77. *Mem:* Can Asn Health, Phys Educ & Recreation (past pres); Can Asn Sports Sci (past pres); fel Am Col Sports Med; Can Physiol Soc; Am Physiol Soc. *Res:* Cardiorespiratory functions at rest and exercise and the effect of physical training through growth and aging on their adaptation to exercise. *Mailing Add:* Sch Phys & Health Educ Queen's Univ Kingston ON K7L 3N6 Can

ANDREW, JAMES F, b Mt Airy, NC, Mar 5, 25; m 51; c 2. EXPERIMENTAL SOLID STATE PHYSICS. *Educ:* Guilford Col, BS, 48; NC State Col, MS, 52; State Univ NY Buffalo, PhD(physics), 62. *Prof Exp:* Weather observer, US Weather Bur, 44-46; instrument specialist, Carter Fabrics Inc, NC, 48-49; asst physics, NC State Col, 49-52; jr engr, Sprague Elec Co, Mass, 52-53; res assoc, State Univ NY Buffalo, 53-62; sr physicist, Thiokol Chem Co, Utah, 62-63; STAFF MEM, LOS ALAMOS SCI LAB, 63- *Mem:* Am Phys Soc. *Res:* Electronic and mechanical properties of carbon base material; physics of solids at high pressures; electronic and thermal properties of plutonium; compatibility of reactor fuels with steel cladding. *Mailing Add:* 198 Navajo Dr Los Alamos NM 87544

ANDREW, KENNETH L, b Wichita, Kans, June 14, 19; m 40; c 3. ATOMIC EMISSION SPECTROSCOPY, FOURIER SPECTROSCOPY. *Educ:* Friends Univ, AB, 40; Johns Hopkins Univ, MA, 42; Purdue Univ, PhD(physics), 51. *Prof Exp:* Head dept physics, Friends Univ, 42-56; chmn dept, Dickinson Univ, 56-57; from assoc prof to prof, 57-89, EMER PROF PHYSICS, PURDUE UNIV, WEST LAFAYETTE, 89- *Concurrent Pos:* Prin investr, NSF Res Grants Atomic Emission Spectroscopy, 58-83, mem comt line spectra of elements, Nat Res Coun, 61-73, chmn, 66-68; consult, Int Astron Union, 64-71, Los Alamos Nat Lab, 65-72 & 77-; mem, Comn 14, Int Astron Union, 71-; prof d'Exchange Labs Aime Cotton, Orsay, France, 68-69; guest scientist, Nat Bur Standards, 86, 88 & 91. *Mem:* Am Phys Soc; fel Optical Soc Am; Am Asn Physics Teachers; Europ Group Atomic Spectroscopists. *Res:* Atomic emission spectroscopy; interferometric measurements; analysis of atomic spectra; atomic energy levels; standard wavelengths; comparison of atomic theory with experiments. *Mailing Add:* 1637 May No 1002 Wichita KS 67213-3584

ANDREW, MERLE M, b St Joseph, Mo, Aug 27, 20; m 50; c 2. APPLIED MATHEMATICS. *Educ:* Univ Nebr, BSEE, 42; Mass Inst Technol, PhD(math), 48. *Prof Exp:* Staff mem, Radiation Lab, Mass Inst Technol, 43-45 & Opers Eval Group, Navy Dept, 48-49; mathematician, Nat Bur Stand, 49-51 & Air Res & Develop Command, US Air Force, 51-54; chief, Math Div, Air Force Off Sci Res, 54-59, dir math sci, 59-70, dir math & info sci, 70-80; RETIRED. *Concurrent Pos:* Lectr, Cath Univ Am, 48-50; guest, Tech Univ, Vienna, Austria, 65-66; prof lectr, Am Univ, 69; mem, Fed Sr Exec Serv, 79- *Mem:* Am Math Soc. *Res:* Electromagnetic theory; differential equations; numerical analysis; computer sciences; scientific research administration. *Mailing Add:* 5914 Walton Rd Bethesda MD 20817-6128

ANDREW, ROBERT HARRY, b Platteville, Wis, Aug 2, 16; m 44; c 5. AGRONOMY. *Educ:* Univ Wis, BA, 38, PhD(agron-genetics), 42. *Prof Exp:* From asst prof to prof, 46-58, EMER PROF AGRON, UNIV WIS-MADISON, 84- *Concurrent Pos:* Vis lectr, Wageningen, Neth, 53-54; vis prof, Portoalegre, Rio Grande Do Sul, Brazil, 68; consult, Padang, W Sumatra, Indonesia, 86, Medan, N Sumatra, 87, Kharkov, Ukraine & Krasnodar, Russia, USSR, 87. *Honors & Awards:* Fel Am Soc Agron; fel Crop Sci Soc Am; fel AAAS. *Mem:* Am Soc Agron; Ecol Soc Am; Am Genetics Asn; Crop Sci Soc Am; AAAS. *Res:* Corn production; sweet corn breeding, genetics; plant ecology. *Mailing Add:* Dept Agron Univ Wis Madison WI 53706

ANDREW, WILLIAM TRELEAVEN, b Lucknow, Ont, Sept 1, 21; m 47; c 5. VEGETABLE CROPS. *Educ:* Univ Alta, BSc, 44; Utah State Col, MS, 49; Mich State Col, PhD(hort), 53. *Prof Exp:* Student asst, Alta Hort Res Ctr, 39-42; lab asst plant path, Univ Alta, 43; tech asst veg crops, Res Br, Can Dept Agr, 44-45; trial ground & receiving supvr veg & flower seeds, BC Seeds Ltd, 45-47; from instr to assoc prof veg crops, Southern Ill Univ, 50-59; prof plant sci, Univ Alta, 59-88; RETIRED. *Concurrent Pos:* Head div hort, Univ Alta, 59-70. *Mem:* Am Soc Hort Sci; Can Soc Hort Sci; Agr Inst Can; Int Soc Hort Sci. *Res:* Flowering and reproduction; growth regulators; moisture temperature and nutritional relationships. *Mailing Add:* Dept Plant Sci 4-10 Agr Forestry Bldg Univ Alta Edmonton AB T6G 2P5 Can

ANDREWS, ARTHUR GEORGE, b Springfield, Mass, Sept 21, 39; m 72; c 3. ORGANIC CHEMISTRY, BIOCHEMISTRY. *Educ:* Am Int Col, BA, 60; Univ Md, MS, 63; NMex State Univ, PhD (chem), 70. *Prof Exp:* Anal chemist, Food & Drug Admin, 63-66; res assoc chem, Univ Calif, 74-76; asst prof, 76-80, ASSOC PROF CHEM & CHMN DEPT, SAGINAW VALLEY STATE COL, 80- *Concurrent Pos:* Res asst, Univ Wis-Madison, 69-71; fel, Univ Trondheim, Norway, 71-74. *Mem:* Am Chem Soc; Sigma Xi. *Res:* Structures; stereochemistry; total synthesis and biosynthesis of naturally occuring polyene compounds, particularly carotenoids. *Mailing Add:* Dept Chem Saginaw Valley State Col University Center MI 48710

ANDREWS, AUSTIN MICHAEL, II, b Arkansas City, Kans, Nov 29, 43; m 61; c 3. ELECTRICAL ENGINEERING, SOLID STATE ELECTRONICS. *Educ:* Okla Univ, BS, 67, ME, 68; Univ Ill, Champaign-Urbana, PhD(elec eng), 71. *Prof Exp:* Mem tech staff elec eng, 71-77, group leader elec eng, 77-79, asst dir electro optics, Sci Ctr, 79-80, mgr focal plane prod, 80-85, chief engr, 85-89, DIR, ELECTRO-OPTICAL CTR, AUTONETICS, ROCKWELL INT, 87- *Mem:* Inst Elec & Electronics Engrs; Sigma Xi. *Res:* Semiconductor devices and materials growth in IV-VI, II-VI and III-V alloys; applied towards infrared detectors used as imaging focal planes; technical management for engineering and production of focal planes. *Mailing Add:* Rockwell Int Electro-Optical Ctr 3370 Miraloma Ave Anaheim CA 92803

ANDREWS, BETHLEHEM KOTTES, b New Orleans, La, Sept 18, 36; m 59; c 2. CHEMISTRY, MATHEMATICS. *Educ:* Tulane Univ, BA, 57. *Prof Exp:* Res chemist durable press, 57-75, res chemist, spec prod res, 76-84, LEAD SCIENTIST, TEXTILE FINISHING CHEM RES, SOUTHERN REGIONAL RES CTR, USDA, 85- *Concurrent Pos:* Am Heart Asn res grant, 57; vis woman scientist, NSF, 78; US deleg, Int Sci Orgn. *Mem:* Am Chem Soc; Am Asn Textile Chemists & Colorists; Fiber Soc; Sigma Xi; Int Sci Orgn. *Res:* Chemical modification and finishing of cotton and cellulose derivatives; finishing systems for knitted cotton; chemistry and catalysis of crosslinking finishes for cellulose; chemistry of formaldehyde release in crosslinked cellulose. *Mailing Add:* 5844 Sylvia Dr New Orleans LA 70124

ANDREWS, BILLY FRANKLIN, b Alamance Co, NC, Sept 22, 32; m 53; c 3. PEDIATRICS. *Educ:* Wake Forest Col, BS, 53; Duke Univ, MD, 57; Am Bd Pediat, cert, 63. *Prof Exp:* Intern, US Army Hosp, Ft Benning, Ga, 57-58; resident pediat, Walter Reed Gen Hosp, Washington, DC, 58-60 & Walter Reed Army Inst Res, 60-61; chief pediat serv, Rodriguez US Army Hosp, Ft Brooke, PR, 61-63; from asst prof to assoc prof pediat, Sch Med, Univ Louisville, 64-68, dir, newborn serv, 64-75, co-dir genetic coun unit, 65-68, PROF PEDIAT, SCH MED, UNIV LOUISVILLE, 68-, DIR COMPREHENSIVE HEALTH CARE CTR FOR HIGH RISK INFANTS & CHILDREN, 68-, CHMN DEPT PEDIAT, 69- *Concurrent Pos:* Consult div maternal & child health, Ky State Dept Health, 66-; civilian consult, US Army, 69-; chief staff, Kosair Children's Hosp, 69-; lectr, J Pediat Found, 72; dir, Community training in pediat, Sch Med, Univ Louisville, 75-77, Pilot Proj in Community Pediat, 76-77; Phi Rho Sigma lectureship, Sch Med, Ind Univ; lectr, Lubbock Pediat Soc, 86; distinguished lectr, Louisville Pediat Soc, 86. *Honors & Awards:* Helen B Fraser Award, 78; Selma Collin Perinatal lectr, Med Sch Ohio, 79; Joseph E Coleman lectr, 81; Munster Pediat lectr, Cork, Ireland, 84; Serv Award, Am Col Osteop Pediatricians, 84. *Mem:* Fel Am Acad Pediat; fel Am Col Physicians; Am Pediat Soc; Soc Pediat Res; fel Royal Soc Med; Am Osler Soc. *Res:* Low birth weight infants; respiratory distress in infants; amniotic fluid studies; infant nutrition; author or coauthor of over 175 publications and abstracts and two books. *Mailing Add:* Dept Pediat Health Sci Ctr Sch Med Univ Louisville Louisville KY 40292

ANDREWS, CECIL HUNTER, b Starkville, Miss, July 26, 32; m 60; c 3. AGRONOMY. *Educ:* Miss State Univ, BS, 54, MS, 58, PhD(agron), 66. *Prof Exp:* Asst prof agron & asst agronomist, 58-70, PROF AGRON, MISS STATE UNIV & AGRONOMIST, MISS AGR & FORESTRY EXP STA, 70- *Concurrent Pos:* Asst agronomist, Chile & Taiwan, 59 & 60; chief of party, Miss State Univ-AID Brazil Contract, 67-68 & Seed Develop Progs in Ecuador, Colombia, Nicaragua, Honduras, Costa Rica, Thailand, Upper Volta, Ghana, Botswana, Sri Lanka, Indonesia & Burma, Senegal, Gambia, 69-88, Burundi, 88-90. *Mem:* Am Soc Agron; Sigma Xi. *Res:* Biochemical and physiological research in seeds of field crops. *Mailing Add:* Box 5267 Miss State Univ Mississippi State MS 39762

ANDREWS, CHARLES EDWARD, b Stratford, Okla, Jan 22, 25; m 46; c 2. INTERNAL MEDICINE. *Educ:* Boston Univ, MD, 49. *Prof Exp:* Instr med, Univ Minn, 55-56; assoc, 56-57,from asst prof to assoc prof, Med Sch, Univ Kans, 57-60; from assoc prof to prof med, WVa Sch Med, 61-63, provost health sci, 61-83, CHANCELLOR, UNIV NEBR MED CTR, 83- *Concurrent Pos:* Chief med serv, Vet Admin Hosp, Kansas City, Mo, 56-61; lectr, Sch Dent, Univ Kans, 58- *Mem:* Am Thoracic Soc; AMA; Fedn Clin Res; Am Col Physicians. *Res:* Pulmonary disease and function. *Mailing Add:* Chancellor Univ Nebr Med Ctr 42nd & Dewey Ave Omaha NE 68105

ANDREWS, CHARLES LAWRENCE, b Atlanta, Ga, Mar 6, 38; m 65; c 2. WILDLIFE ECOLOGY, PARASITOLOGY. *Educ:* Ga State Col, BS, 63; Univ Ga, MS, 66, PhD(wildlife ecol), 69. *Prof Exp:* Res assoc wildlife dis, Sch Vet Med, Univ Ga, 66-69; PROF BIOL, BRENAU COL, 69-, AT DEPT MATH SCI. *Mem:* Wildlife Soc; Am Soc Mammal; Wildlife Dis Asn. *Res:* Cottontail rabbit, gray squirrel and white-tailed deer parasitism; wildlife leptospirosis. *Mailing Add:* 1398 Northwood Circle Gainesville GA 30501

ANDREWS, DANIEL KELLER, b Rockland, Maine, Sept 5, 24; m 50, 77; c 6. POULTRY SCIENCE. *Educ:* Univ Maine, BS, 49; Kans State Univ, MS, 51; Univ Wis, PhD(poultry sci), 63. *Prof Exp:* Supvr, Swift & Co, Pa, 51-52; inspector, US Dept Agr, Iowa, 52-53; asst prof poultry teaching & admin, State Univ NY Agr & Tech Inst Delhi, 53-56; exten poultry specialist, Univ Conn, 56-60; res asst poultry nutrit, Univ Wis, 60-63; exten poultry scientist, Western Wash Res & Exten Ctr, Wash State Univ, 63-90; RETIRED. *Concurrent Pos:* Assoc ed, World Poultry Sci, 82-; Sabbatical leave, NC State Univ, 84, Yemen, 80 & 85. *Mem:* Poultry Sci Asn. *Res:* Effects of arsanilic acid on egg production; extension education for poultrymen and youth; poultry management; fan ventilation for light restricted houses; ahemeral light cycles; molting layers; primary wing feather molt; egg quality; energy efficient light sources for broilers. *Mailing Add:* Western Wash Res & Exten Ctr Wash State Univ Puyallup WA 98371-4998

ANDREWS, DAVID F, b Indianapolis, Ind, Apr 3, 43; Can citizen; m 65; c 3. STATISTICS. *Educ:* Univ Toronto, BSc, 65, MSc, 66, PhD(statist), 68. *Prof Exp:* Lectr statist, Imp Col, Univ London, 68-69; lectr, Princeton Univ, 69-71; asst prof, 71-77, PROF STATIST, UNIV TORONTO, 77- *Concurrent Pos:* Consult, Bell Lab, 69-77; vis asst prof, Univ Chicago, 73; assoc ed, Am Statist Asn, 73-75. *Mem:* Int Statist Inst; Am Statist Asn; Stat Soc Can (pres, 85). *Res:* Robust statistical procedures; graphical methods for data display. *Mailing Add:* Dept Statist Univ Toronto Toronto ON M5S 1A1 Can

ANDREWS, DOUGLAS GUY, b Boston, Eng, Oct 17, 17; Can citizen; m 42; c 3. CHEMICAL ENGINEERING, MECHANICAL ENGINEERING. *Educ:* Univ Cambridge, BA, 39, MA, 43. *Prof Exp:* Exp asst & officer underwater weapons res, Royal Naval Sci Serv, UK, 39-49; engr nuclear reactor & plant design, UK Atomic Energy Authority, 49-57; from assoc prof to prof 57-83, EMER PROF NUCLEAR ENG, UNIV TORONTO, 83- *Concurrent Pos:* Dir, Chem Eng Res Consults Inc, 63-84; pres, Youth Sci Found, Can, 71-73; vpres, Int Coord Comt Presentation Sci, 72-73. *Honors & Awards:* Pro Mundi Beneficio Medal, Brazilian Acad Humane Sci, 76. *Mem:* Am Nuclear Soc; Can Nuclear Soc; Brit Nuclear Energy Soc; fel Brit Inst Mech Eng; fel Royal Soc Arts. *Res:* Hydrodynamics; cavitation; underwater acoustics; screw propulsion; nuclear radiation shielding; nuclear reactor heat transfer; neutron measurement; nuclear reactor safety; radioactive waste management. *Mailing Add:* Dept Chem Eng Univ Toronto Toronto ON M5S 1A4 Can

ANDREWS, EDWIN JOSEPH, b Boston, Mass, Feb 27, 41; m 88; c 3. VETERINARY MEDICINE. *Educ:* Pa State Univ, BS, 64; Univ Pa, VMD, 67, PhD(path), 71. *Prof Exp:* NIH res fel path, Univ Pa Sch Vet Med, 67-71; asst prof comp med, Col Med, Pa State Univ, 71-74; assoc prof vet path, NY State Col Vet Med, 74-77; vpres res & develop, Hancock Labs, Johnson & Johnson, 77-82, mkt dir, Extracorporeal Inc, 82-84, vpres res & develop, Johnson & Johnson Cardiovasc Inc, 84-87; DEAN & PROF VET MED, UNIV PA SCH VET MED, 87- *Concurrent Pos:* Res assoc, Inst Cancer Res, 68-71. *Mem:* Am Vet Med Asn; Am Asn Cancer Res; Am Soc Exp Path; Int Acad Path; Am Col Vet Pathologists; Am Col Lab Animal Med. *Res:* Animal models of human disease, chemical carcinogenesis & tumor immunology. *Mailing Add:* Off Dean Sch Vet Med Univ Pa 3800 Spruce St Rm 110 Philadelphia PA 19104

ANDREWS, EUGENE RAYMOND, b Rockford, Ill, Apr 29, 18; m 48; c 2. ORGANIC CHEMISTRY. *Educ:* Cent YMCA Col, BS, 41. *Prof Exp:* Chief chemist, US Indust Chem Co, 42-57; res chemist, Allied Mills Inc, 58-66, mgr res lab, 66-82; RETIRED. *Mem:* AAAS; Am Chem Soc; Asn Off Anal Chemists. *Res:* Scientific feeding of farm animals, effect of drugs and hormones on such animals. *Mailing Add:* 5810 Nicolet Ave Chicago IL 60631-2429

ANDREWS, FRANK CLINTON, b Manhattan, Kans, May 29, 32; m 64; c 2. CHEMICAL PHYSICS. *Educ:* Kans State Univ, BS, 54; Harvard Univ, AM, 59, PhD(chem physics), 60. *Prof Exp:* NIH fel chem physics, Univ Calif, Berkeley, 60-61; asst prof phys chem, Univ Wis, 61-67, on leave, Theoret Physics Dept, Oxford Univ, 66-67; assoc prof phys chem, 67-74, PROF CHEM, UNIV CALIF, SANTA CRUZ, 74- *Concurrent Pos:* Fulbright scholar, Univ Hull, Eng, 54-55; Sloan fel, 63-67; Danforth assoc, 76-86; vis prof chem, Dartmouth Col, 79-80. *Mem:* Sigma Xi; Asn Humanistic Psychol. *Res:* Statistical mechanics and thermodynamics; general problem-solving; psychological unblocking; role of values in living. *Mailing Add:* Chem Dept Univ Calif Santa Cruz CA 95064

ANDREWS, FRED CHARLES, b Aylesbury, Sask, July 13, 24; nat US; m 44; c 3. MATHEMATICAL STATISTICS. *Educ:* Univ Wash, BS, 46, MS, 48; Univ Calif, PhD(statist), 53. *Hon Degrees:* PhD, Univ Tampere, Finland, 85. *Prof Exp:* Lectr math, Univ Calif, 51-52; res assoc statist, Stanford Univ, 52-54; asst prof math & assoc statistician, Univ Nebr, 54-57; dir statist lab & comput ctr, Univ Ore, 60-69, assoc prof, 57-66, head dept, 73-80, prof, 66-90, EMER PROF MATH, UNIV ORE, 90- *Concurrent Pos:* Fulbright-Hays sr lectr, Univ Tampere, Finland, 69-70; Univ Col, Cork, Ireland, 76-77 & Univ Jordan, 83-84. *Mem:* Fel AAAS; Biomet Soc; Am Statist Asn; Inst Math Statist. *Res:* Statistical theory, including non-parametric inference; statistical computations. *Mailing Add:* 2705 Emerald St Eugene OR 97403

ANDREWS, FRED GORDON, b Los Angeles, Calif, Nov 15, 33; m 60; c 2. SYSTEMATIC ENTOMOLOGY. *Educ:* Los Angeles State Univ, BA, 61; Univ Calif, Riverside, PhD(entom), 72. *Prof Exp:* SYST ENTOMOLOGIST, CALIF DEPT FOOD & AGR, 70- *Concurrent Pos:* Grants, Off Endangered Species, Dept Interior, 75-76 & Bur Land Mgt, 77-78 & 78-79. *Mem:* Pan-Pacific Entom Soc (pres, 75); Coleopterist Soc; Entom Soc Am; Soc Syst Zool; Sigma Xi. *Res:* Systematics and ecology of Lathridiidae; coleoptera of sand dunes; coleoptera associated with fungi. *Mailing Add:* Dept Agr Entomol 1220 N St Sacramento CA 95814

ANDREWS, FREDERICK NEWCOMB, b Boston, Mass, Feb 5, 14; m 38; c 2. ANIMAL PHYSIOLOGY, RESEARCH ADMINISTRATION. *Educ:* Univ Mass, BS, 35, MS, 36; Univ Mo, PhD(physiol reprod), 39. *Hon Degrees:* ScD, Univ Mass, 62; Dr Agr, Purdue Univ, 83. *Prof Exp:* From asst prof to assoc prof, Purdue Univ, 40-49, head dairy dept, 60-62, head animal sci dept, 62-63, prof animal sci, 49-80, dean grad sch & vpres res, 63-80, vpres & gen mgr, Purdue Res Found, 64-80; RETIRED. *Concurrent Pos:* Agr consult, Rockefeller Found, 61-66; mem bd dirs, Indianapolis Ctr for Advan Res, 70-, vpres, 73-80, chmn bd dir, 81-82; consult, Nat Inst Health, 81-83. *Honors & Awards:* Morrison Award, 61. *Mem:* Fel AAAS; Am Asn Anat; fel Am Soc Animal Sci; Am Soc Zool; Am Dairy Sci Asn; Nat Acad Univ Res Admin. *Res:* Physiology of reproduction; endocrinology; nutrition; environmental physiology; growth and development. *Mailing Add:* 691 Sugar Hill Dr West Lafayette IN 47906

ANDREWS, FREDERICK T, JR, b Palmerton, Pa, Oct 6, 26; c 3. TELECOMMUNICATIONS. *Educ:* Pa State Univ, BSEE, 48. *Prof Exp:* Switching res, AT&T Bell Lab, 48-55, supvr, Transmission Systs Develop, 55-58, dept head, 58-62, dir, Transmission Systs Eng, 62-66, dir, Mil Commun Systs Eng, 66-68, dir, Loop Systs Eng, 68-79, exec dir, Switching Systs Eng, 79-83, exec dir, 79-83, exec dir, Technol Systs Planning, AT&T Bell Lab, 83-84; vpres, Technol Systs, Bellcore, 84-90; INDEPENDENT CONSULT, 90- *Honors & Awards:* Inst Elec Electronics Engrs Award, 85; Nat Acad Eng Award, 88. *Mem:* Fel Inst Elec Electronics Engrs; Nat Acad Eng; Inst Elec Electronics Engrs Commun Soc (vpres 82-83 & 84-85). *Res:* Systems engineering and management in the field of telecommunications with emphasis on network evolution and international standardization; author of 14 articles. *Mailing Add:* 25 Clover Hill Lane PO Box 448 Colts Neck NJ 07722

ANDREWS, GEORGE EYRE, b Salem, Ore, Dec 4, 38; m 60; c 3. MATHEMATICS. *Educ:* Ore State Univ, BS & MA, 60; Univ Pa, PhD(math), 64. *Prof Exp:* From asst prof to assoc prof, 64-70, prof math, 70-81, EVAN PUGH PROF MATH, PA STATE UNIV, 81-; ADJ PROF, UNIV WATERLOO, 82- *Concurrent Pos:* vis prof math, Univ Wis, 75-76 & Univ New South Wales, 78-79; NZ Math Soc vis lectr, 79; Guggenheim fel, 82-83; vis fel, Australian Nat Univ, 83; vis prof, Univ Strasbourg, 83; prin lectr, NSF-Conf Bd of the Math Sci regional conf, Tempe, 85; vis Ordway prof, Univ Minn, 88; vis scientist, T J Watson Res Ctr, IBM, 90-91. *Honors & Awards:* Hedrick Lectr, Math Asn Am, 80. *Mem:* Am Math Soc; Edinburgh Math Soc; Math Asn Am; Soc Indust & Appl Math; Australian Math Soc; NZ Math Soc. *Res:* Basic hypergeometric series; partitions; number theory; combinatorics; special functions. *Mailing Add:* Dept Math McAllister Bldg Pa State Univ University Park PA 16802

ANDREWS, GEORGE HAROLD, b Syracuse, NY, July 31, 32; m 55; c 4. MATHEMATICS, NUMERICAL ANALYSIS. *Educ:* Oberlin Col, AB, 52; Univ Mich, AM, 55, PhD(math), 63. *Prof Exp:* From asst prof to assoc prof, 62-73, PROF MATH, OBERLIN COL, 73- *Concurrent Pos:* NSF fac fel, 68-69; consult numerical anal, Am Soc Actuaries, 74-; vis prof, Univ Mich, 75-76; vis scholar, Unv Mich, 82-83. *Mem:* Math Asn Am; Soc Indust & Appl Math; Am Soc Actuaries; Sigma Xi. *Res:* Mathematical statistics; foundations of mathematics; actuarial mathematics; numerical solution of differential equations. *Mailing Add:* Dept Math Oberlin Col Oberlin OH 44074

ANDREWS, GEORGE WILLIAM, b Eau Claire, Wis, Oct 15, 29; m 56; c 3. GEOLOGY, DIATOMS. *Educ:* Univ Wis-Madison, BA, 51, MA, 53, PhD(geol), 55. *Prof Exp:* Geologist, Tech Serv Div, Shell Oil Co, 55-58 & E&P Res Div, Shell Develop Co, 58-59; geologist, US Geol Surv, 59-89; CONSULT, 89- *Concurrent Pos:* Assoc ed, J Bacillaria, 78-84. *Mem:* Int Soc Diatom Res. *Res:* Marine and nonmarine diatoms; paleontology; stratigraphy and paleoecology of Cenozoic strata; petroleum geology. *Mailing Add:* 20 Ridgewood Dr Akron PA 17501

ANDREWS, GLENN COLTON, b Alhambra, Calif, Feb 7, 48; m 72. ORGANIC CHEMISTRY. *Educ:* Univ Calif, Los Angeles, BSc, 70, PhD(org chem), 74. *Prof Exp:* Res scientist org synthesis, Stanford Univ, 74-75; res scientist, 75-77, sr res scientist org synthesis, 77-81, SR RES ADV, PFIZER CENT RES, 81- *Res:* Total synthesis directed toward vitamins and other commercially important natural products. *Mailing Add:* Pfizer Cent Res Eastern Point Rd Groton CT 06340

ANDREWS, GORDON LOUIS, b New Orleans, La, Sept 22, 45; m 74; c 3. AGRICULTURE. *Educ:* Southeastern La Col, BS, 67; La State Univ, MS, 69; Miss State Univ, PhD(entom), 72. *Prof Exp:* Res assoc entom, Miss State Univ, 72-76; area pest mgt specialist, Miss Coop Exten Serv, 76-82; at Dept Entomol, 82-86, assoc entomologist, Delta Br, 86-89, AREA CROP MODEL SPECIALIST, DELTA BR, MISS STATE UNIV, 89- *Res:* Cotton insects; field evaluation and modeling of Heliothis species; optimum pest management trial. *Mailing Add:* Delta Br Exp Sta Box 68 Stoneville MS 38776

ANDREWS, GREGORY RICHARD, b Olympia, Wash, Mar 9, 47; m 67; c 2. COMPUTER SCIENCE. *Educ:* Stanford Univ, BS, 69; Univ Wash, PhD(comput sci), 74. *Prof Exp:* Assoc engr comput sci, Boeing Co, 69-70; asst prof comput sci, Cornell Univ, 74-79; assoc prof, 79-87 PROF COMPUTER SCI, UNIV ARIZ, 87- *Concurrent Pos:* Co-investr grant, NSF, 74-77, prin investr grants, 77-80, 82 & 84-; consult, US Army Electronics Command, 75-; co-prin investr grant, US Army, 80-82, Air Force Off Sci Res, 84-89, NSF, 87-90; vis prof comput sci, Univ Wash, 83-84. *Mem:* Asn Comput Mach. *Res:* Operating systems; concurrent programming languages; protection and security; distributed computing. *Mailing Add:* Dept Computer Sci Univ Ariz Tucson AZ 85721

ANDREWS, HENRY NATHANIEL, JR, b Melrose, Mass, June 15, 10; m 39; c 3. PALEOBOTANY. *Educ:* Mass Inst Technol, BS, 34; Wash Univ, MS, 37, PhD, 39. *Prof Exp:* From instr to prof bot, Wash Univ, 38-64, dean sch, 47-64; prof, 64-75, chmn dept, 64-70, EMER PROF BIOL, UNIV CONN, 75- *Concurrent Pos:* Paleobotanist, Mo Bot Garden, St Louis, 41-64; botanist, US Geol Surv, 50-54, 59-; Guggenheim Mem Found fel, 50-51, 58-59; Fulbright teaching fel, Univ Poona, 60-61; NSF sr fel, Sweden, 64. *Honors & Awards:* Cert of Merit, Bot Soc Am, 66. *Mem:* Nat Acad Sci; Bot Soc Am. *Res:* Carboniferous and Devonian plants; Arctic paleobotany; history of paleobotany. *Mailing Add:* RFD 1 Laconia NH 03246

ANDREWS, HOWARD LUCIUS, b Davisville, RI, Oct 27, 06; m 31; c 3. BIOPHYSICS. *Educ:* Brown Univ, BS, 27, MS, 28, PhD(physics), 31. *Prof Exp:* Instr physics, Brown Univ, 29-34, res assoc psychol, 34-37; assoc physicist, US Pub Health Serv, 37-41, physicist, 41-48, chief nuclear radiation biol sect, Nat Insts Health, 48-61, dept radiation safety, 62-65; asst dir health & safety, PR Nuclear Ctr, 65-67; prof radiation biol & biophysics, Univ Rochester, 67-71; RETIRED. *Concurrent Pos:* Exec secy biol effects of atomic radiation comt, Nat Acad Sci-Nat Res Coun, 59-64; consult, Radiation Safety, 71- *Mem:* Fel Am Phys Soc; Radiation Res Soc; Health Physics Soc (pres, 64-65). *Res:* Biological effects of high energy radiations; radiation health protection. *Mailing Add:* 596 E Shore Rd Jamestown RI 02835

ANDREWS, HUGH ROBERT, b Fredericton, NB, Apr 29, 40; m 71; c 3. EXPERIMENTAL NUCLEAR PHYSICS. *Educ:* Univ NB, BSc, 62; Harvard Univ, AM, 63, PhD(physics), 71. *Prof Exp:* Res officer nuclear physics, 71-82, SR RES SCIENTIST, ATOMIC ENERGY CAN LTD, 83- *Concurrent Pos:* Chmn, Div Nuclear Physics, Can Asn Physicists, 89-90. *Mem:* Sigma Xi; Can Asn Physicists; Can Nuclear Soc; Am Phys Soc. *Res:* Hyperfine interactions and perturbed angular correlations; in-beam gamma ray spectroscopy applied to measurement of short lifetimes and the nature of high spin states in nuclei; atomic physics-heavy ion stopping powers; radioisotope dating using accelerator techniques; public affaris aspects of nuclear energy. *Mailing Add:* Chalk River Labs Sta 49 Chalk River ON K0J 1J0 Can

ANDREWS, JAMES EINAR, b Minneapolis, Minn, Oct 31, 42; m 65, 81; c 1. GEOLOGICAL OCEANOGRAPHY, NAVAL OCEANOGRAPHY. *Educ:* Amherst Col, BA, 63; Univ Miami, PhD(oceanog), 67. *Prof Exp:* Asst prof, Univ Hawaii, 67-74, chmn dept, 73-76, from assoc prof to prog oceanog, 74-81; tech dir, Naval Ocean Res & Develop Activ, Miss, 81-87; SCI DIR, OFF NAVAL RES LONDON, UK, 87- *Concurrent Pos:* Chmn, Hawaii Marine Res, Inc, 75-80; vis prof, Technische Hochschule Aachen, WGermany, 76; vis scientist, Ctr Nat Exploitation Oceans, France, 80. *Honors & Awards:* Humboldt Award, Fed Repub Ger. *Mem:* Fel Geol Soc Am; Am Asn Advan Sci; Europ Geophys Soc; Oceanog Soc. *Res:* Deep sea sediments; sea floor structure and topography; manganese nodules; seafloor mining; ocean resources. *Mailing Add:* Off Naval Res Europe Box 39 FPO New York NY 09510-0700

ANDREWS, JAMES TUCKER, dentistry; deceased, see previous edition for last biography

ANDREWS, JAY DONALD, b Bloom, Kans, Sept 9, 16; m 48; c 2. ZOOLOGY, MARINE ECOLOGY. *Educ:* Kans State Col, BS, 38; Univ Wis, MS, 40, PhD(zool), 47. *Prof Exp:* From asst biologist to assoc biologist, Va Inst Marine Sci, 46-55; prof, 60-83, EMER PROF MARINE SCI, COL WILLIAM & MARY, 83-; SR BIOLOGIST, VA INST MARINE SCI, 55- *Concurrent Pos:* Ed, Proc, Nat Shellfisheries Asn, 57-58; assoc prof marine sci, Univ Va, 60- *Mem:* Hon mem Atlantic Estuarine Res Soc; Am Soc Limnol & Oceanog; Soc Syst Zool; hon mem Nat Shellfisheries Asn (secy-treas, 60-62, vpres, 62-64, pres, 64-66); Am Inst Fishery Res Biologists. *Res:* Marine biology; biological oceanography; ecology of marine invertebrates; shellfish ecology, epizootiology and molluscan diseases. *Mailing Add:* Va Inst Marine Sci Gloucester Point VA 23062

ANDREWS, JOHN EDWIN, b Selkirk, Man, Mar 15, 22; m 48; c 4. GENETICS, PLANT BREEDING. *Educ:* Univ Man, BSA, 49; Univ Minn, MS, 50, PhD(genetics), 53. *Prof Exp:* Cerealist, Rust Res Lab, Univ Man, 47-51; sr cerealist, Res Sta, Alta, 51-60, res dir, Res Br, Man, 60-65; res dir, Res Br, Sask, 65-69, dir res sta, Res Br, Agr Can, Alta, 69-81, dir, Gen Western Res Group, 81-83, sr adv foreign agr, Res Br, Agr Can, 83-86; RETIRED. *Concurrent Pos:* Dir Indo-Canadian Res Proj on Dryland Agr, India, 70-79. *Honors & Awards:* Merit Award, Pub Serv Can, 76. *Mem:* Fel Agr Inst Can; Can Soc Agron. *Res:* Physiology of cold hardiness; winter wheat breeding. *Mailing Add:* 27 Elm Crescent Lethbridge AB T1K 4W8 Can

ANDREWS, JOHN FRANK, b Cave City, Ark, July 10, 30; m 52; c 3. ENVIRONMENTAL & CIVIL ENGINEERING. *Educ:* Univ Ark, BS, 51, MS, 53; Univ Calif, Berkeley, PhD(sanit eng), 64; Am Acad Environ Engrs, dipl. *Prof Exp:* From instr to assoc prof civil eng, Univ Ark, 53-60; proj engr, Univ Calif, Berkeley, 60-63; from assoc prof to prof environ eng, Clemson Univ, 63-74, head dept, 68-74; prof civil eng, Univ Houston, 75-81; PROF ENVIRON SCI & ENG, RICE UNIV, 81- *Concurrent Pos:* Consult, City of Greenwood, SC, 64, Eng Sci, Arcadia, Calif, 67-72, Union Carbide, 69, Bacardi Distilleries, 69, United Aircraft, 70, Pan Am Health Orgn, 70, Int Bus Mach, 71, US Army Environ Hyg Agency, 75-77, Metrop Waste Comn, Minneapolis, 76-78, Shell Develop Co, 78-80, Greeley & Hansen Engrs, 78-80, Weyerhaeuser Co, 80-81, Reynolds, Smith & Hills, 82, Woodlands Develop Corp, 82-83, Gulf Coast Waste Disposal Authority, 84 & Metrop San Dist Greater Chicago, 85-; US ed, Water Res, 74-84; prof eng, Ark, SC, & Tex. *Honors & Awards:* Harrison Prescot Eddy Medal, Water Pollution Control Fedn, 75. *Mem:* Asn Environ Eng Prof (pres, 85); Int Asn Water Pollution Res; Am Water Works Asn; Am Soc Civil Eng; Am Chem Soc; Water Pollution Control Fedn; Am Inst Chem Engrs; AAAS. *Res:* Biological processes for waste treatment; dynamics and control of wastewater treatment plants. *Mailing Add:* Dept Environ Sci & Eng Rice Univ PO Box 1892 Houston TX 77251

ANDREWS, JOHN HERRICK, b Montreal, Que, May 16, 46; m 76; c 3. PLANT PATHOLOGY, AQUATIC BIOLOGY. *Educ:* McGill Univ, BSc, 67; Univ Maine, MS, 69; Univ Calif, Davis, PhD(plant path), 73. *Prof Exp:* Fel plant path, Cambridge Univ, 74-75; fel bot, Univ BC, 75-76; from asst prof to assoc prof, 76-86, PROF PLANT PATH, UNIV WIS-MADISON, 86- *Concurrent Pos:* Leopold Schepp Found fel, 74-75; Nat Res Coun Can fel, 75-76. *Mem:* Am Phytopath Soc; Can Phytopath Soc; AAAS; Am Soc Naturalists. *Res:* Integrated control; aquatic pathology; microbial ecology; biological control. *Mailing Add:* Dept Plant Path 1630 Linden Dr Madison WI 53706

ANDREWS, JOHN PARRAY, b Cambridge, Mass, June 29, 49; m 80. PHARMACOKINETICS, DRUG METABOLISM. *Educ:* Ohio State Univ, BS, 70, PhD(biochem), 74. *Prof Exp:* Fel neurochem, Cornell Univ, 74-77; res assoc chem carcinogenesis, Fox Chase Cancer Ctr, Inst Cancer Res, 77-79; staff scientist drug metabolism, Hazleton Labs Am, 79-81; RES SCIENTIST V BIOANALTICAL CHEM & DRUG METABOLISM, NORWICH-EATON PHARMACEUT, 81- *Mem:* Am Col Toxicol; AAAS; Am Chem Soc; Sigma Xi. *Res:* Biochemical pharmacology; pharmacokinetics; analytical biochemistry; drug metabolism; chemical carcinogenesis; protein chemistry; enzymology. *Mailing Add:* Dept Path/Toxicol Norwich-Eaton Pharmaceut PO Box 191 Norwich NY 13815

ANDREWS, JOHN STEVENS, JR, b Lawrence, Mass, July 12, 27; m 55; c 4. BIOCHEMISTRY. *Educ:* Univ NH, BS, 51, MS, 52; NC State Col, PhD(animal nutrit), 55. *Prof Exp:* Asst res physiol chemist, Univ Calif, Los Angeles, 55-58; instr ophthal res, Harvard Univ, 59-66, assoc, 66-68; asst prof, 68-70, ASSOC PROF OPHTHAL, VANDERBILT UNIV, 70- *Mem:* Am Chem Soc; Am Soc Biol Chemists. *Res:* Abnormal tissue lipid deposition and metabolism; lipid-protein interrelationships; cataractogenesis. *Mailing Add:* Dept Ophthal Vanderbilt Univ Nashville TN 37232

ANDREWS, JOHN THOMAS, b Millom, Eng, Nov 8, 37; m 61; c 1. GEOMORPHOLOGY, QUATERNARY GEOLOGY. *Educ:* Univ Nottingham, BSc, 59, PhD(geomorphol), 65, DSc(geol), 78; McGill Univ, MSc, 61. *Prof Exp:* Res scientist, Can Govt, 61-67; asst prof geol, 68-74, PROF GEOL SCI, UNIV COLO, 74- *Honors & Awards:* Kirk Bryan Award, Geol Soc Am, 73. *Mem:* Geol Soc; Asn Am Geog. *Res:* Glacial chronology and glacioisostatic uplift. *Mailing Add:* Dept Geol Univ Colo Campus Box 250 Boulder CO 80309-0250

ANDREWS, KEN J, b Sandusky, Ohio, Aug 1, 47; m 79; c 4. HOT SPRINGS MICROBIOLOGY, BIOMASS CONVERSION. *Educ:* Purdue Univ, BS, 69; Univ Calif, Berkeley, MA, 71, PhD(microbiol), 74. *Prof Exp:* Fel med micro, Harvard Med Sch, 74-77; instr, 77, asst prof biol, 78-85, ASSOC PROF BIOL, COLO COL, 86- *Concurrent Pos:* Consult biosci, 81- *Mem:* Am Soc Microbiol; Am Inst Biol Sci; Soc Indust Microbiol; Sigma Xi. *Res:* Analysing selected alkaline hot springs (chemically and biologically) with the intent of determining the applicability of using thermophilic organisms in biomass conversion to neutral fuels. *Mailing Add:* Biol Dept Barnes Sci Ctr Colo Col Colorado Springs CO 80903

ANDREWS, LAWRENCE JAMES, b San Diego, Calif, Sept 27, 20; m 44; c 2. CHEMISTRY. *Educ:* Univ Calif, Berkeley, BS, 40; Univ Calif, Los Angeles, AM, 41, PhD(chem), 43. *Prof Exp:* Asst chem, Univ Calif, 40-43; Sharp & Dohme fel & lectr, 43-44; chemist, Tenn Eastman Corp, Oak Ridge, 44-45; from instr to prof, 45-87, chmn dept, 59-62, actg dean, Col Lett & Sci, 62-63, dean, 64-85, fac res lectr, 65, EMER PROF CHEM, UNIV CALIF,

DAVIS, 88- *Concurrent Pos:* Advan Educ Fund fel, 53-54; consult, Kasetsart Univ, Bangkok, 66-68; Fulbright res scholar, Univ Hull, 67-68; chmn, Comn Arts & Sci, Nat Asn State Univs & Land Grant Cols, 73-74. *Mem:* Am Chem Soc; Sigma Xi. *Res:* Mechanisms organic reactions in solution; coordination compounds; molecular complex formation; electrophilic aromatic substitution processes. *Mailing Add:* Dept Chem Univ Calif Davis CA 95616

ANDREWS, LUCY GORDON, b Washington, DC, Mar 27, 41; div; c 2. HUMAN GENETICS, GENETIC COUNSELING. *Educ:* Agnes Scott Col, BA, 63; Univ Ga, PhD(bot), 67. *Prof Exp:* Prof biol, Brenau Col, 67-75, chmn dept, 67-70; genetic counr, Ga Dept Human Resources, 73-76; mem dept psychiat, Sch Med, Emory Univ, 73-74; prof biol, North Ga Col, 76-77; ASST PROF BASIC MED SCI & HUMAN GENETICS, SCH MED, MERCER UNIV, 81- *Concurrent Pos:* Res assoc, USDA, 67-68; adj prof biol, Brenau Col, 75-80. *Mem:* AAAS; Nat Soc Genetic Counselors; Am Soc Human Genetics; fel, Nat Defense Educ Act; NY Acad Sci. *Res:* Nucleic acid relationships during phloem differentiation in Populus deltoides; evaluation of effectiveness of genetic counseling; development of regional genetic counseling programs; case studies in partial aneuploidy; human gene mapping by in situ hybridization. *Mailing Add:* Dept Biomed Scis Mercer Univ Sch Med 1400 Coleman Ave Macon GA 31207

ANDREWS, LUTHER DAVID, b Rogers, Ark, Oct 25, 23; m 54; c 2. MANAGEMENT, ENVIRONMENT. *Educ:* Univ Ark, BS, 52, MS, 53; Univ Mo, PhD(poultry genetics), 66. *Prof Exp:* Mgr, Ark Farmers Poultry Breeding Farm, 53-57; teaching asst poultry, Univ Mo, 62-63; instr, 56-61, asst prof, 61-62 & 63-68, assoc prof, 68-75, PROF POULTRY, DEPT ANIMAL SCI, UNIV ARK, 75- *Mem:* Poultry Sci Asn; Sigma Xi. *Res:* Developing a cage unit to rear broilers; strain comparisons; environmental management; floor systems. *Mailing Add:* Dept Animal & Poultry Sci Univ Ark Fayetteville AR 72701

ANDREWS, MARK ALLEN, b Scranton, Pa, Oct 24, 51; m 74; c 2. ORGANOMETALLIC CHEMISTRY, HOMOGENEOUS CATALYSIS. *Educ:* New Col, BA, 73; Univ Calif, Los Angeles, PhD(chem), 77. *Prof Exp:* Res assoc, Dept Chem, Univ Wis-Madison, 77-79; asst chemist, 79-81, assoc chemist, 81-83, chemist, 83-84, CHEMIST, DEPT CHEM, BROOKHAVEN NAT LAB, 86- *Mem:* Am Chem Soc. *Res:* Synthetic and mechanistic transition metal organometallic chemistry; homogeneous catalysis; metal-mediated reactions of carbohydrates and carbonhydrate model compounds; metal-catalyzed biomass conversion. *Mailing Add:* Dept Chem Brookhaven Nat Lab Upton NY 11973

ANDREWS, MERRILL LEROY, b Albany, NY, Apr 5, 39; m 65; c 2. PLASMA LASER MEDIA. *Educ:* Cornell Univ, BA, 60; Mass Inst Technol, PhD(physics), 67. *Prof Exp:* Res assoc plasma physics, Lab Plasma Studies, Cornell Univ, 67-69, asst prof appl physics, 69-70; asst prof, 70-74, ASSOC PROF PHYSICS & CHMN DEPT, WRIGHT STATE UNIV, 74- *Mem:* Am Phys Soc; Am Asn Phys Teachers; Sigma Xi. *Res:* Far infrared interferometry; electron beam physics; plasma instabilities; plasma laser media; plasma deposition; electron drift velocity. *Mailing Add:* 1031 Wenrick Dr Xenia OH 45385

ANDREWS, MYRON FLOYD, b Huron, SDak, Dec 8, 24; m 48; c 2. VETERINARY PARASITOLOGY. *Educ:* Stanford Univ, AB, 50; Univ Calif, Davis, DVM, 58. *Prof Exp:* Fisheries res biologist, Chesapeake Shellfish Invests, US Fish & Wildlife Serv, MD, 51-52; sr lab technician, Dept Surg, Sch Med, Univ Calif, Los Angeles, 52-54; asst prof vet sci, NDak State Univ, 58-60; dir large animal clin, Norwich Pharmacal Co, 60-61; asst prof, NDak State Univ, 61-65, actg chmn dept, 64-65, chmn, dept vet sci, 65-80, prof, dept vet sci, 65-85; dir, Successways, 85-88; RETIRED. *Concurrent Pos:* Res consult, Norwich Pharmacol Co, 61-63 & Diamond Labs, 63-64; consult vet, NDak Livestock Sanitary Bd, 64-81 & Vet Admin, Fargo, NDak, 79-85. *Mem:* Am Vet Med Asn; Am Asn Vet Parasitol; Wildlife Dis Asn; US Animal Health Asn; Sigma Xi. *Res:* Veterinary parasitology and zoonoses. *Mailing Add:* Vet Sci NDak State Univ Fargo ND 58102

ANDREWS, NEIL CORBLY, b Spokane, Wash, Mar 31, 16; m 43, 70; c 2. THORACIC SURGERY. *Educ:* Univ Ore, BA, 40, MD, 43; Ohio State Univ, MSc, 50. *Prof Exp:* Intern surg, Union Mem Hosp, Baltimore, Md, 43, asst resident, 44; resident, Jefferson Hosp, Roanoke, Va, 44-45; from asst resident to resident, Col Med, Ohio State Univ, 47-50, from instr to prof thoracic surg, 47-70; chmn, Div Community & Postgrad Med, 71-79, chmn, Dept Postgrad Med, 73-81, prof surg, 70-86, EMER PROF SURG, SCH MED, UNIV CALIF DAVIS, 86- *Concurrent Pos:* Chief surg, Ohio Tuberc Hosp, 50-68; consult, Vet Admin Hosp, Chillicothe, 53-70 & Dayton, 54-70; area consult, Vet Admin, 60-70; coordr, Ohio State Regional Med Prog, 66-70 & Area II, Calif Regional Med Prog, 70-73; mem attend staff, Univ Calif, Davis, Med Ctr, 70-89. *Mem:* AAAS; fel Am Col Surg; fel Am Col Chest Physicians; Am Heart Asn; AMA; Sigma Xi; Am Cancer Soc. *Res:* Cancer chemotherapy. *Mailing Add:* Dept Surg PO Box 3007 El Macero CA 95618

ANDREWS, OLIVER AUGUSTUS, b Plymouth, Wis, Dec 15, 31; m 51; c 5. CHEMISTRY, SCIENCE EDUCATION. *Educ:* Univ Wis-Stevens Point, BS, 53; Univ Wis-Madison, MS, 57. *Prof Exp:* Teacher high schs, Wis, 53-59; traveling sci teacher, Mich State Univ & Nat Sci Found, 59-60; from instr to asst prof, 60-65, ASSOC PROF CHEM, UNIV WIS-STEVENS POINT, 65- *Mem:* Nat Educ Asn; Nat Sci Teachers Asn. *Res:* Investigation of learner initiated cues. *Mailing Add:* Dept Chem Univ Wis Stevens Point WI 54481

ANDREWS, PETER BRUCE, b New York, NY, Nov 1, 37; m 85; c 2. MATHEMATICAL LOGIC, INTELLIGENT SYSTEMS. *Educ:* Dartmouth Col, AB, 59; Princeton Univ, PhD(math), 64. *Prof Exp:* From asst prof to assoc prof, 63-79, PROF MATH, CARNEGIE MELLON UNIV, 79- *Mem:* AAAS; Am Math Soc; Asn Symbolic Logic; Asn Comput Mach; Am Asn Artificial Intel; Asn Automated Reasoning. *Res:* Symbolic logic, especially type theory and theorem proving by computer; formalization of mathematics; artificial intelligence. *Mailing Add:* Dept Math Carnegie Mellon Univ Pittsburgh PA 15213

ANDREWS, PETER WALTER, b London, Eng, June 5, 50; m 81; c 3. DEVELOPMENTAL BIOLOGY, CELL BIOLOGY. *Educ:* Univ Leeds, BSc, 71; Oxford Univ, DPhil(genetics), 75. *Prof Exp:* Res fel develop biol, Pasteur Inst, Paris, 74-75; res fel immunobiol, Sloan-Kettering Inst Cancer Res, 76-78; res investr, Wistar Inst, 78-80, res assoc, 80-83, asst prof, 83-90, ASSOC PROF DEVELOP BIOL, WISTAR INST ANAT & BIOL, 91- *Mem:* Soc Develop Biol. *Res:* Regulation of cell differentiation in teratocarcinomas and early mammalian embryos; changes in surface antigen expression; control gene activity during cell differentiation; changes in susceptibility to early mammalian development; viral replication. *Mailing Add:* 3601 Spruce St Philadelphia PA 19104

ANDREWS, RICHARD D, b Mitchell, Nebr, Jan 6, 33; m 55; c 3. ZOOLOGY. *Educ:* Nebr State Teachers Col, Kearney, BA, 58; Iowa State Univ, MS, 60; Univ Ill, Urbana, PhD(vet med sci), 66. *Prof Exp:* Ecologist, Univ Ill, Urbana, 60-63; assoc prof, 66-77, PROF ZOOL, EASTERN ILL UNIV, 77- *Mem:* Wildlife Soc; Am Soc Mammal; Wildlife Dis Asn. *Mailing Add:* Dept Zool Eastern Ill Univ Charleston IL 61920

ANDREWS, RICHARD NIGEL LYON, b Newport, RI, Dec 6, 44; m 69; c 2. ENVIRONMENTAL POLICY. *Educ:* Yale Univ, AB, 66; Univ NC, Chapel Hill, MRP, 70, PhD(environ planning), 72. *Prof Exp:* Budget analyst, US Off Mgt & Budget, 70-72; from asst prof to assoc prof natural resources policy, Sch Natural Resources, Univ Mich, 72-81, chmn, Resource Policy & Mgt Prog, 78-81; dir, Inst Environ Studies, 81-91, PROF ENVIRON POLICY, UNIV NC, CHAPEL HILL, 81-, CHMN, ENVIRON MGT & POLICY PROG, 90- *Concurrent Pos:* Rockefeller fel, 77-78; vis scholar, US Dept State, 81; chief staff mem, Natural Resources Panel, Gov's Comn Future NC, 82-84; consult, NSF, 83-85 & Conserv Found, 84; mem, Bd Environ Studies & Toxicol, Nat Acad Sci, 86-88; chair, Sect X Social Impacts Sci & Eng, AAAS, 89-90; chair, Pub Affairs & Policy Comt, NC Acad Sci, 87-89; Fulbright fel, Vienna Univ Econ, 90; fac, Salzburg Sem, 90. *Mem:* AAAS; Sigma Xi; Asn Pub Policy Anal & Mgt. *Res:* United States environmental policy; environmental policy analysis; solid and hazardous waste policy; science and values in public policy decisions. *Mailing Add:* Inst Environ Studies Univ NC CB 7410 315 Pittsboro St 256 H Chapel Hill NC 27599-7410

ANDREWS, RICHARD VINCENT, b Arapahoe, Nebr, Jan 9, 32; m 54; c 6. PHYSIOLOGY. *Educ:* Creighton Univ, BS, 58, MS, 59; Univ Iowa, PhD(physiol), 63. *Prof Exp:* Instr biol, Creighton Univ, 58-60; instr physiol, Univ Iowa, 60-63; from instr to prof biol, Creighton Univ, 63-70, prof physiol & asst dean med, 70-75, grad dean, 75-85. *Concurrent Pos:* NSF sci fac fel, Univ Iowa, 61-63; Arctic Inst NAm vis investr, Naval Arctic Res Lab, 63-72; vis prof zool, Univ BC, Can, 85-86. *Mem:* Endocrine Soc; Am Physiol Soc; Am Soc Mammologists; Ecol Soc Am; Soc Exp Biol & Med; fel Explorer's Club. *Res:* Regulatory and environmental physiology. *Mailing Add:* Creighton Univ 2500 California Omaha NE 68178

ANDREWS, ROBERT SANBORN, b Minneapolis, Minn, Sept 20, 35; m 60; c 2. GEOPHYSICS, SCIENCE ADMINISTRATION. *Educ:* Univ Minn, Minneapolis, BGE, 58; Univ Wash, MS, 65; Tex A&M Univ, PhD(oceanog), 70. *Prof Exp:* Assoc prof oceanog, Naval Postgrad Sch, 68-80; sr staff officer, Continental Sci Drilling Comt, Nat Acad Sci, 80-85; sr opers officer, Deep Observ & Sampling Earth's Continental Crust, Inc, 85-89; SR STAFF OFFICER, BD RADIOACTIVE WASTE MGT, NAT ACAD SCI, 90- *Concurrent Pos:* Actg dir, Earth Physics Prog, Off Naval Res, 77-80. *Mem:* Am Geophys Union; Geol Soc Am; Sigma Xi; Geothermal Resources Coun; Soc Explor Geophysicists. *Res:* Marine seismic exploration; theoretical seismograms; sedimentary petrology; acoustic and physical properties of marine sediments; hydrographic surveying; crustal tectonics; optical oceanography; antisubmarine warfare; scientific drilling. *Mailing Add:* 6837 Lemon Rd McLean VA 22101-5422

ANDREWS, ROBERT V(INCENT), chemical engineering; deceased, see previous edition for last biography

ANDREWS, ROBIN M, PHYSIOLOGICAL ECOLOGY. *Educ:* Univ Minn, BS, 64; Harvard Univ, MA, 67; Univ Kans, PhD(zool), 71. *Prof Exp:* Fel, Smithsonian Trop Res Inst, 71-72, Int Biol Prog, Grassland Biomed, 73-75; vis asst prof biol, Washington Univ, 75-76; ASSOC PROF BIOL, VA POLYTECH INST & STATE UNIV, 76- *Mem:* Ecol Soc Am; Soc Study Amphibians & Reptiles; Am Soc Naturalists. *Res:* Evolutionary ecology of reptiles; physiological ecology of reptiles. *Mailing Add:* Dept Biol Va Polytech Inst Blacksburg VA 24061

ANDREWS, RODNEY DENLINGER, JR, b Chicago, Ill, May 13, 22. PHYSICAL CHEMISTRY. *Educ:* Princeton Univ, AB, 43, AM, 47, PhD(chem), 48. *Prof Exp:* Asst, Princeton Univ, 43-48, 49-50; phys chemist, Phys Res Lab, Dow Chem Co, 50-58; res assoc, Plastics Res Lab, Mass Inst Technol, 58-61, asst prof mat res lab, Dept Civil Eng, 61-64, asst prof textile div, Dept Mech Eng, 64-65; PROF CHEM, DEPT CHEM & CHEM ENG, STEVENS INST TECHNOL, 65- *Concurrent Pos:* Inst Int Educ fel, Univ Basel, 48-49; ed, Soc Rheol, 57-58. *Mem:* Am Chem Soc; Am Phys Soc; Soc Rheol; Soc Plastics Eng; Fiber Soc. *Res:* Mechanical and optical properties of high polymers in the solid state; viscoelastic properties of materials. *Mailing Add:* Dept Chem Eng & Chem Stevens Inst Technol Castle Point Hoboken NJ 07030

ANDREWS, RONALD ALLEN, b Pontiac, Mich, Mar 1, 40; m 67; c 1. XEROGRAPHY, INTELLIGENT SYSTEMS. *Educ:* Wayne State Univ, BS, 62, PhD(physics), 66; Mass Inst Technol, MS, 76. *Prof Exp:* Fel solid state physics, Wayne State Univ, 66-67; res physicist quantum optics, Naval Res Lab, 67-71, head optical physics br, 71-72, head interaction physics br, 72-76; mgr forward develop area, Wilson Ctr Technol, Xerox Corp, 77-79, mgr, Foward Prod Syst Eng, 79-83; STAFF VPRES & DIR, ADVAN TECHNOL LABS, RCA/GE AEROSPACE, 83- *Concurrent Pos:* Sloan fel, Mass Inst Technol, 75-76. *Mem:* Fel Am Phys Soc; fel Am Optical Soc; sr mem Inst Elec & Electronics Engrs; Sigma Xi. *Res:* Nonlinear optics and parametric phenomenon; product development and engineering; optical waveguides and devices; lasers physics; x-ray lasers; systems engineering; laser-matter interactions; technical program management; xerography; artificial intelligence; digital signal processing; advanced computing architectures; distributed real time processing; intelligent processing systems. *Mailing Add:* G E Adv Technol Labs Moorestown Corp Ctr Rte 38 Moorestown NJ 08057

ANDREWS, RUSSELL S, JR, b Mobile Ala, July 22, 42; m 68. PULP CHEMISTRY. *Educ:* Univ Ala, BSCh, 64, PhD(org chem), 69. *Prof Exp:* Teaching asst, Univ Ala, 64-68; res assoc, Erling Riis Res Lab, Int Paper Co, Mobile, Ala, 69-76, pulp sect leader, 76-77; mgr primary processing develop, Int Paper Co, 77-79, mgr Chem & Energy Recovery, Corp Res Ctr, 79-83, supt Process Eng, Natchez Mill, 83-84, mgr Energy Conserv, NY Off, 84-85, sr coordr, 85-90, STAFF ENGR, PROCESS TECHNOL, INT PAPER CO, MOBILE, 91- *Mem:* Tech Asn Pulp & Paper Indust; Am Chem Soc; Sigma Xi. *Res:* Organic photochemistry; cellulose and lignin chemistry; pulping, bleaching and chemical recovery technology. *Mailing Add:* 6013 Cumberland Rd S Mobile AL 36608

ANDREWS, STEPHEN BRIAN, b McKeesport, Pa, Apr 13, 44; m 69; c 2. CELL BIOLOGY, BIOPHYSICAL CHEMISTRY. *Educ:* Providence Col, BS, 66; Mass Inst Technol, PhD(inorganic chem), 71. *Prof Exp:* Asst prof, 73-79, sr res assoc, physiol & cell biol, Sch Med, Yale Univ, 73-83; res scientist, Spec Expert, Lab Neurobiol, 83-90, CHIEF, ANALYTICAL CELL BIOL, LAB NEUROBIOL, NAT INST NEUROL DIS & STROKE/NIH, 90- *Concurrent Pos:* USPHS fel, Sch Med, Yale Univ, 71-73; Consult, NIH; lectr, Marine Biol Lab. *Mem:* Am Chem Soc; Am Soc Cell Biol; Microbeam Anal Soc. *Res:* Physiology and cell biology of neuron and glia; structure and function of synapses; direct freezing and microanalysis of tissues; analytical electron microscopy. *Mailing Add:* NIH Bldg 36 Rm 2A29 Bethesda MD 20892

ANDREWS, THEODORE FRANCIS, b Atchison, Kans, Aug 17, 17; m 36, 70; c 4. ECOLOGY. *Educ:* Emporia State Univ, BS, 40; Univ Iowa, MS, 42; Ohio State Univ, PhD(zool), 48. *Prof Exp:* From asst prof to prof biol, Emporia State Univ, 48-56, prof & head dept, 57-66; dir sci, Educ Res Coun Am, Ohio, 66-69; actg vpres acad affairs, 76-78, dean, Col Environ & Appl Sci, 69-80, prof, 80-82, EMER PROF BIOL, GOVERNORS STATE UNIV, 82- *Concurrent Pos:* Consult, Biol Sci Curriculum Study, 64-65 & Harvard Univ Nigerian Proj, 65; assoc dir comn undergrad educ biol sci, George Washington Univ, 65-66. *Mem:* AAAS; Nat Asn Biol Teachers (pres, 64); Am Inst Biol Sci; Nat Sci Teachers Asn; Ecol Soc Am. *Res:* Electrophysiology; biological assay of industrial wastes; freshwater biology; limnological trends in artificially impounded waters in Kansas; comparative limnology of Kansas streams; ecology of blueheron Ardea herodias in Kansas; science curriculum development; interdisciplinary science program; environmental science curriculum. *Mailing Add:* 34 Cambre Cr Hot Springs Village AR 71909-3419

ANDREWS, WALLACE HENRY, b Biloxi, Miss, Oct 6, 43. MICROBIOLOGY. *Educ:* Univ Miss, BA, 65, MS, 67, PhD(microbiol), 69. *Prof Exp:* Res microbiologist, Shellfish Sanit Br, Div Food Technol, 69-71, RES MICROBIOLOGIST, FOOD MICROBIOL BR, DIV MICROBIOL, FOOD & DRUG ADMIN, 71- *Concurrent Pos:* Adv, Nat Shellfish Sanit Prog, 73-77; agency expert Salmonella methodoly, 77-; consult, UN Food & Agr Orgn, 83-; gen referee food microbiol, Asn Off Anal Chemists, 83-; Asn Off Anal Chemists rep, Int Dairy Fedn, Int Asn Cereal Chemists, 85- *Mem:* Am Soc Microbiol; Asn Off Anal Chemists; Int Asn Milk Food & Environ Sanitarians Inc. *Res:* Development of sensitive methods for detecting Salmonella and other foodborne enterics; bacteriological standards for foods; recovery of stressed or damaged bacteria. *Mailing Add:* FDA Div Microbiol HFF-234 200 C St SW Washington DC 20204

ANDREWS, WILLIAM ALLEN, b St Louis, Mo, Oct 12, 22; m 55; c 2. CIVIL ENGINEERING. *Educ:* Wash Univ, BS, 43, MS, 48, ScD(civil eng), 54. *Prof Exp:* Jr engr, Curtiss Wright Corp, 43 & US Engrs, 44-46; lectr math, St Louis Univ, 46; instr civil eng, Wash Univ, 47-54; asst prof, NC State Col, 54-55; assoc prof, Wash Univ, 55-65; prof civil eng, Univ Mo- Rolla, 65-88; RETIRED. *Concurrent Pos:* Struct engr, R L Eason & Assocs, 46-50; consult, Steel Joist Inst, 55-65. *Mem:* Am Soc Civil Engrs. *Res:* Structural engineering. *Mailing Add:* Dept Civil Eng Univ Mo Rolla MO 65401

ANDREWS, WILLIAM LESTER SELF, b Lincolnton, NC, Jan 31, 42; m 65; c 2. PHYSICAL CHEMISTRY, SPECTROSCOPY. *Educ:* Miss State Univ, BS, 63; Univ Calif, Berkeley, PhD(phys chem), 66. *Prof Exp:* Asst prof, 66-70, assoc prof, 70-76, PROF PHYS CHEM, UNIV VA, 76- *Concurrent Pos:* Univ Va Ctr Advan Studies sub-grant, 66-68; NSF grants, 68-; A P Sloan fel, 73-75; Petrol Res Fund grants, 75-78, 79-81 & 82-85; Sci Res Coun & Fulbright Res fels, 82-83. *Honors & Awards:* Coblentz Award. *Mem:* Am Chem Soc; Am Phys Soc. *Res:* Infrared and optical spectroscopic studies of complexes, chemical intermediates and molecular ions produced by matrix reactions and photolysis of suitable molecules. *Mailing Add:* Dept Chem Univ Va Charlottesville VA 22901

ANDRIA, GEORGE D, b Hibbing, Minn, Aug 4, 41; m 64; c 5. MATHEMATICAL ANALYSIS, ENERGY CONVERSION. *Educ:* St John's Univ, Minn, BA, 63; St Louis Univ, MS, 65, PhD(math), 68. *Prof Exp:* Asst prof math, Fontbonne Col, 67-68 & Univ Pittsburgh, 68-73; Res mathematician, Bituminous Coal Res, Inc, 74-75, sr mathematician, 76-78, supvr comput servs, 78-80, mgr comput servs & applications, 80-81; DIR, SCIENTIFIC SYST & SERV, AMERICAN GREETINGS, 81-; PRES, SCI SYST. *Concurrent Pos:* Math consult, Bituminous Coal Res, Inc, 72-73. *Mem:* Soc Indust & Appl Math; Int Asn Math & Comput Simulation; Air Pollution Control Asn. *Res:* Scientific computing; numerical analysis; statistics; mathematical modeling and statistical analysis of coal conversion to synthetic gas and coal mining systems; research and marketing systems. *Mailing Add:* Sci Syst 4841 Pin Oak Rd Akron OH 44333

ANDRIACCHI, THOMAS PETER, b Chicago, Ill, May 30, 47. BIOMECHANICS, MECHANICAL ENGINEERING. *Educ:* Univ Ill, Chicago Circle, BS, 69, MS, 71, PhD(mech eng), 74. *Prof Exp:* Asst prof, 74-78, ASSOC PROF & DIR BIOMECH RES, RUSH-PRESBY-ST LUKE'S MED CTR, 78- *Concurrent Pos:* Adj asst prof, Univ Ill, Chicago Circle, 74-; NIH res fel, 75-78 & res career develop award, 78-82, prin investr, 78-82; prin investr, Orthop Res Educ Found, 78-79. *Mem:* Orthop Res Soc; Am Soc Mech Engrs; Am Soc Biomech. *Res:* Biomechanics of the musculoskeletal system and the mechanics of walking; application of biomechanics to the development of internal joint replacement and the treatment of cerebral palsy. *Mailing Add:* Dept Anat Rush Univ Med Col 600 S Paulina St Chicago IL 60612

ANDRICHUK, JOHN MICHAEL, b Downing, Alta, July 4, 26; m 55; c 3. PETROLEUM GEOLOGY, SEDIMENTOLOGY. *Educ:* Univ Alta, BSc, 46, MSc, 49; Northwestern Univ, PhD(geol), 51. *Prof Exp:* Res geologist, Gulf Oil Corp, 51-54; consult geologist, Alex McCoy Assocs, 54-56; CONSULT GEOLOGIST, ANDRICHUK & EDIE, 56- *Honors & Awards:* Pres Award, Am Asn Petrol Geologists, 59; Medal of Merit, Can Soc Petrol Geologists, 59. *Mem:* Geol Soc Am; Soc Econ Paleontologists & Mineralogists; Am Asn Petrol Geologists; Can Geol Asn; Can Soc Petrol Geologists; Asn Prof Engrs. *Mailing Add:* Andrichuk & Edie Third Floor 205 Ninth Ave SE Calgary AB T2G 0R3 Can

ANDRINGA, KEIMPE, b Salatiga, Indonesia, Oct 25, 35; US citizen. LASERS. *Educ:* Delft Univ Technol, MSc, 61. *Prof Exp:* Engr lasers, Laser Advan Develop Ctr, Raytheon Co, 63-66; sr scientist electro-optics, Res Div, 66-74; sr scientist lasers, Exxon Nuclear Corp, 74-81; RES ASSOC LASERS, E I DU PONT DE NEMOURS & CO, INC, 81- *Mem:* Inst Elec & Electronics Engrs. *Res:* High-power laser development; electro-optics; nonlinear optics. *Mailing Add:* Savannah River Labs E I du Pont de Nemours & Co Inc Aiken SC 29803

ANDRIOLE, VINCENT T, b Scranton, Pa, Aug 3, 31; m 56; c 3. INTERNAL MEDICINE, INFECTIOUS DISEASES. *Educ:* Col of Holy Cross, BS, 53; Yale Univ, MD, 57. *Prof Exp:* Intern med, NC Mem Hosp, 57-58, asst resident, 58-59; clin assoc infectious dis, NIH, 59-61; from asst prof to assoc prof, 63-74, PROF INTERNAL MED, SCH MED, YALE UNIV, 74- *Concurrent Pos:* USPHS res fel, Sch Med, Yale Univ, 61-63; clin investr, Vet Admin Hosp, West Haven, Conn, 63-66; estab investr, Am Heart Asn, 66-71. *Mem:* Am Soc Clin Invest; Am Soc Nephrology; Infectious Dis Soc Am; Am Col Physicians; Am Soc Microbiol. *Res:* Host defense mechanisms in the pathogenesis of infectious diseases. *Mailing Add:* 333 Cedar St New Haven CT 06510-8056

ANDRIST, ANSON HARRY, b Omaha, Nebr, Sept 28, 43; m 68; c 3. PHYSICAL ORGANIC CHEMISTRY. *Educ:* Calif State Univ, San Diego, BS, 66; Univ Ill, Urbana, PhD(chem), 70. *Prof Exp:* Assoc chem, Univ Ore, 70-71; res worker, Univ Sheffield, 71; vis lectr, Univ Colo, Boulder, 72-73; from asst prof to assoc prof, 73-80, PROF CHEM, CLEVELAND STATE UNIV, 80- *Mem:* Am Chem Soc; Chem Soc. *Res:* Mechanistic complexities, including reaction stereochemistry of thermal as well as photochemical molecular rearrangements are being probed through structural, kinetic and isotopic labeling studies. *Mailing Add:* Dept Chem Cleveland State Univ Cleveland OH 44115

ANDRLE, ROBERT FRANCIS, b Buffalo, NY, Oct 28, 27; m 53; c 4. BIOGEOGRAPHY, ORNITHOLOGY. *Educ:* Canisius Col, BA, 48; Univ Buffalo, MA, 60; La State Univ, PhD(geog), 64. *Prof Exp:* Asst preparator exhibs, 56-59, cur div biogeog, 59-65, asst dir, 65-72, assoc dir, 72-76, cur, Div Vertebrate Zool, 76-84, RES ASSOC, BUFFALO MUS SCI, 84- *Concurrent Pos:* Res grants, Frank M Chapman Mem Fund, Am Mus Natural Hist, 65, Buffalo Soc Natural Sci & Am Philos Soc, 67 & Int Coun Bird Preserv, 73, ed, The Atlas of Breeding Birds of New York State, 88. *Mem:* Am Inst Biol Sci; Am Ornithologists' Union; Int Coun Bird Preserv, 86. *Res:* Niagara frontier region ornithology; rare and endangered bird species in Mexico, Guatemala, and West Indies. *Mailing Add:* Buffalo Museum Sci Humboldt Pkwy Buffalo NY 14211

ANDROS, GEORGE JAMES, obstetrics & gynecology, for more information see previous edition

ANDRULIS, MARILYN ANN, b Ft Monmouth, NJ, Sept 6, 40; c 2. ACOUSTICS. *Educ:* Univ Mich, BS, 62; Univ Ill, MS, 64; Univ Tex, PhD(mech eng), 68. *Prof Exp:* Consult, Underwater Syst, Inc, 67-68, B-K Dynamics, 69-71 & Tracor Inc, 71-72; res assoc & dir Ocean Acoust Course, Cath Univ Am, 72-73; PRES, ANDRULIS RES CORP, 72- *Concurrent Pos:* Vpres, Mid-Atlantic Res Inst, 73-77; mem bd dirs, Mil Opers Res, 75-79. *Mem:* Acoust Soc Am; Nat Security Indust Asn; Nat Asn Women Fed Contractors. *Res:* Application of underwater acoustics to detection and classification of underwater systems; mathematical modelling. *Mailing Add:* 7315 Wisconsin Ave 650N Bethesda MD 20814

ANDRULIS, PETER JOSEPH, JR, b New York, NY, Apr 16, 40; m 64; c 2. PHYSICAL ORGANIC CHEMISTRY, BIO-ORGANIC CHEMISTRY. *Educ:* Canisius Col, BS, 61, MS, 64; Univ Tex, Austin, PhD(phys org chem), 67. *Prof Exp:* Res asst chem, Canisius Col, 62-64 & Univ Tex, Austin, 64-67; res chemist, US Army Mobility Equip Res & Develop Ctr, 67-69; asst prof chem, Am Univ, 69-71 & Trinity Col, DC, 71-75; dir, chem & biol sci group, 72-82, DIR, BUS DEVELOP, ANDRULIS RES CORP, 82- *Concurrent Pos:* Lectr, NSF High Sch Teachers Inst, Am Univ, 68-69; exec dir, Mid-Atlantic Res Inst, 74- *Mem:* Am Chem Soc; Chem Soc. *Res:* Structure-reactivity correlations; design, synthesis and pharmacological study of biologically important compounds; design and synthesis of cancer chemotherapeutic agents. *Mailing Add:* 58 Conwell Ave No 1 Somerville MA 02144

ANDRUS, JAN FREDERICK, b Washington, DC, Sept 17, 32; m 61; c 2. NUMERICAL ANALYSIS, THEORY OF TIME. *Educ:* Col Charleston, BS, 54; Emory Univ, MA, 55; Univ Fla, PhD(math), 58. *Prof Exp:* Sr mathematician, Ga Div, Lockheed Aircraft Corp, 58-61, math specialist, 61-62; consult mathematician, Huntsville Oper, Comput Dept, Gen Elec Co, 62-63, sub-oper mgr, 63-64, consult mathematician, 64-66; mem tech staff, Northrop-Huntsville, Northrop Corp, 66-68, gen supvr, 68-69, mem sr tech staff, 69-73; assoc prof math, 73-79, PROF MATH, UNIV NEW ORLEANS, 79- *Mem:* Am Math Soc; Soc Indust & Appl Math. *Res:* Optimization; numerical analysis; numerical integration of systems of ordinary differential equations separated into subsystems; numerical solution of two-point boundary-value problems; nonlinear programming; optimal control theory; the theory of time. *Mailing Add:* 237 Dorrington Blvd Metairie LA 70005-3813

ANDRUS, MILTON HENRY, JR, b Omaha, Nebr, Sept 2, 38; m 64; c 2. ORGANIC POLYMER CHEMISTRY, ORGANIC CHEMISTRY. *Educ:* Augustana Col, SDak, BS, 61; Univ Wash, PhD(chem), 67. *Prof Exp:* SR CHEM SPECIALIST, 3M CO, 67- *Mem:* Am Chem Soc. *Res:* Reaction rates and mechanisms of solvolysis of allylic compounds; light sensitive and polymer materials used in printing; synthesis of acrylic monomers and polymers and other organic compounds and polymers. *Mailing Add:* 3M Co 3M Ctr 236-3B-01 St Paul MN 55144-0001

ANDRUS, PAUL GRIER, b Due West, SC, Dec 23, 25; m 48. FIBER OPTICS MICROFABRICATION, ELECTROPHOTOGRAPHY. *Educ:* Univ Wis, PhB, 46. *Prof Exp:* Asst physics, Univ Wis, 46-47; res engr, 47-51, prin physicist, 51-57, asst chief appl physics div, 57-73, sr researcher, 73-76, SR RES PHYSICIST, BATTELLE MEM INST, 76- *Mem:* Soc Imaging Sci & Technol; Inst Elec & Electronics Engrs. *Res:* Fiber optics sensors and couplers; electrostatics; graphic arts; xerography. *Mailing Add:* 9518 Olentangy River Rd Powell OH 43065

ANDRUS, RICHARD EDWARD, b Fulton, NY, Apr 18, 41; m 85; c 3. BRYOLOGY, ECOLOGY. *Educ:* Rensselaer Polytech Inst, BS, 63; State Univ NY, Syracuse, PhD(plant ecol), 74. *Prof Exp:* Policy analyst environ, NY State Dept Environ Conserv, 70-72; asst prof environ studies, 73-76, asst prof, 76-80, ASSOC PROF ENVIRON STUDIES & BIOL SCI, STATE UNIV NY, BINGHAMTON, 80- *Concurrent Pos:* Co-prin investr, NSF grant, 78-81. *Mem:* Am Bryological & Lichenological Soc; Soc Conserv Biol; Soc Wetland Scientists; Torrey Botanical Club. *Res:* Taxonomy, phytogeography and ecology of Sphagnum. *Mailing Add:* Environ Studies Prog State Univ NY Binghamton NY 13902-6000

ANDRUS, WILLIAM DEWITT, JR, b Cincinnati, Ohio, Sept 24, 28; m 56; c 3. ZOOLOGY, CELL PHYSIOLOGY. *Educ:* Oberlin Col, AB, 52; Stanford Univ, PhD(biol), 62. *Prof Exp:* From instr to assoc prof, 60-74, chmn dept, 70-73, PROF ZOOL, POMONA COL, 74- *Mem:* AAAS; Soc Cryobiol; Tissue Cult Asn; Sigma Xi. *Res:* Brine algae and protozoa; cryobiology; mammalian hibernation; tissue culture applications. *Mailing Add:* Biol Dept Pomona Col Claremont CA 91711

ANDRUSHKIW, ROMAN IHOR, b Lviw, Ukraine, May 3, 37; US citizen; c 1. NUMERICAL ANALYSIS. *Educ:* Stevens Inst Technol, BE, 59, PhD(math), 73; Newark Col Eng, MSEE, 64; Univ Chicago, MS, 67. *Prof Exp:* From elec engr to sr elec engr, Weston Instruments & Electronics Co, Schlumberger Inc, 59-64; instr math, Newark Col Eng, 64-66; res asst, Univ Chicago, 66-68; instr, Newark Col Eng, 68-70; from asst prof to assoc prof, 78-84, PROF MATH, NJ INST TECHNOL, 84- *Concurrent Pos:* Union Carbide Corp Scholar, 55-59; consult, Bendix Corp. *Honors & Awards:* NSF Award, 66. *Mem:* Am Math Soc; Soc Indust & Appl Math; Acad Mech; Int Asn Math & Comput Simulation; Shevchenko Sci Soc. *Res:* Spectral theory of non-selfadjoint unbounded operators; finite element analysis with application to parabolic partial differential equations; iterative methods for the solution of linear and nonlinear eigenvalue problems; free boundary problems of the stefan type. *Mailing Add:* Dept Math NJ Inst Technol 323 King Blvd Newark NJ 07102

ANDRYCHUK, DMETRO, b Fisher Branch, Man, Dec 10, 18; nat US; m 44; c 3. SPECTROSCOPY. *Educ:* Univ Man, BS, 41, MA, 42; Univ Toronto, PhD(spectros), 49. *Prof Exp:* Jr res officer spectros, Nat Res Coun, Ottawa, Ont, 49-51; spectroscopist, Diamond Alkali Co, Ohio, 51-57; res scientist, Leeds & Northrup Co, Pa, 57-58; mem tech staff res, Tex Instruments Inc, 58-84; RETIRED. *Mem:* Am Chem Soc; Can Physics Soc. *Res:* Emission, raman, infrared and x-ray spectroscopy; infrared studies of semiconductors; analytical instruments; application of laser and infrared technology to reconnaissance. *Mailing Add:* 422 Terrace Dr Richardson TX 75081

ANDRYKOVITCH, GEORGE, b St Michael, Pa, Jan 1, 41. MICROBIAL PHYSIOLOGY. *Educ:* Univ Pittsburgh, BS, 62, MS, 65; Univ Md, PhD(microbiol), 68. *Prof Exp:* Asst prof, 68-77, ASSOC PROF BIOL, GEORGE MASON UNIV, UNIV VA, 77- *Mem:* AAAS; Am Inst Biol Sci; Sigma Xi; Am Soc Microbiol. *Res:* Regulation and control of the biosynthesis of extracellular polysaccharides of yeasts and bacteria; mechanisms of action of bacteriolytic enzymes elaborated by soil microorganisms. *Mailing Add:* 1735 21st Rd N Arlington VA 22209

ANDY, ORLANDO JOSEPH, b New Britain, Conn, Jan 21, 20. NEUROSURGERY. *Educ:* Ohio Univ, BS, 43; Univ Rochester, MD, 45. *Prof Exp:* Instr neurol surg, Univ & neurol surgeon, Hosp, Johns Hopkins Univ, 52-55; chmn dept, 60-80, PROF NEUROSURG, SCH MED, UNIV MISS, 55- *Concurrent Pos:* USPHS fel, Johns Hopkins Univ, 52-55. *Mem:* Neurosurg Soc Am; Am Asn Neurol Surgeons; AMA; Am Fedn Clin Res; Cong Neurol Surgeons; Sigma Xi. *Res:* Epilepsy. *Mailing Add:* Dept Neurosurg Univ Miss Med Ctr Jackson MS 39216

ANEJA, VINEY P, b Jullunder, India, Nov 21, 48; m 81; c 2. ATMOSPHERIC SCIENCE. *Educ:* Indian Inst Technol, Kanpur, BTech, 71; NC State Univ, Raleigh, MS, 75, PhD(chem eng), 77; Union Col, MBA, 86. *Prof Exp:* Mem sci staff, Brookhaven Nat Lab, 73-75; proj scientist, Northrop Serv, Inc, 76-80; staff chem engr, Gen Elec Co, 80-87; vis assoc prof, 87-90, RES ASSOC PROF ATMOSPHERIC SCI, NC STATE UNIV, RALEIGH, 90- *Concurrent Pos:* Vis prof, Univ Uppsala, Sweden, 79, Jawahar Lal Nehru Univ, India, 80; consult, UN Environ Prog & US Govt; indust reviewer proposals, NSF, US Environ Protection Agency & NASA; tech prog chmn, Int Specialty Conf Environ Impact Natural Emissions, 84; vis scientist, Arrhenius Lab, Sweden, 85; prin investr, US Environ Protection Agency, 87, 88, 89, 90 & 91, USDA Forest Serv, 90-91, Ga Inst Technol, 90-91; mem, comt Global Environ Mgt, Air & Waste Mgt Asn, 89- *Honors & Awards:* Distinguished Serv Award, Air Pollution Control Asn, 84. *Mem:* Am Inst Chem Engrs; Am Chem Soc; Air & Waste Mgt Asn; Am Meteorol Soc; Am Geophys Union. *Res:* Atmospheric chemistry; natural emissions; gas exchange; atmospheric monitoring of trace gases; biospheric-atmospheric interactions; cloud chemistry and physics; photochemical oxidants; transformation and transport of pollutants; climate change; awarded six US patents; author of one book and numerous publications on atmospheric chemistry. *Mailing Add:* Dept Marine Earth & Atmospheric Sci NC State Univ Raleigh NC 27695-8208

ANELLO, CHARLES, b Philadelphia, Pa, Dec 20, 35; m 58; c 2. BIOSTATISTICS, EPIDEMIOLOGY & PUBLIC HEALTH. *Educ:* Towson State Col, BS, 58; Johns Hopkins Univ, ScM, 62, ScD(biostatist), 64. *Prof Exp:* Consult statist, Booz-Allen Appl Res, Md, 64-65; prin investr, Res Anal Corp, Va, 65-67; asst prof biostatist & epidemiol, Sch Hyg & Pub Health, Johns Hopkins Univ, 67-69; DEP DIR OFF EPIDEMIOL & BIOSTATISTICS, CTR DRUG EVAL & RES, FDA, 69- *Concurrent Pos:* Teacher, George Wash Univ. *Honors & Awards:* Merit Award, Food & Drug Admin, 85; Super Serv Award, Pub Health Serv, 87; Distinguished Serv Award, Dept Health & Human Serv, 89. *Mem:* Biomet Soc; Am Statist Soc; Soc Epidemiol Res. *Res:* Competing risk models; application of stochastic processes to medical problems; clinical trial design and analysis; post marketing surveillance. *Mailing Add:* 23 Valerian Ct Rockville MD 20852

ANET, FRANK ADRIEN LOUIS, b Doulcon/Meuse, France, Oct 24, 26; US citizen; m 55. ORGANIC CHEMISTRY, NATURAL MAGNETIC REMANENCE. *Educ:* Univ Sydney, BSc, 49, MSc, 50; Oxford Univ, DPhil(chem), 52. *Prof Exp:* Fel, Nat Res Coun Can, 53-54; from asst prof to prof chem, Univ Ottawa, Can, 54-64; PROF CHEM, UNIV CALIF, LOS ANGELES, 64- *Mem:* Am Chem Soc; Royal Soc Chem; Sigma Xi; fel AAAS. *Res:* Stereochemistry; nuclear magnetic resonance. *Mailing Add:* Dept Chem Univ Calif Los Angeles CA 90024

ANEX, BASIL GIDEON, b Seattle, Wash, May 4, 31; m 59; c 3. SINGLE-CRYSTAL SPECTROSCOPY, ELECTRONIC-STRUCTURE OF MOLECULES. *Educ:* Wesleyan Univ, BA, 53; Univ Wash, PhD(chem), 59. *Prof Exp:* Res assoc quantum chem, Ind Univ, 59-60; from instr to asst prof chem, Yale Univ, 60-67; prof & head dept, NMex State Univ, 67-71; PROF CHEM, UNIV NEW ORLEANS, 71- *Concurrent Pos:* Vis assoc, Comt prof training, Am Chem Soc. *Mem:* Am Chem Soc; fel Am Inst Chem; Sigma Xi. *Res:* Visible, ultraviolet and reflection spectroscopy; electronic properties of highly absorbing crystals; electronic structure of organic molecules, inorganic complexes and charge-transfer complexes; quantum chemistry. *Mailing Add:* Dept Chem Univ New Orleans New Orleans LA 70148-2820

ANFINSEN, CHRISTIAN BOEHMER, b Monessen, Pa, Mar 26, 16; m 41, 79; c 3. PROTEIN CHEMISTRY. *Educ:* Swarthmore Col, BA, 37; Univ Pa, MS, 39; Harvard Univ, PhD(biochem), 43. *Hon Degrees:* DSc, Swarthmore Col, 65, Georgetown Univ, 67, NY Med Col, 69, Univ of Pa, 73, Gustavus Adolphus Col, 75, Brandeis Univ, 76 & Providence Col, 78; MD, Univ Naples, 82, Yeshiva Univ, 82, Adelphi Univ. *Prof Exp:* Asst instr org chem, Univ Pa, 37-39; instr biochem, Harvard Med Sch, 43-45, assoc, 45-47, asst prof, 48-50; chief lab cellular physiol, Nat Heart Inst, 50-52, chief lab cellular physiol & metab, 52-62; prof biochem, Harvard Med Sch, 62-63; chief lab chem biol, Nat Inst Arthritis, Metab & Digestive Dis, 63-81; vis prof biochem, Weizmann Inst Sci, 81-82; PROF BIOL, JOHNS HOPKINS UNIV, 82- *Concurrent Pos:* Am Cancer Soc sr fel, Biochem Div, Med Nobel Inst, Sweden, 47-48; Guggenheim Found fel, 59; vis fel, All Souls Col, 70; hon fel & mem bd gov, Weizman Inst Sci, 62- *Honors & Awards:* Nobel Prize Chem, 72. *Mem:* Nat Acad Sci; Am Soc Biol Chem (pres, 71-72); Am Acad Arts & Sci; Am Philo Soc. *Res:* Structure-function relationship in proteins; genetic basis of protein structure; isolation and characterization of human interferon; protein isolation, synthesis and proteolysis. *Mailing Add:* Dept Biol Johns Hopkins Univ Baltimore MD 21218

ANG, ALFREDO H(UA)-S(ING), b Davao City, Philippines, July 4, 30; m 54; c 3. CIVIL ENGINEERING. *Educ:* Univ Ill, MS, 57, PhD(struct mech), 59. *Prof Exp:* Struct designer, Mulvaney-McMillan Co, 54-55; asst civil eng, 55-57, res assoc, 57-59, from asst prof to assoc prof, 59-65, assoc mem ctr advan study, 64-65, prof civil eng, Univ Ill, Urbana, 65-88; UNIV CALIF, IRVINE, 88- *Concurrent Pos:* NSF res grant, 63-; consult, 62- *Honors & Awards:* Walter L Huber Prize, Am Soc Civil Engrs, 68, State-of-the-Art Award, 73; Freudenthal Medal, 82; Sr Res Award, Am Soc Eng Educ, 83; Newmark Medal, 88. *Mem:* Nat Acad Eng; fel Am Soc Civil Engrs; fel Am Inst Aeronaut & Astronaut; fel Am Soc Mech Engrs; Seismol Soc Am; Earthquake Eng Res Inst; Int Asn Structural Safety & Reliability (pres); Am Soc Eng Educ; Int Asn Structural Mech Reactor Technol; Soc Naval Architects & Marine Engrs. *Res:* Continuum dynamics; non-linear behavioral theories of solid continua and the development of general mathematical approaches to non-linear solids; probability theory in structural mechanics; structural safety and reliability. *Mailing Add:* Dept Civil Eng Univ Calif Irvine CA 92717

ANG, TJOAN-LIEM, b Bogor, Indonesia, Jan 7, 33; m 62; c 2. POLYMER SCIENCE. *Educ:* Inst Technol Bandung, Indonesia, Drs, 62; Univ Akron, MSc, 65, PhD(polymer sci), 75. *Prof Exp:* Lectr org chem, Inst Technol Bandung, 63-71; sr chemist polymer res, Swedlow, Inc, 74-75; sr res chemist, Swedcast Corp, 75-76; sr res chemist polymer appln & prod develop, 76-78, RES ASSOC, POLYMER RES & DEVELOP, AVERY INT CORP, 78- *Mem:* Am Chem Soc. *Res:* Impact modification of glassy polymers; activatable pressure sensitive adhesives; release and anti blocking agents; surfactants; bulk and emulsion polymerization; thermal and radiation curing. *Mailing Add:* 1704 Alhambra Rd S Pasadena CA 91030

ANGEL, AUBIE, b Winnipeg, Man, Can, Aug 28, 35; Can citizen; m 61; c 3. METABOLIC DISEASE, LIPOPROTEIN METABOLISM. *Educ:* Univ Man, BSc, 59, MD, 59; McGill Univ, MSc, 63, FRCP(C), 64. *Prof Exp:* Asst prof, McGill Univ, 65-68; assoc prof, Univ Toronto, 72-81, prof med, 81-90; staff endocrinol, Toronto Gen Hosp, 68-90; PROF & HEAD, DEPT MED, UNIV MAN, 91- *Concurrent Pos:* Staff physician, endocrinol, Royal Victoria Hosp, Montreal, 65-68; scholar med, Med Res Coun, Can, 65-70; proj dir, CIDA grant with INCENSA Costa Rice, 87- *Mem:* Can Soc Clin Invest; Am Soc Clin Invest; Endocrine Soc; Can Soc Endocrinol & Metab; Am Soc Study Obesity; Am Diabetes Asn. *Res:* Regulation of cholesterol storage in adipose tissue and the characterization of lipoprotein-cell interactions; hormonal and nutritional control of lipoprotein binding and metabolism in animals and man. *Mailing Add:* Dept Med Health Sci Ctr Rm GC430 700 William Ave Winnipeg MB R3E 0Z3 Can

ANGEL, CHARLES, b Lockney, Tex, June 10, 23; m 45; c 3. BIOCHEMISTRY. *Educ:* Tex Tech Col, BS, 48, MS, 49; Univ Tex, PhD(biochem), 54. *Prof Exp:* Lab asst org chem, Tex Tech Col, 47-48, fel chem, 48-49; teaching & asst, Med Br, Univ Tex, 49-53; from asst prof to assoc prof physiol chem, Sch Med, Univ Miss, 53-57; chief res lab, Vet Admin Ctr, Gulfport Div, 57-72; prof biochem, Univ Ark Med Ctr, 72-83; res chemist, Vet Admin Hosp, Little Rock, 72-83; EMER PROF, UNIV ARK MED CTR, 87- *Concurrent Pos:* Lectr biochem & psychiat, Sch Med, Tulane Univ, 57-; clin assoc prof anat, La State Univ Med Ctr, 70- *Mem:* Fel AAAS; Am Chem Soc; fel Am Inst Chem; Soc Biol Psychiat; Pavlovian Soc Am; Sigma Xi. *Res:* Biochemical studies in mental diseases; studies relating behavior to physiological and biochemical variables; effects of stress and adrenal hormones on brain protein synthesis and permeability function in the central nervous system. *Mailing Add:* 11 Wheaton Cove Gulfport MS 39503

ANGEL, HENRY SEYMOUR, b London, Eng, Jan 20, 19; nat US; m 41; c 3. ORGANIC CHEMISTRY. *Educ:* Yale Univ, BS, 40; Univ Pa, MS, 47, PhD(org chem), 48. *Prof Exp:* Res chemist, Socony-Vacuum Oil Co, NJ, 40-45; asst instr org chem, Univ Pa, 45-47; sr chemist, Sherwin-Williams Co, 48-52; res chemist, Quaker Oats Co, 53 & Am Cyanamid Co, 53-60; sr scientist, Engelhard Minerals & Chem Corp, 60-83; RETIRED. *Mem:* AAAS; Am Chem Soc. *Res:* Synthesis of organic compounds; synthesis of precious metal-organic compounds. *Mailing Add:* 79 Willow Ave North Plainfield NJ 07060

ANGEL, JOSEPH FRANCIS, b Nablus, Palestine, Aug 31, 40; Can citizen; m 66; c 2. NUTRITION, BIOCHEMISTRY. *Educ:* Ain Shams Univ, Cairo, BSc, 63; Am Univ Beirut, MSc, 65; Univ Toronto, PhD(nutrit), 71. *Prof Exp:* Res asst biochem, Dept Pediat, Augusta Victoria Hosp, Jerusalem, 65-66; asst prof human nutrit, Univ BC, 69-75; assoc prof, 75-83, PROF BIOCHEM, UNIV SASK, 83- *Mem:* Can Soc Nutrit Sci; Can Biochem Soc. *Res:* Energy utilization in pyridoxine deficiency; lipogenesis in meal-fed animals; evolutionary aspects of nutrition; nutrition and metabolic development in mammalia. *Mailing Add:* Dept Biochem Col Med Univ Sask Saskatoon SK S7N 0W0 Can

ANGELAKOS, DIOGENES JAMES, b Chicago, Ill, July 3, 19. ELECTRICAL ENGINEERING, APPLIED PHYSICS. *Educ:* Univ Notre Dame, BS, 42; Harvard Univ, MS, 46, PhD(eng sci), 50. *Prof Exp:* Engr microwaves, Westinghouse Elec, 42-43; instr electronics, Univ Notre Dame, 43-46; teaching fel & res asst eng sci, Harvard Univ, 47-50; asst prof elec eng, Univ Notre Dame, 50-51; from asst prof to assoc prof, 51-61, dir electronics res lab, 64-85, PROF ELEC ENG, UNIV CALIF, BERKELEY, 61- *Concurrent Pos:* Guggenheim Found fel, 57-58; liaison scientist electronics, Off Naval Res, US Embassy, London, Eng, 61-62; mem eval panel, EM Div, Nat Bur Standards, 72-76 & Associateship Prog, Nat Res Coun, 75-; mem, bd dirs, Mod Greek Studies Found, San Francisco State Univ, 85- *Honors & Awards:* Axion Award, Hellenic Am Prof Soc, 78. *Mem:* Sigma Xi; fel Inst Elec & Electronics Engrs. *Res:* Microwave scattering; snow pack measurements; electronics; antennas; microwave communications. *Mailing Add:* Cory Hall Univ Calif Berkeley CA 94720

ANGELAKOS, EVANGELOS THEODOROU, b Tripolis, Greece, July 15, 29; nat US; c 1. PHYSIOLOGY. *Educ:* Nat Univ Athens, BS, 48; Boston Univ, MA, 53, PhD(physiol), 55; Harvard Univ, MD, 59. *Prof Exp:* Asst physiol, Sch Med, Boston Univ, 52-53, from instr to prof, 55-68; chmn dept, 68-83, PROF PHYSIOL & BIOPHYS, GRAD SCH, HAHNEMANN UNIV, 68-, DEAN, 83- *Concurrent Pos:* Med Found fel, 59-60; USPHS res career develop award, 60-68; res assoc biomath, Mass Inst Technol, 59-61; res assoc cardiol Maine Med Ctr, Portland, 59-68; vis scientist, Karolinska Inst, Sweden, 62-63; consult, US Army Natick Labs Environ Med, 63-70; consult, Merck Inst Therapeut Res, 68-73; dir, Biomed Res Inst, Univ Maine, Portland, 72-79. *Mem:* AAAS; fel Am Col Cardiol; fel Am Col Angiol; Microcirc Soc; Am Physiol Soc. *Res:* Physiology and pharmacology of the heart and circulation and autonomic nervous system; electrophysiology of the heart; hypothermia; catecholamines; physiology of emotion and stress; primate physiology; hypertension. *Mailing Add:* Grad Sch Hahnemann Univ Broad Vine MS480 Philadelphia PA 19102-1192

ANGELETTI, RUTH HOGUE, b May 26, 43. EXPERIMENTAL BIOLOGY. *Educ:* Wash Univ, BA, 65, PhD, 69. *Prof Exp:* Res guest, Istituto Superiore di Sanita, Rome, 69-70; postdoctoral fel, Wash Univ, 70-71, res assoc, 73-74; res consult, Lab Cellular Biol, Rome, 71-73; res assoc, Inst Cancer Res, Philadelphia, 74-75; from asst prof to prof, Dept Path & Lab Med, Div Neuropath, Med Sch, Univ Pa, 75-88, dir, Protein Chem Lab, 82-88; PROF & DIR, LAB MACROMOLECULAR ANALYSIS, DEPT DEVELOP BIOL & CANCER, ALBERT EINSTEIN COL MED, 88- *Concurrent Pos:* Res career develop award, NIH, 77-81, mem, Biochem Study Sect, 81-85; lectr peptide synthesis, Albert Einstein Col Med; lectr neurochem-neuropharmacol & develop neurobiol, Univ Pa. *Mem:* NY Acad Sci; AAAS; Am Soc Neurochem; Am Soc Biochem & Molecular Biol; Am Chem Soc; Protein Soc. *Res:* Biochemistry; macromolecular analysis; neurobiology. *Mailing Add:* Dept Develop Biol & Cancer Albert Einstein Col Med 1300 Morris Park Ave Bronx NY 10461

ANGELICI, ROBERT JOE, b Rochester, Minn, July 29, 37; m 60; c 2. INORGANIC CHEMISTRY, ORGANOMETALLIC CHEMISTRY. *Educ:* St Olaf Col, BA, 59; Northwestern Univ, PhD(inorg chem), 62. *Prof Exp:* NSF fel, 62-63; from instr to prof, 63-87, chmn dept, 77-81, DISTINGUISHED PROF CHEM, IOWA STATE UNIV, 87- *Concurrent Pos:* Alfred P Sloan Found fel, 70-72; fel, Royal Soc Guest Res, 87. *Mem:* Am Chem Soc; Royal Soc Chem London. *Res:* Organometallic chemistry of the transition metals; synthesis, characterization, mechanistic studies and catalysts; catalysis. *Mailing Add:* Dept Chem Iowa State Univ Ames IA 50010

ANGELIDES, KIMON JERRY, b Sacramento, Calif, Aug 8, 51. NEUROSCIENCE, CELL BIOLOGY. *Educ:* Lawrence Univ, BA, 73; Univ Calif, Santa Cruz, PhD(chem), 77. *Prof Exp:* NIH postdoctoral fel, Dept Chem, Cornell Univ, 77-79; asst prof biochem, McGill Univ, 79-81; from asst prof to assoc prof biochem, molecular biol & neurosci, Col Med, Univ Fla, 81-86; assoc prof, 86-91, PROF MOLECULAR PHYSIOL & BIOPHYS, NEUROSCI, BIOCHEM & MOLECULAR & CELL BIOL, BAYLOR COL MED, 91- *Concurrent Pos:* Nat Res Coun award, First World Cong, Int Brain Res Orgn, Lausanne, Switz, 82 & 8th Int Biophys Cong, Bristol, UK, 84; vis prof, Nat Ctr Sci Res, Lab Biochem, Univ d'Aix- Marseille, 84 & Nat Inst Physiol Sci, Japan, 84-85; NIH res career develop award, 87-92, res grants, 90-95; mem, Biol Sci Study Sect, NIH, 87-90; Nat Mult Sclerosis Soc res grant, 88-91; Muscular Dystrophy Asn res grant, 90-93. *Mem:* Am Soc Cell Biol; Soc Neurosci; Biophys Soc; Am Chem Soc; Int Soc Toxicol. *Res:* Molecular and cellular basis for signalling in the nervous system and organization of the neuronal membrane and those intra- and extracellular elements that are important in its organization during development; author of numerous technical publications. *Mailing Add:* Dept Molecular Physiol & Biophys Baylor Col Med One Baylor Plaza Houston TX 77030

ANGELINI, ARNALDO M, b Force, Italy, Feb 2, 09; m 37; c 3. ENVIRONMENT, MATHEMATICS & ELECTRICAL ENGINEERING. *Educ:* Univ Rome, Libero Docente(elec eng), 36. *Prof Exp:* Prof elec eng, Univ Rome, 49-73, prof nuclear eng, 57-69; RETIRED. *Concurrent Pos:* Cent mgr, Power Sector, Soc Indust Elec, 44-49, gen mgr, 49-63; gen mgr, Ital Nat Elec Bd, 63-73, chmn, 73-79, hon chmn, 79-; vpres, World Energy Conf, 68-71; hon prof, Polytech Inst NY, 75. *Honors & Awards:* Simon Ramo Gold Medal, Inst Elec & Electronic Engrs, 86, Eng Leadership Recognition, 89. *Mem:* Foreign mem Nat Acad Eng; fel Am Nuclear Soc; fel Inst Elec & Electronic Engrs. *Res:* Energy in general; electric power and nuclear energy; transmission of electric power; application of systems science and engineering; complete reconstruction and large development of the electric power system in central Italy destroyed during the second World War. *Mailing Add:* Enel, Via G B Martini 3 Rome 00198 Italy

ANGELINI, CORRADO ITALO, b Padova, Italy, Aug 3, 41; m 74; c 3. NEUROLOGY. *Educ:* Univ Padova, Italy, Neurol Spec Dipl, 68, MD, 65. *Prof Exp:* Res asst neurol, Mayo Clin, 70-71, res assoc, 71-72, resident, 73; asst prof, Univ Padova, Italy, 73-78; vis asst prof neurol, Univ Calif, Los Angeles, Reed Neurol Res Ctr, 78-79; ASSOC PROF NEUROL, UNIV PADOVA, ITALY, 82-, DIR, NEUROMUSCULAR CTR, 89- *Concurrent Pos:* Prin investr, Muscular Dystrophy Asn, 74-79, sr fel, 78-79; vis assoc prof neurol, Univ Colo, 83; mem med comn, Ital Alpine Club, 85- *Mem:* Am Acad Neurol; Ital Soc Inborn Errors Metab; Int Soc Neuropath; Am Neurol Asn; Soc Neurosci; Ital Alpine Club (vpres, 85-). *Res:* Neuromuscular disorders and muscular dystrophies, inborn errors of metabolism, carnitine deficiency and other lipid storage myopathies; metabolic diseases; glycogenoses and disorders of pyruvate metabolism; multisystemic triglyceride storage disease; energy metabolism and neurological effects of altitude. *Mailing Add:* Noventa Padovana via Cappello 36 Padova 35027 Italy

ANGELINI, PIO, b Ipswich, Mass, May 2, 32; m 62; c 8. FOOD CHEMISTRY. *Educ:* Univ Mass, BS, 54; Mich State Univ, MS, 56, PhD(food sci), 63. *Prof Exp:* Anal chemist, US Army Natick Labs, 62-91; CONSULT, 91- *Mem:* Inst Food Tech; Am Chem Soc; Sigma Xi. *Res:* Analysis of food flavors using gas chromatographic and mass spectrometric techniques. *Mailing Add:* 117 Winter St Ashland MA 01721

ANGELINO, NORMAN J, b Buffalo, NY, May 21, 43; m 83; c 3. ORGANIC CHEMISTRY. *Educ:* State Univ NY, Buffalo, BA, 67. *Prof Exp:* SCIENTIST, NY STATE DEPT HEALTH, ROSWELL PARK MEM INST, 67- *Mem:* Am Chem Soc; AAAS. *Res:* Synthesis of cell-surface related carbohydrate analogs as potential anti-tumor agents; modification of oligonucleotides as anti-sense agents. *Mailing Add:* 666 Elm St Buffalo NY 14263

ANGELL, C A, b Canberra, Australia, Dec 14, 33; m 58; c 3. PHYSICAL CHEMISTRY. *Educ:* Univ Melbourne, BSc, 54, MSc, 56; Univ London, PhD(chem), 61. *Prof Exp:* Lectr chem metall, Univ Melbourne, 62-64; res assoc & fel chem, Argonne Nat Lab, 64-66; from asst prof to assoc prof, 66-70, PROF CHEM, PURDUE UNIV, 71- *Concurrent Pos:* Mem ad hoc comt infra-red transmitting mat, Nat Res Coun, 67-68; mem ed adv bd, J Phys Chem, 74-78; J Chem Phys, 87- *Honors & Awards:* Sigma Xi Fac Res Award, Purdue Univ, 78, McCoy Fac Res Award, 86. *Mem:* Am Chem Soc; Sigma Xi. *Res:* Mass Transport properties of ionic liquids; nature of supercooled liquids and glass transition; spectroscopic studies of coordination states in, and vibrational dynamics of, ionic liquids; geochemical liquids. *Mailing Add:* Dept Chem Ariz State Univ Tempe AZ 85287

ANGELL, CHARLES LESLIE, b July 4, 26; US citizen; m 58; c 2. PHYSICAL CHEMISTRY. *Educ:* Univ Sydney, BSc, 49, MSc, 50; Cambridge Univ, PhD(chem), 55. *Prof Exp:* Res officer, New South Wales Univ Technol, 51-53; fel & res assoc spectros, Univ Mich, 56 & Purdue Univ, 57-58; res scientist, 58-69, group leader spectros, 70-78, SCIENTIST, TARRYTOWN TECH CTR, UNION CARBIDE CORP, 78- *Mem:* Am Chem Soc; Soc Appl Spectros; Coblentz soc (pres 69-71); Royal Soc Chem; Am Soc Testing Mat. *Res:* Structural investigations of molecules by infrared and Raman spectroscopy; adsorbed molecules and catalysis; molecular and bond moments; tautomeric systems; infrared absorption band intensities; solvent effects. *Mailing Add:* Tarrytown Tech Ctr Union Carbide Corp Tarrytown NY 10591

ANGELL, FREDERICK FRANKLYN, b Alto Pass, Ill, July 25, 37; m 64; c 4. PLANT BREEDING, GENETICS. *Educ:* Southern Ill Univ, BS, 60, MS, 61; Univ Wis, PhD(hort, plant path), 65. *Prof Exp:* From asst prof to assoc prof hort, Univ Md, College Park, 70-74; dir res, A L Castle, Inc, 74-85; PROJ LEADER, ASGROW SEED CO, 85-, STA MGR, PAC COAST BREEDING STA, 87- *Mem:* Tomato Genetics Coop; Cucurbit Genetics Coop; Am Soc Hort Sci; Int Hort Soc. *Res:* Vegetable breeding; development of processor tomato, varieties; genetics of tomatoes; especially mutants, gene location and inheritance of disease resistance; direct variety development and evaluation program. *Mailing Add:* Asgrow Seed Co 500 Lucy Brown Lane San Juan Bautista CA 95045

ANGELL, JAMES BROWNE, b Staten Island, NY, Dec 25, 24; m 50; c 2. ELECTRICAL ENGINEERING. *Educ:* Mass Inst Technol, SB & SM, 46, ScD(elec eng), 52. *Prof Exp:* Asst electronics, Mass Inst Technol, 46-52; proj engr, Res Div, Philco Corp, 52-56, mgr solid state circuit res, 56-60; assoc prof, 60-62, PROFELEC ENG, STANFORD UNIV, 62-, ASSOC CHMN DEPT, 69- *Concurrent Pos:* Consult, Philco Corp, 60-62 & Fairchild Semiconductor Div, 63-74; chmn Int Solid-State Circuits Conf, 64-65; mem electronics adv group, US Electronics Command, 64-74 & US Army Sci Adv Panel, 68-75. *Mem:* Fel Inst Elec & Electronics Engrs; Sigma Xi. *Res:* Principles, design and application of solid state devices and integrated circuits; application of integrated circuit technology to bio-medical instrumentation and transducers. *Mailing Add:* Dept Elec Engr Stanford Univ Stanford CA 94305

ANGELL, ROBERT WALKER, b Milwaukee, Wis, Apr 24, 29; m 58. PROTOZOOLOGY, CYTOLOGY. *Educ:* Beloit Col, BS, 58; Univ Chicago, PhD(paleozool), 65. *Prof Exp:* USPHS traineeship, 66; asst prof zool, Univ NC, 66-67; ASSOC PROF ZOOL, UNIV DENVER, 67- *Mem:* Soc Protozool. *Res:* Biological mineralization, especially calcification in invertebrates; biology of Foraminifera. *Mailing Add:* Dept Biol Sci Univ Denver University Park Denver CO 80208

ANGELL, THOMAS STRONG, b Bakersfield, Calif, Jan 8, 42; m 67; c 4. MATHEMATICS. *Educ:* Harvard Univ, AB, 63; Univ Mich, MA, 62, PhD(math), 69. *Prof Exp:* From asst prof to assoc prof, 69-83, PROF MATH, UNIV DEL, 83- *Concurrent Pos:* Vis assoc prof, Ga Inst Technol, 76-77; vis prof, Univ Gottingen, 80-81, Univ NMex, 84-85. *Mem:* Am Math Soc; Soc Indust & Appl Math; Math Asn Am. *Res:* Mathematical theory of optimal control; calculus of variations; ordinary and functional differential equations; scattering theory. *Mailing Add:* Dept Math Sci Univ Del Newark DE 19716

ANGELLO, STEPHEN JAMES, b Haddonfield, NJ, Mar 2, 18; m 44; c 3. PHYSICS, SEMICONDUCTORS. *Educ:* Univ Pa, BS, 39, MS, 40, PhD(physics), 42. *Prof Exp:* Res engr, Westinghouse Elec Co, 42-46; asst prof elec eng, Univ Pa, 46-47, head spec proj, Res Inst, 46-47; res engr, Westinghouse Elec Corp, 47-52, sect mgr, 52-54, eng mgr, Semiconductor Dept, 54-57, consult engr, 57-59, mgr thermoelec proj, 59-60, mgr molecular electronics, 60-62, mgr solid state res & develop, 62-64; prof elec eng, Univ Calif, Santa Barbara, 64-66; consult, Res & Develop Ctr, Westinghouse Elec Corp, 66-82; RETIRED. *Concurrent Pos:* Spec lectr, Duquesne Univ, 47-48. *Mem:* Fel Inst Elec & Electronics Engrs; Am Phys Soc. *Res:* Electrical properties of metal-semiconductor contacts; oxide coated cathodes; Hall effect in semiconductors; low-loss, low-noise crystal rectifier for 3 cm microwave mixers; Hall effect and conductivity of cuprous oxide; silicon rectifier development and thermoelectricity; microelectronics. *Mailing Add:* 112 Woodgate Rd Pittsburgh PA 15235

ANGELO, RUDOLPH J, b Ellwood City, Pa, June 16, 30; m 57; c 1. POLYMER CHEMISTRY. *Educ:* Geneva Col, BS, 51; Univ Fla, MS, 53, PhD(chem), 55. *Prof Exp:* Technician anal, Callery Chem Co, 51; instr, Univ Fla, 51-52, asst, AEC, 52-55; res chemist, Exp Sta, 55-58, Iowa, 58-59, res chemist, Exp Sta, 59-73, res assoc, Exp Sta, E I du Pont de Nemours & Co, Inc, 73-83; CONSULT, 83- *Mem:* Am Chem Soc; Am Inst Chemists; NY Acad Sci. *Res:* Syntheses of vinyl and condensation polymers; cyclic polymerization mechanism of non-conjugated diolefins; film and plastic technology. *Mailing Add:* 723 Blackshire Rd Wilmington DE 19805

ANGELONE, LUIS, b Alliance, Ohio, Oct 3, 19; m 46; c 3. PHYSIOLOGY. *Educ:* Ohio State Univ, BA, 47, MA, 49, PhD(physiol), 52. *Prof Exp:* Asst physiol, Ohio State Univ, 49-52, asst instr, 52; res physiologist, Army Chem Ctr, Md, 52-53; from instr to prof physiol, Wash Univ, 53-65; exec secy heart study sect, 65-67, dep chief, Pac Off, 67-68, asst chief spec progs, Res Grants Review Br, Div Res Grants, 68-71, asst chief referral, 71-74, chief referral, 74-78, CHIEF, REV UNIT, EXTRAMURAL ACTIV PROG, NAT INST ARTHRITIS, DIABETES & DIGESTIVE & KIDNEY DIS, NIH, 78-;

PROF PHYSIOL, WASH UNIV, 65- *Concurrent Pos:* Expert adv, Aberdeen Proving Grounds, Md, 53-54; consult, Army Chem Ctr, 53-59; Nat Bd Dent Exam, 62-67. *Honors & Awards:* Dirs Award, NIH, 78. *Mem:* AAAS; Am Physiol Soc. *Res:* Hematology; cardiovascular physiology; neurophysiology. *Mailing Add:* 5806 Madawaska Rd Bethesda MD 20816

ANGELONI, FRANCIS M, b Butler, Pa, Apr 26, 28; m 52; c 2. ANALYTICAL CHEMISTRY. *Educ:* Pa State Univ, BS, 52, MS, 59, PhD(anal chem), 65. *Prof Exp:* Res asst petrol ref lab, Pa State Univ, 52-61; chief chemist, Appl Sci Lab, Inc, Pa, 61-65; sr scientist, Koopers Co, 65-66; group leader physics & phys chem group, 66-69, mgr anal & res serv sect, Res Dept, 69-78, mgr, res & develop, 78-82, chem prods & process res, 82-83, vpres & dir chem res, 83-88, vpres technol assessment, 88-89; RETIRED. *Mem:* AAAS; Indust Res Inst. *Res:* Thermal methods of analysis; x-ray analysis; petroleum chemistry; instrumental analysis; polymer characterization; organic synthesis; coal tar products; high performance insulating materials; chemical process development. *Mailing Add:* RD 4 Harrison City Rd Export PA 15632

ANGELOPOULOS, EDITH W, b Vienna, Austria, Jan 31, 36; US citizen; m 67. PARASITOLOGY, ENTOMOLOGY. *Educ:* Univ Minn, Minneapolis, BA & BS, 61, MS, 64, PhD(zool), 67. *Prof Exp:* Asst prof, 67-71, ASSOC PROF BIOL, DALHOUSIE UNIV, 71- *Res:* Structural and functional aspects of parasitic flagellates; physiology and ultrastructure of Trypanosomatidae and Trichomonas tenax; microtubules in Trypanosomatidae; scanning and transmission microscopy; freeze etching and histochemical studies of Trichomonas tenax; relation of house dust mites to disease transmission; symbiotic bacteria of Hirudinea; relation of hormones to feline mammary cancer. *Mailing Add:* Dept Biol Dalhousie Univ Halifax NS B3H 3H6 Can

ANGER, CLIFFORD D, b Long Beach, Calif, Nov 15, 34; m 67; c 2. SPACE SCIENCE, IMAGE PROCESSING. *Educ:* Univ Calif, Berkeley, BA, 56, MA, 59, PhD(physics), 63. *Prof Exp:* From asst to assoc prof, 62-74, PROF PHYSICS, UNIV CALGARY, 74-; PRES, ITRES RES LTD CALGARY, 84- *Concurrent Pos:* Mem, Galileo Mission Imaging Sci Team, NASA; prin investr, Satellite Auroral Imaging Instruments; adj prof, Ctr Res Exp Space Sci, York Univ, Toronto, Ont, Can, 88- *Mem:* Am Geophys Union; Can Asn Physicists; Sigma Xi; Am Soc Photogram & Remote Sensing. *Res:* Upper atmosphere physics; satellite, balloon and ground level studies of auroral phenomena. *Mailing Add:* ITRES Res Ltd 110-6815 Eighth St NE Calgary AB T2E 7H7 Can

ANGER, HAL OSCAR, b Denver, Colo, May 24, 20. INSTRUMENTATION, NUCLEAR MEDICINE. *Educ:* Univ Calif, Berkeley, BS, 43. *Hon Degrees:* DSc, Ohio State Univ, 72. *Prof Exp:* Res assoc, Harvard Univ, 43-46; biophysicist, Lawrence Radiation Lab & Donner Lab, Univ Calif, Berkeley, 46-81; CONSULT, 81- *Concurrent Pos:* Guggenheim fel, 66-67. *Honors & Awards:* John Scott Award, 64; Nuclear Med Pioneer Citation, Soc Nuclear Med, 74; Centennial Award, Inst Elec & Electronics Engrs, 84. *Mem:* Inst Elec & Electronics Eng; Soc Nuclear Med. *Res:* Instruments for detecting and measuring radioactive isotopes and for mapping their distribution in patients; invention of the first clinically successful radioisotope camera and other instruments. *Mailing Add:* 2236 Channing Way 2236 Channing Way Berkeley CA 94704-2163

ANGERER, J(OHN) DAVID, b Columbus, Ohio, Mar 9, 42; m 63; c 3. ORGANIC CHEMISTRY, POLYMER CHEMISTRY. *Educ:* Ohio State Univ, BSc, 63; Univ Ill, Urbana, MSc, 65, PhD(org chem), 68. *Prof Exp:* Res chemist, 68-74, develop supvr, 78-81, RES SUPVR, HERCULES INC, 74-, RES ASSOC, 81-, DIR TECHNOL. *Mem:* Sigma Xi; Am Chem Soc. *Res:* Monomer and polymer synthesis; process development of organic reaction mechanisms; chemistry and rheology of polysaccharides. *Mailing Add:* 11 Slashpine Circle Southwood Hockessin DE 19707-9206

ANGERER, LYNNE MUSGRAVE, b Ft Sill, Okla, Dec 7, 44; m 66; c 2. MOLECULAR GENETICS. *Educ:* Ohio State Univ, BSc, 66, MSc, 67; Johns Hopkins Univ, PhD(cell & develop biol), 73. *Prof Exp:* Fel molecular genetics, Calif Inst Technol, 73-78; RES ASSOC, UNIV ROCHESTER, 78- *Concurrent Pos:* Damon Runyan-Walter Winchell fel cancer res, 73-75; NIH fel, 76. *Mem:* Sigma Xi; Am Soc Cell Biol. *Res:* Study of gene expression in the developing sea urchin embryo. *Mailing Add:* 27 Chelmsford Rd Rochester NY 14618

ANGERER, ROBERT CLIFFORD, b Columbus, Ohio, Nov 4, 44; m 66; c 2. DEVELOPMENTAL BIOLOGY, GENE REGULATION. *Educ:* Ohio State Univ, BSc, 66; Johns Hopkins Univ, PhD(cell & develop biol), 73. *Prof Exp:* Res fel molecular biol, Calif Inst Technol, 73-78; from asst prof to assoc prof, 78-89, PROF BIOL, UNIV ROCHESTER, 89- *Concurrent Pos:* Am Cancer Soc fel, 73-78; res career develop award, NIH, 84-89. *Mem:* Am Soc Cell Biol; Soc Develop Biol. *Res:* Early animal development; determination/early differentiation of cell lineages in the sea urchin embryo, studied primarily via assays of gene expression. *Mailing Add:* Dept Biol Univ Rochester Rochester NY 14627

ANGERS, ROCH, b Jonquiere, Que, Nov 16, 34; m 64; c 2. METALLURGY, CERAMICS. *Educ:* Laval Univ, BASc, 59, DSc(metall), 68. *Prof Exp:* Engr, Bell Tel Co Can, 59-62; fel, Univ BC, 68-69; from asst prof to assoc prof, 69-74, PROF MAT, LAVAL UNIV, 74- *Mem:* Am Ceramic Soc; Can Ceramic Soc; Am Powder Metall Inst; Am Soc Metals; Can Inst Metall. *Res:* Mechanical properties of ceramics; applied research on powder metallurgy and rapid solidification technology. *Mailing Add:* Dept Metall Laval Univ Quebec PQ G1K 7P4 Can

ANGEVINE, DANIEL MURRAY, pathology, for more information see previous edition

ANGEVINE, JAY BERNARD, JR, b Boston, Mass, June 29, 28; m; c 3. NEUROANATOMY. *Educ:* Williams Col, BA, 49; Cornell Univ, MA, 52, PhD, 56. *Prof Exp:* Asst histol & embryol, Cornell Univ, 49-51, neuroanat, 52-54, embryol, 54, instr zool, 55-56; asst neuropath, Harvard Med Sch, 56-57, instr, 57-60, assoc anat, 59-64, asst prof, 64-67; assoc prof, 67-70, PROF ANAT, COL MED, UNIV ARIZ, 70-, ASSOC HEAD ANAT, 80- *Concurrent Pos:* USPHS career develop award, 64; mem, Neurol B Study Sect, Nat Inst Neurol Dis & Stroke, 72-74, chmn, 74-76; mem clin teaching fac, Dept Neurol Surg, St Joseph's Hosp & Med Ctr, Phoenix, 74- *Mem:* Sigma Xi; Am Asn Anat; Am Neurol Asn. *Res:* Human and comparative neuroanatomy and neuroembryology. *Mailing Add:* 4015 La Espalda Tucson AZ 85718

ANGEVINE, OLIVER LAWRENCE, b Rochester, NY, Apr 28, 14; m 40; c 3. ACOUSTICS. *Educ:* Mass Inst Technol, BS, 36; Univ Buffalo, MS, 69. *Prof Exp:* Engr, Stromberg Carlson Co, 36-46, chief sound equip engr, 46-51; chief engr, Caledonia Electronics & Transformer Co, 51-60 & Progress Webster Corp, 61-62; mgr new prod, Espey Mfg Co, 62-64; PRES, ANGEVINE ACOUST CONSULT, INC, 64- *Concurrent Pos:* Instr, dept archit, Univ Buffalo, 78-84. *Mem:* Fel Acoust Soc Am; Am Soc Testing & Mat; Nat Coun Acoust Consult (secy, 70-77); Inst Noise Control Eng; Inst Elec & Electronics Engrs. *Res:* Active noise cancellation, especially of the hum of transformers for electric power. *Mailing Add:* Angevine Acoust Consults Inc 1021 Maple St Box 725 E Aurora NY 14052

ANGHAIE, SAMIM, b Mallayer, Iran, Sept 18, 49; US citizen; m 76; c 2. NUCLEAR REACTOR DESIGN, THERMAL HYDRAULICS. *Educ:* Shiraz Pahlavi Univ, Iran, BS, 72, MS, 74; Pa State Univ, PhD(nuclear eng), 82. *Prof Exp:* Asst prof nuclear eng, Ore State Univ, 84-86; vis asst prof, 82-84, ASSOC PROF NUCLEAR ENG, UNIV FLA, 86-, ASSOC DIR SPACE POWER, INNOVATIVE NUCLEAR SPACE POWER & PROPULSION INST, 87- *Concurrent Pos:* Sr res engr & consult, Fla Nuclear Assocs, 82-88; consult, RS&H United Technologies, United Space Boosters, 84 & Exxon Nuclear Co, Licensing & Safety, 85-86; mem, Nuclear Heat Exchanger Comt, Am Soc Mech Engrs, 85-; Norcus appointment, Battelle Pac Northwest Lab, Physics, 86-87; chmn, Am Soc Mech Engrs, Emerging Technol Comt, 88-; mem, Tech Comt, Space Nuclear Power Conf. *Mem:* Am Soc Mech Engrs; Am Inst Aeronaut & Astronaut; Am Nuclear Soc. *Res:* Advanced reactor design; nuclear propulsion; high temperature energy conversion; high temperature materials; thermal hydraulics; single-and-multi-phase flow and heat transfer; applied particle transport; liquid and gaseous nuclear fuels. *Mailing Add:* Innovative Nuclear Space Power & Propulsion Inst Col Eng Univ Fla Gainesville FL 32611

ANGIER, ROBERT BRUCE, b Litchfield, Minn, Mar 24, 17; m 45; c 2. ORGANIC CHEMISTRY. *Educ:* Hamline Univ, BS, 40; Univ Nebr, MS, 42, PhD(org chem), 44. *Prof Exp:* Res chemist, 44-56, res assoc, 56-62, group leader, Lederle Labs Div, Am Cyanamid Co, 62-82; RETIRED. *Mem:* Am Chem Soc. *Res:* Synthetic organic chemistry; folic acid; synthesis of pteroylglutamic acid; pteridine, pyrimidine and purine chemistry; antineoplastic, antifungal and antiviral compounds. *Mailing Add:* 38 Boston Ct Whiting NJ 08759-1707

ANGINO, ERNEST EDWARD, b Winsted, Conn, Feb 16, 32; m 54; c 2. GEOCHEMISTRY, HYDROLOGY & WATER RESOURCES. *Educ:* Lehigh Univ, BS, 54; Univ Kans, MS, 58, PhD(geochem), 61. *Prof Exp:* Instr, Univ Kans, 61-62; asst prof oceanog, Tex A&M Univ, 62-65; res assoc & head div geochem, Kans State Geol Surv, 65-70, assoc dir, 70-72; assoc prof civil eng, 67-71, chmn Dept Geol, 72-86, PROF CIVIL ENG, UNIV KANS, 71-, PROF GEOL, 72-; DIR, WATER RESOURCES CTR, 90- *Concurrent Pos:* Mem US Nat Comt Geochem, Nat Acad Sci-Nat Res Coun, 71-76; assoc ed, Soc Econ Paleontologists & Mineralogists, 73-84; vchmn tech adv comt, Comt Res & Develop, Fed Power Comn, 75-78; mem, NAS-NRC Hydrocarbon res drilling, 84; mem, Irrigation Induced Water Quality Prob, 85-90; NAS-NRC hazardous waste disposal; consult, 62- *Honors & Awards:* Geog feature named in honor, Angino Buttress, Antartica, 67; Dept Defense Medal for Res, 69. *Mem:* Geochem Soc (secy, 70-76); Soc Environ Geochem & Health (pres, 78-79); Sigma Xi; Int Asn Geochem & Cosmochem (treas, 80-); Am Inst Prof Geologists. *Res:* Trace element complexing in natural waters, sediment-water interactions; origin natural hydrogen gas; interactions hazardous wastes & natural materials; water chemistry (quality); environmental geochemistry and geology. *Mailing Add:* Dept Geol Univ Kans Lindley 120 Lawrence KS 66046

ANGLE, WILLIAM DODGE, b Lincoln, Nebr, Jan 28, 26; m 52; c 3. INTERNAL MEDICINE. *Educ:* Univ Nebr, BS, 46, MS, 65; Harvard Univ, MD, 48. *Prof Exp:* Asst prof, 53-59, CLIN ASSOC PROF INTERNAL MED, COL MED, UNIV NEBR, 59- *Res:* Electrocardiography. *Mailing Add:* Clarkson Hosp 44th & Dewey Omaha NE 68105-1018

ANGLEMIER, ALLEN FRANCIS, b Tiffin, Ohio, Sept 10, 26; m 52; c 4. FOOD SCIENCE. *Educ:* Fresno State Col, BS, 53; Ore State Univ, MS, 55, PhD(animal nutrit, physiol), 57. *Prof Exp:* Instr food technol, Ore State Univ, 56-58, from asst prof to prof food sci, 58-89; RETIRED. *Concurrent Pos:* Res grants, NIH, 62-69 & Bur Com Fisheries, 69-71. *Mem:* AAAS; Inst Food Technol; Am Meat Sci Asn. *Res:* Chemistry and physiology of muscle; food protein systems. *Mailing Add:* 1555 NW Hillcrest Corvallis OR 97330

ANGLETON, GEORGE M, b Pontiac, Mich, May 25, 27; m 55; c 5. RADIATION BIOLOGY, BIOMETRICS. *Educ:* Mich State Univ, BS, 49; Univ Rochester, MS, 52; Med Col Va, PhD(biophys, biomet), 61. *Prof Exp:* Asst health physics, Los Alamos Sci Lab, 50-55, staff mem, 55-57; instr biophys, Med Col Va, 61; from asst prof to assoc prof, 62-77, PROF BIOSTATIST & RADIATION BIOL, COLO STATE UNIV, 78-, HEAD BIOMETRY & RADIATION BIOPHYS, COLLAB RADIOL HEALTH LAB, 62-, HEAD COMPUT CTR, VET HOSP, 78- *Concurrent Pos:* Fel biomath, NC State Col, 61-62; prin investr radiation recovery studies & thermoluminescent dosimetry studies, US Air Force Contracts, 65-69 & 74;

consult, Martin Co, 61-62, Los Alamos Sci Lab, 64-65, Datametrics, 67-69, US Air Force Acad, 70-74 & Univ Utah, 76-80. *Mem:* Am Statist Asn; Biomet Soc; Health Physics Soc; Radiation Res Soc. *Res:* Theoretical radiation biology; mathematical biology; computer applications. *Mailing Add:* Colo State Univ 521 Cornell Ave Ft Collins CO 80525

ANGLIN, J HILL, JR, b Iowa Park, Tex, Oct 16, 22; m 48; c 2. ORGANIC BIOCHEMISTRY. *Educ:* York Col, BS, 47; Univ Nebr, MS, 50; Univ Okla, PhD(biochem), 63. *Prof Exp:* Teacher pub schs, NMex, 50-59; asst biochem, 59-62, instr biochem & dermat, 62-63, asst prof, 64-69, assoc prof biochem & dermat, 69-80, ASSOC PROF BIOCHEM & MOLECULAR BIOL, MED SCH, UNIV OKLA, 80- *Mem:* Biophys Soc; Soc Invest Dermat. *Res:* Effect of ultraviolet light on biochemistry of skin; photochemistry; metabolism, especially of photo products, steriochemistry and chemistry of tungsten; affinity chromatography of glyco-proteins. *Mailing Add:* 5701 Terry Oklahoma City OK 73111

ANGONA, FRANK ANTHONY, b Los Angeles, Calif, Apr 15, 20; m 44; c 4. GEOPHYSICS, ACOUSTICS. *Educ:* Univ Calif, Los Angeles, BA, 42, MA, 50, PhD(physics), 52. *Prof Exp:* Res asst, Univ Calif, Los Angeles, 50-52; technologist seismic explor, Socony Mobile Oil Co, 52-53, sr res technologist, 53-70, RES ASSOC, MOBIL RES & DEVELOP CORP, 70- *Mem:* Acoustical Soc Am; Soc Explor Geophys; Sigma Xi (pres, Sci Res Soc Am, 58). *Res:* Transmission of sound in rarified gases; propagation of seismic waves in the earth with seismic and Loggina applications. *Mailing Add:* 1607 Nob Hill Rd Dallas TX 75208

ANGOTTI, RODNEY, b Ellsworth, Pa, Apr 14, 37; m 59; c 2. MATHEMATICS. *Educ:* Univ Pittsburgh, BS, 59, PhD(math), 63. *Prof Exp:* Asst prof math, Univ Akron, 63-64 & State Univ NY Buffalo, 64-67; asst head dept math, 69-71, asst dean, Col Liberal Arts & Sci, 71-83, ASSOC PROF MATH, NORTHERN ILL UNIV, 67- *Mem:* Am Math Soc; Math Asn Am; Ital Math Union. *Res:* Geometric invariant theory; algebraic and topological foundations of geometry. *Mailing Add:* Dept Comput Sci Northern Ill Univ De Kalb IL 60115

ANGRIST, STANLEY W, b Dallas, Tex, June 3, 33; m 55; c 3. RESEARCH ADMINISTRATION. *Educ:* Tex A&M Univ, BSc, 55; Ohio State Univ, MSc, 58, PhD(mech eng), 61. *Prof Exp:* Instr mech eng, Ohio State Univ, 57-62; from asst prof to assoc prof, Carnegie-Mellon Univ, 62-71. prof mech eng, 71-82; columnist, Forbes Magazine, 76-88; STAFF REPORTER, WALL STREET JOUR, 89- *Concurrent Pos:* US Air Force res grant, 62-; NSF lab grant, 63-; Western Elec Fund award, 66-67. *Mailing Add:* 152 Maple Heights Rd Pittsburgh PA 15232

ANGSTADT, CAROL NEWBORG, b Gladwyne, Pa, Oct 23, 35; m 58; c 2. BIOCHEMISTRY, COMPUTER ASSISTED INSTRUCTION SOFTWEAR. *Educ:* Juniata Col, BS, 57; Purdue Univ, West Lafayette, PhD(biochem), 62. *Prof Exp:* ASST PROF BIOCHEM, MED COL & GRAD SCH, HAHNEMANN MED COL, PA, 62-, ASSOC PROF, COL ALLIED HEALTH PROF, 76-, ASST DEAN ADMISSIONS, GRAD SCH. *Concurrent Pos:* Consult, Thomas Jefferson Univ, 76. *Mem:* Chromat Lab Automatic Software; Am Asn Univ Prof; Am Asn Univ Women. *Res:* Radio immunoassay of VIP and substance P; development of CAI programs. *Mailing Add:* Dept Biochem MS 411 Hahnemann Med Col Philadelphia PA 19102

ANGSTADT, ROBERT B, b Kutztown, Pa, Mar 13, 37; m 62; c 2. ANIMAL BEHAVIOR, ANIMAL PHYSIOLOGY. *Educ:* Ursinus Col, BS, 59; Cornell Univ, MS, 61, PhD(ethology), 69. *Prof Exp:* PROF BIOL, LYCOMING COL, 87- *Mem:* Animal Behav Soc. *Res:* Behavior of vertebrates; neural control of behavior. *Mailing Add:* Dept Biol Lycoming Col Williamsport PA 17701

ANGUS, JOHN COTTON, b Grand Haven, Mich, Feb 22, 34; m 60; c 2. MATERIALS SCIENCE ENGINEERING. *Educ:* Univ Mich, BS, 56, MS, 58, PhD(chem eng), 60. *Prof Exp:* Res engr thermoelec mat res, Minn Mining & Mfg Co, 60-63; from asst prof to assoc prof, 63-70, PROF ENG, CASE WESTERN RESERVE UNIV, 70- *Concurrent Pos:* Vis lectr, Univ Edinburgh, 72-73; vis prof, Northwestern Univ, 80-81; NATO sr fel. *Mem:* Am Chem Soc; fel Am Inst Chem Engrs; Electrochem Soc; Mat Res Soc. *Res:* Crystal growth and etching; electrochemistry; diamond growth. *Mailing Add:* Dept Chem Eng Case Western Reserve Univ Cleveland OH 44106

ANGUS, THOMAS ANDERSON, b Toronto, Ont, Sept 19, 15; m 50; c 2. INSECT PATHOLOGY. *Educ:* Univ Toronto, BSA, 49, MSA, 50; McGill Univ, PhD, 55. *Prof Exp:* Res scientist, Can Dept Environ, Can Forestry Serv, 49-70, assoc dir, 70-75, dir, Insect Path Res Inst, 75-78, dep dir, Forest Pest Mgt Inst, 78-80; RETIRED. *Mem:* Soc Invert Path; Can Soc Microbiol; fel Entom Soc Can. *Res:* Bacterial diseases of insects. *Mailing Add:* 24 McCrea St Sault Ste Marie ON P6A 4A1 Can

ANHALT, GRANT JAMES, b Shaunavon, Saskatchewan, Dec 14, 52; Can citizen; m 80; c 2. MEDICINE, IMMUNOLOGY. *Educ:* Univ Manitoba, Winnipeg, MD, 75; Am Bd Dermat, 80; FRCP(C), 80. *Prof Exp:* Resident, internal med, Health Sci Ctr, Winnipeg, 75-77; resident dermat, Univ Mich, 77-79, chief resident, 79-80, fel immunol, 80-81, instr dermat, 81-82; asst prof, 82-87, ASSOC PROF DERMAT, JOHNS HOPKINS UNIV SCH MED, 87- *Concurrent Pos:* Prin investr, Va res assoc, 81-82; prin investr, NIH, 83-, consult, Ad Hoc Study Sect, 87- *Mem:* Am Fedn Clin Res; fel Soc Investigative Dermat; fel Royal Col Phys & Surgeons Can; fel Am Acad Dermat; Dermat Found. *Res:* Immunology in dermatology; autoimmune skin disease and immunologically; mediated mucosal disease; pemphigus; pemphigoid, lichen pianus; benign mucous membrane; pemphigoid, empidermoysis bullosa; lupus. *Mailing Add:* Dept Dermat Johns Hopkins Univ Hosp 909 Dunellen Dr Towson MD 21204

ANJANEYULU, P S R, b Andhra Pradesh, India, Feb 15, 55; m 86. PROTEIN CHEMISTRY, BIO-ORGANIC CHEMISTRY. *Educ:* Andhra Univ, India, BSc, 75, MSc, 77; Indian Inst Tech, Bombay PhD(bio org chem), 84. *Prof Exp:* Proj scientist chem, Indian Inst Technol, 83-84; res assoc, 84-87, RES INSTR BIOCHEM, VANDERBILT UNIV, 87- *Concurrent Pos:* Jr res fel, Cent Food Technol Res Inst, 78-79; grad fel, Indian Inst Technol, Bombay, 79-83. *Mem:* Am Chem Soc; Biophys Soc; Am Soc Protein Chem; Am Soc Biochem & Molecular Biol. *Res:* Development of new photolabels, fluorescent labels, bifunctional spur labels, and cross-linking reagents to study membrane protein structure and function; chemistry of N-hydroxysul fosuccinimide esters, human erythricyte anion exchange channel structure and function. *Mailing Add:* Dept Biochem Vanderbilt Univ Sch Med Nashville TN 37232

ANJARD, RONALD P, SR, b Chicago, Ill, July 31, 35; m; c 4. METALLURGY, MICROELECTRONICS. *Educ:* Carnegie-Mellon Univ, BS, 58; Purdue Univ, MS, 68; Univ State of NY, BS, 78; Univ Wis, PhD, 78; Indiana Univ, AS, 78; T A Edison Col, BA, 79; Columbia Pac Univ, PhD(educ & metal eng), 80-81; Webster Univ, M-CRM. *Prof Exp:* Metall trainee, US Steel, 56-57; metall engr, Crucible Steel, 57-58; process engr, Raytheon Mfg, 58-59; sr engr, Delco Electronics, Gen Motors Co, 59-81; div qual mgr metals, Johnson Matthey Electronic Metals, 81-83; corp dir qual, Kaypro Comput, pres, 83-86; INT CONSULT, ACAD SCREENPRINT TEHCNOL, 86-; COORDR & INTERNAL TECH CONSULT, DIV SPC, GEN DYNAMICS ELECTRONICS & CONVAIR, 89- *Concurrent Pos:* Adj fac, Kokomo Indust Apprentice Prog, 73-81 & Ind Voc Tech Col, 79-81; lectr, Ball State Univ & Chapman Col, 83, Nat Univ, 83, Ala A&M, 84, San Diego Community Col, 84-89 & Univ Southern Calif, 85; Univ La Verne, 82, Univ Cal San Diego, 86, La Jolla Univ, 86-, Union Inst, 87-, Sierra Univ, 87-, Coleman Col, 83-86; bd mem, Dening Users group; rev ed & rev comt, Am Soc Qual Control; chmn, Inst Elec & Electronics Engrs. *Mem:* Int Soc Hybrid Microelectronics; Sigma Xi; Int Electronics Packaging Soc; Am Soc Testing & Mat; Am Ceramic Soc; Inst Elec & Electronic Engrs. *Res:* Theick film materials and testing; solder paste materials and testing; nondestructive testing; computer hardware testing; specialized materials evaluations and tests; manufacture and history of bricks; ancient metallurgy in the Americas and the world; varied leak detection techniques; ancient astronomy & sciences in India and the Americas; author or coauthor of over 750 international technology, business, management, quality, presentations and publications in 16 countries. *Mailing Add:* 10942 Montego Dr San Diego CA 92124

ANKEL, HELMUT K, b Saarbrucken, Germany, Apr 9, 33. BIOCHEMISTRY, IMMUNOLOGY. *Educ:* Univ Marburg, diplom chemiker, 57, PhD(biochem), 60. *Prof Exp:* Nat Res Coun Can fel biochem, 60-61; res assoc, Univ Marburg, 62; NASA fel, Univ Pittsburgh, 63-65; from asst prof to assoc prof, 65-75, PROF BIOCHEM, MED COL WIS, 75- *Concurrent Pos:* vis scientist, Inst Nat de la Sante et de la Recherche Med, Paris, France, 72-75; adj prof biol, Marquette Univ, Milwaukee, 80-; vis scientist, Ger Cancer Res Ctr, Heidelburg, 83-84 & 86. *Honors & Awards:* NIH res career award, 71-76. *Mem:* Am Soc Biol Chem; Soc Complex Carbohydrates; Am Chem Soc; German Chemists Soc; Am Asn Immunol. *Res:* Cell membrane glycoproteins; sugar nucleotides; enzymology; interferon; NK cell action. *Mailing Add:* Dept Biochem Med Col Wis PO Box 26509 Milwaukee WI 53226

ANKEL-SIMONS, FRIDERUN ANNURSEL, b Krofdorf, WGer, Oct 23, 33; m 72; c 2. PRIMATOLOGY, COMPARATIVE ANATOMY. *Educ:* Univ Giessen, Dr rer nat(marine biol), 60; Univ Zurich, Habil(phys anthrop), 68. *Prof Exp:* Res grant primatol, Ger Res Team, Max Planck Inst Brain Res, Univ Giessen, 60-63; res asst & lectr, Inst Anthrop, Univ Zurich, 63-69; instr human anat, Univ Kiel Med Sch, 70-71; assoc cur, Peabody Mus, Yale Univ, 71-78; vis assoc prof anthrop, Univ NC, Chapel Hill, 78-79; vis scholar anthrop, 80-82, RES ASSOC ANAT, DUKE UNIV, DURHAM, NC, 82- *Concurrent Pos:* Res asst paleont, Univ Giessen, 60; Wenner-Gren Found res award, 70; Boise Fund res travel grant, Oxford Univ, 72; lectr human anat, Brown Univ, 72 & Yale Univ, 73-74; Leakey Found res grant, 73-75. *Honors & Awards:* Rolex Enterprise Award. *Mem:* Int Primatol Soc; Soc Vert Paleont. *Res:* Locomotor behavior and comparative anatomy of living and fossil primates and man; investigations of the anatomy of vertebrae, of hands and feet and of the limbs in various extinct primates. *Mailing Add:* 2518 Lanier Pl Durham NC 27705

ANKENBRANDT, CHARLES MARTIN, b Cleveland, Ohio, Aug 20, 39; m 83; c 5. ACCELERATOR PHYSICS, EXPERIMENTAL HIGH ENERGY PHYSICS. *Educ:* St Louis Univ, BS, 61; Univ Calif, Berkeley, PhD(physics), 67. *Prof Exp:* Fel high energy physics, Lawrence Radiation Lab, Univ Calif, Berkeley, 67-68; fel, Brookhaven Nat Lab, 68-70; asst prof physics, Ind Univ, 70-73; PHYSICIST HIGH ENERGY & ACCELERATOR PHYSICS, FERMI NAT ACCELERATOR LAB, 73- *Mem:* Am Phys Soc. *Res:* Experiment and theory for the design and improvement of high-energy particle accelerators and colliding-beam facilities, supervision of such activities. *Mailing Add:* Nat Accelerator Lab PO Box 500 Batavia IL 60510

ANKENEY, JAY LLOYD, b Cleveland, Ohio, June 7, 21; m 46; c 3. MEDICINE. *Educ:* Ohio Wesleyan Univ, BA, 43; Western Reserve Univ, MD, 45. *Prof Exp:* Dir Div Cardiothoracic Surg, Univ Hosp Cleveland, 74-86; from instr to assoc prof thoracic surg, 55-69, chmn Div Cardiothoracic Surg, 74, Actg Dir, Dept Surg, 77-80, PROF SURG, SCH MED, CASE WESTERN UNIV, 69- *Mem:* Am Asn Thoracic Surg; fel Am Col Surgeons; Am Surg Asn; Soc Thoracic Surgeons; Soc Vascular Surg. *Res:* Physiological aspects of cardiovascular and thoracic surgery. *Mailing Add:* Univ Hosps 2074 Abington Rd Cleveland OH 44106

ANKENY, NESMITH CORNETT, b Walla Walla, Wash, Sept 10, 26; m 45; c 3. MATHEMATICS, NUMBER THEORY. *Educ:* Stanford Univ, BS, 48; Princeton Univ, PhD(math), 51. *Prof Exp:* Asst prof math, Johns Hopkins Univ, 52-55; assoc prof, 55-60, PROF MATH, MASS INST TECHNOL, 61-

Concurrent Pos: Res assoc, Inst Advan Study, Princeton, NJ, 51-52; Guggenheim fel, Cambridge Univ, 57-58. *Mem:* Am Math Soc. *Res:* Analytic and algebraic number theory; game theory. *Mailing Add:* Dept Math Rm 2-247 Mass Inst Technol Cambridge MA 02139

ANKRUM, PAUL DENZEL, b Hamlin, Kans, Aug 14, 15; m 40; c 2. ELECTRICAL ENGINEERING. *Educ:* Ind Inst Technol, BS, 35; Ashland Col, AB, 39; Cornell Univ, MS, 44. *Prof Exp:* Asst instr eng drawing, Ashland Col, 35-36; instr elec theory, Ind Inst Technol, 36-38, instr in charge commun eng, 38-42; engr, Magnavox Co, 42; from instr to prof elec eng, 42-82, EMER PROF ELEC ENG, CORNELL UNIV, 82- *Concurrent Pos:* Mem tech staff, Hughes Aircraft Co, 57-58; elec engr, Electronics Lab, Gen Elec Co, 66; consult engr, Semiconductor Prod Div, Gen Elec Co, 73 & Borg Warner Electronics, 79. *Mem:* Sr mem Inst Elec & Electronics Engrs; Am Soc Eng Educ. *Res:* Electronics; physics; basic electronics; electron devices; semiconductor electronics. *Mailing Add:* Phillips Hall Cornell Univ Ithaca NY 14853

ANLAUF, KURT GUENTHER, b WGer, 42; Can citizen. ACID-RAIN, ATMOSPHERIC OXIDANTS. *Educ:* Univ Toronto, BSc, 64; PhD(chem), 69. *Prof Exp:* Fel, Univ Wis, 69-71; res assoc, Univ Laval, 71-72, res prof, 72-73; RES SCIENTIST, ATMOSPHERIC ENVIRON SERV, 73- *Res:* Measurement of atmospheric pollutants, reaction and transport of atmospheric pollutants, development of new techniques for measurements of these species, specifically acid-rain and photochemical smog related chemistry and processes. *Mailing Add:* 29 Four Winds Dr Downsview ON M3J 1K7 Can

ANLYAN, WILLIAM GEORGE, b Alexandria, Egypt, Oct 14, 25; US citizen; m 48; c 5. SURGERY. *Educ:* Yale Univ, BS, 45, MD, 49; Am Bd Surg, cert, 55, Am Bd Thoracic Surg, cert, 56. *Hon Degrees:* DSc, Rush Med Col, 73. *Prof Exp:* Intern & resident, Gen & Thoracic Surg, Duke Univ, 49-55, instr surg, 50-51, assoc, 51-53, from asst prof to assoc prof, 53-61, assoc dean, Sch Med, 63-64, dean, 64-69, assoc provost, 69, vpres health affairs, 69-83, chancellor, 83-87, exec vpres, 87-88, PROF SURG, SCH MED, DUKE UNIV, 61-, EMER CHANCELLOR, 90- *Concurrent Pos:* Markle scholar med sci, Duke Univ, 53-58; consult, numerous orgn, 55- & mem numerous bd dirs & nat comts, 63-; mem, bd dirs, G D Searle & Co, 74-85; mem, bd trustees, NC Sch Sci & Math, 78-85, vchmn, 81-84. *Honors & Awards:* Abraham Flexner Award, Asn Am Med Col, 80. *Mem:* Inst Med-Nat Acad Sci; Soc Univ Surg; Soc Vascular Surg; fel Am Col Surg; AMA; Int Cardiovasc Soc; Soc Clin Surg; Am Heart Asn; Soc Med Adminr(pres, 83-85); Am Surg Asn; AAAS; Sigma Xi. *Res:* Pancreatic physiology; abnormalities in blood clotting; serotonin metabolism and thromboembolic disease; author of numerous articles and publications. *Mailing Add:* Box 3701 Duke Univ Med Ctr Durham NC 27710

ANNAN, MURVEL EUGENE, b Coin, Iowa, July 11, 20; m 45. GENETICS. *Educ:* Univ Nebr, PhD(zool), 54. *Prof Exp:* Instr zool, Univ Nebr, 50-51; from asst prof to prof, 54-81, EMER PROF BIOL, WAGNER COL, 81- *Mem:* Soc Study Evolution; Genetics Soc Am; Sigma Xi. *Res:* Genetics of Drosophilas. *Mailing Add:* 300 N 2nd St Bridgewater VA 22812

ANNAS, RICHARD MORRIS, forest ecology, forestry, for more information see previous edition

ANNAU, ZOLTAN, b Szeged, Hungary, May 20, 36; US citizen; m 60; c 1. BEHAVIORAL TOXICOLOGY. *Educ:* Carleton Univ, Ont, BA, 58; McMaster Univ, MA, 60, PhD(psychol), 64. *Prof Exp:* Res assoc & lectr psychol, Univ Mich, 64-65; fel environ med, 65-66, lectr med psychol, 66-68, asst prof, 68-70, assoc prof environ med, 70-73, asst prof med psychol, Dept Med Psychol, 73-78, assoc prof, 78-80, PROF ENVIRON HEALTH SCI, JOHNS HOPKINS UNIV, 80- *Mem:* Soc Neurosci; fel AAAS; Soc Toxicol; Int Brain Res Orgn. *Res:* Neurobehavioral effects of exposure to toxic environmental agents. *Mailing Add:* Dept Environ Health Sci Johns Hopkins Univ 615 N Wolfe St Baltimore MD 21205

ANNEAR, PAUL RICHARD, b Cedar Rapids, Iowa, Jan 19, 15; m 39; c 3. ASTRONOMY. *Educ:* Drake Univ, BA, 36; Case Western Reserve Univ, MS, 38; Univ Mich, PhD, 49. *Prof Exp:* Tutor physics & astron, Hunter Col, 40-41; from instr to prof math & astron & dir, Burrell Mem Observ, 51-80, EMER PROF MATH & ASTRON, BALDWIN-WALLACE COL, 80- *Mem:* Am Astron Soc; Math Asn Am. *Res:* Astronomical photometry; galactic structure in the constellation of Cygnus; artificial satellite tracking and orbit computation. *Mailing Add:* 66 Barberry Dr Berea OH 44017-1202

ANNESTRAND, STIG A, b Husby, Sweden, Sept 18, 33. TRANSMISSION ENGINEERING. *Prof Exp:* SR TRANSMISSION ENGR, SCECO-EAST, DAMMAM, SAUDI ARABIA, 88- *Mem:* Nat Acad Eng; fel Inst Elec & Electronics Engrs. *Mailing Add:* SCECO-East Rm 2-305W PO Box 5190 Dammam 31422 Saudi Arabia

ANNETT, ROBERT GORDON, b Windsor, Ont, Feb 24, 41; m 64; c 2. BIOCHEMISTRY. *Educ:* Univ Windsor, BSc, 64, PhD(biochem), 68. *Prof Exp:* Asst prof, 68-75, ASSOC PROF BIOCHEM, TRENT UNIV, 75- *Concurrent Pos:* Fel, Univ of Calif, Davis, fel, Univ of Windsor & Wayne State Univ, 81-82. *Mem:* Am Chem Soc; Chem Inst Can; Can Cystic Fibrosis Found. *Res:* Enzymology of Oxalacetic acid tautomerization, fatty acid a-oxidation, furanocoumarin biosynthesis, localization studies and clinical biochemistry; immunochemical determination of Lactate Dehydrogenase Isozyme-1; radioimmunoassay of serum estriol; immuno-assisted determination of Prostatic Acid Phosphatase; Ketone and Bilirubin interference in Creatinine Analyses; serum B-Endorphin levels in chiropractic manipulation; synthesis and enzymology of b-carboxy aspartic acid, spectrophotometric determination of hemoglobin derivatives; effect of sugars on milk protein heat denaturation; rem-sleep deprivation and brain acetycholine enzymology. *Mailing Add:* Dept Chem Trent Univ Peterborough ON K9J 7B8 Can

ANNINO, RAYMOND, b NY, Sept 5, 27; m 50; c 5. ANALYTICAL CHEMISTRY, PHYSICAL CHEMISTRY. *Educ:* Columbia Univ, BA, 50; Okla State Univ, PhD, 56. *Prof Exp:* Asst scientist, Nat Dairy Res Labs, Inc, 50-53; supvr anal res, Westvaco Chlor-Alkali Div, 55-57; assoc prof chem, Northeastern La State Col, 57-60 & Canisius Col, 60-67; sr res scientist, Foxboro Co, Mass, 67-72; prof chem, Canisius Col, 72-81, chmn dept, 77-81; PRIN RES SCIENTIST, FOXBORO CO, MASS, 81- *Concurrent Pos:* Sci adv, Food & Drug Admin, Buffalo Dist, 73-81. *Honors & Awards:* Jacob Schoelkopf Medal, Am Chem Soc, 83. *Mem:* Am Chem Soc; Soc Appl Spectros; Instrument Soc Am. *Res:* Application of signal enhancement techniques to various instrumental methods; design of new chromatographic techniques and instruments; electrochemical methods of analysis. *Mailing Add:* The Foxboro Co Corp Res D33ANO12A Foxboro MA 02035

ANNIS, BRIAN KITFIELD, b Arlington, Mass, Aug 25, 40; m 66; c 2. CHEMICAL PHYSICS. *Educ:* Brown Univ, ScB, 62; Univ Md, PhD(chem physics), 67. *Prof Exp:* RES STAFF MEM CHEM, OAK RIDGE NAT LAB, 68- *Mem:* Am Phys Soc; Sigma Xi. *Res:* Experimental atomic and molecular collision dynamics, gas phase transport properties. *Mailing Add:* 869-A West Outer Dr Oak Ridge TN 37830

ANNIS, JASON CARL, b Minneapolis, Minn, July 11, 30; m 61; c 2. AIR POLLUTION, PARTICULATE TECHNOLOGY. *Educ:* Univ Minn, Minneapolis, BS, 53; Mich Technol Univ, MS, 56; Kans State Univ, PhD(mech eng), 69. *Prof Exp:* Instr mining eng, Mich Technol Univ, 53-55; res fel mech eng, Univ Minn, Minneapolis, 55-59; from instr to assoc prof, Kans State Univ, 59-78; ENG CONSULT, 78- *Concurrent Pos:* Mem, Maj Appliance Consumer Action Panel, 72-88, chmn, 80-84; mem adv comn, US Patent Off, 77-78. *Mem:* Air Pollution Control Asn; Am Indust Hyg Asn; Am Chem Soc; Am Soc Heating, Refrig & Air Conditioning Engrs. *Res:* Atmospheric aerosol measurement; air cleaning device evaluation; fine particle size analysis; air filtration theory; air pollution through process alteration in the grain processing industry. *Mailing Add:* 1325 Hudson Ave Manhattan KS 66502

ANNO, JAMES NELSON, b Niles, Ohio, Feb 6, 34; div; c 3. PHYSICS, NUCLEAR ENGINEERING. *Educ:* Ohio State Univ, BS, 55, MS, 61, PhD(physics), 65. *Prof Exp:* Physicist, Battelle Mem Inst, 55-70; PROF NUCLEAR ENG, UNIV CINCINNATI, 70- *Concurrent Pos:* Consult, Battelle Mem Inst, 70-; pres, Res Dynamics Inc, 77- *Mem:* Am Phys Soc; AAAS; NY Acad Sci. *Res:* Industrial applications of atomic energy, nuclear fusion, heat transfer and fluid flow; lubrication mechanics; radiation effects. *Mailing Add:* Univ Cincinnati Mail Location 163 Cincinnati OH 45221

ANNULIS, JOHN T, b Cincinnati, Ohio, Nov 13, 45; m 69; c 2. MATHEMATICS. *Educ:* Grand Valley State Col, BA, 66; Univ NMex, MA, 68, PhD(math), 71. *Prof Exp:* Asst prof math, Univ Wis-Whitewater, 71-72; asst prof, 72-75, assoc prof, 75-81, PROF MATH, UNIV ARK, MONTICELLO, 81-, HEAD DEPT, 79- *Mem:* Am Math Soc; Math Asn Am; Nat Coun Teachers Math. *Res:* Applications of decompositions of the indentity to problems in vector lattices. *Mailing Add:* Dept Math Univ Ark Monticello AR 71657

ANOUCHI, ABRAHAM Y, b Tel-Aviv, Israel, Oct 3, 30; US citizen; m 58; c 1. ELECTRONIC INSTRUMENTATION, GAS DETECTING SENSORS. *Educ:* Ind Inst Technol, BS, 54; Harvard Univ, PhD, 64. *Prof Exp:* Chief engr & partner, Flow Corp, 64-70; chief engr, Environ Equip Div, EG&G, 70-76; vpres res & eng, Instruments Div, United Technologies, 76-86; PRES, ELPAZ INSTRUMENTS INC, 86- *Concurrent Pos:* Consult. *Mem:* Sr mem Instrument Soc Am; Inst Elec & Electronics Engrs; Int Soc Photoelectronic Engrs. *Res:* Instrumentation technologies using state-of-the-art techniques. *Mailing Add:* AYA Technol 10 Knollwood Dr Pittsburgh PA 15215

ANSARI, ALI, b Hyderabad, India, Dec 29, 34; US citizen; m 67; c 2. CLINICAL CHEMISTRY, LIPID CHEMISTRY. *Educ:* Osmania Univ, India, BSc, 57; Kans State Univ, MS, 63; Tex A&M Univ, PhD(biochem), 66. *Prof Exp:* Res assoc lipid biochem dermat, 66-71, asst prof path, Sch Med, Univ Southern Calif, 71-80, asst clin biochemist, Med Ctr, 70-71, ASSOC PROF PATH, CHAS R DREW POSTGRAD MED SCH, 72-; CLIN CHEMIST, LOS ANGELES COUNTY KING/DREW MED CTR, 71- *Concurrent Pos:* Mem fac, Univ Calif, Los Angeles, 80- *Mem:* Am Chem Soc; Am Asn Clin Chem; AAAS; Am Oil Chem Soc. *Res:* Analytical and clinical chemistry and biochemistry of enzymes, lipids and lipoproteins, and their diagnostic and prognostic application in disease. *Mailing Add:* 15511 Grovehill Lane La Mirada CA 90638

ANSARI, AZAM U, b Hyderabad, India, Sept 28, 40; US citizen. CRITICAL CARE MEDICINE, CARDIOLOGY. *Educ:* Osmania Univ, Hyderabad, India, 63. *Prof Exp:* Instr, Dept Internal Med, Tulane Med Sch, 66-68, St Lukes Hosp, St Paul, 68-70; staff physician, Metrop Mt Sinia Med Ctr, Minneapolis, 72-88; STAFF PHYSICIAN, N MEM MED CTR, ROBBINSDALE, MINN, 73-, FAIRVIEW-SOUTHDALE HOSP; EDINA, MINN, 73-, ABBOTT-NORTHWESTERN HOSP, MINNEAPOLIS, MINN, 75-, FAIRFIELD RIDGES HOSP, BURNSVILLE, MINN, 85- *Concurrent Pos:* Asst vis physician, Tulane Med Serv, Charity Hosp, New Orleans, 67-68; consult internal med, Lallie Kemp Charity Hosp, Independence, La, 67-68; tutor, Vet Admin Hosp, Minneapolis, 70; med & cardiol consult, Minn Armed Forces Ctr, 73- *Mem:* AMA; Fel, Coun Clin Cardiol; fel Am Col Cardiol; fel Am Col Physicians; fel Am Col Chest Physicians; fel Am Col Gastroenterol; fel Am Col Nutrit; fel Am Col Critical Care Med. *Res:* Improvement by research in the diagnosis and management of critical cardiovascular, nutritional and gastrointestinal diseases. *Mailing Add:* Metrop Med Off Bldg 825 S Eighth St Suite 444 Minneapolis MN 55404

ANSARI, GUHLAM AHMAD SHAKEEL, b Barabanki, India, Oct 4, 47; m 74; c 1. ORGANIC CHEMISTRY, BIOCHEMISTRY. *Educ:* Aligarh Muslim Univ, India, BS, 66, MS, 69, MPhil, 70, PhD(chem), 73. *Prof Exp:* Asst prof chem, Aligarh Muslim Univ, India, 73-74; fel, Univ Idaho, 74-76; fel, Univ Tex Med Br, 76-77, res assoc biochem, 78-79, from asst prof to assoc prof, 79-90, PROF BIOCHEM & PATH, UNIV TEX MED BR, GALVESTON, 90- *Mem:* Am Chem Soc; Sigma Xi; Chem Soc Brit. *Res:* Chemistry and biochemistry lipids; biochemical toxicology; biomarkers of chemical exposure. *Mailing Add:* Dept Biochem Univ Tex Med Br Galveston TX 77550

ANSBACHER, RUDI, b Sidney, NY, Oct 11, 34; m 65; c 2. OBSTETRICS & GYNECOLOGY, REPRODUCTIVE BIOLOGY. *Educ:* Va Mil Inst, BA, 55; Univ Va, MD, 59; Univ Mich, MS, 70. *Prof Exp:* Intern, Univ Va Hosp, 59-60; Physician, Richmond Mem Hosp, US Army, 60-62, surg resident, Womack Army Hosp, Ft Bragg, NC, 62-63, resident obstet-gynec, Letterman Gen Hosp, San Francisco, Calif, 63-66, mem staff, US Army Hosp, Ryukyu Islands, Okinawa, 66-68, researcher mil med & allied sci, Walter Reed Army Inst Res, Washington, DC, 68-69; teaching assoc, Dept Obstet & Gynec, Med Ctr, Univ Mich, 69-71, chief family planning & consult serv, Dept Obstet & Gynec & chief clin invest serv, 71-74; asst chief, Dept Obstet & Gynec & chief, Obstet Serv, Brooke Army Med Ctr, Ft Sam Houston, 74-77, chief, Dept Obstet & Gynec, Letterman Army Med Ctr, 77-80; PROF OBSTET & GYNEC & ASST CHMN, UNIV MICH MED CTR, ANN ARBOR, 80- *Concurrent Pos:* clin instr, Dept Obstet & Gynec, Univ Hawaii, 67-68; fel, Ctr Res Reprod Biol, Univ Mich, 69-71, actg chmn gynec, 84-85. *Honors & Awards:* Chmn Award for Clin Res, Armed Forces Dist, Am Col Obstet & Gynec, 70. *Mem:* AMA; Asn Mil Surg US; fel Am Col Obstet & Gynec; Int Fedn Gynec & Obstet; Am Fertil Soc; Am Soc Andrology. *Res:* Reproductive biology, keying on the clinical aspects, using immunologic techniques, especially in the fields of infertility and conception control. *Mailing Add:* Dept Obstet & Gynec MPB Rm D-2206 Univ Mich Med Ctr Ann Arbor MI 48109-0718

ANSBACHER, STEFAN, b Frankfurt-am-Main, Germany, Jan 27, 05; nat US; m 30, 51, 76; c 4. BIOMEDICINE. *Educ:* Univ Frankfurt, BSc, 23; Univ Geneva, MSc, 29, ScD(med chem), 33. *Prof Exp:* Asst, Path Inst Hosp, Univ Geneva, 20-30; res assoc, Med Col, Univ SC, 30-31, Res Lab, Borden Co, 31-36 & Squibb Inst, 37-41; dir res & med rels, Am Home Prod Corp, NY, 41-46; dir nutrit res, Schenley Industs, Cincinnati, 46-47; sci & med consult, 47-70; prof life sci & futurism & dean of men, Ind Northern Grad Sch Prof Mgt, 70-75; TEACHER, FLA PALM BEACH COUNTY SCH SYSTS, 75- *Honors & Awards:* AMA Medal, 36. *Mem:* AAAS; AMA; World Future Soc; Am Humanist Asn. *Res:* Enzymology, hormonology, vitaminology; pharmaceuticals; chemotherapy of tuberculosis; animal diseases; nutrition. *Mailing Add:* 2505 D Lowson Blvd Delray Beach FL 33445

ANSCOMBE, FRANCIS JOHN, b Elstree, Eng, May 13, 18; m 54; c 4. STATISTICS. *Educ:* Cambridge Univ, BA, 39, MA, 43. *Prof Exp:* With Brit Ministry Supply, 40-45 & Rothamsted Exp Sta, Eng, 45-47; lectr math, Cambridge Univ, 48-56; res assoc, Princeton Univ, 53-54, from assoc prof to prof, 56-63; prof, 63-88, EMER PROF STATIST, YALE UNIV, 88- *Concurrent Pos:* Corresp consult, Higher Coun Sci Invests, Madrid, 53; vis assoc prof, Univ Chicago, 59-60. *Honors & Awards:* R A Fisher lectr, Joint Statist Meetings, Cincinnati, 82. *Mem:* Int Statist Inst; fel Inst Math Statist; fel Am Statist Asn; Asn Comput Mach; Royal Statist Soc; Sigma Xi. *Res:* Theory and practice of statistical method; statistical computing. *Mailing Add:* Dept Statist Yale Sta Box 2179 New Haven CT 06520-2179

ANSEL, HOWARD CARL, b Cleveland, Ohio, Oct 18, 33; m 60; c 3. PHARMACY. *Educ:* Univ Toledo, BS, 55; Univ Fla, MS, 57, PhD(pharmaceut sci & chem), 59. *Prof Exp:* Asst prof pharm, Univ Toledo, 59-62; from asst prof to assoc prof, 62-70, head dept, 68-77, PROF PHARM, UNIV GA, 70-, DEAN, COL PHARM, 77- *Mem:* Am Pharmaceut Asn; Am Asn Col Pharm. *Res:* Pharmaceutical sciences; hemolytic effect of various chemical agents; binding of antibacterial preservatives by macromolecules; relationship between bacteriolysis, hemolysis and chemical structure. *Mailing Add:* Col Pharm Univ Ga Athens GA 30602

ANSELL, GEORGE S(TEPHEN), b Akron, Ohio, Apr 1, 34; m 60; c 3. PHYSICAL METALLURGY. *Educ:* Rensselaer Polytech Inst, BMetE, 54, MMetE, 55, PhD(phys metall), 60. *Prof Exp:* Metallurgist, US Naval Res Lab, 57-58; from asst prof to prof, 60-67, dean, 74-84, ROBERT W HUNT PROF METALL ENG, RENSSELAER POLYTECH INST, 67-, CHMN MAT DIV, 69-; PRES, COLO SCH MINES, 84- *Concurrent Pos:* Vis scientist, Ford Sci Lab, 61 & Allegheny Ludlum Steel Corp, 59- *Honors & Awards:* Hardy Gold Medal, Am Inst Mining, Metall & Petrol Engrs, 60; Alfred H Geisler Award, Am Soc Metals, 64; Bradley Stoughton Award, 68; Curtis W McGraw Res Award, Am Soc Eng Educ. *Mem:* Am Inst Mining, Metall & Petrol Engrs; fel Am Soc Metals; fel Metall Soc; Nat Soc Proj Eng. *Res:* Powder metallurgy; mechanical properties of crystals; electron microscopy; dispersion strengthening; diffusion in metals; irradiation damage; magnetic properties. *Mailing Add:* Colo Sch Mines 1500 Illinois St Golden CO 80401

ANSELL, JULIAN SAMUEL, b Portland, Maine, June 30, 22; m 51; c 5. UROLOGY. *Educ:* Bowdoin Col, BA, 47; Tufts Univ, MD, 51; Univ Minn, Minneapolis, PhD(urol), 59. *Prof Exp:* Instr urol, Med Col & mem staff, Univ Hosp, Univ Minn, Minneapolis, 56-59; from asst prof to assoc prof, 59-65, prof & chmn dept, 65-87, PROF, DEPT UROL, UNIV WASH, 59- *Concurrent Pos:* Chief urol, Vet Admin Hosp, Minneapolis, 56-59; urologist in chief, King County Hosp & chief consult, Vet Admin Hosp, Seattle, 59- *Mem:* AAAS; AMA; Am Col Surg; Am Urol Asn. *Res:* Physiology of urinary apparatus; pediatric urology. *Mailing Add:* Dept Urol RL-10 Univ Wash Sch Med Seattle WA 98195

ANSELME, JEAN-PIERRE L M, b Port-au-Prince, Haiti, Sept 22, 36; US citizen; m 60; c 3. ORGANIC CHEMISTRY. *Educ:* St Martial Col, Haiti, BA, 55; Fordham Univ, BS, 59; Polytech Inst Brooklyn, PhD(chem), 63. *Prof Exp:* Res fel org chem, Polytech Inst Brooklyn, 60-63, res assoc, 63, sr instr, 65; NSF fel, Inst Org Chem, Univ Munich, 64-65; from asst prof to assoc prof, 65-69, PROF ORG CHEM, UNIV MASS, BOSTON, 70- *Concurrent Pos:* founder & ed, Org Prep & Procedures, 67-70; founder & ed, Org Prep & Procedures Int, 71-; vis prof, Kyushu Univ, Japan, 72, Univ Miami, 79. *Honors & Awards:* AP Sloan Found fel, 69-71; fel Japan Soc Promotion Sci, 72. *Mem:* Am Chem Soc; Brit Chem Soc; fel Japan Soc Promotion Sci. *Res:* Synthetic organic chemistry; reaction mechanisms; meso-ionic systems; N-nitrenes; diazoalkanes; azides; thionylhydrazines; 3-pyrazolidinones; hydrazinium salts; N-nitrosamines; sulfinyl and sulfenylhydratines; N-nitrosamines; pyrazoles and imidazoles. *Mailing Add:* Dept Chem Univ Mass Harbor Campus Boston MA 02125

ANSELMO, VINCENT C, b New York, NY, July 29, 30; m 54; c 5. TRACE ANALYSIS. *Educ:* Fordham Univ, BS, 51, MS, 59; Univ Kans, PhD(chem), 61. *Prof Exp:* Instr pharm, Fordham Univ, 53-57; asst prof chem, John Carroll Univ, 61-64; from asst prof to prof chem, NMex Highlands Univ, 64-77, chmn dept chem, 74-77; CHIEF, INORG ANALYSIS LAB, TEX AIR CONTROL BD, 83- *Mem:* Fel AAAS; Am Chem Soc. *Res:* Trace analysis of inorganic air pollutants using x-ray fluorescence, ion chromatography or inductively coupled plasma. *Mailing Add:* 2622 W 49 1/2 St Austin TX 78731

ANSELONE, PHILIP MARSHALL, b Tacoma, Wash, Feb 8, 26; m 51; c 1. MATHEMATICAL ANALYSIS. *Educ:* Col Puget Sound, BS, 49, MA, 50; Ore State Univ, PhD(math), 57. *Prof Exp:* Head math anal group, Hanford Atomic Prod Oper, 51-54; res assoc, Johns Hopkins Radiation Lab, 54-58; assoc prof math, Math Res Ctr, US Army, 58-63; chmn dept, 83-87, PROF MATH, ORE STATE UNIV, 64- *Concurrent Pos:* Vis prof, Math Res Ctr, US Army, 66-67 & Mich State Univ, 70-71; Sr Humboldt award, 77-78. *Mem:* Am Math Soc; Math Asn Am; Soc Indust & Appl Math. *Res:* Integral equations; approximation theory. *Mailing Add:* Dept Math Ore State Univ Corvallis OR 97331

ANSEVIN, ALLEN THORNBURG, b Springfield, Ohio, Nov 22, 28; m 61; c 1. BIOPHYSICS. *Educ:* Earlham Col, AB, 51; Univ Pittsburgh, MS, 58, PhD(biophys), 61. *Prof Exp:* Res asst pharmacol, Christ Hosp Int Med Res, 53-56; res assoc biochem, Rockefeller Univ, 61-64; asst physics, 64-70, from asst prof to assoc prof, 65-80, ASSOC PHYSICIST, MD ANDERSON HOSP & TUMOR INST, UNIV TEX, 70-, PROF BIOPHYS, GRAD SCH BIOMED SCI, 80- *Mem:* AAAS; Biophys Soc; Am Chem Soc. *Res:* Virus structure; protein-protein associations; ultracentrifuge methods; nucleoprotein structure; radiation biology; DNA thermal denaturation; sequence-specific protein-DNA complexes. *Mailing Add:* 2127 McArthur Dr Houston TX 77030

ANSHEL, MICHAEL, b New York, NY, Nov 2, 41; m 65; c 2. MATHEMATICS, COMPUTER SCIENCE. *Educ:* Adelphi Univ, BA, 63, MS, 65, PhD(math), 67. *Prof Exp:* Asst prof math, Polytech Inst Brooklyn, 66-67 & Univ Ariz, 67-68; systs analyst, Lambda Corp, 68; from asst prof to assoc prof, 68-78, PROF COMPUT SCI, CITY COL NEW YORK, 79- *Concurrent Pos:* Consult, Princeton-Bethesda, 68 & Mt Sinai Sch Med, 75-80; reviewer, Math Rev, Am Math Soc, 75-; doctoral fac, Grad Ctr, City Univ New York, 73-; fel, Goddard Space Flight Ctr, 82-83; Delphia assoc, 83; AT&T Bell Labs, 86-88. *Mem:* Am Math Soc; Math Asn Am; Ann Symbolic Logic; Sigma Xi; Asn Comput Mach; Inst Elec & Electronics Engrs, Comput Soc; AAAS; Soc Indust Appl Math. *Res:* Group-theoretic decision-problems and their relation to the theory of computation; applications of algebra and logic in the computer sciences and combinatorial mathematics. *Mailing Add:* 1140 Fifth Ave New York NY 10128

ANSHER, SHERRY SINGER, b Washington, DC, Mar 3, 57; m 80; c 1. BIOCHEMISTRY, PHARMACOLOGY. *Educ:* Univ Md, College Park, BS, 78; Johns Hopkins Univ, Md, PhD(biochem & pharmacol), 82. *Prof Exp:* Staff fel, Nat Inst Arthritis, Diabetes & Digestive & Kidney Dis, NIH, 82-85, sr staff fel Biochem, Enzymes & Cellular Metab Sect, 85-87; SR STAFF FEL, LAB BACT TOXINS, FDA, 87- *Mem:* AAAS. *Res:* Investigation and purification of enzymes responsible for detoxation of xenobiotics and carcinogens; mechanism by which compounds inhibiting carcinogenesis in animals through alteration of metabolism may work; role of endogenous methylation inhibitor in gene regulation. *Mailing Add:* 9116 Willow Gate Lane Bethesda MD 20817

ANSON, FRED (COLVIG), b Los Angeles, Calif, Feb 17, 33; m 59; c 2. ELECTROANALYTICAL CHEMISTRY. *Educ:* Calif Inst Technol, BS, 54; Harvard Univ, AM, 55, PhD(chem), 57. *Prof Exp:* From instr to assoc prof, 57-68, PROF CHEM, CALIF INST TECHNOL, 68-, CHMN DIV CHEM & ENG, 84- *Concurrent Pos:* Guggenheim fel, 64; Alfred P Sloan res fel, 65-69; Fulbright scholar, 72; chmn, Gordon Res Conf Electrochem, 70. *Honors & Awards:* David C Grahame Award, Electrochem Soc, Inc, 83; Alexander von Humboldt Award, Fritz-Haber Inst, Berlin, 84; C N Reilley Award, Soc Electroanalytical Chem, 86. *Mem:* Nat Acad Sci; fel AAAS; Am Electrochem Soc; Int Soc Electrochem; Am Chem Soc; Soc Electroanalytical Chem. *Res:* Kinetics of electrode reactions; mechanisms of electrode processes; chemical education. *Mailing Add:* Div Chem & Eng Calif Inst Technol Pasadena CA 91125

ANSPAUGH, BRUCE EDWARD, b Thermopolis, Wyo, Sept 26, 33; m 54. PHYSICS. *Educ:* Nebr Wesleyan Univ, BA, 55; Univ Nebr, MA, 58, PhD(physics), 65. *Prof Exp:* Asst physics, Univ Nebr, 55-64; RES ENGR, JET PROPULSION LAB, CALIF INST TECHNOL, 64- *Mem:* Am Phys Soc. *Res:* Radiation damage; solid state physics. *Mailing Add:* 370 S Allen Ave Pasadena CA 91106

ANSPAUGH, LYNN RICHARD, b Rawlins, Wyo, May 25, 37; m 65; c 2. ENVIRONMENTAL SCIENCES, HEALTH PHYSICS. *Educ:* Nebr Wesleyan Univ, BA, 59; Univ Calif, Berkeley, MBioradiol, 61, PhD(biophys), 63. *Prof Exp:* From biophysicist to sect leader environ sci, 63-82, DIV LEADER, ENVIRON SCI DIV, LAWRENCE LIVERMORE NAT LAB, UNIV CALIF, 82- *Concurrent Pos:* Teacher, Univ Exten, Univ Calif, Berkeley, 66-69; lectr, San Jose State Univ, 75; fac affil, Colo State Univ, 75; sci dir, Nev Test Site Off-Site Radiation Exposure Rev Proj, 79-, Nev Appl Ecol Group, 83-86 & Basic Environ Compliance & Monitoring Proj, 86-; consult, sci adv bd, Environ Protection Agency, 84-85 & Int Atomic Energy Agency, 88-; mem, US deleg to UN Sci Comt Effects of Atomic Radiation, 87- & Nat Coun Radiation Protection, 89- *Mem:* AAAS; fel Health Physics Soc; Soc Risk Anal; Sigma Xi. *Res:* Study of the environmental effects of utilizing geothermal energy resources; experimental study of the resuspension of pollutant aerosols with emphasis on plutonium; calculation of radiation doses from nuclear tests and nuclear reactors. *Mailing Add:* PO Box 2017 Danville CA 94526-7017

ANSPON, HARRY DAVIS, b Washington, DC, Sept 25, 17; m 56; c 2. ORGANIC CHEMISTRY, POLYMER CHEMISTRY. *Educ:* Univ Md, BS, 39, PhD(org chem), 42. *Prof Exp:* Asst, Univ Md, 42-46; res chemist, US Rubber Co, NJ, 46-48; res chemist, Cent Res Labs, Gen Aniline & Film Corp, 48-56, res fel, 56-58; plant chemist, Acetylene Chem Plant, Ky, 58-59; sect leader, Gulf Res & Develop, Res Ctr, Spencer Chem Co, 59-68; mgr prod-process develop, Plastics Dept, USS Chem Div, United States Steel Corp, 68-69, tech dir, 69-76, asst mgr, New Polymer Planning, 76-78; CONSULT PLASTICS & CHEM, 78-, PATENT AGT, 79- *Concurrent Pos:* With Nat Defense Res Coun. *Mem:* Am Chem Soc; Soc Plastics Eng; Plastics & Rubber Inst. *Res:* Hydrogenation and dehydrogenation of pyrethrosin; organic polymers; organic synthesis; plastics; manufacture of plastics. *Mailing Add:* 29 Beaver Sewickley PA 15143-1244

ANSTEY, ROBERT L, b Creston, Iowa, July 15, 21; m; c 2. GEOGRAPHY, CLIMATOLOGY. *Educ:* Univ Nebr, AB, 47, MA, 48; Univ Md, PhD, 57. *Prof Exp:* Geogr, US Bur Census, Washington, DC, 48-49, Off Qm Gen, 49-52 & Natick Labs, US Army, 54-70; from assoc prof to prof geog, Framingham State Col, 70-87; RETIRED. *Concurrent Pos:* Vis lectr, Clark Univ, 57-58 & Mass State Col Framingham, 57-68; consult, Sylvania Elec Prod, Inc, 60-61. *Mem:* Asn Am Geog; Nat Coun Geog Educ. *Res:* Applications of critical data in climatology and military geography to problems of item design, test, issue and storage. *Mailing Add:* 11 Brookdale Rd Natick MA 01760-3144

ANSTEY, THOMAS HERBERT, b Victoria, BC, Dec 27, 17; m 45; c 3. GENETICS, PLANT BREEDING. *Educ:* Univ BC, BSA, 41, MSA, 43; Univ Minn, PhD(plant breeding), 49. *Prof Exp:* Horticulturist, Exp Farm, Agassiz, BC, 46-49, sr horticulturist, 49-53, supt exp sta, Summerland, 53-59, dir res sta, Lethbridge, 59-69, asst dir gen western res br, 69-78, sr adv, Int Res & Develop, Res Br, Cent Exp Farm, Can Dept Agr, Ottawa, 78-82; SR VPRES AGR, INT DEVELOP ASN CAN, INC, 86- *Mem:* Can Soc Hort Sci; fel Agr Inst Can (pres, 70-71); Ont Ins Agr. *Res:* Genetics and breeding for strawberries and broccoli; written and published 100 yr history of research branch, Canada Department Agriculture (1986) 432 pages. *Mailing Add:* 12 Warbonnet Dr Ottawa ON K2E 5M2 Can

ANSUL, GERALD R, b Philadelphia, Pa, July 17, 25; m 58; c 3. ORGANIC CHEMISTRY. *Educ:* Temple Univ, AB, 50; Pa State Univ, PhD(org chem), 54. *Prof Exp:* Res chemist, Marshall Lab, Fabrics & Finishes Dept, 54-58, chem assoc, 58-60, mkt specialist, 60-62, proj supvr, Indust Sales Div, 62-64, tech coord, Wynnewood Automotive & Indust Sales Div, 64-65, Lincolnwood Automotive & Indust Prod, 65-68, mgr insulation sales, Fabrics & Finishes Dept, 68-69, mgr new prod develop & elec sales, Indust Prod Div, 70-71, mgr new prod, 72-76, prod mgr, Corian Bldg Prod, 76-80, prod mgr, Appl Technol Div, Fabrics & Finishes Dept, E I du Pont de Nemours & Co, Inc, 80-85; PRES, BRANDYWINE MKT SERVS, INC, WILMINGTON DEL, 85- *Mem:* AAAS; Am Chem Soc. *Res:* Organosilicon chemistry; urethane chemistry; chemistry of the vinyl dioxolanes; automotive finishes; diallyl phthalate pre-pregs and laminates; wood furniture finishes; adhesives; sealants and paper coatings; polyimide resins; wire enamels; epoxy pre-pregs; rigid and flexible printed circuitry; filled acrylic polymers; industrial hygiene noise and air monitoring instruments. *Mailing Add:* 17 Stonecrop Rd Northminster Wilmington DE 19810

ANTAKLY, TONY, b Nov 9, 51; m 85; c 1. HORMONE ACTION & STEROID PHARMACOLOGY, CELL SPECIFIC GENE REGULATION. *Educ:* Am Univ, Beirut, BSc, 72; Univ Paris, PhD(cell biol), 78; Univ Paris, Sorbonne, DSc, 83. *Prof Exp:* Asst prof anat, McGill Univ, 83-90; ASSOC PROF PATH, UNIV MONTREAL, 91- *Mem:* Endocrine Soc; Am Asn Anatomists; Am Soc Cell Biol; Am Soc Microbiol. *Res:* Molecular mechanisms of steroid hormone action including carcinogenesis, hormone resistance, leukemias; neuroendocrine regulation of pituitary and endocrine cell function. *Mailing Add:* Dept Path Univ Montreal PO Box 6128 Sta A Montreal PQ H3C 3J7 Can

ANTAL, JOHN JOSEPH, b Taylor, Pa, Apr 23, 26; m 55; c 2. NUCLEAR RADIATION PHYSICS, MATERIALS SCIENCE. *Educ:* Univ Scranton, BS, 48; St Louis Univ, MS, 49, PhD(physics), 52. *Prof Exp:* Res physicist, Watertown Arsenal Lab, US Army, 52-53, ord mat res off, 53-62, mat res agency, 62-67, res physicist, Mat & Mech Res Ctr, 67-81, leader, Characterization Res Group, 81-89, RES PHYSICIST, US ARMY MAT TECHNOL LAB, 89- *Concurrent Pos:* Guest assoc physicist, Brookhaven Nat Lab, 53-65; Secy Army res & study fel, 61-62. *Mem:* Am Phys Soc; Am Crystallog Asn; Am Soc Nondestructive Test; AAAS; Mat Res Soc. *Res:* Thermal neutron spectroscopy; neutron and x-ray characterization of materials; digital instrumentation. *Mailing Add:* SLCMT-MRM-S US Army Mat Technol Lab Watertown MA 02172-0001

ANTAL, MICHAEL JERRY, JR, b Monroe, Mich, May 18, 47; m 72; c 2. CARBOHYDRATE THERMOCHEMISTRY. *Educ:* Dartmouth Col, AB, 69; Harvard Univ, MS, 70, PhD(appl math), 73. *Prof Exp:* Res staff, Thermonuclear Weapons Physics Group, Los Alamos Nat Lab, 73-75; asst prof energy sci & appl math, Dept Mech Aerospace Eng, Princeton Univ, 75-81; CORAL INDUST DISTINGUISHED PROF RENEWABLE ENERGY RESOURCES, DEPT MECH ENG, UNIV HAWAII, 82- *Concurrent Pos:* Consult, President's Coun Environ Qual, 77-78, Off Technol Assessment, US Cong, 78-79, Exxon Corp, 79, Solar Eng Group, 79-81, Aerospace Corp, 80, Sao Paulo Elec Ctr, Brazil, 85-; pres, Hawaii Sect, Am Chem Soc, 89. *Mem:* Am Chem Soc; Am Phys Soc; Soc Indust & Appl Math; Am Inst Chem Eng. *Res:* Biomass pyrolysis, emphasizing mechanisms and kinetics, solid and gas phase studies of model compound pyrolysis (carbohydrates, polyols); chemical reactor design; chemical reaction engineering in supercritical fluids. *Mailing Add:* Dept Mech Eng Univ Hawaii 2540 Dole St Honolulu HI 96822

ANTAR, ALI A, b Egypt, Sept 20, 40. ATOMIC & MOLECULAR PHYSICS, OPTICS. *Educ:* Air Shams Univ, BA, 60, Univ Conn, MA, 72, PhD(nuclear physics), 77. *Prof Exp:* Fel, Univ Conn, 77-78, syst prof physics, 78-80; PROF PHYSICS, CENT CONN STATE UNIV, 80- *Mem:* AAAS; Am Asn Physics Teachers; Am Phys Soc; Sigma Xi. *Res:* A research project involving experimental atomic and molecular physics. *Mailing Add:* Phys Dept Cent Conn State Univ New Britain CT 06050

ANTAR, MOHAMED ABDELCHANY, RADIATION BIOLOGY, NUCLEAR CARDIOLOGY. *Educ:* Univ Iowa, MD & PhD(metab radiation biol), 64. *Prof Exp:* PROF MED & CHIEF NUCLEAR MED, CONN VET ADMIN MED CTR, 74- *Mailing Add:* Nuclear Med Dept GB-3176 Cleveland Clin Found One Clinic Ctr 9500 Euclid Ave Cleveland OH 44195

ANTEL, JACK PERRY, b Winnipeg, Man, Oct 25, 45; m 69; c 3. NEUROLOGY, NEUROIMMUNOLOGY. *Educ:* Univ Man, BSc & MD, 69; FRCP(c), 74; Am Bd Psychiat & Neurol, cert neurol, 76. *Prof Exp:* From asst prof to prof neurol, Univ Chicago, 76-86; PROF NEUROL, MCGILL UNIV, 86-, CHMN NEUROL & NEUROSURG, 89- *Concurrent Pos:* Neurologist-in-chief, Montreal Neurol Hosp & Royal Victoria Hosp, 86-; assoc mem microbiol & immunol, McGill Univ, 87- *Mem:* Am Acad Neurol; Am Neurol Asn; Soc Neurosci; Am Asn Immunologists. *Res:* Neuroimmunology; mechanisms whereby immune mediators can induce tissue injury within brain and whereby immune and nervous system constituents can modulate the activity of each other. *Mailing Add:* Montreal Neurol Inst 3801 University St Montreal PQ H3A 2B4 Can

ANTES, HARRY W, b Philadelphia, Pa, June 20, 30; m 53; c 4. FASTENER ENGINEERING, TEACHING. *Educ:* Drexel Inst Tech, BS, 53; Univ Pa, MS, 63; Drexel Univ, PhD, 79. *Prof Exp:* Metallurgist, Frankford Arsenal, 50-68; prof, Drexel Univ, 60-87; dir, res & develop, 79-86, VPRES, SPS TECHNOL, 86- *Concurrent Pos:* Advisor, Adv Tech Ctr Southeast Pa, 80- *Mem:* Am Mgt Asn. *Res:* Mechanical metallurgy; metalworking; materials processing; powder metallurgy; fastener technology; magnetics. *Mailing Add:* SPS Tech Inc SPS Labs Highland Ave Jenkintown PA 19046

ANTHES, JOHN ALLEN, b Janesville, Wis, Aug 18, 13; m 38; c 3. ORGANIC CHEMISTRY, CHEMICAL ENGINEERING. *Educ:* Univ Minn, BChE, 34, PhD(org chem), 39. *Prof Exp:* Asst, Univ Minn, 35-38; res chemist, Union Oil Co, Calif, 39-40; contracting off rep, Ord Dept, US Army, 41-42, asst chief spec mat sect, Manhattan Dist, 43-45; res chemist, Am Cyanamid Co, Conn, 46-48, asst chief chemist, Pa, 48-50, chief chemist, 50-53; proj engr, Dravo Corp, 53-58, asst dir res & develop, 58-65, mgr res, 65-71, sr consult, 71-78; consult engr, 78-80; RETIRED. *Mem:* AAAS; Am Chem Soc; Am Inst Chem Eng; Am Inst Min, Metall & Petrol Eng; NY Acad Sci. *Res:* Ore agglomeration processes; compounds related to perinaphthenone; synthesis of lubricating oil additives and of ion exchange resins; esterification of high boiling alcohols; dehydration of maleic acid; process metallurgy; continuous coking of coal; purification of coke oven gas. *Mailing Add:* Rte 65 Box 279 Arapahoe NC 28510

ANTHES, RICHARD ALLEN, b St Louis, Mo, Mar 9, 44; m 66; c 2. METEOROLOGY. *Educ:* Univ Wis, BS, 66, MS, 67, PhD(meteorol), 70. *Prof Exp:* Res meteorologist, Nat Hurricane Res Lab, Nat Oceanic & Atmospheric Admin, 68-71; from asst prof to prof meteorol, Pa State Univ, 71-81; dir, Atmospheric Anal & Prediction Div, 81-86, DIR, NAT CTR ATMOSPHERIC RES, 86- *Concurrent Pos:* Consult, Atmospheric Sci Lab, White Sands Missile Range, 73-76 & Nat Weather Serv, 73-81; mem storm fury panel, Bd Atmospheric Sci & Climate, Nat Res Coun, 75-81, chair, Panel Mesoscale Res, 84--86, chair, Bd Atmospheric Sci & Climate, Nat Res Coun, 87- *Honors & Awards:* Meisinger Award, Am Meteorol Soc, 80, Charney Award, 87. *Mem:* Am Meteorol Soc; Sigma Xi. *Res:* Numerical modeling of atmospheric phenomena, especially on the mesoscale; hurricane modeling and parameterization of physical processes in models. *Mailing Add:* Nat Ctr Atmospheric Res 1631 Gillespie St Boulder CO 80306

ANTHOLINE, WILLIAM E, b Milwaukee, Wis, July 1, 43; m 65; c 3. PHYSICAL BIOCHEMISTRY. *Educ:* Univ Wis-Madison, BS, 65; Iowa State Univ, PhD(phys chem), 71. *Prof Exp:* Fel phys biochem, Radiation Biol & Biophys Div, Med Col Wis, 72-74; NIH fel & instr phys biochem, Radiation Biol & Biophys Div, Nat Cancer Inst, 74-75, asst prof, 80-86, ASSOC PROF, DEPT RADIOL & MEM, BIOPHYSICS GROUP GRAD PROG, MED COL WIS, 86- *Concurrent Pos:* Fac res specialist, Biochem Sect, Univ Wis-Milwaukee, 75-; lectr, Gordon Res Conf Magnetic Resonance in Biol & Med, 74, 85 & 87. *Res:* Applying physical chemical principles in order to better understand the origin and interaction of paramagnetic metal compounds found in tissues, especially tumor tissue. *Mailing Add:* Nat Biomed ESR Ctr Med Col Wis 8701 W Watertown Plank Rd Milwaukee WI 53226

ANTHONISEN, NICHOLAS R, b Boston, Mass, Oct 12, 33; m 57; c 3. RESPIRATORY PHYSIOLOGY. *Educ:* Dartmouth Col, AB, 55; Harvard Univ, MD, 58; McGill Univ, PhD(exp med), 69. *Prof Exp:* Intern med, NC Mem Hosp, 58-59, jr asst resident, 59-60; sr asst resident, Respiratory Dept, Royal Victoria Hosp, 63-64; demonstr med, McGill Univ, 64-66, asst prof exp med, 69-70, assoc prof exp med, 70-73, prof exp med, 73-75; PROF MED, DEPT MED, UNIV MAN, 75- *Concurrent Pos:* Med Res Coun Can scholar, 69-71. *Mem:* Can Soc Clin Invest; Am Physiol Soc; Can Thoracic Soc; Am Soc Clin Invest; Am Thoracic Soc. *Res:* Chest disease; pulmonary physiology; physiologic aspects of respiratory disease. *Mailing Add:* Med Univ Manitoba 753 McDermot Ave Winnipeg MB R3E 0W3 Can

ANTHONY, ADAM, b Buffalo, NY, Oct 19, 23; div; c 2. PHYSIOLOGY. *Educ:* Univ Buffalo, BA, 43; Marquette Univ, MS, 48; Univ Chicago, PhD(zool), 52. *Prof Exp:* Lab asst comp anat & bact, Univ Buffalo, 43; asst gen zool, parasitol & histol, Marquette Univ, 46-48 & med parasitol, biol sci, endocrinol & vert embryol, Univ Chicago, 49-52; from asst prof to assoc prof zool, 52-61, chmn comt physiol, Col Sci, 69-73, PROF ZOOL, PA STATE UNIV, UNIVERSITY PARK, 61- *Concurrent Pos:* Darbaker Award, 66,68 & 69. *Mem:* Am Physiol Soc; Soc Exp Biol & Med; AAAS. *Res:* Cytochemical aspects of toxication; vascular and neural cytochemistry; cytodiagnosis of neoplasms; scanning microspectrophotometry and interferometry. *Mailing Add:* 418 Mueller Bldg Pa State Univ University Park PA 16802

ANTHONY, DONALD BARRETT, b Kansas City, Kans, Jan 28, 48; m 72; c 2. RESEARCH ADMINISTRATION & TECHNICAL MANAGEMENT. *Educ:* Univ Toledo, BS, 70; Mass Inst Technol, SM, 71, DSc, 74. *Hon Degrees:* DSc, Mass Inst Technol, 74. *Prof Exp:* Asst prof chem eng, Mass Inst Technol, 74-75; supvr coal res, Standard Oil Prod Co, 76-78, mgr, Marine Opers Anal & Control, 78-80, Synfuels Develop, 80-82, vpres & gen mgr, Pfander, 82-85, vpres res develop, 85-87, vpres technol serv & res, Standard Oil Prod Co, 88-90; VPRES TECHNOL, BECHTEL, 90- *Concurrent Pos:* Dir, Sch Chem Eng Pract, Mass Inst Technol, 74-75; vpres, Res & Develop, BPAM, 87-88. *Mem:* Am Chem Soc; Am Inst Chem Engrs; Sigma Xi. *Res:* Synthetic fuels research; systems analysis. *Mailing Add:* 2015 Seven Maples Dr Kingwood TX 77345-1727

ANTHONY, DONALD JOSEPH, b Troy, NY, May 30, 22; m 45; c 5. FISSION & FUSION, ENERGY SYSTEMS. *Educ:* Siena Col, BS, 47; Univ Notre Dame, PhD(physics), 53. *Prof Exp:* Instr chem & math, Siena Col, Univ Notre Dame, 46-47 & physics, 47-52; res assoc, Knolls Atomic Power Lab, 52-56, mgr exp physics, 56-59, reactor physics, 59-68 & operating nuclear plants, 68-74; mem staff, Energy Systs & Technol Div, Gen Elec Co, 74-76, mgr power technol, 76-77, mgr advan energy systs, Energy Systs Progs Dept, 77-80, mgr GE Advan Energy Systs, 80-84, mgr servs mkt develop, GE Nuclear Energy, 84-87; MKT DEVELOP & MGT CONSULT, 87- *Concurrent Pos:* Indust consult, Dept Energy Fusion Power Coord Com, 77-; mem, Oak Ridge Nat Lab Fusion Adv Com, 78-81; mem Univ Chicago Rev Comt Components Technol Div, Argonne Nat Lab, 83-86, Rochester Inst Technol, Dept Physics Planning Coun, 83, Atomic Indust Forum Fusion Comt, 82-84. *Mem:* Am Nuclear Soc; Sigma Xi; Fusion Power Asn. *Res:* Reactor, nuclear physics and energy systems. *Mailing Add:* Three Schuyler Rd Loudonville NY 12211

ANTHONY, ELIZABETH YOUNGBLOOD, b Springfield, Ill, Jan 25, 53; m 79. PETROLOGY, GEOCHEMISTRY. *Educ:* Carleton Col, BA, 75; Univ Ariz, MS, 79, PhD(geol), 86. *Prof Exp:* RES ASSOC GEOL, UNIV ARIZ, 77- *Mem:* Geol Soc Am; Mineral Soc Am; Geochem Soc. *Res:* Igneous petrology and geochemistry of mineralized granites. *Mailing Add:* Dept Geol Sci Univ Tex El Paso TX 79968

ANTHONY, HARRY D, b Fredonia, Kans, Apr 17, 21; m 55; c 1. VETERINARY MEDICINE. *Educ:* Kans State Univ, DVM, 52, MS, 57. *Prof Exp:* Pvt pract, Ill, 52-55; from instr to assoc prof vet med, 55-67, DIR & PROF, KANS VET DIAG LAB, KANS STATE UNIV, 68- *Mem:* Am Asn Vet Lab Diag (pres, 75-). *Res:* Blood parasites of domestic animals; baby pig diseases; respiratory diseases of ruminants; infectious keratitis of the bovine. *Mailing Add:* Dept Vet Med Kans State Univ Manhattan KS 66506

ANTHONY, JOHN WILLIAMS, b Brockton, Mass, Nov 25, 20; m 79; c 3. GEOLOGY, MINERALOGY. *Educ:* Univ Ariz, BS, 46, MS, 51; Harvard Univ, PhD(geol), 65. *Prof Exp:* Mineralogist, Ariz Bur Mines, 46-51; from asst prof to assoc prof, 51-64, actg head dept, 66-67, PROF GEOL, UNIV ARIZ, 64- *Mem:* Fel Geol Soc Am; fel Mineral Soc Am; Soc Econ Geol; Am Crystallog Asn. *Res:* Mineralogy and economic geology of the Southwest; experimental and descriptive mineralogy, crystallography, economic geology. *Mailing Add:* Dept Geol Sci Univ Tex El Paso TX 79968

ANTHONY, LEE SAUNDERS, b Roanoke, Va, Sept 11, 32; m 53; c 2. PHYSICS. *Educ:* Roanoke Col, BS, 53; Va Polytech Inst, MS, 58, PhD(physics), 62; Am Bd Health Physics, cert, 80; Am Bd Radiol, cert, 82. *Prof Exp:* Asst prof, Roanoke Col, 62-65, actg dean, 63-65, chmn, Dept Physics, 62-77, prof physics, 65-79; CONSULT, 79- *Concurrent Pos:* Consult physicist, Roanoke Mem Hosp, 64-80, Physics Asn, 63- *Mem:* Am Phys Soc; Health Physics Soc; Sigma Xi; Am Col Radiol. *Res:* Nuclear, reactor and radiological physics. *Mailing Add:* Phys Assoc 5346 Peters Creek Rd NW Roanoke VA 24019

ANTHONY, LINDA J, b Boston, Mass, Feb 1, 51; m 82; c 2. CHROMATOGRAPHY, INSTRUMENT ANALYSIS & SENSORS. *Educ:* Mt Holyoke Col, AB, 73; Mass Inst Technol, SM, 76, PhD(anal chem), 80. *Prof Exp:* MEM TECH STAFF, ANALYTICAL CHEM RES DEPT, AT&T BELL LABS, 80- *Mem:* Am Chem Soc; Soc Appl Spectros; Sigma Xi. *Res:* Chromatography, especially capillary column technology; use of optical-fiber technology for fabrication of novel capillary columns, trace analysis; miniaturization and sensors; process analytical chemistry. *Mailing Add:* AT&T Bell Labs Rm 1C-259 600 Mountain Ave PO Box 636 Murray Hill NJ 07974-0636

ANTHONY, MARGERY STUART, b New York, NY, Feb 23, 24. BOTANY, RADIATION ECOLOGY. *Educ:* Univ Mich, BS, 45, MS, 46, PhD(bot), 50. *Prof Exp:* Asst bot, Univ Mich, 44-49; from asst prof to assoc prof biol & bot, 49-62, prof biol, Calif State Univ, Chico, 62-83; RETIRED. *Concurrent Pos:* Lectr, NSF Inst, Univ Wyo, 58-62. *Mem:* Ecol Soc Am. *Mailing Add:* PO Box 1469 Gualala CA 95445

ANTHONY, PHILIP JOHN, b New Bremen, Ohio, May 6, 52; m 74; c 2. SEMICONDUCTOR MATERIALS TECHNOLOGY. *Educ:* Univ Dayton, BS, 74; Univ Ill, MS, 75, PhD(physics), 78. *Prof Exp:* mem staff, 78-82, supvr, 82-87, DEPT HEAD, AT&T BELL LABS, 87- *Mem:* Am Phys Soc; Inst Elec & Electronics Engrs; Sigma Xi. *Res:* Low temperature properties of glasses and crystals; electrical and optical properties of semiconductor lasers; photonic switching. *Mailing Add:* 267 White Oak Ridge Rd Bridgewater NJ 08807

ANTHONY, RAYFORD GAINES, b Abilene, Tex, Dec 26, 35; m 59; c 2. CHEMICAL ENGINEERING. *Educ:* Tex A&M Univ, BS, 58, MS, 62; Univ Tex, PhD(chem eng), 66. *Prof Exp:* From asst prof to assoc prof, 66-74, PROF CHEM ENG, TEX A&M UNIV, 74- *Mem:* Am Chem Soc; Am Inst Chem Engrs. *Res:* Methods of calculation for multicomponent distillation; catalytic and reaction engineering; liquefaction of coal. *Mailing Add:* Dept Chem Eng Tex A&M Univ College Station TX 77843

ANTHONY, RONALD LEWIS, b Ft Edward, NY, Sept 7, 38; m 66; c 1. PARASITOLOGY, TROPICAL DISEASES. *Educ:* Susquenna Univ, BA, 61; Univ Kans, PhD(zool), 65. *Prof Exp:* Res assoc int health, 65-66, asst prof, 66-76, ASSOC PROF PATH, SCH MED, UNIV MD, 76-; SR SCIENTIST TROP MED, USNAMRU-2, JAKARTA, IPA AGREEMENT, 90- *Concurrent Pos:* Dir clin immunol lab, Univ Md Med Systs, 71-89; vis scientist, Gorgas Mem Lab, Panama, 77-78; consult, Daho-Lapaz Bolivia, 77; sci adv bd, Gorgas Mem Lab, 86-90. *Mem:* Am Asn Immunologists; Am Soc Microbiol; Am Soc Parasitol; Am Soc Trop Med & Hygiene; Nat Coun Int Health. *Res:* Parasitology; tropical medicine; immunology; naturally acquired immunity in parasitic diseases; diagnostics; antigens; cellular response; field studies; malaria; filariasis; leishmaniasis; trypanosomiasis. *Mailing Add:* USNAMRU-2 Jakarta APO San Francisco CA 96356-5000

ANTHONY, THOMAS RICHARD, b Pittsburgh, Pa, June 27, 41; m 66; c 2. MATERIALS SCIENCE ENGINEERING, INORGANIC CHEMISTRY. *Educ:* Univ Fla, BS, 62; Harvard Univ, MS, 64, PhD(appl physics), 67. *Prof Exp:* Staff metallurgist, Mellon Inst, 62-63; STAFF PHYSICIST, GEN ELEC RES CTR, NY, 67- *Concurrent Pos:* Coolidge fel, Gen Elec, 78. *Honors & Awards:* I R 100 Award, Indust Res Mag, 77 & Nat Acad Eng, 90. *Mem:* Am Soc Metals; Am Inst Mining, Metall & Petrol Engrs; Mat Res Soc. *Res:* Solid state and gaseous diffusion; grain boundaries; segregation in solids; ultracentrifuges; diamond synthesis; artificial intelligence; chemical vapor deposition; laser drilling; diamond crystals are fabricated by a CVD process utilizing hydrogen and hydrocarbons. *Mailing Add:* Gen Elec Res Ctr PO Box 8 Schenectady NY 12308

ANTHONY, W BRADY, b McLinna Co, Tex, Nov 2, 16. RUMINANT NUTRITION. *Educ:* Cornell Univ, PhD(nutrit), 52. *Prof Exp:* Emer prof animal nutrit, Auburn Univ, 53-80. *Concurrent Pos:* Consult nutrit, 80- *Mailing Add:* Rt 1 Box 409 Riesel TX 76682

ANTHONY-TWAROG, BARBARA JEAN, b 1953; m 77; c 2. ASTRONOMY. *Educ:* Univ Notre Dame, BS, 75; Yale Univ, MS, 77, PhD(astron), 81. *Prof Exp:* Instr astron, Univ Tex, Austin, 80-82; asst prof, 82-88, ASSOC PROF ASTRON, UNIV KANS, 88- *Mem:* Am Astron Soc; Int Astron Union; Sigma Xi. *Res:* Stellar photometry in open and globular clusters; white dwarf progenitors; digital image processing. *Mailing Add:* Dept Physics & Astron Univ Kans Lawrence KS 66045-2151

ANTIA, NAVAL JAMSHEDJI, organic chemistry, microbiology, for more information see previous edition

ANTIPA, GREGORY ALEXIS, b San Francisco, Calif, Aug 9, 41; c 2. PROTOZOOLOGY, CELL BIOLOGY. *Educ:* Univ Calif, Berkeley, AB, 63; San Francisco State Univ, MA, 66; Univ Ill, Urbana-Champaign, PhD(zool), 70. *Prof Exp:* NIH fel cell biol, Univ Chicago, 70-71; AEC fel, Argonne Nat Lab, 71-74; asst prof biol, Wayne State Univ, 74-78; from asst prof to assoc prof 78-86, PROF BIOL, SAN FRANCISCO STATE UNIV, 86- *Concurrent Pos:* NSF grant, San Francisco State Univ, 78- *Mem:* Am Micros Soc; Am Soc Cell Biol; Am Soc Zoologists; Electron Micros Soc Am; Soc Protozoologists. *Res:* Temporal events and control of eukaryotic and circadian cell cycles; fine structure; development and phylogeny of ciliates; ciliated protozoa as indicators of water quality and role in activated sludge process. *Mailing Add:* Dept Biol San Francisco State Univ 1600 Holloway Ave San Francisco CA 94132

ANTKIW, STEPHEN, b Passaic, NJ, Sept 27, 22. NUCLEAR PHYSICS. *Educ:* Marietta Col, BS, 47; Univ Del, MS, 53. *Prof Exp:* Physicist x-rays, Atomic & Nuclear Physics Div, Nat Bur Standards, 47-49; physicist nuclear, 49-50, physicist mass spectrometry, 51-53; res physicist nuclear & mass spectrometry, Nucleonics Group, US Geol Surv, 53-54; res physicist, 56-61, sr res physicist nuclear, Schlumberger-Doll Res Ctr, 61-86; CONSULT, 89- *Mem:* Am Phys Soc; Sigma Xi. *Res:* Experimental nuclear physics; gamma ray spectroscopy; nuclear radiation detectors; scintillation counters; neutron induced reactions; neutron transport; coincidence techniques; x-ray shielding and dosimetry; mass spectrometry; nuclear well logging. *Mailing Add:* 104 Blackman Rd Ridgefield CT 06877-4205

ANTLE, CHARLES EDWARD, b East View, Ky, Nov 11, 30; m 53; c 3. MATHEMATICAL STATISTICS, BIOSTATISTICS. *Educ:* Eastern Ky State Col, BS, 54, MA, 55; Okla State Univ, PhD(math), 62. *Prof Exp:* Aerophysics engr, Gen Dynamics/Convair, 55-57; from instr to asst prof math, Univ Mo, Rolla, 57-60; assoc prof, 62-64, PROF STATIST, PA STATE

UNIV, 64- *Concurrent Pos:* Adj prof, Sch Pub Health, Univ Pittsburgh. *Mem:* Am Statist Asn; Inst Math Statist; Royal Statist Soc; Biomet Soc. *Res:* Biostatistics; reliability; Bayesian decision rules; the Weibull Model; stopping rules for clinical trials. *Mailing Add:* Dept Statist Pa State Univ University Park PA 16802

ANTLER, MORTON, b New York, NY, Apr 27, 28; m 50; c 3. TRIBOLOGY, ELECTRICAL CONTACTS. *Educ:* NY Univ, BA, 48; Cornell Univ, PhD(inorg chem), 53. *Prof Exp:* Res chemist, Ethyl Corp, 53-58; supvr phys chem, Borg-Warner Res Ctr, 58-59; adv chemist, Eng Lab, Int Bus Mach Corp, 59-63; dep dir res, Burndy Corp, 63-70; distinguished mem tech staff, Bell Labs, 70-89; PRES & PRIN CONSULT, CONTACT CONSULTS, INC, 89- *Concurrent Pos:* Mem, steering comt, Holm Conf Elec Contacts, Inst Elec & Electronics Engrs; past chmn & USA rep, Adv Group, Int Conf Elec Contacts, 76-; mem, Tech Comt Elec Contacts, Inst Elec & Electronic Engrs; Sci Achievement & Precious Metal Plating Awards, Am Electroplaters & Surface Finishers Soc, 68, 71 & 87; instr, Courses on Elect Contacts & Connectors, Inst Elec & Electronics Engrs & Int Inst Connector & Interconnection Technol; Soc, Ralph Armington & Ragmar Holm Sci Achievement Awards, Inst Elec & Electronics Engrs Chmt Soc, 80, 89 & 90. *Honors & Awards:* Capt Alfred E Hunt Mem Award, Soc Tribologists & Lubrication Engrs, 71; Spec Recognition Award, Int Inst Connector & Interconnection Technol, Inc, 75; Henry J Albert Palladium Medal, Int Precious Metals Inst, 89. *Mem:* Am Chem Soc; Inst Elec & Electronics Engrs; Am Electroplaters & Surface Finishers Soc; Am Soc Testing & Mat; Sigma Xi. *Res:* Inorganic and surface chemistry; friction, wear and lubrication; electric contacts. *Mailing Add:* 821 Strawberry Hill Rd E Columbus OH 43213

ANTMAN, STUART S, b Brooklyn, NY, June 2, 39; m 68; c 2. APPLIED MATHEMATICS, SOLID MECHANICS. *Educ:* Rensselaer Polytech Inst, BS, 61; Univ Minn, MS, 63, PhD(mech), 65. *Prof Exp:* Vis mem, Courant Inst Math Sci, NY Univ, 65-67, from asst prof to assoc prof math, 67-72; PROF MATH, UNIV MD, COLLEGE PARK, 72- *Concurrent Pos:* Sci Res Coun Brit sr vis fel, Oxford Univ, 69-70; co-ed, Springer Tracts in Natural Philos, 72-80; Soc Indust & Appl Math lectr, 73-75; vis prof, Univ de Paris-Sud, 75, Heriot-Watt Univ, Edinburgh, 71, Ecole Polytech, Palaiseau, France, 79, Univ Aut Mex, 81, Univ P&M Curie, Paris, 83, Brown Univ, 78-79, Math Sci Res Inst, Berkeley, 83, Math Res Ctr, Univ Mis, 84, Inst Math Appln, Univ Minn, 85, Univ Bonn, 87; Guggenheim fel, 78-79; consult, Div Appl Math & Comput Sci, NSF, 80-81; mem, US Nat Comt Theoret & Appl Mech, 80-88; prin investr, NSF grants, 72-; assoc ed, Notices of Am Math Soc, 85-87; ed-in-chief, Arch Rational Mech & Anal, 90- *Honors & Awards:* Lester R Ford Award, Math Asn Am, 87. *Mem:* Am Math Soc; Soc Natural Philos; Soc Indust & Appl Math; Math Asn Am; Soc Interaction Mech & Math. *Res:* Nonlinear equations of mechanics; bifurcation and stability theory; theories of rods and shells; nonlinear differential equations. *Mailing Add:* Dept Math Univ Md College Park MD 20742

ANTOGNINI, JOE, plant science, for more information see previous edition

ANTOINE, JOHN EUGENE, b Sheboygan, Wis, May 28, 37. RADIATION ONCOLOGY. *Educ:* Beloit Col, BS, 59; Univ Chicago, MD, 63. *Prof Exp:* Dir, Radiation Res Prog, Nat Cancer Inst, NIH, 85-91; PROF RADIATION MED, LOMA LINDA MED CTR, 91- *Mem:* Fel Am Col Radiol; Am Soc Therapeut Radiation & Oncol; Radiol Soc NAm; Am Soc Clin Oncol; Asn Am Med Cols. *Res:* Radiation oncology related to photon beam therapy. *Mailing Add:* Dept Radiation Med Loma Linda Med Ctr 11234 Anderson St Loma Linda CA 92354

ANTOLAK, ARLYN JOE, b Ladysmith, Wis, Nov 11, 53; m 86. COMPUTATIONAL PHYSICS, NUMERICAL METHODS. *Educ:* Univ Wis, BS, 76; Northwestern Univ, MS, 80 & PhD(appl math), 83. *Prof Exp:* MEM TECH STAFF, SANDIA NAT LAB, 83- *Res:* Charged particle transport theory; numerical modeling; law energy particle physics. *Mailing Add:* 515 Amberina Dr Patterson CA 95363

ANTOLOVICH, STEPHEN D, b Milwaukee, Wis, Dec 29, 39; m 61; c 3. MATERIALS SCIENCE, ENGINEERING. *Educ:* Univ Wis, Madison, BS, 62, MS, 63; Univ Calif, Berkeley, PhD(mat sci), 66. *Prof Exp:* Lectr, Univ Calif, Berkeley, 66-67; res metallurgist, Lawrence Radiation Lab, Calif, 67-68; from assoc prof to prof mat sci, Univ Cincinnati, 68-83; PROF & HEAD DEPT METALL & DIR, FRACTURE & FATIGUE RES LAB, GA INST TECHNOL, 83- *Concurrent Pos:* Grants, Air Force Off Sci Res, 74-, Nat Aeronaut & Space Admin, 76- & NATO, 78-; pres, Metall Res Consult, Inc. *Mem:* Am Soc Metals; Am Soc Mech Engr; Am Soc Testing & Mat; assoc Inst Mining Engrs. *Res:* Fatigue and fracture of metallic materials; mathematical models of fatigue crack propagation and low cycle fatigue; electron microscopy; high temperature low cycle fatigue; composite materials. *Mailing Add:* Dept Mat Sci Ga Inst Technol Atlanta GA 30332-0100

ANTON, AARON HAROLD, b Hartford, Conn, Sept 4, 21; m 47; c 2. PHARMACOLOGY. *Educ:* Univ Conn, BS, 44; Trinity Col, Conn, BS, 52; Yale Univ, PhD(pharmacol), 56. *Prof Exp:* Instr pharmacol, Col Med, Univ Fla, 56-59, asst prof psychiat, 59-65, from assoc prof to prof anesthesiol & pharmacol, Div Anesthesiol, 65-69; PROF ANESTHESIOL & PHARMACOL, SCH MED, CASE WESTERN RESERVE UNIV, 69- *Mem:* Am Soc Pharmacol; Sigma Xi; AAAS; NY Acad Sci; Am Soc Anesthesiologists. *Res:* Biochemical changes in hypertension and mental disease; pharmacological significance of protein binding. *Mailing Add:* Dept Anesthesiol Sch Med Case Western Reserve Univ Cleveland OH 44106

ANTON, DAVID L, b Seattle, Wash, Mar 20, 53; m 75; c 2. ENZYMOLOGY, BIOCATALYSIS. *Educ:* Univ Calif, Berkeley, AB, 75; Univ Minn, PhD(bio chem), 80. *Prof Exp:* Fel biochem, Brandeis Univ, 80-82; res assoc, Harvard Med Sch, 82-83; prin investr enzymol, 83-88, PROJ LEADER, E I DU PONT DE NEMOURS, 88- *Concurrent Pos:* Technol leader, biocatalysis, E

I Du Pont de Nemours, 88. *Mem:* Am Soc Biochem & Molecular Biol; Am Chem Soc; AAAS. *Res:* Design and synthesis of novel enzyme substrate analogs and inhibitors as mechanistic probes and potential pharmaceutical or argicultural chemicals; enzyme mechanisms; use of enzymes as commercial catalysts. *Mailing Add:* Cent Res Exp Sta 328/248 E I Du Pont de Nemours PO Box 80328 Wilmington DE 19880-0328

ANTON, HOWARD, b Philadelphia, Pa, July 27, 39; m 65; c 3. MATHEMATICS. *Educ:* Lehigh Univ, BA, 60; Univ Ill, MA, 63; Polytech Inst Brooklyn, PhD(math), 68. *Prof Exp:* Mathematician, Burroughs Corp, 60-61; lectr math, Hunter Col, City Univ New York, 64-66; assoc prof math, Drexel Univ, 68-83; PRES, ANTON TEXTBOOKS, INC, 83- *Mem:* Am Math Soc; Sigma Xi; Math Asn Am. *Res:* Functional analysis; Banach algebras; distribution theory; approximation theory; linear algebra; calculus; finite mathematics; computer software for linear algebra. *Mailing Add:* 304 Fries Lane Cherry Hill NJ 08003

ANTONACCIO, MICHAEL JOHN, b Yonkers, NY, Mar 6, 43; m 78; c 1. PHARMACOLOGY. *Educ:* Duquesne Univ, BS, 66; Univ Mich, PhD(pharmacol), 70. *Prof Exp:* From sr scientist I, Geigy Pharmaceut to sr scientist II, 70-73, sr staff scientist, 73-75, mgr CV pharmacol, 75-77; dir pharmacol, Squibb Inst Med Res, 77-81; vpres, new drug discovery, Schering-Plough Corp, 81-83; VPRES, CV RES & DEVELOP, BRISTOL-MYERS SQUIBB CO, 83- *Honors & Awards:* Goldblatt Award, 81. *Mem:* Am Soc Exp Ther & Pharmacol; Soc Neurosci; Int Soc Hypertension. *Res:* Central regulation of blood pressure; resin-angiotensin system in the control of blood pressure. *Mailing Add:* Bristol-Myers Squibb Co PO Box 4000 Princeton NJ 08543-4000

ANTONELLI, PETER LOUIS, b Syracuse, NY, Mar 5, 41; div; c 2. TOPOLOGY, GEOMETRY. *Educ:* Syracuse Univ, BS, 63, MA, 65, PhD(math), 67. *Prof Exp:* Asst prof math, Univ Tenn, Knoxville, 67-68; fel, Inst Advan Study, Princeton, 68-70; assoc prof, 70-79, PROF MATH, UNIV ALTA, 79- *Concurrent Pos:* Fel biol, Univ Sussex, Eng, 72-73. *Mem:* Am Math Soc; Can Math Cong; Can Appl Math Soc. *Res:* Mathematical genetics and ecology; geometric diffusion and theoretical developmental biology and non-linear growth mechanics. *Mailing Add:* Dept Math Univ Alta Edmonton AB T6G 2M7 Can

ANTONIADES, HARRY NICHOLAS, b Thessaloniki, Greece, Mar 12, 23; US citizen; m 53; c 2. BIOCHEMISTRY. *Educ:* Nat Univ Athens, BS, 48, PhD(chem), 52. *Prof Exp:* Res assoc biochem, Evangelismos Med Ctr, Athens, Greece, 48-53; vis investr, Univ Lab Phys Chem, Harvard Univ, 53-54; from res asst to res assoc, Ctr Blood Res, 54-56, assoc investr, 56-61; res assoc biol chem, Dept Gynec, Med Sch, 56-63 & Dept Med, 63-66, asst prof biochem, Sch Pub Health, 65-70, ASSOC PROF BIOCHEM, SCH PUB HEALTH, HARVARD UNIV, 70-; SR INVESTR, CTR BLOOD RES, 61- *Concurrent Pos:* Vis assoc prof, Sch Med, Univ Southern Calif, 60; assoc staff med, Peter Bent Brigham Hosp, 61-65; lectr, NATO Advan Study Inst, Stratford-upon-Avon, Eng, 62; vis prof, Med Ctr, Univ Ala, 63; lectr, Cong Int Diabetes Fedn, Toronto, 64, Stockholm, 67 & Buenos Aires, 70; lectr, Cong Int Endocrine Soc, London, 64; vis prof, Inst Physiol, Med Sch, Univ Buenos Aires, 66; spec lectr, 41st Japan Endocrinol Soc, Kyoto, 68; vis prof, Ain Shams Univ Med Sch, Cairo & Alexandria Univ Med Sch, Alexandria, Egypt, 71; assoc biol chem, Harvard Univ, 73- *Honors & Awards:* Eli Lilly Award, Am Diabetes Asn, 62. *Mem:* AAAS; Am Diabetes Asn; Am Soc Biol Chemists; fel NY Acad Sci; Int Soc Thrombosis & Haemostasis. *Res:* Mechanisms of hormone transport and regulation; growth factors and regulation of cell growth; interaction of proteins. *Mailing Add:* Ctr Blood Res & Dept Nutrit Harvard Univ Sch Pub Health 800 Huntington Ave Boston MA 02115

ANTONIAK, CHARLES EDWARD, b Norfolk, Va, Apr 11, 40; m 64; c 2. STATISTICAL ANALYSIS, ACOUSTO-OPTIC ELECTRONICS. *Educ:* Calif Inst Technol, BS, 60; Calif State Univ, San Diego, MS, 63; Univ Calif, Los Angeles, PhD(math), 69. *Prof Exp:* Res physicist, US Navy Electronics Lab, San Diego, 60-61; opers analyst, 63-65, res math, 65-70; asst prof statist, Univ Calif, Berkeley, 70-76; mem Tech Staff, Bell Tel Labs, 76-83; PRIN STAFF ENGR, LOCKHEED-ONT, 83- *Concurrent Pos:* Off Naval Res grant, Univ Calif, Berkeley, 71-77; consult, Lawrence-Berkeley Lab, 74-75; expert witness, Alameda County, Fed Appeals Court, 73. *Mem:* Am Math Soc; Math Asn Am; Inst Elec & Electronics Eng; Inst Math Statist; Am Statist Asn; fel Inst Adv Eng. *Res:* Bayesian, distribution free and sequential applied statistics especially signal detection and data compression; RF spectral analysis using acousto-optic technology. *Mailing Add:* Dept 1-330 Chino Lockheed-Ont PO Box 33 Ontario CA 91761-0033

ANTONIEWICZ, PETER R, b Tarnow, Poland, Feb 5, 36; US citizen; m 61; c 3. THEORETICAL SOLID STATE PHYSICS, SURFACE PHYSICS. *Educ:* NC State Univ, BS, 59; Purdue Univ, MS, 64, PhD, 65. *Prof Exp:* Asst prof, 65-70, ASSOC PROF PHYSICS, UNIV TEX, AUSTIN, 70- *Mem:* AAAS; Am Phys Soc; Am Asn Physics Teachers. *Mailing Add:* Dept Physics Univ Tex Austin TX 78712

ANTONIOU, A(NDREAS), b Cyprus, Mar 3, 38; Can Citizenship; m 64; c 4. ELECTRICAL ENGINEERING. *Educ:* Univ London, BSc, 63, PhD, 66. *Prof Exp:* Mem sci staff, GEC Ltd, Eng, 66; sr sci officer, Post Off Res Dept, Eng, 66-69; mem sci staff, Northern Elec Co, 69-70; from asst prof to prof elec eng, Concordia Univ, 70-83 & chmn dept, 77-83; chmn dept, 83-90, PROF ELEC & COMPUT ENG, UNIV VICTORIA, 83- *Concurrent Pos:* Assoc ed, 83-85, ed, 85-87, Inst Elec & Electronic Engrs Transactions on Circuits & Systs; counr, Asn Prof Engrs of BC, 88-90. *Mem:* Fel Inst Elec Engrs, UK; fel Inst Elec & Electronic Engrs. *Res:* Passive and active networks; active and digital filters; digital signal processing. *Mailing Add:* Dept Elec & Comput Eng Univ Victoria PO Box 3055 Victoria BC V8W 3P6 Can

ANTONIOU, ANTONIOS A, b Thessaloniki, Greece, Aug 9, 23; Can citizen. PHYSICAL CHEMISTRY. *Educ:* Univ Thessaloniki, BSc, 53; Univ Toronto, PhD(phys chem), 58. *Prof Exp:* Res officer capillarity, Div Bldg Res, 59-64, RES OFFICER GAS-SOLID INTERACTION, DIV CHEM, NAT RES COUN CAN, 64- *Mem:* Am Phys Soc; Am Chem Soc; Can Inst Chem. *Res:* Adsorption; capillarity; gas-solid interaction. *Mailing Add:* 44 Pavlou Mela St Thessaloniki TT546-22 Greece

ANTONOFF, MARVIN M, b New York, NY, Nov 29, 30; m 53; c 3. THEORETICAL PHYSICS, SOLID STATE PHYSICS. *Educ:* NY Univ, BS, 52, MA, 53; Cornell Univ, PhD(solid state physics), 62. *Prof Exp:* Staff mem, Gen Elec Co, NY, 55-59; asst physics, Cornell Univ, 59-61; staff mem, Lincoln Lab, Mass Inst Technol, 61-62 & Sperry Rand Res Ctr, 62-65; from asst prof to assoc prof, 65-78, PROF PHYSICS, UNIV MASS, BOSTON, 78-CHMN DEPT, 68- *Concurrent Pos:* Guest scientist, Nat Magnet Lab, Mass Inst Technol, 70-; fel, UK Atomic Energy Res Establishment, Harwell, 71-72; consult, Kennecott Copper Corp, 74-75; vis prof, Tufts Univ, 80- *Mem:* Am Phys Soc; Inst Elec & Electronics Eng. *Res:* Theory of cooperative phenomena in solid state physics, especially ferromagnetism and superconductivity; theory of semiconductor devices; computer-based data acquisition and laboratory automation systems. *Mailing Add:* Dept Physics Univ Mass Harbor Campus Boston MA 02125

ANTONSEN, DONALD HANS, b Weehawken, NJ, Aug 11, 30; m 53; c 2. INDUSTRIAL CHEMISTRY. *Educ:* Davis & Elkins Col, BS, 52; Univ Del, PhD(org chem), 60. *Prof Exp:* Anal chemist, Am Cyanamid Co, 52-54; org chemist, Esso Res & Eng Co, 59 & Sun Oil Co, 60-64; res supvr indust chem, 64-69, appln mgr nickel chem, 69-72, NICKEL CHEM INDUST SALES MGR, INT NICKEL CO, INC, 72- *Mem:* Am Chem Soc; Commercial Develop Asn; Am Inst Chem Eng; NY Acad Sci; Royal Soc Chem. *Res:* Nickel chemicals, salts and catalysts; nickel oxide properties; organometallic chemistry; polymerization; synthetic fibers and films; catalysis; rheology of hydrocarbons; clay mineralogy; precious metals; electrochemistry; pesticides, crude oil production; petroleum products. *Mailing Add:* Int Nickel Co Inc Park 80 W Plaza Two Saddle Brook NJ 07662

ANTONSSON, ERIK KARL, b Milwaukee, Wis, Aug 7, 54; m 85; c 2. COMPUTER-AIDED ENGINEERING DESIGN, COMPUTER VISION. *Educ:* Cornell Univ, BS, 76; Mass Inst Technol, MS, 78, PhD(mech eng), 82. *Prof Exp:* Asst prof mech eng, Univ Utah, 83; asst prof orthop biomech, Harvard Univ, 84; asst prof, 84-90, ASSOC PROF MECH ENG, CALIF INST TECHNOL, 90- *Concurrent Pos:* NSF presidential young investr, 85; assoc ed, Am Soc Mech Engrs J Mech Design, 89- *Mem:* Am Soc Mech Engrs. *Res:* Engineering design; computer aided design and computer aided engineering; representing and manipulating imprecision in preliminary engineering design; design of micro-electrical-mechanical systems; spatial computer vision and optical surface geometry acquisition; photogrammetry. *Mailing Add:* Dept Mech Eng Mail Code 104-44 Calif Inst Technol Pasadena CA 91125

ANTONUCCI, FRANK RALPH, b Auburn, NY, Sept 8, 46; m 68; c 2. ORGANIC CHEMISTRY, ENVIRONMENTAL ANALYSIS. *Educ:* St Michaels Col, BA, 68; Col of Holy Cross, MS, 69; Rensselaer Polytech Inst, PhD(org chem), 73. *Prof Exp:* Sr res chemist, Addressograph-Multigraph Corp, 75-77; group leader, St Regis Paper Co, 77-84; MGR ANALYTICAL SERV, CHAMPION INT CORP, 84- *Mem:* Am Chem Soc; Tech Asn Pulp & Paper Indust. *Res:* spectroscopy and chromatography; environmental analysis. *Mailing Add:* 3 North Dr Washingtonville NY 10992

ANTOSIEWICZ, HENRY ALBERT, b Wollersdorf, Austria, May 14, 25; US citizen; m 58, 70, 80; c 1. MATHEMATICAL ANALYSIS, ORDINARY DIFFERENTIAL EQUATIONS. *Educ:* Univ Vienna, PhD(math, theoret physics), 47. *Prof Exp:* Assoc prof math, Mont State Col, 48-52; assoc prof, Am Univ, 52-55, adj prof, 55-57; assoc ed, Math Rev, 57-58; vis assoc prof, 58-59, assoc prof, 59-61, chmn math dept, 68-77, UNIV PROF MATH, UNIV SOUTHERN CALIF, 61- *Concurrent Pos:* Mathematician, Nat Bur Stand, 52-57; consult, Space Technol Labs, Thompson-Ramo-Wooldridge, Inc, 59-70; consult var financial insts, 74-; mem bd trustees, Math Sci Res Inst, Berkeley, 81-; dir, Beaumont & Co, Newport Beach, 84-88. *Res:* Qualitative theory of ordinary differential equations; applications to finance. *Mailing Add:* Dept Math 1113 Univ Southern Calif University Park Los Angeles CA 90089-1113

ANTZELEVITCH, CHARLES, b Ramat-Gan, Israel, Mar 25, 51; US citizen; m 73; c 2. CARDIAC ELECTROPHYSIOLOGY, CARDIAC ARRHYTHMIAS. *Educ:* City Univ New York, Queens, BA, 73; State Univ New York, Syracuse, PhD(pharmacol), 78. *Prof Exp:* Fel cardiol, Masonic Med Res Lab, 77-80; asst prof pharmacol, Upstate Med Ctr, 80-83; assoc prof pharmacol, State Univ NY Health Sci Ctr, Syracuse, 83-87; res scientist, 80-83, sr res scientist, 84, EXEC DIR & DIR RES, MASONIC MED RES LAB, 84-; PROF PHARMACOL, STATE UNIV NY HEALTH SCI CTR, SYRACUSE, 87- *Concurrent Pos:* Teaching fel, NIH, 73-80, prog dir, minority high sch student res apprentice prog, 84-85, consult, ad hoc study sect, 85, prog dir, training grant, 85-89; prin investr & co-prin, Am Heart Asn & NIH grant , 80-; consult, res peer rev comm, Am Heart Asn, 82-85, mem basic sci coun, 82-; fel, cardiovasc sect, Am Physiol Soc, 84; res comt, Am Heart Asn, 87-90; Gordon K Moe Scholar; mem bd dirs, Clin Med Network, 89-, Cent NY Heart Asn, 89-, Ram Med Res Found, 89-; mem instnl rev bd, Faxton Hosp, 90- *Honors & Awards:* Van Horne Award, 82. *Mem:* Am Heart Asn; AAAS; Am Physiol Soc; Cardiac Electrophysiol Soc; NY Acad Sci; Int Soc Heart Res. *Res:* Electrophysiologic basis of cardiac arrhythmias and mechanisms of action of antiarrhythmic drugs. *Mailing Add:* Masonic Med Res Lab 2150 Bleeker St Utica NY 13501-1787

ANURAS, SINN, b Bangkok, Thailand, Apr 6, 41; m 69; c 2. INTERNAL MEDICINE, GASTROENTEROLOGY. *Educ:* Mahidol Univ, Thailand, dipl sci, 62; Chulalongkorn Univ, Bangkok, MD, 66. *Prof Exp:* Assoc, 74-75, from asst prof to assoc prof med, Univ Iowa, 75-85; PROF MED, TEX TECH UNIV, 85- *Concurrent Pos:* Fel hepatology, Vet Admin Hosp, Boston, 70-72; fel gastroenterol, Univ Iowa, 72-74. *Mem:* Fel Am Col Physicians; Am Fedn Clin Res; Am Gastroenterol Asn; Am Asn For Study Liver Dis. *Res:* Gastrointestinal motility; clinical liver disease; control mechanisms of portal blood flow. *Mailing Add:* Health Sci Ctr Texas Tech Univ Lubbock TX 79430

ANUSAVICE, KENNETH JOHN, b Worcester, Mass, Aug 6, 40. PORCELAINS, CERAMICS. *Educ:* Worcester Polytechnic Inst, BS, 62; Med Col Ga, DMD, 77; Univ Fla, PhD(metall mat eng),77. *Prof Exp:* From asst prof to assoc prof, dept restorative dent, div dental mat, Med Col Ga, 73-83; PROF & CHMN DEPT DENT BIOMAT, COL DENT, UNIV FLA, GAINESVILLE, 83- *Concurrent Pos:* Mem, NIH oral biol & med study sect, 86-90; mem, Am Dent Asn Mat, Instruments & Equip, 90- *Mem:* Am Asn Dent Res; Am Asn Dent Sch; Int Asn Dent Res; Acad Dent Mat; Am Dent Asn. *Res:* Metal-ceramic adherence; thermal stress; thermal compatibility; solidification stress; brazing of nickel-chromium alloys; clinical dental research; optimal design of dental restorations; strengthening of ceramics; ceramic fractology. *Mailing Add:* Dept Dent Biomat Col Dent Univ Fla Gainesville FL 32610-0446

ANVERSA, PIERO, b Parma, Italy, Sept 11, 38; m 68; c 1. EXPERIMENTAL PATHOLOGY, CELL BIOLOGY. *Educ:* Univ Parma, Italy, MD, 65. *Prof Exp:* Asst prof path, Univ Parma, Italy, 66-78; vis asst prof, 72-74, res assoc prof, 76-80, assoc prof, 80-82, PROF PATH, NY MED COL, 82- *Concurrent Pos:* Fel, North Atlantic Treaty Orgn, 71-72. *Mem:* Int Study Group Res Cardiac Metab; Int Soc Stereology; Am Heart Asn. *Res:* Quantitative structure and function of the heart and kidney; light and electron microscopic morphometry. *Mailing Add:* Dept Path NY Med Col Valhalla NY 10595

ANWAR, RASHID AHMAD, b Nakodar, India, Oct 15, 30; m 63. BIOCHEMISTRY. *Educ:* Univ Panjab, WPakistan, BSc, 51, MSc, 52; Mich State Univ, PhD(chem), 57. *Prof Exp:* Lectr pharmaceut chem, Univ Panjab, WPakistan, 52-54; asst biochem, Mich State Univ, 55-57, res instr, 57-60; from res assoc to assoc prof, 63-74, PROF BIOCHEM, UNIV TORONTO, 74- *Concurrent Pos:* Nat Res Coun Can fel, 60-62. *Mem:* AAAS; Am Chem Soc; Am Soc Biol Chem; Can Biochem Soc. *Res:* Proteolytic enzymes; structure of proteins; bacterial cell wall biosynthesis. *Mailing Add:* Dept Biochem Univ Toronto Toronto ON M5S 1A8 Can

ANWAY, ALLEN R, b Cloquet, Minn, Mar 19, 41; m 64; c 3. PHYSICS, ELECTRONICS. *Educ:* Univ Minn, Duluth, BA, 63; Univ Chicago, MS, 65, PhD(physics), 68. *Prof Exp:* Asst prof physics, Chicago State Col, 68-69; asst prof, 69-72, lectr physics, 74-76 & 78-79, electronics consult, Univ Wis-Superior, 83-88; CONSULT, 88- *Concurrent Pos:* Chmn new prod div, Duluth Sci, 73-74; consult, Plaunt & Anderson & Tech Systs, Inc, 74-79. *Mem:* Am Phys Soc; Am Asn Physics Teachers; Sigma Xi. *Res:* Electronic detector of leaks in water mains. *Mailing Add:* 1219 N 21st St Superior WI 54880

ANYSAS, JURGIS ARVYDAS, b Hamburg, Ger, Sept 5, 34; Can citizen. CHEMICAL PHYSICS. *Educ:* Univ Toronto, BASc, 56; Ill Inst Technol, PhD(chem), 66. *Prof Exp:* Geophysicist, Imp Oil Ltd, Can, 56-60; asst prof, 66-77, ASSOC PROF CHEM & CHMN DEPT, DEPAUL UNIV, 77- *Concurrent Pos:* Fel, Ill Inst Technol, 66-67. *Mem:* AAAS; Am Chem Soc. *Res:* Quantum chemistry; environmental chemistry; molecular spectroscopy. *Mailing Add:* Dept Chem DePaul Univ Chicago IL 60614

ANZALONE, LOUIS, JR, b Independence, La, Oct 7, 31; m 53; c 4. PHYTOPATHOLOGY. *Educ:* Southeastern La Col, BS, 54; La State Univ, MS, 56, PhD(plant path), 58. *Prof Exp:* Asst bot bact & plant path, La Agr Exp Sta, La State Univ, Baton Rouge, 54-58, from asst prof to prof bot & plant path, 58-68, prof plant path & supt, Burden Res Ctr Br, 68-77, proj leader dis wheat & grain sorghum, 77-81. *Concurrent Pos:* Mem, Burden Res Found, Sugarcane Variety Comt, Rural Life Mus Comt & Centennial Comt, Exp Sta, La State Univ; Sugarcane Variety Comt; LA State Univ Exp Sta Centennial Comt; Rural Lite Mus Comt. *Mem:* Am Phytopath Soc; Sigma Xi. *Res:* Fungus diseases of ornamental plants; sugar cane breeding and pathology; virus diseases of sugar cane; diseases of winter wheat and grain sorghum. *Mailing Add:* PO Box 404 Coloma CA 95613

AOKI, MASANAO, b Hiroshima City, Japan, May 14, 31; m 62; c 1. ECONOMICS, SYSTEM SCIENCE TIME SERIES. *Educ:* Univ Calif, Los Angeles, PhD(eng), 60. *Hon Degrees:* DSc, Tokyo Inst Technol, 66. *Prof Exp:* prof econ, Inst Social & Econ Res, 81-84; prof syst sci, 60-81, PROF COMPUTER SCI, UNIV CALIF, LOS ANGELES, 84- *Concurrent Pos:* Prof econ, Univ Ill, 71-73. *Honors & Awards:* Fel, Control Systs Soc, Inst Elec & Electronic Engrs; fel, Economet Soc. *Mem:* Inst Elec & Electronic Engrs; Am Econ Asn; Economet Soc; Soc Econ Dynamics & Control. *Res:* Economics; decision analysis. *Mailing Add:* Computer Sci 4731 Boelter Univ Calif 405 Hilgard Ave Los Angeles CA 90024

APAI, GUSTAV RICHARD, II, b Elyria, Ohio, May 27, 46; m 80. SURFACE CHEMISTRY,. *Educ:* Univ Col, BA, 68; Univ Calif, Berkeley, MS, 70, PhD(chem), 77. *Prof Exp:* Commun & electronics officer, US Air Force, 70-74; res chemist, 77-80, SR RES CHEMIST, EASTMAN KODAK CO, NY, 80- *Mem:* Am Phys Soc; Sigma Xi. *Res:* Electronic structure of small metallic clusters and surfaces of bulk materials; surface chemistry studies in heterogeneous catalysis. *Mailing Add:* 110 Bay Knoll Rd Rochester NY 14622

APEL, CHARLES TURNER, b Kearney, Nebr, Apr 28, 31; m 58; c 5. EMISSION SPECTROSCOPY, ANALYTICAL CHEMISTRY. *Educ:* Morehead State Univ, BS, 52; Iowa State Univ, MS, 56. *Prof Exp:* Teaching asst chem, Iowa State Univ, 53-54; res asst phys chem, Ames Lab, Dept Energy, 54-56; anal chemist, Los Alamos Nat Lab, 56-67, emission spectrochemist anal chem, 67-79, Spectrochem Sect leader, 79-86, Spectrochemist, 86-89, ENVIRON SPECTROCHEMIST, LOS ALAMOS NAT LAB, 89- *Mem:* Am Chem Soc; Soc Appl Spectros. *Res:* Emission spectrochemistry with emphasis on the inductively coupled plasma source and the computerization and automation of analytical techniques; special interest in nebulizers; development of fritted disk nebulizer; lasers in analytical chemistry; computer programming and interfacing. *Mailing Add:* 1331 Sage Loop Los Alamos NM 87544

APEL, JOHN RALPH, b Absecon, NJ, June 14, 30; m 56; c 2. WAVES AND INSTABILITIES, REMOTE SENSING. *Educ:* Univ Md, BS, 57, MS, 61; Johns Hopkins Univ, PhD(plasma physics), 70. *Prof Exp:* Phys sci aide, Nat Bur Stand, 55-57; assoc mathematician, Appl Physics Lab, Johns Hopkins Univ, 57-61, sr physicist, 61-70, asst supvr plasma dynamics, 66-70; supvry physicist & dir, Ocean Remote Sensing Lab, Atlantic Oceanog & Meteorol Labs, Nat Oceanic & Atmospheric Admin, Fla, 70-76, Pac Marine Environ Lab, Nat Oceanic & Atmospheric Admin, Wash, 76-81; asst dir, 82-84, CHIEF SCIENTIST, MILTON EISENHOWER RES CTR, APPL PHYSICS LAB, JOHNS HOPKINS UNIV, 84- *Concurrent Pos:* Adj prof physics, Univ Miami, 71-76; consult, NASA, 72-, Dept Defense, 73-, UNESCO/Intergovt Oceanog Comn, 75-80, Int Coun Explor Sea, 78-81; mem comn F, Int Union Radio Sci, 73-, Int Union Comn on Radio Meterol, 75-80 & Sci Comt Oceanic Res, 81-85; expert witness, US House Rep & US Senate, 74; chmn ocean dynamics subcomt, Space Appl Comn, NASA, 73-76; affil prof oceanog & atmospheric sci, Univ Wash, 76-81; sr fel, Joint Inst for Study of Atmosphere & Ocean & Joint Inst for Marine & Atmospheric Res, 77-82; mem, US Nat Comt, Int Union Geod & Geophys, 82-87, Comt Ocean Sci from Shuttle, Off Naval Res, 84-88, adv comt, Inst Naval Oceanog, Univ Corp Atmospheric Res, 88- & Comt Earth Sci, Nat Acad Sci, 88-; lectr, Johns Hopkins Univ, 85-; ed-in-chief, Johns Hopkins APL Tech Digest, 87-89; ed, Rev Geophys, 88-90. *Honors & Awards:* Gold Medal, US Dept Com, 74. *Mem:* Am Phys Soc; Am Geophys Union; Sigma Xi; NY Acad Sci; Explorers Club. *Res:* Fluid and plasma instabilities; waves in fluids; physical oceanography; ocean remote sensing. *Mailing Add:* Appl Physics Lab Johns Hopkins Univ Laurel MD 20722-6099

APELIAN, DIRAN, b Oct 28, 45; m 76; c 2. MATERIALS SCIENCE. *Educ:* Drexel Univ, BS, 68; Mass Inst Technol, ScD, 72. *Prof Exp:* Teaching asst, Mass Inst Technol, 69-72; sr res engr metall, Bethlehem Steel Corp, 72-75; asst prof, 76-79, ASSOC PROF MAT ENG, DREXEL UNIV, PA, 79- *Concurrent Pos:* Vis prof, Katholieke Univ, Belg, 81; consult, var corp, 79- *Mem:* AAAS; Am Inst Minerals Metall & Petrol Eng; Am Soc Metals; Am Inst Chem Eng; Am Powder Metall Inst. *Res:* Field of solidification processing; liquid metal filtration; development of rapid cycle casting technologies such as diffusion solidification; rheocasting; low pressure plasma processing; injection of secondary phases and dissolution kinetics. *Mailing Add:* Provost VP Acad Affairs Worchester Polytech Inst 100 Institute Rd Worchester MA 01609

APELLANIZ, JOSEPH E P, b Caguas, PR, Mar 31, 26; m 50; c 5. PHOTOGRAPHIC CHEMISTRY. *Educ:* Columbia Univ, AB, 48; Rutgers Univ, PhD(chem), 53. *Prof Exp:* Asst, Rutgers Univ, 49-51, res asst, 51; chemist, Stand Oil Co, Ind, 52-55; res specialist, Ansco Div, Gen Aniline & Film Corp, 55-63; res dir, Chemco Photo Prods Inc, 64-80, corp res assoc, 80-85; RETIRED. *Concurrent Pos:* Quality control. *Mem:* Am Chem Soc; Am Soc Quality Control; Soc Photog Sci & Eng; Brit Chem Soc; Royal Photog Soc Gt Brit; Sigma Xi. *Res:* Photography; graphic arts. *Mailing Add:* 361 Gatewood Dr Winston Salem NC 27104-2431

APFEL, ROBERT EDMUND, b New York, NY, Mar 16, 43; m 68; c 2. ACOUSTICS, FLUID PHYSICS. *Educ:* Tufts Univ, BA, 64; Harvard Univ, MA, 67, PhD(appl physics), 70. *Prof Exp:* Res fel acoust, Harvard Univ, 70-71; from asst prof to assoc prof, Yale Univ, 71-81, prof eng & appl sci & chmn, Dept Mech, 81-82, chmn, Dept Mech Eng, 82-86 & Coun Eng, 88-90. *Concurrent Pos:* Independent consult acoust, 74-; pres, Apfel Enterprises, Inc, New Haven, Conn, 87- *Honors & Awards:* A B Wood Medal & Prize, Inst Physics, Gt Brit, 71; Biennial Award, Acoust Soc Am, 76. *Mem:* Fel Acoust Soc Am (vpres-elect, 90); Am Phys Soc; Am Asn Physics Teachers; Am Soc Mech Engrs. *Res:* Study of superheated, supercooled, supersaturated and/or tensilely stressed liquids, possibly irradiated, with emphasis on acoustical techniques for producing phase changes or probing properties of liquids and biological materials; neutron detection with superheated drops. *Mailing Add:* Mason Lab PO Box 2159 Yale Univ New Haven CT 06520

APGAR, BARBARA JEAN, b Tyler, Tex, Mar 4, 36. NUTRITION. *Educ:* Tex Woman's Univ, BA, 57; Cornell Univ, MS, 59, PhD(biochem), 64. *Prof Exp:* RES CHEMIST, SOIL & NUTRIT LAB, US PLANT, 59- *Mem:* Am Soc Animal Sci; Am Inst Nutrit. *Res:* Effect of zinc deficiency on reproduction in the female. *Mailing Add:* US Plant Plant Soil & Nutrit Lab Ithaca NY 14853

APIRION, DAVID, b Petah Tiqwa, Israel, July 17, 35; US citizen; div; c 3. NUCLEIC ACIDS, MICROBIAL GENETICS. *Educ:* Hebrew Univ, Jerusalem, MSc, 60; Glasgow Univ, Scotland, PhD(genetics), 63. *Prof Exp:* Teaching asst genetics, Hebrew Univ, 59-60; asst lectr, Glasgow Univ, 62-63; from asst prof to assoc prof, 65-78, PROF MOLECULAR GENETICS, WASH UNIV, 78- *Concurrent Pos:* Vis scholar, Cambridge Univ, Eng, 73; vis prof, Tel-Aviv Univ, 74 & Hebrew Univ, 80; vis scientist, Hungarian Acad Sci & USSR Acad Sci, 80, Indian Acad Sci, 85 & Soviet Acad Sci, 89; ed, J Bact, 81-89. *Mem:* Genetic Soc Am; Am Soc Microbiol; AAAS; Am Soc Biochem & Molecular Biol; Am Soc Cell Biol; Royal Soc Med. *Res:* Processing and decay of RNA in prokaryotic cells. *Mailing Add:* 408 S Hanley Clayton MO 63105

APLAN, FRANK F(ULTON), b Boulder, Colo, Aug 11, 23; m 55; c 3. METALLURGY, MINERAL ENGINEERING. *Educ:* SDak Sch Mines & Technol, BS, 48; Mont Sch Mines, MS, 50; Mass Inst Technol, ScD(metall), 57. *Hon Degrees:* Minl, Engr Mont Col Mineral Sci & Technol, 68. *Prof Exp:* Mill engr, Climax Molybdenum Co, 50-51; asst prof metall, Univ Wash, 51-53; sr scientist, Kennecott Copper Corp, 57; res & develop group mgr, Mineral Eng, Mining & Metals Div, Union Carbide Corp, 57-67; prof mineral prep & head dept, Pa State Univ, 68-71, chmn, mineral processing sect, 71-77, chmn, metall sect, 73-75, prof, 68-90, DISTINGUISHED PROF METALL & MINERAL PROCESSING, PA STATE UNIV, 90- *Concurrent Pos:* Mem bd dirs, Engr Found, 77-90, chmn, 85-87. *Honors & Awards:* Richards Award, Am Inst Mining, Metall & Petrol Engrs, 78; Eng Found Award, Nat Acad Eng, 89. *Mem:* Nat Acad Eng; Am Chem Soc; Am Inst Chem Engrs; Am Soc Metals; Sigma Xi; Am Filtration Soc; Am Inst Mining Metal & Petrol Engrs; Archaeol Inst Am; Soc Mining Engrs; Metall Soc. *Res:* Mineral process engineering; ore and coal preparation; flotation; flocculation and dispersion; applied surface chemistry; particulate matter systems; hydrometallurgy; pollution control. *Mailing Add:* Mineral Processing Sect Pa State Univ Univ Park PA 16802

APLEY, MARTYN LINN, b Fairbury, Nebr, May 27, 38. INVERTEBRATE ZOOLOGY, AQUATIC ECOLOGY. *Educ:* Kans State Univ, BS, 60; Syracuse Univ, MS, 63, PhD(zool), 67. *Prof Exp:* Fel, Woods Hole Oceanog Inst, 68; fel, Freshwater Biol Lab, Copenhagen Univ, 69; asst prof biol, Brooklyn Col, 69-72, fac res award, 70-71; asst prof biol, Merrimack Col, 75-78; ASSOC PROF BIOL, WESTERN STATE COL, COLO, 78- *Mem:* AAAS. *Res:* Physiological ecology of mollusks, biorhythms and reproductive periodicites neurosecretion and hormonal control mechanisms in invertebrates; limnology, study of high altitude lakes. *Mailing Add:* Dept Sci Western State Col Gunnison CO 81230

APOSHIAN, HURAIR VASKEN, b Providence, RI, Jan 28, 26; m 48; c 3. CELL BIOLOGY, PHARMACOLOGY. *Educ:* Brown Univ, BS, 48; Univ Rochester, MS, 51, PhD(physiol), 54. *Prof Exp:* Asst, Univ Rochester, 48-49; from instr to asst prof pharmacol, Sch Med, Vanderbilt Univ, 54-59; USPHS fel biochem, Stanford Univ, 59-62; assoc prof microbiol, Sch Med, Tufts Univ, 62-67; head dept, 67-72, prof cell biol & pharmacol, Sch Med, Univ Md, Baltimore City, 67-75; PROF CELL & DEVELOP BIOL & CHMN DEPT, COL LIB ARTS & PROF PHARMACOL, COL MED, UNIV ARIZ, 75- *Concurrent Pos:* Vis prof biol, Mass Inst Technol, 83. *Honors & Awards:* Res Award, Am Col Adv Med, 87; Am Soc Toxicol. *Mem:* AAAS; Am Soc Microbiol; Am Soc Biol Chem. *Res:* Development of gene therapy; enzymology of DNA and its precursors; biochemistry of bacterial and animal virus infection; biological chelation; toxicology. *Mailing Add:* Biol Sci West Bldg Rm 308 Univ Ariz Tucson AZ 85721

APOSTOL, TOM M, b Helper, Utah, Aug 20, 23; m 59. MATHEMATICS, STATISTICS. *Educ:* Univ Wash, BS, 44, MS, 46; Univ Calif, Berkeley, PhD(math), 48. *Prof Exp:* Lectr math, Univ Calif, Berkeley, 48-49; C L E Moore instr, Mass Inst Technol, 49-50; from asst prof to assoc prof, 50-62, PROF MATH, CALIF INST TECHNOL, 62- *Concurrent Pos:* Dir, Proj Math, 87- *Mem:* Am Math Soc; Math Asn Am. *Res:* Theory of numbers. *Mailing Add:* Dept Math Calif Inst Technol Pasadena CA 91125

APP, ALVA A, b Bridetown, NJ, Feb 19, 32; m 86; c 3. SOILS & PLANT SCIENCE. *Educ:* Cornell Univ, BS, 53; Rutgers Univ, MS, 55, PhD(soils), 56. *Prof Exp:* Tech adv, Valley Green Mushroom Farms, Pa, 59-61; res assoc, McCollum Pratt Inst, Johns Hopkins Univ, 61-64; assoc seed physiologist, Boyce Thompson Inst, Cornell Univ, 64-69, prog dir cell physiol & virol, 69-76, prog dir nitrogen & crop yields, 77-81; dep dir, Int Ctr Insect Physiol & Ecol, Nairobi Kenya, 81-83; dir agr sci, Rockefeller Found, 83-87; SR SCI ADV, UNITED NATIONS DEVELOP PROG, 87- *Concurrent Pos:* Adj prof biochem, City Univ NY, 73-76; consult, Biochem Prog, US Army Med Res & Develop Command, 75-76 & UN develop progs, 76-; vis scientist, Int Rice Res Inst, Los Banos, Philippines, 76-; adj prof soil sci, Agron Dept, Cornell Univ, 81; bd dirs, Campbell Soup Co, Camden, NJ. *Mem:* Fel AAAS; Am Soc Plant Physiol; sr mem Am Chem Soc; Am Soc Agron; Tissue Cult Asn; Sigma Xi. *Res:* Soil and plant chemistry. *Mailing Add:* UN Develop Prog One UN Plaza New York NY 10017

APPEL, ARTHUR GARY, b Inglewood, Calif, Feb 10, 58; m 87; c 2. URBAN ENTOMOLOGY, ENVIRONMENTAL PHYSIOLOGY. *Educ:* Univ Calif, Los Angeles, BA, 80, Riverside, MS, 82, PhD(entom), 85. *Prof Exp:* ASST PROF ENTOM, AUBURN UNIV, ALA, 85- *Concurrent Pos:* Prin invest, Dept Entom, Auburn Univ; consult, Am Cyanamid Co, 84-, Rainbow Mfg, 85-, Bell S Serv, 86- *Mem:* Entom Soc Am; AAAS; Sigma Xi; Am Soc Zoologists; Animal Behav Soc. *Res:* Cockroach biology, behavior and physiology; urban pest biology and control; distribution and ecology of nonpest cockroaches; environmental physiology of both invertebrates and vertebrates. *Mailing Add:* Dept Entom Auburn Univ Auburn AL 36849-5413

APPEL, DAVID W(OODHULL), b Washington, DC, July 23, 24; m 49; c 3. HYDRAULICS. *Educ:* Lehigh Univ, BS, 47; Univ Iowa, MS, 49, PhD(hydraul), 53. *Prof Exp:* Test engr, Gen Elec Co, 47-48; asst prof, Col Eng, Univ Iowa, 49-56; from assoc prof to prof eng mech, Univ Kans, 56-64; fluid mech engr, Kimberly-Clark Corp, 64-66, sr res scientist, 66-67, sr res assoc res & eng, 67-79; RETIRED. *Concurrent Pos:* Res assoc hydraul, Iowa Inst Hydraul Res, 56-60; consult, 59-64; Nat Sci Found fel, Univ Mich, 62-63. *Honors & Awards:* Ernst Mahler Award, Kimberly-Clark Corp. *Res:* Papermaking and dry forming processes; turbulence and secondary flows. *Mailing Add:* Rte 2 Box 163 Wittenberg WI 54499

APPEL, JAMES B, b New York, NY, Feb 18, 34; m 65. BEHAVIORAL PHARMACOLOGY. *Educ:* Columbia Univ, AB, 55; Ind Univ, PhD(psychol), 60. *Prof Exp:* Asst prof psychol, Dept Psychiat, Sch Med, Yale Univ, 61-66; assoc prof, Dept Psychiat, Univ Chicago, 66-72; prof dept psychol, 72-86, CAROLINA RES PROF, UNIV SC, 86- *Concurrent Pos:* USPHS res scientist develop award, 69-72; consult, NSF, 70-71; vis prof, Med Res Coun Unit, Neural Mechanisms of Behav, Univ Col, London, UK, 78, NIMH, 81-85, Addiction Res Ctr, NIDA, 86; Educ Found Res Award, Univ SC, 89. *Mem:* Fel Am Col Neuropsychopharmacol; fel AAAS; fel Am Psychol Asn; Sigma Xi; Soc Neurosci; Soc Stimulus Properties Drugs (pres, 81-82); Behav Pharmacol Soc. *Res:* Behavioral and neurochemical effects of drugs; learning and conditioning; history and logic of psychology. *Mailing Add:* Dept Psychol Univ SC Columbia SC 29208

APPEL, JEFFREY ALAN, b Cleveland, Ohio, Aug 11, 42; m 65; c 2. ELEMENTARY PARTICLE PHYSICS. *Educ:* Williams Col, AB, 64; Harvard Univ, MA, 65, PhD(physics), 69. *Prof Exp:* Res assoc physics, Nevis Labs, Columbia Univ, 68-70, asst prof, Columbia Univ, 70-74, sr res assoc, Nevis Labs, 74-75; PHYSICIST, FERMI NAT ACCELERATOR LAB, 75- *Mem:* Am Phys Soc. *Res:* Discrete symmetries, resonance phenomena and weak and electromagnetic phenomena. *Mailing Add:* Fermi Lab PO Box 500 Batavia IL 60510

APPEL, KENNETH I, b Brooklyn, NY, Oct 8, 32; m 59; c 3. MATHEMATICS. *Educ:* Queens Col, NY, BS, 53; Univ Mich, MA, 56, PhD(math), 59. *Prof Exp:* Mathematician, Inst Defense Anal, 59-61; asst prof, 61-67, assoc prof, 67-77, PROF MATH, UNIV ILL, URBANA-CHAMPAIGN, 77- *Honors & Awards:* Fulkerson Prize, 79. *Mem:* Am Math Soc; Math Asn Am; Asn Symbolic Logic. *Res:* Combinatorics. *Mailing Add:* Dept Math Univ Ill Urbana IL 61801

APPEL, MAX J, b Giessen, Ger, Dec 13, 29; m 61; c 4. VETERINARY VIROLOGY. *Educ:* Vet Col, Hannover, DVM, 56; Cornell Univ, PhD(vet virol), 67. *Prof Exp:* Res asst vet med, Univ Munich, 57-59; fel, Univ Sask, 59-60; res off, Animal Dis Res Inst, Hull, Que, 61-64; res asst vet virol, Vet Virus Res Inst, 64-67, from asst prof to assoc prof, 67-76, PROF VET VIROL, NY STATE COL VET MED, CORNELL UNIV, 76- *Mem:* Am Vet Med Asn; Am Soc Microbiol. *Res:* Viral diseases of domestic animals. *Mailing Add:* NY State Col Vet Med Cornell Univ Ithaca NY 14853

APPEL, MICHAEL CLAYTON, b Milwaukee, Wis, Nov 19, 47; m 68; c 3. ENDOCRINOLOGY. *Educ:* Western Mich Univ, BS, 70, MA, 73; Univ Minn, PhD(anat), 76. *Prof Exp:* Fel, Univ Mass, 76-78, asst prof anat, 78-81, asst prof path 79-88, ASSOC PROF PATH, MED SCH, UNIV MASS, 88- *Concurrent Pos:* Consult, Bio-Hybrid Technologies & T-cell Scis Inc. *Mem:* Am Diabetes Asn; Europ Asn Study Diabetes; Endocrine Soc. *Res:* Etiology of diabetes mellitus; identify and characterize laboratory animal models of human diabetes; immunology of diabetes. *Mailing Add:* Dept Path Univ Mass Med Sch 55 Lake Ave N Worcester MA 01605

APPEL, STANLEY HERSH, b Boston, Mass, May 8, 33; m 56; c 2. NEUROLOGY. *Educ:* Harvard Univ, AB, 54; Columbia Univ, MD, 60; Am Bd Psychiat & Neurol, dipl, 68. *Prof Exp:* Intern med, Mass Gen Hosp, 60-61; resident neurol, Mt Sinai Hosp, 61-62; res assoc, Lab Molecular Biol, NIH, 62-64; chief res assoc, Sch Med, Univ Pa, 65-66, asst prof, 66-67; assoc neurol, Med Ctr, Duke Univ, 64-65, from assoc prof to prof neurol, 67-77, assoc prof biochem, 68-77, chief div neurol, 69-77; PROF NEUROL & CHMN DEPT, BAYLOR COL MED, 77-, CHMN PROG NEUROSCI, 77-, DIR, JERRY LEWIS NEUROMUSCULAR DIS RES CTR, 77- *Concurrent Pos:* USPHS res career develop award, 65-70. *Mem:* Am Acad Neurol; Am Soc Biol Chem; Am Soc Clin Invest; Am Neurol Asn. *Res:* Molecular neurobiology; neurochemistry; synapse function; muscle membranes and disease. *Mailing Add:* Dept Neurol Baylor Col Med Tex Med Ctr 6501 Fannin NB 302 Houston TX 77030

APPEL, WARREN CURTIS, b Cheyenne, Wyo, Oct 19, 44; m 69; c 1. PHARMACOLOGY, REGULATORY AFFAIRS. *Educ:* Univ Wyo, BSc, 67; Univ Wash, MSc, 70, PhD(pharmacol), 71. *Prof Exp:* Lectr pharmacol, Univ Man, 72-73; SCI ADV, BUR DRUGS, HEALTH & WELFARE CAN, 73- *Concurrent Pos:* Actg chief, Div Gastroenterol & Miscellaneous Drugs, Bur Drugs, 85; WHO Travel Fel, Japan, China, 88. *Mem:* AAAS; NY Acad Sci; Can Soc Clin Pharmacol; Drug Info Asn. *Res:* Mechanisms of drug dependence. *Mailing Add:* Prod Related Div Health & Welfare Can Finance Annex Bldg 267 Ottawa ON K1A 0L2 Can

APPELBAUM, EMANUEL, b Poland, Apr 14, 1894; nat US; m 27; c 2. INTERNAL MEDICINE. *Educ:* Columbia Univ, AB, 16, MD, 18. *Prof Exp:* Instr med, Col Med, NY Univ, 33-43, assoc clin prof, 43-50, assoc clin prof, Postgrad Med Sch, 50-54, assoc prof clin med, 54-57, PROF CLIN MED, MED SCH, NY UNIV, 57- *Concurrent Pos:* Pathologist, NY Health Dept, 21-28, bacteriologist, 28-41, chief div acute infections of cent nervous syst, 41-64; asst vis physician, Bellevue Hosp, 26-38, vis physician, 38-; vis physician, Willard Parker Hosp, 43-55; dir med, Sydenham Hosp, 48-66; consult physician, Long Beach Mem Hosp, 48-, NY Infirmary, 49-, Logan Mem Hosp, 57-, Trafalgar Hosp, 58-, Beth Israel Hosp, Newark, 60- & Sydenham Hosp, 66-; attend physician, Beth David Hosp, 50-61, Univ Hosp, 54- & Doctor's Hosp, 77- *Mem:* Fel AMA; fel Am Pub Health Asn; fel Am Col Physicians; fel NY Acad Sci; fel NY Acad Med. *Res:* Diagnosis and treatment of acute infections of central nervous system; meningitis, encephalitis, poliomyelitis, rabies and tetanus; chemotherapy and antibiotic therapy of bacterial and fungal infections. *Mailing Add:* 40 E 83rd St New York NY 10028

APPELBAUM, JOEL A, b Brooklyn, NY, Dec 30, 41; m 63; c 1. SOLID STATE PHYSICS. *Educ:* City Col New York, BS, 63; Univ Chicago, MS, 64, PhD(physics), 66. *Prof Exp:* Teaching asst physics, Univ Chicago, 63-64; mem tech staff, Bell Labs, NJ, 67; asst prof, Univ Calif, Berkeley, 67-68; MEM TECH STAFF, BELL LABS, 68- *Mem:* Am Phys Soc. *Res:* Solid state physics, particularly electron tunneling, surface physics and magnetism. *Mailing Add:* 100 W Dudley Ave Westfield NJ 07090

APPELGREN, WALTER PHON, b Macomb, Ill, Nov 10, 28; m 61; c 3. HEMATOLOGY, COMPUTER PROGRAMMING. *Educ:* Northwestern Univ, BS, 50; Univ Mich, MS, 53, PhD(microbiol), 67. *Prof Exp:* Res chemist, Continental Can Corp, 54-55; asst plant mgr, Nat Can Corp, 55-57; instr org chem, Sacramento City Col, 57-61; asst prof microbiol, Wis State Univ, Oshkosh, 66-67; from asst prof to assoc prof microbiol, 67-73, dir, Ctr Health Sci, 73-80, chmn, dept microbiol & med technol, 78-83, PROF MICROBIOL, NORTHERN ARIZ UNIV, 74- *Concurrent Pos:* Chmn, Regional Rev, 77-79. *Mem:* Am Soc Microbiol; Am Soc Med Technol. *Res:* Host-parasite relationships in disease; oral microbiology; ribavirin chemotherapy. *Mailing Add:* Dept Biol Northern Ariz Univ Box 5640 Flagstaff AZ 86011-5640

APPELLA, ETTORE, b Castronuovo, Italy, Aug 5, 33; US citizen. IMMUNOLOGY. *Educ:* Univ Rome, MD, 59. *Prof Exp:* Res asst biol, Johns Hopkins Univ, 60-63; vis scientist, Lab Molecular Biol, Nat Inst Arthritis & Metab Dis, 63-64, MED OFFICER IMMUNOL, LAB BIOL, NAT CANCER INST, 67- *Concurrent Pos:* Am Cancer Soc fel, Nat Inst Arthritis & Metab Dis, 64-66. *Mem:* Am Asn Biol Chemist; Am Cancer Soc. *Res:* Biochemistry; protein chemistry. *Mailing Add:* 4112 Aspen St Bethesda MD 20815

APPELMAN, EVAN HUGH, b Chicago, Ill, June 6, 35; m 60; c 2. INORGANIC CHEMISTRY. *Educ:* Univ Chicago, AB, 53, MS, 55; Univ Calif, Berkeley, PhD(chem), 60. *Prof Exp:* From asst chemist to assoc chemist, 60-76, SR CHEMIST, ARGONNE NAT LAB, 76- *Concurrent Pos:* Guggenheim Mem fel, 73-74; vis sr res fel, Brit Sci Res Coun & Univ Oxford, 83-84; Kipping vis, Univ Nottingham, 84. *Honors & Awards:* E O Lawrence Award, US Energy Res & Develop Admin, 76. *Mem:* Fedn Am Scientists; fel AAAS; Am Chem Soc. *Res:* Inorganic reaction kinetics and equilibria in aqueous solution, particularly kinetics of oxidation-reduction reactions; chemistry of the less familiar elements; inorganic fluorine chemistry; chemistry of powerful inorganic oxidants and fluorinating agents. *Mailing Add:* Argonne Nat Lab Bldg 200 9700 S Cass Ave Argonne IL 60439

APPELQUIST, THOMAS, b Emmetsburg, Iowa, Nov 1, 41; m 65; c 2. THEORETICAL PHYSICS. *Educ:* Ill Benedictine Col, BS, 63; Cornell Univ, PhD(theoret physics), 68. *Prof Exp:* Res assoc theoret physics, Stanford Linear Accelerator Ctr, 68-70; from asst prof to assoc prof, Harvard Univ, 70-75; PROF THEORET PHYSICS, YALE UNIV, 75-, CHMN, PHYSICS DEPT, 83- *Concurrent Pos:* AP Sloan Found fel, 74-76; US Sr Scientist Award, Alexander von Humboldt Found, 86- *Mem:* Fedn Am Sci; Am Phys Soc; AAAS. *Res:* Applications of quantum field theory to the interactions of elementary particles. *Mailing Add:* Dept Physics Yale Univ New Haven CT 06520

APPELT, GLENN DAVID, b Yoakum, Tex, Aug 24, 35; m 83; c 1. ETHNOPHARMACOLOGY. *Educ:* Univ Tex, BS, 57, MS, 59; Univ Colo, PhD(pharm), 63. *Prof Exp:* Asst prof pharmacol, Univ Tex, Austin, 63-67; assoc prof, 67-79, asst dean student affairs, 77-82, PROF PHARMACOL, SCH PHARM, UNIV COLO, BOULDER, 79- *Concurrent Pos:* Consult, Pharm Corp Am, Boulder, Colo & Smithkline Beecham Consumer Brands, Clifton, NJ; regist pharmacist; consult pharmacol, Legal & Ins Co. *Mem:* Am Pharm Asn; Am Asn Cols Pharmacy; Sigma Xi; Herb Res Found; Am Soc Pharmacog; fel Am Col Apothecaries. *Res:* Biochemical pharmacology; weight control products; toxicology of natural products; herbal medicines; auth pharmacol textbook. *Mailing Add:* Sch Pharm Univ Colo Boulder CO 80309-0297

APPENZELLER, OTTO, b Czernowitz, Romania, Dec 11, 27; m 56; c 3. NEUROLOGY, MEDICINE. *Educ:* Univ Sydney, MB & BS, 57, MD, 66; Univ London, PhD, 63. *Prof Exp:* Jr resident, Royal Prince Alfred Hosp, Sydney, Australia, 57-58, sr resident, 58-59, med registr, 59-60; clin asst, Nat Hosp Nerv Dis, London, Eng, 61-62; asst prof neurol, Col Med, Univ Cincinnati, 65-67; assoc prof med, 67-70, prof neurol & med, Sch Med, Univ NMex, 70-90; VIS SCIENTIST, LOVELACE MED FOUND, 90- *Concurrent Pos:* Mass Gen Hosp & Harvard Med Sch fel, 63-64; Boston Med Found fel, 65; chief neurol sect, Vet Admin Hosp, Cincinnati, Ohio, 65-67; consult neurologist, Vet Admin Hosp, Albuquerque, NMex, 68; mem res comt sports & phys educ, UNESCO; vis prof neurol, McGill Univ, 77; hon fel, Dept Anat, Univ Col, London, 83. *Mem:* AAAS; AMA; fel Am Acad Neurol; Brit Med Asn; Royal Australasian Col Physicians; Am Col Physicians. *Res:* Clinical neurology; experimental neuropathology; physiology and pathology of autonomic and peripheral nervous systems; neuropathology; biochemistry of peripheral nerves; headache and its mechanism; immunology as applied to the nervous system; sports medicine. *Mailing Add:* Lovelace Med Found Albuquerque NM 87108

APPERSON, CHARLES HAMILTON, b Fairfield, Ala, Feb 19, 25; m 51; c 3. CHEMICAL ENGINEERING. *Educ:* Auburn Univ, BS, 49; WVa COES, MS, 80. *Prof Exp:* Sales engr, Fulbright Labs Div, 49-50; metall engr, US Steel Corp, 50-52; res engr, Chemstrand Res Ctr, Monsanto Co, 52-58, res group leader, 58-63, develop proj supvr, 63-66; res group leader, Union Carbide Tech Ctr, 66-81, RES GROUP LEADER, UNION CARBIDE, RESEARCH TRIANGLE PARK, 81- *Mem:* Am Inst Chem Engrs; Fiber Soc. *Res:* Product/process development and formulations on agricultural chemicals (insecticides and herbicides), polymers and fibers; process design, construction, start-up and technology support for operation. *Mailing Add:* Rhone Polunec PO Box 12014 Attn Beth Rickman Research Triangle Park NC 27709

APPERT, HUBERT ERNEST, PANCREATIC PHYSIOLOGY, GLYCOCONJUGATES. *Educ:* Jefferson Med Col, Toledo, PhD(physiol), 60. *Prof Exp:* ASSOC PROF GASTROINTESTINAL PHYSIOL, MED COL OHIO, TOLEDO, 73- *Mailing Add:* Dept Surg Med Col Ohio Caller Serv No 10008 Toledo OH 43699

APPINO, JAMES B, b Benton, Ill, Apr 27, 31; m 55; c 4. INDUSTRIAL PHARMACY. *Educ:* Purdue Univ, BS, 54, MS, 58, PhD(indust pharm), 60. *Prof Exp:* Sr res scientist, Armour Pharmaceut Co, 60-62; mgr pharm, Baxter Labs, Inc, 62-69; dir pharm res, McNeil Labs, 69-80, GROUP DIR, CHEM & PHARMACEUT DEVELOP, MCNEIL PHARMACEUT, 80- *Mem:* Am Pharm Asn; Acad Pharmaceut Sci; AAAS. *Res:* Product development and dosage form design; analytical development in areas of purity, methods development and stability; biopharmaceutics, including dissolution methodology and physical characterization of drugs; chemical development through pilot plant. *Mailing Add:* 4730 Cheshire Dr Doylestown PA 18901

APPL, FRANKLIN JOHN, b Great Bend, Kans, Dec 23, 37; m 59; c 3. THEORETICAL & APPLIED MECHANICS. *Educ:* Kans State Univ, BS, 60; Univ Ill, MS, 62, PhD(theoret & appl mech), 64. *Prof Exp:* Asst prof, School Aerospace & Mech Eng, 64-69, from assoc prof to prof, 69-78, asst dir, 71-78, ADJ PROF, SCH AEROSPACE MECH ENG, UNIV OKLA, 78-; PRES, APPL ENG CO, 78- *Concurrent Pos:* Res resident, Caterpillar Tractor Co, 66-67. *Honors & Awards:* Ralph R Teetor Award, Soc Automotive Engrs, 68. *Mem:* Am Soc Mech Engrs; Soc Exp Mech; Am Soc Testing & Mat; Soc Automotive Engrs; Am Acad Mech. *Res:* Solid mechanics; design. *Mailing Add:* Appl Eng Co 3503 Charleston Rd Norman OK 73069

APPL, FREDRIC CARL, b Great Bend, Kans, Nov 17, 32; m 54; c 3. MECHANICAL ENGINEERING, APPLIED MATHEMATICS. *Educ:* Carnegie Mellon Univ, BS, 54, MS, 56, PhD(mech eng), 58. *Prof Exp:* Instr mech eng, Carnegie Inst Technol, 57-58, asst prof, 58; res engr drilling, Jersey Prod Res Co, 58-60; from assoc prof to prof, 60-67, JENNINGS PROF MECH ENG, KANS STATE UNIV, 67- *Concurrent Pos:* Consult, Mine Safety Appliances, 57-58, Jersey Prod Res Co, 60-65 & Christensen Diamond Prod Co, 65- *Mem:* Am Soc Mech Engrs; Soc Petrol Engrs. *Res:* Vibration; mechanics; rock mechanics. *Mailing Add:* Mech Eng Dept Kans State Univ Manhattan KS 66502

APPLE, JAY LAWRENCE, b Guilford Co, NC, Jan 8, 26; m 45. GENETICS. *Educ:* NC State Col, BS, 49, MS, 53, PhD(plant path), 55. *Prof Exp:* From res instr to assoc prof, 49-63, dir inst biol sci, 67-71, asst dir acad affairs & res for biol sci, 71-77, assoc dir res & coordr int prog, 77-81, PROF PLANT PATH & GENETICS, NC STATE UNIV, 63-, DIR INT PROG, 81- *Concurrent Pos:* Plant path adv, NC State Univ-US AID Mission, Peru, 63-65, chief-of-party, 65-67; consult int develop, Univ Calif, Berkeley, 72-80; consult pest mgt, Environ Protection Agency, 73-74 & Off Tech Assessment, 78-79 & FAO, 81- *Mem:* Fel AAAS; Am Inst Biol Sci; Am Phytopath Soc; Sigma Xi. *Res:* Genetics of disease resistance in plants; integrated pest management; research administration; international agricultural development. *Mailing Add:* Int Progs Box 7112 Raleigh NC 27695

APPLE, MARTIN ALLEN, b Duluth, Minn, Sept 17, 39; m; c 4. MOLECULAR PHARMACOLOGY, ONCOLOGY. *Educ:* Univ Minn, Minneapolis, AB, 59, MSc, 62; Univ Calif, PhD(biochem), 67. *Prof Exp:* Asst microbiol, Med Sch, Univ Minn, 60-62; asst res biochemist, Sch Med, Univ Calif, San Francisco, 66-68 & Cancer Res Inst, 68-75, lectr biochem, 68-70, asst prof pharmacol & exp therapeut in residence, 70-75, asst prof pharmaceut chem in residence, Sch Pharm, 71-75, assoc res biochemist, Cancer Res Inst, 75-78; pres, Int Plant Res Inst, 78-82; pres, Adytum Int, 82-89; scholar in residence, Nat Health Coun, 90; EXPERT, NSF, 91- *Concurrent Pos:* Consult enzyme, Vet Admin Hosps, 63-69 & Nat Cancer Inst, 73-74; mem bd of regents, Am Col Clin Pharmacol; bd trustees, East-West Ctr, 83-88. *Honors & Awards:* Am Cancer Soc Award, 69. *Mem:* AAAS; Am Soc Microbiol; Am Fedn Clin Res; fel Am Inst Chem; fel Am Col Clin Pharmacol; Am Soc Pharmacol & Exp Therapeut; Sigma Xi (nat bd & chmn, 88-). *Res:* Biochemical and clinical pharmacology; computer aided chemical structure design; biochemical oncology; new drug designs; genetic engineering; agriculture; computer aided chemical structure design. *Mailing Add:* Adytum Int PO Box 2629 San Francisco CA 94126

APPLE, SPENCER BUTLER, JR, b Kansas City, Mo, June 11, 12; m 40; c 3. HORTICULTURE. *Educ:* Agr & Mech Col Tex, BS, 33, MS, 36; Wash State Univ, PhD(hort), 53. *Prof Exp:* From instr to asst prof hort, Agr & Mech Col Tex, 35-41; res asst & exten specialist, Mich State Col, 41-46; exten specialist, Agr & Mech Col Tex, 47-48, assoc prof, 49-50; assoc prof, 50-55, horticulturist & head dept hort, 55-73, prof, 73-75, EMER PROF HORT, ORE STATE UNIV, 75- *Concurrent Pos:* Gen Educ Bd fel, 48-49. *Mem:* Fel AAAS; Am Soc Hort Sci. *Res:* Plant nutrition; physiological problems in vegetable crop production. *Mailing Add:* Dept Hort Ore State Univ 2042 Cordley Hall Corvallis OR 97331-2911

APPLEBAUM, CHARLES H, b Newark, NJ, Nov 26, 42; m 64; c 2. MATHEMATICAL LOGIC. *Educ:* Case Inst Technol, BS, 64; Rutgers Univ, MS, 66, PhD(math), 69. *Prof Exp:* Asst math, Rutgers Univ, 64-69; asst prof, 69-74, ASSOC PROF MATH, BOWLING GREEN STATE UNIV, 74- *Concurrent Pos:* Consult & tech staff mem, Mass Inst Technol Res Estab Corp, 81-88. *Mem:* Math Asn Am; Am Math Soc; Asn Symbolic Logic. *Res:* Theory of recursive equivalence types and recursive functions and their application to various areas of algebra; verification and specification theory in the field of computer science. *Mailing Add:* Dept Math & Statist Bowling Green State Univ Bowling Green OH 43403

APPLEBURY, MEREDITHE L, b Hamilton, Mont, Feb 7, 42; m 76. VISION, RETINAL GENETICS. *Educ:* Univ Wash, BS, 64; Yale Univ, PhD(biochem), 68. *Prof Exp:* Mem tech staff biophys, Bell Labs, Murray Hill, NJ, 70-72; res fel biochem, Max Planck Inst, Gottingen, WGer, 72-75; asst prof biochem, Princeton Univ, 75-82; assoc prof, Dept Biol Sci, Lily, Purdue Univ, 75-82; PROF, DEPT OPHTHAL & VISION SCI, UNIV CHICAGO, 88- *Concurrent Pos:* Fel biochem, Yale Univ, 68-70, res assoc, 68-72; prin investr, Nat Eye Inst, Bethesda, Nat Found March Dimes, 78- *Mem:* Am Chem Soc; AAAS. *Res:* Primary processes of vision; photoreception and transduction; biochemistry of the retina; molecular genetics of retina. *Mailing Add:* Eye Res Labs Univ Chicago 939 E 57 St Chicago IL 60637

APPLEBY, ALAN, b Newcastle, Eng, Apr 25, 37; m 60; c 2. RADIATION CHEMISTRY. *Educ:* Univ Durham, BSc, 58, PhD(radiation chem), 63. *Prof Exp:* Res assoc radiation chem, Univ Durham, 61-63 & Brookhaven Nat Lab, 63-65; sr sci officer, Radiochem Ctr, UK Atomic Energy Authority, Eng, 65-67; asst prof, 67-71, assoc prof, 71-77, PROF RADIATION SCI, RUTGERS UNIV, 77- *Mem:* Am Chem Soc; Sigma Xi. *Res:* Radiation chemistry of water and aqueous systems; radiation dosimetry by magnetic resonance; radiation-induced thermoluminescence. *Mailing Add:* 240 Beechwood Ave Middlesex NJ 08846

APPLEBY, ARNOLD PIERCE, b Formoso, Kans, Oct 24, 35; m 56; c 2. WEED SCIENCE. *Educ:* Kans State Univ, BS, 57, MS, 58; Ore State Univ, PhD(herbicide physiol), 62. *Prof Exp:* Instr agron, 59-62, asst prof, Pendleton br exp sta, 62-63 & Univ, 63-67, assoc prof, 67-72, PROF AGRON, ORE STATE UNIV, 72- *Honors & Awards:* Distinguished Prof, Ore State Univ; Nixon Distinguished Prof. *Mem:* Fel Am Soc Agron; Crop Sci Soc Am; fel Weed Sci Soc Am. *Res:* Agronomic and physiological aspects of herbicides. *Mailing Add:* Dept Crop Sci Ore State Univ Corvallis OR 97331

APPLEBY, JAMES E, b Canton, Ohio, Aug 6, 36. ENTOMOLOGY. *Educ:* Ohio State Univ, BS, 59, MS, 60, PhD(entom), 64. *Prof Exp:* From asst prof to assoc prof, 64-75, PROF ENTOM, ILL NATURAL HIST SURV, 75- *Res:* Life histories and control of insects and mites injurious to trees, shrubs and garden flowers. *Mailing Add:* Dept Econ Entom Univ Ill 163 Nat Resources Bldg Champaign IL 61820

APPLEBY, JOHN FREDERICK, b Houston, Tex, Aug 22, 48; m 77; c 1. PLANETARY ATMOSPHERES. *Educ:* Oberlin Col, BA, 70; Univ Mass, Amherst, MS, 72; State Univ NY, Stony Brook, PhD(planetary astron), 80. *Prof Exp:* Teaching & res asst, Univ Mass, Amherst & State Univ NY, Stony Brook, 70-80; res scientist, Planetary Sci Inst Sci Appln, Inc, 79-81; SR SCIENTIST, JET PROPULSION LAB, CALIF INST TECHNOL, 81- *Concurrent Pos:* Adj asst prof, Suffolk Co Community Col, 77; prin investr, NASA, 81- *Mem:* Am Astron Soc; Sigma Xi. *Res:* Theoretical modeling of the radiative and conective regions of the giant planet atmospheres, including CH_4 non-LTE processors and the effects of multiple aerosol scattering; temperature retrieval problems for Venus. *Mailing Add:* 218 Inverness Pl Glendora CA 91740-3940

APPLEBY, ROBERT H, b Visalia, Calif, Dec 16, 31; m 56; c 4. LOCAL AREA NETWORK COMMUNICATIONS, FIBER OPTIC COMMUNICATIONS. *Educ:* Washington Col, BS, 54; Va Polytech Inst, MS, 60. *Prof Exp:* sr eng assoc, E I du Pont de Nemours & Co, Inc, 60-90. *Concurrent Pos:* US Naval Pilot. *Mem:* Instrument Soc Am; World Futures Soc. *Res:* High speed, high density electronic interconnections. *Mailing Add:* 207 Watts St Durham NC 27701-2036

APPLEGARTH, DEREK A, b London, Eng, July 5, 37; Can citizen; m 84; c 3. GENETICS. *Educ:* Univ Durham, BSc, 58, PhD, 62. *Prof Exp:* Fel, Univ BC, 62-63; asst clin chemist, Vancouver Gen Hosp, 63-66; from asst prof to assoc prof pediat,71-78, PROF PEDIAT, UNIV BC, 78-; DIR BIOCHEM DIS LAB, CHILDREN'S HOSP, VANCOUVER, 69- *Concurrent Pos:* Clin instr, Univ BC, 63-66. *Mem:* Am Soc Human Genetics; fel Can Col Med Genetics; Soc Pediat Path. *Res:* Biochemical diseases of children. *Mailing Add:* Dept Pediat Univ BC Vancouver BC V6H 3V4 Can

APPLEGATE, HOWARD GEORGE, b Philadelphia, Pa, Mar 9, 22. ENVIRONMENTAL SCIENCES. *Educ:* Colo State Univ, BS, 50, MS, 52; Mich State Univ, PhD(physiol), 56. *Prof Exp:* Asst prof floricult, Univ Conn, 56-58; asst biochemist, State Col Wash, 58-60; asst prof plant physiol, Southern Ill Univ, 60-61; assoc prof, Ariz State Univ, 61-62 & Tex A&M Univ, 62-69; assoc prof plant physiol, 69-81, PROF ENVIRON SCI, UNIV TEX, EL PASO, 81- *Mem:* Air Pollution Control Asn; Am Chem Soc; US-Mex Border Health Asn. *Res:* Air pollution. *Mailing Add:* Educ Bldg Univ Tex El Paso TX 79968

APPLEGATE, JAMES EDWARD, b South Amboy, NJ, Aug 28, 42; m 65; c 2. WILDLIFE BIOLOGY. *Educ:* Rutgers Univ, BS, 64; Pa State Univ, MS, 66, PhD(zool), 68. *Prof Exp:* NIH trainee arbovirus ecol, Pa State Univ, 64-66, teaching asst zool, 66-67, USPHS fel parasitol, 67-68; res parasitologist, Naval Med Res Inst, 68-71; asst prof, 71-76, ASSOC PROF WILDLIFE BIOL, RUTGERS UNIV, 76- *Concurrent Pos:* Assoc ed, J Wildlife Mgt. *Mem:* Wildlife Dis Asn; Wildlife Soc; Am Inst Biol Sci. *Res:* Ecology of malaria; social aspects of wildlife management. *Mailing Add:* Dept Forestry Rutgers Univ New Brunswick NJ 08903

APPLEGATE, JAMES KEITH, b Macomb, Ill, Mar 26, 44; m 67; c 4. EXPLORATION GEOPHYSICS, GEOTECHNICAL ENGINEERING. *Educ:* Colo Sch Mines, GPE, 66, MS, 69, PhD(geophys eng), 74. *Prof Exp:* Geophysicist, Marathon Oil Co, 68-70; asst prof geophys, Boise State Univ, 73-77, assoc prof & head geol & geophys dept, 78-80; res assoc prof geophysics & dir explor res lab, Colo Sch Mines, 80-83; PRES, APPLEGATE ASSOC, INC, 83-; SECY, WALDEN & APPLEGATE INC, 85- *Concurrent Pos:* Pres, GeoTechniques, Inc, 77-81 & Terra Concepts Inc, 81-83. *Mem:* Soc Explor Geophysicists; Asn Eng Geologists; Am Assoc of Petrol Geologist. *Res:* High resolution seismic reflection techniques; geothermal exploration; rock and soil properties from geophysical techniques. *Mailing Add:* 105 Flora Way Golden CO 80401

APPLEGATE, LYNN E, b Dayton, Ohio, Apr 3, 41; m 61; c 3. MEMBRANE SEPARATIONS. *Educ:* Univ Dayton, BS, 63; Ind Univ, PhD(chem), 67. *Prof Exp:* Sr tech rep, 67-81, tech consult, 81-86, SR TECH CONSULT, E I DU PONT DE MENOURS & CO, INC, 86- *Honors & Awards:* Outstanding Serv Award, Am Soc Testing & Mat. *Mem:* Am Soc Testing & Mat; Am Chem Soc; Am Inst Chemists; Int Desalination Asn; Water Qual Asn. *Res:* Separations with membranes; purification and or concentration using ultrafiltration and reverse osmosis; water purification; polymer chemistry. *Mailing Add:* 5115 New Kent Rd Wilmington DE 19808

APPLEGATE, RICHARD LEE, b Mt Carmel, Ill, Jan 15, 36; m 57; c 2. ENTOMOLOGY, ZOOLOGY. *Educ:* Univ Southern Ill, BA, 59, MA, 61; SDak State Univ, PhD, 74. *Prof Exp:* Fishery biologist, Va Comn Game & Inland Fisheries, 61-63; fishery res biologist, Bur Sport Fisheries & Wildlife, SCent Reservoir Invests, 63-67; fishery res biologist & unit leader aquatic biol & fisheries, Bur Sport Fisheries & Wildlife, SDak Coop Fishery Unit, SDak State Univ, 67-84; at US Fish & Wildlife Serv, 84-87; COORDR, NAT FISH HATCHERY & TECHNOL CTR, 87- *Concurrent Pos:* Mem, Task Force Team 4 Conserv Aquatic Ecosyst, Int Biol Prog, Nat Acad Sci-Nat Res Coun, 67- *Mem:* Am Fisheries Soc; Am Soc Limnol & Oceanog. *Res:* Aquatic biology and ecology; limnology; fisheries. *Mailing Add:* San Marcos Nat Fish Hatchery & Technol Ctr Rte 1 Box 159-D San Marcos TX 78666

APPLEMAN, DANIEL EVERETT, b Berkeley, Calif, Apr 11, 31; m 67. GEOLOGY, CRYSTALLOGRAPHY. *Educ:* Calif Inst Technol, BS, 53; Johns Hopkins Univ, MA, 54, PhD(geol), 56. *Prof Exp:* Geologist, US Geol Surv, 54-74; crystallogr, Nat Mus Natural Hist, Smithsonian Inst, 74-, chmn, mineral sci dept, 78-; AT DEPT GEOL, GEORGE WASHINGTON UNIV. *Concurrent Pos:* Prof lectr, George Washington Univ, 64- *Mem:* Geol Soc Am; Mineral Soc Am; Am Crystallog Asn; Mineral Soc Gt Brit & Ireland; Am Geophys Union. *Res:* X-ray crystallography; crystal structures of uranium minerals; silicates; geochemistry. *Mailing Add:* 4434 Volta Pl Washington DC 20007

APPLEMAN, GABRIEL, b New York, NY, Dec 10, 22; m 50; c 3. CHEMICAL ENGINEERING. *Educ:* Polytech Inst Brooklyn, BChE, 54. *Prof Exp:* Chem analyst, Foster D Snell, Inc, 42; asst to dir res & develop, Fertilizers & Chem, Ltd, Israel, 49-52; chem engr, Foster D Snell, Inc, 52-53, proj engr, 54-55, asst dir eng, 56, dir eng, 57-59; plant engr, Ansbacher-Siegle Div, Sun Chem Corp, 59-60; proj mgr, Fluor-Singmaster & Breyer, Inc, 60-63; proj engr & gen construct supt, S Pierce Complex & plant engr, Am Agr Chem Co, Div Continental Oil Co, 63-65; prin proj engr, Fluor Corp, Ltd, 65-69; mgr, Eng Div, Makhteshim Chem Works, Ltd, 69-77; PRIN PROJ ENGR, FLUOR CORP, LTD, 77- *Mem:* Am Chem Soc; Am Soc Prof Eng; Am Inst Chem Engrs. *Res:* Chemical process engineering; process evaluation; economic studies; engineering and construction management and supervision; specialties food products; instant coffee; fertilizers; phosphoric acid; spray drying chemical plant design. *Mailing Add:* 1291 S Brass Lantern Dr La Habra CA 90631

APPLEMAN, JAMES R, b Woodward, Okla, Sept 21, 56; m; c 2. BIOCHEMISTRY. *Educ:* Okla State Univ, BS, 78, PhD(biochem), 83. *Prof Exp:* Res assoc, Dept Biochem, Dartmouth Med Sch, Hanover, NH, 83-85; res assoc, 85-88, ASST MEM, DEPT BIOCHEM & CLIN PHARMACOL, ST JUDE CHILDREN'S RES HOSP, MEMPHIS, TENN, 88- *Concurrent Pos:* Guest lectr protein eng, Dept Biochem, Univ Tenn, Knoxville, 91; grant, Nat Inst Gen Med Sci, NIH, 91- *Mem:* AAAS; Am Soc Biochem & Molecular Biol; NY Acad Sci; Sigma Xi; Protein Soc. *Res:* Relationship between the structure of enzymes and their functions utilizing biochemical and biophysical techniques; binding and other interactions of substrates and inhibitors with enzymes and subsequent conformational changes in enzyme-ligand complexes. *Mailing Add:* Dept Biochem & Clin Pharmacol St Jude Children's Res Hosp 332 N Lauderdale Memphis TN 38101

APPLEMAN, M MICHAEL, b Los Angeles, Calif, June 13, 33; m 58; c 3. BIOCHEMISTRY, SCIENCE EDUCATION. *Educ:* Univ Calif, Berkeley, BA, 57; Univ Wash, Seattle, PhD(biochem), 62. *Prof Exp:* Technician, Berkeley Lab, Nat Canners Asn, 57-58; NSF res fel metab, Inst Biochem Res, Argentina, 63-65; actg asst prof biochem, Univ Wash, Seattle, 65-66; from asst prof to assoc prof, 66-76, chmn, dept biol sci, 81-85, PROF BIOCHEM, UNIV SOUTHERN CALIF, 76- *Mem:* AAAS; Am Soc Biochem & Molecular Biol. *Res:* enzymology; biological education. *Mailing Add:* Dept Biol Sci Univ Southern Calif Los Angeles CA 90089-0371

APPLEQUIST, DOUGLAS EINAR, b Salt Lake City, Utah, Oct 29, 30; m 66; c 2. ORGANIC CHEMISTRY. *Educ:* Univ Calif, Berkeley, BS, 52; Calif Inst Technol, PhD(chem), 55. *Prof Exp:* From instr to prof, 55-86, EMER PROF ORG CHEM, UNIV ILL, URBANA-CHAMPAIGN, 86- *Mem:* Am Chem Soc. *Res:* Small-ring compounds; photochemical reactions; bridgehead displacements and eliminations; free-radical reactions. *Mailing Add:* 204 Pell Cir Urbana IL 61801-6637

APPLEQUIST, JON BARR, b Salt Lake City, Utah, Mar 19, 32; m 60; c 5. BIOPHYSICS, PHYSICAL CHEMISTRY. *Educ:* Univ Calif, BS, 54; Harvard Univ, PhD(chem), 59. *Prof Exp:* Asst prof chem, Univ Calif, Berkeley, 58-60 & Columbia Univ, 61-65; assoc prof, 65-68, PROF BIOPHYS, IOWA STATE UNIV, 68- *Concurrent Pos:* Prin investr, Sponsored Res, NIH, 60-, biophys & biophys chem study sect, NIH, 69-73,; secy-treas, Ames Sect, Am Chem Soc, 75, chmn, 76; prof chem courtesy appointment, Iowa State Univ, 86-89. *Res:* Molecular conformations; molecular optical properties; molecular interactions. *Mailing Add:* Dept Biochem & Biophys Iowa State Univ Ames IA 50011

APPLETON, B R, b Pampa, Tex, Nov 24, 37; m 59; c 3. SOLID STATE PHYSICS. *Educ:* Univ Mo, BS, 60; Rutgers Univ, MS, 64, PhD(physics), 66. *Prof Exp:* Mem res staff, Radio Corp Am, 61-63 & Bell Tel Labs Inc, 66-67; sect head, 74-86, div dir, 86-87, ASSOC DIR PHYS RES, OAK RIDGE NAT LAB, 87- *Mem:* Fel Am Phys Soc; fel AAAS; Bohmische Phys Soc; Mat Res Soc. *Res:* Basic ion-solid interactions; alteration of materials properties by ion bombardment and ion implantation techniques. *Mailing Add:* Box 2008 Oak Ridge Nat Lab Oak Ridge TN 37831

APPLETON, JOHN P(ATRICK), b Wolverton, Eng, Dec 17, 34; m 59; c 3. FLUID MECHANICS, APPLIED THERMODYNAMICS. *Educ:* Univ Southampton, BSc, 58, PhD(aeronaut eng), 61. *Prof Exp:* Eng apprentice, Aeronaut Engine Div, Rolls Royce Ltd, Eng, 51-55; res asst aeronaut eng, Univ Southampton, 57-61, lectr, 61-64; sr staff scientist, AC-Electronics Defense Res Labs, Calif, 64-67; from assoc prof to prof, 67-74, SR LECTR MECH ENG, MASS INST TECHNOL, 74-; V PRES RES & DEVELOP, THERMO ELECTRON CORP, 74- *Mem:* Am Phys Soc; Combustion Inst; Am Soc Metals; Tech Asn Pulp & Paper Indust. *Mailing Add:* 38 Sagamore Dr Andover MA 01810

APPLETON, JOSEPH HAYNE, b Collinsville, Ala, Aug 5, 27; m 54; c 6. CIVIL & STRUCTURAL ENGINEERING. *Educ:* Auburn Univ, BCE, 47; Univ Ill, MS, 49, PhD(civil eng), 59. *Prof Exp:* Res engr, US Bur Pub Roads, 49-50; instr civil eng, NC State Col, 50-51; res assoc, Univ Ill, 51-54; struct engr, Ala Cement Tile Co, 54-59; chmn, Div Eng, 67-71, dean eng, 71-78, PROF CIVIL ENG, UNIV ALA, BIRMINGHAM, 59- *Mem:* Am Soc Civil Engrs; Am Concrete Inst: Am Soc Eng Educ; Nat Soc Prof Engrs; Sigma Xi. *Res:* Reinforced and prestressed concrete; structural analysis and design; nature and properties of concrete and dental cements. *Mailing Add:* 4237 Antietam Dr Birmingham AL 35213

APPLETON, MARTIN DAVID, b Bangor, Pa, Feb 11, 17; m 41; c 2. BIOCHEMISTRY. *Educ:* Univ Scranton, BS, 39; Pa State Univ, MS, 41, PhD, 61. *Prof Exp:* Group leader & sr res chemist, Nuodex Prod Co, NJ, 41-42; tech dir & partner, Com & Indust Prod Co, 45-55; from instr to assoc prof, 55-63, PROF CHEM, 63-, CHMN DEPT, 75-, EMER PROF CHEM, UNIV SCRANTON, 88- *Concurrent Pos:* Consult indust chem & biochem; tech dir, Scicon, Inc. *Honors & Awards:* Rosmary Dybwad Int Award, Nat Asn Retarded Children, 70; Spec Citation, Pa Asn Retarded Children, 70. *Mem:* Fel AAAS; Am Chem Soc; fel Am Inst Chem; Sigma Xi. *Res:* Organic synthesis; clinical biochemistry; biochemistry of mental retardation; mechanism of cholera toxin. *Mailing Add:* 8318 Tobin Rd 12 Annandale VA 22003

APPLEWHITE, THOMAS HOOD, b Imperial, Calif, Dec 30, 24; m 45; c 2. ORGANIC CHEMISTRY. *Educ:* Calif Inst Technol, BS, 53, PhD(chem, plant physiol), 57. *Prof Exp:* Res chemist, Dow Chem Co, 56-59; org chemist, Western Utilization Res & Develop Div, Agr Res Serv, USDA, 59-63, head oilseed invest, 63-67; res dir, Pac Veg Oil Corp, 67-69; mgr, Edible Oil Prod Lab, Kraft, Inc, 69-78, dir res serv, Res & Develop, 78-87; CONSULT FATS & OILS, 87- *Concurrent Pos:* Ed, J Am Oil Chemists' Soc, 85-91. *Honors & Awards:* A E Bailey Medal, Am Oil Chemists Soc. *Mem:* Am Chem Soc; Am Oil Chemists Soc (pres, 77-78); Sigma Xi; Inst Food Technologists. *Res:* Enzyme-catalyzed reactions and the chemistry of amino acids; physical chemistry of water-soluble high polymers; chemistry and biochemistry of fats and fatty acids; mechanisms of organic reactions; analytical chemistry; health and nutrition. *Mailing Add:* 1032 Verbena Dr Austin TX 78750

APPLEYARD, EDWARD CLAIR, b Strathroy, Ont, June 22, 34; m 62; c 2. GEOLOGY. *Educ:* Univ Western Ont, BSc, 56; Queen's Univ, Ont, MSc, 60; Cambridge Univ, PhD(petrol), 62. *Prof Exp:* Asst lectr geol, Bedford Col, Univ London, 62-65; asst prof, 65-68, actg chmn dept, 69-70, ASSOC PROF EARTH SCI, UNIV WATERLOO, 68- *Concurrent Pos:* Vis res scientist, Commonwealth Sci & Indust Res Orgn, Australia, 79-80, vis scientist, Geothermal Inst, Univ Auckland, NZ, 86-87. *Mem:* Fel Geol Asn Can; Mineral Asn Can. *Res:* Geochemistry of altered rocks; wall rock alteration; metasomatism; metamorphism of evaporatic deposits; Precambrian geology. *Mailing Add:* Dept Earth Sci Univ Waterloo Waterloo ON N2L 3G1 Can

APPLING, WILLIAM DAVID LOVE, b Chicago, Ill, Sept 13, 34; m 64; c 2. MATHEMATICS. *Educ:* Univ Tex, BA, 55, PhD(math), 58. *Prof Exp:* Instr math, Duke Univ, 60-63; from asst prof to assoc prof, 63-69, PROF MATH, NTEX STATE UNIV, 69- *Mem:* Am Math Soc. *Res:* Real analysis; integral theory; absolute continuity. *Mailing Add:* Univ NTex Denton TX 76203

APPS, MICHAEL JOHN, b Herts, UK, Oct 4, 44; m 68; c 2. ENVIRONMENTAL SCIENCES. *Educ:* Univ BC, BSc, 66, MSc, 71; Univ Bristol, PhD(physics), 72. *Prof Exp:* Fel physics, Simon Fraser Univ, 72-75; res assoc neutron activation analysis, safe-low-power critical exp reactor, Univ Alberta, 76-81; RES SCIENTIST, ENVIRON CAN, 81- *Concurrent Pos:* Vis asst prof, Physics Dept, Simon Fraser Univ, 74-75; adj assoc prof, Fac Pharm, Univ Alberta, 81- *Res:* Application of analytical physics to various problems, including investigation of long-term effects in the terrestrial environment associated with natural radioactive materials; radioanalytical physics techniques of alpha and gamma spectroscopy; neutron activation analysis. *Mailing Add:* 11455 43rd Ave Edmonton AB T6J 0Y2 Can

APRAHAMIAN, ANI, b Beirut, Lebanon, Aug 15, 58; US citizen. NUCLEAR MODELS. *Educ:* Clark Univ, BA, 80, PhD(nuclear chem), 86. *Prof Exp:* Postdoctoral nuclear chem, Lawrence Livermore Nat Lab, 85-88; ASST PROF PHYSICS, UNIV NOTRE DAME, 89- *Mem:* Sigma Xi; Am Phys Soc; Am Chem Soc; AAAS. *Res:* Experimental and theoretical nuclear structure research. *Mailing Add:* Physics Dept Univ Notre Dame Notre Dame IN 46556

APRIL, ERNEST W, b Salem, Mass, Nov 6, 39; m 77; c 2. HUMAN ANATOMY, MUSCLE PHYSIOLOGY. *Educ:* Tufts Univ, BS, 61; Columbia Univ, PhD(anat), 69. *Prof Exp:* ASSOC PROF ANAT & CELL BIOL, COL PHYSICIANS & SURGEONS, COLUMBIA UNIV, 69- *Concurrent Pos:* USPHS grant, 71-; vis asst prof, Sch Med, Univ Miami, 70-71; vis assoc prof, Albert Einstein Col Med, 76-78. *Mem:* Sigma Xi; Am Asn Anat; Am Soc Cell Biol; Biophys Soc Am. *Res:* Investigation, using electron microscopic, x-ray diffraction and physiological methods, of the biophysical aspects of muscle contraction in single fibers by correlation of morphological ultrastructure, physiological processes and physical-chemical phenomena; investigations of anatomical bases of muscular dysfunction, particularly in the upper and lower limbs. *Mailing Add:* Col Physicians & Surgeons Columbia Univ New York NY 10032

APRIL, GARY CHARLES, b New Orleans, La, Jan 5, 40; m 62; c 3. CHEMICAL ENGINEERING. *Educ:* La State Univ, BS, 62, MS, 68, PhD(chem eng), 69. *Prof Exp:* Engr, Sabine River Works, E I du Pont de Nemours & Co, 62-66; grad asst chem eng, La State Univ, 66-69; from asst prof to prof, 69-82, asst dean res & grad studies, 83-89, UNIV RES PROF, UNIV ALA, 82- *Concurrent Pos:* Consult chem process indust; res scholar, Univ Ala, 80, res fel, 84-86; expert witness & prin investr res contracts. *Mem:* Am Inst Chem Engrs; Am Soc Eng Educ. *Res:* Chemicals from woody biomass; modeling natural and reacting systems; transport processes; mathematical modeling; kinetics. *Mailing Add:* Dept Chem Eng Univ Ala PO Box 870203 Tuscaloosa AL 35487-0203

APRIL, ROBERT WAYNE, b Chicago, Ill, Nov 15, 46; div. ECOLOGICAL ASSESSMENT, WATER QUALITY CRITERIA. *Educ:* Univ Chicago, SB, 68; Mass Inst Technol, PhD(anal chem), 72. *Prof Exp:* Asst prof chem, Univ Colo, Denver, 72-74; vis asst prof chem, Colo Sch Mines, 74-77; chemist, 78-81, environ scientist, 81-86, chief, Capacity & Storage Sect, 86-89, CHIEF, CRITERIA BR, US ENVIRON PROTECTION AGENCY, 89- *Res:* Developing variety of tools for ecological quality assessment including water quality criteria, sediment criteria, wildlife criteria and hological criteria. *Mailing Add:* US Environ Protection Agency WH-585 401 M St SW Washington DC 20460

APRILLE, THOMAS JOSEPH, JR, b Boston, Mass, Sept 26, 43. TRANSMISSION SYSTEMS ENGINEERING, SONET ARCHITECTURE. *Educ:* Northeastern Univ, BS, 67; Univ Ill, Urbana, MS, 68, PhD(elec eng), 72. *Prof Exp:* Elec engr, Magnavox Co, Urbana, 68-70; MGR, AT&T BELL LABS, ANDOVER, 72- *Concurrent Pos:* Tech assoc ed, Inst Elec & Electronics Engrs Commun Mag, 80-84; assoc ed, Inst Elec & Electronics Engrs Trans Commun, 81-84; mem, Hist Comt, Inst Elec & Electronics Engrs, 82-, chmn, 90-91; prog evaluator, Accreditation Bd Eng & Technol, 89- *Honors & Awards:* Guillemin-Cauer Award, Inst Elec & Electronics Engrs, Circuits & Systs Soc, 76. *Mem:* Fel Inst Elec & Electronics Engrs. *Res:* Computer analysis of autonomous circuits; development of novel amplifier and equalizer circuits; architecture definition of synchronous telecommunication systems; awarded five patents. *Mailing Add:* AT&T Bell Labs PO Box 369 Andover MA 01810

APRISON, MORRIS HERMAN, b Milwaukee, Wis, Oct 6, 23; m 49; c 2. NEUROCHEMISTRY, NEUROBIOLOGY. *Educ:* Univ Wis, BS, 45, cert, 47, MS, 49, PhD(biochem), 52. *Prof Exp:* Teaching asst physics, Univ Wis, 47-49; tech asst, Inst Paper Chem, Lawrence Col, 49-50; asst path, Univ Wis, 50-51, asst biochem, 51-52; head biophys sect, Galesburg State Res Hosp, Ill, 52-56; from asst prof to assoc prof, Dept Biochem & Psychiat, Ind Univ, 56-64, chief sect neurobiol, Inst Psychiat Res, 69-74, prof biochem, Dept Biochem & Psychiat, Med Ctr, 64-78, exec adminr, 73-74, dir, Inst Psychiat Res, 74-78, DISTINGUISHED PROF NEUROBIOL & BIOCHEM, SCH MED, IND UNIV, INDIANAPOLIS, 78-, CHIEF, SECT APPL & THEORET NEUROBIOL, 78- *Concurrent Pos:* Mem, study sect neuropsychol, NIMH, 70-74 & adv panel, Molecular & Cellular Neurobiol Prog, NSF, 84-86; co-ed, Advances Neurochem, 73-; mem bd overseers, St Meinrad; adv ed, J Biol Psychiat, 68-83, J Neurochem, 72-75, J Comparative & Gen Pharmacol, 74-75, J Develop Psychobiol, 74-77, J Gen Pharmacol, 75-; dep chief ed, J Neurochem, 80-83; mem, Comt Recommendations US Army Sci Res, Bd Physics & Astron, Nat Res Coun, 87-89; mem bd gov, Inst Advan Studies, Ind Univ, 89- *Honors & Awards:* Gold Metal Award Distinguished Res, Soc Biol Psychiat, 75. *Mem:* Biophys Soc; Soc Biol Psychiat; Int Soc Neurochem (secy, 75-79, chmn 79-81); Int Brain Res Orgn; Soc Neurosci; Sigma Xi. *Res:* Neurochemical correlates of behavior; molecular modeling with computers; applied and theoretical neurobiology; development of animal models for studying depression and psychosomatic disease. *Mailing Add:* Inst Psychiat Res Ind Univ Med Ctr 791 Union Dr Indianapolis IN 46202

APRUZESE, JOHN PATRICK, b Hartford, Conn, Nov 2, 48; m 78; c 1. PLASMA PHYSICS, RADIATION TRANSPORT. *Educ:* Yale Univ, BS, 70; Univ Wis Madison, PhD, 74. *Prof Exp:* Lectr & Planetarium dir astron, Univ Wyo, 74-75; post doc res fel, Naval Res Lab, 75-76; physicist, Sci Appln Int Corp, 76-82; RES PHYSICIST & HEAD RADIATION DYNAMICS SECT, NAVAL RES LAB, 82- *Mem:* Am Phys Soc; Am Astron Soc. *Res:* Atomic and plasma kinetics of x-ray lasers; radiation transport and hydrodynamics in dense plasmas; numerical modeling of radiation transport; laser propagaton modeling and phenomenology; optics. *Mailing Add:* Naval Res Lab Plasma Radiation Br Code 4720 Washington DC 20375-5000

APSIMON, JOHN W, b Liverpool, Eng, Aug 5, 35; m 61, 80; c 5. ORGANIC CHEMISTRY. *Educ:* Univ Liverpool, BSc, 56, PhD(org chem), 59. *Prof Exp:* Demonstr org chem, Univ Liverpool, 59-60; fel, Nat Res Coun Can, 60-62; from asst prof to assoc prof, 62-70, PROF ORG CHEM, CARLETON UNIV, ONT, 70-, ASSOC DEAN & DIR RES, 81-, DEAN GRAD STUDIES & RES, 90- *Concurrent Pos:* Sr ed, Can J Chem, 78-83. *Honors & Awards:* Merck Sharpe & Dohme lectr, 74. *Mem:* Am Chem Soc; Chem Inst Can; Royal Soc Chem; AAAS. *Res:* Synthesis, structure and function of natural products chemistry; organic synthesis. *Mailing Add:* Fac Grad Studies & Res 1516 DDT Carleton Univ Ottawa ON K1S 5B6 Can

APT, KENNETH ELLIS, b Bellingham, Wash, Apr 27, 45; m; c 2. NUCLEAR CHEMISTRY. *Educ:* Western Wash Univ, BA, 67; Mass Inst Technol, PhD(nuclear chem), 71. *Prof Exp:* Res asst, Mass Inst Technol, 67-71; fel, Chem & Nuclear Chem Group, Los Alamos Sci Lab, 71-73; staff mem, Environ Studies Group, 73-76, staff mem, Chem & Nuclear Chem Group, 76-78; int nuclear safeguards officer, Int Atomic Energy Agency, Vienna, 78-80; asst assoc dir chem, Earth & Life Sci, 80-84, assoc chem div leader, 84-88, arms control analyst, Ctr Nat Security Studies, 88-91, PROG MGR, ARMS REDUCTION TREATY VERIFICATION, LOS ALAMOS NAT LAB, 91- *Concurrent Pos:* Mem, Tech Comn Natural Fission Reactors, Int Atomic Energy Agency, 77-78. *Mem:* AAAS; Am Chem Soc. *Res:* Low-energy nuclear physics, analytical chemistry, isotope geochemistry, nuclear waste disposal, solid-state radiation dosimetry and atmospheric monitoring; international nuclear safeguards, non-destructive assay, policy analysis; arms control verification. *Mailing Add:* CNSS-C 91-116 Los Alamos Nat Lab MS E550 Los Alamos NM 87545

APT, WALTER JAMES, b Belfield, NDak, Jan 30, 22; m 48; c 3. PLANT PATHOLOGY, NEMATOLOGY. *Educ:* Wash State Univ, BS, 50, PhD(plant path), 58. *Prof Exp:* Agent nematologist, USDA, 55-56, nematologist, 56-63; nematologist, Pineapple Res Inst, 63-73; plant pathologist, 73-80, PROF PLANT PATH, UNIV HAWAII, MANOA, 80- *Concurrent Pos:* Consult nematologist, Hawaiian Sugar Planters Asn, 58-59. *Honors & Awards:* Ciba Geigy Award, 87. *Mem:* Am Phytopath Soc; Soc Nematol; Orgn Trop Am Nematologists. *Res:* Plant parasitic nematodes; control of nematodes; soil fumigation; systematic nematicides; application of nematicides by drip irrigation. *Mailing Add:* 1308 Mokapu Blvd Kailua HI 96734

APTE, CHIDANAND, b Miraj, India. KNOWLEDGE BASED SYSTEMS, BUSINESS APPLICATIONS. *Educ:* Indian Inst Technol, India, BTech, 76; Rutgers Univ, PhD(computer sci), 84. *Prof Exp:* Programmer, Indian Inst Technol Bombay Computer Ctr, 76-78; mem tech staff, RCA Astro-Electronics, 80-82; RES STAFF MEM, INT BUS MACH T J WATSON RES CTR, 84- *Concurrent Pos:* Adj lectr, Polytech Univ, 85; assoc ed, Expert Mag, Inst Elec & Electronics Engrs, Computer Soc, 89- *Mem:* Sr mem Inst Elec & Electronics Engrs; Asn Comput Mach; Am Asn Artificial Intel. *Res:* Knowledge base organizations for intelligent problem solving applications, with an emphasis on knowledge representation and qualitative modeling; exploring use of large-scale knowledge bases in business and marketing applications. *Mailing Add:* 19 Elaine Dr Stamford CT 06902

APTER, NATHANIEL STANLEY, b New York, NY, May 10, 13; m 41, 80; c 2. PSYCHIATRY. *Educ:* Cornell Univ, AB, 33; Univ Buffalo, MD, 38; Chicago Inst Psychoanal, grad, 58. *Prof Exp:* Rotating intern, Beth Israel Hosp, NY, 38-39; psychiat intern, Bellevue Hosp, 39-40; resident neurol, Kings County Hosp, 40-42; asst psychiat, Johns Hopkins Univ, 42-44, house officer, Johns Hopkins Hosp, 42-44; from asst prof to assoc prof & head dept, Univ Chicago, 46-55, Prof lectr psychiat, 55-84; RESIDENT ADJ PROF OCEANOG, NOVA UNIV, 84- *Concurrent Pos:* Chief investr res proj schizophrenia, Manteno State Hosp, 51-69, chief psychiat consult, 69-; attend psychiat, Michael Reese Hosp, 54-73, sr attend & res psychiat, 73-; psychiat consult, State Dept Ment Health, 54-; chief consult, McLean County Ment Health Clin, Bloomington, Ill, 57-68; sr psychiat consult, Ill State Psychiat Inst, 59-71, distinguished consult, 71-78; chief psychiat consult, Tinley Park Ment Health Clin, 70-74, 75-81. *Mem:* Fel AAAS; fel Am Psychiat Asn. *Res:* Schizophrenia; shell development in marine gastropods. *Mailing Add:* Nova Univ Oceanog Ctr 8000 N Ocean Dr Dania FL 33004

APUZZIO, J J, MATERNAL-FETAL MEDICINE. *Educ:* Rutgers Univ, BA, 69; Col Med & Dent, MD, 73. *Prof Exp:* Clin instr, Col Med & Dent NJ, 74-76, from instr to asst prof obstet & gynecol, 76-83; assod prof obstret & gynecol, Univ Med & Dent NJ, 83-86; DIR, HUMAN GENETICS & INFECTIOUS DIS, UNIV HOSP, NEWARK, 83- *Concurrent Pos:* Consult, obstet & gynecol, Vet Admin Hosp, East Orange, NJ, 79-88; attending, Univ Hosp, 79-; consult maternal fetal med, St Michaels Hosp, Newark, 80-88, St Elizabeths Hosp, Elizabeth, NJ, 76-88. *Mem:* Fel Am Col Surgeons; Am Fertility Soc; Am Col Obstetricians & Gynecologists. *Mailing Add:* Dept Obstet & Gynecol Univ Med & Dent NJ Med Sch 185 S Orange Ave Newark NJ 07103-2757

AQUINO, DOLORES CATHERINE, b Chicago, Ill, Oct 12, 49. CHEMISTRY. *Educ:* Ill Inst Technol, BS, 70; Ohio State Univ, PhD(inorg chem), 77. *Prof Exp:* Res chemist, 77-80, SR CHEMIST, EXXON RES & ENG CO, 80-, COORDR TECHNICIAN TRAINING, 81- *Concurrent Pos:* Lectr, Univ Houston, Clear Lake City, 81. *Mem:* Am Chem Soc; Soc Women Engr. *Res:* Synthetic fuels, particularly the chemistry of coal gasification processes; inorganic chemistry, particularly synthesis and characterization of organosilanes and homogeneous catalysis by transition metals. *Mailing Add:* 4600 Beechnut St No 207 Houston TX 77096

ARAB-ISMAILI, MOHAMMAD SHARIF, b Shahrood, Iran, Feb 10, 49; m 75; c 1. COMPUTER & INFORMATION SCIENCES. *Educ:* Abadan Inst Technol, BS, 71; Mass Inst Technol, PhD(chem eng), 78. *Prof Exp:* Systs analyst comput appln, Nat Iranian Oil Co, 71-73; RES ENGR COMPUT AIDED DESIGN, MASS INST TECHNOL, 78- *Concurrent Pos:* Consult, Walden Res Div Abcor, 75-77. *Mem:* Sigma Xi. *Res:* Computer aided design; management information systems. *Mailing Add:* 415 W Aldine Unit 4C Chicago IL 60657

ARABYAN, ARA, b Istanbul, Turkey, Mar 11, 53; US citizen. PARALLEL COMPUTATIONAL METHODS. *Educ:* Tex A&M Univ, BSc, 80; Univ Southern Calif, MSc, 82, PhD(mech eng), 86. *Prof Exp:* Grad asst, Univ Southern Calif, 81-86; mem tech staff, Allied-Signal Garrett Corp, 86; ASST PROF MECH ENG, UNIV ARIZ, 86- *Concurrent Pos:* Consult, Adelberg Labs, 81-85, Phys Educ Dept, Univ Southern Calif, 85-86 & accident invests var law firms, 88-; prin investr, NSF, 86-89 & 88-89, NASA Space Eng Ctr grant & Honeywell, Inc, 90-; panelist, NSF, 89; NSF presidential young investr, 90. *Mem:* Am Soc Mech Engrs. *Res:* Parallel computational methods for simulation and control of complex mechanical systems; distributed control of articulated multijoint biosystems and mechanical linkages. *Mailing Add:* 1431 W Las Lomitas Rd Tucson AZ 85704

ARAGAKI, MINORU, b Honolulu, Hawaii, June 26, 26; m 52; c 3. PLANT PATHOLOGY. *Educ:* Univ Hawaii, BS, 50, MS, 54, PhD(bot), 63. *Prof Exp:* Lab technician, 46-54, jr plant pathologist, 54-61, from asst prof to assoc prof, 61-72, PROF PLANT PATH, UNIV HAWAII, 72- *Mem:* Am Phytopath Soc; Mycol Soc Am. *Res:* Light induced fungal sporulation including inhibitory effects of light; biology of Phytophthora; anatomical changes following invasion by fungal pathogen, particularly host barrier formation. *Mailing Add:* Dept Plant Path Univ Hawaii at Manoa Honolulu HI 96822

ARAI, HISAO PHILIP, b Los Angeles, Calif, Oct 8, 26; m 58. PARASITOLOGY. *Educ:* Univ Calif, Los Angeles, AB, 54, PhD(zool), 60. *Prof Exp:* USPHS fel zool, Univ BC, 60-61; assoc prof biol, Ill State Univ, 61-63; asst prof, Univ Alta, 63-65; assoc prof, 65-77, PROF ZOOL, UNIV CALGARY, 77- *Concurrent Pos:* Assoc mem ctr for zoonoses res, Univ Ill, 62. *Mem:* Am Soc Parasitol; Soc Syst Zool; Can Soc Zool; Sigma Xi. *Res:* Physiological ecology; taxonomy; ecology and physiology of animal parasites. *Mailing Add:* Dept Biol Univ Calgary Calgary AB T2N 1N4 Can

ARAI, KEN-ICHI, b Sapporo, Japan, Oct 18, 42; m 71; c 1. DNA REPLICATION, LYMPHOKINES. *Educ:* Univ Tokyo, MD, 67, PhD, 74. *Prof Exp:* Prin sci, 81-83, DIR, DNAX RES INST, 83- *Concurrent Pos:* Consult prof, Stanford Univ, 87-; prof, Univ Tokyo, 89- *Mem:* Am Asn Biochem & Molecular Biol; Am Asn Immunologists; Japan Molecular Biol Soc. *Res:* T-cell activation and molecular biology of lymphokines; DNA replication. *Mailing Add:* DNAX Res Inst 901 Calif Ave Palo Alto CA 94304

ARAJS, SIGURDS, b Taurkalne, Latvia, Sept 2, 27; US citizen; m 52; c 1. SOLID STATE PHYSICS. *Educ:* Iowa State Univ, BS, 53, PhD, 57. *Prof Exp:* Res technician ultrasonics, Eng Exp Sta, Iowa State Univ, 52-53, asst physics, 53-54, Inst Atomic Res, 54-57; scientist, Bain Lab Fundamental Res,

US Steel Corp, 58-60, from supv scientist to sr scientist, 60-67; chmn dept, 75-83, PROF PHYSICS, CLARKSON UNIV, 83- *Mem:* Fel Am Phys Soc; Europ Phys Soc; Swed Royal Phys Soc; Sigma Xi; foreign mem Latvian Acad Sci. *Res:* Magnetism; transport properties of solids; electronic structures of metals; low temperature phenomena; composite materials; fluid physics; thin films; small particles; electro-magnetorheology. *Mailing Add:* Dept Physics Clarkson Univ Potsdam NY 13676

ARAKAWA, EDWARD TAKASHI, b Honolulu, Hawaii, Apr 8, 29; m 55; c 4. SOLID STATE PHYSICS, OPTICS. *Educ:* Howard Col, BS, 51; La State Univ, MS, 53; Univ Tenn, PhD(physics), 57. *Prof Exp:* Res assoc med physics, Oak Ridge Inst Nuclear Studies, 53-56, SR RES STAFF MEM, OAK RIDGE NAT LAB, 57-, GROUP LEADER, 65- *Concurrent Pos:* Consult, Radiation Effects Res Found, 58-60; guest scientist, Europ Sci Res Orgn, 70. *Mem:* Fel Optical Soc Am; fel Am Phys Soc. *Res:* Dosimetry of nuclear radiation; electron physics; vacuum ultraviolet spectroscopy; soft x-ray spectroscopy; thin film physics; resonance ionization mass spectroscopy; ellipsometry. *Mailing Add:* 4500 S H-160 Oak Ridge Nat Lab PO Box 2008 Oak Ridge TN 37831-6123

ARAKI, GEORGE SHOICHI, b Oakland, Calif, Jan 11, 32; m 58; c 3. BIOLOGY. *Educ:* San Francisco State Col, AB, 57; Stanford Univ, PhD(biol), 64. *Prof Exp:* From asst prof to assoc prof, 62-74, PROF BIOL, SAN FRANCISCO STATE UNIV, 74-; DIR CTR INTERDISCIPLINARY SCI, 74- *Mem:* AAAS; Sigma Xi. *Res:* Comparative invertebrate physiology; cellular physiology; physiology of feeding and digestion in asteroids; human healing and health processes; biofeedback; autogenic training; intracellular digestive processes in asteroids and slime molds. *Mailing Add:* Ten Seadrift Landing Tiburon CA 94920

ARAKI, MASASUKE, b Kyoto, Japan, Mar 13, 50. DEVELOPMENTAL NEUROBIOLOGY, STRUCTURE OF THE RETINA. *Educ:* Kyoto Univ, BSc, 74, MSc, 76, PhD(develop biol), 79; Jichi Med Sch, DMed, 89. *Prof Exp:* Res assoc anat, 78-88, LECTR ANAT, JICHI MED SCH, 88- *Concurrent Pos:* Res fel, Univ BC, 83-84; vis lectr, Shiga Univ Med Sci, 85-86. *Res:* Cellular and molecular mechanism for the development of nervous systems, particularly the neural retina and pineal. *Mailing Add:* Dept Anat Jichi Med Sch Minamikawachi Tochigi 329-04 Japan

ARAKI, MINORU S, AERONAUTICS. *Prof Exp:* EXEC VPRES, LOCKHEED MISSILES & SPACE SYST CO, INC, 88- *Mem:* Nat Acad Eng. *Mailing Add:* Lockheed Missiles & Space Syst Co Inc Dept 100 Bldg 101 PO Box 3504 Sunnyvale CA 94088-3504

ARALA-CHAVES, MARIO PASSALAQUA, b Fafe, Portugal, Dec 2, 39; m 65; c 2. IMMUNOBIOLOGY. *Educ:* Univ Lisbon, MD, 65; Univ Louvain, MS, 67, PhD(transfer factor), 75. *Prof Exp:* Fel res immunobiol, Dept Exp Med, Univ Louvain, 65-67; head dept blood bank, Univ Hosp Luand, 68-70; fel immunobiol, Transplantation Dept, Cent Blood Transfusions, Amsterdam, 70-71; head dept blood bank, Portugese Cancer Inst, Lisbon, 72-75; asst prof immunobiol, Dept Basic & Clin Immunol, Med Univ SC, 75-80; DIR IMMUNOL, INST BIOMED SCI ADEL SALAZAR, UNIV PORTO LARGO MED SCH, PORTUGAL. *Concurrent Pos:* Adj assoc prof immunol & microbiol, Med Univ SC, 80- *Mem:* Brit Soc Immunol; Europ Soc Clin Invest. *Res:* Study of leukocyte and erythrocyte antigens; immunodeficient syndromes; biological aspects of linfoquines; biological aspects and biological characterization of Transfer Factor-d. *Mailing Add:* Inst Biomed Sci Adel Salazar Univ Porto Largo Med Sch Porto 1000 Portugal

ARAMINI, JAN MARIE, b Highland Park, Mich, Sept 29, 55; m 89. HEALTH PSYCHOLOGY. *Educ:* Loyola Univ, Chicago, Ill, BS, 86; Ill Sch Prof Psychol, Chicago, PsyD, 91. *Prof Exp:* Regist nurse, Old Orchard Hosp, 81-87; CLIN PSYCHOLOGIST, N CHICAGO VET AFFAIRS MED CTR, GREAT LAKES NAVAL BASE, USN, 91- *Mem:* Am Psychol Asn; Am Asn Coun & Develop. *Res:* Relationship between chronic fatigue syndrome and replicable personality characteristics profiles on the Minnesota Multiphasic Personality Inventory. *Mailing Add:* 7215 N Hamilton Ave Chicago IL 60645

ARAMS, F(RANK) R(OBERT), b Danzig, Oct 18, 25; nat US; m 52. ELECTRONICS. *Educ:* Univ Mich, BSE, 47; Harvard Univ, ScM, 48; Stevens Inst Technol, MS, 53; Polytech Inst Brooklyn, PhD(electrophysics), 61. *Prof Exp:* Sr staff mem, RCA Corp, 48-56; consult advan tech develop, AIL Div, Eaton Corp, 56-59; head infrared & electrooptics dept, 65-71; V PRES, LNR COMMUN, 71- *Mem:* Fel Inst Elec & Electronics Engrs; Optical Soc Am. *Res:* Satellite communications; low-noise receivers; lasers and coherent optical and infrared systems. *Mailing Add:* LNR Commun Inc 180 Marcus Blvd Hauppauge NY 11788

ARANDA, JACOB VELASCO, b Isabela, Philippines, Dec 29, 42; Can citizen; m 74; c 2. PEDIATRICS, NEONATOLOGY. *Educ:* Manila Cent Univ, MD, 65; McGill Univ, PhD(pharmacol), 75; FRCP(C), 85. *Prof Exp:* Intern med, US Naval Hosp, 64-65; Wash Hosp Ctr, 65-66; resident pediat, State Univ NY Hosp Ctr, 66-68; fel neonatology, Case Western Reserve Univ, Cleveland Metrop Gen Hosp, 68-69; res fel, McGill Univ, Montreal Children's Hosp, 69-71; res fel pharmacol, Med Res Coun Can, 71-74; from asst prof to assoc prof pediat & pharmacol, 74-84, PROF PEDIAT & PHARMACOL, FAC MED, MCGILL UNIV, 84- *Concurrent Pos:* Neonatologist & pediatrician, Royal Victoria Hosp, Jewish Gen Hosp, St Mary's Hosp & Montreal Children's Hosp, 74-; dir, develop pharmacol & perinatal res & Apnea-Sudden Infant Death Syndrome Treat & Res Ctr, Montreal Children's Hosp, 78-; ed-in-chief develop pharmacol & therapeut, Int J Perinatal-Pediat Drug Ther, Karger, Switz, 79- *Honors & Awards:* Queen Elizabeth II Res Scientist Award, Queen Elizabeth II Res Found, 76. *Mem:* Am Acad Pediat; Am Soc Pharmacol & Exp Therapeut; Am Soc Clin Pharmacol & Therapeut; Can Soc Clin Invest; NY Acad Sci; Soc Pediat Res. *Res:* Pharmacology of the fetus and newborn; effects of drugs on control breathing; regulation and drug effect on cerebral blood flow of the newborn; adverse drug effects and utilization in the perinatal period; epidemiology. *Mailing Add:* 2300 Tupper Ave Montreal PQ H3H 1P3 Can

ARANEO, BARBARA ANN, m; c 2. IMMUNOBIOLOGY OF RECOMBINANT VACCINES. *Educ:* Univ Rochester, PhD(microbiol), 76. *Prof Exp:* ASST PROF IMMUNOL, UNIV UTAH, 82- *Mem:* Am Asn Immunologists; Am Asn Pathologists. *Res:* Immunobiology of T-cell ion responsiveness to antigen and new strategies for the development of effective anti-viral vaccines. *Mailing Add:* Dept Path Univ Utah Med Ctr 50 N Medical Dr Salt Lake City UT 84132

ARANOW, RUTH LEE HORWITZ, b Brooklyn, NY, Aug 25, 29; m 50; c 3. PHYSICAL CHEMISTRY. *Educ:* Brooklyn Col, BS, 51; Johns Hopkins Univ, MA, 52, PhD(chem), 57. *Prof Exp:* Jr instr chem, Johns Hopkins Univ, 51-53,; asst, 53-54, inst, McCoy Col, 54-57, staff scientist, Res Inst Advan Studies, MArtin-Marietta Co, 57-69; NIH spec fel, 69-71, res scientist, 69-74, FEL CHEM, JOHNS HOPKINS UNIV, 74-, LECTR, 76- *Mem:* AAAS; Am Phys Soc; Sigma Xi. *Res:* Statistical mechanics; effect of noise on neural membranes; effects of chemicals on behavior. *Mailing Add:* Off Academic Advising Johns Hopkins Univ 3614 Eastwood Dr Baltimore MD 21218

ARATA, DOROTHY, b New York, NY, Mar 8, 28. NUTRITION. *Educ:* Pratt Inst, BS, 48; Cornell Univ, MS, 50; Univ Wis, PhD(biochem), 56. *Prof Exp:* Asst dept foods & nutrit, Cornell Univ, 48-50, technician dept biochem, Med Sch, 50-52, fel, 56-57; asst biochem, Univ Wis, 53-56; from asst prof to prof foods & nutrit, 57-69, assoc dir, Honors Col, 69-71, PROF HUMAN DEVELOP, COL HUMAN MED, MICH STATE UNIV, 69-, ACTG DIR, HONORS COL, 75-, ASST PROVOST, 71- *Concurrent Pos:* Acad admin intern, Am Coun Educ, 68, mem comn educ credit, 74-77; mem, Mich State Selection of Rhodes Scholars, 77 & 78; mem, Col Level Exam Prog, Am Coun Educ, 78-; mem, Mich State Planning Comt Concerning Nat Identification Women in Higher Educ, 78- *Mem:* AAAS; Am Chem Soc; Am Inst Nutrit; Am Inst Chem; Am Asn Higher Educ. *Res:* Biochemical disorders associated with nutritional deficiencies; cellular disruptions in fatty livers induced by threonine deficiency. *Mailing Add:* Off Grad Studies Memphis State Univ Memphis TN 38152

ARAUJO, JOSE EMILIO GONCALVES, soil science, for more information see previous edition

ARAUJO, OSCAR EDUARDO, b Brazil, June 2, 27; nat US; m 50; c 2. PHARMACY. *Educ:* Purdue Univ, BS, 54, MS, 55, PhD(pharm), 57. *Prof Exp:* Asst pharm, Purdue Univ, 54-55; asst prof, Ohio Northern Univ, 57-62; from asst prof to assoc prof, 62-74, PROF PHARM, UNIV FLA, 74- *Mem:* Am Pharmaceut Asn; Am Asn Cols Pharm; Acad Pharmaceut Sci. *Res:* Dermatology and clinical pharmacy. *Mailing Add:* Dept Pharm Univ Fla Med Col J Hills Miller Health Ctr Gainesville FL 32610

ARAUJO, ROGER JEROME, b Fall River, Mass, Mar 13, 34; m 54; c 4. PHYSICAL CHEMISTRY. *Educ:* Bradford Durfee Tech Inst, BS, 56; Brown Univ, PhD(phys chem), 62. *Prof Exp:* RES FEL, CORNING INC, 61- *Mem:* Am Chem Soc; fel Am Ceramic Soc; Mat Res Soc. *Res:* Hydrogen aging of optical fibers; radiation damage in optical fibers. *Mailing Add:* Corning Glass Works Sullivan Park Corning NY 14831

ARAVE, CLIVE W, b Idaho Falls, Idaho, May 12, 31; m 50; c 6. ANIMAL BREEDING. *Educ:* Utah State Univ, BS, 56, MS, 57; Univ Calif, Davis, PhD(genetics), 63. *Prof Exp:* Asst mgr, Lavacre Farms, Calif, 57-59; lab technician genetics, Univ Calif, 59-60; asst prof dairy husb, Chico State Col, 63-65; asst prof, 65-81, ASSOC PROF ADV DAIRY HUSB, UTAH STATE UNIV, 81- *Concurrent Pos:* NSF res grant, New Zealand, 80-81; mem, Animal Care Comt, Am Dairy Sci Asn, 87-, regional proj NC-119, 83-, secy, 86-87 & 88-89, chmn, 87-88 & 90-91, regional proj NCR-131, 84-; consult, Off Int Coop & Develop. *Mem:* Am Dairy Sci Asn; Am Soc Animal Sci. *Res:* Dairy cattle production genetics and management; animal breeding; animal behavior of domesticated species. *Mailing Add:* Animal Dairy Vet Sci Dept Utah State Univ Logan UT 84322-4815

ARBEENY, CYNTHIA M, LIPOPROTEIN METABOLISM. *Educ:* Rutgers Univ, PhD(physiol), 77. *Prof Exp:* ASST PROF MED, ALBERT EINSTEIN COL MED, 82- *Mailing Add:* Squibb Inst Med Res PO Box 4000 Princeton NJ 08543-4000

ARBENZ, JOHANN KASPAR, b Berne, Switz, Aug 27, 18; nat US; m 45; c 3. GEOLOGY. *Educ:* Univ Berne, PhD(geol), 45. *Prof Exp:* Asst geologist, R Helbling Surv Off, 46-47 & Swiss Geol Surv, 47; from asst prof to assoc prof geol, Univ Okla, 48-55; geologist, Shell Oil Co, 55-75, sr staff geologist, 75-83; CONSULT GEOLOGISTS, 83- *Mem:* Geol Soc Am; Am Asn Petrol Geologists; Swiss Geol Soc. *Res:* Structural and petroleum geology. *Mailing Add:* 3964 Wonderland Hill Ave Boulder CO 80304

ARBER, WERNER, b Granichen, Switz, 29; m 66; c 2. MOLECULAR GENETICS. *Hon Degrees:* Dr, Univ Southern Calif, 86, Univ Louis Pasteur, Strasbourg, 88. *Prof Exp:* Asst, Lab Biophys, Univ Geneva, 53-58; res assoc , Dept Microbiol, Univ Southern Calif, 58-59; prof molecular genetics, Univ Geneva, 62-70; vis investr, Dept Molecular Biol, Univ Calif, Berkeley, 70-71; PROF MICROBIOL, UNIV BASEL, 71- *Honors & Awards:* Nobel Prize in Med, 78. *Mailing Add:* Biozentrum Universitat Basel 70 Klingelbergstrasse CH-4056 Basel Switzerland

ARBIB, MICHAEL A, b Eastbourne, Eng, May 28, 40; m 65; c 2. COMPUTER SCIENCES. *Educ:* Univ Sydney, BSc, 61; Mass Inst Technol, PhD(math), 63. *Prof Exp:* Assoc prof elec eng, Stanford Univ, 69-70; chmn dept comput & info sci, 70-75, PROF COMPUT & INFO SCI & ADJ PROF PSYCHOL, UNIV MASS, AMHERST, 70-, DIR CTR SYSTS NEUROSCI, 74- *Res:* Brain modelling; information processing in complex systems, in automata and system theory and in neurophysiology and psychology; social implications of computer science. *Mailing Add:* Univ Mass Amherst MA 01003

ARBITER, NATHANIEL, b Yonkers, NY, Jan 2, 11; m 34, 60; c 6. MINERAL PROCESSING, HYDROMETALLURGY. *Educ:* Columbia Univ, BA, 32. *Prof Exp:* Res metallurgist, Battelle Mem Inst, 43-44 & Phelps Dodge Corp, 44-51; dir res & chief metall, Anaconda Co, Tucson, Ariz, 68-77; res asst eng, 36-43, from assoc prof to prof, 51-69, EMER PROF MINERAL ENG, COLUMBIA UNIV, 77-; ADJ PROF METALL, UNIV UTAH, SALT LAKE CITY, 84- *Concurrent Pos:* Consult, 51-68, 77- *Mem:* Nat Acad Eng; hon mem, Am Inst Mining Engrs; Soc Mining Eng. *Res:* Author of 70 publications in mineral processing; extract metal. *Mailing Add:* 6300 S High Valley Rd Vail AZ 85641

ARBOGAST, RICHARD TERRANCE, b Freeport, Ill, Aug 7, 37; m 58; c 4. INSECT PESTS STORED PRODUCTS, ECOLOGY. *Educ:* Univ Ill, BS, 59; Univ Fla, PhD(entom), 65; Armstrong State Col, BS, 84. *Prof Exp:* RES ENTOMOLOGIST, STORED-PROD INSECTS RES & DEVELOP LAB, AGR RES SERV, USDA, 65- *Mem:* Entom Soc Am; Lepidop Soc; Am Entom Soc; Soc Pop Ecol. *Res:* Ecology and behavior of stored-product insects. *Mailing Add:* 114 Monica Blvd Savannah GA 31419

ARBULU, AGUSTIN, b Lima, Peru, Sept 15, 28; US citizen; c 3. SURGERY, THORACIC SURGERY. *Educ:* San Marcos Univ, Lima, MB, 54, MD, 55. *Prof Exp:* Asst chief surg, Wichita Vet Admin Hosp, Kans, 61-62; from instr to assoc prof, 62-72, PROF SURG, SCH MED, WAYNE STATE UNIV, 72- *Concurrent Pos:* Asst chief surg, Vet Admin Hosp, Allen Park, Mich, 62-66, actg chief surg, 66-67, consult, 67-; fel res coun, Am Heart Asn, 78; assoc, Hutzel & Harper Hosps, Detroit; attend surg, Detroit Gen Hosp; consult chest surg, Grace Hosp, Providence Hosp & Oakwood Hosp; cardiac surgeon, St Joseph Mercy Hosp, Pontiac & St John Hosp, Detroit; gov, Am Col Chest Physicians, 83-88, chmn, Int Comt, 89- *Honors & Awards:* Frederick A Coller Award, Am Col Surgeons, 68 & 72; Cecile Lehman Mayer Res Award, Am Col Chest Physicians, 68, 69 & 70, Regents Award, 72. *Mem:* Am Thoracic Soc; Am Col Chest Physicians; Soc Thoracic Surg; Peruvian Am Med Soc (pres, 73-74); Am Asn Thoracic Surg; Europ Asn Thoracic Surg. *Mailing Add:* Midwest Surg Assn 829 Harper Profl Bldg 4160 John R St Detroit MI 48201

ARCADI, JOHN ALBERT, b Whittier, Calif, Oct 23, 24; m 51; c 7. UROLOGICAL ONCOLOGY, GENERAL ENDOCRINOLOGY. *Educ:* Univ Notre Dame, BS, 47; Johns Hopkins Univ, MD, 50. *Prof Exp:* Nat Cancer Inst fel urol, Brady Urol Inst, Johns Hopkins Hosp, 51-52, 53-55; asst prof, Univ Southern Calif Sch Med, 55-60; res assoc, 57-70, RES PROF BIOL, WHITTIER COL, 70- *Concurrent Pos:* Attend urologist, Murphy Hosp, Whittier, Calif, 55-59 & Presby Community Hosp, 59- *Mem:* Am Asn Anatomists; Endocrine Soc; Am Soc Cell Biol; Am Col Surgeons; Am Urol Asn; fel AAAS. *Res:* Factors associated with the growth and atrophy of the prostate gland; role of the extracellular matrix components including mast cells in prostate growth particularly its role in the production of benign prostatic hyperplasia. *Mailing Add:* PO Box 9220 Whittier CA 90608

ARCAND, GEORGE MYRON, b Ocean Falls, BC, Feb 9, 24; nat US; m 52. ANALYTICAL CHEMISTRY, INORGANIC CHEMISTRY. *Educ:* Calif Inst Technol, BS, 50, PhD(chem), 55. *Prof Exp:* Asst prof chem, Univ Mo, 55-58 & Mont State Col, 58-62; sr scientist, Jet Propulsion Lab, Calif Inst Technol, 62-65, scientist specialist, 65; assoc prof, 65-67, PROF CHEM, IDAHO STATE UNIV, 67-, DEPT CHMN, 87- *Mem:* Am Chem Soc; Sigma Xi. *Res:* Electro-analytical methods; mechanisms of liquid-liquid extraction processes and their use in the study of aqueous systems; inorganic complex ions in aqueous solutions; electrochemistry of batteries; environmental trace analysis. *Mailing Add:* Dept Chem PO Box 8146 Idaho State Univ Pocatello ID 83209

ARCE, A ANTHONY, b San Juan, Puerto Rico, June 13, 23; m 71; c 3. PSYCHIATRY. *Educ:* Washington & Jefferson Col, BS, 42; Med Sch, Temple Univ, MD, 46. *Prof Exp:* Assoc dir psychiat, Grasslands Hosp, 62-67; dir aftercare, NY State Dept Ment Hyg, 68-71; assoc clin prof, Col Physicians & Surgeons, Columbia Univ, 71-76; PROF PSYCHIAT, MED COL, HAHNEMANN UNIV, 76-, CHMN DEPT, 85- *Concurrent Pos:* Vis prof, Univ Mex Res Ctr, 69 & Staff Col, Nat Inst Ment Health, 79-81; dir, Meyer-Manhattan Psychiat Ctr, 71-76 & Hahnemann Community Ment Health Ctr, 76-84. *Honors & Awards:* Warren Williams Award, Am Psychiat Asn, 84. *Mem:* Fel Am Psychiat Asn; fel Am Col Ment Health; fel Am Orthopsychiat Asn. *Res:* Mental health services minorities; administration (planning and development) programs; cross-cultural issues on mental health services and training; homeless mentally ill. *Mailing Add:* Dept Mental Health Sci Hahnemann Med Col 85230 N Broad St Philadelphia PA 19102

ARCE, GINA, b Cuba, Aug 28, 29; nat US. BIOLOGY. *Educ:* George Peabody Col, BA, 48, MA, 49; Vanderbilt Univ, PhD(biol), 56. *Prof Exp:* Teacher high sch, 49-51; instr bot, Vanderbilt Univ, 56-57; assoc prof, 58-69, PROF BOT, CALIF STATE UNIV, FRESNO, 69- *Mem:* Bot Soc Am; Phycol Soc Am. *Res:* Taxonomy of green algae. *Mailing Add:* Dept Biol Calif State Univ Fresno CA 93740

ARCE, JOSE EDGAR, b Lima, Peru, Apr 21, 37; m 61; c 4. GEOPHYSICS, GEOLOGY. *Educ:* San Marcos Univ, Lima, Geol Engr, 60, BSc, 66, PhD(geol), 72. *Prof Exp:* Geologist, Marcona Mining Co, Peru, 60-63, assoc consult geophys, 63-64; assoc prof, San Marcos Univ, Lima, 64-72; independent consult geophysicist & exploration geophysicist, 64-69; prin prof geophys, Univ Eng, Lima, 69-90; CONSULT, 90- *Concurrent Pos:* Chmn, VI Peruvian Geol Cong, 87. *Mem:* Soc Explor Geophys; European Asn Explor Geophys; Geol Soc Peru (pres, 80-81); NY Acad Sci. *Res:* Applied geophysics in mining and groundwater exploration; civil engineering studies with geophysical methods; regular use of electrical, magnetic, seismic, electromagnetic and well logging techniques. *Mailing Add:* Petit Thouars 4380 Miraflores Lima 18 Peru

ARCENEAUX, JOSEPH LINCOLN, b Lafayette, La, Aug 13, 41; m 70; c 1. MICROBIOLOGY, BACTERIAL PHYSIOLOGY. *Educ:* Univ Southwestern La, BS, 63; Univ Tex, Austin, PhD(microbiol), 68. *Prof Exp:* NIH fel, Princeton Univ, 67-70; asst prof, 70-75, ASSOC PROF MICROBIOL, MED CTR, UNIV MISS, 75-, ASST DEAN STUDENT AFFAIRS, 78- *Mem:* AAAS; Am Soc Microbiol. *Res:* Chromosome replication and genetic studies in Bacillus subtitis. *Mailing Add:* Dept Microbiol Univ Miss Med Ctr Jackson MS 39216-4505

ARCESE, PAUL SALVATORE, b West New York, NJ, Mar 14, 30; m 53; c 2. MEDICAL RESEARCH. *Educ:* NY Univ, BA, 51, MS, 58, PhD(endocrinol), 64. *Prof Exp:* Lab technician biochem & endocrinol, Merck Inst Therapeut Res, 54-60, from res asst to res assoc, 60-65; from res assoc to asst dir clin res, 65-74, sr assoc dir clin res, 74-78, DIR CLIN RES, SANDOZ PHARMACEUT, 78- *Mem:* Acad Psychosom Med; Am Asn Study Headache. *Res:* Adrenal steroid metabolism; hormonal effects on erythropoiesis in the hypophysectomized rat. *Mailing Add:* Dept Clin Res Sandoz Pharmaceut Rte 10 Bldg 502 3rd Floor East Hanover NJ 07936

ARCESI, JOSEPH A, b Sayre, Pa, Sept 1, 38; m 78; c 3. ORGANIC CHEMISTRY, POLYMER CHEMISTRY. *Educ:* Gettysburg Col, BA, 60; Univ Del, PhD(org polymer chem), 64. *Prof Exp:* Sr polymer chemist, Eastman Kodak Co, 67-81, asst supt, Synthetic Chem Div, 81-84, div dir, Chem Develop Div, 84-90, DIV MGR, SYNTHETIC CHEM, EASTMAN KODAK CO, 90- *Res:* Polymer preparation and evaluation; photo-polymers; photo-resist. *Mailing Add:* Eastman Kodak Co 343 State Rochester NY 14608

ARCH, STEPHEN WILLIAM, b Los Angeles, Calif, May 15, 42; m 64; c 2. NEUROBIOLOGY. *Educ:* Stanford Univ, AB, 64; Univ Chicago, PhD(biol), 70. *Prof Exp:* Res fel neurobiol, Calif Inst Technol, 70-72; from asst prof to assoc prof, 72-82, PROF BIOL, REED COL, 82- *Concurrent Pos:* Alfred P Sloan Found res fel neurobiol, 73-75; prin investr grants, Med Res Found Ore, Nat Sci Found & NIH; mem neurol B study sect, NIH, 82-86. *Mem:* Soc Gen Physiol; Sigma Xi; Soc for Neuroscience. *Res:* Analysis of biochemical and molecular regulation at the cellular level in identified neuroendocrine cells. *Mailing Add:* Dept Biol Reed Col Portland OR 97202-8199

ARCHAMBEAU, JOHN ORIN, b Maine, Aug 5, 25; c 8. RADIOLOGY. *Educ:* Stanford Univ, AB, 50, MD, 55. *Prof Exp:* Resident path, Univ Chicago, 54-58; fel oncol, Swedish Hosp, 58-60; Fulbright scholar & fel oncol, Curie Found, 60-61; assoc scientist med, Brookhaven Nat Lab, 61-66; dir radiation oncol, Nassau Co Med Ctr, 66-77; chmn radiation oncol, City of Hope, 77-82, dir, Dept Radiation Biol Res, 82-84; DIR RADIATION BIOL, LOMA LINDA UNIV MED CTR, 84- *Concurrent Pos:* Res collabr, Brookhaven Nat Lab, 66-; prof radiol, State Univ NY Stony Brook, 70-77. *Mem:* Radiation Res Soc; fel Am Col Radiol; Am Soc Therapeut Radiol. *Res:* Cell kinetics; radiation therapy; cell kinetics of irradiated tissues. *Mailing Add:* Loma Linda Univ Med Ctr 11234 Anderson St Loma Linda CA 92354

ARCHANGELSKY, SERGIO, b Casablanca, Morocco, Mar 27, 31; m 58; c 2. PALEOBOTANY, PALYNOLOGY. *Educ:* Univ Buenos Aires, Lic en cie, 55, Dr(geol), 57. *Prof Exp:* Prof paleont, Inst Miguel Lillo, Nat Univ Tucuman, 56-61; prof paleobot, La Plata Univ, 61-78; head, paleobot-palynol res unit, Cirgeo, 75-82; RES FEL, BUENOS AIRES NATURAL HIST MUS, ARG, 82- *Concurrent Pos:* Arg Nat Res Coun res fel, 61-; consult, Yacimientos Petroliferos Fiscales, 75-82; distinguished vis prof, Ohio State Univ, Columbus, 84. *Honors & Awards:* Arg Paleontol Asn Award, 78; Arg Geol Soc Award, 87; Arg Acad Exact Phys & Natural Sci Award, 90. *Mem:* Arg Paleont Asn (vpres, 61-63 & 74-75, pres, 63-65 & 67-69); Latin Am Paleobot & Palynology Asn (pres, 74-78); Int Paleontol Asn (vpres, 76-85); corresp mem Bot Soc Am; fel Nat Acad Sci Arg. *Res:* Paleobotany and palynology of the Cretaceous in Argentina; palynology of the Lower Tertiary in Argentina; paleobotany and palynology of Carboniferous and Permian periods in Argentina. *Mailing Add:* Urquiza 1132 Vicente Lopez 1638 Buenos Aires Argentina

ARCHARD, HOWELL OSBORNE, b Yonkers, NY, Mar 25, 29; m 61; c 1. ORAL PATHOLOGY. *Educ:* Rutgers Univ, BSci, 51; Columbia Univ, DDS, 55; Am Bd Oral Path, dipl, 64. *Prof Exp:* Sr asst dent surg, Alaska Native Health Serv, US Pub Health Servs, 55-57; dent intern, Columbia-Presby Med Ctr, NY, 57-58; clin assoc stomatology, Sch Dent & Oral Surg, Columbia Univ, 58-60; resident oral path, Nat Inst Dent Res, USPHS, 60-62; asst pathologist, Armed Forces Inst Path, Walter Reed Army Med Ctr, 62-64; chief diag path, Lab Exp Path, Nat Inst Dent Res, 64-79; PROF ORAL PATH, SCH DENT MED, STATE UNIV NY, STONY BROOK, 87- *Concurrent Pos:* Guest Lectr, Armed Forces Inst Path, Nat Naval Med Ctr, Smithsonian Inst, Univ Ala, Stanford Univ & Georgetown Univ, 64-; consult, Advan Training Oral Path, Am Acad Oral Path, 69; asst ed, J Oral Path, 72-; vis scientist, Inst Dent Res, Univ Ala, 72-78. *Mem:* Am Dent Asn; fel Am Acad Oral Path; Int Asn Dent Res; Am Asn Dent Res; Int Asn Oral Pathologists. *Res:* Clinico-pathologic studies of human oral mucosal diseases; inherited and acquired metabolic diseases affecting the dentition. *Mailing Add:* Dept Oral/Biol/Pathol SUNY Health Sci Ctr Stony Brook Stony Brook NY 11794

ARCHBOLD, NORBERT L, b Ft Wayne, Ind, Apr 7, 30; m 51; c 2. GEOLOGY. *Educ:* Western Reserve Univ, BS, 52; Univ Mich, MS, 56, PhD(geol), 62. *Prof Exp:* Geologist, US Geol Surv, 53-58, Homestake Mining Co, 61-64 & Nev Bur Mines, 64-68; PROF GEOL, WESTERN ILL UNIV, 68- *Concurrent Pos:* Consult to various mining companies. *Mem:* Geol Soc Am; Am Inst Mining, Metall & Petrol Engrs; Soc Econ Geologists. *Res:* Diabase dikes in Canadian shield; mineral deposit exploration; ore deposits of Nevada. *Mailing Add:* Dept Geol Western Ill Univ Macomb IL 61455

ARCHBOLD, THOMAS FRANK, b Ft Wayne, Ind, Aug 24, 33; m 60; c 3. METALLURGICAL ENGINEERING, PHYSICAL METALLURGY. *Educ:* Purdue Univ, BS, 55, MS, 57, PhD(metall), 61. *Prof Exp:* From asst prof to assoc prof 61-77, PROF METALL ENG, UNIV WASH, 77- *Concurrent Pos:* Metall consult, 61- *Mem:* Am Soc Metals; Nat Asn Corrosion Eng; Sigma Xi. *Res:* Diffusion processes; electron microscopy; diffraction; oxidation kinetics; solid state reaction kinetics; metallography; nuclear materials; failure analysis; corrosion. *Mailing Add:* Dept Mat Sci & Eng Univ Wash Seattle WA 98195

ARCHER, CASS L, b Spearman, Tex, June 1, 24; m 53; c 3. MATHEMATICS EDUCATION. *Educ:* Univ Tex, Austin, BS, 50, MEd, 54, MA, 59, PhD(math educ), 67. *Prof Exp:* Teacher pub schs, Tex, 50-58; PROF MATH & HEAD DEPT, ANGELO STATE UNIV, 59- *Concurrent Pos:* Math consult, Tex Educ Agency, 62-64. *Mem:* Am Math Soc; Math Asn Am. *Res:* Mathematics programs of institutions changing from a two-year to a four-year college. *Mailing Add:* 2810 Vista Delarroyo Dr San Angelo TX 76904

ARCHER, DAVID HORACE, b Pittsburgh, Pa, Jan 20, 28; m 50, 76; c 4. PROCESS DEVELOPMENT, PLANT DESIGN. *Educ:* Carnegie Mellon Univ, BS, 48; Univ Del, PhD(chem eng), 53. *Prof Exp:* Instr chem eng, Univ Del, 51-53; from asst prof to assoc prof, Carnegie Mellon Univ, 53-61; mgr, Fuel Cells Systs Technol & Chem Eng, Westinghouse Res Labs, 61-83, MGR PROCESS ENG & CONSULT ENGR, WESTINGHOUSE ENERGY SYSTS, 83- *Concurrent Pos:* Mem, Comt on Nitrogen Oxide, Nat Acad Sci, 76-78. *Mem:* Nat Acad Eng; Am Chem Soc; Combustion Inst; Am Asn Cost Engrs; Sigma Xi; Am Inst Chem Engrs. *Res:* Development of fossil and nuclear fuel processes related to power generation, particularly coal gasification, pyrolysis, fluidized bed combustion, hot gas cleaning, gas turbine corrosion-erosion-deposition; plant design and evaluation; optimum operations; power plant operations, control. *Mailing Add:* 114 Kentzel Rd Pittsburgh PA 15237

ARCHER, DOUGLAS HARLEY, b Saskatoon, Sask, May 20, 25; m 59; c 2. PHYSICS. *Educ:* Univ BC, BA, 47, MS, 48; Harvard Univ, PhD(physics), 53. *Prof Exp:* Staff mem radiowave propagation, Lincoln Lab, Mass Inst Technol, 53-58; staff mem radiowave & auroral physics, Hughes Aircraft Co, 58; staff mem atmospheric physics, Gen Elec Co, 58-71; staff mem atmospheric physics, Mission Res Corp, 71-90; RETIRED. *Mem:* Am Geophys Union. *Res:* Atomic and molecular physics; radiowave propagation; atmospheric physics, especially ionization and deionization of perturbed regions of the atmosphere. *Mailing Add:* 877 Cieneguitas Rd Santa Barbara CA 93110

ARCHER, F JOY, metabolic diseases, for more information see previous edition

ARCHER, JOHN DALE, b Brady, Tex, Mar 10, 23; m 52; c 2. MEDICINE, PHARMACOLOGY. *Educ:* Univ Tex, Austin, BA, 50; Univ Tex, Galveston, MD, 52. *Prof Exp:* Res asst pharmacol, Med Br, Univ Tex, 50-51, res assoc, 51-52, from instr to asst prof, 52-55; staff physician, Student Health Serv, Univ Tex, Austin, 55-57; asst med dir, State Dept Pub Welfare, Tex, 57-58; med officer, Div New Drugs, Bur Med, Food & Drug Admin, Dept Health, Educ & Welfare, Washington, DC, 58-61, dep dir, 61-62, actg dir, 62; staff physician, Med Serv, Vet Admin Ctr, Tex, 62-64; from asst dir to dir drug eval sect, Dept Drugs, 64-72, sr ed jour, AMA, 72-86, contrib ed, 86-87; RETIRED. *Concurrent Pos:* Intern, Brackenridge Hosp, Austin, Tex, 52-53; team physician, Dept Intercollegiate Athletics, Univ Tex, Austin, 55-57. *Mem:* Am Soc Pharmacol & Exp Therapeut; Am Med Writers Asn. *Res:* Evaluation and publication of research and clinical experience with drugs; animal and clinical research in pharmacology and toxicology. *Mailing Add:* 535 N Dearborn St Chicago IL 60610

ARCHER, JUANITA ALMETTA, b Washington, DC, Nov 3, 34; m 58; c 1. INTERNAL MEDICINE. *Educ:* Howard Univ, BS, 56, MS, 58, MD, 65. *Prof Exp:* Intern, Freedmens Hosp, 65-66, resident internal med, 66-69, fel endocrinol, 69-70; with NIH, 70-73; instr, Dept Med, 73-75, asst prof med, 75-77, dir, Endocrine/Metabolic Labs, 77-86, ASSOC PROF MED, HOWARD UNIV HOSP, 77- *Concurrent Pos:* Josiah Macy fac fel, Dept Internal Med, Howard Univ, 74- *Honors & Awards:* Nat Podiatric Asn Award, 86; Moses Wharton Young Res Award, 87. *Mem:* Endocrine Soc; Am Fedn Clin Res. *Res:* Physiological significance of insulin receptors in man. *Mailing Add:* Howard Univ Hosp 2041 Georgia Ave NW Washington DC 20060

ARCHER, LAWRENCE H(ARRY), civil engineering, mathematics; deceased, see previous edition for last biography

ARCHER, ROBERT ALLEN, b Reading, Pa, Apr 3, 36; m 63; c 3. ORGANIC CHEMISTRY. *Educ:* Harvard Univ, AB, 58; Univ Del, MS, 60; Stanford Univ, PhD(chem), 63. *Prof Exp:* Sr org chemist, Chem Res Div, Lilly Res Labs, 64-73, res scientist, 73-78, RES ASSOC, CHEM & CANCER RES DIV, ELI LILLY & CO, 78-, SR RES SCIENTIST, CHEM CANCER & VIROL RES, 84- *Mem:* AAAS; Am Chem Soc; Sigma Xi. *Res:* Immunology; connective tissue; rheumatoid and osteoarthritis; medicinal chemistry. *Mailing Add:* Lilly Res Labs Eli Lilly & Co Indianapolis IN 46285

ARCHER, ROBERT JAMES, b San Francisco, Calif, Jan 17, 26; m 61; c 1. SOLID STATE PHYSICS. *Educ:* Columbia Univ, BA, 51, PhD(phys chem), 54. *Prof Exp:* Mem tech staff, Bell Tel Labs, 54-63; dept head device physics, Hewlett-Packard Assocs, 63-78, dir, solid state lab, Hewlett Packard Labs, Hewlett-Packard Co, 78-82; RETIRED. *Res:* Solid state device research and development in the fields of optoelectronics and microwaves, especially involving III-V compound semiconductors. *Mailing Add:* 165 Fawn Lane Portola Valley CA 94028

ARCHER, ROBERT RAYMOND, b Omaha, Nebr, Sept 8, 28; m 50; c 5. APPLIED MATHEMATICS, ENGINEERING MECHANICS. *Educ:* Mass Inst Technol, SB, 52, PhD(math), 56. *Prof Exp:* Asst prof mech eng, Mass Inst Technol, 56-59; from asst prof to assoc prof, Univ Mass, 59-61; assoc prof eng, Case Western Reserve Univ, 61-66; PROF CIVIL ENG, UNIV MASS, AMHERST, 66- *Concurrent Pos:* Fulbright lectr, Indian Inst Technol, Kanpur, India, 74-75. *Mem:* Sigma Xi; Am Soc Mech Eng. *Res:* Mathematical elasticity; anisotropic stress analysis; continuum mechanical models for study of tree growth stresses and reaction wood mechanics. *Mailing Add:* Dept Civil Eng Univ Mass Amherst MA 01002

ARCHER, RONALD DEAN, b Rochelle, Ill, July 22, 32; m 54; c 4. COORDINATION CHEMISTRY, METAL-CONTAINING POLYMERS. *Educ:* Ill State Univ, BS, 53, MS, 54; Univ Ill, PhD(chem), 59. *Prof Exp:* Asst prof chem, Univ Calif, 59-63; from asst prof to assoc prof, Tulane Univ, 63-66; assoc prof, 66-70, head dept, 77-83, PROF CHEM, UNIV MASS, 70- *Concurrent Pos:* Vis prof, Tech Univ Denmark, 72; consult, Alden Res Found, 73-; vis scientist, Naval Res Lab, Washington, DC, 80; consult, Ventron, 84-85, Monsanto, 85-86; chief chem reader, Educ Testing Serv Advan Placement Prog, 85-88; chair, Am Chem Soc Comn Educ, 87-89; vis prof, Univ Vienna, 87. *Mem:* Fel AAAS; Am Chem Soc; Royal Soc Chem; Sigma Xi. *Res:* Coordination compounds; synthesis; properties; kinetics; biometallic studies; photochemistry; spectroscopy; inorganic polymers. *Mailing Add:* Dept Chem Univ Mass Amherst MA 01003

ARCHER, STANLEY J, b Stephenville, Tex, Aug 13, 44; m 66; c 2. IMMUNOLOGY, ALLERGY. *Educ:* Abilene Christian Col, BS, 66, MS, 68; Univ Tenn, Knoxville, PhD(microbiol), 72. *Prof Exp:* Instr microbiol, Univ Tex, Austin, 72-74; asst prof microbiol, Ariz State Univ, 74-80; DIR RES & DEVELOP, IATRIC CORP, 80- *Mem:* Am Soc Microbiol; AAAS; NY Acad Sci. *Res:* Cellular and molecular events involved in the initiation of an immune response. *Mailing Add:* 21420 S Lindsay Rd Chandler AZ 85249

ARCHER, SYDNEY, b New York, NY, Jan 23, 17; m 46; c 3. MEDICINAL CHEMISTRY. *Educ:* Univ Wis, AB, 37; Pa State Col, MS, 38, PhD(org chem), 40. *Prof Exp:* Procter & Gamble fel, Northwestern Univ, 40-41; Lilly fel, Univ Chicago, 41-42; res chemist, Sun Oil Co, 42-43; res chemist, Sterling-Winthrop Res Inst, 43-60, asst dir chem res, 60-64, dir chem res & develop, 64-68, assoc dir res, 68-73; dean, Sch Sci, 80-85, RES PROF MED CHEM, RENSSELAER POLYTECH INST, 73- *Concurrent Pos:* Consult, Spec Action Off Drug Abuse Prev, 73 & Nat Inst Drug Abuse, 74-; chmn, Med Chem Study Sect, NIH, 76-78. *Honors & Awards:* Award, Am Chem Soc, 68. *Mem:* Am Chem Soc; Am Soc Pharmacol & Exp Therapeut. *Mailing Add:* Res Prof Med Chem Rensselaer Polytech Inst Troy NY 12181-3590

ARCHER, VERNON SHELBY, b Ada, Okla, Dec 5, 39; m 67. ANALYTICAL CHEMISTRY. *Educ:* Univ Okla, BS, 61, PhD(anal chem), 64. *Prof Exp:* Asst prof, 64-74, ASSOC PROF CHEM, UNIV WYO, 74- *Mem:* Am Chem Soc. *Res:* Electrochemistry; fluoride chemistry. *Mailing Add:* 362 N Seventh St Laramie WY 82070-3218

ARCHER, VICTOR EUGENE, b Teigen, Mont, May 21, 22; m 49; c 2. MEDICINE, EPIDEMIOLOGY. *Educ:* Northwestern Univ, BS, 45, MB, 48, MD, 49; Am Bd Prev Med, dipl. *Prof Exp:* Instr physics, Mont State Univ, 43-44; intern clin med, USPHS Hosp, 48-49, instr radiol health, Environ Health Ctr, USPHS, 50-51; radiation safety officer, NIH, 51-55, field investr, Nat Cancer Inst, 56-59; chief, Coffman Res Lab, 59-61; chief epidemiol serv, Occup Health Field Sta, USPHS, 61-72, med dir & epidemiologist, western area lab for occup safety & health, Nat Inst Occup Safety & Health, 72-79; CLIN PROF, MED SCH, UNIV UTAH, 79- *Concurrent Pos:* Clin lectr, Med Sch, Univ Utah, 58-70. *Mem:* Health Physics Soc; Am Pub Health Asn. *Res:* Radiobiology; environmental causes of lung disease and cancer. *Mailing Add:* 4370 Spruce Circle Salt Lake City UT 84124

ARCHIBALD, JAMES, b Bellshill, Scotland, Jan 4, 19; Can citizen; m 48; c 3. VETERINARY MEDICINE. *Educ:* Ont Vet Col, DVM, 49; Univ Toronto, MVSc, 51; Univ Giessen, Dr med vet, 58; FRCVS; Am Col Vet Surgeons, dipl. *Prof Exp:* From asst prof to assoc prof, 49-54, PROF MED VET, DIV SMALL ANIMAL MED & SURG, ONT VET COL, UNIV GUELPH, 54- *Concurrent Pos:* Consult, Defense Res Med Labs, Can, 58; chmn, dept clin studies, Univ Guelph, 63-80. *Honors & Awards:* Order of Ontario, 90. *Mem:* Am Vet Med Asn; Can Vet Med Asn (pres, 63); Am Col Vet Surgeons (pres, 77-78); Commonwealth Vet Asn (sec-treas, 84-91). *Res:* Experimental surgery; liver and tissue transplantation. *Mailing Add:* Dept Clin Studies Ont Vet Col Univ Guelph Guelph ON N1G 2W1 Can

ARCHIBALD, JULIUS A, JR, b New York, NY, May 9, 31; m 54; c 3. SIMULATION & MODELING. *Educ:* NY Univ, BA, 52, MS, 53. *Prof Exp:* Mathematician, Wright Air Develop Ctr, US Air Force, 53-55, comput scientist, Knolls Atomic Power Lab, Ben Elec Co, 55-68, syst specialist, Gen Elec Co, 68-70; assoc prof, PROF, COMPUT SCI, COL ARTS & SCI, STATE UNIV NY, PLATTSBURGH, 70-74. *Concurrent Pos:* Chair, comput sci, Col Arts & Sci, State Univ NY, Plattsburgh, 74-; vis chair, comput sci accreditation comn, 85-87 & 88-89; actg chair math, State Univ NY, Plattsburgh, 88-90. *Mem:* Inst Elec & Electronics Engrs Computer Soc; Asn Comput Mach; Sigma Xi; Am Math Soc; Math Asn Am; Soc Indust & Appl Math. *Mailing Add:* State Univ NY Col Plattsburgh NY 12901-2699

ARCHIBALD, KENNETH C, b White Plains, NY, Aug 1, 27. PHYSICAL MEDICINE. *Educ:* Cornell Univ, MD, 53; St Lawrence Univ, BS, 53. *Prof Exp:* Sr instr phys med, Sch Med, Western Reserve Univ & assoc dir dept phys med & rehab, Univ Hosps Cleveland, 57-58; asst prof phys med, Med Col, Cornell Univ & dir phys med & rehab & asst attend physician, New York Hosp-Cornell Med Ctr, 58-66; from assoc prof to prof phys med & rehab, Sch Med, Temple Univ & dir dept, Episcopal Hosp, 66-69; dir phys med & rehab, Jerd Sullivan Rehab Ctr, Garden Hosp & Presby Hosp of Pac Med Ctr, San Francisco, 69-78; DIR PHYS MED & REHAB, CHILDREN'S HOSP, 74- *Concurrent Pos:* Consult, Muscle Dis Clin, Univ Hosps Cleveland, 58-69;

assoc clin prof orthop surg, Med Ctr, Univ Calif, San Francisco, 70- *Mem:* Am Cong Rehab Med; Am Acad Phys Med & Rehab; Am Asn Electromyography & Electrodiagnosis. *Res:* Physical medicine and rehabilitation; neuromuscular disorders. *Mailing Add:* 2360 Clay St San Francisco CA 94115

ARCHIBALD, PATRICIA ANN, b Olney, Ill, July 18, 34. ALGOLOGY, PHYCOLOGY. *Educ:* Ball State Univ, BS & MA, 61; Univ Tex, Austin, PhD(bot), 69. *Prof Exp:* Teacher high sch, Ind, 59-64; instr biol, Palm Beach Jr Col, 64-66 & Cuyahoga Community Col, 66-67; from asst prof to assoc prof, 69-80, PROF BIOL, SLIPPERY ROCK UNIV, 80- *Concurrent Pos:* Fulbright lectr, Harrow Technol Col, Eng, 62-63; mem, NSF US-Japan Phycol Sem, 75; Int Res & Exchanges Bd exchangee, Czech Acad Sci, 77; Nat Acad Sci exchangee, USSR Acad Sci, 78; coordr, US-Mex phycol sem, 80; assoc sci rev adminr, comp grants prog, Environ Protection Agency, 81-82; Proj-Flora partic, Brazilian Amazon Collection Exped, NY Bot Garden, 84; US-China Exchange Leader, Freshwater Biologist, 88. *Mem:* Phycol Soc Am; Brit Phycol Soc; Int Phycol Soc; Soc Tropical Biol. *Res:* Phycology, systematics morphology and physiology of zoosporic chlorococcales and chlorosarcinales; edaphic algae, distribution, morphology, physiology; culturing of algae. *Mailing Add:* Dept Biol Slippery Rock Univ Slippery Rock PA 16057

ARCHIBALD, RALPH GEORGE, b Sackville, NB, May 23, 01; nat US; m 41; c 1. MATHEMATICS. *Educ:* Univ Man, BA, 22; Univ Toronto, MA, 24; Univ Chicago, PhD(math), 27. *Prof Exp:* Demonstr physics, Univ Man, 21-22; lectr math, Wesley Col, Can, 22-23; from asst prof to assoc prof, Columbia Univ, 27-38; from asst prof to prof, 38-71, EMER PROF MATH, QUEENS COL, CITY UNIV NY, 71- *Mem:* Am Math Soc; Math Asn Am; Am Sci Affil. *Res:* Theory of numbers. *Mailing Add:* 104-15 107th St Ozone Park Queens NY 14417

ARCHIBALD, REGINALD MACGREGOR, b Syracuse, NY, Mar 2, 10; m 48; c 2. PEDIATRIC ENDOCRINOLOGY. *Educ:* Univ BC, BA, 30, MA, 32; Univ Toronto, PhD(path chem), 34, MD, 39. *Prof Exp:* Asst, Univ BC, 30-32; asst, Univ Toronto, 32-33; intern, Hosp for Sick Children, Toronto, 37-38; intern, Toronto Gen Hosp, 39-40; asst resident, Rockefeller Inst Hosp, 41-46; prof biochem & head dept, Johns Hopkins Univ, 46-48; mem staff, 48-59, prof, 59-80, EMER PROF MED, ROCKEFELLER UNIV, 80-, SR PHYSICIAN EMER, ROCKEFELLER UNIV HOSP, 88- *Concurrent Pos:* Nat Res Coun fel, 40-42; vis investr, Rockefeller Univ, 40-43, spec investr, 43-45, assoc, 46, physician, Univ Hosp, 48-59, sr physician, 59-80; vpres, Am Bd Clin Chem, 57-63; mem adv bd, Anal Chem, 57-60. *Mem:* AAAS; Am Chem Soc; Am Soc Biol Chemists; Soc Res Child Develop; Soc Adolescent Med; Lawson Wilkins Pediat Endocrin Soc. *Res:* Non-fermentable carbohydrates of urines; fractionation of urinary steroids; physiological roles of glutamine and other amino acids; measurement of enzymes; use of enzymes in analysis; influences of hormones on enzymes; problems of growth and development of children; skeletal anomalies in children. *Mailing Add:* Rockefeller Univ Hosp 1230 York Ave New York NY 10021

ARCHIBALD, WILLIAM JAMES, b Sydney, NS, Oct 30, 12; m 36; c 3. PHYSICS. *Educ:* Dalhousie Univ, BA, 33, MA, 36; Univ Va, PhD, 38. *Hon Degrees:* DSc, Univ NB, 61. *Prof Exp:* Sterling fel, Yale Univ, 38-39; res scientist, Nat Res Coun, 39-42; from assoc prof to prof, 42-73, dean arts & sci, 55-60, Dr A C Fales Prof theoroe physics & dean freshmen, Dalhousie Univ, 73-78; RETIRED. *Concurrent Pos:* Fel, Univ Va, 51-52; mem, Defense Res Bd, Can, 56-59. *Mem:* Fel Royal Soc Can. *Res:* Ultra centrifuge; theory of molecular weight determination; theoretical physics; nuclear magnetism; field theory. *Mailing Add:* Walnut St Halifax NS B3H 3S4 Can

ARCHIE, CHARLES NEILL, b Coranado, Calif, Feb 22, 48; m 78; c 1. QUANTUM FLUIDS & SOLIDS, MACROSCOPIC QUANTUM TUNNELING. *Educ:* Univ Calif, Davis, BS, 70; Cornell Univ, PhD(physics), 78. *Prof Exp:* Engr, Comn Atomic Energy, Solid Physics & Magnetic Resonance, Ctr Nuclear Studies, Saclay, France, 78-79; PROF, DEPT PHYSICS, STATE UNIV NY, STONY BROOK, 79- *Mem:* Am Phys Soc. *Res:* Very low temperature physics experiments; the production of a new quantum liquid, spin polarized helium 3; the development of devices which exhibit purely quantum properties on a macroscopic scale. *Mailing Add:* Rozerneil Kidlingten Rd Islip Oxford 0X52SS England

ARCHIE, WILLIAM C, JR, b Brownsville, Tex, Mar 1, 44; m 70; c 2. PHYSICAL ORGANIC CHEMISTRY. *Educ:* Duke Univ, BS, 66; Stanford Univ, PhD(chem), 70. *Prof Exp:* Fel chem, Harvard Univ, 70-72; SR RES CHEMIST, EASTMAN KODAK CO, 72- *Mem:* Am Chem Soc. *Res:* Photographic science; use of physical organic chemical methods to elucidate reaction mechanisms in film; stability of photographic color images. *Mailing Add:* 46 Greylock Ridge Pittsford NY 14534-2334

ARCHIMOVICH, ALEXANDER S, b Novosybkov, Ukraine, Apr 23, 1892; US citizen; m 35; c 3. BIOLOGY, AGRONOMY. *Educ:* St Vladimir Univ, Kiev, dipl natural sci, 17; Kiev Polytech Inst, dipl agr sci, 22; T H Shevchenko State Univ, Kiev, MS, 38, PhD(biol), 40. *Prof Exp:* Res botanist, Ukrainian Acad Sci, 19-23; sr specialist, Plant Breeding Sta, Bila Zerkva, Ukraine, 23-33; prof agron, Agr Cols, 24-43; prof plant breeding, Ukrainian Inst, Ger, 45-48; head dept selection sugar beet, Valladolid Selection Sta, Spain, 48-52; prof plant breeding, Ukrainian Tech Inst, NY, 53-64, pres, 55-62; head, Dept Natural Sci & Med, Ukrainian Acad Arts & Sci US, 59-81, pres, 62-70; RETIRED. *Concurrent Pos:* Hon mem, Inst Study USSR, Munich & NY, 59; head chem, Biol & Med Sect, Shevchenko Sci Soc, 74-81. *Mem:* AAAS; Am Soc Sugar Beet Technol; NY Acad Sci; Bavarian Bot Soc; hon mem Am-Ukrainian Vet Med Asn. *Res:* Changes in the world production of grain; USSR's share of world production; methods of producing sugar beet seed in north of Spain; analysis of agriculture in Soviet Union; geography of the field crops in the Ukraine. *Mailing Add:* 155 Mountain Rd Rosendale NY 12472

ARCOS, JOSEPH (CHARLES), b Hungary, Aug 22, 21; US citizen. TOXICOLOGY, ONCOLOGY. *Educ:* Univ Cluj, LChem, 47; Conserv Nat Arts et Metiers, France, ChE, 50; Univ Paris, DSc, 51. *Prof Exp:* Res engr surface chem, Tech Ctr Graphic Indust, 52; asst tech dir develop & prod, Jouan Co, 52-53; res assoc cancer res, McArdle Lab, Univ Wis, 53-57; asst res prof, Cancer Res Lab, Univ Fla, 57-60; from assoc prof to prof med & biochem, Sch Med, 60-68, EMER PROF, SCH MED, TULANE UNIV, 87-; SR SCI ADV, OFF TOXIC SUBSTANCES, US ENVIRON PROTECTION AGENCY, 80- *Concurrent Pos:* Sci ed; ecosystem concept rev. *Honors & Awards:* Fac Res Award, Am Cancer Soc, 72; Creative Advan Environ Sci & Technol Award, Am Chem Soc, 87. *Mem:* Am Soc Biol Chemists; Biophys Soc; Am Asn Cancer Res; Soc Exp Biol & Med; Am Soc Pharmacol & Exp Therapeut; Am Col Toxicol; Europ Asn Cancer Res; fel Am Col Toxicol, 85. *Res:* Structure-activity relationships and molecular geometry of chemical carcinogens; biochemistry of mitochondrial pathways of energy production; conformational changes in biological macromolecules; metabolism of carcinogens; chemical risk assessment; combination carcinogenic effects with complex mixtures; system concept and network analysis of ecocrisis interactions. *Mailing Add:* US Environ Protection Agency TS-778 401 M St Southwest Washington DC 20460

ARD, WILLIAM BRYANT, JR, b Jackson, Tenn, Oct 8, 27; m 50; c 8. PHYSICS. *Educ:* Auburn Univ, BS, 50, MS, 51; Duke Univ, PhD(physics), 55. *Prof Exp:* From asst prof to assoc prof, Univ Ala, 55-59; assoc prof, Univ Fla, 59-62; physicist, Oak Ridge Nat Lab, 62-73; sr res scientist, United Technol Res Ctr, 73-78; STAFF DIR MDC FEL, MCDONNELL DOUGLAS MISSILE SYSTS CO, 78- *Mem:* Fel Am Phys Soc. *Res:* Plasma physics and controlled thermonuclear fusion research; particle beam physics. *Mailing Add:* Dept EBE 3 Bldg 92 PO Box 516 Mc Donnell Douglas Corp St Louis MO 63166

ARDELL, ALAN JAY, b Brooklyn, NY, Mar 24, 39; m 60, 87; c 3. METALLURGY, MATERIALS SCIENCE. *Educ:* Mass Inst Technol, BS, 60; Stanford Univ, MS, 62, PhD(mat sci), 64. *Prof Exp:* Nat Sci Found fel, Cambridge Univ, 64-66; asst prof mat sci, Calif Inst Technol, 66-68; from actg assoc prof to assoc prof, 68-74, PROF MAT, DEPT MAT SCI ENG, UNIV CALIF, LOS ANGELES, 74- *Concurrent Pos:* Fulbright-Hays Res fel, Comn Atomic Energy-Ctr Nuclear Studies, Saclay, France, 74-75. *Mem:* Am Inst Mining, Metall & Petrol Engrs; Electron Micros Soc Am; fel Am Soc Metals; Mat Res Soc. *Res:* Coarsening of precipitates; transmission electron microscopy; mechanical behavior of solids, including aluminum alloys, ion irradiated ordered intermetallic alloys, and creep of metals and alloys. *Mailing Add:* Dept Mat Sci Eng Univ Calif 5732-J BH Los Angeles CA 90024-1595

ARDELL, JEFFREY LAURENCE, CARDIOVASCULAR PHYSIOLOGY. *Educ:* Univ Wash, PhD(physiol & biophys), 80. *Prof Exp:* ASST PROF PHYSIOL, SCH MED, UNIV S ALA, 84- *Mailing Add:* Sch Med Dept Physiol Univ S Ala MSB 3024 Mobile AL 36688

ARDELT, WOJCIECH JOSEPH, b Warsaw, Poland, Apr 5, 39; m 62; c 2. PROTEIN CHEMISTRY, ENZYMOLOGY. *Educ:* Warsaw Univ, Poland, BS, 62, Dr Habilitus, 75; Polish Acad Sci, PhD(biochem), 68. *Prof Exp:* Grad res asst biochem, Inst Rheumatol, Warsaw, Poland, 62-68, postdoctoral fel biochem, 69-70 & biochem/enzymol, 70-71, from asst prof to assoc prof, 71-78; vis prof protein chem, Dept Chem, Purdue Univ, Lafayette, Ind, 79-80, sr res assoc, 81-86; CHIEF CHEMIST PROTEIN CHEM, ALFACELL CORP, BLOOMFIELD, NJ, 87- *Mem:* Sigma Xi; Am Soc Biochem & Molecular Biol; Protein Soc. *Res:* Structure and function of proteins; discovery and characterization of pancreatic elastase II; mechanism of interaction of proteinases with their protein inhibitors; isolation and characterization (including amino acid sequencing) of a novel protein. *Mailing Add:* Alfacell Corp 225 Belleville Ave Bloomfield NJ 07003

ARDEMA, MARK D, b Minneapolis, Minn, Dec 2, 40; m 63; c 2. OPTIMAL CONTROL & DIFFERENTIAL GAMES, SINGULAR PERTURBATIONS & ASYMPTOTIC ANALYSIS. *Educ:* Univ Calif, Berkeley, BS, 63, MS, 64, PhD(mech eng), 74. *Prof Exp:* Res scientist, NASA Ames Res Ctr, 65-80 & 82-86, res asst to dir, 80-82; PROF & CHMN, DEPT MECH ENG, SANTA CLARA UNIV, 86- *Concurrent Pos:* Lectr, Univ Calif, Berkeley, 78, vis prof, 84; vis scholar, Twente Univ Technol, Neth, 82; consult, NASA Ames Res Ctr, 86-; short course instr, Am Inst Aeronaut & Astronaut, 89; assoc ed, J Dynamics & Control, 90-; vchair, Math Control Comt, Int Fedn Automatic Control, 90-93. *Honors & Awards:* Boeing Lectr, Wichita State Univ, 88. *Mem:* Assoc fel Am Inst Aeronaut & Astronaut. *Res:* Optimal control; differential games; singular perturbations; asymptotic analysis; trajectory optimization; aircraft flight dynamics; structural analysis; aircraft conceptual design; performance analysis; author of over 100 publications. *Mailing Add:* Dept Mech Eng Santa Clara Univ Santa Clara CA 95035

ARDEN, DANIEL DOUGLAS, b Bainbridge, Ga, Sept 24, 22; m 43; c 4. EXPLORATION GEOLOGY. *Educ:* Emory Univ, AB, 48, MS, 49; Univ Calif, Berkeley, PhD(paleont), 61. *Prof Exp:* Asst prof geol, Birmingham-Southern Col, 49-51; geologist petrol geol, Standard Oil Co Calif, 54-56; staff geologist, Standard Oil Co Ohio, 56-65; explor adv, Signal Oil & Gas Co, 65-70; PROF GEOL, GA SOUTHWESTERN COL, 70- *Concurrent Pos:* Chief geol consult, Geophys Serv Inc, 71-; mem, Int Comn Hist Geol, 75- *Mem:* Geol Soc Am; Am Geophys Union; Am Inst Prof Geol. *Res:* Geobotanical relationships; stratigraphy of Ocala and Tivola limestones in Georgia. *Mailing Add:* 2127 Broadway St New Orleans LA 70118

ARDEN, SHELDON BRUCE, clinical microbiology; deceased, see previous edition for last biography

ARDIS, COLBY V, JR, b Freeport, NY, May 21, 35; c 3. ENGINEERING MANAGEMENT. *Educ:* Univ Wis, BSCE, 58, MS, 60, PhD(civil eng), 72. *Prof Exp:* Sr engr, Wis Mich Power Co, 60-66; inst dir, Univ Wis, 66-67; civil engr, researcher & mgr, Tenn Valley Authority, 67-76, prog analyst, off gen mgr, 76-79; prof civil eng & chmn dept, Univ Toledo, 79-89; DEAN ENG, SOUTHERN ILL UNIV, EDWARDSVILLE, 89- *Mem:* Am Soc Civil Engrs; Am Soc Eng Educ. *Res:* Design optimization; flood hydrology; technical management; engineering education; project management. *Mailing Add:* Eight Larkmore Dr Glen Carbon IL 62034

ARDITTI, JOSEPH, b Sofia, Bulgaria, May 1, 32; US citizen; div; c 1. PLANT PHYSIOLOGY. *Educ:* Univ Calif, Los Angeles, BS, 59; Univ Southern Calif, PhD(biol, plant physiol), 65. *Prof Exp:* Teaching asst biol , Univ Southern Calif, 60-65, lectr, 65-66; from asst prof to assoc prof, 66-77, PROF BIOL, UNIV CALIF, IRVINE, 77- *Mem:* AAAS; Am Inst Biol Sci; Am Soc Plant Physiol; Bot Soc Am; Am Orchid Soc; Linnean Soc London. *Res:* Plant physiology; plant development; orchid biology; taro. *Mailing Add:* Dept Develop & Cell Biol Univ Calif Irvine CA 92717

AREF, HASSAN, b Alexandria, Egypt, Sept 28, 50; Can citizen; m 74; c 2. FLUID MECHANICS, COMPUTATIONAL SCIENCE. *Educ:* Univ Copenhagen, Denmark, Cand Sci, 75; Cornell Univ, PhD(physics), 80. *Prof Exp:* Asst prof eng, Brown Univ, 80-85; assoc prof, 85-88, PROF FLUID MECH, UNIV CALIF SAN DIEGO, 88- *Concurrent Pos:* Collabr, Los Alamos Nat Lab, 84-; assoc ed, J Fluid Mech, 84-; presidential young investr award, NSF, 85; guest prof, Tech Univ Denmark, 86; chief scientist, San Diego Supercomputer Ctr, 89- *Mem:* Fel Am Phys Soc; Soc Indust & Appl Math. *Res:* Theoretical and computational fluid mechanics; particular emphasis on problems in vortex dynamics and applications of nonlinear dynamics to flow problems. *Mailing Add:* Inst Geophysics & Planetary Physics 0225 Univ Calif San Diego La Jolla CA 92093

ARENA, JAY M, b Clarksburg, WVa, Mar 3, 09; m 31; c 7. PEDIATRICS. *Educ:* WVa Univ, BS, 30; Duke Univ, MD, 32. *Prof Exp:* Intern, Strong Mem Hosp, Rochester, NY, 32 & Johns Hopkins Hosp, 32-33; asst resident & resident, Univ Hosp, Duke Univ, 33-35; instr pediat, Vanderbilt Univ, 36; from asst prof to assoc prof, 36-50, EMER PROF PEDIAT, SCH MED, DUKE UNIV, 50-, EMER PROF COMMUNITY HEALTH SCI, 66-, EMER DIR, DUKE POISON CONTROL CTR, 53- *Concurrent Pos:* Mem ed adv bd, Clin Pediat, Highlight, Emergency Med & Pediat News. *Honors & Awards:* Jacobi Award, Am Acad Pediat, 83. *Mem:* Am Acad Pediat (pres-elect, 70, pres, 71-72); Am Pediat Soc; Am Asn Poison Control Ctrs (pres, 68-70). *Res:* Poisonings and accidents in children. *Mailing Add:* Box 3024 Duke Hosp Durham NC 27710

ARENA, JOSEPH P, b Jersey City, NJ, Oct 2, 54; m; c 1. BIOCHEMICAL PARASITOLOGY. *Educ:* St Johns Univ, BS, 77; Univ Med & Dent, PhD(pharmacol), 86. *Prof Exp:* Mem staff & supv pharmacist, Univ Hosp, Univ Med & Dent, Newark, NJ, 78-81; postdoctoral fel, Dept Physiol, Univ Rochester, NY, 86-89; SR RES BIOLOGIST, BASIC ANIMAL SCI RES, MERCK SHARP & DOHME RES LABS, RAHWAY, NJ, 89- *Concurrent Pos:* Adj asst prof pharmacol & toxicol, Univ Med & Dent, Newark, NJ, 91- *Mem:* AAAS; Biophys Soc; Am Soc Pharmacol & Exp Therapeut. *Res:* Electrophysiology and pharmacology of excitable tissues; neurotransmitter receptors and voltage-gated channels of invertebrates. *Mailing Add:* Biochem Parasitol Dept Merck Sharp & Dohme Res Lab Box 2000 Rm 80T-132 Rahway NJ 07065-0900

ARENAZ, PABLO, b Reno, Nev, Sept 9, 50; m 87; c 1. CARCINOGENESIS-DNA REPAIR, ENZYMOLOGY OF DNA REPAIR. *Educ:* Univ Nev, BS, 72, MS, 76; Wash State Univ, PhD(genetics & cell biol), 81. *Prof Exp:* Fel, Fels Res Inst, Temple Univ Sch Med, 81-84; asst prof, 84-90, ASSOC PROF GENETICS & MOLECULAR BIOL, UNIV TEX, EL PASO, 90- *Concurrent Pos:* Adj asst prof, Tex Tech Univ Sch Med, 87- *Mem:* Am Asn Cancer Res; Am Soc Cell Biol; AAAS; Environ Mutagen Soc. *Res:* Enzymology and molecular biology of DNA repair in eukaryotes; the relationship between aberrant control of DNA repair and carcinogenesis. *Mailing Add:* Biol Dept Univ Tex El Paso TX 79968

ARENBERG, DAVID (LEE), b Baltimore, Md, Nov 14, 27; m; c 2. COGNITIVE PSYCHOLOGY, GERONTOLOGY. *Educ:* Johns Hopkins Univ, AB, 51; Duke Univ, PhD(psychol), 60. *Prof Exp:* Personnel technician, Civil Serv Comm, City of Baltimore, 52-53; res psychologist & chief, Cognition Sect, NIH, 60-89; AT GERONT RES CTR, NAT INST AGING, 89- *Concurrent Pos:* Assoc ed, J Geront, 70-72. *Mem:* Fel Am Psychol Asn; fel Geront Soc. *Res:* Experimental studies of human learning, memory and problem solving with special emphasis on aging and adult development. *Mailing Add:* 13665A Via Aurora Delray Beach FL 33484-1607

AREND, WILLIAM PHELPS, b Utica, NY, Aug 24, 37; m 64; c 3. RHEUMATOLOGY, IMMUNOLOGY. *Educ:* Williams Col, BA, 59; Columbia Univ, MD, 64. *Prof Exp:* Intern, Sch Med, Univ Wash, 64-65, resident, 65-66 & 68-69, fel rheumatol, 69-71, from asst prof to assoc prof med, 71-81; prof med & dir, rheumatol, div, Med Sch, Univ Tex, Houston, 81-82; PROF MED & HEAD, RHEUMATOL DIV, MED SCH, UNIV COLO, 83-, PROF MICRO & IMMUNOL, 88- *Concurrent Pos:* Prin investr res grants, NIH, 71-; chief, arthritis sect, Vet Admin Hosp, Seattle, 71-80; vis scholar, Strangeways Res Lab & Corpus Christi Col, Cambridge, Eng, 80-81; Guggenheim Found fel, 80-81; attend physician, Hermann Hosp, Houston, 81-82 & Univ & Vet Admin Hosp, Denver, 83-; consult, Fitzsimons Army Med Ctr, Denver, 83- *Mem:* Am Soc Clin Invest; Am Asn Immunologists; Am Rheumatism Asn; fel Am Col Physicians; Asn Am Physicians. *Res:* Immunopathology of human autoimmune diseases and the role of monocytes and macrophages, particularly the expression of Fc receptors, production of interleukin-1 and of an interleukin-1 receptor antagonist. *Mailing Add:* Univ Colo Health Sci Ctr Box 115 4200 E Ninth Ave Denver CO 80262

ARENDALE, WILLIAM FRANK, b South Pittsburg, Tenn, Nov 11, 21; m 45; c 3. CHEMICAL PHYSICS. *Educ:* Mid Tenn State Univ, BS, 42; Univ Tenn, MS, 48, PhD(chem & phys), 53. *Prof Exp:* Chemist, Ala Ord Works, 42-44, Tenn Eastman Co, 44-46 & Carbide & Carbon Co, 48-50; chief res, Alpha Div, Thiokol Chem Corp, 51-56, dir res, 56-59, tech dir & asst to gen mgr, 59-64; prof chem & asst dir res inst, Univ Ala, Huntsville, 64-66, dir div natural sci & math, 66-70, PROF CHEM, 70- *Mem:* Fel AAAS; Am Chem Soc; Am Phys Soc; Am Inst Aeronaut & Astronaut; Optical Soc Am. *Res:* Infrared spectroscopy; organometallic compounds; materials properties; organo-selenium chemistry; environmental chemistry; interaction of electromagnetic energy with matter. *Mailing Add:* Univ Ala 1216 Stonehurst Dr Huntsville AL 35801

ARENDS, CHARLES BRADFORD, b Chicago, Ill, June 10, 31; m 55; c 4. PLASTICS CHEMISTRY, PLASTIC MECHANICS. *Educ:* Ill Inst Technol, BS, 52; Univ Wash, PhD(phys chem), 55. *Prof Exp:* From chemist to sr res chemist, 55-74, from res specialist to sr res specialist, 74-84, ASSOC SCIENTIST, DOW CHEM CO, 84- *Res:* Micromechanics of composite plastics; reinforcement of thermoplastics and thermosets; rheology of thermoplastic melts. *Mailing Add:* Dow Chem Co 1702 Bldg Midland MI 48674

ARENDSEN, DAVID LLOYD, b Grand Rapids, Mich, Nov 22, 42; m 66; c 2. MEDICINAL CHEMISTRY. *Educ:* NCent Col, Ill, BA, 65; Northern Ill Univ, MS, 68. *Prof Exp:* Jr res chemist, 69-79, SR RES CHEMIST, ABBOTT LABS, 79- *Mem:* Am Chem Soc; Sigma Xi. *Mailing Add:* Dept 47C Abbott Labs (AP-10) Abbott Park IL 60064

ARENDSHORST, WILLIAM JOHN, b Bloomington, Ind, Aug 25, 44; m 74; c 1. RENAL PHYSIOLOGY, RENAL PATHOPHYSIOLOGY. *Educ:* De Pauw Univ, BA, 66; Ind Univ, PhD(physiol), 70. *Prof Exp:* Res fel physiol, Ind Univ, 70-71; res fel, 71-74, from asst prof to assoc prof, 74-86, PROF PHYSIOL, UNIV NC, CHAPEL HILL, 87- *Concurrent Pos:* NIH res career develop grant, 80-85; mem, sect adv comt & chmn, renal sect, Am Physiol Soc, 87-; mem, Cardiorenal Study Comt, Am Heart Asn, 87- *Mem:* Am Physiol Soc; Am Heart Asn; Am Soc Nephrol; Int Soc Nephrol. *Res:* Mechanisms that control renal hemodynamics and single nephron function in the mammalian kidney; role of the kidney in the pathogenesis of hypertension. *Mailing Add:* 4317 Klein Dr Durham NC 27705

ARENDT, KENNETH ALBERT, b New Britain, Conn, July 24, 25; m 49; c 2. PHYSIOLOGY. *Educ:* Union Col, AB, 49; Boston Univ, AM, 52, PhD(biol), 55. *Prof Exp:* Res assoc, Biol Res Labs, Boston Univ, 53-57; from instr to assoc prof, 57-70, PROF PHYSIOL, SCH MED, LOMA LINDA UNIV, 70- *Mem:* Am Physiol Soc; Microcirculatory Soc; Sigma Xi. *Res:* Microcirculatory physiology. *Mailing Add:* Dept Physiol Loma Linda Univ Loma Linda CA 92354

ARENDT, RONALD H, b Chicago, Ill, Apr 20, 41; m 64; c 2. PHYSICAL CHEMISTRY. *Educ:* Univ Chicago, BS, 62, MS, 65, PhD(phys chem), 68. *Prof Exp:* MEM RES STAFF, GEN ELEC RES & DEVELOP CTR, 68- *Mem:* Am Chem Soc; Electrochem Soc. *Res:* Molten salts; inorganic synthesis; thermodynamics. *Mailing Add:* Gen Elec Res & Develop Ctr One River Rd Schenectady NY 12301

ARENDT, VOLKER DIETRICH, b Berlin, Ger, Sept 18, 34; US citizen; m 66; c 4. POLYMER CHEMISTRY, RUBBER CHEMISTRY. *Educ:* NC State Col, BS, 60; Princeton Univ, MA, 62, PhD(org chem), 64. *Prof Exp:* Textile tech supvr, Duncan Fox Co, Peru, 55-57 & W R Grace Co, Peru, 57; res chemist, 64-73, SR RES SCIENTIST, AM CYANAMID CO, 73- *Mem:* Am Chem Soc. *Res:* Cellulose and carbohydrate chemistry; chemical structure determinations; new analytical methods; synthetic organic and polymer chemistry; structure property relationships; research in elastomers; specialty polyacrylates; polyurethanes; controlled drug release formulations. *Mailing Add:* Eight Monroe Ct RD 4 Princeton NJ 08540

ARENS, JOHN FREDERIC, b New York, NY, July 30, 39. INFRARED ASTRONOMY. *Educ:* Mass Inst Technol, BS, 61; Univ Calif, Berkeley, PhD(physics), 66. *Prof Exp:* Astrophysicist, Goddard Space Flight Ctr, NASA, 66-; AT UNIV CALIF, SPACE SCI LAB. *Concurrent Pos:* Assoc, Nat Acad Sci, 66-69. *Mem:* Am Phys Soc. *Res:* Experimental high energy physics; measurements of cross-sections and polarizations of hadrons; satellite measurements of electrons and protons in the radiation belts and interplanetary space; cosmic ray element and isotope measurements; infrared astronomy photodetectors. *Mailing Add:* Space Sci Lab Univ Calif Berkeley CA 94720

ARENS, MAX QUIRIN, b South Bend, Ind, July 22, 45; m 67; c 2. VIROLOGY. *Educ:* Purdue Univ, BSA, 67; Va Polytech Inst & State Univ, PhD(microbiol), 71. *Prof Exp:* Res assoc, 71-75, instr, 75-76, asst res prof, Inst Molecular Virol, Med Sch, St Louis Univ, 75-80,; assoc res prof, Dept Pediat, St Louis Univ Sch Med, 80-88; ASST RES PROF, DEPT PEDIAT, WASHINGTON UNIV SCH MED, 88- *Mem:* Sigma Xi. *Res:* Mechanism and enzymology of adenovirus DNA replication; specific functions of various enzymes involved in DNA replication. *Mailing Add:* Dept Pediat Washington Univ Med Sch 400 S Kings Hwy St Louis MO 63110

ARENS, RICHARD FRIEDERICH, b Iserlohn, Ger, Apr 24, 19; nat US; m 43; c 1. MATHEMATICS. *Educ:* Univ Calif, Los Angeles, AB, 41; Harvard Univ, AM, 42, PhD(math), 45. *Prof Exp:* Tutor math, Harvard Univ, 42-45, staff asst, Inst Advan Study, 45-47; from asst prof to prof math, 47-89, EMER PROF MATH, UNIV CALIF, LOS ANGELES, 57- *Concurrent Pos:* Mem, Inst Advan Study, 53-54; ed, Pac J Math, 64-; Fulbright fel, India, 83. *Mem:* Am Math Soc; Sigma Xi. *Res:* Analysis of function spaces; topological algebra; dynamics; theoretical physics. *Mailing Add:* 12436 Deerbrook Lane Los Angeles CA 90049

ARENS, YIGAL, b New York, NY, Sept 18, 51; m 74; c 1. HUMAN-COMPUTER INTERFACES, MULTIMEDIA INTERFACES. *Educ:* Hebrew Univ, Israel, BA, 74; Univ Calif, Berkeley, MA, 77, PhD(math), 86. *Prof Exp:* Asst prof computer sci, 83-89, RES SCIENTIST, INFO SCI INST, UNIV SOUTHERN CALIF, 87- *Mem:* Am Asn Artificial Intel; Cognitive Sci Soc; Asn Comput Mach. *Res:* Intelligent interfaces to computer systems; multimedia interfaces; natural language understanding and generation; representation of information in memory. *Mailing Add:* Info Sci Inst Univ Southern Calif 4676 Admiralty Way Marina Del Rey CA 90292

ARENSON, DONALD L, b Chicago, Ill, June 15, 26; m; c 3. APPLIED MATHEMATICS. *Educ:* Ill Inst Technol, BS, 47, MA, 50. *Prof Exp:* PRES, ASSOC CONSULT INT CORP, 76- *Concurrent Pos:* Pres, Basys Inc, 62-72. *Mem:* Inst Mgt Consult; Am Math; Am Inst Aeronaut & Astronaut. *Res:* Engineering mechanics. *Mailing Add:* Assoc Consult Int Corp 626 Grove St Evanston IL 60201-4404

ARENSTORF, RICHARD F, b Hamburg, Ger, Nov 7, 29; US citizen; m 56; c 3. MATHEMATICS. *Educ:* Univ Gottingen, BS, 52, MS, 54; Univ Mainz, Dr rer nat(math), 56. *Prof Exp:* Develop engr, Telefonaktiebelaget L M Ericsson, Ger, 55-56; physicist, Army Ballistic Missile Agency, Redstone Arsenal, Ala, 57-60; staff scientist, Marshall Space Flight Ctr, NASA, 60-69; PROF MATH, VANDERBILT UNIV, 69- *Concurrent Pos:* Asst prof, Univ Ala, 58-59, from assoc prof to prof, 60-69; consult, Math Res Ctr, US Naval Res Lab, Washington, DC, 70-75. *Honors & Awards:* Medal for Except Sci Achievement, NASA, 66. *Mem:* Am Math Soc; Soc Indust & Appl Math. *Res:* Analytic number theory; nonlinear differential equations; celestial mechanics. *Mailing Add:* Dept Math Vanderbilt Univ Nashville TN 37235

ARENTS, JOHN (STEPHEN), b Brooklyn, NY, Dec 11, 26; m 65. PHYSICAL CHEMISTRY. *Educ:* Columbia Univ, AB, 50, AM, 51, PhD(chem), 57. *Prof Exp:* Asst chem, Columbia Univ, 50-56; lectr, 56-57, from instr to assoc prof, 57-79, prof chem, 79-91, EMER PROF CHEM, CITY COL, NY, 91- *Mem:* AAAS; Am Chem Soc; Am Phys Soc. *Res:* Quantum chemistry. *Mailing Add:* 53 Southfield Rd Mt Vernon NY 10552-1337

ARENZ, ROBERT JAMES, b Primghar, Iowa, Aug 20, 24. POLYMER MECHANICS, COMPOSITE MATERIALS. *Educ:* Ore State Univ, BS, 45; St Louis Univ, PhL, 56, MS, 57; Calif Inst Technol, PhD(aeronaut), 64; Alma Col, Calif, STL, 66; Univ Santa Clara, MST, 66. *Prof Exp:* Eng asst, Douglas Aircraft Co, 45-46, mathematician, 46-47, aerodynamicist, 47-50; instr eng mech, Loyola Univ, Los Angeles, 57-58; res asst, Calif Inst Technol, 59-62, res fel, 63-66; vis res scientist, Ernst-Mach-Inst, 66-67; from asst prof to assoc prof mech eng, Loyola Univ, Los Angeles, 67-78, chmn dept, 77-78; prof aerospace eng & dean fac, Parks Aeronaut Col, St Louis Univ, 78-80; prof eng & dean arts & sci, Gonzaga Univ, Spokane, 80-83; Cong fel, Off Technol Assessment, US Cong, Washington, DC, 83-85; PROF MECH ENG, GONZAGA UNIV, SPOKANE, 85- *Concurrent Pos:* Consult, Lockheed Missiles & Space Co, 64-66; vis sr scientist, NASA Jet Propulsion Lab, 69-78; Nat Sci Found grant, Va Polytech Inst, 70. *Mem:* Assoc fel Am Inst Aeronaut & Astronaut; Am Soc Mech Engrs; Soc Exp Mech; Am Acad Mech. *Res:* Solid mechanics, especially experimental and theoretical research in wave propagation in viscoelastic media; structural mechanics; spacecraft meteoroid impact effects; multiaxial mechanical testing of polymeric materials; finite deformation behavior of elastomers. *Mailing Add:* Jesuit House Gonzaga Univ Spokane WA 99258

ARFIN, STUART MICHAEL, b New York, NY, Sept 4, 36; m 64; c 2. BIOCHEMISTRY. *Educ:* City Col New York, BS, 58; Albert Einstein Col Med, PhD(biochem), 66. *Prof Exp:* Res assoc biol, Purdue Univ, 66-69; asst prof biochem, Sch Med, Univ Pittsburgh, 69-71; ASSOC PROF BIOL CHEM, SCH MED, UNIV CALIF, IRVINE, 71- *Mem:* Am Soc Microbiol; Am Soc Biol Chemists; AAAS. *Res:* Biochemical control mechanisms; enzymology. *Mailing Add:* Dept Biol Chem Univ Calif Col Med Irvine CA 92717

ARGABRIGHT, LOREN N, b Nemaha, Nebr, Jan 29, 33; m 61. MATHEMATICS. *Educ:* Nebr State Teachers Col, Peru, BS, 54; Univ Kans, MA, 58; Univ Wash, PhD(math), 63. *Prof Exp:* Instr math, Univ Calif, Berkeley, 63-65; asst prof, Univ Minn, 65-70; assoc prof & chmn dept, Univ Nebr, Lincoln, 70-74; PROF MATH & HEAD DEPT, DREXEL UNIV, 74- *Mem:* Am Math Soc; Math Asn Am. *Res:* Functional analysis; harmonic analysis; group representations. *Mailing Add:* Dept Math Drexel Univ Philadelphia PA 19104

ARGABRIGHT, PERRY A, b Pueblo, Colo, June 6, 29; m 51; c 4. CHEMISTRY. *Educ:* Univ Denver, BS, 51; Univ Colo, PhD (org chem), 56. *Prof Exp:* Res chemist, Shell Develop Co, 51-52; res chemist, Esso Res & Eng Co, NJ, 56-58, group leader, 58-59, sr chemist, 59-62; sr res scientist org synthesis, 62-68, res assoc, 68-82, mgr, 82-86, ASSOC DIR, MARATHON OIL CO, 86- *Honors & Awards:* Gold Medallion, Am Chem Soc, 72. *Mem:* AAAS; Sigma Xi (vpres, Sci Res Soc Am, 63); Am Chem Soc; NY Acad Sci; fel Am Inst Chemists. *Res:* Synthesis and characterization of organic compounds containing N-F bonds; influence of polar-aprotic solvents on rate and course of organic reactions; synthesis of nitrogen heterocycles; preparation and study of polyelectrolytes; chemicals for oil recovery. *Mailing Add:* Echo Village 4453 W Sentinel Rock Terr Larkspur CO 80118

ARGANBRIGHT, DONALD G, b North Platte, Nebr, Dec 27, 39; m 86; c 2. FORESTRY. *Educ:* Iowa State Univ, BS, 62, MS, 64; Univ Calif, Berkeley, PhD(wood sci), 71. *Prof Exp:* Asst specialist, Forest Prod Lab, Univ Calif, 64-70; asst prof wood sci, Univ Calif, Berkeley, 70-74, assoc prof, 74-78, prof, 78-80; DIR WOOD PROCESSING, FOREST PROD LAB, UNIV CALIF, 81- *Concurrent Pos:* Consult, Centre Technique du Bois, France, 75-76; prin investr, Agr Exp Sta, Univ Calif, 70- *Mem:* AAAS; Int Acad Wood Sci & Technol; Forest Prod Res Soc; Soc Wood Sci & Technol; Int Res Group Wood Preservation; Am Soc Testing Mats; Int Union Forestry Res Org; Soc Am Forestry. *Res:* Physical properties and processing of forest products with emphasis on heat and mass transfer; drying and preservative impregnation. *Mailing Add:* Dept Forestry Univ Mass Amherst MA 01003

ARGANBRIGHT, ROBERT PHILIP, b Ft Leavenworth, Kans, Nov 7, 23; m 57; c 2. ORGANIC CHEMISTRY. *Educ:* Univ Kans, BS, 50; Univ Colo, PhD(chem), 56. *Prof Exp:* Chemist, Continental Oil Co, 50-52; chemist, Monsanto Chem Co, 56-60, res specialist, 60-67; res assoc, Petro-Tex Chem Corp, 67-74, res mgr, 74-76; res supvr, M W Kellog Co, 76-82, consult, 83-87, SR SCIENTIST, CHEM RES & LICENSING CORP, 88-; consult, 83-87. *Mem:* Sigma Xi; Catalysis Soc; Am Chem Soc. *Res:* Heterogeneous catalysis; process development. *Mailing Add:* 814 Devonport Seabrook TX 77586

ARGAUER, ROBERT JOHN, b Buffalo, NY, Feb 23, 37; m 64; c 5. ANALYTICAL CHEMISTRY. *Educ:* Canisius Col, BS, 58; Univ Md, PhD(inorg chem), 63. *Prof Exp:* Res chemist, Mobil Oil Co, 63-65; RES CHEMIST, USDA, 65- *Mem:* Am Chem Soc; Entom Soc Am; Sigma Xi. *Res:* Application of light and flame absorption and emission spectro- fluorometry and spectrophotometry, gas chromatography, infrared, mass spectroscopy and radiochemical methods to chemical problems, especially pesticide analysis. *Mailing Add:* 4208 Everett St Kensington MD 20895

ARGENAL-ARAUZ, ROGER, b Leon, Nicaragua, Feb 18, 37; m 65; c 4. GEOMAGNETISM, SEISMOLOGY. *Educ:* Monterrey Inst Technol, Mex, BS, 60; Univ Tex, Austin, MS, 68, PhD(physics), 75. *Prof Exp:* Sci adv geophys, 75-80, DIR, INST SEISMIC RES, 81-; PROF PHYSICS, UNIV NICARAGUA, 60- *Mem:* Soc Centroamericana de Fisica; Soc Nicaraguense de Fisica (pres, 76); Am Geophys Union; Soc Explor Geophysicists. *Res:* Crustal structure in the Nicaraguan region; studies being made through the regional magnetic field anomalies and the seismic characteristics shown by the Seismic Net of Nicaragua formed by sixteen stations; acceleration, attenuation curves for Nicaragua; assessment of seismic risk in Nicaragua. *Mailing Add:* 7731 Hunter Valley Rd Calgary AB T2K 4K9 Can

ARGENZIO, ROBERT ALAN, b Denver, Colo, Apr 27, 41. PHYSIOLOGY, GASTROENTEROLOGY. *Educ:* Colo State Univ, BS, 66, MS, 68; Cornell Univ, PhD(nutrit), 71. *Prof Exp:* Res assoc physiol, Dept Vet Physiol, Cornell Univ, 71-76; physiologist, Nat Animal Dis Ctr, 76-; PROF, DEPT ANAT SCH VET MED, NC STATE UNIV. *Concurrent Pos:* Assoc prof, Dept Vet Physiol, Iowa State Univ, 78- *Mem:* Am Gastroenterol Asn; Am Physiol Soc; Comp Gastroenterol Asn. *Res:* Intestinal absorption; pathophysiology of diarrheal diseases. *Mailing Add:* Dept Anat Sch Vet Med NC State Univ Raleigh NC 27650

ARGERSINGER, WILLIAM JOHN, JR, b Chittenango, NY, Apr 14, 18; m 42; c 3. PHYSICAL CHEMISTRY. *Educ:* Cornell Univ, AB, 38, PhD(phys chem), 42. *Prof Exp:* Asst chem, Cornell Univ, 38-41, instr, 42-44; assoc chemist, Monsanto Chem Co, Ohio, 44-45, group leader, Manhattan Proj, 45-46; from asst prof to prof chem, 46-88, assoc dean fac, 63-70, dean res admin, 70-72, vchancellor res & grad studies, 72-78, dean, Grad Sch, 72-78, EMER PROF CHEM & EMER DEAN GRAD SCH, UNIV KANS, 88- *Concurrent Pos:* Res chemist, Mound Lab, AEC, 51. *Mem:* Fel AAAS; fel Am Inst Chemists; Am Asn Univ Professors; Am Chem Soc; Sigma Xi. *Res:* Surface equilibrium in solutions; coordination compounds; chemical kinetics, cation exchange; thermodynamics of solutions. *Mailing Add:* Dept Chem Univ Kans Lawrence KS 66045-0046

ARGO, HAROLD VIRGIL, b Walla Walla, Wash, Jan 20, 18; m 42; c 4. ASTRONOMY. *Educ:* Whitman Col, AB, 39; George Washington Univ, MA, 41; Univ Chicago, PhD, 48. *Prof Exp:* Jr physicist, Naval Ord Lab, Washington, DC, 42-44; physicist, Los Alamos Nat Lab, 44-; RETIRED. *Mem:* AAAS; Am Geophys Union; Am Phys Soc; Am Astron Soc. *Res:* Experimental nuclear physics; space physics; satellites and probes; solar physics. *Mailing Add:* 301 Potrillo Dr Los Alamos NM 87544

ARGO, PAUL EMMETT, b Los Alamos, NMex, Nov 25, 48; m 83; c 2. IONOSPHERIC PHYSICS, RADIO PROPAGATIONS. *Educ:* Pomona Col, BA, 70; Univ Calif San Diego, MS, 78, PhD(appl physics), 80. *Prof Exp:* Scientist, Naval Ocean Systs Ctr, 71-81; engr, Linkabit, 81-82; STAFF SCIENTIST, LOS ALAMOS NAT LAB, 82- *Mem:* Am Geophys Union; Union Radio Sci Int. *Res:* Radio propagation through ionosphere; effects of ionospheric perturbations on both the propagation channel and communication systems using the channel. *Mailing Add:* Los Alamos Nat Lab MS-D466 Los Alamos NM 87545

ARGON, ALI SUPHI, b Istanbul, Turkey, Dec 19, 30; m 53; c 2. MATERIALS SCIENCE, MECHANICAL ENGINEERING. *Educ:* Purdue Univ, BS, 52; Mass Inst Technol, SM, 53, ScD(mat), 56. *Prof Exp:* Proj engr, High Voltage Eng Corp, 56-58; instr thermodyn, Middle East Tech Univ, Ankara, 59-60; asst prof mat, 60-65, from assoc prof to prof, 65-82, QUENTIN BERG PROF MECH ENG, MASS INST TECHNOL, 82- *Concurrent Pos:* Vis prof polymer physics, Univ Leeds, UK, 71-72. *Honors & Awards:* C R Richards Award, Am Soc Mech Engrs, 76; Hon Fel, Int Cong Fracture, 83. *Mem:* Nat Acad Eng; Am Phys Soc; Am Inst Mining, Metall & Petrol Engrs; fel Am Phys Soc. *Res:* Mechanical behavior of materials; physical mechanisms of deformation and fracture of materials. *Mailing Add:* Rm 1-306 Mass Inst Technol Cambridge MA 02139

ARGOS, PATRICK, b Carbondale, Ill, Nov 24, 42. CRYSTALLOGRAPHY, MOLECULAR BIOPHYSICS. *Educ:* St Louis Univ, BS, 64, MS, 66, PhD(physics), 68. *Prof Exp:* USPHS fel protein crystallog, Sch Med, Wash Univ, 69-71; asst prof physics, Pomona Col, 71-73; res assoc, Purdue Univ, 73-74; assoc prof biophys, Southern Ill Univ, 74-78; ASSOC PROF BIOL, PURDUE UNIV, 78- *Mem:* Am Soc Biol Chemists; Am Crystallog Asn; Biophys Soc; Am Soc Virol; AAAS. *Res:* X-ray crystallographic studies of the structure, function, and evolution of proteins and viruses. *Mailing Add:* European Molecular Biol Lab Postfach 10:2209 Meyerhofstr 1 6900 Heidelberg Germany

ARGOT, JEANNE, b Pocono Lake, Pa, Mar 12, 36. MICROBIOLOGY. *Educ:* Moravian Col, BS, 65; Lehigh Univ, MS, 67, PhD(biol), 69. *Prof Exp:* Biologist, Pharmachem Corp, Pa, 62-66; asst prof biol, Lebanon Valley Col, 69-; AT MARION COL, IND. *Mem:* Am Soc Med Technol; Am Soc Microbiol; Sigma Xi. *Mailing Add:* Indiana Wesleyan Univ 4201 S Washington St Marion IN 46953

ARGOUDELIS, CHRIS J, b Piraeus, Greece, July 30, 29; nat US. ORGANIC CHEMISTRY, BIOCHEMISTRY. *Educ:* Nat Univ Athens, BS, 56; Univ Ill, PhD(food technol), 61. *Prof Exp:* Res asst food technol, 60-61, asst prof, 61-77, ASSOC PROF FOOD SCI, UNIV ILL, URBANA-CHAMPAIGN, 77- *Concurrent Pos:* NIH res grant, 64-71. *Mem:* AAAS; Am Chem Soc; Sigma Xi. *Res:* Chemistry and metabolism of vitamin B6. *Mailing Add:* 580 Bevier Hall Univ Ill 905 S Goodwin Urbana IL 61801

ARGRAVES, W SCOTT, b Jan 30, 56. BIOCHEMISTRY. *Educ:* Univ Conn, BS, 78, PhD(cellular & develop biol), 85. *Prof Exp:* Res asst, Dept Biol Sci & Dept Animal Genetics, Univ Conn, 79-84; postdoctoral fel, La Jolla Cancer Res Found, Calif, 85-87, res assoc, 87-88; staff scientist I, 88-91, STAFF SCIENTIST II, AM RED CROSS, ROCKVILLE, MD, 91-; RES ASST PROF, DEPT ORAL PATH, UNIV MD, 89- *Concurrent Pos:* Postdoctoral fel, Am Chem Soc, 86-88. *Mem:* Soc Develop Biol; Sigma Xi; Am Soc Cell Biol. *Mailing Add:* Dept Biochem J H Holland Lab Am Red Cross 15601 Crabbs Br Way Rockville MD 20855

ARGUS, GEORGE WILLIAM, b Brooklyn, NY, Apr 14, 29; m 55; c 5. SYSTEMATIC BOTANY. *Educ:* Univ Alaska, BS, 52; Univ Wyo, MS; Harvard Univ, PhD(biol), 61. *Prof Exp:* Nat Res Coun Can fel plant taxon, Univ Sask, 61-63, asst prof plant ecol & lectr biol, 63-67, assoc prof, 67-69, cur, W P Fraser Herbarium, 63-69, actg dir, Inst North Studies, 65-66; cur herbarium, Mus Natural Hist, Univ Ore, 69-70; res scientist, Can Forest Serv, 70-72; assoc cur, 72-85, CUR VASC PLANTS, NAT MUS NATURAL SCI, 85-; TAXON ED, FLORA N AM, 86- *Concurrent Pos:* Mem, Nat Res Coun Can, 73-77, chmn plant biol grant selection comt, 75-76; sci authority plants, Int Conv Trade in Endangered Species, 75- *Honors & Awards:* Conserv Award, Fedn Ont Naturalists. *Mem:* Am Soc Plant Taxon; Int Asn Plant Taxon; Can Bot Asn. *Res:* Taxonomy and evolution of Salix in North America; endemism in northwestern Saskatchewan; rare plants of Canada. *Mailing Add:* Bot Div Can Mus Nature Ottawa ON K1P 6P4 Can

ARGUS, MARY FRANCES, b Ironton, Ohio, July 16, 24. ONCOLOGY, TOXICOLOGY. *Educ:* Col Mt St Joseph, BS, 45; Univ Cincinnati, MS, 47; Univ Fla, PhD(biochem), 52. *Prof Exp:* Instr chem, Marygrove Col, 47-49; from asst prof to assoc prof cancer res, Univ Fla, 52-60; from assoc prof to prof biochem & med, Sch Med, Tulane Univ, 60-80; SR SCI ADV, OFF TOXIC SUBSTANCES, US ENVIRON PROTECTION AGENCY, 80- *Honors & Awards:* Joseph Seifter Award, 83. *Mem:* Am Soc Biol Chem; Am Asn Cancer Res; Am Soc Pharmacol & Exp Therapeut; Am Col Toxicol; Soc Toxicol. *Res:* Chemical carcinogenesis; drug metabolism; conformational changes in macromolecules during carcinogenesis; alterations in oxidative metabolism in cardiac hypertrophy; chemical risk assessment. *Mailing Add:* US Environ Protection Agency TS-796 401 M St Southwest Washington DC 20460

ARGY, DIMITRI, b Athens, Greece, Mar 13, 21; US citizen; m 56; c 2. MATERIALS SCIENCE, METALLURGY. *Educ:* Athens Tech Univ, dipl, 46; Aachen Tech Univ, Dr Ing, 55. *Prof Exp:* Res metall engr, Carnegie Inst Technol, 56-58 & E I du Pont de Nemours & Co, Inc, 58-60; sr res metallurgist, Foote Mineral Co, 60-61; chief metallurgist, Hoeganaes Corp, 61-63 & Starlite Indust, Inc, 63-65; sect head, Metals & Controls, Inc, 65-67; chmn dept mech eng, 67-76, PROF MECH ENG, SOUTHEASTERN MASS UNIV, 67- *Mem:* Am Inst Mining, Metall & Petrol Engrs; Am Soc Metals; Nat Soc Prof Engrs; Am Soc Eng Educ; Brit Inst Metals; Sigma Xi. *Res:* Composite materials, properties and evaluation; direct rolling of powders; compaction and sintering mechanism; production of powders by the atomizing process and reduction; alloy development; high temperature oxidation protective coatings; conductivity of molten salts; constitution of the liquid state. *Mailing Add:* Col Eng Southeastern Mass Univ North Dartmouth MA 02747

ARGYRES, PETROS, b Lefkas, Greece, Mar 9, 27; nat US; m 58; c 3. THEORETICAL PHYSICS OF CONDENSED MATTER. *Educ:* Univ Calif, AB, 50, MA, 52, PhD(physics), 54. *Prof Exp:* Res physicist, Westinghouse Res Lab, 54-58 & Lincoln Lab, Mass Inst Technol, 58-67; PROF PHYSICS, NORTHEASTERN UNIV, 67- *Concurrent Pos:* Lectr, Univ Pittsburgh, 55-57; vis assoc prof, Mass Inst Technol, 65-66; vis prof, Ecole Normale Superieure, Univ Paris, 69-70; vis prof, Tech Univ Greece, Athens, 82-83. *Mem:* Fel Am Phys Soc; corresp mem Acad Arts & Sci Greece; AAAS. *Res:* Solid state physics; irreversible phenomena; transport processes; statistical mechanics; spin resonance; many-body theories; equilibrium and transport properties of disordered systems with consideration of electron-electron and electron-phonon interactions and use of formal projection operator techniques. *Mailing Add:* Dept Physics Northeastern Univ Boston MA 02115

ARGYRIS, BERTIE, b Holland, June 27, 30; nat US; m 55. IMMUNOLOGY. *Educ:* Columbia Univ, BA, 51; Brown Univ, MS, 53; Syracuse Univ, PhD, 58. *Prof Exp:* Res assoc exp morphol, Syracuse Univ, 58-62 & immunobiol, 62-68; assoc prof urol, State Univ NY Upstate Med Ctr, 68-71, prof microbiol, 71-85; RETIRED. *Mem:* Am Soc Zool; Soc Study Develop Biol; Am Inst Biol Sci; Am Asn Immunol; Transplantation Soc; fel Am Soc Exp Biol. *Res:* Endocrinology-insulin resistance; experimental morphology-tissue interaction; effects of tumors on mammary glands; transplantation immunity; acquired homograft tolerance; tumor immunology; immunosuppression; cellular interactions during immune response; immunological maturation; neonatal suppressor cells; macrophage immunology; author computer based tutoriols in immunol. *Mailing Add:* 1433 Arboretum Way Burlington MA 01803

ARGYRIS, THOMAS STEPHEN, b Newark, NJ, July 16, 23; m 55. EXPERIMENTAL PATHOLOGY. *Educ:* Rutgers Univ, BS, 48; Brown Univ, PhD, 53. *Prof Exp:* Instr & fel, Harvard Univ, 53-55; from asst prof to prof zool, Syracuse Univ, 55-72; PROF PATH, STATE UNIV NY UPSTATE MED CTR, 72- *Concurrent Pos:* NSF sr fel embryol, Carnegie Inst, 61-62; NIH spec res fel zool, Univ Col, Univ London, 65-66. *Mem:* Fel AAAS; Am Asn Cancer Res; Soc Develop Biol; Am Asn Pathologists; Soc Invest Dermat. *Res:* Wound healing; carcinogenesis; ribosome biogenesis. *Mailing Add:* Dept Pathol 1525 Dolphin Dr Aptes CA 95003

ARGYROPOULOS, STAVROS ANDREAS, b Pireocus, Greece, Dec 15, 45; Can citizen; m 78; c 1. REALTIME SYSTEMS, PROCESS CONTROL. *Educ:* Nat Tech Univ Athens, dipl eng, 69; Grad Sch Indust Studies, dipl statist, 73; McGill Univ, MEng, 77, PhD(metall), 81. *Prof Exp:* Univ res fel metall, 81-86; asst prof, 86-89, ASSOC PROF METALL, UNIV TORONTO, 89- *Concurrent Pos:* Chmn, Computer Applications Comt, Metall Soc, Can Inst Mining & Metall, 90- *Honors & Awards:* C W Briggs Award, Iron & Steel Soc US, 84, 85 & 89, R W Hunt Award, 90. *Mem:* Iron & Steel Soc US; Metall Soc, Can Inst Mining & Metall; Am Soc Qual Control; Instrument Soc Am. *Res:* Transport phenomena in materials processing; mass transfer of ferroalloys in liquid steel; convection in liquid metals. *Mailing Add:* Dept Metall Univ of Toronto 184 College St Toronto ON M5S 1A4

ARGYROS, IOANNIS KONSTANTINOS, b Greece, Feb 20, 56. MATHEMATICS. *Educ:* Univ Ga, PhD(math), 84. *Prof Exp:* PROF MATH, NMEX STATE UNIV, 86- *Mem:* Am Math Soc. *Mailing Add:* Dept Math NMex State Univ Las Cruces NM 88003

ARHELGER, ROGER BOYD, b Green Bay, Wis, May 23, 32; m 53; c 3. MEDICINE, PATHOLOGY. *Educ:* Hamline Univ, BS, 54; Univ Minn, MD, 58. *Prof Exp:* Intern, USPHS Hosp, Boston, Mass, 58-59; resident, Med Ctr, Univ Miss, 59-62, from instr to asst prof path, 62-66; assoc prof path, Univ Tex Med Sch San Antonio, 66-68; assoc prof, Med Ctr, Univ Miss, 68-72; chief surg path & cytol, Fresno Community Hosp, 72-75; chmn dept path, Kern Med Ctr, 75-77; PATHOLOGIST, MISS BAPTIST MED CTR, 77-; ASSOC CLIN PROF, MED CTR, UNIV MISS, 77- *Concurrent Pos:* USPHS res career develop award, 63- *Mem:* Electron Micros Soc Am; Am Soc Exp Path; Am Asn Path & Bact; Int Acad Path; AMA; Sigma Xi. *Res:* Experimental pathology; electron microscopy. *Mailing Add:* Pathology Dept St Marks Hosp 1200 E 3900 South Salt Lake City UT 84124

ARIANO, MARJORIE ANN, b Tokyo, Japan, Feb 13, 51. US citizen. NEUROCHEMISTRY. *Educ:* Univ Calif, Los Angeles, BS, 72, PhD(anat), 77. *Prof Exp:* Asst molecular biol, Univ Southern Calif, 77-80; asst prof anat & neurobiol, 80-86, mem cell biol prog, 82-88, ASSOC PROF ANAAT & NEUROBIOL, UNIV VT, 86- *Concurrent Pos:* Mem bd dirs, Am Heart Asn, 83-87; mem, adv panel, Cell & Molecular Neurobiol, NSF, 85-87; mem, Neurol Sci Study Sect, NIH, 87-91; Res Career Develop Award, Nat Inst Neurol Dis & Stroke, 84-89. *Mem:* Soc Neurosci. *Res:* Elucidation of the specific cellular site of second messenger production in response to dopamine neurotransmitter receptor activation in basal ganglia and sympathetic nervous system in mammals. *Mailing Add:* Dept Anatomy & Neurobiology Given Med Bldg Burlington VT 05405

ARIAS, IRWIN MONROE, b New York, NY, Sept 4, 26; m 53; c 4. MEDICINE, GASTROENTEROLOGY. *Educ:* Harvard Univ, SB, 46; Long Island Col Med, MD, 52; Am Bd Internal Med, dipl. *Prof Exp:* From intern to resident med, Boston City Hosp, 52-54; resident, Boston Vet Admin Hosp, 54-55; assoc, Albert Einstein Col Med, 56-64, assoc prof, 64-67, prof med, 67-; CHMN DEPT PHYSIOL, TUFTS UNIV. *Concurrent Pos:* Res fel, Boston City Hosp, 55-56; NY Heart Asn fel, 56-; instr, NY Heart Asn, 56-; Suiter Mem lectr, 59; attend physician, Albert Einstein Col Med Hosp, 67-; assoc vis prof, Bronx Munic Hosp Ctr; consult, New Rochelle Hosp, Mt Vernon Hosp, USPHS, Pan Am Union & Kellogg Found. *Honors & Awards:* Suiter Mem Lectr, 59; Hon Achievement Award, Am Gastroenterol Asn, 69; Citation in Sci, Univ Recife, 70. *Mem:* AAAS; Am Soc Gastrointestinal Endoscopy; Am Gastroenterol Asn; Am Physiol Soc; Am Asn Study Liver Dis. *Res:* Academic gastroenterology; mechanisms of hepatic excretory function. *Mailing Add:* Dept Physiol Tufts Univ Boston MA 02111

ARIAS, JOSE MANUEL, b Los Angeles, Calif, June 22, 54; m 79; c 4. SEMICONDUCTOR EPITAXIAL GROWTH, ADVANCED INFRARED, MATERIALS & DEVICES. *Educ:* ITESO, Guadalajara, Mex, BS, 78; Univ Calif, Santa Barbara, PhD(phys chem), 84. *Prof Exp:* Teaching asst chem, Univ Calif, Santa Barbara, 79-81, res asst phys chem, 82-84; MEM TECH STAFF ADVAN MAT, SCI CTR, ROCKWELL INT, 85- *Res:* Surface science and solid state physics; research and development of the molecular beam epitaxy growth technology of II-VI semiconductor materials for opto-electronic device applications; author or co-author of 50 scientific papers. *Mailing Add:* 1049 Camino Dos Rios Thousand Oaks CA 91360

ARIEFF, ALEX J, neurology, psychiatry; deceased, see previous edition for last biography

ARIEFF, ALLEN I, MEDICINE. *Educ:* Univ Ill, BS, 60; Northwestern Univ, MS & MD, 64; Am Bd Internal Med, cert, 72, cert nephrology, 76. *Prof Exp:* Rotating intern, Philadelphia Gen Hosp, Univ Pa Serv, 64-65; res med & chief resident, Downstate Med Ctr, State Univ NY, Brooklyn, 67-68; renal fel, Univ Colo, Denver, 68-69; res & educ assoc & clin investr, Wadsworth Vet Admin Med Ctr, Los Angeles, Calif, 70-74; asst prof med & res scientist, Cedar-Sinai Med Ctr, Univ Calif, Los Angeles, 71-74; asst prof med & dir hemodialysis, Vet Admin Med Ctr, Univ Calif, 75-76, assoc prof & dir nephrology sect, 76-83, chief, clin nephrology, 83-86, PROF MED, VET ADMIN MED CTR, UNIV CALIF, 83-, DIR RES & EDUC, GERIAT, 86- *Concurrent Pos:* Res career develop award, Vet Admin, 70-76; sci adv coun, Nat Kidney Found; numerous invited lectrs, 73-91; consult, Food & Drug Admin, 77 & 89, Mobil Oil Co, Dallas, Tex, 85, Abbott Labs, Ill, 87, Vet Admin Coop Studies Sect, 90 & others. *Mem:* Am Soc Nephrology; Int Soc Nephrology; Am Fedn Clin Res; Am Diabetes Asn; Am Physiol Soc; fel Am Col Physicians; Am Soc Neurochem; Am Soc Clin Invest; Am Soc Bone & Mineral Res. *Mailing Add:* Geriat Res Vet Admin Med Ctr Univ Calif 4150 Clement St 111G San Francisco CA 94121

ARIEMMA, SIDNEY, b New York, NY, Apr 9, 22; m 51; c 4. POLYMER CHEMISTRY, ORGANIC CHEMISTRY. *Educ:* Wagner Col, BS, 47; Stevens Inst Technol, MS, 54. *Prof Exp:* Chemist, Armour & Co, 47-50 & Wallerstein Co, 50-52; group leader polymers, Air Reduction Co, Inc, 52-60; sr res scientist, Texaco, Inc, 60-62; mgr polymers, Tex Butadiene & Chem Co, 62-63; mgr natural & synthetic resins, Morningstar Paisley, Inc, 63-64; mgr polymers & coatings, Reichhold Chem Co, 64-66; group leader coatings & finishes, Huyck Corp, 66-69; assoc dir, Polymers, Waxes, Coatings & Rubber Dept, Foster D Snell, Inc, Florham Park, 69-73; PRES & CHIEF EXEC

OFFICER, ARYS LAB, INC, WALDWICK, 73- *Concurrent Pos:* Mem, Int Coun Fabric Flammability. *Mem:* Am Chem Soc; Soc Plastics Engrs; NY Acad Sci. *Res:* Development of petrochemicals, polymers and specialties applied to paper, textiles, paint, polish, adhesives, automotive, leather, moldings, extrusion and casting resins. *Mailing Add:* 29 Skytop Dr Mahwah NJ 07430

ARIES, ROBERT SANCIER, biotechnology, for more information see previous edition

ARIET, MARIO, b Havana, Cuba, July 9, 39; m 60; c 5. SYSTEMS & CHEMICAL ENGINEERING. *Educ:* Univ Fla, BChE, 60, MSE, 62, PhD(chem eng, math), 65. *Prof Exp:* Sr engr, Humble Oil & Refining Co, Tex, 65-67; ASSOC PROF INDUST & SYSTS ENG, UNIV FLA, 67-, PROF COMMUNITY HEALTH & FAMILY MED & DIR COMPUT SCI DIV, 78-,. *Concurrent Pos:* Prof Med, 85- *Mem:* Am Inst Chem Engrs; Asn Comput Mach. *Res:* Computerized analysis of electrocardiagrams; computer simulation of physiochemical systems; computerized closed-loop control of chemical and biological systems; automation of health care delivery systems; management of clinical data base. *Mailing Add:* Dept Med Univ Fla Col Med Gainesville FL 32610

ARIF, BASIL MUMTAZ, b Baghdad, Iraq, April 12, 40; Can citizen; m 75. VIROLOGY, NUCLEIC ACIDS. *Educ:* Queen's Univ, Belfast, UK, BSc, 67; Queen's Univ, Kingston, Ont, MSc, 69, PhD(virol), 71. *Prof Exp:* Sr demonstr microbiol, Queen's Univ, Kingston, Ont, 68-71; res scientist, 72-84, SR RES SCIENTIST VIROL, FOREST RES & MGT INST, 85- *Concurrent Pos:* Vis prof, Lake Superior State Col, 78; prof, Algoma Campus, Laurentian Univ, 79; vis scientist, Univ Cologne, WGermany, 81-83. *Mem:* Am Soc Microbiol; Am Soc Virol; Soc Invert Path. *Res:* Molecular biology of baculoviruses including molecular cloning, mapping and DNA sequencing to elucidate the nature, mode of action and specificity and to examine the potential of improving these visuses as forest pest control agents. *Mailing Add:* Forest Pest Mgt Inst PO Box 490 Sault Ste Marine ON P6A 5M7 Can

ARIF, SHOAIB, b Karachi, Pakistan, May 28, 50; US citizen; m 79; c 5. RESEARCH & DEVELOPMENT PERSONAL CARE HOUSE-HOLD OR I & I PRODUCTS. *Educ:* Univ Karachi, Pakistan, BS, 69, MS, 71. *Prof Exp:* Chemist, Kohimoor Chem Co, Karachi, Pakistan, 71-77, Pvt Formulations, Edison, NJ, 77-78, Dept Water Resources, New York, 78-80; chief chemist, Cameo Inc, Toledo, Ohio, 80-83; assoc dir res & develop, GO-JO Industs, Akron, Ohio, 83-85; chemist res & develop, Sani-Fresh Inc, San Antonio, Tex, 85-87; dir res & develop, Jeri Jacobs Labs, Carrollton, Tex, 87-89; APPLN SPECIALIST, OLIN CORP, CHESHIRE, CT, 89- *Mem:* Am Soc Testing & Mat; Soc Cosmetic Chemists; Chem Specialties Mfrs Asn; Cosmetic Toiletry & Fragrance Asn. *Res:* Developed numerous new consumer products in personal care (cosmetics), house-hold and I and I areas. *Mailing Add:* 350 Knotter Dr Cheshire CT 06410

ARIMOTO, SUGURU, b Hiroshima, Japan, Aug 3, 36; m 65; c 2. ROBOTICS, INFORMATION THEORY. *Educ:* Kyoto Univ, BS Sci, 59; Univ Tokyo, DEng, 67. *Prof Exp:* Engr, Oki Elec Indust Co Ltd, 59-62; assoc prof control eng, Osaka Univ, 68-73, prof robotics, 73-90; res asst appl physics, 62-67, lectr control theory, 67-68, PROF INFO THEORY & ROBOTICS, UNIV TOKYO, 88- *Concurrent Pos:* Vis prof, Tokyo Inst Technol, 88- *Mem:* Fel Inst Elec & Electronics Engrs. *Res:* Theory of learning for skill refinement; information theory; machine intelligence. *Mailing Add:* Fac Eng Univ Tokyo Bunkyo-Ku Tokyo 113 Japan

ARIMURA, AKIRA, b Kobe, Japan, Dec 26, 23; m 57; c 3. ENDOCRINOLOGY, PHYSIOLOGY. *Educ:* Nagoya Univ, BS, 43, MD, 51, PhD(physiol), 57. *Prof Exp:* From asst to instr med, Sch Med, Tulane Univ, 58-61; instr & res assoc physiol, Sch Med, Hokkaido Univ, 61-65; dir, Clin Res Inst Am, 80-87; from asst prof to assoc prof, 65-73, PROF MED, SCH MED, TULANE UNIV, 73- *Concurrent Pos:* J H Brown fel, Sch Med, Yale Univ, 56-58; USPHS res grants, Sch Med, Hokkaido Univ, 63-64 & Sch Med, Tulane Univ, 65-; res consult, Vet Admin Hosp, New Orleans, 65-81; Fulbright Comn Exchange Scholar, 56. *Mem:* AAAS; Endocrine Soc; Am Physiol Soc; Soc Exp Biol & Med; Japan Endocrinol Soc; hon mem Japan Endocrin Soc; NY Acad Sci. *Res:* Experimental and clinical neuroendocrinology. *Mailing Add:* Dir US-Japan Biomed Res Labs Herbert Ctr 3705 Main St Belle Chasse LA 70037

ARIN, KEMAL, b Ankara, Turkey, Jan 3, 41; m 67; c 2. FRACTURE MECHANICS, SOFTWARE DEVELOPMENT. *Educ:* Tech Univ Istanbul, BS & MS, 63; Lehigh Univ, PhD(appl mech), 69. *Prof Exp:* Instr civil eng & appl mech, Tech Univ Istanbul, 65-67; res asst appl mech, Lehigh Univ, 67-69, asst prof, 69-76; staff engr, 76-80, SR ENGR APPL MECH, GEN ELEC CO, 80- *Mem:* Sigma Xi; Am Soc Mech Engrs; Am Acad Mech; Am Soc Eng Educ; Soc Eng Sci. *Res:* Fracture mechanics; fatigue; software development; vibrations; thermoelasticity; plasticity; computer graphics. *Mailing Add:* Gen Elec Co 2266 Pineridge Rd Bldg AEBG Schenectady NY 12309

ARING, CHARLES DAIR, b Dent, Ohio, June 21, 04; m 31; c 2. NEUROLOGY. *Educ:* Univ Cincinnati, BS & MD, 29. *Hon Degrees:* LHD, Univ Cincinnati, 90. *Prof Exp:* Rotating intern, Cincinnati Gen Hosp, 29-30, resident physician, 30-31; resident, Longview State Hosp, Cincinnati, 31; house officer neurol, Boston City Hosp, 32, resident, 33-34; house officer, Children's Hosp, Boston, 33; from instr to assoc prof, Col Med, Univ Cincinnati, 36-46; prof, Col Med, Univ Calif, 46-47, dir neurol serv, 46-47; prof neurol, 47-74, dir neurol serv, 47-76, EMER PROF NEUROL, COL MED, UNIV CINCINNATI, 74- *Concurrent Pos:* Fel neuro-physiol, Yale Univ, 34-35; Rockefeller fel neurol, Madrid, Breslau & Nat Hosp, London, 35-36; attend neurologist, Cincinnati Gen Hosp, 36-45; mem comn neurotropic virus dis, Bd Invest Control Influenza & Other Epidemic Dis. *Mem:* AMA; Am Neurol Asn (pres, 63); Am Psychiat Asn; Asn Res Nerv & Ment Dis (vpres, 47-52). *Res:* Psychiatry; medical administration; medical philosophy. *Mailing Add:* 2444 Madison Rd Cincinnati OH 45208

ARION, DOUGLAS, b New York, NY, Jan 27, 57; m 90. RADIATION PHYSICS, INSTRUMENTATION DESIGN & DEVELOPMENT. *Educ:* Dartmouth Col, AB, 78, Univ Md, MS, 80, PhD(physics), 84. *Prof Exp:* Grad res fel, Goddard Space Flight Ctr, NASA, 82-84; sr scientist, McLean, Va, 84-88, DIV MGR, SCI APPLN INT CORP, ALBUQUERQUE, NM, 88- *Mem:* Am Phys Soc. *Res:* Radiation diagnostics; radiation instrumentation and effects. *Mailing Add:* Sci Appln Int Corp 2109 Air Park Rd SE Albuquerque NM 87106

ARION, WILLIAM JOSEPH, b Cando, NDak, May 31, 40; m 85; c 3. BIOCHEMISTRY. *Educ:* Jamestown Col, BS, 62; Univ ND, 62-65, MS, 64, PhD(biochem), 66. *Prof Exp:* NIH fel biochem, Cornell Univ, 66-68; from asst prof to prof biochem, 68-80, PROF BIOCHEM, DIV NUTRIT SCI & SECT BIOCHEM MOLECULAR CELL BIOL, CORNELL UNIV, 80- *Concurrent Pos:* Vis prof biochem, Univ Stockholm, 76-77. *Mem:* Am Inst Nutrit; NY Acad Sci; Am Soc Biochem & Molecular Biol; Sigma Xi. *Res:* Mechanisms of glucose homeostasis; membrane structure-function; bioenergetics and energy balance. *Mailing Add:* Div Nutrit Sci Cornell Univ 227 Savage Hall Ithaca NY 14853

ARIS, RUTHERFORD, b Bournemouth, Eng, Sept 15, 29; US citizen; m 58. APPLIED MATHEMATICS, CHEMICAL ENGINEERING. *Educ:* Univ London, BSc, 48, PhD(math, chem eng), 60, DSc, 64. *Hon Degrees:* DSc, Univ Exeter, 84 & Clarkson Univ, 85; DEng, Notre Dame Univ, 90. *Prof Exp:* Tech officer, Imperial Chem Industs, 50-55; res fel chem eng, Univ Minn, 55-56; lectr tech math, Univ Edinburgh, 56-58; from asst prof to prof, 58-78, REGENTS' PROF CHEM ENG, UNIV MINN, MINNEAPOLIS, 78- *Concurrent Pos:* Consult, Am Oil Co, Ind, 62-87, Gen Elec Co, 62-64 & Gen Motors, 77-85; Nat Sci Found sr fel, Cambridge Univ, 64-65; Guggenheim fel, 71-72; Lacey lectr, Calif Inst Technol, 70; Hougen Prof, Univ Wis, 79; Fairchild scholar, Calif Inst Tech, 80. *Honors & Awards:* E Harris Harbison Award for Distinguished Teaching, Danforth Found, 69; Alpha Chi Sigma Award for Res in Chem Eng, Am Inst Chem Engrs, 69 & Wilhelm Award, 75; Lewis Award, 81. *Mem:* Nat Acad Eng; Soc Natural Philos; Soc Indust & Appl Math; Soc Scribes & Illum; Am Chem Soc; Mediaeval Acad Am; Am Acad Arts & Sci; fel Inst Math & Applications; Am Inst Chem Engrs. *Res:* Stability and control of chemical reactors; mathematics of chemical kinetics and of diffusion and reaction; mathematical modeling. *Mailing Add:* Dept Chem Eng & Mat Sci Univ Minn Minneapolis MN 55455

ARIYAN, STEPHAN, b Cairo, Egypt, July 30, 41; m 67; c 3. PLASTIC SURGERY, GENERAL SURGERY. *Educ:* Long Island Univ, BS, 62; New York Med Col, MD, 66; Yale Univ, MS, 81. *Prof Exp:* Asst prof surg, 76-79, assoc prof, 79-81, CHMN PLASTIC SURG, SCH MED, YALE UNIV, 79-, PROF SURG, 81- *Concurrent Pos:* Vis prof, Univ Calif, San Diego, 76, Univ Calif, Irvine, 79, Brown Univ, 80; Roswell Park Mem Inst, 80, Nat Naval Med Ctr, 81; consult, Vet Admin Hosp, Conn, 76- *Mem:* NY Acad Sci; Plastic Surg Res Coun; Soc Head & Neck Surgeons; Am Soc Surg Hand; Soc Univ Surgeons; Sigma Xi. *Res:* Cancer immunology, with particular reference to malignant melanoma and head and neck cancer; wound healing; effects of ionizing radiation and chemotherapy; reconstruction following head and neck surgery. *Mailing Add:* 16 Yowagu Ave Branford CT 06405

ARIYAN, ZAVEN S, b Cairo, Egypt, Oct 10, 33; m 66. ORGANIC CHEMISTRY. *Educ:* Univ London, BSc, 57, PhD(org chem), 64. *Prof Exp:* Lectr chem, Medway Col Technol, Eng, 57-59; demonstr org chem, Royal Mil Col Sci, Eng, 60-64; fel, Univ Southern Calif, 64-65; sr res chemist, Uniroyal, Inc, 65-69, coord pharmaceut prog, 69-71, sr group leader pharmaceut chem res, 71-75; mgr, T R Evans Res Ctr, Pharmaceut Diamond Shamrock Corp, 75-79; DIR RES, FAIRMOUNT CHEM CO, INC, 79- *Concurrent Pos:* Grants, Gulbenkian Found, 56-57 & Stauffer Chem Co, 64-65. *Mem:* Am Chem Soc; Brit Chem Soc. *Res:* Chemistry of sulfur compounds. *Mailing Add:* 47 Gallison Dr Murray Hill NJ 07974-2721

ARKELL, ALFRED, b Chicago, Ill, Feb 16, 24; m 43; c 3. ORGANIC CHEMISTRY. *Educ:* Kalamazoo Col, AB, 54; Ohio State Univ, PhD(chem), 58. *Prof Exp:* Asst gen & org chem, Ohio State Univ, 54-55; sr res chemist, Texaco Res Ctr, Texaco Inc, Beacon, NY, 58-82; RETIRED. *Honors & Awards:* Upjohn Prize, 54; Clark MacKenzie Prize, 54; Texaco Res Award, Am Chem Soc, 68. *Mem:* Am Chem Soc; fel Am Inst Chem. *Res:* Synthesis of highly-branched aliphatic compounds; derivatives of acetylene; low temperature infrared spectroscopy; matrix isolation technique; isolation and identification of radicals and unstable compounds. *Mailing Add:* 28 McFarlane Rd Wappingers Falls NY 12590

ARKILIC, GALIP MEHMET, b Sivas, Turkey, Mar 10, 20; US citizen; m 56; c 4. APPLIED MECHANICS. *Educ:* Cornell Univ, BME, 46; Ill Inst Technol, MS, 48; Northwestern Univ, PhD(mech), 54. *Prof Exp:* Develop engr, Miehle Printing Press & Mfg Co, 48-49; res & develop engr, Mech & Chem Indust, Turkey, 49-52; analyst, Miehle Printing Press & Mfg Co, 54-56; asst prof, Pa State Univ, 56-58; assoc prof, 58-63, chmn dept eng mech, 66-69, asst dean, Sch Eng & Appl Sci, 69-74, PROF APPL SCI, GEORGE WASHINGTON UNIV, 63- *Concurrent Pos:* Consult, Beers Co, 62-63, Page Commun Co, 63-64, Bechtel Power Co, 73-75 & Wigman & Cohen Co, 76-77; prin investr, Air Force Off Sci Res, 63-70. *Mem:* Sigma Xi; Am Soc Mech Engrs; Soc Eng Sci; Am Acad Mech. *Res:* Elasticity; plates and shells; mechanics of composite materials. *Mailing Add:* Sch Eng George Washington Univ Washington DC 20006

ARKIN, ARTHUR MALCOLM, b New York, NY, Jan 25, 21; m 43; c 2. PSYCHIATRY, PHYSIOLOGICAL PSYCHOLOGY. *Educ:* Univ Chicago, BS, 42; New York Med Col, MD, 45. *Prof Exp:* Gen intern, Montefiore Hosp, Bronx, 45-46; resident psychiatrist, Bellevue Psychiatric Hosp, New York, 46-47; Menninger Sch Psychiat, Topeka, 47, Brooklyn State Hosp, 48 & Postgrad Ctr Psychother, New York, 48-51; mem headache clin, Montefiore Hosp, Bronx, 50-51; psychiatric consult, Bur Appl Social Res, Columbia Univ, 53-55; asst clin prof psychiat, Albert Einstein Col Med, 55-75; prof psychol, Grad Sch, City Col New York, 75-77; CLIN PROF PSYCHIAT, MT SINAI

SCH MED, 77-; ATTEND PSYCHIATRIST, DEPT PSYCHIAT, MT SINAI HOSP, 77- *Concurrent Pos:* Pvt pract, 49-51; adj attend physician, Montefiore Hosp & Med Ctr, 53-75; psychiat consult, Dept Home Care, Montefiore Hosp, 54-55 & Barrett House, Florence Crittenton League, 54-56; assoc vis psychiatrist, Bronx Munic Hosp Ctr, 56-; adj assoc prof psychol, Grad Sch Arts & Sci, NY Univ, 61- & City Col New York, 66-, ctr staff psychiatrist, 66-; chmn, Comt on Behav Ther, NY Dist Br, Am Psychiat Asn, 74; assoc clin prof, Dept Psychiat, Albert Einstein Col, 75; pres, Biofeedback Soc NY, 78. *Mem:* Am Med Asn; fel Am Psychiat Asn; Soc Psychophysiol Res; Asn Psychophysiol Study Sleep; Asn Advan Psychother. *Res:* Psychophysiology of sleep and dreams; psychotherapy. *Mailing Add:* 60 Horatio St New York NY 10014

ARKIN, GERALD FRANKLIN, b Washington, DC, Sept 16, 42; m 66; c 2. AGRICULTURAL ENGINEERING. *Educ:* Cornell Univ, BS, 66; Univ GA, MS, 68; Univ Ill, PhD(agr eng), 72. *Prof Exp:* Res asst agr eng, Univ Ga, 67-68 & Univ Ill, 68-72; assoc prof, 72-81, prof agr eng, Tex agr exp sta, Tex A&M Univ, 81-; AT CENT TEX RES, BLACKLAND RES CTR. *Concurrent Pos:* Consult comt climate & weather fluctuations & agr prod, Nat Acad Sci, 75-76; prin investr, Statist Reporting Serv, US Dept Agr & Bur Reclamation, US Dept Interior, 81- *Mem:* Am Soc Agr Engrs; Am Soc Agron; Soil Sci Soc Am; Sigma Xi. *Res:* Crop growth simulation modeling to optimize yield and water efficiency. *Mailing Add:* Blackland Res Ctr Cent Tex Res PO Box 6112 Temple TX 76502

ARKIN, JOSEPH, b Brooklyn, NY, May 25, 23; c 4. MATHEMATICS. *Educ:* PhD. *Hon Degrees:* PhD Brantridge School, Eng, 67. *Prof Exp:* PROF & SR LECTR, DEPT MATH, US MIL ACAD, WEST POINT, NY. *Mem:* NY Acad Sci; AAAS; Asn Am Math. *Mailing Add:* 197 Old Nyack Turnpike Spring Valley NY 10977

ARKING, ALBERT, b Brooklyn, NY, Nov 5, 32. PHYSICS. *Educ:* Columbia Univ, AB, 53; Cornell Univ, PhD(physics), 59. *Prof Exp:* Physicist, Artificial Intel, Hughes Aircraft Co, 59; physicist astrophys & atmospheric physics, Goddard Inst Space Studies, New York, 59-74, PHYSICIST, GODDARD SPACE FLIGHT CTR, NASA, GREENBELT, MD, 74- *Concurrent Pos:* Instr, City Col New York, 61-62; from adj asst prof to adj assoc prof, NY Univ, 62-70; vis res assoc, Univ Calif, Los Angeles, 70-71. *Mem:* AAAS; Am Phys Soc; Am Astron Soc; Am Geophys Union; Am Meteorol Soc. *Res:* Atmospheric and nuclear physics; radiative transfer; astrophysics. *Mailing Add:* 11810 Gainsboro Rd Rockville MD 20854

ARKING, ROBERT, b New York, NY, July 1, 36; m 59; c 2. DEVELOPMENTAL GENETICS, GERONTOLOGY. *Educ:* Dickinson Col, BS, 58; Temple Univ, PhD(develop biol), 67. *Prof Exp:* Postdoctoral fel NIH, Univ Va, 67-68; asst prof zool, T H Morgan Sch Biol, Univ Ky, 68-70; res assoc, Develop Biol Labs, Univ Calif, Irvine, 70-75; asst prof, 75-81, ASSOC PROF BIOL, WAYNE STATE UNIV, 81- *Mem:* Soc Develop Biol; Sigma Xi; AAAS; Geront Soc Am; Genetics Soc Am. *Res:* Developmental and physiological genetics of Drosophila; genetic and molecular analysis of delayed senescence and aging mechanisms in Drosophila. *Mailing Add:* Dept Biol Wayne State Univ Detroit MI 48202

ARKINS, JOHN A, b Milwaukee, Wis, Sept 13, 26; m 48; c 3. INTERNAL MEDICINE. *Educ:* Univ Wis, BS, 50, MD, 52. *Prof Exp:* Prof med, Med Col Wis, 70-87, coordr, Clin Student Affairs, 73-87; RETIRED. *Concurrent Pos:* Consult, Vet Admin Hosp, Wood, Wis & Milwaukee Children's Hosp, 65- *Mem:* Am Acad Allergy; Am Thoracic Soc; Am Asn Med Cols; Am Fedn Clin Res; NY Acad Sci. *Res:* Clinical sensitivity to animal danders; induction of homocytotropic antibody in canine models. *Mailing Add:* 10421 E Champagne Dr Sun Lakes AZ 85248

ARKLE, THOMAS, JR, b Colerain, Ohio, Sept 10, 18; m 45; c 2. GEOLOGY. *Educ:* Marietta Col, AB, 40; Ohio State Univ, MS, 50. *Prof Exp:* Teacher high schs, WVa, 40-42 & Ohio, 46-47; asst geol, Ohio State Univ, 48-50; asst geologist, 50-51, asst to dir, 52-58, econ geologist, State Geol Surv, WVa, 58-73; SUPVR COAL RESOURCES STUDY, WVA GEOL ECON SURV, 73- *Mem:* Am Geol Soc; Geol Soc Am; Am Inst Mining Metall Engrs; Am Inst Prof Geol; Sigma Xi. *Res:* Stratigraphy, especially coal bearing strata; economic geology, especially of coal and sandstone; to collect, interpret and collate data for use in the determination of the physical and chemical characteristics of the remainder of coal seams and the associated rocks in West Virginia. *Mailing Add:* 94-179 Hulahe Waipahu HI 96797

ARKLES, BARRY CHARLES, b Philadelphia, Pa, Feb 1, 49; m 67; c 3. POLYMER CHEMISTRY, BIOCHEMISTRY. *Educ:* Temple Univ, BS, 69, PhD(chem), 75. *Prof Exp:* Mgr tech develop, LNP Corp, Beatrice Foods, 69-77; PRES, PETRARCH SYSTS INC, 77- & M ARKLES STEEL INC, 78- *Honors & Awards:* Leo Friend Award, Am Chem Soc, 83; I R 100, Award 84. *Mem:* Am Chem Soc; Mat Res Soc; Soc Plastics Engrs; Soc Mfg Engrs; AAAS. *Res:* Thermoplastic composites; wear and lubrication; silicon chemistry; immobilized biomaterials; protective coatings. *Mailing Add:* 1542 Cooper Dr Ambler PA 19002

ARKO, ALOYSIUS JOHN, b Loski Potok, Slovenia, Yugoslavia, Feb 6, 40; US citizen; m 64; c 3. SOLID STATE PHYSICS, OPTICS. *Educ:* Ill Inst Technol, BS, 62; Northwestern Univ, PhD(physics), 67. *Prof Exp:* Res assoc solid state physics, Northwestern Univ, 67-68; asst physicist, 68-73, physicist, Argonne Nat Lab, 73-87; PHYSICIST, LOS ALAMOS NAT LAB, 87- *Mem:* fel Am Phys Soc. *Res:* Angle-resolved photoemission in bulk and surface studies. *Mailing Add:* Los Alamos Nat Lab MS-K764 Los Alamos NM 87545

ARKOWITZ, MARTIN ARTHUR, b Brooklyn, NY, Apr 17, 35; m 57; c 3. TOPOLOGY. *Educ:* Columbia Univ, BA, 56, MA, 57; Cornell Univ, PhD(math), 60. *Hon Degrees:* MA, Dartmouth Col, 71. *Prof Exp:* Teaching asst, Cornell Univ, 57-60; Off Naval Res assoc, Johns Hopkins Univ, 60-61; instr math, Princeton Univ, 61-63, lectr, 63-64; from asst prof to assoc prof, 64-71, chmn dept, 81-84, PROF MATH, DARTMOUTH COL, 71- *Concurrent Pos:* Dartmouth fac fel, Math Inst, Oxford Univ, 67-68; vis prof, Math Inst, Aarhus Univ, Denmark, 72-73; vis prof, Univ Ariz, 79-80, Max Planck Inst, Bonn, Germany, 85; vis scholar, Univ Penn, 86. *Mem:* Am Math Soc; Math Asn Am. *Res:* Topology. *Mailing Add:* Dept Math & Computer Sci Dartmouth Col Hanover NH 03755

ARLIAN, LARRY GEORGE, b Aspen, Colo, Aug 5, 44; m 66; c 2. PHYSIOLOGY, MEDICAL ENTOMOLOGY. *Educ:* Colo State Univ, BS, 66, MS, 68; Ohio State Univ, PhD(entom), 72. *Prof Exp:* Asst chemist, Great Western Sugar Co, 68-69; from asst to assoc prof biol, 72-82, assoc prof physiol, Sch Med, 80-82, PROF BIOL, PHYSIOL, & BIOPHYS, WRIGHT STATE UNIV, 82- *Concurrent Pos:* NIH grant, 76-79, 81-84 & 85-; res fel, Wright State Univ, 77-79. *Mem:* Sigma Xi; Entom Soc Am; Acarological Soc Am; Am Soc Parasitologists. *Res:* House dust mite allergies; water balance in insects, mites and mammals; scabies biology, culture and host specificity. *Mailing Add:* Dept Biol Sci Wright State Univ Dayton OH 45435

ARLINGHAUS, HEINRICH FRANZ, b Damme, Ger, Mar 11, 58; m 82; c 3. LASER SPECTROSCOPY, DNA SEQUENCING. *Educ:* Westfälische Wilhems Univ, dipl, 82, PhD(physics), 86. *Prof Exp:* Undergrad res asst, Westf06lische Wilhems Univ, Münster, Ger, 81, undergrad teaching asst physics, 81-82, grad res asst physics, 82-84, res scientist, 86; res scientist, Ger Res Asn, 84-86; postdoctoral fel, Argonne Nat Lab, Lawrence Livermore Nat Lab, 86-88; SCIENTIST, ATOM SCI INC, 88- *Concurrent Pos:* Prin investr, Small Bus Innovation Res Prog, Phase I, 89-, Phase II, 90-, Defense Advan Res Projs Agency, Phase I, 91-, NIH, Phase I, 91- *Mem:* Am Vacuum Soc; AAAS; Mat Res Soc. *Res:* Emission process of sputtered particles from various ion- bombarded and laser-irradiated materials; laser spectroscopy; optical damage mechanism; ultra-trace element analysis; depth profile analysis; DNA sequences; 22 publications and 20 conference presentations. *Mailing Add:* Atom Sci Inc 114 Ridgeway Ctr Oak Ridge TN 37830&

ARLINGHAUS, RALPH B, b Newport, Ky, Aug 16, 35; m 57, 68; c 4. BIOCHEMISTRY. *Educ:* Univ Cincinnati, BS, 57, MS, 59, PhD(biochem), 61. *Prof Exp:* NIH fel protein synthesis, Univ Ky, 61-63, Am Cancer Soc fel, 63-65; res chemist, Plum Island Animal Dis Lab, USDA, NY, 65-69; from asst prof to prof biol, Univ Tex M D Anderson Hosp & Tumor Inst, 69-83, chief, sect environ biol, 69-77, chief, sect tumor virol, 77-80, actg head, dept tumor virol, 80-83; dir vaccine develop, Johnson & Johnson Biotechnol Ctr, La Jolla, Calif, 83-86; PROF & CHMN, DEPT MOLECULAR PATH, UNIV TEX M D ANDERSON CANCER CTR, HOUSTON, 86- *Concurrent Pos:* Vis scientist & mem, Scripps Clin & Res Found, La Jolla, Calif, 83-86. *Mem:* AAAS; Am Chem Soc; Am Soc Microbiol; Am Asn Biol Chem; Am Soc Virol. *Res:* Protein biosynthesis and nucleic acid synthesis; animal virus replication at the molecular level; RNA-containing viruses such as Picornaviruses and Murine leukemia viruses; molecular biology of retroviruses and oncogenes. *Mailing Add:* Dept Molecular Path Univ Tex M D Anderson Hosp 1515 Holcombe Blvd Houston TX 77030-4009

ARLINGHAUS, SANDRA JUDITH LACH, b Elmira, NY, April 18, 43; m 66; c 1. MATHEMATICAL GEOGRAPHY, LOCATION THEORY & GEOGRAPHIC. *Educ:* Vassar Col, BA, 64; Wayne State Univ, MA, 76; Univ Mich, PhD(geography), 77. *Prof Exp:* From lectr to asst prof math, Loyala Univ, 79-82; DIR, INST MATH GEOGRAPHY, 84- *Concurrent Pos:* vis asst prof geography, Ohio State Univ, 77-78, lectr math, 78-79; lectr math, Ohio State Univ, 78-79, Loyola Univ, 79-81; lectr math & geog, Univ Mich, 82-83; consult, Univ Mich, 84 & 85, invited lectr, 90. *Mem:* NY Acad Sci; Assoc Am Geographers; Am Math Soc; Am Geog Soc; AAAS; Math Asn Am. *Res:* Geographic location theory; geographic modeling; mathematical modeling; applications of geometry, graphtheory, topology in geography; cartographic analysis; environmental analysis of public transportations systems; fractals and central places; network analysis-algorithms. *Mailing Add:* 2790 Briarcliff Ann Arbor MI 48105-1429

ARLOW, JACOB, b New York, NY, Sept 3, 12; m 36; c 4. PSYCHIATRY. *Educ:* NY Univ, BS, 32, MD, 36. *Prof Exp:* Intern, Harlem Hosp, New York, 36-38; resident neuropsychiatrist, USPHS, 38-39; resident psychiatrist, NY Psychiat Inst, 40-41; assoc psychiatrist, Presby Hosp, New York, 44-52; clin assoc prof, 50-62, CLIN PROF PSYCHIAT, STATE UNIV NY DOWNSTATE MED CTR, 62- *Concurrent Pos:* Vis prof, Albert Einstein Col Med, Mt Sinai Col Med, Columbia Univ & La State Univ; practicing psychiatrist, 42-; ed, Psychoanal Quart. *Honors & Awards:* Int Clin Essay Prize, Brit Psychoanal Soc, 56. *Mem:* Am Psychoanal Asn (pres-elect, 59-60, pres, 60-61); Int Psychoanal Asn (treas, vpres). *Res:* Psychosomatic aspects of arthritis and angina pectoris. *Mailing Add:* Dept Psychiat Columbia Univ 360 W 168th St New York NY 10032

ARMACOST, DAVID LEE, b Santa Monica, Calif, Mar 9, 44. MATHEMATICS. *Educ:* Pomona Col, BA, 65; Stanford Univ, MS, 66, PhD(math), 69. *Prof Exp:* Asst prof, 69-74, assoc prof, 74-81, PROF MATH, AMHERST COL, 81- *Mem:* Am Math Soc; Sigma Xi. *Res:* Topological groups; infinite Abelian groups; locally compact Abelian groups. *Mailing Add:* Dept Math Amherst Col Amherst MA 01002

ARMACOST, WILLIAM L, b Santa Monica, Calif, Oct 6, 41; m; c 2. MATHEMATICS. *Educ:* Pomona Col, BA, 63; Univ Calif, Los Angeles, MA, 65, PhD(math), 68. *Prof Exp:* Asst prof, 68-72, assoc prof, 72-77, PROF MATH, CALIF STATE COL, DOMINGUEZ HILLS, 77- *Mem:* Am Math Soc; Sigma Xi; Math Asn Am. *Res:* Group representation theory and functional analysis; Frobenius reciprocity theorem. *Mailing Add:* 902 Stanford St Santa Monica CA 90403

ARMALY, BASSEM FARID, b Israel, Dec 5, 37; US citizen; m 62; c 2. MECHANICAL ENGINEERING, HEAT TRANSFER. *Educ:* Univ Iowa, BS, 63, MS, 64; Univ Calif, Berkeley, PhD(mech eng), 69. *Prof Exp:* From asst prof to assoc prof, 69-76, PROF MECH ENG, UNIV MO-ROLLA, 76-, CHMN, MECH & AEROSPACE EXP DEPT, 85- *Concurrent Pos:* Prin investr, NSF grants, 72-73, 75-82 & 78-88 & Off Water Res & Technol grant,

74-76; fel, Langley Res Ctr, NASA, 78; Fulbright sr res investr, 79-80. *Mem:* Fel Am Soc Mech Eng; Optical Soc Am; Am Soc Eng Educ; Am Inst Aeronaut & Astronaut; Int Solar Energy Soc. *Res:* Heat and mass transfer; thermophysical properties; solar energy. *Mailing Add:* Dept Mech Eng Univ Mo Rolla MO 65401

ARMALY, MANSOUR F, b Shefa Amer, Palestine, Feb 25, 27. OPHTHALMOLOGY, GLAUCOMA. *Educ:* Univ Iowa, MS, 57. *Prof Exp:* PROF OPHTHAL, GEORGE WASHINGTON UNIV MED CTR, 66-, CHMN DEPT, 70- *Mem:* Am Acad Ophthal & Otolaryngol; Am Asn Ophthal; Am Col Surgeons; AMA; Am Interocular Lens Implant Soc. *Mailing Add:* Dept Ophthal George Washington Univ Med Ctr Washington DC 20037

ARMAMENTO, EDUARDO T, b Lucena City, Philippines, Nov 5, 60; US citizen; m 89. INDUSTRIAL WASTE MANAGEMENT, GENERAL CHEMISTRY & NUCLEAR ENGINEERING. *Educ:* Adamson Univ, Manila, Philippines, BS, 82, MS, 85. *Prof Exp:* Prof chem, Adamson Univ, 83-85, prof nuclear eng, chem eng & indust waste mgt, 83-85; perfume technician, De Lair Inc, NY, 87-88; PLANT & PROD ENGR GLASS FROSTING, MC KAY INT, INC, NY, 88- & RES CHEMIST & ENGR GLASS FROSTING TECHNOL, 88- *Mem:* Soc Glass & Ceramic Decorators; Philippine Inst Chem Engrs. *Res:* New ceramic tile and porcelain etching cream used prior to reglazing; new highly screenable etching paste resulted to new product for glass and mirror decoration. *Mailing Add:* 83-28 Dongan Ave Elmhurst NY 11373

ARMAN, ARA, b Istanbul, Turkey, Sept 12, 30; US citizen; m 63; c 2. CIVIL ENGINEERING. *Educ:* Robert Col, Istanbul, BS, 55; Univ Tex, MS, 56. *Prof Exp:* Asst lab engr, La Dept Hwy, 56-58, head lab engr, 58-61, soil design engr, 61-63; from asst prof to assoc prof civil eng, La State Univ, 63-70, asst dir, Div Eng Res, 65-77, chmn dept, 77-80, assoc dean, Col Eng, 80-87, dir, La Transp Res Ctr, 87-90, PROF CIVIL ENG, LA STATE UNIV, 70-; VPRES, WOODWARD-CLYDE CONSULT, BATON ROUGE, LA, 90- *Concurrent Pos:* Mem maintenance bituminous pavement comt, Hwy Res Bd, Nat Res Coun, Nat Acad Sci, 65, chmn soil-cement stabilization comt, 66; chmn, comt placement of soils, Am Soc Civil Engrs, 76-83; secy, Comt on Geotextiles, Int Soc Soil Mech & Found Eng, 86-90. *Mem:* Am Road Builders Asn (vpres, 70-71, pres, 72-73); Am Soc Eng Educ; Am Soc Testing & Mat; fel Am Soc Civil Engrs. *Res:* Soil mechanics in soil stabilization; organic soils; sampling; pavement design; placement of soils; geotextiles; geomembranes. *Mailing Add:* La Transp Res Ctr La State Univ 4101 Gourrier Ave Baton Rouge LA 70808

ARMANINI, LOUIS ANTHONY, b New York, NY, Feb 11, 30; m 58; c 4. PHYSICAL CHEMISTRY. *Educ:* Manhattan Col, BS, 57. *Prof Exp:* Res chemist, Curtiss-Wright Corp, Pa, 57-59; res chemist, Francis Earle Labs, Mearl Corp, 59-67, group leader interference films, 67-75, asst dir res, 76-90, VPRES, MEARL CORP, 89-, DIR RES, 90- *Mem:* Am Chem Soc; AAAS; Soc Cosmetic Chemists. *Res:* Optical interference pigments made by thin film deposition or controlled crystal growth and the instrumental measurement and optical properties of such films. *Mailing Add:* Mearl Corp 217 N Highland Ave Ossining NY 10562

ARMANT, D RANDALL, b Plainfield, NJ, Nov 13, 51; m 88. DEVELOPMENTAL BIOLOGY, CELL-CELL INTERACTIONS. *Educ:* Va Polytech Inst & State Univ, BS, 74, PhD(develop biol), 80. *Prof Exp:* Res assoc, Worcester Found Exp Biol, 80-82; postdoctoral fel, Dept Biol Chem, Johns Hopkins Univ Sch Med, 82-83 & Dept Biochem & Molecular Biol, Anderson Tumor Inst, Houston, Tex, 83-85; asst prof obstet, gynec & reproductive biol, Beth Israel Hosp, Harvard Med Sch, 85-88; ASSOC PROF OBSTET, GYNEC & ANAT CELL BIOL, WAYNE STATE UNIV SCH MED, 88- *Concurrent Pos:* Nat Res Serv Award, NIH, 82, new investr award, 85. *Mem:* Am Soc Cell Biol; Soc Study Reproduction; Am Fertil Soc; Soc Develop Biol; Soc Cryobiol. *Res:* Biochemistry of early mammalian development; cell adhesion and glycoprotein biosynthesis in mouse embryos; in vitro fertilization and oocyte lembryo cryopreservation; biochemistry of development in the cellular slime mold; role of discoidin and alkaline phosphatase; pattern formation. *Mailing Add:* Dept OB/GYN Mott Ctr Wayne State Univ 275 E Hancock Detroit MI 48201

ARMANTROUT, GUY ALAN, b Salt Lake City, Utah, May 9, 40; m 62; c 9. ELECTRONICS ENGINEERING. *Educ:* Ore State Univ, BS, 62, MS, 64; Purdue Univ, PhD(elec eng), 69. *Prof Exp:* Sr res engr, 68-72, ENG GROUP LEADER RES, LAWRENCE LIVERMORE LAB, UNIV CALIF, 72-, PROJ ENGR. *Concurrent Pos:* Adj prof elec eng, Univ Calif Sch Med, San Francisco, 74-; asst prof appl sci, Univ Calif, Davis, 76- *Mem:* Inst Elec & Electronics Engrs. *Res:* Semiconductor device research and development including gamma radiation detector development and applications and terrestrial photovoltaics basic research and cell development. *Mailing Add:* Lawrence Livermore Nat Lab L-439 7000 East Ave Livermore CA 94550

ARMATO, UBALDO, b Trieste, Italy; m 70; c 2. TUMOR BIOLOGY, GROWTH REGULATION. *Educ:* Univ Padva, Med, 67, Hemat, 69, Internal Med, 75. *Prof Exp:* Prof human anat, Univ Padva, Italy, 69-71, assoc prof, 80-86; PROF HISTOL & EMBRYOL, UNIV VERONA, ITALY, 86- *Concurrent Pos:* Lectr human anat, Univ Padva, Italy, 69-71; consult hemat, Civic Hosp, Este, Italy, 78-87. *Mem:* Tissue Cult Asn; Am Soc Cell Biol; Fedn Am Soc Exp Biol; Europ Tissue Cult Asn. *Res:* Regulation of normal and abnormal growth of neonatal rat hepatocytes and adult human adrenocytes in primary tissue cultures. *Mailing Add:* Inst Anat & Histol Strada Le Grazie Verona 37136 Italy

ARMBRECHT, FRANK MAURICE, JR, b Norfolk, Va, Jan 29, 42; m 65; c 2. CHEMISTRY. *Educ:* Duke Univ, BS, 63; Mass Inst Technol, PhD(chem), 68. *Prof Exp:* Res chemist elastomers chem, E I du Pont de Nemours & Co, Inc, 68-77, process supvr, 77-78, supt res & develop, 78-81, res mgr, 80-82, tech mgr ethylene polymers, 82-87, lab mgr, 84-87, strategic planning consult, 87-90, technol mgr, 91. *Mem:* Am Chem Soc. *Res:* Organometallic chemistry; homogeneous catalysis; elastomers. *Mailing Add:* Du Pont Poly ESL 301-201 Wilmington DE 19880-0301

ARMBRECHT, HARVEY JAMES, b Philadelphia, Pa, March 7, 47; m 71; c 2. GERONTOLOGY. *Educ:* Drexel Univ, BS, 70; Univ Rochester, PhD(biophysics), 74. *Prof Exp:* Fel, Cornell Univ, 74-77; asst prof, 81-85, ASSOC PROF BIOCHEM, MED SCH, ST LOUIS UNIV, 85-, PROF MED, 90-; RES CHEMIST, VET ADMIN HOSP, ST LOUIS, 77- *Mem:* AAAS; Biophys Soc; Gerontol Soc; Am Aging Asn; Am Soc Bone & Mineral Metab; Am Soc Biochem & Molecular Biol. *Res:* Mechanisms of steroid and peptide hormone action; hormonal regulation of calcium metabolism by parathyroid hormone and vitamin B in the elderly and diabetic. *Mailing Add:* Geriatric Ctr 111G-JB St Louis Vet Admin Hosp St Louis MO 63125

ARMBRUSTER, CHARLES WILLIAM, b St Louis, Mo, Mar 24, 37. ORGANIC CHEMISTRY. *Educ:* Univ Notre Dame, BS, 58; Wash Univ, PhD(org chem), 66. *Prof Exp:* From instr to asst prof, 62-66, div sci, 63-67, chmn chem dept, 67-75, ASSOC PROF CHEM, UNIV MO-ST LOUIS, 66- *Mem:* Am Chem Soc; AAAS. *Res:* Physical organic chemistry; biochemical mechanisms. *Mailing Add:* Dept Chem Univ Mo 8001 Nat Bridge Rd St Louis MO 63121

ARMBRUSTER, DAVID CHARLES, b Cincinnati, Ohio, Feb 21, 39; m 62; c 3. INDUSTRIAL CHEMISTRY, ORGANIC POLYMER CHEMISTRY. *Educ:* Xavier Univ, Ohio, BS, 61; Univ Cincinnati, PhD(org chem), 65. *Prof Exp:* Sr res chemist, Rohm and Haas Co, 65-66, res & develop mgr, 66-76; mkt/tech develop mgr, Celanese Chem Co, 76-81; commercial develop mgr, CPS Chem Co, 81-82; PRES, CHEM CONSULT & SPECIALTY PROD, ARMBRUSTER ASSOC, 82- *Mem:* Asn Consult Chemists & Chem Engrs; Am Chem Soc; Commercial Develop Asn; Fedn Soc Coatings Technol; Licensing Execs Soc; Soc Cosmetic Chemists; Soc Plastics Engrs. *Res:* Organic chemistry; heterocycles; polymers; plastics; monomers; emulsion polymerization; radiation curing/photoploymerization; coatings; adhesives; acrylics; epoxies; urethanes; specialty chemicals; water treatment chemicals; applications; marketing; market research; commercial development; technology assessment; technological forecasting; hydrophilic polymers; new products; biotechnology. *Mailing Add:* 43 Stockton Rd Summit NJ 07901

ARMBRUSTER, FREDERICK CARL, b Aurora, Ill, Apr 14, 31; m 54; c 3. FERMENTATION, ENZYMOLOGY. *Educ:* Purdue Univ, BS, 53, MS, 55. *Prof Exp:* res chemist, CPC Int, Inc, 57-61, sect head fermentation res, 61-83, sr sect leader, 83-86; mgr indust enzyme res & develop, Enzyme Biosysts Ltd, 86-90; CONSULT, INDUST ENZYMES, 90- *Mem:* Am Soc Microbiol; Am Chem Soc. *Res:* Fermentation processes; industrial production, purification, and use of enzymes; industrial microbiol enzymes, including strain selection and improved fermentation development and scale-up, through enzyme recovery and application. *Mailing Add:* 5969 Howard Ave La Grange IL 60525-3711

ARMBRUSTER, GERTRUDE D, b Alta, Nov 29, 25; US citizen. FOOD SCIENCE. *Educ:* Univ Alta, BS, 47; Wash State Univ, MS, 50, PhD(food sci), 65. *Prof Exp:* County agent, exten home econ, Wash State Univ, 50-52; asst prof, 52-58, ASSOC PROF FOOD & NUTRIT, CORNELL UNIV, 59-61, 65- *Mem:* Am Home Econ Asn; Inst Food Technologists; Am Soc Hort Sci; Am Dietetic Asn. *Res:* Interrelationship of plant tissue properties and product quality; plant texture; microwave applications. *Mailing Add:* Nutrit 124 Savage Hall Cornell Univ Main Campus Ithaca NY 14853

ARMBRUSTMACHER, THEODORE J, b Owosso, Mich, Jan 11, 38; c 2. MINEROLOGY-PETROLOGY. *Educ:* Univ Iowa, PhD(geol), 65. *Prof Exp:* GEOLOGIST, US GEOL SURVEY, 65- *Mem:* Fel Geol Soc Am; Soc Econ Geologists. *Res:* Nd-Ta-Th commodities; alkaline rock complexes. *Mailing Add:* US Geol Surv Mail Stop 905 Box 25046 Fed Ctr Denver CO 80225

ARMELAGOS, GEORGE JOHN, b Lincoln Park, Mich, May 22, 36; m 91. BIOLOGICAL ANTHROPOLOGY. *Educ:* Univ Mich, BA, 58; Univ Colo, MA, 64, PhD(anthrop), 68. *Prof Exp:* Res assoc, Univ Colo Nubian Exped, 63; instr, Univ Utah, 65-66, asst prof, 67-69; from asst prof to assoc prof anthrop, Univ Mass, Amherst, 69-78, dir univ honors, 74-78, prof anthrop, 78-89; PROF & CHMN, DEPT ANTHROP, UNIV FLA, GAINESVILLE, 89- *Concurrent Pos:* NIH co-prin investr, 67-69; mem adv comt behav & neural sci, NSF Anthropol Panel, 76-78; NSF co-prin investr; pres, Biol Unit, Am Anthrop Asn, 84-87. *Mem:* Am Asn Phys Anthrop; AAAS; Human Biol Coun; Am Anthrop Asn; Sigma Xi; Am Asn Phys Anthrop (pres, 87-89). *Res:* Bone growth and development in archeological populations, especially nutritional stress; paleopathology; hominid evolution; nutrition; skeletal biology and demography of archeological populations; paleopathology human evolution; race; history of physical anthropology. *Mailing Add:* Dept Anthrop Univ Fla 1350 Turlington Hall Gainesville FL 32611

ARMEN, HARRY A, JR, b New York, NY, Feb 4, 40; m 63; c 3. APPLIED MECHANICS. *Educ:* Cooper Union, BCE, 61; New York Univ, MCE, 62, ScD, New York Univ, 64. *Prof Exp:* Head, Appl Mech Br, Grumman Aircraft Eng Corp, 64-70, head, Appl Mech Br, Gruman Aerospace Corp, 70-79, tech specialist, 80-82, head Struct Mech, 82-86, prin engr, 86-88, tech adv, 88-90; CONG FEL, US SENATE, 91- *Concurrent Pos:* Adj prof civil eng & appl mech, Cooper Union, 65-73 & Hofstra Univ, 73-79; Cong Fel, Am Soc Mech Engrs, 91. *Honors & Awards:* Achievement Award, NASA. *Mem:* Am Inst Aeronaut & Astronaut; Am Soc Mech Engrs; Sigma Xi; AAAS. *Res:* Development of methods for the static and dynamic nonlinear analysis of structures; fracture mechanics; crashworthiness evaluation techniques; large displacement and stability analysis of structures; advanced materials development. *Mailing Add:* 27 Jefferson St Glen Cove NY 11542

ARMENAKAS, ANTHONY EMANUEL, b Mytilene, Greece, Aug 23, 24; US citizen; m 50; c 3. CIVIL ENGINEERING. *Educ:* Ga Inst Technol, BS, 50; Ill Inst Technol, MS, 52; Columbia Univ, PhD(appl mech), 59. *Prof Exp:* Instr, Ill Inst Technol, 50-52; sr struct engr, Edwards, Kelcy & Beck, 52-54; lectr, City Col New York, 54-57; assoc prof, civil eng & appl mech, Cooper Union, 58-64; vis assoc prof, Div Eng, Brown Univ, 64-65; prof eng sci, Univ

Fla, 65-67; PROF AEROSPACE ENG, POLYTECH UNIV NY, 67- *Concurrent Pos:* Consult vector eng, Springfield, NJ, 54-59; res assoc, Columbia Univ, 58-62; res consult, Northwestern Univ, 62-65 & Polytech Inst Brooklyn, 62-67; pres, Stress Optics Inc, 70-72; Fulbright lectr to Greece, 72-74; prof & dir, Inst Struct Anal, Nat Tech Univ Athens, 77-84. *Mem:* NY Acad Sci; fel Am Soc Civil Engrs; Am Soc Eng Educ; fel Am Soc Mech Engrs; Am Inst Aeronaut & Astronautics. *Res:* Solid mechanics; vibrations; fracture; shells and structures; author of three books. *Mailing Add:* Dept Mech & Aerospace Eng 333 Jay St Brooklyn NY 11201

ARMENDAREZ, PETER X, b San Pedro, Calif, Sept 7, 30; m 54; c 8. PHYSICAL CHEMISTRY. *Educ:* Loyola Univ Los Angeles, BS, 52; Wash Univ, MS, 54; Univ Ariz, PhD(phys chem), 64. *Prof Exp:* Instr chem, Odessa Col, 58-59; asst prof, Univ Tenn, Martin, 63-65; asst prof physics & phys chem, 65-68, chmn, Div Natural Sci & Math, 74-78, PROF PHYSICS & CHMN DEPT, BRESCIA COL, KY, 68-, CHMN CHEM DEPT, 85- *Concurrent Pos:* Res chemist, US Navy Weapons Support Group, Crane, Ind, 83-84. *Mem:* Am Crystallog Asn; Am Chem Soc; fel Am Inst Chemists. *Res:* Molecular spectra and structure of inorganic complexes; low temperature emission and absorption spectra of chromium complexes; infrared studies and normal coordinate analyses of metal chelate compounds; air-sensitive organomettalic complexes: synthesis and IR characterization. *Mailing Add:* Dept Chem Brescia Col Owensboro KY 42301

ARMENDARIZ, EFRAIM PACILLAS, b Brownsville, Tex, July 9, 38; m 61; c 2. ALGEBRA. *Educ:* Agr & Mech Col, Tex, BA, 60, MS, 62; Univ Nebr, PhD(math), 66. *Prof Exp:* Instr math, Agr & Mech Col, Tex, 62-63 & Univ Nebr, 64-66; asst prof, Univ Tex, 66-67, Univ Southern Calif, 67-68 & Univ Tex, Austin, 68-71; assoc prof, Univ Southwestern La, 71-72; assoc prof, 72-81, PROF MATH, UNIV TEX, AUSTIN, 81- *Mem:* Am Math Soc. *Res:* Ring theory; mathematics. *Mailing Add:* Univ Tex Austin TX 78712

ARMENIADES, CONSTANTINE D, b Thessaloniki, Greece, May 29, 36; US citizen; m 59; c 2. POLYMER CHEMISTRY. *Educ:* Northeastern Univ, BS, 61; Case Inst Technol, MS, 67, PhD(polymer sci), 69. *Prof Exp:* Asst prof, 69-73, assoc prof, 73-78, PROF CHEM ENG, RICE UNIV, 78- *Concurrent Pos:* Master, Will Rice Col, Rice Univ, 76-82; consult polymer properties, E I Du Pont de Nemours & Co, 76-78, Southwestern Bell, 74-79; vis sr lectr, Edinburgh Univ, Scotland, 80-81. *Mem:* Am Inst Chem Engrs; Soc of Plastics Engr; Asn for Res in Vision & Ophthal; Soc Biomat. *Res:* Structure-property relations in synthetic polymers; formulation of polymer composites for application as structural materials; synthesis of five retardant and thermally stable polymers; biocompatibility of synthetic polymers. *Mailing Add:* PO Box 1892 Houston TX 77251

ARMENTANO, LOUIS ERNST, b Brooklyn, NY, Dec 24, 54. RUMINANT NUTRITION. *Educ:* Iowa State Univ, PhD(nutrit physiol), 82. *Prof Exp:* ASST PROF ANIMAL NUTRIT, UNIV WIS, 83- *Mem:* Am Inst Nutrit; Am Dairy Sci Asn. *Res:* Use of by-product feeds; hepatic metabolism; ruminants. *Mailing Add:* Dairy Sci 952 Ansci Bldg Univ Wis Madison WI 53706

ARMENTI, ANGELO, JR, b Bridgeport, Pa, Feb 13, 40. THEORETICAL PHYSICS. *Educ:* Villanova Univ, BS, 63; Temple Univ, MA, 65, PhD(physics), 70. *Prof Exp:* Asst prof physics, Temple Univ, 70-72; from asst prof to assoc prof, 72-79, PROF PHYSICS, VILLANOVA UNIV, 79-, DEAN, UNIV COL, 81- *Mem:* Am Phys Soc; Int Soc Gen Relativity & Gravitation. *Res:* Exact solutions of Einstein's field equations, in particular geodesic motion in the field of compact masses and black holes. *Mailing Add:* Villanova Univ Villanova PA 19085

ARMENTROUT, DARYL RALPH, b Greeneville, Tenn, Apr 6, 42; m 69; c 2. NUCLEAR QUALITY ASSURANCE, PROCUREMENT ENGINEERING. *Educ:* Univ Tenn, Knoxville, BS, 65; PhD(civil eng), 81; Va Polytech Inst, Blacksburg, MS, 68. *Prof Exp:* Prod engr, Humble Oil & Refining Co, 65-66; civil engr, Tenn Valley Authority, 68-74, staff chief, 74-77, asst mgr eng & construct, 77-85, MGR MAT & PROCUREMENT QUAL, TENN VALLEY AUTHORITY, 86- *Concurrent Pos:* Dir, activ under US-China Protocol Hydroelec, Tenn Valley Authority, 80-85. *Mem:* Am Soc Civil Engrs; Nat Soc Prof Engrs. *Res:* Finite element predictions of the behavior of post tensioned reinforced concrete containment vessels specifically the load- deflection behavior of the steel liner and contract interaction; integration of all resources in the planning and development of hydroelectric resources in China. *Mailing Add:* 1908 Windy Oaks Lane Hixson TN 37343

ARMENTROUT, DAVID NOEL, b San Francisco, Calif, Dec 7, 38; m 63; c 3. ANALYTICAL CHEMISTRY. *Educ:* Univ Kans, BS, 61; Cornell Univ, PhD(anal chem), 65. *Prof Exp:* Sr res chemist, 66-80, RES ASSOC ANALYTICAL CHEM, DOW CHEM CO, 80- *Concurrent Pos:* Fel anal chem, NIH, 65-66. *Honors & Awards:* IR-100 Award, 85. *Mem:* Am Chem Soc; Sigma Xi. *Res:* Analytical characterization of polymers. *Mailing Add:* 3707 McKeith Midland MI 48640

ARMENTROUT, STEVE, b Eldorado, Tex, June 19, 30; m 62; c 3. TOPOLOGY. *Educ:* Univ Tex, BA, 51, PhD(math), 56. *Prof Exp:* Teaching asst math, Univ Tex, 52-55; from instr to prof, Univ Iowa, 56-70; PROF MATH, PA STATE UNIV, 70- *Concurrent Pos:* Res fel, Univ Iowa, 62; vis prof, Univ Wis, 66-67. *Mem:* Am Math Soc; Math Asn Am. *Res:* Topology of manifolds; upper semicontinuous decompositions of manifolds. *Mailing Add:* Pa State Univ 207 McAllister Bldg University Park PA 16802

ARMIJO, JOSEPH SAM, b El Paso, Tex, Dec 20, 38; m 58; c 3. NUCLEAR ENGINEERING, TECHNICAL MANAGEMENT. *Educ:* Univ Tex, El Paso, BS, 59; Univ Ariz, MS, 62; Stanford Univ, PhD(mat sci), 69. *Prof Exp:* Qual control engr, Gen Elec Co, Ill, 60; res asst, Los Alamos Sci Lab, NMex, 61; res asst, Univ Ariz, 61-62; res scientist, Vallecitos Nuclear Ctr, Gen Elec Co, Calif, 62-69; mgr plant mat develop, Breeder Reactor Develop Opers, 69-73, mgr fuel chem, 73-75, mgr fuel, mat & mech, 75-76, mgr core mat eng, 76-82, mgr domestic fuel projs, 82-84, mgr nuclear systs technol oper, nuclear energy bus oper, 84-88, PROG GEN MGR SPACE POWER, ASTRO SPACE DIV, GEN ELEC CO, 88- *Honors & Awards:* Indust Res-100 Award, 83; Steinmentz Award, Gen Elec Co, 83. *Mem:* Am Inst Mining, Metall & Petrol Engrs; Sigma Xi; fel Am Nuclear Soc. *Res:* Metallurgical research on stress corrosion of stainless steels in oxidizing media; radiation damage to stainless steels by fast neutrons and fission fragments; light water reactor nuclear fuel and cladding, corrosion and stress corrosion; advanced nuclear reactor design and development. *Mailing Add:* 21053 Canyon View Dr Saratoga CA 95070

ARMINGTON, ALTON, b Everett, Mass, Sept 20, 27; m 52; c 2. PHYSICAL CHEMISTRY, FUEL TECHNOLOGY. *Educ:* Boston Univ, BA, 50; Tufts Univ, MS, 51; Pa State Univ, PhD(fuel technol), 61. *Prof Exp:* Chemist develop, Congoleum-Nairn, 52; phys res, Naval Res Labs, 52-54; asst fuel technol, Pa State Univ, 54-57; from chief chemist to sr chief chemist, 57-81, CHIEF SECT ACOUST & OPTICAL MAT, SOLID STATE CHEM LAB, 81-, CHIEF, ELECTROMAGNETIC MAT DIV, SOLID STATE SCI LAB, ROME AIR DEVELOP COMMAND, DEPT AIR FORCE, 87- *Mem:* Am Chem Soc; Am Asn Crystal Growth. *Res:* Preparation and properties of ultra-pure chemical elements and intermetallic compounds; surface properties; graphite surfaces; crystal preparation. *Mailing Add:* 525 Marrett Rd Lexington Boston MA 02173-7608

ARMINGTON, RALPH ELMER, b Medford, Mass, June 24, 18; m 44; c 4. ELECTRICAL ENGINEERING. *Educ:* Tufts Univ, BS, 40; NY Univ, MS, 42; Pa State Univ, EE, 53; Univ Pittsburgh, PhD(elec eng), 57. *Prof Exp:* Design engr, Transformer Div, Westinghouse Elec Corp, 41-46; asst prof eng res, Pa State Univ, 46-47, from asst prof to assoc prof, 47-61; prof & head dept, Univ Maine, 61-66; lectr, 66-69, PROF ELEC ENG, ILL INST TECHNOL, 69- *Mem:* Inst Elec & Electronics Engrs. *Res:* Education in electrical engineering, especially circuit analysis; electric power systems, electrical contacts and electrical precipitation; educational administration. *Mailing Add:* 6927 Kenton Lincolnwood IL 60646

ARMISTEAD, WILLIS WILLIAM, b Detroit, Mich, Oct 28, 16; m 38, 67; c 3. VETERINARY MEDICINE, MEDICAL EDUCATION. *Educ:* Agr & Mech Col, Tex, DVM, 38; Ohio State Univ, MSc, 50; Univ Minn, PhD(vet med), 55; Am Col Vet Surgeons, dipl, 75. *Prof Exp:* Gen vet practitioner, 38-40; from instr to prof vet med & surg, Agr & Mech Col, Tex, 40-53, dean col vet med, 53-57; dean col vet med, Mich State Univ, 57-74; dean, Col Vet Med, 74-79, vpres agr, Univ Tenn, Knoxville, 79-87; RETIRED. *Concurrent Pos:* Ed, NAm Vet, 50-56 & J Vet Med Educ, 74-80; consult, Southern Regional Educ Bd, 53-56, Surgeon Gen, US Air Force, 60-62 & Tenn Higher Educ Comn, 73-74; collabr, Animal Dis & Parasite Res Div, Agr Res Serv, USDA, 54-65; mem adv bd, Inst Lab Animal Resources, Nat Res Coun, 62-66; mem vet med resident investr selection comt, US Vet Admin, 67-70; mem vet med rev comt, Bur Health Professions Educ & Manpower Training, US Dept Health, Educ & Welfare, 67-71. *Honors & Awards:* Award, Am Vet Med Asn, 77. *Mem:* Nat Inst Med-Nat Acad Sci; Fedn Asns Schs Health Professions (pres, 74-75); Am Vet Med Asn (pres, 57-58); Asn Am Vet Med Col (pres, 64-65, 73-74); NY Acad Sci; hon mem Am Animal Hosp Asn. *Res:* General surgery; tissue transplantation; wound healing. *Mailing Add:* 1101 Cherokee Blvd Knoxville TN 37919

ARMITAGE, IAN MACLEOD, b Sherbrooke, Que, Nov 15, 47; Can citizen; div; c 2. BIOPHYSICAL CHEMISTRY-NMR, BIOCHEMISTRY. *Educ:* Bishop's Univ, BSc, 68; Univ BC, PhD(phys chem), 72. *Prof Exp:* Fel biophys chem, Calif Inst Technol, 73-74; asst prof phys sci, 74-77, asst prof, 77-80, assoc prof biophys & biochem, 80-83, ADJ PROF RES BIOPHYS, BIOCHEM & RADIOL, YALE UNIV, 83- *Concurrent Pos:* Prin investr, NIH & NSF grants, 77-; consult med res div, Am Cyanamid Co, 79-, E I du Pont de Nemours & Co, 85-89; mem, NIH Spec Study Sect, 81-83, NSF Comt Biol Instrumentation, 80-83; consult med res div, Am Cyanamid Co, 79-, NSF Biol Facil Centers, 87-90, NIH metallobiochemical Study Sect, 90-94, Nat Adv Comt, NMR Facil, Madison, 86-, Yale Univ, Sch Med, Admis Comt, 89-; consult, Merck & Co Inc, 86-88. *Mem:* Am Chem Soc; Int Soc Magnetic Resonance; Biophys Soc; NY Acad Sci; AAAS; Soc Magnetic Res Med. *Res:* Application of multinuclear magnetic resonance techniques to the elucidation of macromolecular structure and dynamics; role of metal ions in biology. *Mailing Add:* Dept Molecular Biophys & Biochem Yale Univ PO Box 3333 New Haven CT 06510-3219

ARMITAGE, JOHN BRIAN, b Ripon, Eng, Oct 28, 27; m 54; c 2. ORGANIC POLYMER CHEMISTRY. *Educ:* Univ Manchester, BSc, 48, PhD(chem), 51. *Prof Exp:* Fel, Ohio State Univ, 51-52; sr res fel, Cambridge Univ, 52-55; res chemist, Exp Sta, 55-60, res supvr, 60-61, sr res supvr, Dupont Tex, 61-66, sr res supvr, Plastics Dept, 67-71, res assoc, Plastic Prod & Resins Dept, 71-76, REGULATORY AFFAIRS CONSULT, E I DUPONT DE NEMOURS & CO, 76- *Concurrent Pos:* Sr studentship of 1851 Royal Exhib, 52-54; consult, Chester Water Co, Eng, 54-55. *Mem:* Am Chem Soc; Soc Plastics Eng; fel Royal Soc Chem. *Res:* Synthesis of polyacetylenes; structure of vitamin B12; potato root eel worm chemistry; polymer chemistry; polyolefins; polyamides; fire retardance and toxicity of thermoplastics. *Mailing Add:* 621 Haverhill Rd Wilmington DE 19803

ARMITAGE, KENNETH BARCLAY, b Steubenville, Ohio, Apr 18, 25; m 53; c 3. BEHAVIOR, PHYSIOLOGY. *Educ:* Bethany Col, WVa, BS, 49; Univ Wis, MS, 51, PhD, 54. *Prof Exp:* Asst zool, Univ Wis, 49-52, proj asst bot, 53, instr bot & zool, 54-56; from asst prof to prof zool, 56-75, chmn dept biol, 68-75, chmn dept, 81-88, PROF SYST & ECOL, UNIV KANS, 75- *Concurrent Pos:* Trustee, Rocky Mountain Biol Lab, 70-86 & pres, 85-86; pres, Orgn Biol Field Stas, 88-89. *Honors & Awards:* Antarctic Medal, 68; Baumgartner Distinguised Prof, 87. *Mem:* Fel AAAS; Am Soc Naturalists (treas, 84-86); Ecol Soc Am; Soc Study Evolution; Am Soc Zool. *Res:* Behavioral and physiological ecology; social ecology of marmots. *Mailing Add:* Dept Syst & Ecol Univ Kans Lawrence KS 66045-2106

ARMOR, JOHN N, b Philadelphia, Pa, Sept 14, 44; m 66; c 3. HETEROGENEOUS CATALYSIS ADSORBENTS CERAMICS. *Educ:* Pa State Univ, BS, 66; Stanford Univ, PhD(chem), 70. *Prof Exp:* Asst prof inorg chem, Boston Univ, 70-74; sr res chemist, 74-77, res assoc, Cent Res Ctr, 77-81, sr res assoc, 81-85; PRIN RES ASSOC, AIR PROD & CHEM INC, 89- *Concurrent Pos:* Am Ed J Applied Catalysis. *Mem:* Am Chem Soc; Am Ceram Soc; Mat Res Soc; Catalysis Soc. *Res:* Homo and heterogeneous catalysis using inorganic and organometallic compounds; material synthesis; novel inorganic materials; adsorbents; ceramic materials. *Mailing Add:* Air Prod & Chem Inc Corp Sci & Tech Ctr 7201 Hamilton Blvd Allentown PA 18195

ARMOUDIAN, GARABED, b Houch-Hala, Lebanon, Aug 1, 38; m 62; c 2. ELECTRODYNAMICS, THEORETICAL PHYSICS. *Educ:* Am Univ Beirut, BS, 62; La State Univ, PhD(physics), 66. *Prof Exp:* Instr physics, La State Univ, 65-66; AEC res fel, Columbia Univ, 66-68; ASSOC PROF PHYSICS, SOUTHWESTERN OKLA STATE UNIV, 68- *Mem:* Am Phys Soc; Am Asn Physics Teachers. *Res:* Quantum electrodynamics; theoretical studies on the structure of proton and other elementary particles. *Mailing Add:* Dept Physics Southwestern Okla State Univ 100 Campus Dr Weatherford OK 73096

ARMOUR, EUGENE ARTHUR, b Mercer, Pa, Sept 21, 46; m 68; c 4. ORGANIC CHEMISTRY, PHOTOGRAPHIC CHEMISTRY. *Educ:* Purdue Univ, BS, 68; Ohio State Univ, PhD(org chem), 73. *Prof Exp:* Sr res chemist org chem, 73-80, RES ASSOC, EASTMAN KODAK CO, 80- *Mem:* Soc Photog Scientists & Engrs; Am Chem Soc. *Res:* Materials research and development. *Mailing Add:* 15 Hidden Spring Circle Rochester NY 14616-1922

ARMOUR, JOHN ANDREW, b Montreal, Can, May 22, 37; m 82; c 2. PHYSIOLOGY, ANATOMY. *Educ:* McGill Univ, BSc, 58; Univ Western Ont, MD, 63; Loyola Univ, Chicago, PhD(physiol), 73. *Prof Exp:* Teaching fel physiol, Loyola Univ, 67-73; asst prof, 78-84, PROF PHYSIOL & BIOPHYS, FAC MED, DALHOUSIE UNIV, 84- *Mem:* Am Physiol Soc; Can Physiol Soc; Soc Neurosci; Can Med Asn; Soc Exp Biol & Med; Am Heart Asn. *Res:* Neural control of the heart as it relates to the thoracic autonomic nervous system. *Mailing Add:* Dept Physiol & Biophys Fac Med Dalhousie Univ Halifax NS B3H 4A7 Can

ARMSON, KENNETH AVERY, b Newton Brook, Ont, Feb 19, 27; m 52; c 1. SILVICULTURE, SOIL SCIENCE. *Educ:* Univ Toronto, BScF, 51; Oxford Univ, dipl forestry, 55. *Prof Exp:* Forester, Soils Surv, Res Div, Ont Dept Lands & Forests, 51-52; lectr, Univ Toronto, 52-57, from asst prof to prof forestry, 57-78; spec adv, Ont Ministry Natural Resources,78-80, chief forester, 80-83, exec coordr forest resources, 83-86, prov forester, 86-89; FORESTRY CONSULT, 89- *Honors & Awards:* Gold Medal, Can Inst Forestry, 78. *Mem:* Soc Am Foresters; Soil Sci Soc Am; Can Inst Forestry. *Res:* Forest soils; effect of fertility on tree growth. *Mailing Add:* 446 Heath St E Toronto ON M4G 1B5 Can

ARMSTEAD, ROBERT LOUIS, b Blair, Nebr, Nov 5, 36; m 61; c 2. THEORETICAL PHYSICS. *Educ:* Univ Rochester, BS, 58; Univ Calif, Berkeley, PhD(theoret physics), 65. *Prof Exp:* ASSOC PROF PHYSICS, NAVAL POSTGRAD SCH, 64- *Mem:* Am Phys Soc; Sigma Xi; Am Asn Physics Teachers. *Res:* Theoretical atomic scattering; laser light propagation. *Mailing Add:* Dept Physics Code 61 Ar Naval Postgrad Sch Monterey CA 93940

ARMSTEAD, WILLIAM M, b Meriden, Conn, May 13, 57; m. PHYSIOLOGY. *Educ:* Univ Pa, BA, 79; Tulane Univ, MS, 83, PhD(pharmacol), 85. *Prof Exp:* Postdoctoral fel & instr pharmacol, Tulane Univ, 85-86; postdoctoral fel, 86-88, instr, 88-90, ASST PROF, DEPT PHYSIOL & BIOPHYS, UNIV TENN, MEMPHIS, 90- *Concurrent Pos:* Guest lectr, Pharmacol Dept, NY Med Col, NJ Med Sch, Northeastern Univ, Albany Med Sch, Tex A&M Med Sch, Univ Iowa, Mich State Univ, Neurobiol Dept, Cornell Med Sch, Anat Dept, Univ Ark Med Sch; mem, Stroke Coun, Basic Sci Coun & Circulation Coun, Am Heart Asn; postdoctoral training grant, NIH, 86-87, First award, 89- *Mem:* Am Physiol Soc; Sigma Xi. *Mailing Add:* Dept Physiol Univ Tenn 894 Union Ave Memphis TN 38163

ARMSTRONG, A(RTHUR) A(LEXANDER), JR, b Gastonia, NC, July 13, 21; m 56; c 2. CHEMICAL ENGINEERING. *Educ:* NC State Col, BChE, 47, MS, 49, PhD(chem eng), 57. *Prof Exp:* Chemist, City Gastonia, 46; chem engr, Gen Elec Co, Mass, 47 & Duke Power Co, NC, 49-50; develop engr, Chemstrand Corp, Fla, 53-54; tech supt, Celanese Corp, SC, 54-58; assoc prof chem eng, Univ SC, 58-59; assoc prof chem eng & head radiol lab, Sch Textiles, NC State Col, 59-65; assoc prof chem eng, Univ NMex, 65-68; sr res specialist, 68-71, eng fel, Monsanto Triangle Park Develop Ctr, Inc, 71-80, SR FEL, MONSANTO CO, 80- *Mem:* Am Inst Chem Engrs; Am Soc Eng Educ. *Res:* Mass and heat transfer; applied mathematics; process engineering; radiation-induced processes. *Mailing Add:* 219 Rose St Cary NC 27511

ARMSTRONG, ALFRED RINGGOLD, b Washington, DC, July 6, 11; m 34; c 2. ANALYTICAL CHEMISTRY. *Educ:* Col William & Mary, BS, 32, AM, 34; Univ Va, PhD(chem), 45. *Prof Exp:* Instr, Col William & Mary, 33-35, asst prof chem, 36-42; asst to Dr J. H. Yoe, Nat Defense Res Comt & investigator, Chem Warfare Serv, Univ Va, 44-45; from assoc prof to prof chem Col William & Mary, 45-76; RETIRED. *Honors & Awards:* Thomas Jefferson Award, Col William & Mary, 75. *Mem:* Am Chem Soc. *Res:* Chemistry of sea water; organic analytical reagents. *Mailing Add:* 512 Newport Ave Williamsburg VA 23185-4013

ARMSTRONG, ANDREW THURMAN, b Haslet, Tex, May 26, 35; m 58; c 3. PHYSICAL CHEMISTRY. *Educ:* North Tex State Univ, BS, 58, MS, 59; La State Univ, PhD(chem), 67. *Prof Exp:* Instr chem, West Tex State Univ, 59-61 & La State Univ, 63-66, vis asst prof, 67-68; from asst prof to assoc prof chem, Univ Tex, Arlington, 74-84; PRES, ARMSTRONG FORENSIC LAB INC, 80- *Concurrent Pos:* Chief chemist, Arlington Fire Dept, 78- *Mem:* Am Chem Soc; fel Am Inst Chemists; Am Asn Indust Hygiene; fel Am Acad Forensic Sci; Int Asn Arson Investr. *Res:* Trace analytical techniques for environmental and forensic applications; numerous papers and publications. *Mailing Add:* Armstrong Forensic Lab 330 Loch'n Green Trail Arlington TX 76012

ARMSTRONG, BAXTER HARDIN, b Portland, Ore, Jan 27, 29; m 54; c 4. THERMAL CONDUCTIVITY THEORY, ACOUSTIC EMISSION. *Educ:* Univ Calif, Berkeley, AB, 51, MA, 53, PhD(physics), 56. *Prof Exp:* From assoc res scientist to res scientist, Lockheed Missiles & Space Co, 56-62, staff scientist, 62-63, sr staff scientist & sr mem res lab, 63-64; mem staff, Sci Ctr, IBM Corp, 64-68, sci staff mgr, 69-73, dir, Centro Cientifico de Am Latina, IBM de Mex, 73, sci staff mgr, IBM Sci Ctr, 74-78; vis scholar, Stanford Univ, 78-79, consult prof aeronaut & astronaut, 79-83; mgr eng res, IBM Corp, 83-85; affil prof mat sci & eng, Univ Wash, 85-86; MGR NUMERICALLY INTENSIVE COMPUT APPLNS, PALO ALTO SCI CTR, IBM CORP, 87- *Concurrent Pos:* Asst prof, San Jose State Col, 58-60; guest staff mem, Queen's Univ Belfast, 60-61; vis scholar, Stanford Univ, 78-79, consult prof aeronaut & astronaut, 79-83; assoc ed, J Quant Spectros & Radiative Transfer, 66-86; consult, Syst, Sci & Software Co, 69 & Lockheed Missiles & Space Co, 88-89. *Mem:* Am Phys Soc; Am Asn Physics Teachers; Sigma Xi. *Res:* Theoretical atomic physics; interaction of atoms with radiation fields; atmospheric radiative transfer; thermal conductivity of dielectric crystals; acoustic attenuation theory; acoustic emission from geological materials; phonon interactions. *Mailing Add:* 4339 Miranda Ave Palo Alto CA 94306-3742

ARMSTRONG, CARTER MICHAEL, b Jersey City, NJ, Nov 17, 50; m 72; c 1. PLASMA PHYSICS. *Educ:* Rutgers Univ, AB, 72; Univ MD, PhD(plasma physics), 76. *Prof Exp:* Nat Res Coun assoc laser fusion, Nat Acad Sci, Naval Res Lab, 76-77; asst prof plasma physics, NC State Univ, 77-85; HEAD, FAST WAVE DEVICES SECT, NAVAL RES LAB, 85- *Mem:* Am Phys Soc. *Res:* Intense relativistic electron and ion beams; laser fusion. *Mailing Add:* Naval Res Lab 4555 Overlook Ave SW Washington DC 20375

ARMSTRONG, CLAY M, b Chicago, Ill, Sept 26, 34; m 63; c 3. PHYSIOLOGY, BIOPHYSICS. *Educ:* Rice Univ, BA, 56; Wash Univ, MD, 60. *Prof Exp:* Res assoc biophys, Lab Biophys, NIH, 61-64; hon res asst physiol, Univ Col, Univ London, 64-66; asst prof, Duke Univ, 66-69; assoc prof, Univ Rochester, 69-75, prof, 74-75; PROF PHYSIOL, SCH MED, UNIV PA, 76- *Concurrent Pos:* Mult Sclerosis Soc fel, Neurol Dept, Sch Med, Wash Univ, St Louis, 61; distinguished lectr, Soc Gen Physiologists, 86. *Honors & Awards:* K S Cole Award, Biophys Soc, 75; Bowditch Lectr, Am Physiol Soc, 75. *Mem:* Nat Acad Sci; Biophys Soc; Soc Gen Physiologists (pres, 85-86); Am Physiol Soc; Fedn Am Socs Exp Biol. *Res:* Permeability mechanisms in excitable membranes; excitation-contraction coupling; author of numerous technical publications. *Mailing Add:* Dept Physiol Univ Pa Philadelphia PA 19104-6085

ARMSTRONG, DALE DEAN, b Salina, Kans, Aug 18, 27; m 52; c 3. NUCLEAR PHYSICS. *Educ:* Univ Colo, BS, 57; Univ NMex, MS, 62, PhD(physics), 65. *Prof Exp:* Staff mem physics, 57-80, DEP GROUP LEADER, LOS ALAMOS NAT LAB, 80- *Mem:* Am Phys Soc. *Res:* Design and development of a high intensity neutron source for simulating the radiation damage effects that will be produced in fusion reactor materials. *Mailing Add:* 13839 Elmbrook Dr Sun City West AZ 85375

ARMSTRONG, DANIEL WAYNE, b Ft Wayne, Ind, Nov 2, 49; m 72; c 2. ANALYTICAL CHEMISTRY, COLLOID CHEMISTRY. *Educ:* Washington & Lee Univ, BS, 72; Tex A&M Univ, MS, 74, PhD(chem), 77. *Prof Exp:* Assoc chem, Tex A&M Univ, 77-78; asst prof chem, Bowdoin Col, 78-80; ASST PROF CHEM, GEORGETOWN UNIV, 80- *Concurrent Pos:* Prin investr, Petrol Res Fund, 78-80 & Res Corp, 79-81; consult, E C Jordan Co, Inc, Continental Precious Metals, Inc, Wapora, Inc & Defense Div, Brunswick Corp. *Mem:* Am Chem Soc; AAAS; AAUP. *Res:* Micellar catalysis, environmental chemistry, analytical applications of micelles, chemical evolution, liposome chemistry and chromatographic molecular weight determination of macromolecules. *Mailing Add:* RR 6 Box 263 Rolla MD 65401

ARMSTRONG, DAVID ANTHONY, b Barbados, WI, Aug 27, 30; m 55; c 3. PHYSICAL CHEMISTRY. *Educ:* McGill Univ, BSc, 52, PhD, 55. *Prof Exp:* Brotherton res lectr phys chem, Univ Leeds, 55-57; Nat Res Coun Can fel chem, Univ Sask, 57-58; from asst prof to assoc prof, Univ Alta, 58-68; dean, fac, sci, 84-89, PROF CHEM, UNIV CALGARY, 68- *Mem:* Fel Chem Inst Can; Am Chem Soc; Sigma Xi. *Res:* Radiation chemistry; chemical kinetics. *Mailing Add:* Dept Chem Univ Calgary Calgary AB T2N 1N4 Can

ARMSTRONG, DAVID M(ICHAEL), b Louisville, Ky, July 31, 44; div; c 2. MAMMALOGY, BIOGEOGRAPHY. *Educ:* Colo State Univ, BS, 66; Harvard Univ, MA, 67; Univ Kans, PhD(systs & ecol), 71. *Prof Exp:* From asst prof to assoc prof, 71-85, PROF NAT SCI, 85-, DIR, MUSEUM, UNIV COLO, BOULDER, 88- *Concurrent Pos:* Sci collabr, Nat Park Serv, 71-; ed, Southwestern Naturalist, 76-80 & J Mammal, 81-87; pres-elect, Colo-Wyo Acad Sci, 90-91. *Mem:* Am Soc Mammalogists; AAAS; Ecol Soc Am; Sigma Xi. *Res:* Biogeography and ecology of mammals of the western United States and Middle America; history of evolutionary biology. *Mailing Add:* Univ Colo Natural Sci Prog Boulder CO 80309-0331

ARMSTRONG, DAVID THOMAS, b Kinburn, Ont, Nov 5, 29; m 56; c 3. PHYSIOLOGY, REPRODUCTIVE BIOLOGY. *Educ:* Ont Agr Col, BSA, 51; Cornell Univ, MS, 56, PhD(physiol), 59. *Prof Exp:* Asst agr rep, Ont Dept Agr, 51-54; asst dept animal husb, Cornell Univ, 54-57; res assoc, Dept Biol, Brookhaven Nat Lab, 58-60; from res assoc to assoc anat, Sch Dent Med, Harvard Univ, 60-68; assoc prof, 68-69, PROF PHYSIOL, OBSTET &

GYNEC, UNIV WESTERN ONT, 69-, DIR, MRC GROUP, REPRODUCTIVE BIOL, 79- *Concurrent Pos:* Res career develop award, NIH, 63-; assoc, Med Res Coun Can, 69-; vis prof obstet & gynec, Univ Adelaide, 77-78. *Mem:* Soc Study Reproduction (pres, 77); Endocrine Soc; Am Physiol Soc; Can Physiol Soc; Int Embryo Transfer Soc. *Res:* Pituitary-gonad interrelationships; mechanism of hormone action. *Mailing Add:* Dept Obstet & Gynec Univ Western Ontario London ON N6A 5A5 Can

ARMSTRONG, DON L, b Alhambra, Calif, June 7, 16; m 40; c 2. CHEMICAL ENGINEERING, MARINE SCIENCE. *Educ:* Univ Calif, Los Angeles, AB, 37; Univ Southern Calif, MS, 38, PhD(chem), 42, MChE, 45. *Prof Exp:* Res assoc chem, 41-45; prin chemist, Aerojet-Gen Corp, Gen Tire & Rubber Co, 46-57, mgr chem div, 57-59, dir chem, 59-62, sr scientist, 62-65, mgr corp tech info ctr, 62-64; prof chem & chmn dept, 65-81, DISTINGUISHED SERV PROF, WHITTIER COL, 81- *Concurrent Pos:* Pres, Tech Asst Group, Los Angeles, 64-79. *Mem:* Sigma Xi; Am Chem Soc; Marine Technol Soc. *Res:* Marine chemistry, desalination by reverse osmosis; liquid and solid rocket propellants; technical information retrieval. *Mailing Add:* 20245 Couina Hills Rd Couina CA 91724-3631

ARMSTRONG, DONALD B, b Minneapolis, Minn, June 30, 37; m 64; c 2. ACOUSTICS, CRYSTALLOGRAPHY. *Educ:* Univ Minn, BS, 59, MSEE, 61, PhD(elec eng), 64. *Prof Exp:* Res fel solid state physics, Univ Minn, 61-64; sr engr, Electron Tube Div, Litton Indust, Calif, 64-67, sr scientist, 67-72, prog mgr electrooptic devices, 72-73; dir new prod develop, 73-77, VPRES MFG, CRYSTAL TECHNOL INC, 77- *Mem:* Inst Elec & Electronics Engrs; Sigma Xi. *Res:* Microwave properties of semiconductors; properties of acoustic surface waves on piezoelectric crystals at microwave frequencies. *Mailing Add:* 14 Debbin Lane Belmont CA 94002

ARMSTRONG, DONALD JAMES, b Marshall Co, WVa, Mar 14, 37; m 61, 84; c 1. PLANT HORMONE RESEARCH. *Educ:* Marshall Univ, AB, 59, MA, 61; Univ Wis, Madison, PhD(bot), 67. *Prof Exp:* Proj assoc bot, Univ Wis, Madison, 67-68; Nat Sci Found fel biochem, Univ BC, 68-70; proj assoc bot, Univ Wis, Madison, 70-74; from asst prof to assoc prof, 74-86, PROF, ORE STATE UNIV, 87- *Concurrent Pos:* NIH sr fel, Univ Calif, San Diego, 84-85. *Mem:* Am Soc Plant Physiol; Int Plant Growth Substances Asn; Int Soc Plant Molecular Biol. *Res:* Metabolism and mechanism of action of plant growth substances; plant tissue culture; regulation of plant growth and development. *Mailing Add:* Dept Bot Plant Path Ore State Univ Corvallis OR 97331

ARMSTRONG, EARLENE, b Wilson Co, NC, Apr 11, 47; m 80; c 2. ENTOMOLOGY, MICROBIOLOGY. *Educ:* NC Cent Univ, BS, 69, MS, 70; Cornell Univ, PhD(entom), 75. *Prof Exp:* Grad asst biol, NC Cent Univ, 69-70, instr, 70-71; instr, Southeastern Community Col, 71; asst prof entom, Howard Univ, 75-76; asst prof, 76-81, ASSOC PROF ENTOM, UNIV MD, COLLEGE PARK, 81- *Concurrent Pos:* Consult, Insect Control & Res, Inc, Baltimore, 78-80; invest, Marine Biol Lab, Mass, 76-79. *Mem:* Soc Invert Path; Entom Soc Am; Sigma Xi. *Res:* Protozoan pathogens of insects. *Mailing Add:* Dept Entom Univ Md College Park MD 20742

ARMSTRONG, FRANK BRADLEY, JR, b Brownsville, Feb 15, 28; m 58; c 3. BIOCHEMISTRY. *Educ:* Univ Tex, BS, 50, MA, 53; Univ Calif, PhD(biochem), 59. *Prof Exp:* Fel biochem & genetics, Univ Tex, 59-62; from asst prof to assoc prof genetics & bact, NC State Univ, 62-66, assoc prof biochem, 62-68, dir biotechnol prog, 83-86, UNIV PROF BIOCHEM, 68- *Concurrent Pos:* Career develop award & Sigma Xi res award, 64; NSF panel, 70 & 71; vis prof, Univ Exeter, England, 76. *Mem:* AAAS; Genetics Soc Am; Am Soc Biol Chemists; Am Chem Soc; Am Inst Biol Sci; NY Acad Sci. *Res:* Biochemical studies on Salmonella typhimurium; enzymes and protein chemistry. *Mailing Add:* Dept Biochem NC State Univ Box 7622 Raleigh NC 27695-7622

ARMSTRONG, FRANK CLARKSON FELIX, b New York, NY, June 2, 13; m 47; c 2. GEOLOGY. *Educ:* Yale Univ, BA, 36; Univ Wash, MS, 48; Stanford Univ, PhD(geol), 63. *Prof Exp:* Assayer & engr, Tiblemont Island Mining Co, 36-37; engr & miner, Hollinger Consol Gold Mines Ltd, 37-40; mine geologist, Dome Mines, Ltd, 40-43; geologist strategic minerals invests, 43-45, actg regional geologist, Wash, 45-46, geologist, Wyo, 46, western phosphate deposits, 47 & 48, Idaho, 49 & 50-51, Colo, 51-52, Pac Northwest, 52-54, Idaho, 54-61 & Wyo, 62-64, geologist, US Geol Surv, 43-82; RETIRED. *Concurrent Pos:* Part-time teacher, Ft Wright Col Holy Names, 66. *Mem:* AAAS; fel Geol Soc Am; Soc Econ Geologists; Geol Asn Can. *Res:* Economic and structural geology; metals and nonmetals. *Mailing Add:* US Geol Surv E 2614 40th Spokane WA 99223

ARMSTRONG, GEORGE GLAUCUS, JR, b Houston, Miss, Feb 28, 24; m 48; c 2. PHYSIOLOGY, AEROSPACE MEDICINE. *Educ:* Univ Miss, BA, 48, MS, 50, BS, 52; Univ Ill, MD, 56. *Prof Exp:* Res assoc, Sch Med, Univ Miss, 48-49, instr physiol, 49-52, from asst prof to assoc prof physiol & biophys, 52-64; aerospace technologist, 64-68, chief biomed technol div, 68-70; dep chief, Med Opers Div, Johnson Space Ctr, NASA, 70-72, chief health serv div, 72-77; dir, Health Effects Div, US Environ Protection Agency, 77-80; dir, food ingredient assessment div, USDA, 80-84; RETIRED. *Res:* Cardiovascular reflexes; spaceborne medical systems. *Mailing Add:* PO Box 241 Gilchrist TX 77617

ARMSTRONG, GEORGE MICHAEL, b Madison, Wis, Dec 5, 38; m 60; c 5. BIOLOGY, MOLECULAR BIOLOGY. *Educ:* Univ Wis-Madison, BS, 62; Univ Okla, MNS, 66, PhD(plant physiol), 68. *Prof Exp:* ASSOC PROF BIOL, UNIV WIS, FOND DU LAC, 68- *Mem:* Sigma Xi. *Res:* Phenolic compounds in plants as affected by environmental factors. *Mailing Add:* Dept Biol Univ Wis 400 Campus Dr Fond du Lac WI 54935

ARMSTRONG, HERBERT STOKER, b Toronto, Ont, Nov 23, 15; m 41; c 2. GEOLOGY. *Educ:* Univ Toronto, BA, 38, MA, 39; Univ Chicago, PhD(econ geol), 42. *Hon Degrees:* DSc, McMaster Univ, 67; DUC, Univ Calgary, 72. *Prof Exp:* Asst geol, Univ Toronto, 38-39 & Univ Chicago, 40-41; lectr, McMaster Univ, 41-44, from asst prof to assoc prof, 44-48, prof, 48-62, asst dean arts & sci, 46-48, from assoc dean to dean, 48-62; prof geol, Univ Alta, 62-64, dean sci, 62-63, acad vpres, 63-64; prof geol & pres, Univ Calgary, 64-68, vchancellor, 66-68; dean grad studies, Univ Guelph, 68-80, prof geol, 68-82; RETIRED. *Concurrent Pos:* Chmn, COU Comt Dist Educ, 83-85; consult, Univ Guelph, 82-88, hon fel, 85. *Honors & Awards:* Centennial Medal, 67. *Mem:* Fel Geol Asn Can; fel Royal Soc Can; emer mem Sigma Xi; life mem Can Inst Min & Metal; fel Royal Can Geog Soc; emer mem Can Soc Petrol Geol; Geol Soc Finland. *Res:* Precambrian geology; igneous and metamorphic petrology. *Mailing Add:* 75 Glasgow St N Guelph ON N1H 4W1 Can

ARMSTRONG, JAMES ANTHONY, b Detroit, Mich, Dec 7, 43. POLLUTION CONTROL, CLOUD PHYSICS. *Educ:* Univ Detroit, BME, 66; Northwestern Univ, MS, 70, PhD(mech eng), 73. *Prof Exp:* Coop student engr, Cent Chevrolet Off, Gen Motors Corp, 64-66; res environ engr, 73-81, SR RES ENVIRON ENGR, DENVER RES INST, UNIV DENVER, 81- *Mem:* Sigma Xi; Am Meteorol Soc; Air Pollution Control Asn; AAAS. *Res:* Air pollution investigations, including the measurement and control of particulates and gases from point and non-point sources; cloud physics studies of heterogeneous ice nucleation. *Mailing Add:* PO Box 210 Lima OH 45802

ARMSTRONG, JAMES CLYDE, b Wheelock, Tex, Dec 23, 33; m 58; c 2. CIVIL ENGINEERING, SOIL MECHANICS. *Educ:* Tex A&M Univ, BS, 55, BS, 61, MS, 62, PhD(civil eng), 67. *Prof Exp:* Sr lab asst soil mech, Tex Hwy Dept, 57-59; asst res engr pavement design, Tex Transp Inst, 62-66; asst prof soil mech, Univ Mo-Rolla, 66-69; assoc prof, Univ Ark, 69-71; mem staff, Nat Soil Serv, Arlington, Tex, 75-77; mem staff, Univ Tex, Arlington, 77-89; RETIRED. *Concurrent Pos:* NSF res grant, 68-69; mem, Ford Found Prog Residencies Eng Pract, 70-75. *Mem:* Am Soc Civil Engrs; Nat Soc Prof Engrs; Int Soc Soil Mech & Found Engrs; Am Soc Eng Educr. *Res:* Pore pressure dissipation around piles and pile groups; comparison of particulate mechanics approaches to the action of real solid systems; soil stabilization of active clay soils. *Mailing Add:* PO Box 44 Wheelock TX 77882

ARMSTRONG, JAMES G, b London, Eng, Nov 14, 24; US citizen; m 51; c 4. MEDICAL ADMINISTRATION. *Educ:* Oxford Univ, MA, 49, BM, BCh, 50; Royal Col Physicians & Surgeons Can, cert pediat, 63; FRCP(C). *Prof Exp:* House physician, St Olave's Hosp, London, Eng, 50-51, house surgeon, 51-52; staff physician, Alta Ment Inst, 55-56; resident, St Joseph's Hosp, Toronto, Ont, 56-57; chief resident, Hosp Sick Children, 57-58; physician, Lilly Labs Clin Res, 58-66, sr physician, Lilly Res Labs, 66-68, asst dir med regulatory affairs, 68-69, dir med plans & regulatory affairs, 69-73, dir med liaison, 73-75, clin investr, 75-80; CHMN BD & PRES, ALEXANDER GRANT & ASSOCS INC, 80-; EMER ASST PROF MED, IND UNIV SCH MED, 80- *Concurrent Pos:* Fel pediat, Hosp Sick Children, 57-58; fel pediat med, Univ Toronto, 57-58; instr, Ind Univ, 58-66, asst prof, 66-; staff physician, Dept Pediat, Marion County Gen Hosp, Indianapolis, Ind, 58-79, physician, Tumor Bd, 60-79; mem bd, US Pharmacists, 83- *Mem:* Fel Am Col Physicians; Am Asn Cancer Res; Am Fedn Clin Res; fel Royal Soc Med; Brit Med Asn. *Res:* Pediatric medicine; clinical pediatrics; cancer chemotherapy; medical research administration. *Mailing Add:* 3633 E 71st St Indianapolis IN 46220

ARMSTRONG, JOHN A, b Schenectady, NY, July 1, 34; m 58; c 2. QUANTUM ELECTRONICS, LASER PHYSICS. *Educ:* Harvard Univ, BA, 56, PhD(appl physics), 61. *Hon Degrees:* DSc, State Univ NY, Albany. *Prof Exp:* mem res staff, IBM Corp, 63-76, dir phys sci, Res Ctr, 76-80, corp tech comt, 80-81, mgr mat & technol develop, 81-83, vpres, Logic & Mem, 83-86, dir res, 86-87, vpres & dir res, 87-89, VPRES SCI & TECHNOL, IBM CORP, 89-, MEM CORP MGT BD, 89- *Concurrent Pos:* Mem, Harvard bd Overseers, Nat Adv Comt Semiconductors & policy bd, Nat Nanofabrication Facil Cornell; trustee, Assoc Univs, Inc. *Mem:* Nat Acad Eng; fel Optical Soc Am; fel AAAS; fel Am Phys Soc; Royal Swedish Acad Eng Sci. *Res:* Nonlinear optics, photon statistics; nonlinear optical pulse propagation; picosecond laser pulse techniques; laser spectroscopy of atoms; author of more than 10 publications. *Mailing Add:* PO Box 29 South Salem NY 10590

ARMSTRONG, JOHN BRIGGS, b Toronto, Ont. PEDIATRIC NEUROLOGY, BIOCHEMISTRY. *Educ:* McGill Univ, BSc, 63, MD & CM, 65, PhD(neurochem), 75. *Prof Exp:* Intern med, St Paul's Hosp, Vancouver, BC, 66-67; resident, Montreal Gen Hosp, 67-68; resident neurol, Montreal Neurol Inst, 68-69, res fel neurochem, 69-71; res fel neurochem, Hosp for Sick Children, Toronto, 71-74, chief resident pediat neurol, 74-75; asst prof pediat, Univ Toronto, 75-76; assoc med dir, 76-77, MED DIR, PFIZER CO LTD, 77-; STAFF NEUROLOGIST, MONTREAL CHILDREN'S HOSP, 79- *Concurrent Pos:* Med Res Coun Can fel, 69-73; Muscular Dystrophy Asn Can fel, 73-74, clin fel, 74-75; staff neurologist & res assoc, Hosp for Sick Children, Toronto, 75-76. *Mem:* Can Soc Child Neurol; Am Fedn Clin Res; Am Acad Neurol; Can Neurol Soc. *Res:* Hemodynamic effects of vasodilators in congestive heart failure and hypertension; relationship of therapeutic effect to plasma levels of cyclic antidegenerants. *Mailing Add:* Dept Biol Scis Univ Ottawa Ottawa ON K1N 6N5 Can

ARMSTRONG, JOHN BUCHANAN, b Toronto, Ont, Oct 19, 18; m 44; c 4. INTERNAL MEDICINE. *Educ:* Univ Toronto, MD, 43; McGill Univ, dipl, 51. *Prof Exp:* Sr intern & asst resident med, Royal Victoria Hosp, 46-48; fel, Duke Hosp, 48-49; registr, Post-Grad Med Sch, Hammersmith Hosp, London, 49-50; asst prof physiol & med res & spec lectr med, Univ Man, 50-56; assoc prof pharmacol, Univ Toronto, 56-69; sr lectr pharmacol, Univ Ottawa, 69-75; assoc prof prev med, Univ Toronto, 75-83, sr med consult epidemiol, Ont Ministry Health, 75-83; RETIRED. *Concurrent Pos:* FRCP, Can, 50, Can Life Ins med fel, 50-52; Markle scholar, 52-56; physician,

Winnipeg Munic Hosps, 51-56; chmn & mem panel shock & plasma expanders, Defense Res Bd Can, 53-68; asst physician, Winnipeg Gen Hosp, 54-56; consult, Can Forces Med Coun, 54-60; vol asst, Toronto Gen Hosp, 57-68; exec dir & chief med adv, Can Heart Found, 57-75; pres, Biol Coun Can, 71-72. *Honors & Awards:* Queen Elizabeth's Silver Jubilee Medal, 77. *Mem:* Pharmacol Soc Can (pres, 72); Can Cardiovasc Soc; Can Hypertension Soc; Pharmaceut Soc Can (pres, 72-73); Am Heart Asn; Can Med Asn. *Res:* Clinical cardiovascular physiology, pharmacology and epidemiology. *Mailing Add:* 401 151 Bay St Ottawa ON K1R 7T2 Can

ARMSTRONG, JOHN EDWARD, b Cloverdale, BC, Feb 18, 12; m 37; c 1. QUATERNARY GEOLOGY, GLACIAL GEOLOGY. *Educ:* Univ BC, BASc, 34, MASc, 35; Univ Toronto, PhD(econ geol), 39. *Prof Exp:* Asst geol, Univ Toronto, 35-37; asst geologist, Geol Surv Can, 36-42, from assoc geologist to geologist 43-47, grade 4 geologist, 48-56, geologist in charge BC Off, 56-65, res scientist III, 65-76; chmn geol div, Royal Soc Can, 65-66 & Can Inst Mining & Metall, 67-68; secy gen, Int Geol Cong, 68-76; RETIRED. *Concurrent Pos:* Mem subcomt on Pleistocene, Am Comn Stratig Nomenclature, 54-59; mem steering comt, Int Geol Cong, 76- *Honors & Awards:* Merit Award, Govt Can, 73. *Mem:* Fel Geol Soc Am; fel Royal Soc Can; fel Geol Asn Can. *Res:* Cordilleran field geology; quaternary stratigraphy and geomorphology in the Pacific Northwest. *Mailing Add:* 206-2298 McBain Ave Vancouver BC V6L 3B1 Can

ARMSTRONG, JOHN MORRISON, b Nov 11, 36; US citizen; m 58; c 4. WATER RESOURCES MANAGEMENT, COASTAL ENGINEERING. *Educ:* Univ Mich, Ann Arbor, PhD(eng), 69. *Prof Exp:* Assoc engr systs anal, 64-66, lectr environ & water resource eng, 67-69, asst prof, 69-73, assoc prof environ & water resource eng, Univ Mich, Ann Arbor, 74-; AT DEPT CIVIL ENG, UNIV MICH, 74- *Concurrent Pos:* Dir, Mich Sea Grant Prog, Univ Mich, 70-75 & Great Lakes Resource Mgt Prog & Coastal Zone Lab, 73-; consult, UN, 74-76; mem adv comt effect of offshore develop, NSF, 76-77; mem nat tech comt coastal zone mgt, Am Soc Civil Engrs, 78- *Mem:* Sigma Xi; Am Soc Civil Engrs; AAAS. *Res:* Coastal and ocean resource management; water resource systems analysis; environmental impact analysis; resource policy analysis; engineering economics. *Mailing Add:* 2480 Gale Rd Ann Arbor MI 48105

ARMSTRONG, JOHN WILLIAM, b Roslyn, NY, Mar 15, 48; m 75. ASTRONOMY, SPACE PHYSICS. *Educ:* Harvey Mudd Col, BS, 69; Univ Calif, San Diego, MS, 71, PhD(appl physics), 75. *Prof Exp:* Res assoc astron, Nat Radio Astron Observ, 75-77; MEM TECH STAFF, JET PROPULSION LAB, CALIF INST TECHNOL, 77- *Concurrent Pos:* Mem, Plasma Turbulence Explorer Study Group, NASA, 80; Henry G Booker fel, 20th Union Radio Sci Int Gen Assembly. *Mem:* AAAS; Am Astron Soc; Am Geophys Union. *Res:* Dynamics of interplanetary and interstellar media; wave propagation in random media, turbulence in planetary atmospheres, gravitational wave astronomy. *Mailing Add:* 601 Manzanita Ave Sierra Madre CA 91024

ARMSTRONG, JOSEPH EVERETT, b Rochester, NY, Sept 18, 48; m 71; c 1. MORPHOLOGY & SYSTEMATICS OF FLOWERING PLANTS. *Educ:* State Univ NY Oswego, BA, 70; Miami Univ, MA, 72, PhD(bot), 75. *Prof Exp:* Res fel wood sci, Univ Mo-Columbia, 76-78; from asst prof to assoc prof, 78-89, PROF BOT, ILL STATE UNIV, 90- *Mem:* Bot Soc Am; Am Inst Biol Sci; Sigma Xi; Torrey Botanical Club; Asn Trop Biol. *Res:* Structure, development, and function of flowers in scrophulariaceae, solanaceae, and related families; floral biology and reproduction of primitive, beetle-pollinated, rainforest trees, Myristicaceae Eupomatiaceae; rain forest research in Australia; introduction textbook author. *Mailing Add:* Dept Biol Sci Ill State Univ Normal IL 61761

ARMSTRONG, KENNETH WILLIAM, b Brandon, Man, Sept 4, 35; m 61; c 2. MATHEMATICS. *Educ:* Univ Man, BSc, 57, MSc, 61; McGill Univ, PhD(math), 65. *Prof Exp:* Lectr, Univ Man, 57-58, 59-60, 63-64, from asst prof to assoc prof, 64-69, asst head dept, 66-69; dir bd educ prog, 73-78, ASSOC PROF MATH, UNIV WINNIPEG, 69- *Mem:* Can Math Soc. *Mailing Add:* Dept Math & Statist Univ Winnipeg Winnipeg MB R3B 2E9 Can

ARMSTRONG, LLOYD, JR, b Austin, Tex, May 19, 40; m 65; c 1. ATOMIC PHYSICS. *Educ:* Mass Inst Technol, BS, 62; Univ Calif, Berkeley, PhD(physics), 66. *Prof Exp:* Physicist, Lawrence Radiation Lab, 65-67; sr physicist, Westinghouse Res Labs, Pa, 67-68; res assoc, 68, from asst prof to assoc prof, 69-77, chmn, dept physics & astron, 85-87, PROF PHYSICS, JOHNS HOPKINS UNIV, 77-, DEAN, SCH ARTS & SCI, 87- *Concurrent Pos:* Consult, Westinghouse Res Labs, 69-71; vis fel, Joint Inst Lab Astrophys, Boulder, Co, 78-79; prog off, Nat Sci Found, 81-83; mem, Nat Acad Sci/Nat Res Coun, Comt Recommendations for the US Army Basic Sci Res, 84-87, vchmn, Comt Atomic & Molecular Sci, 84-85 & chmn, 85-88; mem, NSF, Adv Comt Physics, 85-88; mem, adv bd phys rev A, 84-89; mem bd, Physics & Astron, Nat Acad Sci/Nat Res Coun, 90-; mem panel, Nat Measurement Lab, Nat Inst Standards & Technol, 89-, Joint Inst Lab Astrophys, 85- & chmn, 89- *Mem:* Fel Am Phys Soc. *Res:* Theoretical atomic physics, investigation of atomic shell theory through use of group theory and second quantization techniques; many body theory applied to atoms; interaction of atoms with intense laser beams; electron-atom scattering. *Mailing Add:* Dean Sch Arts & Sci Merryman Hall Johns Hopkins Univ Baltimore MD 21218

ARMSTRONG, MARK L, CARDIOVASCULAR DISEASE. *Educ:* Columbia Univ, MD, 51. *Prof Exp:* PROF MED, COL MED, UNIV IOWA, 71- *Concurrent Pos:* Mem coun arteriosclerosis, Am Heart Asn. *Mem:* Am Bd Internal Med; Am Heart Asn; Am Fedn Clin Res; Int Primatol Soc; Am Asn Pathologists. *Res:* Internal medicine. *Mailing Add:* Dept Internal Med Cardiovasc Div Col Med Univ Iowa Iowa City IA 52242

ARMSTRONG, MARTINE YVONNE KATHERINE, MICROBIOLOGY, VIROLOGY. *Educ:* Univ London, MB & BS, 55, MD, 68. *Prof Exp:* RES SCIENTIST EPIDEMIOL, YALE UNIV, 75-, CHMN ANIMAL CARE & USE COMT, 85- *Res:* Parasitology. *Mailing Add:* Sch Med Yale Univ 60 College St PO Box 3333 New Haven CT 06510

ARMSTRONG, MARVIN DOUGLAS, b Wilmington, NC, Apr 15, 18; m 46; c 5. BIOLOGICAL CHEMISTRY. *Educ:* Univ SC, BS, 38; Univ Ill, MS, 39, PhD(org chem), 41. *Prof Exp:* Asst biochem, Med Col, Cornell Univ, 41-42 & 46; from asst res prof to assoc res prof biochem, Col Med, Univ Utah, 46-57; chmn, Dept Biochem, Fels Res Inst, Ohio, 57-77, fels prof biol chem & chief, Biochem Sect, Fels Res Inst, Sch Med, Wright State Univ, 77-80; RETIRED. *Mem:* AAAS; Am Chem Soc; Am Soc Biol Chemists. *Res:* Chemistry and metabolism of the amino acids; metabolism of aromatic compounds; hereditary and metabolic disorders; biochemistry of mental disorders. *Mailing Add:* 1899 Arlington Dr Pocatello ID 83204-5052

ARMSTRONG, NEAL EARL, b Dallas, Tex, Jan 29, 41; m; c 3. ENVIRONMENTAL ENGINEERING, POLLUTION ECOLOGY. *Educ:* Univ Tex, BA, 62, MA, 65, PhD(environ health eng), 68. *Prof Exp:* Res engr, Eng-Sci, Inc, 67-68, asst off mgr & consult sanit engr, 68-70, mgr, Washington Res & Develop Lab, 70-71; assoc prof, 71-79, PROF CIVIL ENG, UNIV TEX, AUSTIN, 79- *Honors & Awards:* Serv Award, Water Pollution Control Fedn, 89. *Mem:* Ecol Soc Am; Am Soc Limnol & Oceanog; Water Pollution Control Fedn; Int Asn Water Pollution Res; Estuarine Res Fedn (vpres, 75-77); Am Soc Civil Engrs; NAm Lake Mgt Soc. *Res:* Water quality management; water quality analysis; water quality modeling. *Mailing Add:* Dept Civil Eng Univ Tex Austin TX 78712-1076

ARMSTRONG, NEIL A, b Wapakoneta, Ohio, Aug 5, 30; m; c 2. AERONAUTICAL ENGINEERING. *Educ:* Purdue Univ, BS, 55. *Hon Degrees:* MS, Univ Southern Calif, 70; DEng, Purdue Univ, 70 & Butler Univ, 72. *Prof Exp:* Joined Lewis Flight Propulsion Lab, NASA, 55 and later aeronaut res pilot, High Speed Flight Sta, Edwards AFB, Calif, selected as astronaut, 62, backup comdr, Gemini V, 65, prime comdr, Gemini VIII, 66, backup comdr, Gemini XI, 66 and Apollo VIII, 68, prime comdr, Apollo XI, 69 and first man on the moon, July 20; prof eng, Univ Cincinnati, 71-79; chmn bd, Cardwell Int Ltd, 80-82; CHMN, CTA INC, 82- *Honors & Awards:* Octave Chanute Award, Am Inst Aeronaut & Astronaut, 62; Presidential Medal for Freedom, 69; Kitty Hawk Mem Award, 69; Hubbard Gold Medal, Nat Geog Soc, 70. *Mem:* Nat Acad Eng; fel Am Inst Aeronaut & Astronaut; fel Soc Exp Test Pilots. *Mailing Add:* PO Box 436 Lebanon OH 45036

ARMSTRONG, P(AUL) DOUGLAS, b New Albany, Ind, Jan 4, 41; m 70; c 2. HETEROCYCLIC CHEMISTRY, MEDICINAL CHEMISTRY. *Educ:* Ind Univ, BS, 63; Univ Iowa, PhD(org med chem), 68. *Prof Exp:* Res assoc chem under Prof John C Sheehan, Mass Inst Technol, 68-69; asst prof chem, Mass Col Pharm, 69-74; assoc prof, Calif Baptist Col, 74-76, chmn div natural sci, 75-76; asst prof chem, Carson-Newman Col, 76-78; chemist & qual control dir, Deltona Corp, 78-81; asst prof chem, Union Univ, 81-83; assoc prof chem, Pikeville Col, 83-85; assoc prof chem, 85-88, PROF CHEM, OLIVET NAZARENE UNIV, 88- *Concurrent Pos:* ACS acad-Indust Polymer Educ Prog, Tenn Eastman Co, 85; Smith Kline Beckman vis res fel, Univ Iowa, 87; consult, Armour Pharmaceut Co, 90- *Mem:* Am Chem Soc; affil Int Union Pure & Appl Chem; Int Soc Heterocyclic Chem. *Res:* Organic synthesis designed to produce novel types of compounds and/or to determine structure-activity relationships for the production of improved drugs. *Mailing Add:* Chem Dept Olivet Nazarene Univ Kankakee IL 60901-0592

ARMSTRONG, PETER BROWNELL, b Syracuse, NY, Apr 27, 39; m 62; c 3. CELL BIOLOGY, DEVELOPMENTAL BIOLOGY. *Educ:* Univ Rochester, BS, 61; Johns Hopkins Univ, PhD(biol), 66. *Prof Exp:* Asst prof zool, Univ Calif, Davis, 66-73; assoc prof, 73-79, PROF ZOOL, UNIV CALIF, DAVIS, 79 - *Concurrent Pos:* Assoc, Clare Hall, Cambridge Univ, 73-74; mem & trustee, Corp Marine Biol Lab, Woods Hole, Mass; regular mem, Cell Biol & Physiol study sect, NIH, 86-90. *Mem:* Am Soc Cell Biol; Soc Develop Biol; Brit Soc Develop Biol. *Res:* Physiological basis for cellular adhesion; mechanisms of morphogenetic movements; mechanisms of intercellular invasion; vitellogenesis; comparative hematology. *Mailing Add:* Dept Zool Univ Calif Davis CA 95616-8755

ARMSTRONG, PHILIP E(DWARD), b St Cloud, Minn, Sept 23, 27; m 51; c 4. PHYSICAL METALLURGY. *Educ:* Iowa State Univ, BS, 50, MS, 52, PhD(metall), 57. *Prof Exp:* Asst, Ames Lab, Iowa State Univ, 50-57; mem staff, 57-90, ASSOC MEM STAFF, LOS ALAMOS NAT LAB, 90- *Mem:* Am Soc Metals; Am Inst Mining, Metall & Petrol Engrs. *Res:* Elastic properties of refractory metals and graphites; mechanical properties of metals and alloys. *Mailing Add:* Los Alamos Nat Labs MS 734 Los Alamos NM 87545

ARMSTRONG, R(ONALD) W(ILLIAM), b Baltimore, Md, May 4, 34; m 58; c 2. MATERIALS SCIENCE, MECHANICAL ENGINEERING. *Educ:* Johns Hopkins Univ, BES, 55; Carnegie Inst Technol, MSc, 57, PhD(metall eng), 58. *Hon Degrees:* MA, Brown Univ, 66. *Prof Exp:* Brit Dept Sci & Indust Res grant, Univ Leeds, 58-59; res metallurgist, Res Labs, Westinghouse Elec Corp, 59-65; assoc prof eng, Brown Univ, 65-68; asst provost, div math phys sci & eng, 85, PROF MAT, UNIV MD, COLLEGE PARK, 68- *Concurrent Pos:* Commonwealth Sci & Indust Res Orgn fel, Australia, 64-65; sr Fulbright-Hays fel, Physics & Eng Lab, Dept Sci & Indust Res, Lower Hutt, NZ, 74; prog mgr, Div Sci Educ Resources Improvement, Directorate Sci Educ, Nat Sci Found, 76-77; NATO Advan Study Inst lectr, Univ Durham, England, 79; US-France coop sci prog lectr, Paris, 80 & Snowmass, Colo, 83; mat sci liaison officer, Europ Sci Off, Off Naval Res, London, 82-84; vis prof, Univ Strathclyde, Scotland, 82-85; vis fel, Clare Hall, Univ Cambridge, Eng, 84; NATO advan study inst contr, Crete, Greece, 85; sabbatical visitor, Cavendish Lab, Univ Cambridge, Eng, 90; liaison scientist, Off Naval Res, Europ Off, London, 91. *Honors & Awards:* Hardy Gold Medal, Am Inst Mining, Metall & Petrol Engrs, 62. *Mem:* Am Inst Mining, Metall & Petrol Engrs; fel Am Soc Metals; Am Cryst Assoc; Am Phys Soc; Mat Res Soc. *Res:* Material deformation and fracture; crystal x-ray diffraction; dislocation model calculations. *Mailing Add:* Mech Eng Dept Univ Md College Park MD 20740

ARMSTRONG, RICHARD LEE, b Seattle, Wash, Aug 4, 37; m 61; c 3. GEOCHRONOMETRY, TECTONICS. *Educ:* Yale Univ, BS, 59, PhD(geol), 64. *Prof Exp:* Actg instr, Yale Univ, 62-63; Nat Sci Found fel, Univ Berne, 63-64; from asst prof to assoc prof, Yale Univ, 70-73; assoc prof, 73-76, PROF GEOL, UNIV BC, 76- *Concurrent Pos:* Guggenheim fel, Canberra, Australia & Pasadena, Calif, 68-69; mem, Working Group 9, Int Geodynamics Comn, 72-80 & Can Nat Comt, 76-80; Killam fel, Univ BC, 79-80; chmn working group on radiogenic isotopes, Int Asn Volcanology & Chem of Earth's Interior, 84-87. *Honors & Awards:* Killam Res Prize, Univ BC, 86; Logan Medal, Geol Asn Can, 90. *Mem:* Geol Soc Am; Geol Asn Can; Geochem Soc; Int Asn Volcanology & Chem of Earth's Interior; Am Geophys Union; Royal Soc Can. *Res:* Geochronometry and regional geology of the Western United States and Canada; lead, strontium and neodymium isotope studies of igneous rocks; major and trace element studies of igneous rocks. *Mailing Add:* Geol Sci Dept Univ BC Vancouver BC V6T 1W5

ARMSTRONG, ROBERT A, b Midland, Ont, July 3, 29; m 56; c 3. SURFACE PHYSICS. *Educ:* Univ Toronto, BA, 52; McGill Univ, MS, 54, PhD(physics), 57. *Prof Exp:* Asst res officer, Nat Res Coun Can, 57-62, assoc res officer, 62-72, sr res officer, 72-90; HON SR SCIENTIST, UNIV OTTAWA, 90- *Mem:* Can Asn Physicists. *Res:* Surface physics; paramagnetic resonance and relaxation; high and low energy electron diffraction; physical adsorption; auger electron spectroscopy; scanning tunneling microscopy; ion scattering spectroscopy. *Mailing Add:* 1523 Chomley Cresc Ottawa ON K1G 0V9 Can

ARMSTRONG, ROBERT BEALL, b Hastings, Nebr, Nov 13, 40; m 66; c 3. PHYSIOLOGY. *Educ:* Hastings Col, BA, 62; Wash State Univ, MS, 70, PhD(exercise physiol), 73. *Prof Exp:* Asst prof biol & health sci, Boston Univ, 73-78; from assoc prof to prof physiol, Oral Roberts Univ, 78-85; prof phys ed, 85-90, RES PROF, UNIV GA, 90- *Concurrent Pos:* Assoc zool, Mus Comp Zool, Harvard Univ, 77-; prin investr, NIH grant, 75- *Mem:* Am Physiol Soc; fel Am Col Sports Med. *Res:* Skeletal muscle; utilization during locomotion and adaptation to exercise. *Mailing Add:* Dept Phys Educ Univ Ga Athens OK 30602

ARMSTRONG, ROBERT G, b Chicago, Ill, Mar 8, 28; m 52; c 2. POLYMER CHEMISTRY, ANALYTICAL CHEMISTRY. *Educ:* Univ Ill, BS, 49. *Prof Exp:* Lab asst, Loyola Univ, Ill, 53-54 & Northwestern Univ, 54-55; asst res chemist, Cent Res & Eng, Continental Can Co, Ill, 55-57, from res chemist to sr res chemist, Corp Res Develop, 57-67; RES SCIENTIST, PHILIP MORRIS INC, 67- *Mem:* AAAS; NY Acad Sci; Am Chem Soc. *Res:* Organic materials; analysis of and determination of physical-chemical properties of materials; high pressure liquid chromatography. *Mailing Add:* 10372 Ashburn Rd Richmond VA 23235-2604

ARMSTRONG, ROBERT JOHN, b Cropsey, Ill, Feb 12, 39. HORTICULTURE, GENETICS. *Educ:* Univ Ill, BS, 62, PhD(hort), 67; Univ Ariz, MS, 63. *Prof Exp:* RES HORTICULTURIST, LONGWOOD GARDENS, 67- *Concurrent Pos:* Adj asst prof, Longwood Grad Prog, Univ Del, 68- *Mem:* Am Soc Hort Sci; Am Genetic Asn; Ecol Soc Am. *Res:* New crops research and breeding of ornamental crops; study of the effects of growth regulators and other factors on new ornamental crops, and the development and maintenance of naturalized areas. *Mailing Add:* 319 W Barnard St West Chester PA 19380

ARMSTRONG, ROBERT LEE, b Xenia, Ohio, July 15, 39; m 63; c 2. BIOCHEMISTRY. *Educ:* Heidelberg Col, BS, 61; Ohio State Univ, MS, 63; Mich State Univ, PhD(biochem), 66. *Prof Exp:* NIH fel, Princeton Univ, 66-68; asst prof biochem, Univ Mich, 68-74; asst prof, 74-80, ASSOC PROF CHEM, ALBION COL, 80- *Concurrent Pos:* Sabbatical MSU, 88-89. *Mem:* Am Soc Microbiol; Sigma Xi; AAAS. *Res:* Spore germination; ribosome biosynthesis; antibiotic resistance; invitro protein syntheses; recombinant DNA. *Mailing Add:* Dept Chem Albion Col Albion MI 49224

ARMSTRONG, ROBIN L, b Galt, Ont, May 14, 35; m 60; c 2. NUCLEAR MAGNETIC RESONANCE, NEUTRON SCATTERING. *Educ:* Univ Toronto, BA, 58, MA, 59, PhD(physics), 61. *Prof Exp:* Nat Res Coun Can & Rutherford Mem fels, 61; from asst prof to assoc prof, 62-71, from assoc chmn dept to chmn dept, 69-82, PROF PHYSICS, UNIV TORONTO, 71-, DEAN FAC ARTS & SCI, 82- *Honors & Awards:* Herzberg Medal, Can Asn Physicists, 74 & Medal of Achievement, 90. *Mem:* Can Asn Physicists; Int Soc Magnetic Resonance; fel Royal Soc Can; Int Soc Magnetic Resonance Med. *Res:* Nuclear spin relaxation studies of molecular motions in gases and liquids; nuclear quadrupole resonance, neutron scattering and thermodynamic studies of lattice dynamics associated with structural phase transitions; in-vivo magnetic resonance imaging and spectroscopy of rodents; magnetic resonance imaging of condensed matter systems. *Mailing Add:* Pres Off Univ New Brunswick PO Box 4400 Fredericton NB E3B 5A3 Can

ARMSTRONG, ROSA MAE, b New Orleans, La, Apr 9, 37; m 58; c 2. HISTOLOGY, BONE PHYSIOLOGY. *Educ:* Southern Univ, BS, 56; Univ Calif, Berkeley, MA, 58; Univ Calif, San Francisco, PhD(anat), 64. *Prof Exp:* Lectr, 67-73, asst prof anat, Sch Med, Univ Calif, San Francisco, 73-81; sr res scientist, Collagen Corp, 81-91; SR RES SCIENTIST, CELTRIX LABS, 91- *Concurrent Pos:* NIH fel, Stanford Univ, 65-67. *Res:* Organ culture; embryonic rudiments; organ culture of human breast tumors; bone growth factors in vivo. *Mailing Add:* Celtrix Labs 2500 Faber Pl Palo Alto CA 94303

ARMSTRONG, THOMAS PEYTON, b Atchison, Kans, Nov 24, 41; m 62; c 1. SPACE PHYSICS. *Educ:* Univ Kans, BS, 62; Univ Iowa, MS, 64, PhD(physics), 66. *Prof Exp:* Res assoc space physics, Univ Iowa, 66-67; res assoc plasma physics, Culham Lab, UK Atomic Energy Auth, Eng, 67-68; from asst prof to assoc prof, 68-74, PROF PHYSICS, UNIV KANS, 74- *Concurrent Pos:* Br chief, Magnetospheric Physics, NASH Hq, 89-90. *Honors & Awards:* NASA Group Achievement for Voyager. *Mem:* AAAS; Am Phys Soc; Am Geophys Union. *Res:* Geomagnetically trapped radiation; solar cosmic ray events; magnetospheric physics; nonlinear waves in plasmas; interplanetary energetic particles and plasmas; Jovian and Saturnian magneto-spheres. *Mailing Add:* Dept Physics Univ Kans Lawrence KS 66044

ARMSTRONG, WILLIAM DAVID, b Clanton, Ala, Jan 15, 44. PHYSICAL CHEMISTRY. *Educ:* Univ Tenn, BSChem, 67; Univ Ga, MS, 70; Brown Univ, PhD(phys chem), 74. *Prof Exp:* Asst chem, Univ Ga, 67-70; res asst chem, Brown Univ, 70-74; res scientist catalysis, Am Cyanamid Co, 74-78; MEM STAFF, HALCON CATALYST INDUSTS, 78- *Mem:* Am Chem Soc; Sigma Xi. *Res:* Development of improved hydrotreating catalysts and reforming catalysts; development of selective oxidation catalyst. *Mailing Add:* 27 Lost Tree Lane Ramsey NJ 07446-1733

ARMSTRONG, WILLIAM LAWRENCE, b Lorain, Ohio, Jan 14, 39; m 75; c 3. ORGANIC CHEMISTRY. *Educ:* Oberlin Col, BA, 60; Univ Rochester, PhD(chem), 66. *Prof Exp:* Asst prof chem, State Univ NY Col Geneseo, 64-65; from asst prof to assoc prof, 65-78, PROF CHEM, STATE UNIV NY COL ONEONTA, 78-, CHMN DEPT, 85- *Mem:* Am Chem Soc; Royal Soc Chem. *Res:* Synthetic organic chemistry. *Mailing Add:* Dept Chem State Univ NY Oneonta NY 13820

ARNAL, ROBERT EMILE, b Orleans, France, Sept 10, 22; nat US; m 54; c 3. GEOLOGY. *Educ:* Univ Poitiers, France, BS, 46; Univ Nancy, France, MS, 48; Univ Southern Calif, PhD(geol), 57. *Prof Exp:* Geophysicist, Cie Gen de Geophysique, France, 48-51; paleontologist, Western Gulf Oil Co, 53-58; asst prof phys sci, San Jose State Col, 58-61, assoc prof geol, 61-69; prof oceanog, Moss Landing Marine Labs, 69-80; MEM STAFF, US GEOL SURV, 80- *Mem:* Fel Geol Soc Am; Am Asn Petrol Geol; Soc Econ Paleontologists & Mineralogists. *Res:* Shallow water oceanography; micropaleontology environmental studies of recent organisms and tertiary fossils. *Mailing Add:* US Geol Surv 170 El Caminito Rd Carmel Valley CA 93924

ARNAOUT, M AMIN, b Sidon, Lebanon, Aug 8, 49; US citizen. NEPHROLOGY, IMMUNOLOGY. *Educ:* Am Univ, Beirut, BS, 69, MD, 74. *Prof Exp:* Resident internal med, Am Univ Hosp, Beirut, 74-76; fel immunol, 76-78, fel nephrology, 78-79, instr, 79-82, asst prof, 82-87, ASSOC PROF PEDIAT, CHILDREN'S HOSP, HARVARD MED SCH, 87- *Mem:* Am Asn Immunologists; Am Soc Nephrology; Int Soc Nephrology; Soc Pediat Res; Am Soc Clin Invest. *Res:* Structure and function of leukosyte membrane molecules involved in adhesion and inflammation. *Mailing Add:* Children's Hosp Renal Div 300 Longwood Ave Boston MA 02115

ARNAS, OZER ALI, b Izmir, Turkey, Sept 1, 36; m 60; c 3. THERMODYNAMICS, HEAT TRANSFER. *Educ:* Robert Col, Istanbul, BSME, 58; Duke Univ, MSME, 61; NC State Univ, PhD(mech eng), 65. *Prof Exp:* From asst prof to assoc prof, 62-68, PROF MECH ENG, LA STATE UNIV, BATON ROUGE, 70- *Concurrent Pos:* Consult, Univ Houston mech eng prog, Univ Costa Rica & consult lectr, Cath Univ Am, 67-69; vis prof, Univ Liege, Belg, 72-73. *Honors & Awards:* Halliburton Educ Found Award, 67. *Mem:* Am Soc Mech Engrs; Am Soc Eng Educ. *Res:* Thermodynamics; heat transfer; direct energy conversion; appropriate technology transfer. *Mailing Add:* Mech Eng Dept San Jose State Univ San Jose CA 95192

ARNASON, BARRY GILBERT WYATT, b Winnipeg, Man, Aug 30, 33; m 60; c 3. IMMUNOLOGY, NEUROLOGY. *Educ:* Univ Man, MD, 57. *Prof Exp:* Intern, Winnipeg Gen Hosp, Man, 56-57, asst resident internal med, 56-58; Nat Mult Sclerosis Soc prin investr, Lab of Dr Pierre Grabar, Inst Sci Res Cancer, Villejuif, France, 62-64; instr, Harvard Med Sch, 64-67, assoc, 67, from asst prof to assoc prof, neurol, 68-76; PROF & CHMN DEPT NEUROL, UNIV CHICAGO, 76-, DIR, BRAIN RES INST, 85- *Concurrent Pos:* Teaching fel neurol, Harvard Med Sch, 58-59 & 61-62, res fel, 59-61; clin & res fel, Lab of Dr Byron H Waksman, 59-61, asst, 64-66, asst neurologist, 66-68, clin fel, Hosp, 61-62; asst resident neurol, Mass Gen Hosp, 58-59, assoc neurologist, 71-76. *Honors & Awards:* Dr Jon Stefansson Mem Prize, 55-56. *Mem:* Am Neurol Asn; Am Acad Neurol; Am Soc Clin Invest; Am Asn Neuropathologists; Am Asn Immunologists. *Res:* Neuroimmunology and multiple sclerosis. *Mailing Add:* Dept Neurol Univ Chicago 5841 S Maryland Ave Chicago IL 60637

ARNAUD, CLAUDE DONALD, JR, b Hackensack, NJ, Dec 4, 29; m 87; c 2. ENDOCRINOLOGY, PHYSIOLOGY. *Educ:* Columbia Col, BA, 51; NY Med Col, MD, 55. *Prof Exp:* Assoc prof med, Mayo Grad Sch Med, 70-74, prof, Mayo Med Sch, 74-77; PROF MED & PHYSIOL, UNIV CALIF, SAN FRANCISCO, 77- *Concurrent Pos:* Head mineral res endocrinol, Mayo Clin, 72-74, endocrine res, 74-77; chief endocrine unit, Vet Admin Med Ctr, San Francisco, 77-89; chief, Div Gerontol, Univ Calif, San Francisco, 89-, dir Ctr Biomed Res on Aging, 89- *Mem:* Am Physiol Soc; Am Soc Biol Chemists; Am Soc Clin Invest; Asn Am Physicians; Endocrine Soc. *Res:* Minerals and bones. *Mailing Add:* Div Gerontol & Geriatric Med 1710 Scott St Third Floor San Francisco CA 94115

ARNAUD, PAUL HENRI, JR, b San Francisco, Calif, Sept 15, 24; m 70; c 2. ENTOMOLOGY. *Educ:* San Jose State Col, AB, 49; Stanford Univ, MA, 50, PhD(biol), 61. *Prof Exp:* Syst entomologist, Bur Entom, Calif Dept Agr, 50-52, 55-56 & 57-59; prin investr med entom, Stanford Univ & Off Surgeon Gen, US Army, Japan, 54-55; entomologist, USDA at US Nat Mus, Washington, DC, 56-57; res entomologist, Calif Acad Sci, 59-61; res fel entom, Am Mus Natural Hist, NY, 61-62; res entomologist, Calif Acad Sci, 63-64, asst cur, 64-65, assoc cur, 65-72, chmn dept, 68-78, CUR ENTOM, CALIF ACAD SCI, 72- *Concurrent Pos:* Mem, Sefton Found Exped, Gulf Calif, 53 & entom expeds, Baja Calif & Mex, 53, 59 & 63; grants, NSF, 60 & 72-80, USDA, 66-68 & Am Philos Soc Penrose Fund, 72 & 76; assoc, Exp Sta, Univ Calif, Albany, 66- & Div Biol Control, Univ Calif, Berkeley, 66- *Mem:* Soc Syst Zool; Entom Soc Am; Am Entom Soc. *Res:* Classification; biology; systematics and distribution of Diptera; zoogeography; evolution. *Mailing Add:* Calif Acad Sci Golden Gate Park San Francisco CA 94118

ARNAUD, PHILIPPE, b Saint-Etienne, France, Mar 30, 35; m 61; c 3. IMMUNOCHEMISTRY, PROTEIN PURIFICATION. *Educ:* Faculte Des Sci, Lyon, France, BS, 55; Univ Claude Bernard, MD, 69; Med Univ SC, PhD(immunol), 77. *Prof Exp:* Resident, Med Univ, Lyon, France, 64-68; instr biochem, Univ Claude Bernard, France, 66-69, asst prof, 70-75, chef de travaux, 75-77; from asst prof to prof immunol, Med Univ SC, 77-84, dir, res & develop, Dept Immunol, 85, dir, Lab Molecular Genetics, 88; vis scientist, Biochem Lab, Nat Cancer Inst, NIH, 85-87; CONSULT, 87- *Concurrent Pos:* Adj prof molecular genetics, Univ Genova, Italy; inter govt personnel act, Nat Cancer Inst, 85 & 86. *Mem:* AAAS; Am Soc Human Genetics; Am Asn Immunologists; Electrophoresis Soc. *Res:* New techniques for protein isolation and purification--microheterogeneity and poly morphism of proteins and their structural basis; effect of acute-phase proteins on the immune response; regulation of acute-phase proteins; gene expression. *Mailing Add:* Dept Microbiol & Immunol Lab Molecular Genetics Med Univ SC 171 Ashley Ave Charleston SC 29425-2230

ARNBERG, ROBERT LEWIS, b San Francisco, Calif, Mar 19, 45; m 67; c 1. NONCOMMUNICATIVE RINGS, HARMONIC ANALYSIS. *Educ:* Univ Calif, Berkeley, AB, 67; Univ Ore, MA, 69. *Prof Exp:* Sr programmer, Consult & Designers, 75-76; teacher math, George Washington Univ, 76-78; systs analyst, Data Transformation Corp, 78-79; programmer-analyst, Nat Econ Res Assoc, 79-80; res fel, Logistics Mgt Inst, 80-88; OR ANALYST, US GEOL SURV, 89- *Concurrent Pos:* Consult, 76-78; book-film reviewer, AAAS, 76-; adj asst prof computer sci-mgt, Univ Col, Univ Md, 88-89. *Mem:* Am Math Soc; Math Asn Am; Soc Indust & Appl Math; AAAS; NY Acad Sci; Int Cong Mathematicians. *Res:* Design and development of scientific and statistical computer applications including energy modelling, logistical modelling and financial modelling; computer systems management; management control analysis of administrative information systems; management and operations of local and circle area networks; optimization theory. *Mailing Add:* 11635 Stoneview Sq No 1B Reston VA 22091

ARNDT, HENRY CLIFFORD, b Caldwell, Idaho, Sept 3, 45; m 69; c 2. ORGANIC CHEMISTRY. *Educ:* Univ Idaho, BS, 67; Univ Wis-Madison, MS, 72, PhD(org chem), 76. *Prof Exp:* Asst res scientist, 72-73, assoc res scientist, 73-76, RES SCIENTIST CHEM, MILES LABS INC, 76- *Mem:* Am Chem Soc. *Res:* Total synthesis of natural products and prostaglandin analogs; new methods in organic synthesis. *Mailing Add:* Miles Labs Inc PO Box 40 Elkhart IN 46515-0040

ARNDT, RICHARD ALLEN, b Cleveland, Ohio, Jan 3, 33; m 53; c 4. PHYSICS. *Educ:* Case Inst Technol, BS, 57; Univ Calif, Berkeley, MA, 62, PhD(physics), 65. *Prof Exp:* Engr, Northrop Corp, Calif, 57-59; physicist, Lawrence Radiation Lab, Univ Calif, 59-67; from asst prof to assoc prof, 67-78, PROF PHYSICS, VA POLYTECH INST & STATE UNIV, 78- *Res:* Elementary particle physics; dispersion relations in nucleon-nucleon scattering; single boson exchange contribution to nucleon-nucleon scattering. *Mailing Add:* 213 Pewter Lane Silver Springs MD 20904

ARNDT, ROGER EDWARD ANTHONY, b New York, NY, May 25, 35; div; c 2. HYDROMECHANICS, HYDRAULIC ENGINEERING. *Educ:* City Col New York, BCE, 60; Mass Inst Technol, SM, 62, PhD, 67. *Prof Exp:* Chemist, Consol Testing Labs, 56-57; res asst hydrodyn, Mass Inst Technol, 60-62; res engr, Allegany Ballistics Lab, 62-63; sr res engr, Lockheed-Calif Co, 63-64; asst prof aerospace eng, 67-71, assoc prof, Pa State Univ, 71-77; DIR & PROF, ST ANTHONY FALLS HYDRAUL LAB, UNIV MINN, 77- *Concurrent Pos:* Consult, Army Aeronaut Res Lab, Ill Inst Technol Res Inst & Bell Aerospace Corp, Richards of Rockford, Yarway Corp, Wylie Labs, Off Naval Res-London, Netherlands Ships Model Basin, Honeywell Corp, Bolt Beranek & Newman, Inc, US Bur Reclamation, Babcock & Wilcox Co, Thermal Systs Inc, ScMed Inc, Indeco, Inc, Metals Selling Corp, Henkel Corp, UN; ASME fel; Assoc fel, AIAA. *Honors & Awards:* First Theodore Ranov Dist Lectr; Lorenz G Straub Award; Alexander Von Humboldt Sr Scientist Award. *Mem:* Am Inst Aeronaut & Astronaut; Acous Soc Am; Am Soc Mech Engrs; Am Soc Civil Engrs; Int Asn Hydraul Res. *Res:* Cavitation in turbulent shear flows both Newtonian and non-Newtonian liquids being considered; hydraulic turbine hydrodynamics and hydropower; aerodynamic noise, especially jet noise and noise induced by other turbulent shear flows. *Mailing Add:* St Anthony Falls Hydraul Lab Mississippi River & 3rd Av SE Minneapolis MN 55414

ARNELL, JOHN CARSTAIRS, b Halifax, NS, Apr 4, 18; m 42, 75; c 3. PHYSICAL CHEMISTRY. *Educ:* Dalhousie Univ, BSc, 39, MSc, 40; McGill Univ, PhD(phys chem), 42. *Prof Exp:* Supt, Defence Res Chem Labs, 49-54, dir sci intel, 55-58, dir plans, 58; sci adv, Royal Can Air Force, 58-63 & Royal Can Navy, 63-64; sci dep chief logistics, eng & develop, Can Forces Hq, Ottawa, 64, sci dep chief tech serv, 64-66; asst dep minister finance, Can Dept Nat Defence, Nat Defence Hq, 66-73; prog consult, Finance Ministry, 73-76, coordr metrication, Govt Bermuda, 75-76; RETIRED. *Mem:* Fel Chem Inst Can; assoc fel Can Aeronaut & Space Inst; Sigma Xi. *Res:* Overall management of planning, programming budgeting and financial management systems for defence. *Mailing Add:* PO Box HM 1263 Hamilton HMFX Bermuda

ARNELL, WALTER JAMES, b Farnborough, Eng, Jan 9, 24; US citizen; m 55; c 3. SYSTEMS & INDUSTRIAL ENGINEERING. *Educ:* Univ London, BSc, 53, PhD, 67; Occidental Col, MA, 56; Univ Southern Calif, MS, 58. *Prof Exp:* Lectr mech eng, City Univ & London Polytech Cols, 48-53; instr gen & aeronaut eng, Univ Southern Calif, 54-59; from asst prof to assoc prof mech eng, Calif State Col, Long Beach, 59-66, chmn dept, 64-65, actg chmn, Div Eng, 64-66, prof mech eng, 66-71, dean, 67-69; researcher, Ctr Eng Res & mem grad affil fac, Dept Ocean Eng, Col Eng, Univ Hawaii, 70-76; PRES & CONSULT, LENRA ASSOCS LTD, 77- *Concurrent Pos:* Dir Cannon Elec Co, Calif, 57-63; consult, Human Factors Inc, 58, Arrowhead Prod, 62-63, Taper Tubes, Inc, 64-65 & Northrup Inst Technol, 69-70; chmn & co-dir, Hawaii Simulation Lab, 71-72; mem bd dirs, Rehab Hosp of the Pac, Honolulu, 75-78; adj prof systems & indust eng, Univ Ariz, 81- *Mem:* Am Inst Aeronaut & Astronaut; Human Factors Soc; Inst Elec & Electronics Engrs; Brit Psychol Soc; Am Psychol Asn; Royal Aeronaut Soc; Ergonomics Soc. *Res:* Man-machine systems; human performance in systems; human factors. *Mailing Add:* 4491 E Ft Lowell Rd Tucson AZ 85712

ARNEMAN, HAROLD FREDERICK, b Mankato, Minn, May 31, 15; c 1. SOILS. *Educ:* Univ Minn, BS, 39, MS, 46, PhD, 50. *Prof Exp:* Soil surveyor, Univ Minn, 35-40; farm credit apprentice, Fed Land Bank, Minn, 40-42; from instr to assoc prof, 46-61, prof, 61-80, EMER PROF SOILS, UNIV MINN, ST PAUL, 80- *Res:* Mineralogical and physical studies of soils. *Mailing Add:* 1403 California Ave Falcon Heights St Paul MN 55108

ARNER, DALE H, b Weissport, Pa, Feb 10, 20; m 45; c 1. WILDLIFE MANAGEMENT. *Educ:* Pa State Univ, BS, 49, MS, 54; Auburn Univ, PhD(wildlife mgt), 59. *Prof Exp:* Hatchery asst, Pa Fish Comn, 41; wildlife field supt, Md Game & Inland Fish Comn, 48-54; biologist, Soil Conserv Serv, USDA, Ala, 57-62; from asst prof to assoc prof wildlife ecol, Miss State Univ, 62-68, prof wildlife, fisheries & zool & head Dept Wildlife & Fisheries, 68-87, emer prof & head dept, 87-; RETIRED. *Concurrent Pos:* Consult, Asplundh, Fish & Wildlife Serv, Environ Protection Agency, Dept Interior, 76-78. *Honors & Awards:* C W Watson Award, Am Fisheries Soc/Wildlife Soc. *Mem:* Wildlife Soc; Audubon & Am Fisheries Soc. *Res:* Use of mechanical equipment, fire fertilizer and seed in utility line right-of-way development for wildlife and reduction of woody vegetation; beaver-pond and duck ecology; wild turkey nutrition. *Mailing Add:* 209 Seville Pl Starkville MS 39759

ARNESON, DORA WILLIAMS, b Fayetteville, Ark, Aug 4, 47; div. ANALYTICAL CHEMISTRY, BIOCHEMISTRY. *Educ:* Univ Mo, BA, 67, PhD(anal chem), 72. *Prof Exp:* Instr anal biochem, Ctr Health Sci, Univ Tenn, Memphis, 72-77; actg chief, Univ Tenn Ctr Health Sci, Memphis, 77-79, assoc dir clin biochem, child develop ctr, 79-83, asst prof biochem & pediat, Col Med, 77-83; SR CHEMIST, MIDWEST RES INST, KANSAS CITY, MO, 84- *Mem:* Am Chem Soc; Am Asn Clin Chem; Soc Inherited Metab Disorders. *Res:* Development of analytical methods applicable to toxicology studies. *Mailing Add:* Midwest Res Inst 425 Volker Blvd Kansas City MO 64110

ARNESON, PHIL ALAN, b Fergus Falls, Minn, May 17, 40; m 63; c 2. PLANT PATHOLOGY. *Educ:* Carleton Col, BA, 62; Univ Wis, PhD(plant path), 67. *Prof Exp:* Plant pathologist, Trop Res Div, United Fruit Co, 67-70; asst prof, 70-76, assoc prof plant path, Cornell Univ, 76-; AT DEPT NUTRIT SERV, CORNELL UNIV. *Mem:* Am Phytopath Soc. *Res:* Orchard replant problems; soil fumigation; nematicides; integrated pest management; computer simulation modeling; pest management curriculum development. *Mailing Add:* Plant Path Cornell Univ Main Campus 334 Plant Sci Bldg Ithaca NY 14853

ARNESON, RICHARD MICHAEL, b Fergus Falls, Minn, Oct 18, 38; m 73. BIOCHEMISTRY. *Educ:* Univ Minn, BA, 60; Duke Univ, PhD(biochem), 71. *Prof Exp:* From instr to asst prof biochem, Ctr Health Sci, Univ Tenn, Memphis, 71-79; biochemist, Oak Ridge Assoc Univs, 79-80; fel, Oak Ridge Nat Lab, 81 & Univ Ga, Athens, 82; Mid-Career Develop fel environ toxicol, Univ Iowa, Iowa City, 82-84; fel, Minneapolis Med Res Found, 85; RETIRED. *Mem:* Sigma Xi. *Res:* Free radical biochemistry. *Mailing Add:* 5440 James Ave S Minneapolis MN 55419-1607

ARNETT, EDWARD MCCOLLIN, b Philadelphia, Pa, Sept 25, 22; m 51, 69; c 5. PHYSICAL ORGANIC CHEMISTRY. *Educ:* Univ Pa, BA, 43, MS, 47, PhD(chem), 49. *Prof Exp:* Asst instr qual, inorg, anal & org chem, Univ Pa, 43-46, 47; res dir, Max Levy & Co, 49-53; asst prof, Western Md Col, 53-55; res fel, Harvard Univ, 55-57; from asst prof to assoc prof gen & org chem, Univ Pittsburgh, 57-64, prof, 64-80; R J REYNOLDS INDUSTS PROF, DUKE UNIV, 80- *Concurrent Pos:* Vis lectr, Univ Ill, 63; adj sr fel, Mellon Inst, 64-; distinguished lectr, Howard Univ, 66 & Guggenheim fel, 68-69; dir, Pittsburgh Chem Info Ctr, 68-; mem adv bd, Petrol Res Fund, 68-; comt chem info, Nat Res Coun, 69-; vis prof, Univ Kent, 70; sr fel, Inst Hydrocarbon Chem, Univ Southern Calif, 80-; A C cope scholar award, Am Chem Soc, 90. *Honors & Awards:* James Flack Norris Award, 77; Pittsburgh Award, 76; Kelly lectr, Purdue Univ, 77; Petrol Chem Award, Am Chem Soc, 85. *Mem:* Nat Acad Sci; Fel AAAS; Am Chem Soc; Royal Soc Chem. *Res:* Acid base behavior of organic compounds; solvent effects in organic chemistry; organic monolayers; stereochemistry of aggregation. *Mailing Add:* Dept Chem Duke Univ Durham NC 27706

ARNETT, JERRY BUTLER, b Pike Co, Ohio, July 19, 38; m 62; c 2. GENERAL ENGINEERING, PHYSICS. *Educ:* Ohio Univ, BS, 60, MS, 63; NMex State Univ, PhD(Physics), 67. *Prof Exp:* Res & teaching asst physics, Ohio Univ, 60-62; res engr, Los Angeles Div, NAm Aviation, 62-64; res & teaching asst physics, NMex State Univ, 64-67; sr res scientist remote sensing & space physics, Gen Dynamics/Ft Worth & Gen Dynamics/Convair, 67-73; staff scientist, Technol Inc, Dayton, Ohio, 73-76; radar engr, BDM Serv Inc, Dayton, 76-77; electronic engr, 77-80, SYSTS STUDY ENGR, US AIR FORCE AERONAUT SYSTS DIV, WRIGHT-PATTERSON AFB, 80- *Concurrent Pos:* Instr, Tex Wesleyan Col, 70 & Tex Christian Univ, 70-71. *Mem:* Sigma Xi; Air Force Asn. *Res:* Innovative systems and applications directed toward near term and future Air Force aeronautical and transatmospheric missions; operational analyses of future weapon systems. *Mailing Add:* ASD/XRM Wright-Patterson AFB OH 45433-6503

ARNETT, ROSS HAROLD, JR, b Medina, NY, Apr 13, 19; m 42; c 8. ENTOMOLOGY. *Educ:* Cornell Univ, BS, 42, MS, 46, PhD(entom), 48. *Prof Exp:* Asst entom, Cornell Univ, 46-48; entomologist, Div Insects, USDA, 48-54; head dept biol, St John Fisher Col, 54-58; from assoc prof to prof biol, Cath Univ, 58-66, head dept, 63-66; prof entom, Purdue Univ, 66-70; res biologist, Tall Timbers Res Sta, 71-73; prof biol, Siena Col, NY, 73-79; prof biol & dir oxycopis, Pond Res Sta, Kanderhook, NY, 79-82; PROF ENTOM, UNIV FLA, GAINESVILLE, 82- *Concurrent Pos:* Asst, State Conserv Dept, NY, 42; instr, US Army Sch Malariol, CZ, 44-45; founder & ed, Coleopterists Bull; exec dir, The Biol Res Inst Am, Inc, 73-80; founder & ed, Insecta Mundi, 85- *Honors & Awards:* Hon Mem, Entom Soc Am, 87. *Mem:* AAAS; Am Soc Zool; Soc Syst Zool; Soc Study Evolution; Entom Soc Am; Am Inst Biol Sci; Nat Asn Biol Teachers; Bot Soc Am. *Res:* Taxonomy of Coleoptera; biology of mosquitoes; Onychophora; revision of world Oedemeridae; experimental systematic biology; desert biology. *Mailing Add:* 2406 NW 47th Terr Gainesville FL 32606

ARNETT, WILLIAM DAVID, b Williamson, WVa, May 2, 40; c 2. ASTROPHYSICS, COMPUTATIONAL PHYSICS. *Educ:* Univ Ky, BS, 61; Yale Univ, MS, 63, PhD(physics), 65. *Prof Exp:* NASA-Nat Res Coun resident res associateship, Goddard Inst Space Studies, New York, 65-67; res fel physics, Kellogg Radiation Lab, Calif Inst Technol, 67-69; asst prof space sci, Rice Univ, 69-71; assoc prof astron & physics, Univ Tex, Austin, 71-74; prof astrophys, Univ Ill, Urbana, 74-76; prof astrophys, Univ Chicago, 76-88; distinguished serv prof, Enrico Fermi Inst, 76-88; PROF ASTROPHYS, UNIV ARIZ, 88- *Concurrent Pos:* Consult, Lawrence Radiation Lab, 66-67; fel, Alfred P Sloan Found, 70-72; comt space astron & astrophys, Nat Acad Sci/Nat Res Coun. *Mem:* Nat Acad Sci; Am Astron Soc; Am Phys Soc; Int Astron Union. *Mailing Add:* Stuart Observ Univ Ariz Tucson AZ 85721

ARNETT, WILLIAM HAROLD, b Louisville, Miss, June 13, 28; m 57; c 3. ECONOMIC ENTOMOLOGY. *Educ:* Miss State Univ, BS, 55, MS, 57; Kans State Univ, PhD(entom), 60. *Prof Exp:* Instr entom, Kans State Univ, 57-60; asst prof & asst entomologist, 60-65, assoc prof & assoc entomologist, 65-74, PROF & ENTOMOL, UNIV NEV, RENO, 74- *Mem:* AAAS; Entom Soc Am; Sigma Xi. *Res:* Biology, ecology and control of agricultural insect pests; forage and animal insects; pollinators; pollination. *Mailing Add:* Col Agr Sect of Entomol Univ Nev Reno NV 89507

ARNFIELD, ANTHONY JOHN, b Nuneaton, Eng, Jan 2, 45; m 69. PHYSICAL GEOGRAPHY, CLIMATOLOGY. *Educ:* Univ Wales, BA, 66; McMaster Univ, MA, 68, PhD(geog), 73. *Prof Exp:* From asst prof to assoc prof, 72-90, PROF GEOG, OHIO STATE UNIV, 90- *Mem:* Royal Meteorol Soc; Am Meteorol Soc; Inst Brit Geographers; Asn Am Geographers; Asn Brit Clomatologists. *Res:* Microclimatology; radiation climatology; investigation of the fluxes of energy, mass and momentum across the earth-atmosphere interface, especially as determined by surface properties; urban surface energy budgets; urban climatology. *Mailing Add:* Dept Geog Ohio State Univ 103 Admin Bldg 190 N Oval Mall Columbus OH 43210-1361

ARNHEIM, NORMAN, b New York, NY, Dec 24, 38; m 68; c 2. MOLECULAR BIOLOGY, EVOLUTION. *Educ:* Univ Rochester, BA, 60, MS, 62; Univ Calif, Berkeley, PhD(genetics), 66. *Prof Exp:* Res biochemist, Dept Biochem, Univ Calif, Berkeley, 67-68; asst prof biochem, 68-74, ASSOC PROF BIOCHEM, STATE UNIV NY, STONY BROOK, 74- *Mem:* Genetics Soc Am; Am Soc Biol Chemists; Am Soc Naturalists. *Res:* Molecular biology; multi-gene evolution. *Mailing Add:* Univ SCalif Los Angeles CA 90089-0371

ARNHEITER, HEINZ, b St Gallen, Switz, May 2, 51. EXPERIMENTAL BIOLOGY. *Educ:* Kantonsschule St Gallen, BA, 70; Univ Zurich, MD, 78. *Prof Exp:* Res fel, Inst Immunol & Virol, Univ Zurich, Switz, 77-81, Lab Molecular Genetics, NIH, 81-82, Fogarty Int fel, Nat Inst Neurol & Commun Dis & Stroke, 82-83, vis assoc, 83-84; oberassistent, Inst Immunol & Virol, Univ Zurich, 84-86; vis scientist, Lab Molecular Genetics, Nat Inst Neurol & Commun Disorders & Stroke, 86-88, CHIEF, VIRAL PATHOGENESIS SECT, LAB VIRAL & MOLECULAR PATHOGENESIS, NAT INST NEUROL DIS & STROKE, NIH, 88- *Concurrent Pos:* Fel, Int Union Against Cancer, 79; res grant, Swiss Nat Sci Found, 84; ad hoc mem, team prog proj grant, Scripps Clin, 89; ad hoc consult, Neurobiol Dis Prog Proj Rev A Comt, 89. *Mem:* Am Soc Cell Biol; Am Soc Microbiol; Am Soc Virol; Am Soc Interferon Res; AAAS. *Mailing Add:* Lab Viral & Molecular Pathogenesis Nat Inst Neurol Dis & Stroke NIH 9000 Rockville Pike Bldg 36 Rm 5D04 Bethesda MD 20892

ARNHOLT, PHILIP JOHN, b Danville, Ill, May 16, 40; m 64; c 2. BOTANY. *Educ:* Eastern Ill Univ, BSEd, 63, MSEd, 67; Univ Nebr, PhD(bot), 73. *Prof Exp:* Teacher biol, Dixon High Sch, Ill, 63-67 & LaSalle-Peru Twp High Sch, Ill, 67-68; asst prof, 71-81, PROF BIOL, CONCORDIA UNIV, WIS, 89-, DIV CHMN BIOL, 85- *Mem:* Bot Soc Am; Nat Asn Biol Teachers; Nat Audubon Soc. *Res:* Morphology and histochemistry of grasses. *Mailing Add:* 10910 San Marino Megnon WI 53092

ARNISON, PAUL GRENVILLE, b Clatterbridge, Eng, Aug 9, 49; Can citizen. MOLECULAR BIOLOGY. *Educ:* McGill Univ, BSc, 70, PhD(plant physiol), 75. *Prof Exp:* Asst prof biol, Northeastern Univ, 75-78; asst prof, Dept Biol, St Francis Xavier Univ, NS, 78-79; asst prof to assoc prof, Dept Biol, Univ New Brunswick, NS, 79-81; supvr agr res, Allied Chem Can, Ltd, 81-85; res dir, 85-90, GEN MGR, PALADIN HYBRIDS, INC, 90- *Mem:* AAAS; Am Soc Plant Physiologists. *Res:* Anther and microspore culture of Brassica Napus, Campestris and Oleracea; microscope specific gene characterization; gene regulation via antisense RNA; use of doubled-haploid hybrids in crop improvement; transformation of plants; Brassica heat shock proteins; development of new hybrid systems. *Mailing Add:* Paladin Hybrids Inc 2475 Don Reid Dr Ottawa ON K1H 8P5 Can

ARNO, STEPHEN FRANCIS, b Seattle, Wash, Oct 1, 43; m 64; c 2. FOREST ECOLOGY. *Educ:* Wash State Univ, BS, 65; Univ Mont, MF, 66, PhD(forestry, plant sci), 70. *Prof Exp:* Park ranger & naturalist, US Nat Park Serv, 63-65; instr forestry, Univ Mont, 67; forester, 70-71, RES FORESTER, US FOREST SERV, 71- *Mem:* Ecol Soc Am; Soc Am Foresters. *Res:* Phytosociology; forest succession; phytogeography; silvicultural use of prescribed fire in western forests; dendrology; effects of fire on forest and range vegetation. *Mailing Add:* Intermountain Fire Sci Lab PO Box 8089 Missoula MT 59807-8089

ARNOFF, E LEONARD, b Cleveland, Ohio, Oct 15, 22; m 48; c 2. APPLIED MATHEMATICS. *Educ:* Western Reserve Univ, BS, 43; Case Inst Technol, MS, 48; Calif Inst Technol, PhD(math), 51. *Prof Exp:* Instr math, Case Inst Technol, 46-48; hydrodynamicist & mathematician, Naval Ord Test Sta, 50-51; mathematician & aeronaut res scientist, Nat Adv Comt Aeronaut, 51-52; assoc prof opers res & asst dir opers res group, Case Inst Technol, 52-61; prin & dir mgt sci, Ernst & Whinney, 60-83; DEAN, COL BUS ADMIN, UNIV CINCINNATI, 83- *Concurrent Pos:* Ed, Mgt Sci, Inst Mgt Sci, 55-70. *Mem:* Fel AAAS; fel Opers Res Soc Am; Math Asn Am; Am Math Soc; Inst Mgt Sci(assoc secy, 62, secy-treas, 63, vchmn, 64, int pres, 68-69). *Res:* Mathematical programming; production and inventory control; development and application of operations research methods, techniques and tools to new problem areas. *Mailing Add:* 80 Carpenters Ridge Cincinnati OH 45241

ARNOLD, ALLEN PARKER, b Oshkosh, Wis, June 25, 23; m 48; c 2. ORGANIC CHEMISTRY. *Educ:* Oberlin Col, AB, 44; Case Inst Technol, MS, 57, PhD(chem), 58. *Prof Exp:* Jr anal chemist, Shell Chem Corp, Calif, 44-45 & 46-47; res assoc org chem, Case Inst Technol, 47-48; res chemist, Cleveland Indust Res, Inc, Euclid, 48-62, asst lab dir, 62-72; group leader res & develop, Lubrizol Corp, 72-84; CONSULT 84- *Concurrent Pos:* Adj asst prof chem, Univ NC, Asheville; fuel additive sem, India, 87. *Mem:* Am Chem Soc; Sigma Xi. *Res:* Organic synthesis, particularly phosphorus chemistry. *Mailing Add:* 44 Wagon Trail Black Mountain NC 28711-2533

ARNOLD, ARTHUR PALMER, b Philadelphia, Pa, Mar 16, 46; m 67; c 2. NEUROBIOLOGY. *Educ:* Grinnell Col, AB, 67; Rockefeller Univ, PhD(neurobiol), 74. *Prof Exp:* Instr psychol, Cent State Univ, Ohio, 68-69; from asst prof to assoc prof, 76-83, PROF PSYCHOL, UNIV CALIF, LOS ANGELES, 83- *Mem:* Soc Neurosci; AAAS. *Res:* Sex differences in the brain; steroid hormone effects on neurons. *Mailing Add:* Dept Psychol Univ Calif Los Angeles CA 90024-1563

ARNOLD, BRADFORD HENRY, b Chehalis, Wash, Oct 14, 16; m 41. PURE MATHEMATICS. *Educ:* Univ Wash, BS, 38, MS, 40; Princeton Univ, PhD(math), 42. *Prof Exp:* Instr math, Purdue Univ, 42-44, 46; major aerodynamicist, Boeing Airplane Co, 44-46; asst prof, Mont State Col, 46-47; from asst prof to prof math, 47-81, asst chmn dept, 72-79, EMER PROF MATH, ORE STATE UNIV, 82- *Concurrent Pos:* Fulbright grants, Iraq, 57-58, China, 62-63 & 69-70. *Mem:* Am Math Soc. *Res:* Topology and abstract algebra. *Mailing Add:* 125 NW 32nd St Corvallis OR 97330

ARNOLD, C(HARLES) W(ILLIAM), b Goree, Tex, June 13, 23; m 51; c 5. PETROLEUM ENGINEERING. *Educ:* Univ Okla, BS, 44; Univ Tex, MS, 50, PhD(chem eng), 54. *Prof Exp:* Prin & teacher pub sch, Tex, 45-46; res engr, Humble Oil & Refining Co, 54-59, res supvr, 59-64, res mgr, 64-76, SR RES ADV, EXXON PROD RES CO, 76- *Mem:* Am Chem Soc; Soc Petrol Engrs; Am Inst Mining, Metall & Petrol Engrs; Sigma Xi. *Res:* Petroleum production research; enhanced oil recovery. *Mailing Add:* 4609 Redstart Houston TX 77035

ARNOLD, CHARLES, JR, b Brooklyn, NY, May 29, 30; m 59; c 6. POLYMER CHEMISTRY. *Educ:* Yale Univ, BS, 52; Purdue Univ, MS, 54, PhD(org chem), 57. *Prof Exp:* From res chemist to sr res chemist, E I du Pont de Nemours & Co, 57-67; MEM STAFF POLYMER CHEM, SANDIA NAT LABS, 67- *Mem:* Am Chem Soc; Soc Advan Mat & Process Engrs. *Res:* Thermal and radiative degradation of polymers; development of novel polyurethanes for encapsulation of electronic components. *Mailing Add:* 3436 Tahoe St Albuquerque NM 87111-5435

ARNOLD, D(ONALD) S(MITH), b Summit Co, Ohio, Sept 14, 20; m 44; c 6. CHEMICAL ENGINEERING. *Educ:* Ohio State Univ, BChE, 42, MSc, 47, PhD(chem eng), 49. *Prof Exp:* Asst, Ohio State Univ, 46-47; instr chem eng, NC State Col, 47-48, asst prof, 48-53, res engr, 51-53; supvr process develop, Nat Lead Co, Ohio, 53-54, head chem dept, 54-59; head high energy fuels sect, Am Potash & Chem Corp, Calif, 59, res mgr, Trona Lab, 59-67, dir cent eng, 67-69, dir process planning & develop, 69-77; CHIEF PROJ ENGR, KERR-McGEE CORP, 77- *Mem:* AAAS; Am Chem Soc; Am Inst Mining, Metall & Petrol Engrs; Am Soc Eng Educ; Am Inst Chem Engrs; Sigma Xi. *Res:* Chemical engineering operations; extraction; distilla- tion; thermodynamics; high energy chemical fuels; uranium refining; process design and development; refining inorganic alkali and borate chemicals. *Mailing Add:* 2005 Briarcliff Bethany OK 73008

ARNOLD, DAVID M, b Falls City, Nebr, Aug 9, 39. ABELTAN GROUP THEORY. *Educ:* Univ Ill, PhD(math), 69. *Prof Exp:* PROF MATH, NEW MEX STATE UNIV, 68- *Mem:* Am Math; Math Asn Am. *Res:* Abeltan group theory. *Mailing Add:* Baylor Univ Waco TX 76798

ARNOLD, DAVID WALKER, b Dundee, Miss, Dec 6, 36; m 58; c 1. CHEMICAL ENGINEERING, OPERATIONS RESEARCH. *Educ:* Univ Miss, BS, 58; Iowa State Univ, MS, 63, PhD(chem eng), 66. *Prof Exp:* Jr engr, Inst Atomic Res, Ames Lab, Atomic Energy Comn, Iowa State Univ, 63-66; process study engr, 66-68, res proj mgr, 68-69, chief process & process study engr, 69-72, mgr process eng, 72-74, dir eng, 74-75, vpres eng, 75-81, sr vpres, 81-87, SR VPRES RES & ENG, MISS CHEM CORP, 87- *Mem:* Am Inst Chem Engrs; Nat Soc Prof Eng. *Res:* Applications of nuclear magnetic resonance techniques to chemical engineering problems; nonlinear optimization; resource allocation and scheduling; fertilizer technology. *Mailing Add:* Dept Chem Eng Univ Ala Box 870202 Tuscaloosa AL 35487

ARNOLD, DEAN EDWARD, b Elmira, NY, Apr 8, 39; m 64; c 2. FRESHWATER FISHERIES, WATER QUALITY. *Educ:* Univ Rochester, AB, 61; Cornell Univ, MS, 65, DPhil(aquatic ecol), 69. *Prof Exp:* Res asst fisheries, Cornell Univ, 63-65, res assoc & asst proj leader warmwater fisheries invests, 65-66, teaching asst limnol, 66-68; asst res limnologist, Great Lakes Res Div, Univ Mich, 69-72; asst prof biol, Pa State Univ, 72-82; asst leader, 72-80, leader, Pa Fishery Res Unit, 80-82, ASSOC LEADER FISHERIES, PA COOP FISH & WILDLIFE RES UNIT, US FISH & WILDLIFE SERV, 82- *Concurrent Pos:* Eng officer, Navy Oceanog Ship, 61-63; consult, Environ Res Goup, Inc, Ann Arbor, Mich, 70-72; assoc ed, Am Fisheries Soc, 75-77 & 84-86; adj asst prof aquatic ecol, Pa State Univ, 82-; chmn adv bd, Sport Fishery Abstracts, 85- *Honors & Awards:* W F Thompson Award, Am Inst Fishery Res Biologists, 74; Qual Performance Award, US Fish & Wildlife Serv, 75. *Mem:* Am Fisheries Soc; Int Asn Theoret & Appl Limnol; Am Inst Fishery Res Biologists; NAm Benthological Soc. *Res:* Hydrobiological effects of acid precipitation; aquatic productivity; effects of manipulations of lakes and streams; lake restoration; pollution, conservation biology and aquatic ecology. *Mailing Add:* Coop Fish & Wildlife Res Unit Ferguson Bldg University Park PA 16802-4399

ARNOLD, DONALD ROBERT, b Buffalo, NY, Mar 1, 35; m 55; c 6. ORGANIC CHEMISTRY, PHOTOCHEMISTRY. *Educ:* Bethany Col, WVa, BS, 57; Univ Rochester, PhD(org chem), 61. *Prof Exp:* Res chemist, Union Carbide Res Inst, New York, 61-70; assoc prof, 70-71, prof chem, Univ Western Ont, 71-79; KILLAM RES PROF, DALHOUSIE UNIV, 79- *Concurrent Pos:* Alfred P Sloan Found fel, 72-74; Chem Inst Can fel, 74; Guggenheim Found fel, 80-81. *Mem:* Am Chem Soc; Chem Inst Can; Inter-Am Photochem Soc (pres). *Res:* Radical ions; heterocyclic compounds; small ring compounds; free radicals; synthetic applications and mechanisms of photochemistry; organic electrochemistry. *Mailing Add:* Dept Chem Dalhousie Univ Halifax NS B3H 4J3 Can

ARNOLD, EMIL, b Luck, Poland, May 20, 32; US citizen; m 65; c 2. SOLID STATE PHYSICS, SEMICONDUCTORS. *Educ:* Dalhousie Univ, BS, 54; Yale Univ, MS, 55, PhD(phys chem), 63. *Prof Exp:* Sr physicist, Gen Instrument Corp, 55-57; sr engr, Gen Tel & Electronics Labs, 57-60; sr physicist, 63-69, sr prog leader, 69-84, PRIN RES SCIENTIST, PHILIPS LABS INC DIV, NAM PHILIPS, 84- *Concurrent Pos:* Vis scientist, Mullard Res Labs, Eng, 74-75. *Mem:* Am Phys Soc; Inst Elec & Electronics Engrs. *Res:* Solid state electronics; transport properties of semiconductors; electrical behavior of semiconductor surfaces; electrochemistry; optoelectronics; semiconductor devices. *Mailing Add:* Philips Labs Inc Box 198 Briarcliff Manor NY 10510

ARNOLD, FRANK R(OBERT), b Winnsboro, La, Oct 20, 10; m 33; c 4. ENGINEERING SCIENCE, FORENSIC ENGINEERING. *Educ:* US Naval Acad, BS, 33, ord Eng, 42; Stanford Univ, MS, 51, PhD(eng mech), 54. *Prof Exp:* US Navy engr, 33-45; teacher & prin high sch, Minn, 46-48; asst nonlinear mech, 49-53 from actg asst prof to assoc prof, 54-66, consult prof, 76-83, lectr mech eng, Stanford Univ, 66-76; pres, Frank R Arnold, Inc, 68-88; RETIRED. *Concurrent Pos:* Consult, Eng Design. *Mem:* Am Soc Mech Engrs; Sigma Xi; US Naval Inst; Am Soc Eng Educ. *Res:* Nonlinear mechanics; thermodynamics; engineering design; naval science; accident analysis theory and applications. *Mailing Add:* Four Grove Ct Portola Valley CA 94028-7635

ARNOLD, FREDERIC G, b Irvington, NJ, Jan 14, 23; m 47; c 2. BIOLOGY, HUMAN PHYSIOLOGY. *Educ:* Montclair State Col, BA, 46, MA, 47; Columbia Univ, EdD(preparation sci teachers), 56. *Prof Exp:* Teacher high schs, NJ, 47-55; prof biol, 56-73, chmn dept, 69-73, PROF INSTR, CURRIC & ADMIN, KEAN COL, NJ, 73-, ASST DEAN EDUC, 80- *Mem:* AAAS; Nat Sci Teachers Asn. *Res:* Body fluids, blood and circulation; science education; desert biology, especially the ecology and plant life of the desert with emphasis on the cacti and succulent plants. *Mailing Add:* Seven Graymoor Rd Livingston NJ 07039

ARNOLD, GEORGE W, b Dayton, Ohio, Nov 24, 23; m 45; c 3. SOLID STATE PHYSICS. *Educ:* Univ Tex, BS, 48, MA, 49. *Prof Exp:* Physicist, Optical Res Lab, Univ Tex, 48-49, Nat Bur Standards, 49-50 & US Naval Res Lab, 50-61; PHYSICIST, SANDIA NAT LABS, 61- *Concurrent Pos:* Fulbright prof, Egypt, 60-61. *Mem:* Am Phys Soc; AAAS; Mat Res Soc. *Res:* Ion implantation and radiation effects studies in inorganic solids. *Mailing Add:* Sandia Nat Labs Div 1112 Albuquerque NM 87185

ARNOLD, HARRY L(OREN), JR, b Owosso, Mich, Aug 7, 12; m 32, 42, 83; c 5. SCIENCE COMMUNICATIONS. *Educ:* Univ Mich, AB, 32, MD, 35, MS, 39. *Prof Exp:* Instr dermat, Med Sch, Univ Mich, 37-39; clin prof med & dermat, Med Sch, Univ Hawaii, 39-69; clin prof dermat, Univ Calif, San Francisco, 69-84; RETIRED. *Concurrent Pos:* Ed, Straub Clin Proc, Straub Clin, Inc, 40-77, Hawaii Med J, Hawaii Med Asn, 41-83 & The Schoch Lett, 75-; vis scholar dermat, Frederick G Novy Found, 75. *Mem:* Fel AAAS; AMA; hon mem Am Acad Dermat (vpres, 65, pres, 75); hon mem Am Dermat Asn (pres, 71). *Mailing Add:* 250 Laurel St No 301 San Francisco CA 94118-2045

ARNOLD, HARVEY JAMES, b Niagara Falls, Ont, Can, Mar 5, 33; US citizen; m 56; c 4. MATHEMATICAL STATISTICS. *Educ:* Queen's Univ, Ont, BA, 51, MA, 52; Princeton Univ, MA, 55, PhD(math), 58. *Prof Exp:* Res asst, Princeton Univ, 57-58; asst prof, Univ Western Ont, 58-60 & Wesleyan Univ, 60-63; assoc prof, Bucknell Univ, 63-67; actg chmn dept, 70-71; assoc prof, 67-70, PROF MATH, OAKLAND UNIV, 70- *Concurrent Pos:* Vis res assoc, Princeton Univ, 62-63; statistician-in-residence, Univ Wis, 72-73. *Mem:* Am Soc Qual Control; Am Statist Asn. *Res:* Nonparametric statistics; design of experiments and statistical applications to the physical and engineering sciences; quality control. *Mailing Add:* Dept Math Sci Oakland Univ Rochester MI 48309-4401

ARNOLD, HENRY ALBERT, b Chewelah, Wash, Feb 18, 14; m 38; c 2. OCEAN ENGINEERING, SCIENCE POLICY. *Educ:* US Naval Acad, BS, 36; Mass Inst Technol, BS & MS, 41. *Prof Exp:* Res officer hydronautics, David Taylor Model Basin, US Navy, 50-53, res officer submarine res & design, 53-58, design officer, Polaris submarine, Spec Proj Off, US Navy, 58-60; asst chief sci, United Aircraft Corp, 61-70; dep dir, AID, Dept of State, 70-73, dir, Off Sci & Technol, 73-79; RETIRED. *Concurrent Pos:* Consult ocean eng, Off of President's Sci Adv, 61-64; tech dir, US Navy Deep Ocean Systs Rev Task Force, 63; sr specialist ocean eng, President's Coun Marine Resources & Eng Develop, 66-68. *Mem:* Sigma Xi. *Res:* Role and application of science and technology to the social and economic development of the lesser developed countries of the world. *Mailing Add:* 1583 Frontier Dr Melbourne FL 32940

ARNOLD, HUBERT ANDREW, b Chicago, Ill, Nov 15, 12. MATHEMATICS. *Educ:* Univ Nebr, AB, 33; Calif Inst Technol, PhD(math), 39. *Prof Exp:* Asst instr math, Calif Inst Technol, 35-39; instr, Univ Minn, 39-40; vis res scholar, Univ Va, 40-41; instr, Princeton Univ, 41-42; from instr to assoc prof, 48-80, EMER ASSOC PROF MATH, UNIV CALIF, DAVIS, 80- *Mem:* Fel AAAS; Math Asn Am; Am Math Soc. *Res:* Differentials in abstract spaces; topological structure of limit sets; numerical analysis and table making. *Mailing Add:* Univ Calif Davis Davis CA 95616

ARNOLD, J BARTO, III, b San Antonio, Tex, Jan 9, 50; m 70; c 3. UNDERWATER ARCHAEOLOGY. *Educ:* Univ Tex, Austin, BA, 71, MA, 73. *Prof Exp:* Res asst, Tex Archaeol Res Lab, Univ Tex, Austin, 71-72; asst marine archaeologist, 72-75, STATE MARINE ARCHAEOLOGIST, TEX ANTIQUITIES COMT, TEX HIST COMN, 75- *Concurrent Pos:* Prog chmn, 9th Conf Underwater Archaeol, 78; bd dir Soc Hist Archaeol, 82-87; mem USS Monitor Tech Adv Comt, 77-90; mem, adv coun Underwater Archaeol, 87-89; mem, Md Gov Adv Comt Underwater Archaeol, 87-89; consult, maritime preservation, Nat Trust Hist Preservation, 79-; USS Monitor Project, 87. *Honors & Awards:* Spec Achievement Award, Soc Prof Archaeologists, 90. *Mem:* Soc Prof Archaeologists (secy-treas, 87-89); Soc Am Archaeol; Soc Hist Archaeol; Archaeol Instit Am. *Res:* Underwater archaeology, excavation of the San Esteban, a wreck of the 1554 New Spain Fleet; surveys and recording shipwrecks in Texas; marine remote sensing surveys. *Mailing Add:* 3610 Crowncrest Dr Austin TX 78759

ARNOLD, JAMES DARRELL, b Emporia, Kans, Apr 21, 37; m 63; c 4. RANGE SCIENCE, CROP PRODUCTION. *Educ:* Panhandle State Univ, BS, 64; SDak State Univ, MS, 66; Tex Tech Univ, PhD(range sci), 73. *Prof Exp:* Res agronomist crop sci, Panhandle State Univ-Okla State Univ & assoc prof agron crop sci, Panhandle State Univ, 66-73; dir forage lab, 76-78, assoc prof, 76-86, ACAD COORDR & PROF CROP SCI, TEX A&I UNIV, 86-, ASST DEAN ACAD AFFAIRS, COL AGR/HOME ECON & PROF, 88- *Mem:* Am Soc Agron; Soc Range Mgt; Crop Sci Soc Am. *Res:* Renewable natural resources; factors relating to range improvement in tropical, subtropical and semi-arid regions. *Mailing Add:* Dept Agr Tex A&I Univ Kingsville TX 78363

ARNOLD, JAMES NORMAN, b Whittier, Calif, Dec 18, 49; m 74; c 2. ENGINEERING HEAT TRANSFER. *Educ:* Univ Calif, Los Angeles, BS, 73, MS, 74, PhD(eng), 78. *Prof Exp:* mem tech staff mech eng, Bell Tel Labs, 78-82, supvr, 82-86, DEPT HEAD, AT&T BELL LABS, 86- *Mem:* Am Soc Mech Eng; Inst Elec & Electronics Engrs. *Res:* Engineering heat transfer with added interest in telecommunications. *Mailing Add:* AT&T Bell Labs 555 Union Blvd Rm 2d-213 Allentown PA 18103

ARNOLD, JAMES RICHARD, b New Brunswick, NJ, May 5, 23; m 52; c 3. PLANETARY SCIENCE, SPACE APPLICATIONS. *Educ:* Princeton Univ, AB, 43, MA, 45, PhD(chem), 46. *Prof Exp:* Asst, Princeton Univ, 43 & Manhattan proj, 43-46; fel, Inst Nuclear Studies, Univ Chicago, 46; Nat Res fel, Harvard Univ, 47; asst prof, Inst Nuclear Studies, Univ Chicago, 48-55; from asst prof to assoc prof, Princeton Univ, 55-58; from assoc prof to prof, 58-83, chmn dept, 60-63, HAROLD C UREY PROF CHEM, UNIV CALIF, SAN DIEGO, 83- *Concurrent Pos:* Dir, Calif Space Inst, 79-89. *Honors & Awards:* Ernest Orlando Lawrence Award, AEC, 68. *Mem:* Nat Acad Sci; Am Acad Arts & Sci; AAAS; Am Chem Soc; Int Acad Astronaut; Meteoritical Soc. *Res:* Cosmic-ray produced nuclides; history of moon and meteorites; cosmochemistry; lunar resources utilization. *Mailing Add:* Dept Chem 0317 Univ Calif San Diego La Jolla CA 92903-0317

ARNOLD, JAMES S(LOAN), b Ely, Nev, Sept 27, 11; m 42; c 2. APPLIED PHYSICS, ACOUSTICS. *Educ:* Univ Calif, Los Angeles, AB, 34; NMex Col Agr & Mech Arts, MS, 51. *Prof Exp:* Prod & tool engr, Trumbull Div, Gen Elec Co, 39-43 & 45-48; assoc physicist, Phys Sci Lab, NMex Col Agr & Mech Arts, 48-53; physicist, Stanford Res Inst, 53-66; sr tech specialist, Reconnaissance Lab, McDonnell Douglas Corp, 66-69; sr group engr, 69-76; RETIRED. *Mem:* Sigma Xi. *Res:* Electronic instrumentation; ultrasonic methods for non-destructive testing; physics of ignition and combustion processes; sonic devices and techniques; photographic and electronic imaging systems for reconnaissance; electronic control systems. *Mailing Add:* 4948 Park Dr Carlsbad CA 92008

ARNOLD, JAMES SCHOONOVER, b Nov 20, 23; c 7. KIDNEY BLOOD FLOW, BONE MORPHOLOGY. *Educ:* Duke Univ, MD, 48. *Prof Exp:* CHIEF, LAB NUCLEAR MED, IRON MOUNTAIN VET ADMIN HOSP, 80- *Concurrent Pos:* Nuclear med physician, 60- *Mem:* Soc Nuclear Med; Soc Clin Path. *Res:* Kidney blood flow & function; bone morphology. *Mailing Add:* Iron Mountain Vet Admin Hosp H St Iron Mountain MI 49801

ARNOLD, JAMES TRACY, b Taiyuanfu, China, Oct 23, 20; US citizen; m 57; c 4. EXPERIMENTAL PHYSICS. *Educ:* Oberlin Col, AB, 42; Stanford Univ, PhD(physics), 54. *Prof Exp:* Lab instr physics, Oberlin Col, 47-48; spec asst, Europ Ctr Nuclear Res, 54-55; res assoc physics, Stanford Univ, 55-57; asst prof, Ore State Col, 57-58; physicist, 58-72, SR SCIENTIST, VARIAN ASSOCS, 72- *Mem:* Sigma Xi. *Res:* High resolution nuclear magnetic resonance spectroscopy, magnetometers, mass spectroscopy, ultrasonic imaging and computer application to instruments. *Mailing Add:* PO Box 314 Los Altos CA 94023

ARNOLD, JESSE CHARLES, b Bowie, Tex, Sept 28, 37; m 59; c 2. APPLIED STATISTICS, MATHEMATICAL STATISTICS. *Educ:* Southeastern State Col, BS, 60; Fla State Univ, MS, 65, PhD(statist), 67. *Prof Exp:* Asst chief exp design, Ctr Dis Control, USPHS, 61-63; asst prof educ res, Fla State Univ, 67-68; assoc prof, 68-74, PROF STATIST & HEAD DEPT, VA POLYTECH INST & STATE UNIV, 74- *Concurrent Pos:* Va State Water Control Bd grant water qual monitoring, Va Polytech Inst & State Univ, 69-71; consult, USPHS, 71, NASA-Langley Res Ctr, 72 & USAID, 72-84; coordr Philippine Nat Nutrit Proj, Va Polytech Inst & State Univ & USAID, 74-84. *Mem:* Biomet Soc(pres, 81); fel Am Statist Asn; Int Statist Inst; Am Statist Asn. *Res:* Theory of estimation; application of statistics to environmental problems; sampling biometrics; international nutrition in the Philippines and Haiti; applications and research in quality control. *Mailing Add:* Dept Statist Va Polytech Inst & State Univ Blacksburg VA 24061

ARNOLD, JIMMY THOMAS, b Crowville, La, May 31, 41; m 67; c 2. POLYNOMIAL & POWER SERIES RINGS. *Educ:* Northeastern La State Col, BS, 63; Fla State Univ, MS, 65, PhD(algebra), 67. *Prof Exp:* From asst prof to assoc prof, 69-76, PROF MATH, VA POLYTECH INST & STATE UNIV, 77- *Mem:* Am Math Soc; Math Asn Am. *Res:* Dimension theory in power series rings and in polynomial rings. *Mailing Add:* Dept Math Va Polytech Inst & State Univ Blacksburg VA 24061

ARNOLD, JOHN MILLER, b St Paul, Minn, Oct 6, 36. ZOOLOGY, EMBRYOLOGY. *Educ:* Univ Minn, BA, 58, PhD(zool), 63. *Prof Exp:* Instr zool, Oberlin Col, 60-61; res fel, Lerner Marine Lab, Am Mus Natural Hist, 63-64; asst prof, Iowa State Univ, 64-66; from asst prof to assoc prof, 66-75, PROF CYTOL, PAC BIOMED RES CTR, UNIV HAWAII, MANOA, 75- *Concurrent Pos:* Mem corp & trustee, Marine Biol Lab, Woods Hole; res assoc, Am Mus Natural Hist, 86. *Mem:* AAAS; Am Soc Zool; Soc Develop Biol; Am Soc Cell Biol; Int Soc Develop Biologists; AAAS. *Res:* Embryological development; biology of cephalopods, first discovered Nautilus embryos. *Mailing Add:* Pac Biomed Res Ctr Univ Hawaii at Manoa Snyder Hall 209A Honolulu HI 96822

ARNOLD, JOHN RICHARD, b Fort McClellan, Ala, Oct 16, 56; m 86. SURFACE SCIENCE, CORROSION & ADHESION SCIENCE. *Educ:* Univ Calif, Santa Barbara, BS, 78; Univ Calif, Irvine, PhD(chem), 84. *Prof Exp:* ENGR RES, BETHLEHEM STEEL CORP, 84- *Mem:* Am Chem Soc; Am Physics Soc. *Res:* Surface analysis; adhesive joint durability in corrosive environments; coated steel products. *Mailing Add:* First Brands Corp Still River 55 Federal Rd Danbury CT 06810-4001

ARNOLD, JOHN RONALD, b Hanford, Calif, June 29, 10; m 33; c 1. ZOOLOGY. *Educ:* Fresno State Col, BA, 32; Univ Calif, MA, 34; Cornell Univ, PhD(zool), 38. *Prof Exp:* Teacher, high sch & jr col, Calif, 33-36; teacher gen zool, Stockton Col, 38-41 & 46-61, chmn sci div, 46-58, dean, 58-61; prof biol, 61-76, EMER PROF BIOL, SONOMA STATE UNIV, CALIF, 76- *Concurrent Pos:* Res assoc, Calif Acad Sci. *Res:* Ornithology; mammalogy; Mexican rodents; mockingbirds. *Mailing Add:* 199 Calistoga Rd Santa Rosa CA 95405

ARNOLD, JONATHAN, b New York, NY, Nov 27, 53; m 80. POPULATION GENETICS, STATISTICAL GENETICS. *Educ:* Yale Univ, BS, 75, MPhil, 78, PhD(statist), 82. *Prof Exp:* Adj instr statist, Rutgers Univ, 81-82; ASST PROF GENETICS, UNIV GA, 82- *Concurrent Pos:* Grant, NSF, 83-86 & Army Res Off, 83-85; adj prof, Emory Univ, 84- *Mem:* Am Statist Asn; Soc Study Evolution; Biometrics Soc; Genetics Soc Am; Inst Math Statist. *Res:* Statistical problems in population genetics, including the analysis of fitness components and the modeling of plant and animal mating systems; statistical analysis of DNA sequence data and models of molecular evolution. *Mailing Add:* Dept Genetics Univ Ga Athens GA 30602

ARNOLD, KEITH ALAN, b Jackson, Mich, Sept 23, 37; m 68; c 2. ORNITHOLOGY. *Educ:* Kalamazoo Col, AB, 59; Univ Mich, MS, 61; La State Univ, PhD(zool), 66. *Prof Exp:* Instr zool, La State Univ, 65-66; from asst prof to assoc prof, 66-78, PROF WILDLIFE SCI, TEX A&M UNIV, 78- *Mem:* Am Ornith Union; Soc Syst Zool; Cooper Ornith Soc; Wilson Ornith Soc; Sigma Xi. *Res:* Taxonomic, ecological and behavioral ornithology; avian ectoparasites; tropical vertebrate ecology. *Mailing Add:* Dept Wildlife Tex A&M Univ College Station TX 77843

ARNOLD, KENNETH JAMES, b Pawtucket, RI, Aug 20, 14; m 39; c 3. MATHEMATICAL STATISTICS. *Educ:* Mass Inst Technol, BS, 37, PhD(math), 41. *Prof Exp:* Asst, Mass Inst Technol, 38-39, 40-41, instr math, 41-43; asst prof, Univ NH, 43-44; sr math statistician, Statist Res Group, Columbia Univ, 44-45; asst prof math, Univ Wis, 45-52; assoc prof statist, 52-62, chmn dept, 63-67, prof, 62-81, EMER PROF STATIST, MICH STATE UNIV, 81- *Concurrent Pos:* Ed, J Soc Indust & Appl Math, 58-69. *Mem:* Math Asn Am; Inst Math Statist (secy-treas, 52-55). *Res:* Statistical theory; computational methods; spherical probability distributions. *Mailing Add:* 12 Cahoon's Hollow Chatham MA 02633

ARNOLD, LESLIE K, b Larned, Kans, Oct 18, 38; m 63; c 2. MATHEMATICS. *Educ:* Rice Univ, BA, 61; Brown Univ, PhD(ergodic theory), 66. *Prof Exp:* Assoc, 66-72, SR ASSOC, DANIEL H WAGNER, ASSOCS, 72- *Mem:* Am Math Soc; Math Asn Am; Sigma Xi. *Res:* Ergodic theory; invariant measures; naval operations analysis; software engineering. *Mailing Add:* 540 Col Dewees Rd Wayne PA 19087

ARNOLD, LUTHER BISHOP, JR, b Duluth, Minn, Aug 9, 07; m 47; c 1. TEXTILE CHEMISTRY. *Educ:* Carleton Col, AB, 29; Harvard Univ, AM, 30, PhD(phys org chem), 33. *Prof Exp:* Asst, Harvard Univ, 33; res chemist, Org Chem Dept, E I du Pont de Nemours & Co, 33-41, res supvr, Rayon Dept, 41-43; asst dir metal lab, Univ Chicago, 43-45; asst to pres, Arthur D Little, Inc, 45-47; proprietor, 47-65, CONSULT, PRES & TREAS, VIKON CHEM CO, 65- *Concurrent Pos:* Gen chmn, Southern Textile Res Conf, 61, treas, 62-78, chmn adv comn, 78-; lectr, Clemson Univ Prof Develop Conf, 77, 83, 85. *Honors & Awards:* Chapin Award, Am Asn Textile Chemists & Colorists, 85. *Mem:* Fel AAAS; Am Chem Soc; Am Asn Textile Chemists & Colorists; Am Soc Test & Mat; Asn Off Anal Chemists. *Res:* Gas reaction kinetics; textile processing; surface active agents and detergents; textile dyeing and bleaching; emulsions; emulsion polymerization; radiochemistry; antimicrobial treatments for textiles, especially mildewproofing; peroxide stabilizer systems for bleaching cellulose fibers in textiles and paper; softeners for textile materials; assistants for dyeing textiles. *Mailing Add:* 1614 Woodland Ave Burlington NC 27215-3530

ARNOLD, MARY B, b Fitchburg, Mass, Sept 29, 24; wid; c 3. PEDIATRIC ENDOCRINOLOGY. *Educ:* Vassar Col, AB, 45; Univ Vt, MD, 50; Am Bd Pediat, dipl, 56. *Hon Degrees:* Brown Univ, MA, 74. *Prof Exp:* Intern, Hartford Hosp, 50-52; asst resident pediat, Babies Hosp, 52-54; asst pediat, Sch Med, Harvard Univ, 55-57; instr, Sch Med, Univ NC, 57-59, asst prof, 59-65; consult biochem, 66-68, DIR PEDIAT OUTPATIENT DEPT & ENDOCRINE CLIN, ROGER WILLIAMS GEN HOSP, 66-, CHMN DEPT PEDIAT, 74- *Concurrent Pos:* Res fel pediat endocrinol, Mass Gen Hosp, Boston, 54-57; lectr med sci, Brown Univ, 66-71; attend pediatrician, Roger Williams Gen Hosp, 66-; consult pediat endocrinol, Women & Infants Hosp RI, 66-; asst physician pediat & attend pediatrician, RI Hosp, 70-, dir pediat endocrinol, 71-; assoc prof pediat & dir pediat endocrinol, prog med, Brown Univ, 71- *Honors & Awards:* Carrbee Award Obstet, Univ Vt, 50. *Mem:* Sigma Xi; Endocrine Soc; Am Fedn Clin Res; AMA; Lawson Wilkins Pediat Endocrinol Soc. *Res:* Endocrine problems of childhood with particular reference to thryoid disorders and disorders of calcium and phosphorus metabolism; fetal endocrinology, and various mechanisms responsible for growth hormone release, including the role of somatostatin and somatomedin in normal growth and development. *Mailing Add:* Dept Pediat Rog William Gen Hosp Providence RI 02908

ARNOLD, PHILIP MILLS, chemical engineering, for more information see previous edition

ARNOLD, R KEITH, forestry, for more information see previous edition

ARNOLD, RALPH GUNTHER, b Potsdam, Ger, Dec 8, 28; Can citizen; m 59; c 4. ENVIRONMENTAL GEOLOGY. *Educ:* Univ Toronto, BASc, 53, MASc, 54; Princeton Univ, PhD(geochem), 58. *Prof Exp:* Asst res off geol, 59-62, assoc res off, 62-67, head geol div, Sask Res Coun, 67-; AT DEPT GEOL SCI, UNIV SASK. *Mem:* Fel Carnegie Inst, 57-59; lectr, Univ Sask, 59, adj prof, 68. Mem: Mineral Soc Am; Soc Econ Geol; Geochem Soc; Mineral Asn Can. *Res:* Phase equilibrium; studies of mineral deposits. *Mailing Add:* 515 North Rd RR 1 Sydney BC V8L 3R9 Can

ARNOLD, RICHARD THOMAS, b Indianapolis, Ind, June 18, 13; m 39; c 2. PHYSICAL ORGANIC CHEMISTRY. *Educ:* Southern Ill State Teachers Col, BEd, 34; Univ Ill, MS, 35, PhD, 37. *Prof Exp:* From asst prof to prof chem, Univ Minn, 37-55, adminr basic sci prog, Alfred P Sloan Found, 55-60; dir res, Mead Johnson & Co, Ind, 60-61, pres res ctr, 61-69; chmn dept, Southern Ill Univ, Carbondale, 70-75, prof chem & biochem, 70-82, EMER PROF, 82- *Concurrent Pos:* Guggenheim fel, 48-49; sci attache, US High Comnr, Ger, 52-53; vis prof chem, Northwestern Univ, 75. *Mem:* Am Chem Soc; Brit Chem Soc. *Res:* Mechanism of organic reactions; stereochemistry and reaction mechanisms. *Mailing Add:* The Georgian Apt 521 422 Davis St Evanston IL 60201

ARNOLD, ROBERT FAIRBANKS, b Gilroy, Calif, July 3, 40; m 66; c 1. MATHEMATICS. *Educ:* Fresno State Col, BS, 63, MA, 64; Univ Calif, Berkeley, PhD(math), 69. *Prof Exp:* From asst prof math to assoc prof, 68-76, chmn dept, 75-77, 88-90, PROF MATH, CALIF STATE UNIV, FRESNO, 76- *Mem:* Math Asn Am. *Mailing Add:* Dept Math Calif State Univ Fresno CA 93740

ARNOLD, ROY GARY, b Lyons, Nebr, Feb 20, 41; m 63; c 2. FOOD SCIENCE, BIOCHEMISTRY. *Educ:* Univ Nebr, BSc, 62; Ore State Univ, MS, 65, PhD(food sci), 67. *Prof Exp:* Proj leader food res, Res & Develop Lab, Fairmont Food Co, Nebr, 62-63; from asst prof to assoc prof, Univ Nebr, Lincoln, 67-74, chmn dept, 73-79, dean & dir, Agr Exp Sta, 80-82, vchancellor, Inst Agr & Natural Resources, 82-87, prof food sci & technol, 74-87; DEAN, COL AGR SCI, ORE STATE UNIV, 87- *Honors & Awards:* William Cruess Award, Inst Food Technologists. *Mem:* AAAS; Inst Food Technologists. *Res:* Flavor chemistry of foods; food lipids. *Mailing Add:* Oregon State Univ Col Agr STAG 126 Corvallis OR 97331

ARNOLD, STEVEN LLOYD, b Springfield, Mass, Apr 14, 49; m 72; c 2. AQUATIC ECOLOGY, ENTOMOLOGY. *Educ:* Johns Hopkins Univ, BA, 71, MA, 72; Cornell Univ, PhD(ecol), 78. *Prof Exp:* asst prof aquatic ecol, State Univ Col, State Univ Ny, 77-82. *Mem:* AAAS; Ecol Soc Am; Entom Soc Am; Entom Soc Can; Freshwater Biol Asn. *Res:* Population dynamics; bioenergetics; natural history and systematics of Sciomyzidae (Diptera) and their aquatic snail prey; realistic simulation of populations; biological control of noxious snails. *Mailing Add:* 20 Fifth Ave New York NY 10011

ARNOLD, WALTER FRANK, b Rochester, NY, Nov 2, 20; m 53; c 1. ENGINEERING, MATERIALS SCIENCE. *Prof Exp:* Field metallurgist, Symington-Gould Corp, Rochester, 42-43; staff metallurgist, Los Alamos Sci Lab, 43-53 & Radiation Lab, 53-60, div head, 60-62, asst head dept mech eng, 62-71, head dept mech eng, 71-80, ASST ASSOC DIR ENG, LAWRENCE LIVERMORE NAT LAB, UNIV CALIF, 80- *Concurrent Pos:* Chmn interagency mech opers group, Atomic Energy Comn, 61- *Mem:* Am Soc Mech Engrs; Soc Mfg Engrs; Am Defense Preparedness Asn. *Res:* Melting and casting methods; nuclear weapons engineering. *Mailing Add:* Electronic Warfare Div Pac Missle Test Ctr EW Code 1153 Point Mugu CA 93042

ARNOLD, WATSON CAUFIELD, b Waco, Tex, Jan 12, 45; m 75; c 2. PEDIATRICS, NEPHROLOGY. *Educ:* Tulane Univ, BA, 66; Univ Tex, Dallas, MD, 70; Am Bd Pediat. *Prof Exp:* Intern pediat, Med Col, Univ Ark, 71, resident, 75; clin fel pediat & nephrology, Med Sch, Univ Calif, San Francisco, 75-76, NIH trainee res fel, 76-78; from asst prof to assoc prof pediat, Sch Med Sci, Univ Ark, 77-90, dir pediat Nephrology, Ark Children's Hosp, 78-90; DIR PEDIAT NEPHROLOGY, COOK FT WORTH CHILDREN'S MED CTR, FT WORTH, TEX, 90- *Concurrent Pos:* Chmn, Sect Nephrology, Am Acad Pediat. *Mem:* Am Acad Pediat; Am Soc Nephrology; Am Soc Pediat Nephrology; Am Fedn Clin Res; Am Fedn Exp Biol; Am Soc Clin Nutrit; Am Inst Nutrit. *Res:* Nutrition and kidney failure in children, including growth and development; emphasis on endocrine aspect of uremia and nutrition and in hypertension. *Mailing Add:* Cook Ft Worth Children's Med Ctr 801 W Seventh Ft Worth TX 76109

ARNOLD, WILFRED NIELS, b Australia, May 29, 36; US citizen; m 61; c 1. BIOCHEMISTRY, HISTORY OF MEDICINE. *Educ:* Univ Queensland, BS, 56; Univ Calif, Los Angeles, MA, 58; Cornell Univ, PhD(plant biochem), 62. *Prof Exp:* Res specialist air pollution, Univ Calif, Riverside, 57-58; res fel, Waite Res Inst, Univ Adelaide, 62-63; res scientist, Commonwealth Sci & Indust Res Orgn, Adelaide, Australia, 63-66; sr fel, Med Sch, Univ Wis, 66-67; asst prof biochem, Med Sch, Wayne State Univ, 67-71; assoc prof, 72-80, PROF BIOCHEM, UNIV KANS MED CTR, KANSAS CITY, 81- *Concurrent Pos:* Lectr, mid Am state univs, 81-82. *Mem:* AAAS; Am Chem Soc; Sigma Xi; Am Soc Biol Chemists. *Res:* Enzymology; structure and function of the cell envelope in yeasts; carbohydrate and amino acid metabolism in higher plants and microorganisms; history and philosophy of science. *Mailing Add:* Med Sch Univ Kans Rainbow Blvd 39th St Kansas City KS 66103-8410

ARNOLD, WILLIAM ARCHIBALD, b Douglas, Wyo, Dec 6, 04; m 29; c 2. PLANT PHYSIOLOGY. *Educ:* Calif Inst Technol, BS, 31; Harvard Univ, PhD(physiol), 35. *Prof Exp:* Sheldon fel biol, Univ Calif, 35-36; Gen Educ Bd fel, Hopkins Marine Sta, Stanford Univ, 36-37, asst, 37-38, res assoc biol, 39-40 & 41, asst prof, 41-46; prin biologist, Oak Ridge Nat Lab, 46-70; RETIRED. *Concurrent Pos:* Rockefeller fel, Inst Theoret Physics, Copenhagen Univ, 38-39; physicist, Princeton Univ, 42; sr physicist, Eastman Kodak Co, NY, 42-44 & Tenn, 44-46; consult, 69-79. *Mem:* Nat Acad Sci; Am Phys Soc; Soc Gen Physiol; Sigma Xi. *Res:* Photosynthesis. *Mailing Add:* 102 Balsam Rd Oak Ridge TN 37830-7823

ARNOLD, WILLIAM HOWARD, JR, b St Louis, Mo, May 13, 31; m 52; c 5. NUCLEAR ENGINEERING. *Educ:* Cornell Univ, AB, 51; Princeton Univ, MA, 53, PhD(physics), 55. *Prof Exp:* Instr physics, Princeton Univ, 55; sr scientist reactor physics, Atomic Power Div, Westinghouse Electric Corp, 55-57, mgr reactor physics design, 57-61; dir nuclear fuel mgt, Nuclear Utility Serv, Inc, 61-62; dep mgr reactor eng, Astronuclear Lab, 62-64, Westinghouse Elec Corp, mgr test systs & opers, 64-66, dep prog mgr NERVA proj, 66-68, proj mgr, 68, mgr, Weapons Dept, Astronuclear Underseas Div, 68-70, mgr eng, 70-72, gen mgr pressurized water reactor systs, 72-80; gen mgr adv energy systs, Westinghouse Hanford Co, 86-89; PRES, LA ENERGY SERV, 89- *Concurrent Pos:* Lectr, Univ Pittsburgh, 58; chmn nuclear propulsion, Inst Aeronaut & Astronaut. *Mem:* Nat Acad Eng; fel Am Nuclear Soc; Am Physics Soc; assoc fel Inst Aeronaut & Astronaut; fel AAAS. *Res:* Cosmic ray and elementary particle physics; reactor physics; nuclear engineering; author of 45 technical publications; awarded two patents. *Mailing Add:* La Energy Serv 600 New Hampshire Ave NW Suite 404 Washington DC 20037-2403

ARNOLDI, ROBERT ALFRED, b New York, NY, Mar 30, 21; m 43; c 3. APPLIED MECHANICS, ACOUSTICS. *Educ:* Stevens Inst Technol, ME, 42; Polytech Inst New York, MS, 50, PhD(appl mech), 53. *Prof Exp:* Anal engr vibrations, Pratt & Whitney Aircraft, 42-46; instr eng physics, Stevens Inst Technol, 46-51; anal engr aerodyn, Res Labs, United Aircraft Corp, 51-68; proj engr Aeroelasticity & Aeroacoust, Pratt & Whitney Aircraft, 68-81; RETIRED. *Concurrent Pos:* Adj prof appl mech, Hartford Grad Ctr, Rensselaer Polytech Inst, 55-70; mem adv subcomt fluid mech, NASA Res & Technol, 69-71. *Mem:* Am Inst Aeronaut & Astronaut. *Res:* Aerolasticity; aeroacoustics. *Mailing Add:* 108 Selden Hill Dr West Hartford CT 06107

ARNOLDI, WALTER EDWIN, b New York, NY, Dec 14, 17; m 42; c 2. MECHANICAL ENGINEERING, APPLIED MATHEMATICS. *Educ:* Stevens Inst Technol, ME, 37; Harvard Univ, MS, 39. *Prof Exp:* Anal engr, Hamilton Standard Div, United Aircraft Corp, 39-44, proj engr, 44-51, systs engr, 51-59, sr tech specialist advance planning, 59-60, chief adv anal, 60-62, head prod res, 62-67, chief div res, 67-70, div tech consult, 70-77; RETIRED. *Concurrent Pos:* Mem subcomt vibration & flutter, Nat Adv Comt Aeronaut, 57-58. *Res:* Vibration of propellers, engines and aircraft; high speed digital computers; aerodynamics and acoustics of aircraft propellers; life support in space vehicles; composite structural materials. *Mailing Add:* 18 Avondale Rd West Hartford CT 06117-1107

ARNOLDY, ROGER L, b LaCrosse, Wis, May 30, 34; m 61; c 5. SPACE PHYSICS. *Educ:* St Mary's Col, Minn, BS, 56; Univ Minn, MS, 60, PhD(physics), 62. *Prof Exp:* Res assoc physics, Univ Minn, 62-64; sr res scientist, Honeywell Res Ctr, Minn, 64-67; assoc prof, 67-78, PROF PHYSICS, UNIV NH, 78-, DIR SPACE SCI CTR, 74- *Mem:* Am Geophys Union. *Res:* Experimentation to measure charged particle radiation in the earth's magnetosphere and in interplanetary space. *Mailing Add:* Dept Physics Univ NH Durham NH 03824

ARNON, DANIEL I(SRAEL), b Poland, Nov 14, 10; nat US; m 40; c 5. BIOCHEMISTRY. *Educ:* Univ Calif, BS, 32, PhD(plant physiol), 36. *Hon Degrees:* Dr, Univ Bordeaux, France. *Prof Exp:* From instr to prof plant physiol, 36-60, biochemist, 58-78, prof cell physiol, 60-78, RES BIOCHEMIST, DEPT PLANT BIOL & EMER PROF CELL PHYSIOL, UNIV CALIF, BERKELEY, 78- *Concurrent Pos:* Guggenheim fel, Cambridge Univ, 47-48 & Hopkins Marine Sta, Stanford Univ, 62-63; Belg-Am Found lectr, Univ Liege, 48; ed, Annual Rev Plant Physiol, 48-55; Fulbright res scholar, Max Planck Inst, Ger, 55-56. *Honors & Awards:* Gold Medal, Univ Pisa, 58; Charles F Kettering Award, Kettering Found & Nat Acad Sci-Nat Res Coun, 63; Nat Medal Sci, President of the US, 73; Newcomb Cleveland Prize, AAAS, 40; Stephen Hales Prize Award, Am Soc Plant Physiologists, 66, Charles Reid Barnes Life Mem Award, 82, Charles F Kettering Award, 84; Finsen Medal, Int Asn Photobiol, 88. *Mem:* Nat Acad Sci; Am Soc Plant Physiologists (pres, 52-53); Am Soc Biol Chemists; Biochem Soc UK; hon mem Span Biochem Soc; Am Chem Soc; Am Acad Arts & Sci; Royal Swed Acad Sci; German Acad Sci Leopoldina; Acad Agr France; Scand Soc Plant Physiologists; Am Soc Photobiol; Am Soc Biochem & Molecular Biol. *Res:* Bioenergetics of photosynthesis. *Mailing Add:* 28 Norwood Ave Berkeley CA 94707

ARNONE, ARTHUR RICHARD, b Syracuse, NY, Sept 25, 42; m 66; c 2. BIOCHEMISTRY. *Educ:* Mass Inst Technol, PhD(phys chem), 70. *Prof Exp:* Fel molecular biol, Med Res Coun Lab, Cambridge, Eng, 70-73; assoc prof, 73-78, PROF BIOCHEM, UNIV IOWA, 78- *Concurrent Pos:* NIH res career develop award, 75. *Mem:* Am Soc Biol Chemists; Am Chem Soc; Am Crystallog Asn; Sigma Xi. *Res:* X-ray crystallographic studies of hemoglobin and other proteins. *Mailing Add:* Dept Biochem Univ Iowa Col Med Iowa City IA 52242

ARNOTT, HOWARD JOSEPH, b Los Angeles, Calif, Mar 9, 28; m 50; c 4. BOTANY, CELL BIOLOGY. *Educ:* Univ Southern Calif, AB, 52, MS, 53; Univ Calif, PhD(bot), 58. *Prof Exp:* Asst bot, Univ Calif, 55-58; asst prof biol, Northwestern Univ, 58-64; fel, cell res inst, Univ Tex, Austin, 64-65, vis assoc prof bot, 65-66, from assoc prof to prof bot, 66-72; prof biol & chmn dept, Univ South Fla, 72-74; prof biol & dean, Col Sci, 74-90, DIR, CTR ELECTRON MICROS, UNIV TEX, ARLINGTON, 85-, ASHBEL SMITH PROF BIOL, 91- *Concurrent Pos:* Adv ed, Protoplasma, 74-; adv ed & consult, J Calcified Tissue Res, 69-78; consult, Alcon Labs, Frito Lay Inc, Tampa Elec Co, many universities; chmn, Gordon Conf Calcium Oxalate, 89; pres, Tex Soc Electron Micros, 88; SW regional dir & bd dirs, Am Soc Plant Physiol, 84-91. *Honors & Awards:* Distinguished Res Award, Univ Tex, Arlington, 84. *Mem:* Am Soc Cell Biol; Bot Soc Am; Am Micros Soc; Sigma Xi; Am Mycological Soc; Am Soc Plant Physiol. *Res:* Biological ultrastructure; development and function of crystals in biological systems; organic crystals in insect viruses and animal reflecting systems; calcium oxalate crystals in plants, animals and fungi; ultrastructural ecology; electron microscopy. *Mailing Add:* Dept Biol Box 19498 Arlington TX 76019

ARNOTT, MARILYN SUE, b Lafayette, Ind, Aug 26, 43. BIOCHEMISTRY, GENETICS. *Educ:* Univ Tex, Austin, BSc, 65; Univ Tex Grad Sch Biomed Sci, Houston, MSc, 67, PhD(biochem), 70. *Prof Exp:* Instr human genetics, Univ Tex Med Br Galveston, 70-74; asst prof biol, Syst Cancer Ctr, Univ Tex M D Anderson Hosp & Tumor Inst, 74-; AT ASN WOMEN SCI. *Mem:* AAAS; Am Chem Soc; Am Soc Human Genetics; Environ Mutagen Soc; Sigma Xi. *Res:* Carcinogen metabolism; chemical carcinogenesis; protein sequence analysis; biochemical genetics. *Mailing Add:* Asn for Women in Sci RD Five Box 740 Boyertown PA 19512

ARNOTT, ROBERT A, b Spencer, WVa, Dec 4, 41; m 63; c 2. ENVIRONMENTAL & HEALTH PROTECTION. *Educ:* WVa Univ, BS, 63; Univ Wis, PhD(phys chem), 68. *Prof Exp:* Asst prof chem, Univ Wis, Oshkosh, 68-75; dir, NSF Environ Assessment Prog, 74-75; mgr, Ambient Air Monitoring Sect, Div Air Pollution Control, Ill EPA, 75-77, mgr, air qual, 77-78; dir air mgt, Wis Dept Natural Resources, 78-80; ASST DIR, COLO STATE DEPT HEALTH, 80- *Concurrent Pos:* Environ intern, German Marshall Fund, 83; mem coun public health consults, NSF, 85. *Mem:* Am Chem Soc; Sigma Xi; Air Pollution Control Asn. *Res:* Environmental trace analysis; environmental regulatory approaches; European waste management. *Mailing Add:* 7823 S Harrison Circle Littleton CO 80122

ARNOTT, STRUTHER, b Larkhall, Scotland, Sept 25, 34;; c 2. MOLECULAR BIOLOGY. *Educ:* Glasgow Univ, BSc, 56, PhD(chem), 60. *Prof Exp:* Scientist, Med Res Coun, biophys unit, King's Col, Univ London, 60-70, dir postgrad studies, dept biophys, col, 67-70; head, Dept Biol Sci, 75-80, PROF BIOL, PURDUE UNIV, 70-, VPRES RES & DEAN, GRAD SCH, 80- *Mem:* Fel Royal Soc Chem; Am Crystallog Asn; Am Soc Biol Chemists; Sigma Xi. *Res:* X-ray diffraction analysis; fibrous polymers; nucleic acids; polysaccharides; proteins; viruses. *Mailing Add:* Univ St Andrews College Gate St Andrews FiFe KY16 9AJ Scotland

ARNOVICK, GEORGE (NORMAN), b Shanghai, China, Aug 22, 25; US citizen; m 56; c 3. COMPUTER SCIENCE, INFORMATION SCIENCE. *Educ:* Univ Southern Calif, BS, 53, MA, 59. *Prof Exp:* Instr high sch, Calif, 54-56; sr systs designer, Rand Corp, Syst Develop Corp, 56-59; consult engr electrodata div, Burroughs Corp, 59-60; mgr systs anal, EDP Div, Radio Corp Am, NJ, 60-61, engr & group leader systs eng div, Data Systs Div, Calif, 61-62; prin scientist & proj engr, Space & Info Systs Div, NAm Aviation, Inc, 62-63; asst to vpres info storage & retrieval systs, Planning Res Corp, 64-65; mgr systs eng, Informatics, Inc, 65-67; dir info systs div, Tracor Inc, 67-68; pres & chmn bd, Digital Industs Inc, 68; PROF COMPUT SCI, CALIF STATE UNIV, CHICO, 68-, DIR, TECH RES CTR, GRAD SCH, 87- *Concurrent Pos:* Res eng assoc, Douglas Aircraft Corp, Calif, 55-56; NATO fels, Int Advan Study Inst on Automatic Documentation, Venice, Italy, 63 & Int Advan Study Inst on Eval Info Retrieval Systs, The Hague, 65; consult, Arnovick Assocs, Consult, 68-80 & Chico Consult Group, 80-; chmn & pres, Calif Microcomput Co, Inc, Chico, 77-80. *Mem:* AAAS; Asn Comput Mach; Am Asn Artificial Intel; Inst Elec & Electronics Engrs Computer Soc; Soc Comput Simulation; Oper Res Soc Am; Inst Mgt Sci; Am Asn Univ Professors. *Res:* Systems architecture; official information systems; management information systems; artificial intelligence software engineering; expert systems, development of case tools for system integration, Interfare requirements for multiple system platforms. *Mailing Add:* Dept Computer Sci Calif State Univ Chico CA 95929

ARNOW, THEODORE, b New York, NY, July 27, 21; m 49; c 2. GEOLOGY. *Educ:* NY Univ, BA, 42; Columbia Univ, MA, 49. *Prof Exp:* Geologist, US Geol Surv, 46-66, Dist Chief, Water Resources Div, 66-86; CONSULT, 86- *Honors & Awards:* Award, US Dept Interior, 84. *Mem:* Geol Soc Am. *Res:* Ground-water geology; Great Salt Lake, Utah. *Mailing Add:* 1064 Hillview Dr Salt Lake City UT 84124

ARNOWICH, BEATRICE, b New York, NY, June 14, 30; div; c 1. PHYSICAL ORGANIC CHEMISTRY. *Educ:* Vassar Col, AB, 51; Univ Mich, MS, 52; NY Univ, PhD(phys-org chem), 57. *Prof Exp:* Asst, NY Univ, 56-57; chemist, Charles Bruning Co, 57-59; sr chemist, Interchem Corp, 59-66, group leader, 66-69; from asst prof to assoc prof, 69-81, PROF CHEM, QUEENSBOROUGH COMMUNITY COL, 81- *Mem:* Fel AAAS; Am Chem Soc; NY Acad Sci. *Res:* Surface chemistry; photoconductivity. *Mailing Add:* Dept Chem Queensborough Community Col Bayside NY 11364

ARNOWITT, RICHARD LEWIS, b New York, NY, May 3, 28. THEORETICAL PHYSICS. *Educ:* Rensselaer Polytech Inst, BS & MS, 48; Harvard Univ, PhD(physics), 53. *Prof Exp:* Res assoc, Radiation Lab, Univ Calif, 52-54; mem, Inst Advan Study, 54-56; from asst prof to assoc prof, Syracuse Univ, 56-59; PROF PHYSICS, NORTHEASTERN UNIV, 59- *Res:* Elementary particle theory; quantum field theory. *Mailing Add:* Dept Physics Tex A&M Univ College Station TX 77843

ARNRICH, LOTTE, b Elberfeld, Ger, Sept 23, 11; US citizen. NUTRITION. *Educ:* Univ Calif, BS, 44, PhD(nutrit), 52. *Prof Exp:* Lectr nutrit, Univ Calif, 48-52, instr, 52-55; from assoc prof to prof, 55-80, EMER PROF NUTRIT, IOWA STATE UNIV, 80- *Mem:* Am Inst Nutrit; Sigma Xi. *Res:* Metabolic interactions of lipids, proteins and vitamins A and E. *Mailing Add:* 777 Cragmont Ave Berkeley CA 94708

ARNS, ROBERT GEORGE, b Buffalo, NY, July 24, 33. NUCLEAR PHYSICS. *Educ:* Canisius Col, BS, 55; Univ Mich, MS, 56, PhD(physics), 60. *Prof Exp:* Res assoc physics, Univ Mich, 59-60, instr, 60; from asst prof to assoc prof physics, State Univ NY Buffalo, 60-64; from assoc prof to prof physics, Ohio State Univ, 64-77; vpres acad affairs, 77-85, PROF PHYSICS, UNIV VT, 77- *Mem:* Am Phys Soc; Hist Sci Soc. *Res:* Nuclear structure; nuclear reactions; gamma and particle-gamma directional correlations; nuclear lifetimes; energy conversion systems; management of technology; history of science; history of technology. *Mailing Add:* Dept Physics Univ Vt Burlington VT 05405

ARNTZEN, CHARLES JOEL, b Granite Falls, Minn, July 20, 41; c 1. CELL BIOLOGY, BIOCHEMISTRY. *Educ:* Univ Minn, BS, 65, MS, 67; Purdue Univ, PhD(cell physiol), 70. *Prof Exp:* NSF fel photosynthesis, C F Kettering Res Lab, 69-70; from asst prof to prof bot, Univ Ill, Urbana, 70-80, plant physiologist, Sci Educ Admin, USDA, 76-80; prof biochem & dir, Dept Energy Plant Res Lab, Mich State Univ, 80-84; dir, Plant Sci & Microbiol, Cent Res & Develop Dept, E I du Pont de Nemours & Co, Inc, 84-87, Agr Biotechnol Res, 87-88; DEP CHANCELLOR & DEAN AGR, TEX A&M UNIV, 88- *Concurrent Pos:* Vis res scientist, Photosynthesis Lab, Europ Orgn Nuclear Res, 76 & Academia Sinica, Beijing, 83; vis prof, Australian Nat Univ, Canberra, 81; chmn, Physiol Sect, Weed Sci Soc Am, 82-83 & Biotechnol Adv Comt to President Reagan's Sci Adv, Nat Acad Sci, 85; mem, var sci comts & adv panels, USDA, NIH, NSF & nat socs, 82- *Honors & Awards:* Charles Albert Shull Award, Am Soc Plant Physiologists, 79. *Mem:* Nat Acad Sci; Nat Acad Sci India; Am Soc Plant Physiologists (pres elect, 84-85, pres, 85-86); Am Soc Cell Biol; Am Biophys Soc; Am Soc Agron; Int Plant Molecular Biol Soc; Weed Sci Soc Am. *Res:* Chloroplast membrane structure; structure-function of photosynthetic membranes; mode of action of photosynthetic herbicides; plant molecular biology. *Mailing Add:* Col Agr & Life Sci Tex A&M Univ College Station TX 77843-2142

ARNTZEN, CLYDE EDWARD, b Crescent City, Ill, Dec 26, 15; m 48; c 2. ORGANIC CHEMISTRY. *Educ:* Univ Ill, BS, 38; Iowa State Col, PhD(org chem), 42. *Prof Exp:* Engr res labs, Westinghouse Elec Corp, 42-45, engr & group leader, 45-46, mgr, Insulation Develop Sect, Mat Eng Dept, 46-48, Insulation Appln Sect, 48-52 & Chem Develop Sect, 52-53, sect mgr chem labs, 53-55, asst to mgr, Mat Eng Depts, 55-58, mgr, 58-62, mgr prog eval & div coord, Res Labs, 62-44, mgr tech serv, Res Labs, 64-76, mgr, Res & Develop Opers, 76-79; RETIRED. *Mem:* Am Chem Soc; Sigma Xi. *Res:* Synthetic organic chemistry; organic resins and plastics; chemical engineering; synthetic resins and plastics; radio; evaluation of technical projects; administration. *Mailing Add:* 5923 Tidewater Dr Jupiter FL 33458

ARNUSH, DONALD, b New York, NY, Feb 7, 36; m 66. THEORETICAL PHYSICS. *Educ:* Mass Inst Technol, SB, 57, PhD(physics), 61. *Prof Exp:* NSF fel, Max Planck Inst Physics & Astrophysics, 61-62; mem tech staff, TWR, 63-72, mgr, plasma physics dept, 72-81, dir, Energy Res Ctr, TRW Energy Tech Div, 81-89, MGR ADVAN TECHNOL, TRW, 90- *Mem:* Am Phys Soc. *Res:* Theory of fundamental particles; engineering physics; electromagnetic wave propagation; ionospheric and fusion related plasma physics; technical management; theoretical physics. *Mailing Add:* R1/1104 TRW 1 Space Park Redondo Beach CA 90278

ARNWINE, WILLIAM CARROL, b McCurtain, Okla, July 28, 29; m 51; c 5. OPERATIONS RESEARCH, INDUSTRIAL ENGINEERING. *Educ:* Okla State Univ, BSIE, 54; Iowa State Univ, MSIE, 62, PhD(opers res, indust eng), 67. *Prof Exp:* Methods engr & group leader indust eng, E I du Pont de Nemours & Co, Iowa, 54-59; from instr to asst prof indust eng & opers res, Iowa State Univ, 59-66; assoc prof, Univ Ark, 66-68; prof & head dept, Univ Nebr, Omaha, 68-69 & NMex State Univ, 69-72; staff specialist, 72-83, PROJECT MGR, ROCKWELL INT, 83- *Concurrent Pos:* Licensed methods-time measurement instr, MTM Asn for Stand & Res, 76-; lectr, Calif State Univ, Fullerton, 75 & Long Beach, 79; work measurement ed, Am Inst Indust Engrs Transactions, 80-85. *Mem:* Am Inst Indust Engrs; Opers Res Soc Am; fel Methods-Time Measurement Asn; Sigma Xi; Am Mgt Asn. *Res:* Allocating multiple warheads; queueing systems; sequencing problems; scheduling models; computer approaches for curve fitting; accuracy of work measurement systems. *Mailing Add:* 508 N Cornell Ave Fullerton CA 92631-2711

ARNY, DEANE CEDRIC, b St Paul, Minn, May 22, 17; m 47; c 5. AGRONOMY. *Educ:* Univ Minn, BS, 39; Univ Wis, PhD(agron, plant path), 43. *Prof Exp:* From instr to prof, 43-84, EMER PROF PLANT PATH, UNIV WIS, MADISON, 84- *Concurrent Pos:* Vis prof, Univ Ife, Nigeria, 66-68, Andalas Univ, Indonesia, 86. *Honors & Awards:* Ruth Allen Award, Am Phytopath Soc, 87. *Mem:* Am Phytopath Soc; Am Soc Agron. *Res:* Diseases of field crops-corn, smallgrains, alfalfa. *Mailing Add:* Dept Plant Path Univ Wis 1630 Linden Dr Madison WI 53706

ARNY, MARGARET JANE, b Madison, Wis, June 18, 48. IMMUNOLOGY. *Educ:* Earlham Col, BA, 71; Sloan-Kettering Inst, PhD(biochem), 79. *Prof Exp:* RES ASST PROF, DEPT OBSTET & GYNECOL, NY UNIV MED CTR, 83- *Mem:* Am Fertil Soc; Soc Study Reproduction; Am Acad Advan Sci; NY Acad Sci. *Res:* Early embryo development and implantation. *Mailing Add:* Nine Sunshine Rd Shelter Island NY 11964

ARNY, THOMAS TRAVIS, b Montclair, NJ, June 2, 40; m 68. ASTROPHYSICS. *Educ:* Haverford Col, BA, 61; Univ Ariz, PhD(star formation), 65. *Prof Exp:* Res assoc astron, Amherst Col, 65-66; asst prof, 66-69, chmn, Five Col Astron Dept, 75-81, ASSOC PROF ASTRON, UNIV MASS, AMHERST, 69- *Concurrent Pos:* NSF grant, 67-71. *Mem:* Am Astron Soc; Royal Astron Soc; Int Astron Union. *Res:* Star formation; interstellar gas dynamics. *Mailing Add:* Dept Astron GRC B Univ Mass Amherst MA 01003

AROESTY, JULIAN MAX, b Rochester, NY, Sept 12, 31; m 71; c 4. CORONARY ANGIOPLASTY, CARDIAC CATHETERIZATION. *Educ:* Cornell Univ, BA, 53; SUNY Syracuse, MD, 60. *Prof Exp:* Instr cardiol, 66-71, ASSOC CLIN PROF MED, HARVARD MED SCH, 71- *Concurrent Pos:* clin asst, Harvard Med Sch, 66-71, clin assoc, 71-86. *Mem:* fel Am Col Physicians; fel Am Col Cardiol; fel Am Heart Assoc; fel Soc Cardiac Angioplasty; fel Am Assoc Adv Sci. *Res:* Cardiogenic shock, valvular and coronary heart disease, permanent pacemaker therapy, coronary angioplasty and valvuloplasty. *Mailing Add:* 333 Longwood Ave Boston MA 02115

AROIAN, LEO AVEDIS, b Holden, Mass, May 10, 07; m 41; c 2. MATHEMATICAL STATISTICS. *Educ:* Univ Mich, AB, 28, MA, 29, PhD, 40. *Prof Exp:* Asst prof math, Colo State Univ, 30-39; instr, Hunter Col, 39-47, asst prof, 48-50; head math sect, Hughes Aircraft Co, 50-59, sr math consult, 59-60; mem tech staff, space technol labs, Thompson-Ramo-Wooldridge, Inc, 60-67, sr staff mathematician, TRW Systs, 67-68; prof indust admin, Union Col, 68-72, res prof admin & mgt, 76-87, emer prof, 72-; RETIRED. *Concurrent Pos:* Rackham fel, Univ Mich, 38-39; res assoc, Mtrop Life Ins Co, 43-44, Statist Lab, Univ Calif, 45; mem Nat Defense Res Comt, 45; vis prof, dept math, Western Ill Univ, 76-77. *Honors & Awards:* Res Award, Am Soc Qual Control, 70. *Mem:* Am Math Soc; Math Asn Am; fel Am Soc Qual Control; fel Am Statist Asn; Inst Math Statist; Sigma Xi. *Res:* High speed computation; numerical methods; mathematical, applied and industrial statistics; reliability theory; sequential analysis; time series; security investments. *Mailing Add:* 32 Union Ave Schenectady NY 12308

ARON, GERT, b Konigsberg, Ger, July 10, 27; US citizen; m 78; c 2. HYDROLOGY, WATER RESOURCES. *Educ:* Univ Iowa, MS, 60; Univ Calif, PhD(hydrol), 69. *Prof Exp:* Flood control engr, Rudolf Goetze Consult, 60-61; hydraul engr, Harza Eng Co, 61-63; res engr hydraul, Univ Calif, Davis, 63-69; from asst prof to assoc prof, 69-81, PROF CIVIL ENG, PA STATE UNIV, UNIV PARK, 81- *Concurrent Pos:* Ground water specialist, UNESCO, 70-73; hydraul eng consult for various firms. *Mem:* Am Soc Civil Eng; Am Water Resources Asn; Am Geophys Union. *Res:* Water resources distribution planning; drought and flood frequencies; urban storm runoff modeling; hydrologic modeling. *Mailing Add:* Dept Civil Eng Pa State Univ Main Campus University Park PA 16802

ARON, WALTER ARTHUR, b Milwaukee, Wis, Aug 30, 21; m 43, 72; c 3. THEORETICAL PHYSICS. *Educ:* Univ Calif, PhD(physics), 51. *Prof Exp:* Physicist, Radiation Lab, Mass Inst Technol, 44-46; physicist, Radiation Lab, Univ Calif, 46-51; res assoc physics, Princeton Univ, 51-55; asst prof, Univ Va, 55-59; theoret physicist, Lawrence Radiation Lab, Univ Calif, Berkeley, 59-64; res physicist, Stanford Res Inst, 64-71; scientist, Sci Applns Inc, 71-83. *Concurrent Pos:* Consult, 83- *Mem:* Sigma Xi; AAAS. *Res:* Particle interactions; data analysis; acoustics; optical transmission. *Mailing Add:* 7 Manor Place Menlo Park CA 94025-3714

ARON, WILLIAM IRWIN, b Brooklyn, NY, June 26, 30; m 61; c 2. BIOLOGICAL OCEANOGRAPHY. *Educ:* Brooklyn Col, BS, 52; Univ Wash, MS, 57, PhD, 60. *Prof Exp:* Biologist, Wash State Dept Fisheries, 52-53, res instr oceanog, Univ Wash, 56-60, res asst prof, 60-61; res supvry biologist, defense res labs, Gen Motors Corp, 61-67; dep head, off oceanog & limnol, Smithsonian Inst, 67-69, dir oceanog & limnol prog, 69-71; dir, Off Ecol & Conserv, 71-78, dir, Off Marine Mammals & Endangered Species, 78-80, DIR, NORTHWEST & ALASKA FISHERIES CTR, NAT OCEANIC & ATMOSPHERIC ADMIN, 80- *Concurrent Pos:* Consult, Vet Corps, US Army, 56-, President's Marine Sci Comn, 68-69 & Nat Water Comn, 69-; mem gen adv comt, Mediter Marine Sorting Ctr, Tunisia, 70-; US del, Coop Invests of the Mediter, 70-; mem comt oceanog biol methods panel, Nat Acad Sci; mem bd dirs, Nat Oceanog Instrumentation Ctr. *Mem:* AAAS; Am Soc Limnol & Oceanog; Am Soc Ichthyologists & Herpetologists; Soc Study Evolution; Sigma Xi. *Res:* Instrumentation for sampling; zoogeography of pelagic fish and plankton; research administration. *Mailing Add:* Northwest-Alaska Fisheries Ctr 7600 Sand Point Bin C15700 Seattle WA 98115

ARONI, SAMUEL, b Kishinew, Romania, May 26, 27; m 56; c 2. STRUCTURES, URBAN PLANNING. *Educ:* Univ Melbourne, BCE, 55; Univ Calif, Berkeley, MS & PhD(struct eng & mech), 66. *Prof Exp:* Lectr civil eng, Univ Melbourne, 55-62; teaching fel & res asst struct eng & mech, Univ Calif, Berkeley, 63-66; assoc prof eng, San Francisco State Col, 66-67; res engr, Am Cement Corp, 67-70; actg dean, Grad Sch Archit & Urban Planning, 74-75 & 83-85, PROF ARCHIT & URBAN PLANNING, UNIV CALIF, LOS ANGELES, 70- *Concurrent Pos:* Consult, Adv Comt Dept Housing & Urban Develop, Nat Acad Sci & Nat Acad Eng, 70-73; mem exec comt, Int Tech Coop Ctr, 71-77; vpres, Urban Innovations Bd, 74-75, chmn, 83-85; mem comt on socioecon effects of earthquake predictions, Nat Acad Sci, 76-78; chmn, Acad Senate, Univ Calif, Los Angeles, 81-82; vpres, Archit Res Ctrs Consortium, 85-86; mem, Adv Comt, NSF, 85-90; mem bd Gov, Ben-Gurion Univ Nyev, Israel, 83- *Honors & Awards:* J James R Croes Gold Medal, Am Soc Civil Engrs, 81. *Mem:* Am Soc Civil Engrs; Am Concrete Inst. *Res:* Cementitious materials; concrete technology; structures; earthquakes; low-cost housing, policy and technology. *Mailing Add:* Grad Sch Archit & Urban Planning Univ Calif Los Angeles CA 90024

ARONIN, LEWIS RICHARD, b Norwood, Mass, Aug 4, 19; m 47; c 2. MATERIALS SCIENCE, PHYSICAL METALLURGY. *Educ:* Mass Inst Technol, BS, 40. *Prof Exp:* Asst res dir precision instruments, Waltham Watch Co, 40-49; staff mem, Metall Proj, Mass Inst Technol, 49-54; dept mgr, Nuclear Metals Inc, 54-66; consult econ eval, Kennecott Copper Ledgemont Lab, 66-67; mat engr res & develop, 67-90, CONSULT ADVAN MAT DEVELOP, ARMY MAT TECHNOL LAB, 90- *Mem:* Am Inst Mining, Metall & Petrol Engrs-Metall Soc; Sigma Xi; Soc Advan Mat & Process Eng; Am Soc Metals; Am Inst Aeronaut & Astronaut. *Res:* Materials research and development in aerospace and nuclear materials with emphasis on refractory metals, beryllium, and advanced composite materials. *Mailing Add:* 20 Ingleside Rd Lexington MA 02173

ARONOFF, GEORGE R, b Peoria, Ill, Mar 6, 50. NEPHROLOGY, CLINICAL PHARMACOLOGY. *Educ:* Indiana Univ, BA, 72, MS, 84 & MD, 75. *Prof Exp:* From asst prof to assoc prof med, Indiana Univ, 80-87; CHIEF, DIV NEPHROL & PROF MED & PHARMACOL,UNIV LOUISVILLE, 87- *Mem:* fel Am Col Physicians; Am Asn Advan Sci; Am Soc Clin Pharmacol & Therapeut; Am Soc Nephrol. *Res:* Drug disposition in patients with renal disease; drug nephrotoxicity. *Mailing Add:* 500 S Floyd St Louisville KY 40292

ARONOFF, SAMUEL, b New York, NY, Feb 27, 15; m 36; c 3. BIOCHEMISTRY. *Educ:* Univ Calif, Los Angeles, AB, 36; Univ Calif, PhD(physico-chem biol), 42. *Prof Exp:* Asst agr, Univ Calif, 41-42, res fel, 42-43; instr chem, Boston Univ, 43-44; res instr, Univ Chicago, 44-46; mem staff, Radiation Lab, Univ Calif, 46-48; from assoc prof to prof bot, Iowa State Univ, 48-60, prof biochem & biophys, 60-69; dean, Grad Sch Arts & Sci & vpres res, Boston Col, 69-71; dean sci, 71-76, PROF CHEM, SIMON FRASER UNIV, 76- *Concurrent Pos:* Sr fel, NSF, 57-58, prog dir molecular biol, 63-64, consult, 64-65. *Mem:* Am Chem Soc; Am Soc Plant Physiol (secy, 62-63, vpres, 64, pres, 65); Am Soc Biol Chemists; Can Soc Plant Physiol. *Res:* Plant biochemistry; photosynthesis; chlorophyll; radiobiochemistry; boron biochemistry. *Mailing Add:* 76 E Vista del Mar Camano Island WA 98292

ARONOFSKY, JULIUS S, b Dallas, Tex, July 29, 21; m 45; c 4. APPLIED MATHEMATICS. *Educ:* Southern Methodist Univ, BS, 44; Stevens Inst Technol, MS, 46; Univ Pittsburgh, PhD(math), 49. *Prof Exp:* Engr, Kellex Corp, 44-45; res engr, Westinghouse Elec Co, 47-49; sr res engr, Magnolia Petrol Co, 49-55, res assoc, 55-57; mgr electronic computer ctr, Socony Mobil Oil Co, Inc, 57-65, mgr opers res dept, 65-68; prof statist & opers res, Univ Pa, 68-70; lectr, 48-57, prof statist & opers res, 70-72, PROF MGT SCI & COMPUT, SOUTHERN METHODIST UNIV, 72-, PROF INDUST ENG & OPER RES, 76- *Concurrent Pos:* Consult, Universal Comput Co, 70- *Honors & Awards:* Lester C Uren Award, Soc Petrol Engrs, 81. *Mem:* Am Soc Mech Eng; Am Inst Mining, Metall & Petrol Eng; Am Phys Soc; Opers Res Soc Am; Inst Mgt Sci. *Res:* Flow of fluids through porous media; elasticity; plastic flow of metals; computer technology; operations research; management science. *Mailing Add:* 5503 Williamstown Rd Dallas TX 75230

ARONOVIC, SANFORD MAXWELL, b New York, NY, June 10, 26; m 57; c 3. ANALYTICAL CHEMISTRY. *Educ:* Columbia Univ, BS, 45; Univ Wis, PhD(chem), 57. *Prof Exp:* Chemist microanal, Lederle Labs Div, Am Cyanamid Co, 47-52, res chemist anal develop, Stamford Res Labs, 57-62; res asst infrared spectros, Univ Wis, 52-57; res anal chemist, Maumee Chem Co, 62-65; res scientist, Union Bag-Camp Paper Corp, 65, res chemist, Union Camp Corp, 65-68; supvr qual control, 69-72, supvr anal serv, 72-76, supvr toxicol, 76-77, coordr environ & regulatory affairs, 77-81, CONSULT, THIOKOL CHEM CORP, 81- *Mem:* AAAS; Am Chem Soc; Coblentz Soc; Sigma Xi. *Res:* Microanalytical chemistry; infrared spectroscopy; instrumental methods of analysis; analytical separations, identification and separations, analytical methods development; gas chromatography. *Mailing Add:* 351 Franklin Ave Princeton NJ 08540

ARONOW, LEWIS, b New York, NY, Mar 28, 27; m 53; c 4. PHARMACOLOGY. *Educ:* City Col New York, BS, 50; Georgetown Univ, MS, 52; Harvard Univ, PhD(pharmacol), 56. *Prof Exp:* Chemist, Lab Chem Pharmacol, Nat Heart Inst, 50-52; fel pharmacol, Harvard Univ, 52-56; from instr to prof, Stanford Univ, 56-76; PROF PHARMACOL & CHMN DEPT, UNIFORMED SERV UNIV HEALTH SCI, 76- *Concurrent Pos:* Vis prof pharmacol & chmn dept, Sch Med Jose Vargas, Univ Cent Venezuela, 62-63; Am Cancer Soc/Eleanor Roosevelt Int Cancer fel, Cambridge Univ, Eng, 70-71; actg chmn dept pharmacol, Stanford Univ, 74-76. *Honors & Awards:* Premio Martin Vegas, Soc Venezolana Dermat, 64; Maloney Lectr, Howard Univ, 77. *Mem:* Am Soc Pharmacol & Exp Therapeut; Am Asn Cancer Res. *Res:* Glucocorticoid hormones; anti-cancer drugs; biochemical and cellular pharmacology; drug metabolism. *Mailing Add:* Dept Pharmacol Uniformed Serv 4301 Jones Bridge Rd Bethesda MD 20814

ARONOW, SAUL, b Brooklyn, NY, Dec 10, 23; m 48; c 3. GEOMORPHOLOGY, SOILS. *Educ:* Brooklyn Col, BA, 45; Univ Iowa, MS, 46; Univ Wis, PhD, 55. *Prof Exp:* Draftsman, Army Map Serv, 45; geologist, US Geol Surv, 48-52; asst geol, Univ Wis, 53-54; PROF GEOL, LAMAR UNIV, 55- *Concurrent Pos:* Consult, Tex Bur Econ Geol, 65-70, Soil Conserv Serv, 65-, Minn Geol Serv, 72-73 & 78-80, Nat Park Serv, 79-, Corps Engrs, 84 & Archeol Res Lab, Tex A&M Univ, 89- *Mem:* AAAS; Am Geol Soc; Sigma Xi; Am Quaternary Asn; Soil Sci Soc Am. *Res:* Geomorphology; glacial geology; ground water; Pleistocene of Gulf Coast region; soils; salt domes; geoarcheology. *Mailing Add:* Dept Geol Box 10031 Lamar Univ Sta Beaumont TX 77710

ARONOW, SAUL, b Brooklyn, NY, Oct 4, 17; m 42; c 6. BIOMEDICAL ENGINEERING, RADIOLOGICAL PHYSICS. *Educ:* Cooper Union, BEE, 39; Harvard Univ, MS, 46, PhD(appl sci), 53. *Prof Exp:* Electronics engr, Harvey Radio Labs, 46-49; from asst physicist to assoc appl physicist, Mass Gen Hosp, 53-68, dir med eng group, 68-72, assoc physicist, Dept Radiol, 72-82, CONSULT PHYSICS, MASS GEN HOSP, 82- *Concurrent Pos:* Prin assoc med, Harvard Med Sch, 54-; lectr, Northeastern Univ, 56- & Mass Inst Technol, 57-82; vis prof, Denmark Tech Univ, 69; consult engr, Mass; pres, Technol Med Res Inst; engr, Proj Hope, Technol Med, Inc, 82-83. *Honors & Awards:* Gano Dunn Medal. *Mem:* Fel Inst Elec & Electronic Engrs; Am Asn Physicists in Med; Soc Nuclear Med; Biomed Eng Soc; Sigma Xi. *Res:* Medical applications of electronic instrumentation; nuclear medicine. *Mailing Add:* 86 Crofton Rd Waban MA 02168

ARONOW, WILBERT SOLOMON, b New York, NY, Oct 30, 31; m 58; c 2. INTERNAL MEDICINE. *Educ:* Queens Col, NY, BS, 53; Harvard Univ, MD, 57. *Prof Exp:* Residency internal med & cardiovascular dis, Michael Reese Hosp & Med Ctr, 57-61; chief cardiol & gen med, US Army Hosp, Ft Chaffee, Ark, 61-63; staff internist, Vet Admin Hosp, E Orange, 63-64; staff cardiologist, Vet Admin Hosp, Long Beach, 64-82, chief, Cardiovasc Sect, 73-82, asst chief med res, 75-80; prof med & chief cardiovasc res, Creighton Univ Sch Med, Omaha, Nebr, 82-84; MED DIR, HEBREW HOSP CHRONIC SICK, BRONX, NY, 84- *Concurrent Pos:* Asst clin prof med, Univ Calif Irvine, 68-72, assoc prof med, 72-75, vchief, Cardiovasc Div & chief cardiovasc res, 74-82, prof med, 75-82 & prof pharmacol & therapeut, 76-82; mem cardiovasc & renal adv comt, US Food & Drug Admin, 73-76; chmn, Cardiovasc & Pulmonary Dis Sect, Am Soc Clin Pharmacol & Therapeut, 73-74 & 75-77; distinguished lectr, Am Col Angiol, 73; consult, US Environ Protection Agency, 73 & 78-83, Calif Air Resources Bd, 73, 78, 80 & 82, NIH, 76 & 80, US Dept Justice Law Enforcement Assistance Admin, 78, Dept Health, WGer, 78, Nat Heart, Lung & Blood Inst, 79, Nat Cancer Inst, 80, Dept Health & Environ Sci, Mont, 80 & Fed Trade Comn, 80-81; Consult, Dept Drugs, AMA, 3rd ed of AMA Drug Eval, 74, 4th ed, 78 & 5t ed, 82 & Nat Ctr Health Statist, 81; pres, Asn Vet Admin Cardiologists, 75-77; vis prof med, Univ Tex Southwestern Med Sch, Dallas, 76, Univ Man & Univ Toronto, 79, Rutgers Med Sch, 83 & pharmacol & therapeut, Texas Tech Univ Sch Med, Lubbock, 83; consult smoking, Am Heart Asn, 80 & mem subcomt, 80-83; guest field ed, J Pharmacol & Exp Therapeut, 81; attend physician, Geront Geriat Soct, Dept Med, Westchester County Med Ctr, Valhalla, NY, 85-87; consult, State NY Dept Health, Off Pub Health, 86; fel, Coun Geriat Cardiol; cardiol consult, Albert Einstein Col Med, 90-; attending physician, Cardiol Div, Bronx Munic Hosp Ctr, 90- *Mem:* Fel Am Col Chest Physicians; Am Soc Clin Pharmacol & Therapeut; fel Am Col Physicians; fel Am Col Cardiol; fel Am Heart Asn; fel Am Geriat Soc; Am Fedn Clin Res; AAAS; mem Am Med Dir Asn; fel Am Col Angiol; fel Coun Geriat Cardiol; Geront Soc Am. *Res:* Cardiovascular diseases; geriatrics. *Mailing Add:* 23 Pebble Way New Rochelle NY 10804

ARONOWITZ, FREDERICK, b New York, NY, July 3, 35. QUANTUM PHYSICS. *Educ:* Polytech Inst Brooklyn, BS, 56; NY Univ, PhD(physics), 69. *Prof Exp:* Physicist, David Taylor Model Basin, 56; consult, Eastern Res Group, 61-62; staff scientist, Honeywell, Inc, 62-80, sect head, 80-83; mgr, Raytheon, 83-84; CHIEF SCIENTIST, ROCKWELL INT, 84- *Honors & Awards:* Prize for Indust Applns of Physics, Am Inst Physics, 83; Elmer A Sperry Award, 84; Kirschner Award, Inst Elec & Electronics Engrs, 88; Michelson Award, Navy League, 90; Technol Achievement Award, Int Soc Optical Eng, 90. *Mem:* Am Phys Soc. *Res:* Quantum electronics with major emphasis of the theory and development of the laser gyroscope. *Mailing Add:* 93 Birch Hill Rd Stow MA 01775

ARONS, ARNOLD BORIS, b Lincoln, Nebr, Nov 23, 16; m 42; c 4. OCEANOGRAPHY. *Educ:* Stevens Inst Technol, ME, 37, MS, 40; Harvard Univ, PhD(phys chem), 43. *Hon Degrees:* AM, Amherst Col, 53; DE, Stevens Inst Technol, 82. *Prof Exp:* Asst, Harvard Univ, 42-43; instr chem, Stevens Inst Technol, 37-40, from asst prof to assoc prof, 46-52; prof, Amherst Col, 68; prof physics, 68-82, EMER PROF, UNIV WASH, 82- *Concurrent Pos:* Res assoc, Oceanog Inst, Woods Hole, 43-72, mem corp, 63-, trustee, 64-; chief sect underwater blast measurements, Oper Crossroads, Navy Bur Ord, 46; Guggenheim fel, 57-58; NSF faculty fel, 62-63; consult, Naval Ord lab, Waterways Exp Sta; mem, Comn Col Physics, 62-68; gov bd, Am Inst Physics, 66-72; baccalaureate degree proj, Asn Am Cols, 82-85; res comt, math, sci & technol educ, Nat Acad Sci-Nat Res Coun, 83-85; adv comt on sci & eng educ, NSF, 84-85. *Honors & Awards:* Honor Medal, Stevens Inst Technol, 68; Oersted Medal, Am Asn Physics Teachers, 72. *Mem:* AAAS; Am Phys Soc; Am Asn Physics Teachers (pres, 67); Am Geophys Union; Nat Sci Teachers Asn. *Res:* Explosion phenomena; development of tourmaline piezoelectric gauges for measurement of explosion pressure waves; physical oceanography; hydrodynamics of rotating systems; science education; cognitive development. *Mailing Add:* Dept Physics Univ Wash Seattle WA 98195

ARONS, JONATHAN, b Philadelphia, Pa, Aug 16, 43; m 74; c 1. THEORETICAL ASTROPHYSICS, THEORETICAL PLASMA PHYSICS. *Educ:* Williams Col, BA, 65; Harvard Univ, AM, 69, PhD(astron), 70. *Prof Exp:* Res assoc astron, Princeton Univ Observ, 70-71; mem astrophys staff, Inst Advan Study, 71-72; from asst prof to assoc prof astron, 72-82, assoc prof physics, 80-82, PROF ASTRON & PHYSICS, UNIV CALIF, BERKELEY, 82- *Concurrent Pos:* Guggenheim Found fel, 80-81; mem, Physics Surv Subcomt, Space & Astrophys Plasmas, Nat Acad Sci, 83-84; vchmn, Astron Dept, Univ Calif, Berkeley, 84-85; Miller Found res prof, 85-86; vchmn & chmn, Astrophys Div, Am Phys Soc, 84-86; mem adv bd, Inst Theoret Physics, 88-91; actg chmn astron dept, Univ Calif, Berkeley, 90-91; mem, Plasma Sci Comt, Nat Acad Sci, 90-93. *Mem:* Am Astron Soc; fel Am Phys Soc; Am Geophys Union; Int Astron Union. *Res:* Astrophysical plasma physics; coherent radiation from plasmas; high energy astrophysics. *Mailing Add:* Dept Astron 601 Campbell Hall Univ Calif Berkeley CA 94720

ARONS, MICHAEL EUGENE, b New York, NY, Mar 29, 39; m 61; c 2. PHYSICS. *Educ:* Cooper Union, BEE, 59; Univ Rochester, PhD(theoret physics), 64. *Prof Exp:* Asst res scientist theoret physics, NY Univ, 64-66, asst prof physics, 66-70; assoc prof, 70-82, chmn dept, 81-87, PROF PHYSICS, CITY COL NY, 82-, DEAN SCI, 88- *Res:* Field theory and quantum electrodynamics; theories of elementary particles, including symmetries and field-theoretic models. *Mailing Add:* Div Sci City Col New York New York NY 10031

ARONSON, A L, b Minneapolis, Minn, Aug 24, 33; m 56; c 3. PHARMACOLOGY, VETERINARY MEDICINE. *Educ:* Univ Minn, BS, 55, DVM, 57, PhD(pharmacol), 63; Cornell Univ, MS, 59. *Prof Exp:* Teaching asst pharmacol, Cornell Univ, 57-58; res fel, Univ Minn, 58-63, res assoc, 63-64; from asst prof to prof, State Univ NY Vet Col, Cornell Univ, 64-80; PROF PHARMACOL & HEAD DEPT ANAT, PHYSIOL SCI & RADIOL, NC STATE UNIV, 80- *Concurrent Pos:* Chmn, Coun Biol & Therapeut Agts, Am Vet Med Asn, 85-86. *Mem:* Am Vet Med Asn; Am Acad Vet Pharmacol & Therapeut (pres-elect, 85-87, pres, 87-89); Soc Toxicol; Am Soc Pharmacol & Exp Therapeut. *Res:* Clinical pharmacology; toxicology. *Mailing Add:* Dept Anat Physiol Sci & Radiol NC State Univ Sch Vet Med Raleigh NC 27606

ARONSON, ARTHUR H, b July 9, 35. METALLURGICAL ENGINEERING. *Educ:* Mass Inst Technol, BS, 58; Rensselaer Polytech Inst, PhD(metall), 65. *Prof Exp:* Pres & chief exec officer, Cold Metal Prod, 80-84; pres & chief oper officer, Lukens Steel, 84-88; EXEC VPRES, ALLEGHENY LUDLUM CORP, 88-, CHIEF OPER OFFICER, 90- *Mem:* Fel Am Soc Metals. *Res:* Metallurgical engineering. *Mailing Add:* Allegheny Ludlum Corp River Rd Brackenridge PA 15104

ARONSON, ARTHUR IAN, b Boston, Mass, July 2, 30; m 56; c 2. MICROBIOLOGY. *Educ:* Univ Chicago, BA, 50; Univ Mass, MS, 53; Univ Ill, PhD(microbiol), 58. *Prof Exp:* Nat Found Polio fel, 58-59; fel, Carnegie Inst, 59-60; from asst prof to assoc prof, 60-67, PROF MICROBIOL, PURDUE UNIV, 67- *Concurrent Pos:* Guggenheim fel, 68-69. *Mem:* AAAS; Am Soc Microbiol; Am Soc Biol Chemists. *Res:* Molecular biology; bacterial spore formation; regulation in bacilli; structure of insect protoxin genes. *Mailing Add:* Dept Biol Sci Purdue Univ Lilly Hall West Lafayette IN 47907

ARONSON, CARL EDWARD, b Providence, RI, Mar 14, 36; m 60; c 2. PHARMACOLOGY, TOXICOLOGY. *Educ:* Brown Univ, AB, 58; Univ Vt, PhD(pharmacol), 66. *Hon Degrees:* MA, Univ Pa, 73. *Prof Exp:* Res technician, Worcester Found Exp Biol, Shrewsbury, Mass, 58-60; instr pharmacol, Sch Med, 67-70, assoc, 70-71, asst prof, Sch Vet Med, 71-73, asst prof pharmacol, Sch Med, 71-75, head labs pharmacol & toxicol, 72-86, ASSOC PROF PHARMACOL & TOXICOL, DEPT ANIMAL BIOL, SCH VET MED, UNIV PA, 73-, ASSOC PROF PHARMACOL, SCH MED, 75- *Concurrent Pos:* Fel pharmacol, Univ Pa, 65-67, Pa Plan scholar, 69-71; Heart Asn Southeastern Pa res fel, 69-72; lectr pharmacol, Div Grad Med Educ, Sch Med, Univ Pa, 65-74, actg assoc dean student affairs, Sch Vet Med, 74-75; vis res scientist, Philadelphia Vet Admin Med Ctr, 86- *Mem:* Sigma Xi; Am Acad Vet Pharmacol & Therapeut (pres, 83-85); Am Acad Vet & Comp Toxicol; Am Soc Pharmacol & Exp Therapeut; Am Vet Med Asn. *Res:* Biochemical pharmacology and toxicology; effects of drugs and hormones on cardiac performance and metabolism; effects of drugs on neurotransmitters and their receptors in the CNS. *Mailing Add:* Dept Animal Biol Lab Pharmacol & Toxicol Sch Vet Med Univ Pa 3800 Spruce St Philadelphia PA 19104

ARONSON, CASPER JACOB, b Canisteo, NY, Sept 1, 16; m 43; c 1. PHYSICS, EXPLOSIVES. *Educ:* Univ Rochester, BS, 38, MS, 39. *Prof Exp:* Asst, Univ Rochester, 38-40; physicist, Carnegie Inst, 40-41 & chief, Magnetic Model Sect, 44-46 & Magnetic Fields Sect, 46-49, sr res assoc, Explosives Res Dept, 49-52, dep chief, Explosion Effects Div, 52-53, asst chief atomic tests, 53-57, chief, Explosion Res Dept, 57-74, actg assoc head chem & explosions res, US Naval Ord Lab, 74-75; scientific asst res & technol, Naval Surface Weapons Ctr, 75 & Ablard Enterprises Inc, 75-76; scientific asst res & technol, VSE Corp, 80-87; RETIRED. *Honors & Awards:* Flemming Award, 55. *Mem:* AAAS; Am Phys Soc; Optical Soc Am; Sigma Xi; NY Acad Sci. *Res:* Ship's magnetic fields; shock waves; effects of explosions. *Mailing Add:* 3401 Oberon St Kensington MD 20895-2935

ARONSON, DONALD GARY, b Jersey City, NJ, Oct 2, 29; m 53; c 3. MATHEMATICS. *Educ:* Mass Inst Technol, SB, 51, SM, 52, PhD(math), 56. *Prof Exp:* Res assoc digital comput lab, Univ Ill, 56-57; from instr to assoc prof, 57-65, PROF MATH, UNIV MINN, MINNEAPOLIS, 65- *Concurrent Pos:* NSF fel, 61-62; Soc Indust & Appl Math vis lectr, 64-66. *Mem:* Am Math Soc; Math Asn Am. *Res:* Partial differential equations. *Mailing Add:* Sch Math Univ Minn Inst Technol Minneapolis MN 55455

ARONSON, HERBERT, b Braunschweig, Ger, Feb 23, 23; US citizen; m 54; c 2. EXPERIMENTAL PSYCHOLOGY. *Educ:* City Col New York, BME, 49; Newark Col Eng, MSME, 54. *Prof Exp:* Liaison engr, Hydropress Co, 49-52; proj engr, Morey Mach Co, 52-53, Curtis Wright and Pratt & Whitney, 53-55, Silent Hoist & Crane Co, 55-57, Sperry Gyroscope Co, 57-59 & Am Mach & Foundry, 59-64; ASSOC PROF MECH ENG, NY INST TECHNOL, 64- *Concurrent Pos:* Eng consult, 60- *Mem:* Soc Automotive Engrs; Am Soc Mech Engrs; Am Soc Safety Engrs. *Res:* Application of the detection theory model to the study of attitudinal factors in temporal integration in vision. *Mailing Add:* 101-06 67th Dr Forest Hills NY 11375

ARONSON, JAMES RIES, b Chicago, Ill, Sept 1, 32; m 63; c 5. SPECTROSCOPY, SCATTERING. *Educ:* Northwestern Univ, BS, 54; Mass Inst Technol, PhD(chem), 58. *Prof Exp:* Res assoc, Mass Inst Technol, 58-59; res scientist, Arthur D Little, Inc, 59-86; res prof appl optics, Fairleigh Dickinson Univ, 87-88; CONSULT, 86- *Mem:* Fel Optical Soc Am. *Res:* Infrared physics and spectroscopy; remote sensing; theory of emittance and reflectance of particulate media; optical constants of anisotropic solids; applications of spectroscopy to earth sciences and pollution; laser spectroscopy; infrared instrumentation; scientific software development. *Mailing Add:* 20 Ridgefield Rd Winchester MA 01890

ARONSON, JAY E, b Boston, Mass, July 19, 53; m 80; c 3. MATHEMATICAL PROGRAMMING. *Educ:* Carnegie-Mellon Univ, BS, 75, MS(elec eng), 76, MS(operations res), 78, PhD(indust admin), 80. *Prof Exp:* Instr operations mgt, Carnegie-Mellon Univ, 78-80; asst prof operations res & eng mgt, 80-84, asst prof mgt info systs, Southern Methodist Univ, 84-87; ASSOC PROF MGT SCI & INFO TECHNOL, COL BUS ADMIN, UNIV GA, 87- *Concurrent Pos:* Consult, 75-; asst prof mgt sci & comput, Southern Methodist Univ, 80- *Mem:* Operations Res Soc Am; Inst Mgt Sci; Am Production & Inventory Control Soc; Math Prog Soc; Am Soc Eng Mgt. *Res:* Large scale mathematical programming; staircase linear programming; planning models; networks; forward algorithms; artificial intelligence; decision support systems; operations research microcomputer applications, Economic Equilibria. *Mailing Add:* Mgt Sci & Info Technol Col Bus Admin Univ Ga Athens GA 30602

ARONSON, JEROME MELVILLE, b Oakland, Calif, May 20, 30; m 55; c 2. EXPERIMENTAL MYCOLOGY. *Educ:* Univ Calif, BA, 52, PhD(bot), 58. *Prof Exp:* Lectr bot, Univ Calif, 57; NIH fel, 58-59; asst prof biol, Wayne State Univ, 59-63; lectr, Univ Calif, Berkeley, 64-66; assoc prof, 66-71, PROF BOT, ARIZ STATE UNIV, 71- *Concurrent Pos:* assoc ed, Exp Mycol, 79- *Mem:* Am Soc Plant Physiol; Bot Soc Am; Mycol Soc Am; Sigma Xi. *Res:* Cellwall biosynthesis, chemistry of fungal cell walls and polysaccharide storage products; biochemical systematics of fungi. *Mailing Add:* Dept Bot MICRD 3 Ariz State Univ Tempe AZ 85287-1601

ARONSON, JOHN FERGUSON, b Philadelphia, Pa. PHYSIOLOGY. *Educ:* Amherst Col, BA, 50; Univ Rochester, PhD(biol), 60. *Prof Exp:* USPHS fel, Dartmouth Med Sch, 60-63; asst prof med, 73-77, STAFF MEM, WISTAR INST ANAT & BIOL, UNIV PA, 77- *Mem:* Soc Gen Physiologists. *Res:* Cell physiology. *Mailing Add:* 601 Childs Ave Drexel Hill PA 19104

ARONSON, JOHN NOEL, b Dallas, Tex, Mar 15, 34; m 58; c 3. BIOCHEMISTRY. *Educ:* Rice Inst, BA; Univ Wis, MS, 55, PhD(biochem), 59. *Prof Exp:* Asst prof chem, Ariz State Univ, 59-65; ASSOC PROF CHEM, STATE UNIV NY ALBANY, 65- *Concurrent Pos:* NIH res fel oncol, Univ Wis, 59; NSF fac fel, Ind Univ, 63-64; Am Soc Microbiol Pres fel, Univ Ill, 64. *Mem:* AAAS; Am Chem Soc; Am Soc Microbiol; Am Soc Biol Chemists; NY Acad Sci; Sigma Xi. *Res:* Biochemistry of bacterial sporulation; structure and function of proteins. *Mailing Add:* Dept Chem State Univ NY Albany NY 12222

ARONSON, M(OSES), b New York, NY, May 2, 19; m 46; c 3. FLIGHT SIMULATION, MATHEMATICAL MODELING. *Educ:* NY Univ, BAE, 40, MAE, 50, MIE, 58. *Prof Exp:* Jr Engr, Langley Aeronaut Lab, Nat Adv Comt Aeronaut, 40-41; sr aerodynamicist, Repub Aviation Corp, 45-49 & 50-51; sr proj engr electronics div, Curtiss-Wright Corp, 50; sr proj engr, Naval Training Equip Ctr, 51-56, br head eng dept, 56-60, div head aerospace syst trainers dept, 60-63, staff res engr, 63-64, head, Visual Simulation Lab, 64-73, head, Electronics Lab, 73-77, head modification eng br, 77-79; CONSULT, ARONSON INDUSTS, 80- *Concurrent Pos:* Mem comt driving simulation, Transp Res Bd, 67-85; adj prof, Univ Cent Fla, 80-81; mem, Am Inst Aeronaut & Astronaut Tech Comt, Flight Simulation, 82-85. *Mem:* Assoc fel Am Inst Aeronaut & Astronaut. *Res:* Conducted training requirements analysis, experimental applications research and preliminary evaluation of potential solutions for operator training simulators and real world visual simulation devices, as well as establishing design parameters. *Mailing Add:* 3705 Wilder Lane Orlando FL 32804-3535

ARONSON, NATHAN NED, JR, b Dallas, Tex, Dec 8, 40; m 64; c 3. BIOCHEMISTRY. *Educ:* Rice Univ, 62; Duke Univ, PhD(biochem), 67. *Prof Exp:* Guest investr cell biol, Rockefeller Univ, 66-68; fel molecular biol, Vanderbilt Univ, 68-69; from asst prof to assoc prof, 69-85, PROF BIOCHEM, PA STATE UNIV, UNIVERSITY PARK, 85- *Concurrent Pos:* Helen Hay Whitney fel, 66-69; fel molecular path, Strange Ways Res Lab, Cambridge, Eng, 76-77; Brit-Am Heart fel, 76-77; Guggenheim fel, 76-77; sabbatical, Univ Calif San Diego, 84. *Mem:* Brit Biochem Soc; Am Soc Biol Chemists. *Res:* Lysosomes; membranes; glycoproteins. *Mailing Add:* 308 Althouse Lab Pa State Univ University Park PA 16802

ARONSON, PETER S, b Brooklyn, NY, Feb 3, 47; m 77; c 2. KIDNEY & MEMBRANE PHYSIOLOGY, NEPHROLOGY. *Educ:* NY Univ, MD, 70. *Hon Degrees:* MA, Yale Univ, 87. *Prof Exp:* assoc prof med & physiol, 81-87, PROF MED & CELLULAR MOLECULAR PHYSIOL, SCH MED, YALE UNIV, 87-, CHIEF NEPHROL, 87- *Concurrent Pos:* Counr, Am Soc Clin Investr, 86-88. *Honors & Awards:* Young Investr Award, Am Soc Nephrol & Am Heart Asn, 85. *Mem:* Am Fed Clin Res; Am Physiol Soc; Asn Am Physicians; Am Soc Nephrol; Soc Gen Physiologists. *Res:* Molecular mechanisms and organic anion transport across plasma membranes of epithelial cells. *Mailing Add:* Dept Med Sch Med Yale Univ 333 Cedar St New Haven CT 06510

ARONSON, RAPHAEL (FRIEDMAN), b Minneapolis, Minn, Feb 13, 28; m 58, 90; c 2. THEORETICAL PHYSICS, REACTOR PHYSICS. *Educ:* Univ Minn, BPhys, 47; Harvard Univ, MA, 48; PhD(physics), 52. *Prof Exp:* Sr scientist, Nuclear Develop Assocs, 51-55; sr supvry scientist, Tech Res Group, Inc, 55-63; sr res scientist, NY Univ, 63-66; prof nuclear eng, 66-73; PROF NUCLEAR ENG & PHYSICS, POLYTECHNIC UNIV, 73- *Mem:* Am Phys Soc. *Res:* Nuclear science; statistical mechanics; transport theory; reactor shielding. *Mailing Add:* Dept Physics Polytech Univ 333 Jay St Brooklyn NY 11201

ARONSON, ROBERT BERNARD, b Seneca Falls, NY, Aug 29, 30; m 55; c 2. BIOCHEMISTRY. *Educ:* Harvard Univ, AB, 52; Boston Univ, AM, 58, PhD(biochem), 62. *Prof Exp:* Res assoc, 60-62, asst chmn med dept, 70-72, from asst scientist to assoc scientist, 62-71, assoc chmn med dept, 72-76, SCIENTIST BIOCHEM, BROOKHAVEN NAT LAB, 71-, DEP CHMN MED DEPT, 76- *Res:* Structure, function and metabolism of collagens. *Mailing Add:* Med Dept Brookhaven Nat Lab Upton NY 11973

ARONSON, RONALD STEPHEN, b Chicago, Ill, Mar 25, 44; m 70; c 3. CLINICAL CARDIOLOGY. *Educ:* Univ Miami, BS, 65; Univ Fla, MD, 69. *Prof Exp:* Intern med, Johns Hopkins Hosp, 69-70, resident, 70-71; fel cardiac electrophysiol, Col Physicians & Surgeons, Columbia Univ, 71-73; fel cardiol, Med Ctr, Duke Univ, 73-75; cardiologist, Andrews AFB, 75-77, dir Cardiovasc Lab, 76-77; from asst prof to assoc prof, 77-89, PROF MED, ALBERT EINSTEIN COL MED, 89- *Concurrent Pos:* Res assoc, Rockefeller Univ, 72-73; adj fac, Cardiac Electrophysiol Lab, 83-; sr investr, NY Heart Asn, 77; estab investr, Am Heart Asn. *Mem:* Am Physiol Soc; Am Soc Clin Invest; fel Am Col Physicians; fel Am Col Cardiol; NY Acad Sci; Am Heart Asn. *Res:* Investigation of the cellular mechanisms that underlie abnormal electrical activity in the heart; electrical measurements in isolated cardiac tissues from animal models of human cardiac disease. *Mailing Add:* Albert Einstein Col Med 1300 Morris Park Ave Bronx NY 10461

ARONSON, SAMUEL HARRY, b Huntington, NY, May 14, 42; m 68; c 2. EXPERIMENTAL HIGH ENERGY PHYSICS. *Educ:* Columbia Univ, AB, 64; Princeton Univ, PhD(physics), 68. *Prof Exp:* Res assoc high energy physics, Enrico Fermi Labs, Univ Chicago, 68-71; asst prof physics, Univ Wis, Madison, 72-77; assoc physicist, 78-80, physicist, Isabelle Proj, Accelerator Dept, 80-82, PHYSICIST, BROOKHAVEN NAT LABS, 82-, ASSOC DEPT CHMN, PHYSICS DEPT, 87- *Mem:* Am Phys Soc. *Res:* Experimental study of fundamental processes involving elementary particles at high energies, using electronic detectors; searcher for weak new intermediate-range forces; research in novel and exotic techniques for particle acceleration. *Mailing Add:* Brookhaven Nat Labs Bldg 510A Upton NY 11973

ARONSON, SEYMOUR, b New Britain, Conn, Jan 23, 29; m 57; c 3. PHYSICAL CHEMISTRY. *Educ:* Yeshiva Col, BA, 50; Polytech Inst Brooklyn, PhD(chem), 56; Univ Pittsburgh, MS(Math), 60. *Prof Exp:* Res chemist, Westinghouse Atomic Power Div, 54-61 & Brookhaven Nat Lab, 61-68; assoc prof, 68-74, chmn dept, 78-81, PROF CHEM, BROOKLYN COL, 74- *Mem:* Am Chem Soc; Sigma Xi. *Res:* Thermodynamic and electrical properties of solids; kinetics of gas-solid reactions. *Mailing Add:* Dept Chem Brooklyn Col Brooklyn NY 11210

ARONSON, STANLEY MAYNARD, b New York, NY, May 28, 22; m 47; c 3. MEDICINE, PUBLIC HEALTH. *Educ:* City Col New York, BS, 43; NY Univ, MD, 47; Brown Univ, MA, 71; Harvard Sch Pub Health, MPH, 81. *Prof Exp:* Asst instr biol, City Col New York, 43; assoc neuropath, Col Physicians & Surgeons, Columbia Univ, 52-54; assoc prof neuropath, State Univ NY Downstate Med Ctr, 54-60, prof path & asst dean, 60-70; chmn, Dept Path, 70-72, dean med, 72-81, PROF MED SCI, BROWN UNIV, 81- *Concurrent Pos:* Res assoc, Mt Sinai Hosp, NY, 51-54; attend neuropathologist, Kingsbrook Jewish Med Ctr, NY, 52-; dir labs & Neuropathologist, Kings County Hosp, 54-70; consult neuropath, NIH, 62-, US Vet Admin, 63- & Brooklyn Hosp, Long Island Col Hosp, RI Hosp, Butler Hosp, Roger Williams Hosp, Mem Hosp, State Univ NY Hosp & Lutheran Med Ctr; lectr, Yale Sch Med, 64-65; pathologist-in-chief, Miriam Hosp, Providence, 70-74; prof lectr, State Univ NY Downstate Med Ctr, 70-; mem, Nat Adv Comn Multiple Sclerosis, 74-75 & Nat Comn Control of Huntington's Dis, 76-77; lectr, Tufts Univ, 78-; vis prof med, Univ New York, 81 & vis prof community med, Dartmouth Med Sch, 82-; pres, Hospice Care, RI, 89-; ed-in-chief, RI Med J, 89- *Mem:* Am Soc Exp Biol & Med; Am Asn Neuropath (pres, 71-72); Am Acad Neurol; Am Neurol Asn; Am Asn Path & Bact; Am Col Epidemiol; Am Pub Health Asn. *Res:* Experimental neuropathology; neuroepidemiology; population studies of vascular and aging disorders of the nervous system; sphingolipidoses. *Mailing Add:* Off Med Affairs Brown Univ Providence RI 02912

ARORA, HARBANS LALL, b Bharthanwala, India, Apr 14, 21; m 51; c 2. ZOOLOGY. *Educ:* Univ Panjab, India, BSc, 44, MSc, 45; Stanford Univ, PhD(fisheries biol), 49. *Prof Exp:* Asst zool, Univ Panjab, India, 44-45; from asst res officer to res officer, Marine Fisheries Inst, Govt India, 49-57; res fel biol, Calif Inst Technol, 57-65; res assoc, Rockefeller Univ, 65-68; from assoc prof to prof biol sci, 68-91, EMER PROF, CALIF STATE COL, DOMINGUEZ HILLS, 91- *Concurrent Pos:* Vis lectr, Univ Panjab, India, 51-53. *Honors & Awards:* First Prize, Am Soc Ichthyologists & Herpetologists, 48. *Mem:* AAAS; Am Soc Ichthyologists & Herpetologists; Am Soc Zoologists; Am Asn Anat; Int Inst Embryol. *Res:* Physiology of learning and behavior in marine organisms and fishes; neuroembryology and experimental biology. *Mailing Add:* Dept Biol Sci Calif State Univ Dominguez Hills CA 90747

ARORA, JASBIR SINGH, b Tarn-Taran, India, Apr 13, 43; US citizen; m 72; c 1. ENGINEERING MECHANICS, STRUCTURAL ENGINEERING. *Educ:* Univ Iowa, PhD(mech), 71. *Prof Exp:* Asst prof struct, GN Eng Col, LDH, India, 64-65; engr, Severud Asn, NY, 67; fel, 71-72, from asst prof to assoc prof mech, 72-81, PROF CIVIL & MECH ENG, UNIV IOWA, 81- *Mem:* Am Soc Civil Engrs; Am Soc Mech Engrs; Am Acad Mech; Am Inst Aeronaut & Astronaut. *Res:* Design optimization, structural dynamics, engineering mechanics. *Mailing Add:* Col Eng Univ Iowa Iowa City IA 52242-1593

ARORA, NARINDER, RESPIRATORY MUSCLES. *Educ:* Univ Ireland, UK, MRCP, 79. *Prof Exp:* Dr, Internal Med, Univ Va, 75-82; PVT PRAC. *Mem:* Fel Am Col Physicians. *Mailing Add:* 308 Tenth St NE Suite 2 Charlottesville VA 22901-5317

ARORA, PRINCE KUMAR, m; c 2. NEUROSCIENCE. *Educ:* Punjab Univ, India, BS, 71, MS, 73; Mich State Univ, PhD(microbiol), 79. *Prof Exp:* Teaching asst & grad res asst, Mich State Univ, East Lansing, 76-78; John E Fogarty vis fel, Immunol Br & Lab Immunodiag, Nat Cancer Inst, 78-82, staff fel, Lab Molecular Genetics & Lab Develop & Molecular Immunity, Nat Inst Child Health & Human Develop, 82-87, guest res, 87-89, SR SCIENTIST, LAB NEUROSCI, NAT INST DIABETES, DIGESTIVE & KIDNEY DIS, NIH, BETHESDA, MD, 89- *Concurrent Pos:* Numerous invited lectrs, 78-91; travel grant, Int Cong Immunol, Am Asn Immunologists, 83; prin investr grant, Off Naval Res Dept Navy, 87-90, Upjohn Corp, Kalamazoo, Mich, 88; co-prin investr, Sect Brain Imaging, Nat Inst Ment Health, 89-91; travel grant, Int AIDS Meeting, Stockholm, Sweden, NIH, 88, Montreal, Can, 89, VII Int AIDS Conf, Florence, Italy, 91. *Honors & Awards:* Cert of Recognition for Promoting Excellence in Sci Achievement, AMA, 90. *Mem:* Sigma Xi; Am Asn Immunologists; NY Acad Sci. *Res:* Author or co-author of over 40 publications. *Mailing Add:* Lab Neurosci Nat Inst Diabetes Digestive & Kidney Dis NIH Bldg 8 Rm 111 Bethesda MD 20892

ARORA, VIJAY KUMAR, b Multan, Pakistan, Nov 13, 45; m 76; c 2. COMPUTER-AIDED DESIGN, MICROELECTRONICS PHYSICS & TECHNOL. *Educ:* Kurukshetra Univ, India, BSc, 65 & MSc, 67; Univ Colo, Boulder, MS, 70 & PhD(physics), 73; Western Mich Univ, 76. *Prof Exp:* Lectr physics, DAV Col, Chandigaru, India, 67-68 & eng physics, Regional Engr Col, Kurukshetra, 68; res & teaching asst eng physics, Univ Colo, Boulder, 68-73, lectr, Univ Colo, Denver, 73-74; asst prof physics, Western Mich Univ, 74-76; from asst prof to prof physics, King Saud Univ, Saudi Arabia, 76-85; vis assoc prof elec eng, Univ Ill, Urbana, 81-82; PROF ELEC ENG, WILKES COL, 85- *Concurrent Pos:* Vis asst prof, Colo Sch Mines, 72-73; guest worker, Nat Bur Standards, 73-74; vis prof, Univ Antwerp, Belgium, 80; vis assoc prof, Univ Ill, Urbana, 81-82; vis scientist, GTE Labs, 84; vis prof, Res Ctr Advan Sci & Technol, Univ Tokyo, 89-91, Dept Elec Eng, Nat Univ Singapore, 91-92. *Mem:* Sr mem, Inst Elec & Electronic Engrs; Am Phys Soc; Indian Phys Soc; Sigma Xi; Am Soc Eng Educ; Nat Geog Soc. *Res:* Electronic transport in quantum; well microstructures and design considerations for high speed devices; process modelling of GaAs-based microdevices; opto-electronic properties of semiconductors; physics of very large-scale integration and very-large-scale integrated circuit. *Mailing Add:* Sch Sci & Eng Wilkes Univ Wilkes-Barre PA 18766

AROTS, JOSEPH B(ARTHOLOMEW), b Waterbury, Conn, Nov 23, 23; m 50; c 3. CHEMICAL ENGINEERING. *Educ:* Northeastern Univ, BS, 44; Wash Univ, St Louis, PhD(chem eng), 50. *Prof Exp:* Sr res chem eng, Res Ctr, Hercules Inc, 50-79, res scientist, 79-86; RETIRED. *Mem:* Nat Asn Corrosion Engrs; Sigma Xi. *Res:* Boiler water treatment chemicals and cooling water additives; failure analysis; corrosion; cost estimating. *Mailing Add:* 721 Ashford Rd Sharpley Wilmington DE 19803

AROYAN, H(ARRY) J(AMES), b Venice, Italy, Dec 10, 21; nat US; m 48; c 5. CHEMICAL ENGINEERING. *Educ:* Univ Pa, BS, 43; Univ Mich, MS, 47, PhD(chem eng), 49. *Prof Exp:* Group supvr, Calif Res Corp, 44-46, 49-64, sr eng assoc, 64-66; mgr planning & develop, Chevron Chem Co, 66-68; VPRES CHEM, CHEVRON RES CO, 68- *Mem:* Am Chem Soc; Am Inst Chem Engrs. *Res:* Petrochemical processes. *Mailing Add:* 66 Seaview Ave San Rafael CA 94901-2362

ARP, ALISSA JAN, b Glendale, Calif, Apr 6, 54; m 86; c 1. ECOLOGICAL PHYSIOLOGY, MARINE INVERTEBRATE PHYSIOLOGY. *Educ:* Sonoma State Univ, BA, 77; Univ Calif, Santa Barbara, MA & PhD(biol), 82. *Prof Exp:* Postdoctoral res biochem, Scripps Inst Oceanog, 84-86; vis scientist marine biol, Moss Landing Marine Lab, 88-89; lectr & res assoc, 86-88, ASSOC PROF BIOL, SAN FRANCISCO STATE UNIV, 89- *Mem:* AAAS; Asn Women Sci; Sigma Xi; Am Soc Zoologists; Soc Conserv Biol. *Res:* Ecological physiology of marine animals; physiological adaptation to low oxygen, high sulfide habitats; function of oxygen transport proteins; hydrothermal vent biology. *Mailing Add:* Dept Biol San Francisco State Univ 1600 Holloway Ave San Francisco CA 94132

ARP, DANIEL JAMES, b Sutton, Nebr, March 14, 54; m 74; c 2. PLANT BIOCHEMISTRY. *Educ:* Univ Nebr-Lincoln, BS, 76; Univ Wis-Madison, PhD(biochem), 80. *Prof Exp:* Res asst, Univ Wis-Madison, 76-80; NATO fel, Univ Erlangen, WGer, 81-82; DEPT BIOCHEM, UNIV CALIF, RIVERSIDE, 82- *Honors & Awards:* Albert Einstein Res Medal. *Mem:* Sigma Xi; Am Chem Soc; Am Micro Chem Sci; Am Asn Adv Sci. *Res:* Plant biochemistry, enzymology, nitrogen fixation and hydrogen metabolism of microorganisms and plant-bacterial symbioses. *Mailing Add:* Lab Nit Fixation Res Ore State Univ 2082 Cordley Corvallis OR 97331-2906

ARP, GERALD KENCH, b Denver, Colo, Dec 6, 47; m 78; c 3. BIOSYSTEMATICS, SURFACE GEOCHEMISTRY. *Educ:* Univ Colo, BA, 70, PhD(plant biosyst), 72. *Prof Exp:* Instr biol, Kans State Univ, 72-73; sr scientist remote sensing, Lockheed Electronics Co, 73-79; SR PRIN RES GEOBOTANIST, ARCO OIL & GAS CO, 79- *Concurrent Pos:* Adj prof biol, Southern Methodist Univ, 81- *Mem:* Asn Petrol Geochem Explor; Ecol Soc Am. *Res:* Geobotany and biogeochemistry for detection of mineral and petroleum deposits; remote sensing natural vegetation of the United States and Mexico, use of surface geochemistry for exploration of petroleum and mineral deposits. *Mailing Add:* Arco Oil & Gas Explor & Prod Res Ctr 2300 W Plano Pkwy Plano TX 75075-8499

ARP, HALTON CHRISTIAN, b New York, NY, Mar 21, 27; m 49, 62, 84; c 4. ASTRONOMY. *Educ:* Harvard Univ, AB, 49; Calif Inst Technol, PhD(astron), 53. *Prof Exp:* Carnegie fel, Mt Wilson & Palomar Observs, 53-55; res assoc, Ind Univ, 55-57; asst astromr, Mt Wilson & Palomar Observ, 57-65, astromr, 65-69, astromr, Hale Observ, 69-81, Mt Wilson & Las Campanas Observ, 81-85; RETIRED. *Concurrent Pos:* Alexander von Humboldt Sr Scientist Award, WGer, 84-85; vis scientist, Max-Planck Inst Astrophys, 85- *Honors & Awards:* Warner Prize, Am Astron Soc, 60; Newcomb Cleveland Prize, AAAS, 61. *Mem:* AAAS; Am Astron Soc; Int Astron Union. *Res:* Globular clusters and globular cluster variable stars; novae; cepheids and extragalactic nebula; peculiar galaxies; formation and evolution of galaxies; quasars, compact galaxies, companion galaxies and the nature of red shifts; cosmology. *Mailing Add:* Max-Planck Inst Astrophys Karl Schwarzschild Strasse No 1 Garching Bei Munchen Germany

ARP, LEON JOSEPH, b Norway, Iowa, June 30, 30; m 55; c 4. INDUSTRIAL & BIOMEDICAL ENGINEERING. *Educ:* Iowa State Univ, BS, 60, MS, 63, PhD(indust educ), 66. *Prof Exp:* From instr to asst prof eng, Iowa State Univ, 60-65; assoc prof, 66-70, prof indust eng & opers res, 70-77, PROF MECH ENG, VA POLYTECH INST & STATE UNIV, 71- *Mem:* Am Soc Eng Educ; Am Inst Indust Engrs; Sigma Xi. *Res:* Design and development of infant respiratory support equipment; extra-corporeal oxygenators; new method for replacement and attachment of detached retinas. *Mailing Add:* Dept Mech Eng Va Polytech Inst & State Univ Blacksburg VA 24061-0238

ARP, VINCENT D, b Grass Valley, Calif, May 9, 30; m 56; c 3. APPLIED MATHEMATICS, SOFTWARE SYSTEMS. *Educ:* Univ Calif, Berkeley, BS, 53, PhD(physics), 59. *Prof Exp:* Physicist, US Nat Bur Stand, 59-89; RETIRED. *Mem:* Am Phys Soc. *Res:* Superconductivity; low temperature physics; fluid dynamics; heat transfer; mechanics. *Mailing Add:* 7837 Fairview Rd Boulder CO 80303

ARPAIA, PASQUALE J, b Brooklyn, NY, July 21, 32. NUMBER THEORY. *Educ:* St Johns Univ, BS, 58 &, MS, 61; Adelphi Univ, PhD(math), 72. *Prof Exp:* PROF & CHMN MATH, ST JOHN FISHER COL, 71- *Mem:* Math Asn Am; Am Math Soc; NY Acad Sci. *Mailing Add:* St John Fisher Col Rochester NY 14618

ARPKE, CHARLES KENNETH, b Beatrice, Nebr, Dec 8, 21; m 50; c 3. SCIENCE ADMINISTRATION. *Educ:* Univ Nebr, AB, 48, MA, 50, PhD(chem), 62. *Prof Exp:* Asst prof air sci, Univ Nebr, 55-59, dep chief, Chem & Mat Br, USAF Rocket Propulsion Lab, 61-63; res assoc, F J Seiler Res Lab, USAF Acad, 63-64, assoc prof chem & dir chem res, 64-66, prof chem & actg head dept chem & physiol, 66-68; chief, Weapons Div, Air Force Armament Lab, 68-69, Technol Div, 69-72 & Weapons Effect Div, 72-74; chief, Anal Div, Armament Develop & Test Ctr, 74-75; publ mgr, Okla State Univ Eng Field Off, Eglin AFB, 76-85; PVT CONSULT, 75- *Concurrent Pos:* Consult environ sci study, Okaloosa-Walton Jr Col. *Mem:* Am Chem Soc; Am Inst Aeronaut & Astronaut; fel Am Inst Chem. *Res:* Smoke obscuration phenomena; obscuration simulation studies; toxicity of rocket propellants; high explosive thermal stability; weapons effects determinations; aeroballistic dispersion phenomena; exterior and terminal ballistics phenomena; computer simulation and modeling studies; library information management systems. *Mailing Add:* 2411 Rocky Shores Dr Niceville FL 32578

ARQUETTE, GORDON JAMES, b North Lawrence, NY, Mar 3, 25; m 48; c 3. PHYSICAL CHEMISTRY. *Educ:* St Lawrence Univ, BS, 47, MS, 49; Cornell Univ, PhD(phys chem), 53. *Prof Exp:* Res assoc, Cornell Univ, 52-56; sr chemist, Cent Res Lab, Air Reduction Co, Inc, 56-57, sect head, 57-58, supvr, Chem Res Div, 58-61, assoc dir, 61-63, dir, 63-67, dir res, 67-68, dir sci & eng & mem operating comt, 68-70; DIR CORP PLANNING DEVELOP & TECHNOL, AIRCO, INC, 70- *Mem:* Am Chem Soc; NY Acad Sci; Am Inst Chemists; Sigma Xi. *Res:* Chemistry and physics of high polymers; phase equilibria; chemical kinetics. *Mailing Add:* 130 Christopher Circle Ithaca NY 14850-1702

ARQUILLA, EDWARD R, b Chicago, Ill, Sept 10, 22; m 49; c 4. PATHOLOGY, IMMUNOLOGY. *Educ:* Northern Ill Univ, BS, 47; Univ Ill, MS, 49; Western Reserve Univ, PhD(anat), 57. *Prof Exp:* Asst Physiol, Univ Ill, 48-49; asst prof path, Univ Southern Calif, 59-61; from asst prof to assoc prof, Univ Calif, Los Angeles, 61-68; PROF PATH & CHMN DEPT, UNIV CALIF, IRVINE-CALIF COL MED, 68- *Concurrent Pos:* USPHS grants, 57-65; Pop Coun grant, 64-65; mem coun, Midwinter Conf Immunologists, 62; mem path training comt, Nat Inst Gen Med Sci, 70- *Mem:* AAAS; Am Asn Immunol; Soc Exp Path; Am Diabetes Asn; Am Asn Path & Bact. *Res:* Immunological aspects of insulin as related to etiology and pathogenesis of diabetes; immunological properties of thyroid stimulating hormones and human chorionic gonadotropin. *Mailing Add:* Dept Path Univ Calif-Calif Col Med Surg 1 Rm 172 Irvine CA 92717

ARRHENIUS, GUSTAF OLOF SVANTE, b Stockholm, Sweden, Sept 5, 22; m 48; c 3. OCEANOGRAPHY. *Educ:* Univ Stockholm, DSc, 53. *Prof Exp:* Geologist, Swed Skagerak Exped,46 & Swed Deep Sea Exped, 47-48; vis asst res oceanogr, 52, asst res oceanogr, 53-55, assoc prof biogeochem, 55-59, asst dir, 67-71, PROF OCEANOG, SCRIPPS INST OCEANOG, UNIV CALIF, 59-, ASSOC DIR, INST PURE & APPL PHYS SCI, 71- *Concurrent Pos:* Vis res fel, Calif Inst Technol, 53-56 & Univ Brussels, 55; vis assoc prof, Int Inst Meterol, Sweden, 57-58; Guggenheim fel, Brussels & Berne, 57-58; chmn, dept earth sci, Univ Calif, La Jolla, 59-61, dir, Space Res Lab, San Diego, 61-63; mem, Lunar Sample Analysis Team, NASA, 70-71 & Asteroid-Comet Mission Study Panel, 71-72; vis comt mem, Max Planck Soc, Ger Fed Repub, 75; Perol Res Fund Award, Am Chem Soc, 60. *Honors & Awards:* Group Achievement Award, Lunar Sci Team, NASA, 73. *Mem:* Am Chem Soc; Geophys Soc; Swed Geol Soc; Swed Acad Sci; Int Acad Astronaut. *Res:* Processes of sedimentation; geochemistry; cosmochemistry; solid state chemistry; oceanography. *Mailing Add:* 2711 Glenwick Pl La Jolla CA 92037

ARRIGHI, FRANCES ELLEN, b Topeka, Kans, July 24, 24. CELL BIOLOGY. *Educ:* Marymount Col, BS, 47; La State Univ, MS, 61; Univ Tex, PhD(zool), 65. *Prof Exp:* Med technologist, Charity Hosp, New Orleans, La, 47-54, Self Mem Hosp, Greenwood, SC, 54 & Sch Med, La State Univ, 54-58; assoc biologist, Univ Tex M D Anderson Hosp & Tumor Inst Houston, 65-; AT DEPT BIOL, UNIV TEX HEALTH SCI CTR. *Mem:* Genetics Soc Am; Am Soc Mammal; Tissue Cult Asn. *Res:* Cell biology, particularly nucleic acid, protein metabolism and chromosome structure. *Mailing Add:* 3036 Albans St Houston TX 77005

ARRINGTON, CHARLES HAMMOND, JR, b Rocky Mount, NC, Dec 23, 20; m 41; c 2. PHYSICAL CHEMISTRY. *Educ:* Duke Univ, BS, 41; Calif Inst Technol, PhD(phys chem), 49. *Prof Exp:* Res chemist, E I du Pont de Nemours & Co, 49-52, res supvr, 52-57, lab dir, 57-67, dir res, 67-74, asst gen dir res, Photog Prod Dept, 74-78, gen dir res, 78-84; RETIRED. *Mem:* AAAS; Am Chem Soc; Am Phys Soc; Sigma Xi. *Res:* Physical chemistry of solutions; photographic science; properties of macromolecules; solid state chemistry. *Mailing Add:* 711 Greenwood Rd Wilmington DE 19807

ARRINGTON, JACK PHILLIP, b Waterloo, Ohio, Dec 28, 42; m 67; c 2. TECHNOLOGY ACQUISITION, RESEARCH MANAGEMENT & DEVELOPMENT. *Educ:* Ohio Univ, BS, 65; Ind Univ, PhD(org chem), 69. *Prof Exp:* Sr res chemist, Spec Assignments Group, Dow Chem Co, 70-71 & org chem res, 71-73, res specialist, Agr Synthesis Lab, 73-79, group leader organic processes, 79-81, res mgr new monomers, Cent Res, 81-83, mgr health & environ sci, 83-84, sr develop mgr, Org Chem Dept, 84-87, sr develop mgr & opers mgr, Chem & Metals Dept, 87-90, MGR TECHNOL ACQUISITION, CHEM & PERFORMANCE PRODS, DOW CHEM USA, 90- *Concurrent Pos:* Chmn, Midland Sect, Am Chem Soc, 81. *Mem:* Am Chem Soc; Sigma Xi. *Res:* Technology assessment and transfer; organic synthesis; polymer synthesis; chemical development; synthesis and structure-activity relationships of agricultural chemicals. *Mailing Add:* 1100 Timber Ave Midland MI 48640

ARRINGTON, LOUIS CARROLL, b Baltimore, Md, July 5, 36; m 60; c 2. POULTRY SCIENCE. *Educ:* Univ Md, BS, 58; Univ Calif, Davis, MS, 61; Mich State Univ, PhD(poultry sci), 66. *Prof Exp:* Res asst, Univ Calif, 58-61 & Mich State Univ, 61-66; from asst to assoc prof, 66-80, PROF POULTRY SCI, UNIV WIS-MADISON, 80- *Mem:* Poultry Sci Asn; World's Poultry Sci Asn. *Res:* Physiology of the domestic fowl; reproductive and environmental physiology of the domestic fowl; artificial insemination of poultry; poultry products technology. *Mailing Add:* Dept Poultry Sci 1675 Observatory Dr Madison WI 53706-1284

ARRINGTON, RICHARD, JR, b Livingston, Ala, Oct 19, 34; m 53; c 4. INVERTEBRATE ZOOLOGY. *Educ:* Miles Col, AB, 55; Univ Detroit, MS, 57; Univ Okla, PhD(zool), 66. *Prof Exp:* Asst prof, 57-61, PROF BIOL, MILES COL, 66-, DEAN, 65- *Concurrent Pos:* Spec instr, Univ Okla, 65-66. *Res:* Insect morphology, particularly taxonomic characters; protozoology involving nucleocytoplasmic studies and their influences on cell division. *Mailing Add:* 1245 Mims SW Birmingham AL 35211

ARRINGTON, WENDELL S, b Brooklyn, NY, Oct 3, 36; c 4. SCIENCE ADMINISTRATION, BIOSTATISTICS. *Educ:* Rensselaer Polytech Inst, BS, 58, PhD(physics), 64; Temple Univ, MBA, 73. *Prof Exp:* NIH trainee biophys, Rensselaer Polytech Inst, 64-65; asst prof physics, State Univ NY Albany, 65-66; head res comput appln, Merck & Co Inc, 66-67; mgr biostatist, 68-72, assoc dir sci comput appln, 72-74, DIR SCI COMPUT APPLN, WYETH LABS, DIV AM HOME PROD, 74- *Mem:* Drug Info Asn; Asn Comput Mach; Sigma Xi. *Res:* Development of new user-oriented computer programming languages and data base technology to maximize cost-effectiveness in the evaluation of biomedical research data; automation and control of biomedical research experiments and related statistical analysis of data. *Mailing Add:* Wyeth Labs Div Am Home Prod 175 E Kenilworth Philadelphia PA 19101

ARROE, HACK, b Denmark, Feb 3, 20; US citizen; m 52; c 3. ATOMIC PHYSICS. *Educ:* Holte Col, Denmark, Cand Art, 38; Copenhagen Univ, PhD, 51. *Prof Exp:* Instr cloud chamber work, Inst Theoret Physics, Denmark, 43; instr atomic spectra, Agr Col, Denmark, 43-47; proj assoc, Univ Wis, 47-50; instr atomic spectra, Agr Col, Denmark, 50-52; proj assoc, Univ Wis, 52-54; prof, Mont State Col, 55-58; head, Physics Div, Denver Res Inst, 58-61; prof, Mont State Col, 61-63; chmn, dept physics, State Univ NY Col Fredonia, 63-78, prof physics, 78-85; RETIRED. *Mem:* Fel Am Phys Soc; Math Asn Am. *Res:* Spectroscopic determination of hyperfine structure and isotope shift. *Mailing Add:* 11 Andrew Ct Fredonia NY 14063

ARROTT, ANTHONY, b Pittsburgh, Pa, Apr 1, 28; m 53; c 4. MAGNETISM. *Educ:* Carnegie Inst Technol, BS, 48, PhD(physics), 54; Univ Pa, MS, 50. *Prof Exp:* Asst prof physics, Carnegie Inst Technol, 55-56; sr scientist, Sci Lab, Ford Motor Co, 56-69; chmn, 77-80, PROF PHYSICS, SIMON FRASER UNIV, 68- *Concurrent Pos:* Guggenheim fel, 63-64; adj prof nuclear eng, Univ Mich, 67-69; mem rev comts, Argonne Univs Assoc, 79-; mem grant selection comt, Nat Sci & Eng Res Coun, Can, 78-81; Japanese Soc Prom Sci fel, Meson Sci Lab, Univ Tokyo, 85. *Mem:* Am Phys Soc; Can Asn Physicists; Am Asn Physics Teachers; Inst Elec & Electronic Engrs; fel Royal Soc Can. *Res:* Magnetism; neutron scattering; electronic structure of metals; liquid crystal elasticity; nuclear engineering; muon spin rotation; molecular bean epitaxy; X-ray photoelectron spectroscopy. *Mailing Add:* Dept Physics Simon Fraser Univ Burnaby BC V5A 1S6 Can

ARROW, KENNETH J, b Aug 23, 21; c 2. OPERATIONS RESEARCH. *Educ:* City Col New York, BS, 40; Columbia Univ, MA, 41, PhD(econ), 51. *Hon Degrees:* Numerous from US & foreign univs, 68-85. *Prof Exp:* Weather officer, US Army Air Force Corp, 42-46; res assoc, Cowles Comn Res Econ, 47-49; asst prof econ, Univ Chicago, 48-49; James Bryant Conant Univ prof, Harvard Univ, 74-79; actg asst prof, econ & statist, 49-50, assoc prof econ & statist, 50-53, exec head, Dept Econ, 53-56, 62-63, prof econ, statist & opers res, 53-68, JOAN KENNEY PROF ECON & OPERS RES, STANFORD UNIV, 79- *Concurrent Pos:* Consult, Rand Corp, 48-; economist, Coun Econ Adv, US Govt, 62; vis prof econ, Mass Inst Technol, 66; fel, Churchill Col, Cambridge, Eng, 63-64, 70, 73 & 86. *Honors & Awards:* Nobel Mem Prize in Econ Sci, 72; John Bates Clark Medal, Am Econ Asn, 57; Marshall lectr, Cambridge Univ, 70; John R Commons Lectr Award, Omicron Delta Epsilon, 73; Tanner lectr, Harvard Univ, 85; Von Neumann Prize, Inst Mgt Sci & Opers Res Soc Am, 86. *Mem:* Nat Acad Sci; fel AAAS; fel Ctr Advan Study Behav Sci; Am Philos Soc; Inst Med; Finnish Acad Sci; Sigma Xi; Economet Soc (vpres, 55, pres, 56); fel Inst Math Statist; Am Statist Asn; Am Econ Asn (pres-elect, 72, pres, 73); Inst Mgt Sci (pres, 62); Am Acad Arts & Sci; Int Soc Inventory Res (pres, 83-); Int Econ Asn (pres, 83-86). *Mailing Add:* Econ & Opers Res Dept Stanford Univ Encino 470 Stanford CA 94305

ARROWOOD, ROY MITCHELL, JR, US citizen. MECHANICAL PROPERTIES MATERIALS, FAILURE ANALYSIS. *Educ:* NC State Univ, BS(mat eng) & BS(geol), 72; Univ Calif, Davis, PhD(mat sci), 81. *Prof Exp:* SR RES ENGR, SOUTHWEST RES INST, 81- *Mem:* Am Soc Metals; Am Ceramic Soc; Am Inst Mining, Metall & Petrol Engrs; Am Soc Mech Engrs. *Res:* Deformation and fracture behavior of metals, ceramics, rocks, and composite materials; effects of chemical and physical environment, loading rate and stress state; ceramic processing. *Mailing Add:* Dept Mech Eng Univ Calif Irvine CA 92717

ARROWSMITH, WILLIAM RANKIN, medicine; deceased, see previous edition for last biography

ARROYAVE, CARLOS MARIANO, ALLERGIES, RHEUMATOLOGY. *Educ:* Univ Mex, PhD(immunol), 69. *Prof Exp:* CLIN IMMUNOLOGIST, COOKS COUNTY HOSP, 85- *Mailing Add:* Cooks County Hosp Attn Miss Turner 749 Winchester Chicago IL 60612

ARROYAVE, GUILLERMO, VITAMIN DEFICIENCIES, NUTRITIONAL FOOD PROGRAMS. *Educ:* Univ Rochester, NY, PhD(biochem), 53. *Prof Exp:* Nutrit consult & prof nutrit, San Diego State Univ, 82-87; NUTRIT CONSULT, 87- *Mailing Add:* 2520 Clairemont Dr No 113 San Diego CA 92117

ARSCOTT, GEORGE HENRY, b Hilo, Hawaii, July 20, 23; m 53; c 2. POULTRY NUTRITION & MANAGEMENT. *Educ:* Ore State Col, BS, 49; Univ Md, MS, 50, PhD(poultry husb), 53. *Prof Exp:* Res asst, Univ Md, 49-50 & 51-53; from asst prof to assoc prof, Ore State Univ, 53-59, actg head dept, 69-70, prof poultry sci, 65-87, head, dept poultry sci, 70-87, EMER PROF POULTRY SCI, ORE STATE UNIV, 88- *Concurrent Pos:* Poultry adv, Consortium Int Develop, Yeman Arab Repub, 81, 83, 85 & 86; poultry specialist, Ore Partners of the Americas, Costa Rica, 89; poultry mgt adv, Int Exec Serv Corps, Dominican Repub, 90 & 91. *Mem:* AAAS; fel Poultry Sci Asn (vpres, 83-84); Am Inst Nutrit; Sigma Xi; World Poultry Sci Asn. *Res:* Nutritional requirements of poultry and evaluation of feedstuffs. *Mailing Add:* 489 NW Witham Dr Corvallis OR 97330

ARSENAULT, GUY PIERRE, b Montreal, Que, Sept 25, 30; m 55; c 2. BIO-ORGANIC CHEMISTRY, ANALYTICAL CHEMISTRY. *Educ:* Univ Toronto, BASc, 53; Ohio State Univ, PhD(org chem), 58. *Prof Exp:* From asst res officer to assoc res officer org chem, Atlantic Regional Lab, Nat Res Coun Can, 60-67; guest dept chem, Mass Inst Technol, 65-66, res chemist, 67-77; PROF CHEM & CHEM ENG, ROYAL MIL COL CAN, 77- *Mem:* Am Chem Soc; Am Soc Mass Spectrometry; Chem Inst Can. *Res:* Structure of fungal and bacterial metabolites; application of mass spectrometry to organic and biological chemistry and to biomedicine and environmental research; transformation of seed oils into petroleum refinery stocks. *Mailing Add:* Dept Chem & Chem Eng Royal Mil Col Can Kingston ON K7K 5L0 Can

ARSENAULT, HENRI H, b Montreal, Que, Sept 29, 37; m 65; c 2. OPTICS. *Educ:* Laval Univ, BSc, 63, MSc, 66, PhD(physics), 69. *Prof Exp:* Scientist x-ray spectros, Aluminium Labs Ltd, 63-64; teacher physics, Col Jonquiere, 64-65; Nat Res Coun Can fel, Inst Optics, Univ Paris, Orsay, 68-70; from asst prof to assoc prof, 70-78, PROF PHYSICS, LAVAL UNIV, 78- *Concurrent Pos:* Vis prof, Stanford Univ, 78-79; asst ed, Optical Eng, 80-; prin investr contracts, Dept Nat Defense, Can. *Mem:* Fel Optical Soc Am; Can Asn Physicists; Inst Elec & Electronics Engrs; fel Int Soc Optical Eng. *Res:* Optical data processing; digital image processing, holography, speckle phenomena; optical computing; neural nets. *Mailing Add:* 2625 Ave Port Royal Port Royal Quebec PQ G1V 1A5 Can

ARSENEAU, DONALD FRANCIS, b St John, NB, Mar 18, 28; m 54; c 10. CELLULOSE CHEMISTRY, TECHNOLOGY TRANSFER. *Educ:* St Francis Xavier Univ, BSc, 50; Laval Univ, DSc(chem), 55. *Prof Exp:* From assoc prof to prof chem, St Francis Xavier Univ, 55-74; prof chem, 74-88, DIR, BRAS D'OR INST, COL CAPE BRETON, 75- *Honors & Awards:* Can Centennial Medal. *Mem:* Am Chem Soc; Forest Prod Res Soc; Chem Inst Can; AAAS; Sci Coun Can. *Res:* Thermal breakdown of wood; differential thermal analysis; thermal degradation of naturally occurring polymers; water quality; utilization of coal for coking; new technologies in community development. *Mailing Add:* PO Box 5300 Sydney NS B1P 6L2 Can

ARSLANCAN, AHMET N, US citizen; m 85; c 1. HEAT TRANSFER PROCESS, MICROELECTRONICS PROCESSING. *Educ:* Carleton Col, BA, 73; Univ Rochester, PhD(chem eng), 80. *Prof Exp:* Res assoc chem eng, Univ Rochester, 74-79; proj engr, Stress Technol Corp, 79-80; control engr, Occidental Chem Corp, 80-83; engr res & develop, 83-88, DIR RES & APPLN, RADIANT TECHNOL CORP, 88- *Concurrent Pos:* Res assoc, Rochester Inst Technol, 72-73. *Mem:* Am Chem Soc; Am Inst Chem Engrs; Int Soc Hybrid Microelectronics; Int Electronics Packaging Soc; Surface Mount Technol Asn. *Res:* CERMET thick film firing; semiconductor packaging processes; polymer thick film curing; hybrid and PWB assembly by reflow soldering methods; other related microelectronics processing applications by infrared technology. *Mailing Add:* 5395 E Hunter Ave Anaheim CA 92807-2054

ARSOVE, MAYNARD GOODWIN, b Lincoln, Nebr, Mar 11, 22; m 44; c 4. MATHEMATICS. *Educ:* Lehigh Univ, BS, 43; Brown Univ, ScM, 48, PhD(math), 50. *Hon Degrees:* DrEd, Saipan Univ, 46. *Prof Exp:* Fulbright scholar, Grenoble Univ, 50-51; from instr to assoc prof, 51-60, PROF MATH, UNIV WASH, 60- *Concurrent Pos:* Guggenheim fel, 57-58; Fulbright fel, Univ Paris, 57-58; NATO fel, Univ Hamburg, 64-65. *Mem:* Am Math Soc. *Res:* Subharmonic function theory; potential theory; complex analysis; theory of bases. *Mailing Add:* Dept Math Univ Wash Seattle WA 98195

ART, HENRY WARREN, b Norfolk, Va, Sept 11, 44; m 66; c 3. FOREST ECOLOGY, ENVIRONMENTAL BIOLOGY. *Educ:* Dartmouth Col, BA, 66; Yale Univ, MPhil, 69, PhD(forest ecol), 71. *Prof Exp:* Res assoc, Williams Col, 70-71, asst prof & asst dir res, Ctr Environ Studies, 71-76, actg dir, 76-79, assoc prof biol, 76-82, asst dir, Ctr Environ Studies, 80-87, chmn dept biol, 81-87, PROF BIOL, WILLIAMS COL, 82-, PREMED ADV, 88-; RES ECOLOGIST, NAT PARK SERV, 88- *Concurrent Pos:* Vis prof environ studies, Univ Vt, 75 & Univ Calif, Santa Barbara, 79-80; vis fel, Corpus Christi Col & Bot Sch, Cambridge Univ, Eng, 87-88. *Honors & Awards:* Quill & Trowel Award, Garden Writers Asn Am. *Mem:* Ecol Soc Am; Brit Ecol Soc; AAAS; Garden Writers Asn Am. *Res:* Biomass and productivity of ecosystems; nutrient cycling; ecosystem dynamics; land use and successional relationships; coastal ecology. *Mailing Add:* Dept Biol Williams Col Williamstown MA 01267

ARTEAGA, LUCIO, b Calatayud, Spain, May 22, 24; Can citizen; m 52; c 2. MATHEMATICS, STATISTICS. *Educ:* Univ Madrid, cert, 56; Dalhousie Univ, MSc, 59; Univ Sask, PhD(math), 64. *Prof Exp:* Statistician, NS Tumor Clin, 57-60; consult, Sask Res Coun, 60-62; asst prof math, Univ Windsor, 62-65; asst prof, Dalhousie Univ, 65-67; asst prof, Univ NC, Charlotte, 67-68; assoc prof, 68-69, PROF MATH, WICHITA STATE UNIV, 69- *Concurrent Pos:* Nat Res Coun Can grants, 64-66. *Mem:* Am Math Soc; Math Asn Am. *Res:* Integral equations; Fourier transforms. *Mailing Add:* 4361 E Clark St Wichita KS 67218

ARTEMIADIS, NICHOLAS, b Constantinople, Turkey, May 17, 17; US citizen; m 63. MATHEMATICAL ANALYSIS. *Educ:* Univ Thessaloniki, MS, 39; French Inst, Athens, dipl, 51; Univ Paris, certs, 54, 55, DSc(math), 57. *Prof Exp:* From asst prof to assoc prof, Univ Wis, Milwaukee, 58-67; assoc prof, Univ Thessoloniki, 60-61; prof math, Southern Ill Univ, Carbondale, 67-74; prof math, Univ Patras, 74-84; RETIRED. *Concurrent Pos:* Vis lectr, Univ Wis, 59 & Univ Chicago, 60; guest, French Govt, Cult Exchange Prog, Univ Paris, 81; study mission, Org Econ Coop Develop, Coop Action Prog, Paris, 81. *Honors & Awards:* Stavropoulos Prize, Greek Embassy Paris, 57; Hon Dipl Distinguished Serv, Greek Math Soc, 88. *Mem:* Am Math Soc; Math Soc France; Greek Math Soc (pres, 85-); Sigma Xi. *Res:* Mathematical and harmonic analysis; author of seven books in real, complex, harmonic and functional analysis. *Mailing Add:* 169 Megalou Alexandrou St 13671 Thrakomakedones Athens Greece

ARTERBURN, DAVID ROE, b Norfolk, Nebr, Dec 29, 39; m 63; c 2. TOPOLOGY. *Educ:* Southern Methodist Univ, BS, 61; NMex State Univ, MS, 63, PhD(functional anal), 64. *Prof Exp:* Asst prof math, Univ Mont, 64-67; ASSOC PROF MATH, NMEX INST MINING & TECHNOL, 67-, CHMN MATH DEPT, 88- *Mem:* Math Asn Am; Sigma Xi. *Res:* Functional analysis; rings of functions; rings of operators. *Mailing Add:* NMex Tech Campus Sta Socorro NM 87801

ARTES, RICHARD H, b Ionia, Iowa, Apr 19, 26; m 52. SPEECH PATHOLOGY. *Educ:* Ariz State Univ, BA, 49; Univ Iowa, MA, 51, PhD(speech path), 67. *Prof Exp:* Clinician speech path, Faribault, Minn Pub Schs, 50-52; PROF SPEECH PATH, BALL STATE UNIV, 52- *Concurrent Pos:* Consult, Vet Admin Hosp, Marion, Ind, 78- *Mem:* Sigma Xi. *Res:* Psychological problems associated with severe language impairment subsequent to brain damage, such as aphasia. *Mailing Add:* 401 N Taft Rd Muncie IN 47304

ARTH, JOSEPH GEORGE, JR, b Rockville Centre, NY, July 2, 45; m 71; c 3. GEOCHEMISTRY. *Educ:* State Univ NY, Stony Brook, BS, 67, MS, 70, PhD(petrol, geochem), 73. *Prof Exp:* Nat Res Coun res assoc isotope geol, Denver, 73-74, GEOLOGIST, US GEOL SURV, 74- *Mem:* Geol Soc Am; Geochem Soc; Am Geophys Union. *Res:* Geochronology, petrology, chemistry and isotopic character of igneous and meta-igneous rock suites to determine their age, origin and role in the earth's crustal history. *Mailing Add:* US Geol Surv 345 Middlefield Rd Menlo Park CA 94025

ARTHAUD, RAYMOND LOUIS, b Cambridge, Nebr, Apr 21, 21; m 47; c 5. ANIMAL SCIENCE. *Educ:* Univ Nebr, BSc, 47, MSc, 49; Univ Mo, PhD(animal breeding), 53. *Prof Exp:* Instr animal husb, Univ Nebr, 49; asst prof beef cattle res sta, Va Polytech Inst, 52-54; animal husbandman beef cattle breeding, Animal & Poultry Husb Res Br, Agr Res Serv, USDA, Univ Nebr, 54-59; prof animal sci, Exten, Univ Minn, 59-80; RETIRED. *Mem:* Am Soc Animal Sci. *Res:* Swine breeding and nutrition; beef cattle breeding and nutrition. *Mailing Add:* 2321 Cedar Rd White Bear Lake MN 55110

ARTHUR, ALAN THORNE, b Evanston, Ill, Feb 26, 42; m 65; c 3. REPRODUCTIVE BIOLOGY, TOXICOLOGY. *Educ:* Hanover Col, BA, 64; Drake Univ, MA, 66; Univ Colo, Boulder, PhD(biol), 71. *Prof Exp:* NIH fel, Worcester Found Exp Biol, 71-73; instr obstet & gynec, M S Hershey Med Ctr, Pa State Univ, 73-77; asst prof med & dir endocrine lab, Hahnemann Med Col & Hosp, Philadelphia, 77-78; mgr reproduction & mutagenesis, dept toxicol, Ciba-Geigy, 78-82, assoc dir reproductive toxicol, 82-83, dir reproductive & genetic toxicol, 83-86, DIR TOXICOL, DEPT TOXICOL-PATH, CIBA-GEIGY, 88- *Concurrent Pos:* Head clin studies, Stamford Lodge, Cheshire, Eng, 86-88. *Mem:* Am Fertil Soc; AAAS. *Res:* Role of proteins in pre- and postimplantation embryonic development; safety evaluation of pharmacologic compounds. *Mailing Add:* Ciba-Geigy 556 Morris Ave Summit NJ 07901

ARTHUR, JAMES ALAN, b Stockton, Calif, June 19, 36; m 58; c 5. POULTRY BREEDING. *Educ:* Univ Calif, Davis, BS, 61, PhD, 65. *Prof Exp:* Pop geneticist, 64-70, mgr appl res, 70-71, RES DIR, HY-LINE INT, 71- *Mem:* Sigma Xi; Genetics Soc Am; Poultry Sci Asn. *Res:* Improvement of hybrid performance in the fowl. *Mailing Add:* Hy-Line Int PO Box 310 Dallas Center IA 50063

ARTHUR, JAMES GREIG, b Hamilton, Can, May 18, 44; m 72; c 2. AUTOMORPHIC FORMS, GROUP REPRESENTATIONS. *Educ:* Univ Toronto, BSc, 66, MSc, 67; Yale Univ, PhD(math), 70. *Prof Exp:* Instr math, Princeton Univ, 70-72; asst prof math, Yale Univ, 72-76; prof math, Duke Univ, 76-79; PROF MATH, UNIV TORONTO, 88- *Concurrent Pos:* Vis mem, Inst Advan Studies, 77-78; mem ed adv bd, Crelles J, 85, Can J Math, 86, J Am Math Soc, 87; Sloan fel, Sloan Found; E W R Stocie fel, Nat Sci & Eng Res Coun Can, 82. *Honors & Awards:* John L Surge Award, Royal Soc Can, 87. *Mem:* Am Math Soc; Can Math Soc; fel Royal Soc Can. *Res:* Trace formula and its applications to group representations, automorphic forms, number theory and algebraic geometry. *Mailing Add:* Dept Math Univ Toronto Toronto ON M5S 1A1 Can

ARTHUR, JETT CLINTON, JR, b Hemphill, Tex, May 31, 18; m 41; c 3. PHYSICAL CHEMISTRY, POLYMER CHEMISTRY. *Educ:* Stephen F Austin State Univ, BA, 39; Univ Tex, MA, 46; Environ Eng Intersoc Bd, dipl, 71. *Prof Exp:* Chemist, Southern Regional Res Ctr, Agr Res Serv, USDA, 41-43 & 46-49, chemist in charge protein prod, 49-52, chemist in charge biochem, 52-56, prin chemist & head radiochem, 56-66, chief chemist, 66-79; BIOMAT CONSULT, 79- *Concurrent Pos:* Abstractor, Chem Abstr, 46-, ed phys org chem, 63-; deleg, Int Atomic Energy Comn, Austria, 62; lectr, Brit Asn Radiol Res, 65; lectr, NATO Advan Study Inst, Wales, 67; deleg, Int Union Pure & Appl Chemists, Japan, 66, Can, 68, Australia, 69, UK, 78, US, 82, Fourth UN Conf Peaceful Uses Atomic Energy, Switz, 71, Pac Chem Eng Cong, Japan, 72, US, 77, Pac Chem Cong, Hawaii, 79 & 89; mem, First Chem Cong-NAm Continent, Mex, 75, Cellucon, Int Conf Cellulose, 84; chmn, res comt, Am Inst Chem Engrs, 77-78, Cellulose, Paper, Textile Div, Am Chem

Soc, 85; vis prof, NWales Inst Higher Educ, UK, 78-86. *Honors & Awards:* Am Chem Soc Southwest Regional Award, 80; Anselme Payen Award in Cellulose, 84; Polymer Sci Pioneer, Polymer News, 85; Herty Medal, 89. *Mem:* Am Chem Soc; Sci Res Soc Am; fel Am Inst Chem Engrs; Am Acad Environ Engrs. *Res:* Radiation chemistry and photochemistry of natural products; physical organic chemistry of natural polymers including cellulose, protein, starch and biotechnology applications; industrial chemistry; applied mathematics. *Mailing Add:* 3013 Ridgeway Dr Metairie LA 70002-5053

ARTHUR, JOHN READ, JR, b Omaha, Nebr, Dec 17, 31; m 54; c 4. SURFACE PHYSICS. *Educ:* Iowa State Univ, BS, 54, PhD(phys chem), 61. *Prof Exp:* Res asst surface chem, Inst Atomic Res, Iowa State Univ, 56-61; mem tech staff, Bell Tel Labs, 61-77; sr proj scientist, phys electronics div, Perkin-Elmer Corp, 77-; AT DEPT ELEC ENG, ORE STATE UNIV. *Honors & Awards:* Morris N Liebmann Award, Inst Elec & Electronics Engrs, 82; IBM Award New Mat, Am Phys Soc, 82; Gaede-Langmuir Award, Am Vac Soc, 88. *Mem:* Am Phys Soc; Am Vacuum Soc (pres, 83). *Res:* Adsorption and surface reaction kinetics on semi-conductors; thin film structures and growth mechanisms; electron diffraction and auger spectroscopy; molecular beam scattering from surfaces; molecular beam epitaxy. *Mailing Add:* Dept Elec Eng Ore State Univ Corvallis OR 97331

ARTHUR, JOHN W, b Superior, Wis, Sept 9, 37; m 61; c 3. AQUATIC TOXICOLOGY, AQUATIC BIOLOGY. *Educ:* Gustavus Adolphus Col, BS, 59; Wash State Univ, MS, 62; Univ Minn, MPH, 64. *Prof Exp:* Aquatic biologist, Upper Miss River Proj, USPHS, Minneapolis, Minn, 64-67; res aquatic biologist bioassays, Duluth, Minn, 67-70 & res team coordr, 70-74, res aquatic biologist 74-78, sta chief stream ecol, Monticello, Minn, 78-86, RES AQUATIC BIOL & TEAM LEADER, WATERSHED ASSESTMENT, US ENVIRON PROTECTION AGENCY, 86- *Concurrent Pos:* Subcomt chmn bioassays, Environ Protection Agency Biol Methods Manual, 73; adv bd Standard Methods for Exam Water, Wastewater, 14th ed, 74-75; 15th ed, 79-80; Planning comt Midwest Conf Environ Technol, 81-85. *Honors & Awards:* Bronze Medal, US Environ Protection Agency, 75. *Mem:* NAm Benthological Soc. *Res:* Aquatic life damage; protective aquatic life criteria evaluation in lab and outdoor experimental streams, watershed assessments. *Mailing Add:* US Environ Protection Agency 6201 Congdon Blvd Duluth MN 55804

ARTHUR, MARION ABRAHAMS, b Port Deposit, Md, Apr 28, 11; m 34; c 3. PLASTICS ENGINEERING, GEOPHYSICS. *Educ:* Haverford Col, BS, 31; Harvard Univ, MS, 32. *Prof Exp:* Seismograph operator, Humble Oil & Refining Co, Houston, 33-34, geophys researcher, 35-42; proj supvr ballistics, Off Sci Res & Develop, Washington, DC, 42-45; res specialist geophys, Exxon Corp USA, Houston, 46-71; PRES, MARION ARTHUR, INC, 71- *Mem:* Emer mem Soc Plastic Engrs (secy-treas, 48-49, pres, 49-50). *Res:* Improvement in the art and science of acrylic embedment. *Mailing Add:* Marion Arthur Inc 1022 Chimney Rock Rd Houston TX 77056

ARTHUR, MICHAEL ALLAN, b Sacramento, Calif, June 25, 48; m 82; c 1. PALEOCEANOGRAPHY, STABLE ISOTOPE GEOCHEMISTRY. *Educ:* Univ Calif, Riverside, BS, 71, MS, 74; Princeton Univ, PhD(geol), 79. *Prof Exp:* Postdoctoral res scientist, Deep Sea Drilling, Scripps Inst Oceanog, 77-79; geologist, Br Oil & Gas Resources, US Geol Surv, 79-81; asst prof geol sci, Univ SC, 81-83; from assoc prof to prof geol oceanog, Grad Sch Oceanog, Univ RI, 83-91; PROF & HEAD GEOSCI, PA STATE UNIV, 91- *Concurrent Pos:* Geologist-WAE, US Geol Surv, 81- *Mem:* Am Asn Petrol Geologists; Am Geophys Union; fel Geol Soc Am; Soc Sedimentary Geol; Geochem Soc; Oceanog Soc. *Res:* Role of the carbon cycle in past climates; causes of global change over the past 140 million years; chemistry of ancient oceans and the sedimentary record; diagenesis of carbonate and organic-carbon rich sediments. *Mailing Add:* Dept Geosci Pa State Univ University Park PA 16802

ARTHUR, PAUL D(AVID), b Washington, DC, Feb 23, 25. MECHANICAL ENGINEERING. *Educ:* Univ Md, BS, 44, MS, 48; Calif Inst Technol, PhD(aeronaut), 52. *Prof Exp:* Engr, Convair, 46-47; instr mech eng, Univ Md, 48-49; res engr hypersonics, Calif Inst Technol, 49-52; oper analyst, USAF, 52-53; Fulbright prof, Iraq, 53-54; syst engr, Ramo-Wooldridge, 54-56; consult, 57-65; vpres, Systs Corp Am, 58-65; prof, Univ Fla, 65-68; prof mech eng, 68-89, EMER PROF MECH ENG, UNIV CALIF, IRVINE, 89- *Concurrent Pos:* Adj prof, Univ Southern Calif, 56-65; vis prof, Technol Univ Delft, 59 & Cairo Univ, 60; Fulbright prof, Univ Naples, 63-64. *Mem:* Am Astronaut Soc; Am Soc Mech Engrs. *Res:* Aerodynamics; orbital mechanics; fluid mechanics. *Mailing Add:* Dept Mech Eng Univ Calif Irvine CA 92717

ARTHUR, RICHARD J(ARDINE), b Springfield, Ohio, Dec 12, 24; m 49. INSTRUMENTATION. *Educ:* Purdue Univ, BS, 48; Univ Ill, MS, 49. *Prof Exp:* Asst math, Univ Ill, 48-51; proj aerodynamicist, Bell Aircraft Corp, 51-55, proj engr inertial instrumentation, 55-57, group leader, 57-61; head dept engr, Sperry Marine Systs Div, Sperry Rand Corp, 61-66, eng mgr, 66-71, dir eng, 71-73; div head, Naval Surface Warfare Ctr, 73-80, dep dept head, 80-86; RETIRED. *Res:* Inertial instrument design; gyro and accelometer design; physics; digital methods applied to instruments; servomechanisms; applied mathematics and electronics; radar systems; collision avoidance systems. *Mailing Add:* 117 Indian Spring Rd Charlottesville VA 22901

ARTHUR, ROBERT DAVID, b Union City, Ind, Oct 3, 42; m 65; c 2. NUTRITION. *Educ:* Purdue Univ, BS, 64, MS, 66; Univ Mo, PhD(agr chem), 70. *Prof Exp:* Fel, Univ Mo, 70-72; asst prof biochem, Miss State Univ, 72-77; from asst prof to assoc prof, 77-88, PROF & CHMN, ANIMAL SCI FOOD & NUTRITION, SOUTHERN ILL UNIV-CARBONDALE, 88- *Concurrent Pos:* Consult, SE Asia-US Feed Grains Coun, 80, Am Soybean Asn, 81 & China-US Feed Grains Coun, 84, 85 & 88. *Mem:* Am Soc Animal Sci; Sigma Xi. *Res:* Study of the factors affecting growth and development in swine with emphasis on nutritional and management practices. *Mailing Add:* Dept Animal Sci Food & Nutrit Southern Ill Univ Carbondale IL 62901

ARTHUR, ROBERT M, b Fond du Lac, Wis, Mar 21, 24, 1924; m 65; c 1. ENVIRONMENTAL HEALTH ENGINEERING. *Educ:* Ripon Col, BA, 49; Northwestern Univ, BS, 53; Harvard Univ, MS, 56; Univ Iowa, PhD(environ health eng), 63. *Prof Exp:* Asst city engr civil eng, City of Fond du Lac, 49-53; asst sanit engr, Chicago Pump Co, 53-55; assoc prof civil eng, Rose Polytech Inst, 56-63, prof & chmn dept biol eng, 63-72; PRES, ARTHUR TECHNOL, 72- *Honors & Awards:* Kermit Fisher Environ Award, Instrument Soc Am, 81. *Mem:* Water Pollution Control Fedn; Am Soc Eng Educ; Instrument Soc Am; Int Asn Water Pollution Res; Am Soc Testing & Mat. *Res:* Research and development in operation of wastewater treatment plants including instrumentation and control; operator training. *Mailing Add:* PO Box 1236 Fond du Lac WI 54936

ARTHUR, ROBERT SIPLE, b Redlands, Calif, Mar 16, 16; m 41, 75; c 1. PHYSICAL OCEANOGRAPHY. *Educ:* Univ Redlands, AB, 38; Univ Calif, Los Angeles, AM, 49, PhD(oceanog), 50. *Prof Exp:* Teacher, high sch, 39-40; asst, Univ Calif, Los Angeles, 40-42, asst prof naval sci, 43-44; assoc oceanog, Scripps Inst Oceanog, La Jolla, 44-46, asst oceanog, 46-48, oceanog, 48-50, from asst prof to prof, 51-63, EMER PROF OCEANOG, SCRIPPS INST OCEANOG, UNIV CALIF, 79- *Mem:* Am Geophys Union; Am Meteorol Soc. *Res:* Physical oceanography. *Mailing Add:* ORD A-030 Scripps Inst Oceanog Univ Calif La Jolla CA 92093

ARTHUR, SUSAN PETERSON, b Berkeley, Calif, June 16, 49; m 74; c 4. BIOSTATISTICS, EPIDEMIOLOGY. *Educ:* Univ Calif, Berkeley, MA, 73; Princeton Univ, PhD(math statist), 79. *Prof Exp:* Statist consult, Epidemiol Div, Col Physicians & Surgeons, Columbia Univ, 75-76; RES SCHOLAR, INT INST APPL SYST ANAL, 79-, STATIST CONSULT, COHERENT LASER MED DIV, 85- *Res:* Applied mathematical statistics and probability theory; clinical trials; uncertainty of oil resource estimates. *Mailing Add:* 1060 Vernier Pl Stanford CA 94305

ARTHUR, WALLACE, b New York, NY, Nov 22, 32; m 60; c 3. NUCLEAR PHYSICS, SPACE PHYSICS. *Educ:* NY Univ, BEngSc & BEE, 57, PhD(physics), 62. *Prof Exp:* Asst physics, NY Univ, 57-61; lectr physics, Rutgers Univ, 61-62; prof physics & chmn dept, 62-73, dean col sci & eng, 73-83, PROF PHYSICS, FAIRLEIGH DICKINSON UNIV, 73- *Mem:* Am Asn Physics Teachers; Am Phys Soc. *Res:* Cosmic ray analyses; atmospheric physics and gas kinetics; low energy experimental nuclear physics; gamma ray lasers. *Mailing Add:* 22 Raymond St Harrington Park NJ 07640

ARTHUR, WILLIAM BRIAN, b Belfast, N Ireland, July 21, 46; m 74; c 4. OPERATIONS RESEARCH, ECONOMICS. *Educ:* Queen's Univ Belfast, BSc, 66; Univ Mich, Ann Arbor, MA, 69; Univ Calif, Berkeley, MA, 73, PhD(opers res), 73. *Prof Exp:* res scholar methodol, Int Inst Appl Systs Anal, 77-82; MORRISON PROF ECON & POP STUDIES, STANFORD UNIV, 83- *Concurrent Pos:* Guggenheim Fel, 88-89; dir, Econ Res Prog, Santa Fe Inst, 88-90. *Honors & Awards:* Schumpeter Prize, Econ, 90. *Mem:* Int Union for Sci Study of Pop; Am Econ Asn. *Res:* Economics; demography. *Mailing Add:* Food Res Inst Stanford Univ Stanford CA 94305-6084

ARTIN, MICHAEL, b June 28, 34; US citizen. MATHEMATICS. *Educ:* Harvard Univ, PhD(math), 60. *Prof Exp:* PROF MATH, MASS INST TECHNOL, 63- *Mem:* Nat Acad Sci. *Res:* Algebra. *Mailing Add:* Dept Math Mass Inst Technol Cambridge MA 02139

ARTIST, RUSSELL (CHARLES), b Francisville, Ind, Jan 5, 11; m 39, 65; c 2. BOTANY. *Educ:* Butler Univ, BS, 32; Northwestern Univ, MS, 34; Univ Minn, PhD(bot, geol), 38. *Prof Exp:* Head dept sci, Amarillo Col, 38-45; head dept natural sci, Westminster Col, 45-47; head dept biol, Abilene Christian Univ, 47-48; prof apologetics, Kolleg Der Gemeinde Christi, Ger, 48-51; prof, David Lipscomb Univ, 53-76, emer prof biol, 76-88; RETIRED. *Res:* Paleoecology and pollen analysis; seedling survival studies. *Mailing Add:* 1507 Blue Springs Dr Franklin TN 37064

ARTMAN, JOSEPH OSCAR, b New York, NY, Apr 22, 26; m 55; c 4. ENGINEERING PHYSICS, SOLID STATE PHYSICS. *Educ:* City Col New York, BS, 44; Columbia Univ, MA, 48, PhD(physics), 53. *Prof Exp:* Asst physics, Columbia Univ, 49-50, asst radiation lab, 50-52; staff mem, Lincoln Labs, Mass Inst Technol, 52-55; res fel, div eng & appl physics, Harvard Univ, 55-58; staff mem, appl physics lab, Johns Hopkins Univ, 58-64; assoc prof, 64-67, PROF ELEC ENG & PHYSICS, CARNEGIE-MELLON UNIV, 67- *Concurrent Pos:* Sr fel, Mellon Inst, 68- *Mem:* Inst Elec & Electronics Engrs; Optical Soc Am; AAAS; Sigma Xi; Am Phys Soc. *Res:* Magnetism; ferromagnetic and paramagnetic resonance; crystal physics; optics; visible and infrared spectroscopy; environmental monitoring; laser doppler velocimetry; chemical physics. *Mailing Add:* Dept Elec Eng Carnegie-Mellon Univ Pittsburgh PA 15213

ARTNA-COHEN, AGDA, b Tartu, Estonia, Oct 19, 30; US citizen; m 67; c 1. INFORMATION SCIENCE, NUCLEAR PHYSICS. *Educ:* McMaster Univ, BSc, 57, PhD(nuclear physics), 61. *Prof Exp:* Nuclear physicist nuclear data proj, Nat Acad Sci-Nat Res Coun, 61-63, nuclear physicist nuclear data proj, Oak Ridge Nat Lab, 64-67, consult, 67-75; res physicist, Naval Res Lab, Washington, DC, 75-85; CONSULT, 85- *Concurrent Pos:* Lectr, USDA Grad Sch, 62-63; assoc ed, Nuclear Data, 65-67. *Mem:* Can Asn Physicists; Am Phys Soc. *Res:* Nuclear structure physics; information retrieval and evaluation; science writings. *Mailing Add:* 8801 Mansion Farm Pl Alexandria VA 22309

ARTRU, ALAN ARTHUR, b Oakland, CA, Apr 30, 49; m 72; c 4. CEREBROSPINAL FLUID DYNAMICS, CEREBRAL BLOOD FLOW & METABOLISM. *Educ:* Univ Calif, Santa Cruz, BA, 71; Med Col Wis, Milwaukee, MD, 75. *Prof Exp:* Intern, Univ Calif, San Francisco, 75-76, res anesthesiol, 76-78; fel neuroanesthesiol, Mayo Clin, Rochester, 78-80, asst consult anesthesiol, 79-80; from asst prof to assoc prof, 80-89, PROF ANESTHESIOL, UNIV WASH, SEATTLE, 89- *Concurrent Pos:* Head educ comt, Soc Neurosurgical Anesthesia & Crit Care, 82-88; comt on Neurosci

& Anesthetic action, Am Soc Anesthesiologists, 85-; vis prof, Univ Oregon dept Anesthesiol, 87; ed lab reports, J Neurosurgical Anesthesiol, 88-; sci adv bd, Asn Univ Anesthetists, 85- *Mem:* Am Soc Anesthesiol; Int Anesthesia Res Soc; Soc Neuroanesthesia & Critical Care; Int Soc Cerebral Blood Flow & Metab; Asn Univ Anesthetists. *Res:* Effects of anesthetics and other medical treatments and drugs on cerebral blood flow, blood volume, and metabolism, cerebrospinal fluid dynamics and pressure; electroencephalogram. *Mailing Add:* Dept Anesthesiol RN-10 Univ Wash Sch Med Seattle WA 98195

ARTUSIO, JOSEPH F, JR, b Jersey City, NJ, Nov 26, 17; m 45; c 6. ANESTHESIOLOGY. *Educ:* St Peter's Col, BS, 39; Cornell Univ, MD, 43; Am Bd Anesthesiol, dipl. *Prof Exp:* Intern, Bellevue Hosp, NY, 43-44; resident anesthesiol, NY Hosp, 46-47; from instr to assoc prof surg, Med Col, Cornell Univ, 47-57, prof anesthesiol in surg, obstet & gynec, 57-67, prof anesthesiol & chmn dept, 67-89, EMER PROF ANESTHESIOL, MED COL, CORNELL UNIV, 89- *Concurrent Pos:* Asst attend anesthesiologist in chg, 48-57, anesthesiologist-in-chief, 57-89, attend anesthesiologist, 89-; ed-in-chief, Clin Anesthesia; mem, Unitarian Serv Comt to Japan, 56. *Mem:* Fel Am Col Anesthesiol; AMA; Am Soc Anesthesiol; Asn Univ Anesthetists; fel NY Acad Med. *Res:* Effect of anesthetic and muscular relaxant agents on the physiology and pharmacology of surgical patients. *Mailing Add:* Dept Anesthesiol Cornell Univ 1300 York Ave New York NY 10021

ARTZT, KAREN, b New York, NY, Sept 4, 42. DEVELOPMENTAL GENETICS. *Educ:* Cornell Univ, AB, 64, PhD(genetics), 72. *Prof Exp:* Fel immunogenetics, Pasteur Inst, Paris, 72-73; res assoc develop biol, Col Med, Cornell Univ, 73-75, asst prof, 75-76; assoc develop genetics, 76-78, assoc mem, Sloan Kettering Inst Cancer Res, 78-86; PROF, UNIV TEX, AUSTIN, 87- *Concurrent Pos:* Mem, Personnel Comt A, Am Cancer Soc, 82-88, Animal Resources Rev Comt, Div Res Resources, Nat Inst Health, 88-; res career develop award, Nat Cancer Inst, 76-81. *Honors & Awards:* Boyer Young Investr Award, Sloan Kettering Inst, 81. *Mem:* AAAS; Genetics Soc Am. *Res:* Molecular biology of mammalian development; genetics of T/t complex. *Mailing Add:* Dept Zool Univ Tex Austin TX 78712-1064

ARUMI, FRANCISCO NOE, b Valparaiso, Chile, Feb 4, 40; m 63; c 2. THERMAL PHYSICS. *Educ:* Univ NC, BS, 62, MS, 64; Univ Tex, PhD(physics), 70. *Prof Exp:* Asst prof physics, Calif State Polytech Col, 64-65; vis prof, Univ Costa Rica, 65-66; regional specialist, Asn Am Univ, 66-67; res assoc physics, 70-72, asst prof archit, 71-75, assoc prof, 75-80, PROF ARCHIT, UNIV TEX, 80-, DIR RES, SCH ARCHIT & DIR, NUMERICAL SIMULATION LAB, 77- *Concurrent Pos:* Fulbright fel, Comn Int Exchange of Persons, 71 & 72. *Mem:* Am Phys Soc; Int Solar Energy Soc; Am Soc Heating & Refrig Engrs; Am Asn Physics Teachers. *Res:* Thermal physics applied to energy analysis in architectural design; thermodynamic modelling of societal systems. *Mailing Add:* Dept Archit Univ Tex Austin TX 78712

ARUNASALAM, VICKRAMASINGAM, b Jaffna, Ceylon, Aug 26, 35; m 68. PHYSICS. *Educ:* Univ Ceylon, BS, 57; Univ Mass, Amherst, MS, 60; Mass Inst Technol, PhD(physics), 64. *Prof Exp:* Asst lectr physics, Univ Ceylon, 57-58; instr physics, Univ Mass, summer 60; res assoc plasma physics, Princeton Univ, 64-67, mem res staff, 67-76, res physicist, 76-80, PRIN RES PHYSICIST, PLASMA PHYSICS LAB, PRINCETON UNIV, 80- *Mem:* Fel Am Phys Soc. *Res:* Plasma physics; quantum theory. *Mailing Add:* Plasma Physics Lab Princeton Univ PO Box 451 Princeton NJ 08543

ARVAN, DEAN ANDREW, b Arta, Greece, Sept 27, 33; US citizen; m 58; c 3. CLINICAL PATHOLOGY, BIOCHEMISTRY. *Educ:* Wilkes Col, BS, 55; Hahnemann Med Col, MD, 59. *Hon Degrees:* MA, Univ Pa, 71. *Prof Exp:* Intern, St Agnes Hosp, Philadelphia, 59-60; resident path, Univ Pa Hosp, 60-64, asst instr, Sch Med, Univ Pa, 60-64, assoc, 64-69, assoc prof path, 69-77, asst dir, Chem Sect, Pepper Lab, Hosp, 67-69, dir, 69-77; PROF & ASSOC CHAIR PATH & DIR, LAB MED DIV, SCH MED & DENT, UNIV ROCHESTER, 77- *Concurrent Pos:* Attend physician, Dept Path, Vet Admin Hosp, Philadelphia, 67-77; consult biochemist, Coatesville Hosp, Pa, 68-77. *Mem:* AAAS; Am Asn Clin Chemists; Am Soc Clin Pathologists. *Res:* Clinical enzymology; use of serum and tissue enzymes in diagnosis; protein chemistry; application of new molecular techniques to diagnosis of disease. *Mailing Add:* Dept Path Univ Rochester Med Ctr Rochester NY 14642

ARVAN, PETER, US citizen. CELL BIOLOGY. *Educ:* Cornell Univ, AB, 77; Yale Univ, MD, 84, PhD(cell biol), 84. *Prof Exp:* Intern med, NC Mem Hosp, 84-85; postdoctoral fel cell biol, Sch Med, Yale Univ, 85-86, resident med, 86-87, fel endocrinol, 87-88; ASST PROF ENDOCRINOL, BETH ISRAEL HOSP, HARVARD MED SCH, 88-, ASST PROF CELL & DEVELOP BIOL, MED SCH, 88- *Concurrent Pos:* Prin investr & consult endocrinol, Beth Israel Hosp, 88- *Mem:* Am Soc Cell Biol; AAAS; NY Acad Sci; Endocrine Soc; Am Thyroid Asn. *Res:* Sorting and targeting of newly-synthesized secretary proteins and peptide hormones. *Mailing Add:* Div Endocrinol Beth Israel Hosp 330 Brookline Ave Boston MA 02215

ARVESEN, JAMES NORMAN, b Portland, Ore, Oct 22, 42; m 65; c 2. MATHEMATICAL STATISTICS, APPLIED STATISTICS. *Educ:* Univ Calif, AB, 64; Stanford Univ, MS, 66, PhD(statist), 68. *Prof Exp:* Mathematician, Daniel H Wagner & Assocs, 66; asst prof statist, Purdue Univ, Lafayette, 68-73; assoc dir sci affairs, Pfizer Pharmaceut, 73-74, group assoc dir sci affairs, 74, dir, Dept Statist & Data Anal, 74-78; pres, Princeton Anal Serv Corp, 78-80; WITH VICKS DIV RES & DEVELOP, 80- *Concurrent Pos:* Consult, Ill Geol Surv, 68-; adj asst prof math statist, Columbia Univ, 73-74; vis assoc prof math, Hunter Col, 74-; sr consult, JJJ Statist Appln, Inc, 75-; vis prof, Grad Bus Sch, NY Univ, 77- *Mem:* Inst Math Statist; Biomet Soc; Am Statist Asn; Inst Mgt Sci; Opers Res Soc Am. *Res:* Robust procedures; regression analysis; biostatistics. *Mailing Add:* Pfizer Pharmaceut 235 E 42nd St New York NY 10017-5755

ARVESON, WILLIAM BARNES, b Oakland, Calif, Nov 22, 34; m; c 3. MATHEMATICS. *Educ:* Calif Inst Technol, BS, 60; Univ Calif, Los Angeles, AM, 63, PhD(math), 64. *Prof Exp:* Mathematician, US Naval Undersea Res Ctr, 60-64; actg asst prof math, Univ Calif, Los Angeles, 65; Benjamin Peirce instr, Harvard Univ, 65-68; lectr, 68-69, assoc prof, 69-73, PROF MATH, UNIV CALIF, BERKELEY, 73- *Concurrent Pos:* Nat Res Coun fel, Brit, 72, sr fel, 81; guest prof, Aarhus Univ, 73-74; Guggenheim fel, 76-77; assoc ed, Duke Math J, 78-87, J Operator Theory, 79-88, prin ed, 88-; Miller Res Prof, Univ Calif, Berkeley, 85-86. *Mem:* Am Math Soc. *Res:* Algebras of operators on Hilbert space; representations of Banach algebras; ergodic and prediction theory. *Mailing Add:* Dept Math Univ Calif Berkeley CA 94720

ARVIDSON, RAYMOND ERNST, b Brooklyn, NY, Jan 22, 48; m 69; c 2. REMOTE SENSING. *Educ:* Temple Univ, BA, 69; Brown Univ, MS, 71, PhD(geol), 74. *Prof Exp:* Res assoc planetary geol, Brown Univ, 73; from asst prof to prof, 74-88, PROF EARTH & PLANETARY SCI DEPT, WASH UNIV, 84-, DEPT CHAIR, 91- *Concurrent Pos:* Fel, McDonnel Ctr Space Sci, Wash Univ, 76-, dir, NASA Planetary Image Facil, 78-; team leader, Viking Lander Imaging Team, 78-82; mem, Space Sci Bd & chmn, Comn Data Mgt & Comput, Nat Acad Sci, 81-; assoc ed, J Geophys Res, 81-84 & Cambridge Univ Press, 83-; mem planetary rev panel, NASA, 81-, chmn, earth observation syst data panel, 84; adv, Mars Inst Planetary Soc, 83-; secy, planetary geol div, Geol Soc Am, 83-85 & planetology div, Am Geophys Union, 84-85; proj scientist, Pilot Planetary Data Syst, 85; chmn working group, Magellan Data Prod Group, 85; interdisciplinary scientist, Mars Observ, 86-; ed, Geol, 88; mem, Mars Mission Coord Implementation Team, US/USSR, 88, Joint Working Group on Solar Syst Explor; Node mgr, Planetary Geosci Node, 89-93; pres-elect, Planetary Sect, Am Geophys Union, 90-91. *Honors & Awards:* Cushing Orator, Am Neurol Asn, 85. *Mem:* Am Geophys Union; Geol Soc Am; Planetary Soc. *Res:* Remote sensing of the surfaces of the terrestrial moons and planets; regional geologic studies of the earth; understanding the relationships between planetary geology records and interior evolution; investigations of the geological histories of Mars and Venus. *Mailing Add:* Dept Earth & Planetary Sci Wash Univ St Louis MO 63130

ARY, DENNIS, b Los Angeles, Calif, July 18, 50; m 83; c 2. PREVENTIVE MEDICINE, BEHAVIORAL MEDICINE. *Educ:* Loyola Univ, BA, 72; Claremont Grad Sch, MA, 75, PhD(psychol), 80. *Prof Exp:* Mgr comput serv, Claremont Grad Sch, 74-77 & instr, grad statist, 76; res assoc, Inst Social Sci & Res, Univ Ore, 81 & Col Health & Phys Educ, 81; RES SCIENTIST BEHAV MED, ORE RES INST, 82- *Mem:* Soc Behav Med. *Res:* Primary prevention of adolescent tobacco and other drug use; Primary prevention of adolescent health and problem behavior with sex, aids, tobacco and other drug use; predictions and risk factors for adolescent tobacco and other drug use; methodological and statistical issues in field research. *Mailing Add:* Ore Res Inst 1899 Williamette St Suite 2 Eugene OR 97401

ARY, THOMAS EDWARD, b Walla Walla, Wash, Apr 30, 50. NEUROPHARMACOLOGY, PSYCHOPHARMACOLOGY. *Educ:* Wash State Univ, BPharm, 73, PhD(pharmaceut sci), 80. *Prof Exp:* ASST PROF, COL PHARM, NDAK STATE UNIV, 89- *Concurrent Pos:* Alta Heritage fel, res asst prof, Univ Nev Sch Med; vis asst prof, Idaho State Univ. *Mem:* Am Asn Col Pharm. *Res:* Smooth muscle excitation-contraction coupling; role of prostaglandins. *Mailing Add:* Col Pharm NDak State Univ Fargo ND 58105-5055

ARYA, ATAM PARKASH, b Panjab, India, June 2, 34; m 61; c 2. NUCLEAR PHYSICS. *Educ:* Univ Rajasthan, India, BS, 53, MS, 55; Univ Panjab, India, MA, 56; Pa State Univ, PhD(physics), 60. *Prof Exp:* Vis res assoc physics, Pa State Univ, 60-62; asst prof, Univ Toledo, 62-64; from asst prof to assoc prof, 64-74, PROF PHYSICS, WVA UNIV, 74- *Mem:* Am Phys Soc; Am Asn Physics Teachers. *Res:* Beta and gamma spectroscopy; neutron inelastic scattering; energy and angular distribution of protons from neutron induced reactions. *Mailing Add:* Dept Physics WVa Univ Morgantown WV 26506

ARYA, SATYA PAL SINGH, b Mavi Kalan, India, Aug 24, 39; c 3. METEOROLOGY, FLUID DYNAMICS. *Educ:* Univ Roorkee, India, BE, 61, ME, 64; Colo State Univ, Ft Collins, PhD(fluid mech), 68. *Prof Exp:* Asst engr civil eng, Irrigation Dept, India, 61-62; lectr, Univ Roorkee, India, 63-65; from res asst to res assoc fluid dynamics, Colo State Univ, Ft Collins, 65-69; from res asst prof to res assoc prof atmospheric sci, Univ Wash, Seattle, 69-76; assoc prof meteorol & geosci, NC State Univ, 76-81, prof meteorol & actg head, 81-83; AT DEPT MARINE-EARTH-ATMOSPHERIC SCI, NC STATE UNIV, 83- *Concurrent Pos:* Vis prof, Ctr Atmospheric Sci, Indian Inst Tech, New Delhi, 83-84. *Mem:* Am Geophys Union; AAAS; Sigma Xi; Am Meteorol Soc. *Res:* Planetary boundary layers; atmospheric turbulence and diffusion; air-sea and air-sea ice interactions; micrometeorology; dispersion in complex terrain/flows. *Mailing Add:* Dept Marine-Earth-Atmospheric Sci NC State Univ PO Box 8208 Raleigh NC 27695-8208

ARZBAECHER, ROBERT, b Chicago, Ill, Oct 28, 31; m 56; c 5. ELECTRICAL ENGINEERING, BIOENGINEERING. *Educ:* Fournier Inst Technol, BS, 53; Univ Ill, MS, 58, PhD(elec eng), 60. *Prof Exp:* Elec engr, Argonne Nat Lab, 54-60; from asst prof to prof elec eng, Christian Bros Col, Tenn, 60-67; from assoc prof to prof, Univ Ill, Chicago Circle, 67-76; prof & chmn dept elec & comput eng, Univ Iowa, 76-87; PROF & DIR, PRITZKER INST MED ENG, ILL INST TECHNOL, 88- *Concurrent Pos:* Mem comt electrocardiography, Am Heart Asn, 71-; prof internal med, Univ Iowa, 78-81. *Mem:* Fel Inst Elec & Electronics Engrs; fel Am Col Cardiol. *Res:* Cardiac electrophysiology; computers in medical research. *Mailing Add:* Pritzker Inst Med Eng IIT Ctr Chicago IL 60616

ARZOUMANIDIS, GREGORY G, b Thessaloniki, Greece, Aug, 16, 36; nat US; m 66; c 2. HOMOGENEOUS CATALYSIS, ORGANOMETALLIC CHEMISTRY. *Educ:* Univ Thessaloniki, BS & MS, 59; Univ Stuttgart, Germany, PhD(inorg chem), 64; Univ Conn, MBA, 79. *Prof Exp:* Sr res chemist, Am Cyanamid Co, Stamford, Conn, 69-72 & Stauffer Chem Co,

Dobbs Ferry, NY, 72-79; RES ASSOC, AMOCO CHEM CO, NAPERVILLE, ILL, 79- *Mem:* Am Chem Soc; Sigma Xi. *Res:* Polyolefin catalysis (Ziegler-Natta). *Mailing Add:* Amoco Chem Co PO Box 3011 Naperville IL 60566

ARZT, SHOLOM, b New York, NY, May 3, 29; m 60; c 1. MATHEMATICS. *Educ:* NY Univ, AB, 46, MS, 48, PhD(math), 51. *Prof Exp:* Rockefeller asst, Courant Inst Math Sci, NY Univ, 46-51; sr mathematician, Appl Physics Lab, Johns Hopkins Univ, 51-54; proj engr, Specialty Electronics & Eng Co, 55-57 & Universal Transistor Prod Corp, 57-58; asst prof, 58-63, assoc prof & head dept, 63-69, PROF MATH, COOPER UNION, 69- *Mem:* Am Math Soc; Math Asn Am; Soc Indust & Appl Math. *Res:* Number theory; analysis. *Mailing Add:* Pinebrook Moretown VT 05660

ASA, SYLVIA L, b New York, NY, May 26, 53; m 76; c 2. ENDOCRINE PATHOLOGY. *Educ:* Univ Toronto, MD, 77, PhD, 90; FRCP(C), 82, FCAP, 82. *Prof Exp:* ASST PROF PATH, UNIV TORONTO-ST MICHAELS' HOSP, 84- *Mem:* Int Acad Path; Am Asn Pathologists; fel Col Am Pathologists; Endocrine Soc. *Res:* Pituitary pathology; pituitary tumor tissue culture; fetal pituitary; hormone production by tumors. *Mailing Add:* Dept Path Univ Toronto-St Michael's Hosp 30 Bond St Toronto ON M5B 1W8 Can

ASAAD, MAGDI MIKHAEIL, b Zagazig, Egypt, Sept 3, 40; US citizen; m 73. PHARMACOLOGY, BIOCHEMICAL PHARMACOLOGY. *Educ:* Cairo Univ, BPharm & Pharm Chem, 62; Univ Houston, PhD(pharmacol), 76. *Prof Exp:* Pharmacist, Zein El Abdine Pharm, Cairo, Egypt, 62-63; pharmacist in charge, 63-71; technician acad res, Dept Pharmacol, Univ Pittsburgh, 71-73; res fel indust res, Squibb Inst Med Res, Princeton, NJ, 76-78, res investr, 78-90; RES LEADER, BRISTOL-MYERS SQUIBB PHARMACEUT RES INST, PRINCETON, NJ, 90- *Mem:* Am Pharmaceut Asn; Am Soc Pharmacol & Exp Therapeut; NY Acad Sci. *Res:* Hypertension; role of central and peripheral nervous system; renin angiotensin system, renal physiology and pharmacology; autonomic physiology and pharmacology; neuronal and receptor functions. *Mailing Add:* Bristol-Myers Squibb Pharmaceut Res Inst PO Box 4000 Princeton NJ 08543-4000

ASADULLA, SYED, b Channapatna, India, June 3, 33; m 59. MATHEMATICS. *Educ:* Cent Col, Bangalore, India, BSc, Hons, 55; Univ Karachi, MSc, 56; Univ Fla, PhD(math), 66. *Prof Exp:* Lectr math, Fed Col, Karachi, Pakistan, 55-56, head dept, 56-57, chmn fac sci, 57-61; interim instr math, Univ Fla, 61-66; asst prof, Miami Univ, 66-68; asst prof, 68-71, ASSOC PROF MATH, ST FRANCIS XAVIER UNIV, 71- *Concurrent Pos:* Coop lectr, Univ Karachi, 57-61. *Mem:* Math Asn Am; Am Math Soc; Can Math Soc. *Res:* Number theory; quadratic forms. *Mailing Add:* St Francis Xavier Univ Box 83 Antigonish NS B2G 1C0 Can

ASAI, DAVID J, b Chicago, Ill, June 7, 53. CYTOSKELETON. *Educ:* Calif Inst Technol, PhD(biochem), 79. *Prof Exp:* Asst res prof, Univ Calif, Santa Barbara, 82-85; asst prof, 85-89, ASSOC PROF BIOL, PURDUE UNIV, 89- *Mem:* Am Soc Cell Biol; Am Soc Biochem & Molecular Biol; Am Soc Develop Biol. *Mailing Add:* Dept Biol Sci Purdue Univ West Lafayette IN 47907

ASAKURA, TOSHIO, b Osaka, Japan, Aug 21, 35; m 67; c 2. BIOCHEMISTRY, HEMATOLOGY. *Educ:* Kyoto Med Col, BS, 56, MD, 60; Univ Tokyo, PhD(biochem), 65; Univ Pa, MA, 74. *Prof Exp:* Intern hosp, Univ Tokyo, 60-61; asst prof biochem, Univ, 65-67; assoc biophysics, Johnson Res Found, 67-69, asst prof, 69-74, assoc prof pediat & biophysics, 74-76, PROF PEDIAT, BIOCHEM & BIOPHYS, UNIV PA, 76- *Concurrent Pos:* Career develop award, NIH, 70-75. *Mem:* AMA; Am Soc Pediat Res; Am Soc Biol Chemists; Am Chem Soc. *Res:* Metabolism of red blood cells; oxygen binding properties of hemoglobin; spin-labeling of hemes and porphyrins; sickle cell disease and other hemoglobinopathies; transgenic mice. *Mailing Add:* Children's Hosp Philadelphia 34th St & Civic Ctr Blvd Philadelphia PA 19104

ASAL, NABIH RAFIA, b Haifa, Israel, Dec 21, 38; m 66. EPIDEMIOLOGY, BIOSTATISTICS. *Educ:* William Jewell Col, AB, 63; Univ Mo, MS, 65; Univ Okla, PhD(epidemiol), 68. *Prof Exp:* Res assoc epidemiol, Univ Mo, 65-66; from instr to asst prof biostatist & epidemiol, 68-72, epidemiologist, 69-72, assoc prof biostatist & epidemiol, 73-78, PROF BIOSTATIST & EPIDEMIOL, SCH HEALTH, UNIV OKLA, 78- *Concurrent Pos:* Consult, Okla Regional Med Prog, 68-; Epidemiol Div, Okla State Health Dept, US Environ Protect Agency, Am Cancer Soc, Cancer Ctr, Fed Aviation Agency & Civil Aeromed Inst. *Mem:* Soc Epidemiol Res; Am Pub Health Asn; Asn Teachers Prev Med. *Res:* Infectious and chronic disease epidemiology; cancer and cerebrovascular disease. *Mailing Add:* Dept Biostatist & Epidemiol 303 Col Health Bldg Univ Okla Health Sci Ctr Box 26901 Oklahoma City OK 73190

ASANO, AKIRA, b Stockton, Calif, Jan 20, 23; m 51; c 2. PHARMACEUTICS. *Educ:* Drake Univ, BS, 44; Univ Minn, MS, 45, PhD(pharmaceut chem), 48. *Prof Exp:* Res assoc, Merck Sharp & Dohme, 48-57; group leader, 57-59, asst dir pharmaceut res, 59-65, dir prod develop, Johnson's Prof Prod Co, 65-67, asst mgr pharmaceut res, 67-77, ASST MGR BIOCHEM & BIOENG RES, JOHNSON & JOHNSON, 77- *Mem:* Am Pharmaceut Asn; Sigma Xi. *Res:* Drug formulation. *Mailing Add:* Johnson & Johnson Res US Rte 1 North Brunswick NJ 08903

ASANO, TOMOAKI, b Tokyo, Japan, Nov 13, 29; m 58. PHYSIOLOGY. *Educ:* Keio Univ, MD, 51, DMS(physiol), 59; Univ Rochester, MS, 55. *Prof Exp:* Asst physiol, Rockefeller Inst Med Res, 55-56 & Sch Med, Keio Univ, 56-57; asst prof physiol, Sch Med, Kanazawa Univ, 57-64; asst prof, 64-68, ASSOC PROF MICROBIOL, LOBUND LAB, NOTRE DAME UNIV, 68- *Mem:* Am Physiol Soc; Soc Exp Biol & Med; Sigma Xi. *Res:* Physiological study of germfree life; carcinogenesis in germfree animals. *Mailing Add:* 217 S Ironwood South Bend IN 46615

ASANUMA, HIROSHI, b Kobe, Japan, Aug 17, 26; m 53; c 2. NEUROPHYSIOLOGY. *Educ:* Keio Univ, Japan, MD, 52; Kobe Med Col, DMedSci, 59. *Prof Exp:* Instr physiol, Kobe Med Col, 53-59; asst prof, Med Sch, Osaka City Univ, 59-61, 63-65; guest investr, Rockefeller Inst, 61-63; from assoc prof to prof physiol, NY Med Col, 65-72; PROF NEUROPHYSIOL, ROCKEFELLER UNIV, 72- *Mem:* Am Physiol Soc; Soc Neurosci; Harvey Soc; Physiol Soc Japan. *Res:* Physiology of mammalian motor system with reference to the function of the pyramidal tract. *Mailing Add:* Dept Motor Physiol Rockefeller Univ 1230 York Ave New York NY 10021

ASARO, ROBERT JOHN, b Brooklyn, NY, Aug 29, 45; m 66; c 2. METALLURGY. *Educ:* Stanford Univ, BS, 67, MS, 69, PhD(mat sci), 72. *Prof Exp:* Res assoc metall, Ohio State Univ, 72-73; staff scientist, Ford Motor Co, 73-75; asst prof, 75-79, ASSOC PROF ENG, BROWN UNIV, 79- *Concurrent Pos:* Consult, US Naval Underwater Systs Command, 80- *Mem:* Am Soc Mining & Metall Engrs. *Res:* Plasticity and fracture of solids; metal corrosion; environmental effects in material behavior; theoretical and experimental studies of the mechanics of crystals. *Mailing Add:* 13 Rustwood Dr Barrington RI 02806

ASATO, GORO, b Mt View, Hawaii, May 29, 31; m 62. ORGANIC CHEMISTRY. *Educ:* Univ Hawaii, BA, 53, MS, 58; Purdue Univ, PhD(org chem), 61. *Prof Exp:* Res chemist, 61-72, group leader, Agr Div, 72-87, MGR, METAB, RESIDUE & ENVIRON CHEM, AM CYANAMID CO, 87- *Mem:* Am Chem Soc. *Res:* Stereochemical and optical rotatory dispersion studies of cyclohexane derivatives; synthesis of pesticides, antifungals and antibacterials; synthesis and infrared spectroscopic studies of transition metal carbonyls; animal health products. *Mailing Add:* Agr Div Am Cyanamid Co Box 400 Princeton NJ 08540

ASATO, YUKIO, b Waipahu, Hawaii, Jan 19, 34; m 69; c 2. MICROBIOLOGY. *Educ:* Univ Hawaii, BA, 57, MS, 66, PhD(microbiol), 69. *Prof Exp:* Microbiologist, State Dept Health, Hawaii, 61-63; asst microbiol, Univ Hawaii, 63-65; res assoc, biol adaptation br, Ames Res Ctr, NASA, 69-71; from asst prof to assoc prof, 71-80, PROF MICROBIOL, SOUTHEASTERN MASS UNIV, 80- *Mem:* AAAS; Am Soc Microbiol; Genetics Soc Am. *Res:* Microbial genetics; biochemical genetics of cyanobacteria. *Mailing Add:* Dept Biol Southeastern Mass Univ North Dartmouth MA 02747

ASAY, KAY HARRIS, b Lovell, Wyo, Nov 20, 33; m 53, 83; c 4. CROP BREEDING, AGRONOMY. *Educ:* Univ Wyo, BS, 57, MS, 59; Iowa State Univ, PhD(crop breeding), 65. *Prof Exp:* Instr high sch, Wyo, 59-61; res assoc crop breeding, Iowa State Univ, 61-65; from asst prof to assoc prof grass breeding, Univ Mo-Columbia, 65-74; RES GENETICIST, AGR RES SERV, USDA, 74- *Res:* Development of improved range grass varieties and related basic studies. *Mailing Add:* 1655 N 1560 E Logan UT 84322

ASBRIDGE, JOHN ROBERT, b Lakeside, Mont, Aug 26, 28; m 54; c 5. SPACE PHYSICS. *Educ:* Mont State Col, BS, 53; Lehigh Univ, MS, 55, PhD(physics), 59. *Prof Exp:* Staff mem, 59-82, SATELLITE PROG MGR, LOS ALAMOS SCI LAB, UNIV CALIF, 82- *Mem:* Am Geophys Union. *Res:* Satellite based study of space environment, the composition of the solar wind, and the earth's magnetosphere and cosmic rays. *Mailing Add:* Los Alamos Nat Lab PO Box 1663 MS D460 Los Alamos NM 87545

ASBURY, JOSEPH G, b Dayton, Ohio, Jan 15, 38; m 68; c 2. PHYSICS. *Educ:* Univ Dayton, BS, 60; Purdue Univ, MS, 62, PhD(physics), 64. *Prof Exp:* Volkswagen fel, Deutsches Elektronen Synchroton, 65-67; from asst physicist to assoc physicist, 67-72, SR ECONOMIST, ARGONNE NAT LAB, 80-, DIR STRATEGIC PLANNING, 85-, DEP LAB DIR, 88- *Concurrent Pos:* Mem, Ill Solid Waste Mgt Task Force, 70-73; mem, Gas Res Inst, Econ Adv Comt, 78-83; mem, Gov Sci Adv Comt, 89- *Mem:* AAAS; Am Phys Soc; Am Water Resources Asn; Sigma Xi. *Res:* Research planning and administration; research and development policy analysis; energy technology assessment. *Mailing Add:* Argonne Nat Lab 9700 S Cass Ave Argonne IL 60439

ASCAH, RALPH GORDON, b Montreal, Que, July 7, 18; m 44; c 3. PHYSICAL CHEMISTRY. *Educ:* McGill Univ, BSc, 39; NY Univ, PhD(chem), 44. *Prof Exp:* Res chemist, Can Indust, Ltd, 43-47; asst prof chem, Pa State Univ, 47-57; assoc prof, 57-80; RETIRED. *Mem:* Am Chem Soc. *Res:* Photochemistry; vapor phase kinetics and low temperature calorimetry; infrared spectroscopy. *Mailing Add:* 223 Westerly Pkwy State College PA 16801

ASCARELLI, GIANNI, b Rome, Italy, Oct 25, 31; US citizen. SPECTROSCOPY, ELECTRONIC TRANSPORT IN SEMICONDUCTORS INSULATORS & LIQUIDS. *Educ:* Univ Rome, Dottore fisica, 55; Mass Inst Technol, PhD(physics), 59. *Prof Exp:* Res assoc physics, Mass Inst Technol, 59, Univ Ill, 59-61; prof incaricato physics, Univ Rome, 61-64; res staff & group leader, Nat Res Coun Rome, 62-64; assoc prof, 64-70, PROF PHYSICS, PURDUE UNIV, 70- *Concurrent Pos:* Fulbright fel, 55 & 91; vis prof, Univ Grenoble, 71-72. *Mem:* Fel Am Phys Soc; AAAS. *Res:* Physics of semiconductors, ionic crystals and silver halides; electronic properties of liquids, particularly electrons injected in insulating liquids, excitoms and Rydberg states. *Mailing Add:* Physics Dept Purdue Univ West Lafayette IN 47907

ASCENSAO, JOAO L, b Maputo, Mozambique, July 6, 48; m 84. HEMATOLOGY, ONCOLOGY. *Educ:* Univ Lisbon Sch Med, MD, 72. *Prof Exp:* Asst prof med, Univ Minn, 81-84; ASSOC PROF MED, NY MED COL, 84- *Concurrent Pos:* Dir CRC, NY Med Col, 86-; assoc dir, Bone Marrow Transplant, NY Med Col, 87-; dir hemat, Westchester Med Ctr, 88- *Mem:* Int Soc Experimental Hemat; Am Soc Hemat; Am Soc Clin Oncol; Am Asn Can Res; Am Fed Clin Res; Europ Soc Med Oncol. *Mailing Add:* Dept Med Div Hemat & Oncol NY Med Col Valhalla NY 10595

ASCENZI, JOSEPH MICHAEL, b Troy, NY, Oct 18, 49; m 71; c 1. MICROBIOLOGY. *Educ:* Univ Dayton, BS, 71, MS, 73; Univ Cincinnati, PhD(microbiol), 77. *Prof Exp:* Res assoc microbiol, Univ Tex, 77-78; sr microbiologist, 78-82, sect head, 82-87, MGR MICROBIOL, JOHNSON & JOHNSON MED, 87- *Concurrent Pos:* Assoc referee, Asn Off Anal Chemists, 78- *Mem:* Am Soc Microbiol; AAAS; Sigma Xi; Soc Indust Microbiol. *Res:* Fatty acid biochemistry; near-ultraviolet radiation effects in bacteria; mode of action of chemo-sterilents and disinfectants. *Mailing Add:* Dept Microbiol Surgikos Inc Arlington TX 76010

ASCH, BONNIE BRADSHAW, m. TUMOR BIOLOGY, CYTOSKELETON. *Educ:* Baylor Col Med, PhD(exp biol), 76. *Prof Exp:* Res assoc, Beth Israel Hosp, Boston, MA, 78-81; asst prof, Harvard Med Sch, 80-81; CANCER RES SCIENTIST, ROSWELL PARK CANCER INST, 81-; ASST RES PROF, STATE UNIV NY, BUFFALO, 87- *Mem:* AAAS; Am Soc Cell Biol; Am Asn Cancer Res. *Res:* Breast cancer; mammary cell biol. *Mailing Add:* Roswell Park Cancer Inst 666 Elm St Buffalo NY 14263

ASCH, HAROLD LAWRENCE, b New York, NY, Nov 6, 43; m 68. MAMMARY CARCINOGENESIS, PANCREATIC CANCER. *Educ:* Cornell Univ, BS, 66; Univ Tex, MS, 71; Rice Univ, PhD(biol), 74. *Prof Exp:* Fel biochem, biochem dept, Baylor Col Med, 73-77, instr, 77-78; actg dir parasitol, Ctr Trop Dis, Univ Lowell, 78-80, assoc dir, 80-81; cancer res scientist cell biol, dept exp path, 81, scientist adminr pancreas cancer, Organ Systs Coord Ctr, 85-89, ASST PROF EXP PATH, ROSWELL PARK MEM INST DIV, STATE UNIV NY, 81- *Concurrent Pos:* Vis scientist, Ain Shams Univ, Cairo, Egypt, 80-81. *Mem:* Am Soc Parasitologists; AAAS; Am Soc Cell Biol. *Res:* Cell biology of mammmary carcinogenesis, especially the roles of retrotransposons and of cytosketal (Paraten) components; cytoskeleton and heat shock proteins in T-cell immonologic activation. *Mailing Add:* Dept Exp Path Roswell Park Cancer Inst Buffalo NY 14263

ASCHBACHER, MICHAEL, b Little Rock, Ark, Apr 8, 44. FINITE GROUP THEORY. *Educ:* Calif Inst Technol, BS, 66; Univ Wis, PhD, 69. *Prof Exp:* Postdoctoral res fel, Univ Ill, 69-70; from asst prof to assoc prof, 72-76, PROF MATH, DEPT MATH, CALIF INST TECHNOL, 76- *Concurrent Pos:* Alfred P Sloan fel, Calif Inst Technol, 72-74. *Mem:* Nat Acad Sci. *Mailing Add:* Math Dept Calif Inst Technol Pasadena CA 91125

ASCHBACHER, PETER WILLIAM, b Ashland, Wis, Apr 20, 28; m 55; c 3. ANIMAL PHYSIOLOGY. *Educ:* Univ Wis, BS, 51, MS, 53, PhD(dairy husb), 57. *Prof Exp:* From asst prof to assoc prof dairy husb, NDak State Univ, 56-64, asst dairy husbandman, 56-61; animal physiologist, 64-78, RES LEADER, METAB & RADIATION RES LAB, AGR RES SERV, USDA, 78- *Concurrent Pos:* Assoc scientist, Agr Res Lab, Atomic Energy Comn 61. *Mem:* AAAS; Am Dairy Sci Asn; Am Soc Animal Sci; Sigma Xi. *Res:* Physiology of farm animals; metabolic fate of xenobiotics in farm animals. *Mailing Add:* Metab/Radiation Res Lab USDA Agr Res Serv PO Box 5674 Fargo ND 58105

ASCHER, EDUARD, b Vienna, Austria, Nov 23, 15; nat US; m 54; c 4. PSYCHIATRY. *Educ:* Wash Univ, BS & MD, 42. *Prof Exp:* Asst prof, 49-73, ASSOC PROF PSYCHIAT, JOHNS HOPKINS UNIV, 73-, PSYCHIATRIST, OUTPATIENT DEPT, JOHNS HOPKINS HOSP, 49- *Concurrent Pos:* Asst clin prof, Sch Med, Univ Md, 64-69, assoc clin prof, 69-; instr, Wash Sch Psychiat, 71-90; consult, off hearings & appeals, Social Security Admin. *Mem:* Fel Am Psychiat Asn. *Res:* Psychotherapy; depressive disorders; expressive behavior; group psychotherapy. *Mailing Add:* 3601 Greenway/Carrollton No 801 Baltimore MD 21218-2439

ASCHER, MARCIA, b New York, NY, Apr 23, 35; m 56. ETHNOMATHEMATICS. *Educ:* Queens Col, NY, BS, 56; Univ Calif, Los Angeles, MA, 60. *Prof Exp:* Comput analyst math, Douglas Aircraft Co, 57-60; specialist tech discipline, Gen Elec Co, 60-61; from asst prof to assoc prof, 61-72, PROF MATH, ITHACA COL, 72- *Concurrent Pos:* Getty Scholar, 87-88. *Mem:* Math Asn Am; Asn Comput Mach. *Res:* Applications of mathematics to anthropology; mathematical ideas of non-literate societies. *Mailing Add:* Dept Math Ithaca Col Ithaca NY 14850

ASCHER, MICHAEL S, b Freeport, Ill, May 17, 42; m 68; c 2. CLINICAL & INFECTIOUS DISEASES, VIROLOGY. *Educ:* Harvard Univ, MD, 68. *Prof Exp:* DEP CHIEF, VIRAL & RICKETTSIAL DIS LAB, CALIF DEPT HEALTH, 85-; LECTR, UNIV CALIF, BERKELEY, 85- *Concurrent Pos:* US Army med res. *Mem:* Am Col Physicians; Infectious Dis Soc; Am Soc Rickettsiology; Am Asn Immunologists; Am Soc Trop & Med Hyg; Am Soc Microbiologists. *Res:* AIDS; infectious disease pathogenesis. *Mailing Add:* Virus Lab Calif Dept Health Serv 2151 Berkeley Way Berkeley CA 94704

ASCHER, ROBERT, b New York, NY, Apr 28, 31; m 50. ANTHROPOLOGY, VISUAL ANTHROPOLOGY. *Educ:* Queens Col, NY, BA, 54; Univ Calif, Los Angeles, MA, 59, PhD, 60. *Prof Exp:* Asst gen anthrop, Univ Calif, Los Angeles, 57-59, instr, Exten Div, 58-60; from asst prof to assoc prof, 60-66, PROF ANTHROP & ARCHAEOL, CORNELL UNIV, 66- *Mem:* AAAS; Am Anthrop Asn; Soc Visual Anthropology. *Res:* Visual anthropology; ethnomathematics. *Mailing Add:* Dept Anthrop 215 Ncgraw Hall Cornell Univ Main Campus Ithaca NY 14853

ASCHNER, JOSEPH FELIX, b Vienna, Austria, Jan 23, 22; US citizen; m 45, 62; c 1. SOLID STATE PHYSICS, SEMICONDUCTORS. *Educ:* Univ Chicago, BS, 43; Univ Ill, PhD(physics), 54. *Prof Exp:* Asst physics, Univ Ill, 50-54; mem tech staff solid state device tech, Bell Tel Labs, 54-62; from asst prof to assoc prof, 62-79, PROF PHYSICS, CITY COL NEW YORK, 80- *Concurrent Pos:* Fac assoc, Boeing Co, 63-66, Gen Atomic Co, 67-68 & Zenith Radio Co, 69. *Mem:* Am Phys Soc; Sigma Xi. *Res:* Alkali halide crystals; solid state diffusion; semiconductor device technology and surfaces; radiation effects; infrared phenomena in semiconductors. *Mailing Add:* Dept Physics City Col New York Convent Ave at 138th St New York NY 10031

ASCHNER, MICHAEL, b Jerusalem, Israel, Nov 11, 55; US citizen; m 79; c 4. NEUROTOXICOLOGY. *Educ:* Univ Rochester, NY, BS, 80, MS, 83, PhD(neurobiol & anat), 85. *Prof Exp:* Postdoctoral fel toxicol, Univ Rochester, 85-87; ASST PROF PHARMACOL & TOXICOL, ALBANY MED COL, 88- *Concurrent Pos:* Teratol Soc young investr award, 85; prin investr, NIH new res serv award, 85-87 & First award, 89-94. *Mem:* Soc Toxicol; Teratology Soc; AAAS; NY Acad Sci. *Res:* Heavy metal, transport across the blood-brain barrier and their effects on homeostasis within the central nervous system employing both in vivo and in vitro techniques. *Mailing Add:* Dept Pharmacol & Toxicol Albany Med Col Albany NY 12208

ASCOLI, GIULIO, b Milano, Italy, Oct 26, 22; nat US; m 50; c 3. EXPERIMENTAL PHYSICS. *Educ:* Mass Inst Technol, PhD(physics), 51. *Prof Exp:* Chem engr nuclear reactor develop, Oak Ridge Nat Lab, 46-47; res asst, Cosmic Rays, Mass Inst Technol, 49-50; from instr to assoc prof, 50-72, PROF PHYSICS, UNIV ILL, URBANA, 72-,. *Mem:* Am Phys Soc. *Res:* Investigation on cosmic radiation. *Mailing Add:* Four Illini Circle Urbana IL 61801

ASCOLI, MARIO, b Guatemala City, Guatemala, May 3, 51; US citizen; m 84. HORMONE ACTION, MOLECULAR ENDOCRINOLOGY. *Educ:* San Carlos Univ, Guatemala, Equi, 71; Vanderbilt Univ, Nashville, PhD(biochem), 75. *Prof Exp:* Postdoctoral fel biochem, M D Anderson Hosp & Tumor Inst, 75-76; res assoc, Biochem Dept, Vanderbilt Univ, 76-77, res instr, 77-78, from asst prof to assoc prof biochem, Med Dept, 78-85; scientist, Pop Coun, 85-88, sr scientist, 89-90; PROF, PHARMACOL DEPT, UNIV IOWA, 90- *Concurrent Pos:* Prin investr, NIH, 78-; ed, Endocrinol, 86-89; Carver sr scientist, Univ Iowa Col Med, 90-95. *Mem:* Sigma Xi; AAAS; Endocrine Soc; Am Soc Cell Biol; Am Soc Biochem & Molecular Biol; Soc Study Reproduction. *Res:* Cellular, biochemical and molecular basis of hormone action. *Mailing Add:* Dept Pharmacol Univ Iowa Col Med Iowa City IA 52242-1109

ASCULAI, SAMUEL SIMON, b Tel Aviv, Israel, Apr 13, 42; m 81; c 3. AGRICULTURAL ECONOMICS. *Educ:* Pace Col, BSc, 65; Rutgers Univ, MSc, 70, PhD(microbiol), 73. *Prof Exp:* Res asst, Ortho Pharmaceut Corp, 65-67, asst scientist, 67-69, assoc scientist, 69-70, scientist, 70-72, sr scientist, 72-74, dir div microbiol, 74-75; mgr res & develop, Monsanto Co, 75-81; pres, Viral Genetics, Inc, 81; vpres, Encore Biol, Inc, 82-87; pres, Firstmiss Seed Co, 88-; Independent consult, Health & Agr Biotech, 87-89; PRES & CHIEF EXEC OFFICER, HYALL PHARMACEUT, INC, 90- *Concurrent Pos:* Adj prof biol, Univ Mo, St Louis, 79-80; pres, Int Genetic Sci, 85- *Mem:* AAAS. *Res:* Antiviral agents immune modulation and reproductive physiology; immunology. *Mailing Add:* Hyall Pharmaceut Inc Eight King St E Suite 202 Toronto ON M5C 1B5 Can

ASEFF, GEORGE V, b Meridian, Miss, May 26, 21; m 46; c 3. METALLURGY, CHEMISTRY. *Educ:* Case Inst Technol, BS, 48; Ga Inst Technol, MS, 51, MS, 63. *Prof Exp:* Sr engr, Lockheed-Ga Co, 51-53, lead engr, 53-58, nuclear specialist, 58-61, scientist, 61-62, proj engr, 62-63, staff & proj engr, 63-65, res scientist & prin investr ultrasonics & metall, 65-70; adminr chem dept, Ga State Univ, 70-71; MGR NOE, MGR ENG FAILURE ANALYSIS, CORPORATE MATRS CONS, LAW ENG, TESTING CO, 71- *Concurrent Pos:* Consult, Photog Assistance Corp, 64- & Minn Mining & Mfg Co, 65- *Mem:* Sr mem AAAS; Am Soc Testing & Mat; sr mem Am Soc Metals. *Res:* Development of materials and material systems for use in a variety of environments, including cryogenic temperature, very high temperature, vibration and nuclear radiation. *Mailing Add:* 1015 Avondale Ave SE Atlanta GA 30312

ASENDORF, ROBERT HARRY, b Philadelphia, Pa, Mar 5, 27; div; c 2. THEORETICAL PHYSICS, INFORMATION SCIENCE. *Educ:* Univ Pa, BA, 47, PhD(physics), 56. *Prof Exp:* Physicist, Selas Corp Am, 52-53; asst instr physics, Bryn Mawr Col, 53-56; physicist, Westinghouse Res Labs, 56-58; sr mem tech staff, Hughes Res Labs, 58-68; with US Govt, 68-73; sr corp scientist, Systs Control, Inc, 73-75; mgr, Imagery & Intel Systs, Ford Aerospace, Western Develop Lab, 75-77; chief engr & dir res & develop, Dalmo-Victor, 77; sr staff engr, Appl Technol, 77-86; SR STAFF SCIENTIST, ESL INC, 86- *Mem:* Inst Elec & Electronics Eng; Am Phys Soc; Sigma Xi; NY Acad Sci; Pattern Recognition Soc. *Res:* Solid state theory; group theory; solid state physics; semiconductors; ferrites; relaxation mechanisms; adaptive pattern recognition; artificial intelligence; radar and sonar signal processing; digital image processing; remote sensing of the environment; computer architecture; electronic warfare systems. *Mailing Add:* 14510 Manuela Rd Los Altos Hills CA 94022

ASENJO, FLORENCIO GONZALEZ, b Buenos Aires, Arg, Sept 28, 26; US citizen; m 60; c 2. MATHEMATICAL LOGIC. *Educ:* La Plata Nat Univ, Lic math, 54, PhD(math), 56. *Prof Exp:* Res mathematician, Lab Testing Mat & Tech Invests, Arg, 48-57, head calculus & statist sect, 57-58; asst prof math, Georgetown Univ, 58-61; assoc prof math, Univ Southern Ill, 61-63; assoc prof, 63-66, PROF MATH, UNIV PITTSBURGH, 66- *Concurrent Pos:* Asst, La Plata Nat Univ, 53-55, instr, 55-57, titular prof, 57-58; Fulbright fel, Univ Lisbon, 70 & 85. *Mem:* Am Math Soc; Asn Symbolic Logic. *Res:* Formalization of internal relations; arithmetic of term-relation numbers; calculus of antinomies; theory of multiplicities; sampling processes; model theory. *Mailing Add:* Dept Math Univ Pittsburgh Pittsburgh PA 15260

ASERINSKY, EUGENE, b New York, NY, May 6, 21; m 42; c 2. PHYSIOLOGY. *Educ:* Univ Chicago, PhD(physiol), 53. *Prof Exp:* Asst physiol, Univ Chicago, 49-52; res assoc, Univ Wash, 53-54; from instr to prof, Jefferson Med Col, 54-76; PROF PHYSIOL & CHMN DEPT, SCH MED, MARSHALL UNIV, 76- *Concurrent Pos:* Vis res scientist, Eastern Pa Psychiat Inst, 61-70; vis prof, NY Col Psychiat, 66- & Cajal Inst, Madrid, 71; adj prof, WVa Univ, 77- *Mem:* Am Physiol Soc; Am Med Writers Asn; sr mem Am Astronaut Soc; Soc Neurosci; Neuroelec Soc. *Res:* Physiology of sleep, eye movements and blinking; electro-oculography; neurophysiology; circulatory and respiratory reflexes; chronaxie; long term effects on synaptic resistance; biomedical application of third derivative of motion. *Mailing Add:* Sch Med Marshall Univ Huntington WV 25701-2901

ASERINSKY, EUGENE, b New York, NY, May 6, 21; m 59; c 3. SLEEP PHYSIOLOGY, ELECTROPHYSIOLOGY. *Educ:* Univ Chicago, PhD(physiol), 53. *Prof Exp:* Res assoc physiol, Sch Fisheries, Univ Wash, 53-54; instr physiol, Dept Physiol, Jefferson Med Col, 54-57, from asst prof to prof physiol, 54-76; prof & chmn physiol & pharmacol, 76-77, prof & chmn physiol, 77-86, EMER PROF PHYSIOL, SCH MED, MARSHALL UNIV, WASH, 87- *Concurrent Pos:* Vis instr, Northern Ill Col Optom, 51-52; vis res scientist, Eastern Pa Psychiat Inst, Philadelphia, Pa, 63-69; vis prof, NY Sch Psychiat, 66-68 & Dept Biophys, Ramón y Cajal Inst, Spain, 71. *Mem:* Am Physiol Soc; Soc Neurosci. *Res:* Physiological parameters of sleep; oculomotor characteristics of rapid eye movement; oculocardiac reflex, chronaxie and CNS changes subsequent to spinal hemisection. *Mailing Add:* 606 Third Ave No 136 San Diego CA 92101

ASFAHL, C RAY, b Enid, Okla, Aug 31, 38; c 4. INDUSTRIAL ENGINEERING, COMPUTER SCIENCE. *Educ:* Okla State Univ, BS, 61; Stanford Univ, MS, 62; Ariz State Univ, PhD(indust eng), 70. *Prof Exp:* Indust engr, Ethyl Corp, 60; engr, Continental Pipeline Co, 61; asst prof indust eng, Ohio Univ, 65-67; assoc prof, 69-74, PROF INDUST ENG, UNIV ARK, 74- *Mem:* Am Inst Indust Engrs; Soc Mfg Engrs; Am Soc Safety Engrs; Am Indust Hyg Asn; Am Soc Eng Educ. *Res:* Industrial safety; robotics; automated systems; reliability of automated systems. *Mailing Add:* Dept Indust Eng Univ Ark Fayetteville AR 72701

ASGAR, KAMAL, b Tabriz, Iran, Aug 28, 22; c 2. BIOENGINEERING. *Educ:* Tech Col Tehran, BA, 45; Univ Mich, Ann Arbor, MS, 48, BS, 50, PhD(mat & metall), 59. *Prof Exp:* From res asst to res assoc, 52-59, from asst prof to assoc prof, 59-66, PROF DENT MAT, SCH DENT, UNIV MICH, ANN ARBOR, 66- *Concurrent Pos:* Mem, Base Metal Alloys Comt, Am Nat Stand Comt & chmn, Casting Investment Comt & Gypsum Subcomt, 73-77; mem consult team, Nat Inst Dent Res; consult, Vet Admin Hosp & USN Dent, 74- *Honors & Awards:* Paul Gibbons Award, Sch Dent, Univ Mich & Souder Award, Int Asn Dent Res, 70; Hollenback Award, Acad Oper Dent, 84; Nikayama Award, 88. *Mem:* Int Asn Dent Res; Am Soc Metallurgists; Microbeam Anal Soc. *Res:* Cast alloys used in dentistry; both noble and base metal alloys; porcelain-metal restorations; dental amalgam; dental investments. *Mailing Add:* Dept Biomat Sch Dent Univ Mich Ann Arbor MI 48109

ASGHAR, KHURSHEED, b India, Aug 16, 40; US citizen; c 3. PHARMACOLOGY, PHARMACY. *Educ:* Univ Panjab, BPharm, 60; Univ Calif, San Francisco, PhD(pharm chem), 66. *Prof Exp:* Res fel, Dept Pharmacol, Univ Calif, San Francisco, 65-67; lectr pharm, Univ Karachi, 67-68; res assoc pharmacol, Univ Chicago, 68-71; spec res fel toxicol, Nat Heart & Lung Inst, NIH, 71-73; pharmacologist, Bur Drugs, Food & Drug Admin, 73-74; mid-level assoc, FDA Career Develop Prog, 74-76; health scientist adminr, Nat Inst Neurol & Commun Disorders & Stroke, NIH, 76-81; health scientist adminr, 81-84, pharmacologist, div preclin res, alcohol, drug abuse & mental health admin, 84-88, CHIEF, EXTRAMURAL POLICY & PROJ REV BR, OFF SCI, NAT INST DRUG ABUSE, 88- *Concurrent Pos:* Teaching & res asst, Dept Pharmaceut Chem, Univ Calif, San Francisco, 61-66. *Res:* Pharmacology; toxicology; drug abuse; policy development and management of biomedical programs. *Mailing Add:* Eight Bouldercrest Ct Rockville MD 20850

ASH, ARLENE SANDRA, b Stamford, Conn, May 15, 46. HEALTH SERVICES RESEARCH, STATISTICS & BIOSTATISTICS. *Educ:* Harvard Univ, BA, 67; Washington Univ, MS, 72; Univ Ill, PhD(math, statist), 77. *Prof Exp:* Instr math, Mindanao State Univ, 67-69; res instr, Dartmouth Col, 76-78; asst prof statist, Boston Univ, 78-84, asst prof math in med, 84-90, ASSOC PROF MATH IN MED, BOSTON UNIV MED SCH, 90- *Concurrent Pos:* Statist consult, Sidney Farber Inst, Boston, 78-80; statist consult in legal disputes, 78-; pres, Boston Chap Am Statist Asn, 82-83; pres, Caucus on Women in Statist, 86. *Mem:* Am Statist Asn; Biomet Soc; Inst Math Statist. *Res:* Experimental design; design optimality; statistical aspects of case-control studies and clinical trials; public health policy; statistical methodology in discrimination litigation. *Mailing Add:* Health Care Res Unit Boston Univ Med Sch 720 Harrison Ave Suite 1102 Boston MA 02118

ASH, J MARSHALL, b New York, NY, Feb 18, 40; m 77; c 3. PURE MATHEMATICS. *Educ:* Univ Chicago, SB, 61, SM, 63, PhD(math), 66. *Prof Exp:* Joseph Fels Ritt instr math, Columbia Univ, 66-69; from asst prof to assoc prof, 69-74, PROF MATH, DEPAUL UNIV, 74- *Concurrent Pos:* Partic harmonic anal conf, Warwick Univ, 68, Williams Col, 78 & Univ Md, 79; Am Math Soc partic, Int Math Cong, 70, 74, 78 & 86; vis prof, Stanford Univ, 77. *Mem:* Am Math Soc; Math Asn Am; Sigma Xi. *Res:* Generalized derivatives of functions of a real variable; multiple trigonometric series; real variable; harmonic analysis; measure theory; singular integrals; Fourier series; analytic number theory. *Mailing Add:* Dept Math DePaul Univ Chicago IL 60614

ASH, KENNETH OWEN, b Provo, Utah, Aug 21, 36; m 56; c 4. PATHOLOGY, CLINICAL CHEMISTRY. *Educ:* Brigham Young Univ, BS, 58, PhD(biochem), 61. *Prof Exp:* Proj leader protein res, Gen Mills, Inc, 61-64; sr res scientist, Honeywell Regulator Co, 64-66; prin scientist, Honeywell, Inc, 66-70, tech dir clin instrumentations dept, 70-71; dir lab, A&M Labs, 71-74; dir, Br Lab, Bio-Sci Labs, 75; DIR CLIN CHEM, MED CTR, UNIV UTAH, 75-, ASSOC DIR CLIN LABS, 77-, DIR CLIN CHEM GROUP, 80-, PROF PATH, 85- *Concurrent Pos:* Lab consult, Vet Admin Hosp, Salt Lake City, 75-; sr vpres, Asn Regional & Univ Pathologists, 84; exec vpres & chief opers officer, Assoc Regional & Univ Pathologists, Inc. *Honors & Awards:* Evans Award, Acad Clin Lab Physicians & Scientists, 89. *Mem:* Am Asn Clin Chemists; Am Soc Clin Pathologists; Nat Acad Clin Biochem; Acad Clin Lab Physicians & Scientists (pres elect, 85, pres, 86). *Res:* Cation flux; hypertension; olfaction; bilirubin; trace metal analysis. *Mailing Add:* Dept Path Univ Utah Med Ctr Salt Lake City UT 84132

ASH, LAWRENCE ROBERT, b Holyoke, Mass, Mar 5, 33; m 60; c 1. PARASITOLOGY, TROPICAL MEDICINE. *Educ:* Univ Mass, BS, 54, MA, 56; Tulane Univ, PhD(parasitol), 60. *Prof Exp:* Asst parasitologist, Univ Hawaii, 60-61; instr parasitol, Tulane Univ, 62-65; med parasitologist, NIH, 65-67; from asst prof to assoc prof infectious & trop dis, 67-75, chmn & assoc dean, 79-84, PROF EPIDEMIOL & INFECTIOUS & TROP DIS, SCH PUB HEALTH, UNIV CALIF, LOS ANGELES, 75- *Concurrent Pos:* Consult, US Naval Med Res Unit 2, 70-; NIH res grant, Univ Calif, Los Angeles, 71-; mem US panel parasitic dis, US-Japan Coop Med Sci Prog, NIH, 72-78, chmn panel, 78-84; mem adv sci bd, Gorgas Mem Inst, 74-76; mem ad hoc study group on parasitic dis, US Army Med Res & Develop Command, 75-79. *Mem:* Am Soc Parasitol; Am Soc Trop Med & Hyg; Royal Soc Trop Med & Hyg; Int Filariasis Asn; Am Soc Clin Pathologists. *Res:* Parasitic diseases of man and animals; biology, pathology and systematics of filariae, metastrongyles, ascarids, spirurids and other helminths; diagnostic methods in parasitic diseases. *Mailing Add:* Pub Health Univ Calif 405 Hilgard Ave Los Angeles CA 90024-1772

ASH, MAJOR MCKINLEY, JR, b Bellaire, Mich, Apr 7, 21; m 47; c 4. PERIODONTICS. *Educ:* Mich State Col, BS, 47; Emory Univ, DDS, 51; Univ Mich, MS, 54. *Hon Degrees:* DrMed, Univ Bern, 75. *Prof Exp:* Instr oral path, Emory Univ, 52-53; teaching fel, Sch Dent, Univ Mich, 53-54; from instr to prof periodont & oral path, 54-69, chmn dept Occlusion, 69-87, PROF PERIODONT, SCH DENT, UNIV MICH, ANN ARBOR, 87-, MARCUS L WARD PROF DENT, 86- *Concurrent Pos:* Attend physician, Vet Admin Hosp, Atlanta, Ga, 52-53; consult, Vet Admin Hosp, Ann Arbor, 61-; pres, Mich Basic Sci Bd, 66-70; consult, Catherine McAuley Health Ctr, Ann Arbor, 86- & Fed Drug Admin, 86- *Mem:* AAAS; Int Asn Dent Res. *Res:* Evaluation of electric and manual toothbrushes; telemetry of intra-oral occlusal forces, pH, muscle forces and jaw movements; evaluation of dental pain thresholds and hypersensitivity; occlusion and temporomandibular joint pathology. *Mailing Add:* 1206 Snyder Ann Arbor MI 48103

ASH, MICHAEL EDWARD, b Detroit, Mich, June 26, 37; m 63; c 2. APPLIED MATHEMATICS, ESTIMATION THEORY. *Educ:* Mass Inst Technol, BS, 59; Princeton Univ, MA, 60, PhD(math), 63. *Prof Exp:* Instr math, Princeton Univ, 62-63; res assoc, Brandeis Univ, 63-64; mem staff celestial mech & gen relativity, Lincoln Lab, Mass Inst Technol, 64-76; STAFF ENGR, INERTIAL INSTRUMENT MODELING & PARAMETER ESTIMATION, CHARLES STARK DRAPER LAB, INC, 76- *Concurrent Pos:* Lectr, Northeastern Univ Grad Eve Sch, 68-71 & Dept Comput Sci, Eve Sch, Boston Univ, 73-87; adj assoc prof, Col Eng, Boston Univ, 86- *Mem:* Am Inst Aeronaut & Astronaut. *Res:* Inertial instruments; filtering and estimation; data acquisition and processing; celestial mechanics. *Mailing Add:* 16 Baskin Rd Lexington MA 02173

ASH, RAYMOND H(OUSTON), b Paterson, NJ, Jan 17, 39; m 60; c 3. CONTROL SYSTEMS, SYSTEMS ANALYSIS. *Educ:* Rensselaer Polytech Inst, BEE, 60, MEE, 65, PhD(elec eng), 69. *Prof Exp:* Engr, Procter & Gamble Co, Ohio, 60-64; from instr to asst prof elec eng, Rensselaer Polytech Inst, 67-69; group leader, Systs Anal Group, 69-75; sect head eng div, 75-78, MGR CONTROL SYSTS TECHNOL, ENG DIV, PROCTER & GAMBLE CO, 78- *Mem:* Inst Elec & Electronic Engrs. *Res:* Real-time process computer control systems applications; state estimation and system identification; optimization; industrial applications of modern control theory. *Mailing Add:* Winton Hill Tech Ctr Proctor & Gamble Eng Cincinnati OH 45224

ASH, ROBERT B, b New York, NY, May 20, 35; m 56. MATHEMATICS. *Educ:* Columbia Univ, BA, 55, BS, 56, MS, 57, PhD(elec eng), 60. *Prof Exp:* From instr to asst prof elec eng, Columbia Univ, 58-62; vis asst prof, Univ Calif, Berkeley, 62-63; assoc prof, 63-71, PROF MATH, UNIV ILL, URBANA-CHAMPAIGN, 71- *Res:* Information theory; probability theory. *Mailing Add:* Univ Ill 1409 W Green St Urbana IL 61801

ASH, ROBERT LAFAYETTE, b Holton, Kans, Dec 27, 41; m 69; c 2. MECHANICAL ENGINEERING. *Educ:* Kans State Univ, BS, 63; Tulane Univ, MS, 66, PhD(mech eng), 68. *Prof Exp:* Asst prof thermal eng, 67-70, from assoc prof to prof eng, 70-79, actg dean, 83-84, EMINENT PROF ENG, MECH ENG & MECH DEPT, OLD DOMINION UNIV, 79-, CHMN, 84- *Concurrent Pos:* NASA grants, 69-72 & 79- & consult, 69-72; sr resident res assoc, Jet Propulsion Lab, Calif Inst Technol, 77-79, Navy grant, 79-84. *Honors & Awards:* Charles T Main Award, Am Soc Mech Engrs, 63. *Mem:* Am Soc Mech Engrs; Am Soc Eng Educ; Am Inst Aeronaut & Astronaut; Soc Natural Philos. *Res:* Thermal sciences; heat transfer; fluid mechanics; thermodynamics; space systems. *Mailing Add:* 1530 Powhatan Ct Norfolk VA 23508

ASH, ROY PHILLIP, b Ft Bragg, NC, July 15, 43; m 77. BIOCHEMISTRY. *Educ:* Va Mil Inst, BS, 65; Univ Wash, PhD(biochem), 78. *Prof Exp:* Instr chem, St Mary's Col, 77-80, asst prof chem, 80-82; at dept chem, SF Austin State Univ, 82-86; asst prof, Pacific Lutheran Univ, Tacoma, Wash, 86-87; instr, Univ Puget Sound, Tacoma, Wash, 87-90; MEM FAC, TACOMA COMMUNITY COL, 90- *Mem:* Am Chem Soc. *Res:* Protein chemistry, crystallography of proteins and related biological molecules. *Mailing Add:* 4806 N 19th St Tacoma WA 98406

ASH, SIDNEY ROY, b Albuquerque, NMex, Nov 25, 28; m 62; c 2. GEOLOGY, PALEOBOTANY. *Educ:* Midland Lutheran Col, BA, 51; Univ NMex, BA, 57, MS, 61; Univ Reading, PhD, 66. *Prof Exp:* Phys sci aid ground water br, US Geol Surv, NMex, 56-58, geologist, 58-61, geologist paleont & stratig br, 61-64; instr natural sci, Midland Lutheran Col, 66-67; asst prof earth sci, 67-69; asst prof geol, Ft Hays Kans State Col, 69-70; assoc prof, Weber State Col, 70-75, chmn dept, 77-83, PROF GEOL, WEBER STATE UNIV, 75- *Concurrent Pos:* US-Australia Coop Sci res grant, Mesozoic plants, 76-77; NSF grant, Mesozoic plants, 79-88; mem, State Adv Coun, Sci & Technol, 87-89. *Honors & Awards:* Govs Medal Sci & Technol, 87. *Mem:* Bot Soc Am; Paleont Soc; Paleont Asn. *Res:* Stratigraphy and paleobotany of the Mesozoic. *Mailing Add:* Dept Geol Weber State Univ Ogden UT 84408-2507

ASH, WILLIAM JAMES, b New York, NY, Nov 3, 31; m 53; c 5. DERMATOGLYPHICS. *Educ:* Cornell Univ, BS, 53, MS, 58, PhD(animal genetics), 60. *Prof Exp:* Res geneticist, Cornell Univ, 59-64; dir res, Crescent Inc, NY, 64-65; mem fac, dept biol, WVa Univ, 65-66; from assoc prof to prof biol, St Lawrence Univ, 66-81; ADV ASSOC INT, INC, 81-; VIS PROF EMBRYOL, STATE UNIV NY, STONY BROOK, 85- *Concurrent Pos:* Vis lectr, Univ RI, 62; Cornell Univ res grant, Europe, 63; prof genetics, Kuwait Univ, Arabian Gulf, 76-78; prog officer, Africa-Asia Sect, NSF, 79-81; prog develop officer, US-Saudi Arabian Joint Comn, Riyadh, 82-83. *Mem:* Am Dermatoglyphics Asn; Int Dermatoglyphics Asn; Sigma Xi; US Power Squadrons. *Res:* Dermatoglyphic genetics; studying qualitative and quantitative dermatoglyphic traits to establish their structural relationships and heritabilities; relationship between dermatoglyphic traits and disease; anthropologic dermatoglyphics. *Mailing Add:* Box 750 Westhampton NY 11977

ASH, WILLIAM WESLEY, b Binghamton, NY, Sept 24, 41; m 63; c 3. SUPERCONDUCTING MAGNETS. *Educ:* Rensselaer Polytech Inst, BS, 62; Cornell Univ, PhD(physics), 67. *Prof Exp:* Res assoc physics, Cornell Univ, 67-68; from instr to asst prof, Princeton Univ, 68-72; STAFF PHYSICIST, STANFORD LINEAR CTR, 72- *Res:* Experimental particle physics with emphasis on electromagnetic interactions (electro and photoproduction and colliding electron-positron physics); detector development. *Mailing Add:* Stanford Linear Accelerator Ctr Box 4349 Stanford CA 94309

ASHBROOK, ALLAN WILLIAM, b Runcorn, Eng, July 18, 29; Can citizen; m 52; c 3. INORGANIC CHEMISTRY, METALLURGICAL CHEMISTRY. *Educ:* Carleton Univ, BSc, 65, PhD(chem), 69. *Prof Exp:* Res chemist metall, Eldorado Nuclear Ltd, 65-69, res supvr, 69-70; head metall chem, Can Govt, 70-75, dir energy res, 75-76; mgr res & develop, 77-80, dir environ & tech, 80-87, VPRES RESOURCES & TECH, EL DORADO NUCLEAR LTD, 87- *Concurrent Pos:* Nat Res Coun Can grant, 69; lectr anal & inorg chem, Carleton Univ, 69-70. *Honors & Awards:* Sherritt Gordon Hydrometall Award, Can Inst Mining & Metall, 78. *Mem:* Am Inst Mining, Metall & Petrol Engrs; Can Inst Mining & Metall; Chem Inst Can; Can Res Mgt Asn; Am Nuclear Soc. *Res:* Manage and direct research and development activities; uranium mining and refining. *Mailing Add:* RR 2 Mountain ON K0E 1S0 Can

ASHBURN, ALLEN DAVID, b Clarkrange, Tenn, Mar 6, 33; m 55; c 3. ANATOMY, PHYSIOLOGY. *Educ:* Tenn Technol Univ, BS, 58, MA, 60; Univ Miss, PhD, 64. *Prof Exp:* From instr to assoc prof, 62-75, PROF ANAT, MED CTR, UNIV MISS, 75- *Mem:* Am Asn Anat. *Res:* Experimental pathology of the cardiovascular system. *Mailing Add:* Dept Anat Sch Med Univ Miss 2500 N State St MS 39216

ASHBURN, EDWARD V, b Pittsburgh, Pa, May 13, 10. PHYSICS, ATMOSPHERIC PHYSICS. *Educ:* Univ Calif-Berkeley, BS, 39; Mass Inst Tech, MS, 40. *Prof Exp:* Physicist, Naval Weapon Ctr, 46-62; res physicist, Lockheed Calif Co, 62-73; RETIRED. *Mem:* Fel Am Phys Soc; fel AAAS; Sigma Xi. *Mailing Add:* 26137 Village 26 Leisure Village Camarillo CA 93012

ASHBURN, WILLIAM LEE, b New York, NY, Jan 18, 33; m 60; c 3. NUCLEAR MEDICINE. *Educ:* Western Md Col, AB, 55; Univ Md, MD, 59; Am Bd Radiol, dipl, 66. *Prof Exp:* Intern, Ohio State Univ Hosp, 59-60; med officer, USPHS Hosp, Savannah, Ga, 60-62, resident radiol, Clin Ctr, NIH, 62-66, chief diag radioisotopes sect, Dept Nuclear Med, 66-68; PROF RADIOL & CHIEF DIV NUCLEAR MED, SCH MED, UNIV CALIF, SAN DIEGO, 68- *Mem:* AMA; Soc Nuclear Med. *Res:* Use of radioisotopes in medical diagnosis; development of improved techniques of radioisotopic organ visualization using computers and other electronic aids with particular emphasis on cardiopulmonary diagnosis and diseases of the central nervous system. *Mailing Add:* Dept Radiol Univ Hosp 225 W Dickenson St San Diego CA 92103

ASHBY, B(ILLY) B(OB), b Gatesville, Tex, Mar 15, 30; m 51; c 5. PROCESS DEVELOPMENT, PROCESS DESIGN. *Educ:* Univ Tex, BS, 52; Univ Mich, MSE, 53, PhD(chem eng), 56. *Prof Exp:* Sr chem engr, Humble Oil & Ref Co, 58-61, staff engr, 61-63, eng supvr, 63-67; dept head, Kawasaki, Japan, 69-70; ENG ASSOC, EXXON CHEM CO AM, 70- *Mem:* Am Inst Chem Engrs. *Res:* Distillation; process design; process simulation and control; computer control of process equipment; environmental control; wastewater treatment; process development. *Mailing Add:* 9344 Meredith Dr Baton Rouge LA 70815

ASHBY, BRUCE ALLAN, b Elmhurst, NY, Mar 28, 22; m 48; c 2. ORGANIC CHEMISTRY. *Educ:* Rutgers Univ, BS, 48, PhD(org chem), 54; Univ Mich, MS, 49. *Prof Exp:* Chemist, Dow Chem Co, Mich, 49-50; instr chem, Rutgers Univ, 50-53; chemist, silicones, Gen Elec Co, 54-60, specialist, 60-66, mgr spec projs, Silicone Prod Div, 66-70, specialist, silicone prod div, 70-84; INDEPENDENT CONSULT, 84- *Mem:* AAAS; emer mem Am Chem Soc; Am Inst Chem; emer mem Sigma Xi. *Res:* Reaction mechanisms; synthetic routes to morphine; agriculturally important sulfur-nitrogen compounds; organo-silicon chemistry; fluorine compounds; organic and organometallic polymers; platinum chemistry. *Mailing Add:* 452 Penn Ave N Forked River NJ 08731-2506

ASHBY, CARL TOLIVER, b Vale, Ark, June 26, 05; m 29; c 2. PHYSICAL CHEMISTRY. *Educ:* Univ Tex, AB, 29, AM, 31, PhD(phys chem), 34. *Prof Exp:* Tutor chem, Univ Tex, 27-32, instr chem, 33-34; res engr, Servel, Inc, 34-43, dir develop, 43-53, chief engr, 53-56; dir eng & pres, Conrad, Inc, Mich, 56-60; dir eng, Norge Div, Borg-Warner Corp, 60-63; adv eng, Arkla Air Conditioning Co, 63-65; res assoc agr-biochem, Univ Ariz, 65-73; RETIRED. *Concurrent Pos:* Consult, 73- *Mem:* Am Chem Soc; Am Soc Heating, Refrig & Air-Conditioning Engrs. *Res:* Absorption refrigeration; low temperature refrigeration; environmental engineering; corrosion; crystallography; hydrothermal growth of crystals. *Mailing Add:* 777 Custer Rd Unit 17-2 Richardson TX 75080-5168

ASHBY, EUGENE CHRISTOPHER, b New Orleans, La, Oct 25, 30; m 52; c 7. ORGANIC CHEMISTRY, INORGANIC CHEMISTRY. *Educ:* Loyola Univ, La, BS, 51; Auburn Univ, MS, 53; Univ Notre Dame, PhD(chem), 56. *Prof Exp:* Res assoc, Ethyl Corp, 56-63; asst prof chem, 63-65, from assoc prof to prof, 65-73, REGENTS' PROF CHEM, GA INST TECHNOL, 73-, DISTINGUISHED PROF, 88- *Concurrent Pos:* Sloan fel, 65-67; Guggenheim fel, 78-79. *Honors & Awards:* Lavoisier Medal, French Chem Soc, 71; Herty Medal, Am Chem Soc, 84. *Mem:* Am Chem Soc. *Res:* Organometallic chemistry; organoaluminum, magnesium, beryllium and boron; complex metal hydrides; organometallic reaction mechanisms; stereochemistry of alkylation and reduction reactions; organic reaction mechanisms; author or coauthor of over 250 research publications. *Mailing Add:* Dept Chem Ga Inst Technol Atlanta GA 30332

ASHBY, JON KENNETH, b Dayton, Ohio, Feb 2, 41; c 2. SPEECH-LANGUAGE PATHOLOGY, AUDIOLOGY. *Educ:* Abilene Christian Univ, BS, 64; La State Univ, MA, 66, PhD(commun disorders), 72. *Prof Exp:* Speech pathologist speech therapy, Hearing & Speech Ctr, Dayton, Ohio, 64; instr & supvr speech & hearing therapy, NTex State Univ, 66-70; PROF & PROG DIR COMMUN DIS, ABILENE CHRISTIAN UNIV, 72- *Concurrent Pos:* Sect chief, WTex Rehab Ctr, 72-73, consult & clinician, 73- *Mem:* Am Speech-Language & Hearing Asn; Int Asn Logopedics & Phoniatrics. *Res:* Aging and communication disorders; gerontology; stuttering and the attitudes toward that problem; conductive hearing impairment in children; history and philosophy of the rehabilitation of the hearing impaired; differential diagnosis of aphasia. *Mailing Add:* Dept Commun Abilene Christian Univ ACU Sta Abilene TX 79699

ASHBY, NEIL, b Dalhart, Tex, Mar 5, 34; div; c 2. THEORETICAL PHYSICS. *Educ:* Univ Colo, BA, 55; Harvard Univ, MA, 56, PhD(physics), 61. *Prof Exp:* Asst prof, 61, Sheldon fel, 61-62, from asst prof to assoc prof, 62-70, PROF PHYSICS, UNIV COLO, BOULDER, 70- *Concurrent Pos:* Chmn, Boulder fac assembly, 80-82, dept physics, Univ Colo, 84-88. *Mem:* Am Phys Soc; Asn Develop Comput-Based Instruct Syst; Int Soc Gen Relativity & Gravitation. *Res:* Transport phenomena; quantum statistical mechanics; relativity. *Mailing Add:* Dept Physics Univ Colo Boulder CO 80309-0390

ASHBY, VAL JEAN, b Kansas City, Mo, June 24, 23; m 48; c 3. APPLIED PHYSICS, ELECTROOPTICS. *Educ:* Univ Kans, BS, 44; Univ Calif, PhD(nuclear physics), 53. *Prof Exp:* Physicist, Lawrence Radiation Lab, Univ Calif, 46-59; physicist, Ramo-Woodridge, 59-61; physicist, Space Technol Labs, 61-62; physicist, Litton Indust Inc, 62-88; RETIRED. *Mem:* Am Phys Soc; Optical Soc Am. *Res:* Signal processing for noise-limited measurement systems; random noise, correlation, filtering and Fourier analyses; multidimensional spatial filtering; geometrical and physical optics; electro-optical systems; nuclear magnetic resonance gyro; ring laser gyro. *Mailing Add:* 19465 Vintage Northridge CA 91324

ASHBY, WILLIAM CLARK, b Duluth, Minn, July 6, 22; m 51; c 3. RECLAMATION, BIOMASS PRODUCTION. *Educ:* Univ Chicago, SB, 47, PhD(bot), 50. *Prof Exp:* Plant physiologist, US Forest Serv, 50-53; res fel, Calif Inst Technol, 53-54; Fulbright res scholar, Sydney, Australia, 54-55; asst prof bot, Univ Chicago, 55-60; assoc prof, 60-69, PROF BOT, SOUTHERN ILL UNIV, CARBONDALE, 69- *Concurrent Pos:* Fac res partic, Argonne Nat Lab, 66 & 70-71; cooperator, USDA Forest Serv; bus mgr, Ecol Soc Am, 69-73. *Mem:* AAAS; Ecol Soc Am (treas, 66-68); Soil Sci Soc Am; Am Soc Agron; Brit Ecol Soc; Am Soc Surface Mining & Reclamation. *Res:* Tree, shrub and herb establishment and growth on surface mines with differing rooting media, grading, existing vegetation, herbicides and mycorrhiza; emphasis on root system development; silver maple production for biomass. *Mailing Add:* Dept Plant Biol Southern Ill Univ Carbondale IL 62901-6509

ASHCRAFT, ARNOLD CLIFTON, JR, b Lexington, Ky, Nov 1, 38; m 59; c 4. ORGANIC CHEMISTRY, POLYMER CHEMISTRY. *Educ:* Univ Cincinnati, BS, 60; Univ Calif, Berkeley, PhD(org chem), 64. *Prof Exp:* MEM RES STAFF ORG & POLYMER CHEM, CHEM & PLASTICS DIV, UNION CARBIDE CORP, 63- *Honors & Awards:* Cert Recognition, NASA, 73. *Mem:* Sigma Xi. *Res:* Synthetic organic chemistry; polymer synthesis; dielectric materials; composite materials. *Mailing Add:* RD 2 Box 2 Hightstown NJ 08520

ASHCROFT, DALE LEROY, b Plainview, Minn, Apr 22, 26; m 49; c 3. ELECTRICAL ENGINEERING. *Educ:* US Naval Acad, BS, 49; Univ Ill, MS, 53. *Prof Exp:* Elec engr, USAF, 49-57, Gen Elec Co, 57-62, Louis Allis Corp, 62-64, Westinghouse Elec Corp, 64-68, US Elec Motors, 68-71, Ward Leonard Elec Co, 71-74; ELEC ENGR, PRINCETON UNIV PLASMA PHYSICS LAB, 74- *Mem:* Inst Elec & Electronics Engrs; Indust Appls Soc. *Res:* Energy conversion and storage; power electronics. *Mailing Add:* Plasma Physics Lab Princeton Univ Box 451 Princeton NJ 08543

ASHCROFT, FREDERICK H, b New York, NY, July 14, 53; m 76. EXPERIMENT & INSTRUMENTATION DESIGN. *Educ:* Univ Mich, Ann Arbor, BSE, 85. *Prof Exp:* CONSULT, FH ASHCROFT, 76-; SR ASSOC RES ENG, UNIV MICH, 78- *Concurrent Pos:* Assoc res eng, Univ Mich, 85- *Mem:* Soc Naval Architects & Marine Engrs; Am Soc Naval Engrs. *Res:* Design and instrumentation of all aspects of hydrodynamic modeling including ships and marine structures. *Mailing Add:* 830 Sunrise Ct Ann Arbor MI 48103-3545

ASHCROFT, NEIL WILLIAM, b London, Eng, Nov 27, 38; US citizen; m 61; c 2. CONDENSED MATTER PHYSICS. *Educ:* Univ NZ, BSc, 58, MScHons, 60, & Dipl (hon); Univ Cambridge, PhD(physics), 64. *Prof Exp:* Res assoc, Univ Chicago, 64-65; from res assoc to assoc prof, Cornell Univ, 65-75, dir, Lab Atomic Solid State Physics, 79-84, prof physics, 75-89, HORACE WHITE PROF PHYSICS, CORNELL UNIV, 90-; DEP DIR, CORNELL HIGH ENERGY SYNCHROTRON SOURCE, 78- *Concurrent Pos:* Sr fel, sci res coun, Cavendish Lab, 73-74; vis fel, Clare Hall, Cambridge

Univ, 73-74 & Churchill Col, 84-85; consult, Los Alamos Nat Lab, 76-; mem panel High Pressure Sci & Technol, Div Mat Sci, US Dept Energy, 82; vchmn, Gordon Conf, High Pressure Physics & Chem, 84, chmn-elect, 86-; mem, NAS Panel, Condensed Matter Physics, 83; actg dir, Cornell High Energy Synchrotron Source, 83-84; adv panel, High Flux Beam Reactor, Brookhaven Nat Lab, 84-; John Simon Guggenheim Mem fel, 84-85; Royal Soc Guest fel, Cambridge, 84-85; vchmn Div Condensed Matter Physics, Am Phys Soc, 85-86; consult, Lawrence Livermore Nat Lab, 85-; chmn, Div Condensed Matter Physics, Am Phys Soc, 86-87; chmn, Gordon Conf Res High Pressures, 86; vis comt mem, Brookhaven Nat Lab, 86; panel, High Temp Superconductivity, Nat Acad Sci, 87. *Mem:* Fel Am Phys Soc; Am Inst Physics; fel AAAS. *Res:* Properties of matter under extreme conditions, theory of disordered systems, theory of classical and quantum liquids, optical properties of solids, theory of the metallic state; numerous articles in physics journals. *Mailing Add:* Clark Hall Cornell Univ Ithaca NY 14853-2501

ASHE, ARTHUR JAMES, III, b New York, NY, Aug 5, 40; m 62; c 2. ORGANIC CHEMISTRY. *Educ:* Yale Univ, BA, 62, MS, 65, PhD(org chem), 66. *Prof Exp:* From asst prof to assoc prof, 66-76, chm, 83-86, PROF CHEM, UNIV MICH, 76- *Concurrent Pos:* A P Sloan fel, 73; vis scholar, Phys Chem Inst, Univ Basel, 74. *Mem:* Am Chem Soc; The Chem Soc. *Res:* Physical organic chemistry; organometallic compounds. *Mailing Add:* Dept Chem Univ Mich Ann Arbor MI 48109

ASHE, JAMES S, b Charlotte, NC, Feb 23, 47; m 84; c 1. SYSTEMATICS, EVOLUTION. *Educ:* Univ NC, BS, 69; Appalachian State Univ, MA, 72; Univ Alta, Edmonton, PhD(systematics), 82. *Prof Exp:* From asst cur to assoc cur insects, Field Mus Natural Hist, 82-88; ASSOC PROF ENTOM & SYSTEMATICS & ECOL, UNIV KANS, LAWRENCE, 88-, DIR ENTOM, 88- *Concurrent Pos:* Asst cur & sessional instr, Univ Alta, 78-79; lectr biol, Field Mus Natural Hist, Chicago, 83-84, adult educ, 85-86, evol biol, Univ Chicago, 84-88; vis cur entom, Am Mus Natural Hist, 83-; div head insects, Field Mus Nat History, 85-88, actg chmn zool, 85, 87 & 88, actg div head zool, 86-87, chmn, Sci Adv Coun, 86-87; naturalist & lectr, Tour-Falkland Island, 88-; sci ed, Fieldiana, 88- *Mem:* Coleopterists Soc (pres & pres-elect, 89-90); Entom Soc Am (secy, 88-90); Soc Syst Zool; Soc Study Evolution; AAAS. *Res:* Systematics and evolution of mushroom inhabiting beetles; evolution, host relationships and ecology of parasitic staphylinid beetles and their rodent hosts. *Mailing Add:* Snow Entom Mus Snow Hall Univ Kans Lawrence KS 66045

ASHE, JOHN HERMAN, b Philadelphia, Pa, Mar 27, 44; div; c 1. NEUROPHYSIOLOGY, SYNAPTIC MECHANISMS. *Educ:* Univ Calif, Riverside, BA, 72, Irvine, PhD(biol), 77. *Prof Exp:* NIH fel neurophysiol, Med Ctr, Univ Calif, San Francisco, 77-78, NIMH fel, 78-79, asst res physiologist, 79-80; from asst prof to assoc prof, 80-90, PROF PHYSIOL PSYCHOL, UNIV CALIF, RIVERSIDE, 90- *Concurrent Pos:* Prin investr, NSF, 81-88, 90-; mem, div res grants study sect behav & neurosci, NIH, 84-87; mem, Ctr Neurobiol Learning & Memory Organized Res Unit, Univ Calif, 84-; NSF adv panel, Neural Mechanisms Behav, 90- *Mem:* Soc Neurosci; Am Physiol Soc; Am Soc Cell Biol; AAAS; Int Brain Res Orgn. *Res:* Electrophysiological and neuropharmacological analysis of synaptic transmission and plasticity in the central and peripheral nervous system; neurophysiological process in learning and memory. *Mailing Add:* Neurosci Prog Dept Psychol Univ Calif Riverside CA 92521

ASHE, WARREN (KELLY), b Halifax, NC, Aug 20, 29; m 51; c 5. MICROBIOLOGY, BIOCHEMISTRY. *Educ:* Howard Univ, BS, 51, MS, 62, PhD, 84. *Prof Exp:* Med biol technician, Nat Inst Dent Res, 54-56, supvr med biol technol, 56-57, from microbiologist to res microbiologist, 57-70, health scientist adminr, 70-71; instr, 71-85, ASST DEAN RES, COL MED, HOWARD UNIV, 71-, ASST PROF, 85- *Concurrent Pos:* Spec consult to secy Dept Health, Educ & Welfare, 68; consult, NIH, 72-, Ohio State Univ, 72-74, Col Dent & dept microbiol, Grad Sch, Southern Ill Univ, 73-74 & Patent Awareness Prog, Atlanta Univ, 74-75; prin investr, biomed res support grant, 72-; mem, Spec Grants Rev Comt, Nat Inst Dent Res, 75-79; mem bd dirs, Am Heart Asn, Dallas, 83-, chmn, 86-88; prin investr, Student Res Asst Prog Grant, 78; prog dir, Howard Univ Ctr Health Sci, Div Res Resources, NIH, 77-81; consult, Nat Cancer Inst, 86- *Mem:* Int Asn Dent Res; Am Asn Lab Animal Sci; Am Soc Microbiol; Soc Exp Biol & Med; Soc Res Adminr; Sigma Xi. *Res:* Virology; immunology; pathology; the role of Herpes simplex viruses in the pathogenesis of chronic diseases, including oral facial ulcerative disorders, autoimmune diseases and vitiligo; research administration. *Mailing Add:* 5051 12th St NE Washington DC 20017

ASHEN, PHILIP, b Brooklyn, NY, Nov 5, 15. INORGANIC CHEMISTRY. *Educ:* City Univ New York, BA, 36; New York Univ, MBA, 57, PhD(chem), 68. *Prof Exp:* Chief chemist, Alco Mfg Corp, 36-48; mgr Chem Div, 48-63, vpres, 63-77, PRES, MW HARDY, & CO, INC, 77- *Concurrent Pos:* Gas reconnaissance officer & lectr chem warfare, US Citizens Defense Corps, 40-45; consult, 50-; lectr, Grad Sch, NY Univ, 54-56. *Mem:* Fel Am Inst Chemists; fel AAAS; NY Acad Sci; Am Chem Soc; Chem & Econ Soc. *Res:* Experimentation with plant life for pharmaceutical and industrial purposes; synthetic organic and natural products chemistry. *Mailing Add:* 2315 Ave I Brooklyn NY 11210

ASHENDEL, CURTIS LLOYD, b St Louis, Mo, June 20,1955; m 84; c 3. CANCER RESEARCH, CELLULAR PHYSIOLOGY. *Educ:* Mich State Univ, BS, 77; Univ Wis, PhD(oncol), 82. *Prof Exp:* Asst prof, 82-88, ASSOC PROF MED CHEM, PURDUE UNIV, 88- *Concurrent Pos:* Consult, NSF Prog Cellular Biosci, 87-91. *Mem:* Am Asn Cancer Res; Am Soc Biochem & Molecular Biol. *Res:* Molecular mechanisms of carcinogenesis and cellular signaling; the roles of oncogene-encoded proteins and protein kinase in normal and altered signal transduction. *Mailing Add:* Dept Med Chem Pharm Bldg Purdue Univ West Lafayette IN 47907

ASHENHURST, ROBERT LOVETT, b Paris, France, Aug 9, 29; div; c 4. APPLIED MATHEMATICS, COMPUTER SCIENCES. *Educ:* Harvard Univ, AB, 50, SM, 54, PhD(appl math), 56. *Prof Exp:* Res assoc, Harvard Comput Lab, 50-56, instr appl math, Harvard Univ, 56-57; from asst prof to assoc prof, 57-65, chmn comt info sci, 69-74, dir, Inst Comput Res, 69-78, PROF APPL MATH, GRAD SCH BUS, UNIV CHICAGO, 65- *Concurrent Pos:* Ed-in-chief, Commun of Asn Comput Mach, 73- *Mem:* Soc Indust & Appl Math; Inst Elec & Electronics Engrs; Asn Comput Mach; Inst Mgt Sci; AAAS. *Res:* Computer and information systems and their application to scientific and management problems. *Mailing Add:* Univ Chicago 1101 E 58th St Chicago IL 60637

ASHER, DAVID MICHAEL, b Chicago, Ill, Nov 10, 37; m 71; c 1. INFECTIOUS DISEASES, PEDIATRICS. *Educ:* Harvard Univ, AB, 59; Harvard Med Sch, MD, 63; Am Bd Pediat, dipl, 75. *Prof Exp:* Intern med, King County Hosp, Seattle, 63-64; jr asst resident pediat, Boston City Hosp, 64-65; jr resident, Mass Gen Hosp, Boston, 65-66; res assoc virol, Nat Inst Child Health & Human Develop, Bethesda, Md, 66-69; Nat Inst Neurol Dis & Stroke spec res fel virol, USA-USSR Health Scientist Exchange, 69-70; res med officer virol, Nat Inst Neurol & Commun Dis & Stroke, NIH, 70-; AT DEPT PEDIAT & ONCOL, JOHNS HOPKINS UNIV. *Concurrent Pos:* Instr pediat, Sch Med, Johns Hopkins Univ, 76. *Mem:* Soc Pediat Res; Infectious Dis Soc Am. *Res:* Persistent viral infections, especially pathogenesis of infections of the nervous system and the urinary tract; viral infections of children with impaired immune defenses. *Mailing Add:* Dept Pediat & Oncol Johns Hopkins Univ Baltimore MD 21205

ASHERIN, DUANE ARTHUR, b Watertown, Wis, Oct 13, 40; m 61; c 4. SYSTEMS APPLICATIONS, ENVIRONMENTAL IMPACT ASSESSMENT. *Educ:* Univ Wis-Stevens Point, BS, 67; Univ Idaho, PhD(wildlife sci), 74. *Prof Exp:* Res assoc wildlife res, Nat Ecol Res Ctr, US Fish & Wildlife Serv, 71-73, res wildlife biologist, Idaho Coop Wildlife Res Unit, 73-77, wildlife data analyst, Ecol Assessment Methods Group, 77-79, group leader, Rapid Assessment Methods Group, Western Energy & Land Use Team, 79-87, sect leader, Resource Problem Anal Sect, Terrestrial Systs Br, 87-90, CHIEF, TECH SERV BR, NAT ECOL RES CTR, US FISH & WILDLIFE SERV, 90- *Concurrent Pos:* Field asst, Dept Natural Resources, Univ Wis, 67; wildlife asst, Idaho Dept Fish & Game, 71. *Mem:* Wildlife Soc; Nat Wildlife Fedn. *Res:* Development of a terrestrial wildlife and habitat characterization process or methodology; prescribed burning effects on nutrition, production, and big game use of key northern Idaho browse species; natural resource applications of geographic information systems. *Mailing Add:* Nat Ecol Res Ctr US Fish & Wildlife Serv 4512 McMurry Ave Ft Collins CO 80525-3400

ASHFORD, ROSS, b Stalwart, Sask, Nov 1, 26; m 53; c 3. AGRONOMY. *Educ:* Univ BC, BSA, 53; Univ Sask, MSc, 59, PhD, 67. *Prof Exp:* Asst economist, Hops Mkt Bd, London, Eng, 47-48; res off, Res Br, Can Dept Agr, 53-60 & 61-66; Nat Res Coun Can scholar, 60-61; prof crop sci, Univ Sask, 66-87; RETIRED. *Concurrent Pos:* Mem Expert Comt weeds, Western Can. *Mem:* Agr Inst Can; Weed Sci Soc Am; Sask Inst Agrologists. *Res:* Chemical control of weeds in field crops; selective action of herbicides. *Mailing Add:* 66 34959 Old Clayburn Rd Abbotsford BC V2F 6W7 Can

ASHFORD, VICTOR AARON, b Bremerton, Wash, Sept 12, 42; m 73; c 3. XRAY IMAGING INSTRUMENTATION, EXPERIMENTAL HIGH ENERGY PHYSICS. *Educ:* Long Beach State Col, BS, 65; Univ Calif, San Diego, MS, 68, PhD(physics), 73. *Prof Exp:* res assoc physics, Fermi Nat Accelerator Lab, 73-76; assoc physicist, Brookhaven Nat Lab, 76-82; res asst, Stanford Linear Accelerator Ctr, 82-86; RES SPECIALIST, UNIV CALIF, SAN DIEGO, 86- *Mem:* Am Phys Soc. *Res:* Particles and fields; physics and society; experimental high energy elementary particle physics; x-ray imaging instrumentation for x-ray protein crystallography. *Mailing Add:* Bio Dept 13-017 Univ Calif-San Diego La Jolla CA 92093

ASHFORD, WALTER RUTLEDGE, organic chemistry, for more information see previous edition

ASHKIN, ARTHUR, b Brooklyn, NY, Sept 2, 22; m 54; c 3. EXPERIMENTAL PHYSICS. *Educ:* Columbia Col, AB, 47; Cornell Univ, PhD(physics), 52. *Prof Exp:* Staff mem, Radiation Lab, Columbia Univ, 42-46; asst, Nuclear Studies Lab, Cornell Univ, 48-52; mem tech staff, Res Dept, 52-63, head dept laser sci, 63-87, MEM TECH STAFF, BELL LABS, 88- *Honors & Awards:* Quantum Electronics Award, Inst Elec & Electronics Eng; Charles Hard Townes Award, Opt Soc Am, 88. *Mem:* Nat Acad Engr; fel Inst Elec & Electronics Eng; fel Am Phys Soc; fel Optical Soc Am. *Res:* Quantum electronics; nonlinear optics; radiation pressure; optical trapping and manipulation of atoms, dielectric particles and biological particles. *Mailing Add:* AT&T Bell Labs 4E-422 Holmdel NJ 07733

ASHLEY, CHARLES ALLEN, b Bronxville, NY, Oct 10, 23; m 48; c 5. PATHOLOGY. *Educ:* Cornell Univ, AB, 44, MD, 47; Univ Ill, MS, 51. *Prof Exp:* Intern, Mary Imogene Bassett Hosp, 47-49; resident path, Presby Hosp, Chicago, 49-52, asst attend pathologist, 52-53; asst chief path, Walter Reed Army Hosp, 53-55; assoc pathologist, 55-67, DIR, MARY IMOGENE BASSETT HOSP, 67- *Concurrent Pos:* Instr, Univ Ill, 52-53; instr path, Columbia Univ, 55-62, assoc clin prof, 69- *Mem:* Am Asn Path; Sigma Xi. *Res:* Structural and chemical studies of muscular contractions; chemical carcinogenesis; pathology of radiation; electron microscopy. *Mailing Add:* Radcliff House Apts C252 1000 Conestoga Rd Bryn Mawr PA 19010

ASHLEY, DOYLE ALLEN, b Collinsville, Ala, May 18, 32; m 58; c 2. AGRONOMY, APPLIED PHYSIOLOGY. *Educ:* Auburn Univ, BS, 54, MS, 57; NC State Univ, PhD(agron), 67. *Prof Exp:* Asst agron, Auburn Univ, 56-58; soil scientist, agr res serv, USDA, 58-67, res soil scientist, Coastal Plains Soil & Water Conserv Res Ctr, 67-69; asst prof, 69-73, assoc prof, 73-78, PROF AGRON, UNIV GA, 78- *Mem:* Am Soc Agron; Soil Sci Soc Am; Crop Sci Soc Am. *Res:* Nutrient and water utilization by field crops; field canopy photosynthesis of crop plants; photosynthate translocation and utilization by crop plants; plant-soil relationships. *Mailing Add:* Dept Agron Univ Ga Athens GA 30602

ASHLEY, HOLT, b San Francisco, Calif, Jan 10, 23; m 47. AERODYNAMICS, AEROELASTICITY. *Educ:* Univ Chicago, BS, 44; Mass Inst Technol, SM, 48, ScD(aeronaut eng), 51. *Prof Exp:* From instr to prof aeronaut eng, Mass Inst Technol, 47-67; PROF AERONAUT & ASTRONAUT, STANFORD UNIV, 67- *Concurrent Pos:* Mem res adv comt aircraft struct, NASA, 58-, mem subcomt aircraft struct; prof aeronaut eng & head dept, Indian Inst Technol, Kampur, India, 64-65; mem struct & mat panel, Adv Group for Aeronaut Res & Develop, 66-69; dir aeronaut & astronaut affiliates prog, Stanford Univ; mem, USAF Sci Adv Bd; distinguished prof aerospace eng, Univ Md, 71-72; div dir, NSF, 73-74; mem bd visitors, USAF Inst Technol, 78-82. *Honors & Awards:* Goodwin Medal, Mass Inst Technol, 53; Struct, Struct Dynamics & Mat Award, Am Inst Aeronaut & Astronaut, 69; 50th Anniversary Medal, Am Meteorol Soc, 71; Wright Bros Lectr, Am Inst Aeronautics & Astronautics, 81- *Mem:* Nat Acad Eng; Am Inst Aeronaut & Astronaut (pres, 73); Am Meteorol Soc; Am Acad Arts & Sci; Am Soc Mech Eng. *Res:* Flutter analysis of wings; structural dynamics; unsteady aerodynamics; application of unsteady aerodynamics to stability of aircraft; wind energy conversion systems; earthquake response of structures. *Mailing Add:* Dept Aeronaut & Astronaut Stanford Univ Stanford CA 94305

ASHLEY, J(AMES) ROBERT, b Kansas City, Mo, Oct 22, 27; m 53; c 3. ELECTRICAL ENGINEERING, ACOUSTICS. *Educ:* Univ Kans, BSEE, 52, MSEE, 56; Univ Fla, PhD(elec eng), 67. *Prof Exp:* Engr klystrons, Sperry Gyroscope Co, Great Neck, NY, 52-53; instr elec eng, Univ Kans, 53-56; sr engr klystron res & develop, Sperry Electron Tube Div, Gainesville, Fla, 56-65; engr cathode ray tubes, Hewlett-Packard, Colo, 67; from assoc prof to prof, Univ Colo, Colorado Springs, 67-77; ENG STAFF CONSULT, SPERRY GYROSCOPE, 77- *Concurrent Pos:* Res & develop consult, Sperry Microwave Electronics Div, 68-72; sci consult, Army Res Off, 72-78; res & develop consult, Koss Corp, 76-80; prof elec eng, Univ Colo, Denver, 77-81. *Mem:* Sr mem Inst Elec & Electronics Engrs; sr mem Simulation Coun; fel Audio Eng Soc. *Res:* Electron tubes; microwave noise measurements; computer aided design procedures; loudspeaker synthesis; loudspeaker crossover networks; loudspeaker placement in listening rooms; theory of musical transients; auditorium psychoacoustics; electronic warfare. *Mailing Add:* 2523 Lake Ellen Lane Tampa FL 33618

ASHLEY, KENNETH R, b Louisville, Ky, Nov 7, 41; m 63; c 2. INORGANIC CHEMISTRY. *Educ:* Southern Ill Univ, BA, 63; Wash State Univ, PhD(chem), 66. *Prof Exp:* Fel inorg chem, Univ Southern Calif, 66-68; asst prof, 68-71, ASSOC PROF CHEM, E TEX STATE UNIV, 71- *Mem:* Am Chem Soc; Sigma Xi. *Res:* Inorganic reaction mechanisms of transition metals. *Mailing Add:* Dept Chem E Tex State Univ Commerce TX 75428

ASHLEY, MARSHALL DOUGLAS, b Portland, Maine, Apr 18, 42; m 65; c 1. FOREST BIOMETRY, REMOTE SENSING. *Educ:* Univ Maine, BS, 65; Purdue Univ, MS, 68, PhD, 69. *Prof Exp:* Forester, Int Paper Co, 65-66; asst prof, 69-75, assoc prof, 75-77, PROF FOREST RESOURCES, UNIV MAINE, ORONO, 77-, ASSOC DIR SCH FOREST RESOURCES, 77- *Mem:* Am Soc Photogram; Soc Am Foresters; Sigma Xi. *Res:* Forest sampling design; application of remote sensors to forest measurement; crop yield from satellites; insect damage from aerial photography. *Mailing Add:* Dept Forest Resources Univ Maine 283 Main St Orono ME 04473

ASHLEY, RICHARD ALLAN, b Wilmington, Del, Mar 20, 41; m 63; c 2. WEED SCIENCE. *Educ:* Univ Del, BS, 63, MS, 65, PhD(biol sci), 68. *Prof Exp:* From asst prof to assoc prof, 68-82, PROF HORT, UNIV CONN, 82- *Concurrent Pos:* Secy-treas, Northwestern Weed Sci Soc, 84-88. *Mem:* Weed Sci Soc Am; Am Soc Plasticult. *Res:* Effect of herbicides on plant-soil water relations and weed interference with crop growth; use of row covers for pest control. *Mailing Add:* Dept Plant Sci U67 Univ Conn Storrs CT 06269-4067

ASHLEY, ROGER PARKMAND, b Portland, Ore, Oct 1, 40; m 64; c 2. ECONOMIC GEOLOGY. *Educ:* Carleton Col, BA, 62; Stanford Univ, PhD(geol), 67. *Prof Exp:* Geologist, 66-79, chief, Br Western Mineral Resources, 80-84, GEOLOGIST, US GEOL SURV, 85- *Mem:* AAAS; Geol Soc Am; Soc Econ Geologists; Am Geophys Union. *Res:* Geology of gold deposits in western United States; detection and mapping of hydrothermal alteration using multispectral imagery; regional geology and ore deposits of Cascade Range, Washington and Oregon. *Mailing Add:* US Geol Surv Mail Stop 901 345 Middlefield Rd Menlo Park CA 94025

ASHLEY, TERRY FAY, b Mayfield, Ky, Aug 24, 42; m. CYTOGENETICS, ZOOLOGY. *Educ:* Duke Univ, BA, 64; Fla State Univ, MS, 67, PhD(genetics), 70. *Prof Exp:* Asst prof bot, Univ Ga, 70-71; fel biol, Univ Lethbridge, 71-73; instr, Univ Calgary, 73-75; adj asst prof, Univ Denver, 75-76; res assoc anat, Med Ctr, Duke Univ, 76-79; PROF DEPT ZOOL, UNIV TENN, 79- *Concurrent Pos:* Fel, Nat Res Coun Can, 71-73; NIH fel, prin investr, 76-79 & 80-83 & 84-87; assoc res prof, Dept Zool, Univ Tenn, 80-; vis scientist, Genetic Inst, Uppsala, Sweden, 87 & 89; consult, Oak Ridge Nat Lab; vis prof, Dept Human Genetics, Yale Univ Sch Med. *Mem:* Am Soc Cell Biol; Genetics Soc Am; Am Genetics Asn; Sigma Xi. *Res:* Molecular cytogenetis of meiosis: factors involved in homologous recognition synapsis and recombination. *Mailing Add:* Dept Human Genetics Yale Univ Sch Med Newtown CT 06510

ASHLEY, WARREN COTTON, b Yorkville, Ill, Dec 25, 04; m 29; c 2. POLYMER CHEMISTRY. *Educ:* Univ Ill, BS, 29, MS, 30, PhD, 35. *Prof Exp:* Org chemist, Eastman Kodak Co, 34; res org chemist, Battelle Mem Inst, 35-38; res org chemist, Pyroxylin Prod, Inc, 38-62; res org chemist, G J Aigner Co, 62-69; INDEPENDENT FORMULATOR, 69- *Honors & Awards:* George B Heckel Award, Fedn Socs Coatings Technol, 45. *Mem:* Am Chem Soc; Fedn Socs Coatings Technol. *Res:* Emulsion and lacquer type and hot melt coatings and adhesives. *Mailing Add:* 656 S Wright St Naperville IL 60540

ASHLOCK, PETER DUNNING, taxonomic entomology; deceased, see previous edition for last biography

ASHMAN, MICHAEL NATHAN, b Baltimore, Md, Oct 15, 40; m 66; c 2. ANESTHESIOLOGY, COMPUTER SCIENCES. *Educ:* Johns Hopkins Univ, BA, 60; Univ Md, MD, 64; Polytech Inst Brooklyn, MS, 69; Am Bd Anesthesiologists, dipl, 70. *Prof Exp:* Instr, 69-71, ASST PROF ANESTHESIOL, SCH MED, UNIV MD, BALTIMORE CITY, 71- *Concurrent Pos:* Fel, Dept Anesthesiol, Columbia-Presby Med Ctr, NY, 67-69; consult, Ctr Educ Comput Develop, Univ Md, Baltimore City, 69-78; chmn, Anesthesiol PSG-37, Am Asn Med Systs & Informatics, 83-; mem, comt Comput Appln, Am Soc Anesthesiolgists, 85- *Mem:* Am Soc Anesthesiologists; fel Am Col Anesthesiologists; Asn Comput Mach; Am Asn Med Systs & Informatics; Asn Advan Med Instrumentation; Soc Comput Simulation. *Res:* Uptake and distribution of inhalation anesthetics; continuous system simulation; continuous system simulation languages; computer appplications in anesthesiology. *Mailing Add:* 7206 Denberg Rd Baltimore MD 21209

ASHMAN, ROBERT F, b Buffalo, NY, Oct 7, 38; c 3. RHEUMATOLOGY, INTERNAL MEDICINE. *Educ:* Wabash Col, BA, 60; Oxford Univ, Eng, MA, 66; Col Surgeons & Physicians, Columbia Univ, MD, 66. *Prof Exp:* PROF RHEUMATOLOGY & MICROBIOL, UNIV IOWA, 80-; STAFF MEM, VET ADMIN HOSP, IOWA CITY, 80- *Mem:* Am Asn Immunologists; Am Rheumatism Asn; Am Soc Clin Invest; Am Asn Physicians. *Mailing Add:* 206 W Park Rd Iowa City IA 52240

ASHMORE, CHARLES ROBERT, b Camden, NJ, Dec 1, 34; m 57; c 4. BIOCHEMISTRY, DEVELOPMENTAL BIOLOGY. *Educ:* Univ Conn, BS, 56, PhD(develop biol), 68; Univ Bridgeport, MS, 61. *Prof Exp:* Assoc prof, 68-80, PROF ANIMAL SCI, UNIV CALIF, DAVIS, 80- *Mem:* Soc Exp Biol & Med; Histochem Soc; Am Asn Animal Sci. *Res:* Hereditary muscular dystrophy; muscle metabolism; muscle growth and development. *Mailing Add:* Dept Animal Sci Univ Calif Davis CA 95616

ASHOK, S, b Coimbatore, India, July 17, 47; m 80; c 1. SEMICONDUCTOR DEVICES, SOLID STATE ELECTRONICS. *Educ:* Univ Madras, India, BE, 68; Indian Inst Technol, MTech, 70; Rensselaer Polytech Inst, NY, PhD(elec eng), 78. *Prof Exp:* Instr & grad asst elec eng, Rensselaer Polytech Inst, NY, 70-78; asst prof, 78-83, assoc prof, 83-87, PROF ENG SCI, PA STATE UNIV, 87- *Concurrent Pos:* Instr, Indust Labor Welfare Asn, India, 68-69; mem staff, Gen Elec Res & Develop Ctr, NY, 73; adj fac, Schenectady Community Col, 75-77; fac fel, Gen Elec Res & Develop Ctr, NY, 82 & Humboldt fel, Inst Semiconductor Electronics, Tech Univ Aachen, WGer, 83; vis scientist, Indian Inst Sci, Bangalore & Uppsala Univ, Sweden, 90. *Mem:* Inst Elec & Electronics Engrs; Am Vacuum Soc; Electrochem Soc; Sigma Xi; NY Acad Sci. *Res:* Semiconductor surface barrier formation and transport; ion-beam and radiation effects on semiconductor devices; photovoltaics; semiconducting and insulating thin films; transparent conductors; ion implantation; semiconductor; surface modification. *Mailing Add:* Dept Eng Sci 130 Hammond Bldg University Park PA 16802

ASHTEKAR, ABHAY VASANT, b Shirpur, India, July 5, 49; m 65. GENERAL RELATIVITY, QUANTUM THEORY. *Educ:* Univ Bombay, India, BSc Hons, 69; Univ Chicago, PhD(physics), 74. *Prof Exp:* Lectr, Univ Clermont, France, 78-80; after 80-84, prof physics, 84-88, DISTINGUISHED PROF PHYSICS, SYRACUSE UNIV, 88- *Concurrent Pos:* Guest scientist, Max Planck Inst Astrophys, Ger, 76-; prin investr, Syracuse Univ, NSF, 80-; Alfred P Sloan res fel, 81-85; prof gravitation, Univ Paris, 83-85; vis prof, Univ Poona, 85-86; coordr quantum gravity prog, NSF Inst Theoret Physics, Santa Barbara, Calif, 86; Mem gov coun, Int Soc Gen Relativity & Gravitation. *Honors & Awards:* First Gravity Prize, Gravity Res Found, 79. *Mem:* Int Soc Math Physics; Int Soc Gen Relativity & Gravitation; Am Phys Soc. *Res:* General relativity; gravitation; quantum field theory; geometric quantization; quantum gravity. *Mailing Add:* Physics Dept Syracuse Univ Syracuse NY 13244-1130

ASHTON, DAVID HUGH, b Rivers, Man, Nov 25, 39; m 66; c 2. FOOD MICROBIOLOGY. *Educ:* Univ Man, BS, 62, MS, 64; NC State Univ, PhD(food sci, microbiol), 67. *Prof Exp:* Nat Acad Sci-Nat Res Coun fel, USDA, 67-69; sr res microbiologist, Campbell Inst Food Res, Campbell Soup Co, 69-72, div head microbiol res, 72-75; Lab mgr microbiol res & develop dept, Hunt-Wesson Foods, Inc, 75-81, lab mgr prod develop, 81-83, assoc dir, 83-84, dir prod develop, 85-89, SR DIR PROD & PROCESS DEVELOP, BEATRICE/HUNT- WESSON INC, 89- *Mem:* Inst Food Technol; Sigma Xi. *Res:* Heat processed microbiology; rapid techniques of Salmonella detection; toxin detection including Clostridium botulinum, Salmonella and Staphylococcus aureus; tomato products, convenience and snack foods, vegetable oils, and peanut butter product and process development. *Mailing Add:* Hunt-Wesson Inc 1645 W Valencia Dr Fullerton CA 92633-3899

ASHTON, FLOYD MILTON, b Indianapolis, Ind, Jan 27, 22; m 42, 86; c 1. PLANT PHYSIOLOGY, WEED SCIENCE. *Educ:* Univ Ill, BS, 47; Ohio State Univ, PhD, 55. *Prof Exp:* Anal chemist, Presto-Lite Battery Co, Inc, 40-42; lab asst, Eli Lilly & Co, 45-46; asst biochemist exp sta, Hawaiian Sugar Planters' Asn, 47-55; instr, 56-58, from asst prof to assoc prof, 58-68, prof, 68-88, EMER PROF BOT, UNIV CALIF, DAVIS, 88- *Concurrent Pos:* NIH fel, 62-63; consult, Nat Acad Sci, 67-68. *Honors & Awards:* Res Award, Weed Sci Soc Am. *Mem:* Am Soc Plant Physiol; fel Weed Sci Soc Am. *Res:* Mode of action of herbicides; weed control. *Mailing Add:* 602 12th St Davis CA 95616

ASHTON, FRANCIS T, ELECTROMICROSCOPY, IMMUNE PROCESSING. *Educ:* Univ Pa, PhD(gen physiol), 70. *Prof Exp:* RES ASSOC, UNIV PA, 77- *Mailing Add:* Dept Anat-Chem 6058 Univ Pa Philadelphia PA 19104

ASHTON, GEOFFREY C, b Croydon, Eng, July 5, 25; nat US; m 51; c 4. GENETICS. *Educ:* Univ Liverpool, BSc, 45, PhD(genetics), 58, DSc, 67. *Prof Exp:* Asst, Univ Toronto, 48-50; sect leader, fermentation res div, Glaxo Labs, Ltd, 51-56; sr sci officer, Farm Livestock Res Ctr, Eng, 56-58; prin sci officer, div animal genetics, Commonwealth Sci & Indust Res Orgn, Australia, 58-64; chmn, Dept Genetics, 65-72, 80-87, from asst vchancellor to vchancellor acad affairs, 72-79, PROF GENETICS, UNIV HAWAII, 64-, DIR, HEALTH INSTRNL RESOURCES UNIT, 86- *Concurrent Pos:* Mem panel blood group scientists, Food & Agric Orgn, UN, 63-68; consult, Joint Comn Rural Reconstruct, Taiwan, 70-74; mem, Health Sci Consortium, 87- *Mem:* Int Soc Myopia Res; Behav Genetics Asn; Am Soc Human Genetics; Am Soc Naturalists; Int Soc Animal Blood Group Res. *Res:* Factors maintaining protein and enzyme polymorphisms; genetic aspects of fertility and disease susceptibility; human behavioral genetics; genetics of myopia; computer assisted instruction in health sciences. *Mailing Add:* Dept Genetics Biomed Sci Bldg Univ Hawaii Honolulu HI 96822

ASHTON, JOSEPH BENJAMIN, b Brownwood, Tex, Sept 17, 30; m 60; c 2. ORGANIC CHEMISTRY, MATHEMATICS. *Educ:* Tex Technol Col, BS, 52; Univ Tex, Austin, PhD(chem), 59. *Prof Exp:* Chemist res & develop lab, Shell Chem Co, Tex, 59-60, chemist, Shell Develop Co, Calif, 60-61, group leader indust chem, Shell Co, Tex, 61-65, proj eval mgr indust chem div, NY, 65-68, asst mgr transportation & supplies, Shell Oil Co, 68-69, mgr chem plant indust chem div, Shell Chem Co, Calif, 69-70, mgr, Dominguez Chem Plant, 70-71, mgr forecasting, Shell Oil Co, Houston, 71-72, mgr elastomers, 72-76, mgr planning & anal & mfg resins & solvents, 76-81, mgr gen facil support, 81-85, ENG ADV, COMPUT SYST PLANNING, MFG & TECHNOL, SHELL OIL CO, 85- *Mem:* Sigma Xi; Am Chem Soc. *Res:* Mechanism of attack of ozone on organic compounds; oxidation of organic compounds; forecasting of energy requirements and development of energy model for interfuel competition. *Mailing Add:* 12625 Memorial 123 Houston TX 77024

ASHTON, JULIET H, CORONARY ARTERY DISEASE, THROMBOSIN. *Educ:* Univ Tex, Dallas, PhD(physiol), 80. *Prof Exp:* ASST PROF MED, HEALTH SCI CTR, UNIV TEX, DALLAS, 83- *Mailing Add:* 6309 Cahoba Dr Ft Worth TX 76135

ASHTON, PETER SHAW, b Boscombe, Eng, June 27, 34; m 58; c 3. BOTANY, FORESTRY. *Educ:* Cambridge Univ, MA, 61, PhD(bot), 62. Prof. *Prof Exp:* Forest botanist, Forest Dept, Brunei, 57-62 & Sarawak, 62-66; lectr, Univ Aberdeen, Scotland, 66-72, sr lectr, 72-78; consult forest botanist, Royal Soc London, 70; consult silvi-cultural botanist, Food Agr Orgn-UN Develop Prog, 66-71; co-dir, UNESCO/Man & Biosphere Prog, 74-75; vis lectr, Univ Malaya, 75; mem, UK Nat Comt Man & Biosphere Prog, 74-75; vis lectr, Univ Malaya, 75; mem, UK Nat Comt Man & Biosphere Prog, 76-78; dir, Arnold Arboretum, Harvard Univ, 78-87; PRES, INT ASN BOT GARDENS & ARBORATOR, 86-; PRES, CTR PLANT CONSERV INC, 88- *Concurrent Pos:* Mem, Bd Govs, Osatom Conser, 89- *Honors & Awards:* Environ Merit Award, Environ Protection Agency, 87. *Mem:* Brit Ecol Soc; Linnean Soc London; Int Soc Trop Ecol; Asn Trop Biol; fel Royal Soc Edinburgh, 77; fel Am Acad Arts & Sci; Bot Soc Am; AAAS. *Res:* Tropical tree taxonomy; tropical forest ecology; tropical tree evolutionary and reproductive biology. *Mailing Add:* Dept Bot Arnold Arboretum 22 Divinity Ave Cambridge MA 02130

ASHUTOSH, KUMAR, b Darbhanga, India, Feb 14, 40; US citizen; m 66; c 3. EMPHYSEMA, COR PULMONALE. *Educ:* Darbhanga Med Col, Bihar, India, MBBS, 63, MD, 66; Am Bd Internal Med, dipl, 72, Am Bd Pulmonary Med, dipl, 74. *Prof Exp:* Fel, pulmonary dis, 72-74, asst prof, 74-81, ASSOC PROF MED, UPSTATE MED CTR, STATE UNIV NY, SYRACUSE, 81-, ASST ATTEND PHYSICIAN, STATE UNIV HOSP, 74- *Concurrent Pos:* Dir respiratory care, respiratory med, Vet Admin Hosp, Syracuse, NY, 77-, chief, pulmonary sect, 80-; prin investr, Res Proj 57207, Am Heart Asn, Syracuse, 80-82 & Res Proj E81-1250-Boehringer, Ingelheim, Ridgefield, Conn, 81-83. *Mem:* Am Thoracic Soc; fel Am Col Chest Physicians; fel Am Col Physicians. *Res:* Acute and chronic respiratory failure in lung disease especially treatment of cor pulmonale; pathogenesis of emphysema and proteolytic antiproteolytic mechanisms in experimental models. *Mailing Add:* Dept Med State Univ NY Upstate Med Ctr 155 Elizabeth Blackwell St Syracuse NY 13210

ASHWELL, G GILBERT, b Jersey City, NJ, July 16, 16; m 42; c 2. BIOCHEMISTRY. *Educ:* Univ Ill, BA, 38, MS, 41; Columbia Univ, MD, 48. *Hon Degrees:* Dr, Univ Paris-Sud, 88. *Prof Exp:* Chemist, Merck & Co, Inc, NJ, 41-44; res fel biochem, Columbia Univ, 48-50; med dir, 50-78, chmn, 78-84, NIH INST SCHOLAR, LAB BIOCHEM & METAB, NIH, 84- *Concurrent Pos:* Chief, Sect Enzymes & Cellular Biochem, Arthritis Inst, NIH & Chief, Lab Biochem & Metab. *Honors & Awards:* Prize for Outstanding Achievement in Field of Med Sci, Gairdner Found, 82; Merck Prize, Am Soc Biol Chemists, 84; Sr Scientist Award, Alexander von Humboldt Found, 89. *Mem:* Nat Acad Sci; Am Soc Biol Chem. *Res:* Pentose metabolism; uronic acid metabolism in bacteria; sugar nucleotide biosynthesis and metabolism; mechanism of polysaccharide biosynthesis; structure and function of carbohydrate moiety of glycoproteins; receptor-mediated endocytosis of asialo glycoprotein. *Mailing Add:* Lab Biochem & Metab NIH Rm 9N105 Bldg 10 Bethesda MD 20892

ASHWIN, JAMES GUY, b Prince Albert, Sask, Nov 18, 26; m 60; c 1. PHYSIOLOGY, PHARMACOLOGY. *Educ:* Univ Sask, BSc, 48, MSc, 50; McGill Univ, PhD(physiol), 53. *Prof Exp:* Lectr physiol, Christian Med Col, India, 53-55; spec lectr, Univ Sask, 56-60; res assoc, Can Heart Found, 60-63; assoc med writer, Sterling-Winthrop Res Inst, 63-66; biologist, Bur Drugs, Dept Health & Welfare, Pub Serv Can, 66-80; RETIRED. *Concurrent Pos:* Fel, Univ Miami, 63; mem, Intervarsity Christian Fel. *Mem:* Pharmacol Soc Can. *Res:* Thrombosis; hemorrhage; histamines; medical writing. *Mailing Add:* 1450 Lexington St Ottawa ON K2C 1R9 Can

ASHWORTH, EDWIN ROBERT, b Evansville, Ind, Feb 12, 27; m 50; c 4. COMPUTER SCIENCE, HIGHER EDUCATION. *Educ:* Purdue Univ, BS, 50, MS, 57; Southern Ill Univ, Carbondale, PhD(higher educ), 72. *Prof Exp:* Supvr indust eng, P R Mallory & Co, 53-55; asst prof indust mgt & eng, Okla State Univ, 57-60; asst prof indust mgt, Auburn Univ, 60-61; dir comput ctr, Tuskegee Inst, 61-63; mgr res & instrnl div, Data Processing & Comput Ctr, 63-67, res assoc, 67-69, asst prof electronic data processing, Sch Tech Careers, 63-85, INSTR DESIGN, SOUTHERN ILL UNIV., CARBONDALE, 69-; DATA PROG SYSTS ANALYST, LOCKHEED MISSILES & SPACE SDI, 85- *Concurrent Pos:* Consult, Carter Oil Co, Okla, 58, USAF, Okla, 59 & Minerva Oil Co, Mo, 63-64; vis prof, Univ Grenoble; Nat Eng Inst Microprocessors fel, 75. *Mem:* Nat Soc Prof Engrs; Asn Comput Mach; Am Inst Indust Engrs; Am Soc Qual Control; Am Soc Eng Educ. *Res:* Graphics; simulation; microprocessors. *Mailing Add:* 715 San Conrado Terr, Apt 7 Sunnyvale CA 94086-3169

ASHWORTH, HARRY ARTHUR, b Rochester, NY, Sept 9, 43; m 66. ANALYTICAL CHEMISTRY. *Educ:* Univ Rochester, BS, 65; Carnegie-Mellon Univ, MS, 67, PhD(physics), 71. *Prof Exp:* ASST PROF PHYSICS, SETON HALL UNIV, 70- *Concurrent Pos:* Vis scientist, Francis Bitter Nat Magnet Lab, 69, Dept Physics, Princeton Univ, 76-77; fel, Sandia Nat Lab, 81; consult, Inverness, Inc, 81-83, Angenics, Inc, 86-88. *Mem:* Am Phys Soc; Sigma Xi; Am Asn Physics Teachers. *Res:* Statistical methods; luminescence. *Mailing Add:* Dept Physics Seton Hall Univ South Orange NJ 07079

ASHWORTH, LEE JACKSON, JR, b Oroville, Calif, Apr 1, 26; m 47; c 6. PLANT PATHOLOGY. *Educ:* Univ Calif, BS, 51, MS, 54, PhD(plant path), 59. *Prof Exp:* Asst, Univ Calif, 51-58; asst prof plant path, Tex A&M Univ, 58-65; res plant pathologist, Cotton Res Sta, USDA, Calif, 65-69; lectr & plant pathologist, Univ Calif, Berkeley, 69-80; WITH SAN JOAQUIN VALLEY AGR, 80- *Concurrent Pos:* Assoc ed, Phytopath, 73-75. *Mem:* Am Phytopath Soc; AAAS; Sigma Xi. *Res:* Soil borne diseases; epidemiology and control of soil borne plant pathogens. *Mailing Add:* San Joaquin Valley Agr 9240 S Riverbend Ave Parlier CA 93648

ASHWORTH, ROBERT DAVID, b Dallas, Tex, Apr 22, 40; m 62, 71; c 3. ANATOMY, NEUROPHYSIOLOGY. *Educ:* NTex State Univ, BS, 63; Univ Tex, Dallas, PhD(anat, neurophysiol), 68. *Prof Exp:* Asst prof anat, Queen's Univ, Ont, 68-69 & Emory Univ, 69-71; ASST PROF ANAT, UNIV TEX HEALTH SCI CTR DALLAS, 71- *Concurrent Pos:* Med Res Coun Can res grant, 69-71; NIH grant, 70-72. *Mem:* Am Asn Anatomists. *Res:* Reflex control of limb musculature; electrical events associated with neurosecretion and hormonal release. *Mailing Add:* 9517 Mossridge Dr Dallas TX 75238

ASHWORTH, T, b Colne, Eng, Sept 18, 40; m 64; c 2. SOLID STATE PHYSICS, THERMODYNAMICS. *Educ:* Univ Manchester, BSc, 61, PhD(physics), 67. *Prof Exp:* Turner & Newall res fel physics, Univ Manchester, 66-68; asst prof, 68-71, assoc prof, 72-76, PROF PHYSICS, SDAK SCH MINES & TECHNOL, 76- *Concurrent Pos:* Sr res fel, Nat Ctr Atmospheric Res, 74-75; res scientist & vis prof physics, Cornell Univ, 85-86. *Mem:* AAAS; Am Phys Soc; Brit Inst Physics. *Res:* Thermal conductivities; specific heats; condensation and coalescence; atmospheric physics. *Mailing Add:* Dept Physics SDak Sch Mines & Technol Rapid City SD 57701

ASHWORTH, WILLIAM BRUCE, JR, b Philadelphia, Pa, June 18, 43; m 80; c 1. SCIENTIFIC REVOLUTION, HISTORY OF ASTRONOMY. *Educ:* Wesleyan Univ, BA, 64; Univ Wis, PhD(hist sci), 75. *Prof Exp:* Vis asst prof hist sci, Univ Wis, 77-78; ASSOC PROF HIST SCI, UNIV MO, KANSAS CITY, 75- *Concurrent Pos:* Consult, Linda Hall Library Sci & Technol, 78- *Mem:* AAAS; Hist Sci Soc; Am Astron Soc; Brit Soc Hist Sci; Renaissance Soc Am. *Res:* Renaissance and 17th century science, especially cosmology, natural history, geology, paleontology, and prehistory. *Mailing Add:* Dept Hist Univ Mo Kansas City MO 64110

ASHY, PETER JAWAD, b Lebanon, Aug 5, 40; US citizen; m 62; c 3. CHEMICAL PHYSICS, MATHEMATICS. *Educ:* Univ Southwestern La, BS, 62; Clemson Univ, MS, 65, PhD(chem physics), 67. *Prof Exp:* Chem engr, textile fiber div, E I du Pont de Nemours & Co, 63-65, res physicist, 67-68; asst prof math, 68-72, chmn dept comput sci, 76-77, ASSOC PROF COMPUT SCI-MATH, FURMAN UNIV, 72-; INT CONSULT, 78- *Concurrent Pos:* Consult, Philips Fibers Res Orgn, 68-, Lockwood Greene Engrs, Daniel Int, Bigelow-Sanford, Dan River, Stone Mfg Co & Fountain Industs; statist consult, Union Carbide Corp, Greenville, 73-75; Enwright assoc, 77-; pres, AMAR Int, Inc; mem exec bd, Southern Fedn, 77-80. *Mem:* Am Inst Chem Engrs; Am Chem Soc; Math Asn Am. *Res:* Fundamental and applied work in thermal reactions; kinetics and sorption; phenomena and intermolecular forces; structure and physical behavior of fibers; thermal diffusivity of gases; quantum theory of surface tension; application of Markov processes; chemical engineering. *Mailing Add:* 329 Pelham Rd Greenville SC 29615

ASIK, JOSEPH R, b Lorain, Ohio, Aug 8, 37; m 59; c 3. AUTOMOTIVE ELECTRONICS, CONTROL SYSTEMS. *Educ:* Case Inst Technol, BS, 59; Univ Ill, Urbana, MS, 61, PhD(physics), 66. *Prof Exp:* Res staff mem phys sci, T J Watson Res Ctr, IBM Corp, 66-69; PRIN STAFF ENGR, PROD & MFG STAFF, FORD MOTOR CO, DEARBORN, 69- *Concurrent Pos:* Lectr elec eng, Lawrence Technol Univ, 78-; chmn, Champion Ignition Conf, 78. *Mem:* Am Phys Soc; Soc Automotive Engrs; sr mem Inst Elec & Electronics Engrs. *Res:* Systems design; ignition systems; electronic engine controls. *Mailing Add:* 776 Great Oaks Dr Bloomfield Hills MI 48304

ASIMAKIS, GREGORY K, b Beaumont, Tex, Sept 23, 47. BIOCHEMISTRY. *Educ:* Rice Univ, BA, 71; Univ Tex Med Br, Galveston, PhD(biochem), 77. *Prof Exp:* Fel, Harvard Univ & Mass Gen Hosp, 77-78; res assoc biochem, Tufts Univ, 78-80; mem fac, biochem div, 80-, AT SURG RES LAB, MED BR, UNIV TEX, GALVESTON. *Mem:* Sigma Xi; Biophys Soc. *Res:* Development of oxidative-phosphorylation in fetal and neotal liver; loss of mitochondrial function in ischemic or pathologic tissues; regulation of mitochondrial ion transport. *Mailing Add:* Surg Res Lab Med Br Univ Tex Galveston TX 77550

ASIMOV, ISAAC, b Petrovichi, Russia, Jan 2, 20; nat US; m 42, 73; c 2. SCIENCE WRITING, BIOCHEMISTRY. *Educ:* Columbia Univ, BS, 39, MA, 41, PhD(enzyme chem), 48. *Prof Exp:* Chemist, Naval Air Exp Sta, Pa, 42-45; from instr to assoc prof, 49-79, PROF BIOCHEM, SCH MED, BOSTON UNIV, 79-; SCI WRITER, 54- *Mem:* Am Chem Soc. *Res:* Kinetics of enzyme inactivation; photochemistry of antimalarials; enzymology of malignant tissues; irradiation of nucleic acids; author of over 400 publications. *Mailing Add:* Ten W 66th St Apt 33A New York NY 10023-6206

ASKANAZI, JEFFREY, b New York, NY, March 23, 51; m 84; c 3. ANESTHESIOLOGY, ARTIFICIAL NUTRITION. *Educ:* Worcester Polytech Inst, BA, 72; Upstate Med Ctr, MD, 75. *Prof Exp:* Intern surg, Columbia Presby Med Ctr, NY, 75-76, fel critical care med, 76-77, resident anesthesiol, 77-79; asst prof anesthesiol, Col Physicians & Surgeons, Columbia Univ, 79-87; ASSOC PROF ANESTHESIOL, ALBERT EINSTEIN COL MED & MONTEFIORE MED CTR, BRONX, NY, 87-, DIR RES CRIT CARE MED, 87- *Concurrent Pos:* Clin dir, Surg Metab Unit, Columbia Presby Med Ctr, 83-87, clin dir parenteral & enteral nutrit, 84-87. *Mem:* Am Soc Anesthesiologists; Am Physiol Soc; Soc Crit Care Med; NY Acad Med; Am Inst Nutrit; Am Soc Clin Nutrit. *Res:* Investigation of the metabolic and pharmacologic effect of artificial nutrition on healthy subjects, malnourished and stressed patients, patients with chronic or acute lung disease and patients with AIDS. *Mailing Add:* Div Crit Care Med Dept Anesthesiol Albert Einstein Col Med Montefiore Med Ctr 111 E 210th St Bronx NY 10467

ASKARI, AMIR, b Ahwaz, Iran, Dec 24, 30; US citizen; m 57; c 2. PHARMACOLOGY, BIOCHEMISTRY. *Educ:* Univ Dubuque, BS, 53; NY Univ, MS, 57; Cornell Univ, PhD(biochem), 60. *Prof Exp:* From instr to prof pharmacol, Med Col, Cornell Univ, 63-75; PROF & CHMN DEPT PHARMACOL & THERAPEUT, MED COL OHIO, 75- *Concurrent Pos:* Res fel pharmacol, Med Col, Cornell Univ, 60-63; USPHS grant, 63. *Mem:* AAAS; Am Soc Biochem & Molecular Biol; Am Soc Pharmacol & Exp Therapeut; Harvey Soc; Biophys Soc. *Res:* Mechanism of ion transport through cell membranes; effects of drugs on membranes. *Mailing Add:* Dept Pharmacol & Therapeut Med Col Ohio CS 10008 Toledo OH 43699-0008

ASKELAND, DONALD RAYMOND, b Spirit Lake, Iowa, Dec 4, 42; m 65; c 1. METALLURGICAL ENGINEERING. *Educ:* Dartmouth Col, BA, 65, MS, 66; Univ Mich, PhD(metall eng), 70. *Prof Exp:* Teaching asst metall eng, Mich State Univ, 68-70; asst prof, 70-76, ASSOC PROF METALL ENG, UNIV MO-ROLLA, 76- *Concurrent Pos:* Res assoc, Gen Motors Corp, 72; partner, Askeland, Kisslinger, & Wolf, Consults, 73- *Mem:* Am Inst Mining, Metall & Petrol Engrs; Am Soc Metals; Am Foundrymen's Soc; Am Welding Soc. *Res:* Effect of casting, welding and metallurgical variables on the structure and behavior of cast iron alloys; general solidification and metals casting; failure analysis. *Mailing Add:* Dept Metall Eng Univ Mo Rolla Rolla MO 65401

ASKENASE, PHILIP WILLIAM, c 2. CLINICAL IMMUNOLOGY, ALLERGY. *Educ:* Yale Univ, MD, 65. *Prof Exp:* PROF MED & PATH, YALE UNIV, 82-, SECT CHIEF, CLIN IMMUNOL & ALLERGY, 85- *Res:* Allergy; immunology; hypersensitivity. *Mailing Add:* Sch Med Yale Univ 333 Cedar St New Haven CT 06510

ASKEW, ELDON WAYNE, b Pontiac, Ill, Aug 23, 42; c 2. BIOCHEMISTRY, PHYSIOLOGY. *Educ:* Univ Ill, BS, 64, MS, 66; Mich State Univ, PhD(nutrition), 69. *Prof Exp:* Biochemist, US Army Med Res & Nutrit Lab, 69-74, Lettermen Army Inst Res, 74-82; asst chief, Dept Clin Invest, Tripler Army Med Ctr, 82-85; DIR, MIL NUTRIT RES, 85- *Mem:* Am Inst Nutrition; Am Col Sports Med; Sigma Xi; Asn Mil Surgeons US. *Res:* Biochemical adaptions of energy metabolism to physical training and the effect of nutrition on physical performance. *Mailing Add:* US Army Res Inst Environ Med Natick MA 01760

ASKEW, RAYMOND FIKE, b Birmingham, Ala, Oct 29, 35; m 81; c 5. PLASMA PHYSICS. *Educ:* Birmingham-Southern Col, BS, 56; Univ Va, MS, 58, PhD(physics), 60. *Prof Exp:* Sr physicist res lab eng sci, Univ Va, 60; from asst prof to assoc prof physics, 60-71, actg dir Nuclear Sci Ctr, 69-71. interim head mech eng, 80-82, dir, 83-86, PROF PHYSICS, AUBURN UNIV, 71-; DIR CTR COM DEVELOP, 87- *Concurrent Pos:* Consult, USAF Systs Command, Eglin AFB, Fla, 72-74. *Mem:* Am Asn Physics Teachers; Am Phys Soc; Soc Physics Students (pres, 85-87). *Res:* Nuclear and atomic physics; gaseous discharges; plasmas; pulsed power; space power systems. *Mailing Add:* 548 Heard Ave Auburn AL 36830

ASKEW, WILLIAM CREWS, b Athens, Ala, Jan 5, 40; m 63; c 3. RADIATION CHEMISTRY, ANALYTICAL CHEMISTRY. *Educ:* Auburn Univ, BS, 62, MS, 63; Univ Fla, PhD(radiolysis of fluorocarbons), 66. *Prof Exp:* Res asst chem eng, Univ Fla, 63-66; asst prof chem eng, Auburn Univ, 67-80; AT DEPT CHEM ENGR, UNIV SALA. *Concurrent Pos:* Proj leader, US AEC res grant, 69- *Mem:* Am Inst Chem Eng. *Res:* Synthesis by radiolysis, identification and purification of perfluoroalkanes. *Mailing Add:* Dept Chem Eng Univ S Ala 307 University Dr Mobile AL 36688

ASKEY, RICHARD ALLEN, b St Louis, Mo, June 4, 33; m 58; c 2. SPECIAL FUNCTION. *Educ:* Wash Univ, AB, 55; Harvard Univ, MA, 56; Princeton Univ, PhD(math), 61. *Prof Exp:* Instr math, Wash Univ, 58-61; instr, Univ Chicago, 61-63; from asst prof to assoc prof, 63-86, GABOR SZEGO PROF MATH, UNIV WIS-MADISON, 86- *Concurrent Pos:* Guggenheim fel, 69-70. *Mem:* Am Math Soc (vpres, 86-87); Math Asn Am; Soc Indust & Appl Math. *Res:* Special functions. *Mailing Add:* Dept Math Univ Wis Madison WI 53706

ASKILL, JOHN, b Kent, Eng, Apr 27, 39; c 2. SOLID STATE PHYSICS, METALLURGY. *Educ:* Univ Reading, BSc, 60 & 61, PhD(physics), 64. *Prof Exp:* Fel, Univ Reading, 64-65; res physicist, Oak Ridge Nat Lab, 65-66; from asst prof to assoc prof, 66-84, PROF PHYSICS, MILLIKIN UNIV, 85-, CHMN DEPT, 66- *Concurrent Pos:* Consult, Metals & Ceramics Div, Oak Ridge Nat Lab, 66-67. *Mem:* Am Asn Physics Teachers; Brit Inst Physics & Phys Soc. *Res:* Diffusion of metals in metals using radioactive tracers. *Mailing Add:* Dept Physics Millikin Univ Decatur IL 62522

ASKIN, RONALD GENE, b Baltimore, Md, Dec 28, 53; m 76; c 2. PRODUCTION CONTROL, REGRESSION ANALYSIS. *Educ:* Lehigh Univ, BS, 75; Ga Inst Technol, MS, 76, PhD(indust & syst eng), 79. *Prof Exp:* Asst indust eng, Ga Inst, Technol, 76-79; from asst prof to assoc prof indust eng, Univ Iowa, 79-85; ASSOC PROF SYSTS & INDUST ENG, UNIV ARIZ, 85- *Concurrent Pos:* Assoc ed, Am Inst Indust Engrs Trans, 81-87, dept educ, 87-; NSF pres young investr, 84. *Honors & Awards:* Am Inst Indust Engrs Transactions Develop & Appln Award, 85; Eugene L Grant Award, 86. *Mem:* Inst Indust Engrs; Oper Res Soc Am; Am Production & Inventory Control Soc; Am Statist Asn; Am Soc Qual Control; Soc Mfg Engrs. *Res:* Production system design, planning and control; manufacturing facility design, multicolinearity and nonnormality in multiple linear regression models; experimental design for parameter optimization; manufacturing planning. *Mailing Add:* Dept Systs & Indust Eng Univ Ariz Tucson AZ 85721

ASKINS, HAROLD WILLIAMS JR, JR, b Chesnee, SC, Sept 3, 39; m 64; c 2. EDUCATIONAL ADMINISTRATION. *Educ:* The Citadel, BS, 61; Clemson Univ, MS, 63; Purdue Univ, PhD(elec eng), 72. *Prof Exp:* Instr elec eng, Clemson Univ, 63-64; mgr, Chesnee Tel Co, SC, 65-66; instr electronics, TEC, Spartanburg, SC, 65-66; res & develop engr elec power, Westinghouse Elec Corp, 71-74; from asst prof to assoc prof, 74-82; PROF ELEC ENG, THE CITADEL, 82-, DEPT HEAD ELEC ENG, 88- *Concurrent Pos:* Electronics engr, Naval Electronic Systs Eng Ctr, Charleston, SC, 76-; lectr, Mod Power Syst Anal Short Course, Auburn Univ, 78-; consult, Nuclear Electromagnetic Pulse Effects, Oak Ridge Nat Lab, 83-86. *Mem:* Inst Elec & Electronics Engrs; Am Soc Eng Educ; Appl Computational Electromagnetics Soc; Sigma Xi. *Res:* Electromagnetic pulse-power system interaction; electromagnetic environmental effect on communication facilities and personnel; lightning and switching surge phenomena; insulation coordination. *Mailing Add:* The Citadel-Mil Col SC Charleston SC 29409

ASLESON, GARY LEE, b St Peter, Minn, June 14, 48; m 71; c 1. ANALYTICAL CHEMISTRY. *Educ:* Gustavus Adolphus Col, BA, 70; Univ Iowa, PhD(anal chem), 75. *Prof Exp:* Teaching asst chem, Univ Iowa, 70-73, res asst, 73-75; asst prof, 75-80, ASSOC PROF CHEM, COL CHARLESTON, 80- *Mem:* Am Chem Soc; Sigma Xi. *Res:* High performance liquid chromatography; gas chromatography; nuclear magnetic resonance spectrometry. *Mailing Add:* Dept Chem Col Charleston Charleston SC 29424

ASLESON, JOHAN ARNOLD, b Stoughton, Wis, Sept 13, 18; m 43; c 3. SOILS. *Educ:* Univ Wis, BS, 42, MS, 47, PhD, 57. *Prof Exp:* From asst prof to asst prof soils, 47-54, asst to dir, Agr Exp Sta, 54-57, from asst dir to assoc dir, 57-61, DIR AGR EXP STA, MONT STATE UNIV, 61-, PROF SOILS, 54-, DEAN COL AGR, 65- *Mem:* Sigma Xi; Am Soc Agron; Soil Sci Soc Am. *Res:* Applied fertilizer research; agronomy and soil science; plant physiology. *Mailing Add:* 2419 Spring Creek Dr Bozeman MT 59715

ASLING, CLARENCE WILLET, b Duluth, Minn, June 17, 13; m 36, 70; c 2. ANATOMY. *Educ:* Univ Kans, AB, 34, AM, 37, MD, 39; Univ Calif, Berkeley, PhD(anat), 47. *Prof Exp:* Asst instr anat, Univ Kans, 34-37; intern, Huntington Mem Hosp, Pasadena, Calif, 39-40; instr anat, Sch Med, Vanderbilt Univ, 40-41; instr, Univ Kans, 41-42, asst prof anat & lectr surg anat, 42-46; res asst, Univ Calif, Berkeley, 44-47, lectr anat, 45-47, from asst prof to prof, 47-66, res assoc, Inst Exp Biol, 51-58, co-chmn dept anat & physiol, 58-62; prof anat, 66-80, EMER PROF ANAT, MED CTR, UNIV CALIF, SAN FRANCISCO, 80- *Concurrent Pos:* Fulbright award, Eng, 53-54; Guggenheim fel, 62-63; guest prof, Univ Geneva, 62-63. *Mem:* Am Asn Anatomists; Anat Soc Gt Brit & Ireland; Teratology Soc; Am Asn Univ Professors. *Res:* Regulation of head form; endocrine and nutritional control of skeletal development; localization of radioactive minerals; teratology and congenital abnormalities. *Mailing Add:* 60 Lomita Ave San Francisco CA 94122

ASMUS, JOHN FREDRICH, b Chicago, Ill, Jan 20, 37; m 63; c 2. LASERS. *Educ:* Calif Inst Technol, BS, 58, MS, 59, PhD(elec eng, physics), 65. *Prof Exp:* Staff assoc high energy fluid dynamics, Gen Atomic Div, Gen Dynamics Corp, 64-67; mem staff spec nuclear effects, Gulf Gen Atomic Inc, 67-69; res staff mem laser technol, Inst for Defense Anal, 69-71; vpres, Sci Applns, Inc, 71-73; RES PHYSICIST, UNIV CALIF, SAN DIEGO, 73- *Concurrent Pos:* Assoc mem spec group on optical lasers, Dir Defense Res & Eng Adv Group Electron Devices, 69-70, mem, 70- *Mem:* Int Inst Conserv; Am Inst Conserv. *Res:* Interaction of high-power laser beams with solid surfaces; laser-induced stress wave propagation and micrometeorite generation; laser excited chemistry; computer image processing in art. *Mailing Add:* Inst Geophys & Planetary Physics Univ Calif San Diego La Jolla CA 92093

ASNER, BERNARAD A, JR, b Oklahoma City, Okla, Dec 28, 33; m 60; c 2. APPLIED MATHEMATICS. *Educ:* Univ Okla, BS, 57; Univ Ala, MA, 64; Northwestern Univ, PhD(eng sci), 68. *Prof Exp:* Mathematician control, NASA, 60-65; assoc prof math, Bogazici Univ, Turkey, 70-72; exchange scientist, Nat Acad Sci, 72-73; assoc prof math & bus mgt, Univ Dallas, Irving, 73-78; res assoc, Systs Dynamics Lab, Marshall Space Flight Ctr, NASA, 78-79; ASSOC PROF MATH, UNIV DALLAS, IRVING, 79- *Concurrent Pos:* Nat Acad Sci res assoc, 78-79. *Mem:* Math Asn Am; Am Math Soc; Soc Indust & Appl Math. *Res:* Degeneracy in functional differential equations and computer science. *Mailing Add:* 8271 Navarre Pkwy Navarre FL 32566

ASNES, CLARA F, b Jersey City, NJ, Aug 31, 46. MICROTUBULES. *Educ:* Boston Univ, PhD(biochem), 75. *Prof Exp:* RES ASSOC CELL BIOL, UNIV ROCHESTER, 82- *Mailing Add:* Dept Biol Univ Rochester Rochester NY 14627

ASOFSKY, RICHARD MARCY, b New York, NY, Sept 25, 33; m 60; c 2. IMMUNOLOGY, PATHOLOGY. *Educ:* State Univ NY, MD, 58. *Prof Exp:* Intern internal med, Kings County Hosp, NY, 58-59; resident, Med Ctr, NY Univ, 59-63; staff assoc immunol, Nat Inst Allergy & Infectious Dis, 63-66, head exp path sect, Lab Germfree Animal Res, 65-70, asst chief, Lab Microbiol Immunity, 70-72, chief, 72-86, CHIEF, EXP PATH SECT, LAB IMMUNOL, NAT INST ALLERGY & INFECTIOUS DIS, 86- *Concurrent Pos:* NIH trainee path, Med Ctr, NY Univ, 59-63. *Honors & Awards:* Arthur S Fleming Award, 71. *Mem:* Am Asn Immunologists; AAAS; NY Acad Sci. *Res:* Sites and control of immunoglobulin synthesis; interaction among lymphocytes in immune responses. *Mailing Add:* Nat Inst Allergy & Infectious Dis-NIH Bethesda MD 20205

ASP, CARL W, b Cleveland, Ohio, Aug 14, 31; m 65; c 2. SPEECH & HEARING SCIENCES, STATISTICS. *Educ:* Ohio State Univ, BS, 63, MA, 64, PhD(speech & hearing sci), 67. *Prof Exp:* PROF SPEECH & HEARING SCI, UNIV TENN, KNOXVILLE, 67- *Concurrent Pos:* Mem field read & site visit team, US Off Educ, 69- *Mem:* AAAS; Acoust Soc Am; Am Speech & Hearing Asn; Audio Eng Soc. *Res:* Psychoacoustics; speech communication; psychophysical experiments; perception of hearing impaired. *Mailing Add:* Dept Audiol & Speech Path Univ Tenn Knoxville TN 37996

ASPDEN, ROBERT GEORGE, b Geneva, Ill, May 22, 27; m 53; c 3. METALLURGY. *Educ:* Univ Ill, BS, 50; Univ Pittsburgh, MS, 57, DScEng, 65. *Prof Exp:* Mat engr, 50-55, res engr, 55-60, supvr engr, 60-65, mgr, 65-68, adv engr, 68-69, mgr, 69-80, ENGR, WESTINGHOUSE ELEC CORP, 80- *Mem:* Am Inst Mining, Metall & Petrol Engrs; Am Soc Metals; Nat Asn Corrosion Engrs. *Res:* High temperature alloys; magnetic materials; corrosion. *Mailing Add:* 2382 Flagstaff Dr Export PA 15632

ASPELIN, GARY B(ERTIL), b Bristol, Conn, Dec 31, 39; m 67; c 2. ORGANIC CHEMISTRY. *Educ:* Brown Univ, AB, 61; Univ Wis, PhD(org chem), 66. *Prof Exp:* Res chemist, Am Cyanamid Co, Bound Brook, NJ, 66-71; res scientist, Johnson & Johnson, New Brunswick, 71-76, group leader tech serv, 76-78, group mgr tech serv, 78-80; group mgr tech serv, Surgikos Inc, Arlington, Tex, 80-90; DIR INT TECH AFFAIRS, JOHNSON & JOHNSON MED INC, ARLINGTON, TEX, 90- *Mem:* Am Asn Textile Chemists & Colorists; Tech Asn Pulp & Paper Indust. *Res:* Cellulosic and synthetic organic chemistry; nonwoven fabrics; film and fabric laminates; disposable hospital packs and gowns; bacterial filtration; cold sterilant solutions; surgical gloves; wound care. *Mailing Add:* 3611 Lake Champlain Dr Arlington TX 76016-3504

ASPER, SAMUEL PHILLIPS, b Oak Park, Ill, July 14, 16; m 42; c 2. MEDICAL ADMINISTRATION. *Educ:* Baylor Univ, AB, 36; Johns Hopkins Univ, MD, 40. *Prof Exp:* House officer med, Johns Hopkins Hosp, 40-41; res fel, Thorndike Mem Lab, Harvard Univ, 41-42 & 46-47; from instr to prof, Sch Med, 47-85, assoc dean, 57-68, vpres med affairs, Johns Hopkins Hosp, 70-73; prof internal med & dean fac med sci, Am Univ Beirut, 73-78, chief of staff, Univ Hosp, 73-78; dep exec vpres, Am Col Physicians, 79-81; EMER PROF MED, SCH MED, JOHNS HOPKINS UNIV, 85- *Concurrent Pos:* Pres, Educ Comn Foreign Med Grads, 82-85, emer pres, 86- *Mem:* Inst Med-Nat Acad Sci; Endocrine Soc (vpres, 66-67); Asn Am Physicians; Am Col Physicians (pres, 69-70, emer pres, 85-); Am Soc Clin Invest. *Res:* Clinical endocrinology; metabolism. *Mailing Add:* PO Box 153 Gibson Island MD 21056

ASPERGER, ROBERT GEORGE, SR, b Detroit, Mich, Oct 1, 37; m 61; c 1. CORROSION ENGINEERING. *Educ:* Eastern Mich Univ, AB, 60; Univ Mich, MS, 63, PhD(chem), 65. *Prof Exp:* Res chemist, Dow Chem Co, 65-75, sr res specialist, 75-78; assoc scientist, Phillips Petrol Co, 78-82; prin investr, Petrolite Corp, 82-88; VCHMN, TECH PRAC COMT, NAT ASN CORROSION ENGRS, 87- *Concurrent Pos:* Chmn & pres, Jay Develop Inc, 80-82. *Honors & Awards:* Distinguished Serv Award, Nat Asn Corrosion Engrs, 91. *Mem:* AAAS; Nat Asn Corrosion Engrs. *Res:* Corrosion inhibitors, gas systems; coordination chemistry; surface characterization using x-ray photoelectron spectroscopy, auger spectroscopy, electron microprobe, transmission electron microscopy and scanning electron microscopy/x-ray energy dispersive spectroscopy; fundamentals of corrosion and its inhibition; inhibitor development; application of oil & gas inhibitors to carbon steet and titanium. *Mailing Add:* 11115 Mills Rd Cypress TX 77429

ASPINALL, GERALD OLIVER, b Chesham Bois, Eng, Dec 30, 24; m 53; c 2. ORGANIC CHEMISTRY. *Educ:* Bristol Univ, BSc, 44, PhD(chem), 48; Univ Edinburgh, DSc(chem), 58. *Prof Exp:* From lectr to sr lectr chem, Univ Edinburgh, 48-63, reader, 63-67; prof & chmn dept, Trent Univ, 67-72; chmn dept, 72-79, PROF CHEM, YORK UNIV, 72-, DISTINGUISHED RES PROF, 88- *Honors & Awards:* C S Hudson Award Carbohydrate Chem, Am Chem Soc, 86; John Labatt Ltd Award, Org/Biochem Res, Can Soc Chem, 87. *Mem:* Am Chem Soc; Chem Inst Can; Royal Soc Chem; Brit Biochem Soc. *Res:* Chemistry of carbohydrates, especially chemical methodology for structure determination of plant and microbial polysaccharides; structural characterization and synthesis of oligosaccharides as serological determinants in bacterial glycolipid and lipopolysaccharide antigens. *Mailing Add:* Dept Chem York Univ 4700 Keele St North York ON M3J 1P3 Can

ASPLAND, JOHN RICHARD, b Leeds, Eng, Oct 22, 36; m 61; c 1. TEXTILE CHEMISTRY. *Educ:* Univ Leeds, BSc, 58, MSc, 60; Univ Manchester, PhD(textile chem), 64. *Prof Exp:* Prof textile chem, Univ Manchester Inst Sci & Technol, 61-66; sr res chemist dye applns res, Southern Dyestuff Co, Martin-Marietta Corp, 66-68, group leader dye applns & eval, 68-74; textile res mgr, Reeves Bros, Inc, 74-82; PROF, TEXTILE CHEM, CLEMSON UNIV, 82- *Concurrent Pos:* Examr, Soc Dyers Colourists, 60-64, chartered colourist; ed, J Coated Fabrics, 84-88. *Mem:* Fiber Soc; Am Asn Textile Chem & Colorists; fel Soc Dyers & Colourists; fel Textile Inst. *Res:* Interactions between dyestuffs and polymeric substances, particularly disperse dyes and hydrophobic fibers and reactive dyes and cellulose; instrumental color measurement and shade sorting; sulfur dyes; flame-resistant fabrics; energy conservation; coated fabrics; carpet dyeing and finishing. *Mailing Add:* Dept Textile Chem Clemson Univ 269 Sirrine Hall Clemson SC 29634-1307

ASPLUND, JOHN MALCOLM, RUMINANT NUTRITION, NUTRITIONAL METABOLISM. *Educ:* Univ Wis-Madison, PhD(nutrit), 60. *Prof Exp:* PROF ANIMAL SCI, UNIV MO, 69- *Mailing Add:* Animal Sci Res Univ Mo Columbia MO 65211

ASPLUND, RUSSELL OWEN, b Lethbridge, Can, May 5, 28; m 51; c 6. BIOCHEMISTRY. *Educ:* Univ Alta, BSc, 49; Utah State Univ, MS, 55; WVa Univ, PhD(biol chem), 58. *Prof Exp:* Asst, Utah State Univ, 49-50 & Univ Utah, 51-52; chemist, exp sta, Utah State Univ, 53-56; asst, WVa Univ, 56-58; from asst prof to assoc prof, 58-65 & 67-72, PROF CHEM, UNIV WYO, 72-73, 75-83 & 85- *Concurrent Pos:* Clayton Found fel, 65-66; vis prof, Univ Calif, Los Angeles, 74, 76 & Brigham Young Univ, 84. *Mem:* AAAS; Am Chem Soc. *Res:* Chemistry of anti-tumor agents; phytotoxins; metal-amino acid complexes. *Mailing Add:* 1824 0RD Univ Wyo Laramie WY 82070

ASPNES, DAVID E, b Madison, Wis, May 1, 39; m 64; c 3. PHYSICS. *Educ:* Univ Wis, Madison, BS, 60, MS, 61; Univ Ill, Urbana, PhD(physics), 65. *Prof Exp:* Res assoc physics, Univ Ill, Urbana, 65-66 & Brown Univ, 66-67; mem tech staff, AT&T Bell Labs, 67-83; DIST RES MGR, BELL COMMUN RES INC, 83- *Concurrent Pos:* Alexander von Humboldt sr scientist award. *Honors & Awards:* Wood Prize, Optical Soc Am, 87. *Mem:* Fel Am Phys Soc; Soc Photooptical Instrumentation Engrs; fel Optical Soc Am; Mat Res Soc; Am Vacuum Soc; Sigma Xi; AAAS. *Res:* Experimental solid state physics; semiconductors; surface physics; optical properties of solids and thin films; ellipsometry; crystal growth. *Mailing Add:* Bell Commun Res Inc 331 Newman Springs Rd Red Bank NJ 07701-7040

ASPREY, LARNED BROWN, b Sioux City, Iowa, Mar 19, 19; m 44; c 7. PHYSICAL CHEMISTRY, INORGANIC CHEMISTRY. *Educ:* Iowa State Univ, BS, 40; Univ Calif, PhD(chem), 49. *Prof Exp:* Chemist, Campbell Soup Co, 40-42, metall lab, Univ Chicago, 44-46 & radiation lab, Univ Calif, 46-49; MEM STAFF, LOS ALAMOS SCI LAB, 49- *Mem:* AAAS; Am Chem Soc; Sigma Xi. *Res:* Physical and inorganic chemistry of actinide and lanthanide elements; fluorine chemistry. *Mailing Add:* 13 Lebanon Ave Las Cruces NM 88005

ASPREY, WINIFRED ALICE, b Sioux City, Iowa, Apr 8, 17. MATHEMATICS, COMPUTER SCIENCES. *Educ:* Vassar Col, AB, 38; Univ Iowa, MS, 43, PhD(math), 45. *Hon Degrees:* Dr Marist Col, Poughkeepsie, NY, 85. *Prof Exp:* Teacher, Brearley Sch, New York, 38-40 & Girls Latin Sch, Chicago, 40-42; asst math, Univ Iowa, 42-45; from instr to prof, Vassar Col, 45-61, chmn, dept math, 58-62, Elizabeth Stillman Williams chair prof math, 61-82, dir comput sci, 77-82; RETIRED. *Concurrent Pos:* IBM Corp indust res fel, 57-58; NSF grant, Univ Calif, Los Angeles, 62, NSF fac fel & vis scholar, 64-65; dir, Acad Year Inst Math, State Dept Educ, NY, 62-64 & Acad Year Inst Comput Math, Vassar Col, spring, 64, 66 & 67; IBM Corp exchange grant, 69-70; vis staff mem & consult, Los Alamos Nat Lab, 72-76; distinguished prof, math & comput sci, Bethany Col, West Va, 83, Marist Col, Poughkeepsie, NY, 85, 88 & 89; prof, math & comput sci, St Louis Univ, Madrid, Spain, 83, 84, 85, 86; consult & lectr, Mount St Mary's Col, Newburgh, NY, 82; vis prof computer sci, Vassar Col, Poughkeepsie, NY, 90. *Mem:* Fel AAAS; Nat Coun Teachers Math; Am Math Soc; Math Asn Am; Sigma Xi; Asn Comput Mach. *Res:* Analysis and topology; families of total oscillators and their derived functions. *Mailing Add:* Box 87 Vassar Col Poughkeepsie NY 12601-6198

ASQUITH, GEORGE BENJAMIN, b Chicago, Ill, June 23, 36; m 63; c 1. PETROLOGY, MINERALOGY. *Educ:* Tex Tech Univ, BS, 61; Univ Wis, Madison, MS, 63, PhD(geol), 66. *Prof Exp:* Instr geol, Univ Wis, Madison, 66; res geologist, Atlantic-Richfield Res Lab, 66-70; from asst prof to assoc prof, 70-81, PROF GEOL, WTEX STATE UNIV, 81- *Mem:* Soc Econ Paleontologists & Mineralogists; Clay Minerals Soc; Sigma Xi. *Res:* Igneous petrology and volcanology; sandstone and carbonate petrology and sedimentation and carbonate diagenesis. *Mailing Add:* 4407 14th St Lubbock TX 79015-4530

ASRAR, JAWED, b India, Oct 1, 49. REACTION-INJECTION MOLDING. *Educ:* Aligarh Univ, India, BS, 69, MS, 71; Moscow Inst Chem Technol, USSR, PhD(chem technol), 78. *Prof Exp:* Teaching fel chem technol, Moscow Inst Chem Technol, USSR, 78-79; res assoc, Duke Univ, Durham, NC, 79-81; res assoc polymer chem & vis lectr gen chem, Univ Lowell, Mass, 81-83; RES SPECIALIST, MONSANTO CO, 83- *Concurrent Pos:* Assoc sci fel, Monsanto Co. *Mem:* Am Chem Soc. *Res:* Synthesis and modification of polyester resins, reinforced plastics, liquid crystalline polymers: polyesters, polyamides, polyamide hydrazides, polyoxadiazoles, polyacylsemicarbazides and ring opening polymerization. *Mailing Add:* 14949 Royal Brook Chesterfield MO 63017

ASSADOURIAN, FRED, b Panderma, Turkey, Apr 13, 15; nat US; m 54; c 2. MATHEMATICS. *Educ:* NY Univ, BS, 35, MS, 36, PhD(math), 40. *Prof Exp:* Instr math, Col Eng, NY Univ, 37-42; assoc prof, Tex Tech Col, 42-44; res engr, Westinghouse Res Labs, Pa, 44-46; develop engr, Fed Telecommun Labs, NJ, 46-51, proj engr, 51-56; sr engr, Radio Corp Am, NY, 56-59, leader tech staff, 59-62, sr staff scientist & head adv transmission, 62-68; from assoc prof to prof elec eng, Pratt Inst, 68-81; PRES, FAHA CONSULTS INC, 83- *Concurrent Pos:* Adj prof elec eng, NJ Inst Technol, 50-57 & Stevens Inst Technol, 68 & 81-82; consult, RCA Astro Electronics, 81-; vis prof elec eng, Pratt Inst, 81- *Mem:* Inst Elec & Electronics Engrs. *Res:* Almost-periodic functions; pulse transformers; equations in gas discharge; microwave transmission; reflex klystron; distortion of amplitude modulation and frequency modulation signals; special microwave components; special problems in communications systems; communications satellite systems. *Mailing Add:* 21 76th St North Bergen NJ 07047

ASSALI, NICHOLAS S, b Lebanon, Apr 7, 16; US citizen. OBSTETRICS. *Educ:* Jesuit Sch, Lebanon, BS, 32; Gymnasio Normal, Brazil, BS, 37; Univ Sao Paulo, MD, 43; Am Bd Obstet & Gynec, cert. *Prof Exp:* Intern med, Univ Hosp, Sao Paulo, 41-42, resident surg & gynec, 42-43; consult obstet & gynec,

Polyclin, Sao Paulo, 44-45; fel obstet & gynec, Bethesda Hosp, Cincinnati, Ohio, 46-47; asst prof obstet, Univ Cincinnati, 48-53, dir clin res & asst attend obstetrician, Dept Obstet, Cincinnati Gen Hosp, 48-53; assoc prof, 53-57, PROF OBSTET & GYNEC, UNIV CALIF, LOS ANGELES, 57-; ATTEND OBSTETRICIAN & GYNECOLOGIST, UNIV CALIF, LOS ANGELES HOSP, 61- *Concurrent Pos:* Consult obstet & gynec, Sao Paulo Health Serv, 44-45; dir, training prog basic sci appl to obstet & gynec, NIH, 57-71; dir, training prog reproductive physiol, USPHS, 59-71; attend obstetrician & gynecologist, Harbor County Gen Hosp, Torrance, Calif, 60-; prof physiol, Univ Calif, Los Angeles, 61-73; mem, Gen Med B Study Sect, NIH, 69-73; mem, Coun Circulation, Am Heart Asn. *Mem:* Am Fedn Clin Res; Am Gynec Soc; Am Heart Asn; Am Physiol Soc; AAAS; Biophys Soc; Soc Gynec Invest (pres, 56); Soc Exp Biol & Med; Sigma Xi; Soc Hist Med Sci. *Res:* Maternal-fetal-neonatal cardiovascular physiology; hypertensive diseases of pregnancy; maternal-fetal exchange mechanisms; author or co-author of over 180 publications. *Mailing Add:* 520 Amalfi Dr Pacific Palisades CA 90272

ASSANIS, DENNIS N, b Athens, Greece, Feb 9, 59; m 84; c 2. THERMAL SCIENCES & SYSTEMS, INTERNAL COMBUSTION ENGINES. *Educ:* Univ Newcastle-Upon-Tyne, UK, BSc, 80; Mass Inst Technol, SM(naval archit & marine eng) & SM(mech eng), 83, PhD(power & propulsion), 85, SM(mgt), 86. *Prof Exp:* Teaching asst, Dept Mech Eng & res asst, Dept Ocean Eng, Mass Inst Technol, 80-82, res asst, Sloan Automotive Lab, 82-85; asst prof, 85-90, ASSOC PROF MECH ENG, UNIV ILL, URBANA-CHAMPAIGN, 90- *Concurrent Pos:* Prin investr over 30 major res projs, Gen Motors, Ford, Chrysler, Cummins, Caterpillar & NSF, Dept Energy & Dept Defense, 86-; consult, Sci Applications Int Corp, 86-88, Adiabatics, Inc, 86-, NASA, 88-89 & Gen Motors, 88-; fac appointee, Energy & Environ Syst Div, Argonne Nat Lab, 87-91; Lilly Endowment teaching fel, 88; NSF presidential young investr, 88; univ scholar, Univ Ill, 91. *Honors & Awards:* Cert Recognition Creative Develop Tech Innovation, NASA, 87; Ralph Teetor Educ Award, Soc Automation Engrs, 87; Gold Medal Award, Am Soc Mech Engrs, 90. *Mem:* Am Soc Mech Engrs; Soc Automotive Engrs; Soc Naval Architects & Marine Engrs; Am Soc Eng Educ; Sigma Xi; Combustion Inst. *Res:* Thermodynamics, fluid mechanics, heat transfer, combustion, and their applications to thermal fluid systems design; physically-based models of internal combustion engine processes and cycles; author of numerous publications. *Mailing Add:* Dept Mech Eng Univ Ill 1206 W Green St Urbana IL 61801

ASSAYKEEN, TATIANA ANNA, b South Orange, NJ, Sept 27, 39; m 80. PHARMACOLOGY, CARDIOLOGY. *Educ:* Trinity Col, DC, BA, 61; Univ Va, PhD(pharmacol), 65. *Prof Exp:* Asst prof surg & pharmacol, 67-73, adj prof surg, Sch Med, Stanford Univ, 73-76; med dir, Astra Chem, Sydney, Australia, 76-80, assoc dir clin res, Astra Pharmaceut Prod, Inc, Framingham, 80-81, vpres & treas, Southboro, 82-83, PRES & TREAS, ASTRA CLIN RES ASSOC, INC, HOPKINTON, MASS, 84- *Concurrent Pos:* USPHS fel, Med Ctr, Univ Calif, San Francisco, 65-67; NIH res grant, 69-76. *Mem:* resigned from all the above Soc Nephrology. *Res:* Renin-angiotensin-aldosterone system; hypertension; catecholamines. *Mailing Add:* Astra Clin Res Assocs Inc 14 Church St Hopkinton MA 01748

ASSENZO, JOSEPH ROBERT, b Boston, Mass, Jan 1, 32; m 60, 90; c 2. BIOSTATISTICS. *Educ:* Northeastern Univ, BS, 54; Harvard Univ, SM, 55; Univ Okla, PhD(biostatist), 63. *Prof Exp:* Asst engr, Camp Dresser & McKee, Mass, 51-54, engr, 54-56, proj engr, 58; instr sanit eng, Univ Okla, 58-61, instr biostatist & sanit eng, 61-63, assoc prof, 63-66; res mgr, Upjohn Co, 66-80, group mgr biostatist & info sci, 80-85, dir proj mgt, 85-90, EXEC DIR, US PHARMACEUT REGULATORY AFFAIRS, UPJOHN CO, 90- *Concurrent Pos:* Consult, Bur Water Resources Res, 58-, Civil Aeromed Res Inst, Fed Aviation Agency, 61-66, med ctr, Univ Okla, 61-66 & pub sch syst, Oklahoma City, 62-66; mem water resources develop prog, Okla, 63-66; adj prof math sci, Oakland Univ, 79-; mem, Human Subj Instnl Review Bd, Western Mich Univ, 81- *Mem:* Fel Am Statist Asn; Asn Comp Mach; Biomet Soc; Drug Info Asn; Am Soc Pharmacol & Therapeut. *Res:* Applications of statistics, biomathematics, epidemiology and computer science and laboratory and medical research and regulations and to the discovery and development of drugs for human use. *Mailing Add:* 7214 Lakeridge Pl Kalamazoo MI 49009-9759

ASSINK, ROGER ALYN, b Holland, Mich, Sept 28, 45; m 67; c 2. POLYMER CHEMISTRY. *Educ:* Mich State Univ, BS, 67; Univ Ill, MS, 69, PhD(chem), 72. *Prof Exp:* MEM STAFF MAT RES, SANDIA LABS, 72- *Mem:* Am Phys Soc; Am Chem Soc. *Res:* Study of the molecular dynamics and structure of polymeric materials by nuclear magnetic resonance spectroscopy. *Mailing Add:* Sandia Labs Div 1812 Albuquerque NM 87185

ASSMUS, ALEXI JOSEPHINE, b New York City, NY, Mar 22, 62. HISTORY OF PHYSICAL SCIENCES, US HISTORY. *Educ:* Stanford Univ, BA, 84; Harvard Univ, MA, 88, PhD(hist sci), 91. *Prof Exp:* Teaching fel, Hist Sci & Physics Dept & Core Prog, Harvard Univ, 86-90; FEL, UNIV CALIF, BERKELEY & BRANDEIS UNIV, 91- *Concurrent Pos:* NSF postdoctoral fel, Univ Calif, Berkeley, 91-92. *Mem:* Am Hist Asn; Hist Sci Soc. *Res:* History of the physical science; development of American science; rise of scientific research in the United States; history of quantum chemistry and quantum theory; American intellechial history; United States history; several publications and presentations. *Mailing Add:* 470 Stephens Hall Univ Calif Berkeley CA 94609

ASSMUS, EDWARD F(ERDINAND), JR, b Nutley, NJ, Apr 19, 31; m 61, 72; c 2. ALGEBRA. *Educ:* Oberlin Col, AB, 53; Harvard Univ, AM, 55, PhD(math), 58. *Prof Exp:* Off Naval Res res assoc, Columbia Univ, 58-59, Ritt instr math, 59-62; lectr, Wesleyan Univ, 62-66; assoc prof, 66-70, PROF MATH, LEHIGH UNIV, 70- *Concurrent Pos:* Consult appl res lab, Sylvania Electronic Systs, 61-71; vis prof, Queen Mary Col, London Univ, 68-69; summer fac, Sandia Lab, 78; vis prof, Univ Ill, Chicago Circle, 81, Univ Rome, 83, Univ Birmingham, Eng, 85 & 87, Inst Math & Applns, Univ Minn, 88. *Mem:* Am Math Soc; Math Asn Am; London Math Soc; Soc Indust & Appl Math; AAAS. *Res:* Homological algebra; algebraic coding theory; combinatorial mathematics. *Mailing Add:* Dept Math Bldg 14 Lehigh Univ Bethlehem PA 18015

ASSOIAN, RICHARD KENNETH, b Englewood, NJ, June, 1954; m 80. CELLULAR BIOCHEMISTRY. *Educ:* Johns Hopkins Univ, BA, 75; Univ Chicago, PhD(biochem), 81. *Prof Exp:* Fel, Nat Cancer Inst, 81-83, staff fel, 83-86; ASST PROF BIOCHEM, COL PHYSICIANS & SURGEONS, COLUMBIA UNIV, 86- *Mem:* Sigma Xi; Am Soc Biochem & Molecular Biol; Am Soc Cell Biol. *Res:* Molecular mechanisms controlling mitosis, particularly the subcellular actions of peptide growth factors and their receptors. *Mailing Add:* Dept Biochem & Molecular Biophys Col Physicians & Surgeons Columbia Univ 630 W 168 St New York NY 10032

ASSONY, STEVEN JAMES, b Czech, Apr 13, 20; nat US; m 48; c 2. ORGANIC CHEMISTRY. *Educ:* Rutgers Univ, BS, 50; Univ Southern Calif, MS, 53, PhD(chem), 57. *Prof Exp:* Analyst, Geigy Co, Inc, 49-50; assoc, Univ Southern Calif, 51-56, res asst, off ord res, 52-55; asst tech dir, Am Latex Prod Corp, 56-61; dir res & develop, Upjohn Co, 61-70, head tech serv, CPR Div, 70-81, mgr tech serv & regulatory affairs, 81-84; RETIRED. *Mem:* Am Chem Soc; The Chem Soc; NY Acad Sci; Sigma Xi. *Res:* Organic and polymer chemistry; polyurethanes and polymers of boron, phosphorus and nitrogen. *Mailing Add:* PO Box 1026 Temecula CA 92390-0013

ASSOUSA, GEORGE ELIAS, b Jerusalem, Israel, Mar 15, 36; US citizen; m 60; c 2. SCIENCE POLICY. *Educ:* Earlham Col, BA, 57; Columbia Univ, MA, 62; Fla State Univ, PhD(exp nuclear physics), 68. *Prof Exp:* Fac mem physics, Earlham Col, 60-63; fel atomic physics, Carnegie Inst Wash, 68-70, sr fac mem atomic physics & radio astron, 70-80, sr fel, 80-81; pres, Partnership Int Inc, 81-86 & Gryphon Technol Investors, 86-87; DIR GEN, TRUST INT DEVELOP & EDUC, 81-; CHIEF SCI & TECHNOL ADV, COATS VIYELLA PLC, 87- *Concurrent Pos:* Consult, Princeton Univ Observ, 71-72; sr consult & vpres sci & educ affairs, Inst Develop & Econ Affairs Serv Inc, 74-; consult, NJ Marine Sci Consortium, 80-81. *Mem:* Am Phys Soc; Am Astron Soc; Am Asn Univ Professors; Int Astron Union. *Res:* Supernova remnant studies at radio frequencies; supernovae and the interstellar medium; beam-foil spectroscopy and heavy ion produced x-ray spectroscopy. *Mailing Add:* Coats Viyella PLC 28 Savile Row London W1X 2DD England

AST, DAVID BERNARD, b New York, NY, Sept 30, 02; m 28; c 1. DENTISTRY, PUBLIC HEALTH ADMINISTRATION. *Educ:* NY Univ, DDS, 24; Univ Mich, MPH, 42. *Hon Degrees:* ScD, Georgetown Univ, 85. *Prof Exp:* Pvt pract, 24-38; asst dir oral hyg, NY State Dept Health, 38-45, chief dent health sect, 45-48, dir bur dent health, 48-66, assoc dir div med serv, 66-70, asst comnr div med care serv & eval, 70-72; RETIRED. *Concurrent Pos:* Dept Health, Educ & Welfare grants, 45-55, 63-64 & 67-69; lectr pub health pract, Columbia Univ, 44-73; spec lectr pub health, Albany Med Col, 50-73; consult, Prof Exam Serv, Am Pub Health Asn, 50-73; mem community health proj rev comt, Dept Health, Educ & Welfare, 63-65. *Honors & Awards:* Gov Alfred E Smith Award, Am Soc Pub Admin, 55; Award, NY State Soc Dent Children, 63; Award, USPHS & Am Dent Asn, 66; Award, NY Acad Prev Med, 66; H Trendley Dean Award, Int Asn Dent Res, 68; Hermann M Biggs Mem Award, NY State Pub Health Asn, 71; John W Knutson Award, Am Pub Health Asn, 82. *Mem:* Am Dent Asn; fel Am Pub Health Asn; fel Am Col Dent; Asn State & Territorial Dent Dirs (pres, 50-51). *Res:* Effectiveness of supplementing a fluoride deficient community water supply with a fluoride compound to bring the fluoride ion concentration up to 1.0 part per million of water to prevent dental caries; public health aspects of malocclusion; administration and medical assistance programs. *Mailing Add:* 954-A Calle Aragon Laguna Hills CA 92653

ASTELL, CAROLINE R, MOLECULAR BIOLOGY. *Educ:* Univ BC, PhD(biochem), 71. *Prof Exp:* From asst prof to assoc prof, 80-90, PROF BIOCHEM, UNIV COLUMBIA, 90- *Mem:* Am Soc Microbiol; Fedn Am Soc Exp Biol. *Res:* Molecular biology; DNA replication and transcription of parvaviruses. *Mailing Add:* Dept Biochem Univ BC 2146 Health Sci Mall Vancouver BC V6T 1Z2 Can

ASTER, ROBERT WESLEY, b Spokane, Wash, Feb 10, 53. PERFORMANCE SIMULATION, PROJECT DESIGN. *Educ:* Stanford Univ, BS & MS, 76. *Prof Exp:* Sr anal photovoltaics, Flat Plate Solar Array Proj, Jet Propulsion Lab, 76-82, sr anal coal mining, Advan Coal Extraction Proj, 82-83, task mgr performance anal, FAA-Radar Weather Proj, 84-89; team leader cost anal, NASA Info Syst Strategic Plan, 89-90; task mgr performance anal, Space Sta Info Syst, 90-91; supvr performance anal, 90, SUPVR PROJ ENG, SYST ENG DIV, JET PROPULSION LAB, 91- *Res:* National authority on photovoltaic economics; development of JPL's information system performance simulation team; developing a project engineering group to ensure JPL flight and instrument projects achieve cost/schedule goals. *Mailing Add:* M/S 601-237 Jet Propulsion Lab 4800 Oak Grove Dr Pasadena CA 91109

ASTERIADIS, GEORGE THOMAS, JR, b Colorado Springs, Colo, May 7, 44; m 77; c 1. HUMAN SEXUALITY. *Educ:* State Univ NY, BS, 66; Purdue Univ, PhD(molecular bio), 71. *Prof Exp:* From asst prof to assoc prof, 71-91, PROF BIOL, PURDUE UNIV, N CENT CAMPUS, 91- *Concurrent Pos:* Fel, Int Coun Sex Educ & Parenthood; infection control consult. *Mem:* Nat Sci Teachers Asn; Am Asn Sex Educr, Counr & Therapists; Asn Practr Infection Control; Am Chem Soc; Am Soc Microbiol. *Res:* Nucleic acid chemistry. *Mailing Add:* Dept Biol Sci Purdue Univ N Cent Campus Westville IN 46391

ASTERITA, MARY FRANCES, b New York, NY. BIOLOGICAL SCIENCES, SCIENCE EDUCATION. *Educ:* Marymount Col, BA, 61; NY Univ, MS, 69; Cornell Univ, PhD(physiol), 73. *Prof Exp:* Lectr physics, Iona Col, 68-69; lectr physiol, Cornell Univ, Med Col, 69-72; teaching fel physiol

& biophys, Yale Univ, Sch Med, 73-75; univ fac prof physiol & pharmacol, Sch Med, Ind Univ, 75-85; VIS PROF, PURDUE UNIV, 80- *Concurrent Pos:* NIH res grants, 69-73 & 73-74; Conn Heart Asn res grant, 74-75; consult, St Anthony's Med Ctr, 84- *Mem:* Biofeedback Soc Am; Am Physiol Soc; Am Soc Exp Biol; Biophys Soc. *Res:* Epithelial transport; self-regulation of the autonomic nervous system; physiological and nutritional changes that take place in the human body as a result of the stress response. *Mailing Add:* Ghent Family Pract 130 Colley Ave Norfolk VA 23510

ASTHEIMER, ROBERT W, b Jersey City, NJ, Oct 16, 22; m 48; c 3. ELECTROOPTICS. *Educ:* Stevens Inst Technol, ME, 44, MS, 49. *Prof Exp:* Staff instr physics & res asst, Stevens Inst Technol, 46-51; res physicist, Naval Ord Lab, 51-53; proj engr, 53-57, dept mgr, 57-60, chief engr, 60-63, tech dir, 63-67, VPRES, BARNES ENG CO, 67- *Mem:* Fel Optical Soc Am; Am Phys Soc. *Res:* Electrooptical systems and components; explosion hydrodynamics; infrared systems. *Mailing Add:* Barnes Eng Co Seven Ellery Lane Westport CT 06880

ASTILL, BERNARD DOUGLAS, b Nottingham, Eng, Feb 11, 25; US citizen; m 55; c 2. BIOCHEMISTRY. *Educ:* Univ Nottingham, BSc, 50, PhD(org chem), 53. *Prof Exp:* Asst anal chemist, Boots Pure Drug Co, Nottingham, Eng, 43-44; chemist, Lab Indust Med, Eastman Kodak, 55-60, sr biol chemist, 60-66, res assoc, 67-72, supvr biochem, Health & Safety Lab, 72-85, dir, Regulatory Affairs, 85-87; CONSULT, BERNARD D ASTILL ASSOC, 87- *Concurrent Pos:* Cerebral Palsy Found grant, Univ Rochester, 52-54, NIH fel, 54-55; clin instr, Dept Community Health & Prev Med, Univ Rochester Med Ctr, 62-76. *Mem:* Am Chem Soc; NY Acad Sci; Soc Toxicol; Am Indust Hyg Asn; Soc Environ Toxicol & Chem. *Res:* Fate of organic chemicals in mammalian and aquatic species; biological monitoring for environmental exposures; biochemical mechanism of toxicity in vitro tests for toxicity; structure-activity relationships in toxicology; environment health and safety. *Mailing Add:* 195 Lyell St Spencerport NY 14559

ASTILL, KENNETH NORMAN, b Westerly, RI, July 16, 23; m 48; c 2. MECHANICAL ENGINEERING. *Educ:* Univ RI, BS, 44; Chrysler Inst Eng, MAE, 46; Harvard Univ, MS, 53; Mass Inst Technol, PhD, 61. *Prof Exp:* Lab engr, Chrysler Corp, 44-47; prof mech eng, 47-90, assoc dean eng, 80-89, EMER PROF, MECH ENG, TUFTS UNIV, 91- *Concurrent Pos:* Instr, Northeastern Univ, 49-54; mech engr, Gen Elec Co, 48-49, Harvard Proj, Pratt & Whitney Aircraft Div, United Aircraft Corp, 51 & Cambridge Air Force Res Ctr, 52-53; consult engr, Sylvania Elec Prod Inc, 54-57, Kaye Instruments, Inc, 57-68, Cohoon & Heasley, 64-69, USMC, US Army-Natick Labs, 67-74, Kaye Instuments, Inc, 70-73, C S Draper Labs, 73- & B-D Electrodyne; vis fel, Univ Sussex, 68; vis prof, Univ Leeds, 76, Foster-Miller Assoc, 79-81, Analytix Inc, 83, Univ Sussex, 83. *Honors & Awards:* Ralph R Teetor Award, 80. *Mem:* Fel Am Soc Mech Engrs; Am Soc Eng Educ. *Res:* Heat transfer and fluid mechanics. *Mailing Add:* Dept Mech Eng Col Eng Tufts Univ Medford MA 02155

ASTLEY, EUGENE ROY, b Alameda, Calif, Dec 5, 26; m 48; c 3. NUCLEAR ENGINEERING, APPLIED PHYSICS. *Educ:* Univ Ore, BS, 48; Ore State Univ, MS, 50. *Prof Exp:* Physicist, Gen Elec Co, Schenectady, NY, 50-55 & Richland, Wash, 55-60, mgr maintenance & indust eng prod reactors, Richland, Wash, 60-65; dir, FFTF Dept, Pac Northwest Div, Battelle Mem Inst, Richland, Wash, 65-69, assoc dir systs & elec, 69-71; dir & vpres res, Exxon Nuclear Co Inc, Bellevue, Wash, 71-79, dir & vpres fuels mfg, Richland, Wash & Lingen, Ger, 79-83; DIR, PRES & CHIEF EXEC OFFICER, SANDVIK SPEC METALS CORP, KENNEWICK, WASH, 83- *Concurrent Pos:* Asst prof, Univ Wash, 58; bd dir, Jersey Nuclear Avco Isotopes, Inc, 72-75; mem, US Coun Energy Awareness. *Mem:* Am Nuclear Soc. *Res:* Instrumentation, reactor physics; industrial engineering, technical management; nuclear reactors and laser isotope enrichment. *Mailing Add:* 2414 Harris Ave Richland WA 99352

ASTLING, ELFORD GEORGE, b Sycamore, Ill, Aug 13, 37; m 71; c 3. METEOROLOGY. *Educ:* Univ Utah, BS, 60; Univ Wis, Madison, MS, 64, PhD(meteorol), 70. *Prof Exp:* Res meteorologist, Nat Environ Satellite Ctr, Environ Sci Serv Admin, 64-66; asst prof meteorol, Fla State Univ, 69-72; asst prof geophys sci, Old Dominion Univ, 72-75; asst prof, 75-80, assoc prof meteorol, Univ Utah, 80-84; RES METEOROLOGIST, US ARMY DUGWAY PROVING GROUND, 84- *Mem:* AAAS; Am Meteorol Soc; Am Geophys Union. *Res:* Applications of satellite technology in synoptic meteorology; mesoscale meteorology, mountain meteorology, remote sensing. *Mailing Add:* 2941 Zenith Circle Univ Utah Salt Lake City UT 84106

ASTON, DUANE RALPH, b Strabane, Pa, Aug 3, 32; m 55; c 6. SOLID STATE PHYSICS. *Educ:* Brigham Young Univ, BS, 55, MS, 57; Temple Univ, PhD(physics), 69. *Prof Exp:* Engr, Gen Dynamics/Convair, 55-56; engr, Boeing Airplane Co, 57-58; assoc scientist, Bettis Atomic, Westinghouse Elec Corp, 58-59; instr physics, Drexel Inst Technol, 59-67; assoc prof, 68-80, PROF PHYSICS, CALIF STATE UNIV, SACRAMENTO, 80- *Mem:* Am Asn Physics Teachers. *Res:* Nuclear reactor calculations; nuclear shielding studies; effects of demagnetization coefficients on type II superconductors; magnetization studies on the strong-coupling Pb-In alloy systems; intermediate and mixed states in type II superconductors. *Mailing Add:* Dept Physics Calif State Univ 6000 J St Sacramento CA 95819

ASTON, RICHARD, b Wilkes-Barre, Pa, Oct 4, 36; m 65; c 3. ELECTROSURGERY, COURSE DEVELOPMENT. *Educ:* Pa State Univ, BS, 61, MS, 64; Ohio State Univ, PhD(med instrumentation), 69. *Prof Exp:* Asst prof elec eng, Rochester Inst Technol, 70-72; staff engr, Hughes Aircraft Co, 72-75; design specialist, Pomona Div, Gen Dynamics, 75-78; assoc prof elec eng, Wilkes Univ, 78-64; ASSOC PROF MED INSTRUMENTATION, PA STATE UNIV, WILKES-BARRE, 84- *Concurrent Pos:* Biol med engr, Proj Hope, Millwood, Va; session chair, Am Soc Eng Educ, 90-93. *Mem:* Inst Elec & Electronics Engrs Eng Med & Biol Soc; Am Soc Eng Educ; Sigma Xi. *Res:* Surgical devices; Impatt power combiners; speech therapy instrumentation; microwave devices; medical instrumentation for hospitals; electrical circuit theory; author of books on medical instrumentation principles, laboratory manual. *Mailing Add:* 33 Barney St Wilkes-Barre PA 18702

ASTON, ROY, b Windsor, Ont, Dec 31, 29; m 73; c 1. PHARMACOLOGY. *Educ:* Univ Windsor, BA, 50; Wayne State Univ, MSc, 54; Univ Toronto, PhD(pharmacol), 58. *Prof Exp:* Res assoc pharmacol, Med Sch, Univ Mich, 58-59; res assoc pharmacol, Sch Med, Wayne State Univ, 59-62, from instr to asst prof, 62-70, assoc prof, 70-80; prof physiol & pharmacol, 80-90, PROF PHARMACOL, DEPT BIOMED SCI, SCH DENT, UNIV DETROIT, MERCY, 90- *Mem:* Am Soc Pharmacol; Pharmacol Soc Can. *Res:* Pharmacodynamics and psychopharmacology of drugs acting on the central nervous system; drug protection against pathological effects of acceleration. *Mailing Add:* Dept Biomed Sci Sch Dent Univ Detroit/Mercy 2985 E Jefferson Ave Detroit MI 48207

ASTON-JONES, GARY STEPHEN, b Havre de Grace, Md, Oct 25, 51; m 81; c 1. BEHAVIORAL NEUROBIOLOGY, NEUROANATOMY. *Educ:* Univ Va, BA, 73; Calif Inst Technol, PhD(neurobiol), 80. *Prof Exp:* Fel, A V Davis Ctr Behav Neurobiol, Salk Inst, 80-82; asst prof psychol, State Univ NY, Binghampton, 82-84; asst prof biol, NY Univ, 84-88; DIR BEHAV & NEUROBIOL LAB & ASSOC PROF DEPT, MENT HEALTH SCI, HAHNEMANN UNIV, 88- *Mem:* AAAS; Soc Neurosci. *Res:* Physiology and anatomy of brain monoaminergic systems. *Mailing Add:* Dept Ment Health Hahnemann Univ MS 543 Broad & Vine St Philadelphia PA 19102-1192

ASTRACHAN, BORIS MORTON, b New York, NY, Dec 1, 31; m 56; c 4. PSYCHIATRY. *Educ:* Alfred Univ, BA, 52; Albany Med Col, MD, 56. *Prof Exp:* Asst dir inpatient psychiat serv, Yale-New Haven Hosp, 63-66; from asst prof to assoc prof psychiat, Sch Med, Yale Univ, 65-71; chief day hosp, Conn Ment Health Ctr, 66-68, dir gen clin div, 68-70, actg dir, Ctr, 70, dir Conn Ment Health Ctr, 71-87; prof psychiat, Sch Med, Yale Univ, 71-90; PROF, HEAD & CHIEF PSYCHIAT, UNIV ILL, CHICAGO, 90- *Concurrent Pos:* NIMH fel, Yale Univ, 61-63; mem res task panel, President's Comn on Ment Health, 77-78; mem adv bd, Alcohol Drug Abuse & Mental Health Admin, 85-86, IBM Ment Health, 90-; mem epidemol & serv comt, NIMH, 87-91, chair, 89-91. *Mem:* AAAS; fel Am Psychiat Asn; Am Pub Health Asn; fel Am Col Psychiatrists; fel Am Asn Psychiat Admin (pres, 90-). *Res:* Organizational group dynamics; schizophrenia, clinical epidemiology and outcome; evaluation research. *Mailing Add:* Dept Psychiat Sch Med Yale Univ New Haven CT 06520

ASTRACHAN, LAZARUS, b New York, NY, Aug 7, 25; m 45; c 2. BIOCHEMISTRY. *Educ:* City Col NY, BS, 44; Yale Univ, PhD(physiol chem), 51. *Prof Exp:* Biochemist, Biol Div, Oak Ridge Nat Lab, 54-62; from asst prof to assoc prof, 62-90, EMER PROF MICROBIOL, SCH MED, CASE WESTERN RESERVE UNIV, 90- *Concurrent Pos:* Fel, McCollum-Pratt Inst, Johns Hopkins Univ, 51-53, Polio Found fel, Sch Hyg & Pub Health, 54. *Mem:* Am Soc Microbiol; Am Soc Biol Chemists; Sigma Xi. *Res:* Biochemistry of nucleic acids; virology; bacterial lipids. *Mailing Add:* Microbiol Dept Case Western Reserve Univ Cleveland OH 44106

ASTRACHAN, MAX, b Rochester, NY, Mar 30, 09; m 31; c 3. MATHEMATICS, STATISTICS. *Educ:* Univ Rochester, AB, 29; Brown Univ, AM, 30, PhD(math), 35. *Prof Exp:* Instr math, Brown Univ, 29-35; from instr to prof math, Antioch Col, 35-50, chmn dept, 42-50; prof statist, Air Force Inst Technol, Wright-Patterson AFB, 49-60, head dept acct & statist, sch bus, 55-60; mathematician, logistics dept, Rand Corp, 60-65; dir educ & training inst, Am Soc Qual Control, 65-67; prof statist, 67-70, prof mgt sci, 70-81, chmn dept, 75-79, EMER PROF MGT SCI, SCH BUS, CALIF STATE UNIV, NORTHRIDGE, 81- *Concurrent Pos:* Instr, exten div, Pa State Col, 41; statistician, Wright Field, Ohio, 44; lectr, Wright-Patterson Grad Ctr, Ohio State Univ, 46-60, Air Inst Technol, 48-49 & Univ Calif, Los Angeles, 61-; opers analyst, Korea & Japan, 53, Eng, 54; consult, Rand Corp, 56-60. *Honors & Awards:* Eugene L Grant Award, 79. *Mem:* Am Statist Asn; Inst Mgt Sci; Am Soc Qual Control. *Res:* Summability of Fourier series by Norlund means. *Mailing Add:* 1870 Kelton Ave No 203 Los Angeles CA 90025

ASTRAHAN, MELVIN ALAN, b Poughkeepsie, NY, Oct 31, 52; m 76; c 3. HYPOTHERMIA, THERMAL PHYSICS. *Educ:* Univ Calif, Irvine, BS, 74; Univ Calif, Los Angeles, MS, 78, PhD(med physics), 81. *Prof Exp:* Res physicist, Vet Admin Res Serv, Brentwood, Calif, 81-84; ASSOC PROF RADIATION ONCOL, UNIV SOUTHERN CALIF, 84- *Honors & Awards:* Silverman Award, Southern Calif Health Physics Soc, 81. *Mem:* Am Asn Physicists Med; NAm Hyperthermia Group. *Res:* Hyperthermia systems for adjunctive use in cancer therapy; treatment planning software for radiation therapy; thermal physics. *Mailing Add:* Dept Radiation Oncol Univ Southern Calif 1441 Eastlake Ave Los Angeles CA 90033

ASTRAHAN, MORTON M, relational data base management systems; deceased, see previous edition for last biography

ASTROLOGES, GARY WILLIAM, b Granby, Que, Oct 22, 49; US citizen. ORGANIC CHEMISTRY. *Educ:* Mass Inst Technol, BS, 71; Univ Ill, Urbana, PhD(org chem), 77. *Prof Exp:* RES CHEMIST, HALOCARBON PROD CORP, 77- *Mem:* Am Chem Soc; AAAS. *Res:* Organofluorine chemistry; hypervalent chemistry, especially involving sulfur and phosphorus. *Mailing Add:* Ariz Chem Co Caller Box 2447 Panama City FL 32402-2447

ASTROMOFF, ANDREW, b Paris, France, Nov 30, 32; US citizen; m 57; c 3. MATHEMATICS. *Educ:* Univ Calif, Berkeley, AB, 54, MA, 58, PhD(math), 63. *Prof Exp:* From asst prof to assoc prof, 62-70, PROF MATH, SAN FRANCISCO STATE UNIV, 70- *Mem:* Math Asn Am. *Res:* Foundations of mathematics; theory of models; general algebraic systems. *Mailing Add:* San Francisco State Univ San Francisco CA 94132

ASTRUC, JUAN, b Utrera, Spain, June 5, 33; m 61; c 7. ANATOMY, NEUROANATOMY. *Educ:* Univ Granada, MD, 57, PhD, 59. *Prof Exp:* From instr to asst prof anat, Sch Med, Univ Granada, 57-61; asst prof, Sch Med, Lit Univ Salamanca, 58-60; from asst prof to assoc prof, Sch Med, Univ

Navarra, 62-67; assoc prof, 67-76, PROF ANAT & NEUROL, MED COL VA, 76- *Concurrent Pos:* Span Govt res fel, 59-60; Ger Govt res fel, 60; March Found res grant, 62-63; USPHS res fel, 63-64 & res grants, 65-67 & 69- *Res:* Influence of the movement of the central nervous system upon the endocrine glands, control of the movement of the eyes; structure and connections in the central nervous system. *Mailing Add:* Dept Anat Va Commonwealth Univ Sch Med UCV Sta Box 565 Richmond VA 23298

ASTRUE, ROBERT WILLIAM, b San Francisco, Calif, Feb 2, 28; m 56; c 5. SOLID STATE PHYSICS. *Educ:* Univ Calif, Berkeley, AB, 51; Univ Wash, MS, 59; Wayne State Univ, PhD(physics), 66. *Prof Exp:* Res physicist, Ford Motor Co Sci Res Lab, 63-66; asst prof, 66-70, assoc prof, 70-77, PROF PHYSICS, HUMBOLDT STATE UNIV, 77- *Mem:* Am Phys Soc; Am Asn Physics Teachers; Sigma Xi. *Res:* Magnetic resonance; ferromagnetic thin films. *Mailing Add:* Dept Physics & Phys Sci Humboldt State Univ Arcata CA 95521

ASTUMIAN, RAYMOND DEAN, b Birmingham, Ala, Oct 3, 56. RELAXATION KINETICS OF MEMBRANE ENZYMES. *Educ:* Univ Tex, PhD(chem & math sci), 83. *Prof Exp:* Staff fel, NIH, 84-88; RES CHEMIST, NAT INST STANDARDS & TECHNOL, 88- *Concurrent Pos:* Vis prof, Bielefeld Univ, Ger, 89-90. *Honors & Awards:* Galvani Prize, Bioelectrochem Soc, 87. *Mem:* Am Soc Biochem & Molecular Biol; Biophys Soc; Am Chem Soc; Bioelectrochem Soc; Bioelectromagnetics Soc. *Res:* Interaction between electromagnetic fields and biomembrane macromolecules, particularly with regard to the kinetics of membrane enzymes in an AC electric field. *Mailing Add:* Biotech Div Nat Inst Standards & Technol Bldg 222 Rm A353 Gaithersburg MD 20899

ASWAD, A ADNAN, b Antioch, Turkey, Dec 6, 33; US citizen; m 62; c 1. INDUSTRIAL & SYSTEMS ENGINEERING. *Educ:* Robert Col, Istanbul, BSc, 55; Univ Mich, Ann Arbor, MSc, 61 & 64, PhD(indust eng), 72. *Prof Exp:* Power plant engr, Damascus Power Co, 55-56; design engr, Mech Construct Co, 56-59; stress anal engr, Int Harvester Co, 61-63; lectr indust eng, 65-72, PROF INDUST & SYSTS ENG & CHMN DEPT, UNIV MICH, DEARBORN, 72- *Concurrent Pos:* Prin researcher, Kellogg Found grant, 75-77; consult, automotive indust, 70- *Mem:* Sr mem Am Inst Indust Engrs; Opers Res Soc Am; Inst Mgt Sci; Am Soc Eng Educ; Soc Mfg Engrs. *Res:* Risk analysis; management of research and development; manufacturing systems; mechanical engineering. *Mailing Add:* Dept Indust Eng Univ Mich 4901 Evergreen Rd Dearborn MI 48128

ASZALOS, ADORJAN, b Szeged, Hungary, May 9, 29; US citizen; m 60; c 2. BIO-ORGANIC CHEMISTRY, CANCER. *Educ:* Tech Univ Budapest, MS, 51; Tech Univ Vienna, PhD(natural prod chem), 61. *Prof Exp:* Plant engr, Hungary, 51-53; supvr, Biochem Res Inst, 53-56; res chemist, Nat Starch & Chem Co, 57-59; sr res scientist, Squibb Inst Med Res, 62-74; head biochem, Frederick Cancer Res Ctr, 74-79; CHIEF CELL BIOL, FOOD & DRUG ADMIN, 79- *Concurrent Pos:* Fel, Rutgers Univ, 61-62, coadj prof, 71-73; res prof, Princeton Univ, 73-75. *Mem:* Am Chem Soc; NY Acad Sci; Intra Sci Found. *Res:* Isolation and structural studies of antibiotics; synthetic organic chemistry in field of carbohydrates; biochemistry; biophysics; cell biology. *Mailing Add:* 7711 Bradley Blvd Bethesda MD 20034

ATACK, DOUGLAS, b Wakefield, Eng, Aug 6, 23; m 50; c 1. PHYSICAL CHEMISTRY. *Educ:* Univ Leeds, BSc, 44, PhD(chem), 46. *Prof Exp:* Res assoc high pressure, PTX Data, Univ Leeds, 46-48; fel critical phenomena, Nat Res Coun Can, 48-50; fel PVI data, Univ Manchester, 50-51; res assoc, Off Naval Res Contract, Univ NC, 51-53; asst prof phys chem, Syracuse Univ, 53-54; head wood & fiber physics div, Pulp & Paper Res Inst Can, 54-67, chmn wood & fiber res dept, 67-69, dir appl physics div, 69-89, dir acad affairs, 85-89; RETIRED. *Concurrent Pos:* Auxillary prof mech eng, McGill Univ, 64-; fel, Asn Pulp & Paper Indust, 75. *Honors & Awards:* Weldon Medal, Can Pulp & Paper Asn, 63 & 81; Walter Brecht Medal, Soc Ger Pulp & Paper Engrs, 82. *Mem:* Am Phys Soc; fel Tech Asn Pulp & Paper Indust; Chem Soc. *Res:* High speed friction phenomena; mechanical properties of polymeric materials. *Mailing Add:* 28 Thornhill Westmount PQ H3Y 2E2 Can

ATAL, BISHNU S, b Kanpur, India, May 10, 33. ACOUSTICS. *Educ:* Univ Lucknow, BSc, 52; Polytech Inst Brooklyn, PhD(elec eng), 68. *Prof Exp:* Lectr acoust, dept elec commun eng, Indian Inst Sci, Bangalore, 57-80; res staff, 81-85, head, res dept, 85-90, HEAD, SPEECH RES DEPT, BELL TEL LABS, 90- *Honors & Awards:* Centennial Medal, Inst Elec & Electronics Engrs; Morris N Liebman Mem Field Award, Inst Elec & Electronics Engrs, 88. *Mem:* Nat Acad Eng. *Res:* Published technical papers in architectural acoustics and speech communication. *Mailing Add:* Dept Speech Res AT&T Bell Labs 600 Mountain Ave Murray Hill NJ 07974

ATALAY, BULENT ISMAIL, b Ankara, Turkey, June 10, 40; US citizen. THEORETICAL PHYSICS, NUCLEAR PHYSICS. *Educ:* Georgetown Univ, BS, 63, MS, 67, PhD(theoret physics), 70; Oxford Univ, MA, 72. *Prof Exp:* Instr area studies, Georgetown Univ, 63-64 & Princeton Univ, 64-65; res assoc nuclear physics, Univ Calif, Berkeley, 68-69; from asst prof to assoc prof, 66-72, PROF & CHMN, MATH SCI & PHYSICS DEPT, MARY WASHINGTON COL, UNIV VA, 74- *Concurrent Pos:* NATO scholar, 60-68; NSF fel, 68-69; Baron Schwarchild fel, 74, 76, 78 & 79; res assoc, Dept Physics, Princeton Univ, 68-69, mem, Inst Advan Study, 74-76; fel, Dept Theoretical Phsyics, Oxford Univ, 72-73. *Mem:* Am Inst Physics. *Res:* Theoretical physics perturbation theory for projected states; nuclear structure and nuclear models. *Mailing Add:* Dept Physics Mary Wash Col Fredericksburg VA 22401

ATALLA, RAJAI HANNA, b Jerusalem, Palestine, Feb 21, 35; nat US; m 63; c 2. CHEMICAL PHYSICS. *Educ:* Rensselaer Polytech Inst, BChE, 55; Univ Del, MChE, 58, PhD(chem eng), 61. *Prof Exp:* Res engr res ctr, Hercules Inc, 60-62, res chemist, 62-68; from assoc prof to prof chem physics & eng, Inst Paper Chem, 68-89; HEAD, DIV CHEM & PULPING RES, FOREST PROD LAB, 89- *Mem:* AAAS; Royal Chem Soc; Am Chem Soc; Tech Asn Pulp & Paper Indust; Sigma Xi. *Res:* Molecular structure and organization in plant cell walls; celluloses, hemi celluloses, and ligrins; vibrational spectroscopy; solid state 13C NHR spectroscopy; thermodynamics and statistical mechanics. *Mailing Add:* Forest Prod Lab One Gifford Pinchot Dr Madison WI 53705-2398

ATALLA, ROBERT E, b Brooklyn, NY, Oct 31, 29. MATHEMATICS. *Educ:* Univ Calif, Los Angeles, AB, 54; Univ Idaho, MA, 62; Univ Rochester, PhD(math), 66. *Prof Exp:* Asst prof, 66-77, ASSOC PROF MATH, OHIO UNIV, 77- *Mem:* Am Math Soc. *Res:* Mathematical analysis and point set topology; applications of topology and functional analysis to matrix summability. *Mailing Add:* Ohio Univ Athens OH 45701

ATALLAH, MIKHAIL JIBRAYIL, b Aleppo, Syria, Apr 19, 53; m 80; c 2. ALGORITHMS, PARALLEL COMPUTATION. *Educ:* Am Univ, Beirut, BS, 75; Johns Hopkins Univ, MS, 80, PhD(elec eng), 82. *Prof Exp:* From asst prof to assoc prof, 82-89, PROF COMPUT SCI, PURDUE UNIV, 89- *Mem:* Inst Elec & Electronics Engrs; Asn Comput Mach; Soc Indust & Appl Math. *Res:* Design and analysis of computer algorithms; parallel computation; computational geometry. *Mailing Add:* Dept Comput Sci Purdue Univ West Lafayette IN 47907

ATAM-ALIBECKOFF, GALIB-BEY, b Paris, France, Jan 1, 23; m 59; c 2. CELLULOSE CHEMISTRY, VISCOSE CHEMISTRY. *Educ:* Univ Paris, Licence Es Sci, 44. *Prof Exp:* Res chemist, French Soc Glycerin, 46-47; chemist, Bact Inst Chile, 47-49; chief chemist, Rayonhil, Chile, 49-57, Von Kohorn Int Corp, 57-66 & Chemtex Inc, 66-67; vpres, Nylonge Corp, 66-78; exec vpres, Sponge Inc, 78-90; EXEC VPRES, NYLONGE CORP, 91- *Mem:* Am Chem Soc; Tech Asn Pulp & Paper Indust. *Res:* Viscose rayon spinning; new products based on viscose and cellulose. *Mailing Add:* Nylonge Corp 1301 Lowell St Elyria OH 44035

ATANASOFF, JOHN VINCENT, b Hamilton, NY, Oct 4, 03; m 26, 49; c 3. PHYSICAL MATHEMATICS, RESEARCH ADMINISTRATION. *Educ:* Univ Fla, BSEE, 25; Iowa State Col, MS, 26; Univ Wis, PhD(theoret physics), 30. *Hon Degrees:* DSc, Univ Fla, Univ Wis 87; ScD, Moravian Col, 81; LittD, Western Md Col, 84, Mt St Mary's Col, 90. *Prof Exp:* Instr math, Iowa State Col, 26-29, assoc prof, 29-30; instr, Univ Wis, 30-42; chief, acoust div, US Naval Ordnance Lab, 42-49, dir, USN Fuze Prog, 51-52; chief scientist, US Army Field Forces, 49-50; vpres, Aerojet Gen Corp, 56-61; PRES, CYBERNETICS, INC, 61- *Concurrent Pos:* Consult, Steward-Warner Corp, 61-63, Honeywell Inc, 67-71 & Control Data Corp, 67-71. *Honors & Awards:* Nat Medal Technol, 90; Comput Pioneer Medal, Inst Elec & Electronics Engrs, 83; Holley Medal, Am Soc Mech Engrs, 85. *Mem:* Bulgarian Acad Sci; Cosmos Club. *Res:* Invention of first electronic digital computer. *Mailing Add:* 11928 E Baldwin Rd Monrovia MD 21770

ATASSI, ZOUHAIR, b Homs, Syria, Dec 20, 34; m 63; c 1. BIOCHEMISTRY, IMMUNOLOGY. *Educ:* Bristol Univ, BSc, 57; Univ Birmingham, MSc, 58, PhD(chem), 60, DSc(chem), 73. *Prof Exp:* Postdoctoral res fel chem, Univ Birmingham, 60-61; res assoc biochem, Albany Med Col & postdoctoral res fel labs protein chem, Div Labs & Res, NY State Dept Health, 62-63; asst prof biochem sch med, State Univ NY Buffalo, 63-68; prof chem, Wayne State Univ, 68-75; prof biochem & immunol, Mayo Med Sch, 75-83; prof biochem, Univ Minn, Minneapolis, 75-83; WELCH CHAIR CHEM, DEPT BIOCHEM, BAYLOR COL, 83- *Concurrent Pos:* Grants, Off Naval Res, 64-70 & 84-90, Nat Inst Arthritis & Metab Dis, 64-, Am Heart Asn, 71-74 & 81-, Nat Inst Allergy & Infectious Dis, 74-, Welch Found, 83- & US Army Med Res & Develop Command, 89-; estab investr, Am Heart Asn, 66-71; immunol consult, Mayo Clin, 75-83, distinguished lectr med sci; pres, Int Symposium on Immunobiol of Proteins; ed in chief, CRC Crit Reviews in Immunol & J Protein Chem. *Honors & Awards:* Harden Medal & Jubilee lect, Biochem Soc, 87. *Mem:* Am Soc Biol Chem; Am Asn Immunol; Brit Biochem Soc; Am Chem Soc; AAAS. *Res:* Protein structure and function; chemistry, biochemistry and molecular and cellular immunology; protein chemistry, immunochemistry and conformation; chemical modification and cleavage of proteins; organic synthesis of peptides; localization and synthesis of protein binding sites. *Mailing Add:* Dept Biochem Baylor Col One Baylor Plaza Houston TX 77030

ATCHESON, J D, b London, Ont, July 27, 17; m 42; c 3. PSYCHIATRY. *Educ:* Univ Western Ont, MD, 41; Univ Toronto, dipl psychiat, 46; FRCP(C), 47; FRCP(C), 72. *Prof Exp:* Resident, Hamilton Gen Hosp, 41-42; surgeon lieutenant comdr, Royal Can Navy, 42-45; registr, Allen Mem Inst Psychiat, McGill Univ, 44-45; teaching assoc, 47-57, from asst prof to prof, 57-83, EMER PROF, FAC MED, DEPT PSYCHIAT, UNIV TORONTO, 83-; SR PSYCHIATRIST, FORENSIC OUT-PATIENT SERV, CLARKE INST PSYCHIAT, UNIV TORONTO, 71- *Concurrent Pos:* Dir, psychiat clin, Juv & Family Court, Toronto, 47-57; consult training schs, Dept Corrections, 49-57, dir treatment serv, 58; supt, Thistletown Children's Psychiat Hosp, 58-69; consult, Eastern Arctic, Dept Nat Health & Welfare, 65- & Ont Correctional Inst, Brampton, 74-; med dir & chief of staff, Thistletown Regional Ctr Children & Adolescents, 69-71; mem, Lieutenant Gov adv rev bd, Ont, 75- *Honors & Awards:* Centennial Medal, Dom Can, 67. *Mem:* Hon fel Am Psychiat Asn; Can Psychiat Asn; Can Med Asn; Am Acad Psychiat & Law. *Res:* Juvenile delinquency; residential care of children; child psychiatry; forensic psychiatry; trans-cultural psychiatry, especially problems of emotional disorder in the Canadian Arctic. *Mailing Add:* 300 Mill Rd Toronto ON M9C 4W7 Can

ATCHISON, F(RED) STANLEY, b Stoddard Co, Mo, Jan 3, 18; m 40. ELECTRONICS, PHYSICS. *Educ:* Southeast Mo State Univ, AB, 38; Univ Iowa, MS, 40, PhD(physics), 42. *Prof Exp:* Instr sci, City HS, Cairo, Ill, 37-38; asst physics, Univ Iowa, 38-42; physicist, US Bur Standards, 42-53; dept head electronics, Naval Ord Lab, Corona, Calif, 53-55, tech dir, 55-68; mem res staff, Inst Defense Anal, 69-80; RETIRED. *Mem:* Am Phys Soc; fel Inst Elec & Electronics Engrs; AAAS; Sigma Xi. *Res:* Nuclear physics; proximity fuse development; guided missile development; microwave radar development; weapons systems analysis. *Mailing Add:* 6197 Palm Ave Riverside CA 92506

ATCHISON, ROBERT WAYNE, b Pratt, Kans, Sept 19, 30; m 52; c 2. VIROLOGY, IMMUNOLOGY. *Educ:* Univ Kans, AB, 52, AM, 55, PhD(virol), 60. *Prof Exp:* Res assoc, 60-62, from asst res prof to assoc prof virol & immunol, 63-69, ASSOC PROF MICROBIOL, GRAD SCH PUB HEALTH, UNIV PITTSBURGH, 69- *Mem:* Am Soc Microbiol. *Res:* Electron microscopy; biochemistry; histochemistry; anatomy; bacteriology. *Mailing Add:* Grad Sch Pub Health Univ Pittsburgh Pittsburgh PA 15261

ATCHISON, THOMAS ANDREW, b Brady, Tex, July 3, 37; m 59; c 3. NUMERICAL ANALYSIS. *Educ:* Univ Tex, BA, 59, MS, 60, PhD(math), 63. *Prof Exp:* Instr math, Howard Payne Col, 59; spec instr, Univ Tex, 61-63; from asst prof to assoc prof, Tex Tech Univ, 63-66; mem staff, LTV Electrosysts, Inc, 66-67; assoc prof math, Tex Tech Univ, 67-72; prof math & head dept, Miss State Univ, 72-79; prof math & chmn dept, 79-89, DEAN SCH SCI & MATH, STEPHEN F AUSTIN STATE UNIV, 89- *Concurrent Pos:* Pres, E Tex Coun Teachers Math, 87-89; mem, Tex Higher Educ Coord Bd Adv Comt Res Prog, 87-; pres, Tex Asn Acad Adminr Math Sci, 89-90. *Mem:* Sigma Xi; Math Asn Am. *Res:* Numerical solutions of differential equations; extrapolation processes; complex analysis. *Mailing Add:* Sch Sci Math Stephen F Austin State Univ Box 13034 Nacogdoches TX 75962

ATCHISON, THOMAS CALVIN, JR, b Fremont, Ohio, June 11, 22; m 50; c 3. MINING ENGINEERING. *Educ:* Princeton Univ, AB, 43. *Prof Exp:* Electronic technician, Palmer Lab, Princeton Univ, 43; physicist, Bur Ships, US Navy Dept, 46-49; supvr physicist, US Bur Mines, 49-70, res dir, Twin Cities Mining Res Ctr, 70-76; RETIRED. *Concurrent Pos:* Vis prof, Dept Civil & Mineral Eng, Univ Minn, 76- *Mem:* Sigma Xi; Int Soc Rock Mech (vpres, 79-); AAAS; Am Planning Asn; World Future Soc. *Res:* Underground space use; mining research; rock mechanics; blasting; lunar surface research. *Mailing Add:* 6512 Warren Ave Edina MN 55439

ATCHISON, WILLIAM FRANKLIN, b Smithfield, Ky, Apr 7, 18; m 47; c 4. MATHEMATICS, COMPUTER SCIENCE. *Educ:* Georgetown Col, AB, 38; Univ Ky, MA, 40; Univ Ill, PhD(math), 43. *Prof Exp:* Asst, Physics Lab, Georgetown Col, 36-38; asst math, Univ Ky, 39-40; asst, Univ Ill, 40-42, instr, 43-44 & 46-48, asst prof, 49-50 & 51-55; asst prof, Harvard Univ, 50-51; res assoc prof, Ga Inst Technol, 55-63, res prof, 63-66; dir comput sci ctr, 66-76, prof, 66-88, EMER PROF COMPUT SCI, UNIV MD, COLLEGE PARK, 88- *Concurrent Pos:* Head prog & coding group, Rich Electronic Comput Ctr, 55-57, dir ctr, 57-66, actg dir sch info sci, 63-64. *Honors & Awards:* Chester Morrill Mem Award, Asn Syst Mgt, 75. *Mem:* Asn Comput Mach; Math Asn Am. *Res:* Digital computation; information science; numerical analysis; algebraic geometry; numerical solution of ordinary differential equations. *Mailing Add:* Dept Comput Sci Univ Md College Park MD 20742

ATCHLEY, ANTHONY A, b Lebanon, Jan 23, 57; m; c 2. PHYSICS, ACOUSTICS. *Educ:* Univ of the South, BS, 79; NMex Tech, MS, 82; Univ Miss, PhD(physics), 84. *Prof Exp:* ASSOC PROF, PHYSICS, NAVAL POST GRAD SCH, 85- *Concurrent Pos:* F V Hunt postdoctoral fel, 85. *Mem:* Am Phys Soc; Acoust Soc Am; Am Asn Physics Teachers. *Res:* Thermoacoustics heat transport. *Mailing Add:* Dept Physics Naval Post Grad Sch Monterey CA 93943

ATCHLEY, BILL LEE, b Cape Girardeau, Mo; m 52; c 3. ENGINEERING MECHANICS. *Educ:* Univ Mo-Rolla, BS, 57, MS, 59; Tex A&M Univ, PhD(struct mech), 65. *Prof Exp:* From instr to prof eng mech, Univ Mo-Rolla, 57-75, dir centennial events, 69-75, asst dean eng, 68-70, assoc dean, 70-75; dean & prof mech engr & mech, Col Eng, WVa Univ, 75-79; pres, 79-, AT SCH ENG, CLEMSON UNIV. *Mem:* Am Soc Eng Educ; Am Soc Civil Engrs; Nat Soc Prof Engrs; Sigma Xi. *Res:* Dynamic properties of engineering materials and the behavior of structures under dynamic loadings. *Mailing Add:* Univ Pac 3601 Pacific Ave Stockton CA 95211

ATCHLEY, WILLIAM REID, b Stilwell, Okla, Sept 6, 42; m 64; c 2. QUANTITATIVE GENETICS, DEVELOPMENTAL GENETICS. *Educ:* Eastern NMex Univ, BS, 64; Univ Kans, MS, 66, PhD(entom), 69. *Prof Exp:* Teaching asst entom, Univ Kans, 64-66, NSF trainee systs & evolutionary biol, 66-69, vis asst prof entom, 70-71; from asst prof to assoc prof biol sci & statist, Tex Tech Univ, 71-77; from assoc prof to prof genetics & entom, Univ Wis-Madison, 77-86; PROF GENETICS & HEAD DEPT, NC STATE UNIV 86-, PROF STATIST. *Concurrent Pos:* Fulbright scholar, Dept Genetics, Melbourne, 69-70; NIH fel, Dept Med Genetics, Univ Wis, 76-77; mem adv panel, NSF Syst Biol Prog, 77-81; res fel, Dept Pop Bio, Australian Nat Univ, 79-81. *Mem:* Soc Syst Zool; Soc Study Evolution; Am Soc Naturalists; Genetic Soc; fel AAAS; Soc Study Evolution; Soc Syst Zool; Am Naturalist (vpres, 87). *Res:* The origin of genetic variability during ontogeny; genetic aspects of growth and morphogenesis; the evolution of developmental processes; the evolutionary aspects of maternal effects. *Mailing Add:* Box 7614 Genetics NC State Univ Main Campus Raleigh NC 27695-7614

ATEMA, JELLE, b Deventer, Neth, Dec 9, 40. SENSORY PHYSIOLOGY. *Educ:* Univ Utrecht, Drs, 66; Univ Mich, PhD(biol), 69. *Prof Exp:* Res assoc, Med Sch & Sch Natural Resources, Univ Mich, 66-70; asst scientist, Woods Hole Oceanog Inst, 70-74; assoc prof, 74-84, PROF, BOSTON UNIV, 84-, DIR, MARINE PROG, 90- *Concurrent Pos:* Res assoc, Univ Hawaii, 76-77; vis prof, Univ Regensburg, WGer, 79-80; trustee, Marine Biol Lab, Woods Hole. *Mem:* Fel AAAS; Am Soc Zoologists; Animal Behav Soc; Soc Neurosci; Asn Chemoreception Sci; Crustacean Soc. *Res:* Neurophysiological and behavioral function of chemoreceptors in aquatic invertebrates and fish, their behavioral and chemical ecology and the effects of pollutants on behavior and chemoreception; physics of odor dispersal; pheromones. *Mailing Add:* Marine Biol Lab Boston Univ Marine Prog Woods Hole MA 02543

ATEN, CARL FAUST, JR, b Lorain, Ohio, Aug 15, 32; m 55; c 4. PHYSICAL CHEMISTRY. *Educ:* Col Wooster, BA, 54; Brown Univ, PhD(phys chem), 59. *Prof Exp:* Chemist, Texaco, Inc, NY, 58-60; asst chem kinetics in shock tubes, Cornell Univ, 60-62; from asst prof to assoc prof, 62-72, PROF PHYS CHEM, HOBART & WILLIAM SMITH COLS, 72- *Mem:* Am Chem Soc. *Res:* Agricultural and food chemistry; analysis of pesticide residues. *Mailing Add:* Hobart Col Geneva NY 14456-3191

ATENCIO, ALONZO C, b Ortiz, Colo, June 24, 29; m 53; c 5. BIOCHEMISTRY. *Educ:* Univ Colo, Boulder, BA, 58; Univ Colo, Denver, MS, 64, PhD(med), 67. *Prof Exp:* Res technician, Sch Med, Univ Colo, Denver, 58-67; ASST PROF BIOCHEM, SCH MED, UNIV NMEX, 70-; ASST DEAN, SCH MED, UNIV NMEX. *Concurrent Pos:* NIH fel chem, Northwestern Univ, 67-70. *Mem:* AAAS; Am Chem Soc; Am Heart Asn; Am Physiol Soc; Int Soc Thrombosis & Haemostasis. *Res:* Hormonal control of protein biosynthesis in mammals using the fibrinogen system as a model. *Mailing Add:* Dept Physiol Univ NMex Sch Med North Campus Albuquerque NM 87131

ATERMAN, KURT, b Bielitz, Sept 9, 13; m 55; c 3. PATHOLOGY. *Educ:* Charles Univ, Prague, MD, 38; Queen's Univ, Belfast, MB & BCh(hons), 42, DSc, 65; Conjoint Bd, London, DCH, 43; Univ Birmingham, PhD, 59. *Prof Exp:* Sr lectr, Univ Birmingham, 50-58; assoc prof path, Dalhousie Univ, 58-61; prof, Woman's Med Col Pa, 61-63, State Univ NY Buffalo & Children's Hosp, 63-67; prof path, Dalhousie Univ & I W Killiam Hosp Children, 67-79; dir, Regional Lab, Dr Everett Chalmers Hosp, NB, Can, 79-86; RETIRED. *Concurrent Pos:* Masaryk Found Stipendium, Commonwealth fel, Harvard & Univ Chicago. *Mem:* Am Asn Path & Bact; Am Soc Exp Path; Path Soc Gt Brit & Ireland; Ger Asn Path; Can Asn Path; Ger Histochem Soc; Am Soc Clin Path. *Res:* Experimental pathology of liver. *Mailing Add:* 5737 Southwood Dr Halifax NS B3H 1E6 Can

ATHANASSIADES, THOMAS J, IMMUNOPATHOLOGY. *Educ:* Tulane Med Sch, MD, 60. *Prof Exp:* ASST PROF PATH, DOWNSTATE MED CTR, STATE UNIV NY, 65- *Mailing Add:* Dept Path Downstate Med Ctr Box 25 Brooklyn NY 11203

ATHANS, MICHAEL, b Drama, Greece, May 3, 37; US citizen. ELECTRICAL ENGINEERING, CONTROL DESIGN. *Educ:* Univ Calif, Berkeley, BS, 58, MS, 59, PhD(eng), 61. *Prof Exp:* Consult, 60-61, mem staff, 61-64, lectr, 63-64, from asst prof to assoc prof, 64-73, consult, Lincoln Lab, 64-76, dir lab info & decision systs, 74-81, PROF ELEC ENG, MASS INST TECHNOL, 73-; CONSULT, ALPHATECH INC, 79- *Concurrent Pos:* Ford Found fel, Mass Inst Technol, 64-66; consult, Bolt, Berannek & Newman, 66-68, Systs Control, Inc, 68-76, Army Mat Command, 68-74, Hamilton Standard, 68-70, Anal Sci Corp, 75-79 & Gen Motors Res Labs, 76; chmn bd, Alphatech, 78-; assoc ed, Automatica. *Honors & Awards:* Eckman Award, 64; Frederick Terman Educ Award, Inst Elec & Electronics Engrs, 69. *Mem:* Fel Inst Elec & Electronics Engrs; Sigma Xi. *Res:* Design of control systems; theory and application of optimal control; command-control and communication systems. *Mailing Add:* Dept Elec Eng Lab Info & Decision Systs Rm 35-406 Mass Inst Technol Cambridge MA 02139

ATHAR, MOHAMMED AQUEEL, b Lucknow, India, Mar 9, 39; Can citizen; m 65; c 2. MEDICAL MICROBIOLOGY, PUBLIC HEALTH. *Educ:* Univ Karachi, BS, 57, MS, 59; Univ Mich, MPH, 63; Univ London, PhD(med microbiol), 69. *Prof Exp:* Lectr microbiol, D J Sci Col, Univ Karachi, 60-62; res asst clin microbiol, Med Ctr, Univ MIch, 63-65; bacteriologist, Holy Cross Hosp, Calgary, Can, 65-66, head, microbiol Div, Dept Labs & Dir, Infection Control, 70-88; HEAD MICROBIOL LAB, CALAGRY DIST HOSP GROUP, 88- *Concurrent Pos:* Res fel, Charing Cross Med Sch, London, Eng, 66-69; adj assoc prof microbiol, Fac Med, Univ, Univ Calgary, 70-; mem bd health, Calgary, 78-80. *Honors & Awards:* Fulbright Scholar, US, 62-65. *Mem:* Am Soc Clin Pathologists; Am Soc Microbiol; Brit Inst Biol; Can Asn Clin Microbiol & Infectious Dis. *Res:* Effects of polyene antibiotics on Candida species; development of resistance in microorganisms to antibiotics; detection and control of legionella species in hopital air conditioners and hot water systems. *Mailing Add:* Rockyview Gen Hosp Microbiol Lab 7007 14th St SW Calgary AB T2V 1P9 Can

ATHELSTAN, GARY THOMAS, b Minneapolis, Minn, Dec 26, 36; m 60; c 2. PHYSICAL MEDICINE & REHABILITATION. *Educ:* Univ Minn, BA, 60, PhD(psychol), 66. *Prof Exp:* Voc rehab counr, Minn Div Voc Rehab, 60-61; res psychologist, Am Rehab Found, 65-68; dir res, Occup Res Div, Inst Interdisciplinary Studies, 68-70; assoc prof, 70-75, PROF PHYS MED & REHAB & PSYCHOL, SCH MED, UNIV MINN, MINNEAPOLIS, 75- *Concurrent Pos:* Consult continuing educ, Am Acad Phys Med & Rehab, 69-70; sect ed, Arch Phys Med & Rehab, 74-81, assoc ed, 77-81; consult psychol, Vet Admin Hosp, Minneapolis, 73- *Honors & Awards:* Licht Award, Am Cong Rehab Med, 82. *Mem:* Am Psychol Asn. *Res:* Development of psychosocial outcome criteria in rehabilitation; psychosocial factors related to rehabilitation success; improved methods to select, train and utilize health professionals. *Mailing Add:* 719 Fifth Ave SE Minneapolis MN 55414

ATHENS, JOHN WILLIAM, b Buhl, Minn, Oct 2, 23; m 49; c 3. INTERNAL MEDICINE, HEMATOLOGY. *Educ:* Univ Mich, BA, 45; Johns Hopkins Univ, MD, 48; Am Bd Internal Med, dipl. *Prof Exp:* Intern med, Peter Bent Brigham Hosp, Boston, Mass, 48-49; asst res, 52-54, from res instr to asst res prof med, 55-62, assoc prof, 62-68, PROF MED & HEAD DIV HEMAT & ONCOL, COL MED, UNIV UTAH, 68- *Concurrent Pos:* AEC fel, Johns Hopkins Hosp, 49-50; Nat Cancer Inst res fel hemat, Col Med, Univ Utah, 54-56; dir outpatient dept, Salt Lake County Gen Hosp, 58-68. *Mem:* Am Fedn Clin Res; Am Col Physicians; Am Soc Clin Invest; Asn Am Physicians; Am Soc Hemat. *Res:* Leukocyte physiology; oncology. *Mailing Add:* Dept Med Hemat Univ Utah Sch Med 50 N Medical Dr Salt Lake City UT 84132

ATHERLY, ALAN G, b Kalamazoo, Mich, Nov 19, 36; m 59; c 2. GENETICS, MOLECULAR BIOLOGY. *Educ:* Western Mich Univ, BS, 60; Univ NC, PhD(biochem), 65. *Prof Exp:* Fel, Dept Biol, Western Reserve Univ, 65-66 & Inst Molecular Biol, Univ Ore, 66-68; from asst prof to assoc prof, 68-70, chmn, 80-90, PROF GENETICS, IOWA STATE UNIV, 76- *Concurrent Pos:* Consult, Eni Chem Am, Princeton, NJ, 86; consult, Pioneer Hibred Int, Johnston, IA, 87; ed, Genetica. *Mem:* Plant Molecular Biol Soc; Am Soc Microbiol. *Res:* Nitrogenase regulation; nodulation of soybeans by rhizobia; phytoph thoria resistance in soybeans. *Mailing Add:* Dept Zool & Genetics Iowa State Univ Ames IA 50011

ATHERTON, BLAIR T, biochemistry, for more information see previous edition

ATHERTON, HENRY VERNON, dairy manufacturing, for more information see previous edition

ATHERTON, ROBERT W, b Auburn, Ala, Dec 20, 46; m 68; c 1. CHEMICAL ENGINEERING. *Educ:* Rice Univ, BA, 69, MChE, 70; Stanford Univ, ChE, 72, PhD, 74. *Prof Exp:* PRES, IN-MOTION TECHNOL, 85- *Honors & Awards:* Donald P Eckman Award, Am Automatic Control Coun, 76. *Mem:* Am Inst Chem Engrs; Soc Indust & Appl Math; Sigma Xi. *Res:* Automatic control of manufacturing systems. *Mailing Add:* 1694 Miller Ave Los Altos CA 94022

ATHEY, ROBERT DOUGLAS, JR, b Washington, DC, Aug 27, 36; m 58, 81; c 3. COLLOID SCIENCE. *Educ:* Univ Md, BS, 64; Univ Del, PhD(org chem), 74. *Prof Exp:* Res chemist, Int Latex & Chem Corp, 64-67; sr res chemist, Scott Paper Co, 67-71 & Gen Tire & Rubber Co, 73-77; mgr proj res, Mellon Inst, 77-80; res assoc, Swedlow, Inc, 80-82; Athey Technol, El Cerrito, Calif, 82-91; CONSULT, 91- *Concurrent Pos:* Instr, Calif State Univ, Fullerton, 81-82, Univ Calif, Berkeley, 83-86 & Calif Polytech State Univ, 91. *Mem:* Am Chem Soc; Sigma Xi; Fedn Soc Coating Technol; Tech Asn Pulp & Paper Indust; Computer Press Asn. *Res:* Tailor-making polymers and formulating through use of organic and colloid chemistry, to assure the polymers utility as coating, adhesives or structural components. *Mailing Add:* Athey Technol Drawer 7 El Cerrito CA 94530-0007

ATHEY, ROBERT JACKSON, b Washington, DC, Mar 7, 25; m 49; c 3. ORGANIC CHEMISTRY. *Educ:* Ga Inst Technol, BS, 48, MS, 50; Univ Wis, PhD(org chem), 54. *Prof Exp:* res chemist, E I du Pont de Nemours & Co, 43-86; RETIRED. *Mem:* Am Chem Soc. *Res:* Spectral investigation of nitryl chloride; preparation of morphine analogs; rubber chemistry; isocyanate chemistry; urethane polymers; fluoro elastomers. *Mailing Add:* 1107 Crestover Rd Graylyn Crest Wilmington DE 19803-3308

ATHOW, KIRK LELAND, b Tacoma, Wash, Jan 22, 20; m 51. PLANT PATHOLOGY. *Educ:* State Col Wash, BSA, 46; Purdue Univ, MS, 48, PhD(plant path), 51. *Prof Exp:* Instr plant path, Purdue Univ, West Lafayette, 49-51, asst plant pathologist, 51-56, from assoc prof bot & plant path, 56-90; RETIRED. *Concurrent Pos:* Mem team, Purdue-Brazilian Proj, 66-68; USAID consult, Brazil, 70, 71, 72 & 73. *Mem:* Am Phytopath Soc. *Res:* Forage crop and soybean diseases; seed treatment; disease resistance. *Mailing Add:* 1216 Farallone Ave Tacoma WA 98466

ATHREYA, KRISHNA BALASUNDARAM, b Madras, India, Dec 12, 39; c 3. PROBABILITY & STOCHASTIC PROCESSES, MATHEMATICAL STATISTICS. *Educ:* Madras Univ, India, BA, 59; Stanford Univ, PhD(math), 67. *Prof Exp:* From asst prof to assoc prof math, Univ Wis-Madison, 68-71; prof math, Indian Inst Sci, Bangalore, 71-79; PROF MATH, IOWA STATE UNIV, 80- *Concurrent Pos:* Assoc ed, Zetschrift fur Wahrscheinlihkeits Theorie, 78- & Letters Probability & Statist, 81-; vis prof, Univ Wis-Madison & Milwaukee, Univ Copenhagen & Australian Nat Univ. *Mem:* Fel Inst Math; fel Indian Acad Sci; fel Inst Math Statist. *Res:* Limit theorems for markov chains, mathematical statistics, branching processes, integral equations, stochastic modelling in applied areas. *Mailing Add:* Dept Math Iowa State Univ 400 Carver Ames IA 50011

ATIA, ALI EZZ ELDIN, b Cairo, Egypt, Aug 10, 41; US citizen; m 64; c 3. MICROWAVE ENGINEERING. *Educ:* Ain Shams Univ, Cairo, MSc, 62; Univ Calif, Berkeley, PhD, 69. *Prof Exp:* Asst prof, Dept Elec Eng & Comput Sci, Univ Calif, Berkeley, 68-69; mem tech staff, COMSAT Labs, 69-76, sr scientist, 76-80, sr dir, Satelite Syst Design, COMSAT Corp, 80-88, VPRES, SYSTS ENG, COMSAT SYSTS DIV, 88- *Mem:* Fel Inst Elec & Electronics Engrs; Sigma Xi; Am Inst Aeronaut & Astronaut. *Res:* Microwave circuits; satellite communications systems; microwave filters and multiplexers; computer aided design. *Mailing Add:* COMSAT 22300 Comsat Dr Clarksburg MD 20871

ATINMO, TOLA, b Lagos, Nigeria, Jan 29, 45; m 71; c 4. NURSING. *Educ:* Unis Ibadan, Nigeria, BS, 70; Cornell Univ, MNS, 73, PhD(nutrit), 75. *Prof Exp:* PROF NUTRIT, UNIV IBADAN, 75- *Concurrent Pos:* Consult, UNICEF, Nigeria, 89-; chmn, Nat Comt Food & Nutrit, 90- *Mem:* Am Inst Nutrit; fel Nutrit Soc Nigeria. *Res:* Food and nutrition policy; community and public health nutrition. *Mailing Add:* Dept Human Nutrit Univ Ibadan Ibadan Nigeria

ATKIN, J MYRON, b New York, NY, Apr 6, 27; m 47; c 3. SCIENCE EDUCATION, SCIENCE POLICY. *Educ:* Col City New York, BS, 47; NY Univ, MA, 48, PhD(sci educ), 56. *Prof Exp:* Teacher biol chem, Ramaz Sch, 48-50; teacher sci, Pub Schs, Great Neck, NY, 50-55; from asst prof to prof sci educ, Univ Ill, 56-79, assoc dean res, 66-70, dean educ, 70-79; dean, 79-86, PROF EDUC, STANFORD UNIV, 79- *Concurrent Pos:* Mem, teacher cert bd, State Ill, 73-76; mem, panel on the high sch, Carnegie Found Advan Teaching, 80-83; vchmn adv comt, directorate sci & eng educ, NSF, 85-86; bd math sci educ, Nat Acad Sci, 85-89. *Mem:* Am Educ Res Asn; Coun Elem Sci Int (pres, 69-70); AAAS; Nat Asn Res Sci Teaching. *Res:* Improvement of science education practice and policy. *Mailing Add:* Sch Educ Stanford Univ Stanford CA 94305-3096

ATKIN, RUPERT LLOYD, b Madison, Ohio, June 7, 18; m 41; c 4. MECHANICAL ENGINEERING. *Educ:* Ohio State Univ, BME, 41. *Hon Degrees:* DSc, Ohio State Univ, 79. *Prof Exp:* With Kelsey-Hayes Co, Mich, 46-64, chief engr, 58-64; mem staff, TRW Inc, 64-74, dir eng chassis components, 74-76, vpres eng chassis components, 76-84; DIR, INT TECH GROUP INC, 84-; DIR, COGSDILL TOOL PROD INC, 84- *Concurrent Pos:* Mem, res adv bd, Col Eng, Univ Detroit, 60-65; mem, indust comt, Col Eng, Univ Mich, 65-83, chmn, 75-77; mem, indust adv comt, Mech Eng Dept, Col Eng, Ohio State Univ, 76-83, mem, steering comt, 79-83. *Honors & Awards:* Engr of Distinction, Engrs Joint Coun; Benjamin G Lamme Medal, Col Eng, Ohio State Univ, 76. *Mem:* Nat Acad Eng; fel Soc Automotive Engrs. *Mailing Add:* 1715 N River Rd Unit 50 St Clair MI 48079

ATKINS, CHARLES GILMORE, b Stambaugh, Mich, July 4, 39; div; c 3. MICROBIAL GENETICS. *Educ:* Albion Col, BA, 61; Eastern Mich Univ, MS, 63; NC State Univ, PhD(genetics), 69. *Prof Exp:* Teaching asst biol, Eastern Mich Univ, 62-63; instr biol, Coe Col, 63-66; from asst prof to assoc prof genetics & microbiol, Ohio Univ, 69-77, assoc prof zool & microbiol, 74-90, assoc dean, Col Osteop Med, 76-90, ASSOC PROF BIOMED SCI, OHIO UNIV, 90- *Concurrent Pos:* Lectr, Cornell Col, 64-65; dir training prog, Appalachian Life Sci Col, 72-74. *Mem:* AAAS; Genetics Soc Am; Am Soc Human Genetics; Sigma Xi. *Res:* Transductional studies and biochemical analysis of Salmonella typhimurium, Salmonella montevideo and hybrids of these species, with emphasis on the isoleucine-valine biosynthetic pathway; recombination in T-even bacteriophage; medical education, preclinical. *Mailing Add:* 356 Irvine Hall Ohio Univ Athens OH 45701

ATKINS, DAVID LYNN, b Wichita Falls, Tex, July 12, 35; m 59. VERTEBRATE ANATOMY, NEUROANATOMY. *Educ:* Univ Tex, Austin, BA, 57; ETex State Univ, MA, 63; Tex A&M Univ, PhD(comp neuroanat), 70. *Prof Exp:* Teacher high schs, Tex, 57-59 & 61-62; asst prof zool, Tarleton State Col, 65-67; asst prof, 70-75, assoc prof zool, 75-81, PROF BIOL, GEORGE WASHINGTON UNIV, 81- *Mem:* Am Soc Zool; Sigma Xi. *Res:* Comparative neuroanatomy; systematics; comparative ontogeny and evolution of the mammalian cerebellum, particularly related to the phylogeny of the major taxa. *Mailing Add:* 2608 Sigmona St Falls Church VA 22046

ATKINS, DON CARLOS, JR, b Denver, Colo, Apr 11, 21; m 47; c 2. INDUSTRIAL CHEMISTRY. *Educ:* Univ Calif, Los Angeles, BA, 43, MS, 48; Chapman Col, DSc, 76. *Prof Exp:* Tech dir, US Chem Milling Corp, 57-62; vpres, Chem & Aerospace Prod, Inc, 62-64; vpres, Wilco Prod, Inc, 64-68; exec vpres, Sanitek Prod, Inc, 68-73; lab dir non-foods chem specialties, Hunt-Wesson Foods, 73-75; PRES, D C ATKINS & SON, 75-, PRES, CONTAINMENT CORP, 87-, VPRES, RESIN SUPPORT SYSTS, 85- *Concurrent Pos:* Staff consult, Solder Removal Co, 70-, Inst Tricology, Inc, 72-, Chemrite Corp, 75- & Clorox Co, 75- *Honors & Awards:* Honor Scroll Award, Am Inst Chemists, 74. *Mem:* Am Inst Chemists; Am Chem Soc; Am Soc Metals; Am Electroplater Soc; Sigma Xi; Am Oil Chemists Soc; Soc Cosmetic Chemists. *Res:* Interaction of chemicals to surfaces with particular emphasis on chemical specialty compounds for consumer and industrial applications; development of cleaning materials, coatings and cosmetics; special expertise in chemical milling and plating of metals onto plastics. *Mailing Add:* 11282 Foster Rd Los Alamitos CA 90720

ATKINS, ELISHA, b Belmont, Mass, Nov 16, 20; m 44; c 5. INTERNAL MEDICINE. *Educ:* Harvard Univ, AB, 42; Univ Rochester, MD, 50. *Prof Exp:* From intern to asst resident, Barnes Hosp, 50-52; instr med, Wash Univ, 54-55; from asst prof to prof, 55-85, EMER PROF MED, YALE UNIV, 85- *Concurrent Pos:* Res fel infectious dis, Wash Univ, 52-54; USPHS res grant, 57-; attend physician, Yale-New Haven Hosp, Conn, 55-; consult, Vet Admin Hosp, West Haven, Conn, 55-; sabbatical, Radcliffe Infirmary, Eng, 62-63; Scripps Clin & Res Found, Calif, 69-70 & Dept Immunol, Yale Univ, 78-79; master, Saybrook Col, Yale Univ, 75-85; exec dir, Habitat Inst Environ, Belmont, Mass, 85- *Honors & Awards:* Bristol Award, Infectious Dis Soc Am, 85. *Mem:* Am Soc Clin Invest; Asn Am Physicians; Infectious Dis Soc Am; Sigma Xi. *Res:* Pathogenesis of fever, especially experimental fevers induced by microbial agents and antigens in specifically sensitized animals. *Mailing Add:* 44 Juniper Rd Belmont MA 01278

ATKINS, FERREL, b West York, Ill, Feb 15, 24; m 55; c 2. MATHEMATICS. *Educ:* Eastern Ill State Col, BS, 45; Univ Ill, MS, 46; Univ Ky, PhD(math), 50. *Prof Exp:* Asst math, Univ Ill, 45-46; instr math, Univ Ky, 46-50; asst prof, Bowling Green State Univ, 50-52; from asst prof to assoc prof math, Univ Richmond, 52-58, chmn dept, 57-58; from asst prof to assoc prof, 58-63, PROF MATH, EASTERN ILL UNIV, 63- *Concurrent Pos:* NSF fac fel, Stanford Univ, 64-65. *Mem:* Math Asn Am; Asn Comput Mach. *Res:* Computer sciences; analysis. *Mailing Add:* Dept Math Eastern Ill Univ Charleston IL 61920

ATKINS, GEORGE T(YNG), b Dallas, Tex, Sept 25, 06; m 29; c 4. CHEMICAL ENGINEERING. *Educ:* Wash Univ, BS, 29. *Prof Exp:* Analytical chemist bur chem & soils, USDA, Tex, 29-30; chem engr tech serv, Humble Oil & Ref Co, 30-55, eng assoc, 56-59; consult chem engr, 59-70; partner, Atkins & McBride, Chem Engrs, 70-72, partner, Atkins, McBride & Owen, Chem Engrs, 72-82; RETIRED. *Mem:* AAAS. *Res:* Process design for oil refinery and petrochemical units; computer programs for physical properties in systems undergoing distillation. *Mailing Add:* 1704 California St Baytown TX 77520

ATKINS, HAROLD LEWIS, b Newark, NJ, Sept 24, 26; m 63; c 4. NUCLEAR MEDICINE, RADIOLOGY. *Educ:* Yale Univ, BS, 48; Harvard Med Sch, MD, 52. *Prof Exp:* Intern med, Grace-New Haven Hosp, 52-53; resident radiol, Hosp Univ Pa, 53-56; instr radiol, Sch Med, Yale Univ, 56-59; from instr to asst prof, Col Physicians & Surgeons, Columbia Univ, 59-63; from assoc scientist to scientist, 63-72, sr scientist, 72-79, RES COLLABR RADIOL NUCLEAR MED, BROOKHAVEN NAT LAB, 79- *Concurrent Pos:* Am Cancer Soc fel, Hosp Univ Pa, 55-56 & Yale-New Haven Med Ctr, 56-57; from adj assoc prof to prof radiol & nuclear med, State Univ NY Stony Brook, 68-; consult, USPHS Hosp, Staten Island, NY, 61-63, Vet Admin Hosp, Northport, 71-, St Charles Hosp, Port Jefferson, 71-, Southside Hosp, Bayshore, 72- & Nassau County Med Ctr, East Meadow, 72- *Mem:* AAAS; Soc Nuclear Med (trustee, 72-76, 78-82, treas, 88-91); Radiol Soc NAm; fel Am Col Radiol. *Res:* Application of radiotracers to problems of renal transplants, gallbladder physiology and lungs. *Mailing Add:* Dept Radiol, Health Sci Ctr State Univ New York Stony Brook NY 11794-8460

ATKINS, HENRY PEARCE, b Birmingham, Ala, Jan 12, 15; m 41; c 2. MATHEMATICS. *Educ:* Cornell Univ, AB, 36; Brown Univ, MSc, 37; Univ Rochester, PhD(math), 47. *Prof Exp:* Instr, Brown Univ, 38-39; asst, Univ Rochester, 39-42, from instr to assoc prof, 42-58, dean of men, 54-58; PROF MATH, UNIV RICHMOND, 58- *Res:* Fractional derivatives of univalent functions and bounded functions. *Mailing Add:* 6705 Lakewood Dr Richmond VA 23229

ATKINS, JAMES LAWRENCE, b York, Pa, Oct 10, 47; m; c 2. NEPHROLOGY. *Educ:* Univ Notre Dame, AB, 69; Univ Md, MD, 75, PhD, 78; Am Bd Internal Med, cert, 78. *Prof Exp:* Med intern, Univ Md Hosp, 75-76, med resident, 76-78, renal fel, 78-79; res fel, Lab Kidney & Electrolytes, NIH, 79-81, sr staff fel, 81-82; ASST PROF MED, F EDWARD HEBERT SCH MED, UNIFORMED SERV UNIV HEALTH SCI, 83- *Concurrent Pos:* Nat res serv award, Nat Inst Arthritis, Metab & Digestive Dis, NIH, 79-81; mem, Libr Comt, Walter Reed Army Inst, 87-; adj asst prof physiol, Sch Med, Univ Md, 90- *Honors & Awards:* Leonard M Hummel Award for Excellence in Internal Med, 75. *Mem:* Fel Am Col Physicians; Am Fedn Clin Res; AAAS; Am Soc Nephrol; Int Soc Nephrol; Am Physiol Soc. *Mailing Add:* Dept Nephrol Walter Reed Army Inst Res Bldg 40 Rm 3035 Washington DC 20307-5100

ATKINS, JASPARD HARVEY, b Boston, Mass, May 9, 26; m 54; c 4. PHYSICAL CHEMISTRY, ANALYTICAL CHEMISTRY. *Educ:* Univ Mass, BS, 50; Rensselaer Polytech Inst, PhD(chem), 55. *Prof Exp:* Anal chemist, US Food & Drug Admin, 50-51; AEC res assoc, Rensselaer Polytech Inst, 53-56; sr scientist, Westinghouse Atomic Power Div, Bettis Atomic Power Lab, 56-57; Portland Cement Asn res fel, Nat Bur Standards, 57-59; proj mgr, Nat Res Corp, 59-61; group leader, Cabot Corp, 61-70, sr res assoc, 70-87; CONSULT, 87- *Mem:* Am Chem Soc; Sigma Xi. *Res:* Physical chemistry measurement; gas-solid interfaces; vacuum techniques; gaseous kinetics; instrumentation. *Mailing Add:* 1536 High St Westwood MA 02090

ATKINS, JOHN MARSHALL, b Charleston, WVa, Mar 2, 41; m 68; c 1. MATHEMATICS. *Educ:* Marshall Univ, BS, 65; WVa Univ, MA, 67, MS, 80; Univ Pittsburgh, PhD(math), 72. *Prof Exp:* Sr teaching fel math, Univ Pittsburgh, 68-73; asst prof math, Bethany Col, 73-80; ASST PROF, WVA UNIV, 80- *Mem:* Math Asn Am; Asn Comput Mach; Sigma Xi. *Mailing Add:* WVa Univ 305 Knapp Hall Morgantown WV 26506

ATKINS, MARVIN C(LEVELAND), b Ballinger, Tex, July 30, 31; m 55; c 2. PHYSICS, NUCLEAR ENGINEERING. *Educ:* Agr & Mech Col, Tex, BS, 52; Univ Ill, MS, 53; Univ Mich, PhD(nuclear sci), 61. *Prof Exp:* Proj engr, Mat Lab, US Air Force, 53-57; group leader, Spec Weapons Ctr, 60-62, chief theoret sect, 62-63; sr scientist, Weapons Lab, 63-64; chief energy effects sect, Res & Develop Div, Avco Corp, Mass, 64-66; asst mgr mat sci dept, Space Systs Div, 66-67, mgr eng tech dept, 67-68; dir environ sci, 68-70; dir, offensive & space systs, Off of Undersecretary of Defense for Res & Eng, 78-83; asst to dep dir sci & technol, Defense Nuclear Agency, Washington, DC, 70-78, dep dir, 83-89; SR VPRES QUAL REV, SCI APPLN INT CORP, MCLEAN, VA, 89- *Mem:* Am Nuclear Soc; Am Inst Aeronaut & Astronaut. *Res:* Shock hydrodynamics and effects; advanced materials; missile system engineering; radiation application and effects; research management. *Mailing Add:* 400 Madison St Apt 1308 Alexandria VA 22314

ATKINS, PATRICK RILEY, b Russellville, Ky, May 20, 42; m 65; c 2. ENVIRONMENTAL HEALTH ENGINEERING. *Educ:* Univ Ky, BS, 64; Stanford Univ, MS, 65, PhD(sanit eng), 68. *Prof Exp:* Asst prof civil eng, Univ Tex, Austin, 68-77; MGR, ALUMINUM CO AM, 77- *Mem:* Air Pollution Control Asn; Am Soc Civil Engrs. *Res:* Air pollutant transport and dispersion; aerosol dynamics; nucleation; rainout and washout processes; air pollution control. *Mailing Add:* 1501 Alcoa Bldg 1501 Alcoa Bldg Pittsburgh PA 15219

ATKINS, PAUL C, b Yonkers, NY, Sept 27, 42; m 63; c 2. ALLERGY, IMMUNOLOGY. *Educ:* Syracuse Univ, BS, 63; NY Med Col, MD, 67; Am Bd Internal Med, cert, 74; Am Bd Allergy & Immunol, cert, 75. *Prof Exp:* Intern, Beth Israel Med Ctr, 67-68; fel allergy & immunol, Univ Pa Hosp, 70-72; chief resident med, Metrop Hosp, NY Med Col, 68-70; allergist & internist med, Madigan Army Med Ctr, 72-74; asst prof med, 74-80, ASSOC PROF MED, ALLERGY & IMMUNOL, SCH MED, UNIV PA, 80- *Concurrent Pos:* Chief, Allergy & Immunology Clin, Univ Pa Hosp, 74-; consult, Vet Admin Hosp, Philadelphia, 74- *Mem:* Am Acad Allergy; Am Fedn Clin Res; Am Thoracic Soc; Am Col Physicians. *Res:* Allergy and immunology; cellular inflammatory response and mediators released during immediate hypersensitivity reactions in man. *Mailing Add:* Dept Allergy & Immunol Univ Pa Sch Med 510 Johnson Pavilion 36th & Hamilton Walk Philadelphia PA 19104

ATKINS, RICHARD ELTON, b Corning, Kans, Feb 10, 19; m 59; c 3. AGRONOMY. *Educ:* Kans State Univ, BS, 41; Iowa State Univ, MS, 42, PhD(crop breeding), 48. *Prof Exp:* Res asst prof, 48-50, assoc prof, 50-60, PROF AGRON, IOWA STATE UNIV, 60- *Mem:* Am Soc Agron; Crop Sci Soc Am. *Res:* Breeding of improved varieties of small grains and grain sorghum for agronomic and disease characters and allied genetic and production studies; crop breeding. *Mailing Add:* Dept Agron 120 Iowa State Univ Ames IA 50011

ATKINS, ROBERT CHARLES, b Norwood, Mass, Aug 19, 44; m 67; c 2. ORGANIC CHEMISTRY. *Educ:* Mass Inst Technol, SB, 66; Univ Wis-Madison, PhD(org chem), 70. *Prof Exp:* Assoc org chem, Columbia Univ, 70-71; asst prof chem, 71-75, ASSOC PROF CHEM, James Madison Univ, 75- *Concurrent Pos:* NIH fel, 70-71. *Mem:* Am Chem Soc. *Res:* Pyrolytic reactions; reactions involving ylides; synthesis of carcinogen metabolites. *Mailing Add:* Dept Chem JAMES MADISON UNIV Harrisonburg VA 22807-0002

ATKINS, ROBERT W, b Belmont, Mass, Nov 30, 17; m 40; c 5. PSYCHIATRY, PSYCHOANALYSIS. *Educ:* Yale Univ, BA, 40; Harvard Univ, MD, 43. *Prof Exp:* Dir, Community Ment Health Ctr Div, Dept Psychiat, Sch Med & Dent, Univ Rochester & Strong Mem Hosp, 68-72, prof psychiat, 70-80; spec consult, Vet Admin Hosp, Canandaigua, NY, 80-86; RETIRED. *Mem:* Fel Am Psychiat Asn. *Res:* Mental illness and unemployment; psychological effects of induced abortion and surgical sterilization in men and women; cerebral metabolic disturbances in hypothyroidism; conceptual thinking in psychiatric patients. *Mailing Add:* 1434 East Ave Rochester NY 14610

ATKINS, RONALD LEROY, b Martinez, Calif, Mar 27, 39; m 64; c 2. ORGANIC CHEMISTRY. *Educ:* Univ Wyo, BS, 66, MS, 68; Univ NH, PhD(org chem), 71. *Prof Exp:* Nat Res Coun res assoc, 71-73, RES CHEMIST, NAVAL WEAPONS CTR, DEPT NAVY, 73- *Mem:* Am Chem Soc; Sigma Xi. *Res:* Synthesis and photochemistry of small ring heterocyclic compounds; synthesis of polynitroaromatic and heterocyclic compounds; synthesis and study of the photostability of dye molecules. *Mailing Add:* 817 Wildflower Ridgecrest CA 93555

ATKINS, WILLIAM M, BIOCHEMISTRY. *Educ:* Col William & Mary, BS, 80; Harvard Univ, AM, 82; Univ Ill, Urbana, PhD(biochem), 88. *Prof Exp:* NIH postdoctoral fel chem, Pa State Univ, 88-91; ASST PROF, DEPT MED CHEM, UNIV WASH, SEATTLE, 91- *Mem:* Fedn Am Socs Exp Biol. *Res:* Biophysical and biotechnical approaches to understand structure-function relationships in the mechanism and regulation of biological macromolecules; author of 20 technical publications. *Mailing Add:* Dept Med Chem Univ Wash BG-20 Seattle WA 98195

ATKINSON, ARTHUR JOHN, JR, b Chicago, Ill, Mar 21, 38; m 50. CLINICAL PHARMACOLOGY. *Educ:* Harvard Univ, AB, 59; Cornell Univ, MD, 63. *Prof Exp:* Intern, internal med, Mass Gen Hosp, 63-64, resident, 64-65; clin assoc, Nat Inst Allergy & Infectious Dis, NIH, 65-67; chief resident internal med, Passavant Mem Hosp, 67-68; fel clin pharmacol, Univ Cincinnati, 68-69, asst prof, 69; vis scientist, Karolinska Inst, 70; from asst prof to assoc prof, 70-76, PROF PHARMACOL & MED, MED SCH, NORTHWESTERN UNIV, 76- *Concurrent Pos:* Mem pharmacol toxicol prog comt, Nat Inst Gen Med Sci, NIH, 75-79, chmn, pharmacol sci prog comt, 84-86. *Honors & Awards:* Rawls Palmer Award, Am Soc Clin Pharmacol & Therapeut, 83; Herry Gold Award, Am Soc Pharmacol & Exp Therapeut, 89. *Mem:* Fel Am Col Physicians; Am Soc Pharmacol & Exp Therapeut; Am Fedn Clin Res; Am Soc Clin Investr; Am Soc Clin Pharmacol & Therapeut. *Res:* Clinical pharmacokinetics; drug metabolism; drug development. *Mailing Add:* Dept Pharmacol & Med Med Sch Northwestern Univ Chicago IL 60611

ATKINSON, BARBARA, CYTOPATHOLOGY, FLOW CYTOMETRY. *Educ:* Jefferson Univ, MD, 74. *Prof Exp:* DIR CYTOPATH, UNIV PA, 78-, ASSOC PROF PATH, 85- *Mailing Add:* Dept Path & Lab Med Med Col Pa 3300 Henry Ave Philadelphia PA 19129

ATKINSON, BURR GERVAIS, b Elizabeth, NJ, Sept 17, 37; div; c 2. BIOCHEMISTRY, DEVELOPMENTAL BIOLOGY. *Educ:* Ohio Univ, BA, 61; Univ Conn, PhD(biochem, biophys), 68. *Prof Exp:* NIH fel, Fla State Univ, 68-71, res assoc develop biochem, 71-72; from asst prof to assoc prof, 72-80, PROF ZOOL, UNIV WESTERN ONT, 80-, CHMN GENETICS, 81- *Concurrent Pos:* Lab instr, Marine Biol Labs, Woods Hole, 66-68, investr, 69; bd dir, Can Fedn Biol Sci. *Mem:* Sigma Xi; Am Soc Biol Chemists; Am Soc Cell Biol; Can Soc Cell Biol (secy, 85-); Can Fedn Biol Sci; Can Soc Genetics. *Res:* Nucleic acid biochemistry; hormone-induced synthesis of nucleic acids; tissue-specific proteins in development; tissue culture; proteins and nucleic acids of genetic machinery; molecular biology; gene expression. *Mailing Add:* Dept Zool Univ Western Ont London ON N6A 5B7 Can

ATKINSON, DANIEL EDWARD, b Pawnee City, Nebr, Apr 8, 21; m 48; c 5. BIOCHEMISTRY. *Educ:* Univ Nebr, BS, 42; Iowa State Col, PhD(biochem), 49. *Hon Degrees:* DSc, Univ Nebr, 75. *Prof Exp:* Res fel, Calif Inst Technol, 49-50; assoc plant physiologist, Argonne Nat Lab, 50-52; from asst prof to assoc prof chem, 52-80, PROF BIOCHEM, UNIV CALIF, LOS ANGELES, 80- *Mem:* Am Chem Soc; Am Soc Plant Physiologists; Am Soc Microbiol; Am Soc Biol Chemists. *Res:* Metabolic regulation and correlation; properties of regulatory enzymes; energy metabolism; pH homeostasis. *Mailing Add:* 3123 Malcolm Ave Los Angeles CA 90034

ATKINSON, EDWARD NEELY, b Shreveport, La, Apr 29, 53; m 81. BIOMATHEMATICS. *Educ:* Rice Univ, BA, 75, MA & PhD(math sci), 81. *Prof Exp:* Res assoc, 81-84, ASST PROF, DEPT BIOMATH, M D ANDERSON HOSP, UNIV TEX, HOUSTON, 84- *Mem:* Soc Indust & Appl Math; Am Math Soc; Am Statist Asn; Inst Math Statist; Biomet Soc. *Res:* Minimization of function whose evaluations are subject to stochastic error; mathematical modelling of neoplastic progression. *Mailing Add:* Dept Biomath Tex Med Ctr M D Anderson Hosp Houston TX 77030

ATKINSON, EDWARD REDMOND, b Boston, Mass, Feb 15, 12; m 44; c 2. ORGANIC CHEMISTRY. *Educ:* Mass Inst Technol, BS, 33, PhD(org chem), 36. *Prof Exp:* Teaching fel chem, Mass Inst Technol, 33-36; instr org chem, Trinity Col, Conn, 36-38; from asst prof to assoc prof chem, Univ NH, 38-51; group leader, Dewey & Almy Chem Co, 51-57; sr chemist, Arthur D Little, Inc, 57-77; INDUST CONSULT, 77- *Concurrent Pos:* Lectr, Northeastern Univ, 51-63 & Boston Univ, 64-65. *Honors & Awards:* Henry A Hill Memorial Award, Am Chem Soc, 81. *Mem:* Am Chem Soc; AAAS. *Res:* Diazonium compounds; history of chemistry; condensation polymers; antiradiation drugs; pharmaceuticals; biphenylene; chemical hazard evaluation. *Mailing Add:* 163 Gray St Amherst MA 01002

ATKINSON, GEORGE FRANCIS, b Toronto, Ont, Feb 25, 32; m 61; c 4. INSTRUMENTATION SPECTROELECTROCHEMISTRY. *Educ:* Univ Toronto, BA, 53, MA, 54, PhD(chem), 60; FRSC(UK), 72; FCIC, 78. *Prof Exp:* Lectr chem, Univ Western Ont, 60-61; from lectr to asst prof, 61-66, ASSOC PROF CHEM, UNIV WATERLOO, 66- *Concurrent Pos:* Mem sci study comt, Ont Curric Inst, 64-68; vis prof, Univ Southampton, 69-70 & 75-76, Univ E Anglia, 88-89. *Mem:* Am Chem Soc; Brit Soc Res Higher Educ; Chem Inst Can; Royal Soc Chem. *Res:* Applications of computers in analysis; instrumentation for electroanalytical and spectrophotometric methods; electrochemistry; continuous analysis and control of process streams; chemical education. *Mailing Add:* Dept Chem Univ Waterloo Waterloo ON N2L 3G1 Can

ATKINSON, GORDON, b Brooklyn, NY, Aug 29, 30; m 76; c 4. PHYSICAL CHEMISTRY, INORGANIC CHEMISTRY. *Educ:* Lehigh Univ, BS, 52; Iowa State Col, PhD(chem), 56. *Prof Exp:* Res assoc, Ames Lab, US AEC, 56-57; instr chem, Univ Mich, 57-61; from asst prof to prof, Univ Md, College Park, 61-71; chmn dept, 71-74, grad dean, 74-78, PROF CHEM, UNIV OKLA, NORMAN, 71- *Concurrent Pos:* Fulbright Fel, 70-71; Regent's award res, 82. *Mem:* AAAS; Am Chem Soc; fel Am Inst Chem; fel NY Acad Sci. *Res:* Electrolyte solutions; conductance; ultrasonic absorption in solution; kinetics and thermodynamics of ion association; structure of aqueous solutions; chemical oceanography; oil field brines. *Mailing Add:* Dept Chem Univ Okla 620 Parrington Oval Norman OK 73019

ATKINSON, HAROLD RUSSELL, b Hamilton, Ont, Sept 25, 37; m 62; c 5. MATHEMATICAL ANALYSIS. *Educ:* Univ Western Ont, BA, 60; Assumption Univ, MSc, 61; Queen's Univ, Ont, PhD(math), 64. *Prof Exp:* Asst prof, 64-70, ASSOC PROF MATH, UNIV WINDSOR, 70- *Mem:* Can Math Soc; Am Math Soc. *Res:* Ordered Banach spaces; spaces of affine continuous functions on compact convex sets; measure theory. *Mailing Add:* Dept Math Univ Windsor Windsor ON N9B 3P4 Can

ATKINSON, JAMES BYRON, ANATOMIC PATHOLOGY, SURGICAL PATHOLOGY. *Educ:* Vanderbilt Univ, MD & PhD(path), 81. *Prof Exp:* ASST PROF GEN PATH, VANDERBILT UNIV, 85- *Mailing Add:* 4032 Albert Dr Nashville TN 37204-4010

ATKINSON, JAMES T N, metallurgical engineering; deceased, see previous edition for last biography

ATKINSON, JAMES WILLIAM, b Pittsburgh, Pa, Mar 15, 42; m 68; c 2. MOLLUSCAN REPRODUCTION & DEVELOPMENT. *Educ:* Kenyon Col, AB, 64; Emory Univ, MS, 66, PhD(biol), 69. *Prof Exp:* From instr to prof natural sci, 68-89, PROF ZOOL, MICH STATE UNIV, 89- *Concurrent Pos:* Adj prof zool, Mich State Univ, 83-89. *Mem:* Am Soc Zoologists; AAAS; Am Malacological Union. *Res:* Analysis of spiralian development with emphasis on the coordination of developmental events in space and time; evolutionary implications of developmental biology; history and philosophy of science; development and evolution of the nervous system in molluscs; history of biology. *Mailing Add:* Dept Zool Mich State Univ East Lansing MI 48824

ATKINSON, JOHN BRIAN, b Louth, Eng, Aug 15, 42; m 70; c 1. MOLECULAR SPECTROSCOPY. *Educ:* Oxford Univ, BA, 63, DPhil(physics), 67. *Prof Exp:* Fel physics, Univ Windsor, 67-69, vis asst prof, 69-72; res assoc, Wayne State Univ, 69-72; from asst prof to assoc prof, 72-87, PROF PHYSICS, UNIV WINDSOR, 87- *Mem:* Can Asn Physicists; Am Phys Soc. *Res:* Study of atomic collision cross-sections and atomic lifetimes using tunable dye lasers; applications of atomic properties to development of tunable lasers; molecular spectroscopy. *Mailing Add:* Dept Physics Univ Windsor Windsor ON N9B 3P4 Can

ATKINSON, JOHN PATTERSON, b Clay Center, Kans, July 4, 43; m 64; c 3. RHEUMATOLOGY, IMMUNOLOGY. *Educ:* Kans Univ, AB, 65, MD, 69. *Prof Exp:* Intern med, Mass Gen Hosp, 69-70, asst resident, 70-71; clin assoc, Lab Clin Ivest, Nat Inst Allergy & Infectious Dis, 71-72, chief clin assoc allergy, 72-73, staff fel, 73-74; NIH res fel, 74-76, asst prof med, 76-78, HEAD DIV RHEUMATOL, SCH MED, WASH UNIV, 76-, ASSOC PROF MED, 78- *Mem:* Am Fedn Clin Res; Am Acad Allergy; Am Asn Immunologists. *Res:* Complement-mediated immune damage; autoimmune hemolytic anemia; hereditary angioedema; receptors on immunocompetent cells; cyclic nucleotides and the immune response. *Mailing Add:* Dept Rheumatol Wash Univ Sch Med 660 S Euclid Ave Box 8045 4566 Scott St Louis MO 63110

ATKINSON, JOSEPH GEORGE, b St Paul, Alta, Mar 9, 35. MEDICINAL CHEMISTRY, PATENT COORDINATOR. *Educ:* Univ Alta, BSc, 57; Mass Inst Technol, PhD(photochem), 62. *Prof Exp:* Group leader, Merck, Sharp & Dohme Can, Ltd, 62-70; res fel, 70-84, SR RES FEL, MERCK FROSST CAN INC, 84- *Mem:* Am Chem Soc; fel Chem Inst Can; Order Chem Quebec. *Res:* Organic photochemistry; synthesis of isotopically labeled organic compounds; synthesis of organometallic compounds and metal hybrides and deuterides; medicinal chemistry; organic synthesis. *Mailing Add:* 1380 Des Pins Montreal PQ H3G 1A8 Can

ATKINSON, KENDALL E, b Centerville, Iowa, Mar 23, 40; m 61; c 2. MATHEMATICS. *Educ:* Iowa State Univ, BS, 61; Univ Wis, Madison, MS, 63, PhD(math), 66. *Prof Exp:* Assoc prof math, Ind Univ, Bloomington, 66-72; PROF MATH, UNIV IOWA, 72- *Concurrent Pos:* vis res fel, Australian Nat Univ, 70-71; Vis prof, Univ New S Wales & Univ Queensland, 88. *Mem:* Soc Indust & Appl Math; Asn Comput Mach. *Res:* Numerical analysis; numerical solution of integral equations; mathematical computer software. *Mailing Add:* Dept Math Univ Iowa Iowa City IA 52242-1466

ATKINSON, LARRY P, b Ames, Iowa, Aug 6, 41; m 66. CHEMICAL OCEANOGRAPHY. *Educ:* Univ Wash, BS, 63, MS, 66; Dalhousie Univ, PhD(oceanog), 73. *Prof Exp:* Res assoc oceanog, Marine Lab, Duke Univ, 66-68; res assoc, Skidaway Inst Oceanog, 72-74, from asst prof to assoc prof oceanog, 74-82; MEM STAFF, DEPT OCEANOG, OLD DOMINION UNIV, NORFOLK, VA. *Mem:* AAAS; Am Geophys Union; Am Soc Limnol & Oceanog. *Res:* Continental shelf oceanography. *Mailing Add:* Dept Oceanog Old Dominion Univ 5215 Hampton Blvd Norfolk VA 24142

ATKINSON, LENETTE ROGERS, b South Carver, Mass, Mar 30, 99; m 28; c 2. PLANT MORPHOLOGY. *Educ:* Mt Holyoke Col, BA, 21; Univ Wis, MA, 22, PhD(bot), 25. *Prof Exp:* Asst bot, Univ Wis, 21-25; Comn Relief Belg fel, Cath Univ Louvain, 25-27; actg asst prof bot, Mt Holyoke Col, 27-28; RES ASSOC BIOL DEPT, AMHERST COL, 46- *Mem:* AAAS; hon mem Am Fern Soc (secy, 63-69); Am Soc Plant Taxon; Bot Soc Am; Int Soc Plant Morphol; Sigma Xi; AAAS. *Res:* Fern gametophyte. *Mailing Add:* Washington Ct Apt 200 900 124th Ave NE Bellevue WA 98005

ATKINSON, MARK ARTHUR LEONARD, b London, Eng, Feb 21, 52; m 84; c 2. CELL BIOLOGY. *Educ:* Oxford Univ, BA, 74, MA, 78, DPhil(cell biol), 79. *Prof Exp:* Fel path, Sch Med, Yale Univ, 79-82; vis assoc cell biol, Nat Heart, Lung & Blood Inst, NIH, 82-87; ASSOC PROF BIOCHEM, UNIV TEX HEALTH CTR AT TYLER, 87- *Concurrent Pos:* Kate Erin res scholar, Linacre Col, Oxford, 77; James Hudson Brown & Alexander B Coxe fel, Yale Univ, 81; estab investr, Am Heart Asn. *Mem:* AAAS; Biophys Soc; Am Soc Cell Biol; Am Soc Biochem & Molecular Biol. *Res:* The structure, function and regulation of non-muscle cell myosins; interrelationships of actomyosin with other cytoskeletal elements. *Mailing Add:* Dept Biochem Univ Texas Health Sci Ctr PO Box 2003 Tyler TX 75710-2003

ATKINSON, PAUL H, b Auckland, NZ, Aug 31, 43; m 67; c 2. CELL BIOLOGY, VIROLOGY. *Educ:* Univ Auckland, PhD(cell biol), 69. *Prof Exp:* PROF DEVELOP BIOL, ALBERT EINSTEIN COL MED, 83- *Concurrent Pos:* Asst dir, Cancer Res Ctr, RECOM, NIH Cell Physiol Study Sect, 77-83, NIH Cell Biol Study Sect, 83-86. *Res:* Specificity of intracellular sorting of viral and membrane glycoproteins especially in endoplasmic reticulum; NMR spectroscopy of glycosylation processing intermediates. *Mailing Add:* Dept Develop Biol & Cancer Albert Einstein Col Med Bronx NY 10461

ATKINSON, RICHARD C, b Oak Park, Ill, Mar 19, 29; m 52; c 1. SCIENCE ADMINISTRATION. *Educ:* Univ Chicago, PhD, 48; Ind Univ, PhD(psychol), 55. *Hon Degrees:* Ten from US univs. *Prof Exp:* Lectr, Appl Math & Statist Labs, Stanford Univ, 56-57; asst prof psychol, Univ Calif, Los Angeles, 57-61; assoc prof psychol, Stanford Univ, 61-65, prof psychol & educ & affil fac mem, Inst Eng & Econ Systs, Sch Eng, 65-80, chmn, dept psychol, 68-73; dir, NSF, 75-80; CHANCELLOR, UNIV CALIF, SAN DIEGO, 80- *Concurrent Pos:* Assoc dir, Inst Math Studies Soc Sci, Stanford Univ, 61-75; vis prof psychol, Univ Mich, 63; dep dir, NSF, 75-76 & 77-80 & actg dir, 76-77; consult, Systs Develop Corp, 58-67, Bell Tel Labs, 61-63, Radio Corp Am, 64-68, Off Comput Activ, NSF, 68-75; ed, J Math Psychol, 63-70; chmn, comt learning & educ process, Soc Sci Res Coun, 66-68 & mem, 63-69; co-dir, Summer Res Conf Learning & Educ Process, US Off Educ, 64; task force on info networks, Educ Comn, Inter Univ Commun Coun, 65-68; dir, Job Corps Reading Inst, Off Econ Opportunities, 66; mem, Math Social Sci Bd, Ctr Advan Study Behav Sci, 68-72, US-USSR Joint Comn Sci & Tech Coop, 77-80, US-Peoples Repub China Joint Comn Sci & Technol Coop, 79-80, Assembly Behav & Soc Sci, Nat Res Coun & Calif Comn Indus Innovation, Exec Dept, State Calif, 81-82, Comn Educ & Pub Policy, Nat Acad Educ, 84-; personality & cognition review comt, Nat Inst Mental Health, 68-71 & chmn, 74-75; comt technol Augmentation Cognition, Smithsonian Inst, 68-70; res adv comt, Children TV Workshop, 71-76; chmn, Math Soc Sci Bd, Ctr Advan Study Behav Sci, 71-73; bd dirs, Am Psychol Asn, 73-76; adv coun, Int Asn Study Attention & Performance, 76-81; intergovt sci, Eng & Technol Adv Panel, Off Sci & Technol Policy & Federal Coun Arts & Humanities, Exec Off Pres, 77-80; Nat Mus Serv Bd, Inst Mus Serv, Dept Educ, 77-80; adv comt, Ctr Advan Study Behav Sci, 81-; bd dirs, Whittier Inst Diabetes & Endocrinol, 81-; pres comt, Nat Medal Sci, 81-84; counlr, NAS, 82-85; adv comt, higher educ policy, Online Comput Libr Ctr, 84-, comt Sci, Eng & Pub Policy, NAS, 84-85, comt nominations, AAAS, 85-; vis comt, dept psychol, Harvard Univ; chmn, Blue Ribbon Panel Info Policy, Implications Archiving Satellite Data, Nat Comn Libr & Info Sci, 83-84, adv bd, Issues Sci & Technol, 88- *Honors & Awards:* Distinguished Res Award, Soc Sci Res Coun, 62; Distinguished Sci Contrib Award, Am Psychol Asn, 77; E L Thorndike Award, Am Psychol Asn, 80; William W Cook Lectr, Univ Mich, 78; Mountain in Antactica named in honor. *Mem:* Inst Med-Nat Acad Sci; Am Acad Arts & Sci; Nat Acad Educ; Am Philos Soc; fel AAAS (pres elec, 88); Psychonomic Soc; Am Psychol Asn (pres, Div Exp Psychol, 74); Soc S Pole. *Res:* Experimental and applied mathematics, particularly with problems of memory and cognition; published over 150 scientific articles; author of 7 books in psychology. *Mailing Add:* Off Chancellor Univ Calif Box 109 La Jolla CA 92093

ATKINSON, RICHARD L, JR, b Petersburg, Va, May 15, 42; m 66; c 3. ENDOCRINOLOGY. *Educ:* Va Mil Inst, BA, 64; Med Col Va, MD, 68. *Prof Exp:* Asst prof, Univ Va, 77-83; assoc prof internal med, Univ Calif-Davis, 83-86; PROF INTERNAL MED, CHIEF DIV CLIN NUTRIT, EASTERN VA MED SCH, 86-, ASSOC CHIEF OF STAFF, RES & DEVELOP, VA MED CTR, HAMPTON, 86- *Concurrent Pos:* Comt, Mil Nutrit Res, Nat Acad Sci, 89-; nutrit study sect, NIH, 90-; pres, NAm Asn, Study Obesity, 90-91. *Mem:* Am Inst Nutrit; Am Soc Clin Nutrit; Endocrine Soc; Am Diabetes Asn; Am Fedn Med Res. *Res:* Clinical nutrition; research in human and animal obesity; regulation of food intake and energy expenditure. *Mailing Add:* Res Serv Va Med Ctr Hampton VA 23667

ATKINSON, ROGER, b Scarborough, UK, July 30, 45; m 88; c 2. GAS KINETICS, ATMOSPHERIC CHEMISTRY. *Educ:* Cambridge Univ, BA, 66, PhD(phys chem), 69. *Prof Exp:* Res fel chem, Nat Res Coun Can, 69-71; res fel, York Univ, Can, 71-72; sr scientist, Shell Res Ltd, UK, 78-79 & ERT, Inc, Calif, 79-80; from asst res chemist to assoc res chemist, 72-78, RES CHEMIST, UNIV CALIF, RIVERSIDE, 80-, PROF, DEPT SOILS & ENVIRON SCIS, 90- *Mem:* Am Chem Soc; Soc Environ Toxicol Chem. *Res:* Kinetics and mechanisms of gas phase reactions of atmospheric interest; radical reactions with organic compounds. *Mailing Add:* Statewide Air Pollution Res Ctr Univ Calif Riverside CA 92521

ATKINSON, RUSSELL H, b Brooklyn, NY, Aug 22, 22; m 44; c 3. PHYSICAL CHEMISTRY, ENGINEERING MANAGEMENT. *Educ:* Polytech Inst Brooklyn, BS, 48, PhD(chem), 52. *Prof Exp:* Jr chemist, Manhattan Proj, Tenn Eastman Corp, 44-46; sr engr, Westinghouse Elec Corp, 51-56, sect mgr metals res group, 56-59, asst res dir, 59-61, mgr parts eng, 61-63, asst mgr, Div Eng, 63-66, mgr parts mfg, 67-69, mgr, Div Eng, 69-76, prod eng mgr, 76-79, dir res & develop, Lamp Div, 79-81; RETIRED. *Concurrent Pos:* Instr, Polytech Inst Brooklyn, 52-54; lectr, Calif State Polytech Univ, 82- *Mem:* Am Illum Eng Soc; Sigma Xi; Soc Advan Mat Processing Eng; Am Soc Metals. *Res:* Chemistry of refractory metals; gas-metal reactions; analysis of refractory metals; vapor deposition reactions; lamp development and materials; composites; material sciences. *Mailing Add:* 313 E Palmyra Orange CA 92666

ATKINSON, SHANNON K C, b Honolulu, Hawaii, Dec 29, 56; m 81; c 1. REPRODUCTIVE BIOLOGY & FERTILITY MANAGEMENT, BREEDING PROGRAMS FOR CAPTIVE WILDLIFE. *Educ:* Univ Hawaii, BSc, 78, MSc, 81; Murdoch Univ, PhD(vet studies), 85. *Prof Exp:* Fel endocrinol & physiol, Vet Sch, Murdoch Univ, 84-86; exp scientist, endocrinol & physiol, Commonwealth Sci & Indust Res Orgn, Australia, 86-88; res affil wildlife res, 89-91, ASSOC RESEARCHER WILDLIFE RES, HAWAII INST MARINE BIOL, UNIV HAWAII, 91- *Honors & Awards:* Mary Walters Mem Bursary, Australian Fedn Univ Women, 83. *Mem:* Am Asn Zool Parks & Aquariums; Soc Study Reproduction; Endocrine & Reproductive Biol Soc (pres, 86-88). *Res:* Breeding strategies and control of fertility in wildlife species. *Mailing Add:* Hawaii Inst Marine Biol MSB No 212 Univ Hawaii 1000 Pope Rd Honolulu HI 96822

ATKINSON, THOMAS GRISEDALE, b Vancouver, BC, Apr 20, 29; m 56; c 3. PLANT PATHOLOGY. *Educ:* Univ BC, BSA, 52; Univ Sask, MSc, 53, PhD(plant physiol), 56. *Prof Exp:* Proj assoc plant physiol, Univ Wis, 56-58; plant pathologist, 58-80, head, Plant Path Sect, 78-80, asst dir, Agr Can Res Sta, Lethbridge, AB, 80-83, DIR, AGR CAN RES STA, WINNIPEG, MAN, 83- *Mem:* Sigma Xi; Can Phytopath Soc. *Res:* Cereal root rots; genetic and microbiological analysis of root rot reaction of cereals; resistance to wheat streak mosaic; disease loss assessment. *Mailing Add:* Agr Can Res Sta 195 Dafoe Rd Winnipeg MB R3T 2M9 Can

ATKINSON-TEMPLEMAN, KRISTINE HOFGREN, b Rockville Centre, NY, Apr 25, 47; m 83; c 3. BIOCHEMISTRY, IMMUNOLOGY. *Educ:* Fla State Univ, BA, 70, MS, 73; Univ Western Ont, PhD(genetics), 80. *Prof Exp:* Res assoc, Univ Western Ont, 72-77; asst prof cell biol, Mount Holyoke Col, 80-81; vis res asst prof biol, Dartmouth Col, 81-83, VIS RES ASSOC PHYSIOL, DARTMOUTH MED SCH, 84- *Concurrent Pos:* Sr scientist; Verax Corp, 81-84. *Mem:* Genetics Soc Can; Am Soc Parasitologists; Am Soc Tropical Med & Hygiene; Sigma Xi. *Res:* Immunocytochemistry of immunoglobulin transport and secretion; immunology of schistosomiasis; cytogenetics and evolution of the schistosome genome. *Mailing Add:* 12 Ten Hills Rd Apt 2 Somerville MA 02145

ATKINSTON, RONALD A, b Nicksburg, MS, May 18, 44. COMPUTER SCIENCE, MATHEMATIC STATISTICS. *Educ:* Univ Ala, PhD(supervision of circular), 77. *Prof Exp:* PROF & CHMN MATH, STILLMAN COL, 68- *Mem:* Am Math Soc. *Mailing Add:* Dept Math Stillman Col PO Drawer 1430 Tuscaloosa AL 35403

ATLAS, DAVID, b Brooklyn, NY, May 25, 24; m 48; c 2. METEOROLOGY, REMOTE SENSING. *Educ:* NY Univ, BSc, 46; Mass Inst Technol, MSc, 51, DSc(meteorol), 55. *Prof Exp:* Chief weather radar br, Air Force Cambridge Res Labs, Bedford, Mass, 48-66; prof meteorol, Univ Chicago, 66-72; dir atmospheric technol div, Nat Ctr Atmospheric Res, Boulder, Colo, 72-73, dir nat hail res exp, 74-75; dir lab atmospheric sci, Goddard Space Flight Ctr, NASA, Greenbelt, MD, 77-84; sr res assoc dept meteorol, Univ Md, 85-87; CONSULT, 84- *Concurrent Pos:* Assoc ed publ, Am Meteorol Soc, 57-74, counr, 61-64 & 72-74; NSF sr fel, Imp Col, London Eng, 59-60; pres interunion comn radio meteorol, Int Radio Sci Union, 69-72; distinguished vis scientist, Jet Propulsion Lab, Calif Inst Technol, 84- & NASA Goddard Space Flight Ctr, 89-; mem US deleg to VII Cong, World Meteorol Orgn, 75; chmn, AAAS, Atmospheric Hydrospheric Sci sect; first remote sensing lectr, Am Meteorol Soc, 91. *Honors & Awards:* Meisinger Award, Am Meteorol Soc, 57, Cleveland Abbe Award, 83; Loeser Award, Air Force Cambridge Res Labs, 57; O'Day Award, 64; Losey Award, Am Inst Aeronaut & Astronuat, 66; Outstanding Leadership Medal, NASA, 82; Symons Mem Medal, Royal Meteorol Soc, 89. *Mem:* Nat Acad Eng; fel Am Geophys Union; fel Royal Meteorol Soc; fel AAAS; fel Am Astronaut Soc; fel Am Meteorol Soc. *Res:* Meteorology and climatology; satellite meteorology; radar meteorology; remote sensing; research management; remote measurement of rainfall both from space and from airborne and surface platforms by radar and other sensors. *Mailing Add:* 7420 Westlake Terr Bethesda MD 20817

ATLAS, RONALD M, b New York, NY, Oct 19, 46; m 70; c 2. MICROBIOLOGY, MICROBIAL ECOLOGY. *Educ:* State Univ NY, Stony Brook, BS, 68; Rutgers Univ, MS, 70, PhD(microbiol), 72. *Prof Exp:* Res assoc, Jet Propulsion Lab, Calif Inst Technol, 72-73; from asst prof to assoc prof, 73-81, PROF BIOL, UNIV LOUISVILLE, 81-; NAT LECTR, SIGMA XI, 81- *Concurrent Pos:* Res grants, Off Naval Res, 73-79, Nat Oceanic & Atmospheric Admin, 75- & Dept Energy, 76-81; Environ Microbiol Comt, Am Soc Microbiol, 81- *Mem:* Am Soc Microbiol; AAAS; Sigma Xi; fel Am Acad Microbiol; Soc Indust Microbiol. *Res:* Microbial degradation of hydrocarbons and other organic pollutants; numerical taxonomy of marine bacteria; effects of pesticides and other organic compounds on microorganisms; ecology of soil and marine microorganisms. *Mailing Add:* 1603 Dunbarton Wynd Louisville KY 40205

ATLAS, STEVEN ALAN, b New York, NY, Sept 3, 46; m 88; c 1. HYPERTENSION, CIRCULATORY PHYSIOLOGY. *Educ:* Johns Hopkins Univ, BA, 68, MD, 71. *Prof Exp:* Intern & resident, New York Hosp, Cornell Med Ctr, 71-73; res assoc pharmacol, Nat Inst Health, 73-76; from instr to assoc prof, 76-82, ASSOC PROF MED, CORNELL UNIV MED COL, 82-; ASSOC ATTEND PHYSICIAN, NEW YORK HOSP, 82-; ASSOC PROF MED, CORNELL UNIV MED COL, 82- *Concurrent Pos:* Asst attending physician, New York Hosp, 77-82; mem, cardiovascular & renal Study Sect, NIH, 84-88; mem, Coun High Blood Pressure Res, Am Heart Asn. *Honors & Awards:* Clin Endocrinol Trust Lectr, Brit Endocrine Soc, 87. *Mem:* Am Soc Clin Invest; Endocrine Soc; Am Heart Asn; Am Soc Hypertension; Int Soc Hypertension; NY Acad Sci; Soc Exp Biol & Med. *Res:* Biochemical endocrinology, pharmacology and physiological role of hormones involved in cardiovascular physiology, with particular expertise on the Renin-Angiotensin-Aldosterone system and on atrial natriuretic factor. *Mailing Add:* Cardiovascular Ctr, NY Hosp Cornell Univ Med Ctr 525 E 68th St New York NY 10021

ATLURI, SATYA N, b Gudivada, India, Oct 7, 45; US citizen; m 72; c 2. FRACTURE MECHANICS & COMPUTATIONAL FLUID DYNAMICS, COMPUTATIONAL SOLID MECHANICS. *Educ:* Andhra Univ, BS, 65; Indian Inst Sci, MS, 66; Mass Inst Technol, DSc, 69. *Hon Degrees:* DSc, Nat Univ Ireland, 89. *Prof Exp:* Res assoc aeronaut & astronaut, Mass Inst Technol, 69-71; asst prof aeronaut & astronaut, Univ Wash, 71-74; from assoc prof to prof mech, 74-79, REGENTS PROF MECH, GA INST TECHNOL, 79-, DIR COMPUT MECH CTR, 80- *Concurrent Pos:* Ed, Int J Comput Mech 86-; Am Inst Aeronaut & Astronaut J, 83-87; prin investr grants, Off Naval Res, US Air Force, Nat Sci Found, Nat Res Coun & many others; chmn Comt Comput Appl Mech, Am Soc Mech Engrs, 81-85; J C Hunsaker prof aeronaut & astronaut, Mass Inst Technol, 90-91. *Honors & Awards:* Aerospace Struct & Mat Award, Am Soc Civil Engrs, 86; Struct, Struct Dynamics & Mat Medal, Am Inst Aeronaut & Astronaut, 88; Monie Ferst Mem Award, Sigma Xi, 88. *Mem:* Fel Am Soc Mech Engrs; fel Am Acad Mech; US Asn Comput Mech; Am Inst Aeronaut & Astronaut. *Res:* Computational mechanics; finite element and boundary element methods; fracture mechanics; dynamin and control; composite materials; author of over 300 articles in archival scientific literature and 16 books. *Mailing Add:* Ctr Comput Mech Ga Tech Atlanta GA 30332-0356

ATNEOSEN, RICHARD ALLEN, b St James, Minn, Sept 11, 34. PHYSICS. *Educ:* Univ Minn, BS, 56, MS, 58; Ind Univ, PhD(physics), 63. *Prof Exp:* Res assoc nuclear chem, Princeton Univ, 63-65; asst res prof physics, Mich State Univ, 65-68; ASSOC PROF PHYSICS & ASTRON, WESTERN WASH STATE UNIV, 68- *Mem:* Am Phys Soc; Am Asn Physics Teachers; Astron Soc Pacific. *Res:* Reaction mechanisms in alpha particle scattering; particle-induced fission; infrared astronomy; acoustics. *Mailing Add:* Dept Physics & Astron Western Wash State Univ 516 High St Bellingham WA 98225

ATNIP, ROBERT LEE, b Bridgeport, Ala, Aug 10, 28; m 51; c 5. ANATOMY. *Educ:* David Lipscomb Col, BA, 51; George Peabody Col, MA, 52; Univ Tenn, PhD(anat), 64. *Prof Exp:* Teacher high schs, Ga & Ala, 52-54; instr biol, Freed-Hardeman Col, 54-59; instr anat & pediat, 66-68, asst prof anat & child develop & res assoc pediat, 68-74, ASSOC PROF ANAT & CHILD DEVELOP, UNIV TENN MED UNITS, 74- *Concurrent Pos:* Nat Inst Child Health & Human Develop fel anat, Col Med, Univ Tenn, 65-66. *Mem:* Am Soc Human Genetics; Sigma Xi. *Res:* Mammalian embryology and experimental teratology; experimentally-produced cleft palate in mice; chromosomal aberrations of malformed mentally retarded children. *Mailing Add:* 1026 Francis Ave Jasper TN 37347-2903

ATOJI, MASAO, b Osaka, Japan, Dec 21, 25; US citizen; m 57; c 3. ELECTRONICS & SEMICONDUCTOR MATERIALS. *Educ:* Shizuoka Univ, BS, 46; Osaka Univ, MS, 48, DSc(chem), 56. *Prof Exp:* Res fel chem, Univ Minn, Minneapolis, 51-56; assoc res, Iowa State Univ, 56-58, asst prof, 58-60; chemist, div chem, Argonne Nat Lab, 60-69, sr chemist & group leader, neutron group, 69-81; sr staff scientist, Airtron Div, Litton Systs, 81-83 & EIC lab, 83-84; mgr, chem lab, Motorola Inc, 84-87; mgr, crystal lab, Northwestern Univ, 88-90; ED, AM CHEM SOC, 90- *Mem:* AAAS; Am Crystallog Asn; Am Chem Soc; fel Am Phys Soc; Phys Soc Japan; Soc Appl Spectroscopy; Am Asn Crystal Growth. *Res:* Crystal and magnetic structures; neutron and x-ray diffraction; crystal growth, fabrication and characterization; semiconductor and electronics materials; AA, thermogravimetric, DSC, thermomechanical and Fourier transform infrared analyses. *Mailing Add:* 702 86th Pl Cowners Grove IL 60516

ATON, THOMAS JOHN, b Miami, Okla, Nov 1, 48; m 71; c 2. SOLID STATE PHYSICS. *Educ:* Rice Univ, BA, 70; Univ Ill, MS, 72, PhD(physics), 76. *Prof Exp:* Instr physics, Princeton Univ, 76-78; asst prof physics, Univ Va, 78-84; SR MEM TECH STAFF, TEX INSTRUMENTS, 84- *Mem:* Am Phys Soc. *Mailing Add:* Tex Instruments MS 369 12840 Hillcrest Rd Dallas TX 75230

ATRAKCHI, AISAR HASAN, b Baghdad, Iraq, Mar 16, 55. ENVIRONMENTAL TOXICOLOGIST, NEUROTOXICOLOGIST. *Educ:* Baghdad Univ, BSc, 78; Univ Missoula, MSc, 82; Univ Calif Davis, PhD(pharmacol & toxicol), 87. *Prof Exp:* Postgraduate researcher, Off Curricular Support, Univ Calif Davis, Sch Med, 85-87; postdoctoral fel free radicals, George Washington Univ, DC, 87-89; Fogarty Int fel electrophysiol, Lab Biophysics, NIH, 89-90; STAFF SCIENTIST ENVIRON TOXICOL, DYNAMAC CORP, 90- *Mem:* Am Soc Pharmacol & Exp Therapeut; Soc Risk Anal. *Res:* Research in pharmacology and physiology of the cardiovascular and central nervous systems; in situ, in vivo, and in vitro preparations; membrane channel activities and kinetics and toxicology. *Mailing Add:* 884 Col Pkwy No 301 Rockville MD 20850

ATREYA, ARVIND, b Bijnor, India, Jan 15, 54; m 83; c 2. COMBUSTION, HEAT & MASS TRANSFER. *Educ:* Indian Inst Technol, New Delhi, BTech, 75; Univ NB, Can, MScE, 77; Harvard Univ, MS, 78, PhD(eng sci), 83. *Prof Exp:* Res asst fire res, Harvard Univ, 78-83; asst prof, 83-87, ASSOC PROF THERMAL SCI, MICH STATE UNIV, 87- *Concurrent Pos:* Mech engr, Nat Inst Standards & Technol, 83-84; NSF presidential young investr award, 86; consult, Hughes, Hubbard & Reed, 88-90 & Hughes Assocs Inc, 89- *Mem:* Combustion Inst; Am Soc Mech Engrs; Int Asn Fire Safety Sci; Sigma Xi. *Res:* Fire safety research including ignition, flame spread, pyrolysis and extinguishment; industrial combustion research to enhance flame radiation and reduce pollutants. *Mailing Add:* Dept Mech Eng A-104 RCE Mich State Univ East Lansing MI 48824

ATREYA, SUSHIL KUMAR, b Apr 15, 46; US citizen; m 70; c 1. ATMOSPHERIC SCIENCES, PLANETARY ATMOSPHERES. *Educ:* Univ Rajasthan, India, BSc, 63, MSc, 65; Yale Univ, MS, 68; Univ Mich, PhD(atmospheric sci), 73. *Prof Exp:* Res fel physics, Univ Delhi, India, 65-66; res assoc physics, Univ Pittsburgh, 73-74; res scientist space physics, 74-78, asst prof, 78-81, assoc prof, 81-87, PROF ATMOSPHERIC & SPACE SCI, UNIV MICH, ANN ARBOR, 87- *Concurrent Pos:* Assoc prof, Univ Paris, 84-85; sci teams, Cassini-Huygens Probe/Gas Chromatograph Mass

Spectrometer, Aerosol Collector Pyrolyzer, Voyager/Ultraviolet Spectrometer, Galileo/Jupiter Entry Probe Mass Spectrometer, Comet Rendezvous Asteroid Flyby Neutral Gas & Ion Mass Spectrometer, Mars '94 & Phobos Projs & Spacelab/Imaging Observ; guest investr, Hubble Space Telescope, Copernicus & Int Ultraviolet Explorer satellites; vis sr res scientist, Imp Col, London, 84; pres, Com Planetary Atmospheres, Int Asn Meteorol & Atmospheric Physics. *Honors & Awards:* Exceptional Sci Accomplishments Award, Voyager UVS Spectroscopy, NASA, Group Achievement Awards, Voyager, 81, 86 & 90. *Mem:* AAAS; Am Geophys Union; Am Astron Soc; Int Astron Union; Comt Space Res; Int Asn Meteorol & Atmospheric Physics. *Res:* Aeronomy, photochemistry, cloud physics, origin and evolution of planetary and satellite atmospheres. *Mailing Add:* Dept Atmospheric Oceanic & Space Sci Space Res Bldg Univ Mich Ann Arbor MI 48109-2143

ATTALLA, ALBERT, b Cuyahoga Falls, Ohio, Sept 29, 31; m 56; c 3. PHYSICAL CHEMISTRY, ANALYTICAL CHEMISTRY. *Educ:* Kent State Univ, BA, 55, MA, 59; Univ Cincinnati, PhD(phys chem), 62. *Prof Exp:* Sr chemist, Corning Glass Works, NY, 62-63; RES SPECIALIST INSTRUMENTAL ANALYSIS, MONSANTO RES CORP, 63- *Res:* Nuclear magnetic resonance and x-ray fluorescence spectroscopy; gas chromatography. *Mailing Add:* 350 Blackstone Dr Centerville OH 45459

ATTALLAH, A M, b Egypt, Feb 2, 44. CELL BIOLOGY, IMMUNOLOGY. *Educ:* George Washington Univ, PhD(genetics), 74. *Prof Exp:* CHIEF IMMUNOL, CTR DRUG & BIOL, NIH, 78-; PROF GENETICS & IMMUNOL, GEORGE WASHINGTON UNIV, 81- *Mailing Add:* Nat Ctr Drugs & Biol NIH Bldg 29 NIH 8800 Rockville Pike Bethesda MD 20892

ATTARDI, GIUSEPPE M, b Yicari, Panerio, Italy, Sept 14, 23. MITOCHONDRIA BIOGENESIS. *Educ:* Univ Padua, MD, 47. *Prof Exp:* Asst prof, Dept Histol & Gen Embryol, Univ Padua, Italy, 47-57; Fulbright fel, Dept Microbiol, Wash Univ, St Louis, 57-59 & Div Biol, Calif Inst Technol, 59-60; sr researcher, Lab Enzymol, Nat Ctr Sci Res & Pasteur Inst, Paris, 61-62; from assoc prof to prof biol, 67-84, GRACE C STEELE PROF MOLECULAR BIOL, CALIF INST TECHNOL, 85- *Concurrent Pos:* Res fel, Inst Cell Res & Genetics, Karolinska Inst, Stockholm, Sweden, 52-55; Guggenheim fel, Molecular Genetics Ctr, Nat Ctr Sci Res, France, 70-71 & 86-87. *Honors & Awards:* Antonio Feltrinelli Int Prize for Med, 89. *Mem:* Nat Acad Sci. *Mailing Add:* Div Biol 156-29 Calif Inst Technol Pasadena CA 91125

ATTAS, ELY MICHAEL, b Montreal, Que, July 18, 52; m 75; c 3. RADIOCHEMICAL ANALYSIS, NUCLEAR SAFEGUARDS TECHNIQUES. *Educ:* McGill Univ, BSc, 73, MSc, 75, PhD(chem & archaeol), 83; Univ Paris XI, Orsay, Doctorat de 3E cycle, 80. *Prof Exp:* Res chemist, Anal Sci Br, 82-90, HEAD, RADIOCHEM ANALYSIS SECT, ANALYSIS SCI BR, AECL RES, WHITESHELL LABS, 90- *Concurrent Pos:* Tech collabr, Chem Dept, Brookhaven Nat Lab, 76-77; lectr, Classics Dept, McGill Univ, Montreal, 76, Dept Etudes Anciennes Modernes, Univ Montreal, 78, Anthrop, Pinawa Sec Sch, Man, 88-89. *Mem:* Chem Inst Can/Can Soc Chem. *Res:* Development of instrumental methods for analysis of radioactive samples; nuclear safeguards instrumentation, including ultraviolet imaging of Cerenkov radiation from used nuclear fuel stored underwater; neutron activation analysis techniques, archaeometry and chemometrics. *Mailing Add:* AECL Res Whiteshell Labs Pinawa MB R0E 1L0 Can

ATTAWAY, CECIL REID, b Fredericksburg, Va, May 19, 40; m 69. MECHANICAL ENGINEERING, MACHINE DESIGN. *Educ:* Clemson Col, BS, 62; Ga Inst Technol, MS, 68, PhD(mach design), 69. *Prof Exp:* Mech engr, Cadre Corp, Ga, 66-67; dir res & develop, Gunter & Cooke Inc, 69-70; res & develop engr, Blue Bell Inc, 70-73; proj engr, Cadre Corp, 73-76; PRIN ENGR, NUCLEAR DIV, UNION CARBIDE CORP, 76- *Mem:* Am Soc Mech Engrs; Nat Soc Prof Engrs. *Res:* Design of state-of-the-art research and development, testing, and production facilities. *Mailing Add:* 8500 Westside Dr Knoxville TN 37909

ATTAWAY, DAVID HENRY, b Sterling, Okla, June 9, 38. BIOCHEMISTRY. *Educ:* Univ Okla, BS, 60, PhD(biochem), 68. *Prof Exp:* Phys oceanogr, US Naval Oceanog Off, 62-65; res assoc org geochem, Marine Sci Inst, Univ Tex, 68-69; chief geochem, State Geol Surv, Univ Kans, 69-71; res chem oceanogr, US Coast Guard Hq, 71-72; PROG DIR BIOTECHNOL & SEAFOOD SCI, 72-, ASST DIR RES, NAT SEA GRANT COL PROG, 86- *Concurrent Pos:* Univ Tex grad sch fel, 68-69; Cong fel, 80-81. *Mem:* AAAS; Marine Technol Soc; Am Chem Soc; Sigma Xi; Inst Food Technol; Am Geophys Union. *Res:* Comparative chemistry of natural products from marine plants and animals; toxic substances of marine origin; organic geochemistry; petroleum in marine environment; marine biotechnology. *Mailing Add:* Nat Sea Grant Col Prog 1335 East-West Hwy Silver Spring MD 20910

ATTAWAY, JOHN ALLEN, b Atlanta, Ga, July 19, 30; m 57; c 4. PLANT BIOCHEMISTRY, RESEARCH ADMINISTRATION. *Educ:* Fla Southern Col, BS, 51; Univ Fla, MS, 53; Duke Univ, PhD(chem), 57. *Prof Exp:* Asst, Off Ord Res, Univ Fla, 53; asst, Duke Univ, 53-54, Off Ord Res & Off Naval Res, 54-55, fel plant biochem, 57; res chemist, Monsanto Chem Co, 57-58; res chemist org chem & biochem, Resources Res Inc, 58-59; res chemist, 59-68, DIR RES, FLA DEPT CITRUS, 68- *Honors & Awards:* Confructa Award, Int Fedn Fruit Juice Producers. *Mem:* Am Chem Soc; Am Soc Plant Physiologists; Inst Food Technologists; Asn Off Anal Chemists; Asn Food & Drug Off. *Res:* Chromatographic analysis; instrumentation; natural products and flavors; organic fluorine compounds; plant growth regulators; organic chemistry. *Mailing Add:* Box 205 Winter Haven FL 33880

ATTEBERY, BILLY JOE, b Bearden, Ark, Dec 26, 27; m 58; c 2. MATHEMATICS. *Educ:* Ark State Teachers Col, BSE, 50; Univ Ark, MA, 54; Univ Mo, PhD, 58. *Prof Exp:* Instr math, Univ Mo, 54-58; from asst prof to assoc prof, Univ Ark, 58-67; assoc prof, 67-68, PROF MATH, LA TECH UNIV, 68- *Mem:* Math Asn Am. *Res:* Probability theory. *Mailing Add:* 3107 Bienville Ave Ruston LA 71270

ATTIE, ALAN D, b New York, NY, June 18, 55. LIPOPROTEINS, CHOLESTEROL. *Educ:* Univ Wis-Madison, BS, 76; Univ Calif, San Diego, PhD(biol), 80. *Prof Exp:* Asst prof, 82-89, ASSOC PROF BIOCHEM, DEPT COMP BIOSCI, UNIV WIS-MADISON, 89- *Concurrent Pos:* Mem Arteriosclerosis Coun, Am Heart Asn; fel res award, Am Asn Study Liver Dis, 80; estab investr award, Am Heart Asn, 87. *Mem:* Am Heart Asn; Am Soc Biochem & Molecular Biol. *Res:* Structure and function of plasma lipoproteins; mutations affecting the structure and expression of apolipoproteins; cholesterol metabolism; animal models for studying lipoprotein and cholesterol metabolism. *Mailing Add:* Dept Biochem Univ Wis 420 Henry Mall Madison WI 53706

ATTIG, THOMAS GEORGE, b Pontiac, Ill, Oct 2, 46; m 68; c 2. INORGANIC CHEMISTRY, ORGANOMETALLIC CHEMISTRY. *Educ:* DePauw Univ, BA, 68; Ohio State Univ, PhD(chem), 73. *Prof Exp:* Teaching asst, Ohio State Univ, 69-70, NSF trainee, 70-73; fel chem, Univ Western Ont, 73-75; asst prof chem, Univ Ky, 75-79; res proj leader, Sohio Res, Standard Oil Co Ohio, 79-85, res assoc, 85, mgr, Process Res, 85-87, mgr process res, Chem Technol, BP Am, 88, MGR DERIVATIVES, BP RES, 88- *Mem:* Am Chem Soc. *Res:* Preparation, structure and reactions of organometallic compounds of the nickel triad; chemistry of transition metal carbonyl complexes; synthesis and stereochemistry of chiral organometallic compounds; catalysis. *Mailing Add:* BP Res 4440 Warrensville Rd Cleveland OH 44128

ATTINGER, ERNST OTTO, b Zurich, Switz, Dec 27, 22; c 3. BIOENGINEERING. *Educ:* Winterthur Cantonal Sch, Switz, BA, 41; Univ Zurich, MD, 48; Drexel Inst Technol, MS, 61; Univ Pa, PhD(biomed eng), 65. *Prof Exp:* Chief res, Dis of Chest, Heilstaette Du Midi, Davos, Switz, 50-52; asst res, Internal Med, Lincoln Hosp, NY, 52-53; asst prof med, Med Sch, Tufts Univ, 56-59; from asst prof to prof physiol, Sch Vet Med, Univ Pa, 61-67; res dir, Res Inst, Presby Hosp, 62-67; prof physiol & chmn biomed eng, Univ Va, 67-90, Forsyth prof appl sci, 84-90; CONSULT, 90- *Concurrent Pos:* Res fels, Cardiopulmonary Lab, Nat Jewish Hosp, Denver, Colo, 43-54, Lung Sta, Boston City Hosp, Tufts Univ, 54- 59 & Res Inst, Presby Hosp, Philadelphia, 59-62. *Mem:* AAAS; Biomed Eng Soc; Inst Elec & Electronics Engrs; fel Biophys Soc; Sigma Xi; Am Physiol Soc. *Res:* Analysis of biological systems; systems analysis with particular emphasis on control hierarchies in biological and social systems; technology assessment in health care. *Mailing Add:* Div Biomed Eng Univ Va Med Ctr Charlottesville VA 22901

ATTIX, FRANK HERBERT, b Portland, Ore, Apr 2, 25; m 59; c 2. RADIOLOGICAL PHYSICS. *Educ:* Univ Calif, Berkeley, AB, 49; Univ Md, MS, 53. *Prof Exp:* Res physicist, NIH, 49-50; res physicist, Nat Bur Standards, 50-57; reactor shielding physicist, ACF Industs, DC, 57-58; res physicist, Naval Res Lab, Washington, DC, 58-76, head dosimetry br, 63-68, consult dosimetry radiation technol div, 69-76; prof, 76-87, EMER PROF MED PHYSICS, UNIV WIS-MADISON, 87-; CONSULT, 87- *Concurrent Pos:* Vis scientist, UK Atomic Energy Res Estab, Harwell, 68-69. *Honors & Awards:* Appl Sci Award, Sigma Xi, 69; Distinguished Sci Achievement Award, Health Physics Soc, 87. *Mem:* Health Phys Soc; Sigma Xi; Am Asn Physicists Med. *Res:* Measurement radiation dose and related quantities; codeveloper of theory of cavity ionization and experimental verification of that theory; invented variable-length free-air ionization chamber; definitive measurement of gamma ray output of radium; measurement of dose by luminescent materials; fast-neutron dosimetry in cancer radiotherapy. *Mailing Add:* Med Physics Dept Univ Wis 1300 University Ave Madison WI 53706

ATTREP, MOSES, JR, b Alexandria, La, Jan 2, 39; m 65. RADIOCHEMISTRY, GEOCHEMISTRY. *Educ:* La Col, BS, 60; Univ Ark, MS, 62, PhD(chem), 65. *Prof Exp:* Res assoc chem, Clark Univ, 65-66; asst prof, 66-74, PROF CHEM, E TEX STATE UNIV, 74- *Mem:* Geochem Soc; Am Chem Soc. *Res:* Analytical chemistry; radiochemistry. *Mailing Add:* Isotope & Nuclear Chem Los Alamos Nat Lab MSJ514 Los Alamos NM 87545-0001

ATTWOOD, DAVID THOMAS, JR, b New York, NY, Aug 15, 41; div; c 3. PHYSICS, ENGINEERING. *Educ:* Hofstra Univ, BS, 63; Northwestern Univ, MS, 64; NY Univ, DSc(plasma physics), 72. *Prof Exp:* Res scientist plasma physics, Gen Appl Sci Labs, 65-68; group leader & physicist plasma physics laser fusion, Lawrence Livermore Nat Lab, 72-83, SR STAFF SCIENTIST & DIR CTR X-RAY OPTICS, LAWRENCE BERKELEY LAB, UNIV CALIF, 83- *Concurrent Pos:* Lectr, dept applied sci, Univ Calif, Davis, 78-83; mem, site rev comt, NIH, NSF, 83 & 84; chmn, prog sub-comt laser fusion & laser-plasma interactions, Conf Laser Eng & Optics, 83; leader, working group short wavelength optics, Conf Free Electron Lasers Extreme Ultraviolet Radiation, 83; vis lectr, USSR, 83; chmn, tech group x-ray & ultraviolet tech, Optical Soc Am, 83-; co-chmn, Topical Conf Coherent Short Wavelength Radiation, 86. *Mem:* AAAS; Am Phys Soc; Sigma Xi; Optical Soc Am. *Res:* Plasma physics, laser fusion, picosecond diagnostic techniques. *Mailing Add:* 320 Garden Creek Pl Danville CA 94526

ATWATER, EDWARD CONGDON, b Rochester, NY, Feb 6, 26; m 50; c 2. INTERNAL MEDICINE, HISTORY OF MEDICINE. *Educ:* Univ Rochester, BA, 50; Harvard Univ, MD, 55; Johns Hopkins Univ, MA, 74. *Prof Exp:* Intern & asst resident, 55-57, chief resident, 59-60, sr instr & asst prof med, 59-69, asst prof hist med, 71-77, ASSOC PROF MED, SCH MED, UNIV ROCHESTER, 69-, ASSOC PROF HIST MED, 77- *Concurrent Pos:* USPHS trainee arthritis & metab dis, Univ Rochester, 57-59; Macy fel, Sch Med, Johns Hopkins Univ, 70-71; from asst physician & assoc physician, Strong Mem Hosp, Rochester, 60-69, sr assoc physician, 69-84, attend physician, 84- *Mem:* Am Rheumatism Asn; Am Asn Hist Med. *Res:* History of American medicine and the medical profession; history of medical education. *Mailing Add:* Dept Med Univ Rochester Med Ctr Rochester NY 14642

ATWATER, HARRY ALBERT, b Boston, Mass, Jan 10, 21; m 58. SOLID STATE PHYSICS. *Educ:* Tufts Col, BS, 40; Harvard Univ, MS, 41, PhD(physics), 57; Boston Univ, MA, 49. *Prof Exp:* Instr physics, Univ Ore, 56-59; assoc prof elec eng, Pa State Univ, 59-65, assoc prof physics, 65-80. *Mem:* Am Phys Soc; Am Asn Physics Teachers; Sigma Xi. *Res:* Electron paramagnetic resonance in solids. *Mailing Add:* 23799 Monterey-Salinas Hwy Apt 21 Salinas CA 93908

ATWATER, NORMAN WILLIS, b Paterson, NJ, Mar 16, 26; m 47; c 3. ORGANIC CHEMISTRY. *Educ:* Rensselaer Polytech Inst, BS, 49; Johns Hopkins Univ, PhD(chem), 53. *Prof Exp:* Dir qual control, G D Searle & Co, 53-76; dir overseas qual control, E R Squibb & Son, 76-88; RETIRED. *Mem:* AAAS; Am Chem Soc. *Res:* Steroids; terpenes. *Mailing Add:* 26 Zion Rd Hopewell NJ 08525

ATWATER, TANYA MARIA, b Los Angeles, Calif, Aug 27, 42; c 1. MARINE GEOPHYSICS, TECTONICS. *Educ:* Univ Calif, Berkeley, AB, 65, Univ Calif, San Diego, PhD(earth sci), 72. *Prof Exp:* Vis res assoc seismol, Univ Chile, 66-67; res assoc paleomagnetism, Stanford Univ, 70-71; asst prof marine geophys, Univ Calif, San Diego, 72-73; Nat Acad Sci exchange scientist, USSR, 73-74; from asst prof to assoc prof marine geophys,Mass Inst Technol, 74-80; PROF, DEPT GEOL SCI, UNIV CALIF, SANTA BARBARA, 80- *Concurrent Pos:* Sloan fel, 75-77; nat lectr, Sigma Xi, 75-76. *Honors & Awards:* Newcomb Cleveland Prize, AAAS, 80; Encourage Award, Asn Woman Geologists, 84. *Mem:* Fel Geol Soc Am; fel Am Geophys Union; AAAS; Am Geol Inst; Asn Women Geologists; Soc Women Geogr. *Res:* Detailed tectonic nature of mid ocean ridges and the mechanisms by which ocean floor is created and modified; plate tectonic reconstructions and their implications for continental geologic structure and history. *Mailing Add:* Dept Geol Sci Univ Calif Santa Barbara CA 93106

ATWELL, CONSTANCE WOODRUFF, b Philadelphia, Pa, Jan 27, 42; m 90; c 2. DEVELOPMENTAL PSYCHOBIOLOGY, SENSORY DEVELOPMENT. *Educ:* Mount Holyoke Col, AB, 63; Univ Calif, Los Angeles, MA, 65, PhD(psychol), 68. *Prof Exp:* Asst prof psychol, Pitzer Col & Claremont Grad Sch, 67-72, assoc prof, 72-77, prof, 77-78; grants assoc, NIH, 78-79; CHIEF, STRABISMUS, AMBLYOPIA & VISION PROCESSING BR, 81-, DEP ASSOC DIR, EXTRAMURAL & COLLABORATIVE PROGS, NAT EYE INST, NIH, 88- *Concurrent Pos:* Res assoc, Fac Med, Univ Col Nairobi, 68-69, lectr child psychol, 69; prin investr res grants, Grant Found, 75-77 & Nat Inst Child Health & Human Develop, NIH, 76-78, chief off clin applications vision res, Nat Eye Inst, 79- *Mem:* AAAS; Am Psychol Asn; Asn Res Vision & Ophthalmol; Asn Women Sci; Women Eye Res; Soc Neurosci. *Res:* Development of the visual system and behavior; development of vestibular responsiveness in human infants; vestibular self-stimulation in normal children. *Mailing Add:* Nat Eye Inst NIH Rm 6A49 Bldg 31 Bethesda MD 20892

ATWELL, WILLIAM ALAN, b Baltimore, Md, Aug 26, 51; m 73; c 2. PRODUCT DEVELOPMENT, CEREAL CHEMISTRY. *Educ:* State Univ NY, BS, 72, MS, 76; Kans State Univ, PhD(grain sci), 78. *Prof Exp:* Teaching asst chem, State Univ NY, Oswego, 73-74; res tech, Lake Ont Environ Labs, 74-75; res asst grain sci, Kans State Univ, 76-78; SR SCIENTIST, PILLSBURY CO, 79- *Concurrent Pos:* Consult, Magnolia Labs, 77-78. *Mem:* Am Asn Cereal Chemists. *Mailing Add:* 650 NE Ione St Spring Lake Park MN 55432

ATWELL, WILLIAM HENRY, b Milwaukee, Wis, Dec 13, 36; m 60; c 4. ORGANIC CHEMISTRY. *Educ:* Marquette Univ, BS, 59, MS, 60; Iowa State Univ, PhD(chem), 64. *Prof Exp:* Res mgr, basic intermediates, 69-72, silicone rubber, 72-75 & elastomers, 75-80, tech dir, basic mat, 81-83, TECH MGR, CERAMICS, DOW CORNING CORP, 83- *Mem:* Am Chem Soc; Sigma Xi; Am Ceramic Soc. *Res:* The use of chemical precursor to produce advanced structural ceramics; development of ceramic fibers an ceramic composites. *Mailing Add:* 1402 Wildwood Midland MI 48640

ATWOOD, CHARLES LEROY, b Decatur, Ill, Dec 24, 51; m 74; c 2. COMPUTER-AIDED ARCHITECTURAL DESIGN. *Educ:* Univ Ill, BSc, 74, MArchit, 76. *Prof Exp:* Spec teaching fel comput appln in archit, Univ Ill, 75-76; asst dir comput serv, Skidmore, Owings & Merrill, 76-81; sr vpres, Hok Comput Serv Corp, 81-84, chief exec officer & pres, 84-87; VPRES & DIR RES & DEVELOP, FORSIGHT RESOURCES, 88- *Honors & Awards:* Prize, Edward C Earl Found, 76; Award, Emerson Found, 76. *Mem:* Asn Comput Mach; Digital Equip Comput Users Soc; Data Processing Mgt Asn; Soc Comput Appln Eng, Planning & Archit. *Res:* Computer applications for architectural design, engineering, interiors and facility management with an emphasis on integration. *Mailing Add:* 7712 N Lucerne Kansas City MO 64151

ATWOOD, DONALD J, b Haverhill, Mass, May 25, 24; m 46; c 2. ELECTRICAL ENGINEERING. *Educ:* Mass Inst Technol, BS, 48, MS, 50. *Hon Degrees:* DEng, Rose Hulman Inst, 87. *Prof Exp:* Dir opers, AC Electronics Div, Gen Motors Corp, 68-70, mgr Indianapolis Opers, Detroit Diesel Div, 70-73, gen mgr, Transp Systs Div, 73-74, gen mgr, Delco Electronics Div, 74-78 & Detroit Div, 78-80, vpres & group execs, 81-83, exec vpres, Delco Electronics Div, 84-87, vchmn bd, 87-89; DEP SECY DEFENSE, DEPT DEFENSE, WASHINGTON, DC, 89- *Concurrent Pos:* Pres, GM Hughes Elec Hughes Electronic Corp, Los Angeles, Calif, 85-89; mem bd dirs, Charles Stark Draper Lab, Automotive Hall Fame, Inc, Nat Bank Detroit. *Mem:* Nat Acad Eng; Am Helicopter Asn; Soc Automotive Engrs; Motor Vehicle Mfrs Asn; Am Inst Aeronaut & Astronaut. *Mailing Add:* Pentagon Rm 3-944 Washington DC 20301

ATWOOD, DONALD KEITH, b Burlington, Vt, June 5, 33; m 54; c 4. CHEMICAL OCEANOGRAPHY. *Educ:* St Michael's Col (Vt), BS, 55; Purdue Univ, PhD(chem), 60. *Prof Exp:* Res engr, Humble Oil & Refining Co, 60-63, sr res chemist, 63-67, sr res specialist, 67-69; assoc prof chem oceanog, Univ PR, Mayaguez; DIR OCEAN CHEM DIV, NAT OCEANIC & ATMOSPHERIC ADMIN, ATLANTIC OCEANOG & METEOROL LABS, 76- *Concurrent Pos:* Subj leader chem oceanog, Coop Invest Caribbean & Adj Regions, 72-76; consult, NSF, 75-76; dir, Intergovt Oceanog Comn Regional Marine Pollution Monitoring Prog Caribbean, 78-86; adj prof, Univ Miami, 77-; fel, Coop Inst Marine & Atmospheric Res, Miami, 90- *Honors & Awards:* Award outstanding contribution investigation of effects of IXTOX-1 blowout in Bay of Campeche, Nat Oceanic & Atmospheric Admin. *Mem:* Am Chem Soc; Sigma Xi; Am Soc Limnol & Oceanog; Am Geophys Union. *Res:* Environmental controls on fisheries; oceanic controls on tropospheric carbon dioxide; ecosystem responses to climate changes. *Mailing Add:* Nat Oceanic & Atmospheric Admin 4301 Rickenbacker Causway Miami FL 33149

ATWOOD, GILBERT RICHARD, b Taunton, Mass, Apr 9, 28; m 55; c 2. PHYSICAL CHEMISTRY. *Educ:* Carnegie Inst Technol, BS, 49; Purdue Univ, MS, 51; Univ Pittsburgh, PhD(chem), 58. *Prof Exp:* From asst to assoc, Mellon Inst, 51-54, from jr fel to fel, 54-58; sr chemist, Koppers Co, Inc, Pa, 58-60, mgr physics & phys chem, 60-63; proj leader, 63-68, from res scientist to sr res scientist, 68-74, RES ASSOC, UNION CARBIDE CORP, 74- *Mem:* AAAS; Am Chem Soc. *Res:* Thermodynamics; phase equilibria; thermometry; cryoscopy; ebulliometry; calorimetry; separation and purification; gas absorption; liquid-liquid extraction and diffusion; fundamental papers and patents in environmental processes, e.g., stack gas scrubbing and removal of PCB's from transformers. *Mailing Add:* Union Carbide Corp Old Saw Mill River Rd Tarrytown NY 10591

ATWOOD, GLENN A, b Rock Rapids, Iowa, Oct 24, 35; m 55; c 5. CHEMICAL ENGINEERING. *Educ:* Iowa State Univ, BS, 57, MS, 59; Univ Wash, PhD(mass transfer), 63. *Prof Exp:* Res asst separation lanthanides, Ames Lab, Iowa, 57-59; instr chem eng, Univ Wash, 59-62; engr, Esso Res & Eng Co, 62-65; from asst prof to prof chem eng, Univ Akron, 65-90, assoc dean, 85-88, actg dean, 88-89; DIR RES & DEVELOP, MIDWEST ORE PROCESSING CO, 90- *Concurrent Pos:* Consult, Khapp Foundry, BF Goodrich, Norton Co, Gen Tire & Rubber Co, Firestone Tire & Rubber Co, Chemstress Consult Co. *Mem:* Am Inst Chem Engrs; Am Soc Eng Educ; Sigma Xi; Nat Soc Prof Engrs. *Res:* Mass transfer; adsorption; process control. *Mailing Add:* Midwest Ore Processing Co Inc 725 Ridgecrest Rd Akron OH 44303-1344

ATWOOD, HAROLD LESLIE, b Montreal, Que, Feb 15, 37; m 59; c 3. PHYSIOLOGY. *Educ:* Univ Toronto, BA, 59; Univ Calif, Berkeley, MA, 60; Glasgow Univ, PhD(zool), 63. *Hon Degrees:* DSc, Glasgow Univ, 79. *Prof Exp:* Res assoc biol, Univ Ore, 62-64; res fel, Calif Inst Technol, 64-65; from asst prof to assoc prof, 65-71, PROF ZOOL, UNIV TORONTO, 71-, PROF & CHMN, DEPT PHYSIOL, 81- *Mem:* Am Physiol Soc; Brit Soc Exp Biol; fel AAAS; fel Royal Soc Can; Soc Neurosci. *Res:* Comparative and neuromuscular physiology; muscular dystrophy. *Mailing Add:* Dept Physiol Med Sci Bldg Univ Toronto Toronto ON M5S 1A8 Can

ATWOOD, JERRY LEE, b Springfield, Mo, July 27, 42; m 64, 84; c 3. INORGANIC CHEMISTRY, PHYSICAL CHEMISTRY. *Educ:* Southwest Mo State Col, BS, 64; Univ Ill, MS, 66, PhD(inorg chem), 68. *Prof Exp:* From asst prof to assoc prof, 68-87, UNIV RES PROF CHEM, UNIV ALA, TUSCALOOSA, 87- *Mem:* Am Inst Chem Eng; Am Chem Soc; Am Crystallog Asn. *Res:* Organometallic chemistry; x-ray crystallography; inclusion chemistry. *Mailing Add:* Dept Chem Univ Ala Box 870336 Tuscaloosa AL 35487-0336

ATWOOD, JIM D, b Springfield, Mo, June 3, 50; m; m 71; c 3. INORGANIC CHEMISTRY, ORGANOMETALLIC CHEMISTRY. *Educ:* Southwest Mo State Univ, BS, 71; Univ Ill, PhD(chem), 75. *Prof Exp:* Fel, US-USSR Prog Coop Catalysis, Cornell Univ, 75-76; assoc dean, fac natural sci & math, 84-85, from asst prof to assoc prof, 77-88, PROF CHEM, STATE UNIV NY, BUFFALO, 89- *Concurrent Pos:* Alfred P Sloan Found fel, 83-85, Alexander von Humboldt Found fel, 84; guest prof, Johann Wolfgang Goethe Univ, Frankfurt, 84. *Mem:* Am Chem Soc. *Res:* Reactivity and mechanisms of organometallic compounds; synthesis of and catalysis by unsaturated transition metal clusters; modeling catalytic reactions; ligand effects; polynuclear complexes; electron transfer. *Mailing Add:* Dept Chem State Univ NY Buffalo NY 14214-3094

ATWOOD, JOHN LELAND, b Walton, Ky, Oct 26, 04; m 72; c 1. ENGINEERING. *Educ:* Hardin Simmons Univ, AB, 26; Univ Tex, BS, 28. *Hon Degrees:* DEng, Stevens Inst Technol, 55, Carnegie-Mellon Univ, 65, Harvey Mudd Col, 79. *Prof Exp:* Jr aeroplane engr, Army Air Corps, Dayton, Ohio, 28-29; design engr, Douglas Aircraft Co, Calif, 30-34; vpres & chief engr, NAm Aviation, Inc, 34-38, vpres & asst gen mgr, 38-41, first vpres, 41, pres, 48-67, chief exec officer, 60-67, chmn bd, 62-67, pres & chief exec officer, NAm Rockwell Corp, 67-70, dir & sr consult, Rockwell Int Corp, 70-77; RETIRED. *Concurrent Pos:* Chmn indust adv comt, Nat Adv Comt Aeronaut, 58. *Honors & Awards:* H H Arnold Trophy, 70; Wright Bros Mem lectr, 76; Wright Bros Mem Trophy. *Mem:* Nat Acad Eng; hon fel Am Inst Aeronaut & Astronaut; Soc Automotive Engrs; Aerospace Indust Asn. *Mailing Add:* PO Box 1587 Vista CA 92085-1587

ATWOOD, JOHN WILLIAM, b Sherbrooke, Que, Feb 23, 41; m 70; c 2. COMPUTER NETWORK PROTOCOLS, DISTRIBUTED OPERATING SYSTEMS. *Educ:* McGill Univ, BEng, 63; Univ Toronto, MASc, 65; Univ Ill, Urbana, PhD(elec eng), 70. *Prof Exp:* Res assoc elec eng, Univ Ill, Urbana, 65-70; asst prof, Univ Toronto, 70-72; asst prof comput sci, 73-75, ASSOC PROF COMPUT SCI, CONCORDIA UNIV, 75- *Concurrent Pos:* Consult, Metrop Toronto Police Comn, 71-72 & Amdahl Corp, 76-79; vis assoc prof comput sci, Univ Tex, Austin, 79-80; vis researcher, Centre Recherche Informatique Montreal, 86- 87. *Mem:* Inst Elec & Electronics Engrs. *Res:* Design, analysis, validation, performance evaluation of high- speed computer network protocols; design, implementation, performance evaluation of distributed operating systems. *Mailing Add:* Dept Comput Sci Concordia Univ 1455 de Maisonneuve Blvd W Montreal PQ H3G 1M8 Can

ATWOOD, KENNETH W, b Cedarview, Utah, Dec 21, 22; m 44; c 3. ELECTRICAL ENGINEERING. *Educ:* Univ Utah, BS, 50, MS, 54, PhD(elec eng), 57. *Prof Exp:* Transmitter engr, Sta KSL-KSL-TV, 49-56; from asst prof to assoc prof elec eng, 56-76, PROF ELEC ENG, UNIV UTAH, 76- *Concurrent Pos:* Sr engr, Hanford Labs, Gen Elec Co, 62-63. *Mem:* Am Soc Eng Educ; Inst Elec & Electronics Engrs. *Res:* Solid state circuitry; digital circuitry design. *Mailing Add:* Dept Elec Eng Univ Utah 1400 E Second St Salt Lake City UT 84112

ATWOOD, LINDA, b Rochester, NY, Nov 13, 46. BIO-ORGANIC CHEMISTRY. *Educ:* Bard Col, BA, 68; Wesleyan Univ, MA, 72, PhD(chem), 74. *Prof Exp:* Teaching & res asst chem, Wesleyan Univ, 70-74; asst prof, 74, assoc prof, 74-84, PROF CHEM, CALIF POLYTECH STATE UNIV, SAN LUIS OBISPO, 84- *Mem:* Am Chem Soc; AAAS. *Res:* Teaching and writing in the areas of pharmacology and nutritional biochemistry. *Mailing Add:* Dept Chem Calif Polytech State Univ San Luis Obispo CA 93407

ATWOOD, MARK WYLLIE, b Fairfield, Iowa, Mar 23, 42; m 62; c 2. ENTOMOLOGY, PHYSIOLOGY. *Educ:* Parsons Col, BS, 62; Univ Wis, Madison, MS, 65, PhD(zool), 68. *Prof Exp:* Asst prof biol, Purdue Univ, Ft Wayne, 67-70; entomologist, Iowa State Dept Agr, 70-77; mem staff, Agricult Div, Am Cyanamid Co, 77-80. *Mem:* AAAS; Am Soc Zool; Cent Plant Bd (pres, 75). *Res:* Excitation in insect visual system and use of visual information in behavior; mechanism of action of pesticides and relation of sub-lethal doses of pesticides to insect behavior. *Mailing Add:* 24 Alpine Lane West Milford NJ 07480-1801

ATYEO, WARREN THOMAS, b Highland Park, Mich, Feb 15, 27; m 55; c 3. ENTOMOLOGY. *Educ:* Western Ill Univ, BS & MS, 53; Univ Kans, PhD(entom), 59. *Prof Exp:* From asst prof to prof entom, Col Agr, Univ Nebr, 58-67, cur mus, 58-67; participant, Grants Assocs Prog, NIH, 67-68; PROF ENTOM & CUR, UNIV GA, 68- *Concurrent Pos:* Res assoc, Field Mus Natural Hist. *Mem:* Soc Study Evolution; Entom Soc Am; Acarological Soc Am; Soc Systematic Zool; Soc Parasitol. *Res:* Acarology; systematics of Acarina: Bdelloidea, Analgoidea. *Mailing Add:* Dept Entom Univ Ga Athens GA 30602

AU, CHI-KWAN, b Macau, China, Jan 21, 46; nat US; m 70; c 2. THEORETICAL PHYSICS. *Educ:* Hong Kong Univ, BSc, 68; Columbia Univ, MA, 70, PhD(physics), 72. *Prof Exp:* Res assoc physics, Univ Ill, Urbana, 72-74; res assoc, Yale Univ, 74-75, lectr, 75; from asst prof to assoc prof, 75-84, PROF PHYSICS, UNIV SC, COLUMBIA, 85- *Concurrent Pos:* Vis asst prof physics, Columbia Univ, 76, 77 & 78; mem, Inst for Theoret Physics, Santa Barbara, 81 & 88; hon lectr, Hong Kong Univ, 81-82; co-investr, 80-85, prin investr, NSF grants, 79-81 & 85-; prin investr, Univ SC fac res & productive scholarship grant, 81-82 & 85-86; vis scientist, Harvard-Smithsonian Ctr Astrophys, Harvard Univ, 82, 83 & 84; partic, Lewis Ctr Physics, Del, 84 & 85. *Honors & Awards:* Univ SC Educ Found Res Award, 85. *Mem:* Am Phys Soc. *Res:* Theoretical astrophysics, neutron star matter; perturbation theory; atomic physics and quantum electrodynamics of simple atomic systems, interpretations of quantum mechanics. *Mailing Add:* Dept Physics & Astron Univ SC Main Campus Columbia SC 29208

AU, TUNG, b Hong Kong, Sept 8, 23; nat US; m 55; c 2. CIVIL ENGINEERING. *Educ:* St Johns Univ, China, BS, 43; Univ Ill, MS, 48, PhD(civil eng), 51; Univ Mich, MSE, 54. *Prof Exp:* Eng construct & teaching, China, 43-47; asst civil eng, Univ Ill, 48-51; struct engr, Harley, Ellington & Day, Inc, Mich, 51-52, 53-55 & Boddy-Benjamin Assocs, Mich, 52-53; asst prof eng mech, Univ Detroit, 55-57; assoc prof, 57-64, PROF CIVIL ENG, CARNEGIE-MELLON UNIV, 64- *Honors & Awards:* State-of-the-Art of Civil Eng Award, Am Soc Civil Engrs, 73; Western Elec Fund Award, Am Soc Eng Educ, 79. *Mem:* Hon mem Am Soc Civil Engrs; Am Soc Eng Educ; Am Asn Cost Engrs; Nat Soc Prof Engrs; Am Soc Testing Mat. *Res:* Engineering system analysis planning and design. *Mailing Add:* Dept Civil Eng Carnegie-Mellon Univ Pittsburgh PA 15213

AUBEL, JOSEPH LEE, b Lansing, Mich, Sept 7, 36; m 67; c 1. PHYSICS. *Educ:* Mich State Univ, BS, 54, PhD(physics), 64. *Prof Exp:* Asst prof, 64-74, ASSOC PROF PHYSICS, UNIV S FLA, 74- *Mem:* Optical Soc Am; Am Asn Physics Teachers; Am Sci Affil. *Res:* Optics; computer-based and individualized competency-based education; semiconductor physics instrumentation. *Mailing Add:* Dept Physics Univ SFla Tampa FL 33620

AUBERTIN, GERALD MARTIN, b Kankakee, Ill, Sept 23, 31; m 57; c 4. WATERSHED MANAGEMENT. *Educ:* Univ Ill, Urbana-Champaign, BS, 58, MS, 60; Pa State Univ, PhD(agron), 64. *Prof Exp:* Asst agron, Univ Ill, 58; res asst, Pa State Univ, 64; res soil physicist & fel, Univ Calif, Riverside, 64-65; res soil scientist, US Forest Serv, Parsons, WVa, 65-76; ASSOC PROF FORESTRY, SOUTHERN ILL UNIV, CARBONDALE, 76- *Concurrent Pos:* Vis prof dept earth resources, Colo State Univ, 75; vis res scientist, Pac Forest Res Ctr, Victoria, BC, 83-86; chair, Qual Assurance Steering Comt, Nat Atmospheric Deposition Prog. *Mem:* Am Soc Agron; Soil Sci Soc Am; Sigma Xi. *Res:* Watershed management; non-point pollution; forestry best management practices; stream water quality; subsurface water movement and soil macropores in forested watersheds; oxygen-salt-moisture and root relations; physical edaphology; soil structure problems; geologic/soil/water quality investigation; acid rain; atmospheric deposition; effect of forest harvesting on soil and water. *Mailing Add:* Dept Forestry Southern Ill Univ Carbondale IL 62901

AUBIN, JANE E, b Cornwall, Ont, Mar 3, 50. CELL BIOLOGY, DIFFERENTIATION. *Educ:* Queen's Univ, BSc, 72; Univ Toronto, PhD(med biophysics), 77. *Prof Exp:* From asst prof to assoc prof, 79-88, PROF CELL BIOL, UNIV TORONTO, 88- *Concurrent Pos:* Oral biol res award, Int Asn Dent Res, 85. *Mem:* Am Soc Cell Biol; Bone & Mineral Res; Int Asn Dental Res; Can Soc Cell Biol; Int Soc Differentiation. *Res:* Osteoblast lineage, differentiation and regulation. *Mailing Add:* Med Res Coun Group Univ Toronto Toronto ON M5S 1A8 Can

AUBORN, JAMES JOHN, b Portland, Ore, Feb 21, 40; m 62; c 2. PHYSICAL CHEMISTRY. *Educ:* Ore State Univ, BS, 61, MS, 62; Univ Utah, PhD(chem), 71. *Prof Exp:* Mem tech staff electrochem, GTE Lab, Inc, Gen Tel & Electronics Corp, Bayside, NY, 71-72 & Waltham, Mass, 72-73, mem tech staff mat sci, 74-75, tech proj mgr electrochem, 75-76; from mem tech staff to distinguished mem tech staff, 76-87, tech supvr undersea lightwave systs, 87-89, DEPT HEAD/PROG MGR, LIGHTWAVE TRANSMISSION SYSTS, AT&T BELL LABS, 89- *Concurrent Pos:* Instr, Northeastern Univ, 74-76; proj mgr electrochem, Off Naval Res, 84; sci tech agent, Dept Defense, 84. *Mem:* Electrochem Soc; Am Chem Soc; AAAS; Am Inst Physics; Sigma Xi; Optical Soc; Int Soc Optical Engrs. *Res:* Undersea fiber optic communications systems; chemistry and physics of inorganic non-aqueous liquids relating to electrochemical power sources; materials science and engineering; high temperature and refractory materials; electrical and optical properties of materials. *Mailing Add:* 121 Mt Horeb Rd Warren NJ 07060

AUBRECHT, DON A(LBERT), chemical engineering, for more information see previous edition

AUBRECHT, GORDON JAMES, II, b Bedford, Ohio, May 2, 43; m 84; c 3. ENERGY ISSUES, PHYSICS PEDAGOGY. *Educ:* Rutgers Univ, BA, 65; Princeton Univ, PhD(theoret physics), 71. *Prof Exp:* Res assoc high energy theory, Ohio State Univ, 70-72; vis asst prof, Inst Theoret Sci, Univ Ore, 72-75; from asst prof to assoc prof, 81-87, PROF PHYSICS, OHIO STATE UNIV, 87- *Concurrent Pos:* Alexander von Humboldt fel, 79-80; gastdozent, Inst Theoret Physics, Univ Karlsruhe, WGer, 79-80; consult, Ctr Fac Develop, Kansas State Univ, 83-86; vis fel, Am Asn Physics Teachers, Univ Md, 85-87; coordr, Conf Teaching Mod Phys, 85-88; chmn, Comt Prof Concerns, Am Asn Physics Teachers, 85-87, Metric Comt, 91-; steering comt, Conf Comput Physics Educ, 87-88; auth. *Honors & Awards:* Woodrow Wilson Fel, 65-66. *Mem:* Am Phys Soc; Am Asn Physics Teachers; Sigma Xi; AAAS. *Res:* Theoretical structure of the electroweak interaction; higher symmetry; chiral group phenomenology; physics pedagogy; alteration of the introductory physics to include more contemporary physics at all levels of education. *Mailing Add:* Dept Physics Ohio State Univ 1465 Mt Vernon Ave Marion OH 43302-5695

AUBREY, DAVID GLENN, b Ft Sill, Okla, Aug 21, 50; m 72; c 3. MARINE SCIENCES. *Educ:* Univ Southern Calif, BS(civil eng) & BS(geol sci), 73; Scripps Inst Oceanog, PhD(oceanog), 78. *Prof Exp:* SR SCIENTIST, WOODS HOLE OCEANOG INST, 78-; DIR, COASTAL RES CTR. *Concurrent Pos:* Consult, 72-; lectr, Mass Inst Technol, 84-85. *Mem:* Am Geophys Union; AAAS. *Res:* Nearshore processes: inner shelf sediment transport; waves and currents; beach erosion; tidal waves; sea-level change; lagoons and estuaries. *Mailing Add:* 204 Woods Hole Rd Falmouth MA 02540

AUCHAMPAUGH, GEORGE FREDRICK, b Chicago, Ill, Jan 19, 39. EXPERIMENTAL NUCLEAR PHYSICS. *Educ:* Univ Ill, BS, 61; Univ Calif, MS, 66, PhD(appl sci), 68. *Prof Exp:* Staff mem neutron physics, Lawrence Livermore Lab, 61-68; STAFF MEM NEUTRON PHYSICS, LOS ALAMOS SCI LAB, 68- *Mem:* Am Phys Soc. *Res:* Neutron physics research with monoenergetic and continuous neutron sources; fission barrier physics; resonance region of sub threshold fission isotopes; nuclear spectroscopy of light elements from R-matrix analysis of neutron cross section data. *Mailing Add:* 519 Rover Blvd Los Alamos NM 87544

AUCHINCLOSS, HUGH, JR, b New York, NY, Mar 15, 49; m; c 3. TRANSPLANT SURGERY. *Educ:* Harvard Univ, MD, 76. *Prof Exp:* From instr to asst prof, 86-90, ASSOC PROF SURG, HARVARD UNIV, 90-; CHIEF PANCREAS TRANSPLANTATION, MASS GEN HOSP, 90- *Concurrent Pos:* Asst surg, 77-85, Mass Gen Hosp, 77-85. *Mem:* Transplantation Soc; Am Soc Transplant Surgeons; Am Asn Immunologists; Am Col Surgeons. *Res:* Transplantation; immunology. *Mailing Add:* Dept Surgery Mass General Hosp Boston MA 02114

AUCHINCLOSS, JOSEPH HOWLAND, JR, b Lawrence, NY, June 28, 21; m 46; c 4. INTERNAL MEDICINE. *Educ:* Yale Univ, BS, 43; Columbia Univ, MD, 45. *Prof Exp:* From instr to assoc prof, 51-71, PROF MED, COL MED, STATE UNIV NY UPSTATE MED CTR, 71- *Mem:* Am Fedn Clin Res; Am Thoracic Soc; Am Physiol Soc; Am Col Physicians; Sigma Xi. *Res:* Cardiopulmonary physiology. *Mailing Add:* 3935 Rippleton Rd Cazenovia NY 13035

AUCHMUTY, GILES, b Dublin, Ireland, 1945; US citizen; m 81. VARIATIONAL METHODS, NUMERICAL ANALYSIS. *Educ:* Australian Nat Univ, BSc, 66; Univ Chicago, SM, 68, PhD(appl math), 70. *Prof Exp:* Res instr math, State Univ NY, Stony Brook, 70-72; from asst prof to assoc prof math, Ind Univ, Bloomington, 72-81; PROF MATH, UNIV HOUSTON, 82- *Concurrent Pos:* Res fel, Math Inst, Oxford Univ, 81; vis mem, Inst Advan Study, Princeton, 89; vis prof, Tex A&M Univ, College Station, 90. *Mem:* Am Math Soc; Soc Indust & Appl Math; Sigma Xi. *Res:* Variational methods; duality theories; applications; numerical analysis; models of rotating stars; thermodynamics; chemical equilibria; reaction diffusion problems. *Mailing Add:* Dept Math Univ Houston Houston TX 77204-3476

AUCHMUTY, JAMES FRANCIS GILES, b Dublin, Eire, June 1, 45. APPLIED MATHEMATICS. *Educ:* Australian Nat Univ, BSc, 65; Univ Chicago, MS, 68, PhD(math), 70. *Prof Exp:* Tutor appl math, Australian Nat Univ, 66; lectr math, State Univ NY Stony Brook, 70-72; from vis asst prof to assoc prof, Ind Univ, 72-81; PROF, UNIV HOUSTON, 81- *Concurrent Pos:* Vis prof, Int Inst Physics & Chem, Belg, 74; fel, Fluid Mech Res Inst, Univ Essex & Brit Sci Res Coun, 75-76; vis assoc prof math, Univ Tex, Austin, 78 & Univ Houston, 79; vis prof, Univ Bonn, 78, Australian Nat Univ, 79, Oxford Univ, 81 & Tex A&M, 90; mem, Inst Adv Studies, 89. *Mem:* Am Math Soc; Soc Indust & Appl Math; Australian Math Soc; Asn of Math Physicists. *Res:* Nonlinear and numerical analysis and applications to astronomy, biology and chemistry; bifurcation theory and variational methods. *Mailing Add:* Dept Math Univ Houston Houston TX 77204-3476

AUCHTER, HARRY A, b St Louis, Mo, Aug 31, 20; m 54; c 2. PHYSICS. *Educ:* Southeast Mo State Col, AB, 42; Univ Iowa, MS, 46. *Prof Exp:* Instr physics, Kearney State Col, 46-47, Bangkok Christian Col, Thailand, 47-50 & Augustana Col (SDak), 51-53; from instr to asst prof & math, 53-58, assoc prof, 58-86, EMER PROF PHYSICS, CARROLL COL, WIS, 86- *Mem:* Am Asn Physics Teachers. *Res:* Adiabatic compressibilities of aqueous solutions using ultrasonics. *Mailing Add:* 347 W College Ave Waukesha WI 53186

AUCLAIR, JACQUES LUCIEN, b Montreal, Que, Apr 2, 23; m 51; c 2. ENTOMOLOGY. *Educ:* Univ Montreal, BSc, 42; McGill Univ, MSc, 45; Cornell Univ, PhD(insect physiol), 49. *Prof Exp:* Asst prof physiol, Univ Montreal, 49-53; res entomologist, Res Lab, Can Dept Agr, 53-64; prof, Dept Bot & Entom, NMex State Univ, 64-67; dir, 67-73, PROF, DEPT BIOL SCI, UNIV MONTREAL, 67- *Mem:* Am Inst Biol Sci; Entom Soc Am. *Res:* General entomology; insect physiology and biochemistry; factors of plant resistance to insects. *Mailing Add:* Dept Biol Sci Univ Montreal C P6128 Montreal PQ H3C 3J7 Can

AUCLAIR, WALTER, b Manchester, NH, Sept 13, 33; m 54; c 3. DAIRY NUTRITION, GENETICS. *Educ:* Univ Conn, BA, 55; NY Univ, MS, 58, PhD(biol), 60. *Prof Exp:* NIH res fel, George Washington Univ, 60-61; from asst prof to assoc prof zool, Univ Cincinnati, 61-66; assoc prof biol, Rensselaer Polytech Inst, 66-77; OWNER, AUCLAIR FARMS, 66- *Concurrent Pos:* Investr, Zool Sta, Naples, Italy, 65-66; trustee & secy, Bermuda Biol Sta Res, Inc; NSF sci consult, India, 68-69; vis assoc prof, Cambridge Univ, England, 74, Rockefeller Univ, 74-75. *Honors & Awards:* Nat Winner, Nat Endowment Soil & Water Conserv, 90. *Mem:* Am Inst Biol Sci; Am Soc Cell Biol; Soc Develop Biol; Am Soc Animal Sci; Am Dairy Sci Asn. *Res:* Differentiation of cilia and other organelles in marine invertebrate embryos; renal hemopoiesis in amphibians; bovine reproductive physiology; dairy nutrition and genetics. *Mailing Add:* RD-1 Box 99 Melrose NY 12121

AUCOIN, PASCHAL JOSEPH, JR, b Houston, Tex, Oct 11, 32; m 57; c 3. ATMOSPHERIC DYNAMICS, FUEL TECHNOLOGY. *Educ:* Rice Univ, BA, 54; Univ Tex, Austin, MA, 56; Univ Calif, Los Angeles, PhD(math), 71. *Prof Exp:* Engr, Exxon Prod Res, 56-62; adv tech staff numerical anal, CCI-Marquardt, 62-71; statistician, Technol Incorp, Houston, 71-72; sci consult math, Digital Resources Corp, Algiers, 72-74; sr scientist, Ford Aerospace Co, 74; engr, Aramco, Saudia Arabia, 74-77; syst analyst, Lemsco, Houston, 77-80; VPRES, O'CONNOR RES, INC, DENVER, 80- *Concurrent Pos:* Geophys consult, C H Burt, Inc, Houston, 77-79, MRO Assoc, Denver, 79-80. *Mem:* Soc Explor Geophysicists; Am Chem Soc; Am Inst Mining, Metall & Petrol Engrs. *Res:* Solutions of non-linear partial differential equations describing physical processes in geophysics, meterology, and oceanography, with attendant numerical analysis formulations and digital computer implementations. *Mailing Add:* 16418 Hickory Knoll Dr Houston TX 77059-5326

AUDESIRK, GERALD JOSEPH, b New Brunswick, NJ, Sept 15, 48; m 70. NEUROETHOLOGY. *Educ:* Rutgers Univ, BA, 70; Calif Inst Technol, PhD(neurobiol), 74. *Prof Exp:* Fel neurobiol, Calif Inst Technol, 74-57; fel, Univ Wash, 76-78, res assoc, 78-79; asst prof neurobiol, Univ Mo-Columbia, 79-; AT DEPT BIOL, UNIV COLO. *Mem:* AAAS; Soc Neurosci; Animal Behav Soc. *Res:* Cellular basis of neurotoxicity of heavy metals especially lead; cellular basis of behavior of invertebrates. *Mailing Add:* Dept Biol Univ Colo 1200 Larimer Denver CO 80204

AUDESIRK, TERESA ECK, b Washington, DC, Jan 28, 50; m 71. NEUROETHOLOGY. *Educ:* Bucknell Univ, BS, 71; Univ Southern Calif, PhD(marine physiol), 76. *Prof Exp:* Res assoc neuroethol, Univ Wash, 76-79; asst prof neuroethol & biol, Univ Mo, 79-; AT BIOL DEPT, DIV NATURAL PHYS SCI, UNIV COLO, DENVER. *Mem:* Soc Neurosci; Sigma Xi; Animal Behav Soc. *Res:* Identifying neurons in simple invertebrate nervous systems which underlie production of behavior; neuronal changes which underlie learning. *Mailing Add:* Biol Dept Box 171 Univ Colo Denver PO Box 173364 Denver CO 80217-3364

AUE, DONALD HENRY, b Columbus, Ohio, June 19, 42; m 64; c 2. ORGANIC CHEMISTRY. *Educ:* Ohio State Univ, BSc, 63; Cornell Univ, PhD(org chem), 67. *Prof Exp:* NSF fel org chem, Columbia Univ, 67-68; ASSOC PROF CHEM, UNIV CALIF, SANTA BARBARA, 68- *Mem:* Am Chem Soc; Brit Chem Soc. *Res:* Synthesis and properties of strained carbocyclic and heterocyclic systems; additions to strained multiple bonds; carbonium ion reactions in solution and gas phase; organic photochemistry; organic synthesis. *Mailing Add:* Dept Chem Univ Calif Santa Barbara CA 93106

AUE, WALTER ALOIS, b Vienna, Austria, Jan 20, 35. ORGANIC CHEMISTRY, ANALYTICAL CHEMISTRY. *Educ:* Univ Vienna, PhD(org chem), 63. *Prof Exp:* Res investr org chem, Western Reserve Univ, 63-65; asst prof anal chem, Univ Mo-Columbia, 65-69, assoc prof chem & res assoc Space Sci Ctr, 69-73; dept chmn, 83-86, PROF CHEM, DALHOUSIE UNIV, 73- *Concurrent Pos:* USDA grant with Dr Billy G Tweedy, 67-70; USPHS grant, 67-70; NSF grant, 70-72; US Environ Protection Agency grant, 72-75; Nat Res Coun Can grant, 73-; Can Defense Res Bd grant, 74-76; Agr Can grant, 74-79; Environ Can grant, 78-80. *Honors & Awards:* Fisher Award, Can, 80. *Res:* Pesticide residue analysis; specific detectors for gas chromatography; support-bonded chromatographic phases. *Mailing Add:* Dept Chem Dalhousie Univ Halifax NS B3H 3J5 Can

AUEL, RAEANN MARIE, electrochemistry, for more information see previous edition

AUER, HENRY ERNEST, b New York, NY, May 20, 38; m 68; c 2. PROTEIN STRUCTURE & STABILITY, PROTEIN FOLDING. *Educ:* Princeton Univ, AB, 60; Harvard Univ, PhD(biochem), 65. *Prof Exp:* Res fel polymers, Weizmann Inst Sci, Israel, 65-67; res fel chem, Cornell Univ, 67-68; asst prof biochem, Sch Med, Univ Rochester, 68-78; assoc prof biochem, Med Col Wis, 78-85; res scientist proteins, Pitman-Moore-Imcera, 85-89; sr res scientist protein pharmaceut, 89-91, ASSOC RES FEL PROTEIN PHARMACEUT, ENZYTECH INC, 91- *Concurrent Pos:* Res fel, Helen Hay Whitney Found, 65-68. *Mem:* Am Chem Soc; Am Soc Biochem & Molecular Biol. *Res:* Protein structure and function; stability and conformational transformations in proteins and enzymes; stabilization of protein pharmaceutical agents in sustained release systems; mechanism of denaturation and refolding of proteins. *Mailing Add:* Enzytech Inc 64 Sidney St Cambridge MA 02139

AUER, JAN WILLEM, b Utrecht, Neth, Apr 10, 42; Can citizen; c 2. MATHEMATICS. *Educ:* McGill Univ, BEng, 64, MSc, 66; Univ Toronto, PhD(math), 71. *Prof Exp:* Lectr, 69-70, asst prof math, 70-76, ASSOC PROF MATH, BROCK UNIV, 76- *Mem:* Am Math Soc; Math Asn Am; Tensor Soc. *Res:* Algebraic topology and differential geometry, specifically in smooth bundles. *Mailing Add:* Brock Univ St Catharines ON L2S 3A1 Can

AUER, LAWRENCE H, b Englewood, NJ, Dec 26, 41; m. ASTROPHYSICS. *Educ:* Haverford Col, BA, 63; Princeton Univ, PhD(astron), 67. *Prof Exp:* Res assoc & fel astron, Joint Inst Lab Astrophys, Univ Colo, 66-68; from asst prof to assoc prof, Yale Univ, 68-77, dir grad studies, Watson Astron Ctr, 74-77; mem staff, Nat Ctr Atomspheric Res, 77-; assoc prof, Pa State Univ, 80-; AT DIV EARTH & SPACE SCI, LOS ALAMOS SCI LAB. *Mem:* Am Astron Soc; Asn Comput Mach; Royal Astron Soc; Sigma Xi. *Res:* Stellar atmospheres; radiative transfer; numerical analysis. *Mailing Add:* Earth & Environ Sci Div Los Alamos Sci Lab Box 1663-MS F665 Los Alamos CA 87545

AUER, PETER LOUIS, b Budapest, Hungary, Jan 12, 28; nat US; m 52; c 4. PLASMA PHYSICS, THEORETICAL PHYSICS. *Educ:* Cornell Univ, BA, 47; Calif Inst Technol, PhD(chem & physics), 51. *Prof Exp:* Fel & res asst, Calif Inst Technol, 48-50; res chemist & physicist, Calif Res & Develop Co, 50-54; physicist, Gen Elec Res Lab, 54-61; head plasma physics dept, Sperry Rand Res Ctr, 62-64; dep dir ballistic missile defense, Advan Res Proj Agency, 64-66; prof, Grad Sch Aerospace Eng, 66-72, dir lab plasma studies, 67-74, PROF MECH & AEROSPACE ENG, CORNELL UNIV, 72- *Concurrent Pos:* Guggenheim fel, 60-61; vis prof, Oxford Univ, 72-73. *Mem:* Fel Am Phys Soc. *Res:* Theoretical physics and chemistry; gaseous electronics. *Mailing Add:* 224 Upson Hall Cornell Univ Ithaca NY 14853

AUERBACH, ANDREW BERNARD, b New York, NY, Jan 23, 48; m 72; c 2. POLYMER CHEMISTRY. *Educ:* Brooklyn Col, BS, 68; City Univ New York, PhD(polymer & org chem), 75; Rutgers Univ, MBA, 80. *Prof Exp:* Res chemist, ITT Rayonier Inc, 74-81, group leader, 79-81; sr res chemist, 81-87, staff chemist, 87-90, RES ASSOC, HOECHST CELANESE, 90- *Concurrent Pos:* Adj asst prof chem, Morris County Community Col, 81-86; dir, Ctr Prof Advan, Course on Eng Plastics, 90- *Mem:* Am Chem Soc; sr mem Soc Plastics Engrs. *Res:* Polymer stabilization; computer databases for plastics formulation; liquid crystals; engineering resins including polyacetals, nylons and polyesters; crosslinking; flame retardants; polymer blends; product development. *Mailing Add:* 23 Orchard Lane Livingston NJ 07039

AUERBACH, ARLEEN D, b New York, NY, May 24, 37; m 58; c 2. HUMAN GENETICS, CYTOGENETICS. *Educ:* William Smith Col, BA, 57; Columbia Univ, MA, 58; NY Univ, PhD(biol), 77. *Prof Exp:* Instr biol, Brearley Sch, NY, 58-59 & Dalton Sch, NY, 71; res fel, 77-78, res assoc genetics, Sloan-Kettering Inst, 78-82; ASST PROF LAB INVEST DERMATOL, ROCKEFELLER UNIV, 82- *Concurrent Pos:* Nat Res Serv fel, Nat Cancer Inst, 78- *Mem:* AAAS; Soc Invest Dermatol; Am Soc Human Genetics; Environ Mutagen Soc; Genetic Toxicol Asn; Tissue Cult Asn. *Res:* Characterization of the molecular defect in Fanconi anemia; somatic cell genetics; molecular genetics and cytogenetics; faniom anemia registry. *Mailing Add:* Rockefeller Univ 1230 York Ave New York NY 10021

AUERBACH, ARTHUR HENRY, b Philadelphia, Pa, Mar 12, 28. PSYCHIATRY. *Educ:* Jefferson Med Col, MD, 51. *Prof Exp:* Resident psychiat, St Elizabeth's Hosp, Washington, DC, 52-53; resident, West Haven Vet Admin Hosp, Conn, 53-54; resident, Hosp, Temple Univ, 56-57; NIMH career develop award, Dept Psychiat, 64-69, ASST PROF PSYCHIAT, SCH MED, UNIV PA, 64- *Concurrent Pos:* Consult, Magee Mem Hosp, Philadelphia, 58-68; NIMH Career Develop Award, Dept Psychiat, 64-69. *Mem:* Am Psychiat Asn; Soc Psychother Res. *Res:* Psychotherapy research. *Mailing Add:* Dept Psychiat Sch Med 133 S 36th St Second Floor Philadelphia PA 19104

AUERBACH, CLEMENS, b Berlin, Ger, Nov 30, 23; US citizen; m 64; c 1. CHEMISTRY. *Educ:* Robert Col, Istanbul, BS, 43; Harvard Univ, AM, 48, PhD(chem), 51. *Prof Exp:* Res fel, Univ Minn, 50-53; asst prof chem, Univ Buffalo, 53-56; from asst chemist to assoc chemist, 56-62, SCIENTIST, DEPT NUCLEAR ENERGY, BROOKHAVEN NAT LAB, 62- *Mem:* Am Chem Soc; Inst Nuclear Mat Mgt. *Res:* Analytical chemistry; analysis of uranium and other reactor materials; nuclear materials safeguards. *Mailing Add:* Bldg 197C Brookhaven Nat Lab Upton NY 11973

AUERBACH, DANIEL J, b New York, NY, Nov 29, 42; m 69; c 3. SURFACE SCIENCE. *Educ:* Univ Chicago, BS, 64, MS, 68, PhD(physics), 71. *Prof Exp:* Vis scientist, Inst Atomic & Molecular Physics, Amsterdam, 71-72; teaching fel physics, Univ Western Ontario, 72-73; sr res fel, James Frank Inst, 73-75; asst prof chem, Johns Hopkins Univ, 76-77; staff mem, 78-83, MGR PHYS SCI, IBM, 83- *Mem:* Fel Am Phys Soc; Am Vacuum Soc; Am Chem Soc; Materials Res Soc. *Res:* Study of the dynamics of gas surface interaction, using a combination of surface science, molecular beam and laser techniques. *Mailing Add:* IBM-K311802 650 Harry Rd San Jose CA 95120-6099

AUERBACH, EARL, b Lodz, Poland; nat US; m 50; c 3. PARASITOLOGY, FOOD SCIENCE. *Educ:* Univ Ill, BS, 46, MS, 47; Northwestern Univ, PhD(biol), 51. *Prof Exp:* Histochemist, Am Meat Inst Found, 51-54, asst to dir, 54-59, chief div histol, 59-64; oper mgr, Polo Food Prod Co, 64-65, dir res & develop, 65-70; dir res & develop, Pronto Food Corp, 70-73; DIR RES & DEVELOP, DAVID BERG & CO, 73-, VPRES CORP AFFAIRS, 80- *Mem:* Am Soc Parasitol; Soc Protozool; Inst Food Technol. *Res:* Parasitic protozoa; cytochemistry and histochemistry; freeze dehydration. *Mailing Add:* David Berg & Co 165 S Watermarket Chicago IL 60608

AUERBACH, ELLIOT H, b New York, NY, July 21, 32; m 71. NUCLEAR PHYSICS, ACCELERATOR PHYSICS. *Educ:* Columbia Univ, AB, 53, AM, 57. *Prof Exp:* Mathematician & physicist, Knolls Atomic Power Lab, Gen Elec Co, 58-61; assoc physics, Brookhaven Nat Lab, 61-64; vis res scientist, Lab Nuclear Sci, Mass Inst Technol, 64-66; staff mem, Physics Dept, 66-82, ALTERNATING GRADIENT SYNCHROTRON DEPT, BROOKHAVEN NAT LAB, 84- *Concurrent Pos:* Vis staff mem, Los Alamos Sci Lab, 65 & 66. *Mem:* Am Phys Soc. *Res:* Nuclear structure and scattering; applications of computers to model and control accelerators. *Mailing Add:* 911-C Brookhaven Nat Lab Upton NY 11973

AUERBACH, IRVING, b Cleveland, Ohio, May 24, 19; m 69. HIGH TEMPERATURE MATERIALS, POLYMER CHEMISTRY. *Educ:* Ohio State Univ, BSc, 42, PhD(chem), 48. *Prof Exp:* Asst org chem, Res Found, Ohio State Univ, 44-46; res assoc, Cleveland Indust Res, 48-50; res assoc, Goodyear Tire & Rubber Co, 50-57; STAFF MEM, SANDIA NAT LABS, 57- *Concurrent Pos:* Res fel, Case Inst Technol, 48. *Mem:* Fel AAAS; Am Chem Soc; fel Am Inst Chem; Am Inst Aeronaut & Astronaut. *Res:* Physical properties of polymers; radiation chemistry and effects of radiation on polymers; graphite and high temperature chemistry. *Mailing Add:* Sandia Nat Labs 1553 PO Box 5800 Albuquerque NM 87185

AUERBACH, ISAAC L, b Philadelphia, Pa, Oct 9, 21; m 76; c 3. INFORMATION SYSTEMS & MANAGEMENT. *Educ:* Drexel Univ, BS, 43; Harvard Univ, MS, 47. *Prof Exp:* Res engr, Univac Div, Sperry Rand Corp, 47-49; mgr defense space & spec prod div, Burroughs Corp, 49-57; pres, Auerbach Assocs Inc, 57-76; chmn & pres, Auerbach Corp Sci & Technol, 57-83; pres, Auerbach Publ, 60-81, chmn, 81-86; PRES, AUERBACH CONSULTS, 76- *Concurrent Pos:* Mem Ctr Strategic & Int Studies, 68-74; mem comt int sci & tech info progs, Nat Acad Sci, 73-78; adj prof mgt, Univ Pa, 74-77; mem tech adv bd, Dept Com & Nat Strategy & Info Ctr, 77-81. *Honors & Awards:* Grand Medal, City Paris, 59. *Mem:* Nat Acad Eng; hon mem Int Fedn Info Processing (pres, 60-65); Am Fedn Info Processing Socs; Info Indust Asn; fel Inst Elec & Electronics Engrs; distinguished fel Brit Comput Soc. *Res:* Computerized information systems; realtime command and control systems; the economic and international value of scientific and technical information and technology transfer. *Mailing Add:* Auerbach Consults 455 Righters Mill Rd Narberth PA 19072

AUERBACH, JEROME MARTIN, b Providence, RI, June 13, 44. PLASMA PHYSICS & FUSION, NUCLEAR ENGINEERING. *Educ:* Brown Univ, ScB, 67; Calif Inst Technol, MS, 68, PhD(eng), 75. *Prof Exp:* Nuclear engr reactor physics, Argonne Nat Lab, 75-76; PHYSICIST LASER FUSION, LAWRENCE LIVERMORE LAB, UNIV CALIF, 76- *Mem:* Am Phys Soc; Am Nuclear Soc. *Res:* Physics of laser-plasma interaction; physics of laser driven thermonuclear fusion. *Mailing Add:* MS L-487 Lawrence Livermore Lab Box 5508 Livermore CA 94550

AUERBACH, LEONARD B, b New York, NY, Aug 11, 29; m 51; c 2. PHYSICS. *Educ:* City Col New York, BS, 51; Univ Ill, MS, 52; Univ Calif, Berkeley, PhD(physics), 62. *Prof Exp:* Res assoc particle physics, Segre Group, Lawrence Radiation Lab, 62-63; res assoc, Univ Pa, 63-65, asst prof, 65-67; ASSOC PROF, TEMPLE UNIV, 67- *Mem:* Am Phys Soc. *Res:* Elementary and experimental high energy particle physics. *Mailing Add:* Dept Physics Temple Univ Philadelphia PA 19122

AUERBACH, MICHAEL HOWARD, b Akron, Ohio, June 24, 43; m 66; c 3. ENHANCED OIL RECOVERY, MINERAL SCALE CONTROL. *Educ:* Mass Inst Technol, BS, 64; Cornell Univ, PhD(anal chem), 69. *Prof Exp:* Sr res chemist, Hydrocarbons & Polymers Div, Monsanto Co, 69-74; res specialist, Monsanto Polymers & Petrochem Co, 74-76; res mgr, 76-82, ASST DIR SPEC CHEM RES & DEVELOP, PFIZER INC, 82- *Concurrent Pos:* Guest lectr, Farleigh Dickinson Univ. *Mem:* Am Chem Soc; Int Erosion Control Asn. *Res:* Enhanced oil recovery polymers; desalination and geothermal antiscalant polyelectrolytes; soil binder gels; chemical applications test development; chemical product market opportunity assessment; shoreline stabilization/erosion control. *Mailing Add:* Two Foot Ct Waterford CT 06385-2712

AUERBACH, OSCAR, b New York, NY, Jan 1, 05; m 32; c 2. MEDICINE. *Educ:* New York Med Col, MD, 29; Am Bd Path, dipl, 44. *Prof Exp:* Pathologist, Sea View Hosp, NY, 32-47; chief lab serv, Halloran Vet Admin Hosp, 47-51; chief lab serv, 52-59, SR MED INVESTR, VET ADMIN HOSP, EAST ORANGE, NJ, 60-, VET ADMIN DISTINGUISHED PHYSICIAN, 78- *Concurrent Pos:* Res fel, Univ Vienna, 32; consult, Richmond Mem Hosp, 38-47, US Vet Admin Hosp, 46-47 & US Naval Hosp, 47-49; vis instr, Sch Med, Wash Univ, 44; assoc prof, New York Med Col, Flower & Fifth Ave Hosps, 49-61, prof, 61-71; prof path, Col Med & Dent, NJ, 71- *Mem:* Am Thoracic Soc; Am Asn Thoracic Surg; fel Am Col Physicians; fel NY Acad Med; hon mem Mex Tuberc Soc. *Res:* Relationship of smoking to lung cancer and other diseases both human and experimental. *Mailing Add:* Dept Path Univ Med & Dent NJ Med Sch 185 S Orange Ave Newark NJ 07103

AUERBACH, ROBERT, b Berlin, Ger, Apr 12, 29; nat; m 50; c 2. DEVELOPMENTAL BIOLOGY, IMMUNOLOGY. *Educ:* Berea Col, AB, 49; Columbia Univ, AM, 50, PhD(zool), 54. *Prof Exp:* Assoc, Biol Div, Oak Ridge Nat Lab, 54-55; fel, Nat Cancer Inst, 55-57; from asst prof to assoc prof, 57-65, PROF ZOOL, UNIV-WIS-MADISON, 65- *Concurrent Pos:* Guggenheim fel, 68; vis prof surg, Harvard Med Sch, 73-74; vis prof cell biol, Southwestern Med Sch, Dallas, 74-75; Rockefeller fel, 74-75; Harold R Wolfe Prof, 84- *Mem:* Soc Develop Biol; Soc Exp Biol & Med; Am Asn Path; Am Asn Immunol; fel AAAS; Am Asn Cancer Res; Soc Anal Cytol. *Res:* Endothelial cell heterogeneity and specificity; ontogeny of immunity; genetics and immunology of granulomatous lung disease; flow cytometry; angiogenesis. *Mailing Add:* Dept Zool Univ Wis Madison WI 53706

AUERBACH, STANLEY IRVING, b Chicago, Ill, May 21, 21; m 54; c 4. RADIATION ECOLOGY, SYSTEM ANALYSIS. *Educ:* Univ Ill, BS, 46, MS, 47; Northwestern Univ, PhD(zool), 49. *Prof Exp:* Asst zool & animal ecol, Northwestern Univ, 47-48; lectr biol, Roosevelt Univ, 50-51, instr, 51-54, asst prof, 50-54; assoc scientist, Oak Ridge Nat Lab, 54-55, health physicist ecol, 54-59, scientist, 55-59, sect chief radiation ecol, 59-70, dir, Ecol Sci Div, 70-72, dir, Environ Sci Div, 72-86, sr staff adv, 86-90; RETIRED. *Concurrent Pos:* Lectr, Univ Tenn, 60; vis res prof, Univ Ga, 64-90; adj res prof, Univ Tenn, 65-90; dir eastern deciduous forest biome, Int Biol Prog, 68-76, vpres US exec comt, 71-; mem bd energy studies & various subcomt, Nat Acad Sci-Nat Res Coun, 74-77; mem sci adv bd, US Environ Protection Agency, 86-, environ adv bd, chief engrs, 89-; mem adv comn Sci & Technol Ctr, NSF, 88- *Mem:* Am Inst Biol Sci; fel AAAS; Ecol Soc Am (secy, 64-69, pres, 71-72); Brit Ecol Soc; Int Union Radioecologists; Health Physics Soc. *Res:* Ecosystem analysis; radioactive waste cycling in terrestrial ecosystems. *Mailing Add:* Environ Sci Div Bldg 1505 Oak Ridge Nat Lab PO Box 2008 Oak Ridge TN 37831-6036

AUERBACH, STEPHEN MICHAEL, b New York, NY, Nov 19, 42; m 78; c 3. CLINICAL PSYCHOLOGY, BEHAVIORAL MEDICINE. *Educ:* Queens Col, CUNY, BA, 65; Fla State Univ, MS, 69, PhD(clin psychol), 71. *Prof Exp:* Asst prof psychol, NMex State Univ, 72-74; from asst prof to assoc prof, Va Commonwealth Univ, 74-82; PROF PSYCHOL & DIR DOCTORAL PROG CLIN PSYCHOL, VA COMMONWEALTH UNIV, 82- *Concurrent Pos:* Adj prof oral & maxillofacial surg, Med Col Va, 89- *Mem:* Fel Am Psychol Asn. *Res:* Psychological stress, coping and stress management; development and evaluation of procedures designed to manage stresses associated with medical treatment and disease and aging. *Mailing Add:* Dept Psychol Va Commonwealth Univ 806 W Franklin St Richmond VA 23284

AUERBACH, VICTOR, b Philadelphia, Pa, July 4, 17; m 46; c 3. SYNTHETIC HIGH POLYMERS, MANAGEMENT OF PATENTS & LICENSING. *Educ:* Brooklyn Col, BA, 40; Polytech Inst New York, PhD(chem), 45. *Prof Exp:* Chemist urea resins, Bakelite Co, 42-43; instr organic chem, Polytechnic Inst New York, 42-46; res assoc thin films, Nat Defense Res Coun, 43-45; group leader reactive resins, Bakelite Co, 45-55; assoc coordr patents, Union Carbide Plastics Co, 55-62, coordr patents, Plastics Div, 62-67, patent mgr chem & plastics, Union Carbide Corp, 67-79, patent mgr, Polyolefins Div, 79-81; CONSULT, 81- *Res:* Thermosetting resins; thermoplastic resins; influence of molecular weight distribution and molecular structure on resin and end-product properties and performance. *Mailing Add:* 1218 Stillman Ave Plainfield NJ 07060-2729

AUERBACH, VICTOR HUGO, b New York, NY, Oct 2, 28; m 56. BIOCHEMISTRY. *Educ:* Columbia Col, AB, 51; Harvard Univ, AM, 55, PhD(biochem), 57. *Prof Exp:* Asst, Col Physicians & Surgeons, Columbia Univ, 45-47; biochemist, Rheumatic Fever Res Inst, 48; asst, Columbia Univ, 48-52; res biochemist, Fleishmann Labs, 52; res assoc pediat & instr physiol chem, Univ Wis, 57-58; dir, enzyme lab & dir res chem, St Christopher's Hosp for Children, 58-85, dept labs, 76-85; asst prof physiol chem & res pediat, 58-64, assoc prof biochem & res pediat, 64-68, res prof, 68-85, SR RES PROF PEDIAT, SCH MED, TEMPLE UNIV, 85- *Mem:* Am Chem Soc; AAAS; Biochem Soc; Am Soc Biol Chem; Am Soc Clin Nutrit; Am Assoc Clin Chem. *Res:* Mammalian amino acid metabolism; control mechanisms such as adaptive enzyme formation and negative feedback regulation of enzyme synthesis; inborn errors of metabolism and genetic disorder in man; clinical chemistry. *Mailing Add:* 1244 Hoffman Rd Ambler PA 19002-5033

AUERBACK, ALFRED, b Toronto, Ont, Sept 20, 15; nat US; m 42; c 3. PSYCHIATRY. *Educ:* Univ Toronto, MD, 38. *Prof Exp:* Intern, Toronto Gen Hosp, 38-39; resident med, French Hosp, Calif, 39-40; resident neurol & psychiat, Univ Calif Hosp, 40-42; psychiatrist, Sheppard Pratt Hosp, Md, 42-43; clin asst, Med Sch, 43-46, clin instr, 46-53, from asst clin prof to clin prof, 53-83, EMER CLIN PROF PSYCHIAT, UNIV CALIF, SAN FRANCISCO, 83- *Concurrent Pos:* Consult, Vet Admin, 46-52; chmn, Ment Health Adv Bd, San Francisco, 66-70; pvt practr psychiat; mem, Adv Comt, Div Rehab, Calif Dept Pub Health, 71-76; consult, Nat Adv Coun Alcohol & Alcohol Abuse, 73-77. *Honors & Awards:* Royer Award, Univ Calif Bd Regents, 66. *Mem:* Fel Am Psychiat Asn (vpres, 66-67); fel Am Col Psychiatrists; hon fel Royal Col Psychiatrists; hon fel Royal Australian & NZ Col Psychiatrists. *Res:* Psychosomatic relationships; community and social aspects of psychiatry; alcoholic rehabilitation; marital and sex counseling. *Mailing Add:* 1000 Northpoint St San Francisco CA 94109

AUERSPERG, NELLY, b Vienna, Austria, Dec 13, 28; Can citizen; m 55; c 2. CANCER, CELL BIOLOGY. *Educ:* Univ Wash, MD, 55; Univ BC, PhD(cell biol), 68. *Prof Exp:* Intern med, Vancouver Gen Hosp, BC, 55-56, res asst tissue cult, 59-60; cytologist, Cytol Lab, Vancouver, 60-65; mem staff, Cancer Res Ctr, 68-78, from assoc prof to prof zool, 68-77, PROF ANAT, UNIV BC, 78- *Concurrent Pos:* Terry Fox res scientist. *Mem:* Am Asn Cancer Res; Am Soc Cell Biol; Tissue Cult Asn; Can Soc Cell Biol (pres, 78); Int Soc Differentiation; Sigma Xi. *Res:* Regulation of phenotypic expression in malignant tumors. *Mailing Add:* Dept Anat Univ BC Vancouver BC V6T 1W5 Can

AUFDEMBERGE, THEODORE PAUL, b Winnipeg, Man, Feb 1, 34; US citizen; m 60; c 4. PHYSICAL GEOGRAPHY, REGIONAL GEOGRAPHY. *Educ:* Concordia Teachers Col, BSc, 56; Wayne State Univ, MA, 64; Univ Mich, Ann Arbor, PhD(geog), 71. *Prof Exp:* Prin & teacher, Immanual Lutheran Sch, Brownton, Minn, 56-58; teacher sci & english, Trinity Lutheran Sch, Mt Clemens, Mich, 58-64; instr geog, 64-66, asst prof, 66-70, assoc prof, 70-81, PROF GEOG, CONCORDIA COL, ANN ARBOR, 81- *Mem:* Creation Res Soc; Nat Coun Geog Educ. *Res:* Spatial distributions of economic activities particularly agricultural & manufacturing activities in Anglo-America; physical geography of Middle-America and glacial surging applied to Pleistocene chronology. *Mailing Add:* Concordia Col 4090 Geddes Rd Ann Arbor MI 48105

AUFDERHEIDE, KARL JOHN, b Minneapolis, Minn, Aug 17, 48; m 69; c 1. CELL BIOLOGY, DEVELOPMENTAL BIOLOGY. *Educ:* Univ Minn, BS, 70, MS, 72, PhD(cell biol), 74. *Prof Exp:* Res assoc cell biol, Ind Univ, Bloomington, 74-77; instr, Univ Iowa, 77-79; asst prof, 79-86, ASSOC PROF, TEX A&M UNIV, 86- *Concurrent Pos:* NIH fel, 75-77; sr scientist, A&M Inst Develop Biol, 83-85; vis scholar, Ind Univ, 86. *Mem:* Soc Protozoologists; AAAS. *Res:* Intracellular development and differentiation; intracellular pattern formation in ciliate protozoa; developmental genetics of exocytotic organelles; formation and positioning of cytoskeletal arrays in the cell cortex; mechanisms of clonal cellular aging; molecular biology of the expression and differentiation of serotype genes. *Mailing Add:* Dept Biol Texas A&M Univ College Station TX 77843-3258

AUFDERMARSH, CARL ALBERT, JR, b Cincinnati, Ohio, Nov 9, 32; m 55; c 4. COMPOSITES, ENGINEERING PLASTICS. *Educ:* Univ Cincinnati, BS, 54; Yale Univ, MS, 55, PhD(chem), 58. *Prof Exp:* Res chemist, E I du Pont de Nemours & Co, 58-82; MAT ENGR, SCHLUMBERGER WELL SERV, 83- *Mem:* Am Chem Soc; Am Inst Chemists; Soc Advan Mat & Process Eng. *Res:* Polymer chemistry; organic synthesis; isocyanates and polyurethanes; fluoroelastomers; advanced composites; engineering plastics. *Mailing Add:* 3919 Marlowe Houston TX 77005-2045

AUFENKAMP, DARREL DON, mathematics, computer science; deceased, see previous edition for last biography

AUFFENBERG, WALTER, b Dearborn, Mich, Feb 6, 28; m 49; c 4. HERPETOLOGY. *Educ:* Stetson Univ, BS, 51; Univ Fla, MS, 53, PhD(biol), 56. *Prof Exp:* Cur vert paleont, Charleston Mus, 53; asst prof biol, Univ Fla & assoc cur vert paleont, Fla State Mus, 56-59; assoc dir, Biol Sci Curric Study, Univ Colo, 59-63; chmn dept natural sci, 63-73, CUR HERPET, FLA STATE MUS, 73- *Concurrent Pos:* Res assoc, Univ Colo Mus, 59-63 & NY Zool Soc, 69-; consult, AID, India, 62 & -64 & NSF, India, 67. *Honors & Awards:* 1st Prize Honorarium, Am Soc Ichthyol & Herpet, 53. *Mem:* Fel AAAS; Am Soc Ichthyologists & Herpetologists; Am Soc Naturalists; Am Soc Zool; Sigma Xi. *Res:* Behavior; field biology of large reptiles; biosystematics of fossil and living land tortoises of the world. *Mailing Add:* Fla State Mus Univ Fla Gainesville FL 32611

AUGENLICHT, LEONARD HAROLD, b New York, NY, Aug 16, 46; m 70; c 1. CANCER, COLON. *Educ:* State Univ NY Binghamton, BA, 67; Syracuse Univ, PhD(biol), 71. *Prof Exp:* Trainee path, Med Sch, Temple Univ, 71-74; assoc mem, Sloan-Kettering Cancer Ctr, 74-83; RES SCIENTIST, DEPT ONCOL, MONTEFIORE MED CTR, 83-; ASSOC PROF, DEPT MED & CELL BIOL, ALBERT EINSTEIN MED SCH, 84- *Concurrent Pos:* NSF & NASA predoctoral fel. *Mem:* Am Asn Cancer Res. *Res:* Euraryotic gene expression; control of cell proliferation; gene expression in transformation. *Mailing Add:* Dept Oncol Montefiore Med Ctr 111 E 210th St Bronx NY 10467

AUGENSEN, HARRY JOHN, b Chicago, Ill, July 18, 51; m. ASTRONOMY, PHYSICS. *Educ:* Elmhurst Col, BA, 73; Northwestern Univ, MS, 74, PhD(astron), 78. *Prof Exp:* lectr astron, Northwestern Univ, 78-80; ASSOC PROF PHYSICS, WIDENER UNIV, 80- *Concurrent Pos:* Vis observer, Cerro Tololo Inter-Am Observ, 76-77 & 81; lectr astron, Swarthmore Col,80- *Honors & Awards:* Harlow Shapley Lectr, 81-82 & 87-88; Goddard Summer Fac Fel, NASA, 82, 83. *Mem:* Am Astron Soc; Royal Astron Soc; Sigma Xi. *Res:* High-velocity stars; galactic structure; planetary nebulae. *Mailing Add:* Dept Physics Widener Univ Chester PA 19013

AUGENSTEIN, BRUNO (WILHELM), b Germany, Mar 16, 23; nat US; m 50; c 3. PHYSICS, AERONAUTICS. *Educ:* Brown Univ, ScB, 43; Calif Inst Technol, MS, 45. *Prof Exp:* Supvr, Aerophys Lab, NAm Aviation, Inc, 46-48; asst prof aeronaut, Purdue Univ, 48-49; consult scientist, Lockheed Missiles & Space Co, Calif, 58-59, sci adv, 59-61, dir adv planning, 61; asst dir defense res & eng, US Dept Defense, 61-65, spec asst intel reconnaissance to dir, 63-65, res adv, Inst Defense Anal, 65-67; assoc phys scientist, 49-53, phys scientist, 53-56, asst head sci staff, 56-57, sr staff, 56-58, vpres, 67-71, chief scientist, 71-72, consult, 72-76, SR SCIENTIST, RAND CORP, 76- *Concurrent Pos:* Consult, Nat Acad Sci, 65-, US Dept Defense, 78 & Off Sci Technol Policy, 78-; assoc chmn bd regents, Nat Libr Med; head, US Navy Med Rev Comt, 75; dir, Xerad, 81-83. *Mem:* AAAS; Am Phys Soc; Am Inst Aeronaut & Astronaut; Am Geog Soc. *Res:* Research and development administration; applied physics; weapon systems; operational and systems analysis; strategic policies; policy analysis; astronautics. *Mailing Add:* 1144 Tellem Dr Pacific Palisades CA 90272

AUGENSTEIN, MOSHE, b Baltimore, Md, Nov 27, 47; m 70; c 5. SOFTWARE SYSTEMS. *Educ:* Brooklyn Col, BS, 69; NY Univ, MS, 71, PhD(opers res), 75. *Prof Exp:* PROF COMPUTER SCI, BROOKLYN COL, 72- *Concurrent Pos:* Consult, 78- *Mem:* Asn Comput Mach. *Res:* Data structures; microcomputers; graphic display of data structures; author of several publications. *Mailing Add:* Computer Sci Brooklyn Col Brooklyn NY 11210

AUGHENBAUGH, NOLAN B(LAINE), b Akron, Ohio, July 29, 28; m 59; c 3. GEOLOGICAL & GEOTECHNICAL ENGINEERING, WASTE DISPOSAL. *Educ:* Purdue Univ, BSCE, 55, PhD(civil eng), 63; Univ Mich, MS, 59. *Prof Exp:* From instr to asst prof civil eng, Purdue Univ, 59-65; from assoc prof to prof geol eng, Univ Mo, Rolla, 66-83; dean, Sch Mineral Eng, Univ Alaska, Fairbanks, 83-84; PROF GEOL ENG, UNIV MISS, 85-; CONSULT, 60- *Concurrent Pos:* Mem, Mo Land Reclamation Comn, 77-; deleg leader, People-to-People Tour to Europe on Underground Space Use, 78; lab asst geol, Purdue Univ, 54-55; chmn dept mining, petrol & geol eng, Univ Mo, Rolla, 70-80; Fulbright scholar, NZ; chmn, Miss Hazardous Waste Tech Siting Comt. *Honors & Awards:* Mountain & Rock Group named in honor, Anatarctica. *Mem:* Asn Eng Geologists; Soc Mining Engrs; Am Soc Civil Engrs; Sigma Xi; Am Underground Space Asn; Nat Soc Prof Engrs. *Res:* Agregate quality and degradation, evaluating laboratory tests with physical properties; mechanics of rock fracture; coal mine roof stability; subsidence; effects of humidity and temperature on rocks; properties and durability of shale; basic engineering properties of clay and shales; flexibility of compacted clays; durability of shales; photoelastic analysis of earth minerals; general soil and rock mechanics; author of more than 50 technical publications. *Mailing Add:* Geol Dept Univ Miss University MS 38677

AUGL, JOSEPH MICHAEL, b Pasching, Austria, Jan 8, 32; m 62; c 1. ORGANIC CHEMISTRY. *Educ:* Univ Vienna, PhD(org chem), 59. *Prof Exp:* Asst prof org chem, Univ Vienna, 59-61; sr res organometallic chemist, Sohio Res Ctr, Ohio, 61-63, proj leader, 63; org chem, Melpar Inc, 63-64, res leader, 64-68; RES CHEMIST, NAVAL ORD LABS, 68- *Honors & Awards:* Capt Walter S Diehl Award, Sci & Eng Symp, 78. *Mem:* Am Chem Soc; Soc Adv Mat & Process Eng. *Res:* Synthesis of heterocyclic compounds; transition metal coordination compounds; synthesis of high temperature stable polymers; composite materials; material aging phenomena. *Mailing Add:* 606 W Nettle Tree Rd Sterling VA 22170-4724

AUGOOD, DEREK RAYMOND, b London, Eng, Mar 7, 28; nat US; m 55; c 2. PHYSICAL CHEMISTRY, RESEARCH ADMINISTRATION. *Educ:* Univ London, BSc, 49, PhD(phys org), 52; Cambridge Univ, PhD(chem eng), 55. *Prof Exp:* Harwell fel, UK Atomic Energy Res Estab, 54-56; engr, E I du Pont de Nemours & Co, 56-62; head sect prod develop & mgr, Res & Develop Lab, Kaiser Chem, 72-89; AT SCI CERT SYSTS INC. *Honors & Awards:* Moulton Medal, Brit Inst Chem Engrs, London, 55; Neal Ricc Award, Int Briquetting Assoc, 87. *Res:* Homolytic aromatic substitution; isotope separation; nylon intermediates; polyethylene, fluorocarbon and vinyl butyral polymers; isocyanates and urethanes; research and development project evaluation; manufacture and performance of extruded dielectric cables; non-bauxitic alumina processes; alumina reduction plant processes, industrial hygiene, carbon product, waste product utilization, environmental and risk assessment studies; author of over 50 publications. *Mailing Add:* Sci Cert Systs Inc 1611 Telegraph Ave Oakland CA 94612

AUGSBURGER, LARRY LOUIS, b Baltimore, Md, July 1, 40; m 65; c 4. DOSAGE FORM DESIGN, INDUSTRIAL PHARMACY. *Educ:* Univ Md, BS, 62, MS, 65, PhD(pharm), 67. *Prof Exp:* Sr res scientist, Johnson & Johnson, 67-69; asst prof, 69-74, ASSOC PROF PHARMACEUT, SCH PHARM, UNIV MD, 74-, DIR, PHARMACEUT GRAD PROG, 76- *Concurrent Pos:* Consult, FMC Corp, 75-76 & 81-82, Mallincordt, Inc, 79-80, E R Squibb & Sons, 79-80 & Pharmakinetics, Inc, 80-81. *Mem:* Am Pharmaceut Asn; Acad Pharmaceut Sci; Am Asn Col Pharm; Fine Particle Soc; Sigma Xi. *Res:* Instrumentation of table presses and capsule filling machines; interplay of formulation and process variables on the performance and manufacturability of solid dosage forms; drug release from dosage forms. *Mailing Add:* Dept Pharmaceut Univ Md Sch Pharm 20 N Pine St Baltimore MD 21201

AUGSPURGER, CAROL KATHLEEN, US citizen. ECOLOGY, EVOLUTION. *Educ:* Bowling Green State Univ, BS, 63; Univ Mich, MS, 75, PhD(biol), 78. *Prof Exp:* ASST PROF BOT, UNIV ILL, URBANA, 78- *Mem:* Ecol Soc Am; Soc Study Evolution; Asn Trop Biol; AAAS. *Res:* Evolutionary ecology of tropical plant populations; reproductive strategies and reproductive synchrony; phenological patterns and environmental cues; life history and demography of semelparous plants; plant-animal coevolution. *Mailing Add:* Dept Plant Biol Univ Ill 505 S Goodwin Ave Urbana IL 61801

AUGSTKALNS, VALDIS ANSIS, b Rezekne, Latvia, Mar 31, 39; US citizen; m 86; c 1. CHEMICAL ENGINEERING, POLYMER CHEMISTRY. *Educ:* Univ Mass, BSChE, 61, MSChE, 63. *Prof Exp:* Res engr polymers, E I Du Pont de Nemours & Co, 62-63, 65-75; res chem engr polymers & pyrotech, Chem Res & Develop Labs, Edgewood Arsenal, US Army, Md, 63-65; sr res engr polymers, 75-79, RES ASSOC, E I DU PONT DE NEMOURS & CO, INC, 79- *Concurrent Pos:* NSF coop grad fel, Univ Mass, 61-62. *Mem:* Am Inst Chem Engrs; AAAS; Sigma Xi. *Res:* Nylon and polyester polymerization, degradation and processing; computer modeling of these. *Mailing Add:* E I du Pont de Nemours & Co Inc Wash Lab PO Box 1217 Parkersburg WV 26102

AUGUST, JOSEPH THOMAS, b Whittier, Calif, Sept 19, 27; m 51; c 3. BIOCHEMISTRY, MOLECULAR BIOLOGY. *Educ:* Stanford Univ, AB, 51, MD, 55. *Prof Exp:* Intern med, Los Angeles County Gen Hosp, 54-55; resident, Royal Infirmary, Univ Edinburgh, 55-56; resident & fel, Harvard Med Sch, 56-59; asst prof, Sch Med, Stanford Univ, 59-61; assoc prof, Sch Med, NY Univ, 61-63; from asst prof to prof molecular biol, Albert Einstein Col Med, 63-76, chmn & dir, Dept Molecular Biol & Div Biol Sci, 72-76; PROF & DIR, DEPT PHARMACOL & EXP THERAPEUT, SCH MED, JOHNS HOPKINS UNIV, 76- *Concurrent Pos:* Res fel, Am Heart Asn, 57-59; career scientist grant, Health Res Coun New York City, 62-71; Guggenheim fel, 71-72; vis prof biochem, Oxford Univ, 71-72; Merkle Scholar Med Sci, 60-65; Siegfried Ullman Prof Molecular Biol, 68-76; Johnson & Johnson Focused Giving Award, 82-87. *Mem:* Am Soc Biol Chemists; Am Chem Soc; Am Soc Microbiol; Harvey Soc; Am Soc Pharmacol & Exper Therap. *Res:* Identification, purification and characterization of eukaryotic cell surface protein; immunopharmacology. *Mailing Add:* Dept Pharmacol Johns Hopkins Sch Med 725 N Wolfe St Baltimore MD 21205

AUGUST, LEON STANLEY, b New Orleans, La, Sept 30, 26; m 52. NUCLEAR SCIENCE, RADIATION DAMAGE. *Educ:* La State Univ, BS, 50, PhD(physics), 57; Tulane Univ, MS, 52. *Prof Exp:* res physicist, US Naval Res Lab, 57-86; CONSULT PHYSICIST, 86- *Concurrent Pos:* Mem cyclotron group, Univ Calif, Los Angeles, 65-66. *Mem:* Am Phys Soc; Sigma Xi; AAAS; Inst Elec & Electronics Engrs. *Res:* Radiation damage experiments in semiconductors; space dosimetry, neuron gamma, electron, proton, and heavy-ion dosimetry and spectroscopy; microdosimetry and ion-track theory as applied to semiconductor; accelerator physics. *Mailing Add:* 6920 Baylor Dr ALexandria VA 22307-1703

AUGUSTEIJN, MARIJKE FRANCINA, b Rotterdam, Neth, Mar 30, 46; m 70. THEORETICAL PHYSICS. *Educ:* Tech Univ Delft, Ing, 70; Ohio Univ, PhD(physics),80. *Prof Exp:* Instr physics & math, Haags Montessori Lyceum, 70-75; res fel med physics, Inst Med & Math, Ohio Univ, 80-82; AT DEPT COMPUT SCI, UNIV COLO, 82- *Concurrent Pos:* Vis asst prof, Math Dept, Ohio Univ, 81-82. *Mem:* Asn Comput Mach; Am Asn Artificial Intel; Inst Elec & Electronics Engrs. *Res:* Nonlinear classical mechanics; oscillatory systems; mathematical physics; intelligent training systems. *Mailing Add:* Dept Comput Sci Univ Colo PO Box 7150 Colorado Springs CO 80933-7150

AUGUSTIN, JORG A L, b Heerbrugg, Switz, Jan 5, 30; US citizen; m 62. FOOD SCIENCE, PLANT BIOCHEMISTRY. *Educ:* Swiss Fed Inst Technol, dipl agrotech, 55; Univ Ill, MS, 57; Mich State Univ, PhD(food sci), 64. *Prof Exp:* Res assoc, Hero Konserven, Switz, 57-60; asst plant supt food processing, Libby, McNeil & Libby, 60; res asst food sci, Mich State Univ, 61-64; food technologist food res div, Armour & Co, 64-68, from asst sect head to sect head meat prod develop, 66-68; assoc res prof agr biochem, Br Exp Sta, 68-75, GROUP LEADER & PROF BACT & BIOCHEM, UNIV IDAHO 75-, IN CHG, FOOD RES CTR, 75- *Mem:* Inst Food Technol; Am Chem Soc; Am Asn Cereal Chemists; Sigma Xi. *Res:* Process and product development and improvement of foods, especially canned foods; studies on thermal destruction of bacterial spores; sugar beet storage; nutrient composition of fresh, stored and processed potatoes, legumes and cereal products; vitamin analysis. *Mailing Add:* Food Res Ctr Univ Idaho Moscow ID 83843

AUGUSTINE, JAMES ROBERT, b Mascoutah, Ill, Jan 27, 46; m 70; c 3. NEUROANATOMY. *Educ:* Millikin Univ, BA, 68; St Louis Univ, Mo, MS, 70; Univ Ala, Birmingham, PhD(anat), 73. *Prof Exp:* From instr to asst prof anat, Sch Med & Sch Dent, Univ Ala, Birmingham, 73-76; asst prof, 76-86, ASSOC PROF ANAT, SCH MED, UNIV SC, COLUMBIA, 86- *Concurrent Pos:* Scholarly exhib, Mus Royal Col Surgeons, London, Eng. *Honors & Awards:* Stefan Mironescu Res Award, 85. *Mem:* Am Asn Anatomists; AAAS; Cajal Club. *Res:* Primate neuroanatomy; immunocytochemistry; tissue culture. *Mailing Add:* Dept Anat Univ SC Sch Med Columbia SC 29208

AUGUSTINE, NORMAN R, b Colo, July 27, 35; m; c 2. AERONAUTICAL ENGINEERING. *Educ:* Princeton Univ, BScE, 57, MScE, 59. *Hon Degrees:* Dr Eng, Rensselaer Polytech Inst & Western Md Col, DSc, Univ Colo. *Prof Exp:* Eng exec, Douglas Aircraft, 58-65; off secy of defense, 65-70; vpres advan progs, Vought Missiles & Space Co, 70-73; asst secy, Army Res & Develop, 73-74; under secy army, 75-77; staff mem, 77-82, pres, Denver Aerospace, 82-85, sr vpres Coro, 85, exec vpres, 85-86, pres & chief oper officer, 86-87, CHIEF EXEC OFFICER, MARTIN MARIETTA CORP, 87-, CHMN, 88- *Concurrent Pos:* Chmn, Defense Policy Adv Comt Trade; trustee, Int Acad Astronaut; mem bd dirs, Phillips Petrol Co, Procter & Gamble Corp, Riggs Nat (Bank) Corp, Ethics Resource Ctr, Atlantic Coun, Alliance to Save Energy & Wolf Trap Found; mem, President's Nat Security Telecommun Adv Comt, Chief Naval Opers Exec Panel & numerous others. *Honors & Awards:* Distinguished Serv Medal, Dept Defense, Meritorious Serv Medal; James Forrestal Mem Award, Nat Security Indust Asn; Mil Astronaut Trophy, Am Astronaut Soc; Goddard Medal, Am Inst Aeronaut & Astronaut; Gold Medal, Am Defense Preparedness Asn, Knowles Award & John C Jones Award; Nat Eng Award, Am Asn Eng Soc; Carlton Award, Inst Elec & Electronics Engrs; Bernard Baruch Lectr, Nat Defense Univ. *Mem:* Nat Acad Eng; fel Inst Elec & Electronics Engrs; Int Acad Astronaut; hon fel Soc Tech Commun; hon fel Am Inst Aeronaut & Astronaut; fel Am Astronaut Soc. *Mailing Add:* Martin Marietta Corp 6801 Rockledge Dr Bethesda MD 20817

AUGUSTINE, PATRICIA C, b Lanham, Md, June 7, 37; m 58; c 3. COCCIDIOSIS, CELL CULTURE. *Educ:* Univ Md, BS, 59, MS, 76, PhD(poultry sci), 80. *Prof Exp:* Microbiologist, NIH, 59-61; MICROBIOLOGIST, USDA, 65- *Concurrent Pos:* Adj prof, Univ Md, 83- *Mem:* Am Soc Parasitologists; Poultry Sci Asn; Soc Exp Biol & Med. *Res:* Mechanisms and modification of host cell invasion by coccidia; employing tissue culture and hybridoma technology also in in vivo studies; pathophysiologic effects of coccidial infection in turkeys. *Mailing Add:* Protozoan Dis Lab USDA Bldg 1040 Rm 101 BARC-East Beltsville MD 20705

AUGUSTINE, ROBERT LEO, b Omaha, Nebr, Nov 15, 32; m 57; c 2. CATALYSIS ORGANIC CHEMISTRY. *Educ:* Creighton Univ, 54; Columbia Univ, MA, 55, PhD(chem), 57. *Prof Exp:* From instr to asst prof chem, Univ Tex, 57-61; from instr to assoc prof, 61-69, PROF CHEM, SETON HALL UNIV, 69-, CHMN, CHEM DEPT, 88- *Honors & Awards:* Joseph P Hyman Award, Am Chem Soc, 88; Paul N Rylander Award, Org Reactions Catalysis Soc, 90. *Mem:* Am Chem Soc; NY Acad Sci; Brit Chem Soc; Org Reactions Catalysis Soc (dir, 75-77, chmn, 84-86). *Res:* Synthetic and mechanistic applications of catalytic hydrogenation, oxidation and related reactions; characterization of surface of dispersed metal catalysts; synthetic application of catalytic reactions. *Mailing Add:* Dept Chem Seton Hall Univ South Orange NJ 07079

AUGUSTINE, ROBERTSON J, b Grand Rapids, Mich, Jan 28, 33; m 56; c 3. HEALTH PHYSICS. *Educ:* Univ Mich, BS, 55, MS, 57, MPH, 59, PhD(environ health), 62. *Prof Exp:* Health physicist, Univ Mich, 56-57, res assoc radiol health, 57-62; asst officer-in-charge, SE Radiol Health Lab, USPHS, 62-65; tech dir, Appl Health Physics, Inc, 65-66; chief environ radioactivity sect, Nat Ctr Radiol Health, USPHS, 66-68, dep dir div environ radiation, Bur Radiol Health, 68-70; SCIENTIST DIR, OFF RADIATION PROG, ENVIRON PROTECTION AGENCY, 70- *Mem:* Health Physics Soc; Am Pub Health Asn; Sigma Xi. *Res:* Radiological health; radiation protection and measurement. *Mailing Add:* 15412 Hannans Way Rockville MD 20853

AUGUSTYN, JOAN MARY, b Buffalo, NY. BIOCHEMISTRY. *Educ:* D'Youville Col, BA, 62; State Univ NY Buffalo, PhD(biochem), 69. *Prof Exp:* Trainee steroid biochem, Worcester Found Exp Biol, Shrewsbury, Mass, 69-71; RES BIOCHEMIST, RES SERV, VET ADMIN HOSP, ALBANY, 71-, CLIN CHEMIST, LAB SERV, 79-; RES ASST PROF ATHEROSCLEROSIS, ALBANY MED COL, 76- *Concurrent Pos:* Mem, Am Heart Asn. *Mem:* Tissue Cult Asn; Sigma Xi; AAAS; Am Asn Clin Chem. *Res:* Biochemical mechanisms involved in atherosclerotic heart disease. *Mailing Add:* Three Cornsilk St Watervliet NY 12189

AUKERMAN, LEE WILLIAM, b North Grove, Ind, June 7, 23; m 62; c 4. SOLID STATE PHYSICS. *Educ:* Purdue Univ, BS, 49, MS, 53, PhD(physics), 58. *Prof Exp:* Asst, Purdue Univ, 49-58; prin physicist, Battelle Mem Inst, 58-63; MEM TECH STAFF, AEROSPACE CORP, LOS ANGELES, 63- *Mem:* AAAS; Am Phys Soc; Sigma Xi. *Res:* Semiconductors and insulators; radiation damage; electrical and optical properties; infrared detectors. *Mailing Add:* 516-21 Manhattan Beach CA 90266

AUKLAND, JERRY C, b Macedonia, Iowa, Oct 7, 31; m 51; c 4. SIGNAL PROCESSING DEVICES, REMOTE SENSING. *Educ:* Iowa State Univ, BS, 59. *Prof Exp:* Res engr, Autonetics Div, NAm Aviat, Inc, 59-63, sr res physicist, 63-64, res specialist, 64-66; dept mgr sensor systs & instrumentation, Space Gen, 66-68; dir eng microwave sensor systs, Spectran, Inc, 68-71; group leader, Electronics Div, Rockwell Int, 71-78, mgr advan prods, 78-80, mgr microwave develop, 80-84; RES & DEVELOP MGR, HEWLETT-PACKARD, SAN JOSE, CALIF, 84- *Concurrent Pos:* Dir, Region 6, Inst Elec & Electronic Engrs, 91 & 92. *Honors & Awards:* Centennial Award, Inst Elec & Electronics Engrs; fel Inst Advan Eng. *Mem:* Inst Elec & Electronics Engrs. *Res:* Parametric amplifiers and their applications at microwave frequencies; millimeter semiconductor devices and parametric interactions; earth resource applications; pollution detection; surface acoustic wave devices; Rf and microwave semiconductor devices and IC's. *Mailing Add:* Hewlett-Packard 350 W Trimble Rd San Jose CA 95131

AUKRUST, EGIL, b Lom, Norway, June 16, 33; US citizen; m 62. METALLURGY, PHYSICAL CHEMISTRY. *Educ:* Norweg Inst Technol, DiplomEng, 57, DrEng, 60. *Prof Exp:* Res assoc metall, Pa State Univ, 60-62; res engr process metall, Jones & Laughlin Steel Corp, 81-84, sr res engr, 63-65, res supvr steelmaking, 65-66, res supvr ironmaking, 66-69, asst dir process metall, 69-71, dir process metall, 71-71, gen mgr res, 74-81, sr dir technol, 81-84; SR TECH DIR, LTV STEEL, 84- *Mem:* Am Inst Mining, Metall & Petrol Engrs; Indust Res Inst; Brit Metals Soc; Asn Iron & Steel Engrs; Am Iron & Steel Inst. *Res:* Extractive metallurgy and process control; phase relationship of oxide systems; thermodynamics of liquid salts. *Mailing Add:* 1700 Grandview Ave Pittsburgh PA 15211

AUKSMANN, BORIS, b Tartu, Estonia, Mar 12, 27; Can citizen; m 55. ENGINEERING DESIGN. *Educ:* Univ BC, BASc, 55; Calif Inst Technol, MS, 58, ME, 59, PhD(mech eng), 64. *Prof Exp:* Design engr, Can Industs Ltd, 55-56 & H A Simons, Consult Engr, 56-57; res engr, Ingersoll-Rand Co, 59-61; asst mech eng, Calif Inst Technol, 61-64, asst prof eng design, 64-70; CONSULT MECH ENG, 70- *Mem:* Am Soc Mech Engrs; Sigma Xi. *Res:* Journal bearings; planetary gear trains; mechanical seals; engineering design; accident reconstruction and analysis. *Mailing Add:* 1816 Anita Crest Dr Arcadia CA 91506

AULD, DAVID STUART, b Newton, NJ, Jan 8, 37; m 61; c 2. BIOPHYSICS, BIOLOGICAL CHEMISTRY. *Educ:* Lehigh Univ, BA, 60, MS, 62; Cornell Univ, PhD(chem), 67. *Prof Exp:* Am Cancer Soc res fel biol chem, 67-69, assoc, 69-70, from asst prof to assoc prof biol chem, 70-83, ASSOC PROF PATH, HARVARD MED SCH, 83-; ASSOC MED, BRIGHAM & WOMENS HOSP, 67- *Concurrent Pos:* Res fel, Cancer Div, NIH, 69-70, spec fel, 70-71; consult, Monsanto Res Corp, 75-77; bd tutors biochem sci, Harvard Univ, 77- *Mem:* Am Chem Soc; Am Soc Exp Biol Chemists; Sigma Xi; NY Acad Sci. *Res:* Mechanism of enzyme action; relationship of diseased states and abnormal enzyme activities. *Mailing Add:* 29 Railroad Ave Bedford MA 01730

AULD, EDWARD GEORGE, b Chilliwack, BC, Apr 27, 36; m 60; c 4. NUCLEAR PHYSICS. *Educ:* Univ BC, BASc, 59, MASc, 61; Univ Southampton, PhD(nuclear physics), 64. *Prof Exp:* Athlone fel, 61, sci officer, Rutherford High Energy Lab, eng, 63-64, sr sci officer, 64-66; assoc prof, 66-83, PROF PHYSICS, UNIV BC, 83-, DIR ENG PHYSICS, 81-, PROF ENG, 76- *Concurrent Pos:* Vpres, Univ BC Res Enterprizes, technol transfer & commercialization. *Mem:* Am Asn Physics Teachers; Can Asn Physicists. *Res:* Particle physics, especially proton-antiproton interactions; nuclear physics, especially deuteron interactions; high energy physics; meson resonance and pion inelastic interactions; cyclotron magnetic design for the TRIUMF negative ion cyclotron. *Mailing Add:* Dept Physics Univ BC Vancouver BC V6T 1W5 Can

AULD, PETER A MCF, b Toronto, Ont, Feb 5, 28; m 51; c 2. MEDICINE, PEDIATRICS. *Educ:* Univ Toronto, BA, 48; McGill Univ, MD, 52; FRCP(C), 57. *Prof Exp:* Instr pediat, Harvard Med Sch, 59-60; res assoc, McGill Univ, 60-62; from asst prof to prof pediat, 62-74, PROF PEDIAT & PERINATAL MED IN OBSTET & GYNEC, MED COL, CORNELL

UNIV, 74- *Concurrent Pos:* Res fel pediat, Harvard Med Sch, 57-59; career investr, NY Health Res Coun, 62- *Mem:* Fel Am Acad Pediat; Perinatal Res Soc; Soc Pediat Res; Am Pediat Soc. *Res:* Cardiology; cardio-pulmonary physiology; neonatal pulmonary disease. *Mailing Add:* Dept Pediat Cornell Univ Med Col New York NY 10021

AULENBACH, DONALD BRUCE, b Berwick, Pa, Mar 7, 28; m 52; c 4. ENVIRONMENTAL ENGINEERING. *Educ:* Franklin & Marshall Col, BS, 50; Rutgers Univ, MS, 52, PhD(chem, sanit), 54. *Prof Exp:* Chemist-bacteriologist, Del Water Pollution Comn, 54-60; from asst prof to assoc prof, 60-65, chmn environ eng curric, 67-71, PROF ENVIRON ENG, RENNSSELAER POLYTECH INST, 73-, EMER PROF, 90- *Concurrent Pos:* Consult, Gen Elec Co, 65-; invited lectr, Peoples Rupub China, 85. *Mem:* Am Chem Soc; Health Physics Soc; Water Pollution Control Fedn; Am Water Works Asn; Asn Environ Eng Prof; Sigma Xi; Nat Soc Prof Eng. *Res:* Treatment of laundromat wastes; chemical nutrients in natural waters; fate of chemical nutrients in ground water; thermoluminescent dosimetry; effects of man's works on environment; Lake George ecosystem; removal of heavy metals in wastewaters; leachate. *Mailing Add:* 24 Valencia Lane Clifton Park NY 12065

AULERICH, RICHARD J, b Detroit, Mich, Mar 12, 36; m 63; c 3. ANIMAL SCIENCE, FURBEARING ANIMAL BIOLOGY. *Educ:* Mich State Univ, BS, 58, MS, 64, PhD(poultry sci), 67. *Prof Exp:* Technician, Fur Animal Proj, 62-67, from asst prof to prof poultry sci, 67-80, PROF ANIMAL SCI, MICH STATE UNIV, 80- *Concurrent Pos:* Mem subcomt furbearer nutrit, Nat Res Coun, 67, univ coordr, live animal res, 87- *Mem:* Sigma Xi; Am Soc Animal Sci. *Res:* Nutrition, physiology, toxicology and management of furbearing animals. *Mailing Add:* Dept Animal Sci Mich State Univ East Lansing MI 48824

AULETTA, CAROL SPENCE, b Princeton, NJ, Apr 14, 42; m 65; c 2. ACUTE & OCULAR TOXICOLOGY. *Educ:* Rutgers Univ, BA, 64. *Prof Exp:* prof anat & physiol Mercer County Community Col, 67-75; ASSOC DIR TOXICOL, BIO DYNAMICS, INC, 64- *Concurrent Pos:* Consult, 70-75. *Mem:* Soc Toxicol. *Mailing Add:* Bio Dynamics Inc Mettlers Rd East Millstone NJ 08873

AULICK, LOUIS H, POST-BURN HYPERMETABOLISM, INFECTION. *Educ:* Ind Univ, PhD(physiol), 74. *Prof Exp:* ASSOC PROF SURG & PHYSIOL, SCH MED, MARSHALL UNIV, 84- *Res:* Temperature regulation. *Mailing Add:* Dept Surg Sch Med Marshall Univ Huntington WV 25204

AULL, CHARLES EDWARD, b US, Sept 1, 27. MATHEMATICS. *Educ:* Columbia Univ, BS, 49; Univ Ore, MS, 53; Univ Colo, PhD(math), 62. *Prof Exp:* Asst math, Univ Ore, 52-53; instr, Univ Ariz, 53-55; instr, Univ Colo, 55-62; asst prof, Kent State Univ, 62-65; assoc prof, 65-68, PROF MATH, VA POLYTECH INST & STATE UNIV, 68- *Mem:* Am Math Soc; Math Asn Am. *Res:* Topological spaces; separation and base axioms; covering properties and sequences; algebra, especially ideal theory; analysis, especially generalized variations. *Mailing Add:* Va Polytech Inst & State Univ Blacksburg VA 24061

AULL, FELICE, b Vienna, Austria, Aug 12, 38; US citizen; m 62; c 1. PHYSIOLOGY. *Educ:* Columbia Univ, AB, 60; Cornell Univ, PhD(physiol), 64. *Prof Exp:* Physiologist, Radiobiol Lab, Bur Com Fisheries, NC, 64-65; from instr to asst prof, 66-72, ASSOC PROF PHYSIOL, SCH MED, NY UNIV, 72- *Concurrent Pos:* USPHS fel physiol, Med Col, Cornell Univ, 65-66. *Mem:* NY Acad Sci; Soc Gen Physiol; Am Physiol Soc; Biophys Soc; Assoc Women Sci. *Res:* Anion transport. *Mailing Add:* Dept Physiol & Biophys NY Univ Sch Med 550 First Ave New York NY 10016

AULL, JOHN LOUIS, b Newberry, SC, May 7, 39; m 62; c 2. ENZYMOLOGY. *Educ:* Univ NC, Chapel Hill, AB, 64; NC State Univ, PhD(biochem), 72. *Prof Exp:* Chemist, NC Dept Agr, 62-63; teaching asst, NC State Univ, 68-69, res asst, 69-72, NIH fel, 72-73; Am Cancer Soc fel, Univ SC, 73-74; from asst to assoc prof chem, 74-85, PROF CHEM, AUBURN UNIV, 85-, ACTG DEPT HEAD, 86- *Concurrent Pos:* Line officer, US Navy, 64-67. *Mem:* Am Soc Biol Chemists & Molecular Biologists; Am Chem Soc; Sigma Xi. *Res:* Interactions of potential chemotherapeutic compounds with enzymes; enzyme mechanisms and metabolic regulation of enzymes; protein structures; function relationships; chemical modification of proteins. *Mailing Add:* Dept Chem Auburn Univ Auburn AL 36830

AULL, LUTHER BACHMAN, III, b Greenwood, SC, Mar 1, 29; m 50; c 4. SCIENCE ADMINISTRATION, NUCLEAR & RADIATION PHYSICS. *Educ:* US Mil Acad, BS, 50; Univ Va, PhD(physics), 59. *Prof Exp:* Physicist nuclear weapon effects, Defense Atomic Support Agency, 59-63; physicist nuclear physics, US Army Res Off, 63-67; tech asst nuclear weapon effects, Defense Res & Eng, Off Secy Defense, 67-70; planning consult nuclear weapon progs, Combined Opers Planning Group, Union Carbide 70-73, mgr planning & spec tasks gas centrifuge develop, Nuclear Div-Separation Systs Div, 73-81, mgr, tech integration off, advan gas centrifuge prog, Nuclear Div, 81-84; MED PHYSICIST, SELF MEM HOSP, 85- *Mem:* Sigma Xi; Am Phys Soc; Am Asn Physicists Med. *Res:* Planning and coordinating continuing development program for the gas entrifuge for use in uranium isotope separation. *Mailing Add:* 1711 Hwy 34 W Ninety Six SC 29666

AULSEBROOK, LUCILLE HAGAN, b Houston, Tex, Dec 31, 25; m 52; c 4. ANATOMY. *Educ:* Univ Tex, Austin, BA, 46, MA, 47; Univ Ark, Little Rock, PhD(anat), 66. *Prof Exp:* Instr zool, Univ Tex, Austin, 48; res assoc cancer res, M D Anderson Hosp Cancer Res, Houston, Tex, 48-50; teaching asst zool, Univ Wis, Madison, 50-53; instr biol & chem, Univ Ark, Little Rock, 57-60, teaching asst anat, Sch Med, 60-64; instr, 69-71, ASST PROF ANAT, SCH MED, VANDERBILT UNIV, 71-, ASSOC PROF, SCH NURSING, 73- *Concurrent Pos:* USPHS fel, Sch Med, Vanderbilt Univ, 66-69. *Mem:* Am Asn Anat; Soc Neurosci; Sigma Xi. *Res:* Hypothalamic-endocrine interrelations, particularly neural regulation of releasing factors involved in reproduction, parturition and lactation. *Mailing Add:* Cell Biol Med Ctr N Vanderbilt Univ Sch Med Nashville TN 37232-2175

AULT, ADDISON, b Boston, Mass, July 3, 33; m 58; c 5. ORGANIC CHEMISTRY. *Educ:* Amherst Col, AB, 55; Harvard Univ, PhD(chem), 60. *Prof Exp:* Asst prof chem, Grinnell Col, 59-61; resident res assoc, Argonne Nat Lab, 61-62; from asst prof to assoc prof, 62-70, PROF CHEM & CHMN DEPT, CORNELL COL, 70- *Concurrent Pos:* NSF sci fac fel, Pa State Univ, 68-69; consult & evaluator, NCent Asn Cols & Schs, 72-; tour speaker, Am Chem Soc, 79-81, 84, 85. *Mem:* Am Chem Soc; AAAS; Brit Chem Soc. *Res:* Structure determination of natural products; reaction mechanisms; physical organic chemistry; nuclear magnetic resonance; sterochemistry; enzyme kinetics; chemical education; author of laboratory manuals and textbooks for organic chemistry. *Mailing Add:* Dept Chem Cornell Col Mt Vernon IA 52314

AULT, GEORGE MERVIN, b Pomeroy, Iowa, Aug 23, 21; m 64; c 2. METALLURGY, MATERIALS ENGINEERING. *Educ:* Iowa State Univ, BS, 43. *Prof Exp:* Engr & head, Mat Res Sect, Lewis Res Ctr, NASA, 47-55, chief, 55-59, asst chief, Mat & Struct Div, 59-64, assoc chief, 64-70, dir, Power & Mat, 70-75, dir energy progs directorate, 75-81; RETIRED. *Concurrent Pos:* Mem mat adv bd, Nat Acad Sci & Eng, 67-70. *Honors & Awards:* Except Sci Achievement Medal, NASA, 68, Except Serv Medal, 76, Distinquished Serv Medal, 80. *Mem:* Fel Am Soc Metals; Am Inst Mining, Metall & Petrol Engrs; AAAS. *Res:* Superalloys; refractory metals; ceramics and cermets; composites; thermal and mechanical fatigue; space power; terrestrial energy technology for propulsion and electric power generation. *Mailing Add:* 25920 Myrtle Ave Olmsted Falls OH 44138

AULT, N(EIL) N(ORMAN), b Dunkirk, Ohio, Mar 31, 22; m 46; c 3. CERAMICS. *Educ:* Ohio State Univ, BCerE, 47, MSCerE, 48, PhD(ceramic eng), 50. *Prof Exp:* Res assoc eng exp sta, Ohio State Univ, 48-50; res ceramic engr, Norton Co, 50-57, asst dir res & develop, 57-64, dir res, Refractories & Protective Prod Div, 64-70, tech dir, Indust Ceramics Div, 70-78, res mgr technol, Indust Ceramics Div, 78-80, Tech Dir, 81-87; CONSULT, 87- *Concurrent Pos:* Chair, Mat Tech Adv Comt, US Dept Com. *Honors & Awards:* Distinguished Ceramist of New Eng, Am Ceramic Soc, 76. *Mem:* Am Ceramic Soc; Nat Inst Ceramic Engrs. *Res:* Materials science; special refractories for high temperatures. *Mailing Add:* 52 Holden St Holden MA 01520

AULT, WAYNE URBAN, b Monroe Co, Mich, Jan 20, 23; m 47; c 4. GEOCHEMISTRY. *Educ:* Wheaton Col, BA, 50; Columbia Univ, MA, 56, PhD(geochem), 57. *Prof Exp:* Asst geochem, Lamont Geol Observ, 50-57; res geochemist, US Geol Surv, 57-62; sr res scientist, Isotopes, Inc, 62-64, dir tech representation, 64-68; assoc prof, Wheaton Col, 68-69; assoc prof chem, 69-70, actg chmn dept, 69, assoc prof geol, 70-72, prof geol, King's Col, NY, 72-, chmn dept, 70-; AT DEPT GEOL, NYACK COL; PROF GEOL, MEMPHIS STATE UNIV, 85- *Concurrent Pos:* Asst prof, Shelton Col, 52-54; asst prof, Nyack Col, 64-72, assoc prof geol, 72- *Mem:* Fel Geol Soc Am; Am Geophys Union; Am Nuclear Soc; Int Asn Geochem & Cosmochem; Sigma Xi. *Res:* Isotope geology; mass spectrometry; vacuum fusion; age determination; volcanology; volcanic fluids; subsurface tracing; radioisotope applications; environmental geology. *Mailing Add:* 595 Watson Memphis TN 38111

AUMAN, JASON REID, b High Point, NC, Feb 5, 37; m 58; c 2. ASTROPHYSICS. *Educ:* Duke Univ, BS, 59; Northwestern Univ, PhD(astron), 65. *Prof Exp:* Res assoc astron, Princeton Univ, 64-68; from asst prof to assoc prof, 68-77, PROF ASTRON, UNIV BC, 77- *Mem:* Am Astron Soc; Int Astron Union. *Res:* Structure and evolution of stars; atmospheres of late-type stars; infrared opacity of water vapor at elevated temperatures; nucleii of galaxies. *Mailing Add:* Dept Geophys & Astron Univ BC 2075 Westbrook Pl Vancouver BC V6T 1W5 Can

AUMANN, GLENN D, b Elbowoods, NDak, June 17, 30; m 50; c 4. ECOLOGY, ETHOLOGY. *Educ:* Wis State Univ, Eau Claire, BS, 57; Univ Wis, MS, 58, PhD(ecol), 64. *Prof Exp:* Asst prof biol, Gogebic Community Col, 57-61; NIH fel electron micros, Sch Med, Univ Wis, 64-65; assoc prof, Univ Houston, 65-70, chmn dept, 67-76, assoc dean, 76-81, interim dean, 82-83, assoc provost, 84-85, assoc vpres res, 85-90, PROF BIOL, UNIV HOUSTON, 70-, DEAN, NATURAL SCI & MATH, 90- *Mem:* AAAS. *Res:* Vertebrate population control mechanisms. *Mailing Add:* Col Natural Sci & Math Univ Houston Houston TX 77204-5502

AUMANN, HARTMUT HANS-GEORG, b Frankfurt, Ger, Nov 18, 40; US citizen; m 70; c 3. SPACE SCIENCE, ASTRONOMY. *Educ:* Univ Idaho, BS, 62; Univ Okla, MS, 64; Rice Univ, PhD(space sci), 70. *Prof Exp:* Res scientist physics, Esso Prod Res Co, 64-66; SR SCIENTIST SPACE SCI, JET PROPULSION LAB, CALIF INST TECHNOL, 70-, IPAC. *Concurrent Pos:* IRAS proj scientist, Jet Propulsion Lab, 76- *Res:* Remote sounding of earth and planetary atmospheres in the infrared. *Mailing Add:* Mail Stop 100-22 4800 Oak Grove Dr Calif Inst Technol Pasadena CA 91109

AUMANN, ROBERT JOHN, b Frankfurt-am-Main, Germany, June 8, 30; US & Israeli citizen; m 55; c 5. GAME THEORY, MATHEMATICAL ECONOMICS. *Educ:* City Col New York, BS, 50; Mass Inst Technol, SM, 52, PhD(math), 55. *Prof Exp:* Res asst, dept math, Princeton Univ, 54-56; from instr to assoc prof, 56-68, PROF MATH, HEBREW UNIV, JERUSALEM, 68- *Concurrent Pos:* Res assoc, Princeton Univ, 60-61; vis prof, Yale Univ, 64-65, Univ Calif, Berkeley, 71, Stanford Univ, 75-76 & 80-81, Univ Louvain, 72, 78, 84. *Honors & Awards:* Harvey Prize, Israel Inst Technol, 83. *Mem:* Nat Acad Sci; fel Econometric Soc; foreign hon mem Am Acad Arts & Sci. *Res:* Interactive decision making (game theory); economics and general equilibrium theory; interactive information theory (repeated games). *Mailing Add:* Inst Math Hebrew Univ Jerusalem 91904 Israel

AUMENTO, FABRIZIO, marine geology, volcanology, for more information see previous edition

AUNE, JANET, b St Louis, Mo. BIOLOGY. *Educ:* St Louis Univ, PhD(biol), 68. *Prof Exp:* VPRES ACAD SERV, HEALTH SCI CTR, UNIV TEX, 81- *Mailing Add:* c/o Sheila Bradley Univ Tex Southwestern Med Ctr 5323 Harry Hines Blvd Dallas TX 75235-9023

AUNE, KIRK CARL, b Winona, Minn, Nov 5, 42; m 64; c 2. PHYSICAL BIOCHEMISTRY. *Educ:* Univ Minn, Minneapolis, BCh, 64; Duke Univ, PhD(biochem), 68. *Prof Exp:* Mem staff biochem, Brandeis Univ, 68-70; asst prof, Ohio State Univ, 70-73; assoc prof biochem, Baylor Col Med, 73-86; ASSOC DEAN, OFF INFO SYSTS, UNIV NC, CHAPEL HILL, 86- *Concurrent Pos:* Am Cancer Soc res fel, 68-70; NIH career develop award, 73 & 75. *Mem:* AAAS; Am Chem Soc; Am Soc Biol Chemists; Sigma Xi. *Res:* Physical chemistry of macromolecules; energetics of interaction in complex biological structures such as the ribosomal subunits from ultracentrifugal techniques. *Mailing Add:* Off Info Systs Rm 76 MacNider Bldg Univ NC Med Sch Chapel Hill NC 27599-7045

AUNG, TAING, b Hmawbe, Yangon, Myanmar, July 15, 44; Can citizen. CYTOGENETICS, PLANT BREEDING. *Educ:* Univ Rangoon, Burma, BSc, 65, MSc, 72; Univ Wales, UK, MSc, 75, PhD(cytogenesis), 78. *Prof Exp:* Lect demonstrator forest ecol, Bot Dept, Univ Rangooon, 70-72, jr lectr genetics statist, 78-80; fel, Agr Bot Dept, Univ Col wales, Aberystwyth, 81-84; res assoc, Plant Sci Dept, Univ Alta, 84-87; asst prof gen genetics, plant breeding & cytogenetics, Plant Soil & Insect Sci Dept, Univ Wyo, 87-90; RES SCIENTIST, AGR CAN RES STA, WINNIPEG, MAN, 90- *Concurrent Pos:* Wyo state rep, Regional Tech Comt on Plant Germplasm, Western US, 87-90. *Mem:* Genetics Soc Can; Am Soc Agron; Crop Sci Soc Am. *Res:* Transfer of desirable genes from related and unrelated species of wheat, oat and barley into the cultivated forms through cytogenetic techniques. *Mailing Add:* Agr Can Res Sta 195 Dafoe Rd Winnipeg MB R3T 2M9 Can

AUNGST, BRUCE J, b Pottsville, Pa, Nov 22, 52; m 80; c 3. PHARMACOKINETICS, PHARMACEUTICS. *Educ:* Pa State Univ, BS, 74, MS, 77; State Univ NY, Buffalo, PhD(pharmaceut), 81. *Prof Exp:* prin scientist, E I Du Pont de Nemours & Co, 81-90; SR SCIENTIST, DUPONT MERCK, 91- *Mem:* Am Asn Pharmaceut Scientists. *Res:* Drug metabolism; drug delivery; biopharmaceutics. *Mailing Add:* Dupont Merck PO Box 80400 Wilmington DE 19880-0400

AUNON, JORGE IGNACIO, b Havana, Cuba, Sept 2, 42; US citizen; m 65; c 4. BIOMEDICAL ENGINEERING. *Educ:* George Washington Univ, BS, 67, MS, 69, ScD(eng), 72. *Prof Exp:* Med engr, George Washington Univ Hosp, 69-72; res assoc neurol, Children's Hosp, Washington, DC, 72-73; from asst prof to prof elec eng, Purdue Univ, West Lafayette, 73-87; PROF & HEAD DEPT ELEC ENG, COLO STATE UNIV, FT COLLINS, 88- *Honors & Awards:* Cert of Recognition, Inst Elec & Electronics Engrs, 77. *Mem:* Fel Inst Elec & Electronics Engrs. *Res:* Electroencephalography research; evoked potentials; computer processing of physiological data. *Mailing Add:* Dept Elec Eng Colo State Univ Ft Collins CO 80523

AURAND, HENRY SPIESE, JR, b Columbus, Ohio, Feb 24, 24; m 46; c 2. UNDERWATER ACOUSTICS. *Educ:* US Mil Acad, BS, 44; State Univ Iowa, MS, 47. *Prof Exp:* Design engr ballistic missiles, Douglas Aircraft Co, 54-56; mem tech staff weapon syst, Tech Mil Planning Oper, Gen Elec, 57-63; mgr eng develop dept, Electronics Div, Gen Mill Inc, 63; sr res engr ocean acoustics, Lockheed Missiles & Space Co, 63-69; phys sci adminr, Off Naval Res, Washington, 69-71; supvr physicist Ocean Acoust, Naval Undersea Ctr, San Diego, 71-84; RETIRED. *Mem:* Acoust Soc Am; Am Geophys Union; Am Soc Civil Engr. *Res:* Large scale computers and sensors in ocean acoustics. *Mailing Add:* 734 Camino Solana Beach CA 92075

AURAND, LEONARD WILLIAM, b Shamokin Dam, Pa, Feb 5, 20; m 43; c 3. BIOCHEMISTRY. *Educ:* Pa State Univ, BS, 41, PhD(biochem), 49; Univ NH, MS, 47. *Prof Exp:* From res asst prof to res assoc prof animal indust, NC State Univ, 49-60, prof food, 60-88; RETIRED. *Mem:* Inst Food Technol; Am Inst Nutrit. *Res:* Off-flavors in milk; enzyme chemistry; oxidation of lipids. *Mailing Add:* 236 Food Sci NC State Univ Raleigh NC 27650

AURBACH, GERALD DONALD, b Cleveland, Ohio, Mar 24, 27; m 60; c 2. MEDICINE, ENDOCRINOLOGY. *Educ:* Univ Va, BA, 50, MD, 54. *Prof Exp:* Intern med, New Eng Ctr Hosp, 54-55, assoc, 58-59; resident, Boston City Hosp, 55-56; res assoc, Nat Inst Arthritis, Metab & Digestive Dis, 59-61, mem, Sect Endocrinol, 61-65, mem, Endocrinol Study Sect, 67-69, chief, Mineral Metab Sect, 65-73, CHIEF, METAB DIS BR, NAT INST ARTHRITIS, METAB & DIGESTIVE DIS, 73- *Concurrent Pos:* Res fel med, New Eng Ctr Hosp, 56-58; asst, Sch Med, Tufts Univ, 56-58, instr, 58-59; USPHS fel, Nat Inst Arthritis & Metab Dis, 56-59; secy-treas, Int Cong Endocrinol, 72; vol ed sect endocrinol, Handbook Physiol, Am Physiol Soc; assoc ed, Endocrine Soc, coun, 83-, pres, 88-; mem ed comt, Am Soc Clin Invest, 71-; ed in chief, Vitamins and Hormones; Burroughs Wellcome vis prof, Royal Soc Med, 89; Bley Stein vis prof, Univ Southern Calif, 86; vis prof, Mayo Clinic, 90; mem exec comt, Int Soc Endocrinol. *Honors & Awards:* John Horsley Mem Prize, Univ Va, 60; Andre Lichtwitz Prize, 68; Gordon Wilson Lectr, Am Clin & Climatol Asn, 73; William F Neuman Award, Am Soc Bone & Mineral Res, 81; Gairdner Found Int Award, 83; Edwin B Astwood Lectureship Award, Endocrine Soc, 85; Distinguished Serv Medal, USPHS, 88. *Mem:* Nat Acad Sci; Am Fedn Clin Res; Endocrine Soc (vpres, 78-79, pres, 88-89); Am Soc Biol Chem; Asn Am Physicians; Am Soc Clin Invest; Int Soc Endocrinol. *Res:* Endocrinology; biochemistry. *Mailing Add:* Metab Dis Br Nat Inst Diabetes & Digestive & Kidney Dis NIH Bldg 10 Rm 9C101 Bethesda MD 20892

AURELIAN, LAURE, b Bucharest, Rumania, June 17, 39; m 64. VIROLOGY. *Educ:* Tel Aviv Univ, MSc, 62; Johns Hopkins Univ, PhD(microbiol), 66. *Prof Exp:* Res asst microbiol, Tel Aviv Univ, 60-62; from instr to asst prof, Johns Hopkins Med Inst, 66-74, assoc prof microbiol & biophys, 74-; PROF, DEPT PHARMACOL & EXP THERAPEUT, DEPT MICROBIOL & IMMUNOL, SCH MED, UNIV MD; PROF, DEPT BIOPHYS & COMP MED, JOHNS HOPKINS. *Concurrent Pos:* Res fel, Sch Med, Johns Hopkins Univ, 65-66. *Honors & Awards:* Georgian Republican & Tbilisi Sci Soc of Oricologists, Georgia, USSR; Premlo XXIV Casali, 90; Pro Loco Bronte Edizione Speciale Medicina, Catania, Sicily. *Mem:* AAAS; Am Asn Cancer Res; Soc Exp Biol & Med; Am Soc Microbiol; Am Asn Immunol; D A Boyes Soc Gynaecol Oncol. *Res:* Biological, biochemical and biophysical properties of viruses; basis of host-range restrictions; role of viruses in cancer. *Mailing Add:* Dept Pharmacol & Exp Therapeut Univ Md Sch Med 10 S Pine St Baltimore MD 21201

AUSICH, RODNEY L, b Casper, Wyo, Oct 28, 53; m 86. PLANT TISSUE CULTURE, PLANT MOLECULAR BIOLOGY. *Educ:* Univ Wyo, BS, 76; Ind Univ, MA, 78, PhD(biol), 80. *Prof Exp:* Res Biol, 80-84, staff res scientist, Amoco Corp, 84-89, RES SUPVR, AMOCO TECHNOL CO, 89- *Mem:* Am Soc Plant Physiologists; AAAS. *Res:* Regulation of gene activity in plant, algal and yeast cells. *Mailing Add:* Amoco Res Ctr PO Box 400 Naperville IL 60566

AUSICH, WILLIAM IRL, b Kewanee, Ill, Feb 2, 52; m 73; c 3. INVERTEBRATE PALEONTOLOGY, PALEOECOLOGY. *Educ:* Univ Ill, BS, 74; Ind Univ Bloomington, AM, 76, PhD(geol, biol), 78. *Prof Exp:* Instr geol, Indiana Univ, Bloomington, 78; asst prof, Wright State Univ, 78-82, assoc prof, 82-84; assoc prof, 84-90, PROF, GEOL SCI DEPT, OHIO STATE UNIV, 90- *Honors & Awards:* Schuchert Award, Paleont Soc, 90. *Mem:* Paleont Soc; Int Palaeont Soc; Sigma Xi. *Res:* Paleoecology, functional morphology, taxonomy, and evolution of Paleozoic crinoids; Paleozoic community paleoecology; lower Mississippian paleontology. *Mailing Add:* Geol Sci Dept Ohio State Univ 125 S Oval Mall Columbus OH 43210

AUSIELLO, DENNIS ARTHUR, b Chelsea, Mass, Sept 12, 45. MEDICINE. *Educ:* Harvard Col, BA, 67; Univ Pa, MD, 71; Am Bd Internal Med, dipl, 74; Am Bd Nephrology, dipl, 78. *Prof Exp:* Intern med, Mass Gen Hosp, 71-72, asst resident, 72-73, clin fel med-nephrology, 75-76, asst med, 76-82, asst physician, 82-84, assoc physician, 85-88; clin fel med, Harvard Med Sch, 71-73, res fel, 75-76, from instr to asst prof, 76-85, ASSOC PROF MED, HARVARD MED SCH, 85-; CHIEF, RENAL UNIT, MASS GEN HOSP, 84-, PHYSICIAN & CHIEF, 88- *Concurrent Pos:* Surgeon, Pub Health Serv & staff assoc, Lab Kidney & Electrolyte Metab, Nat Heart & Lung Inst, NIH, 73-75; NIH res fel med-nephrology, Mass Gen Hosp, 76-77, attend physician, Med Serv & visit, Renal Serv, 76-; lectr & sect leader renal pathophysiol, Harvard Med Sch, 76-; res career develop award, NIH, 77-82; assoc ed, Am J Physiol, 87- *Mem:* Am Soc Nephrology; Int Soc Nephrology; Am Fedn Clin Res; Am Physiol Soc; Am Soc Clin Invest; Asn Am Physicians. *Mailing Add:* Harvard Med Sch & Renal Unit Mass Gen Hosp 149 13th St 8th Floor Charlestown MA 02129

AUSKAPS, AINA MARIJA, b Raiskums, Latvia, Sept 2, 21; US citizen. DENTISTRY, BIOCHEMISTRY. *Educ:* Univ Latvia, MD, 44; Ludwic Maximilian Univ, Munich, DDS, 45; Harvard Univ, DMD, 55. *Prof Exp:* Res asst physiol & dent, Sch Med, Yale Univ, 49-52; res assoc nutrit & biochem, Harvard Univ, 52-61; CONSULT, 86- *Mem:* Fel AAAS; fel Am Acad Dent Sci. *Res:* Studies on bone and tooth metabolism using phosphorus-32; nutrition; fluoride metabolism. *Mailing Add:* 104 Perkins St Jamaica Plain MA 02130-4303

AUSLANDER, BERNICE LIBERMAN, b Brooklyn, NY, Nov 21, 30; m 50; c 2. MATHEMATICS. *Educ:* Columbia Univ, AB, 51; Univ Chicago, MS, 54; Univ Mich, PhD(math), 63. *Prof Exp:* Asst prof, Wellesley Col, 65-71; ASSOC PROF MATH, UNIV MASS, BOSTON, 71- *Concurrent Pos:* Fel, Radcliffe Inst Independent Study, 63-65. *Mem:* Am Math Soc; Math Asn Am. *Res:* Algebra, particularly the theory of commutative rings. *Mailing Add:* 16 Everett Ave Newton Centre MA 02159

AUSLANDER, JOSEPH, b New York, NY, Sept 10, 30; m 57; c 2. MATHEMATICS. *Educ:* Mass Inst Technol, SB, 52; Univ Pa, PhD(math), 57. *Prof Exp:* Asst prof math, Carnegie Inst Technol, 57-60; mem staff, Res Inst Advan Study, 60-62; assoc prof, 62-66, PROF MATH, UNIV MD, COLLEGE PARK, 66- *Concurrent Pos:* Res assoc, Yale Univ, 64-65; vis sr lectr, Imp Col, Univ London, 67-68. *Mem:* AAAS; Am Math Soc; Math Asn Am. *Res:* Topological dynamics; ergodic theory. *Mailing Add:* Univ Md College Park MD 20742

AUSLANDER, LOUIS, b Brooklyn, NY, July 12, 28; m 77; c 3. MATHEMATICS. *Educ:* Columbia Univ, BA, 49, MA, 50; Univ Chicago, PhD, 54. *Prof Exp:* Instr, Yale Univ, 53-55; asst prof math, Univ Pa, 56-57; from asst prof to assoc prof, Ind Univ, 57-62; prof, Purdue Univ, 62-64 & Belfer Grad Sch Sci, Yeshiva Univ, 64-65; DISTINGUISHED PROF MATH, GRAD SCH & UNIV CTR, CITY UNIV NEW YORK, 65- *Concurrent Pos:* NSF fel, Inst Advan Study, 55-57; vis assoc prof, Yale Univ, 60-62; vis prof, Univ Calif, Berkeley, 63-64; Guggenheim fel, 71-72. *Mem:* Am Math Soc. *Res:* Structure of solvmanifolds and harmonic analysis on solvable Lie groups and solvmanifolds. *Mailing Add:* Dept Math Grad Sch & Univ 33 W 42nd St New York NY 10036

AUSLANDER, MAURICE, b Brooklyn, NY, Aug 3, 26; m 50; c 2. MATHEMATICS. *Educ:* Columbia Univ, BA, 49, PhD(math), 54. *Prof Exp:* Instr math, Univ Chicago, 53-54 & Univ Mich, 54-56; NSF fel, Inst Advan Study, 56-57; from asst prof to assoc prof, 57-63, PROF MATH, BRANDEIS UNIV, 63- *Mem:* Fel AAAS; fel Am Acad Arts & Sci; fel Royal Norweg Acad Arts & Sci; Sigma Xi. *Res:* Homological, associative and communatative algebra. *Mailing Add:* Dept Math Brandeis Univ Waltham MA 02254

AUSMAN, ROBERT K, b Milwaukee, Wis, Jan 31, 33. MEDICAL ADMINISTRATION, SURGERY. *Educ:* Marquette Univ, MD, 57. *Prof Exp:* Intern, Univ Minn, 57-58; dir, Health Res, Inc, 61-69; mem staff, NY State Dept Health, Roswell Park Mem Inst, 69; dep dir, Fla Regional Med Prog, 69-70; dir clin res, 70-73, VPRES CLIN RES, BAXTER LABS, 73- *Concurrent Pos:* Res fel, Univ Minn, 58-61; Damon Runyon res fel, 59-61;

sci consult, Northern Eng Co, 61-; assoc clin prof surg, Med Col Wis. *Mem:* AAAS; Am Asn Cancer Res; Am Geriat Soc; Am Soc Clin Oncol; Sigma Xi. *Res:* Medical systems and data processing; cancer chemotherapy; public health statistics; nutrition; computer science. *Mailing Add:* Box 3538 RFD Long Grove IL 60047

AUSPOS, LAWRENCE ARTHUR, b Anoka, Minn, Aug 27, 17; m 45; c 2. CHEMISTRY. *Educ:* Univ Portland, BS, 40; Univ Notre Dame, MS, 41, PhD(org chem), 43. *Prof Exp:* Sr lab asst chem, Univ Portland, 39-40; lab asst, Univ Notre Dame, 40-41; instr, Holy Cross Sem, 41-42; mem staff, Pioneering Res Div, Rayon Tech Div, 43-50, MEM STAFF, TEXTILE FIBRE DEPT, E I DU PONT DE NEMOURS & CO, 50- *Mem:* Am Chem Soc. *Res:* Synthetic plastics and fibers; preparation of fiberforming polymers; preparation of dibutyl benezenes. *Mailing Add:* 2536 Blackwood Rd Wilmington DE 19810

AUST, CATHERINE COWAN, b Atlanta, Ga, Apr 26, 46; m 67; c 1. MATHEMATICS, UNIVERSAL ALGEBRA. *Educ:* Univ Ga, BS, 68; Emory Univ, PhD(math), 73. *Prof Exp:* From instr to asst prof math, Ga Inst Technol, 72-75; assoc prof, 80-85, PROF MATH, CLAYTON STATE COL, 85- *Mem:* Am Math Soc; Math Asn Am; Am Math Asn of Two Year Cols; Asn Women in Math. *Mailing Add:* Sch Arts & Sci Clayton State Col Morrow GA 30260

AUST, J BRADLEY, b Buffalo, NY, Sept 8, 26; m 49; c 6. SURGERY. *Educ:* Univ Buffalo, MD, 49; Univ Minn, MS, 57, PhD(surg), 58; Am Bd Surg, dipl, 58; Am Bd Thoracic Surg, dipl, 65. *Prof Exp:* Res asst, Univ Minn, 53-57, from instr to prof surg, 57-66, coordr cancer res, 60-66; PROF SURG & CHMN DEPT, UNIV TEX MED SCH, SAN ANTONIO, 66- *Concurrent Pos:* Am Cancer Soc scholar, 57-62; surg consult, Minn State Prison, 57-63 & Anoka State Hosp, 63-66. *Mem:* Soc Univ Surg; Soc Surg Oncol (pres, 88); fel Am Col Surg; Western Surg Asn (pres, 89); Am Surg Asn (vpres, 86); Sigma Xi. *Res:* Cancer immunity; homotransplantation; surgical oncology. *Mailing Add:* Dept Surg Univ Tex Health Sci Ctr San Antonio TX 78284

AUST, KARL T(HOMAS), b Toronto, Ont, Aug 9, 25; nat US; m 50; c 4. PHYSICAL METALLURGY, MATERIALS SCIENCE. *Educ:* Univ Toronto, BASc, 46, MASc, 48, PhD(phys metall), 50. *Prof Exp:* Asst metallurgist, US Metals Refining Co, 46-47; res metallurgist, Kaiser Aluminum & Chem Corp, 50-52; res assoc metall, Johns Hopkins Univ, 52-55; res metallurgist, Gen Elec Co, 55-67; chmn, Mat Res Ctr, 69-74, PROF METALL & MAT SCI, UNIV TORONTO, 67- *Concurrent Pos:* Guggenheim fel, Ecole des Mines, Paris, France, 62-63; spec lectr, Nat Inst Nuclear Sci & Technol, Paris, 63; chmn phys metall, Gordon Res Conf, 72; vis prof, Univ Hawaii, 74 & Univ Kyoto, 75; vis scientist, Ecole des Mines, St Etienne, France, 76; vis prof, Univ Saarlandes, 81; vis prof, Univ Kyoto, 81 & 88; vis prof, Univ Gottingen, 87; Gauss prof, Göttigen Acad Sci, 87. *Honors & Awards:* Mathewson Gold Medal, Am Inst Mining, Metall & Petrol Engrs, 61; Hoffman Mem Award, Lead Develop Asn, 71; Japan Soc Promotion Sci Award, 75 & 88; Yamada Sci Found Award, 81; Can Metal Physics Medal, 89. *Mem:* Fel Am Soc Metals; Am Inst Mining, Metall & Petrol Engrs; Hist Metall Soc; Sigma Xi. *Res:* Structure and properties of grain boundaries in metals; plastic deformation and annealing phenomena in metals; structure and properties of metals and metallic glasses; triple line defects and nanocrystals. *Mailing Add:* Dept Metall-Mat Sci Univ Toronto Toronto ON M5S 1A4 Can

AUST, RICHARD BERT, b Oklahoma City, Okla, Aug 29, 38; m 68; c 2. CHEMICAL ENGINEERING, PHYSICAL CHEMISTRY. *Educ:* Carnegie Inst Technol, BS, 60; Univ Ill, MS, 62, PhD(chem eng), 64. *Prof Exp:* Sr res chemist, 64-69, RES ASSOC, EASTMAN KODAK CO, 69- *Mem:* SPIE; Sigma Xi. *Res:* Properties of matter; preparation and evaluation of thin film optical coatings. *Mailing Add:* 150 Mt Airy Dr Rochester NY 14617

AUST, STEVEN DOUGLAS, b South Bend, Wash, Mar 11, 38; m 72; c 2. BIOCHEMISTRY, TOXICOLOGY. *Educ:* Wash State Univ, BS, 60, MS, 62; Univ Ill, PhD(dairy sci), 65. *Prof Exp:* Prof biochem, Mich State Univ, 67-87, dir, Ctr Study Active Oxygen in Biol & Med, 80-87; PROF CHEM & BIOCHEM, NUTRIT & FOOD SCI, UTAH STATE UNIV, 87-, DIR, BIOTECHNOL CTR. *Concurrent Pos:* USPHS fel, Karolinska Inst, Sweden, 66; Ministry Agr & Fisheries NZ fel, Ruakura Agr Res Ctr, Hamilton, NZ, 75-76; consult, Nat Ctr Dis Control & Environ Protection Agency Sci Adv Bd; mem toxicol study sect, NIH, 79-83; comnr, Mich Toxic Substance Control Comn, 79-83, chmn, 81-83. *Mem:* Soc Toxicol; Am Soc Photobiol; AAAS; Am Soc Biol Chem; Am Soc Pharmacol & Exp Therapeut; Am Chem Soc; Soc Ecotoxicol & Environ Safety. *Res:* Mixed function oxidation of drugs, the peroxidation of lipids, toxicity of halogenated aromatic hydrocarbons; biological oxidation of environmental pollutants. *Mailing Add:* Biotechnol Ctr Utah State Univ Logan UT 84322-4430

AUSTEN, K(ARL) FRANK, b Akron, Ohio, Mar 14, 28; m 59; c 4. INTERNAL MEDICINE, IMMUNOLOGY. *Educ:* Amherst Col, BA, 50; Harvard Univ, MD, 54; Am Bd Internal Med, dipl; Am Bd Allergy, dipl, 65. *Prof Exp:* Intern med, Mass Gen Hosp, Boston, 54-55, asst resident, 55-56, sr resident, 58-59, chief resident, 61-62, asst, 62-63, asst physician, 63-66, chief, pulmonary unit, 64-66; captain, US Army Med Corp Div Immunochem, Walter Reed Army Inst Res, 56-58; asst, 61, instr, 61-62, assoc, 62-64, from asst prof to prof, 65-72, THEODORE BEVIER BALES PROF MED, MED SCH, HARVARD UNIV, 72-; CHMN, RHEUMATOLOGY & IMMUNOL DIV, BRIGHAM & WOMEN'S HOSP, BOSTON, 80- *Concurrent Pos:* USPHS fel, Nat Inst Med Res, Eng, 59-61, USPHS res career develop award, NIH, 66-66; mem, Comt Drug Reactions, AMA, 63-66, sci & ed coun, Allergy Found Am, 64-, Comt Collab Res Transplantation & Immunol, NIH, 65-68, Nat Comn Arthritis & Related Musculoskeletal Dis; physician-in-chief, Robert B Brigham Hosp, 66-80; physician, Peter Bent Brigham Hosp, 66-80; mem bd dirs, Arthritis Found & chmn res comt; Am Asn Immunol rep, Nat Res Coun; mem, Am Bd Allergy & Immunol & chmn, res & develop comt; vchmn, Immunol & Microbiol Interdisciplinary Cluster, President's Biomed Res Panel, 75-76; chmn, Allergy & Immunol Res Comt, Nat Int Allergy & Infectious Dis, NIH, 76-79, mem, Adv Comt to Dir, 86-90 & Task Force Allergy & Immunol, 89-90. *Honors & Awards:* Geigy Prize, Int League Against Rheumatism, 73; Squibb Award, Infectious Dis Soc Am, 73; Lila Gruber Award, Am Acad Dermat, 75; Marion L Sulzberger Award, Am Dermat Soc Allergy, 79; Waterford Biomed Sci Award, 82; Distinguished Serv Award, Int Asn Allergol & Clin Immunol, 88. *Mem:* Nat Acad Sci; Inst Med-Nat Acad Sci; fel Am Acad Allergy (pres, 81-82); Am Asn Immunologists (pres, 77-78); fel Am Acad Arts & Sci; Int Soc Immunopharmacol; Am Fedn Clin Res; Am Soc Clin Invest; Asn Am Physicians (pres, 89-90); Brit Soc Immunol; Am Soc Pharmacol & Exp Therapeut; Am Soc Exp Path; Am Rheumatism Asn; Transplantation Soc; fel Am Col Physicians. *Res:* Basic and clinical immunology. *Mailing Add:* Brigham & Women's Hosp 75 Francis St PBB2 Boston MA 02115

AUSTEN, W(ILLIAM) GERALD, b Akron, Ohio, Jan 20, 30; m 61; c 4. SURGERY. *Educ:* Mass Inst Technol, BS, 51; Harvard Univ, MD, 55; Am Bd Surg, cert, 62; Am Bd Thoracic Surg, cert, 64. *Hon Degrees:* DHH, Univ Akron, 80; DSc, Univ Athens, Greece, 81, Univ Mass, 85. *Prof Exp:* Intern, Mass Gen Hosp, Boston, 55-56, asst resident surg, 56-59; sr registr surg, Kings Col Hosp, London, Eng, 59; hon sr registr, Thoracic Unit, Gen Infirmary, Univ Leeds, 59; chief resident, E Surg Serv, Mass Gen Hosp, 60-61; clin surgeon, NIH, 61-62; assoc, 63-65, from assoc prof to prof, 65-74, EDWARD D CHURCHILL PROF SURG, MED SCH, HARVARD UNIV, 74-; CHIEF SURG SERV, MASS GEN HOSP, 69-, SURGEON-IN-CHIEF, 89- *Concurrent Pos:* Teaching fel surg, Med Sch, Harvard Univ, 60-61; chief surg cardiovasc res unit, Mass Gen Hosp, 63-69, vis surgeon, 66-; Markle scholar acad med, 63-68; numerous vis prof, US & foreign; mem, Coun Cadiovasc Surg, Am Heart Asn, 69-76, chmn, 74-76; mem, Ad Hoc Policy Adv Bd Coronary Surg, NIH, 72-74, Heart & Lung Prog Proj Comt, 73-76; mem bd trustees, Mass Inst Technol, 72-, chmn, 74-84; mem bd gov, Am Col Surgeons, 80-83; consult, Univ Mass Hosp, 76-; dir, Am Bd Thoracic Surg, 84-, bd dirs, 84-90; mem bd dirs, Found Biomed Res, 88. *Honors & Awards:* Gold Heart Award, Am Heart Asn, 80; Louis Mark Mem Lectr Award, Am Col Chest Physicians, 81. *Mem:* Inst Med-Nat Acad Sci; fel Am Acad Arts & Sci; Soc Univ Surg (secy, 67-70, pres, 72-73); Am Surg Asn (secy, 79-84, pres, 85-86); Am Heart Asn (pres, 77-78); Am Asn Thoracic Surg (vpres, 87-88, pres, 88-89); Am Col Surgeons; AAAS; Am Asn Univ Professors. *Res:* Cardiovascular physiology. *Mailing Add:* Dept Surg Mass Gen Hosp Boston MA 02114

AUSTENSON, HERMAN MILTON, b Viscount, Sask, Nov 28, 24; m 52; c 3. AGRONOMY. *Educ:* Univ Sask, BSA, 46, MSc, 48; State Col Wash, PhD, 51. *Prof Exp:* Asst agronomist, Wash State Univ, 51-53; asst prof, State Univ NY Col Agr, Cornell Univ, 53-54; asst agronomist, Wash State Univ, 54-59, assoc agronomist, 59-66; assoc prof crop sci, 66-69, head Dept Crop Sci & Dir Crop Develop Ctr, 75-83, PROF CROP SCI, UNIV SASK, 69- *Mem:* Am Soc Agron; fel Agr Inst Can; Can Soc Agron (pres, 85-86). *Res:* Ecology and production of grain and forage crops; seed quality. *Mailing Add:* Dept Crop Sci Univ Sask Saskatoon SK S7N 0W0 Can

AUSTERN, BARRY M, b New York, NY, Nov 2, 42; m 68; c 1. ANALYTICAL CHEMISTRY, BIOCHEMISTRY. *Educ:* Columbia Univ, AB, 63; NY Univ, MS, 65; Univ Mass, PhD(biochem), 69. *Prof Exp:* Res chemist, US Dept Interior, 69-70; RES CHEMIST, US ENVIRON PROTECTION AGENCY, 70- *Concurrent Pos:* Abstractor, Chem Abstracts Serv, 66-77. *Honors & Awards:* Bronze Medal, Environ Protection Agency, 79. *Mem:* Am Chem Soc; Sigma Xi; Soc Appl Spectros; Am Soc Mass Spectrometry. *Res:* Analytical chemistry and biochemistry of wastewaters and effluents; chromatography; mass spectroscopy. *Mailing Add:* Risk Reduction Eng Lab Environ Protection Agency Cincinnati OH 45268

AUSTERN, NORMAN, b New York, NY, Feb 23, 26; m 64; c 4. THEORETICAL NUCLEAR PHYSICS. *Educ:* Cooper Union, BSEE, 46; Univ Wis, PhD(physics), 51. *Prof Exp:* Asst, Univ Wis, 46-48; AEC fel physics, Cornell Univ, 51-52, res assoc, 52-54; staff scientist, Comput Facility, AEC & math inst, NY Univ, 54-55; from asst prof to assoc prof physics, 56-70, PROF PHYSICS, UNIV PITTSBURGH, 70- *Concurrent Pos:* Fulbright res scholar, Univ Sydney, 57-58, Fulbright lectr, 68-69; NSF sr fel, Inst Theoret Physics, Copenhagen Univ, 61-62; Vis Prof, Univ Wash, Seattle, 76-77; Vis Fel, Japan Soc Prom Sci, 84- *Mem:* Am Phys Soc; AAAS; Sigma Xi. *Res:* Nuclear reactions and structure, especially direct reaction theory. *Mailing Add:* Dept Physics Univ Pittsburgh Pittsburgh PA 15260

AUSTIC, RICHARD EDWARD, b Ithaca, NY, Apr 10, 41; m 63; c 2. NUTRITION. *Educ:* Cornell Univ, BS, 63; Univ Calif, Davis, PhD(nutrit), 68. *Prof Exp:* Res assoc nutrit, 68-70, asst prof animal nutrit, 70-75, ASSOC PROF ANIMAL NUTRIT, CORNELL UNIV, 75- *Mem:* AAAS; Poultry Sci Asn; Am Inst Nutrit; World Poultry Sci Asn; NY Acad Sci. *Res:* Poultry nutrition; amino acid requirements and interactions; hyperuricemia and gout; nutrition and embryonic development. *Mailing Add:* Dept Poultry & Avian Sci Cornell Univ NYS Col Agr & Life Sci 200 Rice Hall Ithaca NY 14853

AUSTIN, ALFRED ELLS, b Waltham, Mass, Oct 26, 20; m 43; c 2. PHYSICAL INORGANIC CHEMISTRY, SOLID STATE SCIENCE. *Educ:* Alfred Univ, BA, 42; Yale Univ, PhD(chem), 44. *Prof Exp:* Res chemist, Corning Glass Works, 44-47; res engr, 47-56, div consult, 56-60, fel, Battelle Mem Inst, 60-82; RETIRED. *Concurrent Pos:* At AEC, 44. *Mem:* Am Chem Soc; Am Crystallog Asn; Am Phys Soc; Electrochem Soc; fel Am Inst Chemists. *Res:* Solid state chemistry and physics; crystallography; nonaqueous electrochemistry. *Mailing Add:* 2189 Dart Ave Apt B Largo FL 34640

AUSTIN, ARTHUR LEROY, b Vancouver, BC, July 21, 29; US citizen; m 59; c 3. ENERGY SYSTEMS, MECHANICAL ENGINEERING. *Educ:* Univ Calif, Berkeley, 56, MS, 57, Dr Eng(mech eng), 62. *Prof Exp:* Instr mech eng, Univ Calif, Berkeley, 58-62; proj engr mat res, Lawrence Livermore Lab, 62-65, assoc div leader advan weapons, 65-69; assoc prof mech eng, Univ Calif,

Davis, 69-71; mem sr res staff energy systs, Lawrence Livermore Lab, 71-74, proj mgr geothermal energy, 74-78, sr staff long range planning, 78-80, prog leader, Treaty Verifications Technol Develop, 80-; RETIRED. *Concurrent Pos:* Consult, Lawrence Livermore Lab, 58-60; mem, Synthetic Fuels Panel, Off Sci & Technol, 72-73 & Geothermal Adv Comt, Dept Energy, 73-78. *Honors & Awards:* Mat Testing Methods Award, Am Soc Testing & Mat, , 62. *Mem:* Am Soc Mech Eng. *Res:* Geothermal energy; material response to high loading rates; dynamics; mechanical vibrations; continuum mechanics; mechanical design. *Mailing Add:* Mech Eng Dept L-123 PO Box 808 Livermore CA 94550

AUSTIN, BERT PETER, b Utica, NY, July 26, 46; m 66; c 2. ANATOMICAL SCIENCE, CELL BIOLOGY. *Educ:* Hobart Col, BA, 68; State Univ NY Upstate Med Ctr, Syracuse, PhD(anat), 73. *Prof Exp:* Fel, Orthop Res Labs, Vet Admin Hosp, Syracuse, 73-75; asst prof, 75-83, ASSOC PROF ANAT, SCH DENT, MARQUETTE UNIV, 83- *Concurrent Pos:* NIH fel, 73, NIH Biomed Res Develop grant, investr, 77-80. *Mem:* Sigma Xi; Am Asn Anatomists. *Res:* Structural alterations in an experimentally induced arthritis; osseous reactions to dental materials. *Mailing Add:* Sch Dent Marquette Univ 604 N 16th St Milwaukee WI 53233

AUSTIN, BILLY RAY, b Atkins, Ark, July 25, 40; m 62; c 3. INTEGRAL THEORY. *Educ:* Ark Col, BA, 62; La State Univ, MS, 64; Univ Miss, PhD(math), 76. *Prof Exp:* Instr math, Christian Bros Col, 64-65; from inst to assoc prof, 65-82, PROF & CHAIR MATH, UNIV TENN, MARTIN, 82- *Mem:* Sigma Xi; Math Asn Am; Am Math Soc; Nat Coun Teachers Math. *Res:* Real variables, integration theory. *Mailing Add:* Dept Math Univ Tenn Martin TN 38238-5049

AUSTIN, CARL FULTON, b Oakland, Calif, July 18, 32; m 53; c 3. MINERALOGY, EXPLOSIVES. *Educ:* Univ Utah, BS, 54, MS, 55, PhD(geol eng), 58. *Prof Exp:* Geologist, NMex Bur Mines & Mineral Resources, 58-61; GEOLOGIST, US NAVAL WEAPONS CTR, 61- *Concurrent Pos:* Owner/operator, Cedarsage Farm & Golden Jubilee Mining. *Honors & Awards:* L T E Thompson Award, 82. *Mem:* Am Inst Mining, Metall & Petrol Eng; Sigma Xi. *Res:* Exploration geochemistry, geothermal deposits, lined-cavity and other impulsive loading phenomena in brittle solids; fracture and penetration of rock and concrete; sub-sea floor construction; geochemistry. *Mailing Add:* US Naval Weapons Ctr Sci Res Off Code 2607 China Lake CA 93555

AUSTIN, CHARLES WARD, b Seattle, Wash, Nov 5, 32; m 55; c 4. MATHEMATICS. *Educ:* Univ Wash, BS, 54, MS, 60, PhD(math), 62. *Prof Exp:* Actg instr math, Univ Wash, 61-62; asst prof, Univ Colo, Boulder, 62-66; assoc prof, 66-71, chmn dept, 67-73, PROF MATH, CALIF STATE UNIV, LONG BEACH, 71-, ASSOC DEAN SCH LETT, 73- *Mem:* Am Math Soc; Math Asn Am. *Res:* Topological algebra, especially semigroups and continuous functions on them. *Mailing Add:* Calif State Univ Long Beach CA 90840

AUSTIN, DANIEL FRANK, b Paducah, Ky, May 18, 43. SYSTEMATIC BOTANY, ETHNOBOTANY. *Educ:* Murray State Univ, BA, 66; Wash Univ, AM, 69, PhD(biol), 70. *Prof Exp:* From asst prof to assoc prof, 70-78, PROF BOT, FLA ATLANTIC UNIV, 78- *Concurrent Pos:* Vis res assoc bot, Inst Pesquisas Norte, Brazil, 69; nat sci trainee, Wash Univ, 69-70; adj asst prof bot, Fla Int Univ, 73; res grants, Fla Atlantic Univ, Fla Int Univ Joint Ctr Urban & Environ Issues, 74-75, 78-79 & Off Endangered Species, Dept Interior, 78-81; adj prof bot, Univ South Fla, 79-; mem, Food & Agr Orgn, UN, 81-84; res grant, Fla Dept Transp, 85. *Mem:* Asn Trop Biol; Soc Econ Bot; Torrey Bot Club; Int Asn Plant Taxonomists. *Res:* Systematic, evolutionary and ethnobotanical studies of the Morning Glory family (Convolvulaceae); ecosystem dynamics in southern Florida. *Mailing Add:* Dept Biol Sci Fla Atlantic Univ Boca Raton FL 33431

AUSTIN, DONALD FRANKLIN, b Forsyth, Mont, Nov 1, 37; m 58; c 2. MEDICINE, EPIDEMIOLOGY. *Educ:* Univ Ore, BS, 62, MD, 65, MS, 71; Univ Calif, Berkeley, MPH(pub health), 70; Am Bd Prev Med, dipl, 72. *Prof Exp:* Cancer control officer, Ky State Dept, USPHS, 66-68, res scientist viral oncol, Commun Dis Ctr, 68-69; med dir field screening, Health Testing Serv, Inc, 70; med epidemiologist chronic dis, 70-71, emergency med serv, 71-72, chief, Cardiovasc Control Unit, 72-73, actg chief, 73, med epidemiologist, 73-74, chief, Calif Tumor Registry, 74-75, asst head, Resource Cancer Epidemiol Ctr, 75-76, CHIEF, RESOURCE CANCER EPIDEMIOL CTR, CALIF STATE DEPT HEALTH, 76- *Concurrent Pos:* Epidemiol consult respiratory devices, Tecna Corp, Nat Heart-Lung Inst, 71-72; epidemiol consult Adventist health study, Loma Linda Univ, 74-; consult cancer, Univ Southern Calif, 75-, San Francisco Regional Tumor Found, 75-76 & Ad Hoc Group Evaluation, Hospice Prog, Nat Cancer Inst, 78-; epidemiol consult breast cancer, Stanford Res Inst, 75-; res collabr cancer, Univ Calif, Berkeley, Sch Pub Health, 75- & 76- & Children's Hosp, San Francisco, 76- *Mem:* Am Soc Prev Oncol; Am Pub Health Asn; Soc Epidemiol Res. *Res:* Epidemiology of cancer and other chronic diseases on evaluation of public health programs. *Mailing Add:* 1918 University Ave Suite 3D Berkeley CA 94204

AUSTIN, DONALD GUY, b Chicago, Ill, Sept 24, 26. MATHEMATICS. *Educ:* Univ Ill, BS, 47; Northwestern Univ, MA, 50, PhD(math), 51. *Prof Exp:* Asst, Northwestern Univ, 47-49; sr mathematician, US Air Force Proj, Univ Chicago, 51; instr math, Syracuse Univ, 51-55; asst prof, Ohio State Univ, 55-56; NSF fel, Yale Univ, 57-58; assoc prof, Univ Miami, 59-61; assoc prof, 61-66, PROF MATH, NORTHWESTERN UNIV, 66- *Mem:* Am Math Soc. *Res:* Mathematical analysis; measure theory; theory of probability. *Mailing Add:* Dept Math Northwestern Univ Evanston IL 60208

AUSTIN, DONALD MURRAY, b Memphis, Tenn, Aug 11, 38; m 61; c 2. APPLIED MATHEMATICS. *Educ:* Memphis State Univ, BS, 64; Iowa State Univ, PhD(physics), 69. *Prof Exp:* Postdoctoral fel physics, Lawrence Berkeley Lab, Univ Calif, Berkeley, 69-70, computer scientist, 70-80; mathematician math & computer sci, Dept Energy, 80-90; EXEC DIR COMPUTER SCI, UNIV MINN, 90- *Concurrent Pos:* Mem, Gov Task Force on Minn Comput Indust, 90-91; vchmn, Supercomput Activ Group, Soc Indust & Appl Math, 91- *Mem:* AAAS; Soc Indust & Appl Math. *Res:* Computational aspects of science, engineering and mathematics; massively parallel computation and graphics/visualization techniques for displaying results of scientific computation. *Mailing Add:* Univ Minn AHPCRC 1100 Washington Ave S Minneapolis MN 55415

AUSTIN, FAYE CAROL, b Philadelphia, Pa, Dec 3, 44; c 2. IMMUNOLOGY, CANCER BIOLOGY. *Educ:* Pa State Univ, BS, 62, MS, 64; George Washington Univ, PhD(microbiol), 78. *Prof Exp:* Sr chemist to res assoc, Biosci Labs, 67-70; scientist, Meloy Labs, 72-74, sr scientist, 74-76, prin scientist, 76; chemist, Lab Viral Carcinogenesis, Nat Cancer Inst, NIH, 76-79, microbiologist, Lab Cellular & Molecular Biol, 79-81, prog dir cellular immunol, Immunol Prog, 81-87, chief, Cancer Immunol Br, 85-90, ASSOC DIR, EXTRAMURAL RES PROG, DIV CANCER BIOL DIAG & CTRS, NAT CANCER INST, NIH, 90- *Mem:* Sigma Xi; Am Asn Immunologists; Am Asn Cancer Res. *Res:* Virus-augmented skin test antigens specific for human melanoma or breast cancer; virus-augmented tumor antigens as vaccines in mice; therapeutic and diagnostic potential of monoclonal antibodies against human breast cancer. *Mailing Add:* Suite 642 Exec Plaza S Nat Cancer Inst NIH Bethesda MD 20892

AUSTIN, FRED, b Vienna, Austria, Apr 3, 36; US citizen; m 57; c 2. COMBAT ANALYSIS, STRUCTURAL & CONTROL ANALYSIS. *Educ:* Cooper Union, BME, 57; NY Univ, MME, 60, PhD(mech eng), 68. *Prof Exp:* Engr, Curtiss Wright Corp, 57, Control Instrument Co, Burroughs Corp, 57-58 & Sperry Gyroscope Co, 58-62; SR STAFF SCIENTIST, GRUMMAN AEROSPACE CORP, BETHPAGE, 62- *Concurrent Pos:* Adj asst prof, NY Univ; lectr, Hofstra Univ, 74-81. *Mem:* Sigma Xi; Am Soc Mech Engrs. *Res:* Spacecraft structural dynamics and controls; combat analysis; structural optimization; dynamics of rotating elastic satellites; development of theory and computer program for automatically controlling fixed-wing end helicopter aircraft during dogfights; damping spacecraft structural vibration. *Mailing Add:* Five Zinnia Ct Commack NY 11725

AUSTIN, GEORGE M, b Philadelphia, Pa, May 10, 16; m 42; c 4. NEUROPHYSIOLOGY. *Educ:* Lafayette Col, AB, 38; Univ Pa, MD, 42; McGill Univ, MSc, 51; Am Bd Neurol Surg, dipl. *Prof Exp:* Assoc neurosurg, Univ Pa, 51-54, asst prof, 54-57; prof & head div, Med Sch, Univ Ore, 57-68; prof neurosurg & chief sect, Loma Linda Univ, 68-81; NEUROL SURGEON, SANTA BARBARA COTTAGE HOSP, 81-; CLIN PROF, UNIV SOUTHERN CALIF, 81- *Concurrent Pos:* USPHS spec fel, Dept Theoret Chem, Cambridge Univ; chief neurosurg, Philadelphia Gen Hosp & Lankenau Hosp. *Mem:* Am Physiol Soc; Am Col Surgeons; AMA; Soc Univ Surgeons; Am EEG Soc. *Res:* Brain edema and water flux in nerve cells; mechanisms of spasticity in spinal cord and brain lesions; new techniques and measurement of cerebral blood flow in conditions of cerebrovascular disease and brain edema. *Mailing Add:* 2320 Bath St No 301 Santa Barbara CA 93105

AUSTIN, GEORGE STEPHEN, b Roanoke, Va, Mar 24, 36; m 60; c 2. GEOLOGY. *Educ:* Carleton Col, BA, 58; Univ Minn, MS, 62; Univ Iowa, PhD(geol), 71. *Prof Exp:* Teacher geol, Col of St Thomas, 62-67; stratigrapher, Minn Geol Surv, 67-69; clay mineralogist, Ind Geol Surv, 71-74; indust minerals geologist, 74-76, dep dir, 76-88, SR INDUST MINERALS GEOLOGIST, NMEX BUR MINES & MINERAL RES, 88- *Mem:* Soc Econ Paleontologists & Mineralogists; Clay Mineral Soc; Am Inst Mining & Petrol Engrs; Geol Soc Am. *Res:* The study of industrial minerals, especially clay and shales, potash, perlite, fluorite and coal. *Mailing Add:* NMex Bur Mines & Mineral Res Campus Sta Socorro NM 87801

AUSTIN, GEORGE T(HOMAS), b Salem, Ill, June 6, 14; m 37; c 3. CHEMICAL ENGINEERING. *Educ:* Univ Ill, BS, 36; Purdue Univ, PhD(chem eng), 43. *Prof Exp:* Jr engr, Socony-Vacuum Oil Co, Kans, 36-39; fel, Purdue Univ, 39-43, instr eng sci & mgr war training, 40-41; chief chem engr, Mallinckrodt Chem Works, Mo, 43-47; assoc prof, 47-49, chmn dept, 49-75, PROF CHEM ENG, WASH STATE UNIV, 49-, CHEM ENGR, COL ENG RES DIV, 63- *Concurrent Pos:* Vis prof, Indian Inst Sci, India, 54-56; vis prof, Chonnam Nat Univ, Kwangju, Korea, 80-81 & 82-83. *Mem:* Am Inst Chem Engrs; Am Chem Soc. *Res:* Instruments; gas adsorption; organic and inorganic syntheses; computers. *Mailing Add:* 1880 Landis Pl Pullman WA 99163

AUSTIN, HOMER WELLINGTON, b Greenlee, Va, Aug 7, 44;; m 77; c 2. MATHEMATICAL EDUCATION. *Educ:* James Madison Univ, BS, 66; Univ Wyoming, MS, 68; Univ Virginia, PhD(math), 75. *Prof Exp:* PUb teacher math, Rockbridge County, Va, 66; grad asst math, Univ Wyo, 66-67; from instr to assoc prof math sci, James Madison Univ, 76-83; from instr to assoc prof, 67-83, PROF MATH DEPT, SALISBURY STATE UNIV, 83- *Mem:* Math Asn Am. *Mailing Add:* Math Sci Dept Salisbury State Univ Salisbury MD 21801

AUSTIN, JAMES BLISS, physical chemistry, metallurgy; deceased, see previous edition for last biography

AUSTIN, JAMES HENRY, b Cleveland, Ohio, Jan 4, 25; m 48; c 3. NEUROLOGY. *Educ:* Harvard Univ, MD, 48; Brown Univ, AB, 61. *Prof Exp:* Intern med, Boston City Hosp, 48-49, asst resident neurol, 49-50; asst resident, Neurol Inst NY, 53-55; assoc neurol, Med Sch, Univ Ore, 55-57, from asst prof to assoc prof, 57-67; chmn dept, 75-83, PROF NEUROL, SCH MED, UNIV COLO, 67- *Concurrent Pos:* Univ fel neuropath, Col Physicians & Surgeons, Columbia Univ, 53; Kenny Found fel, 58-63; Commonwealth Fund fel, 62-63. *Mem:* Am Acad Neurol; Asn Res Nervous & Ment Dis; Am Asn Neuropath; Am Neurol Asn. *Res:* Hypertrophic neuritis; recurrent polyneuropathy; metachromatic leukodystrophy; globoid leukodystrophy; genetically-determined neurological diseases; sulfatase enzymes; cerebrovascular diseases; aging of the brain; biogenic amines; chance and creativity in medical research. *Mailing Add:* Dept Neurol Univ Colo Sch Med Denver CO 80262

AUSTIN, JAMES MURDOCH, b Dunedin, NZ, May 25, 15; US citizen; m 41; c 2. METEOROLOGY. *Educ:* Univ NZ, BA, 35, MA, 36; Mass Inst Technol, ScD, 41. *Prof Exp:* Asst meteorologist, Apia Observ, West Samoa, 37-39; asst, 40-41, from asst prof to prof, 41-79, dir summer session, 56-83, EMER PROF METEOROL, MASS INST TECHNOL, 79- *Concurrent Pos:* Consult, Army Air Forces, Washington, DC, 43-45; expert consult, Joint Res & Develop Bd, 47-48; consult, AEC, 47-53 & US Navy, 55-62; vis lectr, Harvard Univ, 55-79; mem adv coun air pollution emergencies, Dept Pub Health, Mass, 69-77. *Mem:* Fel Am Acad Arts & Sci; fel Am Meteorol Soc. *Res:* Development of quantitative forecasting methods; growth of cumulus clouds; climatology; meteorological aspects of air pollution. *Mailing Add:* 100 Keyes Rd Unit 421 Concord MA 01742

AUSTIN, JOHN H(ENRY), b Washington, DC, Feb 22, 29; m 56; c 4. SANITARY ENGINEERING. *Educ:* Syracuse Univ, BCE, 51; Mass Inst Technol, SM, 53; Univ Calif, Berkeley, PhD(sanit eng), 63. *Prof Exp:* Asst sanit engr, Taft Ctr, USPHS, Ohio, 55-57; sanit engr, Adv to Ministry Health, Int Coop Admin, Vietnam, 57-59; sr asst sanit engr, San Francisco Bay Study, USPHS, Calif, 59; assoc air pollution, Sch Pub Health, Univ Calif, Berkeley, 59-60, lectr radiol eng, 63, res engr, Col Eng, 60-62; from asst prof to assoc prof sanit eng, Univ Ill, Urbana, 63-69; head dept environ systs eng, Clemson Univ, 73-77, prof, 69-76; environ engr, USAID, 81-84; US NUCLEAR REGULATORY COMN, 72- *Concurrent Pos:* HEW grant, 64-67; US Dept Interior Fed Water Qual Admin grants, 69-70; Environ Protection Agency grants, 71-; vis prof civil eng, Univ Md, 76-78; consult, Maxima Corp, 78-, vpres, 78-81. *Mem:* Am Water Works Asn; Water Pollution Control Fedn; Sigma Xi. *Res:* Water and wastewater treatment plant operations. *Mailing Add:* 1895 Milboro Dr Potomac MD 20854

AUSTIN, JOSEPH WELLS, b Snowville, Utah, Mar 10, 30; m 54; c 5. ANIMAL SCIENCE. *Educ:* Utah State Univ, BS, 58; Univ Tenn, MS, 61; Tex A&M Univ, PhD(physiol reprod), 67. *Prof Exp:* Res asst radiation, Tex A&M Univ, 62-65, from asst radiobiologist to assoc radiobiologist, 65-67; asst prof biol, Ill Wesleyan Univ, 67-74; area livestock specialist, Coop Exten Serv, 74-; AT DEPT ANIMAL & DAIRY & VET SCI, UTAH STATE UNIV. *Mem:* Am Soc Animal Sci; Sigma Xi. *Res:* Mammalian reproduction. *Mailing Add:* 22613 Elm Ave Torrance CA 50505

AUSTIN, LEONARD G(EORGE), b London, Eng, Oct 5, 29; m 51, 84; c 3. METALLURGICAL ENGINEERING, FUEL TECHNOLOGY. *Educ:* Univ London, BSc, 50; Pa State Univ, PhD(fuel technol), 61. *Prof Exp:* Asst chemist cent labs, Southeastern Gas Bd, Eng, 52-54; gen engr fuel technol subdiv, Chief Engrs Dept, Cent Elec Generating Bd, Eng, 54-55, asst engr, 55-57; from instr to prof fuel sci, Pa State Univ, 57-67; prof chem eng, NC State Univ, 67-68; prof mat sci, 68-77, prof fuels & mineral eng, 77-89, EMER PROF FUELS & MINERAL ENG, PA STATE UNIV, 89- *Concurrent Pos:* Vis prof chem eng, Durban Univ, 72-73; ed, Powder Technol; vis prof metallurgy, Concepcion Univ Chile, 86-87; vis prof chem eng, Tech Univ Delft, 89, Neth, 89; vis prof mineral processing, Univ BC, Can, 89, 90; fel chem eng, Univ Manchester, Inst Sci & Technol, UK, 90; vis scientist, Commonwealth, Sci & Indus Res Orgn, Australia, 91. *Honors & Awards:* Distinguished Visitors Award, Commonwealth Sci & Indust Res Orgn, SAfrica, 72; A M Gaudin Award, Soc Mining Eng, Am Inst Mining, Metall & Petrol Engrs, 84, P W Nicholls Award, 87; Babcock Power Award, Inst Energy UK, 86. *Mem:* Am Inst Mining, Metall & Petrol Engrs; fel Inst Energy (London); Fine Particle Soc. *Res:* Boiler slagging; comminution; powder technology. *Mailing Add:* Dept Mineral Eng Pa State Univ University Park PA 16802

AUSTIN, MAX E, b Pine Grove, Pa, July 17, 33; m 53; c 4. HORTICULTURE. *Educ:* Univ RI, BS, 55, MS, 60; Mich State Univ, PhD(hort), 64. *Prof Exp:* Jr res asst hort, Univ RI, 57-60; res assoc, Mich State Univ, 60-64, dist agt, Coop exten serv, 64-65; from assst prof to assoc prof, Va Polytechnic Inst & State Univ, 65-72; head dept, 72-85, PROF HORT, COASTAL PLAIN EXP STA, UNIV GA, 79- *Concurrent Pos:* Exchange prof, Kagoshima Univ, Japan, 81; vpres, Int Affairs Div, Am Soc Hort, Sci, 89. *Honors & Awards:* Tifton Sigma Xi Outstanding Res Award; Japanese Ministry Educ Fel Award. *Mem:* Sr mem Am Soc Hort Sci; Am Pomol Soc; Int Hort Soc; fel Am Soc Hort Sci. *Res:* Rabbiteye blueberry breeding, production, harvest including fertility, pruning, breeding and evaluation of Southern highbush blueberries. *Mailing Add:* Dept Hort Coastal Plain Exp Sta Univ Ga Box 748 Tifton GA 31793-0748

AUSTIN, OLIVER LUTHER, JR, ornithology; deceased, see previous edition for last biography

AUSTIN, PAUL ROLLAND, b Monroe, Wis, Dec 9, 06; m 34; c 4. ORGANIC CHEMISTRY. *Educ:* Univ Wis, AB, 27; Northwestern Univ, MS, 29; Cornell Univ, PhD(org chem), 30. *Prof Exp:* Asst chem, Cornell Univ, 28-30; Nat Res fel, Univ Ill, 30-32; res chemist, Rockefeller Inst Technol, 32-33; res chemist, Chem Dept, E I du Pont de Nemours & Co, 33-36, group leader, 36-39, asst lab dir, 39-49, dir res div, Electrochem Dept, 49-59, mgr patent & licensing div, Int Dept, 59-66; counsel, Tech Serv Div, 67-73, ADJ PROF MARINE CHEM, COL MARINE STUDIES, UNIV DEL, 73- *Mem:* fel AAAS; Am Chem Soc. *Res:* Marine polymer chemistry; chitin isolates, their separation, characterization, derivatives and applications; particularly solvents, filaments, chiroptical properties and use in animal feed supplements. *Mailing Add:* 2327 W 18th St Wilmington DE 19806

AUSTIN, PAULINE MORROW, b Kingsville, Tex, Dec 18, 16; m 41; c 2. METEOROLOGY. *Educ:* Wilson Col, BA, 38; Smith Col, MA, 39; Mass Inst Technol, PhD(physics), 42. *Hon Degrees:* ScD, Wilson Col, 64. *Prof Exp:* Computer, Radiation Lab, Mass Inst Technol, 41-42, mem staff, 42-45, res staff, 46-53; lectr, Wellesley Col, 53-55; sr res assoc, Mass Inst Technol, 56-79; RETIRED. *Concurrent Pos:* Mem Comn II, Int Sci Radio Union. *Mem:* Fel Am Meteorol Soc. *Res:* Radar scattering cross sections; weather radar; propagation of electromagnetic waves in the atmosphere; precipitation physics. *Mailing Add:* 100 Keyes Rd Unit 421 Concord MA 01742

AUSTIN, ROBERT ANDRAE, b Wilmington, Del, Sept 26, 38; m 68; c 2. PAPER CHEMISTRY. *Educ:* Bucknell Univ, BS, 61; Univ Mass, MS, 67, PhD(chem), 68. *Prof Exp:* develop chemist, E I du Pont de Nemours & Co, 68-74; res chemist, Mead Corp, 74-77, sr res chemist, 77-78, tech dir Opas, 78-87, PROJ MGR, MEAD CORP, 87- *Mem:* Am Chem Soc. *Res:* Conformation of substituted indans by nuclear magnetic resonance; building products from polymers; carbonless copy papers; paper coatings. *Mailing Add:* Mead Adv Concepts Team 3020 Newmark Dr Miamisburg OH 45342

AUSTIN, ROSWELL W(ALLACE), b Taunton, Mass, July 28, 20; m 54; c 3. ENVIRONMENTAL OPTICS, REMOTE SENSING. *Educ:* Mass Inst Technol, SB, 42. *Prof Exp:* Develop engr, Gen Eng Lab, Gen Elec Co, 42-53; from assoc dir to dir, 53-87, EMER DIR VISIBILITY LAB, UNIV CALIF, SAN DIEGO, 87-, RES ENGR, SCRIPPS INST OCEANOG, 53- *Mem:* Optical Soc Am; Soc Photo-Optical Instrumentation Engrs. *Res:* Optical properties of natural environment and instrumentation for their measurement; optical oceanography; remote sensing of the oceans. *Mailing Add:* 952 Amiford Dr San Diego CA 92107

AUSTIN, SAM M, b Columbus, Wis, June 6, 33; m 59. NUCLEAR PHYSICS. *Educ:* Univ Wis, BS, 55, MS, 57, PhD(physics), 60. *Prof Exp:* Res assoc physics, Univ Wis, 60; NSF fel, Oxford Univ, 60-61; asst prof, Stanford Univ, 61-65; from assoc prof to prof physics, Mich State Univ, 65-70, assoc dir, Cyclotron Lab, 76-79, res dir, 83-85, chmn, dept physics & astron, 80-83, co-dir, Nat Superconducting Cyclotron Lab, 85-91, UNIV DISTINGUISHED PROF PHYSICS, MICH STATE UNIV, 90-, DIR, NAT SUPERCONDUCTING CYCLOTRON LAB, 89- *Concurrent Pos:* Sloan res fel, 63-66; vis scientist, Neils Bohr Inst, 70; guest prof, Univ Munich, 72-73; collaborateur etranger, CEN Saclay, 79-80; vis scientist, Lab Rene Bernas, Orsay, 80; mem, subcomt adv comt physics to Rev Nuclear Physics Labs, NSF, 79, Argonne Univs Asn Spec Comt Elec Accelerator, 81-84, nuclear sci adv comt, NSF & Dept Energy, 81-83, nuclear sci adv comt Panel Electron Accelerators, 83, Panel Solar Neutrino Prob, 85 & adv comt physics, NSF, 83-85; vis staff mem, Los Alamos Nat Lab, 81-; vchair, Div Nuclear Physics, Am Phys Soc, 81-82, chair, 82-83; Am Phys Soc Coun, 86-89; AAU vis comt, physics dept, Brookhaven Nat Lab, 86-91; ed, Phys Reviews C, 88-; mem dirs rev comt, Nuclear Sci Div, Lawrence Berkeley Lab, 89-91; mem rev comt, Physics Div, Argonne Nat Lab, 89- *Mem:* Fel Am Phys Soc; fel AAAS; Am Asn Physics Teachers; Fedn Am Scientists. *Res:* Experimental study of the structure of nuclei, of nuclear reaction mechanisms and of the two body force; nuclear astrophysics; nitrogen fixation. *Mailing Add:* Cyclotron Lab Mich State Univ East Lansing MI 48824

AUSTIN, STEVEN ARTHUR, b Stanford, Calif, Jan 25, 48; m 87. COAL GEOLOGY. *Educ:* Univ Wash, BS, 70; San Jose State Univ, MS, 71; Pa State Univ, PhD(geol), 79. *Prof Exp:* Teaching asst geol, Earth & Mineral Sci, Pa State Univ, 76, res asst, 77-79; CHMN, GEOL DEPT, INST CREATION RES, 79- *Concurrent Pos:* Geologist, Keymar Resources Inc, 83-86. *Mem:* Geol Soc Am; Am Asn Petrol Geologists; Soc Econ Paleontologists & Mineralogists. *Res:* Modern and ancient catastrophic geologic processes, especially those that form sedimentary strata; processes at Mount St Helens and application to Grand Canyon and origin of coal. *Mailing Add:* Inst Creation Res PO Box 2667 El Cajon CA 92071

AUSTIN, T LOUIS, JR, ENGINEERING. *Prof Exp:* PRES, BROWN & ROOT. *Mem:* Nat Acad Eng. *Mailing Add:* Brown & Root 4100 Clinton Dr Houston TX 77020

AUSTIN, THOMAS HOWARD, b Mt Pleasant, Tex, Aug 12, 37; m 57; c 2. ORGANIC CHEMISTRY. *Educ:* Cent State Col, Okla, BS, 61; Okla State Univ, PhD(org chem), 65. *Prof Exp:* Asst chem, Okla State Univ, 61-65; res chemist, Jefferson Chem Co, Texaco, Inc, 65-87; RES CHEMIST, ARCO CHEM CO, 87- *Mem:* Am Chem Soc. *Res:* Nucleophilic reactions of organophosphorus esters; catalytic reactions of petrochemical derivatives; application research and product development of isocyanates polyisocyanurates and polyurethanes. *Mailing Add:* Arco Chem Co 3801 Westchester Pike Newton Square PA 19073

AUSTIN, TOM AL, b Fort Worth, Tex, Nov 21, 43; m 63; c 6. GROUNDWATER HYDROLOGY, CONTAMINATE HYDROLOGY. *Educ:* Tex Tech Univ, BSCE, 67, Utah State Univ, MS, 69; PhD(civil eng), 71. *Prof Exp:* Civil engr, Forest Serv, US Dept Agr, 67-69; systs engr, Tex Water Develop Bd, 71-72; from asst prof to assoc prof civil eng, 72-79, PROF HYDROL, IOWA STATE UNIV, 79-; DIR, IOWA STATE WATER RESOURCES RES INST, 83- *Concurrent Pos:* Exec comt, Irrig & Drainage Div, Am Soc Civil Engrs, 84-87; bd dir, Univs Coun Water Resources, 87-90; dir, Iowa State Water Resources Res Inst, 83-90. *Mem:* Am Soc Civil Engrs; Am Water Resources Asn; Int Water Resources Asn; Nat Asn Water Inst Dirs. *Res:* Contaminate transport of organic chemicals in groundwater aquifers; fundamental studies on absorption of pesticides in groundwater aquifer materials. *Mailing Add:* 376 Town Eng Bldg Iowa State Univ Ames IA 50011

AUSTIN, WALTER J, b St Louis, Mo, Feb 6, 20; m 49; c 2. STRUCTURAL ENGINEERING. *Educ:* Rice Univ, BS, 41; Univ Ill, MS, 46, PhD(civil eng), 49. *Prof Exp:* Struct designer, Chicago Bridge & Iron Co, 42-46; from asst prof to assoc prof civil eng, Univ Ill, 49-60; prof, 60-87, chmn dept, 63-64 & 77-82, EMER PROF CIVIL ENG, RICE UNIV, 87- *Concurrent Pos:* Mem exec comt, Column Res Coun, 76-82. *Honors & Awards:* Moisseiff Award, Am Soc Civil Engrs, 58. *Mem:* Am Soc Civil Engrs; Struct Stability Res Coun. *Res:* Structural analysis and design; numerical methods; plates; buckling; metal compression member design. *Mailing Add:* Dept Civil Eng Rice Univ PO Box 1892 Houston TX 77251

AUSTIN, WILLIAM W(YATT), b Vicksburg, Miss, May 29, 15; m 39; c 4. METALLURGY, MATERIALS & ENGINEERING SCIENCE. *Educ:* Birmingham-Southern Col, BS, 35; Vanderbilt Univ, MS, 39, PhD(metall), 48. *Prof Exp:* Inspector, Gulf States Steel Corp, Ala, 35-36; res chemist &

opers foreman, Swann & Co, Birmingham, 36-38; fel metall & chem, Vanderbilt Univ, 39-41; inspection foreman & supvr metall labs, Consol Vultee Aircraft Corp, Nashville, 41-45; sr res metallurgist, Southern Res Inst, 45-52; assoc prof & head, Dept Mech Eng, NC State Univ, 52-54, prof mineral industs, 54-70, prof mat eng & head dept, 70-79; METALL CONSULT, RALEIGH, NC, 79- Concurrent Pos: Mem exten fac, Univ Ala, 49-51. Mem: Am Inst Mining, Metall & Petrol Engrs; Am Soc Metals Int; Am Foundrymen's Soc; Am Soc Eng Educ. Res: Metallurgy of cast iron; phase distribution in alloy steels; direct reduction of iron ores; nondestructive testing and failure analysis; metallurgical factors in metal cutting; substitutes for manganese in steel making; metallic corrosion; metallurgy of nuclear reactor materials; accident reconstruction. Mailing Add: 3017 Rue Sans Famille Raleigh NC 27607

AUSTON, DAVID H, b Toronto, Ont, Nov 14, 40; m 62; c 2. LASERS, OPTICAL SPECTROSCOPY. Educ: Univ Toronto, BA, 62, MA, 63; Univ Calif, Berkeley, PhD(elec eng), 69. Prof Exp: Mem tech staff physics res, AT&T Bell Labs, Inc, 69-87, dept head, 82-87; PROF ELEC ENG & APPL PHYSICS, COLUMBIA UNIV, 87-, DEAN, SCH ENG & APPL SCI. Concurrent Pos: Chair, Dept Elec Eng, Columbia Univ, 91- Honors & Awards: R W Wood Prize, Optical Soc Am, 85; Quantum Electronics Award, Inst Elec & Electronics Engrs, 90, Morris E Leeds Award, 91. Mem: Nat Acad Sci; Inst Elec & Electronic Engrs; Lasers & Electro Optics Soc; Optical Soc Am; Am Phys Soc. Res: Optical and electronic properties of materials and devices using ultrashort optical and electronic pulses; femto second lasers; far-infrared and optical spectroscopy; nonlinear optics. Mailing Add: Sch Eng & Appl Sci Columbia Univ Seeley W Mudd Bldg New York NY 10027

AUSTRIAN, ROBERT, b Baltimore, Md, Apr 12, 16; m 63. INTERNAL MEDICINE, BACTERIAL GENETICS. Educ: Johns Hopkins Univ, AB, 37, MD, 41; Am Bd Internal Med, dipl. Hon Degrees: DSc, Hahnemann Med Col & Hosp, 80, Philadelphia Col Pharmacy & Sci, 81, Univ Pa, 86. Prof Exp: Asst med, Sch Med, Johns Hopkins Univ, 42-43, instr, 43-47; res assoc microbiol, Col Med, NY Univ, 47-48; instr med, Sch Med, Johns Hopkins Univ, 49-52; from assoc prof to prof med, Col Med, State Univ NY Downstate Med Ctr, 52-62; John Herr Musser prof res med & chmn dept, 62-86, EMER PROF, SCH MED, UNIV PA, 86- Concurrent Pos: Vis physician, Univ Div Med Serv, Kings County Hosp, 52-62 & Univ Pa Hosp, 62-; consult, Brooklyn & Maimonides Hosps, 53-62 & Univ Pa Hosp, 62-; vis scientist, Dept Microbial Genetics, Pasteur Inst, 60-61; trustee, Johns Hopkins Univ, 63-69; Tyndale vis lectr & prof, Col Med, Univ Utah, 64; mem study sect A, Allergy & Immunol, NIH, 65-69, mem bd sci counr, Nat Inst Allergy & Infectious Dis, 67-70, chmn, 69-70; mem subcomt streptococci & pneumococci, Int Comn Bact Nomenclature; mem comt meningococcal infections, Comn Acute Respiratory Dis, Armed Forced Epidemiol Bd, 66-72. Honors & Awards: USA Typhus Commission Medal, 47; Maxwell Finland Lectureship Award, Infectious Diseases Soc Am, 74; Albert Lasker Clin Med Res Award, 78; James D Bruce Mem Award, Am Col Physicians, 79; Bristol Award, Infectious Dis Soc Am, 86. Mem: Nat Acad Sci; Asn Am Physicians; Am Soc Clin Invest; master Am Col Physicians; Am Philos Soc; Infectious Dis Soc Am (pres, 71). Res: Infectious disease; pneumococcal transformation reactions; pneumococcal vaccines. Mailing Add: Dept Res Med Univ Pa Sch Med Philadelphia PA 19104-6088

AUSUBEL, FREDERICK MICHAEL, b New York, NY, Sept 2, 45. MOLECULAR GENETICS. Educ: Univ Ill, Urbana, BS, 66; Mass Inst Technol, PhD(biol), 72. Prof Exp: Asst prof, 75-80, assoc prof biol, Harvard Univ, 80-; AT DEPT GENETICS, HARVARD MED SCH. Res: Molecular genetic analysis of nitrogen-fixation; somatic cell genetics of plant cell tissue cultures. Mailing Add: 271 Lake Ave Newton MA 02161

AUSUBEL, JESSE HUNTLEY, b New York, NY, Sept 27, 51. EMISSION PROJECTION, TECHNOLOGICAL EVOLUTION. Educ: Harvard Univ, AB, 74; Columbia Univ, MBA, 77, MIA, 77. Prof Exp: Resident fel, Nat Acad Sci, 77-79; res scholar, Int Inst Appl Systs Anal, 79-81; staff officer, Nat Res Coun, 81-83; spec asst to pres, 83-85, DIR PROG OFF, NAT ACAD ENG, 85- Concurrent Pos: Consult, Nat Sci Bd, NSF, 83-84 & Exec Off, Rockefeller Univ, 84- Mem: AAAS. Res: Technological change; science and technology policy; carbon dioxide; climate; atmospheric and ocean policy; environment; energy; impact assessment; general systems theory; long economic cycles. Mailing Add: 2130 North St Washington DC 20037

AUTENRIETH, JOHN STORK, b Weehawken, NJ, Dec 28, 16; wid. ORGANIC CHEMISTRY. Educ: City Col New York, BS, 37; Mass Inst Technol, PhD(org chem), 41. Prof Exp: Res chemist, Exp Sta, Hercules Powder Co, 41-44, tech serv rep, 44-48, tech sales rep, 48-53, dist mgr, Synthetics Dept, 53-56, indust sales mgr, 56-59, mgr mkt serv, 59-61, mgr hard resin develop, Pine & Paper Chem Dept, 61-63, sales mgr resins, 63-65, mgr sales develop, 65, mgr mkt develop, 65-67, mgr tech serv pine chem, Pine & Paper Chem Dept, 67-76, mgr prod safety, Org Dept, Hercules Inc, 76-77; RETIRED. Mem: Sigma Xi. Res: Identification of aromatic hydrocarbons; research on development of synthetic resins; synthetic rosin ester resins; paints; varnishes; enamels; lacquers; adhesives; printing inks; floor tiles. Mailing Add: 39 Paschall Rd Shell Burne Wilmington DE 19803-4943

AUTH, DAVID C, b Akron, Ohio, Dec 5, 40; m 64; c 3. PHYSICS, ELECTRICAL ENGINEERING. Educ: Cath Univ Am, AB, 62; Georgetown Univ, MS, 66, PhD(physics), 69. Prof Exp: Teaching asst physics, Univ Fla, 62-63; aerospace technologist, Cape Kennedy, NASA, 63; physicist, Harry Diamond Labs, Dept Army, 63-67; res asst physics, Georgetown Univ, 67-68, Dept Defense res assoc, 68-69; from asst prof to prof elec eng, 69-85, adj prof bioeng, 78-85, AFFIL PROF, UNIV WASH, 85-; DIR BIOPHYS INT, SQUIBB CORP, 85- Concurrent Pos: Dir new ventures, Squibb Corp, 82-85. Mem: Inst Elec & Electronics Engrs; Am Asn Univ Professors; Am Heart Asn; fel Am Soc Laser Med & Surg. Res: Acoustics; laser interactions; endoscopic instruments for control of gastrointestinal bleeding; laser surgery; ophthalmic holography; medical instruments; fiberoptics; electrooptics; bioengineering; angioplasty. Mailing Add: Hart Technol 2515 140th Ave NE Bellevue WA 98005

AUTHEMENT, RAY PAUL, b Chauvin, La, Nov 19, 29; m 51; c 2. MATHEMATICS. Educ: Southwestern La Inst, BS, 50; La State Univ, MS, 52, PhD, 56. Prof Exp: Instr math, La State Univ, 52-56; assoc prof, McNeese State Col, 56-57; assoc prof, 57-60, vpres, 66-73, PRES, UNIV SOUTHWESTERN LA, 73-, PROF MATH, 60- Concurrent Pos: Vis prof, Univ NC, 62-63. Mem: Math Asn Am. Res: Algebraic number theory; matrix theory; Galois theory. Mailing Add: USL Drawer 41008 Lafayette LA 70504-1008

AUTIAN, JOHN, b Philadelphia, Pa, Aug 20, 24; m 62; c 1. PHARMACY. Educ: Temple Univ, BS, 50; Univ Md, MS, 52, PhD(pharm chem), 54. Prof Exp: Instr pharm, Univ Md, 52-53, Franklin Sq Hosp Sch Nursing, Md, 53 & Sinai Hosp Sch Nursing, Md, 54; asst prof, Sch Pharm, Temple Univ, 54-56, Sch Pharm, Univ Md, 56-57 & Col Pharm & Rackham Sch Grad Study, Univ Mich, 57-60; assoc prof pharm, Col Pharm, Univ Tex, 60-67, dir drug-plastic res lab, 61-67; PROF PHARMACEUT & DENT & DIR, MAT SCI TOXICOL LAB, COL DENT & COL PHARM, UNIV TENN, MEMPHIS, 67-, DEAN, COL PHARM, 75-, PROF ORAL PATH, 80- Concurrent Pos: Mem, Nat Formulary Adv Panel, 57; fel, Am Found Pharmaceut Educ; consult, Clin Ctr, NIH, 60- Mem: AAAS (vpres, 62); Am Pharmaceut Asn; Am Chem Soc. Res: Toxicology; plastic materials and their use in pharmacy and medicine. Mailing Add: 6997 Fords Station Rd Memphis TN 38138-1513

AUTON, DAVID LEE, b Lexington, Ky, April 1, 39; m 63; c 2. PHYSICS. Educ: Univ Chicago, BS, 62, SM, 64, PhD(physics), 69. Prof Exp: Sr scientist, Analytic Serv, Inc, 69-75; OPER RES PHYSICIST, DEFENSE NUCLEAR AGENCY, 75- Res: Radiation transport, fallout; civil defense; civil protection; biological response; strategic studies. Mailing Add: HQ Defense Nuclear Agency RARP 6801 Telegraph Rd Alexandria VA 22310-3398

AUTOR, ANNE POMEROY, b Prince George, BC, Jan 26, 35; div; c 2. MOLECULAR BIOLOGY. Educ: Univ BC, BA, 56, MSc, 57; Duke Univ, PhD(biochem), 70. Prof Exp: Res asst biochem, Duke Univ, 61-65; fel, Univ Mich, 70-72; res assoc, Univ Iowa, 72-73, from asst prof to prof pharmacol, 73-83; PROF, DEPT PATH, UNIV BC, 83- Concurrent Pos: Damon Runyon fel, 70; Nat Found Basil O'Connor starter grant, 74; Nat Inst Child Health & Human Develop res career develop award, 75; consult study sects, NIH, 76- Mem: Am Soc Biochem & Molecular Biol; Am Soc Pharmacol & Exp Therapeut; Soc Toxicol; Biophys Soc; Int Acad Path; Oxygen Soc. Res: Biology and toxicology of endothelial cells; oxygen and oxygen radical toxicity; biological generation of oxygen radical; enzyme antioxidants, superoxide dismutase glutothioneperoxidase; regulation and transport of mitochondrial proteins; cell culture. Mailing Add: Dept Path Univ BC 2211 Westbrook Mall Vancouver BC V6T 2B5 Can

AUTREY, ROBERT LUIS, b Indio, Calif, Feb 24, 32; m 55, 79; c 5. ORGANIC CHEMISTRY, TECHNICAL MANAGEMENT. Educ: Reed Col, AB, 53; Harvard Univ, AM, 55, PhD(org chem), 58. Prof Exp: Instr chem, Reed Col, 57-58; from instr to asst prof org chem, Univ Rochester, 59-65; lectr, Harvard Univ, 65-67; assoc prof chem, Ore Grad Ctr, 67-72, chmn fac, 70-72; secy & treas, Chiron Press, Inc, 71-77; exec vpres, Baleking Systs, 78-80; mgr systs develop, Heidelberg Eastern, Inc, 81-82; MGR RES & DEVELOP, NERCO, INC, 83- Concurrent Pos: NSF fel, Imp Col, Univ London, 58-59; asst ed, Am Chem Soc Jour, 62-65. Mem: AAAS. Res: Organic chemical reactions of synthetic utility; structure and synthesis of natural products and related substances. Mailing Add: NERCO Inc 500 NE Multnomah Suite 1500 Portland OR 97232-2045

AUVIL, PAUL R, JR, b Charleston, WVa, Aug 4, 37; m 60; c 2. THEORETICAL HIGH ENERGY PHYSICS, THEORETICAL CONDENSED MATTER PHYSICS. Educ: Dartmouth Col, BA, 59; Stanford Univ, PhD(physics), 62. Prof Exp: NSF fel physics, Imperial Col, London, 62-64; res assoc, Northwestern Univ, 64-65, asst prof, 65-68, actg chmn, Dept Physics & Astron, 78, dir, Integrated Sci Prog, 79-82, assoc chmn physics, 82-88, ASSOC PROF PHYSICS, NORTHWESTERN UNIV, 68-, ASSOC VPRES RES & ASSOC DEAN GRAD SCH, 88- Concurrent Pos: Vis scientist, CERN, Switz, 69-70. Mem: Am Phys Soc; Sigma Xi; NY Acad Sci. Res: Theory of elementary particles, Coulomb corrections, current algebra, symmetry breaking and quark-gluon models; superconductor superlattice structures; Josephson radiation; transition temperatures. Mailing Add: Dept Physics Northwestern Univ Evanston IL 60208

AUXIER, JOHN A, b Paintsville, Ky, Oct 7, 25; m 48; c 2. HUMAN RADIOBIOLOGY. Educ: Berea Col, AB, 51; Vanderbilt Univ, MS, 52; Ga Tech, PhD(nuclear eng), 72. Hon Degrees: PhS, Berea Col, 87. Prof Exp: Head, Radiobiol Lab, Dept Physics & Eng, Univ Tex, 52-55; dir, Appl Health Physics & Safety Div, Union Carbide Nuclear, 55-82; dir, Appl Sci Lab, Gulf Nuclear Corp, 83-85; dir, Radiation Sci Lab, 85-89, NUCLEAR SCI CORP MGR, IT CORP, 89- Concurrent Pos: Ed, Health Physics, J, 58-77; mem, Nat Coun Radiation Protection & Measurements, 73- & Staff President's Comn Accident Three Mile Island, 79; consult, Radiation Protection & Environ Assessment, 77-; corp scientist, Erc Corp, 83. Mem: Fel Health Physics Soc; Am Acad Health Physics (pres, 91); Int Radiation Protection Asn; Soc Risk Assessment. Res: Radiation measurements; instrument development; environmental assessments for radiation and radionuclides; human radiobiology; author of 100 publications. Mailing Add: Rte 1 Box 303 Lenoir City TN 37771

AUYONG, THEODORE KOON-HOOK, b Honolulu, Hawaii, Jan 18, 25; m 70. PHARMACOLOGY. Educ: Univ Mo, Kansas City, BSc, 54, MSc, 55; Univ Mo, Columbia, PhD(pharmacol), 62. Prof Exp: Res assoc pharmacol, Univ Mo, Kansas City, 55-56; instr, Univ Mo, Columbia, 62-63; asst prof, 63-72, ASSOC PROF PHARMACOL, SCH MED, UNIV N DAK, 72- Concurrent Pos: Consult community physicians on drug interactions. Mem: Sigma Xi; Int Soc Biochem Pharmacologists. Res: Isolation of pharmacologically active extracts of plants and determination of its mechanism of action; structure-function relationships of drugs; mechanism of renal toxicity by lithium. Mailing Add: 3614 11th Ave N Grand Forks ND 58201

AVADHANI, NARAYAN G, b Honavar, India, Jan 28, 41; m 72; c 1. BIOCHEMISTRY, MOLECULAR GENETICS. *Educ:* Karnatak Univ, India, BSc, 61; Univ Bombay, PhD(biochem), 69. *Prof Exp:* Res assoc physiol, Univ Ill, Urbana, 69-70, res asst prof, 70-72; res investr, 72-73, from asst prof to assoc prof, 73-82, PROF BIOCHEM, UNIV PA, 82- *Concurrent Pos:* Consult, Vet Admin Hosp, Coatesville, Pa, 73- *Mem:* Am Chem Soc; Brit Biochem Soc; Am Soc Biol Chem. *Res:* Biogenesis of mitochondria with special reference to the information contents and the expression of the mitochondrial genome in normal and tumor cells. *Mailing Add:* Dept Animal Biol Univ Pa 3800 Spruce St Philadelphia PA 19104-6008

AVAKIAN, PETER, b Tabriz, Iran, May 15, 33; US citizen; m 57; c 2. POLYMER PHYSICS. *Educ:* Univ Rochester, BS, 55; Mass Inst Technol, PhD(physics), 60. *Prof Exp:* Res physicist, Lab Insulation Res, Mass Inst Technol, 60; RES PHYSICIST, CENT RES & DEVELOP DEPT, E I DU PONT DE NEMOURS & CO, 61- *Concurrent Pos:* Fulbright fel, Univ Stuttgart, 60-61. *Mem:* AAAS; fel Am Phys Soc; Optical Soc Am; Sigma Xi. *Res:* Morphology, structure, molecular motions, and structure-property relationships in polymers; dielectric and optical properties of polymers, organic crystals and insulating inorganic crystals; electronic excitation energy transfer in organic crystals. *Mailing Add:* Cent Res & Develop Dept E I du Pont de Nemours & Co Wilmington DE 19880-0356

AVAULT, JAMES W, JR, b East St Louis, Ill, May 20, 35; m 66; c 2. BIOLOGY. *Educ:* Univ Mo, BS, 61; Auburn Univ, MS, 63, PhD(fisheries), 66. *Prof Exp:* Res asst fish parasitol, Auburn Univ, 61-63 & biol weed control, 63-66; from asst prof to assoc prof, 66-75, PROF FISHERIES, SCH OF FORESTRY, WILDLIFE & FISHERIES, LA STATE UNIV, BATON ROUGE, 75- *Concurrent Pos:* Ed, J World Maricult Soc, 70-86; consult, govt Philipines, 77, Dominican Republic, 78, Sierra Rutile Ltd, Sierra Leone, 87- *Mem:* Am Fisheries Soc; World Maricult Soc (secy-treas, 70-71, vpres, 74, pres, 75); Int Asn Astacology (pres, 74-). *Res:* Fish culture; marine fisheries; ichthyology; fishery biology; author of numerous publications in various journals. *Mailing Add:* 1058 Oak Hills Pkwy Baton Rouge LA 70810

AVCIN, MATTHEW JOHN, JR, b Pittsburgh, Pa, June 8, 43; m 75. ECONOMIC GEOLOGY, STRATIGRAPHY. *Educ:* Lafayette Col, BA, 65; Univ Ill, Urbana, MS, 69, PhD(geol), 74. *Prof Exp:* Res asst geol, Ill State Geol Surv, 69-73; chief coal sect geol, Iowa State Geol Surv, 73-90; GEOLOGIST, COLEMAN CO, WICHITA, 90- *Mem:* Soc Econ Paleontologists & Mineralogists; Paleont Res Inst; Sigma Xi; AAAS. *Res:* Coal research and the practical applications of stratigraphy, biostratigraphy and paleobotany. *Mailing Add:* 202 N Rock Rd Apt 1401 Wichita KS 67206

AVEGEROPOULOS, G, b Phthiotis, Greece, Jan 25, 34; US citizen; m 64; c 3. RUBBER COMPOUNDING & PROCESSING, PROCESS ENGINEERING & DESIGN. *Educ:* Nat Univ Athens, Greece, MA, 61; Wayne State Univ, Detroit, MS, 67, PhD(theoret physics), 68. *Prof Exp:* Postdoctoral statist thermodynamics, Univ Ga, 68; sr res scientist polymer physics, Firestone CEnt Res, 68-78; tech dir polymer physics, Standard Prod Co, 79-83; dir technol polymer eng, Gen Corp-EED, 83-87; DIR TECHNOL RES & DEVELOP & QUAL, RM ENG PROD-INTERTECH, 87- *Concurrent Pos:* Consult process eng, 77-88; chmn, Akron Sect, Am Chem Soc, 78. *Mem:* Am Phys Soc; AAAS; Soc Rheology; Am Inst Chemists; Am Inst Chem Engrs; Am Soc Testing Mat. *Res:* Morphology of elastomers and elastomeric blends; physical characterization of polymers and blends; laminated reinforced composites with continuous and discontinuous fibers; mechanical modeling of molecules and engineering systems; antivibration systems; molded and extruded product studies. *Mailing Add:* RM Eng Prod Intertech PO Box 5205 Charleston SC 29406

AVE LALLEMANT, HANS GERHARD, b Benkulen, Indonesia, May 2, 38; m 66; c 2. STRUCTURAL ANALYSIS, TECTONOPHYSICS. *Educ:* State Univ Leiden, BSc, 60, MSc, 64, PhD(struct petrol), 67. *Prof Exp:* Res staff geologist, Yale Univ, 67-70; from asst prof to assoc prof, 70-81, PROF GEOL, RICE UNIV, 81- *Concurrent Pos:* Vis assoc prof, State Univ NY, Stony Brook, 76-77; assoc ed technophysics, 80-; vchmn, Dept Geol & Geophys, Rice Univ, 80-83, chmn, 86-89; vis prof, Univ Pierre, Mary Curie, Paris, 84-85. *Honors & Awards:* Basic Res Award, US Nat Comn Rock Mech, 81. *Mem:* Am Geophys Union; fel Geol Soc Am; Am Asn Petrol Geologists; corresp mem Royal Dutch Acad Arts & Sci. *Res:* Structural analysis in fold and thrust belts and wrench fault zones in Oregon, Alaska and Venezuela. *Mailing Add:* Dept Geol & Geophys Rice Univ PO Box 1892 Houston TX 77251-1892

AVELSGAARD, ROGER A, b Minneapolis, Minn, Jan 5, 32; m 63. ALGEBRA. *Educ:* Univ Minn, Minneapolis, BChE, 54, MA, 63; Univ Iowa, PhD(math), 69. *Prof Exp:* Teaching assoc math, Univ Minn, 59-63; asst prof, 64-66, ASSOC PROF MATH, BEMIDJI STATE UNIV, 69- *Mem:* Am Math Soc; Math Asn Am. *Res:* Cohomology of non-associative algebra. *Mailing Add:* Rte 1 Box 274 Solway MN 56678

AVEN, MANUEL, b Tallinn, Estonia, Dec 25, 24; nat US; m 47; c 2. MATERIALS SCIENCE. *Educ:* Univ Pittsburgh, BS, 51, PhD(phys chem), 55. *Prof Exp:* Chemist, Mercy Hosp, 50-51; res asst, Univ Pittsburgh, 51-55, lectr, 54; res phys chemist, Lamp Develop Lab, Cleveland, Ohio, 55-59, phys chemist, Res Lab, Schenectady, NY, 59-68, mgr luminescence br, Gen Elec Res & Develop Ctr, 68-72, mgr phys chem lab, 72-82, mgr mat labs, Gen Elec Res & Develop Ctr, 82-86, phys chemist, 86-88, mgr power syst prog, 88-89, MGR MAT RES PROG, GEN ELEC CO, SCHENECTADY, 89- *Mem:* Am Chem Soc; Am Phys Soc; Mat Res Soc. *Res:* Managing research in material science and solid state physics. *Mailing Add:* 38 Forest Rd Burnt Hills NY 12027

AVEN, RUSSELL E(DWARD), b Water Valley, Miss, Feb 18, 23; m 54; c 1. CHEMICAL ENGINEERING. *Educ:* Univ Miss, BSE, 44, MA, 50; Univ Tenn, PhD, 63. *Prof Exp:* Lab chemist, E I du Pont de Nemours & Co, Tex, 46-48; develop engr, Oak Ridge Nat Lab, Tenn, 51-56; assoc prof, 56-66, PROF CHEM ENG, SCH ENG, UNIV MISS, 66- *Mem:* AAAS; Am Chem Soc; Am Inst Chem Engrs; Sigma Xi. *Res:* Heat transfer; nuclear engineering. *Mailing Add:* Dept Chem Eng Sch Eng Univ Miss University MS 38677

AVENI, ANTHONY, b New Haven, Conn, Mar 5, 38; m 59; c 2. ASTRONOMY. *Educ:* Boston Univ, AB, 60; Univ Ariz, PhD(astron), 65. *Prof Exp:* From instr to asst prof, 63-68, assoc prof, 68-73, PROF ASTRON, COLGATE UNIV, 74-, R B COLGATE PROF ASTRON & ANTHROP, 88- *Concurrent Pos:* NSF grants, 66-71; vis assoc prof, Univ SFla, 71; consult, Ferson Optical Div, Bausch & Lomb. *Mem:* AAAS; Am Astron Soc. *Res:* Galactic structure; star formation; works on observational evidence relating to formation of stars in low mass primary condensations; astroarcheology of Mesoamerica; studies on orientation of buildings in Maya and Zapotec zone of Mexico/Peru. *Mailing Add:* Dept Physics & Astron Colgate Univ Hamilton NY 13346

AVENS, JOHN STEWART, b Geneva, NY, Mar 23, 40; m 63; c 3. FOOD MICROBIOLOGY & SAFETY. *Educ:* Syracuse Univ, BS, 62; Colo State Univ, MS, 69, PhD(animal sci), 72. *Prof Exp:* Res technician hemat, NY Univ Med Ctr, 62-63; res scientist food microbiol, Syracuse Univ Res Corp, 63-65; sr microbiologist, Food & Drug Admin, Buffalo, 65-67; from instr to assoc prof, 67-82, PROF FOOD SCI & ANIMAL SCI, COLO STATE UNIV, 82- *Concurrent Pos:* Consult, Agr & Food Indust, 77- *Mem:* Inst Food Technologists; Coun Agr Sci & Technol; Sigma Xi; Nat Registry Microbiologists. *Res:* Improvement of microbiological analytical methodology for food; microbiology related to food safety in food production and processing and food service establishments; coauthor of textbook on poultry science and production; author of chapter in textbook on food safety; efficacy and safety testing of animal pharmaceuticals. *Mailing Add:* Dept Food Sci & Human Nutrit Colo State Univ Ft Collins CO 80523

AVENT, JON C, b Billings, Mont, July 4, 34. GEOLOGY. *Educ:* Univ Colo, BA, 56; Univ Wash, Seattle, MS, 62, PhD(phys stratig), 65. *Prof Exp:* From asst prof to assoc prof, 65-77, PROF GEOL, CALIF STATE UNIV, FRESNO, 77- *Mem:* Geol Soc Am; Am Asn Petrol Geol. *Res:* Cenozoic stratigraphy and structure; trace element distribution in basalt flows. *Mailing Add:* Dept Geol Calif State Univ Fresno CA 93740

AVENT, ROBERT M, b Jacksonville, Fla, Feb 24, 42; div; c 2. OCEANOGRAPHY. *Educ:* Jacksonville Univ, BA, 67; Fla State Univ, MS, 70, PhD(oceanog), 73. *Prof Exp:* Oceanographer, Harbor Br Found, 73-77 & Tex Instruments, Inc, 78; marine biologist, Tex Parks & Wildlife Dept, 79; oceanographer, Nat Oceanic & Atmospheric Admin/Nat Marine Fisheries Serv, 79-81; oceanographer, Bureau Land Mgt, 81-82, OCEANOGRAPHER, MINERALS MGT SERV, US DEPT INTERIOR, 82- *Concurrent Pos:* Adj prof oceanog, Fla Inst Technol, 73-74. *Mailing Add:* Minerals Mgt Serv 1201 Elmwood Park Blvd New Orleans LA 70123-2394

AVERA, FITZHUGH LEE, b Pocahontas, Ark, July 4, 06; m 50; c 1. FOOD SCIENCE, CHEMISTRY. *Prof Exp:* Sales mgr, Skippy CPC Int Inc, 26-38, chief chemist, 38-43, dir res, Best Foods Peanut Butter Div, 46-69; CONSULT CHEM ENGR, 69- *Concurrent Pos:* Mem, Food & Agr Orgn, UN survey West Indies; bd dirs, Driwater Inc, 90. *Mem:* Nat Soc Prof Engrs; Am Oil Chem Soc; NY Acad Sci; Sigma Xi. *Res:* Edible fats and oils; quantal chemical analysis; food products; numerous industrial process patents. *Mailing Add:* 1809 Yale Dr Alameda CA 94501

AVERBACH, B(ENJAMIN) L(EWIS), b Rochester, NY, Aug 12, 19; m 47. METALLURGY. *Educ:* Rensselaer Polytech Inst, BS, 40, MS, 42; Mass Inst Technol, ScD(metall), 47. *Prof Exp:* Chief metallurgist, US Radiator Corp, NY & Mich, 43-45; metallurgist, Gen Elec Co, NY, 45; asst, 45-47, from asst prof to assoc prof, 47-60, PROF METALL, MASS INST TECHNOL, 60- *Concurrent Pos:* Past pres, Int Cong Fracture; deleg, Int Inst Welding. *Honors & Awards:* Am Welding Soc Prize, 45; Howe Medal, Am Soc Metals, 49. *Mem:* Am Inst Mining, Metall & Petrol Engrs; fel Am Soc Metals; Am Phys Soc; Am Crystallog Soc; fel Inst Metals. *Res:* Physical metallurgy; solid state physics; x-ray diffraction; neutron diffraction; phase transformations; thermodynamics; fracture correlations; structure of liquids; magnetic materials; theory of alloys. *Mailing Add:* Dept Mat Sci & Eng Rm 13- 5001 Mass Inst of Technol Cambridge MA 02139

AVERETT, JOHN E, b Coleman, Tex, Apr 19, 43; m 63; c 2. SYSTEMATIC BOTANY. *Educ:* Sul Ross State Univ, BS, 66, MA, 67; Univ Tex, Austin, PhD(bot), 70. *Prof Exp:* from asst prof to prof biol, Univ Mo-St Louis, 70-88; RES DIR, NAT WILDFLOWER RES CTR, 88- *Concurrent Pos:* Res assoc, Mo Bot Garden, 70-88; ed rep, Bot Soc Am, 81; assoc ed, Phytochemical Bull, 76-79; ed, Wildflower, 88-; Alexander von Humboldt fel. *Mem:* Am Soc Plant Taxon; Bot Soc Am. *Res:* Biosystematics and chemosystematics; research in Onagraceae and Solanaceae; native plant propagation; revegetation. *Mailing Add:* Nat Wildflower Res Ctr 2600 FM 973 N Austin TX 78725

AVERILL, BRUCE ALAN, b Bucyrus, Ohio, May 19, 48; m 69, 86; c 2. INORGANIC CHEMISTRY, BIOLOGICAL CHEMISTRY. *Educ:* Mich State Univ, BS, 69; Mass Inst Technol, PhD(inorg chem), 73. *Prof Exp:* Res asst, Los Alamos Sci Lab, 69-70; fels biochem, Brandeis Univ, 73-74 & Univ Wis, 74-76; from asst prof to assoc prof chem, Mich State Univ, 76-82; FROM ASSOC PROF TO PROF CHEM, UNIV VA, 82- *Concurrent Pos:* A P Sloan fel, 81-85; vis prof, Univ Sydney, Australia, 89. *Mem:* Am Chem Soc; Royal Soc Chem; AAAS; Sigma Xi. *Res:* Role of transition metal ions in biology; synthesis of novel metal complexes; electrically conducting solids. *Mailing Add:* Dept Chem, Univ Virginia McCormick Rd Charlottesville VA 22901

AVERILL, FRANK WALLACE, b Anniston, Ala. QUANTUM MECHANICS, APPLIED MATHEMATICS. *Educ:* Univ Fla, BS, 67, PhD(physics), 71. *Prof Exp:* Res assoc quantum mech, Univ Fla, 72; res assoc, Northwestern Univ, 72-73; PROF MATH & PHYSICS, JUDSON COL, 73- *Concurrent Pos:* Consult, Oak Ridge Nat Lab, 81- *Mem:* Math Asn Am; Am Phys Soc. *Res:* Developing computer programs which produce approximate solutions to Schrödinger's equation for large electronic systems. *Mailing Add:* 1151 N State St Elgin IL 60120

AVERILL, SEWARD JUNIOR, b Punxsutawney, Pa, Jan 5, 21; m 43; c 3. RUBBER CHEMISTRY. *Educ:* Kent State Univ, BS, 42. *Prof Exp:* Instr chem, Williams Col, Mass, 43-45; res chemist, B F Goodrich, 46-49, mgr raw mat develop, 59-70, sr develop engr, 70-76, mgr gen chem lab, 76-80; RETIRED. *Concurrent Pos:* Consult, Sprague Elec Co, 43-45. *Res:* Rubber reinforcement and natural rubber technical characterization; rubber compounding; organo metallics; synthesis and reactions of vinylidene cyanide; plasticizers for vinyl polymers. *Mailing Add:* 71 Colony Dr Hudson OH 44236

AVERILL, WILLIAM ALLEN, b Albuquerque, NMex, Sept 10, 48; m 72. CHEMICAL METALLURGY, METALLURGICAL ENGINEERING. *Educ:* NMex Inst Mining & Technol, BS(metall eng) & BS(ceramic eng), 71, MS, 73; Univ Utah, PhD(metall), 76. *Prof Exp:* Metall technician, Hughes Tool Co, Houston, 70; res asst metall, NMex Inst Mining & Technol, 71-73 & Univ Utah, 73-76; asst prof metall eng, Colo Sch Mines, 76-81; MEM STAFF, SANDIA NAT LABS, 81- *Concurrent Pos:* Consult, Kaman Sci, Inc, Colorado Springs, 76-77 & Rocky Flats Plant, Rockwell Int, 78-81. *Mem:* Am Inst Mining, Metall & Petrol Engrs; Sigma Xi; Am Soc Metals. *Res:* Kinetics of hydrometallurgical operations; process control of hydrometallurical systems; nuclear fuel reprocessing; corrosion in solar systems; electrodeposition phenomena; materials extraction from brines. *Mailing Add:* 31789 Robinson Hill Rd Golden CO 80403

AVERITT, PAUL, economic geology; deceased, see previous edition for last biography

AVERRE, CHARLES WILSON, III, b Puerto Castilla, Honduras, June 3, 32; US citizen; m 55; c 2. PLANT PATHOLOGY. *Educ:* NC State Univ, BS, 55, MS, 60; Purdue Univ, PhD(plant path), 63. *Prof Exp:* Asst plant pathologist, Subtrop Exp Sta, Univ Fla, 63-67; asst prof plant path, Ga Exp Sta, 67-68; PROF PLANT PATH, NC STATE UNIV, 68- *Concurrent Pos:* Consult, Int Biol Ser, Inc, 73, Catie-MIP Turrialba, Costa Rica; ed, Fungicide & Nematicide Tests, Am Phytopath Soc, 74-79. *Mem:* Am Phytopath Soc. *Res:* Extension; diseases of vegetable crops; disease control systems; evaluation of fungicides and nematicides; plant disease diagnoses. *Mailing Add:* Dept Plant Path NC State Univ Raleigh NC 27695-7616

AVERS, CHARLOTTE J, cell biology; deceased, see previous edition for last biography

AVERY, CHARLES CARRINGTON, b Syracuse, NY, July 22, 33; m 61; c 4. FOREST HYDROLOGY. *Educ:* Utah State Univ, BS, 61; Duke Univ, MF, 63; Univ Wash, PhD(hydrol), 72. *Prof Exp:* Forester, Wenatchee Nat Forest, US Forest Serv, 61-64, forest hydrologist, Med Bow Nat Forest, 64-66, res hydrologist, Rocky Mountain Forest Exp Sta, 68-74; assoc prof, 74-80, PROF FORESTRY, NORTHERN ARIZ UNIV, 80- *Mem:* Am Geophys Union; Soc Am Foresters; Sigma Xi; Am Inst Hydrol; NY Acad Sci. *Res:* The relationships between forest practices and water quantity and quality. *Mailing Add:* Box 22514 Flagstaff AZ 86002

AVERY, DONALD HILLS, b Hartford, Conn, May 7, 37; m 63; c 3. ACCIDENT RECONSTRUCTION. *Educ:* Mass Inst Technol, BS, 59, ScD(metall), 62; Brown Univ, MA, 69. *Prof Exp:* Staff engr, Div Sponsored Res, Mass Inst Technol, 62-63, res assoc metall, 63-65,; from asst prof to assoc prof, 66-74, PROF ENG, BROWN UNIV, 74- *Concurrent Pos:* Consult, Chomerics Inc, 62-66, Brunswick Corp, 65-, Gillette Safety Razor Co, 66-, Advan Res Projs Agency, 67-69 & Armorflite, 73-; pres, Strathmore Res Corp, 63-70; Ford fel, 65-66; lectr hist mat, RI Sch Design, 67-68; vis prof, Univ Capetown, 74, 76, 79, 82 & 83; vis prof, Univ Rhodesia, 76; pres, ATS, 80; div res, armor flite group, 72-81; expert witness, State & Fed Courts, 63- *Honors & Awards:* H M Howe Medal, Am Soc Metals, 64. *Mem:* AAAS; Am Soc Metals; Am Inst Mining, Metall & Petrol Engrs; Soc Automotive Eng. *Res:* Mechanical, physical and powder metallurgy; crystal plasticity; fatigue; texture hardening; superplasticity; history of technology wear; composite materials; polymer processing. *Mailing Add:* 45 Jenny Lane Barrington RI 02806

AVERY, GORDON B, b Beirut, Lebanon, Dec 10, 31; US citizen; m 54; c 3. MEDICINE, EMBRYOLOGY. *Educ:* Harvard Univ, AB, 53; Univ Pa, MD, 58, PhD(embryol), 59; Am Bd Pediat, dipl. *Prof Exp:* Dir clin res ctr, Children's Hosp DC, 66-71, DIR NEWBORN NURSERY, CHILDREN'S HOSP DC, 63-, PROF CHILD HEALTH, SCH MED, GEORGE WASHINGTON UNIV, 71- *Concurrent Pos:* Grants, Gen Res Support, 63-65, Children's Bur, 65-67, NSF, 65-67 & Clin Res Ctr, 66-73; consult, Wash Hosp Ctr, Bethesda Naval Hosp & Nat Inst Child Health & Human Develop, 64-; assoc prof pediat, Sch Med, George Washington Univ, 67-71; mem comt ment retardation res & training, NIH, 69- *Honors & Awards:* Leavonson Prize, 57; Pepper Prize, 58. *Mem:* AMA; Am Acad Pediat; Transplantation Soc; Soc Pediat Res; Am Pediat Soc. *Res:* Pediatrics; neonatology; pulmonology; nutrition. *Mailing Add:* Dept Child Health Develop Childrens Hosp Nat Med Ctr 111 Mich Ave Washington DC 20010

AVERY, HOWARD S, b Canyon, Colo, Apr 18, 06. METALLURGY & PHYSICAL METALLURGICAL ENGINEERING, MINING ENGINEERING. *Educ:* Va Polytech Inst, BS, 27, EM, 28. *Prof Exp:* Teacher, Roanoke, Va Pub Sch, 31-34; res metallurgist, Abex Corp, 34-71; CONSULT ENGR, 71- *Honors & Awards:* Lincoln Gold Metal Award, Am Welding Soc, 50 & 52, Adams Lectr, 63. *Mem:* Am Welding Soc; fel Am Soc Metals; Am Inst Mining Metall & Petrol Engrs; Am Soc Testing & Mat. *Res:* Metallurgy; hard facing alloys; heat resisting alloys; wear resistant alloys; high temperature alloys. *Mailing Add:* 69 Alcott Rd Mahwah NJ 07430

AVERY, JAMES KNUCKEY, b Holly, Colo, Aug 6, 21; m 50; c 3. EMBRYOLOGY, ANATOMY. *Educ:* Univ Kansas City, DDS, 45; Univ Rochester, BA, 48, PhD(anat), 52. *Prof Exp:* Res assoc & instr, Univ Rochester, 52-54; from asst prof to assoc prof dent, Sch Dent, 54-63, assoc prof anat, Sch Med, 61-71, chmn, Dept Oral Biol, 77, PROF DENT, SCH DENT, UNIV MICH, ANN ARBOR, 63-, PROF ANAT, SCH MED, 71- *Concurrent Pos:* Dir, Dent Res Inst, Univ Mich & res & educ consult, Ann Arbor Vet Admin Hosp, 63-; consult dent training comt, NIH, 64-68, Sci Sessions of Am Dental Assoc, 60-70, Coun on Dent Educ & Accreditation, 75-80. *Honors & Awards:* Acad Dent Med Award; Pulp Biol IADR Award. *Mem:* Fel AAAS (chmn sect dent, 75-76); fel Am Col Dent; Int Asn Dent Res (pres, 74-75); Am Asn Anat; Electron Micros Soc Am; Sigma Xi. *Res:* Embryology of teeth, jaws, and face, comparative and human; histology and histochemistry of oral and related structures; in vivo and in vitro study of developing organs and structures; cytology of formative cells by means of light, phase, interference and electron microscopy. *Mailing Add:* Rm 5213 Univ Mich Sch Dent Ann Arbor MI 48109

AVERY, MARY ELLEN, b Camden, NJ, May 6, 27. MEDICINE, PEDIATRICS. *Educ:* Wheaton Col, BA, 48; Johns Hopkins Univ, MD, 52; Am Bd Pediat, cert, 59. *Hon Degrees:* MA, Harvard Univ, 74; DSc, Wheaton Col, 74, Univ Mich, 75, Med Col Pa & Trinity Col, 76, Albany Med Col, 77, Med Col Wis & Radcliffe Col, 78, Russel Sage Col, 83; LHD, Emmanuel Col, 79. *Prof Exp:* From asst prof to assoc prof pediat, Johns Hopkins Univ, 61-69; prof, Montreal Children's Hosp, McGill Univ, 69-74; physician-in-chief, Children's Hosp, Boston, 74-85; THOMAS MORGAN ROTCH PROF PEDIAT, HARVARD MED SCH, 74- *Concurrent Pos:* Res fel pediat, Harvard Med Sch, 57-59; Nat Inst Neurol Dis & Blindness spec trainee, 57-60; fel, Sch Med, Johns Hopkins Univ, 59-60; USPHS res grants, 61-; Markle scholar med sci, 61-66; pediatrician in chg newborn nurseries, Johns Hopkins Hosp, 61-69; consult, US Dept Health, Educ & Welfare, 64; mem, Med Res Coun Can, 70-74, adv comt, NIH & Robert Wood Johnson Found; bd dirs, AAAS, 88-; consult, Food & Drug Admin, 89-; chair, Fedn Pediat Orgn, 90-; mem, Comt Health & Human Rights, Inst Med, 90- *Honors & Awards:* Nat Medal of Sci, 91; Wall-Copeland Lectr, Georgetown Univ, 81; Arthur McElfresh Lectr, St Louis Univ, 81; John Howland Lectr, Johns Hopkins Univ, 81, Anna Baetjer Lectr, 91; Theodore L Badger Lectr, Harvard Med Sch, 83; Harry Bawkin Mem Lectr, 84; Edward Livingston Trudeau Medal, Am Lung Asn, 84; Charles McNeil Lectr, Royal Col Physicians, Scotland, 88; Donald F Eagan Mem Lectr, New Orleans, 90. *Mem:* Inst Med-Nat Acad Sci; fel Am Acad Pediat; Soc Pediat Res (pres, 72); Am Pediat Soc (pres-elect, 89, pres, 90); Am Physiol Soc; fel Am Acad Arts & Sci; fel AAAS; Am Thoracic Soc; fel Am Col Chest Physicians; fel Royal Col Physicians Can. *Res:* Respiratory problems of the newborn infant; pulmonary surfactant; control of breathing. *Mailing Add:* Children's Hosp 300 Longwood Ave Boston MA 02115

AVERY, ROBERT, b Ft Worth, Tex, Sept 7, 21. THEORETICAL & REACTOR PHYSICS. *Educ:* Univ Ill, BS, 43; Univ Wis, MS, 47, PhD(physics), 50. *Prof Exp:* Mem, Naval Reactor Div, Argonne Nat Lab, 50-56, sect head, theoret reactor physics sect, Reactor Eng Div, 56-63, dir, Appl Physics Div, 63-73, dir, Reactor Analysis & Safety Div, 73-86, SR PHYSICIST, ARGONNE NAT LAB, 86- *Mem:* Nat Acad Eng; fel Am Nuclear Soc; fel Am Phys Soc. *Res:* Nuclear reactor theory. *Mailing Add:* Reactor Analysis & Safety Div Argonne Nat Lab 9700 S Cass Ave Argonne IL 60439

AVERY, SUSAN KATHRYN, b Detroit, Mich, Jan 5, 50; m 72; c 1. ATMOSPHERIC DYNAMICS. *Educ:* Mich State Univ, BS, 72; Univ Ill, MS, 74, PhD(atmospheric sci), 78. *Prof Exp:* Asst prof atmospheric dynamics elec eng, dept elec eng, Univ Ill, 78-83; ASSOC PROF ELEC ENG, UNIV COLO, 85-, ASSOC DEAN RES & GRAD EDUC, 90- *Concurrent Pos:* Mem, Int Comn Meteorol Upper Atomosphere, Working Group Tides in the Mesosphere & Lower Thermosphere, 81-86; fel, Coop Inst Res Environ Sci, 83-; mem adv panel, NSF, 85-88; mem, Comt Solar-Terrestrial Res, Nat Res Coun, 87-90, Comt Educ & Human Res, Am Geophys Union, 88-, Comt Middle Atmosphere, Am Meteorol Soc, 90-93, Steering Comt, Equatorial Middle Atmosphere Dynamics & Mesosphere-Lower Thermosphere Network, Int STEP Prog, 90-; mem-at-large, Int Union Radio Sci, Nat Res Coun, 91-93, mem, Comns F & G; bd trustees, Univ Corp Atmospheric Res, 91-93. *Mem:* Am Meteorol Soc; Am Geophys Union; Inst Elec & Electronics Engrs; Sigma Xi; Union Radio Sci Int. *Res:* Dynamics of the mesosphere and stratosphere; unifying observational analyses and theoretical studies; modeling large scale atmospheric waves; ground based measurement techniques to observe the atmosphere. *Mailing Add:* Elec & Computer Eng Univ Colo Campus Box 425 Boulder CO 80309

AVERY, WILLIAM HINCKLEY, b Ft Collins, Colo, July 25, 12; m 38; c 2. PHYSICAL CHEMISTRY, PHYSICS. *Educ:* Pomona Col, AB, 33; Harvard Univ, AM, 35, PhD(photochem), 37. *Prof Exp:* Asst chem, Harvard Univ, 34-39; res chemist, Shell Oil Co, 39-43; res assoc & chief propellants div, Nat Defense Res Comt, Allegany Ballistics Lab, 43-46; res assoc, A D Little, Inc, 46-47; res assoc, Appl Physics Lab, 47-50, supvr, Launching & Propulsion, 50-54, Res & Develop, 54-60 & Appl Res, 60-73, asst dir explor develop, Appl Physics Lab, 73-78, supvr, Aeronaut Div, 61-80, DIR, OCEAN ENERGY PROGS, JOHNS HOPKINS UNIV, 73- *Concurrent Pos:* Mem res adv comt air breathing propulsion systs, NASA; mem advan systs panel, Hwy Res Bd, 70-; mem, Sr Adv Comt, Cent Energy & Environ Res, Univ Puerto Rico, 77-84. *Honors & Awards:* President's Cert Merit, 47; Hickman Medal, 50; Sir Alfred C Egerton Gold Medal, Combustion Inst, 71. *Mem:* Am Chem Soc; Am Inst Aeronaut & Astronaut; Combustion Inst; AAAS; Int Hydrogen Energy Asn; Soc Naval Architects & Marine Engrs. *Res:* Photochemistry; spectroscopy; molecular structure; chemical kinetics rocket propellants and internal ballistics; ramjet propulsion; combustion; urban transportation; ocean thermal energy conversion. *Mailing Add:* 724 Guilford Ct Silver Spring MD 20901-3218

AVGEROPOULOS, GEORGE N, b Greece, Jan 25, 34; m 64; c 1. PHYSICAL CHEMISTRY, QUANTUM CHEMISTRY. *Educ:* Nat Univ Athens, MA, 61; Wayne State Univ, MSc, 67, PhD(quantum chem), 68. *Prof Exp:* Chief res chemist, Michaelides Food Indust, Greece, 61-62; asst chem, Wayne State Univ, 62-68; eng scientist, Cent Res Labs, Firestone Tire & Rubber Co, 68-72, polymer physicist-rheologist, 72-78; tech dir, Standard

Prod Co, 78-83; dir technol, Diversitech-Gen Corp, 83-87; RES & DEVELOP DIR, RM ENGINEERED PROD INC, 87- *Mem:* Am Chem Soc; Am Phys Soc; Soc Rheol; AAAS; Am Inst Chemists; Am Soc Testing Mat; Soc Automotive Engr; Soc Mining Engr; Am Inst Chem Engr. *Res:* Applied mathematics; rheology of polymers and polymeric blends; mechanical properties and characterization of polymers; morphology of heterogeneous blends; mathematical models for the study of synthesis and characterization of polymers; processing of polymers, extrusion moldings & coating. *Mailing Add:* RM Engineered Products Intertech PO Box 5205 North Charleston SC 29406

AVIADO, DOMINGO MARIANO, b Manila, Philippines, Aug 28, 24; m 53; c 4. TOXICOLOGY, PHARMACOLOGY. *Educ:* Univ Pa, MD, 48. *Prof Exp:* From asst instr to instr pharmacol, Sch Med, Univ Pa, 48-50, assoc, 50-52, from asst prof to assoc prof, 52-65, prof, 65-77; sr dir biomed res, Allied Chem Corp Med Affairs, 77-80; PRES, ATMOSPHERIC HEALTH SCI INC, 80- *Concurrent Pos:* NIH res fel, 48-50; Guggenheim Found fel, 62-63; assoc ed, Circulation Res, 58-62; treas sect pharmacol, Int Union Physiol Sci, 59-; adj prof pharmacol, NJ Col Med & Dent, 77-83; mem clear air sci adv comt, Environ Protection Agency, 78-82; ed, Inhalation Toxicol & Path Sect, Environ Path & Toxicol, 79-82. *Mem:* AAAS; AMA; Am Soc Clin Pharmacol & Therapeut; Am Soc Pharmacol & Exp Therapeut; Am Physiol Soc. *Res:* Respiratory and circulatory reflexes; physiology and pharmacology of pulmonary circulation. *Mailing Add:* Atmospheric Health Sci Inc PO Box 307 152 Parsonage Hill Rd Short Hills NJ 07078

AVIGAD, GAD, b Jerusalem, Israel, Apr 30, 30; nat US; m 65; c 1. BIOLOGICAL CHEMISTRY. *Educ:* Hebrew Univ Jerusalem, MSc, 55, PhD(biochem), 58. *Prof Exp:* Asst biochem, Hadassah Med Sch, Hebrew Univ Jerusalem, 54-57, instr, 57-59; res fel, Med Col, NY Univ, 59-61; sr lectr, Hebrew Univ Jerusalem, 61-65, assoc prof, 65-69; assoc prof, 70-79, PROF BIOCHEM, RUTGERS UNIV & R W JOHNSON MED SCH, 79- *Concurrent Pos:* Vis scientist, Carlsberg Labs, Copenhagen, Denmark, 58; Jane Coffin Childs Mem Fund fel, 60-61; vis asst prof, Albert Einstein Col Med, 64, vis assoc prof, 67-70, vis res scientist, 88; vis prof, Parana State Univ, Brazil, 72. *Honors & Awards:* Israel Sci Prize, 58. *Mem:* Soc Gen Microbiol; NY Acad Sci; Am Chem Soc; Am Soc Microbiol; Am Soc Biol Chem; Soc Complex Carbohydrates. *Res:* Carbohydrate chemistry; biosynthesis and degradation of polysaccharides and complex glycosides; mechanisms of enzyme action; sugar metabolism. *Mailing Add:* Dept Biochem Univ Med & Dent NJ R W Johnson Med Sch PO Box 101 Piscataway NJ 08854-5635

AVIGAN, JOEL, b Warsaw, Poland, Jan 8, 20; m 48; c 3. LIPIDS, NUCLEIC ACIDS. *Educ:* Hebrew Univ, Israel, MSc, 43; McGill Univ, PhD(biochem), 54. *Prof Exp:* Res chemist, Israel Res Coun, 50-52; vis scientist, Bio-Med Res, Lab Metab, 55-64, RES CHEMIST, LAB CELLULAR METAB, NAT HEART, LUNG, & BLOOD INST, BIO-MED RES, 64- *Concurrent Pos:* Vis fel, Hadassah Med Sch, Hebrew Univ, 63; vis scientist, Lab Chem Enzymol, Shell Res Ltd, Eng, 63; mem coun arteriosclerosis, Am Heart Asn. *Mem:* Am Soc Biochem & Molecular Biol; Am Heart Asn Coun Arteriosclerosis. *Res:* Biochemistry of sterols, fatty acids and metabolism of plasma lipoproteins in relation to atherosclerosis; molecular mechanisms of hormonal effects on mammalian cells. *Mailing Add:* Bldg 10 5N311 NIH Bethesda MD 20892

AVIGNONE, FRANK TITUS, III, b New York, NY, May 9, 32; m 54; c 2. ELEMENTARY PARTICLE PHYSICS. *Educ:* Ga Inst Technol, BS, 60, MS, 62, PhD(physics), 65. *Prof Exp:* From asst prof to assoc prof physics, 65-73, PROF PHYSICS & ASTRON, UNIV SC, 73- *Res:* Experimental elementary particle physics. *Mailing Add:* Dept Physics Univ SC Columbia SC 29208

AVILA, CHARLES FRANCIS, b Taunton, Mass, Sept, 17, 06; m 34; c 2. MILITARY CAMERA DESIGN. *Educ:* Harvard Univ, BS, 29. *Hon Degrees:* LLD, Univ Mass, 63. *Prof Exp:* Eng mgt, Boston Edison Co, 29-60, pres & chmn, 60-71; dir, Liberty Mutual Ins Co, 60-70, Liberty Mutual Fire Ins Co, 69-78, Chas T Main, Inc, 77-84, C T M Ltd, Bermuda, 80-84; RETIRED. *Concurrent Pos:* Chief exec officer, Boston Edison Co, 60-70; dir, Boston Edison Co, 60-75, Shawmut Bank Boston, 58-79, Shawmut Corp, 65-79, John Hancock Mutual Life Ins Co, 64-74, Raytheon Co, 62-78. *Honors & Awards:* Edison Medal, Inst Elec & Electronic Engrs, 68. *Mem:* Nat Acad Eng; Inst Elec & Electronics Engrs. *Res:* Study and design or redesign of various plant and equipment of a major utility including management. *Mailing Add:* 272 Atlantic Ave Swampscott MA 01907-1638

AVILA, JESUS, b Madrid, Spain, Dec 12, 45; m 80; c 1. MICROTUBULE PROTEINS. *Educ:* Univ Complutense Madrid, Master, 67, PhD(chem), 71. *Prof Exp:* Res fel, Molecular Biol Lab, NIH, Bethesda, 72-75; mem staff res, Inst Biol Develop (CSIC), 75-76; head Microtub Lab, Molecular Biol Ctr (CSIC-UAM), 76-84, head struct unit, 84-86, dir, 86-88; br pres, Nat Agency Perspective Eval, 88-90; SCI COORDR, SCI INVEST, 90- *Mem:* Am Soc Cell Biol; Int Union Biol Sci. *Res:* Structure and function of microtubule proteins. *Mailing Add:* Dept Differentiation & Morphogen Ctr Molec Biol Univ Autonoma Fac Sci Cantoblanco Madrid 28049 Spain

AVILA, VERNON LEE, b Segundo, Colo, Apr 5, 41; c 4. COMPARATIVE ENDOCRINOLOGY, ANIMAL BEHAVIOR. *Educ:* Univ NMex, BS, 62; Northern Ariz Univ, MA, 66; Univ Colo, PhD(biol), 73. *Prof Exp:* Teacher, Los Lunas High Sch, 62-65 & Valley High Sch, 65-71; lectr biol, Univ Colo, 72-73; asst prof, 73-76, ASSOC PROF BIOL, SAN DIEGO STATE UNIV, 76- *Concurrent Pos:* Sci adv, Goodyear Publ Co, 74-77; sci consult, Scott Foresman & Co Publ, 75-; expert consult & health scientist adminr, Nat Cancer Inst, NIH, 78-79; admin fel, Col Sci, San Diego State Univ, 80-81; prof biol & asst dean, Col Natural Sci, Univ PR, Rio Piedras, 85-87; expert consult & health scientist adminr, Div Res Resources, NIH, 87-88. *Mem:* AAAS; Am Soc Zoologists; Animal Behav Soc; Am Inst Biol Sci. *Res:* Endocrinological aspects of behavior in teleost fishes; behavior in aquatic frogs; hormonal aspects of behavior in vertebrates. *Mailing Add:* Dept Biol San Diego State Univ 500 Campanile Dr San Diego CA 92182-0057

AVILES, JOSEPH B, b Atlantic City, NJ, Aug 21, 27; m 59; c 2. THEORETICAL PHYSICS. *Educ:* Rutgers Univ, BS, 50; Johns Hopkins Univ, PhD, 58. *Prof Exp:* Theoret physicist & head, Radiation Matter Interactions Br, US Naval Res Lab, 56-87; CONSULT PHYSICIST, 87- *Mem:* Am Phys Soc. *Res:* Radiation-matter interactions; nuclear physics. *Mailing Add:* 710 W Tantallon Dr Ft Washington MD 20744

AVILES, RAFAEL G, b Mexico City, Mex, April 28, 52; m 82; c 2. PHYSICAL CHEMISTRY, POLYMER CHEMISTRY. *Educ:* Univ Guanajuato, Mex, BChemE, 74; Univ Nev, Reno, PhD(phys chem), 79. *Prof Exp:* Res asst phys chem, Univ Nev, Reno, 75-79; fel, Cornell Univ, 79-80; sr scientist coatings res, Rohme & Haas Co, 80-84, res sect mgr, indust coatings res, 84-87, res sect mgr, trade sales coatings res, 87-89, RES MGR, FORMULATION CHEM, ROHM & HAAS CO, 89- *Concurrent Pos:* Adj prof, Philadelphia Col Textiles & Sci, 84- *Mem:* Am Chem Soc. *Res:* Kinetics and photophysics of aromatic polymers; gas phase kinetics; laser initiated reactions; energy transfer; structure-property relationships in polymeric systems; design of latex polymers. *Mailing Add:* Rohm and Haas Co 727 Norristown Rd Spring House PA 19477

AVINS, JACK, b New York, NY, Mar 18, 11; m 43; c 2. TELEVISION ENGINEERING. *Educ:* Columbia Col, AB, 32; Polytech Inst Brooklyn, MEE, 49. *Prof Exp:* Mem staff, Radio Corp Am Labs, 46-55, mgr int indust serv lab, Consumer Electronics Div, RCA Corp, 55-56, res appln lab, 57-64, staff engr, 64-76; CONSULT, 76- *Mem:* Fel Inst Elec & Electronics Engrs. *Res:* Design of radio and television receivers; design of integrated circuits. *Mailing Add:* 178 Herrontown Rd Princeton NJ 08540

AVIOLI, LOUIS, b Coatesville, Pa, Apr 13, 31; m 55; c 5. NUCLEAR MEDICINE, ENDOCRINOLOGY. *Educ:* Princeton Univ, AB, 53; Yale Univ, MD, 47. *Prof Exp:* From intern to resident med, NC Mem Hosp, 57-59; clin investr, NIH, 59-61; from instr to asst prof med, Seton Hall Col Med, 61-66, dir isotope lab & asst sci dir clin res ctr, 61-66; asst prof, 66-70, PROF MED, MED SCH, WASH UNIV, 70-; CHIEF ENDOCRINOL & METAB, JEWISH HOSP ST LOUIS, 66- *Concurrent Pos:* Grants, Nat Inst Arthritis & Metab Dis, 60-67, AEC, 62-67 & NIH, 65-70; attend physician, Jersey City Med Ctr, 62-66. *Mem:* AAAS; Soc Nuclear Med; Radiation Res Soc; Am Nuclear Soc; Am Fedn Clin Res. *Res:* Metabolism; radioactive isotopes. *Mailing Add:* Jewish Hosp 216 S Kings Hwy St Louis MO 63110

AVIRAM, ARI, b Braila, Romania, Aug 18, 37; m 59; c 3. ORGANIC CHEMISTRY, CHEMICAL PHYSICS. *Educ:* Israel Inst Technol, BSc, 65; NY Univ, MSc, 71, PhD(theoret chem), 75. *Prof Exp:* Synthetic chemist pharmaceut, Ayerst Lab, Can, 67-68; RES STAFF MEM ORG & THEORET CHEM, T J WATSON RES LAB, IBM CORP, 68- *Mem:* Am Chem Soc; Am Physics Soc. *Res:* The ability of molecules to perform device functions; molecular rectifiers and molecular memory elements; theoretical and synthetic aspects. *Mailing Add:* TJ Watson Res Ctr PO Box 218 Yorktown Heights NY 10598

AVIS, KENNETH EDWARD, b Elmer, NJ, June 3, 18; m 43; c 3. PHARMACEUTICS, PARENTERAL TECHNOLOGY. *Educ:* Philadelphia Col Pharm, BSc, 42, MSc, 47, DSc(pharm), 56. *Prof Exp:* From instr to assoc prof pharm, Philadelphia Col Pharm, 46-61; from assoc prof to prof, 61-77, vchmn dept, 68-77, dir Div Parenteral Medications, 72-83, Goodman prof pharmaceut, 68-88, chmn dept, 83-88, EMER PROF PHARMACEUT, COL PHARM, UNIV TENN MED UNIV, 88- *Concurrent Pos:* Consult, Nat Cancer Inst, 66-67, Food & Drug Admin, 68-, Vet Admin Hosp, Memphis & over 50 pharmaceut co, 60-; dir, Parenteral Drug Asn, 60-77, pres, 68-69; mem, Nat Formulary Comt on Specifications, 70-74; mem, nat coord comt on large vol parenterals, US Pharmacopeia & Food & Drug Admin. 72-76; consult, Pan-Am Health Orgn, 79 & 83. *Honors & Awards:* Parenteral Drug Asn Res Award, 71, 74, & 79; Schaufus Parenteral Technol Achievement Award, 76. *Mem:* Am Soc Hosp Pharmacists; Am Pharmaceut Asn; Am Asn Cols Pharm; fel Acad Pharmaceut Sci; Parenteral Drug Asn (pres, 68 & 69); fel Am Asn Pharm Scientists, 86- *Res:* Sterile pharmaceutical products; formulation, evaluation of effects of environmental and processing factors on sterile products; radiopharmaceuticals; use of principles of aerobiology to evaluate potential for contamination of containers and systems. *Mailing Add:* Rm 214 Col Pharm Univ Tenn Med Univ 26 S Dunlop Memphis TN 38163

AVISE, JOHN C, b Grand Rapids, Mich, Sept 19, 48. EVOLUTIONARY GENETICS. *Educ:* Univ Mich, BS, 70; Univ Tex, MA, 71; Univ Calif, Davis, PhD(genetics), 75. *Prof Exp:* From asst prof to prof, 75-87, RES PROF GENETICS, UNIV GA, 87- *Mem:* Nat Acad Sci; Soc Study Evolution; Genetics Soc Am; Am Ornith Union. *Res:* Ecology and evolution of natural populations. *Mailing Add:* Dept Genetics Univ Ga Athens GA 30602

AVITZUR, BETZALEL, b Haifa, Israel, May 7, 25; m 55; c 4. MECHANICAL ENGINEERING. *Educ:* Israel Inst Technol, BSc & Dipl Ing, 49; Univ Mich, MS, 56, PhD(mech eng), 60. *Prof Exp:* Res & develop engr, Ministry of Defence, Israel, 49-54; dept mgr indust eng, Vulcan Foundries, Israel, 54-55; res & develop engr, Micrometrical Develop Corp, Mich, 56-58; res engr sci lab, Ford Motor Co, 59-61; sr lectr mech eng, Israel Inst Technol, 61-64; assoc prof, 64-68, PROF METALL & MAT SCI, LEHIGH UNIV, 68-, DIR INST METAL FORMING, 70- *Concurrent Pos:* Guest lectr, Univ Mich, 61. *Mem:* Am Soc Mech Engrs; Am Soc Metals; Am Soc Mfg Engrs. *Res:* Design of mechanical instruments; metal behavior undergoing gross plastic deformations. *Mailing Add:* 817 N 31st St Allentown PA 18104

AVIZIENIS, ALGIRDAS, b Kaunas, Lithuania, July 8, 32; US citizen; m 62; c 1. COMPUTER SCIENCE. *Educ:* Univ Ill, BS, 54, MS, 55, PhD(elec eng), 60. *Prof Exp:* Res engr, Jet Propulsion Lab, Calif Inst Technol, 55-56; staff engr, Barnes & Reinecke, Inc, Ill, 58-60; sr res engr, Jet Propulsion Lab, Calif Inst Technol, 60-68, mem tech staff, 68-76; asst prof eng, 62-68, assoc prof comput sci, 68-76, PROF ENG & APPL SCI, UNIV CALIF, LOS

ANGELES, 76- *Concurrent Pos:* Mem adv panel comput sci & comput eng, NSF, 71- *Honors & Awards:* Apollo Achievement Award, NASA, 69. *Mem:* Asn Comput Mach; Inst Elec & Electronics Engrs. *Res:* Information systems engineering; digital computer system architecture; theory of digital arithmetic; fault-tolerant and self-repairing digital systems. *Mailing Add:* Dept Comput Sci Univ Calif Boelter Hall 3732 Los Angeles CA 90024

AVIZONIS, PETRAS V, b Lithuania, Aug 17, 35; US citizen; m 59; c 2. LASERS, PHYSICAL OPTICS. *Educ:* Duke Univ, BS, 57; Univ Del, MS, 59, PhD(phys chem), 62. *Prof Exp:* Chemist, Sun Oil Co, Pa, 61-62; sr proj scientist, 62-68, tech dir laser div, 68-71, TECH DIR, ADVAN RADIATION TECHNOL OFF, US AIR FORCE WEAPONS LAB, 71- *Honors & Awards:* Arthur S Flemming Award, Civil Serv Comn, Washington, DC, 71. *Mem:* Am Phys Soc; Am Optical Soc. *Res:* High energy lasers; optical systems; exited state chemistry and physics. *Mailing Add:* 25802 Prairestone Dr Laguna Hills CA 92653

AVNER, BARRY P, b Chicago, Ill, June 19, 44. MONOCLONAL ANTIBODIES. *Educ:* State Univ NY, Buffalo, PhD(biochem pharmacol), 70. *Prof Exp:* Asst vpres sci affairs & dir res, Summa Med Corp, Albuquerque, NMex, 81-85; dir Antibody Proj, Biotherapeutics, Inc, 85-88; DIR IMMUNOL ONCOL LAB, BAPTIST HOSP MIAMI, 88- *Mem:* NY Acad Sci; Am Soc Pharmacol & Exp Therapeut. *Res:* Technology assessment and acquisition; monoclonal antibodies. *Mailing Add:* Baptist Hosp Miami 8900 N Kendall Dr Miami FL 33176-2197

AVNER, ELLIS DAVID, b Pittsburgh, Pa, June 12, 48; m 70; c 2. PEDIATRIC NEPHROLOGY, DEVELOPMENTAL BIOLOGY. *Educ:* Princeton Univ, AB, 70; Univ Pa, MD, 75. *Prof Exp:* Resident pediat, Children's Hosp, Boston, Mass, 75-78, clin fel nephrology, 78-79, res fel, 79-80; asst prof pediat, 80-86, ASSOC PROF PEDIAT, SCH MED, UNIV PITTSBURGH, 86- *Concurrent Pos:* Scholar, Woodrow Wilson Nat Fel, 70; staff nephrologist, Children's Hosp, Pittsburgh, 80- & dir, organ cult lab, 81-; med adv, Nat Kidney Found, Western Pa, 81-; prin investr, NIH, 87- *Honors & Awards:* Basil O Connor Res Award, 82; New Investr Res Award, NIH, 84-87. *Mem:* Int Soc Pediat Nephrol (counr, 86-); Soc Pediat Res (counr, 88-); Am Soc Cell Biol; Am Fedn Clin Res; Soc Pediat Res. *Res:* Use of tissue and organ culture models to study normal and abnormal mammalian kidney development; pathophysiology of renal cystic maldevelopment. *Mailing Add:* Div Nephrology Box C5371 Childrens Hosp Med Ctr 4800 Sand Point Way NE Seattle WA 98105

AVONDA, FRANK PETER, b New York, NY, Nov 26, 24; m 56; c 4. ORGANIC CHEMISTRY. *Educ:* City Col New York, BS, 48; Columbia Univ, AM, 49; Ohio State Univ, PhD(chem), 53. *Prof Exp:* Res assoc org fluorides, Duke Univ, 53-54; res chemist, Hooker Electrochem Co, 54-56; instr org chem, Univ Tampa, 56-58; asst prof, Univ Md, 58-59 & Seton Hall Univ, 59-61; proj leader res & develop, Maxwell House Div, Gen Foods Corp, 61-62; prof Am indust technol, Ger Employees Acad, 62-63; sr res chemist, Allied Chem Corp, NJ, 63-64; assoc prof chem, Northern State Col, 64-65; prof, Nicholls State Col, 65-67; chmn div sci & math, Southern State Col, SDak, 67-68; prof chem & asst dean arts & sci, Delgado Col, 68-69; PROF CHEM, NICHOLLS STATE UNIV, 69-, HEAD DEPT, 76- *Concurrent Pos:* NSF grant, 70-72. *Mem:* Am Chem Soc. *Res:* Synthesis and chemistry of organic fluorine compounds. *Mailing Add:* PO Box 1041 Thibodaux LA 70301-1041

AVOURIS, PHAEDON, b Athens, Greece, June 16, 45; US citizen; m 76; c 1. CHEMISTRY METAL SURFACES. *Educ:* Aristotelian Univ, Greece, BS, 68; Mich State Univ, PhD(phys chem), 74. *Prof Exp:* Fel, Univ Calif, Los Angeles, 75-77; res assoc, AT&T Bell Labs, 77-78; mgr, 78-84, MGR CHEM PHYSICS, IBM RES DIV, THOMAS J WATSON RES CTR, 84- *Mem:* Fel Am Phys Soc; Am Chem Soc; Mat Res Soc; Am Vacuum Soc. *Res:* Experimental and theoretical studies of the electronic structure and reactivity of atoms and molecules on semiconductor and metal surfaces; mechanisms and dynamics of particle/photon-surface interactions; scanning tunneling microscopy; charge and energy transfer processes at surfaces. *Mailing Add:* IBM Thomas J Watson Res Ctr PO Box 218 Yorktown Heights NY 10598

AVRETT, EUGENE HINTON, b Atlanta, Ga, Oct 28, 33; m 61; c 2. ASTROPHYSICS. *Educ:* Ga Inst Technol, BS, 57; Harvard Univ, PhD(physics), 62. *Prof Exp:* PHYSICIST, SMITHSONIAN ASTROPHYS OBSERV, 62- *Concurrent Pos:* Lectr, Harvard Univ, 64- *Res:* Radiative transfer in stellar atmospheres; spectral line formation; applied mathematics. *Mailing Add:* Astron Observ Harvard Univ Cambridge MA 02138

AVRIN, WILLIAM F, b Pomona, Calif, June 5, 53. SUPERCONDUCTING TECHNOLOGIES, MAGNETIC MEASUREMENTS. *Educ:* Univ Calif, San Diego, BA, 75; Cornell Univ, MS, 79, PhD(physics), 91. *Prof Exp:* Res physicist, Standard Oil, Ohio, 83-85 & Quantum Design, Inc, 85-87; RES PHYSICIST, QUANTUM MAGNETICS, INC, 87- *Concurrent Pos:* Prin investr, NSF, 90. *Res:* New measurement instruments involving magnetic sensors, superconductivity and superconducting quantum interference devices: optically pumped magnetometers, Josephson-junction voltage standards, seismometers, nuclear magnetic resonance and nondestructive inspection. *Mailing Add:* Quantum Magnetics Inc 11578 Sorrento Valley Rd No 30 San Diego CA 92121

AVRUCH, JOSEPH, b New York, NY, Apr 13, 41. MEDICINE. *Educ:* Brooklyn Col, BS, 61; Wash Univ, MD, 65; Am Bd Internal Med, dipl, 75. *Prof Exp:* Intern, Ward Med, Barnes Hosp, St Louis, Mo, 65-66, asst resident, 68-69, chief resident, 71-72; res fel biol chem & med, Harvard Med Sch & Mass Gen Hosp, Boston, 69-71; asst med, Mass Gen Hosp, 72-75, asst physician, 75-80, assoc physician, 80-85; from asst prof to assoc prof, 72-88, PROF MED, HARVARD MED SCH, 88-; CHIEF, DIABETES UNIT, MASS GEN HOSP, 79- & PHYSICIAN, 86- *Concurrent Pos:* Clin assoc, Metab Study Sect, Geront Res Ctr, NIH, 66-68; instr med, Sch Med, Wash Univ, 71-72; investr, Howard Hughes Med Inst, 74-89; lectr, 74-; mem, Comt Res & Prog Comt, Am Diabetes Asn, 79-82; assoc ed, J Clin Invest, 82-86; vis scientist, Molecular Biol Dept, Mass Gen Hosp & Dept Genetics, Harvard Med Sch, 85-86; vchmn, comt res, Mass Gen Hosp, 88-91, chmn, Subcomt Res Bldgs, 88-, dir, Endocrine Div, 89- & chmn, comt res, 91-; spec lectr, Swed Diabetes Asn, 89 & 33rd Gen Assembly, Japan Diabetes Soc, 90. *Mem:* AAAS; Am Diabetes Asn; Am Soc Clin Invest; Endocrine Soc; Am Soc Biol Chemists; Asn Am Physicians. *Res:* Insulin action; hormonal control of protein phosphorylation; author of numerous publications. *Mailing Add:* Dept Med Mass Gen Hosp E Bldg 149 8th Floor 13th St Charlestown MA 02129

AWAD, ALBERT T, b Cairo, Egypt, Nov 25, 28; m 58; c 3. NATURAL PRODUCTS CHEMISTRY, PHARMACOGNOSY. *Educ:* Cairo Univ, BSc, 52, MSc, 60; Ohio State Univ, PhD(pharmacog), 66. *Prof Exp:* instr & researcher pharmacog, Col Pharm, Cairo Univ, 56-61; teacher & res asst, Col Pharm, Ohio State Univ, 61-66; assoc prof, 66-71, PROF PHARMACOG, COL PHARM, OHIO NORTHERN UNIV, 71- *Concurrent Pos:* Lederle pharm fac award, 67; Gilford instrument apparatus grant, 68; NSF col sci improvement grant, 69-73. *Mem:* Am Asn Col Pharm; Am Soc Pharmacog; Acad Pharmaceut Sci; Marine Technol Soc; Sigma Xi. *Res:* Medicinal plant alkaloids and glycosides; study of active principles in medicinal plants, including the screening, isolation and characterization of these co nstituents which might possess certain pharmacological and clinical activities in treatment. *Mailing Add:* Col Pharm Ohio Northern Univ Ada OH 45810

AWAD, ATIF B, US citizen; m 74; c 4. MEMBRANES, LIPIDS. *Educ:* Ain Shams Univ, Cairo, BS, 59, MS, 66; Rutgers Univ, PhD(nutrit), 74. *Prof Exp:* Fel biochem, Univ Iowa, 74-76; asst prof biochem, Kirksville Col Osteop Med, 76-80, assoc prof, 80-85; ASSOC PROF NUTRIT & DIR, CLIN NUTRIT PROG, SCH HEALTH RELATED PROFESSIONALS, STATE UNIV NY, BUFFALO, 85- , ASSOC PROF BIOCHEM, SCH MED, 87- *Concurrent Pos:* Prin investr, Kirksville Col Osteop Med, NIH, 78-80, Am Osteop Asn, 81-83, USDA, 81-84; adj asst prof, Northeast Mo State Univ, 76-80, adj assoc prof, 80-85. *Mem:* Am Inst Nutrit; Am Diet Asn; Am Oil Chemists Soc; Am Soc Biol Chemists. *Res:* The effect of dietary fats on structure and function of mammalian cell membranes; the role of saturated fatty acids in the inhibition of hormone-stimulated lipolysis in adipose tissue; nutrition and cancer. *Mailing Add:* Nutrit Prog 301 Parker Hall State Univ NY 3435 Main St Buffalo NY 14214

AWAD, ELIAS M, b Latakia, Syria, Oct 06, 34; US citizen; m 70; c 3. EXPERT SYSTEMS, STRUCTURED SYSTEMS DESIGN. *Educ:* Geneva Col, BA, 56; Tulsa Univ, MBA, 58; Northwestern Univ MA, 66; Univ Ky, PhD(personnel info), 75. *Prof Exp:* Asst prof computer systs, Rochester Inst Tech, 60-65; asst prof info systs, DePaul Univ, 67-75; prof info systs, Ball State Univ, 75-77 & Fla Int Univ, 77-82; PROF INFO SYSTS, UNIV VA, 82- *Concurrent Pos:* Consult, Awad & Assoc Inc, 67-; personnel dir, First Nat Bank SMiami, 78-82, dir, 83-88; vis prof, Dartmouth Col, 86-87. *Mem:* Asn Comput Mach; Data Processing Mgt Asn. *Res:* Author of 19 textbooks on management information systems, system design and database design; published over 20 papers and presented over 50 papers in National Conferences on computer systems. *Mailing Add:* McIntire Sch Com Univ Va Charlottesville VA 22903

AWAD, ESSAM A, b Cairo, Egypt, Mar 5, 32; m 65; c 4. PHYSICAL MEDICINE & REHABILITATION. *Educ:* Cairo Univ, MD, 57; Univ Minn, PhD(phys med rehab), 64. *Prof Exp:* From asst prof to assoc prof, Temple Univ, 65-67; lectr, Med Sch, Al-Azhar Univ, 67-68; med dir, phys med & rehab dept, United Hosp, St Paul, Minn, 80-86; instr, 64, prof, 69-80, CLIN PROF PHYS MED & REHAB, UNIV MINN, MINNEAPOLIS, 80-; PROF & CLIN CHIEF, DEPT PHYS MED & REHAB, UNIV MINN HOSP, 87- *Honors & Awards:* Gold Medal Award, Am Acad Phys Med & Rehab, 66. *Mem:* Am Cong Rehab Med; NY Acad Sci; Am Acad Phys Med. *Res:* Neuromuscular diseases; electromyography and nerve conduction studies. *Mailing Add:* Dept Phys Med & Rehab Univ Hosp Minneapolis MN 55455

AWAD, WILLIAM MICHEL, JR, b Shanghai, China, Nov 5, 27; US citizen; m 57; c 2. INTERNAL MEDICINE, HEMATOLOGY. *Educ:* Manhattan Col, BS, 50; State Univ NY, MD, 54; Univ Wash, PhD(biochem), 65. *Prof Exp:* Intern med, Kings County Hosp, Brooklyn, NY, 54-55; asst resident path, St Vincent's Hosp, Manhattan, 55-56; asst resident med, Bronx Munic Hosp Ctr, 56-57; fel med, Univ Minn Hosps, 57-58; fel cardiol, New York Hosp, 58-59; physician, NY Tel Co, 59-60; Damon Runyon fel, 60-63; Nat Cancer Inst fel, Univ Wash, 63-64; from asst prof to assoc prof, 65-78, dir PhD-MD prog, 73-88, PROF MED, UNIV MIAMI, 78-, PROF BIOCHEM & MOLECULAR BIOL, 79- *Mem:* Am Soc Biol Chem; Am Asn Oncol Soc; Am Chem Soc; Biochem Soc; Am Soc Hemat. *Res:* Protein and peptide chemistry; biosynthesis of methyl containing metabolites; macrophage function. *Mailing Add:* Dept Med & Biochem & Molecular Biol Univ Miami Sch Med R-123A PO Box 016960 Miami FL 33101

AWADALLA, FAROUK TAWFIK, b Kalyolria, Egypt, Jan 20, 51; m 81; c 3. RECOVERY OF METALS, HYDROMETALLURGY. *Educ:* Cairo Univ, BSc, 73, MSc, 81; Laval Univ, Que, PhD(metall eng), 87. *Prof Exp:* Asst res extractive metall, Dept Metall, Laval Univ, 82-86, postdoctoral, 86-88; postdoctoral, Extractive Metall, Dept Metall, Univ Idaho, Moscow, 88. *Concurrent Pos:* Res asst, Nat Res Ctr, Cairo, Egypt, 74-81. *Mem:* Electrochemical Soc; Can Inst Mining & Metall. *Res:* Extraction of metals from rock by ion-exchange, solvent extraction, precipitation or crystallization methods; recovery of mineral salts from sea-water; detoxification of asbestos fibers; ultrasound leaching; treatment of waste water; biosorption. *Mailing Add:* Canmet Ottawa ON Can

AWADALLA, NABIL G, b Cairo, Egypt, Jan 10, 45; m 71; c 1. STRESS ANALYSES, MATERIAL PROPERTIES. *Educ:* Cairo Univ, BS, 65; Duke Univ, MS, 70, PhD(civil eng), 72. *Prof Exp:* Engr anal & design, Dravo Corp, 72; sr engr, Duke Power Co, Inc, 72-81; STAFF ENGR RES & DEVELOP,

E I DU PONT DE NEMOURS & CO, INC, 81- *Mem:* Am Soc Civil Engrs. *Res:* Effects of long term radiation on stainless steels from the mechanical properties standpoint; analyses of critical reactor systems subjected to random vibratory excitation in a seismic event. *Mailing Add:* Westinghouse-Savanna Co Aikes SC 29808

AWAPARA, JORGE, b Arequipa, Peru, Dec 15, 18; nat US; m 61; c 1. BIOCHEMISTRY. *Educ:* Mich State Univ, BS, 41, MS, 42; Univ Southern Calif, PhD(biochem), 47. *Prof Exp:* Biochemist, Univ Tex M D Anderson Hosp & Tumor Inst, 48-57, assoc biochemist, 57-74; assoc prof biol, 57-62, PROF BIOCHEM, RICE UNIV, 62- *Mem:* AAAS; Am Soc Biol Chemists. *Res:* Intermediary metabolism of amino acids; metabolism of amino acids in the brain; mechanism of enzymatic decarboxylation of amino acids in the brain; sulfur metabolism. *Mailing Add:* 3426 Deal St Houston TX 77025

AWASTHI, YOGESH C, b July 13, 39; Indian citizen; m 61; c 4. BIOLOGICAL CHEMISTRY. *Educ:* Univ Lucknow, India, BS, 57, MS, 59, PhD(chem), 67. *Prof Exp:* Jr res fel plant chem, 59-60, from sr sci asst to jr sci officer, 60-70, sr sci officer plant chem, Nat Bot Gardens, Lucknow, India, 70-76; sr res assoc, 74-76, from asst prof to assoc prof, 76-84, PROF HUMAN GENETICS, UNIV TEX MED BR GALVESTON, 84- *Concurrent Pos:* Fel biol sci, Purdue Univ, 68-70; res assoc hemat, City of Hope Nat Med Ctr, Duarte, Calif. *Mem:* Am Soc Biol Chemists; Am Soc Human Genetics; Biochem Soc London. *Res:* Biochemical genetics of Tay-Sach's disease and other lysosomal storage diseases; red cell metabolism with particular Glutathione metabolism and defense mechanisms against xenobiotics and oxidative stress; chemoprevention of cancer; protein chemistry and enzymology. *Mailing Add:* Dept Human Biol Chem & Genetics Univ Tex Med Br Galveston TX 77550

AWBREY, FRANK THOMAS, b Carlsbad, NMex, Oct 16, 32; m 56; c 3. BIOLOGY. *Educ:* Univ Calif, Riverside, AB, 60; Univ Tex, MA, 63, PhD(bioacoust), 65. *Prof Exp:* from asst prof to assoc prof, 64-79, PROF BIOL, SAN DIEGO STATE UNIV, 79- *Concurrent Pos:* NSF res grant, 65-67 & 76-; sr res fel, Hubbs/SeaWorld Res Inst; mem bd dir, Porpoise Rescue Found. *Mem:* AAAS; Soc Study Evolution; Am Soc Ichthyologists & Herpetologists; Sigma Xi; Acoust Soc Am. *Res:* Biological significance of vocalization in amphibians and marine mammals; effects of noise on animals. *Mailing Add:* Dept Biol San Diego State Univ San Diego CA 92182-0057

AWRAMIK, STANLEY MICHAEL, b Lynn, Mass, Aug 11, 46; m; c 1. PALEOBIOLOGY. *Educ:* Boston Univ, AB, 68; Harvard Univ, PhD(geol), 73. *Prof Exp:* Res fel biol, Harvard Univ, 73-74; from lectr to assoc prof, 80-85, PROF GEOL, UNIV CALIF, SANTA BARBARA, 85- *Mem:* Fel AAAS; Sigma Xi; Soc Econ Paleontologists & Mineralogists; Paleont Soc; Int Soc Study Origin of Life. *Res:* Precambrian microbial evolution; role of cyanobacteria in sedimentary constructions; Phanerozoic algal-invertebrate interactions; early earth history; stromatolites. *Mailing Add:* Dept Geol Sci Univ Calif Santa Barbara CA 93106

AWSCHALOM, MIGUEL, nuclear physics; deceased, see previous edition for last biography

AXE, JOHN DONALD, b Denver, Colo, Sept 6, 33; m 63. CHEMICAL PHYSICS. *Educ:* Univ Denver, BS, 55; Univ Calif, Berkeley, PhD(chem), 60. *Prof Exp:* NSF res fel, Dept Physics, Johns Hopkins Univ, 61-62; mem res staff, Thomas J Watson Res Ctr, Int Bus Mach Corp, 62-69; MEM RES STAFF, BROOKHAVEN NAT LAB, 69-, ASSOC DIR, 90- *Honors & Awards:* Warren Award, Am Crystallog Soc, 73. *Mem:* Fel Am Phys Soc. *Res:* Neutron spectroscopy; lattice dynamics; structural phase transformations in solids; x-ray scattering; surface structure. *Mailing Add:* Brookhaven Nat Lab Bldg 510A Upton NY 11973-5000

AXE, WILLIAM NELSON, organic chemistry; deceased, see previous edition for last biography

AXEL, LEON, b Lakewood, NJ, Nov 1, 47; c 1. RADIOLOGY, MEDICAL PHYSICS. *Educ:* Syracuse Univ, BS, 67; Princeton Univ, PhD(astrophys), 71; Univ Calif, San Francisco, MD, 76. *Prof Exp:* From asst prof to assoc prof, 81-90, PROF RADIOL, UNIV PA, 90- *Concurrent Pos:* Vis prof, Stanford Univ, 88. *Mem:* Radiol Soc N Am; Soc Magnetic Resonance in Med; Soc Magnetic Resonance Imaging; Am Asn Physicists Med; Asn Univ Radiologists; Am Col Radiol. *Res:* Magnetic resonance imaging, particularly for the study of blood flow and the heart. *Mailing Add:* Dept Radiol Pendergrass Diag Radiol Res Lab Univ Pa Philadelphia PA 19104-6086

AXEL, RICHARD, b Brooklyn, NY, July 2, 46; m 75. MOLECULAR BIOLOGY. *Educ:* Columbia Univ, AB, 67; Johns Hopkins Sch Med, MD, 70. *Prof Exp:* Res assoc, NIH, 72-74; resident, Dept Path, Col Physicians & Surgeons, Columbia Univ, 70-71, fel, Inst Cancer Res, 71-72, asst prof, Dept Path, 74-78, prof, Dept Path & Biochem, 78-83, actg dir, Inst Cancer Res, 83-85, HIGGINS PROF BIOCHEM & PATH, COLUMBIA UNIV, 83-, INVESTR, HOWARD HUGHES MED INST, 84- *Concurrent Pos:* Vis fel, Dept Path, Columbia Univ, 71-72; sci adv bd, Israel Cancer Res Fund, 84, Cancer Res Inst, 86-; asst ed, Cell, 76-, Neuron, 87; young scientist award, Passano Found, 79. *Honors & Awards:* Alan T Waterman Award, 82; Eli Lilly Award, 83; Harvey Lectr, Rockefeller Univ, 83; NY Acad Sci Award Biol & Med Sci, 84; Chipperfield Lectr, Mass Inst Technol 86; Chiron Lectr, Univ Calif, Berkeley, 87; Richard Lounsbery Award, Nat Acad Sci, 89. *Mem:* Nat Acad Sci; Am Acad Arts & Sci. *Res:* Control of gene expression in normal and transformed cells; published numerous articles in various journals. *Mailing Add:* Howard Hughes Med Inst Columbia Univ 722 W 168th St New York NY 10032

AXELRAD, ARTHUR AARON, b Montreal, Que, Dec 30, 23; m 60; c 3. HEMATOLOGY & IMMUNOLOGY, CELL BIOLOGY. *Educ:* McGill Univ, BSc, 45, MD & CM, 49, PhD(anat), 54. *Prof Exp:* Demonstr zool, McGill Univ, 43-44, comp anat, 44-45, med histol, 46 & exp morphol, 52-54; asst prof anat & histol, 56-57, assoc prof med biophys, 57-66, head, Div Histol, 66-85, PROF ANAT, UNIV TORONTO, 66-; CONSULT, DEPT MED & HEMAT, MT SINAI HOSP, TORONTO, 81- *Concurrent Pos:* Nat Cancer Inst Can fel, 50-57 & Inst Cell Res & Genetics, Karolinska Inst, Sweden, 54-56; scientist & head, Subdiv Immunogenetics & Cytol, Ont Cancer Inst, Toronto, 57-68; mem, Inst Med Sci, Univ Toronto, 67-; mem, res adv group, Nat Cancer Inst Can, 68-74; Chercheur assoc, Nat Ctr Sci Res, Inst Pasteur, 76-77; Michael Sela fel, Weizmann Inst Sci, Israel, 77. *Honors & Awards:* Thomas W Eadie Medal, Royal Soc Can, 88. *Mem:* Asn Cancer Res; Can Soc Cell Biol; Int Soc Exp Hemat; Am Soc Hemat; Can Asn Anatomists; fel Royal Soc Can. *Res:* Leukemia, genetics and virology; gene control of hemopoietic stem and progenitor cell proliferation and differentiation. *Mailing Add:* Dept Anat Univ Toronto Toronto ON M5S 1A8 Can

AXELRAD, D(AVID) R(OBERT), b Vienna, Austria, Apr 24, 10; Brit citizen; m 47; c 1. MICROMECHANICS, RHEOLOGY. *Educ:* Vienna Tech Univ, Dipl Ing, 36, DEngSc, 63; Univ Sydney, MEngSc, 59. *Prof Exp:* Sr lectr mech eng, Univ Sydney, 54-64; prof, 64-70, THOMAS WORKMAN PROF MECH ENG, MCGILL UNIV, 70- *Concurrent Pos:* Alexander von Humboldt fel, Inst Theoret Physics, Aachen, Ger, 63; vis prof, Univ Ill, Urbana, 64; mem Can Nat Comt, Int Union Theoret & Appl Mech, 66-70. *Mem:* Am Soc Mech Engrs; Brit Inst Mech Engrs; Soc Rheol; Soc Eng Sci; NY Acad Sci. *Res:* Response behavior of solid matter to external fields; theoretical analysis; random theory of deformation of solids; thermodynamic aspects; experimental work concerned with stress holographic interferometry; x-ray holography. *Mailing Add:* Dept Mech Eng McGill Univ Box 6070 Sta A Montreal PQ H3C 3G1 Can

AXELRAD, GEORGE, b Trutnov, Czech, Oct 3, 29; US citizen; m 62; c 2. ORGANIC CHEMISTRY. *Educ:* City Col New York, BS, 54; Univ Kans, PhD(org chem), 60. *Prof Exp:* Fel chem, Ohio State Univ, 60-61; from asst prof to assoc prof, 61-81, PROF CHEM, QUEENS COL, 81-, CHMN DEPT, 80- *Mem:* Am Chem Soc; Royal Soc Chem. *Res:* Synthetic organic chemistry; reaction mechanisms. *Mailing Add:* Dept Chem & Biochem Queens Col Flushing NY 11367-0904

AXELROD, ABRAHAM EDWARD, b Cleveland, Ohio, June 10, 12; m 39; c 2. BIOCHEMISTRY. *Educ:* Western Reserve Univ, BA, 33, MA, 36; Univ Wis, PhD(biochem), 39. *Prof Exp:* Commercial Solvents fel, Univ Wis, 39-40, Rockefeller Found fel, 40-42; res chemist, Western Pa Hosp, Pittsburgh, 42-51; assoc prof chem, Univ Pittsburgh, 46-51, lectr nutrit, Sch Pub Health, 50-51; assoc prof biochem, Western Reserve Univ, 51-54; prof, Sch Med, Univ Piitsburgh, 54-87, actg chmn dept, 64-66, assoc dean, 65-70, EMER PROF BIOCHEM, SCH MED, UNIV PITTSBURGH, 87- *Mem:* AAAS; Am Soc Biol Chem; Am Inst Nutrit. *Res:* Relationships of vitamins and enzymes; physiological studies in nutritional deficiencies; environmental factors and bacterial virulence; mode of action of vitamins; role of vitamins in antibody formation; chemistry of skin; skin transplantation. *Mailing Add:* Five Bayard Rd Apt 608 Pittsburgh PA 15213

AXELROD, ARNOLD RAYMOND, b Cleveland, Ohio, Jan 1, 21; m 43; c 3. HEMATOLOGY. *Educ:* Ohio Univ, AB, 41; Wayne Univ, MD, 44; Am Bd Internal Med, dipl, 51. *Prof Exp:* Intern, Billings Hosp, Chicago, 44-45; first asst resident med, Detroit Receiving Hosp, 48-49; instr clin med, 49-51, asst prof med, 51-53, from asst prof to assoc prof clin med, 53-70, PROF MED, COL MED, WAYNE STATE UNIV, 70-, CHMN DEPT MED, SINAI HOSP, 74- *Concurrent Pos:* Teaching fel hemat, Col Med, Wayne Univ, 45-48. *Mem:* Am Fedn Clin Res; Cent Soc Clin Res; fel Am Col Physicians (gov, 82); Int Soc Hemat; Am Soc Hemat. *Mailing Add:* Dept Med Sinai Hosp 6767 W Outer Dr Detroit MI 48235

AXELROD, BERNARD, b New York, NY, Oct 16, 14; m 34; c 2. BIOCHEMISTRY. *Educ:* Wayne Univ, BS, 35; George Washington Univ, MS, 39; Georgetown Univ, PhD(biochem), 43. *Hon Degrees:* PhD(sci), Purdue Univ, 89. *Prof Exp:* Chemist, USDA, Washington, DC, 38-43 & Calif, 43-50, chief enzyme sect, Western Regional Res Lab, Bur Agr & Indust Chem, 52-54; assoc prof biochem, 54-58, head dept, 65-75, PROF BIOCHEM, PURDUE UNIV, WEST LAFAYETTE, 58- *Concurrent Pos:* Sr res fel, Calif Inst Technol, 50-52; NSF sr fel, Carlsberg Lab, Copenhagen, 60-61; NSF sr res fel, Univ Calif, Santa Cruz, 70-71; adj prof biochem, Sch Med, Indiana Univ, 71. *Mem:* Am Chem Soc; Am Soc Biochem & Molecular Biol; Am Soc Plant Physiol. *Res:* Enzymology; molecular biology; plant biochemistry; lipoxygenase. *Mailing Add:* Dept Biochem Purdue Univ West Lafayette IN 47907

AXELROD, DANIEL, b Brooklyn, NY, Mar 15, 48. BIOPHYSICS, PHYSICS. *Educ:* Brooklyn Col, BS, 68; Univ Calif, Berkeley, PhD(physics), 74. *Prof Exp:* NIH fel biophys, Cornell Univ, 74-77; asst prof & asst res scientist, 77-81, assoc prof & assoc res scientist, 81-87, PROF & RES SCIENTIST, DEPT PHYSICS & BIOPHYS RES DIV, UNIV MICH, 87- *Mem:* Biophys Soc. *Res:* Biophysics of membranes and cell surfaces; development of nerve/muscle synapses; fluorescence spectroscopy. *Mailing Add:* Biophys Res Div 2200 Bonisteel Blvd Ann Arbor MI 48109

AXELROD, DANIEL ISAAC, b Brooklyn, NY, July 16, 10; m 85; c 1. PALEOBOTANY. *Educ:* Univ Calif, AB, 33, MA, 36, PhD(paleobot), 38. *Prof Exp:* Technician, Calif Forest & Range Exp Sta, 34, jr forester, 36; Nat Res fel, US Nat Mus, 39-41; asst cur, Carnegie Inst, 41-42 & 46; from asst prof to prof geol, Univ Calif, Los Angeles, 46-68; PROF BOT, UNIV CALIF, DAVIS, 68- *Concurrent Pos:* John Simon Guggenheim fel, 52-53. *Honors & Awards:* Fel Medal Cal Acad Sci, 81, Int Medal, Palaeobot Soc, 85; Sir Albert Charles Seward Mem Lectr, India, 86; Paleont Soc Medal, 90. *Mem:* Fel Geol Soc Am; Paleont Soc; Soc Study Evolution; AAAS; fel Am Acad Arts & Sci. *Res:* Tertiary paleobotany; evolution of vegetation; paleoecology; paleoclimate; plate tectonics and angiosperm evolution; evolution of Madrean-Tethyan Sclerophyll vegetation; biogeography. *Mailing Add:* Dept Botany Univ Calif Davis CA 95616

AXELROD, DAVID, HEALTH ADMINISTRATION. *Prof Exp:* Comnr, NY State Dept Health, 69-91; RETIRED. *Mem:* Inst Med-Nat Acad Sci. *Mailing Add:* 98 Terrace Ave Albany NY 12203

AXELROD, DAVID E, b Chicago, Ill, Aug 25, 40; m 72; c 1. MICRO & MOLECULAR BIOLOGY, GENETICS. *Educ:* Univ Chicago, BS, 62; Univ Tenn, PhD(radiation biol), 67. *Prof Exp:* Nat Cancer Inst fel develop biol & cancer, Albert Einstein Col Med, 68-70; asst prof, 70-75, ASSOC PROF BIOL, RUTGERS UNIV, 76- *Concurrent Pos:* Prin investr, USPHS Res Grant, Rutgers Univ, 86-89; prin investr, NSF Res Grant, 75-77, 78-80 & 80-83; res biologist, Nat Heart & Lung Inst, NIH, 76, 77; prin investr, NJ Cancer Comt Res Grant, 84-88, Nat Cancer Inst, 86-90, Coun Tobacco Res, 90-92. *Mem:* Am Soc Cell Biol; Am Soc Microbiol; Genetics Soc Am; Cell Kinetics Soc. *Res:* Molecular cell genetics, oncogenes and anti-oncogenes. *Mailing Add:* Waksman Inst Rutgers Univ PO Box 759 New Brunswick NJ 08855-0759

AXELROD, JULIUS, b New York, NY, May 30, 12; m 38; c 2. BIOCHEMICAL PHARMACOLOGY. *Educ:* City Col New York, BS, 33; NY Univ, MS, 41; George Washington Univ, PhD(chem pharmacol), 55. *Hon Degrees:* DSc, Univ Chicago, 66, NY Univ & Med Col Wis, 71; LLD, George Washington Univ, 71, City Col NY, 72 & Med Col Pa, 74; Dr, Univ Panama, 72, Tel Aviv Univ, 86, Ripon Col, 84. *Prof Exp:* Chemist, Lab Indust Hyg, Inc, NY, 35-45; res assoc, Res Div, Goldwater Mem Hosp, Welfare Island, NY Univ, 45-49; biochemist, Nat Heart Inst, 49-55, chief sect pharmacol, Lab Clin Sci, 55-84, GUEST WORKER, LAB CELL BIOL, NIMH, 84- *Concurrent Pos:* Lectr, NIH, 67; Claude Bernard prof, Univ Montreal, 69. *Honors & Awards:* Nobel Prize in Med Physiol, 70; Otto Loewi Mem Lectr, Med Sch, NY Univ, 64; Gardner Award, 67; Albert Einstein Award, Yeshiva Univ, 71; Parkinson Lectr, Columbia Univ, 71; Hodge Lectr, Rochester Univ, 71. *Mem:* Sr mem Inst Med-Nat Acad Sci; AAAS; fel Am Acad Arts & Sci; hon mem Am Psychophysiol Asn; Am Soc Pharmacol & Exp Therapeut; foreign mem Royal Soc; foreign mem German Acad Sci; Int Brain Res Orgn. *Res:* Biochemical mechanisms of drug and hormone action; drug and hormone metabolism; enzymology. *Mailing Add:* 10401 Grosvenor Pl Rockville MD 20852

AXELROD, LLOYD, b Brooklyn, NY, July 29, 42; div; c 1. ENDOCRINOLOGY, METABOLISM. *Educ:* Princeton Univ, AB, 63; Harvard Med Sch, MD, 67. *Prof Exp:* Intern med, Peter Bent Brigham Hosp, 67-68, jr resident, 68-69, res fel, 69-70, asst med, 70-72; resident, Mass Gen Hosp, 70-71, clin & res fel, 71-72, chief resident, 73; from instr to asst prof med, 73-83, ASSOC PROF MED, HARVARD MED SCH, 83-; CHIEF, JAMES HOWARD MEANS FIRM, MASS GEN HOSP, 89- *Concurrent Pos:* Res fel med, Harvard Med Sch, 71-72; Daland fel, Am Philos Soc, 74-77; asst med, Mass Gen Hosp, 74-78, asst physician, 78-80, assoc physician, 81-89, physician, 89-; mem, US Med Deleg to People's Repub China, 75; chief md unit, Mass Eye & Ear Infirmary, 77-85. *Mem:* Am Diabetes Asn; Am Fedn Clin Res; NY Acad Sci; AAAS; Endocrine Soc. *Res:* Role of prostaglandins in the regulation of lipolysis and ketogenesis; prostaglandin production by adipose tissue in vitro and in vivo; role of prostaglandins in the pathogenesis of diabetic microvascular disease and diabetic ketoacidosis; role of Omega-3 fatty acids in patients with diabetes mellitus. *Mailing Add:* Mass Gen Hosp Boston MA 02114

AXELROD, NORMAN NATHAN, b New York, NY, Aug 26, 34; m 75; c 2. OPTICS, OPTICAL ENGINEERING. *Educ:* Cornell Univ, AB, 54; Univ Rochester, PhD(physics, optics), 59. *Prof Exp:* Aerospace scientist, Goddard Space Flight Ctr, Nat Aeronaut & Space Admin, 59-60; res fel, Univ London, 60-61; asst prof physics, Univ Del, 61-65; mem tech staff, Bell Tel Labs, 65-72; PRES, NORMAN N AXELROD ASSOC, 72- *Concurrent Pos:* Univ res found grant, Univ Del, 63-64, Off Naval Res grant, 64-65; dir, World Resources Develop Co, 71-; consult, Gen Elec, 73-, Recognition Equipment Inc, 74-, Timex, 74-, Konishiroku, 76-, Picker, 77-, Wall St J, 78-, Timken Research, 78-, Polychrome, 79-, ITT, 80-, Thatcher Glass, 80-, Am Cyanamid Co, 81-, Torrington, 81, Perkin-Elmer, 82, Comput Scan, 82, IBM, 83, Becton-Dickenson, 83, Firestone, 84, RCA, 84, Johnson & Johnson, 85 & Emhart, 85, Gen Foods, 86, Celanese, 87-, Medtronic, 88-, Conrail, 89, Bausch & Lomb, 89- *Mem:* Inst Elec & Electronics Engr; fel AAAS; NY Acad Sci; Am Phys Soc; Optical Soc Am. *Res:* Optical-electronic and laser measurement, control, and display systems and techniques; dynamic materials characterization; 0503406lxattern recognition; non-contact defect detection, metrology, and alignment; applications to industrial, consumer and professional products. *Mailing Add:* Norman N Axelrod Assocs 445 E 86th St New York NY 10028

AXELSON, MARTA LYNNE, b Lewiston, Idaho, March 5, 53. FOOD RELATED BEHAVIOR. *Educ:* Fla State Univ, BS, 75; Univ Tenn, Knoxville, PhD(food sci), 79; State Univ NY at Albany, MS, 91. *Prof Exp:* Res asst, Univ Tenn, Knoxville, 75-79; fac intern, Univ Md, College Park, 81, assoc prof food sci, 79-87; RES, STATE UNIV NY, 88- *Mem:* Inst Food Technologists; Soc Nutrit Educ. *Res:* Food related behavior of elderly people living independently, incidence and duration of breast feeding among women, food classification systems and dietary fiber intake. *Mailing Add:* PO Box 2 Sch Bus State Univ NY Albany NY 12063

AXEN, DAVID, b Brackendale, BC, Aug 6, 38; m 65. NUCLEAR PHYSICS. *Educ:* Univ BC, BASc, 60, PhD(physics), 65. *Prof Exp:* Nat Res Coun Can-NATO res fel physics, 65-68, asst prof, 68-74, ASSOC PROF PHYSICS, UNIV BC, 74- *Res:* A polarized helium-3 ion source. *Mailing Add:* Dept Physics Univ BC 2075 Westbrook Pl Vancouver BC V6T 1W5 Can

AXEN, KENNETH, b New York, NY, Mar 23, 43. PULMONARY PHYSIOLOGY. *Educ:* New York Univ, PhD(biomed eng), 72. *Prof Exp:* Asst prof, 79-85, ASSOC PROF PHYSIOL, INST REHAB MED, NEW YORK UNIV, 85- *Mailing Add:* 114 Garfield Pl Brooklyn NY 11215

AXEN, UDO FRIEDRICH, b Siegburg, Ger, Sept 10, 35; m 62; c 2. ORGANIC CHEMISTRY. *Educ:* Univ Bonn, Dipl Chem, 61, Dr rer nat(org chem), 63. *Prof Exp:* Res assoc org chem, Mass Inst Technol, 64-66; head res, 74-79, res mgr, 79-81, group mgr, 81-83, dir, 83-85, SCIENTIST, UPJOHN CO, 66-, VPRES, RES & DEVELOP, JAPAN, 85- *Mem:* Am Chem Soc. *Res:* Structure elucidation of natural products; saponines; macrolides; prostaglandin synthesis. *Mailing Add:* Upjohn Co 7942-24-1 Kalamazoo MI 49001-3298

AXENROD, THEODORE, b New York, NY, Aug 27, 35; m 61; c 1. ORGANIC CHEMISTRY. *Educ:* NY Univ, BA, 56, PhD(chem), 61. *Prof Exp:* Res fel, Univ Wis, 60-61; from asst prof to assoc prof, 61-72, PROF CHEM & CHMN DEPT, CITY UNIV NEW YORK, 72- *Mem:* Am Chem Soc; Brit Chem Soc. *Mailing Add:* Dept Chem City Univ NY Convent Ave & 138th St New York NY 10031

AXFORD, ROY ARTHUR, b Detroit, Mich, Aug 26, 28; m 54; c 3. NUCLEAR ENGINEERING, THEORETICAL PHYSICS. *Educ:* Williams Col, BA, 52; Mass Inst Technol, SB, 52, SM, 55, ScD(nuclear eng), 58. *Prof Exp:* Supvr theoret physics group, Atomics Int Div, NAm Aviation, Inc, 58-60; assoc prof, Tex A&M Univ, 60-62, prof, 62-63; assoc prof, Northwestern Univ, 63-66; assoc prof, 66-68, PROF NUCLEAR ENG, UNIV ILL, URBANA-CHAMPAIGN, 68- *Concurrent Pos:* Consult, United Nuclear Corp, 63, Los Alamos Sci Lab, 63- & Argonne Nat Lab, 64-66. *Mem:* Am Nuclear Soc; Am Soc Mech Engrs; Am Inst Aeronaut & Astronaut. *Res:* Theoretical reactor physics; transport theory; reactor heat transfer; thermoelasticity; nuclear reactor design; plasma physics. *Mailing Add:* Dept Nuclear Eng Univ Ill 103 S Goodwin Ave Urbana IL 61801

AXLER, DAVID ALLAN, b Philadelphia, Pa, Feb 6, 42; m 65; c 1. VIROLOGY. *Educ:* Pa State Univ, BS, 63; Hahnemann Med Col, MS, 66, PhD(virol), 69. *Prof Exp:* Asst prof vet path, Ohio State Univ, 70-73; prin virologist, Battelle Mem Inst, 73-75; assoc prof microbiol, 75-79, PROF & CHMN, DEPT MICROBIOL, IMMUNOL & PATH, PA COL PODIATRIC MED, 79-, VPRES RES & DEAN FOR EDUC AFFAIRS, 84- *Mem:* AAAS; Am Soc Microbiol; Tissue Cult Asn; Sigma Xi. *Res:* Mechanism of carcinogenesis induced by viruses and development of methods for extraction and in vitro assay in tumor specific transplantation antigens; in vitro and in vivo assays of antifungal agents; susceptibility to infection in the diabetic. *Mailing Add:* 716 Suffolk Rd Jenkintown PA 19046

AXLEY, JOHN HAROLD, b Butternut, Wis, June 2, 15; m; c 3. SOILS. *Educ:* Univ Wis, BA, 37, PhD(soil chem), 45. *Prof Exp:* Mem staff, Riverside Exp Sta, Univ Calif, 46; agronomist, Zeloski Potato Farm, 47; assoc prof soils, 48-68, PROF AGRON, UNIV MD, COLLEGE PARK, 68- *Concurrent Pos:* Chmn phosphorus & lime work group, Regional Soil Nitrogen Proj & mem soil test work group, Northeast Soil Res Comt. *Mem:* Am Soc Agron; Soil Sci Soc Am; Potato Asn Am; Sigma Xi. *Res:* Response of Maryland tobacco to liming; clay separation and identification by a density gradient method; chemistry and phytotoxicity of arsenic in soils--arsenic soil test correlation; technique for determining phototoxicity of water extracts of plants; nitrogen removal from sewage waters by soils and plants. *Mailing Add:* 50 Cedar Rd Port Republic MD 20676

AXTELL, DARRELL DEAN, b Grants Pass, Ore, July 27, 44; m 69; c 1. INORGANIC CHEMISTRY, CHEMICAL EDUCATION. *Educ:* Linfield Col, BA, 67; Ore State Univ, PhD(chem), 73. *Prof Exp:* Fel chem, Univ Ariz, 73-74; fel & lectr chem, Tex A&M Univ, 75-77; asst prof chem, Eastern Mont Col, 77-; AT UNITED WORLD COL,. *Mem:* Am Chem Soc; Am Asn Univ Professors; Sigma Xi. *Res:* Inorganic coordination chemistry; chemical education; self-paced instruction; microforms. *Mailing Add:* Dept Chem St Martins Col Lacey WA 98503

AXTELL, JOHN DAVID, b Minneapolis, Minn, Feb 5, 34; m 57; c 3. PLANT BREEDING. *Educ:* Univ Minn, BS, 57, MS, 65; Univ Wis, PhD(genetics), 67. *Prof Exp:* Res asst genetics, Univ Wis, 59-65, res assoc, 67; from asst prof to prof, 67-82, LYNN DISTINGUISHED PROF AGRON, PURDUE UNIV, 82- *Concurrent Pos:* Proj leader, USAID res grants, 72-79, 79- & 82; Alexander von Humboldt award, 76; sci liaison officer, USAID & Int Crops Res Inst Semi-Arid Tropics, Hyderabad, India, 84-86; chmn, Agr Sci Sect, Nat Acad Sci, 87-90; mem, Expert Panel USAID Biotechnol Res, Nat Acad Sci, 90, Sci Coun Plant Gene Expression Ctr, 90 & Comt Plant Sci, 90. *Honors & Awards:* Crop Sci Award, Crop Sci Soc Am, 77. *Mem:* Nat Acad Sci; Genetics Soc Am; Am Soc Agron; Am Genetic Asn; Crop Sci Soc Am; AAAS. *Res:* Controlling elements in maize; paramutation in maize; chemical paramutagenesis; mutagenesis in higher plants; regulation of gene action in higher plants; plant genetics and breeding; genetic improvement of protein quality in cereals. *Mailing Add:* Dept Agron Purdue Univ 1150 Lilly Hall Sci Bldg West Lafayette IN 47907-1150

AXTELL, RALPH WILLIAM, b Norfolk, Nebr, Apr 20, 28; m 67; c 1. VERTEBRATE ZOOLOGY. *Educ:* Univ Tex, BA, 53, MA, 54, PhD, 58. *Prof Exp:* Res sci zool, Univ Tex, 55-57; instr biol, ETex State Col, 57-58; asst prof, Sul Ross State Col, 58-60; asst prof zool, 60-65, assoc prof biol, 65-70, chmn fac biol sci, 65-67, PROF BIOL, SOUTHERN ILL UNIV, EDWARDSVILLE, 70- *Concurrent Pos:* Herpet ed, Copeia, Am Soc Ichthyologists & Herpetologists, 68- *Honors & Awards:* Student Paper Award, Am Soc Ichthyol & Herpet, 57. *Mem:* Fel AAAS; Am Soc Ichthyologists & Herpetologists; Soc Study Evolution; Soc Syst Zool; Am Inst Biol Sci. *Res:* North American herpetology and zoogeography; lizard genus Holbrookia; lizard family Iguanidae; reptilian evolution. *Mailing Add:* Dept Biol Sci Southern Ill Univ Edwardsville IL 62026

AXTELL, RICHARD CHARLES, b Medina, NY, Aug 4, 32; m 61; c 2. MEDICAL ENTOMOLOGY. *Educ:* State Univ NY Albany, BS, 54, MS, 55; Cornell Univ, PhD(entom), 62. *Prof Exp:* Vis instr zool, Col Agr, Univ Philippines, 55-56; from asst prof to assoc prof, 62-68, PROF ENTOM, NC STATE UNIV, 68- *Concurrent Pos:* Consult, Nat Develop Co, Philippines, 55-56; prin investr, USPHS Grant, 63-70; dir, NIH med entom training grant, 66-73; Sigma Xi res award, NC State Univ, 69; Off Naval Res contract, 70-75; dir dept com, Nat Oceanic & Atmospheric Admin sea grant proj, 70-76; prin investr, NIH grant, 83- *Mem:* AAAS; Entom Soc Am; Am Mosquito Control Asn (vpres, 80, pres, 82); Acarological Soc Am. *Res:* Medical and veterinary entomology; biology and control of Diptera; mosquito control; arthropod sensory structures and behavior; biological control; poultry pest management. *Mailing Add:* Dept Entom NC State Univ Raleigh NC 27695-7613

AXTHELM, DEON D, b Hallam, Nebr, Mar 6, 20; c 3. WATER MANAGEMENT. *Educ:* Univ Nebr, BS, 41, BS, 48, MA, 65. *Prof Exp:* Agr exten asst, Coop Exten Serv, Clay County, Nebr, 48-49; self employed, 50-55; asst prof irrig, Coop Exten Serv, 56-65, from assoc prof to prof, 66-80, emer prof agr eng & water resources, Univ Nebr, Lincoln, 80-; WATER CONSULT, 80- *Concurrent Pos:* Mem water develop comt, Great Plains Coun, 68-, mem feedlot pollution res comt, 69-71; consult, Com Water Measurement Mfg Firms, Demonstration & Sales. *Mem:* Am Soc Agr Engrs. *Res:* Groundwater management and related administrative-political systems. *Mailing Add:* 3731 S Glenstone No 198 Springfield MO 65804

AXTMANN, ROBERT CLARK, b Youngstown, Ohio, Feb 25, 25; m 49; c 3. APPLIED PHYSICS, CHEMICAL ENGINEERING. *Educ:* Oberlin Col, AB, 47; Johns Hopkins Univ, PhD(phys chem), 50. *Prof Exp:* Instr org chem, Johns Hopkins Univ, 48-49; res physicist, Ballistic Res Lab, Aberdeen Proving Ground, Md, 50; res physicist, E I du Pont de Nemours & Co, 50-53, res supvr reactor physics, 53-54, area tech supvr reactor physics & eng, 55-56, sr res supvr chem physics, 57-59; assoc prof chem eng, 59-65, chmn prog nuclear studies, 60-68, chmn coun environ studies, 70-73, PROF CHEM ENG, PRINCETON UNIV, 65- *Concurrent Pos:* Vis fel, Israel AEC, 64 & Nat Comn Nuclear Energy, Mex, 69; mem, NJ Comn Radiation Protection, 66-70; mem vis comt, Brookhaven Nat Lab, 68-71; vis scientist, Dept Sci & Indust Res, Govt NZ, 74 & 79-80 & Inst InvestS Elec, Mex, 84-85; mem, Rev Comt, Argonne Nat Lab, 75-79; mem Adv Comt Geothermal Energy, US Dept Energy, 76-80 & Adv Comt Reactor Safeguards, US Nuclear Regulatory Comn, 81-85. *Mem:* AAAS; Am Phys Soc; Am Chem Soc; Am Inst Chem Eng. *Res:* Nuclear fusion technology; geothermal energy; radiation chemistry; nuclear reactor physics; Mossbauer effect; nuclear magnetic resonance; environmental effects of geothermal power; desilication of geothermal fluids. *Mailing Add:* 346 Burnt Hill Rd Skillman NJ 08558-9408

AYALA, FRANCISCO JOSE, b Madrid, Spain, Mar 12, 34; m 85; c 2. GENETICS, EVOLUTION. *Educ:* Univ Madrid, BS, 55; Lit Univ Salamanca, STL, 60; Columbia Univ, MA, 63, PhD(genetics), 64. *Hon Degrees:* Dr, Univ Leon, 82, Univ Madrid, 86 & Univ Barcelona, 86. *Prof Exp:* Res assoc genetics, Rockefeller Inst, 64-65; prof biol, Providence Col, 65-67; asst prof, Rockefeller Univ, 67-71; assoc prof, Univ Calif, Davis, 71-74, prof genetics, 74-87, dir, Inst Ecol, 77-81, PROF BIOL SCI, UNIV CALIF, IRVINE, 87- *Concurrent Pos:* Guggenheim fel, 77-78; assoc ed, Molecular Evolution & Paleobiol; mem, Nat Adv Coun, Nat Inst Gen Med Sci; mem exec comt, Sci Adv Bd, Environ Protection Agency; Fulbright fel, 79 & 81; assoc ed, Biol & Philos; chmn, Sect Population Biol, Evolution & Ecol, Nat Acad Sci, 83-86; mem, Comn Life Sci, Nat Res Coun, 84-; chmn, Bd Basic Biol, Nat Res Coun, 85-; Coun, Nat Acad Sci, 86-89; bd dir, Am Asn Adv Sci, 89-; mem, Adv Comm Ed Hum Res, Nat Sci Found, 89- *Honors & Awards:* Medal, Col France, 79; W E Key Award, Am Gen Asn, 85; Sci Freedom & Responsibility Award, Amer Asn Adv Sci, 87. *Mem:* Nat Acad Sci; Am Acad Arts & Sci; AAAS; Genetics Soc Am; Soc Study Evolution (vpres, 72, pres, 79-80); Am Philos Soc; hon mem Fedn Yugoslavian Genetics Socs; Am Philos Soc; hon mem Genetics Soc Spain; foreign mem Royal Acad Sci Spain. *Res:* Evolution; population genetics, fitness of natural and experimental populations; reproductive isolation and origin of species; biochemical variation and molecular evolution; philosophy of science. *Mailing Add:* Dept Ecol & Evolutionary Biol Univ Calif Irvine CA 92717

AYARS, JAMES EARL, b Penns Grove, NJ, Jan 28, 43; m 69; c 2. ENGINEERING, AGRICULTURE. *Educ:* Cornell Univ, BSAgE, 65; Colo State Univ, MS, 73, PhD(eng), 76. *Prof Exp:* Asst prof eng, Univ Md, 76-80; RES AGR ENGR, AGR RES SERV, USDA, 80- *Mem:* Sigma Xi; Am Soc Agr Engrs; Am Soc Agron. *Res:* Groundwater hydrology; irrigation water management; drainage water reuse; drainage system design and management. *Mailing Add:* PO Box 26341 Fresno CA 93729

AYCOCK, BENJAMIN FRANKLIN, b Washington, DC, May 18, 22; m 48; c 5. TEXTILE CHEMISTRY. *Educ:* Univ NC, BS, 42; Univ Ill, MS, 45, PhD(org chem), 47. *Prof Exp:* Asst, Nat Defense Res Comt, Univ Ind, 44-45; res assoc, Explosives Res Lab, Carnegie Inst Technol, 45; from instr to asst prof chem, Univ Wis, 47-50; chemist, Rohm and Haas Co, Pa, 50-59, sect head, Redstone Arsenal Res Div, 59-69; head, Chem Dept, Burlington Idust Res Ctr, 69-72, sr scientist, 72-80, mgr res & develop, 80-84; RETIRED. *Mem:* Am Chem Soc; Swiss Chem Soc; Am Asn Textile Chemists & Colorists. *Res:* Chemistry of natural products; nitrogeneous resins; plasticizers; propellant chemistry; textile chemistry. *Mailing Add:* 2310 N Elm St Greensboro NC 27408

AYCOCK, MARVIN KENNETH, JR, b Warrenton, NC, Nov 12, 35; m 60; c 2. PLANT BREEDING, PLANT GENETICS. *Educ:* NC State Univ, BS, 59, MS, 63; Iowa State Univ, PhD(plant breeding), 66. *Prof Exp:* Teacher, Bear Grass High Sch, 59-60; from asst prof to prof plant breeding, 66-87, PROF & CHMN, DEPT AGRON, UNIV MD, COLLEGE PARK, 87- *Honors & Awards:* Res Award, Am Soc Agron, 82. *Mem:* Am Soc Agron; Crop Sci Soc Am; Am Genetic Asn. *Res:* Effects of a male-sterile cytoplasm in Nicotiana tabacum and inbreeding depression in Medicago sativa; developing new and improved varieties of Maryland tobacco. *Mailing Add:* Dept Agron Univ Md College Park MD 20742

AYCOCK, ROBERT, b Lisbon, La, Dec 23, 19; m 41; c 2. PHYTOPATHOLOGY. *Educ:* La State Univ, BS, 40, NC State Col, MS, 42, PhD(plant path), 49. *Prof Exp:* Assoc plant pathologist, 49-55, Clemson Col; res assoc prof, NC State Univ, 55-61, prof plant path, 61-84, head dept, 73-84; RETIRED. *Concurrent Pos:* Ed-in-chief, Phytopathology, Am Phytopath Soc, 70-72. *Mem:* Fel Am Phytopath Soc (pres, 75-76). *Res:* Plant disease diagnosis and diseases of ornamental crops. *Mailing Add:* Dept Plant Path Box 7616 NC State Univ Raleigh NC 27607

AYDELOTTE, MARGARET BEESLEY, b London, Eng, June 1, 34; m 61; c 3. BIOCHEMISTRY. *Educ:* Cambridge Univ, BA, 56, MA, 60, PhD(physiol), 61. *Prof Exp:* Demonstr physiol, Cambridge Univ, 59-63; assoc prof biol, Tarkio Col, Mo, 64-68; fel anat, Col Med, Univ Iowa, 69-70, asst prof, 71-75; assoc prof biol, Franklin Col, 75-78; asst prof anat, Sch Med, Ind Univ, 78-81; asst prof, 81-89, ASSOC PROF BIOCHEM, COL MED, RUSH UNIV, 89- *Concurrent Pos:* Vis physiologist, Univ Calif, San Francisco, 61. *Honors & Awards:* Carol Nachman Prize in Rheumatol, 87. *Mem:* Am Asn Anatomists; Tissue Cult Asn; AAAS; Am Soc Cell Biol; Orthop Res Soc. *Res:* Morphological and biochemical studies of connective tissues; cartilage matrix synthesis and degradation by articular chondrocytes. *Mailing Add:* 418 Greenleaf Ave Wilmette IL 60091

AYDELOTTE, MYRTLE K, b May 31, 17; m 56; c 2. NURSING. *Educ:* Univ Minn, BS, 39, MA, 48, PhD(educ admin & educ psych), 55. *Hon Degrees:* DS, Univ Nebr, 81. *Prof Exp:* Head nurse, Charles T Miller Hosp, St Paul, 39-41; surgical teaching supvr, St Mary's Hosp, Sch Nursing, Minn, 41-42; asst chief nurse, Army Nurse Corps, 26th Gen Hosp, Europ Theatre, 42-45, chief nurse, 52nd Sta Hosp, 45; instr, Univ Minn, 45-49; assoc chief nursing, Res Nursing, VA Hosp, Iowa, 63-64, chief nursing, Nursing Res, 64; exec dir, Am Nurses Asn, Kans, 77-81; clin prof, Yale Univ, 83-85; dir, dean & prof, Col Nursing, 49-57, prof, 57-62 & 64-76, dir nursing, 68-76, PROF & EMER DEAN, UNIV IOWA, 76-; ACTG EXEC DIR, CTR FOR NURSING INNOVATION, NEW HAVEN, CONN, 83- *Concurrent Pos:* Vis prof, Univ Ill, Col Nursing, 82-; mem, Nat Acad Sci Comt Study of Health Care Resources, Va, 75-77, Comn Nursing Servs, Am Nurses Asn, 76-77, Nat Comn Nursing, 80-83. *Honors & Awards:* Luther Christman Award, 77; Elizabeth Soule lectr, Univ Wash, 81. *Mem:* Inst Med-Nat Acad Sci; Am Acad Nursing; Am Nurses Asn. *Res:* Structural arrangements and financial mechanisms for delivery of nursing services; author of two articles and 48 publications. *Mailing Add:* 408 Montclair Park 201 N First Ave Iowa City IA 52245

AYDLETT, LOUISE CLATE, b Elizabeth City, NC, Aug 04, 41. OPERATIONS RESEARCH. *Educ:* Duke Univ, AB, 62; E Carolina Univ, MA, 66. *Prof Exp:* PROF MATH, COL ALBERMARLE, 64- *Mem:* Math Asn Am. *Mailing Add:* Col Albermarle Elizabeth City NC 27909

AYE, MAUNG TIN, b Rangoon, Burma, Aug 9, 41; Can citizen; m; c 2. EXPERIMENTAL HEMATOLOGY, CELL BIOLOGY. *Educ:* Univ Cambridge, Eng, BChir, MB & MA, 66; Univ Toronto, PhD(med sci), 74, FRCP, 81. *Prof Exp:* House physician & surgeon, Hereford Gen Hosp, UK, 66-67; sr house officer, St Thomas' Hosp, London, 67-68; resident hemat & internal med, Ottawa Gen Hosp, 68-70; ASSOC PROF & SCHOLAR MED RES COUN, DEPT MED & PHYSIOL, 74- *Concurrent Pos:* Fel Med Res Coun, Ont Cancer Inst, 70-74, scholar, 76-81, grant, 74-; Nat Cancer Inst Can grant, 77-81; chmn comt human experimentation, Ottawa Gen Hosp, 77- *Mem:* Am Soc Hemat; Can Soc Hemat; Can Soc Cell Biol. *Res:* Regulation of cell proliferation and differentiation in mammalian hemopoietic systems. *Mailing Add:* 50 Whitemarl Dr Rockliffe Park ON K1L 8J6 Can

AYEN, RICHARD J(OHN), b Loyal, Wis, June 8, 38; m 63; c 2. CHEMICAL ENGINEERING. *Educ:* Univ Wis, BS, 60; Univ Ill, MS, 61, PhD(chem eng), 64. *Prof Exp:* Asst prof chem eng, Univ Calif, Berkeley, 64-68, consult, 65-67; sr res engr, Stauffer Chem Co, 68-70, supvr, Richmond Res Ctr, 70-74, sect mgr, 74-79, sr sect mgr, 79-81, dept mgr, Eastern Res Ctr, 81-88, DIR CHEM PROCESS DEVELOP, CHEM WASTE MGT, STAUFFER CHEM CO, 88- *Mem:* Am Inst Chem Engrs. *Res:* Catalyzed reactions of nitrogen oxides; gas-solids fluidization; scaleup of processes for sulfur and phosphorus-containing compounds; chemical waste treatment. *Mailing Add:* 5N745 Jens Jensen Lane St Charles IL 60175

AYENGAR, PADMASINI (MRS FREDERICK ALADJEM), b Bangalore, India, July 31, 24; m 57; c 1. BIOCHEMISTRY, MICROBIOLOGY. *Educ:* Travancore Univ, India, BSc, 44; Univ Wash, St Louis, PhD, 53. *Prof Exp:* Fel enzymechem, Enzyme Inst, Univ Wis, 53-54; res assoc biochem, City of Hope Med Ctr, Calif, 54-56; vis scientist, Nat Inst Arthritis & Metab Dis, NIH, Md, 57; res assoc immunochem, Dept Med Microbiol, 58-76, RES ASSOC MICROBIOL, DEPT MICROBIOL, SCH MED, UNIV SOUTHERN CALIF, 76- *Mem:* Am Chem Soc; Am Soc Biol Chem. *Res:* Microbial fermentation; growth factors and inhibitors; biosynthesis and metabolism of phosphodiester of serine and ethanolamine; enzymes; purification; properties and mechanism of action. *Mailing Add:* Dept Microbiol Sch Med Univ Southern Calif Los Angeles CA 90033

AYENI, BABATUNDE J, b Lagos, Nigeria, May 29, 54; m 78; c 4. RESERVOIR ENGINEERING, STATISTICS. *Educ:* Univ Southwestern La, BS, 80, MS, 81, PhD(statistics), 84. *Prof Exp:* Grad asst, petroleum eng, Univ Southwestern Univ, La, 81-84; from asst prof to actg dir tech, Southwestern Univ, New Orleans, 85-88, asst prof tech, 88; MEM STAFF, 3M CO, ST PAUL, MINN. *Mem:* Math Asn Am; Am Statist Asn; Nat Soc Prof Eng. *Res:* Petroleum reserves analysis; property evaluation; pressure transient analysis; forecasting; estimation in a mixture of distribution. *Mailing Add:* 3M Co Bldg 549-25-03 St Paul MN 55144-1000

AYER, DARRELL, (JR), pathology; deceased, see previous edition for last biography

AYER, HOWARD EARLE, b Brookings, SDak, July 28, 24; m 47; c 4. INDUSTRIAL HYGIENE. *Educ:* Univ Minn, BChE, 48; Harvard Univ, SM, 55. *Prof Exp:* Engr, Div Occup Health, US Pub Health Serv, 48-64; asst dir field studies, Nat Inst Occup Safety & Health, 64-72; PROF ENVIRON HYG, UNIV CINCINNATI, 72- *Concurrent Pos:* Chmn, Am Conf Govt Indust Hygienists, 71-72. *Honors & Awards:* Meritorious Serv Medal, Pub Health Serv, 72. *Mem:* Am Indust Hyg Asn; Health Physics Soc; Brit Occup Hyg Soc; Am Pub Health Asn. *Res:* Behavior of aerosols in environment and health effects; methods of measurement which relate to these effects on worker health. *Mailing Add:* Kettering Lab Univ Cinn Environ Health 056 107 Ketter Cincinnati OH 45267-0056

AYER, WILLIAM ALFRED, b Sackville, NB, July 4, 32; m 54; c 6. ORGANIC CHEMISTRY. *Educ:* Univ NB, BSc, 53, PhD(chem), 56. *Prof Exp:* Res assoc, Harvard Univ, 57-58; from asst prof to assoc prof, 58-67, PROF CHEM, UNIV ALTA, 67- *Concurrent Pos:* Fel, Alfred P Sloan Found, 65-67; vis prof, Japan Soc Prom Sci, 76; org chem ed, Can J Chem, 76-83, sr ed, 83-88; Job vis prof, Mem Univ Nfld, 83. *Honors & Awards:* Merck Sharp & Dohme Award, Chem Inst Can, 70; John Labatt Award, Chem Inst Can, 81; Van Cleave Lectr, Univ Sask, Regina, 85. *Mem:* Sr mem Am Chem Soc; fel Chem Inst Can; fel Royal Soc Can; sr mem Can Soc Chem (pres, 88-89). *Res:* Structural and synthetic studies of naturally occurring substances. *Mailing Add:* Dept Chem Univ Alta Edmonton AB T6G 2G2 Can

AYERS, ALVIN DEARING, b Alvin, Tex, Apr 27, 09; m 39; c 2. SOIL SCIENCE, PLANT PHYSIOLOGY. *Educ:* Univ Ariz, BS, 32, MS, 34; Univ Calif, PhD(soil sci), 39. *Prof Exp:* Asst agr chemist, Exp Sta, Univ Ariz, 32; chemist, Salt River Valley Water Users Asn, Ariz, 34 & Asn Lab, Calif, 35; asst plant nutrit, Univ Calif, 36-37; agent, Salinity Lab, USDA, 38-41, asst chemist, 41-42; chemist, Calif Inst Technol, 42-43; supvr safety & personnel, 43-45; assoc chemist, Salinity Lab, USDA, 45-48, chemist, 48-50, agriculturist field comt, Ark White & Red Basins Inter-Agency Comt, Okla, 51-53, soil scientist, US Salinity Lab, 53-59, asst dir, Europ Res Off, Agr Res Serv, Italy, 59-62, dir, Far Eastern Regional Res Off, 62-66; soil & water mgt specialist, Agr & Rural Develop Serv, War on Hunger, AID, 66-70 & Off Agr & Fisheries Tech Assistance Bur, 70-73; CONSULT, 73- *Mem:* Fel Soil Sci Soc Am; Am Soc Agron. *Res:* Agricultural research administration; soil salinity and alkali; salt tolerance of plants; soil moisture-plant relationships; soil chemistry. *Mailing Add:* 5621 Del Rio Ct Cape Coral FL 33904

AYERS, ARTHUR RAYMOND, b San Diego, Calif, Sept 28, 47; m 84; c 3. PLANT MOLECULAR BIOLOGY, PLANT DISEASE RESISTANCE. *Educ:* Univ Calif, San Diego, BA, 69; Univ Colo, Boulder, PhD(molecular cell develop biol), 75. *Prof Exp:* Guest researcher biomass conversion, Swedish Forest Prods Res Labs, Stockholm, Sweden, 75-77; res asst phytopath, dept plant path, Univ Mo, Columbia, 77-79, & plant cell cult, Kans State Univ, Manhattan, 79-80; asst prof cell & develop biol, Harvard Univ, 81-86; DIR GENETIC ENG TECHNOL PROG, CEDAR CREST COL, ALLENTOWN, PA, 86- *Concurrent Pos:* Consult, Vital Technol Inc, Cambridge, 83-85; consult Agr Diag Assoc Inc, Cinnaminson, NJ, 86-; assoc ed, J Pa Acad Sci, 88- *Mem:* Am Soc Plant Physiologists; Am Phytopath Soc; Sigma Xi; Phytochem Soc NAm. *Res:* Molecular basis of plant disease resistance using monoclonal antibodies to identify pathogen and host molecules that mediate race-cultivar specificity; immunodiagnostics for plant pathogens. *Mailing Add:* Genetic Eng Cedar Crest Col Allentown PA 18104

AYERS, CARLOS R, b Oakvale, WVa, Apr 2, 32; m 58; c 3. INTERNAL MEDICINE. *Educ:* Lincoln Mem Univ, BS, 53; Univ Va, MD, 58. *Prof Exp:* Intern med, Univ Va, 58-S9; resident univ affil hosp, Univ Utah, 61-62; resident, 63-64, from instr to assoc prof internal med, 64-75, PROF INTERNAL MED, SCH MED, UNIV VA, 75-; DIR, VA HEART RES LAB, 69- *Concurrent Pos:* Fel cardiol, Univ Hosp, Univ Va, 62-63; mem coun high blood pressure res, Am Heart Asn; Va Heart Asn res prof cardiol, 68- *Mem:* Am Fedn Clin Res; fel Am Col Physicians. *Res:* Peripheral vascular disease; hypertension; aldosterone metabolism; renin; angiotensin metabolism. *Mailing Add:* Hypertension Unit Univ Va Sch Med Univ VA Hosp Box 146 Charlottesville VA 22908

AYERS, CAROLINE LEROY, b Augusta, Ga, Oct 30, 41; m 63; c 2. PHYSICAL CHEMISTRY, SCIENCE EDUCATION. *Educ:* Univ Ga, BS, 62, PhD(phys chem), 66. *Prof Exp:* Asst prof chem, Portland State Col, 66-67; from asst prof to assoc prof, 67-84, chmn dept, 83-88, PROF CHEM, E CAROLINA UNIV, 84- *Mem:* Am Chem Soc; Sigma Xi. *Res:* Electron paramagnetic resonance studies of gamma irradiated solids, particularly triphenylmethyl derivatives and organo-metallic compounds; chemical education. *Mailing Add:* Dept Chem ECarolina Univ Greenville NC 27858-4353

AYERS, JACK DUANE, b Nampa, Idaho, Feb 25, 41; m 63; c 3. PHYSICAL METALLURGY. *Educ:* Univ Wash, BS, 66; Carnegie-Mellon Univ, PhD(metall & mat sci), 70. *Prof Exp:* NATO fel, Univ Oxford, 70-71; METALLURGIST, NAVAL RES LAB, 71- *Mem:* Am Inst Mining, Metall & Petrol Engrs Metall Soc; Am Soc Metals. *Res:* Kinetics and crystallography of solid state phase transformations; solidification processing of metals with the objective of improving their physical properties. *Mailing Add:* Code 6320 Naval Res Lab Washington DC 20375-5000

AYERS, JERRY BART, b Atlanta, Ga, Apr 17, 39; m 71; c 2. SCIENCE EDUCATION. *Educ:* Oglethorpe Univ, BS, 60; Univ Ga, MEd, 66, EdD(sci educ), 67. *Prof Exp:* Assoc physicist, Union Carbide Nuclear-Oak Ridge Nat Labs, 60-61; instr chem, Lenoir Rhyne Col, 64-65; asst prof sci educ, Univ Ga, 67-70; PROF SCI EDUC & ASSOC DEAN EDUC, TENN TECHNOL UNIV, 70- *Mem:* Am Chem Soc; Nat Sci Teachers Asn. *Res:* Preparation of improved teaching materials for use at the college level in training teachers; evaluation of science teaching; scientific information retrieval. *Mailing Add:* 580 Pleasant Hill Dr Cookeville TN 38501

AYERS, JOHN E, b Indianapolis, Ind, Dec 20, 41; m 88; c 5. PLANT PATHOLOGY, PLANT GENETICS. *Educ:* Purdue Univ, BS, 63, MS, 65; Pa State Univ, PhD(genetics), 69. *Prof Exp:* From asst prof to assoc prof, 69-80, PROF PLANT PATH, PA STATE UNIV, UNIVERSITY PARK, 80- *Mem:* Am Phytopath Soc. *Res:* Nature and inheritance of disease resistance in plants. *Mailing Add:* Dept Plant Path Pa State Univ Buckhout Lab Rm 211 University Park PA 16802

AYERS, JOSEPH LEONARD, JR, b Long Beach, Calif, Nov 14, 47. NEUROPHYSIOLOGY, COMPARATIVE PHYSIOLOGY. *Educ:* Univ Calif, Riverside, AB, 70; Univ Calif, Santa Cruz, PhD(biol), 75. *Prof Exp:* Res asst neurophysiol, Univ Calif, Santa Cruz, 70-75; fels, Nat Ctr Sci Res, Marseille, France, 75-76 & Univ Calif, San Diego, 76-78; ASST PROF BIOL,

NORTHEASTERN UNIV, 78- *Mem:* Soc Neurosci; AAAS; Int Union Physiologists. *Res:* Neurophysiology, neural control at locomotion, neuroethology, cellular mechanisms underlying behavior, pattern generation, computer applications in neurosciences. *Mailing Add:* Dept Biol Northeastern Univ 360 Huntingten Ave Boston MA 02115

AYERS, ORVAL EDWIN, b Grant, Ala, Mar 29, 32; m 52; c 3. ORGANIC CHEMISTRY. *Educ:* Berea Col, Ky, BA, 55; Auburn Univ, MS, 58; Univ Ala, Tuscaloosa, PhD(org chem), 73. *Prof Exp:* Chemist qual control, Nat Cash Regist Co, Dayton, Ohio, 55-56; RES CHEMIST PROPELLANT RES, US ARMY MISSILE COMMAND, REDSTONE ARSENAL, ALA, 59- *Honors & Awards:* Army Res & Develop Achievement Awards, Dept Army, 75. *Mem:* Am Chem Soc (treas, 75-76). *Res:* Development and evaluation of solid propellants for rocket motors; application of high energy lasers to laser-induced chemical syntheses and propulsion techology. *Mailing Add:* 7805 Martha Dr SE Huntsville AL 35802-2458

AYERS, PAUL WAYNE, b Winter Garden, Fla, Oct 18, 36; m 63. PHYSICAL ORGANIC CHEMISTRY. *Educ:* David Lipscomb Col, BA, 60; Univ Ga, MS, 62, PhD(chem), 66. *Prof Exp:* Asst prof chem, Portland State Col, 66-67; asst prof, 67-73, ASSOC PROF CHEM, E CAROLINA UNIV, 73- *Mem:* Am Chem Soc. *Res:* Free radical rearrangements and oxidation processes. *Mailing Add:* Dept Chem ECarolina Univ Greenville NC 27834

AYERS, RAYMOND DEAN, b Ossining, NY, July 11, 40; m 69. MUSICAL ACOUSTICS. *Educ:* Calif Inst Technol, BS, 63, MS, 64; PhD(mat sci), 71. *Prof Exp:* from asst prof to assoc prof, 67-79, PROF PHYSICS & ASTRON, CALIF STATE UNIV, LONG BEACH, 79- *Mem:* Am Asn Physics Teachers; Acoust Soc Am. *Res:* Acoustics of early wind instruments; acoustical imaging; experimental solid state physics. *Mailing Add:* Dept Physics & Astron Calif State Univ 1250 Bellflower Blvd Long Beach CA 90840

AYERS, WILLIAM ARTHUR, b Highgrove, Calif, Dec 22, 24; m 51; c 2. MICROBIOLOGY. *Educ:* Univ Calif, Los Angeles, AB, 49; Rutgers Univ, MS, 51; Univ Wis, PhD(bact), 54. *Prof Exp:* Asst bact, Rutgers Univ, 49-51 & Univ Wis, 51-54; asst prof, Univ NH, 54-59; microbiologist, Brooklyn Bot Garden, 59-62; microbiologist, USDA, 62-84; RETIRED. *Mem:* Am Soc Microbiol. *Res:* Soil microbiology; ecology and physiology of soilborne plant pathogenic fungi. *Mailing Add:* 11472 Chelsea Ct Fredericksburg VA 22407

AYKAN, KAMRAN, inorganic chemistry; deceased, see previous edition for last biography

AYLER, MAYNARD FRANKLIN, b Tacoma, Wash, Oct 15, 22; m 45; c 2. MINE-ASSISTED RESIDUAL OIL RECOVERY. *Educ:* Colo Sch Mines, E of M, 45, MS, 63. *Prof Exp:* Geologist, US Bur Reclamation, 45-47; Calif Co, 47-52; instr mining, Colo Sch Mines, 58-63; mining engr, US Bur Mines, 60-77; MINING CONSULT, 52- *Concurrent Pos:* Chief, Libyan Geol Surv, USAID & Govt Libya, 64-66; instr geol, Univ Md, 65-66; pres, Hydrocarbon Mining Co, 86- *Mem:* Am Asn Petrol Geologists. *Res:* Developing and patenting a safe system and needed equipment for mine-assisted mobile residual oil recovery; recipient of 2 US and 3 Canadian patents. *Mailing Add:* 1315 Normandy Rd Golden CO 80401

AYLES, GEORGE BURTON, b Prince Albert, Sask, Nov 28, 45. FISH GENETICS, AQUACULTURE. *Educ:* Univ BC, BSc, 67, MSc, 69; Univ Toronto, PhD(zool), 72. *Prof Exp:* Res scientist, 72-80, regional planning officer, 80-82, dir res, 82-86, DIR RES, FISHERIES HABITAT MGT, CENT ARCTIC REGION, 86- *Concurrent Pos:* Adj prof, dept zool, Univ Man, 77- *Mem:* Aquacult Asn Can; Can Soc Zoologists; Can Genetics Soc; Am Genetics Soc. *Res:* Fish genetics and fish culture; management techniques of rainbow trout culture in small lakes and in tanks using recirculated water and waste heat; quantitative genetics of rainbow trout and Arctic char. *Mailing Add:* Dept Fisheries & Oceans 501 University Crescent Winnipeg MB R3T 2N6 Can

AYLESWORTH, THOMAS GIBBONS, b Valparaiso, Ind, Nov 5, 27; m 49; c 2. ZOOLOGY. *Educ:* Ind Univ, AB, 50, MS, 53; Ohio State Univ, PhD(sci educ), 59. *Prof Exp:* Asst prof sci educ, Mich State Univ, 57-62; lectr sci & sr ed, Current Sci, Wesleyan Univ, 62-65; sr ed sci bks, Doubleday & Co, 65-80; ed in chief, Bison Books Ltd, 80-86; RETIRED. *Concurrent Pos:* Vis fac mem, Ohio State Univ, 62, Wis State Univ, Whitewater, 64, Fairfield Univ, State Univ NY, 80 & Western Conn State Univ, 81-; New Eng ed, Am Biol Teacher, 62-64. *Mem:* Nat Asn Biol Teachers; Nat Asn Sci Writers; Nat Sci Teachers Asn; Nat Asn Res Sci Teaching; NY Acad Sci. *Res:* Critical thinking and its applications in the field of teaching; uses of printed materials in the teaching of science; author of 42 science trade books. *Mailing Add:* 48 Van Rensselaer Stamford CT 06902

AYLING, JUNE E, b US citizen. BIOCHEMISTRY, PHARMACOLOGY. *Educ:* Univ Calif, Berkeley, BS, 63, PhD(biochem), 66. *Prof Exp:* Fel biochem, Univ Calif, 66-67; Max Planck Inst, Munich, 67-69; asst prof, Univ Calif, Los Angeles Sch Med, 69-76; assoc prof biochem, Univ Tex, San Antonio, 76-81; PROF PHARMACOL, UNIV S ALA, MOBILE, 81- *Concurrent Pos:* Fel, Am Cancer Soc, 67-69; prin investr res grant, Nat Cancer Inst, NIH, Dept Health & Human Resources, 74-87, Robert A Welch Found, 78-81, Nat Inst Gen Med Sci, 80-, Nat Inst Neurol & Commun Dis & Stroke, 84- & Ciba Geigy, 85- *Mem:* Am Soc Biol Chemists; Am Chem Soc; AAAS; Am Soc Pharmacol & Exp Therapeut; Soc Neurosci. *Res:* Mechanism of enzyme action; coenzyme catalysis; structure function relationship; physical and chemical properties of enzymes; regulation of enzyme activity; role of tetrahydrobiopterin in regulation of neurotransmitter biosynthesis. *Mailing Add:* Pharmacol Dept Col Med Univ S Ala Mobile AL 36688

AYLOR, DONALD EARL, b Hunt, NY, Dec 22, 40; m 64; c 2. BIOPHYSICS, MICROMETEOROLOGY. *Educ:* State Univ NY, Stony Brook, BES, 64, MES, 66, PhD(mech eng), 70. *Prof Exp:* Asst scientist, 69-75, chief scientist biophysics, dept ecol & climat, 76-83, CHIEF SCIENTIST, DEPT PLANT PATH & ECOL, CONN AGR EXP STA, 84- *Concurrent Pos:* Mem comt aerobiol, Nat Res Coun, 76-80; Guggenheim fel, 80. *Mem:* Am Meteorol Soc; Am Phytopath Soc; fel AAAS. *Res:* Turbulent dispersion of particles and gases in the atmosphere and of solutes in rivers; plant disease epidemiology; noise propagation outdoors; mechanics of plant cells. *Mailing Add:* Dept Plant Path & Ecol PO Box 1106 New Haven CT 06504

AYLOR, JAMES HIRAM, b Charlottesville, Va, May 30, 46; m 73; c 2. ELECTRICAL ENGINEERING, COMPUTER ENGINEERING. *Educ:* Univ Va, BSEE, 68, MSEE, 71, PhD(elec eng), 77. *Prof Exp:* Res engr, 73-78, PROF ELEC ENG, UNIV VA, 78- *Concurrent Pos:* Consult, Gen Elec Co, 74, Alert Commun Corp, & Va Highway & Transp Res Coun, 77-; vis scientist, Fed Sys Div, IBM, 82-83. *Mem:* Inst Elec & Electronics Engrs; Inst Elec & Electronics Engrs Comput Soc; AAAS. *Res:* Computer and microprocessor applications; pattern recognition; rehabilitation engineering; very large scale integration systems; test technology. *Mailing Add:* Dept Elec Eng Univ Va Charlottesville VA 22903-2442

AYNARDI, MARTHA WHITMAN, CELL & MOLECULAR BIOLOGY, GENETICS. *Educ:* Carnegie Mellon Univ, PhD(intermediate filaments), 82. *Prof Exp:* ASST PROF BIOL, PA STATE UNIV, 84- *Mailing Add:* 1512 Concord Rd Wyomissing PA 19610-1106

AYO, DONALD JOSEPH, b Bourg, La, Apr 1, 34; m 58; c 2. AGRONOMY, HORTICULTURE. *Educ:* La State Univ, BS, 56, MS, 58, PhD, 64. *Prof Exp:* Asst prof hort, La State Univ, 58; from instr to prof plant sci, 58-67, head dept agr, 66-68, dean div sci, 68-69, dean div life sci & technol, 69-77, ALCEE FORTIER DISTINGUISHED HONOR PROF PLANT SCI, NICHOLLS STATE UNIV, 67-, VPRES, 71-, PROVOST, 75- *Concurrent Pos:* Asst ed, Nat Asn Col Teachers Am J, 63-65. *Mem:* Am Soc Hort Sci; Am Soc Agron; Soil Sci Soc Am; NY Acad Sci; Am Inst Biol Sci. *Res:* Dormancy and rest periods in decidious plants. *Mailing Add:* Nicholls State Univ Thibodaux LA 70310

AYOUB, CHRISTINE WILLIAMS, b Cincinnati, Ohio, Feb 7, 22; m 50; c 2. MATHEMATICS. *Educ:* Bryn Mawr Col, AB, 42; Radcliffe Col, AM, 43; McGill Univ, MA, 44; Yale Univ, PhD(math), 47. *Prof Exp:* Fel, Inst Advan Study, Off Naval Res, 47-48; instr math, Cornell Univ, 48-51; from instr to assoc prof, 53-69, PROF MATH, PA STATE UNIV, 69- *Concurrent Pos:* Res asst, Radcliffe Col, 51-52; NSF sci fac fel, 66-67; vis prof, Univ Frankfurt, 66-67 & Univ Harwich, 79-80. *Mem:* Am Math Soc; Math Asn Am; Asn Women Mathematicians. *Res:* Theory of groups and normal chains; modern algebra; constructive algebra. *Mailing Add:* 120 Ridge Ave State College PA 16801

AYOUB, ELIA MOUSSA, b Haifa, Palestine, Apr 12, 28; m 54; c 5. PEDIATRICS. *Educ:* Am Univ Beirut, BS, 49, MD, 53; Am Bd Pediat, dipl. *Prof Exp:* Intern, Am Univ Beirut, 53; physician in-chg, Qaisumah Hosp, Beirut, 53-56; resident pediat, Univ Wis, 56-57; from instr to prof, Univ Minn, 59-69; PROF PEDIAT, COL MED, UNIV FLA, 69- *Concurrent Pos:* Fel, Col Med, Univ Minn, 57-58, res fel, 58-59; Helen Hay Whitney res fel, 60-63; guest investr, Rockefeller Univ, 63-65; estab investr, Helen Hay Whitney Res Found, 63-68. *Mem:* AAAS; Am Soc Microbiol; Soc Pediat Res; Am Asn Immunol; Am Acad Pediat. *Mailing Add:* Dept Pediat Infect Dis Univ Fla Col Med Box J 296 JHMHC Gainesville FL 32610

AYOUB, MAHMOUD A, b Cairo, Egypt, Jan 1, 42; US citizen. OCCUPATIONAL SAFETY & HEALTH. *Educ:* Cairo Univ, BS, 64; Tex Tech Univ, MS, 69, PhD(indust eng), 70. *Prof Exp:* Field engr, Egypt, 64-65; instr civil eng, Cairo Univ, 65-66; res asst indust eng, Tex Tech Univ, 67-69; asst prof, 71-75, assoc prof, 75-80, PROF INDUST ENG, NC STATE UNIV, 80- *Concurrent Pos:* Assoc dir syst safety eng, Nat Inst Occup Safety & Health, 71-; adj asst prof environ sci, Univ NC, Chapel Hill, 73-; prin investr, NIH biomed sci grant, 74-75, NSF, 74-76 & NC Dept Transp, 75-76; participant, Pres Comt on Employ of Handicapped, 79; consult, Reynolds Metals Co, 79-80 & Hanes Knitwear, 80-; dir develop support mat for training naval aviation personnel, US Navy, 80- *Honors & Awards:* Phil Carrol Award, Am Inst Indust Engrs, 80. *Mem:* Am Inst Indust Engrs; Human Factors Soc; Ergonomics Res Soc. *Res:* Evaluating manual lifting hazards and developing recommended solutions; defining an acceptable solution to the tendonitis problem; simulation modeling. *Mailing Add:* Indust Sci Box 7906 NC State Univ Main Campus Raleigh NC 27695-7906

AYOUB, MOHAMED MOHAMED, b Tanta, Egypt, Feb 17, 31; US citizen; m 69; c 2. INDUSTRIAL ENGINEERING. *Educ:* Cairo Univ, BS, 53; Univ Iowa, MS, 55, PhD(indust eng), 64. *Prof Exp:* Indust engr, Maytag Co, Iowa, 56-58, planning engr, 58-61; from asst prof to prof indust eng, 61-69, prof indust eng & statist, 69-76, prof, 76-78, HORN PROF INDUST ENG & BIOMED ENG, TEX TECH UNIV, 78- *Honors & Awards:* Paul Fitts Award, Human Factors Soc, 75; David Baker Research Award, Am Inst Indust Engrs, 81. *Mem:* Am Inst Indust Eng; Am Indust Hyg Asn; Human Factors Soc; Ergonomics Res Soc; Am Soc Eng Educ; Sigma Xi. *Res:* Ergonomics, with particular interest in occupational biomechanics and work physiology; contributions of these areas to occupational safety and health. *Mailing Add:* Dept Indust Eng Tex Tech Univ Lubbock TX 79409

AYOUB, RAYMOND G DIMITRI, b Sherbrooke, Can, Jan 2, 23; m 50; c 2. MATHEMATICS. *Educ:* McGill Univ, BSc, 43, MSc, 46; Univ Ill, PhD(math), 50. *Prof Exp:* Lectr math, McGill Univ, 47-49; Pierce instr, Harvard Univ, 50-52; from asst prof to prof, 66-84, EMER PROF MATH, PA STATE UNIV, 84- *Mem:* Am Math Soc; Math Asn Am. *Res:* Analytic theory of numbers; theory of algebraic numbers. *Mailing Add:* Pa State Univ University Park PA 16802

AYRAL-KALOUSTIAN, SEMIRAMIS, b Istanbul, Turkey; m 69. ORGANIC SYNTHESIS, ORGANIC PHOTOCHEMISTRY. *Educ:* Am Univ, Beirut, BS, 66; Univ Calif, Los Angeles, PhD(org chem), 73. *Prof Exp:* Asst prof org chem, Haigazian Col, Beirut, 72-73; asst prof chem & thermodynamics, Lebanese Univ, 73-74; res assoc, Rockefeller Univ, 75-78, asst prof org chem, 78-82; sr res chemist, 82-88, PROJ LEADER, AM CYANAMID CO, 88- *Concurrent Pos:* Teaching assoc, Univ Calif, Los Angeles, 66-69, res assoc, 69-71; res asst, Imp Col Sci & Technol, London, 72. *Mem:* Am Chem Soc. *Res:* Synthesis, photolysis and thermolysis of organic compounds; investigations in the photochemistry of epoxy and cyclopropyl ketones; synthesis of fructose bisphophate analogs as potential antidiabetic agents; synthesis of immunostimulants and peptide mimics. *Mailing Add:* Am Cyanamid Co Med Res Div Pearl River NY 10965

AYRES, DAVID SMITH, b Boston, Mass, June 14, 39; m 65. EXPERIMENTAL HIGH ENERGY PHYSICS. *Educ:* Williams Col, BA, 61; Univ Calif, Berkeley, MA, 63, PhD(physics), 68. *Prof Exp:* US Peace Corps lectr physics, Univ Nigeria, 63-65; res asst physics, Univ Calif, Berkeley, 65-68, res assoc, Lawrence Berkeley Lab, 68-69; res assoc, 69-70, from asst physicist to physicist, 71-84, SR PHYSICIST, ARGONNE NAT LAB, 84- *Mem:* Fel Am Phys Soc; AAAS. *Res:* Planning, execution, and analysis of experiments in elementary particle physics, with emphasis on hadron-hadron colliding beams at very high energies, nucleon decay, and other underground experiments. *Mailing Add:* Argonne Nat Lab Bldg 362 Argonne IL 60439

AYRES, GILBERT HAVEN, b Upland, Ind, Aug 29, 04; m 26; c 2. ANALYTICAL CHEMISTRY. *Educ:* Taylor Univ, AB, 25; Univ Wis, PhD(chem), 30. *Prof Exp:* Instr chem, Taylor Univ, 25-27; asst, Univ Wis, 27-30, instr, 30-31; from asst prof to assoc prof, Smith Col, 31-47; from assoc prof to prof, 47-74, chmn dept chem, 50-52, grad adv, 57-61 & 65-69, EMER PROF CHEM, UNIV TEX, AUSTIN, 74- *Mem:* AAAS; Am Chem Soc. *Res:* Spectrophotometric methods of analysis; analytical chemistry of platinum elements. *Mailing Add:* 3307 Perry Lane Austin TX 78731-5330

AYRES, JAMES WALTER, b Boise, Idaho, Apr 14, 42; m 64; c 2. PHARMACOKINETICS, BIOPHARMACEUTICS. *Educ:* Idaho State Univ, BS, 65; Univ Kans, PhD(org med chem), 70. *Prof Exp:* PROF PHARMACOKINETICS & BIOPHARMACEUT, ORE STATE UNIV, 70- *Concurrent Pos:* Scholar pharmacokinetics & biopharmaceut, Univ Mich, 77; scholar pharmaceut, Glaxo, UK, 90. *Honors & Awards:* Inst Food Technol Indust Achievement Award, 82. *Mem:* Acad Pharmaceut Sci; fel Am Pharmaceut Asn; Am Asn Cols Pharm; fel Am Asn Pharmaceut Sci. *Res:* Drug product formulation and evaluation; biopharmaceutics; pharmacokinetics. *Mailing Add:* Ore State Univ Sch Pharm Corvallis OR 97331

AYRES, JOHN CLIFTON, b Beckemeyer, Ill, Apr 17, 13; m 35; c 2. FOOD SCIENCE. *Educ:* Ill State Normal Univ, BEd, 36; Univ Ill, MS, 38, PhD(microbiol), 42. *Prof Exp:* Teacher, Pub Schs, Ill, 31-34 & 36-41; res microbiologist, W S Merrell Co, Ohio, 42-43; food microbiologist, Res Div, Gen Mills, Inc, Minn, 43-46; from asst prof to prof bact, Iowa State Univ, 46-54, prof food technol, 54-67; fac mem, 67, head dept & chmn div, 67-73, prof food sci, 73-75, D W BROOKS DISTINGUISHED PROF AGR, FOOD SCI, UNIV GA, 75- *Concurrent Pos:* Mem adv comt, Qm Food & Container Inst Armed Forces, 51-62 & Iowa Agr Adjust Ctr, 56-57; assoc ed, Food Technol, 54-57 & Appl Microbiol, 58-63; mem, NIH Toxicol Study Sect, 60-64; mem food technol comt, Nat Acad Sci-Nat Res Coun, 62-70, mem food protection comt & chmn subcomt food microbiol, 65-, mem microbiol subcomt mil personnel supplies & comt Salmonella, Div Biol & Agr; US mem comt food microbiol & Hyg Sect, Int Asn Microbiol Socs, 64-66; mem environ health training study sect, USPHS, 66-70; Inst Food Technol rep, Int Comt Food Sci & Technol; mem, Citizens' Comn Sci, Law & Food Supply, 73-76; US deleg, Int Union Food Sci & Technol, 77-82. *Honors & Awards:* Achievement Award, Inst Am Poultry Indust, 66; Nicholas Appert Medalist, Inst Food Technol, 72; Inst Food Tech Int Award, 80. *Mem:* Am Soc Microbiol; fel Inst Food Technol (pres, 76-77); Soc Appl Bact; Am Meat Sci Asn; Inst Am Poultry Indust. *Res:* Microbiology of meats, poultry and eggs; ecology of food spoilage microorganisms; psychrophilic flora of flesh foods; aflatoxin and other mycotoxins in foods. *Mailing Add:* 970 Old Powder Springs Mableton GA 30606

AYRES, KATHLEEN N, CELL & DEVELOPMENTAL BIOLOGY. *Educ:* Univ Chicago, PhD(cell & develop biol), 76. *Prof Exp:* Peace Corps, 62-65; fel, Ohio State Univ, 77-78; res assoc, Univ Chicago, 78-80 & Univ Ill Med Ctr, 82-85; CONSULT, 85- *Mailing Add:* 6708 Meadowcrest Dr Downers Grove IL 60516

AYRES, ROBERT ALLEN, b Dallas, Tex, Jan 3, 46. SHEET METAL FORMABILITY. *Educ:* Colo Sch Mines, BS, 67; Univ Minn, MS, 69; Mich Technol Univ, PhD(metall eng), 74. *Prof Exp:* Phys Sci Asst, Redstone Arsenal, 69-71; sr res scientist, Gen Motors Res Labs, 74-80, staff res scientist, 80-85. *Concurrent Pos:* Group leader, Metal Deformation Group, GM Labs, 78-85. *Honors & Awards:* Arch T Colwell Award, Soc Automotive Engr, Inc, 80. *Mem:* Am Soc Metals. *Res:* Cleavage fracture; dislocation dynamics; high purity metals; thermo-mechanical strengthening of steel; sheet metal formability. *Mailing Add:* Gen Motors Corp CPC Hq 3001 Van Dyke Ave Rm 232-25 Warren MI 48090-9020

AYRES, ROBERT U, b Plainfield, NJ, June 29, 32; m 54; c 1. ENGINEERING, PUBLIC POLICY. *Educ:* Univ Chicago, BS, 54; Univ Md, MS, 56; Univ London, PhD(math physics), 58. *Prof Exp:* Res assoc physics, Univ Md, 58-60, G C Dewey Corp, New York, 60-61 & Grad Sch Sci, Yeshiva Univ, 61-62; mem res staff, Hudson Inst, 62-67; vis scholar, Resources for the Future, 67-68; vpres, Int Res & Technol Corp, 69-75, chmn, 76; vpres & dir, Delta Res Corp, 76-78; chmn, Vari-Flex Corp, 69-87; dep prog leader, Int Inst Appl Sci Anal, 86-90; PROF ENG & PUB POLICY, CARNEGIE-MELLON UNIV, 79- *Concurrent Pos:* Mem, Hwy Res Bd

Comt on New Transp Systs & Technol, 71-, Nat Acad Sci Comt Technol & Water, Comt Alternatives for Reduction Chlorofluorocarbon Emissions, Nat Mat Adv Bd Comt Tech Aspects Strategi & Critical Mat; prin consult, UN Statist Off, 74-78. *Mem:* Soc Hist Technol; Am Econ Asn; fel AAAS; Asn Environ Resource Economists. *Res:* Resource and environmental economics and technology; technology and economics systems modelling and forecasting; robotics and computer integrated manufacturing. *Mailing Add:* Dept Eng & Pub Policy Schenley Park Pittsburgh PA 15213

AYRES, STEPHEN MCCLINTOCK, b Elizabeth, NJ, Oct 29, 29; m 55; c 3. INTERNAL MEDICINE. *Educ:* Gettysburg Col, BA, 51; Cornell Univ, MD, 55. *Prof Exp:* Dir cardiopulmonary lab, St Vincent's Hosp, New York, 63-73; prof med, Med Sch Univ Mass, 73-75; prof int med & chmn dept, St Louis Univ Sch Med, 75-85; DEAN, MED COL VA, 85- *Concurrent Pos:* Chmn, Pulmonary Dis Adv Comt, 76 & 79-80. *Honors & Awards:* Phoenix Award, Found Critical Care Med, 83. *Mem:* Soc Critical Care Med (pres, 79); Am Lung Asn; Am Soc Clin Invest; Asn Am Prof; Am Col Cardiol. *Mailing Add:* Va Commonwealth Univ Sch Med MCV Sta Box 565 Richmond VA 23298

AYRES, WESLEY P, b Los Angeles, Calif, Sept 26, 24; m 48; c 2. PHYSICS. *Educ:* Fresno State Col, BS, 51; Stanford Univ, MS, 53, PhD(physics), 54. *Prof Exp:* Sr engr, Electronic Defense Lab, Sylvania Elec Prod, Gen Tel & Electronics Corp, 54-56; vpres eng, Melabs, Inc, 56-69; dir systs eng, Bus Equip Div, SCM Corp, 69-73, mgr consult, 73-75; INDEPENDENT MGT CONSULT, 75- *Mem:* Am Phys Soc; Inst Elec & Electronics Engrs; Sigma Xi. *Res:* Application of solid state physics at microwave frequencies to achieve new devices; study and design of microwave systems; study and development of systems engineering techniques to office equipment research and development. *Mailing Add:* 12760 Camino Medio Ln Los Altos Hills CA 94022

AYROUD, ABDUL-MEJID, b Aleppo, Syria, Nov 24, 26; nat Can; m 54; c 3. CHEMISTRY. *Educ:* Univ Lille, ChE, 49; French Sch Paper Making, Papermaking E, 50; Univ Grenoble, PhD(chem), 53. *Prof Exp:* Head res dept, French Sch Paper Making, 53-57; chem engr, Pulp & Paper Res Inst Can, 57-58; sr res chemist, Res Ctr, Consol Paper Corp, Ltd, 58-60, leader chem pulping, 60-66, sect chief, Res & Develop Ctr, 66-67, mgr chem pulping & bleaching, Res & Develop Ctr, Consol-Bathurst Ltd, 67-71, asst dir res & tech serv, 71-84, DIR RES, CONSUL-BATHURST LTD, 84- *Concurrent Pos:* Lectr, Univ Laval & Univ Que. *Mem:* Tech Asn Pulp & Paper Indust; Can Pulp & Paper Asn; Res Mgt Asn. *Res:* Applied chemistry; wood chemistry; pulp and paper technology; pollution abatement; administration. *Mailing Add:* 31 Dr Rigby Lachute PQ J8H 4A2 Can

AYUSO, KATHARINE, b New York, NY, June 25, 18. FOOD SCIENCE. *Educ:* St Lawrence Univ, BS, 40. *Prof Exp:* Asst chief chemist, Great Atlantic & Pac Tea Co, 42-65; assoc res dir, Foster D Snell, Inc, Booz Allen & Hamilton, 65-73; dir food sci, Rosner-Hixson Labs, 73-78; dir, Food Prod Develop, Herbert V Shuster, Inc, 78-85; RETIRED. *Mem:* Am Inst Chemists; Am Asn Cereal Chemists; Inst Food Technologists; Am Chem Soc; Am Asn Candy Technologists. *Res:* New product development, product reformulation, utilization of by-products; development of foods for the future; consumer evaluation studies. *Mailing Add:* Eight Plympton St Cambridge MA 02138-6606

AYVAZIAN, L FRED, b Ordu, Turkey, Oct 3, 19; US citizen; m 47; c 3. INTERNAL MEDICINE. *Educ:* Columbia Col, BA, 39; NY Univ, MD, 43. *Prof Exp:* Fel med, Sch Med, NY Univ, 44-45; res fel, Thorndike Med Lab, Boston City Hosp, Mass, 47-48; fel, Harvard Med Sch, 47-48; investr sterilization blood plasma, Goldwater Mem Hosp, Welfare Island, NY, 53-54; asst chief med serv, Vet Admin Hosp, NY, 54-61; med dir, Will Rogers Hosp & O'Donnell Mem Res Labs, 61-71; CHIEF CHEST SERV, VET ADMIN HOSP, 71-; PROF MED, NJ MED SCH, NEWARK, 71- *Concurrent Pos:* Assoc clin prof, Sch Med, NY Univ, 61-71, prof clin med, Med Ctr, 77- *Mem:* Harvey Soc; NY Acad Sci; fel Am Col Physicians; Am Thoracic Soc; Fedn Clin Res. *Res:* Tuberculosis and pulmonary diseases; sterilization of blood plasma; uric acid metabolism; hematologic changes in arterial oxygen desaturation states secondary to pulmonary disease; big ACTH in lung cancer. *Mailing Add:* Vet Admin Med Ctr East Orange NJ 07019

AYYAGARI, L RAO, b Mogallu, India, Dec 30, 40; US citizen; m 72; c 2. DNA REPLICATION, BIOLOGY EDUCATION. *Educ:* Bombay Univ, India, BSc, 60, MSc, 65; Loyola Univ Chicago, MS, 69, PhD(biochem), 72. *Prof Exp:* Postdoctoral fel, Univ Calif, Davis, 72-77; res scientist, Mich Cancer Found, 77-79; instr chem, Univ Mich, Dearborn, 79-82; from asst prof to assoc prof, 83-91, PROF BIOL, LINDENWOOD COL, ST CHARLES, 91- *Concurrent Pos:* Vis res fel, Dept Biochem & Biophys, Wash Univ Med Sch, 90- *Mem:* AAAS; Am Arachnological Soc; Int Soc Chem Ecol; Nat Asn Biol Teachers. *Res:* Pheromones in invertebrates; DNA replication, especially fidelity in yeasts; radioactive waste disposal. *Mailing Add:* 16084 Meadow Oak Chesterfield MO 63017

AYYANGAR, KOMANDURI M, b Andhra, India, Jan 13, 40; m 70; c 2. PHYSICS, MEDICAL PHYSICS. *Educ:* Andhra Univ, India, BSc, 58, MSc, 60, PhD(nuclear physics), 65. *Prof Exp:* Res fel nuclear physics, Andhra Univ, India, 60-66; res fel med physics, Physics Res Lab, Mass Gen Hosp, Boston, MA, 66-69; sci officer, Div of Radiol Protection, Bhabha Atomic Res Ctr, Bombay, India, 70-74; trainee, 75, from asst prof to assoc prof, 76-90, PROF MED PHYSICS, DEPT RADIATION THER & NUCLEAR MED, THOMAS JEFFERSON UNIV HOSP, PHILADELPHIA, 90- *Concurrent Pos:* Physicist radiation ther, Bryn Mawr Hosp, Pa, 76-77, Burlington County Med Hosp, NJ, 77-81. *Mem:* Am Asn Physicists in Med; Sigma Xi; Am Col Med Physics. *Res:* Nuclear medicine; radiation physics; radiation therapy physics; nuclear cardiology; radiation dosimetry. *Mailing Add:* Dept Radiation Ther & Nuclear Med Thomas Jefferson Univ Hosp 1025 Walnut St Philadelphia PA 19107

AYYASWAMY, PORTONOVO S, b Bangalore, India, Mar 21, 42; US citizen; m 75; c 2. APPLIED MECHANICS, MECHANICAL ENGINEERING. *Educ:* Univ Mysore, BE, 62; Columbia Univ, MS, 65, ME, 67; Univ Calif, Los Angeles, PhD(eng), 71. *Prof Exp:* Postdoctoral fel geophysics, Inst Geophys & Planetary Phys, 71-72, postdoctoral scholar mech eng, Univ Calif, Los Angeles, 72-74; from asst prof to assoc prof, 74-87, PROF MECH ENG, UNIV PA, 87. *Concurrent Pos:* Consult, Nat Air Oil Burner Co, Inc, Philadelphia, 76- & Combustion Unlimited Inc, Jenkintown, Pa, 78-; electric power res inst grant, 75-77; NSF grant, 78; NIH grants, 81-, Battelle, NC, 84-; US deleg, Nuclear reactor safety to People's Repub of China, 85; key note lectr, Nat Heat & Mass Transfer Conf, India, 85; invited lectr Condensation Theory, Cavendish Labs, Univ Cambridge, Eng, 86; invited lectr Condensation Theory, Royal Inst Technol, Stockholm, Sweden, 86; invited lectr Condensation Theory, Danish Ctr Appl Math & Mech, Lyngby, Denmark, 86; mem, eval panel for award-NSF, 87, mem, panelist Workshop Combustion, Sandia Nat Labs, 88. *Mem:* Am Soc Mech Engrs; Am Nuclear Soc; Radiation Res Soc. *Res:* Stability of fluid motion, natural convection flow and stability, two-phase flows condensation, evaporation, combustion, plasma heat transfer and stability; nuclear reactor safety analysis, solar energy heat collection and utilization; bioheat transfer. *Mailing Add:* Dept Mech Eng SEAS Univ Pa Philadelphia PA 19104

AZAD, ABDU F(ARHANG), b Iran, Nov 9, 42; US citizen; m 71; c 1. PARASITOLOGY, TROPICAL MEDICINE. *Educ:* Univ Tehran, DrPhar, 66, MPH, 69; Johns Hopkins Univ, PhD(parasitol), 76. *Prof Exp:* Fel, 76-78, from asst prof to assoc prof, 78-89, PROF MICROBIOL, DEPT MICROBIOL, SCH MED, UNIV MD, 90- *Concurrent Pos:* Team-leader, Iran-WHO, 68-69; vis scientist USSR, Nat Acad Sci, 71; mem field team, Ethiopia, Burma, Pakistan & Iran, 76-78; consult, WHO, 76-78 & Med Serv Consult, 79-; mem expert comt rodent-borne dis, WHO, 73-77; mem NIH Alaska-hemorraghic fever proj, Univ Md, 81. *Mem:* AAAS; Am Soc Trop Med & Hyg. *Res:* Vector biology; transmission dynamics of arthropods-borne diseases; host immunological resposes to insect bitos; host-parasite interactions; reactivation of latent infection by artificial stress. *Mailing Add:* Dept Microbiol & Immunol Sch Med Univ Md 660 W Redwood St Baltimore MD 21201

AZAD, HARDAM SINGH, b Abohar, India, Aug 10, 38; m 65; c 2. ENVIRONMENTAL ENGINEERING, CHEMICAL ENGINEERING. *Educ:* Punjab Univ, BA, 60; Kans State Univ, BS, 64; Univ Mo-Columbia, MS, 65; Univ Mich, PhD(environ eng), 68. *Prof Exp:* Water resources engr, Irrigation Dept, Punjab Govt, 58-62; res asst, Univ Mo, 64-65; teaching fel & res assoc, Univ Mich, 65-68; supvr pollution control res & develop lab & mgr water pollution control plant, Abbott Labs, 68-70; sr res group leader, Monsanto Biodize Systs, Inc, 70-71; indust develop mgr, Monsanto Envirochem Systs Inc, 71-72; mgr environ control, McKee Corp, 72-75; dir environ projs, 75-77, dir eng, 77-81, GEN MGR, HOUSTON ENG DIV, NUS CORP, 81- *Mem:* Water Pollution Control Fedn; Am Inst Chem Engrs. *Res:* Biochemical wastewater treatment processes; activated carbon adsorption; chemical oxidations; ecological and eutrophication investigation; water renovation and reuse; recovery and recycling in industrial pollution control. *Mailing Add:* 8730 Memorial Dr Houston TX 77024

AZAM, FAROOQ, b Lahore, Pakistan. MICROBIAL ECOLOGY. *Educ:* Panjab Univ, BSc, 61, MSc, 63; Czech Acad Sci, Prague, CSc(microbiol), 68. *Prof Exp:* Res assoc anal biochem, West Regional Labs, Coun Sci & Indust Res, Pakistan, 63-65; fel biochem, State Univ NY Stony Brook, 68-69; res biologist, Scripps Inst Oceanog, Univ Calif, San Diego, 69-73, res biologist marine microbial ecol, Inst Marine Resources, PROF MARINE BIOL, SCRIPPS INST OCEANOG, UNIV CALIF, LA JOLLA, 88- *Honors & Awards:* Rosenstiel Award, Marine Sci, 84. *Mem:* AAAS; Am Soc Microbiol; Am Soc Limnol Oceanog. *Res:* Biochemistry; role of bacteria in marine food webs; transport of sugars and amino acids in marine bacteria; metabolic regulation in marine bacteria. *Mailing Add:* Scripps Inst Oceanog Univ Calif La Jolla CA 92093

AZAR, HENRY A, b Heliopolis, Egypt, Dec 21, 27; US citizen; m 60; c 2. PATHOLOGY. *Educ:* Am Univ Beirut, BA, 48, MD, 52; Am Bd Path, dipl, cert anat path, 58, cert clin path, 74, cert hematol, 76. *Prof Exp:* Asst prof path, Am Univ Beirut, 58-60; from asst prof to assoc prof, Col Physicians & Surgeons, Columbia Univ, 60-70; prof path & dir surg path, Univ Kans Med Ctr, Kansas City, 70-72; chief lab serv, 72-83, CHIEF ANAT PATH, VET JAMES A HALEY HOSP, TAMPA, 83-; PROF PATH, COL MED, UNIV S FLA, 72- *Concurrent Pos:* Assoc dir path, Francis Delafield Hosp, 68-70; assoc attend pathologist, Presby Hosp, 68-70. *Mem:* Col Am Path; Harvey Soc; Int Acad Path; Arthur Purdy Stout Soc Surg Path. *Res:* Lymphomas; myeloma; human tumor xenografts. *Mailing Add:* Dept Path Univ SFla Col Med Tampa FL 33620

AZAR, JAMAL J, b Tripoli, Lebanon, sept 19, 37; US citizen; m 62; c 2. STRUCTURAL ANALYSIS, PETROLEUM ENGINEERING. *Educ:* Univ Okla, BS, 61, MS, 62, PhD(eng sci), 65. *Prof Exp:* Sr res engr, Okla Univ Res Inst, 61-65; asst prof struct dynamics & vehicle struct, 65-68, assoc prof mech & aerospace eng, 68-73, PROF PETROLEUM ENG & DIR DRILLING RES PROJS, UNIV TULSA, 74- *Concurrent Pos:* Consult, var companies & govt orgns; int lectr Drilling Eng. *Mem:* Soc Petroleum Engrs. *Res:* Drilling-bit mechanics; Wellsore hydraulics; drillstring mechanics; Wellsore mechanics. *Mailing Add:* Dept Petroleum Eng Univ Tulsa Tulsa OK 74104

AZAR, MIGUEL M, b Cordoba, Arg, Oct 21, 36; m 60; c 6. IMMUNOLOGY, PATHOLOGY. *Educ:* Nat Col Dean Funes, Arg, BA, 53; Cordoba Nat Univ, MD, 58; Univ Tenn, PhD(path), 65. *Prof Exp:* Assoc physician, Children's Hosp, Cordoba, 58-60; intern, Mercy Hosp, Des Moines, Iowa, 60-61; resident pediatrician, Med Units, Univ Tenn, 61-63, res assoc, 66-67; asst prof lab med & dir blood bank, 69-70, assoc prof lab med & dir clin lab, 70-75, PROF LAB MED, DIR LABS & DIR GRAD STUDIES LAB MED & PATH, VET ADMIN HOSPS, SCH MED, UNIV MINN, MINNEAPOLIS, 75- *Concurrent Pos:* Nat Inst Allergy & Infectious Dis spec fel immunol, Sch Med, Univ Minn, Minneapolis, 67-69; vis prof path, Harvard Med Sch, 87-89; USA rep, Allergy & Immunol Divs, Asn Latin Am Pediat. *Mem:* Am Asn Immunol; Soc Exp Biol & Med; Am Soc Exp Path; Asn Latin Am Pediat. *Res:* Pediatric pathology; immunopathology. *Mailing Add:* Vet Admin Hosp Lab Med Minneapolis MN 55417-2300

AZARI, PARVIZ, b Baku, Russia, Feb 3, 30; US citizen; m 55; c 5. BIOCHEMISTRY. *Educ:* Univ Calif, AB, 55; Univ Nebr, MS, 58, PhD(chem), 61. *Prof Exp:* Asst prof chem & physiol, 63-70, assoc prof biochem, 70-75, PROF BIOCHEM, COLO STATE UNIV, 75- *Concurrent Pos:* USPH fel, 61-63, grant, 64- *Mem:* Am Soc Biol Chem; Sigma Xi. *Res:* Structure and biological function of iron-transferrins; biochemistry of cataract. *Mailing Add:* 1825 Essex Dr Ft Collins CO 80521

AZARNOFF, DANIEL LESTER, b Brooklyn, NY, Aug 4, 26; m 51; c 3. CLINICAL PHARMACOLOGY. *Educ:* Rutgers Univ, BS, 47, MS, 48; Univ Kans, MD, 55. *Prof Exp:* Res assoc, Univ Kans, 49-52, intern, 55-56; asst prof med, Sch Med, St Louis Univ, 60-62; from asst prof med to assoc prof med & dir clin pharmacol study unit, 64-68, prof med & pharmacol, 68-73, distinguished prof med & pharmacol, Med Ctr, Univ Kans, 73-78; sr vpres, 78-79, pres res & develop, G D Searle & Co, 79-85; PROF PATH & CLIN PROF PHARMACOL, NORTHWESTERN UNIV MED SCH, 78-; PRES, D L AZARNOFF ASSOC, 86- *Concurrent Pos:* USPHS res fel med, Univ Kans, 56-58; Nat Inst Neurol Dis & Blindness spec trainee, Sch Med, Wash Univ, 58-60; Markle scholar, 62-; Burroughs-Wellcome scholar clin pharmacol, 64-69; Fulbright scholar, Karolinska Inst, Sweden, 68; mem drug res bd, Nat Res Coun, mem adv coun & pharmacol-toxicol prog adv comt, Nat Inst Gen Med Sci & mem comt on revision, US Pharmacopoeia; ed, NY Drug Interactions & Year Bk Drug Ther; clin prof med, Univ Kans Med Sch. *Mem:* Inst Med Nat Acad Sci; fel Am Col Physicians; Brit Pharmacol Soc; Am Soc Clin Nutrit; Am Soc Pharmacol & Exp Therapeut; fel NY Acad Sci; Am Asn Pharmaceut Scientists. *Res:* Lipid metabolism; drug development. *Mailing Add:* DL Azarnoff Assoc 400 Oyster Blvd Suite 325 South San Francisco CA 94080

AZAROFF, LEONID VLADIMIROVICH, b Moscow, Russia, June 19, 26; nat US; m 46, 71. X-RAY CRYSTALLOGRAPHY. *Educ:* Tufts Univ, BS, 48; Mass Inst Technol, PhD(crystallog), 54. *Prof Exp:* Asst res engr, Raytheon Mfg Co, 41-44; asst, Mass Inst Technol, 50-52; assoc, 52-53; assoc physicist, Armour Res Found, 53-55, res physicist, 55-56, sr scientist, 56-57; from assoc prof to prof metall eng, Ill Inst Technol, 57-66; PROF PHYSICS & DIR INST MAT SCI, UNIV CONN, 66- *Concurrent Pos:* Guest physicist, Brookhaven Nat Lab, 61-64. *Mem:* Am Phys Soc; Am Crystallog Asn; Mineral Soc Am; Am Inst Min, Metall & Petrol Eng; Am Soc Eng Educ. *Res:* Physics of metals; x-ray diffraction; soft x-ray spectroscopy; liquid-crystal polymers. *Mailing Add:* Dept Physics Univ Conn Main Campus 4-46 2152 Hillside Rd Storrs CT 06268

AZBEL, MARK, b Poltava, USSR, May 12, 32; Israel citizen; m 67; c 2. RANDOM SYSTEMS. *Educ:* Kharkov Univ, USSR, MA, 53, PhD, 55; Inst Phys Problems, Moscow DSc(physics), 57. *Prof Exp:* Prof, physics, Moscow State Univ, Union Soviet Socialist Repub, 65-72; prof, physics, 73-87, CHMN, THEORET PHYSICS, TEL-AVIV, UNIV ISRAEL, 88. *Concurrent Pos:* Consult, AT&T Bell Labs, IBM, 78-; chmn, theor physics, Univ Lousanne, Switz, 81-82; adj prof, Univ Pa, Philadelphia, 80- *Honors & Awards:* Lomonosov Prize; Landau Prize; Christopher Award. *Mem:* Israeli Phys Soc; fel Am Phys Soc; Europ Phys Soc; NY Acad Sci. *Res:* Predictions of cyclotron resonance in metals of the Kantor set-type spectrum; theory of the helix coil transition in DNA and of resonance tunneling in random systems. *Mailing Add:* Sch Physics & Astron Tel-Aviv Univ Tel-Aviv Israel

AZBELL, WILLIAM, b Manito, Ill, Jan 1, 06; m; c 3. PHYSICS. *Educ:* Ill State Norm Univ, BS, 33; Univ Ill, AM, 34. *Prof Exp:* Pub sch instr, Ill, 34-41; instr physics, Exten Div, Univ Ill, 41-43, Army Specialized Training Prog, Ball State Teachers Col, 43-44, Navy V-12, DePauw Univ, 44-45 & Univ Ill, 45-46; asst prof, Bradley Univ, 46-52; ASSOC PROF PHYSICS & HEAD DEPT, WARTBURG COL, 52- *Mem:* Am Asn Physics Teachers. *Res:* Atomic physics. *Mailing Add:* 502 Fourth St SW Waverly IA 50677

AZEN, EDWIN ALLAN, b Pittsburgh, Pa, March 13, 31; c 4. HEMATOLOGY, MEDICINE. *Educ:* Univ Chicago, BA, 51; Univ Pittsburgh, MD, 55. *Prof Exp:* Asst prof, 63-69, assoc prof, 69-74, PROF MED, UNIV WIS-MADISON, 74-, PROF MED GENETICS, 81- *Mem:* AAAS; Am Fedn Clin Res; Am Soc Hematol; Am Soc Human Genetics. *Res:* Genetic research on human salivary proteins; genetic polymorphisms of salivary proteins using electrophoretic and recombinant DNA techniques. *Mailing Add:* Dept Med Hemat Univ Wis Hosp Madison WI 53706

AZHAR, SALMAN, REPRODUCTIVE ENDOCRINOLOGY, HORMONE METABOLISM. *Educ:* La, Kampur Univ, India, PhD(biochem), 73. *Prof Exp:* BIOCHEMIST, VET ADMIN MED CTR, 80- *Mailing Add:* GRECC 182B Vet Admin Hosp 3801 Miranda Ave Palo Alto CA 94304-9991

AZIZ, KHALID, b Pakistan, Sept 29, 36; Can citizen; m 62; c 2. CHEMICAL ENGINEERING, PETROLEUM ENGINEERING. *Educ:* Univ Mich, BSE, 55; Univ Alta, BSc, 58, MSc, 61; Rice Univ, PhD(chem eng), 66. *Prof Exp:* Jr design eng, Massey-Ferguson, 55-56; distribution eng, 58-59, Karachi Gas Co, chief eng, 62-63; from instr to asst prof petrol eng, Univ Alta, 60-62; from asst prof to prof chem eng, Univ Calgary, 65-82; assoc dean res, Sch Earth Sci, 83-86, PROF PETROL ENG, STANFORD UNIV, 82-, CHMN PETROL ENG, 86- *Concurrent Pos:* Vis scientist, Inst Francais Petrol, 70; vis scholar, Stanford Univ, 71; Killam resident fel, Univ Calgary, 77; mem, US Nat Comt World Petrol Cong, 83-; dir, Golden Gate Sect Soc Petrol Eng, 86- *Honors & Awards:* Gold Medal, Asn Prof Engrs Alta, 58; Cedric K Ferguson Cert Award, Soc Petrol Engrs, 79; Reservoir Eng Award, Soc Petrol Eng, 87. *Mem:* Fel Chem Inst Can; Soc Petrol Engrs; Sigma Xi; Am Inst Chem Engrs; Soc Indust & Appl Math. *Res:* Multiphase flow in reservoirs; pipeline transportation of oil and gas; gas technology; reservoir simulation. *Mailing Add:* 112 Peter Coutts Circle Stanford CA 94305-2170

AZIZ, PHILIP MICHAEL, b Toronto, Ont, Mar 12, 24. PHYSICAL CHEMISTRY. *Educ:* Univ Toronto, BASc, 46, MA, 47, PhD(phys chem), 50. *Prof Exp:* Head theoret corrosion sect, Aluminum Labs Ltd, Can, 49-51; res assoc, Inst Study Metals, Univ Chicago, 52; head theoret corrosion sect, Aluminum Labs, Ltd, Can, 53-59; mgr opers res & tech comput div, Systs Dept, Aluminum Co Can, Ltd, 59-70; corp systs mgr, Alcan Aluminum Corp, 70-78, chief systs officer, Alcan Aluminum Ltd, 78-85; RETIRED. *Mem:* Fel Chem Inst Can. *Res:* Mechanisms of the corrosion and oxidation of aluminum and its alloys; operations research; technical computing. *Mailing Add:* 154 Cheverie St Oakville ON L6J 6C3 Can

AZIZ, RONALD A, b Toronto, Ont, Sept 27, 28; m 59; c 3. CHEMICAL PHYSICS. *Educ:* Univ Toronto, BA, 50, MA, 51, PhD(physics), 55. *Prof Exp:* Lectr physics, Royal Mil Col, Can, 55-57; asst prof, Univ Windsor, 57-58; from asst prof to assoc prof, 58-67, assoc dean grad studies, Fac Sci, 76-81, PROF PHYSICS, UNIV WATERLOO, 67- *Concurrent Pos:* Heinemann Found scholar, 65-66. *Mem:* Fel Am Phys Soc; Can Asn Physicists. *Res:* Low temperature physics; solidified inert gas solids; intermolecular potentials; thermodynamic and transport properties. *Mailing Add:* Dept Physics Univ Waterloo Waterloo ON N2L 3G1 Can

AZIZI, SAYED AUSIM, b Kabul, Afghanistan, March 13, 54; US citizen/Afghanistan. NEUROPHYSIOLOGY, ANATOMY. *Educ:* Univ Tex, BS, 76; Univ Tex Health Sci Ctr, PhD(physiol), 81. *Prof Exp:* Instr physiol, Univ Tex Health Sci Ctr, Dallas, 77-78; INSTR NEUROBIOL, SOUTHWESTERN MED SCH, 78-, LECTR ADVAN NEUROBIOL, 79-, FEL PHYSIOL, 81- *Concurrent Pos:* Lectr, Sch Allied Health, Univ Tex Health Sci Ctr, 78- *Mem:* Soc Neurosci; AAAS; Sigma Xi. *Res:* Cerebro-cerebellar communication systems, specifically defining pathways that carry visual and auditory information from cerebral cortex to CBM; anatomy of brain stem connections with the cerebellum; motor cortical and brain stem control of cardiovascular function. *Mailing Add:* 607 Woodcrest Lane No 135 Arlington TX 76010

AZOFF, EDWARD ARTHUR, b Chicago, Ill, Oct 4, 45. MATHEMATICS. *Educ:* Univ Chicago, BS, 67; Univ Mich, 69, PhD(math), 72. *Prof Exp:* Res assoc math, Univ Ga, 72-73, vis asst prof, 73-74; asst prof, Univ Iowa, 74-75; from asst prof to assoc prof, 75-84, PROF MATH, UNIV GA, 85- *Concurrent Pos:* Res assoc, NSF, 73-74, prin investr, 75, co-prin investr, 76-85. *Mem:* Am Math Soc. *Res:* Operator theory; operator algebras; measurable selections. *Mailing Add:* Univ Ga Athens GA 30602

AZPEITIA, ALFONSO GIL, b Madrid, Spain, Feb 22, 22; nat US. MATHEMATICS. *Educ:* Univ Madrid, BA, 39, MA, 46, PhD, 52. *Prof Exp:* Res assoc, Nat Res Coun, Spain, 52-55; res assoc, Brown Univ, 55-57; from instr to prof math, Univ Mass, Amherst, 57-65; PROF MATH, UNIV MASS, BOSTON, 65- *Concurrent Pos:* Fulbright lectr, 69-70. *Mem:* Am Math Soc; Math Asn Am; Soc Indust & Appl Math; Opers Res Soc Am. *Res:* Theory of functions; linear and mathematical programming. *Mailing Add:* 650 Huntingten Ave Boston MA 02115

AZUMAYA, GORO, b Yokohama, Japan, Feb 26, 20; m 50; c 1. ALGEBRA. *Educ:* Tokyo Univ, BS, 42; Nagoya Univ, PhD(math), 49. *Prof Exp:* From lectr to asst prof math, Nagoya Univ, 45-53; prof, Hokkaido Univ, 53-68; vis prof, 65-66, PROF MATH, IND UNIV, BLOOMINGTON, 68- *Concurrent Pos:* Fel math, Off Naval Res, Yale Univ, 56-58; NSF fel, Northwestern Univ, 58-59; vis prof, Univ Mass, 64-65; vis prof, Univ Munich, 75-76; fel math, ETH Zurich, 83-84. *Honors & Awards:* Chunichi Cultural Prize, Japan, 49. *Mem:* Math Soc Japan; Am Math Soc. *Res:* Ring theory; number theory; homological algebra. *Mailing Add:* Ind Univ Bloomington IN 47405

AZZAM, RASHEED M A, b Elminia, Egypt, Mar 9, 45; m 74; c 6. ELLIPSOMETRY, POLARIMETRY. *Educ:* Cairo Univ, Egypt, BSc, 67; Univ Nebr, Lincoln, PhD(elec eng), 71. *Prof Exp:* Fel elec eng, Univ Nebr, Lincoln, 72-74, assoc prof eng & med, 74-79; from assoc prof to prof, 79-82, DISTINGUISHED PROF ELEC ENG, UNIV NEW ORLEANS, 82- *Concurrent Pos:* Researcher eng & med, Med Ctr, Univ Nebr, Omaha, 74-79; prin investr, NSF grant, Univ New Orleans, 81-83 & 86-89; sr Fulbright scholar, Univ Provence, 85-86; topical ed, J Optical Soc Am, 85-89. *Mem:* Fel Optical Soc Am; Soc Photo-Optical Instrumentation Engrs; Sigma Xi. *Res:* Surface science and thin-film optics; electrical engineering. *Mailing Add:* Dept Elec Eng Univ New Orleans Lakefront New Orleans LA 70148

AZZARO, ALBERT J, NEUROLOGY. *Educ:* Wheeling Jesuit Col, BS, 65; WVa Univ, PhD(pharmacol), 70. *Prof Exp:* Instr high sch biol & chem, Bishop Donahue High Sch, WVa, 65-66; postdoctoral fel pharmacol, Health Sci Ctr, Univ Colo, Denver, 70-71; from asst prof to prof, Dept Neurol & Pharmacol/Toxicol, 71-83, PROF, DEPT NEUROL, PHARMACOL/TOXICOL & BEHAV MED/PSYCHIAT, HEALTH SCI CTR, WVA UNIV, MORGANTOWN, 83-; DIR CLIN PSYCHOPHARMACOL LAB SERV, CHESTNUT RIDGE HOSP, MORGANTOWN, WVA, 89- *Concurrent Pos:* Grant reviewer, NSF, 75-85, Nat Inst Drug Abuse, 75 & US-Israel Binat Sci Found, 88- *Mem:* Sigma Xi; Soc Neurosci; AAAS; Am Soc Pharmacol & Exp Therapeut; Am Acad Neurol; NY Acad Sci. *Res:* Role of antidepressant monoamine oxidase inhibitors in the regulation of central monoaminergic transmission. *Mailing Add:* Dept Neurol WVa Univ Health Sci Ctr Morgantown WV 26506

B

BAAD, MICHAEL FRANCIS, b Hillsdale, Mich, Jan 5, 41; m 65, 79; c 1. PLANT TAXONOMY, ECOLOGY. *Educ:* Univ Mich, BS, 63, MS, 66; Univ Wash, PhD(bot), 69. *Prof Exp:* assoc prof, 69-80, PROF BIOL SCI, CALIF STATE UNIV, SACRAMENTO, 80- *Mem:* Ecol Soc Am; Am Soc

Plant Taxon; Int Soc Plant Taxon; Soc Study Evolution; Am Inst Biol Sci; Sigma Xi. *Res:* Taxonomy of the Caryophyllaceae; pollination ecology; plant succession; genecology; floristics of Central Sierra Nevada. *Mailing Add:* Dept Biol Sci Calif State Univ Sacramento CA 95819

BAADSGAARD, HALFDAN, b Minneapolis, Minn, Apr 16, 29; Can citizen; m 57; c 5. GEOCHEMISTRY. *Educ:* Univ Minn, BS, 51; Swiss Fed Inst Technol, PhD, 55. *Prof Exp:* Sr res assoc, Univ Minn, 55-57; from asst prof to assoc prof, 57-66, PROF GEOCHEM, UNIV ALTA, 66- *Concurrent Pos:* NATO exchange prof, Copenhagen Univ, 64; vis fel, Res Sch Earth Sci, Australian Nat Univ, 82-83. *Mem:* Geochem Soc; Am Geophys Union; fel Geol Asn Can. *Res:* Methods of geologic dating; geochronology; trace elements in geochemical research. *Mailing Add:* Dept Geol Univ Alta Edmonton AB T6G 2E3 Can

BAALMAN, ROBERT J, b Grinnell, Kans, Oct 28, 39; m 64; c 3. BOTANY, ECOLOGY. *Educ:* Ft Hays Kans State Univ, BS, 60, MS, 61; Univ Okla, PhD(bot), 65. *Prof Exp:* Teaching asst bot, Ft Hays Kans State Col, 60-61; teacher, High Sch, Kans, 61-62; teaching asst bot, Univ Okla, 62-64; from asst prof to assoc prof biol, 65-74, PROF BIOL, CALIF STATE UNIV, HAYWARD, 74- *Mem:* AAAS; Ecol Soc Am; Am Forestry Asn; Am Inst Biol Sci. *Res:* Plant succession; fire as a successional factor; seed dissemination and germination; saline plant communities in inland and coastal areas; plant competition. *Mailing Add:* Dept Biol Calif State Univ Hayward CA 94542

BAARDA, DAVID GENE, b Newton, Iowa, Apr 23, 37; m 62; c 3. ORGANIC CHEMISTRY, ENVIRONMENTAL CHEMISTRY. *Educ:* Cent Col, Iowa, BA, 59; Univ Fla, MS, 60, PhD(org chem), 62. *Prof Exp:* Res asst chem, Univ Fla, 62-63; chemist, Texaco, Inc, 63-64, sr chemist, 64-65; from asst prof to assoc prof, 65-74, PROF CHEM, GA COL, 74- *Concurrent Pos:* Danforth assoc, 70. *Mem:* Am Chem Soc. *Res:* Use of laboratory investigation and computerized data handling to teach chemistry at the undergraduate level. *Mailing Add:* Dept Chem Ga Col Milledgeville GA 31061

BAARS, DONALD LEE, b Oregon City, Ore, May 27, 28; m 48; c 3. STRATIGRAPHY, SEDIMENTARY PETROLOGY. *Educ:* Univ Utah, BS, 52; Univ Colo, Boulder, PhD(geol), 65. *Prof Exp:* Explor geologist, Shell Oil Co, 52-61; res geologist, Continental Oil Co, 61-62; assoc prof geol, Wash State Univ, 65-68; assoc prof, 68-70, prof geol, Ft Lewis Col, 70-80; consult petroleum geologist, 80-88; RES PETROL GEOLOGIST, KANS GEOL SURV, 88- *Concurrent Pos:* Proj dir, NSF Res Grant, 68-; dir short course for col teachers proj, NSF, 70-71. *Mem:* Am Asn Petrol Geol; fel Geol Soc Am; Soc Econ Paleont & Mineral. *Res:* Paleozoic stratigraphy of Colorado Plateau; carbonate sedimentation and sedimentary petrology; early Paleozoic paleotectonic history of the ancestral Rockies and related tectonic system; basement and early Paleozoic paleotectonic history of southern midcontinent. *Mailing Add:* Kans Geol Surv Campus W 1930 Constant Ave Lawrence KS 66047-2598

BAASEL, WILLIAM DAVID, b Chicago, Ill, Mar 18, 32; m 60, 90; c 3. CHEMICAL ENGINEERING. *Educ:* Northwestern Univ, BS, 54, MS, 56; Cornell Univ, PhD(chem eng), 62. *Prof Exp:* Asst prof chem eng, Clemson Col, 59-62; asst prof, 62-66, assoc prof, 66-70, PROF CHEM ENG, OHIO UNIV, 70- *Concurrent Pos:* Ford Found res eng practice, Dow Chem Co, Mich, 65-66, consult, 66-68; mem, Nat Sci Found Summer Inst, Mich Technol Univ, 63, US Environ Protection Agency, 78-80 & Coop Agreement, 80-83; actg chmn chem eng, Ohio Univ, 88-90; instr, Petaling Jaya Community Col, Petaling Jaya Malaysia. *Honors & Awards:* Fritz & Dolores Russ Res Award, 81. *Mem:* AAAS; Am Inst Chem Engrs; Am Soc Eng Educ; Sigma Xi; Am Asn Univ Professors; Air & Waste Mgt Asn. *Res:* Environmental assessment, air pollution abatement, acid rain; plant design; author of three books. *Mailing Add:* Dept Chem Eng Ohio Univ Athens OH 45701

BAASKE, DAVID MICHAEL, b Milwaukee, Wis, Feb 15, 47; div; c 2. PHARMACEUTICAL ANALYSIS, BIOANALYTICAL CHEMISTRY. *Educ:* Carroll Col, BS, 69; DePauw Univ, MA, 71; Purdue Univ, PhD(pharmaceut anal), 76. *Prof Exp:* Res fel pharmaceut anal, Drug Dynamics Inst, Univ Tex, 76-77; res investr, Arnar-Stone Labs, Subsid Am Hosp Supply, 77-80, sr res investr, 80-82, group leader, 82-85, sect head, Am Crit Care, Am Hosp Supply, 85-89; sr res supvr, Dupont Pharmaceut, E I Dupont de Nemours & Co, 89-91, SR RES SUPVR, DUPONT MERCK PHARMACEUT CO, 91- *Mem:* Am Chem Soc; Am Asn Pharmaceut Scientists; NY Acad Sci; Sigma Xi. *Res:* Characterization of new drug substances and the stability-indicating analytical methods for use in the evaluation of pharmaceutical dosage forms and their compatibility with excipients, diluents and medical devices; preparation of regulatory submissions. *Mailing Add:* 3109 Stone Pl Newark DE 19702

BABA, ANTHONY JOHN, b Bethleham, Pa, May 2, 36; m 59; c 6. RADIATION PHYSICS. *Educ:* Georgetown Univ, BS, 57. *Prof Exp:* Physicist, Nat Bur Standards, 58-60; RES PHYSICIST, HARRY DIAMOND LABS, 60-, SUPVR PHYSICIST. *Concurrent Pos:* Lectr, Trinity Col, DC, 63-69; assoc mem adv group electron devices, Dept Defense, 72-79. *Mem:* Sr mem Inst Elec & Electronics Engrs. *Res:* Chemical vapor depositions for silicon integrated circuit applications; transient radiation effects on integrated circuits; thermal radiation effects on materials. *Mailing Add:* Harry Diamond Labs 2800 Powder Mill Rd Adelphi MD 20783

BABA, PAUL DAVID, b Elizabeth, NJ, Dec 15, 32; m 59; c 2. CERAMICS. *Educ:* Rutgers Univ, BS, 54, PhD(ceramics), 60; Golden Gate Univ, MBA, 80. *Prof Exp:* Mem tech staff, Bell Tel Labs Inc, 60-65; head ferrite res group, Ampex Corp, 65-68, mgr ferrite res sect, 68-72, mgr ferrite mat dept, 72-78, gen mgr mat & devices group, 78-80, dir bus mgt, Magnetic Tape Div, 80-81, dir res & develop eng, 81-84, dir, Media Technol Lab, 84-91; DIR, DATA TAPES, SANTA CLARA FAC, 91- *Mem:* Inst Elec & Electronics Engrs; KERAMOS; Soc Photo-Optical Instrumentation Engrs. *Res:* Electronic materials, devices such as ferromagnetics, ferroelectrics, insulators, semiconductors, and magnetic tapes; optical recording media. *Mailing Add:* 142 Coronado Ave San Carlos CA 94070

BABAD, HARRY, b Vienna, Austria, Mar 12, 36; US citizen; m 65; c 2. ORGANIC CHEMISTRY, NUCLEAR CHEMISTRY. *Educ:* Polytech Inst Brooklyn, BS, 56; Univ Ill, MS & PhD(org chem), 61. *Prof Exp:* Fel org chem, Mass Inst Technol, 61-62; res assoc, Univ Chicago, 62-63; asst prof, Univ Denver, 63-67; group leader new prod res, Ott Chem Co, 67-69, head sect res & develop, 69-74; mgr, Chem Technol Lab, Atlantic Richfield Hanford Co, 74-77; MGR, RES DEPT, ROCKWELL HANFORD OPER, 77-, PRIN SCIENTIST, BASALT WASTE ISOLATION PROJ, 80- *Concurrent Pos:* Consult, Colo Int, 74- & Argent Chem, 79- *Mem:* AAAS; Am Chem Soc; NY Acad Sci; fel Am Inst Chemists. *Res:* Areas of chemical and environmental sciences aimed at nuclear waste treatment, immobilization and isolation as well as aspects of chemical processing of nuclear materials; multidisciplinary technical integration. *Mailing Add:* 2540 Cordoba Ct Rockwell Hanford Opers Richland WA 99352

BABAYAN, VIGEN KHACHIG, b Armenia, Jan 1, 13; nat US; m 42; c 3. FOOD SCIENCE, LIPID CHEMISTRY. *Educ:* NY Univ, BA, 38, PhD, 43. *Prof Exp:* Chemist, E F Drew & Co, 38-40; res chemist, Warwick Chem Co, 40-41; dir res & develop, Ridbo Labs, 41-47; dir labs, Theobald Industs, 47-48; dir res & develop, E F Drew & Co, 48-61, vpres & dir labs, Drew Chem Corp, 61-65; asst prof med, Sch Med, Ind Univ, 69-77; vpres & dir res & develop & qual control, 65-80, VPRES SCI & TECHNOL, STOKELY-VAN CAMP, INC, 80-; ASSOC BIOCHEM SURG, HARVARD MED SCH, 80- *Concurrent Pos:* Mem res & develop tech comts, Res & Develop Assocs, Am Asn Clin Nutrit, Am Soc Parenteral & Enterial Nutrit, Inst Food Technol & Am Oil Chem Soc. *Honors & Awards:* Glycerin Asn Award, 64. *Mem:* Am Chem Soc; Am Oil Chem Soc; Inst Food Technol; Am Asn Cereal Chemists; Am Asn Candy Technologists. *Res:* Foods, canned and frozen; edible oils; fatty derivatives; lipids; diet-nutrition; medical specialties; lipids for nutritional and dietetic use; medium chain triglyceredes; structured lipids and their use in medical specialties. *Mailing Add:* 178 Beethoven Ave Waban MA 02168

BABB, ALBERT L(ESLIE), b Vancouver, BC, Nov 7, 25; US citizen; m 72; c 3. CHEMICAL ENGINEERING, NUCLEAR ENGINEERING. *Educ:* Univ BC, BASc, 48; Univ Ill, MS, 49, PhD(chem eng), 51. *Prof Exp:* Res engr, Rayonier, Inc, Wash, 51-52; from asst prof to assoc prof, Univ Wash, 52-60, chmn, Nuclear Eng Group, 56-65, dir, Nuclear Reactor Labs, 60-77, chmn, Dept Nuclear Eng, 65-81, actg chmn, Dept Chem Eng, 85, actg chmn, Dept Nuclear Eng, 84-86, PROF CHEM ENG & NUCLEAR ENG, UNIV WASH, 60- *Concurrent Pos:* Consult, Puget Sound Bridge & Dry Dock Co, Seattle, 58-65, R W Beck & Assocs, 63-67, Swed Hosp, Seattle, 65-67, Atomic Energy Comn, 65-70, Kidney Dis Br, USPHS & Nat Inst Arthritis & Metab Dis, 66-, Seattle Res Ctr, Battelle Mem Inst, 69- & Physio Control Corp, 68-70; mem bd dirs and consult, Thermo-Dynamics Inc, 67-72, Sweden Freezer, 72-78, Biomedics, 78-81 & Renal Devices, Inc, 79-81. *Honors & Awards:* Dialysis Pioneer Award, Nat Kidney Found. *Mem:* Nat Acad Eng; Inst Med-Nat Acad Sci; fel Am Inst Chem Engrs; fel Am Nuclear Soc. *Res:* Diffusion in liquids; separations processes in nuclear energy field; nuclear reactor analysis; bioengineering; biomedical engineering. *Mailing Add:* Dept Nuclear Eng Univ Wash Seattle WA 98195

BABB, DANIEL PAUL, b Red Wing, Minn, Aug 1, 39. INORGANIC CHEMISTRY. *Educ:* Mankato State Col, BA, 63; Univ Idaho, PhD(inorg chem), 68. *Prof Exp:* Asst prof chem, Kearney State Col, 67-68; asst scientist, Inorg Chem Inst, Univ Gottingen, 68-69; asst prof, Va Polytech Inst, 70-72; assoc prof, 75, asst dean, Col Sci, 86-90, PROF CHEM, MARSHALL UNIV 72-, CHMN DEPT, 90- *Concurrent Pos:* Vis assoc prof, Univ Idaho, 77-78. *Mem:* Am Chem Soc. *Res:* Fluorine and inorganic synthesis chemistry. *Mailing Add:* Dept Chem Marshall Univ Huntington WV 25701

BABB, DAVID DANIEL, b Saugus, Mass, Sept 20, 28; m 49. ELECTROMAGNETISM. *Educ:* Mass Inst Technol, BS & MS, 50, MS, 58. *Prof Exp:* Student engr, Philco Corp, 47-49; proj engr, Radio Div, Bendix Corp, 49-52; res asst atomic physics, Mass Inst Technol, 55-58; sr nuclear engr, Gen Dynamics, Ft Worth, 58-61; sr res physicist, Dikewood Corp, NMex, 61-66, head, Physics Div, 66-68, physics prog mgr, 68-69, prog mgr theoret res, 69-70; pres, Albuquerque Res Assocs, Inc, 70-73; sr res physicist, Dikewood Corp, NMex, 73-78; eng mgr, Babb Sound Corp, Tex, 78-80; SR ELEC ENGR, HOLMES & NARVER INC, NEV, 80- *Mem:* Am Phys Soc; Inst Elec & Electronics Engrs; Sigma Xi. *Res:* Nuclear weapons effects, including electromagnetic pulse, neutron and gamma transport and x-ray vulnerability; air and space borne reactor hazards, including reactor shielding and space radiation; atomic frequency standards; radio design; lasers, loudspeakers, electrical power distribution. *Mailing Add:* 4424 St Andrews Circle Las Vegas NV 89107

BABB, ROBERT MASSEY, b Durham, NC, Apr 24, 38; m 66; c 1. ORGANIC CHEMISTRY, SAFETY & HEALTH. *Educ:* Univ NC, Chapel Hill, BS, 60; Univ Utah, PhD(org chem), 69. *Prof Exp:* Jr chemist, Merck Sharp & Dohme Res Labs, NJ, 60-63; res chemist, Res Ctr, Burlington Industs, Inc, NC, 68-72; group leader, Dyestuffs & Chem Div, 72-75, staff chemist, 75-76, group leader, 76-77, safety supt, Agr Div, 77-87, SR STAFF CHEMIST, CIBA-GEIGY CORP, 88- *Mem:* Am Chem Soc; Am Soc Safety Engrs; Am Inst Chem Eng. *Res:* Reactions of cyclopropenes and methylenecyclopropanes; organophosphorus chemistry; textile chemistry; pesticides and process development; chemical permeation of protective clothing; chemical spill mitigation; emergency modeling of gaseous chemical releases. *Mailing Add:* 1336 Tara Blvd Baton Rouge LA 70806

BABBITT, DONALD GEORGE, b Detroit, Mich, Feb 24, 36; m 64; c 2. MATHEMATICAL ANALYSIS. *Educ:* Univ Detroit, BS, 57, Univ Mich, MA, 58, PhD(math), 62. *Prof Exp:* From asst prof to assoc prof, 62-73, PROF MATH, UNIV CALIF, LOS ANGELES, 73- *Concurrent Pos:* Mem, Inst Advan Study, 70-71; managing ed, Pacific J Mathematics, 79- *Mem:* AAAS; Am Math Soc. *Res:* Study of mathematical structures which arise in theoretical physics. *Mailing Add:* Dept Math Univ Calif Los Angeles CA 90024

BABBOTT, FRANK LUSK, JR, b New York, NY, Feb 6, 19; m 50; c 3. EPIDEMIOLOGY. *Educ:* Amherst Col, BA, 47; State Univ NY, MD, 51; Harvard Univ, MPH, 53, MS, 54; Am Bd Prev Med, dipl. *Prof Exp:* Instr epidemiol, Sch Pub Health, Harvard Univ, 54-55, assoc, 55-58; asst prof prev med, Sch Med, Univ Pa, 58-62; vis fel epidemiol, Pub Health Lab Serv, London, Eng, 62-63; assoc prof epidemiol, Col Med, Univ Vt, 63-91; RETIRED. *Concurrent Pos:* Vchmn, Am Bd Prev Med, 74-79. *Mem:* Fel Am Pub Health Asn; fel Am Col Prev Med; Am Epidemiol Soc; Asn Teachers Prev Med. *Mailing Add:* Dept Med Univ Vt Col Med Burlington VT 05401

BABBS, CHARLES FREDERICK, b Toledo, Ohio, July 6, 46; m 81; c 3. CARDIOVASCULAR PHYSIOLOGY, HEAT THERAPY FOR CANCER. *Educ:* Yale Univ, BA, 68; Baylor Col Med, MD, 74 & MS, 75; Purdue Univ, PhD(pharmacol), 77. *Prof Exp:* Instr anat, Baylor Col Med, 71-73; instr biol sci & res assoc biomed eng, 74-80, ASSOC RES SCHOLAR BIOMED ENG, PURDUE UNIV, 80-; INSTR FAMILY MED, IND UNIV SCH MED, 80- *Concurrent Pos:* Fel, Coun Circulation, Am Heart Asn. *Honors & Awards:* James R MacKenzie Award, Univ Asn Emergency Med, 84. *Mem:* Shock Soc; Nat Soc Med Res; Asn Advan Med Instrumentation; Am Heart Asn. *Res:* Cardiopulmonary resuscitation, pathophysiology of tissue ischemia and reperfusion; free radicals in medicine and biology; heat therapy for cancer; drug effects upon all of the foregoing; computer modeling of complex biological systems. *Mailing Add:* William A Hillenbrand Biomed Eng Ctr Purdue Univ A A Potter Bldg Rm 204 West Lafayette IN 47907

BABCOCK, ANITA KATHLEEN, b Tucson, Ariz, Sept 2, 48. COMPUTER MODELING & SIMULATION. *Educ:* Pomona Col, BS, 70; State Univ NY, Buffalo, PhD(theoretical biophysics), 76. *Prof Exp:* Res asst, Mich State Univ, 72-73 & State Univ NY, Buffalo, 73-76; staff fel, Nat Cancer Inst, NIH, 76-79; res assoc, Georgetown Univ, 79-80; PHYSICIST, VITRO CORP, 80- *Concurrent Pos:* Vchmn, Vitro Tech J, 88 - *Res:* Computer systems development, simulation and modeling; applied mathematics; network analysis; statistics; interactive graphics information systems development; statistical analysis of sampling plans using Monte Carlo simulation methods; parameter estimation using modeling package; covariance analysis of Kalmen filters; test tolerance derivation. *Mailing Add:* Vitro Corp Bldg 12 Rm 2109 14000 Georgia Ave Silver Spring MD 20906-2972

BABCOCK, BYRON D(ALE), b Wilmington, Del, Mar 17, 31; m 61; c 2. CHEMICAL ENGINEERING. *Educ:* Univ Del, BChE, 52; Univ Wis, MS, 53, PhD(chem eng), 55. *Prof Exp:* Res supvr, Sabine River Works, 61-62, asst supt, 62-64, tech supt, Victoria Plant, 64-68, lab supt, Washington Works, 68-74, dir planning & develop, Mexico City, 75-76, mgr mfg & res, 77-87, PROG MGR, E I DU PONT DE NEMOURS & CO, 87- *Mem:* Am Chem Soc; Am Inst Chem Eng. *Mailing Add:* 738 Taunton Rd Tavistock Wilmington DE 19803

BABCOCK, CLARENCE LLOYD, physics; deceased, see previous edition for last biography

BABCOCK, DALE F, b Minneapolis, Kans, Oct 5, 06; m 30; c 2. NUCLEAR ENGINEERING. *Educ:* Kans State Univ, Pillsburg, BS, 24; Univ Ill, PhD(chem), 29. *Prof Exp:* Phys chemist, Dupont Co, 29-45, dir chem eng, 45-50, Nuclear engr, 50-71; RETIRED. *Mem:* Fel Am Nuclear Soc; Am Chem Soc. *Res:* Chemical separations; nylon spinning and drawing; major contributions in design, construction and operation of nuclear reactors, including Hanford, Washington reactors and Savannah River reactors. *Mailing Add:* Cokesbury Village No 95 Hockessin DE 19707-1504

BABCOCK, DANIEL LAWRENCE, b Philadelphia, Pa, Nov 25, 30; m 80; c 3. ENGINEERING MANAGEMENT, SYSTEMS ENGINEERING. *Educ:* Pa State Univ, BS, 52; Mass Inst Technol, MS, 53; Univ Calif, Los Angeles, PhD(eng), 70. *Prof Exp:* Chemist & tech writer, Dow Corning Corp, Mich, 56-59; tech ed, Chem Propulsion Info Agency, Johns Hopkins Univ, 59-62; supvr proj engr & sr specialist, Space Div, NAm Rockwell Corp, 63-69; assoc prof, 70-78, PROF ENG MGT, UNIV MO-ROLLA, 78- *Concurrent Pos:* Exec dir, 79-89, emer exec dir, Am Soc Eng Mgt, 89- *Mem:* Am Soc Eng Educ; fel Am Soc Eng Mgt; Am Soc Quality Control; Soc Am Mil Engrs. *Res:* engineering and technology management as an educational discipline; military logistics and systems acquisition; quality assurance and reliability; author of one book. *Mailing Add:* Dept Eng Mgt Univ Mo 214 Eng Mgt Bldg Rolla MO 65401

BABCOCK, DONALD ERIC, b Canton, Ohio, Nov 10, 07. ELECTRO ORGANIC CHEMISTRY. *Educ:* Ohio State Univ, BS, 31, MS, 33, PhD(chem), 35. *Prof Exp:* CONSULT, REPUB STEEL CORP, 31- *Concurrent Pos:* Consult prof engr; tech consult adv, UN. *Mem:* Am Chem Soc; hon fel Am Soc Metals; Am Inst Mining Metall & Petrol Engrs. *Mailing Add:* D E Babcock Assoc 29611 W Oakland Rd Bay Village OH 44140

BABCOCK, ELKANAH ANDREW, b Elizabeth, NJ, Nov 23, 41; m 64; c 1. STRUCTURAL GEOLOGY, REMOTE SENSING. *Educ:* Union Col, NY, BS, 63; Syracuse Univ, MS, 65; Univ Calif, Riverside, PhD(geol), 69. *Prof Exp:* Asst prof geol, Univ Alta, 69-75; dir, 75-80, VPRES, NATURAL RESOURCES DIV, ALTA RES COUN, 80- *Mem:* AAAS; Geol Asn Can; Can Soc Petrol Geologists. *Res:* Structural geology of the Salton Trough, Southern California; terrain analysis and geologic mapping applications of remote sensing; study of fracture phenomena in rock. *Mailing Add:* 500 Booth St Ottawa ON K1A 0E4 Can

BABCOCK, GEORGE, JR, b NJ, Apr 26, 16; m 52; c 3. PHARMACOLOGY, CLINICAL IMMUNOLOGY. *Educ:* Rutgers Univ, BA, 42, MSc, 43; Western Reserve Univ, MD, 47; Am Bd Allergy & Immunol, cert, 75. *Prof Exp:* Intern, East Orange Gen Hosp, 47-48; mem res staff, Schering Corp, 48-52, asst dir clin res, 52-54, assoc dir, 55-60, dir, Med Res Div, 60-63, vpres, 65-80, DIR SCI AFFAIRS, IVES LABS, INC, 64-, SR VPRES, 80- *Mem:* Fel Am Col Allergists; fel Am Geriat Soc; Am Acad Allergy; Sigma Xi; fel Am Col Nutrit. *Res:* Allergy; endocrinology; clinical pharmacology. *Mailing Add:* 55 Hilltop Terr Kinnelon NJ 07405

BABCOCK, GERALD THOMAS, b Minneapolis, Minn, Feb 9, 46; m 76; c 2. BIOPHYSICAL CHEMISTRY, PHYSICAL CHEMISTRY. *Educ:* Creighton Univ, BS, 68; Univ Calif, Berkeley, PhD(chem), 73. *Prof Exp:* Fel chem, Univ Calif, Berkeley, 73-74; NIH fel biochem, Rice Univ, 74-76; from asst prof to assoc prof chem, 76-84, PROF CHEM, MICH STATE UNIV, 84-, CHMN, DEPT CHEM, 90- *Concurrent Pos:* Vis prof, Col de France, 90. *Honors & Awards:* Phillips Lectr, Haverford Col, 90. *Mem:* Am Chem Soc; Biophys Soc; Fedn Am Soc Exp Biol. *Res:* Physical chemistry of bioenergetic reactions; electron transfer kinetics in photosynthesis associated with water oxidation; cytochrome oxidase-structure and function. *Mailing Add:* Dept Chem Mich State Univ East Lansing MI 48824

BABCOCK, HORACE W, b Pasadena, Calif, Sept 13, 12; m 40, 58; c 3. ASTRONOMY. *Educ:* Calif Inst Technol, BS, 34; Univ Calif, PhD(astron), 38. *Hon Degrees:* DSc, Univ Newcastle-upon-Tyne, 65. *Prof Exp:* Asst, Lick Observ, Univ Calif, 38-39; instr, McDonald Observ, Univ Chicago & Univ Tex, 39-41; res assoc, Radiation Lab, Mass Inst Technol, 41-42; res assoc, Rocket Proj, Calif Inst Technol, 42-46; staff mem to dir, Mt Wilson & Palomar Observ, 46-74; dir observ site-surv & construct, Las Campanas Observ, Chile, Carnegie Inst Wash, 63-78; CONSULT, 78- *Concurrent Pos:* Counr, Am Astron Soc, 52-53 & Nat Acad Sci, 73-76. *Honors & Awards:* US Navy Bur Ord Develop Award, 46; Draper Medal, Nat Acad Sci, 57; Eddington Medal, Royal Astron Soc, 58, Gold Medal, 70; Bruce Gold Medal, Astron Soc of the Pac, 69. *Mem:* Nat Acad Sci; Am Astron Soc; Am Acad Arts & Sci; Int Astron Union; Am Philos Soc. *Res:* Rotation and mass distribution of spiral galaxy in Andromeda (M31); stellar spectroscopy; diffraction gratings; adaptive optics; astronomical instrumentation; magnetic fields of sun and stars; telescope design (engineering). *Mailing Add:* Observ Carnegie Inst Wash 813 Santa Barbara St Pasadena CA 91101

BABCOCK, LYNDON ROSS, (JR), b Detroit, Mich, Apr 8, 34; m 57; c 4. ENVIRONMENTAL MANAGEMENT, ENVIRONMENTAL EDUCATION. *Educ:* Mich Technol Univ, BS, 56; Univ Wash, MS, 58, PhD(environ eng), 70. *Prof Exp:* Chem engr polymers, Shell Chem Co, 58-67; from assoc prof to prof, Environ & Occup Health Sci Prog, Sch Pub Health, Univ Ill, 70-79, dir, 79-84, assoc dean, 84-85. *Concurrent Pos:* Consult, Ill Inst Natural Resources, 70-71 & 76- & Oak Ridge & Argonne Nat Labs, 71-75; consult & proj mgr, Indust Environ Res Lab, US Environ Protection Agency, 72-80; sr lectr, Fulbright-Hays Prog, Turkey & India, 75-76, Mexico, 86-87; mem, Nat Air Conserv Comn, 77-82, World Health Orgn, 85, Los Alamos Nat Lab, 90-, Interam Develop Bd, 90- *Mem:* Air & Waste Mgt Asn; Nat Asn Environ Prof. *Res:* Environmental management especially in developing countries; environmental education, air quality characterization, control and health effects; land use; urban transportation, environmental policy and analysis. *Mailing Add:* Pub Health EOHS MC922 UIC Box 6998 Chicago IL 60680

BABCOCK, MALIN MARIE, b Juneau, Alaska, Oct 31, 39; c 2. TOXICOLOGY. *Educ:* Ore State Univ, BS, 62; Univ Alaska, MS, 69. *Prof Exp:* Arctic physiologist, Arctic Health Res Ctr, HEW, 66-69; FISHERY RES BIOLOGIST HISTOL & PHYSIOL, AUKE BAY LAB, NAT MARINE FISHERIES SERV, 69- *Mem:* Am Fisheries Soc; Am Inst Fishery Res Biologists; Crustacean Soc; Nat Shellfisheries Asn. *Res:* Histopathology and physiology of fish and invertebrates exposed to stress and pollutants. *Mailing Add:* Nat Marine Fisheries Serv Auke Bay Lab Box 210155 Auke Bay AK 99821-0155

BABCOCK, PHILIP ARNOLD, b Clinton, Mass, May 11, 32; m 59; c 2. PHARMACOGNOSY, BOTANY. *Educ:* Mass Col Pharm, BS, 54, MS, 59; Univ Iowa, PhD(pharm), 62. *Prof Exp:* Asst prof, 62-66, ASSOC PROF PHARMACOG, COL PHARM, BUSCH CAMPUS, RUTGERS UNIV, 66- *Concurrent Pos:* Dir, Scholastic Standing & Student Recs. *Mem:* Acad Pharmaceut Sci; Am Pharmaceut Asn; Am Soc Pharmacog; Soc Econ Bot. *Res:* Biosynthesis of drugs in plants; tissue culture of medicinal plants; chemotaxonomy. *Mailing Add:* Dept Pharm Rutgers Univ New Brunswick NJ 08903

BABCOCK, ROBERT E, b Chicago, Ill, Oct 10, 37; m 59; c 2. THERMODYNAMICS. *Educ:* Univ Okla, BS, 59, MChE, 62, PhD(heat transfer), 64. *Prof Exp:* Sr res engr, Esso Prod Res Co, 64-65; from asst prof to assoc prof, 65-77, asst dean res, 69-77, DIR WATER RESOURCES RES CTR & PROF CHEM ENG, UNIV ARK, FAYETTEVILLE, 77-, RES COORD OFF, 71- *Mem:* Am Inst Chem Engrs; Am Inst Mining, Metall & Petrol Engrs; Am Soc Eng Educ. *Res:* Transport properties; irreversible thermodynamics; heat transfer; fluid flow; petroleum reservoir engineering. *Mailing Add:* Dept Chem Eng Univ Ark Main Campus Fayetteville AR 72701

BABCOCK, ROBERT FREDERICK, b Ft Wayne, Ind, Feb 26, 30; m 53; c 2. ANALYTICAL CHEMISTRY, ENVIRONMENTAL ANALYSIS. *Educ:* DePauw, BA, 51; Ind Univ, PhD(anal chem), 58. *Prof Exp:* From proj chemist to sr proj chemist, Am Oil Co, Whiting, Ind, 57-74; res chemist, Stand Oil Co (Ind), 74-82; SR RES CHEMIST, AMOCO CORP, 82- *Mem:* Am Chem Soc; Am Soc Testing Mat; Sigma Xi. *Res:* Development of analytical methods; spectrophotometry; environmental analytical research. *Mailing Add:* 26 W 270 Durfee Rd Wheaton IL 60187

BABCOCK, WILLIAM EDWARD, b Buhl, Idaho, Oct 15, 22; m 45; c 4. VETERINARY MEDICINE. *Educ:* State Col Wash, BS, 44, DVM, 45; Ore State Col, MS, 51. *Prof Exp:* Assoc vet, Dept Vet Med, Ore State Col, 49-63; res vet, Agr Res Dept, Pfizer, Inc, 63, mgr, 63-65, asst dir, 65-68-; RETIRED. *Mem:* Am Vet Med Asn; Poultry Sci Asn; Am Asn Avian Path. *Res:* Poultry diseases. *Mailing Add:* 87 Allendale Pl Terre Haute IN 47802

BABCOCK, WILLIAM JAMES VERNER, ecology, science education; deceased, see previous edition for last biography

BABEL, FREDERICK JOHN, b Traverse City, Mich, Oct 22, 11; m 39. FOOD MICROBIOLOGY. *Educ:* Mich State Univ, BS, 35; Purdue Univ, MS, 36; Iowa State Univ, PhD(dairy bact), 39. *Prof Exp:* Res chemist & bacteriologist, NAm Creameries, Inc, Minn, 39-40; res assoc dairy bact, Iowa State Univ, 40-43, asst prof, 43-44, res assoc prof, 44-47; res assoc prof, 47-50, prof, 50-78, EMER PROF DAIRY & FOOD MICROBIOL, PURDUE UNIV, LAFAYETTE, 78- *Concurrent Pos:* Consult, C Hansen's Lab, Milwaukee & Kraft, Inc, Glenview, Ill. *Honors & Awards:* Paul-Lewis Pfizer Award, 61. *Mem:* Am Dairy Sci Asn; Int Asn Milk & Food Sanitarians. *Res:* Microbiology and enzymology of dairy products; bacteriophage; flavor of cultured dairy products; milk-coagulating enzymes; dairy plant sanitation. *Mailing Add:* Food Sci Dept Purdue Univ Smith Hall West Lafayette IN 47907-1160

BABER, BURL B, JR, b Wellston, Okla, June 15, 28; m 49; c 3. FUEL TECHNOLOGY & PETROLEUM ENGINEERING. *Educ:* Okla State Univ, BS, 51. *Prof Exp:* Prod tester, Int Havester Co, Ind, 51-53; from res engr to sr res engr, Southwest Res Inst, 53-60, sect mgr, Appl Lubrication Sect, 60-69, asst dir, Div 02, 69-89; RETIRED. *Mem:* Am Soc Lubrication Eng; Am Soc Testing Mat. *Res:* Applied lubrication of gears and bearings, primarily as used in aircraft and aerospace applications; use of methanol fluid (M85) in various current engine designs as observed from active fleet tests.. *Mailing Add:* PO Box 9 Kingsland TX 78639

BABERO, BERT BELL, b St Louis, Mo, Oct 9, 18; m 50; c 2. PARASITOLOGY. *Educ:* Univ Ill, BS, 49, MS, 50, PhD, 57. *Prof Exp:* Med parasitologist, USPHS, Alaska, 50-53; asst parasitol, Univ Ill, 54-57; prof zool & head dept, Ft Valley State Col, 57-59; prof, Southern Univ, 59-60; lectr, Fed Emergency Sci Scheme, Lagos, Nigeria, 60-62; parasitologist, Col Med, Univ Baghdad, Iraq, 62-65; from assoc prof to prof, 65-85, EMER PROF BIOL SCI, UNIV NEV, LAS VEGAS, 87- *Concurrent Pos:* Fulbright-Hayes fel, US Dept State, Iraq, 63-65; La State Univ fel, 68; prof, Grambling State Univ, 87-89. *Mem:* Soc Syst Zool; Soc Protozool; Am Soc Parasitol; Wildlife Dis Asn; Am Micros Soc. *Res:* Animal and human parasitology; ascariasis; diphyllobothriasis; echinococcosis; zoonotic diseases; author of approximately 100 scientific publications. *Mailing Add:* Dept Biol Sci Univ Nev Las Vegas NV 89154

BABICH, HARVEY, b Brooklyn, NY, Mar 19, 47. POLLUTION BIOLOGY, IN VITRO ASSAYS. *Educ:* Yeshiva Col, BA, 68; Long Island Univ, MS, 71; NY Univ, PhD(biol), 76. *Prof Exp:* Adj asst prof gen biol, NY Univ, 76-79, sr res scientist, 80-84; sr staff scientist, Environ Law Inst, 79-80; SR RES ASSOC, ROCKEFELLER UNIV, 84- *Concurrent Pos:* Adj asst prof, NY Univ, 84- *Mem:* AAAS; Am Soc Microbiol. *Res:* In vitro cytotoxicity testing using cultured fish cells; microbial ecology and heavy metal pollution. *Mailing Add:* Rockefeller Univ 1230 York Ave New York NY 10021

BABICH, MICHAEL WAYNE, b Milwaukee, Wis, Sept 23, 45; m 73; c 2. PHYSICAL INORGANIC CHEMISTRY, SOLID STATE CHEMISTRY. *Educ:* Univ Wis-Madison, BS, 67; Univ Nev, Reno, PhD(inorg chem), 74. *Prof Exp:* Teaching fel chem, Univ Nev, Reno, 70-74; res assoc, Iowa State Univ-Ames Lab, US Dept Energy, 74-76; vis asst prof, Ohio Wesleyan Univ, 76-77 & Univ Del, 77-78; from asst prof to assoc prof, 71-85, chmn chem prog, 85-87, PROF & HEAD DEPT CHEM, FLA INST TECHNOL, 87-, CO-CHAIR BIOCHEM PROG, 88- *Concurrent Pos:* Prin investr, NSF-ISEP, 80-83, NSF-CIP, 82-84, Am Chem Soc-PRF, 84-86, Am Chem Soc-PRF-SRF, 85, Fla Solar Energy Ctr, 87-90. *Mem:* Am Chem Soc; Am Crystallog Asn; Sigma Xi; fel Am Inst Chemists. *Res:* Solid phase reaction kinetics; crystallography of inorganic complexes; synthetic inorganic chemistry; thermal analysis; DSC and TG. *Mailing Add:* Dept Chem Fla Inst Technol Melbourne FL 32901-6988

BABINEAU, G RAYMOND, b New York, NY, Apr 8, 37; m 60; c 3. PSYCHIATRY. *Educ:* Bowdoin Col, BA, 59; Harvard Med Sch, MD, 63. *Prof Exp:* Chief ment hyg psychiat, US Army Hosp, Berlin, Ger, 67-70; asst prof & sr assoc psychiat, Univ Health Serv, Univ Rochester, 70-, chief, Ment Health Sect, 73-; PVT PRACT. *Mem:* Am Psychiat Asn; Am Col Health Asn. *Res:* Psychological problems of university students. *Mailing Add:* 428 White Spruce Blvd Rochester NY 14623

BABIOR, BERNARD M, b Los Angeles, Calif, Nov 10, 35; m 61; c 2. BIOCHEMISTRY, HEMATOLOGY. *Educ:* Univ Calif, San Francisco, MD, 59; Harvard Univ, PhD(biochem), 65. *Prof Exp:* Asst prof med, Harvard Med Sch, 69-71, assoc prof, 71-72; from assoc prof to prof med, Sch Med, Tufts Univ, 72-86; HEAD DIV BIOCHEM, DEPT MOLECULAR & EXP MED & STAFF PHYSICIAN, DEPT MED, SCRIPPS CLIN RES FOUND, 86- *Mem:* AAAS; Am Soc Biol Chem; Am Soc Clin Invest. *Res:* Mechanism of action of coenzyme B12; granulocyte physiology. *Mailing Add:* Scripps Clin-Res Found La Jolla CA 92037

BABISH, JOHN GEORGE, b Johnstown, Pa, Dec 10, 46; div; c 2. TOXICOLOGY, EPIDEMIOLOGY. *Educ:* Pa State Univ, BS, 68; Cornell Univ, PhD(food chem & toxicol), 76. *Prof Exp:* Res assoc, Pesticide Res Lab, Pa State Univ, 68-71; toxicologist, Food & Drug Res Lab, 76-78; asst prof toxicol, Dept Prev Med, 78-85, ASSOC PROF, DEPT PHARMACOL, CORNELL UNIV, 85- *Mem:* Soc Toxicol; AAAS. *Res:* Control of cytochrome p-450 expression and pharmacokinetics. *Mailing Add:* Dept Pharmacol Cornell Univ Ithaca NY 14850

BABITCH, JOSEPH AARON, b Detroit, Mich, July 14, 42; m 64; c 2. NEUROCHEMISTRY. *Educ:* Univ Mich, BS, 65; Univ Calif, Los Angeles, PhD(biol chem), 71. *Prof Exp:* from asst prof to assoc prof, 73-83, PROF CHEM, TEX CHRISTIAN UNIV, 83- *Concurrent Pos:* Fel, US Nat Mult Sclerosis Soc, 71-73; dir chem of behav prog, Tex Christian Univ, 75-83; NIH res career develop grant, 78-83. *Mem:* Am Soc Neurochem; Soc Neurosci; Am Soc Biochem & Molecular Biol; Am Soc Cell Biol. *Res:* Structure and function of brain synapses; calcium binding proteins. *Mailing Add:* Dept Chem Tex Christian Univ Ft Worth TX 76129

BABIUK, LORNE A, b Canora, Sask, Jan 25, 46; m 73; c 2. VIROLOGY. *Educ:* Univ Sask, BSA, 67, MSc, 69, PhD(virol), Univ Brit Columbia, 72. *Hon Degrees:* DSc, Univ Sask, 87. *Prof Exp:* Post doctoral virol, Univ Toronto, 72-73; from asst prof to assoc prof, 73-79, PROF VIROL, UNIV SASK, 79-; ASSOC DIR RES, VET INFECTIOUS DIS ORGN, 84- *Concurrent Pos:* Consult, Molecular Genetic, Inc, 80-85, Genentech Inc, 81-85; chmn, Med Res Coun, Biohazards Comt, 86-, strategic grants biotechnol, Nat Sci Eng Res Coun Can, 85-88. *Mem:* Am Soc Microbiol; Can Soc Microbiol (vchmn, 80-82, chm, 82-84); Am Soc Virol. *Res:* Virus cell interactions molecular level and host response to virol pathogenesis, Bovine Rotavirus, Coronavirus and herpes virus; immune responses and modulation by lymphokines. *Mailing Add:* 245 E Pl Saskatoon SK S7N 2Y1 Can

BABLANIAN, ROSTOM, b Addis-Ababa, Ethiopia, Nov 24, 29; US citizen; m 55; c 1. VIROLOGY, BIOLOGY. *Educ:* Am Univ Cairo, BS, 53, BA, 55; NY Univ, MS, 59, PhD(cell biol), 64. *Prof Exp:* Res assoc virol, Rockefeller Univ, 64-65; from asst prof to assoc prof, 66-81, PROF MICROBIOL & IMMUNOL, STATE UNIV NY DOWNSTATE MED CTR, 81- *Mem:* Am Soc Microbiol. *Res:* Mechanism of animal virus-induced cell damage. *Mailing Add:* Dept Microbiol & Immunol Downstate Med Ctr Brooklyn NY 11203

BABLER, JAMES HAROLD, b Evanston, Ill, June 14, 44. SYNTHETIC ORGANIC CHEMISTRY. *Educ:* Loyola Univ, Ill, BS, 66; Northwestern Univ, PhD(chem), 71. *Prof Exp:* From asst prof to assoc prof, 70-85, PROF CHEM, LOYOLA UNIV, CHICAGO, 85- *Mem:* AAAS; Am Chem Soc; Sigma Xi. *Res:* Natural products; new synthetic methodology; chemicals controlling insect behavior; allylic rearrangements; synthesis of retinoids and carotenoids. *Mailing Add:* Loyola Univ Dept Chem 6525 N Sheridan Rd Chicago IL 60626

BABOIAN, ROBERT, b Watertown, Mass, Nov 17, 34; m 59; c 4. ELECTROCHEMISTRY, CORROSION. *Educ:* Suffolk Univ, BS, 59; Rensselaer Polytech Inst, PhD(chem), 64. *Prof Exp:* Ford Found fel, Univ Toronto, 64-65, sr res assoc, 65-66; mem tech staff, 66-68, HEAD LAB, TEX INSTRUMENTS INC, 68- *Concurrent Pos:* Chmn res comt, Nat Asn Corrosion Eng, 81-84; chmn panel mats sci, Nat Bur Standards, 83-84, review panel mats effects, Nat Acid Precipitation Assessment Prog, 84-86, Automotive Corrosion Prev Comt, 84-86; mem bd, Gen Mats Coun, Soc Automotive Eng, 84-86; fel, Texas Instruments, 76, sr fel, 80, prin fel, 88; distinguished vis prof, Cal State Univ, Chico, 88. *Mem:* Am Chem Soc; Electrochem Soc; Sigma Xi; Am Soc Testing & Mat; Nat Asn Corrosion Eng; Soc Automotive Eng; Fed Mats Soc. *Res:* Electrochemical behavior of metals in electrolyte solutions; corrosion properties of materials; processes for surface treatment of metals, electrocleaning, electropolishing, electromachining; ionic interactions in molten salts. *Mailing Add:* Advan Develop Texas Instruments Inc Mail Sta 10-13 Attleboro MA 02703

BABOOLAL, LAL B, mass transfer kinematics, cloud physics, for more information see previous edition

BABRAUSKAS, VYTENIS, b Ravensburg, WGermany, June 4, 46; US citizen; div. FIRE PROTECTION ENGINEERING, INSTRUMENT DEVELOPMENT. *Educ:* Swathmore Col, AB, 68; Univ Calif, Berkeley, MS, 72, PhD(fire protection eng), 76. *Prof Exp:* Civil eng, US Army Corps Engrs, 69-71; asst res specialist, Univ Calif, Berkeley, 72-76; fire protection engr, 77-82, group head, Mats Fire Properties, 82-85, FLAMMABILITY & TOXICITY MEASUREMENT, NAT BUR STANDARDS, 85- *Mem:* Combustion Inst; Nat Fire Protection Asn; Soc Photo-Optical Instrumentation Engrs. *Res:* Development of new engineering test methods for flammability and combustion toxicity; studies of furniture flammability, pool fires, rate of heat release and smoke production; modeling of room fires. *Mailing Add:* Nat Bur Standards Bldg 224 A363 Gaithersburg MD 20899

BABROV, HAROLD J, b New York, NY, Apr 21, 26; m 52; c 4. MOLECULAR SPECTROSCOPY, ATMOSPHERIC CHEMISTRY & PHYSICS. *Educ:* Univ Calif, Los Angeles, AB, 50, MA, 51; Univ Pittsburgh, PhD(physics), 59. *Prof Exp:* Physicist, US Naval Ord Test Sta, 50; asst, Inst Geophys, Univ Calif, 51; physicist, Knolls Atomic Power Lab, Gen Elec Co, 51-54; asst, Univ Pittsburgh, 54-60; sr physicist, Control Instrument Div, Warner & Swasey Co, NY, 60-63, chief physicist, 63-68; assoc prof physics, Ind Univ, South Bend, 68-72; mem tech staff, Rockwell Int Space Div, 72-79; mem tech staff, Aerospace Corp, 79-85; mem tech staff, Rockwell Int, 85-88; RETIRED. *Concurrent Pos:* Cottrell grant, Res Corp, 69. *Mem:* Fel Optical Soc Am; Am Phys Soc; Sigma Xi. *Res:* Infrared physics; intensity spectra of small molecules and atoms, especially the broadening of spectral lines and determination of transition probabilities; applications of intensity spectroscopy to problems of radiant energy transfer. *Mailing Add:* 9881 Cheshire Ave Westminster CA 92683

BABSON, ARTHUR LAWRENCE, b Orange, NJ, Mar 3, 27; m 50; c 2. CLINICAL CHEMISTRY. *Educ:* Cornell Univ, BA, 50; Rutgers Univ, MS, 51, PhD(physiol, biochem), 53. *Prof Exp:* Res assoc, Radiation Res Lab, Univ Iowa, 53-54; sr res assoc biochem, Warner-Lambert Res Inst, 54-66, dir, Dept Diag Res, 66-76, vpres res & develop, Gen Diag Div, 76-80; PRES, BABSON RES LABS, 80- *Honors & Awards:* Bernard F Gerault Mem Award, Am Asn Clin Chemists, 75. *Mem:* AAAS; Am Asn Clin Chemists; Am Chem Soc; NY Acad Sci; Can Soc Clin Chemists. *Res:* Blood coagulation; diagnostic enzymology; immanoassay. *Mailing Add:* Babson Res Labs Old Mill Rd Chester NJ 07930

BABU, SURESH PANDURANGAM, b Secunderabad, India, July 15, 41; US citizen; m 70. CHEMICAL ENGINEERING, BIOFUELS CONVERSION. *Educ:* Osmania Univ, Hyderabad, India, BS, 62; Indian Inst Technol, Kharagpur, India, MS, 63; Ill Inst Technol, PhD(gas eng), 71. *Prof Exp:* Sr res fel coal conversion, Regional Res Labs, Coun Sci & Indust Res, Hyderabad, India, 63-65; asst prof mineral processing, WVa Univ, 71-73; res suprv, 73-77, mgr process develop, 79-84, assoc dir process develop, 84-87, dir process res, 88-90, DIR, LICENSING & COMMERCIALIZATION,

INST GAS TECHNOL, 90- Concurrent Pos: Vis expert, UN Indust Develop Orgn, Vienna, Austria, 81, USAID, Bandung, Indonesia, 82 & UN Develop Progs, India, 85; task leader, Int Energy Agency, Thermal Gasification Proj, 89-91. Mem: Am Inst Chem Engrs; Am Chem Soc; Lic Exec Soc. Res: Development of substitute fuel production processes from fossil fuels and renewable resources and process modeling. Mailing Add: Inst Gas Technol 3424 S State St Chicago IL 60616

BABU, UMA MAHESH, b Mysore City, India, Mar 29, 47; m 75; c 1. DIAGNOSTICS, BIOLOGICALS. Educ: Univ Mysore, India, BSc, 68; Univ Nebr, PhD(biochem), 74. Prof Exp: Asst immunol, Univ Man, Can, 74-77; res assoc, Thomas Jefferson Univ, Philadelphia, 77-82, asst prof biochem, 82-84; sr scientist, 84-86, Mgr Diag, Pitman-Moore Inc, 86-87; prin scientist, 87-90, GROUP MGR, IMMUNOBIOL RES INST, ANNANDALE, NJ, 90- Concurrent Pos: Adj asst prof biochem, Thomas Jefferson Univ, Philadelphia, 84- Mem: Am Asn Immunologists; NY Acad Sci; Sigma Xi. Res: Diagnostics and immunoassays; application of immunochemistry in diagnostics; hybridoma technology; protein conjugation; vaccine developments; application of biotechnology in general medicine; analysis of feasibility of new projects and techniques. Mailing Add: Immunobiol Res Inst Box 999 RT 22 E Annandale NJ 08801-0999

BABUSKA, IVO MILAN, b Prague, Czech, Mar 22, 26. NUMERICAL ANALYSIS, APPLIED MATHEMATICS. Educ: Tech Univ Prague, PhD(civil eng), 51; Czech Acad Sci, PhD(math), 55, DSc(math), 60. Prof Exp: Res scientist math, Math Inst, Czech Acad Sci, 50-69; RES PROF MATH, UNIV MD, 69- Concurrent Pos: Alexander von Humboldt sr scientist award, 77. Honors & Awards: Czechoslovak State Award, Math, 68. Mem: Am Math Soc; Soc Indust Appl Math. Res: Numerical analysis of partial differential equations; applied mathematics related to continuum theory. Mailing Add: Inst Phys Sci & Technol Univ Md College Park MD 20742-2431

BACA, GLENN, b Socorro, NMex, Nov 22, 43. PHYSICAL INORGANIC CHEMISTRY, ELECTROCHEMISTRY. Educ: NMex State Univ, BS, 65, MS, 67, PhD(inorg chem), 69. Prof Exp: Teaching asst chem, NMex State Univ & Rose-Hulman Inst Technol, 65-69; asst prof chem, Mercer County Community Col, 69-70; ASSOC PROF CHEM, ROSE-HULMAN INST TECHNOL, 70- Mem: Sigma Xi; Am Chem Soc. Res: The determination of transference numbers in non-aqueous solvents in conjunction with the development and construction of batteries in water-like solvents. Mailing Add: 202 Capri Rd Mesilla Park NM 88047

BACA, OSWALD GILBERT, b Belen, NMex, July 6, 42; m 63; c 2. MICROBIOLOGY, BIOLOGY. Educ: Univ NMex, BS, 66, MS, 69; Univ Kans, PhD(microbiol), 73. Prof Exp: Fel biochem, Friedrich Miescher Inst, Basel, Switz, 73-75; res assoc biochem, Univ Minn, Minneapolis, 75-76; ASST PROF MICROBIOL, UNIV NMEX, 76- Concurrent Pos: Prin invest, NSF, Univ NMex, 77-79; prin investr, NSF, Univ NMex, 77- Mem: Am Soc Microbiol; AAAS; Am Soc Rickettsiology & Rickettsial Dis; Fedn Am Soc Exp Biol. Res: Host parasite interactions; protein synthesis. Mailing Add: Dept Biol & Microbiol Univ NMex Albuquerque NM 87131

BACANER, MARVIN BERNARD, b Chicago, Ill, Mar 18, 23; m 48; c 4. PHYSIOLOGY, MEDICINE. Educ: Boston Univ, MD, 53. Prof Exp: Intern path, Boston City Hosp, Mass, 53-54; resident internal med, Cambridge City Hosp, 54-55 & Mt Sinai Hosp, New York, 55-56; res fel cardio-pulmonary physiol, Stanford Univ, 57-59; res physician biophys & physiol, Donner Lab, Lawrence Radiation Lab, Univ Calif, Berkeley, 59-61; PROF PHYSIOL, UNIV MINN, MINNEAPOLIS, 61- Concurrent Pos: Percy Klingenstein fel cardiol, Mt Sinai Hosp, New York, 56-57; USPHS res fel, Med Sch, Stanford Univ, 57-59; Burroughs-Wellcome fel, Rambam Hosp, Haifa, Israel, 68-69. Honors & Awards: William & Dorothy Fish Kerr Award, San Francisco Heart Asn, 57. Mem: Am Physiol Soc. Res: Metabolic determinants of heart performance; development of bretylium tosylate antiarrhythmic drug; in-vivo methods of measuring intestinal circulation; x-ray microanalysis of deep frozen tissue with the scanning electron microscope; electron optical imaging of deep-frozen muscle; voltage clamp studies of antifibrillatory drugs. Mailing Add: Dept Physiol Univ Minn Med Sch Minneapolis MN 55455

BACASTOW, ROBERT BRUCE, b Kearny, NJ, June 4, 30; m 76; c 4. NATURAL CARBON CYCLE. Educ: Mass Inst Technol, BS, 52, MS, 54; Univ Calif, Berkeley, PhD(physics), 63. Prof Exp: Asst prof physics, Univ Calif, Riverside, 65-69, assoc res physicist, San Diego, 69-70; assoc specialist, 74-76, SPECIALIST, SCRIPPS INST OCEANOG, 76- Mem: Am Geophys Union; AAAS; Sigma Xi. Res: Modeling the natural carbon cycle in the world oceans; uptake of anthropogenic carbon dioxide by the oceans. Mailing Add: Scripps Inst Oceanog La Jolla CA 92093

BACCANARI, DAVID PATRICK, b Wilkes-Barre, Pa, Apr 15, 47; m 68; c 2. ENZYMOLOGY. Educ: Wilkes Col, BS, 68; Brown Univ, PhD(biochem), 72. Prof Exp: Fel, Brown Univ, 72-73; SR RES MICROBIOLOGIST, WELLCOME RES LAB, 73- Mem: Am Soc Microbiol; Am Soc Biol Chem; Am Asn Cancer Res. Res: Comparative enzymology and its relationship to chemotherapy. Mailing Add: Div Molecular Genetics & Microbiol Wellcome Res Labs Res Triangle Park NC 27709

BACCEI, LOUIS JOSEPH, b Torrington, Conn, Dec 16, 41; m 64; c 4. POLYMER CHEMISTRY. Educ: Cent Conn State Col, BS, 63; Univ Md, PhD(org chem), 70. Prof Exp: Res & develop chemist monomer polymers, Am Cyanamid Co, 69-72; res & develop chemist, 72-73, sr chemist, 74-75, res assoc, 75-78, MGR ANAEROBIC CHEM, LOCTITE CORP, 78- Mem: Am Chem Soc. Res: Monomer and polymer synthesis; study of structure-property relationships; free radical chemistry. Mailing Add: 60 Oak Ridge Rd Unionville CT 06085

BACCHETTI, SILVIA, b Rome, Italy, Sept 20, 39; m 68. VIROLOGY, CANCER RESEARCH. Educ: Univ Rome, Laurea, 61. Prof Exp: Fel, Univ Rome, 61-64, researcher, 64-65; fel, Ont Cancer Inst, 66-68; vis scientist, Argonne Nat Lab, 68-70; asst prof, Univ Leiden, Holland, 70-74; from asst prof to assoc prof, 75-82, PROF, MCMASTER UNIV, 82- Concurrent Pos: Res scholar, Nat Cancer Inst, Can, 76-82, res assoc, 82-85, Terry Fox cancer res scientist, 85- Res: Mechanisms of mammalian cell transformation by DNA tumor virus; induction of genetic damage by viral oncogenes; characterization of viral functions involved in DNA metabolism. Mailing Add: Dept Path Rm 4H30 McMaster Med Ctr Hamilton ON L8N 3Z5 Can

BACCHUS, HABEEB, b Triumph, Brit Guiana, Oct 15, 28; US citizen; m 56; c 6. INTERNAL MEDICINE, ENDOCRINOLOGY. Educ: Howard Univ, BS, 47; George Washington Univ, MS, 48, PhD(physiol), 50, MD, 54; Am Bd Internal Med, dipl, 63 & 74, cert endocrinol & metab, 75. Prof Exp: Res assoc physiol, George Washington Univ, 51-54, asst res prof, 54-57, asst clin prof, 57-59; prin investr, Res Labs Biochem, Providence Hosp, Washington, DC, 60-65, dir metab lab & sr attend physician internal med, 65-69, res mem dept path, 65-67, chief res div, 67-69; assoc prof med, 69-79, PROF MED, SCH MED, LOMA LINDA UNIV, 79-; ASSOC CHIEF INTERNAL MED, RIVERSIDE GEN HOSP, 69-; CHIEF ENDOCRINOL & METAB, 84-, CHIEF MED, 87- Concurrent Pos: Clin assoc, Nat Cancer Inst, 57-59; coordr health serv, Am Univ, 68-69; med consult, Vet Admin, 61-; assoc clin pathologist, Riverside Gen Hosp, 70-80. Mem: Am Col Physicians; Endocrine Soc; Am Fedn Clin Res; Am Physiol Soc. Res: Pituitary-adrenal axis; ascorbic acid adrenal hormone formation and breakdown; ascorbic acid and intermediary metabolism; adrenogenital syndromes; metabolic diseases; Marfan's syndrome; connective tissue diseases; hypertensive vascular disease; experimental hypertension; glycopeptide markers in malignant neoplastic disorders; disorders of calcium metabolism; vitamin D hormones; thyroid stimulating immunoglobulins; insulin resistance phenomena. Mailing Add: Riverside Gen Hosp 9851 Magnolia Ave Riverside CA 92503

BACH, DAVID RUDOLPH, b Shimonoseki, Japan, Apr 24, 24; US citizen; m 48, 75; c 4. NUCLEAR PHYSICS, REACTOR PHYSICS. Educ: Univ Mich, BA, 48, MS, 50, PhD(physics), 55. Prof Exp: Res assoc physics, Knolls Atomic Power Lab, Gen Elec Co, NY, 55-58; from assoc prof to prof nuclear eng, Univ Mich, Ann Arbor, 64-79; res physicist, 79-84, CONSULT, LOS ALAMOS NAT LAB, 84- Concurrent Pos: Consult, Knolls Atomic Power Lab, 65-69, Lawrence Livermore Nat Lab, 86- Mem: Am Phys Soc. Res: Experimental nuclear and reactor physics; neutron physics, especially pulsed neutron research; plasma physics, especially laser interaction with dense plasmas; inertial confinement fusion. Mailing Add: 3142 Grove St Ventura CA 93003

BACH, ERIC, b Chicago, Ill, Nov 30, 52. NUMBER-THEORETIC ALGORITHMS, COMPLEXITY THEORY. Educ: Univ Mich, BA, 74; Univ Calif, Berkeley, PhD(computer sci), 84. Prof Exp: Asst prof, 84-89, ASSOC PROF COMPUTER SCI, UNIV WIS-MADISON, 89- Concurrent Pos: NSF presidential young investr, 86; vis assoc prof, Univ Paris-Sud, Orsay, 89; mem, Spec Interest Group Automata & Computability Theory, Asn Comput Mach. Mem: Asn Comput Mach; Am Math Soc; Math Asn Am. Res: Design and analysis of algorithms for algebraic and number-theoretic problems; computer algebra; cryptography; pseudo-random number generation. Mailing Add: Dept Computer Sci Univ Wis Madison WI 53706

BACH, FREDERICK L, b Corning, NY, Feb 5, 21; m 47; c 2. ORGANIC CHEMISTRY. Educ: Univ Md, BS, 43; NY Univ, MS, 57, PhD(chem), 63. Prof Exp: Sr res chemist, Lederle Labs Div, 49-71, dir Tech Regulatory Affairs, Am Cyanamid, 71-78; RETIRED. Concurrent Pos: Lectr org chem, NY Univ, 63-73. Mem: AAAS; Am Chem Soc; Brit Chem Soc; Brit Inst Regulatory Affairs. Res: Tuberculostatic agents; tropical diseases; cardiovascular drugs; hypocholesteremic agents; free radical chemistry; computer solutions of photochemical mechanisms. Mailing Add: 44 Sunrise Dr Montvale NJ 07645-1030

BACH, HARTWIG C, b Kiel, Ger, May 24, 30; m 57. POLYMER CHEMISTRY. Educ: Kiel Univ, BS, 54, MS, 57, PhD(org chem), 59. Prof Exp: Res assoc fels, org chem, Univ SC, 59-60 & Fla State Univ, 60-61; res specialist polymer chem, Chemstrand Res Ctr, Inc, 61-69, sr res specialist nylon res, 69-71, sci fel nylon res, 71-81, sr sci fel adv technol, Monsanto Co, 81-89; RETIRED. Mem: Am Chem Soc. Res: Organic chemistry; catalysis; polymer chemistry; synthetic fibers; solution and melt spinning. Mailing Add: 11416 High Springs Rd Pensacola FL 33534

BACH, JOHN ALFRED, b Detroit, Mich, July 26, 35; m 57; c 3. MICROBIOLOGY. Educ: Mich State Univ, BS, 58, MS, 61, PhD(microbiol), 63. Prof Exp: Sr microbiologist, Bristol Labs, NY, 63-67; ASSOC SCIENTIST, UPJOHN CO, 67- Mem: Am Soc Microbiol. Res: Physiology of bacterial endospores; antibiotics; chemotherapy; fluorescent antibody techniques; fermentation microbiology. Mailing Add: 3300 Woodstone Kalamazoo MI 49007

BACH, MARILYN LEE, b Lynn, Mass, Apr 24, 37; m 58; c 3. SCIENCE POLICY. Educ: Simmons Col, BS, 58; NY Univ, PhD(biochem), 66. Prof Exp: Fel oncol, Univ Wis, 66-67, proj assoc, Dept Genetics, 67-68, res assoc, 68-70, from asst prof to assoc prof, Dept Pediat, 70-78; fel sci policy, Brookings Inst, 81-82; ASSOC PROF, DEPT LAB MED/PATH & HEALTH SERV RES CTR, UNIV MINN, MINNEAPOLIS, 78-; ASSOC PROF, DIV HEALTH SERV RES & POLICY, SCH PUB HEALTH, 82- Concurrent Pos: Vis prof, Univ Leiden, Netherlands, 73; mem, Nat Adv Allergy & Infectious Dis, 74-77 & Basil O'Connor Starter Res Grant Panel, Nat Found, 74-77; sci liaison, Off Med Applications Res, 79-80; spec asst to dir, Nat Inst Allergy & Infectious Dis, NIH, 80-81; exec dir, Minnesota Coun on Biotechnol, 85-87. Honors & Awards: Gold Medal Award, Am Inst Chemists, 58; Fac Res Award, Am Cancer Soc, 71. Mem: Transplantation Soc; Am Asn Immunologists; AAAS. Res: Relationships between academia, industry and government, focusing on work in recombinant DNA, hybridomas, and vaccine development. Mailing Add: 644 Goodrich Ave St Paul MN 55105

BACH, MICHAEL KLAUS, b Stuttgart, Ger, Oct 2, 31; US citizen; m 54; c 2. BIOCHEMISTRY, IMMUNOLOGY. *Educ:* Queens Col (NY), BS, 53; Univ Wis, MS, 55, PhD, 57. *Prof Exp:* Res plant biochemist, Union Carbide Chem Co, 57-60; biochem res assoc, 60-70, sr res scientist, 70-81, DISTINGUISHED SCIENTIST, UPJOHN CO, 81- *Concurrent Pos:* Vis lectr med, Harvard Med Sch, 69-70; allergy/immunol study sect, NIH, 82-86; vis foreign scientist, Karolinka Inst, Stockholm, 86-87. *Mem:* Am Asn Immunol; Am Soc Biol Chemists & Molecular Biol. *Res:* Mediators of anaphylaxis: immuno-biochemistry of their release and production, their structure and pharmalologic interactions; cellular aspects of immunoglobulin E-mediated tissue damage; leukotrienes: their biosythesis, metabolism, actions and pharmacologic interventions. *Mailing Add:* 2115 Frederick Ave Kalamazoo MI 49001

BACH, RICARDO O, b Ulm, Ger, Dec 25, 17; US citizen; m 45; c 3. BIOCHEMISTRY. *Educ:* Univ Zurich, PhD(phys chem), 42. *Prof Exp:* Dir res electrometall & inorg chem, Meteor Est Met, Buenos Aires, Arg, 42-54; res investr, Cent Res Labs, Am Smelting & Ref Co, NJ, 54-59; dir inorg res, Lithium Corp Am, Inc, 59-67, mgr res dept, 67-82; PRES, BACH ASSOCS, INC, 83- *Mem:* Am Chem Soc; Am Inst Mining, Metall & Petrol Engrs; fel Am Inst Chem; Sigma Xi. *Res:* Extractive metallurgy of lithium; biochemistry of antiviral compounds. *Mailing Add:* 44 Old Post Rd Lake Wylie SC 29710

BACH, SHIRLEY, b Williston Park, NY, Nov 22, 31; m 54; c 2. BIOMEDICAL ETHICS. *Educ:* Queens Col, NY, BS, 53; Univ Wis, PhD(org chem), 57. *Prof Exp:* from instr to assoc prof, 74-81, PROF NATURAL SCI, WESTERN MICH UNIV, 81- *Mem:* Am Chem Soc; AAAS. *Res:* Chemical reactions in monomolecular layers; natural product chemistry; human serum complement genetics. *Mailing Add:* Dept Gen Sci Western Mich Univ Kalamazoo MI 49008

BACH, WALTHER DEBELE, JR, b Pensacola, Fl, Apr 11, 39; m 61; c 3. ATMOSPHERIC DYNAMICS. *Educ:* Ga Inst Technol, BS, 61; Univ Okla, MS, 66, PhD(meteorol), 70. *Prof Exp:* Res meteorologist, Res Triangle Inst, 69-82; METEOROLOGIST, US ARMY RES OFF, 83- *Concurrent Pos:* Vis assoc prof, NC State Univ, 81-82; vis fac assoc, Savannah River Lab, 82; consult meteorologist, 82- *Mem:* Am Meteorol Soc. *Res:* Manager of basic research program in dynamics of the planetary boundary layer for Army Research Office. *Mailing Add:* 68 Kimberly Dr Durham NC 27707

BACHA, JOHN D, b Minneapolis, Minn, Aug 21, 41; m 66; c 4. ORGANIC CHEMISTRY, DISTILLATE & RESIDUAL PETROLEUM FUELS. *Educ:* St Mary's Col (Minn), BA, 63; Case Inst Technol, PhD(chem), 67. *Prof Exp:* Res chemist, Gulf Res & Develop Co, 67-72, sr res chemist, 72-78, res assoc, 78-85; sr res chemist, 85-87, SR RES ASSOC, CHEVRON RES & TECHNOL CO, 87- *Mem:* Am Chem Soc; Am Inst Mining, Metall & Petrol Engrs; Metall Soc; AAAS; Am Carbon Soc; Am Soc Test & Mat. *Res:* Free radical chemistry; carbonium ion chemistry and aromatic substitution; metal catalysis of organic reactions; solvent effects upon organic reactions; monomers for high performance polymers; carbon products and their precursors; distillate and residual petroleum fuels. *Mailing Add:* Chevron Res & Technol Co PO Box 1627 Richmond CA 94802-0627

BACHA, WILLIAM JOSEPH, JR, b US, Mar 1, 30; m 53, 79; c 2. PARASITOLOGY. *Educ:* Long Island Univ, BS, 51; NY Univ, PhD(biol), 59. *Prof Exp:* Asst biol, NY Univ, 51-54 & 56-58; lectr, City Col New York, 58-59; from instr to assoc prof, 59-70, actg dean, 69-70, chmn dept biol & dir grad prog, 71-74, PROF BIOL, RUTGERS UNIV, CAMDEN, 70- *Mem:* Sigma Xi; Am Soc Parasitol; Am Micros Soc; AAAS. *Res:* Biology; life cycles, behavior and development of trematodes. *Mailing Add:* 311 N Fifth St Rutgers Univ Camden NJ 08102

BACHARACH, MARTIN MAX, b Munich, Germany, Dec 18, 25; nat US. BIOLOGY, GENETICS. *Educ:* Rutgers Univ, BS, 50; Univ Wis, MS, 52, PhD(poultry genetics), 55. *Prof Exp:* Asst, Univ Wis, 50-55, assoc, 55-57; cytologist, Tumor Res, Vet Admin Hosp, Hines, Ill, 57-58; instr biol, NCent Col, Ill, 59; geneticist, Honegger Farms Co, Inc, Ill, 59-60; res nutritionist & librn, Dawe's Labs, Inc, Chicago, 60-62; info res scientist, Squibb Inst Med Res, NJ, 62-64; mem staff, Med Serv Dept, Sandoz, Inc, NJ, 64; dir sci info, Cortez F Enloe, Inc, NY, 65-66; supvr info serv, William D McAdams, Inc, New York, 66-82 CONSULT 82- *Res:* Immuno-genetics of species and species hybrids; developmental genetics; physiological reactions of normal and malignant cells; literature research. *Mailing Add:* 390 Prospect Ave Hackensack NJ 07601-2547

BACHE, ROBERT JAMES, b 1938; m; c 3. CORONARY ARTERY DISEASE, ISCHEMIC HEART DISEASE. *Educ:* Harvard Univ, MD, 64. *Prof Exp:* PROF MED, UNIV MINN, 76- *Mem:* Am Soc Clin Invest; Ctrl Soc Clin Res; Asn Univ Cardiologists; Am Physiol Soc; fel Am Col Cardiol; Am Fedn Clin Res; Asn Am Physicians. *Res:* Cardiovascular physiology. *Mailing Add:* Univ Minn Hosp Minneapolis MN 55455

BACHELIS, GREGORY FRANK, b Los Angeles, Calif, Mar 9, 41; m; c 1. ANALYSIS OF ALGORITHMS, PARALLEL ALGORITHMS. *Educ:* Reed Col, BA, 62; Univ Ore, MA, 63, PhD(math), 66. *Prof Exp:* Asst, Univ Ore, 62-63 & 64-66; asst prof math, State Univ NY Stony Brook, 66-70; vis assoc prof, Kans State Univ, 70-71; assoc prof math, 71-79, PROF MATH, WAYNE STATE UNIV, 79- *Concurrent Pos:* Programmer & analyst, Ford Motor Co, 82-83; adj prof computer sci, Wayne State Univ, 88- *Mem:* Am Math Soc; Math Asn Am; Asn Comput Mach; Soc Indust & Appl Math. *Mailing Add:* Dept Math Wayne State Univ Detroit MI 48202

BACHELOR, FRANK WILLIAM, b New York, NY, Nov 2, 28; m 55; c 3. ORGANIC CHEMISTRY. *Educ:* Univ Calif, Berkeley, BS, 51; Mass Inst Technol, PhD(org chem), 60. *Prof Exp:* Asst chemist, Merck & Co, Inc, NJ, 51-54; NSF fel, Inst Org Chem, Munich Tech Univ, 60-61; res assoc org chem, Johns Hopkins Univ, 61-63; from asst prof to assoc prof org chem, Univ Calgary, 63-89; AT FWB CHEM CONSULT LTD, 89- *Mem:* Am Chem Soc; Brit Chem Soc; fel Chem Inst Can. *Res:* Structure and synthesis of natural products; organic reaction mechanisms; chemistry of coal and asphaltenes. *Mailing Add:* FWB Chem Consult Ltd 339 50th Ave SE No 118 Calgary AB T2G 2B3 Can

BACHENHEIMER, STEVEN LARRY, b Chicago, Ill, Sept 19, 45; m 67, 86; c 2. ANIMAL VIROLOGY. *Educ:* Univ Ill, Urbana, BS, 67; Univ Chicago, PhD(microbiol), 72. *Prof Exp:* From res assoc to lectr animal virol, Columbia Univ, 72-74; fel, Rockefeller Univ, 74-75; asst prof, 75-80, ASSOC PROF ANIMAL VIROL, DEPT MICROBIOL & IMMUNOL, UNIV NC CHAPEL HILL 80- *Concurrent Pos:* Damon Runyon fel, 72-74; NIH res fel. *Mem:* Am Soc Microbiol; AAAS; Am Soc Virol. *Res:* Synthesis and post-transcriptional modification of herpes simplex virus transcripts; replication and maturation of herpes simplex virus DNA. *Mailing Add:* Dept Microbiol & Immunol Univ NC Chapel Hill NC 27599

BACHER, ANDREW DOW, b Ithaca, NY, Sept 5, 38; m 62. NUCLEAR PHYSICS. *Educ:* Harvard Univ, AB, 60; Calif Inst Technol, PhD(nuclear physics), 67. *Prof Exp:* Res fel physics, Calif Inst Technol, 66-67; NSF fel, Lawrence Radiation Lab, Univ Calif, 67-68, res physicist, 68-70; from asst prof to assoc prof, 70-77, PROF PHYSICS, IND UNIV, BLOOMINGTON, 77- *Mem:* Am Phys Soc. *Res:* Nuclear astrophysics. *Mailing Add:* Dept Physics Ind Univ Bloomington IN 47401

BACHER, FREDERICK ADDISON, b Northampton, Mass, July 18, 15; m 44, 54; c 3. PHYSICAL CHEMISTRY. *Educ:* Harvard Univ, SB, 36; Rutgers Univ, MSc, 45. *Prof Exp:* Asst chemist, Plastics Dept, E I du Pont de Nemours & Co, Mass, 36-38; chemist, Merck Sharp & Dohme Res Labs, 39-57, mgr new prod control, 57-70, dir pharmaceut anal, 70-80; RETIRED. *Mem:* AAAS; Am Chem Soc; Sigma Xi. *Res:* Stability-indicating assays for drug formulations; purity of organic compounds; analytical methods; bioavailability of drug dosage forms. *Mailing Add:* 666 W Germantown Pike Apt 520S Plymouth Meeting PA 19462

BACHER, ROBERT FOX, b Loudonville, Ohio, Aug 31, 05; m 30; c 2. HIGH ENERGY PHYSICS, NUCLEAR ENERGY. *Educ:* Univ Mich, BS, 26, PhD(physics), 30. *Hon Degrees:* ScD, Univ Mich, 48; LLD, Claremont Grad Sch, 75. *Prof Exp:* Nat res fel physics, Calif Inst Technol, 30-31 & Mass Inst Technol, 31-32; Lloyd fel, Univ Mich, 32-33; instr, Columbia Univ, 34-35; from instr to prof, Cornell Univ, 35-49, dir nuclear studies lab, 46; res assoc radiation lab, Mass Inst Technol, 41-45; head, Exp Physics Div, Atomic Bomb Proj, Los Alamos Lab, 43-44, head bomb physics div, 44-45; mem, US AEC, 46-49; chmn div physics, math & astron, 49-62, provost, 62-70, prof, 49-76, EMER PROF PHYSICS, CALIF INST TECHNOL, 76- *Concurrent Pos:* Sci Adv to US Rep UN AEC, 46-49; Mem, President's Sci Adv Comt, 53-55, 57-60 & Naval Res Adv Comt, 57-62; trustee, Assoc Univs, 46, 62-66, 70-73, Rand Corp, 50-60, Carnegie Corp, 59-76 & Univ Res Assoc, 65-75, chmn, 70-73, pres, 73-74; bd fels, Claremont Univ Ctr, 71-88. *Honors & Awards:* Medal for Merit. *Mem:* Nat Acad Sci; AAAS; Am Phys Soc (pres, 64); Am Philos Soc; Int Union Pure & Appl Physics (vpres, 66-69, pres, 69-72). *Res:* High energy physics; photoproduction of mesons; nuclear physics; atomic energy. *Mailing Add:* 300 Hot Springs Rd No 12 Montecito CA 93108

BACHMAN, BONNIE JEAN WILSON, b Mt Pleasant, Iowa, Sept 20, 50; m 70, 82; c 1. MECHANICS, ENGINEERING & SURFACE PHYSICS. *Educ:* Ill Benedictine Col, BS, 79; Rutgers Univ, MS, 87. *Prof Exp:* Lab supvr, Norplex Div, UOP, 79-80; mem tech staff, AT&T Bell Labs, 81-91; RES ASSOC, SOLAR ENERGY RES INST, 91- *Concurrent Pos:* Chair, Plastics Anal Div, Soc Plastics Engrs, 85, chair- elect, Eng Properties & Struct Div, 91. *Honors & Awards:* Outstanding Achievement Award, Soc Plastics Engrs, 91. *Mem:* Soc Plastics Engrs (treas, 90, 2nd vpres, 91; NAm Thermal Anal Soc (secy, 85-87, vpres, 88, pres, 89); Am Chem Soc; Am Phys Soc; Mat Res Soc. *Res:* Developing polymers and their processing conditions for electronic, microelectronic and optoelectronic applications; property-structure relationships of high temperature polymers; polymer-metal interface studies. *Mailing Add:* 9627 Mountain Ridge Pl Boulder CO 80302-9336

BACHMAN, GEORGE, b New York, NY, Jan 17, 29. MATHEMATICS. *Educ:* NY Univ, BEE, 50, MS, 52, PhD(math), 56. *Prof Exp:* From instr to asst prof math, Rutgers Univ, 57-60; from asst prof to assoc prof, 60-65, PROF MATH, POLYTECH INST BROOKLYN, 65- *Honors & Awards:* Distinguished Teacher Award, Polytech Inst NY, 74. *Mem:* Am Math Soc; Math Asn Am; Math Soc France; Indian Math Soc; Can Math Cong; Sigma Xi. *Res:* Functional analysis; algebraic number theory; topological measure theory. *Mailing Add:* Dept Math Polytech Inst NY 333 Jay St Brooklyn NY 11201

BACHMAN, GEORGE O, b Wichita, Kans, Aug 4, 20. ROCKY MOUNTAIN, DESERT REGIONS. *Educ:* Univ NMex, BA, 48. *Prof Exp:* Geologist, US Geol Surv, 48-78; independent consult, 78-88; RETIRED. *Mailing Add:* 4008 Hannett Ave NE Albuquerque NM 87110

BACHMAN, GEORGE S(TRICKLER), b Lebanon, Pa, Jan 6, 15; m 50. CERAMICS. *Educ:* Pa State Col, BS, 36, MS, 38; Univ Mich, BSE, 43; Univ Ill, PhD, 47. *Prof Exp:* Ceramic engr, Owens-Ill Glass Co, 37-42; res ceramist, Pittsburgh Plate Glass Co, 47-52, dir, Res Fiberglass Div, 52-57, asst dir res, Gen Ceramics Corp, 57-62, mfr thermoelements, 62-69, proprietor patent referencing serv, 69-80; RETIRED. *Mem:* Am Chem Soc; Am Ceramic Soc; Am Inst Chem. *Res:* Glass compositions; thermoelectric materials. *Mailing Add:* Redfield Village 16 B 3 Edison NJ 08837

BACHMAN, GERALD LEE, b Alton, Ill, Oct 6, 32; m 57; c 2. ORGANIC CHEMISTRY. *Educ:* Univ Ill, BS, 54; Wash Univ, PhD(org chem), 63. *Prof Exp:* Res chemist biochem, 56-58, res chemist explor synthesis, 63-80, sr res group leader, 80-84, res mgr, 84-86, mgr res & develop, 86-87, RES DIR, MONSANTO CO, 87- *Mem:* Am Chem Soc. *Res:* Chemistry of carbenes, particularly phenylcarbene; synthesis of heterocyclic compounds; amino acid chemistry; asymmetric hydrogenation; protein purification. *Mailing Add:* 1627 Dunmorr Des Peres MO 63131-3830

BACHMAN, HENRY L(EE), b New York, NY, Apr 29, 30; m 51; c 3. ENGINEERING MANAGEMENT. *Educ:* Polytech Univ, BEE, 51, MEE, 54. *Prof Exp:* Elec engr, Wheeler Labs, Inc, 51-61, asst chief engr, 62-68, pres, 68-70; prod line dir radio navig, Hazeltine Corp, 70-71, dir qual & logistics eng, 71-75, vpres qual & customer serv, 76-78, vpres opers, 79-89, vpres eng, 86-89, VPRES MKT PLANNING, HAZELTINE CORP, 89- *Concurrent Pos:* Bd dir, Inst Elec & Electronics Engrs; bd trustees, Polytechnic Univ. *Honors & Awards:* Centennial Medal, Inst Elec & Electronics Engrs. *Mem:* Fel Inst Elec & Electronics Engrs (exec vpres, 84, treas 85, pres-elect, 86, pres, 87); fel AAAS. *Res:* Antennas and microwave components; manufacturing technology. *Mailing Add:* Five Brandy Rd Huntington NY 11743

BACHMAN, KENNETH CHARLES, b West New York, NJ, Mar 1, 22; m 51; c 3. PHYSICAL CHEMISTRY, FUEL SCIENCE. *Educ:* Rutgers Univ, BS, 44, MS & PhD(phys chem), 50. *Prof Exp:* Chemist, Benzol Prod Co, 44 & 46; asst, Rutgers Univ, 46-49, res asst, 49-50; res chemist, M W Kellogg Co, 51-55; res chemist, Esso Res & Eng Co, Standard Oil Co NJ, 55-66, sr chemist, 66-70, res assoc, Exxon Res & Eng Co, Exxon Corp, Linden, 70-84; RETIRED. *Honors & Awards:* Arch T Colwell, Soc Automotive Eng, 66. *Mem:* Am Chem Soc; Sigma Xi. *Res:* Cause of and control methods for ignitions induced by static discharges during fuel transfer, such as truck loading and aircraft fueling; methods for control of automotive exhaust emissions, particularly sulphates; combustion, particularly residual fuels. *Mailing Add:* 404 Wells St Westfield NJ 07090

BACHMAN, MARVIN CHARLES, b Decorah, Iowa, Feb 8, 21; m 44; c 3. MICROBIOLOGY. *Educ:* Luther Col, BS, 42; Okla State Univ, MS, 47. *Prof Exp:* Chemist, E I du Pont de Nemours & Co, 42-44; teaching fel, Okla State Univ, 46-47; microbiologist, Com Solvents Corp, 47-51, dir microbiol res, 51-56, lab res, 56-62, dir res, 62-75; dir res & develop, IMC Chem Group, Int Min & Chem Corp, 75-79, dir life sci, 79-82; RETIRED. *Mem:* Am Soc Microbiol; Soc Chem Indust; Am Soc Testing & Mat; Agr Res Inst. *Mailing Add:* 2931 Placita Nublada Green Valley AZ 85614

BACHMAN, PAUL LAUREN, photo chemistry, photoimaging systems; deceased, see previous edition for last biography

BACHMAN, RICHARD THOMAS, b Los Angeles, Calif, May 10, 43; m 67; c 2. GEOACOUSTICS, GEOTECHNOLOGY. *Educ:* San Diego State Univ, AB, 66. *Prof Exp:* OCEANOGR, NAVAL OCEAN SYSTS CTR, 67- *Mem:* Fel Geol Soc Am. *Res:* Measurement of physical properties of marine sediments and rocks which control elastic wave propagation; generalization and prediction of acoustic properties; integration of sediment acoustics with general studies of oceanic sound propagation. *Mailing Add:* Dept Immunol IMM3 La Mesa CA 91942

BACHMAN, WALTER CRAWFORD, b Pittsburgh, Pa, Dec 24, 11; wid; c 2. HYDRODYNAMICS, THERMODYNAMICS. *Educ:* Lehigh Univ, BS, 33, MS, 35. *Prof Exp:* Asst instr mech eng, Lehigh Univ, 35-36; engr, Fed Shipbuilding & Dry Dock Co, US Steel Corp, 36; marine engr, Gibbs & Cox, Inc, 36-70, chief engr, 58-70, vpres, 63-70; RETIRED. *Concurrent Pos:* Mem comt eng, Am Bur Shipping, 59-; mech eng adv comt, Norwich Univ, 64- *Mem:* Nat Acad Eng; fel Am Soc Mech Engrs; Soc Naval Archit & Marine Engrs; Am Soc Naval Engrs; Nat Soc Prof Engrs. *Res:* Ship design; development of propulsion systems and advanced experimental power plants for naval and merchant ships, including the SS United States and various unusual ship types and marine vehicles. *Mailing Add:* Wayside Short Hills NJ 07078

BACHMANN, BARBARA JOYCE, b Ft Scott, Kans, May 16, 24; div. MICROBIAL GENETICS. *Educ:* Baker Univ, AB, 45; Univ Ky, MS, 47; Stanford Univ, PhD(microbiol), 54. *Prof Exp:* Assoc bact, Univ Calif, 53-56; res assoc biochem, Columbia Univ, 57-58; res assoc microbiol, Yale Univ, 58-64; asst prof, Sch Med, NY Univ, 64-68; lectr microbiol, Sch Med, 68-76, sr res scientist, dept human genetics, 82-87, CUR, E COLI GENETIC STOCK CTR, 70-, SR RES SCIENTIST, DEPT BIOL, YALE UNIV, 87- *Concurrent Pos:* Res asst, Oak Ridge Nat Lab; ed, Neurospora Newslett, 63-76; lectr, dept biol, Yale Univ, 76- *Honors & Awards:* J R Porter Award, Am Soc Microbiol. *Mem:* AAAS; Am Soc Microbiol; Genetics Soc Am. *Res:* Linkage map of Escherichia coli. *Mailing Add:* Dept Biol Yale Univ PO Box 6666 New Haven CT 06511-7444

BACHMANN, FEDOR W, b Zurich, Switz, May 23, 27; US & Swiss citizen; m 57; c 1. INTERNAL MEDICINE, HEMATOLOGY. *Educ:* Gym Zurich, Matura, 46; Univ Zurich, Switz, MD. *Prof Exp:* Resident, Univ Zurich Med Sch, 55-61; trainee enzym, Wash Univ, St Louis, Mo, 61-64, asst prof med, 64-69; assoc prof med, Rush Med Sch, Chicago, 69-73; dir med res, Schering-Plough, US, Lucerne, 73-76; PROF HEMAT, MED SCH, UNIV LAUSANNE, 77-, PROVOST, 87- *Concurrent Pos:* Mem, Sci Coun, Swiss Cancer Inst, 80-88 & Coun Thrombosis, Am Heart Asn, 80-; chmn, Int Comt Thrombosis & Hemostatis, 84-86. *Mem:* Am Heart Asn; Am Col Physicians; Am Soc Hemat; Am Physiol Soc; Cent Soc Clin Res. *Res:* Hematology, particularly hemostasis, thrombosis and fibriolysis; author of over 200 publications. *Mailing Add:* Hemat Div Lausanne Univ Med Ctr Lausanne 1011 Switzerland

BACHMANN, JOHN HENRY, JR, b Akron, Ohio, Nov 27, 44; m 82; c 1. CARBON BLACK, RUBBER COMPOUNDING. *Educ:* Ohio State Univ, BS, 66; Univ Chicago, MS, 69. *Prof Exp:* Teacher chem & math, Revere Local Sch Dist, Revere High Sch, Bath, Ohio, 68-72; Compounder, Firestone Tire & Rubber Co, 72-76, mgr chem labs, 76-83; MGR RES CTR, SID RICHARDSON CARBON & GASOLINE CO, 83- *Mem:* Am Chem Soc; Am Chem Soc Rubber Div; Am Indust Hyg Asn. *Res:* Carbon black and rubber compounding. *Mailing Add:* Sid Richardson Carbon & Gasoline Co 4825 N Freeway Ft Worth TX 76106

BACHMANN, KENNETH ALLEN, b Columbus, Ohio, Mar 21, 46; m 69. PHARMACOLOGY. *Educ:* Ohio State Univ, BS, 69, PhD(pharmacol), 73. *Prof Exp:* ASSOC PROF PHARMACOL, COL PHARM, UNIV TOLEDO, 73- *Mem:* AAAS; NY Acad Sci; Sigma Xi. *Res:* Pharmacokinetic aspects of oral anticoagulant-drug interactions; influence of environmental chemical exposure and disease upon oral anticoagulant disposition. *Mailing Add:* 308 E Sixth St Perrysburg OH 43551

BACHMANN, ROGER WERNER, b Ann Arbor, Mich, Dec 11, 34; m 60; c 2. LIMNOLOGY. *Educ:* Univ Mich, 56, PhD(zool), 62; Univ Idaho, MS, 58. *Prof Exp:* Res zoologist, Univ Calif, Davis, 62-63; from asst prof to assoc prof, 63-71, PROF LIMNOL, IOWA STATE UNIV, 71- *Concurrent Pos:* Assoc ed, Ecol & Ecol Monogr, 77-82; mem, Iowa Gov Sci Adv Coun, 77-82; ed, J Iowa Acad Sci, 83- *Mem:* AAAS; Am Soc Limnol & Oceanog; Ecol Soc Am; Am Fisheries Soc. *Res:* Biological productivity of lakes; aquatic mineral cycles. *Mailing Add:* Dept Animal Ecol Iowa State Univ Ames IA 50011

BACHNER, FRANK JOSEPH, b Brookline, Mass, Feb 2, 40; m 66; c 2. PHYSICAL METALLURGY. *Educ:* Mass Inst Technol, SB, 61, SM, 63, PhD(phys metall), 66. *Prof Exp:* Consult electronic mat group, Lincoln Lab, Mass Inst Technol, 66, mem staff, 68-75, asst group leader, 75-77, assoc group leader, 77-; DIR MICROELECTRONICS, MICROELECTRONIC & MAT CTR, POLAROID CORP. *Mem:* Inst Elec & Electronics Engrs. *Res:* Thin films for microelectronics; silicon charge-coupled devices for imaging and sampled-data signal processing; GaAs microwave devices. *Mailing Add:* 121 Forest Ave West Newton MA 02165

BACHOP, WILLIAM EARL, b Youngstown, Ohio, Aug 31, 26; m 58; c 2. GROSS ANATOMY, DEVELOPMENTAL BIOLOGY. *Educ:* Western Reserve Univ, AB, 50; Ohio State Univ, MSc, 58, PhD(zool), 63. *Prof Exp:* Asst prof biol, Univ Omaha, 63-65; training prog fel develop biol res, Sch Med, Univ Wash, 65-69; asst prof zool, Clemson Univ, 69-73; asst prof & actg chmn, Nat Col Chiropractic, 74-75, assoc prof, 76-80, chmn dept, 76-86, PROF ANAT, NAT COL CHIROPRACTIC, LOMBARD, 80- *Mem:* Am Anatomists Asn; Soc Social Studies Sci; Am Asn Clin Anatomists. *Res:* Anatomy of human lumbar spine; yolk sac syncytium of bony fishes. *Mailing Add:* Dept Anat Nat Col Chiropractic 200 E Roosevelt Rd Lombard IL 60148

BACHRACH, ARTHUR JULIAN, b New York, NY, Mar 20, 23; m 54; c 2. EXPERIMENTAL PSYCHOLOGY. *Educ:* Col City New York, BS, 47; Western Reserve Univ, MA, 50; Univ Va, PhD(educ & psychol), 52. *Prof Exp:* From instr to assoc prof neurol & psychiat, Sch Med, Univ Va, 50-62, dir, div clin & med psychol, 50-58, dir, div behav sci, 58-62; prof psychol & chmn dept, Ariz State Univ, 62-69; chmn, Dept Behav Sci, Naval Med Res Inst, 69-79, dir, Environ Stress Prog Ctr & chmn, Sci Psychophysiol, 79-87; CONSULT, 87- *Concurrent Pos:* Adj prof med psych, Uniformed Serv Univ Health Sci, 78-87; chmn, bd adv, Col Oceaneering, 83-87; mem, bd adv, Op Raleigh, US, 84-88. *Honors & Awards:* Award for Sci, Underwater Soc Am, 73; Greenstone Award Contrib Diving Safety, Nat Asn Underwater Instrs, 83; Shilling Award, Undersea & Hyperbaric Med Soc, 86. *Mem:* Undersea Med Soc (pres, 76-77); fel Am Psychol Asn; fel AAAS; Am Littoral Soc; Human Factors Soc; Royal Soc Med. *Res:* Stress physiology and behavior under conditions of environmental extremes of cold, heat, and underwater; effects of protection equipment such as garments and masks. *Mailing Add:* PO Box 3600 Taos NM 87571

BACHRACH, HOWARD L, b Faribault, Minn, May 21, 20; m 43; c 2. MOLECULAR VIROLOGY, SUB-UNIT VACCINES. *Educ:* Univ Minn, BA, 42, PhD(biochem), 49. *Prof Exp:* Res asst, Explosives Res Lab, Nat Defense Res Comt Proj, Carnegie Inst Technol, 42-45; res asst, Univ Minn, 45-49; biochemist, Foot & Mouth Dis Res Mission, USDA, Denmark, 49-50; res biochemist, Biochem & Virus Lab, Univ Calif, Berkeley, 50-53; chief scientist & head biochem & phys invest, 53-81, res chemist, 81-85, CONSULT-COLLABR, PLUM ISLAND ANIMAL DIS CTR, USDA, 85- *Concurrent Pos:* Chemist, Jos Seagram & Co, Lawrenceburg, Ind, 42; consult, Walter Reed Army Inst Res, 81-84, Am Chem Soc Spencer Awards Comt, 83-85, Nat Acad Sci, 84-91, Off Tech Assessment, US Cong, 84-85, Nat Cancer Inst, 84-87, Inst Biosci & Technol, Tex A&M Univ, 87-89 & var industs, 83- *Honors & Awards:* Theobald Smith Lectr, Am Soc Microbiol, 81; Newcomb Cleveland Prize, AAAS, 82; Kenneth A Spencer Medal, Am Chem Soc, 82; Nat Medal Sci, President Ronald Reagan, 83; Alexander von Humboldt Award, 83. *Mem:* Nat Acad Sci; fel NY Acad Sci; hon mem Am Col Vet Microbiologists; Am Chem Soc; Am Soc Virol. *Res:* Structure and function of animal viruses; viral replication strategies; cloned viral protein vaccines; immunology; immunization with a capsid protein of foot and mouth disease virus; structure and molecular biology of viruses; first production through gene splicing of an effective vaccine against any disease of animals or humans; first purification and electron microscopic identification of poliovirus. *Mailing Add:* USDA Plum Island Animal Dis Ctr PO Box 848 Greenport NY 11944

BACHRACH, JOSEPH, b Ger, Feb 9, 18; US citizen; m 58. CARBOHYDRATE CHEMISTRY. *Educ:* Queen's Univ (Ont), BA, 44, MA, 45; Purdue Univ, PhD(bio-org-chem), 50. *Prof Exp:* Asst prof chem, Concord Col, 50-53; from asst prof to assoc prof, Univ Ill, Chicago Circle, 53-66; prof chem, Northeastern Ill Univ, 66-89, chmn dept, 77-88, emer prof, 89-; RETIRED. *Mem:* Am Chem Soc; Sigma Xi. *Res:* Carbohydrates; hemicelluloses. *Mailing Add:* 6744 Trumbull Lincolnwood IL 60645

BACHRACH, ROBERT ZELMAN, b Worcester, Mass, June 28, 42; m 69; c 5. SOLID STATE PHYSICS. *Educ:* Mass Inst Technol, BS, 64; Univ Ill, Urbana, MS, 65, PhD(physics), 69. *Prof Exp:* Mem tech staff physics, Bell Labs, 69-73; MEM RES STAFF PHYSICS, XEROX PALO ALTO RES CTR, 73- *Concurrent Pos:* Consult prof, Stanford Univ, 78- *Mem:* Fel Am Phys Soc; Am Vacuum Soc; Am Optical Soc; Mat Res Soc; AAAS. *Res:* Studies of metals and semiconductors using electron spectroscopies and other techniques; molecular beam epitaxy for integrated optics. *Mailing Add:* 4179 Oak Hill Ave Palo Alto CA 94306

BACHUR, NICHOLAS R, SR, b Baltimore, Md, July 21, 33; m 52; c 3. BIOCHEMISTRY, PHARMACOLOGY. *Educ:* Johns Hopkins Univ, AB, 54; Univ Md, MD & PhD(biochem), 61. *Prof Exp:* Res asst biochem, McCollum-Pratt Inst, Johns Hopkins Univ, 51-55; instr biochem, Univ Md, 61, intern med, Univ Hosp, 61-62; res biochemist, Lab Clin Biochem, Nat Heart Inst, 62-65; chief, lab clin biochem, Baltimore Cancer Res Ctr, Nat Cancer Inst, 66-82; Dir Res, UNIV MD CANCER CTR, 82-87, PROF, 82- *Concurrent Pos:* Actg dir, Univ Md Cancer Ctr, 85-87. *Honors & Awards:* Commendation Medal, USPHS, 80. *Mem:* AAAS; AMA; Am Fedn Clin; Fedn Am Soc Exp Biol. *Res:* Am Soc Pharmacol & Exp Therapeut; Am Asn Cancer Res. Res: Biochemical mechanisms of pathological processes; enzymology. *Mailing Add:* Cancer Res Ctr Univ Md 655 W Baltimore St Baltimore MD 21201

BACHVAROFF, RADOSLAV J, b Blagoevrad, Bulgaria, Nov 5, 35; US citizen; m 65; c 4. IMMUNOGENETICS, HISTOCOMPATIBILITY ANTIGENS. *Educ:* Med Acad Sofia, Bulgaria, MD, 59. *Prof Exp:* Resident, Alexandrov Hosp, Sofia, 59-61; res assoc immunol, Inst Microbiol & Epidemiol, Sofia, 61-63; guest investr biochem & immunol, Immunol Lab, Nat Inst Allergy & Infectious Dis, NIH, 63-64 & biochem, Rockefeller Univ, 64; Eleanor Roosevelt fel cancer res, Inst Biol & Med Chem, Acad Med Sci, Moscow, 65-66; res assoc immunol, Biochem Lab, Pasteur Inst, Paris, 67-69, Rockefeller Univ, 69-70; res assoc immunol & transplantation, Med Ctr, NY Univ, 71-73, asst prof, 73-77; from assoc prof to prof, depts surgery & path, State Univ NY, Stony Brook, 77-85; res internal med, 86-88, chief resident internal med, 88-90, NEPHROLOGY FEL, NASSAU COUNTY MED CTR, 90-, ATTEND PHYSICIAN & CLIN INSTR, FLUSHING HOSP MED CTR, 90- *Mem:* Am Asn Immunologists; Transplantation Soc; Soc Exp Biol & Med; NY Acad Sci; Harvey Soc; AAAS; Am Col Physicians. *Res:* Transplantation biology; lymphocyte differentiation in the immune response; histocompatibility testing; immunogenetics; nephrology; kidney and bone marrow transplantation. *Mailing Add:* Flushing Hosp Med Ctr 45th Ave at Parsons Blvd Flushing NY 11355

BACHVAROVA, ROSEMARY FAULKNER, b Brookline, Mass, Oct 2, 38; div; c 3. EMBRYOLOGY, MOLECULAR BIOLOGY. *Educ:* Radcliffe Col, BA, 61; Rockefeller Univ, PhD(molecular embryol), 66. *Prof Exp:* Res fel, 70-72, from instr to assoc prof anat, 73-90, PROF CELL BIOL, MED COL, CORNELL UNIV, 90- *Mem:* Soc Develop Biol; Sigma Xi. *Res:* Oogenesis and early embryological development, especially in mammals. *Mailing Add:* Dept Biol Cornell Univ Med Col 1300 York Ave New York NY 10021

BACHYNSKI, MORREL PAUL, b Bienfait, Sask, July 19, 30; m 59. PLASMA PHYSICS. *Educ:* Univ Sask, BEng, 52, MSc, 53; McGill Univ, PhD(physics), 55. *Prof Exp:* Res assoc, Eaton Electronics Lab, McGill Univ, 55; mem sci staff, Res Labs, RCA Corp, 55-57, assoc labs dir, 58-59, dir microwave & plasma physics labs, 59-65, dir res, RCA Ltd, 65-72, dir res & develop, 72-75, vpres res & develop, RCA Ltd, 75-76; FOUNDER & PRES, MPB TECHNOLOGIES INC, 77- *Concurrent Pos:* Pres, Asn Sci, Eng & Tech Community of Can, 74-75; mem, Sci Coun Can. *Honors & Awards:* David Sarnoff Gold Medal, 77; Prix Sci du Quebec, 77; Achievement in Physics Medal, Can Asn Physicists, 87. *Mem:* Fel Inst Elec & Electronics Engrs; fel Am Phys Soc; fel Can Aeronaut & Space Inst; fel Royal Soc Can (vpres, 81-82); Can Asn Physicists (pres, 68-69); hon mem Eng Inst Can. *Res:* Electromagnetic wave propagation; microwave optics; plasma physics; geophysics and space physics; plasma and laser technology, including controlled fusion and communications. *Mailing Add:* MPB Technol Inc 1725 N Service Rd Trans-Can Hwy Dorval PQ H9P 1J1 Can

BACH-Y-RITA, PAUL, b New York, NY, Apr 24, 34; m 77; c 4. NEUROPHYSIOLOGY, REHABILITATION. *Educ:* Nat Univ Mex, MD, 59. *Prof Exp:* Res physiologist dept physiol, Sch Med, Univ Calif, Los Angeles, 54-56, jr res pharmacologist, Dept Biophys & Nuclear Med, 60; sci transl & writer, Ctr Sci & Tech Doc Mex, UNESCO, 55-57, rural pub health physician, 58-59; intern, Presby Med Ctr, San Francisco, 60-61; sr res mem, Pac Med Ctr, Smith-Kettlewell Inst Visual Sci, 63-79, assoc dir, 68-79; prof visual sci, Univ Pac, 69-79; chief, Rehab Med Serv, Martinez Vet Admin Med Ctr, 79-; prof, depts human physiol & phys med & rehab & vchmn, dept phys med & rehab, Univ Calif, Davis, 79-; chmn, 83-85, PROF, REHAB MED, CLIN SCI CTR, UNIV WIS, 83- *Concurrent Pos:* USPHS ment health trainee, Univ Calif, Los Angeles, 59-60, fel, Dept Clin Neurophysiol, Univ Freiburg, 62-63 & res career award, 63-73; Bank of Am-Giannini Found fel, Ctr Study Nerv Physiol & Electrophysiol, Nat Ctr Sci Res, Paris, 61-62; resident physician med & rehab, Stanford Univ, 77-79; vis prof, Univ de Pisa, 70-71, Univ Autonoma Metropolilana, Mex, 75-76; vis scientist, Karolinska Inst, Stockholm, 89-90. *Honors & Awards:* Silver Hektoen Medal, AMA, 72; Franceschetti-Liebrecht Prize, Ger Ophthal Soc, 74. *Mem:* AAAS; Am Physiol Soc; NY Acad Sci; Asn Res Vision & Ophthal; Int Rehab Med Asn. *Res:* Brain control of eye movement; tactile vision substitution based on brain plasticity; helping blind persons learn to see by means of images from a television camera delivered to the skin; brain plasticity applied to the development of rehabilitation procedures for stroke and head injury. *Mailing Add:* Clin Sci Ctr Univ Wis E3-350 600 Highland Ave Madison WI 53705

BACK, BIRGER BO, b Copenhagen, Denmark, Mar 26, 46; m 73; c 3. EXPERIMENTAL NUCLEAR PHYSICS. *Educ:* Univ Copenhagen, MS, 70, PhD(nuclear physics), 74. *Prof Exp:* Res asst, Univ Copenhagen, 70-71; vis staff mem, Los Alamos Sci Lab, 71-72; fel, Univ Copenhagen, 72-77; vis staff mem, Los Alamos Sci Lab, 77; asst physicist, 77-81, PHYSICIST, NUCLEAR PHYSICS, ARGONNE NAT LAB, 81- *Mem:* Am Phys Soc. *Res:* Nuclear fission and heavy ion reactions. *Mailing Add:* Argonne Nat Lab 9700 S Cass Argonne IL 60439

BACK, KENNETH CHARLES, b Wharton, NJ, Nov 17, 25; m 50, 81; c 9. MEDICAL SCIENCES. *Educ:* Muhlenberg Col, BS, 51; Univ Okla, MS, 54, PhD(med sci), 57. *Prof Exp:* Sr scientist, Warner-Lambert Res Inst, 57-58; pharmacologist, Pitman-Moore Co, 58-60; pharmacologist-toxicologist, Wright-Patterson AFB, 60-66, supvry pharmacologist-toxicologist & chief toxicol br, 6570th Aerospace Med Res Labs, 66-82; prof prev med & biomet, Uniformed Serv Univ Health Sci, Bethesda, Md, 82-87. *Concurrent Pos:* Adj prof biol sci, Wright State Univ, 73-; mem, Am Conf Govt Indust Hygienists. *Mem:* Sigma Xi; Soc Toxicol; Am Soc Pharmacol & Exp Therapeut; NY Acad Sci. *Res:* General pharmacology; intermediary metabolism; aerospace toxicology; propellant toxicology. *Mailing Add:* Int Toxicol Consults Abbey Lodge E Cliff Whitby North Yorks Y022-4JT England

BACK, MARGARET HELEN, b Ottawa, Ont, Dec 7, 29; m 54; c 2. PHYSICAL CHEMISTRY. *Educ:* Acadia Univ, BSc, 50; McGill Univ, MSc, 54, PhD(chem), 59. *Prof Exp:* Fel chem, Nat Res Coun Can, 59-61; res assoc, 61-65, from asst prof to assoc prof, 65-77, PROF CHEM, UNIV OTTAWA, 77- *Concurrent Pos:* Nat Res Coun Can res grants, 63-; Ont Res Found grant, 64-67; Natural Sci & Eng res grants, Res Coun Can, 79- *Mem:* Fel Chem Inst Can. *Res:* Kinetics of gas phase reactions; formation of carbon from hydrocarbons; reactions of carbon. *Mailing Add:* Dept Chem Univ Ottawa Ottawa ON K1N 6N5 Can

BACK, NATHAN, b Philadelphia, Pa, Nov 30, 25; m 51; c 5. PHARMACOLOGY. *Educ:* Pa State Univ, BS, 48; Philadelphia Col Pharm, MS, 52, DSc(pharmacol), 55. *Prof Exp:* Asst admin & tech orgn med labs, Israel, 48-49; biochemist, Wyeth Inst, 50-51, pharmacologist, 51-52; res asst pharmacol, Philadelphia Col Pharm, 52-54, instr, 54-55; cancer res scientist, Roswell Park Mem Inst, 55-58, sr cancer res scientist, 58-66; PROF BIOCHEM PHARMACOL, STATE UNIV NY BUFFALO, 64- *Concurrent Pos:* Res fel, State Univ NY Buffalo, 56-58, from asst prof to prof pharmacol, Sch Pharm, 56-64, prof, Grad Sch Arts & Sci, 64-, from actg chmn to chmn dept biochem pharmacol, 67-71; vis prof & actg dir sch pharm, Fac Med, Hebrew Univ, Jerusalem, 69-71, vis prof, 71-; exec ed, Pharmacol Res Commun; chmn, State Univ Task Force on Israel Progs; UN expert pharmacol, 75-; UN Sci Expert, Jerusalem, Israel, 76-77; vis prof, Weizman Inst Sci, 85-86. *Honors & Awards:* E K Frey Medal, 70. *Mem:* Fel AAAS; Am Soc Hemat; Int Soc Hemat; Fedn Am Soc Exp Biol; fel Am Soc Clin Pharmacol & Therapeut; fel Royal Soc Med; fel NY Acad Sci; Am Soc Exp Pharmacol. *Res:* Chemotherapy of cancer; alkylating agents; physiology and pathology of blood and blood-forming organs; pharmacology of fibronolytic agents and anticoagulants; extracorporeal circulation; thrombo-embolic phenomena; mechanisms of drug resistance; shock mechanisms; vasoactive polypeptides; atherosclerosis; biochemical pharmacology; angiogenesis; Kallikrein-Kininogen-Kinin system; proteases and protease inhibitors. *Mailing Add:* H347 Hochstetter State Univ NY N Campus Buffalo NY 14260

BACK, ROBERT ARTHUR, b Delhi, Ont, Aug 13, 29; m 54; c 2. CHEMICAL KINETICS. *Educ:* Univ Western Ont, BSc, 50, MSc, 51; McGill Univ, PhD(chem), 53. *Prof Exp:* Dewar res fel natural philos, Univ Edinburgh, 53-56; Nat Res Coun Can fel chem, McGill Univ, 56-59; from assoc res officer to sr res officer, 59-71, prin res officer photo chem & kinetics, 71-87, GUEST WORKER, NAT RES COUN CAN, 88- *Mem:* Fel Chem Inst Can. *Res:* Kinetics of gas-phase free-radical and atomic reactions; photochemistry; mercury photosensitization; reactions of diimide; radiation chemistry of gases; isotope separation; photochemistry with lasers; spectroscopy. *Mailing Add:* Steacie Inst Molecular Sci Nat Res Coun Can Ottawa ON K1A 0R9 Can

BACK, WILLIAM, b East St Louis, Ill, Aug 9, 25; m 50; c 4. HYDROGEOLOGY. *Educ:* Univ Ill, AB, 48; Univ Calif, MS, 55; Harvard Univ, MPA, 56; Univ Nev, Reno, PhD(chem hydrogeol), 69. *Prof Exp:* Asst, Ill State Geol Surv, 45-48; asst geol, Univ Calif, 49-50; geologist, Calif, 50-54, GEOLOGIST, WATER RESOURCES DIV, US GEOL SURV, VA, 54- *Concurrent Pos:* Instr hist & Eng, Fla Mil Acad, 47; asst geol, Harvard Univ, 56-57; lectr, Am Univ, 57-59; vis scholar, Desert Res Inst, Nev, 66-68; continuing FAO adv to Israel, 67; USAID adv to Pakistan, 71; UNDP consult, Costa Rica, 71, Bolivia, 77, Turkey, 85; Dept Interior adv to Poland, 74; prin investr, NSF grant, Mexico, 75-77 & US-Spain Proj, 84-; adj prof, George Washington Univ, 75- *Honors & Awards:* O E Meinzer Award, Geol Soc Am, 73; Distinguished Lectr in Hydrogeol, Geol Soc Am, 78-79. *Mem:* Int Asn Hydrogeologists; Geol Soc Am; Am Geophys Union; Geochem Soc. *Res:* Isotopes and geochemistry of ground water; regional ground water systems; Karst processes. *Mailing Add:* 4100 Nellie Custis Dr Arlington VA 22207

BACKER, DONALD CHARLES, b Plainfield, NJ, Nov 9, 43; m; c 1. ASTRONOMY. *Educ:* Cornell Univ, BEP, 66, PhD(astron), 71; Univ Manchester, MSc, 69. *Prof Exp:* Res asst, Nat Radio Astron Observ, 71-73; res assoc, Goddard Space Flight Ctr, NASA, 73-75; RES ASTRONR, RADIO ASTRON LAB, UNIV CALIF, BERKELEY, 75-, PROF ASTRON DEPT, 90- *Concurrent Pos:* Mem vis comt, Nat Radio Astron Observ, 78-81. *Mem:* Am Astron Soc; AAAS; Int Astron Union; Union Radio Sci Int. *Res:* Pulsars; interstellar scattering; extragalactic radio sources. *Mailing Add:* Radio Astron Lab 601 Campbell Hall Univ Calif Berkeley CA 94720

BACKER, RONALD CHARLES, b Newark, NJ, May 3, 43; c 5. TOXICOLOGY. *Educ:* Univ Ariz, BA, 64, PhD(pharmaceut chem), 70; Am Bd Forensic Toxicol, dipl. *Prof Exp:* Asst chief forensic chem, Milwaukee Health Dept, 69-74; asst toxicol, Off Chief Med Examr, State Md, 74-76; chief toxicol, Off Chief Med Examr, State W Va, 76-88; BUR CHIEF TOXICOL, SCI LAB DIV, STATE HEALTH & ENVIRON DEPT, ALBUQUERQUE, NMEX, 88-; ASST PROF PATH, UNIV NMEX SCH MED, ALBUQUERQUE, NMEX, 88- *Concurrent Pos:* Instr toxicol, dept path, div forensic path, Univ Md, Baltimore City, 74-76; supvr toxicol, Cent Labs Assoc Md Pathologists, Ltd, 74-76; lectr toxicol, Sch Pub Health & Hyg, Johns Hopkins Univ, 74-76. *Mem:* Soc Forensic Toxicologists; Am Acad Forensic Sci. *Res:* Improvement and development of analytical procedures for drugs from biological specimens. *Mailing Add:* State Health & Environ Dept Sci Lab Div 700 Camino-Salud NE PO Box 4700 Albuquerque NM 87196-4700

BACKMAN, KEITH CAMERON, b Utica, NY, Feb 22, 47; m 79; c 1. MOLECULAR BIOLOGY, GENETICS. *Educ:* Univ Chicago, BS, 69; Harvard Univ, PhD(biophys), 77. *Prof Exp:* Helen Hay Whitney fel, Univ Calif, San Francisco, 77-79 & Mass Inst Technol, 79-80; Charles A King mem trust fel, Mass Inst Technol, 80-82; sr scientist, 82-85, VPRES RES & DEVELOP, BIOTECHNICA INT, INC, 85- *Concurrent Pos:* Consult, Genentech, Inc, 78-81. *Mem:* AAAS. *Res:* Genetic engineering of microorganisms for the purpose of altering their physiology toward the production of useful and valuable compounds. *Mailing Add:* Biotechnica Int Inc 85 Bolton St Cambridge MA 02140

BACKMAN, PAUL ANTHONY, b Shrewsbury, Eng, Nov 7, 44; US citizen. PHYTOPATHOLOGY. *Educ:* Univ Calif, Davis, BS, 66, PhD(plant path), 70. *Prof Exp:* Fel mycotoxins, NC State Univ, 70; res plant pathologist, US Dept Agr, 70-71; ASSOC PROF PLANT PATHOLOGY, AUBURN UNIV, 71- *Concurrent Pos:* Assoc ed, Plant Dis Reporter, 74-77; consult, Off Technol Asessment, US Cong, 78 & Food & Agr Orgn, UN, 78-79; assoc ed, Peanut Sci, 81- *Mem:* Am Phytopath Soc; AAAS; Sigma Xi; NY Acad Sci. *Res:* Control of diseases of peanuts and soybeans, including loss estimation, optimization of pesticide programs, seed bacterization and post harvest diseases; sugar beet and peach diseases. *Mailing Add:* Dept Bot Plant Path & Microbiol Auburn Univ Auburn AL 36849

BACKOFEN, WALTER A, b Bockville, Conn, Dec 8, 25. MATERIALS SCIENCE ENGINEERING, METALLURGY & PHYSICAL METALLURGICAL ENGINEERING. *Educ:* Mass Inst Technol, SB, 48, ScD, 50. *Prof Exp:* Prof, 50-80, EMER PROF ENG, MASS INST TECHNOL, 80- *Mem:* Fel Am Soc Metals. *Res:* Social history with a numerical taxonomic bias. *Mailing Add:* Methodist Hill Lebanon NH 03766

BACKUS, CHARLES E, b Wadestown, WVa, Sept 17, 37; m 57; c 3. SOLAR ENERGY, ELECTRICAL ENGINEERING. *Educ:* Univ Ohio, BSME, 59; Univ Ariz, MS, 61, PhD(nuclear eng). *Prof Exp:* Res asst nuclear eng, Univ Ariz, 62-65; supvr systs dynamics, Westinghouse Astronuclear Lab, 65-68; from asst prof to assoc prof, 68-76, PROF ENG, ARIZ STATE UNIV, 76-, ASST DEAN, COL ENG & APPL SCI, 79-, DIR, CTR FOR RES, 80- *Concurrent Pos:* Vis staff mem, Los Alamos Sci Lab, 69-; dir, Advan Energy Conversion Short Course, 69-; consult to various industs and govt; dir, Photovoltaics Short Course, 69- *Mem:* Am Nuclear Soc; Am Soc Eng Educ; fel Inst Elec & Electronics Engrs; Am Soc Mech Engrs; AAAS. *Res:* Sunlight concentration onto solar cells; photovoltaics; solar thermal power generation; solar energy utilization. *Mailing Add:* 9619 S 157 Pl Gilbert AZ 85234

BACKUS, GEORGE EDWARD, b Chicago, Ill, May 24, 30; m 61; c 3. GEOPHYSICS. *Educ:* Univ Chicago, BS, 48, MS, 50 & 54; PhD(physics), 56. *Prof Exp:* Physicist, Proj Matterhorn, Princeton Univ, 57-58; from asst prof to assoc prof math, Mass Inst Technol, 58-60; assoc prof geophys, 60-62, PROF GEOPHYS, UNIV CALIF, SAN DIEGO, 62- *Concurrent Pos:* Guggenheim Mem fel, 63-64 & 70-71; mem comt on sci & pub policy, Nat Acad Sci, 70-73, mem report rev comt, 73-76, chmn Day fund selection comt, 75-78, co-chmn, Int Working Group on Magnetic Field Satellites, 83- *Honors & Awards:* Gold Medal, Royal Astron Soc; John Adam Fleming Medal, Am Geophy Union. *Mem:* Nat Acad Sci; fel Am Geophys Union; Royal Astron Soc; Soc Indust & Appl Math; Am Acad Arts & Sci. *Res:* Origin of geomagnetic field; normal modes of elastic-gravitational oscillation of the earth; general theory of geophysical inverse problems; lower mantle electrical conductivity; fluid motion in earth's core; analysis of satellite magnetic data. *Mailing Add:* Inst Geophys & Planetary Phys Univ Calif San Diego La Jolla CA 92093-0225

BACKUS, JOHN, b Philadelphia, Pa, Dec 3, 24; c 2. FUNCTIONAL PROGRAMMING. *Educ:* Columbia Univ, BS, 49, AM, 50. *Hon Degrees:* Doctor Univ, Univ York, England, 85; DSc, Univ Ariz, 88; Dr Hon Causa Use, Univ de Nancy I, France, 89. *Prof Exp:* Programmer, Pure & Appl Sci Depts, IBM Corp, 50-53, mgr, Prog Res Dept, 54-58, res staff mem, Thomas J Watson Res Ctr, 59-63, IBM fel, 63-64, IBM FEL, ALMADEN RES CTR, IBM CORP, SAN JOSE, 65-, MGR FUNCTIONAL PROG, 73- *Concurrent Pos:* Mem, Working Group 2.2, 65-, Working Group 2.8, Int Fedn Info Processing, 88; vis prof, Univ Calif, Berkeley, 80. *Honors & Awards:* W W McDowell Award, Inst Elec & Electronics Engrs, 67; Nat Medal Sci, 75; A M Turing Award, Asn Comput Mach, 77; Harold Pender Award, Univ Pa, 83. *Mem:* Nat Acad Sci; Nat Acad Eng; Asn Comput Mach; fel Am Acad Arts & Sci. *Res:* Programming languages, their syntax and semantics; functional programming and its algebra of programs; program optimization. *Mailing Add:* 91 St Germain Ave San Francisco CA 94114

BACKUS, JOHN (GRAHAM), physics; deceased, see previous edition for last biography

BACKUS, JOHN KING, b Buffalo, NY, May 22, 25; m 50; c 4. PHYSICAL CHEMISTRY. *Educ:* Hamilton Col, AB, 47; Cornell Univ, MS, 50, PhD(physics, math), 52. *Prof Exp:* Res chemist, Procter & Gamble Co, 52-53; res chemist, O-Celo Dept, Gen Mills, Inc, 53-56, res supvr, 56-61; res specialist, Mobay Corp, 62-63, group leader, 64-67, sr group leader, 67, coordr regulatory affairs, Ind Hyg Med Serv, 71- 77, mgr, 67-90; CONSULT, 90- *Concurrent Pos:* Chmn, Gordon Conf Chem & Physics of Cellular Mat, 70; comt Toxicol & Indust Health, Int Isocyante Inst, 74-80, ad hoc comt Criteria Doc, 74 & 78; bd dirs, Western Pa Safety Coun, 75-90, SPE Div of Thermoplastic Mat & Forms 76-79; chmn, Tech Conf, Urethane Div, SPI, 77, Urethane Div Flammability Comm, 73, Isocyanate Shipping Comm, 74. *Mem:* AAAS; NY Acad Sci; Sigma Xi. *Res:* Physical polymer chemistry; flammability; degradation; molecular structure; health effects; government regulation; safety; polymer physics; author and co-author of numerous publications; patents. *Mailing Add:* 9441 Katherine Dr Allison Park PA 15101

BACKUS, MILO M, b Ill, May 3, 32; m 52; c 5. GEOPHYSICS. *Educ:* Mass Inst Technol, BS, 52, PhD(geophys), 56. *Prof Exp:* From res physicist to vpres, Geophys Serv Inc, 56-74; consult, 74-75; prof, 75-80, WALLACE E PRATT PROF GEOPHYS, UNIV TEX, AUSTIN, 80- *Honors & Awards:* Conrad Schlumberger Award, Europ Asn Explor Geophys, 76; Maurice Ewing Medal, Soc Explor Geophys, 90. *Mem:* Soc Explor Geophysicists (vpres, 77, pres, 80); Europ Asn Geophysicists; Am Asn Petrol Geologists. *Res:* Petroleum exploration. *Mailing Add:* 6606 Mesa Dr Sci Univ Tex PO Box 7909 Austin TX 78731

BACKUS, RICHARD HAVEN, b Rochester, NY, Dec 5, 22; m 49; c 3. MARINE BIOLOGY. *Educ:* Dartmouth Col, BA, 47; Cornell Univ, MS, 48, PhD(ichthyol), 53. *Prof Exp:* Supvry res assoc oceanog, Cornell Univ, 51-52; res assoc marine biol, 52-59, marine biologist, 59-63, chmn dept biol, 70-74, SR SCIENTIST, WOODS HOLE OCEANOG INST, 63- *Mem:* AAAS; Ecol Soc Am. *Res:* Biology of marine vertebrates, especially mesopelagic fishes; marine biogeography. *Mailing Add:* 244 Woods Hole Rd Falmouth MA 02540

BACLAWSKI, LEONA MARIE, b Akron, Ohio, Dec 26, 45. ORGANIC CHEMISTRY. *Educ:* Univ Akron, BS, 67, PhD(chem), 74. *Prof Exp:* Fel org chem, State Univ NY Col Environ Sci & Forestry, Syracuse Univ, 74; fel chem, 75, lab dir org chem, Bucknell Univ, 75-77; MEM STAFF, MONSANTO AKRON RES CTR, 77- *Mem:* Sigma Xi; Am Chem Soc. *Res:* Diaziridine chemistry; 1,3-dipolar addition; heterocyclic chemistry; carbanion synthesis; rubber chemicals; thermal analysis. *Mailing Add:* 137 Quaker Ridge Dr Akron OH 44313

BACON, ARTHUR LORENZA, b West Palm Beach, Fla, Sept 22, 37. PROTOZOOLOGY, EMBRYOLOGY. *Educ:* Talladega Col, AB, 61; Howard Univ, MS, 63, PhD, 67. *Prof Exp:* Instr protozool, Howard Univ, 63-65, res assoc, 65-67, asst prof, 67-68; PROF BIOL, TALLADEGA COL, 80-, CHMN, DEPT BIOL & DIV NATURAL SCI & MATH, 69-, DIR, BIOMED RES, 73- *Concurrent Pos:* Postdoctoral fel, Univ Miami, 68-69; presentor, President's Cancer Panel, 84; consult, Div Res Resources, NIH, Nat Inst Gen Med Sci. *Honors & Awards:* Armstrong Award for Creative Ability, 59. *Mem:* Soc Protozoologists; Am Soc Zoologists; AAAS; Sigma XI. *Res:* Studies of the effects of certain food and drug additives on development in sea urchins and protozoan morphology. *Mailing Add:* Dept Biol Talladega Col Talladega AL 35160

BACON, CHARLES WILSON, b Bradenton, Fla; m 68; c 2. MYCOLOGY, BOTANY. *Educ:* Clark Col, BS, 65; Univ Mich, PhD(bot), 72. *Prof Exp:* Res assoc biochem, 71-72; RES MICROBIOLOGIST, AGR RES SERVS, USDA & ASST PROF PLANT PATH, UNIV GA, ATHENS, 72- *Mem:* Mycological Soc Am; Am Soc Microbiol; Agron Soc Am; Ga Asn Plant Pathologists. *Res:* Fungal physiology and biochemistry; mycotoxicology and plant pathology. *Mailing Add:* Toxicol & Mycotoxin Res Unit PO Box 5677 Athens GA 30613

BACON, DAVID W, b Peterborough, Ont, Sept 12, 35; m 63; c 2. CHEMICAL ENGINEERING, STATISTICS. *Educ:* Univ Toronto, BASc, 57; Univ Wis, MS, 62, PhD(statist), 65. *Prof Exp:* Comput analyst, Can Gen Elec Co, Ont, 57-61; comput programmer, Math Res Ctr, Univ Wis-Madison, 61-63; statist & math group leader, Du Pont Can, Ont, 65-67; assoc prof chem eng, 68-73, dean, Fac Appl Sci, 80-90, PROF CHEM ENG, QUEEN'S UNIV, ONT, 73- *Concurrent Pos:* Vis assoc prof, Univ Wis-Madison, 69-70; vis lectr, Univ Sydney, 76-77; chmn, Nat Comt Deans Eng & Appl Sci, 82-83 & Comt Ont Deans Eng; vis scholar, Stanford Univ, 85-86. *Honors & Awards:* Shewell Award, Am Soc Qual Control, 84. *Mem:* Fel Chem Inst Can; Can Soc Chem Eng; Royal Statist Soc; Statist Soc Can; Can Res Mgt Asn; fel Am Statist Asn. *Res:* Statistical process control; mathematical modeling of processes; design of experimental programs; on-line optimization. *Mailing Add:* 368 Chelsea Rd Kingston ON K7M 3Z9 Can

BACON, EDMOND JAMES, b Chidester, Ark, Jan 10, 44; m 73. AQUATIC ECOLOGY. *Educ:* Southern State Col, Ark, BS, 66; Univ Ark, MS, 68; Univ Louisville, PhD(biol), 73. *Prof Exp:* From teaching asst to res asst zool, Univ Ark, 66-68; technician radiation, Univ Louisville, 68-70, res asst water resources, 72-74; asst prof, 74-80, ASSOC PROF BIOL, UNIV ARK, MONTICELLO, 80- *Concurrent Pos:* Environ consult, Aquatic Control, Inc & Dames & Moore, 72-74; ichthyologist, Water Resources Lab, Univ Louisville, 73-74; proj engr, E D'Appolonia Consult Engrs, 74. *Mem:* Am Soc Limnol & Oceanog; Int Soc Limnol; Am Fisheries Soc; Brit Freshwater Biol Asn; assoc Sigma Xi. *Res:* Environmental deterioration appraisal and the impact of costly restoration efforts; natural history of lower vertebrates of Arkansas and primary productivity, water quality and limiting factors in oxbow lakes. *Mailing Add:* Dept Biol Univ Ark Monticello AR 71655

BACON, EGBERT KING, b Sault Ste Marie, Mich, Aug 20, 00; m 26; c 1. CHEMISTRY. *Educ:* Univ Mich, BS, 22, MS, 23, PhD(chem), 26. *Prof Exp:* Asst quantitative anal, Univ Mich, 20-22, asst gen chem, 22-26; instr chem, Brown Univ, 26-30; from instr to prof, 30-66, chmn dept, 63-64, chmn div sci, 56-57, EMER PROF CHEM, UNION UNIV, NY, 66- *Concurrent Pos:* City chemist, Schenectady, NY, 41-51; lectr, Albany Col Pharm, 45-46, prof, 68-78. *Mem:* AAAS; Am Chem Soc. *Res:* Diffusion potentials; equivalent weight of gelatin; viscosity of gelatin; analytical chemistry; chemical history. *Mailing Add:* 1964 Eastern Pkwy Schenectady NY 12309-6229

BACON, FRANK RIDER, b Wilton, Iowa, Apr 28, 14; m 38; c 3. PHYSICAL INORGANIC CHEMISTRY. *Educ:* Iowa State Univ, BS, 35; Washington Univ, MS, 37. *Prof Exp:* Asst chem, Washington Univ, 35-37; res chemist, Owens-Ill Glass Co, 37-58; chief surface chem, Glass Container Div, Owens-Ill, Inc, 58-82; RETIRED. *Mem:* Am Chem Soc; fel Am Ceramic Soc; Sigma Xi. *Res:* Chemical durability of glass; light transmission of glass; strength of glass. *Mailing Add:* 3153 Darlington Rd Toledo OH 43606

BACON, GEORGE EDGAR, b New York, NY, Apr 13, 32; m 56; c 3. PEDIATRIC ENDOCRINOLOGY. *Educ:* Wesleyan Univ, BA, 53; Duke Univ, MD, 57; Univ Mich, MS, 67. *Prof Exp:* Intern pediat, Duke Hosp, 57-58; resident, Columbia Presby Med Ctr, 61-63; prof & chair, dept pediat, Tex Tech Univ, Lubbock, 86-89; from instr to assoc prof, 63-74, PROF PEDIAT, UNIV MICH, ANN ARBOR, 74-; DIR MED EDUC & RES, BUTTERWORTH HOSP, GRAND RAPIDS, 90- *Concurrent Pos:* Fel physiol, Univ Mich, Ann Arbor, 65, fel pharmacol, 65-67; pediat endocrinologist, Univ Pittsburgh, 70. *Mem:* Soc Pediat Res; Endocrine Soc; Am Pediat Soc; Am Fedn Clin Res; Am Acad Pediat; Pediat Endocrin Soc; Am Diabetes Asn. *Res:* Adrenal disorders, particularly modes of therapy and metabolic effects of treatment in congenital adrenal hyperplasia; diabetes mellitus. *Mailing Add:* Dept Med Educ Butterworth Hosp 100 Michigan NE Grand Rapids MI 49503

BACON, LARRY DEAN, b Topeka, Kans, Jan 24, 38; m 69. GENETICS, IMMUNOGENETICS. *Educ:* Kans State Univ, BS, 61, MS, 67, PhD(genetics), 69. *Prof Exp:* Instr path, NY Med Col, 69-72; asst prof microbiol, Sch Med, Wayne State Univ, 72-77; IMMUNOGENETICIST, USDA, SEA, REGIONAL POULTRY RES LAB, 78- *Concurrent Pos:* NIH fel, NY Med Col, 69-71. *Mem:* Fedn Am Soc Exp Biol. *Res:* Immunological responses to sex-linked histocompatibility antigens in chickens; genetics of autoimmune disease and the relationship of alloantigens to autoimmunity, tumor regression and disease resistance. *Mailing Add:* USDA SEA 3606 E Mt Hope Rd East Lansing MI 48823

BACON, LYLE C(HOLWELL), JR, b Los Angeles, Calif, June 23, 31. ELECTRONICS. *Educ:* Pomona Col, BA, 53; Stanford Univ, MS, 54, PhD(elec eng), 58. *Prof Exp:* Mem tech staff, Gen Elec Microwave Lab, 58-62; sr engr, Electromagnetic Technol Corp, 62-66; mem staff, Develco, Inc, 66-78; res scientist, 78-86, ADVAN DECISION SYSTS, LOCKHEED PALO ALTO RES LAB, 86- *Mem:* AAAS; Inst Elec & Electronics Engrs; Sigma Xi. *Res:* Microwave electronics; active and passive microwave devices; electromagnetic field theory; systems and circuit analysis; antennas and propagation; computer analyses; expert systems; image processing. *Mailing Add:* 205 Golden Oak Dr Portola Valley CA 94025

BACON, MARION, b Hemingford, Nebr, May 26, 14; m 50. MEDICAL MICROBIOLOGY. *Educ:* Iowa State Univ, BSc, 50; Wash State Univ, MS, 52, PhD(bact), 56. *Prof Exp:* Asst entomol, Wash State Univ, 50-52; med entomologist, Ctr for Dis Control, USPHS, 52-53; asst bact, Wash State Univ, 53-56; from instr to asst prof microbiol, Univ Nebr, 56-59; from assoc prof to emer prof biol, Eastern Wash Univ, 59-79; RETIRED. *Mem:* Am Soc Microbiol; Sigma Xi. *Res:* Diseases of nature communicable to man; medical entomology; bacterial indicators of fecal pollution. *Mailing Add:* 2808 Bancroft Missoula MT 59801

BACON, MERLE D, b Dacoma, Okla, Nov 28, 31; m 51; c 1. ENGINEERING MECHANICS, AERONAUTICAL ENGINEERING. *Educ:* Wichita State Univ, BS, 54; Univ Okla, MS, 61, PhD(eng mech), 66. *Prof Exp:* Liaison engr, Beech Aircraft Corp, 51-53; structural engr, 53-54; USAF, 54-, proj engr, Armament Ctr, Elgin AFB, 54-55, interceptor pilot, 57-60, from instr to assoc prof eng mech, USAF Acad, 61-63 & 65-69, Air Liaison Off, Vietnam, 69-70; chief, Laser Effects Br, Laser Div, Air Force Weapons Lab, 70-75; comdr & assoc prof eng mech, Frank J Seiler Res Lab, USAF Acad, 75-79, dir fac res, 79-81; dir training, Bechtel Corp, Saudi Arabia, 81-84; proj mgr, Nat Educ Corp, Newport Beach, Calif, 84-86; EXEC VPRES, INT EDUC FOUND, 86- *Mem:* Am Inst Aeronaut & Astronaut; Soc Exp Stress Anal; Am Soc Eng Educ. *Res:* Structural dynamics; aircraft structures; stress analysis. *Mailing Add:* Int Educ Found 1500 Garden of the Gods Rd Colorado Springs CO 80907

BACON, OSCAR GRAY, b Sanger, Calif, Nov 8, 19; m 45; c 2. ENTOMOLOGY. *Educ:* Fresno State Col, AB, 41; Univ Calif, MS, 44, PhD(entom), 48. *Prof Exp:* Field aide, Div Fruit Insects, Bur Entom & Plant Quarantine, USDA, 41-43; assoc div entom, 46-47, sr lab technician, 47-48, instr econ entom & jr entomologist, 48-50, from asst prof to assoc prof econ entom, 50-63, chmn dept entom, 67-74, PROF ECON ENTOM & ENTOMOLOGIST, EXP STA, UNIV CALIF, DAVIS, 63- *Mem:* Entom Soc Am. *Res:* Economic entomology; biology and control of insect pests of field crops; small seeded legumes for seed; field corn and grain sorghums; insect pests of potato; insect pests of cereal grains. *Mailing Add:* 615 Cordova Pl Davis CA 95616

BACON, ROBERT ELWIN, b Lansdowne, Pa, Apr 24, 34; m 57; c 2. PHOTOGRAPHIC SCIENCE. *Educ:* Univ Mich, BSChem, 56, BSChE, 56; Mass Inst Technol, PhD(phys chem), 60. *Prof Exp:* Sr chemist, 60-64, res assoc, 65-73, lab head res, 73-79, sr lab head, 73-84, SR TECH ASSOC, EASTMAN KODAK CO, 84- *Concurrent Pos:* Fel, Dept Mat Sci, Stanford Univ, 70-71. *Mem:* Soc Imaging Sci & Technol; Asn Col Entrepreneurs. *Res:* Physical and chemical aspects of photographic response in silver halides; process engineering and research; internal corporate venturing. *Mailing Add:* Six Marks Hill Lane Rochester NY 14617

BACON, ROGER, b Cleveland, Ohio, Apr 16, 26; m 72; c 2. MATERIALS SCIENCE. *Educ:* Haverford Col, AB, 51; Case Inst Technol, MS, 53, PhD, 56. *Prof Exp:* Res physicist, 55-62, group leader carbon fiber develop, Union Carbide Corp, 62-86; SR RES ASSOC, AMOCO PERFORMANCE PROD, INC, 86- *Mem:* Am Phys Soc; Am Carbon Soc; Soc Advan Mat & Process Eng. *Res:* Studies of structure and properties of graphite crystals, whiskers and fibers; experimental demonstration of extremely high strength and stiffness inherent in oriented graphite structures; development of processes for producing high performance carbon fibers from organic precursors; development of carbon-carbon composites. *Mailing Add:* Amoco Performance Prod Inc Res & Develop Ctr 4500 McGinnis Ferry Rd Alpharetta GA 30202-3944

BACON, VINTON WALKER, b Estelline, SDak, Dec 21, 16; m 40; c 4. CIVIL ENGINEERING, WATER POLLUTION CONTROL. *Educ:* Univ Calif, Berkeley, BS, 40; Am Acad Environ Engrs, dipl, 56. *Prof Exp:* Engr, East Bay City Sewage Disposal Surv, Berkeley, 40-41; designer, Los Angeles County Sanit Dist, 41-43 & 46; office engr, Orange County Sewerage Surv, Calif, 46-49; exec officer, Calif State Water Pollution Control Bd, 50-56; exec secy, NW Paper & Pulp Asn, Tacoma, Wash, 56-62; gen supt, Metrop Sanit Dist Greater Chicago, 62-70; prof, 70-82, EMER PROF CIVIL ENG, UNIV WIS, MILWAUKEE, 82- *Concurrent Pos:* Consult civil eng; chmn, Wis Gov Solid Waste Recycling Task Force, 71-76; mem, Wis Solid Waste Recycling Authority, 74-76. *Honors & Awards:* Awards, Am Soc Civil Engrs, 56, 67. *Mem:* Nat Acad Eng; Am Pub Health Asn; Water Pollution Control Fedn; Am Water Works Asn. *Res:* Originator deep tunnel project. *Mailing Add:* 2616 143 Rd Pl SE Mill Creek WA 98012

BACON, WILLIAM EDWARD, b Dansville, NY, Dec 12, 17; c 5. INORGANIC CHEMISTRY, PHYSICAL ORGANIC CHEMISTRY. *Educ:* St Lawrence Univ, BS, 42; Ind Univ, Bloomington, MA, 50; Kent State Univ, PhD(inorg chem), 67. *Prof Exp:* Chemist, E I du Pont de Nemours & Co, 43-47; res supvr chem, Lubrizol Corp, 50-61; SR RES FEL CHEM, LIQUID CRYSTAL INST, KENT STATE UNIV, 67- *Concurrent Pos:* Consult, Gen Tire & Rubber Co, Akron, 72; prin investr, NSF, 74-78, US Army Res Off, 75 & Air Force Off Sci Res, 76. *Mem:* Am Chem Soc; AAAS. *Res:* Structure and properties of thermotropic and lyotropic liquid crystalline substances; study of chemical reactions within these media; comparison of kinetics of isomerizations, polymerizations and hydrolysis in anisotropic and isotropic solvents. *Mailing Add:* 1010 S Willow St Kent OH 44240

BACOPOULOS, NICHOLAS G, b Athens, Greece, Mar 13, 49; US citizen; m 80. NEUROPHARMACOLOGY, RECEPTOR REGULATION. *Educ:* Cornell Col, BA, 71; Univ Iowa, PhD(pharmacol), 76. *Prof Exp:* Fel pharmacol, Yale Univ, 76-79; asst prof pharmacol & psychiat, Med Sch, Dartmouth Col, 79-83; asj assoc prof pharmacol & psychiat, 83-88, DIR, DEPT CANCER & NEUROSCI, PFIZER CENT RES, 83- *Concurrent Pos:* Prin investr res grant, NIMH, 81-83; ad hoc mem, Neurol Sci Initial Rev Group, NIH, 81; mem Tech Adv Bd, Chicago Consortium Psychiat Res. *Mem:* Am Soc Pharmacol & Exp Therapeut; Soc Exp Biol & Med; Soc Neurosci; NY Acad Sci; Sigma Xi; Col Int Neuropsychopharm; Am Asn Cancer Res. *Res:* Regulation of dopamine receptors in mammalian brain; mechanisms of drug-induced tolerance and supersensitivity; mechanism of acute and chronic effects of antipsychotic drugs. *Mailing Add:* CNS Res Pfizer Cent Res Eastern Paint Rd Groton CT 06385

BACSKAI, ROBERT, b Ujpest, Hungary, Feb 17, 30; US citizen; m 56; c 2. ORGANIC CHEMISTRY, POLYMER CHEMISTRY. *Educ:* Eotvos Lorand Univ, Budapest, dipl, 52. *Prof Exp:* Res chemist, Plastics Res Inst, Budapest, Hungary, 52-56; res assoc polymer chem, Princeton Univ, 57-60; from res chemist to sr res chemist, 60-68, sr res assoc polymer chem, 68-89, RES SCIENTIST, CHEVRON RES & TECHNOL CO, 89- *Mem:* Am Chem Soc. *Res:* Ferrocene chemistry; hydrogen bonding; water soluble, stereoregular, cationic, anionic and free radical polymers; nuclear magnetic resonance of polymers; lactam polymerization; high activity catalysts; corrosion inhibitors. *Mailing Add:* Chevron Res & Technol Co 100 Chevron Way Richmond CA 94802

BACUS, JAMES NEVILL, b Ft Worth, Tex, Jan 8, 48; m 70; c 3. BACTERIOLOGY, FOOD SCIENCE. *Educ:* Univ Evansville, BA, 70; Univ Wis, MS, 72, PhD(bact), 73. *Prof Exp:* Res asst microbiol, Food Res Inst, 70-73; res scientist processed meats, Swift & Co, 74-75; vpres prod develop, ABC Res Corp, 75-85; vpres, Qual Sausage, 85-87; PRES, DIVERSITECH, INC, 87- *Mem:* Inst Food Technologists; Am Soc Microbiol; Am Meat Sci Asn; Am Meat Inst; Am Chem Soc. *Res:* Meat and poultry, especially formulation of processing and characteristics. *Mailing Add:* 922 NW 45th Terr Gainesville FL 32605

BACUS, JAMES WILLIAM, b Alton, Ill, Apr 21, 41; m 62; c 2. BIOENGINEERING, PHYSIOLOGY. *Educ:* Mich State Univ, BS, 64; Ill Inst Technol, 64-66; Univ Ill, PhD(physiol), 71. *Prof Exp:* Sci asst radiation biol, Argonne Nat Lab, 64-66; asst biomed engr, Bioeng Dept & chief automation, Sect Clin Hemat, Rush-Presby-St Luke's Med Ctr, 70-76, dir med automation res unit, 76-; AT CELL ANALYSIS SYSTS INC. *Concurrent Pos:* Consult, Electronics Res Div, Corning Glass Works, 71-73. *Mem:* AAAS; Pattern Recognition Soc; Inst Elec & Electronics Engrs; Asn Advan Med Instrumentation; Sigma Xi. *Res:* Automatic pictorial pattern recognition; medical image processing and classification. *Mailing Add:* 826 S Lincoln Hinsdale IL 60521

BACZEK, STANLEY KARL, b New Bedford, Mass, Dec 30, 46; m 69; c 2. POLYMER SCIENCE, PHYSICAL CHEMISTRY. *Educ:* Southeastern Mass Univ, BS, 69; Univ Mass, MS, 75, PhD(polymer sci), 77. *Prof Exp:* Jr scientist phys chem, Firestone Tire & Rubber Co, 69-73; tech asst polymer sci, Univ Mass, 73-77; sr res engr, Electrochem, Diamond Shamrock Corp, 77-80; mem staff, 80-87, DIR PHYS/ANALYTICAL SCI, LUBRIZOL CORP, 87- *Concurrent Pos:* Vis scientist, Oak Ridge Nat Labs, 76-77. *Mem:* Am Chem Soc; Am Phys Soc; AAAS; Sigma Xi; Creation Res Soc. *Res:* Physical chemistry and engineering of ion-exchange polymers; semi-crystalline polymer deformation studies using rheo-optical techniques; solution properties of polymers; lubrication chemistry; rheology. *Mailing Add:* 6940 Mildon Dr Painesville OH 44077-9701

BADA, JEFFREY L, b San Diego, Calif, Sept 10, 42; c 2. COSMOGEOCHEMISTRY, GEOCHRONOLOGY. *Educ:* San Diego State Univ, BS, 65; Univ Calif, San Diego, PhD(chem), 68. *Prof Exp:* Instr chem, Univ Calif, San Diego, 68-69; res fel geophys & environ chem, Harvard Univ, 69-70; asst prof oceanog, 70-74, assoc prof, 74-80, PROF MARINE CHEM, SCRIPPS INST OCEANOG, UNIV CALIF, SAN DIEGO, 80- *Concurrent Pos:* Res grants, NSF, 70-71, 73-75, 75-79 & 88-; Petrol Res Fund, 70-73, 84-87, NIH, 77-85; Alfred P Sloan res fel, 75-79; ONR, 75-82,

86- *Mem:* AAAS; Am Chem Soc. *Res:* Organic chemistry of natural waters, sediments and fossils; stability of organic compounds on the primitive earth; geochemical implications of the kinetics of reactions involving biologically produced organic compounds; biochemistry of fishes. *Mailing Add:* Scripps Inst Oceanog Univ Calif San Diego La Jolla CA 92093-0212

BADALAMENTI, ANTHONY FRANCIS, b Bronx, NY, Feb 2, 43; m 68; c 1. MATHEMATICAL MODELLING OF BEHAVIORAL PHENOMENA, APPLYING PSYCHOLOGICAL MODELS TO ARTIFICIAL INTELLIGENCE. *Educ:* Manhattan Col, BS, 64; Stevens Inst Technol, MS, 67; Brooklyn Polytech Inst, PhD(math), 70. *Prof Exp:* Mem tech staff, Bell Tel Labs, 64-70; asst prof math & computer sci, Fairleigh Pickinson Univ, 70-72; mem tech staff, Gen Res Corp, 72-74; dir revenue modeling, Western Union Tel Corp, 74-75; RES SCIENTIST, NATHAN KLINE INST, 76- *Mem:* Am Math Soc; Asn Comput Mach; Soc Indust & Appl Math; Soc Psychoanal Psychother; NY Acad Sci. *Res:* Quantifying the study of psychotherapy and of human emotional communication; research in number theory; modelling for artificial intelligence. *Mailing Add:* 46 Charles St Apt 2A Westwood NJ 07675

BADASH, LAWRENCE, b Brooklyn, NY, May 8, 34. HISTORY OF MODERN PHYSICS, SCIENCE & SOCIETY. *Educ:* Rensselaer Polytech Inst, BS, 56; Yale Univ, PhD(hist sci), 64. *Prof Exp:* Instr hist sci, Yale Univ, 64-65, res assoc, 65-66; from asst prof to assoc prof, 66-79, PROF, HIST SCI, UNIV CALIF, SANTA BARBARA, 79- *Concurrent Pos:* NATO fel, Cambridge Univ, 65-66, prin investr, NSF grant, 69-70, 90-92; dir sem, Global Security & Arms Control, Univ Calif, 83 & 86; Guggenheim fel, 84-85; chair, Div Hist Physics, Am Phys Soc, 88-89. *Mem:* Hist Sci Soc; fel Am Phys Soc; fel AAAS; Fedn Am Scientists. *Res:* History of radioactivity and nuclear physics; history of the nuclear arms race and arms control; science and society; science and social responsibility. *Mailing Add:* Dept Hist Univ Calif Santa Barbara CA 93106

BADDER, ELLIOTT MICHAEL, b Philadelphia, Pa, Feb 7, 43; div; c 1. GENERAL SURGERY, VASCULAR SURGERY. *Educ:* Univ Pa, BA, 63; Thomas Jefferson Univ, MD, 67. *Prof Exp:* Asst prof surg, Pa State Univ, 75-77; asst chief surg, Mercy Hosp, Baltimore, 77-83, actg chief, 83; asst prof, 77-79, ASSOC PROF SURG, UNIV MD HOSP, 80-; CHIEF SURG, MERCY HOSP, BALTIMORE, 84- *Concurrent Pos:* Inst gen surg, Frances Delafield Hosp, New York, NY, 72; asst attend surgeon, Vanderbilt Clin, Presby Hosp, New York, 72; co-investr, NIH Grant, Cardiovasc Determinants of Catecholamine Release, 76-83. *Mem:* Am Fedn Clin Res; Asn Acad Surg; fel Am Col Surgeons; Am Asn Endocrine Surgeons; Soc Univ Surgeons. *Res:* Catecholamines under conditions of stress. *Mailing Add:* 301 St Paul Pl Baltimore MD 21202

BADDILEY, JAMES, b Manchester, Eng, May 15, 18; m 44; c 1. MICROBIAL BIOCHEMISTRY. *Educ:* Univ Manchester, BSc, 41, PhD(chem), 44, DSc, 53; Univ Cambridge, ScD, 86. *Hon Degrees:* DSc, Heriot-Watt, Edinburgh, 79, Bath, 86. *Prof Exp:* Res fel chem, Imp Chem Industs, Pembroke Col, Cambridge, 45-49; staff mem biochem, Lister Inst Prev Med, Univ London, 49-54; prof org chem, Newcastle upon Tyne Univ, 54-77, prof chem microbiol, 77-83; RETIRED. *Concurrent Pos:* Rockefeller fel, Mass Gen Hosp, Harvard Med Sch, 54; Karl Folkers vis prof, Univ Ill, 62; head, Sch Chem, Newcastle upon Tyne Univ, 68-78, dir, Microbiol Chem Res Lab, 77-83; sr fel sci & eng res coun, Univ Cambridge, 81-84; fel, Pembroke Col, Cambridge, 81-84, emer fel, 84- *Honors & Awards:* Meldola Medal, Royal Soc Chem, 47, Corday-Morgan Medal, 52, Tilden Medal, 59, Pedler Medal, 78; Leeuwenhock lectr, Royal Soc, 67, Davy Medal, 74. *Mem:* Fel Royal Soc UK; Royal Soc Chem; hon mem Am Soc Biochem & Molecular Biol; fel Royal Soc Edinburgh; Biochem Soc; Soc Gen Microbiol. *Res:* Chemical structure and synthesis of nucleotide coenzymes as ATP, coenzyme A; discovery, isolation, structure of bacterial cell-wall polymers including the teichoic acids and lipoteichoic acids in gram positive bacteria; biosynthesis and function of these polymers. *Mailing Add:* Dept Biochem Univ Cambridge Tennis Court Rd Cambridge CB2 1QW England

BADDING, VICTOR GEORGE, b Buffalo, NY, Dec 15, 35; m 61; c 2. ORGANIC CHEMISTRY. *Educ:* Canisius Col, BS, 57; Univ Notre Dame, PhD(org chem), 61. *Prof Exp:* Fel, Univ Wis, 61-62; res chemist, Silicones Div, Union Carbide Corp, 62-65; from asst prof to assoc prof, 65-90, PROF CHEM, MANHATTAN COL, 90- *Concurrent Pos:* Fel, State Univ NY Buffalo, 65; sabbatical leave, Princeton Univ, 79 & Polytech Univ, 87. *Mem:* Am Chem Soc; Royal Soc Chem; Sigma Xi; NY Acad Sci. *Res:* Chemistry of acetals; polymer stereochemistry; intermolecular interactions in polymer blends. *Mailing Add:* Dept Chem Manhattan Col Bronx NY 10471

BADDOUR, RAYMOND F(REDERICK), b Laurinburg, NC, Jan 11, 25; m 54; c 3. CHEMICAL ENGINEERING. *Educ:* Univ Notre Dame, BS, 45; Mass Inst Technol, MS, 49, ScD, 51. *Prof Exp:* Asst dir, Eng Practice Sch, Tenn, 48-49; from asst prof to prof, 51-63, head dept, 69-76, dir, Environ Lab, 70-76, LAMMOT DU PONT PROF CHEM ENG, MASS INST TECHNOL, 73- *Concurrent Pos:* P C Reilley lectr, Univ Notre Dame, 65; co-founder & dir, Amgen, Inc, Enterprise Mgt Corp, & Tekmat, Inc. *Mem:* Fel Am Acad Arts & Sci; NY Acad Sci; Am Chem Soc; fel Am Inst Chem Engrs; fel Am Inst Chem. *Res:* Chemical reactions in plasmas; diffusion in adsorptive solids; preparation and properties of selected sorbents and membranes; chromatography; ion exchange. *Mailing Add:* Dept Chem Eng Rm 66-440 Mass Inst Technol Cambridge MA 02139-3561

BADE, MARIA LEIPELT, b Hamburg, Ger, Dec 13, 25; US citizen; m 49; c 1. BIOCHEMISTRY. *Educ:* Univ Nebr, BS, 51, MS, 54; Yale Univ, PhD(biochem), 60. *Prof Exp:* Fel biochem, Harvard Univ, 60-62, fel, Sch Med, 63-64; USPHS fel, Mass Inst Technol, 64-65, res assoc, 65-67; asst prof biol, 67-72, ASSOC PROF BIOL, BOSTON COL, 72- *Mem:* AAAS; Am Chem Soc; Am Soc Zool; Am Physiol Soc; Am Soc Biol Chem; Sigma Xi. *Res:* Biochemistry and structure of arthropod cuticle; water quality; human mycoses. *Mailing Add:* Four Bowser Rd Lexington MA 02173

BADE, WILLIAM GEORGE, b Oakland, Calif, May 29, 24; m 52; c 6. MATHEMATICS. *Educ:* Calif Inst Technol, BS, 45; Univ Calif, Los Angeles, MA, 48, PhD(math), 51. *Prof Exp:* Instr math, Univ Calif, 51-52 & Yale Univ, 52-55; from asst prof to assoc prof, 55-64, PROF MATH, UNIV CALIF, BERKELEY, 64- *Concurrent Pos:* NSF sr fel, 58-59. *Mem:* Am Math Soc. *Res:* Functional analysis. *Mailing Add:* Dept Math Univ Calif Berkeley CA 94720

BADEER, HENRY SARKIS, b Mersine, Turkey, Jan 31, 15; nat US; m 48; c 2. MEDICAL PHYSIOLOGY. *Educ:* Am Univ Beirut, MD, 38. *Prof Exp:* Asst physiol, Am Univ Beirut, 38-41, instr, 41-45, from adj prof to asst clin prof, 45-51, from assoc prof to prof, 51-65, actg head dept, 51-56, chmn dept, 56-65; vis prof physiol, State Univ NY Downstate Med Ctr, 65-67; prof physiol & pharmacol, 67-76, prof physiol, 76-91, EMER PROF PHYSIOL, SCH MED, CREIGHTON UNIV, 91- *Concurrent Pos:* Rockefeller res fel physiol, Harvard Med Sch, 48-49; vis prof, Univ Iowa, 57-58; actg chmn dept physiol & pharmacol, Creighton Univ, 71-72. *Honors & Awards:* Golden Apple, 75. *Mem:* AAAS; Am Physiol Soc; Int Soc Heart Res. *Res:* Hypothermia on heart; ventricular fibrillation after coronary occlusion; myocardial oxygen uptake; cardiac hypertrophy and hemodynamics; author text. *Mailing Add:* Div Physiol Creighton Univ Sch Med Omaha NE 68178-0224

BADEN, ERNEST, b Berlin, Ger, Feb 2, 24; nat US. ANATOMIC PATHOLOGY, ORAL PATHOLOGY. *Educ:* Univ Algiers, PhB, 43; Univ Paris, Sorbonne, Lic ès Let, 46, Chirugien-Dentiste, 47; Odont Sch France, Laureat, 47; NY Univ, DDS, 50, MS, 64; Univ Geneva, MD, 63; Univ Toulouse, DDS, 78; Am Bd Oral Path, dipl, 60; Am Bd Oral Surg, dipl, 61; Am Bd Path Anat, dipl, 68, Am Bd Oral Med, dipl, 81. *Prof Exp:* Asst stomatol, Univ Hosp Cochin, Univ Paris, 45-46 & St Louis Univ Hosps, 46-47; intern oral surg, Mt Sinai Hosp, New York, 50-51, asst dentist, 51-59; from asst prof to prof path, Fairleigh Dickinson Univ, 57-79, chmn dept, 57-73, assoc scientist path, Sch Dent, Oral Health Res Ctr, 79-90; CLIN PROF ORAL MED, COL DENT, NY UNIV, 81- *Concurrent Pos:* Clin asst oral surg, City Hosp New York, 52-58; resident, Queens Gen Hosp, 56-57; vis asst prof, Albert Einstein Col Med, 59-69; consult, Hackensack Gen Hosp, 62-72; resident, Francis Delafield Hosp Div, Columbia Med Ctr, 64, fel path, 65 & 67-68, fel path, Presby Hosp Div, 65 & 66; res path, Univ Miami & Jackson Mem Hosp, 67-68; consult oral path, Holy Name Hosp, Teaneck, NJ, 67-, Out-Patient Dent Clin, USPHS, New York, 68-, Englewood Hosp, 70-, St Joseph Hosp, Paterson, 72-, Bergen Pines County Hosp, 78- & Lutheran Med Ctr, Brooklyn, 78-; asst clin prof path, Col Physicians & Surgeons, Columbia Univ, 68-72, adj assoc prof, 73; consult surg, Harlem Hosp Ctr, 68-72, consult path, 68-; consult path, Cath Med Ctr, Brooklyn & Queens, 74-; vis prof, Dept Anatomic Path, Fac Med, Univ Paul Sabatier, Toulouse, France, 77-78; chief, Sect Oral Med & Oral Path, Dept Dent & consult, Dept Path, Lutheran Med Ctr, Brooklyn & Sect Oral Med & Oral Path, Dept Dent, Bergen Pines County Hosp, 78- *Mem:* Fel Am Acad Oral Path; Am Dent Asn; fel Col Am Pathologists; Am Acad Oral Med; Int Acad Path. *Res:* Experimental oral pathology and surgery; pathobiology of neoplasia; pathology of salivary glands; oral manifestations of systemic diseases; oral medicine; lymphomas. *Mailing Add:* Dept Oral Med Col Dent NY Univ 421 First Ave New York NY 10010

BADEN, HARRY CHRISTIAN, b Brooklyn, NY, Oct 22, 23; m 47; c 2. ANALYTICAL CHEMISTRY. *Educ:* Col Educ Albany, BA, 48, MA, 49; Rensselaer Polytech Inst, PhD(chem), 55. *Prof Exp:* Teacher, High Sch, NY, 48-50; asst chem, Rensselaer Polytech Inst, 50-52, asst chem physics, 52-54; anal chemist, 54-58, sr anal chemist, 58-64, tech assoc, Photog Technol Div, Eastman Kodak Co, Rochester, 65-83; IND CONSULT, 84- *Mem:* Am Chem Soc; Sigma Xi. *Res:* Explosive oxidation of boron hydrides; instrumental analysis; infrared spectra; gas chromatography; photographic chemistry. *Mailing Add:* 205 Curtice Park Webster NY 14580

BADEN, HOWARD PHILIP, b Boston, Mass, Feb, 23, 31; m 81; c 3. DERMATOLOGY. *Educ:* Harvard Univ, AB, 52; Harvard Med Sch, MD, 56. *Prof Exp:* Intern med, Peter Bent Brigham Hosp, Boston, 56-57; asst resident dermat, Mass Gen Hosp, Boston, 59-60, clin fel, 60-61; assoc dermatologist, Mass Gen Hosp, 76-80; res assoc, 62-63, instr, 63-64, assoc, 64-67, from asst prof in assoc prof, 67-75, mem fac; Ctr Human Genetics, 71-75, PROF DERMAT, HARVARD MED SCH, 75-; DERMATOLOGIST, MASS GEN HOSP, 80- *Concurrent Pos:* NIH fel, Res Lab, Harvard Med Sch, Dept Dermat, Mass Gen Hosp & Grad Dept Biochem, Brandeis Univ, 61-62; clin assoc, Mass Gen Hosp, Boston, 62-64, asst, 64-67, from asst dermatologist to dermatologist, 67-75; mem, Gen Med A Study Sect, NIH, 69-73; Am Cancer Soc-Eleanor Roosevelt int cancer fel, Galton Lab, Univ Col, Univ London, 70-71; ed, Progress in Dermat, 71-74; assoc ed, J Soc Invest Dermat, 72-77. *Mem:* Am Fedn Clin Res; Soc Invest Dermat; Am Soc Human Genetics; AAAS; Am Soc Clin Invest; Fedn Am Soc Exp Biol. *Res:* Differentiation of epidermal tissues with emphasis on the structural proteins; diagnosis and management of genetic disorders of keratinization. *Mailing Add:* Dept Dermat Harvard Med Sch Mass Gen Hosp 32 Fruit St Boston MA 02114

BADENHOP, ARTHUR FREDRICK, food science, biochemistry; deceased, see previous edition for last biography

BADENHUIZEN, NICOLAAS PIETER, b Zaandam, Netherlands, June 14, 10; Can citizen; m; c 6. BOTANY. *Educ:* Univ Amsterdam, DSc, 38. *Prof Exp:* Asst bot, Univ Amsterdam, 31-39; geneticist & biochemist, Tobacco Exp Sta, Java, 39-42; biochemist, Royal Netherlands Yeast Factory, 46-50; prof bot & head dept, Univ Witwatersrand, 50-60; head dept, 61-71, prof, 61-75, EMER PROF BOT, UNIV TORONTO, 75- *Concurrent Pos:* Teacher sec sch, Netherlands, 35-39. *Honors & Awards:* Thomas Burr Osborne Gold Medal, Am Asn Cereal Chemists, 69. *Mem:* Can Bot Asn; fel Royal Netherlands Acad Sci. *Res:* Fine structure and physiology of cells in relation to starch granule production and composition. *Mailing Add:* 159 Russell Hill Toronto ON M4V 2S9 Can

BADER, HENRY, b Warsaw, Poland, Apr 26, 20; nat US; m 50; c 2. ORGANIC CHEMISTRY. *Educ:* Univ Strasbourg, Ing Chem, 40; Univ Paris & Univ Montpellier, LSc, 42; Univ London, PhD(chem), 48, Imperial Col, dipl, 48. *Prof Exp:* Res chemist, Beecham Res Labs, Eng, 48-50 & May & Baker, Ltd, 50-51; group leader & sr chemist, Dominion Tar & Chem Co, Can, 51-53; sr res assoc, Ortho Res Found, Raritan, NJ, 53-61 & Am Cyanamid Co, Bound Brook, 61-63; dir labs, Aldrich Chem Co, 63-66; sr scientist, Polaroid Corp, 66, group leader, 67-69, sr group leader, 70-77, res fel, 77-82; RES ASSOC, DANA FARBER CANCER INST, BOSTON, MA, 82- *Concurrent Pos:* Expert, UN Chem Indust Develop Orgn, Cuba, 82. *Mem:* Am Chem Soc; Chem Soc. *Res:* Synthetic organic chemistry; acetylenes and heterocyclics; chemotherapy; process research. *Mailing Add:* 19 Maplewood Ave Newton MA 02159

BADER, HERMANN, b Furtwangen, Ger, Nov 23, 27; m 58; c 9. CIRCULATION, MEMBRANE TRANSPORT. *Educ:* Univ Munich, MD, 55. *Prof Exp:* Asst, Physiol Inst, Univ München, 54-56, med apprentice, Med Outpatient Dept, 56-57, Ger Res Asn scholar, Physiol Inst, 57-58, sci asst, 60-61; instr, Dept Physiol, Vanderbilt Univ, 59-60, asst prof, 60 & 63-66; sci asst, Physiol Inst, Univ Würzburg, 61-63; res fel, Biochem Inst, Univ Strassburg, Frankreich, 62; assoc prof, Dept Pharmacol & Toxicol, Sch Med, Univ Miss, 66-70, prof, 70-72; PROF & HEAD, DEPT PHARMACOL & TOXICOL, UNIV ULM, 72- *Concurrent Pos:* Proctor & chmn senate, Univ Ulm, 73, dean, Fac Theoret Med, 74-75, prodekan, 75-77, chmn, Ctr Biol & Theoret Med, 74-79, dean, Fac Theoret Med, 79-83. *Mem:* Ger Physiol Soc; Ger Soc Biol Chem; Ger Soc Biophysics; Biophys Soc; Am Soc Pharmacol & Exp Therapeut; Ger Pharmacol Soc. *Res:* Pharmacology. *Mailing Add:* Neue Welt 6 Vohringen-Illerberg 7917 Germany

BADER, KENNETH L, b Carroll, Ohio, May 4, 34; m 55; c 2. AGRONOMY, ACADEMIC ADMINISTRATION. *Educ:* Ohio State Univ, BSc, 56, MSc, 57, PhD(agron), 60. *Prof Exp:* Asst agron, Ohio State Univ, 56-57, from instr to prof agron, 57-72, actg asst dean col agr & home econ, 63-64, asst dean col agr, 64-68, from assoc dean students to dean students, 68-72; prof agron & vchancellor student affairs, Univ Nebr, Lincoln, 72-76; CHIEF EXEC OFFICER, AM SOYBEAN ASN, 76- *Concurrent Pos:* Am Coun Educ fel acad admin, 67-68. *Mem:* Am Soc Agron; Sigma Xi; Agr Coun Am. *Res:* Grass physiology and turf management; business administration. *Mailing Add:* 1635 Trotting Trail Rd Chesterfield MO 63017

BADER, RICHARD FREDERICK W, b Kitchener, Ont, Oct 15, 31; m 58; c 3. THEORETICAL CHEMISTRY. *Educ:* McMaster Univ, BSc, 53, MSc, 55; Mass Inst Technol, PhD(chem), 57. *Prof Exp:* Fel chem, Mass Inst Technol, 58, Univ Cambridge, UK, 58-59; from asst prof to assoc prof, Univ Ottawa, 59-63; assoc prof, 63-66, PROF CHEM, McMASTER UNIV, 66- *Concurrent Pos:* A P Sloan res fel, Sloan Found, 64-66; E W R Steacie mem fel, Nat Res Coun Can, 67-69; Guggenheim Mem Found fel, 79-80. *Mem:* Royal Soc Can; Chem Inst Can; Am Phys Soc. *Res:* Development of a theory that quantifies the chemical concepts of atoms, bonds, structure and reactivity, and is based upon the properties of the distribution of electronic charge in a molecule. *Mailing Add:* Dept Chem McMaster Univ Hamilton ON L8S 4M1 Can

BADER, ROBERT SMITH, b Falls City, Nebr, June 18, 25; m 48; c 4. EVOLUTIONARY BIOLOGY. *Educ:* Kans State Col, BS, 49; Univ Chicago, PhD(paleozool), 54. *Prof Exp:* From instr to asst prof biol sci, Univ Fla, 52-56; from asst prof to prof zool, Univ Ill, 56-68; DEAN COL ARTS & SCI, UNIV MO, ST LOUIS, 68- *Mem:* Soc Vert Paleont; Soc Study Evolution; Am Soc Zool; Am Soc Mammal. *Res:* Vertebrate evolution; variation in rodent dentition. *Mailing Add:* 4609 Shores Dr St Louis MO 63125

BADER, SAMUEL DAVID, b New York, NY, Feb 4, 47; m 71; c 2. PHYSICAL CHEMISTRY, SOLID STATE PHYSICS. *Educ:* Univ Calif, Berkeley, BS, 67, PhD(chem), 74. *Prof Exp:* Appointee superconductivity, 74-76, asst physicist, 76-79, physicist, 79-87, sr physicist, 90, GROUP LEADER, MAT SCI, ARGONNE NAT LAB, 87- *Concurrent Pos:* Assoc ed Applied Physics Lett; adv ed, J Magnetism & Magnetic Mat. *Mem:* Fel Am Phys Soc; Am Vacuum Soc; AAAS. *Res:* Experimental surface science; electron spectroscopies; surface magnetism. *Mailing Add:* Mat Sci Div Argonne Nat Lab Bldg 223 Argonne IL 60439

BADERTSCHER, ROBERT F(REDERICK), b Cincinnati, Ohio, June 4, 22; m 46; c 4. AERONAUTICAL & ASTRONAUTICAL ENGINEERING. *Educ:* Ohio State Univ, BAAE & MSc, 50. *Prof Exp:* Engr, Aeroprod Div, Gen Motors Co Corp, 50-52; engr, Columbus Labs, Battelle Mem Inst, 52-60, chief aerospace mech res group, 60-74, prog mgr, 74-76, assoc dir, 76-78, proj mgr, Battelle Advan Systs Lab, 79-; RETIRED. *Mem:* Assoc fel Am Inst Aeronaut & Astronaut; Am Defense Preparedness Asn. *Res:* General aeropsace vehicle design and development problems; aircraft and missile systems analysis; armored combat vehicle systems assessment and related technologies; technology assessment. *Mailing Add:* 218 Kenbrook Dr Worthington OH 43085

BADGER, ALISON MARY, b Croyden, Eng, Nov 25, 35; US citizen; m 61; c 2. IMMUNOBIOLOGY. *Educ:* Univ London, BSc, 58; Boston Univ, PhD(microbiol), 72. *Prof Exp:* Res asst microbiol, United Fruit Co, Boston, 58-64; res asst biochem, Blood Res Inst, Boston, 66-68; res assoc microbiol, Sch Med, Boston Univ, 68-73, from instr to asst prof, 73-77, asst prof surg, 75-77, assoc prof microbiol, 77-79; sr investr, Smith Kline & French, 79-82; ASST DIR, SMITHKLINE BEECHAM PHARMACEUT, PHILADELPHIA, 82-, RES FEL, 90- *Mem:* Am Asn Immunologists. *Res:* Cellular immunology. *Mailing Add:* Smithkline Beecham Pharmaceut 1500 Spring Garden St King of Prussia PA 19406

BADGER, GEORGE FRANKLIN, b Everett, Mass, May 14, 07; m 34; c 3. BIOSTATISTICS. *Educ:* Mass Inst Technol, BS, 29; Johns Hopkins Univ, MPH, 32; Univ Mich, MD, 38. *Prof Exp:* Assoc epidemiologist, Dept Health, Detroit, Mich, 29-34; assoc biostatist, Johns Hopkins Univ, 38-45, asst prof biostatist, 45-46; from assoc prof to prof, 46-72, dir dept biomed, 63-69, EMER PROF BIOSTATIST, SCH MED, CASE WESTERN RESERVE UNIV, 72- *Concurrent Pos:* Consult to Secy War, 42-46; mem comn acute respiratory dis, Armed Forces Epidemiol Bd, 42-44 & 53-60; mem adv comt epidemiol & biomet, NIH, 58-62, human ecol study sect, 63-65; med adv comt clin invest, Nat Found, 59-61. *Mem:* Am Epidemiol Soc (pres, 61). *Res:* Epidemiology of communicable diseases; the family as an epidemiological unit; statistical evaluation of clinical and laboratory procedures. *Mailing Add:* 750 Spafford Oval Sagamore Hills OH 44067

BADGER, RODNEY ALLAN, b Fullerton, Calif, Feb 7, 43; m 66. ORGANIC CHEMISTRY. *Educ:* Ore State Univ, BA & BS, 64; Univ Calif, Berkeley, MS, 66, PhD(chem), 68. *Prof Exp:* Asst chemist, Univ Hawaii, 68-69; PROF CHEM, SOUTHERN ORE STATE COL, 69- *Concurrent Pos:* Vis assoc prof chem, Univ BC, 76-77; vis prof chem, Ore State Univ, 83-84; indust consult, 83- *Mem:* Am Chem Soc; Sigma Xi. *Res:* Isolation and synthesis of natural products; investigation of new synthetic methods; isolation and structure determination of pharmacologically-active compounds. *Mailing Add:* Dept Chem Southern Ore State Col Ashland OR 97520

BADGETT, ALLEN A, b Van Nuys, Calif, Sept 8, 43; m 68; c 3. BIOCHEMICAL GENETICS, HUMAN GENETICS. *Educ:* Humboldt State Univ, AB, 67, MS, 68; Utah State Univ, PhD(genetics), 73. *Prof Exp:* ASSOC PROF GENETICS, SOUTHWESTERN OKLA STATE UNIV, 73- *Concurrent Pos:* Instnl res grant, Southwestern Okla State Univ, 74-80. *Mem:* Genetics Soc Am. *Res:* Regulation of gene activity and how it relates to biochemical and cytogenetic parameters. *Mailing Add:* Dept Biol Sci Southwestern Okla State Univ 100 Campus Dr Weatherford OK 73096

BADGLEY, FRANKLIN ILSLEY, b Mansfield, Ohio, Dec 20, 14; m 43; c 2. METEOROLOGY. *Educ:* Univ Chicago, BS, 35; NY Univ, MS, 49, PhD(meteorol), 51. *Prof Exp:* Chemist, Swift & Co, 36-42; meteorologist, Trans World Airlines, Inc, 47; from instr to assoc prof, 50-67, PROF METEOROL, UNIV WASH, 67-, PROF ATMOSPHERIC SCI, 73-, DEPT CHMN, 77- *Concurrent Pos:* Assoc dir, Quarternary Res Ctr, Univ Wash, 73-76. *Mem:* AAAS; Am Meteorol Soc; Am Geophys Union; Sigma Xi. *Res:* Atmospheric turbulence. *Mailing Add:* 13749 41st Ave NE Seattle WA 98125

BADHWAR, GAUTAM D, b New Delhi, India, Nov 8, 40; US citizen; m 70; c 2. RADIATION PHYSICS, CROP MONITORING USING SATELLITE DATA. *Educ:* Agra Univ, BS, 59; Univ Rochester, PhD(physics), 67. *Prof Exp:* Asst prof physics, Univ Rochester, 67-72; mem staff, 72-90, HEAD, SPACE RADIATION, NASA JOHNSON SPACE CTR, HOUSTON, 90- *Concurrent Pos:* Lectr, Univ Houston, 76-90. *Res:* Radiation monitoring; astronaut health; crop identification; orbital monitoring. *Mailing Add:* Code SN3 NASA Johnson Space Ctr Houston TX 77058

BADIN, ELMER JOHN, b Canonsburg, Pa, Aug 21, 17. PHYSICAL ORGANIC CHEMISTRY. *Educ:* Cooper Union, BChE, 42; Princeton Univ, MA, 44, PhD(org chem), 45. *Prof Exp:* Res assoc combustion chem, Princeton Univ, 45-46, instr chem, 46-50; sr res chemist, Celanese Corp Am, 56-59; prof chem & chmn dept, Union Col, NJ, 59-63; combustion chemist, 63-65; sr res chemist, Cities Serv Oil Co, 65-72; combustion chemist, 72-74; staff mem, Mitre Corp, 74-76; combustion chemist, 76-78; mgr combustion chem, Tenn Valley Authority, 78-82; COMBUSTION CHEM BK, 82- *Concurrent Pos:* Consult, Wesco Paint Co, 45-47; assoc, US Navy Proj Squid, Princeton Univ, 45-47; researcher, Univ Calif, Berkeley, 49-50. *Mem:* Am Chem Soc; Combustion Inst; Sigma Xi. *Res:* Inorganic-organic-physical organic chemistry; combustion-oxidation; petroleum and polymer chemistry; coal science and energy; synthesis and chemical mechanisms. *Mailing Add:* 803 Blue Spring Rd Princeton NJ 08540-1637

BADLER, NORMAN IRA, b Los Angeles, Calif, May 3, 48; m 68; c 2. HUMAN MOVEMENT, COMPUTER GRAPHICS. *Educ:* Univ Calif, Santa Barbara, BA, 70; Univ Toronto, MS, 71, PhD(comput sci), 75. *Prof Exp:* From asst prof to assoc prof, 74-86, PROF COMPUT SCI, UNIV PA, 87- CHAIRPERSON, 90- *Concurrent Pos:* Prin investr, NSF grants, 75- & ARO grants, NASA, 82-; consult, 76-; vchair, Spec Interest Group on Comput Graphics & Interactive Tech, Asn Comput Mach, 79-82; sr ed, Comput Graphics & Image Processing, 81-90; mem, Comput Soc, Inst Elec & Electronics Engrs, assoc ed, Comput Graphics & Appln, 84-90. *Mem:* Asn Comput Mach (vchmn, 79-82); Inst Elec & Electronics Engrs; Asn Comput Ling; Am Asn Artificial Intelligence; Cognitive Sci Soc. *Res:* The computer representation, simulation, animation, analysis and description of human movement; the understanding and processing of movement concepts in natural languages; interactive and graphics system design; archaeological site reconstruction. *Mailing Add:* Dept Comput & Info Sci Philadelphia PA 19104-6389

BADONNEL, MARIE-CLAUDE H M, b France, 39. TISSUE TYPING, TEACHING. *Educ:* Univ Geneva, Switzerland, PhD(exp path), 75. *Prof Exp:* ASST PROF PATH, UNIV MASS MED CTR, 78-, ASSOC DIR, TISSUE TYPING LAB/BLOOD BANK, 87- *Mem:* Am Soc Histocompatibility & Immunogenetics; Am Asn Path; AAAS; Am Heart Asn. *Mailing Add:* Univ Mass Med Ctr HLA Blood Bank 55 Lake Ave N Worcester MA 01655

BADOYANNIS, HELEN LITMAN, b New York, NY, Apr 16, 59; m 89. MOLECULAR NEUROBIOLOGY, DEVELOPMENTAL NEUROBIOLOGY. *Educ:* Smith Col, AB, 81; NY Med Col, MS, 88, PhD(neurosci), 91. *Prof Exp:* Teaching asst psychol, Smith Col, 81-82; teaching assoc psychol, Univ Mass, 81-82; teaching asst histol, NY Med Col, 86 & 87, teaching asst neurosci, 88 & 89; POSTDOCTORAL FEL NEUROBIOL, STATE UNIV NY, STONY BROOK, 91- *Mem:* Soc Neurosci; Soc Cell Biol; AAAS; Am Women Sci. *Res:* Molecular and cellular neurobiology; examining various growth factors and cell-cell contact alone or in combination for their ability to alter catecholaminergic phenotypic expression in the neural crest derived PCIZ cell line. *Mailing Add:* 61 Hastings Dr Stony Brook NY 11790

BADR, MOSTAFA, b Cairo, Egypt, Oct 5, 50; nat US; m 82. PHARMACOLOGY, TOXICOLOGY. *Educ:* Cairo Univ, Egypt, BS, 73, MS, 78; Univ Louisville, PhD(pharmacol), 83. *Prof Exp:* Postdoctorate fel pharmacol, Univ Louisville, 83-84, Univ NC, Chapel Hill, 84-86; asst prof, 87-90, ASSOC PROF PHARMACOL, UNIV MO, KANSAS CITY, 90- *Concurrent Pos:* Prin invest, NIH, 88- *Mem:* Am Soc Pharmacol Experimental Ther; Soc Toxicol; Soc Experimental Biol & Med; NY Acad Sci. *Res:* Investigating mechanisms of chemical-induced hepatotoxicity and effects of metabolic disease on liver cell damage. *Mailing Add:* M3-115 2411 Holmes St Kansas City MO 64108

BADRE, ALBERT NASIB, b Beirut, Lebanon, Aug 15, 45; US citizen; m 69; c 1. INFORMATION SCIENCE. *Educ:* Univ Iowa, BA, 68; Univ Mich, Ann Arbor, MA, 71, PhD(behav sci), 73. *Prof Exp:* Soc worker, New York City Dept Soc Serv, 68-69; asst res, Ment Health Res Inst, Univ Mich, 70-71, res asst comput applns, 71-73; asst prof, 73-79, ASSOC PROF, SCH INFO & COMPUT SCI, GA INST TECHNOL, 79- *Mem:* AAAS; Am Asn Univ Prof; Am Soc Info Sci; Am Educ Res Asn; Am Psychol Asn; Sigma Xi. *Res:* The study, design, and analysis of man-computer problem-solving, information-processing, learning, pattern recognition, memory and decision processing systems. *Mailing Add:* Sch Info & Comput Sci Ga Inst Technol Atlanta GA 30332

BAE, JAE HO, b Taegu, Korea, Dec 11, 39; m 68; c 3. CHEMICAL ENGINEERING. *Educ:* Seoul Nat Univ, BSE, 62; Univ Fla, MSE, 64, PhD(chem eng), 66. *Prof Exp:* Engr, Cheil Sugar Indust Co, 62-63; res engr, Gulf Res & Develop Co, 66-73, sr res engr, 73-78, res assoc, 78-81, dir chem recovery processes, 81-85; SECT SUPVR, MOBILITY CONTROL, CHEVRON OIL FIELD RES CO, 85- *Concurrent Pos:* Lectr chem eng, Carnegie-Mellon Univ, 67-68. *Mem:* Am Inst Chem Engrs; Soc Petrol Engrs. *Res:* Chemical engineering thermodynamics and reaction kinetics; thermal recovery of crude oil; underground oxidation of hydrocarbons; enhanced oil recovery by surfactant flooding; alkaline flooding; polymer flooding. *Mailing Add:* PO Box 446 La Habra CA 90631

BAECHLER, CHARLES ALBERT, b Lima, Ohio, Dec 15, 34; m 57; c 3. MEDICAL & PHARMACEUTICAL INFORMATION, INDUSTRIAL & MANUFACTURING ENGINEERING. *Educ:* Ohio State Univ, BScEduc, 57; Univ Toledo, MS, 64; Wayne State Univ, PhD(med microbiol), 71. *Prof Exp:* Teacher biol & chem, Libby High Sch, Toledo, Ohio, 60-63; teacher chem, Sylvania High Sch, 63-64; microbiol res specialist, Parke-Davis & Co, 64-71; res assoc, Sch Med, Wayne State Univ, 71-74, asst prof physiol, 74-78; microbiologist-tech serv, 78-80, appln develop engr, Merck & Co, Inc, 80-85, mgr ed & labeling, 85-87, MGR MED INFO MGT SYSTS, MERCK SHARP & DOHME, 87- *Concurrent Pos:* Lectr, Univ Toledo Community Col, 64; consult electron micros, Parke-Davis & Co, 71-78; ed adv bd, Pharmaceut Technol, 83- *Mem:* Soc Exp Biol & Med; Sigma Xi. *Res:* Function, structure and developmental aspects of blood and cardiovasculature interactions and microbial interactions at the ultrastructural level; technical services-vaccine and biological development; development of computer systems for scientific and industrial laboratory and production applications. *Mailing Add:* 250 Hampshire Dr Chalfont PA 18914

BAECHLER, RAYMOND DALLAS, b Brooklyn, NY, Mar 11, 45; m 67; c 2. ORGANIC CHEMISTRY. *Educ:* Fordham Univ, BS, 66; Princeton Univ, MA, 68, PhD(chem), 71. *Prof Exp:* Instr, Princeton Univ, 71-73; teaching fel, Miami Univ, 73-74; from asst prof to assoc prof, 74-82, chmn, dept chem, 83-86, PROF CHEM, RUSSELL SAGE COL, 82-, CHMN, DEPT CHEM, 90- *Mem:* Am Chem Soc. *Res:* Elucidation of reaction mechanisms; correlations between structure and reactivity; conformational processes of bond torsion and pyramidal inversion; silicon, phosphorus and sulfur chemistry; thermal rearrangements; unstable intermediates. *Mailing Add:* Dept Chem Russell Sage Col Troy NY 12180

BAEDECKER, MARY JO, b Richmond Ky, Nov 6, 41. ORGANIC GEOCHEMISTRY. *Educ:* Vanderbilt Univ, BS, 64; Univ Ky, MS, 67; George Wash Univ, PhD(geochem), 85. *Prof Exp:* Chemist, Inst Geophys & Planetary Physics, Univ Calif, Los Angeles, 68-73; RES CHEMIST, WATER RESOURCES DIV, US GEOL SURV, RESTON, VA, 74- *Concurrent Pos:* Adj prof, George Washington Univ, 87- *Honors & Awards:* Super Serv Award, Dept Interior, 88, Meritorious Serv Award, 91. *Mem:* Geol Soc Am; Geochem Soc; Am Geophys Union; Am Chem Soc; Nat Well Water Asn. *Res:* Organic-inorganic interactions in aquifers and recent sediments; degradation of natural and man-made organic materials and compounds in sub-surface environments; analytical methods for analyses of organic and inorgainc compounds. *Mailing Add:* 431 Nat Ctr US Geol Surv Reston VA 22092

BAEDECKER, PHILIP A, b East Orange, NJ, Dec 19, 39; m 66; c 1. NUCLEAR CHEMISTRY, GEOCHEMISTRY. *Educ:* Ohio Univ, BS, 61; Univ Ky, MS, 64, PhD(chem), 67. *Prof Exp:* Res assoc nuclear chem, Mass Inst Technol, 67-68; asst res chemist, Univ Calif, Los Angeles, 68-73, asst prof in residence, 70-71; chief, Anal Chem Br, 81-86, chmn, Effects on mat Task group, Nat Acid Precipitation Assessment Prog, 86-90, RES CHEMIST, US GEOL SURV, 74- *Mem:* AAAS; Am Chem Soc; Meteoritical Soc; Sigma Xi. *Res:* Radiochemistry; trace element geochemistry; activation analysis. *Mailing Add:* Mail Stop 923 US Geol Surv Reston VA 22092

BAEDER, DONALD L(EE), b Cleveland, Ohio, Aug 23, 25; m 52; c 5. CHEMICAL ENGINEERING, CHEMISTRY. *Educ:* Baldwin-Wallace Col, BA, 49; Carnegie Inst Technol, BS, 51. *Prof Exp:* Group head process res, Esso Res & Eng Co, 53-55, asst sect head, explor & fundamental res, 55-56, sect head, 56-58, asst dir process res div, 58-62, dir, cent basic res lab & spec proj unit, 62-63; mgr res & develop, Humble Oil & Ref Co, 64-66, mgr plastics div, Enjay Chem Co Div, 66-68; vpres corp & govt res, Esso Res & Eng Co, 69-75; exec vpres res & develop, Occidental Petrol Corp, 75-77; pres, Hooker Chem Co, 77-81; exec vpres sci & technol, Occidental Petrol Corp, 81-84; CHMN, BAEDER NERUDA INTERESTS INC, 85- *Concurrent Pos:* Mem div sci & technol, Am Petrol Inst. *Mem:* Am Chem Soc; fel Am Inst Chem Engrs; Sigma Xi. *Res:* Kinetics of thermal and catalytic reactions; radiation initiation of free radical reactions. *Mailing Add:* 279 Sugarberry Circle Houston TX 77024

BAEKELAND, FREDERICK, b Bronxville, NY, Sept 3, 28; m 50; c 2. PSYCHIATRY. *Educ:* Columbia Univ, BS, 51; Yale Univ, MS, 54, MD, 58; State Univ NY Downstate Med Ctr, DMSc(psychiat), 67. *Prof Exp:* From instr to asst prof, 64-71, ASSOC PROF PSYCHIAT, STATE UNIV NY DOWNSTATE MED CTR, 71- *Mem:* Asn Psychophysiol Study Sleep; fel Am Psychiat Asn. *Res:* Sleep: effects of drugs, physiology, dream content, requirements; alcoholism; correlates of treatment outcome; psychology of art. *Mailing Add:* 155 W 68th St Apt 28A New York NY 10023

BAENZIGER, NORMAN CHARLES, b New Ulm, Minn, Sept 23, 22; m 44; c 4. PHYSICAL CHEMISTRY. *Educ:* Hamline Univ, BS, 43; Iowa State Col, PhD(phys chem), 48. *Prof Exp:* Assoc chemist, Manhattan Proj & AEC, Iowa State Col, 44-48; instr phys chem, 46-48; fel, Mellon Inst, 48-49; from asst prof to assoc prof, 49-57, PROF PHYS CHEM, UNIV IOWA, 57- *Mem:* Am Chem Soc; Am Crystallog Asn. *Res:* X-ray crystallography; intermetallic compounds; metal-olefin complexes. *Mailing Add:* Dept Chem Univ Iowa Iowa City IA 52240

BAER, ADELA (DEE), b Blue Island, Ill, Apr 4, 31; c 2. GENETICS. *Educ:* Univ Ill, BS, 53; Purdue Univ, 54-56; Univ Calif, Berkeley, PhD(genetics), 63. *Prof Exp:* From asst prof to prof biol, San Diego State Univ, 62-83, chmn dept, 75-78; ADJ PROF, DEPT ZOOL, ORE STATE UNIV, CORVALLIS, 84- *Concurrent Pos:* Fulbright prof genetics, Univ Malaya, 67-68; res scientist, Int Ctr Med Res, Inst Med Res, Malaysia, 71-72; vis prof genetics, Univ Calif, Berkeley, 75 & 81; vis prof, Ctr Theoret Studies, Indian Inst sci, Bangalore, India, 78 & 89; instr, Ore State Univ, 84- *Mem:* Am Soc Human Genetics; Sigma Xi. *Res:* Human ecological and biochemical genetics. *Mailing Add:* 42220 Marks Ridge Rd Sweet Home OR 97386

BAER, ADRIAN DONALD, b St Louis, Mo, Nov 3, 42; m 64; c 3. SOLID STATE PHYSICS. *Educ:* Univ Denver, BS, 64; Stanford Univ, MS, 65, PhD(elec eng), 71. *Prof Exp:* Res assoc physics, Mont State Univ, 71-73 & Stanford Univ, 73; res physicist, 74-79, OPER RES ANALYST, NAVAL WEAPONS CTR, 79- *Mem:* Sigma Xi; Oper Res Soc Am. *Res:* Measuring various electron emission and optical spectra and utilizing these spectra to study electronic states at surfaces and interfaces, and in thin films. *Mailing Add:* 417 Cisco St Ridgecrest CA 93555

BAER, DONALD RAY, b Warren, Ohio, Nov 12, 47; w; c 2. SURFACE SCIENCE, SOLID STATE PHYSICS. *Educ:* Carnegie-Mellon Univ, BS, 69; Cornell Univ, PhD(physics), 74. *Prof Exp:* Res assoc physics, Mat Res Lab & Dept Physics, Univ Ill, Urbana, 74-76; res scientist, 76-80, SR RES SCIENTIST, SURFACE SCI GROUP, MAT DEPT, PAC NORTHWEST DIV, BATTELLE MEM INST, 80- *Concurrent Pos:* Vis res fel, Dept Mat Sci & Eng, Univ Surrey, Eng, 84-85; adj fac chem, Wash State Univ. *Mem:* Am Phys Soc; Am Soc Testing & Mat; Electrochem Soc; Minerals Metals & Mat Soc; AAAS; Am Sci Affil. *Res:* Surface and interface science; oade surfaces; high temperature corrosion; amorphous materials; surface reactivity; aqueous corrosion; auger electron spectroscopy; x-ray photo-electrons spectroscopy; ion beam materials analysis; metals physics; low temperature physics. *Mailing Add:* Battelle NW Labs, Mat Dept Box 999 Mail Stop K2-57 Richland WA 99352

BAER, DONALD ROBERT, b Paterson, NJ, Apr 20, 28; m 52; c 3. RESEARCH ADMINISTRATION. *Educ:* Cornell Univ, BA, 48; Univ Mich, MS, 50, PhD(org chem), 52. *Prof Exp:* Res chemist dyes, Fluorine Chem, E I du Pont de Nemours & Co, Inc, 52-57, res supvr dyes, 57-62, div head dyes-textile chem, 62-75, tech supt, 75-85; RETIRED. *Mem:* Am Chem Soc. *Res:* Fluorochemicals; dyes; textile chemicals; specialty chemicals processes; ornithology. *Mailing Add:* 2002 Dogwood Lane Foulk Woods Wilmington DE 19810

BAER, ERIC, b Nieder-Weisel, Germany, July 18, 32; US citizen; m 56; c 2. POLYMER PHYSICS & CHEMISTRY. *Educ:* Johns Hopkins Univ, MA, 53, PhD(chem eng), 57. *Prof Exp:* Res engr, Polychem Dept, E I Du Pont de Nemours & Co, 57-60; asst prof chem eng, Univ Ill, 60-62; assoc prof, 62-66, prof-in-chg polymer sci & eng, 62-66, chmn, Dept Macromolecular Sci, 67-78,; dean, 78-83, PROF, CASE INST TECHNOL, 83-; PROF ENG, CASE WESTERN RESERVE UNIV, 66- *Concurrent Pos:* Indust consult ed, Polymer Eng & Sci. *Honors & Awards:* McGraw Award, Am Soc Eng Educ, 68; Int Award, Soc Plastics Engrs, 80; Borden Award, Am Chem Soc, 81. *Mem:* Am Inst Chem Engrs; Am Soc Mech Engrs; Am Chem Soc; fel Am Phys Soc; fel Am Inst Chemists; Sigma Xi. *Res:* Physical behavior and structure of polymers; engineering design for plastics; reactions on polymer crystal surfaces; properties of polymeric materials under high pressure and at cryogenic temperatures; nucleation phenomena in phase changes; heat transfer during condensation. *Mailing Add:* Dept Macromolecular Sci Case Western Reserve Univ Cleveland OH 44106

BAER, FERDINAND, b Dinkelsbuhl, Ger, Aug 30, 29; US citizen; m; c 4. METEOROLOGY, NUMERICAL WEATHER PREDICTION. *Educ:* Univ Chicago, AB, 50, MS, 54, PhD(geophys sci), 61. *Prof Exp:* Asst atmospheric physics, Univ Ariz, 55-56; asst meteorol, Univ Chicago, 56-61; from asst prof to assoc prof atmospheric sci, Colo State Univ, 61-71; prof atmospheric sci, Univ Mich, 72-77; dir meteorol prog, Univ Md, 77-79, chmn, Dept Meteorol, 79-87, dir coop inst climate studies, 84-89, PROF METEOROL, UNIV MD, COLLEGE PARK, 77- *Concurrent Pos:* Vis res fel, Geophys Fluid Dynamics Lab, Princeton Univ, 68-69; consult, McGraw Hill Info Systs, 71-79; vis prof, Stockholm Univ, 74-75 & Frie Univ, Berlin, 75; mem rep, Univ Corp for Atmospheric Res, 77-90; mem comt on atmospheric sci, Nat Acad Sci, 80-82; mem, Bd Atm Sci & Climate, Nat Res Coun, 82-85; trustee, Univ Corp Atmospheric Res, 85-91. *Mem:* Fel Am Meteorol Soc; Am Geophys Union; Fel Royal Meteorol Soc; Meteorol Soc

Japan; Can Meteorol & Oceanog Soc. *Res:* Dynamics of the atmosphere; expansion of atmospheric space variables in series of orthogonal polynominals; numerical analysis of atmospheric space fields; normal mode analysis; initialization. *Mailing Add:* Dept Meteorology Univ Md College Park MD 20742

BAER, GEORGE M, b London, Eng, Jan 12, 36; US citizen; m 60; c 2. EPIDEMIOLOGY, VIROLOGY. *Educ:* Cornell Univ, DVM, 59; Univ Mich, MPH, 61. *Prof Exp:* Practicing vet, 59-60; vet epidemiologist, NY State Health Dept, 61-63; actg chief rabies invests lab, Commun Dis Ctr, Ga, 63-64, actg chief, Southwest Rabies Invests Lab, NMex, 64-66, chief lab, 66-69, CHIEF VIRAL ZOONOSES BR, CTR DIS CONTROL, USPHS, 69- *Concurrent Pos:* Rabies consult, Pan Am Health Orgn, Mex, 66-69. *Mem:* Am Vet Med Asn; Am Pub Health Asn; NY Acad Sci. *Res:* Veterinary epidemiology, especially zoonotic diseases such as rabies, brucellosis and tuberculosis. *Mailing Add:* 1476 Peachtree Battle Ave Atlanta GA 30327

BAER, HANS HELMUT, b Karlsruhe, Ger, July 3, 26; Can citizen; m 56; c 2. ORGANIC CHEMISTRY, CARBOHYDRATE CHEMISTRY. *Educ:* Karlsruhe Tech Univ, cand chem, 47; Univ Heidelberg, dipl chem, 50, Dr rer nat(chem), 52. *Prof Exp:* Res assoc chem, Max Planck Inst Med Res, Heidelberg, 52-57; vis scientist biochem, Univ Calif, Berkeley, 57-59; vis scientist chem, Nat Inst Arthritis & Metab Dis, 59-61; assoc prof, 61-65, chmn dept, 69-75, PROF CHEM, UNIV OTTAWA, 65- *Honors & Awards:* C S Hudson Award, Am Chem Soc, 75. *Mem:* Am Chem Soc; Chem Inst Can; Soc Ger Chemists. *Res:* Chemistry of natural products, especially carbohydrates; general synthetic organic chemistry. *Mailing Add:* Dept Chem Univ Ottawa Ottawa ON K1N 9B4 Can

BAER, HAROLD, b New York, NY, Oct 3, 18; m 46; c 2. ALLERGY, IMMUNOLOGY. *Educ:* Brooklyn Col, BA, 38; Columbia Univ, MA, 40; Harvard Univ, MA, 42, PhD(chem), 43. *Prof Exp:* From asst to res assoc, Col Physicians & Surgeons, Columbia Univ, 43-50; from asst prof to assoc prof, Med Sch, Tulane Univ, 50-60; chief sect allergenic prod, Div Biologies Stand, NIH, 60-71, chief lab bact prod, 71-74, dir allergenic prod, Bur Biologies, Food & Drug Admin, 74-88; RETIRED. *Concurrent Pos:* Instr, City Col New York, 48-50. *Mem:* AAAS; Sigma Xi; Can Soc Allergy & Clin Immunol; Am Acad Allergy. *Res:* Allergen chemistry and standardization of allergenic extracts; delayed contact sensitivity. *Mailing Add:* 20500 Highland Hall Dr Gaithersburg MD 20879-4003

BAER, HELMUT W, b Hunan, China, Aug 16, 39; US citizen; m; c 2. NUCLEAR PHYSICS. *Educ:* Franklin & Marshall Col, BA, 61; Univ Mich, Ann Arbor, PhD(physics), 67. *Prof Exp:* Res asst physics, Univ Mich, Ann Arbor, 67-69; res asst, Univ Colo, Boulder, 69-71; physicist, Lawrence Berkeley Lab, Univ Calif, 71-74; asst prof physics, Case Western Reserve Univ, 74-78; STAFF PHYSICIST, LOS ALAMOS SCI LAB, UNIV CALIF, 78- *Mem:* Fel Am Phys Soc. *Res:* Experimental nuclear research at low, medium and high energies. *Mailing Add:* Los Alamos Sci Lab MP-4 MS H845 PO Box 1663 Los Alamos NM 87545

BAER, HERMAN, b Uetikon, Switz, Sept 11, 33; US citizen; m; c 3. MEDICAL MICROBIOLOGY. *Educ:* Univ Basel, Switz, MD, 60; Am Acad Med Microbiol, dipl med bact, 69. *Prof Exp:* Fel, Inst Med Microbiol, Sch Med, Univ Zurich, 60-62, head diag unit, 62-63; fel, Div Infectious Dis, Dept Med, Med Sch, Univ Pittsburgh, 63-64, fel, Dept Bact, Div Natural Sci, 65-66; asst prof, 66-70, ASSOC PROF IMMUNOL & MED MICROBIOL, COL MED, UNIV FLA, 70-, ASSOC PROF PATH, 74-, DIR CLIN MICROBIOL LAB, 66- *Mem:* Am Soc Microbiol. *Res:* Role of hand contamination of personnel in the epidemiology of gram negative nosocomal infections; role of airborne contamination in orthopedic surgery. *Mailing Add:* Col Med Univ Fla Gainesville FL 32611

BAER, HOWARD A, b Milwaukee, Wis, Nov 4, 57. PHENOMENOLOGY, COMPUTER SIMULATIONS. *Educ:* Univ Wis, BS, 79, MS, 81, PhD(physics), 84. *Prof Exp:* Researcher theoret particle physics, Europ Orgn Nuclear Res, 84-85, Argonne Nat Lab, 85-87; asst res scientist, 87-90, ASST PROF THEORET PARTICLE PHYSICS, FLA STATE UNIV, 90- *Mem:* Am Phys Soc. *Res:* Theoretical physics of elementary particles; interface of experiment and theory; phenomenology; search for supersymmetry; new quarks and leptons; perturbative quantum chromodynamics; computer simulation of particle scattering. *Mailing Add:* Physics Dept Fla State Univ Tallahassee FL 32306

BAER, JAMES L, b Van Wert, Ohio, Oct 6, 35; m 61; c 4. PALEOECOLOGY, STRUCTURAL GEOLOGY. *Educ:* Ohio State Univ, BS, 57; Brigham Young Univ, MS, 62, PhD(paleoecol), 68. *Prof Exp:* Asst prof geol, Northeast La State Col, 68-69; asst prof, 69-74, ASSOC PROF GEOL, BRIGHAM YOUNG UNIV, 74- *Mem:* Geol Soc Am; Paleont Soc. *Res:* Paleoecology of ancient lake sediments; structural evolution of western United States; continental drift; geochemistry of sediments. *Mailing Add:* Dept Geol Brigham Young Univ Provo UT 84602

BAER, JOHN ELSON, b Cleveland, Ohio, Apr 25, 17; m 47; c 4. PHARMACOLOGY, DRUG METABOLISM. *Educ:* Swarthmore Col, AB, 38; Univ Pa, MS, 40, PhD(org chem), 48. *Prof Exp:* Chemist, Med Sch, NY Univ, 43-46; instr chem, Haverford Col, 47-48; asst prof, Carleton Col, 48-51; res assoc, Pharmacol Sect, Sharp & Dohme Div, Merck & Co, 51-58, dir pharmacol chem, Merck Inst, 58-69, assoc dir, Merck Inst Therapeut Res, 69-71, sr dir, 71-76, exec dir drug metab, 76-81, staff executive, Merck Sharp & Dohme Res Labs, 81-84; RETIRED. *Honors & Awards:* Albert Lasker Spec Award, Lasker Found, 75. *Mem:* AAAS; Am Soc Pharmacol & Exp Therapeut; Am Soc Nephrology; Am Soc Clin Chemists. *Res:* Chemical methods for analysis of compounds in biological systems; pharmacology of diuretics and saluretics. *Mailing Add:* 517 Tennis Ave Ambler PA 19002

BAER, LEDOLPH, b Monroe, La, Nov 21, 29; m 57; c 3. PHYSICAL OCEANOGRAPHY. *Educ:* La Polytech Inst, BS, 50; Tex Agr & Mech Col, MS, 55; NY Univ, PhD(ocean waves), 62. *Prof Exp:* Meteorologist, Gulf Consult, Tex, 55-56; res specialist, Lockheed Missiles & Space Co, 59-62, res & develop scientist, Lockheed-Calif Co, 62-66, mgr ocean sci dept, Lockheed Ocean Lab, 66-74; dir oceanog serv, Nat Oceanic & Atmospheric Admin, 74-78, dir policy & long range planning, 78-80, oceanog sci adv & mgr coastal waves prog, 80-82, spec asst, off chief scientist, 82-89, CHIEF, PHYS OCEANOG DIV, NAT OCEAN SERV, NAT OCEANIC & ATMOSPHERIC ADMIN, 89- *Mem:* AAAS; Am Meteorol Soc; Am Geophys Union. *Res:* Surface wave forecasting; physical oceanography; oceanographic statistics; quality assurance of oceanic data and products; synoptic oceanography. *Mailing Add:* Nat Oceanic & Atmospheric Admin N/OMAI 6001 Executive Blvd Rockville MD 20852

BAER, NORBERT SEBASTIAN, b New York, NY, June 6, 38; m 59; c 2. ENVIRONMENTAL SCIENCE, MATERIALS DEGRADATION. *Educ:* Brooklyn Col, BSc, 59; Univ Wis, MSc, 62; NY Univ, PhD(phys chem), 69. *Prof Exp:* Physicist, Control Instrument Div, Warner & Swasey Co, 62-63; researcher sci bks, Sci Lib, Bk Div, Time Inc, 63-64; lectr chem, Queensborough Community Col, 67-68, asst prof, 68-69; from instr to prof, 69-86, actg dir admin, 78-79, HAGOP KEVORKIAN PROF CONSERV, INST FINE ARTS, NY UNIV, 86- *Concurrent Pos:* Ed adv & assoc ed, Studies Conserv, 71-87; mem exec comt, Nat Conserv Adv Coun, 72-79; co-chmn, Conserv Ctr, Inst Fine Arts, NY Univ, 75-83; chmn ad hoc vis comt, Conserv Anal Lab, Smithsonian Inst, 77-78; chmn, Adv Comt, Nat Archives & Records Serv, 80-; chmn, Comt Conserv Historic Stone Bldg & Monuments, Nat Mat Adv Bd, Nat Acad Sci, 80-82; US exec ed, Butterworth Series Conserv Arts, 79-; John Simon Guggenheim Mem Found fel, 83-84; tech adv, Comt Preserv Hist Rec, Nat Mat Adv Bd, Nat Acad Sci, 85-86; mem, Nat Mat Adv Bd, Nat Acad Sci, 86-, comt Natural Disasters, 88-; mem, vis comt, Dept Objects Conserv, Metropoliton Mus Art, 80-, chmn, 85-90; chmn, rev panel progs dry deposition & mat effects, Environ Protection Agency, 85; US coordr, conserv workshops Indo-US subcomn educ & cult, 78-; mem, book preserv technol adv panel, Off Technol Assessment, 87-88; comn Sistine Chapel, 87- *Mem:* Am Chem Soc; fel Am Inst Chemists; sr mem, Instrument Soc Am; Air Pollution Control Asn; Sigma Xi. *Res:* Application of physiochemical methods to the examination and preservation of artistic and historic works; effects of air pollution on materials; physical chemistry. *Mailing Add:* Conserv Ctr Inst Fine Arts 14 E 78th St New York NY 10021

BAER, PAUL NATHAN, b New York, NY, Oct 13, 21; m 50; c 1. PERIODONTOLOGY. *Educ:* Brooklyn Col, BA, 42; Columbia Univ, DDS, 45; Am Bd Periodont, dipl, 56. *Prof Exp:* Asst prof periodont, Univ Southern Calif, 55-56; peridontist, Nat Inst Dent Res, 56-70, chief adolescent periodont unit, 70-73; PROF PERIODONT & CHMN DEPT, SCH DENT MED, STATE UNIV NY STONY BROOK, 73- *Concurrent Pos:* Vis assoc prof, Grad Sch Dent, Boston Univ; assoc ed, J Dent Res; consult, Long Island Jewish-Hillside Med Ctr Hosp, Vet Admin Northport Hosp, Nassau County Med Ctr & Sagamore Childrens Ctr. *Honors & Awards:* William J Gies Award, 80. *Mem:* Am Dent Asn; Am Acad Periodont; Int Asn Dent Res; hon mem Colombia Soc Periodont. *Res:* Periodontal disease. *Mailing Add:* Dept Periodont State Univ NY Sch Dent Med Stony Brook NY 11794-8703

BAER, RALPH NORMAN, b Flushing, NY, Jan 2, 48. UNDERWATER ACOUSTICS, ACOUSTIC PROPAGATION. *Educ:* Rensselaer Polytech Inst, BS, 68, MS, 70, PhD(appl math), 74. *Prof Exp:* Mathematician, 74-80, SUPVR MATHEMATICIAN, US NAVAL RES LAB, 80- *Concurrent Pos:* Sci officer, Off Naval Res, 90- *Honors & Awards:* R Bruce Lindsay Award, Acoust Soc Am. *Mem:* Acoust Soc Am; Soc Indust & Appl Math; Am Math Soc; Math Asn Am. *Res:* Basic research in underwater acoustic propagation, scattering, and signal processing; acoustic computer-based model development. *Mailing Add:* US Naval Res Lab Code 5163 Washington DC 20375

BAER, RICHARD, b Aug 29, 53. MICROBIOLOGY. *Educ:* Rutgers Univ, BA, 76, PhD(microbiol), 81. *Prof Exp:* Postdoctoral fel, Lab Molecular Biol, MRC, Cambridge, Eng, 80-83 & 84-87, Mem Sloan-Kettering Cancer Ctr, NY, 83-84; ASST PROF MICROBIOL, UNIV TEX SOUTHWESTERN MED CTR, DALLAS, 87- *Concurrent Pos:* Damon Runyon/Walter Winchell Cancer Fund fel, 80-82 & Lady Tata Mem Trust fel, 82-85; jr fac res award, Am Cancer Soc, 90-92. *Mailing Add:* Dept Microbiol Univ Tex Southwestern Med Ctr 5323 Harry Hines Blvd Dallas TX 75235-9048

BAER, ROBERT LLOYD, b New York, NY, Jan 29, 31; m 76; c 2. MECHANICAL ENGINEERING. *Educ:* Clarkson Col Technol, BME, 52; Rensselaer Polytech Inst, MSME, 58. *Prof Exp:* Engr, Nuclear Div, Pratt & Whitney Aircraft Div, United Aircraft Corp, 54-57; sr engr, 57-59; eng specialist, Nuclear Div, Martin Co, 59-61, proj engr, 61-63; chief reactor systs, Hittman Assoc, Inc, 63-67, dept mgr, 67-70; sr proj mgr, Atomic Energy Comn, 71-74; sect leader, 74-76, BR CHIEF, NUCLEAR REGULATORY COMN, 76- *Concurrent Pos:* Lectr, Catholic Univ, 63-76. *Mem:* Am Soc Mech Engrs; Am Nuclear Soc. *Res:* Nuclear power plant design, development and operation; safety, systems and economic analysis of large central station power plants; heat transfer and flow analysis of nuclear reactor cores. *Mailing Add:* Off Nuclear Regulatory Res Nuclear Regulatory Comn Washington DC 20555

BAER, ROBERT W, b Minneapolis, Minn, Nov 8, 48; m 79; c 2. CARDIOVASCULAR & CORONARY PHYSIOLOGY. *Educ:* Carleton Col, BA, 71; Johns Hopkins Univ, PhD(physiol), 79. *Prof Exp:* Res fel physiol, Cardiovasc Res Inst, Univ Calif, San Francisco, 79-82; asst prof, 82-89, ASSOC PROF PHYSIOL, KIRKSVILLE COL OSTEOP MED, 89- *Mem:* Am Physiol Soc; Am Heart Asn. *Res:* Control of coronary blood flow and myocardial oxygen delivery; transmural distribution of myocardial blood flow; cardiac function during respiratory maneuvers. *Mailing Add:* Dept Physiol Kirksville Col Osteop Med Kirksville MO 63501

BAER, RUDOLF L, b Strasbourg, Alsace-Lorraine, July 22, 10; nat US; m 41; c 2. DERMATOLOGY, IMMUNOLOGY. *Educ:* Univ Frankfurt, Basel, MD, 34; Am Bd Dermat, dipl, 40. *Hon Degrees:* Dr Med, Univ Munich, WGer, 81. *Prof Exp:* Teacher dermat allergy, Med Sch, Columbia Univ, 39-46, instr dermat & syphil, 46-48, asst clin prof, Postgrad Med Sch & Hosp, 48; asst prof clin dermat & syphil, Postgrad Med Sch, 49-50, from assoc prof to prof, 50-61, chmn dept, 61-81, PROF DERMAT, SCH MED, NY UNIV, 61- *Concurrent Pos:* Consult, Med Adv Panel on Hazardous Substances, US Dept Health, Educ & Welfare, 61, Nat Inst Allergy & Infectious Dis, 68; chmn Dermat Sect, AMA, 66, bd, Dermat Found, 74-78; founder & mem bd trustees, Rudolf L Baer Found Skin Dis, Inc, 75-; mem comt cutaneous dis, Armed Forces Epidemiol Bd, 67-72; consult, Surgeon Gen, US Army, 67; mem, Int Comt Dermat, 67-82, pres, 72-77; consult, Food & Drug Admin, 69; mem comt on revision, US Pharmacopeia, 70; mem, Am Bd Dermat, 64-72, pres, 70-; O'Leary lectr, Mayo Clin, Rochester, Minn, 71; Robinson lectr, Univ Md, 72; Barrett Kennedy Mem lectr, SCent Dermat Soc, 73; Duhring lectr, Pa Acad Dermat, 74; Bluefarb lectr, Chicago Dermat Soc, 75; Dome lectr, Am Acad Dermat, 76; F C Novy vis scholar, Univ Calif, Davis, 78. *Honors & Awards:* Dohi Mem Medal, Japanese Dermat Soc, 65; Hellerstrom Medal-Karolinska Inst, Sweden, 70; Stephen Rothman Medal, Soc Invest Dermat, 73; Alfred Marchionini Gold Medal, XV Int Cong of Dermat, 77; Gold Medal, Am Acad Dermat, 78; M H Samitz lectr, Univ Pa, 79; Nomland-Carney lectr, Univ Iowa, 79; Barrett Kennedy Mem lectr, La Dermat Soc, 80; A Harvey Neidoff Lectr, Am Asn Clin Immunol Allergy, 85-; Morrow-Miller-Taussig Mem Lectr, San Francisco Dermat Soc, 61; Leo Ritter von Zumbusch Mem Lectr, Univ Munich, 67; Ferdinand von Hebra Lectr, Austrian Soc Dermat & Venerology, 88. *Mem:* AAAS; fel Am Acad Allergy; fel & hon mem Am Acad Dermat (pres, 74-75); fel Am Col Allergists (vpres, 54); hon mem Am Dermat Asn (pres, 77); hon mem Soc Invest Dermat (pres, 64); hon mem Int Dermat Socs; Brazilian Nat Acad Med. *Res:* Immunology and allergy cutaneous aspects; cutaneous aspects; cross-sensitization; photobiology of the skin; biology of fungous infections of the skin; Langerhans cells; contact sensitization. *Mailing Add:* Dept Dermat NY Univ Sch Med 530 First Ave New York NY 10016

BAER, THOMAS STRICKLAND, b Huntington, WVa, May 24, 42; m 64; c 2. NUCLEAR & ENVIRONMENTAL ENGINEERING. *Educ:* US Naval Acad, BS, 64; Univ Cincinnati, MS, 71, PhD, 73. *Prof Exp:* Sta engr, Metrop Edison Co, 73-74; vpres & gen mgr, Protective Packaging Inc, 74-76; vpres, Nuclear Eng Co, 76-81; mgr health physics-environ safety, Formerly Utilized Sites Remedial Action Prog, Bechtel Nat, Inc, 81-83, proj mgr, Hazardous Chem Waste Dept, 83-85, proj mgr, 85-86; SR VPRES, US ECOL INC, 86- *Mem:* Am Nuclear Soc; Health Physics Soc; Soc Am Mil Engrs; Am Soc Eng Mgt. *Res:* Evaluation and development of methods to treat, handle, transport, and dispose of hazardous chemical and radioactive wastes. *Mailing Add:* US Ecol Inc 621 Summit Lake Ct Knoxville TN 37922

BAER, TOMAS, b Zurich, Switz, Aug 27, 39; US citizen; m 62; c 3. CHEMICAL PHYSICS, CHEMICAL DYNAMICS. *Educ:* Lawrence Univ, BA, 62; Wesleyan Univ, MA, 64; Cornell Univ, PhD(chem), 69. *Prof Exp:* Fel chem, Northwestern Univ, 69-70; from asst prof to assoc prof, 70-79, PROF CHEM, UNIV NC, CHAPEL HILL, 79- *Concurrent Pos:* J S Guggenheim fel, Univ Paris, 76-77; vis prof, Univ Paris, 79; ed bd, Int J Mass Spectrom Ion Proc, 86-; organizing comt, Symp Synchroton Radiation, 87; exec comt, Advan Light Source Synchroton, 87- *Mem:* Am Chem Soc; fel Am Phys Soc; Am Soc Mass Spectrometry. *Res:* Photoion-photoelectron coincidence spectroscopy used to study the dissociation rates of energy selected ions and the role of ion internal energy in ion-molecule reactions; use of synchrotron radiation and multiphoton ionization. *Mailing Add:* Dept Chem Univ NC Chapel Hill NC 27599-3290

BAER, WALTER S, b Chicago, Ill, July 27, 37; m 59; c 2. TELECOMMUNICATIONS, SCIENCE ADMINISTRATION. *Educ:* Calif Inst Technol, BS, 59; Univ Wis, PhD(physics), 64. *Prof Exp:* Mem tech staff, Bell Tel Labs, 64-66; White House fel, Off of Vice President, 66-67; staff mem, Off Sci & Technol, Exec Off of President, 67-69; consult & sr scientist, Rand Corp, 70-78, dir energy policy prog, 78-81; DIR ADVAN TECHNOL, TIMES MIRROR CO, 81- *Concurrent Pos:* Mem comput sci & eng bd, Nat Acad Sci, 69-72; consult, maj US corps & govt agencies, 70-; mem cable TV adv comt, Fed Commun Comn, 72-74; dir, Aspen Cable Workshop, 72-73; mem adv coun, Aspen Prog Commun & Soc, 74-; mem adv comt, Univ Calif, Los Angeles, communications law prog; mem energy mgt bd, City of Los Angeles, adv coun, Electric Power Res Inst. *Honors & Awards:* Preceptor Award, Broadcast Indust Conf, 74. *Mem:* AAAS; Am Phys Soc; Inst Elec & Electronics Engrs; Int Inst Commun. *Res:* Science technology and public policy; telecommunications; technical management. *Mailing Add:* Rand Corp 1700 Main St Santa Monica CA 90406

BAERG, DAVID CARL, b Dinuba, Calif, June 23, 38; m 62; c 2. MEDICAL ENTOMOLOGY. *Educ:* Univ Calif, Davis, BS, 64, MS, 65, PhD(entom), 67. *Prof Exp:* Res asst mosquito res, Univ Calif, Davis, 64-67; res entomologist, Gorgas Mem Lab, 67-76; clin asst prof & assoc mem grad fac trop med & med parasitol, La State Univ Med Ctr, New Orleans, 73-76; environ health inspection, CZ Health Bur, 76-78; ENVIRON & ENERGY CONTROL OFFICER, OFF CANAL IMPROV, PANAMA CANAL COMN, 78- *Mem:* Audubon Soc; Animal Welfare Inst. *Res:* Mosquito biology and ecology; transmission, development and chemotherapy of malarias. *Mailing Add:* PAC 0636 PO Box 37301 Washington DC 20013

BAERG, WILLIAM, b Fayetteville, Ark, Oct 16, 38; m 60, 72; c 3. ELECTRONICS ENGINEERING, MATERIALS SCIENCE ENGINEERING. *Educ:* Stanford Univ, BS, 60, PhD(chem eng), 65. *Prof Exp:* NSF fel eng & appl math, Harvard Univ, 65-66; sr scientist, chem lab, Aeronutronic Div, Philco-Ford Corp, 66-68; private res, 68-69; vpres, Linviron Corp, 69-72; sr staff engr, 72-80, PRIN ENGR, INTEL CORP, 80- *Mem:* Inst Elec & Electronics Engrs. *Res:* Acoustics and turbulence; plasma physics; reverse osmosis and module design; home water purification devices; semiconductor technology. *Mailing Add:* Intel Corp SC9-03 2250 Mission College Blvd Santa Clara CA 95052-8125

BAERNSTEIN, ALBERT, II, b Birmingham, Ala, Apr 25, 41; m 62; c 2. MATHEMATICAL ANALYSIS. *Educ:* Cornell Univ, AB, 62; Univ Wis, MA, 64, PhD(math), 68. *Prof Exp:* Cost analyst, Prudential Ins Co, 62-63; instr math, Wis State Univ, Whitewater, 66-68; asst prof, Syracuse Univ, 68-72; assoc prof, 72-74, PROF MATH, WASH UNIV, 74- *Concurrent Pos:* Fulbright-Hays sr res scholar, Imp Col, Univ London, 76-77. *Mem:* Am Math Soc. *Res:* Complex analysis and related areas. *Mailing Add:* Dept Math Wash Univ St Louis MO 63130

BAERTSCH, RICHARD D, b Robbinsdale, Minn, Mar 14, 36; m 60; c 2. ELECTRICAL ENGINEERING. *Educ:* Univ Minn, BSEE, 59, MSEE, 61, PhD(elec eng), 64. *Prof Exp:* ELEC ENGR, GEN ELEC RES & DEVELOP CTR, 65- *Mem:* Sr mem Inst Elec & Electronics Engrs. *Res:* Noise in semiconductor devices; semiconductor physics and detectors; surface charge transport; integrated circuit design and process. *Mailing Add:* Gen Elec Res & Develop Ctr PO Box 8 Schenectady NY 12301

BAERWALD, JOHN E(DWARD), b Milwaukee, Wis, Nov 2, 25; m 48, 75; c 3. TRANSPORTATION & TRAFFIC ENGINEERING. *Educ:* Purdue Univ, BSCE, 49, MSCE, 50, PhD(civil eng), 56. *Prof Exp:* Instr, surv lab, Valparaiso Univ, 47-48; asst traffic eng, Purdue Univ, 49-52, res engr, 52-55; from asst prof to prof, transp & traffic eng, 55-83, dir Hwy Traffic Safety Ctr, 61-83, EMER PROF, UNIV ILL, URBANA-CHAMPAIGN, 83- *Concurrent Pos:* Consult transp & traffic engr, 52-; mem, Traffic Conf Exec Comt, Nat Safety Coun, 62-81, bd dirs, 75-80; chmn adv comt ambulance design criteria, Nat Acad Eng, 68-69; tech adv, tech comt traffic & safety, Pan Am Hwy Cong, 68-74; mem group 3 coun, oper & mgt serv & maintenance facil, Transportation Res Bd, Nat Res Coun-Nat Acad Sci, 69-73; mem, Bd Trustees, Champaign-Urbana, 73-83, chmn transit dist, 75-83; mem, tech adv comt, Ill Transportation Study Comn, 77- *Honors & Awards:* Award, Inst Transp Engrs, 53, Matson Award, 88. *Mem:* Inst Transp Engrs (int pres, 70); Am Soc Civil Engrs; Am Road & Transp Builders Asn (pres, 79); Sigma Xi. *Res:* Traffic safety; traffic planning and administration; transportation and traffic engineering; traffic accident reconstruction. *Mailing Add:* RR 2 Box 927 Santa Fe NM 87505

BAES, CHARLES FREDERICK, JR, b Cleveland, Ohio, Dec 2, 24; m 48; c 3. SEPARATION SCIENCE. *Educ:* Rutgers Univ, BSc, 46; Univ Southern Calif, MSc, 48, PhD(phys inorg chem), 50. *Prof Exp:* Fel metal, Columbia Univ, 50-51; res chemist, Oak Ridge Nat Lab, 51-87, CONSULT, 88- *Concurrent Pos:* Vis prof chem, Col William & Mary, 72-73. *Mem:* Am Chem Soc; Sigma Xi. *Res:* Solution chemistry; thermodynamics; solvent extraction chemistry; environmental chemistry; high temp aqueous chem. *Mailing Add:* 102 Berwick Dr Oak Ridge TN 37830

BAETKE, EDWARD A, b Marinette, Wis, March 3, 37; m 58; c 2. ENVIRONMENTAL ANALYSIS. *Educ:* St Norbert Col, BS, 59; Northwestern Univ, MS, 62. *Prof Exp:* Res chemist, Central Soya, 60-62, The Ansul Co, 62-70; LAB DIR, CORY LABS INC, 70- *Mem:* Am Chem Soc. *Mailing Add:* 4000 Irving St Marinette WI 54143-1045

BAETZ, ALBERT L, b Cleveland, Ohio, Dec 25, 38; m 89; c 1. CLINICAL CHEMISTRY. *Educ:* Purdue Univ, BS, 61; Iowa State Univ, MS, 63, PhD(anal chem), 66. *Prof Exp:* RES CHEMIST, NAT ANIMAL DIS CTR, AGR RES SERV, USDA, 66- *Mem:* AAAS; Am Chem Soc; Am Asn Vet Physiologists & Pharmacologists; Sigma Xi. *Res:* Purification and characterization of protein antigens from Listeria Monocytogenes in order to elucidate their roles in animal disease. *Mailing Add:* Nat Animal Dis Ctr PO Box 70 Ames IA 50010

BAETZOLD, ROGER C, b Warsaw, NY, Feb 26, 42; m 64; c 4. PHYSICAL CHEMISTRY, CHEMICAL PHYSICS. *Educ:* Univ Buffalo, BA, 63; Univ Rochester, PhD(chem), 67. *Prof Exp:* Sr res chemist, 66-71, RES ASSOC, EASTMAN KODAK CO, NY, 71- *Mem:* Am Chem Soc. *Res:* Experimental and theoretical kinetics of chemical reactions; molecular orbital calculations; solid state chemistry. *Mailing Add:* 1188 Clarkson Parma TL Rd Brockport NY 14420-9417

BAEUMLER, HOWARD WILLIAM, b Buffalo, NY, Sept 29, 21. MATHEMATICS, ELECTRICAL ENGINEERING. *Educ:* State Univ NY teachers Col Buffalo, BS, 43; Univ Buffalo, MA, 50; Ohio State Univ, PhD(elec eng), 64. *Prof Exp:* Instr math & sci, Park Sch, Buffalo, 43-44; instr math, Univ Buffalo, 46-53; asst prof, Marshall Univ, 54-55; asst instr, Ohio State Univ, 55-56, res assoc elec eng, Res Found, 56-61, instr, univ, 59-64; ASSOC PROF MATH, OLD DOMINION UNIV, 65- *Concurrent Pos:* Lectr, Va High Sch Vis Scientist Prog, 66-67. *Mem:* AAAS; Am Soc Eng Educ; Asn Comput Mach; Inst Elec & Electronics Engrs; Math Asn Am. *Res:* Electromagnetic field theory; numerical methods, computation and analysis; Boolean algebra and Boolean matrices; electronic and mechanical switching circuits; symbolic binary and ternary logic; computer science, programming languages and non-numerical applications. *Mailing Add:* Dept Math Old Dominion Univ 5215 Hampton Blvd Norfolk VA 23508

BAEZ, ALBERT VINICIO, b Puebla, Mex, Nov 15, 12; nat US; m 35; c 3. PHYSICS, ENVIRONMENTAL EDUCATION & ETHICS. *Educ:* Drew Univ, BA, 33; Syracuse Univ, MA, 35; Stanford Univ, PhD, 50. *Hon Degrees:* Dr, Open Univ, Gt Brit, 74. *Prof Exp:* Instr physics & math, Morris Jr Col, 36-38; instr math, Drew Univ, 38-40; from instr math to prof physics, Wagner Col, 40-44; instr, Stanford Univ, 44-45, actg instr math, 45-46, res asst physics, 46-49; physicist, Aeronaut Lab, Cornell Univ, 49-50; prof physics, Univ Redlands, 51-56, physicist, Film Group, Phys Sci Study Comt, 58-60; assoc prof physics, Harvey Mudd Col, 60-61; dir div sci teaching, UNESCO, 61-67; consult, sci & technol educ, UN, NY & UNESCO, Paris, 67-74; assoc, Lawrence Hall Sci, Univ Calif, Berkeley, 74-78; chmn 79-84, EMER CHMN, COMN EDUC, INT UNION CONSERV NATURE & NATURAL RESOURCES, GLAND, SWITZ, 84-; PRES, VIVAMOS MEJOR/USA, 86- *Concurrent Pos:* Prof, Col Arts & Sci, Univ Baghdad & head, UNESCO Tech Asst Mission, 51-52; vis prof, Stanford Univ, 56-58; hon res assoc, Dept

Physics, Harvard Univ, 70-71; sci dir, Physics Loop Film Proj, Encycl Britannica, 67-75; consult int sci educ, NSF, 70-; vchmn, Comt Teaching Sci, Int Coun Sci Unions, 70-74, chmn, 74-78; vis prof, Open Univ, Gt Brit, 71-72; mem, Bd Int Orgns & Progs, Nat Acad Sci, 74-76; proj head, Inst Elec Electronics, Algeria, 76-78. *Mem:* Fel AAAS; Am Phys Soc; Optical Soc Am; Am Asn Physics Teachers; Nat Sci Teachers Asn. *Res:* Absolute intensity of x-ray radiation; x-ray optics and microscopy; x-ray optical images; holography; science and environmental education; instrumentation for x-ray astronomy. *Mailing Add:* 58 Greenbrae Boardwalk Greenbrae CA 94904

BAEZ, SILVIO, b July 6, 15; US citizen; m 50; c 3. PHYSIOLOGY. *Educ:* Nat Univ Paraguay, BS, 35, MD, 42. *Prof Exp:* Res assoc, Cornell Univ, 48-52, asst prof, 52-58; asst prof anesthesiol, NY Univ, 58-61; assoc prof, 61-69, PROF ANESTHESIOL, ALBERT EINSTEIN COL MED, 69- *Concurrent Pos:* Fel physiol, Int Training Admin, Cornell Univ, 44, res fel med, Med Col, 45-48; hon prof, Univ Montevideo, 68. *Mem:* AAAS; Am Heart Asn; Am Physiol Soc; Harvey Soc; NY Acad Sci. *Res:* Peripheral vascular physiology; shock; hypertension; adaptation; anesthesiology. *Mailing Add:* RD 1 Box 170 Yorktown Heights NY 10598

BAG, JNANANKUR, b Calcutta, India, Oct 17, 47. BIOCHEMISTRY, MOLECULAR BIOLOGY. *Educ:* Univ Calcutta, India, PhD(biochem), 73. *Prof Exp:* Asst prof, Mem Univ Newfoundland, 81-85; ASSOC PROF MOLECULAR BIOL & GENETICS, UNIV GUELPH, 85- *Mem:* Am Soc Cell Biol; Can Biochem Soc; Can Cell Biol Soc. *Res:* Regulation of gene expression in enkanyotes. *Mailing Add:* Dept Molecular Biol & Genetics Univ Guelph Guelph ON N1G 2W1 Can

BAGASRA, OMAR, b Bagasra, India, Oct 9, 48; US citizen; c 2. CELLULAR & CLINICAL IMMUNOLOGY, IMMUNOPATHOLOGY. *Educ:* Karachi Univ, Pakistan, BSc, 68, MSc, 70; Univ Louisville, Ky, PhD(microbiol & immunol), 79; Autonomous City Univ, Juarez, Mex, MD, 85. *Prof Exp:* Chief biochemist, Sind Flour Mills Co, Karachi, Pakistan, 68-72; med technologist, Clark Co Mem Hosp, Jeffersonville, Ind, 74-79; teaching fel, dept clin microbiol, Div Infectious Dis, Albany Med Col, NY, 80; sr instr & dir, Treponemal Res Sect, Dept Med, Hahnemann Med Col, 80-82, from asst prof to assoc prof, Dept Pathol & Lab Med, 86-87; DIR, MOLECULAR RETROVIROL, DIV INFECTIOUS DIS, THOMAS JEFFERSON UNIV, 91- *Concurrent Pos:* Grad student asst scholar, dept microbiol & immunol, Sch Med, Univ Louisville, 76-79; instrnl grant, Hahnemann Univ, 81; Am Cancer Soc grant, 82-83; adj asst prof pediat & immunol, St Christopher's Hosp & Sch Med, Temple Univ, Philadelphia, 85- *Mem:* Am Asn Immunologists; AAAS; fel Sigma Xi; Am Asn Immunol. *Res:* Immunobiology of syphilitic hamsters; immunobiology of congenital syphilis in Syrian hamsters; immunobiology of experimental alcoholism in rat model; in vitro cloning of tumors in soft agarose and its use in chemotherapy; preparation of monoclonal antibodies to various tumor and cellular antigens; effect of Cis-platinum, cyclophosphamide and irradiation on the immune system; preparation of HTLV-III/LAV neuratizers as a possible alternate to vaccine. *Mailing Add:* Dept Med Thomas Jefferson Univ 1020 Locust Dr Rm 342 Philadelphia PA 19107

BAGBY, FREDERICK L(AIR), b Salt Lake City, Utah, Aug 2, 20; m 43; c 3. AEROSPACE & MECHANICAL ENGINEERING. *Educ:* Univ Utah, BS, 42. *Prof Exp:* Prod design engr, Curtis-Wright Corp, Columbus, 42-46; res engr, 46-48, asst div chief, 48-55, div chief, 55-60, assoc mgr, mech eng dept, 60-66, mgr, 66-70, asst dir, Columbus Div, 70-76, dir, Advan Concepts Lab, 76-78, SR PROG MGR, BATTELLE MEM INST, 78- *Mem:* AAAS; Am Soc Mech Engrs; Am Inst Aeronaut & Astronaut; Sigma Xi. *Res:* Aeronautical sciences; astronautics; fluid mechanics; combat and tactical vehicle engineering; thermodynamics; high temperature technology; propulsion technology; combustion physics; aircraft and missile engineering; military weapons technology; inter and multi disicplinary program management; applications of technology to contemporary social problems. *Mailing Add:* Columbus Labs 505 King Ave Columbus OH 43201

BAGBY, GREGORY JOHN, METABOLIC REGULATION DURING INFECTIONS. *Educ:* Wash State Univ, PhD(phys educ), 76. *Prof Exp:* PROF PHYSIOL, LA STATE UNIV MED CTR, 75- *Mem:* Am Physiol Soc; Am Col Sports Med; Shock Soc; Am Heart Asn; Endotoxin Soc. *Res:* Regulation of carbohydrate/lipid metabolism during infection by macrophage derived cytokines and its impart on host defense. *Mailing Add:* Physiol Dept La State Univ Med Ctr 1901 Perdido St New Orleans LA 70112

BAGBY, GROVER CARLTON, b Summit, NJ, July 15, 42; m 67; c 2. HEMATOLOGY, ONCOLOGY. *Educ:* Pomona Col, BS, 64; Baylor Col Med, MD, 68. *Prof Exp:* Resident med, 70-72, fel hematol, 74-76, instr, 75-76, from asst prof to assoc prof, 76-79, PROF MED & HEAD DIV HEMAT & ONCOL, ORE HEALTH SCI UNIV, 85- *Concurrent Pos:* Prin investr, Vet Admin, 76-81, 82-85 & NIH, 84-87; vis prof, Good Samaritan & St Joseph's Hosp, Phoenix, 78; dir res, E E Osgood Leukemia Ctr, 79-; clin investr, Vet Admin Med Ctr, 80-83; mem , Hemat Study Sect, NIH, 84-88, (chmn, 88). *Honors & Awards:* Med Res Found Discovery Award; Kendall Award. *Mem:* Am Soc Clin Invest; fel Am Col Physicians; Am Soc Hematol; Int Soc Exp Hematol. *Res:* Cellular and molecular regulation of growth and differentiation of hematopoietic cells in normal humans and in patients with leukemia, preleukemia, and other disorders of marrow function. *Mailing Add:* Med Res Vet Admin Med Ctr 3710 S US Vet Admin Hosp Rd Portland OR 97201

BAGBY, JOHN P(ENDLETON), optical engineering, space science, for more information see previous edition

BAGBY, JOHN R, JR, b Aurora, Mo, Mar 3, 19; m 43; c 3. PUBLIC HEALTH, BIOLOGY. *Educ:* Univ Ark, BS, 54, MS, 55; Emory Univ, PhD(parasitol), 62. *Prof Exp:* Malaria control aid, USPHS, 46-51, info officer, Econ Stabilization Agency, 51-53, vector control specialist, 55-59, asst to chief tech br, 59-62, asst chief tech br, 62-63 & 64-66, dep chief tech br, 63-64, dept dir, Nat Commun Dis Ctr, 66-69; dir Inst Environ Health & head dept microbiol, Colo State Univ, 69-84; dir, Div Environ Health & Epidemiol, 84-88, DIR, MO DEPT HEALTH, 89- *Concurrent Pos:* Consult, NASA, Jet Propulsion Lab & Am Inst Biol Sci. *Honors & Awards:* Walter F Snyder Award, Environ Health, 83. *Mem:* NY Acad Sci; Nat Environ Health Asn; Am Soc Trop Med & Hyg; Sigma Xi. *Res:* Parasitology; entomology; bacteriology; serology; control of vectors; reservoirs; etiologic agents of communicable diseases; health effects of dioxin; cancer epidemiology. *Mailing Add:* 5315 Foxfire Lane Lohman MO 65053

BAGBY, MARVIN ORVILLE, b Macomb, Ill, Sept 27, 32; m 57; c 2. ORGANIC CHEMISTRY, NATURAL PRODUCTS CHEMISTRY. *Educ:* Western Ill Univ, BS & MS, 57. *Prof Exp:* Org chemist, Nat Ctr Agr Utilization Res, Agr Res Serv, 57-75, res leader, 75-81, mgr, Northern Agr Energy Ctr, 81-85, RES LEADER, OIL CHEM RES, USDA, 85- *Concurrent Pos:* Tech adv, technologies for fiber uses, Sci & Educ Admin, USDA, 75-81, technologies for indust uses of plant & animal prods (hydrocarbon plants), 80-81; team leader, utilization of fibrous mat, Bilateral Res & Educ Proj, Am Chem Soc/Egyptian Nat Res Ctr, 77; mem, Secy Agr Task Force Critical Mat, 83-85; postdoctoral res assoc grant, 90. *Honors & Awards:* Cert Merit, Am Paper Ind, 88; ARS Technol Transfer Award, 90; Domestic Mkt Award, Am Soybean Asn, 90. *Mem:* AAAS; Am Chem Soc; Am Oil Chem Soc; Tech Asn Pulp & Paper Indust; NY Acad Sci; Am Soc Agr Eng. *Res:* Natural products; lipids, especially isolation, characterization, and reactions of fatty acids and sugars; annual plants for pulp and paper raw materials; plant biomass for energy and strategic materials; hydrocarbon plants; vegetable oils as alternative fuels; industrial materials from soybean oil and other seed oils; alternative industrial crops. *Mailing Add:* USDA 1815 N University St Peoria IL 61604

BAGBY, ROLAND MOHLER, IMMUNOFLUORESCENCE, THREE DIMENSIONAL RECONSTRUCTION. *Educ:* Univ Ill, Urbana, PhD(physiol & biophys), 64. *Prof Exp:* PROF PHYSIOL, UNIV TENN, 66- *Mailing Add:* Dept Zool Univ Tenn Knoxville TN 37996

BAGCHI, AMITABHA, b Calcutta, India, June 12, 45; m 74. SOLID STATE PHYSICS, SURFACE PHYSICS. *Educ:* Calcutta Univ, BSc, 64; Univ Calif, San Diego, MS, 67, PhD(physics), 70. *Prof Exp:* Res assoc physics, Univ Ill, Urbana, 70-71; res assoc surface physics, James Franck Inst, Univ Chicago, 71-73; res fel physics, Battelle Mem Inst, Columbus, Ohio, 73-74; asst prof physics, Univ Md, 74-80; dept physics, Zerox Corp, 80-; at AT&T Bell Labs; ASST PROF, DEPT MECH ENG, CLEMSON UNIV, 90- *Mem:* Am Phys Soc; Am Vacuum Soc. *Res:* Electronic properties of metal surfaces including chemisorption and photoemission and optical reflectance from clean and adsorbate-covered surfaces; excitation spectrum of He3-He4 mixtures. *Mailing Add:* Dept Mech Eng Clemson Univ 3118 Riggs Hall Clemson SC 29634-0921

BAGCHI, MIHIR, b Ranchi, India, Nov 28, 38; m 72; c 3. CELL BIOLOGY, OPHTHALMOLOGY. *Educ:* Univ Bihar, BS, 59; Univ Ranchi, MS, 62; Univ Vt, PhD(zool), 69. *Prof Exp:* Res asst biol, Ciba Res Ctr, India, 63-65; res assoc, Oakland Univ, 63-75; asst prof, 75-81, ASSOC PROF ANAT, SCH MED, WAYNE STATE UNIV, 81- *Concurrent Pos:* NIH spec fel, 73-75; NIH grants, 76-91. *Mem:* Am Soc Cell Biol; Sigma Xi; Am Asn Anatomists; Asn Res Vision & Ophthal. *Res:* Biochemistry and morphology of mammalian lens in culture. *Mailing Add:* Dept Anat Wayne State Univ Sch Med Detroit MI 48201

BAGCHI, PRANAB, b Calcutta, India, Jan 14, 46. COLLOID CHEMISTRY, SURFACE CHEMISTRY. *Educ:* Jadavpur Univ, India, BSc, 65; Univ Southern Calif, PhD(phys chem), 70. *Prof Exp:* Sr res chemist, 70-80, RES ASSOC, EASTMAN KODAK CO, 80- *Mem:* Am Chem Soc; AAAS; NY Acad Sci; The Chem Soc; Fedn Am Scientists. *Res:* Investigation of the mechanism of the stability of lyophobic dispersions in nonaqueous media and in the presence of nonionic surfactants; long chain nonionic molecules and polymeric nonionic molecules, both experiment and theory considered; surface colloidal properties and stability of silver salts in nonaqueous media; interfacial behavior of model polymer colloids; interactions of immunoproteins at latex/water interface. *Mailing Add:* Res Lab Eastman Kodak Co Rochester NY 14650

BAGCHI, SAKTI PRASAD, b Tatanagar, India, Dec 26, 31; US citizen; m 59; c 2. NEUROSCIENCE, PHARMACOLOGY. *Educ:* Univ Calcutta, BS, 51, MS, 54, PhD(biochem), 59. *Prof Exp:* Res assoc biochem, Okla State Univ, 59-60; jr res biochemist, Sch Med, Univ Calif, San Francisco, 60-62; res assoc, Kinsmen Lab of Neurol Dis, Univ BC, 63-66; res biochemist, Psychiat Res Unit, Dept Pub Health, Univ Hosp, 66-73; SR RES SCIENTIST, NY STATE DEPT MENT HYG, NATHAN KLINE INST, 73- *Mem:* AAAS; Int Soc Neurochem; Int Brain Res Orgn; Am Soc Neurochem; Soc Neurosci. *Res:* Neuropharmacology; psychotropic drug action; biochemistry of mental disorders. *Mailing Add:* Nathan Kline Inst Bldg 37 Orangeburg NY 10962

BAGDASARIAN, ANDRANIK, b Tehran, Iran, Dec 5, 35; m 69; c 3. BIOCHEMISTRY. *Educ:* Univ Tehran, BS, 60, DrPharm, 61; Univ Louisville, PhD(biochem), 67. *Prof Exp:* Fel protein chem, Calif Inst Technol, 67-68; fel enzymol, Syntex Res, Palo Alto, 68-70; res fel plasma proteases, Mass Gen Hosp, Harvard Med Sch, 70-73; res assoc plasma proteases, Univ Pa, 73-75, res asst prof biochem, dept med, 75-78; sr res scientist, Hyland Div, Travenol Labs, Inc, 78-80; res mgr, protein res labs, 80-83, ASSOC DIR RES, HYLAND THERAP LABS, 84- *Concurrent Pos:* NIH award studies prekallikrein activation, 71; Nat Cancer Inst & Am Cancer Soc grant kinin pathway normal & malignant tissue, 75. *Mem:* Sigma Xi; Am Chem Soc; Am Soc Biol Chemists. *Res:* Purification and characterization of proteases and protease inhibitors involved in the kallikrein-kinin, coagulation and fibrinolysis systems of human plasma, platelets and normal and malignant tissue; hemoglobin-structure/function relationship; stroma free hemoglobin as a blood substitute. *Mailing Add:* 1227 Calle Estrella San Dimas CA 91773

BAGDON, ROBERT EDWARD, b Newark, NJ, Sept 18, 27; m 56; c 2. PHARMACOLOGY. *Educ:* Upsala Col, BS, 49; Univ Chicago, PhD, 55. *Prof Exp:* Chemist, Ciba Pharmaceut Prod Inc, 49-52; pharmacologist, Am Cyanamid Co, 56-57; sr pharmacologist, Hoffmann-La Roche, Inc, 57-67; HEAD TOXICOL, TOXICOL & PATH SECT SANDOZ PHARMACEUT DIV, SANDOZWANDER, INC, 67-; PRES, ROBERT EDWARD BAGDON, INC. *Mem:* Am Chem Soc; NY Acad Sci; Am Soc Pharmacol & Exp Therapeut; Soc Toxicol; Soc Exp Biol & Med. *Res:* Toxicology; mechanism; drug action. *Mailing Add:* Robert Edward Bagdon Inc 21 Tilden Ct Livingston NJ 07939

BAGDON, WALTER JOSEPH, b Camden, NJ, Apr 22, 38; m 62; c 3. TOXICOLOGY, PHARMACOLOGY. *Educ:* Philadelphia Col Pharm, BS, 59; Temple Univ, MS, 61, PhD(pharmacol), 64. *Prof Exp:* Sr res toxicologist, 64-68, res fel, 68-74, sr res fel, 74-79, dir, toxicol & path, 79-86, SR INVESTR, MERCK INST THERAPEUT RES, 86- *Mem:* AAAS; Soc Toxicol; Am Pharmaceut Asn; Sigma Xi; Dipl Am Bd Toxicol. *Res:* Safety evaluation of current and potential therpeutic agents; evaluation of clinical tests concerning normal body function of animals. *Mailing Add:* Merck Inst Therapeut Res West Point PA 19486

BAGG, THOMAS CAMPBELL, b Philadelphia, Pa, Oct 28, 17; m 42; c 3. INFORMATION SYSTEMS, PHOTO OPTICAL. *Educ:* Lafayette Col, BS, 39. *Prof Exp:* Lab instr, Univ Pa, 39-41; jr physicist electronic ord, Carnegie Inst Washington, 41; physicist electronic ord, 41-52, physicist infrared & ultraviolet sources detectors, 52-56, SYSTS ENGR & CONSULT HIGH DENSITY INFO STORAGE, NAT INST STANDARDS & TECHNOL, 56- *Mem:* Fel Nat Micrographics Asn; Inst Elec & Electronics Engrs; Soc Photog Scientists & Engrs; Am Phys Soc; Am Soc Info Scientists; fel AAAS. *Res:* High density information storage and retrieval systems using photooptical techniques and associated image quality and standards. *Mailing Add:* Nat Inst Standards & Technol Rm A 51 Technol Gaithersburg MD 20899

BAGGENSTOSS, ARCHIE HERBERT, b Richardton, NDak, Apr 13, 08; m 34; c 3. PATHOLOGY. *Educ:* Univ NDak, AB, 30, BS, 31; Univ Cincinnati, BM, 33, MD, 34; Univ Minn, MS, 38. *Prof Exp:* Consult path anat, 38-52, head sect exp & anat path, 55-68, prof path, 52-75, EMER PROF PATH, MAYO GRAD SCH MED, UNIV MINN, 75-, SR CONSULT, MAYO CLIN, 68- *Mem:* Fel AMA; fel Am Soc Clin Path; Am Asn Path & Bact; Am Gastroenterol Asn; Am Asn Study Liver Dis (pres, 68-69). *Res:* Pathologic anatomy; liver; pancreas; gastrointestinal tract. *Mailing Add:* 2100 Valkyrie Dr Rochester MN 55901

BAGGERLY, LEO L, b Wichita, Kans, Mar 13, 28; m 66; c 5. PHYSICS. *Educ:* Calif Inst Technol, BS, 51, MS, 52, PhD(physics), 56. *Prof Exp:* Sr res engr, Jet Propulsion Lab, Calif, 55-56; Fulbright lectr physics, Univ Ceylon, 56-59; from asst prof to prof, Tex Christian Univ, 59-69; assoc prog dir, NSF, 69, prog dir, 70-71; vis fel, Cornell Univ, 71-72; prof physics, Calif State Col, Bakersfield, 72-75; vis prof physics, Pomona Col, 75-76; vis prof, Harvey Mudd Col, 76-77; sr consult, Milco Int, 76-77; SR MEM TECH STAFF, TRW INC, 77- *Concurrent Pos:* Sr scientist, LTV Res Ctr Div, Ling-Temco-Vought Corp, 63-64. *Mem:* AAAS; Am Asn Physics Teachers; Sigma Xi. *Res:* Electromagnetic properties of materials; science education; radiation physics. *Mailing Add:* 2218 Grand Ave Claremont CA 91711-2210

BAGGETT, BILLY, b Oxford, Miss, Oct 23, 28; m 90; c 5. BIOCHEMICAL ENDOCRINOLOGY. *Educ:* Univ Miss, BA, 47; St Louis Univ, PhD(biochem), 52. *Prof Exp:* Instr biochem, Harvard Med Sch, 52-57; asst, Mass Gen Hosp, 52-57; from asst prof to prof pharmacol & biochem, Univ NC, Chapel Hill, 57-69; chmn dept, 69-78, PROF BIOCHEM, MED UNIV SC, 69-, PROF OBSTET & GYNEC, 80- *Concurrent Pos:* USPHS sr res fel, Univ NC, Chapel Hill, 57-62; consult, Res Triangle Inst, 66-71, NIH, 80-82. *Mem:* AAAS; Endocrine Soc; Am Soc Biol Chemists; Am Chem Soc; Am Asn Cancer Res; Soc Study Reproduction; Am Fertility Soc. *Res:* Metabolism and actions of steroid hormones. *Mailing Add:* Dept Obstet & Gynecol Med Univ SC 171 Ashley Ave Charleston SC 29425

BAGGETT, JAMES RONALD, b Boise, Idaho, Apr 24, 28; m 51; c 2. HORTICULTURE. *Educ:* Univ Idaho, BS, 52; Ore State Col, PhD(hort), 56. *Prof Exp:* From asst prof to assoc prof, 56-71, PROF HORT, ORE STATE UNIV, 71- *Mem:* Am Soc Hort Sci. *Res:* Genetics and breeding of disease resistance in vegetables. *Mailing Add:* Dept Hort Ore State Univ Corvallis OR 97331

BAGGETT, LAWRENCE W, b Morehead, Miss, Mar 3, 39; m 79; c 2. MATHEMATICS. *Educ:* Davidson Col, BS, 60; Univ Wash, Seattle, PhD(math), 66. *Prof Exp:* From asst prof to assoc prof math, 66-77, dept chmn, 84-87, PROF MATH, UNIV COLO, BOULDER, 77- *Concurrent Pos:* NSF grant, 67-88. *Mem:* Am Math Soc. *Res:* Topological groups and their representations; ergodic theory. *Mailing Add:* Dept Math Univ Colo Box 426 Boulder CO 80309

BAGGETT, LESTER MARCHANT, b Oakland, Calif, Nov 28, 20; m 51; c 2. NUCLEAR PHYSICS, HYDRODYNAMICS. *Educ:* Rhodes Col, BA, 43; Ga Inst Technol, MS, 48; Rice Univ, PhD(physics), 51. *Prof Exp:* Assoc div leader, Los Alamos Sci Lab, 51-84; RETIRED. *Mem:* Am Phys Soc. *Res:* Energy levels of light nuclei. *Mailing Add:* 996 Nambe Pl Los Alamos NM 87544

BAGGETT, MILLICENT (PENNY), b Buffalo, NY, Sept 15, 39; div. DATA BASE SYSTEMS. *Educ:* Antioch Col, BS, 61; Univ Md, PhD(physics), 69. *Prof Exp:* Res assoc, Univ Heidelberg, 69-71; sr programmer, Imperial Col, 72-73; asst prof comput prog, Purdue Univ, 73-77; systs analyst, Calculon Corp, 78-80; comput analyst, Brookhaven Nat Lab, 80-90; STAFF PHYSICIST, SSC LAB, 90- *Mem:* Asn Comput Mach; Asn Women Sci. *Res:* Design and administration of data bases for research and development information on superconducting magnets. *Mailing Add:* SSC Lab MS 1002 2550 Beckleymeade Ave Dallas TX 75237

BAGGETT, NEIL VANCE, b DeRidder, La, Aug 12, 38; m 65. PHYSICS. *Educ:* Tulane Univ, BS, 60; Univ Md, PhD(physics), 69. *Prof Exp:* Physicist heat transfer & nuclear physics, Nat Bur Standards, 60-66; res assoc high energy physics, Inst Hochenergiephysik, Univ Heidelberg, 66-71; res asst high energy physics, Imp Col, London, 71-73; proj assoc high energy physics, Purdue Univ, 73-77; physicist, Div High Energy Physics, Brookhaven Nat Lab, US Dept Energy, 77-80, physicist & spec asst to assoc dir high energy physics, 80-90; PHYSICIST DIRECTORATE, SUPER CONDUCTING SUPER COLLIDER LAB, 89- *Concurrent Pos:* Physicist, Div High Energy Physics, US Dept Energy, 77-80. *Mem:* Am Phys Soc; AAAS. *Res:* Experimental high energy physics; science communications; scientific computing; weak decays; neutrino interactions; hadronic interactions. *Mailing Add:* SCSCL 2550 Beckley Meade Ave Dallas TX 75237

BAGGI, DENIS LOUIS, b Zurich, Switz, Feb 10, 45; c 2. COMPUTER SCIENCE, ELECTRICAL ENGINEERING. *Educ:* Swiss Fed Inst Technol, Dipl Ing elec eng, 69; Polytech Inst, New York, MS, 71; Univ Calif, Berkeley, PhD(elec eng, comput sci), 74. *Prof Exp:* Instr comput sci, Univ Calif, Berkeley, 73-74; asst prof elec eng & comput sci, Polytech Inst New York, 74-76; consult speech synthesis, Bell Tel Labs, Murray Hill, 75-77; asst prof comput sci, Queens Col Ny, 87-; AT DEPT CENT RES & COMPUT SCI, METTLER INSTRUMENTE AG, 88- *Concurrent Pos:* Actg res prof, US Air Force, Rome, NY, 77-80; vpres, XI Comput Corp, Irvine, Calif; pres, Ventur Tech, SA, Lupano, Switz; vis prof, Univ Calif, Berkeley, 85 & 88. *Mem:* Inst Elec & Electronics Engrs; Asn Comp Mach; Swiss Soc Comp Music. *Res:* Software engineering and reliability; artificial intelligence applied to music composition; modeling of social systems; speech synthesis; expert systems. *Mailing Add:* Via Casagrande 12 CH-6932 Breganzona Switzerland

BAGGOT, J DESMOND, b Ireland, Oct 11, 39; m 65; c 2. CLINICAL PHARMACOKINETICS, ANTIMICROBIOL THERAPY. *Educ:* Nat Univ Ireland, Dublin, BSc, 62; Vet Col Ireland, Dublin, 66; Nat Univ Ireland, Dublin, MVM, 68; Ohio State Univ, PhD(biochem pharmacol), 71, DSc, 80. *Prof Exp:* Prof clin pharmacol & chmn, Vet Pharmacol & Toxicol, 81-86, RES PROF CLIN PHARMACOL, UNIV CALIF, DAVIS, 86-; DIR, IRISH EQUINE CTR, JOHNSTOWN, COUNTY KILDARE, IRELAND, 86- *Concurrent Pos:* Vet consult, Alza Corp, Palo Alto, Calif, 84- *Mem:* Am Soc Pharmacol & Exp Therapeut; Controlled Release Soc; fel Am Col Clin Pharmacol; Europ Asn Vet Pharmacol & Toxicol. *Res:* Drug bioavailability and disposition kinetics in animals and humans; influence of disease conditions on drug disposition and dosage; species variations in drug response. *Mailing Add:* Irish Equine Ctr Johnstown County Kildare Ireland

BAGGOTT, JAMES PATRICK, b Milwaukee, Wis, Jan 19, 41; m 66; c 7. BIOCHEMISTRY. *Educ:* Marquette Univ, BS, 62; Johns Hopkins Univ, PhD(biochem), 66. *Prof Exp:* Sr instr, 68-71, asst prof, 71-78, asst dean student affairs, 78-84, ASSOC PROF BIOCHEM, HAHNEMANN UNIV SCH MED, 78- *Concurrent Pos:* Fel biol chem, Univ Utah, 66-68. *Mem:* Am Chem Soc. *Res:* medical education; development and evaluation of computer based learning. *Mailing Add:* Dept Biol Chem Hahnemann Univ Sch Med Philadelphia PA 19102-1192

BAGINSKI, F(RANK) C(HARLES), b Holyoke, Mass, Oct 6, 23; m 51; c 6. CHEMICAL ENGINEERING. *Educ:* Worcester Polytech Inst, BS, 44; Yale Univ, DEng, 52. *Prof Exp:* Res chem engr, Monsanto Chem Co, 46-47; lab asst units oper lab, Yale Univ, 48-49; res engr, E I du Pont de Nemours & Co Inc, 51-53, res supvr polymer mfg, 53-56, res supvr cellophane mfg, 56-68, from plant supt polyethylene film mfg res & develop to tech supt, Spruance Film Plant, 58-66, prod mgr polyolefin films mfg, 66-68, mfg supt, 68-71, tedlar plant supt, Yerkes Plant, 71-84; RETIRED. *Mem:* AAAS; Am Chem Soc. *Res:* Research and development in films manufacturing in polyvinyl fluoride, cellophane and polyethylene films. *Mailing Add:* 123 Meadow View Lane Buffalo NY 14221

BAGLEY, BRIAN G, b Racine, Wis, Nov 20, 34; m 59; c 3. ELECTRONIC & PHOTONIC MATERIALS SCIENCE. *Educ:* Univ Wis-Madison, BS, 58, MS, 59; Harvard Univ, AM, 64, PhD(appl physics), 68. *Prof Exp:* Mem res staff, Univ Wis, 59-60; metallurgist, Ladish Co, Wis, 61; mem tech staff, Bell Labs, Murray Hill, NJ, 67-83, MEM TECH STAFF, BELL COMMUN RES, RED BANK, NJ, 84- *Mem:* Am Phys Soc; Mat Res Soc; Am Vacuum Soc. *Res:* Preparation and properties of high transition temperature superconductors in both bulk and thin film forms; glasses for photonic applications. *Mailing Add:* Bell Commun Res 331 Newman Spring Rd Rm 3X-257 Red Bank NJ 07701-7040

BAGLEY, CLYDE PATTISON, b Shreveport, La, Dec 19, 51; m 75; c 2. ANIMAL SCIENCE & NUTRITION, AGRONOMY. *Educ:* La State Univ, BS, 72, MS, 75; Va Tech Univ, PhD(animal sci), 78. *Prof Exp:* Res asst animal sci, La State Univ, 74-75; asst prof, WLa Exp Sta, 78-80; assoc prof animal sci, 80-88, RESIDENT DIR ANIMAL SCI, ROSEPINE RES STA, LA STATE UNIV, 80-, PROF, 80- *Mem:* Am Soc Animal Sci; Am Forage & Grassland Coun; Am Soc Agron; Sigma Xi; Coun Agr Sci Technol. *Res:* Beef cattle production using existing forage resources; evaluation of systems to optimize beef cattle performance, using high quality forages. *Mailing Add:* N Mississippi Res & Extension Ctr PO Box 456 Verona MS 38879

BAGLEY, GEORGE EVERETT, b Corry, Pa, Feb 14, 33; m 56; c 4. POLYMER CHEMISTRY. *Educ:* Houghton Col, BS, 54; Univ Pa, PhD(phys chem), 59. *Prof Exp:* Res chemist, Armstrong World Industs, Inc, 58-69, res suprv, 69-70, sr res scientist, 70-77, res assoc, 77-89, SR PRIN SCIENTIST, ARMSTRONG WORLD INDUSTS, INC, 90- *Mem:* Am Chem Soc; Soc Plastics Engrs; Soc Photog Scientists & Engrs; Rad Tech Int; Oil & Color Chemists Asn; Am Soc Testing & Mat. *Res:* Inorganic synthesis and inorganic polymers; polyvinyl chloride technology; radiation curable polymers; elastomers. *Mailing Add:* 2425 Mayfair Dr Lancaster PA 17603-4135

BAGLEY, JAY M(ERRILL), b Koosharem, Utah, Oct 14, 25; m 47; c 5. HYDRAULIC ENGINEERING. *Educ:* Utah State Univ, BS, 52, MS, 53; Stanford Univ, PhD(hydraul eng), 64. *Prof Exp:* Engr, Irrig Equip Co, Ore & Colo, 52-53; field engr, Farm Improv Co, 53-54; gen mgr, West Irrig Co, 54; asst prof civil & irrig eng, 54-60, assoc prof dept civil eng & Utah Water Res Lab, 60-66, PROF DEPT CIVIL ENG & DIR UTAH WATER RES LAB, UTAH STATE UNIV, 66- *Concurrent Pos:* Consult, Sprinkler Irrig Assocs, 55, Plantacoes do Mucoso, Portuguese West Africa, 55, Foreign Agr Serv, 56, Gunnison Irrig Co, 57, Resources Co, 64, United Park City Mines Co, Atlantic Refining Co, Anaconda Co, Ralph M Parsons Co, US Agency Int Develop, Nat Oceanic & Atmospheric Admin, US Dept Agr & various assignments to Africa, Vietnam & China. *Mem:* Am Soc Civil Engrs; Nat Soc Prof Engrs; Int Asn Sci Hydrol; Am Geophys Union; Sigma Xi. *Res:* Water resource development; irrigation and drainage; hydrology; fluid mechanics; watershed and river basin modeling; water resources planning methodology; hydrologic processes; extending utility of water supplies; legal and institutional impediments to effective water use. *Mailing Add:* Utah Water Res Lab Utah State Univ Logan UT 84322

BAGLEY, JOHN D(ANIEL), b Chicago, Ill, Oct 2, 35; m 58; c 5. COMPUTER SCIENCES. *Educ:* Univ Notre Dame, BSEE, 57; Univ Ill, Urbana-Champaign, MS, 58; Univ Mich, AM, 63, PhD(commun sci), 67. *Prof Exp:* Mem staff, Systs Develop Div, IBM Corp, 59, Advan Systs Develop Div, 59-61 & 65-67 & Res Div, NY, 67-76, adv programmer, Gen Systs, 76-77; SUPVR SYSTS ENG, INDIAN HILL LAB, BELL TEL LABS, 77- *Mem:* Inst Elec & Electronics Engrs; Asn Comput Mach. *Res:* Adaptive systems; simulation of biological systems; digital computer operating system and architecture; voice and graphical input to digital computers; digital computer simulation of continuous systems; distributed computer systems; computer networking; unix operating system. *Mailing Add:* AT&T Bell Labs Rm 1B-224 2000 N Naperville Rd Naperville IL 60566-7033

BAGLEY, ROBERT WALLER, b Wesson, Miss, Apr 14, 21; m 44; c 3. MATHEMATICS. *Educ:* Univ Mich, BS, 47; Tulane Univ, MS, 49; Univ Fla, PhD(math), 54. *Prof Exp:* Asst prof math, Univ Ky, 54-55; assoc res scientist, Lockheed Missile Systs, 55-58; prof math, Miss Southern Univ, 58-60 & Univ Ala, 60-63; assoc prof, 63-64, PROF MATH, UNIV MIAMI, 64- *Concurrent Pos:* Consult, Grad Sch Bus, Stanford Univ, 57. *Mem:* Am Math Soc; Math Asn Am. *Res:* Topological groups. *Mailing Add:* Dept Math Univ Miami Coral Gables FL 33146

BAGLI, JENANBUX FRAMROZ, b Bombay, India, Sept 25, 28; m 56; c 2. PHARMACEUTICAL CHEMISTRY. *Educ:* Univ Bombay, BSc, 49 & 51; Univ London, PhD(pharmaceut chem), 55. *Prof Exp:* Fel, Johns Hopkins Univ, 55-59; Nat Res Coun Can fel, Laval Univ, 59-60; sr res scientist, 60-69, sr res adv, 69-79, ASSOC DIR CHEM, WYETH-AYERST RES, 79- *Mem:* Am Chem Soc; Royal Soc Chem; fel Can Inst Chem. *Res:* Elucidation of structure and stereochemistry of natural products; mechanism of the organic reactions and the biogenesis of the steroids and terpenoids; mechanism and pharmacological profile of drug molecules. *Mailing Add:* Chem Res Dept Wyeth Ayerst Res CN 8000 Princeton NJ 08540

BAGLIN, RAYMOND EUGENE, JR, b Hartford, Conn, Aug 26, 44; m 66; c 2. FISHERIES SCIENCE. *Educ:* Univ Hartford, BA, 66; Univ Ark, MS, 68; Univ Okla, PhD(fisheries), 75. *Prof Exp:* Fishery biologist, US Army Corps Engrs, 69-71; from aquatic ecologist to sr aquatic ecologist, Off Environ Planning, NY State Pub Serv Comn, 72-74; FISHERY BIOLOGIST, NAT MARINE FISHERIES SERV, US DEPT COM, 74- *Mem:* Am Fisheries Soc; Am Inst Fishery Res Biologists. *Res:* Reproduction, migration, age and growth, and stock structure of fishes. *Mailing Add:* Nat Marine Fisheries Serv Box 21668 Juneau AK 99802

BAGLIO, JOSEPH ANTHONY, b New York, NY, May 16, 39; m 60; c 2. PHYSICAL CHEMISTRY. *Educ:* Rutgers Univ, PhD(phys chem), 65. *Prof Exp:* Chemist, Atomic Energy Comn, 60; chemist, David Sarnoff Res Ctr, Radio Corp Am Inc, 61; PRIN MEM TECH STAFF SOLID STATE CHEM & THERMODYNAMICS, GTE LABS INC, 80- *Concurrent Pos:* Dupont teaching fel; NSF fel; NIH fel. *Mem:* Am Chem Soc; Electrochem Soc; Sigma Xi. *Res:* Development of catalysts for removal of sulfur dioxide, hydrogen sulfide and carbon sulfide compounds from plant stacks; development of amorphous and crystalline materials for magnetic bubble memory applications; precious metal replacement in materials for the communications industry; plasma chemistry, luminescent and catalytic materials research; correlation of crystal structure with solid state properties; x-ray diffraction and knudsen cell mass spectrometry; investigation of semiconductors for applications in liquid-junction solar cells; physics and chemistry of III-IV semiconductor surfaces for applications in metal insulator semiconductor field effect transistor devices; solid state chemistry of high temperature superconductors; thermodynamics of high temperature materials; plasma-enhanced chemical vapor deposition of diamond films. *Mailing Add:* GTE Labs Inc 40 Sylvan Rd Waltham MA 02254

BAGLIONI, CORRADO, b Rome, Italy, Aug 1, 33; m 57; c 3. BIOCHEMICAL GENETICS, IMMUNOLOGY. *Educ:* Univ Rome, MD, 57. *Prof Exp:* Vis lectr biochem, Mass Inst Technol, 59-61, asst prof genetics, 61-62; Ital Nat Res Coun intern genetics, Biophys Lab, 62-66; assoc prof biol, Mass Inst Technol, 67-73; prof & chmn dept, 73-75, assoc dean, Sch Health Sci, 85-88, LEADING PROF BIOL, STATE UNIV NY, ALBANY, 76- *Concurrent Pos:* Mem dir, Ital Asn Biophys & Molecular Biol, 64-66; mem, Immunol & Allergy Study Sect, NIH, 67-71; mem, Cell & Develop Biol Study Sect, Am Cancer Soc, 81-85; Rockfeller & Foggarty scholar. *Mem:* Europ Molecular Biol Orgn. *Res:* Genetic control of protein structure; action of suppressor genes in Drosophila; human abnormal hemoglobins; mechanism of assembly of hemoglobin and immunoglobulin peptide chains; genetic basis of antibody variability; regulation of histone synthesis; mechanisms of action of interferon and tumor neurosis factor. *Mailing Add:* Bi-126 State Univ NY 1400 Washington Ave Albany NY 12222

BAGLIVO, JENNY ANTOINETTE, b New York, NY, Nov 7, 48. MATHEMATICS. *Educ:* Fordham Univ, BA, 70; Syracuse Univ, MA, 72, MS, 76, PhD(math), 76. *Prof Exp:* Asst prof, 76-81, ASSOC PROF MATH, FAIRFIELD UNIV, 81- *Mem:* Am Math Soc; Math Asn Am. *Res:* Topology; transformation groups. *Mailing Add:* Dept Math Boston Col Carney Hall 318 Chestnut Hill MA 02167

BAGNALL, LARRY OWEN, b Enumclaw, Wash, Dec 19, 35; m 58; c 3. AGRICULTURAL ENGINEERING. *Educ:* Wash State Univ, BS, 57; Cornell Univ, PhD(agr eng), 67. *Prof Exp:* Trainee, Allis-Chalmers, 57-58, eng scientist, 58-66, res engr, 66-68; from asst prof to assoc prof, 69-81, PROF AGR ENG, UNIV FLA, 81- *Mem:* Am Soc Agr Engrs; Aquatic Plant Mgt Soc; Sigma Xi; Am Soc Eng Educ. *Res:* Aquatic plant harvesting, processing, utilization; forage harvesting, processing; aquaculture. *Mailing Add:* Dept Agr Eng Univ Fla Gainesville FL 32611

BAGNALL, RICHARD HERBERT, b Hazel Grove, PEI, Feb 23, 23; m 52; c 4. PLANT PATHOLOGY. *Educ:* Prince of Wales Col, jr col dipl, 44; McGill Univ, BSc, 46, MSc, 49; Univ Wis, PhD, 56. *Prof Exp:* Agr scientist, 46, res scientist, Plant Path Lab, 46-85, RES SCIENTIST, PLANT VIROL, CAN AGR RES STA, NB, 85- *Mem:* Am Phytopath Soc; Potato Asn Am. *Res:* Epidemiology, resistance, serology of potato viruses. *Mailing Add:* Can Agr Res Sta PO Box 20280 Fredericton NB E3B 4Z7 Can

BAGNARA, JOSEPH THOMAS, b Rochester, NY, July 26, 29; m 55. EMBRYOLOGY, ENDOCRINOLOGY. *Educ:* Univ Rochester, BA, 52; Univ Iowa, PhD, 56. *Prof Exp:* Lab instr, Univ Iowa, 52-56, res asst, 55; from instr to prof zool, 56-64, dept head gen biol, 80-83, PROF BIOL SCI & ANAT, UNIV ARIZ, 64- *Concurrent Pos:* Fulbright res scholar, Univ Paris, 64 & Zool Sta, Naples, 69-70; managing ed, Am Zoologist, Am Soc Zool, 71-75; res scholar, Japan Soc Prom Sci, 79; ed in chief, Pigment Cell Res, 86- *Honors & Awards:* Myron Gordon Award, 86. *Mem:* Am Soc Zool; Am Asn Anat; Soc Develop Biol; Am Soc Cell Biol. *Res:* Experimental embryology of amphibia; endocrinology; physiology of pigmentation; growth factors; cell culture. *Mailing Add:* Dept Anat Univ Ariz Tucson AZ 85724

BAGNELL, CAROL A, b Avenel, NJ, July 10, 52. ENDOCRINOLOGY, IMMUNOHISTOCHEMISTRY. *Educ:* Glassboro State Col, BA, 74; WVa Univ, Morgantown, MS, 76; Med Col Ga, PhD(endocrinol), 83. *Prof Exp:* Res fel, Dept Anat, Univ Hawaii, 83-84, asst researcher, 84-85, assoc researcher, 85-88; ASST PROF, DEPT ANIMAL SCI, RUTGERS UNIV, 88- *Mem:* Sigma Xi; Endocrine Soc; Soc Study Reproduction. *Res:* Paracrine control of follicular development and ovulation; immunohistochemical localization of the peptide hormone relaxin in rat and pig ovarian tissue; changes in the protease, plasminogen activator, during follicular development. *Mailing Add:* Dept Animal Sci Rutgers Univ New Brunswick NJ 08903-0231

BAGSHAW, JOSEPH CHARLES, b Niagara Falls, NY, Sept 2, 43; m 71; c 2. GENE STRUCTURE & FUNCTION. *Educ:* Johns Hopkins Univ, BA, 65; Univ Tenn, PhD(biochem), 69. *Prof Exp:* From asst prof to assoc prof biochem, Sch Med, Wayne State Univ, 71-84; PROF & HEAD, DEPT BIOL & BIOTECHNOL, WORCESTER POLYTECH INST, 84- *Concurrent Pos:* Res fel, Harvard Med Sch-Mass Gen Hosp, Boston, 70-71; dir, Worcester Consortium PhD Prog Biomed Sci, 84- *Mem:* AAAS; Am Soc Biol Chemists; Am Soc Cell Biol. *Res:* Transcriptional regulation in eukaryotes; gene structure and expression; biochemical aspects of development and differentiation; cloning and characterization of histone and tubulin genes. *Mailing Add:* Dept Biol & Biotechnol Worcester Polytech Inst 100 Institute Rd Worcester MA 01609

BAGSHAW, MALCOLM A, b Adrian, Mich, June 24, 25; m 48; c 3. RADIOTHERAPY, RADIOBIOLOGY. *Educ:* Wesleyan Univ, BA, 46; Yale Univ, MD, 50. *Prof Exp:* From instr to assoc prof, 55-69, PROF RADIOL, SCH MED, STANFORD UNIV, 69-, CHMN DEPT, 72-, DIR DIV RADIOTHER, 60- *Concurrent Pos:* Consult, Vet Admin Hosp, Palo Alto, 60- *Mem:* Am Col Radiol; Asn Univ Radiol; Radiol Soc NAm; Radiation Res Soc; Am Soc Therapeut Radiol; Am Asn Cancer Res. *Mailing Add:* Radiol/Oncol Stanford Univ Sch Med Stanford Univ Hosp Stanford CA 94305

BAGUS, PAUL SAUL, b New York, NY, Nov 19, 37; m 81; c 1. SOLID STATE PHYSICS, QUANTUM CHEMISTRY. *Educ:* Univ Chicago, BS & MS, 58, PhD(physics), 65. *Prof Exp:* Fel, Argonne Nat Lab, 62-64, resident res assoc physics, Solid State Sci Div, 64-66; assoc prof, Fac Sci, Univ Paris, 66-67; res assoc physics, Univ Chicago, 67-68; PROF STAFF MEM, INT BUS MACH RES LAB, 68- *Concurrent Pos:* Consult, Lockheed Missiles & Space Corp, 62-63; US sr scientist, Alexander von Humboldt Found, 80-81. *Mem:* Fel Am Phys Soc; Am Vacuum Soc. *Res:* Ab initio calculation of the electronic structure of atoms and molecules to determine properties of atoms, molecules, solids and surfaces; application to the theory of x-ray photo-electron spectroscopy and related matters; use of cluster models to determine the nature of the interaction of adsorbates with surfaces. *Mailing Add:* IBM Res Lab K31/802 650 Harry Rd San Jose CA 95120-6094

BAGWELL, ERVIN EUGENE, b Honea Path, SC, Jan 24, 36; m 58; c 2. PHARMACOLOGY. *Educ:* Furman Univ, BS, 58; Med Univ SC, MS, 60, PhD, 63. *Prof Exp:* From asst prof to assoc prof pharmacol, 65-75, PROF PHARMACOL, MED UNIV SC, 75- *Mem:* Am Soc Pharmacol & Exp Therapeut. *Res:* Autonomic and cardiovascular drugs; electropharmacology of antiarrhythmic drugs. *Mailing Add:* Dept Pharmacol Med Univ SC 171 Ashley Ave Charleston SC 29425

BAHADUR, BIRENDRA, b Gorakhpur, India, July 1, 49; m 70; c 2. LIQUID CRYSTALS, DISPLAYS. *Educ:* Gorakhpur Univ, BSc, 67, MSc, 69, PhD(physics), 76. *Prof Exp:* Res fel physics, Gorakhpur Univ, India, 69-76, asst prof, 76-77; sr sci officer res & develop, Nat Phys Lab, New Delhi, India, 77-81; vpres & dir res & develop, Data Images Inc, Ottawa, 81-85; MGR RES

& DEVELOP, LITTON SYST CAN, 85- *Mem:* Inst Physics, UK; Optical Soc Am; Soc Phys Chem France; Soc Info Displays. *Res:* Developed manufacturing process & technology for LCDS; author of over 60 research papers, review articles & three books on LCDS and liquid crystals. *Mailing Add:* Litton Data Images 1283 Algoma Rd Ottawa ON K1B 3W7 Can

BAHADUR, RAGHU RAJ, b Delhi, India, Apr 30, 24; m 50; c 2. MATHEMATICAL STATISTICS. *Educ:* Univ Delhi, BA, 43, MA, 45; Univ NC, Chapel Hill, PhD(math statist), 50. *Prof Exp:* Instr statist, Univ Chicago, 50-51, asst prof, 54-56; prof, Indian Coun Agr Res, 51-52; asst prof math statist, Columbia Univ, 52-53; prof statist, Indian Statist Inst, 56-61; assoc prof, 62-64, PROF STATIST, UNIV CHICAGO, 65- *Mem:* Inst Math Statist; Int Statist Inst; Nat Inst Sci India; Indian Soc Agr Statist; Sigma Xi. *Res:* Theory of statistical tests and decisions; approximations to classical distributions; large sample theory. *Mailing Add:* 5500 Southshore Dr Chicago IL 60637

BAHAL, NEETA, b Philadelphia, Pa, May 10, 65. PEDIATRICS, CRITICAL CARE. *Educ:* Philadelphia Col Pharm & Sci, BS, 88; Med Col Va, PharmD, 90. *Prof Exp:* Pharmacist-intern, Children's Hosp, Philadelphia, 84-89; pharmacist, Med Col Va Hosp, 88-90. *Concurrent Pos:* Postdoctoral fel pediat pharmacother, Wexner Inst Pediat Res, Children's Hosp, Columbus, 90-92; mem, Educ Comt, Am Col Clin Pharm. *Mem:* Am Col Clin Pharm; Fel Am Soc Hosp Pharmacists. *Res:* Pharmacokinetics in critically ill infants and children; gastroenterology and pharmacy education in US; infectious disease pharmacotherapy. *Mailing Add:* 650 Timber Dr Wayne PA 19087

BAHAL, SURENDRA MOHAN, b India, July 1, 35; m 61; c 2. PHYSICAL PHARMACY. *Educ:* Univ Bombay, BS, 54, BS, 56, MS, 59; Temple Univ, PhD(phys pharm), 65. *Prof Exp:* Instr pharm, Dept Chem Technol, Univ Bombay, 58-61; res pharmacist, 64-72, supvr oral liquids & topical prod unit, Wyeth Labs, Inc, 73-88, res assoc, Oral Parenteral & Topical Prod, E I Dupont de Nemours & Co, Inc, 88-90, SR GROUP LEADER, LIQUIDS & TOPICAL PROD, DUPONT MERCK PHARMACEUTICAL CO, 91- *Honors & Awards:* Lunsford-Richardson Pharm Award, Eastern Region, Richardson Merrell Co, Inc, 65; Pharmaceut Res Discussion Group Award, 70. *Mem:* Am Asn of Pharm Scientists. *Res:* Excretion kinetics of drugs as modified by protein binding and renal tubular secretion; antiinfective agents; powder caking; antacid technology; effect of sugars on penicillin degradation kinetics; stabilization of pharmaceutical formulations; drug stabilization patents; inhalation aerosol technology. *Mailing Add:* 650 Timber Dr Wayne PA 19087

BAHAM, ARNOLD, b Folsom, La, Oct 7, 43; m 68; c 2. DAIRY NUTRITION. *Educ:* La State Univ, Baton Rouge, BS, 66, MS, 68; Auburn Univ, PhD(dairy nutrit), 71. *Prof Exp:* Asst dairy sci, La State Univ, 66-68; asst dairy nutrit, Auburn Univ, 68-71; asst prof agr, Southeastern La Univ, 71-74; asst prof, 74-79, ASSOC PROF DAIRY SCI, LA STATE UNIV, 79- *Concurrent Pos:* Transp chmn, Am Dairy Sci Asn Meeting, 81; pres, Noba-Atlantic-LABC Affil Breeders, 82; secy fedn coun, Atlantic-LABC-Eastern AI, 85; mgr, La Animal Breeders Coop, 77- *Mem:* Am Dairy Sci Asn; Nat Asn Animal Breeders. *Res:* Nutrition and reproduction of dairy animals. *Mailing Add:* 2288 Gourrier Baton Rouge LA 70820

BAHAR, EZEKIEL, b Bombay, India, May 23, 33; US citizen; m 57; c 3. ELECTROMAGNETIC PROPAGATION, REMOTE SENSING. *Educ:* Israel Inst Technol, BSc, 58, MSc, 60; Univ Colo, PhD(wave propagation), 64. *Prof Exp:* Teaching asst elec eng, Israel Inst Technol, 58-60, instr, 60-62; res assoc electromagnetic model studies, Univ Colo, 62-64, asst prof elec eng, 64-67; from assoc prof to prof, 67-80, REGENTS PROF, DURHAM PROF ELEC ENG & GEORGE HOLMES DISTINGUISHED PROF, UNIV NEBR, LINCOLN, 80- *Concurrent Pos:* Res grants, Adv Res Projs Agency, 62-67 & Off Naval Res & NSF, 66-67, 72-78 & 87-; Environ Sci Serv Admin, US Dept Com, 70-71, Army Res Off, 76-79, 82-86 & 87-, Air Force, 81-83, NSF, 85- *Mem:* Fel Inst Elec & Electronics Engrs; Int Union Radio Sci. *Res:* Electromagnetic theory; propagation in general multimode waveguide structures; electromagnetic modelling techniques; synthesis of waveguide junctions; propagation in inhomogeneous anisotropic media; scattering, depolarization and diffraction by irregular media; generalized field transform techniques for nonuniform boundary value problems; scattering depolarization by random rough surfaces; remote sensing. *Mailing Add:* Dept Elec Eng Univ Nebr Lincoln NE 68588

BAHAR, LEON Y, b Turkey, Apr 4, 28; US citizen; m 65; c 3. APPLIED MECHANICS. *Educ:* Robert Col, Istanbul, BS, 50; Lehigh Univ, MS, 59, PhD(mech), 63. *Prof Exp:* Instr mech, Lehigh Univ, 57-63; asst prof eng, City Univ New York, 63-66; assoc prof, 66-79, PROF APPL MECH, DREXEL UNIV, 79- *Concurrent Pos:* Consult, Data Processing Div, IBM Corp, NY, 65, United Engrs & Constructors, Inc, 74-75, Gulf Res & Develop Corp & Western Res & Develop, 74, INA Corp, 76, George M Ewing Co, 76, Hudson Int, 90. *Mem:* Am Soc Mech Engrs; Am Acad Mech; Sigma Xi; fel NY Acad Sci. *Res:* Dynamics and stability of constrained dynamical systems and interaction with control theory. *Mailing Add:* Dept Mech Eng & Mech Col Eng Drexel Univ Philadelphia PA 19104

BAHARY, WILLIAM S, b Kermanshah, Iran, Jan 20, 36; US citizen; m 79. POLYMER PHYSICAL CHEMISTRY. *Educ:* Harvard Univ, AB, 57; Columbia Univ, AM, 58, PhD(chem), 61. *Prof Exp:* Sr res chemist, Tex-US Chem Co, 61-68; vis asst prof chem, Fairleigh Dickinson Univ, 68-73; res scientist, Stevens Inst Technol, 73-75, adj asst prof chem, 75-79; consult chemist, Wallace Clark Co, 75-79; eng supvr, Duracell, Inc, 79-88; RES SCIENTIST, UNILEVER, INC, 89- *Mem:* AAAS; Am Chem Soc. *Res:* Structure and properties of synthetic and biological macromolecules; viscoelasticity of concentrated polymer solutions; structure and function of polysaccarides of dental plaque; chemistry of biomimetic blood anticoagulation; battery research and technology; characterization, including light scattering GPC and EM of solutions of polymers and biopolymers; materials science and process engineering. *Mailing Add:* 291 N Middletown Rd Pearl River NY 10965

BAHCALL, JOHN NORRIS, b Shreveport, La, Dec 30, 34; m 66; c 3. ASTROPHYSICS. *Educ:* Univ Calif, Berkeley, BA, 56; Univ Chicago, MA, 57; Harvard Univ, PhD(physics), 61. *Prof Exp:* Res fel physics, Ind Univ, 60-62; res fel physics, Calif Inst Technol, 62-63, sr res fel, 63-65, from asst prof to assoc prof theoret physics, 65-70; mem, 68-70, PROF ASTROPHYSICS, INST ADVAN STUDY, 71- *Concurrent Pos:* Sloan found fel, 68-71; vis lectr, Dept Astron, Princeton Univ, 71-; mem physics adv panel, NSF, 72-75; comt mem, Div High Energy Astrophys, Am Astron Soc, 72-74; chmn, Sect Astron, Nat Acad Sci, 80-83, Astron & Astrophysics Surv Comt, 89-; Regent's fel, Smithsonian Inst, 86-89; mem, Astrophysics Mgt Opers Working Group, 87. *Honors & Awards:* Warner Prize, Am Astron Soc, 70; Shulamet Goldhaber Lectr, Tel Aviv Univ, 87; James Arthur Prize Lectr, Harvard-Smithsonian Ctr Astrophys, 88. *Mem:* Nat Acad Sci; Am Acad Arts & Sci; fel Am Phys Soc; Am Astron Soc (pres-elect, 89-90, pres, 90-92). *Mailing Add:* Sch Natural Sci Inst Advan Study Olden Lane Princeton NJ 08540

BAHCALL, NETA ASSAF, b Tel-Aviv, Israel, Dec 16, 42; m 66; c 3. ASTROPHYSICS, ASTRONOMY. *Educ:* Hebrew Univ, Jerusalem, BSc, 63; Weizmann Inst Sci, Israel, MSc, 65; Tel-Aviv Univ, PhD(astrophys), 70. *Prof Exp:* Res asst astrophys, Calif Inst Technol, 65-66, res fel, 70-71; res assoc, Princeton Univ Observ, 71-74, res staff mem, 74-75, res astronr, 75-79, sr res astronr,; off head sci prog, Space Telescope Sci Inst, 79-90; PROF, DEPT ASTROPHYS SCI, PRINCETON UNIV, 90- *Mem:* Am Astron Soc. *Res:* Galactic and extra-galactic astronomy; optical properties of galaxies and clusters of galaxies; optical identification and properties of x-ray sources. *Mailing Add:* Dept Astro-Phys Sci Princeton Univ Payton Hall Princeton NJ 08544

BAHE, LOWELL W, b Sycamore, Ill, Jan 30, 27; div; c 3. PHYSICAL CHEMISTRY. *Educ:* Purdue Univ, BS, 49; Princeton Univ, AM, 51, PhD(chem), 52. *Prof Exp:* Res chemist, Allis-Chalmers Mfg Co, 53-57; from asst prof to assoc prof phys chem, 57-65, PROF PHYS CHEM, UNIV WIS-MILWAUKEE, 65- *Mem:* Sigma Xi; Am Chem Soc. *Res:* Electrolytic solutions in aqueous and nonaqueous solvents; thermodynamics of solutions; dielectric constants. *Mailing Add:* 2106 E Newton Milwaukee WI 53211

BAHILL, A(NDREW) TERRY, b Washington, Pa, Jan 31, 46; m 71; c 2. EXPERT & MODELING PHYSIOLOGICAL SYSTEMS, SCIENCE OF BASEBALL. *Educ:* Univ Ariz, BSEE, 67; San Jose State Univ, MSEE, 70; Univ Calif, Berkeley, PhD(elec eng & comput sci), 75. *Prof Exp:* Asst prof elec eng & bioeng, Carnegie-Mellon Univ, 76-81, assoc prof, 81-84; PROF SYSTS & INDUST ENG, UNIV ARIZ, 84- *Concurrent Pos:* Asst prof neurol, Sch Med, Univ Pittsburgh, 77-84. *Mem:* Sr mem Inst Elec & Electronics Engrs; Sigma Xi; Soc Neurosci. *Res:* Experiments and modeling analysis of human head and eye movements, with applications to the science of baseball; computer techniques for validating expert system knowledge bases. *Mailing Add:* Systs Ind US & Eng Univ Ariz Tucson AZ 85721

BAHL, INDER JIT, b Punjab, India, Jan 27, 44; US citizen; m 68; c 2. MICROWAVE TECHNOLOGY, MICROWAVE INTEGRATED CIRCUITS. *Educ:* Punjab Univ, India, BS, 65; Birla Inst Technol & Sci, India, MS, 67, MS, 69; Indian Inst Technol, PhD(elec eng), 75. *Prof Exp:* Res engr, Indian Inst Technol, India, 69-78; res assoc microwave, Univ Ottowa, Can, 79-81; res scientist, Defense Res Estab, Can, 81; EXEC SCIENTIST, ITT GALLIUM ABSENIDE TECHNOL CTR, 81- *Concurrent Pos:* Assoc ed, Int J Microwave & Millimeter-Wave Computer-Aided Eng, 90- *Mem:* Fel Inst Elec & Electronics Engrs; Electromagnetics Acad. *Res:* Advanced state-of-the-art in the areas of microwave and millimeter-wave integrated circuits including antennas; author of five books and numerous publications. *Mailing Add:* 6825 Wood Haven Rd NW Roanoke VA 24019

BAHL, LALIT R(AI), b Shekhupura, India, July 30, 43; m 68; c 3. COMMUNICATIONS & COMPUTER SCIENCE. *Educ:* Indian Inst Technol, Kharagpur, BTech, 64; Univ Ill, Urbana-Champaign, MS, 66, PhD(elec eng), 68. *Prof Exp:* RES STAFF MEM COMPUT SCI, 68-, MGR NATURAL LANG SPEECH RECOGNITION, IBM CORP RES CTR. *Concurrent Pos:* Adj assoc prof, dept elec eng & comput sci, Columbia Univ, 70-75. *Mem:* Inst Elec & Electronics Engrs. *Res:* Coding theory for reliable transmission; information theory; speech recognition. *Mailing Add:* IBM Corp Watson Res Ctr Div Natural Lang Speech Recognition PO Box 218 Yorktown Heights NY 10598

BAHL, OM PARKASH, b Lyallpur, Punjab, India, Jan 10, 27; m 52; c 3. CHEMISTRY. *Educ:* Punjab Univ, India, MSc, 50; Univ Minn, PhD, 62. *Prof Exp:* Lectr chem, Arya Col, Ludhiana, India, 50-52 & Govt Col, 52-57; res assoc biochem, Univ Minn, 62-63 & Univ Calif, 63-64; Am Cancer Soc career investr & Dernham fel, Univ Southern Calif, 64-65, asst prof biochem, 65-66; from asst prof to assoc prof, State Univ NY, 66-71, Prof, Dept Chem, 71-74, prof & dir, Div Cell & Molecular Biol, 74-78, chmn Dept Biol Scis, 76-83, PROF, DEPT BIOL SCIS, STATE UNIV NY, BUFFALO, 76- *Concurrent Pos:* Nat Found res grant, 65-66; Am Cancer Soc career develop award, 65-66, 69-82; USPHS res grant, 66-92. *Honors & Awards:* Schoellkopf Award, Am Chem Soc, 78. *Mem:* Am Chem Soc; Am Soc Biol Chemists; Brit Chem Soc; NY Acad Sci; Endocrine Soc; AAAS; Am Soc Cell Biol. *Res:* Proteins and enzymes; polysaccharides; glycoproteins; hormones and receptors. *Mailing Add:* Dept Biol Sci 347 Cooke Hall State Univ NY Buffalo Amherst Campus Buffalo NY 14260

BAHLER, THOMAS LEE, b Walnut Creek, Ohio, Feb 1, 20; m 48; c 2. ZOOLOGY. *Educ:* Col Wooster, BA, 43; Univ Wis, PhD(zool), 49. *Prof Exp:* From res asst to teaching asst, Univ Wis, 44-49; from asst prof to assoc prof, 49-59, prof zool & physiol, 59-73, PROF BIOL, UTAH STATE UNIV, 73- *Concurrent Pos:* NSF fac fel, 58-59. *Res:* Physiology of parasites; physiological effects of insecticides; physiology of eosinophils. *Mailing Add:* Dept Zool Physiol Utah State Univ Logan UT 84322-5305

BAHN, ARTHUR NATHANIEL, b Boston, Mass, Jan 5, 26; m 84; c 5. MICROBIOLOGY, DENTAL RESEARCH. *Educ:* Boston Univ, AB, 49; Univ Kans, MA, 52; Univ Wis, PhD(bact), 56. *Prof Exp:* Asst bacteriologist, Kans State Bd Health, 51-52; asst bact, Univ Wis, 55-66; instr, Col Med, Univ Ill, 56-69; asst prof, Sch Dent, Northwestern Univ, 59-63; assoc prof bact, Sch Dent & assoc prof microbiol, Sch Med, 63-71; PROF & SECT HEAD SCH DENT MED, SOUTHERN ILL UNIV, ALTON, 71- *Concurrent Pos:* Proj asst, Univ Wis, 52-56; mem, Am Bd Med Microbiol, 70-; consult, Food & Drug Admin Oral Cavities Prod Panel, 74-; adj prof oral path, Med, Univ SC. *Mem:* Fel AAAS; fel Am Soc Microbiol; Int Asn Dent Res. *Res:* Resistance to infection; immunochemistry, oral bacterial enzymes; oral streptococci and actinomycetes; experimental endocarditis; viruses. *Mailing Add:* 1453 E 56th St Chicago IL 60637

BAHN, EMIL LAWRENCE, JR, b Cape Girardeau, Mo, Sept 6, 24; m 49; c 2. INORGANIC CHEMISTRY, NUCLEAR CHEMISTRY. *Educ:* Mo Sch Mines, BSChE, 46; Washington Univ, MSChE, 49, MA, 61, PhD(chem), 62. *Prof Exp:* Prof chem, Southeast Mo State Col, 62-73, head dept, 62-73; RETIRED. *Mem:* AAAS; Am Chem Soc; Am Nuclear Soc; Health Physics Soc; Soc Appl Spectros. *Res:* X-ray fluorescence; trace elements emitted from coal burning; utilization of drinking water and waste water sludges. *Mailing Add:* Rte 1 Box 141 Cape Girardeau MO 63701

BAHN, GILBERT S(CHUYLER), b Syracuse, NY, Apr 25, 22; wid; c 1. ENGINEERING, MATHEMATICAL MODELING. *Educ:* Columbia Univ, BS, 43; Rensselaer Polytech Inst, MS, 65; Columbia Pacific Univ, PhD, 79. *Prof Exp:* Chem engr, sci prog, chem dept, Gen Elec Co, 46-48, develop engr, Thermal Power Systs Div, 48-53; sr thermodyn engr, Marquardt Aircraft Co, 53-54, from res specialist to res consult, Marquardt Corp, 54-70; eng specialist, Hampton Tech Ctr, Planning Res Corp, 70-88; RETIRED. *Concurrent Pos:* Chmn, Nat Conf Performance of High Temperature Systs, 59-68; ed, Pyrodynamics, 63-69; mem performance standardization working group, JANNAF Interagency Propulsion Comt, 66-84, thermochem working group, 67-72; free-lance hist res proj, 88- *Mem:* Am Soc Mech Engrs; Combustion Inst. *Res:* High-temperature thermodynamics; kinetics, and thermal power performance analysis; documentation; environmental research with remotely-sensed digital imagery; statistical inference of undefined populations; analysis of aberrant engineering data. *Mailing Add:* 4519 N Ashtree St Moorpark CA 93021

BAHN, ROBERT CARLTON, b Newark, NY, July 24, 25; m 49; c 4. PATHOLOGY. *Educ:* Univ Buffalo, MD, 47; Univ Minn, PhD(path), 53; Am Bd Path, dipl path anat, 53, cert clin path, 54 & cert neuropath, 59. *Prof Exp:* Asst to staff, Mayo Clin, 53; sr asst surgeon, NIH, 54-55; CONSULT SECT PATH ANAT, MAYO CLIN, 56-, PROF PATH, 69- *Concurrent Pos:* Instr grad sch, Univ Minn & Mayo Med Sch, 57-; from asst prof to prof path, Mayo Med Sch, 59-69, chmn-vpres, Mayo Comput Comt, 65-76. *Honors & Awards:* Heinrich Leonard Award, Univ Buffalo, 47. *Mem:* Am Heart Asn; Am Physiol Soc; Int Acad Path; Am Asn Pathologists; Inst Elec & Electronics Engrs Computer Soc; Am Math Asn. *Res:* Cytophysiology of hypothalmus and pituitary; human and experimental pituitary neoplasms; medical applications of digital computers; biomedical digital image reconstruction and analysis. *Mailing Add:* Dept Path Mayo Clin Rochester MN 55905

BAHNER, CARL TABB, b Conway, Ark, July 14, 08; m 31; c 3. ORGANIC CHEMISTRY, BIOMEDICAL ENGINEERING. *Educ:* Hendrix Col, AB, 27; Univ Chicago, MS, 28; Southern Baptist Theol Sem, ThM, 31; Columbia Univ, PhD(chem eng), 36. *Prof Exp:* Asst prof chem & head dept physics, Union Univ, Tenn, 36-37; prof chem, Carson-Newman Col, 37-73, res coordr, 67-73; from assoc prof to prof chem, 73-78, Walters State Community Col, 78-81; prof chem, Bluefield Col, 79-91; RETIRED. *Concurrent Pos:* Consult, Tenn Valley Authority, 42-45, Carbide & Carbon Chem Corp, 48-79 & Oak Ridge Inst, 50-55. *Mem:* AAAS; Am Chem Soc; Am Asn Cancer Res; fel Am Inst Chem. *Res:* Reactions of aliphatic nitro compounds and of organic halides; amines; heterocyclic nitrogen compounds; protein extraction; immersion freezing solutions; hydrotropic solutions; organic complexes of radioisotopes; cancer chemotherapy; nitrates in saliva. *Mailing Add:* PO Box 549 Jefferson City TN 37760

BAHNFLETH, DONALD R, b West Chicago, Ill, July 16, 27; m 51; c 3. APPLIED ENGINEERING PROBLEM SOLVING, SYSTEM ANALYSES. *Educ:* Univ Ill, BSME, 52, MSME, 56. *Prof Exp:* Res asst prof, Univ Ill, 55-60; ed, Heating/Piping/Air Conditioning, 60-70, ed dir, 70-71; tech dir, 71-76, exec vpres, 76-84, PRES, ZBA, INC, 84- *Concurrent Pos:* Mgt rep, Automated Procedures Eng Consults. *Honors & Awards:* Distinguished Serv Award, Am Soc Heating, Refrig & Air Conditioning Engrs, 76; First Outstanding Achievement Award, Nat Air Filtration Asn, 90. *Mem:* Nat Soc Prof Engrs; fel Am Soc Heating Refrig & Air Conditioning Engrs; Air Pollution Control Asn; Sigma Xi; Am Consult Engrs Coun; Automated Procedures Eng Consults. *Res:* Applied research identifying thermal and hydraulic system deficiencies in large scale central steam, high temperature water and chilled water generation and distribution systems. *Mailing Add:* 7466 Glenover Dr Cincinnati OH 45236-2114

BAHNG, JOHN DEUCK RYONG, b Bookchung, Korea, Mar 21, 27; US citizen; m 61; c 3. ASTRONOMY. *Educ:* St Norbert Col, BS, 50; Univ Wis, MS, 54, PhD(astron), 57. *Prof Exp:* Res assoc astron, Princeton Univ, 57-62; asst prof, 62-66, chmn dept, 75-77, ASSOC PROF ASTRON, NORTHWESTERN UNIV, 66-, DIR, DEARBORN OBSERV, 75- *Mem:* Am Astron Soc; Royal Astron Soc; Int Astron Union; Sigma Xi. *Res:* Photoelectric spectrophotometry of stars; stellar structure and evolution. *Mailing Add:* Dearborn Observ Northwestern Univ Evanston IL 60208

BAHNIUK, EUGENE, b Weirton, WVa, Mar 10, 26; m 77; c 5. BIOMECHANICS. *Educ:* Case Inst Technol, BSME, 50, MS, 61; Case Western Reserve Univ, PhD(eng), 70. *Prof Exp:* Develop engr, NY Air Brake Co, 50-54; proj engr, Lear Siegler Inc, 54-56; supvr, Borg-Warner Corp, 56-61; res & develop mgr, Weatherhead Co, 61-69; assoc prof eng, 69-70, assoc prof mech eng, 71-76, ASST PROF ORTHOP SURG, CASE WESTERN RESERVE UNIV, 71-, PROF MECH ENRG, 77- *Concurrent Pos:* Res assoc, Vet Admin, Cleveland, 76-; fel, NIH; fel, NSF; vis prof & fel, NASA Langley. *Honors & Awards:* Award of Merit, Am Soc Testing & Mat; Venor Award. *Mem:* Am Soc Biomechanics; fel Am Soc Testing & Mat; Sigma Xi; Int Soc Ski Safety. *Res:* Biomechanics of bone; sports trauma; cervical spinal cord trauma; loosening of orthopedic implants; mathematical modeling of the human neuromuscular system; mechanical design. *Mailing Add:* Glennan Bldg University Circle Cleveland OH 44106

BAHNS, MARY, MICROBIOLOGY, MOLECULAR BIOLOGY. *Educ:* Univ Va, PhD(cell biol), 85. *Prof Exp:* ASST PROF BIOL, BRIAR CLIFF COL, 83-; ASST PROF SCI EDUC, TEX CHRISTIAN UNIV, 86. *Mailing Add:* 712 S Pitt St Alexandria VA 22314

BAHR, CHARLES CHESTER, b Orange, Tex, Apr 2, 58; m 89. SURFACE CHEMISTRY, SURFACE MORPHOLOGY. *Educ:* Tex A&M Univ, BS, 80; Univ Calif, Berkeley, PhD(phys chem), 86. *Prof Exp:* Mem tech staff, AT&T Eng Res Ctr, 86-88, MEM TECH STAFF, AT&T BELL LABS, 88- *Mem:* Am Phys Soc. *Res:* Dynamics of surface chemical reactions to develop novel and technologically important materials in electronic and optical applications. *Mailing Add:* AT&T Bell Labs Rm 1C-402 600 Mountain Ave Murray Hill NJ 07974-0636

BAHR, DONALD WALTER, b Chicago, Ill, Dec 13, 27; m 60; c 2. COMBUSTION TECHNOLOGY, GAS TURBINE ENGINE TECHNOLOGY. *Educ:* Univ Ill, BS, 49; Ill Inst Technol, MS, (chem eng) & MS (gas technol), 51. *Prof Exp:* Aeronaut res scientist, jet engine combustion, NACA Lewis Flight Propulsion Lab, Cleveland, 51-54; first lieutenant, Environ Health Lab, Wright Patterson AFB, Ohio, 54-56; combustion chem engr, GE Aircraft Engines, 56-59, mgr aerothermochem, 59-60, mgr combustion appl res, 60-62, mgr propulsion equip eng, GE Missile & Space Div, 62-68, mgr, Combustion & Emission Control, 68-86, MGR, COMBUSTION & HEAT TRANSFER TECHNOL, GE AIRCRAFT ENGINES, 86- *Concurrent Pos:* Chmn, Aircraft Engine Emissions Comt, Aerospace Industs Asn, 71-; vchmn, Ad Hoc Panel Jet Engine Hydrocarbon Fuels, NASA, 73-76; tech rep, Ad Hoc Comt Aircraft Fuel Conserv, Nat Acad Eng, 75-78; indust deleg, Comt Aircraft Engine Emissions, Int Civil Aviation Orgn, 78-; mem, Aircraft Fuel, Lubricant & Equip Res Comt, Coord Res Coun, Inc, 78-; vchmn, Combustion & Fuels Comt, Am Soc Mech Engrs, 85-87, chmn, 87-89; mem bd governors, cent sect, Combustion Inst, 86- *Honors & Awards:* Perry T Egbert Award for Eng Achievement, Gen Elec Co, 82; Air Breathing Propulsion Award, Am Inst Aeronaut & Astronaut, 83. *Mem:* Nat Acad Eng; assoc fel Am Inst Aeronaut & Astronaut; Combustion Inst; Am Soc Mech Engrs. *Res:* Design and development of advanced combustion systems for gas turbine engines with performance and durability, reduced pollutant emissions and enhanced fuels flexibility. *Mailing Add:* 6576 Branford Ct Cincinnati OH 45236

BAHR, GUSTAVE KARL, b Chicago, Ill, Mar 7, 29; m 61; c 2. RADIOLOGICAL PHYSICS, NUCLEAR MEDICINE. *Educ:* Xavier Univ, Ohio, BS, 51; Univ Cincinnati, MS, 59, PhD(physics), 64; Am Bd Radiol, dipl, 66. *Prof Exp:* Instrumentation engr, Keleket Instrument Co, Ohio, 51-53; exp physicist, Ohmart Corp, 53-54; res asst appl mech, Univ Cincinnati, 54-56; tech engr, Gas Turbine Div, Gen Elec Co, 56-58; radiol physicist, 58-59, from instr to asst prof, 59-70, ASSOC PROF RADIOL, COL MED, UNIV CINCINNATI, 70- *Concurrent Pos:* Adj prof physics, Xavier Univ Ohio, 65-; lectr eng physics, Eve Col, Univ Cincinnati, 66-; consult, Bur Dis Prev & Environ Control, Nat Ctr Radiol Health, 66-, Christian Holmes Hosp & Good Samaritan Hosp, Cincinnati, 68- *Mem:* AAAS; Am Asn Physicists in Med; Am Col Radiol. *Res:* Application of digital computer techniques to radiation therapy planning and nuclear medicine diagnostic techniques; application of linear programming to radiation therapy; application of stochastic simulation of radiation to radiologic physics problems. *Mailing Add:* 2743 Bentley Anderson Turnpike Cincinnati OH 45244

BAHR, JAMES THEODORE, b East Orange, NJ, Nov 19, 42; m 66; c 3. AGRICULTURAL & FOOD CHEMISTRY, BIOPHYSICS. *Educ:* Mass Inst Technol, BS, 64; Univ Pa, PhD(biophys), 71. *Prof Exp:* Res assoc chem, Univ Ariz, 71-76; sr res biologist, Mobil Chem Co, 76-80, assoc biol & crop chem, 80-81; res mgr, Agrochem Div, Rhone-Poulenc Inc, 81-85; mgr, plant growth regulators, 85-88, MGR HERBICIDE DISCOVERY PROJS, FMC CORP, 88- *Res:* Regulation of photosynthetic carbon metabolism; regulation and mechanism of ribulose diphosphate carboxylase-oxygenase; cyanide-insensitive respiration in plant mitochondria; yield enhancement and plant growth regulators in soybeans; herbicidal agrochemicals. *Mailing Add:* FMC Corp PO Box 8 Princeton NJ 08543

BAHR, JANICE M, b LaCrosse, Wis. ANIMAL SCIENCE PHYSIOLOGY. *Educ:* Viterbo Col, BA, 64; Univ Ill, Urbana, MSc, 68, PhD(physiol), 74. *Prof Exp:* Instr biol, Viterbo Col, LaCrosse, Wis, 68-70; endocrinol trainee, Dept Physiol, 70-72, from asst prof to assoc prof, 74-83, PROF ANIMAL SCI & PHYSIOL, UNIV ILL, URBANA, 83- *Concurrent Pos:* NSF, NIH & USDA res grants, 74-; NIH vis scientist, Univ Krakow, Poland, 77; asst ed, Biol Reproduction, 77-81; mem, Prog Site Visit Team, NIH, 78 & 89-90, Panel Rev Competitive Grants Regulatory Biol, NSF, 81-84, Develop Comt, Soc Study Reproduction, 83-86, bd dirs, 84-87, Panel Animal Sci Competitive Grants Prog, USDA, 87, Animal Welfare Comt, Endocrine Soc, 86-90, Reproductive Biol Study Sect, NIH, 88-92 & Panel Rev Progs Biol, Behav & Social Sci, NSF, 89-90; chair, Sci Sessions Endocrine Soc, Soc Study Reproduction & Poultry Sci Asn, 80-; assoc, Ctr Advan Study, 90. *Mem:* Fel AAAS; Am Physiol Soc; Am Soc Animal Sci; Endocrine Soc; Soc Study Fertil; Poultry Sci Asn; Soc Study Reproduction; Sigma Xi; World Poultry Sci Asn. *Res:* Author of more than 150 technical publications. *Mailing Add:* 102 Animal Genetics Lab Univ Ill 1301 W Lorado Taft Dr Urbana IL 61801

BAHR, KARL EDWARD, b Rochester, NY, Dec 11, 33; m 52; c 2. AUTOMATION EQUIPMENT DESIGN, PROCESS DEVELOPMENT. *Educ:* Clarkson Univ, BS, 59 & MS, 71. *Prof Exp:* Dir res & develop, Talon Div, Textron, 72-79; chief engr, Copier Div, Nashua Corp, 80-82; mgr mfg eng, Data Prod Corp, 83-86; vpres eng, Hollins Automation, Esterline Corp, 86-89; ENG MGR, SIERRA RES & TECHNOL, 89- *Concurrent Pos:* Lectr, Ohio State Univ, 71-72; instr, Hesser Col, 82-83; cosult, DBA Automation Technol, 82-83. *Mem:* Am Soc Mech Engrs; Inst Elec & Electronics Engrs; Air Pollution Info Comput Systs. *Res:* Plastics processes; wave soldering and reflow technologies. *Mailing Add:* Sierra Res & Technol 59 Power Rd Westford MA 01886

BAHR, LEONARD M, JR, b Baltimore Md, June 17, 40; m 79; c 1. SYSTEMS ECOLOGY. *Educ:* Univ Md, BS, 63; Univ Richmond, MS, 68; Univ Ga, PhD(zool & ecol), 74. *Prof Exp:* Res asst, Chesapeake Biol Lab, Univ Md, 63-66; res assoc, Ctr Wetland Resources, 73-75, asst prof, 75-79, assoc prof coastal ecol, 79-84, PUBL, CITY FITNESS MAG, DEPT MARINE SCI, LA STATE UNIV, 87- *Res:* Coastal systems ecology, especially as relating to resource management; energetics of ecosystems; identification of the functional roles of organisms within ecosystems, particularly estuarine epifauna, including oyster reef inhabitants. *Mailing Add:* PO Box 94004 Baton Rouge LA 70806

BAHR, THOMAS GORDON, b La Crosse, Wis, Apr 17, 40; m 60; c 1. LIMNOLOGY. *Educ:* Univ Idaho, BS, 63; Mich State Univ, MS, 66, PhD(limnol, physiol), 68. *Prof Exp:* Asst prof zool, Colo State Univ, 68-70; from asst dir to dir, Inst Water Res, Mich State Univ, 70-78; DIR, WATER RESOURCES RES INST, N MEX STATE UNIV, 78- *Mem:* AAAS; Am Soc Limnol & Oceanog. *Res:* Applied limnology; water management in arid regions; nutrient cycling; waste water management; dynamics of toxic materials in natural systems. *Mailing Add:* 1728 Pomona Dr Las Cruces NM 88001

BAI, ZHI DONG, b Leting Cty, China, Nov 27, 43; m 72; c 2. STATISTICS. *Educ:* China Univ Sci & Technol, BS, 68, PhD (statist), 82. *Prof Exp:* Assoc prof math, China Univ Sci & Technol, 82-84; RES ASSOC STATIST, DEPT MATH, UNIV PITTSBURGH, 84- *Mem:* Chinese Soc Math; Chinese Soc Probability & Statist. *Res:* Density and nonparametric regression estimation; limiting theorems; model selection; multivariate analysis; nonparametric discrimination and prediction; spectral analysis of random matrices and signal processing; author of 80 papers. *Mailing Add:* 801 W Aaron Dr State College PA 16803

BAIAMONTE, VERNON D, b Greeley, Colo, Apr 29, 34; m 58; c 4. PHYSICAL CHEMISTRY. *Educ:* Colo State Col, AB, 59, MA, 60; Univ Ind, PhD(phys chem), 66. *Prof Exp:* Instr chem, Ball State Univ, 60-62; fel phys chem, Los Alamos Sci Lab, 66-67; assoc prof, 67-69, PROF CHEM, MO SOUTHERN STATE COL, 70-, HEAD DEPT, 69- *Mem:* Am Chem Soc. *Res:* Kinetic spectroscopy, especially the fast reactions initiated by flash photolysis and detected by spectroscopic methods. *Mailing Add:* Dept Chem Mo Southern State Col Joplin MO 64801

BAIARDI, JOHN CHARLES, b Brooklyn, NY, Feb 9, 18; m 43; c 2. HEMATOLOGY, PHYSIOLOGY. *Educ:* St Francis Col, NY, BS, 40; Brooklyn Col, MA, 43; NY Univ, PhD(biol), 53. *Prof Exp:* Instr biol, St Francis Col, NY, 42-43, asst prof & chmn dept, 46-48; assoc prof, St John's Univ, NY, 48-54; prof biol, Long Island Univ, 54-70, chmn dept, 54-62, assoc dean sci, 58-62, vpres & provost, 62-67, vchancellor, 67-70; pres, Affil Cols & Univs, Inc, 70-82; dir, NY Ocean Sci Lab, Montauk, 70-82; RETIRED. *Concurrent Pos:* Mem med adv bd, Cooley's Anemia Found, NY, vchmn, 63-75. *Mem:* Fel NY Acad Sci; Harvey Soc. *Res:* Research administration; blood and hemopoietic organs of mammals, fish and invertebrates. *Mailing Add:* 9930-4 Eagles Pt Circle Port Richey FL 34668

BAICH, ANNETTE, b Chicago, Ill, Mar 2, 30; m 50. BIOCHEMISTRY. *Educ:* Roosevelt Univ, BS, 51; Univ Ore, MS, 54, PhD(chem), 60. *Prof Exp:* Asst chem, Univ Ore, 51-54, instr, 55-56, asst, 56-57; lab technician, Agr Exp Sta, Univ Calif, 54-55; fel biochem, Ore State Univ, 60-61; fel microbiol, Rutgers Univ, 61-62; asst prof biochem, Ore State Univ, 63-69; assoc prof, 69-74, PROF BIOCHEM, SOUTHERN ILL UNIV, EDWARDSVILLE, 74-, CHMN BIOL DEPT, 84- *Concurrent Pos:* NIH career develop award, 65-69. *Mem:* AAAS; Am Chem Soc; Am Soc Biol Chemists. *Res:* The relation and response of an organism to its environment and how this relationship may be explained in chemical terms. *Mailing Add:* Dept Biol Sci Southern Ill Univ Edwardsville IL 62025

BAIDINS, ANDREJS, b Dauguli, Latvia, Dec 16, 30; US citizen; m 60; c 2. COLLOID & SURFACE CHEMISTRY. *Educ:* Franklin & Marshall Col, BS, 54; Rutgers Univ, PhD(phys chem), 58. *Prof Exp:* CHEMIST, RES DIV, CHEM & PIGMENTS DEPT, E I DU PONT DE NEMOURS & CO, INC, 58- *Concurrent Pos:* Abstractor, Chem Abstr Serv, 59-81. *Mem:* AAAS; Am Chem Soc. *Res:* Kinetics of decomposition of light and heat sensitive compounds; physical and chemical properties of inorganic crystal surfaces and solid-liquid interfaces. *Mailing Add:* Jackson Lab 2104 E I Du Pont de Nemours & Co Inc Deepwater NJ 08023

BAIE, LYLE FREDRICK, b Rockford, Ill, Dec 14, 42; m 69; c 4. EXPLORATION GEOLOGY, OCEANOGRAPHY. *Educ:* Univ Notre Dame, BA, 64; Tex A&M Univ, MS, 67, PhD(oceanog), 70. *Prof Exp:* Geologist explor, Texaco, Inc, 70-74; staff geologist, 74-77, geol supvr, Cent Gulf Mexico Area, 77-78, MGR GEOL RES EXPLOR, CITIES SERV CO, 78- *Honors & Awards:* Sr Geol Assoc, Cities Serv Co, 77. *Mem:* Am Asn Petrol Geologists. *Mailing Add:* 2310 Ivy Glen Dr Houston TX 77077

BAIER, JOSEPH GEORGE, b New Brunswick, NJ, Feb 12, 08; m 29; c 3. IMMUNOBIOLOGY. *Educ:* Rutgers Univ, BS, 28, MS, 29; Univ Wis, PhD(zool), 32. *Prof Exp:* Asst zool, Rutgers Univ, 28-29; asst, Univ Wis, 29-32, from instr to prof, 32-68, dean, Col Lett & Sci, 56-66, Michael F Guyer prof, 69-75, EMER MICHAEL F GUYER PROF ZOOL, UNIV WIS-MILWAUKEE, 75-; PRES & TREAS, TIME RESTORATION, INC, 78- *Concurrent Pos:* Dir, Am Watchmakers Inst, 79-85, second vpres, 80-82, sem instr, 81- & first vpres, 82-83. *Mem:* Fel AAAS; Am Soc Zoologists; fel Linnaean Soc London. *Res:* Immunotaxonomy; quantitative studies on precipitins; electronic instrumentation; photoelectric densitometry; horological research. *Mailing Add:* 10624 N 24th Pl Phoenix AZ 85028

BAIER, ROBERT EDWARD, b Buffalo, NY, Oct 31, 39; m 61; c 2. SURFACE CHEMISTRY, BIOMATERIALS. *Educ:* Cleveland State Univ, BES, 62; State Univ NY Buffalo, PhD(biophys), 66; Environ Eng Intersoc Bd, dipl. *Prof Exp:* Res asst, Buffalo Gen Hosp, 59-60; nucleonics engr, Electromech Res Lab, Repub Steel Corp, 61-62; res chemist & Nat Res Coun-Nat Acad Sci fel surface chem, Naval Res Lab, 66-68; prin physicist, Cornell Aeronaut Lab, 68-71, head chem sci sect, 71-74; mem tech staff, Calspan Corp, 74-77, staff scientist, 77-84; res prof & dir biomat, Ctr Advan Technol, Health-Care Instruments & Devices Inst, 85-89, CO-DIR, NSF INDUST/UNIV COOP RES CTR BIOSURFACES, STATE UNIV NY, BUFFALO, 88-, ASSOC PROF BIOMAT, 89- *Concurrent Pos:* Guest worker, Nat Heart & Lung Inst, 68; consult, Roswell Park Mem Inst, NY, 70-89; mem, Adv Bd, Prog Biomed Eng, Clemson Univ, 70-82; chmn, Columbia Univ Biomat Sem, 71-72; adj assoc prof chem eng, Cornell Univ, 71-76; res assoc prof biophys, State Univ NY, Buffalo, 71-, Dent Mat, 75-88, res prof biophys sci, 83-, assoc prof oral path, 84-88; consult biomat develop & environ quality control; vpres eng, B/A/I/E/R Inc, 81- *Honors & Awards:* Union Carbide Chem Award, Am Chem Soc, 71; Clemson Award Basic Res, Soc Biomat, 83; Distinguished Serv Award, Soc Appl Spectros, 87; Award Innovation Med Devices, 87. *Mem:* Soc Biomat; Am Soc Artificial Internal Organs; Biophys Soc; Am Chem Soc; Am Soc Cell Biol. *Res:* Interdisciplinary studies in surface chemistry and physics of materials at liquid-gas, solid-gas and solid-liquid interfaces, including biomedical and environmental systems. *Mailing Add:* Ctr Biosurfaces 110 Parker Hall State Univ NY Buffalo NY 14214

BAIER, WILLIAM H, b Chicago, Ill, Oct 1, 22; m 58. SOLID MECHANICS, ELECTRONICS. *Educ:* Ill Inst Technol, BS, 44 & 49, MS, 50, PhD(mech eng), 58. *Prof Exp:* Instr mech, Ill Inst Technol, 47-49; designer fluid power, Hannifin Corp, 49-51; design engr, Askania Regulator Co, 51-53; mgr space systs, IIT Res Inst, 53-63; mgr aerospace mech, Emerson Elec Co, 64-66; dir res, Thor Power Tool Co, 66-68; dir res & eng, Cenco Instruments Corp, 68-76; vpres res & eng, Cent Sci Co, 76-78; dir eng, Fitzpatrick Co, 78-83; LECTR, DU PAGE COL, 83- *Concurrent Pos:* Lectr, Ill Inst Technol, 55-57; pres, DuPage Inst Res & Educ, 69-; mem, bd standardization, Am Soc Mech Eng, 70- *Mem:* Am Inst Physics; Am Soc Testing & Mat; Am Vacuum Soc; Am Soc Mech Engrs; Am Inst Aeronaut & Astronaut. *Res:* Lunar surface petrographic analyzers; physics educational instrumentation; in-space rendezvous mechanisms and systems; vacuum systems; portable tools for industrial and extraterrestrial use; fluid power mechanisms and systems; fluidized beddryers. *Mailing Add:* 3624 Creekwood Ct Downers Grove IL 60515

BAIER, WOLFGANG, b Liegnitz, Ger, Dec 10, 23; nat US; m 50; c 2. AGRICULTURAL METEOROLOGY. *Educ:* Hohenheim Agr Univ, Dipl, 49, Dr Agr, 52; Univ Pretoria, MSc, 64. *Prof Exp:* Officer in charge agrometeorol sect, Agr Res Inst Highveld Region, SAfrican Dept Agr, 55-64; res scientist agrometeorol sect, Plant Res Inst, Agr Can, 64-69, chief sect, Res Br, 69-73, head agrometeorol res & serv, Agrometeorol Sect, Chem & Biol Res Inst, 73-78, head agrometeorol sect & asst dir, Land Resource Res Inst, Agr Can, 78-90; RETIRED. *Concurrent Pos:* Secy, Can Comt Agr Meteorol; pres, Comn Agr Meteorol, World Meteorol Orgn, 71- *Honors & Awards:* Patterson Medal, Can, 75. *Mem:* Am Meteorol Soc; Can Meteorol Soc; Int Soc Biometeorol. *Res:* Problems related to the impact of weather and climate on agricultural crops and production; development and application of computer models to evaluate plant-weather relationships. *Mailing Add:* 55 Pemberton Crescent Nepean ON K2G 4N5 Can

BAIERLEIN, RALPH FREDERICK, b Springfield, Mass, Dec 19, 36; m 65; c 2. COSMOLOGY. *Educ:* Harvard Univ, AB, 58; Princeton Univ, PhD(gen relativity), 62. *Prof Exp:* Instr physics, Harvard Univ, 62-65; from asst prof to prof, 66-77, CHARLOTTE AYRES PROF PHYSICS, WESLEYAN UNIV 77-, PHYSICIST, 69- *Concurrent Pos:* Res assoc, Smithsonian Astrophys Observ, 65-66. *Mem:* Am Phys Soc. *Res:* Star formation and turbulence. *Mailing Add:* Dept Physics Wesleyan Univ Middletown CT 06457

BAIG, MIRZA MANSOOR, b Jabulpar, India, June 16, 42; m 72; c 2. BIOCHEMISTRY. *Educ:* Univ Karachi, BS, 60, MS, 62; State Univ NY Buffalo, PhD(biol), 69. *Prof Exp:* Lectr biol, Jamia Col, Karachi, 62-64; fel chem, Fla State Univ, 69-70; fel internal med & biol chem, Univ Mich, Ann Arbor, 70-71; fel, dept biochem & pediat, Col Med, Univ Fla, 72-75, asst prof pediat, 75-78, asst prof med, 78-; ASSOC PROF PATH, COL MED, KING SAUD UNIV. *Mem:* Am Inst Nutrit; Am Soc Biol Chemists; Soc Complex Carbohydrates; Am Soc Clin Nutrit. *Res:* Chemistry and biology of complex carbohydrates; glycoproteins and glycosamynoglycans; nutrition. *Mailing Add:* Dept Pathol King Saud Univ Col Med PO Box 2925 Riyadh Saudi Arabia

BAILAR, BARBARA ANN, b Monroe, Mich, Nov 24, 35; m 66; c 2. STATISTICS. *Educ:* State Univ NY Albany, AB, 56; Va Polytech Inst, MS, 65; Am Univ, PhD(statist), 72. *Prof Exp:* Math statistician, Bur Census, 58-72, chief, Res Ctr Measurement Methods, 72-79; assoc dir, Statist Standards & Methodology, 79-87; EXEC DIR, AM STATIST ASN, 88- *Concurrent Pos:* Chmn rev comt for grant to Am Statist Asn from NSF, 75-76; chmn dept math & statist, USDA Grad Sch, 75-87. *Mem:* Fel Am Statist Asn; Pop Asn Am; Int Asn Surv Statisticians; Am Asn for Pub Opinion Res; AAAS; Int Statist Inst; Int Asn Survey Samples (pres, 90-91). *Res:* Effect of different methods of measurement on final published census statistics. *Mailing Add:* Am Statist Asn 1429 Duke St Alexandria VA 22314-3402

BAILAR, JOHN CHRISTIAN, JR, b Golden, Colo, May 27, 04; m 31, 76; c 2. INORGANIC CHEMISTRY. *Educ:* Univ Colo, AB, 24, AM, 25; Univ Mich, PhD(org chem), 28. *Hon Degrees:* DSc, Univ Colo, 59, Univ Buffalo, 59 & Lehigh Univ, 73; LHD, Monmouth Col, 83. *Prof Exp:* Asst chem, Univ Mich, 26-28, from instr to prof, 28-72, secy dept, 37-54, EMER PROF CHEM, UNIV ILL, URBANA, 72- *Concurrent Pos:* With Nat Defense Res Comt, 44; treas, Int Union Pure & Appl Chem, 63-71; vis prof, Univ Colo, Ariz, Wyo, Sao Paulo, Brazil, Kyushu, Japan, Wash State, Guanajuato, Mexico & Colo Col. *Honors & Awards:* Chem Educ Award, Am Chem Soc, 61, Priestley Medal, 64, Distinguished Serv Award, 72; John R Kuebler Award, Alpha Chi Sigma, 62; Frank P Dwyer Medal, Chem Soc NSW, 65; Alfred Werner Gold Medal, Swiss Chem Soc, 66; Midwest Award, 71; Distinguished Service Award Advan Inorganic Chem, Am Chem Soc, 72; J Heyrovsky Medal, Czech Acad Sci, 78. *Mem:* Am Chem Soc (pres, 59); hon fel India Chem Soc, 74; fel Am Inst Chemists; Sigma Xi; hon mem Chem Soc Japan, 85. *Res:* Complex inorganic ions, including stereochemistry; polymerization through coordination; valence stabilization through coordination; selective catalyses by complex compounds. *Mailing Add:* 354 Noyes Lab Univ Ill Urbana IL 61803

BAILAR, JOHN CHRISTIAN, III, b Urbana, Ill, Oct 9, 32; m 66; c 4. RESEARCH ADMINISTRATION, BIOSTATISTICS. *Educ:* Univ Colo, BA, 53; Yale Univ, MD, 55; Am Univ, PhD, 73. *Prof Exp:* Field investr cancer res, Nat Cancer Inst, 56-62, head demog sect, 62-70, dir, Third Nat Cancer Surv, 67-70; dir Res Serv, US Vet Admin, 70-72; dep assoc dir cancer control, Nat Cancer Inst, 72-74, sr consult coop studies & ed-in-chief, jour, 74-80; prof biostatist, Harvard Sch Pub Health, 80-87; PROF EPIDEMIOL & BIOSTATIST, MCGILL UNIV FAC MED, 88- *Concurrent Pos:* Lectr, Sch Med, Yale Univ, 59-; assoc ed, J Nat Cancer Inst, 64-66; instr, Grad Sch, USDA, 66-76; assoc ed, Cancer Res, 68-72; mem, Task Force Cancer of Prostate & Bladder, 68-70 & Clin Cancer Training Grants Comt, 69-73; vis prof, State Univ NY Buffalo, 74-82; instr, George Washington Univ, 75-80; vis prof, Harvard Univ, 76-79; ed-in-chief, Jour Nat Cancer Inst, 74-80; statist consult, New Eng J of Med, 80-; pres, Coun Biol Ed, 87-88; McArthur fel, 90-95. *Mem:* Fel Am Statist Asn; Biomet Soc; Coun Biol Eds; Soc Risk Anal; Int Statist Inst. *Res:* Administration of medical research; cancer epidemiology; randomized clinical trials; development of related statistical methods and theory; risk assessment and environmental epidemiology. *Mailing Add:* Dept Epidemiol & Biostatist McGill Univ 1020 Pine Ave West Montreal PQ H3A 1A2 Can

BAILDON, JOHN DAVID, b Johnstown, Pa, Nov 14, 43; m 68; c 2. CONVEXITY, STARSHAPED SETS. *Educ:* Lafayette Col, BS, 65; Rutgers Univ, MS, 67; State Univ NY Binghamton, PhD(math), 71. *Prof Exp:* from instr to asst prof, 70-77, ASSOC PROF MATH, PA STATE UNIV, WORTHINGTON SCRANTON CAMPUS, 77- *Concurrent Pos:* Vis assoc prof math, Univ Okla, 85-86; reviewer, NSF, 91- *Mem:* Sigma Xi; Am Math Soc; Math Asn Am. *Res:* Studies in convexity on visibility problems concerning conditions under which sets will be star shaped or unions of star shaped sets; light open maps on 2-manifolds. *Mailing Add:* Worthington Scranton Campus Pa State Univ Dunmore PA 18512

BAILE, CLIFTON AUGUSTUS, III, b Warrensburg, Mo, Feb 8, 40; m 60; c 2. NUTRITION, PHYSIOLOGY. *Educ:* Cent Mo State Col, BS, 62; Univ Mo, PhD(nutrit), 64; Univ Pa, MA, 79. *Prof Exp:* NIH fel, Sch Pub Health, Harvard Univ, 64-66, from instr to asst prof nutrit, 66-71; sr investr res & develop, Animal Health Prod Div, Smith Kline & French Labs, 71-73, mgr neurobiol res, 73-75; from assoc prof to prof nutrit, Dept Clin Studies, Sch Vet Med, Univ Pa, 75-82; DIST FEL & DIR RES & DEVELOP, AM SCI DIV, MONSANTO CO, 82- *Concurrent Pos:* Lectr, Univ Pa & res assoc, Monell Chem Senses Ctr, 71-75; adj assoc prof, Pa State Univ, 71-81, adj prof, 81- & adj prof, Wash Univ, 82- *Honors & Awards:* Nutrit Res Award, Am Feed Mfgrs Asn & Am Dairy Sci Asn, 79; Animal Growth & Develop Award, Am Soc Animal Sci, 89. *Mem:* Am Physiol Soc; Am Inst Nutrit; Am Dairy Sci Asn; Am Soc Animal Sci. *Res:* Regulatory mechanisms for control of food intake and energy balance. *Mailing Add:* Monsanto Co 700 Chesterfield Village Pkwy St Louis MO 63198

BAILES, GORDON LEE, b Greenwood, SC, Apr 18, 46; m 72; c 3. ARTIFICIAL INTELLIGENCE, SOFTWARE ENGINEERING. *Educ:* Clemson Univ, BS, 68, MS, 69, PhD(math sci), 72. *Prof Exp:* Asst prof, 72-77, chmn & assoc prof, 77-83, CHMN & PROF COMPUTER SCI, ETENN STATE UNIV, 83- *Concurrent Pos:* Tech chair, Computer Sci Conf, Asn Comput Mach, 88-89, abstr chair, 89-91 & 90-92, publicity chair, 89-91; team chair, Computer Sci Accreditation Comn, 89- *Mem:* Asn Comput Mach; Inst Elec & Electronics Engrs; Am Asn Artificial Intel; Spec Interest Group Artificial Intel; Spec Interest Group Computer Sci Educ. *Res:* Authored a book on computer organization and assembly language. *Mailing Add:* ETenn State Univ Box 23 830A Johnson City TN 37601

BAILEY, ARTHUR W, b Chilliwack, BC, July 22, 38; m 62; c 3. RANGE SCIENCE. *Educ:* Univ BC, BSA, 60; Ore State Univ, MS, 63, PhD(range mgt), 66. *Prof Exp:* From asst prof to assoc prof, 66-77, PROF PLANT SCI, UNIV ALTA, 77- *Concurrent Pos:* Grants, Can Dept Agr, 66-, Alta Agr Res Trust, 67-82, Nat Res Coun Can, 69-78 & Alta Farming Future, 79-; vis prof, Tex Tech Univ, 72-83 & Univ Natal, SAfrica, 81-82, Univ Witwatersrand, SAfrica, 88-89. *Mem:* Soc Range Mgt; Agr Inst Can. *Res:* Range ecology; brush control; rangeland fire; grazing effects on vegetation. *Mailing Add:* Dept Plant Sci Univ Alta Edmonton AB T6G 2M7 Can

BAILEY, BRUCE M(ONROE), b Stoneham, Mass, Nov 5, 28; m 51; c 3. MECHANICAL ENGINEERING. *Educ:* Worcester Polytech Inst, BS, 51; Northeastern Univ, MS, 75. *Prof Exp:* Aeronaut res scientist, Nat Adv Comt Aeronaut, Langley Field, Va, 51-53; sr mech engr, Arthur D Little, Inc, 53-59; sr scientist, Air Prod, Inc, 59-60; chief mech engr, Lab Nuclear Sci, Mass Inst Technol, 60-89; RETIRED. *Concurrent Pos:* Lectr, Northeast Univ, 64-65; mech engr consult. *Mem:* Am Soc Mech Engrs; Am Vacuum Soc. *Res:* Low-temperature thermodynamics applications; high-vacuum techniques; gas separation processes. *Mailing Add:* 41 Woodland St Sherborn MA 01770

BAILEY, BYRON JAMES, b Oklahoma City, Okla, Apr 5, 34; m 57; c 5. OTOLARYNGOLOGY. *Educ:* Univ Okla, BA, 55, MD, 59. *Prof Exp:* Surg intern, Univ Calif, Los Angeles, 59-60, asst resident gen surg, 60-61, resident head & neck surg, 61-62; assoc prof, 64-70, PROF OTOLARYNGOL, WEISS PROF & CHMN DEPT, UNIV TEX MED BR GALVESTON, 70- *Concurrent Pos:* USPHS fel, Univ Calif, Los Angeles, 62-64; NIH res grant, 67-; bd dirs, Am Coun Otolaryngol, 73- & Am Bd Otolaryngol, 76-; mem, Residency Rev Comt Otolaryngol, 73-79; consult, Fed Drug Admin Panel Classification of Ear, Nose & Throat Devices, 77-80; chmn, Commun Sci Study Sect, NIH, 77-79. *Mem:* AAAS; Soc Univ Otolaryngol (pres, 76-77); fel Am Acad Ophthal & Otolaryngol; fel Am Col Surg; fel Am Soc Head & Neck Surg; Sigma Xi. *Res:* Laryngeal reconstruction; tracheal transplantation. *Mailing Add:* Dept Otolaryngol Univ Tex Med Br Galveston TX 77550

BAILEY, CARL LEONARD, b Grafton, NDak, Aug 2, 18; m 42; c 2. PHYSICS. *Educ:* Concordia Col, BA, 40; Univ Minn, MA, 42, PhD(physics), 47. *Prof Exp:* Assoc scientist, Off Sci Res & Develop proj, Univ Minn, 42-43; scientist, Los Alamos Lab, Univ Calif, 43-46; with AEC, 46; PROF PHYSICS, CONCORDIA COL, 47- *Mem:* Am Phys Soc; Sigma Xi. *Res:* Development of electrostatic generators and ion sources; efficiency of nuclear reactions; neutron scattering; charged particle scattering. *Mailing Add:* 425 Valley Ave Moorhead MN 56560

BAILEY, CARL WILLIAMS, III, b New Haven, Conn, Feb 7, 41; m 63; c 2. WOOD CHEMISTRY, PAPER CHEMISTRY. *Educ:* State Univ NY Col Environ Sci & Forestry & Syracuse Univ, BS, 63, MS, 65, PhD(pulp & paper), 68. *Prof Exp:* From res chemist to sr res chemist, 68-73, group leader res lignin chem, 73-74, group leader res tall oil chem, 74-79, SECT LEADER RES, CHARLESTON RES CTR, WESTVACO CORP, 79- *Mem:* Am Oil Chem Soc. *Res:* Preparation of derivatives of rosin and fatty acids; characterization of these products and finding commercial areas where they can be utilized. *Mailing Add:* Westvaco Corp PO Box 70848 Charleston Heights SC 29405-0848

BAILEY, CARROLL EDWARD, b Lewiston, Maine, Jan 14, 40; m 75; c 2. SOLID STATE PHYSICS. *Educ:* Bates Col, BS, 62; Dartmouth Col, MA, 64, PhD(physics), 68. *Prof Exp:* Nat Res Coun-Naval Res Lab res assoc physics, US Naval Res Lab, 68-70; PROF PHYSICS, UNIV PR, MAYAGUEZ, 70- *Mem:* Am Phys Soc; Am Asn Physics Teachers; Sigma Xi. *Res:* Electron spin resonance and electron-nuclear double resonance in the study of defect centers in solids and phase transitions. *Mailing Add:* Dept Physics Col Arts & Sci Univ PR Mayaguez PR 00708

BAILEY, CATHERINE HAYES, b New Brunswick, NJ, May 9, 21. POMOLOGY. *Educ:* Douglass Col, BA, 42; Rutgers Univ, PhD(fruit breeding), 57. *Prof Exp:* Tech asst, Agr Exp Sta, Rutgers Univ, 48-54, res assoc, 54-57, asst prof pomol, 57-66, assoc res prof, 66-72, prof, 72-80. *Concurrent Pos:* Exchange scientist, Res Sta, Can Dept Agr, BC, 65-66. *Mem:* Am Soc Hort Sci; Am Pomol Soc; Torrey Bot Club; Int Soc Hort Sci. *Res:* Fruit breeding; breeding for disease resistance; peaches, apricots, apples; inheritance of season of ripening in progenies from certain early ripening peach varieties and selections. *Mailing Add:* 71 Mountain Terr Bristol VT 05443

BAILEY, CECIL DEWITT, b Ayers, Miss, Oct 25, 21; m 42; c 2. DYNAMICS. *Educ:* Miss State Univ, BS, 51; Purdue Univ, MS, 54, PhD(trapezoidal plates), 62. *Prof Exp:* Instr B-47 bomber sch, USAF, 51-52, eng mech & aircraft structure, Air Force Inst Technol, 54-56, asst prof, 56-58, chief aerospace res, Air Force Systs Command Off, Langley Res Ctr, NASA, 59-63, aeronaut engr, Hq Air Force Systs Command, 64-65, assoc prof mech, Air Force Inst Technol, 66-67; from assoc prof to prof, 67-85, EMER PROF AERONAUT & ASTRONAUT ENG, OHIO STATE UNIV, 85- *Concurrent Pos:* Dir, USAF-ASEE Summer Fac Res Prog, Wright-Patterson AFB, Ohio, 76-78. *Mem:* Soc Exp Stress Anal; Am Soc Eng Educ; Am Acad Mech; Soc Natural Philos. *Res:* Alternatives to the theory of differential equations; systems dynamics; particles, rigid and deformable body mechanics; dynamics of thermally stressed structures. *Mailing Add:* Dept Aeronaut Eng Ohio State Univ Columbus OH 43210

BAILEY, CHARLES BASIL MANSFIELD, b Vancouver, BC, Feb 10, 30; m 57; c 3. ANIMAL PHYSIOLOGY, NUTRITION. *Educ:* Univ BC, BSA, 54, MSA, 56; Univ Reading, PhD(animal physiol), 59. *Prof Exp:* Res scientist animal physiol, Can Dept Agr, 59-90; RETIRED. *Concurrent Pos:* Nat Res Coun Can fel, 59-60. *Honors & Awards:* Can Asn Animal Breeders Medal, 83. *Mem:* Can Soc Animal Sci; Brit Nutrit Soc. *Res:* Renal physiology, particularly urinary calculi formation in cattle; effect of diet on muscle protein synthesis in cattle. *Mailing Add:* 4235 Frederick Rd RR 2 C-18 Armstrong BC V0E 1B0 Can

BAILEY, CHARLES EDWARD, b Johnson City, Tenn, May 25, 32; m 58; c 3. PHYSICAL CHEMISTRY. *Educ:* Univ Tenn, BS, 53, MS, 55, PhD(phys chem), 58. *Prof Exp:* Sr physicist, Polymer Intermediates Dept, Savannah River Lab, E I du Pont de Nemours & Co, Inc, 58-69, sr res supvr comput appln div, 69-74, staff chemist, Environ Anal Planning Div, Savannah River Lab, 74-90; RETIRED. *Mem:* Am Chem Soc; Am Nuclear Soc. *Res:* Environmental management; nuclear reactor safety; environmental sciences. *Mailing Add:* 563 Coker Spring Rd Aiken SC 29801

BAILEY, CHARLES LAVON, b Sterling, Okla, June 26, 42; m 63; c 2. MEDICAL ENTOMOLOGY. *Educ:* Okla State Univ, BS, 65, MS, 66, PhD(entom), 68. *Prof Exp:* Chief entom br, US Army Med Res & Develop Command, 68-70; res entomologist, SEATO Med Res Lab, 71-75; res entomologist, Walter Reed Inst Res, 70-71 & 75-; DEP COMDR RES, US ARMY MED RES INST INFECTIOUS DIS, 88- *Mem:* Am Soc Trop Med Hyg; Entom Soc Am; Am Mosquito Control Asn. *Res:* Mosquito biology, physiology, and overwintering ecology as they relate to disease transmission. *Mailing Add:* Dep Comdr US Army Med Res Inst Infectious Dis Ft Detrick MD 21702-5011

BAILEY, CLAUDIA F, TELEOST NEURULATION, ENDOCYTOSIS. *Educ:* Bryn Mawr Col, PhD(develop biol), 70. *Prof Exp:* ASSOC PROF ZOOL, UNIV ARK, 70- *Mailing Add:* 1008 Lakeside Dr Fayetteville AR 72701

BAILEY, CLETA SUE, b Bartlesville, Okla, Oct 8, 45; m 75. VETERINARY MEDICINE. *Educ:* Okla State Univ, BS, 67, DVM, 70; Univ Calif, Davis, PhD(comp path), 77, Am Col Vet Internal Med, dipl. *Prof Exp:* Intern small animal med & surg, Univ Calif, Davis, 70-71, resident neurol & neurosurg, 71-73, asst prof, 78-84, ASSOC PROF NEUROL & NEUROSURG, UNIV CALIF, DAVIS, 84- *Concurrent Pos:* Gen res support fel & grant, Univ Calif, Davis, 74-76; Gianni Found fel, 76-77; biomed res support grant, 76-77; Nat Inst Neurol & Commun Dis & Stroke fel, 77-78. *Mem:* Am Vet Med Asn; Am Vet Neurol Asn. *Res:* Cutaneous innervation of canine and feline. *Mailing Add:* Dept Surg Univ Calif Davis CA 95616

BAILEY, CURTISS MERKEL, b Cleveland, Ohio, Aug 31, 27; m 54; c 4. ANIMAL BREEDING. *Educ:* Univ Wis, BS, 52, PhD(genetics, animal husb), 60; Tex A&M Univ, MS, 54. *Prof Exp:* From asst prof to assoc prof, 60-71, prof & geneticist, 71-90, EMER PROF ANIMAL SCI, UNIV NEV, RENO, 90- *Concurrent Pos:* Ed-in-chief, J Animal Sci. *Mem:* Am Genetic Asn; Am Soc Animal Sci; Sigma Xi. *Res:* Selection responses; crossbreeding; bull beef concept; total life-cycle beef systems. *Mailing Add:* Animal Sci Dept Univ Nev Reno NV 89557

BAILEY, DANA KAVANAGH, b Clarendon Hills, Ill, Nov 22, 16. GEOPHYSICS, SYSTEMATIC BOTANY. *Educ:* Univ Ariz, BS, 37; Oxford Univ, BA, 40, MA, 43, DSc, 67. *Prof Exp:* Observ asst, Steward Observ, Univ Ariz, 33-37; astronr, Hayden Planetarium, Am Mus Natural Hist, New York, 37; physicist, Anarctic Serv, US Dept Interior, 40-41; proj engr, Rand, Douglas Aircraft Co, Calif, 46-48; consult to chief, Cent Radio Propagation Lab, Nat Bur Standards, 48-55; sci dir, Page Commun Engrs, Inc, DC, 55-59; consult, Space Environ Lab, Environ Res Labs, Nat Oceanic & Atmospheric Admin, 59-78; RES ASSOC, UNIV COLO MUS, 72- *Concurrent Pos:* Mem, Hayden Planetarium-Grace Eclipse exped, Peru, 37 & US Antarctic exped, 40-41; chmn, Int Study Group Ionospheric Radio Progagation, Int Radio Consult Comt, Int Telecommun Union, 56-; hon res assoc, Rhodes Univ, 70-71. *Honors & Awards:* US Antarctic Serv Medal, 47. *Mem:* Am Astron Soc; fel Am Phys Soc; Am Geophys Union; fel Royal Geog Soc; fel Royal Astron Soc; Sigma Xi. *Res:* Solar terrestial relationships; cosmic ray physics; ionosphere; radio wave propagation; phytogeography; dendrochronology; systematic and evolutionary aspects of gymnosperms. *Mailing Add:* 624 Pearl St No 403 Boulder CO 80302-5073

BAILEY, DAVID GEORGE, b New York, NY, Feb 21, 40; m 61; c 3. LEATHER CHEMISTRY, WASTE TREATMENT. *Educ:* Univ Del, BS, 63, MS, 64; Univ Vt, PhD(biochem), 67. *Prof Exp:* Res chemist, 67-80, SUPR RES CHEMIST, AGR RES SERV, USDA, PA, 80- *Mem:* Am Chem Soc; Am Leather Chemists Asn. *Res:* Chemistry of conversion of rawhides to leather; anaerobic treatment of tannery waste. *Mailing Add:* 600 E Mermaid Lane Philadelphia PA 19118-2551

BAILEY, DAVID NELSON, b Anderson, Ind, June 21, 45. CLINICAL TOXICOLOGY, LABORATORY MEDICINE. *Educ:* Indiana Univ, BS, 67; Yale Univ, MD, 73. *Prof Exp:* Clin fel, Dept Lab Med, Yale Univ, 73-75; asst resident clin path, Yale-New Haven Hosp, 75-76, chief resident, 76-77; from asst prof to assoc prof, 77-86, PROF PATH, DEPT PATH, UNIV CALIF, SAN DIEGO, 86-, CHMN, 88- *Concurrent Pos:* Clin instr, Dept Lab Med, Yale Univ, 76-77; head, div lab med, Univ Calif, San Diego, 83-89, vchmn, 85-86, actg chmn, 86-88. *Mem:* AAAS; Am Chem Soc; Am Asn Clin Chem; Acad Clin Lab Physicians & Scientist (pres, 88-89); Asn Path Chmn. *Res:* Percutaneous absorption of drugs in vivo utilizing the hairless mouse as experimental model; clinical pharmacology of drugs of abuse; epidemiology of drug overdose. *Mailing Add:* Dept Pathol Univ Calif H720T 225 Dickinson St San Diego CA 92103

BAILEY, DAVID NEWTON, b Pittsburgh, Pa, Dec 1, 41; m 61; c 3. ANALYTICAL CHEMISTRY. *Educ:* Juniata Col, BS, 63; Mass Inst Technol, PhD(anal chem), 68. *Prof Exp:* Asst prof anal chem, Gustavus Adolphus Col, 68-71; from asst prof to assoc prof anal chem, Lebanon Valley Col, 71-80; PROF CHEM, ILL WESLEYAN UNIV, 80- *Concurrent Pos:* Consult, State Farm Ins Co. *Mem:* Sigma Xi; Am Chem Soc. *Res:* Drug analysis; computer applications to chemistry. *Mailing Add:* Dept Chem Ill Wesleyan Univ Bloomington IL 61701

BAILEY, DAVID TIFFANY, b Olney, Tex, Aug 26, 42; m 66. ORGANIC CHEMISTRY, BIOCHEMISTRY. *Educ:* Univ Colo, Boulder, BA, 64; Iowa State Univ, PhD(org chem), 68. *Prof Exp:* NIH res fel & assoc chem, Yale Univ, 68-69; from asst prof to prof chem, Calif State Univ, Fullerton, 69-79; prof chem, Flathead Valley Community Col, 80-; MGR NAT PROD RES, HAUSER CHEM RES INC. *Mem:* AAAS; Am Chem Soc. *Res:* Organic chemistry and biosynthesis of natural products. *Mailing Add:* PO Box 4989 Boulder CO 80306-4989

BAILEY, DENIS MAHLON, b St Louis, Mo, Aug 14, 35; m 61; c 4. ORGANIC CHEMISTRY. *Educ:* Washington Univ, BA, 57, PhD(org chem), 61; Rensselaer Polytech Inst, MBA, 85. *Prof Exp:* NIH fel, Stanford Univ, 62-63; res assoc, 63-65, group leader, 65-68, sr res chemist, 68-74, sect head, 74-77, sr res assoc, 74-77, dir med chem, 77-82, vpres chem, 82-84, EXEC VPRES DRUG DISCOVERY, STERLING WINTHROP RES INST, 84- *Mem:* Am Chem Soc; AAAS; Am Soc Pharmacol & Exp Therapeut; Soc Magnetic Resonance Med. *Res:* Medicinal chemistry; cardiovascular and metabolic diseases; central nervous system, analgesic, antiinflammatory and antiinfective agents; diagnostic imaging agents; Ed-in-chief, Annual reports in Medicinal Chemistry. *Mailing Add:* Sterling Res Group Nine Great Valley Pkwy Malvern PA 19355-1314

BAILEY, DONALD ETHERIDGE, b Moore Co, NC, May 4, 31; m 54; c 2. SCIENCE EDUCATION, BIOLOGY. *Educ:* Univ NC, BS, 53, MEd, 58, EdD(sci educ), 62. *Prof Exp:* Teacher pub schs, NC, 53-54 & 56-60; prof sci educ, 61-69, dean gen col, 69-90, PROF SCI EDUC, E CAROLINA UNIV, 90- *Mem:* Nat Sci Teachers Asn. *Res:* Incidence and prevalence of misconceptions of matters scientific in secondary schools of North Carolina. *Mailing Add:* 214 York Rd Greenville NC 27858

BAILEY, DONALD FOREST, b Cliffside, NC, Apr 10, 39; m 61; c 2. MATHEMATICS. *Educ:* Wake Forest Univ, BS, 61; Vanderbilt Univ, MA, 63, PhD(math), 65. *Prof Exp:* From asst prof to assoc prof math, ECarolina Univ, 65-69; assoc prof, 69-78, PROF MATH, CORNELL COL, 78- *Mem:* Math Asn Am; Am Math Soc. *Res:* Iteration techniques for locating fixed points of continuous mappings. *Mailing Add:* 15103 Circle Oak St San Antonio TX 78232

BAILEY, DONALD LEROY, b Sterling, Okla, Feb 22, 39; m 61; c 2. ENTOMOLOGY. *Educ:* Okla State Univ, BS, 61, MS, 62, PhD(entom), 64. *Prof Exp:* Entomologist, Walter Reed Army Inst Res, Washington, DC, 64-66; ENTOMOLOGIST, USDA, 66- *Mem:* Entom Soc Am. *Res:* Mass rearing of insects; insect nutrition; mosquito and house fly pathology; biological control of snails. *Mailing Add:* Ceportodo 544 Tuxtla Gutierrez Mexico

BAILEY, DONALD LEROY, b Benzonia, Mich, July 29, 22; m 43. ORGANIC CHEMISTRY. *Educ:* Mich State Col, BS, 43, MS, 44; Pa State Univ, PhD(org chem), 49. *Prof Exp:* Res chemist, Linde Co, 48-55, res supvr, 55-56, asst mgr res, 56-57; mgr res, NY, 57-64, res & develop, Calif, 64-69, from asst to assoc dir, 69-78, CONSUL CHEMS & PLASTICS, SILICONES DIV, UNION CARBIDE CORP, 78- *Honors & Awards:* Schoelkopf Medal, Am Chem Soc, 68. *Mem:* AAAS; Am Chem Soc. *Res:* organosilicon chemistry; silane monomers and silicone polymers; processes for semiconductor grade silicon; synthesis and applications of silicon chemicals. *Mailing Add:* 208 W 19th St Traverse City MI 49684

BAILEY, DONALD WAYNE, b Hutchinson, Kans, Feb 28, 26. DEVELOPMENTAL GENETICS, IMMUNOGENETICS. *Educ:* Univ Calif, AB, 49, PhD(genetics), 53. *Prof Exp:* Fel, USPHS, Jackson Lab, 53-55, lab fel, 55-56, staff scientist, 56-57; asst prof, Univ Kans, 57-59; geneticist, Lab Aids Br, NIH, 59-61; assoc res geneticist, Univ Calif, San Francisco, 61-67; staff scientist, Jackson Lab, 67-70, sr staff scientist, 70-90; RETIRED. *Mem:* Genetics Soc Am. *Res:* Development of methodologies that use recombinant-inbred and bilineal congenic strains of mice to analyze complexly inherited traits, such as morphogenesis and disease resistance. *Mailing Add:* Jackson Lab 600 Main St Bar Harbor ME 04609

BAILEY, DONALD WYCOFF, b Emory, Va, Dec 17, 33; m 58; c 3. PHYSIOLOGY. *Educ:* Vanderbilt Univ, BA, 55; Emory Univ, MS, 56, PhD(biol), 58. *Prof Exp:* Prof biol & head dept sci, Tift Col, 58-62; PROF BIOL, WESTERN KY UNIV, 62- *Mem:* Am Soc Zoologists; Sigma Xi. *Res:* Chromatography of iodinated compounds; histochemical localizations of alkaline phosphatase; effects of testicular function on alloxan diabetes; physiology of exercise; lipid heat increment; metabolic cycles. *Mailing Add:* 1750 Normal Dr Bowling Green KY 42101

BAILEY, DUANE W, b Moscow, Idaho, Sept 22, 36; m 59; c 3. COMPUTER NETWORKS. *Educ:* Wash State Univ, BA, 57; Univ Ore, MA, 59, PhD(math), 61. *Hon Degrees:* MA, Amherst Col, 71. *Prof Exp:* Instr math, Yale Univ, 61-63; from asst prof to assoc prof, 63-71, chmn dept, 67-72, PROF MATH, AMHERST COL, 71- *Concurrent Pos:* Vis prof, Univ Texas, Austin, 84-85. *Mem:* Am Math Soc; Math Asn Am; London Math Soc; Asn Comput Mach; Soc Indust Appl Math. *Res:* Banach and topological algebras. *Mailing Add:* 115 Van Meter Dr Amherst MA 01002

BAILEY, EDWARD D, b Port Clinton, Pa, Sept 22, 31. ANIMAL BEHAVIOR. *Educ:* Mont State Univ, BS, 58, MS, 60; Pa State Univ, PhD(zool), 63. *Prof Exp:* Instr zool, Mont State Univ, 63-64; asst prof, Ont Agr Col, 64-69; ASSOC PROF ZOOL, UNIV GUELPH, 69- *Mem:* Fel AAAS; Animal Behav Soc; Am Orinthologist Union. *Res:* Social behavior in wild mammals and birds; development of communication and early learning. *Mailing Add:* Dept Zool Univ Guelph Guelph ON N1G 2W1 Can

BAILEY, EVERETT MURL, JR, b Big Spring, Tex, Mar 24, 40; m 61; c 2. VETERINARY TOXICOLOGY. *Educ:* Tex A&M Univ, DVM, 64; Iowa State Univ, MS, 66, PhD(physiol), 68; Am Bd Vet Toxicol, dipl, 72. *Prof Exp:* Staff pathologist, US Army Inst Environ Med, 68-70; asst prof, 70-75, assoc prof, 75-81, PROF PHYSIOL & PHARMACOL, COL VET MED, TEX A&M UNIV, 81- *Concurrent Pos:* Vis prof, Food & Drug Admin, 81; chmn, IRB, 88- *Honors & Awards:* Comnr Spec Citation, FDA. *Mem:* Am Vet Med Asn; Am Acad Vet & Comp Toxicol; Am Asn Vet Physiol & Pharmacol; Am Bd Vet Toxicol (pres, 79-82). *Res:* Chemical restraining agents in horses; experimental surgery; toxicology; anesthesiology; toxic plants; treatment of intoxications; pesticides. *Mailing Add:* Dept Physiol & Pharmacol Tex A&M Univ Col Vet Med College Station TX 77843-4466

BAILEY, F WALLACE, b Britt, Iowa, Nov 6, 29; m 50; c 3. PHYSICAL CHEMISTRY, POLYMER PROCESSING. *Educ:* Univ Okla, BS, 53, MS, 54. *Prof Exp:* Chemist, Silicones Div, Union Carbide Corp, 54-61; CHEMIST, RES & DEVELOP DEPT, PHILLIPS PETROL CO, 61- *Mem:* Soc Plastics Engrs. *Res:* Product development and application research of polymers. *Mailing Add:* 87F PRC Phillips Petrol Co Bartlesville OK 74004

BAILEY, FRED COOLIDGE, b Claremont, NH, Oct 5, 25; m 48; c 2. MECHANICS, ENGINEERING. *Educ:* Mass Inst Technol, SB, 48, SM, 49. *Prof Exp:* Res engr, Caterpillar Tractor Co, 49-52; actg tech dir, comt ship steel & ship struct design, Nat Acad Sci, 52-55; pres, Lessells & Assocs, 55-66 & Teledyne Mat Res Co, 66-76; pres, Teledyne Eng Serv, Teledyne Inc, 76-87, group exec, 83-87, consult, 87-89; RETIRED. *Honors & Awards:* Linnard

Prize, Soc Naval Archit & Marine Engrs, 71. *Mem:* Fel Soc Exp Stress Anal (treas, 61-65, vpres, 65-67, pres, 67-69); Am Soc Mech Engrs; Am Welding Soc; Soc Naval Archit & Marine Engrs. *Res:* Theoretical and experimental stress analysis; metals engineering; applied mechanics. *Mailing Add:* 48 Coolidge Ave Lexington MA 02173

BAILEY, FREDERICK EUGENE, JR, b Brooklyn, NY, Oct 8, 27; m 79. PHYSICAL CHEMISTRY. *Educ:* Amherst Col, AB, 48; Yale Univ, MS, 50, PhD(chem), 52. *Prof Exp:* Asst, Yale Univ, 50-52; res chemist, 52-59, group leader, 59-61, asst dir res & develop dept, Chem Div, 61-69, tech mgr calendering, flooring & rec prod dept, Plastics Div, 64-65, tech mgr vinyl resins, 65-68, mgr mkt res, 69-71, SR RES SCIENTIST POLYMER SCI, UNION CARBIDE CORP, 71- *Concurrent Pos:* Lectr, Kanawha Valley Grad Ctr, WVa Univ, 59-60 & Morris Harvey Col, 61-62, 65-66; chmn, Gordon Res Conf Polymers, 72 & 84, Gordon Res Conf Forms, 90 & 92; adj prof chem, Marshall Univ, 75-; prof chem eng, WVa Col Grad Studies, 81-; chmn, Div Polymer Chem, Am Chem Soc, gen secy, Macromolecular Secretariat, 78, counr, 78-89, exec-officer counr, 90-, bd dirs, 90-93. *Honors & Awards:* Chem Pioneers Award, Am Inst Chem, 87; Sci Achievement Award, Kanawha Valley Sect, Am Chem Soc, 88. *Mem:* Fel AAAS; Am Chem Soc; Am Phys Soc; fel NY Acad Sci; fel Am Inst Chem; Am Chem Soc. *Res:* Polymer chemistry and physics; polymer morphology-property relationships and polymer synthesis, polyurethanes; author of twenty-one publications, three books, sixty-eight technical articles and sixty patents. *Mailing Add:* 848 Beaumont Rd Charleston WV 25314

BAILEY, FREDRIC NELSON, b Toledo, Ohio, Jan 5, 32; div; c 5. CONTROL SYSTEMS, ROBOTICS. *Educ:* Purdue Univ, BSEE, 53; Univ Mich, MSE, 60, PhD(elec eng), 64. *Prof Exp:* Asst res engr, Cooley Electronics Lab, Univ Mich, 57-62, assoc res eng, 62-64; from asst prof to assoc prof, 64-90, PROF ELEC ENG, UNIV MINN, MINNEAPOLIS, 90- *Concurrent Pos:* Consult, Unisys Corp, 66-84, Cimcorp Inc, 85-88, Graco Inc, 89; vis assoc prof, Inst Automation, Tech Univ Warsaw, 73, 74 & 77, Univ Budapest, 80; assoc ed, Trans Automatic Control, 82-84; vis scientist, Dept Comput Sci, Tech Univ Madrid, 86, 87 & 88. *Mem:* Sr mem Inst Elec & Electronics Engrs; Soc Indust & Appl Math; Int Fedn Automatic Control. *Res:* Analog and digital control systems; computer aided design of control systems; industrial applications of control in manufacturing; machine tools and robotics. *Mailing Add:* Dept Elec Eng Univ Minn 200 Union St SE Minneapolis MN 55455

BAILEY, GARLAND HOWARD, b Giatto, WVa, Nov 6, 90; m 20; c 1. IMMUNOLOGY. *Educ:* WVa Univ, BS, 15; Johns Hopkins Univ, MD, 20, PhD, 21; Harvard Univ, MA, 65. *Prof Exp:* Instr path, Univ Wis, 21-22; instr epidemiol, Harvard Univ, 22-23, prev med & hyg, 23-25; assoc immunol, Johns Hopkins Univ, 25-26, assoc prof, 26-46, instr serol, Army Med Off, 43-45; MED & RES CONSULT BACT & IMMUNOL, 46- *Mem:* Am Soc Microbiol; Am Pub Health Asn; fel Royal Soc Health; NY Acad Sci. *Res:* Fatigue and susceptibility to infection; cold, influenza, diphtheria and pneumococcus immunity; antisera for pneumonia; respiratory immunity in rabbits; effects of antiseptics on nasal flora; accessory etiological factors of infection; measles; heterophile antigen and antibody; infectious mononucleosis; anaphylaxis; bacterial polysaccharides; organ specificity; the myxoma virus in rabbits; experimental cancer. *Mailing Add:* 378 Toll Gate Rd Warwick RI 02886

BAILEY, GEORGE WILLIAM, b Monmouth, Ill, May 14, 30; m 53; c 4. ANALYTICAL CHEMISTRY. *Educ:* Monmouth Col, BS, 52; Univ Ill, PhD(chem), 55. *Prof Exp:* Anal chemist, Esso Res & Eng Co, 55-56 & Borg-Warner Corp, 56-58; microscopist, Esso Res Labs, Humble Oil & Refining Co, 58-72, microscopist, Exxon Res & Develop Labs, Exxon Co, 72-87. *Mem:* Am Chem Soc; Electron Micros Soc Am; Microbeam analysis Soc. *Res:* Optical and electron microscopy; electron diffraction; electron probe analysis; thermal analysis; x-ray diffraction and spectrography. *Mailing Add:* Box No 108 Baton Rouge LA 71414

BAILEY, GEORGE WILLIAM, b Des Moines, Iowa, Oct 30, 33; m 58, 77; c 4. ENVIRONMENTAL QUALITY. *Educ:* Iowa State Univ, BS, 55; Purdue Univ, MS, 58, PhD(soil chem, mineral), 61. *Prof Exp:* Soil surveyor, Soil Conserv Serv, 55-56; NIH res fel, Purdue Univ, 61-64; soil-pesticide chemist, Div Water Supply & Pollution Control, USPHS, 64-66; soil phys chemist, Fed Water Pollution Control Admin, US Dept Interior, 66-70, Fed Water Qual Admin,; supvry res soil phys chemist, Southeastern Environ Lab, US Environ Protection Agency, 70-74, supvry res soil phys chemist, 74-76, assoc dir rural lands res, 76-85, RES SOIL PHYS CHEMIST, ENVIRON RES LAB, US ENVIRON PROTECTION AGENCY, ATHENS, GA, 85- *Honors & Awards:* Bronze Medal, Environ Protection Agency. *Mem:* AAAS; fel Am Soc Agron; fel Soil Sci Soc Am; Am Chem Soc; Sigma Xi. *Res:* Soil pesticide chemistry; water quality; clay and soil mineralogy; mathematical modeling and simulation of the transformation, transport and fate of pesticides, plant nutrients, hazardous wastes and toxics from agricultural and rural land; fate and transport of metals in the environment; surface chemistry of metal interactions; redox reactions at mineral surfaces. *Mailing Add:* Environ Res Lab US EPA Athens GA 30613-7799

BAILEY, GLENN CHARLES, b Pekin, Ill, Aug 21, 30; m 58; c 5. SOLID STATE PHYSICS. *Educ:* Univ Ill, BS, 52; Univ Wis, MS, 54; Cath Univ Am, PhD(physics), 65. *Prof Exp:* Physicist, Caterpillar Tractor Co, Ill, 55-57; res physicist, condensed matter & radiation sci div, US Naval Res Lab, 57-85; physicist, 85-87, CONSULT, SACHS/FREEMAN ASSOC, LANDOVER, MD, 87- *Mem:* Am Phys Soc. *Res:* Microwave interactions in materials and non-equilibrium superconductivity; ferromagnetic resonance of thin films and single crystals; radiation damage in semiconductors, ferrites and thin films; x-ray diffraction and fluorescent analysis; residual stresses in metals. *Mailing Add:* 1915 Taylor Ave Ft Washington MD 20744

BAILEY, GORDON BURGESS, b Worcester, Mass, Feb 13, 34; m 58; c 3. BIOCHEMISTRY. *Educ:* Brown Univ, BA, 56; Univ Mass, MA, 61; Univ Fla, PhD(biochem), 66. *Prof Exp:* Vis prof biochem, Field Staff, Rockefeller Found, 66-73; dep dir, Anemia & Malnutrition Res Ctr, Fac Med, Chiang Mai Univ, Thailand, 73-76; PROF BIOCHEM, MOREHOUSE SCH MED, 76- *Concurrent Pos:* Vis prof, Mahidol Univ, Bangkok, 66-73; consult human reproduction unit, WHO, 75-76. *Mem:* AAAS; Am Soc Cell Biol; Am Soc Microbiol; Am Soc Parasitol; Soc Protozool. *Res:* Motility and cytopathogenicity of Entamoeba. *Mailing Add:* Dept Biochem Morehouse Sch Med 720 Westview Dr Atlanta GA 30310

BAILEY, HAROLD EDWARDS, b Salt Lake City, Utah, July 26, 06; m 33. PHARMACOGNOSY. *Educ:* Univ Calif, Los Angeles, AB, 30, PhD(microbiol), 35. *Prof Exp:* Teaching fel bot, Univ Calif, Los Angeles, 31-34; asst forester, Nat Park Serv, 35-38, jr park naturalist, 39-42; instr bot, Univ Tenn, 38; res asst, Parke Davis & Co, 42-47; asst prof biol sci & pharmacog, 47-53, assoc prof biol sci, 53-58, prof pharmacog, 59-74, EMER PROF PHARMACOG, WAYNE STATE UNIV, 74- *Mem:* AAAS; Am Pharmaceut Asn; Am Soc Pharmacog; Soc Econ Bot; Acad Pharmaceut Sci. *Res:* Antibiotics; physiology of fungi; chemotaxonomy. *Mailing Add:* 4727 Second Detroit MI 48201-1244

BAILEY, HAROLD STEVENS, b Springfield, Mass, Apr 18, 22; m 46; c 5. PHARMACEUTICAL CHEMISTRY. *Educ:* Mass Col Pharm, BS, 44, MS, 48; Purdue Univ, PhD(pharmaceut chem), 51. *Prof Exp:* Asst chem, Mass Col Pharm, 46-48; instr pharm, Purdue Univ, 50-51; asst prof pharmacol, 51-52, from assoc prof to prof pharmaceut chem, 52-73, dean acad affairs, 61-73, dean grad sch, 65-76, vpres acad affairs, 73-85, VPRES EMER & DISTINGUISHED PROF HIGHER EDUC, SDAK STATE UNIV, 85- *Concurrent Pos:* Ed, Pharm Sect, SDak J Med & Pharm, 53-61. *Mem:* Fel AAAS; Sigma Xi; fel Am Found Pharm Educ. *Res:* Application of biochemical research to pharmacology and radiation biology; dental research. *Mailing Add:* Admin Bldg PO Box 2201 Brookings SD 57007-2726

BAILEY, HARRY HUDSON, soil science; deceased, see previous edition for last biography

BAILEY, HERBERT R, b Denver, Colo, Nov 2, 25; m 51; c 5. MATHEMATICS. *Educ:* Rose Polytech Inst, BS, 45, 46; Univ Ill, MS, 47; Purdue Univ, PhD(math), 55. *Prof Exp:* Instr math, Gen Motors Inst, 47-49; mathematician, US Naval Ord Plant, Ind, 49-55 & Denver Res Ctr, Ohio Oil Co, 56-62; assoc prof math, Colo State Univ, 62-66; prof & chmn dept, 66-75, vpres acad affairs & dean fac, 75-77, PROF MATH, ROSE-HULMAN INST TECHNOL, 77- *Mem:* Math Asn Am. *Res:* Differential difference equations; fluid flow and diffusion. *Mailing Add:* Math Dept Rose-Hulman Inst Technol 5500 Wabash Ave Terre Haute IN 47803

BAILEY, IAN L, b Melbourne, Australia, Dec 22, 40; m 63; c 2. VISION SCIENCE, ILLUMINATING ENGINEERING. *Educ:* Victorian Col Optom, LOSc, 62; Univ Melbourne, B App Sc, 63; City Univ London, FBOA HD, 67; Ind Univ, MS, 71. *Prof Exp:* Lectr optom, Victorian Col Optom, Univ Melbourne, 68-74; sr res officer, Nat Vision Res Inst Australia, 75-76; PROF OPTOM, UNIV CALIF, BERKELEY, 76- *Concurrent Pos:* Gov, Asn Blind, Melbourne, Australia, 75; rep, comt vision, Nat Acad Sci/Nat Res Coun, Am Acad Optom, 85-; chair, Low Vision Sect, Am Acad Optom, 85-86. *Honors & Awards:* Glen A Fry Award, Am Acad Optom, 86. *Mem:* Am Acad Optom; Am Optom Asn; Asn Res Vision & Ophthal. *Res:* Low vision; clinical assessment of vision; visual ergonomics; visual optics. *Mailing Add:* Sch Optom Univ Calif Berkeley CA 94720

BAILEY, J EARL, b State College, Miss, Apr 16, 33; m 58; c 2. AERONAUTICS, ASTRONAUTICS. *Educ:* Miss State Univ, BS, 55, MS, 58. *Prof Exp:* Sr aerophys engr, Gen Dynamics/Ft Worth, 58-62; assoc prof, 63-81, PROF AEROSPACE ENG, UNIV ALA, TUSCALOOSA, 81-, DIR, FLIGHT DYNAMICS LAB, 77- *Concurrent Pos:* NSF fac fel, Mass Inst Technol, 66-69; sabbatical leave, Boeing Com Airplane Co, 85-86. *Mem:* Am Inst Aeronaut & Astronaut; Soc Automotive Engrs. *Res:* Flight dynamics; automatic control of aircraft and space vehicles; aeroelasticity; flight simulation. *Mailing Add:* Dept Aerospace Eng Univ Ala Box 2901 Tuscaloosa AL 35487

BAILEY, JACK CLINTON, b Austin, Tex, Aug 10, 36; m 61; c 2. ENTOMOLOGY. *Educ:* Tex A&M Univ, BS, 59; Miss State Univ, MS, 65, PhD(entom), 67; Am Registry Prof Entomologists, cert. *Prof Exp:* Jr entomologist, Tex Agr Exp Sta, Weslaco, 59-63; res asst entom, Miss State Univ, 63-67; res entomologist, Soybean Insects, Agr Res Serv, USDA, 67-71 & res leader quarantine insects res, 71-77, res entomologist, cotton host plant resistance, 78-89; CONSULT, 89- *Concurrent Pos:* Adj assoc prof entom & assoc mem grad coun, Miss State Univ, 68- *Mem:* Sigma Xi; Entom Soc Am. *Res:* Biological control both of plants and insects; pest management systems. *Mailing Add:* Agr Res Serv USDA PO Box 225 Stoneville MS 38776

BAILEY, JAMES ALLEN, b Chicago, Ill, May 14, 34; m 59; c 2. UNGULATE MANAGEMENT. *Educ:* Mich Technol Univ, BS, 56; State Univ NY Col Forestry, MS, 58, PhD(zool), 66. *Prof Exp:* Res assoc ecol, Ill Natural Hist Surv, 64-68; asst prof wildlife biol, Univ Mont, 68-69; from asst prof to assoc prof, 69-77, PROF WILDLIFE BIOL, COLO STATE UNIV, 80- *Mem:* Wildlife Soc. *Res:* Population dynamics; ungulate behavior, foraging efficiency; habitat requirements of mountain ungulates. *Mailing Add:* Dept Fishery & Wildlife Biol Colo State Univ Ft Collins CO 80523

BAILEY, JAMES E(DWIN), b Great Falls, Mont, Feb 28, 44; m 66; c 1. CHEMICAL ENGINEERING. *Educ:* Rice Univ, BA, 66, PhD(chem eng), 69. *Hon Degrees:* Dr, Univ Brussel, 88. *Prof Exp:* Engr, Shell Develop Co, Calif, 69-71; from asst prof to prof chem eng, Univ Houston, 71-80, assoc dean fac res, 76-78; PROF CHEM ENG, DEPT CHEM ENG, CALIF INST TECHNOL, PASADENA, 80-, CHEVRON PROF, 89- *Concurrent Pos:* Distinguished biotechnol lectr, Univ Fla, 88; distinguished vis lectr, Univ

Alta, 88. *Honors & Awards:* Allan T Colburn Award, Am Inst Chem Eng, 79, Prof Progress Award, 87; Curtis W Mc Graw Award, Am Soc Eng Educ, 83; Marvin J Johnson Award, Am Chem Soc, 90; Cert of Recognition, NASA, 90; John C & Florence W Holtz Lectr, Johns Hopkins Univ, 90; Merck Distinguished Lectr, Rutgers Univ, 91. *Mem:* Nat Acad Eng; Am Inst Chem Engrs; Sigma Xi; Am Chem Soc; fel AAAS. *Res:* Enzyme and microbial reaction engineering; catalytic reactor dynamics. *Mailing Add:* Dept Chem Eng 210-41 Calif Inst Technol Pasadena CA 91125

BAILEY, JAMES EDWARD, b Detroit, Mich, Jan 16, 42; m 66; c 2. INDUSTRIAL ENGINEERING. *Educ:* Wayne State Univ, BS, 64, MS, 66, PhD(indust eng), 75. *Prof Exp:* Mfg engr, Gen Motors Corp, 65-67; staff engr, Nat Bank Detroit, 67-69; instr, Wayne State Univ, 69-72, dir co-op educ, 72-74; PROF INDUST ENG, ARIZ STATE UNIV, 74- *Concurrent Pos:* Mem bd & actg chief exec officer, Kurta Corp, 81. *Mem:* Am Inst Indust Engrs (vpres); Inst Mgt Sci; Soc Info Mgt. *Res:* Concurrent engineering; production control; data systems management; data systems analysis; expert systems. *Mailing Add:* Dept Indust Eng Ariz State Univ Tempe AZ 85287

BAILEY, JAMES L, b Tiffin, Ohio, Mar 21, 30; m 52; c 3. APPLIED MATHEMATICS. *Educ:* Heidelberg Col, BS, 52; Mich State Univ, MS, 54, PhD(math), 58. *Prof Exp:* Instr math, Mich State Univ, 58; asst prof, Case Inst Technol, 58-63; assoc prof, 63-73, PROF MATH, UNIV TOLEDO, 73- *Mem:* Math Asn Am. *Res:* Mechanics. *Mailing Add:* 2423 Middlesex Toledo OH 43606

BAILEY, JAMES STEPHEN, b Paris, Tex, Oct 3, 42. ELECTRICAL ENGINEERING, PHYSICS. *Educ:* Univ Tex, BS, 65, MS, 68, PhD(elec eng), 71; Southern Methodist Univ, MS, 82. *Prof Exp:* Aerosysts engr analog comput, Gen Dynamics Corp, 64-66; res engr microwave res, Collins Radio Co, 67-68; eng specialist high frequency technol, Vought Corp, 73-78; CONSULT DEFENSE INDUST, 78- *Concurrent Pos:* Res scientist & adj prof, Strategic Defense Initiate Prog, Univ Tex, Arlington, 85-89; proj engr advan avionics, A-12 Navy Stealth Aircraft Prog, Gen Dynamics, Ft Worth, 89-91. *Mem:* Inst Elec & Electronics Engrs; Am Inst Physics. *Res:* Microwave devices; electromagnetic wave propagation and antenna theory; radar systems; electronic warfare systems. *Mailing Add:* PO Box 399 Palo Pinto TX 76072

BAILEY, JAMES STUART, b Chicago, Ill, Dec 23, 21; m 45. GEOLOGICAL OCEANOGRAPHY. *Educ:* Univ Wash, BS, 50; Northwestern Univ, MS, 53; Univ Ariz, PhD(geol), 55. *Prof Exp:* Explor geologist, Calif Explor Co, Standard Oil Co, Calif, 55-60; independent consult geol, 60-64; geol oceanogr, US Naval Oceanog Off, 64-67; res oceanogr fisheries, Bur Com Fisheries, Dept Interior, 67-68; sr eng scientist electro-optics, TRW Systs, 68-70; PROF DIR GEOG, OFF NAVAL RES, DEPT NAVY, 70- *Concurrent Pos:* Geol oceanogr, Spacecraft Oceanog Proj, US Naval Oceanog Off, 65-67; mem, Am Cong Surv & Mapping Marine Surv & Mapping Comt, 75-78. *Honors & Awards:* Sustained Superior Achievement Award, US Naval Oceanog Off, 67. *Mem:* Am Geophys Union; Am Soc Photogram. *Res:* Research on marine geology and coastal processes including sea floor geology and structure; paleostratigraphic and structural reconstructions. *Mailing Add:* 6007 Williamsburg Blvd Arlington VA 22207

BAILEY, JOHN ALBERT, b Liverpool, Eng, June 8, 37; m 63. METALLURGY, METALLURGICAL ENGINEERING. *Educ:* Univ Col Swansea, Wales, BSc, 60, PhD(metall), 63. *Prof Exp:* Res asst metall, Imp Chem Industs, Ltd, Eng, 56-57 & Univ Col Swansea, Wales, 57-63; from asst prof to assoc prof mech eng, Ga Inst Technol, 63-67; assoc prof, 67-72, PROF MECH ENG, NC STATE UNIV, 72-, HEAD DEPT MECH & AERO ENG, 84- *Mem:* Brit Inst Metall; fel Inst Metallurgists; Soc Mfg Engrs; Metals Soc. *Res:* Properties of materials at elevated temperatures; mechanics of metal machining; wear, surface integrity in machining; friction. *Mailing Add:* Dept Mech & Aerospace Eng NC State Univ Raleigh NC 27695-7910

BAILEY, JOHN CLARK, b Newport, Vt, Aug 25, 42; m 67; c 4. SURFACE CHEMISTRY, ELECTROCHEMISTRY. *Educ:* Cornell Univ, BA, 64; Duke Univ, PhD(phys chem), 69. *Prof Exp:* Asst, Manchester Univ, 69-71; asst prof chem, Univ NC, Charlotte, 74-75 & Swarthmore Col, 75-76; TECH ASSOC BATTERY PRODS, UNION CARBIDE CORP, 76- *Mem:* Electrochem Soc. *Res:* Battery research and development. *Mailing Add:* Eveready Battery Co PO Box 45035 Westlake OH 44145

BAILEY, JOHN MARTYN, b Hawarden, Eng, May 13, 29; US citizen; m 63; c 3. BIOCHEMISTRY. *Educ:* Univ Wales, BSc, 49, PhD(biochem), 52. *Hon Degrees:* DSc, Univ Wales, 70. *Prof Exp:* Res assoc, Iowa State Univ, 54-55; res assoc physiol chem, Johns Hopkins Univ, 55-56, res assoc biochem, Dept Surg, Med Sch, 56-59; from asst prof to assoc prof, 59-69, PROF BIOCHEM, SCH MED, GEORGE WASHINGTON UNIV, 69- *Concurrent Pos:* Fel chem, Prairie Regional Lab, Nat Res Coun Can, 52-54; USPHS career develop award, 63-73; consult, Div Res Grants, USPHS, 69- & NSF, 74-; fel coun arteriosclerosis, Am Heart Asn, 69-, prog comnr, 75-77; vis prof biochem, Univ Miami Sch Med, 71, 78 & 85; pres, Convirex Inc. *Mem:* Am Heart Asn; AAAS; Am Soc Biol Chem; Soc Exp Biol & Med. *Res:* Lipid and lipoprotein metabolism in cultured mammalian cells, genetic disorders, essential fatty acids, prostaglandins and thromboxanes; experimental atherosclerosis; immune cell functions; anti-viral drugs. *Mailing Add:* Dept Biochem George Washington Univ Sch Med Washington DC 20037

BAILEY, JOSEPH RANDLE, b Fairmont, WVa, Sept 17, 13; m 46; c 2. ZOOLOGY. *Educ:* Univ Mich, AB, 35, PhD(zool), 40; Haverford Col, MA, 37. *Prof Exp:* Int exchange fel, Brazil, 40-41 & fel, 41-42; from instr to prof, 46-83, EMER PROF ZOOL, DUKE UNIV, 83- *Concurrent Pos:* Fulbright lectr, Sao Paulo Univ, 61-62; Guggenheim Mem Found fel, 53-54; Fulbright sr scholar, James Cook Univ NQueensland, Townsville, 71. *Mem:* Soc Study Amphibians & Reptiles; Herpetologists League; AAAS; Am Soc Ichthyol & Herpet (pres, 72). *Res:* Systematics; life history and distribution of herpetology and ichthyology; zoogeography. *Mailing Add:* Dept Zool Duke Univ Durham NC 27706

BAILEY, JOSEPH T(HOMAS), b Clemson, SC, Oct 16, 37; m 62. CERAMICS ENGINEERING. *Educ:* Clemson Univ, BS, 59, MS, 60; Ohio State Univ, PhD(ceramic eng), 66. *Prof Exp:* Metallurgist, Savannah River Lab, E I du Pont de Nemours & Co, 60; proj engr, Chattanooga, 60-64, res supvr oxide mat, 66-74, tech dir, Tech Ceramic Prod Div, 74-81, dept mgr, Tech Ceramic Resources Dept, 81-82, dept mgr, Ceramic Mat Dept, 83-89, VPRES RES & DEVELOP, I & E SECTOR, 3M CORP, ST PAUL, 89- *Concurrent Pos:* Mem, indust adv comt, dept ceramic eng, Ohio State Univ. *Mem:* Fel Am Ceramic Soc; Nat Inst Ceramic Engrs; Sigma Xi; Int Soc Hybrid Microelectronics. *Res:* Electronic materials; ceramic substrates; mechanical ceramics; polymeric inorganic composites; technical management; ceramic fibers; sol-gel ceramics. *Mailing Add:* 9800 Janero Ct Mahtomedi MN 55115

BAILEY, KINCHEON HUBERT, JR, b Zebulon, NC, Dec 21, 21; m 48; c 5. RADIO FREQUENCY COMMUNICATION, MICROWAVES. *Educ:* US Mil Acad, West Point, BS, 45; Pa State Univ, MEE, 67; NC State Univ, EdD, 75. *Prof Exp:* Instr electronic eng, 67-77, HEAD, ELEC ENG DEPT, WAKE TECH COMMUNITY COL, 78- *Concurrent Pos:* Consult, John Wiley Publ, 70- & MacMillan, McGraw-Hill Publ, 80- *Honors & Awards:* Wiley Award, Am Soc Eng Educ, 88. *Mem:* Inst Elec & Electronics Engrs; Am Soc Eng Educ. *Res:* Radio frequency communication; microwaves. *Mailing Add:* 701 Currituck Dr Raleigh NC 27609

BAILEY, LEO L, b Blossom, Tex, Jan 17, 22; m 45; c 4. HORTICULTURE, ENTOMOLOGY. *Educ:* Tex A&M Univ, BS, 43; Sam Houston State Teachers Col, MS, 49; La State Univ, PhD(hort), 54. *Prof Exp:* Instr voc agr, Red River County Voc Sch, Tex, 46-48; instr hort, Sam Houston State Teachers Col, 48-49; from asst prof to prof, Tex A&I Univ, 49-74, for student adv, 65-73, teacher agr, 75-88; RETIRED. *Honors & Awards:* Hon Agr Educ Award, Am Soc Hort Sci. *Mem:* Am Soc Hort Sci. *Res:* Maturity test for vegetable crops; propagation of ornamental plants; teaching techniques. *Mailing Add:* 431 W King Kingsville TX 78363

BAILEY, LEONARD CHARLES, b Jersey City, NJ, Feb 5, 36; m 64; c 2. PHARMACEUTICAL CHEMISTRY. *Educ:* Fordham Univ, BS, 57; Rutgers Univ, MS, 65, PhD(pharmaceut sci), 69. *Prof Exp:* Sci drug analyst, Ortho Pharmaceut Corp, 69-73; ASST PROF PHARMACEUT CHEM, RUTGERS UNIV, NEW BRUNSWICK, 73- *Mem:* Am Pharmaceut Asn; Am Chem Soc; Sigma Xi. *Res:* Application of modern chromatographic techniques to the analysis of drugs in formulations and in biological fluids. *Mailing Add:* 48 Fieldstone Dr Somerville NJ 08876

BAILEY, LEONARD LEE, b Takoma Park, Md, Aug 28, 42; m 66. CARDIO-THORACIC SURGERY. *Educ:* Columbia Union Col, BS, 64; Loma Linda Univ, MD, 69. *Prof Exp:* ASST PROF SURG, ASST PROF PEDIAT & DIR PEDIAT CARDIAC SURG, MED CTR, LOMA LINDA UNIV, 76- *Mem:* Am Col Surgeons. *Res:* Transplant whole organ grafting, such as heart, lung and kidney. *Mailing Add:* Dept Cardio-Thoracic Surg Loma Linda Univ Loma Linda CA 92350

BAILEY, LESLIE EDGAR, b Los Angeles, Calif, Oct, 25, 31. PHARMACOLOGY. *Educ:* Utah State Univ, BS, 59; Univ Utah, PhD(pharmacol & biophys), 66. *Prof Exp:* Asst pharmacologist, Riker Labs, 59-60; asst statistician, Thiokol Chem Corp, 60-61; postdoctoral pharmacol, Univ Man, 66-68, from asst prof to assoc prof, 68-76; from assoc prof to prof, Dalhousie Univ, 76-80; PROF PHARMACOL, UNIV HAWAII, 80- *Concurrent Pos:* Res fel, Can Heart Found, 68-71, res scholar, 71-76, mem, Sci Subcomt, 74-80, chmn, Personnel Support Comt, 75-79, vchmn, Sci Res Comt, 77-80; counr, Pharmacol Soc Can, 76-79 & Western Pharmacol Soc, 83-85; mem, Prog Comt, Am Soc Pharmacol & Exp Therapeut, 76-84 & Pulmonary & Cardiovasc Study Sect, 77; field ed, J Pharmacol & Exp Therapeut, 77 & 84-; Med Res Coun vis prof, Univ BC, 79. *Mem:* Am Soc Pharmacol & Exp Therapeut; Int Soc Heart Res; Can Cardiovasc Soc. *Res:* Author of numerous technical publications. *Mailing Add:* Dept Pharmacol Sch Med Univ Hawaii 1960 East-West Rd Honolulu HI 96822

BAILEY, LORAINE DOLAR, b St Vincent, Wis, Sept 5, 36; Can citizen; m 62; c 3. SOIL CHEMISTRY, PLANT NUTRITION. *Educ:* Univ Man, BSA, 65, MSc, 67; Univ Guelph, PhD(soil-plant relationship), 71. *Prof Exp:* Res scientist soil-plant relationship, 66-78, HEAD PLANT & SOIL SCI RES, BRANDON RES STA, AGR CAN, 78- *Mem:* Can Soc Soil Sci; Int Soc Soil Sci; Am Soc Agron; Am Soc Soil Sci; Agr Inst Can; Can Soc Agron. *Res:* Nitrogen fixation by legumes and the value of fixed nitrogen to cereals; placement of phosphorus with crop seeds and the residual value of fertilizer phosphorus; sulpher status of chernozemic soils and methods of increasing soil sulphate-sulphurs. *Mailing Add:* Brandon Res Sta Agr Can Box 610 Brandon MB R7A 5Z7 Can

BAILEY, LYNN BONNETTE, b Charleston, SC, Aug 25, 48; m 70. HUMAN NUTRITION. *Educ:* Winthrop Col, BA, 70; Clemson Univ, MS, 72; Purdue Univ, PhD(nutrit), 75. *Prof Exp:* Res asst nutrit, Clemson Univ, 70-73; res instr, Purdue Univ, 73-75; ASST PROF NUTRIT, UNIV FLA, 77- *Concurrent Pos:* Assoc, Dept Nutrit, Purdue Univ, 75-76; affil asst prof, Col Health Related Professions, Univ Fla, 77- *Mem:* Assoc Am Inst Nutrit; Am Dietetic Asn; Inst Food Technologists; Sigma Xi. *Res:* Evaluation of the etiology of nutritional anemia in low-income adolescent and elderly people. *Mailing Add:* 403 Food Sci Bldg Univ Fla Gainesville FL 32611

BAILEY, MARION CRAWFORD, b Carrollton, Miss, Oct 26, 37; m 63; c 2. ELECTROMAGNETICS, ANTENNAS. *Educ:* Miss State Univ, BS, 64; Univ Va, MS, 67; Va Polytech Inst & State Univ, PhD(eng), 72. *Prof Exp:* RES ENGR ELECTROMAGNETICS, NASA, 64- *Concurrent Pos:* Vis prof, Univ Miss, 78-79. *Honors & Awards:* Sci Achievement Medal, NASA. *Mem:* Inst Elec & Electronics Engrs. *Res:* Antennas; microwaves. *Mailing Add:* Seven Cortez Ct Hampton VA 23666

BAILEY, MAURICE EUGENE, b Portsmouth, Ohio, Mar 14, 16; m 37, 74; c 4. SCIENCE EDUCATION, MINING. *Educ:* Ohio Wesleyan Univ, AB, 37; Purdue Univ, MS, 39, PhD, 41. *Prof Exp:* Res chemist, Purdue Univ Res Found, 40-41; res chemist, Allied Chem Corp, 41-55, mgr appln res, 55-59, tech liaison, 59-63, chem planning, 63-68, dir res, 68-70; dir mining technol & chmn, Div Sci & Technol, Pikeville Col, 71-81, prof chem, 70-87; dir coal develop, Ky dept Energy, 81-82, ENVIRON ENGR, KY PARTNERS, 88- *Concurrent Pos:* Consult, Ky Coun Pub Higher Educ, 75. *Mem:* AAAS; Am Chem Soc. *Res:* Author of 85 publications. *Mailing Add:* 100 Poplar St Pikesville KY 41501

BAILEY, MICHAEL JOHN, b Philadelphia, Pa, Oct 16, 53. COMPUTER GRAPHICS, COMPUTER AIDED DESIGN. *Educ:* Purdue Univ, BSME, 75, MSME, 76, PhD(comput graphics), 79. *Prof Exp:* Tech staff, Sandia Nat Labs, 79-81; asst prof comput graphics, Purdue Univ, 81-85; dir advan develop, Megatek Corp, 85-89; MGR SCI VISUALIZATION, SAN DIEGO SUPERCOMPUTER CTR, 89- *Concurrent Pos:* Sr instr, Interactive Comput Systs, 81- *Honors & Awards:* Ralph Teetor Award, Soc Automotive Engrs, 83. *Mem:* Asn Comput Mach; Am Soc Mech Eng; Nat Comput Graphis Asn. *Res:* Creation and use of high performance computer graphics displays to solve scientific visualization problems, particularly in the physical sciences. *Mailing Add:* San Diego Supercomputer Ctr PO Box 85608 San Diego CA 92186-9784

BAILEY, MILTON, b New York, NY, May 20, 17; m 54; c 1. INDUSTRIAL CHEMISTRY, LEATHER CHEMISTRY. *Educ:* City Col NY, BA, 40, MS, 49; Pratt Inst, cert, 51. *Prof Exp:* Ed newspaper & educr, Adj Gen Off, US Dept Army, 41-43; chem supt, Ruderman, Inc, NY, 46-52; leather chemist, US Naval Supply Res & Develop Facil, NJ, 52-66, footwear specialist, 66-67; leather chemist & opers res analyst, US Naval Clothing Res & Develop Unit, Mass, 67; leather chemist, footwear specialist & phys scientist, US Navy Clothing & Textile Res Facil, 67-89; SELF-EMPLOYED, 89- *Concurrent Pos:* Lectr, City Col NY, 51-52 & NY Community Col, 53-61; arbitrator, Am Arbit Asn, 73-; chmn elec hazard & conductive safety shoe comt, Am Nat Standards Inst, 73-; chmn, safety & traction for footwear, Am Soc Testing & Mat, 78-82. *Mem:* Am Leather Chem Asn; Am Soc Testing & Mat. *Res:* Impregnation of leather and protein fibers; coloring of hair and protein fibers; instrumentation; leather and footwear technology; chemical specialties; static electricity; conductive rubber; personnel protection asbestos fibers; traction; slips and falls; athletic and general footwear; walkways; environmental health; human factors. *Mailing Add:* 18 Bayfield Rd Wayland MA 01778

BAILEY, MILTON (EDWARD), b Shreveport, La, June 7, 24; m 58. FOOD CHEMISTRY. *Educ:* Tulane Univ, BS, 49; La State Univ, MS, 53, PhD(biochem), 58. *Prof Exp:* Res assoc, La State Univ, 53-58; asst prof animal husb, 58-60, assoc prof food sci & nutrit, 60-69, PROF FOOD SCI & NUTRIT, UNIV MO-COLUMBIA, 69- *Concurrent Pos:* Fulbright-Hays fel, Agr Res Serv, USDA, 74-75 & Univ Col Cork, Ireland, 75-76. *Mem:* Am Chem Soc; Inst Food Technologists. *Res:* Muscle physiology and biochemistry; flavoring constituents of foods; enzymology. *Mailing Add:* Dept Food Sci & Nutrit Univ Mo 21 Agr Bldg Columbia MO 65211

BAILEY, NORMAN SPRAGUE, biology; deceased, see previous edition for last biography

BAILEY, ORVILLE TAYLOR, b Jewett, NY, May 28, 09. PATHOLOGY. *Educ:* Syracuse Univ, AB, 28; Albany Med Col, MD, 32. *Prof Exp:* Rotating intern, Albany Hosp, NY, 32-33; house officer path, Peter Bent Brigham Hosp, Boston, 33-34; resident, Children's Hosp, 34-35; from instr path to asst prof path, Harvard Med Sch, 35-51; prof neuropath, Sch Med, Ind Univ & chief neuropath, Larue Carter Mem Hosp, 51-59; prof neurol, 59-70, prof neurol surg, 70-72, prof neuropath, 72-77, EMER PROF NEUROPATH, ABRAHAM LINCOLN SCH MED, UNIV ILL COL MED, 77-, DISTINGUISHED PROF PATH, RUSH MED COL, 77- *Concurrent Pos:* Guggenheim fel, Univ Cambridge, 46-47; resident, Peter Bent Brigham Hosp, 35-37, assoc pathologist, 40-43, assoc pathologist neuropath, 46-51; jr fel, Soc Fels, Harvard Univ, 37-40; lectr, Lowell Inst, 40; mem, Off Sci Res & Develop, 41-46; neuropathologist & vchmn, Neurol Inst, Children's Hosp, 47-51; chmn deleg & vpres, US Int Cong Neuropath, 59-67. *Mem:* Am Asn Neurol Surgeons; AMA; Am Asn Pathologists; Sigma Xi. *Res:* Dural sinus thrombosis; alloxan diabetes; myelin degeneration; brain tumors; radioactive materials in brain; pediatric neuropathology; hydrocephalus. *Mailing Add:* PO Box 2517 Chicago IL 60690

BAILEY, PAUL BERNARD, mathematics, for more information see previous edition

BAILEY, PAUL TOWNSEND, b Sydney, Australia, Nov 14, 39; m 65; c 2. PHYSICS. *Educ:* Univ New Eng, Australia, BSc, 62; Mass Inst Technol, ScD(physics), 66; Harvard Bus Sch, PMD, 73. *Prof Exp:* Sr res physicist, Monsanto Co, 66-69; group leader, Electronic Prod Div, 69-72, mgr com develop, 72-75, mgr New Ventures, 75-81; venture capitalist, Advent Ltd, 82-84; VENTURE CAPITALIST, BARING BROTHERS HAMBRECHT & QUIST LTD, LONDON, 84- *Mem:* Am Phys Soc. *Mailing Add:* Baring Brothers Hambrecht & Quist Ltd 140 Park Lane London W1Y England

BAILEY, PHILIP SIGMON, b Chickasha, Okla, June 9, 16; m 41, 73; c 3. ORGANIC CHEMISTRY. *Educ:* Okla Baptist Univ, BS, 37; Univ Okla, MS, 40; Univ Va, PhD(org chem), 44. *Prof Exp:* Asst chem, Univ Okla, 37-40; chemist, Halliburton Oil Well Cementing Co, Okla & Kans, 40-41; asst under contract, Off Sci Res & Develop, 42-45; from asst prof to prof, Univ Tex, Austin, 45-84, emer prof chem, 84; RETIRED. *Concurrent Pos:* Fulbright res award, Univ Karlsruhe, WGer, 53-54; consult. *Mem:* Am Chem Soc. *Res:* Reactions of ozone with organic compounds. *Mailing Add:* PO Box 4827 Lago Vista TX 78645

BAILEY, R L, b Trenton, Tenn, Sept 14, 16; m 44; c 1. AVIAN PHYSIOLOGY, VETERINARY ANATOMY. *Educ:* Tenn State Univ, BS, 42; Iowa State Univ, MS, 46, PhD(avian physiol, vet anat), 50. *Prof Exp:* Instr animal husb, Tenn State Univ, 42-45; instr poultry husb, WVa State Col, 46-48, asst prof, Agr & Tech Col, NC, 53-58; prof, Agr & Tech Col, NC, 53-58; prof poultry husb, Grambling Col, 58-76, head dept agr, 63-83, prof biol sci, 76-83; RETIRED. *Mem:* Am Inst Biol Sci; Poultry Sci Asn; World Poultry Sci Asn. *Res:* Reproduction in the male animal. *Mailing Add:* 201 Adams Ave Grambling LA 71245

BAILEY, R V, b Hanna, Wyo, Sept 16, 32. ECONOMIC GEOLOGY. *Educ:* Univ Wyo, BS, 56. *Prof Exp:* Geologist uranium geol, Utah Int, 59-64; geologist coal geol, Page T Jenkins, Geologist, 64-65; dist geologist uranium, Union Carbide Corp, 65-68; independent consult geol, 68-71; pres, Aquarius Resources Corp, 71-73; pres, Power Resources Corp, 73-79; PRES, ASPEN EXPLORATION CORP, 80- *Mem:* Soc Econ Geol; Soc Mining Engrs; Am Asn Petrol Geologists; Can Inst Mining. *Mailing Add:* 4161 S Quebec Denver CO 80237

BAILEY, RAYMOND VICTOR, b Strong, Ark, Nov 22, 23; m 43; c 3. CHEMICAL ENGINEERING. *Educ:* La Polytech Inst, BS, 44; La State Univ, MS, 48, PhD(chem eng), 49. *Prof Exp:* Asst chem, La Polytech Inst, 41-44; chemist, Cities Serv Ref Corp, 44-45; asst chem eng, La State Univ, 46-48; assoc prof, Univ Miss, 48-51; head dept, Eng Res Inst, 53-72, asst dean eng & grad prog, 70-73, PROF CHEM ENG, TULANE UNIV, 51-, ASSOC DEAN ENG, 73- *Mem:* Sr mem Am Chem Soc; Am Inst Chem Engrs; Am Soc Eng Educ. *Res:* Distillation heat transmission; fluid dynamics. *Mailing Add:* Dept Chem Eng Tulane Univ La New Orleans LA 70118

BAILEY, RICHARD ELMORE, b Cleveland, Ohio, Nov 4, 29; m 53; c 4. MEDICINE. *Educ:* Stanford Univ, BA, 51, MD, 55. *Prof Exp:* Trainee diabetes & metab dis, Sch Med, Stanford Univ, 57-59; trainee steroid biochem training prog, Clark Univ & Worcester Found Exp Biol, 60-61; dir res training prog diabetes & metab, Med Sch, Univ Ore, 61-71, asst prof, 61-66, assoc prof med, 66-73; assoc prof med, Col Med, Univ Utah, 73-86. *Concurrent Pos:* Fel diabetes, Sch Med, Univ Southern Calif, 56-57; NASA fac fel award, 69; USPHS spec res fel, Case Western Reserve Univ, 69-70; asst prof biochem, Case Western Reserve Univ, 69-70. *Honors & Awards:* Ayerst-Squibb Res Award, Endocrine Soc, 69; Estab Investigatorship Award, Am Heart Asn, 70. *Mem:* Endocrine Soc; Am Diabetes Asn; Am Fedn Clin Res; AMA; Am Physiol Soc. *Res:* Metabolic diseases; hypertension; endocrinology and diabetes. *Mailing Add:* 880 E 9400 S Sandy UT 84094

BAILEY, RICHARD HENDRICKS, b Rocky Mt, NC, Oct 30, 46; m 72; c 3. INVERTEBRATE PALEONTOLOGY. *Educ:* Old Dominion Univ, BS, 68; Univ NC, MS, 71, PhD(geol), 73. *Prof Exp:* Asst prof, 72-80, ASSOC PROF GEOL, NORTHEASTERN UNIV, 80- *Mem:* Soc Econ Paleontologists & Mineralogists. *Res:* Cenozoic molluscan communities from Atlantic Coastal Plain; systematic and evolutionary studies of Cenozoic Mollusca; stratigraphy and depositional history of Avalonian rocks of Eastern Massachusetts. *Mailing Add:* Dept Geol Northeastern Univ Boston MA 02115

BAILEY, ROBERT BRIAN, b Champaign, Ill, Oct 22, 50. SOLID STATE ELECTRONICS. *Educ:* Mass Inst Technol, BS, 72; Univ Calif, Berkeley, PhD(physics), 79. *Prof Exp:* Res assoc, Lawrence Berkeley Lab, 79; mem tech staff, Electron Res Ctr, Rockwell Int, Anaheim, 79-81; MEM TECH STAFF, ELECTRON RES LAB, AEROSPACE CORP, EL SEGUNDO, CALIF, 81- *Mem:* Am Phys Soc. *Res:* Processing and characterization of thin film integrated circuits fabricated on insulating substrates and in radiation effects on semiconductor devices. *Mailing Add:* 6500 Falconbridge Rd Chapel Hill NC 27514-8610

BAILEY, ROBERT CLIFTON, b Richmond, Va, Mar 29, 41; m 65; c 2. BIOMETRICS, STATISTICS. *Educ:* Randolph-Macon Col, BS, 62; Iowa State Univ, MS, 64; Emory Univ, PhD(biomet), 72. *Prof Exp:* Res asst physics, Iowa State Univ, 62-64; assoc prof statist, Morris Brown Col, 67-73; math statistician, Food & Drug Admin, 73-75; statistician & head biomet, Naval Med Res Inst, 75-80; statistician, Off Water Regulations & Standards, Environ Protection Agency, 80-88; STATIST ADV TO DIR, BUR HEALTH STANDARD QUAL, HEALTH CARE FINANCING ADMIN, 88- *Concurrent Pos:* Statist consult, Asn Off Anal Chemists, 74-75; mem prof adv comt, Food & Drug Admin, 75-78; mem human relations comt, Naval Med Res Ins, Bethesda, Md, 79-80; guest scientist, 84-91; lectr decision sci, George Mason Univ, 85; chair, Am Statist Asn Rep to AAAS, 81-86, co-chair, 86-; chair comt reps to AAAS for the comt of pres of Statist Soc, 87-; secy statist sect, 91-95. *Mem:* Fel AAAS; Am Statist Asn; Biomet Soc; Sigma Xi; Soc Math Biol; Inst Math Statist; Math Asn Am. *Res:* Mathematical and statistical models for data analysis of health care and medical outcomes and design of sampling studies and systems; environmetrics. *Mailing Add:* 6507 Divine St McLean VA 22101

BAILEY, ROBERT E, b Lafayette, Ind, Dec 28, 31; m 53; c 2. NUCLEAR ENGINEERING. *Educ:* Univ Ill, BS, 55, MS, 56; Purdue Univ, PhD(nuclear eng), 63. *Prof Exp:* Asst chem engr, Argonne Nat Lab, 56-59; from instr to assoc prof nuclear eng, Purdue Univ, 59-71, prof, 71-80. *Concurrent Pos:* Ford Found res specialist, Atomics Int Div, NAm Aviation, Inc, 64-65; mem adv comt civil defense, Nat Acad Sci. *Mem:* AAAS; Am Nuclear Soc; Am Soc Eng Educ. *Res:* Nuclear kinetics and controls, sociotechnical problems and interdisciplinary studies. *Mailing Add:* 5104 Sansom Ct Columbus OH 43220

BAILEY, ROBERT LEROY, b Carrollton, Miss, Oct 16, 40; m 64; c 2. BIOMETRICS, FORESTRY. *Educ:* Miss State Univ, BS, 62; Univ Ga, MS, 68, PhD(biomet), 72. *Prof Exp:* Math statistician, USDA Forest Serv, 71-74; biometrician, Weyerhaeuser Co, 74-76, res mgr, 76-78, sr biometrician, 78-80; MEM STAFF, SCH FOREST RESOURCES, UNIV GA, 80- *Concurrent Pos:* Assoc ed, Forest Sci, Soc Am Foresters, 76- *Mem:* Biomet Soc; Soc Am

Forestry; Am Statist Asn; Am Forestry Asn. *Res:* Mathematical and statistical models as applied to forest growth, yield and inventory; experimental design in forestry and biology. *Mailing Add:* Dept Forestry Univ Ga Athens GA 30602

BAILEY, RONALD ALBERT, b Winnipeg, Man, June 2, 33; m 61. COORDINATION CHEMISTRY, MOLTEN SALTS. *Educ:* Univ Man, BSc, 56, MSc, 57; McGill Univ, PhD(radiochem), 60. *Prof Exp:* NATO sci fel, Univ Col, Univ London, 60-61; from asst prof to assoc prof, 61-71, PROF INORG CHEM, RENSSELAER POLYTECH INST, 71- *Mem:* Am Chem Soc; Brit Chem Soc. *Res:* Preparation and characterization of coordination complexes; chemistry of metal compounds in melter salt solutions especially electrochemistry and electroplating. *Mailing Add:* Dept Chem Rensselaer Polytech Inst Troy NY 12180

BAILEY, ROY ALDEN, b Providence, RI, July 28, 29; m 58; c 3. MINERALOGY & PETROLOGY, GEOLOGY. *Educ:* Brown Univ, AB, 51; Cornell Univ, MSc, 54; Johns Hopkins Univ, PhD, 78. *Prof Exp:* GEOLOGIST, US GEOL SURV, 53-55, 57- *Concurrent Pos:* Fulbright scholar, Auckland Univ, NZ, 56; coodinator US Geol Surv, Volcano Hazards Prog, 80-83; vis sci, NZ Geol Surv, 89. *Mem:* Geol Soc Am; Mineral Soc Am; Am Geophys Union; Sigma Xi; Int Asn Volcanology. *Res:* Volcanology, petrology of igneous rocks, volcano tectonics, geothermal resources, Jeniez Mountains, New Mexico, Long Valley Caldera, eastern California, New Zealand volcanism. *Mailing Add:* US Geol Survey MS-910 345 Middlefield Rd Menlo Park CA 94025

BAILEY, ROY HORTON, JR, b Maxton, NC, Nov 7, 21; m 47; c 3. PHYSICAL ORGANIC CHEMISTRY. *Educ:* Univ NC, BS, 48, PhD(chem), 58. *Prof Exp:* Asst anal chem, Univ NC, 48-53; asst prof chem, King Col, 53-57; asst prof & chmn dept, 57-63; from asst prof to assoc prof chem, Clemson Univ, 63-88; RETIRED. *Mem:* Am Chem Soc. *Res:* Heterocyclic reaction kinetics; analytical instrumentation. *Mailing Add:* 105 Carolus Dr Clemson SC 29631

BAILEY, SAMUEL DAVID, b Cedar Falls, Iowa, July 14, 15; m 42; c 1. PHYSICAL CHEMISTRY. *Educ:* Univ Northern Iowa, BA, 37; Univ Iowa, MS, 40, PhD(phys chem), 42. *Prof Exp:* From instr to asst prof chem, Univ Northern Iowa, 41-42 & 46-48; res chemist, Rohm and Haas Co, 42-46; head phys chem sect, Smith Kline & French Lab, 48-51; dir, Pioneering Res Lab, Army Natick Res & Develop Labs, 51-74, Food Sci Lab, 74-81 & Sci & Advan Technol, 81; RETIRED. *Mem:* AAAS; Am Chem Soc; Am Inst Chemists. *Res:* Physical and optical properties of matter; research administration; radiation and flavor chemistry; photochemistry; food chemistry; pollution abatement; food nutrition and microbiology; behavioral sciences. *Mailing Add:* 901 N Abrego Dr Green Valley AZ 85614

BAILEY, STURGES WILLIAMS, b Waupaca, Wis, Feb 11, 19; m 49; c 2. CLAY MINERALOGY, CRYSTALLOGRAPHY. *Educ:* Univ Wis, BA, 41, MA, 48; Cambridge Univ, PhD(physics), 54. *Prof Exp:* From instr to prof x-ray crystallog, Dept Geol, 51-76, chmn dept geol & geophys, 68-71, ROLAND D IRVING PROF MINERAL, DEPT GEOL & GEOPHYS, UNIV WIS-MADISON, 76- *Concurrent Pos:* Ed, Clays & Clay Minerals, 65-69; chmn joint comt on nomenclature, Int Union Crystallog & Int Mineral Asn, 70-76. *Honors & Awards:* Neil Miner Award, Am Asn Geol Teachers. *Mem:* Clay Minerals Soc (vpres, 70-71, pres, 71-72); Mineral Soc Am (vpres, 72-73, pres, 73-74); Int Asn Study Clays (pres, 75-78); Geol Soc Am; Am Crystallog Asn. *Res:* Potassium feldspars; layer silicates. *Mailing Add:* Dept Geol & Geophys Univ Wis 1215 W Dayton St Madison WI 53706

BAILEY, SUSAN GOODMAN, b Washington, DC, Aug 5, 40; m 65; c 2. ENVIRONMENTAL SCIENCES. *Educ:* Vassar Col, BA, 62; Iowa State Univ, MS, 65; Emory Univ, PhD(inorg chem), 69. *Prof Exp:* Instr chem, Ga State Col, 65, Northern Va Community Col, 73-74 & Marshall Adult Ctr, 86-87; chemist, Ctr Dis Control, 69; tutor chem, 87-88; CHEMIST, ASCI CORP, 89- *Res:* Environmental remediation work; quality assurance audits of environmental research laboratories. *Mailing Add:* 6507 Divine St McLean VA 22101

BAILEY, THOMAS DANIEL, b Birmingham, Ala, July 15, 45; m 66; c 1. ORGANIC CHEMISTRY. *Educ:* Univ Montevallo, BS, 69; Univ Ala, PhD(chem), 74. *Prof Exp:* Res chemist, Reilly Tar & Chem Corp, 75-79, assoc dir res, 80-83; dir com develop, 84-89, VPRES CORP DEVELOP, REILLY INDUST, 89- *Mem:* Am Chem Soc. *Res:* Synthesis of heterocyclic compounds; synthesis and chemistry of heterocyclic-N-oxides; process development. *Mailing Add:* Reilly Labs 1500 S Tibbs Ave PO Box 41076 Indianapolis IN 46241

BAILEY, THOMAS L, III, b Newnan, Ga, Dec 24, 23; m 55; c 2. CHEMICAL PHYSICS. *Educ:* Carnegie Inst Technol, BS, 49; Univ Chicago, SM, 50, PhD(chem), 53. *Prof Exp:* Asst res elec eng, 53-54, asst res prof, 54-57, from assoc prof to prof, 57-87, EMER PROF PHYSICS, CHEM & ELEC ENG, UNIV FLA, 87- *Concurrent Pos:* Guggenheim fel, 59-60. *Mem:* Fel Am Phys Soc. *Res:* Molecular and atomic structure; experimental mass spectroscopy; collisions of positive ions, negative ions and electrons with atoms and molecules; applied physics. *Mailing Add:* Dept Physics Univ Fla Gainesville FL 32611

BAILEY, VERNON LESLIE, JR, b St Louis, Mo, Apr 27, 41; m 65; c 2. PLASMA PHYSICS. *Educ:* Northwestern Univ, BSME, 63; Univ Wash (plasma physics), 68. *Prof Exp:* Tech staff mem fluid mech, Sandia Corp, 68-70; dep dept mgr theoret groups plasma physics, Physics Int Co, 70-82; STAFF PHYSICIST, PULSE SCI INC, SAN LEANDRO, 82- *Mem:* Am Phys Soc. *Res:* Application of relativistic electron beams to controlled thermonuclear fusion (steady state tokamaks, low aspect ratio tokamaks, linear systems, tandem mirror and inertial confinement), ion acceleration, and inductive storage. *Mailing Add:* 2047 Westbrook Lane Livermore CA 94550

BAILEY, WAYNE L, b Woodbury, NJ, Dec 12, 42. AEROSPACE ENGINEERING. *Educ:* Rutgers Univ, BS, 65; Univ Ariz, PhD(physics), 71. *Prof Exp:* Prof, physics, NJ State Univ, 72-77; SR SYSTS ANALYST, PHYSICS, TELEDYNE BROWN ENG, 77- *Mem:* Am Phys Soc; Am Astronaut Soc; Am Inst Astronaut & Aeronaut; Royal Astronaut Soc. *Mailing Add:* 12003 Pulaski Pike Toney AL 35773

BAILEY, WILFORD SHERRILL, b Somerville, Ala, Mar 2, 21; m 42; c 4. VETERINARY PARASITOLOGY. *Educ:* Auburn Univ, DVM, 42, MS, 46; Johns Hopkins Univ, ScD(hyg), 50. *Hon Degrees:* DHH, Auburn Univ, 84. *Prof Exp:* From instr to assoc prof, Sch Vet Med, Auburn Univ, 42-48, head prof, 50-62, res prof parasitol & assoc dean grad sch & coordr res, 62-66, vpres acad affairs, 66-69, vpres acad & admin affairs, 69-72, pres, 83-84; health scientist adminr, Nat Inst Allergy & Infectious Dis, 72-74; prof path & parasitol, 74-84, PROF, AUBURN UNIV, 84- *Concurrent Pos:* NSF sci fac fel, 59; mem, Nat Res Coun, 63-69; mem comt vet med educ & res, Nat Acad Sci, 68-70; mem, Nat Adv Allergy & Infectious Dis Coun, NIH, 71-72 & 75-78. *Mem:* Am Soc Parasitol (pres, 71); Am Vet Med Asn; Am Soc Trop Med & Hyg (pres, 77); World Asn Advan Vet Parasitol. *Res:* Life cycle and immunity to Hymenolepis nana; immunology and pathology of Cooperia punctata and ruminant stomach worms; toxic hepatitis in dogs; life cycle and pathology of Spirocerca lupi and its relationship to neoplasia; prevalence of Pneumocystis carinii in dogs and cats; biology and life cycle of Cryptosporidium in man and animals. *Mailing Add:* 778 Moores Mill Rd Auburn AL 36830

BAILEY, WILLIAM C, b Jacksonville, Fla, Aug 4, 39; m 63; c 3. PULMONARY DISEASES. *Educ:* Washington & Lee Univ, BA, 61; Tulane Univ, MD, 65. *Prof Exp:* Dir, Bur Tuberculosis Control, City New Orleans Health Dept, 70-73; asst prof med, Sch Med, Tulane Univ, 72-73; asst prof, 73-75, assoc prof, 75-79, PROF MED, SCH MED, UNIV ALA, BIRMINGHAM, 79- *Concurrent Pos:* Med dir, Bur Commun Dis, Jefferson County Health Dept, 73-78; assoc chief staff educ, Vet Admin Med Ctr, Birmingham, Ala & chief pulmonary dis med sect, 73-; tuberculosis coordr, Ala Dept Public Health, Birmingham, 73-; consult, Bur Commun Dis, Jefferson County Health Dept, 78- *Mem:* Am Col Chest Physicians; Am Col Physicians; Am Thoracic Soc; Int Union Against Tuberculosis. *Res:* Clinical pulmonary disease; occupational pulmonary disease; tuberculosis control. *Mailing Add:* Univ Ala Univ Hosp Univ Ala Birmingham Med Ctr Birmingham AL 35294

BAILEY, WILLIAM CHARLES, b Albany, NY, Aug 21, 39. NUCLEAR MAGNETIC RESONANCE. *Educ:* Siena Col, BS, 63; State Univ NY Albany, MS, 66, PhD(physics), 71. *Prof Exp:* Res assoc physics, State Univ NY, Albany, 71-80; fel mat sci & eng, Univ Pa, 80-82; ASST PROF PHYSICS, KEAN COL, 83- *Concurrent Pos:* J Corbett fel physics, Atomic Energy Comn, State Univ NY, Albany, 71-73, fel, Inst Study of Defects in Solids, 75-80, instr phys sci, Educ Opportunities Prog, 75-80. *Mem:* Am Phys Soc; Am Asn Physics Teachers. *Res:* Ion diffusion in superionic conducting solids such as sodium beta alumina. *Mailing Add:* Kean Col NJ Union NJ 07083-9982

BAILEY, WILLIAM FRANCIS, b Jersey City, NJ, Dec 8, 46. ORGANIC CHEMISTRY. *Educ:* St Peter's Col, NJ, BS, 68; Univ of Notre Dame, PhD(chem), 73. *Prof Exp:* Post doc assoc chem, Univ NC, 73 & Yale Univ, 73-75; from asst prof to assoc prof, 75-85, PROF CHEM, UNIV CONN, 85- *Concurrent Pos:* Fac summer fel, Univ Conn, 76; co-chmn, Int Conf Conformational Anal, 81; vis prof, Nat Polytech Inst Mex, 81, Oxford Univ, 83. *Mem:* Am Chem Soc; Sigma Xi. *Res:* Synthetic methods; molecular structure and energetics; group I organometallic chemistry; chemistry of organolithium compounds. *Mailing Add:* Dept Chem Univ Conn Storrs CT 06268

BAILEY, WILLIAM JOHN, b East Grand Forks, Minn, Aug 11, 21; m 49; c 3. POLYMER CHEMISTRY, ORGANIC CHEMISTRY. *Educ:* Univ Minn, BChem, 43; Univ Ill, PhD(org chem), 46. *Prof Exp:* Asst org chem, Univ Ill, 43, asst War Prod Bd prog, 43-46; Little fel, Mass Inst Technol, 46-47; from asst prof to assoc prof org chem Wayne State Univ, 47-51; PROF ORG CHEM, UNIV MD, 51- *Concurrent Pos:* Welch Found lectr, 70; consult, Am Cyanamid Co, Goodyear Tire & Rubber Co, Phillips Fibers, BASF Wyandotte Corp, King Industs, Naval Surface Weapons Lab, Hydron Labs, Nat Starch & Chem Co & Armstrong World Indust; chmn, Gordon Res Conf Org Reactions, 60, vpres, 1st Chem Cong NAm, 75; mem, Fel Selection Comt, NSF, 65-68; mem, Nat Res Coun Elastomers Adv Comt, US Army Natick Lab; chmn, Comt Macromolecules, Nat Res Coun, 68-75; nat rep, Macromolecular Div, Int Union Pure & Appl Chem, 68-81; Rauscher Mem lectr, 76; co-chmn, US-Japan Macromolecular Symposium, 80; chmn bd, Am Chem Soc, 79 & 81, chmn, Div Polymer Chem; pres, Chem Soc, Wash, 61. *Honors & Awards:* Res Award, Fatty Acid Producers, 55; Chem Soc Wash Serv Award, 69, Hillebrand Prize, 85; Hon Scroll Award, DC Inst Chemists, 75; Outstanding Achievement Award, Univ Minn, 76; Am Chem Award in Polymer Chem, 77; Gosset Award, NC State Univ, 83; Mobay Award lectr, Charleston Col; Am Chem Soc Award Appl Polymer Chem, 86; Serv Award, Am Chem Soc, 85. *Mem:* AAAS; Am Chem Soc (pres, 75); Am Oil Chemists Soc. *Res:* Cyclic dienes; pyrolysis of esters and unsaturated compounds; polypeptides; phosphorus compounds; spiro and ladder polymers; biodegradable polymers; monomers that expand on polymerization; mechanism of thermal decomposition of vinyl polymers; free radical ring-opening polymerization. *Mailing Add:* Dept Chem Univ Md College Park MD 20742

BAILEY, WILLIAM T, b Buffalo, NY, Apr 4, 36; m 83; c 2. MATHEMATICS. *Educ:* State Univ NY Buffalo, BA, 59, MA, 62, EdD(math educ), 71. *Prof Exp:* Instr statist, State Univ NY Buffalo, 60-64; PROF MATH, STATE UNIV NY COL BUFFALO, 64- *Concurrent Pos:* Mem Nat Coun Teachers Math. *Mem:* Math Asn Am; Sch Sci & Math Asn. *Res:* Group reaction to and productivity on a mathematical task involving productive thinking. *Mailing Add:* Dept Math Southern Univ NY 1300 Elmwood Ave Buffalo NY 14222

BAILEY, ZENO EARL, b Frisco City, Ala, Aug 9, 21; m 56; c 2. BOTANY, GENETICS. *Educ:* Auburn Univ, BS, 50, MS, 53; Ohio State Univ, PhD(agr educ), 55; ETex State Univ, MS, 61. *Prof Exp:* Teacher high sch, Ala, 50-53; instr biol, Snead Jr Col, 55-56; assoc prof agr educ, ETex State Univ, 56-60, assoc prof biol, 61-67; prof, Livingston Univ, 67-69; prof bot, Eastern Ill Univ, 69-87; RETIRED. *Concurrent Pos:* Tex State Univ study grant, 65-67. *Mem:* Hon mem Ill Acad Sci; Am Genetic Asn. *Res:* Phases of the life history of the boat-tailed grackle. *Mailing Add:* Bot Dept Eastern Ill Univ Charleston IL 61920

BAILEY-BROCK, JULIE HELEN, b Birmingham, Eng; US citizen; m 70. INVERTEBRATE ZOOLOGY, POLYCHAETE BIOLOGY. *Educ:* Univ Wales, BSc, 64, PhD(zool), 68. *Prof Exp:* From asst prof to assoc prof 69-85, PROF ZOOL, UNIV HAWAII, 85- *Concurrent Pos:* Vis scholar, dept zool, Univ Wash, Seattle, 73-74; vis researcher, Lizard Island Res Sta, 77, Univ Malaysia, 77, Marine Lab, Univ Guam, 81; res assoc, Bernice P Bishop Museum, Honolulu, 83- *Honors & Awards:* Fulbright-Hays Res Abroad Award, SE Asia & E Australia, 77. *Mem:* Sigma Xi; Am Soc Zoologists. *Res:* Polychaete communities of coral reefs; polychaete fauna of Hawaii & West Pacific Islands; deep sea polychaetes; Serpulidae and Spirorbidae; polychaetes as pollution indicators; coral reef bryozoans; benthic community development on artificial reefs benthic communities near sewage outfalls in Hawaii. *Mailing Add:* Dept Zool Univ Hawaii 2538 The Mall Honolulu HI 96822

BAILEY-SERRES, JULIA, b Redwood City, Calif, July 9, 58. PLANT MOLECULAR BIOLOGY, PLANT GENETICS. *Educ:* Univ Utah, BS, 82; Univ Edinburgh, PhD(plant biol), 86. *Prof Exp:* Res assoc, Univ Edinburgh, 84-86, postdoctoral res assoc, 86; postdoctoral res assoc, 86-90, ASST PROF GENETICS, DEPT BOT & PLANT SCI, UNIV CALIF, RIVERSIDE, 90- *Concurrent Pos:* Fels & res grants, several insts & schs, 86- *Mem:* Am Soc Plant Physiologists; Maize Genetics Coop. *Res:* Plant response to environmental stress; molecular genetic analysis of the response of maize to hypoxia and ozone stress; analysis of isozyme expression in tissues of maize. *Mailing Add:* Dept Bot & Plant Sci Univ Calif Riverside CA 92521

BAILIE, MICHAEL DAVID, b South Bend, Ind, Sept 27, 36; m 58; c 2. PEDIATRICS, PHYSIOLOGY. *Educ:* Ind Univ, Bloomington, AB, 59, MA, 60, PhD(physiol), 66; Ind Univ, Indianapolis, MD, 64. *Prof Exp:* From asst prof to prof human develop & physiol, Dept Human Develop, Col Human Med, Mich State Univ, 70-79 & assoc chmn, 78-79; PROF PEDIAT & CHMN, UNIV CONN HEALTH CTR, 79- *Concurrent Pos:* Fel, Univ Tex Southwestern Med Sch Dallas, 68-70. *Mem:* Am Fedn Clin Res; Am Soc Nephrology; Int Soc Nephrology; Soc Pediat Res; Am Physiol Soc. *Res:* Renal physiology, specifically, control of renal hemodynamics; relation of renin-angiotensin system to renal hemodynamics; intra renal actions and formation of angiotensin II; development of renal function; renal metabolism. *Mailing Add:* Dept Pediat Univ Conn Sch Med Farmington CT 06032

BAILIE, RICHARD COLSTEN, b Chicago, Ill, July 4, 28; m 63; c 4. CHEMICAL ENGINEERING. *Educ:* Ill Inst Technol, BS, 51; Wayne State Univ, MS, 57; Iowa State Univ, PhD(chem eng), 65. *Prof Exp:* Jr engr, Victor Chem Works, 51-53; res & develop chem engr, US Rubber Co, Joliet Arsenal, 53; instr chem eng, Wayne State Univ, 56-57; partic, Int Sch Nuclear Sci & Eng, Argonne Nat Lab, 57-58; asst prof nuclear eng, Kans State Univ, 58-61, assoc prof, 61-65; PROF CHEM ENG, WVA UNIV, 65- *Mem:* Am Inst Chem Engrs; Am Chem Soc; Am Soc Eng Educ; Sigma Xi. *Res:* Gasification, including combustion and pyrolysis in fluidized beds; kinetics of solid-gas reactions. *Mailing Add:* Box 4214 Morgantown WV 26506

BAILIE, WAYNE E, b Bondurant, Iowa, Aug 28, 32; m 55; c 3. VETERINARY MICROBIOLOGY. *Educ:* Kans State Univ, BS & DVM, 57, PhD(path), 69; Am Col Vet Microbiologists, cert, 80. *Prof Exp:* Vet, self-employed, Nebr, 57-64; instr path, Kans State Univ, 64-69; assoc prof, SDak State Univ, 69-70 & Ahmadu Bello Univ, Nigeria, 70-72; assoc prof infectious diseases, 71-81, PROF LAB MED, KANS STATE UNIV, 81- *Mem:* Am Vet Med Asn; Am Soc Microbiol. *Res:* Infectious diseases of domestic animals; taxonomy of unnamed species of bacteria associated with domestic animals. *Mailing Add:* Col Vet Med Kans State Univ Manhattan KS 66506

BAILIN, GARY, b New York, NY, Apr 2, 36; m; c 2. BIOCHEMISTRY. *Educ:* City Col New York, BS, 58; Adelphi Univ, PhD(biochem), 65. *Prof Exp:* Res asst, Adelphi Univ, 59-61; res assoc protein chem, Inst Muscle Dis, 64-74, asst prof biochem, Mt Sinai Sch Med, 74-77; asst prof, 77-79, ASSOC PROF BIOCHEM, UNIV MED & DENT NJ-SCH OSTEOP MED, 79- *Concurrent Pos:* Adj asst prof, Brooklyn Col, 68-75. *Mem:* Am Soc Biochemists & Molecular Biologists; Biophys Soc. *Res:* Regulation of muscle contraction by muscle and membrane proteins; correlation of structure and function of muscle contractile and membrane proteins. *Mailing Add:* Dept Molecular Biol Univ Med & Dent NJ-Sch Osteop Med 401 S Central Plaza Stratford NJ 08084

BAILIN, LIONEL J, b New York, NY, Oct 28, 28; m 58; c 4. PHYSICAL INORGANIC CHEMISTRY. *Educ:* NY Univ, BA, 49; Polytech Inst Brooklyn, MS, 52; Tulane Univ, PhD(inorg chem), 58. *Prof Exp:* Develop chemist, Am Cyanamid Co, 51-52; phys chemist chem warfare, US Army Edgewood Arsenal, Md, 52-54; chemist pigment technol, E I Du Pont de Nemours & Co, 57-62; res scientist, 62-73, staff scientist, Chem Dept, 74-78, SR STAFF SCIENTIST, PALO ALTO RES LAB, LOCKHEED MISSILES & SPACE CO, 78- *Mem:* Am Chem Soc; Am Soc Metals; fel Am Inst Chemists; Sigma Xi; Electrochem Soc Am. *Res:* Materials science and engineering; inorganic and modified inorganic materials; protective coatings technology, coatings for anticorrosion applications on metals; reactions at polymer-metal interfaces. *Mailing Add:* 740 Crane Ave Foster City CA 94404

BAILIT, HOWARD L, b Boston, Mass, Dec 17, 37. PUBLIC HEALTH. *Educ:* Tufts Dent Sch, DMD, 62; Harvard Univ, PhD, 67. *Prof Exp:* from asst prof to prof, Dept Behav Sci & Community Health, Univ Conn Health Ctr, 67-82; prof & head, Dept Health Admin, Sch Pub Health & Med, Columbia Univ, 82-86; VPRES MED POLICY & PROGS, AETNA LIFE & CASUALTY, 86- *Concurrent Pos:* Mem, Study Sect, Nat Ctr Health Serv Res, 78-82 & 86-, Nat Affairs Comt, Am Asn Dent Res, 78-82, Future Dent Comn, Am Dent Asn, 83 & Comt Cost Mgt Systs, Inst Med Nat Acad Sci, 87-; consult, Rand Health Ins Exp, 79-86 & Off Qual Assurance, Am Dent Asn, 83- *Mem:* Nat Acad Sci; Inst Med-Nat Acad Sci; Am Col Dentists. *Res:* 98 published articles in national reference journals; focus of work is on health services, research and policy. *Mailing Add:* Aetna Life & Casualty 151 Farmington Ave MC1A Hartford CT 06156

BAILLARGEON, VICTOR PAUL, b Dover, NH, Apr 4, 58; m 83. RESEARCH ADMINISTRATION. *Educ:* Univ NH, BS, 80; Colo State Univ, PhD(chem), 85. *Prof Exp:* Res chemist, Humko Chem Div, Witco Corp, 85-88; sr res assoc, 88-89, dir, Int Res & Develop, 89-91, MGR KNOWLEDGE TRANSFER, BUCKMAN LABS INT, 91- *Mem:* Am Chem Soc; Tech Asn Pulp & Paper Indust; Int Mgt Coun; Am Mgt Asn. *Res:* Corporate knowledge base management. *Mailing Add:* Buckman Labs Int 1256 N McLean Blvd Memphis TN 38108

BAILLIE, ANDREW DOLLAR, b Dollar, Scotland, Nov 20, 12; m 46; c 1. STRATIGRAPHY. *Educ:* Univ Man, BSc, 49, MSc, 50; Northwestern Univ, PhD(geol), 53. *Prof Exp:* Geologist, Mines Br, Man, 50-53; Can Gulf Oil Co, 53-57; from regional geologist to res geologist, Brit Am Oil Co, 57-70; dir res geol, 70-72, mgr geol serv, Gulf Oil Can, 72-78; RETIRED. *Concurrent Pos:* Vis prof basin studies, Univ Alta, 77-85; consult, int training basin anal, 78-90. *Honors & Awards:* Spencer Gold Medal, Univ Man, 53; Medal Merit, Alta Soc Petrol Geol, 55. *Mem:* Hon mem Am Asn Petrol Geologists; fel Geol Soc Am; hon mem Soc Econ Paleont & Mineral; fel Geol Asn Can; hon mem Can Soc Petrol Geol. *Res:* Sedimentary basin analysis and oil occurrence. *Mailing Add:* 917 Rideau Rd Calgary AB T2S 0S3 Can

BAILLIE, DONALD CHESLEY, b Toronto, Ont, Apr 3, 15; m 57; c 2. ACTUARIAL MATHEMATICS. *Educ:* Univ Toronto, BA, 35, MA, 36. *Prof Exp:* Lectr, Univ Toronto, 38-41, asst prof math, 45-58, assoc prof, 58-80; CONSULT, ECKLER PARTNERS LTD, 80- *Concurrent Pos:* Consult actuary, various plans in fraternal socs, 45-88. *Honors & Awards:* Triennial Prize, Soc Actuaries, 47. *Mem:* Assoc Soc Actuaries; fel Can Inst Actuaries. *Res:* Dynamics of pension funding; actuarial model building. *Mailing Add:* 20 Wychwood Pk Toronto ON M6G 2V5 Can

BAILLIE, PRISCILLA WOODS, b Buffalo, NY, Jan 18, 35; m 53. ESTUARINE ECOLOGY, MICROALGAL ECOLOGY. *Educ:* Univ Hartford, BS, 74; Univ Conn, MS, 78, PhD, 83. *Prof Exp:* PRIN INVESTR, MARINE & FRESH WATER RES SERV, 83- *Mem:* Aquatic Plant Mgt Soc; NAm Lake Mgt Soc; Sigma Xi. *Res:* Eutrophication problems associated with perturbations of estuarine, lacustrine, and wetlands ecosystems. *Mailing Add:* 276 State St Guilford CT 06437

BAILLIE, THOMAS ALLAN, b Isle-of-Islay, Scotland, May 30, 48; m 74; c 2. DRUG METABOLISM, MASS SPECTROMETRY. *Educ:* Univ Glasgow, Scotland, BSc, 70, PhD(chem), 73; Univ London, Eng, MSc, 78. *Prof Exp:* Fel, Karolinska Inst, Sweden, 73-75; lectr, Royal Postgrad Med Sch, Eng, 75-78; asst prof pharmaceut, Univ Calif, San Francisco, 78-81; asst prof, 81-83, ASSOC PROF MED CHEM, UNIV WASH, 83- *Concurrent Pos:* Sci adv, Food & Drug Admin, San Francisco, 80-81. *Mem:* Royal Soc Chem; Am Soc Mass Spectrometry; Am Chem Soc. *Res:* Application of stable isotope labelling techniques and mass spectrometry to studies of pathways of drug metabolism and their relationship to drug-induced toxicities. *Mailing Add:* Dept Med Chem BG-2D Univ Wash Seattle WA 98175

BAILLOD, CHARLES ROBERT, b Milwaukee, Wis, Mar 21, 41; c 3. CIVIL ENGINEERING, ENVIRONMENTAL ENGINEERING. *Educ:* Marquette Univ, BCE, 63; Univ Wis, MSCE, 65, PhD(civil eng), 68. *Prof Exp:* From asst prof to assoc prof civil eng, 68-78, PROF CIVIL & ENVIRON ENG & DIR, ENVIRON ENG CTR WATER & WASTE MGT, MICH TECHNOL UNIV, 82- *Concurrent Pos:* Vis prof civil eng, Univ Wis-Madison, 77-78, Rogaland Regional Col, Stavanger, Norway, 83. *Mem:* Am Soc Civil Engrs; Am Inst Chem Engrs; Water Pollution Control Fedn; Asn Environ Eng Professors; Am Acad Environ Engrs. *Res:* Biological waste treatment; hazardous waste destruction processes, wet oxidation. *Mailing Add:* Environ Eng Mich Technol Univ Houghton MI 49931

BAILY, EVERETT M, b Twin Falls, Idaho, June 9, 38; m 61; c 2. RESEARCH & DEVELOPMENT, MANUFACTURING. *Educ:* Univ Idaho, BS, 61, MS, 64; Stanford Univ, PhD(network theory), 68. *Prof Exp:* Instr elec eng, Univ Idaho, 61-64, asst prof, 64-70, assoc prof, 70-74; mem staff reliability, design eng & mfr, 74-84, prod eng mgr, 84-87, SR MFR DEVELOP ENGR, BOISE DIV, HEWLETT-PACKARD, 87- *Concurrent Pos:* Am Soc Eng Educ & Dow Chem Co Pac Northwest faculty award, 70. *Mem:* Inst Elec & Electronics Engrs; Am Soc Eng Educ. *Res:* Remote sensing; control systems; printer mechanisms. *Mailing Add:* Hewlett-Packard MS 514 PO Box 15 Boise ID 83707

BAILY, NORMAN ARTHUR, b New York, NY, July 2, 15; m 40; c 2. MEDICAL PHYSICS, RADIOBIOLOGY. *Educ:* St John's Univ, NY, BS, 41; NY Univ, MA, 43; Columbia Univ, PhD(physics), 52. *Prof Exp:* Res scientist, Columbia Univ, 46-52; sci adv, US Air Force, 52-54; prin cancer res scientist & chief physicist, Dept Radiation Ther, Roswell Park Mem Inst, 54-59; from assoc clin prof to prof in residence radiol, Univ Calif, Los Angeles, 59-68; prof radiol & physics & dir div radiol sci, Emory Univ, 67-68; prof, 68-88, EMER PROF RADIOL, UNIV CALIF, SAN DIEGO, 88- *Concurrent Pos:* NASA grants, Univ Calif, San Diego, 68-73 & 73-75, AEC grant, 74-75 & NIH grant, 75-78; radiation physicist, Marine Biol Lab, Woods Hole, Mass, 46-52; assoc radiol, Sch Med, Univ Buffalo, 54-59; asst res prof biophys, Roswell Park Div, 57-59; lectr chem, Canisius Col, 57-59; mem sci adv bd, US Air Force, 63-74; consult, Rand Corp, 67-71, US Naval Hosp, San Diego, 68- & Vet Admin Hosp, San Diego, 71-; vis scientist, Europ Orgn Nuclear Res, 70; vis prof, Hebrew Univ, Jerusalem, 72; mem NIH radiation study sect, 75-; assoc ed, Med Physics, 76-; Henry Goldberg prof biomed eng, Technion, Israel, 80; vis prof imaging sci, Korea Advan Inst Sci & Technol, 82; mem, Comprehensive Panel, Am Bd Health Physics. *Honors & Awards:* Norman A Baily Student Res Award, Am Asn Physicists Med, 86. *Mem:* AAAS; Am Asn Physicists Med; fel Am Col Radiol; Am Phys Soc; Radiation Res Soc; Radiol Soc NAm; Am Endocurietherapy Soc; fel Am Asn Physics; Am Col Med Physics. *Res:* Quantitation of diagnostic radiological procedures; dose reduction in diagnostic radiology; microdosimetry; radiobiological modeling. *Mailing Add:* Radiol 0632 Univ Calif San Diego 9500 Gilman Dr La Jolla CA 92093-0632

BAILY, WALTER LEWIS, JR, b Waynesburg, Pa, July 5, 30; c 1. MATHEMATICS. *Educ:* Mass Inst Technol, SB, 52; Princeton Univ, MA, 53, PhD(math), 55. *Prof Exp:* Instr math, Princeton Univ, 55-56 & Mass Inst Technol, 56-57; from asst prof to assoc prof, 57-63, PROF MATH, UNIV CHICAGO, 63- *Concurrent Pos:* Mathematician, Bell Tel Labs, 57-58; NSF sr fel, 65-66. *Mem:* Am Math Soc; Math Soc Japan. *Res:* Algebraic groups; modular forms; arithmetic properties of automorphic forms. *Mailing Add:* Dept Math Univ Chicago 1118-32 E 58th St Chicago IL 60637

BAILYN, MARTIN H, b Conn, June 24, 28. PHYSICS. *Educ:* Williams Col, BA, 48; Harvard Univ, PhD, 56. *Prof Exp:* From asst prof to assoc prof, 58-66, PROF PHYSICS, NORTHWESTERN UNIV, 66- *Mem:* Am Phys Soc. *Res:* Magnetism in metals. *Mailing Add:* Dept Physics Northwestern Univ Evanston IL 60208

BAIN, BARBARA, b Montreal, Que, May 8, 32. EXPERIMENTAL MEDICINE. *Educ:* McGill Univ, BSc, 53, MSc, 57, PhD(exp med), 65. *Prof Exp:* Res asst hemat, Royal Victoria Hosp, 65-68; demonstr, 65-67, lectr, 67-68, ASST PROF EXP MED, McGILL UNIV, 68-; RES ASSOC HEMAT, ROYAL VICTORIA HOSP, 68- *Concurrent Pos:* Med Res Coun Can scholar, 66-71, grant, 70- *Mem:* Tissue Cult Asn; Transplantation Soc; Can Soc Immunol; Can Soc Cell Biol; Am Soc Cell Biol. *Res:* Lymphocyte transformation to blast cells in vitro, using allogenic leukocytes and other stimuli; application of mixed leukocyte reaction to histocompatibility testing; mechanisms of blast cell transformation. *Mailing Add:* Dept Zool Univ Western Ontario 201 Huron St London ON N6A 5B7 Can

BAIN, GORDON ORVILLE, b Lethbridge, Alta, Nov 1, 26; m 51; c 5. PATHOLOGY. *Educ:* Univ Alta, BSc, 49, MD, 51. *Prof Exp:* From asst prof to assoc prof, 56-67, prof path, Univ Alta, 67-80, Chmn Dept, 70-80; RETIRED. *Concurrent Pos:* dir, Dept Lab, Cross Cancer Inst, 76- *Mem:* Can Asn Path; Royal Col Physicians & Surgeons of Can. *Res:* Hematological pathology; immunopathology; oncology. *Mailing Add:* 13007 63rd Ave Edmonton AB T6H 1R9 Can

BAIN, JAMES ARTHUR, b Langdon, NDak, May 22, 18; m 47; c 2. BIOCHEMICAL PHARMACOLOGY. *Educ:* Univ Wis, BS, 40, PhD(physiol), 44. *Prof Exp:* Asst, Univ Wis, 40-44; res assoc psychiat & pharmacol, Med Sch, Univ Ill, 47-50, from asst prof to assoc prof, 50-54; chmn, Dept Pharmacol, 57-62, dir, Div Basic Health Sci, 60-76, prof pharmacol, 54-88, exec assoc dean, sch med, 65-88, EMER PROF PHARMACOL, EMORY UNIV, 88- *Concurrent Pos:* Consult, gov & indust. *Mem:* AAAS; Asn Cancer Res; Am Chem Soc; Am Soc Pharmacol & Exp Therapeut; Soc Exp Biol & Med. *Res:* Biochemical basis of drug action; epilepsy; carcinogenesis; neurochemistry. *Mailing Add:* Dept Pharmacol Sch Med Emory Univ Atlanta GA 30322-0309

BAIN, LEE J, b Newkirk, Okla, Jan 11, 39; m 62; c 5. MATHEMATICAL STATISTICS. *Educ:* Okla State Univ, BS, 60, MS, 62, PhD(math), 63. *Prof Exp:* From asst prof to assoc prof, 63-73, PROF MATH, UNIV MO-ROLLA, 73- *Concurrent Pos:* Co-editor, Commun in Statist, Series E, 81- *Mem:* Fel Am Statist Asn. *Res:* Statistical inference concerning life-testing distribution and reliability problems. *Mailing Add:* Dept Math Univ Mo-Rolla Rolla MO 65401

BAIN, ROGER J, b Kenosha, Wis, Mar 29, 40; m 66; c 2. GEOLOGY. *Educ:* Univ Wis, BS, 62, MS, 64; Brigham Young Univ, PhD(geol), 68. *Prof Exp:* Asst prof geol, Univ RI, 67-68 & Univ Va, 68-70; asst prof, 70-74, ASSOC PROF GEOL, UNIV AKRON, 74- *Concurrent Pos:* Res grants, 68-70, 71, 73, 75, 76-81, 83, 90-91; consult, NAm Explor, 70. *Mem:* Int Asn Sedimentologists; Soc Econ Paleont & Mineral; Am Asn Petrol Geologists; Geol Soc Am. *Res:* Sedimentary geology; sedimentology; carbonate petrology; paleoecology; paleoenvironments. *Mailing Add:* Dept Geol Univ Akron Akron OH 44325

BAIN, WILLIAM MURRAY, b Indianapolis, Ind, Dec 28, 28; m 64; c 2. MICROBIAL PHYSIOLOGY. *Educ:* Ind Univ, AB, 51, MA, 53, PhD(bact), 59. *Prof Exp:* From asst prof to assoc prof, 59-75, PROF BACT, UNIV MAINE, 75- *Mem:* AAAS; Am Soc Microbiol. *Res:* Endogenous metabolism of bacteria; survival of bacteria, terrestial and aquaric bacteria. *Mailing Add:* Dept Microbiol Univ Maine Orono ME 04473

BAINBOROUGH, ARTHUR RAYMOND, b Can, Mar 8, 18; m 45; c 2. PATHOLOGY. *Educ:* Univ Western Ont, BA, 43, MD, 46; McGill Univ, MSc, 51. *Prof Exp:* Pathologist & dir labs, Med Hat Gen & Galt Hosps, 51-55; pathologist & dir Labs, St Michael's Gen & Lethbridge Munic Hosps & Lethbridge Cancer Clin, 51-80-83, chief med staff, Lethbridge Munic Hosp, 80-82; RETIRED. *Mem:* Fel Am Soc Clin Pathologists; Can Asn Pathologists (pres, 65-66); Int Acad Path. *Res:* Atherosclerosis; histopathology of prostate, bladder and lungs. *Mailing Add:* 3610 Spruce Dr Lethbridge AB T1K 4P6 Can

BAINBRIDGE, KENNETH TOMPKINS, b Cooperstown, NY, July 27, 04; m 31, 69; c 3. NUCLEAR PHYSICS. *Educ:* Mass Inst Technol, BS, 25, MS, 26; Princeton Univ, AM, 27, PhD(physics), 29. *Hon Degrees:* AM, Harvard Univ, 42. *Prof Exp:* Nat res fel physics, 29-31; fel, Bartol Res Found, 31-33; Guggenheim fel, 33-34; from asst prof to prof, 34-46, chmn dept, 53-56, Leverett Prof, 61-75, EMER LEVERETT PROF PHYSICS, HARVARD UNIV, 75- *Concurrent Pos:* Wagner Inst lectr, 40; div leader & mem steering comt, Radiation Lab, Mass Inst Technol, 40-43; div leader, mem steering comt & dir first atom bomb test, Los Alamos, NMex, 43-45; mem, Solvay Cong Chem, 47; chmn, Task Force Acad Planning Sci, Iran/Harvard Univ-Reza Shah Kabir Univ, 75. *Honors & Awards:* Levy Medal, Franklin Inst, 33. *Mem:* Nat Acad Sci; Am Acad Arts & Sci; fel Am Phys Soc. *Res:* Nuclear physics; mass-spectroscopy, mass-spectrometry; photoelectric effect; radar development; high sensitivity caesium-oxygen-silver photocell; amplification of photocell currents by secondary emission; electromagnetic mercury pump; linear direct current motor; change of decay rates of radioactive nuclei. *Mailing Add:* Five Nobscot Rd Weston MA 02193-1146

BAINE, WILLIAM BRENNAN, b Washington, DC, Aug 10, 45; m 69; c 1. INFECTIOUS DISEASE, MICROBIAL PATHOGENESIS. *Educ:* Princeton Univ, AB & cert, 66; Vanderbilt Univ, MD, 70. *Prof Exp:* Intern med, Cleveland Metrop Gen Hosp, 70-71, resident, 71-72; Epidemic Intel Serv officer, Bact Dis Div, Epidemiol Prog, Ctr Dis Control, Atlanta, 72-74; resident internal med, Parkland Mem Hosp, Dallas, 74-75; fel infectious dis, Univ Tex Health Sci Ctr at Dallas, 75-77; med epidemiologist, Bact Dis Div, Epidemiol Prog, Ctr Dis Control, Atlanta, 77-79, Epidemiol Prog Off, 79-81; asst prof, Southwestern Med Sch & Southwestern Grad Sch Biomed Sci, 81-88, ASSOC PROF INTERNAL MED & MICROBIOL, UNIV TEX SOUTHWESTERN MED CTR DALLAS, 81- *Concurrent Pos:* Guest, Lab Epidemiol & Biostatist, Instituto Superiore di Sanità, Rome, Italy, 79-81. *Mem:* Fel Am Col Epidemiol; Am Fedn Clin Res; Am Soc Microbiol; fel Infectious Dis Soc Am; Soc Epidemiol Res. *Res:* Bacterial phospholipase C; microbial toxinology; applied epidemiology and biostatistics of acute and infectious diseases. *Mailing Add:* Dept Internal Med Univ Tex Southwestern Med Ctr Dallas TX 75235-8859

BAINER, ROY, agricultural engineering; deceased, see previous edition for last biography

BAINES, A D, b Toronto, Ont, July 17, 34; m; c 2. PHYSIOLOGY, CLINICAL BIOCHEMISTRY. *Educ:* Univ Toronto, MD, 59, PhD(path chem), 65; FRCP(c), 88. *Prof Exp:* Assoc prof path chem, 68-72, PROF CLIN BIOCHEM, UNIV TORONTO, 72-, DEPT CHMN, 88-; BIOCHEMIST-IN-CHIEF, TORONTO GEN HOSP, 84-, TORONTO HOSP, 87- *Concurrent Pos:* Res fel, Dept Med, Univ NC, Chapel Hill, 65-67; res fel, Dept Physiol, AEC, France, 67-68; mem, Inst Biomed Eng & lectr, Dept Med, Univ Toronto, 72-; assoc physician, Toronto Gen Hosp, 72-; prin, New Col, Univ Toronto, 74- & Dept Clin Biochem, Oxford Univ, Eng, 80-81. *Res:* Normal and abnormal renal physiology; cellular transport mechanisms; histochemistry. *Mailing Add:* Toronto Gen Hosp 200 Elizabeth St Toronto ON M5G 2C4 Can

BAINES, KIM MARIE, b Halifax, NS, Oct 6, 60. ORGANOSILICON CHEMISTRY, ORGANOGERMANIUM CHEMISTRY. *Educ:* St Mary's Univ, BSc, 82; Univ Toronto, PhD(chem), 87. *Prof Exp:* Postdoctoral fel chem, Dortmund Univ, 87-88; ASST PROF CHEM, UNIV WESTERN ONT, 88- *Honors & Awards:* John C Polanyi Prize, Govt Ont, 88. *Mem:* Am Chem Soc; Chem Inst Can; Soc Chem. *Res:* Synthesis of novel organosilicon and organogermanium compounds; germasilenes; siladigermiranes; silylgermanes. *Mailing Add:* Dept Chem Univ Western Ont London ON N6A 5B7 Can

BAINES, RUTH ETTA, b Red Oak, NC. EDUCATIONAL OUTCOMES ASSESSMENT, ALLIED HEALTH & NURSING ACCREDITATION. *Educ:* Spelman Col, Atlanta, BA, 57; NY Univ, MA, 71; State Univ NY, Albany, EdD, 90. *Prof Exp:* Asst dir, Phys Ther Dept, NY Univ Med Ctr, 68-69, res assoc, 69-74; dir, interdisciplinary clin educ, Sch Allied Health Prof, State Univ NY, Stony Brook, 74-76, asst chmn & asst prof, Dept Allied Health Resources, 76-77, chmn & assoc prof, 77-80; assoc chancellor health sci, 80-82, asst vchancellor, 82-88, ASST PROVOST ALLIED HEALTH & NURSING, CENT ADMIN, STATE UNIV NY, 88- *Mem:* Fel Am Soc Allied Health Prof; Am Phys Ther Asn. *Res:* Health professional accreditation reviews; curriculum and instruction in post-secondary health professional programs. *Mailing Add:* State Univ NY State Univ Plaza S300 Albany NY 12246

BAINES, WILLIAM DOUGLAS, b Edmonton, Alta, Feb 11, 26; m 50; c 4. FLUID MECHANICS, HEAT TRANSFER. *Educ:* Univ Alta, BSc, 47; Univ Iowa, MS, 48, PhD(hydraul), 50. *Prof Exp:* Asst prof civil eng, Mich State Univ, 50-51; res officer mech eng, Nat Res Coun Can, 51-59; assoc prof, 59-65, dean grad studies, 70-71, chmn dept, 71-76, dir cont studies, 85- 86, PROF MECH ENG, UNIV TORONTO, 65- *Res:* Viscous flow; turbulence; flow processes of paper manufacture; heat and mass transfer. *Mailing Add:* Dept Mech Eng Univ Toronto Toronto ON M5S 1A4 Can

BAINS, MALKIAT SINGH, b Montgomery, India, Jan 29, 32; m 66; c 3. PHYSICAL INORGANIC CHEMISTRY. *Educ:* Panjab Univ, India, BSc, 55, MSc, 57; Univ London, PhD(inorg chem), 59; Atomic Energy Estab, Trombay, Bombay, cert radiation nuclear chem, 63. *Prof Exp:* Nat Res Coun Can fel, Univ Western Ont, 59-61; lectr inorg chem, Panjab Univ, India, 61-65; reader, Marathwada Univ, India, 65-66; Nat Acad Sci-Nat Res Coun resident res assoc inorg free radicals & electron spin resonance, Agr Res Serv, USDA, 66-68; assoc prof, 68-73, PROF GEN & INORG CHEM, SOUTHERN UNIV, 73- *Concurrent Pos:* New Delhi Univ Grants Comn grant, 64-66. *Mem:* Am Chem Soc; Sigma Xi; The Chem Soc; life mem Indian Chem Soc. *Res:* Coordination chemistry of metal alkoxides and metal carboxylates; flame retardation; inorganic free radicals in solution; application of nuclear magnetic resonance and electron spin resonance to biomedical problems; synthesis of transition metal compounds; metalloporphysins. *Mailing Add:* 5742 Louis Prima E Dr New Orleans LA 70128

BAINTON, CEDRIC R, ANESTHESIOLOGY. *Educ:* Oberlin Col, BA, 53; Univ Rochester, MD, 58; Am Bd Anesthesiol, cert, 67. *Prof Exp:* Intern, Strong Mem Hosp, 58-60; surgeon, Heart Dis Control Prog, Seattle Health Dept, USPHS, 60-62; resident anesthesia, Univ Calif, San Francisco, 62-66, asst clin prof anesthesia, 66-74, from assoc prof to prof in residence, 74-85, PROF ANESTHESIA & PHYSIOL & CHIEF ANESTHESIA SERV, GEN HOSP, UNIV CALIF, SAN FRANCISCO, 85-, CHMN, DEPT ANESTHESIA, 86- *Concurrent Pos:* Fel anesthesia, Cardiovasc Res Inst, Univ Calif, San Francisco, 63-65; guest referee ed, Am J Physiol, 67 & J Appl Physiol, 67-; NIH career develop award, 69-74; clin investr, Vet Admin Med Ctr, 74-77, asst chief anesthesiol serv & dir, Anesthesia Res Labs, 77-85, chief inhalation ther, 79-80 & 83-85, dir, Anesthesia Clin Blood-Gas Lab, 77-85; consult, Children's Hosp, Palo Alto, Calif, 75-78; chmn, Sci Session, Am Soc Anesthesiologists, 76 & Am Physiol Soc, 77 & 79-80, Sci Adv Bd, Asn Univ Anesthetists, 89-90. *Mem:* Am Heart Asn; Am Soc Anesthesiologists; Am Physiol Soc; Asn Univ Anesthetists. *Res:* Author of more than 60 technical publications. *Mailing Add:* Dept Anesthesia San Francisco Gen Hosp 1001 Potrero Ave Rm 35-50 San Francisco CA 94110

BAINTON, DOROTHY FORD, b Magnolia, Miss, June 18, 33; m 59; c 3. PATHOLOGY, INTERNAL MEDICINE. *Educ:* Millsaps Col, BS, 55; Tulane Univ, MD, 58; Univ Calif, San Francisco, MS, 66. *Prof Exp:* Intern med, Strong Mem Hosp, Univ Rochester, 58-59, resident internal med, 59-60; resident hemat, Sch Med, Univ Wash, 60-62; from asst prof to assoc prof, 66-86, PROF PATH & CHMN, SCH MED, UNIV CALIF, SAN FRANCISCO, 86- *Concurrent Pos:* NIH career develop award, 70-75; mem, Hemat Study Sect, NIH, 76-78, Subcomt Res Needs Hemat, Nat Inst Arthritis, Metab, Digestive Dis, 78-81, comt, 78-81, Spec Grants Rev Comt, Nat Inst Arthritis, Diabetes, Digestive & Kidney Dis, 85-87; mem coun, Am Asn Pathologists, 77-80; mem coun, Am Soc Cell Biol, 77-80, Nat Pub Policy Comt, 81-84, publ comt, 84-87; vis prof cell biol, Univ Tex, Dallas, 83; lectr, Royal Micros Soc, Eng, 83; NIH merit award, 86-; mem sci adv bd, Armed Forces Inst Path, 89-; mem, Vascular Biol Res Study Comt, Am Heart Asn, 89-; mem Health Sci Policy Comt, Inst Med-Nat Acad Sci, 90- *Honors & Awards:* Phyllis Bodel Med Lectr, Yale Univ, 79. *Mem:* Inst Med-Nat Acad Sci; Am Soc Cell Biol; AAAS; Am Soc Hemat; Am Soc Histochemists & Cytochemists; Am Asn Pathologists (pres, 90-91); Am Asn Physicians; Am Heart Asn. *Res:* Hematology and cell biology; leukocyte maturation and function using the light and electron microscopes, histochemistry and cell fractionation procedures. *Mailing Add:* Dept Path Sch Med Univ Calif San Francisco CA 94143-0506

BAINUM, PETER MONTGOMERY, b St Petersburg, Fla, Feb 4, 38; m 68; c 1. SPACECRAFT SYSTEMS, STABILITY THEORY. *Educ:* Tex A&M Univ, BS, 59; Mass Inst Technol, SM, 60, Catholic Univ Am, PhD(aerospace eng), 67. *Prof Exp:* Asst engr, Naval Supersonic Lab, Mass Inst Technol, 59-60; sr engr, Martin Marietta Corp, 60-62; staff engr, Space Systs Ctr, IBM Fed Syst Div, 62-65; sr staff engr, Appl Physics Lab, Johns Hopkins Univ, 65-69; assoc prof, Howard Univ, 69-73, dir grad studies, dept mech eng, 74-84, prof aerospace eng, 73-90, GRAD PROF, HOWARD UNIV, 76-, DISTINGUISHED PROF, AEROSPACE ENG, 90- *Concurrent Pos:* Consult, Appl Physics Lab, Johns Hopkins Univ, 69-72; NASA-Am Soc Eng Educ fac fel, Goddard Spaceflight Ctr, NASA, 70-71; prin investr res grants, NASA, 71-, USAF, 88-; vpres & res consult, WHF & Assoc, Inc, 77-86; Outstanding Res, Howard Univ, 80-81. *Honors & Awards:* Teetor Award, Soc Automotive Engrs, 71; Dirk Brouner Award, Am Astronaut Soc, 90. *Mem:* Fel Am Astronaut Soc; assoc fel Am Inst Aeronaut & Astronaut; fel Brit Interplanetary Soc; Sigma Xi; Int Acad Astronaut; German Soc Aviation & Spaceflight. *Res:* Analysis of passive and active satellite control systems, as applied to proposed large space structural systems where both shape and orientation control are required. *Mailing Add:* Dept Mech Eng Howard Univ Washington DC 20059

BAIR, EDWARD JAY, b Ft Collins, Colo, June 30, 22; m 58. PHYSICAL CHEMISTRY, ANALYTICAL CHEMISTRY. *Educ:* Colo Agr & Mech Col, BS, 43; Brown Univ, PhD(chem), 49. *Prof Exp:* Chemist, Manhattan Proj, Tenn Eastman Corp, 44-46; res assoc chem, Univ Wash, 49-54; from instr to assoc prof, 54-65, PROF CHEM, IND UNIV, BLOOMINGTON, 65- *Mem:* Am Chem Soc; Am Phys Soc; Faraday Soc. *Res:* Photochemistry; spectroscopy; kinetics of atmospheric and combustion reactions. *Mailing Add:* Dept Chem Ind Univ Bloomington IN 47401

BAIR, HARVEY EDWARD, b Williamsport, Pa, June 6, 36; m 58; c 4. THERMAL PHYSICAL PROPERTIES OF MATERIALS. *Educ:* Dickinson Col, BS, 58; Pa State Univ, MS, 62. *Prof Exp:* Res training fel, Gen Elec Res Lab, 62-64; assoc chemist, Appl Physics Lab, Johns Hopkins Univ, 64-65; assoc mem tech staff, 65-75, MEM TECH STAFF, AT&T BELL LABS, 75- *Concurrent Pos:* Guest fac mem, Polymer Res Inst, Polytech Univ, NY, 90-91. *Honors & Awards:* Mettlar Award in Thermal Anal, NAm Thermal Anal Soc, 87. *Mem:* Fel NAm Thermal Anal Soc (pres, 84); fel Am Phys Soc; Am Chem Soc. *Res:* Application of thermoanalytical techniques to characterize the structure and behavior of polymeric materials, especially multicomponent blends, reactive polymer systems, and materials containing additives or contaminants. *Mailing Add:* AT&T Bell Labs 600 Mountain Ave Murray Hill NJ 07974-2070

BAIR, JOE KEAGY, b Massillon, Ohio, Mar 10, 18; m 41; c 1. NUCLEAR PHYSICS. *Educ:* Rice Inst, BA, 40. *Prof Exp:* Asst physics, Columbia Univ, 40-41; physicist, Naval Ord Lab, 41-47 & Fairchild Eng & Airplane Co, 47-51; Physicist, Oak Ridge Nat Lab, 51-80; RETIRED. *Concurrent Pos:* Civilian technician, Naval Mine Modification Unit, 44-45. *Mem:* Am Phys Soc. *Res:* Low energy nuclear physics. *Mailing Add:* 200 W Fairview Rd Oak Ridge TN 37830

BAIR, KENNETH WALTER, b Detroit, Mich, Mar 20, 48; m 69, 79; c 1. SYNTHETIC ORGANIC CHEMISTRY. *Educ:* Wayne State Univ, BS, 70, MS, 73; Brandeis Univ, PhD(org chem), 76. *Prof Exp:* Lab technician clin chem, Clin Chem Lab, Harper Hosp, 69-72; res assoc enzyme chem, Dept Internal Med, Hutzel Hosp, 72-73; res assoc org chem, Wayne State Univ, 73; res assoc org chem, Brandeis Univ, 73-76; Damon Runyon-Walter Winchell res fel, Mass Inst Technol, 76-78; SR RES CHEMIST, BURROUGHS-WELLCOME CO, 78- *Mem:* Am Chem Soc; The Chem Soc. *Res:* Synthesis of medicinally important compounds, development of new synthetic methods. *Mailing Add:* Org Chem Burroughs 3030 Cornwallis Rd Research Triangle Park NC 27709-4416

BAIR, THOMAS DE PINNA, b New York, NY, Mar 19, 22; m 84; c 2. PARASITOLOGY, PHYSIOLOGY. *Educ:* DePauw Univ, AB, 46; Ind Univ, MA, 47; Univ Ill, PhD(zool, physiol), 51. *Prof Exp:* Asst zool, Ind Univ, 47; asst zool & physiol, Univ Ill, 47-49, vet parasitol, 49-50; instr biol, Utica Col, Syracuse Univ, 50-53; instr physiol, Albany Med Col, 53-56; from asst prof to prof zool, Calif State Univ, Los Angeles, 72-88; RETIRED. *Concurrent Pos:* Vis asst prof, Calif Col Med, 58-64; lectr & consult, Gormac Polygraph Inst, 59-; sr lectr, Univ Southern Calif, 64-; Fulbright prof, Nangrahar Med Fac, Jalalabad, Afghanistan, 70-71 & India, 77-78; fel trop med, La State Univ Sch Med, USPHS, fel physiol, Univ Calif, Los Angeles. *Mem:* Indian Soc Parasitol; AAAS; Am Soc Parasitol; Sigma Xi. *Res:* Comparative physiology; parasitology; history and philosophy of science; physiology of parasites. *Mailing Add:* Journey's End Dr La Canada CA 91012

BAIR, THOMAS IRVIN, b Grier City, Pa, Feb 12, 38; m 66. POLYMER CHEMISTRY, PHYSICAL ORGANIC CHEMISTRY. *Educ:* Pa State Univ, BS, 60; Univ Wis, PhD(electrophilic substitution), 66. *Prof Exp:* Res chemist, 65-69, sr res chemist, 69-72, supvr res & develop, 72-77, RES ASSOC, E I DU PONT DE NEMOURS & CO, 77- *Mem:* Am Chem Soc; Sigma Xi; NY Acad Sci. *Res:* Cyclopropanol chemistry; nuclear magnetic resonance spectroscopy of cyclopropane compounds; electrophilic substitution at saturated carbon; synthesis and characterization of condensation polymers; fiber technology; rubber reinforcing materials; adhesives and finishes. *Mailing Add:* Four Little Leaf Ct Foulk Woods Wilmington DE 19810

BAIR, WILLIAM J, b Jackson, Mich, July 14, 24; m 52; c 3. RADIOBIOLOGY, RADIOLOGICAL HEALTH. *Educ:* Ohio Wesleyan Univ, BA, 49; Univ Rochester, PhD(radiation biol), 54. *Prof Exp:* Nat Res Coun-AEC fel, Univ Rochester, 49-50, res assoc, 50-54; biol scientist, Hanford Labs, Gen Elec Co, Richland, Wash, 54-56; mgr inhalation toxicol sect, Biol Dept, Battelle Mem Inst, 56-68, mgr Biol Dept, 68-74, dir life sci prog, 73-75, mgr biomed & environ res prog, 75-76, mgr environ health & safety res prog, 76-86, MGR LIFE SCI CTR, PAC NORTHWEST LABS, BATTELLE MEM INST, 86- *Concurrent Pos:* Lectr radiation biol, Joint Ctr Grad Study, Richland, Wash, 55-; mem subcomt on inhalation hazards, Comt on Path Effects of Atomic Radiation, Nat Acad Sci, 57-64; Japan AEC invited lectr, Nat Radiol Health Inst, Chiba, 69, Radiation Effects Res Found, Hiroshima, 89; chmn, task force, Int Comn Radiol Protection, 70-79, ad hoc comt hot particles, 74-75 & task group respiratory tract model, 84-; mem, sci comt radionuclides, Int Coun Radiol Protection, 70-77, comt internal radiation, 73-, Nat Coun Radiation Protection & Measurements, 74-, comt basic radiation protection, 75- & bd dirs, 76-82, Presidents Comn Accident Three Mile Island, 79-80 & regional steering comt eruption Mt St Helens, 80-; consult to Adv Comt on Reactor Safeguards, Nuclear Regulatory Comn, 71-87; chmn, Transuranium Tech Group, Am Inst Biol Sci-AEC- Energy Res & Develop Admin, 72-75; mem ad hoc comt on hot particles of subcomt biol effects of ionizing radiation, Nat Acad Sci-Nat Res Coun, 74-76 & vchmn, comt on biol effects of ionizing radiation IV alpha radiation, 85-88; chmn, Marshall Islands radiol adv group, Dept Energy, 78-81; mem, Argonne Univ Asn Review Comt, 77-80; lectr, SAfrican Asn Physicists in Med & Biol, Pretoria, 80; invited lectr, NChina Inst Radiation Protection, 84; guest scientist, USSR Meeting Med Consequences Chernobyl Accident, 88. *Honors & Awards:* E O Lawrence Mem Award, US AEC, 70. *Mem:* Fel AAAS; Am Soc Exp Biol & Med; Radiation Res Soc; Health Physics Soc (pres-elect, 83-84, pres, 84-85); Sigma Xi; NY Acad Sci; Soc Occup & Environ Health. *Res:* Radiation biology of inhaled radionuclides; health and environmental effects of transuranium elements; establishment of radiation protection criteria and standards for radionuclides. *Mailing Add:* 102 Somerset Richland WA 99352

BAIRD, ALBERT WASHINGTON, III, electromagnetics, engineering physics, for more information see previous edition

BAIRD, ALFRED MICHAEL, b Philadelphia, May 29, 48; m 70; c 2. TARGET SIGNATURE ANALYSIS. *Educ:* St Joseph's Col, BS, 70; Dartmouth Col, PhD(physics), 77. *Prof Exp:* Lead scientist, Analytics Inc, 78-83; lead engr, Raytheon Co, 83-85; DIV DIR, ANALYTICS INC, 85- *Mem:* Optical Soc Am. *Res:* Analysis and simulation of electromagnetic signatures of military targets and performance evaluation of autonomous acquisition sensors. *Mailing Add:* 2418 Rocky Shores Dr Niceville FL 32578

BAIRD, BARBARA A, b Decatur, Ill, June 8, 51; m 79; c 3. MEMBRANE BIOCHEMISTRY. *Educ:* Knox Col, BA, 73; Cornell Univ, MS, 75, PhD(chem), 79. *Prof Exp:* Fel immunol, Nat Cancer Inst, 78-80; from asst prof to assoc prof, 80-91, PROF CHEM, CORNELL UNIV, 91- *Honors & Awards:* Lamport Award Young Investr Biophys, NY Acad Sci, 87. *Mem:* Am Soc Biol Chemists; Biophys Soc; Am Chem Soc. *Res:* Biophysical chemistry of immunological systems; molecular details of the structure and function of the cell surface receptor for immunoglobulin E from a mammalian cell line. *Mailing Add:* Box 555 Lab Chem Cornell Univ Ithaca NY 14853-1301

BAIRD, CHARLES ROBERT, b Windsor, NS, Nov 2, 35; m 62; c 6. CONTROL SYSTEMS THEORY, OPTIMAL SYSTEMS THEORY. *Educ:* Tech Univ NS, BEng, 57; Univ BC, MASc, 62; Univ NB, PhD(elec eng), 74. *Prof Exp:* From asst prof to assoc prof, 74-79 & 81-86, PROF & HEAD, ELEC ENG DEPT, TECH UNIV NS, 86- *Concurrent Pos:* Dir & prin investr, Vehicle Safety Res Team, TUNS for Transp Can, 75-; adj prof elec eng, US Navy Postgrad Sch, 79-81; pres, Synergetic Eng Assoc Inc, 84-; mem bd examiners, Asn Prof Eng NS, 89-92. *Mem:* Soc Automotive Engrs; Am Asn Automotive Med. *Res:* Identification and estimation theory applied to optimal control systems; knowledge based intelligent controllers; vehicle safety research. *Mailing Add:* 6748 Quinpool Rd Halifax NS B3L 1C3 Can

BAIRD, CRAIG RISKA, b Woodland, Utah, May 6, 39; m 63; c 6. ENTOMOLOGY, WILDLIFE BIOLOGY. *Educ:* Utah State Univ, BS, 67, MS, 70; Wash State Univ, PhD(entom), 73. *Prof Exp:* Sr sci aide entom, Wash State Univ, 72-73, pest mgt specialist, Dept Entom, 73-74; asst prof, 74-80, PROF ENTOM & EXTEN ENTOMOLOGIST, UNIV IDAHO, 80- *Mem:* Entom Soc Am; Entom Soc Can; Soc Vector Ecologists. *Res:* Life history and biology of rodent and rabbit botflies; livestock parasites and control; alfalfa seed production and pest management; leafcutting bee management; ecology of wildlife ectoparasites; hops insects and pest management. *Mailing Add:* Univ Idaho SW Idaho Res & Exten Ctr Parma ID 83660-9637

BAIRD, D C, b Edinburgh, Scotland, May 6, 28; m 54; c 4. LOW TEMPERATURE PHYSICS. *Educ:* Univ Edinburgh, BSc, 49; Univ St Andrews, PhD(physics), 52. *Prof Exp:* Lectr, 52-54, from asst prof to assoc prof, 54-65, head dept, 72-78, PROF PHYSICS, ROYAL MIL COL CAN, 65-, DEAN SCI, 80- *Concurrent Pos:* Defence Res Bd Can res grant, 58- *Res:* Low temperature physics; magnetic properties of superconductors. *Mailing Add:* Dept Physics Royal Mil Col Kingston ON K7L 2W3 Can

BAIRD, DAVID MCCURDY, geology, for more information see previous edition

BAIRD, DONALD, b Pittsburgh, Pa, May 12, 26; m 48; c 2. VERTEBRATE PALEONTOLOGY. *Educ:* Univ Pittsburgh, BS, 48; Univ Colo, MS, 49; Harvard Univ, PhD(biol), 55. *Hon Degrees:* MS, Wagner Free Inst Sci, 90. *Prof Exp:* Asst geologist, Pa Topog & Geol Surv, 47-48; cur, Univ Mus, Univ Cincinnati, 49-51; asst cur vert paleont, Mus Comp Zool, Harvard Univ, 54-57; from asst cur to assoc cur, Mus Natural Hist, Princeton Univ, 57-67, cur, 67-73, dir, 73-88; RETIRED. *Concurrent Pos:* Res assoc, Am Mus Natural Hist, 64-, Acad Natural Sci, Philadelphia, 84-, Carnegie Mus Natural Hist, 84-; pres, Princeton Jr Mus, 68-70. *Mem:* Soc Vert Paleont; Paleont Soc; Hist Earth Sci Soc; Soc Hist Archaeol; Brit Palaeont Asn; Sigma Xi. *Res:* Carboniferous, Triassic and Cretaceous reptiles, amphibians and footprints. *Mailing Add:* Four Ellsworth Terr Pittsburgh PA 15213

BAIRD, DONALD HESTON, b Milwaukee, Wis, July 3, 21; m 51; c 2. PHYSICS & CHEMISTRY OF THIN FILMS. *Educ:* Haverford Col, BS, 43; Harvard Univ, MA, 50, PhD(phys chem), 50. *Prof Exp:* Asst, Manhattan Proj, 43-46; MEM TECH STAFF, GEN TEL & ELECTRONICS LABS, INC, 50- *Mem:* Am Phys Soc; Inst Elec & Electronics Engrs; AAAS. *Res:* Fundamental development of materials for thin film electroluminescent services, for magnetic bubble memories and other magnetic devices and for various types of opto-electronic devices. *Mailing Add:* GTE Labs Inc 31 Locke Rd Newton MA 02168

BAIRD, GORDON CARDWELL, b Rochester, NY, Oct 6, 46. HISTORICAL GEOLOGY. *Educ:* Earlham Col, BA, 69; Univ Nebr-Lincoln, MS, 71; Univ Rochester, PhD(geol), 75. *Prof Exp:* Res assoc geol sci, State Univ NY Binghamton, 75-76; asst curator, Field Mus Natural Hist, 76-; AT DEPT GEOL, STATE UNIV NY, FREDONIA. *Mem:* AAAS; Paleontological Soc; Soc Econ Paleontologists & Mineralogists; Sigma Xi. *Res:* Paleoecology-sedimentary geology; animal-sediment relationships in ancient deposits in order to do paleocommunity-paleoenvironmental reconstructions. *Mailing Add:* Dept Geol State Univ NY Fredonia NY 14063

BAIRD, HENRY W, III, pediatrics, neurology; deceased, see previous edition for last biography

BAIRD, HERBERT WALLACE, b Morganton, NC, Sept 2, 36; m 58; c 2. STRUCTURAL CHEMISTRY. *Educ:* Berea Col, BA, 58; Univ Wis, PhD(chem), 63. *Prof Exp:* Asst chem, Univ Wis, 59-60; from asst prof to assoc prof, 63-75, PROF CHEM, WAKE FOREST UNIV, 75- *Mem:* Am Chem Soc; Am Crystallog Asn. *Res:* X-ray diffraction structure analysis. *Mailing Add:* PO Box 849 Walkerton NC 27051-0849

BAIRD, JACK R, b Colchester, Ill, May 1, 31; m 54; c 3. ELECTRICAL ENGINEERING. *Educ:* Univ Ill, BS, 58, MS, 59, PhD(elec eng), 63. *Prof Exp:* Res assoc elec eng, Univ Ill, 63-64, asst prof, 64-68; from assoc to prof elec eng, Univ Colo, Boulder, 68-80; PRES, LED SYSTEMS INC, BOULDER, 80- *Mem:* Inst Elec & Electronics Engrs. *Res:* Millimeter and submillimeter wave length generation, transmission, detection and optical communication systems. *Mailing Add:* 150 Manhattan Dr Boulder CO 80303

BAIRD, JACK VERNON, b Grand Island, Nebr, July 7, 28; m 49; c 3. SOIL SCIENCE. *Educ:* Univ Nebr, BS, 49, MS, 51; Wash State Univ, PhD(soil sci), 55. *Prof Exp:* Exten agronomist soil mgt, Univ Ill, 58-60; gen agron, Kans State Univ, 61-64; exten agronomist soil sci, 64-71, actg head dept, 70-71, EXTEN PROF SOIL SCI, NC STATE UNIV, 71- *Mem:* Am Soc Agron; Soil Sci Soc Am. *Res:* Soil fertility problems in North Carolina; proper fertilizer usage and soil management. *Mailing Add:* Dept Soil Sci NC State Univ Box 7619 Raleigh NC 27695-7619

BAIRD, JAMES CLYDE, b Montgomery, Ala, Oct 26, 31; m 59; c 1. PHYSICAL CHEMISTRY. *Educ:* Stanford Univ, BS, 53; Rice Inst, PhD(chem), 59. *Prof Exp:* Fel chem, Harvard Univ, 58-60; res chemist, Calif Res Corp, 60-62; from asst prof to assoc prof chem, 62-71, PROF CHEM & PHYSICS, BROWN UNIV, 71- *Concurrent Pos:* Res assoc, Univ Calif, Berkeley, 60-62; vis fel, 65-67; vis fel, Joint Inst Lab Astrophys, Univ Colo, 70-71. *Mem:* Am Phys Soc. *Res:* Atomic and molecular physics and chemistry; radio frequency spectroscopy. *Mailing Add:* Dept Chem Brown Univ Providence RI 02912

BAIRD, JAMES KERN, b Pittsburgh, Pa, Aug 24, 41; m 67; c 1. RADIATION CHEMISTRY, CRYSTAL GROWTH. *Educ:* Yale Univ, BS, 63; Harvard Univ, AM, 65, PhD(chem physics), 69. *Prof Exp:* Captain & dep test group coordr, Test Command, Defense Atomic Support Agency, Sandia Base, US Army, NMex, 69-70; physicist, Oak Ridge Nat Lab, Tenn, 70-81; mgr, radiochem unit, Knolls Atomic Power Lab, Schenectady, NY, 81-82; chmn dept, 82-90, PROF CHEM, UNIV ALA, HUNTSVILLE, 82- *Concurrent Pos:* Adj assoc prof, dept physics, Univ Kans, Lawrence, 77-80; prin sci advisor, consortium for mat develop in space, Univ Ala, 85- *Honors & Awards:* Cert of Achievement, Test Command, Defense Atomic Support Agency, 70. *Mem:* Am Phys Soc; Am Chem Soc; Radiation Res Soc; Mat Res Soc. *Res:* Theoretical chemistry with emphasis on problems in diffusion, radiation chemistry, plasma chemistry and crystal growth. *Mailing Add:* Dept Chem Univ Ala Hunstville AL 35899

BAIRD, JAMES LEROY, JR, b Bridgeport, Conn, Aug 5, 34; m 56; c 3. COMPARATIVE PHYSIOLOGY. *Educ:* Tufts Univ, BS, 56; Univ Minn, MS, 59; Univ Conn, PhD(zool), 64. *Prof Exp:* Res entomologist, US Forest Serv, 59; instr biol, Conn Col, 62-63 & zool, Univ Conn, 63-64; asst prof, Lafayette Col, 64-71; actg dean studies, 71; assoc prof biol & head dept, Rochester Inst Technol, 71-74; fac assoc student affairs, 74-75; DIR, AVERY POINT CAMPUS, UNIV CONN, 75- *Mem:* Entom Soc Am; Am Soc Zoologists. *Res:* Comparative physiology of insect flight; insect neuromuscular biology; biological control systems; arthropod behavior. *Mailing Add:* Pres/Chancellor Univ Conn Avery Point Groton CT 06340

BAIRD, JOHN JEFFERS, b North English, Iowa, Jan 1, 21; m 45; c 1. NEUROEMBRYOLOGY. *Educ:* Iowa State Teachers Col, AB, 48; Univ Iowa, MS, 53, PhD, 57. *Prof Exp:* Teacher high sch, Iowa, 48-54; instr, Univ Iowa, 54-56; from asst prof to prof biol embryol, Calif State Col Long Beach, 56-67, chmn, Dept Biol, 61-67, assoc dean acad planning, 67-73; dep dean 73-78; prof biol, educ progs & resources, 67-78, prof biol, 78-84, actg chmn dept, 81-82, EMER PROF, CALIF STATE UNIV, LONG BEACH, 84- *Mem:* AAAS; Am Soc Zool; Sigma Xi. *Res:* Peripheral control of development of motor and sensory elements of central nervous system. *Mailing Add:* Dept Biol 1250 Bellflower Blvd Long Beach CA 90840

BAIRD, KENNETH MACCLURE, b Hawaiking Fu, China, Jan 23, 23; Can citizen; m 51; c 4. PHYSICS. *Educ:* Univ New Brunswick, BSc, 43; Univ Bristol, PhD(physics), 53. *Prof Exp:* Prin res scientist, Nat Res Labs, 43-48 & 50-81; RETIRED. *Mem:* Optical Soc Am. *Res:* Optics; interferometry. *Mailing Add:* 25 Rothwell Dr Gloucester ON K1J 7G5 Can

BAIRD, MALCOLM BARRY, b Wilkes Barre, Pa, Feb 4, 43; m 64; c 4. PHYSIOLOGY, BIOCHEMICAL PHARMACOLOGY. *Educ:* Wilkes Col, AB, 64; Univ Del, MA, 68, PhD(genetics), 70. *Prof Exp:* Olsen Mem fel, 70-72, res scientist, 72-82, SR RES SCIENTIST, MASONIC MED RES LAB, 82- *Concurrent Pos:* Vis prof, Syracuse Univ, 73-75 & 80. *Mem:* Genetics Soc Am; Am Asn Cancer Res; AAAS; Am Aging Asn. *Res:* Genetics of development and mechanisms of aging; chemical carcinogenesis; metabolism of xenobiotics via mixed-function oxidase activity; xenobiology; modulation of mutagenesis and carcinogenesis by retinoids. *Mailing Add:* Masonic Med Res Lab 2150 Bleecker St Utica NY 13503

BAIRD, MALCOLM HENRY INGLIS, b London, Eng, July 2, 35; Can & Brit citizen; m 65; c 2. CHEMICAL ENGINEERING. *Educ:* Univ Glasgow, BSc, 57; Univ Cambridge, PhD(chem eng), 60. *Prof Exp:* Res engr, Can Industs Ltd, 60-63; teaching fel chem eng, Univ Edinburgh, 63-65, lectr, 65-67; assoc prof, 67-73, chmn dept, 82-85, PROF CHEM ENG, MCMASTER UNIV, 73- *Concurrent Pos:* Assoc ed, Can J Chem Eng, 69-73, 74-76 & 81-84; vis fel, Univ Queensland, Australia, 73-74; vis prof, Univ Bradford, UK, 85; assoc ed, Chem Eng Res & Design, 90- *Mem:* Can Soc Chem Eng. *Res:* Mass transfer and fluid mixing processes such as mixing, solvent extraction and gas absorption; oscillation flow phenomena; resonant drops and bubbles; vortex rings in liquids; hydrometallurgy. *Mailing Add:* Chem Eng Dept McMaster Univ Hamilton ON L8S 4L7 Can

BAIRD, MERTON DENISON, b Ft Wayne, Ind, Sept 9, 40; m 67; c 2. ORGANIC CHEMISTRY. *Educ:* Purdue Univ, BS, 62; Univ Wis-Madison, PhD(org chem), 69. *Prof Exp:* From asst prof to assoc prof, 68-80, chmn dept, 74-77, PROF CHEM, SHIPPENSBURG UNIV, 80- *Mem:* Am Chem Soc. *Res:* Synthesis of organic compounds related to natural products; applications of nuclear magnetic resonance and computational methods to conformational analysis of molecules and correlations of structure and reactivity; synthesis of silicon containing anticonvulsants. *Mailing Add:* Dept Chem Shippensburg Univ Shippensburg PA 17257

BAIRD, MICHAEL JEFFERSON, b New Orleans, La, Oct 21, 39; m 67; c 4. CATALYST EVALUATIONS. *Educ:* Tex A&M Univ, BS, 63, PhD(phys chem), 71; Harvard Univ, MS, 79. *Prof Exp:* Process engr, E I Du Pont, Chattanooga, 63-64; res chemist, US Bur Mines, Minneapolis, 71-74; supvry chem engr, Pittsburgh Energy Res Ctr, US Dept Energy, 74-79; sr res chemist, Res & Develop, Ashland Oil, 79-80; STAFF RES ENGR, HYDROTREATING PROCESS DEVELOP, AMOCO OIL, NAPERVILLE ILL, 80- *Concurrent Pos:* Adj prof chem eng, Ill Inst Technol, 81; Univ Ill, Chicago, 82 & 84. *Mem:* Am Chem Soc. *Res:* Pilot plant evaluations of hydrotreating and hydrocracking catalysts; catalyst preparation of new hydrotreating catalysts; catalyst regeneration processes. *Mailing Add:* Amoco Res Ctr Mail Code H-4 Box 400 Naperville IL 60566

BAIRD, NORMAN COLIN, b Montreal, Que, May 22, 42; c 1. PHYSICAL CHEMISTRY. *Educ:* McGill Univ, BSc, 63, PhD(phys chem), 67. *Prof Exp:* Robert A Welch fel, Univ Tex, 66-68; asst prof to assoc prof, 68-78, PROF PHYS CHEM, UNIV WESTERN ONT, 78- *Res:* Quantum-mechanical investigations of the structure, energy and chemical bonding in molecules. *Mailing Add:* Dept Chem Univ Western Ont London ON N6A 5B9 Can

BAIRD, PATRICIA A, b Rochdale, Eng; Can citizen; m 64; c 3. MEDICAL GENETICS. *Educ:* McGill Univ, BS, 59, MD & CM, 63. *Prof Exp:* Instr pediat, Univ BC, 68-72, from asst prof to prof med genetics, Dept Med Genetics, 72-89, head, 79-89, actg dir, Ctr Molecular Genetics, 82-87; head med genetics, Grace & Children's Hosps, BC, 81-89; CHAIR, FED ROYAL COMN NEW REPRODUCTIVE TECHNOLOGIES, 89-; VPRES, CAN INST ADVAN RES, 91- *Concurrent Pos:* Med consult, Health Surveillance Registry BC, 77-90; mem, Bd Gov, Univ BC, 84-90; vpres, Can Col Med Geneticists, 84-86; mem, Nat Consult Comt Epidemiol Ment Retardation, Can, 84-87; mem, Nat Adv Bd Sci & Technol, Fed govt, 87-; co-chair, Nat Forum Sci & Technol Councils, 91; mem ethics panel, Int Pediat Asn. *Mem:* Am Soc Human Genetics; Genetics Soc Can; Royal Col Physicians Can. *Res:* Use of large population-based registry of genetic and handicapping conditions to provide insights and information useful in clinical counselling and program planning; natural history of genetic disorders; documentation of the occurrence of conditions associated with various birth defects in a non-random way; biological background and mechanisms responsible for some handicapping disorders. *Mailing Add:* Dept Med Genetics Univ BC 226-6174 Univ Blvd Vancouver BC V6T 1Z3 Can

BAIRD, QUINCEY LAMAR, b Hoschton, Ga, Mar 22, 32; m 56; c 4. REACTOR SAFETY. *Educ:* Berry Col, AB, 52; Emory Univ, MS, 54; Vanderbilt Univ, PhD(physics), 58. *Prof Exp:* Teacher pvt sch, Tenn, 52-53; physicist, Argonne Nat Lab, 58-63; engr Nuclear Technol Div, Idaho Opers Off, US AEC, 63-66; sr res scientist, Pac Northwest Lab, Battelle Mem Inst, 66-70; sr res scientist, Wadco Corp, 70-74, PRIN ENGR & TECH ASSOC, WESTINGHOUSE-HANFORD CO, 74- *Mem:* Am Phys Soc; Am Nuclear Soc. *Res:* Experimental reactor physics related to power reactor design and operation; nuclear engineering aspects of system design; liquid metal and fast breeder reactor safety, design requirements and operations analysis. *Mailing Add:* Westinghouse-Hanford Co Hanford Eng Dev Lab, Bldg 337 Richland WA 99352

BAIRD, RAMON CONDIE, b Ogden, Utah, May 25, 29; m 61; c 4. PHYSICS, ELECTRICAL ENGINEERING. *Educ:* Brigham Young Univ, BS, 55, MS, 57; Univ Colo, PhD(physics), 65. *Prof Exp:* Proj leader, mm-wave interferometry, Microwave Physics Sect, Radio Physics Div, Nat Bur Standards, 57-66, sect chief antenna measurements, antenna systs metrol, Electromagnetic Fields Div, 66-87, DIV CHIEF, NAT INST STANDARDS & TECHNOL, 87- *Concurrent Pos:* Chmn sub-comt measurements & instrumentation & mem comt radiation hazards, Am Nat Standards Inst, 70-; chmn US Comn Electromagnetic Metrol, Int Union Radio Sci, 78- *Mem:* Am Nat Standards Inst; Int Union Radio Sci; Inst Elec & Electronics Engrs. *Res:* Millimeterwave measurements of the speed of light and refractive indices of gases; near-field antenna measurement techniques; metrology for satellite communication systems. *Mailing Add:* Electromagnetic Fields Div 72300 Nat Inst Standards & Technol Boulder CO 80303

BAIRD, RICHARD LEROY, b Sioux City, Iowa, Aug 15, 31; m 59; c 2. POLYMER CHEMISTRY. *Educ:* Reed Col, BA, 53; Univ Calif, Los Angeles, PhD, 58. *Prof Exp:* Instr chem, Yale Univ, 58-60, asst prof, 60-63, lectr, 63-64; res chemist, Cent Res Dept, E I Du Pont de Nemours & Co, 64-69, Elastomers Dept, 69-73, develop supvr, 73-74, div head, Elastomers Dept, 74-79, res mgr, Polymer Prod Dept, 79-80, tech mgr, 80-81. *Mem:* Am Chem Soc. *Res:* Physical organic chemistry; coordination chemistry. *Mailing Add:* Eight Aldrich Way Wilmington DE 19807

BAIRD, RONALD C, b New Albany, Ind, Apr 1, 36; m 62; c 2. EDUCATION ADMINISTRATION, ECOLOGY. *Educ:* Yale Univ, BS, 58; Univ Tex, MA, 65; Harvard Univ, PhD, 69. *Prof Exp:* Asst cur, Mus Comp Zool, 68-69; from asst prof to assoc prof marine sci, Univ SFla, 70-78; vpres & exec vpres, 79-84, PRES, SCHUSTER CORP, 85-; DIR CORP RELS, WORCESTER POLYTECH INST, 91- *Concurrent Pos:* Dir res, Geo-Marine, Inc, 76-79; courtesy prof marine sci, Univ SFla, 79-; trustee, Worcester Hort Soc, 89-; vchmn, Nat Sea Grant Adv panel, 90- *Mem:* AAAS; Am Soc Limnol & Oceanog; Am Soc Naturalists; Am Soc Eng Educ; Am Soc Zoologists; Am Soc Ichthyologists & Herpetologists. *Res:* Biology of deep-sea ecosystems, oceanic ichthyology and coastal resource management. *Mailing Add:* Worcester Polytech Inst 100 Institute Rd Worcester MA 01609-2280

BAIRD, RONALD JAMES, b Toronto, Ont, May 3, 30; m 55; c 3. CARDIOVASCULAR SURGERY. *Educ:* Univ Toronto, MD, 54, BSc, 56, MSurg, 64; FRCS(C), 59, cert cardiovasc & thoracic surg, 64. *Prof Exp:* Dir surg res, Univ & Toronto Gen Hosp, 72-77, cardiovasc surg, 77-88; PROF SURG, UNIV TORONTO, 72- *Concurrent Pos:* Res assoc, Ont Heart Found, 64-; dir, Can Heart Found, 67- & Ont Heart Found, 67-; assoc ed, Can J Surg, 67-; mem, Med Res Coun Can, 67-; chief cardiovasc surg, Toronto Western Hosp, 72-77. *Honors & Awards:* Royal Col Surg Medal, 69. *Mem:* Can Cardiovasc Soc (secy-treas, 70-73); fel Am Col Surg; Int Cardiovasc Soc (vpres, 73-75); Am Surg Asn; Soc Univ Surg; Int Soc Cardiovasc Surg (pres, 88-89). *Res:* Cardiovascular surgery and physiology; surgery of the ischemic heart and the ischemic limb. *Mailing Add:* Toronto Gen Hosp 101 Col St Suite 12-220 Eaton Tower Toronto ON M5G 1L7 Can

BAIRD, TROY ALAN, b Pomona, Calif, Dec 29, 54; m 79. EVOLUTION OF SOCIAL BEHAVIOR IN VERTEBRATES, REPRODUCTIVE ECOLOGY OF VERTEBRATES. *Educ:* Calif State Univ, San Diego, BS, 78, MS, 80; Univ BC, PhD(zool), 89. *Prof Exp:* Postdoctoral res assoc, Marine Sci Ctr, Ore State Univ, Hatfield, 88-89; ASST PROF BIOL, ZOOL & ANIMAL BEHAV, CENT STATE UNIV, 89- *Mem:* Am Soc Zoologists; Am Soc Ichthyologists & Herpetologists; Sigma Xi. *Res:* Evolution of social systems; adaptive significance of group living, individual use of space, reproductive patterns and both intra- and inter-specific social interactions. *Mailing Add:* Dept Biol Univ Cent Okla 100 N University Dr Edmond OK 73034

BAIRD, WILLIAM C, JR, b McKeesport, Pa, Sept 23, 33; m 61; c 2. ORGANIC CHEMISTRY, CATALYSIS. *Educ:* Pa State Univ, BS, 55; Univ Iowa, PhD(chem), 59. *Prof Exp:* Am Chem Soc-Petrol Res Fund fel, Swiss Fed Inst Technol, 59-60; NSF fel, La State Univ, 60-61; chemist catalysis; Exxon Res & Eng Co, 61-68, res assoc, 68-72, res assoc, Exxon Res & Develop Lab, 72-75, SR RES ASSOC, EXXON RES & DEVELOP LAB, 75- *Mem:* Am Chem Soc. *Res:* Organometallic chemistry; homogeneous catalysis; heterogeneous catalysis; refining processes. *Mailing Add:* 5905 Bennington Ave Baton Rouge LA 70808

BAIRD, WILLIAM MCKENZIE, b Philadelphia, Pa, Mar 23, 44; m 69; c 3. CHEMICAL CARCINOGENESIS. *Educ:* Lehigh Univ, BS, 66; Univ Wis, PhD(oncol), 71. *Prof Exp:* Fel, Inst Cancer Res, London, 71-73; asst prof, Wistar Inst, 73-77, assoc prof, 78-80; assoc prof, 80-82, PROF MED CHEM, PURDUE UNIV, 82-, DIR, PURDUE CANCER CTR, 86-, GLENN L JENKINS PROF MED CHEM, 89- *Concurrent Pos:* Fac partic, biochem prog, Purdue Univ, 80- & Purdue Univ Cancer Ctr, 80-; adv comt, biochem & chem carcinogenesis, Am Cancer Soc, 83-86; mem, chem path study sect, NIH, 86-90. *Mem:* Am Asn Cancer Res; AAAS; Am Chem Soc; Genetic Toxicol Asn; Am Soc Biochem & Molecular Biol; Am Soc Biol Chemists. *Res:* Metabolism of polycyclic aromatic hydrocarbons; interactions of chemical carcinogens with nucleic acids; use of cells in culture for studies on carcinogenesis; tumor promotion and environmental carcinogenesis; chemoprevention of cancer; genetic toxicology. *Mailing Add:* Dept Med Chem & Pharmacog Pharm Bldg Sch Pharm & Pharm Sci Purdue Univ West Lafayette IN 47907

BAISDEN, CHARLES ROBERT, b Logan, WVa, Apr 23, 39; m 62; c 3. MEDICINE, PATHOLOGY. *Educ:* WVa Univ, AB, 61, MD, 65. *Prof Exp:* Intern med, WVa Univ Hosp, 65-66, resident path, 66-68; resident, Baptist Mem Hosp, Memphis, Tenn, 68-70; asst prof lab med, Sch Med, Johns Hopkins Univ & pathologist, Johns Hopkins Hosp, 70-79; asst prof path, 84, DIR CLIN LABS, MED COL GA HOSP & CLINS, 79- *Concurrent Pos:* Path consult, Vet Admin Hosp, Perry Point, Md, 70-79; dir clin labs, Franklin Sq Hosp, Baltimore, 72-76; consult, Vet Admin Hosp, Augusta, Ga, 79- *Mem:* Am Soc Clin Path. *Res:* Types of bisalbuminemia and methods of detection. *Mailing Add:* Dept Path Med Col Ga Augusta GA 30912

BAISDEN, PATRICIA ANN, b Montgomery, WVa, July 10, 49; m 78. FISSION PROPERTIES, HEAVY-ION REACTIONS. *Educ:* Fla State Univ, BS, 71, PhD(chem), 75. *Prof Exp:* Res assoc chem oceanog, Univ Puerto Rico, 71; res chemist spectros, Procter & Gamble, 75-76; staff scientist, 76-81, GROUP LEADER NUCLEAR CHEM, LAWRENCE LIVERMORE NAT LAB, 81- *Concurrent Pos:* Consult, Int Atomic Energy Agency, 80- *Mem:* Am Chem Soc; Am Phys Soc; Sigma Xi. *Res:* Measurement of heavy element fission properties using both chemical and on-line techniques; solution of chemistry of lanthanide and actinide elements; heavy ion collisions leading to complete or incomplete fusion. *Mailing Add:* Nuclear Chem Div L-234 Lawrence Livermore Nat Lab PO Box 808 Livermore CA 94550

BAISH, JAMES WILLIAM, b Carlisle, Pa, Nov 2, 57. BIOHEAT TRANSFER. *Educ:* Bucknell Univ, BSME, 79; Univ Pa, MSE, 83, PhD (mech eng & appl mech), 86. *Prof Exp:* Edison engr, Space Div, Gen Elec, 79-81; ASST PROF, BUCKNELL UNIV, 86- *Concurrent Pos:* NSF presidential young investr, 90. *Mem:* Am Soc Mech Engrs; Inst Elec & Electronics Engrs; Am Soc Eng Educ. *Res:* Heat and mass transport in biological tissues; transport in multiphase media; mathematical modeling of tree structures; control theory; application of engineering to medicine and biology. *Mailing Add:* Dept Mech Eng Bucknell Univ Lewisburg PA 17837

BAISTED, DEREK JOHN, b London, Eng, Oct 5, 34; m 60; c 2. BIOCHEMISTRY. *Educ:* Exeter Univ, BSc, 57, PhD(org chem), 60. *Prof Exp:* USPHS trainee steroid biochem, Clark & Worcester Found Exp Biol, 60-62; NSF fel, Ind Univ, 62-63; NIH fel, 63-64, USPHS career develop award, 65-70, from asst prof to assoc prof, 65-81, PROF BIOCHEM, ORE STATE UNIV, 81- *Mem:* Fedn Am Soc Exp Biol; Royal Soc Chem; Am Soc Plant Physiologists. *Res:* Biosynthesis and metabolism of lipids; membrane-associated enzymes. *Mailing Add:* Prof Biochem Dept Biochem & Biophys Ore State Univ Corvallis OR 97331

BAITINGER, WILLIAM F, JR, b Bridgeton, NJ, Nov 24, 35; m 58; c 3. TEXTILE CHEMISTRY. *Educ:* Albright Col, BS, 58; Princeton Univ, MA, 62; PhD(chem), 64. *Prof Exp:* Chemist, Am Cyanamid Co, 58-60, res chemist, 63-66, group leader textile chem, 66-71, sr res scientist, 71-74; mgr fire retardant res, Cotton, Inc, 74-78, assoc dir prod safety res, 78-79, dir, 80-84; VPRES TECH, WESTEX INC, 84- *Mem:* Am Chem Soc; Am Soc Testing & Mat; Am Asn Textile Chemists & Colorists. *Res:* Synthesis of aromatic and aliphatic light stabilizers; effects of ultra-violet light on polymers and organic compounds; chemistry of textile wet processing. *Mailing Add:* Westex Inc 2845 W 48th Pl Chicago IL 60632

BAIZER, JOAN SUSAN, b New York, NY, Sept 12, 46. NEUROPHYSIOLOGY, VISION. *Educ:* Bryn Mawr Col, BA, 68; Brown Univ, MS, 70, PhD(psychol), 73. *Prof Exp:* Fel neurophys, Lab Neurobiol, NIMH, 73-76; ASSOC PROF PHYSIOL, STATE UNIV NY BUFFALO, 76-, ASSOC PROF NEUROBIOL, DEPT PHYSIOL. *Concurrent Pos:* Nat Inst Neurol & Commun Disorders & Stroke fel, 73-75; NIMH fel, 75-76; Nat Eye Inst res grant, 77-80. *Mem:* Asn Res Vision & Ophthal; Soc Neurosci; Asn Women in Sci. *Res:* Neurophysical vision system; prestriate cortex; eye movements. *Mailing Add:* Div Neurobiol Dept Physiol State Univ NY 4234 Ridge Lea Rd Amherst NY 14226

BAIZER, MANUEL M, electrochemical synthesis, for more information see previous edition

BAJAJ, JAGMOHAN, b New Delhi, India, Mar 12, 54; US citizen; m 83; c 2. SEMICONDUCTORS, INFRARED IMAGING. *Educ:* Univ Delhi, India, BSc, 73, MSc, 75; Northwestern Univ, Evanston, PhD(solid state physics), 81. *Prof Exp:* MEM TECH STAFF, ROCKWELL INT SCI CTR, 81- *Mem:* Am Phys Soc. *Res:* HgCdTe alloy semiconductor based materials and device physics for development of infrared imaging. *Mailing Add:* Rockwell Int Sci Ctr 1049 Camino Dos Rios Thousand Oaks CA 91360

BAJAJ, PREM NATH, b Ferozepur City, India, Oct 6, 32; m 56; c 5. MATHEMATICAL ANALYSIS, TOPOLOGY. *Educ:* Punjab Univ, India, BA, 51, MA, 54; Case Western Reserve Univ, MS, 67, PhD(math), 68. *Prof Exp:* Teacher, 51-52; lectr, RKSD Col, Kaithal, 54-57 & DAV Col, Jullundur, 57-65; vis asst prof, Case Western Reserve Univ, 68; asst prof, 68-72, ASSOC PROF MATH, WICHITA STATE UNIV, 72- *Concurrent Pos:* Mem, Vishveshvaranand Vedic Res Inst. *Mem:* Math Asn Am; Indian Math Soc. *Res:* Ordinary differential equations; dynamical systems; topology. *Mailing Add:* Dept Math Wichita State Univ Wichita KS 67208

BAJAJ, RAM, b Sukkur, Pakistan, May 15, 46; m 74; c 3. METALLURGY, RADIATION EFFECTS. *Educ:* Indian Inst Technol, Bombay, BTech, 68; Univ Mo-Rolla, MS, 70; Iowa State Univ, PhD(metall), 75. *Prof Exp:* FEL ENGR MAT SCI, BETTIS ATOMIC POWER LAB, WESTINGHOUSE ELEC CORP, 74- *Concurrent Pos:* Mem task group, Nat Alloy Develop Prog, Dept Energy, 76- *Mem:* Metall Soc. *Res:* Alloy development; mechanical properties of high temperature alloys; radiation damage; refractory materials; macro and micro analytical techniques; electron microscopy; corrosion; ceramics. *Mailing Add:* 886 Freightner Rd RD 7 Greensburg PA 15601

BAJAJ, S PAUL, b Punjab, Pakistan, Nov 12, 46; m; c 1. BIOCHEMICAL HEMOSTASIS. *Educ:* Univ Minn, PhD(biochem), 74. *Prof Exp:* Asst prof med, Univ Calif, San Diego, 78-85; assoc prof, 85-89, PROF MED & BIOCHEM, MED CTR, ST LOUIS UNIV, 89- *Concurrent Pos:* Estab investr, Am Heart Asn. *Mem:* Am Soc Hemat; Int Soc on Thrombosis & Hemostasis; Am Heart Asn; Am Soc Biochem & Molecular Biol. *Res:* Biochemical regulation of thrombosis. *Mailing Add:* Med Ctr St Louis Univ St Louis MO 63104

BAJARS, LAIMONIS, b Viipuri, Finland, June 24, 08; US citizen; m 47; c 2. CHEMICAL ENGINEERING. *Educ:* Univ Latvia, MS, 35, ChE habil, 42; NY Univ, MChE, 56. *Prof Exp:* Asst & instr unit oper & heat econ, Univ Latvia, 31-39, assoc prof, 39-44; lectr physics & chem, Int Refugee Orgn Tech Sch, Ger, 45-48; eng scientist, NY Univ, 52-55; res chemist, Petrotex Chem Corp Div, FMC Corp, 56-60, group leader, 60-63, res supvr, 63-66, res assoc, 66-69, eng supvr, 69-71, consult, 71-; RES ASSOC & SECY, CATALYSIS RES CORP, 74- *Concurrent Pos:* Tech mgr & plant supvr, Methyl Chem Corp, Latvia, 39-41; adj assoc prof, NY Univ, 57- *Mem:* Am Chem Soc; Am Soc Eng Educ. *Res:* Catalysis; dehydrogenation; oxydation; petrochemicals; thermodynamics; industrial heat economy; combustion; boiler feed water. *Mailing Add:* Three Harrison Lane Princeton Junction NJ 08550

BAJCSY, RUZENA K, b Czech, May 28, 33; c 2. COMPUTER SCIENCE. *Educ:* Slovak Tech Univ, Bratislava, PhD(elec eng), 67; Stanford Univ, PhD(comput sci), 72. *Prof Exp:* Comput engr, Comput Ctr, Slovak Tech Univ, Bratislava, 62-64, asst prof comput sci, 64-67; res assoc, Stanford Univ, 67-72; asst prof, 72-77, ASSOC PROF COMPUT SCI, UNIV PA, 77- *Concurrent Pos:* NSF res grant, 74. *Mem:* Asn Comput Mach; Inst Elec & Electronics Engrs. *Res:* Artificial intelligence; scene analysis; pattern recognition of outdoor scenes; analysis of biological images. *Mailing Add:* Dept Comput Sci Univ Pa Philadelphia PA 19104

BAJEMA, CARL J, b Plainwell, Mich, May 25, 37; m 59, 78; c 4. HUMAN ECOLOGY, BIOLOGICAL ANTHROPOLOGY. *Educ:* Western Mich Univ, BS, 59, MA, 61; Mich State Univ, PhD(zool), 63. *Prof Exp:* Instr sci, Grand Rapids Pub Sch Syst, Mich, 59-60; asst zool, Mich State Univ, 62-63; asst prof biol, Mankato State Col, 63-64; from asst prof to assoc prof, 64-72, PROF BIOL, GRAND VALLEY STATE UNIV, 72- *Concurrent Pos:* Pop Coun sr fel demog & pop genetics, Univ Chicago, 66-67; res assoc pop studies, Ctr Pop Studies, Harvard Univ, 67-73, vis prof anthrop, 74-75. *Mem:* Hist Sci Soc; Forest Hist Soc; Am Inst Biol Sci; Nat Asn Biol Teachers; Sigma Xi. *Res:* Measurement of natural selection in human populations; interactions between genetic and cultural systems for adapting to the environment; history of evolutionary thought; teaching of human ecology and evolution; environmental history of Michigan forests. *Mailing Add:* Dept Biol Grand Valley State Univ Allendale MI 49401

BAJER, ANDREW, b Czestochowa, Poland, Jan 3, 28; US citizen; m 51; c 2. CELL BIOLOGY. *Educ:* Jagiellonian Univ, MA, 49, PhD(cytol), 50, DSc, 56. *Prof Exp:* Asst cytol, Jagiellonian Univ, 48, assoc prof, 56-63; assoc prof, Polish Acad Sci, 63-64; assoc prof, 64-69, PROF BIOL, UNIV ORE, 69- *Concurrent Pos:* NIH career develop award, 67-72. *Honors & Awards:* Sci Award, Alfred Jurzykowski Found, 74. *Mem:* AAAS; Am Soc Cell Biologists; Bot Soc Fund; Int Soc Cell Biol; Am Sci Film Asn; Sigma Xi. *Res:* Physiology of cell division and mechanism of chromosome movements; mitosis and meiosis, especially microcinematrographic technique on plant endosperm; developed technique permitting study of the same cell with light microscopy and electron microscopy. *Mailing Add:* Dept Biol Univ Ore Eugene OR 97403

BAJIKAR, SATEESH S, b Mangalore, India, Sept 24, 65. MICROELECTRONICS PROCESSING, MICROMACHINING & MICROMECHANICS. *Educ:* Univ Poona, BE, 87; Univ Louisville, MS, 88; Univ Wis-Madison, MS, 90. *Prof Exp:* Teaching asst chem, Univ Louisville, 87-89; res asst, 89-91, TEACHING ASST PHYSICS, UNIV WIS-MADISON, 91- *Mem:* Am Inst Chem Engrs. *Res:* Deep ultraviolet and x-ray resist technology for micromachining application and the application of the micromachining process to the development of vacuum microelectronics and field emission devices. *Mailing Add:* 2302 University Ave No 229 Madison WI 53705

BAJOREK, CHRISTOPHER HENRY, b Tel Aviv, Palestine, Nov 26, 43; US citizen; m 65; c 2. APPLIED PHYSICS, MAGNETISM. *Educ:* Calif Inst Technol, BS, 67, MS, 68, PhD(elec eng), 71. *Prof Exp:* Res staff mem & var mgt positions, Thomas J Watson Res Ctr, IBM, Yorktown Heights, NY, 71-81, dir storage systs & technol, Almaden Ctr, San Jose, Calif, 81-83, dir technol develop, GPD, San Jose, Calif, 83-86, dir Rochester storage prod, ABS, Rochester, Minn, 87-90, DIR TECH STRATEGY DEVELOP, IBM HQ, ARMONK, NY, 90- *Concurrent Pos:* Mem, Tech Acad, IBM. *Mem:* Sigma Xi; fel Inst Elec & Electronics Engrs. *Res:* Thin film magnetic recording heads and storage media; magnetic bubble devices and materials; ferromagnetic resonance; thin film magnetic materials for general use; bipolar semiconductor devices and materials; semiconductor packaging design and materials. *Mailing Add:* 120 Clover Way Los Gatos CA 95032

BAJPAI, PRAPHULLA K, b Charkhari, India, Sept 24, 36; m 66. PHYSIOLOGY, IMMUNOLOGY. *Educ:* Agra Univ, BVSc, 58, MVSc, 60; Ohio State Univ, MSc, 63, PhD(avian physiol), 65. *Prof Exp:* Vet surgeon, Govt India, 58; instr physiol, 64-66, asst prof physiol & immunol, 66-70, assoc prof, 70-77, PROF PHYSIOL & IMMUNOL, UNIV DAYTON, 77- *Concurrent Pos:* Res assoc & fel, Reprod Endocrinol Prog, Dept Path, Univ Mich, Ann Arbor, 72-73; adj assoc prof, 75-77, adj prof, Sch Med & Sch Sci & Eng, Wright State Univ, 77- *Mem:* Am Fedn Clin Res; Am Physiol Soc; Soc Study Reprod; Fedn Am Socs Exp Biologists; Soc Biomaterials. *Res:* Physiology and immunology of heart, reproductive system and bone. *Mailing Add:* Dept Biol Univ Dayton Dayton OH 45469

BAJPAI, RAKESH KUMAR, b Kanpur, India, Dec 24, 50; US citizen; m 77; c 2. BIOTECHNOLOGY, FERMENTATION TECHNOLOGY. *Educ:* Kanpur Univ, BSO, 69; Indian Inst Technol, MTech, 72, PhD(chem eng), 76. *Prof Exp:* Lectr chem eng, Indian Inst Technol, New Delhi, 76-79; wissenschaftlicher mitarbeiter, Microbiol Inst, Swiss Fed Inst Technol, 78 & Inst Gärungsgewerbe & Biotechnol, West Berlin, 79-82; asst prof, 82-86, ASSOC PROF CHEM ENG, UNIV MO, COLUMBIA, 86- *Mem:* Am Chem Soc; Am Inst Chem Engrs. *Res:* Study of growth and product formation in bioprocesses; design and scale-up of bioreactors; hydrodynamics in bioreactors; shear sensitive organisms; fermentation; production of biological insecticides and pharmaceuticals. *Mailing Add:* Chem Eng Dept Univ Mo 1030 Eng Bldg Columbia MO 65211

BAJZAK, DENES, b Cegled, Hungary, Dec 25, 33; Can citizen; m 62; c 2. REMOTE SENSING, FOREST ENGINEERING. *Educ:* Univ BC, BSF, 58, MF, 60; Syracuse Univ, PhD(photogram), 67. *Prof Exp:* Forest res off, Forestry Br, Can Dept Northern Affairs & Nat Resources, 61-62; res scientist, Forest Res Lab, Can Dept Forestry, Nfld, 66-69; prof remote sensing & measurements, 69-71, PROF ENG & APPL SCI, MEM UNIV NFLD, 71- *Mem:* Am Soc Photogram; Can Inst Forestry; Sigma Xi; Can Remote Sensing Soc. *Res:* Land and forest site classification research in Newfoundland; application of remote sensing techniques for forest insect detection, bird and animal census and mapping of thermal pollution; the effect of flooding in reservoirs of hydroelectric power development in Labrador; slow-mater equivalent using multi-homrel radar. *Mailing Add:* Dept Eng Mem Univ Nfld St John's NF A1B 3X5 Can

BAJZER, WILLIAM XAVIER, b Cleveland, Ohio, June 17, 40; m 63; c 4. ORGANIC CHEMISTRY, MEDICINAL CHEMISTRY. *Educ:* Case Inst Technol, BS, 63; Ohio State Univ, MS, 66, PhD(org chem), 68. *Prof Exp:* Chemist, 63-65, res chemist, 69-70, supvr chem, 70-73, develop mgr, 73-75, res sect mgr, 75-77, sr res specialist, 77-81, PROCESS ENG SECT MGR, DOW CORNING CORP, 81- *Mem:* Am Chem Soc; NY Acad Sci; Am Pharmaceut Asn; Sigma Xi. *Res:* Chemistry of fluorinated compounds; methods of fluorine introduction in organic molecules; synthesis and chemistry of fluoroalkylsilicon compounds and polymers derived therefrom; fluoroalkyl triazine chemistry; biocompatible polymers and elastomers; flexible silicone contact lens; biologically active compounds containing silicon; polymer bound drugs; organofunctional silicones and specialty silane intermediates for silicone polymer preparation. *Mailing Add:* 4201 McKeith Rd Midland MI 48640

BAK, CHAN SOO, b Chun-nam, Korea, June 15, 36; US citizen; m 66; c 2. LIQUID CRYSTALS, HOLOGRAMS. *Educ:* Seoul Nat Univ, BS, 59; Univ Pittsburgh, PhD(solid state physics), 70. *Prof Exp:* Communs officer, Korean Air Force, 60-64; res assoc liquid crystals, State Univ NY, Stony Brook, 70-72, Univ Mass, 72-73 & Temple Univ, 73-76; sr physicist res & develop, Spectro-Systs Inc, 76-80; MEM STAFF, HUGHES RES LAB, MALIBU, CALIF, 80- *Concurrent Pos:* Lectr, George Washington Univ & Northern Va Community Col, 78-80. *Mem:* Am Phys Soc; Sigma Xi; Soc Info Display; Int Soc Optical Eng. *Res:* Liquid crystals; critical phenomena in liquid mixtures; laser light scattering study; pyroelectricity in polymers; geothermal energy; silicon on insulators; holographic optical elements; artificial neural networks; CRT phosphors; low temperature ceramics by Sol-Gel. *Mailing Add:* Hughes Res Lab 3011 Malibu Canyon Rd Malibu CA 90265

BAK, DAVID ARTHUR, b Yankton, SDak, Feb 6, 39; div; c 2. PHYSICAL CHEMISTRY, ORGANIC CHEMISTRY. *Educ:* Augustana Col, BA, 61; Kans State Univ, PhD(org chem), 66. *Prof Exp:* Res assoc org chem, Mich State Univ, 65-66; from asst prof to assoc prof, 66-77, chmn dept, 74-85, PROF CHEM, HARTWICK COL, 77- *Concurrent Pos:* Am Chem Soc-Petrol Res Fund res grant, 67-69; vis prof, NTex State Univ, 77-78; sr res scientist, AKZO, 87-88; res assoc, Hoffman-LaRoche, 90. *Mem:* Am Chem Soc. *Res:* Reactions of cyclooctatetraenyl dianion; electron transfer reactions; small ring ketone photochemistry; synthetic organic electrochemistry. *Mailing Add:* Dept Chem Hartwick Col Oneonta NY 13820

BAK, PER, b Bronderslev, Denmark, Dec 8, 47; c 3. PHASE TRANSITIONS, DYNAMICAL SYSTEMS. *Educ:* Tech Univ Denmark, PhD(physics), 74; Univ Copenhagen, DPhil, 82. *Prof Exp:* Res assoc physics, Brookhaven Nat Lab, 74-76; asst prof, Nordita, Copenhagen, 76-78; vis scientist, T J Watson Res Ctr, IBM, 78-79; assoc prof physics, Univ Copenhagen, 79-83;

SCIENTIST, BROOKHAVEN NAT LAB, 83-, HEAD CONDENSED MATTER THEORY GROUP, 84- *Res:* Theoretical condensed matter physics; statistical mechanics; dynamical systems; theory of chaos; incommensurate structures; quantum mechanical tunneling; surface transitions. *Mailing Add:* Dept Physics Brookhaven Nat Lab Upton NY 11973

BAKAC, ANDREJA, b Varazdin, Croatia, Yugoslavia. CATALYSIS BY TRANSITION METALS. *Educ:* Univ Zagreb, BS, 68, MS, 72 & PhD(chem), 76. *Prof Exp:* Res assoc, 76-79, from asst chemist to assoc chemist, 79-84, CHEMIST AMES LAB, IOWA STATE UNIV, 84- *Mem:* Am Chem Soc; AAAS; Sigma Xi. *Res:* Kinetic and mechanistic studies of chemical and photochemical reactions of transition metal complexes, organometallics and free radicals. *Mailing Add:* 1432 Breckinridge Ct Ames IA 50010

BAKAL, ABRAHAM I, b Bagdad, Iraq, July 5, 36; US citizen; m 66; c 2. FOOD SCIENCE, CHEMICAL ENGINEERING. *Educ:* Israel Inst Technol, BSc, 62, MSc, 64; Rutgers Univ, NB, PhD(food sci), 70. *Prof Exp:* Tech head food & chem engr, Israel Indust Adv Sta, 63-66; asst prof food sci, Univ Sask, 67-68; assoc res dir food consult, Booz, Allen & Hamilton, Inc, 70-73; tech dir food & cosmetics, Centro Indust Exp Para la Exportacion, Venezuelan Corp Develop, 73-75; prin, Booz, Allen & Hamilton, Inc, 75-78; PRES, ABIC INT CONSULTS, INC, 78- *Concurrent Pos:* Vis scientist, Nat Ctr Sci Res, France, 65; res asst, Rutgers Univ, NB, 68-70. *Mem:* NY Acad Sci; Inst Food Technologists. *Res:* Product development, process development, mostly in the food and nutrition field with emphasis on system innovation. *Mailing Add:* Ten Stafford Rd Parsippany NJ 07054

BAKANOWSKI, STEPHEN MICHAEL, b Washington, DC, May 28, 45; m 67; c 5. MICROWAVE PHYSICS, THERMOGRAPHY. *Educ:* Rutgers Univ, BA, 66; Temple Univ, PhD(physics), 75. *Prof Exp:* NSF fel, Temple Univ, 75-76; asst prof physics, Calif State Univ, Fullerton, 76-77; PHYSICIST, GEN ELEC CO, 77- *Concurrent Pos:* Vis prof, Physics Dept, Univ Louisville, 78- *Mem:* Am Phys Soc. *Res:* Thermodynamic and transport properties of metallic magnetic systems; apply principles of microwave physics and develop experimental techniques to expand knowledge of microwave oven operation; apply engineering mechanics principles to reduce noise and vibrations in consumer products. *Mailing Add:* AP 35-1101 Gen Elec Co Louisville KY 40225

BAKAY, LOUIS, b Pozsony, Hungary, June 18, 17; US citizen; m 54; c 2. NEUROSURGERY, NEUROPHYSIOLOGY. *Educ:* Univ Budapest, MD, 41. *Prof Exp:* Asst prof surg, Univ Budapest, 45-47; clin asst neurosurg, Serafimer Hosp, Stockholm, Sweden, 47-48; instr neurosurg, Harvard Med Sch, 53-61; PROF SURG, SCH MED, STATE UNIV NY BUFFALO, 61- *Concurrent Pos:* Res fel, Harvard Univ, 48-50; clin assoc, Mass Gen Hosp, Boston, 52-61; chief neurosurg serv, E J Meyer Mem Hosp, Buffalo Gen Hosp & Children's Hosp, Buffalo, 61-87. *Mem:* Am Col Surg; Harvey Cushing Soc; Am Acad Neurol. *Res:* Surgery of pituitary tumors and cervical spine; cerebral circulation and metabolism; hydrodynamics of cerebrospinal fluid; blood-brain barrier; radioactive isotopes in nervous system; cerebral edema; head injuries; medical history. *Mailing Add:* 152 Bryant St Buffalo NY 14222

BAKELMAN, ILYA J(ACOB), b Leningrad, USSR, Nov 30, 28; US citizen; m 62; c 1. DIFFERENTIAL GEOMETRY. *Educ:* Leningrad State Univ, Master degree, 51, PhD(math), 54. *Hon Degrees:* DSc, Math & Physics, Leningrad State Pedag Univ (Inst), 60. *Prof Exp:* Asst prof math, Leningrad Technol Inst, 51-55; from asst prof to prof math, Leningrad State Pedag Univ(Inst), 55-78; vis prof math, Sch Math, Univ Minn, 79-81; PROF MATH, TEX A&M UNIV, 81- *Concurrent Pos:* Vis prof, Moscow State Univ, 61, Leningrad State Univ, 62 & 64 & Univ Bonn, 80 & 81; NSF grants, 80-82, 83-84 & 85-87; vis mem, Inst Higher Sci Studies, France, 82 & 84, Math Sci Res Inst, Berkeley, Calif, 83 & 86 & Inst Math & Applns, Univ Minn, 85; mem, Inst Advan Study, Princeton, 87-88; pres, Conf Diffential Geometry, 67 & 69. *Mem:* AAAS; NY Acad Sci; Am Math Soc. *Res:* Analysis, functional analysis and geometry; global problems of existence; uniqueness, non-uniqueness and stability of solutions for non-linear partial differential equations; differential geometry; calculus of variations; theory of convex bodies and functions; elasticity-plasticity. *Mailing Add:* Dept Math Tex A&M Univ Col Sta TX 77843

BAKER, A LEROY, b Hagerstown, Md, Oct 9, 39; m 60; c 2. BIOCHEMISTRY, CHROMATOGRAPHY. *Educ:* Bridgewater Col, BA, 61; Va Polytech Inst, PhD(biochem), 64. *Prof Exp:* Res assoc, 78-88, SR RES SCI, ELI LILLY & CO, 88- *Mem:* Am Soc Biochem & Molecular Biol. *Mailing Add:* Lilly Corp Ctr Eli Lilly & Co Indianapolis IN 46285

BAKER, ADOLPH, b Russia, Nov 15, 17; US citizen; m 42; c 3. PHYSICS. *Educ:* City Col New York, BA, 38, MS, 39; Polytech Inst Brooklyn, BME, 46; NY Univ, MS, 49; Brandeis Univ, PhD(physics), 64. *Prof Exp:* Stress analyst, Repub Aviation Corp, NY, 46-47; sr engr, Ranger Aircraft Engines, 47-48; mem staff, Int Bus Mach Corp, 48-49; sr engr, Raytheon Co, Mass, 49-55; mgr airborne digital comput develop, Radio Corp Am, 55-59; prof, 63-86, EMER PROF PHYSICS & APPL PHYSICS, UNIV LOWELL, 86- *Concurrent Pos:* Consult, Radio Corp Am, Mass, 59-64; sr Fulbright-Hays scholar, USSR, 74-75. *Mem:* Am Phys Soc. *Res:* Scattering theory; high energy electron scattering from nuclei; potential scattering theory; digital computers; optical image enhancement. *Mailing Add:* Seven Gage Rd Wayland MA 01778

BAKER, ALAN GARDNER, b Inyokern, Calif, Dec 29, 47; m 75; c 1. ENGINEERING PHYSICS. *Educ:* Rensselaer Polytech Inst, BS, 69, MS, 72, PhD(physics), 75. *Prof Exp:* Res assoc phase transitions, Physic Dept, Montana State Univ, 74-77; scientist instrument res & develop for nuclear reactor safety testing, Idaho Nat Eng Lab, EG&G Idaho, Inc, 77-81; sr scientist, process & control instrument res & develop, Leeds & Northrup, 81-84; PROD DEVELOP ENGR, APPARATUS DIV, EASTMAN KODAK, 84- *Mem:* Am Inst Physics; Am Asn Physics Teachers; Sigma Xi; Instrument Soc Am. *Res:* Applied physics and system engineering of digital devices and computer based data processing systems. *Mailing Add:* 4047 Buffalo Rd Rochester NY 14624

BAKER, ALAN PAUL, b Saltsburg, Pa, Aug 6, 38; m 61; c 2. PRODUCT ACQUISITION & LICENSING. *Educ:* Philadelphia Col Pharm, BSc, 60; Hahnemann Med Col, PhD(biochem), 64. *Prof Exp:* Damon Runyon Mem Fund fel cancer res, Royal Univ Umea, Sweden, 64-66; fel enzymol, Brandeis Univ, 66-67; sr chemist, Wyeth Labs, 67-68; sr scientist, 68-71, sr investr, 71-79, asst dir pharmacol, 79-81, assoc dir new compound eval, 81-84, dir, R & D Licensing, Smith Kline & French Labs, 84-86; ASST VPRES, LICENSING & NEW BUS PLANNING, WYETH-AYERST LABS, 87- *Mem:* Am Chem Soc; Licensing Exec Soc; Am Soc Biol Chemists. *Res:* Structure and function of myeloperoxidase; uterine peroxidase; eosinophils; mast cells; electron spin resonance of flavin enzymes; microbial cell wall synthesis; mucopolysaccharides; immunopharmacology; respiratory pharmacology; regulation of biosynthesis and secretion of respiratory mucus; reaginic antibody production. *Mailing Add:* Licensing & New Business Planning Wyeth Ayerst Labs PO Box 8299 Philadelphia PA 19101-8299

BAKER, ALLEN JEROME, b Buffalo, NY, Oct 19, 36; m 61; c 4. ENGINEERING, FLUID MECHANICS. *Educ:* Union Col, BS, 58; State Univ NY Buffalo, MSc, 68, PhD(eng sci), 70. *Prof Exp:* Engr, Linde Div, Union Carbide Corp, 58-64; lectr eng sci, State Univ NY Buffalo, 65-70; prin scientist res, Bell Aerospace Div, Textron, 70-74; vis prof mech eng, Old Dominion Univ, 74-75; assoc prof, 75-80, PROF ENG SCI, UNIV TENN, KNOXVILLE, 80- *Concurrent Pos:* Digital analyst, Bell Aerospace Div, Textron, 66-70; pres, Computational Mech Corp, 75-; IBM professorship, Univ Tenn, 83-86, Chancellor's res scholar, 88. *Mem:* Am Inst Aeronaut & Astronaut; Am Acad Mech; Am Soc Eng Educ. *Res:* Computational fluid dynamics, finite element analysis, turbulent flows; large scale fluid dynamics computer codes; parallel processing. *Mailing Add:* 310 Perkins Hall Knoxville TN 37996-2030

BAKER, ANDREW NEWTON, JR, b Wellington, Kans, Oct 21, 28; m 51; c 2. AERONAUTICAL & ASTRONAUTICAL ENGINEERING, PHYSICS. *Educ:* Univ Rochester, BS, 49; Univ Calif, Los Angeles, MS, 51, PhD(physics), 54. *Prof Exp:* Assoc physics, Univ Calif, Los Angeles, 52-54; mem tech staff, Bell Tel Labs, 54-60; head solid state lab, Lockheed-Calif Co, 60-63, mgr phys sci, 64-66; Sloan exec fel, Stanford Univ, 66-67; dep chief engr, 68-70, dir res, 70-72, dir develop planning, 72-73, dir res, 73-77, dir res & planning, 77-85, dep dir technol, 85-87, dir Eng-Aero Systems, Lockheed Corp, 87-90, ENG CONSULT, 91- *Mem:* Am Phys Soc; Am Inst Aeronaut & Astronaut; Sigma Xi. *Res:* Infrared spectroscopy; hydrogen bonds; semiconductors and semiconductor devices; electro-optics and ferroelectrics; titanium metallurgy and corrosion; CAD/CAM; low obscurables technology; aircraft safety. *Mailing Add:* 12324 Woodley Ave Granada Hills CA 91344

BAKER, ARTHUR A, b New Britain, Conn, Oct 31, 97; m 25; c 1. GEOLOGY. *Educ:* Yale Univ, PhB, 19, PhD, 31. *Prof Exp:* Admin geol, US Geol Surv, 53-56, assoc dir, 56-69, spec asst to dir, 69-72, Rep, Dept Interior, US Bd on Geog Names, 53-72, mem geol staff, 21-72; RETIRED. *Mem:* Am Asn Petrol Geologists; Geol Soc Am. *Mailing Add:* 5201 Westwood Dr Bethesda MD 20816

BAKER, BARTON SCOFIELD, b Paint Bank, Va, Oct 4, 41; m 64. AGRONOMY. *Educ:* Berea Col, BS, 64; WVa Univ, MS, 66, PhD(agron), 69. *Prof Exp:* From asst prof to prof agron, 69-87, dir, Allegheny Highlands Proj, 78-80, DIR, DIV PLANT & SOIL SCI, 87- *Mem:* Am Soc Agron; Am Forage & Grassland Coun. *Res:* Influence of environmental conditions on the growth and composition of forage crops. *Mailing Add:* Dept Plant Sci WVa Univ Box 6108 Morgantown WV 26506-6108

BAKER, BERNARD RAY, b Wheatland, Wyo, Dec 9, 32; div; c 2. SURFACE CHEMISTRY. *Educ:* Univ Denver, BS, 54; Northwestern Univ, PhD(chem), 58. *Prof Exp:* Instr chem, Northwestern Univ, 57-58 & Univ Nev, 58-60; from instr to asst prof, Univ Rochester, 60-65; res assoc, Brookhaven Nat Lab, 65-66; RES CHEMIST, CTR TECHNOL, KAISER ALUMINUM & CHEM CORP, 66- *Concurrent Pos:* Sect ed, Chem Abstr, 62-63. *Mem:* Am Soc Testing & Mat. *Res:* Kinetics and mechanisms of inorganic reactions; electrochemistry; surface chemistry; chemistry of thin films; metal finishing; development of new surface finishes for aluminum including anodizing, chemical coatings and electroplating; general surface chemistry of aluminum and aluminum oxides. *Mailing Add:* 459 Sanchez St San Francisco CA 94114

BAKER, BERNARD S, b Philadelphia, Pa, June 26, 36; m 59; c 4. RESEARCH ADMINISTRATION, TECHNICAL MANAGEMENT. *Educ:* Univ Pa, BS, 57, MS, 59; Ill Inst Technol, PhD(chem eng), 69. *Prof Exp:* Sr scientist electrochem, Lockheed Missiles & Space Co, 60-62 & Electrochem Corp, 62; from mgr energy conversion res to dir basic sci, Int Gas Technol, 62-70; pres, Energy Res Corp, 70-91. *Concurrent Pos:* Chmn steering comt, Intersoc Energy Conversion Eng Conf, 72. *Mem:* Electrochem Soc; Am Inst Chem Engrs. *Res:* Energy conversion research, especially electrochemical power sources such as fuel cells and batteries. *Mailing Add:* Energy Res Corp Three Great Pasture Rd Danbury CT 06810

BAKER, BRENDA SUE, b Oakland, Calif, Dec 19, 48; m 82; c 2. COMPUTER SCIENCE. *Educ:* Radcliffe Col, BA, 69; Harvard Univ, MS, 70, PhD(appl math), 73. *Prof Exp:* Res fel comput sci, Harvard Univ, 73-74; mem tech staff comput sci, Bell Labs, 74-78; asst prof comput & commun sci, Univ Mich, 78-79; mem staff, Bell Labs, 80-86; RETIRED. *Concurrent Pos:* Vis lectr, Univ Calif, Berkeley, 77-78. *Mem:* Asn Comput Mach; AAAS. *Res:* Theoretical computer science. *Mailing Add:* 281 Timber Dr Berkeley Heights NJ 07922

BAKER, BRYAN, JR, b Grenada, Miss, Feb 24, 23; m 46; c 1. ANIMAL HUSBANDRY, ANIMAL PHYSIOLOGY. *Educ:* Miss State Univ, BS, 47, BS, 48, MS, 52; Univ Ill, PhD(animal sci), 55. *Prof Exp:* Teacher voc agr, Biggersville High Sch, 48-51; asst animal husb, Miss State Univ, 51-52; asst animal sci, Univ Ill, 52-55; assoc prof animal husb, Va Polytech Inst, 55-56; assoc prof, 56-60, PROF ANIMAL SCI, MISS STATE UNIV, 60-, HEAD DEPT, 78- *Mem:* AAAS; Am Soc Animal Sci; Sigma Xi. *Res:* Effect of environmental factors on physiology of reproduction of sheep, cattle and swine. *Mailing Add:* 11 Talley Ho Dr Mississippi State MS 39759

BAKER, CARL GWIN, b Louisville, Ky, Nov 27, 20; m 49, 75; c 2. RESEARCH ADMINISTRATION. *Educ:* Univ Louisville, AB, 41, MD, 44; Univ Calif, MA, 49. *Hon Degrees:* DSc, 80, Univ Louisville. *Prof Exp:* Intern, Milwaukee County Gen Hosp, 44-45; physician, US Navy, 45-46; Childs fel, Pub Health Serv, 46-48; sr asst surgeon to med dir, Nat Cancer Inst, 49-55; asst to dir intramural res, NIH, 56-58; asst dir, Nat Cancer Inst, 58-61, actg sci dir, 60-61, assoc dir prog, 61-67, sci dir etiology, 68-69, dir, 69-72; ASST SURGEON GEN, PUB HEALTH SERV, 70- *Concurrent Pos:* Spec fel, Nat Cancer Inst, 49; assoc ed, Jour Nat Cancer Inst, 54-55; spec lectr, Sch Med, Georgetown Univ, 54-64; mem subcomt amino acids, Comt Biochem, Nat Res Coun, 56-58; mem planning comt, Nat Cancer Conf, 62-64; mem ed adv bd, Cancer, 65-73; dir, Am Cancer Soc, 69-72; vpres, Tenth Int Cancer Cong, 70; mem sci adv comt, Ludwig Inst Cancer Res, 71-85; pres & sci dir, Hazelton Labs, 72-73; mem bd dirs, Am Asn Cancer Res, 73-76; consult, Res Admin, 73-75; dir prog policy staff, Health Resources Admin, HEW, 75-77; med dir, Ludwig Inst Cancer Res, Switz, 77-82, emer med dir, 82-, consult, 85-87; adj instr, Columbia Union Col, 88-, Univ Md Univ Col, 91- *Honors & Awards:* Meritorious Serv Medal, Pub Health Serv, 66. *Mem:* Am Chem Soc, Div Biol Chem (secy, 54-57); Soc Exp Biol & Med; Am Soc Biol Chem; Am Asn Cancer Res; Sigma Xi. *Res:* Enzymatic resolution of amino acids; cancer research; systems management and research administration; systems analysis; networking techniques were developed and used for planning, obtaining funds for and operating major cancer research programs. *Mailing Add:* 19408 Charline Manor Rd Olney MD 20832

BAKER, CARLETON HAROLD, b Utica, NY, Aug 2, 30; m 63; c 2. CARDIOVASCULAR PHYSIOLOGY, MICROCIRCULATION. *Educ:* Syracuse Univ, BA, 52; Princeton Univ, MA, 54, PhD(biol), 55. *Prof Exp:* Asst instr biol, Princeton Univ, 52-54, asst res biol, 54-55; from asst prof to prof physiol, Med Col Ga, 55-67; prof physiol & biophys, Sch Med, Univ Louisville, 67-71; dep dean res grad affairs, 80-82, PROF PHYSIOL & BIOPHYS & CHMN DEPT, COL MED, UNIV S FLA, TAMPA, 71- *Mem:* AAAS; Am Heart Asn; Am Physiol Soc; Microcirc Soc; Soc Exp Biol & Med; Shock Soc. *Res:* Adrenal cortex; blood volume regulation; cardiovascular system; microcirculation. *Mailing Add:* Dept Physiol & Biophys Col Med Univ S Fla Box 8 Tampa FL 33612

BAKER, CHARLES EDWARD, b Manchester, Ky, Nov 8, 31; m 52; c 4. PHYSICAL CHEMISTRY. *Educ:* Berea Col, BA, 53; Univ Fla, PhD(phys chem), 60. *Prof Exp:* Chemist, Res Div, Monsanto Chem Co, Ohio, 53-54; asst, Univ Fla, 56-60; sr res engr, Gen Dynamics/Convair, Calif, 60-62; aeronaut res scientist, Lewis Res Ctr, NASA, 62-84; RETIRED. *Mem:* AAAS; Am Chem Soc; Sigma Xi. *Res:* Atmospheric chemistry; low-energy scattering of negative-ion beams; thermal conductivities of ordinary and isotopically substituted polar gases; diffusion of polar gases; recombination of atoms; combustion fundamentals; alternate fuels. *Mailing Add:* 510 Wyleswood Dr Berea OH 44017

BAKER, CHARLES P, b Leominster, Mass, Aug 2, 10; m 53. ACCELERATOR DESIGN. *Educ:* Denison Univ, BA, 33; Cornell Univ, MA, 35, PhD(physics & math), 40. *Prof Exp:* Res assoc physics, Cornell Univ, 40-42 & 46-47, asst prof, 47-49; res assoc physics, Purdue Univ, 42-43; engr physics, Manhattan Dist, Los Alamos Nat Lab, 43-46; physicist, Brookhaven Nat Lab, 50-75, consult, 75-80; RETIRED. *Mem:* Fel Am Phys Soc; AAAS. *Res:* Neutron velocity spectrometer; cyclotron ion source; Los Alamos Baker experiment; critical assemblies cyclotron activities including design construction sector-focussing; interdisciplinary liasion. *Mailing Add:* Fairhaven C011 7200 Third Ave Sykesville MD 21784

BAKER, CHARLES RAY, b Pine Bluff, Ark, May 22, 32; m 52; c 1. APPLIED MATHEMATICS, ENGINEERING. *Educ:* Univ Southwestern La, BS, 57; Univ Calif, Los Angeles, MS, 63, PhD(eng), 67. *Prof Exp:* Staff engr, Land-Air, Inc, Holloman Air Develop Ctr, NMex, 57-59; engr aerospace systs div, Bendix Corp, Mich, 59-60, sr engr electrodynamics div, Calif, 60-67, eng res specialist, 67-68; assoc prof statist, 68-73, PROF STATIST, UNIV NC, CHAPEL HILL, 73- *Mem:* Inst Elec & Electronics Engrs; Inst Math Statist; Am Math Soc. *Res:* Statistical communication theory; probability theory; stochastic processes and applications to communication systems. *Mailing Add:* Dept Statist Univ NC Chapel Hill NC 27599-3260

BAKER, CHARLES TAFT, b Carbondale, Ill, Aug 29, 39; m 62; c 3. ELECTROCHEMISTRY, EPOXY POLYMERS. *Educ:* Southern Ill Univ, BA, 61; Univ Ill, MS, 64, PhD(chem), 66. *Prof Exp:* Mem tech staff, Mat Sci & Res Lab, Cent Res, 66-70, sect leader, 70-72, mgr qual reliability assurance, Anal Serv Lab, 72-84, mgr qual reliability assurance, Eng Labs, Semiconductor Group, 84-87, SR MEM TECH STAFF/VLSI QRA ENG MGR, TEX INSTRUMENTS INC, 87- *Res:* Semiconductor device material reliability evaluation and qualification; electromigration; metal migration; plating processes and organic polymer encapsulation compounds; analytical characterization techniques; ion-selective electrodes; anodization of (Hg, Cd)Te; low-temperature kinetics of battery electrodes. *Mailing Add:* 7631 El Pensador Dr Dallas TX 75248

BAKER, CHARLES WESLEY, b Binghamton, NY, May 29, 45; m 67; c 2. FERMENTED BEVERAGES, SENIOR MANAGEMENT. *Educ:* Northeastern Univ, BA, 67; Purdue Univ PhD(biochem), 73. *Prof Exp:* Asst prof cereal chem, NDak State Univ, 73-77; mgr res & develop, Pabst Brewing Co, 77-80; mgr tech ctr, Jos Schlitz Brewing Co, 80-82; dir tech ctr, Stroh Brewery Co, 82-86, dir tech serv, 86-90; VPRES SCI AFFAIRS, SUGAR ASN INC, 90- *Mem:* Am Soc Brewing Chemists (pres, 86-88); Am Asn Cereal Chemists; Master Brewers Asn of Am; Asn Official Anal Chemists. *Res:* Effect of protein structure and geometry on foam and foam stability; development and refinement of analytical methodology for fermented malt beverages and juice products. *Mailing Add:* Sugar Asn Inc 1101 15th St NW Suite 600 Washington DC 20005-5076

BAKER, D(ALE) B(URDETTE), b Bucyrus, Ohio, Sept 19, 20; m 47; c 3. CHEMICAL ENGINEERING. *Educ:* Ohio State Univ, BChE, 42, MS, 48. *Hon Degrees:* PhD, Ohio State Univ, 86. *Prof Exp:* Chemist supvr, E I du Pont de Nemours & Co, 42-45; asst ed, 46-50, assoc ed, 51-57, dir chem abstr, 58-86, EMER DIR, AM CHEM SOC, 86- *Concurrent Pos:* Mem bd, Nat Fedn Sci Abstr & Indexing Serv; mem planning & steering comt, abstr bd, Int Coun Sci Unions; comt int sci & tech info progs, Nat Acad Sci, 73-; mem bd visitors, grad sch libr & info sci, Univ Pittsburgh, 76-82 & Case-Western Reserve Univ. *Honors & Awards:* Austin Patterson-Crane Award, Am Chem Soc, 79, Herman Skolnik Award, 86. *Mem:* Nat Acad Sci; AAAS; Am Chem Soc; Am Inst Chem Engrs; Am Soc Info Sci. *Res:* Chemical literature, documentation, communication, research, development and planning. *Mailing Add:* Chem Abstrs Serv Am Chem Soc PO Box 3012 Columbus OH 43210

BAKER, D(ONALD) JAMES, b Long Beach, Calif, Mar 23, 37; m 68. PHYSICAL OCEANOGRAPHY. *Educ:* Stanford Univ, BS, 58; Cornell Univ, PhD(physics), 62. *Prof Exp:* Res assoc oceanog, Univ RI, 62-63; NIH fel biophys, Univ Calif, Berkeley, 63-64; res fel geophys fluid dynamics, Harvard Univ, 64-66, asst prof oceanog, 66-70, assoc prof phys oceanog, 70-73; from res assoc prof to res prof oceanog, Univ Wash, 73-79, sr oceanog, Appl Physics Lab, 75-79, group leader deep-sea physics, Pac Marine Environ Lab, Univ Wash-Nat Oceanic & Atmospheric Admin, 77-79, chmn dept oceanog, Univ Wash, 79-81, dean col ocean & fishery sci, 81-83, prof, Sch Oceanog, 79-86, adj prof, Dept Atmospheric Sci, 79-86; PRES, JOINT OCEANOG, INC, 83- *Concurrent Pos:* Vis scholar, Woods Hole Oceanog Inst, 70; chmn, Joint Polar Exp Panel, Nat Res Coun, 73-79, co-chmn, Int Southern Ocean Studies Prog, 74-84 & mem, Climate Dynamics Panel, US Global Atmospheric Res Prog, Nat Acad Sci, 76-78; consult, Polar Exp, World Meteorol Orgn, 75-79; Univ Wash/Nat Oceanic & Atmospheric Admin & Univ Hawaii/Nat Oceanic & Atmospheric Admin sr fels, Joint Insts Study of Atmospheres & Oceans & Marine & Atmospheric Res, 77-83; mem bd gov, Joint Oceanog Inst, Inc, 79-83; mem, Ocean Studies Bd, Nat Res Coun, 79-, Space Sci Bd, 84-87, Climate Res Comt, 79-, Space & Earth Sci Adv Comt, NASA, 82-86, Earth Syst Sci Comt, 83-87, environ panel, Navy Res Adv Comt, 83-85, Comt Global Change, Nat Res Coun, 87-, Panel on Climate & Global Change, Nat Oceanic & Atmospheric Admin, 87-; distinguished vis scientist, Jet Propulsion Lab, Calif Inst Tech, 82-; officer, Joint Sci Comt, World Climate Res Prog, 87-; co-chmn, World Ocean Circulation Exp Sci Steering Group, 88-; chmn, Comt Ocean Processes & Climate, Intergov Oceanog Comn, 89-, Panel to Rev Earth Observing Syst, Nat Res Coun, 89-90; mem bd dirs, Coun Ocean Law, 89- *Mem:* Am Geophys Union; Am Asn Univ Profs; Fel AAAS; Am Meteorl Soc; Sigma Xi; Marine Technol Soc; The Oceanog Soc (pres, 88-). *Res:* Physical oceanography, physics of large-scale ocean circulation and climate; ocean instrumentation; ocean measurements from satellites, research management; co-awardee of one patent; author of one book. *Mailing Add:* Joint Oceanog Inst Inc 1755 Massachusetts Ave Suite 800 Washington DC 20036

BAKER, DALE E, b Marble Hill, Mo, May 5, 30; m 49; c 3. AGRONOMY, SOIL CHEMISTRY. *Educ:* Univ Mo, BS, 57, MS, 58, PhD(soils), 60. *Prof Exp:* Agronomist, NE Exp Sta, Duluth, Minn, 60-61; from asst prof to assoc prof, 61-70, PROF SOIL CHEM, PA STATE UNIV, 70-, DIR, SOIL & ENVIRON CHEM LAB, 74- *Concurrent Pos:* Consult, Amax, Inc, 72-75 & Food & Drug Admin, Dept Health, Educ & Welfare, 74; pres, Land Mgt Decisions, Inc, 89- *Mem:* Fel Am Soc Agron; fel Soil Sci Soc Am; fel AAAS; Sigma Xi. *Res:* Physical and colloidal chemistry of soils, soil testing and genetic control of physiological processes in plants in relation to ion uptake, environmental quality and production of field crops; land management. *Mailing Add:* 1429 Harris St State College PA 16801

BAKER, DANIEL NEAL, b Postville, Iowa, Nov 10, 48; m 71. SPACE PLASMA PHYSICS. *Educ:* Univ Iowa, BA, 69, MS, 73, PhD(physics), 74. *Prof Exp:* Res assoc physics, Univ Iowa, 74-75; res fel, Calif Inst Technol, 75-77; staff mem space physics, Los Alamos Nat Lab, Univ Calif, 77-81, leader, Space Plasma Physics Group, 81-87; DIR, LAB EXTRATERRESTRIAL PHYSICS, NASA, 87- *Concurrent Pos:* Fel, dept physics, math & astron, Calif Inst Technol, 75-77; prin investr, NASA, basic energy sci, Dept Energy & Inst Geophys Progs, Univ Calif, 81-; chmn, Data Systs Users Working Group, NASA, 82-90; mem, comt on solar & space physics, Nat Acad Sci, 84-87 & Bd Atmospheric Sci, Climate Panel on Long-term Observ, Nat Res Coun, 85-88 & mgt & opers comt, NASA; assoc ed, Geophys Res Letters, 85-88; mem, comt on data mgt & comput, Nat Acad Sci, 85-89; organizer, Coord Data Analysis Workshop Prog, NASA-ESA, 85-; US mem comn D, Int Asn Geomagnetism & Aeronomy, 86-; US rep, Int Sci Comt Solar-Terrestrial Physics Res, 87-; consult, Los Alamos Nat Lab, Univ Calif, 87-; mem, Geospace Environ Modeling Steering Comt, NSF, 88-; Space Sci & Appln Adv Comt, NASA, 88-; proj scientist, Small Explorer Prog, NASA, 88- & US Solar Terrestrial Energy Prog, 89-; mem, external adv comt, Boston Univ, Univ Md; magnetospheric secy, Am Geophys Union, 88-90. *Mem:* Am Geophys Union; Sigma Xi; AAAS. *Res:* Magnetospheric physics, energetic particle and cosmic ray physics, solar-terrestrial relationships; space plasma physics; study of planetary magnetospheres; solar planetary relationships; plasma astrophysics. *Mailing Add:* Lab Extraterrestrial Physics Code 690 Goddard Space Flight Ctr NASA Greenbelt MD 20771

BAKER, DAVID BRUCE, b Akron, Ohio, May 29, 36; m 60; c 1. PLANT PHYSIOLOGY. *Educ:* Heidelberg Col, BS, 58; Univ Mich, MS, 60, PhD(bot), 63. *Prof Exp:* Asst prof bot, Rutgers Univ, 63-66; from asst prof to assoc prof biol, 66-76, DIR WATER QUAL LAB, HEIDELBERG COL, 76- *Concurrent Pos:* NSF fel, 63-64. *Mem:* Am Soc Plant Physiol. *Res:* Effects of agricultural land use on water quality; large scale agri-eco system research. *Mailing Add:* Water Qual Lab Heidelberg Col Tiffin OH 44883

BAKER, DAVID H, b DeKalb, Ill, Feb 26, 39; m 57, 81; c 3. ANIMAL NUTRITION. *Educ:* Univ Ill, BS, 61, MS, 63, PhD(animal nutrit), 65. *Prof Exp:* Res asst nutrit, Univ Ill, 61-65; sr scientist, Greenfield Labs, Eli Lilly & Co, Ind, 65-67; from asst prof to assoc prof, 67-74, PROF NUTRIT, UNIV

ILL, URBANA-CHAMPAIGN, 74- *Honors & Awards:* Res Award, Am Soc Animal Sci, 71; Nutrit Res Award, Am Feed Mfrs Asn, 73; Paul A Funk Award, 77; Merck Award, 77; Broiler Res Award, 83; Gustav Bohstedt Award, 85; Borden Award, Am Inst Nutrit, 86; Distinguished Serv Award, USDA, 87. *Mem:* Am Soc Animal Sci; Poultry Sci Asn; Am Inst Nutrit. *Res:* Amino acid nutrition and metabolism; trace minerals. *Mailing Add:* Dept Animal Sci Univ Ill Urbana IL 61801

BAKER, DAVID H, b Concord, NH, Aug 25, 25; m 52; c 3. PEDIATRICS, RADIOLOGY. *Educ:* Boston Univ, MD, 51. *Prof Exp:* From instr to asst prof pediat & radiol, NY Hosp-Cornell Med Ctr, 54-59; asst prof pediat & radiol, 59-64, assoc prof radiol, 64-68, PROF RADIOL, COL PHYSICIANS & SURGEONS, COLUMBIA UNIV, 68- *Concurrent Pos:* Consult, New York Foundling Hosp, 56-, Roosevelt Hosp, 57-, Montefiore Hosp & USPHS Hosp, Staten Island, 62- *Mem:* Soc Pediat Radiol; fel Am Acad Pediat; Am Col Radiol; Asn Univ Radiol; Am Roentgen Ray Soc. *Res:* Pathophysiologic states in children involving the gastrointestinal, cardiovascular, bronchovascular and urologic systems. *Mailing Add:* Presby Hosp 622 W 168th St New York NY 10032

BAKER, DAVID KENNETH, b Glasgow, Scotland, Oct 2, 23; nat US; m 47; c 1. PHYSICS. *Educ:* McMaster Univ, BSc, 46; Univ Pa, PhD, 53. *Prof Exp:* Instr physics, Univ Pa, 52-53; prof, Union Col, 53-65; prof personnel & univ rels, Res & Develop Ctr, Gen Elec Co, 65-67; vpres & dean, St Lawrence Univ, 67-76; pres, Harvey Mudd Col, 76-88; RETIRED. *Mem:* Am Phys Soc; Am Asn Physics Teachers. *Res:* Semiconductors and solid state physics. *Mailing Add:* 3088 Fairway Woods Catlonia Trace Sanford NC 27330

BAKER, DAVID THOMAS, b Granite Falls, Minn, June 27, 25; m 64; c 3. NUCLEAR PHYSICS, SCIENCE EDUCATION. *Educ:* US Mil Acad, BS, 46; Purdue Univ, MS, 51; Am Univ, MS, 71. *Prof Exp:* US Army, 46-70, instr nuclear physics & elec eng, US Mil Acad, 51-54, asst prof nuclear physics, 54-55, res & develop officer, nuclear weapons, Army Artil & Guided Missile Off, Ft Sill, Okla, 56-58, nuclear weapons effects, Atomics Off, Off Chief Res & Develop, Hq Dept Army, 60-64, dir Nuclear Weapons Assembly Dept, Army Sch Europe, Ger, 65-66, develop officer, Nuclear Weapons Req Div, US Army Mobility Equip Res & Develop Ctr, 69-71; MASTER INSTR PHYSICS, CULVER MIL ACAD, 71- *Mem:* Am Asn Physics Teachers. *Res:* Nuclear weapons, their development, testing, effects and employment. *Mailing Add:* Box 45 Culver Mil Acad Culver IN 46511

BAKER, DAVID WARREN, b Great Falls, Mont, Nov 9, 39; m 62; c 3. STRUCTURAL GEOLOGY. *Educ:* Mass Inst Technol, BS, 61; Swiss Fed Inst Technol, Dipl Nat Sci, 64; Univ Calif, Los Angeles, PhD(geol), 69. *Prof Exp:* Asst prof geol sci, Univ Ill, 70-76: RES GEOLOGIST, GULF RES, 76- *Concurrent Pos:* Res geophysicist, Inst Geophys & Plantetary Physics, Univ Calif, Los Angeles, 69-70; res geologist, Dept Geol & Geophys, Yale Univ, 70. *Mem:* Am Geophys Union; Swiss Mineral & Petrog Soc; Geol Soc Am. *Res:* X-ray analysis of preferred orientation in experimentally and naturally deformed rocks; deep-seated thrust zones; seismic velocity anisotropy in the upper mantle; plate and regional tectonics; remote sensing and lineament analysis, in situ stress. *Mailing Add:* PO Box 906 Monarch MT 59463

BAKER, DENNIS JOHN, b Detroit, Mich, July 25, 40; m 70; c 5. COMMUNICATION, SIMULATION. *Educ:* Univ Detroit, BS, 63; Univ Mich, PhD(physics), 73. *Prof Exp:* Res physicist, US Army Tank Plant, 66; RES PHYSICIST, NAVAL RES LAB, 67- *Concurrent Pos:* Consult, Gen Dynamics, 68. *Res:* Distributed simulation; communication network design; software systems; co-originator of the linked cluster architecture for mobile, broadcast radio networks. *Mailing Add:* US Naval Res Lab Code 5521 4555 Overlook SW Washington DC 20375-5000

BAKER, DON H(OBART), JR, b Lawrence, Kans, Mar 15, 24; m 45; c 2. EXTRACTIVE & CHEMICAL METALLURGY. *Educ:* Univ Ariz, BS, 47, MS, 49, EMet, 60. *Prof Exp:* Plant metallurgist, Am Smelting & Ref Co, 48-49; metallurgist, Titanium Projs, US Bur Mines, 49-54, supvr pure metals projs, 54-60, phys sicentist, DC, 60-62; coordr mineral beneficiation proj, Reno Metall Res Ctr, 62-67; prof extractive metall, Univ Waterloo, 67-69; prof, NMex Inst Mining & Technol, 69-; RETIRED. *Concurrent Pos:* Dir, NMex State Bur Mines, 69-73; coordr aluminum projs & pilot plants, US Bur Mines, 73-77, dir proj develop for geothermal mineral recovery, 76-77; adj prof metall, NMex Inst Mining & Technol, 78-80; several consult contracts; pres, Don H Baker & Assoc, 81-86; pres & mgr, Albany Titanium Inc, Albany, Ore, 84-85. *Honors & Awards:* Flemming Award, 58. *Mem:* Am Inst Mining, Metall & Petrol Engrs. *Res:* Extractive metallurgy and physical chemistry; high temperature electrochemical phenomena; production of high purity titanium by metallic sodium reduction of titanic halide; pilot plant design and construction; mineral recovery from geothermal fluids and plant designs. *Mailing Add:* PO Box 707 Socorro NM 87801

BAKER, DON ROBERT, b Salt Lake City, Utah, Apr 6, 33; m 54; c 4. ORGANIC CHEMISTRY. *Educ:* Sacramento State Col, AB, 55; Univ Calif, PhD(chem), 59. *Prof Exp:* Sr res chemist, Stauffer Chem Co, 59-72, res assoc, 72-74, supvr, 74-85, sr res assoc, 85-87; SR RES ASSOC, ICI AMERICAS INC, 87- *Concurrent Pos:* Counr, Am Chem Soc, 70-; ed, Calif Chemists Alert, 86-, Synthesis & Chem of Agrochem, 87, 91. *Mem:* Am Chem Soc; Plant Growth Regulator Soc; Am Orchid Soc. *Res:* Organic synthesis; agricultural chemistry; microbiology; computer applications; mineralogy; 100 patents. *Mailing Add:* 15 Muth Dr Orinda CA 94563

BAKER, DONALD GARDNER, b St Paul, Minn, July 20, 23; m 53; c 1. MICROCLIMATOLOGY. *Educ:* Univ Chicago, prof cert, 44; Univ Minn, BS, 49, MS, 51, PhD(soils), 58. *Prof Exp:* Asst soils, 49-51, 53-58, from instr to assoc prof, 58-69, PROF SOILS, UNIV MINN, ST PAUL, 69- *Concurrent Pos:* Weather forecaster, 44-46 & 51-52. *Honors & Awards:* Spec Serv Award, Nat Weather Serv, Nat Oceanic & Atmospheric Admin, 86. *Mem:* Fel AAAS; Am Meteorol Soc; Am Geophys Union; Soil Sci Soc Am; Royal Meteorol Soc, London; fel Am Soc Agron. *Res:* Agricultural meteorology and climatology. *Mailing Add:* Dept Soil Sci Univ Minn Inst Agr St Paul MN 55108

BAKER, DONALD GRANVILLE, b Toronto, Ont, Oct 19, 24; m 52; c 2. PHYSIOLOGY. *Educ:* Univ Toronto, BA, 51, MA, 52, PhD(physiol), 55. *Prof Exp:* Res assoc, Banting & Best Dept Med Res, Univ Toronto, 55-62, lectr, Sch Hyg, 58-64, from asst prof to assoc prof physiol & spec lectr zool & radiobiol, 59-64; radiobiologist, Biol Div, Brookhaven Nat Lab, 64-68; radiobiologist, Claire Zellerbach Saroni Tumor Inst, Mt Zion Hosp & Med Ctr, 68-77; ASSOC PROF, DIV RADIATION ONCOL, UNIV VA HOSP, 77- *Mem:* Radiation Res Soc; NY Acad Sci; Am Soc Therapeut; Am Asn Cancer Res; Am Sci Affil. *Res:* Radiation injury; radiation biology; hyperthermia anti tumor biology. *Mailing Add:* Dept Radiol Div Radiation Oncol Univ Va Hosp Charlottesville VA 22908

BAKER, DONALD ROY, b Norfolk, Va, May 8, 27; m 48; c 2. ORGANIC GEOCHEMISTRY, PETROLEUM GEOLOGY. *Educ:* Calif Inst Technol, BS, 50; Princeton Univ, PhD(geol), 55. *Prof Exp:* Instr petrol & chem geol, Northwestern Univ, 54-56; sr res geologist, Denver Res Ctr, Marathon Oil Co, 56-66; assoc prof, 66-72, chmn dept, 77-80, PROF GEOL, RICE UNIV, 72- *Mem:* Fel Geol Soc Am; Am Asn Petrol Geol; Soc Econ Paleontologists & Mineralogists; Geochem Soc; Am Geophys Union; Europ Asn Org Geochemists; Int Asn Geochem & Cosmochem. *Res:* Organic and stable isotope geochemical research with focus on problems related to petroleum evolution and sedimentary petrology. *Mailing Add:* Dept Geol Rice Univ PO Box 1892 Houston TX 77251

BAKER, DORIS, b Pt Marion, Pa, Nov 16, 21. ANALYTICAL CHEMISTRY, CEREAL CHEMISTRY. *Educ:* Univ Md, BS, 48. *Prof Exp:* Chemist grain & oil seed technol, 48-52, res chemist, 52-74, res chemist & proj leader Food Composition Methodology, USDA, 74-85; RETIRED. *Mem:* Am Asn Cereal Chemists; Am Chem Soc. *Res:* Analytical chemistry as applied to nutrients in foods; methodology for determining composition of food fiber and its physical and chemical relationship to other food nutrients. *Mailing Add:* 3413 Dunnington Rd Beltsville MD 20705-3225

BAKER, DUDLEY DUGGAN, III, b Seguin, Tex, Feb 1, 36; m 59; c 3. UNDERWATER ACOUSTICS. *Educ:* Univ Tex, BS, 57, MA, 58. *Prof Exp:* Res scientist assoc, Defense Res Lab, 59-67, head, Electroacoust Div, 67-76, asst dir, 76-81, head acoust measurement div, 81-84, RES ASSOC, APPL RES LABS, UNIV TEX, AUSTIN, 84- *Mem:* Acoust Soc Am; Sigma Xi. *Res:* Underwater acoustical measurements, especially related to sonar transducer testing. *Mailing Add:* Box 8029 University Sta Austin TX 78712

BAKER, DURWOOD L, b Algona, Iowa, June 16, 19; m 45; c 4. VETERINARY MEDICINE. *Educ:* Iowa State Univ, DVM, 43. *Prof Exp:* From instr to prof vet med, 47-89, from asst dean to assoc, Col Vet Med, 64-89, EMER PROF, IOWA STATE UNIV, 89- *Mem:* Am Vet Med Asn. *Res:* Small animal medicine and surgery. *Mailing Add:* 809 16th St Ames IA 50010

BAKER, DWIGHT DEE, symbiology, nitrogen-fixation, for more information see previous edition

BAKER, EARL WAYNE, b Lewistown, Mont, Sept 26, 28; m 49; c 3. ORGANIC CHEMISTRY. *Educ:* Mont State Univ, BS, 52; Johns Hopkins Univ, MA, 62, PhD(org chem), 64. *Prof Exp:* Jr chemist, Lago Oil & Transport Co, Ltd, Aruba, 52-60; chemist, Stand Oil Co, NJ, 53-60; instr chem, Johns Hopkins Univ, 60-63, Petrol Res Found fel, 63-64; res fel petrol chem, Mellon Inst, 64-70; assoc prof chem, Univ Pittsburgh, 69-70; prof chem & head dept, Northeast La Univ, 70-76; dean col sci, 77-83, PROF CHEM, FLA ATLANTIC UNIV, 77- *Concurrent Pos:* Am Chem Soc grant & lectr, Carnegie-Mellon Univ, 67-69; mem adv panel, Joint Oceanog Inst for Deep Earth Sampling, 68-83; mem, US Nat Comt Geochem, Nat Acad Sci, 76-79; vchmn, Conf Org Geochem, Gordon Res Conf, 77-78, chmn, 79-80; secy-treas, Div Geochem, Am Chem Soc, 79-, vchmn, 80, chmn, 81 & alt counr, 82-; ed-in-chief, Org Geochem, 81- *Mem:* AAAS; Am Chem Soc; NY Acad Sci. *Res:* Synthesis, reactions and spectrometry of the transition metal porphyrins and related materials; biogeochemistry of tetrapyrrole pigments. *Mailing Add:* Col Sci Fla Atlantic Univ Boca Raton FL 33431

BAKER, EDGAR EUGENE, JR, b Visalia, Calif, Oct 12, 13; m 38; c 1. MEDICAL MICROBIOLOGY. *Educ:* Univ Calif, Los Angeles, AB, 35, MA, 37, PhD(microbiol), 41. *Prof Exp:* Res assoc, Hooper Found, Univ Calif, 42-46; assoc path & bact, Rockefeller Inst, 46-49; assoc prof, 49-52, PROF MICROBIOL & HEAD DEPT, SCH MED, BOSTON UNIV, 52- *Mem:* Am Soc Microbiol; Am Asn Immunol; Soc Exp Biol & Med. *Res:* Coccidioidomycosis; immunology of plague; antigenic structure of enterobacteriaceae; antibiotics. *Mailing Add:* Dept Microbiol Boston Univ Sch Med Boston MA 02118

BAKER, EDGAR GATES STANLEY, b Peotone, Ill, June 7, 09; m 35, 88; c 3. BIOLOGY. *Educ:* DePauw Univ, AB, 31; Stanford Univ, PhD(biol), 43. *Prof Exp:* From instr to asst prof zool, Wabash Col, 32-38, actg head dept, 38-39; asst biol, Stanford Univ, 39-42; head biol sect, Del Mar Col, 46; asst prof biol, Cath Univ Am, 46-50; instr, Cath Sisters Col, 47-50; from assoc prof to prof, 50-74, head dept, 51-70, EMER PROF ZOOL, DREW UNIV, 74- *Concurrent Pos:* Consult surv physiol sci, Am Physiol Soc, 52-54; NSF sci fac fel, 57-58; chief reader biol, Adv Placement Prog, Col Entrance Exam Bd, 64-67; staff biologist, Comn on Undergrad Educ in Biol Sci, 67-68. *Mem:* Fel AAAS; Am Soc Zool; Sigma Xi. *Res:* Physiology of protozoan populations; bacteria-free cultures; effects of nutrition on rate of growth. *Mailing Add:* 1029 W Walnut St Brownstown IN 47220

BAKER, EDWARD THOMAS, b Chicago, Ill, May 21, 45; m 70; c 3. OCEANOGRAPHY, MARINE GEOLOGY. *Educ:* Univ Notre Dame, BS, 67; Univ Wash, MS, 69, PhD(oceanog), 73. *Prof Exp:* OCEANOGR, PAC MARINE ENVIRON LAB, NAT OCEANIC & ATMOSPHERIC ADMIN, DEPT COM, 76-; AFFIL ASSOC PROF, DEPT OCEANOG, UNIV WASH, 85- *Mem:* AAAS; Am Geophys Union; Oceanog Soc. *Res:* Geological and geochemical oceanography; composition, distribution and transport of particulate matter in the ocean; submarine hydrothermal processes. *Mailing Add:* NOAA/PMEL 7600 Sandpoint Wy NE Seattle WA 98115-0070

BAKER, EDWARD WILLIAM, b Porterville, Calif, Dec 29, 14; m 35; c 1. ACAROLOGY. *Educ:* Univ Calif, BS, 36, PhD(entom), 38. *Prof Exp:* Mem staff, Mex Fruit-Fly Lab, Bur Entom & Plant Quarantine, Agr Res Serv, USDA, 39-44, entomologist, Div Insect Identification, 44-53, Div Insects, Entom Res Br, Agr Res Serv, 53-58, res entomologist acarologist, Insect Identification & Parasite Introd Labs, Entom Res Div, 58-87; RETIRED. *Concurrent Pos:* Vis lectr, Univ Md, 53 & Ohio State Univ, 62. *Mem:* Entom Soc Am; Am Acarology Soc; Acarology Soc India. *Res:* Acarina; biology and taxonomy of mites; plant feeding mites of importance to agriculture. *Mailing Add:* Systematic Entom Lab Bldg 046 BARC W ARS USDA Beltsville MD 20705

BAKER, ELIZABETH MCINTOSH, b Washington, DC, Sept 30, 45; div. INVERTEBRATE ZOOLOGY. *Educ:* George Washington Univ, BA, 67; Univ Mich, MS, 68; Univ Va, PhD(develop biol), 73. *Prof Exp:* Vis asst prof, 75, electron micros technician biol, Univ NC, Charlotte, 75-77; adj prof, Sacred Heart Col, 77-78; CHMN, DEPT BIOL, BELMONT ABBEY COL, 78-, CHMN MATH & NATURAL SCI DIV, 81- *Res:* Insectan prothoracic gland development and function. *Mailing Add:* Dept Biol Belmont Abbey Col Belmont NC 28012

BAKER, FLOYD B, physical chemistry, for more information see previous edition

BAKER, FRANCIS EDWARD, JR, b Baltimore, Md, Nov 9, 44; m 66; c 3. STATIC MAGNETIC FIELD ANALYSIS. *Educ:* Va Polytech Inst & State Univ, BS, 67, MS, 69, PhD(elec eng), 71; Mass Inst Tech, SM, 89. *Prof Exp:* Instr elec eng, Va Polytech Inst & State Univ, 70-71; asst dir sci & technol, chief Naval Opers Exec Panel; head, Surface Warfare Anal Off, 84-88, HEAD, COMBAT ENGR & ASSIGNMENT DIV, NAVAL SURFACE WARFARE CTR, 89- *Concurrent Pos:* Vis assoc prof physics, US Naval Acad, Md, 80. *Mem:* Sigma Xi; Inst Elec & Electronics Engrs; Am Soc Naval Engr. *Res:* Static and low frequency electromagnetic field phenomena; electroacoustic phenomena; circuit analysis; electric power system design. *Mailing Add:* 1501 Kingsway Dr Gambrills MD 21054

BAKER, FRANCIS TODD, b Chicago, Ill, Feb 22, 42; m 64; c 3. NUCLEAR PHYSICS. *Educ:* Miami Univ, AB, 63, MA, 64; Univ Mich, PhD(physics), 70. *Prof Exp:* Asst prof, Carroll Col, Wis, 66-68 & St Lawrence Univ, 70-71; res assoc, Univ Mich, 71 & Rutgers Univ, 71-74; asst prof, 74-79, ASSOC PROF PHYSICS, DEPT PHYSICS & ASTRON, UNIV GA, 79- *Mem:* Am Phys Soc. *Res:* Experimental nuclear physics, particularly nuclear structure and reactions. *Mailing Add:* Dept Physics Univ Ga Athens GA 30602

BAKER, FRANK, b Dallas, Tex, Feb 28, 36; m 60; c 3. PSYCHONEUROIMMUNOLOGY. *Educ:* Vanderbilt Univ, BA, 58; Northwestern Univ, MA, 62, PhD(psychol), 64. *Prof Exp:* Lectr psychol, Northwestern Univ, 61-62; asst prof social psychol, Lehigh Univ, 63-65, psychiat, Harvard Med Sch, 65-74; prof psychol, social & prev med, med sch, State Univ NY, Buffalo, 74-85; prof & chmn, dept behav sci, 85-87, PROF HEALTH PSYCHOL ENVIRON HEALTH SCI, JOHNS HOPKINS SCH HYG & PUB HEALTH, 87- *Concurrent Pos:* Dir, Div Community Psychiat, State Univ NY Med Sch, Buffalo, 74-85; vis lectr psychiat, Harvard Med Sch, 81-82. *Mem:* Sigma Xi; fel Am Psychol Asn; Am Pub Health Asn; AAAS; fel Am Orthopsychiat Asn; Am Sociol Asn; Soc Behav Med. *Res:* Work stress and immunomodulation; psychosocial factors and quality-of-life of cancer patients; case management systems and community adjustment of the severely mentally ill; new techniques for health program evaluation. *Mailing Add:* 3029 St Paul St Baltimore MD 21218

BAKER, FRANK HAMON, b Stroud, Okla, May 2, 23; m 46; c 4. ANIMAL SCIENCE. *Educ:* Okla Agr & Mech Col, BS, 47, MS, 51, PhD(animal nutrit), 54. *Prof Exp:* County agr agent, Del County, Okla, 47-48; instr high sch, Okla, 49-50; asst animal husb, Okla Agr & Mech Col, 51-53; instr, Kans State Col, 53-54, asst prof, 54-55; assoc prof, Univ Ky, 55-58; exten livestock specialist, Okla State Univ, 58-62; exten animal scientist, USDA, Washington, DC, 62-66; prof animal sci & chmn dept, Univ Nebr, Lincoln, 66-74; prof & dean agr, 74-79, Int Agr Prog officer, 80-81, EMER PROF AGR, OKLA STATE UNIV, 81- *Concurrent Pos:* Dir, Int Stockmens Sch & US Progs, Winrock Int, Morrilton, Ark, 81-88, sr assoc, 88-; mem, Int Rev & Study Teams, Ecuador, 78 & Botswana, 81. *Honors & Awards:* Distillers Feed Res Coun Award, 64; Indust Serv Award, Am Soc Animal Sci. *Mem:* Am Soc Animal Sci (pres, 73-74); Am Meat Sci Asn; fel AAAS; Am Inst Biol Sci; Coun Agr Sci & Technol (pres, 79); fel Am Soc Animal Sci. *Res:* Animal production. *Mailing Add:* Winrock Int R 3 Box 376 Morrilton AR 72110-9537

BAKER, FRANK SLOAN, JR, b Brownwood, Tex, May 20, 21; m 42; c 3. ANIMAL HUSBANDRY, ANIMAL NUTRITION. *Educ:* Agr & Mech Col Tex, BS, 42; Univ Fla, MSA, 57. *Prof Exp:* From asst animmal husbandman to assoc animal husbandman, NFla Exp Sta, Univ Fla, Quincy, 45-63, animal husbandman & prof animal sci, Agr Res & Educ Ctr, Inst Food & Agr Sci, 63-89; RETIRED. *Mem:* Am Soc Animal Sci; Am Dairy Sci Asn. *Res:* Production of beef cattle; beef cattle nutrition and finishing. *Mailing Add:* PO Box 470 Quincy FL 32351

BAKER, FRANK WEIR, b Anderson, Ind, Nov 22, 38; m 64; c 2. ORGANIC CHEMISTRY. *Educ:* Col Wooster, BA, 60; Univ Chicago, MS, 62, PhD(org chem), 66. *Prof Exp:* Res chemist, Miami Valley Labs, Procter & Gamble Co, 66-71, Ivorydale Tech Ctr, 71-75, assoc dir prod develop, Winton Hill Tech Ctr, 75-80, dir res, Miami Valley Labs, 80-85, DIR, RES & DEVELOP, SHARON WOODS TECH CTR, PROCTER & GAMBLE CO, 85- *Mem:* AAAS; Am Chem Soc; Am Soc Photobiol; Sigma Xi. *Res:* Photochemical stability of biologically active compounds; toxicology; biological effects of environmental contaminants. *Mailing Add:* 952 Springbrook Dr Cincinnati OH 45224

BAKER, FRANK WILLIAM, b Kittanning, Pa, Oct 17, 35; m 58; c 3. ELECTROCHEMISTRY, FINISHING TECHNOLOGY. *Educ:* Grove City Col, BS, 57. *Prof Exp:* Scientist phys chem, Allegheny Ludlum Steel Corp, 57-59; develop engr process develop, 59-73, sect head, 73-78, SR TECH SUPVR PROCESS METALL, ALUMINUM CO AM, 78- *Mem:* Am Electroplaters Soc; Electrochem Soc; Sigma Xi. *Res:* Metal finishing including mechanical, chemical, electrochemical and applied finishing treatments; aluminum reduction processes. *Mailing Add:* Alcoa Tech Ctr Aluminum Co Am Alcoa Center PA 15069

BAKER, FREDERICK CHARLES, b North Shields, Eng, July 30, 48; m 78; c 2. ENZYMOLOGY, INSECT BIOCHEMISTRY. *Educ:* Univ Newcastle-Upon-Tyne, Eng, BSc, 69; NC State Univ, PhD(biochem), 74. *Prof Exp:* Res fel plant biosynthesis, Dept Chem, Univ Glasgow, Scotland, 74-77; res fel insect biosynthesis, 77-78, SR RES BIOCHEMIST, BIOCHEM DEPT, ZOECON CORP, PALO ALTO, CALIF, 78- *Mem:* Am Soc Biol Chemists; Phytochem Soc Europe; AAAS. *Res:* Lipid biochemistry of bacteria, plants, and insects particularly fatty acid metabolism in ruminant bacteria and sesquiterpenoid biosynthesis (phytoalexins and juvenile hormones) in plants and insects; enzyme purification; juvenile hormone titer determinations. *Mailing Add:* Dept Biochem & Insect Metab Sandoz Crop Protection Corp 975 California Ave Palo Alto CA 94304

BAKER, GEORGE ALLEN, b Robinson, Ill, Oct 31, 03; m 30; c 2. MATHEMATICAL STATISTICS. *Educ:* Univ Ill, BS, 26, PhD(math statist), 29. *Prof Exp:* Assoc statistician, USPHS, 29; Milbank Mem Fund fel, Columbia Univ, 29-30; statistician, State Dept Health, NY, 30-31; head dept math, Shurtleff Col, 31-34; prof & head dept, Miss Woman's Col, 34-36; with consumers purchase study, Bur Home Econ, USDA, 36-37; from instr math & jr statistician to prof math & statistician, 37-74, EMER PROF MATH, UNIV CALIF, DAVIS, 71- *Concurrent Pos:* Fac Res Lectr, 56. *Mem:* Am Math Soc; Math Asn Am; fel Inst Math Statist; Am Statist Asn; Biomet Soc (vpres, 50); fel AAAS. *Res:* Application of field trials and growth curves; selection, prediction and transformation of data; random sampling from nonhomogeneous populations; taste-testing; factor analysis; applications to astronomy, physics and education. *Mailing Add:* Dept Math Univ Calif Davis CA 95616

BAKER, GEORGE ALLEN, JR, b Alton, Ill, Nov 25, 32; m 90; c 3. STATISTICAL MECHANICS, APPLIED MATHEMATICS. *Educ:* Calif Inst Technol, BS, 54; Univ Calif, Berkeley, PhD(physics), 56. *Prof Exp:* NSF fel, Columbia Univ, 56-57; mem staff, Los Alamos Sci Lab, 57-66; physicist, Brookhaven Nat Lab, 66-71, sr physicist, 71-75; MEM STAFF, LOS ALAMOS NAT LAB, 75- *Concurrent Pos:* Assoc res physicist, Univ Calif, San Diego, 61-62; vis prof, Kings Col, Univ London, 64-65, Univ Nice, 70 & Cornell Univ, 71-72, Univ New S Wales, 85; assoc ed, J Math Physics & J Statist Physics; assoc group leader, 76-81, Los Alamos Fel, Los Alamos Nat Lab, 83-; vis physicist, serv physic theory, Saclay Ctr Nucl Study; vis lectr, Princeton Univ, 83. *Mem:* Fel Am Phys Soc; Int Asn Math Physics. *Res:* Statistical mechanics; mathematical methods of theoretical physics; quantum theory; field theory; nuclear physics. *Mailing Add:* T-11 MS-B262 Los Alamos Nat Lab Los Alamos NM 87545

BAKER, GEORGE SEVERT, b Chicago, Ill, Aug 2, 27; m 59; c 2. SOLID STATE PHYSICS. *Educ:* Purdue Univ, BS, 50; Univ Ill, PhD(physics), 57. *Prof Exp:* Res assoc metall, Univ Ill, 56-57; asst prof physics, Univ Utah, 57-62; tech specialist, Aerojet Gen Corp, Calif, 62-67; asst dean, Col Eng & Appl Sci, 72-76, PROF, DEPT MAT, UNIV WIS-MILWAUKEE, 67-, CHMN DEPT, 78- *Mem:* Am Soc Metals. *Res:* Defect properties of crystals; mechanical properties; refractory metals; forging; manufacturing processes. *Mailing Add:* Univ Wis Milwaukee WI 53201

BAKER, GEORGE THOMAS, III, b Waterbury, Conn, Sept 10, 40. BIOMEDICAL, GERONTOLOGY. *Educ:* Univ Conn, BA & BS, 65; Univ RI, MS, 69; Univ Miami, PhD(physiol & biophys), 71. *Prof Exp:* Res assoc biochem, Cancer Res Inst, New Eng Deaconess Hosp & Harvard Med Sch, Boston, 65-66; instr anat & physiol, Registr Nurse Prog, RI Hosp, Univ RI, 66-67; res physiologist, Naval Submarine Med Ctr, Groton, Conn, 67-69; res assoc, dept physiol & biophys, Univ Miami, 69-70, instr gen physiol, 71-73; vis prof biochem genetics, Univ Zurich, 73-75; vis scientist molecular biol, Max Planck Inst, Tubingen, Germany, 75-76; dir med & physiol, Inst Aging, Drexel Univ, 76-82; dir biomed, Ctr Aging, Univ Md, College Park, 82-86; PRES, BAKER & ASSOCS, 86- *Concurrent Pos:* Consult, space re-entry div, Gen Elec, Valley Forge, Pa, 79-84, GTE Serv Corp, Stamford, Conn, 82- & Milnor Fenwick, 83-; mem adv bd, nat community educ proj, Nat Coun Aging, 82-84 & technol & aging, Am Soc Aging, 82-85; consult, Nat Inst Aging/NIH, 86-87, Bessmer Carraway Med Ctr, Bessmer Ala, 86-, Montgomery Gen Hosp, Md, 86-87, Travelers Ins, Off Technol Assessment, US Cong, Am Asn Homes Aging, 87, Nat Oceanic & Atmospheric Admin, 89-; guest researcher, Gerontol Res Ctr, Nat Inst Aging/NIH, 89-; pres, Nathan Shock Aging Res Found. *Honors & Awards:* Swiss Am Found Award, 73; fel, Gerontol Soc Am, 78. *Mem:* Geront Soc (vpres, 80-81); AAAS; Sigma Xi; Am Aging Asn (pres, 85-); NY Acad Sci; Am Asn Advan Health Care Res. *Res:* Molecular-genetic mechanisms of aging,; impact of technology on older Americans health care issues and delivery systems; biomarkers of aging development and marketing of products and services for mature individuals. *Mailing Add:* 14628 Carna Dr Silver Spring MD 20907

BAKER, GLADYS ELIZABETH, b Iowa City, Iowa, July 22, 08. BOTANY. *Educ:* Univ Iowa, AB, 30, MS, 32; Washington Univ, PhD(mycol), 35. *Prof Exp:* Asst bot, Washington Univ, 35-36; instr biol, Hunter Col, 36-40; from instr to prof plant sci, Vassar Col, 40-63, chmn dept, 48-60; vis prof, 61-62, prof, 63-73, EMER PROF BOT, UNIV HAWAII, 73- *Concurrent Pos:* Vassar Col fac fel, Stanford Univ, 45 & Univ Calif, 54. *Mem:* AAAS; Mycol Soc Am; Sigma Xi; Brit Mycol Soc. *Res:* Cytology and morphology of myxomycetes, lichens and basidiomycetes; cytogenetics of imperfect fungi; distribution of fungi. *Mailing Add:* 154 Casa Del Sol 11411 N 91st Ave Peoria AZ 85345

BAKER, GLEN BRYAN, b Watrous, Sask, July 26, 47; m 75. NEUROCHEMISTRY, ANALYTICAL CHEMISTRY. *Educ:* Univ Sask, BSP, 70, MSc, 72, PhD(biol psychiat), 74. *Prof Exp:* Fel, Med Res Coun Neuropharmacol Unit, Univ Birmingham, UK, 74-77; from asst prof to assoc prof, 77-85, hon asst prof, 79-81, HON PROF, FAC PHARM & PHARMACEUT SCI, PROF PSYCHIAT, DEPT PSYCHIAT, UNIV ALTA, 85- *Concurrent Pos:* From asst dir to assoc dir, 79-90, co-dir, Neurochem Res Unit, Univ Alta, 90- *Mem:* Can Col Neuropsychopharm; Int Soc Neurochem; Pharmacol Soc Can; European Soc Neurochem; West Pharmacol Soc; Soc Neurosci. *Res:* Biochemical bases of psychiatric disorders; modes of action of antidepressant, neuroleptic and antianxiety drugs; development of assays for biogenic amines, psychotropic drugs and their metabolites. *Mailing Add:* Neurochem Res Unit Dept Psychiat MacKenzie Ctr Univ Alta Edmonton AB T6G 2B7 Can

BAKER, GRAEME LEVO, b Kalispell, Mont, Mar 7, 25; m 49; c 3. ANALYTICAL CHEMISTRY. *Educ:* Mont State Univ, BSc, 47, MS, 53, PhD(chem), 59. *Prof Exp:* Instr chem, Mont State Col, 49-51; assoc state feed control chemist, State Dept Agr, Mont, 51-52; asst chem, Mont State Col, 52-53, from instr to prof, 53-68; chmn dept, 68-78, PROF CHEM, UNIV CENT FLA, 68- *Mem:* Am Chem Soc; Sigma Xi. *Res:* Lipids of insect origin; protective coatings and corrosion. *Mailing Add:* 914 Arabian Lane Libby MT 59923-9537

BAKER, GRIFFIN JONATHAN, b Marion, Ill, July 19, 17; m 47; c 2. ENTOMOLOGY. *Educ:* Univ Ill, BSc, 40. *Prof Exp:* Asst, United Fruit Co, Honduras, 40-42 & Univ Ill, 42-43; asst proj chemist, Standard Oil Co (Ind), 46-50, proj chemist, 50-59; with McLaughlin, Gormley, King & Co, 59-70, mgr tech serv, 70-71, mgr res & develop, 71-82; RETIRED. *Mem:* Entom Soc Am. *Res:* Control of the banana thrips; product development of insecticidal aerosols; space and residual sprays; stock sprays and general garden sprays. *Mailing Add:* 6312 St John Ave Edina MN 55424

BAKER, HAROLD LAWRENCE, b Ogden, Utah, Jan 14, 18; m 46; c 2. FOREST ECONOMICS. *Educ:* Utah State Agr Col, BS, 39; Univ Calif, Berkeley, MS, 42, PhD(forest econ), 65. *Prof Exp:* Asst forestry res, Sch Forestry, Univ Calif, Berkeley, 40-42, asst forestry instr & res, 46-49; forest economist, Pac Southwest Forest & Range Exp Sta, US Forest Serv, 49-58; land economist, Hawaii State Land Study Bur, 58-74, dir, 64-74, prof forest econ & mem grad fac, 61-74, prof resource econ, 74-84, EMER PROF RESOURCE ECON, DEPT AGR & RESOURCE ECON, UNIV HAWAII, 84- *Concurrent Pos:* Forest econ consult, US Forest Serv, 61-74; resource economist, Water Resources Res Ctr, Univ Hawaii, 64-74, land econ consult, Am Factors, 64; forest & land use econ consult, UN Food & Agr Orgn, Philippine Govt, 66 & Malaysian Govt, 69-74; land econ consult, Govt Am Samoa, 72; res economist & res econ specialist, Coop Educ Serv, Hawaii Inst Trop Agr & Human Resources, 74-84, emer res econ & res economist specialist, 84- *Mem:* Soc Am Foresters. *Res:* Forest mensuration, management and policy; agricultural policy; production and land economics; land tenure; regional economic analyses; economics of resource allocation and development; land use planning; land use controls. *Mailing Add:* 423 Puamamane St Honolulu HI 96821

BAKER, HAROLD NORDEAN, b Iowa City, Iowa, May 18, 43; c 4. BIOCHEMISTRY, ATHEROSCLEROSIS. *Educ:* Lamar Univ, BS, 65; Tulane Univ, PhD(org chem), 70. *Prof Exp:* Robert A Welch Found fel, Univ Tex M D Anderson Hosp, Houston, 69-71, res assoc protein struct, 71; res assoc, Baylor Col Med, 71-72, from instr to asst prof, 72-76; asst prof, Univ Minn, 76-78; asst prof lipoproteins in atherosclerosis, La State Univ Med Ctr, 78-87; RES SCIENTIST, ABBOTT LAB, 87- *Concurrent Pos:* Dir, Lipid Res Clin, Core Lab, 72-76. *Mem:* Am Chem Soc; AAAS; Am Heart Asn. *Res:* Protein structure; metabolic basis of atherosclerosis; lipid transport; development of immunochemical assays. *Mailing Add:* Abbott Lab Dept 93Y Bldg AP8B Abbott Park IL 60064

BAKER, HAROLD WELDON, b Lincoln, Nebr, Jan 23, 31; m 57; c 2. ANALYTICAL CHEMISTRY. *Educ:* Nebr Wesleyan Univ, BA, 54; Purdue Univ, MS, 56; Univ Iowa, PhD, 61. *Prof Exp:* Instr chem, Butler Univ, 56-57; from assoc prof to prof, Parsons Col, 59-62; from asst prof to assoc prof, 62-74, PROF CHEM, SCH PHARM, TEMPLE UNIV, 74- *Mem:* Am Chem Soc. *Res:* Atomic absorption and fluorescence; trace analysis of metals; air and water pollution; heavy metal toxicity of life systems; pharmaceutical analysis; clinical chemistry. *Mailing Add:* Dept Chem Temple Univ 3307 N Broad St Philadelphia PA 19140-5010

BAKER, HERBERT GEORGE, b Brighton, Eng, Feb 23, 20; m 45; c 1. EVOLUTION, ECOLOGY. *Educ:* Univ London, BSc, 41, PhD, 45. *Prof Exp:* Res chemist & plant physiologist, Hosa Res Labs, 40-45; lectr bot, Univ Leeds, 45-54; sr lectr, Univ Col of Gold Coast, 54-55, prof & head dept, 55-57; dir bot garden, 57-69, assoc dir, 69-74, assoc prof, 57-60, PROF BOT, UNIV CALIF, BERKELEY, 60- *Concurrent Pos:* Res fel, Carnegie Inst, 48-49; assoc ed, Evolution, 56-59, 62-65 & Ecology, 63-66; res prof, Miller Inst, 66-67 & 88; mem bd gov, Orgn Trop Studies, 64-76; distinguished econ botanist, Soc Econ Bot. *Honors & Awards:* Merit Award, Am Bot Soc. *Mem:* Am Bot Soc (vpres, 77, pres, 79); Soc Study Evolution (secy, 67-69, pres, 69); Ecol Soc Am; Int Asn Bot Gardens (vpres, 64-69); fel AAAS; fel Am Acad Arts & Sci; fel Asn Trop Biol; Am Philos Soc; hon mem Brit Ecol Soc. *Res:* Ecology and evolution of higher plants, especially on reproductive biology; palynology; cytogenetics; history of biology; general tropical botany; nectar and pollen chemistry; weed evolution. *Mailing Add:* Dept Integrative Biol Univ Calif Berkeley CA 94720

BAKER, HERMAN, b New York, NY, Jan 22, 26; m 52; c 2. METABOLISM, NUTRITION. *Educ:* City Col New York, BS, 46; Emory Univ, MS, 48; NY Univ, PhD(metab), 56; Am Bd Nutrit, dipl, 68. *Prof Exp:* Assoc prof med, 60-70, PROF MED & PREV MED & DIR DIV NUTRIT & PREV MED, UNIV MED & DENT NJ, 70- *Concurrent Pos:* Mem, Coun Nutrit & pres Prev Med Sect, Pan Am Med Asn. *Honors & Awards:* Hans Selye Award, 79. *Mem:* Soc Exp Biol & Med; Am Soc Clin Nutrit; Pan Am Med Asn; fel Am Col Nutrit. *Res:* Vitamin metabolism; nutrition and analysis. *Mailing Add:* Dept Prev Med Pub Health UMD NJ Med Sch 185 S Orange Ave Newark NJ 07103

BAKER, HOUSTON RICHARD, b Pittsfield, Mass, Aug 27, 40; m 66; c 2. PHARMACOLOGY, PHYSIOLOGY. *Educ:* Harvard Col, BA, 63; Ohio State Univ, PhD(physiol), 69. *Prof Exp:* Res asst physiol, Physiol Inst, Univ Saarlandes, 69-70; res assoc cryobiol, Blood Res Lab, Am Nat Red Cross, 70-74; bus exec, Am Soc Plant Physiologists, 74-77, EXEC OFFICER, AM SOC PHARMACOL & EXP THERAPEUT, 77- *Concurrent Pos:* Ed, Newsletter, Am Soc Plant Physiologists, 74-77 & Pharmacologist, 77-; publ, What's New in Plant Physiol, 76- *Mem:* Am Soc Pharmacol & Exp Therapeut; Coun Eng & Sci Soc Execs; Soc Cryobiol; Am Soc Plant Physiologists; fel AAAS. *Res:* Muscle contraction; electrophysiology; freezing injury and mitigation; plant hardiness; scientific society management; scientific journal publishing and management. *Mailing Add:* 9411 Warfield Rd Gaithersburg MD 20882

BAKER, HOWARD CRITTENDON, b Lexington, Ky, Sept 4, 43; m 67. THEORETICAL PHYSICS. *Educ:* Berea Col, BA, 65; Washington Univ, St Louis, MA, 67, PhD(physics), 72. *Prof Exp:* From asst prof to prof physics, Berea Col, 72-90; HEAD, DEPT PHYSICS, BUTLER UNIV, 90- *Concurrent Pos:* Vis asst prof physics, Univ Conn, 74-75; NSF fel, Oakridge Nat Lab, 79; vis res fel, Johns Hopkins Univ, 80; mem tech staff, Rocketdyne Div, Rockwell Int, 85-86. *Mem:* Am Phys Soc; Am Asn Physics Teachers. *Res:* Field theory quantum measurement, decay theory and nonlinear optics. *Mailing Add:* Physics Dept Box 209 Butler Univ Indianapolis IN 46208

BAKER, JAMES ADDISON, b Eugene, Ore, Aug 20, 22; m 46; c 1. COMPUTER SCIENCE. *Educ:* Pomona Col, AB, 44. *Prof Exp:* Head math & comput, Lawrence Berkeley Lab, Univ Calif, Berkeley, 52-58; head data systs div, Broadview Res Corp, 58-61; head math & comput group, Lawrence Berkeley Lab, Univ Calif, Berkeley, 61-84; consult, Sci Comput Systs Corp, 84-88, Silicon Mgt Group, 88-90; RETIRED. *Concurrent Pos:* Consult, Comt Uses of Comput, Nat Acad Sci, 63; lectr elec eng, comput sci, indust eng, opers res & transp eng, Univ Calif, Berkeley, 63-; vis scholar, Europ Orgn Nuclear Res, Geneva, 74-75; mem, Am Nat Standards Comt, X3 Info Processing. *Mem:* Asn Comput Mach; Inst Elec & Electronics Engrs Comput Soc. *Res:* Simulation of communications systems; programming systems; mass storage systems; network analysis. *Mailing Add:* 131 Avenida Dr Berkeley CA 94708

BAKER, JAMES BERT, b Bernice, La, Feb 7, 39; m 60; c 2. FOREST SOILS, SILVICULTURE. *Educ:* Univ Ark, Monticello, BSF, 61; Duke Univ, MF, 62; Miss State Univ, PhD(forest soils), 70. *Prof Exp:* Res forester silvicult, 62-71, res soil scientist forest soils, 71-78, SUPVRY RES FORESTER, SOUTHERN FOREST EXP STA, USDA FOREST SERV, 78- *Mem:* Soc Am Foresters; Sigma Xi. *Res:* Uneven-aged management; development of silvicultural systems; timber management; multi-resource management. *Mailing Add:* USDA Forest Serv PO Box 3516 Monticello AR 71655

BAKER, JAMES E, b Columbus, Ohio, Apr 16, 31; m 58; c 2. SCIENTIFIC ADMINISTRATION, SPEECH RECOGNITION. *Educ:* US Naval Acad, BS, 54; Okla State Univ, BS & MS, 62, PhD(elec eng), 71. *Prof Exp:* Dir, Satellite Commun Prog, USAF, 74-77, dir elec systs, hq, Systs Command Prog, 77-79, Comdr, Geophysics Lab, 79-81, dir, Off Sci Res, 81-84; PROF & HEAD ELEC ENGR, OKLA STATE UNIV, 84- *Concurrent Pos:* US rep, Res Panel Physics & Electronics, NATO, 81-84. *Mem:* Inst Elec & Electronics Engrs; Sigma Xi; Nat Soc Prof Engrs. *Res:* Speech processing for recognition by machines; scientific disciplines. *Mailing Add:* 2809 N Husband St Stillwater OK 74075

BAKER, JAMES EARL, plant physiology, for more information see previous edition

BAKER, JAMES GILBERT, b Louisville, Ky, Nov 11, 14; m 38; c 4. OPTICAL PHYSICS, ASTRONOMY. *Educ:* Univ Louisville, AB, 35; Harvard Univ, MA, 36, PhD(astron, astrophys), 42. *Hon Degrees:* ScD, Univ Louisville, 48. *Prof Exp:* Res fel, 42-45, dir, Optical Res Lab, 43-45, assoc prof, Harvard Observ, 46-48, res assoc, 49-62, ASSOC, HARVARD OBSERV, HARVARD UNIV, 62- *Concurrent Pos:* Lowell lectr, 40; res assoc, Lick Observ, 48-60; consult optical physics & aerial photog, US Air Force, 49-57, mem sci adv bd, 52-57; chmn, US Nat Comn, Int Comn Optics, 56-59, vpres, 59-62; assoc, Ctr Astrophys, Harvard & Smithsonian Observs, 66-; consult, optical physics & astron, Aerospace Corp, 66-75, photog physics, Polaroid Corp, 66-; trustee, The Perkin Fund, 70- *Honors & Awards:* Presidential Medal of Merit, 47; Exceptional Civilian Serv Award, USAF, 57; Adolph Lomb Medal, 42; Magellanic Medal, Am Philos Soc, 53; Elliott Cresson Medal, Franklin Inst, 62; Frederick Ives Medal, Optical Soc Am, 65; Alan Gordon Award, Soc Photooptical Instrumentation Engrs, 76, Gold Medal, 78; Int Lens Design Conf Award, 90; Fraunhofer Award, Optical Soc Am, 91. *Mem:* Nat Acad Sci; Nat Acad Eng; Am Astron Soc; fel Optical Soc Am (pres, 60); Am Acad Arts & Sci; Am Philos Soc. *Res:* Instrumentation and optical design; astrophysics. *Mailing Add:* 14 French Dr Bedford NH 03102

BAKER, JAMES HASKELL, b Ft Worth, Tex, Sept 8, 40. MARINE ZOOLOGY. *Educ:* Tex Christian Univ, BA, 62, MS, 65; Univ Houston, PhD(biol), 75. *Prof Exp:* Mus technician, Smithsonian Oceanog Sorting Ctr, 67; res assoc aquatic biol, TCU Res Found, Tex Christian Univ, 68-70; SR RES SCIENTIST AQUATIC BIOL, SOUTHWEST RES INST, 73- *Mem:* AAAS; Am Soc Limnol & Oceanog; Asn Meiobenthologists; NY Acad Sci; Soc Syst Zool. *Res:* Ecology and systematics of marine invertebrates. *Mailing Add:* 2011 Singleton Houston TX 77006-4455

BAKER, JAMES LEROY, b Chamberlain, SDak, Apr 25, 44; m 65; c 2. PHYSICAL CHEMISTRY, AGRICULTURAL ENGINEERING. *Educ:* SDak Sch Mines & Technol, BS, 66; Iowa State Univ, PhD(phys chem), 71. *Prof Exp:* Asst chem, Iowa State Univ 66-71, agr engr, 71-73, asst prof, 73-76, assoc prof, 76-, PROF AGR ENG, IOWA STATE UNIV. *Mem:* Am Chem Soc; Am Soc Agr Engrs; Am Soc Agron; Soil Sci Soc Am. *Res:* Effects of farm management practices on the efficient use of pesticides and nutrients by agriculture, with emphasis on losses to the environment. *Mailing Add:* Dept Agr Eng Iowa State Univ 102 Davidson Ames IA 50011

BAKER, JAMES ROBERT, b Urbana, Ill, Apr 20, 41; m 66; c 3. ENTOMOLOGY. *Educ:* NC State Univ, BS, 66, MS, 68; Univ Kans, PhD(entom), 72. *Prof Exp:* Exten entomologist, NC Agr Exten Serv, NC State Univ 72-81. *Mem:* Entom Soc Am; Sigma Xi. *Res:* Mealybugs infesting ornamental plants in North Carolina, identification and control. *Mailing Add:* Rte 4 Box 82C Apex NC 27502

BAKER, JEFFREY JOHN WHEELER, b Montclair, NJ, Feb 2, 31; m 55; c 4. DEVELOPMENTAL BIOLOGY. *Educ:* Univ Va, BA, 53 & MS, 59. *Prof Exp:* Supvr qual control, Libby, McNeill & Libby Food Co, Inc, 53-54; teacher biol, Mt Hermon, Mass, 54-62; lectr sci, Wesleyan Univ, 62-66; vis assoc prof biol, George Washington Univ, 66-68; prof, Univ PR, 68-69; sr fel, sci soc prog, 75-82, LECTR SCI, WESLEYAN UNIV, 69- *Concurrent Pos:* Staff biologist & educ dir, Comn Undergrad Educ Biol Sci, 66-68; vis prof biol, Washington Univ, St Louis, Mo, 85; scientist scholar, res dept religious studies, Charlottesville, Va. *Mem:* AAAS; Am Inst Biol Sci; Nat Asn Sci Writers; Soc Sci Study Sex. *Res:* Amphibian development; religion and science; science writing. *Mailing Add:* Box 205 Ivy VA 22945

BAKER, JOFFRE B, b Alameda, Calif, Dec 23, 47. CARDIOVASCULAR RESEARCH. *Educ:* Univ Calif, San Diego, BA, 71; Univ Hawaii, PhD(biochem), 77. *Prof Exp:* Postdoctoral fel, Dept Med Microbiol, Univ Calif, Irvine, 77-79; from asst prof to assoc prof, Dept Biochem, Univ Kans, 79-88; sr scientist, 88-90, ACTG DIR, DEPT CARDIOVASC RES, GENENTECH, INC, 90- *Concurrent Pos:* NIH & Am Heart Asn grants, 80-88; NIH res career develop award, 83-88. *Mem:* Am Soc Cell Biol; Am Soc Biochem & Molecular Biol. *Mailing Add:* Dept Cardiovasc Res Genentech Inc 480 Pt San Bruno Blvd S San Francisco CA 94080

BAKER, JOFFRE BERNARD, b Alameda, Calif, Dec 23, 47; m 70; c 1. CELL BIOLOGY. *Educ:* Univ Calif, San Diego, BA, 71; Univ Hawaii, PhD(biochem), 77. *Prof Exp:* Res asst, Dept Biochem, Univ Hawaii, 76; fel, Dept Microbiol, Univ Calif, Irvine, 77-79; ASST PROF, DEPT BIOCHEM, UNIV KANS, 79- *Mem:* Am Soc Cell Biol; Sigma Xi. *Res:* Interactions of protein hormones with cells, cellular regulation of protease activity, control of proliferations of eukaryotic cells, and biochemistry of cell membranes. *Mailing Add:* Dept Biochem Univ Kans Lawrence KS 66045

BAKER, JOHN ALEXANDER, b Stratford, Ont, Apr 11, 39; m 65; c 2. ANALYSIS & FUNCTIONAL ANALYSIS. *Educ:* Univ Sask, BA, 60, MA, 63; Univ Waterloo, PhD(math), 68. *Prof Exp:* Instr math, Can Serr Col Royal Roads, 64-65; lectr, Univ Sask, 65-66; from asst prof to assoc prof, 68-82, PROF MATH, UNIV WATERLOO, 82- *Mem:* Can Math Soc; Math Asn Am. *Res:* Analytic aspects of the theory of functional equations; regularity properties; stability and techniques of solution involving differential equations; distribution theory and functional analysis. *Mailing Add:* Dept Pure Math Univ Waterloo Waterloo ON N2L 3G1 Can

BAKER, JOHN BEE, b Clarksdale, Miss, Mar 30, 27; m 56; c 2. PLANT PHYSIOLOGY, WEED SCIENCE. *Educ:* Miss State Col, BS, 50; Univ Wis, MS, 51, PhD(bot), 55. *Prof Exp:* Asst prof plant path & asst pathologist weed control, 53-59, assoc prof bot & plant path, 59-66, PROF BOT & PLANT PATH, LA STATE UNIV, BATON ROUGE, 66-, CHMN DEPT, 90- *Mem:* Am Soc Plant Physiol: Weed Sci Soc Am. *Res:* Mechanism of action of herbicides and growth regulators; autecology of weeds; weed control in rice. *Mailing Add:* Dept Plant Path La State Univ Baton Rouge LA 70803

BAKER, JOHN CHRISTOPHER, b Warren, Ohio, Feb 25, 52; m 79; c 1. LARGE ANIMAL INTERNAL MEDICINE. *Educ:* Ohio State Univ, BS, 73, MS, 75, DVM, 80; Univ Minn, PhD(vet med), 84; Am Col Vet Internal Med, dipl, 85. *Prof Exp:* Asst prof, 84-88, ASSOC PROF, MICH STATE UNIV, 88- *Mem:* Am Vet Med Asn; Res Workers Animal Dis. *Res:* Bovine respiratory disease, specifically the epidemiology pathogenesis of respiratory syncytial virus infection. *Mailing Add:* Vet Clin Ctr Mich State Univ East Lansing MI 48824-1314

BAKER, JOHN DAVID, b Columbus, Ohio, Aug 9, 49; m 80; c 2. RADIOCHEMISTRY, SEPARATIONS CHEMISTRY. *Educ:* Northern Ill Univ, BS, 72; Univ Idaho, MS, 80. *Prof Exp:* Assoc chemist, Allied Chem Corp, 73-78; group leader, Exxon Nuclear Corp, 78-80; scientist, 80-83, sr scientist, 83-88, SCI SPECIALIST, EG&G, IDAHO, INC, 88- *Mem:* Am Chem Soc. *Res:* Applying chemical separation techniques, such as high performance liquid chromatography solvent extraction to the rapid radiochemical separation of short lived fission products; basic solvent extraction processes; automating chemical separations; radioisotope production. *Mailing Add:* EG&G Idaho Inc TRA 604 PO Box 1625 Idaho Falls ID 83415-7111

BAKER, JOHN KEITH, b San Antonio, Tex, Dec 1, 42; m 65; c 2. MEDICINAL CHEMISTRY, PHYSICAL CHEMISTRY. *Educ:* Univ Tex, Austin, BS, 66; Univ Calif, San Francisco, PhD(pharmaceut chem), 70. *Prof Exp:* From asst prof to assoc prof 70-79, PROF MED CHEM, SCH PHARM, UNIV MISS, 79- *Concurrent Pos:* Sr fel, Nat Inst Health, 85-86. *Mem:* Am Pharmaceut Asn; Am Chem Soc. *Res:* Application of spectroscopic methods in molecular pharmacological studies; thermodynamic studies of the interaction of adenosine triphosphate; drug metabolism; high pressure liquid chromatography; quantitative drug analysis. *Mailing Add:* Dept Med Chem Sch of Pharm Univ Miss University MS 38677

BAKER, JOHN P, b Aledo, Tex, Jan 31, 23; m 44; c 3. EQUINE NUTRITION & MANAGEMENT. *Educ:* Univ Ill, PhD(animal sci), 60. *Prof Exp:* PROF ANIMAL SCI, UNIV KY, 67- *Concurrent Pos:* Assoc prof, Animal Sci, Univ Idaho, 60-67. *Mem:* Equine Nutrit & Physiol Soc; Am Inst Nutrit; Am Soc Animal Sci. *Res:* Equine nutrition and exercise physiology. *Mailing Add:* Dept Animal Sci 905 Garrigus Bldg Univ Ky Lexington KY 40546-0215

BAKER, JOHN ROWLAND, b Hadleigh, Essex, Eng, Sept 26, 34. DENTISTRY, BIOCHEMISTRY. *Educ:* Univ Reading, Eng, BSc Hons, 56, PhD(physiol chem), 59. *Hon Degrees:* MA, Trinity Col, Dublin, 67. *Prof Exp:* Postdoctoral fel, Dept Biochem & Biophys, Univ Calif, Davis, 59-61; res assoc, Queen Charlotte's Hosp, London, Eng, 62-63; lectr biochem, Trinity Col, Univ Dublin, Ireland, 63-72; assoc prof, 72-80, PROF DENT, UNIV ALA, BIRMINGHAM, 80-, PROF BIOCHEM, 82- *Concurrent Pos:* USPHS int fel, 68; sabbatical leave, Dept Pediat, Univ Chicago, 68-70; sr scientist, Diabetes Res & Training Ctr, Univ Ala, Birmingham; comt mem, Brit Biochem Soc, 70-72. *Mem:* Biochem Soc; Sigma Xi; assoc Royal Inst Chem; Soc Complex Carbohydrates; Fedn Am Socs Exp Biol. *Res:* Problems of proteoglycan structures and biosynthesis; structure of proteoglycans from other connective tissues; interaction of proteoglycans with other connective tissue macromolecules; immunological changes in proteoglycans in disease. *Mailing Add:* Dept Dent & Biochem Univ Ala Rm 712 Diabetes Hosp Birmingham AL 35294

BAKER, JOHN WARREN, b El Paso, Tex, Aug 24, 36; m 70. FUNCTIONAL ANALYSIS, TOPOLOGY. *Educ:* Hardin-Simmons Univ, BS, 58; Univ Tex, Austin, MA, 65, PhD(math), 68. *Prof Exp:* Asst prof math, Fla State Univ, 68-73; asst prof, 73-75, ASSOC PROF MATH, KENT STATE UNIV, 75- *Concurrent Pos:* Undergrad coordr, Dept Math, Kent State Univ, 74-76; colloquium chmn math sci, Kent State Univ, 81- *Mem:* Asn Comput Mach; Math Asn Am. *Res:* Computer algebra; spaces of continuous functions; projections in Banach spaces; zero-dimension topological spaces. *Mailing Add:* Dept Math Kent State Univ Main Campus Kent OH 44242

BAKER, JOSEPH WILLARD, b Luray, Va, Sept 15, 24; m 52; c 2. ORGANIC CHEMISTRY, PROCESS DEVELOPMENT. *Educ:* Bridgewater Col, BA, 45; Univ Va, MS, 49, PhD(org chem), 52. *Prof Exp:* Instr, Univ Va, 51-52; res chemist, 52-56, res group leader, 56-72, sr res group leader, Monsanto Co, 72-85,; CONSULT, 85- *Mem:* Am Chem Soc; Sigma Xi. *Res:* Organophosphorus chemistry; agricultural chemicals; food and fine chemicals; pharmaceuticals; bacteriostats; plasticizers; functional fluids; amino acids, ketones and alcohols; organic heterocycles; catalytic reductions and chemical process development; antioxidants; pharmaceuticals. *Mailing Add:* 421 Greenleaf Dr St Louis MO 63122-4451

BAKER, JUNE MARSHALL, b Napton, Mo, Nov 11, 22; m 47; c 3. ANALYTICAL CHEMISTRY, INSTRUMENTAL CHEMISTRY. *Educ:* Mo Valley Col, BA, 44; Ohio State Univ, MSc, 50; Univ Mo, PhD(chem), 55. *Prof Exp:* Instr chem, Mo Valley Col, 47-49; analyst, Univ Mo, 52-55; asst prof, Tenn Polytech Univ, 55-56; assoc prof, Eastern Ill State Col, 56-57; assoc prof, 57-65, PROF CHEM, AUBURN UNIV, 65- *Concurrent Pos:* Consult, Orradio Industs & Auburn Res Found. *Mem:* Am Chem Soc. *Res:* Ion-exchange method for study of chemical reactions; determinations for chromium; trace element contamination; spectrophotometry. *Mailing Add:* 950 Terrace Acres Dr Auburn AL 36830

BAKER, KAY DAYNE, b Escalante, Utah, Jan 31, 34; m 55; c 3. METEOROLOGY, ELECTRICAL ENGINEERING. *Educ:* Univ Utah, BS, 56, MS, 57, PhD(elec eng), 66. *Prof Exp:* Res asst, Upper Air Res Lab, Univ Utah, 54-57, res engr, 57-62, asst dir lab, 62-69, dir lab, 69-70; dir, Space Sci Lab, 70-79, dept head, Elec Engr, 72-77, dir, Ctr Atmospheric & Space Sci, 81-83, PROF ELEC & PHYSICS, UTAH STATE UNIV, 70- *Concurrent Pos:* Mem Comn 3, Inst Sci Radio Union, 67- *Honors & Awards:* Alexander von Humboldt Found sr scientist award. *Mem:* Am Geophys Union. *Res:* Development of measuring techniques and investigations of the upper atmosphere of the earth, with special emphasis on ionospheric auroral measurements. *Mailing Add:* Dept Physics Utah State Univ Logan UT 84322-4415

BAKER, KENNETH FRANK, b Ashton, SDak, June 3, 08; m 44. PHYTOPATHOLOGY, SOIL MICROBIOLOGY. *Educ:* State Col Wash, BS, 30, PhD(plant path), 34. *Prof Exp:* Asst plant path, State Col Wash, 30-34; Nat Res fel bot, Univ Wis, 34-35; jr pathologist, Div Forest Path, USDA, 35-36; assoc pathologist, Exp Sta, Pineapple Producers Coop Asn, 36-39; from asst plant pathologist & asst prof to plant pathologist, Exp Sta & prof plant path, Univ Calif, Los Angeles, 39-60; prof & plant pathologist, 61-75, EMER PROF PLANT PATH & EMER PLANT PATHOLOGIST, UNIV CALIF, BERKELEY, 75- *Concurrent Pos:* Supvry technician, Div Forest Path, USDA, 34; mem, Bot Exped, SAm, 38-39 & Cent Am, 57, Comt Biol Control Agr Bd, Nat Res Coun, 58-65; ed, Phytopathology, 58-60, Ann Rev Phytopath, 62-77 & Plant Dis Reporter, 72-77; Fulbright res scholar, Univ South Australia, 61-62; collaborator, USDA, Corvallis, Ore, 77-; courtesy prof, Ore State Univ, Corvallis, 77- *Mem:* Fel Am Phytopath Soc; Mycol Soc Am; Brit Mycol Soc; Brit Asn Appl Biol; Netherlands Soc Plant Path; Australasian Plant Path Soc; Brit Soc Plant Pathologists. *Res:* Storage decays of apples; diseases of ornamental plants; seedborne pathogens; heat therapy; soil steaming; biological control of root pathogens; history of plant pathology. *Mailing Add:* Hort Crops Res Lab USDA 3420 NW Orchard Ave Corvallis OR 97330

BAKER, KENNETH L(EROY), b Inwood, Ind, Feb 3, 19; m 45; c 2. MECHANICAL ENGINEERING. *Educ:* Purdue Univ, BSME, 42. *Prof Exp:* Mech engr, US Naval Surface Weapons Ctr, Silver Spring, 43-79; RETIRED. *Res:* Naval ordnance. *Mailing Add:* 9104 49th Ave College Park MD 20740

BAKER, KENNETH MELVIN, b Harrisburg, Pa, Oct 30, 47; m; c 1. EXPERIMENTAL BIOLOGY. *Educ:* Lebanon Valley Col, BS, 69; Temple Univ, MD, 74. *Prof Exp:* Med residency, Temple Univ Hosp, Philadelphia, Pa, 74-77, clin instr & chief med resident, 77-78; cardiol fel & res fel hypertension & pharmacol, Univ Va Hosp, Charlottesville, 78-81; from asst prof to assoc prof med, Milton S Hershey Med Ctr, Pa, 81-87; asst dir, Cardiovasc Non-Invasive Lab, 81-86, ASSOC, DEPT CARDIOL, GEISINGER MED CTR, DANVILLE, PA, 81-, STAFF SCIENTIST, WEIS CTR RES, GEISINGER CLIN, 87-; ASSOC PROF MED, THOMAS

JEFFERSON SCH MED, PHILADELPHIA, PA, 89- *Concurrent Pos:* Emergency Rm physician, Germantown Hosp, Charlottesville, Va, 77-78; co-investr & prin investr, var insts, 78-; mem, Sci Res Comt, 84-88 & Hazardous Waste Comt, 90-; fel, Coun High Blood Pressure Res, Am Heart Asn, 90, estab investr, Asn, 91 & mem, Coun Basic Sci. *Mem:* Am Soc Cell Biol; fel Am Col Cardiol; assoc mem Am Col Physicians; AMA; Am Heart Asn; Am Fedn Clin Res; Am Physiol Soc. *Mailing Add:* Geisinger Clin 26-11 Weis Ctr Res Danville PA 17822

BAKER, KIRBY ALAN, b Boston, Mass, June 17, 40; m 66, 80. MATHEMATICS. *Educ:* Harvard Univ, AB, 61, PhD(math), 66. *Prof Exp:* Ford Found res fel math, Calif Inst Technol, 66-68; from asst prof to assoc prof, 68-76, PROF MATH, UNIV CALIF, LOS ANGELES, 76- *Mem:* Am Math Soc; Math Asn Am; Asn Comput Mach. *Res:* Lattice theory and partial order; algebraic systems. *Mailing Add:* Dept Math Univ Calif Los Angeles 405 Hilgard Ave Los Angeles CA 90024

BAKER, LEE EDWARD, b Springfield, Mo, Aug 31, 24; m 48; c 2. PHYSIOLOGY, ELECTRICAL ENGINEERING. *Educ:* Univ Kans, BS, 45; Rice Univ, MS, 60; Baylor Col Med, PhD(physiol), 65. *Prof Exp:* Design engr, Radio Corp Am, 46-47; consult engr, 47-51; chief engr & part owner, Radio Sta KDKD, Mo, 53-54; owner & mgr, Radio Sta KOKO, 54-55; from instr to asst prof elec eng, Rice Univ, 55-65; from asst prof to assoc prof physiol, Baylor Col Med, 65-75; prof elec eng & dir biomed eng prof, 75-82, ROBERT L PARKER SR PROF ELEC & DIR BIOMED ENG PROG, UNIV TEX, AUSTIN, 82- *Concurrent Pos:* Consult, Sci Res Ctr, Int Bus Mach Corp, 66-67 & Southwest Res Inst, 82-84. *Mem:* Am Inst Elec & Electronics Eng; Asn Advan Med Instrumentation; Biomed Eng Soc; Am Physiol Soc; NY Acad Sci. *Res:* Use of electrical impedance for the measurement of physiological events. *Mailing Add:* Dept Biomed Eng Univ Tex Eng Sci Bldg 610 Austin TX 78712

BAKER, LENOX DIAL, b DeKalb, Tex, Nov 10, 02; m 33, 67; c 2. ORTHOPAEDIC SURGERY. *Educ:* Duke Univ, MD, 34; Am Bd Orthop Surg, dipl. *Prof Exp:* Athletic trainer, Univ Tenn, 25-29, lab asst zool, 27-29; athletic trainer, Duke Univ, 29-33; intern orthop, Johns Hopkins Hosp, 33-34, intern surg, 34-35, asst resident orthop surg, 35-36, resident, 36-37, asst instr, Sch Med, 35-37; from asst to prof orthop, 37-72, orthopedist, 37-72, dir, Div Phys Ther, 43-62, EMER PROF ORTHOP, SCH MED, DUKE UNIV, 72- *Concurrent Pos:* Orthop surgeon, Crippled Children's Div, NC Bd Health, 37-74, Voc Rehab Div, NC Dept Pub Instr, 37-74; orthopedist, Lincoln Hosp, 37-74; vis orthopedist, Watts Hosp, 37-74; med dir, Lenox Baker Children's Hosp, 49-72; mem gov cabinet, Secy Human Resources, State of NC, 72-; orthop consult various hosps, found, sanitariums & gov agencies. *Mem:* Am Med Asn; fel Am Acad Orthop Surg; Am Orthop Asn; Int Cerebral Palsy Soc. *Res:* Osteomyelitis; arthritis; fractures; cerebral palsy. *Mailing Add:* Dept Orthop Duke Univ Sch Med Box 3706 Durham NC 27710

BAKER, LEONARD MORTON, b Medford, Mass, Oct 2, 34; m 58; c 3. ORGANIC CHEMISTRY, POLYMER CHEMISTRY. *Educ:* Harvard Univ, AB, 56; Mass Inst Technol, PhD(org chem), 59. *Prof Exp:* Chemist, Plastics Div, 59-62, proj scientist, 62-64, group leader, 64-69, assoc dir res & develop chem & plastics, 69-74, dir, 74-77, vpres res & develop, 77-84, CORP DIR, UNION CARBIDE CORP, 84-, CORP VPRES TECHNOL, 84- *Concurrent Pos:* Mem gov bd, Coun Chem Res, 89- *Mem:* Nat Acad Sci; Am Chem Soc; Am Inst Chem Engrs; Indust Res Inst; NY Acad Sci; Coun Chem Res. *Res:* Catalysis; process research and development; polymer applications development. *Mailing Add:* Union Carbide Corp 39 Old Ridgebury Rd Danbury CT 06817-0001

BAKER, LEONARD SAMUEL, geodesy, for more information see previous edition

BAKER, LOUIS, JR, b Chicago, Ill, Dec 31, 27; m 51; c 4. PHYSICAL CHEMISTRY. *Educ:* Ill Inst Technol, BS, 49, MS, 51, PhD(chem), 55. *Prof Exp:* Aeronaut res scientist, Lewis Lab, Nat Adv Comt Aeronaut, 54-58; assoc chemist, 58-70, SR CHEMIST, ARGONNE NAT LAB, 70- *Honors & Awards:* E O Lawrence Award, US AEC, 73. *Mem:* Am Chem Soc; Am Nuclear Soc. *Res:* Reactor safety; chemical kinetics and heat transfer. *Mailing Add:* 760 61st St Downers Grove IL 60516

BAKER, LOUIS COOMBS WELLER, b New York, NY, Nov 24, 21; m 64; c 2. STRUCTURAL INORGANIC CHEMISTRY, HETEROPOLY COMPLEXES. *Educ:* Columbia Univ, AB, 43; Univ Pa, MS, 47, PhD(chem), 50. *Hon Degrees:* Dr, Georgetown Univ, 88. *Prof Exp:* Asst instr chem, Univ Pa, 43-50, instr, 45-50, assoc, Johnson Found, 50-51; from asst prof to assoc prof & head inorg div, Boston Univ, 51-62; chmn dept, 62-84, PROF CHEM, GEORGETOWN UNIV, 62- *Concurrent Pos:* Co-proj dir, Off Prod Res & Develop, 43-46; instr, Pa Area Cols, 45-49; external examr for doctorate, Univ Calcutta, 56-; Guggenheim fel, 60-; fel, Wash Acad Sci, 64-; consult, Various Petrochem Co, 57-; John Wiley & Sons, 68- & Nat Res Coun, 69-; lectr, Tour Univs, Romanian Ministry Educ, 73; Univs & Res Insts, USSR Acad Sci, 73; Plenary lectr, XV Int Conf on Coord Chem, 73; chmn, Comt Recommendations US Army Basic Sci Res, Nat Acad Sci, 74-; consult sci educ, Ferdowsi Univ, Iran, 75-77; mem comt, Accreditation Cols & Univs, Middle States Asn, 75- *Honors & Awards:* Tchugaev Medal, USSR Acad Sci, 73. *Mem:* Am Chem Soc; Sigma Xi (past pres). *Res:* Structures, properties and syntheses of heteropoly molybdates, tungstates and vanadates; coordination complexes; electron exchange and delocalization, magnetochemistry; high thermal efficiency engines; multinuclear nuclear magnetic resonance; synthetic inorganic chemistry; academic administration. *Mailing Add:* Dept Chem Georgetown Univ Washington DC 20057

BAKER, MALCHUS BROOKS, JR, b Alton, Ill, June 19, 41; m 70; c 3. HYDROLOGY, FORESTRY. *Educ:* Southern Ill Univ, BS, 64; Yale Univ, MF, 65; Univ Minn, PhD(forestry), 71. *Prof Exp:* HYDROLOGIST WATERSHED MGT, ROCKY MOUNTAIN FOREST & RANGE EXP STA, 69- *Mem:* Am Geophys Union; Sigma Xi; Soc Am Foresters; Am Water Resources Asn. *Res:* Quantify the effects of vegetation manipulation upon stream flow, water quality and sediment yield. *Mailing Add:* Forestry Sci Lab Ariz State Univ Tempe AZ 85287-1304

BAKER, MARY ANN, b Los Angeles, Calif, Oct 11, 40; m 82. MAMMALIAN PHYSIOLOGY, NEUROSCIENCES. *Educ:* Univ Redlands, BA, 61; Univ Calif, Santa Barbara, MA, 64; Univ Calif, Los Angeles, PhD(anat), 68. *Prof Exp:* Bank of Am-Giannini Found res fel, 68-69; NIH trainee physiol & biophys, Univ Wash, 69-70; asst prof, Sch Med, Univ Southern Calif, 70-73, assoc prof physiol, 73-76; assoc prof, 76-81, PROF BIOMED SCI, UNIV CALIF, RIVERSIDE, 81- *Mem:* Am Physiol Soc. *Res:* Mammalian thermoregulation and water balance. *Mailing Add:* Div Biomed Sci Univ Calif Riverside CA 92502

BAKER, MAX LESLIE, b Batesville, Ark, Aug 4, 43; m 69; c 3. RADIOBIOLOGY, BIOPHYSICS. *Educ:* Ark Col, BA, 65; Univ Ark, Little Rock, MS, 67, PhD(physiol, biophys), 70. *Prof Exp:* Instr radiol, 69, instr radiobiol, 70-71, asst prof radiobiol, 71-77, assoc prof radiol, 77-86, PROF RADIOL, UNIV ARK, LITTLE ROCK, 86-, PROF PHYSIOL & BIOPHYS, MED SCI, 88- *Concurrent Pos:* Fel biol, Univ Tex M D Anderson Hosp & Tumor Inst Houston, 69-70. *Mem:* Radiation Res Soc; Sigma Xi. *Res:* radiobiological basis of radiation therapy; radioprotective drugs; medical lasers. *Mailing Add:* 19 Greenview Circle Sherwood AR 72116

BAKER, MERL, b Cadiz, KY, July 11, 24; m 46; c 2. MECHANICAL ENGINEERING, ENGINEERING MANAGEMENT. *Educ:* Univ Ky, BS, 45; Purdue Univ, MS, 48, PhD, 52. *Prof Exp:* Asst, Purdue Univ, 46-48; from asst prof mech eng to prof, Univ Ky, 48-63, exec dir res found, 53-63, res dir, Overseas prog, 56-63; dean of fac, sch mines & metall, Univ Mo-Rolla, 63-64, chancellor, 64-73; spec asst to pres, Univ Mo Syst, 74-77; lab wide coordr energy conserv, Oak Ridge Nat Labs, 77-79, energy mgt res specialist, Oak Ridge Nat Lab, 79-82; provost prof eng, 82-85, DIR PRO CAREER ENHANCEMENT & PROF ENG, UNIV TENN, 82- *Honors & Awards:* Award, Am Soc Heat, Refrig & Air-Conditioning Engrs, 59. *Mem:* Am Soc Mech Engrs; Am Soc Eng Educ; Am Soc Eng Mgt; Sigma Xi. *Res:* Engineering management and management of technology; refrigeration; air conditioning; heat transfer; energy policy, legal and ethical issues in engineering. *Mailing Add:* Univ Tenn Chattanooga TN 37402

BAKER, MICHAEL ALLEN, b Toronto, Ont, Jan 24, 43; m 67; c 2. HEMATOLOGY, CANCER. *Educ:* Univ Toronto, MD, 66; FRCP(C), cert clin haematol, 71 & internal med, 74; Am Bd Internal Med, dipl & cert hemat, 72. *Prof Exp:* Intern, Toronto Gen Hosp, 66-67; asst resident internal med, Mt Sinai Hosp, New York, 67-69, chief resident med & clin resident hemat, 69-70, chief resident hemat, 70-71, res fel, 70-72; clin hematologist & dir, Div Hemat Oncol, Toronto Western Hosp, 72-84; assoc med, 72-73, from asst prof to assoc prof, 73-81, PROF MED, UNIV TORONTO, 81-, ASSOC, INST MED SCI, SCH GRAD STUDIES, 74-; DIR, ONCOL PROG & DEPT HEMAT, TORONTO GEN HOSP, 84- *Concurrent Pos:* Asst med, Mt Sinai Sch Med, 67-69, instr, 69-72; Med Res Coun Can fel, 71-72; assoc, Ont Cancer Treatment & Res Found, 73- *Mem:* Am Fedn Clin Res; Am Soc Hemat; Can Soc Hemat; Can Soc Clin Invest. *Res:* Studies of cell membrane changes in acute and chronic leukemia; clinical trials of treatment for leukemia. *Mailing Add:* Toronto Gen Hosp 200 Elizabeth St Toronto ON M5G 1L7 Can

BAKER, MICHAEL HARRY, b Roanoke, Va, Oct 25, 16; m 40; c 3. APPLIED CHEMISTRY, CHEMICAL ECONOMICS. *Educ:* Pratt Inst, ChE, 38; Va Polytech Inst, BS, 39, MS, 40; Univ Md, cert, 42. *Prof Exp:* Owner, Chem Prod Distrib Co, 37-40; plant-prod chem engr, Joseph Seagram Sons Co, 40-42; field develop engr, Davison Chem Co, 42-47; head chem prod eval, Gen Mills Inc, 47-51; PRES, CHEM/SERV INC, 51-, CHMN, 80- *Concurrent Pos:* Ed, Minn Chemist, 60-62; consult, Vols for Int Tech Asst; ed, Chem Distribr, 78-80. *Honors & Awards:* Honor Medal, Minn Fedn Eng Socs, 63; Honor Award, Minn Indust Chemist's Forum, 68; Albert Einstein Award, Am Technicon Soc, 85. *Mem:* AAAS; Am Inst Chemists; Am Chem Soc; Inst Food Technol; Am Inst Chem Eng. *Res:* Application of raw materials of a chemical nature in the production of food products, paints, paper, cosmetics and related industries; chemical economics. *Mailing Add:* 2012 Girard Ave S Minneapolis MN 55405

BAKER, NEAL KENTON, b Boston, Mass, Mar 20, 45; m 68. SOLAR PHYSICS, METEOROLOGY. *Educ:* Harvard Univ, AB, 67; Pa State Univ, PhD(astron), 75. *Prof Exp:* Mem tech staff solar physics, 68-71, mem techstaff optics, 75-79, proj engr, 79-83, SYST DIR, AEROSPACE CORP, 83- *Mem:* Am Astron Soc; Am Soc Photogram; Sigma Xi; Optical Soc Am. *Res:* Solar radio astronomy-microwave bursts; solar flares; image processing; meteorology. *Mailing Add:* 22468 Paul Revere Dr Woodland Hills CA 91364

BAKER, NOME, b Los Angeles, Calif, July 19, 27; m 50; c 4. BIOCHEMISTRY. *Educ:* Univ Calif, Los Angeles, AB, 49; Univ Calif, Berkeley, PhD(physiol), 52. *Prof Exp:* Sr instr biochem, Sch Med, Western Reserve Univ, 52-56; chief biochemist radioisotope res, Vet Admin Ctr, Los Angeles, 56-74, chief tumor-lipid res, 75-; ADJ PROF, DEPT MED & CRUMP INST MED ENG, MED CTR, UNIV CALIF, LOS ANGELES. *Concurrent Pos:* Prin scientist, Radioisotope Unit, Vet Admin Hosp, Ohio, 52-56; asst clin prof biochem, Sch Med, Univ Calif, Los Angeles, 56-70, adj prof med, 71-; Multiple Sclerosis Found fel, 61-62. *Mem:* AAAS; Am Soc Biol Chem. *Res:* Carbohydrate, fat metabolism; diabetes; tracer kinetics; cancer-host metabolism; lipid autoxidation. *Mailing Add:* Crump Inst Med Eng Med Ctr Univ Calif 982 Santa Barbara Rd Berkeley CA 94707

BAKER, NORMAN FLETCHER, veterinary parasitology, for more information see previous edition

BAKER, NORMAN HODGSON, b Fergus Falls, Minn, Oct 23, 31; m 76. ASTROPHYSICS. *Educ:* Univ Minn, BA, 52; Cornell Univ, PhD(physics), 59. *Prof Exp:* Vis fel, Max Planck Inst, Munich, 59-61; res fel, NASA Goddard Inst, 61-64; res assoc, NY Univ, 64-65; from asst prof to assoc prof, 65-71 chmn, dept astron, 72-76, PROF ASTRON, COLUMBIA UNIV, 71- *Concurrent Pos:* Mem, Inst F Advan Study, 62-63; co-ed, Astron J, 67-79, 79-84, Bulletin Am Astron Soc, co-ed, 69-79 & ed, 79-83. *Mem:* Am Phys Soc; Am Astron Soc; Royal Astron Soc; Int Astron Union. *Res:* Stellar structure and evolution; theory of stellar pulsation; convection in stars. *Mailing Add:* Dept Astron Columbia Univ New York NY 10027

BAKER, PATRICIA J, IMMUNONEPHROLOGY. *Educ:* Univ Ill, PhD(microbiol), 77. *Prof Exp:* ASST PROF MED, UNIV WASH, 82- *Mailing Add:* Dept Med Div Nephrol Univ Wash Seattle WA 98195

BAKER, PAUL, JR, b Ashland, Ky, Feb 10, 21; m 48; c 3. ASTROPHYSICS. *Educ:* Washington & Lee Univ, BA, 42; US Mil Acad, BS, 45; NC State Univ, MS, 52; Univ Denver, PhD(physics), 66. *Prof Exp:* Proj officer nuclear propulsion, Directorate Requirements, Hq, US Air Force, 52-56, chief tech br, Hartford Area, US AEC, 56-61, from instr to prof physics, US Air Force Acad, 61-67, head dept, 64-65, res assoc, 65-66, dir fac res, 66-67, chief tech div, Directorate Space, Pentagon, 67-71, prog mgr, Defense Advan Res Proj Agency, Va, 71-75; consult, 75-76, res projs mgr, 76-85, SR SAFEGUARDS SCIENTIST, NUCLEAR REGULATORY COMN, 85- *Res:* Cosmic rays. *Mailing Add:* 4404 Random Ct Annandale VA 22003

BAKER, PAUL THORNELL, b Burlington, Iowa, Feb 28, 27; m 49; c 4. BIOLOGICAL ANTHROPOLOGY. *Educ:* Univ NMex, BA, 51; Harvard Univ, PhD(anthrop), 56. *Prof Exp:* Res phys anthropologist, Environ Protection & Res Div, US Army Qm Res & Develop Ctr, 52-57; res assoc, Biophys Lab, Pa State Univ, 57-58, from asst prof to prof anthrop, 58-68, actg head, Dept Sociol & Anthrop, 64-65, prof, Dept Anthrop, 64-80, exec officer, Anthrop Br, 65-68, actg head, Dept Anthrop, 68-69, head dept, 80-85, Evan Pugh prof, 81- 87, EVAN PUGH EMER PROF ANTHROP, PA STATE UNIV, UNIVERSITY PARK, 88- *Concurrent Pos:* Fulbright lectr, Brazil, & res scholar, Peru, 62; US Army grant, 62-68; mem, NIH Behav Sci Fels Rev Comt, 65-69; NATO sr sci fel, Oxford Univ, 68; prog dir, US Int Biol Prog Biol Human Pop at High Altitude, 65-74, mem, US Exec Comt Int Biol Prog, 70-74; Nat Inst Gen Med Sci grant, 69-79; NSF grant, 71-74; mem, Int Coun Sci Unions Comn Predictive World Ecosysts, 71-72; ed, Am Anthropologist, 73-76; Guggenheim fel, 74-75; chmn, US Man & Biosphere Prog 12 Directorate, 75-77; consult, UNESCO, 75 & 79-, UN Environ Prog, 79; Nat Inst Child Health & Human Develop grant, 75-77; mem, Adv Coun Bd Trustees, Wenner-Gren Found Anthrop Res, Inc, 80-84; chmn, US Man & Biosphere Prog, 83-85; chair, Anthrop Sect, Nat Acad Sci, 84-87 & US Deleg, Int Cong Anthrop & Ethnol Sci, 88. *Honors & Awards:* Distinction in Soc Sci Award, Pa State Univ, 77; Huxley Mem lectr & medallist, Royal Anthrop Inst, Gt Brit & Ireland, 82; Gorjanovic-Krambergeri Medal, Univ Zagreb, Yugoslavia. *Mem:* Nat Acad Sci; fel Am Anthrop Asn; Am Asn Phys Anthropologists (pres, 69-71); Int Asn Human Biologists (pres, 80-); fel AAAS; Human Biol Coun (pres, 74-77); Int Union Anthrop & Ethnol Sci (vpres, 88-90). *Res:* Adaptation to high altitude; biological effects of modernization in human populations; environmental health. *Mailing Add:* Dept Anthrop Pa State Univ University Park PA 16802

BAKER, PETER C, b San Francisco, Calif, Feb 14, 33; c 4. NEUROBIOLOGY, EMBRYOLOGY. *Educ:* Univ Calif, Berkeley, AB, 56, MA, 61, PhD(zool), 66. *Prof Exp:* Instr biol, City Col San Francisco, 66-67; from asst prof to assoc prof, 67-70, PROF BIOL, CLEVELAND STATE UNIV, 74- *Concurrent Pos:* Lectr genetics, Dept Biol, San Francisco State Col, 66-67. *Mem:* AAAS; Sigma Xi; AAAS; Am Soc Zool. *Res:* Maturation of indoleamines in the mouse brain; mammals; drug affects on the maturing brain. *Mailing Add:* Dept Biol Cleveland State Univ Cleveland OH 44115

BAKER, PHILIP SCHAFFNER, chemistry; deceased, see previous edition for last biography

BAKER, PHILLIP JOHN, b East Chicago, Ind, Aug 21, 35; m 67; c 2. MICROBIOLOGY, IMMUNOLOGY. *Educ:* Ind State Univ, BA, 60; Univ Wis, MS, 62, PhD(bact), 65. *Prof Exp:* Res asst bact, Univ Wis, 60-65; MICROBIOLOGIST, LAB MICROBIAL IMMUNITY, NIH, 70- *Concurrent Pos:* USPHS fel immunol, Lab Immunol, NIH, 65-67, staff fel microbiol & immunol, Lab Germfree Animal Res, 67-69, sr staff fel, Lab Microbial Immunity, 69-70, res microbiologist, 70-76, head microbiol & immunol sect, 76-; chmn, Immunol Div, Am Soc Microbiol, 83-85. *Mem:* AAAS; Am Asn Immunologists; Reticuloendothelial Soc; Am Soc Microbiol; Sigma Xi. *Res:* Antibody formation at cellular level, particularly microbial antigens or microbial products; initiation of antibody synthesis and tolerance to microbial polysaccharide antigens; factors related to virulence of bacteria; regulation of antibody responses; suppressor T cells; genetics of the immune response. *Mailing Add:* LIG IRP NIAID NIH Twinbrook II 12241 Parklawn Dr Rockville MD 20852

BAKER, R RALPH, b Houston, Tex, Aug 31, 24; wid; c 6. PLANT PATHOLOGY, MYCOLOGY. *Educ:* Colo Agr & Mech Col, BS, 48, MS, 50; Univ Calif, Berkeley, PhD(plant path), 54. *Prof Exp:* PROF BOT & PLANT PATH, COLO STATE UNIV, 54- *Concurrent Pos:* Assoc ed, Phytopath, 61-63 & 76-79, sr ed, 79-81; mem comt control soil borne pathogens, Nat Acad Sci, 62-; asst dir biol res, Res Found, 62-64; vis prof, Univ Calif, Berkeley, 63-64; chmn comt planning manned orbital res & lab exp in space biol, Am Inst Biol Sci, 65-68, chmn Study Group II, 69-70, comt study role of the lunar receiving lab in the post Apollo biol & biomed activities; mem plant sci comt, Comn Undergrad Educ Biol Sci, 66, partic workshop for teaching biol to nonsci majors, Bar Harbor, 68; NSF sr fel & vis prof, Cambridge Univ, 68-69; Lunar Sample Review Bd, 70-73; mem bd govs, Am Inst Biol Sci, 71-75, mem educ comt, 74-; partic, USA/Repub of China Coop Sci Prog, 75; chmn fungal antagonists group, Fungal Biocontrols Workshop, Am Inst Biol Sci-Environ Protect Agency, Univ Ark, 75; mem space processing payload adv subcomt, NASA, 75-76; NSF partic, US-USSR Coop Symp, 76; Sigma Xi lectr, 76-78; mem study group recombinant DNA, Environ Protect Agency, 77; mem Adv Comt concerned with treatment of returned Mars (NASA-JPL) samples; mem IR4-Biorational Workshop; sr ed, Can J Microbiol, 87-; assoc ed, Crop Protection, 87-; mem, Subcomt Biol Control, Exp Stas Comt Organized Policy, 88- *Honors & Awards:* Florists Mutual Award, 59; Pennock Distinguished Serv Award, 65; Apollo Prog Group Achievement Award, NASA, 73, Group Achievement Award, US-USSR Biol Satellite Mission, 76-; Sci Achievement Medal, Colo State Univ, 76. *Mem:* Fel AAAS; fel Am Phytopath Soc. *Res:* Ecology of soil microorganisms; physiology of sexual reproduction in fungi; mathematical biology; ornamental pathology; space biology; biological control. *Mailing Add:* Dept Bot & Plant Path Colo State Univ Ft Collins CO 80521

BAKER, RALPH ROBINSON, b Baltimore, Md, Dec 30, 28; m 53; c 4. SURGERY. *Educ:* Johns Hopkins Univ, AB, 50, MD, 54. *Prof Exp:* From instr to prof, 62-75, PROF ONCOL, SCH MED, JOHNS HOPKINS UNIV, 75- *Concurrent Pos:* Am Cancer Soc adv clin fel, 65-68; consult, US Vet Admin Hosp, Perry Point & Loch Raven, Md, 64-68. *Mem:* Am Surg Asn; Am Asn Thoracic Surg; Soc Surg Alimentary Tract; Soc Univ Surg; Asn Acad Surg. *Res:* Academic surgery; cancer and transplantation research. *Mailing Add:* Johns Hopkins Univ 601 N Broadway Baltimore MD 21205

BAKER, RALPH STANLEY, b LeRaysville, Pa, Aug 1, 27; m 54; c 2. WEED SCIENCE, COTTON. *Educ:* Univ Del, BS, 57; Purdue Univ, MS, 59, PhD(plant physiol), 61. *Prof Exp:* RESEARCHER WEED SCI, DELTA BR, MISS AGR & FORESTRY EXP STA, 61- *Mem:* Weed Sci Soc Am; Sigma Xi; Coun Agr & Sci Technol. *Res:* Weeds, herbicides and cultural methods for controlling weeds in cotton. *Mailing Add:* Delta Br Exp Sta Box 197 Stoneville MS 38776-0197

BAKER, RALPH THOMAS, b Penticton, BC, Apr 10, 53; m 78; c 2. HOMOGENEOUS CATALYSIS. *Educ:* Univ BC, BSc, 75; Univ Calif, Los Angeles, PhD(chem), 80. *Prof Exp:* Lab asst, BC Govt Lab, 75; teaching asst, Univ Calif, Los Angeles, 75-76; fel, Pa State Univ, 80-81; RES CHEMIST, E I DU PONT DE NEMOURS & CO, INC, 81- *Mem:* Am Chem Soc. *Res:* Metalloborane and carborane synthesis; metal atom synthesis; homogeneous catalysis; multinuclear nuclear magnetic resonance; early transition metal diorganophosphide synthesis; early-late phosphide-bridged heterobimetallic synthesis; fluoroorganometallic synthesis; transition metal-catalyzed hydroboration. *Mailing Add:* Cent Res & Develop Exp Sta E I du Pont de Nemours & Co Inc Wilmington DE 19880-0328

BAKER, RAYMOND MILTON, b Kansas City, Mo, Nov 7, 40; m 78; c 2. GENETICS, CELL BIOLOGY. *Educ:* Yale Univ, BS, 62; Univ Calif, Berkeley, PhD(biophys), 69. *Prof Exp:* Lectr med biophys, Univ Toronto, 71-73, asst prof, 73-74; asst prof biol, Ctr Cancer Res, Mass Inst Technol, 74-; AT DEPT EXP THERAPEUT, GCDC, ROSWELL PARK MEM INST. *Concurrent Pos:* USPHS fel med biophys, Univ Toronto-Ont Cancer Inst, 69-70; staff scientist, Ont Cancer Inst, 71-74 & Hosp Sick Children, Toronto, 73-74. *Mem:* AAAS; Biophys Soc; Genetics Soc Am; Am Soc Cell Biologists. *Res:* Somatic cell genetics, particularly concerning membrane mutants and mutagenesis; molecular genetics, particularly recombination and repair. *Mailing Add:* 123 Wesley Ave Buffalo NY 14214

BAKER, REES TERENCE KEITH, b Coventry, UK, Nov 24, 38; m 63; c 2. KINETICS, CATALYSIS. *Educ:* Liverpool Polytech, ARIC, 63, ARC, 64; Univ Wales, PhD(chem), 66, DSc, 78. *Prof Exp:* Sr scientific res officer, Atomic Energy Res Estab, 68-75; sr res assoc, Exxon Res & Eng Co, 75-86; PROF, CHEM ENG DEPT, AUBURN UNIV, ALA, 86- *Concurrent Pos:* Hougen prof, Univ Wis, Madison, 88. *Honors & Awards:* Pettinos Award, Am Carbon Soc, 87. *Mem:* Am Carbon Soc; Am Chem Soc. *Res:* Examination of various aspects of reaction involving gas and solid systems, with the emphasis on the catalytic behavior of metals and metal oxides during carbon deposition and carbon gasification. *Mailing Add:* 86 Lee Rd 827 Opelika AL 36801

BAKER, RICHARD A, b Enumclaw, Wash, Mar 13, 34; m 55; c 3. ELECTRICAL ENGINEERING. *Educ:* Wash State Univ, BS, 56; NY Univ, MEE, 59; Univ Calif, Berkeley, PhD(elec eng), 67. *Prof Exp:* Mem tech staff, Bell Tel Labs, 56-60; sr proj engr, AC Spark Plug Div, Gen Motors Corp, 60-61; from asst prof to assoc prof, 61-77, PROF ELEC ENG, WASH STATE UNIV, 77- *Concurrent Pos:* Guest prof, Brown Boveri Res Ctr, Baden, Switz, 80-81, Asea Brown Boveri, 87-88. *Mem:* Inst Elec & Electronics Engrs; Sigma Xi. *Res:* System theory and control theory, principally, stability analysis; power systems. *Mailing Add:* SW 210 Spruce St Pullman WA 99163

BAKER, RICHARD DEAN, b Hot Springs, SDak, June 9, 13; m 46; c 1. CHEMISTRY. *Educ:* SDak Sch Mines & Technol, BS, 36; Iowa State Col, PhD(phys chem), 41. *Prof Exp:* Asst chem, Iowa State Col, 37-41; res assoc, 42-43; res chemist, US Gypsum Co, Ill, 41-42; sr scientist, Los Alamos Sci Lab, 43-45, group leader, 45-56, div leader, 56-79; assoc dir, 79-81, CONSULT TO DIR, NAT SECURITY PROG, 81- *Mem:* Am Chem Soc; Am Soc Metals. *Res:* Atomic energy; lime and building materials; low pressure carburization. *Mailing Add:* 1999 Juniper S Los Alamos NM 87544

BAKER, RICHARD FRELIGH, b Westfield, Pa, Feb 14, 10; m 39, 84; c 1. BIOPHYSICS. *Educ:* Pa State Col, BS, 32, MS, 33; Univ Rochester, PhD(physics), 38. *Prof Exp:* Asst physics, Univ Minn, 37-38; fel physiol, Col Physicians & Surgeons, Columbia Univ, 39-40; assoc anat, Johns Hopkins Univ, 40-41; physicist, Radio Corp Am Labs, 41-47; from asst prof to assoc prof exp med, 47-59, PROF MICROBIOL, UNIV SOUTHERN CALIF, 59- *Concurrent Pos:* Sr res fel, Calif Inst Technol, 53-56, Commonwealth Fund fel, 57-, res assoc, 68- *Honors & Awards:* Electron Micros Award, Electron Microscope Soc, 85. *Mem:* Electron Micros Soc Am; Sigma Xi. *Res:* Mass spectrometry; biological electron microscopy; photoelectric properties of cadmium in the Schumann region; membrane structure; normal and sickle human red cells; scanning and transmission electron microscopy. *Mailing Add:* 526 Mudd Bldg Sch Med Univ Southern Calif Los Angeles CA 90033

BAKER, RICHARD GRAVES, b Merrill, Wis, June 12, 38; m 61; c 3. PALYNOLOGY, QUATERNARY GEOLOGY. *Educ:* Univ Wis-Madison, BA, 60; Univ Minn, Minneapolis, MS, 64; Univ Colo, Boulder, PhD(geol), 69. *Prof Exp:* Res assoc, Ctr Climatic Res, Univ Wis-Madison, 69-70; from asst prof to assoc prof, 70-82, PROF GEOL, UNIV IOWA, 82- *Mem:* AAAS; Geol Soc Am; Am Quaternary Asn; Am Asn Stratig Palynologists; Ecol Soc Am. *Res:* Quaternary palynology and plant macrofossil analysis in the Rocky Mountains and Midwest. *Mailing Add:* Dept Geol Univ Iowa Iowa City IA 52242

BAKER, RICHARD H(OWELL), b Long Beach, Calif, Jan 5, 21; m 47; c 2. ELECTRONICS ENGINEERING. *Prof Exp:* Prototype engr, Lockheed Aircraft Corp, 41-43; res engr, Northrop Aircraft Corp, 47-48, chief electronic comput opers, 48-50; proj engr, Bill Jack Sci Instrument Co, 50-51; electronic syst consult, 51-68, mgt consult, 58-76, CONSULT, 76- *Concurrent Pos:* Chief engr, Digital Control Systs, Inc, 51-53; lectr eng & head dept, US Navy Elec Eng & Radio Mat Sch, Tex; consult, Stanford Res Inst, 53-, head electronics indust res, 54-58; consult to chief sig off, US Army, 53; chmn bd dirs & vpres, ABL, Inc, 62-68. *Mem:* Sigma Xi. *Res:* Electronic computer and control systems; communications system design; electronics management and marketing; magnetic recording technique; system architect, video-graphic data systems and hardware design. *Mailing Add:* PO Box 24245 San Diego CA 92124

BAKER, RICHARD H, b Hayfield, Minn, Sept 14, 36; m 61; c 2. GENETICS. *Educ:* Univ Ill, BS, 59, MS, 62, PhD(zool, genetics), 65. *Prof Exp:* Res assoc mosquito genetics, Univ Ill, 65; res assoc, 65-66, asst prof mosquito genetics, 66-69, assoc prof int med, Sch Med, Univ Md, Baltimore, dir, Pakistan Med Res Ctr, Lahore, Pakistan, 69-79, prof & dir, Int Health Prog, Univ Md, Baltimore, 82-; FLA MED ENTOM LAB, UNIV FLA. *Concurrent Pos:* Consult, WHO, 65. *Mem:* Genetics Soc Am; Entom Soc Am; Am Soc Trop Med & Hyg; Am Genetic Asn; Am Mosquito Control Asn; Sigma Xi. *Res:* Mosquito genetics and cyto-genetics. *Mailing Add:* Fla Med Entom Lab 200 Ninth St SE Vero Beach FL 32962

BAKER, RICHARD WILLIAM, b Boreham Wood, Eng, Aug 18, 41; m 68. PHYSICAL CHEMISTRY, POLYMER CHEMISTRY. *Educ:* London Univ, BS, 63, DIC, 65, PhD(phys chem), 66. *Prof Exp:* Sr chemist, Amicon Corp, 66-68; res fel, Polymer Res Inst, Brooklyn Polytech Inst, 68-70; sr scientist, Alza Corp, 70-74; dir res, Bend Res Inc, 74-; AT MEMBRANE TECHNOL & RES. *Mem:* Royal Soc Chem; AAAS; Am Chem Soc. *Res:* Physical chemistry of synthetic polymer membranes as applied to separation processes such as ultrafiltration and reverse osmosis, and the application of membranes to sustained release drug-delivery systems. *Mailing Add:* Membrane Technol & Res 1360 Willow Rd No 103 Menlo Park CA 94025-1516

BAKER, ROBERT ANDREW, b Lakewood, Ohio, Sept 8, 25; m 47. ENVIRONMENTAL SCIENCES. *Educ:* NC State Univ, BS, 49; Villanova Univ, MS, 55 & 58; Univ Pittsburgh, DSc(environ health eng), 69. *Prof Exp:* Chem engr, Atlantic Refining Co, 49-57; sr staff eng environ sci, Franklin Inst, 57-64; sr fel, Mellon Inst Sci, 64-70; dir environ sci, Teledyne Brown Eng, 71-73; REGIONAL RES HYDROLOGIST & CHIEF, GULF COAST HYDRO-SCI CTR, US GEOL SURV, 73- *Honors & Awards:* Award of Merit, Am Soc Testing & Mat, 68, Max Hecht Award, 70 & Award Sensory Res, 73; Eng of Distinction, Eng Joint Coun, 70; Buswell-Porges Award, Inst Advan Sanitation Res, 72; Larry Cecil Environ Award, Am Inst Chem Engrs, 84; Res Award, AWWA, 85. *Mem:* Am Chem Soc; Am Water Works Asn; fel Am Soc Testing & Mat; fel Am Inst Chem Engrs; Int Soc Water Pollution & Control. *Res:* Direction of interdisciplinary research involving various aspects of the hydrological cycle with personal emphasis on trace organic contaminants in the environment. *Mailing Add:* US Geol Surv J C Stennis Space Ctr NSTL Station MS 39529

BAKER, ROBERT CARL, b Newark, NY, Dec 29, 21; m 44; c 5. FOOD SCIENCE. *Educ:* Cornell Univ, BS, 43; Pa State Univ, MS, 49; Purdue Univ, PhD(food sci), 58. *Prof Exp:* Asst prof poultry exten, Pa State Univ, 46-49; from asst prof to assoc prof, 49-56, dir food sci & mkt, 69-75, PROF FOOD SCI, CORNELL UNIV, 57-, CHMN, DEPT POULTRY & AVIAN SCI, 79- *Concurrent Pos:* Inst Am Poultry Industs food res award, 59; mem, Comt Animal Prod, Nat Acad Sci, 65-68. *Mem:* AAAS; Sigma Xi; Poultry Sci Asn; Inst Food Technologists. *Res:* Bacteriology; microbiology of eggs; new poultry meat and egg products; new fish products; biochemistry. *Mailing Add:* Dept Poultry 112 Rice Hall Cornell Univ Ithaca NY 14853

BAKER, ROBERT CHARLES, b Blackduck, Minn, June 17, 30; m 54; c 6. ECOLOGY, SCIENCE EDUCATION. *Educ:* Bemidji State Col, BS, 54; Univ NDak, MSEd, 55; Ore State Univ, MS, 63. *Prof Exp:* Teacher high sch, 55-56; instr sci & math, Lab Sch, 56-58, actg prin, 59-62, asst prof sci educ, 63-65, Minn State Col Bd fel, 65-66; asst prof sci educ, 66-67, dir continuing educ, 67-68, assoc prof, 67-81, PROF BIOL & SCI EDUC, BEMIDJI STATE UNIV 81- *Concurrent Pos:* Adv, State Curric Environ Educ Kindergarten to 12th grade, 70- *Mem:* AAAS; Am Inst Biol Sci; Nat Soc Stud Educ; Nat Asn Biol Teachers; Nat Sci Teachers Asn. *Res:* Environmental pollution, especially solutions to land use problems and waste disposal; statistical evaluation of general biology program, traditional versus experimental; biological and attitudinal analysis of Upper Mississippi, especially Iowa-Itasca. *Mailing Add:* Dept Biol Bemidji State Univ Bemidji MN 56601

BAKER, ROBERT DAVID, b Chicago, Ill, July 7, 29; div; c 4. EPITHELIAL PHYSIOLOGY. *Educ:* Iowa Wesleyan Col, BS, 54; Univ Iowa, MS, 57, PhD(physiol), 59. *Prof Exp:* From instr to asst prof, 59-64, ASSOC PROF PHYSIOL, UNIV TEX MED BR GALVESTON, 64- *Concurrent Pos:* USPHS grants, 60-64 & 68-; Liberty Muscular Dystrophy Res Asn grant, 61-64. *Mem:* AAAS; Am Physiol Soc; Biophys Soc. *Res:* Intestinal absorption; transport of materials across biological membranes. *Mailing Add:* Dept Physiol & Biophys Univ Tex Med Br Galveston TX 77550

BAKER, ROBERT DONALD, b Chico, Calif, Dec 7, 27; m 58; c 1. FOREST MANAGEMENT, REMOTE SENSING. *Educ:* Univ Calif, BS, 51, MF, 52; State Univ NY, PhD, 57. *Prof Exp:* From asst prof to prof forest mgt, Stephen F Austin State Univ, 56-74; PROF FOREST MGT, TEX A&M UNIV, 75- *Concurrent Pos:* Nat dir, Am Soc Photogram, 75-77. *Mem:* Fel Soc Am Foresters; emer mem Am Soc Photogram. *Res:* Forest mensuration; remote sensing; geographic information systems. *Mailing Add:* Dept Forest Sci Tex A&M Univ College Station TX 77843-2135

BAKER, ROBERT FRANK, b Weiser, Idaho, Apr 9, 36; m 65; c 2. MOLECULAR BIOLOGY, GENETICS. *Educ:* Stanford Univ, BS, 59; Brown Univ, PhD(biol), 66. *Prof Exp:* Fel microbial genetics, Stanford Univ, 66-68; from asst prof to assoc prof, 68-83, PROF MOLECULAR BIOL, UNIV SOUTHERN CALIF, 83- *Concurrent Pos:* Vis prof, Harvard Med Sch, 75-76; genetics study sect, NIH, 77-79; dir Molecular Biol Div, Univ Southern Calif, 78-80, mem Comprehensive Cancer Ctr, 84- *Mem:* Am Soc Microbiol; Am Soc Zoologists; Sigma Xi. *Res:* Gene regulation in eucaryotic cells; molecular oncology. *Mailing Add:* Molecular Biol Univ Southern Calif Los Angeles CA 90089-1340

BAKER, ROBERT G, b Kindersley, Sask, Dec 17, 18; m 42; c 4. BIOPHYSICS, NUCLEAR PHYSICS. *Educ:* Univ Sask, BA, 50, MA, 52; Univ Toronto, PhD(physics), 60. *Prof Exp:* Res officer radiol, Nat Res Coun Can, 52-54; physicist radiation ther, Ont Cancer Found, 54-58; physicist isotopes, Ont Cancer Inst, 60-68; asst prof med biophys, Univ Toronto, 62-68; chief physicist, Ont Cancer Found, Ottawa Civic Hosp, 68-81; RETIRED. *Mem:* Can Asn Physicists; Soc Nuclear Med. *Res:* Medical biophysics; instrumentation and application of isotopes to diagnosis and treatment of disease, especially cancer. *Mailing Add:* Six Morningside Community New Hamburg ON N0B 2G0 Can

BAKER, ROBERT GEORGE, b Streator, Ill, Sept 1, 40; m 64; c 2. NEUROBIOLOGY. *Educ:* NCent Col, BA, 62; Univ Ill, PhD(pharmacol), 67. *Prof Exp:* Fel pharmacol, Univ Ill, Nat Inst Neurol Dis & Blindness, 67-68; USPHS spec fel neurobiol, Inst Brain Res, Univ Tokyo, 68-69; asst prof, Dept Neurobiol, Inst Biomed Res, Am Med Asn, 69-70; asst prof physiol & biophysics, Div Neurobiol, Univ Iowa, 70-73, assoc prof, 73-76; PROF PHYSIOL & BIOPHYS, MED CTR, NY UNIV, 76- *Concurrent Pos:* Vis prof, Physiol du Travail Ergonomie Conservatone Nat des Artsetmetiels, Paris, France, 71, Reg Primate Res Ctr, Univ Wash, 73; consult, Commmun Sci Study Sect, NIH, 76-80. *Mem:* Fedn Am Scientists; AAAS; Soc Neurosci; Am Physiol Soc; Int Brain Res Orgn. *Res:* Physiological organization of eye movements with emphasis on synaptic mechanisms in the brainstem and cerebellum of vertebrates. *Mailing Add:* Dept Physiol & Biophys Med Ctr NY Univ 550 First Ave New York NY 10016

BAKER, ROBERT HENRY, b Central City, Ky, June 14, 08; m 32; c 2. ORGANIC CHEMISTRY. *Educ:* Univ Ky, BS, 29, MS, 31; Univ Wis, PhD(org chem), 40. *Hon Degrees:* ScD, Univ Ky, 68. *Prof Exp:* From instr to asst prof chem, Univ Ky, 31-41; from asst prof to assoc prof, 41-49, from asst dean to assoc dean, 49-64, prof chem, 49-76, dean grad sch, 64-76, EMER PROF CHEM & EMER DEAN GRAD SCH, NORTHWESTERN UNIV, 76- *Concurrent Pos:* Mem exec comt grad deans, African-Am Inst, 64-; vis comt grad sch, Vanderbilt Univ, 68- *Mem:* AAAS; Am Chem Soc; Asn Grad Schs (pres, 71). *Res:* Synthesis, stereochemistry and the mechanism of organic reactions; hydrogenation. *Mailing Add:* 1700 Bronson Way No 122 Kalamazoo MI 49009

BAKER, ROBERT HENRY, JR, b Lexington, Ky, Apr 5, 34; m 57; c 3. COMPUTER SCIENCES. *Educ:* Trinity Col, Conn, BS, 56; Ind Univ, PhD(biochem), 60. *Prof Exp:* NSF fels, Univ Wis, 60-62; res assoc phys & anal chem, Upjohn Co, 61-66, sect head comput systs, Res Comput Ctr, 66-69, mgr comput systs res & develop, 69-72, mgr admin, Comput Ctr, 72-79, mgr, Future Technologies Res & Develop, 79-90; RETIRED. *Mem:* Am Chem Soc. *Res:* Computer and telephone systems resource management. *Mailing Add:* 5147 Colony Woods Kalamazoo MI 49009

BAKER, ROBERT J(ETHRO), JR, b Dayton, Ohio, June 10, 38; m 61; c 3. CERAMIC ENGINEERING, CHEMISTRY. *Educ:* Univ Ill, MS, 63, PhD(ceramic eng), 64; Mass Inst Technol, SM, 72. *Prof Exp:* Res scientist, Hanford Labs, Gen Elec Co, 64-65; res scientist, Pac Northwest Labs, Battelle Mem Inst, 65-66, mgr mat process develop, 66-67, mgr, Battelle Develop Corp, Washington/Ohio, 67-74; FOUNDER, VENTURE RESOURCES INT, OHIO, 74-; PRES, RATHBONE PROD, 77- *Mem:* Sigma Xi; Am Ceramic Soc; Nat Inst Ceramic Engrs; Newcomen Soc. *Res:* Materials and process research, development and application; high temperature materials; characterization of powders, microstructure analysis and correlation with properties and behavior. *Mailing Add:* 2901 Berwick Blvd Columbus OH 43209

BAKER, ROBERT JOHN, b Cold Lake, Alta, June 10, 38; m 60; c 2. PLANT BREEDING, QUANTITATIVE GENETICS. *Educ:* Univ Sask, BSA, 61, MSc, 62; Univ Minn, PhD(genetics), 66. *Prof Exp:* Res scientist plant breeding, Can Dept Agr, 66-78; SR RES SCIENTIST, UNIV SASK, 78- *Mem:* Crop Sci Soc Am; Agr Inst Can; fel Am Soc Agron; fel Crop Sci Soc Am. *Res:* Methods of improving quantitative characteristics in self-pollinated species. *Mailing Add:* Dept Crop Sci Univ Sask Saskatoon SK S7N 0W0 Can

BAKER, ROBERT NORTON, b Inglewood, Calif, Mar 25, 23; m 71. MEDICINE. *Educ:* Park Col, BA, 44; Univ Southern Calif, MD, 50. *Prof Exp:* Lab asst comp anat, Univ Mo, 42-43; asst biochem, Univ Southern Calif, 46-48, asst anat, 49-50, instr neuroanat, 50-53; from instr to assoc prof med neurol, Med Ctr, Univ Calif, Los Angeles, 53-70; prof neurol, Univ Nebr Med Ctr, Omaha, 70-72, chief neurol serv, Creighton-St Joseph Hosp, 74-76, PROF NEUROL & PATH, UNIV NEBR & CREIGHTON COL MED, 72-, CLIN PROF NEUROL, UNIV NEBR, 77- *Concurrent Pos:* Chief neurol sect, Vet Admin Ctr, Gen Med & Surg Hosp, 53-54 & 56-70; Vet Admin liaison, Neurol A Study Sect, NIH, 71-75. *Mem:* AMA; Am Acad Neurol; Am Asn Neuropath; Am Neurol Asn. *Res:* Experimental neuropathology; electron microscopic studies of cerebral vascular disease especially cerebral edema, hypertensive encephalopathy and cerebral embolism; clinical and pathological studies of cerebral vascular diseases and their metabolic aspects. *Mailing Add:* Univ Nebr Sch Med Good Samaritan Hosp Kearny NE 68847

BAKER, ROLLIN HAROLD, b Cordova, Ill, Nov 11, 16; m 39; c 3. ZOOLOGY. *Educ:* Univ Tex, BA, 37; Tex A&M Univ, MS, 38; Univ Kans, PhD(zool), 48. *Prof Exp:* Asst, Tex A&M Univ, 38; field biologist, State Coop Wildlife Res Unit, Tex, 38-39; wildlife biologist, State Game, Fish & Oyster Comn, 39-43; vis investr, Rockefeller Inst, 44; asst instr zool, Univ Kans, 46-48, from instr to assoc prof & asst cur mammals, Mus Natural Hist, 48-55; prof zool, Fish & Wildlife & Dir Mus, Mich State Univ, 55-83; CUR, PRAIRIE EDGE MUS, EAGLE LAKE, TEX, 83- *Concurrent Pos:* Leader, Mus Expeds, Latin Am, 50-82. *Honors & Awards:* Fogelsanger award, Shippensburg State Univ, 72. *Mem:* AAAS; Am Soc Mammal; Wildlife Soc; Soc Study Evolution; Soc Syst Zool. *Res:* Mammalian taxonomy, distribution and ecology; author of 160 publications on vertebrate zoology mostly mammals. *Mailing Add:* 302 N Strickland St Eagle Lake TX 77434

BAKER, RONALD G, b Erie, Pa, Mar 21, 39; m 61; c 5. HISTORY OF MATH, PROGRAMS IN BASIC. *Educ:* Edinboro Univ, Pa, BS, 60; State Univ NY, Albany, MD, 65; State Univ NY, Buffalo, DEduc, 71. *Prof Exp:* PROF MATH & CHMN, DEPT MATH & COMPUT SCI, EDINBORO UNIV PA, 85- *Mem:* Math Asn Am; Data Processing Mgt Assoc. *Res:* History of mathematics. *Mailing Add:* Dept Math & Comput Sci Edinboro Univ Pa Edinboro PA 16444

BAKER, SAMUEL I, b Cannelton, Ind, Nov 5, 34; m 62; c 1. RADIATION PHYSICS. *Educ:* Ind Univ, BS, 56; Univ Ill, MS, 58; Ill Inst Technol, PhD(physics), 67. *Prof Exp:* Appointee reactor training, Int Sch Nuclear Sci & Eng, 58-60; physicist, Argonne Nat Lab, 60-62; res assoc nuclear physics, Univ Iowa, 67 & Cyclotron Br, US Naval Res Lab, 67-69; res physicist, IIT Res Inst, 69-72; physicist, Fermi Nat Accelerator Lab, 72-; AT SSC LAB. *Mem:* AAAS; Am Phys Soc; Inst Elec & Electronics Engrs; Sigma Xi; Am Nuclear Soc. *Res:* Non-linear response of thallium-activated sodium iodide to gamma rays and x-rays measurement of nuclear lifetimes by the Doppler-Shift attenuation method; fast neutron spectrometry; radioactivation by 400 GeV protons. *Mailing Add:* SSC Lab 2550 Beckleymeade Ave Mail Stop 1071 Dallas TX 75237

BAKER, SAUL PHILLIP, b Cleveland, Ohio, Dec 7, 24. GERIATRICS, CARDIOLOGY. *Educ:* Case Inst Technol, BS, 45; Ohio State Univ, MSc, 49, MD, 53, PhD(physiol), 57; Case Western Univ, Sch Law, JD, 81. *Prof Exp:* Asst physiol, Ohio State Univ, 48, 50 & 52; health educ & field coordr, Greater Cleveland Tuberc X-ray Surv, 49; intern, Cleveland Metrop Gen Hosp, 53-54; asst vis staff physician, Dept Med, Baltimore City & Johns Hopkins Hosps, 54-56; sr asst resident physician, Dept Med, Billings Hosp & Univ Chicago Hosps, 56-57; asst prof med, Chicago Med Sch, 57-62; pvt pract geriatrics, cardiol & internal med, 62-70; Medicare med consult, Gen Am Life Ins Co, 70-71; PVT PRACT GERIAT, CARDIOL & INTERNAL MED, 72- *Concurrent Pos:* Cent Ohio Heart Asn res fel physiol, 52-53; sr asst surgeon, Geront Res Ctr, Nat Inst Aging, NIH, 54-56; Nat Heart Inst res grants, 58-62; Chicago Heart Asn res grants, 58-62; assoc attend physician & physician-in-chg, Chicago Med Sch Med Serv, Cook County Hosp, 57-62; res assoc, Hektoen Inst Med Res, 58-62; assoc prof med, Grad Sch Med, Cook County Hosp, 58-62; consult internal med & cardiol, Bur Disability Determination, Old-Age & Survivor's Ins, Social Security Admin, 63-; consult internal med & cardiol, Ohio Bur Worker's Compensation, 63-; mem comt older people, Welfare Fedn Cleveland, 63-70; head dept geriatrics, St Vincent Charity Hosp, 64-67; mem adv comt, Sr Adult Div, Jewish Community Ctr, 64-; consult internal med, City of Cleveland, 64-; cardiovasc consult, Fed Aviation Admin, 73-; consult internal med & cardiol, State of Ohio, 74-; mem proj comt aging & comt older people, Fedn Community Planning, 74-77; partic, 19th Int Physiol Cong & 5th Int Cong Gerontol; fel coun arteriosclerosis, Am Heart Asn. *Mem:* Fel AAAS; fel Geront Soc Am; fel Am Geriat Soc; Am Heart Asn; fel Am Col Cardiol; Am Soc Law & Med; Sigma Xi; Am Pub Health Asn; Am Physiol Soc; Soc Exp Biol & Med; Am Fedn Clin Res; Am Diabetes Asn. *Res:* Effects of heparin activated clearing factor or lipoprotein lipase; blood lipid enzymes; physiology of aging; immunophysiology of serum lipoproteins associated with atherosclerosis; lipid metabolism; effects of thiouracil-cholesterol atherogenic diets; basal metabolism; functional cell mass; thyroid function in aging. *Mailing Add:* 6803 Mayfield Rd PO Box 24246 Cleveland OH 44124

BAKER, STEPHEN DENIO, b Durham, NC, Nov 30, 36; m 62; c 2. PHYSICS. *Educ:* Duke Univ, BS, 57; Yale Univ, MS, 59, PhD(physics), 63. *Prof Exp:* Res assoc & lectr, 63-66, from asst prof to assoc prof, 66-73, PROF PHYSICS, RICE UNIV, 73- *Mem:* Am Phys Soc; Am Asn Physics Teachers; AAAS. *Mailing Add:* Dept Physics Rice Univ Houston TX 77251-1892

BAKER, STEPHEN PHILLIP, b Auburn, NY, Nov 20, 48; m; c 1. BIOCHEMICAL PHARMACOLOGY. *Educ:* Univ Mass, BA, 71; Univ Aston, Birmingham, Eng, PhD(biochem pharmacol), 76. *Prof Exp:* Res technician, Dept Pharmacol, Sch Med, Univ Rochester, 71-72; res asst, Dept Pharmacol, Univ Aston, Birmingham, Eng, 72-73; res assoc, Dept Pharmacol, Univ Miami, 76-79; from asst prof to assoc prof, 79-89, PROF, DEPT PHARMACOL & INTERIM CHMN, DEPT PHARMACOL & THERAPEUT, COL MED, UNIV FLA, GAINESVILLE, 89- *Concurrent Pos:* Prin investr, NIH, 80-91, Am Heart Asn, 80-92 & Am Fedn Aging Res, Inc, 85; consult, Brain Inst, Mt Sinai Med Ctr & Univ Miami, 80 & NIH, 88; mem, Peer-Rev Comt, Am Heart Asn, 82-87, Cardiovasc & Renal Rev Group, NIH, 88 & Neurol Sci Study Sect, NIH, 89. *Mem:* Am Soc Pharmacol & Therapeut; Soc Neurosci. *Res:* Author of numerous technical publications; awarded one US patent. *Mailing Add:* Dept Pharmacol & Therapeut Univ Fla JHMHC Box J-267 Gainesville FL 32610

BAKER, SUSAN PARDEE, b Atlanta, Ga, May 31, 30; m 51; c 3. EPIDEMIOLOGY, INJURY CONTROL. *Educ:* Cornell Univ, BA, 51; Johns Hopkins Univ, MPH, 68. *Prof Exp:* PROF HEALTH POLICY & MGT, SCH HYG & PUB HEALTH, JOHNS HOPKINS UNIV, 68- *Concurrent Pos:* Res assoc, Off Chief Med Exam Md, 68-81; mem bd dirs, Baltimore Safety Coun, 69-74; adj asst prof prev med, Sch Med, Univ Md, 79-; chmn, Nat Accident Sampling Syst Rev Bd, 76-80; co-chmn, Nat Symp Injury Control, 81; vchmn trauma comt, Nat Res Coun, 84-85. *Honors & Awards:* Prince Bernhard Medal for Dissertation in Traffic Med, 74; Stone Lectr, Am Trauma Soc, 85. *Mem:* Am Asn Automotive Med (pres, 74-75); Am Trauma Soc (bd dirs, 72-); Am Pub Health Asn. *Res:* Injury epidemiology and prevention; burns; driver and pedestrian fatalities; alcohol; drowning; carbon monoxide poisoning; emergency medical services; evaluation; falls in the elderly; injury severity scoring. *Mailing Add:* Sch Hyg & Pub Health Johns Hopkins Univ Baltimore MD 21205

BAKER, THEODORE PAUL, b Abington, Pa, Dec 15, 49; m 73; c 1. COMPUTER SCIENCE. *Educ:* Cornell Univ, AB, 70, PhD(comput sci), 74. *Prof Exp:* Asst prof comput sci, Fla State Univ, 73-79; assoc prof comput sci, Univ Iowa, 79-80; MEM FAC, DEPT COMPUT SCI, FLA STATE UNIV, 80- *Mem:* Asn Comput Mach; Soc Indust & Appl Math; Sigma Xi. *Res:* Computational complexity; relative computability; theory of computation; formal languages. *Mailing Add:* Dept Math Fla State Univ Tallahassee FL 32306

BAKER, THOMAS, neuropharmacology, neurotoxicology, for more information see previous edition

BAKER, THOMAS IRVING, b La Rue, Ohio, Sept 28, 31; m 53; c 4. MICROBIOLOGY, BIOCHEMISTRY. *Educ:* Kent State Univ, BS, 54; Ohio State Univ, MS, 56; Western Reserve Univ, PhD(microbiol), 65. *Prof Exp:* Assoc chemist fermentation lab, Northern Utilization Res & Develop Div, USDA, 57-60; asst prof, 67-74, ASSOC PROF MICROBIOL, SCH MED, UNIV NMEX, 74- *Concurrent Pos:* Res fel microbiol, Western Reserve Univ, 65-67. *Mem:* Genetics Soc Am; Am Soc Microbiol; AAAS. *Res:* Microbiol genetics, specifically gene-enzyme interrelationships; molecular genetics; DNA repair. *Mailing Add:* Dept Microbiol Univ NMex Sch Med Albuquerque NM 87131-5276

BAKER, TIMOTHY D, b Baltimore, Md, July 4, 25; m 51; c 3. PREVENTIVE MEDICINE. *Educ:* Johns Hopkins Univ, BA, 48, MPH, 54; Univ Md, MD, 52; Am Bd Prev Med, dipl, 59. *Prof Exp:* Clinician, Med Care Clin, Johns Hopkins Univ, 53-54; NY State actg dist health officer, Geneva, 54-56; from asst chief to actg chief health div, Tech Coop Mission, India, 56-58; asst prof prev med, State Univ NY Upstate Med Ctr, 58-59; prof int health & pub health admin, asst dean curric & assoc dir dept int health, 59-72, PROF INT HEALTH, OCCUP HEALTH & HEALTH POLICY & MGT, JOHNS HOPKINS UNIV, 72- *Concurrent Pos:* Intern, Univ Baltimore Hosp & pub health resident, NY State, 52-54; NY State Syracuse Dist Health Officer, 58-59; consult health manpower, Taiwan, Thailand, Indonesia & US Dept Health, Educ & Welfare; consult health planning, Brazil, Saudi Arabia, El Salvador, Burma & Sri Lanka; consult med educ, Peru & Saudi Arabia; vpres & dir, Univ Assocs Inc, dir, Intermed Inc, 72-74; asst secy-treas, Am Bd Prev Med; treas & dir, Pan Am Health Educ Found. *Mem:* AAAS; fel Am Pub Health Asn; Am Soc Trop Med & Hyg; Am Col Prev Med. *Res:* Health manpower and planning; medical education; international health and population dynamics; public health administration; medical care; epidemiology. *Mailing Add:* Sch Hyg & Pub Health Johns Hopkins Univ 615 N Wolfe St Baltimore MD 21205

BAKER, VICTOR RICHARD, b Waterbury, Conn, Feb 19, 45; m 67; c 2. GEOMORPHOLOGY, PLANETARY GEOLOGY. *Educ:* Rensselaer Polytech Inst, BS, 67; Univ Colo, PhD(geol), 71. *Prof Exp:* Geophysicist hydrol, US Geol Surv, 67-69; city geologist urban geol, Boulder, Colo, 69-71; asst prof geol sci, Univ Tex, Austin, 71-76, assoc prof, 76-81; PROF GEOSCI, UNIV ARIZ, TUCSON, 81-, PROF PLANETARY SCI, LUNAR & PLANETARY LAB, 82- & REGENTS PROF, DEPT GEOSCI & DEPT PLANETARY SCI, 88- *Concurrent Pos:* Res scientist, Bur Econ Geol, Univ Tex, 73; mem, NASA Viking Orbital Imaging Team, 77-79; vis res fel, Australian Nat Univ, 79-80; Fulbright-Hays sr res scholar, 79-80; mem, Nat Res Coun Comt & Panels, 79-; prin investr, NASA grants, 73- , NSF grants, 78-85, 88-, US Dept Interior grants, 80-86; consult, NASA, 82, US Dept Energy, 83- & US Bur Reclamation, 89-; mem Magellan Space Mission Team, 83-; vis res fel, Nat Instit Hydrology, India, 87-88; vis fel, Univ Adelaide, Australia, 88; chmn, Planetary Geol Div, Geol Soc Am, 86, Quaternary Geo & Geomorphol Div, 87, Counr, 90-; mem, Panel Global Surficial Geofluxes, Nat Res Coun, 88-, Global Change Comt Working Group on Solid Earth Processes, 90- & Panel Sci Responsibility & Conduct of Res, 90-; assoc ed, J Geophys Res, 89-91; chair-elect, Sect Geol & Geog, AAAS, 91- *Mem:* Fel Geol Soc Am; Am Geophys Union; fel AAAS; Am Quaternary Asn; Int Asn Geomorphologists. *Res:* Geomorphology; paleohydrology; quaternary geology; natural hazards; geology of Mars and Venus; fluvial geomorphic studies in western United States, Australia, India, Israel and South America; flood geomorphology; philosophy of earth and planetary sciences. *Mailing Add:* Dept Geosci Univ Ariz Tucson AZ 85721

BAKER, WALTER L(OUIS), b Earlton, NY, Aug 7, 24; m 44; c 3. ELECTRONICS ENGINEERING, ACOUSTICS. *Educ:* Clarkson Col Technol, BEE, 44; Pa State Univ, MS, 54, EE, 55. *Prof Exp:* Tech asst, Tenn Eastmn Corp, 44-45; sr engr, Philco Corp, 45-49; from res assoc to assoc prof, Pa State Univ, 49-82, sr mem grad fac, 65-82; PRES, BAKER ENG CO, 80- *Concurrent Pos:* Consult, John I Thompson & Co, 52-68, Sparton Elec Corp, 60-61, HRB-Singer Corp, 58-67, US Marine Corp, 65, DSI, 79-84, Woods Hole Oceanog Inst, 80-, Vitro Corp, 81, SRM, 84-86 & BB&N, Inc, 86- *Honors & Awards:* Navy Meritorious Pub Serv Citation, US Navy, 75. *Mem:* Sr mem Inst Elec & Electronics Engrs; NY Acad Sci; Acoustical Soc Am; Sigma Xi. *Res:* Underwater acoustics; signal processors; data processing; computer systems; weapon system feasibility studies; methodology and digital simulation. *Mailing Add:* 22 Oliver Hazard Perry Rd Portsmouth RI 02871

BAKER, WALTER WOLF, b Philadelphia, Pa, Oct 29, 24; m 52; c 3. NEUROPHARMACOLOGY. *Educ:* Franklin & Marshall Col, BS, 48; Univ Iowa, MS, 50; Jefferson Med Col, PhD(pharmacol), 53. *Prof Exp:* Res asst, Radiation Res Lab, Univ Iowa, 48-50; admin asst & pharmacologist, Wyeth Inst Appl Biochem, 50-52; from instr to assoc prof pharmacol, 53-60, assoc prof neuropharmacol, 60-70, PROF PSYCHIAT, JEFFERSON MED COL, 70-, PROF PHARMACOL, 73-; DIR NEUROPHARMACOL, EASTERN PA PSYCHIAT INST, 63- *Mem:* AAAS; Am Soc Neurochem; Am Soc Pharmacol & Exp Therapeut; Am Col Clin Pharmacol; Am Col Neuropsychopharmacol. *Res:* Electron microscopy of muscle; cerebral evoked potentials; central autonomic regulation; chemically-induced tremors; psychomotor epilepsy; local action of drugs on brain electrical activity; neurotransmitter regulatory mechanisms in caudate nucleus and hippocampus. *Mailing Add:* Dept Pharmacol Jefferson Med Col 1025 Walnut St Philadelphia PA 19107

BAKER, WARREN J, b Fitchburg, Mass, Sept 5, 38; m 62; c 2. CIVIL ENGINEERING, SOIL MECHANICS. *Educ:* Univ Notre Dame, BSCE, 60, MSCE, 62; Univ NMex, PhD(civil eng), 66. *Prof Exp:* Proj engr, C E Williams & Assoc, Inc, 59-62; res asst engr, Air Force Shock Tube Facility, 62-65, res assoc engr, 65-66; asst prof civil eng, Univ Detroit, 66-67, assoc prof, 67-, acad vpres, 76-; PRES, CALIF POLYTECH STATE UNIV. *Concurrent Pos:* Lectr, Univ NMex, 65-66; consult, Woodward, Clyde & Assoc, 62-65, E H Wang Civil Eng Res Facility, 67- & Off Civil Defense, Dept Defense Mem Instrumentation panel, Defense Atomic Support Agency, 63-65. *Mem:* Nat Soc Prof Engrs; Am Soc Civil Engrs; Am Soc Eng Educ; Int Soc Soil Mech & Found Eng. *Res:* Soil dynamics; foundation vibrations; shock loading response of earth masses; wave propagation in non-linear materials; foundation engineering. *Mailing Add:* Off Pres Chancellor Calif Polytech State Univ San Luis Obispo CA 93407

BAKER, WILBER WINSTON, b Ponca City, Okla, Nov 2, 34; m 57; c 4. BIOCHEMISTRY, CELL BIOLOGY. *Educ:* Grinnell Col, AB, 56; Iowa State Univ, MS, 56; Ore State Univ, PhD(biochem), 64. *Prof Exp:* Fel biol chem, Harvard Med Sch, 64-66; res assoc biol, Cancer Res Inst, Philadelphia, 66-68; ASST PROF BIOL, VILLANOVA UNIV, 68- *Mem:* Am Chem Soc; Am Inst Biol Sci. *Res:* Chemical embryology; biochemical genetics; isozymes. *Mailing Add:* Dept Biol Villanova Univ Villanova PA 19085

BAKER, WILFRED E(DMUND), b Baltimore, MD, Jan 3, 24; m 47; c 6. MECHANICAL ENGINEERING, PHYSICS. *Educ:* Johns Hopkins Univ, BE, 43, MSE, 49, DE, 58. *Prof Exp:* Jr instr mech eng, Johns Hopkins Univ, 46-47, instr, 47-49; physicist, Ballistics Res Labs, Aberdeen Proving Ground, 49-61; prin develop engr, Aircraft Armaments, Inc, 61-64; mgr eng dynamics sect, Dept Mech Sci, Southwest Res Inst, 64-72, inst scientist, 72-84; PRES, WILFRED BAKER ENG, 84- *Concurrent Pos:* Consult, Aberdeen Proving Ground, 61-63; Armed Serv Explosive Safety Bd, 68-70; chmn joint serv hazards evaluation comn, Joint Army, Navy, NASA & Air Force Safety & Environ Protection Working Group, 75-78. *Mem:* Assoc fel Am Inst Aeronaut & Astronaut; fel Am Soc Mech Engrs; Sigma Xi; Soc Am Mil Engrs. *Res:* Reaction of structures to transient loads; damping of vibrations; explosion hazards evaluation; explosive effects research; design of motion simulators; instrumentation for dynamic effects; scale modeling of physical effects; design of explosion-resistant structures. *Mailing Add:* 8103 Broadway Suite 102 PO Box 6477 San Antonio TX 78209

BAKER, WILLIAM KAUFMAN, b Portland, Ind, Dec 2, 19; m 44; c 3. GENETICS. *Educ:* Col of Wooster, BA, 41; Univ Tex, MS, 43, PhD(genetics), 48. *Hon Degrees:* DSc, Col of Wooster, 84. *Prof Exp:* Nat Res Coun fel, Univ Tex, 46-48; asst prof zool, Univ Tenn, 48-52; from biologist to sr biologist, Oak Ridge Nat Lab, 51-55; assoc prof zool, Univ Chicago, 55-59, chmn, dept biol, 68-72, prof biol, 59-76; chmn dept, 77-80, prof biol, 80-85, EMER PROF, UNIV UTAH, 85- *Concurrent Pos:* NSF sr fel, 63-64; co-ed, Am Naturalist, 65-70; NIH spec fel, 72-73. *Mem:* Am Soc Naturalists (secy, 62-64, vpres, 76); Genetics Soc Am (pres, 80); Soc Study Evolution. *Res:* General genetics and development of Drosophila. *Mailing Add:* Rte 9 Box 86WB Santa Fe NM 87505

BAKER, WILLIAM OLIVER, b Chestertown, Md, July 15, 15; m 41; c 1. PHYSICAL CHEMISTRY, POLYMER SCIENCE. *Educ:* Wash Col, BS, 35; Princeton Univ, PhD(phys chem), 38. *Hon Degrees:* Twenty-four from US and foreign univs, 57-90. *Prof Exp:* Mem tech staff, Bell Labs, 39-51, asst dir chem & metall res, 51-54, dir res phys sci, 54-55, vpres res, 55-73, pres, AT&T Bell Labs, 73-79, chmn, 79-80; RETIRED. *Concurrent Pos:* mem, Nat Sci Bd, 60-66; Mem, President's Sci Adv Comt, 57-60; mem Sci Info Coun, NSF, 58-61, chmn, 59-61; consult, US Dept Defense, 58-71, spec asst sci & technol, 63-73; mem adv bd, Mil Personnel Supplies, Nat Res Coun, 58-78, chmn, 64-78, mem, Comn Sociotech Systs, 63-70, mem, Comt Phys Chem, Div Chem & Chem Technol, 63-70; mem, President's Foreign Intel Adv Bd, 59-77 & 81-90; mem sci adv bd, Nat Security Agency, 60-76; consult, Panel Opers Eval Group, US Navy, 60-62 & Nat Sci Bd, 60-66; mem bd visitors, US Air Force Systs Command, 62-73; mem bd visitors, Tulane Univ, 63-80; mem, Liaison Comt Sci & Technol, Libr Cong, 63-73, Comt Sci & Technol, US Chamber of Com, 63-66, Coun Trends & Perspective, 66-74; mem sci adv bd, Robert A Welch Found, 68-; mem bd regents, Nat Libr Med, 69-73; chmn adv tech panel, Nat Bur Stand & Nat Acad Sci-Nat Res Coun, 69- 78; mem coun, Nat Acad Sci, 69-72 & Inst Med, 73-75; mem mgt adv coun, Oak Ridge Nat Lab, 70-, chmn, Andrew W Mellon Found, 70-90, emer chmn, 90-, trustee Rockefeller Found, 60-90, chmn, 80-90, emer chmn, 90-; mem, Nat Comn Libr & Info Sci, 71-75; mem, Nat Policy Panel on Future UN Role in Sci & Technol, UN Asn, 72-, Nat Coun Educ Res, Nat Inst Educ, 73-75, Energy Res & Develop Adv Coun, Energy Policy Off, 73-75, Comn on Critical Choices for Americans, 73-75, Proj Independence Adv Comt, Fed Energy Admin, 74-75, Nat Cancer Adv Bd, 74-80, Governor's Comn Evaluate Capital Needs NJ, 74-75, Steering Comt, President's Food & Nutrit Study & Comn Int Rels, Nat Acad Sci, 75; mem gov bd, Nat Enquiry Into Scholarly Commun, 75-; bd overseers, Col Eng & Appl Sci, Univ Pa, 75-; consult, Nat Security Agency, 76-; vchmn, President's Comt Sci & Technol, Off Sci & Technol Policy, 76-77, Nat Comt Excellence in Educ, 81-84. *Honors & Awards:* Honor Scroll, Am Inst Chemists, 62, Gold Medal, 75; Perkin Medal, 63; Priestley Medal, Am Chem Soc, 66, Charles Lathrop Parsons Award, 76, J Willard Gibbs Medal, 78; Edgar Marburg Award, 67; Indust Res Inst Medal, 70; Frederik Philips Award, Inst Elec & Electronics Engrs, 72; Procter Prize, Sigma Xi, 73; James Madison Medal, Princeton Univ, 75; Mellon Inst Award, 75; Franklin Inst Fahrney Medal, 77; Thomas Alva Edison Medal, 87. *Mem:* Nat Acad Sci; Am Philos Soc; Am Acad Arts & Sci; Am Inst Chemists; Am Phys Soc; Nat Acad Eng; Am Chem Soc; Sigma Xi. *Res:* Study of solid state materials and macromolecules; dielectric and dynamic mechanical properties of crystals and glasses; information processing; author of approximately 95 papers; holder of 13 patents. *Mailing Add:* AT&T Bell Labs Spring Valley Rd Morristown NJ 07960

BAKER, WILLIE ARTHUR, JR, b San Antonio, Tex, Nov 7, 33; m 54; c 2. INORGANIC CHEMISTRY. *Educ:* Tex Col Arts & Indust, BS, 55; Univ Tex, PhD(chem), 59. *Prof Exp:* From asst prof to prof chem, Syracuse Univ, 59-70, chmn dept, 66-70; PROF CHEM & DEAN GRAD SCH, UNIV TEX, ARLINGTON, 70-, VPRES ACAD AFFAIRS, 72- *Mem:* AAAS; Am Chem Soc; The Chem Soc; Sigma Xi. *Res:* Coordination compounds; magnetic properties; spectra structure. *Mailing Add:* 2014 Greencove Dr Arlington TX 76012

BAKER, WINSLOW FURBER, b Brockton, Mass; m 67; c 2. HIGH ENERGY PHYSICS, ELEMENTARY PARTICLE PHYSICS. *Educ:* Bowdoin Col, AB, 50; Columbia Univ, MA, 53, PhD(physics), 57. *Hon Degrees:* DSc, Allegheny Col, 75. *Prof Exp:* Res assoc physics, Columbia Univ, 57-59; physicist, Brookhaven Nat Lab, 59-64, Europ Orgn Nuclear Res, 64-69; PHYSICIST, FERMI LAB, UNIV RES ASN, 69- *Mem:* Am Phys Soc. *Res:* Strong interactions at high energies; backward scattering, baryon exchange, elastic scattering, total cross sections; direct photons. *Mailing Add:* Fermi Lab PO Box 500 Batavia IL 60510

BAKER-COHEN, KATHERINE FRANCE, b Winnipeg, Man, Mar 1, 28; US citizen; m 57. ZOOLOGY. *Educ:* Antioch Col, BA, 47; Columbia Univ, MA, 52, PhD(zool), 57. *Prof Exp:* USPHS fel biochem, Columbia Univ, 57-59; fel physiol, 60-61, interdisciplinary prog fel, 61-62, instr anat, 62-67, ASST PROF ANAT, ALBERT EINSTEIN COL MED, 67- *Mem:* Am Asn Anatomists; Am Soc Zoologists. *Res:* Neurohistochemistry; biochemical embryology; thyroid development and function in fish. *Mailing Add:* Dept Anat Albert Einstein Col Med 1300 Morris Park Ave Bronx NY 10461

BAKERMAN, SEYMOUR, b Brooklyn, NY, May 26, 24; m 54; c 3. BIOPHYSICS, PATHOLOGY. *Educ:* NY Univ, BA, 49; Purdue Univ, MS, 52, PhD(chem), 57; Western Reserve Univ, MD, 59. *Prof Exp:* From isntr to assoc prof path, Sch Med, Univ Kans, 60-68; prof, Med Col Va, 68-78; PROF & CHMN DEPT PATH & LAB MED, E CAROLINA UNIV, 78- *Mem:* Soc Exp Path; Am Chem Soc; Geront Soc; Sigma Xi. *Res:* Molecular biophysics; molecular structure of collagen and membranes and alterations with age and disease; computer diagnosis of disease. *Mailing Add:* Dept Path Sch Med E Carolina Univ Greenville NC 27858

BAKHIET, ATEF, b El Minia, Egypt, Apr 21, 41; Can citizen; m 68; c 2. MAINTENANCE, QUALITY CONTROL. *Educ:* El Minia Univ, BSc, 66; Turin Univ, Italy, dipl, 70. *Prof Exp:* Maintenance supt, Egyptian Co Metall Construct, 66-71; plant supvr, Oxygen Plant, Egyptian Co & Steel & Iron, 71-74; proj engr, Que Iron & Titanum, Soul, Que, 76-83; maintenance consult, Ministry Recreation, Hunting & Fishing, Que, 83-85; mgr mfg servs, Simonds Industs Inc, 85-87, mgr eng & mfg servs, 87-90; PVT CONSULT, 90- *Mem:* Order Eng Quebec; Standards Coun Can; Indust Admin Clubs Can; Asn Can Mfg. *Res:* General maintenance and quality control. *Mailing Add:* 588 Des Mouettes Granby PQ J2H 1S6 Can

BAKHRU, HASSARAM, b Rohri, India, June 2, 37; c 2. NUCLEAR & SOLID STATE PHYSICS. *Educ:* Banaras Hindu Univ, India, BS, 58, MS, 60; Calcutta Univ, PhD(nuclear physics), 64. *Prof Exp:* Fel, Heavy Ion Accelerator, Yale Univ, 65-69, res staff fel nuclear physics, 65-66, res assoc & assoc dir, 66-69; asst prof, 70-73, PROF PHYSICS, STATE UNIV NY ALBANY, 73-, DIR, ACCELERATOR LAB, 72- *Concurrent Pos:* USAEC res grant, 71- *Honors & Awards:* Frederick Gardner Cottrell Res Award, Res Corp, 71. *Mem:* Am Phys Soc. *Res:* Nuclear reactions and spectroscopy; new isotopes; heavy ion and neutron induced reactions; x-ray spectroscopy; modification of surface sensitive properties of metals using MeV ions; research on fatigue, corrosion, wear and hardness of metals, analysis using microbeams, materials characterization using MeV ions. *Mailing Add:* Dept Physics Nuclear Accelerator Lab State Univ NY Albany NY 12222

BAKHSHI, NARENDRA NATH, b New Delhi, India, July 27, 28; Can citizen; m 51; c 4. CHEMICAL ENGINEERING. *Educ:* Benares Hindu Univ, BSc, 46, MSc, 48; Va Polytech Inst, PhD(chem eng), 57. *Prof Exp:* Lectr chem, Yuvraj Dutta Col, India, 48-49 & Dayanand, India, 49-50; asst, Nat Phys Lab, 50-53; process develop engr, Hydrocarbon Res, Inc, 56-59; process engr, Air Prod & Chem, Inc, Pa, 59-60; sr chem engr, Air Prod Ltd, Eng, 60-63; sr process engr, Brit Oxygen Co, Ltd, 63-64; from asst prof to assoc prof, 64-74, PROF CHEM & CHEM ENG, UNIV SASK, 74- *Concurrent Pos:* Spec lectr chem eng, Univ Sask, 63-64. *Mem:* Fel Chem Inst Can; Can Soc Chem Eng; Asn Prof Engrs Sask; Am Chem Soc; Am Inst Chem Engrs. *Res:* Heterogeneous catalysis; chemical reactor design; biomass pyrolysis; upgrading of biomass-derived oils to fuels and chemicals. *Mailing Add:* Dept Chem & Chem Eng Univ Sask Saskatoon SK S7N 0W0 Can

BAKHSHI, VIDYA SAGAR, b India, Sept 23, 39; m; c 2. APPLIED MATHEMATICS. *Educ:* Univ Delhi, BA, 59, MA, 63; Ore State Univ, PhD(math), 68. *Prof Exp:* Lectr math, Kurukshetra Univ, India, 63-65; teaching asst, Ore State Univ, 65-68; assoc prof, 68-75, PROF MATH, VA STATE COL, 75- *Mem:* Math Asn Am. *Res:* Fluid dynamics and mathematical seismology; two and three dimensional boundary layer problems. *Mailing Add:* Dept Math Va State Univ Petersburg VA 23803

BAKIS, RAIMO, b Tartu, Estonia, May 7, 33; US citizen; m 65; c 2. PHYSICS. *Educ:* Sterling Col, BA, 54; Kans State Univ, MS, 56, PhD(physics), 59. *Prof Exp:* Assoc physicist res ctr, 59, staff physicist, 59-60, res staff mem speech recognition, 60-61, res staff mem pattern recognition, 61-66, res staff mem speech processing res lab, Ruschlikon ZH, Switz, 66-68, RES STAFF MEM, IMAGE PROCESSING, RES CTR, INT BUS MACH CORP, 68- *Mem:* AAAS. *Res:* Speech processing--automatic recognition of continuous speech. *Mailing Add:* 204 Ridge Top Lane Brewster NY 10503-9389

BAKISH, ROBERT, b Sofia, Bulgaria, Jan 5, 26; US citizen; m 59; c 1. PHYSICAL METALLURGY. *Educ:* Columbia Univ, BSMetEng, 52; Yale Univ, MEng, 53, DEng(metall), 55. *Prof Exp:* Sr metallurgist, Sprague Elec Co, Mass, 55-57; consult phys metall, Rare Metals Div, Ciba Corp, Switz, 57-59; vpres res & develop, Alloyd Electronics Corp, Mass, 59-62; exec vpres, Electronics & Alloys, Inc, NJ, 62-64; PRES, BAKISH MAT CORP, 64- *Concurrent Pos:* Lectr, Rensselaer Polytech Inst, 56-57, Univ Calif, Los Angeles, 64-68 & Am Welding Soc, 66-; consult, Repub Foil, Inc, Conn, 60-, vpres res & develop, 66-70; chmn, Int Conf Electron & Ion Beams in Sci & Technol, 63-83; adj prof, Fairleigh Dickinson Univ, 64-80; dir, St Croix Corrosion Installation, 70-80; pres, Int Desalination Environ Asn, 78-81; chmn, Electron Beam Melting & Refining State of the Art, 83-89, Laser vs Electron Beam, 85-89. *Mem:* Inst Mining, Metall & Petrol Engrs; Am Welding Soc; Nat Asn Corrosion Engrs; Am Vacuum Soc; Am Soc Metals. *Res:* Corrosion; electron beam technology; metals processing; various publications regarding electron beam and vacuum technology and desalination; over 90 publications including contribution & editor to 24 volumes dealing with electron beam technology. *Mailing Add:* 171 Sherwood Pl PO Box 190 Englewood NJ 07631

BAKKE, JEROME E, b Mapes, NDak, July 14, 31; m 54; c 2. BIOCHEMISTRY. *Educ:* NDak State Univ, BS & MS, 56; Univ NDak, PhD(biochem), 60. *Prof Exp:* Chemist, Lignite Res Lab, US Bur Mines, NDak, 55-58; res biochemist, Lipid Sect, Fundamental Food Res Dept, Gen Mills, Inc, Minn, 60-61, Natural Prod Sect, 61-63, Explor Food Res Dept, 63-64; asst prof biochem, NDak State Univ, 68-73; RES BIOCHEMIST RADIATION & METAB LAB, USDA, 64- *Concurrent Pos:* Adj prof biochem, NDak State Univ, 73- *Mem:* Int Soc Study Xenobiotics. *Res:* Natural product isolation, identification and synthesis; lipid metabolism; food research; insecticide metabolism in the animal. *Mailing Add:* 301 15th Ave N Fargo ND 58201

BAKKEN, AIMEE HAYES, b Camden, NJ, Oct 18, 41; m 63; c 1. CELL BIOLOGY, DEVELOPMENTAL BIOLOGY. *Educ:* Univ Chicago, BA, 63; Univ Iowa, PhD(zool), 70. *Prof Exp:* Res asst reprod biol, Dept Obstet & Gynec, Univ Chicago, 63-66; investr, Biol Div, Oak Ridge Nat Lab, 70-72; res assoc biol, Yale Univ, 72-73; asst prof, 73-78, ASSOC PROF ZOOL, UNIV WASH, 79- *Concurrent Pos:* NSF res grant, 74-82, 89-, NIH grant, 80- *Mem:* AAAS; Am Soc Cell Biol. *Res:* Regulation of RNA genes; correlation of chromosome and nucleolar structure and function in eukaryotic germ cells, embryonic cells and somatic cells, using molecular and cytogenetic techniques. *Mailing Add:* Dept Zool NJ-15 Univ Wash Seattle WA 98195

BAKKEN, ARNOLD, b Antelope, Mont, Feb 19, 21; m 47; c 2. ZOOLOGY. *Educ:* Univ Mont, BA, 43; Univ Wis, MS, 49, PhD(zool, mammal, ecol), 53. *Prof Exp:* Teaching & res zool & anat, Reed Col, 52-53; teaching, 53-55, from assoc prof to prof, 55-83, EMER PROF BIOL, UNIV WIS, EAU CLAIRE, 83- *Concurrent Pos:* Fulbright grant, Norway, 51-52; NSF award, 58. *Mem:* AAAS; Am Inst Biol Sci; Animal Behav Soc; Am Soc Mammal; Ecol Soc Am; Sigma Xi. *Res:* Behavior and interrelationships between fox squirrels and grey squirrels in mixed populations; population problems of Norwegian lemmings; microclimate and behavior. *Mailing Add:* 500 Big Flat Rd Missoula MT 59801

BAKKEN, EARL E, b Minneapolis, Minn, 1924. RESEARCH ADMINISTRATION. *Educ:* Univ Minn, BEE, 48. *Hon Degrees:* DSc, Univ Minn, 88, Tulane Univ, 88. *Prof Exp:* Partner, Medtronic Inc, 49-57, chief exec officer & chmn bd, 57-76, sr chmn bd, 76-89, DIR, MEDTRONIC INC, 89- *Concurrent Pos:* Chmn bd dirs, Archaeus Proj & Bakken, pres bd dirs, Pavek Mus Broadcasting, mem bd dirs, Children's Heart Fund, Med Graphics Corp & N Hawaii Community Hosp. *Honors & Awards:* Eng for Gold Award, Nat Soc Prof Engrs, 84; Centennial Medal, Inst Elec & Electronics Engrs, 84; Distinguished Serv Award, NAm Soc Pacing & Electrophysiol, 85; Officer in the Order of Orange-Nassau, Neth, 89. *Mem:* Nat Acad Eng; fel Inst Elec & Electronics Engrs; fel Instrument Soc Am; Am Antiquarian Soc; Asn Advan Med Instrumentation; assoc mem NAm Soc Pacing & Electrophysiol; hon fel Am Col Cardiol. *Res:* Bioengineering. *Mailing Add:* Medtronic Inc 7000 Central Ave NE Minneapolis MN 55432

BAKKEN, GEORGE STEWART, b Denver, Colo, Feb 17, 43; m 73; c 2. PHYSIOLOGICAL ECOLOGY. *Educ:* NDak State Univ, BS, 65; Rice Univ, MA, 67, PhD(physics), 70. *Prof Exp:* Res assoc physics, Rice Univ, 69-70; fel physiol ecol biophys, Mo Bot Garden, 70-71, Univ Mich, 71-73; res assoc, Div Biol Sci, 73-75, from asst prof to assoc prof, 75-83, PROF LIFE SCI, IND STATE UNIV, 83- *Mem:* AAAS; Am Inst Biol Sci; Am Ornithol Union; Ecol Soc Am; Brit Ecol Soc; Am Soc Zoologists. *Res:* The study of energy and mass flow between organisms and their physical environment and the consequences for physiology, ecology, behavior and evolution; biological, physical and mathematical methods are employed. *Mailing Add:* Dept Life Sci Ind State Univ Terre Haute IN 47809

BAKKER, CORNELIS BERNARDUS, b Rotterdam, Holland, Jan 6, 29; US citizen; m 55; c 3. PSYCHIATRY. *Educ:* State Univ Utrecht, MD, 52. *Prof Exp:* Intern, Clin of Rotterdam, Holland, 52-53; intern, Sacred Heart Hosp, Spokane, Wash, 53-54; resident psychiat, Eastern State Hosp, Medical Lake, 54-56; resident, Psychiat Clin, State Univ Utrecht, 56-57; resident, Med Sch, Univ Mich, Ann Arbor, 57-59, instr, 59-60, res assoc ment health res inst, 58-60; from instr to assoc prof, Univ Wash, 60-72, dir adult psychiat inpatient serv, Univ Hosp, 61-68, actg chmn psychiat, Sch Med, 69-70, dir adult prog, Univ Hosp, Univ Wash, 68-79, prof psychiat, Sch Med, 72-79; prof, Dept Psychol, Univ Ill Col Med, Peoria, 79-84; MED DIR, DEPT PSYCHIAT, SACRED HEART MED CTR, SPOKANE, WASH, 84-; CLIN PROF, DEPT PSYCHIAT & BEHAV SCI, UNIV WASH SCH MED, SEATTLE, WASH, 85- *Concurrent Pos:* Fulbright & Comt Int Exchange Persons grant, 53; Fogart sr fel, Univ Leuven, Belg, 77-78; psychiat consult, Soc See Hearing & Appeals, 63-79; Ketchikan Community Ment Health Ctr, 72-77. *Honors & Awards:* Significant Achievement Award, Am Psychiat Asn, 75. *Mem:* Am Psychiat Asn; Am Col Psychiat. *Res:* Behavior change techniques; human territoriality. *Mailing Add:* Dept Psychiat Sacred Heart Med Ctr TAF-C9 Spokane WA 99220-4045

BAKKER, GERALD ROBERT, b Chicago, Ill, May 16, 33; m 57; c 2. ORGANIC CHEMISTRY. *Educ:* Calvin Col, AB, 55; Univ Ill, PhD(chem), 59. *Prof Exp:* Res assoc chem, Univ Kans, 58-59; from asst prof to prof, Earlham Col, 59-71, consult teaching & learning, 75-77, actg provost, 88-89, PROF CHEM, EARLHAM COL, 71-, ASSOC ACAD DEAN, 86-88 & 89- *Mem:* AAAS; Am Chem Soc; Sigma Xi. *Res:* Physical organic chemistry; ferritin and transferrin interactions. *Mailing Add:* Dept Chem Earlham Col Richmond IN 47374

BAKKER, JAAP JELLE, b Kerkrade, Netherlands, Dec 12, 29; Can citizen; m 56; c 5. CIVIL & TRANSPORTATION ENGINEERING. *Educ:* Glasgow Univ, BSc, 54; Imp Col Sci & Technol, Univ London, dipl, 55; Purdue Univ, MSCE, 57. *Prof Exp:* Instr graphics, Purdue Univ, 55-57; design engr, City of Moose Jaw, Sask, 57-58, city engr, 58-59; from asst prof to assoc prof, 59-76, PROF TRANSP, UNIV ALTA, 76- *Concurrent Pos:* Transp consult to City of Edmonton, 60-77; univ rep, Transp Res Bd, Nat Acad Sci-Nat Res Coun, 71- *Mem:* Can Soc Civil Engrs. *Res:* Traffic and transit; long-range transit planning; transportation prediction and performance; geometric design. *Mailing Add:* Dept Civil Eng Univ Alta Edmonton AB T6G 2M7 Can

BAKKER-ARKEMA, FREDERIK WILTE, b Groningen, Netherlands, Aug 18, 32; m 57; c 3. AGRICULTURAL ENGINEERING. *Educ:* Mich State Univ, BS, 59, MS, 61, PhD(agr eng), 64. *Prof Exp:* Asst, 57-63, from asst prof to assoc prof, 63-71, PROF AGR ENG, MICH STATE UNIV, 71- *Mem:* Am Soc Agr Engrs. *Res:* Bioengineering; mass and heat transfer in biological products; drying and processing of biological products; digital computer techniques in agriculture. *Mailing Add:* Dept Agr Eng Mich State Univ 101 Farrell Hall East Lansing MI 48824

BAKKUM, BARCLAY W, b Manchester, Iowa, Apr 10, 57; m. CHIROPRACTIC. *Educ:* Wartburg Col, BA, 78; Palmer Col Chiropractic, DC, 81; Univ Ill, Chicago, PhD(anat), 91. *Prof Exp:* Chiropractor, Breitbach Chiropractic Off, Charles City, Iowa, 81-85; INSTR ANAT, NAT COL CHIROPRACTIC, 89- *Mem:* Soc Neurosci. *Res:* Clinical anatomy of the region of the intervertebral foramen including gross anatomical, histological and in vivo imaging techniques. *Mailing Add:* Anat Dept Nat Col Chiropractic 200 E Roosevelt Rd Lombard IL 60148

BAKOS, GUSTAV ALFONS, b Trnava, Czech, Aug 2, 18; Can citizen; m 62; c 1. ASTRONOMY. *Educ:* Comenius Univ, Bratislava, BA, 42; Univ Toronto, MA, 53, PhD(astron), 60. *Prof Exp:* Astronr, Smithsonian Astrophys Observ, 59-61; asst prof astron, Northwestern Univ, 61-65; PROF PHYSICS, UNIV WATERLOO, 65- *Mem:* Am Astron Soc; Royal Astron Soc Can; Int Astron Union. *Res:* The effects of soft non-perturbative self-consistent fragmentation of strings and gluon flux tubes on superstring and hadron mass spectra. *Mailing Add:* Dept Physics Univ Waterloo Waterloo ON N2L 3G1 Can

BAKOS, JACK DAVID, JR, b Robins, Ohio, Nov 7, 40; m 71; c 2. STRUCTURAL MECHANICS, TRANSPORTATION. *Educ:* Univ Akron, BSCE, 63; WVa Univ, MSCE, 65, PhD(civil eng), 67. *Prof Exp:* Eng aide, Hwy Eng Dept, City of Akron, Ohio, 60-62; instr civil eng, WVA Univ, 64-67; Assoc prof, 69-80, PROF & CHMN, CIVIL ENG DEPT, YOUNGSTOWN STATE UNIV, 80- *Concurrent Pos:* Consult, Republic Steel Corp & Commercial Shearing, Inc, 77- *Honors & Awards:* Watson Merit Award, 83; Outstanding Civil Engr, Am Soc Civil Engrs, 85. *Mem:* Am Soc Civil Engrs; Nat Soc Prof Engrs; Sigma Xi; Am Soc Eng Educ. *Res:* Highway impact attenuators; low density concrete; freeze-thaw protection of concrete. *Mailing Add:* Dept Civil Eng Youngstown State Univ Youngstown OH 44555

BAKSAY, LASZLO ANDRAS, b Budapest, Hungary, July 22, 45; m 83; c 2. INSTRUMENTATION, DETECTOR DEVELOPMENT. *Educ:* Rheinisch-Westfalische Technische Hochschule Aachen, WGer, dipl physics, 72, Dr rer nat, 78, Staats examen, 79. *Prof Exp:* Researcher, Centre Europ Recherche Nucleaire, 72-75; sci collabr, Rheinisch-Westfälisch Technische Hochschule Aachen, WGer, 75-78; asst prof physics, Northeastern Univ, 78-82; staff scientist, Univ Calif, Berkeley, 82-83; assoc prof physics, Univ Dallas, 83-85; assoc prof physics, Union Col, 85-90; guest scientist, Brookhaven Nat Lab, 86-90; ASSOC PROF, UNIV ALA, 91- *Concurrent Pos:* Vis scientist, Stanford Univ, 78-82; rev panelist, NSF, 87-, US Dept Ed, 87-; prin investr, NSF, 87-, Res Corp, 87-88; vis prof, Hungarian Acad Sci, 88-; hon mem, Maifhe. *Mem:* Am Asn Physics Teachers; Am Phys Soc; Europ Phys Soc; NY Acad Sci; Soc Col Sci Teachers; Ger Phys Soc; hon mem Int Asn Physics Studies. *Res:* Experimental elementary particle physics; L3 experiment at LEP, CERN; instrumentation for high energy physics; astronomy; high pressure research; superconducting super collider. *Mailing Add:* Dept Physics & Astron Univ Ala Tuscaloosa AL 35487t

BAKSHI, PRADIP M, b Baroda, India, Aug 21, 36; m 67; c 2. THEORETICAL PHYSICS. *Educ:* Univ Bombay, BSc, 55; Harvard Univ, AM, 57, PhD(physics), 62. *Prof Exp:* Res fel physics, Harvard Univ, 63; res physicist, Air Force Cambridge Res Labs, 63-66; sr res assoc physics, Brandeis Univ, 66-68, assoc prof, 68-70, assoc prof, 70-75, PROF PHYSICS, BOSTON COL, 75- *Mem:* Am Phys Soc. *Res:* New techniques in mathematical physics; quantum field theory; plasma physics. *Mailing Add:* Dept Physics Boston Col Chestnut Hill MA 02167

BAKSHI, TRILOCHAN SINGH, b Lonavla, India, Sept 13, 25; Can citizen; div; c 3. ECOLOGY. *Educ:* Univ Bombay, BSc, 47; Univ Saugar, MSc, 49; Wash State Univ, PhD(bot), 58. *Prof Exp:* Asst prof bot, Birla Col, India, 50-52; lectr, Univ Delhi, 52-53; head dept biol, SNat Col, India, 54; asst bot, Wash State Univ, 54-58; fel ecol & anat, Univ Sask, 58-59; syst botanist & ecologist, Ministry Nat Resources, Sierra Leone, WAfrica, 60-63; lectr bot, Univ Ghana, 63-64; from assoc prof to prof biol, Notre Dame Univ, BC, 64-73; PROF BIOL, ATHABASCA UNIV, 73- *Concurrent Pos:* Fulbright travel grantee, 54-58; Nat Res Coun Can study grants plant ecol southeast BC; vis prof bot, Univ WIndies, 78, Indira Gandhi Nat Open Univ, India, 86-87, Univ of South Pacific, Fiji, 87; Lady Davis vis prof agr eng, Technion-Israel Inst Technion, 81-82; vis prof, distance educ, Deakin Univ, Australia, 91- *Mem:* Int Soc Plant Morphol; Indian Bot Soc; Int Asn Ecol; Int Coun Distance Educ. *Res:* Ecology of vascular plants; environmental education; distance education. *Mailing Add:* Dept Biol Athabasca Univ Box 10000 Athabasca AB T0G 2R0 Can

BAKSI, SAMARENDRA NATH, b Rajshahi, India, Dec 28, 40; US citizen; m 71; c 2. BIOCHEMICAL PHARMACOLOGY, ENDOCRINOLOGY. *Educ:* Bihar Vet Col, India, BVSc, 60; Univ Mo-Columbia, MS, 67, PhD(endocrinol), 71. *Prof Exp:* Vet surg, Govt Bihar, India, 60-66; res asst endocrinol, Univ Mo-Columbia, 67-71; res fel, Worcester Found Exp Biol, 71-72; vis fel environ toxicol, NIH, 72-74; res assoc, Univ Tex Med Br Galveston, 75-76; res assoc pharmacol, Sch Med, Tex Tech Univ, 76-79; RES ASSOC PHYSIOL, HEALTH SCI CTR, TEX TECH UNIV, 80-; ASST PROF, DEPT VET COMP ANAT, PHARMACOL & PHYSIOL, WASH STATE UNIV, 88- *Mem:* Endocrine Soc; Am Soc Pharmacol & Exp Therapeut; Am Physiol Soc; Am Soc Bone & Mineral Res; Am Col Toxicol. *Res:* Endocrine aspects of calcium metabolism; pharmacology of calcemic hormones; regulation of vitamin D metabolism; neurotransmitter physiology; dietary calcium and hypertension Angiotensin II receptor bindings. *Mailing Add:* Dept Vet Comp Anat Pharmacol & Physiol Wash State Univ Pullmann WA 99164

BAKULE, RONALD DAVID, b Chicago, Ill, Nov 19, 36; m 68; c 2. TECHNOLOGY PROTECTION. *Educ:* Ill Inst Technol, BS, 57; Cornell Univ, PhD(phys chem), 62. *Prof Exp:* Chemist, 62-73, proj leader, 73-80, res sect mgr, 80-87, RES PATENT LIAISON, ROHM & HAAS CO, 87- *Mem:* Am Chem Soc. *Res:* Polymer chemistry. *Mailing Add:* Res Labs Rohm & Haas Co Spring House PA 19477

BAKUN, ANDREW, b Tacoma, Wash, Apr 20, 39; c 2. PHYSICAL OCEANOGRAPHY, FISHERIES. *Educ:* Univ Wash, BS, 62, MS, 68. *Prof Exp:* Staff scientist oceanog, res vessel, R V Anton Bruun, US Prog Biol Int Indian Ocean Exped, 62-64; oceanogr environ Pac salmon, Nat Marine Fisheries Serv, Nat Oceanic & Atmospheric Admin, Seattle, Wash, 68-70; oceanogr fishery-environ processes, 70-78, TASK LEADER FISHERY/ENVIRON MODELLING & FORECASTING, PAC ENVIRON GROUP, SOUTHWEST FISHERIES CTR, NAT MARINE FISHERIES SERV, NAT OCEANIC & ATMOSPHERIC ADMIN, 78- *Res:* Environmental processes affecting fisheries in eastern boundary current regions; upwelling, drift, distribution of properties and organisms; development of practical environmental inputs to fishery management models. *Mailing Add:* Nat Marine Fisheries PO Box 831 Monterey CA 93942

BAKUN, WILLIAM HENRY, b Seattle, Wash, Dec 23, 41; m 70; c 2. EARTHQUAKE PREDICTION. *Educ:* Univ Wash, BS, 64; Univ Calif, Berkeley, MA, 66, PhD(geophys), 70. *Prof Exp:* RES GEOPHYSICIST, US GEOL SURV, 73- *Concurrent Pos:* Chief scientist, Parkfield Earthquake Prediction Proj, 85- *Mem:* Seismol Soc Am; Am Geophys Union. *Res:* Earthquake prediction in California; work in California and the People's Republic of China on earthquake magnitude and seismic scales and source mechanics; earthquake source mechanics. *Mailing Add:* Eight Patricia Pl Menlo Park CA 94025

BAKUS, GERALD JOSEPH, b Thorpe, Wis, Dec 5, 34; m 53; c 2. MARINE ECOLOGY. *Educ:* Calif State Univ, Los Angeles, BA, 55; Univ Mont, MA, 57; Univ Wash, PhD(zool), 62. *Prof Exp:* Asst prof biol, Calif State Univ Northridge, 61-62; asst prof & asst cur, Allan Hancock Collections, Univ Southern Calif, 62-67; chief biologist, Tetra Tech, Inc, Calif, 76-79; ASSOC PROF BIOL & CUR SPONGES, ALLAN HANCOCK FOUND, UNIV SOUTHERN CALIF, 67-, RES FEL, INST MARINE & COASTAL STUDIES, 81- *Concurrent Pos:* NSF & Cocos Found grants, Fanning Island Exped, 63; NSF grant, 64-66; NIH biomed sci grant, 68-70; staff officer, Div Biol & Agr, Nat Acad Sci, 69-70; various environ contracts, 74-; mem, Mirex Comt, Environmental Protection Agency, 71-72; chmn comt marine biol, Orgn Trop Studies, 72-; prin biologist, Tetra Tech Inc, Calif, 74-75, consult, 75-; hon mem, Great Barrier Reef Comt, Univ Queensland, Australia, 76; Off Naval Res grant, 84-87; mem, Indo-US Marine Bioactive Substances Prog, 84- *Mem:* Fel AAAS; Ecol Soc Am. *Res:* Natural toxins as chemical defense mechanisms in marine organisms; bioactive substances; biology of sponges. *Mailing Add:* Dept Biol Sci 0371 Univ Southern Calif University Park Los Angeles CA 90089-0371

BAKUZIS, EGOLFS VOLDEMARS, b Kuldiga, Latvia, May 27, 12; US citizen; m 40; c 6. FOREST ECOLOGY. *Educ:* Univ Latvia, ForestE, 35; Univ Minn, PhD(forestry), 59. *Prof Exp:* Instr forest mgt, Sch Forestry, Acad Agr Jelgava, 40-41; asst forest growth & yield, Eberswalde Forest Acad, Ger, 44-45; lectr forest mensuration, valuation & mgt, Baltic Univ, Ger, 47-50; res fel forest ecol, Sch Forestry, 56-59, res assoc, 59-61, asst prof forest synecol, 61-63, from assoc prof to prof, 63-82, EMER PROF FOREST ECOSYSTS, SCH FORESTRY, UNIV MINN, ST PAUL, 82- *Concurrent Pos:* Assoc mem, Forest Res Inst Latvia, 36-44; res grant, Univ Minn, 60-63; NSF travel grant, Int Union Forest Res Orgn Cong, Vienna, Austria, 61, Munich, Ger, 67 & Gainesville, Fla, 71; Int Bot Cong, Seattle, Wash, 69. *Mem:* Fel AAAS; Ecol Soc Am; Soc Am Foresters; Soil Sci Soc Am; Am Inst Biol Sci; Sigma Xi. *Res:* Forest measurements and silviculture; forest ecology, particularly theoretical aspects of forest ecosystems; growth and yield studies; systems ecology. *Mailing Add:* Univ Minn Forestry 314 Green Hall St Paul MN 55108

BAL, ARYA KUMAR, b Jorhat, India, Aug 24, 34; c 1. CELL BIOLOGY. *Educ:* Univ Calcutta, BSc, 51, MSc, 53, PhD(bot), 60. *Prof Exp:* Lectr bot, Bangabasi Col, India, 54-58; res asst cytol, Carnegie Inst Genetics Res Unit, 58-60; lectr bot, Bangabasi Col, India, 60-62; fel biol, Brown Univ, 62-65; res assoc anat, McGill Univ, 65-66; res assoc biol, Univ Montreal, 66-69; from asst prof to assoc prof, 69-77, PROF BIOL, MEM UNIV NFLD, 77- *Concurrent Pos:* Nat Res Coun Can grants, 68- *Mem:* Am Soc Cell Biol; Can Soc Cell Biol; Electron Micros Soc Am; Micros Soc Can. *Res:* Cell biology of symbiotic nitrogen fixation; cell biology of the potato wart disease. *Mailing Add:* Dept Biol Mem Univ Nfld St John's NF A1B 3X9 Can

BALA, SHUKAL, b New Delhi, India, Nov 27, 51. PARASITE IMMUNOLOGY, NUTRITION IMMUNOLOGY. *Educ:* Univ Delhi, BSc, 73, MSc, 75; All India Inst Med Sci, PhD, 81. *Prof Exp:* Jr res fel immunol filariasis, All India Inst Med Sci, 75-76, sr res fel, 76-81, res officer immunol filariasis & brain tumor, 81-86, sr res officer, molecular biol HSV, HPV cancer cervix, 86-87; res assoc bact-immunol, Ala State Univ, Montgomery, 87-88; res assoc nutrit-immunol, US Dept Agr, 88-90; MICROBIOLOGIST PARASITE-IMMUNOL, FOOD & DRUG ADMIN, 91- *Mem:* Assoc mem Am Inst Nutrit; Sigma Xi; Indian Immunol Soc; Indian Asn Med Microbiologists; Indian Asn Pathologists & Microbiologists. *Res:* Opportunistic parasites in patients with AIDS; cryptosporidium; toxoplasmosis; pneumocystis; giardia; leishmaniasis. *Mailing Add:* Food & Drug Admin HFD-530 5600 Fishers Lane Rockville MD 20857

BALABA, WILLY MUKAMA, b Uganda, May 5, 53. TRIBOMECHANICAL PROPERTIES OF POLYMERS, POLYMERS IN ELECTRONIC PACKAGING. *Educ:* Makerere Univ, Kampala, BS, 76; Univ Wash, MS, 80; Wash State Univ, PhD(eng sci), 84. *Prof Exp:* Spec asst, Chem Dept, Makerere Univ, 76-78; res asst, Forestry Dept, Univ Wash, 78-80, mat sci & eng, Wash State Univ, 80-84; scientist, Surface & Polymer Div, 85-86, sr scientist, Packaging Div, 86-89, STAFF SCIENTIST, SURFACE TECHNOL DIV, ALCOA TECH CTR, 89- *Res:* Application of polymers in electronic packaging and in tribological settings. *Mailing Add:* Alcoa Tech Ctr Alcoa Center PA 15069

BALABAN, MARTIN, b New York, NY, Oct 5, 30; m 50; c 3. ZOOLOGY. *Educ:* Univ Chicago, BA, 50, PhD(biopsychol), 59. *Prof Exp:* USPHS fel neurochem, Univ Ill, 59-61; USPHS fel neuroembryol, Univ Wash, St Louis, 61-62, res assoc, 61-64; from asst prof to assoc prof, 64-71, PROF ZOOL, MICH STATE UNIV, 71- *Mem:* AAAS; Am Soc Zoologists. *Res:* Behavioral biology; neuroembryology; ontogeny of behavior; ontogeny of central nervous system development. *Mailing Add:* Dept Zool Mich State Univ 203 Natural Sci Bldg East Lansing MI 48824

BALABAN, ROBERT S, b Los Angeles, Calif, May 9, 53; m. HEART & KIDNEY PHYSIOLOGY, BIOENERGETICS. *Educ:* Duke Univ, PhD(physiol & pharmacol), 79. *Prof Exp:* Staff fel, 81-85, res physiologist, 85-88, CHIEF, LAB CARDIAC ENERGETICS, NAT HEART, LUNG & BLOOD INST, NIH, 88- *Concurrent Pos:* NATO postdoctoral fel, Oxford, Eng. *Mem:* Soc Gen Physiologists; Am Physiol Soc; Biophys Soc; Soc Magnetic Resonance Med. *Mailing Add:* Nat Heart Lung & Blood Inst NIH Bldg 1 Rm B3-07 Bethesda MD 20892

BALABANIAN, NORMAN, b New London, Conn, Aug 13, 22; m; c 4. ELECTRICAL ENGINEERING. *Educ:* Aleppo Col, Syria, Sci Dipl, 42; Syracuse Univ, BEE, 49, MEE, 51, PhD(elec network synthesis), 54. *Prof Exp:* Instr English, Aleppo Col, Syria, 42-43; from instr to assoc prof elec eng, 49-61, chmn elec & computer eng, 83-90, PROF ELEC ENG, SYRACUSE UNIV, 61- *Concurrent Pos:* Mem tech staff, Bell Labs, 56; consult ed, Allyn & Bacon Elec Eng Serv, 59-74; dir, Syracuse Electronics Corp, 63-65; mem develop lab, Int Bus Mach Corp, 63; ed, Trans Circuit Theory, 63-65, Trans on Commun, 76 & Technol & Soc Mag, 76 & 79-86; vis prof, Univ Colo, 65 & Univ Calif, Berkeley, 65-66; consult, UNESCO, 66; mem vis comt, Engrs Coun for Prof Develop, 67-70; specialist in tech teacher training, UNESCO, Nat Polytech Inst, Mex, 69-70; sr Fulbright fel, Univ Zagreb, Yugoslavia, 74-75; dir eng & pub affairs, Syracuse Univ, 75-77; consult, UNESCO, Paris, 76; Nat Inst Elec & Electronics Engrs, Algeria, 77-78; vis scholar, Prog Sci Technol & Soc, Mass Inst Technol, 90-91. *Honors & Awards:* Centennial Award, Inst Elec & Electronics Engrs, 84. *Mem:* Fel AAAS; fel Inst Elec & Electronics Engrs; Am Soc Eng Educ; Inst Elec & Electronics Engrs Soc Implications Technol (pres, 89-91). *Res:* Electric network theory; network synthesis; linear network analysis; programmed learning; engineering and public affairs; telecommunications policy; technology and society; active network design. *Mailing Add:* Dept Elec & Comput Eng Syracuse Univ Syracuse NY 13244-1240

BALACHANDRAN, KASHI RAMAMURTHI, b Madras, India, June 16, 41; m 69; c 2. OPERATIONS RESEARCH, APPLIED PROBABILITY. *Educ:* Univ Madras, India, BE, Hons, 62; Univ Calif, MS, 64; Univ Calif, Berkeley, PhD(opers res), 68; Inst Mgt Acct, CMA, 76. *Prof Exp:* Field engr, Neyveli Lignite Corp, India, 62; designer mech eng, Qual Casting Systs, 63; mem opers res staff planning, Gen Mills, Inc, 63; opers res analyst planning & designing, Nat Cash Regist Co, 67-68; asst prof mgt sci, Univ Wis-Milwaukee, 68-72; assoc prof mgt sci, Ga Inst Technol, 72-76; assoc prof bus, Univ Ky, 76-79; PROF BUS, NY UNIV, 79-; MGT CONSULT, COST MGT SYSTS. *Concurrent Pos:* Res fel, Univ Wis, 69; mem, Govr's Comn Educ, Wis, 69-70; assoc ed, J Oper Res Soc India, 75-84; PhD prog adv, NY Univ, 84,; assoc ed, J Acct, Auditing & Finance. *Mem:* Inst Mgt Sci; Opers Res Soc Am; Oper Res Soc India; Am Acct Asn; Inst Mgt Acct. *Res:* Individual and collective optima in queues; operations research applications; cost allocations and efficient usage of scarce resources; transfer pricing under uncertainty; cost management systems for new manufacturing environemnt. *Mailing Add:* Stern Sch Bus 432 Tisch Hall NY Univ New York NY 10003

BALAGNA, JOHN PAUL, b Florence, Colo, Aug 19, 20; m 49; c 2. ANALYTICAL CHEMISTRY, RADIOCHEMISTRY. *Educ:* Holy Cross Col, Colo, AB, 41. *Prof Exp:* Chemist, Chem Warfare Serv, US Army, 43-44; staff mem anal chem, Manhattan Dist, Corps Engrs, 44-47; staff mem radiochem, Los Alamos Nat Lab, 47-74, assoc group leader, 74-86; RETIRED. *Concurrent Pos:* Consult, 66-69. *Mem:* AAAS; Inst Elec & Electronics Engrs; Am Chem Soc. *Res:* Design of ultra low level gas filled proportional counters; laboratory scale electromagnetic isotope separation. *Mailing Add:* 223 Rio Bravo Dr Los Alamos NM 87544

BALAGOT, REUBEN CASTILLO, b Philippines, July 28, 20; nat US; m 46; c 4. ANESTHESIOLOGY. *Educ:* Univ Philippines, BS, 41, MD, 44; Am Bd Anesthesiol, dipl, 55. *Prof Exp:* Resident, Res & Educ Hosps, Ill, 49-50; clin instr, 52-54, from asst prof to prof anesthesiol, Univ Ill Med Ctr, 54-75, chmn dept, 69-75; PROF ANESTHESIOL & CHMN DEPT, CHICAGO MED SCH, 75-, CHMN DEPT ANESTHESIOL, DOWNEY VET ADMIN HOSP, 75- *Concurrent Pos:* Res fel anesthesiol, Res & Educ Hosps, Ill, 51; asst head div anesthesiol, Res & Educ Hosps, 54-; consult, Coun Pharmacol, AMA, 55 & 66-, Coun Drugs, 63 & 67-; chmn dept anesthesiol, Grant Hosp, 57-, Ill Masonic Hosp, 66-67, Presby-St Luke's Hosp, 67- & Vet Admin Hosp, Hines, 71-; chmn, Dept Anesthesiology, Cook County Hosp, 81- *Honors & Awards:* Outstanding Filipino Res Award, Philippine Govt. *Mem:* AAAS; AMA; Am Soc Anesthesiol; Am Fedn Clin Res; NY Acad Sci. *Res:* Physiology and pharmacology in anesthesiology. *Mailing Add:* 859 Evergreen Glen Ellyn IL 60137

BALAGURA, SAUL, b Cali, Colombia, Jan 11, 43; US citizen; m 64. NEUROSURGERY, PHYSIOLOGICAL PSYCHOLOGY. *Educ:* Univ Valle, Colombia, MD, 64; Princeton Univ, MA, 66, PhD(psychol), 67, Am Bd Neurol Surgeon, 83. *Prof Exp:* Asst prof psychol, Univ Chicago, 67-71; assoc prof, Univ Mass, Amherst, 71-73; resident neurosurg, Downstate Med Ctr, Brooklyn, 73-76; resident neurosurg, Albert Einstein Sch Med, Yeshiva Univ, 76-80; assoc prof clin, Univ Ill, Champaign, 83-88; PVT NEUROSURG, VICTORIA, TEX, 88- *Concurrent Pos:* NIMH res grant, 67-71. *Mem:* Fel Am Phycol Asn; Am Physiol Soc; Soc Neurosci; Cong Neurol Surg; Am Asn Neurol Surg; fel Am Psychol Asn. *Res:* Neurophysiological and endocrine regulation of feeding behavior; recovery of function after brain injury; feedback systems involved in the control of food intake. *Mailing Add:* 115 Medical Dr Victoria TX 77904

BALAGURU, PERUMALSAMY N, b Tamil Nadu, India, Mar 26, 47; m 74; c 2. STRUCTURAL ANALYSIS & DESIGN, REINFORCED & PRESTRESSED CONCRETE. *Educ:* Univ Madras, BE Hons, 68; Indian Inst Sci, ME, 70; Univ Ill, Chicago, PhD(struct eng), 77. *Prof Exp:* Assoc lectr civil eng, Univ Madras, 70-73; from asst prof to assoc prof, 77-87, PROF CIVIL ENG, RUTGERS, STATE UNIV NJ, 88- *Concurrent Pos:* Assoc grad dir, Rutgers, State Univ NJ, 80-86; consult, DuPont de Nemours & Co, 87-89 & Allied Signal, Inc, 87-; vis prof, Northwestern Univ, 90. *Mem:* Am Concrete Inst; Am Soc Civil Engrs. *Res:* Reinforced and prestressed concrete; new construction materials; construction management; computer aided analysis and design; fiber reinforced cement composites and ferrocements; author of 105 publications. *Mailing Add:* Rutgers State Univ NJ Box 909 Piscataway NJ 08855-0909

BALAJEE, SHANKVERM R, b Chikkaballupur, India, Feb 3, 39; m 69; c 2. METALLURGICAL ENGINEERING. *Educ:* Univ Nagpur, India, BSC, 58, MSc, 60; Univ Minn, MS, 68; Univ Wis, PhD(metall eng), 71. *Prof Exp:* Sr tech asst mineral technol, Indian Bur Mines, Nagpur, India, 61-62, sr ore dressing off, 62-65; res engr, 71-72, SR RES ENGR METALL ENG, RES DEPT, INLAND STEEL CO, 75- *Concurrent Pos:* Guest lectr, Purdue Univ, 81. *Mem:* Am Inst Mining, Metall & Petroleum Engrs; Soc Mining Engrs; Am Soc Metals. *Res:* Process metallurgy dealing with benefication and agglomereter areas dealing with steel plant. *Mailing Add:* Inland Steel Res Labs 3001 E Columbus Dr E Chicago IN 46312

BALAKRISHNAN, A V, b Palghat, India, Dec 4, 22; US citizen; m 52; c 4. ENGINEERING. *Educ:* Univ Madras, BSc, 43, MA, 47; Univ Southern Calif, AM & MS, 50, PhD(math), 54. *Prof Exp:* Sr engr, Radio Corp Am, NJ, 54-56; asst prof math, Univ Southern Calif, 56-57; asst prof math, Univ Calif, Los Angeles, 57-59; sr staff engr systs & res, Space Technol Labs, Inc, 59-61; assoc prof eng, 61-62, prof, 62-65, prof math & eng, 65-76, chmn dept syst sci, 69-76, PROF MATH & PROF ENG & APPL SCI, UNIV CALIF, LOS ANGELES, 76- HEAD INFO SYSTS DIV, DEPT ENG, 64- *Concurrent Pos:* Consult, NASA, 64; consult ed, Syst Sci Series, McGraw-Hill Book Co, 64; chief ed, J Comput & Syst Sci; mem, US Comn 6, Union Radio Sci Int, 63; chmn, Int Symposium Info Theory, 66. *Mem:* Am Math Soc; Inst Math Statist; fel Inst Elec & Electronics Engrs. *Res:* Communication theory; control theory; systems optimization theory. *Mailing Add:* Dept Syst Sci Univ Calif Los Angeles CA 90024

BALAKRISHNAN, NARAYANA SWAMY, b Madras, India. NUCLEAR MAGNETIC RESONANCE SPECTROSCOPY, CHEMICAL KINETICS. *Educ:* Univ Madras, BS, 65; Indian Inst Technol, MS, 67; Univ Hawaii, PhD(chem), 78. *Prof Exp:* Instr, Univ Hawaii, 74-81; res scientist & tech sales dir, Indust Anal Lab, Inc, 82-; AT DEPT CHEM, UNIV GUAM. *Concurrent Pos:* Vis prof, Hawaii Loa Col, 80-81. *Mem:* Am Chem Soc. *Res:* Rotational dynamics and geometries in the liquid state; nuclear magnetic resonance in liquid crystal solvents to probe the details of the liquid state. *Mailing Add:* Dept Chem Univ Guam PO Box 5034 Mangilao GU 96923

BALAKRISHNAN, NARAYANASWANY, b Tamil Nadu, India, May 2, 56; Can citizen; m 85. ORDER STATISTICS, STATISTICAL INFERENCE. *Educ:* Univ Madras, India, BSc, 76, MSc, 78; Indian Inst Technol, Kanpur, India, PhD(statist), 82. *Prof Exp:* Lectr statist, Indian Inst Technol, Delhi, 81-82; postdoctoral fel statist, Univ Man, Winnipeg, Can, 83 & Univ Guelph, Can, 84-85; postdoctoral fel statist, McMaster Univ, Hamilton, Can, 82-83 & 84, univ res fel, 85-86, asst prof, 86-89, ASSOC PROF STATIST, MCMASTER UNIV, HAMILTON, CAN, 89- *Concurrent Pos:* Assoc ed, Selected Tables in Math Statist, 90- & Commun Statist, 90-; consult & proprietor, Statpro Consult, 91- *Mem:* Int Statist Inst; Inst Math Statist; Am Statist Asn; Statist Soc Can; Am Soc Qual Control. *Res:* Order statistics; inference for life-testing and reliability models; robust inference; classification theory; statistical quality control; characterizations of statistical distributions; logistic model and inference; regression problems; censoring and truncation. *Mailing Add:* Dept Math & Statist McMaster Univ Hamilton ON L8S 4K1 Can

BALAKRISHNAN, V K, b India; US citizen. COMBINATORIAL & DISCRETE MATHEMATICS. *Educ:* Univ Wis, MS, 65; State Univ NY, Stony Brook, PhD(math), 70. *Prof Exp:* From asst prof to assoc prof math, 70-81, PROF MATH, UNIV MAINE, 81- *Concurrent Pos:* Vis fel, Yale Univ, 77-78; assoc dean, Univ Maine, 80-81. *Mem:* Am Math Soc; Soc Indust & Appl Math; Oper Res Soc Am; Asn Comput Mach. *Res:* Graph Theory network flows combinatorial optimization and linear programming; writer of books on discrete mathematics and discrete optimization. *Mailing Add:* Math Dept Univ Maine Orono Neville Hall Orono ME 04469

BALAM, BAXISH SINGH, b Kukar Pind, India, Aug 16, 30; US citizen; m 65; c 3. ANALYTICAL CHEMISTRY, ENVIRONMENTAL CHEMISTRY. *Educ:* Punjab Univ, India, BSc, 51, MSc, 53; Ohio State Univ, PhD(soil chem), 65. *Prof Exp:* Agr asst soil chem, Dept Agr, Punjab, 53-55; asst prof chem, Col Agr, Univ Nagpur, 55-56 & Univ Jabalpur, 56-58; from asst prof to assoc prof, Punjab Univ, India, 58-66; res assoc agron, Ohio State Univ, 66; adv chem, Int Atomic Energy Agency, UN, 66-68; PROF CHEM, MISS VALLEY STATE UNIV, 69- *Concurrent Pos:* Consult, Govt Iran, Ministry Agr & Rural Develop, 77-78; Fac Res Intern, Environ Protection Agency, 87-89. *Mem:* Am Soc Agron; Soil Sci Soc Am; Am Chem Soc. *Res:* Availability, plant uptake and interactions of soil and fertilizer phosphates; fertilizer technology; assessment of fertilizer requirements; radioisotopes in agricultural research; partial acidulation of rock phosphate; soil salinity and alkalinity; radioactive waste management. *Mailing Add:* 123 Bluebird Dr Greenwood MS 38930

BALAMUTH, DAVID P, b New York, NY, Nov 4, 42; m 69. NUCLEAR PHYSICS. *Educ:* Harvard Col, AB, 64; Columbia Univ, PhD(physics), 68. *Prof Exp:* Res assoc physics, Univ Pa, 68-70, asst prof, 70-74; prog officer for nuclear physics, NSF, 74-75; assoc prof, 75-80, PROF PHYSICS, UNIV PA, 80- *Mem:* Fel Am Phys Soc; AAAS; Sigma Xi. *Res:* Nuclear structure and nuclear reaction mechanisms in light nuclei; study of formation and decay of continum states using coincidence techniques; use of computers in experimental nuclear physics data acquisition and analysis; spectroscopy of nuclei far from stability. *Mailing Add:* Dept Physics E1 Univ Pa Philadelphia PA 19104-6396

BALANDRIN, MANUEL F, b Havana, Cuba, Oct 5, 52, US citizen; m; c 1. ECOLOGICAL BIOCHEMISTRY, CHEMOTAXONOMY. *Educ:* Univ Ill, BS, 77; PhD(pharmacog), 82. *Prof Exp:* Scientist natural prod chem, 82-83, coordr phytochem, 83-88, SR RES SCIENTIST NAT PROD CHEM, NATIVE PLANTS, INC, 88. *Concurrent Pos:* Consult, Standard Oil Ohio, 82-85; prin investr, Dept Energy, 83; US Dept Agr, 83-85, Nat Sci Found, 84 & 86. *Mem:* Am Chem Soc; Am Pharmaceut Asn; Am Soc Pharmacog; Int Soc Chem Ecol; Acad Pharmaceut Res & Sci. *Res:* Isolation and structure elucidation of biologically active natural plant products used as chemotherapeutic and insect control agents; quinolizidine alkaloids, steroidal and triterpenoid saponins, terpenoids, tannins. *Mailing Add:* Natural Prod Sci Inc Univ Res Park 420 Chipeta Way Salt Lake City UT 84108

BALANIS, CONSTANTINE A, b Trikala, Thessalia, Greece, Oct 29, 38; m; c 2. ELECTRICAL ENGINEERING. *Educ:* Va Polytech Inst, BSEE, 64; Univ Va, MEE, 66; Ohio State Univ, PhD(elec eng), 69. *Prof Exp:* Engr, Langley Res Ctr, NASA, Hampton, Va, 64-70; from vis assoc prof to assoc prof elec eng, WVa Univ, 70-76, prof, 76-83; DEPT ELEC ENG, COL ENG & APPL SCI, ARIZ STATE UNIV, 83-, DIR, TELECOMMUNICATIONS RES CTR, 88- *Concurrent Pos:* Asst prof lectr, George Wash Univ, 68-70; consult, Nat Radio Astron Observ, 72-74 & Naval Air Test Ctr 76-, Nat Inst Occup Safety & Health, 80-81, Southwest Res Inst, 81-84, Motorola Inc, 84-, Loral Syst Group, 86-88 & Gen Dynamics, 86; grants, NASA, Dept Trans, Fed Aviation Admin, NSF, Naval Air Test Ctr, Army Res Off, Energy Res Develop Admin, Dept Energy, Motorola Inc, Boeing Helicopters, IBM, McDonnell Douglas, Silorsky Aircraft, Rockwell Int, Gen Dynamics, US Bur Mines; assoc ed, Inst Elec & Electronics Engrs Trans Geoscience & Remote Sensing, Inst Elec & Electronics Engrs Trans Antennas Propagation. *Honors & Awards:* Individual Achievement Award, Region 6, Inst Elec & Electronics Engrs. *Mem:* Fel Inst Elec & Electronics Engrs; Am Soc Eng Educ; Sigma Xi. *Res:* high and low frequency asymptotic electromagnetic techniques, electromagnetic wave radiation and scattering; electromagnetic wave propagation in microwave integrated circuit lines; reconstructive imaging, tomography. *Mailing Add:* Dept Elec Eng Ariz State Univ Tempe AZ 85287-7206

BALAS, EGON, b Cluj, Romania; US citizen; m; c 2. MATHEMATICAL PROGRAMMING. *Educ:* Bolyai Univ, Romania, Diploma Licentiae econ; Univ Brussels, DScEc; Univ Paris, DrSci(math). *Prof Exp:* Lectr, Inst Econ Sci & Planning, Bucharest, 49-58; res assoc, Inst Econ Res Romanian Acad, 56-58; proj engr, Designing Inst Forestry & Timer Indust, 59-61, head, Math Prog Group, 62-64; head, Math Prog Sect, Ctr Math Statist Romanian Acad, 64-65; res fel, Int Computation Ctr, Rome, 66; Ford distinguished res prof, 67-68, PROF INDUST ADMIN & APPL MATH, CARNEGIE-MELLON UNIV, 68-, GSIA ALUMNI CHAIR PROF, 80-, UNIV PROF, 90- *Concurrent Pos:* Vis prof opers res, Univ Toronto & Stanford Univ, 67 & vis prof systs eng, Fed Energy Admin, 76; assoc ed, Opers Res, 67-, Annals Opers Res, 83- & Naval Res Logistics Quart, 79-89, ed assoc, Europ J Oper Res, 76-; var res grants, NSF, Off Naval Res, NATO & Air Force Off Sci Res, 67-; mem, Comt Recommendations US Army Basic Res, Nat Res Coun, 77-80; consult, var indust co & govt agencies. *Mem:* Opers Res Soc Am; Soc Indust & Appl Math; Math Prog Soc; Am Math Soc; Inst Mgt Sci. *Res:* Integer programming; combinatorial optimization; network models; polyhedral theory; projection methods; scheduling and sequencing; distribution and location theory; author or co-author of over 140 publications; developer of scheduling system for steel rolling. *Mailing Add:* 104 Maple Heights Rd Pittsburgh PA 15232

BALASINSKI, ARTUR, b Warsaw, Poland, June 30, 57; m 83; c 1. PROCESS-RELATED DEFECTS IN SILICON DIOXIDE. *Educ:* Warsaw Univ Technol, MS, 80, PhD(elec eng), 86. *Prof Exp:* Res scientist elec eng, Warsaw Univ Technol, 87-88, asst prof, 88-89; POSTDOCTORAL ASSOC ELEC ENG, YALE UNIV, 89- *Mem:* Electrochem Soc. *Res:* Investigation of radiation and hot-electron effects in metal-oxide-semiconductor capacitors and transistors; properties of dielectric layers (silicon dioxide, diamond like carbon, etc); charge trapping; studies of work function difference; experimental work involving device fabrication and measurements. *Mailing Add:* 256 Park St 7 New Haven CT 06511-4711

BALASKO, JOHN ALLAN, b Morgantown, WVa, July 16, 41; m 63; c 3. AGRONOMY, CROP PHYSIOLOGY. *Educ:* WVa Univ, BS, 63, MS, 67; Univ Wis, Madison, PhD(agron), 71. *Prof Exp:* Asst prof agron, 70-74, assoc prof, 74-78, PROF AGRON, WVA UNIV, 78- *Mem:* Crop Sci Soc Am; Am Soc Agron; Am Forage & Grassland Coun. *Res:* Carbohydrate metabolism in perennial forage plants; forage management and forage quality. *Mailing Add:* Div Plant & Soil Sci WVa Univ Morgantown WV 26506-6108

BALASSA, LESLIE LADISLAUS, b Bacsfoeldvar, Yugoslavia, Sept 6, 03; US citizen; m 63; c 1. PHYSIOLOGICALLY ACTIVE ANIMAL TISSUE EXTRACTS. *Educ:* Univ Vienna, PhD(org chem & chem eng), 26. *Prof Exp:* Res chemist, E I DuPont de Nemours, 28-35, res mgr, 36-44; res dir, US Finishing Co, Providence, RI, 44-48; exec vpres, Aula Chem Inc, 49-57; mgr, textile color develop labs, Geigy Chem, 57-59; pres, 60-80, VCHMN, LESCARDEN INC, 80- *Concurrent Pos:* Pres, Barrington Industs, Inc, 50-68, Balchem Corp, 67-72 & Barinco Corp, 68-85; consult, Arde Barinco Assocs, 71-85; pres, Balassa Res Assoc, 80-; pres & C E O, HFM Inc, 87- *Mem:* Am Chem Soc; NY Acad Sci; AAAS. *Res:* Animal tissue extract (Catrix) effective in the treatment of arthritis, psoriasis and certain types of human cancer; medicated skin creams and lotions; injectable solutions, capsules and tablets for oral administration; firefighting and fire suppression with cellulose fiber dispersions and hydrated fibrous mats, agrochemicals and processes. *Mailing Add:* Shore Dr Blooming Grove NY 10914

BALASUBRAHMANYAN, VRIDDHACHALAM K, b Tiruchendoor, India, Nov 11, 26; m 54; c 2. PHYSICS, ASTROPHYSICS. *Educ:* Calcutta Univ, MSc, 48; Bombay Univ, PhD(cosmic rays), 61. *Prof Exp:* Fel cosmic rays, Tata Inst Fundamental Res, 50-62; resident res assoc, Nat Res Coun-Nat Acad Sci, 62-65; astrophysicist cosmic rays, Goddard Space Flight Ctr, NASA, 65-; ASTROPHYSICIST, RADIOL DEPT, JOHNS HOPKINS UNIV. *Mem:* Am Phys Soc; Am Geophys Union. *Res:* Cosmic rays, solar physics, interstellar medium, nucleosynthesis. *Mailing Add:* Radiol Dept Johns Hopkins Univ 720 Rutland Ave Baltimore MD 21205

BALASUBRAMANIAN, KRISHNAN, b Bangalore, India, Apr 10, 56; m 82; c 1. THEORETICAL CHEMISTRY, RELATIVISTIC QUANTUM CHEMISTRY. *Educ:* Birla Inst Technol & Sci, MSc, 77; Johns Hopkins Univ, MA, & PhD, 80. *Prof Exp:* Res assoc chem phys, Lawrence Berkeley Lab, Univ Calif, Berkeley, 80-82, vis lectr chem, 83; asst prof 83-86, ASSOC PROF CHEM, ARIZ STATE UNIV, 87- *Concurrent Pos:* Alfred P Sloan fel, Sloan Foundation, 84-; panelist, Air Force Off Sci Res, Chem Sci Panel, Nat ACad Sci, 88- *Mem:* NY Acad Sci; Am Chem Soc. *Res:* Theoretical and computer chemistry; relativistic quantum chemistry; chemical applications of group theory and graph theory; artificial intelligence. *Mailing Add:* Dept Chem Ariz State Univ Tempe AZ 85287-1604

BALAZOVICH, KENNETH J, b Detroit, Mich, Dec 21, 54; m. PEDIATRICS. *Educ:* Wayne State Univ, BS, 79; Univ Miami, PhD(human anat & cell biol), 84. *Prof Exp:* Res technician, Wayne State Univ, 80-81; teaching asst, Sch Med, Univ Miami, 83; res fel, 84-87, ASST RES SCIENTIST, DEPT PEDIAT, MED SCH, UNIV MICH, 87- *Concurrent Pos:* First award, NIH, 88- & 91- *Mem:* Am Soc Cell Biol; Electron Micros Soc Am; NY Acad Sci; Sigma Xi; Am Soc Hemat; Am Fedn Clin Res. *Mailing Add:* Dept Pediat Univ Mich M7510 Med Sci Res Bldg 1 Ann Arbor MI 48109

BALAZS, ENDRE ALEXANDER, b Budapest, Hungary, Jan 10, 20; nat US; m 45; c 2. BIOCHEMISTRY. *Educ:* Univ Budapest, MD, 43. *Hon Degrees:* MD, Royal Univ Uppsala, 67. *Prof Exp:* Res asst histol, biol & embryol, Univ Budapest, 40-43, asst prof, 43-44, asst prof pharmacol, 45; dir, Lab Biol Anthrop, Mus Natural Hist, Budapest, 45-47; assoc exp histol, Karolinska Inst, Sweden, 48-51; assoc dir, Retina Found-Int Biol & Med Sci, 51-62, pres, 62-63 & 65-68, vpres, 63-65, dir dept connective tissue res, 62-69; res dir, Dept Connective Tissue Res, Boston Biomed Res Inst, 69-76, exec dir, 74-75; MALCOLM P ALDRICH PROF OPHTHAL, COL PHYSICIANS & SURGEONS, COLUMBIA UNIV, 76-, DIR RES, 76-; PRES, BIOMATRIX INC, 81- *Concurrent Pos:* Vis scientist, Swed Inst, 47; instr, Harvard Med Sch, 51-56, lectr, 56-, mem bd freshman adv, 57-61; Mass Eye & Ear Infirmary fel ophthal, 51-56, biochemist, 56-6S; co-ed-in-chief, Exp Eye Res, 62-; Guggenheim fel, 68-69; vis scientist, Dept Tumor Biol, Karolinska Inst, Sweden, 68-69; vis prof chem, Univ Salford, Eng, 69-76; mem, Coun Biol Educ. *Honors & Awards:* Friedenwald Award, Asn Res Vision & Ophthal, 63. *Mem:* Reticuloendothelial Soc; fel Geront Soc; Asn Res Vision & Ophthal; fel NY Acad Sci; Int Soc Eye Res (pres 74-78). *Res:* Biochemistry and physiology of connective tissues, especially of the eye and joint; molecular biology of the intercellular matrix; physical biochemistry of glycosaminoglycans. *Mailing Add:* Matrix Biol Inst 65 Railroad Ave Ridgefield NJ 07657

BALAZS, LOUIS A P, b Tsing-Tao, China, July 31, 37; US citizen. THEORETICAL HIGH ENERGY PHYSICS. *Educ:* Univ Calif, Berkeley, BA, 59, PhD(physics), 62. *Prof Exp:* Mem, Sch Math, Inst Advan Study, 62-63; res fel, Calif Inst Technol, 63-64; vis prof, Tata Inst Fundamental Res, India, 64-65; asst prof, Univ Calif, Los Angeles, 65-70; assoc prof, 70-78, PROF PHYSICS, PURDUE UNIV, 78- *Concurrent Pos:* Alfred P Sloan Found fel, 67-69; visitor, Imp Col, London, 68-69; Fermi Nat Accel Lab, 76 & Univ Pierre et Marie Curie, Paris, 79 & 81. *Mem:* Fel Am Phys Soc. *Res:* Light and heavy hadron mass-spectrum calculations in strong-interaction physics in which open hadronic channels are taken into account non-perturbatively from the beginning, along with the usual closed valence-quark channels; studies of baryonium and other multiquark states. *Mailing Add:* Dept Physics Purdue Univ Sch Sci West Lafayette IN 47907

BALAZS, NANDOR LASZLO, b Budapest, Hungary, July 7, 26; nat US; m; c 1. THEORETICAL PHYSICS. *Educ:* Univ Budapest, MSc, 48; Univ Amsterdam, PhD(physics), 51. *Prof Exp:* Fel, Dublin Inst Advan Study, 51-52; Nat Res Coun Can fel, 52; mem, Inst Advan Studies, Princeton Univ, 53; vis assoc prof physics, Univ Ala, 53, assoc prof, 53-54; sr theoret physicist, Am Optical Co, 55-56; res assoc, Enrico Fermi Inst Nuclear Studies, Chicago, 56-59; mem res staff, Plasma Physics Lab, Princeton Univ, 59-61; PROF PHYSICS, STATE UNIV NY STONY BROOK, 61- *Concurrent Pos:* Consult, Gen Atomics, La Jolla, Calif, 59, Am Optical Co, Mass, 60 & Dept Chem, Univ Calif, San Diego, 62; NSF grants, 62-64 & 64-; consult comt col physics, 63 & 65, Gen Tel & Electronics, Bayside, NY, 63-65 & Laser Inc, 66-67; vis prof, Univ Wash, 65, Dublin Inst Advan Study, 67, 70 & 71, Univ Cambridge, 68, Oxford Univ, 72, Max-Planck-Inst Nuclear Physics, Heidelberg, 76 & 78, Inst Theor Phys, Tech Univ Munich, 77-79 & 81, & 89 Centre d'Etudes Nucleaire, Saclay, 80 & 84, Inst Nuclear Physics, Kernforschungsanlage, Julich, 81 & Inst Phys Nucleire, Orsay, 84; State of NY distinguished fac res fel, 68; Exchange visit, Nat Acad Sci, Budapest, 80, NSF, 81, 82 & 84; vis prof Theor Physics Serv, CEN-Saclay & Inst Nuclear Physics, Orsay, 84-88 & 86-90. *Honors & Awards:* Schrödinger Mem Lectr, Zürich & Dublin, 76; J L Synge Jubilee Lectr, Dublin, 82. *Mem:* Am Phys Soc. *Res:* Statistical mechanics; relativity theory; quantum mechanics. *Mailing Add:* Dept Physics State Univ NY Stony Brook NY 11790

BALAZS, TIBOR, b Sarbogard, Hungary, Mar 1, 22; US citizen; m 49; c 1. TOXICOLOGY. *Educ:* Univ Agr Sci, Hungary, Vet Surg, 45, DMV, 49; Pazmany Peter Univ, MPharm, 48. *Prof Exp:* Asst prof pharmacol, Univ Agr Sci, Hungary, 46-50; sr biologist, Nat Inst Pub Health, Hungary, 50-56; res toxicologist, Ayerst Lab Div, Am Home Prod Co, Quebec, 57-59; res toxicologist, Food & Drug Directorate, Can Dept Nat Health & Welfare, 59-63; res pharmacologist, Lederle Labs Div, Am Cyanamid Co, 63-67, group leader toxicol, 67-69; asst dir toxicol, Smith Kline & French Labs, Pa, 69-71; chief Drug Toxicol Br, Div Drug Biol Ctr Drugs & Biologics, Food & Drug Admin, 71-88; RETIRED. *Concurrent Pos:* Adj prof, Dept Pharmacol, Sch Med, Howard Univ, Washington, DC. *Honors & Awards:* Arnold J Lehman Award, Soc Toxicol. *Mem:* Vet Med Asn; Soc Toxicol; Am Soc Pharmacol & Exp Therapeut; Soc Toxicol Path; fel Acad Vet & Comparative Toxicol; Acad Toxicol Sci. *Res:* Drug toxicology; pharmacology; experimental pathology. *Mailing Add:* Drug Toxicol Br DDB PRT Food & Drug Admin HFN 414 Washington DC 20204

BALBES, RAYMOND, b Los Angeles, Calif, Dec 14, 40; m 62; c 2. MATHEMATICS. *Educ:* Univ Calif, Los Angeles, BS, 62, MA, 64, PhD(math), 66. *Prof Exp:* Asst math, Univ Calif, Los Angeles, 62-66; from asst prof to assoc prof, 66-78, PROF MATH, CHMN DEPT & COORDR SEC MATH, UNIV MO-ST LOUIS, 78- *Concurrent Pos:* Consult microcomput graphics. *Mem:* Math Asn Am; Am Math Soc. *Res:* Lattice theory, particularly distributive lattices and Boolean algebras; projective and injective distributive lattices; order sum of distributive lattices; representation theory for prime and implicative semi-lattices. *Mailing Add:* Dept Math & Comput Sci Univ Mo St Louis 8001 Natural Bridge St Louis MO 63121

BALBINDER, ELIAS, b Warsaw, Poland, Jan 22, 26; US citizen; m 55; c 2. GENETICS. *Educ:* Univ Mich, BS, 49; Ind Univ, PhD(cytogenetics), 57. *Prof Exp:* Res assoc & instr zool, Univ Pa, 56-57; res assoc genetics, Carnegie Inst, 57-60; Am Cancer Soc fel biol, Univ Calif, La Jolla, 60-63; vis investr Fulbright travel grant, Univ Buenos Aires, 63; from asst prof to prof genetics, Syracuse Univ, 63-76; dir genetics & carcinogenesis res, AMC, Cancer Res Ctr, 76-82; AT DEPT BIOCHEM, BIOPHYS & GENETICS, UNIV COLO HEATLH SCI CTR, 82- *Concurrent Pos:* Adj prof microbiol, Univ Colo Med Sch, 76-; mem, NSF genetics panel, 77-79; vis prof Fulbright fel, Univ Los Andes, Bogota, Colombia, 81. *Mem:* Fel AAAS; Am Soc Microbiol; Genetics Soc Am; Environ Mutagen Soc. *Res:* Evolution of tryptophan synthetase; mutation and recombination in bacteria; fine structure genetics; regulation of gene expression; carcinogenesis. *Mailing Add:* Dept Biochem Biophys & Genetics Univ Colo Health Sci Ctr 4200 E Ninth Ave B-121 Denver CO 80262

BALBONI, EDWARD RAYMOND, b Springfield, Mass, Dec 7, 30; m 54. INSECT PHYSIOLOGY. *Educ:* Boston Univ, AB, 57, AM, 58; Univ Mass, PhD(entom), 62. *Prof Exp:* From instr to asst prof, 62-69, ASSOC PROF BIOL SCI, HUNTER COL, 69- *Concurrent Pos:* NSF grant, 65-70. *Mem:* AAAS; Entom Soc Am; Am Soc Zool; NY Acad Sci; Sigma Xi. *Res:* Insect flight muscle metabolism. *Mailing Add:* Bio Sci Dept Hunter Col 695 Park Ave New York NY 10021-5024

BALCER-BROWNSTEIN, JOSEFINE P, b Augsburg, Germany, Mar 19, 48; m; c 1. ENVIRONMENTAL PHYSIOLOGY, PUBLIC HEALTH. *Educ:* Univ Calif, Santa Barbara, PhD(environ physiol), 77. *Prof Exp:* assoc res coordr, Sch Med, La State Univ, 79-86; EVALUATOR, CTRS FOR DIS CONTROL, DIV CHRONIC DIS, COMMUNITY HEALTH PROM BR, 90- *Mem:* Am Pub Health Asn; Am Soc Training & Develop; Nat Soc Performance & Instr. *Mailing Add:* 5014 Brookside Court Alpharetta GA 30201

BALCERZAK, MARION JOHN, b Baltimore, Md, Oct 28, 33; m 58; c 4. MECHANICAL ENGINEERING. *Educ:* Univ Detroit, BME, 56; Northwestern Univ, Evanston, MSME, 58, PhD(mech eng), 61. *Prof Exp:* Res engr, R C Ingersol Res Ctr, Borg Warner Corp, 60-62; res engr, Gard, Inc, 62-64, dir appl mech, 65-68, tech dir, 69-76, vpres & gen mgr, 77-80, pres & dir res & develop Gard Inc, GATX Corp, 80-84; group vpres technol group, Chamberlain Mfg Corp, Unit Duchossois Indust Inc, 84-90; FOUNDER, M T BALCERZAK & ASSOCS, 90- *Honors & Awards:* Charles T Main Award, Am Soc Mech Engrs, 56. *Mem:* Am Soc Mech Engrs; Am Inst Aeronaut & Astronaut; Am Soc Testing & Mat; Am Mgt Asn. *Res:* Management of research and development subsidiary which provides advanced design, analysis and application engineering services to industry and the government. *Mailing Add:* 2750 Crabtree Northbrook IL 60062-3460

BALCERZAK, STANLEY PAUL, b Pittsburgh, Pa, Apr 27, 30; m 53; c 8. INTERNAL MEDICINE, HEMATOLOGY. *Educ:* Univ Pittsburgh, BS, 53; Univ Md, Baltimore, MD, 55. *Prof Exp:* Instr med, Univ Chicago, 59-60; consult hemat & med, Walter Reed Army Inst, 60-62; from instr to asst prof med, Univ Pittsburgh, 62-67; assoc prof, 67-71, PROF MED, OHIO STATE UNIV, 71-, DIR, DIV HEMAT-ONCOL, 69- & DEP DIR, COMPREHENSIVE CANCER CTR, 84- *Concurrent Pos:* Consult hemat, Dayton Vet Admin Hosp, 67- & Wright Patterson Air Force Hosp, 67- *Mem:* Am Col Physicians; Cent Soc Clin Res; Am Fedn Clin Res; Am Soc Hemat; AMA; Am Soc Clin Oncol; Am Asn Cancer Res. *Res:* Iron metabolism; oxygen transport; abnormal hemoglobins; tumor immunology; clinical trials in new drug development; magnetic resonance imaging and spectroscopy as applied to sickle cell disease and cancer. *Mailing Add:* Div Hemat Oncol N-1021 Ohio State Univ Hosp 410 W Tenth Ave Columbus OH 43210

BALCH, ALAN LEE, b Pottstown, Pa, Aug 21, 40; m 64; c 3. INORGANIC CHEMISTRY. *Educ:* Cornell Univ, BA, 62; Harvard Univ, PhD(chem), 67. *Prof Exp:* Asst prof chem, Univ Calif, Los Angeles, 66-70, from asst prof to assoc prof, 70-77, PROF CHEM, UNIV CALIF, DAVIS, 77- *Mem:* Chem Soc London; Am Chem Soc. *Res:* Synthesis, structure and reactions of transition metal complexes; electrochemistry; reactions of metalloporphyrius; oxidation catalysts. *Mailing Add:* Dept Chem Univ Calif Davis CA 95616

BALCH, ALFRED HUDSON, b Manhattan, Kans, May 22, 28; m 53; c 3. GEOPHYSICS. *Educ:* Stanford Univ, BS, 50; Colo Sch Mines, DSc(geophys), 64. *Prof Exp:* Asst petrol eng, Stanford Univ, 50-52; seismic computist, Phillips Petrol Co, 55-57 & Chevron Oil Co div, Stand Oil Co Calif, 57-58; asst, Colo Sch Mines, 58-63; adv res geophysicist, Marathon Oil Co, 63-71; assoc prof geophys, Colo Sch Mines, 71-74; MEM STAFF, US GEOL SURV, DENVER, COLO, 74- *Mem:* AAAS; Soc Explor Geophys; Am Geophys Union; Sigma Xi; Europ Asn Explor Geophys. *Res:* Wave propagation; geophysical data analysis techniques from random noise theory. *Mailing Add:* PO Box 129 Golden CO 80401

BALCH, CHARLES MITCHELL, b Milford, Del, Aug 24, 42; m 66; c 4. SURGICAL ONCOLOGY, IMMUNOLOGY. *Educ:* Univ Toledo, BS, 63, Columbia Univ, MD, 67. *Prof Exp:* Intern surg, Med Ctr, Duke Univ, 67-68; prog specialist, Gen Clin Res Ctr Br, NIH, 68-70; resident gen surg, Med Ctr, Univ Ala, Birmingham, 70-71 & 73-75; fel immunol, Scripps Clin & Res Found, 71-73; from asst prof to assoc prof surg, Med Ctr, Univ Ala, Birmingham, 75-81, prof, 81-, assoc prof microbiol, 81-; HEAD, DIV SURG & CHMN, DEPT GEN SURG, M D ANDERSON CANCER CTR, 86- *Concurrent Pos:* Scientist, Comprehensive Cancer Ctr, 75-; Am Cancer Soc jr fac fel, 76-79. *Mem:* Soc Univ Surgeons; Asn Acad Surg; Am Asn Immunologists; Am Asn Cancer Res. *Res:* Human cellular immunobiology; tumor immunology; adjunctive therapy for neoplast disease. *Mailing Add:* M D Anderson Cancer Ctr 1515 Holcombe Blvd Houston TX 77030

BALCH, DONALD JAMES, b Hanover, NH, July 15, 22; m 46; c 2. ANIMAL BREEDING. *Educ:* Univ NH, BS, 48, MS, 52; Va Polytech Inst, PhD(animal breeding), 62. *Prof Exp:* Instr animal husb, Univ NH, 50-51; from instr to prof, 52-85, dir Morgan Horse Farm, 72-85, EMER PROF ANIMAL SCI, UNIV VT, 85- *Mem:* Am Soc Animal Sci. *Mailing Add:* Underhill Burlington VT 05401

BALCH, WILLIAM E, b Corvallis, Ore, Feb 22, 49. MOLECULAR BIOLOGY, BIOCHEMISTRY. *Educ:* Portland State Univ, BSc, 71; Univ Ill, PhD(microbiol), 79. *Prof Exp:* Res asst & res assoc, Dept Biol, Portland State Univ, 71-73 & Dept Microbiol, Univ Ill, 73-79; NIH postdoctoral scholar, Dept Biochem, Stanford Univ, 79-83; from asst prof to assoc prof, Dept Molecular Biophys & Biochem, Yale Univ, 83-88; ASSOC MEM, DEPT CELL & MOLECULAR BIOL, SCRIPPS RES INST, 88- *Mem:* Fedn Am Socs Exp Biol. *Res:* In vitro reconstitution of vesicular trafficking in the secretory pathway; biochemical and molecular characterization of vesicle fission and fusion events regulating endoplasmic reticulum to Golgi protein transport; author of 27 technical publications. *Mailing Add:* Dept Molecular & Cellular Biol Scripps Res Inst Mail Drop IMM 11 10666 N Torrey Pines Rd La Jolla CA 92037

BALCHUM, OSCAR JOSEPH, b Detroit, Mich, Nov 9, 17; m 49; c 2. INTERNAL MEDICINE, PULMONARY DISEASES. *Educ:* Wayne Univ, BS, 38, MS, 42, MD, 43; Univ Colo, PhD, 53. *Prof Exp:* Pulmonary fel, Cardio-Pulmonary Lab, Nat Jewish Hosp, Denver, Colo, 50-53; intern, Med Ctr, Univ Colo, 53-54, resident internal med, 56-57; instr med & asst dir, Clin Radioisotope Ctr, Vanderbilt Univ Hosp, 57-59; asst prof, 59-61, Hastings assoc prof, 61-67, HASTINGS PROF MED, SCH MED, UNIV SOUTHERN CALIF, 68- *Concurrent Pos:* Am Heart Asn res fel cardiol, Univ Colo, 54-56; dir pulmonary dis sect, Los Angeles County-Univ Southern Calif Med Ctr, 69-82, dir pulmonary med sect, 70-82. *Mem:* AAAS; Am Thoracic Soc; fel Am Col Cardiol; fel Am Col Physicians; fel Am Col Chest Physicians. *Res:* Diseases of the chest and cardiovascular system; pulmonary physiology; lung cancer; photodynamic therapy and diagnostic imaging bronchoscopy of bronchial cancers; diagnostic imaging to detect small and even invisible bronchial cancers, particularly in individuals with positive sputum cytology and negative chest X-rays; photodynamic therapy of early stage bronchial cancer lesions and obstructing endobiomedical cancer lesions employing hematoporphyrin derivative and red laser light. *Mailing Add:* 826 Garfield Ave South Pasadena CA 91030-2809

BALCZIAK, LOUIS WILLIAM, b Duluth, Minn, Feb 11, 18; m 46; c 1. EDUCATIONAL STATISTICS. *Educ:* Duluth State Teachers Col, BS, 40; Univ Minn, AM, 49, PhD(ed statist), 53. *Prof Exp:* Instr chem, Duluth State Teachers Col, 41-44 & 46-47 & Duluth Br, Univ Minn, 47-48; from asst prof to assoc prof, 50-63, PROF CHEM, MANKATO STATE UNIV, 63- *Concurrent Pos:* Mem, NSF Summer Insts, 59-65. *Mem:* Am Chem Soc. *Res:* Inorganic chemistry; nuclear chemistry. *Mailing Add:* 1114 Anderson Dr Mankato MN 56001-4704

BALD, KENNETH CHARLES, b Rock Island, Ill, Oct 27, 30; m 65; c 4. PHYSICAL CHEMISTRY. *Educ:* St Ambrose Col, AB, 52; St Louis Univ, PhD(phys chem), 62. *Prof Exp:* Chemist, Callery Chem Co, 52-57; teaching fel, St Louis Univ, 57-62; res chemist, M W Kellogg Co, 62-69; res chemist, Riegel Prod Corp, 69-82, mgr coating develop, 82-; SR PROD TECHNOLOGIST, JAMES RIVER GRAPHICS. *Mem:* Am Chem Soc; Tech Asn Pulp & Paper Indust; Am Soc Qual Eng. *Res:* Development of electrochemical processes for industrial applications; development of electrophotographic products, cured release products, and other specialty coated papers; development of coated printable decorative products; quality assurance decorative products. *Mailing Add:* James River Graphics 28 Gaylord St South Hadley MA 01075

BALDA, RUSSELL PAUL, b Oshkosh, Wis, June 14, 39; m 87; c 3. ANIMAL ECOLOGY. *Educ:* Univ Wis, Oshkosh, BS, 61; Univ Ill, MS, 63, PhD(zool), 67. *Prof Exp:* From asst prof to prof, 66-88, REGENT'S PROF BIOL, NORTHERN ARIZ UNIV, 88- *Mem:* Ecol Soc Am; Am Ornith Union; Cooper Ornith Soc (secy, 75-76); Wilson Ornith Soc; Asn Study Animal Behav; Sigma Xi. *Res:* Spatial memory in seed caching birds; social behavior of birds. *Mailing Add:* Dept Biol Sci Northern Ariz Univ Box 5640 Flagstaff AZ 86011

BALDAUF, GUNTHER H(ERMAN), b Munich, Ger, May 20, 23; nat US; m 48; c 4. CHEMICAL ENGINEERING, INDUSTRIAL CHEMISTRY. *Educ:* Mass Inst Technol, 44, MS, 46, DSc, 49. *Prof Exp:* Chem engr, Godfrey L Cabot, Inc, 45-47 & Ecusta Paper Div, Olin Mathieson Chem Corp, 49-59; res mgr, Allied Paper Corp, 59-63; consult, Nadelman-Baldauf & Assocs, Inc, 63-65; mgr cent tech serv, Simpson Paper Co, 65-86; PAPER CONSULT, 86- *Mem:* Am Soc Qual Control; Tech Asn Pulp & Paper Indust; Paper Mgt Asn. *Res:* Adhesives; pulp and paper manufacture; paper coating; process and quality control; copier technology. *Mailing Add:* 609 Fleetwood Dr Modesto CA 95350

BALDER, JAY ROYAL, b Bridgeport, Conn, Feb 4, 41; m; c 2. CHEMICAL ENGINEERING. *Educ:* Univ Del, BChE, 63; Univ Calif, Berkeley, PhD(chem eng), 67. *Prof Exp:* CONSULT MGR, ENG EVAL, ENG SERV DIV, E I DU PONT DE NEMOURS & CO, INC, 85- *Mem:* Am Inst Chem Engrs. *Mailing Add:* 803 Rock Lane Newark DE 19711

BALDERRAMA, FRANCISCO E, internal medicine, for more information see previous edition

BALDESCHWIELER, JOHN DICKSON, b Elizabeth, NJ, Nov 14, 33; c 3. BIOCHEMISTRY, CHEMICAL PHYSICS. *Educ:* Cornell Univ, BChemE, 56; Univ Calif, Berkeley, PhD(phys chem), 59. *Prof Exp:* Lectr chem, Harvard Univ, 60, instr, 60-62, asst prof, 62-65; assoc prof, Stanford Univ, 65-67, prof, 67-71; dep dir, Off Sci & Technol, Washington, DC, 71-73; PROF CHEM & CHMN DIV CHEM & CHEM ENG, CALIF INST TECHNOL, 73- *Concurrent Pos:* Consult, Ballistic Res Labs, Aberdeen Proving Grounds, Md, 60-, Monsanto Chem Co, Mo & Res Systs, Inc, Mass; Alfred P Sloan Found fel, 62-66; mem, Army Sci Adv Panel, 66-69; mem, President's Sci Adv Comt, 69-; vis scientist, Nat Cancer Inst, 73; consult, NSF, Nat Security Coun, Nat Cancer Inst & Merck Sharp & Dohme, 73-; chmn, Bd Dirs, Vestar Res Inc; vis prof, E T H, Zurich, Switz, 85; Biophys Chem Study Sect, NIH, 84-86. *Honors & Awards:* Am Chem Soc Award, 67; Am Acad Arts & Sci Award, 72. *Mem:* Nat Acad Sci; Am Inst Physics; Am Chem Soc; Am Acad Arts & Sci, 72; Am Philos Soc. *Res:* Molecular structure and spectroscopy, including nuclear magnetic resonance, nuclear spectroscopy, mass spectroscopy, x-ray diffraction and applications of these methods to biological systems. *Mailing Add:* Div Chem & Chem Eng Calif Inst Technol Pasadena CA 91125

BALDESSARINI, ROSS JOHN, b Sept 30, 37; m; c 2. PHARMACOLOGY, PSYCHIATRY. *Educ:* Johns Hopkins Univ, MD, 63. *Hon Degrees:* MA, Harvard Univ, 77. *Prof Exp:* PROF PSYCHIAT & NEUROSCI, MAILMAN RES CTR, MCLEAN HOSP, 77- *Honors & Awards:* Efron Res Prize, Am Col Neuropsychopharmacol. *Mem:* Soc Neurosci; Am Col Neuropsychopharmacol; Am Soc Pharmacol & Exp Therapeut; Soc Neurochem; Am Psychiat Asn. *Res:* Pharmacology of antipsychotic drugs and of dopamine system in the brain. *Mailing Add:* Mailman Res Ctr McLean Hosp Belmont MA 02178

BALDINI, JAMES THOMAS, b Paterson, NJ, Jan 21, 27; m 51; c 4. NUTRITION, DERMATOLOGY. *Educ:* Rutgers Univ, BS, 47; Purdue Univ, MS, 49, PhD(nutrit), 51; Univ Del, MBA, 69. *Prof Exp:* Jr asst nutrit, Purdue Univ, 47-51; from res biologist to clin res monitor, Stine Lab, E I du Pont de Nemours & Co, 51-69; med serv mgr, White Labs Div, Schering Corp, 69-70, dir sci affairs, Schering Labs, 70-71, dir prof serv, 71- 80, dir prof & med serv, Int Div, Schering-Plough, Inc, 80-85; MGT & BIOMED CONSULT SERV, 85- *Mem:* Am Col Allergy & Immunol. *Res:* Toxicology; pharmacology; virology; biochemistry and low protein diets in nutrition; amino acid nutrition; endocrinology; adrenal ascorbic acid; estrogen and calcium utilization; energy metabolism; appetite control; dermatology; allergy; antibiotics. *Mailing Add:* 4285 Crested Butte Run Onondaga NY 13215

BALDINO, FRANK, JR, b Passaic, NJ, May 13, 53; m 76; c 2. ELECTROPHYSIOLOGY, MOLECULAR NEUROBIOLOGY. *Educ:* Muhlenberg Col, BS, 75; Temple Univ, PhD(pharmacol), 79. *Prof Exp:* Teaching fel pharmacol, Sch Med, Rutgers Univ, 79-80, instr, 80-81; PRIN SCIENTIST NEUROBIOL, E I DU PONT DE NEMOURS, 81- *Concurrent Pos:* Adj asst prof, Dept Pharmacol, Sch Med, Temple Univ, 81- *Mem:* AAAS; Soc Neurosci. *Res:* Multidisciplinary approach to the function and regulation of neuropeptides in the central nervous system; trophic influences on gene expression of hypothalamic peptides. *Mailing Add:* 145 Brandywine Pkwy West Chester PA 19380

BALDOCCHI, DENNIS D, b Antioch, Calif, Apr 12, 55; m 84; c 1. CANOPY MICROMETEOROLOGY, BIOPHYSICAL ECOLOGY. *Educ:* Univ Calif, Davis, BS, 77; Univ Nebr, Lincoln, MS, 79, PhD(bioenviron eng), 82. *Prof Exp:* Postdoctoral fel, Oak Ridge Assoc Univ, 82-83, biometeorologist, 83-86; PHYS SCIENTIST, ATMOSPHERIC TURBULENCE DIFFUSION DIV, NAT OCEANIC & ATMOSPHERIC ADMIN, 86- *Concurrent Pos:* Mem, Comt Agr Forest Meteorol, Am Meteorol Soc, 85-88 & Comt Biometeorol, 90-93; vis scientist, Meteorol Dept, Univ Stockholm, 90. *Mem:* Am Meteorol Soc; Am Geophys Union; Sigma Xi. *Res:* Experimental and theoretical studies on the physical, biological and chemical processes affecting trace gas exchange between vegetation and the atmosphere. *Mailing Add:* NOAA-ATDD PO Box 2456 Oak Ridge TN 37831-2456

BALDONI, ANDREW ATELEO, b Pekin, Ill, May 16, 16; m 42; c 2. ORGANIC CHEMISTRY. *Educ:* Bradley Univ, BS, 47; Univ Notre Dame, PhD(chem), 51. *Prof Exp:* Chemist, Control Lab, Am Distilling Co, 37-42; res chemist & group leader, Ringwood Chem Corp, 50-56, dir res, 57; asst dir res, Morton Chem Co, 58-64, dir tech serv, 64-67; dir res, Simoniz Co, 67-70; dir corp res, Morton Int, Inc, 70-72, dir res & develop dyes, 72-74, dir new prod res, 74-79, vpres res, Morton Chem Co, 79-85; RETIRED. *Mem:* Am Chem Soc. *Res:* Organic research. *Mailing Add:* 344 Hilkert Ct Crystal Lake IL 60014-5814

BALDRIDGE, ROBERT CRARY, b Herington, Kans, Jan 9, 21; m 43; c 5. BIOCHEMISTRY. *Educ:* Kans State Univ, BS, 43; Univ Mich, MS, 48, PhD(biol chem), 51. *Prof Exp:* Instr biol chem, Univ Mich, 51-53; from asst prof to prof biochem, Sch Med, Temple Univ, 53-70, assoc dean grad sch, 65-70; dean, Col Grad Studies, Thomas Jefferson Univ, 70-81, PROF BIOCHEM, JEFFERSON MED COL, 70- *Mem:* Am Soc Biol Chem. *Res:* Amino acid metabolism; biochemical genetics; histidinemia; steroid hormone receptors. *Mailing Add:* Dept Biochem Thomas Jefferson Univ 1020 Locust St Philadelphia PA 19107

BALDRIDGE, WARREN SCOTT, b Lyons, NY, Jan 7, 45; m 74; c 1. IGNEOUS PETROLOGY, REGIONAL GEOLOGY. *Educ:* Hamilton Col, AB, 67; Calif Inst Technol, MS, 70, PhD(geol), 78. *Prof Exp:* Res asst geophysics, Mass Inst Technol, 69-72; fel, 78-80, GEOLOGIST, LOS ALAMOS NAT LAB, 80- *Concurrent Pos:* Fulbright fel exp petrol, Univ Gottingen, 67-68; vis assoc prof, Hebrew Univ, Jerusalem, 84-85. *Mem:* Sigma Xi; fel Geol Soc Am; Am Geophys Union. *Res:* Petrologic and geochemical studies of mafic volcanic rocks; mantle xenoliths; petrogenesis of rift zone igneous rocks; evolution and regional tectonic studies of Rio Grande rift. *Mailing Add:* Earth & Environ Sci Div MS D462 Los Alamos Nat Lab Los Alamos NM 87545

BALDUCCI, LODOVICO, b Borgonovo, Italy, Apr 7, 44; nat US; m 71; c 1. GERIATRICS, NUTRITION. *Educ:* Catholic Univ Sch Med Rome, Italy, MD, 68. *Prof Exp:* Resident med, Univ Miss, 73-76, fel med & oncol, 76-78, asst prof, 78-84, assoc prof med, 84-87; ASSOC PROF MED, UNIV S FLA, 87- *Concurrent Pos:* Staff oncologist, Jackson Vet Admin, Jackson, Miss, 78-87; chief hemat & oncol, Buy Pines Vet Admin, Fla, 87- *Mem:* Am Assoc Cancer Res; Am Soc Hemat; Am Soc Clin Oncol; Am Col Physicians; Am Cancer Soc. *Res:* Cancer in the elderly, geriatric oncology; prostate cancer; hemopoiesis and cancer. *Mailing Add:* Bay Pines Vet Admin Bay Pines FL 33504

BALDUS, WILLIAM PHILLIP, b Milwaukee, Wis, Mar 11, 32; m 56; c 5. INTERNAL MEDICINE, GASTROENTEROLOGY. *Educ:* Marquette Univ, BS, 54, MD, 57; Univ Minn, MS, 64. *Prof Exp:* Intern, Boston City Hosp, 57-58; resident, Univ Minn Hosps, 58-61; instr, 65-68, asst prof, 68-78, ASSOC PROF GASTROENTEROL, MAYO GRAD SCH MED, UNIV MINN, 78- *Mem:* Am Gastroenterol Asn; fel Am Col Physicians; Am Fedn Clin Res. *Res:* Renal failure and disturbances of water and electrolyte metabolism in liver disease. *Mailing Add:* Dept Med Univ Minn Mayo Grad Sch Med Rochester MN 55901

BALDUZZI, PIERO, b Voghera, Italy, Mar 22, 28; m 56; c 2. MICROBIOLOGY, VIROLOGY. *Educ:* Univ Pavia, MD, 52. *Prof Exp:* Asst genetics, Univ Pavia, 52-53; res in microbiol, Pharmaceut Co, Italy, 56-58; asst, Univ Florence, 58-62; from asst prof to assoc prof, 62-84, PROF MICROBIOL, SCH MED, UNIV ROCHESTER, 84- *Concurrent Pos:* Fel, Univ Rochester, 59-60; sr fel, Nat Cancer Inst, Imp Cancer Res Fund Lab, London, Eng, 73-74. *Res:* Tumor viruses; genetics of RNA tumor viruses. *Mailing Add:* Dept Microbiol Univ Rochester Med Sch 601 Elmwood Rochester NY 14642

BALDWIN, ARTHUR DWIGHT, JR, b Boston, Mass, Oct 28, 38; m 62; c 2. GEOLOGY. *Educ:* Bowdoin Col, BA, 61; Univ Kans, MA, 63; Stanford Univ, PhD(geol), 67. *Prof Exp:* Assoc prof, 66-79, dep dir affairs, Inst Environ Sci, 77-79, prof & chmn dept, 79-87, PROF GEOL, MIAMI UNIV, 87- *Mem:* Geol Soc Am; AAAS; Nat Water Well Asn; Sigma Xi. *Res:* Problems related to finding sufficient water, both in quantity and in quality, to satisfy our nation's increasing needs. *Mailing Add:* Dept Geol Miami Univ Oxford OH 45056

BALDWIN, ARTHUR RICHARD, b Palmyra, Mich, Feb 20, 18; m 42; c 3. CHEMISTRY. *Educ:* Stetson Univ, BS, 40; Univ Pittsburgh, PhD(biochem), 43. *Prof Exp:* Lab asst chem, Univ Pittsburgh, 40-42, Swift & Co fel, 43, Nutrit Found fel, 44; sect leader food & lipid chem, Corn Prod Refining Co, 44-54; dir res, Cargill, Inc, 54-64, vpres & exec dir res, 64-83, secy, Cargill Long Rauge Planning Comt, 68-83; CONSULT, 83- *Concurrent Pos:* Ed, J Am Oil Chem Soc; mem space applns bd, Nat Acad Eng, 73 & 78; dir, Agribus Coun, 71-; mem, Joint Res Coun, Bd Int Food & Agr Develop, 78- & Joint Coun Food & Agr Sci, USDA, 78-; chmn, World Conf Vegetable Food Proteins, 78 & World Conf Emerging Technol on Fats & Oils; ed, Nine World Conf Proceedings; mem, Res Adv Comt of AID, 82-89; bd trustees, Pan Am Sch Agr, 82-; chmn, Minn Coun Biotechnol, 85-87 & Minn Comt Sci & Technol, 87. *Honors & Awards:* Normann Medal, Ger Oil Sci Soc, 51; Alton E Bailey Award, Am Oil Chem Soc, 63; Chevreul Medal, Groupement Technique des Corps Gras, 65; A Richard Balwin Award, Am Oil Chem Soc, 81. *Mem:* Am Chem Soc; Am Oil Chem Soc; Inst Food Technologists; Am Asn Cereal Chem. *Res:* Purification of inositol and phytates; fat and oil analyses recoveries and processing; development of food products; fat digestibilities and metabolism; compounding hydrogenated shortening; preparation of synthetic lipids; carbohydrate oxidations; new packaged food products; plant breeding; feed and animal nutrition research; drying oils and protection coatings; plant biology; agricultural research management. *Mailing Add:* 4854 Thomas Ave S Minneapolis MN 55410

BALDWIN, BARRETT S, JR, b Bakersfield, Calif, July 22, 21; m 43; c 2. AERONAUTICS, ASTRONAUTICS. *Educ:* Univ Calif, BS, 43; Stanford Univ, MS, 51, PhD(aeronaut, astronaut), 62. *Prof Exp:* Teaching asst physics, Stanford Univ, 48-51; res scientist, Nat Adv Comt Aeronaut, 51-58; RES SCIENTIST, NASA, 58- *Mem:* Am Phys Soc; Am Inst Aeronaut & Astronaut; Am Geophys Union. *Res:* Wind tunnel wall interference in airfoil testing; behavior of radiating gas flows; application of aeronautical technology to interpretation of meteor phenomena; computational fluid dynamics and turbulence modeling. *Mailing Add:* 2468 Whitney Dr Mountain View CA 94043

BALDWIN, BERNARD ARTHUR, b Chicago, Ill, Sept 7, 40; c 2. PHYSICAL CHEMISTRY, MOLECULAR SPECTROSCOPY. *Educ:* La Verne Col, BA, 62; San Diego State Col, MS, 65; Univ Calif, Santa Barbara, PhD(phys chem), 68. *Prof Exp:* Asst chem, San Diego State Col, 62-64, asst phys chem, 63-65; asst chem, Univ Calif, 65-66, asst phys chem, 66-67; SR RES CHEMIST, PHILLIPS PETROL CO, 68- *Honors & Awards:* Walter D Hodson Award, 77; Captain Alfred E Hunt Award, 87. *Mem:* Am Soc Lubrication Engrs; Soc Core Analysts; Soc Petrol Engr. *Res:* Perturbation effects on the population and depopulation processes of electronically excited molecules; chemical character of natural adsorbates and mechanism for adsorption of molecular species on solid surfaces; chemical characterization of surfaces, adsorbed molecular species and their reaction products; mechanism(s) responsible for boundary lubrication of metals; petrophysical properties of fluids in porous media; NMR imaging of fluids in porous media; 36 technical publications. *Mailing Add:* Phillips Petrol Co 103 GB Bartlesville OK 74004

BALDWIN, BERNELL ELWYN, b Angwin, Calif, Jan 21, 24; m 60. NEUROPHYSIOLOGY. *Educ:* Columbia Union Col, BA, 56; George Washington Univ, MA, 58, PhD, 63. *Prof Exp:* Instr physiol, 63-67, asst prof physiol & biophys, 67-69, asst prof physiol, Sch Med, 69-77, ASSOC PROF APPL PHYSIOL, SCH HEALTH, LOMA LINDA UNIV, 77- *Concurrent Pos:* Mem staff, Appl Physiol Lab, Wildwood Sanitarium & Hosp. *Res:* Effects of combined exercise, metabolic alteration and stress control on human lipid profiles and on blood pressure; frontal lobe contribution to autonomic nervous system dynamics and feeding behavior. *Mailing Add:* Appl Physiol Lab Wildwood Sanitarium & Hosp Wildwood GA 30757

BALDWIN, BREWSTER, b Brooklyn, NY, Nov 10, 19; m 46; c 4. GEOLOGY. *Educ:* Williams Col, Mass, AB, 41; Columbia Univ, AM, 44, PhD(geol), 51. *Prof Exp:* Jr geologist, US Geol Surv, 42-44; instr geol, Univ SDak, 47-51; econ geologist, Bur Mines, NMex, 51-58; from asst prof to prof geol, Middlebury Col, 58-89,; RETIRED. *Concurrent Pos:* Mem, Coun Educ in Geol Sci, 67-70 & New Eng Sect, Correlation Stratig Units NAm. *Mem:* Geol Soc Am; Soc Econ Paleontologists & Mineralogists; Nat Asn Geol Teachers. *Res:* Sedimentology of Taconics; environmental geology; curriculum philosophy; rates of sedimentation. *Mailing Add:* 17 Buttolph Dr Middlebury VT 05753

BALDWIN, CHANDLER MILNES, b Oakland, Calif, Nov 22, 51; m 82; c 2. ELECTRONICS ENGINEERING, ELECTROMAGNETISM. *Educ:* UCLA, BS, 74, MS, 75, PhD(eng), 79. *Prof Exp:* Engr develop, Menorex Corp, 78-79, sr engr advan thin film, 79-81; prin engr process develop, Exxon Enterprises, 81-82; prin engr media develop, 83-85, mgr, 85-89, DIR DEVELOP, CENSTOR CORP, 89- *Honors & Awards:* Grad Res Award, Electrochem Soc, 78. *Mem:* Inst Elec Electronics Engrs Magnetics Soc. *Res:* Development and characterization of glassy BeF2 based materials; magnetic recording industry primarily thin film device development. *Mailing Add:* Censtor Corp 530 Race St San Jose CA 95126

BALDWIN, DAVID ELLIS, b Bradenton, Fla, June 12, 36; m 61; c 2. PLASMA PHYSICS. *Educ:* Mass Inst Technol, BSc, 58, PhD(physics), 62. *Prof Exp:* Res assoc physics, Stanford Univ, 62-64 & Theory Div, Culham Lab, Eng, 64-66; from asst prof to assoc prof appl sci, Yale Univ, 66-70; res physicist, Lawrence Livermore Lab, 70-78, dep theory prog leader, 78-84, from dep assoc dir to assoc dir, 84-88; dir, Inst Fusion Studies & prof physics, Univ Tex, Austin, 88-91; ASSOC DIR MAGNETIC FUSION ENERGY, LAWRENCE LIVERMORE NAT LAB, 91- *Mem:* Fel Am Phys Soc. *Res:* Theoretical plasma physics; controlled thermonuclear fusion. *Mailing Add:* Lawrence Livermore Nat Labs Livermore CA 94550

BALDWIN, DAVID HALE, b Manchester, Conn, Dec 4, 36; m 59; c 3. METALLURGY. *Educ:* Univ Fla, BME, 59, MSE, 61, PhD, 66. *Prof Exp:* Asst prof metall eng, Univ Fla, 66-77; ENGR, MAT & PROCESSES LAB, GEN ELEC CO, 77- *Mem:* Am Soc Metals; Am Inst Mining, Metall & Petrol Engrs; Sigma Xi. *Res:* Mechanical properties of superalloys. *Mailing Add:* 6036 Loma Prieta Dr San Jose CA 95123-3931

BALDWIN, ELDON DEAN, b Ashley, Ohio, Sept 23, 39; m 64; c 2. AGRICULTURAL ECONOMICS. *Educ:* Ohio State Univ, BS, 63; Univ Ill, MS, 67, PhD(agr econ), 70. *Prof Exp:* Agr statistician, Bur Census, US Dept Com, 63-65; res asst agr econ, Univ Ill, 65-69; asst prof econ, Eastern Ill Univ, 69-70 & Miami Univ, 70-74; asst prof, 74-80, ASSOC PROF AGR ECON, OHIO STATE UNIV, 80- *Mem:* Am Agr Econ Asn. *Res:* Analyze and model structures and performance variables for grain markets; electronic marketing of livestock, technical feasibility, application and its impact on price and marketing costs and performance. *Mailing Add:* Dept Agr Econ Ohio State Univ 2120 Fyffe Rd Columbus OH 43210

BALDWIN, EWART MERLIN, b Pomeroy, Wash, May 17, 15; m 42; c 2. GEOLOGY. *Educ:* State Col Wash, BS, 38, MS, 39; Cornell Univ, PhD(geol), 43. *Prof Exp:* Asst, State Col Wash, 38-39 & Cornell Univ, 40-43; from asst geologist to geologist, Ore Dept Geol & Mining Industs, Portland, 43-47; from asst prof to prof, Univ Ore, 47-80, emer prof geol, 81; RETIRED. *Concurrent Pos:* Fulbright award, Dacca, EPakistan, 59-60; Arnold vis prof, Whitman Col, 81. *Mem:* Fel Geol Soc Am; Am Asn Petrol Geologists; Paleont Soc; Sigma Xi. *Res:* Oregon coast range stratigraphy; Coos Bay coal fields; Hambaek coal field of Korea; geology of Oregon and parts of Washington and Idaho; regional geology of Oregon, in particular the Oregon Coast Range and Klamath Mountains. *Mailing Add:* 1020 E 18th Ave Apt 3 Eugene OR 97403

BALDWIN, GEORGE C, b Denver, Colo, May 5, 17; m 44, 52; c 3. PHYSICS. *Educ:* Kalamazoo Col, BA, 39; Univ Ill, MA, 41, PhD(physics), 43. *Prof Exp:* Instr physics, Univ Ill, 43-44; res physicist, Gen Elec Co, 44-55, physicist, Aircraft Nuclear Propulsion Dept, 55-57; mgr, Argonaut Reactor, Argonne Nat Lab, 57-58; appl physicist, Adv Tech Labs, Gen Elec Co, 58-65, sr physicist, 65-67; mem staff, Los Alamos Sci Lab, 77-87; from adj prof to prof nuclear eng & sci, 61-77, EMER PROF NUCLEAR ENG & SCI, RENSSELAER POLYTECH INST, 77-; PRES, BALDWIN SCI INC, 87- *Concurrent Pos:* Vis mem staff, Los Alamos Sci Lab, 75-77; vis lectr, Naval Res Lab, 88. *Mem:* AAAS; fel Am Phys Soc. *Res:* Nuclear photo-effect; high energy radiation and neutron physics; nuclear instrumentation; reactor kinetics and physics; nuclear and ion propulsion; particle accelerators; stimulated emission devices; electron collision spectrometry of molecules; nuclear lasers. *Mailing Add:* 1016 Calle Bajo Santa Fe NM 87501

BALDWIN, GEORGE L(EE), communication engineering, computer science, for more information see previous edition

BALDWIN, HENRY IVES, b Saranac Lake, NY, Aug 23, 96; wid; c 4. FORESTRY. *Educ:* Yale Univ, BA, 19, MF, 22, PhD(bot), 31. *Prof Exp:* Asst to dean, Yale Forest Sch, 22-23; exec secy, Conn Forestry Asn, 23; res asst forest sci, Swed Forestry Exp Sta, 23-24; chief forest invest, Brown Co, Berlin, NH, 24-32; prof forestry, Pa State Col, 32-33; res forester, NH Forestry Comn, 33-62; consult, NH State Planning Bd, 62-65; prof bot & ecol, Franklin Pierce Col, 65-75; consult, forestry, Soc Protection NH Forests, 75-80 & Audubon Soc, NH, 75-85; RETIRED. *Concurrent Pos:* Consult, Nat Resources Planning Bd, 41-42, Fed Reserve Bank, Boston, 48-49 & Foreign Agr Orgn-UN. *Mem:* Fel Am Acad Arts & Sci; fel Soc Am Foresters; fel AAAS; Ecol Soc Am; Am Inst Biol Sci; Nat Audubon Soc. *Res:* Physiology of tree seed; provenance tests of Norway spruce, scotch pine, European larch, red spruce and jack pine; seed testing; local flora, timberline research; natural distribution of rare trees. *Mailing Add:* Ctr Rd RFD 2 Box 465 Hillsboro NH 03244

BALDWIN, HOWARD WESLEY, b Winnipeg, Man, June 11, 28; m 58; c 3. INORGANIC CHEMISTRY. *Educ:* Univ Sask, BA, 50, MA, 51; Univ Chicago, PhD(chem), 59. *Prof Exp:* Asst prof, 59-67, asst dean, Fac Grad Studies, 69-78, ASSOC PROF CHEM, UNIV WESTERN ONT, 69-, ASST PROVOST & UNIV RES OFFICER, 78- *Mem:* Chem Inst Can. *Res:* Kinetic and isotopic studies of reactions of complex ions in solution. *Mailing Add:* Dept Chem Univ Western Ont London ON N6A 5B9

BALDWIN, JACK NORMAN, microbiology, for more information see previous edition

BALDWIN, JACK TIMOTHY, b Abilene, Tex, Oct 6, 45; m 67. COMPUTER SCIENCE. *Educ:* Univ Tex, Austin, BA, 67, PhD(comput sci), 76. *Prof Exp:* Programmer-technician, Int Bus Mach, 67-69; teaching asst comput sci, Univ Tex, Austin, 69-72; asst prof comput sci, Univ Nebr, Omaha, 75-80; ASSOC PROF, DEPT MATH, UNIV ILL, CHICAGO CIRCLE, 80- *Mem:* Asn Comput Mach. *Res:* Programming techniques; artificial intelligence; computer assisted instruction. *Mailing Add:* Dept Math Box 4348 Univ Ill Chicago Circle Chicago IL 60680

BALDWIN, JAMES GORDON, b May 24, 45; m 68. PLANT NEMATOLOGY, PLANT PATHOLOGY. *Educ:* Bob Jones Univ, BS, 67; NC State Univ, MS, 70, PhD(plant path, bot), 73. *Prof Exp:* Asst prof biol & bot, Bryan Col, 73-74; res assoc plant nematol, NC State Univ, 74-75; nematologist, Fla Dept Agr, 75-78; PROF NEMATOL, UNIV CALIF, 78-, CHAIR, DEPT NEMATOL, 90- *Concurrent Pos:* USDA fel, NC State Univ, 74-75, AID grant, 75-; Calif Dept Agric grant, 85-87; NSF grants, 85-88, 89-90, 84-85, 86, 87. *Mem:* Orgn Trop Nematologists; Soc Nematol; Sigma Xi; Soc Syst Zool; Helminthological Soc Wash. *Res:* The fine structure, phylogenetic analysis and systematics of plant parasitic nematodes with emphasis on the Heteroderidae. *Mailing Add:* Dept Nematol Univ Calif Riverside CA 92521

BALDWIN, JAMES W(ARREN), JR, b Peoria, Ill, June 29, 29; m 58; c 2. CIVIL ENGINEERING, MATERIALS. *Educ:* Univ Ill, BS, 51, MS, 56, PhÐ(mech), 61. *Prof Exp:* Res asst theoret & appl mech, Univ Ill, 55-56, instr, 57-60; assoc prof, 60-66, chmn dept, 67-72, PROF CIVIL ENG, UNIV MO-COLUMBIA, 66-, CHMN DEPT, 88- *Concurrent Pos:* Comt mem, Hwy Res Bd, Nat Acad Sci-Nat Res Coun; sr design engr, Rust Eng, 65-66; vis prof, Univ Warwick, Coventry, Eng, 72. *Mem:* Am Concrete Int; Am Soc Civil Engrs; Am Soc Eng Educ; Sigma Xi. *Res:* Fatigue, fracture and field testing of structures; structure monitoring; motor vehicle roll over; failure analysis. *Mailing Add:* Dept Civil Eng Univ Mo Columbia MO 65211

BALDWIN, JOHN CHARLES, b Ft Worth, Tex, Sept 23, 48. EXPERIMENTAL BIOLOGY. *Educ:* Harvard Univ, BA, 71; Stanford Univ, MD, 75; Am Bd Internal Med, cert; Am Bd Surg, cert; Am Bd Thoracic Surg, cert. *Hon Degrees:* MA, Yale Univ, 89. *Prof Exp:* Clin fel med, Med Sch, Harvard Univ, 75-77, clin fel surg, 77-81; asst prof & physician specialist cardiovasc surg, Sch Med, Stanford Univ, 84-85, asst prof & dir, Transplantation Res Lab, 85-87; PROF SURG & CHIEF CARDIOTHORACIC SURG, SCH MED & CARDIOTHORACIC SURGEON-IN-CHIEF, NEW HAVEN HOSP, YALE UNIV, 88- *Concurrent Pos:* Prin investr, Stanford Univ Med Ctr, 81-88, Datascope Corp, 85-87 & 88-90, Cystic Fibrosis Found, 86-88 & Upjohn Corp, 87-; attend surgeon cardiovasc & thoracic surg serv, Stanford Univ Med Ctr, 84-87, head, Heart-Lung Transplant Prog, 85-87; mem, Res Grant Peer Rev Subcomt, Am Heart Asn, 84-87, Prog Comt, Int Soc Heart Transplantation, 84-85 & 87, Comt Organ Transplantation, Pan Am Med Asn, 85-, Clin Res Comt Ad Hoc Res Grant Rev, Cystic Fibrosis Found, 86- & Adv Comt Issues, Am Soc Transplant Surgeons, 89-; assoc ed, J Appl Cardiol, 85-; chmn, Comt Heart Transplantation, Am Soc Transplant Surgeons, 86-89, mem, 90-93, chmn, Subcomt Heart Transplantation, 89-; Royal Col Surgeons Eng traveling lectr, 88. *Honors & Awards:* Medaille de la Ville de Bordeaux, French Thoracic Soc, 87. *Mem:* Am Asn Thoracic Surg; fel Am Col Angiol; fel Am Col Cardiol; fel Am Col Chest Physicians; fel Am Col Physicians; fel Am Col Surgeons; Am Heart Asn; AMA; Am Organ Transplant Soc; Am Physiol Soc. *Res:* Author of more than 350 technical publications. *Mailing Add:* Cardiovasc Surg Yale Univ 333 Cedar St PO Box 3333 New Haven CT 06510

BALDWIN, JOHN E, b Berwyn, Ill, Sept 10, 37; m 61; c 3. PHYSICAL ORGANIC CHEMISTRY. *Educ:* Dartmouth Col, AB, 59; Calif Inst Technol, PhD, 63. *Prof Exp:* From instr to assoc prof chem, Univ Ill, 62-68; prof chem, Univ Ore, 68-84, dean, Col Arts & Sci, 75-80; PROF CHEM, SYRACUSE UNIV, 84- *Concurrent Pos:* Alfred P Sloan res fel, 66-68; Guggenheim Mem Found fel, 67; consult, Stauffer Chem Co & Off Sci & Technol; mem bd ed, Org Reactions & Chem Rev; Alexander von Humboldt Found sr US scientist award, 74-75; Jubileums prof, Chalmers Univ Technol, Göteborg, Sweden, 90. *Mem:* Am Chem Soc. *Res:* Stereochemistry; reaction mechanisms; molecular rearrangements; cycloadditions; stereochemical definition of reaction paths for hydrocarbon molecular rearrangements. *Mailing Add:* Dept Chem Syracuse Univ Syracuse NY 13244-4100

BALDWIN, JOHN THEODORE, b Danbury, Conn, Aug 14, 44; m 65; c 1. MATHEMATICAL LOGIC. *Educ:* Mich State Univ, BSc, 66; Univ Calif, Berkeley, MS, 67; Simon Fraser Univ, PhD(math), 71. *Prof Exp:* Res assoc math, Mich State Univ, 71-73; asst prof, 73-77, ASSOC PROF MATH, UNIV ILL, CHICAGO CIRCLE, 77- *Mem:* Am Math Soc; Asn Symbolic Logic; Math Asn Am. *Res:* Model theory; universal algebra. *Mailing Add:* Dept Math Univ Ill Chicago IL 60680

BALDWIN, KATE M, b Seattle, Wash, July 12, 40. CELL COMMUNICATIONS, CELL MEMBRANES. *Prof Exp:* Asst prof, 72-84, ASSOC PROF ANAT, COL MED, HOWARD UNIV, 84- *Mem:* Am Soc Cell Biol; Am Asn Anatomists. *Mailing Add:* Dept Anat Howard Univ Col Med 520 W St NW Washington DC 20059

BALDWIN, KEITH MALCOLM, b Buffalo, NY, May 25, 28; m 51; c 4. PHYSICS, ELECTRONICS. *Educ:* Mich State Univ, BS, 50; Univ Maine, MS, 55. *Prof Exp:* Assoc scientist, Raytheon Mfg Co, 55-57; assoc scientist, Res & Adv Develop Div, Avco Corp, 57-59, from sr scientist to sr proj scientist, 59-63; assoc prof physics, Mich Technol Univ, 63-86; PRES, KMB/TECH, 86- *Mem:* AAAS; Am Asn Physics Teachers; Sigma Xi. *Res:* Analysis of properties of ionized gas by microwave, optical and ultrasonic diagnostic techniques; microwave transmission through ionized air during missile re-entry; electronic instrumentation. *Mailing Add:* KMB/Tech 1700 College Ave Houghton MI 49931

BALDWIN, KENNETH M, b Leominster, Mass, Apr 25, 42. MUSCLE PHYSIOLOGY. *Educ:* Univ Iowa, PhD(exercise sci), 70. *Prof Exp:* PROF PHYSIOL & BIOPHYS, UNIV CALIF, IRVINE, 78- *Concurrent Pos:* Sr assoc dean, Col Med. *Mem:* Am Physiol Soc; Am Col Sports Med; Am Heart Asn. *Mailing Add:* Dept Physiol Univ Calif Irvine CA 92717

BALDWIN, LIONEL V, b Beaumont, Tex, May 30, 32; m 55; c 4. CHEMICAL ENGINEERING. *Educ:* Univ Notre Dame, BS, 54; Mass Inst Technol, SM, 55; Case Inst Technol, PhD(chem eng, phys chem), 59. *Prof Exp:* Res engr, Nat Adv Comt Aeronaut, Ohio, 55-59; unit head, NASA, 59-61; assoc prof eng, Colo State Univ, 61-64, actg dean, 64-65, dean col eng, 65-84; PRES, NAT TECHNOL UNIV, 84- *Concurrent Pos:* Off Naval Res, NIH & NASA eng res grants; asst engr, E I du Pont de Nemours & Co, Tex, 51-53; jr engr, W R Grace & Co, NY, 54. *Mem:* Am Soc Mech Engrs; Am Inst Chem Engrs; Am Inst Aeronaut & Astronaut; Am Soc Eng Educ. *Res:* Engineering science; turbulent fluid flow and diffusion; heat transfer in rarefied gas flows; low density, high energy plasma research. *Mailing Add:* Pres Nat Technol Univ PO Box 700 Ft Collins CO 80522

BALDWIN, MARK PHILLIP, b Schenectady, NY, Mar 4, 58; m 82. RESEARCH ADMINISTRATION. *Educ:* State Univ NY, Albany, BS, 80, MS, 81; Univ Wash, MS, 85, PhD(atmospheric sci), 87. *Prof Exp:* POSTDOCTORAL ASSOC, NORTHWEST RES ASSOC, 87- *Mem:* Am

Meteorol Soc. *Res:* Dynamics of stratosphere and trosposphere; observational studies of wave propagation, wave-mean flow interaction and sudden strastospheric warning; relation of polar strastospheric flow to tropospheric and tropical flow. *Mailing Add:* PO Box 3027 Bellevue WA 98009

BALDWIN, RANSOM LELAND, b Meriden, Conn, Sept 21, 35; m 57; c 3. ANIMAL SYSTEMS, RUMINANT NUTRITION. *Educ:* Univ Conn, BS, 57; Mich State Univ, MS, 58, PhD(biochem), 62. *Prof Exp:* Asst prof animal husb, 62-70, chmn dept, 78-81, PROF ANIMAL SCI, UNIV CALIF, DAVIS, 70- *Concurrent Pos:* Guggenheim Mem Found fel, 68-69; assoc ed, J Nutrit, 76-79 & 84-; Fulbright fel, UK, 83-84. *Honors & Awards:* Am Feed Mfrs Award, 70; Borden Award, 80. *Mem:* AAAS; Am Dairy Sci Asn; Am Inst Nutrit; Am Soc Animal Sci. *Res:* Ruminant digestion; physiology of lactation; nutritional energetics; mechanisms and quantitative aspects of regulation of animal and tissue metabolism; computer simulation modeling of animal systems. *Mailing Add:* Dept Animal Sci Univ Calif Davis CA 95616

BALDWIN, RICHARD H(AROLD), b Mich, Aug 7, 27. CHEMICAL ENGINEERING. *Educ:* Ill Inst Technol, BS, 50. *Prof Exp:* Chem engr, Res Dept, Standard Oil Co, Ind, 53-54, asst proj engr, 54-57, proj chem engr, 57-59, group leader, 59-61; group leader, Res Dept, Amoco Chem Corp, 61-65, sect leader, 65-67, asst div dir, 67-69, dir, 69-77, mgr, Chem Intermediates Process Design & Econ Anal, 77-81, MGR PROCESS ENG, ENG DEPT, AMOCO CHEM CORP, 81- *Mem:* Sigma Xi. *Res:* Chemical engineering fundamentals; fluid mixing. *Mailing Add:* 310 Euclid Ave Winnetka IL 60680

BALDWIN, ROBERT CHARLES, b Oakland, Calif, Sept 13, 42; c 2. RISK ASSESSMENT, ANALYTICAL CHEMISTRY. *Educ:* Univ Calif, Davis, BS, 64; Stanford Univ, PhD(org chem), 72. *Prof Exp:* Chemist anal chem, Vacuum Anal Lab, Lawrence Livermore Lab, 65-66; trainee toxicol, Univ Calif, San Francisco, 72-74; toxicologist inhalation toxicol, Flammability Res Ctr, Univ Utah, 74-77; sr inhalation toxicologist, Rohm & Haas Co, 77-91; DIR, MAMMALIAN TOXICOL, SRI INT, 91- *Res:* General toxicology; inhalation toxicology; safety evaluation. *Mailing Add:* SRI Int 333 Ravenswood Ave Menlo Park CA 94025-3493

BALDWIN, ROBERT LESH, b Madison, Wis, Sept 30, 27; m 65; c 2. PHYSICAL BIOCHEMISTRY. *Educ:* Univ Wis, BA, 50; Oxford Univ, PhD, 54; Univ Wis, postdoc, 54-55. *Prof Exp:* From asst prof to assoc prof biochem, Univ Wis, 55-59; assoc prof, 59-64, PROF BIOCHEM, STANFORD UNIV, 64- *Concurrent Pos:* Rhodes Scholar, 50-53, Guggenheim fel, 58-59, NIH sr fel, 63-64; mem, USPHS Fel Comt Biochem & Molecular Biol, 69-71; mem, NSF Adv Panel Biochem & Biophys, 74-76; vis prof, Col de France, Paris, 71; NIH Adv Panel Molecular & Cell Biophysics, 84-88; Sabbatical, Institut Pasteur, Dept Molecular Biol, Paris, 71-72, MRC Lab Molecular Biol, Cambridge, Eng, 78-79. *Mem:* Nat Acad Sci; Am Soc Biol Chemists; Biophys Soc; Am Chem Soc; Am Acad Arts & Sci; Protein Soc. *Res:* Mechanism of protein folding; author of numerous articles and publications. *Mailing Add:* Dept Biochem Stanford Univ Stanford CA 94305

BALDWIN, ROBERT RUSSEL, b Chicago, Ill, Nov 15, 16; m 44; c 2. BIOCHEMISTRY. *Educ:* DePauw Univ, AB, 38; Iowa State Univ, PhD(plant chem), 43. *Prof Exp:* Instr chem, Iowa State Col, 39-43, assoc chemist, Manhattan Dist Proj, 43-46; biophys sect head, Gen Foods Corp, NY, 46-48, dir biochem lab, 48-56 & chem res lab, 56-61; res supvr, Continental Baking Co, 61-66, assoc dir res, ITT Continental Baking Co, 66-74; vpres & dir res & develop, Morton Frozen Foods, Del Monte Frozen Foods, 74-81; RETIRED. *Concurrent Pos:* Mem, Nat Res Coun Comt Civil Defense, 56. *Mem:* AAAS; fel Am Inst Chem; fel Inst Food Technol; Am Asn Cereal Chem; NY Acad Sci; Am Chem Soc. *Res:* Physico-chemical studies; carbohydrates, fats and proteins; toxicology; quality control; product development. *Mailing Add:* 105 Wilson Ct Charlottesville VA 22901-2941

BALDWIN, ROBERT WILLIAM, b Huddersfield, Mar 27, 27; m 52; c 1. TUMOR IMMUNOLOGY, CLINICAL IMMUNOLOGY. *Educ:* Univ Birmingham, UK, BSc, 48, PhD, 51; Royal Col Pathologists, UK, MRC Path, 66; FRC Path, 73. *Prof Exp:* DIR ONCOL, CANCER RES LABS, UNIV NOTTINGHAM, UK, 60-; GIBBS SR RES FEL ONCOL, CANCER RES CAMPAIGN LABS, 71- *Concurrent Pos:* Prof tumour biol Oncol, Cancer Res Campaign Labs, Univ Nottingham, UK, 72-; chmn, Europ Orgn Res, Treatment of Cancer, Tumour Immunol, 72-75; counr, Sci Adv Bd, NY Cancer Res Inst Inc, 72-89 & MRC Leukaemic Steering Comt, 82-90. *Mem:* Am Asn Cancer Res; Am Asn Immunologists; Europ Asn Cancer Res (pres, 73-79); Royal Col Pathologists; Royal Inst Biol. *Res:* Basic immunology; immune related disorders; allergy; monoclonal antibodies; immunoconjugates; drug targeting. *Mailing Add:* Cancer Res Campaign Labs Univ Nottingham Nottingham NG7 2RD England

BALDWIN, ROGER ALLAN, b Decatur, Ill, June 2, 31; m 54; c 2. ORGANIC CHEMISTRY, FUEL SCIENCE. *Educ:* Millikin Univ, BS, 53; Mich State Univ, MS, 56, PhD(chem), 59. *Prof Exp:* Asst, Mich State Univ, 53-59; sr res chemist, Am Potash & Chem Corp, 59-60, res proj chemist, 60-63, group leader, 63-64, head org sect, 64-69; group leader, 69-78, MGR PROCESS CHEM, KERR MCGEE CORP, 78- *Mem:* Am Chem Soc; The Chem Soc. *Res:* Phosphorus and boron chemistry; azide chemistry; thiophene and sulfur chemistry; solvent extraction; coal conversion. *Mailing Add:* Tech Ctr Kerr McGee Corp PO Box 25861 Oklahoma City OK 73125

BALDWIN, RONALD MARTIN, b Argentina, June 6, 47; US citizen; m 77. RADIOCHEMISTRY. *Educ:* Univ Calif, Berkeley, BS, 69, PhD(org chem), 74. *Prof Exp:* Proj chemist, 74-76, res scientist, 76-79, sr res scientist, 79, proj mgr, 79-82, DEVELOP MGR, MEDI-PHYSICS, INC, 82- *Mem:* Am Chem Soc; Soc Nuclear Med; Sigma Xi; AAAS. *Res:* Development of derivatives of short-lived radioisotopes for use as radiodiagnostic agents in nuclear medicine; iodine 123 labeled amines for brain imaging and iodine 123 labeled fatty acids for heart imaging, and 99 Tc complexes. *Mailing Add:* Vet Admin Res Ctr 151 950 Campbell Ave West Haven CT 06516

BALDWIN, THOMAS O, b Ironton, Mo, Sept 18, 38; m 61; c 2. SOLID STATE PHYSICS. *Educ:* Univ Mo, BS, 59, MS, 61, PhD(physics), 64. *Prof Exp:* Res physicist, Solid State Div, Oak Ridge Nat Lab, 64-69; assoc prof physics, Southern Ill Univ, Edwardsville, 69-72, chmn dept physics, 71-74, prof physics, 72-80; MGT CONSULT, BOULDER CONSULT GROUP, 80- *Concurrent Pos:* Consult, Oak Ridge Nat Lab, 69-72. *Mem:* Am Chem Soc; Am Asn Physics Teachers. *Res:* Debye-Waller factors; expansion coefficients in ionic crystals; anomalous transmission in perfect metal crystals; x-ray studies of defects and radiation damage in crystals. *Mailing Add:* PO Box 3320 Boulder CO 80307

BALDWIN, THOMAS OAKLEY, b Jackson, Miss, June 3, 47; m 67, 78; c 3. BIOCHEMISTRY, MOLECULAR BIOLOGY. *Educ:* Univ Tex, Austin, BS, 69, PhD(zool), 71. *Prof Exp:* Fel biochem, Harvard Univ, 72-75; asst prof biochem, Univ Ill, Urbana, 75-81; assoc prof, 81-86, PROF BIOCHEM, TEX A&M UNIV, COLLEGE STATION, 86- *Concurrent Pos:* Mem corp & instr physiol, Marine Biol Lab, 77- *Mem:* AAAS; Biophys Soc; Am Chem Soc; Am Soc Biochem & Molecular Biol; Am Soc Photobiol; Am Soc Microbiol. *Res:* Structure and function of proteins; structure of bacterial luciferase; analysis of the structures of mutant forms of bacterial luciferase; effects of protein structure upon the rate of enzyme inactivation and degradation in vivo; mechanism of protein folding and assembly of multimeric structures. *Mailing Add:* Dept Biochem & Biophys Tex A&M Univ College Station TX 77843-2128

BALDWIN, VANIAH HARMER, JR, physical chemistry, for more information see previous edition

BALDWIN, VIRGIL CLARK, JR, b Salt Lake City, Utah, Jan 15, 40; m 64; c 8. GROWTH & YIELD PREDICTION OF PINE STANDS, ECOLOGICAL MODELING. *Educ:* Univ Utah, BS, 67; Colo State Univ, MS, 69; Duke Univ, PhD(forest biomet), 75. *Prof Exp:* Res scientist, Maritimes Forest Res Ctr, Can Forestry Serv, 72-75; asst prof forestry, Wash State Univ, 75-78; RES FORESTER, SOUTHERN FOREST EXP STA, US FOREST SERV, 78-; INSTR STATIST, NORTHWESTERN STATE UNIV LA, 81- *Concurrent Pos:* Consult, Ouzinkie Native Corp, 76; prin investr, US Dept Energy grant, 78 & US Bur Indian Affairs grant, 78; sabbatical, Univ Ga, 90-91. *Mem:* Soc Am Foresters; Am Forestry Asn. *Res:* Modeling physiological processes; modeling forest stand growth and yield in terms of weight and volume for selection of optimum management practices under specified conditions; silviculture and ecology of forest stands. *Mailing Add:* Southern Forest Exp Sta 2500 Shreveport Hwy Pineville LA 71360

BALDWIN, W GEORGE, b Winnipeg, Man, May 3, 38; m 65; c 2. INORGANIC CHEMISTRY. *Educ:* Univ Man, BSc, 59, MSc, 61; Univ Melbourne, PhD(chem), 66. *Prof Exp:* Fel, Univ Adelaide, 66 & Royal Inst Technol, Sweden, 66-68; from asst prof to assoc prof, 68-84, PROF INORG CHEM, UNIV MAN, 84- *Mem:* Chem Inst Can. *Res:* Thermodynamic and kinetic studies of complex ions. *Mailing Add:* Dept Chem Univ Man Winnipeg MB R3T 2N2 Can

BALDWIN, WILLIAM RUSSELL, b Danville, Ind, July 29, 26; m 47; c 1. OPTOMETRY, EDUCATION ADMINISTRATION. *Educ:* Pac Univ, BS, 49, OD, 51; Ind Univ, MS, 56, PhD(phys optics), 64. *Hon Degrees:* LHD, New Eng Col Optom, 82. *Prof Exp:* Dir, Optom Clin, Ind Univ, 59-63; dean, Col Optom, Pac Univ, 63-69; pres, New Eng Col Optom, 69-79; dean, Col Optom, Univ Houston, 79-90; PRES, RIVER BLINDNESS FOUND, 90- *Concurrent Pos:* Pvt pract optom, Beech Grove & Bloomington, Ind, 51-63; mem bd dirs, Am Optom Found, 70-74; mem bd, Nat Soc Prevent Blindness, 88- *Mem:* Am Optom Asn; Asn Schs & Cols Optom (pres, 73-75); Am Acad Optom; AAAS. *Res:* Mechanisms of refractive change of the eye; myopia and physiological and behavioral correlates; optometric education. *Mailing Add:* Col Optom Univ Houston 4901 Calhoun Houston TX 77004

BALDWIN, WILLIAM WALTER, b Ft Wayne, Ind, July 27, 40. MICROBIOLOGY, BIOCHEMISTRY. *Educ:* Ind Cent Col, AB, 62; Ind Univ, MSEd, 67, PhD(microbiol), 73. *Prof Exp:* ASSOC PROF MICROBIOL, MED SCH, IND UNIV NORTHWEST, 73- *Mem:* AAAS; Am Soc Microbiol; Can Soc Microbiol. *Res:* Antitumor agents; cellular energy metabolism and osmolarity responses. *Mailing Add:* Ind Univ Northwest Med Sch 3400 Broadway Gary IN 46408

BALDY, MARIAN WENDORF, b Cleveland, Ohio, June 18, 44; m 65. GENETICS, ENOLOGY. *Educ:* Univ Calif, Davis, AB, 65, PhD(genetics), 68. *Prof Exp:* Res assoc biochem, Med Sch, Univ Ore, Portland, 69-70, NIH fel, 70; asst prof plant sci, Calif State Univ, Chico, 71-73; inst agr, Butte Community Col, 73; from asst prof to assoc prof plant sci, Calif State Univ, Chico, 73-80, spec asst to dean, Sch Agr & Home Econ, 80-83, PROF AGR, CALIF STATE UNIV, CHICO, 80- *Concurrent Pos:* Winemaker, Butte Creek Vineyards, Inc, 72-76; dir, Univ Hons Prog, Calif State Univ, Chico, 90- *Mem:* Soc Wine Educrs; Nat Col Hons Coun. *Mailing Add:* Sch Agr Calif State Univ Chico CA 95929-0310

BALDY, RICHARD WALLACE, b Madena, Calif, Jan 15, 42; m 65. AGRICULTURE. *Educ:* Univ Calif, Davis, BS, 64, PhD(plant physiol), 68. *Prof Exp:* Sloan fel biol, Reed Col, 68-70; PROF POMOL & VITICULTURE, CALIF STATE UNIV, CHICO, 70- *Mem:* Am Soc Hort Sci; Appl Econ Entomology. *Res:* Integrated crop and pest management. *Mailing Add:* Dept Agr Calif State Univ Chico CA 95929

BALE, HAROLD DAVID, b Fargo, NDak, Oct 3, 27; m 54; c 4. X-RAY CRYSTALLOGRAPHY. *Educ:* Concordia Col, BA, 50; Univ NDak, MS, 53; Univ Mo, PhD, 59. *Prof Exp:* Phys sci aide, US Bur Mines, 50-53; instr physics, Univ NDak, 53-55 & Univ Mo, 55-59; from asst prof to assoc prof, 59-70, actg chmn dept, 63-67, PROF PHYSICS, UNIV NDAK, 70- *Concurrent Pos:* Fel, Univ Mo, 69-70; vis prof, Univ Mo, 78 & 84. *Res:* Small angle x-ray scattering. *Mailing Add:* PO Box 8008 Grand Forks ND 58202

BALENTINE, J DOUGLAS, b Greenville, SC, Sept 24, 37. NEUROPATHOLOGY. *Educ:* Med Univ SC, MD, 63. *Prof Exp:* PROF PATH, MED UNIV SC, 75-, VCHMN DEPT, 79- *Mem:* Am Asn Neuropathologists (pres, 86-);; Int Soc Neuropath; Am Asn Pathologists; Int Acad Path; Undersea Med Soc. *Mailing Add:* Dept Path & Neuropath Med Univ SC 171 Ashley Ave Charleston SC 29425

BALES, BARNEY LEROY, b Amarillo, Tex, Jan 4, 39. BIOPHYSICS. *Educ:* Univ Colo, BS, 62, PhD(physics), 68. *Prof Exp:* Staff scientist biophys, Nat Inst Sci Res, Venezuela; assoc prof, 75-80, PROF PHYSICS, CALIF STATE UNIV, NORTHRIDGE, 80- *Mem:* Am Phys Soc; Sigma Xi. *Res:* Structure and dynamics of the biological membrane and of liquid crystals; micelles. *Mailing Add:* Dept Physics Calif State Univ Northridge CA 91330

BALES, CONNIE WATKINS, b Morristown, Tenn, Oct 21, 54; m 75. BIOLOGY OF AGING, CLINICAL NUTRITION. *Educ:* Univ Tenn, Knoxville, BS, 76, PhD(nutrit sci), 81. *Prof Exp:* Therapeut Nutritionist, St Mary's Med Ctr, Knoxville, Tenn, 76-77; res asst, Univ Tenn, Knoxville, 77-81; asst prof nutrit foods, Univ Tex, Austin, 81-; PROF NUTRIT, CTR AGING, DUKE UNIV. *Mem:* Am Dietetic Asn; Am Geront Soc; AAAS; Am Home Econ Asn; Am Col Nutrit. *Res:* Relationships between nutrition and aging, including prolongation of lifespan by dietary restriction; protein and mineral status of the elderly; trace mineral metabolism. *Mailing Add:* Ctr Aging Duke Univ Med Ctr Box 3003 Durham NC 27710

BALES, HOWARD E, b Ohio, Dec 22, 12; m 37; c 2. SPECTROSCOPY, RESEARCH ADMINISTRATION. *Educ:* Wilmington Col, BS, 34. *Prof Exp:* Teacher, pub schs, Ohio, 34-43; chemist, Nat Cash Register Co, 43-46, x-ray diffractionist, 46-49, spectroscopist, 49-53; spectroscopist, C F Kettering Res Lab, 53-57, res physicist, 57-62, staff scientist, 62-64; instr biol & asst dir, Col Sci & Eng, Wright State Campus, Miami Univ-Ohio State Univ, 64-68, from assoc dir to dir, Ohio State Tech Serv, 68-70, assoc dir res & develop & Ed, Wright State Univ Res News, Wright State Univ, 70-77, dir tech & bus serv & assoc dir res serv, 77-78; RETIRED. *Concurrent Pos:* Fel & vis scientist, Ohio Acad Sci. *Mem:* AAAS; Am Chem Soc; Am Phys Soc; Soc Appl Spectros; Am Crystallog Asn. *Res:* Application of instrumental methods to research allied with the study of photosynthesis and nitrogen fixation. *Mailing Add:* 247 Edison Blvd Xenia OH 45385

BALESTRINI, SILVIO J, b Fresno, Calif, June 8, 22; m 48; c 4. PHYSICS. *Educ:* Univ Calif, Berkeley, BA, 48, PhD(physics), 54. *Prof Exp:* Staff mem mass spectros, Los Alamos Sci Lab, 54-61, prin investr nuclear criticality studies, 61-71, PRIN INVESTR, ISOTOPIC SEPARATIONS ON-LINE & FISSION YIELDS, LOS ALAMOS NAT LAB, 70-, PRIN INVESTR, ISOTOPIC SEPARATIONS OFF-LINE, 77- *Concurrent Pos:* Invited researcher deleg by II Phys Inst, Univ Giessen with Ger Res Asn grant, Laue-Langevin Inst, Grenoble, France, 76-77; consult, Los Alamos Nat Lab, Isotope Servs, Inc. *Mem:* Am Phys Soc; Sigma Xi. *Res:* Mass spectroscopy; nuclear critical assembly studies; fission yield determinations by on-line isotopic separation; off-line isotopic separations; 235U assays by delayed neutron detection. *Mailing Add:* 113 Grand Cayon Dr Los Alamos NM 87544

BALFOUR, HENRY H, JR, b Jersey City, NJ, Feb 9, 40; m 67; c 3. PEDIATRICS, VIROLOGY. *Educ:* Princeton Univ, BA, 62; Columbia Univ Col Physicians & Surgeons, MD, 66. *Prof Exp:* Attend pediatrician, Wright-Patterson AFB, Ohio, 68-70; from asst prof to assoc prof, 72-79, PROF LAB MED & PATH & PEDIAT, UNIV MINN, 79-, DIR, DIV CLIN MICROBIOL, 74- *Concurrent Pos:* prin investr, NIH grant, 76-, Ctr Dis Control grant, 77-79 & Wellcome Found grant, 78-, NIH NIAID-Funded Aids Clin Trials Univ, MN, 87-; proj dir NIH grant, 77-79. *Mem:* Am Soc Microbiol; Soc Exp Biol & Med; Soc Pediat Res; Infectious Dis Soc Am; Am Pediat Soc. *Res:* Pathogenesis and treatment of herpes virus diseases and AIDS. *Mailing Add:* Health Sci Ctr Box 437 UMHC Harvard E River Rd Minneapolis MN 55455

BALFOUR, WALTER JOSEPH, b Hilton, Scotland, May 1, 41; Can citizen; m 74; c 2. MOLECULAR SPECTROSCOPY, PHYSICAL CHEMISTRY. *Educ:* Aberdeen Univ, BSc, 63, DSc, 85; Mc Master Univ, PhD(chem), 67. *Prof Exp:* Nat Res Coun Can fel physics, 67-69; from asst prof to assoc prof, 69-81, PROF CHEM, UNIV VICTORIA, BC, 81- *Concurrent Pos:* Vis scientist, Nat Sci Res Ctr, Orsay, France, 75-76, Inst Physics, Univ Stockholm, Sweden, 82-83 & Nat Res Coun Can, 89-90. *Mem:* Can Asn Physicists; Can Inst Chem. *Res:* Spectroscopy and structure of diatomic and small polyatomic molecules. *Mailing Add:* Dept Chem Univ Victoria Victoria BC V8W 3P6 Can

BALFOUR, WILLIAM MAYO, b Pasadena, Calif, Nov 26, 14; m 39; c 4. PHYSIOLOGY. *Educ:* Univ Minn, MD, 40, MS, 48. *Prof Exp:* Intern internal med, Univ Rochester, 39-40, resident path, 40-42; consult, Mayo Clin & Mayo Found, 48-56; asst prof physiol, 57, res assoc, 59-64, assoc prof, 62-66, dean student affairs, 68-70, vchancellor student affairs, 68-70, PROF PHYSIOL & CELL BIOL, SCH MED, UNIV KANS, 66- *Concurrent Pos:* Fel internal med, Mayo Found, 42-48, instr, Univ Minn, 50-56; USPHS fel, Univ Kans, 57-59. *Mem:* AAAS; Am Diabetes Asn; AMA; Am Physiol Soc; NY Acad Sci. *Res:* Cerebral energy metabolism; cellular structure and function of brain; endocrine physiology. *Mailing Add:* 1505 University Dr Lawrence KS 66045

BALGOOYEN, THOMAS GERRIT, b Grand Haven, Mich, July 5, 43; c 1. ORNITHOLOGY. *Educ:* San Jose State Univ, BA, 66; MA, 67; Univ Calif, Berkeley, PhD(zool), 72. *Prof Exp:* Fel ornith, Univ Calif, Berkeley, 72-74; ASSOC PROF BIOL, SAN JOSE STATE UNIV, 75- *Mem:* Am Ornithologists Union; Cooper Ornith Soc; Wilson Ornith Soc; Sigma Xi. *Res:* Adaptive strategies of raptorial birds in terms of time and energy; interactions of avian behavior and ecology. *Mailing Add:* Dept Biol Sci San Jose State Univ San Jose CA 95192

BALIGA, B SURENDRA, b Bangalore, India, Jan 12, 35; US citizen; m 64; c 2. BIOCHEMISTRY, NUTRITION. *Educ:* Poona Univ, BSc Hons, 56, MSc, 58; Indian Inst Sci, Bangalore, PhD(biochem), 63. *Prof Exp:* Fel biochem, Indian Inst Sci, Bangalore, 63-64 & Med Ctr, Tufts Univ, Boston, 65-66; res assoc biochem nutrit, Mass Inst Technol, 67-68; ASSOC PROF BIOCHEM NUTRIT & DIR NUTRIT LAB, MED CTR, UNIV S ALA, 78- *Mem:* Am Soc Biol Chemists; Am Inst Nutrit. *Res:* Regulatory mechanisms in eucaryotic systems. *Mailing Add:* Dept Pediat Univ S Ala Mobile AL 36617

BALIGA, JAYANT, b Madras, India, Apr 28, 48; m 75; c 1. SEMICONDUCTOR DEVICES, SEMICONDUCTOR TECHNOLOGY. *Educ:* Indian Inst Technol, Madras, BTech, 69; Rensselaer Polytech Inst, NY, MS, 71, PhD(elec eng), 74. *Prof Exp:* Mem staff power devices, 74-80, proj leader gallium arsenide power devices, 80-81, mgr, power device unit, 81-83, MGR, HV DEVICE PROG, GEN ELEC RES LABS, NY, 83- *Concurrent Pos:* Adj prof, Rensselaer Polytech Inst, NY, 74-80; mem, Electron Mat Comt, Am Inst Mech Eng, 78-, Int Electron Devices Meeting, Inst Elec & Electronics Engrs, 82-; UN consult, Cent Electron Eng Res Inst, 79-80; Coolidge Mem Found fel. *Mem:* Fel Inst Elec & Electronics Engrs; Electrochem Soc; Sigma Xi. *Res:* Investigation of the physics and technology of power semiconductor devices and the development of new power switching with improved performance for power electronics; author or co-author of over 150 publications, and contributor to seven books. *Mailing Add:* 2612 Bembridge Dr Raleigh NC 27613

BALIN, ARTHUR KIRSNER, b Philadelphia, Pa, June 17, 48. DERMATOLOGY, AGING RESEARCH. *Educ:* Northwestern Univ, BA, 70; Univ Pa, MD, 74, PhD(biochem), 76, Am Bd Internal Med, dipl, 79, Am Bd Dermat, dipl, 80, Am Bd Path, Dermatopathology, dipl, 81. *Prof Exp:* Resident internal med, Hosp Univ Pa, 76-78; resident dermat, Sch Med, Yale Univ, 79-80, res dermatopathology, 80-81; ASST PROF & ASSOC PHYSICIAN, ROCKEFELLER UNIV, NY, 81- *Concurrent Pos:* Assoc scientist aging res, Wistar Inst Anat & Biol, 76-78; adj asst prof dermat, Sch Med, Cornell Univ, 81-; asst attend physician, New York Hosp, 81-; fel Am Acad Dermat; fel Col Am Pathologists; fel Am Soc Dermatopathology; fel Am Soc Dermatologic Surg. *Honors & Awards:* Wilton M Earle Award, Tissue Cult Asn, 75. *Mem:* Am Chem Soc; AMA; Am Geriat Soc; Geront Soc; Tissue Cult Asn. *Res:* Biology of aging; molecular biology, clinical; photobiology, radiation biology; oxygen toxicity, free radical biology, cell cycle, cell culture; dermatology. *Mailing Add:* Arthur Kirsner Balin 2129 Providence Ave Chester PA 19103

BALINSKI, MICHEL LOUIS, b Geneva, Switz, Oct 6, 33; US citizen; m 57; c 2. APPLIED MATHEMATICS. *Educ:* Williams Col, BA, 54; Mass Inst Technol, MS, 56; Princeton Univ, PhD(math), 59. *Hon Degrees:* MA, Yale Univ, 78. *Prof Exp:* Instr math, Princeton Univ, 59-61, lectr & res assoc, 61-63; assoc prof econ, Univ Pa, 63-65; assoc prof, 65-69, prof math, Grad Ctr, City Univ NY, 69-77; prof admin sci, Yale Univ, 78-80; prof, Inst August Comte, Paris, France, 80-81 & Ecole Polytech, 81-82; prof appl math, Statist & Econ, State Univ NY, Stony Brook, 83-89; DIR RES, NAT CTR SCI RES, LAB ECON ECOLE POLYTECH, PARIS, FRANCE, 83-; DIR, LABORATOIRE D'ECONOMETRIE, 89-; CO-DIR, PHD PROG MODELS & MATH METHODS ECON, UNIV PARIS ONE-ECOLE POLYTECHIQUE, 89- *Concurrent Pos:* Consult, Mathematica, 59-74, Rand Corp, 60-66 & Mobil Oil Res Labs, 66-69; lectr, Univ Mich Summer Eng Conf, 62-69; IBM Corp World Trade Corp prof fel, Paris, 69-70; mem opers res coun adv body, Mayor, City of New York, 67-70; founding ed-in-chief, Math Programming and Math Programming Studies, 70-80; consult, Off Radio-diffusion Television, France, 71; vis prof, Swiss Fed Inst Technol, Lausanne, 72-73 & Univ Grenoble, 74-75; chmn syst & decision sci area, Int Inst Appl Systs Anal, Laxenburg, Austria, 75-77; sr consult econ, 78-80; chmn, Math Prog Soc, 86-89. *Honors & Awards:* Lanchester Prize, Opers Res Soc Am, 65; Lester R Ford Award, Math Asn Am, 75; Spec Serv Award, Math Prog Soc, 82. *Mem:* Math Prog Soc; Soc Indust Appl Math; Math Asn Am; Opers Res Soc Am. *Res:* Mathematical programming; combinatorial applied mathematics; decision and system sciences; mathematical methods in the social sciences. *Mailing Add:* Laboratoire d'Econométrie Ecole Polytech One rue Descartes Paris 75005 France

BALINSKY, DORIS, b Frankfurt, Germany, Dec 3, 34; c 2. CANCER RESEARCH, NUTRITION. *Educ:* London Univ, PhD(biochem), 59. *Prof Exp:* Prof biochem, Iowa State Univ, 81-85; prog adminr, USDA, 85-88; prog dir, NIH, 88-91; AT NAT HEART LUNG & BLOOD INST, NIH, 91- *Mem:* Am Asn Cancer Res; Am Soc Biol Chemists. *Mailing Add:* 10A10 Westwood Bldg Rm 554 Nat Heart Lung & Blood Inst Bethesda MD 20892

BALINT, FRANCIS JOSEPH, b Johnstown, Pa, Mar 2, 32; m 56; c 6. MATHEMATICS. *Educ:* Indiana Univ Pa, BS, 54; Univ Pittsburgh, MS, 59. *Prof Exp:* Customs engr, Int Bus Mach Corp, 54-55; programmer, Gulf Res & Develop Co, 55-60, opers leader, 60-62, sr proj engr, 63-65; chief, Mgt & Prog Br, US Weather Bur, 65-66, chief, Mgt & Planning Br, 66-70, chief, Mgt Systs Div, 70-72, dir, Off Info & Mgt Serv, 78-84, DEP DIR, OFF MGT & COMPUT SYSTS, NAT OCEANIC & ATMOSPHERIC ADMIN, 72-; DIR, OFF INFO SYSTS, US DEPT COM, 84- *Concurrent Pos:* Instr, Indiana State Col Pa, 62; mem exec bd, Share, Inc, 68-69, mgr, Basic Systs Div, 72-73; Nat lectr, Asn Comput Mach. *Honors & Awards:* US Dept Com Silver Medal, 71. *Mem:* NY Acad Sci; Asn Comput Mach; AAAS. *Res:* Application of supercomputers for scientific and engineering problem solutions. *Mailing Add:* 2400 Jackson Pkwy Vienna VA 22180

BALINT, JOHN ALEXANDER, b Budapest, Hungary, Feb 11, 25; m 49; c 2. MEDICINE. *Educ:* Univ Cambridge, BA, 45, MB, BCh, 48; FRCP(UK), 76. *Prof Exp:* Fel gastroenterol, Col Med, Univ Cincinnati, 58-59; fel med, Sch Med, Johns Hopkins Univ, 59-60; asst prof med & gastroenterol, Med Col Ala, 60-63; assoc prof, 63-68, chmn, Dept Med, 81-88, PROF MED & GASTROENTEROL, ALBANY MED COL, 68-, RICHARD T BEEBE PROF MED, 88- *Concurrent Pos:* NIH trainee lipid biochem, Sinai Hosp, Baltimore, 59-60. *Mem:* Am Fedn Clin Res; Am Physiol Soc; Am Soc Clin

Invest; Brit Med Asn; fel Royal Soc Med; Sigma Xi; Am Gastroenterol Asn; Am Asn Study Liver Dis; Am Clin & Climat Asn. *Res:* Lipid metabolism in relation to fat absorption and intrahepatic phospholipid metabolism; lipid and protein disorders in hereditary lipid storage diseases. *Mailing Add:* Dept Gastroenterol Albany Med Col Albany NY 12208

BALINTFY, JOSEPH L, b Budapest, Hungary, May 21, 24; US citizen; m 83; c 2. OPERATIONS RESEARCH. *Educ:* Univ Tech Sci, Budapest, dipl eng, 47, dipl econ, 48; Johns Hopkins Univ, DEng, 62. *Prof Exp:* Adj prof mech, Univ Tech Sci, 52-56; prof, Tulane Univ, 60-70; prof mgt sci, Univ Mass, 70-87; AT CONVERSE COL, 87- *Mem:* Operations Res Soc Am; Inst Mgt Sci; Math Prog Soc. *Res:* Mathematical programming models of food consumption decision support systems optimal diets. *Mailing Add:* Dept Bus & Econ Studies Converse Col Spartanburg SC 29301

BALIS, JOHN ULYSSES, b July 7, 32; m; c 1. LUNG PATHOLOGY. *Educ:* Nat Univ Athens, MD, 57. *Prof Exp:* PROF PATH, COL MED, UNIV SFLA, 78- *Mem:* Am Asn Pathologists; Int Acad Path. *Res:* Lung surfactant system. *Mailing Add:* Dept Path Univ SFla Col Med Box 11 Tampa FL 33612

BALIS, MOSES EARL, b Philadelphia, Pa, June 19, 21; m 45; c 2. BIOCHEMISTRY. *Educ:* Temple Univ, BA, 43; Univ Pa, MA, 47, PhD(org chem), 49. *Prof Exp:* Assoc, 49-54, assoc mem, 54-65, mem, 65-87, chief div cell metab, Sloan-Kettering Inst Cancer Res, 71-87; prof biochem, Sloan-Kettering Div, Cornell Univ, 66-87; CONSULT, 87- *Concurrent Pos:* Assoc prof biochem, Sloan-Kettering Div, Cornell Univ, 54-56, chmn dept, 69-72; chmn biochem unit, Sloan-Kettering Inst Cancer Res, 69-71, 77-79, mem ed adv bd, Cancer Res, 70-73, assoc ed, 69-81; partic, Nat Cancer Planning Prog, 72; mem adv comt pathogenesis, Am Cancer Soc; mem rev comt, Nat Large Bowel Cancer Proj, 76-81. *Honors & Awards:* USPHS Res Career Award, 63. *Mem:* Am Chem Soc; Am Soc Biol Chem; Harvey Soc; Am Asn Cancer Res. *Res:* Biochemistry of differentiation and malignancy; regulation of enzyme function; purine metabolism; action of antimetabolites; genetic defects; mechanisms of carcinogenesis. *Mailing Add:* 146 Byrdcliffe Rd Woodstock NY 12498-1026

BALISH, EDWARD, b Scranton, Pa, Apr 23, 35; m 61; c 1. MICROBIOLOGY, BIOCHEMISTRY. *Educ:* Univ Scranton, BS, 57; Syracuse Univ, MS, 59, PhD(microbiol), 63. *Prof Exp:* Asst, Syracuse Univ, 57-60, res assoc, 62-63; resident res assoc, Argonne Nat Lab, 63-65; res scientist-microbiologist, Oak Ridge Assoc Univs, 66-70; assoc prof, 70-77, PROF MED MICROBIOL & SURG, MED SCH, UNIV WIS-MADISON, 78- *Mem:* AAAS; Am Soc Microbiol; Radiation Res Soc; Can Soc Microbiol; Int Soc Human & Animal Mycol; Sigma Xi. *Res:* Microbial biochemistry; enzymology; gnotobiology; mechanism of pathogenicity of fungi, particularly Candida albicans. *Mailing Add:* Med Sch Univ Wis 179 Med Sci Bldg Rm 185 Madison WI 53706

BALK, PIETER, b Breukelen, Neth, Dec 27, 24; m 53; c 3. MICROELECTRONICS, SEMICONDUCTOR DEVICE TECHNOLOGY. *Educ:* State Univ, Utrecht, MSc, 53; Free Univ, Amsterdam, PhD(chem physics), 57. *Prof Exp:* Res fel, Nat Res Coun, Ottawa, Can, 57-58; res staff mem, Int Bus Mach Res Ctr, NY, 59-72; prof semiconductor elec, Tech Univ, Aachen, 72-87; DIR & PROF SEMICONDUCTOR ELEC, DELFT INST MICROELECTRONICS & SUBMICRON TECHNOL, TECH UNIV, 88- *Mem:* Fel Inst Elec & Electronics Engrs; Electrochem Soc. *Res:* Physics and technology of mos devices and III-V devices. *Mailing Add:* DIMES Tech Univ PO Box 5053 Delft 2600 GB Netherlands

BALKA, DON STEPHEN, b South Bend, Ind, Aug 14, 46; m 69; c 2. MATHEMATICS. *Educ:* Mo Valley Col, BS, 68; Ind Univ, MS, 70; St Francis Col, MS, 71; Univ Mo, Columbia, PhD(math educ), 74. *Prof Exp:* Jr high sch teacher math, Plymouth Community Sch Corp, 68-69; sr high sch teacher math, Union-North United Sch Corp, 69-71; lectr math educ, Univ Mo, Columbia, 73-74; from asst prof to assoc prof, 74-87, PROF MATH, ST MARY'S COL, 88- *Concurrent Pos:* Math consult, Creative Publ, Inc. 74-, Ind Dept Pub Instr, 74-75; assoc fac, Ind Univ, South Bend, 75-; math consult, Didax Educational Resources, 87-, Cuisenaire Co Am, 87- *Honors & Awards:* Maria Piata Award. *Mem:* Nat Coun Teachers Math; Sch Sci & Math Asn; Nat Educ Asn; Math Asn Am. *Res:* Mathematics manipulatives in the classroom; creativity in mathematics; handheld calculators in classrooms; problem solving; analysis of Indiana high school math contest results. *Mailing Add:* Dept Math St Mary's Col Notre Dame IN 46556

BALKANY, THOMAS JAY, b Miami, Fla; m 70; c 1. OTOLOGY. *Educ:* Univ Miami, MD, 72; Am Bd Otolaryngol, cert. *Prof Exp:* Resident otolaryngol, 73-76, chief resident, 76-77, ASST PROF OTOL, MED CTR, UNIV COLO, 77-, DIR SECT PEDIAT OTOLARYNGOL, 72-, VCHMN, DEPT OTOLARYNGOL. *Concurrent Pos:* Chmn, Dept Otolaryngol, Denver Children's Hosp, 79- *Mem:* Fel Am Acad Otolaryngol; Soc Univ Otolaryngologists; fel Am Col Surgeons; fel Int Col Surgeons. *Res:* Electrical cochlear prosthesis for the profoundly deaf; etiology, diagnosis and treatment of otitis media; developmental delays in Down's Syndrome children due to hearing loss. *Mailing Add:* Sch Med Dept Otolaryngol Univ Miami PO Box 016960 Miami FL 33101

BALKE, BRUNO, b Braunschweig, Ger, Sept 6, 07; US citizen; m 35; c 4. PHYSIOLOGY. *Educ:* Acad Phys Educ, Ger, dipl, 31; Univ Berlin, MD, 36; Univ Leipzig, Dr med habil, 45. *Prof Exp:* Consult sports med, Univ Berlin, 36-42; head dept performance physiol, Med Sch Ger Army Mt Corps, 42-45; physiologist, Sch Aviation Med, US Air Force, 50-60; chief biodynamics br, Civil Aeromed Res Inst, Fed Aviation Agency, 60-64; prof, 64-73, EMER PROF PHYS EDUC & PHYSIOL, UNIV WIS-MADISON, 73-; DIR ASPEN CARDIO-PULMONARY REHAB UNIT, 73- *Concurrent Pos:* Assoc prof, Air Univ, 58-60; prof, Col Med, Univ Okla, 60-; attend physician, Vet Admin Hosp, 60-; ed, J Med & Sci in Sports, 69-73; consult, Colo Govt Coun Phys Fitness, Health & Tennis Corp Am, Longevity Ctr, Santa Barbara

& Health Enhancement & Lifestyle Med Inst, Palm Springs, Calif. *Mem:* Soc Exp Biol & Med; Aerospace Med Asn; Am Physiol Soc; Am Col Sports Med (pres, 65). *Res:* Medical and physiological research applied to human stress tolerance under condition of exercise, climate, altitude and high speed flying; role of exercise in cardiac rehabilitation. *Mailing Add:* PO Box 1354 Basalt CO 81621

BALKE, NELSON EDWARD, b Edwardsville, Ill, Mar 28, 49. PLANT PHYSIOLOGY, WEED SCIENCE. *Educ:* Univ Ill, BS, 71; Purdue Univ, MS, 74, PhD(plant physiol), 77. *Prof Exp:* asst prof, 77-83, ASSOC PROF AGRON, UNIV WIS-MADISON, 83- *Mem:* Am Soc Plant Physiologists; Weed Sci Soc Am; Am Soc Agron; Crop Sci Soc Am; Coun Agr Sci & Technol; AAAS. *Res:* Absorption and metabolism of organic molecules by plant cells; enzymatic detoxication of xenobiotics by plants. *Mailing Add:* 259 Moore Hall Univ Wis 1575 Linden Dr Madison WI 53706

BALKISSOON, BASDEO, b Trinidad, WI, July 29, 22; US citizen; m 53; c 2. CARDIOVASCULAR PHYSIOLOGY. *Educ:* Univ London, Inter BA, 46; Howard Univ, BS, 48, MD, 52, MS, 56, PhD(physiol & surg), 64; Am Board Surg, dipl, 1959. *Prof Exp:* Resident surg, 53-58, asst prof, 59-62, ASST PROF SURG & PHYSIOL, COL MED, HOWARD UNIV, 65-; PVT PRACT GEN SURG, 65- *Mem:* Fel Am Col Surgeons; NY Acad Sci; AMA; Nat Med Asn; Pan Am Med Asn; Am Heart Asn. *Res:* Production and cardiovascular effects of heart blocks in dogs. *Mailing Add:* 1248 Monroe St NE Washington DC 20017-2596

BALKWILL, DAVID LEE, b Sheboygan, Wis, Jan 19, 49. MICROBIOLOGY. *Educ:* Univ Wis-Madison, BS, 71; Pa State Univ, MS, 73, PhD(microbiol), 77. *Prof Exp:* Asst prof, dept microbiol, Univ NH, 77-82; from asst prof to assoc prof, 82-90, PROF, DEPT BIOL SCI, FLA STATE UNIV, 90- *Mem:* Am Soc Microbiol; Electron Microscopy Soc Am; AAAS. *Res:* Natural ecology of aquifer and other subsurface sediment; physiological and genetic characteristics of micro-organisms in subsurface sediments. *Mailing Add:* Dept Biol Sci Fla State Univ Tallahassee FL 32306-2043

BALL, BILLIE JOE, b Louin, Miss, Apr 8, 29; m 54; c 2. ELECTRICAL ENGINEERING. *Educ:* Miss State Univ, BS(chem eng) & BS(elec eng), 56, MS, 59; Tex A&M Univ, PhD(elec eng), 62. *Prof Exp:* Engr, Westinghouse Elec Corp, 55 & Reynolds Metals Co, 56 & 57; instr elec eng, Miss State Univ, 57-59; from instr to assoc prof, Tex A&M Univ, 59-63; assoc prof, 63, PROF ELEC ENG, MISS STATE UNIV, 80-, HEAD DEPT, 64- *Mem:* Inst Elec & Electronics Engrs; Am Soc Eng Educ; Nat Soc Prof Engrs; Instrument Soc Am. *Mailing Add:* Box 2848 Mississippi State MS 39762

BALL, BILLY JOE, b Crowell, Tex, Nov 29, 25; m 47; c 1. MATHEMATICS. *Educ:* Univ Tex, BA, 48, PhD(math), 52. *Prof Exp:* Instr pure math, Univ Tex, 49-52; actg asst prof math, Univ Va, 52-53, asst prof, 53-59; head dept math, Univ Ga, 63-69, from assoc prof to prof math, 59-86; RETIRED. *Mem:* Am Math Soc; Math Asn Am. *Res:* Point set topology. *Mailing Add:* 3304 Glen Rose Dr Austin TX 78731

BALL, CARROLL RAYBOURNE, b Leakesville, Miss, Oct 11, 25; div; c 3. ANATOMY. *Educ:* Univ Miss, BA, 47, MS, 48, PhD(anat), 63. *Prof Exp:* Asst zool, Univ Miss, 46-48; asst, Duke Univ, 48-50, instr, 50-51; instr micros & gross anat, Med Sch, WVa Univ, 51-57; asst prof biol, Univ Southern Miss, 57-60, instr anat, 62-63; from asst prof to assoc prof, 63-71, PROF ANAT, SCH MED, UNIV MISS, 71- *Mem:* Am Asn Anat; Am Soc Exp Path; Soc Exp Biol & Med; Sigma Xi. *Res:* Experimental pathology; nutrition; cardiovascular disease; hepatic liposis. *Mailing Add:* Dept Anat Univ Miss Sch Med Jackson MS 39216

BALL, DAVID RALPH, b Swansea, Wales, July 20, 40; m 64; c 3. INDUSTRIAL ORGANIC CHEMISTRY, FUEL SCIENCE. *Educ:* Univ Leeds, Eng, BSc, 64, PhD(coal chem), 69. *Prof Exp:* Chem engr & group leader, Brit Acheson Electrodes Ltd, 68-73; SCIENTIST INDUST GRAPHITE RES & DEVELOP, PARMA TECH CTR, UNION CARBIDE CORP, 73- *Mem:* Brit Inst Fuel; Soc Chem Indust. *Res:* Preparation, composition, structure, chemical reactivity, carbonisation and pyrolysis of fossil fuel derivatives. *Mailing Add:* Parma Tech Ctr Box 6116 Union Carbide Corp Cleveland OH 44101

BALL, DEREK HARRY, b London, Eng, Sept 1, 31; m 65; c 3. CARBOHYDRATE CHEMISTRY. *Educ:* Bristol Univ, BSc, 53; Queen's Univ, Ont, PhD(chem), 56. *Prof Exp:* Res fel carbohydrate chem, Div Appl Biol, Nat Res Coun Can, 56-58; fel catalysis, State Univ NY Col Ceramics, Alfred Univ, 59-60; res chemist, Org Chem, 60-72, head org chem group, Pioneering Res Lab, 72-74, res chemist, Food Sci Lab, 74-83, HEAD, BIOTECHNOL BR, SCI & ADV TECH LAB, US ARMY NATICK RES & DEVELOP CTR, 83- *Mem:* Am Chem Soc; fel Brit Chem Soc. *Res:* Selective substitution of carbohydrates; branched chain sugars; sucrochemistry; H-1 and C-13 nmr of carbohydrates; liquid chromatographic methods for the analysis of free sugars in foods. *Mailing Add:* Sci Advan Technol Lab US Army Natick Res & Develop Ctr Natick MA 01760-5020

BALL, DONALD LEE, b Kalamazoo, Mich, June 9, 31; m 69. CHEMISTRY. *Educ:* Kalamazoo Col, AB, 53; Brown Univ, PhD(chem), 56. *Prof Exp:* Proj assoc chem, Univ Wis, 56-57; sr res chemist res labs, Gen Motors Corp, Mich, 57-67; chief phys chem prog, Chem Res Directorate, 67-70, prog mgr chem energetics, 70-73, DIR CHEM & ATMOSPHERIC SCI, AIR FORCE OFF SCI RES, 73- *Mem:* Am Chem Soc; Sigma Xi. *Res:* Physical and inorganic chemistry. *Mailing Add:* AFOSR-NC Bolling AFB Bldg 410 Washington DC 20332

BALL, DOUGLAS, b Cheyenne, Wyo, Mar 5, 20; m; c 4. GEOLOGY. *Educ:* Colo Sch Mines, PE(EE Petrol), 43. *Prof Exp:* PRES, BALL ASN, 57-, INTERSTATE SERV CORP, 60- *Concurrent Pos:* Assoc prof, Clear Creek Co Metal Mining Asn, 88-90. *Mem:* Fel Geol Soc Am; Soc Petrol Engrs; Am Asn Petrol Geologists. *Mailing Add:* PO Box I Boulder CO 80306

BALL, EDWARD D, b Syracuse, NY, Mar 15, 50. HEMATOLOGY, IMMUNOLOGY. *Educ:* Case Western Reserve Univ, MD, 76. *Prof Exp:* Instr med, Case Western Reserve Univ, 79-81; instr, 81-82, asst prof, 82-87, ASSOC PROF MED & MICROBIOL, SCH MED, DARTMOUTH UNIV, 87- *Mem:* Am Soc Hemat; Am Asn Immunologists; AAAS; Am Fedn Clin Res. *Res:* Tumor antigens; bone marrow transplantation; leukemia biology. *Mailing Add:* Dept Microbiol Sch Med Dartmouth Univ Hanover NH 03756

BALL, EDWARD JAMES, b Ashtabula, Ohio, May 6, 48; m 71; c 2. TRANSPLANTATION IMMUNOLOGY, TCELL ANTIGEN RECEPTOR. *Educ:* Cleveland State Univ, BS, 70; Incarnate Word Col, San Antonio, MS, 75; Univ Cincinnati, (microbiol), 79. *Prof Exp:* From instr to asst prof internal med transp, Univ Tex Southwestern Med Sch Dallas, 85-88; DIR TRANSP IMMUNOL, HISTOCOMPATIBILITY LAB, DAYTON COMMUNITY BLOOD CTR, 88- *Mem:* Am Asn Immunologists; Am Soc Histocompatibility & Immunogenetics. *Res:* Human major histocompatibility complex (HLA) D-region polymorphism and recognition by alloreactive and antigen-specific T cells; polymorphism and structure of human T cell antigen reception. *Mailing Add:* Community Blood Ctr Dayton Regional Tissue Bank 349 S Main St Dayton OH 45402

BALL, FRANCES LOUISE, b Murfreesboro, Tenn, Oct 6, 24. ORGANIC CHEMISTRY, ELECTRON MICROSCOPY. *Educ:* Mid Tenn State Col, BS, 45; Vanderbilt Univ, MS, 48. *Prof Exp:* Jr chemist, E I du Pont de Nemours & Co, 45; instr high sch, Tenn, 45-46; instr chem, Limestone Col, 48-49; chemist, Oak Ridge Gaseous Diffusion Plant, Oak Ridge Nat Lab, Union Carbide Corp,49-68, physicist, Molecular Anat Prog, 68-75, res staff mem, Anal Chem Div, 75-83; RETIRED. *Mem:* Electron Micros Soc Am (secy, 75-84); Am Chem Soc. *Res:* Beckmann rearrangement of oximes; oxidation of single crystals of copper; structure and reactions of thin metal films. *Mailing Add:* 1128 E Northfield Blvd Murfreesboro TN 37130

BALL, FRANK JERVERY, b Charleston, SC, Jan 29, 19; m 50; c 4. ORGANIC CHEMISTRY, WOOD CHEMISTRY. *Educ:* Univ of the South, BS, 41; Univ Rochester, PhD(org chem), 44. *Prof Exp:* Res chemist, 44-46; group leader, 46-53; dir, Charleston Res Ctr, Westvaco Corp, 53-75, dir, Tyrone Res Lab, 58-60, assoc corp res dir, 75-84; RETIRED. *Concurrent Pos:* Pres, Empire State Pulp Res Assocs, 74-78. *Mem:* Am Chem Soc; Tech Asn Pulp & Paper Industs; Can Pulp & Paper Asn; Am Soc Magnesium Res. *Res:* Sulfate lignin chemicals; fatty and rosin derived surfactants and chemicals from tall oil; activated carbon; pulping, bleaching and chemical recovery; air pollution control; nutrition. *Mailing Add:* Four Atlantic St Charleston SC 29401

BALL, GENE V, b Rivesville, WVa, June 28, 31. INTERNAL MEDICINE. *Educ:* WVa Univ, BS, 57; Vanderbilt Univ, MD, 59; Am Bd Internal Med, dipl; Am Bd Rheumatol, dipl. *Prof Exp:* Intern, Cincinnati Gen Hosp, 59-60; resident, Hosp Univ Pa, 60-61; resident, Jackson Mem Hosp, 61-64; pract internal med, Univ Miami, 64-65; from instr to assoc prof, 65-71, PROF INTERNAL MED, MED CTR, UNIV ALA, BIRMINGHAM, 71-, CHIEF, RHEUMATOL INPATIENT SERV, UNIV HOSP, 79- *Concurrent Pos:* Fel, Jackson Mem Hosp, 61-64; assoc ed, Arthritis & Rheumatism, 75-80; mem, Subspecialty Sect Rheumatol, Am Bd Internal Med; mem, Arthritis Adv Coun, Food & Drug Admin, 89-92; mem, Rheumatology Sect, MKSAP, Am Col Physicians; consult, Accreditation Coun Grad Med Educ. *Mem:* Fel Am Col Physicians; Am Rheumatism Asn. *Res:* Gout; Saturnine gout; systemic lupus erythematosus. *Mailing Add:* Dept Internal Med Univ Ala Med Ctr Birmingham AL 35294

BALL, GEORGE A(PPLETON), engineering, applied physics, for more information see previous edition

BALL, GEORGE EUGENE, b Detroit, Mich, Sept 25, 26; m 49; c 2. SYSTEMATICS. *Educ:* Cornell Univ, AB, 49, PhD, 54; Univ Ala, MS, 50. *Prof Exp:* From asst prof to assoc prof entomol, 54-65, dept chmn, 73-83, PROF ENTOM, UNIV ALTA, 65- *Honors & Awards:* Gold Medal, Entom Soc Can, 80. *Mem:* Soc Syst Zoologists; Soc Study Evolution; Royal Entom Soc London. *Res:* Systematic entomology, especially classification, way of life and geographical distribution of beetles of the family Carabidae. *Mailing Add:* Dept Entom Fac Agr Univ Alta Edmonton AB T6G 2E2 Can

BALL, GEORGE WILLIAM, b Warren, Pa, Apr 3, 41; m 78, 84; c 2. SOFTWARE ENGINEERING, PROGRAMMING LANGUAGES. *Educ:* Union Col, NY, BS, 63; Syracuse Univ, MS, 65, PhD(math), 72; Rochester Inst Technol, MS, 88. *Prof Exp:* ASSOC PROF MATH, ALFRED UNIV, 68- *Mem:* Asn Comput Mach. *Res:* Computer science education; compiler construction. *Mailing Add:* Div Math & Comput Sci Alfred Univ Alfred NY 14802

BALL, GORDON CHARLES, b Calgary, Alta, Oct 18, 42; m 65; c 4. NUCLEAR STRUCTURE, NUCLEAR CHEMISTRY. *Educ:* Univ Alta, HonBSc, 64; Univ Calif, Berkeley, PhD(nuclear chem), 68. *Prof Exp:* Asst res officer, Exp Nuclear Physics, Chalk River Nuclear Labs, Atomic Energy Can Ltd, 68-70, assoc res officer, 71-83, SR RES OFFICER, EXP NUCLEAR PHYSICS, CHALK RIVER LABS, AECL RES, 84- *Concurrent Pos:* Vis scientist, Univ Liverpool, 86. *Mem:* Am Phys Soc; Can Asn Physicists. *Res:* Nuclear spectroscopy, gamma decay, lifetime measurements, exotic transfer reactions; precise nuclear mass determinations, heavy-ion reaction mechanisms, nuclear astrophysics, nuclei at high spin, accelerator mass spectrometry, atomic physics. *Mailing Add:* Chalk River Labs AECL Res Chalk River ON K0J 1J0 Can

BALL, GREGORY FRANCIS, b Washington, DC, May 6, 55. NEUROETHOLOGY, BEHAVIORAL NEUROENDOCRINOLOGY. *Educ:* Columbia Univ, BA, 77; Rutgers Univ, PhD(psychobiol), 83. *Prof Exp:* Postdoctoral fel behav neuroendocrinol, Rockefeller Univ, New York, NY, 83-86, asst prof animal behav, 86-88, asst dir animal behav, Field Res Ctr, Millbrook, NY, 86-88; asst prof psychol, Boston Col, Chestnut Hill, Mass, 88-91; ASST PROF PSYCHOL, JOHNS HOPKINS UNIV, BALTIMORE, MD, 91- *Concurrent Pos:* Adj asst prof, Rockefeller Univ, New York, NY, 89. *Mem:* Soc Neurosci; Animal Behav Soc; Int Soc Neuroethology; Am Ornithologists Union. *Res:* Interaction of hormones, brain and behavior in birds; steroid hormones regulate the production and acquisition of birdsong; environmental stimuli regulate seasonal reproduction in birds; hormones influence the development of sex differences in the brain. *Mailing Add:* Dept Psychol Johns Hopkins Univ Charles & 34th St Baltimore MD 21218

BALL, HAROLD JAMES, b Beaver Dam, Wis, Mar 13, 19; m 60; c 1. INSECT PHYSIOLOGY. *Educ:* Univ Wis, BA, 43, MA, 46, PhD(entom), 51. *Prof Exp:* Asst prof naval sci & tactics, Univ SC, 46; from asst prof to assoc prof entom, 51-60, PROF ENTOM, UNIV NEBR, LINCOLN, 60- *Mem:* Entom Soc Am. *Res:* Insect photoreceptors; insecticide resistance. *Mailing Add:* Insectary Bldg Univ Nebr Lincoln NE 68583-0721

BALL, IRVIN JOSEPH, b Peoria, Ill, Oct 17, 44; m 77; c 3. WATERFOWL ECOLOGY. *Educ:* Utah State Univ, BS, 66; Univ Minn, MS, 71, PhD(wildlife), 73. *Prof Exp:* From asst prof to assoc prof, Zool Dept, Wash State Univ, 74-79; ASST LEADER, MONT COOP WILDLIFE RES UNIT, UNIV MONT, MISSOULA, 79- *Mem:* Wildlife Soc. *Res:* Ecology, behavior and management of gallinaceous birds and waterfowl; habitat use, movements, population dynamics, nesting ecology and effects of exploitation. *Mailing Add:* Mont Coop Wildlife Res Unit 107 Health Sci Univ Mont Missoula MT 59812

BALL, JAMES BRYAN, b Portland, Ore, Oct 2, 32; m 60; c 2. NUCLEAR CHEMISTRY. *Educ:* Ore State Col, BS, 54; Univ Wash, PhD, 58. *Prof Exp:* Resident res assoc, 58-59, dir, Holifield Heavy Ion Res Facil, 74-83, PHYSICIST, OAK RIDGE NAT LAB, 59-, DIR, PHYSICS DIV, 83- *Concurrent Pos:* Mem, nuclear sci adv comt, Dept Energy-NSF, 80-82 & 90-91; chmn, Div Nuclear Physics, Am Phys Soc, 90-91. *Mem:* AAAS; Am Phys Soc. *Res:* Experimental studies of reaction mechanisms and validity of nuclear models in medium energy nuclear reactions. *Mailing Add:* 110 Berwick Dr Oak Ridge TN 37830

BALL, JAMES STUTSMAN, b Reno, Nev, Sept 13, 34; m 57; c 1. PHYSICS. *Educ:* Calif Inst Technol, BS, 56; Univ Calif, Berkeley, PhD(physics), 60. *Prof Exp:* Res assoc physics, Lawrence Radiation Lab, Univ Calif, Berkeley, 60, asst res physicist, La Jolla, 60-63, asst prof physics, Los Angeles, 63-68; assoc prof, 68-72, PROF PHYSICS, UNIV UTAH, 72- *Mem:* Am Phys Soc. *Res:* Elementary particle physics; structure of elementary particles based on quantum-chromodynamics. *Mailing Add:* Dept Physics Univ Utah Salt Lake City UT 84112

BALL, JOHN ALLEN, b Ravenna, Nebr, Dec 8, 35; m 55; c 4. RADIO ASTRONOMY, ASTROPHYSICS. *Educ:* Univ Nebr, BSEE, 57; Harvard Univ, PhD(astron), 69. *Prof Exp:* Staff mem, Haystack Observ, Mass Inst Technol, 64-69; res fel, 69-73, res assoc & lectr, 73-77, dir radio astron facil & lectr, Harvard Col Observ, 77-84; CHIEF TELESCOPE OPERS, MIT HAYSTACK OBSERV, 84- *Honors & Awards:* Rumford Prize, Am Acad Arts & Sci, 71. *Mem:* Inst Elec & Electronics Engrs; Am Astron Soc; Int Astron Union; Am Asn Physics Teachers; Sigma Xi. *Res:* Studies of the interstellar medium and circumstellar clouds using radio-frequency atomic and molecular spectral lines; studies of circumstellar molecular masers using very-long-baseline interferometry; data-acquisition and data-reduction techniques in radio astronomy. *Mailing Add:* 85 Oak Hill Rd Harvard MA 01451

BALL, JOHN SIGLER, b Hamlin, Tex, Aug 2, 14; m 38; c 6. PETROLEUM CHEMISTRY. *Educ:* Tex Tech Col, BS, 34, MS, 36. *Prof Exp:* Jr chem engr petrol & oil-shale exp sta, US Bur Mines, 38-42, asst chemist, 42-43, assoc petrol chemist, 43-44, chemist, 44-45, head petrol chem & refining sect & shale oil res & anal sect, 45-56, chief br petrol, Region III, 56-58, projs coordr petrol res, 58-63, dir, Am Petrol Inst Projs, 48-63, res dir, Bartlesville Energy Res Ctr, US Dept Energy, 63-78; consult, 78-88; RETIRED. *Mem:* Fel AAAS; fel Am Inst Chemists; Am Chem Soc; Soc Petrol Engrs. *Res:* Sulfur and nitrogen compounds of petroleum; composition of shale oil; analytical methods for petroleum. *Mailing Add:* 5934 Cornell Dr Bartlesville OK 74006

BALL, JOSEPH ANTHONY, b Washington, DC, June 4, 47; m 82. OPERATOR & SYSTEM THEORY. *Educ:* Georgetown Univ, BS, 69; Univ Va, MS, 70, PhD(math), 73. *Prof Exp:* From asst prof to assoc prof, 73-82, PROF MATH, VA TECH, 82- *Mem:* Am Math Soc; Soc Indust & Appl Math. *Res:* Matrix and operator valued functions; interpolation. *Mailing Add:* Dept Math Va Polytech Inst & State Univ Blacksburg VA 24601

BALL, KENNETH STEVEN, b Philadelphia, Pa, Sept 10, 60; m 83; c 1. TURBULENCE, COMPUTATIONAL FLUID DYNAMICS & HEAT TRANSFER. *Educ:* Lehigh Univ, BSME, 82; Drexel Univ, MSME, 84, PhD(mech eng), 87. *Prof Exp:* Postdoctoral res assoc appl math, Ctr Fluid Mech, Brown Univ, 87-89; ASST PROF MECH ENG, UNIV TEX, AUSTIN, 89- *Mem:* Am Soc Mech Engrs; Am Phys Soc. *Res:* Computational fluid dynamics and heat transfer; hydrodynamic stability; transition; turbulence; buoyancy induced flow; thermal radiation. *Mailing Add:* Dept Mech Eng Univ Tex Austin TX 78712

BALL, LAURENCE ANDREW, b York, Yorkshire, Eng, July 9, 44; m 68, 87; c 2. ANIMAL VIROLOGY, INTERFERON ACTION. *Educ:* Oxford Univ, Eng, BA, 66, MA, 69, Phd(biochem), 69. *Prof Exp:* Fel biochem, Univ Wis-Madison, 69-71; staff scientist, Nat Inst Med Res, Mill Hill, London, Eng, 72-74; from asst prof to assoc prof microbiol, Univ Conn, Storrs, 74-79; from assoc prof to prof biochem, Univ Wis-Madison, 79-87; PROF MICROBIOL, UNIV ALA, BIRMINGHAM, 87- *Concurrent Pos:* Res career develop award, NIH, 78, mem, Study Sect Virol, 80-83; assoc ed, Acad Press, Inc, 80-83 & J Interferon Res, Liebert, 80-83. *Mem:* Am Soc Microbiol; Am Soc Virol; Am Soc Biochem & Molecular Biol. *Res:* Control of gene expression in animal cells infected with viruses; molecular biology of pox viruses. *Mailing Add:* Microbiol Dept Univ Ala Birmingham UAB Sta Birmingham AL 35294

BALL, LAWRENCE ERNEST, b Akron, Ohio, Nov 11, 35; m; c 1. POLYMER CHEMISTRY, ORGANIC CHEMISTRY. *Educ:* Univ Akron, BS, 57, MS, 58, PhD(polymer org chem), 61. *Prof Exp:* Proj assoc, Standard Oil Co, 61-64, proj leader, 64-68, res assoc Polymer Synthesis, 68-89, RES SCIENTIST, RES LAB, BP RES, 89- *Mem:* Soc Petrol Engrs. *Res:* Polymer synthesis; structure and property relations in polymers specialty; acrylonitrile; nylon; water soluble polymers; 27 patents; 7 publications. *Mailing Add:* 4288 W Bath Rd Akron OH 44333

BALL, M ISABEL, b Elmendorf, Tex, June 1, 29. BIO-ORGANIC CHEMISTRY. *Educ:* Our Lady of the Lake Univ, BA, 50; Univ Tex, MA, 63, PhD, 69. *Prof Exp:* Teacher parochial sch, Okla, 52-54; from instr to asst prof chem, Our Lady of the Lake Univ, 54-65; res assoc, Clayton Found Biochem Inst, Univ Tex, 65-68; PROF CHEM, OUR LADY OF LAKE UNIV, 75-, CHMN DEPT, 68-, DIR, DIV SCI & MATH, 73-, DEAN, COL ARTS & SCI, 80- *Concurrent Pos:* AEC res partic, Oak Ridge Nat Lab, 69. *Mem:* Am Chem Soc; Nat Sci Teachers Asn. *Res:* Synthesis and assay of analogs of metabolites; studies of Aspergillus niger using purine analogs and derivatives; metabolic changes caused by phthalic acid esters in mice and bacteria. *Mailing Add:* Dept Chem 411 SW 24th St San Antonio TX 78285

BALL, MAHLON M, b Lawrence, Kans, Apr 7, 31; m 53; c 2. GEOLOGY, GEOPHYSICS. *Educ:* Univ Kans, BSc, 53, MSc, 58, PhD(geol), 60; Univ Birmingham, MSc, 59. *Prof Exp:* Geologist, Shell Develop Co, 59-62, res geologist, 62-63, sr geologist, Shell Oil Co, 63-65, div geologist, Shell Develop Co, 65-66, res assoc, 66-67; assoc prof geophys, 67-70, prof geophys, Rosenstiel Sch Marine & Atmospheric Sci, Univ Miami, 70-75, res geophysicist, Inst Marine Sci, 67-75; geophysicist, Woods Hole Oceanographic Inst, 75-89, GEOPHYSICIST, PETROL GEOL DIV, US GEOL SURV, DENVER, 89- *Concurrent Pos:* Mem geol adv comt, US Air Force Weapons Res, 73; mem site surv comt, Int Prog Ocean Drilling, 74. *Mem:* Geol Soc Am; Soc Econ Paleontologists & Mineralogists; Am Geophys Union; Soc Explor Geophys. *Res:* Recent carbonate sediments; marine geophysical investigations of the Bahamas, Caribbean and equatorial Atlantic Ocean. *Mailing Add:* Petrol Geol Div US Geol Surv Box 25046 MS 911 Denver CO 80225

BALL, MARY UHRICH, b Pittsburg, Kans, May 26, 44; m 62; c 2. QUANTITATIVE GENETICS. *Educ:* Trinity Univ, BS, 66; Tex A&M Univ, MS, 72, PhD(genetics), 74. *Prof Exp:* Asst prof zool-entom, Auburn Univ, 74-85; ASST PROF BIOL, CARSON-NEWMAN COL, 85- *Mem:* Am Genetic Asn; Genetics Soc Am. *Res:* Genetics of Spodoptera science education exigua. *Mailing Add:* Carson-Newman Col Box 2044 Jefferson City TN 37760

BALL, MICHAEL OWEN, b Washington, DC, July 20, 50; m; c 3. OPERATIONS RESEARCH. *Educ:* Johns Hopkins Univ, BES, 72, MSE, 73; Cornell Univ, PhD(oper res), 77. *Prof Exp:* Mem tech staff comput networks, Bell Labs, 76-78; from asst prof to assoc prof, 78-90, PROF OPER RES, COL BUS & MGT & SYST RES CTR, UNIV MD, 90- *Mem:* Opers Res Soc Am; Math Prog Soc; Inst Elec & Electronics Engrs; Soc Indust & Appl Math. *Res:* Operations research especially in the areas of computer and transportation networks. *Mailing Add:* Col Bus & Mgt Univ Md College Park MD 20742

BALL, RAIFORD MILL, b Bryan, Tex, Nov 16, 41; m 62; c 3. SOLID STATE PHYSICS, SCIENCE EDUCATION. *Educ:* Trinity Univ, BS, 64; Tex A&M Univ, MS, 71, PhD(physics), 75. *Prof Exp:* Physics teacher, Auburn City Sch, Ala, 75-85; assoc prof physics & phys sci, E Tenn State Univ, 85-87; instr, Morristown Col, Tenn, 87-88; ASSOC PROF PHYS, E TENN STATE UNIV, 87- *Concurrent Pos:* Teaching asst, Physics Dept, Tex A&M Univ, 69-75; adj asst prof, physics dept, Auburn Univ, Ala, 80-85. *Mem:* Am Asn Physics Teachers; Am Phys So. *Mailing Add:* 344 Summitt Heights Dr Jefferson City TN 37760

BALL, RALPH WAYNE, b Los Angeles, Calif, Oct 14, 25. MATHEMATICS. *Educ:* Pomona Col, BA, 47; Univ Southern Calif, MA, 52; NMex State Univ, MS, 63, PhD(math), 64. *Prof Exp:* Teacher, Los Angeles City Schs, 49-59; secy curric coord, Inyo County Schs, Calif, 59-61; instr math, NMex State Univ, 61-64; assoc prof, Appalachian State Univ, 64-66; prof math, 66-85, EMER PROF MATH, NMEX INST MINING & TECHNOL, 85- *Mem:* Am Math Soc; Math Asn Am. *Res:* Group theory, particularly infinite symmetric groups. *Mailing Add:* Dept Math NMex Inst Mining & Technol Socorro NM 87801

BALL, RICHARD WILLIAM, b Streator, Ill, Aug 16, 23. MATHEMATICS. *Educ:* Univ Ill, BA, 44, MA, 45, PhD(math), 48. *Prof Exp:* Instr math, Univ Wash, 48-54; from asst prof to assoc prof, 54-60, PROF MATH, AUBURN UNIV, 60- *Mem:* Am Math Soc; Math Asn Am; Sigma Xi. *Res:* Theory of groups, semigroups. *Mailing Add:* Auburn Univ 768 Hollon Auburn AL 36830

BALL, RUSSELL MARTIN, b Chicago, Ill, Oct 28, 27; m 55; c 3. REACTOR PHYSICS, NUCLEAR PHYSICS. *Educ:* Univ Ill, BS, 49, MS, 50; Univ Va, PhD, 70. *Prof Exp:* Process engr, Elec Storage Battery Co, 46-47; asst, Univ Ill, 47-50; chief engr, Nuclear Instrument & Chem Corp, 50-55; with Oak Ridge Sch Reactor Technol, 55; SUPVR, BABCOCK & WILCOX, 55- *Concurrent Pos:* Lectr, Univ Wis. *Mem:* Am Phys Soc; Am Asn Physics Teachers; Inst Elec & Electronics Engrs; Am Nuclear Soc. *Res:* Isometric transitions; radiation detectors and computers; industrial application of nucleonics and isotopes; nuclear instrumentation; process control; computers. *Mailing Add:* PO Box 10935 Lynchburg VA 24506

BALL, STANTON MOCK, b Lawrence, Kans, Aug 20, 33; m 55; c 2. GEOLOGY. *Educ:* Univ Kans, BS, 56, MS, 58, PhD(geol), 64. *Prof Exp:* Geologist, Kans State Geol Surv, 57-64; geologist, Pan Am Petrol Corp, 64-70, res scientist, Okla, 70-74, RES SCIENTIST, AMOCO PRODS CO, TEX, 74- *Mem:* Geol Soc Am; Am Asn Petrol Geologists; Soc Econ Paleont & Mineral; Sigma Xi. *Res:* Stratigraphy and carbonate petrology. *Mailing Add:* 4608 Waterway Ft Worth TX 76137-1530

BALL, WILBUR PERRY, b Briggsdale, Colo, Jan 2, 23; m 48; c 3. AGRICULTURE. *Educ:* Colo State Univ, BS, 48, MEd, 53; Iowa State Univ, PhD(agr), 56. *Prof Exp:* Teacher pub sch, Colo, 48-53; instr agr mech, Iowa State Univ, 53-56; consult agr educ, Stanford Univ Contract, Int Coop Admin, Philippines, 56-58; from asst prof to assoc prof agr educ, Calif State Univ, Fresno, 58-68, grad coordr, Sch Agr & Sch Supvr Teacher Educ, 65-77, prof agr & educ, 78-84, PROF INT AGR & EDUC, CALIF STATE UNIV, FRESNO, 85- *Concurrent Pos:* Consult rural educ, Calif State Univ, Fresno, contract, US AID, Sudan, 61-63; vis prof agr educ, Ohio State Univ, 67-68; res assoc, Nat Ctr Voc & Tech Educ, 67-68; instr hist, AIMS Community Col, Greeley, Colo. *Mem:* Assoc Am Soc Agr Eng; Soc Int Agr. *Res:* Agricultural engineering; international agriculture; author of ten books about Colorado history. *Mailing Add:* PO Box 392 Eaton CO 80615

BALL, WILLIAM DAVID, b Newark, NJ, July 22, 37; m 75; c 3. DEVELOPMENTAL BIOLOGY, PROTEIN SECRETION. *Educ:* Univ Wis, BS, 58, MS, 61; Univ Chicago, PhD(zool), 65. *Prof Exp:* Proj asst zool, Univ Wis, 58-61; res assoc biochem, Univ Ill, 65; res assoc, Univ Wash, 65-67, asst prof zool, 67-73, res assoc zool, 73-75; assoc prof oral biol, Sch Dent, Univ Colo, 75-78; ASSOC PROF ANAT, COL MED, HOWARD UNIV, 78- *Concurrent Pos:* NIH res grant, 69-72 & 83-91; Cystic Fibrosis Found grant, 77-79, NSF grant, 79-82. *Mem:* Am Soc Cell Biologists; Am Soc Zoologists; Soc Develop Biol; Int Asn Dent Res; Am Asn Anatomists. *Res:* Epithelio-mesenchymal interactions in development of rat salivary glands; formation of specific proteins in embryonic tissues; secretory proteins as markers for cellular differentiation in salivary glands. *Mailing Add:* Dept Anat Col Med Howard Univ 520 W St NW Washington DC 20059

BALL, WILLIAM E(RNEST), b Emporia, Kans, May 26, 30; m 53; c 2. INTELLIGENT DATABASE SYSTEMS. *Educ:* Wash Univ, BS, 52, DSc(chem eng), 58. *Prof Exp:* Instr, Wash Univ, 54-56; appl sci rep, Int Bus Mach Corp, 57; mathematician, Monsanto Co, 58-62; chmn dept appl math & comput sci, 70-74, assoc prof, 62-67, PROF COMPUT SCI, WASH UNIV, 67- *Concurrent Pos:* prin, AutoComp Comput Consult, 76-; dir, Ctr Intelligent Computer Systs. *Mem:* Am Asn Artificial Intel; Asn Comput Mach; Inst Elec & Electronics Engrs; Sigma Xi. *Res:* Graphical understanding; automatic scanning and understanding of maps and mechanical drawings; knowledge-based information systems. *Mailing Add:* Dept Comput Sci Box 1045 Wash Univ One Brookings Dr St Louis MO 63130-4899

BALL, WILLIAM HENRY WARREN, b Grand Falls, Nfld, Nov 6, 21; US citizen; m 48. ELECTRICAL ENGINEERING. *Educ:* NS Tech Col, BSc, 44; Univ Toronto, MASc, 46; Cornell Univ, PhD(math, mech), 57. *Prof Exp:* Demonstr elec eng, Univ Toronto, 46-47; from instr to asst prof, Univ Buffalo, 47-52; instr, Cornell Univ, 52-57; mem tech staff, Bell Tel Labs, Inc, 57-68; adj prof, Newark Col Eng, 59-68; assoc prof, 68-75, PROF ELEC ENG, NJ INST TECHNOL, 75- *Concurrent Pos:* Consult, NIKE Weapons Systs, Bell Tel Labs, Inc, 59-68; Found Advan Grad Studies Eng res grant, Newark Col Eng, 69-70. *Mem:* AAAS; sr mem Inst Elec & Electronics Engrs. *Res:* Noise in feedback control systems; solar and thermal noise in tracking systems; sea noise in underwater systems. *Mailing Add:* 24 Brookfield Way Morristown NJ 07960

BALL, WILLIAM JAMES, JR, b New York, NY, Sept 5, 42. BIOCHEMISTRY. *Educ:* San Diego State Univ, BS, 66, MS, 68; Univ Calif, Los Angeles, PhD(biochem), 72. *Prof Exp:* Fel, Sloan-Kettering Cancer Ctr, 72-75; instr, Baylor Col Med, 75-77; asst prof, 77-84, ASSOC PROF, COL MED, UNIV CINCINNATI, 84- *Concurrent Pos:* Investr, Am Heart Asn; immunol trainee, Columbia Univ, 78. *Mem:* NY Acad Sci; Biophys Soc; Am Soc Biochem & Molecular Biol. *Res:* Mechanisms of enzyme regulation; utilization of immunological techniques to study sodium potassium, adenosine triphosphatase-ligand interactions; isolation of and characterization of monoclonal antibodies. *Mailing Add:* Dept Pharmacol & Cell Biophys Col Med Univ Cincinnati 231 Bethesda Ave ML 575 Cincinnati OH 45267

BALL, WILLIAM PAUL, b San Diego, Calif, Nov 16, 13; m 41; c 2. NUCLEAR PHYSICS, RADIATION SCIENCE. *Educ:* Univ Calif, Los Angeles, AB, 40; Univ Calif, Berkeley, PhD(physics), 53. *Prof Exp:* High sch teacher math & phys educ, Montebello Unified Sch Dist, 41-42; instr math & physics, Santa Ana Army Air Base, 42-43; physicist, Radiation Lab, Univ Calif, 43-58 & Ramo Wooldridge Corp, 58-59; sr scientist & mgr nuclear technol staff, Hughes Aircraft Co, 59-64; sr scientist engr space & nuclear physics, TWR Systs & Energy, 64- 83; staff engr, Hughes Aircraft Co, 83-86; RETIRED. *Concurrent Pos:* Bd dirs, S Dist, Los Angeles Chapter ARC, 79-86, Torrance, Calif Area Chamber of Commerce, 78-84; Consult, 86- *Mem:* AAAS; Am Phys Soc; Am Nuclear Soc; Am Inst Aeronaut & Astronaut; NY Acad Sci; Sigma Xi. *Res:* Space and nuclear radiation environments and lasers and their effects on properties and responses of materials, sensors, electronics and systems. *Mailing Add:* 209 Via El Toro Redondo Beach CA 90277

BALLAL, S K, b Trichur, Kerala, India, Dec 31, 38; m; c 2. BOTANY, PHYSIOLOGY. *Educ:* Univ Madras, BS, 59, MS, 61; Univ Tenn, PhD(bot), 64. *Prof Exp:* Asst bot, Univ Tenn, 61-64; from asst prof to assoc prof biol, 65-74, admin asst to chancellor, 89, PROF BIOL, TENN TECHNOL UNIV, 74- *Concurrent Pos:* Shell Merit fel, Stanford Univ, 70. *Mem:* AAAS; Bot Soc Am; Am Inst Biol Sci. *Res:* Plant growth regulations. *Mailing Add:* Dept Biol Tenn Tech Univ Cookeville TN 38505

BALLAM, JOSEPH, b Boston, Mass, Jan 2, 17; m 38; c 2. PHYSICS. *Educ:* Univ Mich, BS, 39; Univ Calif, PhD, 51. *Prof Exp:* Physicist, US Navy Dept, 41-45; asst physics, Univ Calif, 45-46, physicist, Radiation Lab, 46-48, asst physics, 48-51; asst, Princeton Univ, 51-52, instr, 52-53, res assoc, 53-56; from assoc prof to prof, Mich State Univ, 56-61; assoc prof, 61-64, assoc dir res div, Linear Accelerator Ctr, 61-82, prof, 64-87, EMER PROF PHYSICS, STANFORD UNIV, 87- *Concurrent Pos:* Res collabr, Brookhaven Nat Lab, 56-; Ford Found fel, Europ Orgn Nuclear Res, 60-61; Guggenheim Found fel,

71-72; coun fel sci res, Imperial Col, London, 80; vis prof, Polytechnic Sch, Paris, 81. *Mem:* Fel Am Phys Soc. *Res:* Elementary particles; cosmic rays; high energy physics. *Mailing Add:* Stanford Linear Accelerator Ctr Stanford Univ PO Box 4349 Stanford CA 94305

BALLANTINE, C S, b Seattle, Wash, Aug 27, 29; m 82. MATHEMATICS. *Educ:* Univ Wash, BS, 53; Stanford Univ, PhD(math), 59. *Prof Exp:* Asst math, Stanford Univ, 56-58; actg instr, Univ Calif, 58-59, instr, 59-60; from instr to assoc prof, 60-72, PROF MATH, ORE STATE UNIV, 72- *Mem:* Am Math Soc; Sigma Xi. *Res:* Matrix theory. *Mailing Add:* Dept Math Ore State Univ Corvallis OR 97331-4605

BALLANTINE, DAVID STEPHEN, b New York, NY, Dec 26, 22; m 49; c 7. ENVIRONMENTAL CHEMISTRY, RADIATION TECHNOLOGY. *Educ:* St Francis Col, BS, 43; Univ Notre Dame, PhD(chem), 52. *Prof Exp:* Jr chemist, Clinton Lab, 44-46; res chemist, Kellex Corp, 49-51; chemist, Brookhaven Nat Lab, 51-67; sr scientist, 67-75, ATMOSPHERIC SCIENTIST, US DEPT ENERGY, 75- *Mem:* Am Chem Soc; Sigma Xi. *Res:* Atmospheric science; environmental pollution; applied radiation chemistry. *Mailing Add:* 66 Webster No 104 Washington DC 20011

BALLANTINE, LARRY GENE, b Webster City, Iowa, July 6, 44; m 73; c 2. AGRICULTURAL CHEMISTRY. *Educ:* Iowa State Univ, BS, 66; Univ Mass, PhD(soil chem), 71. *Prof Exp:* Teacher chem, Essex Agr & Tech Inst, 71-73; asst prof soil chem, State Univ NY Agr & Tech Col, Cobleskill, 73-74; environmentalist, Ciba-Geigy Corp, 74-80, mgr environ invests, 80-; HAZLETON LABS, MADISON, WI. *Mem:* Am Chem Soc; Am Soc Agron; Sigma Xi; Soc Environ Chem & Toxicol. *Res:* Environmental research; emphasis on pesticide usage. *Mailing Add:* Hazleton Labs 3301 Kinsmen Blvd PO Box 7545 Madison WI 53707

BALLANTYNE, BRYAN, b Halifax, Eng, Sept 11, 34; m 67; c 4. TOXICOLOGY. *Educ:* Univ Leeds, BSc, 58, MB & ChB, 61, PhD(histochem), 68, MD, 71, DSc, 78; Royal Col Pathologists, MRCPath, 71, FRCpath, 73; Am Col Toxicol Sci, dipl, 82. *Prof Exp:* House surgeon, Gen Infirmary, Leeds Univ, 61-62; house physician, Bradford Royal Infirmary, 62-63; lectr anat & hist, Sch Med, Univ Leeds, 63-68; sr med officer, Chem Defense Estab, 68-78; DIR TOXICOL, UNION CARBIDE CORP, 78- *Concurrent Pos:* Mem, Sci Subcomt Pesticides, Mininstry Agr, Fisheries & Food, 69-78, Comt Experts Pesticides, Coun Europe, 76-78, Intergovn Maritime Consult Orgn, 77-; ed, Forensic Toxicol, 74, Current Approaches Toxicol, 77, Respiratory Protection, 81; adj prof toxicol, Univ Pittsburgh & Univ WVa, 81-; ed, Perspecher Basic & Appl Toxicol, 88, Clin & Exp Toxicol Cyanides, 87. *Mem:* Royal Col Physicians; Soc Toxicol; Europ Soc Toxicol; Am Acad Occup Med; Am Occup Med Asn; Fac Occup Med; fel Am Acad Clin Toxicol; Am Col Toxicol; Sigma Xi; Brit Pharmacol Soc. *Res:* Acute toxicity; primary irritancy; ophthalmic toxicology; application of histochemical methods; toxicology of cyanides and cyanogens; evaluation of potential human health hazards from chemicals; clinical toxicology. *Mailing Add:* 871 Chappell Rd Charleston WV 25304

BALLANTYNE, DAVID JOHN, b Victoria, BC, May 1, 31; m 62; c 3. HORTICULTURE, PLANT PHYSIOLOGY. *Educ:* Univ BC, BCom, 54; Wash State Univ, MS, 57; Univ Md, PhD(hort), 60. *Prof Exp:* Nat Res Coun Can fel, 60-61; res officer plant physiol, Exp Farm, Can Dept Agr, 61-63; asst prof, 62-68, ASSOC PROF PLANT PHYSIOL, UNIV VICTORIA, BC, 68- *Mem:* Am Soc Plant Physiol; Am Soc Hort Sci; Can Soc Plant Physiol. *Res:* Effect of air pollutants on plant metabolism; photosynthesis, growth and development of Rhododendron. *Mailing Add:* Dept Biol Univ Victoria Victoria BC V8W 2Y2 Can

BALLANTYNE, DONALD LINDSAY, b Peking, China, Nov 8, 22; m 63; c 3. TRANSPLANTATION BIOLOGY, MICROSURGERY. *Educ:* Princeton Univ, AB, 45; Cath Univ Am, MS, 48, PhD(biol), 52. *Prof Exp:* Parasitologist, E R Squibb & Sons, 51-53 & Ill State Psychopath Inst, 53-54; from asst prof to assoc prof, 62-75, res assoc tissue transplantation, 54-62, prof, 75-90, EMER PROF EXP SURG, INST RECONSTRUCT PLASTIC SURG, NY UNIV MED CTR, 90- *Concurrent Pos:* Dir, Microsurgical Res & Training Labs, NY Univ Med Ctr, 78- *Honors & Awards:* Edmund Lyon Mem Lectr, Nat Inst Deaf, Rochester Inst Technol. *Mem:* AAAS; Transplantation Soc; Sigma Xi; Found Sci & Handicapped (pres, 84-86). *Mailing Add:* Dept Surg NY Univ Med Ctr 550 First Ave New York NY 10016

BALLANTYNE, GARTH H, b Mineola, NY, June 4, 51; m 75; c 5. COLON & RECTAL SURGERY, GASTROINTESTINAL PHYSIOLOGY. *Educ:* Harvard Col, AB, 73; Col Physicians & Surgeons Columbia Univ, MD, 77. *Prof Exp:* ASST PROF SURG, YALE UNIV SCH MED, 84-; CHIEF GEN SURG, VET ADMIN CTR, WEST HAVEN, CONN, 84- *Mem:* Asn Acad Surg; Am Soc Colon & Rectal Surgeons; Am Asn Gastrointestinal Endoscopists; Sigma Xi. *Res:* Mechanisms of release of putative colonic hormones and their modulation of proximal gastrointestinal function; downstaging colorectal concers by endoscopic surveillance of high risk populations. *Mailing Add:* One Sunset Beach Bd Branford CT 06405

BALLANTYNE, JOSEPH M(ERRILL), b Tucson, Ariz, Dec 16, 34; m 61; c 8. SEMICONDUCTOR OPTICAL DEVICES. *Educ:* Univ Utah, BS & BSEE, 59; Mass Inst Technol, SM, 60, PhD(elec eng), 64. *Prof Exp:* Asst elec eng, Mass Inst Technol, 60-61, staff mem & consult, Lincoln Lab, 61, res asst & staff mem, Lab Insulation Res, 62-63; asst prof, 64-68, assoc prof elec eng, 68-71, actg dir, Nat Res & Resource Facil, Submicron Struct, 77-78, dir, Sch Elec Eng, 80-84, PROF ELEC ENG, CORNELL UNIV, 71-, VPRES, RES & ADVAN STUDIES, 84- *Concurrent Pos:* NSF sr fel, Stanford Univ, 70-71; consult, Mat Technol Inc, 62-63, Gen Tel & Electronics Lab, 65, Int Bus Mach Corp, 66, 78 & 79, Sandia Corp, 67-70, Cayuga Assocs, 68-69, Battelle Mem Inst, 68-69 & 76-78, Eastman Kodak & Schlumberger, 81-, McDonald Douglas Astron Ctr, 88; mem, adv comt, Sherman Fairchild Lab, Lehigh Univ, 78-; external adv comt, Elec Eng Dept, Univ Del, 79-; policy bd, Nat Nanofabrication Facil, Cornell Univ, 78-; Mit Whitney Fel, Schlumberger Fel, 59-63; Sr Fel, NSF, 70-71; Fel Inst Elec & Eletronics Engrs, 85. *Mem:* Am Phys Soc; Inst Elec & Electronics Engrs; Am Vacuum Soc; Asn Inst Mech Engrs; Metall Soc Am Inst Mining Metall & Petrol Engrs. *Res:* Ferroelectricity; infrared and optical properties of solids; quantum electronics; magnetic thin films; far infrared spectroscopy; solid state electronic devices; optical semiconductor devices; microfabrication technology. *Mailing Add:* Dir Res Found Cornell Univ Main Campus 312 Day Hall Ithaca NY 14853

BALLARD, HAROLD NOBLE, b Little Rock, Ark, Feb 11, 19; m 45, 60; c 4. ATMOSPHERIC PHYSICS. *Educ:* Univ Tex, El Paso, BS, 48; Tex A&M Univ, MS, 50. *Prof Exp:* Res physicist, NMex Inst Mining & Technol, 50; asst prof physics, Univ Tex, El Paso, 50-54; physicist, Los Alamos Sci Lab, 54-57; res physicist, Univ Tex, El Paso, 57-64; res physicist, Atmospheric Sci Lab, US Army Electronics Command, White Sands Missile Range, 64-79; RES PHYSICIST, UNIV TEX, EL PASO, 81- *Concurrent Pos:* Mem, Joint Comt Space Environ Forecasting, 68-; assoc prof & lectr, physics, Univ Tex, El Paso, 70. *Mem:* Am Phys Soc; Am Meteorol Soc; Am Geophys Union; AAAS. *Res:* Composition and thermal structure of stratosphere; multiple instrument balloon-borne experiments; development of rocket and balloon-borne atmospheric sensory instruments. *Mailing Add:* 239 El Puente St El Paso TX 79912

BALLARD, JAMES ALAN, b Rockingham, NC, Aug 13, 29; m 53; c 4. SEISMIC STRATIGRAPHY. *Educ:* Univ NC, BS, 53, MS, 59, PhD(oceanog), 79. *Prof Exp:* Geophysicist oil explor, Independent Explor Inc, 53-54; oceanographer geol & geophysic oceanog, Naval Oceanog Off, 59-64, supvr, 64-68, res scientist, 68-76; res scientist marine geophys, Naval Ocean Res & Develop Activ, 76-79, prog mgr, 79-88; prin scientist oceanog/geophys, MAR, Inc, 88-90; sci consult, Environ Protection Agency, 90-91; CONSULT, 91- *Concurrent Pos:* Lectr, Univ NC, 56-58; tech adv, Md Geol Surv, 75-76; tech prog chmn, US Navy Soc Explor Geophysicists Joint Tech Symp, 78, 80, 82, 84, 86 & 90. *Mem:* Soc Explor Geophysicists; Sigma Xi. *Res:* Development, installation and operation of a broad band marine bore hole seismograph station; instrumentation, techniques and applications of seismic stratigraphy to naval problems; geohazards analysis. *Mailing Add:* 1520 Fourth Ave Picayune MS 39466

BALLARD, KATHRYN WISE, b Waverly Hills, Ky, June 10, 30. CARDIOVASCULAR PHYSIOLOGY. *Educ:* Howard Univ, BS, 51; Univ Mich, MS, 53; Western Reserve Univ, MS, 59; Univ Southern Calif, PhD(physiol), 67. *Prof Exp:* Res asst physiol, Western Reserve Univ, 59 & Univ Southern Calif, 59-67; NIH fel, Sch Med, Univ Southern Calif, 67-68; trainee & vis scientist, Dept Pharmacol, Karolinska Inst, Sweden, 68-70, fel, 70; assoc res physiologist, dept med, Univ Calif, Los Angeles & Vet Admin Wadsworth Hosp Ctr, 78-81; asst prof physiol, Sch Med, 71-79, res staff, Cardiovasc Res Lab, Univ Southern Calif, 70-87; HEALTH SCIENTIST ADMIN, NAT HEART LUNG & BLOOD INST, 87- *Mem:* AAAS; Microcirc Soc; Am Heart Asn; Am Physiol Soc. *Res:* Effects of various stimuli on nutritive and total blood flow in skeletal muscle and skin; vascular and metabolic responses in canine adipose tissue; effects of sympathetic nerve stimulation and of catecholamines on circulation; microvascular hemodynamics during aging; measurement of microcirculatory intravascular pressures during aging; measurement of microvascular reactivity in aging. *Mailing Add:* Rev Br NHLBI 5333 Westbard Ave Rm 550 Westwood Bldg Bethesda MD 20814

BALLARD, KENNETH J, b Highland Park, Ill, Jan 22, 30; m 56; c 2. PHARMACY. *Educ:* Univ Calif, Berkeley, BA, 50, BSc, 54, PharmD, 55. *Prof Exp:* Staff pharmacist, USPHS Hosp, San Francisco, 55-57; sr asst pharmacist, Div Biologics Standards, Lab Blood & Blood Prod, NIH, 57-58; asst prof pharm, Sch Pharm & Grad Div Arts & Sci, Univ Buffalo, 58-60; assoc prof, Col Pharm, Northeastern Univ, 60-64; res assoc, Allergan Pharmaceut, 64-66; dir pharm serv, Student Health Serv, Univ Calif, Los Angeles, 66-70; ASST CLIN PROF PHARM & DIR CLIN PHARM SERV, LOS ANGELES COUNTY/UNIV SOUTHERN CALIF MED CTR, 70- *Mem:* Am Pharmaceut Asn; Am Soc Hosp Pharmacists. *Res:* Developing innovations in delivery of clinical pharmacy services; education of clinical pharmacists; electronic data processing of pharmacy drug and information systems; drug product development. *Mailing Add:* Dept Pharm Los Angeles County 1200 N State St Los Angeles CA 90033

BALLARD, LEWIS FRANKLIN, b Mooresville, NC, Dec 17, 34; m 61; c 3. POLLUTION CONTROL INSTRUMENTATION. *Educ:* NC State Univ, BS, 58; Duke Univ, MS, 67, PhD(elec eng), 69. *Prof Exp:* Rocket engr, Thiokol Chem Corp, 58-59; nuclear engr, Res Triangle Inst, 59-65, staff scientist, 65-69, supvr anal chem & physics sect, 69-73; pres, Nutech Corp, 73-85; CONSULT, 85- *Mem:* Inst Elec & Electronics Engrs; Am Phys Soc. *Res:* Electrical and optical properties of insulators, semiconductors, and organometallics; design of radiation detectors; applications of nuclear techniques in industry and highway engineering; ambient air monitoring instrumentation; gas chromatography, mass spectrometry. *Mailing Add:* 108 Maybank Ct Durham NC 27713

BALLARD, MARGUERITE CANDLER, b Atlanta, Ga, June 14, 20; m 73. INTERNAL MEDICINE, HEMATOLOGY. *Educ:* Vassar Col, BA, 42; Emory Univ, MS, 43, MD, 48. *Prof Exp:* Bacteriologist, US War Dept, 43-44; intern & med resident, Md Gen Hosp, 48-51; exam physician, Lockheed Aircraft Corp, 51-52; hemat fel, Emory Clin, 52-54; comn officer, Ctr Dis Control, USPHS, 54-85; CLIN INSTR & ASSOC DEPT MED, EMORY UNIV SCH MED, 54- *Concurrent Pos:* Auth, Lab Techniques in Central of Anticoaqulent Therapy, 63, Atlas of Blood Cell Morphol, 87; consult & maj contr, Third Ed Am Soc Hemat Slide Bank, 90. *Mem:* AAAS; Int Soc Hemat; AMA; Am Pub Health Asn; Am Soc Hemat. *Res:* Standardization of hematological laboratory procedures; development and presentation of lectures and bench training; photomicrographer of blood cells in health and disease for publication of Atlas. *Mailing Add:* 3092 Argonne Dr NW Atlanta GA 30305-1950

BALLARD, NEIL BRIAN, b Mankato, Minn, Jan 26, 38. ZOOLOGY, PARASITOLOGY. *Educ:* Mankato State Col, BS, 59; Colo State Univ, MS, 65, PhD(zool), 68. *Prof Exp:* Assoc prof, 68-80, PROF BIOL, MANKATO STATE UNIV, 80- *Concurrent Pos:* Brown-Hazen Fund grant, 69-70. *Mem:* Soc Protozool; Am Soc Parasitol; Wildlife Disease Asn. *Res:* Host-specificity and immunity phenomena with regard to coccidian infections in rodents. *Mailing Add:* Dept Biol Mankato State Univ Mankato MN 56001

BALLARD, RALPH CAMPBELL, b Washington, DC, Jan 4, 26; m 51, 80; c 2. ENTOMOLOGY, PHYSIOLOGY. *Educ:* George Washington Univ, BS, 51; Ohio State Univ, MS, 53; Rutgers Univ, PhD(entom), 56. *Prof Exp:* Assoc prof biol, Pac Union Col, 56-58; from assoc prof to prof biol, 58-88, dir toxicol grad prog, 77-88, EMER PROF BIOL, SAN JOSE STATE UNIV, 88- *Concurrent Pos:* Environ toxicol fel, Univ Calif, Davis, 75-76. *Mem:* AAAS; Am Indust Hyg Asn; Entom Soc Am; Am Chem Soc; Soc Risk Anal; Col Toxicol. *Res:* Toxicology; general physiology. *Mailing Add:* 127 Altura Vista Los Gatos CA 95030

BALLARD, ROBERT D, b Wichita, Kans, June 30, 42; m 91; c 1. PLATE TECTONICS. *Educ:* Univ Calif, Santa Barbara, BS(chem & geol), 65; Univ Hawaii, BS(oceanog & marine geol), 66; Univ Southern Calif, BS(marine geol & geophysics), 67; Univ RI, PhD(marine geol & geophysics), 74. *Hon Degrees:* DSc, Clark Univ, 86; Univ Rhode Island, 86; Southern Mass Univ, 86; Long Island Univ, 87; Univ of Bath, Eng, 88; Tufts Univ, 90. *Prof Exp:* Res assoc ocean eng, Woods Hole Oceanog Inst, 69-74, asst scientist geol & geophysics, 74-76, assoc scientist, 76-78, assoc scientist ocean eng dept, 78-79 & 80-83, SR SCIENTIST & HEAD, DEEP SUBMERGENCE LAB, OCEAN ENG DEPT, WOODS HOLE OCEANOG INST, 83- *Concurrent Pos:* Vis scholar, Geol Dept, Stanford Univ, 79-80, consult prof, 80-81; pvt consult, Benthos, Inc, Mass, 82-83; consult, Dep Chief Naval Operations for Submarine Warfare, 84-, Marine Bd, Nat Res Coun, Comn on Eng & Tech Systs, 84-87; dir, Ctr Marine Explor, Woods Hole Oceanog Inst, 87-; founder, Jason Found Educ. *Honors & Awards:* Newcomb Cleveland Prize, AAAS, 81; Cutty Sark Award, Sci Dig, 82; Westinghouse Award, AAAS, 90; William Proctor Award, Sigma Xi, 90; Am Geol Inst Award, 90. *Mem:* Geol Soc Am; Marine Technol Soc; Explorers Club; Am Geophys Union. *Res:* Development of advanced deep water technology aimed at better understanding of the ocean floor with special emphasis on the Mid-Ocean Ridge; deep diving submersibles. *Mailing Add:* Woods Hole Oceanog Inst Water St Woods Hole MA 02543

BALLARD, STANLEY SUMNER, b Los Angeles, Calif, Oct 1, 08; m 35, 74; c 2. PHYSICS, OPTICS. *Educ:* Pomona Col, AB, 28; Univ Calif, MA, 32, PhD(physics), 34. *Hon Degrees:* DSc, Pomona Col, 74. *Prof Exp:* Asst physics, Dartmouth Col, 28-30; asst, Univ Calif, 30-34, res fel, 34-35; from instr to asst prof, Univ Hawaii, 35-41; prof physics & chmn dept, Tufts Univ, 46-54; res physicist, Scripps Inst, Univ Calif, 54-58; prof physics, 58-78, chmn dept physics & astron, 58-71, dir div phys & math sci, 68-71, distinguished serv prof physics, 78-81, EMER PROF PHYSICS, UNIV FLA, 82- *Concurrent Pos:* Consult exp sta, Hawaiian Sugar Planters Asn, 37-40; collabr, Hawaii Agr Exp Sta & res assoc, Hawaii Nat Park, 39-41; consult physicist, Polaroid Corp, Mass, 46-54 & Baird Assocs, Inc, 46-51; US deleg int comn optics, Int Union Pure & Appl Physics, Delft, 48, London, 50, Madrid, 53, Cambridge, Mass, 56, Stockholm, 59, Munich, 62, Paris, 66, Reading, Eng, 69, Calif, 72, Czech, 75 & Madrid, 78 & Graz, Austria, 81, vpres, 48-56, pres, 56-59, chmn, US Nat Comt, 48-56; consult, Rand Corp, 52-53, physicist, 53-54, consult, 54-71; mem vision comt, Armed Forces-Nat Res Coun, exec secy, 56-59; consult, Inst Sci & Technol, Univ Mich, 57-65, Bendix Systs Div, 57-59, Perkin-Elmer Corp, 60-62 & Atlantic Missile Range, 60-63; consult astrionics div, Aerojet-Gen Corp, 61-66; mem bd dirs, Am Inst Physics, 62-65 & 67-73; mem phys sci div, Nat Res Coun, 63-65; consult, Douglas Advan Res Labs, 66-70, Aerospace Corp, 67-70 & Honeywell Radiation Ctr, 69. *Honors & Awards:* Pegram Award, Am Phys Soc, 81; Oersted Medal, Am Asn Physics Teachers, 85. *Mem:* AAAS (secy physics sect, 60-67, vpres, 68); fel Am Phys Soc; fel Optical Soc Am (pres, 63); Am Asn Physics Teachers (pres, 68-69); fel Brit Phys Soc; fel Soc Photo-Optical Instrumentation Engrs. *Res:* Spectroscopy; optical and infrared instrumentation; properties of optical materials. *Mailing Add:* Dept Physics Univ Fla Gainesville FL 32611-2085

BALLAS, ZUHARI K, ALLERGY IMMUNOLOGY. *Educ:* Am Univ Beirut, Lebanon, MD, 74. *Prof Exp:* Asst prof, 80-86, ASSOC PROF MED, UNIV IOWA, 86- *Mailing Add:* Dept Internal Med Univ Iowa Iowa City IA 52242

BALLATO, ARTHUR, b Astoria, NY, Oct 15, 36; m 59; c 2. ELECTROPHYSICS, ELECTRONICS ENGINEERING. *Educ:* Mass Inst Technol, SB, 58; Rutgers Univ, MS, 62; Polytech Inst Brooklyn, PhD(electrophys), 72. *Prof Exp:* Electronics engr, 58-86, RES PHYS SCIENTIST, US ARMY ELECTRONICS TECHNOL & DEVICES LAB, NJ, 86- *Concurrent Pos:* Adj prof, Monmouth Col, NJ, 83-; tech adv group TC-49, Int Electrotech Comn, 83-; nat lectr, Inst Elec & Electronics Engrs, Sonics & Ultrasonics Group, 84-85. *Honors & Awards:* C B Sawyer Mem Award, Sawyer Res Prod Inc, 78; H Jocobs Award, 85. *Mem:* Sigma Xi; Am Phys Soc; Inst Elec & Electronics Engrs; Inst Elec Engrs, London. *Res:* Piezoelectric crystal resonators and filters; crystal physics; stacked crystal filters; surface acoustic waves; stress effects in crystal plates; temperature compensated acoustic devices; frequency control devices. *Mailing Add:* 150 Atlantic Ave Long Branch NJ 07740

BALLENGER, JAMES CAUDELL, b Raleigh, NC, Mar 2, 44; m 69; c 2. PSYCHIATRY, CLINICAL PSYCHOPHARMACOLOGY. *Educ:* Univ NC, AB, 66; Duke Univ, MD, 70. *Prof Exp:* Intern internal med, Duke Univ Med Ctr, 70-71; resident psychiat, Harvard Med Sch, Mass Gen Hosp, Boston, 71-74; pvt pract adult gen psychiat, Boston, Mass, 74-76; staff psychiatrist, Nat Naval Med Ctr, Bethesda, Md, 76-77; ward chief, 3 West Clin Res Unit, Inst Mental Health, 77-79; dir inpatient serv, 79-81, prof & dir res behav med & psychiat, Univ Va Med Ctr & Western State Hosp, 79-, head, div psychopharmacology, 81-; CHMN DEPT PSYCHIAT & BEHAV SCI & DIR, INST PYSCHIAT, MED UNIV SC, 88- *Concurrent Pos:* Mem & co-chmn, publ comt, Am Col Neuropsychiat Pharmacol, 87. *Mem:* Fel Am Psychiat Asn; Soc Biol Psychiat; fel Am Col Psychiatrists. *Res:* Clinical psychological and biological interfaces in patients with spectrum of psychiatric problems, especially anxiety, manic-depressive illness, schizophrenia, alcoholism and personality disorders; relationship between biological factors and manifest psychopathology; biological and psychological relationships in normals; clinical psychopharmacology. *Mailing Add:* Dept Psychiat & Behav Sci Med Univ SC 171 Ashley Ave Charleston SC 29425

BALLENTINE, ALVA RAY, b Irmo, SC, Feb 19, 31. PHYSICAL ORGANIC CHEMISTRY, SYNTHETIC ORGANIC CHEMISTRY. *Educ:* Univ SC, BS, 50, MS, 52, PhD(org chem), 56. *Prof Exp:* Asst prof chem, Presby Col, SC, 55-56; from asst prof to assoc prof, 56-68, PROF CHEM, THE CITADEL, 68- *Mem:* Am Chem Soc; Sigma Xi. *Res:* Reaction mechanisms; organic synthesis. *Mailing Add:* Dept Chem The Citadel Charleston SC 29409

BALLENTINE, LESLIE EDWARD, b Wainwright, Alta, Apr 19, 40; c 2. QUANTUM MECHANICS. *Educ:* Univ Alta, BSc, 61, MSc, 62; Cambridge Univ, PhD(physics), 66. *Prof Exp:* Fel, Theoret Phys Inst, Univ Alta, 65-67; from asst prof to assoc prof physics, 67-73, PROF PHYSICS, SIMON FRASER UNIV, 73- *Mem:* Can Asn Physicists; Am Phys Soc. *Res:* Foundations of quantum mechanics; dynamical chaos and quantum mechanics. *Mailing Add:* Dept Physics Simon Fraser Univ Burnaby BC V5A 1S6 Can

BALLENTINE, ROBERT, b Orange, NJ, Mar 6, 14; m 68. BIOCHEMISTRY, MICROBIAL ECOLOGY. *Educ:* Princeton Univ, AB, 37, PhD(biol), 40. *Prof Exp:* Proctor fel, Princeton Univ, 40-41; Nat Res fel zool, Rockefeller Inst, 41-42; lectr, Columbia Univ, 42-43, instr, 43-48; scientist, Brookhaven Nat Lab, 48-49; assoc prof, 49-81, EMER PROF BIOL, JOHNS HOPKINS UNIV, 81- *Concurrent Pos:* Guggenheim fel, Calif Inst Technol, 47; assoc res specialist, Rutgers Univ, 48-49. *Mem:* Fel AAAS; Am Chem Soc; Soc Gen Physiol; Inst Elec & Electronics Engrs Computer Soc; Asn Computer Mach. *Res:* Computer information science; radioisotope assay; inorganic ion accumulation; surface colonization by bacteria in estuarine environments. *Mailing Add:* Dept Biol Johns Hopkins Univ Baltimore MD 21218

BALLENTINE, RUDOLPH MILLER, b Ballentine, SC, Aug 16, 41; m 79; c 4. HOLISTIC MEDICINE. *Educ:* Duke Univ, BS, 62, MD, 67. *Prof Exp:* From instr to asst clin prof psychiat, La State Univ Med Sch, 71-72; PVT PRACT PSYCHOSOMATIC MED, CTR HOLISTIC MED, 74-; DIR COMBINED THER, HIMALAYAN INST, 75- *Concurrent Pos:* Pres & adminr, Himalayan Inst, 81- *Honors & Awards:* Ebert Mem Lectr, Univ Ill Col Pharm, 78. *Mem:* Am Inst Homeopath; NY Acad Sci; AMA. *Res:* Holistic medicine; writings on yoga, nutrition and holistic medicine. *Mailing Add:* Himalayan Inst RR 1 PO Box 400 Honesdale PA 18431

BALLERINI, ROCCO, b Pittsburgh, Pa, June 24, 58; m 89; c 1. EXTREME VALUE THEORY, TIME SERIES ANALYSIS. *Educ:* Pa State Univ, BS, 80; Colo State Univ, MS, 83, PhD(statist), 85. *Prof Exp:* Asst prof, 85-90, ASSOC PROF STATIST, UNIV FLA, 90- *Mem:* Am Statist Asn; Inst Math Statist; Am Geophys Union. *Res:* Extreme value theory; time series analysis; geophysical applications. *Mailing Add:* 484 Little Hall Dept Statist Univ Fla Gainsville FL 32611

BALLESTERO, THOMAS P, b Saratoga Springs, NY, Aug 10, 53; m 79; c 3. GROUNDWATER, HYDROLOGIC SIMULATION. *Educ:* Pa State Univ, BS, 75 & MS, 77; Colo State Univ, PhD(civil eng), 81. *Prof Exp:* Inst civil eng, Pa State Univ, 77-78; div mgr, Simons, Li & Assoc, 80-83; ASST PROF CIVIL ENGR, UNIV NH, 83-, DIR, WATER RESOURCES RES CTR, 86- *Concurrent Pos:* Pvt consult, 77- *Mem:* Am Geophys Union; Am Inst Hydrol; Am Water Resources Asn; Am Water Works Asn; Nat Water Well Asn. *Res:* Computer simulation of hydrologic phenomena and field measurement of parameters of such models; new methodologies for field measurements in which the areas of interest include groundwater, flow and bioremediation. *Mailing Add:* Civil Eng Dept Univ NH 236 Kingsbury Hall Durham NH 03824

BALLEW, DAVID WAYNE, b Mangum, Okla, Aug 20, 40; m 60; c 1. NUMBER THEORY. *Educ:* Univ Okla, BA, 62, MA, 64; Univ Ill, PhD(math), 69. *Prof Exp:* Asst math, Univ Okla, 62-64 & Univ Ill, 64-67; asst prof, 67-75, dept head, 81-86, PROF MATH, SDAK SCH MINES & TECHNOL, 75- *Mem:* Am Math Soc; Math Asn Am; Am Soc Eng Educ. *Res:* Extensions of results in commutative ring theory to results in noncommutative ring theory, especially orders and separable algebras; application of computers to number theory. *Mailing Add:* 125 Pawn Ridge Macomb IL 61455

BALLHAUS, WILLIAM F(RANCIS), b San Francisco, Calif, Aug 15, 18; m 44; c 4. AERONAUTICAL & ASTRONAUTICAL ENGINEERING. *Educ:* Stanford Univ, BS, 40, ME, 42; Calif Inst Technol, PhD(aerodyn, math), 47. *Prof Exp:* Chief engr, Northrop Corp, 53-57, vpres & gen mgr, Nortronics Div, 57-61, exec vpres & dir corp, 61-64; pres & dir, 65-82, chief exec off, 83-84, CONSULT, BECKMAN INSTRUMENTS, INC, 84- *Concurrent Pos:* Mem, Nat Adv Comt Aeronaut, 54-57; mem tech adv panel aeronaut, Off Secy Defense, 54-60; adv, Off Critical Tables, Nat Acad Sci. *Mem:* Nat Acad Eng; fel Am Inst Aeronaut & Astronaut. *Mailing Add:* 150 S El Camino Dr No 200 Beverly Hills CA 90212

BALLHAUS, WILLIAM FRANCIS, JR, b Los Angeles, Calif, Jan 28, 45; m 83; c 4. COMPUTATIONAL AERODYNAMICS. *Educ:* Univ Calif, Berkeley, BS, 67, MS, 68, PhD(mech eng), 71. *Prof Exp:* dir astron, NASA, Ames Res Ctr, 71-78, res scientist computational aerodyn, US Army Aviation Res Develop Comn & Ames Res Ctr, 71-79, chief, Appl Computational

Aerodyn Br & Ames Res Ctr, NASA & US Army Aviation Res & Develop Comn, 79-84, dir, Ames Res Ctr, 84-89 & actg assoc adminr, Aeronaut & Space Technol, 88-89; PRES, CIVIL SPACE & COMMUN, MARTIN MARIETTA, 90- *Honors & Awards:* Lawrence Sperry Award, Am Inst Aeronaut & Astronaut, 80; Arthur S Flemming Award, 80. *Mem:* Nat Acad Eng; fel Am Inst Aeronaut & Astronaut (pres, 88-89); Int Acad Astronaut; Int Asn Astronaut. *Res:* Computational simulation of fluid flows about aerodynamic configurations; transonic regime; applied mathematics. *Mailing Add:* Civil Space & Commun Martin Marietta PO Box 179 Mail Stop DC 8000 Denver CO 80201

BALLIF, JAE R, b Provo, Utah, July 5, 31; m 53; c 7. SOLAR PHYSICS, TERRESTRIAL PHYSICS. *Educ:* Brigham Young Univ, BS, 53; Univ Calif, Los Angeles, MA, 61, PhD(atmospheric physics), 62. *Prof Exp:* From asst prof to assoc prof, 62-70, PROF PHYSICS, BRIGHAM YOUNG UNIV, 70-, DEAN PHYS & MATH SCI, 74- *Concurrent Pos:* Provost & acad vpres, Brigham Young Univ, 80-89. *Mem:* Am Geophys Union. *Res:* Atmospheric airglow, particularly nocturnal emission of sodium and hydroxyl emissions; interplanetary magnetic field structure, its solar origin and terrestrial effects. *Mailing Add:* 1790 N 1500 E Provo UT 84057

BALLINA, RUDOLPH AUGUST, b New York, NY, Nov 30, 34; m 67; c 5. ORGANIC CHEMISTRY. *Educ:* St John's Univ, NY, BS, 56, MS, 61, PhD(org chem), 65. *Prof Exp:* Anal chemist, US Treas Dept, 59-60; prod develop chemist, 65-77, ASST DEPT HEAD, TOMS RIVER CHEM CORP, 77- *Mem:* Am Chem Soc. *Res:* Natural products, specifically peptides and polypeptides; organic synthesis; dyestuffs. *Mailing Add:* 30 Smith Rd Toms River NJ 08753

BALLING, JAN WALTER, b Louisville, Ky, Apr 8, 34. ANIMAL BEHAVIOR. *Educ:* Univ Louisville, BA, 56, PhD(radiation biol), 64; Purdue Univ, MS, 57. *Prof Exp:* Instr biol, Univ Louisville, 60-64; dir labs, Cent State Hosp, Louisville, 64-65; asst prof biol, George Peabody Col, 65-66; PROF BIOL, CALIF STATE COL, PA, 66- *Mem:* Am Soc Parasitol; Am Nuclear Soc; Am Soc Ichthyologists & Herpetologists; Animal Behav Soc; Sigma Xi. *Res:* Turtle ethology; operant behavior of rats; trematode systematics; chemical protection of cold-blooded vertebrates against radiation damage; operant behavior of chronic schizophrenics. *Mailing Add:* PO Box 576 California PA 15419

BALLINGER, CARTER M, b Lewistown, Mont, Mar 9, 22; m 47; c 5. ANESTHESIOLOGY. *Educ:* Univ Iowa, BA, 43, MD, 45; Univ Pa, MMedSci, 61; Am Bd Anesthesiol, dipl, 55. *Prof Exp:* Intern, Fresno County Gen Hosp, Calif, 45-46; asst resident surg, Crile Vet Admin Hosp, Cleveland, Ohio, 48-50; resident anesthesiol, Columbia Univ, 51-53, instr, 53-54; chmn div anesthesiol, Univ Utah, 54-73, anesthesiologist-in-chief, 65-73, assoc prof anesthesiol, 54-76; prof anesthesiol & chmn dept, NJ Med Sch, Col Med & Dent, 76-78; PROF ANESTHESIOL, SCH MED, UNIV COLO, 79- *Concurrent Pos:* Asst attend, Presby Hosp, NY & adj attend, Valley Hosp, Ridgewood, NJ, 53-54; chmn div anesthesiol, Salt Lake County Hosp, Utah, 54-65; consult, Salt Lake Vet Admin Hosp, 54-75, Shriners Hosp, 57-72, Primary Childrens Hosp, 67-76 & McKay Hosp, Ogden, Utah, 70-75; secy-treas, Biennial Western Conf Anesthesiol, 61-65, pres, 65-67, prog chmn, 67-69; NIH spec fel neurophysiol, Univ Wis, 62-63; pres, Guedel Mem Anesthesia Libr, San Francisco, Calif, 68-70; sr staff mem anesthesiol, Martland Hosp, Newark, 76-78; sr staff anesthesiol, Univ Hosp, Denver, 79- *Mem:* Sigma Xi; AMA; Am Soc Anesthesiol; Asn Univ Anesthetists; Int Asn Study Pain. *Res:* Antiemetic and anesthetic agents in dogs and man; blood transfusion and intravenous therapy; mechanism of action anesthetic agents; treatment of chronic pain; training of general practitioners in anesthesiology; development of a disposable anesthesia bacterial filter system; continuing medical education in anesthesiology. *Mailing Add:* Univ Colo Med Ctr Box B-113 4200 E Ninth Ave Denver CO 80262

BALLINGER, PETER RICHARD, b London, Eng, Oct 15, 32; US citizen; m 59, 70; c 3. ORGANIC CHEMISTRY. *Educ:* Univ London, BSc, 54, PhD(org chem), 57. *Prof Exp:* Fel phys org chem, Cornell Univ, 57-58, instr chem, 58-59; res chemist petrol prod, Calif Res Corp, 59-64, sr res chemist, 64-67, SR RES ASSOC PETROL PROD, CHEVRON RES CO, 67- *Concurrent Pos:* AEC fel, 57-59. *Mem:* NY Acad Sci; Am Chem Soc. *Res:* Chemistry of air pollution; composition of automobile exhaust; mechanism of detergency; detailed analysis of hydrocarbon fuels. *Mailing Add:* 1310 Mariposa St Richmond CA 94804

BALLINGER, ROYCE EUGENE, b Burkburnett, Tex, Feb 21, 42. EVOLUTIONARY ECOLOGY, VERTEBRATE ZOOLOGY. *Educ:* Univ Tex, Austin, BA, 64; Tex Tech Univ, MS, 67; Tex A&M Univ, PhD(ecol), 71. *Prof Exp:* Res asst ecol, Ctr Biol Nat Systs, Wash Univ, 66-67; instr biol, Angelo State Univ, 67-71, from asst prof to assoc prof, 71-76; assoc prof ecol, Sch Biol Sci, Univ Nebr-Lincoln, 76-82, vdir, Sch Life Sci, 77-82, prof ecol & dir, 82-90, PROF, SCH BIOL SCI, UNIV NEBR-LINCOLN, 90- *Concurrent Pos:* NSF grant, 72-76, 78-80; Dept Interior Water Resources grant, 77-79; panel mem, NSF, 78-79; bd gov, Ctr Great Placing Studies, 80-88; ecol ed, Copeia, Am Soc Ichthyologists & Herpetologists, 84-87; proj specialist, Chinese Provincial Univ Dev Proj, Guilin, Peoples Repub China, 88. *Mem:* Ecol Soc Am; AAAS; Am Soc Ichthyologists & Herpetologists; Am Soc Naturalists; Sigma Xi. *Res:* Evolutionary ecology of life history and reproductive strategies of lizards; ecology of natural selection and the importance of demographic and competitive environments to ecological adaptations; reptile thermoregulation; prairie ecology. *Mailing Add:* 710 Cottonwood Dr Lincoln NE 68801

BALLINGER, WALTER ELMER, b Burlington, NJ, Jan 22, 26; m 62; c 3. HORTICULTURE. *Educ:* Rutgers Univ, BS, 52; Mich State Univ, MS, 55, PhD(hort), 57. *Prof Exp:* Res asst, Mich State Univ, 54-57; from asst prof to prof hort, NC State Univ, 57-87, coordr acad affairs, Dept Hort Sci, 73-87; RETIRED. *Concurrent Pos:* Guest lectr, Peach Cong, Verona, Italy, 65 & Univ Berlin, 65. *Mem:* Am Soc Hort Sci. *Res:* Pre- and post-harvest physiology of blueberries and grapes, especially anthocyanin characterization, physical and chemical measurement of quality and maintenance of quality. *Mailing Add:* 5612 Winthrop Dr Raleigh NC 27612

BALLINGER, WALTER F, II, b Philadelphia, Pa, May 16, 25; m 53, 80; c 3. SURGERY. *Educ:* Univ Pa, MD, 48. *Prof Exp:* From asst to assoc prof surg, Jefferson Med Col, 56-64; assoc prof, Sch Med, Johns Hopkins Univ, 64-67; Bixby prof surg & chmn dept, 67-78, PROF SURG, SCH MED, WASH UNIV, 78- *Concurrent Pos:* Markle scholar acad med, 61-66; surgeon-in-chief, Barnes & Allied Hosps, 67-78, surgeon, 78- *Mem:* Am Col Surg; Soc Univ Surg; Soc Clin Surg; Soc Vascular Surg. *Res:* Gastrointestinal, vascular and metabolic research. *Mailing Add:* Wash Univ Sch Med Barnes Hosp 4989 Barnes Hospital Plaza St Louis MO 63110

BALLMANN, DONALD LAWRENCE, b Dayton, Ohio, Apr 25, 27; m 72. GEOLOGY. *Educ:* St Joseph's Col, Ind, AB, 54; Univ Ill, BS, 55, MS, 56, PhD(geol), 59. *Prof Exp:* Asst prof geol, St Joseph's Col, Ind, 56-63, dir geol field camp, 56-60, dir undergrad res, 58-60, asst dir develop, 61-62, exec asst to pres, 62-63, assoc prof geol, 63-72, acad dean, 63-68, dir develop found & govt rels, 68-72; proj mgr to assoc, Dames & Moore, 73-82; sr licensing adv, Battelle, 83-89, tech admin, off Low-Level Waste Technol, 89-90, GEOSCI MGR, CONN LOW-LEVEL WASTE PROJ, BATTELLE, 91- *Mem:* Am Asn Petrol Geol; Geol Soc Am; Am Inst Prof Geologists; Soc Econ Paleontologists & Mineralogists; Sigma Xi; AAAS; Am Nuclear Soc. *Res:* Areal geology; geomorphology; historical geology; engineering geology; soil mechanics. *Mailing Add:* 74 Village Lane Windsor CT 06095

BALLOU, CLINTON EDWARD, b King Hill, Idaho, June 18, 23; m 49; c 2. BIOCHEMISTRY. *Educ:* Ore State Col, BS, 44; Univ Wis, MS, 48, PhD(biochem), 50. *Prof Exp:* USPHS res fels, Univ Edinburgh, 50-51 & Univ Calif, Berkeley, 51-52; asst res biochemist, 52-55, from asst prof to assoc prof biochem, 55-61, chmn dept, 64-68, PROF BIOCHEM, UNIV CALIF, BERKELEY, 61- *Concurrent Pos:* NSF sr fel, 61-62; Guggenheim fel, Kyoto & Paris, 68-69; vis scientist, Basel Inst Immunol, 75-76, Imp Col London & Nat Ctr Sci Res, Grenoble, 82-83. *Honors & Awards:* Claude Hudson Award, 81. *Mem:* Nat Acad Sci; AAAS; Am Chem Soc; Am Soc Biol Chemists. *Res:* Structure and biosynthesis of complex polysaccharides, lipids and lipopolysaccharides and the role these substances play in the biochemistry of the cell envelope; glycoproteins. *Mailing Add:* Div Biochem & Molecular Biol Barker Hall Univ Calif Berkeley CA 94720

BALLOU, DAVID PENFIELD, b New Britain, Conn, Aug 25, 42; m 72; c 2. BIOCHEMISTRY, ENZYMOLOGY. *Educ:* Antioch Col, BS, 65; Univ Mich, MS, 67, PhD(chem), 71. *Prof Exp:* Asst res biochemist, 71-72, from instr to assoc prof biochem, 72-87, PROF BIOCHEM, UNIV MICH, ANN ARBOR, 87- *Concurrent Pos:* Consult, Henry Ford Hosp, Detroit, 72- *Mem:* AAAS; Am Soc Biol Chemists. *Res:* Study of oxygen activation in biological systems; reactions of oxygen with flavins, nonheme iron proteins and cytochromes employing computerized rapid kinetics, absorption, fluorescence and magnetic resonance spectroscopy; development of instrumentation for the above. *Mailing Add:* 1321 Moringside Ann Arbor MI 48103-2524

BALLOU, DONALD HENRY, b Chester, Vt, Mar 28, 08; m 33; c 1. MATHEMATICS. *Educ:* Yale Univ, AB, 28; Harvard Univ, AM, 31, PhD(math), 34. *Prof Exp:* Instr, Harvard Univ, 33-34; instr math, Ga Inst Technol, 34-35, asst prof, 35-42; from asst prof math to prof, 42-64, Beman prof, 64-71, Charles A Dana prof, 71-73, head dept, 54-68, CHARLES A DANA EMER PROF MATH, MIDDLEBURY COL, 73- *Concurrent Pos:* Vis assoc prof, Yale Univ, 51-52. *Mem:* Am Math Soc; Math Asn Am. *Res:* Laplace transforms; roots of polynomial equations. *Mailing Add:* 27 Weybridge St Middlebury VT 05753

BALLOU, DONALD POLLARD, b Atlanta, Ga, Sept 30, 40; m 69; c 2. MANAGEMENT SCIENCE. *Educ:* Harvard Col, BA, 62; Univ Mich, MA, 63, PhD(math), 68. *Prof Exp:* Res assoc math, Brown Univ, 69-70; asst prof math, State Univ Ny, Albany, 70-76, lectr mgt sci, 76-81, chmn, Mgt Sci & Info Systs Dept, 87-88, ASSOC PROF MGT SCI, STATE UNIV NY, ALBANY, 81- *Concurrent Pos:* Prin or Co-prin investr, Nat Sci Found, 76-78, US Dept Transp, 77-79, 79-80 & 83-85, USCG, 87-88. *Mem:* Asn Comput Mach; Math Asn Am; Inst Mgt Sci; Sigma Xi. *Res:* Mathematical modelling of non-linear wave phenomona, transportation flows, production systems and decision making; quality of information systems; management information systems. *Mailing Add:* Three Rita Ct Delmar NY 12054

BALLOU, JOHN EDGERTON, b King Hill, Idaho, Oct 25, 25; m 52; c 4. INHALATION TOXICOLOGY. *Educ:* Ore State Col, BS, 50; Univ Rochester, PhD(toxicol), 69. *Prof Exp:* Biol scientist, Hanford Atomic Prod Oper, Gen Elec Co, 51-64; biol scientist, Pac Northwest Lab, Battelle Mem Inst, 65-69, staff scientist, 69-88; RETIRED. *Mem:* Am Chem Soc; Health Physics Soc. *Res:* Absorption, distribution and retention of inhaled radioisotopes. *Mailing Add:* 1628 Butternut Richland WA 99352

BALLOU, NATHAN ELMER, b Rochester, Minn, Sept 28, 19; m 45, 73; c 2. SEPARATIONS SCIENCE, MASS SPECTROSCOPY. *Educ:* Minn State Teachers Col, BS, 41; Univ Ill, MS, 42; Univ Chicago, PhD(chem), 47. *Prof Exp:* Asst chem, Univ Ill, 41-42; asst, Metall Lab, Univ Chicago, 42-43; asst, Clinton Labs, Oak Ridge, 43-44, assoc chemist, 45-46; res chemist, Hanford Eng Works, Wash, 44-45; assoc, Radiation Lab, Univ Calif, 47-48; head nuclear chem br, Naval Radiol Defense Lab, 48-69; sr res mgr, 69-87, LEAD SCIENTIST, PAC NORTHWEST LABS, BATTELLE MEM INST, 87- *Concurrent Pos:* Head chem dept, Belgian Nuclear Ctr, 59-60; mem comt nuclear sci, Nat Acad Sci-Nat Res Coun, chmn subcomt radiochem, 62-66; mem exec comt, Div Nuclear Chem & Technol, Am Chem Soc, 69-77. *Mem:* AAAS; Am Chem Soc. *Res:* Fission process; radiochemical techniques; ultrasensitive analytical techniques. *Mailing Add:* Battelle Northwest Richland WA 99352

BALLUFFI, R(OBERT) W(EIERTER), b Bayshore, NY, Apr 18, 24; m; c 3. PHYSICAL METALLURGY. *Educ:* Mass Inst Technol, BS, 47, DSc, 50. *Prof Exp:* Sr engr, Sylvania Elec Prod, Inc, 50-54; res assoc, Columbia Univ, 54; prof mat sci & eng, Univ Ill, 54-64 & Cornell Univ, 64-78; PROF MAT

SCI & ENG, MASS INST TECHNOL, 78- *Honors & Awards:* David Adler Lectr, Am Phys Soc, 89; Von Hippel Award, Mat Res Soc, 90. *Mem:* Nat Acad Sci; fel Am Inst Mining, Metall & Petrol Engrs; Metall Soc; Brit Inst Metals; fel Am Acad Arts & Sci; fel Am Phys Soc. *Res:* Diffusion in metals; radiation damage; crystal imperfections. *Mailing Add:* Dept Mat Sci & Eng Mass Inst Technol Cambridge MA 02139

BALLY, ALBERT WALTER, b The Hague, Holland, April 21, 25. STRUCTURAL, PETROLEUM. *Educ:* Univ Zurich, PhD(geol), 52. *Prof Exp:* Chief geologist, Shell Oil Co, 77-81; PROF GEOL, RICE UNIV, 81- *Mem:* Fel Geol Soc Am; Am Asn Petrol Geologists. *Mailing Add:* Dept Geol Rice Univ PO Box 1892 Houston TX 77251

BALMAIN, KEITH GEORGE, b London, Ont, Aug 7, 33; m 58. ELECTROMAGNETICS. *Educ:* Univ Toronto, BASc, 57; Univ Ill, Urbana-Champaign, MS, 59, PhD(elec eng), 63. *Prof Exp:* Res asst elec eng, Univ Ill, Urbana-Champaign, 61-63; from asst prof to assoc prof, 63-73, PROF ELEC ENG, UNIV TORONTO, 73- *Concurrent Pos:* Chmn, Div Eng Sci, Univ Toronto, 85-87, Res Bd, 87-90. *Mem:* Inst Elec & Electronics Engrs; Can Asn Physicists. *Res:* Antennas in plasmas; broadband antennas; spacecraft electrostatic charging and discharging; electromagnetic compatibility. *Mailing Add:* Univ Toronto Ten King's College Rd Univ Toronto Toronto ON M5S 1A4 Can

BALMER, CLIFFORD EARL, b Brownstown, Pa, Sept 17, 21; m 44; c 3. ORGANIC CHEMISTRY, POLYMER CHEMISTRY. *Educ:* Albright Col, BS, 47. *Prof Exp:* Org chemist, Pioneer Res Labs, US Army Qm Depot, 47-53; chemist, Appln Labs, Rohm & Haas Co, 53-59, tech sales, 59-68, tech sales rep, Resins Dept, 68-70, tech sales rep, Indust Chem Dept, 70-73, tech sales rep, Plastics Int Dept, 73-78, tech sales rep, Plastics Dept, 78-84; RETIRED. *Mem:* Soc Plastics Engrs; Am Chem Soc. *Res:* Development of plasticizers for polyvinyl chloride. *Mailing Add:* PO Box 562 Eight Lord Jeffrey Dr Amherst NH 03031-0562

BALMER, ROBERT THEODORE, b Chelsea, Mich, Nov 26, 38; m 61; c 3. THERMODYNAMICS, RHEOLOGY. *Educ:* Univ Mich, BSE, 61, BSE, 64, MSE, 63; Univ Va, ScD, 68. *Prof Exp:* NATO vis prof res, Univ Naples, Italy, 68-69; from asst prof to assoc prof, 69-79, PROF TEACHING & RES, UNIV WIS-MILWAUKEE, 79- *Concurrent Pos:* Dept chair, Mech Eng Dept, Univ Wis-Milwaukee, 72-74, 89-, eng res coordr, Grad Sch, 84-87, dir, CAE Lab, 90-; vis prof, Tech Univ, Budapest, Hungary, 90. *Honors & Awards:* Teetor Award, Soc Automotive Engrs, 66. *Mem:* Am Soc Mech Engrs; Soc Rheology; Am Soc Eng Educ. *Res:* Authored an engineering thermodynamics textbook (1990) and over 50 research papers in the areas of fluid flow, bioengineering, thermodynamics, history of technology and engineering education; developed nonequilibrium thermodynamic theory of static charge generation in flowing fluids. *Mailing Add:* Mech Eng Dept Univ Wis-Milwaukee Shorewood WI 53211

BALMFORTH, DENNIS, b Pudsey, Eng, Apr 29, 30; m 56; c 3. CHEMISTRY. *Educ:* Univ Leeds, BSc, 54, PhD(colour chem, dyeing), 61. *Prof Exp:* Res chemist, James Anderson & Co, Ltd, Scotland, 54-56; chemist, Courtaulds Ltd, Eng, 57-58 & 61-63; supvr dyeing res, Chemstrand Co, 63-67; DIR, CHEM & ENVIRON RES & SERV DEPT, INST TEXTILE TECHNOL, 67- *Mem:* Royal Inst Chem; Brit Soc Dyers & Colourists. *Res:* Color chemistry; fine structure of fibrous materials and its influence on dyeing and finishing behavior. *Mailing Add:* Chem Inst Textile Tech Box 391 Charlottesville VA 22902

BALOG, GEORGE, b Gunn, Wyo, Oct 12, 28; m 51; c 4. LASERS. *Educ:* Univ Wyo, BS, 50; Ill Inst Technol, MS, 54, PhD(chem), 58. *Prof Exp:* mem staff, Los Alamos Sci Lab, 56-90; RETIRED. *Mem:* Am Chem Soc. *Res:* Polarographic studies in non-aqueous solvents; refractory materials, specifically the thermionic emission microscopy; high temperature x-ray studies; chemical lasers; dye lasers and laser dye development. *Mailing Add:* Four Erie Lane Pajarito Acres Los Alamos NM 87544

BALOGH, CHARLES B, b Beled, Hungary, Sept 13, 29; US citizen; m 54; c 1. MATHEMATICS. *Educ:* Eotvos Lorand Univ, Budapest, Ms, 54; Ore State Univ, PhD(math), 65. *Prof Exp:* Res fel, Astrophys Observ, Budapest, 54-56; instr math, Jr Col, 57-58 & Univ Portland, 59-61; asst, Ore State Univ, 61-64; asst prof, 64-68, ASSOC PROF MATH, PORTLAND STATE UNIV, 68- *Mem:* Am Math Soc; Soc Indust & Appl Math. *Res:* Asymptotic solutions of differential equations; special functions; computer languages. *Mailing Add:* Dept Comput Sci Portland State Univ Portland OR 97207

BALOGH, KAROLY, b Budapest, Hungary, Sept 24, 30; m 55; c 3. PATHOLOGY, HISTOCHEMISTRY. *Educ:* Med Univ Budapest, MD, 54. *Prof Exp:* Resident path anat, Univ Budapest, 54-56; dir otolaryngol path, Mass Eye & Ear Infirmary, 62-68; asst prof path, Sch Med, Harvard Univ, 68; from assoc prof to prof, Sch Med, Boston Univ, 68-75, chief path, Univ Hosp, 68-74; sr staff mem, Mallory Inst Path, 74-75; chief path, Worcester City Hosp, 75-77; prof, Sch Med, Univ Mass, 75-77; PATHOLOGIST, NEW ENG DEACONESS HOSP, 77-; ASSOC PROF PATH, HARVARD MED SCH, 77- *Concurrent Pos:* Rockefeller fel, Sch Med, Tulane Univ, 57-58; USPHS trainee path, Mass Gen Hosp, Boston, 58-61. *Mem:* Am Asn Path & Bact; Int Acad Path; Histochem Soc; Ger Path Soc. *Res:* Pathological anatomy; pediatric pathology; bone pathology; enzyme histochemistry; histochemistry of calcified tissues; histochemistry of the inner ear. *Mailing Add:* New Eng Deaconess Hosp 185 Pilgrim Rd Boston MA 02215

BALON, EUGENE KORNEL, b Orlova, Czechoslovakia, Aug 1, 30; Can citizen; c 1. EPIGENETICS, EVOLUTIONARY BIOLOGY. *Educ:* Charles Univ, Prague, RNDr, 53, CSc(zool), 62. *Prof Exp:* Res Biologist, Lab Ichthyol, Bratislava, 54-67; fish biologist fishery sci, UN Develop Prog, Food & Agr Orgn UN, Res Inst, Chilanga, Zambia, 67-71; assoc prof, 72-76, PROF ZOOL, UNIV GUELPH, ONT, 76- *Concurrent Pos:* Vis prof,

Forschungsinstitut Senckenberg, Frankfurt, 76, 79, 84; vis fel, JLB Smith Inst of Ichthyology, Grahamstown, 86, 87; res assoc, Royal Ont Mus, Toronto, 72-; ed-in-chief, Int Jour Environ Biol of Fishes, 76-; Hugh Kelly fel, 87. *Mem:* Am Soc Zool; Am Soc Ichthyologists & Herpetologists; Am Fisheries Soc; Can Soc Zoologists; Europ Ichthyol Union; Sigma Xi. *Res:* Ecology, developmental and evolutionary biology; patterns and mechanisms of fish reproduction, early life history, epigenesis, and their applications to evolution and aquaculture; philosophy and metatheoretical aspects of science; comparative ontogeny and epigenetics. *Mailing Add:* Dept Zool Inst Ichthyol Guelph ON N1G 2W1 Can

BALOUN, CALVIN H(ENDRICKS), b Chicago, Ill, Apr 14, 28; m 51. CORROSION, MATERIALS SCIENCE. *Educ:* Univ Cincinnati, ChE, 54, MS, 57, PhD(metall eng), 62. *Prof Exp:* Instr metall eng, Univ Cincinnati, 60-61; from asst prof to assoc prof, 61-71, chmn dept, 72-77, PROF CHEM ENG, OHIO UNIV, 71- *Concurrent Pos:* Corrosion consult, 64-; vis prof metall eng, Univ Cincinnati, 65 & 68; vis fac mem, Esso Res & Eng Co, 68. *Mem:* Nat Asn Corrosion Engrs; Am Inst Mining, Metall & Petrol Engrs; Am Soc Metals; Am Soc Testing & Mat; Am Inst Chem Engrs. *Res:* Corrosion science; corrosion engineering; physical metallurgy. *Mailing Add:* Dept Chem Eng Ohio Univ Main Campus Athens OH 45701

BALOW, JAMES E, b Wabasha, Minn, Nov 6, 42. NEPHROLOGY, INTERNAL MEDICINE. *Educ:* Col St Thomas, BS, 64; Univ Minn, MD, 68. *Prof Exp:* CLIN DIR, PEDIAT METAB BR, NAT INST DIABETES, DIGESTIVE & KIDNEY DIS, NIH, 88- *Mem:* Am Soc Nephrology; Am Asn Immunobiol; Nat Kidney Found; Am Col Physicians. *Mailing Add:* Pediat Metab Br Nat Inst Diabetes Digestive & Kidney Dis NIH Bldg 10 Rm 9N222 Bethesda MD 20892

BALOWS, ALBERT, microbiology, infectious diseases, for more information see previous edition

BALPH, DAVID FINLEY, animal behavior, ecology; deceased, see previous edition for last biography

BALPH, MARTHA HATCH, b Boston, Mass, May 27, 43; wid; c 2. ORNITHOLOGY, ANIMAL BEHAVIOR. *Educ:* Wellesley Col, BA, 65; Univ Wyo, MS, 69; Utah State Univ, PhD(biol), 75. *Prof Exp:* Mus technician & biol illusr ichthyol, Nat Mus Natural Hist Smithsonian Inst, Washington, DC, 67-70; res asst prof, 75-80, RES ASSOC PROF FISHERIES & WILDLIFE, UTAH STATE UNIV, 80- *Concurrent Pos:* Biol consult, White River Oil Shale Proj, 74-77; F M Chapman grants, Am Mus Natural Hist, NY, 76-77 & 78-79; second vpres, Western Bird Banding Asn, 77-78, first vpres, 78-79, pres, 79-81; bd dirs, Cooper Ornith Soc, 84-87. *Honors & Awards:* Carnes Award, Am Ornithologists Union, 79. *Mem:* Am Ornithologists Union; Cooper Ornith Soc; Animal Behav Soc; Wilson Ornith Soc; Wildlife Soc; Sigma Xi. *Res:* Avian social behavior and behavioral ecology, especially communication and social organization in winter flocking passerines; avian ontogeny. *Mailing Add:* Dept Fisheries & Wildlife UMC 5210 Utah State Univ Logan UT 84322

BALSANO, JOSEPH SILVIO, b Kenosha, Wis, Apr 19, 37; m 60; c 5. POPULATION BIOLOGY, SCIENCE EDUCATION. *Educ:* Marquette Univ, BS, 59, MS, 62, PhD(biol), 68. *Prof Exp:* From instr to assoc prof biol, Dominican Col, Wis, 61-68; asst prof, Marquette Univ, 68-69; assoc prof, 74-84, PROF BIOL SCI, UNIV WIS-PARKSIDE, 84- *Mem:* AAAS; Am Inst Biol Sci; Am Soc Ichthyologists & Herpetologists; Nat Sci Teachers Asn; Sigma Xi. *Res:* Ecological genetics of fish populations; use of electrophoretic analyses and tissue transplantation to assay variation in bisexual/unisexual complexes of molly fishes. *Mailing Add:* Dept Biol Sci Univ Wis-Parkside Kenosha WI 53141

BALSBAUGH, EDWARD ULMONT, JR, b Harrisburg, Pa, Jan 12, 33; m 68; c 2. ENTOMOLOGY. *Educ:* Lebanon Valley Col, BS, 55; Pa State Univ, MS, 61; Auburn Univ, PhD(entom), 66. *Prof Exp:* Entomologist, Bur Plant Indust, Pa Dept Agr, 58-62; assoc prof entom, SDak State Univ, 65-76; PROF ENTOM, N DAK STATE UNIV, 76- *Concurrent Pos:* Consult, Am Mus Nat Hist, 82, US Army Corps Engrs, 90; ed, Schafer-Post Series NDak Insects, 76- *Mem:* Coleopterists Soc; Entom Soc Am; Sigma Xi. *Res:* Taxonomy, biology, ecology, evolution and zoogeography of Coleoptera, especially Chrysomelidae. *Mailing Add:* Dept Entom NDak State Univ Fargo ND 58102

BALSLEY, BEN BURTON, b Santa Monica, Calif, Apr 23, 32; m 56; c 3. ATMOSPHERIC PHYSICS, IONOSPHERIC PHYSICS. *Educ:* Calif State Polytech Univ, BSc, 57; Univ Colo, MSc, 64, PhD(elec eng), 67. *Prof Exp:* Physicist & engr ionospheric studies, Jicamarca Observ, Peru, 60-69; supvry physicist atmospheric studies, Aeronomy Lab, Nat Oceanic Atmospheric Admin, Boulder, Colo, 69-90; SR RES FEL, COOP INST RES ENVIRON SCI, UNIV COLO, BOULDER, 85- *Concurrent Pos:* Vchmn orgn comt, Int Symp Equatorial Aeronomy, 75-; adj prof elec eng, Univ Colo & Geophys Inst, Univ Alaska. *Honors & Awards:* Gold Medal, US Dept Com, 74. *Mem:* Sigma Xi; Am Meteorol Soc; Am Geophys Union; Int Union Radio Sci. *Res:* Investigations of neutral atmospheric dynamics in the troposphere, stratosphere and mesosphere by radar techniques; similar studies of small-scale ionospheric irregularity structure in the E and F regions at both equatorial and auroral latitudes. *Mailing Add:* Coop Inst Res Environ Sci Univ Colo PO Box 216 Boulder CO 80309-0216

BALSTER, CLIFFORD ARTHUR, b Monmouth, Iowa, Feb 27, 22; m 43; c 4. GEOLOGY. *Educ:* Iowa State Univ, BS, 48, MS, 50. *Prof Exp:* Geologist, Pure Oil Co, 51-53; div stratigrapher, 53-54; stratigrapher, Am Stratig Co, 54-56, mgr Northern Div, 56-60; group leader carbonate res, Continental Oil Co, 60-61; area geologist, Soil Conserv Serv, USDA, 61-67; from asst prof to assoc prof geol, Mont Col Mineral Sci & Technol, 67-74; consult geologist, 74-78; explor mgr, Burlington Northern, 78-79; pres, Cana Corp, 81-88; CONSULT GEOLOGIST, 79- *Concurrent Pos:* Asst prof, Ore State Univ, 61-67; res petrol geologist, Mont Bur Mines & Geol, 67-74. *Mem:* Fel Geol Soc Am; Am Asn Petrol Geol; Am Inst Prof Geologists; Sigma Xi. *Res:* Petrology of sedimentary rocks, especially carbonate rocks; geology of soils and soil genesis, particularly weathering and geomorphology; energy mineral resources and supplies. *Mailing Add:* HC 55 Box 545 Fishtail MT 59028

BALSTER, ROBERT L(OUIS), b St Cloud, Minn, Oct 12, 44; m 66; c 1. BEHAVIORAL PHARMACOLOGY. *Educ:* Univ Minn, BA, 66; Univ Houston, PhD(psychol), 70. *Prof Exp:* Teaching fel psychiat & pharmacol, Univ Chicago, 70-72; asst prof psychiat, Sch Med, Duke Univ, 72-73; from asst prof to assoc prof, 73-84, dir, grad studies, 80- 87, PROF PHARMACOL & TOXICOL, MED COL VA, VA COMMONWEALTH UNIV, 84- *Concurrent Pos:* Prin investr, Environ Protection Agency, Nat Inst Environ Health Sci & Nat Inst Drug Abuse, 76-; pres, Int Study Group Invest Drugs as Reinforcers, 79-81; mem, Drug Adv Comt, Food & Drug Admin, 80-84, chmn, 83-84; chmn, Drug Abuse Clin & Behav Res Rev Comt, Nat Inst Drug Abuse, 85-89; bd dirs, Comt Probs Drug Dependence, 87-; pres, Psychopharmacol Div, Am Psychol Asn, 89-90. *Mem:* Fel Am Psychol Asn; Soc Stimulus Properties Drugs (pres, 84-85); fel Am Col Neuropsychopharmacol; Am Soc Pharmacol & Exp Therapeut; Behav Pharmacol Soc. *Res:* Effects of drugs of abuse, particularly phencyclidine and volatile solvents, on learned behavior of animals; drug self-administration and drug discrimination; behavioral toxicology of inhalants. *Mailing Add:* Dept Pharmacol & Toxicol Med Col Va Commonwealth Univ Richmond VA 23298

BALTAY, CHARLES, b Budapest, Hungary, Apr 15, 37; US citizen; m 61; c 5. NUCLEAR PHYSICS. *Educ:* Union Col, BS, 58; Yale Univ, MS, 59, PhD(physics), 63. *Prof Exp:* Res physicist, Yale Univ, 63-64; from instr to assoc prof, 64-72, dir, Nevis Labs, 79-88, PROF PHYSICS, COLUMBIA UNIV, 72-; EUGENE HIGGINS PROF PHYSICS, YALE UNIV, 88- *Concurrent Pos:* Alfred P Sloan fel, 66-68; CERN fel, 80. *Mem:* fel Am Phys Soc. *Res:* Elementary particle physics. *Mailing Add:* Physics Dept Yale Univ New Haven CT 06511

BALTAZZI, EVAN S, b Izmir, Turkey, Apr 11, 19; US citizen; m 45; c 3. ORGANIC CHEMISTRY, MATERIALS SCIENCE ENGINEERING. *Educ:* Univ Paris, DSc(phys sci), 49; Oxford Univ, PhD(org reaction mechanisms), 54. *Prof Exp:* Head org chem res, French Nat Res Ctr, 46-59; group leader org synthesis, Nalco Chem Co, 59-61; mgr org res, IIT Res Inst, 61-63; dir org res, Multigraphics Develop Ctr, Addressograph Multigraph Corp, 63-66, dir phys & chem res, 66-70, dir appl sci, 70-77; PRES, EVANEL ASSOCS, 77- *Concurrent Pos:* Brit Coun Exchange scholar, Oxford Univ, 52-54; Can Nat Res Coun fel, 55-56; chmn, Gordon Res Conf Photoconductors, 76; mem, Comt US currency, Nat Res Coun, 85-87. *Honors & Awards:* Distinguished Serv Award in Sci, 65. *Mem:* Sr mem Am Chem Soc; fel Chem Soc; fel Soc Chem Indust; sr mem French Chem Soc; sr mem Soc Photog Scientists & Engrs; Inst Elec & Electronics Engrs; Tech Asn Pulp & Paper Indust. *Res:* Heterocyclic chemistry; synthetic methods; biologically active compounds; photosensitive compounds; photocopying and duplicating materials and systems. *Mailing Add:* 825 Greengate Oval Sagamore Hills OH 44067-2311

BALTENSPERGER, ARDEN ALBERT, b Kimball, Nebr, Dec 25, 22; c 4. PLANT BREEDING. *Educ:* Univ Nebr, BS, 47, MS, 49; Iowa State Col, PhD(agron), 58. *Prof Exp:* Jr agronomist field crop res, Agr Exp Sta, Agr & Mech Col, Tex, 52-54; instr agron, Iowa State Col, 54-56, res assoc, 56-58; asst plant breeder sorghum res, Univ Ariz, 58-59, assoc prof & assoc agronomist grass breeding, 59-62, prof & agronomist, 62-63; head dept, 63-77, PROF AGRON, DEPT AGRON & HORT SCI, NMEX STATE UNIV, 63- *Mem:* Fel Am Soc Agron (pres, 90). *Res:* Grass, corn and sorghum breeding and production. *Mailing Add:* Dept Agron & Hort Sci NMex State Univ PO Box 3Q Las Cruces NM 88003

BALTENSPERGER, DAVID DWIGHT, b Kimball, Nebr, Dec 28, 53; m 76; c 3. FORAGE LEGUME BREEDING. *Educ:* Nebr Wesleyan Univ, BS, 76; Univ Nebr, MS, 78; NMex State Univ, PhD(agron), 80. *Prof Exp:* asst prof, 81-86, ASSOC PROF AGRON, UNIV FLA, 86- *Mem:* Am Soc Agron; Crop Sci Soc Am; Soc Range Mgt; Sigma Xi. *Res:* Forage legume breeding; cultivar development including Trifolium Medicago and Alysicarpus species. *Mailing Add:* 4502 Ave I Scotts Buff NE 69361

BALTHASER, LAWRENCE HAROLD, b Camden, NJ, Feb 19, 37; m 67; c 1. GEOLOGY. *Educ:* Rutgers Univ, BA, 60; Ind Univ, Bloomington, MA, 63, PhD(geol), 69. *Prof Exp:* Asst prof geol, Southampton Col, Long Island Univ, 67-69, assoc prof, 69-76, PROF PHYSICS, CALIF STATE POLYTECH COL, SAN LUIS OBISPO, 77- *Concurrent Pos:* E R Cumings award, Ind Univ, 70; map draftsman, Sun Oil Co, field asst, NJ Agr Exp Sta & teaching asst, Indiana Univ, 75- *Mem:* Soc Econ Paleont & Mineral. *Res:* Sedimentary petrology and paleoecology of carbonate rocks; paleocurrent analysis of Franciscan and Great Valley sequence rocks. *Mailing Add:* Dept Physics Calif Poly State Univ San Luis Obispo CA 93407

BALTHAZOR, TERRELL MACK, b Hays, Kans, Aug 25, 49; m 68; c 1. ORGANIC CHEMISTRY. *Educ:* Ft Hays Kans State Col, BS, 71; Univ Ill, PhD(org chem), 75. *Prof Exp:* SR RES CHEMIST ORG CHEM, MONSANTO CO, 75- *Mem:* Am Chem Soc. *Res:* Polyvalent iodine chemistry; synthesis of plant growth hormones and their derivatives. *Mailing Add:* Monsanto Via 800 N Lindbergh Blvd St Louis MO 63137

BALTIMORE, DAVID, b New York, NY, Mar 7, 38; m 68; c 1. VIROLOGY, BIOCHEMISTRY. *Educ:* Swarthmore Col, BA, 60; Rockefeller Inst, PhD(biol), 64. *Hon Degrees:* DSc, Swarthmore Col, 76; DSc, Mt Holyoke Col, 87. *Prof Exp:* Fel microbiol, Mass Inst Technol, 63-64; fel molecular biol, Albert Einstein Col Med, 64-65; res assoc virol, Salk Inst Biol Studies, 65-68; assoc prof microbiol, 68-72, Am Cancer Soc Prof Microbiol, 73-84, PROF BIOL, MASS INST TECHNOL, 72-; PRES, ROCKEFELLER UNIV, 90- *Concurrent Pos:* Bd govs, Weizmann Inst Sci, Israel, 79-; bd mem, Life Sci Res Found, 81-; co-chmn, Comt Nat Strategy AIDS, Nat Acad Sci & Inst Med, 86; bd mgrs, Swarthmore Col; NIH ad hoc prog adv comt complex genomes; dir, Whitehead Inst, 82-90. *Honors & Awards:* Nobel Prize, Physiology & Med, 75; Gustav Stern Award Virol, 70; Warren Trienniel Prize, Mass Gen Hosp, 71; Eli Lilly Award, 71; US Steel Found Award, Molecular Biol, Nat Acad Sci, 74. *Mem:* Nat Acad Sci; Inst Med; AAAS; Am Soc Microbiol; Am Soc Biol Chem; Pontif Acad Sci; Am Acad Arts & Sci; for mem Royal Soc Eng. *Res:* Protein and nucleic acid synthesis of RNA animal viruses, especially poliovirus and the RNA tumor viruses; immunoglobulin gene expression. *Mailing Add:* Rockefeller Univ 1230 York Ave New York NY 10021-6399

BALTRUSAITIS, ROSE MARY, b Saginaw, Mich, Aug 28, 50. ELEMENTARY PARTICLE PHYSICS. *Educ:* Northwestern Univ, BA, 71; Princeton Univ, PhD(physics), 77. *Prof Exp:* res assoc physics, Fermi Nat Accelerator Lab, 77-80; mem staff, Stanford Linear Accelerator Ctr, Stanford Univ, 80-; AT PHYSICS DEPT, UNIV UTAH. *Mem:* Am Phys Soc. *Mailing Add:* 5259 University Dr Santa Barbara CA 93111

BALTZ, ANTHONY JOHN, b Indianapolis, Ind, Mar 10, 42; m 68; c 2. THEORETICAL NUCLEAR PHYSICS. *Educ:* Spring Hill Col, BS, 66; Case Western Reserve Univ, MS, 68, PhD(physics), 71. *Prof Exp:* From res assoc to assoc physicist, 71-76, PHYSICIST, BROOKHAVEN NAT LAB, 76- *Mem:* Am Phys Soc; fel AAAS. *Res:* Theoretical nuclear and particle physics. *Mailing Add:* Dept Physics Brookhaven Nat Lab Upton NY 11973

BALTZ, HOWARD BURL, b St Louis, Mo, July 27, 30; m 68; c 3. APPLIED STATISTICS, MANAGEMENT SCIENCES. *Educ:* Baylor Univ, BBA, 55, MS, 56; Okla State Univ, PhD(econ statist), 64. *Prof Exp:* Instr statist, Baylor Univ, 57-59, asst dir Bur Bus & Econ Res, 58-59; teaching asst, Okla State Univ, 59-62; asst prof statist & dir Comput Ctr, Midwestern Univ, 62-64; asst prof, Univ Tex, Austin, 64-68; assoc prof, 68-76, PROF STATIST, UNIV MO-ST LOUIS, 76- *Concurrent Pos:* Coordr, Sci. Quant Mgt Sci Group, Univ Mo, 68-73. *Mem:* Am Statist Asn; Am Inst Decision Sci. *Res:* Applied forecasting. *Mailing Add:* Dept Bus Admin Univ Mo St Louis 8001 Natural Bridge St Louis MO 63121

BALTZER, OTTO JOHN, b St Louis, Mo, Dec 24, 16; m 81; c 4. PHYSICS. *Educ:* Washington Univ, AB, 38, MS, 39, PhD(physics), 41. *Prof Exp:* Res assoc radiation lab, Mass Inst Technol, 41-45; asst dir & supvr, Defense Res Lab, Tex, 45-56; pres & tech dir, Textran Corp, 56-62, prin scientist, Tracor, Inc, 62-80; CONSULT, 81- *Mem:* Am Phys Soc. *Res:* Electromagnetic propagation; guided missile research and development; precision timing; navigation. *Mailing Add:* 11805 Arabian Trail Austin TX 78759

BALTZER, PHILIP KEENE, b Quincy, Mass, Feb 11, 24; m 44; c 4. PHYSICS. *Educ:* Northeastern Univ, SB, 52; Mass Inst Technol, SM, 55; Rutgers Univ, PhD, 63. *Prof Exp:* Draftsman, Fore River Shipyard, Mass, 46-47; student engr, Charles T Main, Inc, 48-52; res staff magnetic mat, Digital Comput Lab, Mass Inst Technol, 52-55; mem res staff, RCA Sarnoff Res Labs, 55-67, dir res, RCA Labs, Tokyo, 67-72, staff scientist, 72-76, group head, LSI Systs Design & Appln, RCA Sarnoff Res Labs, 76-86; RETIRED. *Mem:* Sigma Xi. *Res:* Solid state physics; magnetic materials. *Mailing Add:* 536 Rosedale Rd Princeton NJ 08540

BALUDA, MARCEL A, b France, June 30, 30; nat US; c 3. VIROLOGY, ONCOLOGY. *Educ:* Univ Pittsburgh, BS, 51, MS, 53; Calif Inst Technol, PhD(virol, chem), 56. *Prof Exp:* Res assoc pediat, City of Hope Med Ctr, 56-62, sr virologist, Biol Dept, 62-66; PROF VIRAL ONCOLOGY, UNIV CALIF, LOS ANGELES, 66- *Concurrent Pos:* Dep dir, Jonsson Comprehensive Cancer Ctr, 81- *Mem:* Am Soc Microbiol; Am Asn Cancer Res; AAAS. *Res:* Tumor viruses; leukemia; oncogenesis and cellular differentiation. *Mailing Add:* Dept Path Univ Calif Sch Med Los Angeles 405 Hilgard Ave Los Angeles CA 90024

BALWANZ, WILLIAM WALTER, b Glencoe, Ohio, Mar 9, 13; m 36; c 3. ENGINEERING, PHYSICAL SCIENCE. *Educ:* George Washington Univ, BEE, 41; Univ Md, MSEE, 49. *Prof Exp:* Radio engr, Naval Air Sta, Anacostia & Patuxent Rivers, 42-46; electronic engr, Naval Res Lab, 46-74, consult phys sci, 74-79; PRES, MATTOX, INC, 77- *Concurrent Pos:* Lectr, Univ Md, 51-65 & George Washington Univ, 52-64. *Mem:* Am Inst Aeronaut & Astronaut; Inst Elec & Electronics Engrs; Am Inst Physics; Combustion Inst; Sigma Xi. *Res:* Chemically generated plasmas, free stream turbulent flow and space simulation. *Mailing Add:* PO Box 1073 Oak Grove VA 22443

BALWIERCZAK, JOSEPH LEONARD, b Staten Island, NY, Apr 4, 54. CARDIOVASCULAR PHARMACOLOGY. *Educ:* State Univ NY, Buffalo, BS, 76, PhD(biochem pharmacol), 83. *Prof Exp:* Fel, Dept Pharmacol & Cell Biophys, Sch Med, Univ Cincinnati, 83-85, res assoc, 85-87; sr scientist, 87-90, RES SCIENTIST II, DEPT CARDIOVASC RES, PHARMACEUT DIV, CIBA-GEIGY CORP, 91- *Mem:* Am Soc Pharmacol & Exp Therapeut; NY Acad Sci. *Res:* Ion channel activity and regulation; receptor classification: receptor binding, in vitro organ studies; receptor activated messenger systems: cyclic nucleotides, inositol phosphates. *Mailing Add:* Cardiovasc Pharmacol Dept Ciba-Geigy Corp Pharmaceut Div 556 Morris Ave Summit NJ 07901

BALZHISER, RICHARD E(ARL), b Elmhurst, Ill, May 27, 32; m 51; c 4. ENERGY RESEARCH, ENVIRONMENTAL RESEARCH. *Educ:* Univ Mich, BSE, 55, MSE, 56, PhD(chem eng), 61. *Prof Exp:* prof chem eng, Univ Mich, 61-71, chmn chem eng, 70-71; asst dir, Off Sci & Technol, 71-73; dir, Fossil Fuel & Adv Syst, 73-79, vpres res & develop, 79- 83, sr vpres res & develop, 83-87, exec vpres res & develop, 87- 88, PRES-CHIEF EXEC OFFICER, ELEC POWER RES INST, 88- *Concurrent Pos:* Consult, Allegany Ballistics Lab, 61-62 & E I du Pont de Nemours & Co, Inc, 67-70; White House fel, 67-68; consult prof, Stanford Univ, 74-79; mem adv bd, Univ Calif, Berkeley, 74-77, Argonne Nat Lab, 74-80, Oak Ridge Nat Lab, 76-79, Cal Tech Energy Comt, 76-, Gas Res Inst, 79- & Inst Energy Anal, 80-

Honors & Awards: Charles M Schwab Mem lectr, Am Iron & Steel Inst. *Mem:* AAAS; Am Inst Chem Engrs; Am Inst Mining, Metall & Petrol Engrs. *Res:* Heat transfer; thermodynamics studies in liquid metal systems; direct research in generation, delivery and use of electrical energy including environmental and economic consideration; electric power. *Mailing Add:* Elec Power Res Inst PO Box 10412 Palo Alto CA 94303

BAMBARA, ROBERT ANTHONY, b Chicago, Ill, Jan 3, 49; m 74; c 2. BIOCHEMISTRY, MOLECULAR BIOLOGY. *Educ:* Northwestern Univ, BA, 70; Cornell Univ, PhD(biochem), 74. *Prof Exp:* Fel biochem, Stanford Univ, 74-76; from asst prof to assoc prof oncol biochem, 77-89, PROF ONCOL BIOCHEM, UNIV ROCHESTER, 89- *Honors & Awards:* Faculty Res Award, Am Cancer Soc. *Mem:* Am Asn Biochem & Molecular Biol. *Res:* Mechanism of DNA synthesis in procaryotes and eukaryotes; role of DNA polymerases and DNA replication-associated proteins; inhibition of DNA synthesis by anti-tumor drugs; hormonal regulation of gene expression. *Mailing Add:* Cancer Ctr & Dept Biochem Univ Rochester Box 607 Rochester NY 14642

BAMBENEK, MARK A, b Watertown, SDak, Nov 22, 34; m 57; c 2. ANALYTICAL CHEMISTRY. *Educ:* Col St Thomas, BS, 56; Univ Iowa, MS, 59, PhD(chem), 61. *Prof Exp:* Asst org photochem, Res Dept, Minneapolis-Honeywell Regulator Co, 56-57; res asst, Univ Iowa, 57-59; res assoc, Cornell Univ, 61-62; res assoc, Univ Pittsburgh, 62-65; from asst prof to assoc prof, 65-75, PROF CHEM, ST MARY'S COL, INC, 75- *Mem:* Am Chem Soc. *Res:* Organic reagents for analysis; ion exchange; optical methods of analysis. *Mailing Add:* Dept Chem St Mary's Col Notre Dame IN 46556-5001

BAMBERGER, CARLOS ENRIQUE LEOPOLDO, b Offenbach, Ger, Mar 26, 33; m 59; c 3. PHYSICAL INORGANIC CHEMISTRY. *Educ:* Univ Buenos Aires, Licenciado, 57; Dr en Quimica, 58. *Prof Exp:* Chemist, Nat Atomic Energy Comn, Arg, 57-60, head lab, 60-63, head div beryllium chem, 63-66; chemist, 66-80, chem div, 80-85, SR STAFF MEM, CHEM DIV, OAK RIDGE NAT LAB, 85- *Concurrent Pos:* Int Atomic Energy Agency fel, 61-63. *Honors & Awards:* Martin-Marietta Energy Systs Inventor Award, 85. *Mem:* Am Chem Soc; Sigma Xi; Am Ceramic Soc; fel AAAS. *Res:* Beryllium chemistry; preparation of high purity compounds, solvent extraction, stability of complexes; molten (fluorides) salts physical chemistry; actinide chemistry in molten salts; inorganic chemistry for the development of thermochemical cycles for hydrogen production; nuclear waste chemistry; synthesis of advanced ceramics. *Mailing Add:* Chem Div Oak Ridge Nat Lab PO Box 2008 Oak Ridge TN 37831-6119

BAMBERGER, CURT, b Stettin, Ger, Jan 8, 00; nat US; m 30; c 2. ANTHRAQUINONE INTERMEDIATES & DYES. *Educ:* Univ Würzburg, PhD(chem), 23. *Prof Exp:* Asst, Univ Würzburg, 23-25; res chemist, I G Farbenindust AG, Works Elberfeld & Leverkusen, 25-39; res & develop chemist, Union Chimique Belge, Brussels, 39-40, Nat Aniline Div, Allied Chem & Dye Corp, 41-42 & Patent Chem, Inc, 43-68; sr res chemist, Morton Chem Co Div, Morton-Norwich Prod Inc, 68-71; CHEM CONSULT, 71- *Mem:* Am Chem Soc; fel Am Inst Chemists. *Res:* Inorganic salts and organic pesticides; intermediates and dyestuffs of the anthraquinone series; fast wool dyes; vat dyes; fast pigments; oil soluble colors dyestuffs for plastics; fluorescent dyes of the anthraquinone series. *Mailing Add:* 1050 George St Apt 11-L New Brunswick NJ 08901

BAMBERGER, JUDY, b 1952. SOFTWARE SYSTEMS. *Educ:* Univ Wis, BS, 74; Univ NColo, MS, 79; Univ Calif, Los Angeles, MS, 85. *Prof Exp:* MEM TECH STAFF, SOFTWARE ENG INST, CARNEGIE MELLON UNIV, 87- *Concurrent Pos:* Chair, Joint Users Group, Asn Comput Mach, 86-87. *Mem:* Asn Comput Mach. *Mailing Add:* Software Eng Inst Carnegie Mellon Univ Pittsburgh PA 15213

BAMBURG, JAMES ROBERT, b Chicago, Ill, Aug 20, 43; m 70, 85; c 2. BIOCHEMISTRY, NEUROSCIENCES. *Educ:* Univ Ill, Urbana, BS, 65; Univ Wis-Madison, PhD(biochem), 69. *Prof Exp:* From res asst biochem to res assoc physiol chem, Univ Wis-Madison, 65-69; fel biochem, Med Sch, Stanford Univ, 69-71; from asst prof to assoc prof, 71-81, interim chmn dept, 82-85 & 88-89, assoc dirneurol growth develop, 86-90, PROF BIOCHEM, COLO STATE UNIV, 81-, DIR NEUROL GROWTH DEVELOP, 90- *Concurrent Pos:* Acad coordr, Grad Prog Cellular & Molecular Biol, Colo State Univ, 75-76; vis scientist, MRC Lab Molecular Biol, Univ Postgrad Med Sch, Cambridge, Eng, 78-79; J S Guggenheim Mem fel, 78-79; chmn elect grad prog cellular & molecular biol, Colo State Univ, 82; vis scientist, M R C Cell Biophys, London, England & sr int fel, Fogarty Ctr, 85-86; W Evans vis fel, Univ Otago, Dunedin, NZ, 91. *Mem:* Sigma Xi; Am Chem Soc; Int Soc Neurochem; Am Soc Cell Biol; Am Soc Biol Chem & Molecular Biol. *Res:* Structural and functional role of microtubules, microfilaments and other proteins on the processes of nerve growth, cytoplasmic transport, secretion and mitosis. *Mailing Add:* Dept Biochem Colo State Univ Ft Collins CO 80523

BAME, SAMUEL JARVIS, JR, b Lexington, NC, Jan 12, 24; m 56; c 3. SPACE PHYSICS. *Educ:* Univ NC, BS, 47; Rice Univ, AM, 49, PhD(physics), 51. *Prof Exp:* Mem staff, Space Physics Group, 51-80, LAB FEL, SPACE PLASMA GROUP, LOS ALAMOS NAT LAB, 80- *Concurrent Pos:* Prin investr, Int Sun Earth Explorer Mission & Ulysses Int Solar Polar Mission. *Mem:* Fel AAAS; fel Am Phys Soc; fel Am Geophys Union; Am Astron Soc. *Res:* Light particle reactions; neutron physics; particles and radiations in space. *Mailing Add:* MS D438 Los Alamos Nat Lab Los Alamos NM 87545

BAMFORD, ROBERT WENDELL, b Glendale, Calif, July 13, 37. ECONOMIC GEOLOGY, GEOCHEMISTRY. *Educ:* Univ Wash, BSc, 59; Stanford Univ, PhD(geol geochem), 70. *Prof Exp:* Res chem engr by-prod chem res, Ga Pac Corp, 60-63; res geologist geol & geochem res, 70, chief site geologist ore deposit eval, 71, sr res geologist geol & geochem, Kennecott

Explor, Inc, 72-77; sr geochemist & proj mgr geochem res, Earth Sci Lab, Univ Utah Res Inst, 77-80; CONSULT, 80- *Concurrent Pos:* Consult specializing in applns & technique develop for mineral & geothermal explor, 77- *Mem:* Soc Econ Geologists; Geol Soc Am; Mineral Soc Am; Soc Mining Engrs; Am Inst Mining & Metall Engrs. *Res:* Development of ore deposit and geothermal system empirical models, chemical and mineralogical, and of new or refined techniques in exploration geochemistry. *Mailing Add:* 2315 26th Ave E Seattle WA 98112

BAMFORTH, BETTY JANE, b New Britain, Conn, Jan 20, 23. ANESTHESIOLOGY. *Educ:* Bates Col, BS, 44; Boston Univ, MD, 47. *Prof Exp:* Intern, Mt Auburn Hosp, 48; resident, Univ Wis Hosp, 51; asst prof anesthesiol, Sch Med, Univ Okla, 51-54; assoc prof, 54-64, actg chmn dept, 69-71, PROF ANESTHESIOL, SCH MED, UNIV WIS, MADISON, 64-, ASST DEAN EDUC ADMIN, 73- *Concurrent Pos:* Consult, Madison Vet Admin Hosp, 54-; mem anesthetic & respiratory adv comt, Food & Drug Admin, 73- *Mem:* Am Soc Anesthesiol; Int Anesthesia Res Soc; Asn Univ Anesthesiol. *Res:* Medical education. *Mailing Add:* Univ Hosps Clins Univ Wis 600 Highland Ave Madison WI 53792

BAMFORTH, STUART SHOOSMITH, b White Plains, NY, Oct 23, 26; m 52; c 2. PROTOZOOLOGY. *Educ:* Temple Univ, AB, 51; Univ Pa, AM, 54, PhD(zool), 57. *Prof Exp:* Asst instr biol, Univ Pa, 51-57; from instr to assoc prof, 57-72, PROF BIOL, NEWCOMB COL, TULANE UNIV, 72- *Mem:* Am Soc Zoologists; Soc Study Evolution; Soc Protozool; Am Soc Limnol & Oceanog; Sigma Xi. *Res:* Ecology of protozoa and plankton. *Mailing Add:* Dept Biol Newcomb Col Tulane Univ New Orleans LA 70125

BAMRICK, JOHN FRANCIS, b Rockwell, Iowa, Dec 15, 26; m 56; c 3. BIOLOGY, GENETICS. *Educ:* Loras Col, BS, 53; Marquette Univ, MS, 56; Iowa State Univ, PhD(zool), 60. *Prof Exp:* From instr to assoc prof biol, 55-70, PROF BIOL, LORAS COL, 70- *Mem:* AAAS; Soc Invert Path; Sigma Xi. *Res:* Disease resistance; insect pathology. *Mailing Add:* Dept Biol Loras Col Dubuque IA 52001

BAN, STEPHEN DENNIS, b Gary, Ind, Dec 16, 40; m 66; c 3. TRANSPORT PHENOMENA, PROCESS RESEARCH. *Educ:* Rose Polytech Inst, BSME, 62; Case Inst Technol, MSME, 64, PhD(fluid, thermal & aero sci eng), 67. *Prof Exp:* Res engr, Battelle Mem Inst, 67-68, sr res engr, 68-69, sr proj leader, 69-70, assoc chief, fluid & gas dynamics div, 70-71, chief, 71-77; dir & vpres res & develop, Develop & Testing Lab, Bituminous Mat Co, 77-80; vpres, 81-85, sr vpres res & develop opers, 85-86, exec vpres, 86-87, PRES & CHIEF EXEC OFFICER, GAS RES INST, CHICAGO, 87- *Concurrent Pos:* Adj prof mech eng, Rose Hulman Inst Technol, 80- *Mem:* Am Inst Aeronaut & Astronaut; Sigma Xi. *Res:* Fluid mechanics; heat transfer in gases and liquids; singular parabolic differential equations; boundary layer mechanics; applying research and development to real industrial processes; gas utilization technology, cooperative research management. *Mailing Add:* 8600 W Bryn Mawr Gas Res Inst Chicago IL 60631

BAN, THOMAS ARTHUR, b Budapest, Hungary, Nov 16, 29; Can citizen; m 63; c 1. PSYCHOPHARMACOLOGY, PSYCHIATRIC NOSOLOGY. *Educ:* Med Univ Budapest, MD, 54; McGill Univ, dipl, 60. *Prof Exp:* Demonstr, McGill Univ, 60-63, lectr, 64-65, asst prof, 65-70, assoc prof psychiat, 71-76, dir div psychopharmacol, Dept Psychiat, 70-76; PROF PSYCHIAT, VANDERBILT UNIV, 76- *Concurrent Pos:* Sr psychiatrist, Douglas Hosp, 60-61; sr res psychiatrist, 61-65, asst dir res, 66-70, chief res serv, 70-72; psychiat res consult, Hopital des Laurentides, 63-72, Lakeshore Gen Hosp, 67- & St Marys Hosp, 71- *Honors & Awards:* McNeil Award, Can Psychiat Asn. *Mem:* AAAS; Am Psychiat Asn; fel Am Col Neuro-Psychopharmacol; Collegium Int Neuro-Psychopharmacologicum; Am Psychopath Asn. *Res:* Psychopharmacology and the classification of psychiatric disorders; composite diagnostic evaluation of psychiatric disorders. *Mailing Add:* A-2215 Med Ctr N Dept Psychiat Vanderbilt Univ Nashville TN 37232

BAN, VLADIMIR SINISA, b Bjelovar, Yugoslavia, Sept 7, 41; US citizen; m; c 2. SOLID STATE SCIENCE. *Educ:* Univ Zagreb, dipl Eng, 64; Pa State Univ, PhD(solid state sci), 69. *Prof Exp:* Res asst solid state sci, Mat Res Lab, Pa State Univ, 64-69; mem tech staff solid state sci, Mat Res Lab, RCA Lab, Princeton, NJ, 69-79, group head, Videodisc Mat & Diag, 79-84; EXEC VPRES, EPITAXX INC, PRINCETON, 84- *Concurrent Pos:* Vis fel, Imp Col, Univ London, 72. *Mem:* Electrochem Soc; Am Crystal Growth Soc; Sigma Xi. *Res:* Chemical vapor deposition of semiconducting films; fundamental research on thermodynamics and kinetics of vapor deposition of semiconductors and the subsequent optimization of their synthesis; laser solid interactions; high temperature mass spectrometry; videodisc materials; polymer surfaces; fiber optic components. *Mailing Add:* 1061 The Great Rd Princeton NJ 08540

BANAKAR, UMESH VIRUPAKSH, b Solapur, India, Dec 4, 56; m 84; c 1. PHARMACEUTICAL TECHNOLOGY, DRUG PRODUCT DEVELOPMENT. *Educ:* Bombay Univ, India, BPharm, 78; Duquesne Univ, Pittsburgh, PhD(pharmaceut sci), 85. *Prof Exp:* Prod chemist trainee, Roussel Pharmaceut Ltd, Bombay, 78-79; lectr pharmaceut technol, J N Med Col, India, 80; instr-in-charge, Sch Pharm, Duquesne Univ, Pittsburgh, 81-84; assoc prof pharmaceut sci, Sch Pharm, Creighton Univ, 85-90; ASSOC PROF PHARMACEUT TECHNOL & DIR RES INST, ST LOUIS COL PHARM, 90- *Concurrent Pos:* Book rev ed, BioPharm Mfg J, 87; expert, WHO, 87-; mem, Int Exec Serv Corps, 89- *Mem:* Am Asn Pharmaceut Scientists; Sigma Xi; Am Pharmaceut Asn; Am Asn Cols Pharm. *Res:* Controlled release technology; pharmaceutical product development; dissolution technology; bioavailability-bioequivalence; transdermal drug delivery; prolonged release pharmaceuticals; biopharmaceutics; novel drug delivery systems; pharmaceutical technology; technology transfer through education. *Mailing Add:* St Louis Col Pharm 4588 Parkview Pl St Louis MO 63110

BANASIAK, DENNIS STEPHEN, b Chicago, Ill, June 3, 50; m 73. ORGANIC CHEMISTRY. *Educ:* DePaul Univ, BS, 72; Iowa State Univ, PhD(org chem), 77. *Prof Exp:* Res chemist, Phillips Petrol Co, 77-81, MKT DEVELOP MGR, PHILLIPS CHEM CO, 82- *Mem:* Am Chem Soc. *Res:* Organometallic chemistry, organosilicon chemistry, biotechnology, enzymatic chemistry, agricultural chemicals. *Mailing Add:* 8477 N Del Mar Fresno CA 93711-6020

BANASIK, ORVILLE JAMES, b Wales, NDak, Nov 17, 19; m 47; c 4. BIOCHEMISTRY. *Educ:* NDak State Univ, BS, 43, MS, 46. *Prof Exp:* Lab asst & fel, NDak State Univ, 46-47, asst cereal technologist, 47-58, assoc cereal technologist, 58-63, assoc prof cereal technol, 63-70, prof cereal chem & technol, 70-85, chmn dept, 71-85; RETIRED. *Mem:* Am Asn Cereal Chem; Am Soc Brewing Chem; Sigma Xi; Am Chem Soc. *Res:* Cereal research, especially barley; development of testing methods and basic research; durum wheat and pasta quality research. *Mailing Add:* 2406 Tenth St N Fargo ND 58201

BANASZAK, LEONARD JEROME, b Milwaukee, Wis, Feb 1, 33; m; c 4. CRYSTALLOGRAPHY, BIOCHEMISTRY. *Educ:* Univ Wis, Madison, BS, 55; Loyola Univ, Chicago, MS, 60, PhD(biochem), 61. *Prof Exp:* Asst prof physiol & biophys, 66-71, assoc prof biol chem, 71-75, PROF BIOL CHEM, SCH MED, WASH UNIV, 75- *Concurrent Pos:* USPHS fels biochem, Ind Univ, 61-63 & Med Res Coun Lab Molecular Biol, Eng, 63-65; USPHS career develop award, 69-74. *Mem:* AAAS; Am Chem Soc; Am Crystallog Asn. *Res:* Structure of enzymes and the relationship of these structures to biological function. *Mailing Add:* Dept Biol Chem Wash Univ Sch Med St Louis MO 63110

BANAUGH, ROBERT PETER, b Los Angeles, Calif, Oct 27, 22; m 46; c 8. COMPUTER SCIENCES, GEOPHYSICS. *Educ:* Univ Calif, Berkeley, AB, 46, MA, 52, PhD(eng sci), 62. *Prof Exp:* High sch teacher, Calif, 47-50; mathematician, Calif Res & Develop Corp, 52-54; aeronaut engr, Pioneer Industs, Nev, 54-55; theoret physicist, Lawrence Radiation Lab, Univ Calif, 55-62; head appl math group, Ventura Div, Northrop Corp, 62-64; dir, Comput Ctr, 64-72, chmn dept, 67-80, PROF COMPUT SCI, UNIV MONT, 64- *Concurrent Pos:* Consult, Air Force Shock Tube Facility, Univ NMex, 65-66, Boeing Co, 65-68, Appl Theory, Inc, 70-76 & USDA, Forest Serv, 75-; vis prof, Col Natural Resources, Univ Calif, Berkeley, 78-79; vis prof, Dept Comput Sci, Univ Wollongong, Wollongong, Australia, 85-86. *Mem:* AAAS; Am Geophys Union; Seismol Soc Am. *Res:* Computer based methods of analysis of biological and physical systems; educational uses of computers in science; computer modeling; computational methods in linear and nonlinear ware interaction phenomena; quantitative ecology. *Mailing Add:* Dept Comput Sci Univ Mont Missoula MT 59801

BANAVAR, JAYANTH RAMARAO, b Bangalore, Karnataka, India, Oct 8, 53; m 82. DISORDERED SYSTEMS, POROUS MEDIA. *Educ:* Bangalore Univ, India, BSc Hons, 72, MSc, 74; Univ Pittsburgh, MS, 75, PhD(physics), 78. *Prof Exp:* Res assoc physics, Univ Chicago, 78-81; mem tech staff, Bell Labs, 81-83; mem prof staff, Schlumberger-Doll Res, 83-88; ASSOC PROF PHYSICS & MAT RES, PA STATE UNIV, 88- *Mem:* Am Phys Soc. *Res:* Theoretical condensed matter physics; disordered systems and porous media; statistical mechanics. *Mailing Add:* Dept Physics Davey Lab Pa State Univ University Park PA 16802

BANAY-SCHWARTZ, MIRIAM, b Sibiu, Rumania, Oct 9, 29; US citizen; m 51; c 2. BIOCHEMISTRY, ORGANIC CHEMISTRY. *Educ:* Hebrew Univ, Jerusalem, MSc, 54, PhD(biochem), 61. *Prof Exp:* Instr biochem, 62-68, ASST PROF BIOCHEM, ALBERT EINSTEIN COL MED, 68-; SR RES SCIENTIST, NY STATE RES INST NEUROCHEM & DRUG ADDICTION, 68- *Mem:* Am Soc Neurochem; Int Soc Neurochem; NY Acad Sci; Am Inst Chemists. *Res:* Electron transport; intermediate metabolism. *Mailing Add:* Ctr Neurochem The Nathan S Kline Inst Psychiat Res Orangeburg NY 10962

BANCHERO, J(ULIUS) T(HOMAS), b New York, NY, June 18, 14; m 38; c 3. PROCESS DESIGN. *Educ:* Columbia Univ, AB, 33, BS, 35, ChE, 36; Univ Mich, PhD(chem eng), 50. *Prof Exp:* Chem engr prod dept, Carbide & Carbon Chem Corp, 36-38; instr, Univ Detroit, 38-41; from instr chem eng to prof, Univ Mich, 43-59; prof & head dept, 59-79, EMER PROF CHEM ENG, UNIV NOTRE DAME, 81- *Mem:* Am Chem Soc; Am Inst Chem Engrs; Am Soc Eng Educ. *Res:* Kinetics of epoxide-alcohol reactions; chemical engineering process design. *Mailing Add:* 202 S Jacob St South Bend IN 46615-1002

BANCHERO, NATALIO, b Lima, Peru, June 6, 35. CARDIOVASCULAR RESPIRATORY PHYSIOLOGY. *Educ:* San Marcos Univ, MD, 62. *Prof Exp:* PROF PHYSIOL, SCH MED, UNIV COLO, 77- *Mem:* Am Physiol Soc; Int Soc Oxygen Transport Tissues; Microcirulatory Soc; Soc Exp Biol & Med. *Mailing Add:* Dept Physiol Univ Colo Health Sci Ctr C240 4200 E Ninth Ave Denver CO 80262

BANCROFT, DENNISON, b Newton, Mass, Oct 3, 11; m 39, 68; c 3. PHYSICS. *Educ:* Amherst Col, AB, 33; Harvard Univ, PhD(physics), 39. *Prof Exp:* Res assoc geophys, Harvard Univ, 36-41; physicist, David W Taylor Model Basin, US Navy Dept, 41-42; asst prof physics, Princeton Univ, 46-47; from asst prof to assoc prof, Swarthmore Col, 47-59; prof physics & chmn dept, 59-74, EMER PROF PHYSICS, COLBY COL, 74- *Concurrent Pos:* Consult, Los Alamos Sci Labs, 55-56. *Res:* Velocity of sound in nitrogen at high pressures; elastic constants of solids at high pressures and high temperatures; stress-strain relations under explosive load; thermodynamic properties of gases. *Mailing Add:* Rte One Box 279 Brooksville ME 04617

BANCROFT, GEORGE MICHAEL, b Saskatoon, Sask, Apr 3, 42; m 67; c 2. PHYSICAL INORGANIC CHEMISTRY. *Educ:* Univ Man, BSc, 63, MSc, 64; Univ Cambridge, PhD(chem), 67. *Hon Degrees:* MA & ScD, Univ Cambridge, 70. *Prof Exp:* Demonstr & teaching fel chem, Univ Cambridge, 67-70; from asst prof to assoc prof, 70-74, PROF CHEM, UNIV WESTERN ONT, 74-, CHMN, 86- *Concurrent Pos:* Prin investr, NASA, 70-72; E W Steacie mem fel, Nat Res Coun Can, 73-76; vis investr, Phys Sci Lab, Univ Wis, 75-76; mem, Int Comn Applns Mossbauer Effect, 75-; Guggenheim fel, 82-83. *Honors & Awards:* Harrison Mem Prize, Brit Chem Soc, 72; Meldola Medal, Royal Inst Chem, 72; Rutherford Medal, Royal Soc Can, 80; Alcan Lectr Award, Can Soc Chem, 90. *Mem:* The Chem Soc; Chem Inst Can; Am Chem Soc; fel Royal Soc Can. *Res:* High resolution photoelectron spectroscopy of inorganic and organometallic gases using resonance lamps and synchrotron radiation; adsorption and leaching of ions from minerals and glasses with particular application to nuclear waste disposal; Mossbauer spectroscopy of iron and tin organometallic compounds, and iron containing silicate minerals. *Mailing Add:* Dept Chem Univ Western Ont London ON N6A 5B8 Can

BANCROFT, HAROLD RAMSEY, b Meridian, Miss, May 3, 32; m 59; c 2. ENTOMOLOGY. *Educ:* Miss State Univ, BS, 58, MS, 59, PhD(entom), 62. *Prof Exp:* Asst prof, 62-66, ASSOC PROF BIOL, MEMPHIS STATE UNIV, 66- *Concurrent Pos:* Consult entomologist, Memphis State Univ. *Mem:* Entom Soc Am; Am Registry Prof Entomologists. *Res:* Insect resistance to pesticides; pest control research. *Mailing Add:* Dept Biol Memphis State Univ Memphis TN 38152

BANCROFT, JOHN BASIL, b Vancouver, BC, Can, Dec 31, 29; m 54; c 4. BIOLOGY. *Educ:* Univ BC, BA, 52; Univ Wis, PhD(plant path), 55. *Prof Exp:* from asst prof to prof bot & plant path, Purdue Univ, 55-70; prof & head, John Innes Inst, 70-73; prof plant sci & chmn dept, 73-78, DEAN SCI & FAC SCI, UNIV WESTERN ONT, 78- *Mem:* Am Phytopath Soc; fel Royal Soc Can; Sigma Xi. *Res:* Biological and biophysical characteristics of certain simple plant viruses and their assembly. *Mailing Add:* Dept Plant Sci Univ Western Ont London ON N6A 3B7 Can

BANCROFT, LEWIS CLINTON, b Reading, Mass, Jan 1, 29; m 51, 62; c 8. ELECTRICAL ENGINEERING. *Educ:* Princeton Univ, BSE, 50, MSE, 52. *Prof Exp:* Res mgt, Nuclear Reactor Instrumentation, Microwave Heating, Photopolymer & Silver Halide Chem, E I DuPont De Nemours & Co, 51-83, mgr, Patents & Contracts, Photosystems & Electronic Prods, 83-85; RETIRED. *Concurrent Pos:* Sayre Fel, Princeton Univ, 51. *Mem:* Sigma Xi. *Res:* Physics and chemistry of magnetic recording materials, especially chromium dioxide. *Mailing Add:* 12 Red Oak Rd Wilmington DE 19806

BANCROFT, RICHARD WOLCOTT, b Denver, Colo, Mar 28, 16; m 50. AEROSPACE PHYSIOLOGY. *Educ:* Stanford Univ, BA, 38; Univ Southern Calif, MS, 47, PhD(physiol), 51. *Prof Exp:* Res asst aviation physiol, US Air Force Sch Aerospace Med, 47-48, physiologist, 50-68, chief appl physiol br, 68-72; RETIRED. *Mem:* Fel Aerospace Med Asn; Am Physiol Soc; Soc Exp Biol & Med; assoc mem Royal Soc Med London. *Res:* Aerospace physiology; respiration; circulation. *Mailing Add:* 7709 Prospect Pl La Jolla CA 92037

BAND, HANS EDUARD, b Vienna, Austria, Oct 14, 24; US citizen; m 48; c 3. PHYSICS, ATOMIC & MOLECULAR PHYSICS. *Educ:* Harvard Univ, AB, 47; Boston Univ, AM, 53. *Prof Exp:* Elec engr, Gen Elec Co, 48-52; mem staff, Lincoln Lab, Mass Inst Technol, 52-53; lit analyst, Gillette Safety Razor Co, 53-56; sr res engr, Melpar, Inc, 56-57; opers analyst, Tech Opers, Inc, 57-58; staff physicist, Pickard & Burns, Inc, 58-61; sr res scientist, Raytheon Co, 61-63; optical physicist, Concord Radiance Lab, Utah State Univ, 63-66; prin res scientist, Avco Everett Res Lab, Mass, 66-67; physicist, Army Mat & Mech Res Ctr, 68-80, electronics engr, US Army Electronics Command, 80-83; electronics engr, USAF, 83-87, Defense Logistics Agency, 87-89; RETIRED. *Mem:* Am Phys Soc; Am Sci Affil. *Res:* Optical and radiation physics; atomic spectroscopy; radio communication; solid state physics; materials science and electronics engineering. *Mailing Add:* 22 Reeves Rd Bedford MA 01730-1335

BAND, PIERRE ROBERT, b Paris, France, June 23, 35; Can citizen; m 60; c 4. MEDICAL ONCOLOGY. *Educ:* Col Stanislas Montreal, BA, 57; Univ Montreal, MD, 62; FRCP(C), 68. *Prof Exp:* Assoc prof, Dept Med, Univ Alta, 70-76; dir clin res oncol, Inst Cancer Montreal, 76-; prof med, Univ Montreal, 76-; AT CANCER CONTROL AGENCY BC. *Concurrent Pos:* McEachern fel, Can, 64-65; NIH fel, 67-68; prin investr, Eastern Coop Oncol Group, 71-; mem, Breast Cancer Task Force, Combination Chemother Subcomt, NIH, 73-74; prin investr, Europ Orgn Res & Treat Cancer, 73-; chmn, Working Group Lung Cancer, Cancer Res Coord Comt, Can, 77- *Mem:* Am Asn Cancer Res; Am Soc Clin Oncol; Int Asn Study Lung Cancer. *Res:* Phase I-II-III investigation of antineoplastic agents; epidemiology of lung cancer in uranium workers. *Mailing Add:* Cancer Control Agency BC 600 W Tenth Ave Vancouver BC V5Z 4E6 Can

BAND, RUDOLPH NEAL, b Oakland, Calif, Aug 2, 29; m 55. ZOOLOGY. *Educ:* Univ San Francisco, BS, 52, MS, 54; Univ Calif, Berkeley, PhD(zool), 59. *Prof Exp:* NIH fel, Amherst Col, 58-59; from instr to asst prof, 59-63; from asst prof to assoc prof, 63-69, PROF ZOOL, MICH STATE UNIV, 69- *Concurrent Pos:* USPHS spec res fel, physiol lab, Cambridge Univ, 65-66 & Dept Zool, Univ Calif, Berkeley, 72-73. *Mem:* Soc Protozool; Am Soc Parasitol; Am Soc Cell Biologists. *Res:* Biochemistry and physiology of protozoan organelles; locomotion; phagocytosis; physiology and biochemistry of amoebae. *Mailing Add:* Dept Zool Mich State Univ East Lansing MI 48824

BAND, WILLIAM, b Birkenhead, UK, Aug 27, 06; m 31. PHYSICS. *Educ:* Univ Liverpool, BS, 25, MS, 27, DSc(physics), 46. *Prof Exp:* Lectr, physics, Liverpool Univ, 26-28; prof & chmn physics, Yenching Univ, Peking, China, 29-46; res assoc, Univ Chicago, 46-49; prof & chmn physics, Wash State Univ, 49-71; EMER PROF PHYSICS, WASH STATE UNIV, 71-, EMER CHMN, 71- *Concurrent Pos:* Physicist, Stanford Res Inst, 56-64. *Mem:* Inst Physics, London; Am Phys Soc; Sigma Xi. *Res:* Theory of relativity, thermo-electric and thermomagnetic effects; quantum statistics and low temperature phenomena; foundations of quantum mechanics; superoptic electromagnetic wave packets and John Bells theorem. *Mailing Add:* NW 915 Fisk Pullman WA 99163

BAND, YEHUDA BENZION, b Munich, Ger, Dec 1, 46; US citizen; m 70; c 4. ATOMIC PHYSICS, QUANTUM PHYSICS. *Educ:* Cooper Union, BS, 68; Univ Chicago, MS, 70, PhD(physics), 73. *Prof Exp:* Fel chem physics, Univ Chicago, 73-75; asst scientist physics, Argonne Nat Lab, 75-78; PROF, DEPT CHEM, BEN-GURION UNIV, 78- *Concurrent Pos:* Res scientist, Allied-Signal Inc; Barecha Fel, outstanding young scientists, 78-81. *Honors & Awards:* Allied-Signal Inventor Award. *Mem:* Am Inst Physics; Am Phys Soc. *Res:* Collision theory; charge exchange processes; dissociation of molecules; surface phenomena; many body physics; laser physics; nonlinear optics; transport processes. *Mailing Add:* Dept Chem Ben-Gurion Univ Beer-Sheva Israel

BANDA, SIVA S, b Vijayawada, India, Mar 28, 51; US citizen; c 1. CONTROL SYSTEMS. *Educ:* Regional Eng Col, Warangal, India, BS, 74; Indian Inst Sci, Bangalore, MS, 76; Wright State Univ, Ohio, MS, 78; Univ Dayton, PhD(aerospace eng), 80. *Prof Exp:* AEROSPACE ENGR, FLIGHT CONTROL, FLIGHT DYNAMICS LAB, WRIGHT PATTERSON AFB, OHIO, 81- *Concurrent Pos:* Adj assoc prof elec eng & adj asst prof math, Wright State Univ, 78- *Mem:* Am Inst Aeronaut & Astronaut. *Res:* Robust multivariable control theory and application to modern aircraft and highly flexible large space structures. *Mailing Add:* WL/FIGC Wright Patterson AFB OH 45433

BANDEEN, WILLIAM REID, b Escanaba, Mich, Oct 11, 26; m 60; c 3. METEOROLOGY, SPACE SCIENCE. *Educ:* US Mil Acad, BS, 48; NY Univ, MS, 55. *Prof Exp:* Meteorologist, Meteorol Br, NASA, 59-65, head Planetary Radiations Br, 65-72, asst chief, Lab Meteorol & Earth Sci, 72-74, chief, Atmospheric & Hydrospheric Appln Div, 74-77, assoc chief, Lab Atmospheric Sci, 77-86, assoc dir, Space & Earth Sci, Goddard Space Flight Ctr, 86-89; RETIRED. *Mem:* Fel Am Meteorol Soc; fel AAAS; fel Am Geophys Union. *Res:* Radiation budget of the earth from remotely sensed measurements from satellites; remote sensing of atmospheric parameters from space; radiative transfer in the atmosphere. *Mailing Add:* 12 Brinkwood Rd Brookeville MD 20833

BANDEL, HANNSKARL, b Dessau, Ger, May 3, 25. STRUCTURAL ENGINEERING. *Educ:* Univ Berlin, DEng, 52. *Prof Exp:* Partner, Severud Eng, 54-80; VPRES, DRC CO, QUEENS, NY, 81- *Mem:* Nat Acad Eng. *Mailing Add:* Skyhigh Lane Canadensis PA 18325

BANDEL, HERMAN WILLIAM, b Great Falls, Mont, Sept 23, 16; m 45. PLASMA PHYSICS. *Educ:* Univ Calif, Berkeley, AB, 47, PhD(physics), 54. *Prof Exp:* Sr engr, Appl Physics Sect, Electron Defence Lab, Sylvania Elec Prod, Inc, 54-57, adv res physicist, Microwave Physics Lab, 57-63; res scientist, Lockheed Missiles & Space Co, 63-76; RETIRED. *Mem:* Am Phys Soc. *Res:* Discharge through gases. *Mailing Add:* 1753 Mayflower Ct Mountain View CA 94040

BANDEL, VERNON ALLAN, b Baltimore, Md, May 8, 37; m 59; c 2. AGRONOMY, PLANT PHYSIOLOGY. *Educ:* Univ Md, BS, 59, MS, 63, PhD(agron, soil fertil), 65. *Prof Exp:* Asst prof soils, 64-69, assoc prof, 69-77, PROF SOILS, UNIV MD, COLLEGE PARK, 77-, EXTEN SOILS SPECIALIST, 64- *Mem:* Fel Am Soc Agron; Soil Sci Soc Am; Coun Soil Testing & Plant Anal; Am Forage & Grassland Coun. *Res:* Soil fertility; nutrient balance in plants; micro-nutrient fertilization of corn, soybeans; nitrogen fertilization of corn and small grains; no-tillage and conventional tillage corn fertilization; soil test-plant response; nutrient management. *Mailing Add:* Dept Agron Univ Md College Park MD 20742

BANDER, MYRON, b Belzyce, Poland, Dec 11, 37; US citizen; m 67. THEORETICAL PHYSICS. *Educ:* Columbia Univ, BA, 58, MA, 59, PhD(physics), 62. *Prof Exp:* NSF fel, 62-63; res assoc theoret physics, Linear Accelerator Ctr, Stanford Univ, 63-66; assoc prof physics, 66-72, dean Sch Phys Sci, 80-86, PROF PHYSICS, UNIV CALIF, IRVINE, 72- *Concurrent Pos:* Sloan Found fel, 67-69. *Mem:* Fel Am Phys Soc. *Res:* Elementary particle physics. *Mailing Add:* Dept Physics Univ Calif Irvine CA 92717

BANDES, DEAN, b New York, NY, Jan 16, 1944; m 68; c 2. MATHEMATICS, SOFTWARE SYSTEMS. *Educ:* Williams Col, BS, 65; Brandeis Univ, MA, 67, PhD(math), 74. *Prof Exp:* Asst prof math, Univ Mass, Boston, 74-76; mem staff, Parke Math Labs, 76-79; mem staff, LTX Corp, 79-88; MEM STAFF, CREDENCE SYSTS CORP, 88- *Res:* Topology-knot theory; computer combinatorics; partitions, transportation problem; ray-tracing; real-time control. *Mailing Add:* 225 Cypress St Newton Centre MA 02159-2226

BANDES, HERBERT, b New York, NY, May 23, 14; m; c 3. ELECTROCHEMISTRY. *Educ:* Univ Mich, BS, 35, MS, 36, PhD(electrochem), 38. *Prof Exp:* Res chemist, Alrose Chem Co, RI, 39; plant chemist, Cohn & Rosenberger, Inc, 40; physicist, Brooklyn Naval Shipyard, Bur Ord, 40-43; electrochemist, Kellex Corp, NY, 43-44; sr engr, Sylvania Elec Prod, Inc, 44-50, mgr chem lab, Cent Res Labs, 50-55, chief engr, Semiconductor Div, 55-58; mem sr staff & dir res, Western Div, Arthur D Little, Inc, 58-63; dir res, Eitel-McCullough, Inc, Calif, 63-65; chief engr, Electronics Dept, Hamilton Stand Div, United Aircraft Corp, 65-69; dean admin affairs, 69-84, ADJ PROF, MANCHESTER COMMUNITY COL, 84- *Mem:* Am Chem Soc; Electrochem Soc; Sigma Xi. *Res:* Kinetics of electrode reactions; mechanism of electrodeposition; solid state chemistry. *Mailing Add:* 35 London Rd Hebron CT 06248

BANDI, WILLIAM R, analytical chemistry; deceased, see previous edition for last biography

BANDICK, NEAL RAYMOND, b Orange, Calif, Mar 29, 38; m 63; c 2. HUMAN PHYSIOLOGY, MEDICAL EDUCATION. *Educ:* Univ Calif, Davis, BS, 60; Trinity Univ, MS, 65; Univ Mich, EdD(physiol), 70. *Prof Exp:* Instr physiol, Univ Mich, 69-70; assoc prof, 70-81, PROF BIOL, ORE COL EDUC, WESTERN ORE STATE COL, 81- *Mem:* Am Physiol Soc. *Res:* Cardiovascular physiology; visco-elastic and contractile properties of arterial walls during hypothermia. *Mailing Add:* Dept Sci & Math Western Ore State Col Monmouth OR 97361

BANDLER, JOHN WILLIAM, b Jerusalem, Nov 9, 41; Can citizen; c 2. ELECTRICAL ENGINEERING. *Educ:* Imperial Col, Univ London, BSc(eng), 63, DIC & PhD, 67. *Hon Degrees:* DSc, Univ London, 76. *Prof Exp:* Engr, elec eng, Mullard Res Labs, Eng, 66-67; lectr & fel, Univ Man, 67-69; from asst prof to prof, McMaster Univ, 69-74, coordr group sumulation, optimization & control, 73-83, chmn dept, 78-79, dean fac eng, 79-81, PROF ELEC & COMPUT ENG, MCMASTER UNIV, 74-, PRES, OPTIMIZATION SYSTS ASSOCS INC & DIR, OPTIMIZATION SYST RES LAB, 83-; PRES, BANDLER RES INC, 89- *Concurrent Pos:* Assoc Ed, IEEE transactions on microwave theory and techniques, 69-74, Guest Ed, Inst Elec & Electronics Engrs, 74. *Mem:* Fel Inst Elec & Electronics Engrs; fel Inst Elec Engrs, UK; Asn Prof Engrs, Can; fel Royal Soc Can. *Res:* Integrated circuits and systems; computer-aided design; optimization methods; microwave circuits; engineering design; power systems; CAD/CAE software systems; optimization with tolerances; computer methods; electronic circuit fault diagnosis; author of more than 230 technical papers. *Mailing Add:* Dept Elec & Comput Eng McMaster Univ Hamilton ON L8S 4L7 Can

BANDMAN, EVERETT, b New York, NY, June 29, 47; c 1. MOLECULAR BIOLOGY. *Educ:* City Col New York, BS, 69; Univ Calif, Berkeley, PhD(molecular biol), 74. *Prof Exp:* Res fel surg, Mass Gen Hosp, 74-76; fel biochem, Harvard Med Sch & fel, Shriner's Burns Inst, 74-76; fel zool, 76-78, asst res zoologist, Univ Calif, Berkeley, 78-81, asst prof, 81-86, ASSOC PROF, UNIV CALIF, DAVIS, 86- *Concurrent Pos:* Res technician biochem, Sloan Kettering Inst, 69; USPHS traineeship, 69-74. *Honors & Awards:* Nat Res Serv Award, NIH, 76. *Mem:* AAAS; Am Soc Cell Biol; Tissue Cult Asn. *Res:* Biochemical control of the regulation of cell growth in tissue culture; muscle protein isozymes during development. *Mailing Add:* Dept Food Sci & Technol Univ Calif Davis CA 95616

BANDONI, ROBERT JOSEPH, b Weeks, Nev, Nov 9, 26; Can citizen; m 56; c 1. MYCOLOGY. *Educ:* Univ Nev, BSc, 53; Univ Iowa, MSc, 56, PhD(mycol), 57. *Prof Exp:* Instr & asst prof bot, Univ Wichita, 57-58; PROF MYCOL, UNIV BC, 58- *Concurrent Pos:* Gertrude S Burlingham fel mycol, NY Bot Garden, 57. *Mem:* AAAS; Mycol Soc Am; Brit Mycol Soc; Japanese Mycol Soc. *Res:* Studies of leaf-decay fungi; taxonomic studies of lower Basidiomycetes. *Mailing Add:* Univ BC Bot Dept 6270 University Blvd No 3529 Vancouver BC V6T 2B1 Can

BANDURSKI, ROBERT STANLEY, b Chicago, Ill, May 11, 24; m 44; c 1. PLANT BIOCHEMISTRY. *Educ:* Univ Chicago, BS, 46, PhD(plant physiol), 49. *Hon Degrees:* DSc, Copernicus Univ, Poland, 87. *Prof Exp:* Instr bot, Univ Chicago, 48-49; Nat Res Coun fel biol, Calif Inst Technol, 49-50, res fel, 50-52, sr res fel, 52-53; asst biochem, Mass Gen Hosp, 53-54; from assoc prof to prof, 54-63, DISTINGUISHED PROF BOT, MICH STATE UNIV, 63- *Mem:* Fedn Biol Chem; Am Soc Plant Physiol. *Res:* Plant pigments; respiratory enzymes; auxins; chemistry of plant hormones and enzymes of sulfur metabolism. *Mailing Add:* Dept Bot Mich State Univ East Lansing MI 48824

BANDY, ALAN RAY, b Indiahoma, Okla, Sept 13, 40; m 63; c 4. ATMOSPHERIC, PHYSICAL & ANALYTICAL CHEMISTRY. *Educ:* Okla State Univ, BS & MS, 64; Univ Fla, PhD(chem), 68. *Prof Exp:* Res assoc chem, Univ Md, College Park, 68-70; from asst prof to assoc prof, Old Dom Univ, 70-75; assoc prof, 75-82, PROF CHEM, DREXEL UNIV, 82- *Mem:* Am Chem Soc; Am Geophhys Union; AAAS. *Res:* Photochemistry of trace atmospheric constituents; atmospheric chemistry of atmospheric sulfur and nitrogen compounds. *Mailing Add:* Dept Chem Drexel Univ Philadelphia PA 19104

BANDY, PERCY JOHN, b Mexico City, Mex, Aug 13, 27; Can citizen; m 52; c 3. ZOOLOGY. *Educ:* Univ BC, BA, 52, MA, 55, PhD(zool), 65. *Prof Exp:* Regional wildlife biologist, 58-62, wildlife res biologist, 62-73, asst dir, BC Fish & wildlife Br, 73-88; RETIRED. *Concurrent Pos:* Chmn, Nat Adv Comt Land Capability Classification for Wildlife, 66; chmn, BC Wildlife & Recreation Comt, Can Land Inventory, 66-71: hon lectr fac agr, Univ BC, 67-71, res assoc inst animal resource ecol, 69-71; ed, Fish & Wildlife Publ, 70-73, prog eval, 81-87. *Res:* Wildlife management research of various types including population dynamics, diseases, parasites and nutrition. *Mailing Add:* 4580 Torquay Dr Victoria BC V8N 3L7 Can

BANE, GILBERT WINFIELD, b San Diego, Calif, Dec 11, 31; m 70; c 3. MARINE ECOLOGY, BIOLOGICAL OCEANOGRAPHY. *Educ:* San Jose State Col, BA, 54; Cornell Univ, MS, 61, PhD(vert zool), 63. *Prof Exp:* From jr scientist to sr scientist, Inter-Am Trop Tuna Comn, 55-58; scientist, Starkist Foods, Inc, 59-60; lab asst vert zool, Conserv Dept, Cornell Univ, 60-63; asst prof marine biol, Univ PR, Mayaguez, 63-65 & Univ Calif, Irvine, 65-69; assoc prof, Southampton Col, LI Univ, 69-73; chmn marine & natural sci, St Francis Col, Maine, 73-75; dir marine sci & environ studies, Univ NC, Wilmington, 75-80; dir marine sci wetland res, La Univ, 80-88; PROF BIOL, KODIAK COL, 88- *Concurrent Pos:* Danforth assoc, Danforth Found, 68. *Mem:* Am Fisheries Soc; Am Soc Ichthyol & Herpet; Am Soc Mammal; Am Soc Limnol & Oceanog; Am Soc Zool. *Res:* Biology and population studies of marine fishes, especially sharks; environmental stresses on coastal fishes, especially offshore petroleum operations and coastal nuclear facilities; fishes of southern Mexico. *Mailing Add:* Dept Biol Kodiak Col 117 Benny Benson Dr Kodiak AK 99615

BANE, JOHN MCGUIRE, JR, b Kalamazoo, Mich, May 3, 46; m 70; c 3. PHYSICAL OCEANOGRAPHY. *Educ:* Western Mich Univ, BS, 70; Fla Atlantic Univ, ME, 71; Fla State Univ, PhD(phys oceanog), 75. *Prof Exp:* From asst prof to assoc prof phys oceanog & physics, 76-86, PROF PHYS OCEANOG, GEOL & PHYSICS, UNIV NC, CHAPEL HILL, 87- *Mem:* Am Geophys Union; Am Meteorol Soc; Oceanog Soc. *Res:* Theoretical and observational studies of western boundary current dynamics and ocean circulation; air-sea interaction. *Mailing Add:* Curric Marine Sci Univ NC Chapel Hill NC 27599-3300

BANERJEE, AMIYA KUMAR, b Rangoon, Burma, May 3, 36; nat US; m 65; c 2. BIOCHEMISTRY, MOLECULAR BIOLOGY. *Educ:* Univ Calcutta, BS, 55, MS, 58, PhD(biochem), 64. *Hon Degrees:* DSc, Univ Calcutta, 70. *Prof Exp:* Res assoc molecular biology, Albert Einstein Col Med, 66-69; asst mem, 70-73, assoc mem, 74-80, MEM MOLECULAR BIOL, ROCHE INST MOLECULAR BIOL, 81- *Mem:* Am Soc Biol Chemists; Am Soc Microbiol; AAAS; Harvey Soc. *Res:* Biosynthesis, structure and function of viral and cellular macromolecules; regulation of eucaryotic gene expression. *Mailing Add:* Dept Molecular Biol Cleveland Clin Found Res Inst 9500 Euclid Ave Cleveland OH 44195-5175

BANERJEE, CHANDRA MADHAB, b Calcutta, India, Aug 28, 32; m 66; c 4. RESPIRATORY PHYSIOLOGY. *Educ:* Univ Calcutta, MB & BS, 55; Med Col Va, PhD(physiol), 67. *Prof Exp:* Clin asst med, Calcutta Nat Med Col, 55-56, house surgeon venereal dis, 56-57; med officer leprosy, Cent Leprosy Teaching & Res Inst, India, 57-58; rotating intern, St Vincent's Hosp, NY, 59; intern, St Mary's Hosp, 59-60; staff scientist respiratory physiol, Hazleton Labs Inc, Va, 67-68; from asst prof to assoc prof physiol & anesthesiol, Jefferson Med Col, Pa, 68-74; PROF PHYSIOL, SCH MED, SOUTHERN ILL UNIV, 74- *Concurrent Pos:* Fel & asst, Dept Med, Med Col Va, 60-66. *Mem:* AAAS; Am Physiol Soc; NY Acad Sci; Am Heart Asn. *Res:* Air pollution; myocardial infarction; pulmonary edema; right duct lymph; cardiovascular physiology. *Mailing Add:* Dept Physiol Sch Med Southern Ill Univ Carbondale IL 62901

BANERJEE, DEBENDRANATH, b Calcutta, India, Sept 10, 35; m 70; c 1. ANIMAL & PLANT PHYSIOLOGY. *Educ:* Univ Calcutta, BSc, 57, MSc, 59; Univ Nebr, PhD(biol), 72. *Prof Exp:* Teaching fel, 72-74, res assoc, 76-78, assoc investr, 78-87, INVESTR, NY BLOOD CTR, 87- *Mem:* Am Soc Cell Biol; NY Acad Sci. *Res:* Cellular mechanisms governing the biosynthesis, intracellular movement, secretion and regulation of secretory proteins; studies involve small laboratory animals, a variety of biochemical techniques, electron microscopy, gene cloning and the use of monoclonal antibodies. *Mailing Add:* Membrane Biochem New York Blood Ctr 310 E 67th St New York NY 10021

BANERJEE, DIPAK KUMAR, b Pakshi, Bangladesh, Jan 5, 47; m 79; c 1. GLYCOPROTEINS, BIOMEMBRANES. *Educ:* Univ Calcutta, BSc, 66, MSc, 68, PhD(biochem), 76. *Prof Exp:* Sr sci asst, 69-75, jr sci officer, Inst Nuclear Med & Allied Sci, 75-79; postdoctoral fel, Univ Md Med Sch, 79-82; biochemist, NIH, 83-86; ASSOC PROF, UNIV PR Sch Med, 86- *Concurrent Pos:* Jr & sr res fel, Christian Med Col Hosp, 72-75; postdoctoral fel, HIH, 82-83. *Mem:* Am Soc Complex Carbohydrates; Biochem Soc London; Am Soc Biochem & Molecular Biol; NY Acad Sci; Sigma Xi. *Res:* Angiogenesis; capillary proliferation. *Mailing Add:* Dept Biochem & Nutrit Univ Sch Med GPO Box 5067 San Juan PR 00936-5067

BANERJEE, KALI SHANKAR, b Dacca, Bangladesh, Sept 1, 14; m; c 2. MATHEMATICAL STATISTICS. *Educ:* Univ Calcutta, BA, 35, MA, 37, PhD(statist), 50. *Prof Exp:* Sr asst statistician, Cent Sugarcane Res & Develop, Pura Bihor, India, 42-44, statistician, 46-51; statist officer, Govt of Orissa, India, 44-46; dep dir, State Statist Bur, WBengal, 57-62, 63-64 & 66, additional dir, 66-67; vis assoc prof, Cornell Univ, 62-63; vis prof, 65-66, prof, Kans State Univ, 68-69; additional dir eval develop & planning dept, West Bengal, India, 67-68; prof statist, Univ Del, 69-74, H Rodney Sharp prof, 74-80; mem fac, Dept Math & Comput Sci, Univ Md, Baltimore County Campus, 80-86; RETIRED. *Concurrent Pos:* Fulbright travel assistance, 62-63; NSF grants, 64-66, 68-69 & 70-71. *Honors & Awards:* Distinguished Scientist Award, Sigma Xi, 75. *Mem:* Int Statist Inst; fel AAAS; fel Royal Statist Soc; fel Am Statist Asn; fel Inst Math Statist. *Res:* Weighing designs; index numbers and quadratic forms; design of experiments in general; one hundred research publications and five published monographs. *Mailing Add:* 66 Eight Ave Kings Park NY 11754

BANERJEE, MIHIR R, b Calcutta, India, Oct 24, 27. GENETICS CELLULAR & MOLECULAR BIOLOGY, MAMMARY GLAND RESEARCH. *Educ:* Univ Calcutta, BSc, 49, MSc, 51, PhD(zool), 59. *Prof Exp:* Lectr zool, City Col, Calcutta, India, 53-59; res fel, Cancer Res Lab, Univ Calif, Berkeley, 60-62, Dept Exp Path & Biol, City of Hope Med Ctr, Duarte, Calif, 62; vis res scientist, Div Biol, Inst Cancer Res, Fox Chase, Philadelphia, Pa, 63-64; vis scientist, Carcinogenesis Area, Nat Cancer Inst, NIH, 64-67; assoc prof zool, Sch Life Sci, 67-71, PROF GENETICS, CELLULAR & MOLECULAR BIOL, SCH BIOL SCI, UNIV NEBR, LINCOLN, 71- *Concurrent Pos:* Fulbright scholar, 60-63; grants, USPHS-NIH, 68-78, 78-81, 79-82, 83-86 & 86-90, HEW-NIH, 68-78, Univ Res Coun, 83, 84, 86 & 87; chmn, Cell Biol & Genetics Sect, Sch Life Sci, Univ Nebr, Lincoln, 74-77; mem, Prog Comt, Tissue Cult Asn, 74-76; mem, Chem Path Study Sect, NIH, 75 & 81, Breast Cancer Task Force, Nat Cancer Inst, 79-84; mem, Cell & Molecular Biol Prog, NSF, 78-; UNESCO consult, Cell & Molecular Biol Prog, Dept Zool, Univ Calcutta, India, 83-84. *Mem:* Am Asn Cancer Res; Am Soc Biol Chemists; Am Soc Cell Biol; Tissue Cult Asn. *Res:* Molecular biology of hormone regulated development of the mammary gland and breast cancer; culture model of the whole mammary organ; development of neoplastic transformation model in the whole mammary organ in vitro; cloning of the milk-protein genes. *Mailing Add:* Tumor Biol Lab Sch Biol Sci Univ Nebr 201 Lyman Hall Lincoln NE 68588-0342

BANERJEE, MUKUL RANJAN, b Dacca, India, Jan 3, 37; US citizen; m 68; c 2. RESPIRATORY PHYSIOLOGY. *Educ:* Univ Calcutta, BVSc, 57; La State Univ, PhD(physiol), 64. *Prof Exp:* Res assoc, Ind Univ, Bloomington, 64-65, from instr to asst prof, 66-72; NIH res fel, Univ Fla, 73-74; assoc prof, Tenn State Univ, 75-76; assoc prof, 77-79, PROF PHYSIOL, MEHARRY MED COL, 79- *Mem:* Am Physiol Soc; Int Soc Biometeorol; Sigma Xi. *Res:* Respiratory physiology; environmental physiology. *Mailing Add:* 1005 18th Ave N Nashville TN 37208

BANERJEE, PRASANTA KUMAR, b Feb 1, 41; m 71; c 3. COMPUTATIONAL MECHANICS, GEOTECHNICAL ENGINEERING. *Educ:* Jadavpur Univ, India, BCE, 63, Southampton Univ, UK, PhD (civil eng), 70. *Prof Exp:* Prin sci officer, bridges, Dept Transp London, 70-73; sr lectr geotechnol eng, Univ Wales, 73-80; PROF CIVIL ENG, STATE UNIV NY, BUFFALO, 80-; PROJ DIR, NASA PROGS, 83- *Concurrent Pos:* Prin invest, Nat Sci Found, 80- *Mem:* Inst Civil Engrs London; Am Soc Civil Engrs; Am Soc Mech Engrs. *Res:* Development of boundary element method in engineering mechanics, development of computational mechanics, mechanical behavior of materials, dynamic as well as seismic behavior of foundations, specifically piled foundations. *Mailing Add:* Dept Civil Eng State Univ Ny, Buffalo Buffalo NY 14260

BANERJEE, PRITHVIRAJ, b Khartoum, Sudan, July 17, 60; Indian citizen; m 84. FAULT TOLERANT PARALLEL ARCHITECTURES, VLSI CAD TOOLS. *Educ:* Indian Inst Technol, India, BTech, 81; Univ Ill, Urbana-Champaign, MS, 82; PhD(elec Eng), 84. *Prof Exp:* Res asst, Coord sci lab, 81-84, ASST PROF, ELEC ENG, UNIV ILL, URBANA-CHAMPAIGN, 85- *Concurrent Pos:* Consult, Westinghouse Corp, Baltimore, 86-87, Jet Propulsion Lab, Pasedena, 88-, Res Triangle Inst, NC, 88-; prog comt, Int Fault-Tolerant Computing Symp, Japan, 88, organizing comt, Chicago, 89. *Honors & Awards:* Press India Gold Medal, 81. *Mem:* Inst Elec & Electronics Engrs; Asn Comput Mach. *Mailing Add:* Dept Elec Eng Univ Ill Urbana Campus 1406 W Green St Urbana IL 61801

BANERJEE, R L, b Bengal, India, Mar 1, 33; m 66; c 1. SOLID STATE PHYSICS. *Educ:* Univ Calcutta, BSc, 53, MSc, 56; Univ Paris, Sorbonne, DSc(physics), 66. *Prof Exp:* Jr lectr physics, Univ Calcutta, 57; res fel, Bengal Eng Col, Calcutta, 57-58; crystallog, Univ Madrid, 58-60; from jr researcher to full researcher physics, Univ Paris, Sorbonne, 60-67; sr lectr, Univ WI, 67-69; asst prof, 69-70, assoc prof, 71-77, PROF PHYSICS, UNIV MONCTON, 78- *Concurrent Pos:* Nat Res Coun Can Res grants, 69-76. *Mem:* Fr Soc Mineral & Crystallog. *Res:* Clay-mineral contents of Indian soils; polymorphic transitions in single crystals; thermal diffuse scattering of x-rays from molecular crystals; lattice dynamics; Compton scattering of x-rays; structures and properties of metals. *Mailing Add:* Dept Physics Univ Moncton Moncton NB E1A 3E9 Can

BANERJEE, RANJIT, b Calcutta, India, June 6, 48; US citizen; m 84; c 2. MOLECULAR ONCOLOGY, MOLECULAR VIROLOGY. *Educ:* Univ Calcutta, India, BS Hons, 69, MS, 71; NY Univ, MS, 79, PhD(genetics), 81. *Prof Exp:* Res fel cell & molecular biol, Dept Path, Sch Med, Temple Univ, 81-83; res scientist, Inst Cancer Res, Col Physicians & Surgeons, Columbia Univ, 83-86; ASST PROF, DEPT NEOPLASTIC DIS, MT SINAI SCH MED, 86- *Mem:* Am Asn Cancer Res; AAAS; Am Soc Microbiol; NY Acad Sci. *Res:* Molecular biology of gene expression in normal and transformed cells; effect of cofactors in carcinogenesis and viral gene expression; role of transcription factors and cytokines in regulation of human immunodeficiency virus and hepatitis B virus gene expression. *Mailing Add:* Dept Biochem Box 1020 Mt Sinai Sch Med One Gustave L Levy Pl New York NY 10029

BANERJEE, SANJAY KUMAR, b Khartoum, Sudan, Feb 24, 58; m 83; c 1. SEMICONDUCTOR DEVICE & PROCESS PHYSICS, LOW TEMPERATURE SEMICONDUCTOR CRYSTAL GROWTH. *Educ:* Indian Inst Technol, Kharagpur, BTech (Hons), 79; Univ Ill, Urbana-Champaign, MS, 81, PhD (elec), 83. *Prof Exp:* Res asst elec, Coord Sci Lab, Univ Ill, 79-83; mem tech staff, Very Large Scale Integration Circuits Design Lab, Tex Instruments, 83-87; asst prof, 87-90, ASSOC PROF ELEC & COMPUTER ENG, UNIV TEX, AUSTIN, 90- *Concurrent Pos:* Vis asst prof, Elec Eng, Univ Ill, 83; NSF presidential young investr, 88; consult, Corp Res & Develop, Tex Instruments, 88. *Mem:* Sr mem Inst Elec & Electronic Engrs; Am Phys Soc; Mat Res Soc; Am Soc Eng Educ. *Res:* Semiconductor physics, devices, and processes; device modeling; very low temperature plasma-enhanced and laser-enhanced semiconductor crystal growth; 3-D integrated circuits using polysilicon; drams; author of over 100 publications; awarded four US patents. *Mailing Add:* Elec & Computer Eng Rm 404B ENS Univ Tex Austin TX 78712-1084

BANERJEE, SIPRA, b Calcutta, West Bengal, Feb 20, 39; US citizen; m 65; c 2. ENVIRONMENTAL CARCINOGENESIS, CANCER. *Educ:* Presidency Col, Calcutta, India, BS, 58; Calcutta Univ, MS, 60, PhD(biochem), 65. *Prof Exp:* Fel biochem, Dept Biochem, Albert Einstein Col Med, 66-69; res biochemist, NY Med Col, 69-70; fel, Hoffmann LaRoche Inc, 72-73; assoc res scientist, 74-78, asst prof, 78-85, assoc res prof, 85-87; ASSOC STAFF, CLEVELAND CLIN FOUND, 88- *Mem:* Am Asn Cancer Res; NY Acad Sci; Am Soc Biol Chemists. *Res:* Biochemical studies of molecular mechanism of chemical carcinogenesis with the emphasis on the metabolic pathways, enzymes, DNA damage and repair, and oncogenes. *Mailing Add:* Dept Molecular Biol Cleveland Clin Found Res Inst 9500 Euclid Ave Cleveland OH 44195-5108

BANERJEE, SUBIR KUMAR, b Jamshedpur, India, Feb 19, 38; m 63, 90; c 3. GEOMAGNETISM, ROCK MAGNETISM. *Educ:* Univ Calcutta, BSc, 56; Indian Inst Technol, Kharagpur, MTech, 59; Univ Cambridge, PhD(geophys), 63. *Hon Degrees:* DSc, Univ Cambridge, 83. *Prof Exp:* Res physicist magnetism of ferrites, Mullard Res Labs, Redhill, Eng, 63-64; sr res assoc rock magnetism, Dept Geophys & Planetary Physics, Univ Newcastle, 64-66, univ lectr geophys, 66-69; sr staff scientist geophys & magnetic mat, Dept Mat Sci & Eng, Franklin Inst Res Labs, Philadelphia, 69-71; assoc prof geophys, 71-74, PROF GEOPHYS, UNIV MINN, MINNEAPOLIS, 74- *Concurrent Pos:* Consult, Atomic Energy Res Estab, Eng, 65-69; mem res staff, Ampex Corp & res assoc, Stanford Univ, 67-68; lectr, Univ Pa, 69-71; vis sr res assoc, Lamont-Doherty Geol Observ, Columbia Univ, 70-71; adj prof Middle-Eastern & Islamic Studies, Univ Minn, Minneapolis, 76-; vis prof geophys, Stanford Univ & Univ Calif, Berkeley, vis scholar, Off for Hist of Sci & Technol, 77-78; vis prof geophys & vis scholar hist sci & technol, Univ Calif, Berkeley, 84-85; chmn, Int Asn Geomagnetism & Aeronomy's Working Group on Rock Magnetism, 83-87; pres, Geomag & Paleomag Sect, Am

Geophys Union, 86-88; mem, Adv Panel Earth Sci, Nat Sci Found. *Mem:* Fel Am Geophys Union. *Res:* Rock magnetism-magnetism of natural and synthetic minerals relevant to geomagnetism and paleomagnetism; magnetic records of climatic and environmental change; geomagnetic field fluctuations; correlations between climate and geomagnetism; continental drift and sea floor spreading; magnetism of lunar samples, meteorites, recording tapes, ferrites and permanent magnets; history of medieval Arabic-Islamic science. *Mailing Add:* Dept Geol & Geophys Univ Minn 310 Pillsbury Dr SE Minneapolis MN 55455-0219

BANERJEE, SURATH KUMAR, b W Bengal, India, Nov 3, 38; m 70; c 1. CLINICAL CHEMISTRY. *Educ:* Calcutta Univ, India, BSc, 69 & MS, 61, Jadavpur Univ, PhD(chem), 66. *Prof Exp:* ASSOC PROF PATH, WAYNE STATE UNIV, 80-; CHIEF CHEM, VET MED CTR, 83- *Concurrent Pos:* Res assoc, Purdue Univ, 69-70, Univ Ariz, 67-69 & 70-74; res asst prof, dept Internal Med & Biochem, Univ Ariz, 74-80. *Mem:* Am Soc Biochem & Molecular Biol; Am Asn Clinical Chemists; Am Chem Soc. *Res:* Contractile protein chemistry; cardiac myosin gene expression; drug effects on myosin genes; heart nuclear thyroid hormone receptors. *Mailing Add:* Dept Chem Lab Serv Vet Med Ctr Southfield & Outer Dr Allen Park MI 48101

BANERJEE, SUSHANTA KUMAR, b Ranchi, India, Sept 25, 27; m 52; c 2. AGRONOMY, ENVIRONMENTAL HEALTH. *Educ:* Patna Univ, BS, 48; Tex A&M Univ, MS, 57, PhD(plant & soil sci), 64; Univ Pittsburgh, MPH, 80. *Prof Exp:* Owner & mgr, Liluah Dairy Farm, India, 48-50; asst storage & transp off, Govt WBengal, 50-55, subdiv agr off, 55-57; supt agr, Govt India, 57-60; instr biol, Mitchell Col, Conn, 64-67; from asst prof to assoc prof, 67-72, chmn, Biol Sci Dept, 71-73, PROF BIOL SCI & PUB HEALTH, POINT PARK COL, 72- *Concurrent Pos:* Mem, Bd Dir, Allegheny County Environ Coalition, Pittsburgh, Pa, 72-75; Dir, Point Park Col Chapter Inst Human Ecol Southwestern Pa, 72-76; Chmn, Bd Dir, India Asn, Pittsburgh, Pa, 73; Pres, Bengali Asn, Pittsburgh, Pa, 81-82; Vis prof, Visva Bharati Univ, Santiniketan, West Bengal, India, 88. *Mem:* Soc for Environ Geochem & Health. *Res:* Soil fertility and plant nutrition; response of rice crop to fertilizers; protective effects of chemicals on radiation damage of plants. *Mailing Add:* Dept Natural Sci & Technol Point Park Col Pittsburgh PA 15222

BANERJEE, UMESH CHANDRA, b Aligarh, India, July 15, 37; US citizen; div; c 1. ENVIRONMENTAL BIOLOGY & POLLUTION, TISSUE CULTURE. *Educ:* Muslim Univ, India, BSc, 58, MSc, 61; Univ Mass, Amherst, MS, 66; Harvard Univ, PhD(biol), 73. *Prof Exp:* Res assoc electron micros, dept biol, Yale Univ, 66-68; electron microscopist lunar res, NASA Proj, Harvard Univ, 68-69 & 71-73, res assoc & fel biol, 73-74; asst prof biol, Boston Univ, 74-75; lectr, Univ Mass, Boston, 75-78; asst prof & dir, Electron Microscope Ctr, NTex State Univ, Denton, 78-81; ASSOC PROF BIOL & ENVIRON BIOL, NC CENT UNIV, DURHAM, 81- *Concurrent Pos:* Dir, Boston Univ Herbarium, 74-75, Benjamin Harris Herbarium, NTex State, Univ, 78-81 & Tissue Cult Labs, NC Cent Univ, 81-; fel biol, Harvard Univ, 75-78; asst dir, Electron Microscope & Immunol Ctr, NC Cent Univ, 81- *Mem:* Bot Soc Am; Am Asn Stratig Palynology; Indian Palynological Soc; Int Asn Aerobiol; Electron Micros Soc Am. *Res:* Biotechnology of guayule rubber, a native species of North America; allergenic pollen grains and spores; electron microscopy, transmission, scanning electron microscopy and X-ray analysis; tissue culture; origin of corn; medicinal plants; systematic biology. *Mailing Add:* Five Rollingview Ct Rollingwood Develop Durham NC 27713-9344

BANERJEE, UTPAL, b Howrah, India, Aug 4, 42; m 69; c 1. PARALLEL PROCESSING. *Educ:* Calcutta Univ, India, BSc, 61, MSc, 63; Carnegie-Mellon Univ, MS, 67, PhD(math), 70; Univ Ill, Urbana-Champaign, MS, 76, PhD(comput sci), 79. *Prof Exp:* Asst prof math, Univ Cincinnati, 69-75; prin analyst, Large Info Systs Div, Honeywell Corp, Phoenix, 79-81; mem res staff, Fairchild Advan Res Lab, Calif, 81-82; consult, Sunnyvale Develop Div, Control Data Corp, 82-89; SR RES, INTEL CORP, 89- *Concurrent Pos:* Vis sr software engr, Ctr Supercomputing Res & Develop, Univ Ill, Urbana-Champaign, 86- *Mem:* Am Math Soc; Math Asn Am; Asn Comput Mach; Inst Elec & Electronics Engrs; Sigma Xi. *Mailing Add:* Intel Corp MS NW 1-18 2801 Northwestern Pkwy Santa Clara CA 95051

BANERJI, RANAN BIHARI, b Calcutta, India, May 5, 28; m 54; c 2. ENGINEERING, COMPUTER SCIENCE. *Educ:* Patna Univ, BSc, 47; Univ Calcutta, MSc, 49, DPhil(physics), 56. *Prof Exp:* Vis asst prof eng res, Pa State Univ, 53-56; lectr radio eng, Univ Calcutta, 56; maintenance engr comput sect, Indian Statist Inst, 56-58; res assoc opers res, Case Inst Technol, 58-59; asst prof eng, Univ NB, 59-61; from asst prof to prof, Case Western Reserve Univ, 61-73; prof comp sci, Temple Univ, 73-81; PROF COMP SCI, ST JOSEPH'S UNIV, 81- *Concurrent Pos:* Res asst, Linguistic Data Processing Seminar, Ind Univ, 64; consult, Smith Electronics, Ohio, 62, RCA, 67, GE, 73, Analytics, Inc, 80, Chase Financial, 90 & Philips Res, 90; vis prof, Univ Paris, 77, Univ Calcutta, 76, Univ Vienna, 85 & 91 & Univ Sydney, 90. *Mem:* Asn Comput Mach; sr mem Inst Elec & Electronics Engrs; Math Asn Am; Am Assoc Artificial Intelligence. *Res:* Statistics of radio wave fluctuations; information theory and coding; automata and linguistics; artificial intelligence. *Mailing Add:* 7612 Woodlawn Ave Melrose Park PA 19126-1428

BANERJI, SHANKHA K, b Lucknow, India, Jan 25, 36; m 64; c 3. CIVIL ENGINEERING, ENVIRONMENTAL ENGINEERING. *Educ:* Univ Calcutta, BE, 57; Univ Ill, Urbana, MS, 62, PhD(sanit eng), 65. *Prof Exp:* Asst engr, Hindustan Steel Ltd, India, 57-60; asst prof sanit eng, Univ Ill, Urbana, 65-66; asst prof, Univ Del, 66-70, res grant, Water Resources Ctr, 67-68; consult engr, Kidde Consults, Inc, 70-75; PROF CIVIL ENG, UNIV MO-COLUMBIA, 75- *Concurrent Pos:* Consult, Stauffer Chem Co, Del, 66-67; Kent Country Sewer Study Comt, 67; Del Pub Serv Comn, 68-70; consult, Kidde Consults Inc, Baltimore, Md, 75-81, R W Booker & Assoc, St Louis, Mo, US Corp Engrs, St Louis, Mo, & Shell Engr & Assoc, Columbia, Mo, 75. *Mem:* Int Asn Water Pollution Res; Am Soc Civil Engrs; Am Water Works Asn; Water Pollution Control Fedn; Sigma Xi. *Res:* Biological treatment of waste water and stream pollution control; water quality control facility design; water quality deterioration in distribution systems; sludge stability analysis; waste stabilization; pond effluent polishing. *Mailing Add:* 2301 Ridgefield Rd Columbia MO 65203

BANERJI, TAPAN KUMAR, b Calcutta, India, July 12, 40; m 74; c 1. ENDOCRINOLOGY, MICROANATOMY. *Educ:* Univ Calcutta, BSc Hons, 61, MSc, 63, PhD(endocrinol), 70. *Prof Exp:* Asst prof biol sci, Himachal Pradesh Univ, Simla, India, 72-74; fel neuroendocrinol, Waisman Ctr, Univ Wis-Madison, 74-77; res assoc, 77, fac assoc, 77-78, from asst prof to prof, 78-87, PROF ANAT, UNIV TEX MED BR, GALVESTON, 87- *Honors & Awards:* S P Basu Mem Medal, Zool Soc, 68. *Mem:* Fel Zool Soc; NY Acad Sci; Am Asn Anatomists; Int Brain Res Org. *Res:* Neuroendocrinology, comparative endocrinology, biogenic amines, ultrastructure and rhythmic body functions; reproduction biology, psychopharmacology. *Mailing Add:* Dept Anat Sch Med Univ Tex Med Br Galveston TX 77550

BANES, ALBERT JOSEPH, b McKeesport, Pa, Aug 26, 47; m 68; c 2. MICROBIOLOGY, BIOCHEMISTRY. *Educ:* Lehigh Univ, BA, 69; Univ Richmond, MS, 71; Va Commonwealth Univ, PhD(microbiol), 75. *Prof Exp:* Fel microbiol, Duke Univ, 74-76; res assoc, Va Commonwealth Univ, 76-77; res assoc biochem, 77-78, ASSOC PROF SURG, UNIV NC, 78- *Concurrent Pos:* Asst dir res, J C Burn Ctr, Univ NC, 78-, mem, Dent Res Ctr, 78- *Mem:* Fedn Am Soc Exp Biol; Am Burn Asn; Am Asn Dent Res; Am Soc Bone & Mineral Res. *Res:* Developmental biology and biochemistry; identification of the molecular mechanisms underlying disease, scar, mechanical. *Mailing Add:* Dept Surg Univ NC 253 Clin Sci Bldg 229H Chapel Hill NC 27514

BANEY, RONALD HOWARD, b Alma, Mich, Nov 26, 32; m 55; c 2. INDUSTRIAL CHEMISTRY. *Educ:* Alma Col, BS, 55; Univ Wis, PhD(inorg chem), 60. *Prof Exp:* Group leader, 60-74, assoc scientist, 74-83, RES SCIENTIST, DOW CORNING CORP, 83- *Concurrent Pos:* Vis indust scientist, Nagoya Univ, 86-87. *Mem:* Am Chem Soc; Sigma Xi; Am Ceramic Soc. *Res:* Physical organic chemical investigations, especially organosilicon chemistry; development of new silicone materials; ceramic materials; superconductors. *Mailing Add:* Dow Corning Corp Midland MI 48640

BANFIELD, ALEXANDER WILLIAM FRANCIS, b Toronto, Ont, Mar 12, 18; m 42; c 3. ENVIRONMENTAL IMPACT ASSESSMENT, MAMMALOGY. *Educ:* Univ Toronto, BA, 42, MA, 46; Univ Mich, PhD, 52. *Prof Exp:* Mus asst Royal Ont Mus Zool & Paleont, 45-46; mammalogist, Nat Parks Serv, 46-47; chief mammalogist, Can Wildlife Serv, Dept Northern Affairs & Nat Resources, 48-57, chief zoologist, Nat Mus Can, 57-63, dir Mus Nat Sci, Ottawa, 64-69; prof biol sci & environ studies, Brock Univ, 69-79, dir, Inst Urban & Environ Studies, 75-80, emer prof biol sci & environ studies, 80-; RETIRED. *Concurrent Pos:* Mem, Can Comt, Int Biol Prog; hon res fel, Univ Edinburgh, 75-76; mem, Can comt, Man & Biosphere Prog; pres, Rangifer Assocs Environ Consult, 79-85. *Mem:* Fel Arctic Inst NAm; Am Soc Mammal. *Res:* Status, ecology and utilization of the barren-ground caribou; ecology; mammalian systematics; zoogeography of mammals. *Mailing Add:* 37 Yates St St Catharines ON L2R 5R3 Can

BANFIELD, WILLIAM GETHIN, b Hartford, Conn, Mar 2, 20; m 44; c 3. EXPERIMENTAL PATHOLOGY. *Educ:* RI State Col, BS, 41, MS, 43; Yale Univ, MD, 46; Am Univ, Washington, DC, JD, 74. *Prof Exp:* Am Cancer Soc fel, Yale Univ, 49-52, from instr to asst prof path, 52-54; pathologist, Nat Cancer Inst, NIH, 54-80; RETIRED. *Concurrent Pos:* Assoc pathologist, Grace-New Haven Community Hosp, 52-54. *Res:* Connective tissues; viruses. *Mailing Add:* 15715 Avery Rd Rockville MD 20855

BANG, NILS ULRIK, b Copenhagen, Denmark, Sept 24, 29; m 60; c 3. MEDICINE, HEMATOLOGY. *Educ:* Copenhagen Univ, MD, 55. *Prof Exp:* Intern, Munic Hosp, Copenhagen, 55-56; resident, Univ Hosp, Univ of Copenhagen, 57; resident cardiol, Mem Ctr Cancer & Allied Dis, Med Col, Cornell Univ, 57-58; NIH res trainee enzymol, Sch Med, Wash Univ, 59-61; asst prof med, NY Hosp-Cornell Med Ctr, 62-66; assoc prof, 66-72, PROF MED, SCH MED, IND UNIV, INDIANAPOLIS, 72-, PROF PATH, 80-; CHIEF HEMAT SERV, LILLY LAB CLIN RES, WISHARD MEM HOSP, 66- *Concurrent Pos:* Spec fel enzymol, Munic Hosp, Copenhagen, 56; res fel, Sloan-Kettering Inst Cancer Res, 58-S9; fel couns thrombosis, stroke & arteriosclerosis, Am Heart Asn. *Mem:* AAAS; Harvey Soc; Am Fedn Clin Res; fel Am Col Cardiol; Am Hemat Soc; Sigma Xi. *Res:* Biochemistry of blood coagulation and platelet function. *Mailing Add:* 1001 W Tenth St Indianapolis IN 46202

BANGDIWALA, ISHVER SURCHAND, b Surat, India, Jan 9, 22; US citizen; m 47; c 2. STATISTICS METHODOLOGY, APPLIED RESEARCH. *Educ:* Bombay, BSc, 43, MSc, LLB, JD, 46; Univ NC, MS, 50, PhD(exp statist, math statist), 58. *Prof Exp:* Head statist sect & legal adv, K A Pandit, Actuary, India, 44-48; head statist dept, Agr Exp Sta, Univ PR, 52-58; dir res & consult statistician, 58-66, prof, 66-82, EMER PROF STATIST & RES, UNIV PR, RIO PIEDRAS, 82- *Concurrent Pos:* Prof & consult, Inst Statist, Univ PR, 53-66, Col Agr & Mech Arts, 59-82, Med Sch, 69-82. *Mem:* Fel Royal Statist Soc; Inst Math Statist; Biomet Soc; fel Int Statist Inst; emer Sigma Xi; Int Asn Survey Statisticians; Int Asn Off Statist. *Res:* Statistical methodology; sampling; experimental designs; educational and social science research; medical science research; evaluation design and statistical analysis. *Mailing Add:* Calle Cisne 792 URB Dos Pinos Rio Piedras PR 00923-2792

BANGE, DAVID W, b St Louis, Mo, Apr 29, 45. GRAPH THEORY. *Educ:* St Loui Univ, BS, 67; Colo State Univ, MS, 69 &, PhD(math), 71. *Prof Exp:* PROF MATH, UNIV WIS, 70- *Mem:* Am Math Soc. *Mailing Add:* Dept Math Univ Wis-La Crosse 1725 State St La Crosse WI 54601

BANGERT, JOHN T(HEODORE), b Chicago, Ill, Jan 8, 19; m 50; c 2. COMMUNICATION ENGINEERING. *Educ:* Univ Mich, BS, 42; Stevens Inst Technol, MS, 47. *Prof Exp:* Dir, Carrier Transmission Lab, AT&T Bell Labs, Inc, 68-75, dir, Transmission Technol Lab, 62-68 & 76-79, dir, Transmission Develop Planning Lab, 79-81; RETIRED. *Mem:* Fel Inst Elec & Electronics Engrs. *Res:* Transient and steady state analysis and synthesis of active and passive networks; time domain equalization; time varying networks; computer aided design; pulse code modulation systems; coaxial cable systems; international standards. *Mailing Add:* 75 Dascomb Rd Andover MA 01810

BANGERTER, ROGER ODELL, b Salt Lake City, Utah, Mar 9, 39; m 65; c 6. PHYSICS. *Educ:* Univ Utah, BA, 63; Univ Calif, Berkeley, PhD(physics), 69. *Prof Exp:* Physicist elem particle physics, 69-74, inertial confinement fusion, Lawrence Livermore Lab, 74-81, Los Alamos Nat Lab, 81-, PHYSICIST HEAVY ION FUSION/INERTIAL FUSION, LAWRENCE BERKELEY LABS. *Mem:* Am Phys Soc. *Res:* Inertial confinement fusion; specifically, the use of heavy ion accelerators for inertial confinement fusion. *Mailing Add:* Lawrence Berkeley Labs Heavy Ion Fusion/Inertial Fusion Bldg 47/12 Berkeley CA 94720

BANGHART, FRANK W, b Michael, Ill, Oct 30, 23; m 46; c 2. BIOSTATISTICS. *Educ:* Quincy Col, AB, 49; Drake Univ, MA, 50, EdD(statist), 57. *Prof Exp:* Instr psychol, Quincy Col, 49-52; asst instr ed, Univ Va, 55-56, instr, 56-57, from asst prof to assoc prof biostatist, 57-65; PROF EDUC MGT SYSTS, FLA STATE UNIV, 65- *Concurrent Pos:* Prin investr, Off Naval Res, 57-61; Air Force Syst Command Hqs, 57-; mem adv comt new educ media, US Health, Educ & Welfare Dept, 61- *Mem:* AAAS; Am Statist Asn; Biomet Soc; NY Acad Sci. *Res:* Statistical methodology, research designs. *Mailing Add:* Dept Educ Leadership Fla State Univ Tallahassee FL 32306

BANGLE, RAYMOND, JR, pathology, for more information see previous edition

BANGS, LEIGH BUCHANAN, b Marblehead, Mass, Oct 9, 36; m 59; c 3. SURFACE CHEMISTRY. *Educ:* Colby Col, BA, 58; Mass Inst Technol, SM, 62, PhD(metall), 65. *Prof Exp:* Spec assignments chemist, Dow Chem Co, 64-66, res chemist, Dow Chem USA, 66-68, res specialist, 68-81, nat serv mgr, Div Dow Chem, Photovolt Corp, prod mgr, Dow Inst & Reagents, 82-83; bus mgr, Seradyn Diag Inc, 85-88; PRES, BANGS LAB INC, 88- *Mem:* Am Chem Soc; Sigma Xi; AAAS. *Res:* Surface chemistry; colloid chemistry; uniform latex particles applications; diagnostic chemistry. *Mailing Add:* Bangs Lab Inc 979 Keystone Way Carmel IN 46032

BANHOLZER, WILLIAM FRANK, US citizen. DIAMOND. *Educ:* Marquette Univ, BS, 79; Univ Ill, MS, 81, PhD(chem eng), 83. *Prof Exp:* Staff scientist, 83-89, MGR, ADVAN INORG MAT LAB, GEN ELEC CORP RES & DEVELOP, 89- *Mem:* Am Vacuum Soc; Mat Res Soc. *Res:* Surface reactions and material science; catalysis of silicon reaction; chemical vapor deposition; diamond films; author of 38 technical publications. *Mailing Add:* 30 Olde Coach Rd Glenville NY 12302

BANICK, WILLIAM MICHAEL, JR, b Scranton, Pa, Feb 6, 32; m 56; c 3. ANALYTICAL CHEMISTRY. *Educ:* Kings Col, BS, 53; Univ Ill, PhD(chem), 57. *Prof Exp:* Anal proj chemist, 57-65, sr res chemist, 66-68, GROUP LEADER, AM CYANAMID CO, 68- *Mem:* Am Chem Soc. *Res:* Analytical chemistry of basic intermediates; rubber chemicals; brighteners; ultraviolet absorbers; antioxidants, especially acid-base behavior, thin-layer and liquid chromatography. *Mailing Add:* 119 W High St Somerville NJ 08876-2108

BANIGAN, THOMAS FRANKLIN, JR, b Dover, NJ, Sept 22, 20; m 46; c 1. POLYMER CHEMISTRY, ORGANIC CHEMISTRY. *Educ:* Univ Notre Dame, BS, 42, MS, 43, PhD(org chem), 46. *Prof Exp:* Chemist, Sinclair Refining Co, Ind, 46-48; res chemist, rubber res sta, USDA, 48-53; sr res chemist, Tide Water Oil Co, 53-57; chemist, Arthur D Little, Inc, 57-61; vpres & dir res, Ben Holt Co, 62-65; tech dir, Pilot Chem Co, 65-68; mgr chem res, Avery Prod Corp, 68-70; tech dir, Pilot Chem Co, 70-73; assoc, Chem Res & Develop Co, 74-76; VPRES, ANVER BIOSCI DESIGN, 77- *Mem:* Am Chem Soc; Consult Chemists Asn. *Res:* Natural resources; detoxification of plant materials, adhesive and detergent chemicals; protective coatings. *Mailing Add:* 403 Bougainvillea Lane Glendora CA 91740-2667

BANIK, NARENDRA LAL, b Ganganagar, India, Jan 2, 38; m 68; c 2. NEUROCHEMISTRY. *Educ:* Univ Calcutta, BSc, 59; Univ London, MSc, 66, PhD(biochem), 70. *Prof Exp:* Res asst biochem, Inst Psychiat, London, 61-65; res asst, Charing Cross Hosp Med Sch, London, 65-71; lectr neurochem, Inst Neurol, London, 71-74; res assoc neurol, Sch Med, Stanford Univ, 74-76; res assoc, 76-77, asst prof neurol & biochem, 77-80, ASSOC PROF NEUROL, MED UNIV SC, 80- *Mem:* Biochem Soc London; Soc Neurosci; AAAS; Int Soc Neurochem; Am Soc Neurochem; Am Soc Biol Chem. *Res:* Brain development process of myelination and dissolution of the myelin membrane in demyelinating diseases; metabolism and function of membranes and isolated myelin-forming cells. *Mailing Add:* Dept Neurol Med Univ SC 171 Ashely Ave Charleston SC 29425

BANISTER, ERIC WILTON, b Barrow-in-Furness, Lancashire, Eng, May 18, 32; Can citizen; m 60; c 3. ERGONOMICS, SPORTS MEDICINE. *Educ:* Manchester Univ, Eng, BSc, 53; Loughborough Univ, Eng, DPE, 54; Univ BC, Can, MPE, 62; Univ Ill, PhD(phys ed & physiol), 64. *Prof Exp:* Asst prof physiol exercise, Dept Phys Ed, Univ BC, 64-67; vis prof res, Dept Naval & Aviation Med, Karolinska Inst, Stockholm, 69-70, Dept Physiol, Sch Med, Univ Hawaii, 75-76 & Royal Postgrad Med Sch, Univ London, Eng, 76; asst prof physiol exercise, Phys Develop Studies, 67-69, assoc prof physiol exercise, Kinesiology, 69-72, chmn physiol exercise, 1st Dept Kinesiology NAm, 70-82, dir physiol exercise, Sch Kinesiology, 87-88, PROF PHYSIOL EXERCISE, KINESIOLOGY, SIMON FRASER UNIV, 72- *Concurrent Pos:* Prin investr, Submissions to Can Natural Sci & Eng Res Coun Funded, 67-; exec, Can Asn Sport Sci, 78-81, pres, 79-80; coun mem, Can Sports Med Coun, 79-82, vpres, 80-81; ed, Can J Appl Sport Sci, Can Asn Sport Sci, 80-83; ed-in-chief, J Sports Med, Training & Rehab, Gordon, Breach Harwood Sci Publ, NY, USA, 89-; guest ed, Int J Sports Med, Ger, Vol II, Suppl 2, 90- *Mem:* Can Asn Sport Sci (pres, 79-80); fel Am Asn Sports Med; AAAS; Undersea Med Soc. *Res:* Researcher and writer in contemporary health issues, mathematically modeling human physical performance and topics which include occupational health and safety, ergonomics and toxicology of the silviculture industry and optimization of the athletic potential. *Mailing Add:* Sch Kinesiology Fac Appl Sci Simon Fraser Univ Burnaby BC V5A 1S6 Can

BANISTER, JOHN ROBERT, physics, for more information see previous edition

BANITT, ELDEN HARRIS, b Red Wing, Minn, Apr 7, 37; m 59; c 2. ORGANIC CHEMISTRY. *Educ:* St Olaf Col, BA, 59; Univ Wis, PhD(org chem), 64. *Prof Exp:* Fel, Univ Calif, Berkeley, 64-65; res specialist, 65-75, sr res specialist, Riker Labs, 75-89, DIV SCIENTIST, 3M PHARMACEUT, 89- *Mem:* Am Chem Soc; Int Soc Heterocylic Chem; NY Acad Sci. *Res:* Synthetic medicinal chemistry. *Mailing Add:* 1181 Woodhill Dr Woodbury MN 55125

BANK, ARTHUR, b New York, NY, Apr 20, 35; m; c 2. HEMATOLOGY. *Educ:* Harvard Univ, MD, 60. *Prof Exp:* PROF MED & GENETICS, COLUMBIA UNIV, 70- *Concurrent Pos:* Dir hemat & clin path, Presby Hosp. *Mem:* Am Soc Hemat; Am Assn Physicians; Am Soc Biol Chemists. *Res:* Globin gene regulation; gene transfer and gene therapy. *Mailing Add:* 701 W 168th St Columbia Univ New York NY 10032

BANK, HARVEY L, b Brooklyn, NY, Feb 13, 43; m 65; c 3. CRYOBIOLOGY, CELL BIOLOGY. *Educ:* Hunter Col, City Univ New York, BA, 65; Oak Ridge Grad Sch Biomed Sci, Univ Tenn, PhD(biophysics), 72. *Prof Exp:* Technician marine biol, Osborn Labs Marine Biol, 65-67; sr technician develop biol, Rockefeller Univ, 67-68; fel, Anat Dept, Duke Univ, 72-73; assoc, 73-74, from asst prof to assoc prof path, Med Univ SC, 74-90; EXEC DIR, BEACON LIGHT CTR, 90- *Concurrent Pos:* Guest investr, Nat Inst Environ Health Sci, 72-73; Smith, Kline & French fel, 73-75; prin investr, Anal Scanning Electron Micros Facil, Med Univ SC, 74-81, crybiol div, 74-; consult, Environ Protection Agency, 79-81; ed, Int J Risk Assessment, 82-84; dir res Cryolife Inc, 85-87. *Mem:* Soc Cryobiol; Sigma Xi; Am Soc Cell Biol; Biophys Soc; Electron Micros Soc Am; AAAS; Tissue Culture Asn. *Res:* Methods to cryogenically preserve embryos, granulocytes, veins, ligaments and islets of langerhans; freeze-fracture and cytochemistry of membrane structure and function; reaction of women to IUDs. *Mailing Add:* Dept Path & Lab Med Univ SC Charleston SC 29425

BANK, NORMAN, b New York, NY, Oct 19, 25; m 54; c 2. NEPHROLOGY. *Educ:* NY Univ, AB, 49; Columbia Univ, MD, 53. *Prof Exp:* From instr to assoc prof med, Sch Med, NY Univ, 60-71; assoc prof, 71-73, PROF MED, ALBERT EINSTEIN SCH MED, 73- *Concurrent Pos:* Life Ins Med Res Found res fel, New Eng Med Ctr, Tufts Univ, 57-59; assoc attend physician, NY Univ Med Ctr, 68-71, attend physician, 71-; chief nephrology, Montefiore Hosp, 71- *Mem:* Am Soc Clin Invest; Asn Am Physicians; Am Physiol Soc; Am Fedn Clin Res; Am Soc Nephrology. *Res:* Regulation of acid secretion by kidney; renal concentrating defects in hypercalcemia; renal function in pyelonephritis and mercury poisoning; regulation of sodium transport; renal function in diabetes. *Mailing Add:* Montefiore Hosp Dept Nephrol 111 E 210th St Bronx NY 10467

BANK, SHELTON, b New York, NY, Jan 28, 32; m 57; c 2. ORGANIC & ENVIRONMENTAL CHEMISTRY. *Educ:* Brooklyn Col, BS, 54; Purdue Univ, PhD(phys chem), 60. *Prof Exp:* Res fel, Harvard Univ, 60-61; res chemist, Esso Res & Eng Co, 61-64, sr res chemist, 64-66; assoc prof chem, 66-72, PROF CHEM, STATE UNIV NY, ALBANY, 72-, PROF ENVIRON HEALTH & TOXICOL, GRAD SCH PUB HEALTH SCI, 85- *Concurrent Pos:* Consult, Col Chem Consults, Am Chem Soc, 72-, Tour Speaker, 84- *Mem:* Am Chem Soc; Chem Soc Brit. *Res:* Clay catalysed organic reactions; solid state NMR; chemistry of environmental pollutants; nuclear magnetic resonance method in environmental chemistry; multinuclear NMR spectroscopy. *Mailing Add:* Dept Chem State Univ NY Albany NY 12222

BANK, STEVEN BARRY, b New York, NY, Mar 14, 39; m 72; c 1. MATHEMATICS. *Educ:* Columbia Univ, AB, 59, AM, 60, PhD(math), 64. *Prof Exp:* PROF MATH, UNIV ILL, URBANA-CHAMPAIGN, 64- *Concurrent Pos:* Prin investr, Nat Sci Found grant, 76-91. *Mem:* Am Math Soc. *Res:* Pure mathematics; ordinary differential equations and complex analysis. *Mailing Add:* Dept Math Univ Ill Urbana IL 61801

BANKER, GARY A, b Seattle, Wash, July 1, 46. CELLULAR NEUROBIOLOGY, DEVELOPMENTAL NEUROBIOLOGY. *Educ:* Univ Wash, BS, 68; Univ Calif, Irvine, PhD(neurobiol), 73. *Prof Exp:* Prof neurosci, Albany Med Col, 76-89; PROF NEUROSCI, UNIV VA, 89- *Res:* Neurosciences. *Mailing Add:* Dept Neurosci Univ Va Med Ctr MR-4 Box 5148 Charlottesville VA 22908

BANKER, GILBERT STEPHEN, b Tuxedo Park, NY, Sept 12, 31; m 56; c 4. PHARMACY. *Educ:* Albany Col Pharm, Union Univ (NY), BS, 53; Purdue Univ, MS, 55, PhD(pharm), 57. *Prof Exp:* From asst prof to assoc prof pharm, 57-64, PROF INDUST PHARM, PURDUE UNIV, WEST LAFAYETTE, 64-, HEAD DEPT INDUST & PHYS PHARM, 66- *Concurrent Pos:* Consult, G D Searle, 76-, Richardson Merrell, 74- & FMC Corp, 79- *Honors & Awards:* Lederle Fac Res Award, 61; Am Pharm Asn Indust Pharm Award, 71. *Mem:* Fel AAAS; Am Pharmaceut Asn; Am Chem Soc; NY Acad Sci; fel Acad Pharmaceut Sci (vpres, 70-71). *Res:* Quantitative evaluations of pharmaceutical unit operations; applications of radioactive tracer techniques to the evaluation of pharmaceutical products; emulsion and suspension rheology; applications of synthetic polymers to pharmaceutical product development and controlled drug and chemical agent release. *Mailing Add:* 308 Harvard St SE Minneapolis MN 55455

BANKERT, RALPH ALLEN, b York, Pa, Jan 19, 18; m 46; c 2. ORGANIC CHEMISTRY. *Educ:* Gettysburg Col, BA, 40; Pa State Col, MS, 41, PhD(org chem), 44. *Prof Exp:* Res chemist, Res Ctr, B F Goodrich Co, 43-50; RES CHEMIST, RES CTR, HERCULES INC, 50- *Mem:* Am Chem Soc. *Res:* Antioxidants; preparation of aliphatic ketimines; reactions of N-monochloramines; terpenes; amine chemistry; peroxides; polymers. *Mailing Add:* 28 The Strand New Castle DE 19720-4826

BANKERT, RICHARD BURTON, b St Louis, Mo, Apr 22, 40; m 70; c 3. VETERINARY MEDICINE. *Educ:* Gettysburg Col, BA, 62; Univ Pa, VMD, 68, PhD(immunol), 73. *Prof Exp:* CANCER RES SCIENTIST IMMUNOL, ROSWELL PARK MEM INST, 73-, ASSOC CHIEF, MOLECULAR IMMUNOL, 83- *Concurrent Pos:* Res fel, Sch Vet Med, Univ Pa, 68-70 & 70-73; res prof, Dept Microbiol/Immunol & Molecular Biol, State Univ NY, Buffalo; mem, Immunobiology Study Sect, NIH, 84-88, NIH Reviewers Reserve, 88-92. *Mem:* Am Asn Immunologists; AAAS; Sigma Xi. *Res:* Membrane events associated with lymphocyte function; molecular genetics of the expression and regulation of the B-cell repertoire; human tumor cell biology. *Mailing Add:* Dept Molecular Immunol Roswell Park Mem Inst 666 Elm St Buffalo NY 14263

BANKO, WINSTON EDGAR, wildlife ecology, for more information see previous edition

BANKOFF, S(EYMOUR) GEORGE, b New York, NY, Oct 7, 21. CHEMICAL ENGINEERING. *Educ:* Columbia Univ, BS, 40, MS, 41; Purdue Univ, PhD(chem eng), 52. *Prof Exp:* Chem engr, Sinclair Refining Co, East Chicago, 41-42; chem engr, E I du Pont de Nemours & Co, Ala, 42-43; res assoc, Manhattan Project, Univ Chicago, 43-44; chem engr, Hanford Eng Works, Washington, 44-45, NJ, 45-48; from asst prof to prof & head dept, Rose Polytech Inst, 48-59; prof chem eng, 59-73, WALTER P MURPHY PROF CHEM, MECH & NUCLEAR ENG, NORTHWESTERN UNIV, 73- *Concurrent Pos:* NSF fel, Calif Inst Technol, 58-59; Guggenheim fel, 66-67; Shell vis prof, Imp Col, Univ London, 67; Fulbright intercountry exchange lectr, Scandinavia & Israel, 67; consult, Los Alamos Nat Lab, Argonne Nat Lab & various co; chmn, Energy Eng Coun, Northwestern Univ, 75-79; vis res scientist, Centre d'Etudes Nucleaires, Comn Atomic Energy, Grenoble, France, 80; fel, Int Ctr Heat & Mass Transfer, Belgrade, Yugoslavia, 83; mem, Organizing Comt, Int Symp Multi-phase Transport, 86 & US Team, Seminar Two-Phase Flow, 88; co-chmn, US Sci Comt, Int Heat Transfer Conf, 87-; chmn, Heat Transfer & Energy Conversion Div, Am Inst Chem Engrs, 87. *Honors & Awards:* Wolfson Lectr, Sch Chem Eng, Technion, Israel, 70; Max Jacob Mem Award, 87. *Mem:* Am Nuclear Soc; fel Am Soc Mech Engrs; fel Am Inst Chem Engrs. *Res:* Two-phase flow and heat transfer; nuclear reactor safety; vapor explosions. *Mailing Add:* Dept Chem Eng Northwestern Univ Evanston IL 60208

BANKOWSKI, RAYMOND ADAM, b Chicago, Ill, Feb 4, 14; m 40; c 2. VETERINARY MEDICINE. *Educ:* Mich State Col, DVM, 38; Univ Calif, MS, 40, PhD(comp path), 46. *Prof Exp:* Asst, Univ Calif, Berkeley, 38-40; assoc, Agr Exp Sta, 40-42, asst prof vet sci & asst vet, 46-52, assoc prof & assoc vet, 52-57, prof vet med & vet, 57-78, EMER PROF, EXP STA, UNIV CALIF, DAVIS, 78- *Concurrent Pos:* Collabr bur animal indust, USDA, Mex, 48-49; pres, Conf Vet Lab Diag, 59-60; mem bd dirs, Am Col Vet Microbiologists, 66-70, pres, 69-70; mem virus study sect, NIH, 68-70; consult animal & plant health inspection serv, USDA, 68-85, consult, Plum Island Animal Dis Res Ctr, 78-81. *Honors & Awards:* Mark Morris Found Award, 61; Am Vet Med Asn Res Award, 71; Vet Med Res Award, 71; K F Meyer Gold Cane Award, 85; E P Pope Diagnosticians Award, 85; Fulbright Award, 57 & 58. *Mem:* Fel Am Acad Microbiol; Am Vet Med Asn; Am Asn Avian Path; Sigma Xi. *Res:* Virus diseases, especially vesicular and myxovirus groups; diagnosis; prevention and control of diseases of animals. *Mailing Add:* VM-EPM Sch Vet Med Univ Calif Davis CA 95616

BANKS, DALLAS O, b Campbell, Nebr, Oct 21, 28; m 52; c 3. APPLIED MATHEMATICS. *Educ:* Ore State Univ, BS, 50, MS, 52; Carnegie Mellon Univ, PhD(math), 59. *Prof Exp:* Math asst, Gen Elec Co, Wash, 52-54; from asst prof to assoc prof math, 59-70, PROF MATH, UNIV CALIF, DAVIS, 70- *Concurrent Pos:* Grant, Air Force Off Sci Res, 62-65; consult, Lawrence Livermore Lab, 78- *Mem:* Math Asn Am; Am Math Soc; Soc Indust & Appl Math. *Res:* Differential equations of engineering and physics; mathematical modeling in engineering. *Mailing Add:* Dept Math Univ Calif Davis CA 95616

BANKS, DONALD JACK, b Sentinel, Okla, July 11, 30; m 52; c 2. BOTANY, PHYTOPATHOLOGY. *Educ:* Okla State Univ, BS, 53, MS, 58; Univ Ga, PhD(bot), 63. *Prof Exp:* Instr agron, Auburn Univ, 58-60; asst prof biol, Stephen F Austin State Col, 63-66; RES GENETICIST, PLANT SCI RES DIV, AGR RES SERV, USDA, 66-; PROF AGRON, OKLA STATE UNIV, 77- *Concurrent Pos:* Assoc prof agron, Okla State Univ, 68-77. *Mem:* Bot Soc Am; Am Soc Plant Taxon; Am Soc Agron; Crop Sci Soc Am; Sigma Xi; AAAS; fel Am Peanut Res & Educ Soc; Crop Sci Soc Am; Carribean Food Crop Soc; Soc Econ Bot. *Res:* Plant taxonomy, agrostology, cytology, genetics and breeding; speciation in genus Paspalum; studies in the Paspalum setaceum Michaux complex; breeding, genetics, cytology, speciation and evolution of peanuts. *Mailing Add:* Dept Agron Okla State Univ Stillwater OK 74078-0507

BANKS, EPHRAIM, b Norfolk, Va, Apr 21, 18; wid; c 2. INORGANIC CHEMISTRY. *Educ:* City Col New York, BS, 37; Polytech Inst Brooklyn, PhD(chem), 49. *Prof Exp:* Jr metallurgist, NY Naval Shipyard, 41-46; res fel, Polytech Inst Brooklyn, 46-49, res assoc, 49-50, from instr to assoc prof, 50-58, actg head, 68-71, head dept chem, 71-76, prof, 58-87, EMER PROF, INORG CHEM, POLYTECH UNIV, 87- *Concurrent Pos:* Consult, Westinghouse Elec Corp, 57-63, Gen Motors Corp, 65-70 & Mallinckrodt Chem Works, 67-73; fel, Weizmann Inst, 63-64; NSF fac fel, 71-72. *Mem:* Fel AAAS; Am Chem Soc; Am Phys Soc; Electrochem Soc; fel NY Acad Sci. *Res:* Solid state chemistry and physics; crystal chemistry; luminescence; semiconductors; magnetic materials and structures; Mossbauer spectroscopy; superconducting compounds. *Mailing Add:* Dept Chem Polytech Univ Brooklyn NY 11201

BANKS, GRACE ANN, b Philadelphia, Pa, Apr 1, 42. PHYSICAL CHEMISTRY. *Educ:* Chestnut Hill Col, BS, 72; Univ NC, Chapel Hill, MS, 75, PhD(chem), 78. *Prof Exp:* Teacher physics, chem, Notre Dame High Sch, Easton, Pa, 70-73; teaching asst chem, Univ NC, 73-75, res asst, 75-77; asst prof, St Augustine's Col, 77-79; ASSOC PROF CHEM & PHYS, CHESTNUT HILL COL, 79- *Mem:* Am Chem Soc; Nat Soc Teachers Asn. *Res:* Nuclear magnetic resonance spin lattice relaxation in low temperature organic solids. *Mailing Add:* Chestnut Hill Col Philadelphia PA 19118

BANKS, HARLAN PARKER, b Cambridge, Mass, Sept 1, 13; m 39; c 2. PLANT MORPHOLOGY & ANATOMY, PALEOBOTANY. *Educ:* Dartmouth Col, AB, 34; Cornell Univ, PhD(bot), 40. *Prof Exp:* Instr bot, Dartmouth Col, 34-36; asst, Cornell Univ, 37-39, instr, 39-40; from instr to assoc prof biol, Acadia Univ, 40-47; asst prof bot, Univ Minn, 47-49; assoc prof, 49-50, head dept bot, 52-61, prof bot, 50-77, Liberty Hyde Bailey prof bot, 77-78, EMER PROF BOT & EMER LIBERTY HYDE BAILEY PROF BOT, CORNELL UNIV, 78- *Concurrent Pos:* Fulbright res scholar, Univ Liege, 57-58; Guggenheim fel, 63-64; fel, Clare Hall, Cambridge, 68-; lectr, Am Inst Biol Sci Vis Biologists Prog, 69-70 & 71-73; mem, Paleont Res Inst. *Honors & Awards:* Int Medal, Paleont Asn India, 91. *Mem:* Nat Acad Sci; Bot Soc Am (secy pro-tem, 52-53, treas, 64-67, vpres, 68, pres, 69); Torrey Bot Club; Int Orgn Paleobot (vpres, 64-69, pres, 69-75); foreign mem Geol Soc Belg. *Res:* Paleobotany, origin and early evolution of land plants (Devonian and other early Paleozoic Periods). *Mailing Add:* Plant Biol 228 Plant Sci Bldg Cornell Univ Ithaca NY 14853

BANKS, HAROLD DOUGLAS, b Brooklyn, NY, Apr 28, 43; m 67; c 2. ORGANIC CHEMISTRY. *Educ:* Brooklyn Col, BS, 63, MA, 65; Cornell Univ, PhD(org chem), 70. *Prof Exp:* Assoc prof chem, Univ Bridgeport, 70-81; assoc prof chem, Atlanta Univ, 81-86; RES CHEMIST, US ARMY CHEM RES, DEVELOP & ENG CTR, ABERDEEN PROVING GROUND, 86- *Concurrent Pos:* Am Chem Soc Petrol Res Fund grant, Univ Bridgeport, 70; sabbatical leave, Nat Inst Arthritis, Metabolism, & Digestive Dis, NIH, 79-80. *Mem:* Am Chem Soc. *Res:* Synthesis of bicyclo(2.2.2) octanes, and substituted piperidines. *Mailing Add:* US Army Chem Res Develop & Eng Ctr Aberdeen Proving Ground MD 21010-5423

BANKS, HARVEY OREN, b Chaumont, NY, Mar 29, 10; m 34; c 3. CIVIL ENGINEERING, WATER RESOURCES. *Educ:* Syracuse Univ, BSCE, 30; Stanford Univ, MS, 55; Am Acad Environ Engrs, dipl. *Prof Exp:* From asst state engr to state engr, State of Calif, 53-56, dir water resources, 56-60; from vpres to chmn, Leeds, Hill & Jewett, Inc, Consult Engrs, San Francisco, 61-69; pres, Harvey O Banks, Consult Engr, Inc, Belmont, Calif, 70-76 & water resources div, Camp Dresser & McKee Inc, 77-82; CONSULT CIVIL ENG, 82- *Concurrent Pos:* Chmn bd dirs, Systs Assocs, Inc, 70-76; mem water adv panel to dir, Calif Water Resources; mem bd dirs, Belmont County Water Dist, 72-; pres, water resources div, Camp Dresser & Mckee Inc, 77-82. *Honors & Awards:* George Arents Pioneer Medal, Syracuse Univ, 61; Royce J Tipton Award, Am Soc Chem Engrs, 73, Julian Hinds Award, 76. *Mem:* Nat Acad Eng; fel Am Consult Engrs Coun; hon mem Am Water Works Asn; Water Pollution Control Fedn; hon mem Am Soc Civil Engrs. *Mailing Add:* Three Kittie Lane Belmont CA 94002

BANKS, HENRY H, b Boston, Mass, Mar 9, 21; m 45; c 3. ORTHOPEDIC SURGERY. *Educ:* Harvard Univ, BA, 42; Tufts Univ, MD, 45; Am Bd Orthop Surg, cert, 56. *Prof Exp:* Geogr,Peter Brent Brigham Hosp, 53-70; PROF ORTHOP SURG & CHMN DEPT, SCH MED, TUFTS UNIV, 70-, ASSOC DEAN AFFIL HOSPS, 72- *Concurrent Pos:* Asst clin prof orthop surg, Harvard Med Sch, 65-70; head div orthop, Peter Bent Brigham Hosp, 68-70; surgeon-in-charge pediat orthop surg, New Eng Med Ctr Hosps, 70-71; dir orthop surg, Boston City Hosp, 70-75, chmn surg serv, 61-73; consult orthop surg, numerous hosps. *Mem:* Am Orthop Asn; Am Acad Orthop Surg; Int Soc Orthop Surg & Traumatology; Orthop Res Soc; Am Acad Cerebral Palsy (past pres); Am Bd Orthop Surg(secy-treas, 75-78, pres, 78-79). *Mailing Add:* 1819 N River Rd St Clair MI 48079

BANKS, JERRY, b Birmingham, Ala, Apr 23, 39; m 67; c 1. INDUSTRIAL ENGINEERING, OPERATIONS RESEARCH. *Educ:* Univ Ala, BS, 61, MS, 63; Okla State Univ, PhD(indust eng), 66. *Prof Exp:* Asst prof indust eng, 65-66, ASSOC PROF INDUST & SYSTS ENG, GA INST TECHNOL, 70- *Mem:* Am Inst Indust Engrs; Opers Res Soc Am. *Res:* Inventory theory. *Mailing Add:* Dept Indust Eng Ga Inst Technol Atlanta GA 30332

BANKS, KEITH L, IMMUNOPATHOLOGY. *Educ:* Wash State Univ, PhD(exp path), 72. *Prof Exp:* ASSOC PROF VET IMMUNOL, WASH STATE UNIV, 77- *Res:* Chronic virus infections. *Mailing Add:* Dept Vet Microbiol & Path Wash State Univ Pullman WA 99164

BANKS, NORMAN GUY, b Philadelphia, Pa, Feb 19, 40; m; c 2. GEOLOGY, VOLCANOLOGY. *Educ:* NMex Inst Mining & Technol, BS, 62; Univ Calif, San Diego, MS, 65, PhD(geol), 67. *Prof Exp:* Geologist, US Geol Survey, ore deposits, Ariz & Wash, 67-78, Volcanoes-Hawaii, 78-85, Mt St Helens, 80, Pagan Mariana Islands, 81, Indonesia, 83, Papua New Guinea & Mayon, Philippines, 83-84, GEOLOGIST, US GEOL SURVEY, CENT & SAM, 85- *Mem:* Soc Econ Geologists; Int Asn for Genesis Ore Deposits; AAAS; Geol Soc Am; Int Asn Volcanology & Chem Earths Interior; Am Geophys Union; Northwest Mining Asn. *Res:* Geologic mapping; economic geology; petrology; geochemistry; volcanology, domestic and international; petrology of prophyry copper deposits, Arizona and Washington. *Mailing Add:* Cascades Volcano Observ US Geol Surv 5400 MacArthur Blvd Vancouver WA 98661

BANKS, PETER MORGAN, b San Diego, Calif, May 21, 37; m 60, 83; c 4. SPACE PLASMA SCIENCE, ACTIVE EXPERIMENTS IN SPACE. *Educ:* Stanford Univ, MS, 60; Pa State Univ, PhD(physics), 65. *Prof Exp:* Res asst physics, Pa State Univ, 63-65; res assoc aeronomy, Inst Space Aeronomy, Brussels, Belgium, 65-66; from asst prof to prof electrophys, Univ Calif, San Diego, 66-76; prof & head, Dept Physics, Utah State Univ, 76-81; prof elec eng & dir, Space, Telecommun & Radiosci Lab, Stanford Univ, 81-90; DEAN & PROF, COL ENG, UNIV MICH, ANN ARBOR, 90- *Concurrent Pos:* fel, Pa State Univ, 81; prin investr, Vehicle Charging & Potential Exp, NASA, 78-84, Shuttle Electrodynamic Tether Syst, 85-; mem, Space Sci Adv Comt, 83-87; chmn, Task Force on Sci Uses of Space Sta, NASA, 83-87; mem bd dirs, Ctr Space & Applns Technol, Fairfax, Va, 89-, Indust Technol Inst, Ann Arbor, Mich, 90- & Mich Instrnl Television Network, 91-; mem sci bd, Innovare, SRL, Naples, Italy, 89-; mem adv group, Aspen Global Change Inst, 90- *Honors & Awards:* Appleton Prize, Royal Soc London, 78; Space Sci Award, Am Inst Aeronaut & Astronaut, 81; Distinguished Pub Serv Medal, NASA, 86; Nicolet Lectr, Am Geophys Union, 91. *Mem:* AAAS; fel Am Geophys Union; Int Sci Radio Union; Soc Automotive Engrs. *Res:* Structure of the upper atmosphere; aeronomy; magnetospheric physics. *Mailing Add:* Off Dean Col Eng Univ Mich 2104 EECS Ann Arbor MI 48109

BANKS, PHILIP ALAN, b Sentinel, Okla, Jan 12, 52; div; c 2. WEED SCIENCE, SOIL-HERBICIDE INTERACTIONS. *Educ:* Okla State Univ, BS, 74, MS, 76; Tex A&M Univ, PhD(agron), 78. *Prof Exp:* Instr agron, Tex A&M Univ, 77-78; from asst prof to prof agron, Univ Ga, 79-90; PRES, MARATHON AGR & ENVIRON CONSULT INC, 90- *Concurrent Pos:* Lectr, Fed Rural Univ Pernambues, Brazil, 87. *Honors & Awards:* Outstanding Young Scientist Award, Weed Sci Soc Am, 88. *Mem:* Weed Sci Soc Am; Am Soc Agron. *Res:* Weed biology, herbical persistence in soils; evaluation of general weed control practices in agronomic crops. *Mailing Add:* Marathon Agr & Environ Consult Inc 3001-A Majestic Ridge Las Cruces NM 88001

BANKS, PHILIP OREN, b Palo Alto, Calif, Jan 31, 37. GEOLOGY, GEOCHEMISTRY. *Educ:* Mass Inst Technol, BS, 58; Calif Inst Technol, PhD(geochem), 63. *Prof Exp:* Asst prof, 63-80, ASSOC PROF GEOL, CASE WESTERN RESERVE UNIV, 68- *Mem:* AAAS; Geol Soc Am; Mineral Soc Am; Am Geophys Union; Sigma Xi. *Res:* Geochronology; isotope geochemistry. *Mailing Add:* Dept Geol Case Western Reserve Univ Cleveland OH 44106

BANKS, RICHARD C, b Green Bay, Wis, Aug 19, 40; m 65; c 2. ORGANIC CHEMISTRY. *Educ:* Col Idaho, BS, 63; Ore State Univ, PhD(org chem), 68. *Prof Exp:* Asst prof org chem, 68-75, assoc prof, 75-77, PROF CHEM, BOISE STATE UNIV, 78-, CHMN DEPT, 88- *Mem:* Am Chem Soc; Sigma Xi. *Res:* Isolations and identification problem on terpenes and terpene derivatives; synthesis of stable isotope labeled metabolites. *Mailing Add:* 1910 College Blvd Boise ID 83725

BANKS, RICHARD CHARLES, b Steubenville, Ohio, Apr 19, 31; m 67; c 2. ORNITHOLOGY, MAMMALOGY. *Educ:* Ohio State Univ, BS, 53; Univ Calif, Berkeley, MS, 58, PhD(zool), 61. *Prof Exp:* NSF grant & res biologist, Birds & Mammals, Calif Acad Sci, 61-62; cur & ed, San Diego Nat Hist Mus, 62-66; chief bird sect, Bird & Mammal Labs, 66-71, dir, 71-73, staff specialist, Div Wildlife Res, 73-78, ZOOLOGIST, BIOL SURV SECT, NERC, US FISH & WILDLIFE SERV, 78- *Concurrent Pos:* Res assoc, Dept Vert Zool, Smithsonian Inst, 67- *Mem:* Am Ornithologists Union (secy, 69-73, vpres, 86-87); Am Soc Mammal; Cooper Ornith Soc; Wilson Ornith Soc (2nd vpres, 87-89, 1st vpres, 89-). *Res:* Systematics, distribution, molt and hybridization of American birds; endangered species; biology of introduced species. *Mailing Add:* US Fish & Wildlife Serv Nat Mus Natural Hist Washington DC 20560

BANKS, ROBERT B(LACKBURN), b Wichita, Kans, Oct 12, 22; m 60; c 2. FLUID MECHANICS. *Educ:* Northwestern Univ, BSCE, 47, MS, 49; Univ Calif, PhD(fluid mech), 51; Univ London, DIC, 52. *Hon Degrees:* DTech, Asian Inst Technol, 84. *Prof Exp:* Res engr, Univ Calif, 49-51; develop engr fluid mech res, Infilco, Inc, 52-54; lectr civil eng, Northwestern Univ, 54-55, assoc prof civil eng & sci eng, 55-59, chmn sci-eng comt, 55-57 & dept civil eng 56-58, prof eng sci, 59-61, 56-58, asst dean res & grad studies, 60-61; prof eng & dir res, Asian Inst Technol, Bangkok, 61-63; prof eng sci & dean eng, Univ Ill, Chicago, 63-67; prog adv sci & technol, Ford Found, Mexico, 67-76; prof eng, Nat Autonomous Univ Mex, 67-76; pres, Asian Inst Technol, Bangkok, 76-83; exec officer, Int Serv for Nat Agr Res, The Hague, 83-85; CONSULT, 85- *Mem:* Am Soc Eng Educ; AAAS; Am Geophys Union. *Res:* Fluid mechanics; biomathematics; environmental engineering; scientific and engineering education; international education. *Mailing Add:* 2420 Torrey Pines Rd Apt A-101 La Jolla CA 92037

BANKS, ROBERT L(OUIS), chemical engineering; deceased, see previous edition for last biography

BANKS, ROBERT O, RENAL HEMODYNAMICS, ATRIAL NATRIURETIC FACTOR. *Educ:* State Univ NY, Buffalo, PhD(physiol), 68. *Prof Exp:* ASSOC PROF PHYSIOL, COL MED, UNIV CINCINNATI, 70- *Mailing Add:* Dept Physiol & Biophys Univ Cincinnati Col Med 231 Bethesda Ave Cincinnati OH 45267-0576

BANKS, ROBERT R(AE), b New York, NY, Oct 19, 25; m 61; c 2. CHEMICAL ENGINEERING, METALLURGY. *Educ:* Columbia Univ, BS, 44, MS, 46, PhD(chem eng), 50. *Prof Exp:* Assoc chem engr, Armour Res Found, 50-51, res metallurgist, 51-53; sr engr, Air Reduction Co, Inc, 53-54, sect head, 54-56, supvr, 56-59, proj engr, Air Reduction Sales Co, 59-60, engr, 60-63, supvr, Air Reduction Co, Inc, 63-67, mgr develop, Pittsburgh Metall Co Div, 67-69, mem staff, Airco Alloys, NY, 69-76; mgr res & develop, Wheelabrator-Frye Inc, 76-86; CONSULT ENG, 86- *Mem:* Am Chem Soc; Am Inst Chem Engrs; Iron & Steel Soc; Am Soc Metals; Air & Waste Mgt Asn; Am Filtration Soc. *Res:* Ferroalloys; piloting of smelting processes; electron beam equipment; cryogenic engineering; brazing, high vacuum equipment; inorganic chemical processes; air pollution control equipment. *Mailing Add:* 1817 Tilton Dr Pittsburgh PA 15241

BANKS, THOMAS, b New York, NY, Apr 19, 49; m 69; c 2. STRING THEORY, QUANTUM FIELD THEORY. *Educ:* Reed Col, BA, 69; Mass Inst Technol, PhD(physics). *Prof Exp:* Fel physics, Tel Aviv Univ, Israel, 73-75, lectr, 75-77, sr lectr, 77-80, from assoc prof to prof, 80-85; prof physics, Univ Calif, Santa Cruz, 86-89; PROF PHYSICS, RUTGERS UNIV, 89- *Concurrent Pos:* Mem, Inst Advan Study, 76-78, Einstein fel, 83-84; vis prof, Stanford Univ, 82-86. *Mem:* Am Phys Soc. *Res:* Theory of elementary particles; quantum field theory; renormalization group; string theory; quantum cosmology. *Mailing Add:* Rutgers Univ Piscataway CA 08855-0849

BANKS, WILLIAM JOSEPH, JR, b Port Chester, NY, Nov 3, 38; m 60; c 2. VETERINARY HISTOLOGY, VETERINARY GROSS ANATOMY. *Educ:* Calif State Polytech Univ, Kellogg/Voorhis, BS, 64; Colo State Univ, MS, 66, PhD(anat), 68, DVM, 80. *Prof Exp:* Res chemist, Aerojet Gen Corp, Calif, 63; asst prof zool, Calif State Col, Fullerton, 67-68; asst prof zool, Calif State Polytech Col Kellogg Voorhis, 68; vet anat, Vet Col, Univ Sask, 68-70, assoc prof, 70; from assoc prof to prof anat, Col Vet Med & Biomed Sci, Colo State Univ, 71-86, prof radiol & radiation & biol, 81-86; assoc dean acad prog, La State Univ, 86-90; ASSOC DEAN ACAD PROGS, DEPT VET MED, TEX A&M UNIV, 90- *Concurrent Pos:* Med Res Coun Can res grant, 69-71. *Mem:* Am Vet Med Asn; Am Asn Anat; Am Asn Vet Anat. *Res:* Chondrogenic and osteogenic processes; fracture repair. *Mailing Add:* Dept Vet Med Tex A&M Univ College Station TX 77843-0000

BANKS, WILLIAM LOUIS, JR, b Paterson, NJ, Mar 25, 36; m 65; c 1. BIOCHEMISTRY, NUTRITION. *Educ:* Rutgers Univ, BS, 58, PhD(physiol, biochem), 63; Bucknell Univ, MS, 61. *Prof Exp:* Clin lab officer, US Air Force Sch Aerospace Med, 63-65; from asst prof to prof biochem, 65-74, co-dir Cancer Ctr, 74-88, PROF BIOCHEM/MOLECULAR BIOPHYS & SURG, MED COL VA, 74-, DIR, NUTRIT SCI CTR, 88- *Concurrent Pos:* Lectr, St Mary's Univ, Tex, 64-65. *Mem:* Am Asn Cancer Res; Am Chem Soc; Am Inst Nutrit; NY Acad Sci. *Res:* Hydrazine drug toxicology; protein nutrition; protein and nucleic acid metabolism; diet, nutrition and cancer. *Mailing Add:* Dept Biochem & Molecular Biophys Med Col Va Richmond VA 23298-0614

BANKS, WILLIAM MICHAEL, b New York, NY, Jan 10, 14; m 43; c 1. ZOOLOGY. *Educ:* Univ Iowa, BA, 42; Ohio State Univ, MA, 49, PhD, 53. *Prof Exp:* From assoc prof to prof biol, Agr & Technol Col NC, 52-56; assoc prof, Grambling Col, 56-58; prof & chmn dept, Clark Col, 58-61; from asst prof to assoc prof, 61-67, PROF ZOOL, HOWARD UNIV, 70- *Mem:* AAAS; Am Micros Soc; Am Soc Zool; Entom Soc Am; Am Inst Biol Soc. *Res:* Taxonomy; life cycles and nutritional requirements of trematodes; invertebrate physiology; carbohydrases; composition of hemolymph; cardiology of cockroaches; characterization and biological effects of bacterial toxins. *Mailing Add:* 4207 Blagdon Ave Washington DC 20011

BANK-SCHLEGEL, SUSAN PAMELA, CELL BIOLOGY, TUMOR BIOLOGY. *Educ:* Northwestern Sch Med, PhD(microbiol), 73. *Prof Exp:* SR STAFF FEL, NAT CANCER INST, 80- *Mailing Add:* Div Lung Dis Airways Dis Br Nat Heart Lung & Blood Inst, NIH Westwood Bldg Rm 6A15 Bethesda MD 20892

BANKSON, DANIEL DUKE, b Moscow, Idaho, June 27, 56; m 82; c 1. CANCER PREVENTION, CLINICAL CHEMISTRY. *Educ:* Univ BC, BSc, 78; Mass Inst Technol, SM, 82, PhD(nutrit), 85. *Prof Exp:* Res asst nutrit, Univ BC, 78 & Mass Inst Technol, 79-82; sr res asst nutrit, Tufts Univ, 82-85, res assoc, 85-86; fel clin chem, NC Mem Hosp, 86-88; res fel endocrin, Univ NC, 88-89; ASSOC CANCER PREV, FRED HUTCHINSON CANCER RES CTR, 89- *Concurrent Pos:* Dipl, Am Bd Clin Chem, 90. *Honors & Awards:* Young Investr Award, Acad Clin La Physicians & Scientists, 87; Travel Award, Am Asn Clin Chem, 87. *Mem:* Am Asn Clin Chem (treas, 90-92); Fedn Am Socs Exp Biol; Am Inst Nutrit; Am Soc Clin Nutrit; Am Asn Cancer Res; Nat Acad Clin Biochem. *Res:* Cancer prevention research: how nutrition may help prevent cancer; role of vitamin A in health and disease. *Mailing Add:* 1124 Columbia St E-601 Seattle WA 98104

BANKSTON, CHARLES A, JR, b Los Angeles, Calif, Feb 4, 35; m 56; c 2. HEAT TRANSFER, FLUID MECHANICS. *Educ:* Univ NMex, BSME, 57, MS, 59, ScD(eng), 65. *Prof Exp:* GROUP LEADER, LOS ALAMOS SCI LAB, UNIV CALIF, 58- *Concurrent Pos:* Adj prof, Univ NMex, 68- *Mem:* Am Soc Mech Engrs; Int Solar Energy Soc. *Res:* Heat transfer and fluid mechanics, especially turbulent boundary layer and shear flows with heat transfer and applications to solar energy for heating and cooling. *Mailing Add:* 14401 Alene Ct Albuquerque NM 87123

BANKSTON, DONALD CARL, analytical chemistry, computer science; deceased, see previous edition for last biography

BANN, ROBERT (FRANCIS), b New Baden, Ill, June 30, 21; m; c 4. ORGANIC CHEMISTRY. *Educ:* Univ Ill, BS, 43; Rutgers Univ, MS, 59. *Prof Exp:* Sect chief chemist, 56-65, chief chemist, 65-68, gen supt dyes mfg, 68-73, mgr dyes & chem intermediates prod, Dyes Dept, Org Chem Div, 73-76, PLANT MGR, AM CYANAMID CO, 76- *Mem:* Am Chem Soc; Am Asn Textile Chemists & Colorists. *Res:* Azo dyes and intermediates; triphenylmethane dyes; dyes for synthetic fibers. *Mailing Add:* 119 Meadow Lane Marietta OH 45750-1345

BANNA, SALIM M, physical chemistry, electron spectroscopy; deceased, see previous edition for last biography

BANNAI, EICHI, b Tokyo, Japan, Feb 07, 46. COMBINATORICS MATHEMATICS. *Educ:* Tokyo Univ, PhD(math), 74. *Prof Exp:* PROF MATH, OHIO STATE UNIV, 74- *Mem:* Am Math Soc; Japanese Math Soc. *Mailing Add:* Dept Math Ohio State Univ Main Campus 100 Math Bldg Columbus OH 43210

BANNAN, ELMER ALEXANDER, b Wilmington, Del, Sept 5, 28; m 58; c 3. MICROBIOLOGY. *Educ:* Univ Mich, BS, 58, MS, 60. *Prof Exp:* Microbiologist, 60-73, mgr dermal safety testing, 73-87, MGR QUAL ASSURANCE, PROCTER & GAMBLE, CO, 88- *Mem:* Am Soc Microbiol; Soc Indust Microbiol; Soc Qual Assurance. *Res:* Bacterial physiology, the ultra structure of bacterial spores; antimicrobial cleansing products; skin degerming; human skin response to cleaning products. *Mailing Add:* Ivorydale Tech Ctr 5299 Spring Grove Ave Cincinnati OH 45217

BANNAN, MARVIN WILLIAM, botany, for more information see previous edition

BANNARD, ROBERT ALEXANDER BROCK, b Edmonton, Alta, Apr 28, 22; m 46; c 2. ORGANIC CHEMISTRY. *Educ:* Queen's Univ, BSc, 45, MSc, 46; McGill Univ, PhD(chem), 49. *Prof Exp:* Jr res officer, Atomic Energy Proj, Nat Res Coun Can, 49-50, asst res officer, Div Appl Chem, 50-51; Sci Serv Officer, Defense Res Estab Ottawa, Defense Res Bd, 52-86; RETIRED. *Mem:* Am Chem Soc; fel Chem Inst Can; Royal Soc Chem; fel Royal Inst Chem. *Res:* Stereochemistry; organophosphorus chemistry; heterocyclics; alicyclic compounds; gas-liquid chromatographic separations and analysis. *Mailing Add:* 152 Robertlee Dr PO Box 219 Carp ON K0A 1L0 Can

BANNER, DAVID LEE, b St Joseph, Mo, Jan 7, 42; m 86; c 2. PHYSICS. *Educ:* Purdue Univ, BS, 64; Univ Ill, MS, 66, PhD(physics), 72. *Prof Exp:* Res physicist mat studies, B-Div, Lawrence Livermore Lab, Univ Calif, 72, group leader, 73-75, proj mgr lab liaison with Air Force, D-Div, 75-77, asst assoc prog leader laser fusion, Y Div, 77-81; mgr, Advan Technol, FMC Corp, San Jose, Calif, 81-82; group leader, Appl Technol, Z Div, Lawrence Livermore Nat Lab, 82-90; HEAD PHYSICS SECT DEPT RES & ISOTOPES, IAEA, VIENNA, AUSTRIA, 90- *Mem:* Am Phys Soc. *Res:* Weak interactions of elementary particles; material properties of materials under high pressure and high rates of strain; inertial confinement fusion using lasers as the driver. *Mailing Add:* Lawrence Livermore Nat Lab PO Box 808 L-366 Livermore CA 94550

BANNERMAN, DOUGLAS GEORGE, b Calgary, Alta, Can, Feb 22, 17; US citizen; m 42; c 4. TEXTILE CHEMISTRY. *Educ:* Beloit Col, BS, 39; Colo Col, MA, 41; Mass Inst Technol, PhD(chem), 45. *Prof Exp:* Instr chem, Colo Col, 39-41 & Mass Inst Technol, 42-45; asst prof, Colo Col, 45-46; res chemist, E I du Pont De Nemours & Co, 45-56, merchandising mgr, 56-60, mkt develop mgr, 60-78; adv toxic substances, US Environ Protection Agency, 78-82; ENVIRON CONSULT, NEMA, 83- *Mem:* Am Chem Soc. *Res:* Organic research in heterocyclic chemistry; high polymer chemistry; synthetic fiber research; textile development; synthetic fiber paper; toxic substances regulation. *Mailing Add:* 2101 L St NW Nema Suite 300 Washington DC 20037-1526

BANNERMAN, JAMES KNOX, b Houston, Tex, Sept 11, 46. PHYSICAL CHEMISTRY. *Educ:* Houston Baptist Univ, BA, 68; Baylor Univ, PhD(chem), 73. *Prof Exp:* PRES, BANNERMAN & ASSOC, SPECIALTY CHEM & MINERALS, 84- *Mem:* Am Chem Soc; Am Inst Chemists; Mineral Soc Am; Clay Minerals Soc. *Res:* Clay mineral syntheses; clay surface modification; water-soluble polymer chemistry. *Mailing Add:* 12905 Trail Hollow Dr Houston TX 77079-3707

BANNING, JON WILLROTH, b Akron, Ohio, Oct 17, 47; m 71; c 2. REGULATORY AFFAIRS. *Educ:* Otterbein Col, BS, 69; Miami Univ, MS, 74; Univ Cincinnati, PhD(pharmacol), 77. *Prof Exp:* Res assoc, Col Pharm, Ohio State Univ, 78-82; asst prof pharm, Col Pharm & Allied Health Prof, Wayne State Univ, 82-88; REGULATORY AFFAIRS SPECIALIST, BEECHAM LABS, 88- *Concurrent Pos:* Instr pharmacol, Col Pharm, Ohio State Univ, 82- *Mem:* Sigma Xi; NY Acad Sci; AAAS; Soc Exp Biol & Med; Am Soc Pharmacol & Exp Therapeutics; Soc Toxicol. *Res:* Characterize the autonomic, cardiovascular, central nervous system, and neuromuscular pharmacology of isoquinoline alkaloids isolated form plants; experiments for medicinal chemists to elucidate direct and indirect mechanisms of action for catecholimidazolines, imidates and their stereoisomers on alpha and beta adrenoceptors and histamine receptors in vitro. *Mailing Add:* 501 Fifth St PO Box 398707 Cincinnati OH 45239-8707

BANNING, LLOYD H(AROLD), b St Lawrence, SDak, Feb 21, 09; m 35; c 1. METALLURGY. *Educ:* SDak Sch Mines & Technol, BS, 31. *Prof Exp:* Assayer, Canyon Corp, 35-38; Gilt Edge Mining Co, 38-39; chemist, Bald Mountain Mining Co, 39-42; mine mgr, Va Hardwood Lumber Co, 42; metallurgist, Bur Mines, Md, 42-47; asst plant supt, Titanium Dioxide Plant, Chem & Pigment Co, 47-48; metallurgist, Bur Mines, US Dept Interior, 48-52, proj coordr, Pyometallurgic Lab, 52-65, Scrap Utilization Lab, 65-67, metallurgist, 67-74; RETIRED. *Mem:* Am Inst Mining, Metall & Petrol Engrs; Sci Res Soc Am. *Res:* Recovery of copper, nickel, chromium and titanium from low-grade ores and tin, copper and iron from municipal incinerator residues. *Mailing Add:* 1214 NE Springwood Dr Albany OR 97321

BANNISTER, BRIAN, b Gateshead, Eng, Mar 10, 26; m 57. ORGANIC CHEMISTRY. *Educ:* Oxford Univ, BA, 47, BSc, 47, MA & DPhil(org chem), 50. *Prof Exp:* Nuffield Found res fel, Dyson Perrins Lab, Univ Oxford, 50-52; Alumni Res Found fel, Univ Wis, 52-53, NSF fel, 53-54; proj leader, dept chem, 54-73, SR SCIENTIST, UPJOHN CO, 73- *Mem:* Am Chem Soc; The Chem Soc. *Res:* Degradative and synthetic investigations in field of natural products, with major emphasis on antibiotics. *Mailing Add:* Dept Infectious Dis Res Upjohn Co Henrietta St Kalamazoo MI 49001

BANNISTER, BRYANT, b Phoenix, Ariz, Dec 2, 26; m 51; c 2. DENDROCHRONOLOGY. *Educ:* Yale Univ, BA, 48; Univ Ariz, MA, 53, PhD(anthrop), 60. *Prof Exp:* Lab asst dendrochronology, Univ Ariz, 51-53, res asst, 53-54; cur archaeol collections, 54-59, from instr to prof dendrochronology, 59-89, dir, Lab Tree Ring Res, 64-82, asst dean, Col Earth Sci, 71-72, actg dean, 72, assoc dean, 72-82, EMER DIR, LAB TREE RING RES, UNIV ARIZ, 82-, EMER PROF DENDROCHRONOLOGY, 89- *Concurrent Pos:* Ed, Tree-Ring Bull, 58-69; collabr, Nat Park Serv, 60-; res assoc, Mus Northern Ariz, 60-77. *Mem:* Fel AAAS; fel Am Anthrop Asn; Soc Am Archaeol; Tree-Ring Soc. *Res:* Derivation and application of archaeological tree-ring dates. *Mailing Add:* Lab Tree-Ring Res Univ Ariz Tucson AZ 85721

BANNISTER, PETER ROBERT, b Guilford, Conn, Aug 26, 38; m 58; c 4. PHYSICS, MATHEMATICS. *Educ:* Univ Conn, BA, 61, MS, 65. *Prof Exp:* RES PHYSICIST ELECTROMAGNETICS, NAVAL UNDERWATER SYSTS CTR, NEW LONDON LAB, 61- *Mem:* Inst Elec & Electronics Engrs; Int Sci Radio Union. *Res:* Extremely low frequency electromagnetic wave propagation and radiation from various antenna configurations; determining new methods for measuring the electrical properties of the earth. *Mailing Add:* Code 3411 Naval Underwater Systs Ctr Ft Trumbull New London CT 06320

BANNISTER, ROBERT GRIMSHAW, b Terre Haute, Ind, Oct 27, 25; m 56; c 4. TEXTILE CHEMISTRY, TEXTILE EQUIPMENT. *Educ:* Rose Polytech Inst, BS, 46; Univ Ill, MS, 48, PhD(org chem), 51. *Prof Exp:* Chemist anal chem, Com Solvents Corp, 47-48; res chemist textile chem, 51-53, patent chemist, 53-63, sr patent specialist, 63-79, PATENT ASSOC TEXTILE CHEM, E I DU PONT DE NEMOURS & CO, INC, 79- *Mem:* Am Chem Soc. *Res:* Fiber forming polymers; fiber structure; industrial fibers; textile machinery; chemical and textile patents. *Mailing Add:* 2103 Largo Rd Wilmington DE 19803-2307

BANNISTER, THOMAS TURPIN, b Orange, NJ, Apr 20, 30; m 53; c 2. BIOLOGY. *Educ:* Duke Univ, BS, 51; Univ Ill, MS, 53, PhD(biophys), 58. *Prof Exp:* Res asst, Duke Univ, 48-51 & Univ Ill, 51-53, 55-58; instr biol, 58-60, from asst prof to assoc prof, 61-69, chmn dept biol, 69-75, PROF RADIATION BIOL & BIOPHYS, UNIV ROCHESTER, 69- *Concurrent Pos:* NIH res fel, Nat Cent Sci Res, France, 62-63. *Mem:* Am Soc Plant Physiol; Biophys Soc. *Res:* Photosynthesis; photochemistry of chlorophyll and energy transfer; plant and cell physiology; biophysics; ecology of algae. *Mailing Add:* Dept Biol Univ Rochester Rochester NY 14627

BANNISTER, WILLIAM WARREN, b Terre Haute, Ind, Feb 24, 29; m 52; c 3. PHYSICAL ORGANIC CHEMISTRY & COMBUSTION CHEMISTRY, FIRE FIGHTING TECHNOLOGIES. *Educ:* Purdue Univ, BS, 53, PhD(org chem), 61. *Prof Exp:* Chemist, Miami Valley Labs, Procter & Gamble Co, Ohio, 60-65; asst prof org chem, Tenn Tech Univ, 65-66; asst prof, Univ Cincinnati, 66-67; PROF CHEM, UNIV LOWELL, 67- *Concurrent Pos:* Res grants, Sigma Xi, 66, NSF, 67 & 68, US Navy, 75, 76 & 79-80, Environ Protection Agency, 77 & 78, USAF/AFESC, 85- *Mem:* Am Chem Soc; Sigma Xi. *Res:* Structure-property correlations of organic compounds; acidities of organic compounds; organic reaction mechanisms; environmental chemistry with regard to inland, coastal and ocean waterways; fire initiation mechanisms. *Mailing Add:* Dept Chem Univ Lowell Lowell MA 01854

BANNON, JAMES ANDREW, b Philadelphia, Pa, June 17, 53; m 76; c 2. CLINICAL RESEARCH, POST-MARKETING SURVEILLANCE. *Educ:* Philadelphia Col Pharm & Sci, BSc, 76, PharmD, 81. *Prof Exp:* Clin instr, Philadelphia Col Pharm & Sci, 80-83; clin res assoc, 81-82, asst dir clin res, 82-86, VPRES CLIN RES, PHILADELPHIA ASN CLIN TRIALS, 86-; CLIN ASST PROF, PHILADELPHIA COL PHARM & SCI, 83- *Concurrent Pos:* Clin pharmacist, Temple Univ Hosp, 76-79, clin instr, Temple Univ Sch Pharm, 76-79. *Mem:* Am Soc Clin Pharmacol & Therapeut; Am Soc Hosp Pharmacists; Am Heart Asn; Drug Info Asn. *Res:* The planning, conduct and analysis of large-scale post-marketing drug surveillance studies (up to 75,000 patients) to treat patients not well studied in the phases of clinical drug development (e.g. elderly). *Mailing Add:* 150 Radnor-Chester Rd Suite D 200 St David's PA 19087

BANNON, ROBERT EDWARD, b Schenectady, NY, Feb 27, 10; m 39; c 1. OPHTHALMOLOGY, OPTOMETRY. *Educ:* Columbia Univ, BS, 34; Dartmouth Med Sch, OD, 40. *Hon Degrees:* DOS, Chicago Col Optom, 50 & Mass Col Optom, 57. *Prof Exp:* Fel med sch, Dartmouth Col, 35-36, clin fel, 36-40, from instr to asst prof physiol optics, 40-47; assoc optom, Columbia Univ, 47-49, asst prof, 49-50; res optometrist bur visual sci, 50-58, CONSULT, REICHERT SCI INSTRUMENTS, 58- *Concurrent Pos:* Res ed, Am J Optom, 47-51; vis lectr, Mass Col Optom, 57-; assoc res prof ophthal, Sch Med, State Univ NY Buffalo, 67-73. *Honors & Awards:* Am Acad Ophthal Award, 66. *Mem:* AAAS; assoc Optical Soc Am; fel Am Acad Optom (vpres, 50-52); Asn Res in Vision & Ophthal; Nat Soc Prev Blindness; hon mem Am Optom Asn; Am Acad Optom. *Res:* Refraction; aniseikonia; physiologic optics. *Mailing Add:* Leica Sci Instrument Co Box 123 Buffalo NY 14240

BANNON, ROBERT PATRICK, b Joliet, Ill, May 3, 27; m 51; c 5. PHYSICAL SEPARATIONS, ENERGY CONSERVATION. *Educ:* Univ Ill, BS, 50; Univ Mich, MS, 51. *Prof Exp:* Engr, Process Design, 51-67, group leader, Refinery Technol, Martinez, Calif, 67-70, SR ENGR, STAFF ENGR & SR STAFF ENGR, ENG ADV, SHELL OIL CO HEAD OFF, 70- *Concurrent Pos:* Shell Oil Co rep to Fractionation Res, Inc, 85- *Mem:* Am Inst Chem Engrs. *Res:* Physical separations, primarily distillation, and energy conservation in refinery and chemical processes; 19 patents. *Mailing Add:* 810 Thornwick Dr Houston TX 77079

BANOVITZ, JAY BERNARD, immunochemistry, for more information see previous edition

BANSCHBACH, MARTIN WAYNE, b Glen Cove, NY, July 5, 46; m 68; c 2. BIOCHEMISTRY, LIPID BIOCHEMISTRY. *Educ:* Susquehanna Univ, BA, 68; Va Polytech Inst & State Univ, MS, 71, PhD(biochem), 72. *Prof Exp:* NIH trainee, Univ Wis-Madison, 72-74, NIH fel, 74-75; asst prof biochem, Sch Med, La Univ, Shreveport, 75-80; assoc mem, Sch Grad Studies, La Univ

Med Ctr, 75-80; ASSOC PROF BIOCHEM, OKLA COL OSTEOP MED, TULSA, 80-, DIR, MOLECULAR MED RES CTR, 81- *Concurrent Pos:* Special lectr biochem, Centenary Col, La, 77-80. *Mem:* AAAS; Am Chem Soc; Soc Exp Biol & Med; NY Acad Sci; Sigma Xi. *Res:* Lipid metabolism; role of membrane lipids in regulating calcium transport; role of prostaglandins in the regulation of cellular activity and the induction of human disease; calcium metabolism in patients with cystic fibrosis. *Mailing Add:* Okla Col Osteop 1111 W 17th St Tulsa OK 74107

BANSE, KARL, b Koenigsberg, Ger, Feb 20, 29. BIOLOGICAL OCEANOGRAPHY, HYDROBIOLOGY. *Educ:* Univ Kiel, PhD(oceanog zool), 55. *Prof Exp:* From asst prof to prof, 60-66, PROF OCEANOG, UNIV WASH, 66- *Concurrent Pos:* Mem, Sci Comt Oceanic Res Working Groups on Oceanog Methods 63-72; mem, comt Oceanog, Nat Acad Sci Nat Res Coun, 64-70. *Mem:* AAAS; Am Soc Limnol & Oceanog; Marine Biol Asn UK; fel Marine Biol Asn India; German Zool Soc; Am Geophys Union. *Res:* Marine plankton production; Arabian Sea; remote sensing. *Mailing Add:* Sch Oceanog WB-10 Univ Wash Seattle WA 98195

BANSIL, ARUN, b New Delhi, India, Sept 16, 48; m; c 3. CONDENSED MATTER PHYSICS. *Educ:* Univ Delhi, BSc, 67; Stony Brook Univ, MA, 68; Harvard Univ, PhD(physics), 74. *Prof Exp:* From asst prof to assoc prof, 76-87, PROF PHYSICS, NORTHEASTERN UNIV, 87- *Concurrent Pos:* Scientific consult, Netherland Energy Res Found, Holland, 87-; Tampere, Univ Technol, Finland, 89- *Mem:* Fel Am Phys Soc. *Mailing Add:* Dept Physics Northeastern Univ Boston MA 02115

BANSIL, RAMA, b Bilaspur, India, Oct 29, 47; US citizen; m 72; c 3. POLYMER PHYSICS, STATISTICAL MECHANICS. *Educ:* Delhi Univ, India, BSc Hons, 67, MSc, 69; Univ Rochester, PhD(physics), 75. *Prof Exp:* Res assoc, Harvard Univ, 74-75 & Mass Inst Technol, 75-76; res asst prof, 76-77, asst prof, 77-84, ASSOC PROF PHYSICS, BOSTON UNIV, 84- *Concurrent Pos:* Asst prof, Dept Physiol, Boston Univ Sch Med, 76- *Mem:* Am Phys Soc; Am Chem Soc. *Res:* Physics of gels and polymers using light scattering techniques to study diffusion and phase separation kinetics related phenomena; computer simulation of gels; bio-macromolecule structure, phase transitions in membranes and biological gels. *Mailing Add:* Physics Dept Boston Univ Boston MA 02215

BANTA, JAMES E, b Tucumcari, NMex, July 1, 27; m 50; c 2. EPIDEMIOLOGY. *Educ:* Marquette Univ, MD, 50; Johns Hopkins Univ, MPH, 54; Am Bd Prev Med, dipl, 60. *Prof Exp:* Intern, Hosp Good Samaritan, US Navy, Los Angeles, Calif, 50-51, epidemiologist, US Naval Med Sch, 51-52 & Streptococcal Dis Res Unit, Naval Training Ctr, Md, 52-53, prev med officer, Third Marine Div, Far East, 54-55, sr investr virol sect, Naval Res Inst, Md, 55-57, chief virol sect, US Naval Med Res Unit, Egypt, 57-59, chief mil med, Naval Dispensary, Va, 59-60; chief coronary artery dis unit, USPHS, 60-62, dir ecol field sta, Univ Mo, Columbia, 62-63, dir & dep dir, Med Prog Div, Peace Corps, 63-65, head spec int progs, Off Int Res, NIH, 65-68; assoc prof, Sch Med, Georgetown Univ, 68-69, clin assoc prof, 69-70; prof pub health, Univ Hawaii, Manoa, 70-73; dep dir, Off Health, US AID, 73-75; PROF EPIDEMIOL & DEAN, SCH PUB HEALTH & TROP MED, TULANE UNIV, 75- *Concurrent Pos:* Lectr, US Naval Med Sch, 56-57; mem subcomt criteria & methodology, Comt Epidemiol, Am Heart Asn, 61-63; clin asst prof, Univ Mo, 62-68; chronic dis adv, WHO-Pan-Am Health Orgn, 69-70; fel Philadelphia Col Physicians; fel Am Col Epidemiol. *Mem:* Fel AAAS; fel Am Pub Health Asn; fel Am Col Prev Med; fel Am Heart Asn; NY Acad Sci. *Res:* Research epidemiology in infectious and chronic disease; epidemiologic studies ischemic heart disease; hypertension; environmental health; public health operations; international health. *Mailing Add:* 146 G St SW Washington DC 20024

BANTA, MARION CALVIN, b Marshall, Tex, July 19, 34; m 61; c 2. PHYSICAL CHEMISTRY, ANALYTICAL CHEMISTRY. *Educ:* NTex State Univ, BS, 56, MS, 57; Univ Tex, Austin, PhD(chem), 63. *Prof Exp:* Res chemist, Tracor, Inc, Tex, 60-61; sr res scientist, Esso Prod Res Co, 62-66; asst prof chem, ETex Baptist Col, 66-68; asst prof, 68-77, ASSOC PROF CHEM, SAM HOUSTON STATE UNIV, 77- *Mem:* Am Chem Soc. *Res:* Electrochemical kinetics; kinetics of reaction in solution. *Mailing Add:* Dept Chem Sam Houston State Univ Huntsville TX 77341

BANTA, WILLIAM CLAUDE, b Long Beach, Calif, Nov 13, 41; m 65; c 3. MARINE ZOOLOGY. *Educ:* Univ Calif, Berkeley, BA, 63; Univ Southern Calif, PhD(biol), 69. *Prof Exp:* Res fel, Div Invert Paleont, Smithsonian Inst, 69-70; ASSOC PROF BIOL, AM UNIV, 70- *Concurrent Pos:* Assoc ed, Chesapeake Sci, 75-; res prof, Univ Sydney, Australia, 77; vis prof, US Naval Acad, 86; fel, Am Soc Eng Educ, David Taylor Lab, Annapolis, 87, 88. *Mem:* NY Acad Sci; Sigma Xi; Am Micros Soc; Am Soc Zoologists. *Res:* Recent cheilostome and ctenostome Bryozoa; fouling; salt marshes. *Mailing Add:* Dept Biol Amern Univ Washington DC 20016

BANTER, JOHN C, b Marion, Ind, Dec 25, 31; m 55; c 4. PHYSICAL CHEMISTRY. *Educ:* DePauw Univ, BA, 54; Western Reserve Univ, MS, 56, PhD(chem), 57. *Prof Exp:* Sr res chemist, Oak Ridge Nat Lab, Union Carbide Nuclear Corp, 57-66; assoc prof chem, 66-69, PROF CHEM, FLA ATLANTIC UNIV, 69- *Concurrent Pos:* Consult, Metals & Ceramics Div, Oak Ridge Nat Lab, 67-69, Van Prod Corp, Chem Form Div, KMS Industs, Int Bus Mach Corp & City of Boca Raton, Fla, 68-69. *Mem:* Am Chem Soc; Electrochem Soc; Sigma Xi. *Res:* Kinetics and mechanisms of metallic corrosion; anodization of metals; solid state electrochemistry; thin oxide films; spectroscopy applied to the study of thin dielectric films. *Mailing Add:* 6624 Burningwood Dr No 267 Boca Raton FL 33433-4797

BANTLE, JOHN ALBERT, b Detroit, Mich, Jan 20, 46; m 69; c 2. CELL BIOLOGY, DEVELOPMENTAL BIOLOGY. *Educ:* Eastern Mich Univ, BA, 68, MS, 70; Ohio State Univ, PhD(zool), 73. *Prof Exp:* Fel cell biol, Anat Dept, Univ Colo Med Ctr, 73-76; assoc prof zool, 80-87, ASST PROF CELL BIOL, DEPT CELL MOLECULAR & DEVELOP BIOL, OKLA STATE UNIV, 76-, PROF ZOOL, 87- *Concurrent Pos:* NSF grant, 77-81, March Dimes Birth Defects Found, 81-, USDA Coop Agreement, 84- *Honors & Awards:* MASUA Hon Lectr, 87. *Mem:* Am Soc Cell Biol; Am Soc Testing & Mat; Soc Environ Toxicologists & Chemists. *Res:* Regulation and expression of the animal genome; interrelationships of nuclear and cytoplasmic RNA populations; biogenesis of messenger RNA; mutagenesis and teratogenesis; Brucella antigen expression in E coli. *Mailing Add:* Dept Zool Okla State Univ Stillwater OK 74078

BANTTARI, ERNEST E, b Bismarck, NDak, Aug 2, 32. PLANT PATHOLOGY. *Educ:* Univ Minn, BS, 54, MS, 59, PhD(plant path), 62. *Prof Exp:* From asst prof to assoc prof, 63-73, PROF PLANT PATH, UNIV MINN, ST PAUL, 73- *Mem:* Am Phytopath Soc. *Res:* Plant virology. *Mailing Add:* Dept Plant Path Univ Minn St Paul 1991 Upper Buford Cr 495 Borlaugh Hall St Paul MN 55108

BANUCCI, EUGENE GEORGE, b Racine, Wis, June 5, 43; m 66; c 3. ORGANIC POLYMER CHEMISTRY. *Educ:* Beloit Col, BA, 65; Wayne State Univ, PhD(org chem), 70. *Prof Exp:* Staff chemist, Polymer Plastic Div, Gen Elec Corp, 70-75, tech coordr, Phys Chem Lab, Res & Develop Ctr, 75-76, mgr advan develop, Lexan Prod Dept, Plastics Div, 77-78, mgr, Arnox Prod Sect, 78-80; AT AM CYANAMID CO, 80- *Mem:* Am Chem Soc. *Res:* Heterocyclic chemistry; cycloaddition reactions; condensation polymers; high temperature materials; cyanopolymers. *Mailing Add:* 106 Adams Lane New Canaan CT 06840

BANUS, MARIO DOUGLAS, b Geneva, Switz, Nov 11, 21; US citizen; m 43, 68; c 4. MARINE ECOLOGY, ENVIRONMENTAL CHEMISTRY. *Educ:* Mass Inst Technol, SB, 44, PhD(inorg chem), 49. *Prof Exp:* Sr res chemist, Metal Hydrides, Inc, 49, asst dir chem res lab, 49-51, assoc dir, 51-52, dir, 52-55, dir res & develop labs, 55-60; mem sr staff chem, Lincoln Lab, Mass Inst Technol, 60-71; mem sr staff chem, B U Marine Prog, Marine Biol Lab, Woods Hole, 71-73; scientist, Marine Ecol Div, Sigma Environ Sci, PR Nuclear Ctr, 73-76, consult, 76-79; consult, Banus Enterprises, 79-84; RETIRED. *Concurrent Pos:* Prof marine sci, Univ SC, 88- *Mem:* Sigma Xi; AAAS; Am Chem Soc. *Mailing Add:* PO Box 296 Seabrook SC 29940-0296

BANVILLE, BERTRAND, b Bic, Rimouski Co, Can, Feb 19, 31; m 54; c 3. PHYSICS. *Educ:* Univ Montreal, BSc, 53, MSc, 55, PhD(physics), 60. *Prof Exp:* Lectr physics, Univ Montreal, 57-59; assoc res officer health physics, Atomic Energy Can, Ltd, 59-64; asst prof, 64-66, ASSOC PROF PHYSICS, UNIV MONTREAL, 66- *Mem:* AAAS; Pattern Recognition Soc; Can Asn Physicists; Can Standards Asn. *Res:* Nuclear physics instrumentation; health physics; biomedical engineering. *Mailing Add:* Dept Physics Univ Montreal Case Postale 6128 Montreal PQ H3C 3J7 Can

BANVILLE, MARCEL, b St Octave, Que, June 29, 33; m 62; c 4. THEORETICAL PHYSICS. *Educ:* Univ Montreal, BSc, 56; Univ BC, MSc, 59, PhD(physics), 65. *Prof Exp:* Asst prof, 63-79, PROF PHYSICS, UNIV SHERBROOKE, 79- *Concurrent Pos:* Nat Sci & Eng Res Coun Can res grants, 78- *Mem:* Can Asn Physicists. *Res:* Optical properties of solids. *Mailing Add:* Dept Physics Univ Sherbrooke Sherbrooke PQ J1K 2R1 Can

BANWART, GEORGE J, b Algona, Iowa, Sept 15, 26; m 55; c 2. FOOD MICROBIOLOGY. *Educ:* Iowa State Univ, BS, 50, PhD(bact), 55. *Prof Exp:* Asst prof food technol, Univ Ga, 55-57; head egg prod sect, USDA, 57-62; assoc prof bact, Purdue Univ, 62-65; res microbiologist, Plant Indust Sta, USDA, Md, 65-69; prof food microbiol, 69-88, EMER PROF, OHIO STATE UNIV, 88- *Mem:* Inst Food Technologists. *Res:* Salmonella in foods; methodology of Salmonella; methods to destroy Salmonella; egg products; food spoilage; public health organisms. *Mailing Add:* Dept Microbiol Ohio State Univ Columbus OH 43210-1292

BANWART, WAYNE LEE, b West Bend, Iowa, Jan 9, 48; m 70; c 3. SOILS. *Educ:* Iowa State Univ, BS, 69, MS, 72, PhD(soil fertil), 75. *Prof Exp:* Res asst soil fertil, Iowa State Univ, 69-72, res assoc, 72-75; from asst prof to assoc prof soils, 75-84, PROF SOIL SCI, UNIV ILL, 84- *Honors & Awards:* George D Scarseth Award, Am Soc Agron, 74. *Mem:* Fel Am Soc Agron; Soil Sci Soc Am; fel Nat Asn Cols & Teachers Agr. *Res:* Factors affecting ionic balance of plants as it related to plant growth and yield; sorption of organic compounds by soils, including herbicides; organic transformations of sulfur in soils and other natural systems; characterization of natural organics in soil rhizosheres; effect of acid rain on agricultural crops. *Mailing Add:* S-512 Turner Univ Ill 1102 S Goodwin Urbana IL 61801

BANWELL, JOHN G, b Pegu, Burma, Aug 2, 30; US citizen; m 61; c 4. MEDICINE, PHYSIOLOGY. *Educ:* Oxford Univ, BA, 53, MD, 56, DM, 63. *Prof Exp:* Resident, Middlesexon Hosp, London, 56-59; fel, Johns Hopkins Univ Hosp, 59-61, asst prof, Dept Med, 66-72; prof med & chief gastroenterol, Dept Med, Med Ctr, Univ Ky, 72-81; PROF MED, GASTROENTEROL DIV, CASE WESTERN RESERVE UNIV, 81- *Mem:* Am Fedn Clin Res; Am Gastroenterol Asn; Am Col Physicians. *Res:* Intestinal function, scentern and absorptive junction; plant lectins. *Mailing Add:* Gastroenterol Div Lakeside Hosp Case Western Reserve Univ Cleveland OH 44106

BANZETT, ROBERT B, b Englewood, NJ, Aug 16, 47. RESPIRATORY PHYSIOLOGY. *Educ:* Univ Calif, Davis, PhD(physiol), 74. *Prof Exp:* Asst prof, 80-85, ASSOC PROF PHYSIOL, SCH PUB HEALTH, HARVARD UNIV, 85- *Mem:* Am Physiol Soc. *Res:* Respiratory sensation; avian respiration. *Mailing Add:* Dept Environ Health Sch Pub Health Harvard Univ 665 Huntington Ave Boston MA 02115

BAO, QINGCHENG, b Jinan, China, Feb 15, 48; m 74; c 2. MICROCHEMICAL SENSOR & INSTRUMENTATION, OPTICAL ENGINEERING. *Educ:* Tsing Hua Univ, Beijing, China, BS, 70; Semiconductor Inst, Beijing, China, MS, 82; Chinese Univ Technol & Sci, PhD(solid state physics), 86. *Prof Exp:* Mfg engr semiconductor, Changchun Semiconductor Device Co, China, 70-79; res asst laser diode, Semiconductor

Inst, Beijing, China, 79-82; res assoc laser spectroscopy, Changchun Physics Inst, China, 82-86; postdoctoral mod optics, Chem Dept, Univ Rochester, NY, 86-88; RES SPECIALIST DESIGN & MFG, DEPT CHEM, ARIZ STATE UNIV, 88- Concurrent Pos: Pres, Silverdow Corp, 90-, prin investr, 91-; chief scientist, Endetec Ltd, 91- Mem: Am Vacuum Soc. Res: Non-linear optics; high density excitation; laser holographic technology and spectroscopy; acoustic generation; dynamic chemistry; surface science; thin film technology; device design and processing development. Mailing Add: Dept Chem Ariz State Univ Tempe AZ 85287-1604

BAPATLA, KRISHNA M, b Guntur, India, Feb 18, 36; nat US; m 67; c 1. PHARMACEUTICS, PHARMACEUTICAL CHEMISTRY. Educ: Andhra Univ, India, BPharm, 55, MPharm, 59; Univ Southern Calif, PhD(pharmaceut chem), 66; Fairleigh-Dickinson Univ, MBA, 76. Prof Exp: Lectr pharm, Andhra Univ, India, 59-60; asst, Univ Southern Calif, 64-66; asst prof pharm, SDak State Univ, 66-69; res assoc pharm, State Univ Ny, Buffalo, 70-71; sr scientist, Warner-Lambert Res Inst, 71-76; section head, Cooper Labs, 76-78; ASST DIR, ALCON LABS, 78- Concurrent Pos: Res fel pharmaceut chem, Cancer Chemother Nat Serv Ctr, Univ Ariz, 66; Syntex Res Ctr, Calif, 69-70. Honors & Awards: Gold Medal. Mem: Am Pharm Asn; fel Am Inst Chem; Sigma Xi; Am Asn Pharmaceut; Parenteral Drug Asn; Regulatory Affairs Professionals Soc. Res: Synthesis of organic medicinal compounds; biopharmaceutics; research and development of new drug products; regulatory affairs; project management. Mailing Add: Alcon Labs Inc 6201 S Freeway Ft Worth TX 76134

BAPTIST, JAMES (NOEL), b Shelbyville, Ill, June 6, 30. MICROBIAL BIOCHEMISTRY. Educ: Case Inst Technol, BS, 52; Univ Ill, PhD(biochem), 57. Prof Exp: Chemist, Phillips Petrol Co, Okla, 52-54; Damon Runyon Fund fel enzymatic study omega oxidation, Univ Mich, 57-59; biochemist, W R Grace & Co, Md, 59-63; microbiologist, Int Minerals & Chem Co, Ill, 63-65; self-employed, Fla, 65-68; res assoc biol, Univ Tex M D Anderson Hosp & Tumor Inst, 68-69; from asst to assoc biologist, 69-77; SELF-EMPLOYED, 77- Mem: Am Chem Soc; Am Soc Microbiol; Sigma Xi. Res: Isozymes; metabolism of chemical carcinogens; protein purification methods; aquatic biology. Mailing Add: 2620 Lake Victoria Dr El Paso TX 79936-3107

BAPTIST, VICTOR HARRY, b Redlands, Calif, Mar 26, 23; m 46; c 2. PHYSICAL BIOCHEMISTRY. Educ: Northwestern Univ, BS, 48, PhD(phys biochem), 52. Prof Exp: Control chemist, 42-43 & 46-48; res assoc protein chem, Northwestern Univ, 51-52 & Univ Iowa, 52-53; res chemist in chg chem control, Don Baxter, Inc, 53-63, control adminr, 63-68; MGR RES ANALYSIS, MAX FACTOR & CO, 68- Mem: Am Chem Soc; Soc Cosmetic Chemists; Soc Appl Spectros; Am Soc Testing Mat; Coblentz Soc; Sigma Xi. Res: Methods of analysis; solar energy; making potable water from sea water; levitation. Mailing Add: 2636 N Brighton Burbank CA 91504

BAPTISTA, JOSE ANTONIO, b Azores, Portugal, Jan 30, 39; Can citizen; m 67; c 1. CARBOHYDRATE ANALYSIS, OLIGONUCLEOTIDE SYNTHESIS. Educ: Univ Porto, Portugal, Bachelor, 64; Univ Coimbra, Portugal, Master, 70; Columbia P Univ, PhD(anal chem), 85. Prof Exp: Researcher org chem, Fac Sci, Univ Porto, Portugal, 70-80; physico-chem specialist anal chem, Ludwig Inst Cancer Res, 80-88; RES ASSOC ANALYTIC BIOCHEM, FAC MED, UNIV TORONTO, CAN, 88- Concurrent Pos: Dir, Anal Biochem Lab, Carbohydrate Res Ctr, Fac Med, Univ Toronto, 88-91. Res: Separation and analysis of mixtures of natural and synthetic products; purification of chemical and biochemical substances; quantitative and qualitative analysis of carbohydrates of biological specimens using HPLC and GC techniques. Mailing Add: Dept Med Genetics Med Sci Bldg Univ Toronto Toronto ON M5S 1A8 Can

BAQUET, CLAUDIA, CANCER RESEARCH. Educ: Meharry Med Col, MD, 77; Johns Hopkins Univ, MPH, 83. Prof Exp: Pvt gen pract, Los Angeles, 79-80, St Louis, Mo, 80-82; parochial sch physician, Sch Health Prog, City St Louis, 80-82; dir, Cancer Prog & Epidemiologist/Cancer Control Specialist, Provident Hosp Inc, 83-84; expert, 84-87, chief spec pop studies br, 87-89, ASSOC DIR, CANCER CONTROL SCI PROG DIV CANCER PREV & CONTROL, NAT CANCER CTR, 89- Concurrent Pos: Consult, Mo Div Health, Bur Chronic Dis, St Louis, 81-82, Charles R Drew Postgrad Med Sch, Los Angeles, 83-84, Hanes Group, Winston-Salem, NC, 83-84, Richmond Community Hosp, 83-84, Provident Hosp, 84; mem, Nat Adv Comt on Cancer in Minorities, Am Cancer Soc, 82-86, Nat Subcomt Cancer in Econ Disadvantaged Patient, 85-86; bd dirs, Pub Educ Comt, Am Cancer Soc, 83-86; chairperson, Community Med & Pub Health Sect, Nat Med Asn, 85-, fac, Cancer Prev Fel Prog, Nat Cancer Inst, 86- Honors & Awards: W S Quinland Prize, Meharry Med Col, 76; ONI Award, Int Black Womens Health Cong, 88. Mem: Am Soc Clin Path; Am Pub Health Asn; Am Soc Prev Oncol; Nat Med Asn. Res: Preventive oncology; nutrition as an etiologic and preventive factor and cancer; public education for cancer; epidemiologic methods; breast cancer in young women; aged populations and cancer; cancer in racial/ethnic minorities; numerous technical publications. Mailing Add: Nat Cancer Inst DCPC NIH Exec Plaza N Rm 243 9000 Rockville Pike Bethesda MD 20892

BARABAS, SILVIO, water pollution, environmental health, for more information see previous edition

BARACH, JOHN PAUL, b New York, NY, Dec 21, 35; m 57; c 2. LIVING STATE PHYSICS, PLASMA PHYSICS. Educ: Princeton Univ, BA, 57; Univ Md, PhD(physics), 61. Prof Exp: From asst to assoc prof, 61-73, dir undergrad study, 79-83, PROF PHYSICS, VANDERBILT UNIV, 73- Mem: Am Phys Soc; Sigma Xi. Res: Experiments on biomagnetism; electromagnetic shock tubes; magnetic interactions; spectroscopy; function of nerves; computer modeling of non-linear waves. Mailing Add: Dept Physics Box 1807 Sta B Vanderbilt Univ Nashville TN 37235

BARAC-NIETO, MARIO, b Bogotá, Colombia, DE, Jan 1, 40; m 64; c 2. RENAL FUNCTION, EXERCISE & FITNESS. Educ: Univ Javeriana, Bogot01, MD, 62; Univ Rochester, NY, PhD(physiol), 68. Prof Exp: Teaching asst physiol, Univ Javeriana, Bogotá, Colombia, 59, instr biol, 60-61; res asst biol, Univ los Andes, Bogotá, Colombia, 60-62; from asst prof to prof physiol, Univ Valle, Cali, Colombia, 60-80; assoc prof, 80-85, ASSOC PROF & SR SCIENTIST, UNIV ROCHESTER, NY, 85- Concurrent Pos: Fel, Rockefeller Found, 63-65; dept med, Buswell, Univ Rochester, 67-68, Monroe County chap, Am Heart Asn, 66-67; vis scientist, Venezuela Inst Sci Res, 74-75; vis prof, Max Planck Inst Biophys, Frankfurt, 77-78, dept physiol, Med Col Wis, 78-80. Mem: Am Physiol Soc; Am Soc Nephrol; Am Col Sports Med; Am Soc Renal Biochem & Metab; NY Acad Sci. Res: Renal trasport and metabolism of nutrients; influence of nutrition on renal function, body composition and physical work capacity as influenced by development and aging. Mailing Add: Dept Physiol Univ Rochester 601 Elm Ave Box 642 Rochester NY 14642n

BARACOS, ANDREW, b Calgary, Alta, May 10, 25; m 51; c 3. CIVIL ENGINEERING, GEOLOGICAL ENGINEERING. Educ: Univ Alta, BSc, 47, MSc, 49. Prof Exp: Prof civil eng, 67-78, head geol eng dept, 78-83, PROF GEOL ENG, UNIV MAN, 83- Concurrent Pos: Vis prof, Chiang Mai Univ, Thailand, 86. Mem: Fel Eng Inst Can; Can Geotech Soc. Res: Geological engineering environmental impact; overburden stabilization for mining clays. Mailing Add: Dept Civil Eng Geol Eng Bldg Univ Man Winnipeg MB R3T 2N2 Can

BARAFF, GENE ALLEN, b Washington, DC, Dec 27, 30; m 58; c 3. SOLID STATE PHYSICS. Educ: Columbia Univ, AB, 52; NC State Col, MS, 56; NY Univ, PhD(physics), 61. Prof Exp: Reactor physicist, Astra Assocs, Inc, 56-58; PHYSICIST, SOLID STATE THEORY, BELL TEL LABS, 61- Mem: Fel Am Phys Soc. Res: Electronic properties of metals, semimetals, semiconductor surfaces and point defects. Mailing Add: Bell Labs Rm 1D-333 Murray Hill NJ 07974-2070

BARAGAR, WILLIAM ROBERT ARTHUR, b Brandon, Man, Mar 7, 26; m 55; c 2. GEOLOGY, PETROLOGY. Educ: Univ BC, BASc, 50; Queen's Univ, Ont, MSc, 52; Columbia Univ, PhD, 59. Prof Exp: Geologist, Falconbridge Nickel Mines Ltd, 52-54; TECH OFF & GEOLOGIST, GEOL SURV CAN, 57- Concurrent Pos: Assoc ed, Canadian J Earth Science; chmn, publ comt, Geol Asn Can. Mem: Geol Soc Am; Geol Asn Can; Mineral Asn Can; Mineral Soc Am. Res: Petrology; petrology of basic igneous rocks; petrochemistry of the Troodos ophiolite, Cyprus. Mailing Add: Geol Survey Can Geol Survey Bldg 615 Booth St Ottawa ON K1A 0E8 Can

BARAJAS, LUCIANO, b Santa Cruz de Tenerife, Canary Islands, Spain, Oct 10, 33; m 60; c 3. PATHOLOGY. Educ: Univ Madrid, MD, 56. Prof Exp: Rotating intern med, Alexian Bros Hosp, Elizabeth, NJ, 57-58; res path, Malmonides Hosp, Brooklyn, NY, 58-59; res path, Barnes Hosp, Wash Univ, 59-61, USPHS trainee exp path, 59-60 & Nat Cancer Inst trainee, 60-61; from instr to asst prof path, Sch Med, 61-69, assoc prof zool, Univ, 69-74, PROF PATH, UNIV CALIF, LOS ANGELES & HARBOR HOSP, TORRANCE, 74- Concurrent Pos: NIH spec fel, 66-69. Mem: Am Soc Cell Biol; Am Asn Anat; Soc Exp Biol & Med; Am Asn Pathologists; Int Acad Path. Res: Normal and pathologic anatomy of the kidney. Mailing Add: Dept Path Harbor Hosp 1000 W Carson St Torrance CA 90509

BARAK, ANTHONY JOSEPH, b Petersburg, Nebr, July 18, 22; m 50; c 5. BIOCHEMISTRY. Educ: Creighton Univ, BS, 48, MS, 50; Univ Mo, PhD(biochem), 53. Prof Exp: From instr to assoc prof, 53-77, PROF BIOCHEM, UNIV NEBR MED CTR, 77- Concurrent Pos: Biochemist, Vet Admin Hosp, Omaha, 56- Res: Alcohol metabolism; liver disease. Mailing Add: Dept Med Univ Nebr Med Ctr 600 42nd St Omaha NE 68198

BARAK, EVE IDA, b New York, NY, Jan 28, 48; m 70, 90. CELL BIOLOGY. Educ: Brown Univ, AB, 68; Rockefeller Univ, PhD(cell biol), 74. Prof Exp: Postdoctoral fel, Sch Med, Wash Univ, 74-78, Univ Ala, Birmingham, 78-79; asst prof anat, Univ Ala, Birmingham, 79-82, tumor biol, Univ Tex Syst Cancer, 83-86; asst prog dir, 86-89, ASSOC PROG DIR CELL BIOL, NSF, 89- Mem: Am Soc Cell Biol; Soc Complex Carbohydrates. Res: Cell biology; microbiology; biochemistry. Mailing Add: NSF Rm 321 1800 G St NW Washington DC 20550

BARAKAT, HISHAM A, b Sept 22, 43; m 69; c 2. BIOCHEMISTRY, MICROBIOLOGY. Educ: Am Univ Beirut, BSc, 65, MSc, 67; Univ Mass, PhD(microbiol), 72. Prof Exp: Instr sci, Teacher-Training Inst, Siblin, Lebanon, 67-68; fel biochem, Baylor Col Med, 72-73; res assoc, 73-75, from asst prof to assoc prof, 75-86, PROF BIOCHEM, MED SCH, E CAROLINA UNIV, 87- Mem: Soc Exp Biol & Med; Am Physiol Soc; Am Heart Asn; Am Diabetes Asn. Res: Metabolism of plasma lipoproteins in obese subjects and subjects with diabetes mellitus. Mailing Add: Dept Biochem Sch Med E Carolina Univ Greenville NC 27834

BARALD, KATE FRANCESCA, b May 11, 45; US citizen; m 71; c 1. NEUROBIOLOGY, DEVELOPMENTAL BIOLOGY. Educ: Bryn Mawr Col, AB, 67; Univ Wis-Madison, MS, 69, PhD(oncol immunol), 74. Prof Exp: NIH fel neurobiol, Cornell Univ, 75; MDA fel biol & neurobiol, Univ Calif, San Diego, 75-78; NIH fel biol, Stanford Univ, 78-81; asst prof, 81-90, ASSOC PROF ANAT & CELL BIOL, MED SCH, UNIV MICH, ANN ARBOR, 90- Concurrent Pos: Mem, Develop Biol Panel, NSF, 83-86, Develop Neurosci Panel, 90-; consult, univs & indust. Mem: Soc Neurosci; Am Soc Cell Biol; Am Soc Develop Biol; Sigma Xi; AAAS. Res: Developmental neurobiology and molecular biology; cell-cell interactions in the formation of specific connections in the nervous system; neural crest; development and regeneration of muscle. Mailing Add: Children's Med Res Found Univ Sydney PO Box 61 Camperdown N S W 2050 Australia

BARAM, PETER, immunology, microbiology, for more information see previous edition

BARAN, GEORGE ROMAN, b Philadelphia, Pa, Nov 3, 50. DENTAL MATERIALS. *Educ:* Univ Mich, BSE, 72, MSE, 73, PhD(mat sci eng), 76. *Prof Exp:* Res fel, Free Univ Berlin, 77-78; ASSOC PROF DENT, SCH DENT, TEMPLE UNIV, PA, 78- *Concurrent Pos:* Res fel, Alexander von Humboldt Found, 77; consult, Dentsply Int, 79- *Mem:* Am Asn Dent Res; Am Ceramic Soc; Am Asn Dent Sch. *Res:* Properties of casting alloys that may be used as alternatives to high gold content alloys in dentistry; enameling. *Mailing Add:* Dept Oper Dent Sch Dent Temple Univ 3223 N Broad St Philadelphia PA 19140

BARAN, JOHN STANISLAUS, b Chicago, Ill, Feb 7, 29; m 54; c 2. ORGANIC CHEMISTRY, BIOLOGICAL CHEMISTRY. *Educ:* Univ Chicago, BA, 51, MSc, 53; Univ Wis, PhD(chem), 55. *Prof Exp:* Res investr, G D Searle & Co, Inc, 56-59, sr res investr, 59-66, res fel, 66-88, SR RES FEL, G D SEARLE & CO, INC, 89- *Concurrent Pos:* Fel, Univ Wis, 56; vis lectr, Univ Southern Calif, 68. *Mem:* Am Chem Soc; AAAS; NY Acad Sci. *Res:* Medicinal chemistry; biological chemistry; synthetic, organic and natural products. *Mailing Add:* G D Searle & Co Box 5110 Chicago IL 60680

BARANCZUK, RICHARD JOHN, b Hrubieszow, Poland, Nov 15, 43; US citizen; m 67; c 2. ENDOCRINOLOGY, CLINICAL CHEMISTRY. *Educ:* Loyola Univ, BS, 65; Northern Ill Univ, MS, 67; Med Col Ga, PhD(endocrinol), 72. *Prof Exp:* Co-investr obstet & gynec, Univ Kans Med Ctr, 73-76, asst prof, 75-76, res assoc endocrinol, 76-77; INSTR BIOCHEM, AVILA COL, 76-; DIR ENDOCRINOL & LAB SCI DIR CLIN CHEM, PHYSICIANS REF LAB, 76-; LAB DIR, BIOMED RES LABS, INC, 81- *Concurrent Pos:* Fel, Ford Found, 71-73; fel, USPHS, 71-73; instr, Univ Kans Med Ctr, 73-75; prin investr, SBIR Awards & Res Contract, Nat Cancer Inst, 81- *Mem:* Endocrine Soc; Soc Study Reproduction. *Res:* Reproductive endocrinology, clinical chemistry. *Mailing Add:* 10127 Horton Overland Park KS 66207

BARANGER, ELIZABETH UREY, b Baltimore, Md, Sept 18, 27; m 51; c 3. THEORETICAL PHYSICS. *Educ:* Swarthmore Col, BA, 49; Cornell Univ, MA, 51, PhD, 54. *Prof Exp:* Res fel, Calif Inst Technol, 53-55; res fel, Univ Pittsburgh, 55-58, from asst prof to prof, 58-69; res assoc, Mass Inst Technol, 69-73; assoc dean, 73-78, dean grad studies in arts & sci, 78-89, PROF PHYSICS, UNIV PITTSBURGH, 73-, ASSOC PROVOST GRAD STUDIES, 89- *Mem:* Fel Am Phys Soc; fel AAAS. *Res:* Theoretical nuclear physics. *Mailing Add:* Dept Physics Univ Pittsburgh Pittsburgh PA 15260

BARANGER, MICHEL, b Le Mans, France, July 31, 27; nat US; m 51; c 3. THEORETICAL PHYSICS. *Educ:* Ecole Normale Superieure, Paris, 45-49; Cornell Univ, PhD(physics), 51. *Prof Exp:* Res assoc physics, Cornell Univ, 51-53; res fel, Calif Inst Technol, 53-55; res physicist, Carnegie Inst Technol, 55-56, from asst prof to assoc prof, 56-64, prof, 64-69; PROF PHYSICS, MASS INST TECHNOL, 69- *Concurrent Pos:* NSF sr fel, Univ Paris, 61-62. *Mem:* Fel Am Phys Soc. *Res:* Nuclear physics; plasma spectroscopy; quantum electrodynamics. *Mailing Add:* Dept Physics Rm 6-312 Mass Inst Technol Cambridge MA 02139

BARANKIEWICZ, JERZY ANDRZEJ, b Warsaw, Poland, Jul 27, 42; c 2. PURINE BIOCHEMISTRY, NUCLEOTIDE METABOLISM. *Educ:* Univ Warsaw & Univ Lodz, MS, 68; Inst Biochem & Biophys, Polish Acad Sci, PhD(biochem), 73. *Prof Exp:* Jr asst biochem, Polish Acad Sci, 68-69, asst lectr, 69-73, adj, 73-75; postdoctoral fel, Univ Alta Cancer Res Unit, 75-77; adj, Polish Acad Sci, 77-81; sr scientist, Dir Immunol, Res Inst, Hosp for Sick Children, Can, 81-88. *Concurrent Pos:* Postdoctoral fel, Univ Alta, 75. *Honors & Awards:* Awad of Pres, Polish Acad Sci, 77. *Mem:* Polish Biochemical Soc; Polish Bot Soc; Europ Acad Sci & Humanities; Europ Soc Nuclear Methods in Agr; NY Acad Sci; Europ Soc Purine Pyrimidine Res. *Res:* Purine and pyrimidine; anticancer drugs and immunomodulation; comparative and evolutionary biochemistry; experimental chemotherapy; cardiovascular and cerebrovascular research. *Mailing Add:* Gensia Pharmaceut Inc 11025 Roselle St San Diego CA 92121

BARANOWSKI, RICHARD MATTHEW, b Utica, NY, Mar 1, 28; m 51; c 3. ENTOMOLOGY. *Educ:* Syracuse Univ, BA, 51; Univ Conn, MS, 53, PhD, 59. *Prof Exp:* From asst entomologist to entomologist, Sub Trop Exp Sta, 58-84, CTR DIR, TROP RES & EDUC CTR, UNIV FLA, 84- *Concurrent Pos:* Tech adv, Animal Plant Health Inspection Serv, Nat Biol Control Prog, USDA, 80-82. *Mem:* Entom Soc Am; Int Org Biol Control. *Res:* Economic entomology; biology of hemiptera; biological control. *Mailing Add:* Dept Entom Univ Fla Gainesville FL 32611

BARANOWSKI, ROBERT LOUIS, BIOCHEMISTRY, PATHOPHYSIOLOGY. *Educ:* Univ Ill, Chicago, PhD(human physiol), 79. *Prof Exp:* RES ASST RENAL PHYSIOL & PROF MED, UNIV UTAH, 83- *Mailing Add:* Dept Med Div Nephrol Vet Admin Med Ctr 500 Foothill Blvd 11H Salt Lake City UT 84148

BARANOWSKI, TOM, b Brooklyn, NY, Dec 3, 46; m; c 2. HEALTH PSYCHOLOGY, HEALTH EDUCATION. *Educ:* Princeton Univ, AB, 68; Univ Kans, MA, 70, PhD(social psychol), 74. *Prof Exp:* Res scientist, Health Care Study Ctr, Battelle Mem Inst, 74-76; asst prof health educ, WVa Univ Med Ctr, 76-80; assoc prof prev med, Univ Tex Med Br-Galveston, 80-90; PROF PEDIAT & DEP DIR, GA PREV INST, 90- *Concurrent Pos:* Prin investr, Nat Heart, Lung & Blood Inst, 78-80, 83-, Moody Found, 81-84, Nat Cancer Inst, 86-87. *Mem:* Am Pub Health Asn; Am Psychol Asn; Am Heart Asn; Soc Behav Med; Asn Hypertension Black Pop. *Res:* Behavioral determinants of cardiovascular disease risk factors; behavioral change for cardiovascular risk reduction; psychosocial aspects of infant feeding practices; ethnic group differences in health related behaviors; measurement of health related behaviors and cognitions. *Mailing Add:* Ga Prev Inst Med Col Ga HB 5040 Augusta GA 30912-3710

BARANSKI, MICHAEL JOSEPH, b Wheeling, WVa, Dec 8, 46; m 69; c 1. PLANT TAXONOMY, ECOLOGY. *Educ:* West Liberty State Col, BS, 68; NC State Univ, PhD(bot), 74. *Prof Exp:* Instr bot, NC State Univ, 73-74; asst prof, 74-80, ASSOC PROF BIOL, CATAWBA COL, 81- *Mem:* Assoc Sigma Xi; Am Soc Plant Taxonomists; Bot Soc Am; Am Inst Biol Sci; Ecol Soc Am. *Res:* Taxonomy and ecology of woody plants; floristics and natural area analysis; vegetation analysis. *Mailing Add:* Dept Biol Catawba Col Salisbury NC 28144

BARANWAL, KRISHNA CHANDRA, b Bahadurpur, India, Jan 1, 36; m; c 2. POLYMER SCIENCE. *Educ:* Univ Allahabad, BSc, 57, MSc, 59; Indian Inst Technol, Kharagpur, MTech, 60; Univ Akron, PhD(polymer sci), 65. *Prof Exp:* DIR, TIRE RES, UNIROYAL GOODRICH RES & DEVELOP CTR, 65- *Mem:* Am Chem Soc; World Future Soc; Soc Plastics Engrs. *Res:* Basic understanding and physical, mechanical and dynamic properties of polymers; extensive and execellent research and development management experience; technology assessment and strategic planning. *Mailing Add:* Uniroyal Goodrich Res Ctr 3360 Gilchrist Rd Mogadore OH 44260

BARANY, FRANCIS, b Afula, Israel, Apr 4, 57. PROTEIN ENGINEERING. *Educ:* Univ Ill, Chicago Circle, BA, 76; Rockefeller Univ, PhD(microbiol), 81. *Prof Exp:* Fel, Rockefeller, Univ, 81-82,John Hopkins Univ Sch Med, 82-85; asst prof, 85-90, ASSOC PROF MICROBIOL, CORNELL UNIV MED COL, 90- *Concurrent Pos:* Adj asst prof, Rockefeller Univ, 86-, adj assoc prof, 90- *Mem:* Sigma Xi; Am Soc Microbiol; Fed Am Soc Exp Biol. *Res:* Using protein engineering, Tag I mutants with alatered cleavage properties are being constructed,these mutant proteins are being characterized via "star" reaction, NMR spectroscopy and x-ray crystallography; developed ligase chain reaction (LCR) for detection of single base genetic diseases using cloned thermostable ligase. *Mailing Add:* Dept Microbiol Box 62 1300 York Ave Cornell Univ Med Col New York NY 10021

BARANY, GEORGE, b Budapest, Hungary, Feb 19, 55; US citizen; m 86; c 2. PEPTIDE SYNTHESIS, ORGANOSULFUR CHEMISTRY. *Educ:* Rockefeller Univ, PhD(biochem), 77. *Prof Exp:* USPHS fel peptide synthesis, Rockefeller Univ, 78-80; from asst prof to assoc prof, 80-91, PROF CHEM, UNIV MINN, 91- *Concurrent Pos:* Searle Scholar, 82; USPHS res career develop award, 82-87; consult, peptide synthesis. *Mem:* Sigma Xi; AAAS; Am Soc Mass Spectrometry; NY Acad Sci; Am Chem Soc. *Res:* Peptide synthesis, particularly the solid-phase method; protecting groups for organic fuctionalities with an emphasis on the concept of orthogonality; chemistry of thiols, disulfides and polysulfanes; biochemistry of garlic. *Mailing Add:* Dept Chem Univ Minn 207 Pleasant St SE Minneapolis MN 55455

BARANY, JAMES W, b South Bend, Ind, Aug 24, 30; m 60; c 1. INDUSTRIAL ENGINEERING. *Educ:* Univ Notre Dame, BS, 53; Purdue Univ, MS, 58, PhD(indust eng), 61. *Prof Exp:* Jr engr, Bendix Aviation Corp, 53-54, prod liaison engr, 55-56; from instr to assoc prof, 58-69, PROF INDUST ENG, PURDUE UNIV, 69-, ASSOC HEAD DEPT, 70- *Concurrent Pos:* Instr training prog, Western Elec Co, Inc, 61-; res grants, NSF, 61-62, NIH, 63- & Easter Seal Found, 64-; consult, Int Coop Admin, China Productivity & Trade Ctr, 61-62. *Mem:* Fel Am Inst Indust Engrs; Human Factors Soc; Am Soc Eng Educ. *Res:* Human factors engineering pertaining to the assessment of the degree of handicap of hemiplegic patients; the application of job design principals for rehabilitation. *Mailing Add:* 101 Andrew Pl Apt 201 Lafayette IN 47906

BARANY, KATE, m 49; c 2. MYOSIN LIGHT CHAIN PHOSPHORYLATION. *Educ:* Goethe Univ, PhD(phys chem), 59. *Prof Exp:* PROF PHYSIOL & BIOPHYS, DEPT PHYSIOL & BIOPHYS, COL MED, UNIV ILL, 74- *Concurrent Pos:* Mem, Basic Sci Coun, Am Heart Asn. *Mem:* Biophys Soc; Am Physiol Soc; Am Heart Asn. *Res:* Physical-chemical characterization of proteins; physiology of muscle contraction; protein phosphorylation. *Mailing Add:* Dept Physiol & Biophys M/C 901 Col Med Univ Ill Chicago IL 60612

BARANY, MICHAEL, b Budapest, Hungary, Oct 29, 21; nat US; m 49; c 2. BIOCHEMISTRY. *Educ:* Univ Budapest, MD, 51, PhD(biochem, physiol), 56. *Prof Exp:* Lectr gen biochem, Inst Biochem, Univ Budapest, 51-56; intermediate scientist protein chem, Weizman Inst, 56-57; assoc mem muscle biochem, Max Planck Inst, 58-60; mem contractile proteins, Inst Muscle Dis, Inc, 61-74; PROF BIOL CHEM, UNIV ILL COL MED, 74- *Mem:* Am Soc Biol Chem; Am Physiol Soc; Am Chem Soc; Biophys Soc. *Res:* Contractile proteins of muscle; actin and myosin, isolation, characterization and mechanism of their interaction; muscle contraction; protein phosphorylation in smooth muscle hereditary muscular dystrophy; nuclear magnetic resonance of intact muscle and live human patients. *Mailing Add:* Dept Biol Chem M/C 536 Chicago Col Med Univ Ill Chicago IL 60612

BARANY, RONALD, b Bronx, NY, Mar 11, 28; m 60; c 4. PHYSICAL CHEMISTRY. *Educ:* City Col New York, BS, 49; NY Univ, MS, 51; Yale Univ, PhD(chem), 55. *Prof Exp:* Phys chemist thermochem measurements, US Bur Mines, 55-67; sr math analyst, Kaiser Engrs Div, Kaiser Indust Corp, 67-72; sr engr, Bechtel Corp, 72-81, sr analyst, 81-86; sr analyst, Lawrence Livermore Nat Lab, 87-89; RETIRED. *Mem:* Am Chem Soc; Asn Comput Mach. *Res:* Thermochemistry and thermodynamics; data processing; theoretical chemistry. *Mailing Add:* 6553 Chabot Rd Oakland CA 94618

BARASCH, GUY ERROL, b Baltimore, Md, Aug 28, 37; m 77; c 2. ATMOSPHERIC PHYSICS. *Educ:* Johns Hopkins Univ, AB, 58, PhD(physics), 65. *Prof Exp:* Staff scientist atmospheric physics, Los Alamos Sci Lab, Univ Calif, 65-69; sci specialist weapon physics, Santa Barbara Opers, EG&G, Inc, 69-71; staff scientist atmospheric physics, 71-77, GROUP LEADER, ATMOSPHERIC NUCLEAR WEAPONS EFFECTS GROUP, LOS ALAMOS SCI LAB, UNIV CALIF, 78- *Mem:* Sigma Xi; Optical Soc Am. *Res:* Physics of atmospheric nuclear detonations; light emission and propagation in highly dosed air; production and dispersal of pollutants due to nuclear, chemical and natural sources. *Mailing Add:* 2805 Glade Vale Way Vienna VA 22180

BARASCH, WERNER, b Breslau, Ger, May 27, 19; nat US. PLASTICS CHEMISTRY. *Educ:* Univ Calif, AB, 47; Mass Inst Technol, SM, 49; Univ Colo, PhD(org chem), 52. *Prof Exp:* Asst chem, Mass Inst Technol, 47-49 & Univ Colo, 49-52; res chemist, Olin Industs, Inc, 52-56; chief chemist resins, US Plywood Corp, 56-62; dir chem res, Plastomeric, Inc, 62-67 & Becton, Dickinson of Calif, Inc, 67-79; CONSULT, 79- *Concurrent Pos:* Study supvr, Harvard Univ, 49; consult, 79- *Mem:* Am Chem Soc; Sigma Xi. *Res:* Carbohydrate chemistry; mechanisms of organic reactions; polymers. *Mailing Add:* 23049 Santa Cruz Hwy Los Gatos CA 95030

BARASH, MOSHE M, b Lwow, Poland, Apr 5, 22; m 49; c 3. MACHINE DESIGN. *Educ:* Israel Inst Technol, BSc & Dipl Ing, 47, BSc, 50; Univ Manchester, PhD(mech eng), 58. *Prof Exp:* Sci officer mech eng, Israel Ministry Defence, 48-51, sect head, 51-53; sr engr, ATA Textile Co, Israel, 53-55; lectr mech eng, Univ Manchester, 56-63; PROF INDUST ENG, PURDUE UNIV, 63- *Concurrent Pos:* Consult eng, Pergamon Press, 60-62; contrib ed, Machinery, 65-70. *Mem:* Am Soc Mech Engrs; sr mem Am Inst Indust Engrs; Am Soc Eng Educ; Sigma Xi. *Res:* Materials processing; kinematics and dynamics of machines; synthesis of mechanisms; theory and effect of metal working processes on machined parts; computerized automation; reliability of systems; manufacturing science. *Mailing Add:* 124 Pawnee Dr West Lafayette IN 47906

BARASH, PAUL G, b Brooklyn, NY, Feb 22, 42; m 67; c 2. ANESTHESIOLOGY. *Educ:* City Col New York, BA, 63; Univ Ky, MD, 67. *Prof Exp:* From instr to asst prof, 72-77, ASSOC PROF ANESTHESIOL, SCH MED, YALE UNIV, 78- *Concurrent Pos:* Attend physician, Yale New Haven Hosp, 73-74, asst clin dir anethesiol, 74-, dir surg intensive care unit, 74- *Mem:* Int Anesthesia Res Soc; fel Am Col Anesthesiol; Am Soc Anesthesiol; Soc Comput Med. *Res:* Critical care medicine; cardiovascular and respiratory physiology; pharmacology of cocaine; computer applications in medicine. *Mailing Add:* Yale-New Haven Hosp Yale Univ Sch Med 867 Robert Treat Ext Orange CT 06477

BARATOFF, ALEXIS, b France, May 23, 37; US citizen; m 61; c 2. THEORY OF TUNNELING & SUPERCONDUCTIVITY. *Educ:* Mass Inst Technol, BS, 59; Cornell Univ, PhD(theoret physics), 64. *Prof Exp:* Res fel appl physics, Harvard Univ, 64-66; asst prof physics, Brown Univ, 66-71; res physicist, IFF KernForschungsanlage Julich, WGer, 71-73; RES PHYSICIST, IBM RES LAB, ZURICH, 73- *Concurrent Pos:* Ed bd, Europhysics News, 69-88. *Mem:* Am Phys Soc; European Phys Soc; German Phys Soc; Swiss Phys Soc. *Res:* Electronic properties of metals semiconductors and surfaces; superconductivity; tunneling and transport theory; scanning tunneling microscopy and extensions. *Mailing Add:* IBM Forschungs Lab Zurich Saumerstrasse 4 Rueschlikon CH 8803 Switzerland

BARATTA, ANTHONY J, b Bayonne, NJ, Dec 24, 45; m 68; c 2. RADIATION DAMAGE, ENGINEERING MANAGEMENT. *Educ:* Columbia Univ, BA & BS, 68; Brown Univ, MS, 70, PhD(physics), 78. *Prof Exp:* Proj engr, Div Naval Reactors, US Dept Energy, 69-71, proj officer, 71-74, refueling engr, 74-75, proj mgr, 74-78; res assoc, US Naval Res Lab, 78; from asst prof to assoc prof, 78-89, PROF NUCLEAR ENG, PA STATE UNIV, 89- *Concurrent Pos:* Consult, US Dept Energy, 79-80, US Naval Res Lab, 79-, Lom Tech, 85-90, Vollian Enterprises, 90- & Escom, Johannesberg, 88-; vis prof, Vrije Univ, Amsterdam, 86. *Mem:* Am Nuclear Soc; Am Phys Soc; Inst Elec & Electronics Engrs. *Res:* Neutron transport; nuclear reactor safety analysis; thermal hydraulic analysis; severe accident analysis; radiation damage to semiconductor devices and materials. *Mailing Add:* 231 Sackett University Park PA 16802

BARATTA, EDMOND JOHN, b Somerville, Mass, June 22, 28; m 53; c 1. ANALYTICAL CHEMISTRY, RADIOCHEMISTRY. *Educ:* Northeastern Univ, BSc, 53. *Prof Exp:* Chemist, Control Lab, Shell Oil Co, 53-56; chemist, Petrol Lab, Naval Supply Depot, US Navy, 57-59; sr prof biol & med, Nat Lead Co, Inc, 59-61, chief anal serv, 61-68; chief anal qual control serv, Bur Radiol Health, USPHS, 68-70; chief radiation off, Environ Protection Agency, 71-72; chief radionuclide sect, 72-85, NAT EXPERT RADIOACTIVITY, WINCHESTER ENG & ANALYSIS CTR, FOOD & DRUG ADMIN, 85- *Mem:* Am Chem Soc; Health Physics Soc; fel Asn Off Anal Chem. *Res:* Development, testing and standardization of methods for various radionuclides in environmental media; quality control for radionuclides in environment from fallout, nuclear reactors and fuel processing plants; radiopharmaceutical methodology investigations. *Mailing Add:* Winchester Eng & Anal YSIS Ctr 109 Holton St Winchester MA 01890

BARATZ, ROBERT SEARS, b New London, Conn, July 15, 46; m 75; c 2. DEVELOPMENTAL BIOLOGY. *Educ:* Boston Univ, AB, 69, MD, 87; Northwestern Univ, DDS, 72, PhD(anat & cell biol), 75; Am Bd Oral Med, dipl, 88. *Prof Exp:* Nat Inst Dent Res fel, 72-74; instr anat, Med Sch, Northwestern Univ, 74-76; asst prof anat, Sch Grad Dent, Boston Univ, 76-83, asst res prof oral &maxillofacial surg, 77-86, asst res prof dermat, Sch Med, 83-86; resident internal med, Carney Hosp, Boston, 87-91; ASST RES PROF ORAL MED, TUFTS UNIV, 87-; PHYSICIAN & RES SCIENTIST, BOSTON VA MED CTR, 88- *Concurrent Pos:* Staff dentist, Harvard Community Health Plan, 81-83; consult US Air Force, 83-84 & Am Dent Asn, 85-; indust consult Biomed Devices, 82-; med dir, Thermedics Inc, 85-87. *Mem:* Am Soc Cell Biol; AAAS; Int Asn Dent Res; Am Dent Asn; Soc Develop Biol; Am Med Assn. *Res:* Keratinization, cell differentiation, organ culture, tooth development and epithelio-mesenchymal interactions; oral mucosa in health and disease; carcinogenesis; biomedical devices; wound healing; allergic responses to biomedical materials. *Mailing Add:* Boston VA Dept Med 159 Bellevue St Newton MA 02158

BARB, C RICHARD, b Washington, DC, July 19, 48; m 70; c 3. REPRODUCTIVE PHYSIOLOGY, NEUROENDOCRINOLOGY. *Educ:* Univ Md, BS, 75; Univ Ga, MS, 80, PhD(neuroendocrin), 85. *Prof Exp:* Biol lab tech, Beltsville Agr Res Ctr, USDA, Agr Res Serv, Beltsville, Md, 70-76; biol lab tech, 76-85, RES PHYSIOLOGIST, RUSSELL RES CTR, USDA,

AGR RES SERV, ATHENS, GA, 85- *Concurrent Pos:* Vis scientist, Inst Animal Husb & Animal Behav, Fed Agr Res Ctr, Mariensee, Fed Repub Ger, 86 & Univ Hohenheim, Stuttgart, Fed Repub Ger, 86; adj asst prof, Animal & Dairy Sci Dept, Univ Ga, 88- *Mem:* Am Soc Animal Sci; Soc Study Reproduction; Soc Neurosci. *Res:* Opioid peptide regulation of pituitary hormone secretion changes during different physiological states; identified site of action of these peptides on modulating pituitary hormone secretion. *Mailing Add:* Russell Res Ctr Animal Physiol Unit USDA-Agr Res Serv PO Box 5677 Athens GA 30613

BARBA, WILLIAM P, II, b Philadelphia, Pa, Sept 25, 22; m 47; c 4. PEDIATRICS. *Educ:* Princeton Univ, BA, 44. *Prof Exp:* Intern, Pa Hosp, 46-47; Univ Pa, Md, 46; pediat resident, Res & Educ Hosp, Univ Ill, 49-51, instr pediat, 51-52; instr, Sch Med, Temple Univ, 52-54, assoc attend pediatrician, 57-61; dir med educ, Babies Unit, United Hosps, Newark, NJ, 61-64; actg asst dean, 65-67, assoc dean, 67-69, actg dean, 69-71, ASSOC PROF PEDIAT, SCH MED, TEMPLE UNIV, 64-, ASSOC VPRES, 71- *Concurrent Pos:* Mem adv comt, Child Guid Clin, 62-64; assoc attend pediatrician, St Christopher's Hosp Children, 64- *Mem:* Asn Am Med Cols; AAAS. *Res:* Pediatrics and medical education; systems of providing health care. *Mailing Add:* 326 Summit Ave Jenkintown PA 19046

BARBACCI, MARIO R, b Lima, Peru, Aug 28, 45; US citizen. SOFTWARE ENGINEERING, COMPUTER ARCHITECTURE. *Educ:* Univ Eng, Lima, BS, 66, EE, 68; Carnegie-Mellon Univ, PhD(comput sci), 74. *Prof Exp:* Res assoc, Sch Computer Sci, Carnegie-Mellon Univ, 74-77, res computer scientist, 77-82, dir, Technol Explor Dept, Software Eng Inst, 85-86, SR RES COMPUTER SCIENTIST, SCH COMPUTER SCI, CARNEGIE-MELLON UNIV, 82-, PROJ LEADER, SOFTWARE ENG INST, 86- *Concurrent Pos:* Consult, Naval Res Lab, 75-77, Digital Equipment Corp, 76-77 & Bendix Corp, 78-80; assoc ed hardware, Comput Reviews, Asn Comput Mach, 79-84, assoc ed software, 86-; consult, ITT, 82-84; consult, Wright Patterson Labs, USAF, 82-84; chmn, IFIP Working Group 10.2, 79-84; bd gov, Inst Elec & Electronics Engrs Computer Soc, 86-90, vpres tech activ, 90- & Outstanding Contribr Award, 89 & 90; dir, Real-time Distrib Systs Prog, Software Eng Inst, Carnegie-Mellon Univ, 91- *Honors & Awards:* Recognition Serv Award, Asn Comput Mach, 87. *Mem:* Fel Inst Elec & Electronics Engrs; Asn Comput Mach; Sigma Xi. *Res:* Computer architecture; design automation; programming languages and environments; real-time and distributed systems. *Mailing Add:* Software Eng Inst Carnegie Mellon Univ Pittsburgh PA 15213

BARBACK, JOSEPH, b Buffalo, NY, Oct 27, 37; m; c 1. MATHEMATICS. *Educ:* Univ Buffalo, AB, 59; Rutgers Univ, MS, 61, PhD(math), 64. *Prof Exp:* Asst prof math, Univ Pittsburgh, 63-65 & State Univ NY Buffalo, 65-67; vis asst prof, Ariz State Univ, 67-68; NIH fel biostatist, Stanford Univ, 68-69; assoc prof math, 69-72, PROF MATH, STATE UNIV NY COL BUFFALO, 72- *Mem:* Asn Symbolic Logic; Am Math Soc. *Res:* Recursive function theory; theory of computation; isols. *Mailing Add:* Dept Math State Univ NY Col Buffalo Buffalo NY 14222

BARBAN, STANLEY, b New York, NY, Mar 16, 21; m 50; c 2. BIOCHEMISTRY. *Educ:* City Col New York, BS, 43; Univ Mich, MS, 49; Wash Univ, PhD(bact & biochem), 52. *Prof Exp:* Bacteriologist, Dept Health, Syracuse, NY, 50; instr bact, Med Col, State Univ NY, 52-53; biochemist, Nat Inst Allergy & Infectious Dis, 54-78, SCIENTIST-ADMINR, OFF RECOMBINANT DNA ACTIV, NAT INST ALLERGY & INFECTIOUS DIS, NIH, 78- *Concurrent Pos:* USPHS fel, 53-54. *Mem:* Am Soc Microbiol; Am Soc Biol Chem. *Res:* Tissue cell culture metabolism; bacterial metabolism; enzymology; experimental chemotherapy; molecular biology of tumor viruses; recombinant DNA. *Mailing Add:* 6603 Pyle Rd Bethesda MD 20817

BARBAREE, JAMES MARTIN, b Union Springs, Ala, June 23, 40; m 61; c 3. BACTERIOLOGY. *Educ:* Univ Southern Miss, BS, 63, MS, 64; Univ Ga, PhD(bact), 67. *Prof Exp:* Instr biol, Univ Southern Miss, 63-64; chief microbiol, US Army William Beaumont Gen Hosp, 67-69; asst prof, Univ W Fla, 69-71; scientist microbiol, 71-74; sr scientist, 74-83, SCIENTIST (DIR GRADE) MICROBIOL & CHIEF BACT RESPIRATORY DIS EPIDEMIC INVEST LAB, USPHS, CTR DIS CONTROL, 83- *Concurrent Pos:* Consult, Pen Bay Clin & Anal Lab, 69-71; bacteriologist, Sacred Heart Hosp, 69-71; instr, DeKalb Community Col, 76, Oglethorpe Univ, 78-79; pres, Southeastern br, Am Soc Microbiol, 85; Int lectr, 88-90, Am Soc Microbiol Found lectr, 88; adj prof, Ga Tech, 88-90. *Honors & Awards:* CDC Mackel Award, 88 & 90. *Mem:* Am Soc Microbiologists; Sigma Xi. *Res:* Preservation and stability tests for bacteria and biologic reagents; laboratory methods for support of epidemic investigations of Legionnaires' disease and Whooping Cough; lyophilization/cryopreservation. *Mailing Add:* Ctr Dis Control 1-B360 1600 Clifton Rd NE Atlanta GA 30333

BARBARO, RONALD D, b Brooklyn, NY, Aug 5, 33; m 56; c 4. ENVIRONMENTAL SCIENCES. *Educ:* Providence Col, BA, 56; Univ RI, MS, 60; Rutgers Univ, PhD(environ sci), 66. *Prof Exp:* Technician, Gen Dynamics/Elec Boat Div, 59; lit scientist, Stamford Res Lab, Am Cyanamid Corp, 60; chief recreation-water qual unit, Southeast Water Lab, Fed Water Pollution Control Admin, US Dept Interior, 66-68; scientist adminr extramural prog br, Off Manpower Develop, Nat Air Pollution Control Admin, 68-70; vpres & dir, Princeton Aqua Sci, Inc, 70-72; PROF CHEM & ENVIRON SCI, NORTHERN VA COMMUNITY COL, 74- *Concurrent Pos:* Mgr air & water qual progs, Glass Container Mfrs Inst, 72; pres, GBC Assocs Inc-Environ Consult, 72- *Mem:* Am Chem Soc; Water Pollution Control Fedn; Sigma Xi. *Res:* Administration of programs with scientific orientation; water pollution control equipment design, development; pollution control consulting; educational program development; hazardous waste management; patent water filtration. *Mailing Add:* 7036 Lee Park Ct Falls Church VA 22042

BARBAS, JOHN THEOPHANI, b Kyrenia, Cyprus; US citizen; m 70; c 2. FREE RADICAL CHEMISTRY. *Educ:* Bob Jones Univ, BS, 66; Univ Ga, PhD(org chem), 71. *Prof Exp:* Res assoc chem, Fla State Univ, 70-72; prof, Lane Col, Jackson, Tenn, 72-78, Univ Montevallo, Ala, 79-82; PROF CHEM, VALDOSTA STATE COL, GA, 82- *Concurrent Pos:* Fulbright scholar, 66; vis res prof, Fla State Univ, 73-76. *Mem:* Am Chem Soc. *Res:* Free radical mechanisms in solution and on surfaces; electron transfer reactions; photochemistry; chemically induced dynamic nuclear polarization. *Mailing Add:* 2616 Quincy Circle Valdosta GA 31602-1591

BARBAT, WILLIAM FRANKLIN, geology; deceased, see previous edition for last biography

BARBE, DAVID FRANKLIN, b Webster Springs, WVa, May 26, 39; m 65; c 2. ELECTRONIC DEVICES, INTEGRATED CIRCUITS. *Educ:* WVa Univ, BS, 62, MS, 64; Johns Hopkins Univ, PhD(elec eng), 69. *Prof Exp:* Instr elec eng, WVa Univ, 62-65; engr res & develop, Westinghouse Elec Corp, 69-71; br head, Naval Res Lab, 71-79; asst for electronics & phys sci, Off Asst Secy Navy for res, Eng & Systs, 79-83, dir submarine & anti-submarine systs, 83-85; assoc dir, 85-87, EXEC DIR, ENG RES CTR, UNIV MD, 87- PROF ELEC ENG, 85- *Honors & Awards:* VHSIC Pioneer Award, 87. *Mem:* Fel Inst Elec & Electronics Engrs; Sigma Xi. *Res:* Charge coupled devices for low-light level imaging; infrared imaging and signal processing; very large scale integrated circuits. *Mailing Add:* Exec Dir Eng Res Ctr Univ Md College Park MD 20742

BARBEAU, EDWARD JOSEPH, JR, b Toronto, Ont, June 25, 38; m 61; c 2. PURE MATHEMATICS. *Educ:* Univ Toronto, BA, 60, MA, 61; Univ Newcastle, PhD(semialgebras), 64. *Prof Exp:* Lectr math, Univ Newcastle, 63-64; asst prof, Univ Western Ont, 64-66; from asst prof to assoc prof, 67-88, PROF MATH, UNIV TORONTO, 88- *Concurrent Pos:* Fel, Ont Inst Studies Educ. *Mem:* Am Math Soc; Math Asn Am; Can Math Soc; Can Soc Hist & Philos Math; Nat Coun Teachers Math. *Res:* History of analysis (17th and 18th centuries); elementary number theory; optimization. *Mailing Add:* Dept Math Univ Toronto Toronto ON M5S 1A1 Can

BARBEE, R WAYNE, b Greensboro, NC, Feb 6, 56; m 81. CARDIOLOGY. *Educ:* Univ NC, Wilmington, BS, 77, BA, 78; Univ Fla, Gainesville, MS, 81; La State Univ, New Orleans, PhD(physiol), 86. *Prof Exp:* Res asst, Dept Physiol, Univ Fla, 79-80, teaching asst, 80-81; postdoctoral fel, Dept Hypertension Res, 86-88, staff res scientist, 88-90, STAFF RES SCIENTIST, CARDIOL SECT, DEPT MED, ALTON OCHSNER MED FOUND, 90-; ASST PROF, DEPT PHYSIOL, MED SCH, LA STATE UNIV, 88- *Concurrent Pos:* Sigma Xi res award, 85-86; innovative res am grant award, 85-86; adj asst prof, Dept Physiol, Tulane Sch Med, 88-; mem, Safety Comt, Alton Ochsner Med Found, 88-90, chmn, Animal Care & Use Comt, 89; mem, Coun Circulation & Res Peer Rev Comt, Am Heart Asn, 89- *Mem:* Am Fedn Clin Res; Am Heart Asn; Am Physiol Soc; AAAS; Int Soc Heart Res; Sigma Xi. *Res:* Skeletal muscle efficiency and acid-base metabolism; cardiovascular changes during shock and diabetes; atrial natriuretic peptides; endothelial derived contracting and relaxing factors; author of various publications. *Mailing Add:* Div Res Alton Ochsner Med Found 1520 Jefferson Hwy New Orleans LA 70121

BARBEHENN, ELIZABETH KERN, b Washington, DC, June 21, 33; m 57; c 3. BIOCHEMISTRY. *Educ:* Cornell Univ, BS, 55; Wash Univ, PhD(biochem pharmacol), 74. *Prof Exp:* Staff fel, NIH, 74-80, expert, 80-85; PHARMACOLOGIST, DIV METAB, FOOD & DRUG ADMIN, 85- *Mem:* Fed Am Soc Exp Biol. *Res:* Metabolism of retina; micromethods of analysis; enzymatic assays. *Mailing Add:* Div Metabolism Food & Drug Admin 5600 Fishers Lane HFD-510 Rockville MD 20857

BARBEHENN, KYLE RAY, b Cumberland, Md, Dec 24, 28; m 56; c 3. VERTEBRATE ECOLOGY. *Educ:* Rutgers Univ, BS, 50; Cornell Univ, MS, 52, PhD(vert zool), 55. *Prof Exp:* Biologist, Nat Res Coun, 57-58; res fel, NIMH, 59-62; prin investr, NSF grant, 62-64; lectr biol, Univ Pa, 64-65; dir field ctr, Smithsonian Inst, 65-68; res fel, Ctr for Biol Natural Systs, Wash Univ, 68-74; PROJ MGR, OFF SPEC PESTICIDE REV, US ENVIRON PROTECTION AGENCY, 74- *Mem:* AAAS; Am Soc Mammal; Ecol Soc Am; Am Soc Naturalists; Wildlife Soc; Sigma Xi. *Res:* Intraspecific and interspecific behavioral relationships among small mammals and their relevance to population dynamics; urban rat ecology; management of vertebrate pest species by ecological methods; pharmacodynamics of pesticides. *Mailing Add:* Kile Ray Barbehenn Bethesda MD 20817

BARBER, ALBERT ALCIDE, b Providence, RI, July 13, 29; m 56; c 2. CELL BIOLOGY. *Educ:* Univ RI, BS, 50, MS, 52; Duke Univ, PhD, 58. *Prof Exp:* Assoc, Duke Univ, 57-58; from instr to assoc prof zool, 58-68, PROF ZOOL, UNIV CALIF, LOS ANGELES, 68-, ASSOC VCHANCELLOR RES, 71- *Concurrent Pos:* Consult, NIH, 73-75. *Mem:* Am Soc Biol Chemists; AAAS; Am Soc Zoologists; Am Physiol Soc. *Res:* Cell membranes; aging and antioxidant activity. *Mailing Add:* Chancellor's Off Univ Calif 405 Hilgard Ave Los Angeles CA 90024

BARBER, BRIAN HAROLD, MOLECULAR IMMUNOLOGY. *Educ:* Univ Toronto, PhD(med biophys), 74. *Prof Exp:* ASSOC PROF IMMUNOL, DEPT IMMUNOL, UNIV TORONTO, 76- *Res:* Viral vaccines; histocompatibility antigens. *Mailing Add:* Dept Immunol Univ Toronto Med Sci Bldg Toronto ON M5S 1A8 Can

BARBER, DONALD E, b Harrisburg, Pa, Apr 1, 31; div; c 3. HEALTH PHYSICS, INDUSTRIAL HYGIENE. *Educ:* Dickinson Col, BS, 53; Univ Mich, MPH, 59; PhD(environ health), 61. *Prof Exp:* Health physicist, Elec Boat Div, Gen Dynamics Corp, 54-55; asst prof environ health, Univ Mich, 61-66; assoc prof, 66-74, head, Environ & Occup Health Div, 81-89, PROF ENVIRON HEALTH, SCH PUB HEALTH, UNIV MINN, MINNEAPOLIS, 74- *Concurrent Pos:* Chmn, Sci Comt, 45, Nat Coun Radiation Protection & Measurements, 83-89; cert health physicist, cert safety prof. *Mem:* Health Physics Soc; Am Indust Hyg Asn. *Res:* Radiation dosimetry; environmental radioactivity. *Mailing Add:* Sch Pub Health Box 197 UMHC 420 Delaware St SE Minneapolis MN 55455

BARBER, EUGENE DOUGLAS, b Philadelphia, Pa, Jan 20, 43; m 65; c 1. TOXICOLOGY, GENETIC. *Educ:* Drexel Univ, BS, 65; Univ Pa, PhD(biochem), 70; Am Bd Toxicol, dipl, 87. *Prof Exp:* NIH fel lipid chem, Univ Mich, 70-72; SR RES CHEMIST, EASTMAN KODAK CO, 72-, MGR GENETIC TOXICOL, 78- *Mem:* AAAS; Environ Mutagen Soc; Genetic Toxicol Asn. *Res:* Metabolism of foreign compounds; mutagenicity of chemicals; composition and function of biological membranes; in vitro percutaneous absorption studies; toxicity of phthalate esters. *Mailing Add:* Eastman Kodak Co Kodak Park Bldg 320 Rochester NY 14650

BARBER, EUGENE JOHN, b Kit Carson, Colo, Jan 8, 18; m; c 4. PHYSICAL INORGANIC CHEMISTRY. *Educ:* Univ Nev, BS, 40; Univ Wash, PhD(chem), 48; Oakridge Sch Reactor Technol, dipl, 59. *Prof Exp:* Jr analyst, US Bur Mines, 40-41; res chemist, Div War Res, Columbia Univ, 42-45; res chemist, Union Carbide Corp, 48-58, sr develop consult chemist, 59-78, sr res scientist, Nuclear Div, 78-86; CORP FEL, MARTIN MARIETTA ENERGY SYSTS, 87- *Mem:* AAAS; Am Chem Soc; Sigma Xi. *Res:* Thermo-dynamics of nonaqueous solutions; gas phase corrosion; chemistry of fluorine and uranium. *Mailing Add:* PO Box 476 Kingston TN 37763

BARBER, FRANKLIN WESTON, b New York, NY, July 3, 12; m 37; c 1. FOOD SCIENCE. *Educ:* Aurora Col, BS, 34; Univ Wis, MS, 42, PhD(agr bact), 44. *Prof Exp:* Lab technician, H P Hood & Sons, Boston, 37-40; instr bact, Univ Wis, 40-44; asst, Golden State Co, Ltd, San Francisco, 44-45; new prod develop, Res & Develop Div, Kraftco Corp, 45-46, bacteriologist & head dept bact res, 46-53, sr scientist & chief div fundamental res, 53-57, assoc mgr fundamental res, Ill, 58-60, div res dir, 60-64, asst mgr patents & regulatory compliance, 64-66; from mgr to dir regulatory compliance, 66-75; CONSULT, FOOD REGULATIONS & QUAL CONTROL, 75- *Concurrent Pos:* Chmn, Food & Agr Orgn-WHO Joint Expert Comt Milk Hyg, Geneva, 59; mem expert adv panel environ health, WHO, 59-78; secy-treas, Fla Asn Milk, Food & Environ Sanitarians, 81- *Mem:* Am Soc Microbiol; Am Dairy Sci Asn; Inst Food Technologists; hon mem Int Asn Milk, Food & Environ Sanitarians (pres, 58-59). *Res:* Dairy bacteriological research; factors affecting accuracy of the phosphatase test; dissociation in lactobacilli; coliform bacteria in ice cream; psychrophilic bacteria in dairy products; milk plate count media; millipore filter techniques; patent and regulatory problems in dairy and food products. *Mailing Add:* 464 Bouchelle Dr No 305 New Smyrna Beach FL 32169-5414

BARBER, GEORGE ALFRED, b New York, NY, May 7, 25. BIOCHEMISTRY. *Educ:* Rutgers Univ, AB, 51; Columbia Univ, PhD(plant physiol, chem), 55. *Prof Exp:* Jr asst res biochemist, Univ Calif, Berkeley, 55-57; asst biochemist, Conn Agr Exp Sta, 57-59; biochemist, Stanford Res Inst, 59-60; asst res biochemist, Univ Calif, Berkeley, 60-65; assoc prof biochem, Univ Hawaii, 65-68; PROF BIOCHEM, OHIO STATE UNIV, 68- *Mem:* Am Soc Biol Chemists. *Res:* Intermediary metabolism of higher plants. *Mailing Add:* Dept Biochem Ohio State Univ 484 W 12th Ave Columbus OH 43210

BARBER, GEORGE ARTHUR, b Chicago, Ill, Dec 6, 29; m 52; c 4. SCIENTIFIC DRILLING MANAGEMENT, MINERAL EXPLORATION. *Educ:* Univ Ariz, BS, 51. *Hon Degrees:* ME, Univ Ariz, 66. *Prof Exp:* Field geologist, Anaconda Co, 51-66, chief geologist, 66-71, explor mgr, Prime Metals Div, 71-73, vpres geol & explor, 73-74, vpres geol & technol, 74-77; vpres explor & geol, Anaconda Minerals-Atlantic Richfield Co, 77-82, vpres minerals adv, 82-85; pres & chief exec officer, Deep Observ & Sampling Earths Continental Crust, Inc, 85-89; INDEPENDENT RESOURCE ANALYST, 89- *Concurrent Pos:* Film ed, Geophys Film Comt, Nat Acad Sci-Nat Res Coun, 81-86, mem comt global & int geol, 82-85, mem Panel Mineral Res, 83-87 & mem comt hydrocarbon res drilling, 87-; mem, Prog Comt, Am Mining Cong, 82-85 & Resources for the future Policy Adv Comt, 83- *Mem:* Soc Econ Geologists (vpres, 84, pres, 90-); Geol Soc Am; Am Asn Petrol Geologists; Am Geophys Union; Soc Mining Engrs; Int Asn Genesis Ore Deposits. *Res:* Organization and management of national continental scientific drilling program; world wide exploration for base and precious metal, energy resources, and non-metallic minerals. *Mailing Add:* 5800 Morning Glory Lane Littleton CO 80128-2708

BARBER, J(AMES) C(ORBETT), b Ruskin, Ga, Nov 28, 16; m 46; c 3. CHEMICAL ENGINEERING. *Educ:* Ga Inst Technol, BS, 38. *Prof Exp:* Chief plant chem control br, Tenn Valley Authority, 38-63, staff chem engr & off mgr, Off Agr & Chem Develop, 63-76; CONSULT, JAMES C BARBER & ASSOCS, 76- *Concurrent Pos:* Mem comt med & biol effects environ pollutants, Nat Res Coun, 76-77. *Mem:* Fel Am Chem Soc; Am Inst Chem Engrs. *Res:* Development of processes for production of phosphate fertilizers by the electric furnace method; environmental control for fertilizer operations; development of new fertilizer production processes; elemental phosphorus production; biomass utilization. *Mailing Add:* Suite 115 Courtview Towers Florence AL 35630-5434

BARBER, JAMES RICHARD, b High Wycombe, UK, Apr 15, 42; c 2. CONTACT MECHANICS, THERMOELASTICITY. *Educ:* Univ Cambridge, BA, 63, MA, 67 & PhD(eng), 68. *Prof Exp:* Tech asst, British Rail, 63-69; lectr mech eng, Univ Newcastle Upon Tyne, 69-81; reader solid mech, 81; assoc prof, 81-85, PROF MECH ENG & APPL MECH, UNIV MICH, 85- *Mem:* Inst Mech Engrs. *Res:* Elasticity and thermoelasticity, particularly in systems where thermal distortion affects the static or sliding contact of elastic bodies; engine-generator brakes, seals, heat transfer between contacting bodies, thermally driven fracture; solidification of castings. *Mailing Add:* Dept Mech Eng & Appl Mech Univ Mich Ann Arbor MI 48109-2125

BARBER, JOHN CLARK, forest genetics, for more information see previous edition

BARBER, JOHN THRELFALL, b Lancaster, Eng, Mar 13, 37; m 61; c 2. PLANT PHYSIOLOGY. *Educ:* Univ Liverpool, BSc, 58 & 59, PhD(plant physiol), 62. *Prof Exp:* Res assoc, Lab Cell Physiol, Growth & Develop, Cornell Univ, 62-69; asst prof, 69-71, ASSOC PROF BIOL, TULANE UNIV, 71- *Mem:* Sigma Xi; Am Soc Plant Physiol; Am Mosquito Control Asn. *Res:* Plant proteins in relation to growth, development and morphogenesis; ecology of mucilaginous plant seeds. *Mailing Add:* Dept Biol Tulane Univ New Orleans LA 70118

BARBER, M(ARK) R, b Wellington, NZ, July 23, 31; US citizen. ELECTRONICS. *Educ:* Univ Auckland, BSc, 54, BE, 55; Cambridge Univ, PhD(eng), 59. *Prof Exp:* Res engr, Naval Res Lab, NZ, 59-61; MEM TECH STAFF SEMICONDUCTOR ELECTRONICS, BELL LABS, 61- *Mem:* Fel Inst Elec & Electronics Engrs. *Res:* Electrocardiography; medical electronics; high power electron guns; processing signals from underwater acoustic arrays; solid state microwave switching networks; tunnel diode amplifiers and frequency converters; microwave solid state oscillators; semiconductor memories; integrated circuits; automated test equipment for integrated circuits. *Mailing Add:* AT&T Bell Labs 555 Union Blvd Allentown PA 18103

BARBER, MICHAEL JAMES, b UK, Aug 8, 50. METALLOENZYMES. *Educ:* Univ Kent, BSc, 72, MSc, 73; Univ Sussex, DPhil(biochem), 76. *Prof Exp:* Teaching fel biochem, Univ Sussex, 75-77; lectr phys sci, Open Univ, 76-77; res assoc biochem, Duke Univ Med Ctr, 77-83; asst prof, 83-87, ASSOC PROF BIOCHEM, COL MED, UNIV S FLA, 87- *Mem:* Am Soc Biol Chemists. *Res:* Structure and function of metalloenzymes with special emphasis on those containing molybdenum; mechanism of action of nitrate reductase; use of spin-labels in biochemistry. *Mailing Add:* Dept Biochem MS MDC7 Univ South Fla Col Med Tampa FL 33612

BARBER, PATRICK GEORGE, b Santa Barbara, Calif, Dec 14, 42; m 67; c 2. PHYSICAL CHEMISTRY. *Educ:* Stanford Univ, BS, 64; Cornell Univ, PhD(phys chem), 69. *Prof Exp:* Res technician radiol chem, US Naval Radiol Defense Lab, 63; res assoc chem, Duke Univ, 69-71; asst prof chem & math, Daniel Campus, Southside Va Community Col, 71-75, chmn div arts & sci, 73-75, prof chem, 75-78; assoc prof chem, 78-85, PROF CHEM & DIR CHEM PROG, DEPT NATURAL SCI, LONGWOOD COL, 85- *Concurrent Pos:* NSF prof develop grant, Duke Univ, 77; summer fac fel, Lewis Res Ctr, NASA, 76 & 77, Langley Res Ctr, 83, 84, 87 & 88, res scientist, 84-85 & 87-91; summer fac fel, David W Taylor Naval Ship Res & Dev Ctr, 81 & 83; researcher crystal growth, Spain, 90, liquid crystal res, Halle, Ger, 91. *Mem:* AAAS; Am Chem Soc; Am Crystallog Asn; Sigma Xi; Electrochem Soc; fel Am Inst Chem; fel Cert Prof Chemist; Am Asn Crystal Growth. *Res:* Crystallographic studies on derivatives of Gramicidin-S; structure and properties of liquid crystalline compounds; crystal growth in gels; alkaline-zinc battery for electric car; computer generated displays and computer software; conformation of polymers in the solid state; crystal growth of semiconductor crystals, etches for semiconductor crystals; x-ray crystallography. *Mailing Add:* Rte 2 Box 29B Keysville VA 23947

BARBER, ROBERT CHARLES, b Sarnia, Ont, Apr 20, 36; m 62; c 3. PHYSICS. *Educ:* McMaster Univ, BSc, 58, PhD(physics), 62. *Prof Exp:* Fel physics, McMaster Univ, 62-65; from asst prof to assoc prof, 65-75, PROF PHYSICS, UNIV MAN, 75-, HEAD DEPT, 87- *Concurrent Pos:* Sabbatical leave, Sch Physics, Univ Minn, Minneapolis, 71-72; mem, Int Union Pure & Appl Physics, Comn C13 on Atomic Massas & Fundamental Constants, 72-75, secy, 75-78, Comn C2 on Symbols, Units, Nomenclature, SUN-AMCO, secy, 78-84, chmn, 86-90, mem, 90-93. *Mem:* Can Asn Physicists; Can Asn Univ Teachers. *Res:* Determination of atomic mass differences by high resolution mass spectrometer. *Mailing Add:* Dept Physics Univ Man Winnipeg MB R3T 2N2 Can

BARBER, ROBERT EDWIN, b Long Beach, Calif, July 8, 32; m 56; c 4. ORGANIC RANKINE ENGINES, LOW SPECIFIC TURBINE DESIGN & ANALYSIS. *Educ:* Ore State Univ, BS, 57; Rensselaer Polytech Inst, MS, 60. *Prof Exp:* Res engr, United Aircraft Res Labs, 57-61 & Sundstrand Aviation, Denver, 61-66; VPRES & FOUNDER, BARBER-NICHOLS ENG CO, 66- *Mem:* Soc Automotive Engrs. *Res:* Cycle analysis and thermodynamics of organic rankine engines; design of low specific speed turbines for solar and geothermal engines. *Mailing Add:* 6325 W 55th Ave Arvada CO 80002

BARBER, SAUL BENJAMIN, b Somerville, Mass, Sept 3, 20; m 55; c 3. ZOOLOGY, INVERTEBRATE NEUROPHYSIOLOGY. *Educ:* RI State Col, BS, 41; Yale Univ, PhD(zool), 54. *Prof Exp:* Instr zool, RI State Col, 46-48; instr biol, Williams Col, 52-54; instr zool, Smith Col, 54-55; res assoc, Narragansett Marine Lab, Univ RI, 55-56; from asst prof to prof, 56-85, chmn, Biol Dept, 65-72, 74-76 & 78-80, assoc dean, Col Arts & Sci, 76-77, dean pro-tem, 77-78, EMER PROF BIOL, LEHIGH UNIV, 85- *Concurrent Pos:* NIH res contract; Nat Inst Neurol Dis & Blindness spec res fel zool, Oxford, 63-64. *Mem:* AAAS; Am Soc Zoologists; Electron Micros Soc Am; Sigma Xi; Am Asn Univ Professors. *Res:* Chemoreception and proprioception in the horseshoe crab; fish sound production; physiology of insect flight and muscle. *Mailing Add:* 11 W Yarmouth Rd Yarmouth Port MA 02675-1946

BARBER, SHERBURNE FREDERICK, b Nunda, NY, Oct 25, 07; m 44; c 3. MATHEMATICS. *Educ:* Univ Rochester, AB, 29, AM, 30; Univ Ill, PhD(math), 33. *Prof Exp:* Nat res fel, Johns Hopkins Univ, 33-34 & Princeton Univ & Inst Advan Study, 34-35; assoc math, Univ Iowa, 35-37; tutor, City Col New York, 37-40, from instr to prof math, 40-76, from asst dean to dean, Col Lib Arts & Sci, 53-71; RETIRED. *Mem:* Am Math Soc. *Res:* Cremona and birational transformations; algebraic geometry. *Mailing Add:* Five Yorkshire Ave Stony Brook NY 11790

BARBER, STANLEY ARTHUR, b Wolseley, Can; Mar 29, 21; nat; m 50; c 2. SOIL CHEMISTRY, SOIL FERTILITY. *Educ:* Univ Sask, BSA, 45, MSc, 47; Univ Mo, PhD(soils), 49. *Hon Degrees:* LLD, Univ Sask, 86. *Prof Exp:* Instr soils, Univ Sask, 45-47; from asst prof to assoc prof, 49-59, PROF SOILS, PURDUE UNIV, WEST LAFAYETTE, 59- *Concurrent Pos:* Vis prof, Univ Calif, Berkeley, 61 & Univ Adelaide, 68-69; NSF sr fel, 68-69; vis prof, Cornell Univ, 79; Can Indust Ltd distinguished vis lectr, Univ Sask, 87. *Honors & Awards:* Soil Res Award, Am Soc Agron, 74, Agron Res Award, 83, Agron Achievement Award, 84; Bouyoucos Distinguished Career Award, Soil Sci Soc Am, 85; Alexander von Humboldt Award, Alexander von Humboldt Found, Ger, 86. *Mem:* Nat Acad Sci; fel Am Soc Agron; fel Soil Sci Soc Am; Int Soc Soil Sci; Am Soc Plant Physiol. *Res:* Mechanisms for the movement of nutrients through the soil to the plant root; influence of the plant root on the soil; soil nutrients availability and crop growth. *Mailing Add:* Dept Agron Purdue Univ West Lafayette IN 47907-1150

BARBER, THOMAS KING, b Highland Park, Mich, Sept 26, 23; m 47; c 3. DENTISTRY. *Educ:* Mich State Col, BS, 45; Univ Ill, BS, 47, DDS & MS, 49. *Prof Exp:* Instr pedodontics, Marquette Univ, 50-51; from instr to prof, Univ Ill, 51-69; assoc dean, 78-80, 81-83, PROF PEDIAT, SCH MED & PEDIAT DENT, SCH DENT, UNIV CALIF, LOS ANGELES, 69- *Concurrent Pos:* Consult, Coun Dent Educ, Am Dent Asn, Am Fund Dent Health & Educ Testing Serv; actg dean, Univ Calif, Los Angeles, 80-81. *Honors & Awards:* Hon mem Award, Peruvian Dent Soc, 68; Hon mem Award, French Soc Dent Children, 70; Award of Excellence, Am Soc Dent Children, 81. *Mem:* Fel Am Col Dent; Am Dent Asn; Am Soc Dent for Children; Am Acad Pedodontics. *Res:* Pedodontics, especially growth and dental development of children and minor orthodontics in children; educational research in dentistry. *Mailing Add:* Univ Calif Sch Dent Los Angeles CA 90024

BARBER, THOMAS LYNWOOD, b Dothan, Ala, Feb 4, 34; div; c 3. SCIENCE EDUCATION, ANIMAL DISEASE RESEARCH. *Educ:* Auburn Univ, DVM, 58; Cornell Univ, MS, 61, PhD, 69. *Prof Exp:* Res vet, Plum Island Animal Dis Ctr, 58-69; vet med officer agr res, Arthoropod-Borne Animal Dis Res Lab, Agr Res Serv, USDA, 69-85; CONSULT, 85- *Concurrent Pos:* Fac affil, Colo State Univ, 73-88. *Mem:* US Animal Health Asn; Am Vet Med Asn. *Res:* Arthropod borne diseases of animals; bluetongue virus vaccine development for sheep and cattle. *Mailing Add:* 7855 Armadillo Trail Evergreen CO 80439

BARBER, WALTER CARLISLE, nuclear physics; deceased, see previous edition for last biography

BARBER, WILLIAM AUSTIN, b Brooklyn, NY, Oct 2, 24; m 52; c 7. PHYSICAL CHEMISTRY. *Educ:* Holy Cross Col, BS, 49; Cornell Univ, PhD(phys chem), 52. *Prof Exp:* Res assoc, Cornell Univ, 52-53; phys res chemist, Am Cyanamid Co, 53-60, sr res chemist, 60-62 & 69-85, group leader res div, 62-69; RETIRED. *Mem:* Am Chem Soc. *Res:* Clay-water systems; soap-hydrocarbon gelation; polymers; organometallic chemistry; fused salts; fuel cells; heterogeneous catalysis. *Mailing Add:* 42 Simsbury Rd Stamford CT 06905-2431

BARBER, WILLIAM J, MICROCIRCULATION. *Educ:* Univ Ky, PhD(biol sci), 76. *Prof Exp:* Assoc prof physiol, Med Col Wis, 80-90; ASSOC RES PROF, CTR BIOMED ENG, UNIV KY, 90- *Mailing Add:* Tobacco & Health Res Inst Univ Ky Cooper & University Dr Lexington KY 40546-0236

BARBERO, GIULIO J, b Mt Vernon, NY, Oct 13, 23; m 47; c 6. PEDIATRICS, GASTROENTEROLOGY. *Educ:* Univ Maine, BS, 43; Univ Pa, MD, 47. *Prof Exp:* Sr physician, dir div gastroenterol & cystic fibrosis res group, Children's Hosp Philadelphia, 59-67; prof pediat & chmn dept, Hahnemann Med Col, 67-73; prof pediat & chmn dept, Sch Med, Univ Mo-Columbia, 73-90; RETIRED. *Concurrent Pos:* Chmn, gen med & sci adv coun, Nat Cystic Fibrosis Res Found, 71-74; mem, Coun Heart, Lung & Blood Inst, NIH. *Mem:* Soc Pediat Res; Am Acad Pediat; Am Gastroenterol Asn; Am Psychosom Soc. *Res:* Cystic fibrosis; gastrointestinal disturbances of children; psychosomatic aspects of gastrointestinal disorders of childhood. *Mailing Add:* Dept Pediat Univ Mo Columbia MO 65201

BARBERY, GILLES A, mineral engineering; deceased, see previous edition for last biography

BARBIN, ALLEN R, b Beaumont, Tex, July 3, 34; m 54; c 2. MECHANICAL ENGINEERING. *Educ:* Lamar State Col, BS, 55; Tex A&M Univ, MS, 59; Purdue Univ, PhD(mech eng), 61. *Prof Exp:* Asst prof mech eng, Lamar State Col, 56-58; assoc prof, 61-67, PROF MECH ENG, AUBURN UNIV, 67- *Concurrent Pos:* Consult, Reactor Div, Oak Ridge Nat Lab, 63- *Mem:* Assoc mem Am Soc Mech Engrs; Sigma Xi. *Res:* Turbulent flow in tubes; jet flows; forced turbulent boundary layers; fluid jet amplifiers. *Mailing Add:* Rte 1 Box 160 Buna TX 77612

BARBO, DOROTHY M, b River Falls, Wis, May 28, 32. NUTRITION. *Educ:* Asbury Col, Wilmore, Ky, AB, 50; Univ Wis, Madison, MD, 54. *Hon Degrees:* DSc, Asbury Col, 81. *Prof Exp:* Instr obstet & gynec, Med Col Wis, 62-66, asst prof, 66-67; assoc prof obstet & gynec, Ludhiana Christian Med Col, Punjab, India, 68-72, actg chmn dept, 70, chief gynec oncol, 69-72, chief family planning, 70-72; coordr & clin dir, Med Col Pa, 73-77, dir gynec oncol, 75-85, assoc prof obstet & gynec, 72-78, dir, Ctr Mature Women, 83-91, prof obstet & gynec, 88-91; PROF OBSTET & GYNEC, UNIV NMEX, SCH MED, 91- *Concurrent Pos:* Sr clin trainee obstet & gynec, USPHS CA Control, 63-65; mem bd dirs, Ludhiana Christian Med Col, NY, 73-, Am Cancer Soc, Philadelphia Div, 80-86 & Serv Master Co LP, Chicago, 82-; consult, Vet Admin Hosp, Philadelphia, 75-85; examr, Am Bd Obstet & Gynec, 84-; mem, Dept Health Human Serv, Fed Drug Admin Fertil & Maternal Health Drug Adv Comn, 86-91. *Mem:* Am Col Obstet & Gynec; Am Col Surgeons; AMA; Am Soc Colposcopy & Cervical Path; Am Women's Med Asn; NAm Menopause Soc. *Res:* Gynecological oncology in the areas of prevention, screening techniques and early diagnosis; gynecological surgery and care of the post-menopausal woman with emphasis on menopausal problems; prevention, early diagnosis and treatment of post-menopausal osteoporosis. *Mailing Add:* Dept Obstet/Gynec Univ NMex Sch Med 2211 Lomas Blvd NE Albuquerque NM 87131-5286

BARBORAK, JAMES CARL, b Moulton, Tex, Sept 8, 41; m 69. MECHANISTIC ORGANOMETALLIC CHEMISTRY. *Educ:* Univ Tex, Austin, BS, 63, PhD(chem), 68. *Prof Exp:* Res scientist, Uniroyal, Inc, 68-69; fel, Princeton Univ, 69-70, instr chem, 70-72; from asst prof to assoc prof chem, 72-80, PROF CHEM, UNIV NC, GREENSBORO, 80- *Mem:* Am Chem Soc; Sigma Xi. *Res:* Organic synthesis of strained ring systems; organometallic chemistry applied to organic synthesis; rearrangements of olefin systems induced by iron carbonyl catalysts and the mechanism of these rearrangements. *Mailing Add:* Dept Chem Univ NC Greensboro Greensboro NC 27412

BARBORIAK, JOSEPH JAN, b Slovakia, Feb 19, 23; nat US; m 56; c 3. BIOCHEMICAL PHARMACOLOGY, ALCOHOLISM. *Educ:* Inst Technol, Bratislava, BS, 44; Swiss Fed Inst Technol, ScD(nutrit biochem), 53. *Prof Exp:* Res assoc, Swiss Fed Inst Technol, 47-53; res assoc, Yale Univ, 54-58; group leader, Mead Johnson & Co, 59-61; assoc prof, 62-71, PROF PHARMACOL & TOXICOL, MED COL WIS, 71-, PROF PREV MED, 79- *Concurrent Pos:* Chief biochem sect, Res Serv, Vet Admin Ctr, 61- *Mem:* AAAS; Am Soc Pharmacol & Exp Therapeut; Am Inst Nutrit; Am Acad Allergy & Immunol; Soc Exp Biol & Med; Am Col Epidemiol; fel Am Col Clin Nutrit; fel Soc Behav Med; Am Col Epidemiol. *Res:* Lipid metabolism; pharmacology of alcoholism; nutrition; epidemiology of cardiac risk factors. *Mailing Add:* Zablocki VA Med Ctr 500 W Nat Ave Milwaukee WI 53295

BARBOUR, CLYDE D, b New York, NY, Oct 30, 35; m 62; c 2. ICHTHYOLOGY. *Educ:* Stanford Univ, AB, 58; Tulane Univ, PhD(zool), 66. *Prof Exp:* Asst prof biol, Univ Utah, 66-72 & Tuskegee Inst, 74-75; vis scientist, Mus Zool, Univ Mich, 72-73; vis asst prof, Miss State Univ, 73-74; asst prof biol, Tuskegee Inst, 74-75; ASSOC PROF BIOL, WRIGHT STATE UNIV, 75- *Mem:* AAAS; Am Soc Ichthyologists & Herpetologists; Soc Study Evolution; Soc Syst Zool. *Res:* Ichthyology, including systematics, zoogeography and ecology. *Mailing Add:* Dept Biol Sci Wright State Univ Dayton OH 45435

BARBOUR, MICHAEL G, b Jackson, Mich, Feb 24, 42; m 63; c 3. ECOLOGY, BOTANY. *Educ:* Mich State Univ, BSc, 63; Duke Univ, PhD(bot), 67. *Prof Exp:* From asst prof to assoc prof, 67-76, PROF BOT, UNIV CALIF, DAVIS, 76- *Concurrent Pos:* NSF grants, Study Coastal Vegetation, 69-71, Study Beach Vegetation, 73-75 & Study Montane Forests, 89-91; NSF/Int Biol Prog grant desert vegetation in NAm & SAm, 70-73; Univ Calif sea grant mgt dune vegetation, 74-76; fed & state grants to study salt marshes, 77-80, grasslands and meadows, 84-87; Fullbright fel, Univ Adelaide, Australia, 64; Guggenheim fel, 78; vis prof bot, Hebrew Univ, Israel, 79-80, La State Univ, 83. *Mem:* Am Inst Biol Sci; Ecol Soc Am; Brit Ecol Soc. *Res:* Physiological ecology; ecological life histories of coastal, desert and mountain plant species; community ecology of California vegetation. *Mailing Add:* Dept Bot Univ Calif Davis CA 95616

BARBOUR, MICHAEL THOMAS, b Stockbridge, Mich, Sept 21, 47; m 69; c 2. DEVELOPMENT OF BIOASSESSMENT APPROACHES. *Educ:* Eastern Mich Univ, BS, 70, MS, 72. *Prof Exp:* Biol tech, US Fish & Wildlife Serv, 70-72; instr zool, Eastern Mich Univ, 72; assoc biologist, NALCO Chem Co, 72-74; assoc ecologist, NUS Corp, 74-77; SR SCIENTIST, EA ENG, SCI & TECHNOL, 77- *Concurrent Pos:* Treas, Atlantic Estuarine Res Soc, 86-88; chmn, Exec Comt, N Am Benthological Soc, 89-90. *Mem:* Estuarine Res Fedn; Soc Toxicol & Chem; N Am Benthological Soc. *Res:* Development of bioassessment approaches for evaluating pollutant effects on indigenous aquatic communities; concept of an integrated assessment utilizing habitat quality; physicochemical measurements and biological condition indices. *Mailing Add:* EA Eng Sci & Technol Inc 15 Loveton Circle Sparks MD 21152

BARBOUR, ROGER WILLIAM, b Morehead, Ky, Apr 5, 19; m 38; c 3. VERTEBRATE ZOOLOGY. *Educ:* Morehead State Univ, BS, 38; Cornell Univ, MS, 39, PhD(vert zool), 49. *Hon Degrees:* DSc, Morehead State Univ, 83. *Prof Exp:* Instr zool, Morehead State Col, 39-40 & 46-47 & Western Ky State Col, 42-43; dir nature educ, Oglebay Inst, 49-50; from instr to prof zool, Univ Ky, Lexington, 50-84; RETIRED. *Concurrent Pos:* Vis prof, Bandung Technol Inst, 57-59; res grants, NIH, 60-62, 62-65 & 77-79. *Honors & Awards:* Wildlife Publ Award, Wildlife Soc, 74. *Mem:* Am Soc Mammal; Am Soc Ichthyol & Herpet; Wildlife Soc; Am Ornith Union; Sigma Xi. *Res:* Ecology, movements, life history, and distribution of vertebrates, particularly amphibians, reptiles, and mammals; author of ten books. *Mailing Add:* 4880 Tates Creek Pike Lexington KY 40515

BARBOUR, STEPHEN D, b St Louis, Mo, Oct 7, 42; m 67; c 2. GENETICS, BIOCHEMISTRY. *Educ:* Temple Univ, AB, 64; Princeton Univ, MA, 66, PhD(biol), 67. *Prof Exp:* Jane Coffin Childs Fund fel, 67-69; assoc prof, 69-80, SR RES ASSOC MICROBIOL, CASE WESTERN RESERVE UNIV, 80- *Concurrent Pos:* Sabbatical, Inst Molecular Biol, Paris, 75-76; NIH career develop award, 70-75. *Mem:* Am Soc Microbiol; Am Gen Soc. *Res:* Mechanism of genetic recombination; DNA replication and repair; control of enzyme synthesis. *Mailing Add:* Dept Pediat Infectious Dis Butterworth Hosp 100 Michigan Ave NW Grand Rapids MI 49503

BARBOUR, WILLIAM E, JR, b Evanston, Ill, Nov 1, 09; m 50; c 2. APPLIED PHYSICS, ELECTRICAL ENGINEERING. *Educ:* Mass Inst Technol, SB, 33. *Prof Exp:* Consult elec instrumentation, 33-36; engr, Raytheon Co, 36-39 & Boston Edison Co, 39-41; pres & chmn nuclear instrumentation & chem, Tracerlab, Inc, 45-56; pres nuclear controls, Controls for Radiation, Inc, 57-59; pres & chmn, Magnion, Inc, 60-64; dir, Alloyd Gen Corp, 66-66; consult, Barbour & Assocs, Inc, 66-76; RETIRED. *Concurrent Pos:* Mem, Engrs Joint Coun, Int Nuclear Cong, 54-56, New Eng Coun Comt Patent Revision, AEC, 55 & Southern Regional Educ Bd Nuclear Develop, 55-56; observer, Atoms for Peace Cong, Geneva, 55; exec mgr, Asn Nuclear Instrument Mfr, Inc, 66-72; dir, QSC Indust, Inc, 71-80; dir, Gen Aircraft Corp, 75-77. *Mem:* Am Nuclear Soc; Inst Elec & Electronics Engrs; Asn Advan Med Instrumentation. *Res:* Instrumentation; applied nuclear physics; high magnetic fields; superconductivity; nuclear medicine; x-ray, beam electronics and related vacuum technology. *Mailing Add:* Box 460 Concord MA 01742

BARBUL, ADRIAN, b Bucharest, Romania, Jan 11, 50; US citizen; m; c 2. SURGERY. *Educ:* City Col New York, BS, 69; Gen Sch Med, Bucharest, Romania, MD, 74; Am Bd Surg, dipl, 83. *Prof Exp:* Surg resident, Albert Einstein Col Med, 74-76, NIH res trainee trauma & burns, 76-78; surg resident, Sinai Hosp, Baltimore, 78-82; asst prof, 83-87, ASSOC PROF, DEPT SURG, JOHNS HOPKINS UNIV, 87-; ASST SURGEON-IN-CHIEF, SINAI HOSP, BALTIMORE, 82- *Concurrent Pos:* Mem, Comt Infections, Sinai Hosp, 83-, Tissue Comt, 85-, Nutrit Care Comt, 86-88, chmn, 89-, mem, Animal Care Comt, 88-, Inst Rev Bd, 88-; prin investr, NIH, 87-92, Johnson & Johnson Patient Care, Inc, 88-90; rep to Nat Asn Biomed Res, Soc Univ Surgeons, 90-91, counr- at-large, 91-93. *Honors & Awards:* Mead-Johnson Excellence of Res Award, 77; Resident Res Award, Asn Acad Surg, 82. *Mem:* Fel Am Col Surgeons; Soc Univ Surgeons; Am Soc Parenteral & Enteral Nutrit; Asn Acad Surg; Am Inst Nutrit; Am Soc Clin Nutrit; fel Am Col Nutrit; AAAS; NY Acad Sci; Soc Crit Care Med. *Res:* Studies on nutritional modification of the deleterious effects of trauma and tumors on protein metabolism; wound healing and immunity. *Mailing Add:* Dept Surg Sinai Hosp 2401 W Belvedere Ave Baltimore MD 21215

BARCELLONA, WAYNE J, b Chicago, Ill, Sept 22, 40; m 70; c 2. CELL BIOLOGY. *Educ:* Univ Southern Calif, AB, 62, MS, 65, PhD(biol), 70. *Prof Exp:* Fel biol, Sect Cell Biol, M D Anderson Hosp & Tumor Inst, Univ Tex, 70-72; res assoc, Univ Tex Med Br, 72-73; asst prof, 73-79, chmn dept, 81-87, ASSOC PROF BIOL, TEX CHRISTIAN UNIV, 79- *Mem:* Am Soc Androl; AAAS; Am Inst Biol Sci; Am Soc Cell Biol; Soc Study Reproduction. *Res:* Reproductive biology; differentiation; control and regulation of cell division and mammalian spermatogenesis. *Mailing Add:* Dept Biol Tex Christian Univ Ft Worth TX 76129-0000

BARCELO, RAYMOND, b Montreal, PQ, July 14, 31; m 54; c 4. PHYSIOLOGY, BIOCHEMISTRY. *Educ:* Univ Montreal, BSc, 50, MD, 57; Univ Pa, DSc(biochem), 60; FRCPC, 62. *Prof Exp:* Intern, Hosp Necker, Paris, France, 60-61; asst prof, 62-70, ASSOC PROF MED, UNIV MONTREAL, 71-, CHIEF SERV NEPHROLOGY, MAISONNEUVE HOSP, 64- *Concurrent Pos:* Head dept, Clin & Res Lab, Maisonneuve Hosp, 67-70. *Mem:* Am Soc Clin Res; Can Soc Clin Res; Can Soc Immunol; Can Soc Nephrol; Int Soc Nephrology. *Res:* Protein biochemistry; immunochemical studies of serum and urinary proteins; plasma seromucoids in various inflammatory conditions; perfusion of isolated kidney; diuretics; antibiotics; antihypertensive drugs. *Mailing Add:* Dept Nephrology Maisonneuve Hosp Montreal PQ H3C 3J7 Can

BARCELONA, MICHAEL JOSEPH, b Chicago, Ill, Aug 20, 49; m 72; c 3. GROUNDWATER GEOCHEMISTRY. *Educ:* St Mary's Col, Minn, BA, 71; Northeastern Univ, MS, 74; Univ PR, PhD(marine chem), 77. *Prof Exp:* Environ sect head, OMNI Res, Inc, 73-75; instr chem oceanog, Univ PR, 75-76; res assoc environ eng, Keck Lab Environ Eng Sci, 77 & Calif Inst Technol, 78-79; assoc prof sci, 79-80, AQUATIC CHEM SECT HEAD, WATER SURV DIV, ILL DEPT ENERGY & NATURAL RESOURCES, 80- *Concurrent Pos:* Vis res assoc, Ocean Chem Lab-Atlantic Ocean Marine Lab, Nat Oceanog & Atmospheric Admin, 76; consult, Ga Environ Res Lab, US Environ Protection Agency, 80 & Environ Processes Br, 82; assoc prof sci, Calif Inst Technol, 79-80. *Mem:* Am Chem Soc; Sigma Xi; Am Soc Limnol & Oceanog; Nat Water Well Asn. *Res:* Chemistry of environmental processes in natural waters; sediment geochemistry. *Mailing Add:* Dept Chem Western Mich Univ Kalamazoo MI 49008

BARCHAS, JACK D, b Los Angeles, Calif, Nov 2, 35. BEHAVIORAL NEUROCHEMISTRY. *Educ:* Pomona Col, BA, 56; Yale Univ, MD, 61. *Prof Exp:* Intern psychiat, Pritzker Sch Med, Univ Chicago, 61-62; res assoc, NIH, 62-64; res prof, Stanford Univ, 64-67, Nancy Friend Pritzer prof psychiat, 67-89; ASSOC DEAN NEUROSCI, SCH MED, UNIV CALIF, LOS ANGELES, 90-, PROF PSYCHIAT BEHAV SCI, 90- *Mem:* Inst Med-Nat Acad Sci; Am Soc Neurochem; Am Soc Pharmacol; AAAS; Am Chem Soc. *Mailing Add:* Dept Neurosci Univ Calif Sch Med 10833 Le Conte Ave Los Angeles CA 90024-1722

BARCHET, WILLIAM RICHARD, b Baltimore, Md, June 24, 43; m 68; c 2. WIND ENERGY, CLOUD PHYSICS. *Educ:* Drexel Inst Technol, BS, 66; Colo State Univ, MS, 68, PhD(atmospheric sci), 71. *Prof Exp:* Asst prof physics, Univ Northern Ariz, 70-71; vis asst prof meteorol, Univ Wis-Madison, 71-72, asst prof, 72-78; SR RES SCIENTIST EARTH SCI, BATTELLE PAC NORTHWEST LAB, 79- *Mem:* Am Meteorol Soc. *Res:* Atmospheric processes affecting the fate of pollutants; cloud modeling, microphysics and weather modification; wind energy resource assessment and data analysis. *Mailing Add:* 308 Scot St Richland WA 99352

BARCHFELD, FRANCIS JOHN, b Pittsburgh, Pa, Sept 12, 35; m 58; c 1. MECHANICAL ENGINEERING. *Educ:* Carnegie-Mellon Univ, BS, 58. *Prof Exp:* Engr, US Steel Corp, 58-60; engr, Hagan Controls Corp, 60-61; dir control res, Jones & Laughlin Steel Corp, 61-84; MGR, LTV, 84- *Mem:* Asn Iron & Steel Engrs; Instrument Soc Am; Am Soc Nondestructive Testing. *Res:* Design of automatic inspection equipment. *Mailing Add:* 1788 Farr's Garden Path Westlake OH 44145

BARCILON, ALBERT I, b Alexandria, Egypt, July 21, 37; m 61; c 2. DYNAMIC METEOROLOGY. *Educ:* McGill Univ, BEngPhysics, 60; Harvard Univ, MA, 61, PhD(fluid dynamics), 65. *Prof Exp:* Res assoc, Meteorol Dept, Mass Inst Technol & Harvard Univ, 65-66; scientist, Space Sci Lab, Gen Elec Co, Philadelphia, 66-68; asst prof meteorol & res assoc, 68-72, assoc prof, 72-80, PROF METEOROL, FLA STATE UNIV, 81- *Concurrent Pos:* Adj asst prof, Physics Dept, Drexel Inst & Villanova Univ, 66-68; consult, Naval Res Lab, Washington, DC, 70-; liaison scientist, Off Naval Res, London, 75-77. *Mem:* Am Meteorol Soc. *Res:* Dynamics of intense atmospheric vortices; geophysical fluid dynamics, internal gravity waves, nearshore circulations; turbulence; nonlinear baroclinic waves; moist flow over topography. *Mailing Add:* Geophys Fluid Dynam Inst Fla State Univ Tallahassee FL 32306

BARCILON, VICTOR, b Alexandria, Egypt, Oct 10, 39; m 61; c 1. APPLIED MATHEMATICS, FLUID DYNAMICS. *Educ:* McGill Univ, BSc, 59; Harvard Univ, AM, 60, PhD(math), 63. *Prof Exp:* Res assoc meteorol & oceanog, Harvard Univ, 63-64; fel meteorol, Univ Oslo, 64-65; res assoc, Mass Inst Technol, 65-66, asst prof appl math, 66-69; assoc prof math, Univ Calif, Los Angeles, 69-72; PROF MATH, DEPT GEOPHYS SCI, UNIV CHICAGO, 72- *Concurrent Pos:* Ed, J Soc Indust & Appl Math, 75- *Mem:* Soc Indust & Appl Math; Royal Astron Soc; Am Geophys Union. *Res:* Inverse problems of geophysics; dynamics of rotating/stratified flows. *Mailing Add:* Dept Geophys Scis Univ Chicago 5734 Ellis Ave Chicago IL 60637

BARCKETT, JOSEPH ANTHONY, b New Bedford, MA, Oct 8, 55; c 1. MATHEMATICAL STATISTICS, ANALYSIS & FUNCTIONAL ANALYSIS. *Educ:* Southeastern Mass Univ, BA, 77; Ariz State Univ, MA, 79. *Prof Exp:* Qual statistician, Sippican Ocean Systs, 80-83; qual eng, Roger Corp, 84-85, sr qual eng, 85-86, mgr qual assurance, Flexible Interconnections Div, 86-89, TECH QUAL CONTROL IMPLEMENTATION MGR, MICRO-INTERCONNECTIONS DIV, ROGERS CORP, 89- *Concurrent Pos:* Staff instr, Ariz State Univ, 84. *Mem:* Am Math Asn; Am Soc Qual Control. *Res:* Functional analysis. *Mailing Add:* Micro-Interconnections Div Rogers Corp PO Box 700 Chandler AZ 85244

BARCLAY, ALEXANDER PRIMROSE HUTCHESON, b Glasgow, Scotland, Apr 19, 13; Can citizen; m 38; c 2. ELECTRICAL ENGINEERING, COMMUNICATIONS ENGINEERING. *Educ:* McMaster Univ, BA, 35; Cornell Univ, MSc, 36. *Prof Exp:* Engr radio, Can Westinghouse Ltd, 36-38; engr commun, Northern Elec Ltd, 38-41; mgr radar, Res Enterprises Ltd, 41-45; gen mgr, Philips Electronics Indust Ltd, 45-71; gen mgr res, Ryerson Appl Res Ltd, 71-73; mgr elec, Can Post, 73-78; RETIRED. *Concurrent Pos:* Fel, Postal Eng, 78; assoc prof Eng, Ont. *Mem:* Inst Elec & Electronics Engrs. *Res:* Communications; general electronics; microwave, radar; closed circuit television and mobile radio. *Mailing Add:* Valley View Apts No 913 240 Scarlet Rd York Toronto ON M6N 4X4 Can

BARCLAY, ARTHUR S, b Minneapolis, Minn, Aug 5, 32. PLANT TAXONOMY. *Educ:* Univ Tulsa, BS, 54; Harvard Univ, MA, 58, PhD(biol), 59. *Prof Exp:* Res botanist, New Crops Res Br, USDA, 59-72, Med Plant Resources Lab, 72-78, Econ Bot Lab, 78-79; RETIRED. *Res:* Evolution; economic botany; cancer. *Mailing Add:* 8455 Greenbelt Rd No 101 Greenbelt MD 20770-2528

BARCLAY, BARRY JAMES, molecular biology, biochemistry, for more information see previous edition

BARCLAY, FRANCIS WALTER, b Vancouver, BC, Apr 11, 31; m 55; c 3. THERMALHYDRAULICS, REACTOR PHYSICS. *Educ:* Univ BC, BA, 55. *Prof Exp:* Shift supvr reactor opers, 58-63, reactor physicist opers, 63-66, res officer reactor physics & econ studies, 66-76, res officer exp thermalhydraul, 76-80, RES OFFICER ANAL YTIC THERMALHYDRAUL, ATOMIC ENERGY CAN, LTD, 80- *Res:* Analytical thermalhydraulics; validation of the two-fluid thermalhydraulics code ATHENA against experimental data; use of the code, primarily for reactor accident analysis calculations. *Mailing Add:* Box 313 Pinawa MB R0E 1L0 Can

BARCLAY, JACK KENNETH, b July 27, 38; c 2. CARDIOVASCULAR PHYSIOLOGY, SKELETAL MUSCLE PHYSIOLOGY. *Educ:* Univ Mich, PhD(physiol), 69. *Prof Exp:* FROM ASST PROF TO PROF PHYSIOL, UNIV GUELPH, 71- *Concurrent Pos:* Vis prof physiol, Monash Univ, 77-78; vis prof physiol, Mich State Univ, 84-85. *Mem:* Am Physiol Soc; Can Physiol Soc; Sigma Xi; NY Acad Sci; fel Am Col Sports Med. *Res:* muscle fatigue; local blood flow regulation; skeletal muscle function and physiology. *Mailing Add:* Sch Human Biol Col Biol Sci Univ Guelph Gordon St Guelph ON N1G 2W1 Can

BARCLAY, JAMES A(LEXANDER), b Barre, Vt, May 7, 18; m 45, 85; c 4. CHEMICAL ENGINEERING. *Educ:* Kalamazoo Col, BA, 39. *Prof Exp:* Asst chem, Brown Univ, 39-41, res chemist, 41-42; res chemist, Naval Res Lab, Washington, DC, 42-43, Manhattan Dist, Columbia Univ, 43-44 & Sharples Corp, Philadelphia, 44-47; from asst prof to assoc prof chem eng, Cath Univ Am, 47-66, head dept, 57-63; chem res engr, Col Park Metall Res Ctr, 66-73, CHEM ENGR, DIV MINERALS & MAT SCI, US BUR MINES, 73- *Mem:* Am Chem Soc; Am Inst Chem Engrs; Electrochem Soc. *Res:* Reactions between gases and solids; dielectric properties of polar solids; electrodeposition from fused salts. *Mailing Add:* 7314 Piney Branch Rd Takoma Park MD 20912

BARCLAY, JOHN ARTHUR, b Beach, NDak, Apr 30, 42; m 69; c 5. LOW TEMPERATURE PHYSICS, SOLID STATE PHYSICS. *Educ:* Univ Notre Dame, BS, 64; Univ Calif, Berkeley, PhD(chem-physics), 69; Monash Univ, Australia, dipl tertiary educ, 72. *Prof Exp:* Fel physics, Australian Inst Nuclear Sci & Eng, Monash Univ, 69-71, lectr, Monash Univ, 71-75, sr lectr, 76-77; staff scientist physics, Los Alamos Sci Lab, 77-85; ASTRONAUT TECH CTR, 85- *Concurrent Pos:* Consult-vis scientist, Los Alamos Sci Lab, 75-76. *Mem:* Am Phys Soc; Sigma Xi; Am Soc Med Eng. *Res:* Refrigeration; magnetism of solids. *Mailing Add:* Astronaut Tech Ctr 5800 Cottage Grove Rd Madison WI 53716

BARCLAY, LAWRENCE ROSS COATES, b Wentworth, NS, Oct 24; 28; m 50; c 3. ORGANIC CHEMISTRY. *Educ:* Mt Allison Univ, BSc, 50, MSc, 51; McMaster Univ, PhD(org chem), 57. *Prof Exp:* Asst prof chem, Mt Allison Univ, 51-55; vis lectr, McMaster Univ, 55-56; from assoc prof to prof, 56-67, CARNEGIE PROF CHEM & HEAD DEPT, MT ALLISON UNIV, 67- *Concurrent Pos:* Grants, Am Chem Soc Petrol Res Fund, 59-61, Res Corp, 60 & Nat Res Coun Can. *Mem:* Chem Inst Can. *Res:* Orthodiquaternary aromatic compounds; solvolysis hindered benzyl compounds; chemistry of crowded and sterically hindered aromatics; photochemistry of aryl-nitro, nitroso, and halo compounds; hindered aryl radicals; autoxidation of biomembranes; free radical rearrangements. *Mailing Add:* Dept Chem Mt Allison Univ Sackville NB E0A 3C0 Can

BARCLAY, ROBERT, JR, b Mt Vernon, NY, Apr 1, 28. INDUSTRIAL ORGANIC CHEMISTRY, POLYMER CHEMISTRY. *Educ:* Cornell Univ, AB, 48; Univ Md, PhD(chem), 57. *Prof Exp:* Patent chemist, Barrett Div, Allied Chem Corp, 48-51; res chemist, Am Cyanamid Co, 51-52; res chemist, Union Carbide Corp Plastics Div, 56-60, proj scientist, 60-69; sr res chemist, Chem Div, Thiokol Corp, 69-71, res specialist, 71-73, res scientist, 73-75, sr res scientist, 75-79; sr process develop specialist, Hydrocarbon Res Inc, 79-81, sect head, 81-86; CONSULT, AMOCO PERFORMANCE PROD, INC, 86- *Mem:* Am Chem Soc. *Res:* Hydrogenation and hydrocracking processes; photocurable polymers; condensation polymers and monomers. *Mailing Add:* Six Berrywood Dr Trenton NJ 08619-1906

BARCLAY, W(ILLIAM) J(OHN), b Corvallis, Ore, Nov 18, 12; m 39; c 2. ELECTRICAL ENGINEERING. *Educ:* Ore State Col, BS, 39; Stanford Univ, EE, 41, PhD, 49. *Prof Exp:* Asst prof elec eng, Stanford Univ, 42-45; res engr, Loran Pro, 45-46, assoc prof, 47; instr, Heald Eng Col, 47-48; asst prof, Ore State Col, 49-54; res engr, 54, assoc tech dir, Eng Res Dept, 57-58, assoc prof, 54-58, PROF ELEC ENG, NC STATE UNIV, 58- *Concurrent Pos:* Field engr, Western Elec Co, 55; res engr, Autonetics Div, NAm Aviation Co, 59; partic radio propagation course, Cent Radio Propagation Lab, Nat Bur Standards, 62. *Mem:* Sr mem Inst Elec & Electronics Engrs; Sigma Xi. *Res:* Electronic instrumentation and measurements; microwave theory, devices and components; radiation; propagation; antennas. *Mailing Add:* Elec/Comput Eng Dept NC State Univ 7911 Raleigh NC 27695

BARCLAY, WILLIAM R, b Golden, BC, May 25, 19; m 44; c 2. MEDICINE. *Educ:* Univ BC, BA, 41; Univ Alta, MD, 45. *Prof Exp:* Instr biochem, Univ Alta, 42; mem, Dept Nat Health & Welfare, Can, 46-47; asst, Dept Med, Univ Chicago, 48-49; mem, Dept Nat Health & Welfare, Can, 49-51; instr med, Univ Chicago, 51-52, asst prof chest dis, 52-56, assoc prof, 56-64, prof med, 64-70; dir sci activities & asst exec vpres, 70-75, VPRES, AMA & ED, J AMA, 75- *Concurrent Pos:* Palmer sr fel, Univ Chicago, 54-56; instr surg, Univ Alta, 50-51; bd mem, Nat Tuberc & Respiratory Dis Asn, 70; chmn tuberc panel, US-Japan Med Sci Prog, 70; mem adv comt to dir, NIH, 70. *Mem:* Am Thoracic Soc (pres, 63-64); fel NY Acad Sci; Am Col Chest Physicians; Am Soc Clin Invest. *Res:* Metabolic studies of tubercle bacillus. *Mailing Add:* Dept English Elmhurst Col 190 Prospect Elmhurst IL 60126

BAR-COHEN, YOSEPH, b Baghdad, Irag, Sept 3, 47; US citizen; m; c 2. NONDESTRUCTIVE EVALUATION SPECIALIST, COMPOSITE MATERIALS. *Educ:* Hebrew Univ, Jerusalem, Israel, BS, 71, MS, 73, PhD(physics), 79. *Prof Exp:* Consult nondestructive eval, Israel Aircraft Co, 71-72, sr specialist, 72-79; postdoctoral fel nondestructive eval, NRC Award, Air Force Mat Lab, 79-80; sr physicist nondestructive eval, Systs Res Lab, Dayton, Ohio, 80-83; prin specialist nondestructive eval, McDonnell Douglas Corp, Long Beach, 83-91; MEM TECH STAFF, JET PROPULSION LAB, PASADENA, 91- *Concurrent Pos:* Comt mem, Aerospace & Sonics, Am Soc Nondestructive Testing, 81-; adj prof, Mane Dept, Univ Calif, Los Angeles, 89- *Mem:* Am Soc Physics; Acoust Soc Am; Am Soc Nondestructive Testing. *Res:* Nondestructive evaluation technology for characterization of defects and determination of properties of aerospace material and structures; ultrasonic polar backscattering and leaky lamb wave for nondestructive evaluation of composite materials; author of 80 publications; granted six patents. *Mailing Add:* 3721 Fuchsia St Seal Beach CA 90740

BARCOS, MAURICE P, b Medellin, Colombia, Jan 13, 35; US citizen; m; c 3. PATHOLOGY. *Educ:* McGill Univ, BS, 55; Dalhousie Univ, MD, 62; Univ Chicago, PhD(biophysics), 75. *Prof Exp:* Res assoc biophysics, Univ Chicago, 63-65, resident & Seymour Coman fel path, 65-69, instr & fel hematopath, 69-70; asst prof path, Univ Southern Calif Sch Med, 70-73; asst clin prof, 74-80, cancer res pathologist, 73-81, assoc chief path, 81-90, ACTG CHIEF PATH, ROSWELL PARK CANCER INST, 90-; ASSOC CLIN PROF PATH, STATE UNIV NY, BUFFALO, 80- *Concurrent Pos:* Fel hematopath, Univ Southern Calif, 70-72; dir clin path lab, East Los Angeles Child & Youth Clin, 72; mem, Cancer Regional Studies Rev Comt, 82-84, Cancer Clinical Invest Rev Comt, 84-86, Nat Path Panel Lymphoma Clin Studies, NIH, 80-88; chmn, Path Comt, Cancer & Leukemia Group B, 79-, Sci Rev Comt, Roswell Park Cancer Inst, 88- *Mem:* Sigma Xi; Am Asn Cancer Res; Int Acad Path; Am Soc Hematology. *Res:* Neoplastic hematopathology; immunologic phenotyping of lymphomas. *Mailing Add:* Dept Path Roswell Park Cancer Inst Elm & Carlton Sts Buffalo NY 14263

BARCUS, WILLIAM DICKSON, JR, b Mineola, NY, Mar 19, 29; m 55; c 2. MATHEMATICS. *Educ:* Mass Inst Technol, SB, 50; Oxford Univ, DPhil(math), 55. *Prof Exp:* Instr, Princeton Univ, 55-56; from instr to asst prof, Brown Univ, 56-61; assoc prof, 61-66, PROF MATH, STATE UNIV NY, STONY BROOK, 66- *Concurrent Pos:* Vis sr res fel, Jesus Col, Oxford Univ, 65-66 & Math Inst, 68-69. *Mem:* Am Math Soc. *Res:* Algebraic topology. *Mailing Add:* Dept Math State Univ NY Stony Brook Main Campus Stony Brook NY 11794

BARCZAK, VIRGIL J, b Toledo, Ohio, Nov 29, 31; m 59; c 4. MINERALOGY, GEOLOGY. *Educ:* Univ Mich, BS, 58, MS, 59; Oklahoma City Univ, MBA, 66. *Prof Exp:* Petrographer, Ceramic Div, Champion Spark Plug Co, 59-64; res petrographer, Kerr McGee Corp, 64-77, sr proj anal chemist, 77-80, sr staff res chemist, 80-83, prin mineralogist, Tech Ctr, 83-91; RETIRED. *Honors & Awards:* Purdy Award, Am Ceramic Soc, 66. *Mem:* Am Ceramic Soc; fel Geol Soc Am; Mineral Soc Am; Mineral Asn Can; Nat Inst Ceramic Engrs; Soc Mining Engrs. *Res:* Determinative mineralogy; instrumental techniques applicable to quantitative determination of minerals and chemical compositions; general geology; ceramic technology; crystal chemistry; geochemistry; management. *Mailing Add:* 2500 NW 109th St Oklahoma City OK 73120

BARD, ALLEN JOSEPH, b New York, NY, Dec 18, 33; m 57; c 2. ELECTROANALYTICAL CHEMISTRY, PHYSICAL CHEMISTRY. *Educ:* City Col New York, BS, 55; Harvard Univ, PhD(chem), 58. *Hon Degrees:* Dr, Univ Paris, 86. *Prof Exp:* Anal chemist, Gen Chem Co Res Lab,

55; prof, 58-80, JACK S JOSEY PROF CHEM, UNIV TEX, AUSTIN, 80-, NORMAN HACKERMAN-WELCH REGENTS CHAIR CHEM, 85- *Concurrent Pos:* Consult, E I du Pont de Nemours & Co, Solar Energy Res Inst & Air Force Off Sci Res; Fulbright prof, Univ Paris, 73-74; Sherman Mills Fairchild scholar, Calif Inst Technol, 77; centenary lectr, Royal Soc Chemists, UK, 88; co-chmn, chem sci & technol bd, Nat Acad Sci; ed-in-chief, J Am Chem Soc. *Honors & Awards:* Harrison Howe Award, Am Chem Soc, 80, Willard Gibbs Award, 87; Carl Wagner Mem Award, Electrochem Soc, 81, Olin-Palladium Medal, 87; Bruno Breyer Mem Medal, Royal Australian Chem Inst, 84; Fisher Award, Am Chem Soc, 84; Charles N Reilley Award, Soc Electroanal Chem, 84; Iddles Lectr, Univ NH, 85; Baker Lectr, Cornell Univ, 87; Priestly Lectr, Pa State Univ, 88; Woodward Vis Lectr, Harvard Univ, 88; Mack Mem Award Lectr, Howard Univ, 88; W Heinlen Hall Lect Ser, Bowling Green State Univ, 89; Hill Mem Lectr, Duke Univ, 90; Mary E Kapp Lectr, Va Commonwealth Univ, 90; Venable Lectr, Univ SC, Chapel Hill, 90. *Mem:* Nat Acad Sci; Am Acad Arts & Sci; Am Chem Soc; Int Soc Electrochem; AAAS; Int Union Pure & Appl Chem (vpres, 90-91). *Res:* Electro-organic chemistry; kinetics and mechanisms of electrode reactions; semiconductor electrodes for solar energy conversion, electrogenerated chemiluminescence; electrochemical techniques and instrumentation; photoelectrochemistry; electroanalytical chemistry. *Mailing Add:* Dept Chem Univ Tex Austin TX 78712

BARD, CHARLETON CORDERY, b Chicago, Ill, Feb 2, 24; m 46; c 4. PHOTOGRAPHIC CHEMISTRY. *Educ:* Univ Chicago, PhB & BS, 46, MS, 48, PhD, 50. *Prof Exp:* Instr, Wilson Jr Col, 47-49; org chemist, Photog Technol Div, Eastman Kodak Co, 49-68, supvr, Processing Chem Sect, 68-86; RETIRED. *Concurrent Pos:* Consult, Image Conserv. *Mem:* Am Chem Soc; Soc Photog Sci & Eng; Int Color Coun; fel Soc Imaging Sci & Technol; Soc Motion Picture & TV Engrs. *Res:* Infrared spectroscopy and molecular structure; physical organic chemistry; conservation of images; environmental chemistry; photographic processing chemistry. *Mailing Add:* 74 Cornwall Ln Rochester NY 14617

BARD, DAVID S, b Columbua, Ohio, May 31, 35; m 81; c 4. GYNECOLOGIC ONCOLOGY. *Educ:* J B Stetson Univ, BS, 57; Columbia Col Physicians & Surgeons, MD, 61. *Prof Exp:* Clin assoc surg, Nat Cancer Inst, NIH, 63-65; intern & jr resident surg, Univ Hosp, Columbus, Ohio, 61-63; residency obstet & gynec, Boston Hosp Women, 65-68; fel gynec oncol, M D Anderson Hosp, Houston, Tex, 69-70; asst prof, Col Med, Univ Fla, 70-76; assoc prof, Med Sch, Univ Tenn, 76-78; PROF GYNEC ONCOL, COL MED, UNIV ARK, 78- *Concurrent Pos:* Consult, Little Rock Vet Admin Hosp, 76- *Mem:* Soc Gynec Oncologists; fel Am Col Obstet & Gynec; Int Soc Obstet & Gynec Path. *Res:* Biological behavior of ovarian neoplasmas and other neoplasmas of female genital tract. *Mailing Add:* Dept Obstet-Gynec 4301 W Markham St Sch Med Univ Ark Little Rock AR 72205

BARD, ENZO, b Luis Palacios, Santa Fe, Argentina, Oct 18, 38; m 65; c 3. MOLECULAR BIOLOGY, MICROBIOLOGY. *Educ:* La Plata Nat Univ, Argentina, BPharm, 60, MSc, 63, PhD(biochem), 72. *Prof Exp:* Head instr inorg chem, Sch Agron, La Plata Nat Univ, 65-66, head instr biochem, Sch Exact Sci, 68-72; fel molecular biol, Inst Cancer Res, Fox Chase Md, 72-74; vis prof molecular biol, dept biol, Univ Ottawa, Ont, 75-77; res asst prof cell biol, NY Univ Med Ctr, 77-81; ASST PROF PATH, SCH MED, HEALTH SCI CTR BROOKLYN, STATE UNIV NY, 82-, SCH GRAD STUDIES, 83- *Concurrent Pos:* Head, vaccine control sect, Biol Inst, Ministry Health, Buenos Aires, Argentina, 60-62 & Div Biol Control, 63-67; res consult cell biol, NY Univ Med Ctr, 75; dir clin chem, State Univ Hosp, State Univ NY, Health Sci Ctr Brooklyn, 81-83, dir clin virol, 81-87. *Mem:* AAAS; Am Soc Cell Biol; Am Soc Microbiol; NY Acad Sci (vchmn, biol sci sect, 88, chmn, 89-90); Sigma Xi. *Res:* Regulation of gene expression during adaptative differentiation in Leishmania; identification of antigens for possible use as immunogens and for diagnosis; production of these antigens using recombinant DNA technology. *Mailing Add:* Dept Path Health Sci Ctr Brooklyn State Univ NY 450 Clarkson Ave Box 25 Brooklyn NY 11203-2098

BARD, EUGENE DWIGHT, b Blackwell, Okla, Nov 8, 28; m 50; c 4. SCIENCE EDUCATION. *Educ:* Okla A&M Col, BS, 50, MS, 54; Univ Northern Colo, EdD(sci educ), 74. *Prof Exp:* Teacher phys sci, Planeview Pub Sch, 54-55; teacher phys sci, Pueblo Pub Sch, 55-65, dept head sci, 58-65, curric specialist math-sci, 63-65; instr chem, Southern Colo State Col, 65-66, from instr to assoc prof physics, 66-74; assoc prof, 74-78, PROF PHYSICS, UNIV SOUTHERN COLO, 78- *Mem:* Am Asn Physics Teachers. *Res:* Development and evaluation of non-traditional teaching materials and techniques. *Mailing Add:* 3500 St Clair Ave Pueblo CO 81005

BARD, GILY EPSTEIN, b Berlin, Ger, Sept 14, 24; nat US; m 45; c 2. PLANT ECOLOGY. *Educ:* Hunter Col, BA, 45; Cornell Univ, MS, 47; Rutgers Univ, PhD(bot), 51. *Prof Exp:* Res asst forest soils, Cornell Univ, 45-46, res asst bot, 46-47; res fel plant physiol, Brooklyn Bot Garden, 47-48; asst biol, Rutgers Univ, 48-51; lectr bot, 51-53, from instr to asst prof, 53-70, ASSOC PROF BIOL, LEHMAN COL, 70- *Concurrent Pos:* Res assoc, Columbia Univ, 53-54; ed, Torrey Bot Club Bull, 70-75. *Mem:* Am Bryol Soc; Bot Soc Am; Ecol Soc Am; Torrey Bot Club (treas, 55-56, pres, 78). *Res:* Mineral nutrition of forest species; secondary succession; vegetational dynamics of natural areas. *Mailing Add:* Herbert H Lehman Col Bronx NY 10468

BARD, JAMES RICHARD, b Chelsea, Okla. PHYSICAL CHEMISTRY, SPECTROSCOPY. *Educ:* Univ Okla, BS, 59; Univ Ark, MS, 64; Univ Mo, Kansas City, PhD(chem), 77. *Prof Exp:* Res chemist, Shell Chem Co, 62-67 & Vet Admin, 67-70; PROF CHEM, PENN VALLEY COMMUNITY COL, 70- *Mem:* Am Chem Soc; Sigma Xi. *Res:* Infrared and Raman spectra of polyatomic molecules; application of magnetic resonance to structure and equilibria of complex ions in mixed solvents; enzyme kinetics; electrochemistry; development of industry related laboratory experiments. *Mailing Add:* Penn Valley Community Col Kansas City MO 64111

BARD, JOHN C, b Huron, SDak, Apr 7, 25; m 47; c 3. FOOD SCIENCE. *Educ:* Huron Col, SDak, BS, 49; Iowa State Univ, MS, 50. *Prof Exp:* Technologist, Meat Prod Control, Oscar Mayer & Co, 51-52, head dept, 52-58, gen dept head, 58-61, dir res, Meat Prod & Packaging, 61-75, vpres res, 75-78, gen mgr, Sci Protein Labs, 78-83; RETIRED. *Concurrent Pos:* Mem sci adv comt, Am Meat Inst, 66-; mem liaison panel, Food & Nutrit Bd, Nat Acad Sci, coun mem, 74, vchmn coun, 75- *Mem:* Inst Food Technologists; Am Chem Soc. *Res:* Product, package and processing developments of application to meat industry, particularly an improved control of quality, costs and sanitation. *Mailing Add:* 4318 Winnequah Rd Monona WI 53716

BARD, MARTIN, b New York, NY, Dec 3, 42; m; c 2. GENETICS. *Educ:* City Col NY, BS, 65; Univ Calif, Berkeley, PhD(genetics), 71. *Prof Exp:* Fel genetics, Univ Sheffield, 71-73, State Univ NY, Albany, 73-74 & Univ Calif, Santa Cruz, 74-75; asst prof, 75-79, ASSOC PROF BIOL, IND UNIV-PURDUE UNIV, INDIANAPOLIS, 79- *Concurrent Pos:* Merck Sharp & Dohme Res Lab, 84-87, NIH, 86-88 & Eli Lilly & Co, 88- *Mem:* Am Soc Microbiol; AAAS. *Res:* Molecular genetics; sterol synthesis, cloning and regulation in Saccharomyces cerevisiae and candida albicans. *Mailing Add:* Dept Biol Ind Univ-Purdue Univ Indianapolis IN 46205

BARD, RAYMOND CAMILLO, b New Britain, Conn, Aug 26, 18; m; c 5. MICROBIOLOGY. *Educ:* City Col New York, BS, 38; Ind Univ, MA, 47, PhD(bact), 49; Am Bd Microbiol, dipl. *Prof Exp:* Med technologist, St John's Hosp, New York, 38-40 & St Joseph's Hosp, 40-42; from instr to asst prof bact, Ind Univ, 49-53; head microbiol sect, Smith Kline & French Labs, 53-56, assoc dir res & develop, 56-60; res dir, Nat Drug Co, 60-62; prof dent, Univ Ky, 62-65, prof cell biol, 65-67, dir res, Sch Dent, 62-63, asst dean, 63-64, asst vpres res & exec dir univ res found, 64-67; vpres, Col, 67-72, actg dean 68-72, emer prof cell & molecular biol & dean, Sch Allied Health Sci, Med Col Ga, 72-88; RETIRED. *Concurrent Pos:* Assoc prof, Hahnemann Med Col, 54-62 (part-time). *Mem:* Fel AAAS; fel Am Acad Microbiol; Am Soc Microbiol; fel Am Soc Allied Health Profs. *Res:* Microbiological metabolism; chemotherapy; dental research. *Mailing Add:* 422 Waverly Dr Augusta GA 30912-3121

BARD, RICHARD JAMES, b Mt Kisco, NY, July 19, 23; m 46; c 3. PHYSICAL CHEMISTRY. *Educ:* Univ Mich, BS, 44, MS, 48, PhD(chem), 51. *Prof Exp:* Instr colloid & surface chem, Univ Mich, 50-51; mem staff, Phys Chem Res, 51-56; group leader uranium chem group, Los Alamos Sci Lab, 53-73, leader phys chem & metall group, 73-82, asst div leader nuclear mat, 82-83; RETIRED. *Mem:* Am Chem Soc; Sigma Xi. *Res:* Physical and inorganic chemistry; gas-solid reaction kinetics; uranium processing; development of coated nuclear fuel particles; studies of pyrolytic carbons. *Mailing Add:* 975 Nambe Loop Los Alamos NM 87544

BARDACH, JOHN E, b Vienna, Austria, Mar 6, 15; US citizen; m 47. AQUATIC ECOLOGY, FISH BIOLOGY. *Educ:* Queen's Univ, Ont, BA, 46; Univ Wis, MSc & PhD(zool), 49. *Prof Exp:* From instr to asst prof biol, Iowa State Teachers Col, 49-53; from asst prof to prof fisheries zool, Univ Mich, 53-71; dir, Hawaii Inst Marine Biol, Univ Hawaii, 71-77; res assoc & asst dir acad affairs, East West Ctr, Resource Syst Inst, 77-85; RETIRED. *Concurrent Pos:* Dir, Fisheries Res Prog, Bermuda Govt, 55-58; adv, Royal Cambodian Govt, 58-59; ecol adv, UN Mekong Comt; sr vis fel sci, OEEC, 61; independent investr, Int Indian Ocean Exped, 64; rep, Nat Acad Sci Coun, Pac Sci Asn, 78-; res fel, Bellagio, 79; adj prof geog & oceanog, Univ Hawaii, 78-85. *Mem:* Fel AAAS; Ecol Soc Am; Am Soc Limnol & Oceanog; Am Soc Zoologists; Am Fisheries Soc. *Res:* Aquaculture; physiology and behavior of fishes; coral reef fishes; the senses of fishes; fisheries development; natural resources ecology; resource management. *Mailing Add:* 2979 Kalakaua Ave Honolulu HI 96815

BARDACK, DAVID, b New York, Apr 11, 32; m 60; c 2. VERTEBRATE PALEONTOLOGY, ICHTHYOLOGY. *Educ:* Columbia Univ, AB, 54, AM, 55; Univ Kans, PhD(zool), 63. *Prof Exp:* Asst zool, Columbia Univ, 55-56; vert paleontologist & ichthyologist, Am Mus Natural Hist, 56-58; instr zool, Hunter Col, 57-58; asst, Univ Kans, 59-62, curatorial asst vert paleont, Mus Natural Hist, 63, res assoc, 63-64; from asst prof to assoc prof, 64-75, assoc dean, Grad Col, 75-81, PROF BIOL SCI, UNIV ILL, CHICAGO CIRCLE, 75- *Concurrent Pos:* Res assoc, Field Mus Natural Hist. *Mem:* Fel AAAS; Soc Vert Paleont; Am Soc Ichthyologists & Herpetologists; Am Soc Zoologists; Paleont Soc. *Res:* Anatomy and evolution of living and fossil paleoniscoid and teleostean fishes. *Mailing Add:* Dept Biol Sci Univ Ill Box 4348 Chicago IL 60680

BARDANA, EMIL JOHN, JR, b New York, NY, May 21, 35; m 60; c 2. ALLERGY, IMMUNOLOGY. *Educ:* Georgetown Univ, BS, 57; McGill Univ, MD, CM, 61. *Prof Exp:* Intern gen med, Univ Calif Med Ctr, San Francisco, 61-62; gen med officer, Bremerton Naval Hosp, US Navy, Wash, 62-64, US Marine Corps, S Vietnam, 65; from resident med to instr med allergy immunol, Univ Ore Health Sci Ctr, 65-69; res trainee, Dept Allergy & Immunol Clin, Nat Jewish Hosp & Res Ctr, Denver, 69-71; from asst prof to assoc prof med, 71-79, PROF MED, VCHMN DEPT & HEAD DIV ALLERGY & CLIN IMMUNOL, ORE HEALTH SCI UNIV, 80- *Concurrent Pos:* Allergy Found Am training grant, 68-70; Am Acad Allergy travel grant, 69; dir, Am Bd Allergy & Immunol, 89-94. *Mem:* Fel Am Col Physicians; fel Am Acad Allergy & Immunol; Am Thoracic Soc; fel Am Col Allergy & Immunol; Am Col Legal Med; fel Am Col Chest Physicians. *Res:* Development of better understanding of the immunopathogenesis of mycotic hypersensitivity disorders of the lung; establishment of sensitive radioassays which will permit their earlier diagnosis; occupational lung diseases. *Mailing Add:* Ore Health Sci Univ 3181 SW Sam Jackson Park Rd L329 Portland OR 97201-3098

BARDASIS, ANGELO, b New York, NY, Nov 26, 36. SOLID STATE PHYSICS. *Educ:* Cornell Univ, AB, 57; Univ Ill, MS, 59, PhD(physics), 62. *Prof Exp:* From asst prof to assoc prof, 63-88, PROF PHYSICS, UNIV MD, COLLEGE PARK, 88- *Concurrent Pos:* NSF fel, 61-63. *Mem:* Am Phys Soc. *Res:* Superconductivity; helium three; helium four; semiconductors. *Mailing Add:* Dept Physics Univ Md College Park MD 20742

BARDASZ, EWA ALICE, b Warsaw, Poland; US citizen. CHEMICAL ENGINEERING, SURFACE CHEMISTRY. *Educ:* Warsaw Tech Univ, MS, 67; Case Western Reserve Univ, PhD(chem eng), 74. *Prof Exp:* Asst, Case Western Reserve Univ, 71-73; engr, Exxon Res & Eng Co, 73-74, res engr, 74-77, sr res engr chem eng, 77-79, sr staff engr, 79-81, eng assoc, 81-; AT UNION CAMP CORP. *Mem:* NY Acad Sci; Am Inst Chem Engrs; AAAS; Am Chem Soc. *Res:* Lubrication, absorption, corrosion prevention, monolayers and thin films; interfaces in composite materials; adhesion; catalysis. *Mailing Add:* 6900 Weatherby Dr Mentor OH 44060-8408

BARDAWIL, WADI ANTONIO, b Mexico, May 13, 21; nat US; m 47; c 6. PATHOLOGY. *Educ:* Nat Univ Mex, MD, 46; Am Bd Path, dipl. *Prof Exp:* Instr path, Univ Vt, 50-52; asst, Med Sch, Harvard Univ, 52-53, instr, 53-57; from asst prof to assoc prof path, Tufts Univ, 57-71, from asst prof to prof obstet & gynec, 57-73, prof path, 71-73, chmn dept, 73-77; assoc head dept, 80-83, PROF PATH, UNIV ILL, COL MED AT CHICAGO, 77- *Concurrent Pos:* Pathologist, Robert Breck Brigham Hosp & asst, Peter Bent Brigham Hosp, 54-55; res assoc, Med Sch, Harvard Univ, 57-59; dir dept path & med res, St Margaret's Hosp, 57-74. *Mem:* Am Soc Exp Path; fel Col Am Pathologists; AMA. *Res:* Gynecological and obstetrical research; regression of trophoblast and possible immune reactions involved; pathogenesis of collagen diseases and pathogenesis of vascular pathology. *Mailing Add:* Dept Path 1853 Polk St Chicago IL 60612

BARDEEN, JAMES MAXWELL, b Minneapolis, Minn, May 9, 39; m 68. THEORETICAL ASTROPHYSICS. *Educ:* Harvard Univ, AB, 60; Calif Inst Technol, PhD(physics), 65. *Prof Exp:* Res assoc physics, Calif Inst Technol, 65; res physicist, Univ Calif, Berkeley, 66; from asst prof to assoc prof astron, Univ Wash, 66-72; from assoc prof to prof physics, Yale Univ, 74-76; PROF PHYSICS, UNIV WASH, 76- *Concurrent Pos:* NSF res grants, 67-76; Sloan fel, 68-72; Fairchild fel, Calif Inst Technol, 77. *Mem:* AAAS; Am Phys Soc; Am Astron Soc; Int Astron Union. *Res:* Astrophysics of black holes and compact x-ray sources; stability of disks; dynamics of spiral structure in galaxies. *Mailing Add:* Dept Physics Univ Wash Seattle WA 98195

BARDEEN, JOHN, physics; deceased, see previous edition for last biography

BARDEEN, WILLIAM A, b Washington, Pa, Sept 15, 41; m 61; c 2. PARTICLE PHYSICS. *Educ:* Cornell Univ, AB, 62; Univ Minn, Minneapolis, PhD(physics), 68. *Prof Exp:* Res assoc physics, State Univ NY Stony Brook, 66-68; mem, Inst Advan Study, 68-69; from asst prof to assoc prof physics, Stanford Univ, 69-75; PHYSICIST, FERMI NAT ACCELERATOR LAB, 75- *Concurrent Pos:* Vis scientist, Europ Coun Nuclear Res, 71-72, Max-Planck Inst, Munich, 77, 86, Res Inst Fundamental Physics, 85; Sloan Found fel, 71,; Alexander von Humboldt Found sr US scientist award, 77; Guggenheim Found fel, 85. *Mem:* Am Phys Soc. *Res:* Theoretical particle physics and quantum field theory. *Mailing Add:* Theory Group Fermilab MS 106 PO Box 500 Batavia IL 60510

BARDELL, DAVID, b Braintree, Eng; US citizen. VIROLOGY. *Educ:* City Univ New York, BA, 66; Univ NH, MS, 68, PhD(microbiol), 72. *Prof Exp:* Fel microbiol, Harvard Univ, 72-73; res assoc, Univ Wis, 73-75; from asst prof to assoc prof, 75-86, PROF BIOL, KEAN COL NJ, 86- *Res:* Effect of adenoviruses on host cell metabolism; adenovirus induced cytopathology; survival and transmission of herpes simplex virus Type 1; history of microbiology. *Mailing Add:* Dept Biol Kean Col NJ Union NJ 07083

BARDELL, EUNICE BONOW, b Milwaukee, Wis, Feb 8, 15; m 72. PHARMACY. *Educ:* Univ Wis, BS, 38, MS, 49, PhD(pharm), 52. *Prof Exp:* From instr to prof pharm, 48-73, EMER PROF PHARM & HEALTH SCI, UNIV WIS-MILWAUKEE, 73- *Mem:* Am Pharmaceut Asn; Am Inst Hist Pharm. *Res:* History of pharmacy. *Mailing Add:* 1539 N 51st St Milwaukee WI 53208

BARDEN, JOHN ALLAN, b Providence, RI, Oct 30, 36; m 58; c 2. HORTICULTURE. *Educ:* Univ RI, BS, 58; Univ Md, MS, 61, PhD(hort), 63. *Prof Exp:* Instr hort, Univ Md, 63; from asst prof to assoc prof, 63-81, PROF HORT, VA POLYTECH INST & STATE UNIV, 82- *Mem:* Fel Am Soc Hort Sci (vpres educ, 84-85); Int Soc Hort Sci; Sigma Xi. *Res:* Horticultural physiology; fruit tree physiology with emphasis on apple production intensification, including size controlling rootstocks, tree spacing and photosynthetic efficency as influenced by various factors. *Mailing Add:* Dept Hort Va Polytech Inst & State Univ Blacksburg VA 24061-0327

BARDEN, LAWRENCE SAMUEL, b Little Rock, Ark, Nov 12, 42; m 64; c 1. FOREST ECOLOGY. *Educ:* Hendrix Col, BA, 64; Univ Maine, Orono, MS, 68; Univ Tenn, Knoxville, PhD(ecol), 74. *Prof Exp:* Park ranger, Nat Park Serv, 68-70; from asst prof to assoc prof biol, 70-88, PROF UNIV NC, CHARLOTTE, 88- *Concurrent Pos:* Grants, Found Univ NC, Charlotte, 75-89, Highlands Biol Sta, 76-79 & Southern Regional Educ Bd, 78-81 & NSF, 80. *Mem:* Ecol Soc Am; Southern Appalachian Bot Club. *Res:* Tree species replacement in old-growth forest; effects of fire on forest succession; competition of Lonicera Japonica with ground flora; ecology of Microstegiuan Vimineum. *Mailing Add:* Dept Biol Univ NC Charlotte Charlotte NC 28223

BARDEN, NICHOLAS, b Rotherham, Eng, July 8, 46; Can citizen; c 4. NEUROENDOCRINOLOGY. *Educ:* Aberdeen Univ, BSc, 67; Sussex Univ, PhD(biochem), 70. *Prof Exp:* Fel dept physiol, 70-73, from asst prof to assoc prof, 73-82, PROF NEUROENDOCRINOL, LAVAL UNIV, 82- *Concurrent Pos:* Sr mem, Med Res Coun Group, 73-86; vis scientist, NIH, 80-81, Duke Univ Med Ctr, 81. *Mem:* Endocrine Soc; Int Soc Neuroendocrinol; Can Soc Biochem; Soc Neurosci. *Res:* Control of neuropeptide gene expression; molecular genetics of affective disorders. *Mailing Add:* Le Centre Hospitalier de l'Univ Laval Ste-Foy PQ G1V 4G2 Can

BARDEN, ROLAND EUGENE, b Powers Lake, NDak, Sept 11, 42; m 67; c 3. BIOCHEMISTRY, SURFACTANT CHEMISTRY. *Educ:* Univ NDak, BS, 64; Univ Wis, Madison, MS, 66, PhD(biochem), 69. *Prof Exp:* Fel biochem, Case Western Reserve Univ, 68-70, sr instr, 70-71; from asst prof to prof Chem & Biochem, Univ Wyo, 71-89, dept head, 80-83, assoc dean arts & sci, 83-84, assoc vpres Acad Affairs, 84-89; VPRES ACAD AFFAIRS, MOORHEAD STATE UNIV, 89- *Concurrent Pos:* NIH fel, Case Western Reserve, 68-70; res career develop award, Nat Inst Gen Med Sci, Univ Wyo, 76-80. *Mem:* Am Chem Soc; Am Soc Biol Chemists. *Res:* Enzyme chemistry, lipid enzymology, affinity labels, physical properties of detergent solutions, micelles, organic geochemistry. *Mailing Add:* Moorhead State Univ Academic Affairs Moorhead MN 56563

BARDIN, CLYDE WAYNE, b McCamey, Tex, Sept 18, 34. BIOMEDICAL RESEARCH. *Educ:* Rice Univ, BA, 57; Baylor Univ, MS & MD, 62. *Hon Degrees:* Dr, Univ Caen, France, 90. *Prof Exp:* Intern & asst resident med, NY Hosp-Cornell Med Ctr, New York, 62-64; clin assoc, Endocrinol Br, Nat Cancer Inst, 64-67, sr investr, 67-70; from assoc prof to prof med, Milton S Hershey Med Ctr, Pa State Univ, 70-78, chief, Div Endocrinol, 70-78, assoc physiol, Dept Physiol, 70-78; VPRES, POPULATION COUN, 78-, DIR, CTR BIOMED RES, 78- *Concurrent Pos:* Ayerst travel award, 65; assoc ed, J Nat Cancer Inst, 69-70; consult physician, Lebanon Vet Admin Hosp, Pa, 70-78; mem, Contract Rev Panel Fertil Regulating Methods, Eval Br, Nat Inst Child Health & Human Develop, NIH, 71-76, Contract Rev Panel Contraceptive Develop Br, 73-76; mem, Task Force Pop Control: Use of Antihormones, 72-73, Sperm Maturation, 74-75; mem, Nat Prostate Cancer Task Force, Nat Cancer Inst, 73-78; coun mem, Endocrine Soc, 76-79, Soc Study Reproduction, 79-82, Am Soc Andrology, 84-87, exec coun, Int Soc Andrology, 81-85; Josiah Macy Jr Found scholar, 76-77; chmn, Int Comt Contraceptive Res, 78-; adj prof, Rockefeller Univ, New York, 78-; Wynd vis prof endocrinol & metab, Univ Rochester, NY, 81; co-chmn, Gorden Res Conf on Hormone Action, 84-85; adj prof med, Cornell Univ Med Col, 85- *Honors & Awards:* Commander of the Order of the Lion, Finland, 83; Serono Award for Outstanding Contrib in the Field of Male Reproductive Endocrinol, Am Soc Andrology, 84; David Rabin Lectr, Vanderbilt Med Sch, 85; Gregory Pincus Mem Lectr, Laurentian Hormone Conf, 86; Shirley & Maurice Saltzman Med Lectr, Mt Sinai Med Ctr, 87; Transatlantic Medalist, Brit Endocrine Socs, 88; Moritmer B Lipsett Mem Lectr, NIH, 90; Ramon Guiteras Lectr, Am Urol Asn, 90; Joji Ishigami Lectr, Kobe, Japan, 91. *Mem:* Inst Med-Nat Acad Sci; fel AAAS; Am Asn Physicians; Am Fedn Clin Res; Am Physiol Soc; Am Soc Andrology (vpres, 87-88, pres, 88-89); Am Soc Clin Invest; NY Acad Sci; Endocrine Soc; Soc Study Reproduction. *Res:* Reproductive medicine; endocrinology; contraceptive development; author of over 400 publications. *Mailing Add:* Ctr Biomed Res Population Coun 1230 York Ave New York NY 10021

BARDIN, RUSSELL KEITH, b Fresno, Calif, Mar 22, 32; m 65; c 2. INSTRUMENTATION PHYSICS. *Educ:* Calif Inst Technol, BS, 53, PhD(physics), 61. *Prof Exp:* Res fel, Calif Inst Technol, 61-62; res assoc, Columbia Univ, 62-67; res scientist physics, 67-74, staff scientist, 74-84, SR STAFF SCIENTIST, LOCKHEED PALO ALTO RES LAB, 84- *Mem:* Am Phys Soc. *Res:* Applications of nuclear and electronic techniques to instrumentation specialized to high-speed pulse measurement, signal and data processing systems; specialized high-speed pulse measurement, signal and data processing systems. *Mailing Add:* Lockheed Palo Alto Res Lab Bldg 251 Dp 91-50 Palo Alto CA 94304-1191

BARDIN, TSING TCHAO, b Tsintao, China, July 31, 38; US citizen; m 65; c 2. NUCLEAR PHYSICS. *Educ:* Univ London, BS & ARCS, 61; Columbia Univ, PhD(physics), 66. *Prof Exp:* Res asst physics, Columbia Univ, 62-66; asst prof, Haverford Col, 66-67; assoc, Univ Pa, 67; consult, 68-87, RES SCIENTIST, PALO ALTO RES LAB, LOCKHEED CORP, 87- *Mem:* Am Phys Soc. *Res:* Experimental physics in nuclear spectroscopy and nuclear structure studies and application to solid state material sciences. *Mailing Add:* 120 Bear Gulch Dr Fortola Valey CA 94028

BARDO, RICHARD DALE, b Garner, Iowa. QUANTUM CHEMISTRY. *Educ:* Univ Wyo, BS, 66; Iowa State Univ, PhD(phys chem), 73. *Prof Exp:* Res assoc, dept chem, Univ Calif, Irvine, 73-77; RES CHEMIST, NAVAL SURFACE WEAPONS CTR, 77- *Mem:* Am Phys Soc. *Res:* Electronic structure calculations on explosive molecules; calculations of chemical reaction rates and electronic spectra at high pressure. *Mailing Add:* 10413 Tullymore Dr Hyattsville MD 20783

BARDOLIWALLA, DINSHAW FRAMROZE, b Bombay, India, Jan 19, 45; m 71; c 2. POLYMER CHEMISTRY, PLASTICS ENGINEERING. *Educ:* Univ Bombay, BS, 65, 68; Univ Lowell, MS, 71, PhD(polymer sci), 74; Rutgers Univ, MBA, 86. *Prof Exp:* Polaroid fel, Dept Chem, Univ Lowell, 71-74; res chemist polymer chem, Am Cyanamid Co, Stamford, Conn, 74-80, sr res chemist, 80-; mgr polymer chem, Diamond Shamrock Corp, Morristown, NJ, 82-85, mgr develop, 85-87; tech dir, 87-88, VPRES RES & TECHNOL, OAKITE PROD INC, BERKELEY HEIGHTS, NJ, 88- *Mem:* Am Chem Soc; Soc Plastic Engrs; Am Mgt Asn; Nat Coil Coaters Asn; Chem Specialties Mfrs Asn. *Res:* Rheslogy modifiers for paints, associative thickness, adhesives, corrosion protection coatings for automotive, aircrafts and aerospace applications; pretreatment of plastics prior to painting and decoration, plastics recycling; synthetic coolants and lubricants for metalworking operations; flame retardant acrylic polymers. *Mailing Add:* Three Darlene Ct Randolph NJ 07869-2948

BARDON, MARCEL, b Paris, France, Sept 16, 27; US citizen; m 64; c 3. PHYSICS ADMINISTRATION. *Educ:* Univ Paris, dipl, 52; Columbia Univ, MA, 56, PhD(physics), 61. *Prof Exp:* Instr physics, Columbia Univ, 59-63, res assoc & asst dir, Nevis Labs, 63-64, assoc dir, 64-66, sr res assoc, 64- 70, dep dir, 66-70; prog dir, Phys Sect, Nat Sci Found, 70-71, head physics sect, 71-75, dep div dir physics, 75-77, dir, physics div, 77-79, asst dir, 82-85; sci-officer, US deleg UNESCO, 79-81; dep asst secy gen sci, NATO, 86-88; DIR, DIV PHYSICS, NAT SCI FOUND, 88- *Mem:* Fel AAAS; fel Am Phys Soc; Sigma Xi. *Res:* Intermediate and high energy physics. *Mailing Add:* 1051 Waverly Way McLean VA 22101

BARDOS, DENES I(STVAN), b Budapest, Hungary, Feb 24, 38; US citizen; m 56. BIOMATERIALS, METALLURGY. *Educ:* Univ Ill, BS, 62, MS, 63, PhD(metall eng), 66. *Prof Exp:* Res assoc magnetism, Argonne Nat Lab, 65-66, appointee, 66-67, metallurgist, 67-69; chief metallurgist & res dir, Ft Wayne Metals, Inc, 69-72; dir res, 72-82, SR VPRES SCI AFFAIRS, ZIMMER USA, INC, 82- *Concurrent Pos:* Am Soc Metals Scholar. *Mem:* Am Soc Metals; Soc Biomat; Ortho Res Soc. *Res:* Theory of alloying behavior and magnetic phenomena; crystal chemistry of intermetallic compounds; wire technology; superalloys; orthopaedic implant materials; composite materials for total joint prostheses; development of hot isostatically pressed cobalt alloy for surgical implants; titanium alloys for medical applications; ceramics for surgical implants. *Mailing Add:* Smith & Nephew Richards 1450 Brooks Rd Memphis TN 38116

BARDOS, THOMAS JOSEPH, b Budapest, Hungary, July 20, 15; nat US; m 51. MEDICINAL CHEMISTRY. *Educ:* Tech Univ Budapest, ChE, 38; Univ Notre Dame, PhD(chem), 49. *Prof Exp:* Chem engr, Vacuum Oil Co, 38-46; res fel, Biochem Inst, Univ Tex, 48-51; res chemist, Armour & Co, 51-55, sect head, 55-59, lab head, 59-60; PROF MED CHEM, STATE UNIV NY BUFFALO, 60- *Concurrent Pos:* Consult, Roswell Park Mem Inst, 60- & Walter Reed Army Inst Res, 79-81. *Honors & Awards:* Ebert Prize, Am Pharmaceut Asn, 71; Schoellkopf Medal, Am Chem Soc, 74. *Mem:* Fel AAAS; Am Chem Soc; fel NY Acad Sci; fel Acad Pharmaceut Sci; Am Soc Biol Chemists; hon mem Hungarian Acad Sci. *Res:* Organic synthesis; isolation and identification of natural growth factors and inhibitors; biosynthetic pathways; design of antimetabolites; chemotherapy; synthesis of anticancer agents; modification of nucleic acids. *Mailing Add:* Dept Medicinal Chem State Univ NY Buffalo NY 14260

BARDSLEY, CHARLES EDWARD, b Newport, RI, Apr 23, 21; m 46; c 1. SOIL CHEMISTRY, PLANT PHYSIOLOGY. *Educ:* Univ RI, BS, 48; Miss State Univ, MS, 50, PhD(soils), 59. *Prof Exp:* Soil conservationist, Soil Conserv Serv, USDA, 50-52, soil scientist, Res Serv, 52-61; assoc prof soils, Clemson Univ, 61-68; res assoc, Am Can Co, 68-72; vpres, Agritec Co, 72-75; prin investr, Mallinckrodt, Inc, 75-80, res assoc prod develop, 80-82, mgr, speciality agr technol, 83-; RETIRED. *Mem:* Am Soc Agron; Soil Sci Soc Am; Weed Sci Soc Am; AAAS; Entomol Soc Am. *Res:* Pesticide and fertilizer formulation; synthetic soils for horticulture and forestry; environmental fate and behavior of pesticides and plant nutrients. *Mailing Add:* 1113 Cotita Dr Guatier MS 39553

BARDSLEY, JAMES NORMAN, b Ashton-U-Lyne, Eng, May 10, 41; US citizen; m 65; c 3. ATOMIC & MOLECULAR COLLISIONS, SHORT PULSE LASERS. *Educ:* Cambridge Univ, BA, 61, MA, 65; Manchester Univ, PhD(physics), 65. *Prof Exp:* Lectr physics, Univ Manchester, 65-69; asst prof physics, Univ Tex, Austin, 69-70; prof physics, Univ Pittsburgh, 70-87; ASSOC DIV LEADER, LAWRENCE LIVERMORE NAT LAB, 87- *Concurrent Pos:* Vis fel, Joint Inst Lab Astrophys, Boulder, 74-75 & 85-86; vis astrom, Observatoire Paris, Meudon, 81; prog dir, NSF, 84-85; secy, Int Conf Physics Electronic & Atomic Collisions, 89-95. *Mem:* Fel Am Phys Soc. *Res:* Atomic and molecular collisions; physics of short-pulse lasers; gas discharges; pulsed-power systems; plasma processing; effluent gas cleaning; ion-surface interactions. *Mailing Add:* L-296 Lawrence Livermore Nat Lab Livermore CA 94550

BARDUHN, ALLEN J(OHN), b Seattle, Wash, Aug 24, 18; m 46; c 2. CHEMICAL ENGINEERING. *Educ:* Univ Wash, BS, 40, MS, 42; Univ Tex, PhD(chem eng), 55. *Prof Exp:* Jr engr, Tide Water Oil Co Assocs, Calif, 41-44, engr, 44-47; sr engr, 47-49; prof chem eng, Syracuse Univ, 54-86; RETIRED. *Concurrent Pos:* UNESCO expert chem eng, Indust Univ Santander, 69-70; consult various industs & govt; vis prof, Shiraz Tech Inst, Iran, 77-78. *Mem:* Am Chem Soc; Am Inst Chem Engrs. *Res:* Thermodynamics and kinetics of gas hydrate systems and their use in desalting sea water and waste water; growth rate of ice crystals. *Mailing Add:* Dept Chem Eng Syracuse Univ Hinds Hall Syracuse NY 13244

BARDWELL, GEORGE, b Denver, Colo, Jan 6, 24; m 46; c 5. MATHEMATICS, STATISTICS. *Educ:* Univ Colo, BS, 44, MS, 49, PhD(math), 61, MAA, 65; Bowdoin Col & Mass Inst Technol, certs, 45. *Prof Exp:* Res technician, Philadelphia Navy Yard, 46-47; instr statist, Univ Colo, 48-49; res assoc, Bur Bus & Social Res, Univ Denver, 49-54, asst prof statist & res, 54-61, assoc prof & chmn dept, 61-62, joint assoc prof math & statist, 63-71, prof math, 71-90; RETIRED. *Concurrent Pos:* Gulf Oil Co fac res grant, 61; consult, indust firms & govt agencies; impartial labor arbitrator, 62-; mem, Fed Mediation & Conciliation Serv. *Mem:* Am Math Asn; Am Statist Asn; Inst Math Statist. *Res:* Mathematical analysis of discrete probability distributions; sampling theory; social and economic research. *Mailing Add:* 2201 S Harrison Denver CO 80210

BARDWELL, JENNIFER ANN, b Saskatoon, Sask, Oct 3, 57; m 81. PHYSICAL CHEMISTRY. *Educ:* Univ Sask, BSc, 78; Univ Western Ont, PhD(chem), 83. *Prof Exp:* Res fel, Univ Toronto, 83-85; res assoc, 85-86, RES OFFICER, NAT RES COUN CAN, 86- *Mem:* Am Chem Soc; Chem Inst Can; Electrochem Soc. *Res:* Spectroscopic and electrochemical properties of thin solid films. *Mailing Add:* Inst Microstruct Sci Nat Res Coun Can Ottawa ON K1A 0R9 Can

BARDWELL, JOHN ALEXANDER EDDIE, b Appin, Ont, Dec 25, 21; m 54; c 3. PHYSICAL CHEMISTRY. *Educ:* Univ Western Ont, BA, 44, MSc, 46; McGill Univ, PhD(chem), 48; Oxford Univ, DPhil, 50. *Prof Exp:* Spec lectr chem, Univ Sask, 50-51, from asst prof to prof chem, 51-84, asst dean arts & sci, 61-67, univ secy, 68-74, asst to pres, 67-84, dir res admin, 78-84; RETIRED. *Mem:* Chem Inst Can. *Res:* Polymers; combustion of hydrocarbons; radiation chemistry. *Mailing Add:* Box 97 Cochin SK S0M 0L0 Can

BARDWELL, STEVEN JACK, b Denver, Colo, Dec 2, 49; m 68; c 2. PLASMA PHYSICS, CONTINUUM MECHANICS. *Educ:* Swarthmore Col, BA, 71; Univ Colo, PhD(physics), 76. *Prof Exp:* Dir plasma physics, Fusion Energy Found, 77-80; VPRES, ADVAN TECHNOL ENTERPRISES INC, 77- *Concurrent Pos:* Ed Int J Fusion Energy, 77- *Mem:* Am Phys Soc. *Res:* Self-organized and nonlinear structure in plasmas and fluids with application to fusion research and meteorology. *Mailing Add:* 18 Calumet Ave Hastings-on-Hudson NY 10706

BARDWICK, JOHN, III, b Harvey, Ill, Mar 10, 31; m 54; c 3. PHYSICS. *Educ:* Purdue Univ, BS, 53; Univ Mich, MS, 59, PhD(physics), 64. *Prof Exp:* Res assoc, Univ Mich, 64, from instr to asst prof physics, 64-79, assoc prof, 69-80. *Mem:* Am Phys Soc. *Res:* Accelerator design and development; experimental nuclear physics. *Mailing Add:* 1206 Harbrooke Ann Arbor MI 48103

BARE, BARRY BRUCE, b South Bend, Ind, Apr 24, 42; c 2. FOREST MANAGEMENT, FOREST VALUATION. *Educ:* Purdue Univ, BSF, 64, PhD(quant methods), 69; Univ Minn, MS, 65. *Prof Exp:* Res asst, Univ Minn, 64-65; res asst, Purdue Univ, 66 & 68, instr forest financial mgt, 68-69; from asst prof to assoc prof forest resources, 69-82, PROF FOREST MGT & QUANTITATIVE ANAL YSIS, UNIV WASH, 83-, CHMN QUANTITATIVE ECOL & RESOURCE MGT, INTERDISCIPLINARY GRAD PROG, 90- *Mem:* Soc Am Foresters; Inst Mgt Sci; Opers Res Soc Am. *Res:* Application of mathematical programming to forest production processes; application of financial management principles to forestry-oriented problems; forest taxation; forest biometry. *Mailing Add:* Col Forest Resources Univ Wash Seattle WA 98195

BARE, CHARLES L, b Okla City, Okla, Apr 22, 32; m 53; c 2. FUEL TECHNOLOGY, PETROLEUM ENGINEERING. *Educ:* Univ Okla, BS, 54. *Prof Exp:* Engr, Mobil Oil, 54-61; res scientist, Conoco Inc, 61-67, mgr comput applns, 67-71, mgr planning & admin, 71-80, oper mgr, Far East & Africa, 80-84; dir & gen mgr planning & admin, Conoco Ltd, London, 84-90, MGR BUS DEVELOP & INDUST RELS, CONOCO INC, 90- *Concurrent Pos:* Dir, Soc Petrol Engrs, UK, 87- *Mem:* Soc Petrol Engrs (treas, 77-78 & pres, 79); Soc Petrol Engrs UK; Am Inst Mining Engrs. *Mailing Add:* Conoco Inc PO Box 2197 DU 3024 Houston TX 77252

BARE, THOMAS M, b Lancaster, Pa, Nov 24, 42; m 67; c 3. SYNTHETIC ORGANIC CHEMISTRY & MEDICINAL CHEMISTRY, CENTRAL NERVOUS SYSTEM RESEARCH. *Educ:* Pa State Univ, BS, 64; Mass Inst Technol, MS, 66, PhD(org chem), 68. *Prof Exp:* Sr res chemist, Lakeside Labs div, Colgate-Palmolive Co, 69-75; sr chemist, Morton Chem Co Div, Morton-Norwich Prod, Inc, 75-76; PRIN RES CHEMIST, BIOMED RES DEPT, ICI PHARMACEUT GROUP, 76- *Mem:* Am Chem Soc; Sigma Xi; Int Soc Heterocyclic Chem. *Res:* Synthesis and design of novel therapeutic agents for diseases of the central nervous system; natural products chemistry. *Mailing Add:* Med Chem Dept ICI Pharmaceut Group Wilmington DE 19897

BAREFIELD, EDWARD KENT, b Hopkinsville, Ky, Feb 26, 43; m 66; c 2. INORGANIC CHEMISTRY, ORGANOMETALLIC CHEMISTRY. *Educ:* Western Ky Univ, BS, 65; Ohio State Univ, PhD(chem), 69. *Prof Exp:* Assoc, Cent Res Dept, E I du Pont de Nemours & Co, Inc, 69-70; asst prof chem, Univ Ill, Urbana, 70-76; ASSOC PROF CHEM, GA INST TECHNOL, 76- *Mem:* Am Chem Soc; Sigma Xi. *Res:* Inorganic and organometallic synthesis; homogeneous catalysis. *Mailing Add:* 1622 Montcliff Ct Decatur GA 30333

BAREFOOT, ALDOS CORTEZ, JR, b Angier, NC, Feb 25, 27; m 49; c 3. WOOD SCIENCE & TECHNOLOGY. *Educ:* NC State Univ, BS, 50; MWT, 51; Duke Univ, DF(wood anat), 58. *Prof Exp:* Qual control engr, Henry County Plywood Corp, Va, 51-52; asst exp statistician, NC State Univ, 52-54, technologist, 54, asst prof, 54-59; forestry adv, Int Coop Admin, Pakistan, 59-61; chief of party, Reforestation Proj, Sri Lanka, 82-84; assoc prof wood technol, 62-68, leader wood prod exten, 71-75, head, div univ studies, 75-82, prof wood technol, 68-86, prof univ studies, 75-86, EMER PROF, NC STATE UNIV, 86- *Concurrent Pos:* Fulbright-Hays fel, Univ Oxford, 73-74; consult, 86- *Mem:* Int Asn Wood Anatomists; Forest Prod Res Soc; fel Brit Inst Wood Sci; Tree-Ring Soc; Sigma Xi. *Res:* Wood anatomy; tropical woods; statistical quality control; dendrochronology. *Mailing Add:* 3401 Hampton Rd Raleigh NC 27607-3131

BAREIS, DAVID W(ILLARD), b Irondequoit, NY, Feb 9, 22; m 47; c 4. CHEMICAL ENGINEERING. *Educ:* Univ Rochester, BS, 43; Mass Inst Technol, MS, 47. *Prof Exp:* Assoc chem warfare develop lab, Mass Inst Technol, 43-44 & dept chem eng, 46-47; assoc dept metall, US Bur Ships Proj, 44-46; group leader chem processing group, Nuclear Eng Dept, Brookhaven Nat Lab, 47-53; supvr separations chem group, Atomic Energy Res Dept, NAm Aviation, Inc, 53-56, supvr process develop group, Reactor Develop Dept, Atomics Int Div, 56-58, res specialist, 58-60; res scientist, Nuclear Syst Div, Marquardt Corp, 60-61, mgr nuclear res dept, Astro Div, 61-63 & nuclear prod, Power Syst Div, 63-64; consult, 64-65; staff engr, Agbabian Assocs Inc, 65-67, tech adminr, 67-70, prin engr, 70-72; exec vpres, ATSA Calif, Inc, 82-86; CONSULT ENGR, REGIST NUCLEAR ENGR, CALIF, 86- *Mem:* AAAS; Am Nuclear Soc. *Res:* Energy conversion; dynamic systems; direct conversion systems; nuclear reactor, radioisotope and chemical heat sources; salt water conversion; power generation; radioisotope propulsion; radioisotope sterilization; nuclear systems design. *Mailing Add:* 31912 Kingspark Ct West Lake Village CA 91361

BAREIS, DONNA LYNN, b Abington, Pa, May 1, 54; m 81. MEDICAL DEFENSE, CHEMICAL WARFARE AGENTS. *Educ:* Pa State Univ, BS, 75; Duke Univ, PhD(pharmacol), 79. *Prof Exp:* Res assoc, Lab Clin Sci, Sect Pharmacol, NIH, 79-81; prin investr pharmacol, US Army Med Res Inst Chem Defense, Aberdeen Proving Ground, 81-82, mgr Super-fund Occup Health Proj, US Army Corp Engrs, Washington, DC, 82-83; SR SCI, DIV MGR & ASST VPRES, MED CHEM DEFENSE, SCI APPLN INT CORP,

JOPPA, MD, 83- *Mem:* AAAS; Sigma Xi; Am Chem Soc; Soc for Risk Anal. *Res:* Mechanisms of toxicity of chemical warfare agents and antidotes, worker health and safety procedures for handling hazardous materials; biochemical mechanisms of receptor activation. *Mailing Add:* 8805 Blue Sea Dr Columbia MD 21046

BAREISS, ERWIN HANS, b Schaffhausen, Switz, May 10, 22; US citizen; m 60; c 2. APPLIED MATHEMATICS, COMPUTER SCIENCES. *Educ:* Schaffhausen Teachers Col, BA, 43; Univ Zurich, PhD(math), 50; Lehigh Univ, MS, 52. *Prof Exp:* Mem staff, C H Knorr, Switz, 43-45; instr math, Schaffhausen Teachers Col & Winterthur Tech Col, 45-51; mathematician, Naval Res & Develop Ctr, Washington, DC, 52-54, specialist math physics, 54-56, consult numerical anal, 56-57; consult appl math, Argonne Nat Lab, 57-63; vis lectr, 69-70, dir tech comput, 88-89, PROF ELEC ENG & COMPUT SCI, APPL MATH & ENG SCI & NUCLEAR ENG, NORTHWESTERN UNIV, 70- *Concurrent Pos:* Janggen-Pohen fel, 49-50; Baldwin Res fel, 51-52; lectr, Univ Md, 55-56; sr mathematician, Argonne Nat Lab, 63-76; vis lectr, Harvard Univ, 64; chmn comput sci prog, Northwestern Univ, 77-88. *Mem:* Am Math Soc; Sigma Xi; Soc Indust & Appl Math; Swiss Math Soc. *Res:* Function theory of a hypercomplex variable; theoretical and applied mechanics; applied mathematics and numerical analysis; mathematical structure and methods of transport theory; design of CRD comupter. *Mailing Add:* Dept Elec Eng & Comput Sci Northwestern Univ Evanston IL 60208

BAREL, MONIQUE, b Boulogne, France, Mar 8, 52; m 74; c 2. IMMUNOLOGY. *Educ:* Univ Sci, Paris, MSc, 73, PhD(biochem), 87. *Prof Exp:* Sr researcher, 82-86, MEM STAFF, NAT INST HEALTH & MED RES, ST-ANTOINE, PARIS, 87- *Mem:* Am Asn Immunologists. *Res:* Structure and function of gp140, the receptor for c3d fragment of complement and for Epstein-Barr virus, expressed on human B lymphocytes. *Mailing Add:* Centre INSERM Hospival St-Antoine Paris 75012 France

BARENBERG, ERNEST J(OHN), b Herndon, Kans, Apr 9, 29; m 53; c 6. CIVIL ENGINEERING. *Educ:* Kans State Univ, BS, 53; Univ Kans, MS, 58; Univ Ill, PhD(civil eng), 65. *Prof Exp:* Engr, Cessna Aircraft Co, 53; instr eng drawing, Univ Kans, 55-56, from instr to asst prof civil eng, 56-60; from res asst to res assoc, 60-65, from asst prof to assoc prof, 64-71, assoc head civil eng, 80-85, PROF CIVIL ENG, UNIV ILL, URBANA-CHAMPAIGN, 71- *Concurrent Pos:* Mem, Hwy Res Bd, Nat Acad Sci-Nat Res Coun; indust consult, 58-; consult, Nat Technol Comt Pozzolanic Pavements, 65-, US Army Corps Engrs, 71-74, US Air Force, 71-75 & Fed Aviation Admin, 74-; Paul Kent fac travel award, 78. *Mem:* Am Soc Civil Engrs; Am Soc Eng Educ; Nat Soc Prof Engrs; Am Soc Testing & Mat. *Res:* Structural engineering; pavement behavior analysis and pavement design; paving materials. *Mailing Add:* Dept Civil Eng Univ Ill Urbana Campus 208 N Romine St Urbana IL 61801

BARENBERG, SUMNER, b Boston, Mass, May 11, 45; m 70. BIOMEDICAL & PACKAGING MATERIALS, ENGINEERING PLASTICS. *Educ:* Case Inst Technol, BS, 67; Case Western Reserve Univ, MS, 70, PhD(macromolecular sci), 76; Northwestern Univ, MM, 91. *Prof Exp:* Chief operating officer, Cambridge Chem Co, Inc, 70-74; biomat proj dir, Cleveland Clin Found, 75-76; fac, Univ Mich, Ann Arbor, 77-80; bus mgr, E I du Pont & Co, Inc, 80-88; dir res, Baxter Health Care, 88-90; VPRES & CHIEF TECH OFFICER, PCA(TENNECO), 91- *Concurrent Pos:* Consult, 70-; adj staff scientist, Dept Path, Cleveland Clin, 76-83; prin investr, NIH grants, 76-80; mem, bd dirs, Soc Biomat, 81-; asst ed, J Biomed Mat Res, 82- & J Bioactive & Compatible Polymers, 86-; chmn, Nat Res Coun Comt Biomat. *Mem:* Sigma Xi; Soc Biomat; NY Acad Sci; AAAS; Am Chem Soc; Am Phys Soc; Am Soc Artificial Intern Organs; Intern Soc Artificial Organs. *Res:* Biomedical materials; engineering plastics; membranes; polymer structure property relationships; interfacial phenomena; biotechnology; surface science; materials science engineering. *Mailing Add:* 222 E Chestnut St Chicago IL 60611

BARENDREGT, RENE WILLIAM, b Lethbridge, Alta, Mar 28, 50. SOIL MICROMORPHOLOGY. *Educ:* Univ Lethbridge, BSc, 71; Univ Delft, Neth, MSc, 74; Queen's Univ, PhD(geomorphol & soils), 77. *Prof Exp:* Asst prof geog & geol, San Francisco State Univ, 77-81; res scientist irrigation eng, Agr Can, 81-82; ASSOC PROF GEOG & GEOL, UNIV LETHBRIDGE, 82- *Mem:* Can Soc Agr Eng; Can Quaternary Geologists Asn (secy, 85-); Int Quaternary Asn; Am Soc Photogrammetric Engrs; Can Geophys Union; Geol Asn Can. *Res:* Quaternary environments with special reference to glaciated environments in western Canada and the Arctic; quaternary correlation and dating using paleomagnetism; holocene geomorphic processes; landscapes of Alberta and geology of western Canada. *Mailing Add:* Dept Geog & Geol Univ Lethbridge 4401 University Dr Lethbridge AB T1K 3M4 Can

BARER, RALPH DAVID, b Bruk, Austria, July 8, 22; Can citizen; m 50; c 5. PHYSICAL METALLURGY. *Educ:* Univ BC, BASc, 45; Mass Inst Technol, SM, 48. *Prof Exp:* Observer, McKinnon Indust, Ltd, 45-46; metallurgist, Polymer Corp, 46-47; instr phys metall, Univ BC, 48-50; metallurgist, Consol Mining & Smelting Co, 50-52; sect head mat eng, Defence Res Estab Pac, 52-88; RETIRED. *Honors & Awards:* Int Galvanizers Award, 61. *Mem:* Am Soc Metals; BC Asn Prof Engrs. *Res:* Metal failures; problems associated with marine boilers; applications of non-destructive testing; marine corrosion. *Mailing Add:* 2123 Sandowne Rd Victoria BC V8R 3J2 Can

BARES, WILLIAM ANTHONY, b Beach, NDak, Apr 1, 35; m 54; c 4. ELECTRICAL ENGINEERING, BIOENGINEERING. *Educ:* Univ NDak, BSEE, 59, MSEE, 61; Stanford Univ, Engr, 65; Univ Wyo, PhD(elec eng), 68. *Prof Exp:* Instr elec eng, Univ NDak, 59-61; design engr, Data Systs Div, Int Bus Mach Corp, 61-62, assoc engr, Gen Prod Div, 62-65, staff engr, 66-67; prof elec eng, Univ NDak, 67-78, chmn dept, 74-76; PROF ELEC & ELECTRONICS ENG & CHMN DEPT, N DAK STATE UNIV, 78- *Concurrent Pos:* IBM fel, Stanford Univ, 63-65; Off Naval Res grant, 68-77. *Mem:* Nat Soc Prof Engrs; Inst Elec & Electronics Engrs. *Res:* Development of computer circuits, memory and control systems; simulation, life systems modeling. *Mailing Add:* RR 3 Box 118 Fargo ND 58104

BARESI, LARRY, b Glendale, Calif; m 75; c 2. BIOTECHNOLOGY BIOINSTRUMENTATION. *Educ:* Cal State Univ, BA, 69, MS, 72; Univ Cal, PhD(environ nutrit), 78. *Prof Exp:* Sr scientist, biotechnol group, 80-82, MEM TECH STAFF, BIOTECHNOL GROUP, JET PROPULSION LAB, 82- *Concurrent Pos:* Postgrad researcher I & II, Univ Col, Los Angeles, 72-78; res assoc, Dept Microbiol, Univ Ill, 78-80. *Mem:* Sigma Xi; Am Soc Microbiol. *Mailing Add:* 25523 Meadow Mont Valencia CA 91355-1942

BARFIELD, BILLY JOE, b Logansport, La, Oct 8, 38; m 65; c 2. HYDROLOGY, SEDIMENTOLOGY. *Educ:* Tex A&M Univ, BS(civil eng) & BS(agr eng), 61, cert meteorol, 62, PhD(agr eng), 68. *Prof Exp:* Res asst agr eng, Tex A&M Univ, 64-68; from asst prof to assoc prof, 68-77, PROF AGR ENG, UNIV KY, 77-, DIR KY WATER RESOURCES RES INST, 88- *Concurrent Pos:* Vis agr engr, Snake River Res Ctr, Agr Res Serv, USDA, Idaho, 74-75. *Honors & Awards:* Outstanding Young Researcher, Am Soc Agr Engrs, 78. *Mem:* Am Soc Agr Engrs. *Res:* Sedimentology of surface mined lands; channel erosion; non point source pollution control; reparion vegetation impacts on pollution; frost protection. *Mailing Add:* Dept Agr Eng Univ Ky Lexington KY 40506

BARFIELD, MICHAEL, b Charleston, SC, Oct 14, 34; m 60; c 3. PHYSICAL CHEMISTRY. *Educ:* San Diego State Col, BA, 57; Univ Utah, PhD(chem), 62. *Prof Exp:* Eng chemist, Gen Dynamics/Astronaut, Calif, 57-59; res fel, Harvard Univ, 62-63; asst prof chem, Univ SFla, 63-65; from asst prof to assoc prof, 65-72, PROF CHEM, UNIV ARIZ, 72- *Concurrent Pos:* Sr Fulbright-Hays scholar, Australia, 76-77; US Ed, Magnetic Resonance Chem, 90. *Honors & Awards:* US Sr Scientist Award, Alexander von Humboldt Found, 86-87. *Mem:* Sigma Xi; Am Chem Soc; Int Soc Magnetic Resonance. *Res:* Application of molecular quantum mechanics to problems in nuclear magnetic resonance spectroscopy and molecular structures. *Mailing Add:* Dept Chem Univ Ariz Tucson AZ 85721

BARFIELD, ROBERT F(REDRICK) (BOB), b Thomaston, Ga, Feb 8, 33; m; c 2. MECHANICAL ENGINEERING. *Educ:* Ga Inst Technol, BME, 56, MSME, 58, PhD(mech eng), 65. *Prof Exp:* Preliminary design engr, AiRes Mfg Co Div, Garrett Corp, Calif, 57-58; from instr to asst prof mech eng, Ga Inst Technol, 58-65; corp staff engr, Thomaston Mills Corp, Ga, 65-67; from assoc prof to prof mech eng, 67-82, DEAN ENG, UNIV ALA, 83- *Concurrent Pos:* Fac fel, NSF, 62-63; vis prof & consult to fac eng, Kabul Univ, Afghanistan, 63, mech eng & chmn dept, Univ Petrol & Minerals, Saudi Arabia, 71-73; E I Dupont fel, Mech Eng, 64, Am Soc Eng Educ/NASA fel 67-68; sr adv, Shiraz Tech Inst, Iran, 75-77; cultural specialist, Dept State, Univ Jordan, 81. *Mem:* Am Soc Eng Educ; fel Am Soc Mech Engrs; fel Am Soc Prof Engrs; Sigma Xi. *Res:* Thermal and fluid sciences; energy conversion. *Mailing Add:* Col Eng Univ Ala Box 870200 Tuscaloosa AL 35487-0200

BARFIELD, WALTER DAVID, b Gainesville, Fla, Nov 25, 28. APPLIED THEORETICAL PHYSICS, COMPUTATIONAL PHYSICS. *Educ:* Univ Fla, BS, 50, MS, 51; Rice Univ, PhD(physics), 61. *Prof Exp:* Asst, Theoret Physics Div, Los Alamos Sci Lab, 51-53, mem staff, 53-63; mem tech staff, Res Eng Support Div, Inst Defense Analysis, 63-67; staff mem, Los Alamos Sci Lab, 68-90, GUEST SCIENTIST, LOS ALAMOS NAT LAB, 90- *Mem:* Am Phys Soc. *Res:* Detonations, nuclear weapons effects; radiation transfer; opacities; fluid dynamics; nuclear physics; numerical methods; gamma ray spectroscopy. *Mailing Add:* 4647 Ridgeway Dr Los Alamos NM 87544

BARFORD, ROBERT A, b Philadelphia, Pa, Feb 27, 36; m; c 2. CHROMATOGRAPHY. *Educ:* Temple Univ, BS, 65. *Prof Exp:* RES LEADER, AGR RES SERV, USDA, 85- *Mem:* Am Chem Soc; AAAS. *Res:* The structure and function of food biopolymers and their aggregates; development of rapid cost effective methods for detection of toxic residues in foods. *Mailing Add:* Agr Res Serv USDA 600 E Mermaid Ln Philadelphia PA 19118

BARFUSS, DELON WILLIS, b Glendale, Calif, Dec 20, 42. RENAL PHYSIOLOGY. *Educ:* Univ Ariz, Tucson, PhD(human physiol), 75. *Prof Exp:* Res asst prof human physiol, Univ Ala, 75-85; ASST PROF HUMAN PHYSIOL, GA STATE UNIV, 85- *Mailing Add:* Dept Physiol-Nephrol Univ Ala Med Ctr LHR 646 Birmingham AL 35294

BARGELLINI, P(IER) L(UIGI), b Florence, Italy, Feb 7, 14; nat US; m 41; c 3. ELECTRICAL ENGINEERING. *Educ:* Univ Florence, Dipl, 34; Polytech Inst Turin, Dr Ing, 37; Cornell Univ, MEE, 49. *Prof Exp:* Engr, Italoradio, Italy, 37-41; head spec tests lab, Fivre, 41-44; researcher, Microwave Res Ctr, Ital Nat Res Coun, 47-50; from instr to assoc prof, Moore Sch Elec Eng, Univ Pa, 50-68; sr scientist, Comsat Labs, 68-83; ADJ PROF ELEC ENG, UNIV PA, 87- *Concurrent Pos:* Consult, RCA Corp, 53- & Aerospace Corp, 62-; lectr, Inst Humanistic Studies for Execs, Univ Pa, 57; vis prof, Pahlavi Univ, Iran, 66-67; consult, Comsat Labs, 84- *Honors & Awards:* Columbus Award, Int Commun, 75; Columbus Gold Medal, Genoa, Italy, 87. *Mem:* Fel Inst Elec & Electronics Engrs; assoc fel Am Inst Aeronaut & Astronaut; Sigma Xi. *Res:* Theory of communication; electronics; satellite and space communications. *Mailing Add:* PO Box 256S Clarksburg MD 20871

BARGER, A(BRAHAM) CLIFFORD, b Greenfield, Mass, Feb 1, 17; m 43; c 3. PHYSIOLOGY, MEDICINE. *Educ:* Harvard Univ, AB, 39, MD, 43. *Hon Degrees:* DSc, Univ Cincinnati, 79. *Prof Exp:* Res asst, Fatigue Lab, Harvard Univ, 38-41, instr, 48-49, assoc, 50-52, from asst prof to prof, 53-63, chmn dept, 74-76, Robert Henry Pfeiffer Prof, 63-87, EMER PROF PHYSIOL, HARVARD MED SCH, 87- *Concurrent Pos:* Med house officer, Peter Bent Brigham Hosp, 43, asst resident, 45, asst med, 46-53, assoc staff mem, 53-58, consult physiol, 58-; res fel physiol, Harvard Med Sch, 46-47; assoc ed, Circulation Res, 63-66; mem bd sci counr, Nat Cancer Inst, 69-74, chmn, 73-74; mem fedn bd, Fedn Am Socs Exp Biol, 69-73; pres, Harvard Apparatus Found, 70- *Honors & Awards:* Ray G Daggs Award, Am Physiol Soc. *Mem:* Nat Inst Med; Am Physiol Soc (pres, 70-71); Am Acad Arts & Sci. *Res:* Cardiovascular-renal physiology; congestive failure; hypertension. *Mailing Add:* Dept Physiol Harvard Med Sch Boston MA 02115

BARGER, JAMES DANIEL, b Bismarck, NDak, May 17, 17; m 52, 71, 80; c 4. PATHOLOGY. *Educ:* Univ NDak, AB & BS, 39; Univ Pa, MD, 41; Univ Minn, MS, 49. *Prof Exp:* Dir labs, Pima County Hosp, Tucson, Ariz, 49-50 & Maricopa County Hosp, Phoenix, 50-51; med dir, Southwest Blood Bank Ariz, 51-64; dir clin path, Sunrise Hosp, 64-69, chmn, Dept Path, 64-81; PROF LAB MED & PATH, UNIV NEV, RENO, 81- *Concurrent Pos:* Dir labs, Good Samaritan Hosp, Ariz, 51-63; asst clin prof path, Univ Nev, Reno, 74-; quondam chmn, Dept Path, Sunrise Hosp, 81-; consult, US Vet Admin, USPHS Indian Hosp & St Luke's Hosp, Phoenix; distinguished practr med, NAP, 83. *Mem:* AAAS; Am Soc Clin Path; Am Asn Pathologists; sr mem Am Soc Qual Control; AMA; Col Am Pathologists (gov, 66-71, secy-treas, 71-79, vpres, 79-81, pres, 81-83); corresp mem Europ Acad Humanities, Arts & Sci. *Res:* Blood group serology and clinical biochemistry as related to hematologic diseases; quality control methods for categorical procedures in medicine. *Mailing Add:* 3186 S Maryland Pkwy Las Vegas NV 89109

BARGER, JAMES EDWIN, b Manhattan, Kans, Dec 28, 34; m 57; c 4. ACOUSTICS. *Educ:* Univ Mich, BS, 57; Univ Conn, MS, 60; Harvard Univ, MA, 62, PhD(appl physics), 64. *Prof Exp:* Sr consult appl physics, 64-68, VPRES, BOLT BERANEK & NEWMAN, INC, 68-, CHIEF SCIENTIST, 75- *Mem:* Fel Acoust Soc Am. *Res:* Underwater sound; acoustic cavitation in water; sonar development; medical ultra sonics. *Mailing Add:* Bolt Beranek & Newman Inc 50 Moulton St Cambridge MA 02138

BARGER, SAMUEL FLOYD, b Butler County, Pa, July 12, 36. COMBINATORICS MATHEMATICS, GRAPH THEORY. *Educ:* Clarion Univ, BS, 60; Univ Minn, MS, 68 &, PhD(math), 70. *Prof Exp:* PROF MATH, YOUNGSTOWN STATE UNIV, 70- *Mem:* Am Math Soc; Math Asn Am. *Mailing Add:* Dept Math & Comput Sci Youngstown State Univ 410 Wick Ave Youngstown OH 44555

BARGER, VERNON DUANE, b Curllsville, Pa, June 5, 38; m 67; c 3. THEORETICAL HIGH ENERGY PHYSICS. *Educ:* Pa State Univ, BS, 60, PhD(physics), 63. *Prof Exp:* Res assoc theoret physics, 63-65, from asst prof to prof physics, 65-83, J H VAN VLECK PROF & DIR, INST ELEM PARTICLE PHYSICS RES, UNIV WIS-MADISON, HILLDALE PROF, 87- *Concurrent Pos:* Vis prof, Univ Hawaii, 70, 79 & 82, Univ Durham, 83; vis scientist, CERN, Switz & Rutherford Labs, Eng, 72 & SLAC, Stanford, Calif, 75; Guggenheim fel, 72; sr vis fel, Brit Sci-Eng Res Coun, 83-84. *Mem:* Fel Am Phys Soc. *Res:* Fundamental particle theory; elementary particle physics. *Mailing Add:* Dept Physics Univ Wis Madison WI 53706

BARGERON, CECIL BRENT, b Provo, Utah, May 8, 43; m 66; c 5. PHYSICS, SURFACE SCIENCE. *Educ:* Brigham Young Univ, BS, 67; Univ Ill, MS, 68, PhD(physics), 71. *Prof Exp:* Teaching asst physics, Univ Ill, 67-68, res asst, 68-71; PHYSICIST, APPL PHYSICS LAB, JOHNS HOPKINS UNIV, 71- *Concurrent Pos:* NSF fel, 67-71; Woodrow Wilson hon fel. *Mem:* Am Phys Soc. *Res:* Surface physics and chemistry, electrochemistry (corrosion), infrared laser damage to biological tissue; light scattering from biological and related systems; high temperature physics. *Mailing Add:* Appl Physics Lab Johns Hopkins Univ Johns Hopkins Rd Laurel MD 20723-6099

BARGERON, LIONEL MALCOLM, JR, b Savannah, Ga, Nov 4, 23; m 50; c 3. PEDIATRICS, CARDIOLOGY. *Educ:* Univ Ala, BS, 48; Med Col Ala, MD, 52; Am Bd Pediat, dipl & cert cardiol. *Prof Exp:* Intern, Jefferson-Hillman Hosp, Birmingham, Ala, 52-53, resident, 53; asst resident, Babies Hosp, Columbia-Presby Med Ctr, NY, 54-55, sr resident, 55; instr pediat, 56-58, asst prof, 58-59, assoc prof pediat cardiol, 60-66, PROF PEDIAT, SCH MED, UNIV ALA, 66-, ASSOC PROF SURG, 77- *Concurrent Pos:* USPHS res fel pediat cardiol, 55-57. *Mem:* Fel Am Acad Pediat. *Res:* Pediatric cardiology. *Mailing Add:* Univ Ala Hosp Birmingham Univ Ala Univ Sta Birmingham AL 35294

BARGHUSEN, HERBERT RICHARD, b Englewood, NJ, July 9, 33; m 57; c 2. PALEOZOOLOGY, ANATOMY. *Educ:* Lafayette Col, AB, 55; Univ Chicago, PhD(paleozool), 60. *Prof Exp:* Asst prof zool, Smith Col, 61-64; instr anat, Stritch Sch Med, Loyola Univ, Ill, 64-67; from asst prof to prof anat & oral anat, Col Med & Col Dent, Univ Ill Med Ctr, 67-87; RETIRED. *Concurrent Pos:* NSF fel, 60-61; res grants, Loyola Univ, Ill, 64-67 & Univ Ill, 67-70; NSF grant, 73-75; vis prof anat, Univ Chicago, 78-79. *Mem:* AAAS; Soc Vert Paleont. *Res:* Biomechanics of vertebrate feeding mechanism; evolutionary origin of mammals and the mammalian feeding mechanism. *Mailing Add:* 1166 E 53rd St Chicago IL 60615

BARGMANN, ROLF ERWIN, b Glueckstadt, Ger, May 13, 21; nat; m 49; c 4. MATHEMATICAL STATISTICS. *Educ:* Univ NC, PhD(math statist), 57. *Prof Exp:* Head dept statist, Inst Educ Res, Frankfurt, 52-55; asst, Psychomet Lab, Univ NC, 55-57; from assoc prof to prof statist, Va Polytech Inst, 58-61; mgr appl math & info sci, Systs Develop div, Thomas J Watson Res Ctr, Int Bus Mach Corp, 61-65; PROF STATIST, UNIV GA, 65- *Concurrent Pos:* Vis lectr, Columbia Univ, 63-64, Univ Paris, 65, Univ Ill, 66, Univ Erlangen, 78 & Univ Dortmund, 80; assoc ed, J Statist Comp. *Mem:* Fel AAAS; fel Am Statist Asn; Economet Soc; Psychomet Soc; Sigma Xi. *Res:* Multivariate analysis; statistical computation. *Mailing Add:* Dept Statist Univ Ga Athens GA 30601

BARGMANN, VALENTINE, mathematical physics; deceased, see previous edition for last biography

BARGON, JOACHIM, b Wiesbaden, Ger, Apr 13, 39; m 63; c 3. PHYSICAL CHEMISTRY. *Educ:* Darmstadt Tech Univ, BS, 61, MS, 65, PhD(physics), 68. *Prof Exp:* Staff mem spectroscopy, Ger Plastics Inst, Darmstadt, 65-69; fac mem org chem, Calif Inst Technol, 69-70; staff mem gen chem, res div, Yorktown, NY, IBM Corp, 70-71, staff mem, 71-74, mgr radiation chem, 74-76, dep mgr chem, San Jose Res Lab; AT INST PYHS CHEM, UNIV BONN. *Concurrent Pos:* Consult chem & plastics indust. *Mem:* Ger Phys Soc; Am Phys Soc; Am Chem Soc. *Res:* Materials science, new polymers, polymer modification, lithography, reaction mechanisms in radiation, free radical and organo-mettalic chemistry using spectroscopic techniques. *Mailing Add:* Univ Bonn Inst Phys Chem Wegelerstr 12 Bonn 1 5300 Germany

BARHAM, WARREN SANDUSKY, b Prescott, Ark, Feb 15, 19; m 40; c 4. PLANT BREEDING, HORTICULTURE. *Educ:* Univ Ark, BSA, 41; Cornell Univ, PhD, 50. *Prof Exp:* Asst, Fruit & Truck Exp Sta, Univ Ark, 41-42, supt, 45-46; asst, Cornell Univ, 46-49; from asst prof to assoc prof hort, NC State Col, 49-58; dir raw mat res & seed prod, Basic Veg Prod, Inc, 58-76; head Hort Sci Dept, 76-80, PROF HORT, TEX A&M UNIV, 76- *Mem:* Fel Am Soc Hort Sci (pres, 81); Am Genetic Asn; Am Inst Biol Sci; AAAS. *Res:* Plant physiology, genetics and breeding; onions, watermelons, cantaloupes, tomatoes and garlic; disease and insect resistance; quality characteristics; seed production problems. *Mailing Add:* Hort Sci Dept Tex A&M Univ College Station TX 77843

BARHYDT, HAMILTON, b New Haven, Conn, Nov 4, 28; m 52, 78; c 2. ELECTROOPTICS, INFRARED. *Educ:* Yale Univ, BS, 50; Cornell Univ, PhD(eng physics), 54. *Prof Exp:* Mem tech staff, Systs Eng Lab, 554-59, sect head, Infrared Systs Dept, Res & Develop Div, 59-62, mgr, Advan Systs Res Dept, 62-67, asst mgr, Space & Tactical Electrooptical Lab, 67-76, chief scientist electrooptical sensors, Strategic Systs Div, Gulver City, 76-80, CHIEF SCIENTIST, SPACE SENSORS DIV, ELECTROOPTICAL & DATA SYSTS GROUP, HUGHES AIRCRAFT CO, EL SEGUNDO, CALIF, 80- *Mem:* Soc Photo-Optical Instrumentation Engrs. *Res:* Infrared and electrooptical systems for military applications; space borne surveillance; night vision; missile guidance. *Mailing Add:* 3037 Grass Valley Hwy No 8084 Auburn CA 95603

BARI, ROBERT ALLAN, b Brooklyn, NY, Sept 3, 43; m 66; c 2. SOLID STATE PHYSICS, NUCLEAR ENGINEERING. *Educ:* Rutgers Univ, AB, 65; Brandeis Univ, PhD(physics), 70. *Prof Exp:* Physicist solid state physics, Lincoln Lab, Mass Inst Technol, 69-71; asst physicist, Brookhaven Nat Lab, 71-73; vis asst prof physics, State Univ NY Stony Brook, 73-74; SR PHYSICIST, NUCLEAR ENERGY SCI, BROOKHAVEN NAT LAB, 74- *Concurrent Pos:* Prin investr, NSF res grant, 73-74; adj assoc prof mech eng, Manhattan Col, 78-; adj prof nuclear eng, Polytech Inst NY, 78- *Mem:* Am Phys Soc; Am Nuclear Soc. *Res:* Nuclear reactor safety analysis; condensed matter physics; reliability and risk assessment. *Mailing Add:* Brookhaven Nat Lab Upton NY 11973

BARI, WAGIH A, PATHOLOGY. *Prof Exp:* DIR HEMAT PATH, ST JOHN'S MERCY MED CTR, 68- *Mailing Add:* Dept Path St John's Mercy Med Ctr 615 S New Ballas Rd St Louis MO 63141

BARIA, DORAB NAOROZE, b Bombay, India, Apr 5, 42. HAZARDOUS WASTE MANAGEMENT, LOW-RANK COAL UTILIZATION. *Educ:* Univ Bombay, BChem Eng, 64; Northwestern Univ, MS, 68, PhD(chem eng), 71. *Prof Exp:* Postdoctoral assoc chem eng, Ames Lab, Iowa State Univ, 72-73; from asst prof to prof, Univ NDak, 73-85; PROF CHEM ENG, UNIV MINN, DULUTH, 85- *Mem:* Am Inst Chem Engrs; Am Soc Eng Educ; Nat Soc Prof Engrs; Sigma Xi. *Res:* Beneficiation of low-rank coals, removal of sodium, water and sulfur; flue gas desulfurization; hazardous waste management. *Mailing Add:* Chem Eng Dept Univ Minn Duluth MN 55812

BARIC, LEE WILMER, b Carlisle, Pa, Nov 19, 32; m 56; c 3. MATHEMATICS. *Educ:* Dickinson Col, BSc, 56; Lehigh Univ, MSc, 62, PhD(math), 66. *Prof Exp:* Instr math, Lafayette Col, 58-63; mathematician, McCoy Electronics, Pa, 63-64; from asst prof to assoc prof, 64-75, chmn dept, 74-77, PROF MATH, DICKINSON COL, 75- *Concurrent Pos:* Consult, McCoy Electronics, 64-90. *Mem:* Am Math Soc. *Res:* Schauder bases in Banach spaces; summability theory. *Mailing Add:* Dept Math Dickinson Col Carlisle PA 17013

BARICA, JAN M, b Varin, Czech, Feb 10, 33; m 60; c 3. LIMNOLOGY, AQUATIC ECOLOGY. *Educ:* Univ Bratislava, MSc, 57; Univ Prague, PhD(ecol), 67; Charles Univ, DSc, 88. *Prof Exp:* Vis scientist, Inst Limnol, Univ Freiburg, 68-69; res scientist fish ecol, Freshwater Inst, Winnipeg, 69-80; res mgr aquatic ecol, 80-86, SR SCIENTIST, LAKES RES BR, NAT WATER RES INST, BURLINGTON, ONT, 87- *Concurrent Pos:* UN expert, UN Develop Prog, SChina Sea Fisheries Prog, 75; Humboldt fel, Federal Republic Germany. *Mem:* Int Soc Limnol; Soc Can Limnologists; Asn Great Lakes Res; Water Studies Inst; Can Asn Water Pollution Res. *Res:* Characterization of hypereutrophic ecosystems; instability of algal (cyanophyte) blooms, their collapses and ecological impact; predictive models for summer and winter fish kills in hypereutrophic lakes; manipulation of algal bloom composition by selective nutrient additions; ammonia contamination; Great Lakes areas of concern rehabilitation. *Mailing Add:* Nat Water Res Inst PO Box 5050 Burlington ON L7R 4A6 Can

BARIE, PHILIP S, b Buffalo, NY, Aug 18, 53; m 81; c 2. SURGICAL CRITICAL CARE. *Educ:* Boston Univ, MD, 77 Am Bd Surg, dipl, 85 & 87. *Prof Exp:* Instr physiol, Albany Med Col, 80-81; asst prof surg, Cornell Univ Med Col, 84-89, ATTEND SURGEON & DIR, SURG INTENSIVE CARE UNIT, NY HOSP-CORNELL MED CTR, 84-, ASSOC PROF SURG, CORNELL UNIV MED COL, 89- *Concurrent Pos:* Consult surg, Cath Med Ctr, NY, 85- *Mem:* Fel Am Col Surgeons; NY Acad Sci; Asn Acad Surg; Surg Infection Soc; Soc Critical Care Med; Am Physiol Soc; Shock Soc; Am Asn Surg Trauma; Soc Univ Surgeons; Societe Internationale de Chirurgie. *Res:* Pathophysiology respiratory failure, pulmonary transvascular fluid and protein exchange. *Mailing Add:* Dept Surg Med Ctr New York Hosp-Cornell Univ 525 E 68th St New York NY 10021

BARIE, WALTER PETER, JR, b Pittsburgh, Pa, Oct 28, 26; m 64; c 1. ORGANIC POLYMER CHEMISTRY. *Educ:* Duquesne Univ, BS, 50, MS, 51; Pa State Univ, PhD(chem), 54. *Prof Exp:* Asst, Duquesne Univ, 50-51; res chemist, Gulf Res & Develop Co, 54-67, sr res chemist, 67-75, res assoc, 75-83; tech mgr resins, Sterling Div, Reichold Chem, Inc, 83-87; RETIRED. *Concurrent Pos:* Consult, 87- *Mem:* AAAS; Am Chem Soc. *Res:* Monomers; electrical-electronic coatings, adhesives and potting compounds. *Mailing Add:* 213 Amherst Rd Glenshaw PA 15116

BARIEAU, ROBERT (EUGENE), b Lindsay, Calif, Mar 12, 15; m 46 & 64; c 1. PHYSICAL CHEMISTRY. *Educ:* Univ Calif, BS, 37, PhD(chem), 40. *Prof Exp:* Asst chem, Univ Calif, 37-40, instr, 40-41, investr, Nat Defense Res Comt, 41-44; res chemist, Sharples Corp, Philadelphia, 44-45 & Calif Res Corp, 45-62; res chemist, Helium Res Ctr, 62-71; RES PROF PHYSICS, WTEX STATE UNIV, 73- *Concurrent Pos:* Res fel, Calif Inst Technol, 49-50. *Mem:* AAAS; Am Chem Soc; Sigma Xi. *Res:* X-ray and electron diffraction; electron microscopy; analytical chemistry by x-ray fluorescence and x-ray absorption; thermodynamics of galvanic cells; visible light scatterings; crystal structures; atomic emission and absorption spectroscopy; thermodynamics of gases and gas mixtures; cryogenics; wind energy conversion systems. *Mailing Add:* PO Box 781 Canyon TX 79015-0781

BARIL, ALBERT, JR, b New Orleans, La, Aug 19, 26; m 49; c 6. RESEARCH ADMINISTRATION. *Educ:* Loyola of the South, BS, 49; Tulane Univ, MS, 52. *Prof Exp:* Dir mfg photog, Kalvar Corp, La, 62-66; dir spec prod res, Allied Paper Co, Mich, 66-67; res physicist textiles, Mach Develop, Agr Res Serv, USDA, 67-73, res leader, New Systs Res, 73-74, chief res admin, Cotton Textile Processing Lab, Southern Regional Res Ctr, 74-88; RETIRED. *Mem:* Electrostatic Soc Am; Am Soc Mech Engrs; Sigma Xi; Am Asn Textile Chemists & Colorists; Inst Elec & Electronics Engrs. *Res:* New systems for textile processing, particularly, integrated system, producing yarn continuously from a supply of fibers by one machine; electrostatic manipulation of fibers; electrostatic spinning of yarns; agri-particulates analysis and control. *Mailing Add:* 6300 Ackel St No 293 Metairie LA 70003

BARIL, EARL FRANCIS, b Claremont, NH, Apr 23, 30; m 65. BIOCHEMISTRY. *Educ:* St Anselms Col, AB, 59; Univ Houston, MS, 61; Univ Conn, PhD(biochem), 66. *Prof Exp:* NIH fel biochem, Univ Wis, 66-69; asst prof pharmacol, Med Sch, Duke Univ, 69-72; sr scientist, 72-86, PRIN SCIENTIST, WORCESTER FOUND EXP BIOL, 86-; PROF PHARMACOL, MED SCH, UNIV MASS, 84-, PROF BIOCHEM, 86- *Concurrent Pos:* Exchange scientist to Poland, Nat Cancer Inst, 77; mem, Physiol Chem Study Sect, NIH, 84-88. *Mem:* AAAS; Am Asn Cancer Res; Am Chem Soc; Am Soc Biol Chemists; Biophys Soc; Protein Soc; fel Am Inst Chem. *Res:* Regulatory control mechanisms; enzymology of DNA replication in eucaryotes. *Mailing Add:* Worcester Found Exp Biol Shrewsbury MA 01545

BARILE, DIANE DUNMIRE, b Beaver Falls, Pa, July 1, 40; m 64; c 5. ENVIRONMENTAL SYSTEMS ANALYSIS. *Educ:* Univ Fla, BS, 62; Fla Inst Technol, MS, 76. *Prof Exp:* Teacher biol sci, Broward Co, Fla, 62-63; grad asst, sci educ, Univ Miami, 63-64; ENVIRON PLANNER, CITY OF PALM BAY, FLA, 76-; EXEC DIR, MARINE RESOURCES COUN, 84-; ADJ FAC URBAN PLANNING, FLA INST TECHNOL, 76-, GRAD FAC COASTAL MGT, 85- *Concurrent Pos:* Dir, Initiative to Manage Fla Indian River Lagoon, 77-78, implementation of local manatee proj prog, Nat Oceanic & Atmospheric Admin, 77-79; dir, Coastal Mgt Prog, Fla Inst Technol, 83-84. *Honors & Awards:* Cauler Award Spec Recognition. *Mem:* Am Planning Asn; Am Water Resources Asn. *Res:* Environmental and urban systems; analysis of hydrological, biological and physical parameters translated into programs for enhancement of natural function for sound resource management for man's use; public awareness and policy implimentation. *Mailing Add:* Marine Resources Coun 730 E Strawbridge Ave Melbourne FL 32901

BARILE, GEORGE CONRAD, b New York, NY, Jan 7, 48; m 71; c 2. ORGANIC CHEMISTRY, CATALYSIS. *Educ:* City Col NY, BS, 70; NY Univ, PhD(org chem), 75. *Prof Exp:* Res photochem, Univ Calif, Los Angeles, 75-77; RES CHEMIST CATALYSIS, MOBIL CHEM CO, 77- *Mem:* Am Chem Soc. *Res:* Photochemistry and catalysis for the production of organic chemicals. *Mailing Add:* Mobil Chem Co Res & Develop PO Box 240 Edison NJ 08817

BARILE, MICHAEL FREDERICK, b Bound Brook, NJ, Jan 9, 24; m 46; c 4. MICROBIOLOGY. *Educ:* Univ Ga, BS, 49; Univ Mich, MS, 51, PhD(bact), 54. *Prof Exp:* Instr microbiol, Univ Mich, 54; bacteriologist, 406th Gen Med Lab, 54-56; supvry bacteriologist, Med Lab, Tokyo, Japan, 56-57; bacteriologist, NIH, 58-62, res bacteriologist, 62-68, chief sect Mycoplasma, 68-72, CHIEF LAB MYCOPLASMA, 72- *Honors & Awards:* Presidential Citation Award, Int Organ Mycoplasmol, 84. *Mem:* AAAS; Am Acad Microbiol; Tissue Cult Asn; Int Soc Mycoplasm. *Res:* Non-gonococcal urethritis; penile lesions caused by herpesvirus, S pallidum and H ducreyi; immunization against neonatal tetanus; tetanus toxin neutralization test; diphtheria toxin for the Schick test; M pneumoniae as cause of primary typical pneumonia; etiology of recurrent aphthous stomatitis, rheumatoid arthritis and Wegener's granulomatosis; pathogenicity of Mycoplasma pneumoniae, M hominis, M genitalium, and Ureaplasma urealyticum; Mycoplasma-induced septic arthritis in patients and experimentally-induced septic arthritis in chimpanzees by M hominis, M pneumoniae and U urealyticum; M genitalium induced urogenital infections in chimpanzees and in new and old world monkeys; development of an acellular M pneumoniae vaccine containing virulence/immunogenic attachment, ciliotoxic and leukocyte recruitment factors; potency assays to evaluate M pneumoniae vaccines; other mycoplasma virulence factors; specific IgA proteases of U urealyticum; characteristics of U urealyticum urease mycoplasma lipoglycans; monoclonal antibodies to surface components of M hominis and M pneumoniae; sialic acid-containing glycoconjugates and sulfated glycolipids receptor sites for M pneumoniae; guinea pig, monkey and chimpanzee animal models to study mycoplasma disease; ocular pathology induced in hamsters and rats by Spiroplasma mirum, SMCA; cause of epidemic caprine keratoconjunctivitis contamination of cell culture. *Mailing Add:* Div Bact Prod Ctr Biol Eval Res Bldg 29 Rm 420 Bethesda MD 20892

BARILE, RAYMOND CONRAD, b New York, NY, June 30, 36; m 59; c 2. PHYSICAL CHEMISTRY, INORGANIC CHEMISTRY. *Educ:* Manhattan Col, BS, 58; Fordham Univ, MS, 60, PhD(phys chem), 65. *Prof Exp:* Asst chem, Fordham Univ, 58-59; asst, 59-60, from instr to asst prof, 60-68, ASSOC PROF CHEM, MANHATTAN COL, 68- *Concurrent Pos:* Res assoc, Brookhaven Nat Lab, 66-67; fels, NSF & Atomic Energy Comn. *Mem:* Am Chem Soc. *Res:* Kinetics and mechanism of reactions of transition metal complexes, electron transfer, formation reaction and characterization of polymers. *Mailing Add:* Dept Chem Manhattan Col Riverdale NY 10471

BARINGER, J RICHARD, b Columbus, Ohio, Jan 19, 35; c 3. NEUROLOGY, PATHOLOGY. *Educ:* Ohio State Univ, BSc, 55; Western Reserve Univ Sch Med, MD, 59. *Prof Exp:* Intern, Mass Gen Hosp, 59-60, resident, 60-61, asst resident, 64-65, fel, 66-69; asst prof, 69-75, PROF NEUROL & PATH, UNIV CALIF, SAN FRANCISCO, 75-; CHIEF NEUROL, VET ADMIN HOSP, SAN FRANCISCO, 69- *Concurrent Pos:* Vchmn, Dept Neurol, Univ Calif, San Francisco, 74-; chmn sci adv comt, Nat Multiple Sclerosis Soc, 76-78; mem path A study sect, NIH, 76-; Fulbright grant, 77-; med investr award, Vet Admin, 71-75. *Honors & Awards:* Med Investr Award, Vet Admin, 71-75; Sr US Scientist Award, Alexander von Humboldt Found, 77. *Mem:* Am Asn Neuropathologists; Am Soc Cell Biologists; Electron Micros Soc Am; Am Neurol Asn; Am Soc Clin Invest. *Res:* Viral diseases of the central nervous system; mechanisms of viral persistence; slow virus diseases of the nervous system. *Mailing Add:* Dept Neurol Univ Utah Med Ctr Salt Lake City UT 84132

BARIOLA, LOUIS ANTHONY, b Lake Village, Ark, July 7, 32; m 59; c 2. ECONOMIC ENTOMOLOGY. *Educ:* Univ Ark, Fayetteville, BSA, 58, MS, 62; Tex A&M Univ, PhD(entom), 69. *Prof Exp:* Res assoc entom, Univ Ark, Fayetteville, 59-62; entomologist, Entom Res Div, Tex, 62-69, RES ENTOMOLOGIST, WESTERN COTTON RES LAB, AGR RES SERV, USDA, ARIZ, 69- *Mem:* Entom Soc Am; Sigma Xi. *Res:* Ecology of cotton insects; control of cotton insects using alternative methods, such as non-insecticides--cultural or biological methods; biology of cotton insects in the arid West, with major emphasis on the pink bollworm, cotton leafperforator, cotton bollworm and lygus bugs. *Mailing Add:* Western Cotton Res Lab 4135 E Broadway Rd Phoenix AZ 85040

BARISAS, BERNARD GEORGE, JR, b Shreveport, La, July 16, 45; m 81. CELL SURFACE BIOPHYSICS, MICROCALORIMETRY. *Educ:* Univ Kans, BA, 65; Oxford Univ, BA, 67, MA, 83; Yale Univ, MPhil, 69, PhD(chem), 71. *Prof Exp:* Rhodes scholar, Oxford Univ, 67-69; NIH fel, Yale Univ, 71-72; NIH fel, Univ Colo, 73-75, lectr, 75; assoc prof biochem, Sch Med, St Louis Univ, 75-81; assoc prof, 81-87, PROF CHEM & MICROBIOL, COLO STATE UNIV, 87-, ASSOC DEAN, COL NAT SCI, 90- *Concurrent Pos:* Prin investr, NIH, NSF, March of Dimes, Am Heart Asn, 75-; mem, Midwest & Northwest Regional Rhodes Scholar Select Comts, 76-81, 82-; res career develop award, NIH, 78; Fulbright sr fel, G06ttingen, 85. *Mem:* Am Soc Biol Chemists; Am Asn Immunologists; Biophys Soc; Am Chem Soc; Sigma Xi. *Res:* Laser optical techniques for analysis of cell surface molecular motions; fluorescence photobleaching recovery; molecular rotation in membranes; reaction microcalorimetry and differential scanning calorimetry of biological systems; cellular immunology. *Mailing Add:* Dept Chem Colo State Univ Ft Collins CO 80523M

BARISH, BARRY C, b Omaha, Nebr, Jan 27, 36; m 60; c 2. HIGH ENERGY PHYSICS. *Educ:* Univ Calif, Berkeley, AB, 57, PhD(physics), 62. *Prof Exp:* Physicist, Radiation Lab, Univ Calif, 62-63; res fel physics, 63-66, from asst prof to assoc prof, 66-72, PROF PHYSICS, CALIF INST TECHNOL, 72- *Mem:* Fel Am Phys Soc. *Res:* Experimental elementary particle physics; studies of neutrino interactions at high energies; production of short lived weakly decaying states by hadrons; investigation of high energy electron-positron annihilations; non-accelerator physics. *Mailing Add:* Dept Physics Calif Inst Technol 256-48 Pasadena CA 91125

BARISH, LEO, b New Bedford, Mass, Jan 19, 30; m 52; c 2. MICROSCOPY, TEXTILE CHEMISTRY. *Educ:* Southeastern Mass Univ, BS, 52; Univ Lowell, MS, 54. *Prof Exp:* Cost-finder, Dartmouth Finishing Co, Mass, 51-52; res fel, Am Asn Textile Chemists & Colorists, 53-54; sr res assoc, Fabric Res Labs, Inc, 56-72, sr res assoc, 72-79, SR MICROSCOPIST, ALBANY INT RES CO, MANSFIELD, 79- *Mem:* Electron Micros Soc Am; Fiber Soc. *Res:* Optical and scanning electron microscopy; fiber structure; polymer morphology; scientific photography; textile mechanics; literature searches; fire resistant fabrics evaluation and development; stereoscopy; optical analysis; composite structures. *Mailing Add:* Two Pole Plain Rd Sharon MA 02067

BARISH, ROBERT JOHN, b Newark, NJ, May 14, 46. RADIOLOGICAL PHYSICS, MEDICAL PHYSICS. *Educ:* NY Univ, BS, 68, MEng, 70; Univ London, PhD(med physics), 76; Am Bd Radiol, cert; Am Bd Health Physics, cert. *Prof Exp:* Clin instr radiol physics, State Univ NY, Downstate Med Ctr, 70-73; res student, Inst Cancer Res Royal Marsden Hosp, 73-76; asst prof radiol physics, 76-81, ASSOC PROF, MED CTR, NY UNIV, 81- *Concurrent Pos:* Consult physicist, Maimonides Med Ctr, Coney Island Hosp Affil, 71-73, Booth Mem Med Ctr, 80-82, Hackensack Hosp, 81-83, Westchester Sq Hosp, 82-85, Cabrini Med Ctr, 83-85, Mary Immaculate Hosp, 81- *Mem:* Am Asn Physicists Med; Health Physics Soc. *Res:* Radiation dosimetry; physics applications in radiology. *Mailing Add:* Cancer Inst Mary Immaculate Hosp 152-11 89th Ave Jamaica NY 11432

BARK, LAURENCE DEAN, b Chanute, Kans, Mar 18, 26; m 64; c 3. AGRICULTURAL METEOROLOGY. *Educ:* Univ Chicago, BSc, 48, MSc, 50; Rutgers Univ, PhD(agron), 54. *Prof Exp:* Asst, Rutgers Univ, 51-54; bioclimatologist, US Weather Bur, 55; assoc prof physics & assoc climatologist, Agr Exp Sta, Kans State Univ, 56-67, prof climat & climatologist, 67-90; RETIRED. *Honors & Awards:* Nat Weather Serv Centennial Medallion. *Mem:* AAAS; Am Meteorol Soc; Am Soc Agron; Crop Sci Soc Am. *Res:* Relation of weather to the life processes of plants, animals and humans. *Mailing Add:* 1741 Fairview Ave Manhattan KS 66502

BARKA, TIBOR, b Debrecen, Hungary, Mar 31, 26; US citizen; m. EXPERIMENTAL MEDICINE. *Educ:* Debrecen Univ, MD, 50. *Prof Exp:* Asst, Inst Histol & Embryol, Med Univ Budapest, 51-54; res assoc, Inst Exp Med, Hungarian Acad Sci, 54-56; res assoc, Inst Cell Res & Genetics, Karolinska Inst, Sweden, 56-58; res assoc, 58-62, asst attend pathologist, 62-64, assoc attend pathologist, Mt Sinai Hosp, 64-80, prof anat & chmn dept, 67-86, PROF ANAT, MT SINAI SCH MED, 86-, PROF PATH, 66- *Concurrent Pos:* Ed-in-chief, J Histochem & Cytochem, 65-73; prof path, Mt Sinai Sch Med, 66- *Mem:* Histochemical Soc; Am Asn Path; Am Soc Cell Biol; Am Asn Anat. *Res:* Histochemistry; experimental pathology; cell biology. *Mailing Add:* Dept Anat Mt Sinai Sch Med Box 1007 New York NY 10029

BARKAI, AMIRAM I, b June 20, 36; m; c 2. NEUROTRANSMITTERS, DRUG EFFECTS. *Educ:* Hebrew Univ, Israel, PhD(physiol), 70. *Prof Exp:* RES SCIENTIST & ASSOC PROF PSYCHIAT, NY STATE PSYCHIAT INST, 72- *Concurrent Pos:* Med entomol; pub health serv grants. *Mem:* Am Soc Neurochem; Am Soc Pharmacol & Exp Therapeut. *Res:* Neuroscience, neurochemistry, neuropharmacology. *Mailing Add:* Dept Psychiat NY State Psychiat Inst Columbia Univ 722 W 168th St New York NY 10032

BARKALOW, DEREK TALBOT, b New York, NY, May 25, 51. MOLECULAR BIOLOGY. *Educ:* Univ Wis, Madison, BS, 73; Rutgers Univ, MS, 74, PhD(physiol), 78. *Prof Exp:* Res asst physiol, Marine Biol Lab, Mass, 75-78; asst prof biol, 78-87, ASSOC PROF BIOL, STETSON UNIV, 87-, CHAIRPERSON, 90- *Concurrent Pos:* Res fel cell physiol, Sch Med, Wash Univ, 81; res fel cell biol, Princeton Univ, 85. *Mem:* AAAS; Am Soc Zoologists; Am Soc Photobiol; Am Inst Biol Sci; Hist Sci Sco. *Res:* Membrane transport; intestinal absorption; cell physiology. *Mailing Add:* Dept Biol Stetson Univ De Land FL 32720-3756

BARKALOW, FERN J, b Aug 25, 60. CELL ADHESION, EXTRACELLULAR MATRIX BIOLOGY. *Educ:* Rutgers Univ, BA, PhD(microbiol), 88. *Prof Exp:* Teaching asst introductory genetics, Biol Dept, State Univ NY, Binghamton, 82; POSTDOCTORAL FEL, DEPT MOLECULAR BIOL, PRINCETON UNIV, 88- *Concurrent Pos:* Nat Cancer Inst nat res serv award, 90-92. *Mem:* Am Soc Cell Biol; AAAS. *Res:* Author of eight technical publications. *Mailing Add:* Dept Molecular Biol Princeton Univ Moffett Hall Princeton NJ 08544

BARKAN, P(HILIP), b Boston, Mass, Mar 29, 25; wid; c 2. MECHANICAL ENGINEERING DESIGN, MANUFACTURABILITY. *Educ:* Tufts Univ, BS, 46; Univ Mich, MS, 48; Pa State Univ, PhD(mech eng), 53. *Prof Exp:* Asst prof diesel engine res, Pa State Univ, 48-51; proj engr, Gen Elec Co, 53-66, mgr mech eng & circuit interruption, Eng Res Power Transmission Div, 66-77; PROF MECH ENG, STANFORD UNIV, 77- *Concurrent Pos:* Pres, Philip Barkan, Inc, 78- *Honors & Awards:* Steinmetz Medal, Gen Elec Co, 73. *Mem:* Nat Acad Eng; Am Soc Mech Engrs; fel Inst Elec & Electronics Engrs; Soc Mfg Eng; Sigma Xi. *Res:* Transient behavior of high speed mechanisms; design synthesis; fluid dynamic transient phenomena; electric power protection devices; high current phenomena; design for manufacturability. *Mailing Add:* Dept Mech Eng Design Div Stanford Univ Stanford CA 94305

BARKATE, JOHN ALBERT, b Sulphur, La, Dec 4, 36; m 64; c 4. MICROBIOLOGY, FOOD SCIENCE. *Educ:* Northwestern State Univ, MS, 63; La State Univ, PhD(food microbiol), 67. *Prof Exp:* Dir, microbiol dept & asst dir cent res, Ralston Purina Co, 79-; AT USDA. *Mem:* Am Soc Microbiol; Inst Food Technologists; Int Asn Milk, Food & Environ Sanitarians, Inc; Am Soc Cereal Chemist; Sigma Xi. *Res:* Microbiological research and analysis. *Mailing Add:* USDA-ARS-SRRC PO Box 19687 New Orleans LA 70179

BARKAUSKAS, ANTHONY EDWARD, b Seattle, Wash, Jan 30, 46; m 70; c 3. GRAPH THEORY. *Educ:* Whitman Col, AB, 68; Duke Univ, MA, 70, PhD(math), 72. *Prof Exp:* From asst prof to assoc prof, 72-82, PROF MATH, UNIV WIS-LA CROSSE, 82- *Mem:* Am Math Soc; Math Asn Am; Soc Indust & Appl Math. *Res:* Graph theory, specifically dominating sets, graph reconstruction and graph labeling. *Mailing Add:* 2103 Cass St La Crosse WI 54601

BARKE, HARVEY ELLIS, b Plymouth, Mass, Nov 20, 17; m 41; c 5. ECONOMIC ENTOMOLOGY, BOTANY. *Educ:* Univ Mass, BS, 39, MS, 43; Univ Ga, PhD(entomol), 70. *Prof Exp:* Head res plant breeding, Arnold-Fisher Rose Co, Mass, 40-43; lab instr basic bot & crypt, Univ Mass, 46-47; instr bot & entom, Long Island Agr Tech Inst, 47-56; from asst prof to assoc prof, 56-62, prof biol, 64-78, coordr biol technol, 70-74, chmn, Dept Biol Sci, 74-78, EMER PROF BIOL, STATE UNIV NY AGR & TECH COL, FARMINGDALE, 78-; CONSULT IPM, HORT & LECTR, 80- *Concurrent Pos:* Consult, Am Rose Soc, 60-70; consult rosarian, 65-; mem adj fac biol, C W Post Col, 78-79; dir res & employer training, Flower Time Inc, 78-80, consult & adv, 80-88. *Mem:* Entom Soc Am; Weed Sci Soc Am; Am Inst Biol Sci; Sigma Xi. *Res:* Screening new pesticides for agriculture and ornamental horticulture; turf insect and disease control, weed control; house and tropical plants; eriophyid mites in agriculture and horticulture; peach silver mite. *Mailing Add:* 11 Stoddard St Plymouth MA 02360

BARKELEW, CHANDLER H(ARRISON), b Fresno, Calif, Oct 12, 19; m 47; c 3. CHEMICAL ENGINEERING, PROCESS CONTROL. *Educ:* Pomona Col, BA, 41; Univ Calif, Berkeley, PhD(chem), 44. *Prof Exp:* Engr chem eng, Shell Develop Co, 45-85; RETIRED. *Concurrent Pos:* Lectr, Univ Calif, 70- *Mem:* Am Chem Soc; Am Inst Chem Engrs; Instrument Soc Am; Soc Comput Simulation; Am Inst Chemists. *Res:* Reaction engineering; control theory; thermodynamics; simulation; chemical kinetics. *Mailing Add:* 1110 Spyglass Dr Eugene OR 97401

BARKER, ALFRED STANLEY, JR, b Vancouver, BC, Dec 18, 33; m 55; c 3. EFFICIENT USE OF ENERGY, SOLAR & WIND ENERGY SOURCES. *Educ:* Univ BC, BA, 55, MSc, 57; Univ Calif, PhD(physics), 62. *Prof Exp:* Mem tech staff physics, AT&T Bell Labs, 62-79; vis prof physics, Simon Fraser Univ, 79-81; PROF PHYSICS, TRINITY WESTERN UNIV, 81- *Concurrent Pos:* Vis lectr, NSF, 69-70; vis prof physics, Univ Southern Calif, 70; consult, Energy Study Panel, Am Phys Soc, 74-; adv, USAF Study Panel, 75-; adj prof, Energy Res Inst, Simon Fraser Univ, 80-85. *Mem:* Fel Am Phys Soc. *Res:* Lattice vibrations in solids; alternate energy sources (solar, wind, fuel cells, energy conservation). *Mailing Add:* 6680 248th St Aldergrove BC V0X 1A0 Can

BARKER, ALLEN VAUGHAN, b McLeansboro, Ill, Aug 15, 37; m 67; c 2. PLANT PHYSIOLOGY, SOIL SCIENCE. *Educ:* Univ Ill, BS, 58; Cornell Univ, MS, 59, PhD(agron), 62. *Prof Exp:* Fel soil sci, NC State Univ, 62-64; asst prof plant physiol, Univ Mass, 64-66 & Mich State Univ, 66-67; asst prof, 67-70, assoc prof plant physiol, 70-76, PROF PLANT & SOIL SCI, UNIV MASS, AMHERST, 76- *Concurrent Pos:* Plant physiologist, Agr Res Serv, USDA, 66-67. *Honors & Awards:* Dow Chem Co Environ Qual Res Award, 75; Marion Meadows Award, Am Soc Hort Sci, 77. *Mem:* Crop Sci Soc Am; Am Soc Agron; Soil Sci Soc Am; Am Soc Plant Physiol; Am Soc Hort Sci. *Res:* Ammonium toxicity in plants; nitrate and heavy metal accumulation in vegetables. *Mailing Add:* Dept Plant & Soil Sci Univ Mass Amherst MA 01002

BARKER, BARBARA ELIZABETH, HEMATOLOGY, LYMPHOCYTES. *Educ:* Univ RI, PhD(cell biol), 65. *Prof Exp:* ASSOC DIR HEMAT, RI HOSP, 56-; ASSOC DIR PATH, BROWN UNIV, 74-; ASSOC DIR PATH, UNIV RI, 81- *Res:* Immunology. *Mailing Add:* Dept Spec Hemat RI Hosp Providence RI 02903

BARKER, BEN D, b Burlington, NC, Dec 19, 31; m; c 3. DENTISTRY. *Educ:* Davidson Col, BS, 54; Univ NC, DDS, 58; Duke Univ, MED. *Prof Exp:* Prog dir, W K Kellogg Found, 75-81; from instr to assoc prof fixed prosthodontics, Univ NC, Chapel Hill, 58-65, assoc prof prev dent & dent sci, 65-69, asst dean, 68-69, prof & assoc dean, 69-75, prof, Sch Dent, 81- 89, PROF, DEPT DENT ECOL, SCH DENT, UNIV NC, CHAPEL HILL, 89- *Concurrent Pos:* Mem, Coun Dent Educ, Am Dent Asn, 71-, Coun Int Rels, 79-83, Bd Health Care Serv, Nat Acad Sci, 87-, Comt Estab Res Agenda Aging & Comt Health Promotion & Disability Rev, 88-90; vis prof, New Cross Hosp, London, 72-73. *Mem:* Inst Med-Nat Acad Sci; fel Am Col Dentists; fel Int Col Dentists. *Mailing Add:* Sch Dent 310 Braver Hall Univ NC CB 7450 Chapel Hill NC 27599-7450

BARKER, C(ALVIN) L R, b Dallas, Tex, Nov 25, 30; m 52; c 2. MECHANICAL ENGINEERING. *Educ:* Univ Tex, Austin, BS, 53; Calif Inst Technol, MS, 54, PhD(mech eng), 58. *Prof Exp:* Lab res asst, Defense Res Lab, Univ Tex, 53; res engr, Jet Propulsion Lab, Calif Inst Technol, 54-58; sr propulsion engr, Gen Dynamics/Ft Worth, 58-60; assoc prof, 60-63, assoc dean eng, 74-80, PROF MECH ENG, UNIV TEX, ARLINGTON, 63- *Mem:* Asn Comput Mach; Am Soc Mech Engrs. *Res:* Thermodynamics; gas dynamics; fluid mechanics. *Mailing Add:* Mech Eng Dept Box 19023 Univ Tex Arlington TX 76019

BARKER, CLYDE FREDERICK, b Salt Lake City, Utah, Aug 16, 32; m 56; c 4. SURGERY, IMMUNOLOGY. *Educ:* Cornell Univ, BA, 54, MD, 58; Am Bd Surg, dipl, 65. *Prof Exp:* Intern, Hosp Univ Pa, 58-59, from asst resident to chief resident surg, 59-64; assoc, 64-68, from asst prof to assoc prof, 68-73, PROF SURG, SCH MED, UNIV PA, 73-, ASSOC MED GENETICS, 66-, CHMN, DEPT SURG, 83- *Concurrent Pos:* Fel, Harrison Dept Surg Res, Sch Med, Univ Pa, 59-64, Am Cancer Soc fel, 61-62, USPHS fel med genetics, 65-66; Hartford fel vascular surg, Hosp Univ Pa, 64-65; Markle Found scholar acad med, 68; attend surgeon & assoc chief sect vascular surg, Hosp Univ Pa, 66-82, chief sect renal transplantation, 69-, chief sect vascular surg, 82-, chief surg, 83-; chief surg, Hosp Univ Pa, 83- *Mem:* Transplantation Soc; Soc Univ Surgeons; Soc Vascular Surg; Am Surg Asn; Soc Clin Surg. *Res:* Gastrointestinal and cardiovascular physiology; transplantation biology; immunologically privileged sites; microvascular transplantation surgery; transplantation of pancreatic islets; clinical renal transplantation; immunological aspects of diabetes and pathogenesis of diabetes. *Mailing Add:* Three Coopertown Rd Haverford PA 19041

BARKER, COLIN G, b Plymouth, Eng, Aug 3, 39; m 65; c 2. ORGANIC GEOCHEMISTRY. *Educ:* Oxford Univ, BA, 62, DPhil(geol), 65. *Prof Exp:* Res fel geol, Univ Tex, Austin, 65-67; sr res chemist, Esso Prod Res Co, Tex, 67-69; from asst prof to assoc prof, 69-76, PROF GEOSCI, UNIV TULSA, 76-, CHMN, GEOSCI DEPT, 87- *Concurrent Pos:* Consult, Amoco Prod Res Co, 70-, Sun Oil Co, 74-75, Agrico, 74, Rockwell Int, 75, CONOCO, 76- & Gulf Oil Co, 77; Res Corp grant; NASA grants; NSF grant, 73-; assoc ed, Geochem & Cosmochem Acta, 78-85 & Bull Am Asn Petrol Geologists, 80-86; distinguished lectr, Am Asn Petrol Geologists, 80-81; Dept Energy grants, 81-86. *Honors & Awards:* Matson Award, Am Asn Petrol Geologists, 78 & 82; Esso Distinguished Lectr, Univ Sydney, Australia, 85. *Mem:* Geochem Soc; Europ Asn Org Geochemists; Am Asn Petrol Geologists. *Res:* Generation and migration of petroleum; analysis of inorganic gases in minerals from igneous rocks and ore deposits; pyrolysis fluid inclusions. *Mailing Add:* Dept Geosci Univ Tulsa Tulsa OK 74104

BARKER, DANIEL STEPHEN, b Waltham, Mass, Feb 27, 34; div; c 2. GEOLOGY. *Educ:* Yale Univ, BS, 56; Calif Inst Technol, MS, 58; Princeton Univ, PhD(geol), 61. *Prof Exp:* Res assoc geol, Cornell Univ, 61-62; res assoc, Yale Univ, 62-63; from asst prof to assoc prof, 63-72, PROF GEOL, UNIV TEX, AUSTIN, 72- *Concurrent Pos:* Fulbright-Hays res fel, 74. *Mem:* Geol Soc Am; Mineral Soc Am; Am Geophys Union. *Res:* Igneous and metamorphic petrology; volcanology. *Mailing Add:* Dept Geol Sci Univ Tex Austin TX 78713-7909

BARKER, DAVID LOWELL, b Price, Utah, May 3, 41; m 64; c 1. NEUROBIOLOGY. *Educ:* Calif Inst Technol, BS, 63; Brandeis Univ, PhD(biochem), 69. *Prof Exp:* Res fel neurobiol, Harvard Med Sch, 69-71; asst prof biol, Univ Ore, 71-80; mem fac, dept zool, Univ Iowa, 80-; AT MARINE SCI CTR, ORE STATE UNIV. *Mem:* Soc Neurosci; AAAS. *Res:* Identification of neurotransmitters in individual invertebrate neurons; analysis of biochemical control mechanisms that regulate neurotransmitter metabolism and electrical activity in single cells and in interconnected neurons generating motor patterns. *Mailing Add:* 1171 Rickover Lane Foster City CA 94404

BARKER, DEE H(EATON), b Salt Lake City, Utah, Mar 28, 21; m 45; c 5. CHEMICAL ENGINEERING, EDUCATION. *Educ:* Univ Utah, BS, 48, PhD(chem eng), 51. *Prof Exp:* Res engr, Eng Res Lab, E I du Pont de Nemours & Co, 51-54, engr, 54-59; prof, 59-90, EMER PROF CHEM ENG, BRIGHAM YOUNG UNIV, 90- *Concurrent Pos:* Fulbright fel, Chonnam Nat Univ, Kuang Ju, Korea, 80-81. *Mem:* Fel Am Inst Chem Engrs; Am Soc Eng Educ. *Res:* Heat transfer; fluid flow; particle dynamics; automatic control and process dynamics; atmospheric pollution; dust and mist collection. *Mailing Add:* 1398 Cherry Lane Provo UT 84604

BARKER, DONALD YOUNG, b Glenboro, Man, Apr 4, 25. PHARMACY. *Educ:* Univ Man, BS, 49; Purdue Univ, MS, 53, PhD(pharm), 55. *Prof Exp:* Instr pharm, Univ Man, 49-53; PROF PHARM, UNIV PAC, 60- *Mem:* Am Pharmaceut Asn; Soc Cosmetic Chem; Royal Soc Health; Sigma Xi. *Res:* Pharmaceutical education; product development; dermatological vehicles; emulsion technology. *Mailing Add:* Sch Pharm 751 W Brookside Stockton CA 95207

BARKER, EARL STEPHENS, b Salt Lake City, Utah, Sept 21, 20. INTERNAL MEDICINE. *Educ:* Univ Utah, BA, 41; Univ Pa, MD, 45. *Prof Exp:* Asst instr pharmacol, 48-49, from asst instr to instr, 51-53, assoc, 53-55, asst prof, 55-61, ASSOC PROF MED, SCH MED, UNIV PA, 61-, DIR, DIAG CLIN, 70- *Concurrent Pos:* Godey-Seger fel med, Univ Pa, 49-51; NIH res fel, 51-53; asst attend physician, Univ Hosps, 51-55, attend physician, 55-; investr, Am Heart Asn, 55-60. *Mem:* Fel Am Col Physicians; Am Fedn Clin Res; Am Soc Clin Invest. *Res:* Renal physiology; tubular transport mechanisms; acid-base balance; electrolyte excretion; oxygen metabolism of kidney. *Mailing Add:* Dept Med Renal Sect Univ Pa 36th & Spruce St Philadelphia PA 19104

BARKER, EDWIN STEPHEN, b Santa Fe, NMex, Oct 22, 40; m 62, 89; c 5. PLANETARY ASTRONOMY. *Educ:* NMex State Univ, BS, 62; Univ Kans, MA, 64; Univ Tex, Austin, PhD(astron), 69. *Prof Exp:* Staff astron planetary, Univ Tex McDonald Observ, 69-75; staff astron Copernicus, Princeton Univ, 75-77; staff astron, Univ Tex McDonald Observ, 77-78, res astron planetary, 78-79, res staff, 79-88, SUPT, UNIV TEX MCDONALD OBSERV, 88- *Mem:* Am Astron Soc; Int Astron Union. *Res:* Planetary atmospheres; cometary spectroscopy and spectrophotometry; space astronomy: Copernicus observations of planets; reflectance spectroscopy of outer planets, satellites and asteroids. *Mailing Add:* Univ Tex McDonald Observ PO Box 1337 Ft Davis TX 79734

BARKER, FRANKLIN BRETT, b Shawnee, Okla, Nov 12, 23; m 51; c 1. RADIOCHEMISTRY. *Educ:* Univ Okla, BA, 44; Univ NMex, PhD(chem), 54. *Prof Exp:* Asst chemist, Cities Serv Oil Co, 44-45, control chemist, 45-46; mem staff, Los Alamos Sci Lab, 47-51; res asst, Univ NMex, 51-54; chemist, US Geol Surv, Colo, 54-57, res chemist & proj chief, 57-61; sr scientist, Bettis Atomic Power Lab, Westinghouse Elec Corp, 61-67, fel scientist, 67-78, adv scientist, 78-89; RETIRED. *Mem:* AAAS; Am Chem Soc; Am Nuclear Soc; Am Inst Chemists; Sigma Xi. *Res:* Mass spectrometry of reactor materials; automatic control and data acquisition; computer programming; radiochemical analysis; gamma-ray spectroscopy; nuclear fuel burn-up measurement; reactor plant radiochemistry; environmental radiochemistry. *Mailing Add:* 3347 Forest Rd Bethel Park PA 15102

BARKER, FRED, b Seekonk, Mass, Nov 4, 28; m 61; c 3. GEOLOGY. *Educ:* Mass Inst Technol, BS, 50; Calif Inst Technol, MS, 52 PhD, 54; Harvard, post-doc, 57. *Prof Exp:* GEOLOGIST, US GEOL SURV, 54- *Concurrent Pos:* Vis res fel, Univ Witwatersrand, 74. *Mem:* Fel Geol Soc Am; Am Geophys Union. *Res:* Petrology and geochemistry: trondhjemites, igneous rocks of accreted terranes; geology of Precambrian rocks; origin of granitic batholiths. *Mailing Add:* US Geol Surv Box 25046 Stop 913 Denver CO 80225

BARKER, GEORGE EDWARD, b Detroit, Mich, Apr 3, 28; m 52; c 3. CHEMICAL ENGINEERING. *Educ:* Univ Mich, BSE, 50, MSE, 51, PhD(chem eng), 52. *Prof Exp:* Chem engr, Monsanto Co, 52-55 & Process Res Sect, 57-59, res group leader, 59-62, Monsanto fel, 62-70, SR MONSANTO FEL, MONSANTO CO, 70- *Mem:* Am Chem Soc; Am Inst Chem Engrs. *Res:* Chemical reaction kinetics; process instrumentation and control; automation by computer; mammalian cell culture. *Mailing Add:* Monsanto Co St Louis MO 63167

BARKER, GEORGE ERNEST, organic chemistry; deceased, see previous edition for last biography

BARKER, HAL B, b Palmer, Tenn, Feb 5, 25; m 52; c 2. ANIMAL PHYSIOLOGY, ANIMAL NUTRITION. *Educ:* Tenn Polytech Inst, BS, 47; Iowa State Col, MS, 49; Ala Polytech Inst, PhD, 59. *Prof Exp:* Assoc prof dairying, La Tech Univ, 49-58, head, Dept Animal Indust, 53-64, dean, Sch Agr & Forestry, 64-71, prof animal husb, 58-88, dean, Col Life Sci, 71-88, EMER PROF, LA TECH UNIV, 88- *Mem:* AAAS; Am Soc Animal Sci; Am Dairy Sci Asn. *Res:* Milkfat replacements in rations for dairy calves; physiology of reproduction. *Mailing Add:* 2802 Lakeview Dr Ruston LA 71270

BARKER, HAROLD CLINTON, b Akron, Ohio, Aug 6, 22; m 44; c 3. ORGANIC CHEMISTRY, POLYMER CHEMISTRY. *Educ:* Univ Akron, BS, 44; Ohio State Univ, MSc, 48, PhD(org chem), 51. *Prof Exp:* Res chemist, E I du Pont de Nemours Inc, 51-54, sales technologist, 54-59, indust specialist, 59-71, sales develop plastic pipe, 71-74, mkt develop polyimide plastics, 74-78, mkt develop polyvinyl butyral sheeting, 78-82; RETIRED. *Mem:* Sigma Xi. *Res:* Organic free radical chemistry; degradation of polymers present; sales development of polyolefin plastics; market development of glass reinforced thermoplastics. *Mailing Add:* 3204 Tanya Dr Delwynn Wilmington DE 19803

BARKER, HAROLD GRANT, b Salt Lake City, Utah, June 10, 17; m 49; c 2. SURGERY. *Educ:* Univ Utah, AB, 39; Univ Pa, MD, 43; Am Bd Surg, dipl. *Prof Exp:* Intern, Hosp Univ Pa, 43-44, asst instr pharmacol, Med Sch, 46-47, asst instr surg, 47-51, instr, 51-52, assoc surg, 52-53; from asst prof surg to assoc prof, 53-68, PROF CLIN SURG, COL PHYSICIANS & SURGEONS, COLUMBIA UNIV, 68- *Concurrent Pos:* Res fel pharmacol, Univ Pa, 46-47, Harrison res fel surg, 47-51 & Runyon res fel, 51-52; asst resident surg, Hosp, Univ Pa, 47-51, sr resident, 51-52, asst attend surgeon, 52-53; from asst attend surgeon to attend surgeon, Presby Hosp, 53-, dir med affairs, 74-82. *Mem:* Am Surg Asn; Soc Univ Surgeons; Soc Surg Alimentary Tract; Am Col Surgeons; Halsted Soc. *Res:* Surgical physiology of gastrointestinal tract. *Mailing Add:* One Forest Ave Rye NY 10580

BARKER, HORACE ALBERT, b Oakland, Calif, Nov 29, 07; m 33; c 3. BIOCHEMISTRY. *Educ:* Stanford Univ, AB, 29, PhD(chem), 33. *Hon Degrees:* ScD, Western Reserve Univ, 64; Dr, Munich Univ, 90. *Prof Exp:* Asst zool, Univ Chicago, 30-31; Nat Res Coun fel, Hopkins Marine Sta, Pacific Grove, 33-35; Gen Educ Bd fel, Technol Univ Delft, 35-36; instr soil microbiol, Univ Calif & jr soil microbiologist, Agr Exp Sta, 36-41, asst prof & asst soil microbiologist, 41-45, assoc soil microbiologist, 45-46, prof, 46-51, prof plant biochem, 51-59, prof biochem, 59-75, microbiologist, 46-75, chmn dept biochem, 62-64, EMER PROF BIOCHEM, AGR EXP STA, UNIV CALIF, BERKELEY, 75- *Concurrent Pos:* Guggenheim fel, 41-42 & 62; assoc ed, Ann Rev Microbiol; dir, Ann Rev, Inc, 46-62. *Honors & Awards:* Sugar Res Award, 45; Carl Neuberg Medal, 59; Borden Award, 62; F G Hopkins Medal, Brit Biochem Soc, 67; Nat Medal Sci, 68. *Mem:* Nat Acad Sci; Am Soc Microbiol; Am Chem Soc; Am Soc Biol Chem; Brit Biochem Soc; Am Acad Arts & Sci. *Res:* Soil microbiology; physiology and biochemistry of microorganisms; vitamin B-12 coenzymes. *Mailing Add:* H A Barker Hall Univ Calif Berkeley CA 94720

BARKER, JAMES CATHEY, b Mt Pleasant, Tenn, Oct 30, 45. WATER POLLUTION ABATEMENT SYSTEMS. *Educ:* Univ Tenn, BS, 67, MS, 69, PhD(agr eng), 73. *Prof Exp:* Res asst, Univ Tenn, 67-73; res assoc, Univ Ga, 73-74; exten asst prof, 74-79, exten assoc prof, 79-85, PROF AGR WASTE MGT, NC STATE UNIV, 85- *Concurrent Pos:* Engr consult, RI Dept Environ Mgt, 83, Neuhoff Farms, Inc, 83-84, John A Lut & Son, Inc, 84; Nat Fertilizer Serv, 87 & Shimar Farms, 87, US Feed Grains Coun, 89, Am Asn Swine Practitioners, 89, Agrocarne, SA, 90 & Phillips Dairy Farm, 90. *Honors & Awards:* Aerovent Young Exten Worker Award, Am Soc Agr Engrs, 84. *Mem:* Am Soc Agr Engrs; Water Pollution Control Fedn; Coun Agr Sci & Technol. *Res:* Developing educational programs and fostering the advancement of technology in the areas of agricultural waste management and pollution abatement systems; land application and utilization of wastes; preservation of environmental quality. *Mailing Add:* NC State Univ Box 7625 Raleigh NC 27695-7625

BARKER, JAMES J(OSEPH), b Elmhurst, NY, Apr 15, 22; m 45, 79; c 3. CHEMICAL ENGINEERING, NUCLEAR ENGINEERING. *Educ:* NY Univ, BChE, 43, MChE, 50, DESc(chem eng), 59. *Prof Exp:* Jr engr, Kellex Corp, 43-45 & Hydrocarbon Res Inc, 47-48; sewer cleaner, Roto-Rooter, 49; engr, Atomic Energy Div, H K Ferguson Co, 49-52; mgr process dept, Walter Kidde Nuclear Labs, Inc, 52-57; assoc prof nuclear eng, NY Univ, 57-62; CONSULT ENGR, 59- *Concurrent Pos:* Consult, Res Div, W R Grace & Co, 67, AiRes Mfg, 79-80 & 83-85 & Atlantic Richfield Corp, 80-82; prof physics, C W Post Col, Long Island Univ, 67-70; chmn dept chem eng, Manhattan Col, 75-76; CSULB, SJSU, De Anza Community Col, 86-88; prin engr, Westinghouse Hanford Co, 89- *Mem:* AAAS; Am Nuclear Soc; Am Inst Chem Engrs; Am Chem Soc; NY Acad Sci; Am Phys Soc; Sigma Xi. *Res:* Mass transfer in agitated liquids; isotope separation; radioisotopes; nuclear engineering; mathematics. *Mailing Add:* 1508 Arbor St Richland WA 99352-3963

BARKER, JANE ELLEN, b Bangor, Maine, June 21, 35. DEVELOPMENTAL BIOLOGY, HEMATOLOGY. *Educ:* Univ Maine, BA, 57; Wellesley Col, MA, 59; Univ Wis, PhD(zool), 67. *Prof Exp:* Asst investr, Inst Med Res, Putnam Mem Hosp, 69-73; sr staff fel, Clin Hemat Br, NIH, 73-78, sr investr, 78-80; STAFF SCIENTIST, JACKSON LAB, 80- *Concurrent Pos:* Fel develop genetics, Univ Wis, 66-67; fel develop hemat, Jackson Lab, 67-69. *Mem:* AAAS; Soc Develop Biol; Am Soc Hemat; Am Soc Cell Biol. *Res:* Embryonic development of mouse hematopoietic system; cellular and molecular defects in mice with hereditary anemias; gene transfer. *Mailing Add:* Jackson Lab Bar Harbor ME 04609

BARKER, JEFFERY LANGE, b New York, NY, Jan 29, 43; m 70; c 2. NEUROBIOLOGY, NEUROPHARMACOLOGY. *Educ:* Harvard Col, BA, 64; Boston Univ, MD, 68. *Prof Exp:* Intern surg, Boston Univ Hosp, 68-69; res assoc neurobiol, Nat Inst Neurol Dis & Stroke, 69-72, spec fel, 72-73, med officer neurobiol, Nat Inst Child Health & Human Develop, 73-76, med officer, lab neurol & commun dis & stroke, 76-, CHIEF MED DIR, PUBLIC HEATH SERV, LAB NEUROPHYSIOL, NAT INST NEUROL & COMMUN DISORDER & STROKE, NIH. *Concurrent Pos:* NSF grants res consult neurobiol, 75- *Mem:* AAAS; Soc Gen Physiologists; Soc Neurosci; Am Physiol Soc; Am Soc Neurochem. *Res:* Physiological roles of peptides and neurotransmitters in neuronal function; cellular mechanisms of neuropharmacologic agents, such as anesthetics and convulsants; biophsyical mechanisms of neuronal pacemaker activity; function of rapidly transported proteins in axons. *Mailing Add:* Lab Neurophysiol Nat Inst Neurol Dis & Stroke NIH Bldg 36 Rm 2C02 Bethesda MD 20892

BARKER, JOHN L, JR, b New York, NY, Sept 4, 37; m 59; c 2. EARTH SCIENCES, REMOTE SENSING. *Educ:* Johns Hopkins Univ, AB, 58; Univ Chicago, MS, 62, PhD(phys chem), 67. *Prof Exp:* Res assoc chem, Univ Chicago, 67-68; asst prof, Univ Md, College Park, 68-72; radiochemist, US Geol Surv, 72-73; PHYS SCIENTIST, GODDARD SPACE FLIGHT CTR, NASA, 74-, NASA LANDSAT ASSOC, PROJECT SCI, 81- *Concurrent Pos:* Nat Acad Sci sr fel, 73 & 74. *Mem:* Am Chem Soc. *Res:* Earth resources; applications of remote sensing from satellites; evaluation and improvement of digital image processing of imagery from scanners and linear array sensors; identifying requirements for future satellite sensors; renewable resources; satellite performance simulation; spectral analysis for global habitability. *Mailing Add:* 9102 Tuckahoe Lane Adelphi MD 20783-1438

BARKER, JOHN ROGER, b Sewanee, Tenn, Oct 17, 43; m 72. CHEMICAL KINETICS. *Educ:* Hampden-Sydney Col, BS, 65; Carnegie-Mellon Univ, MS, 69, PhD(chem), 70. *Prof Exp:* Res assoc chem kinetics, Univ Wash, 69-71; res assoc, Brookhaven Nat Lab, 71-73; sr res assoc, Yeshiva Univ, 73-74; CHEMIST CHEM KINETICS, SRI INT, 74- *Concurrent Pos:* Asst ed, Int J Chem Kinetics, 74-76. *Mem:* Am Chem Soc; AAAS; Sigma Xi. *Res:* Gas phase chemical kinetics; inter- and intra-molecular energy transfer processes; laser-induod chemistry and photophysics. *Mailing Add:* Chem Kinetics SRI Int Menlo Park CA 94025

BARKER, JUNE NORTHROP, b Milwaukee, Wis, June 29, 28; m 51. PHYSIOLOGY. *Educ:* Univ Rochester, BS, 52; Duke Univ, MA, 54, PhD(physiol), 56. *Prof Exp:* Instr physiol, Med Sch, Duke Univ, 57-58; from instr to asst prof, Jefferson Med Col, 58-64; RES ASSOC, DEPT CHEM, RUTGERS UNIV. *Mem:* AAAS; Am Physiol Soc; Biomed Eng Soc; Am Soc Neurochem. *Res:* Developmental and cerebrovascular physiology. *Mailing Add:* RD 2 Box 227 Barbertown Rd Frenchtown NJ 08825

BARKER, KENNETH LEROY, b Columbus, Ohio, July 15, 39; m 64; c 1. BIOCHEMISTRY, ENDOCRINOLOGY. *Educ:* Ohio State Univ, BS, 60, MS, 62, PhD(biochem), 64. *Prof Exp:* Res asst dairy sci, Ohio State Univ, 61-64; res assoc obstet & gynec, Univ Kans, 65-66; from instr to res prof obstet & gynec & from instr to prof biochem, Univ Nebr Med Ctr, Omaha, 66-81, dean grad studies & res, 79-81; PROF & CHMN BIOCHEM & PROF OBSTET & GYNEC, TEX TECH UNIV HEALTH SCI CTR, LUBBOCK, 81- *Concurrent Pos:* Fel biochem, Univ Kans, 65-66; Pop Coun res grant, 67; Nat Inst Child Health & Human Develop res grant, 67-; NIH res career develop award, 68-73; guest scientist, Biol Div, Oak Ridge Nat Lab, 70; Nat Acad Sci exchange fel to Inst Molecular Biol, Moscow, USSR, 73; mem, Reproductive Biol Study Sect, NIH, 75-79 & chmn, 85-87. *Mem:* Endocrine Soc; Soc Gynec Invest; Am Physiol Soc; Soc Study Reproduction; Am Soc Biochem Molecular Biol; Am Chem Soc. *Res:* Investigations of hormonal control mechanisms in the mammalian uterus, translational and transcriptional control of uterine, protein synthesis and enzyme activities. *Mailing Add:* Dept Biochem Tex Tech Univ Health Sci Ctr Lubbock TX 79430

BARKER, KENNETH NEIL, b Spring Valley, Ohio, Mar 25, 37; m 57; c 3. PHARMACY ADMINISTRATION, HOSPITAL PHARMACY. *Educ:* Univ Fla, BS, 59, MS, 61; Univ Miss, PhD(pharm admin), 70. *Prof Exp:* Residency, Hosp Pharm, Univ Fla, 61; proj coordr, Hosp Systs Res, Sch Pharm, Univ Miss, 67-70; dir admin res, US Pharmacopeial Conv, Inc, 70-72; assoc dir res inst & assoc prof pharm admin, Sch Pharm, Northeast La Univ, 72-75; CHMN, DEPT PHARM ADMIN & ALUMNI PROF PHARM ADMIN, AUBURN UNIV, 75- *Concurrent Pos:* Mem vis scientist prog, Am Asn Col Pharm, 70-72; consult & ed dir, Health Care Facil Serv, Health Resources Admin, Dept Health, Educ & Welfare, 70-73; mem adv panel hosp pharm res needs, Am Soc Hosp Pharmacists, 73; proj dir & chmn, Nat Coord Comt Large Volume Parenterals, 72-80; consult, Bd Trustees, US Pharmacopeial Conv, Inc, 74-75. *Honors & Awards:* Res Award, Am Soc Hosp Pharmacists, 73, 85, 87 & Whitney Award, 81. *Mem:* Acad Pharmaceut Sci; Am Asn Col Pharm; Am Soc Hosp Pharmacists; Am Pharmaceut Asn. *Res:* Socioeconomics of health care with emphasis on pharmacy services; hospital medication systems, errors; co-originator of unit-dose concept; organizational research; utilization of pharmacists: design of hospital pharmacy facilities; future of pharmacy. *Mailing Add:* Dept Pharm Auburn Univ Main Campus Auburn AL 36849

BARKER, KENNETH REECE, b Roaring River, NC, Feb 1, 32; m 58; c 2. PLANT PATHOLOGY. *Educ:* NC State Col, BS, 56, MS, 59; Univ Wis, PhD(plant path), 61. *Prof Exp:* Asst prof, Univ Wis, Madison, 61-66; assoc prof, 66-71, PROF PLANT PATH, NC STATE UNIV, 71- *Concurrent Pos:* Ed-in-chief, Jour Nematol, 75-78. *Honors & Awards:* Fel, Am Phytopath Soc; Fel, Soc Nematologists. *Mem:* Am Phytopath Soc; Soc Nematologists (vpres, 78-79, pres, 79-80). *Res:* Ecology and physiology of plant parasitic nematodes and their interaction with other plant pathogens. *Mailing Add:* Dept Plant Path NC State Univ Raleigh NC 27695-2716

BARKER, LAREN DEE STACY, b Honolulu, Hawaii, Mar 1, 42; m 64; c 4. VERTEBRATE PHYSIOLOGY. *Educ:* Univ Minn, BS, 64, MS, 66; Pa State Univ, PhD(physiol), 69. *Prof Exp:* Res asst dairy husb, Univ Minn, 64-66; assoc prof, 69-76, PROF BIOL, SOUTHWEST STATE UNIV, 76- *Mem:* Soc Study Reproduction; Am Inst Biol Sci; Brit Soc Study Fertil. *Res:* Physiology of the male reproductive system; epididymal physiology; immunoreproduction. *Mailing Add:* Dept Math & Sci Southwest State Univ Marshall MN 56258

BARKER, LEROY N, b Brigham City, Utah, Oct 18, 28; m 56; c 5. AGRONOMY, PLANT BREEDING. *Educ:* Utah State Univ, BS, 53, MS, 57; Univ Wis, PhD(agron), 64. *Prof Exp:* Plant breeder & res sta mgr, Asgrow Seed Co, Wis, 60-65; from asst prof to assoc prof, 65-73, PROF AGRON, CALIF STATE UNIV, CHICO, 73- *Concurrent Pos:* Vis prof, US AID, Univ Wis Proj, Univ Ife, Nigeria, 68-70; plant breeder & consult, Rice Researchers, Inc, 74- *Mem:* Am Soc Agron; Crop Sci Soc Am. *Res:* Breeding of cowpeas, peppers and rice; agronomic research on cowpeas and tomatoes; basic genetic studies on cowpeas; field crop yield and fertilizer trials and genetic studies; variety development, agronomic research and extension service on rice. *Mailing Add:* Dept Agr Calif State Univ Chico Chico CA 95929

BARKER, LEWELLYS FRANKLIN, b Baltimore, Md, Sept 9, 33; m 64; c 3. INFECTIOUS DISEASES, IMMUNOLOGY. *Educ:* Princeton Univ, BA, 55; Johns Hopkins Univ, MD, 59, MPH, 90- *Prof Exp:* Intern, Johns Hopkins Univ Hosp, Baltimore, Md, 59-60; resident, Bellevue Hosp, New York, NY, 60-62; med officer virol, Div Biol Standards, NIH, Bethesda, 62-72; dep dir virol, Bur Biologics, Food & Drug Admin, Rockville, Md, 72-73, dir blood & blood prod, 73-78; vpres blood serv, Am Red Cross, 78-81, vpres, 81-84, sr vpres, 84-90, chief med officer, health serv, 88-90. *Concurrent Pos:* Counr, Int Soc Blood Transfusion, 75-90; bd mem, Nat Health Coun, 81-90. *Mem:* Nat Acad Sci; Am Asn Blood Banks; Am Epidemiol Soc; NY Acad Sci; Am Asn Immunologists; Am Soc Microbiol; Nat Res Coun; Nat Health Coun (pres, 85-87). *Res:* Medical virology; detection of viruses transmissible by blood transfusion; prevention of infectious diseases. *Mailing Add:* Am Red Cross 4007 Laird Pl Chevy Chase MD 20815

BARKER, LOUIS ALLEN, b Charleston, WVa, Nov 23, 41; m 66; c 1. PHARMACOLOGY, NEUROCHEMISTRY. *Educ:* WVa Univ, BSc, 64; Tulane Univ, PhD(pharmacol), 68. *Prof Exp:* Grant researcher biochem, Queens Col, NY, 70-72, adj asst prof psychol, 72; from instr to assoc prof pharmacol, Mt Sinai Med Sch, 72-82; assoc prof, 82-86, PROF PHARMACOL, LA STATE UNIV MED CTR, 86- *Concurrent Pos:* Nat Inst Neurol Dis & Stroke fel, NY State Inst Basic Res Ment Retardation, 68-70. *Mem:* Am Soc Pharmacol & Exp Therapeut; Shock Soc. *Mailing Add:* Dept Pharmacol La State Univ Med Ctr New Orleans LA 70112

BARKER, LYNN MARSHALL, b Florence, Ariz, Apr 4, 28; m 64; c 2. APPLIED PHYSICS. *Educ:* Univ Ariz, BS, 54, MS, 55. *Prof Exp:* Staff mem, Sandia Labs, 55-74; sr staff consult, Terra Tek, Inc, 74-81; distinguished mem tech staff, Sandia Labs, 81-90; PRES, VALYN INT, 90- *Concurrent Pos:* Consult, Sandia Labs, 90- *Mem:* Am Phys Soc; Soc Exp Mech; Am Soc Testing & Mat; Am Soc Metals Int. *Res:* Shock waves in condensed matter; originator of VISAR laser interferometer instrumentation; in fracture mechanics originator of Am Standards Testing Mat-standardized chevron-notched short rod test method for measuring fracture toughness; eight patents. *Mailing Add:* 13229 Circulo Largo NE Albuquerque NM 87112-3771

BARKER, MARY ELIZABETH, b 1930; m 75. LIVER CELL BIOLOGY. *Educ:* Univ Calif, Berkeley, PhD(zool), 68. *Prof Exp:* Asst prof biol, Rosary Col, Ill, 68-72; staff physiologist, Lawrence Berkeley Lab, 76-82; ASST RES CELL BIOL, UNIV CALIF, SAN FRANCISCO, 82-, CO-DIR, LIVER CTR MICROS CORE FACIL, 82- *Mem:* Am Soc Cell Biol. *Res:* Immunocytochemistry; role of growth factors in control of liver regeneration; intracellular pH and ion fluxes in regenerating liver cells. *Mailing Add:* Vet Admin Med Ctr 151 E 4150 Clement St San Francisco CA 94121

BARKER, MORRIS WAYNE, b Astoria, Ore, May 31, 45; m 66; c 2. APPLIED FISHERIES MANAGEMENT. *Educ:* Ore State Univ, BS, 72, MS, 74; Univ Wash, PhD(fisheries), 79. *Prof Exp:* Fish biologist III, 78-79, fish biologist IV, 79-80, ASST CHIEF I, DEPT FISHERIES, WASH STATE, 80- *Concurrent Pos:* Vis lectr, Univ Alaska, 77-78; consult, Peter Pan Seafoods & Nichiro Pac Ltd, 78. *Mem:* Pac Fisheries Biologists; Am Inst Fishery Res Biologists. *Res:* Fishing gear efficiencies and economics; population dynamics of sport exploited near shore ground fish, including fishing and natural mortality rates; growth rate estimates; tagging studies. *Mailing Add:* 8608 Ninth Way SE Lacey WA 98503

BARKER, NORVAL GLEN, b Lincoln, Nebr, Aug 6, 25; m 45; c 3. NUTRITIONAL BIOCHEMISTRY. *Educ:* Univ Nebr, BS, 47, MS, 49, PhD(chem), 50. *Prof Exp:* Sect leader chem res, Gen Mills, Inc, 50-58; VPRES RES, SANDOZ NUTRIT, 58- *Mem:* Am Chem Soc; Inst Food Technologists. *Res:* Therapeutic nutritional products; institutional foods. *Mailing Add:* RR 3 Box 121 Annandale MN 55302

BARKER, PAUL KENNETH, b Leeds, Eng, Nov 26, 49. STELLAR SPECTROSCOPY, STELLAR WINDS. *Educ:* Univ Manchester, Eng, BSc, 72; Univ Colo, Boulder, PhD(astro-geophysics), 79. *Prof Exp:* fel, Univ Western Ont, 79-82, res assoc, Dept Astron, 82-; AT DEPT PHYSICS, YORK UNIV. *Mem:* Am Astron Soc; AAAS. *Res:* Spectroscopic observations of early-type stars; dynamics and geometrical configuration of circumstellar envelopes; magnetic stars; infuence of rotation and magnetic fields upon circumstellar dynamics. *Mailing Add:* Dept Physics York Univ 4700 Keele St North York ON M3J 1P3 Can

BARKER, PETER EUGENE, b Ithaca, NY; m 83; c 1. CANCER RESEARCH, CYSTIC FIBROSIS. *Educ:* Cornell Univ, BA, 73; Univ Tex, PhD(cell & molecular biol), 81. *Prof Exp:* ASST PROF MED GENETICS, UNIV ALA, BIRMINGHAM, 85- *Concurrent Pos:* Postdoctoral fel, Yale Univ, 80-85. *Mem:* Sigma Xi; fel Muscular Dystrophy Asn. *Res:* Human genetics; molecular cell biology; cancer genetics. *Mailing Add:* Lab Med Genetics 316 Sparks Ctr Univ Ala Birmingham AL 35294

BARKER, PHILIP SHAW, b Queretaro, Mex, Aug 2, 33; c 3. ENTOMOLOGY, INSECT TOXICOLOGY. *Educ:* Univ Calif, Berkeley, MSc, 60; McGill Univ, PhD, 65. *Prof Exp:* RES SCIENTIST, AGR CAN, 65- *Mem:* Entom Soc Am; Entom Soc Can; Sigma Xi. *Res:* Control of insects and mites in stored cereals; effect of freezing temperatures on fumigants; life histories of mites; statistical programs for calculators. *Mailing Add:* Res Sta Agr Can 195 Dafoe Rd Winnipeg MB R3T 2M9 Can

BARKER, RICHARD CLARK, b Bridgeport, Conn, Mar 27, 26; m 48; c 2. APPLIED PHYSICS, ELECTRONICS. *Educ:* Yale Univ, BE, 50, ME, 51, PhD(elec eng), 55. *Prof Exp:* From instr to assoc prof, 52-74, PROF APPL SCI, YALE UNIV, 74-, DIR UNIV RES CONTRACTS, 50- *Concurrent Pos:* Consult, Naval Ord Lab, 62-68, Sandia Corp, 63-68, AMP Inc, 72-, Echlin Mfg Co, 76- & Rockwell Int, 78- *Honors & Awards:* Alexander von Humboldt Found Sr Am Scientist Award, 75. *Mem:* AAAS; fel Inst Elec & Electronics Engrs; Am Phys Soc. *Res:* Basic theory of nonlinear magnetic devices; physics of magnetic materials; ferromagnetic resonance in metals; electron tunneling in semiconductors; semimetals; superconductors; metals; semiconductor device physics. *Mailing Add:* Elec Eng Yale Univ 525 Becton Ctr PO Box 2157 Yale Sta New Haven CT 06520

BARKER, RICHARD GORDON, b Rochester, NY, Feb 8, 37; m 57; c 3. ORGANIC CHEMISTRY. *Educ:* Hamilton Col, BA, 58; Lawrence Univ, MS, 60, PhD(org chem), 63. *Prof Exp:* Res scientist, Union Bag-Camp Paper Corp, 62-69, group leader, Union Camp Corp, 69-72, sect leader, 72-74, dir res & develop projs, 74-79, lab dir corp res & develop, 79-85, CORP DIR RES & DEVELOP, UNION CAMP CORP, 85- *Concurrent Pos:* Mem indust liaison comt, Forest Prod Lab, 74-; chmn, Res Steering Comt, Empire State Paper Res Inst; chmn, Res Adv Comt, Inst Paper Chem; pres, ESPRA; trustee, Miami Univ, Ohio, Univ Maine. *Mem:* Am Chem Soc; fel Tech Asn Pulp & Paper Indust; Soc Res Adminr. *Res:* Chemical wood pulping processes; bleaching processes; pulp and paper. *Mailing Add:* Union Camp Corp Box 3301 Princeton NJ 08543-3301

BARKER, ROBERT, b Bedlington, Eng, Sept 21, 28; m 55; c 2. BIOCHEMISTRY. *Educ:* Univ BC, BA, 52, MA, 53; Univ Calif, Berkeley, PhD(biochem), 58. *Prof Exp:* Atlas Powder Co fel chem, Washington Univ, 58-59; vis scientist, NIH, 59-60; from asst prof to assoc prof biochem, Univ Tenn, 60-64; from assoc prof to prof, Univ Iowa, 64-74, assoc dean med, 70-74; prof biochem & chmn dept, Mich State Univ, 74-80; dir, Div Biol Sci, 80-83, vpres res & advan studies, 83-84, PROVOST, CORNELL UNIV, 84- *Concurrent Pos:* Mem, Biochem Training Comt, Nat Inst Gen Med Sci, NIH, 65-69, chmn, 69-70, mem, Med Scientist Training Comt, 70-73, mem, Cellular & Molecular Basis Dis Rev Comt, 77-78; vis prof, Duke Univ, 70-71; chmn biochem test comt, Nat Bd Med Examr, 73-76; mem, Merit Rev Bd Basic Sci, Vet Admin, 73-78, chmn, 76-78; mem, Exec Comt, Nat Bd Med Examr, 77-80; chmn, Panel Res Needs Basic Biomed Sci, Nat Acad Sci-Nat Res Coun, 79-84 & mem, Comt Nat Needs Biomed & Behav Res Personnel, 79-87; mem bd dirs, Corning Glass Works, 87-; mem adv coun, Oak Ridge Nat Lab, 88- *Mem:* AAAS; Am Soc Biol Chemists; Am Chem Soc. *Res:* Effects of configuration and substitution on reactions of carbohydrates; solution structures of mono and oligosaccharides. *Mailing Add:* 19 Spruce Lane Ithaca NY 14850

BARKER, ROBERT EDWARD, JR, b Bonifay, Fla, Oct 8, 30; m 54; c 2. POLYMER PHYSICS. *Educ:* Univ Ala, BS, 52, MS, 54, PhD(physics), 60. *Prof Exp:* High sch teacher, Fla, 50-51; physicist, US Navy Mine Defense Lab, 54 & 55; engr, Hayes Aircraft Corp, Ala, 56 & 57, scientist, 58; from instr to asst prof physics, Univ Ala, 58-60; physicist, Gen Elec Res Lab, 60-67; chmn eng sci prog, 69-73, ASSOC PROF MAT SCI, UNIV VA, 67- *Concurrent Pos:* Consult, Hayes Aircraft Corp, 56-59; mem exec bd & mem tech adv comt, Conf Elec Insulation & Dielectric Phenomena, Nat Acad Sci. *Mem:* Am Phys Soc; Am Asn Physics Teachers; Am Chem Soc; Metric Asn. *Res:* Thermomechanical properties of solids; radiation damage of polymers; electron spin resonance; wave motion; thermodynamics; thermal conductivity; diffusion; ionic mobility; glass transition phenomena in polymers. *Mailing Add:* Dept Mat Sci Univ Va Thorton Hall B-120 Charlottesville VA 22903

BARKER, ROBERT HENRY, b Washington, DC, Aug 17, 37; div; c 3. SCIENCE POLICY, POLYMER CHEMISTRY. *Educ:* Clemson Col, BS, 59; Univ NC, PhD(org chem), 63. *Prof Exp:* From instr to asst prof org chem, Tulane Univ, 62-67; assoc prof textiles & chem, Clemson Univ, 67-74, J E Sirrine prof textile & polymer chem, 74-84; sci adv & chief staff, Hon Judd Gregg, US House Rep, 83-88; ADV TECH & GOVT AFFAIRS, AM FIBER MFRS ASN, 88- *Concurrent Pos:* Consult, Fiber Indust, Inc, 81-83; Am Chem Soc Cong Sci fel, 81-82; Nat Res Coun-Nat Acad Sci Eval Panel, Ctr Fire Res, Nat Bur Standards, 76-82. *Mem:* Am Chem Soc; Am Asn Textile Chem & Colorists; Fiber Soc. *Res:* Chemistry of fibers; polymer pyrolysis and combustion; textile flammability; science policy. *Mailing Add:* 3516 S Ninth St Arlington VA 22204

BARKER, ROY JEAN, b Norborne, Mo, July 9, 24; m 48; c 2. ENTOMOLOGY. *Educ:* Univ Mo, BS, 48; Univ Ill, PhD(entom), 53. *Prof Exp:* Asst entom, Univ Ill, 48-50 & Ill State Natural Hist Surv, 50-53; entomologist, E I Du Pont de Nemours & Co, 53-55; sr insect physiologist, Pioneering Lab, USDA, Md, 55-64; entomologist, Res Labs, Rohm and Haas Co, Pa, 64-67; res entomologist, USDA, 67-79; PVT CONSULT, 80- *Mem:* Am Chem Soc. *Res:* Pesticide trouble shooting; insect physiology and biochemistry; pheromones for pest control. *Mailing Add:* 4620 N Calle Altivo Tucson AZ 85718-5810

BARKER, SAMUEL BOOTH, b Montclair, NJ, Mar 3, 12; m 34. PHYSIOLOGY, PHARMACOLOGY. *Educ:* Univ Vt, BS, 32; Cornell Univ, PhD(physiol), 36. *Hon Degrees:* ScD, Univ Ala, 79, Univ Vt, 84. *Prof Exp:* Asst physiol, Med Col, Cornell Univ, 37-40; from instr to asst prof physiol, Col Med, Univ Tenn, 41-44; from asst prof to assoc prof, Col Med, Univ Iowa, 44-52; prof pharm, Med Col & Sch Dent, Med Ctr, Univ Ala, 52-62; prof, Col Med, Univ Vt, 62-65; dir grad studies, 65-70, prof, 65-76, dean, Grad Sch, 70-78, DISTINGUISHED PROF PHYSIOL & BIOPHYS, UNIV ALA, BIRMINGHAM, 76-, EMER DEAN, GRAD SCH, 78- *Concurrent Pos:* Fel med, Cornell Univ, 38-41; Krichesky fel, Univ Calif, Los Angeles, 51-52, USPHS spec fel, 61-62, career res award, 62-65; pres, Am Thyroid Asn, 71. *Mem:* AAAS; Am Soc Exp Pharmacol & Therapeut; Am Physiol Soc; Soc Exp Biol & Med; Endocrine Soc. *Res:* Fat, carbohydrate and protein metabolism; tissue metabolism; effect of endocrines on metabolism; thyroid and iodine. *Mailing Add:* Dept Physiol Univ Ala Birmingham AL 35294

BARKER, SAMUEL LAMAR, b Pine Bluff, Ark, Sept 22, 42; m 63; c 2. RADIOPHARMACEUTICAL CHEMISTRY, CLINICAL CHEMISTRY. *Educ:* Henderson State Col, BS, 64; Univ Ark, MS, 66; Purdue Univ, PhD(bionucleonics), 69. *Prof Exp:* Res scientist, 69-72, sr res scientist, 72-73, sect head radiopharmaceut res, 73-75, Squibb mgt develop prog, 75-77, dir parenteral mfg, 77-78, ASSOC DIR SQUIBB INST & DIR DIAG RES & DEVELOP, E R SQUIBB & SONS, 78- *Concurrent Pos:* USPHS radiol health fel, Univ Ark, 64-65 & Purdue Univ, 66-69. *Mem:* Soc Nuclear Med; Am Asn Clin Chemists; Clin Radioassay Soc; Parenteral Drug Asn. *Res:* Chemistry of technetium complexes and radioimmunoassay methodology as applied to medical diagnosis. *Mailing Add:* Bristol Myers Squibb Co PO Box 4000 Princeton NJ 08543-4000

BARKER, S(HIRLEY)HUGH, zoology, for more information see previous edition

BARKER, WILEY FRANKLIN, b Santa Fe, NMex, Oct 16, 19; m 43; c 3. MEDICINE. *Educ:* Harvard Univ, BS, 41, MD, 55; Am Bd Surg, dipl. *Prof Exp:* Intern & resident surg, Peter Bent Brigham Hosp, 44-46; asst, 49-51, clin instr, 51-54, from asst prof to assoc prof, 54-63, PROF SURG, SCH MED, UNIV CALIF, LOS ANGELES, 63-, CHIEF, DIV GEN SURG, 69- *Concurrent Pos:* Jr assoc, Peter Bent Brigham Hosp, 48-49; resident surg, Vet Admin Hosp, Los Angeles, 49-51, attend physician, 54-; chief gen surg sect & asst chief gen surg serv, Wadsworth Vet Admin Hosp, 51-54; mem bd trustees, Am Bd Surg, 64-70; chief staff, Sepulveda Vet Admin Ctr, 48- *Mem:* Fel Am Col Surg; Soc Univ Surg; Soc Vascular Surg; Am Surg Asn; Soc Clin Surg. *Res:* Clinical surgical aspects of arteriosclerosis, gastroenterological disease, and aseptic technique. *Mailing Add:* 13216 Dobbins Pl W Los Angeles CA 90000

BARKER, WILLIAM ALFRED, b Los Angeles, Calif, May 9, 19; m 41, 85; c 5. PHYSICS. *Educ:* Yale Univ, BA, 41; Calif Inst Technol, MS, 48; St Louis Univ, PhD(physics), 52. *Prof Exp:* Instr, St Louis Univ, 52-55, from asst prof to prof, 55-64; prof, 64-69, chmn dept, 69-73, prof, 73-88, EMER PROF PHYSICS, UNIV SANTA CLARA, 88- *Concurrent Pos:* Res theoret physics, Swiss Fed Inst Technol, 53-55; consult, Argonne Nat Lab, 58-63, McDonnell Aircraft Corp, 60-, Hewlett-Packard, NASA-Ames, 68-77. *Mem:* Am Phys Soc; Am Asn Physics Teachers. *Res:* Nuclear orientation; quantum electronics; solid state; photoelectron spectroscopy; atmospheric physics; interaction physics; cosmology; acoustics; theory of the magnetic monopole; radioactive half lives, fusion. *Mailing Add:* Altran Corp 540 Weddell Dr Suite 8 Sunnyvale CA 94089

BARKER, WILLIAM GEORGE, b Stratford, Ont, Apr 2, 22; m 46; c 4. PLANT PHYSIOLOGY. *Educ:* Univ Western Ont, BSc, 48, MSc, 49; Univ Mich, PhD(bot), 53. *Prof Exp:* Lectr bot, Univ Western Ont, 51-52; asst prof, Ont Col, 52-54; plant physiologist, Coto Res Sta, United Fruit Co, Costa Rica, 54-55; plant physiologist, Vining G Dunlap Labs, La Lima, Honduras, 55-59 & Cornell Univ, Ithaca, NY, 59-60; res officer, Can Dept Agr, 60-64; from assoc prof to prof biol sci, Univ Man, 64-72, dir biol teaching unit, Div Biol Sci, 70-72; prof genetics & chmn dept, Univ Guelph, 72-83, prof bot, 83-87; RETIRED. *Mem:* Bot Soc Am; Can Bot Asn; Can Soc Plant Physiol. *Res:* Plant tissue culture; activity of the pith region of 50 year old basswood stem; in vitro potato tuberization; banana growth and propagation; low bush blueberry; adventitious root development of impatiens; stomatal behavior of drought stressed citrus. *Mailing Add:* 136 West Mount Rd Guelph ON N1H 3J4 Can

BARKER, WILLIAM HAMBLIN, II, b Albuquerque, NMex, Aug 10, 48. OPERATIONS RESEARCH, PURE MATHEMATICS. *Educ:* Univ Calif, Santa Barbara, BA, 70; Stanford Univ, MS, 72, PhD(math), 75. *Prof Exp:* Res asst math, Ames Res Ctr, NASA, 69-72, consult, 73; mathematician & asst, Stanford Univ, 70-75; mathematician, 75-80, SR ASSOC, DANIEL H WAGNER ASSOCS, 80- *Mem:* Am Math Soc; Acoust Soc Am; Soc Indust & Appl Math. *Res:* Theory of search; theory of surveillance; complex analysis, conformal mappings; calculus of variations, hyperelliptic Reimann surfaces. *Mailing Add:* Box 492 Naalehu HI 96772

BARKER, WILLIAM T, b Larned, Kans, Aug 3, 41; m 62; c 2. SYSTEMATIC BOTANY. *Educ:* Kans State Teachers Col, BA, 63, MA, 67; Univ Kans, PhD(bot), 68. *Prof Exp:* Assoc prof, 68-80, PROF BOT, NDAK STATE UNIV, 80- *Mem:* Am Soc Plant Taxon; Int Asn Plant Taxon; Soc Range Mgt. *Res:* Floristic studies of the United States prairies and plains; grazing management. *Mailing Add:* Dept Bot NDak State Univ Fargo ND 58102

BARKER, WINONA CLINTON, b New York, NY, Sept 5, 38; m 59; c 2. MOLECULAR EVOLUTION. *Educ:* Conn Col, BA, 59; Univ Chicago, PhD(physiol), 66. *Prof Exp:* Sr res technician physiol, Univ Chicago, 65-66, trainee comt math, 66-67, res assoc med, 67-68; assoc dir res, 73-87, VPRES, NAT BIOMED RES FOUND, 87- *Concurrent Pos:* Staff scientist, Atlas Protein Sequence & Struct, 68-; lectr, Sch Med, Georgetown Univ, 72-, assoc mem grad fac, 74-; NIH res grant, 75-88; proj dir, Protein Identification Resource, 84-; mem, task grop, comt Data Sci & Technol. *Mem:* Biophys Soc (secy, 79-87); Am Soc Biol Chemists; Am Soc Cell Biol; AAAS; Protein Soc. *Res:* Analysis of domains in proteins; evolutionary history, function, genetic basis, role in human development, differentiation and oncogenesis; computer analysis of related proteins; protein and nucleic acid sequence databases. *Mailing Add:* Nat Biomed Res Found Georgetown Univ Med Ctr 3900 Reservoir Rd NW Washington DC 20007

BARKEY, KENNETH THOMAS, b Auburn, Wash, Dec 29, 16; m 42; c 2. ORGANIC CHEMISTRY. *Educ:* Univ Wash, BS, 38, MS, 40; Mass Inst Technol, PhD(org chem), 43. *Prof Exp:* Oceanog chemist, Univ Wash, 36-38; supvr, Eastman Kodak Co, 43-79, sr tech assoc, 75-79; RETIRED. *Mem:* Am Chem Soc. *Res:* Cellulose esters; plasticizers; polyesters; air pollution abatement. *Mailing Add:* 3976 Rte 364 E Lake Rd Canandaigua NY 14424

BARKIN, STANLEY M, b New York, NY, July 19, 26; m 55; c 2. MATERIALS SCIENCE. *Educ:* Brooklyn Col, BS, 50; Polytech Inst Brooklyn, MS, 55, PhD(polymer chem), 57. *Prof Exp:* Anal chemist, Lucius Pitkin Inc, 48-50; control chemist, Davis & Geck, Inc, 50-53; res asst, Polymer Inst Brooklyn, 53-57; res chemist, Tex-US Chem Co, 57-58; mgr, Ethicon, Inc, 58-63; sect head, Colgate-Palmolive Co, 63-70; sect supvr, US Steel Corp, 70-75; ASSOC DIR, NAT MAT ADV BD, NAT ACAD SCI, 75-

Concurrent Pos: Adj prof, Dept Math, Seton Hall Univ, 61-70; lectr, Dept Chem, George Wash Univ, 76-; adj prof, Col Arts & Sci, Marymount Univ, 88- *Mem:* Am Chem Soc; AAAS. *Res:* Solution properties of macromolecules; characterization techniques for polymers. *Mailing Add:* Nat Acad Sci 2101 Constitution Ave NW Washington DC 20418

BARKLEY, DWIGHT G, b Indiana, Pa, Nov 13, 32; m 59; c 2. HORTICULTURE. *Educ:* Pa State Univ, BS, 55; Va Polytech Inst, MS, 63, PhD(seed germination), 69. *Prof Exp:* Instr hort, Va Polytech Inst, 63-68; from asst prof to assoc prof agr, 68-71, PROF AGR, EASTERN KY UNIV, 71-, CHMN DEPT, 77- *Mem:* Am Soc Agron; Crop Sci Soc Am. *Res:* Microclimate and microenvironment as they affect germination of grass and woody plant seeds; nursery crop production; turf-grass management. *Mailing Add:* Dept Agr Eastern Ky Univ Richmond KY 40475

BARKLEY, JOHN R, b Oswego, NY, Aug 22, 38; div; c 3. POLYMER COMPOSITES & STRUCTURE, FINE PARTICLES. *Educ:* LeMoyne Col, BS, 60; Ohio Univ, MS, 63, PhD(physics), 66. *Prof Exp:* Res physicist, Cent Res Dept, 66-81, SR RES ASSOC, ELECTRONICS DEPT, E I DU PONT DE NEMOURS & CO, INC, 81- *Res:* Low loadings of inorganic particulates that affect the physical and electrical properties of polymer composites and topography of polymer films; understanding relationships among process, structure and functioning of polyester films. *Mailing Add:* E I du Pont de Nemours & Co Inc PO Box 89 Circleville OH 43113

BARKLEY, LLOYD BLAIR, b Ellwood City, Pa, Mar 10, 25; m 50; c 1. ORGANIC CHEMISTRY. *Educ:* Pa State Univ, BS, 47; Univ Pittsburgh, MS, 50, PhD(org chem), 52. *Prof Exp:* Res chemist, Monsanto Chem Co, 52-56, res group leader, 56-61; dir res & vpres, Pa Indust Chem Corp, 62-69; mgr res, Newport Div, Tenneco Inc, 69-74; mem res staff, Southern Resins, Inc, 74-80. *Mem:* Am Chem Soc. *Res:* Carbon to carbon condensations; steroids; polymers. *Mailing Add:* 166 Hardwood Rd Pelzer SC 29669-9177

BARKLEY, THEODORE MITCHELL, b Modesto, Calif, May 14, 34; m 81. SYSTEMATIC BOTANY. *Educ:* Kans State Univ, BS, 55; Ore State Univ, MS, 57; Columbia Univ, PhD(bot), 60. *Prof Exp:* Instr biol, Occidental Col, 60-61; PROF BOT & CUR HERBARIUM, KANS STATE UNIV, 61- *Mem:* Am Soc Plant Taxon; Bot Soc Am. *Res:* Monographic studies of the genus Senecio; floristics of the Central Great Plains. *Mailing Add:* Herbarium Div Biol Kans State Univ Manhattan KS 66506

BARKMAN, ROBERT CLOYCE, b Massillon, Ohio, Oct 10, 42; m 65; c 2. COMPARATIVE PHYSIOLOGY. *Educ:* Wittenburg Univ, BA, 64; Univ Cincinnati, MS, 66, PhD(zool), 69. *Prof Exp:* Teaching asst biol, Univ Cincinnati, 64-66, 68-69; from asst prof to assoc prof biol, 69-82, PROF BIOL, SPRINGFIELD COL, 82- *Concurrent Pos:* Res biologist, US Environ Res Lab; fel NASA. *Mem:* Am Soc Zoologists; Am Inst Biol Sci; Sigma Xi; Fedn Am Scientists. *Res:* Development of culture methods for all life stages of the Atlantic silverside, Menidia; varying temperature, salinity and plankton on growth of otolith daily rings of marine fish; new ways to teach science. *Mailing Add:* Biol Dept Springfield Col 263 Alden St Springfield MA 01109

BARKO, JOHN WILLIAM, b Detroit, Mich, June 28, 47; m 68; c 1. AQUATIC ECOLOGY. *Educ:* Mich State Univ, BS, 72, PhD(bot), 75. *Prof Exp:* Teaching asst biol, 72-75, asst prof bot, Mich State Univ, 75; RES BIOLOGIST, WATERWAYS EXP STA, US ARMY CORPS ENGRS, 75- *Mem:* Int Asn Aquatic Vascular Plant Biologists; Am Inst Biol Sci; Am Soc Limnol & Oceanog; Ecol Soc Am. *Res:* Physiological ecology of vascular aquatic plants; ecosystem metabolism (fresh water). *Mailing Add:* Waterways Exp Sta US Army Corps Engrs 3909 Halls Ferry Rd Vicksburg MS 39180

BARKS, PAUL ALLAN, b Ft Morgan, Colo, May 28, 36; m 60; c 3. ORGANIC CHEMISTRY, CHEMICAL EDUCATION. *Educ:* Grinnell Col, BA, 58; Iowa State Univ, PhD(org chem), 63. *Prof Exp:* Instr chem, Hamline Univ, 63-64; asst prof, St Norbert Col, 64-68; asst prof, Monmouth Col, Ill, 68-71; HEAD DEPT CHEM, NORTH HENNEPIN COMMUNITY COL, 71- *Concurrent Pos:* NSF undergrad instr equip grant, 65-, res participation for chem teachers grant, 65-67; vis prof, Univ Ky, 80 & Macalester Col, 83-88. *Mem:* Am Chem Soc. *Res:* Photochemical reactions of organic chemicals and natural products, particularly amino acids; effectual teaching; manuscript reviewer for preparation of chemistry textbooks. *Mailing Add:* Dept Chem North Hennepin Community Col Minneapolis MN 55445

BARKSDALE, CHARLES MADSEN, b San Diego, Calif, June 20, 47. PHARMACOKINETICS, DRUG METABOLISM. *Educ:* Univ Calif San Diego, BA(chem) & BA(math), 70; Columbia Pac Univ, MS, 82, PhD(chem), 83. *Prof Exp:* Res neurochemist, Dept Psychiat, VA Hosp, Madison, Wis, 82-88; res assoc, Dept Psychiat, Univ Wis-Madison, 85-86, sr res assoc, 86-87, res scientist, 87-88; SCIENTIST PHARMACOKINETICS & DRUG METABOLISM, PARKE-DAVIS PHARMACEUT RES DIV, WARNER LAMBERT CO, 88- *Concurrent Pos:* NIH postdoctoral res fel, Dept Psychiat, Univ Wis-Madison, 83-86. *Mem:* Endocrine Soc; Soc Neurosci; Int Soc Psychoneuroendocrinol; Asn Tropical Biol; AAAS; fel Royal Entom Soc London. *Res:* Development of radioimmunoassays in both academic and industrial positions; psychoneuroendocrine research in plasma, CSF, and urine of human and research animals; RIA development of drugs, hormones and other substances; author of various publications. *Mailing Add:* Parke-Davis PDM Rm 358-S 2800 Plymouth Rd Ann Arbor MI 48105-2430

BARKSDALE, JAMES BRYAN, JR, b Blytheville, Ark, Dec 29, 40; m 72. MATHEMATICS. *Educ:* Univ Ark, BA, 64, MS, 66, PhD(math), 69. *Prof Exp:* Instr high sch, 63-64; asst prof, 68-73, PROF MATH, WESTERN KY UNIV, 73- *Concurrent Pos:* Mu Alpha Theta vol lectr, 69- *Mem:* AAAS; Math Asn Am; Am Math Soc. *Res:* Analysis in normed vector spaces, especially research involving integrals and differentials of vector functions. *Mailing Add:* Dept Math Western Ky Univ Bowling Green KY 42101

BARKSDALE, LANE W, b Emporia, Va, Nov 23, 14; m. MICROBIOLOGY. *Educ:* Univ NC, AB, 38, MA, 40; NY Univ, PhD(microbiol), 53. *Prof Exp:* Chief bact, 406 Med Gen Lab, Tokyo, 46-49; fel, Pasteur Inst, 53-54; from asst prof to assoc prof microbiol, 54-69, PROF MICROBIOL, SCH MED, NY UNIV, 69- *Concurrent Pos:* Vis prof, Osaka, 60-61; Guggenheim fel, 60-61; chmn subcomt Corynebacterium & Related Organisms, Int Comt Systematic Bact, 74- *Mem:* Am Soc Microbiol; Soc Gen Physiol; Harvey Soc; NY Acad Med; Am Acad Microbiol. *Res:* Biology of Corynebacterium, Mycobacterium, Nocardia group and its viruses. *Mailing Add:* Dept Microbiol NY Univ Sch Med 550 First Ave Med Sci Bldg 147 New York NY 10016

BARKSDALE, RICHARD DILLON, b Orlando, Fla, May 2, 38; m 62; c 1. CIVIL ENGINEERING, GEOTECHNICAL ENGINEERING. *Educ:* Ga Inst of Technol, BCE, 62, MS, 64, PhD(civil eng), 66. *Prof Exp:* Spec lectr civil technol, Southern Technol Inst, 58-60; from asst prof to assoc prof, 65-76, PROF CIVIL ENG, GA INST TECHNOL, 76- *Concurrent Pos:* Mem comt mechanics of earth masses & layered systs, Hwy Res Bd, Nat Acad Sci-Nat Res Coun, chmn comt strength & deformation characteristics of pavement sect, 72-78, mem, 72-; vpres & consult, Soil Systs Inc, 74-80; mem, Comt on Placement Improvement of Soils & Pavement Design, Am Soc Civil Engrs. *Honors & Awards:* Norman Medal, Am Soc Civil Eng. *Mem:* Int Soc Soil Mech & Found Eng; Am Soc Civil Engrs. *Res:* Stress distribution in earth masses; soil-structure interaction; development of a rational pavement design procedure. *Mailing Add:* Sch Civil Eng Ga Inst Technol Atlanta GA 30332-0355

BARKSDALE, THOMAS HENRY, b Trenton, NJ, Nov 7, 32; m 62; c 2. PLANT PATHOLOGY. *Educ:* Rutgers Univ, BS, 54; Cornell Univ, PhD(plant path), 59. *Prof Exp:* Res asst, Cornell Univ, 54-58; res plant pathologist, US Army Biol Labs, 59-66; RES PLANT PATHOLOGIST, AGR RES SERV, USDA, 66- *Mem:* Am Phytopath Soc; Sigma Xi; Am Soc Hort Sci. *Res:* Vegetable diseases and breeding for disease resistance. *Mailing Add:* PGGI Veg Lab USDA Agr Res Ctr Beltsville MD 20705

BARKWORTH, MARY ELIZABETH, b Marlborough, UK, Aug 5, 41; Can citizen. PLANT TAXONOMY. *Educ:* Univ BC, BSc, 61; Western Wash Univ, MEd, 70; Wash St Univ, PhD(bot), 75. *Prof Exp:* res scientist Taxon Can Grasses, Biosysts Res Inst Agr Can, 75-79; asst prof, 79-85, dir, Intermountain Herbarium, 80-82 & 84, ASSOC PROF BIOL, UTAH STATE UNIV, 82- *Mem:* Int Asn Plant Taxon; Am Soc Plant Taxonomists; Sigma Xi; AAAS; Linnean Soc London. *Res:* Taxonomy of North American Gramineae; evolutionary relations within the Gramineae, particularly members of Stipeae and Triticeae; application of numerical and chemical methods to taxonomic problems in angiosperms. *Mailing Add:* Dept Biol Utah State Univ Logan UT 84322-5305

BARLAS, JULIE S, b Eugene, Ore, Sept 15, 44; m 74. COMPUTER SECURITY, COMPUTER-AIDED SOFTWARE ENGINEERING TOOLS. *Educ:* Univ Ore, BA, 65; Univ Pittsburgh, MLS, 67; Harvard Univ, MS, 85. *Prof Exp:* Librn educ, Harvard Univ, 67-70, appl sci, 70-86; MEM TECH STAFF COMPUTER SCI, MITRE CORP, 86- *Res:* Supervision of computer security aspects of software acquired on contract by the federal government; computer-aided software engineering tools; formal verification of software; user interfaces. *Mailing Add:* Mitre Corp Burlington Rd Bedford MA 01730

BARLETTA, MICHAEL ANTHONY, b Brooklyn, NY, July 15, 42; m 66; c 2. PHARMACOLOGY, TOXICOLOGY. *Educ:* St John's Univ, BS, 65, MS, 70; NY Med Col, PhD(pharmacol), 73. *Prof Exp:* LECTR PHARMACOL, USPHS PHYSICIANS ASST PROG, 77-; ASSOC PROF, COL PHARM, ST JOHN'S UNIV, 73- *Concurrent Pos:* Consult pharmacologist, Queens Gen Med Ctr, 75- & Cath Med Ctr Brooklyn & Queens, 76- *Mem:* Am Col Clin Pharmacol; Am Pharmaceut Asn; Am Soc Hosp Pharmacists. *Res:* Cardiovascular pharmacology; establishing animal models for myocardial infarction, ischemic coronary vascular disturbances, atherosclerosis; testing known and unknown drugs against these models in an effort to diminish or prevent these conditions. *Mailing Add:* Dept Pharm St John's Univ Grand Cent & Utopia Jamaica NY 11439

BARLOW, ANTHONY, b Southport, Eng, June 25, 38; m 61; c 3. POLYMER CHEMISTRY. *Educ:* Univ Birmingham, BSc, 59, PhD(phys chem), 62. *Prof Exp:* Proj leader polymer struct, Res Div, US Indust Chem Co, 65-68, asst mgr res div, 68-75, mgr Spec Prod Res Div, 75-89, MGR PROJ DEVELOP, US INDUST CHEM CO, 90- *Mem:* Am Chem Soc; Inst Elec & Electronics Engrs. *Res:* Polymer degradation by pyrolysis; radiation effect on polyethylene and ethylene copolymers; polyethylene structure and its relation to physical properties and synthesis conditions; crosslinking and fire retardacy of polymer compounds; adhesives and coatings development; development of insulation and semiconductive compounds for high voltage insulation. *Mailing Add:* 9363 Long Lane Cincinnati OH 45231

BARLOW, CHARLES F, b Mason City, Iowa, Nov 10, 23; m 53; c 3. MEDICINE. *Educ:* Univ Chicago, SB, 45, MD, 47. *Prof Exp:* Intern, Johns Hopkins Hosp, 47-48; jr asst resident, Boston Children's Hosp, 48-49; resident neurol, Clins Univ Chicago, 51-53, instr, 53-55, asst prof, 55-63; PROF NEUROL, HARVARD MED SCH, 63- *Concurrent Pos:* Emer chief neurologist, Children's Hosp, Med Ctr, Boston, emer dir, Ment Retardation Res Prog; consult, Peter Bent Brigham Hosp, Boston Hosp Women & Beth Israel Hosp, Boston. *Mem:* Am Acad Neurol; Am Asn Neuropath; Am Neurol Asn. *Res:* Clinical neurology of children; blood-brain barrier with isotope labeled compounds; neuropathology. *Mailing Add:* Dept Neurol Harvard Med Sch Boston MA 02115

BARLOW, CLYDE HOWARD, BIOMEDICAL SPECTROSCOPY, BIOINORGANIC CHEMISTRY. *Educ:* Ariz State Univ, PhD(chem), 73. *Prof Exp:* MEM FAC, DEPT CHEM, EVERGREEN STATE COL, 81- *Mailing Add:* Evergreen State Col Olympia WA 98505

BARLOW, EDWARD J(OSEPH), b East Orange, NJ, Sept 5, 20; m 46; c 3. ELECTRONICS, COMPUTER SCIENCE. *Educ:* Cooper Union, BEE, 42. *Prof Exp:* Mem staff, Rand Corp, 48-58, dir, Interdiv projs, 58-60; vpres & opers gen mgr, Aerospace Corp, 60-68; group vpres int, Varian Assocs, 68-69, group vpres instruments, 69-76, group pres, instruments, 76-78, vpres, res & develop, 78-84; RETIRED. *Concurrent Pos:* Mem tech capabilities panel, President's Sci Adv Comt, 54, adv panel aeronaut, Off Secy Defense, 59; chmn aerospace vehicle panel, Air Force Sci Adv Bd, 60; mem large launch vehicle planning group, NASA Dept Defense, 61; mem report review comt, Nat Acad Sci, 74- *Honors & Awards:* Inst Radio Engrs Award, 49. *Mem:* Nat Acad Eng. *Res:* General engineering. *Mailing Add:* 26073 Mulberry Lane Los Altos Hills CA 94022

BARLOW, GEORGE, b Springfield, Mass, Dec 13, 26; m 51; c 2. PHYSIOLOGY. *Educ:* Syracuse Univ, AB, 50; Princeton Univ, MS, 52, PhD, 53. *Prof Exp:* Instr physiol, Col Med, Univ Tenn, 53-57, asst prof, 57-63, instr clin physiol, 53-55, from asst prof to assoc prof, 55-63, asst dir clin physiol labs, 57-63; assoc prof, 64-66, PROF BIOL, HEIDELBERG COL, 66- *Concurrent Pos:* Nat Acad Sci-Nat Res Coun travel award, Int Cong Physiol Sci, Buenos Aires, Arg, 58; Rockefeller Found travel award, Latin Am, 59; vis prof, Univ Valle, Colombia, 60-61; consult, John Gaston Hosp, Tenn. *Mem:* AAAS; Am Physiol Soc; Sigma Xi. *Res:* Adrenal cortex, water and electrolyte metabolism; capillary permeability; circulation and renal physiology. *Mailing Add:* Dept Biol Heidelberg Col Tiffin OH 44883

BARLOW, GEORGE WEBBER, b Long Beach, Calif, June 15, 29; m 55; c 3. ANIMAL BEHAVIOR, ICHTHYOLOGY. *Educ:* Univ Calif, Los Angeles, AB, 51, MA, 55, PhD(zool), 58. *Prof Exp:* From asst prof to assoc prof zool, Univ Ill, Champaign, 60-66; assoc prof, 66-70, Miller prof, 70-71, PROF ZOOL, UNIV CALIF, BERKELEY, 70- *Concurrent Pos:* NIMH fel, Max Planck Inst Physiol Behav, Ger, 58-60; mem, NSF Adv Panel Psychobiol, 65-68; Am Inst Biol Sci vis biologist, 69; US rep, Int Ethology Comt, 69-75; fel, Calif Acad Sci, 76; resident scholar, Ctr Interdisciplinary Res, Univ Bielefeld, 77-78; ed, Cambridge Studies Behav Biol; consult ed, J Comp Psychol. *Mem:* Fel AAAS; Am Soc Zoologists; Am Soc Ichthyologists & Herpetologists; fel Animal Behav Soc (pres elect, 77, pres, 79); Sigma Xi. *Res:* Lab and field studies on social behavior of New World cichlid fishes; eco-ethology of coral-reef fishes; nature of patterned movements; biology of social systems. *Mailing Add:* Dept Integrative Biol Univ Calif Berkeley CA 94720

BARLOW, JAMES A, JR, b Englewood, NJ, Sept 4, 23; m 46; c 2. PHYSICAL GEOLOGY. *Educ:* Middlebury Col, BA, 49; Univ Wyo, MA, 50, PhD(geol), 53. *Prof Exp:* Geologist, Ohio Oil Co, 52-53 & Forest Oil Corp, 53-57; GEOLOGIST, BARLOW & HAUN, INC, 57- *Mem:* Fel Geol Soc Am; Am Asn Petrol Geologists. *Res:* Oil and gas exploration, stratigraphy and structure. *Mailing Add:* 1420 S David St Casper WY 82601

BARLOW, JOEL WILLIAM, b Burbank, Calif, May 2, 42; m 65; c 2. CHEMICAL ENGINEERING, POLYMER ENGINEERING. *Educ:* Univ Wis, BS, 64, MS, 65, PhD(chem eng), 70. *Prof Exp:* Fel mat & thermodyn, Wash Univ, 68-70; res engr plastics, Union Carbide Corp, Bound Brook, NJ, 70-73; from asst prof to prof, 73-84, Z D Bonner prof, 84-90, CULLEN PROF CHEM ENG, UNIV TEX, AUSTIN, 90- *Concurrent Pos:* Asst chmn, Dept Chem Eng, Univ Tex, Austin, 86-90, grad adv chem eng, 90- *Mem:* Am Chem Soc; Am Inst Chem Engrs; Soc Plastics Engrs. *Res:* Polymer processing and physics; thermodynamics of polymer blends; polymer flammability characterization; reaction injection molding; selective laser sintering. *Mailing Add:* Chem Eng Dept Univ Texas Austin TX 78712

BARLOW, JOHN LESLIE ROBERT, clinical medicine, for more information see previous edition

BARLOW, JOHN SUTTON, b Raleigh, NC, June 10, 25; m 50; c 3. NEUROPHYSIOLOGY, BIOPHYSICS. *Educ:* Univ NC, BS, 44, MS, 48; Harvard Univ, MD, 53. *Prof Exp:* Asst neurol, 57-61, NEUROPHYSIOLOGIST, NEUROL SERV, MASS GEN HOSP, 61- *Concurrent Pos:* Clin & res fel neurol, Mass Gen Hosp, 53-57; Nat Inst Neurol Dis & Stroke spec trainee, 58-61, res career develop award, 62-71, res grant, 62-; asst, Harvard Med Sch, 57-61, res assoc, 61-69, prin res assoc, 69-78, sr res assoc, 79-; res assoc, Mass Inst Technol, 54-64, res affil, 64-; res assoc, Sch Med, Univ Calif, Los Angeles, 66; mem, Neurol Study Sect, NIH, 66-70; consult ed, EEG Clin Neurophysiol, 70-85; mem, Panel Neurocommunication & Biophys, Int Brain Res Orgn; mem, Panel Rev Neurol Devices, Fed Drug Admin, 74-76; Fogarty sr int fel, 79; vis scientist, Nat Prog Comt Scholar Communication with People's Repub of China, 88, sci visitor, USSR, 81, 82 & 88, NAS-USSR Acad Exchange Prog. *Honors & Awards:* US Sr Scientist Award, Alexander von Humboldt Found, Ger, 79. *Mem:* Am EEG Soc (pres, 75-76); Am Neurol Asn; Am Acad Neurol; Am Geophys Union; Soc Neurosci. *Res:* Electrical activity of the nervous system in man; computer design and development; medical geography and geophysics; animal navigation. *Mailing Add:* Mass Gen Hosp Boston MA 02114

BARLOW, JON CHARLES, b Jacksonville, Ill, Oct 31, 35; m 81; c 5. ORNITHOLOGY. *Educ:* Knox Col, Ill, BA, 57; Univ Kans, MA, 60, PhD(zool), 65. *Prof Exp:* Res chemist, Corn Prod Refining Co, 57-58; teaching asst zool, Univ Kans, 58-62 & 63-64; field assoc mammal, Am Mus Natural Hist, NY, 62-63; asst prof biol, Rockhurst Col, 64-65; head, Dept Ornith, Royal Ont Mus, 65-76; from asst prof to assoc prof, 65-80, PROF ZOOL, UNIV TORONTO, 80-; CUR, DEPT ORNITH, ROYAL ONT MUS, 65- *Concurrent Pos:* Counr, Am Ornithologists Union, 80-83; chmn bd sci, Chihuahuan Desert Res Inst, 84-86; mem bd dirs, Metrop Toronto Zool Soc, 85-; ed, Wilson Bull, 78-84. *Mem:* Fel AAAS; fel Am Ornithologists Union; Cooper Ornith Soc (vpres, 83-85); Am Soc Zoologists; Animal Behav Soc; Wilson Ornith Soc (vpres, 85-87); Soc Can Ornith (pres-elect 87-89). *Res:* Avian systematics and mammalian ecology and zoogeography; avian behavior, especially of vireos; introduced passer vocalizations evolutionary genetics. *Mailing Add:* Dept Ornith Royal Ont Mus Toronto ON M5S 2C6 Can

BARLOW, MARK OWENS, b Salt Lake City, Utah, July 24, 46; m 74; c 7. MISSILE GUIDANCE, HARDWARE DESIGN. *Educ:* Univ Utah, BS, 71, ME, 72. *Prof Exp:* Missile guidance engr, Boeing, Aerospace Co, 72-78; software engr, TRW, 78-81; elec engr, Diasonics Inc, 81-84; elec engr, Cordin Co, 84-90; HEFU ENGR, HERCULES INC, 90- *Concurrent Pos:* Consult, Cordin Co, 90- *Res:* ECM resistant high altitude guidance. *Mailing Add:* 6750 N 6500 W American Fork UT 84003-9735

BARLOW, RICHARD EUGENE, b Galesburg, Ill, Jan 12, 31; m 56; c 4. MATHEMATICAL STATISTICS. *Educ:* Knox Col, BA, 53; Univ Ore, MA, 55; Stanford Univ, PhD, 60. *Prof Exp:* Adv res engr math, Sylvania Electronic Defense Lab, Gen Tel & Electronics Corp, 57-60, mem tech staff, Res Labs, 61-62; mem tech staff, Inst Defense Anal, 60-61; assoc prof opers res & statist, 63-69, PROF OPERS RES & STATIST, UNIV CALIF, BERKELEY, 69- *Concurrent Pos:* Dir opers res ctr, Univ Calif, Berkeley; consult, Rand Corp, Calif, 63-69; vis prof statist, Fla State Univ, 75-76; assoc ed, J Inst Math Statist, 75- *Mem:* Fel Inst Math Statist; fel Am Statist Asn. *Res:* Statistics; probability theory and its applications, especially mathematical theory of reliability. *Mailing Add:* IEOR Dept Univ Calif Berkeley CA 94720

BARLOW, ROBERT BROWN, JR, b Trenton, NJ, July 31, 39; m 61; c 3. NEUROPHYSIOLOGY. *Educ:* Bowdoin Col, AB, 61; Rockefeller Univ, PhD(life sci), 67. *Prof Exp:* Asst prof, 67-71, assoc prof, 71-77, PROF SENSORY SCI, SYRACUSE UNIV, 77- *Concurrent Pos:* Trustee, Marine Biol Lab, Woods Hole, Mass, 72- *Honors & Awards:* Sigma Xi Fac Res Award. *Mem:* Asn Res Vision & Ophthal; Soc Neurosci; Sigma Xi. *Res:* Processing of contrast and intensity information by the eye and the brain; neurophysiology of circadian rhythms in the visual system of Limulus; efferent control of sensory processing. *Mailing Add:* Inst Sensory Res Syracuse Univ Merrill Lane Syracue NY 13210-5290

BARLOZZARI, TERESA, b Masmassamartana, Italy, May 23, 54. EXPERIMENTAL BIOLOGY. *Prof Exp:* SR SCIENTIST, BASF BIORES CORP. *Mailing Add:* BASF Biores Corp 195 Albany St Cambridge MA 02139

BARMACK, NEAL HERBERT, b New York, NY, Aug 23, 42; m 64; c 2. NEUROBIOLOGY, NEUROPHYSIOLOGY. *Educ:* Univ Mich, BS, 63; Univ Rochester, PhD, 70. *Prof Exp:* Asst lectr psychol, Univ Rochester, 68-69; from res assoc to sr res assoc neurophysiol, Dept Ophthal, assoc scientist, 75-80, sr scientist, Neurol Sci Inst, Good Samaritan Hosp & Med Ctr, Portland, 80-81; assoc prof biol sci, Univ Conn, 81-82; AT R S DOW NEUROL SCI INST, 82- *Mem:* Soc Neurosci; Am Physiol Soc; Asn Res Vision; Int Brain Res Orgn; Nat Eye Inst. *Res:* Neural control of eye movements; plasticity of reflexive eye movements; the cellular and biochemical basis of cerebellar modulation of reflex function. *Mailing Add:* Neurol Sci Inst 1120 NW 20th Ave Portland OR 97209

BARMAN, SUSAN MARIE, b Joliet, Ill, Aug 28, 49. NEUROPHYSIOLOGY, AUTONOMIC PHYSIOLOGY. *Educ:* Loyola Univ, Chicago, BS, 71, PhD(physiol), 76. *Prof Exp:* Res assoc, 75-77, from instr to asst prof, 77-84, ASSOC PROF NEUROPHYSIOL, MICH STATE UNIV, 84- *Concurrent Pos:* Sci consult, study sect, NIH, 84- *Mem:* Soc Neurosci; Am Physiol Soc; AAAS; Sigma Xi. *Res:* Neural control of circulation; identification of areas of brain which are involved in control of blood pressure and heart rate. *Mailing Add:* Dept Pharmacol & Toxicol Mich State Univ East Lansing MI 48824-1317

BARMATZ, MARTIN BRUCE, b Los Angeles, Calif, May 25, 38; m 61; c 2. ACOUSTICS. *Educ:* Univ Calif, Los Angeles, BA, 60, MA, 62, PhD(physics), 66. *Prof Exp:* Asst prof physics in residence, Univ Calif, Los Angeles, 66-67; mem tech staff, Condensed State Physics Res Dept, Bell Tel Labs 67-78; MEM TECH STAFF, PHYS ACOUST & CONTAINERLESS SCI GROUP, JET PROPULSION LAB, 78- *Mem:* Acoust Soc Am; Am Phys Soc. *Res:* Critical point phenomena; sound velocity, sound attenuation, sound dispersion measurement and nonlinear acoustics. *Mailing Add:* Jet Propulsion Lab 4800 Oak Grove Dr Pasadena CA 91109

BARMBY, DAVID STANLEY, b Hull, Eng, Mar 4, 28; m 57; c 3. PHYSICS. *Educ:* Univ Leeds, BSc, 49, PhD(physics), 54. *Prof Exp:* Physicist, Wool Industs Res Asn, Eng, 54-57; physicist, Sun Co, Inc, 57-66, chief petrol processing sect, Basic Res Div, 66-70, chief explor petrol res, Corp Res Div, 70-75, mgr technol assessment, 75-84, dir Advan Technol, 84-88, TECH CONSULT, SUN CO, INC, CORP HQ, RADNOR, PA, 89- *Mem:* Am Phys Soc; Inst Elec & Electronics Engrs; Catalyst Soc. *Res:* Solid state physics and catalysis; paraffin wax structures and phase transitions; correlations between physical properties and structure of solids; infrared spectroscopy of polymers; fuel technology; alternative energy technology. *Mailing Add:* 1237 Hunt Club Lane Media PA 19063

BARMBY, JOHN G(LENNON), b Lawrence, Mass, Oct 22, 22; m 55; c 2. SYSTEMS ENGINEERING. *Educ:* Mass Inst Technol, BS, 44; Middlebury Col, AB, 44; George Washington Univ, MBA, 51; Am Univ, PhD(pub admin), 57. *Prof Exp:* Aeronaut res scientist, Langley Lab, NASA, 44-48; aeronaut res engr, Bur Aeronaut, US Navy Dept, 48-57, asst br head, 57-59; mem tech staff, Systs & Opers Anal, Weapons Systs Eval Div, Inst Defense Anal, 59-62; mem staff systs anal, Bellcomm, Inc, Am Tel & Tel Co, 62-64; sr staff assoc, Ill Inst Technol Res Inst, 64-71; sr tech adv nat security, US Gen Acct Off, 71-90; ENG & MGT CONSULT, 90- *Concurrent Pos:* Adj prof, Univ Va, 69-81. *Mem:* Assoc fel Am Inst Aeronaut & Astronaut; Am Soc Pub Admin; Inst Elec & Electronics Engrs. *Res:* Engineering management; systems analysis and research and development management of space and weapons programs; trade-off studies of the technical performance of various systems with cost, schedule and reliability factors. *Mailing Add:* 924 Fairway Dr NE Vienna VA 22180

BARMORE, FRANK E, b Manhattan, Kan, June 20, 38; m 67; c 2. PHYSICS. *Educ:* Wash State Univ, BS, 60; Univ Wis-Madison, MS, 63, PhD(physics), 73. *Prof Exp:* Sr technician, Thule Field Sta, Greenland, Am Geog Soc, 61-62; teaching asst, Dept Physics, Univ Wis-Madison, 66; asst prof natural sci & actg chmn interdisciplinary div, Milton Col, 70-73; proj assoc, Dept Physics, Univ Wis-Madison, 73-74; asst prof physics, Mid-East Tech Univ, Ankara, 74-76; res assoc, Dept Physics, Univ Calgary, 76-77; proj scientist, Ctr Res Exp Space Sci, York Univ, 77-78; lectr, 77-83, asst prof, 83-86, ASSOC PROF, DEPT PHYSICS, UNIV WIS-LA CROSSE, 86- *Mem:* Optical Soc Am; AAAS; Am Geophys Union; Am Asn Physics Teachers; Sigma Xi; Soc Archit Historians. *Res:* Minor atmospheric constituents by ground-based spectroscopic techniques; modern optics; application of high resolution-luminosity interferometers to studies in astronomy and aeronomy; archaeoastronomy. *Mailing Add:* Dept Physics Cowley Hall Univ Wis La Crosse La Crosse WI 54601

BARNA, GABRIEL GEORGE, b Mar 13, 46; Can citizen. ELECTROCHEMISTRY. *Educ:* McGill Univ, BSc, 68, PhD(chem), 73. *Prof Exp:* MEM TECH STAFF CHEM, TEX INSTRUMENTS INC, 72- *Mem:* Am Chem Soc; Electrochem Soc; Sigma Xi. *Res:* Passive display technology; liquid crystal and electrochromic displays; fuel cells. *Mailing Add:* 1613 Morning Star Terr Richardson TX 75081

BARNAAL, DENNIS E, b Sacred Heart, Minn, Jan 5, 36; m 62; c 3. SOLID STATE PHYSICS. *Educ:* Univ Minn, BS, 58, MS, 62, PhD(physics), 65. *Prof Exp:* From asst prof to assoc prof, 64-74, PROF PHYSICS, LUTHER COL, 74-, HEAD DEPT, 86- *Concurrent Pos:* Consult, Kaman Sci Corp, 80, Am Inst Prof Educ, 77-82, Oper Rec Inc, 79-81. *Mem:* Am Phys Soc; Am Asn Physics Teachers. *Res:* Pulsed nuclear magnetic resonance in solids; ice physics; wind energy research. *Mailing Add:* Dept Physics Luther Col Decorah IA 52101

BARNABEO, AUSTIN EMIDIO, b New York, NY, Feb 7, 33; m 56; c 2. ORGANIC CHEMISTRY, POLYMER CHEMISTRY. *Educ:* Queens Col, BS, 54; Brooklyn Col, MS, 60. *Prof Exp:* Chemist, 56-62, proj scientist, 62-69, res scientist, 69-81, SR RES SCIENTIST, UNION CARBIDE CORP, 81- *Mem:* Am Chem Soc; Sigma Xi. *Res:* Polyhydroxyethers; epoxies; polyesters; polyethylene for wire and cable applications; cellular products. *Mailing Add:* 533 Spring Valley Dr Bridgewater NJ 08807

BARNABY, BRUCE E, b Milwaukee, Wis, Sept 24, 29; m 57; c 6. PHYSICS, ELECTRICAL ENGINEERING. *Educ:* DePaul Univ, BS, 51; Univ Notre Dame, PhD(physics), 60. *Prof Exp:* Res physicist, Eitel-McCullough, Inc, 60-63, proj engr, 63-65; sr proj engr, Eimac Div, Varian Assocs, 65-67; staff mem, Los Alamos Sci Lab, 67-70; DIV SUPVR, TUBE DEVELOP DIV, SANDIA LABS, 70- *Mem:* Am Soc Testing & Mat; Am Nuclear Soc. *Res:* Electron physics technology, especially use of technology in industry. *Mailing Add:* PO Box 129 Peralta NM 87042

BARNARD, ADAM JOHANNES, b Murraysburg, SAfrica, June 9, 29; m 57; c 4. PLASMA PHYSICS. *Educ:* Univ SAfrica, BSc, 49, MSc, 51; Glasgow Univ, PhD(physics), 54. *Prof Exp:* Sr lectr, Univ Natal, 54-59; from asst prof to assoc prof, 59-68, PROF PHYSICS, UNIV BC, 68- *Mem:* Am Phys Soc; Can Asn Physicists. *Res:* Spectroscopic investigation of plasmas. *Mailing Add:* Dept Physics Univ BC 2075 Westbrook Pl Vancouver BC V6T 1W5 Can

BARNARD, ANTHONY C L, b Birmingham, Eng, Apr 30, 32; m 64; c 2. COMPUTER SCIENCE. *Educ:* Univ Birmingham, Eng, BSc, 53, PhD(nuclear physics), 57, DSc(physics), 74, Univ Ala, Birmingham, MBA, 79. *Prof Exp:* Res physicist, Res Lab, Assoc Elec Industs, Ltd, 56-58; res assoc physics, Univ Iowa, 58-60; from instr to asst prof, Rice Univ, 60-66; mgr physics dept, Int Bus Mach Sci Ctr, 66-68; assoc prof biomath, 68-71, chmn dept comput & info sci, 72-78, assoc vpres acad affairs, 79-84, PROF COMPUT & INFO SCI, UNIV ALA, BIRMINGHAM, 71-, ASSOC PROF PHYSICS, 76-, DEAN GRAD SCH, 84- *Mem:* Am Phys Soc; Asn Comput Mach; Data Processing Mgt Asn; Inst Elec & Electronics Engrs Comput Soc. *Res:* Medical applications of physics and computer science. *Mailing Add:* 511 University Ctr Univ Sta Birmingham AL 35294

BARNARD, DONALD ROY, b Santa Ana, Calif, June 7, 46; m 67; c 2. PEST-MANAGEMENT, POPULATION ECOLOGY OF ARTHROPODS. *Educ:* Calif State Univ, BS, 69, MA, 72; Univ Calif-Riverside, PhD(entom), 77. *Prof Exp:* Res entomologist, 79-85, SUPVRY RES ENTOMOLOGIST, AGR RES SERV, USDA, 85- *Concurrent Pos:* Consult, US AID, Somali Dem Repub, 81-83; World Health Orgn, 82-85, Dept Defense, Armed Forces Pest Mgt Bd, 85-, Repub S Africa, Karoo Paralysis Proj, 88- *Mem:* Entom Soc Am; Entom Soc Can; Ecol Soc Am. *Res:* Integrated pest management and population ecology of arthropod pests of domestic livestock. *Mailing Add:* 4403 NW 77th Terr Gainesville FL 32606

BARNARD, GARLAND RAY, b Victoria, Tex, Apr 12, 32; m 54; c 2. ACOUSTICS. *Educ:* Univ Tex, BS, 57, MA, 60. *Prof Exp:* ASSOC DIR RES, APPL RES LAB, UNIV TEX, AUSTIN, 56- *Mem:* Fel Acoust Soc Am; Sigma Xi. *Res:* Underwater acoustics; basic and applied research. *Mailing Add:* 3307 Stardust Dr Austin TX 78757

BARNARD, JERRY LAURENS, b Pasadena, Calif, Feb 27, 28; m 49; c 3. MARINE ZOOLOGY. *Educ:* Univ Southern Calif, AB, 49, MS, 50, PhD(zool), 53. *Prof Exp:* Instr biol sci, Calif State Polytech Col, 52-53; investr, US Air Force Arctic res grant, Univ Southern Calif, 53-59; res assoc, Beaudette Found, 59-64; CUR CRUSTACEA, NAT MUS NATURAL HIST, SMITHSONIAN INST, 64- *Mem:* Marine Biol Asn UK. *Res:* Taxonomy and ecology of marine Amphipoda; estuarine and coastal shelf biology. *Mailing Add:* Div Crustacea NHB-163-W323 Nat Mus Natural Hist Washington DC 20560

BARNARD, JOHN WESLEY, b St Thomas, Ont, Dec 26, 46, Canadian; m 71; c 1. HEALTH PHYSICS, CRITICALITY SAFETY. *Educ:* McMaster Univ, BSc, 69; Univ Man, PhD, 76. *Prof Exp:* Res scientist, Man Inst Cell Biol, Man Cancer Treatment & Res Found, 77-80; HEALTH PHYSICIST, RADIATION APPL RES BR, WHITESHELL NUCLEAR RES ESTAB, ATOMIC ENERGY CAN LTD, 80- *Concurrent Pos:* Fel, Dept Physics, Univ Man, 76-77; asst prof, Dept Internal Med, Univ Man, 77-80. *Mem:* Health Physics Soc. *Res:* Nuclear physics; health physics, including radiation dose control; external dosimetry, shielding and critical safety assessment; biosphere modeling for nuclear waste management and safety assessment; industrial radiation process development. *Mailing Add:* Box 1215 Lac du Bonnet MB R0E 1A0 Can

BARNARD, KATHRYN E, b Omaha, Nebr, Apr 16, 38. NURSING. *Educ:* Univ Nebr-Omaha, BS, 60; Boston Univ, MS, 62; Univ Wash-Seattle, PhD(nursing), 72. *Hon Degrees:* DSc, Univ Nebr, 90. *Prof Exp:* Actg instr nursing, Univ Nebr-Omaha, 60-61; actg instr nursing, Univ Wash & Seattle, 63-65; assoc prof nursing, 65-69, PROF NURSING, UNIV WASH, 72- *Concurrent Pos:* Prin investr or co-investr grants, 72-; vis prof, Sch Nursing, Univ Ala, 81; mem bd dirs, Nat Ctr Clin Infant Progs, 81-; adj prof psychol, Univ Wash, 85-, assoc dean acad progs, 87-; mem, Adv Group Prev Res, NIMH, 90- *Honors & Awards:* Jessie M Scott Award, Am Nurses Asn, 82; Martha May Eliot Award, Am Pub Health Asn, 83. *Mem:* Inst Med-Nat Acad Sci; Am Nurses Asn; Am Asn Univ Professors; Am Asn Ment Deficiency; fel Am Acad Nursing; Am Pub Health Asn; World Asn Infant Psychiat & Allied Disciplines. *Mailing Add:* 11508 Dunland Ave NE Univ Wash Seattle WA 98125

BARNARD, R JAMES, KINESIOLOGY. *Educ:* Kent State Univ, BS, 59, MA, 63; Univ Iowa, PhD, 68. *Prof Exp:* Teacher, Hawthorne, Calif, 59-61; postdoctoral fel, Sch Med, Univ Calif, Los Angeles, 68-71, asst res cardiologist, 71-73, assoc res cardiologist, 73-79, assoc prof, Dept Kinesiol, 73-79, res cardiologist, Sch Med, 79-88, PROF KINESIOL, UNIV CALIF, LOS ANGELES, 79-, PROF MED, DIV CLIN NUTRIT, SCH MED, 88- *Concurrent Pos:* Muscular Dystrophy Asn fel, 70-71; res career develop award, USPHS, 75-80; mem bd dirs & dir res, Nathan Pritkin Res Found, 81-88; mem, Coun Circulation, Am Heart Asn. *Mem:* Am Col Sports Med (vpres, 74-75); Am Physiol Soc; Am Heart Asn; Am Diabetes Asn. *Res:* Exercise physiology; mechanisms responsible for adaptations associated with exercise and diet; regulation of glucose transport in skeletal muscle; value of diet and exercise as preventive measures for degenerative diseases. *Mailing Add:* Dept Kinesiol/Med Univ Calif 405 Hilgard Ave Los Angeles CA 90024-1568

BARNARD, ROBERT D(ANE), JR, b Chicago, Ill, Mar 17, 29; m 49; c 4. ELECTRICAL ENGINEERING, COMPUTER ENGINEERING. *Educ:* Polytech Inst Brooklyn, BEE, 52, MEE, 55; Case Inst Technol, PhD(solid state physics), 59. *Prof Exp:* Instr elec eng, Polytech Inst Brooklyn, 53-55 & Case Inst Technol, 55-59; mem tech staff, Bell Tel Labs, Inc, 59-65; assoc prof, 61-62, PROF ENG, WAYNE STATE UNIV, 65- *Mem:* Inst Math Soc; Am Math Soc; Inst Elec & Electronics Engrs; Soc Indust & Appl Math. *Res:* Systems and computer science; applied mathematics. *Mailing Add:* Col Eng Wayne State Univ Detroit MI 48202

BARNARD, ROGER W, b One, Ohie, Sept 20, 42. COMPLEX ANALYSIS. *Educ:* Kent State Univ, BS, 65 & MS, 68; Univ Md, PhD(math), 71. *Prof Exp:* PROF MATH TEX TECH UNIV, 73- *Mem:* Am Math Soc; Soc Indust & Appl Math. *Mailing Add:* Dept Math Tex Tech Univ Lubbock TX 79409

BARNARD, WALTHER M, b Hartford, Conn, May 30, 37. MINERALOGY, GEOCHEMISTRY. *Educ:* Trinity Col, Conn, BS, 59; Dartmouth Col, AM, 61; Pa State Univ, PhD(mineral), 65. *Prof Exp:* Asst prof geol, 64-70, assoc prof, 70-77, actg chmn, 86-87, PROF GEOL, STATE UNIV NY COL FREDONIA, 77-, CHMN, 87- *Concurrent Pos:* Res Corp res grant, 65-66; State Univ NY Res Found res grants-in-aid, 65-66, 67-68, 69-70 & 72-73; res grant, Energy & Resources Develop Agency, 75-77; res consult, Res Corp, Univ Hawaii, 81-82; collab res grant, SUNY Buffalo, 87-89. *Mem:* Geochem Soc; AAUP; Am Chem Soc; fel Geol Soc Am; Mineral Soc Am. *Res:* Environmental geochemistry; variability, controls and potential applications of chemistry of fumarolic condensates of Kilauea volcano, island of Hawaii. *Mailing Add:* Dept Geoscis State Univ NY Fredonia NY 14063

BARNARD, WILLIAM SPRAGUE, b Hillsboro, Ill, May 20, 25; m 48, 75; c 3. PHYSICAL CHEMISTRY. *Educ:* Harvard Univ, AB, 47; Princeton Univ, PhD(phys chem), 51. *Prof Exp:* With res dept, Nat Lead Co, 50-53; group leader res div, Chicopee Mfg Corp, 53-57, dir woven prod, 57-64, dir res & develop, 64-69, mem bd dirs, 79-80, vpres res, 69-80, vpres sci affairs, 80-90; RETIRED. *Concurrent Pos:* Chmn exec comt, Textile Res Inst. *Mem:* Am Chem Soc; Am Asn Textile Chemists & Colorists. *Res:* Textile chemistry and resin technology; fibers and finishes; non-woven fabric technology. *Mailing Add:* 326 Ridgeview Rd Princeton NJ 08540

BARNARTT, SIDNEY, b Toronto, Ont, July 31, 19; nat US; m 43; c 2. ELECTROCHEMISTRY, CORROSION. *Educ:* Univ Toronto, BA, 41, MA, 42, PhD(chem), 44. *Prof Exp:* Asst, Univ Toronto, 41-43; res chemist, Westinghouse Res Labs, 46-56, adv chemist, 56-62, mgr electrochem sect, 62-64; res chemist, Fundamental Res Lab, US Steel Corp, 64-72; dir electrochem res, Technicon Instruments Corp, 72; adv scientist, Westinghouse Bettis Atomic Power Lab, 72-84; consult chemist, 84-85; RETIRED. *Concurrent Pos:* Instr, Carnegie Inst Technol, 54. *Honors & Awards:* Bronze Medal, Brit Asn Advan Sci, 41. *Mem:* Am Chem Soc; Electrochem Soc; fel Am Inst Chemists; Nat Asn Corrosion Eng. *Res:* Adsorption; batteries; electrodeposition; electrode kinetics; corrosion. *Mailing Add:* 1358 Hillsdale Dr Monroeville PA 15146

BARNAWELL, EARL B, b Maryville, Tenn, Nov 26, 22; m 46; c 2. COMPARATIVE ENDOCRINOLOGY. *Educ:* Univ Calif, Berkeley, AB, 51, MA, 54, PhD(endocrinol, zool), 64. *Prof Exp:* Exp animal biologist, Univ Calif, Berkeley, 54-60; asst prof, Univ Nebr, Lincoln, 64-69, dir inst cellular res, 66-71, assoc prof, Sch Biol Sci, 69-91; RETIRED. *Mem:* Sigma Xi; Am Soc Zool. *Res:* Physiology and comparative endocrinology; organ culture. *Mailing Add:* 3400 J St Lincoln NE 68510

BARNEKOW, RUSSELL GEORGE, JR, b Cleveland, Ohio, Jan 26, 32; m 57; c 4. BACTERIOLOGY, BACTERIAL PHYSIOLOGY. *Educ:* Miami Univ, AB, 55; Kans State Univ, MS, 61, PhD(bact), 67. *Prof Exp:* From asst prof to assoc prof bact, Univ Mo-Kansas City, 62-76, from asst dean to assoc dean, Sch Grad Studies, 68-76, lectr, Sch Med, 75-76; DEAN GRAD SCH & PROF LIFE SCI, SOUTHWEST MO STATE UNIV, 76- *Mailing Add:* Dept Biol SW Mo State Univ 901 S Nat Springfield MO 65804

BARNER, HENDRICK BOYER, b Seattle, Wash, Feb 23, 33; m 61; c 3. SURGERY. *Educ:* Univ Wash, BS, 54, MD, 57; Am Bd Surg, dipl, 67; Am Bd Thoracic Surg, dipl, 68. *Prof Exp:* Instr surg, Univ Rochester, 64-65; from instr to assoc prof, 65-73, PROF SURG, ST LOUIS UNIV, 73-, DIR, CARDIOTHORACIC SURG RES, 87-, ASSOC DIR, DIV CARDIOTHORACIC SURG, 87- *Concurrent Pos:* USPHS res fel, 64-65, fel cardiovasc surg, 65-66. *Mem:* Am Col Surg; AMA; Am Surg Asn; Am Asn Thoracic Surg; Soc Univ Surg; Soc Thoracic Surg. *Res:* Cardiovascular surgery and physiology. *Mailing Add:* Dept Surg 3635 Vista PO Box 15250 St Louis MO 63110

BARNERJEE, SHAILESH PRASAD, b June 22, 40; m; c 2. NEUROCHEMISTRY, NEUROSCIENCE. *Educ:* Univ Toronto, PhD(pharmacol), 71. *Prof Exp:* PROF MED, DEPT PHARMACOL, SCH MED, CITY UNIV NEW YORK. *Mem:* Am Soc Pharmacol & Therapeut; Am Soc Neurosci. *Res:* Neurotransmitter receptor regulation. *Mailing Add:* Dept Pharmacol City Univ New York 138th St & Convent Ave New York NY 10031

BARNES, AARON, b Shenandoah, Iowa, May 9, 39; m 62; c 2. SPACE PLASMA PHYSICS. *Educ:* Univ Chicago, SB, 61, SM, 62, PhD(physics), 66. *Prof Exp:* Res fel, 66-67, RES SCIENTIST, NASA-AMES RES CTR, 67- *Concurrent Pos:* Vis scientist, Max-Planck Inst Physics & Astrophysics, 73; prin investr plasma analyzers, Pioneer 10 & 11 & Pioneer Venus Orbiter, NASA-Ames Res Ctr, 80- *Honors & Awards:* Except Sci Achievement Medal, NASA, 82; Sackler Distinguished lectr, Tel Aviv Univ, 86. *Mem:* Am Phys Soc; Am Geophys Union; Am Astron Soc; AAAS; Sigma Xi; Int Astron Union; Astron Soc Pac. *Res:* Theoretical and observational investigation of the physics of interplanetary plasma and magnetic field; development of the theory of hydromagnetic waves and turbulence in collisionless plasmas; hydromagentic waves and turbulence. *Mailing Add:* 245-3 NASA Ames Res Ctr Moffett Field CA 94035

BARNES, ALLAN MARION, b Stirling City, Calif, Dec 8, 24; m 62; c 2. MEDICAL ENTOMOLOGY. *Educ:* San Diego State Col, AB, 53; Univ Calif, PhD(med entom), 63. *Prof Exp:* Vector control specialist, Calif State Health Dept, 55-67; chief ecol & control unit, 67-70, from asst chief to actg chief, 70-75, CHIEF PLAGUE SECT, VECTOR BORNE DIS DIV, CTR INFECTIOUS DIS, CTR DIS CONTROL, 75- *Concurrent Pos:* Consult, Indonesian Ministry Health, 68 & 88; affil prof entom & affil prof grad sch, Colo State Univ; mem, Expert Comt on Plague, WHO, 79 & Expert Panel Vector Biol & Control; consult plague prob, WHO & Pan Am Health Orgn, Venezuela, 71, Brazil, 74, Bolivia, 78, Peru, 79, 81, 84, Ecuador, 81, 84, Burma, 82, Saudi Arabia, 84 & Peoples Repub China, 90; secy, Int Nouthwest Conf Dis in Nature Commun to Man, 80, vpres, 81, pres, 82. *Mem:* AAAS; Wildlife Dis Asn; Ecol Soc Am; Am Soc Trop Med & Hyg; Soc Vector Ecologists. *Res:* Research on the ecology, epidemiology and control of bubonic plague, tularemia, Colorado tick fever and other vector-borne diseases with emphasis on the dynamics of interactions between host-ectoparasite-pathogen complexes. *Mailing Add:* Ctr Dis Control Plague Br PO Box 2087 Ft Collins CO 80522

BARNES, ALLEN LAWRENCE, b Wooster, Ohio, Sept 28, 32; m 58; c 3. PETROLEUM & MECHANICAL ENGINEERING. *Educ:* Ohio State Univ, BPetE, 55; Pa State Univ, MS, 60; Okla State Univ, PhD(mech eng), 68. *Prof Exp:* Engr, Magnolia Petrol Co, 55-57 & 57-58; res asst petrol eng, Pa State Univ, 58-59; res engr, Sinclair Res, Inc, 59-65, sect head reservoir & recovery eng, 66-68; teaching asst petrol eng, Okla State Univ, 65-66; tech asst, Explor & Prod Dept, Ashland Oil & Refining Co, 68-69, mgr tech serv, 69-71, energy specialist planning dept, Ashland Oil Inc, 71-73; MGR RESERVOIR RES ENG, ATLANTIC RICHFIELD CO, 73- *Mem:* Soc Petrol Engrs; Opers Res Soc Am. *Res:* New secondary recovery methods of economically extracting oil from natural reservoirs; use of heating techniques for the extraction of mineral wealth; mathematical simulation of hydraulic petroleum reservoir and network problems; risk analysis and economic analyses; managing production research programs. *Mailing Add:* Mgr Reservoir Res Eng Atlantic Richfield Co 2300 Plano Pkwy Plano TX 75075

BARNES, ARNOLD APPLETON, JR, b Charleston, WVa, June 10, 30; m 60; c 3. METEOROLOGY, ATMOSPHERIC PHYSICS. *Educ:* Princeton Univ, AB, 52; Mass Inst Technol, MSc, 57, PhD(meteorol), 62. *Prof Exp:* Asst meteorol, Mass Inst Technol, 56-61; res scientist, Meteorol Develop Lab, Tech Br, 61-62, res scientist, Meteorol Res Lab, Upper Atmosphere Br, 62-63, res scientist, Meteorol Lab, 63-71, res scientist, Weather Radar Br, 71-74, chief, Convective Cloud Physics Br, Meteorol Div, Air Force Cambridge Res Labs, 74-82, Chief, Cloud Physics Br, 82-87, SR SCIENTIST, ATMOSPHERIC SCI DIV, AIR FORCE GEOPHYS LAB, HANSCOM AFB, 82- *Concurrent Pos:* Consult, Allied Res Corp, 59-60; Dept Defense team leader, Nat Storm Prog, 83- *Mem:* AAAS; Am Meteorol Soc; Am Geophys Union. *Res:* Large scale dynamics of the atmosphere; transport of water vapor; numerical weather prediction; radar meteor trail winds and densities; radar acoustic low level temperature soundings; weather radar; cloud physics; aircraft icing; cirrus clouds; millimeter wave precipitation attenuation; mesoscale modelling; lighting and atmospheric electricity. *Mailing Add:* 32 Bradyll Rd Weston MA 02193-1711

BARNES, ASA, JR, b Cape Girardeau, Mo, Jan 30, 33; m 57; c 4. PATHOLOGY, HEMATOLOGY. *Educ:* Univ Ky, BA, 55; Yale Univ, MD, 59. *Prof Exp:* Intern, Univ Hosps, Cleveland, Ohio, 59-60; resident anat path, Grace-New Haven Community Hosp, New Haven, Conn, 63-65; resident clin path, Walter Reed Gen Hosp, Washington, DC, 65-67, asst chief, 67; blood prog officer, US Army, Vietnam, 67-68; pathologist, Armed Forces Inst Path, Washington, DC, 69-70; prof path, Sch Med, Univ Mo-Columbia, 70-77; chief clin path, Ellis Fischel Cancer Hosp, 74-76; sr scientist, Cancer Res Ctr, 74-76; CLIN PROF PATH, COL MED, UNIV CALIF, IRVINE, 77-; ASSOC PATHOLOGIST & DIR BLOOD BANK, MEM HOSP MED CTR, LONG BEACH, 77- *Honors & Awards:* US Maj Gary Wratten Award, Asn Mil Surg, 69. *Mem:* AMA; Am Soc Clin Path; Asn Mil Surg; Am Asn Blood Banks; Col Am Pathologists. *Res:* Blood banking; morphologic hematology; clinical and anatomical pathology. *Mailing Add:* Dept Path 2801 Atlantic Ave Long Beach CA 90801

BARNES, BRUCE HERBERT, b Minneapolis, Minn, May 28, 31; m 53; c 3. COMPUTER SCIENCE, MATHEMATICS. *Educ:* Mich State Univ, BS, 53, MS, 58, PhD(math), 60. *Prof Exp:* Teaching asst math, Mich State Univ, 56-58, res instr comput sci, 58-60; sr res engr, Jet Propulsion Labs, 60-61; asst prof math, Pa State Univ, 61-65, assoc prof, 65-68; vis assoc prof, Univ Iowa, 68-69; assoc prof comput sci, Pa State Univ, 69-74; prog dir, Comput Sci Sect, NSF, 74-86; VPRES, SOFTWARE PRODUCTIVITY CONSORTIUM, 86- *Mem:* Am Math Soc; Asn Comput Mach; Inst Elec & Electronics Engrs Comput Soc. *Res:* Automata theory; numerical solution of ordinary differential equations; software engineering, especially decision tables and computer science education. *Mailing Add:* Rte 1 Box 323G Leesburg VA 22075

BARNES, BURTON VERNE, b Bloomington, Ill, Nov 4, 30; m 57; c 3. FORESTRY, BOTANY. *Educ:* Univ Mich, BSF, 52, MF, 53, PhD(forestry), 59. *Prof Exp:* Forester, Region VI, US Forest Serv, 53-59, res forester, Intermountain Forest & Range Exp Sta, 59-63; from asst prof to assoc prof, 63-70, PROF FORESTRY, SCH NATURAL RESOURCES, UNIV MICH, ANN ARBOR, 70-, FOREST BOTANIST, BOT GARDEN, 67- *Concurrent Pos:* NSF fel, Ger, 63-64; Danforth assoc, 69-; mem staff, Syst-Geobot Inst, Univ Gottingen, 70. *Honors & Awards:* Distinguished Serv Award, Univ Mich, 67. *Mem:* Ecol Soc Am; Soc Am Foresters. *Res:* Forest ecology; dendrology; genecology; forest genetics; silviculture; biosystematics of Populus and Betula. *Mailing Add:* Sch Natural Resources Ann Arbor MI 48109-1115

BARNES, BYRON ASHWOOD, b Peoria, Ill, Mar 14, 27; m 48; c 2. PHARMACOLOGY. *Educ:* St Louis Col Pharm, BS, 51; Univ Fla, MS, 53, PhD(pharmacol), 54. *Prof Exp:* Lab instr pharm, St Louis Col Pharm, 49-51; instr, Clin Lab, Sch Aviation Med, US Air Force, Gunter Air Force Base, 54-57; head dept pharmacol & physiol, 57-70, dean, 70-89, EMER DEAN PHARM, ST LOUIS COL, 89- *Concurrent Pos:* Ed, Pharmaceut Trends; ed-in-chief, Rx Sat Continuing Educ Pharmacist Series. *Mem:* Am Pharmaceut Asn; Am Asn Cols Pharm; Royal Soc Health; Sigma Xi. *Res:* Continuing pharmacy education; drug abuse. *Mailing Add:* 10232 Richview Dr St Louis MO 63127

BARNES, C(ASPER) W(ILLIAM), JR, b Leesburg, Fla, Nov 24, 27; m 47. ELECTRICAL ENGINEERING. *Educ:* Univ Fla, BE, 50; Stanford Univ, PhD(elec eng), 54. *Prof Exp:* Engr, Sylvania Microwave Tube Lab, 54-56; sr res engr, Stanford Res Inst, 56-67; assoc prof, 67-69, PROF ENG, UNIV CALIF, IRVINE, 69- *Mem:* Am Phys Soc; Inst Elec & Electronics Engrs. *Res:* Digital signal processing; image processing; applied mathematics. *Mailing Add:* Dept Elec Eng Univ Calif Irvine CA 92717

BARNES, CARL ELDON, b Clovis, NMex, June 9, 36; m 54; c 2. AGRONOMY, FIELD CROPS. *Educ:* NMex State Univ, BS, 59, MS, 60. *Prof Exp:* Asst miller, Western Wheat Qual Lab, 60-61; res asst crops, Kans State Univ, 61-63; asst agronomist, Colo State Univ, 63-65; asst agronomist, 65-68, asst prof agron, 68-75, ASSOC PROF AGRON, NMEX STATE UNIV, 75-, SUPT CROPS RES, SOUTHEASTERN BR EXP STA, 68- *Mem:* Am Soc Agron. *Res:* Crop production research; cotton breeding; water requirements. *Mailing Add:* 1904 Briscoe Ave Artesia NM 88210

BARNES, CHARLES ANDREW, b Toronto, Ont, Dec 12, 21; nat US; m 50; c 2. NUCLEAR ASTROPHYSICS, NUCLEAR PHYSICS. *Educ:* McMaster Univ, BA, 43; Univ Toronto, MA, 44; Cambridge Univ, PhD(physics), 50. *Prof Exp:* Res physicist, Can-Brit Atomic Energy Proj, 44-46; instr physics, Univ BC, 50-53, asst prof, 55-56; res fel, Calif Inst Technol, 53-55; sr res fel, 56-57, assoc prof, 57-62, PROF PHYSICS, CALIF INST TECHNOL, 62- *Concurrent Pos:* NSF sr fel, Niels Bohr Inst, Copenhagen, Denmark, 62-63; guest prof, Nordita, Copenhagen, Denmark, 73-74; Alexander von Humboldt sr fel, Max Planck Inst, Heidelberg & Univ Münster, Ger, 87-88. *Mem:* Fel Am Phys Soc; AAAS. *Res:* Astrophysical nuclear reactions; nuclear structure physics; application of nuclear and accelerator techniques to other sciences and technologies. *Mailing Add:* Div Physics Math & Astron Calif Inst Technol Pasadena CA 91125

BARNES, CHARLES DEE, b Carroll, Iowa, Aug 17, 35; m 57; c 4. NEUROPHYSIOLOGY. *Educ:* Mont State Col, BS, 58; Univ Wash, MS, 61; Univ Iowa, PhD(physiol), 62. *Prof Exp:* Postdoctoral fel, Univ Calif, 62-64; from asst prof to assoc prof physiol, Ind Univ, 64-71; prof life sci, Ind State Univ, Terre Haute, 71-75; prof & chmn, physiol, Tex Tech Univ Sch Med, 75-83; PROF & CHMN VCAPP DEPT, WASH STATE UNIV, 84- *Concurrent Pos:* Vis scientist, Inst Physiol, Univ Pisa, 68-69; prof physiol, Sch Med, Ind Univ, 71-75. *Honors & Awards:* Career Develop Award, USPHS, 67. *Mem:* Int Brain Res Orgn; Am Physiol Soc; Am Soc Pharmacol & Exp Therapeut; Radiation Res Soc; Soc Neurosci; fel AAAS. *Res:* Neurophysiology, neuropharmacology and radiation effects on the nervous system; interaction of different levels of the central nervous system, particularly brain stem-spinal cord. *Mailing Add:* NW 1235 State St Pullman WA 99163

BARNES, CHARLES M, b Rising Star, Tex, July 21, 22; m 43; c 4. VETERINARY MEDICINE, MEDICAL PHYSICS. *Educ:* Tex A&M Univ, DVM, 44; Univ Calif, PhD(comp path), 57. *Prof Exp:* Pvt pract, 44-47; livestock inspector, Comn Eradication Foot & Mouth Dis, USDA, 47-50; base vet, USAF, 50-52, res radiobiologist, Hanford Atomic Prod Oper, Gen Elec Co, 52-54, life sci proj officer, Aircraft Nuclear Propulsion Off, US AEC, 56-60, safety officer, Systs Nuclear Auxiliary Power Prog, 60-62, pathologist, Armed Forces Inst Path, 62-66, dir vet med serv, Air Proving Ground Ctr & Spec Air Warfare Ctr, Eglin AFB, Fla, 66-68; mgr radiol health & pub health ecol, L B Johnson Space Ctr, NASA, 68-71, mgr health appln off, 71-78, mgr radiation protection, space shuttle, 78-82; regional supvr Cent Am & Columbia, Animal Plant Health Inspection Serv, USDA, 82-85; RETIRED. *Concurrent Pos:* Mem, Nat Comt Radiation Protection & Measurements, 60-66; mem subcomt livestock damage, Adv Comt Civil Defense, Nat Acad Sci-Nat Res Coun, 61-62; prof comp path, Grad Sch Biomed Sci; chmn, bd dirs, Int Vet Med Found, 73-, mgr, animal geriat res, 85- *Mem:* Am Vet Med Asn. *Res:* Remote sensing; radiation toxicology; radiation biology; geographic pathology; use of aircraft and spacecraft remote sensor data in application to public health. *Mailing Add:* Rte 2 Box 245C Rising Star TX 76471

BARNES, CHARLES WINFRED, b Oklahoma City, Okla, Oct 21, 34; m 58; c 4. OBLIQUE-SLIP FAULTING, FORCED-DRAPED FOLDING. *Educ:* Univ Okla, BS, 57; Univ Idaho, MS, 62; Univ Wis, PhD(geol), 65. *Prof Exp:* Asst prof geol, Eastern NMex Univ, 65-68; from asst prof to assoc prof geol, 68-74, chair, Dept Geol, 79-84, PROF GEOL, NORTHERN ARIZ UNIV, 74-, ASSOC DEAN SCI, 84- *Concurrent Pos:* Res prof, Univ Wyo, 66-68; prin investr, NSF, 68-76 & NASA, 83-86; chair fac, Northern Ariz Univ, 73-74; sr geologist, Berge Explor, 77-79; sci preceptor, Nat Col Hons Coun, 77; fac assoc, Danforth Found, 78; res geologist, US Geol Surv, 80-81. *Mem:* Geol Soc Am; Sigma Xi; assoc Am Mus Natural Hist; Hist Earth Sci Soc; Nat Asn Geol Teachers. *Res:* Dynamics of planetary surfaces; energetics of planets; science administration; science education; history of geology. *Mailing Add:* 250 N Circle Dr Flagstaff AZ 86001

BARNES, CHARLIE JAMES, b Greenville, NC, Aug 23, 30; m 57; c 1. PHYSICAL CHEMISTRY, ORGANIC CHEMISTRY. *Educ:* Va Union Univ, BS, 51; Va State Col, MS, 57; Howard Univ, PhD, 59. *Prof Exp:* Phys & org chemist, Res & Develop Dept, US Naval Propellant Plant, 59-63; ANALYTICAL CHEMIST, FOOD & DRUG ADMIN, 63- *Mem:* Am Chem Soc. *Res:* Development of pesticide reference standards; stability of pesticide materials; analysis of food additives; drugs in edible animal tissue. *Mailing Add:* 2203 Piermont Dr Ft Washington MD 20744

BARNES, CHRISTOPHER RICHARD, b Nottingham, UK, Apr 20, 40; Can citizen; m 61; c 3. PALEONTOLOGY. *Educ:* Birmingham Univ, UK, BSc, 61; Univ Ottawa, Can, PhD(geol), 64. *Prof Exp:* NATO res fel geol, Univ Wales, 64-65; from asst prof to assoc prof & chmn dept, Univ Waterloo, 65-81; prof & head geol, Mem Univ Nfld, 81-87; dir, Gen Sedimentary & Marine Geosci Br, Geol Surv Can, 87-89; DIR, CTR EARTH & OCEAN RES, UNIV VICTORIA, 89- *Concurrent Pos:* Lectr, Geol Asn Can, 78; chmn, Coun Chmn Can Earth Sci Dept, 83-85; group chmn, Earth Sci & Interdisciplinary, Natural Sci & Eng Res Coun Can, 87-90. *Honors & Awards:* Bancroft Award, Royal Soc Can, 82. *Mem:* Fel Royal Soc Can; Geol Asn Can (pres, 83-84); Can Soc Petrol Geologists; Geol Soc Am; Paleont Soc; Paleont Asn; Can Geosci Coun (pres, 79); Acad Sci Royal Soc Can (vpres, 87-88, pres, 90-93). *Res:* Evolution; biostratigraphy; paleoecology and paleobiogeography of lower Paleozoic conodonts. *Mailing Add:* Ctr Earth & Ocean Res Univ Victoria PO Box 3055 Victoria BC V8W 3P6 Can

BARNES, CLIFFORD ADRIAN, b Goldendale, Wash, Oct 14, 05; m 38; c 3. OCEANOGRAPHY. *Educ:* Univ Wash, BS, 30, PhD(chem), 36. *Prof Exp:* Res chemist, Fuels Div, Battelle Mem Inst, 36-40; assoc phys oceanogr, US Coast Guard, 40-46; oceanogr, US Navy Hydrographic Off, Washington, DC, 46-47; from assoc prof to prof oceanog, 47-73, actg chmn dept, 64-65, EMER PROF OCEANOG, UNIV WASH, 73- *Concurrent Pos:* Oceanogr, Exped to Aleutian Islands USN, 33 & oceanog cruise of US Coast Guard cutter Chelan, 34; officer in chg, Oceanog Surv Subsect JTF-1, Oper Crossroads, 46 & Beauford Sea Exped, 50, 51, 52; consult, President's Sci Adv Comt, 60 & 61; consult, Off Sci & Technol, 62-64. *Mem:* Am Geophys Union; Am Chem Soc; Am Soc Limnol & Oceanog; Arctic Inst NAm. *Res:* Ocean currents; chemistry of sea water; sea ice and icebergs; oceanography of arctic and inshore waters; river water at sea. *Mailing Add:* Sch Oceanog Univ Wash Seattle WA 98195

BARNES, CRAIG ELIOT, b Berkeley, Calif, June 17, 55. ORGANOMETALLIC CHEMISTRY, CATALYSIS. *Educ:* Harvey Mudd Col, BS, 77; Stanford Univ, PhD(chem), 82. *Prof Exp:* Postdoctoral fel res, Alexander von Humbolt Found, 82-83 & NATO, 83-84; asst prof, 84-90, ASSOC PROF CHEM, UNIV TENN, KNOXVILLE, 90- *Mem:* Am Chem Soc. *Res:* Roles transition metals play in both homogeneous and heterogeneous catalytic systems; techniques of solid state nuclear magnetic resonance, x-ray crystallography and extended x-ray absorption fine structure spectroscopy to study known catalytic systems and develop new ones. *Mailing Add:* Dept Chem Univ Tenn Knoxville TN 37996-1600

BARNES, DAVID FITZ, b Boston, Mass, May 23, 21; m 66; c 2. GEOPHYSICS. *Educ:* Harvard Univ, BS, 43, MA, 55. *Prof Exp:* Asst marine geophys, Woods Hole Oceanog Inst, 43-49; geophysicist, US Geol Surv, 49-54 & Air Force Cambridge Res Ctr, 56-58; geophysicist, 58-90, EMER GEOPHYSICIST, US GEOL SURV, 90- *Mem:* Am Geophys Union; Soc Explor Geophys; fel Arctic Inst NAm; Geol Soc Am. *Res:* Application of geophysics to Arctic research; Alaskan regional gravity; precision gravimetry; Alaskan aeromagnetics; permafrost; perennially frozen lakes; infrared luminescence. *Mailing Add:* US Geol Surv Br Geophys 345 Middlefield Rd MS 989 Menlo Park CA 94025

BARNES, DAVID KENNEDY, b Concordia, Kans, Apr 23, 23; m 42; c 4. ORGANIC CHEMISTRY. *Educ:* Olivet Col, BS, 43; Ind Univ, AM, 44, PhD(chem), 47. *Hon Degrees:* DSc, Olivet Col, 78. *Prof Exp:* Off Sci Res & Develop contract, 43-44, asst, Ind, 46; sr chemist, Process Res Dept, Stanolind Oil & Gas Co, 47-53; with Dacron process develop, Textile Fibers Dept, 53-54, Dacron plant tech, 54-55, group supvr, 55-57, sr supvr, Nylon Plant Technol, 57-59, supt, 59-60, asst mgr, Seaford Nylon Plant, 61-62, prod mgr, Nylon Mfg Div, 63-66, dir mfg, Orlon-acetate-lycra Div, 66-67, with Indust & Biochem Dept, 67-69, asst gen mgr indust chem dept, 69-75, vpres & gen mgr, Energy & Mat Dept, 75, vpres textile fibers, 77-81, exec vpres & dir, 81-87, DIR, E I DU PONT DE NEMOURS & CO, 88- *Mem:* Am Chem Soc; Soc Chem Indust; Sigma Xi. *Res:* Anti-malarials synthesis; diphenylmethanes; azlactones; hydrocarbon synthesis process; petrochemicals; synthetic textile fibers. *Mailing Add:* 35 Southridge Dr Kennett PA 19348-2714

BARNES, DEREK A, b Sussex, Eng, Sept 10, 33; US citizen; m 58; c 2. PLASMA PHYSICS, FLUID DYNAMICS. *Educ:* Oxford Univ, BA, 56 & 57, MA, 60, PhD(plasma physics), 63. *Prof Exp:* Theoret physicist, UK Atomic Energy Authority, 58-60; res scientist, Courant Inst Math Sci, NY Univ, 60-61; mem tech staff, Bell Tel Labs, Inc, 61-64; assoc prof, 64-66, chmn dept, 66-75, prof physics, 66-80, PROF MATH & COMPUT SCI, MONMOUTH COL, 80- *Concurrent Pos:* Consult, Aeronaut Res Assocs Princeton, Inc, 64-68 & Corp Comput Systs, Inc, 81- *Mem:* AAAS; Am Phys Soc; Am Inst Aeronaut & Astronaut. *Res:* Laminar to turbulent transition and jet impingement; macro processors. *Mailing Add:* Dept Comput Sci Monmouth Col West Long Branch NJ 07764

BARNES, DONALD KAY, b Minneapolis, Minn, Jan 15, 35; div; c 3. GENETICS, PLANT BREEDING. *Educ:* Univ Minn, BS, 57, MS, 58; Pa State Univ, PhD(genetics), 62. *Prof Exp:* Res geneticist, Regional Pasture Res Lab, Agr Res Serv, USDA, Pa, 62-63, res geneticist, Fed Exp Sta, PR, 63-65 & Plant Indust Sta, Md, 65-68, RES GENETICIST, USDA, 68-; PROF AGRON & PLANT GENETICS, UNIV MINN, ST PAUL, 68- *Concurrent Pos:* Res geneticist, USDA, 68- *Mem:* Fel Am Soc Agron; fel Crop Sci Soc Am. *Res:* Alfalfa genetics; breeding for disease and insect resistance; breeding for improved nitrogen fixation potential; alfalfa germplasm evaluation and coordination. *Mailing Add:* Dept Agron & Plant Genetics Univ Minn St Paul MN 55108

BARNES, DONALD MCLEOD, b New Lisbon, Wis, Sept 25, 21; m 49; c 2. VETERINARY PATHOLOGY, MICROBIOLOGY. *Educ:* Univ Minn, BSc, 53, DVM, 55, PhD(vet med), 63; Am Col Vet Path, dipl. *Prof Exp:* Instr vet diag, 55-63, from asst prof to assoc prof, 63-72, PROF VET PATH, UNIV MINN, ST PAUL, 72- *Mem:* Wildlife Disease Asn; Am Vet Med Asn; Sigma Xi. *Res:* Wildlife disease. *Mailing Add:* Vet Diag Lab Univ Minn St Paul MN 55108

BARNES, DONALD WESLEY, b Norfolk, Va, June 28, 44; m 69; c 2. TOXICOLOGY, IMMUNOTOXICOLOGY. *Educ:* Univ Richmond, BS, 66, Med Col Va, PhD(pharmacol), 71. *Prof Exp:* Grad asst pharmacol, Med Col Va, 70-71; asst prof, 71-82, ASSOC PROF PHARMACOL, E CAROLINA UNIV, 82- *Concurrent Pos:* External expert, consult & reviewer, Environ Protection Agency, 79. *Mem:* Int Soc Study Xenobiotics; Reticuloendothelial Soc; Sigma Xi; Am Soc Exp Pharmacol & Therapeut. *Res:* Biochemical and immunological mechanisms for the inhibition of drug metabolizing enzymes by immunomodulators; environmental toxicology: drinking water contaminant effects on immune systems and mixed-function oxidases. *Mailing Add:* Dept Pharmacol Sch Med E Carolina Univ Greenville NC 27858

BARNES, EARL RUSSELL, b Bennettsville, SC, May 2, 42; m 66. MATHEMATICS. *Educ:* Morgan State Col, BS, 64; Univ Md, PhD(math), 68. *Prof Exp:* Res mathematician, IBM Watson Res Ctr, 68-88; PROF INDUST ENG, GA INST TECH, 88- *Concurrent Pos:* Adj assoc prof, Columbia Univ. *Mem:* Am Math Soc; Soc Indust & Appl Math; Inst Elec & Electronics Engrs. *Res:* Variational analysis and differential equations. *Mailing Add:* 205 Hamden Trace Atlanta GA 30331

BARNES, EDWIN ELLSWORTH, b Kinzua, Ore, Sept 1, 29; m 52; c 3. ANALYTICAL CHEMISTRY, LIQUID CHROMATOGRAPHY. *Educ:* Univ Puget Sound, BS, 51; Univ Wash, PhD(chem), 61. *Prof Exp:* Chemist, Hooker Chem Corp, 51-53; res chemist, Am Marietta Co, 60-62; prof spec phys chem, Weyerhaeuser Co, 62-68, sci specialist, 68-77, scientist 5, 77-90; RETIRED. *Mem:* Am Chem Soc. *Res:* Properties of forest products and synthetic polymers; thermal analysis. *Mailing Add:* 17158 Sixth Pl SW Seattle WA 98166

BARNES, EUGENE MILLER, JR, b Versailles, Ky, Dec 9, 43; m 69. BIOCHEMISTRY, BIOLOGY. *Educ:* Univ Ky, BS, 65; Duke Univ, PhD(biochem), 70. *Prof Exp:* asst prof, 71-76, assoc prof, 77-85, PROF BIOCHEM, PHYSIOL, MOL BIOPHYS & NEUROSCI, BAYLOR COL MED, 86- *Concurrent Pos:* Am Cancer Soc fels, Nat Heart & Lung Inst, 69-70 & Roche Inst Molecular Biol, Nutley, NJ, 70-71; NIH res career develop award, 75-80; Vis prof, Univ Louis Pasteur, Strasbourg, France, 79-80; NIH Neurol Dis Prog Proj Rev B Comt, 86-90. *Mem:* AAAS; Am Soc Biol Chem; Am Soc Microbiol. *Res:* Membrane transport; mechanics and energetics of transport in bacterial membrane vesicles; membrane interactions and transport of neurotransmitter and neuromodulator substances. *Mailing Add:* Dept Biochem Baylor Col Med Tex Med Ctr Houston TX 77030

BARNES, FRANK STEPHENSON, b Boulder, Colo, July 31, 32; m 55; c 2. ELECTRICAL ENGINEERING, PHYSICS. *Educ:* Princeton Univ, BS, 54; Stanford Univ, MS, 56, PhD(elec eng), 58. *Prof Exp:* Fulbright prof elec eng, Col Eng, Baghdad, 57-58; physicist, US Nat Bur Standards, Boulder, 59-62; assoc prof, 59-65, chmn dept, 64-81, PROF ELEC ENG, UNIV COLO, BOULDER, 65- *Concurrent Pos:* Educ activ bd, Inst Elec & Electronics Engrs, 75-; bd dirs, Eng & Info, 85-; consult, Electronic Tire Co, Columbine

Ventures. *Honors & Awards:* Curtis McGraw Res Award, Am Soc Eng Educ, 65; Stearns Award, 80; Centennial Award, Inst Elec & Electronic Engrs; Fel Am Soc Lasers & Med. *Mem:* Fel Inst Elec & Electronics Engrs; Am Phys Soc; fel AAAS; Am Soc Educ Advan; Bioelectromagnetics Soc. *Res:* Applications of electromagnetic fields, ultrasonics and lasers to medicine and biology; electron devices. *Mailing Add:* Dept Elec Eng Univ Colo Boulder CO 80309

BARNES, FREDERICK WALTER, JR, b Cleveland, Ohio, Mar 3, 09; m 40; c 2. MEDICINE, NATURE OF HUMAN INTERACTION. *Educ:* Yale Univ, BA, 30; Johns Hopkins Univ, MD, 34; Columbia Univ, PhD(chem), 43. *Prof Exp:* Intern med, Johns Hopkins Hosp, 34-35, pediat, 35-36; resident, Children's Hosp, Boston, 36-37, asst physician, 37-38; asst biochem, Col Physicians & Surgeons, Columbia Univ, 40-41; asst prof pediat, Col Med, Univ Cincinnati, 42-46, biol chem, 42-46; assoc prof med & physiol chem, Sch Med, Johns Hopkins Univ, 46-62; PROF MED SCI, DIV MED SCI, BROWN UNIV, 62- *Concurrent Pos:* Vis researcher, Oxford Univ, Eng, 70. *Mem:* Am Soc Biol Chem; Soc Pediat Res; Am Heart Asn; fel Am Col Physicians; Physicians for Social Responsibility; Am Lung Asn. *Res:* Protein regeneration and interaction after harmful stimuli; development immunologic theory; detoxification dysentery vaccine; evolution of the immune system. *Mailing Add:* 229 Medway Providence RI 02906

BARNES, G RICHARD, b Milwaukee, Wis, Dec 12, 22. SOLID STATE PHYSICS. *Educ:* Univ Wis, BA, 48; Dartmouth Col, MS, 49; Harvard Univ, PhD(physics), 52. *Prof Exp:* Prog dir, Solid State Physics, Nat Sci Lab, 88-90; prof, 56-88, EMER PROF PHYSICS, IOWA STATE UNIV, 90- *Honors & Awards:* Alexander von Humboldt Sr Scientist Award, 75. *Mem:* Am Phys Soc. *Mailing Add:* 4925 Utah Dr Ames IA 50011

BARNES, GARRETT HENRY, JR, b Akron, Ohio, Mar 7, 26; m 57; c 3. ORGANIC CHEMISTRY. *Educ:* Case Inst Technol, BS, 51; Pa State Univ, PhD(chem), 54. *Prof Exp:* Dow-Corning fel, Mellon Inst, 54-66; res chemist, Dow Corning Corp, 66-68, sr res chemist, 68-75, proj engr, 76-81; CONSULT, 81- *Res:* Organosilicon compounds. *Mailing Add:* 1325 S Portofino No 306 Sarasota FL 34242

BARNES, GENE A, b Topeka, Kans, Feb 9, 35; m 68; c 2. SOLID STATE PHYSICS. *Educ:* Calif Inst Technol, BS, 56; Univ Ill, Urbana-Champaign, MS, 57; Univ Ore, PhD(physics), 67. *Prof Exp:* Sr res engr, Autonetics Div, NAm-Rockwell Corp, 60-63; PROF PHYSICS, CALIF STATE UNIV, SACRAMENTO, 67- *Concurrent Pos:* Visitor, Fac Math Studies, Univ Southhampton, Eng, 79. *Mem:* Am Asn Physics Teachers; Am Phys Soc; Sigma Xi. *Res:* Photovoltaics; electrical conductivity in high magnetic fields; solar optics. *Mailing Add:* Dept Physics-Astron Calif State Univ Sacramento CA 95819

BARNES, GEORGE, b Denver, Colo, May 27, 21; m 44; c 3. SCIENCE EDUCATION. *Educ:* Pomona Col, AB, 42; Univ Colo, MS, 49; Ore State Univ, PhD(physics), 55. *Prof Exp:* Instr physics, Pomona Col, 43; jr physicist, US Naval Res Lab, 43-44; asst physics, Univ Colo, 44-45, instr eng math, 45-47; teacher, Taft Jr Col, 47-48; from asst prof to assoc prof physics, Linfield Col, 48-55; asst prof, Humboldt State Col, 55-57; from assoc prof to prof, 57-83, EMER PROF PHYSICS, UNIV NEV, RENO, 83- *Mem:* Am Asn Physics Teachers; Sigma Xi. *Res:* The physics of biological systems; the physics of heat engines. *Mailing Add:* Dept Physics Univ Nev Reno NV 89577

BARNES, GEORGE EDGAR, b San Antonio, Tex, Jan 27, 43; m 60; c 2. CARDIOVASCULAR PHARMACOLOGY, OCULAR PATHOPHYSIOLOGY-PHARMACOLOGY. *Educ:* Southwest Tex State Univ, BS, 67, MA, 69; Univ Tex Med Sch, San Antonio, PhD(pharmacol), 74. *Prof Exp:* Am Heart Asn fel, Univ Miss, 73-75, instr physiol, Med Ctr, 74-75; asst prof pharmacol, Col Med, Tex A&M Univ, 75-85; scientist II & III, 85-90, PRIN SCIENTIST, ALCON LABS, 91-; Am Heart Asn fel, Univ Miss, 73-75. *Concurrent Pos:* co-investr, NIH grant, 77-; prin investr, Am Heart Asn & NIH grant. *Mem:* Am Heart Asn; Microcirculatory Soc; Soc Exp Med & Biol; Sigma Xi; Am Soc Pharmacol Exp Therapeut; Asn Res Vision & Opthal. *Res:* Pathophysiology of cerebral edema; basic mechanisms of edema formation to provide better understanding of how to treat cerebral or brain edema; synthesis and testing of blood pressure control drugs; cardiovascular physiology; pathophysiology of degenerative diseases of the eye; testing of ophthalmic drugs. *Mailing Add:* Alcon Labs Degenerative Dis 6201 S Freeway Ft Worth TX 76134

BARNES, GEORGE LEWIS, b Detroit, Mich, Aug 21, 20; m 47; c 4. ENTOMOLOGY. *Educ:* Mich State Univ, BS, 48, MS, 50; Ore State Univ, PhD(plant path), 53. *Prof Exp:* Res asst, Ore State Univ, 50-53; res assoc, Ohio State Univ, 53-55; plant pathologist, Olin Mathieson Chem Corp, 55-58; from asst prof to prof, 58-86, EMER PROF PLANT PATH, OKLA STATE UNIV, 86- *Concurrent Pos:* Plant pathologist, Agr Res Serv, USDA, 58-61; consult, Allergy Labs, Inc, 75-82. *Mem:* Sigma Xi; Am Phytopath Soc. *Res:* Fungicides; mycology; entomology; fungus physiology; peanut mold fungi; diseases of pecans and alfalfa; insect pathology. *Mailing Add:* 424 N Donaldson Dr Stillwater OK 74075

BARNES, GERALD JOSEPH, b Beech Grove, Ind, May 7, 36; m 57; c 2. AUTOMOTIVE ENGINEERING. *Educ:* Purdue Univ, BS, 58, MS, 60, PhD(mech eng), 63. *Prof Exp:* Sr res engr combustion, Gen Motors Res Labs, 62-70, sr res engr catalysts, 70-76, emissions engr, 76-78, STAFF ENGR AUTO EMISSION CONTROL, ENVIRON ACTIV STAFF, GEN MOTORS CORP, 78- *Mem:* Soc Automotive Engrs; Combustion Inst. *Res:* Combustion in spark ignition and diesel engines; emission control; catalytic control of auto emissions. *Mailing Add:* 405 Applehill Lane Rochester MI 48306

BARNES, GLOVER WILLIAM, b Birmingham, Ala, Sept 7, 23; m 52; c 2. IMMUNOLOGY, BACTERIOLOGY. *Educ:* Univ Akron, BSc, 49; Univ Buffalo, MA, 56, PhD(bact, immunol), 62. *Prof Exp:* Asst cancer res scientist, Roswell Park Mem Inst, 55-58; res assoc bact & immunol & asst prof path, State Univ NY Buffalo, 61-63, from instr to asst prof path, 63-69; assoc prof, 69-76, PROF UROL & LECTR MICROBIOL, UNIV WASH, 76-; RES CONSULT IMMUNOL, VA MASON RES CTR, 78- *Concurrent Pos:* Res consult, Millard Fillmore Hosp, 60-64. *Mem:* Sigma Xi; AAAS; Soc Study Reproduction; Am Soc Primatologists; Am Fertil Soc. *Res:* Immunologic studies on the organ-specific character and function of the male reproductive system. *Mailing Add:* 3415 S McClellan St Seattle WA 98144

BARNES, H VERDAIN, b Borger, Tex, Nov 20, 35; m 57; c 4. ADOLESCENT MEDICINE, ENDOCRINOLOGY. *Educ:* McMurry Col, Abilene, Tex, BA, 58; Yale Univ, BD, 61; Vanderbilt Sch Med, Nashville, Tenn, MD, 65. *Prof Exp:* Med intern, Vanderbilt Hosp, 65-66; asst resident, Johns Hopkins Hosp, 68-69, fel adult & pediat endocrinol & Metab, 69-71, chief resident med, 71-72, asst prof med & pediat, 72-75, dir adolescent med, 72-75, asst prof radiol, 73-75; assoc prof med & pediat & dir introd clin med & adolescent med, Univ Iowa Hosps & Clins, 75-78; prof med & pediat & chmn dept internal med, Col Med, Univ Tenn, Chattanooga, 78-83; PROF MED & PEDIAT & CHMN DEPT INTERNAL MED, SCH MED, WRIGHT STATE UNIV, 83- *Concurrent Pos:* Vis prof at numerous univs; mem comp II comt, Nat Bd Med Examrs, 77-83; mem, Adolescent Training Grant Guidelines Comt, Maternal & Child Health, Health Educ & Welfare, 78-79; mem, Coun Socs, Am Col Physicians, 79-; ed-in-chief, J Adolescent Health Care, 79-; Wyeth vis prof, 86-87; chmn, Am Col Physicians Coun Med Soc, Am Col Physicians Bd of Regents. *Honors & Awards:* Schweikert Award Distinguished Clin Res, Johns Hopkins Sch Med, 71; Bella Schick Lectr, Brookdale Hosp, New York, 77; Robert Ward Mem Lectr, Los Angeles Children's Hosp, Univ Southern Calif, 84; Mead Johnson Award, Soc Adolescent Med, 87; Adele Hofmann Award, Am Acad Pediat, 90. *Mem:* Am Fedn Clin Res; fel Soc Adolescent Med (pres, 67-68 & 77-78); Am Diabetes Asn; fel Am Col Physicians; fel Am Acad Pediat; Endocrine Soc; Asn Profs Med; AAAS. *Res:* Adolescent growth and development; adolescent endocrinology; therapy of hyperthyroidism; thyroid dysfunction in puberty; thyroid hormone physiology; medical education. *Mailing Add:* PO Box 927 Dayton OH 45401-0927

BARNES, HERBERT M, b Paulding, Ohio, Mar 27, 18; m 45; c 2. ANIMAL SCIENCE. *Educ:* Ohio State Univ, BSc, 40. *Prof Exp:* assoc prof & exten specialist animal sci, Ohio State Univ, 46-86; RETIRED. *Mem:* Am Soc Animal Sci. *Res:* Swine production; nutrition; physiology; swine husbandry. *Mailing Add:* Animal Sci Bldg Ohio State Univ 2029 Fyffe Rd Columbus OH 43210

BARNES, HOWARD CLARENCE, b Terre Haute, Ind, May 28, 12; m 32; c 3. ELECTRICAL ENGINEERING. *Educ:* Rose Polytech Inst, BSEE, 34. *Hon Degrees:* DEng, Rose Polytech Inst, 77. *Prof Exp:* Elec engr, Wheeling Steel Corp, 36-39 & Ohio Power Co, 39-44; from engr to asst vpres, Am Elec Power Serv Corp, 44-75, vpres power & environ systs, Charles T Main, Inc, 75-77, chief engr, 77-78, sr vpres, 77-82, dir, 78-82; RETIRED. *Concurrent Pos:* Vpres, Charles T Main, NY & Charles T Main Int; dir, Main Engenharia, Brazil; vis prof, Mass Inst Technol; lectr, Univ Tex, Chalmers Univ, Univ Denmark & Ohio State Univ; consult, ERDA; mem & vpres, Int Conf Large Elec Systs High Tension; vpres & dep chief engr, Am Elec Power Serv Corp, 39-44. *Honors & Awards:* Habirshaw Award, 80. *Mem:* Nat Acad Eng; fel AAAS; fel Inst Elec & Electronics Engrs; NY Acad Sci. *Mailing Add:* 209 Lehuerta Green Valley AZ 85614

BARNES, HOYT MICHAEL, b Long Beach, Calif, June 9, 43; m 64; c 4. WOOD SCIENCE & TECHNOLOGY, WOOD PRESERVATION. *Educ:* La State Univ, BSF, 65, MS, 68; State Univ NY, PhD(wood prod eng), 73. *Prof Exp:* From asst prof to assoc prof, 71-85, PROF WOOD SCI & TECHNOL, FOREST PROD UTILIZATION LAB, MISS STATE UNIV, 85- *Concurrent Pos:* Consult wood preserv, 81-; vis prof dept biol, Imperial Col Sci & Technol, London, 88-89; mem, Int Res Group Wood Preserv. *Mem:* Forest Prod Res Soc; Soc Wood Sci & Technol; Am Wood-Preservers' Asn; Am Railway Eng Asn; Railway Tie Asn. *Res:* Wood preservation; wood modification; wood physics; biodeterioration of wood; treated wood products; fire retardants. *Mailing Add:* Forest Prod Utilization Lab PO Drawer FP Mississippi State MS 39762

BARNES, HUBERT LLOYD, b Chelsea, Mass, July 20, 28; m 50; c 2. HYDROTHERMAL PROCESSES. *Educ:* Mass Inst Technol, BS, 50; Columbia Univ, PhD(geol), 58. *Prof Exp:* Res geologist, Peru Mining Co, 50-52; lectr geol, Columbia Univ, 52-54; fel, Geophys Lab, Carnegie Inst, 56-60; from asst prof to prof, 60-90, DISTINGUISHED PROF GEOCHEM, PA STATE UNIV, UNIVERSITY PARK, 90-, DIR ORE DEPOSITS RES SECT, 69- *Concurrent Pos:* Guggenheim fel, Geochem Inst, Univ Goettingen, 66-67; vis prof, Univ Heidelberg, 74; exchange scientist, USSR Acad Sci, 74; chmn, US Nat Comt Geochem, Nat Res Coun Geophys Res Bd, 76-80; bd dirs, Sci Systs, Inc, 69-; mem, US Geodynamics comt, 77-80, counr, Soc Econ Geol, 80-82; vis prof, Academia Sinica, China, 83; mem, US Nat Comt Geol, Nat Res Coun, 83-86; mem, US Dept Energy Mat Rev, 85-88; distinguished vis prof, Univ Sydney, 87; chmn, NATO Advan Study Inst, Geochm Hydrothermal Ore Deposits, Spain, 87. *Honors & Awards:* Lindsley lectr, Soc Econ Geol; Sr Humboldt Prize, Univ Munich, 87-88; Davidson Mem lectr, St Andrews, 71; Crosby lectr, Mass Inst Technol, 83. *Mem:* AAAS; Geochem Soc (pres, 82-85); fel Soc Econ Geol; fel Geol Soc Am; Am Geophys Union; fel Mineral Soc Am. *Res:* Geochemistry of hydrothermal processes in geothermal systems and in the formation of ore deposits. *Mailing Add:* 235 Deike Bldg Pa State Univ University Park PA 16802

BARNES, IRA LYNUS, b Los Angeles, Calif, Mar 13, 28; m 46; c 2. ANALYTICAL CHEMISTRY. *Educ:* Ind Univ, BS, 55; Univ Hawaii, PhD(anal chem), 63. *Prof Exp:* Assoc prof chem, Hawaii Inst Geophysics, Univ Hawaii, 65-69; res chemist, 69-73, sect chief, 73-78, chief, Inorg Anal Res Div, 78-79, RES CHEMIST, NAT BUR STANDARDS, 80- *Concurrent Pos:* Titular mem comn atomic weights, Int Union Pure & Appl Chem, 75-, secy subcomn assessment isotopic abundance, 75- *Honors & Awards:* Silver Medal, US Dept Com, 76. *Mem:* Am Chem Soc; Am Soc Mass Spectrometry; AAAS; Geochem Soc; Soc Appl Spectros. *Res:* Systematic redetermination of the atomic weights of polynuclidic elements, with the development of mass spectrometry and the determination of precise, accurate isotopic ratios in extremely small samples. *Mailing Add:* A-23 Physics Nat Bur Standards Gaithersburg MD 20899

BARNES, IVAN, geochemistry; deceased, see previous edition for last biography

BARNES, JAMES ALFORD, b Charlotte, NC, Aug 20, 44; m 66; c 2. LITHIUM BATTERIES, ELECTROCHEMISTRY. *Educ:* Davidson Col, BS, 66; Univ NC, Chapel Hill, PhD(inorg chem), 70. *Prof Exp:* Teaching assoc inorg chem, Southampton Univ, Eng, 70-71; NIH res assoc, Univ SC, 71-72; vis asst prof chem, Western Md Col, 72-73; assoc prof, Austin Col, Sherman, Tex, 73-83; RES CHEMIST ELECTROCHEM, NAVAL SURFACE WARFARE CTR, 82- *Concurrent Pos:* Res chemist phys chem, Shell Develop Corp, 76; inorg chem, Tex Instruments, 80, environ, Southwest Res Inst, 81 & inorg chem, Naval Res Lab, 81-82. *Mem:* Am Chem Soc; Electrochem Soc; Royal Soc Chem; AAAS. *Res:* Lithium batteries; electrochemical power sources. *Mailing Add:* Naval Surface Warfare Ctr Code R33 Silver Spring MD 20903-5000

BARNES, JAMES ALLEN, b Denver, Colo, Dec 7, 33; m 59; c 3. PHYSICS. *Educ:* Univ Colo, BS, 56, PhD(physics), 66; Stanford Univ, MS, 58; Denver Univ, MBA, 79. *Prof Exp:* Physicist, Atomic Frequency & Time Interval Stand Sect, 56-65, from actg sect chief to sect chief, 65-67, chief Time & Frequency Div, Nat Bur Stand, 67-81; dir, Res & Develop, Austron Inc, 82-88; RETIRED. *Concurrent Pos:* Mem comn, Int Sci Radio Union; mem study group VII, Int Radio Consultative Comt; consult, comn 31, Int Astron Union, 72- *Honors & Awards:* Silver Medal Award, US Dept Com, 65, Gold Medal Award, 75. *Mem:* Fel Inst Elec & Electronics Engrs. *Res:* Atomic and molecular spectroscopy; noise limitations of devices; statistics. *Mailing Add:* 3772 Lakebriar Dr Boulder CO 80304

BARNES, JAMES MILTON, b Ypsilanti, Mich, July 5, 23; m 49. PHYSICS. *Educ:* Eastern Mich Univ, BS, 48; Mich State Univ, MS, 50, PhD(physics), 55. *Prof Exp:* From asst prof to assoc prof, 55-62, head dept, 61-74, PROF PHYSICS, EASTERN MICH UNIV, 62- *Mem:* AAAS; Am Asn Physics Teachers; Acoust Soc Am; Nat Sci Teachers Asn. *Res:* Ultrasonics in glass by optical methods. *Mailing Add:* 4872 Whitman Circle Ann Arbor MI 48103

BARNES, JAMES RAY, b Woodland, Calif, Feb 16, 40; m 62; c 3. AQUATIC ECOLOGY. *Educ:* Brigham Young Univ, BS, 63; Ore State Univ, MS, 66, PhD(zool), 72. *Prof Exp:* Asst prof, 69-73, ASSOC PROF ZOOL, BRIGHAM YOUNG UNIV, 73- *Concurrent Pos:* Adj prof, Kellogg Biol Sta, Mich State Univ, 75. *Mem:* AAAS; Ecol Soc Am; Am Soc Limnol & Oceanog; NAm Benthol Soc; Orgn Inland Biol Field Sta. *Res:* Terrestrial litter decomposition in lakes and streams; annual reproductive cycles of marine and freshwater invertebrates. *Mailing Add:* Dept of Zool Brigham Young Univ Provo UT 84602

BARNES, JEFFREY LEE, GLOMERLONEPHRITIS. *Educ:* Univ Md, PhD(path), 78. *Prof Exp:* ASST PROF PATH, HEALTH SCI CTR, UNIV TEX, 78- *Mailing Add:* Dept Med Univ Tex Health Sci Ctr 7703 Floyd Curl Dr San Antonio TX 78284

BARNES, JOHN DAVID, b Allentown, Pa, Aug 12, 39; m 65. POLYMER PHYSICS. *Educ:* DePauw Univ, BA, 60; Univ Md, MS, 65; Cath Univ Am, PhD(physics), 72. *Prof Exp:* Physicist ceramics, Eastern Res Ctr, US Bur Mines, College Park, Md, 61-64; PHYSICIST POLYMER PHYSICS, POLYMERS DIV, NAT INST STANDARDS & TECHNOL, 64- *Mem:* Am Phys Soc; Mat Res Soc; Am Crystallog Asn. *Res:* Permeation, diffusion, and solubility of gases, vapors and liquids in polymers; x-ray diffraction and scattering from polymers. *Mailing Add:* Nat Inst Standards & Technol Polymers Bldg Rm B210 Gaithersburg MD 20899

BARNES, JOHN FAYETTE, b Santa Cruz, Calif, Jan 28, 30; m 55; c 2. ATOMIC PHYSICS. *Educ:* Univ Calif, Berkeley, BA, 51; Univ Denver, MS, 52; Univ NMex, PhD(physics), 63, MMS, 81. *Prof Exp:* Staff mem, Los Alamos Nat Lab, 53-68, asst group leader, 68-71, alt group leader, 71, group leader, 71-77, asst theoret div leader, 76-77, assoc theoret div leader, 77-80, dep theoret div leader, 80-81, dep assoc dir energy, 81-82, dep assoc dir physics & math,82-83, appl theoret physics div leader, 83-85, sr scientist, actg group leader, 85-87, LAB ASSOC, LOS ALAMOS NAT LABORATORY, 87- *Concurrent Pos:* Adj instr, Univ NMex, 85-; coordr Math & Tech Progs, 87-; adj instr, Santa Fe Community Col, 87- *Mem:* Am Phys Soc. *Res:* Thomas-Fermi theory; atomic structure; solid state physics; hydrodynamics; equation of state and opacity research for applied problems. *Mailing Add:* 2213 Calle Cacique Santa Fe NM 87505

BARNES, JOHN MAURICE, b Washington, DC, Apr 22, 31; m 57; c 5. PLANT PATHOLOGY. *Educ:* Univ Md, BS, 54; Cornell Univ, MS, 57, PhD(plant path), 60. *Prof Exp:* Opers analyst, Opers Res Off, Johns Hopkins Univ, 60-63; plant scientist, Hazleton Labs, TRW Systs, 63-67; plant pathologist, Coop State Res Serv, 67-75, coordr, Corn Blight Info Ctr, 71-72, pesticide coordr, Off Secy, 75-77, pesticide impact assessment coordr, Sci & Educ Admin, 77-80, PRIN PLANT PATHOLOGIST, COOP STATE RES SERV, USDA, 80- *Concurrent Pos:* Mem, acid rain competitive grant mgt; co-ed, Appl Bot, Crop Sci, Acad Press; cert merit, USDA, 72, 84, 86 & 88. *Mem:* Am Phytopath Soc; Sigma Xi; Soc Nematologists; AAAS. *Res:* Coordination and policy planning relative to agricultural pest control. *Mailing Add:* Coop State Res Serv USDA 901 D St SW Washington DC 20250-2200

BARNES, KAREN LOUISE, b Cleveland, Ohio, Feb 24, 42; m 73. NEUROPHYSIOLOGY, NEUROANATOMY. *Educ:* Mt Holyoke Col, AB, 63; Univ Mich, Ann Arbor, AM, 65; Case Western Reserve Univ, PhD(exp psychol), 70. *Prof Exp:* Asst prof psychol, John Carroll Univ, 69-73; res psychologist med res, Cleveland Vet Admin Hosp, 73-75; asst prof neurosurg & biomed eng, Sch Med, Case Western Reserve Univ, 75-83, adj assoc prof biomed eng & neurol, 83-90; STAFF DOCTOR, NEUROL & BRAIN & VASC RES CLEVELAND CLIN FOUND, 76-, HEAD, SECT NEUROL RES, 80- *Concurrent Pos:* Mem, Prog Advan Training Biomed Res Mgt, Sch Pub Health, Harvard Univ, 89. *Mem:* Soc Neurosci; Am Heart Asn; Am Physiol Soc; Am Acad Neurol; Sigma Xi. *Res:* Neurohormonal mechanisms in cardiovascular control and hypertension; neuronal actions of angiotensin peptides and substance P in pathways of the dorsomedial medulla that subserve cardiovascular function. *Mailing Add:* Res Inst 9500 Euclid Ave Cleveland OH 44195-5070

BARNES, LARRY D, PROTEIN-LIGAND INTERACTION. *Educ:* Univ Calif, Los Angeles, PhD(biochem), 70. *Prof Exp:* ASSOC PROF BIOCHEM, HEALTH SCI CTR, UNIV TEX, 76- *Res:* Metabolic regulation by nucleotides. *Mailing Add:* Dept Biochem Health Sci Ctr Univ Tex 7703 Floyd Curl Dr San Antonio TX 78284-7760

BARNES, LAWRENCE GAYLE, b Portsmouth, Va, Dec 26, 45; m 67; c 1. VERTEBRATE PALEONTOLOGY, MAMMALOGY. *Educ:* Univ Calif, Berkeley, BA, 67, MA, 69, PhD(paleont), 72. *Prof Exp:* Mus asst paleont, Univ Calif, Berkeley, 67-69, mus preparator, 69-72, teaching asst, 71-72; assoc cur, 72-75, CUR VERT PALEONT, NATURAL HIST MUS LOS ANGELES COUNTY, 75- *Concurrent Pos:* Res assoc, Mus Paleont, Univ Calif, Berkeley, 74- *Mem:* Soc Vert Paleont; Am Soc Mammal; Soc Syst Zool; Soc Marine Mammal. *Res:* Paleontology of fossil marine vertebrates, mainly marine mammals of Tertiary age. *Mailing Add:* Natural Hist Mus of Los Angeles County 900 Exposition Blvd Los Angeles CA 90007

BARNES, MARTIN MCRAE, b Calgary, Alta, Aug 3, 20; US citizen; wid; c 4. ECONOMIC ENTOMOLOGY. *Educ:* Univ Calif, Berkeley, BS, 41; Cornell Univ, PhD(entom), 46. *Prof Exp:* Jr entomologist, 46-49, from asst entomologist to assoc entomologist, 49-61, ENTOMOLOGIST, UNIV CALIF, RIVERSIDE, 61-, PROF ENTOM, 63- *Concurrent Pos:* Chmn, Dept Entom, Univ Calif, Riverside, 88. *Mem:* Fel AAAS; Entom Soc Am; Can Entom Soc; Mex Soc Entom. *Res:* Biology, ecology and management of insects and mites of deciduous fruit and nut orchards and vineyards. *Mailing Add:* Dept Entom Univ Calif Riverside CA 92521

BARNES, MARY WESTERGAARD, b Champaign, Ill, May 20, 27; m 50; c 2. PHYSICAL CHEMISTRY, MATERIALS SCIENCE. *Educ:* Swarthmore Col, BA, 48; Pa State Univ, PhD(phys chem), 66. *Prof Exp:* Res asst gen chem, Mass Inst Technol, 48-50; from tech asst to tech staff assoc electron devices, Bell Tel Labs, 52-56; chemist, Nat Bur Standards, 56-60; proj assoc water struct, 67-69, proj assoc nuclear waste disposal, 77-81, PROJ COORDR, MAT RES LAB, PA STATE UNIV, 81- *Concurrent Pos:* Fel, Max Planck Inst Phys Chem, Gottingen, WGer, 66-67. *Mem:* Am Chem Soc; Mat Res Soc; Am Ceramic Soc; Geochem Soc. *Res:* Physical chemistry: interactions between nuclear waste and rock; water and groundwater; chemistry of cement and its application to waste disposal; statistical thermodynamics; materials for electron devices; phase chemistry of calcium aluminate silicate hydrates; calcium phosphates. *Mailing Add:* 208 Mat Res Lab Pa State Univ University Park PA 16802

BARNES, PAUL RICHARD, b San Francisco, Calif, May 1, 36; m 64; c 3. SPACE PHYSIOLOGY. *Educ:* San Francisco State Univ, BA, 60, MA, 63; Univ Kans, PhD(biochem & physiol), 69. *Prof Exp:* Asst microbiol, Sch Med, Univ Kans, 69-71; asst pharmacol, Sch Med, Univ Md, 71-72; asst biochem, Univ Southern Calif, 72-73; lectr, 73-75, from asst prof to assoc prof, 75-83, PROF BIOL, SAN FRANCISCO STATE UNIV, 83-; CONSULT, AMES BIOMED DIV, NASA, 81- *Mem:* Am Physiol Soc; Am Chem Soc; Sigma Xi. *Res:* Human physiological changes associated with space flight; alterations in body fluid composition and volumes; electrolyte-balance; hormone levels; blood pressure control mechanisms in the absence of gravity. *Mailing Add:* 1724 Holly San Bruno CA 94066

BARNES, PETER DAVID, b Garden City, NY, Oct 10, 37; m 61; c 3. EDUCATION, NUCLEAR & PARTICLE PHYSICS. *Educ:* Univ Notre Dame, BS, 59; Yale Univ, MS, 60, PhD(physics), 65. *Prof Exp:* Fel nuclear physics, Niels Bohr Inst, Copenhagen, Denmark, 64-66 & Los Alamos Sci Lab, 66-68; assoc prof, 68-78, PROF PHYSICS, CARNEGIE-MELLON UNIV, 78- *Concurrent Pos:* NATO fel, 65-66; vis scientist, Los Alamos Sci Lab, 73-74 & Centre Etude Nucleaire, France, 78-79; mem, Nuclear Sci Adv Comt, USA, 77-80, chmn electromagnetic interactions, 81-82; assoc ed, Phys Rev C, 80-82. *Mem:* Fel Am Phys Soc; Sigma Xi. *Res:* Strong electromagnetic and weak interactions in atomic, nuclear and elementary particle systems; kaonic, antiprotonic and sigma atoms, few nucleon and complex nuclei, lamda hypernuclei, antiproton, kaon physics. *Mailing Add:* Dept Physics Carnegie-Mellon Univ Pittsburgh PA 15213

BARNES, RALPH CRAIG, b Summerfield, Ohio, May 4, 14; m 86; c 1. PUBLIC ENTOMOLOGY, PUBLIC HEALTH. *Educ:* Ohio Univ, BS, 39; NC State Col, MS, 41. *Prof Exp:* Eng aide. Commun Dis Ctr, 42-43, asst sanitarian, 43-44, sr asst sanitarian, 44-47, sr asst scientist, 47-48, scientist, 48-52, sr scientist, 52-56, scientist dir, 56-70, regional prog dir, Ctr Dis Control, 70-74, dir div prev, Public Health Serv, US Dept Health, Educ & Welfare, Reg VIII, 74-78; RETIRED. *Concurrent Pos:* Part-time instr biol, Metrop State Col, Denver, 78-83. *Mem:* Entom Soc Am; Am Pub Health Asn. *Res:* Ecology of mosquitoes; medical entomology; communicable disease control. *Mailing Add:* 185 Grape St Denver CO 80220

BARNES, RALPH W, b New York, NY, Nov 28, 36; m; c 2. ELECTRICAL ENGINEERING. *Educ:* Duke Univ, BSEE, 58, PhD, 69; Univ Pa, MSE, 65. *Prof Exp:* Design engr, Radio Corp Am, Camden & Moorestown, NJ, 58-62; sr engr, Motorola, Chicago, 63-64; instr elec eng, Duke Univ, 64-65, NSF trainee, 65 & 66, teaching intern, 65-66, trainee, Div Biomed Eng, 66-69; res assoc, Sonic Lab, 69, res instr ultrasound, 70, from res asst prof to res assoc prof med sonics, 70-84, PROF NEUROL, BOWMAN GRAY SCH MED, 85 -; DIR RES, AUTREC, WINSTON-SALEM, NC, 84 - *Concurrent Pos:* NC Heart Asn res grant; consult, NIH & NSF; prin investr grants & contracts, NIH. *Honors & Awards:* Small Bus Innovative Res Award, NIH. *Mem:* Am Inst Ultrasound in Med; Inst Elec & Electronics Engrs; Sigma Xi. *Res:* Ultrasound; diagnostic data processing; biomedical instrumentation; industrial instrumentation; research and development. *Mailing Add:* Dept Neurol Bowman Gray Sch Med 300 S Hawthorne Winston-Salem NC 27103

BARNES, RAMON M, b Pittsburgh, Pa, Apr 24, 40; m 69. ANALYTICAL CHEMISTRY. *Educ:* Ore State Univ, BS, 62; Columbia Univ, AM, 63; Univ Ill, Urbana-Champaign, PhD(anal chem), 66. *Prof Exp:* Spectrochemist, NASA Lewis Res Ctr, Ohio, 67-68, mat engr, 68-69; assoc anal chem, Iowa State Univ, 68-69; asst prof, 69-75, assoc prof, 75-80, PROF ANAL CHEM, UNIV MASS, AMHERST, 80- *Concurrent Pos:* Lectr chem, Baldwin-Wallace Col, 67-68; ed & publ, ICP Info Newslett; chmn, Winter Conf Plasma Spectrochem, 80, 82, 84, 86, 88, 90 & 92; consult, Phillips Petrol & other insts. *Mem:* Am Chem Soc; Soc Appl Spectros; Spectros Soc Can; Am Soc Test & Mat; Optical Soc Am; Sigma Xi; Royal Soc Chem; Soc Environ Geochem & Health; fel AAAS; Int Asn Bioinorganic Scientists; Am Soc Mass Spectrometry. *Res:* Chemical reactions in high energy spectroscopic discharges, including sparks and radio frequency, induction heated plasmas for spectrochemical analysis; applications of atomic and mass spectroscopy with plasma sources for spectrochemical analyses and chromatographic detection. *Mailing Add:* Dept Chem 102 Lederle GRC Towers Univ Mass Amherst MA 01003-0035

BARNES, RICHARD N, b Washington, DC, Dec 10, 28; m 57; c 2. ECOLOGY. *Educ:* Univ Calif, Berkeley, AB, 52; Univ Calif, Davis, MA, 57, PhD(zool), 62. *Prof Exp:* Instr biol, Sacramento State Col, 57-62; from asst prof to assoc prof, 62-72, PROF BIOL, BEREA COL, 72- *Mem:* Sigma Xi; Am Soc Limnol & Oceanog; Ecol Soc Am. *Res:* Limnology; aquatic ecology. *Mailing Add:* Dept of Biol Berea Col Berea KY 40404

BARNES, ROBERT DRANE, b New Orleans, La, Apr 3, 27; m 53; c 3. INVERTEBRATE ZOOLOGY. *Educ:* Davidson Col, BS, 49; Duke Univ, PhD, 53. *Hon Degrees:* DSc, Davidson Col. *Prof Exp:* Instr zool, Duke Univ, 52-53 & Smith Col, 53-54; res assoc biol, Rice Inst, 54-55; from asst prof to assoc prof, 55-63, chmn dept, 65-70, PROF BIOL, GETTYSBURG COL, 63- *Concurrent Pos:* Mem, Foreign Currency Prog, Res Coun Biol Studies, Smithsonian Inst, 71-74; trustee, Bermuda Biol Sta, 79- *Honors & Awards:* Lindback Found Award, 82. *Mem:* AAAS; Am Soc Zool; Crustacean Soc; Marine Biol Asn UK; Int Soc Reef Studies. *Res:* Invertebrate zoology. *Mailing Add:* Dept Biol Gettysburg Col Gettysburg PA 17325-1486

BARNES, ROBERT F, b Estherville, Iowa, Feb 6, 33; m 55; c 4. AGRONOMY, CROP SCIENCE. *Educ:* Estherville Jr Col, AA, 53; Iowa State Univ, BS, 57; Rutgers Univ, MS, 59; Purdue Univ, PhD(agron), 63. *Prof Exp:* Res agronomist, Ind, USDA, 59-70, res agronomist, Pa, 70-75, dir, US Regional Pasture Res Lab, 70-75, staff scientist forage, range, plant & entom sci, Nat Prog Staff, Agr Res Serv, 75-79, assoc regional adminr, Agr Res Serv, Southern Region, 79-82, regional adminr, 82-86; EXEC VPRES, AM SOC AGRON, CROP SCI SOC AM & SOIL SCI SOC AM, 86- *Concurrent Pos:* From instr to assoc prof, Purdue Univ, 59-70; adj prof, Pa State Univ, 70-76; participant, Int Grassland, Cong, Helsinki, Finland, 66, Surfer's Paradise, Australia, 70, Moscow, USSR, 73, Leipzig, E Ger Dem Repub, 77, Lexington, Ky, 81 & Kyoto, Japan, 85. *Honors & Awards:* Medallion Award, Am Forage Grassland Coun, 81. *Mem:* Fel AAAS; fel Am Soc Agron; Am Soc Animal Sci; Am Forage & Grassland Coun; fel Crop Sci Soc Am; Soil Sci Soc Am. *Res:* Forage, pasture, range and turfgrass breeding, production and utilization; crop physiology; development and application of forage evaluation and utilization techniques; biochemistry of forage and range plants; agricultural research priorities, program planning and leadership; administration of and writing for agronomy, crop science and soil science. *Mailing Add:* Am Soc Agron 677 S Segoe Rd Madison WI 53711

BARNES, ROBERT KEITH, b Connersville, INd, Apr 25, 25; m 57; c 3. INDUSTRIAL CHEMICALS. *Educ:* Purdue Univ, BS, 49; Univ Idaho, MS, 51; Univ Wash, PhD(chem), 55. *Prof Exp:* Technol mgr oxide-derivatives & amines & assoc dir res & develop dept, Ethylene Oxide & Glycols Div, Union Carbide Corp, 55-86; RETIRED. *Concurrent Pos:* Consult, 86- *Mem:* Am Chem Soc. *Res:* Alkylene oxide monomer and polymer research; agricultural chemicals and intermediates; management of pesticide research, screening, formulations and product and process development; alkylamines processes; glyoxal technology; heterogeneous catalysis. *Mailing Add:* 8 Bridlewood Rd Charleston WV 25314

BARNES, ROBERT LEE, b Niagara Falls, NY, Aug 1, 35. PHARMACEUTICALS, MATERIALS SCIENCE. *Educ:* Hamilton Col, BA. 57; Pa State Univ, MS, 60, PhD(chem', 63. *Prof Exp:* Res fel chem. Harvard Univ, 63-65; staff scientist, Aerospace Div, Res Ctr, Gen Precision, Inc, Little Falls, 65-69, res mgr chem dept, Singer-Gen Precision Kearfott Res Ctr, 69-74; mgr inorg chem, US Testing Co, Hoboken, NJ, 74-76; mgr anal labs, Booz, Allen & Hamilton, Foster D Snell Div, Florham Park, NJ, 76-80; mgr, Chem & Anal Serv, Case Consult Labs, Inc, Whippany, NJ, 80-82; sect head chromatography, E R Squibb, 82-84, dept head, Chem Control, 84-87, dir prod qual control, 87-89, DIR WORLDWIDE TECH SERV, E R SQUIBB, NEW BRUNSWICK, NJ, 89- *Mem:* Am Chem Soc; AAAS; Pharmaceut Qual Control. *Res:* Electron deficient compounds, especially boron hydrides and metal alkyls; mechanism of inorganic reactions; synthesis of inorganic materials; organometallics; adhesives; chemiluminescence; analytical chemistry; pharmaceutical analysis; environmental analysis; thermal analysis, materials research. *Mailing Add:* 48 High Ridge Rd Randolph NJ 07869

BARNES, RODERICK ARTHUR, b Madison, Wis, June 21, 20; m 44, 66; c 2. ORGANIC CHEMISTRY. *Educ:* Univ Wis, BS, 39; Univ Minn, PhD(org chem), 43. *Prof Exp:* Asst chem, Univ Minn, 39-42, instr & fel, 44-45; instr, Columbia Univ, 45-47; asst prof, Rutgers Univ, 47-53, from assoc prof to prof, 53-63; prof, Fla Atlantic Univ, 63-67; coordr res & grad prog, Engenharia Mil Inst, 70-75; PROF, FED UNIV RIO DE JANEIRO, 75- *Concurrent Pos:* NSF sr fel, 57-58; Conselho Nacional de Pesquisas fel, Brazil, 57-58; Fulbright lectr, Sao Paulo, 62-63; vis prof, Brazil, 63-65; temp prof, Univ Rio de Janeiro, 67. *Mem:* Am Chem Soc; Pharmaceut Soc Japan. *Res:* Biogenesis of natural products; reaction mechanisms; alkaloids; determination of structure of new natural products; chemical transformations of abundant natural products; synthesis of natural products and analogues. *Mailing Add:* CCS-NPPN Fed Univ Rio De Janeiro Rio de Janeiro 21941 Brazil

BARNES, RONNIE C, b Union City, Tenn, Sept 10, 41; m 66; c 1. ASTROPHYSICS. *Educ:* Vanderbilt Univ, BA, 63; Ind Univ, AM, 66, PhD(astrophys), 68. *Prof Exp:* From instr to asst prof physics & astron, Univ Mo-Columbia, 68-75; chmn, Div Natural Sci, 81-83, DIR, M D ANDERSON PLANETARIUM, LAMBUTH COL, 75-, PROF ASTRON, 85- *Concurrent Pos:* Summer fac fel, George C Marshall Spaceflight Ctr, NASA, 79-80. *Mem:* Am Astron Soc; fel Royal Astron Soc; Astron Soc Pac. *Res:* Close binary systems; cosmology. *Mailing Add:* Dept Sci Lambuth Col Jackson TN 38301

BARNES, ROSS OWEN, b Wahroonga, Australia, May 18, 46; m 67; c 2. MARINE & SEDIMENTARY GEOCHEMISTRY. *Educ:* Andrews Univ, BA, 67; Univ Calif, San Diego, PhD(earth sci), 73. *Prof Exp:* Res chemist, Univ Calif, San Diego, 73-74; res assoc, 74-77, asst res prof, 77-78, ASSOC RES PROF MARINE SCI, MARINE STA, WALLA WALLA COL, 78- *Mem:* AAAS; Geol Soc Am; Am Geophys Union; Geochem Soc; Am Soc Limnol & Oceanog. *Res:* Marine biogeochemical cycles of carbon, nitrogen and sulfur; helium flux from the earth's crust; circulation of crustal waters; specialized in-situ pure water filtration. *Mailing Add:* 104 Harbor Lane Anacortes WA 98221

BARNES, THOMAS GROGARD, b Blanket, Tex, Aug 14, 11; m 38; c 4. PHYSICS. *Educ:* Hardin-Simmons Col, AB, 33; Brown Univ, ScM, 36. *Hon Degrees:* ScD, Hardin-Simmons Col, 60. *Prof Exp:* Asst physics, Brown Univ, 35-36; teacher high sch, Tex, 36-38; instr math & physics, Col Mines & Metall, Univ Tex, El Paso, 38-43; elec engr, Div Phys War Res, Duke Univ, 43-45; dean, Grad Sch, Inst Creation Res, El Cajon, Calif, 81-83; dir Schellenger Res Labs, 55-65, from asst prof to prof, 45-81, EMER PROF PHYSICS, UNIV TEX, EL PASO, 81-; VPRES, GEO-SPACE RES & EXPLOR FOUND, 85- *Concurrent Pos:* Vis fel, Brown Univ, 52-53; consult physicist, US Army Res Off, Durham, 63; consult, Res Adv Proj Div, Globe Universal Sci, 65-77. *Mem:* Am Asn Physics Teachers; fel Creation Res Soc (pres, 73-77). *Res:* Electromagnetism; acoustics; atmospheric physics; electronic instrumentation; relativity. *Mailing Add:* Dept Physics Univ Tex 500 W University Ave El Paso TX 79968-0151

BARNES, VIRGIL EVERETT, b Chehalis, Wash, June 11, 03; m 32; c 3. GEOLOGY. *Educ:* State Col Wash, BS, 25, MS, 27; Univ Wis, PhD(geol), 30. *Prof Exp:* Cur dept geol, Univ Wis, 27-29; res fel, Am Petrol Inst, 30-31 & US Geol Surv, 33-35; geologist, Bur Econ Geol, Univ Tex, 35-77, assoc dir, 61-68, prof dept geol sci, 61-77, SR RES SCI, UNIV TEX, AUSTIN, 73-, EMER PROF GEOL, 77- *Concurrent Pos:* NSF grants study of origin & compos of tektites, 60-75; mem, Am Comn Stratig Nomenclature, 70-76; dir, Tektite Res, 60- *Honors & Awards:* Barringer Award, Meteocritical Soc, 89. *Mem:* Fel Geol Soc Am; fel Mineral Soc Am; Asn Petrol Geologists; Geochem Soc; Soc Econ Paleontologists & Mineralogists; fel AAAS. *Res:* Cambrian, Ordovician and Devonian stratigraphy; geophysics; mineralogy; world wide occurrence of tektites; Lunar glass; meteorites; economic geology of Texas. *Mailing Add:* Bur Econ Geol Box X Univ Sta Austin TX 78713-7508

BARNES, VIRGIL EVERETT, II, b Galveston, Tex, Nov 2, 35; m 57, 70; c 4. HIGH ENERGY PARTICLE PHYSICS. *Educ:* Harvard Univ, AB, 57; Cambridge Univ, PhD(physics), 62. *Prof Exp:* Res assoc, Brookhaven Nat Lab, 62-64, from asst physicist to assoc physicist, 64-69; assoc prof, 69-79, asst dean, Sch Sci, 74-78, PROF PHYSICS, PURDUE UNIV, WEST LAFAYETTE, 79- *Concurrent Pos:* Spokesman, Dept Energy, High Energy Physics Contract, Task D. *Honors & Awards:* Perkin Elmer Prize, Harvard Univ, 56. *Mem:* Am Phys Soc; NY Acad Sci; AAAS; Sigma Xi. *Res:* Experimental elementary particle physics, currently collaborator on collider detector at Fermilab, 1.8 trillion eV antiproton-proton collisions; author or coauthor of over 110 publications. *Mailing Add:* 801 N Salisburg Lafayette IN 47906

BARNES, WALLACE EDWARD, mathematics; deceased, see previous edition for last biography

BARNES, WAYNE MORRIS, b Riverside, Calif, Sept 30, 47; m 74; c 2. BIOCHEMISTRY, MOLECULAR BIOLOGY. *Educ:* Univ Calif, Riverside, AB, 69; Univ Wis-Madison, PhD(biochem), 74. *Prof Exp:* Fel, MRC Lab Molecular Biol, Cambridge, Eng, 75-77; ASST PROF BIOCHEM, WASHINGTON UNIV SCH MED, 77- *Concurrent Pos:* Fel, Am Cancer Soc, 75-76 & jr fac res award, 78-81; NSF fel, 76-77. *Mem:* Sigma Xi. *Res:* DNA sequence and mechanism of action of genetic control signals. *Mailing Add:* 223 Renaldo Ct Chesterfield MO 63017-2210

BARNES, WILFRED E, b Oak Park, Ill, June 3, 24; m 46; c 2. ALGEBRA. *Educ:* Univ Chicago, BS, 49, MS, 50; Univ BC, PhD(math), 54. *Prof Exp:* From instr to prof math, Wash State Univ, 54-66; head dept, 66-82, PROF MATH, IOWA STATE UNIV, 66- *Mem:* Am Math Soc; Math Asn Am; Sigma Xi. *Res:* Algebra; theory of rings and ideals. *Mailing Add:* Dept Math Iowa State Univ Ames IA 50011

BARNES, WILLIAM CHARLES, b Chicago, Ill, Dec 12, 34; m 67; c 1. GEOLOGY. *Educ:* Colo Sch Mines, Geol E, 56; Univ Wyo, MS, 58; Princeton Univ, PhD(geol), 63. *Prof Exp:* Geologist, Humble Oil & Ref Co, 58-59; asst prof geol, Ore State Univ, 63-67; geologist, US Geol Surv, 67-68; asst prof, 68-77, ASSOC PROF GEOL, UNIV BC, 77- *Mem:* Geol Soc Am; Geol Asn Can. *Res:* Sedimentology; organic geochemistry of recent marine and lacustrine sediments; overthrust fault mechanics. *Mailing Add:* Dept Geol Univ BC 2075 Westbrook Pl Vancouver BC V6T 1W5 Can

BARNES, WILLIAM GARTIN, b Leavenworth, Kans, June 13, 27; c 3. MICROBIOLOGY, BIOLOGY. *Educ:* Univ Kans, BA, 59, PhD(microbiol), 68; Univ Kansas City, MA, 60. *Prof Exp:* Tech rep pharmaceut, Winthrop Labs, Kansas City, 53-58; instr biol, Kansas City Jr Col, 60-63; instr microbiol, Sch Dent, Univ Mo, 64-65, asst prof, 65-69; chief microbiologist, Gen Hosp, Kansas City, 69-74; dir microbiol, Vet Admin Hosp, 74-83; from asst prof to assoc prof path, Sch Med, Univ Kans, 79-82; dir microbiol, Truman Med Ctr, Kansas City, MO, 83-88; PROF PATH, SCH MED, UNIV MO, KANSAS CITY, 83-; DIR MICROBIOL, TRUMAN MED CTR, KANSAS CITY, MO, 83- *Concurrent Pos:* Consult microbiologist. *Honors & Awards:* C Herrick Award, Eli Lilly Lab, 68; Sci Exhibit Award, Am Acad Dermat, 71; Gold Award, Am Soc Clin Pathologists, 72. *Mem:* Am Soc Microbiol; fel Am Acad Microbiol; fel Infectious Dis Soc Am. *Res:* Medical-clinical microbiology; host-parasite relationships. *Mailing Add:* 6308 W 67th Terr Overland Park KS 66204

BARNES, WILLIAM SHELLEY, b Wallingford, Conn, Feb 15, 47; m 70; c 2. FOOD MUTAGENS. *Educ:* Marietta Col, Ohio, BA, 69; Univ Mass, Amherst, PhD(bot), 80. *Prof Exp:* Res demonstr genetics, Univ Wales, UK, 78-79; res fel carcinogenesis, BC Cancer Res Ctr, Vancouver, 80-81; sr res fel, Am Health Fedn, Valhalla, NY, 81-82, assoc, 82-84; asst prof, 84-87, ASSOC PROF BIOL, CLARION UNIV, PA, 87- *Concurrent Pos:* Consult, Am Health Found, 84-86. *Mem:* Environ Mutagenesis Soc; AAAS. *Res:* Dietary origin of cancer; carcinogens formed in meat during cooking, their metabolism and genotoxic effects in vivo. *Mailing Add:* Dept Biol Clarion Univ Clarion PA 16214

BARNES, WILLIAM WAYNE, b Amsterdam, Ohio, May 26, 27; m 50; c 3. MEDICAL ENTOMOLOGY, INSECT TOXICOLOGY. *Educ:* Ohio State Univ, BSc, 50, MSc, 58, PhD(insect toxicol), 64. *Prof Exp:* Entomologist, US Army, 52-70, chief med entom div, 66-70; chief environ biol br, Tenn Valley Auth, 70-; RETIRED. *Mem:* Am Mosquito Control Asn. *Res:* Medical acarology; insect pathology; insect ecology; epidemiology of arthropod-borne diseases. *Mailing Add:* 706 Riverview Dr Florence AL 35630

BARNESS, LEWIS ABRAHAM, b Atlantic City, NJ, July 31, 21; m; c 7. NUTRITION, PEDIATRICS. *Educ:* Harvard Univ, AB, 41, MD, 44. *Hon Degrees:* MA, Univ Pa, 71. *Prof Exp:* From instr to prof pediat, Med Sch, Univ Pa, 51-72; prof pediat & chmn dept, Col Med, UNIV SFLA, TAMPA, 72-88; VIS PROF PEDIAT, UNIV WIS-MADISON, 87- *Concurrent Pos:* Res fel, Children's Med Ctr, Boston, 47-50. *Honors & Awards:* Borden Award Nutrit, Am Acad Pediat, 72, Jacobi Award, 91; Noer Distinguished Professor, Univ S Fla, 80; Goldberger Award Clin Nutrit, AMA, 84. *Mem:* AAAS; Soc Pediat; Am Pediat Soc; Am Acad Pediat; Am Inst Nutrit; Am Col Nutrit. *Res:* Liver necrosis; nitrogen requirements of infants. *Mailing Add:* Dept Pediat Waisman Ctr Univ Wis Madison WI 53705

BARNET, ANN B, b Chicago, Ill; m 53; c 3. PEDIATRIC NEUROLOGY. *Educ:* Sarah Lawrence Col, AB, 51; Harvard Univ, MD, 55. *Prof Exp:* Trainee ment health, Seizure Unit, Children's Med Ctr, Boston, Mass, 52; asst pediat, US Army 130th Field Hosp, Heidelberg, Ger, 56-57; clin dir, Child Develop Clin, Cambridge, Mass, 58-61; res assoc exp psychophysiol, Walter Reed Army Inst Res, 61-73; asst to assoc prof, 67-84, PROF NEUROL & CHILD HEALTH & DEVELOP SCH MED, GEORGE WASHINGTON UNIV, 84-, DIR EEG RES LAB & EVOKED RESPONSE LAB, DEPT NEUROL, CHILDREN'S HOSP, NAT MED CTR, 65- *Concurrent Pos:* Guest physician, Nerve Clin, Heidelberg Univ, 57; asst physician, Wrentham State Sch, Mass, 58; fel pediat, Mass Gen Hosp, Boston, 58-61; prin investr, Army Med Res & Develop Command Contract, 61-64; res assoc, Wash Sch Psychiat, 61-65; consult, Dept Neurol, Walter Reed Army Hosp, 61-65 & Children's Bur Proj 237, DC, 64-66; deleg, Conf Young Deaf Child, Toronto, Ont, 64 & Int Cong Physiol, Moscow, USSR, 66; USPHS grant, 64-78; mem, Int Elec Response Audiometry Study Group; NIMH res scientist develop award, 70-75; W T Grant Found res grant, 73-75; mem, NIH Commun Sci Study Sect, 73-77; consult, Va Comn Visually Handicapped, 77-80, United Nat Develop Prog, 77-83, & Food Drug Admin, 78-83; Nat Found, March of Dimes grants, 78-; Children's Hosp Res Found grant, 81-84; pres, The Family Place, Inc; proj dir, Better Babies, 83-85. *Mem:* AAAS; Soc Neuroscience. *Res:* Research on development of sensory and perceptual processes in normal and abnormal infants and young children using the method of computer analysis of electroenchephalographic responses to sensory stimuli; research in methods of reducing the incidence of low birth weight infants born to high risk women. *Mailing Add:* Childrens Nat Med Ctr 111 Michigan Ave NW Washington DC 20010

BARNET, HARRY NATHAN, b San Diego, Calif, Apr 30, 23; m 46; c 2. BIOCHEMISTRY. *Educ:* San Diego State Col, AB, 47; Univ Southern Calif, MS, 50, PhD(biochem), 53. *Prof Exp:* Asst, Med Sch, Univ Southern Calif, 49-53; res assoc, Scripps Metab Clin, 53-57; from asst prof to prof chem, US Int Univ, 57-75; prof chem, Palomar Col, 75-90; RETIRED. *Mem:* AAAS; Am Chem Soc; Sigma Xi. *Res:* tracer carbohydrate metabolism; serum protein chemistry; erythrocyte metabolism; blood storage. *Mailing Add:* 6063 La Jolla Blvd La Jolla CA 92037

BARNETT, ALLEN, b Newark, NJ, May 5, 37; m 65; c 2. NEUROPHARMACOLOGY. *Educ:* Rutgers Univ, BS, 59; State Univ NY Buffalo, PhD(pharmacol), 65. *Prof Exp:* Pharmacologist, Hoffman-La Roche Inc, 65-66; PHARMACOLOGIST, SCHERING CORP, 66- *Mem:* AAAS; Am Chem Soc; Am Soc Pharmacol & Exp Therapeut; sr mem Acad Pharmaceut Sci; Am Pharmaceut Asn. *Res:* Central nervous system dopamine receptors; analgesics. *Mailing Add:* Dept Biol Res Schering-Plough Res 60 Orange St Bloomfield NJ 07003

BARNETT, ALLEN M, b Oklahoma City, Okla, June 20, 40; m 68; c 3. ELECTRICAL ENGINEERING. *Educ:* Univ Ill, Urbana, BSEE, 62, MS, 63; Carnegie Inst Technol, PhD(elec eng), 66. *Prof Exp:* Proj engr, Electronics Lab, Gen Elec Co, 66-69, physicist, Res & Develop Ctr, 69-76; pres, XCITON Corp, 71-75; dir, Inst Energy Conversion, Univ Del, 76-79; gen mgr, Astro Power Div, Astro Systs, Inc, 83-89; PRES, ASTROPOWER, INC, 89-, PROF ELEC ENG, UNIV DEL, 76- *Concurrent Pos:* Consult, Technol & Mktg Assessment, 75- *Honors & Awards:* IR-100 Awards, 70, 73, 74 & 79. *Mem:* Am Phys Soc; Inst Elec & Electronics Engrs; Int Solar Energy Soc. *Res:* Solar energy, solar cells, especially development of record energy conversion efficiencies for thin-film solar cells; invention of thin film crystalline silicon solar cells on low cost substrates; energy storage; energy systems; energy conservation; development of planar processing technology for compound semiconductors. *Mailing Add:* Two Polaris Dr N Star Newark DE 19711

BARNETT, AUDREY, b Somerset, Pa, May, 1933. GENETICS, PROTOZOOLOGY. *Educ:* Wilson Col, AB, 55; Ind Univ, MA, 57, PhD(genetics), 62. *Prof Exp:* Asst, Ind Univ, 55-58; asst prof biol, Western Col, 61-63; NIH vis fel, Princeton Univ, 63-65; asst prof, Bryn Mawr Col, 65-70; ASSOC PROF ZOOL, UNIV MD, COLLEGE PARK, 71- *Concurrent Pos:* NIH spec res fel, 68-69; vis scientist, Argonne Nat Lab, 68-70; guest worker, NIH & Nat Inst Arthritis, Metab & Digestive Dis, 77-78. *Mem:* AAAS; Soc Protozool; Genetics Soc Am; Sigma Xi. *Res:* Genetics of mating types and circadian rhythm of mating type reversals in Paramecium multimicronucleatum; cell surface antigens. *Mailing Add:* Dept Zool Univ Md College Park MD 20742

BARNETT, BILL B, b Pendelton, Ore, Jan 14, 46. VIROLOGY, IMMUNODIAGNOSTICS. *Educ:* Wash State Univ, BS, 68; Utah State Univ, PhD(biochem), 75. *Prof Exp:* Radiochemist, US Atomic Energy Comn, 68-71; NIH trainee, Utah State Univ, 71-75; sr scientist, Becton-Dickinson, Inc, 75-77; PROF, UTAH STATE UNIV, 77- *Concurrent Pos:* Tech adv, Int Atomic Energy Agency, Vienna, 81- *Mem:* Am Soc Microbiol; Sigma Xi; Soc Exp Biol & Med; AAAS. *Res:* Arena viruses; bunyaviruses; detection procedures for viruses and antibodies; vaccine development; monoclonal antibodies and their application to virology; antiviral drugs. *Mailing Add:* Dept Biol UMC 55 Utah State Univ Logan UT 84322

BARNETT, BOBBY DALE, b Elm Springs, Ark, Aug 12, 27; m 48. POULTRY NUTRITION. *Educ:* Univ Ark, BSA, 50, MS, 54; Univ Wis, PhD(biochem, poultry husb), 57. *Prof Exp:* Asst poultryman, 56-58, assoc poultryman, 58-59, PROF POULTRY SCI & HEAD DEPT, CLEMSON UNIV, 59- *Concurrent Pos:* Coop res tour, USDA, Washington, DC. *Mem:* Poultry Sci Asn. *Res:* Dietary toxins; nutrition-disease relationships. *Mailing Add:* Dept Poultry Sci Clemson Univ Clemson SC 29631

BARNETT, CHARLES JACKSON, b Chattanooga, Tenn, Feb 15, 42; m 70; c 2. ORGANIC CHEMISTRY. *Educ:* Vanderbilt Univ, BA, 64; Rice Univ, PhD(org chem), 69. *Prof Exp:* Sr org chemist, 69-75, res scientist, 75-79, SR RES SCIENTIST, LILLY RES LABS, ELI LILLY & CO, 80- *Mem:* Am Chem Soc. *Res:* Heterocyclic synthesis; new synthetic methods; asymmetric synthesis; process research. *Mailing Add:* 7540 N Pennsylvania St Indianapolis IN 46240

BARNETT, CLARENCE FRANKLIN, physics; deceased, see previous edition for last biography

BARNETT, CLAUDE C, b Woodinville, Wash, Nov 8, 28; m 51; c 2. THEORETICAL PHYSICS. *Educ:* Walla Walla Col, BS, 52; Wash State Univ, MS, 56, PhD(physics), 60. *Prof Exp:* Instr, 57-58, from asst prof to assoc prof, 58-64, PROF PHYSICS, WALLA WALLA COL, 64-, CHMN DEPT, 61- *Mem:* Am Phys Soc; Am Asn Physics Teachers; Sigma Xi. *Res:* Electroluminescence; quantum hydrodynamics; relativity; bio-electronics; radar; many-body problem; modelling and simulation. *Mailing Add:* Dept Physics 704 SW Evans College Place WA 99324

BARNETT, DAVID, b Arlington, Wash, Sept 4, 40; m 65. GEOMETRY. *Educ:* Univ Wash, BS, 64, MS, 66, PhD(math), 67. *Prof Exp:* Asst prof, Western Wash State Col, 67; asst prof, 67-70, assoc prof, 70-76, PROF MATH, UNIV CALIF, DAVIS, 76- *Concurrent Pos:* NSF res grant, Univ Calif, Davis, 68- *Mem:* Math Asn Am. *Res:* Combinatorial structure of convex polytopes, particularly problems dealing with the relationships between the numbers of faces of different dimensions of convex polytopes. *Mailing Add:* Dept Math Univ Calif Davis CA 95616

BARNETT, DAVID M, b Washington, DC, Nov 16, 39; m 72; c 2. APPLIED MECHANICS, MATERIALS SCIENCE. *Educ:* Rice Univ, BA, 61, MS, 63; Stanford Univ, PhD(mat sci), 67. *Prof Exp:* NATO fel, Inst Theoret Physics, Clausthal Tech Univ, 67-68; staff scientist, Sci Labs, Ford Motor Co, Mich, 68-69; from asst prof to assoc prof, 69-78, PROF MAT SCI & APPL MECH, STANFORD UNIV, 78- *Concurrent Pos:* Vis scientist, Physics Inst, Oslo Univ, Norway, 74; consult, Sandia Labs, Calif, 77-80, IBM, 83- & Intel, 84. *Res:* Defects in anisotropic elastic solids; surface waves in anisotropic elastic and piezoelectric solids; non-destructive testing; mechanics of thin films. *Mailing Add:* Dept Mat Sci & Eng Stanford Univ Stanford CA 94305-2205

BARNETT, DON(ALD) R(AY), b Morton, Tex, Oct 19, 35. HUMAN GENETICS, BIOCHEMICAL GENETICS. *Educ:* Univ Tex, Austin, BA, 67; Univ Tex Med Br Galveston, MA, 69, PhD(human genetics), 71. *Prof Exp:* Res scientist assoc biochem genetics, Dept Zool, Univ Tex, Austin, 62-63 & 65-68; res asst, Dept Genetics, Rockefeller Inst, 63-65; res scientist, Univ Tex Med Br & Grad Sch Biomed Sci, Galveston, 68-71, from instr to assoc prof biochem genetics,71-82; DEPT CELLULAR & STRUCT BIOL, UNIV TEX HEALTH SCI CTR, SAN ANTONIO, 82- *Concurrent Pos:* Nat Found March of Dimes & Cystic Fibrosis Found grants, Univ Tex Med Br Galveston, 74-76, USPHS grant, 74-88; mem, res comt, Nat Cystic Fibrosis Found, 75-79; pres, Tex Genetics Soc, 89. *Honors & Awards:* Award for

Excellence in Res, Sigma Xi, 72. *Mem:* AAAS; Am Soc Human Genetics; Soc Cellular Biol; Sigma Xi. *Res:* Biochemical basis for inherited disease; genetic control of protein and DNA structure and function; med genetics education. *Mailing Add:* Dept Cellular & Struct Biol Univ Tex Health Sci Ctr 7703 Floyd Curl Dr San Antonio TX 78284-7762

BARNETT, DOUGLAS ELDON, b Walnut Ridge, Ark, Feb 11, 44; m 73. ENTOMOLOGY. *Educ:* Ark State Univ, BSE, 67; Northwestern State Univ, MS, 70; Univ Ky, PhD(entom), 74. *Prof Exp:* Lab asst biol, Northwestern State Univ, 67-68; sci instr, Maynard Pub Schs, 68-69; instr biol, Paragould Pub Schs, Ark, 69-70; researcher systs, Dept Entom, Univ Ky, 70, surv entomologist, 71-76; AGRICULTURIST-ENTOMOLOGIST, ANIMAL & PLANT HEALTH INSPECTION SERV, USDA, 76- *Concurrent Pos:* Chmn sect E, Entom Soc Am. *Mem:* Entom Soc Am; Sigma Xi. *Res:* Revision of various genera in the Cicadellidae; taxonomic methods, cave ecology, pest detection. *Mailing Add:* Rm 665 Plant Protection Fed Ctr Hyattsville MD 20782

BARNETT, EUGENE VICTOR, b New York, NY, Mar 28, 32; m 58; c 4. BIOLOGY, MEDICINE. *Educ:* Univ Buffalo, BA, 55, MD, 56; Am Bd Internal Med, dipl, 63. *Prof Exp:* Asst resident med, Rochester Med Ctr, 59-60; sr instr med & microbiol, Sch Med & Dent, Univ Rochester, 62-63, asst prof med & sr instr microbiol, Med Ctr, 63-65; assoc prof, 65-70, PROF MED, CTR HEALTH SCI, UNIV CALIF, LOS ANGELES, 70- *Concurrent Pos:* USPHS trainee med & immunol, 60-61; fel rheumatic dis, Med Res Coun Rheumatism Ctr, Taplow, Eng, 61-62; fel med & microbiol, Univ Rochester, 62-63; dir immunol sect, Rochester Gen Hosp, 63-65. *Mem:* Am Fedn Clin Res; Am Rheumatism Asn; Am Asn Immunol; Am Soc Clin Invest; Am Col Physicians. *Res:* Immunology; internal medicine. *Mailing Add:* Med Ctr Univ Calif Los Angeles CA 90024

BARNETT, GORDON DEAN, materials science, physical chemistry; deceased, see previous edition for last biography

BARNETT, GUY OCTO, b Chula Vista, Calif, Sept 18, 30; m 58; c 3. INFORMATION SCIENCE & SYSTEMS, INTELLIGENT SYSTEMS. *Educ:* Vanderbilt Univ, BA, 52; Harvard Univ, MD, 56. *Prof Exp:* Med resident, Peter Bent Brigham Hosp, 56-61; clin assoc, Nat Heart Inst, 58-60; estab investr, Am Heart Assoc, 61-67; PHYSICIAN, MASS GEN HOSP, 79-; PROF MED, HARVARD MED SCH, 80- *Concurrent Pos:* lectr elec eng, Mass Inst Technol, 72-; bd dirs, Am Med Informatics Asn, 84- *Mem:* Inst Elec & Electronics Engrs; Fel Inst Med; Asn Comput Mach; Biomed Eng Soc; Am Col Physicians. *Res:* Computer applications in patient care; computer-aided instruction in medical education. *Mailing Add:* Lab Comput Sci Mass Gen Hosp Boston MA 02114

BARNETT, H J M, b Newcastle-on-Tyne, Eng, Feb 10, 22; Can citizen; m 46; c 4. NEUROLOGY. *Educ:* Univ Toronto, MD, 44; FRCP(C), 52. *Prof Exp:* Assoc prof neurol, Univ Toronto, 67-69; prof clin neurol sci & chmn dept, Univ Western Ont, 69-; JOHN P ROBARDS RES INST. *Mailing Add:* John P Robards Res Inst 100 Perth Dr PO Box 5015 London ON N6A 5K8 Can

BARNETT, HENRY LEWIS, b Detroit, Mich, June 25, 14; m 40; c 2. PEDIATRICS, PEDIATRIC NEPHROLOGY. *Educ:* Wash Univ, BS & MD, 38; Am Bd Pediat, dipl. *Prof Exp:* Instr pediat, Sch Med, Wash Univ, 41-43; from asst prof to assoc prof, Med Col, Cornell Univ, 46-55; prof & head dept, 55-69, assoc dean clin affairs, 69-72, Univ prof, 72-81, EMER PROF PEDIAT, ALBERT EINSTEIN COL MED, 81-; MED DIR, CHILDREN'S AID SOC, 81- *Concurrent Pos:* Vis pediatrician, Bronx Munic Hosp, Ctr, 55-; consult, Appleton-Century-Crofts, 81- *Honors & Awards:* Mead-Johnson Award, 49; David Hume Award, 77; John Howland Award & Medal, 84; John P Peters Award, 88. *Mem:* Asn Am Physicians; Am Soc Clin Invest; Am Pediat Soc (pres, 81-82); Am Soc Pediat Nephrology; Soc Ped Res, (pres, 59-60). *Res:* Physiology of infants and children with particular reference to kidney physiology and electrolyte metabolism as affected by growth; pediatric clinical epidemiology; pediatric and medical education. *Mailing Add:* Children's Aid Soc 150 E 45th St New York NY 10017

BARNETT, HERALD ALVA, b Orlando, WVa, Oct 2, 23; m 47; c 2. SPECTROSCOPY. *Educ:* Salem Col, BS, 44; WVa Univ, MS, 50. *Prof Exp:* Sect supvr anal chem, chromatography & spectrometry, Tech Ctr, US Steel Corp, 50-84, res mgr, Tech Serv, 82-86; RETIRED. *Mem:* Am Chem Soc; Am Soc Testing & Mat; Coblentz Soc; Soc Appl Spectros. *Res:* Applications of mass spectrometry; absorption and fluorescence spectrophotometry; atomic absorption spectrometry; applications of gas chromatography analytical chemistry, x-ray and optical emission spectrometry to coal-chemical recovery processes, steel-making processes and air and water pollution abatement studies. *Mailing Add:* 4133 Dundee Dr Murrysville PA 15668

BARNETT, JAMES P, b Mena, Ark, May 19, 35; m 65; c 2. PLANT PHYSIOLOGY, FORESTRY. *Educ:* La State Univ, BSF, 57, MF, 63; Duke Univ, DF, 68. *Prof Exp:* RES FORESTER, SOUTHERN FOREST EXP STA, US FOREST SERV, 57- *Mem:* Soc Am Foresters; Am Soc Plant Physiol. *Res:* Seed dormancy and physiology in the genus Pinus; artificial regeneration of forest stands; vegetation control; silviculture of Pinus. *Mailing Add:* 3783 S Loop Pineville LA 71360

BARNETT, JEFFREY CHARLES, b Philadelphia, Pa, Oct 24, 46; m; c 1. MATHEMATICS EDUCATION. *Educ:* Shippensburg State Col, BSEd, 67; Bucknell Univ, MS, 69; Pa State Univ, PhD(math educ), 74. *Prof Exp:* Instr math, Neshaminy High Sch, 69-70 & Williamsport Area Community Col, 70-71; asst math educ, Pa State Univ, 71-74; asst prof math, Northern Ill Univ, 74-78; from asst prof to assoc prof, 78-86, PROF MATH, FT HAYS STATE UNIV, 86- *Concurrent Pos:* NSF grant co-researcher, Northern Ill Univ, 77-79; mem, Nat Coun Tachers Math. *Mem:* Nat Coun Teachers Math; Am Educ Res Asn; Math Asn Am; Int Group Psychol Math Educ. *Res:* Mathematics education, particularly problem solving, use of manipulatives and the role of visual abilities; teaching heurntics. *Mailing Add:* Col Educ Univ Wis Whitewater WI 53190

BARNETT, JOHN BRIAN, b Cardston, Alberta, Apr 17, 45; US citizen; m 66; c 2. IMMEDIATE HYPERSENSITIVITY, IMMUNOTOXICOLOGY. *Educ:* Mont State Univ, BS, 67, MS, 69; Univ Louisville, PhD(microbiol), 73. *Prof Exp:* From asst prof to assoc prof, 75-88, PROF IMMUNOL, UNIV ARK MED SCI, 88- *Concurrent Pos:* Asst prof, Dept Pharmacol & Interdisciplinary Toxicol, Univ Ark, 79-81, assoc prof, 85-90, assoc dean res, 88-, prof, 91- *Mem:* Am Asn Immunologists; Am Soc Microbiol; Sigma Xi; Soc Toxicol. *Res:* Extent and mechanism of the immunomodulating activity of natural and synthetic retinoid (vitamin A) compounds; effect on pesticides such as Chlordane, and other environmental contaminants, on developing immune responses. *Mailing Add:* Dept Microbiol & Immunol Univ Ark Med Sci 4301 W Markham Little Rock AR 72205-7199

BARNETT, JOHN DEAN, b Payson, Utah, May 21, 30; m 54; c 6. HIGH PRESSURE, CRITICAL PHENOMENA. *Educ:* Univ Utah, BA. 54, PhD(physics). 59. *Prof Exp:* Assoc prof, 58-65, PROF PHYSICS, BRIGHAM YOUNG UNIV, 65- *Mem:* Am Phys Soc; Am Asn Physics Teachers. *Res:* Experimental high pressure electron spin resonance; x-ray diffraction; electrical and thermal studies relating to phase transformation mechanisms and critical phenomena. *Mailing Add:* 621 E Sagewood Ave Provo UT 84604

BARNETT, JOHN WILLIAM, b Gadsden, Ala, Sept 17, 41; m 62; c 2. TOXICOLOGY, ENTOMOLOGY. *Educ:* Auburn Univ, BS, 62, MS, 64, PhD(entom), 69. *Prof Exp:* Asst prof entom, Univ Fla, 69-70; exten entomologist, Auburn Univ, 70-72; res specialist, Ciba-Geigy Corp, 73-75, toxicologist, 75-77, sr toxicologist, 78-79, mgr toxicol, 79-80, dir, Safety Evals, 80-89, EXEC DIR, SAFETY EVALS, CIBA-GEIGY CORP, 89- *Mem:* Entom Soc Am; AAAS; Soc Toxicol; Soc Environ Toxicol & Chem; Sigma Xi. *Res:* Animal toxicology; insect physiology and biochemistry; pesticide development; agricultural pest management. *Mailing Add:* 3202 Cabarrus Dr Greensboro NC 27407

BARNETT, KENNETH WAYNE, b Mobile, Ala, Apr 8, 40; m 67; c 2. INORGANIC CHEMISTRY, POLYMER CHEMISTRY. *Educ:* Univ Tex, Arlington, BS, 63; Univ Wis, PhD(inorg chem), 67. *Prof Exp:* Res chemist, Shell Develop Co, 67-70; asst prof chem, Univ Mo-St Louis, 70-76, assoc prof, 76-79; AT ASHLAND CHEM CO, 79- *Mem:* Am Chem Soc; Am Foundrymen's Soc. *Res:* Organometallic chemistry of the transition elements; organic reactions catalyzed by transition metals; inorganic photochemistry; polymer chemistry; adhesives and coatings; foundry technology. *Mailing Add:* Ashland Chem Co PO Box 2219 Columbus OH 43216

BARNETT, LELAND BRUCE, b Los Angeles, Calif, Aug 21, 35; m 61; c 3. ZOOLOGY, ECOLOGY. *Educ:* Brigham Young Univ, BA, 62, MS, 64; Univ Ill, PhD(zool), 68. *Prof Exp:* From asst prof to assoc prof, 68-78, PROF BIOL, WAYNESBURG COL, 78- *Mem:* Ecol Soc Am; Am Inst Biol Sci. *Res:* Ecology; streams; birds; botany; genetics. *Mailing Add:* Dept Biol Waynesburg Col Waynesburg PA 15370

BARNETT, LEWIS BRINKLEY, b Lexington, Ky, Jan 29, 34; m 56; c 4. PHYSICAL BIOCHEMISTRY. *Educ:* Univ Ky, BS, 55; Univ Iowa, MS, 57, PhD(biochem), 59. *Prof Exp:* NSF fel phys chem, Univ Wis, 59-60, univ res fel, 60-61; Am Heart Asn advan res fel, Van't Hoff Lab, Neth, 61-63; asst prof, 63-67, ASSOC PROF BIOCHEM & NUTRIT, VA POLYTECH INST & STATE UNIV, 67-, ASST DEAN, COL ARTS & SCI, 84- *Concurrent Pos:* Mem, Blacksburg town coun, 86- *Mem:* Am Chem Soc; Am Soc Biol Chem; Sigma Xi. *Res:* Protein chemistry. *Mailing Add:* 126 Williams Hall Va Polytech Inst & State Univ Blacksburg VA 24061

BARNETT, MARK, b Amityville, NY, Mar 1, 57; m 87; c 1. COMPUTATIONAL FLUID DYNAMICS. *Educ:* Univ Cincinnati, BS, 79, MS, 81, PhD(aerospace eng), 84. *Prof Exp:* Aerospace engr, NASA Ames Res Ctr, Univ Cincinnati, 77-79; assoc res engr, 84-87, RES ENGR, UNITED TECHNOL RES CTR, 87- *Concurrent Pos:* Treas, Am Inst Aeronaut & Astronaut, Conn Sect, 86-; adj prof, Hartford Grad Ctr, Rensellear Polytech Inst, 88- *Mem:* Am Inst Aeronautics & Astronaut; Sigma Xi. *Res:* Computational fluid dynamics; viscous flows in turbomachinery; strong viscous/inviscid interactions in internal and external flows; boundary layer theory. *Mailing Add:* United Tech Res Ctr E Hartford CT 06108

BARNETT, MICHAEL PETER, b London, Eng, Mar 24, 29; m 61; c 3. LITIGATION & COMPUTING, PROGRAMMING LANGUAGES. *Educ:* King's Col, London, BSc, 48, PhD(chem), 52. *Prof Exp:* Teaching fel chem, King's Col, London, 52-53; sr gov res fel solid state phys, Royal Radar Estab, 53-55; head appl sci dept comput, IBM, UK Ltd, 55-57; res assoc chem, Chem Dept, Univ Wis, 57-58; res assoc physics, Dept Physics, Mass Inst Technol, 58-60, assoc prof, 60-64; reader comput sci, Inst Comput Sci, Univ London, 64-65; mgr comput sci, Data Processing Div, RCA Graph Syst Div, 65-69; dir res & develop comput sci, H W Wilson Co, 69-75; prof libr sci, Libr Ach, Columbia Univ, 75-77; PROF COMPUT SCI, SCI DEPT, BROOKLYN COL, CITY UNIV NEW YORK, 77- *Concurrent Pos:* Dir, Coop Comput Lab, Mass Inst Technol, 61-64; adj lectr, Libr Sch, Columbia Univ, 69-75; chmn off automation sect, Asn Comput Mach, 81-82, mem SIGBd, 82-85, nat lectr, 85- *Mem:* Asn Comput Mach. *Res:* Quantum theoretical molecular structure calculations; econometric simalation of river systems; text and natural language processing; electronic publishing; symbolic algebraic manipulation; computer graphics; educational technology; neurophysiological modelling; artifical intelligence. *Mailing Add:* Comput & Info Sci Dept Brooklyn Col 2118 Ingersoll Hall Bedford Ave & Ave H Brooklyn NY 11210-2816

BARNETT, NEAL MASON, b Lafayette, Ind, May 31, 37; m 60; c 2. PLANT PHYSIOLOGY. *Educ:* Purdue Univ, BS, 59; Duke Univ, PhD(bot), 66. *Prof Exp:* Res assoc plant physiol, Purdue Univ, 65-67; asst prof, 68-75, ASSOC PROF PLANT PHYSIOL, UNIV MD, COLLEGE PARK, 75- *Concurrent Pos:* NIH fel, Mich State Univ & AEC Plant Res Lab, 67-68. *Mem:* AAAS; Am Soc Plant Physiol; Am Inst Biol Sci. *Res:* Biochemistry of water stress in plants; plant cell wall protein; nitrogen assimilation. *Mailing Add:* Dept Bot Univ Md College Park MD 20742

BARNETT, ORTUS WEBB, JR, b Pine Bluff, Ark, July 2, 39; m 64; c 3. PLANT VIROLOGY, PLANT PATHOLOGY. *Educ:* Univ Ark, BSA, 61, MS, 65; Univ Wis, PhD(plant path), 69. *Prof Exp:* Sr sci officer, Scottist Crop Res Inst, 68-69; from asst prof to assoc prof, 69-78, PROF PLANT PATH & PHYSIOL, CLEMSON UNIV, 78- *Concurrent Pos:* Collabr, USDA, 69-; joint appointment, Dept Microbiol, 72-; assoc ed, Phytopath, 80-82, sr ed, 85-87; Fulbright travel Award, Waite Agr Res Inst, Australia, 82-83; mem plant virus subcomt, Int Comt Taxon Viruses, 87- *Mem:* Am Phytopath Soc (secy, 86-89, vpres, 90, pres-elect, 91, pres, 92); Int Soc Plant Path; Am Soc Virol. *Res:* White clover viruses; virus purification and serology; plant virus inactivation in vitro; relationships of plant viruses by complementary DNA probes; plant virus epidemiology. *Mailing Add:* Dept Plant Path & Physiol Clemson Univ Clemson SC 29634-0377

BARNETT, PAUL EDWARD, b Jerryville, WVa, Mar 19, 36; m 59; c 2. FOREST GENETICS. *Educ:* WVa Univ, BS, 64; Univ Tenn, MS, 67, PhD(genetics), 72. *Prof Exp:* Res asst forest genetics, Forestry Dept, Univ Tenn, 64-69; botanist forest physiol, Tenn Valley Authority, 69-73, staff forester forest genetics, 73-74, staff forester & proj leader forest genetics, Div Forestry, Fisheries & Wildlife Develop, 74-79, staff forester, Forest & Biomass Harvesting Opers, 79-88; SCIENTIST, BECHTEL NAT INC, 89- *Mem:* Sigma Xi. *Res:* Genetics and breeding of forest trees with superior growth and form; geographic and altitudinal variation; establishment of seed production areas and hardwood plantation establishment and growth; investigation of equipment and methods to improve quantity and efficiency of recovery in forest harvesting operations; harvesting biomass plantations. *Mailing Add:* Bechtel Nat Inc 151 Lafayette Dr Oak Ridge TN 37831-0350

BARNETT, R(ALPH) MICHAEL, b Gulfport, Miss, Jan 25, 44. SUPERSYMMETRY, QUANTUM CHROMODYNAMICS. *Educ:* Antioch Col, BS, 66; Univ Chicago, PhD(physics), 71. *Prof Exp:* Fel physics, Univ Calif, Irvine, 72-74; res fel, Harvard Univ, 74-76; res assoc physics, Stanford Linear Accelerator Ctr, 76-83; vis physicist, Inst Theoret Physics, Univ Calif, Santa Barbara, 83- 84; staff scientist, 84-89, SR SCIENTIST, LAWRENCE BERKELEY LAB, 90- *Concurrent Pos:* Head, Particle Data Group, Lawrence Berkeley Lab; vpres, Contemp Physics Educ Proj. *Mem:* Am Phys Soc. *Res:* Theoretical analysis of searches for supersymmetry; Higgs bosons and other new particles at the Superconducting Super Collider and at other accelerators; quantum chromodynamics and electro-weak theories of elementary particles; solenoidal detector collaboration; detector requirements for observation of physics processes. *Mailing Add:* Lawrence Berkeley Lab Univ Calif Bldg 50A Rm 3115 Berkeley CA 94720

BARNETT, RONALD DAVID, b Texarkana, Ark, Nov 20, 43; m 61; c 3. AGRONOMY, PLANT GENETICS. *Educ:* Univ Ark, BS, 65, MS, 68; Purdue Univ, PhD(genetics, plant breeding), 70. *Prof Exp:* PROF AGRON, UNIV FLA, 70- *Mem:* Am Soc Agron; Crop Sci Soc Am; Am Genetic Asn; AAAS; Coun Agr Sci & Technol. *Res:* Plant breeding and genetics with emphasis on varietal development in wheat, triticale, rye and oats. *Mailing Add:* N Fla Res & Educ Ctr Univ Fla Rte 3 Box 4370 Quincy FL 32351

BARNETT, RONALD E, b Pueblo, Colo, Dec 26, 42; m 63; c 1. PHYSICAL ORGANIC CHEMISTRY, ENZYMOLOGY. *Educ:* Univ Colo, BA, 65; Brandeis Univ, PhD(biochem), 69. *Prof Exp:* Res assoc biochem, Stanford Univ, 68-69; asst prof chem, Univ Minn, Minneapolis, 69-76; assoc prof biochem & nutrit, Va Polytech Inst & State Univ, 76-80; PRIN SCIENTIST, GEN FOODS, 80- *Concurrent Pos:* NSF fel, 68-69; Merck Co Found Faculty develop award, 69; NIH res grant, 70- *Mem:* Fedn Am Soc Exp Biol; Am Chem Soc. *Res:* Carbonyl and acyl group reactions with diffusion limited rate determining steps; models for enzymatic reactions; conformational changes in enzymes; mechanisms of active transport of cations through cell membranes. *Mailing Add:* Gail Borden Res Ctr Gail Borden Dr Syracuse NY 13204-1436

BARNETT, RONALD E, b Pueblo, Colo, Dec 26, 42. ENTOMOLOGY. *Educ:* Univ Colo, BA, 65; Brandeis Univ, PhD(biochem), 68. *Prof Exp:* NSF postdoctoral fel, Stanford Univ, 68-69; from asst prof to assoc prof chem, Univ Minn, 69-76; assoc prof biochem, Va Polytech Inst & State Univ, 76-80; res scientist, Cent Res, Gen Foods, 80-82, sr scientist, 82-84, prin scientist, 84-86; dir, Biochem Sci, Nutrasweet Co, 86-88; group mgr ingredient res, Pepsico, 88-89; DIR, FUNDAMENTAL SCI, BORDEN INC, 89- *Concurrent Pos:* Vis prof biochem, State Univ NY, 80-82. *Mem:* Am Chem Soc; Am Soc Biochem & Molecular Biol; Am Asn Cereal Chemists. *Res:* Author of 39 publications in the areas of enzymes, membrane enzymes and model systems for enzymes; 46 patents on high intensity sweeteners, sweetener inhibitors and food ingredients. *Mailing Add:* Gail Borden Res Ctr One Gail Borden Dr Syracuse NY 13204

BARNETT, SAMUEL C(LARENCE), b Chatsworth, Ga, May 10, 22; m 50; c 1. MECHANICAL ENGINEERING. *Educ:* Ga Inst Technol, BIE, 48, MS, 56, PhD(mech eng), 62. *Prof Exp:* From asst prof to prof mech eng, Ga Inst Technol, 56-80, asst dir sch mech eng, 62-68, assoc dean undergrad div, 68-72; mgr educ assistance prog, Inst Nuclear Power Opers, Atlanta, Ga, 80-90; RETIRED. *Mem:* Am Soc Mech Engrs. *Res:* Heat transfer; fluid mechanics; thermodynamics; power production; air conditioning; refrigeration. *Mailing Add:* 1938 Gotham Way NE Atlanta GA 30324-4817

BARNETT, SCOTT A, THIN FILM PHYSICS, SURFACE SCIENCE. *Educ:* Univ Ill, MS, 76, PhD(Metallurgy), 82. *Prof Exp:* Res asst, Univ Ill, 76-82; vis sci, Univ Ill, 82-85; Linköping Univ, 85-86; ASST PROF MAT SCI, NORTHWESTERN UNIV, 86- *Mem:* Am Vacuum Soc. *Res:* Deposition and properties of thin films by techniques such as molecular beam epitaxy and sputter deposition. *Mailing Add:* Dept Mat Sci Northwestern Univ Evanston IL 60208

BARNETT, STANLEY M(ARVIN), b New York, NY, Apr 8, 36; m 59; c 1. BIOCHEMICAL & SEPARATIONS ENGINEERING, WASTE MINIMIZATION. *Educ:* Columbia Univ, BA, 57, BS, 58; Lehigh Univ, MS, 59; Univ Pa, PhD(chem eng), 63. *Prof Exp:* Chem engr, Elec Boat Div, Gen Dynamics Corp, 59-61; engr, Process Res Div, Esso Res & Eng Co, 63-64 & Plastics Technol Ctr, Shell Chem Co, 64-69; PROF CHEM ENG, UNIV RI, 69-, CHMN DEPT, 88- *Concurrent Pos:* Mem adv bd, RI Solid Waste Mgt Corp, 74-; chmn, Food Pharmaceut & Bioeng Div, Am Inst Chem Engrs, 83. *Mem:* Am Inst Chem Engrs; Inst Food Technologists; Soc Indust Microbiologist; Am Chem Soc; N Am Membrane Soc. *Res:* Separations; alternate sources of chemicals, food and energy; waste minimization. *Mailing Add:* Dept Chem Eng Univ RI Kingston RI 02881-0806

BARNETT, STOCKTON GORDON, III, b East Orange, NJ, July 18, 39; m 66. GEOLOGY. *Educ:* Dartmouth Col, BA, 61; State Univ Iowa, MS, 63; Ohio State Univ, PhD(geol), 66. *Prof Exp:* Asst prof geol, 66-73, assoc prof, 66-77, PROF EARTH SCI, STATE UNIV NY COL PLATTSBURGH, 77- *Concurrent Pos:* State Univ NY Res Found grants-in-aid, 67-71. *Mem:* Soc Econ Paleont & Mineral; Paleont Soc; Sigma Xi. *Res:* Air pollution; advanced wood fuel combustion; lake management; application of biometrical techniques to the study of evolution of invertebrate fossils, particularly conodonts; paleoecology; carbonate petrology; studies of hydrology, earth crustal movements, and biomass combustion. *Mailing Add:* 15480 Overton Dr Beaverton OR 97006

BARNETT, THOMAS BUCHANAN, b Lewisburg, Tenn, May 27, 19; m 44; c 3. MEDICINE, PHYSIOLOGY. *Educ:* Univ Tenn, BA, 44; Univ Rochester, MD, 49. *Prof Exp:* Instr internal med, Univ Rochester, 51-52; from instr to assoc prof, 52-64, PROF INTERNAL MED, SCH MED, UNIV NC, CHAPEL HILL, 64- *Concurrent Pos:* Vis prof & researcher, Copenhagen Univ, 66-67, 75-76 & 82-83. *Mem:* Am Thoracic Soc; Am Col Physicians; Am Clin & Climat Asn; Am Fedn Clin Res; Am Physiol Soc. *Res:* Pulmonary diseases and clinical pulmonary physiology; disorders of control of respiration. *Mailing Add:* Sch Med Bldg 229H Rm 724 Box 7020 Univ NC Chapel Hill NC 27599-7020

BARNETT, WILLIAM ARNOLD, b Boston, Mass, Oct 30, 41; div. ECONOMIC STATISTICS, MATHEMATICAL STATISTICS. *Educ:* Mass Inst Technol, BS, 63; Univ Calif, Berkeley, MBA, 65; Carnegie-Mellon Univ, PhD(statist), 74. *Prof Exp:* Res engr, Rocketdyne Div, Rockwell Int Corp, 63-69; res economist, Bd Govrs, Fed Reserve Syst, 73-81; PROF ECON, UNIV TEX, AUSTIN, 82- *Concurrent Pos:* Res assoc, Univ Chicago, 77-80; prin investr, NSF, 77-80; vis lectr, Univ Aix-Marsielle, 79; vis prof, Johns Hopkins Univ, 80; ed, J Econometrics, 80; assoc ed, J Bus & Econ Statist, 81-; ed, Cambridge Univ Press. *Mem:* Sigma Xi; Am Statist Asn; Econometric Soc; Am Econ Asn; Fedn Am Scientists. *Res:* Development of asymptotic theory of inference with nonlinear models; demand theory and empirical systems of consumer demand functions; construction of monetary asset demand and aggregation model; monetary theory and policy. *Mailing Add:* Dept Econ Wash Univ Campus Box 1208 St Louis MO 63130-4899

BARNETT, WILLIAM EDGAR, b Kempner, Tex, Aug 10, 34; m 54; c 3. MOLECULAR BIOLOGY. *Educ:* Southwestern Univ, Tex, BS, 56; Northwestern Univ, MS, 57; Fla State Univ, PhD(biol genetics), 61. *Prof Exp:* Instr genetics, Fla State Univ, 61; USPHS fel, Oak Ridge Nat Lab, 61-63, biologist, 63-69, sci dir genetics & develop biol, 69-79; prof, Biomed Sci, Univ Tenn, 79-86; RETIRED. *Concurrent Pos:* Vis lectr, Univ Tenn, 62-67, prof, 67-; dir, Grad Sch Biomed Sci, 77- *Mem:* AAAS; Genetics Soc Am; Am Soc Biol Chemists. *Res:* Genetic translational apparatus of cellular organelles. *Mailing Add:* Rte 3 Box 575 Harriman TN 37748

BARNETT, WILLIAM OSCAR, b Tuscola, Miss, Sept 20, 22; m 47; c 3. SURGERY. *Educ:* Univ Miss, BS, 44; Univ Tenn, MD, 46; Univ Md, 51; Am Bd Surg, dipl, 56. *Prof Exp:* Intern, Tampa Gen Hosp, Fla, 46-47; resident surg, Lutheran Hosp, Baltimore, Md, 50-53; from assoc prof to prof surg, Sch Med, Univ Miss, 67-89; CONSULT, NAT CONTINENT OSTOMY CTRS, NATMED ENTERPRISES, 89- *Concurrent Pos:* Mem, Miss Comn Hosp Care, 62-69, vchmn, 67-68, chmn, 68-69; vis lectr surg, Med Ctr, La State Univ, 67, Sch Med, Emory Univ, 67, Med Col SC, 68 & Lubbock Med Soc, 74; chmn, Am Col Surg, 70- *Mem:* Am Surg Asn; Soc Univ Surg; Soc Surg Alimentary Tract; Int Soc Surg; Am Col Surg. *Res:* Shock in late gangrenous bowel obstruction; pre-operative management of strangulation obstruction. *Mailing Add:* 1609 Pasadena Ave S No 2A St Petersburg FL 33707

BARNEY, ARTHUR LIVINGSTON, b Homer, Minn, Apr 4, 18; m 42; c 4. POLYMER CHEMISTRY. *Educ:* Middlebury Col, BS, 38; Syracuse Univ, MS, 40; Purdue Univ, PhD(org chem), 43. *Prof Exp:* Asst chem, Syracuse Univ, 38-40 & Purdue Univ, 40-41; res chemist, Cent Res Dept, E I du Pont de Nemours & co, Inc, 42-61, res assoc, Elastomer Chem Dept, 61-62, res mgr, Polymer Prod Dept, Elastomer Div, 62-81; RETIRED. *Mem:* Am Chem Soc. *Res:* Organic synthesis; fluorine chemistry; chemistry of high polymers; catalysis. *Mailing Add:* 104 Peirce Rd Wilmington DE 19803

BARNEY, CHARLES WESLEY, b Brewster, NY, Apr 17, 15; m 43; c 3. FOREST ECOLOGY, SILVICULTURE. *Educ:* Syracuse Univ, BS, 38; Univ Vt, MS, 39; Duke Univ, DF(root growth), 47. *Prof Exp:* Asst agr aide, Soil Conserv Serv, USDA, 41; from asst prof to prof, 47-85, head dept forest mgt & utilization, 53-65, EMER PROF SILVICULTURE, COLO STATE UNIV, 85- *Mem:* Soc Am Foresters; Soil Conserv Soc Am; Am Soc Photogram & Remote Sensing; Forest Hist Soc; Sigma Xi. *Res:* Arid zone forestry; seed physiology. *Mailing Add:* 1001 W Mulberry Ft Collins CO 80521

BARNEY, CHRISTOPHER CARROLL, b Sept 20, 51; m 73; c 2. TEMPERATURE REGULATION, WATER BALANCE. *Educ:* Wright State Univ, Dayton, Ohio, BS, 73; Ind Univ, Bloomington, PhD(physiol), 77. *Prof Exp:* Assoc physiol & pharmacol, Univ Fla, 79-89; asst prof, 80-86, ASSOC PROF BIOL, HOPE COL, 86- *Concurrent Pos:* Postdoctoral traineeships, NIH, 73-79; vis asst prof pharmaceut biol, Univ Fla, 84; mem, coun undergrad res, Am Physiol Soc; Rol Award, Nat Inst Diabetes & Digestive & Kidney Dis, NIH, 87-89; Rev Awards, Nat Sci Found, 88, 89 & 90. *Mem:* Sigma Xi; AAAS. *Res:* Regulatory physiology; heat stress and thirst; hypertension and temperature regulation; endocrine/metabolic effects of food and water deprivation. *Mailing Add:* Dept Biol Peale Sci Ctr Hope Col Holland MI 49423

BARNEY, DUANE LOWELL, b Topeka, Kans, Aug 3, 28; m 50; c 2. ELECTROCHEMISTRY. *Educ:* Kans State Univ, BS, 50; Johns Hopkins Univ, MA, 51, PhD(chem), 53. *Prof Exp:* Res chemist, Gen Elec Co, 53-66, mgr battery tech lab, Battery Bus Sect, 66-68, mgr eng, 68-72, mgr, Battery Bus Sect, 72-74, gen mgr, Home Laundry Eng Dept, 74-78; assoc dir, Chem Eng Div, 78-84, SR ENGR, ARGONNE NAT LAB, 78- *Mem:* Electrochem Soc; AAAS; Am Chem Soc. *Res:* Battery technology; corrosion; materials development; fission product chemistry. *Mailing Add:* 4970 Sentinel Dr Apt 205 Bethesda MD 20816

BARNEY, GARY SCOTT, b Monroe, Utah, Sept 17, 42; m 63; c 4. PHYSICAL INORGANIC CHEMISTRY. *Educ:* Brigham Young Univ, BS, 64, PhD(inorg chem), 70. *Prof Exp:* RES STAFF CHEMIST, ATLANTIC RICHFIELD HANFORD CO, 68- *Mem:* Am Chem Soc. *Res:* Plutonium process chemistry; development of chemical processes for conversion of radioactive wastes from nuclear fuel processing to forms which can be safely stored. *Mailing Add:* Rte 1 Box 5336 Richland WA 99352-9766

BARNEY, JAMES EARL, II, b Rossville, Kans, Sept 1, 26; m 46; c 2. ANALYTICAL CHEMISTRY. *Educ:* Univ Kans, BS, 46, PhD(anal chem), 50. *Prof Exp:* Mem staff, AEC, Oak Ridge, 46-47; res chemist, Stand Oil Co, Ind, 50-54, group leader, 55-60; res scientist, Spencer Chem Co, 60-62; from sr chemist to prin chemist, Midwest Res Inst, 62-65, head anal chem sect, 65-69; sect mgr, 69-74, sr sect mgr, 74-76, dept mgr, 76-77, mgr admin, Toxicol Dept, Stauffer Chem Co, 77-88; MGR TOXICOL ADMIN, CIBA-GEIGY, FARMINGTON, CONN, 88- *Concurrent Pos:* Instr, Purdue Univ, 51-53; vis lectr, Univ Kans, 61. *Mem:* Am Chem Soc. *Res:* Pesticide residue analysis; gas chromatography; applied spectroscopy; spectrophotometry; determination of anions; physical testing of polymers; rheology of gluten; spot tests; complex ions; analysis of spices; analysis of medical gases; air pollution. *Mailing Add:* 64 Cold Spring Rd Avon CT 06001

BARNGROVER, DEBRA ANNE, b Denver, Colo. HIGH DENSITY MAMMALIAN CELL CULTURE, RECOMBINANT PROTEIN PRODUCTION. *Educ:* Colo State Univ, BS, 77; Cornell Univ, PhD(nutrit biochem), 82. *Prof Exp:* Postdoctoral fel, Mass Inst Technol, Cambridge, 81-83; asst dir, Elf Biorecherches, Toulouse, France, 83-85; staff scientist, Integrated Genetics, Inc, 85-88; SR SCIENTIST, GENZYME CORP, 88- *Concurrent Pos:* Adj prof chem eng, Worcester Polytech Inst, 86- *Mem:* Am Chem Soc; Am Inst Nutrit; Int Soc Pharmaceut Eng. *Res:* Scale-up and production of therapeutic proteins using mammalian cells; bioreactor technology; automated control and monitoring systems for pilot plant fermentations; modeling of cellular energy metabolism; serum-free media. *Mailing Add:* Dept Bioeng Genzyme Corp One Mountain Rd Framingham MA 01701

BARNHARD, HOWARD JEROME, b New York, NY, July 18, 25; m 47; c 3. MEDICINE, RADIOLOGY. *Educ:* Univ Miami, BS, 44; Med Col SC, MD, 49; Am Bd Radiol, dipl, 55. *Prof Exp:* Intern, US Naval Hosp, Charleston, SC, 49-50; resident radiol, Roper Hosp, 50-51; radiologist, US Naval Hosp, Quantico, Va, 52-53; resident radiol, Roper Hosp, 53-54; from instr to asst prof radiol, Med Ctr, Univ Ark, 54-59; asst prof, Hahnemann Med Col, 59-60; chmn dept, 60-73, dir planning, 73-77, PROF RADIOL, MED CTR, UNIV ARK, LITTLE ROCK, 60- *Concurrent Pos:* Teaching fel radiol, Med Col SC, 53-54; attend radiologist, Vet Admin Hosp, Little Rock, Ark, 54-59, consult, 60- *Mem:* Radiol Soc NAm; fel Am Col Radiol; Asn Univ Radiol; Am Roentgen Ray Soc; Sigma Xi. *Res:* Osseus system; medical computer applications. *Mailing Add:* Dept Radiol Univ Ark Med Ctr Little Rock AR 72205

BARNHARDT, ROBERT ALEXANDER, b Jenkins Twp, Pa, Sept 21, 37; m 61. TEXTILE SCIENCE. *Educ:* Philadelphia Col Textiles & Sci, BSTE, 59; Inst Textile Technol, MS, 61; Univ Va, EdD, 73. *Prof Exp:* Assoc prof textile mat, Philadelphia Col Textiles & Sci, 61-64, chmn, Dept Textiles, 64-66; dean, Inst Textile Technol, 66-75, vpres res & educ, 75-87; DEAN COL TEXTILE, NC STATE UNIV, 87- *Mem:* Nat Coun Textile Educ; Am Soc Eng Educ. *Res:* Research projects related to textile science including raw material research, energy and environmental research, flammability research. *Mailing Add:* 114 Windy Rush Lane Cary NC 27511-9758

BARNHART, BARRY B, b Dallastown, Pa, Oct 7, 36; m 58; c 2. CURRICULUM DEVELOPMENT, COMPUTERS. *Educ:* Lebanon Valley Col, BS, 58; Univ NH, MS, 68. *Prof Exp:* Teacher chem, Palurgra Aneot High Sch, 58-66 & Manheim Twp High Sch, 66-71; PROF CHEM, HARRISBURG AREA COMMUNITY COL, 71- *Concurrent Pos:* Dir, Regional Computer Resource Ctr, 84-; pres, Computer Consult Serv, 84-; adj prof, Clarion State Univ Pa Grad Sch, 84-; PSTA fel, 90. *Mem:* Am Chem Soc; Asn Supv & Curric Develop. *Res:* Microcomputer impact on delivery strategies and learning by objective mastery; electrolytic initiation of cold fusion. *Mailing Add:* 3300 Cameron St Rd Harrisburg PA 17110

BARNHART, BENJAMIN J, b Winchester, Ind, July 27, 35; m 57; c 2. GENETICS. *Educ:* Ind Univ, AB, 58, AM, 59; Johns Hopkins Univ, ScD(biochem), 62. *Prof Exp:* Teaching asst zool, Ind Univ, 58-59; NIH trainee biochem, Johns Hopkins Univ, 62-63; Nat Inst Allergy & Infectious Dis fel, Lab Genetics, Univ Brussels, Belg, 63-64; WITH LIFE SCI DEPT, MIDWEST RES INST. *Concurrent Pos:* Nat Acad Sci-Nat Res Coun travel grant, Int Biophys Cong, Austria, 66. *Mem:* Am Soc Microbiol; fel Am Inst Chemists; Am Soc Cell Biologists. *Res:* Mutagenesis; microbial genetics; mammalian cell genetics; cell biology. *Mailing Add:* 8808 Waxwing Terr Gaithersburg MD 20879

BARNHART, CHARLES ELMER, b Windsor, Ill, Jan 25, 23; m 46, 73; c 3. ANIMAL NUTRITION. *Educ:* Purdue Univ, BS, 45; Iowa State Univ, MS, 48, PhD(animal nutrit), 54. *Prof Exp:* Instr, Iowa State Univ, 47-48; instr, 48-50, from asst prof to assoc prof, 50-57, assoc dir agr exp sta, 62-67, assoc dean col agr, 66-69, PROF ANIMAL HUSB, UNIV KY, 57-, DIR AGR EXP STA, 67-, DEAN & DIR COL AGR, 69- *Concurrent Pos:* pres, Southern Assoc Agr Sci, 82-83. *Mem:* Am Soc Animal Sci. *Res:* Swine management, production, breeding and nutrition research. *Mailing Add:* 1017 Turkey Foot Rd Lexington KY 40502

BARNHART, CYNTHIA, b West Point, NY, July 23, 59; m. NETWORK OPTIMIZATION, DECOMPOSITION TECHNIQUE. *Educ:* Univ Vt, BS, 81; Mass Inst Technol, MS, 85, PhD(transp & opers res), 88. *Prof Exp:* Sch engr, Bechtel Inc, 81-84; ASST PROF OPERS RES, GA INST TECHNOL, 88- *Concurrent Pos:* Mem coun, Transp Sci Sect, Opers Res Soc Am & Inst Mgt Sci, 90; panelist, Nat Res Coun, 90; NSF presidential young investr, 90. *Mem:* Opers Res Soc Am; Inst Mgt Sci. *Res:* Apply operations research methods to large-scale network flow problems arising in transportation communications and productions; linear programming. *Mailing Add:* Sch Indust & Systs Eng Ga Inst Technol Atlanta GA 30332-0205

BARNHART, DAVID M, b Wenatchee, Wash, May 28, 33; m 54; c 3. PHYSICAL CHEMISTRY. *Educ:* Univ Ore, BS, 57, MS, 59; Ore State Univ, PhD(chem), 64. *Prof Exp:* Fel, Harvey Mudd Col, 64-65; assoc prof, 65-76, PROF PHYSICS, EASTERN MONT COL, 76- *Mem:* Am Phys Soc; Am Crystallog Asn; Am Chem Soc. *Res:* Molecular structure studies by electron and x-ray diffraction. *Mailing Add:* Dept Phys Sci Eastern Mont Col Billings MT 59101

BARNHART, JAMES LEE, MAGNETIC RESONANCE IMAGING. *Educ:* Kans State Univ, PhD(physiol), 71. *Prof Exp:* ADJ ASSOC PROF, UNIV CALIF, SAN DIEGO, 78- *Res:* Physiology; biochemistry. *Mailing Add:* Radiol Res Lab H 756 Univ Hosp UCSD 225 W Dickinson St San Diego CA 92103

BARNHART, JAMES WILLIAM, b Wadena, Minn, Apr 23, 35; m 56; c 3. BIOCHEMISTRY. *Educ:* Columbia Union Col, BA, 57; Univ Md, PhD(biochem), 63. *Prof Exp:* CHEMIST, DOW CHEM CO, 62- *Res:* Biochemistry and methodology of lipids and their relationship to atherosclerosis; drug metabolism; trace drug analysis; bioavailability of drugs. *Mailing Add:* Marion Merrell Dow Inc Box 68470 Indianapolis IN 46268-0470

BARNHART, MARION ISABEL, b Webb City, Mo, Sept 23, 21. PHYSIOLOGY. *Educ:* Univ Mo, AB, 44, PhD, 50; Northwestern Univ, MS, 46. *Prof Exp:* Instr zool, Univ Mo, 46-49; from instr to assoc prof, 50-67, PROF PHYSIOL, SCH MED, WAYNE STATE UNIV, 67- *Mem:* AAAS; Am Soc Hemat; Am Physiol Soc; Soc Exp Biol & Med; Am Acad Neurol; Am Asn Blood Banks. *Res:* Cell physiology of microvasculature, liver, bone marrow, spleen and blood in relation to hemostasis, thrombosis and fibrinolysis; inflammation; pathophysiology of platelets and hemolytic anemias; scanning electron microscopy applied to medicine; transfusion medicine. *Mailing Add:* 540 E Canfield Detroit MI 48201

BARNHILL, MAURICE VICTOR, III, b Rocky Mount, NC, Mar 16, 40. THEORETICAL PARTICLE PHYSICS. *Educ:* Univ NC, Chapel Hill, BS, 62; Stanford Univ, MS, 65, PhD(physics), 67. *Prof Exp:* Res assoc physics, Univ Va, 67-68; from asst prof to assoc prof, 68-87, PROF PHYSICS, UNIV DEL, 87- *Mem:* Am Phys Soc; Am Asn Physics Teachers; AAAS. *Res:* High energy theory; bag models, including electromagnetic form factors and recoil, and in fermion mass matrices. *Mailing Add:* Dept Physics Univ Del Newark DE 19716

BARNHILL, ROBERT E, b Lawrence, Kans, Oct 31, 39; m 63. MATHEMATICAL ANALYSIS. *Educ:* Univ Kans, BA, 61; Univ Wis, MA, 62, PhD(math), 64. *Prof Exp:* From asst prof to assoc prof, 64-71, PROF MATH, UNIV UTAH, 71- *Mem:* Soc Indust & Appl Math; Asn Comput Mach. *Res:* Numerical analysis; computer aided geometric design. *Mailing Add:* Dept Computer Sci & Eng Col Eng & Appl Sci Ariz State Univ Tempe AZ 85207-5406

BARNHISEL, RICHARD I, b Peru, Ind, Mar 1, 38; m 58; c 1. SOIL MINERALOGY. *Educ:* Purdue Univ, BS, 60; Va Polytech Inst, MS, 62, PhD(soil mineral), 65. *Prof Exp:* Asst prof soil mineral, 64-69, assoc prof, 69-77, PROF SOIL MINERAL, UNIV KY, 77- *Mem:* Am Soc Agron; Soil·Sci Soc Am; Clay Minerals Soc. *Res:* Formation and stability of hydroxy-aluminum interlayers in clay minerals; weathering of primary minerals in clay formation; reclamation methods of surface-mined coal spoils. *Mailing Add:* 3134 Montavesta Rd Lexington KY 40502

BARNOFF, ROBERT MARK, b Punxsutawney, Pa, Aug 28, 26; m 54; c 4. CIVIL ENGINEERING. *Educ:* Pa State Univ, BS, 51, MS, 56; Carnegie Inst Technol, PhD(civil eng), 66. *Prof Exp:* Bridge designer, Gannett, Fleming, Corddrg & Carpenter, 53-54; from instr to assoc prof, 55-70, PROF CIVIL ENG, PA STATE UNIV, 70-, HEAD, 80- *Concurrent Pos:* NSF fel, Carnegie Inst Technol, 63-; consult eng, 56-; mem trans res bd, Nat Acad Sci. *Mem:* Fel Am Soc Civil Engrs; Am Soc Eng Educ; Am Concrete Inst; Am Soc Testing & Mat; Sigma Xi. *Res:* Static and fatigue strength of structures; structural behavior of prestressed concrete. *Mailing Add:* 606 Nimitz Ave State College PA 16801

BARNOSKI, MICHAEL K, b Williamsport, Pa, Aug 19, 40; m 63; c 2. SOLID STATE PHYSICS. *Educ:* Univ Dayton, BSEE, 63; Cornell Univ, MS, 65, PhD(solid state physics), 68. *Prof Exp:* Asst res engr, Res Inst, Univ Dayton, 63; res asst microwave electronics, Cornell Univ, 63-65, infrared spectros, 65-68; prin sr res engr, Honeywell Radiation Ctr, 68-69; mem tech staff, Hughes Aircraft Co, 69-78, sect head, Optical Device Sect, 75, mgr optical circuits prog, Hughes Res Labs, 78-79, dir, TRW technol res ctr, 79-83; pres-chief exec officer, Plesseor Optronics Inc, 84-90; CHIEF EXEC OFFICER, UPSTATE TOOL INC, 90- *Concurrent Pos:* Consult, 83-84. *Mem:* Optical Soc Am. *Res:* Infrared spectra of solids; Fourier transform spectroscopy; coherence properties of radiation; microwave electronics; opto-electronic devices; fiber optics; integrated optics; lasers. *Mailing Add:* PO Box 1470 Pacific Palisades CA 90272

BARNOTHY, JENO MICHAEL, b Kassa, Hungary, Oct 28, 04; nat US; m 38. NUCLEAR PHYSICS, ASTROPHYSICS. *Educ:* Royal Hungarian Univ, PhD(physics), 33. *Prof Exp:* From instr to prof physics, Royal Hungarian Univ, 35-48; prof, Barat Col, 48-53; res physicist, Nuclear Instrument & Chem Corp, 53-55; OWNER & TECH DIR, FORRO SCI CO, 55-; PRES, BIOMAGNETIC RES FOUND, 60- *Honors & Awards:* Medal, Royal Hungarian Acad Sci, 39; Eotvos Medal, 48. *Mem:* Am Astron Soc; Am Phys Soc; Biophys Soc; Ger Astron Soc; Int Astron Union. *Res:* Constructed the first cosmic ray telescope; developed the FIB cosmological theory and the gravitational lens explanation of quasars and pulsars. *Mailing Add:* 833 Lincoln St Evanston IL 60201

BARNS, ROBERT L, b Wichita, Kans, Mar 11, 27; m 55; c 1. CRYSTALLOGRAPHY. *Educ:* Wichita State Univ, BS, 49, MS, 51. *Prof Exp:* Res chemist, Found Indust Res, Wichita State Univ, 50-55; mem tech staff, Phys Chem Res Dept, 55-60, mem tech staff, Crystal Chem Res Dept, 60-83, MEM TECH STAFF, OPT MAT RES DEPT, BELL LABS, 83- *Mem:* Am Crystallog Asn; Am Asn Crystal Growth; AAAS. *Res:* Combustion; electrical contacts; crystal growth; imperfections in crystals; crystal characterization; x-ray diffraction; petrography; superconductivity; semiconductors. *Mailing Add:* 63 Martins Lane Berkeley Heights NJ 07922

BARNSLEY, ERIC ARTHUR, b Birmingham, Eng, July 31, 34; c 2. MICROBIAL PHYSIOLOGY. *Educ:* Univ London, BSc, 60, PhD(biochem), 63. *Prof Exp:* Assoc prof, 77-80, PROF BIOCHEM, MEM UNIV, 80- *Mem:* Biochem Soc; Can Soc Microbiologists. *Res:* Regulation of the degradation of polycyclic armatic hydrocarbons by micro-organisms. *Mailing Add:* Dept Biochem Mem Univ St John's NF A1B 3X9 Can

BARNSTEIN, CHARLES HANSEN, b Newton, Wis, June 25, 25; m 54; c 3. PHARMACEUTICAL CHEMISTRY. *Educ:* Univ Wis, BSc, 52, MSc, 55, PhD(pharm), 60. *Prof Exp:* From instr to assoc prof pharm, Idaho State Col, 56-63; assoc prof, Univ Wis-Milwaukee, 63-70; assoc dir, Nat Formulary, Am Pharmaceut Asn, 70-75; sr scientist, 75-78, DEP DIR, DRUG STANDARDS DIV, US PHARMACOPEIAL CONV, INC, 78- *Mem:* AAAS; Am Chem Soc; Am Pharmaceut Asn; Acad Pharmacuet Sci. *Res:* Organic analytical chemistry; physical pharmacy and chemistry; chemical kinetics. *Mailing Add:* 16900 Freedom Way Rockville MD 20853

BARNTHOUSE, LAWRENCE WARNER, b Cincinnati, Ohio, Oct 19, 46. ECOLOGY. *Educ:* Kenyon Col, AB, 68; Univ Chicago, PhD(biol), 76. *Prof Exp:* RES ASSOC AQUATIC ECOL, OAK RIDGE NAT LAB, 76- *Mem:* AAAS; Ecol Soc Am; Am Soc Limnol & Oceanog; Sigma Xi. *Res:* Application of ecological theory to environmental impact assessment. *Mailing Add:* 1069 W Outer Dr Oak Ridge TN 37830-8635

BARNUM, DENNIS W, b Portland, Ore, June 30, 31; div; c 2. CHEMISTRY. *Educ:* Univ Ore, BA, 53, MA, 55; Iowa State Univ, PhD(chem), 57. *Prof Exp:* Chemist, Shell Develop Co, 57-64; from asst prof to assoc prof, 64-75, PROF CHEM, PORTLAND STATE UNIV, 75- *Mem:* Am Chem Soc. *Res:* Coordination chemistry; geochemistry; soils. *Mailing Add:* Dept Chem Portland State Univ PO Box 751 Portland OR 97207

BARNUM, DONALD ALFRED, b Ont, June 20, 18; m 42; c 3. VETERINARY SCIENCE, MICROBIOLOGY. *Educ:* Ont Vet Col, DVM, 41; Univ Toronto, DVPH, 49, DVSc, 52. *Prof Exp:* Bacteriologist, Ont Dept Health Labs, 41-43; lectr bact, 45-47, from asst prof to prof, 47-84, head, Dept Vet Bact, 64-69, chmn, Dept Vet Microbiol & Immunol, 69-80, EMER PROF BACT, ONT VET COL, UNIV GUELPH, 84- *Mem:* Am Soc Microbiol; fel Am Acad Microbiol; Can Vet Med Soc; Am Vet Med Asn; Path Soc Gt Brit & Ireland. *Res:* Veterinary bacteriology and mycology; bacterial infections of animals, especially of the mammary glands of dairy cattle and gastrointestinal tract of the young. *Mailing Add:* 28 Colborn Guelph ON N1G 2M5 Can

BARNUM, EMMETT RAYMOND, b Rushville, Nebr, May 4, 13; m 41; c 1. ORGANIC CHEMISTRY. *Educ:* Nebr State Teachers Col, BS, 37; Univ Nebr, MS, 39, PhD(org chem), 42. *Prof Exp:* Asst, Univ Nebr, 37-38; res supvr, 52-75, CHEM CONSULT, SHELL DEVELOP CO, 77- *Mem:* Am Chem Soc. *Res:* Organic synthesis; lubrication and industrial oil; sulfur compounds to reduce wear in lubricating oils; synthetic lubricants; polymers; petrochemicals; chemical intermediates; polymers for contact lens. *Mailing Add:* 6100 Estates Dr Oakland CA 94611

BARNUM, JAMES ROBERT, b Palo Alto, Calif, Feb 19, 44; m 68; c 2. RADAR ENGINEERING, OVER-THE-HORIZON RADAR. *Educ:* Stanford Univ, BS, 65, MS, 67, Eng, 69, PhD(elec eng), 70. *Prof Exp:* Res asst elec eng, Stanford Univ, 65, res assoc radiosci, 70; res engr, 70-74, sr res engr, 74-81, prog mgr, 81-89, staff scientist, 89-91, PRIN SCIENTIST, SRI INT, 91- *Mem:* sr mem Inst Elec & Electronics Engrs. *Res:* High frequency backscatter from sea and land; ionospheric propagation; polarization of radio waves; antenna theory; wide-aperature antennas; detection of aircraft, ships, and storms at sea by backscattering of high frequency radio waves; ocean surveillance; multiple-sensor interaction; HF-radar systems engineering and development. *Mailing Add:* SRI Int Bldg 301-77 Menlo Park CA 94025

BARNWELL, FRANKLIN HERSHEL, b Chattanooga, Tenn, Oct 14, 37; m 59; c 1. COMPARATIVE PHYSIOLOGY, BIOLOGICAL RHYTHMS. *Educ:* Northwestern Univ, BA, 59, PhD(biol sci), 65. *Prof Exp:* Instr biol sci, Northwestern Univ, 64, res assoc, 65-67; asst prof, Univ Chicago, 67-70; from asst prof to assoc prof, 70-84, PROF ECOL, UNIV MINN, MINNEAPOLIS, 84-, HEAD, DEPT ECOL, EVOLUTION & BEHAV, 86-

Concurrent Pos: Mem fac, Orgn Trop Studies, Costa Rica, 66, 69-71 & 85; bd dirs, Orgn Trop Studies, 85-; gov bd, Nat Conferences Undergrad Res, 90- *Mem:* Am Soc Zoologists; Crustacean Soc (secy, 91-); Int Soc Chronobiol; Soc Res Biol Rhythms; fel Linnean Soc. *Res:* Persistent daily and tidal rhythms; invertebrate behavior; biochemical systematics, zoogeography and functional morphology of marine crustacea. *Mailing Add:* Dept Ecol Evolution & Behav Univ Minn 318 Church St SE Minneapolis MN 55455

BARNWELL, THOMAS OSBORN, JR, b Charleston, SC, Mar 12, 47; m 72; c 1. HYDROLOGY & WATER RESOURCES, COMPUTER SCIENCES. *Educ:* Clemson Univ, BS, 69, MS, 71. *Prof Exp:* Sanit engr, Surveillance & Anal Div, Region IV, 71-77, CIVIL ENGR, ASSESSMENT BR, ENVIRON RES LAB, OFF RES & DEVELOP, US ENVIRON PROTECTION AGENCY, 77- *Mem:* Am Soc Civil Engrs; Int Asn Water Pollution Res & Control; Water Pollution Control Fedn; Am Geophys Union; Int Soc Econ Modeling. *Res:* Water quality modeling and management technology; nonpoint source pollution control; expert systems. *Mailing Add:* US Environ Protection Agency Environ Res Lab Col Sta Rd Athens GA 30613-7799

BAROCZY, CHARLES J(OHN), b New York, NY, Apr 21, 24; div; c 4. MECHANICAL ENGINEERING. *Educ:* Pratt Inst, BME, 47; Harvard Univ, MS, 48. *Prof Exp:* Design engr, Ebasco Serv, Inc, 48-50; develop engr anal & develop, Gen Elec Co, 50-57; self employed, 57-58; res specialist, Atomics Int Div, NAm Aviation, Inc, 58-70, mem tech staff, NAm Rockwell Corp, 70-73; reactor engr, US Atomic Energy Comn, 73-74; staff engr, Gen Atomic Co, 74-83; CONSULT, 84. *Mem:* Am Soc Mech Engrs. *Res:* Steam generators; heat exchangers; heat transfer and fluid flow in two phase flow, boiling and condensing. *Mailing Add:* 13735 Franciscan Dr Sun City West AZ 85375

BARON, ARTHUR L, b New York, NY; m 65; c 2. MATERIALS SCIENCE ENGINEERING, ORGANIC CHEMISTRY. *Educ:* Columbia Univ, BA, 57; Purdue Univ, MS, 59; Univ Hamburg, WGermany, PhD(chem), 63. *Prof Exp:* Res chemist, Exxon Corp, 63-66; group leader, Celanese Corp, 66-72; dir, res & develop, Mobay Chem Corp, 72-84; vpres res & develop, Pennwalt Corp, 84-89; VPRES RES & DEVELOP, ATOCHEM NAM, 90- *Mem:* Am Chem Soc; WGerman Chem Soc; Soc Plastics Engrs; Soc Chem Indust; Asn Res Dirs. *Res:* Polymer science and technology; thermoplastic, thermosetting and elastomeric materials; coatings and adhesives; composites. *Mailing Add:* 30 Randwood Dr Getzville NY 14068

BARON, DAVID ALAN, b Chicago, Ill, Aug 14, 51; m 83; c 2. PHYSIOLOGY & PHARMACOLOGY. *Educ:* Univ Chicago, BA, 72, PhD(anat), 79. *Prof Exp:* Postdoctoral fel path, Med Univ SC, 79-82; res scientist pharmacol, Drug Sci Found, 82-84; asst prof pharmacol, Med Univ SC, 84-89; RES SCIENTIST PATH, SEARLE RES & DEVELOP, 89- *Mem:* AAAS; Am Soc Cell Biol; Electron Micros Soc Am. *Res:* Quantitative morphologic analysis of cardiovascular and gastrointestinal cell physiology and pharmacology; electron microscopy of target organs in drug toxicology studies. *Mailing Add:* Dept Path Searle Res & Dev 4901 Searle Pkwy Skokie IL 60077

BARON, FRANK A, organic chemistry; deceased, see previous edition for last biography

BARON, HAZEN JAY, b Detroit, Mich, March 12, 34; m 54; c 2. CARIES RESEARCH, PERIODONTAL DISEASE. *Educ:* Wayne State Univ, BS, 55; Northwestern Univ, DDS, 58, PhD(path), 68. *Prof Exp:* Lectr path, Northwestern Univ, 62-64; asst dir dent therapeut, Warner-Lambert Co, 64-69, dir dent sci, 69-80; dir dent care, 80-87, VPRES DENT RES & DEVELOP, JOHNSON & JOHNSON CO, 87- *Concurrent Pos:* Consult, Am Dent Asn, 72- *Mem:* Am Dent Asn; Am Asn Dent Res; Am Acad Oral Path; Am Acad Periodont. *Res:* Consumer oriented oral hygiene products designed to prevent dental caries and periodontal disease. *Mailing Add:* Johnson & Johnson 501 George St New Brunswick NJ 08903

BARON, JEFFREY, b Brooklyn, NY, July 10, 42; m 65; c 3. TOXICOLOGY, CHEMICAL CARCINOGENESIS. *Educ:* Univ Conn, BS, 65; Univ Mich, PhD(pharmacol), 69. *Prof Exp:* Res fel biochem, Southwestern Med Sch, Univ Tex, Dallas, 69-71, res asst prof biochem & pharmacol, 71-72; from asst prof to assoc prof, 72-80, PROF PHARMACOL, UNIV IOWA, 80- *Concurrent Pos:* Res Career Develop Award, Nat Inst Arthritis, Diabetes, Digestive & Kidney Dis, NIH, 75-80; mem, chem path study sect, NIH, 83-87, environ health sci rev comt, Nat Inst Environ Health Sci, 90-94. *Mem:* AAAS; Am Soc Pharmacol & Exp Therapeut; Am Soc Biochem & Molecular Biol; Am Asn Cancer Res; Soc Toxicol; Int Soc Study Xenobiotics. *Res:* Biochemical pharmacology and toxicology of xenobiotic metabolism and chemical carcinogenesis; immunohistochemistry of drug- and carcinogen-metabolizing enzyme systems; histopathologic toxicology. *Mailing Add:* Dept Pharmacol Univ Iowa 2-270 Bowen Sci Bldg Iowa City IA 52242

BARON, JUDSON R(ICHARD), b New York, NY, July 28, 24; m 49; c 2. AERONAUTICAL ENGINEERING. *Educ:* NY Univ, BAeroE, 47; Mass Inst Technol, SM, 48, ScD, 56. *Prof Exp:* Engr, Chance Vought Aircraft, Stratford, Conn, 47; res staff, supersonic lab, Mass Inst Technol, 48-52, asst, dept aeronaut, 52-56, from asst prof to prof, 56-89, EMER PROF & SR LECTR AERONAUT & ASTRONAUT, MASS INST TECHNOL, 89- *Concurrent Pos:* Mem, USAF Sci Adv Bd, 88- *Mem:* Assoc fel Am Inst Aeronaut & Astronaut; Sigma Xi. *Res:* Gas dynamics; aerothermodynamics; computational fluid dynamics. *Mailing Add:* Dept Aeronaut & Astronaut 37-467 Mass Inst Technol Cambridge MA 02139

BARON, LOUIS SOL, b New York, NY, Jan 2, 24; m 61; c 1. MICROBIAL GENETICS. *Educ:* City Col New York, BS, 47; Univ Ill, MS, 48, PhD(bact), 51. *Prof Exp:* Asst bact, Univ Ill, 49-52; bacteriologist, 52-56, CHIEF DEPT BACT IMMUNOL, DIV IMMUNOL, WALTER REED ARMY INST RES, 56- *Concurrent Pos:* Squibb & Sons fel, 51; President's fel, Soc Bact, 57; Army rep, Genetics Study Sect, USPHS, 58-; spec lectr, Dept Microbiol, Sch

Med, George Wash Univ, 61; res fel, Stanford Univ, 61-62; mem subcomt maintenance of genetic stocks, NSF, 64-68, chmn, 68-72; ed, J Bact, 65-70 & Proc Soc Exp Biol Med, 65-72; mem genetics fac, Grad Prog, NIH, 67-69; mem adv comt, Am Type Cult Collection, 69-; prof lectr, Sch Med, Georgetown Univ, 69-75; Coun biol ed. *Mem:* Am Soc Microbiol; Am Asn Immunologists; Biophys Soc; Am Soc Molecular Biol & Biochem; Am Acad Microbiol. *Res:* Bacterial genetics, virulence and immunology; enteric diseases; molecular biology. *Mailing Add:* Dept Bact Immunol Walter Reed Army Inst Res Washington DC 20307

BARON, MELVIN L(EON), b New York, NY, Feb 27, 27; m 50; c 2. APPLIED MECHANICS, STRUCTURAL ENGINEERING. *Educ:* City Col New York, BCE, 48; Columbia Univ, MS, 49, PhD(appl mech), 53. *Prof Exp:* Struct designer, Corbett-Tinghir Co, 49-50; res assoc civil eng, Columbia Univ, 51-53, asst prof, 53-57; chief engr, 57-60, assoc partner, 60-64, PARTNER & DIR RES, APPL MECH & STRUCT ENG, WEIDLINGER ASSOCS, CONSULT ENGR, 64- *Concurrent Pos:* From adj asst prof to adj assoc prof, Columbia Univ, 57-61, adj prof, 61-; mem US Nat Comt Theoret & Appl Mech, 74-82; chmn adv panel, Eng Mech Prog, NSF, 77-78. *Honors & Awards:* Spirit of St Louis Jr award, Am Soc Mech Engrs, 58; J James Croes Medal, 63, Am Soc Civil Engrs, Walter L Huber Res Prize, 66, Arthur M Wellington Prize, 69 & Nathan M Newmark Medal, 77; Except Pub Serv Medal, Defense Nuclear Agency, US Dept Defense, 85. *Mem:* Nat Acad Eng; AAAS; fel Am Soc Civil Engrs; fel Am Soc Mech Engrs; NY Acad Sci. *Res:* Applied mechanics and mathematics; mathematical methods and computer technology; shock propagation in elastic and inelastic media; theory of vibrations; theory of structures; nuclear weapons effects; hydrodynamics; sound radiation from submerged structures. *Mailing Add:* 3801 Hudson Manor Terr Apt 6H Bronx NY 10463

BARON, ROBERT, b Chicago, Ill, May 16, 32; div; c 2. PHYSICS OF INFRARED DETECTORS, CURRENT TRANSPORT IN SEMICONDUCTORS. *Educ:* Univ Chicago, BA, 52, MS, 55, PhD(physics), 62. *Prof Exp:* SR SCIENTIST, SOLID STATE PHYSICS, HUGHES RES LABS, 62- *Mem:* Am Inst Physics; Am Phys Soc; Inst Elec & Electronics Engrs. *Res:* Behavior of insulators and semi-insulators under high injection current conditions; physics of ion-implanted layers in semiconductors; physics of current transport and materials for extrinsic infrared detectors. *Mailing Add:* Hughes Res Lab MS RL63 3011 Malibu Canyon Rd Malibu CA 90265

BARON, ROBERT WALTER, b Winnipeg, Man, Aug 6, 47; m 70; c 3. IMMUNOLOGY, PARASITOLOGY. *Educ:* Univ Man, BSc, 68, MSc, 71; McGill Univ, PhD(immmunoparasitol), 76. *Prof Exp:* Res assoc immunol, Univ Man, 76-78; RES SCIENTIST IMMUNOCHEM, AGR CAN, 78- *Mem:* Am Soc Parasitologists; Can Asn Advan Vet Parasitol. *Res:* Characterization of immune response to arthropods; development of vaccines for protection of livestock against arthropod infestation. *Mailing Add:* Lethbridge Res Ctr PO Box 3000 Main Lethbridge AB T1J 4B1 Can

BARON, SAMUEL, b New York, NY, July 27, 28; m 49; c 5. EXPERIMENTAL ONCOLOGY, INTERFERON. *Educ:* New York Univ, BA, 48, MD, 52. *Prof Exp:* Intern, Montefiore Hosp, New York, 53; asst virol, Univ Mich, Ann Arbor, 55; sect head cell virol, NIH, Bethesda, Md, 68-75; CHMN, DEPT MICROBIOL, MED BR, UNIV TEX, GALVESTON, 75- *Concurrent Pos:* Ed, Tex Reports, Med Br, Univ Tex, 77-86 & med microbiol, Churchill Livingston Inc, New York, NY, 91- *Honors & Awards:* USPHS Meritorious Serv Award, NIH, 70. *Mem:* Am Asn Immunologists; Soc Exp Biol & Med; Am Soc Microbiol; Infectious Dis Soc Am; Asn Med Sch Microbiol Chmn; Soc Gen Microbiol; Sigma Xi. *Res:* Host defenses against viruses and cancer; virology; cancer; interferon; immunity. *Mailing Add:* Dept Microbiol Univ Tex Med Br Galveston TX 77550

BARON, SEYMOUR, b New York, NY, Apr 5, 23; m 50; c 2. ENGINEERING MANAGEMENT, RESEARCH DEVELOPMENT. *Educ:* Johns Hopkins Univ, BS, 44, MS, 47; Columbia Univ, PhD(chem eng), 50. *Prof Exp:* Lab researcher, US Indust Chem Co, 44-47; chief engr, Burns & Roe, Inc, 50-64, vpres, 64-75, sr vpres, 75-76, sr corp vpres eng & technol, 76-84; ASSOC DIR, BROOKHAVEN NAT LAB, 84- *Concurrent Pos:* Adj prof, Columbia Univ & Polytech Inst Brooklyn, 66-68; mem, NJ Comn Radiation Protection, 78-; mem bd trustees, Argonne Univ Asn, 76-, bd liaison, Reactor & Safety Anal, 76- & Ad Hoc Comt, 80-; lectr, Univ Fla, 80- *Mem:* Nat Acad Eng; fel Am Soc Mech Engrs; Am Nuclear Soc; fel Am Soc Nuclear Engrs; Am Inst Chem Engrs; fel AAAS. *Res:* Energy conversion; advanced power technology; electric power plant design; net energy analysis; economics; transfer of technology to developing countries. *Mailing Add:* Brookhaven Nat Lab Upton NY 11973

BARON, WILLIAM ROBERT, b Providence, RI, May 22, 47; m 72. HISTORICAL CLIMATOLOGY, HISTORICAL NATURAL DISASTER IMPACT ASSESSMENT. *Educ:* Allegheny Col, BA, 69; Univ RI, MA, 70; Univ Maine, Orono, PhD(interdisciplinary), 80. *Prof Exp:* Res assoc quaternary studies, Univ Maine, 78-82; sr res assoc, Ctr Colo Plateau Studies, 83-91; res fel hist climat, Old Sturbridge Village, Mass, 90; CUR, HIST CLIMATE RECORDS OFF, NORTHERN ARIZ UNIV, 88-, ASST PROF HIST & QUATERNARY STUDIES, 89- *Concurrent Pos:* Adj res assoc, Inst Quaternary Studies, Univ Maine, 83-91. *Mem:* Asn Am Geographers; Am Soc Environ Hist; Agr Hist Soc; Am Hist Soc; Orgn Am Historians. *Res:* Reconstruction of historical world climates using instrument and qualitative data; environmentally induced stress on society; environmental and earth science records from historical data. *Mailing Add:* Hist Climate Records Off Northern Ariz Univ Box 5613 Flagstaff AZ 86011

BARONA, NARSES, b Cali, Colombia, Feb 14, 32; m 58; c 2. CHEMICAL ENGINEERING. *Educ:* Univ Valle, Colombia, ChE, 53; Carnegie Inst Technol, MSc, 56; Univ Houston, PhD(chem eng), 64. *Prof Exp:* Prof chem eng, Univ Valle, Colombia, 53-61, secy, Sch Chem Eng, 56-57, head dept, 58-61; RES ENGR & CONSULT, ETHYL CORP, 63- *Concurrent Pos:* Hon prof, Cent Univ Ecuador, 60; prof, Am Inst Chem Engrs, 83- *Mem:* Am Chem Soc; fel Am Inst Chem Engrs; Colombian Inst Chem Engrs; Peruvian Inst Chem Engrs. *Res:* Reaction kinetics; mass transfer; process design and evaluation; mass transfer with chemical reaction; mathematical modeling and simulation of chemical processes; physical properties of materials; batch processing (chemical engineering). *Mailing Add:* 2947 Myrtle Ave Baton Rouge LA 70806

BARONDES, SAMUEL HERBERT, b Brooklyn, NY, Dec 21, 33; m 63; c 2. PSYCHIATRY, NEUROBIOLOGY & CELL BIOLOGY. *Educ:* Columbia Col, AB, 54; Columbia Univ, MD, 58. *Prof Exp:* From intern to asst resident, Peter Bent Brigham Hosp, 58-60; clin assoc, NIH, 60-63; resident, McLean & Mass Gen Hosps, 63-66; from asst to assoc prof psychiat & molecular biol, Albert Einstein Col Med, 66-69; prof psychiat, Univ Calif, San Diego, 69-86; PROF PSYCHIAT & NEUROSCI, UNIV CALIF, SAN FRANCISCO, 86-, CHMN, DEPT PSYCHIAT, 86- *Concurrent Pos:* NIH fel, 60-63; teaching fel psychiat, Harvard Med Sch, 63-66; NIH career develop award, 66-69; mem alcoholism & alcohol probs rev comt, NIMH, 67-70; mem neurobiol rev comt, NSF, 70-73; mem neurobiol merit rev bd, Vet Admin Cent Off, 72-75; mem bd dirs, Found Fund for Res in Psychiat, 74-76; assoc ed, J Neurobiol, 70-77; Fogarty Int scholar-in-residence, NIH, 79-; scholar rev comt, McKnight Found, 76-; bd sci advs, La Jolla Cancer Res Found, 84-; coun mem, Am Soc Cell Biol, 81-84; bd dir, McKnight Endowment Fund Neurosci, 86-; dir, Langley Porter Psychiat Inst, Univ Calif, San Francisco, 86-; sci adv comn, Charles E Culpeper Found, 87- *Honors & Awards:* J Elliott Royer Award, 90. *Mem:* Inst Med-Nat Acad Sci; Am Soc Cell Biol; Soc Neurosci; Am Soc Biol Chemists; Psychiatric Res Soc; fel Am Col Neuropsychopharm; fel AAAS. *Res:* Endogenous lectins; cell adhesion; genetics of mental illness. *Mailing Add:* Dept Psychiat Univ Calif San Francisco CA 94143-0984

BARONDESS, JEREMIAH A, b New York, NY, June 6, 24; m 52; c 1. INTERNAL MEDICINE. *Educ:* Johns Hopkins Univ, MD, 49; Am Bd Internal Med, dipl, 56 & 75. *Hon Degrees:* DSc, Albany Med Col, 78. *Prof Exp:* Intern, Johns Hopkins Hosp, 49-50, asst med, 50-51; res fel, Dept Virol, Sch Med, Univ Pa, 51-53; asst resident med, New York Hosp, 53-54, chief resident, 54-55; instr med, Med Col, Cornell Univ, 54-59, from asst prof to assoc prof clin med, 59-71, clin prof med, 71-79, prof clin med, 79-86, assoc chmn, Dept Med, 82-90; PRES, NY ACAD MED, 90- *Concurrent Pos:* Attend physician, NY Hosp, 71-, chief, Pvt Med Serv, 73-82; vis prof med, Univ PR, 72, Univ Ill, 74, Univ Va, 76, Mayo Med Sch & Found, 78, Univ Iowa, 79, Univ Tex, Houston, 86-91, NY Med Col, 91; vchmn bd gov, Am Col Physicians, 72-73, chmn, 73-74; trustee, Johns Hopkins Univ, 77-; chief consult, Med Necessity Prog, Nat Blue Cross/Blue Shield Asn, 83-86; Irene F & I Roy Psaty distinguished prof clin med, Med Col, Cornell Univ, 86-89, William T Foley distinguished prof, 89-90, adj prof, 90- *Honors & Awards:* Alfred E Stengel Award, Am Col Physicians, 83. *Mem:* Inst Med-Nat Acad Sci; Am Clin & Climat Asn; master Am Col Physicians (pres-elect, 77-79, pres, 78-79); Am Fedn Clin Res; fel AAAS; Am Pub Health Asn; Asn Am Physicians; NY Acad Sci; Int Soc Internal Med; fel Royal Soc Med London. *Res:* Medical education and training. *Mailing Add:* NY Acad Med Two E 103rd St New York NY 10029

BARONE, FRANK CARMEN, b Syracuse, NY, July 5, 49; div; c 2. NEUROSCIENCE, PHYSIOLOGY. *Educ:* Syracuse Univ, BA, 73, PhD(biopsychol), 78. *Prof Exp:* USPHS trainee & res asst, Syracuse Univ, 73-77, res assoc, Brain Res Lab, 77-82, asst prof psychol, 78-82; SCIENTIST & ASSOC FEL, SMITH KLINE BEECHAM PHARMACEUT, 82- *Concurrent Pos:* Vis researcher, NSF Joint US-Repub China Res Proj, 77-; res assoc, NIH grant, 78-82; ed assoc sci journals, Brain Res Publ, Inc, 78-82; adj asst prof physiol, Temple Univ Sch Med, 87. *Mem:* Soc Neurosci; Am Physiol Soc; Am Gastroenterol Asn. *Res:* Neurophysiology and behavior; gastrointestinal pharmacology; cerebrovascular diseases; development of animal models of CNS ischemia/CNS trauma in order to discover drugs that can be useful as therapy in man. *Mailing Add:* Smith Kline Beecham Pharmaceut PO Box 1539 King of Prussia PA 19406-0939

BARONE, JOHN A, b Dunkirk, NY, Aug 30, 24; m 47. ORGANIC CHEMISTRY. *Educ:* Univ Buffalo, BA, 44; Purdue Univ, MS, 48, PhD(chem), 50. *Prof Exp:* Asst chem, Purdue Univ, 47-48; from instr to assoc prof, 50-62, dir, NSF In-Serv Inst, 61-68, dir res & grad sci, 63-66, vpres planning, 66-70, PROF CHEM, FAIRFIELD UNIV, 62-, PROVOST, 70- *Concurrent Pos:* Dir, Jesuit Res Coun Am, 63-70, chmn, 68-70; consult, Conn Regional Med Prog, 70-76; proj mgr, Housing & Urban Develop-New Rural Soc Contracts, 72-76; Dir, Blue Cross-Blue Shield Conn, 73; mem, Statewide Health Coord Coun, 79-87, chmn, 83-87; mem, Conn Health & Educ Facil Auth, 87- *Mem:* AAAS; Am Chem Soc; Sigma Xi. *Res:* Heterocyclic and medicinal chemistry; organohalogen compounds; academic and research administration; resource management. *Mailing Add:* Fairfield Univ Fairfield CT 06430-7524

BARONE, JOHN B, US citizen. ECOLOGY. *Educ:* Kent State Univ, BS, 72, Univ Calif, Davis, PhD(ecol), 80. *Prof Exp:* Environ engr, Firestone Tire & Rubber Co, 72-73; environ specialist, Ohio Environ Protection Agency, 73-74; res asst, Crocker Nuclear Lab, Univ Calif, Davis, 74-79, lectr meteorol, Land, Air & Water Resources Dept, 79-80; proj mgr, 81-87, TECH DIR, WESTON, 87- *Mem:* Air Pollution Control Asn. *Res:* Ambient air monitoring and emission testing; collection techniques and analysis of fine aerosols; statistical analysis of environmental data; evaluation of transport and disperson of air pollutants; meteorological monitoring, measurement and data analysis; micro and mesoscale meteorological studies. *Mailing Add:* Weston One Weston Way Bldg 5 West Chester PA 19380

BARONE, LEESA M, b Fitchburg, Mass, Feb 15, 58. MOLECULAR BIOLOGY. *Educ:* Smith Col, BA, 80; Boston Univ, PhD(biochem), 87. *Prof Exp:* Postdoctoral fel osteoarthritis, Med Ctr, Tufts Univ, 87-88; postdoctoral fel, Children's Hosp & Med Ctr, Univ Mass, 88-90, INSTR BONE & CARTILAGE DEVELOP, MED CTR, UNIV MASS, 90- *Concurrent Pos:* Grant, Orthop Res & Educ Found, 88-90. *Mem:* Am Soc Cell Biol; Am Soc Bone & Mineral Res; Orthop Res Soc. *Mailing Add:* Dept Cell Biol Univ Mass Med Ctr 55 Lake Ave N Worchester MA 01655

BARONE, LOUIS J(OSEPH), b Brooklyn, NY, Oct 10, 29; m 59; c 2. PRINCIPAL LICENSING ENGINEERING. *Educ:* Polytech Inst Brooklyn, BChE, 51, NY Univ, MS, 58, MChE, 58. *Prof Exp:* Chem engr, res & develop, Cent Res Lab, Allied Chem & Dye Corp, 51-54; sr res engr, 54-72, sr proj engr, 72-79, PRIN LICENSING ENGR, STAUFFER CHEM CO, 80- *Mem:* Am Chem Soc; Am Inst Chem Engrs. *Res:* Plant design; process development. *Mailing Add:* 23 Biltom Rd White Plains NY 10607

BARONE, MILO C, b Throop, Pa, Dec 4, 41; m 82. ANIMAL PHYSIOLOGY. *Educ:* Univ Scranton, BS, 63; John Carroll Univ, MS, 65; St Bonaventure Univ, PhD(physiol), 68. *Prof Exp:* ASST PROF BIOL, FAIRFIELD UNIV, 68- *Mem:* AAAS; NY Acad Sci. *Res:* Reptilian hematology; seasonal variations in serum proteins and other blood properties in Pseudemys and Chrysemys turtles. *Mailing Add:* Dept Biol Fairfield Univ Fairfield CT 06430

BARONET, CLIFFORD NELSON, b Quebec, Que, Oct 17, 34; m 60; c 1. MECHANICAL ENGINEERING. *Educ:* Laval Univ, BS, 59, MS, 61; Rensselaer Polytech Inst, PhD(appl mech), 66. *Prof Exp:* From instr to assoc prof mech eng, Laval Univ, 66-70; dir mech div, Que Indust Res Ctr, 70-71, dir mech & mat div, 71-76, dir res & develop, 76-82, vpres opers, Montreal, 82-85, EXEC VPRES, QUE INDUST RES CTR, 88- *Honors & Awards:* Ralph Teetor Award, Soc Automotive Engrs, 70. *Mem:* Soc Mfr Eng; Can Soc Mech Engrs. *Res:* Machine design. *Mailing Add:* Scheminde Montreal 8475 Christophe Colomb Ottawa ON K1A 0R6 Can

BAROSS, JOHN ALLEN, b San Francisco, Calif, Aug 27, 41; m 67. MARINE MICROBIOLOGY. *Educ:* San Francisco State Univ, BS, 63, MA, 65; Univ Wash, PhD(marine microbiol), 72. *Prof Exp:* Res asst marine microbiol, Col Fisheries, Univ Wash, 66-70; res assoc, 70-77, ASST PROF & SR RESEARCHER, MARINE MICROBIOL, ORE STATE UNIV, 77- *Mem:* Am Soc Microbiol; AAAS; Am Indust Microbiol; Sigma Xi; Audubon Soc. *Res:* Effects of temperature, pressure and salinity on the physiology and molecular biology of inshore and deepsea marine bacteria; incidence, pathogenicity and possible significance of genetic exchange among in-shore marine vibrio populations, including the human pathogen, Vibrio parahaemolyticus. *Mailing Add:* Dept Oceanog Univ Wash Seattle WA 98195

BAROUDY, BAHIGE M, b Beruit, Lebanon, July 1, 50; nat US. MOLECULAR VIROLOGY, RECOMBINANT DNA. *Educ:* Am Univ, Beirut, BSc, 72; Georgetown Univ, PhD(biochem), 78. *Prof Exp:* Res asst, dept biochem, Georgetown Univ, 74-78; vis fel scientist, Lab Biol Viruses, Nat Inst Allergy & Infectious Dis, NIH, 79-81; fel, div molecular genetics, Lombardi Cancer Ctr, Georgetown Univ Med Ctr, 82; vis assoc scientist, Lab Molecular Oncol, Nat Cancer Inst, NIH, 82-83, Lab Infectious Dis, Nat Inst Allergy & Infectious Dis, 83-85; res assoc prof, Div Molecular Virol & Immunol, Dept Microbiol, Georgetown Univ, 85-89; DIR, DIV MOLECULAR VIROL, JAMES N GRANBLE INST MED RES, 89- *Concurrent Pos:* Guest scientist, Lab Immunopath, Nat Inst Allergy & Infectious Dis, NIH, 85-87; assoc mem, Lombardi Cancer Res Ctr, Georgetown Univ Med Ctr, 86-89. *Mem:* Am Soc Biochem & Molecular Biol; Am Soc Microbiol; Am Soc Virol; Sigma Xi; Am Soc Study Liver Dis. *Res:* Molecular virology of retroviruses with emphasis on murine leukemic viruses and human immunodeficiency virus; molecular virology of delta hepatitis and its interaction with hepatitis B and the molecular virology of hepatitis C virus. *Mailing Add:* 2141 Auburn Ave Cincinnati OH 45219

BAROUKI, ROBERT, b Beirut, Lebanon, Apr 7, 57; French citizen; m 83; c 2. GENE REGULATION IN THE LIVER, HORMONE & DRUG ACTION. *Educ:* Univ Paris, MD, 83, PhD(pharmacol), 83, Habilitation, 91. *Prof Exp:* Postdoctoral molecular biol, Sch Med, Johns Hopkins, 83-85; RES SCIENTIST MOLECULAR BIOL, UNIT 99, NAT INST HEALTH & MED RES, CRETEIL, FRANCE, 86- *Mem:* Am Soc Cell Biol; French Soc Endocrinol. *Res:* Hormonal regulation of the promoter of aspartate aminotransferase gene: glucocorticoid, CAMP and insulin effects; DNA-protein interactions; physiopathological regulation of the human isoenzymes in the liver and serum. *Mailing Add:* Nat Inst Health & Res Unit 99 Hosp Henri Mondor Creteil 94010 France

BARPAL, ISAAC RUBEN, b Argentina, Feb 21, 40; c 3. RESEARCH ADMINISTRATION. *Educ:* Calif State Polytech Univ, BS, 67; Univ Calif, Santa Barbara, MS, 68, PhD(elec eng), 70. *Prof Exp:* Sr engr, Minicars Inc, 68-70; teaching, res, Univ Calif, Santa Barbara, 70-71; engr mgr, Westinghouse Transp, 71-74, gen mgr, Brazil, 74-80; pres, Westinghouse Int, Latin Am, 80-87; VPRES, WESTINGHOUSE SCI & TECHNOL CTR, 87- *Concurrent Pos:* Mem, Indust Sci & Technol, Nat Sci Found, 88- *Mem:* Fel Inst Elec & Electronics Engrs; Soc Am Eng. *Mailing Add:* Westinghouse Sci & Technol Ctr 1310 Beulah Rd Pittsburgh PA 15235-5098

BARR, ALFRED L, b Rig, WVa, Jan 17, 33. AGRICULTURAL ECONOMICS, AGRICULTURAL ADMINISTRATION. *Educ:* WVa Univ, BS, 55; Univ Okla, MS, 57; Okla State Univ, PhD(agr econ), 60. *Prof Exp:* Agr economist, Res Serv, USDA, WVa, 60-61; agr economist, WVa Univ, 61-70, chmn, Div Animal Vet Sci, Col Agr, 70-80; ASSOC DIR, WVa AGR FORESTRY EXP STA, 80- *Mem:* Prof Agr Econ Asn. *Mailing Add:* WVa Agr & Forestry Exp Sta PO Box 6108 Rm 1164-AS Morgantown WV 26506

BARR, ALLAN RALPH, b Ft Worth, Tex, Aug 13, 26; m 52. MEDICAL ENTOMOLOGY. *Educ:* Southern Methodist Univ, BS, 48; Johns Hopkins Univ, ScD(parasitol), 52. *Prof Exp:* Instr med, econ entom & parasitol, Univ Minn, 52-55; asst prof gen & med entom, Univ Kans, 55-58; supvr vector res, Bur Vector Control, State Dept Health, Calif, 58-66; assoc entomologist, Mosquito Control Res-Fresno Proj, 66-67, assoc prof, 67-70, PROF INFECTIOUS & TROP DIS, UNIV CALIF, LOS ANGELES, 70- *Concurrent Pos:* Vis prof, Fac Med, Univ Singapore, 62-63; vis scientist, Inst Med Res, Kuala Lumpur, 75. *Mem:* Entom Soc Am; Am Soc Trop Med & Hyg; Am Mosquito Control Asn. *Res:* Systematics and biology of mosquitoes. *Mailing Add:* Div Infectious & Trop Dis Univ Calif Sch Pub Health Los Angeles CA 90024

BARR, B(ILLIE) GRIFFITH, b Birmingham, Ala, Sept 19, 23; m 45; c 3. MECHANICAL ENGINEERING. *Educ:* Univ Ala, BS, 48; Univ Kans, MS, 61. *Prof Exp:* Cost Control engr, Procter & Gamble Mfg Co, Kans, 48-50, instr engr, 50-52, tech engr, 52-56, power plant mgr, 56-60, tech construct engr, 60-62; assoc prof, 62-76, assoc dir, Ctr Res, 62-69, PROF MECH ENG, UNIV KANS, 76-, EXEC DIR, CTR RES, INC, 69-, ASST DEAN RES ADMIN, 70- *Concurrent Pos:* Consult, Lawrence Paper Co, Kans, 65-; Ling-Tempco-Vought Inc, Tex, 66- *Mem:* Am Soc Mech Engrs. *Res:* Research coordination; technical information transfer to aid industrialization; automation; industrial management; mechanical design; use of case studies in engineering education. *Mailing Add:* Dept Mech Eng Univ Kans Lawrence KS 66045

BARR, CHARLES (ROBERT), b Altoona, Pa, Mar 21, 22; m 43; c 4. PHOTOGRAPHIC CHEMISTRY. *Educ:* St Vincent Col, BS, 46. *Prof Exp:* Res chemist, Remington-Rand, Inc, 46-48; res chemist, Eastman Kodak Co, 48-57, sr res chemist, 57-62, res assoc, 62, asst to div head, Color Photog Div, 62-65, head exp color photog lab, 65-75, dir info serv, 75-84; RETIRED. *Mem:* Am Chem Soc; Soc Photog Sci & Eng. *Res:* Synthesis of organic compounds for use in color photography; color and constitution of indoaniline dyes; new color photographic products and processes. *Mailing Add:* 18410 129th Ave Sun City West AZ 85375

BARR, CHARLES E, b Chicago, Ill, Sept 20, 29; m 56; c 2. CELL PHYSIOLOGY. *Educ:* Iowa State Col. BS. 54; Univ Calif, Berkeley, PhD(plant physiol), 61. *Prof Exp:* Trainee biophys, Sch Med, Univ Md, 61-63, instr, 63-64; asst prof biol, Duquesne Univ, 64-68; assoc prof, 68-74, PROF BIOL, STATE UNIV NY COL BROCKPORT. 74- *Mem:* Sigma Xi. *Res:* Electrophysiology and ion transport in plant cells; origin of the resting potential in Nitella. *Mailing Add:* Dept Biol Sci State Univ Col Brockport NY 14420

BARR, CHARLES RICHARD, b Dakota, Ill, May 16, 32. BIOCHEMISTRY. *Educ:* NCent Col, BA, 54; Mich State Univ, MS, 57, PhD, 60. *Prof Exp:* Asst Mich State Univ, 54-58; asst prof, Dubuque Univ, 58-62; asst prof, 62-66, assoc prof chem, 66-72, PROF CHEM, AUSTIN COL, 72- *Mem:* Am Chem Soc. *Res:* Isolation and synthesis of biologically active compounds; binding of metals in enzymes; Mossbauer spectroscopy. *Mailing Add:* Austin Col Sherman TX 75091-1177

BARR, DAVID JOHN, b Evansville, Ind, Mar 5, 39; m 66; c 1. REMOTE SENSING, GEOGRAPHIC INFORMATION SYSTEMS. *Educ:* Purdue Univ, MSCE, 64, PhD(geol eng), 68. *Prof Exp:* CHMN, DEPT GEOL ENG, UNIV MO, ROLLA, 67-, PROF, GEOL ENG, 72- *Concurrent Pos:* Dir, Mo Mining & Mineral Resources, Res Inst, 80-88; asst vchancellor for academic comput, Univ Mo, Rolla, 86-87. *Mem:* Am Soc Civil Engrs; Am Soc Photogrammetry & Remote Sensing; Asn Eng Geol; Soc Mining Engrs; Nat Soc Prof Engrs. *Res:* Use of expert systems and geographic information systems for thematic map creation and compilation; remote sensing and digital image processing for minerals; exploration and/or environmental site evaluation. *Mailing Add:* Dept Geol Eng 129 McNutt Hall Univ Mo Rolla Rolla MO 65401

BARR, DAVID ROSS, b Madison, Wis, Aug 21, 32; m 57; c 2. MATHEMATICAL STATISTICS. *Educ:* Miami Univ, BA & MA, 54, MS, 57; Univ Iowa, PhD(math), 64. *Prof Exp:* Programmer, Appl Math Lab, Eglin Air Force Base, US Air Force, 55-57, instr math, US Air Force Acad, 59-62, asst prof, 62, res math statistician, Aerospace Res Labs, 64-68; ASST PROF MATH, USAF INST TECHNOL, 68- *Mem:* Inst Math Statist; Am Statist Asn; Math Asn Am. *Res:* Power function of the likelihood ratio test and ranking and selection problems of the uniform distribution and related distribution. *Mailing Add:* Dept Math USAF Inst Technol Wright-Patterson OH 45433

BARR, DAVID WALLACE, b Toronto, Ont, Jan 1, 43. EVOLUTIONARY BIOLOGY, PHYLOGENY. *Educ:* Univ Toronto, BSc, 65; Cornell Univ, PhD(entom), 69. *Prof Exp:* Curatorial asst, Royal Ont Mus, 68-69, asst cur, 69-73, assoc cur, 73-76, assoc prof in-charge, 76-80, cur-in-charge entom, 80-84, assoc dir curatorial, 84-89, CUR-IN-CHARGE INVERTEBRATE ZOOL, ROYAL ONT MUS, 89- *Concurrent Pos:* Asst prof zool, Univ Toronto, 73-78, assoc prof, 78- *Res:* Evolution and comparative biology of Mollusca. *Mailing Add:* Invertebrate Zool Royal Ont Mus 100 Queen's Park Toronto ON M5S 2C6 Can

BARR, DENNIS BRANNON, b Bonham, Tex, Apr 23, 43; m 65; c 3. PHYSICS, ELECTRON MICROSCOPY. *Educ:* Tex A&M Univ, BS, 65, MS, 67, PhD(physics), 71. *Prof Exp:* Physicist electron microscopy, Army Mat & Mech Res Ctr, 71-73; res physicist, 73-76, SR RES PHYSICIST ELECTRON MICROSCOPY, EASTMAN CHEM DIV, 76- *Mem:* Electron Microscopy Soc Am. *Res:* Industrial microscopy; polymer physics, polymer morphology. *Mailing Add:* 5329 or 5333 Kiowa St Kingsport TN 37664

BARR, DONALD EUGENE, b June 30, 34; US citizen; m 56; c 2. CHEMISTRY. *Educ:* Elizabethtown Col, BS, 56; Bucknell Univ, MS, 61; Univ Mass, PhD(chem), 65. *Prof Exp:* Mem staff, Bucknell Univ, 61; res chemist, Chem Res Div, Armstrong Cork Co, 65-69; res specialist new imaging processes res, GAF Corp, 69-71, mgr photoimaging, 71-74, mgr, Raw Mat Qual Assurance Dept, 74-76, staff mgr photog prod, 76-78, tech dir res & develop, 78-80; adv engr, IBM Corp, 80, mgr, 81-84, sr eng mgr mat sci, 84-86, AREA MGR & MAT ENG, IBM CORP, 86- *Mem:* Am Chem Soc; Soc Photog Sci & Eng. *Res:* Seven-azabenzonorbornadiene synthesis; polyvinyl chloride polymerization, characterization and stabilization; high resolution nuclear magnetic resonance, polyvinyl chloride and model compounds; photosensitive compounds and polymeric coatings for microelectronics industry; electronic packaging (computers). *Mailing Add:* 1508 Drexel Dr Vestal NY 13850

BARR, DONALD JOHN STODDART, b Surrey, Eng, Sept 18, 37; Can citizen; m 61; c 2. MYCOLOGY. *Educ:* McGill Univ, BSc, 60, MSc, 62; Univ Western Ont, PhD(mycol, phycol), 65. *Prof Exp:* Res scientist mycol, Plant Res Inst, 65-73, SR RES SCIENTIST MYCOL, BIOSYSTS RES INST, CAN DEPT AGR, 73- *Mem:* Mycol Soc Am (pres, 90-91); Can Phytopath Soc; Int Soc Evolutionary Protistology (pres, 90-). *Res:* Fungal parasites of freshwater algae; taxonomy, ecology and physiology of the Mastigomycotina, particularly the Chytridiomycetes; virus transmission by zoosporic fungi; beneficial fungi in the rumen. *Mailing Add:* Biosysts Res Inst Can Dept Agr Ottawa ON K1A 0C6 Can

BARR, DONALD R, b Durango, Colo, Dec 10, 38; m 58; c 2. MATHEMATICAL STATISTICS. *Educ:* Whittier Col, BA, 60; Colo State Univ, MS, 62, PhD(statist), 65. *Prof Exp:* Instr math. Colo State Univ, 63-65; asst prof, Wis State Univ, Oshkosh, 65-66; vpres, Evaluation Technol Inc, 87-89; from asst prof to assoc prof opers res, 66-77, prof opers res & statist, 77-87, PROF MATH, NAVAL POSTGRAD SCH, 90- *Concurrent Pos:* Consult, Litton Mellonics, 67-77, Braddock, Dunn & McDonald, Inc, 71-76, Dynamic Sci, Inc, 77-80, ORI, Inc, 81-87; vis prof, Ministry Nat Defense, Thailand, 70, Tamkang Col, Taiwan, 75; liaison scientist, Office Naval Res, London, 82-84. *Mem:* Am Statist Asn; Int Test & Eval Asn; Am Defense Preparedness Asn. *Res:* Statistical theory; sequential decision theory; reliability growth and classification models; test and evaluation, data analysis. *Mailing Add:* 1495 Prescott Ave Monterey CA 93940-1647

BARR, DONALD WESTWOOD, b Worcester, Mass, May 6, 32; m 54; c 4. NUCLEAR CHEMISTRY. *Educ:* Univ Mass, BS, 54; Univ Calif, Berkeley, PhD(nuclear chem), 57. *Prof Exp:* Staff mem, Los Alamos Nat Lab, 57-71, assoc group leader, 71-72, dep group leader, 72-82, fel, 81-82, assoc div leader, 82-84, div leader, 84-90; RETIRED. *Honors & Awards:* E O Lawrence Award, Dept Energy, 80. *Mem:* Am Chem Soc. *Res:* Neutron physics and radiochemistry. *Mailing Add:* PO Box 647 Jemez Springs NM 87025

BARR, FRANK T(HOMAS), b Elwood, Ind, July 1, 10; m 36; c 3. ENERGY, CHEMICAL ENGINEERING. *Educ:* Wash Univ, BS, 29, MS, 31; Univ Ill, Urbana, PhD(chem eng), 34. *Prof Exp:* Asst chem, Wash Univ, 29-31; chem engr, Univ Ill, 31-34; instr, Ill Inst Technol, 34-36; chem engr, Exxon Res & Eng Co, 36-52, sr eng assoc, 52-69, mem staff planning & coord, Exxon Nuclear Co, 69-73, sr assoc, 70-73; CONSULT ENERGY, 73- *Concurrent Pos:* Consult, Off Sci Res & Develop, Manhattan Dist, 41-45 & Wash State Energy Policy Co, 74-75; chmn energy task force, San Diego Chamber Com, 76-78. *Mem:* Fel Am Inst Chem Engrs; Am Nuclear Soc; Am Chem Soc. *Res:* Hydrogenation of oil and coal; hydrocarbon synthesis from CO and H2; hydrogen and synthesis gas manufacture; synthetic fuels; energy supply and conversion; atomic energy and nuclear technology; petroleum processing; coking. *Mailing Add:* 39951 Pierce Dr Three Rivers CA 93271

BARR, FRED S, b Benhams, Va, Jan 22, 26; m 50; c 2. MEDICAL MICROBIOLOGY. *Educ:* Univ Tenn, BS, 52; East Tenn State Univ, MA, 57. *Prof Exp:* Prod asst, S E Massengill Co, Tenn, 52-53, res bacteriologist, 53-60, res assoc microbiol & biochem, 60-70, sect head biopharmaceut res, 70-71; MGR MICROBIOL SECT, BEECHAM LABS, 71- *Mem:* AAAS; Am Soc Microbiol; Am Soc Trop Med & Hyg; Am Chem Soc. *Res:* Antibiotic isolation, identification and purification; antibiotic combinations in human and animal medications; microbiological diagnostic reagents for clinical use; drug levels in body tissues and fluids; drug metabolism and absorption-excretion kinetics. *Mailing Add:* 1760 Overhill Rd Bristol VA 24201

BARR, GEORGE E, b Milwaukee, Wis, Oct 16, 37; wid. APPLIED MATHEMATICS. *Educ:* Univ Wis, Madison, BA, 58; Ore State Univ, PhD(theoret physics), 64. *Prof Exp:* Jr res assoc math physics, Brookhaven Nat Lab, 62-63; MEM TECH STAFF APPL MATH, SANDIA LAB, 65- *Mem:* Soc Indust & Appl Math; Am Math Soc. *Res:* Special functions; theory of diffraction; fluid dynamics of viscous media; nuclide transport in geologic media. *Mailing Add:* Sandia Lab Orgns 6312 Bldg 823 Rm 4021 Albuquerque NM 87185

BARR, HARRY, automotive engineering; deceased, see previous edition for last biography

BARR, HARRY L, b Bethesda, Ohio, Feb 15, 22; m 44; c 4. REPRODUCTIVE PHYSIOLOGY, GENETICS. *Educ:* Ohio State Univ, BSc, 54, MS, 55, PhD(genetics), 60. *Prof Exp:* From instr to asst prof physiol, 55-67, assoc prof, 67-71, PROF DAIRY SCI, OHIO STATE UNIV, 71-, SPECIALIST, COOP EXTEN SERV, 67- *Concurrent Pos:* Vis prof, Cornell Univ, 70. *Mem:* Am Dairy Sci Asn; Am Soc Animal Sci. *Res:* Reproductive physiology and population genetics. *Mailing Add:* Dairy Sci Ohio State Univ Main Campus 116 Plumb Hall Columbus OH 43210

BARR, JAMES K, physical chemistry, research adminstration, for more information see previous edition

BARR, JOHN BALDWIN, b Niagara Falls, NY, Nov 8, 32; m 54; c 4. PHYSICAL CHEMISTRY. *Educ:* Univ Buffalo, BA, 54; Univ Mich, MS, 56; Pa State Univ, PhD(phys chem), 61. *Prof Exp:* Sr chemist, Corhart Refractories Div, Corning Glass Works, NY, 60-62; sr res chemist, 62-71, res scientist, Carbon Prod Div, 71-82, sr res scientist, Specialty Polymers & Composites Div, Res Lab, Union Carbide Corp, Parma, 82-86; sr res scientist, Parma, Ohio, 87-90, SR RES SCIENTIST, AMOCO PERFORMANCE PRODS INC, ALPHARETTA, GA, 90- *Mem:* Am Chem Soc; Sigma Xi; Am Carbon Soc; NAm Thermal Anal Soc; Soc Advan Mat & Process Eng. *Res:* Carbon fibers; refractory compounds; carbon and graphite chemistry; boron halide reactions; high temperature chemistry. *Mailing Add:* Amoco Performance Prod Inc 4500 McGinnis Ferry Rd Alpharetta GA 30202-3944

BARR, JOHN TILMAN, organic chemistry, for more information see previous edition

BARR, KEVIN PATRICK, b London, Eng, May 4, 44; m. HEAT TRANSFER SYSTEMS ENGINEERING. *Educ:* Univ Santa Clara, BS, 66; Univ Kans, MS, 68, PhD(physics), 76. *Prof Exp:* PRELIMINARY DESIGN ENGR, AIRES LOS ANGELES DIV, 77- *Concurrent Pos:* Instr physics, Univ Kans, 72-73, physics & math, Ottawa Univ, Kans, 75-76. *Mem:* Am Phys Soc; Am Soc Mech Engrs; Sigma Xi; Am Inst Aeronaut & Astronaut. *Res:* Heat transfer and cryogenic systems and components; new energy sources including solar systems and fuel cells; aerospace; life support and thermal management systems. *Mailing Add:* 23026 Dana Ct Torrance CA 90501

BARR, LAWRENCE DALE, b Adelanto, Calif, Nov 5, 30; m 51; c 6. ASTRONOMY ENGINEERING, MECHANICAL ENGINEERING. *Educ:* Univ Wis, BS, 57; Rensselaer Polytech Inst, MS, 60. *Prof Exp:* Develop engr & dept mgr plastics, Monsanto Corp, 57-64; eng mgr, Celanese Corp, 64-66; gen mgr res & develop, Container Corp Am, 66-68; sr engr 4m telescope, Kitt Peak Nat Observ, 68-70, eng mgr astron, Nat Optical Astron Observ, 70-90; CONSULT, 90- *Concurrent Pos:* Consult engr mult mirror telescope, Univ Ariz, Smithsonian Inst, 73-76, 3.6m Canada-France-Hawaii telescope & Univ Calif 10 meter telescope, 3m IR telescope facil, Keck Observ. *Mem:* Optical Soc Am. *Res:* Large telescope design, most recently for the 8 meter and 15 meter National New Technology Telescopes which are an outgrowth of earlier Next Generation Telescope studies; instrumentation for astronomy and diffraction grating ruling engines; large optics fabrication. *Mailing Add:* 1167 N Winstel Blvd Tucson AZ 85716

BARR, LLOYD, b Chicago, Ill, Dec 27, 29. PHYSIOLOGY. *Educ:* Univ Chicago, BS, 54; Univ Ill, MS, 56, PhD(physiol), 58. *Prof Exp:* Instr physiol, Univ Mich, 58-61, from asst prof to assoc prof, 61-65; prof, Med Col Pa, 65-70; PROF PHYSIOL & BIOPHYS, UNIV ILL, URBANA-CHAMPAIGN, 70- *Concurrent Pos:* Spec fel, USPHS, 63-64; vis prof, Physiol Inst, Ud Saarlandes, 63-64, 72, Baumfield Marine Lab, State Univ NY, Stony Brook, 77. *Mem:* Fel AAAS; Soc Gen Physiol; NY Acad Sci; Biophys Soc; Sigma Xi. *Res:* Electrophysiology; cellular transport processes; kinetics and thermodynamics. *Mailing Add:* Univ Ill 407 S Goodwin 524 Burrill Hall Urbana IL 61801

BARR, MARTIN, b Philadelphia, Pa, Nov 11. 25; m 51; c 4. PHYSICAL PHARMACY. *Educ:* Temple Univ, BSc, 46; Philadelphia Col Pharm, MSc, 47; Ohio State Univ, PhD, 50. *Prof Exp:* Asst pharm, Ohio State Univ, 47-49, instr, 50; from asst prof to prof, Philadelphia Col Pharm, 50-61; chmn dept, Wayne State Univ, 61-64, dean, Col Pharm, 64-72, vis lectr pharm, Col Med, 64-72, vpres spec assignments, 72-78, vpres health affairs & dep provost, 78-82, dean, Col Pharm, 82-87; EXEC VPRES, CORP BUS & MED DEVELOP, MICH HEALTH CARE CORP, 87- *Mem:* Am Pharmaceut Asn. *Res:* Clays; ion exchange; emulsions; ointments; colloids; product development; aerosols; health care delivery systems. *Mailing Add:* 32355 Susanne Dr Franklin MI 48025-1120

BARR, MASON, JR, b Washington, DC, June 15, 35. PEDIATRICS, TERATOLOGY. *Educ:* Haverford Col, BA, 57; George Washington Univ, MD, 61. *Prof Exp:* Rotating intern, 61-62, resident, 62-64, from instr to assoc prof, 66-83, PROF PEDIAT, UNIV MICH, 83-, PROF PATH, 87-, PROF OBSTET, 88- *Concurrent Pos:* Fel embryol, Univ Mich, 66-68; fel teratology, Thomas Jefferson Univ, 68-69; vis prof path, Children's Hosp Mich, 77-78; rev ed, 79-83, sect ed, J Teratology, 83-; coun mem, Teratology Soc, 84-87. *Mem:* Soc Pediat Res; Teratology Soc; Am Acad Pediat; Soc Pediat Path. *Res:* Prenatal pathology; teratology; birth defects rehabilitation. *Mailing Add:* Dept Pediat Univ Mich D1109 MPB Box 0718 Ann Arbor MI 48109

BARR, MICHAEL, b Philadelphia, Pa, Jan 22, 37; m 64; c 3. MATHEMATICS. *Educ:* Univ Pa, AB, 59, PhD(math), 62. *Prof Exp:* Instr, Columbia Univ, 62-64; from asst prof to assoc prof, Univ Ill, Urbana, 64-68; assoc prof, 68-72, PROF MATH, MCGILL UNIV, 72- *Concurrent Pos:* Guest, Res Inst Math, Swiss Fed Inst Technol, 67 & 75-76; res assoc, Math Inst, Univ Fribourg, 70-71; vis prof, Univ Pa, 89-90. *Mem:* Math Asn Am; Can Math Soc. *Res:* Categorical algebra, especially the study of those aspects of mathematical structures describable in terms of mapping properties; applications to theoretical computer science. *Mailing Add:* Dept Math McGill Univ 805 Rue Sherbrooke Ouest Montreal PQ H3A 2K6 Can

BARR, MURRAY LLEWELLYN, b Belmont, Ont, June 20, 08; m 34; c 4. GENETICS, NEUROSCIENCES. *Educ:* Univ Western Ont, BA, 30, MD, 33, MSc, 38; FRCP(C), 64; FACP, 65; FRCOG, 72. *Hon Degrees:* LLD, Queen's Univ, Ont, 63, Univ Toronto, 64, Univ Alta, 67, Dalhousie Univ, 68 & Univ Sask, 73; Dr Med, Univ Basel, 66; DSc, Univ Western Ont, 74. *Prof Exp:* From instr to prof anat, 36-52, prof micros anat & head dept, 52-64, head, Dept Anat, 64-67, prof, 64-79, EMER PROF ANAT, UNIV WESTERN ONT, 79- *Concurrent Pos:* Med officer, Royal Can Air Force, 39-45. *Honors & Awards:* Award, Soc Obstet & Gynec Can, 56; Borden Award, 57; Flavelle Medal, Royal Soc Can, 59; Charles Mickle Award, Univ Toronto, 60; Ortho Medal, Am Soc Study Sterility, 62; Medal, Am Col Physicians, 62; Joseph P Kennedy Jr Found Award, 62; Gairdner Found Award Merit, 63; Papanicalaou Award, Am Soc Cytol, 64; F N G Starr Medal, Can Med Asn, 67; Maurice Goldblatt Cytol Award, Int Acad Cytol, 68. *Mem:* Fel Royal Soc; Asn Res Nerv & Ment Dis; Am Soc Human Genetics; Can Neurol Soc; fel Royal Soc Can. *Res:* Neuroanatomy; cytogenetics; mental deficiency. *Mailing Add:* 452 Old Wonderland Rd London ON N6K 3R2 Can

BARR, NATHANIEL FRANK, b Union City. NJ, Dec 28, 27; m 56; c 3. RADIATION CHEMISTRY. *Educ:* Queens Col, NY, BS. 50; Columbia Univ, PhD(chem). 54. *Prof Exp:* Asst, Columbia Univ, 51-54; res assoc, Brookhaven Nat Lab, 54-56; asst, Div Biophys, Sloan Kettering Inst, 56-61; biophysicist, US AEC, 61-62, chief, Radiol Physics & Instrumentation Br, Div Biol & Med, 62-67, tech adv to asst gen mgr res & develop, 67-69, asst dir, Div Biol & Med, 69-75; sci adv, Div Biomed & Environ Res, 75-78, mgr, Health & Environ Risk Anal Prog, 78-83, BIOPHYSICIST, US DEPT

ENERGY, 83- *Concurrent Pos:* Asst prof, Med Sch, Cornell Univ & lectr, Columbia Univ, 58-61; prof, Southeastern Univ, 80-81. *Mem:* Am Chem Soc; Radiation Res Soc (pres, 71-72); Health Physics Soc; Pattern Recognition Soc; AAAS; NY Acad Sci. *Res:* Dosimetry and properties of ionizing radiation; risk analysis. *Mailing Add:* Off Health Environ Res/ER-72 US Dept Energy Washington DC 20545

BARR, RICHARD ARTHUR, b Southport, NY, Mar 12, 25; m 61; c 2. PLANT PHYSIOLOGY, PHYTOCHEMISTRY. *Educ:* Univ Vt, BS, 50, MS. 55; Cornell Univ, PhD(plant physiol), 63. *Prof Exp:* Res assoc plant physiol, Cornell Univ, 61-63, asst prof, 64-66; asst prof biol, Univ Mo-St Louis, 66-68; assoc prof, 68-72, PROF BIOL, SHIPPENSBURG UNIV, 72- *Res:* Growth of intact plants from cell suspensions; tissue culture; somatic cell fusion. *Mailing Add:* 201 Franklin Sci Ctr Shippensburg Univ Shippensburg PA 17257

BARR, RITA, b Riga, Latvia, Sept 30, 29. PLANT PHYSIOLOGY. *Educ:* Northern Ill Univ, BS, 54, MS, 56; Purdue Univ, PhD(genetics), 60. *Prof Exp:* Fel genetics, 60-62, res asst plant physiol, 63-65, ASSOC BIOL SCI, PURDUE UNIV, 65- *Mem:* Sigma Xi; Am Soc Plant Physiol. *Res:* Electron transport reactions of isolated plasma membranes in plants; transmembrane ferricyanide reduction in cultured carrot cells; plant growth regulator effects upon plasma membrane redox and concomitant proton secretion. *Mailing Add:* Dept Biol Sci Purdue Univ Lafayette IN 47907

BARR, ROBERT ORTHA, JR, b Lima, Ohio, Jan 4, 40; m 64; c 2. CONTROL SYSTEMS, COMPUTER SCIENCE. *Educ:* Univ Mich, BSE, 61, MSE, 62, PhD(info & control eng), 66. *Prof Exp:* Res asst optimal control, Univ Mich, 65-66; asst prof systs sci, 66-69, assoc prof elec eng & systs sci, 69-87, ASSOC CHAIR ELEC ENG, MICH STATE UNIV, 87- *Mem:* Inst Elec & Electronics Engrs; Soc Indust & Appl Math. *Res:* Interactive computing procedures for optimal control problems; control and optimization applications to environmental systems. *Mailing Add:* Dept Elec Eng Mich State Univ East Lansing MI 48824-1226

BARR, ROGER COKE, b Jacksonville, Fla, Feb 21, 42; m 67; c 2. BIOMEDICAL ENGINEERING, COMPUTER SCIENCE. *Educ:* Duke Univ, BS, 64, PhD(elec eng), 68. *Prof Exp:* Res assoc, 68-69, from asst prof to assoc prof, 69-79, PROF BIOMED ENG, DUKE UNIV, 79-, ASSOC PROF PEDIAT, 72- *Mem:* Am Heart Asn; Am Col Cardiol; Inst Elec & Electronics Engrs; Sigma Xi. *Res:* Electrocardiology; electrophysiology of the heart; digital computing. *Mailing Add:* 4710 Oak Hill Rd Chapel Hill NC 27514

BARR, RONALD DUNCAN, b Cardross, Scotland, May 28, 43; m 70; c 2. HEMATOLOGY, ONCOLOGY. *Educ:* Univ Glasgow, Scotland, MB & ChB, 66, MD, 77. *Prof Exp:* Res scholar hemat, Univ Glasgow, 67-68; lectr internal med, Univ Nairobi, Kenya, 70-72; lectr hemat, Univ Aberdeen, Scotland, 73-77; assoc prof, 77-81, PROF PEDIAT, MCMASTER UNIV, CAN, 81-, PROF PATH & MED, 85- *Concurrent Pos:* Vis scientist, Nat Cancer Inst, NIH, 74-76. *Mem:* Fel Royal Col Physicians; Royal Col Pathologists; fel Am Col Physicians; Am Soc Hemat; Int Soc Exp Hemat. *Res:* Interaction of cellular and humoral mechanisms in the regulation of normal hemopiesis with particular reference to the roles of lymphocyte subpopulations and their ontogenesis. *Mailing Add:* 3-41 Charlton West Hamilton ON L8P 2C2 Can

BARR, RONALD EDWARD, b St Louis, Mo, Sept 9, 36; m 64; c 4. BIOPHYSICS, BIOENGINEERING. *Educ:* St Louis Univ, BS, 58, MS, 63, PhD(physics), 66. *Prof Exp:* Res asst, St Louis Univ, 61-65; res fel biophys, Space Sci Res Ctr, Univ Mo-Columbia, 65-66, asst prof radiol, 66-74, assoc prof ophthalmol, 74-77, assoc dean, Grad Sch, 77-79; ASSOC PROVOST, OHIO UNIV, 79- *Concurrent Pos:* NIH spec fel, Cambridge Univ, 69-70. *Mem:* AAAS; Am Phys Soc; Nat Coun Univ Res Admin; Soc Univ Patent Adminr. *Res:* Oxygen transfer properties of the cornea and other ocular tissues; development of long term stability for oxygen, pH and pCO2 electrodes; research and graduate education administration. *Mailing Add:* Admin Bldg 213 Ariz State Univ Tempe AZ 85287-2103

BARR, SUMNER, b Worcester, Mass, July 12, 38; m 61; c 2. METEOROLOGY. *Educ:* Univ Mass, BS, 60; Mass Inst Technol, SM, 65; Univ Utah, PhD(meteor), 69. *Prof Exp:* Weather officer, US Air Force, 60-63; scientist meteorol, GCA Corp, 64-68; asst prof physics, Drexel Univ, 69-72; group leader atmospheric physics, 77-81, assoc group leader atmospheric sci, 81-84, STAFF MEM, LOS ALAMOS SCI LAB, 72-, PROJ MGR, ATMOSPHERIC STUDIES & OIL SHALE, 84- *Mem:* Am Meteorol Soc; Am Geophys Union. *Res:* Physics of transport and diffusion in the atmospheric boundary layer over non-homogeneous surfaces. *Mailing Add:* 605 Meadow Lane Los Alamos NM 87544

BARR, SUSAN HARTLINE, b Findlay, Ohio, June 23, 42; m 66; c 2. BIOLOGICAL EDITING, RADIATION BIOLOGY. *Educ:* Bowling Green State Univ, BA, 64; Univ Mich, Ann Arbor, MS, 65, PhD(radiation biol), 67. *Prof Exp:* Sci asst, Argonne Nat Lab, Ill, 67-69, tech ed, 80-83, managing tech ed, 83-90, USER PROG ADMINR, ARGONNE NAT LAB, 90- *Concurrent Pos:* Mem, Coun Biol Ed. *Mem:* Soc Tech Commun. *Res:* Carcinogenesis; biophysics; radiation biology; toxicology. *Mailing Add:* Advan Photon Source Argonne Nat Lab 9700 S Cass Ave Argonne IL 60439-4833

BARR, TERY LYNN, b Renova, Pa, Feb 9, 39; m 59; c 4. CHEMICAL PHYSICS, SURFACE CHEMISTRY. *Educ:* Univ Va, BS, 60; Univ SC, MS, 62; Univ Ore, PhD(chem), 68. *Prof Exp:* Res chemist, Union Oil Co, 63-64 & Shell Develop Co, 69-71; vis asst prof chem, Univ Calif, Berkeley, 71; asst prof, Harvey Mudd Col, 71-73; asst prof, Claremont Men's Col, 73-74; sr res chemist, 74-80, res specialist, Corp Res Ctr, UOP, Inc, 80-84, assoc res scientist, Signal Res Inc, 84-85; PROF SURFACE SCI, DEPT MAT & LAB SURFACE SCI, UNIV WIS MILWAUKEE, 85- *Concurrent Pos:* Res assoc, Univ Wash, 68-69; consult, R&D Assoc, 72-74; vis prof chem, Harvey Mudd Col, 74-75; adj assoc prof, 78-89, adj prof chem, Univ Ill, Chicago Circle, 79-86; vis prof, Univ Messina, Italy, 83; vis prof, Inst Physics, Univ Uppsala, Sweden, 85; consult, Signal Res Ctr Inc, 85-, S C Johnson Co Res Ctr, 86- *Mem:* Sigma Xi; Am Phys Soc; Am Vacuum Soc; Soc Electron Spectros (pres, 75-); Am Soc Testing & Mat; Mats Res Soc. *Res:* X-ray photoelectron spectroscopy; surface chemistry and physics; quantum mechanics; spectroscopic characterization of materials; catalysts, composites, and thin films; corrosion science; oxides, high Tc superconductivity, ion beam deposition. *Mailing Add:* Dept Mat Col Eng & Appl Sci Univ Wis Milwaukee WI 53201

BARR, THOMAS ALBERT, JR, b Chattanooga, Tenn, Aug 18, 24; m 49; c 5. PLASMA PHYSICS. *Educ:* Univ Chattanooga, BS, 47; Vanderbilt Univ, MS, 50, PhD(physics), 53. *Prof Exp:* Mem staff, Vanderbilt Univ, 50-51; asst prof physics, Univ Ga, 51-56; supvy res physicist, high energy lasers, US Army Missile Res & Develop Command, Redstone Arsenal, 56-81; RES PROF, DEPT PHYSICS, UNIV ALA, HUNTSVILLE, 81- *Concurrent Pos:* Govt liaison rep, Plasma Phenomena Comt, AIAA, 60. *Mem:* Fel AAAS; Am Phys Soc; assoc fel Am Inst Aeronaut & Astronaut; Sigma Xi; fel Optical Soc Am. *Res:* Investigation of the properties of metal membranes; x-ray and neutron dosimetry; basic and applied research on partially and fully ionized gases; chemical and high power gas lasers. *Mailing Add:* Dept Physics Univ Ala Huntsville AL 35899

BARR, WILLIAM A(LEXANDER), b Boston, Mass, Jan 9, 21; m; c 4. MECHANICAL ENGINEERING. *Educ:* St John's Col, Md, BA, 42; Univ Va, BSME, 48; Va Polytech Inst & State Univ, MS, 51. *Prof Exp:* Lab technician plastic testing, Libbey-Owens-Ford Glass Co, 43; instr mech eng, Va Polytech Inst, 48-50, asst prof, 50-51; develop engr, Metal Prod Div, Koppers Co, 51-53, staff mech engr, 54; sr engr, Kaiser Aluminum & Chem Co, 55-56; plant engr, Crown Cork & Seal, 57, mgr equip eng, 57-58; ASSOC PROF MECH ENG, US NAVAL ACAD, 58- *Mem:* Nat Soc Prof Engrs. *Res:* Heat transfer; design and evaluation of mechanical equipment; waste disposal problems. *Mailing Add:* Gibson Island MD 21056

BARR, WILLIAM FREDERICK, b Oakland, Calif, Oct 20, 20; m 46; c 3. ENTOMOLOGY. *Educ:* Univ Calif, BS, 45, MS, 47, PhD(syst entom), 50. *Prof Exp:* Teaching asst entom, Univ Calif, 45-47; asst entomologist, Exp Sta, Univ Idaho, 47-53, asst prof entom, 48-53, assoc prof & assoc entomologist, 53-58, prof & entomologist, 58-82, head, dept entom, 78-82; RETIRED. *Concurrent Pos:* NSF fac fel, 59-60. *Mem:* Entom Soc Am; Pac Coast Entom Soc; Soc Syst Zool; Royal Entom Soc London. *Res:* Systematic entomology; distribution and biology of Coleoptera; biological control of weeds; insects affecting range plants. *Mailing Add:* Dept Entom Univ Idaho Moscow ID 83843

BARR, WILLIAM J, b Reading, Pa, Feb 5, 19. ENERGY CONVERSION. *Educ:* Princeton Univ, AB, 39; Columbia Univ, MA, 40. *Prof Exp:* Instr math, Polytech Inst Brooklyn, 41-42 & Brooklyn Col, 42; mathematician, Naval Res Lab, 42-46; head analysis group, Exp, Inc, 46-50; tech asst to dir Proj Squid, Princeton Univ, 51; mathematician, Res Div, Detroit Controls Corp, 52, asst res dir, 53-55, res dir, 55-57, asst to pres, 57-58; mgr physics res, Am Stand Inc, 58-60 & opers res, 60-72; mgr mfg systs develop, Am Can Co, 72-73; CONSULT, MATHEMATICA INC, 73- *Concurrent Pos:* Consult, Mathtech, 86- *Mem:* Am Math Soc; Am Chem Soc; Opers Res Soc Am. *Res:* Thermochemistry; interior ballistics; thermodynamics; combustion; propulsion. *Mailing Add:* 127 Westerly Rd Princeton NJ 08540

BARR, WILLIAM LEE, b Wilbur, Wash, Feb 17, 25; m 52; c 3. ATOMIC & MOLECULAR PHYSICS. *Educ:* Univ Wash, BS, 50; Univ Calif, PhD(physics), 57. *Prof Exp:* PLASMA PHYSICS, LAWRENCE LIVERMORE NAT LAB, 57- *Mem:* Am Phys Soc. *Res:* Atomic spectroscopy; plasma physics. *Mailing Add:* Lawrence Livermore Lab L-644 Univ Calif PO Box 808 Livermore CA 94551

BARRACK, CARROLL MARLIN, b Baltimore, Md, Mar 15, 27; m 49; c 3. ELECTRICAL ENGINEERING, OPERATIONS RESEARCH. *Educ:* Johns Hopkins Univ, BE, 50, DE, 56, MS, 65. *Prof Exp:* Sr design engr, Aircraft Armaments Inc, 54-57; res assoc res div, Electronic Commun Inc, 57-58; mgr, electronics dept, Miller Res Corp, Md, 58-59; prin develop engr, AAI Corp, 59-68; SR STAFF ENGR, APPL PHYSICS LAB, JOHNS HOPKINS UNIV, 68- *Concurrent Pos:* Treas, E Coast Conf Aerospace & Navig Electronics, 58-61, vchmn, 61-62, chmn, 62-63; vchmn, Balto Sect, Inst Elec & Electronics Engrs, 67-68, chmn, 68-69. *Mem:* Inst Elec & Electronics Engrs; Sigma Xi. *Res:* Operations analysis in radar and missile systems; system analysis in electronic simulation, electronic warfare and digital systems; electromagnetic backscattering; applied statistics; experimental design. *Mailing Add:* Appl Physics Lab Johns Hopkins Rd Laurel MD 20707-6099

BARRACK, EVELYN RUTH, NORMAL & ABNORMAL PROSTATE GROWTH. *Educ:* Johns Hopkins Univ, PhD(pharmacol), 75. *Prof Exp:* ASST PROF UROL, SCH MED, JOHNS HOPKINS UNIV, 80-; ASSOC DIR RES, BRADY UROL INST, JOHNS HOPKINS HOSP, 80- *Res:* Nuclear matrix. *Mailing Add:* Dept Urol Johns Hopkins Univ Sch Med 600 N Wolfe St Marburg 115 Baltimore MD 21205

BARRACLOUGH, CHARLES ARTHUR, b Vineland, NJ, July 13, 26; m 52; c 2. ENDOCRINOLOGY. *Educ:* St Joseph's Col, Pa, BS, 47; Rutgers Univ, MS, 51, PhD(zool), 52. *Prof Exp:* Asst, Rutgers Univ, 50-53; jr res anatomist, Sch Med, Univ Calif, Los Angeles, 53-54, from instr to asst prof, 54-61; assoc prof, 62-65, actg chmn dept, 71-72, PROF PHYSIOL, SCH MED, UNIV MD, BALTIMORE CITY, 65-, DIR, CTR STUDIES REPRODUCTION, 84- *Concurrent Pos:* Spec res fel, Cambridge Univ, 61-62; spec res fel, Univ Milan, 69-70; consult res div, Vet Admin Hosp, Calif, 59-61; mem reproductive biol study sect, NIH, 66-68 & 70-74, chmn, 73-74. *Honors & Awards:* Res Award, Soc Study Reproduction, 84, Carl Hartman Award, 90. *Mem:* Fel AAAS; Soc Exp Biol Med; Int Brain Res Orgn; Am Physiol Soc; Soc Study Reproduction (dir, 71-73); Soc Neurosci; Endocrine Soc. *Res:* Reproductive physiology; neuroendocrinology. *Mailing Add:* Dept Physiol Sch Med Univ Md Baltimore MD 21201

BARRACO, ROBIN ANTHONY, b Detroit, Mich, Mar 24, 45. NEUROCHEMISTRY. *Educ:* Georgetown Univ, BA, 66; Wayne State Univ, PhD(physiol), 71. *Prof Exp:* Res assoc, Wayne State Univ, 70-71, asst prof physiol, 71-80; CONSULT, 73- *Concurrent Pos:* Mem task force, Off Drug Abuse & Alcoholism, Gov State of Mich, 72-73. *Res:* Behavioral neurochemistry; psychopharmacology of memory and learning; developmental neurobiology. *Mailing Add:* Dept Physiol Wayne State Univ Med Sch 540 E Canfield 6213 Scott Hall Detroit MI 48201

BARRADAS, REMIGIO GERMANO, b Hong Kong, Oct 23, 28; Can citizen; m 53; c 3. PHYSICAL CHEMISTRY. *Educ:* Univ Liverpool, BSc, 52; Royal Col Sci & Technol, Scotland, dipl, 53; Univ Ottawa, PhD(electrochem), 60. *Prof Exp:* Govt chemist, Govt Lab, Hong Kong, 53-56; chemist inspection serv, Chem Div, Dept Nat Defence, Can, 56-57; demonstr phys chem, Univ Ottawa, 57-60; res fel chem, Royal Mil Col, Can, 60-62; from asst prof to assoc prof, Univ Toronto, 62-68; PROF CHEM, CARLETON UNIV, ONT, 68- *Mem:* Electrochem Soc; Chem Inst Can; fel Royal Inst Chem; Brit Polarographic Soc. *Res:* Theoretical electroanalytical chemistry; electrochemistry of transition metal ion complexes including photocatalytic and spectroscopic studies; battery systems and related power sources; photogalvanic cells; surface electrochemistry; electrocrystallization; electro-organic chemistry. *Mailing Add:* Dept Chem Carleton Univ Ottawa ON K1S 5B6 Can

BARRALL, EDWARD MARTIN, II, b Louisville, Ky, Dec 8, 34; m 77; c 3. ANALYTICAL CHEMISTRY. *Educ:* Univ Louisville, BS, 55, MS, 57; Mass Inst Technol, PhD(anal chem), 61. *Prof Exp:* Instr anal chem, Univ Ga, 57-58; asst, Mass Inst Technol, 58-60; mem part-time staff, Dewey & Almy Chem Co Div, W R Grace Corp, 60-61; anal chemist, Chevron Res Co, Va, 61-69; res chemist, 69-80, mgr org & polymer anal, Gen Prod Div, 80-85, MGR, MAT LAB, SYSTS STORAGE PROD DIV, IBM CORP, 85- *Concurrent Pos:* Mem part-time staff, E I du Pont de Nemours & Co, Inc, 57-59; vis prof, Univ Conn, 74-75; thesis adv, Univ Conn, 75-, Univ Southern Miss, 76- & Univ Calif, Berkeley, 79-81. *Honors & Awards:* Mettler Award, NAm Thermal Anal Soc, 73. *Mem:* AAAS; fel NAm Thermal Anal Soc; Int Confedn Thermal Anal. *Res:* Thermal methods of analysis; gas chromatography; polymer characterization; thermodynamics of mesophase transitions; physics of polymers; thermal properties of organic compounds; thermal properties of cross linked systems. *Mailing Add:* Dept E40-028 Systs Storage Prod Div IBM Corp 5600 Cottle Rd San Jose CA 95193

BARRALL, RAYMOND CHARLES, b Philadelphia, Pa, Jan 4, 30; m 53; c 5. RADIATION PHYSICS, HEALTH PHYSICS. *Educ:* Western NMex Univ, BS, 55; Univ Rochester, MS, 57. *Prof Exp:* Radiation physics fel, Brookhaven Nat Lab, 57; supv health physicist, Mare Island Naval Shipyard, 57-59; chief health physicist, IIT Res Inst, 59-65; dir health physics & safety, Stanford Univ, 65-75; dir res ctr, King Faisal Spec Hosp, 75-81, asst dir, Cancer Ther Inst, 81-83; DIR RADIATION PROTECTION, UNIV ILL, CHICAGO, 84- *Concurrent Pos:* AEC radiol physics fel, 56-57. *Mem:* Am Soc Testing & Mat; Am Asn Physicists in Med; Health Physics Soc; Am Nuclear Soc. *Res:* Radiation measurements, especially neutron cross-sections, flux and fluence with activation and moderator techniques; evaluation of radioactive impurities in radiopharmaceuticals; nuclear technology and applications. *Mailing Add:* 8001 Winter Circle Downers Grove IL 60516

BARRAN, LESLIE ROHIT, b Berbice, Guyana, July 29, 39; Can citizen; m 61; c 1. AGRICULTURAL MICROBIOLOGY. *Educ:* McGill Univ, BSc, 63, MSc, 65; Mich State Univ, PhD(biochem), 69. *Prof Exp:* RES SCIENTIST AGR MICROBIOL, PLANT RES CTR, AGR CAN, 71- *Concurrent Pos:* Coordr, Network of Can Nitrogen Fixation Researchers. *Mem:* Am Soc Microbiol. *Res:* Molecular genetics of nitrogen fixing bacteria. *Mailing Add:* 81 Renfrew Ottawa ON K1S 1Z6 Can

BARRANCO, SAM CHRISTOPHER, III, b Beaumont, Tex, Nov 17, 38; m; c 2. RADIOBIOLOGY, CANCER. *Educ:* Tex A&M Univ, BS, 60, MS, 62; Johns Hopkins Univ, PhD(cellular radiobiol), 69. *Prof Exp:* Asst prof surg, M D Anderson Hosp & Tumor Inst, 71-72; assoc prof biol, 72-78, ASSOC DIR BASIC & PRECLIN RES, CANCER CTR & PROF BIOL, UNIV TEX MED BR GALVESTON, 78- *Concurrent Pos:* NIH fel, M D Anderson Hosp & Tumor Inst, 69-71, Damon Runyon res grant, 69-71; NIH cancer grants, Univ Tex Med Br Galveston, 72-, 78-81 & 85-90. *Mem:* Am Soc Cell Biol; Am Asn Cancer Res; Cell Kinetics Soc; Tissue Cult Asn; Soc Anal Cytol. *Res:* Normal and tumor cell kinetics; cancer chemotherapy; radiobiology. *Mailing Add:* Radiation Therapy Univ Tex Med 310 Gail Borden Bldg Galveston TX 77550

BARRANGER, JOHN ARTHUR, b Baltimore, Md, Aug 5, 45; m 69; c 3. BIOCHEMISTRY, PEDIATRICS. *Educ:* Loyola Col, Baltimore, BS, 67; Univ Southern Calif, PhD(biochem), 73, MD, 74. *Prof Exp:* Intern, Univ Minn, 74-75, resident, 75-76; clin assoc, 76-77, CHIEF, SECT CLIN INVEST, NAT INST NEUROL & COMMUN DIS, NIH, 77-, CHIEF, MOLECULAR & MED GENETICS SECT, DEVELOP & METAB NEUROL BR, NAT INST NEUROL & COMMUN DIS & STROKE, 82-, ASSOC CHIEF, DEVELOP & METAB NEUROL BR, 83- *Concurrent Pos:* Res trainee, Univ Md, 67-69; res fel, Univ Southern Calif, 69-72, teaching asst, 72-74; clin assoc, develop & metab neurol, NIH, 76-77; clin staff, Clin Ctr, NIH, 76-78; Phillips Aull scholar; ed, Biochem Med. *Honors & Awards:* Arthur S Fleming Award, 85. *Mem:* Sigma Xi; AAAS; NY Acad Sci; Soc Inherited Metabolic Dis; Am Soc Biol Chemists. *Res:* Inborn errors of metabolism; molecular genetics; enzymology; lysosomal storage disorders. *Mailing Add:* Develop Metabolic Neurol Br NIH Bldg 10 Rm 4N-248 Bethesda MD 20205

BARRANGER, JOHN P, b Worcester, Mass, June 26, 30; m 56; c 8. INSTRUMENTATION. *Educ:* George Washington Univ, BS, 57; Purdue Univ, MS, 60; Case Western Reserve, PhD(eng), 69. *Prof Exp:* Engr, ITT Fed, 57-60 & Delco Electronics, 60-63; RES ENGR, LEWIS CTR, NASA, 63- *Concurrent Pos:* Consult to various orgn, 70- *Mem:* Inst Elec & Electronics

Engrs; Am Soc Nondestructive Testing; British Inst Nondestructive Testing. *Res:* Magnetics; magnetic measurements; material losses; network theory, eddy current and capacitance sensors; jet engine instrumentation; high temperature technology; strain gages. *Mailing Add:* 22324 Sharon Lane Fairview Park OH 44126

BARRANTE, JAMES RICHARD, b Torrington, Conn, Apr 30, 38; m 65; c 3. PHYSICAL CHEMISTRY. *Educ:* Univ Conn, BA, 60; Harvard Univ, MA, 62, PhD(chem), 64. *Prof Exp:* Sr res chemist, Olin-Mathieson Chem Corp, Conn, 64-66; asst prof phys chem, 66-69, assoc prof chem, 69-76, PROF PHYS CHEM, SOUTHERN CONN STATE UNIV, 76- *Concurrent Pos:* Instr, Eve Sch, Southern Conn State Univ, 65-66; consult, Macdermid Corp, 81-83. *Mem:* Am Chem Soc. *Res:* Nuclear magnetic resonance studies of inorganic polymers; x-ray crystal structure analysis; theoretical chemistry. *Mailing Add:* Dept Chem Southern Conn State Univ 501 Crescent St New Haven CT 06515

BARRAR, RICHARD BLAINE, b Dayton, Ohio, Oct 12, 23; m 47; c 4. MATHEMATICS. *Educ:* Univ Mich, MS, 48, PhD(math), 53. *Prof Exp:* Res assoc, Harvard Univ, 51-52; head theoret dept, McMillan Lab, 52-54; res physicist, Hughes Aircraft Co, 54-57; head, Antenna Group, Hoffman Lab, Inc, 57-58; sr opers res scientist, Syst Develop Corp, 58-66, sr scientist, 66-68; PROF MATH, UNIV ORE, 68- *Mem:* Am Math Soc; Soc Indust & Appl Math; Inst Elec & Electronics Engrs; Asn Comput Mach; Am Astron Soc. *Res:* Electromagnetic theory; existence theorems for partial differential equations; system analysis; celestial mechanics. *Mailing Add:* Dept Math Univ Ore Eugene OR 97403

BARRAS, DONALD J, b St Martinsville, La, Jan 11, 32; m 61; c 3. PHYSIOLOGY, BIOCHEMISTRY. *Educ:* Univ Southwestern La, BS, 54; Univ Miss, MS, 64; Miss State Univ, PhD(physiol), 70. *Prof Exp:* Teaching assoc physiol, Univ Miss, 62-64; res assoc biol warfare, Univ Okla, 64-65; asst prof physiol, Nicholls State Col, 65-67; res assoc, Miss State Univ, 67-70; prof biol, Norman Col, 70; assoc prof, 70-78, PROF PHYSIOL, TROY STATE UNIV, 78-, CHMN, DEPT BIOL SCI, 76-; COLLAB RES EXPERT BIOL CONTROL, AGR RES SERV, USDA, 70- *Mem:* Sigma Xi. *Res:* Biochemical and physiological studies of parasitoids in the use of biological control of harmful insects. *Mailing Add:* Dept Biol Sci Troy State Univ Main Campus Troy AL 36082

BARRAS, STANLEY J, b New Orleans, La, Jan 4, 36. FOREST ENTOMOLOGY. *Educ:* La State Univ, BS, 59, MS, 61; Univ Wis, PhD(entom), 65. *Prof Exp:* Res entomologist, Southern Forest Exp Sta, US Forest Serv, 65-75, proj leader, 75-76; FIDR staff, Forest Serv, 76-80, ASST DIR, SOUTHERN FOREST EXP STA, USDA, 80- *Mem:* Entom Soc Am; Soc Am Foresters; Sigma Xi. *Res:* Fungal repositories and associated microorganisms of bark beetles; interactions between bark beetle symbiotic fungi; ecology of symbiotic associations in pine phloem. *Mailing Add:* Southern Forest Exp Sta 701 Loyola Ave Rm T-10210 USPSB New Orleans LA 70113

BARRAT, JOSEPH GEORGE, b New Haven, Conn, May 30, 22; m 48; c 3. PLANT PATHOLOGY. *Educ:* RI State Col, BS, 48; Univ RI, MS, 51; Univ NH, PhD(bot, plant path), 58. *Prof Exp:* Exten specialist entom & plant path, State Agr Exten Serv, Univ RI, 48-50; plant pathologist, Nursery Improv Prog, Dept Hort, State Dept Agr, Washington, 50-55; state exten specialist plant path, Coop Exten Serv, WVa Univ, 58-70, assoc prof, 67-70, supt exp farm, 70-80, prof plant path, 70-88, State Exten Specialist Plant Path, 80-88; RETIRED. *Mem:* Am Phytopath Soc. *Res:* Stone and pome fruit disease and insect control; virus diseases of stone and pome fruits. *Mailing Add:* Rte 1 Box 714 Shepherdstown WV 25443

BARRATT, MICHAEL GEORGE, b London, Eng, Jan 26, 27; m 52; c 5. PURE MATHEMATICS. *Educ:* Oxford Univ, BA, 48, MA, 52, DPhil, 52; Manchester Univ, MSc, 68. *Prof Exp:* Jr lectr math, Oxford Univ, 50-52; fel, Oxford Univ, 52-56; lectr, Brasenose Col, 55-59; sr lectr, Manchester Univ, 59-63, reader, 63-64, prof pure math, 64; PROF MATH, NORTHWESTERN UNIV, 72- *Concurrent Pos:* Vis asst prof, Princeton Univ, 56-67; prof, Univ Chicago, 63-64 & Northwestern Univ, 69-70. *Mem:* Am Math Soc. *Res:* Algebraic topology, particularly homotopy theory; homological algebra; topology of manifolds. *Mailing Add:* Dept Math Northwestern Univ Evanston IL 60208

BARRATT, RAYMOND WILLIAM, b Holyoke, Mass, May 4, 20; div; c 2. MICROBIAL GENETICS. *Educ:* Rutgers Univ, BSc, 41; Univ NH, MSc, 43; Yale Univ, PhD(microbiol), 48. *Prof Exp:* Asst plant path & hort, Univ NH, 43-44; res assoc, Crop Protection Inst & asst plant pathologist, Conn Agr Exp Sta, 44-46; res assoc biol, Stanford Univ, 48-53, res biologist & actg asst prof, 53-54; asst prof bot, Dartmouth Col, 54-57, prof, 58-62, prof biol, 62-70, lectr microbiol, Med Sch, 61-70, chmn dept biol sci, 65-69; PROF BIOL & DEAN SCH SCI, HUMBOLDT STATE UNIV, 70-, DIR FUNGAL GENETICS STOCK, 65- *Concurrent Pos:* USPHS spec fel, 61-62. *Mem:* Am Inst Biol Sci; Genetics Soc Am; Sigma Xi. *Res:* Genetics and biochemistry of microorganisms. *Mailing Add:* Dept Biol Humboldt State Univ Arcata CA 95521

BARREKETTE, EUVAL S, b New York, NY, Feb 18, 31; m 56; c 3. ELECTROOPTICS, APPLIED MECHANICS. *Educ:* Columbia Univ, AB, 52, BS, 53, MS, 56, PhD(appl mech), 59. *Prof Exp:* Asst prof civil eng, Columbia Univ, 59-60; mem tech staff, IBM Corp, 60-63, tech asst to dir res, 63, mgr tech studies, 64-65, explor systs, 65-66, electrooptical technol, 67-71, prog dir advan eng, Corp Staff, 71-72, mgr advan technol, Res Div, 73-76, asst dir appl res, 76-81, asst dir comput sci, 81-84, prog dir optical technol, Corp Staff, 84-86; ENG CONSULT, 86- *Concurrent Pos:* Adj assoc prof, Columbia Univ, 60-70. *Mem:* Am Inst Aeronaut & Astronaut; Am Soc Civil Eng; Optical Soc Am; Soc Eng Sci; Inst Elec & Electronics Engrs; Sigma Xi; Nat Soc Prof Engrs. *Res:* Optical information processing; laser applications; microwave acoustics; computer printer and display technologies; semiconductor lasers. *Mailing Add:* 90 Riverside Dr New York NY 10024

BARRER, DANIEL EDWARD, b Oelwein, Iowa, Aug 24, 49; m 77; c 2. SYNTHETIC ORGANIC & NATURAL PRODUCTS CHEMISTRY, PATENT CHEMIST. *Educ:* Loras Col, BS, 71; John Carroll Univ, MS, 74. *Prof Exp:* Process develop chemist, Gen Elec Co, 73-74; res chemist, Lubrizol Corp, 74-79, proj mgr, 79-83, prof res chemist, 83-89; PATENT CHEMIST, 89- *Mem:* Am Chem Soc. *Res:* Synthetic lubricants and lubricant additives; fuel conserving motor oil additives; synthetic overbased detergents for motor oil and specialty chemical applications; awarded two US patents. *Mailing Add:* 755 E 258th St Euclid OH 44132

BARRERA, CECILIO RICHARD, b Rio Grande City, Tex, Nov 30, 42; m 70; c 2. MICROBIAL PHYSIOLOGY. *Educ:* Univ Tex, Austin, BA, 65, MA, 67, PhD, 70. *Prof Exp:* Fel biochem, Clayton Found Biochem Inst, Univ Tex, Austin, 70, res assoc, 70-75; asst prof, 75-81, ASSOC PROF BIOL, NMEX STATE UNIV, 81- *Concurrent Pos:* Served on Various Eval Panels, NIH & NSF. *Mem:* Am Soc Microbiol; AAAS; Sigma Xi; Soc Advan Chicanos & Native Americans Sci (secy). *Res:* Enzyme isolation and characterization; regulation of enzyme activity; microbiol ultrastructure. *Mailing Add:* Dept Biol NMex State Univ PO Box 30001 Las Cruces NM 88003

BARRERA, FRANK, b Havana, Cuba, Nov 7, 17; m 49; c 5. PHYSIOLOGY. *Educ:* Belen Col, BS, 34; Univ Havana, MD, 41. *Prof Exp:* Instr med, Sch Med, Univ Havana, 46-47; from instr to assoc prof physiol, Sch Med, Temple Univ, 55-73, from asst prof to assoc prof med, 63-73; chief Respiratory Serv, Am Hosp, 73-85; RETIRED. *Concurrent Pos:* Chief clin & cardio-respiratory physiol dept, Inst Cardiovasc & Thoracic Surg, Cuba, 51-60. *Honors & Awards:* Tamayo Award, 52-59; Farinas Award, 58. *Mem:* Am Physiol Soc; Sigma Xi. *Res:* Cardiopulmonary physiology. *Mailing Add:* 7095 Sunset Dr Miami FL 33143

BARRERA, JOSEPH S, b Brandon, Vt, Mar 13, 41; m 63; c 2. SOLID STATE PHYSICS. *Educ:* Harvey Mudd Col, BS, 62; Carnegie Inst Technol, MS, 63, PhD(solid state elec eng), 66. *Prof Exp:* Res scientist, Hewlett-Packard Labs, 67-74, mgr, Res & Develop Sect, Hewlett-Packard Assoc, 74-80; dir gas opers, 80-85, VPRES & GEN MGR, HARRIS MICROWAVE SEMICONDUCTORS, 85- *Mem:* Inst Elec & Electronics Engrs. *Res:* Fluctuation studies in space charge and recombination limited, single and double injection devices; solid state microwave bulk oscillator; III-V microwave devices; field effect transistors; Gunn microwave diode devices; Ga As digital and microwave ICs. *Mailing Add:* Harris Microwave Semiconductors 1530 McCarthy Blvd Milpitas CA 95035

BARRERAS, RAYMOND JOSEPH, b Albuquerque, NMex, June 8, 40; m 59; c 5. ORGANIC CHEMISTRY. *Educ:* Univ NMex, BS, 61; Mich State Univ, PhD(chem), 66. *Prof Exp:* From asst prof to assoc prof chem, Tuskegee Inst, 66-74; prof natural sci & head div, Navajo Community Col, Tsaile, Ariz, 74-78; prof biochem & dir tutorial prog, Sch Med, Morehouse Col, 78-80; prof med educ & dir admis, 80-84, SPEC ASST TO DEAN & RES RESOURCES COORDR, MOREHOUSE SCH MED, 84- *Concurrent Pos:* Res Corp Cottrell grant, 67-68; consult, Los Alamos Sci Lab, Univ Calif. *Mem:* AAAS; Am Chem Soc; Royal Soc Chem. *Res:* Synthesis of important organic intermediates; stereochemistry; reaction mechanisms; photochemistry; use of computers in education; medical data bases. *Mailing Add:* 720 Westview Dr SW Atlanta GA 30310

BARRERE, CLEM A, JR, b Bradford, Pa, Jan 5, 39; m 69. CHEMICAL ENGINEERING. *Educ:* Yale Univ, BEng, 60; Rice Univ, PhD(chem eng), 65. *Prof Exp:* Res group leader, 65-69, sr staff engr, 69-70, DIR TECH SERV, CONTINENTAL OIL CO, 70- *Mem:* Am Inst Chem Engrs. *Res:* Thermodynamics; phase behavior of vapor-liquid and vapor-solid systems; adsorption; chromatography; absorption. *Mailing Add:* 5430 Lynbrock Houston TX 77056

BARRETO, ERNESTO, b Colombia, SAm, Nov 9, 34; US citizen; m 60; c 2. ELECTRODYNAMICS, FLUID DYNAMICS. *Educ:* NY Univ, BA, 58, MS, 60. *Prof Exp:* Res assoc physics, Dept Indust Med, NY Univ-Bellevue Med Ctr, 59-60; sci dir physics, Marks Polarized Corp, 60-64; proj engr, Curtiss-Wright Corp, 64-69; SR RES ASSOC, ATMOSPHERIC SCI RES CTR, STATE UNIV NY, ALBANY, 69- *Concurrent Pos:* Consult, Mobil Res Labs & US Coast Guard, 72-; mem comt hazardous mat, Nat Res Coun, 74-76; comt mem, Subcomn IV, Ions-Aerosols-Radioactivity, Int Comn Atmospheric Elec, 74- *Mem:* Am Phys Soc; Am Geophys Union; Am Inst Aeronaut & Astronaut; Elec Soc Am. *Res:* Transformation of stored electrical energy into heat in electrical discharges; oil transportation hazards, lightning studies, charged aerosol physics. *Mailing Add:* Atmospheric Sci Res Ctr 1400 Washington Ave Albany NY 12222

BARRETT, ALAN H, radio astronomy; deceased, see previous edition for last biography

BARRETT, ANTHONY GERARD MARTIN, b Exeter, Eng, Mar 2, 52; m 73; c 2. HETEROCYCLIC NATURAL PRODUCT CHEMISTRY. *Educ:* Imp Col London, BSc, 73, PhD(org chem) & DIC(org chem), 75. *Prof Exp:* From lectr to sr lectr org chem, Imp Col, 75-83; prof chem, Northwestern Univ, 83-90; PROF CHEM, COLO STATE UNIV, 90- *Concurrent Pos:* Consult, W R Grace, 80, G D Searle, 84 & B F Goodrich, 85, NIH Medicinal Chem, 86, Amoco, 88; Camille & Henry Dreyfus Teacher Scholar Award, 87; Japan Soc Promotion of Sci Fel, 89. *Honors & Awards:* Meldola Medal, Royal Soc Chem, London, 80, Harrison Medal, 82; Arthur C Cope Scholar Award, Am Chem Soc, 86; Corday Morgan Medal, Royal Soc Chem, London, 86. *Mem:* Royal Soc Chem, London; Am Chem Soc. *Res:* The development of novel synthetic methods and their application to the total synthesis of bioactive natural products; design of host guest ensembles; discovery of versatile organometallic reagents; total synthesis. *Mailing Add:* Dept Chem Colo State Univ Ft Collins CO 80523

BARRETT, BRUCE RICHARD, b Kansas City, Kans, Aug 19, 39; m 79. THEORETICAL NUCLEAR PHYSICS. *Educ:* Univ Kans, BS, 61; Stanford Univ, MS, 64, PhD(physics), 67. *Prof Exp:* Res fel physics, Weizmann Inst Sci, Israel, 67-68; Andrew Mellon res fel, Univ Pittsburgh, 68-69, res assoc, 69-70; from asst prof to assoc prof, 70-76, PROF PHYSICS, UNIV ARIZ, 76- *Concurrent Pos:* Alfred P Sloan Found res fel, 72-76; Alexander von Humboldt fel, 76-77; Fundamental Onderzoek der Materie fel, Netherlands, 80; Alexander von Humboldt sr US scientist award, 83-84; prog dir, Theoret Physics Div, NSF, 85-87. *Honors & Awards:* Alexander von Humboldt Sr US Scientist Award. *Mem:* Fel Am Phys Soc; Sigma Xi. *Res:* Structure of finite nuclei; particle-hole and particle-particle states with realistic nuclear forces; effective interaction calculations with realistic nuclear forces; exact reaction matrix calculations; real nuclear three-body forces; microscopic theory of nuclear collective motion; interacting Boson model of nuclear collective properties. *Mailing Add:* Dept Physics Bldg 81 Univ Ariz Tucson AZ 85721

BARRETT, C BRENT, b Angora, Maine, Aug 30, 55. PHARMACOLOGY. *Educ:* Univ Maine, BS, 77; Boston Col, MS, 83, PhD(biol), 88. *Prof Exp:* POSTDOCTORAL FEL CELL BIOL & CHEM, UNIV COLO HEALTH SCI CTR, 88- *Concurrent Pos:* Nat res serv award, NIH, 89. *Mem:* Am Soc Cell Biol. *Mailing Add:* Dept Pharmacol Univ Colo Health Sci Ctr Box C-236 4200 E Ninth Ave Denver CO 80262

BARRETT, CHARLES SANBORN, b Vermillion, SDak, 02; m 28; c 1. MATERIALS SCIENCE, PHYSICAL METALLURGY. *Educ:* Univ SDak, BS, 25; Univ Chicago, PhD(physics), 28. *Hon Degrees:* MA, Oxford Univ, Eng. *Prof Exp:* Asst physicist, Naval Res Lab, 28-32; metallurgist, Metall Res Lab, Carnegie Inst Technol, 32-46, lectr metall, 32-41, assoc prof, 41-44, prof, 45-46; res prof, Inst Study Metals, 46-71, EMER RES PROF, UNIV CHICAGO, 71-; EMER SR RES PROF & ADJ PROF PHYSICS, UNIV DENVER, 71- *Concurrent Pos:* Nat Res Coun, 48-65; vis prof, Univ Birmingham, 52, Univ Denver, 62, Stanford Univ, 63, Univ Va, 68, 70 & Ga Inst Technol, 73; George Eastman prof, Oxford Univ, 65-66; co-ed, Advances in X-Ray Anal, 70-; adj prof, Univ Denver, 70- *Honors & Awards:* Mathewson Gold Medal, Am Inst Mining, Metall & Petrol Engrs, 34, 44 & 50 & Hume Rothery Award, 75; Clamer Medal, Franklin Inst, 50; Howe Medal, Am Soc Metals, Sauveur Medal & Gold Medal; Heyn Medal, Ger Metall Soc; Gold Medal, Japan Inst Metals, 76. *Mem:* Emer mem Nat Acad Sci; fel Am Phys Soc; fel Am Inst Mining, Metall & Petrol Engrs; hon mem Am Soc Metals; emer mem Japan Inst Metals. *Res:* Preferred orientations in metals and alloys; x-ray equipment and methods; structure, deformation and transformation of solids; crystallography at low temperatures; stress analysis by X-ray diffraction in metals and composites; metallurgical research; editing; author. *Mailing Add:* Eng Div Univ Denver Denver CO 80208

BARRETT, DENNIS, b Philadelphia, Pa, Feb 13, 36; m 64. DEVELOPMENTAL BIOLOGY, BIOCHEMICAL GENETICS. *Educ:* Univ Pa, AB, 57; Calif Inst Technol, PhD(biochem), 63. *Prof Exp:* USPHS res fel biochem, Univ Calif, Berkeley, 63-65; asst prof zool, Univ Calif, Davis, 65-73; asst prof biol sci, 73-77, ASSOC PROF BIOL SCI, UNIV DENVER, 77- *Concurrent Pos:* Spec consult, US AID Mission, Tunisia, 69; instr physiol, Marine Biol Lab, Woods Hole, 72-76; vis prof, Univ Edinburgh, 83-84. *Mem:* AAAS; Am Soc Zoologists; Soc Develop Biol; NY Acad Sci. *Res:* Developmental biochemistry; control of gene expression in development; substrate specificity of proteases. *Mailing Add:* Dept Biol Sci Univ Denver Denver CO 80208-0178

BARRETT, EAMON BOYD, mathematics, artificial intelligence, for more information see previous edition

BARRETT, EDWARD JOSEPH, b New York, NY, July 4, 31; c 2. ORGANIC CHEMISTRY. *Educ:* Fordham Univ, BA, 53; Columbia Univ, MA, 60, PhD(chem), 62. *Prof Exp:* From asst prof to assoc prof chem, 64-70, chmn dept, 68-73, dean & assoc provost, 74-77, PROF CHEM, HUNTER COL, 71- *Mem:* AAAS; Am Chem Soc; NY Acad Sci; Royal Soc Chem. *Res:* Design of atomic and molecular models. *Mailing Add:* Dept Chem Hunter Col 695 Park Ave New York NY 10021

BARRETT, ELLEN FAYE, b Ontario, Ore, Nov 17, 44; m 68; c 2. NEUROBIOLOGY & NEUROSCIENCES, PHYSIOLOGY. *Educ:* Univ Wash, BS, 66, PhD(physiol & biophysics), 72. *Prof Exp:* NIH fel neurophysiol, Med Sch, Univ Colo, 72-73; fel neurobiol, Harvard Med Sch, 73; asst prof neurobiol & zool, Univ Iowa, 74; from asst prof to assoc prof, 78-90, PROF PHYSIOL & BIOPHYS, UNIV MIAMI, 90- *Concurrent Pos:* Prin investr, NIH grant, 75-93; co-prin investr, Nat Parkinson Found grant, 78-79. *Honors & Awards:* Javits Neuroscience Investr Award, NIH, 86. *Mem:* Am Physiol Soc; AAAS; Soc Neurosci. *Res:* Mechanisms of transmitter release from nerve terminals; electrical properties of myelinated axons and motor nerve terminals. *Mailing Add:* Dept Physiol & Biophys Sch Med Univ Miami PO Box 016430 Miami FL 33101

BARRETT, GARY WAYNE, b Princeton, Ind, Jan 3, 40; m 69; c 1. ECOLOGY. *Educ:* Oakland City Col, Ind, BS, 61; Marquette Univ, MS, 63; Univ Ga, PhD(zool), 67. *Hon Degrees:* DSc, Oakland City Col, 87. *Prof Exp:* Asst prof biol, Drake Univ, 67-68; from asst prof to prof zool, Miami Univ, 68-86, actg dir, Inst Environ Sci, 70-71, dep dir, 71-77, dep dir res, 77-81, coordr environ educ, 76-77, CO-DIR, ECOL RES CTR, UNIV MIAMI, 77-, DISTINGUISHED PROF ECOL, 86- *Concurrent Pos:* NSF grants, 70, 71, 75, 78-79, 80-84, 85 & 89-93; adv, Citizens Task Force on Environ Protection, 71-72; Environ Protection Agency Grant, 80; dir ecol prog, NSF, 81-83; Earthwatch grants, 86, 87 & 88. *Honors & Awards:* Award of Distinction, Inst Environ Sci, 77. *Mem:* Fel AAAS; Ecol Soc Am; Am Soc Mammal; Wildlife Soc; Am Inst Biol Sci; Sigma Xi; Am Soc Naturalists; Asn Ecosyst Res Ctrs; Int Asn Ecol; Int Soc Behav Ecol. *Res:* Pesticide stresses on total ecosystems; mammalian population regulation and dynamics; species diversity in nature; bioenergetics of small mammal populations; importance of ecotone areas in relation to agricultural productivity. *Mailing Add:* Dept Zool Miami Univ Oxford OH 45056

BARRETT, HARRISON HOOKER, b Springfield, Mass, July 1, 39; m 59; c 2. MEDICAL INSTRUMENTATION, OPTICS. *Educ:* Va Polytech Inst, BS, 60; Mass Inst Technol, SM, 62; Harvard Univ, PhD(appl physics), 69. *Prof Exp:* From res scientist to prin res scientist, Res Div, Raytheon Co, 62-74; assoc prof, 74-76, PROF, RADIOL DEPT & OPTICAL SCI CTR, UNIV ARIZ, 76- *Concurrent Pos:* Alexander von Humboldt award. *Mem:* Am Phys Soc; fel Optical Soc Am. *Res:* Instrumentation for radiology and nuclear medicine; medical ultrasound; optical data processing. *Mailing Add:* Dept Radiol Univ Ariz Col Med Tucson AZ 85724

BARRETT, IZADORE, b Vancouver, BC, Oct 4, 26; m 58; c 4. MARINE FISHERIES. *Educ:* Univ BC, BA, 47, MA, 49; Univ Wash, PhD, 80. *Prof Exp:* Biologist, Trout Hatcheries, BC Game Comn, 52-56; scientist fishery biol, Inter-Am Trop Tuna Comn, 56-61, sr scientist tuna physiol, 61-67, ed, Bull, 64-65; chief fishery biologist, Inst Fishery Develop, Food & Agr Orgn, Chile, 67-69; asst dir, Fishery-Oceanog Ctr, 70-72, dep dir, 72-77, dir, 77-88, DIR SCI & RES, SOUTHWEST REGION, NAT MARINE FISHERIES SERV, NAT OCEANIC & ATMOSPHERIC ADMIN, US DEPT COM, 88- *Concurrent Pos:* Marine fisheries adv, Govt Chile, Food & Agr Orgn of UN, 69-70; res assoc, Scripps Inst Oceanog, 77- *Mem:* AAAS; Am Soc Ichthyologists & Herpetologists; Am Inst Fishery Res Biologists (vpres, 73-74); Soc Marine Mammal; Western Soc Naturalists. *Res:* Management and administration of marine fisheries research. *Mailing Add:* Southwest Fisheries Sci Ctr PO Box 271 La Jolla CA 92038

BARRETT, J(AMES) CARL, b Portsmouth, Va, Dec 28, 46; m 68; c 3. CANCER, CELL BIOLOGY. *Educ:* Col William & Mary, BS, 69; Johns Hopkins Univ, PhD(biophys chem), 74. *Prof Exp:* Res assoc chem carcinogenesis, Johns Hopkins Univ, 74-77; NAT INST ENVIRON HEALTH SCI, 77-; CHIEF, LAB MOLECULAR CARCINOGENESIS, 87- *Concurrent Pos:* Adj prof, Univ NC, 78- *Mem:* Am Asn Cancer Res; Am Chem Soc; Tissue Culture Asn; AAAS. *Res:* Cellular and molecular mechanisms of carcinogenesis with particular emphasis on chemical carcinogenesis, the relationship between carcinogenesis and mutagenesis, tumors suppressor genes and oncogenes. *Mailing Add:* Nat Inst Environ Health Sci PO Box 12233 Research Triangle Park NC 27709

BARRETT, JAMES E, b Camden, NJ, Aug 9, 42; m 62; c 3. DRUG ABUSE, NEUROCHEMISTRY. *Educ:* Pa State Univ, PhD, 72. *Prof Exp:* PROF PSYCHIAT PHARMACOL & MED PSYCHOL, UNIFORMED SERV UNIV HEALTH SCI, 79- *Concurrent Pos:* Consult, NASA Life Sci Space Lab; ed, Advances Behav Pharmacol & J Exp Anal Behav. *Mem:* Soc Neurosci; AAAS; Behav Pharmacol Soc; Fed Am Soc Exp Biol; Am Soc Pharmacol & Exp Therapeut; Am Col Neuropsychopharmacol. *Res:* Behavioral and neurochemical effects of drugs; particularly those used in treating anxiety and drugs of abuse; also in dynamic interactions between behavior and neurochemistry. *Mailing Add:* Dept Psychiat Uniformed Serv Univ Health Sci 4301 Jones Bridge Rd Bethesda MD 20814-4799

BARRETT, JAMES MARTIN, b Chippewa Falls, Wis, July 4, 20; m 47; c 5. PROTOZOOLOGY. *Educ:* Marquette Univ, BS, 47, MS, 49; Univ Ill, PhD(zool), 53. *Prof Exp:* Asst zool, Marquette Univ, 47-48; asst, Univ Ill, 48-50; from instr to prof biol, Marquette Univ, 51-86; RETIRED. *Mem:* Soc Protozool; Am Soc Zool; Sigma Xi. *Res:* Biology of protozoa. *Mailing Add:* Dept Biol Marquette Univ 1311 W Wisconsin Ave Milwaukee WI 53233

BARRETT, JAMES PASSMORE, b Atlanta, Ga, June 20, 31; m; c 2. FORESTRY. *Educ:* NC State Univ, BS, 54; Duke Univ, MF, 57, PhD(forest mensuration), 62. *Prof Exp:* Forest researcher, US Forest Serv, 57-62; from asst prof forestry to assoc prof, 62-75, PROF FOREST BIOMET, UNIV NH, 75- *Mem:* Soc Am Foresters. *Res:* Applications of mathematical and statistical techniques in forestry. *Mailing Add:* Dept Forestry Univ NH Durham NH 03824

BARRETT, JAMES THOMAS, b Iowa, May 20, 27; m 49, 67; c 4. IMMUNOLOGY, MICROBIOLOGY. *Educ:* Univ Iowa, BA, 50, MS, 51, PhD, 53. *Hon Degrees:* MD, Univ Repub Uruguay. *Prof Exp:* Asst prof microbiol, Sch Med, Univ Ark, 53-57; from asst prof to assoc prof, 57-69, PROF MICROBIOL, SCH MED, UNIV MO-COLUMBIA, 69- *Concurrent Pos:* NIH spec fel, Sweden, 63-64; sabbatical, Dept Immunol, Med Microbiol Inst, Fac Med, Gothenberg Univ, 70-71; Nat Acad Sci exchange scientist, Romanian Acad Sci, 71; sabbatical, Inst Med Microbiol, Vet Fac, Ludwig Maximilians Univ, Munich & NIH Fogarty sr int fel, 77-78 & Univ Murcia, Spain, 86-87; Fulbright scholar, Univ Repub Montevideo, 84; vis prof, Univ Autonoma of Guadalajara, Mex, 86- *Mem:* Am Asn Immunol; Am Soc Microbiol; Sigma Xi. *Res:* Immunological nature of proenzyme-enzyme pairs; spider venoms and ultrasound. *Mailing Add:* M258 Med Sci Bldg Univ Mo Columbia MO 65212

BARRETT, JERRY WAYNE, b Marshall, Tex, Apr 29, 36; m 59; c 2. HISTORY & PHILOSOPHY OF SCIENCE, SCIENCE EDUCATION & POLICY. *Educ:* ETex Baptist Col, BS, 61; Baylor Univ, PhD(chem), 68. *Prof Exp:* From asst prof to prof chem, Samford Univ, 66-76; sr lectr chem, Hong Kong Baptist Col, 77-84, head dept, 80-84, dean sci, 82-88, ACAD VPRES, HONG KONG BAPTIST COL, 88- *Mem:* Am Chem Soc; Sigma Xi; Hong Kong Chem Soc; fel Royal Soc Chem. *Res:* Electrodeposition, corrosion, electrodissolution and chemical plating of metals; electroanalytical methods; electrode kinetics and electrode reaction mechanism studies; philosophy of science. *Mailing Add:* Hong Kong Baptist Col 224 Waterloo Rd Kowloon Hong Kong

BARRETT, JOHN HAROLD, b Springfield, Mo, Oct 9, 26; m 52; c 1. THEORETICAL SOLID STATE PHYSICS. *Educ:* Rice Inst, BS, 48, MA, 50, PhD(physics), 52. *Prof Exp:* Eli Lilly fel physics, Mass Inst Technol, 52-53; asst prof, NC State Col, 53-56 & La State Univ, 56-59; physicist, Oak Ridge Nat Lab, 59-78, sr res staff mem, 78-90; RETIRED. *Concurrent Pos:* Consult, 90- *Mem:* Am Phys Soc; AAAS. *Res:* Particle-solid interactions. *Mailing Add:* 827 W Outer Dr Oak Ridge TN 37831-8413

BARRETT, JOHN NEIL, b Rochester, Minn, June 1, 43; m 68; c 2. NEUROSCIENCES & NEUROBIOLOGY, PHYSIOLOGY. *Educ:* St Mary's Col, Minn, BA, 65; Univ Wash, PhD(physiol & biophysics), 72. *Prof Exp:* Res assoc neuroanat, Med Ctr, Univ Colo, 72-73; fel neurobiol, Harvard Med Sch, 73 & Univ Iowa, 74; from asst prof to assoc prof, 74-85, PROF PHYSIOL & BIOPHYS, UNIV MIAMI, 85- *Concurrent Pos:* Prin investr, NIH grants, 75-93; co-prin investr, Nat Parkinson Found grant, 78-79 & 85-88; Javits Neurosci Investr Award. *Mem:* Soc Neurosci. *Res:* Trophic control of dopaminergic and cholinergic mammalian central neurons; neuronal cell culture; neuronal membrane electrical properties; physiology. *Mailing Add:* Dept Physiol & Biophys Univ Miami PO Box 016430 Miami FL 33101

BARRETT, JOSEPH JOHN, b Scranton, Pa, Mar 11, 36; m 67; c 2. QUANTUM OPTICS, MOLECULAR SPECTROSCOPY. *Educ:* Univ Scranton, BS, 58; Fordham Univ, MS, 60, PhD(physics), 64. *Prof Exp:* Res asst physics, Fordham Univ, 62-64; sr scientist physics, Perkin-Elmer Corp, 64-70; sr physicist, 70-80, mgr laser instrumentation, 80-85, SR RES ASSOC, ALLIED-SIGNAL CORP, 85- *Honors & Awards:* Williams-Wright Award, 86. *Mem:* Optical Soc Am; Am Phys Soc; Sigma Xi. *Res:* Lasers for spectroscopic applications with emphasis on the generation and analysis of coherent Raman radiation; laser imaging research. *Mailing Add:* Allied-Signal Inc 101 Columbia Rd Box 1021 Morristown NJ 07962-1021

BARRETT, LIDA KITTRELL, b Houston, Tex, May 21, 27; wid; c 3. MATHEMATICS. *Educ:* Rice Inst, BA, 46; Univ Tex, MA, 49; Univ Pa, PhD(math), 54. *Prof Exp:* Mathematician, Schlumberger Well Surv Corp, 46-47; teacher math, Tex State Col Women, 47-48; res mathematician, Defense Res Lab, Univ Tex, 49-50; asst instr math, Univ Pa, 53-54; instr, Univ Conn, 55-56; lectr, Univ Utah, 56-61; from assoc prof to prof math, Univ Tenn, Knoxville, 61-80, head dept, 73-80; prof math & assoc provost, Northern Ill Univ, 80-87; PROF MATH & STATIST & DEAN, COL ARTS & SCI. MISS STATE UNIV, 87- *Concurrent Pos:* Vis lectr, Univ Wis, 59-60; consult, Oak Ridge Nat Lab, 64-75. *Mem:* Am Math Soc; Math Asn Am (pres, 89-90); Soc Indust & Appl Math; Sigma Xi. *Res:* Point set topology; applications of topology in metallurgy. *Mailing Add:* Col Arts & Sci Mississippi MS 39762

BARRETT, LOUIS CARL, b Murray, Utah, Jan 23, 24; m 47; c 4. MATHEMATICS. *Educ:* Univ Utah, BS, 48, MS, 51, PhD(appl math), 56. *Prof Exp:* Instr math, Univ Utah, 53-56; assoc prof math & physics, Ariz State Univ, 56-57; from assoc prof to prof math, SDak Sch Mines & Technol, 57-65, head dept, 60-65; prof & chmn dept, Clarkson Col Technol, 65-67; prof & head dept, 67-72, PROF MATH, MONT STATE UNIV, 72- *Concurrent Pos:* Res mathematician, Holloman Air Develop Ctr, NMex, 55, math consult, 56; res mathematician, US Naval Ord Test Sta, Calif, 57, 58, 63-64, consult, 63-64; lectr, Ariz State Univ, 57 & NSF Inst, SDak, 59; mem anal panel, NSF Modules & Monographs Undergrad Math & Its Appln Proj, Educ Develop Ctr, Newton, Mass, 78. *Mem:* Am Math Soc; Math Asn Am; Soc Indust & Appl Math. *Res:* Eigenvalue problems; differential equation theory; boundary value problems; optimization problems; Ordinary & Interface Strum-Liouville Systems; linear algebra; statistics; probability, mathematical physics and analysis; calculus variations; advanced engineering mathematics. *Mailing Add:* Dept Math 1721 S Willson Bozeman MT 59715

BARRETT, O'NEILL, JR, b Baton Rouge, La, Mar 21, 29; m 52; c 3. HEMATOLOGY, ONCOLOGY. *Educ:* La State Univ, BS, 49, MD, 53; Baylor Univ, MSc, 58. *Prof Exp:* Clin investr hemat & asst dir basic sci course, Walter Reed Army Inst Res, 59-60, chief gen med & hemat, Madigan Gen Hosp, 60-62, dep chief dept med, Letterman Gen Hosp San Francisco, Calif, 63-68, chief dept med, Tripler Gen Hosp, Honolulu, 68-71, chief dept med, Walter Reed Gen Hosp, 71-77; prof med, Col Med, Univ S Fla, 73-77, prof comprehensive med & chmn dept, 74-77; undergrad distinguished adj prof pharm & grad prof med, 77-80, asst dean acad affairs, Sch Med, 80-87, PROF MED, UNIV SC, 80-, CHMN, DEPT MED, 87- *Concurrent Pos:* Clin asst prof, Sch Med, Univ Calif, San Francisco, 63-68; clin prof, Sch Med, Univ Hawaii, 69-71. *Mem:* Fel Am Col Physicians; Am Soc Hemat; Am Fedn Clin Res; Am Soc Clin Oncol; AMA; fel Am Col Pharm Ther. *Mailing Add:* Univ S Carolina Sch Med Five Richland Med Park Columbia SC 29203

BARRETT, PAUL HENRY, b Petaluma, Calif, Dec 4, 22; m 48; c 3. PHYSICS. *Educ:* Mont State Col, BS, 44; Univ Calif, PhD(physics), 51. *Prof Exp:* Res assoc physics, Cornell Univ, 51-53; asst prof, Syracuse Univ, 53-55; assoc prof, 55-66, PROF PHYSICS, UNIV CALIF, SANTA BARBARA, 66- *Mem:* Am Phys Soc. *Res:* Solid state physics. *Mailing Add:* Dept Physics Univ Calif Santa Barbara CA 93106

BARRETT, PETER FOWLER, b Port Dover, Ont, May 27, 39; m 65; c 2. INORGANIC CHEMISTRY. *Educ:* Queen's Univ, Ont, BSc, 61, MSc, 62; Univ Toronto, PhD(inorg chem), 65. *Prof Exp:* Nat Res Coun overseas fel, 65-67; from asst prof to assoc prof, 67-84, actg chmn dept chem, 75-76, chmn, 83-86, PROF INORG CHEM, TRENT UNIV, 84-, ASSOC DEAN SCI, 88- *Mem:* Chem Inst Can. *Res:* Kinetics and mechanisms of reactions of transition metal carbonyl complexes; energy storage by supercooling of liquid salt hydrates. *Mailing Add:* Dept Chem Trent Univ Peterborough ON K9J 7B8 Can

BARRETT, PETER VAN DOREN, b Los Angeles, Calif, Oct 22, 34; m 59; c 3. INTERNAL MEDICINE, GASTROENTEROLOGY. *Educ:* Univ Calif, Los Angeles, BA, 56; Harvard Med Sch, MD, 60; Am Bd Internal Med, dipl & cert internal med & gastroenterol. *Prof Exp:* From intern to asst resident med, Mass Gen Hosp, Boston, 60-62; clin assoc, NIH, 62-65; resident, Mass Gen Hosp, Boston, 65-66; asst prof med, Sch Med, Univ Calif, Los Angeles, 67-71; assoc chief, Div Gastroenterol, Harbor Gen Hosp, Torrance, 67-75; assoc prof, 71-79, PROF MED, SCH MED, UNIV CALIF, LOS ANGELES, 79-; DIR MED EDUC, ST MARY MED CTR, LONG BEACH, CALIF, 75- *Concurrent Pos:* Fel gastroenterol, Wadsworth Vet Admin Hosp, Los Angeles, 66-67; NIH res grants, 67-76. *Mem:* Am Gastroenterol Asn; fel Am Col Physicians; Am Fedn Clin Res. *Res:* Hepatic physiology and bile pigment metabolism. *Mailing Add:* St Mary Med Ctr Univ Calif Los Angelea Sch Med Long Beach CA 90801

BARRETT, RICHARD JOHN, b West Pittston, Pa, Mar 28, 45; m 69; c 2. TECHNICAL ASSESSMENT, NUCLEAR REACTOR SAFETY. *Educ:* Univ Scranton, BS, 67; Univ Va, PhD(physics), 73. *Prof Exp:* Res assoc, Case Western Reserve Univ, 72-75; mem staff, Theoret Div, Los Alamos Nat Lab, 75-79, Systs Anal Div, 79-82; NUCLEAR ENG, US NUCLEAR REGULATORY COMN, 82- *Concurrent Pos:* Res prog analyst, Off Energy Res, US Dept Energy, 80- *Mem:* Am Nuclear Soc. *Res:* Formal analytical techniques to evaluate the technical, economic and safety performance of advanced energy technologies; identify areas of needed research. *Mailing Add:* 7501 Mill Run Dr Derwood MD 20855

BARRETT, RICHARD JOHN, b Newark, NJ, Oct 10, 49; m 71; c 2. PHARMACOLOGY. *Educ:* Villanova Univ, BS, 71; Fairleigh Dickinson Univ, MS, 79; Univ Houston, PhD(pharmacol), 82. *Prof Exp:* Asst scientist I, Hoffmann La Roche Inc, 71-75; lab supvr, Hoechst-Roussell Pharmaceut Div, Am Hoechst Co, 75-79; teaching training, Univ Houston, 79-82; res pharmacologist, Stuart Pharmaceut Div, ICI Am, Inc, 82-85; sr res biologist, 85-87, RES ASSOC, A H ROBINS CO, 87- *Concurrent Pos:* Adj asst prof pharmacol & toxicol, Philadelphia Col Pharm & Sci, 82-85; mem coun, High Blood Pressure Res & Kidney & Cardiovasc Dis, Am Heart Asn. *Mem:* Am Soc Pharmacol & Exp Therapeut; Soc Neurosci; Am Soc Hypertension; Am Heart Asn. *Res:* Cardiovascular pharmacology and physiology; renal and neural actions of drugs in the therapy of cardiovascular diseases. *Mailing Add:* Dept Pharmacol A H Robins Res Labs PO Box 26609 Richmond VA 23261-6609

BARRETT, ROBERT, b Los Angeles, Calif, Jan 4, 14; m 38; c 1. NEUROPHYSIOLOGY. *Educ:* Univ Calif, Los Angeles, BA, 34, MA, 65, PhD(zool), 69. *Prof Exp:* Tutor & instr sociol, Harvard Univ, 34-36; mem, NSF Cetacean Exped, Tierra del Fuego, 68; NIMH fel, Brain Res Inst, 69-70, dir, 70-85, EMER DIR, DEPT ENG & SCI, EXTEN & LECTR BIOL, UNIV CALIF, LOS ANGELES, 85- *Mem:* Am Inst Biol Sci; Soc Neurosci; Am Soc Zool. *Res:* Sensory physiology, particularly in areas of heat sensing, olfaction and taste; comparative neurophysiology. *Mailing Add:* 498 Drexel Dr Santa Barbara CA 93103

BARRETT, ROBERT EARL, physical chemistry, polymer chemistry, for more information see previous edition

BARRETT, ROLIN F(ARRAR), b White Sulphur Springs, WVa, Aug 25, 37; m 60; c 2. MECHANICAL ENGINEERING. *Educ:* NC State Univ, BS, 59, MS, 62, PhD(mech eng), 65. *Prof Exp:* Consult engr, L E Wooten & Co, 59-60; from instr to assoc prof mech eng, NC State Univ, 62-76, prof & asst admin dean res, 76-80; CONSULT, 80- *Concurrent Pos:* Shell Oil fel. *Honors & Awards:* Centennial Award, Am Soc Mech Engrs; Ralph Teetor Award, Soc Automotive Engrs. *Mem:* Soc Automotive Engrs; Am Inst Aeronaut & Astronaut; Am Soc Mech Engrs; Sigma Xi; Am Soc Metals; Am Soc Testing & Mat. *Res:* High speed mass transportation; turbine engine diagnostic techniques; mechanical design; transportation; automobile engineering; highway safety. *Mailing Add:* 3701 National Dr Suite 202 Raleigh NC 27612

BARRETT, SPENCER CHARLES HILTON, b Bushey, Hertfordshire, Eng, June 7, 48; m 73; c 2. BOTANY, POPULATION BIOLOGY. *Educ:* Univ Reading, UK, BSc Hons, 71; Univ Calif, Berkeley, PhD(bot), 77. *Prof Exp:* Weed biologist, Commonwealth Develop Corp, 69-70; lectr ecol, Univ Calif, Santa Cruz, 76; PROF BOT, UNIV TORONTO, 77- *Concurrent Pos:* Chmn & mem Pop Biol Grants Panel, Res Coun Can, 84-; E W R Steacie mem Fel, 88-89; assoc ed, J Evolution. *Mem:* Soc for Study Evolution; Brit Ecol Soc; Am Soc Plant Taxon; Asn for Trop Biol. *Res:* Ecological genetics and reproductive biology of plants; evolution and ecology of weeds; systematics of Pontederiaceae; biology of the water hyacinth; conservation biology. *Mailing Add:* Dept Bot Univ Toronto Toronto ON M5S 3B2 Can

BARRETT, STUART WILLIAM, chemical engineering, for more information see previous edition

BARRETT, TERENCE WILLIAM, biophysics, for more information see previous edition

BARRETT, WALTER EDWARD, b Stamford, Conn, May 26, 21; m; c 4. MEDICAL SCIENCES GENERAL, TOXICOLOGY. *Educ:* Columbia Univ, AB, 44; Princeton Univ, MA, 57, PhD, 58. *Prof Exp:* From asst pharmacologist to sr pharmacologist, Ciba Pharm Co, 44-61, assoc dir gen pharmacol, 62-64, head, 64-66, from asst dir macrobiol to dir pharmacol, 66-69; dir pharmacol, Parke, Davis & Co, 69-70; clin res assoc, Sandoz Res Inst, Sandoz Pharmaceut Corp, 71-72, head short term res & develop, 72-73, dir short term res & develop, Pharmaceut Res & Develop, 74-83, dir med res, 83-86; PRES, WEB INT RES CORP, 87- *Concurrent Pos:* Mem med adv bd, Coun High Blood Pressure Res, Am Heart Asn; mem comt, Am Soc Clin Pharmacol & Therapeut. *Mem:* Am Soc Pharmacol; NY Acad Sci; Sigma Xi; Am Soc Clin Pharmacol & Therapeut. *Res:* Mechanism of action of autonomic drugs; cardiovascular drugs; clinical pharmacology and research; biopharmaceutics. *Mailing Add:* Four Hoffman Ct Convent Station NJ 07961

BARRETT, WILLIAM AVON, b Central City, Nebr, Nov 8, 30; m 76; c 2. COMPUTER SCIENCE, ELECTRICAL ENGINEERING. *Educ:* Univ Nebr, BS, 52, MS, 53; Univ Utah, PhD(physics), 57. *Prof Exp:* Mem tech staff devices, Bell Tel Labs, 57-63; asst prof physics, Muhlenberg Col, 63-66; assoc prof elec eng, Lehigh Univ, 66-74; staff mem comput, Hewlett-Packard Co, 74-84; DIR SOFTWARE, LASA INDUST, INC, 86- *Concurrent Pos:* Consult, Bell Tel Labs, 64-74; RCA Astro-Electronics, 71, Comput Ctr, Lehigh Univ, 72 & Int Tel & Tel Corp, 73. *Mem:* Inst Elec & Electronics Engrs; Asn Comput Mach. *Res:* LR(1) parser error recovery; fault detection and diagnosis; memory test systems; magnetic memory models. *Mailing Add:* 1164 Hyde Ave San Jose CA 95129

BARRETT, WILLIAM JORDAN, b Harrison, Ga, Nov 7, 16; wid; c 4. ANALYTICAL CHEMISTRY. *Educ:* Mercer Univ, AB, 36, MA, 37; Univ Fla, PhD(chem), 50. *Prof Exp:* Instr chem, Univ Fla, 47-50; head, Anal Chem Sect, 50-67, Anal & Phys Chem Div, 67-79, dir, applied sci res, Southern Res Inst, 79-82; ADMIN OFFICER, ALA ACAD SCI, 82- *Mem:* Am Chem Soc; AAAS. *Res:* Electroanalytical chemistry; polarography; spot reactions; air pollution; water pollution; military defense. *Mailing Add:* 216 Woodbury Dr Birmingham AL 35216-2547

BARRETT, WILLIAM LOUIS, b Phoenix, Ariz, June 3, 33; m 76; c 3. SOLID STATE PHYSICS, NUCLEAR ENGINEERING. *Educ:* Univ Idaho, BS, 56; Univ Wash, MS, 61, PhD(physics), 68. *Prof Exp:* Res engr, Boeing Co, Wash, 61-64; asst prof, 68-71, ASSOC PROF PHYSICS, WESTERN WASH UNIV, 71-, CHMN DEPT PHYSICS & ASTRON, 75- *Mem:* AAAS; Am Inst Physics; Am Phys Soc. *Res:* Ferromagnetic resonance in metals; crystal growth; nuclear rocket concepts. *Mailing Add:* Phys & Astron Western Wash Univ Bond Hall No 152 Bellingham WA 98225

BARRETT-CONNOR, ELIZABETH L, b Evanston, Ill, April 8, 35; m 65; c 3. EPIDEMIOLOGY. *Educ:* Mt Holyoke Col, BA, 56; Med Col, Cornell Univ, MD, 60; London Sch Hyg & Trop Med, dipl, 65. *Hon Degrees:* DSc, Mt Holyoke Col, 85. *Prof Exp:* Instr med, Univ Miami, 65-68, asst prof, 68-70; from asst prof to assoc prof, 70-81, PROF MED & COMMUNITY MED, UNIV CALIF, SAN DIEGO, 81-, CHMN, DEPT COMMUNITY & FAMILY MED, 82- *Concurrent Pos:* Consult, Nat Inst Med, 78, Med Res Labs, Dept Defense, 80-83; comt mem, Nat Inst Allergies & Infectious Dis, 78-83; adv mem, Walter Reed Army Inst Res, 79-83, Nat Ctr Health Statist, 83-87. *Honors & Awards:* Living Legacy Award, 84. *Mem:* Inst Med-Nat Acad Sci; Royal Soc Health; Soc Epidemiol Res; Infectious Dis Soc Am; Am Heart Asn; Am Epidemiol Soc; Am Diabetes Asn; Nat Diabetes Adv Bd; Am Venereal Dis Asn. *Res:* Epidemiology of tropical medicine, tuberculosis, nosocomial infection, sickle cell anemia, cardiovascular disease risk factors and diabetes mellitus in adults; geriatrics. *Mailing Add:* Dept Community & Family Med M-007 Univ Calif San Diego CA 92093

BARRETTE, DANIEL CLAUDE, b Berthierville, Que, Apr 22, 43; m 67. RUMINANT FEEDING & HUSBANDRY, SWINE FEEDING. *Educ:* Univ Montreal, BA, 62, DMV, 66; Univ Laval, MSc, 70. *Prof Exp:* Asst prof physiol, fac agri sci & nutrit, Univ Laval, Que, 70-72; asst prof nutrit, 72-78, assoc prof nutrit & animal sci, Fac Vet Med, 77-78, PROF NUTRIT & ANIMAL SCI, UNIV MONTREAL, 88- *Concurrent Pos:* Lectr, fac agr sci & nutrit, Univ Laval, 72-81; vpres, coun animal prod, Que Dept Agr, 83-85, pres, 85-87; mem expert comt pet food, Can Vet Med Asn, 85- *Mem:* Nutrit Today Soc; Am Asn Bovine Practrs; Asn Am Vet Med Col; French Can Asn Advan Sci. *Res:* Evaluating metabolic and zootechnic effects of different levels of alimentary fiber and protein on growth, gestation and lactation in Beagle dogs; dog and cat feeding; small animal feeding and nutrition. *Mailing Add:* Fac Vet Med CP 5000 St Hyacinthe PQ J2S 7C6 Can

BARRETTE, JEAN, b Montreal, Que, May 1, 46. NUCLEAR STRUCTURE. *Educ:* Univ Montreal, BSc, 67, MSc, 68, PhD(nuclear physics), 74. *Prof Exp:* Postdoctoral fel nuclear physics, Max Planck Inst Nuclear Physics, Heidelberg, Ger, 74-76; asst physicist, Brookhaven Nat Lab, 76-78, assoc physicist, 78-80, physicist, 80-82; engr-physicien nuclear physics, Commissariat a L'Energie Atomique, France, 82-87; PROF PHYSICS, MCGILL UNIV, 87-, DIR, FOSTER RADIATION LAB, 88- *Mem:* Am Phys Soc; Can Asn Physics. *Res:* Nucleus-nucleus reactions & heavy-ion physics with particular interest in the study of reaction mechanism at intermediate and relativistic bombarding energies. *Mailing Add:* Foster Radiation Lab McGill Univ 3610 University St Montreal PQ H3A 2B2 Can

BARRICK, DONALD EDWARD, b Tiffin, Ohio, Nov 7, 38; m 61; c 2. ELECTRICAL ENGINEERING. *Educ:* Ohio State, BSEE & MSc, 61, PhD(elec eng), 65. *Prof Exp:* Res assoc electromagnetic theory, electrosci lab, Ohio State Univ, 63-65; fel, Battelle-Columbus Labs, 65-69; adj asst prof, Ohio State Univ, 69-77; CHIEF SEA STATE STUDIES PROG, ENVIRON RES LABS, NAT OCEANIC & ATMOSPHERIC ADMIN, 77- *Concurrent Pos:* Mem Comn II, Int Sci Radio Union, 69-77. *Mem:* AAAS; Inst Elec & Electronics Engrs. *Res:* Electromagnetic theory; radar systems; scattering from rough surfaces; radar signal processing; antenna and phased array theory; radar cluttering and clutter rejection; radar scattering theory. *Mailing Add:* 632 Lakeview Way Redwood City CA 94062

BARRICK, ELLIOTT ROY, b Worthington, Minn, Feb 7, 15; m 45; c 3. ANIMAL SCIENCE. *Educ:* Okla State Univ, BS, 38; Purdue Univ, MS, 41, PhD(animal nutrit), 47. *Prof Exp:* Asst prof animal husb, Purdue Univ, 47-48; assoc prof, Univ Tenn, 48-49; assoc prof animal indust, NC State Univ, 49-54, head animal husb sect, 54-67, prof animal sci, 54-80; RETIRED. *Mem:* Am Soc Animal Sci; Am Dairy Sci Asn. *Res:* Animal production and nutrition. *Mailing Add:* 5310 Old Stage Rd Raleigh NC 27603

BARRICK, P(AUL) L(ATRELL), b Villa Grove, Ill, Aug 22, 14; m 47; c 3. CHEMICAL ENGINEERING, ORGANIC CHEMISTRY. *Educ:* Univ Ill, BS, 35; Cornell Univ, PhD(org chem), 39. *Prof Exp:* Asst org chem, Cornell Univ, 35-38; res chemist, Exp Sta, E I DuPont de Nemours & Co, 39-48; prof chem eng, Univ Colo, Boulder, 48-85; RETIRED. *Mem:* Am Chem Soc; Soc Plastics Engrs; Am Inst Chem Engrs; Am Soc Eng Educ. *Res:* Synthetic polymers; catalysis; cryogenics; phase equilibria. *Mailing Add:* Dept Chem Eng Univ Colo Box 424 Boulder CO 80309

BARRIE, LEONARD ARTHUR, b Que. WET & DRY REMOVAL PROCESSES, GAS POLLUTANT TRANSFORMATION AEROSOLS. *Educ:* Queen's Univ, Can, BSc, 70; Univ Toronto, MSc, 72; Johann Wolfgang Von Goethe Univ, WGer, PhD(atmospheric chem), 75. *Prof Exp:* SR RES SCIENTIST, ATMOSPHERIC ENVIRON SERV, CAN, 75- *Concurrent Pos:* Mem, comt global change prob, Can Royal Soc, 85-, comt atmospheric chem & radioactiv, 86- & Comn Atmospheric Chem & Global Pollution, 87- *Mem:* Am Geophys Union; Can Meteorol & Oceanog Soc. *Res:* Transport, transformation and removal of atmospheric trace constituents over regional and global scale; arctic air chemistry; aerosol chemistry; pollutant removal by clouds; gas pollutant transformation; aerosols. *Mailing Add:* Atmospheric Environ Serv 4905 Dufferin St Downsview ON M3H 5T4 Can

BARRIE, ROBERT, b Newmains, Scotland, Sept 19, 27; m 54. THEORETICAL PHYSICS. *Educ:* Univ Glasgow, BSc, 49, PhD(physics), 54. *Prof Exp:* From sci officer to sr sci officer, Servs Electronics Res Lab, Baldock, Herts, Eng, 53-57; from instr to assoc prof, 57-64, PROF PHYSICS, UNIV BC, 64- *Res:* Electronic properties of solids. *Mailing Add:* Dept Physics Univ BC 6224 Agricultural Rd Vancouver BC V6T 1Z1

BARRIENTOS, CELSO SAQUITAN, b Cavite, Philippines, Jan 9, 36; m 59; c 2. METEOROLOGY, OCEANOGRAPHY. *Educ:* Feati Univ, Philippines, BSc, 60; NY Univ, MSc, 65, PhD(meteorol), 69. *Prof Exp:* Asst res scientist geophys sci lab, NY Univ, 62-66; res scientist, Isotopes, Inc, 66-68 & Geophys Sci Lab, NY Univ, 68-69; res materologist, Tech Develop Lab, Nat Weather Serv, 69-81, SUPVRY PHYS SCIENTIST, ENVIRON DATA & INFO SERV, NAT OCEANIC & ATMOSPHERIC ADMIN, 81- *Mem:* AAAS; Am Meteorol Soc; Am Geophys Union; foreign mem Royal Meteorol Soc; Meteorol Soc Japan. *Res:* Hurricane dynamics and structures; tropical meteorology; atmospheric diffusion; turbulence; micrometeorology; ocean waves; air-sea interactions; marine environmental predictions; Great Lakes wind-wave forecasting; oil spill trajectory forecasts; environmental monitoring and management; strategic petroleum reserve. *Mailing Add:* 16609 Cutlass Drive Rockville MD 20853

BARRIGA, OMAR OSCAR, b Santiago, Chile, Mar 1, 38; m 60; c 2. PARASITOLOGY, IMMUNOBIOLOGY. *Educ:* Univ Chile, BA, 58, DVM, 63; Univ Ill, PhD(parasitol-immunol), 72. *Prof Exp:* Small animal vet, 63; asst prof parasitol, Sch Med, Univ Chile, 64-71, prof immunol, Grad Sch Med & assoc prof parasitol, Schs Med & Pub Health, 72; asst prof parasitol, Sch Vet Med, Univ Pa, 73-80; ASSOC PROF & MEM GRAD FAC, DEPT VET PREV MED, OHIO STATE UNIV, 80- *Concurrent Pos:* Ed, J Chilean Soc Vet Med, 66-67; int consult immunoparasitol, Grad Sch, Porto Alegre, Brazil, 77-78; vis prof immunoparasitol, Mex, 82 & Dominican Repub, 84, Costa Rica, 88, Morocco, 89, Brazil, 90. *Mem:* Am Asn Vet Parasitologists; Am Asn Vet Immunologists; Am Asn Parasitologists; Am Soc Trop Med & Hyg. *Res:* Immunity to parasitic infections; modifications of the immune response by parasitic infections; immunodiagnosis and vaccination of parasitic infections; immunological control of arthropod infestations. *Mailing Add:* Dept Vet Prev Med Ohio State Univ 1920 Coffey Rd Columbus OH 43210

BARRINGER, DONALD F, JR, b Cleveland, Ohio, June 3, 32; m 58; c 4. METABOLISM. *Educ:* Denison Univ, BS, 54; Ohio State Univ, PhD(org chem), 60. *Prof Exp:* Res chemist, E I du Pont de Nemours & Co, 59-60; RES CHEMIST, AM CYANAMID CO, 60- *Mem:* Am Chem Soc. *Res:* Metabolism of pesticides and veterinary drugs in plants, animals and the environment. *Mailing Add:* Seven Stonicker Dr Lawrenceville NJ 08648

BARRINGER, WILLIAM CHARLES, b Cleveland, Ohio, Feb 28, 34; c 4. INFORMATION SCIENCE & SYSTEMS. *Educ:* Denison Univ, BS, 56; NY Univ, MS, 64, PhD(photochem), 68. *Prof Exp:* Develop chemist, 57-75, group leader preformulation chem, 75-78, SR RES CHEMIST, LEDERLE LAB DIV, AM CYANAMID CO, 78- *Concurrent Pos:* Prof chem, The King's Col, 69-71. *Mem:* Am Chem Soc; Am Asn Pharmaceut Scientists; NY Acad Sci. *Res:* The relationship between the physical and chemical properties of chemical compounds and their performance in pharmaceutical formulations; stability evaluation of potential market products. *Mailing Add:* Am Cyanamid Co Lederle Lab N Middletown Rd Pearl River NY 10965

BARRINGTON, A(LFRED) E(RIC), b Vienna, Austria, Mar 22, 21; wid; c 1. ELECTRONICS, APPLIED PHYSICS. *Educ:* London Univ, BSc, 47, PhD(elec eng), 50. *Prof Exp:* Design engr vacuum tubes, Machlett Labs, Conn, 50-52; sr sci officer microwave tubes, serv electronics res lab, Brit Admiralty, 52-54; Harwell res fel & lectr elec eng, Queen Mary Col, London, 55-57; lectr appl physics, Harvard & res fel physics, Cambridge Electron Accelerator, 57-60; dept mgr vacuum physics, Varian Assoc, Calif, 60-62; mgr ion physics dept, Geophys Corp Am, 62-66; head astrophys br, instrumentation lab, electronics res ctr, NASA, 66-68; div chief, 68-70; CHIEF, SAFETY & ENVIRON TECHNOL DIV, RES & SPECIAL PROGS ADMIN, JOHN A VOLPE TRANSP SYSTS CTR, DEPT TRANSP, 70- *Res:* Microwave tubes; exploding wires; electron accelerators; ion pumps; mass spectrometers; ultra-high vacuum; analytical instruments. *Mailing Add:* John A Volpe Nat Transp Syst Ctr Kendall Sq Cambridge MA 02142

BARRINGTON, DAVID STANLEY, b Boston, Mass, Sept 4, 48. BOTANY. *Educ:* Bates Col, BS, 70; Harvard Univ, PhD(biol), 75. *Prof Exp:* asst prof, 77-83, CUR PRINGLE HERBARIUM, UNIV VT, 74-, ASSOC PROF BOT, 83- *Mem:* Bot Soc Am; Am Soc Plant Taxonomists; Am Fern Soc; Soc Study Evolution; AAAS. *Res:* Systematics and evolution of tropical American ferns; anatomy of the fossil tree fern stems of the Cyatheaceae and Dicksoniaceae. *Mailing Add:* Dept Bot Univ Vt Burlington VT 05405

BARRINGTON, GORDON P, b Dayton, Wis, Apr 14, 23; m 48; c 1. AGRICULTURAL ENGINEERING. *Educ:* Univ Wis, BS, 49, MS, 51. *Prof Exp:* From assoc prof to prof agr eng, 52-77, prof agr & life sci, 77-88, EMER PROF AGR & LIFE SCI, UNIV WIS-MADISON, 88- *Mem:* Am Soc Agr Engrs. *Res:* Forage seeding and harvesting machinery, forage preservation; separation of protien from plant materials. *Mailing Add:* 26 Shadow Woods Lane Waupaca WI 54981

BARRINGTON, NEVITT H J, b St Catharines, Ont, Can, Jun 1, 08; m 35; c 2. COMMUNICATION, INNOVATION. *Educ:* Univ Toronto, Can, BA Sc, 41, McGill Univ, Montreal, Can, 45. *Hon Degrees:* LLD, Concordia Univ, Montreal, Can, 84. *Prof Exp:* Res develop eng, Zavod Elektropribor,

Leningrad, USSR, 32-33; mfg eng, Northern Elec Co, Montreal, Can, 34-39; comt syst eng, Can Pac & Defense Comn, 39-44; exec eng, RCA Int Div, 44-47; Int consult, LM Ericsson Telephone Co, Sweden, 47-60; Nat consult, Royal Comn Govt Orgn, Can, 60-65; consult dir, Ont Develop Corp, Can, 63-76; CONSULT & LECTR, COMMUN & INNOVATION, 76- *Concurrent Pos:* Lectr hist & philos sci, Sr George Williams Col, Que, 43; vis prof, Sch Indust Design, Carleton Univ, Concordia Univ, Que & Univ Stockholm, Sweden, 69-; lectr, UNESCO, NATO, 69- *Mem:* Fel Royal Soc Arts UK; Inst Elec Engrs UK; Engrs Inst Can; AAAS; Soc Gen Systs Res; Sigma Xi; Inst Elec & Electronics Engrs. *Res:* Material, mental, and social effects of human artifacts; hardware products and information software considered as human communication media. *Mailing Add:* Two Clarendon Ave Apt 207 Toronto ON M4V 1H9 Can

BARRINGTON, RONALD ERIC, b Toronto, Ont, Aug 1, 31; m 56; c 3. PHYSICS. *Educ:* Univ Toronto, BA, 54, MA, 55, PhD(physics), 57. *Prof Exp:* Sci officer, Defence Res Telecommun Estab, 57-60 & 63-65 & Norweg Defence Res Estab, 60-62; sect head plasma physics, 65-71, dir propagation res, 71-78, DIR GEN RADAR & COMMUN TECHNOL RES, COMMUN RES CTR, 78- *Concurrent Pos:* Exec mem, Assoc Comt Space Res, 67- *Mem:* Can Asn Physicists. *Res:* Ionospheric physics; electromagnetic wave propagation; plasma physics. *Mailing Add:* Commun Res Ctr Box 11490 Sta H Ottawa ON K2H 8S2 Can

BARRIOS, EARL P, b New Orleans, La, Dec 14, 17; m 46. HORTICULTURE. *Educ:* La State Univ, PhD(bot, hort), 60. *Prof Exp:* Chief, Nebr Potato Develop Div, State Dept Agr & Univ Nebr, 48-54; sales mgr agr chem, P V Fertilizer Co, 54-57; from instr to asst prof, 58-64, assoc prof, 64-77, PROF HORT, LA STATE UNIV, BATON ROUGE, 77- *Mem:* Am Soc Hort Sci. *Res:* Breeding and genetics of horticultural crops, especially vegetables. *Mailing Add:* 1420 Aster Baton Rouge LA 70802

BARRNETT, RUSSELL JOFFREE, anatomy; deceased, see previous edition for last biography

BARROIS, BERTRAND C, b Princeton, NJ, Nov 10, 52. DEFENSE. *Educ:* Mass Inst Technol, BS, 73; Caltech, PhD(physics), 79. *Prof Exp:* RES STAFF, INST DEFENSE ANALYSIS, 78- *Mem:* Am Phys Soc. *Res:* Strategic warfare, arms control, air defense. *Mailing Add:* Inst Defense Analysis SED 1801 N Beauregard St Alexandria VA 22311

BARRON, ALMEN LEO, b Toronto, Ont, Jan 19, 26; nat US; m 49; c 1. MICROBIOLOGY. *Educ:* Ont Agr Col, Univ Toronto, BSA, 48, MSA, 49; Queen's Univ (Can), PhD(bact), 53. *Prof Exp:* Mem fac, State Univ NY Buffalo, 54-74, prof microbiol, 68-74; prof microbiol & chmn dept microbiol & immunol, Col Med, Univ Ark Med Sci, Little Rock, 74-91; RETIRED. *Concurrent Pos:* Fulbright scholar, 64; Commonwealth Fund travel fel, 64; Fight for Sight travel award, 72; vis prof, Hebrew Univ Hadassah Med Sch, Israel, 72 & Kaohsiung Med Sch, Taiwan, 82; consult microbiol, Little Rock Vet Admin Hosp, 74-91. *Honors & Awards:* Student Am Med Asn Golden Apple Award, 74-75, 76-77. *Mem:* Am Acad Microbiol; fel Infectious Dis Soc Am; Am Soc Microbiol; Am Asn Immunologists; Am Veneral Dis Asn. *Res:* Chlamydia group, biology and immunology, animal models for study of genital tract infections. *Mailing Add:* 14000 Rivercrest Dr Little Rock AR 72212

BARRON, BRUCE ALBRECHT, b New York, NY, Nov 3, 34; m 75; c 1. OBSTETRICS & GYNECOLOGY. *Educ:* Allegheny Col, AB, 55; Yale Univ, MPH, 60, PhD(biomet), 65; NY Univ, MD, 71. *Prof Exp:* Res asst, Dept of Surg, Bellevue Med Ctr, NY Univ, 56-57; biochemist, Walter Reed Army Inst Res, 57-59; res assoc biomet, Rockefeller Univ, 65-67, asst prof, 67-74; intern, Dept Med, Bellevue Med Ctr, 71-72, asst res obstet & gynec, 72-73, chief res, 73-74; ASSOC PROF OBSTET & GYNEC, COL PHYSICIANS & SURGEONS, COLUMBIA UNIV, 74-; SR ATTEND PHYSICIAN, SLOAN HOSP WOMEN, PRESBY MED CTR, 74- *Concurrent Pos:* Consult, Biomed Div, Population Coun, 65-71; chmn & bd dir, New York City Health & Hosp Corp, 80- *Mem:* AAAS; Biomet Soc; Am Statist Asn; Harvey Soc; Sigma Xi. *Res:* Mathematical models in carcinogenesis. *Mailing Add:* Columbia-Presby Med Ctr 161 Ft Washington Ave New York NY 10032

BARRON, CHARLES IRWIN, b Chicago, Ill, July 21, 16; m 45; c 4. AEROSPACE MEDICINE. *Educ:* Univ Ill, BS, 40, MD, 42. *Prof Exp:* Surgeon, US Dept Army Qm Depot, Ill, 48-50; flight surgeon, Lockheed-Calif Co, 50-51, sr exam physician, 51-53, med dir, 53-79. *Concurrent Pos:* Lectr, Univ Southern Calif, 54-, assoc clin prof, 63-70, clin prof, 70-; med examr, Fed Aviation Agency, 55-; mem comt hearing & bio-acoustics, Nat Res Coun, 55-57; lectr, Univ Calif, Los Angeles, 55, assoc clin prof, Sch Pub Health, 63-74; chmn adv coun, Civil Air Surgeon, 60-75; lectr, Ohio State Univ, 60, vis prof, 63; chmn res adv comt advan biotechnol & human res, NASA, 63- *Honors & Awards:* Arnold D Tuttle Award, 63; Jeffries Award, Am Inst Aeronaut & Astronaut, 68. *Mem:* AMA; fel Indust Med Asn; Am Inst Aeronaut & Astronaut; fel Aerospace Med Asn (pres, 63-64); Civil Aviation Med Asn (vpres, 55-59). *Res:* Industrial and aerospace medicine; life sciences and human factors engineering; education in aerospace safety; basic and applied research in aviation physiology, pathology and toxicology; environmental medicine. *Mailing Add:* 19303 Itasca St Northridge CA 91324

BARRON, EDWARD J, b Sudan, Tex, June 2, 27; m 53; c 3. CLINICAL BIOCHEMISTRY. *Educ:* Univ Wash, BS, 50, PhD(biochem), 59. *Prof Exp:* Clin chemist, Group Health Co-op Puget Sound, 50-55; NIH trainee biochem, Univ Calif, Davis, 60-62; assoc dir, Virginia Mason Res Ctr, 62-68, ASSOC DIR LABS, THE MASON CLIN, 68- *Mem:* Am Chem Soc; Am Asn Clin Chemists; NY Acad Sci; Sigma Xi. *Res:* Lipid chemistry; metabolism of lipids; biosynthesis of fatty acids. *Mailing Add:* 19528 38 NE Seattle WA 98155

BARRON, EMMANUEL NICHOLAS, b Lodi, Calif, Oct 14, 49. OPTIMAL CONTROL THEORY, HAMILTON-JACOBI EQUATIONS. *Educ:* Univ Ill Chicago, BS, 70; Northwestern Univ, MS, 72 & PhD(math), 74. *Prof Exp:* Asst prof math, Ga Inst Technol, 74-80; PROF MATH, LOYOLA UNIV, CHICAGO, 80- *Concurrent Pos:* Mem tech staff, AT&T Bell Labs, 78-80. *Mem:* Soc Indust & Appl Math. *Res:* Optimal control theory and non-linear partial differential equations; stochastic processes and its applications to financial economics. *Mailing Add:* Dept Math Sci Loyola Univ Chicago IL 60626

BARRON, ERIC JAMES, b Lafayette, Ind, Oct 26, 51; m 82; c 2. PALEOCLIMATOLOGY, PALEOCEANOGRAPHY. *Educ:* Fla State Univ, BS, 73; Univ Miami, MS, 76, PhD(geophys), 80. *Prof Exp:* Fel, Nat Ctr Atmospheric Res, 80-81; scientist, 81-85; assoc prof marine geol & geophys, Univ Miami, 85-86; DIR EARTH SYST SCI CTR & ASSOC PROF GEOSCI, PA STATE UNIV, 86- *Concurrent Pos:* Co-ed-chief, Paleogeog, Paleoclimat & Palaeocol Geol, Geol Soc Am, 85-; chmn, Int Geol Correl Prog, 82-87, Climate Variations Comt, Am Meteorol Soc, 88-, Paleoceanog Comt, Am Geophys Union, 87-88. *Mem:* Am Geophys Union; Geol Soc Am; Soc Econ Paleontologists & Mineralogists; Am Meterol Soc; Am Asn Petrol Geologists; Asn Am Geographers. *Res:* Global change, specifically numerical models of the climate system and the study of climate change throughout Earth history. *Mailing Add:* Nat Ctr Atmospheric Res PO Box 3000 Boulder CO 80307

BARRON, EUGENE ROY, b Somerset, Pa, Sept 10, 41; m 63; c 2. SYNTHETIC FIBERS, TIRES. *Educ:* Univ Md, BS, 63, PhD(chem), 67. *Prof Exp:* SR RES ASSOC, E I DU PONT DE NEMOURS & CO, INC, 67- *Res:* Nylon, Kevlar aramid and Dacron polyester for industrial applications such as tires, mechanical rubber goods, industrial fabrics, ropes and sewing thread. *Mailing Add:* 1111 Dardel Dr Wilmington DE 19803

BARRON, JOHN ARTHUR, b Omaha, Nebr, Oct 22, 47. GEOLOGY, PALEONTOLOGY. *Educ:* Univ Calif, Los Angeles, BS, 69, PhD(geol), 74. *Prof Exp:* GEOLOGIST, US GEOL SURV, 74- *Concurrent Pos:* Comt mem, Int Geol Correlations Prog, UNESCO, 76-; Int Symposium Living & Fossil Diatoms, 76, 80, 82, 84, 86, 88, 90. *Honors & Awards:* Schuchert Award Paleont, 86. *Mem:* Geol Soc Am; Soc Econ Petrol Mineralogists; Am Geophys Union. *Res:* Marine diatom and silicoflagellate biostratigraphy and paleoecology of the Cenozoic. *Mailing Add:* P&S Br 345 Middlefield Rd Menlo Pk CA 94025

BARRON, JOHN ROBERT, b Niagara Falls, Ont, Dec 23, 32; m 55; c 1. ENTOMOLOGY, TAXONOMY. *Educ:* McGill Univ, BSc, 61, MSc, 62; Univ Alta, PhD(entom, taxon), 69. *Prof Exp:* Tech asst taxon, McGill Univ, 58-61; res asst exten entom, Univ Alta, 62-68; RES SCIENTIST, BIOSYSTEMATICS RES INST, CAN DEPT AGR, 69- *Concurrent Pos:* Vis prof, Univ Laval, 72- *Mem:* Entom Soc Am; Soc Study Evolution; Soc Syst Zool; Entom Soc Can. *Res:* Taxonomy of Gryllus of North America; taxonomy and zoogeography of Cleroidea of the world, particularly the family Trogositidae; taxonomy and genetics of Ichneumonidae, particularly of North America. *Mailing Add:* Biosystematics Res Ctr Can Dept Agr Cent Exp Farm Ottawa ON K1A 0C6 Can

BARRON, KEVIN D, b St John's, Nfld, Apr 21, 29; US citizen; m 56; c 2. MEDICINE, NEUROLOGY. *Educ:* Dalhousie Univ, MD & CM, 52. *Prof Exp:* Instr neurol, Col Physicians & Surgeons, Columbia Univ, 57-59; assoc neurol & psychiat, Med Sch, Northwestern Univ, 59-61, from asst prof to prof, 61-69; PROF NEUROL & CHMN DEPT & PROF PATH, ALBANY MED COL, 69- *Concurrent Pos:* USPHS fel neuropath, Montefiore Hosp, NY, 56-59; USPHS grants, 60-; Nat Multiple Sclerosis Soc grant, 61-63, 79-81. *Mem:* Am Neurol Asn; Am Soc Neurochem; Asn Univ Prof Neurol; fel Am Acad Neurol; Am Asn Neuropath. *Res:* Histochemistry and cytology; electron microscopy of the nervous system; neuroenzymology, particularly hydrolytic enzymes; neuropathology. *Mailing Add:* Dept Neurol Albany Med Col Albany NY 12208

BARRON, RANDALL F(RANKLIN), b Many, La, May 16, 36; m 58; c 4. MECHANICAL ENGINEERING. *Educ:* La Polytech Inst, BSME, 58; Ohio State Univ, MSc, 61, PhD (mech eng), 64. *Prof Exp:* From instr to asst prof mech eng, Ohio State Univ, 58-65; assoc prof mech eng, 65- 70, dir eng res, Col eng, 76-80, dir eng res & grad studies, 80-89, PROF MECH ENG, LA TECH UNIV, 70- *Concurrent Pos:* Consult, CryoVac, Inc, Ohio, 63-65; mem, Cryogenic Eng Conf Comt, 77-83; consult, Riley-Beaird, 66-86. *Honors & Awards:* Res Award, Sigma Xi, 71. *Mem:* AAAS; Am Soc Mech Engrs; Am Soc Eng Educ. *Res:* Heat transfer; thermodynamics; design in cryogenic engineering systems; noise control. *Mailing Add:* Dept Mech Eng La Tech Univ Ruston LA 71272-0046

BARRON, RONALD MICHAEL, b Windsor, Ont, July 12, 48; m 69; c 2. COMPUTATIONAL FLUID DYNAMICS, TRANSONIC AERODYNAMICS. *Educ:* Univ Windsor, BA, 70, MS, 71; Carleton Univ, PhD(appl math), 74. *Prof Exp:* Postdoctoral fel, 74-75, from asst prof to assoc prof, 75-84, PROF MATH, UNIV WINDSOR, 84- *Concurrent Pos:* Vis scientist, Dept Aeronaut & Astronaut, Stanford Univ, 81; dept head, Dept Math & Statist, Univ Windsor, 86-89, dir, Fluid Dynamics Res Inst, 87-89; vis prof, UGC-DSA Ctr Fluid Mech, Bangalore Univ, India, 89-90. *Mem:* Assoc mem Am Inst Aeronaut & Astronaut; Can Aeronaut & Space Inst; Am Acad Mech; Can Appl Math Soc. *Res:* Navier-Stokes; transonic full potential; Euler solvers; grid system for finite difference calculations; design problems. *Mailing Add:* Dept Math & Statist Univ Windsor Windsor ON N9B 3P4 Can

BARRON, SAUL, b New York, NY, Feb 24, 17; m 41; c 3. THERMODYNAMICS, PHYSICAL CHEMISTRY. *Educ:* Lafayette Col, BS, 41; Ohio State Univ, MS, 48, PhD(chem eng), 54. *Prof Exp:* Architect & engr, Bur Ships, US Navy, Pa, 41-43; mech engr, US Govt, Wright-Patterson AFB, 46-51; design engr, Martin-Marietta Corp, Md, 54-56; sr scientist, Avco Mfg Corp, Mass, 56-57; dir res, Thiokol Chem Corp, Pa, 58-59 & Bell Aerosysts Co, NY, 60-64; prof chem, State Univ NY, Buffalo, 64-85; RETIRED. *Concurrent Pos:* Consult, Marine Trust Co, NY, 64- & Conax Corp, 65-; vis prof, Inst Chem, Hebrew Univ, Israel, 82; consult, 85- *Mem:* Am Chem Soc; NY Acad Sci; Sigma Xi. *Res:* Oxide ceramics; volume and phase chemical equilibria, especially critical region, using multi-component mixtures; chemical thermodynamics; thermal effects upon polymers. *Mailing Add:* 249 Troy Del Way Buffalo NY 14221

BARROW, EMILY MILDRED STACY, b Washington, DC, July 2, 27; div; c 3. BIOCHEMISTRY. *Educ:* Meredith Col, AB, 50; Univ NC, Chapel Hill, AM, 52, PhD(physiol), 63. *Prof Exp:* Grad asst bot, 50-52, res technician path, 53-56, res asst path, 56-59, asst physiol, 59-62, res assoc path, 62-69, from assoc prof to prof, 69-82, EMER PROF PATH, UNIV NC, CHAPEL HILL, 82- *Res:* Biochemical genetics of blood coagulation disorders. *Mailing Add:* 209 Barclay Rd Chapel Hill NC 27516

BARROW, GORDON M, b Vancouver, BC, Nov 13, 23; m 57. PHYSICAL CHEMISTRY. *Educ:* Univ BC, BASc, 46, MASc, 47; Univ Calif, PhD(chem), 50. *Prof Exp:* Instr chem, Northwestern Univ, 51-52, asst prof, 53-57, assoc prof, 58-59; prof & head dept, Case Western Reserve Univ, 59-69; adj prof, Dartmouth Col, 69-70; PROF, ROYAL ROADS MILITARY COLLEGE, 83- *Mem:* 380-4586. *Mailing Add:* Dept Chem Royal Rds Military Col Victoria FMG BC V0S 1B0 Can

BARROW, JAMES HOWELL, JR, b West Point, Ga, June 28, 20; m 50. PROTOZOOLOGY, PARASITOLOGY. *Educ:* Emory Univ, BA, 43; Yale Univ, PhD(zool), 51. *Prof Exp:* Instr, Emory Univ, 43-44; asst, Yale Univ, 44-47; lectr biol, Albertus Magnus Col, 47-48; prof biol & chmn dept, Huntingdon Col, 48-57; chmn dept, 57-, PROF BIOL, HIRAM COL, 57-, DIR BIOL STA, 70- *Concurrent Pos:* Ford Found fel, Cambridge Univ, 53-54; vis prof protozool, biol sta, Univ Mich, 56-64. *Mem:* Wildlife Disease Asn; Soc Protozool; Am Micros Soc; AAAS; Sigma Xi. *Res:* Trypanosomes of poikilothermic vertebrates; haemosporidians of waterfowl and invertebrates; avian behavior and parasitism. *Mailing Add:* Dept Biol Hiram Col PO Box 1808 Hiram OH 44234

BARROW, THOMAS D, b San Antonio, Tex, Dec 27, 24; m 50; c 4. PETROLEUM GEOLOGY, MINERAL EXPLORATION. *Educ:* Univ Tex, BS, 45, MA, 48; Stanford Univ, PhD(geol), 53. *Prof Exp:* Jr geologist, Humble Oil & Ref Co, Calif, 51-55, area explor geologist, 55-59, div explor geologist, La, 59-61, regional explor mgr, 62-64, exec vpres & dir, Esso Explor, Inc, NY, 64-65, dir, Humble Oil & Ref Co, Houston, 65-72, pres, 70-72, sr vpres, 67-70; sr vpres, Exxon Corp, 72-78; chmn & chief exec officer, Kennecott Copper Corp, 78-81; vchmn, Standard Oil Co, Ohio, 81-85; CHMN, GEOQUEST TECHNOL, 90-; SR CHMN, GEOQUEST INT, 90- *Concurrent Pos:* Mem, Oceanog Adv Comt, 68; mem, Sea Grant Adv Panel, 69; comn natural resources, Nat Res Coun, 73-78; comn phys sci, math & nat resource, Nat Res Coun, 84-87; Task Force US Energy Policy, 74; Bd Earth Sci, Nat Res Coun, 82-84; mem, Comt Eng Int Enterprise, Nat Acad Eng, 90-91. *Mem:* Nat Acad Eng; Am Geophys Union; Am Geog Soc; Am Asn Petrol Geologists; Am Inst Mining & Metall Engrs; Geol Soc Am; AAAS. *Res:* Petroleum exploration. *Mailing Add:* 4605 Post Oak Pl No 207 Houston TX 77027-9728

BARROWMAN, JAMES ADAMS, b Edinburgh, Scotland, June 4, 36; m 64; c 3. GASTROENTEROLOGY. *Educ:* Edinburgh Univ, BSc, 58, MBChB, 61; Univ London, MRCP, 64, PhD(physiol), 66; FRCP(C), 78; FRCP, 81. *Prof Exp:* From lectr to sr lectr physiol, London Hosp Med Col, 63-71, from lectr to sr lectr med, 71-75; prof med, Mem Univ Nfld, 80, asst dean res, 81; CONSULT, 81- *Mem:* Brit Soc Gastroenterol; Brit Physiol Soc. *Res:* Physiology of intestinal lymph in relation to digestion and absorption; trophic action of gastrointestinal hormones on the alimentary tract. *Mailing Add:* Gastroenterol Dept Health Sci Ctr Prince Phillip Driveway St John's NF A1B 3V6 Can

BARROWS, AUSTIN WILLARD, b Addison, Vt, July 3, 37; m 60; c 4. APPLIED PHYSICS. *Educ:* Univ Vt, BA, 60, MS, 61; Univ Ky, PhD(physics), 65. *Prof Exp:* Res assoc nuclear physics, Univ Ky, 65; res physicist, US Army Nuclear Defense Lab, 66-67, br chief, 67-70; asst to dir admin, US Army Ballistic Res Lab, 70-71; BR CHIEF APPL PHYSICS ADMIN, COMBUSTION & PROPULSION BR, BALLISTIC RES LABS, 71- *Mem:* Am Phys Soc; AAAS; Combustion Inst; Am Defense Preparedness Asn; Sigma Xi. *Res:* The phenomena of combustion of propellants and interior ballistics of guns and rockets. *Mailing Add:* 3103 Whitefield Rd Churchville MD 21028-1304

BARROWS, EDWARD MYRON, b Detroit, Mich, Aug 8, 46; m 79; c 2. BEHAVIORAL ECOLOGY, ENTOMOLOGY. *Educ:* Univ Mich, BS, 68; Univ Kans, PhD(entom), 75. *Prof Exp:* Asst prof, 75-81, ASSOC PROF BIOL, GEORGETOWN UNIV, 81- *Mem:* AAAS; Entom Soc Am; Animal Behavior Soc; Ecol Soc Am; Soc Study Evol; Orgn Trop Studies. *Res:* Reproductive strategies in insects; behavior ecology and evolution of insects in urban ecosystems. *Mailing Add:* Dept Biol Georgetown Univ Washington DC 20057

BARROWS, JOHN FREDERICK, b Detroit, Mich, Dec 26, 28; m 51; c 3. AERODYNAMICS. *Educ:* Univ Mich, BSME, 51, MSE, 59, PhD(mech eng), 62. *Prof Exp:* Proj engr, Allison Div, Gen Motors Corp, 51-58; lectr fluid mech & thermodyn, Univ Mich Dearborn Ctr, 60-62; asst prof mech eng, Cornell Univ, 62-67; sr engr, Res Div, Carrier Corp, 67-70, chief engr, 70-78, asst dir, Res Labs, 78-85; PROF & DIR, MFG ENG PROG, SYRACUSE UNIV, 85- *Mem:* Am Soc Mech Engrs; Soc Mfg Engrs; Am Soc Eng Educ. *Res:* Aerodynamics; applied thermodynamics; fluid mechanics; design for manufacturing. *Mailing Add:* Col Eng Syracuse Univ Syracuse NY 13205

BARRUETO, RICHARD BENIGNO, b Guatemala City, Guatemala, Feb 16, 28; US citizen; m 54; c 3. BIOCHEMISTRY, ORGANIC CHEMISTRY. *Educ:* Norm Sch, Guatemala, AB, 46; Eastern Nazarene Col, BS, 52; Boston Univ, MA, 54. *Prof Exp:* Res chemist, Sch Med, Boston Univ, 52-54; asst biochemist, US Qm Res & Eng Command, US Army, Mass, 54-60, res biochemist, Army Inst Environ Med, 60-62; bioscientist, Biotech Sect, Aerospace Div, Boeing Co, 62-65, sr bioscientist, 65-75; PRES, CARL F MILLER & CO, INC, 75- *Mem:* Am Sci Affil; Am Inst Chemists; Sigma Xi. *Res:* Life support and protection problems associated with man in a space vehicle; high altitude chamber; space-suit-vehicle integration; closed environment toxicological problems; optimizing man-machine interface in missile systems from psycho-physiological viewpoint. *Mailing Add:* 5155 NE Laurelcrest Lane Seattle WA 98105

BARRY, ARTHUR JOHN, b Buffalo, NY, Mar 11, 09; m 36; c 3. ORGANIC CHEMISTRY, PHYSICAL CHEMISTRY. *Educ:* State Univ NY Col Forestry, BS, 32, PhD(org chem), 36. *Prof Exp:* Actg prof org chem & carbohydrate chem, State Univ NY Col Forestry, 36-37; res chemist, Dow Chem Co, 37-47; res group supvr, 47-50, assoc dir res, 50-65; dir chem res, Dow Corning Corp, 65-70, res consult, 70-74; RETIRED. *Concurrent Pos:* Vis scholar dept chem, Univ Ariz, 74-83, adj prof, 83- *Mem:* AAAS; Am Chem Soc. *Res:* Preparation of cellulose derivatives; plastics; reactions in liquid ammonia, metallo-organic chemistry; synthesis of organosilicon compounds; investigation of polysiloxane resins, fluids and elastomers. *Mailing Add:* 9102 N Riviera Dr Tucson AZ 85737-7427

BARRY, ARTHUR LELAND, b Spokane, Wash, Aug 2, 32; m 55; c 4. MICROBIOLOGY, BACTERIOLOGY. *Educ:* Gonzaga Univ, BS, 55; St Lukes Sch Med Technol, MT, 55; Wash State Univ, MS, 57; Ohio State Univ, PhD(microbiol), 62. *Prof Exp:* Chief microbiologist, Clin Lab Med Group, Los Angeles, 64-66; med microbiologist, Los Angeles County Gen Hosp, 66-68; asst clin prof microbiol, Univ Calif-Calif Col Med, 67-68; lectr internal med & path, 68-75, prof clin microbiol, Dept Med & Path, Sch Med, 75-82, RESIDENT, CLIN MICROBIOL INST, UNIV CALIF, DAVIS, 82- *Concurrent Pos:* NIH trainee clin microbiol, Univ Wash Hosp, 62-64; sr ed, Am J Med Technol, 68-74; mem ed bd, Antimicrobial Agents & Chemother, J Clin Microbiol, Current Microbiol & Diag Microbiol & Infectious Dis; dir, Microbiol Labs, Sacremento Med Ctr, 68-82, Clin Microbiol Inst, 82- *Mem:* Am Soc Med Technol; Am Soc Microbiol; Infectious Dis Soc Am. *Res:* Methods for rapid identification of bacterial pathogens; antimicrobic susceptibility tests; quality control in microbiology. *Mailing Add:* Clin Microbiol Inst PO Box 947 Tualatan OR 97062

BARRY, B(ENJAMIN) AUSTIN, b Newburgh, NY, July 23, 17. CIVIL ENGINEERING. *Educ:* Cath Univ, AB, 37; Rensselaer Polytech Inst, BCE, 42; NY Univ, MCE, 49. *Prof Exp:* From instr to assoc prof, 43-70, head dept, 62-71, prof civil eng, 70-88, EMER PROF CIVIL ENG, MANHATTAN COL, 88- *Concurrent Pos:* Consult to obtain secure water supply, Rongai Village, USAID, 77. *Honors & Awards:* Surv & Mapping Award, Am Soc Civil Engrs, 70; Hon Mem, Am Soc Civil Engrs, 87; WS Dix Award, Am Cong Surv & Mapping. *Mem:* Am Soc Civil Engrs; hon mem Am Cong Surv & Mapping (pres, 61-62); Am Soc Eng Educ (vpres, 69-71). *Res:* Surveying and mapping; photogrammetry; transportation engineering. *Mailing Add:* Dept Civil Eng Manhattan Col Bronx NY 10471-4098

BARRY, CORNELIUS, b Canandaigua, NY, Aug 23, 34; m 59. ENTOMOLOGY. *Educ:* St John Fisher Col, BS, 56; Univ Md, MS, 62, PhD(entom), 64. *Prof Exp:* Asst zool & parasitol, Univ Md, 56-59; teacher, Md Bd Educ, 59-61; asst insect physiol, Univ Md, 61-62, NSF res asst, 62-64; res entomologist, US Army Biol Labs, 64-68; asst prof int med, Sch Med, Univ Md, 68-74; prof, Inst Microbiol, Fed Univ Rio de Janeiro, 75-78, prof, Inst Biophys, 78-80. *Mem:* AAAS; Am Soc Trop Med & Hyg; Sigma Xi; Entom Soc Am. *Res:* Medical entomology; insect physiology and biochemistry; invertebrate tissue culture. *Mailing Add:* 209 Brackenwood Ct Timonium MD 21093

BARRY, DON CARY, b Los Angeles, Calif, Jan 4, 41; m 67. ASTRONOMY. *Educ:* Univ Southern Calif, BS, 62; Univ Ariz, PhD(astron), 67. *Prof Exp:* Asst prof, 67-71, ASSOC PROF ASTRON, UNIV SOUTHERN CALIF, 71-, CHMN DEPT, 74- *Mem:* AAAS; Am Astron Soc; Int Astron Union; Astron Soc Pac. *Res:* Astronomical spectroscopy; photoelectric photometry; spectral quantification. *Mailing Add:* Dept Astron Univ Southern Calif 01340 University Park Los Angeles CA 90089

BARRY, EDWARD GAIL, b Butte, Mont, May 4, 33; m 58, 84; c 3. GENETICS. *Educ:* Dartmouth Col, AB, 55; Stanford Univ, PhD(biol), 61. *Prof Exp:* USPHS trainee genetics, Yale Univ, 60-62; from asst prof to assoc prof bot, 62-75, PROF BOT, UNIV NC, CHAPEL HILL, 75- *Mem:* Genetics Soc Am; Genetical Soc Gt Brit. *Res:* Microbial genetics; cytogenetics of Neurospora. *Mailing Add:* Dept Biol 203 Univ NC Chapel Hill Wilson Hall 046A Chapel Hill NC 27514

BARRY, GUY THOMAS, organic chemistry, for more information see previous edition

BARRY, HENRY F, b Detroit, Mich, June 25, 23; m 47; c 6. CHEMICAL ENGINEERING, PHYSICAL CHEMISTRY. *Educ:* Stanford Univ, BS, 50; Univ Mich, MS, 52, MBA, 78. *Prof Exp:* Sr engr, Standard Oil Co, Ind, 52-60; dir spec chem, Haviland Prod Co, 60-62; res assoc chem, Climax Molybdenum Co, 62-65, supvr chem res, 65-67, dir chem res, 67-75, dir chem develop, 75-82; VPRES, SHATTUCK CHEM CO, 83- *Mem:* Am Chem Soc; Am Soc Lubrication Engrs; Nat Soc Prof Engrs; Am Inst Chem Engrs; Soc Plastics Engrs. *Res:* Solid lubricants; catalysts; petroleum processing; engineering research; molybdenum chemistry; environmental sciences. *Mailing Add:* 5337 W Iliff Dr No 101 Denver CO 80227-3791

BARRY, HERBERT, III, b New York, NY, June 2, 30. PSYCHOPHARMACOLOGY. *Educ:* Harvard Col, BA, 52; Yale Univ, MS, 53, PhD(psychol), 57. *Prof Exp:* From instr to asst prof psychol, Yale Univ, 58-61; asst prof, Univ Conn, 61-63; res assoc prof pharmacol, 63-70, PROF ANTHROP, SCH DENT MED, 70-, PROF PHYCOL & PHARMACOL, UNIV PITTSBURGH, 87- *Concurrent Pos:* NIMH fel, Yale Univ, 57-59; res sci develop award, NIMH, 67-77; consult, Nat Inst Alcohol Abuse & Alcoholism, 72-76; managing ed, Psychopharmacol, 74-84, field ed, 85- *Honors & Awards:* Distinguished Scientist Award, Soc Stimulus Properties & Drugs, 86. *Mem:* Am Psychol Asn; Am Anthrop Asn; Soc Cross-Cult Res (pres, 73-74); Am Soc Pharmacol & Exp Therapeut; Soc Stimulus Properties Drugs (pres, 80-81); Am Col Neurophsychopharmacol. *Res:* Psychopharmacology, testing effects of drugs, especially alcohol and other sedatives; cross-cultural research, correlating child training practices with other customs; relationships of birth order and sharing parent's first name to personality characteristics. *Mailing Add:* Sch Dent Med Univ Pittsburgh 552 N Neville St Apt 83 Pittsburgh PA 15261

BARRY, JAMES DALE, b Washington, DC, Feb 8, 42; m 64; c 2. SPACE PHYSICS, LASERS. *Educ:* Univ Calif, Los Angeles, BS, 64, PhD(space physics), 69; Calif State Col, Los Angeles, MS, 66. *Prof Exp:* Asst high energy physics, Univ Calif, Los Angeles, 64-65, res asst space physics, 65-68, res geophysicist, 68-69; laser scientist, Air Force Atomics Lab, 69-73, tech dir, Space Laser Commun, 73-75, sr scientist, Tech Div, Air Force Space & Missiles Orgn, 75-76, sr scientist, Space Laser Commun, Air Force Space & Missiles Systs Orgn, 76-78; sr scientist, Space Commun, Hughes Aircraft Co, 78-81, sr scientist, Electro-Optical Div, 81-86, PROG MGR, ELECTRO-OPTICAL/COMPUTER DATA PROCESSING, HUGHES AIRCRAFT CO, 86- *Mem:* Am Geophys Union; Inst Electronics Engrs; Sigma Xi. *Res:* Interplanetary particles and fields; ionosphere; magnetosphere; wave-particle interactions; stimulated emission; gaseous lasers; space laser communications; laser design and theory, medical applications of lasers; author of over seventy-five technical articles. *Mailing Add:* Hughes Aircraft Co EDSG/E51/B222 PO Box 902 El Segundo CA 90245

BARRY, JOAN, b New York, NY, Sept 17, 53; m; c 3. CARDIOVASCULAR RESEARCH. *Educ:* Univ Calif, Los Angeles, BA, 78. *Prof Exp:* Res assoc, Div Cardiol, Univ Calif, Los Angeles Med Ctr, 80-83; RES ASSOC CARDIOVASC DIV, BRIGHAM & WOMEN'S HOSP, 83-, ASSOC ISCHEMIA LAB SCIENTIST, DEPT MED, 87-; RES ASSOC, HARVARD MED SCH, 87- *Concurrent Pos:* Comt to Enhance Cardiac Patient Family Support Groups, Am Heart Asn, Greater Los Angeles Affiliate, 83-83, Pub Educ Forum Comt, 82-83; Family Health Promo Prog, Boston Univ Sch Med, 83. *Mem:* Am Heart Asn. *Res:* The utilizing of ambulatory electrocardiographic monitors in the detection of transient myocardial ischemia and pharmacologica investigations aimed at the treating of silent ischemia; author and coauthor of numerous articles books and papers on coronary artery disease, its investigation, and treatment. *Mailing Add:* Cardiovasc Div Harvard Univ Sch Med 75 Francis St Boston MA 02115

BARRY, KEVIN GERARD, b Newton, Mass, May 12, 23; m 42; c 4. INTERNAL MEDICINE. *Educ:* Georgetown Univ, MD, 49; Am Bd Internal Med, dipl, 56. *Prof Exp:* Intern, Walter Reed Gen Hosp, Washington, DC, US Army, 49-50, mem staff internal med, Soldiers Home Hosp, 50-51 & 98th Gen Hosp, Munich, Ger, 51-52, resident internal med, Walter Reed Gen Hosp, 53-56, batallion surgeon, 43rd Combat Engrs, 52-53, chief med serv, Army Hosp, NY, 56-58; instr med, Sch Med, Georgetown Univ, 58-63, clin asst prof, 63-69, assoc prof, 69-76; PRES & CLIN DIR, BIO-MED INC, 76- *Concurrent Pos:* Res internist, Walter Reed Army Inst Res, 58-61, chief dept metab, Inst & chief med sect, Hosp, 61-65, dir div med, Inst, 65-66; dir med educ & chief, Renal & Metab Sect, Wash Hosp Ctr, 66-69; chief, DC Gen Hosp, Med Div, Georgetown Univ, 69-71; med dir, Morris Cafritz Mem Hosp, 71-74. *Mem:* Fel Am Col Physicians; fel Am Col Clin Pharmacol & Chemother; AMA; Am Fedn Clin Res; Am Soc Artificial Internal Organs. *Res:* Renal, water and electrolyte metabolism; prevention and therapy of acute renal failure in man and animals; pathogenesis of acute renal failure in animals; peritoneal membranes used as absorptive surfaces. *Mailing Add:* 1295 Lavalle Dr Davidsonville MD 21035

BARRY, MICHAEL ANHALT, b Atlanta, Ga, July 11, 53; m 77; c 2. REGENERATION, COMPARATIVE NEUROBIOLOGY. *Educ:* Univ Pa, BA, 75; Univ Hawaii, MS, 78; Univ Del, PhD(life & health sci), 85. *Prof Exp:* Postdoctoral fel neurosci, Albert Einstein Col Med, 85-88; ASST PROF NEUROSCI, SCH DENT MED, UNIV CONN HEALTH CTR, 88- *Mem:* Am Soc Zoologists; Soc Neurosci; Asn Chemoreception Sci; AAAS. *Res:* Response and regeneration of taste system following nerve injury in mammals; comparative neurobiology of auditory, lateral line, and gustatory systems; fish neurobiology; anatomy; electrophysiology; behavior. *Mailing Add:* Dept Biostruct & Function Univ Conn Health Ctr Farmington CT 06030

BARRY, MICHAEL LEE, b Helena, Mont, Oct 11, 35; m 61; c 2. ELECTRONICS ENGINEERING. *Educ:* Colo Sch Mines, PRE, 57; Univ Calif, Berkeley, PhD(chem eng), 65. *Prof Exp:* Res engr, Esso Res & Eng Co, 57-59; mem tech staff, Fairchild Res & Develop, Fairchild Camera & Instrument Corp, 65-72; dir eng, Advan Memory Systs, Intersil, 72-77; prod line mgr, Synertek, 77-79; prod line mgr, Nat Semiconductor, 79-80; dir, VLSI Res Lab, Fairchild Res Ctr, Schlumberger, Ltd, 80-87; vpres, Res & Develop, 87-89, VPRES & GEN MGR, VITELIC SEMICONDUCTOR CORP, 89- *Mem:* Inst Elec & Electronics Engrs. *Res:* Electrochemistry of elemental phosphorus; vapor-phase deposition of dielectric films; semiconductor processing. *Mailing Add:* 1221 Dana Ave Palo Alto CA 94301

BARRY, ROGER DONALD, b Columbus, Ohio, Nov 26, 35; m 57; c 4. ORGANIC CHEMISTRY, BIOCHEMISTRY. *Educ:* Univ Cincinnati, BS, 57, PhD(org chem), 60. *Prof Exp:* Res fel, Ohio State Univ, 60-61, from instr to asst prof obstet & gynec, Col Med, 61-66; PROF CHEM & HEAD DEPT, NORTHERN MICH UNIV, 66- *Concurrent Pos:* Comnr, Accrediting Bur

Med Lab Schs, 69- *Mem:* Am Chem Soc; Sigma Xi. *Res:* Organic synthesis; spectroscopy; steroid chemistry and biochemistry; gas chromatography; analytical organic chemistry; chemical education. *Mailing Add:* Dept Chem Northern Mich Univ Marquette MI 49855-5343

BARRY, ROGER GRAHAM, b Sheffield, Eng, Nov 13, 35. CLIMATOLOGY, SNOW & ICE. *Educ:* Univ Liverpool, BA, 57; McGill Univ, MSc, 59; Univ Southampton, PhD(geog), 65. *Prof Exp:* Lectr geog, Univ Southampton, 60-66; res scientist climat, Geog Br, Dept Energy, Mines & Resources, Ont, 66-67; lectr geog, Univ Southampton, 67-68; assoc prof, 68-71, PROF GEOG, UNIV COLO, BOULDER, 71-, FEL, COOP INST RES ENVIRON SCI, 81- *Concurrent Pos:* Mem glaciol panel, Comt Polar Res, Nat Acad Sci, 73-77; dir, World Data Ctr-A for Glaciol, 76-; JS Guggenheim fel, 82-83; mem, comt arctic ocean info systs, marine bd, Nat Acad Sci, 85-87; mem, Polar Res Bd, Nat Acad Sci, 88- *Honors & Awards:* Honors Award, Asn Am Geogr, 86. *Mem:* Am Meteorol Soc; Am Quaternary Asn; Asn Am Geogr; Int Mountain Soc (secy, 81-); Am Geophys Union. *Res:* Climates of arctic and mountain regions; climatic change, interrelations between sea ice and climate, snow cover and sea ice data. *Mailing Add:* Coop Inst Res Environ Sci Box 449 Univ Colo Boulder CO 80309

BARRY, RONALD A, b Twin Falls, Idaho, Feb 19, 50; m. IMMUNOLOGY. *Educ:* Col Idaho, BS, 72; Univ Md, MD, 73; Navy Med Technol Sch, MT, 74; Wash State Univ, MS, 81, PhD, 83. *Prof Exp:* Clin bact instr, Nat Naval Med Ctr, 74; clin microbiologist, Navy Med Hosp, Guam, 75-76, San Diego, Calif, 76-77; teaching asst, Wash State Univ, 78-82, res asst, 79-81; res fel, Univ Calif, San Francisco, 83-86; RES IMMUNOLOGIST/ MICROBIOLOGIST, VET ADMIN MED CTR, PORTLAND, ORE, 86-; ASST PROF, DEPT MICROBIOL & IMMUNOL, ORE HEALTH SCI UNIV, PORTLAND, 89- *Concurrent Pos:* Immunol instr, Sch Nursing, Ore Health Sci Univ, Portland, Ore, 88; res rep, Libr Comt, Vet Admin Med Ctr, Portland, Ore, 88-, mem, Inst Animal Care & Use Subcomt, 88-89, chmn, 89- *Mem:* Am Soc Clin Pathologists; Am Soc Microbiologists; Am Soc Virol; Am Asn Immunologists; AAAS. *Res:* Immune T cell phenotypes in the expression of immunity and delayed-type hypersensitivity to Listeria monocytogenes; T cell phenotypes in antilisterial immunity. *Mailing Add:* Immunol Res 151R Vet Admin Med Ctr Bldg 101 Rm 529 Portland OR 97207

BARRY, RONALD EVERETT, JR, b Great Barrington, Mass, Jan 4, 47; m 69; c 2. MAMMALOGY, VERTEBRATE ECOLOGY. *Educ:* Univ Conn, BA, 68; Ind State Univ, MA, 74; Univ NH, PhD(zool), 78. *Prof Exp:* Res asst biochem, Health Ctr, Univ Conn, 68-73; instr biol, Bates Col, 77-78; teaching fel, W Va Univ, 79-80; asst prof zool, Unity Col, 80-84; ASSOC PROF BIOL, FROSTBURG STATE UNIV, 84- *Concurrent Pos:* Fel & fac mem mammal, Mountain Lake Biol Sta, Univ Va, 80; prin invest, NSF Cause grant, 83-84. *Mem:* AAAS; Am Soc Mammalogists; assoc mem Sigma Xi. *Res:* Habitat selection in small mammals; gastrointestinal morphology of small mammals; orientation, navigation and spatial resource partitioning in deer mice (peromyscus); competition in small mammals. *Mailing Add:* Dept Biol Frostburg State Univ Frostburg MD 21532

BARRY, SUE-NING C, b Shanghai, China, Nov 5, 32; US citizen; m 59. CELL PHYSIOLOGY. *Educ:* Barat Col, BA, 55; Univ Md, PhD(zool), 61. *Prof Exp:* Asst instr physiol, 61, asst prof, 62-68, assoc prof histol, 68-75, PROF ANAT, UNIV MD, 75-, ACTG CHMN, 87- *Concurrent Pos:* Res fel histol, Univ Md, 61-62. *Mem:* AAAS; Am Soc Cell Biol; Am Soc Zool. *Res:* Carbohydrate metabolism of fresh water protozoan astasia longa; effects of sugar metabolism in streptococcus salivarius on oral tissues; effects of systemic flouride on tooth formation. *Mailing Add:* Dept Anat Univ Md Sch Dent Baltimore MD 21201

BARRY, WILLIAM EUGENE, b Staten Island, NY, Aug 29, 28; m 54; c 5. MEDICINE. *Educ:* Villanova Univ, BS, 50; NY Med Col, MD, 54; Am Bd Internal Med, dipl, 61, cert hemat, 74. *Prof Exp:* Asst resident internal med & resident, Health Ctr, Ohio State Univ, 57-59, sr asst resident hemat & chief resident, 59-61; from instr to assoc prof, 61-72, PROF MED, MED CTR, TEMPLE UNIV, 72- *Concurrent Pos:* Am Col Physicians Brower travelling scholar, 65; res fel, Div Hemat, Univ Wash, 69-70. *Mem:* Am Fedn Clin Res; fel Am Col Physicians; Am Soc Hemat. *Res:* Clincial hematology; iron metabolism. *Mailing Add:* Temple Univ Hosp 3400 N Broad St Philadelphia PA 19140

BARRY, WILLIAM JAMES, b Hudson, NY, May 14, 48; m 72; c 1. TERRESTRIAL ECOLOGY, ENVIRONMENTAL IMPACT ASSESSMENT. *Educ:* Cornell Univ, BS, 70; Mich State Univ, MS, 71, PhD(zool), 74. *Prof Exp:* Asst prof zool, Conn Col, 75-80; PRIN SCIENTIST, NORMANDEAU ASSOCS. *Concurrent Pos:* Consult environ, 81- *Mem:* Animal Behav Soc; Am Soc Mammalogists; The Wildlife Soc; Ecol Soc Am. *Res:* The effect of photoperiod, temperature and other environmental factors on food hoarding behavior in small mammals, especially deermice; the study of the ecological relationships of small mammals on tidal marshes; environmental impacts on wildlife. *Mailing Add:* Normandeau Assocs Inc 25 Nashua Rd Bedford NH 03110

BARS, ITZHAK, b Izmir, Turkey, Aug 31, 43; US citizen; m 67; c 2. ELEMENTARY PARTICLE PHYSICS. *Educ:* Robert Col, BS, 67; Yale Univ, MPhil, 69, PhD(physics), 71. *Prof Exp:* Res assoc physics, Univ Calif, Berkeley, 71-73; asst prof physics, Stanford Univ, 73-75; from asst prof to assoc prof physics, Yale Univ, 75-83; PROF PHYSICS, UNIV SOUTHERN CALIF, 83- *Concurrent Pos:* Fel, A P Sloan Found, 76-80; Vis prof, Harvard Univ, 78, Inst Advan Study, 79 & 90, Inst Theoret Physics, Univ Calif, Santa Barbara, 86; jr fac fel, Yale Univ 78; prin investr, Univ Southern Calif, 83-; outstanding jr investr, Dept Energy, 83. *Mem:* Am Phys Soc; NY Acad Sci. *Res:* Formulated, developed and applied symmetry and supersymmetry principles in unified gauge theories, composite models of quarks and leptons, nuclear supersymmetries, feeble forces, superstring and supermembrane theories. *Mailing Add:* Dept Physics Univ Southern Calif Los Angeles CA 90089-0484

BARSCH, GERHARD RICHARD, b Berlin, Ger, June 22, 27; m 59; c 3. SOLID STATE PHYSICS. *Educ:* Tech Univ Berlin, MS, 51; Univ Gottingen, PhD(theoret physics), 55. *Prof Exp:* Mem sci staff theoret physics, Battelle Inst, Ger, 56-62; sr res assoc appl physics, 62-64, assoc prof solid state sci, 64-70, PROF PHYSICS, PA STATE UNIV, 70- *Mem:* Sigma Xi. *Res:* Lattice dynamics of anharmonic properties and crystal defects; nonlinear crystal elasticity; high pressure physics. *Mailing Add:* 245 Madison St State College PA 16801

BARSCHALL, HENRY HERMAN, b Berlin, Ger, Apr 29, 15; nat US;. NUCLEAR SCIENCE. *Educ:* Princeton Univ, AM, 39, PhD(physics), 40. *Hon Degrees:* Dr Natural Sci, Marburg (WGer), 82. *Prof Exp:* Instr physics, Princeton Univ, 40-41 & Univ Kans, 41-43; mem staff, Los Alamos Sci Lab, NMex, 43-46; from asst prof to prof physics, 46-73, chmn dept, 51, 54, 56-57 & 63-64, Bascom prof physics, med physics & nuclear eng, 73-87, EMER PROF PHYSICS, MED PHYSICS & NUCLEAR ENG, UNIV WIS-MADISON, 87- *Concurrent Pos:* Assoc Div Leader, Los Alamos Sci Lab, 51-52 & Lawrence Livermore Lab, 71-73; continuing consult, Lawrence Livermore Lab; assoc ed, Rev Mod Physics, 51-53 & Nuclear Physics, 59-72; vis prof, Univ Calif, Davis, 71-73; ed, Phys Rev C, 72-87; mem, Assembly Math & Phys Sci, Nat Res Coun, 80-83; mem gov bd, Am Inst Physics, 83-88; coun, Am Physics Soc, 83-86. *Honors & Awards:* T W Bonner Prize, Am Phys Soc, 65. *Mem:* Nat Acad Sci (chmn physics sect, 80-83); Am Nuclear Soc; Am Asn Physicists in Med; fel Am Phys Soc; fel Am Acad Arts & Sci. *Res:* Nuclear physics; fast neutrons; neutron physics; applications of neutron physics to radiotherapy and fusion technology. *Mailing Add:* 1150 University Ave Univ Wis Madison WI 53706

BARSDATE, ROBERT JOHN, b Richmond Hill, NY, Sept 4, 34; m 59; c 2. GEOCHEMISTRY. *Educ:* Allegheny Col, BS, 59; Univ Pittsburgh, PhD(geochem), 64. *Prof Exp:* from asst prof to assoc prof, 63-72, PROF MARINE SCI, UNIV ALASKA, 72- *Concurrent Pos:* Vis prof, Swiss Fed Inst Water Resources, 73-74. *Mem:* AAAS; Am Soc Limnol & Oceanog; Geochem Soc; fel Arctic Inst NAm; Soc Int Limnol; Sigma Xi. *Res:* Biologically mediated movement of trace metals in natural waters. *Mailing Add:* Box 80174 College AK 99708

BAR-SHAVIT, RACHEL, b Hadera, Israel, Dec 1, 53. GROWTH FACTORS. *Educ:* Bar-Ilan, Israel, PhD(biochem), 83. *Prof Exp:* RES ASSOC, JEWISH HOSP ST LOUIS, 85- *Mailing Add:* Hadassah Med Sch Hebrew Univ PO Box 1172 Jerusalem 91010 Israel

BAR-SHAVIT, ZVI, CELL BIOLOGY. *Educ:* Weizmann Inst, Israel, PhD(membrane res), 81. *Prof Exp:* PROF CELL BIOLOGY, RES ASST & LECTR, DIV CELL BIOL, SCH DENT MED, WASH UNIV, 83- *Mailing Add:* Hubert H Humphrey Ctr Exp Med & Cancer Res Hebrew Univ Hadassah Med Sch Jerusalem 91010 Israel

BARSHAY, JACOB, b New York, NY, Apr 7, 40; m 64; c 1. NUMBER THEORY. *Educ:* Princeton Univ, AB, 61; Brandeis Univ, MA, 63 & PhD(math), 66. *Prof Exp:* From instr to asst prof math, Northeastern Univ, 65-70; vis prof math, Univ Aarhus, Denmark, 70-71, Univ Genoa, Italy, 70-72; asst prof, 72-77, ASSOC PROF MATH, CITY UNIV NEW YORK, 77-, CHMN, MATH DEPT, 86- *Mem:* Am Math Soc; Math Assoc Am. *Res:* Commutative ring theory. *Mailing Add:* Dept Math NAC Bldg Rm 8133 City Col NY New York NY 10031

BARSIS, EDWIN HOWARD, b Brooklyn, NY, June 28, 40; div; c 2. APPLIED PHYSICS, ELECTRONICS. *Educ:* Cornell Univ, BEP, 63, MS, 65, PhD(appl physics), 67. *Prof Exp:* Mem tech staff, Device Studies Div, Livermore, Calif, 6871, supvr, 71-78, supvr, Advan Develop Div, 78-79; mgr, Electronics Subsysts Dept, 79-86, DIR, COMPUTER SCI & MATH, SANDIA NAT LABS, ALBUQUERQUE, 86-, DIR, ENG SCI, 89- *Concurrent Pos:* Consult, Air Force Weapons Lab. *Mem:* Am Phys Soc. *Res:* Ionic conductivity; dielectric relaxation; defects in insulators; radiation-induced optical properties; piezoresistivity; optical computing; parallel computing; computer architectures; intelligent machines; computer science. *Mailing Add:* 1538 Catron Ave SE Albuquerque NM 87123

BARSKE, PHILIP, b Fairfield, Conn, Jan 12, 17; m 48; c 1. WILDLIFE MANAGEMENT & WETLANDS. *Educ:* Univ Conn, BS, 40; Univ Mich, MS, 43; Union Grad Sch, Ohio PhD(appl ecol), 76. *Hon Degrees:* DSc, Univ Bridgeport, 77. *Prof Exp:* Game biologist, Conn State Bd Fish & Game, 42-43 & 46; FIELD CONSERVATIONIST, WILDLIFE MGT INST, WASHINGTON, DC, 46-, ENVIRON CONSULT, WETLANDS SPECIALTY. *Concurrent Pos:* Consult environ systs to several major indust concerns, 70-; mem, Gov Coun Environ Qual, Conn. *Honors & Awards:* Am Motors Award, 63; Cert of Recognition, Wildlife Soc, 67, John Pearce Mem Award, 75. *Mem:* Wildlife Soc; Soil Conserv Soc Am; Nat Asn Environ Prof; Soc Am Foresters; Soc Wetland Scientists; Natl Asn State Wetland Mgrs. *Res:* Applied habitat development; wildlife areas and natural areas development and wetland evaluation and development. *Mailing Add:* 200 Audubon Lane Fairfield CT 06430

BARSKY, BRIAN ANDREW, b Montreal, Que, Sept 17, 54. COMPUTER GRAPHICS. *Educ:* McGill Univ, DCS, 73, BSc, 76; Cornell Univ, MS, 78; Univ Utah, Salt Lake City, PhD(comput sci), 81. *Prof Exp:* Asst prof, 81-86, ASSOC PROF COMPUT SCI, UNIV CALIF, BERKELEY, 86- *Concurrent Pos:* Presidential young investr, NSF, 83; consult, 81-; adj asst prof, Univ Waterloo, Ont, 82-; prin investr, NSF grants, 82- & MICRO grant, 84-85; vis researcher, Nat Advan Sch Telecommun, Paris, 85-86; fac develop award, IBM, 85. *Mem:* Asn Comput Mach; Nat Comput Graphics Asn; Inst Elec & Electronics Engrs; Can Man-Comput Commun Soc; Soc Indust & Appl Math. *Res:* Interactive three-dimensional computer graphics, concentrating on various aspects of the realistic image synthesis problem; computer aided geometric design and modelling concentrating on mathematical techniques for curve and surface representation; development of the Beta-splire representation and Liang-Barsky clipping algorithms; computer aided geometric design and modelling. *Mailing Add:* Comput Sci Div Univ Calif Berkeley CA 94720

BARSKY, CONSTANCE KAY, b Newark, NJ, Nov 3, 44; m 74. POLYMER SCIENCE, GLASS CHEMISTRY. *Educ:* Denison Univ, BS, 66; Washington Univ, PhD(geochem), 75. *Prof Exp:* Res assoc geochem petrol, Univ Mo-Columbia, 71-77; advan scientist, Owens-Corning Fiber Glass Corp, 77-78, supvr mat anal lab, 78-81, supvr, indust operating div prod develop lab, 81-85, supvr, inorg & surface technol lab, 85-86; ADMIN MGR, DEPT CHEM, OHIO STATE UNIV, 87-, ASST TO DEAN, COL MATH & PHYS SCI, 90- *Mem:* Sigma Xi; Elec Micros Soc Am; Microbeam Anal Soc; Am Women Sci. *Res:* Applications of electron beam microanalysis to problems in glass chemistry, metallurgy and environmental analysis; scanning electron microscopy of glass composites; major and trace element geochemistry of igneous rocks applied to problems of magma origin and evolution; diversification projects related to inorganic and surface technology of glass forming; development of polymeric coatings for continuous and chopped fibrous glass products used as reinforcements; applications of chemical microanalysis and microscopy to glass composites and metals. *Mailing Add:* 221 Elm St Granville OH 43023

BARSKY, JAMES, b Regina, Sask, Nov 28, 25; US citizen; m 57; c 2. INFORMATION SCIENCE. *Educ:* Univ Sask, BA, 46; Univ Toronto, MA, 51; Northwestern Univ, PhD(biochem), 58. *Prof Exp:* Asst biochem, Univ Sask, 46-48; fel physiol chem, Univ Calif Col Med, Ment Health Prog & Vet Admin Ctr, Los Angeles, 58-60; mgr dept biochem, Res Div, Ethicon, Inc, 60-63; biol sci ed, Acad Press Inc, 63- 67, exec ed, 67-68, vpres ed, 68-73, sr vpres, 73-78, pres, 78-87; sr vpres, Harcourt Brace Jovanovich, 81-87; RETIRED. *Concurrent Pos:* Doctoral fel, Univ Calif, Los Angeles; mem adv panel, NSF Off Sci Info Serv, 75 & adv comt, Nat Comn Libr & Info Sci, 75-87; mem bd dirs, Copyright Clearance Ctr, 77-87; chmn, Academic Press, 85. *Mem:* AAAS; Am Chem Soc; Am Asn Clin Chem; Biochem Soc, UK; NY Acad Sci; Sigma Xi. *Res:* Enzymology; scientific technical, medical and scholarly publications. *Mailing Add:* 53 Rosemont Terr West Orange NJ 07052

BARSOM, JOHN M, MATERIALS SCIENCE ENGINEERING. *Educ:* Univ Pittsburgh, BS, 60, MS, 62, PhD(mech eng), 69. *Prof Exp:* Sr res engr, Pittsburgh Plate Glass Indust, 60-67; sr res consult & chief mat, Technol Div, 67-83, SR CONSULT, US STEEL CORP, 83- *Concurrent Pos:* Adj prof, Civil Eng Dept, Univ Pittsburgh; chmn, Comt E-24 on Fracture of Mat, Am Soc Testing Mat, Transp Strut Subcomt, Am Iron & Steel Inst; mem, Comt E-9, Am Soc Testing Mat, subcomt F-704 Hydrogen Embrittlement, Ship Res Comt, Maritime Transp Res Bd, adv mem steel indust coun, Advan Steel Bridge Technol, Am Inst Steel Construct Subcomt on Brittle Fracture. *Honors & Awards:* Award of Merit, Am Soc Testing & Mat, 77. *Mem:* Fel Am Soc Metals; fel Am Soc Testing & Mat. *Res:* Material selection; prediction of behavior of fabricated components in structures; fracture and fatigue prevention; structural reliability; failure analysis; accident reconstruction; numerous publications. *Mailing Add:* US Steel Corp 600 Grant St Rm 755 Pittsburgh PA 15230

BARSS, WALTER MALCOMSON, b Corvallis, Ore, Jan 29, 17; nat Can; m 47; c 1. PHYSICS. *Educ:* Univ BC, BA, 37, MA, 39; Purdue Univ, PhD(physics), 42. *Prof Exp:* Asst, Univ BC, 38-39 & Purdue Univ, 40-42; res physicist, Nat Res Coun Can, 42-52 & Atomic Energy Can Ltd, 52-64; assoc prof physics, Univ Victoria, 64-82; RETIRED. *Concurrent Pos:* Can Asn Physicists; Acoust Soc Am. *Res:* Acoustics, especially reverberation problems in rooms and in underwater sound using laboratory models. *Mailing Add:* 3930 Cherrilee Crescent Victoria BC V8N 1R9 Can

BARSTON, EUGENE MYRON, b Chicago, Ill, Aug 7, 35; m 62; c 2. APPLIED MATHEMATICS. *Educ:* Calif Inst Technol, BS, 57, MS, 58; Stanford Univ, PhD(physics), 64. *Prof Exp:* Math physicist, Stanford Res Inst, 63-65; assoc res scientist, Courant Inst Math Sci, 65-67, from asst prof to assoc prof math, NY Univ, 67-72; assoc prof, 72-77, PROF MATH, UNIV ILL, CHICAGO CIRCLE, 77- *Mem:* Soc Indust & Appl Math. *Res:* Stability theory; fluid mechanics; plasma physics. *Mailing Add:* Dept Math Univ Ill Chicago IL 60680

BARSTOW, DAVID ROBBINS, b Middletown, Conn, Aug 5, 47; m 70; c 2. AUTOMATIC PROGRAMMING. *Educ:* Carleton Col, BA, 69; Stanford Univ, MS, 71, PhD(comput sci), 77. *Prof Exp:* J W Gibbs instr comput sci, Yale Univ, 77-79, asst prof, 79-80; prog leader software res, Schlumberger-Doll Res, Ridgefield, Conn, 80-87, sci adv, 87-89; SCI ADV, SCHLUMBERGER LAB COMPUTER SCI, 89- *Mem:* Asn Comput Mach; Comput Soc-Inst Elec & Electronics Engrs; Am Asn Artificial Intel. *Res:* Applications of artificial intelligence techniques to software engineering; automatic programming; programming environments; models of parallel computation. *Mailing Add:* Schlumberger Lab Computer Sci PO Box 200015 Austin TX 78720-0015

BARSTOW, LEON E, b Union City, Pa, Dec 14, 40; div; c 5. ORGANIC CHEMISTRY, BIOCHEMISTRY. *Educ:* Edinboro State Col, BS, 62; Syracuse Univ, PhD(chem), 66. *Prof Exp:* Asst prof chem, Robert Col, Istanbul, 66-69; res assoc, 69-72, RES ASST PROF BIOCHEM, UNIV ARIZ, 72- *Concurrent Pos:* NIH postdoctoral fel, Univ Ariz, 71-72; consult, Vega Eng, 72-75; pres & gen mgr, Vega Biochem, 74-79, pres & chief exec officer, Vega Biotechnologies, Inc, 79-85; pres & chief exec officer, Protein Technol, Inc. *Honors & Awards:* IR-100 Award, 82; Award of Excellence, Soc Tech Commun, 84. *Mem:* Am Chem Soc; AAAS; Sigma Xi. *Res:* Preparation of semi-synthetic analogs of proteins for structure-function studies; synthesis of peptide hormones with novel biological activity; development of peptide substrates for various enzymes. *Mailing Add:* Protein Technol Inc 1665 E 18th St Suite 106 Tucson AZ 85719

BART, GEORGE RAYMOND, b Oak Park, Ill; m 61; c 2. MATHEMATICAL PHYSICS, ELEMENTARY PARTICLE PHYSICS. *Educ:* Loyola Univ, Chicago, BS, 61; Ill Inst Technol, PhD(theoret physics), 70. *Prof Exp:* Physicist & mgr crystal contracts, Electronics & Res Ctr, Victor-Comptometer Corp, 63-66; resident assoc, High Energy Physics Div,

Argonne Nat Lab, 67-69; asst prof physics, Loyola Univ, Chicago, 69-70; asst prof physics & math, Mayfair Col, 70-74; assoc prof, 74-81, chmn, Nat Sci Dept, 81-86, PROF PHYSICS & MATH, HARRY S TRUMAN COL, 81-, CHMN PHYSICS SCI & ENG DEPT, 86- *Concurrent Pos:* Fac res prog, accelerator res fac div, Argonne Nat Lab, 79- *Mem:* Am Phys Soc; Am Asn Physics Teachers. *Res:* Theories of elementary particles; null-plane quantum field theories; S-matrix theory; dispersion relations; analysis of non-linear equations; soliton theory; integral equations. *Mailing Add:* Dept Nat Sci Truman Col 1145 W Wilson Ave Chicago IL 60640

BART, ROGER, chemical engineering, for more information see previous edition

BARTA, ALICE, biochemistry, for more information see previous edition

BARTA, OTA, b Ostrava, Czech, Aug 18, 31; US citizen; m 56; c 2. VETERINARY IMMUNOLOGY. *Educ:* Univ Vet Med, Brno, Czech, MVDr, 55, CSc, 63; Univ Guelph, PhD(immunolo microbiol), 69. *Prof Exp:* Asst prof animal hyg, Univ Vet Med, Brno, Czech, 55-61; scientist animal hyg, Cent Res Inst Animal Husb, Prague-Uhrineves, Czech, 61-64 & Inst Vet Med Res, Brno, 64-69; from asst prof immunol to assoc prof, Dept Microbiol, Okla State Univ, 69-75; from assoc prof to prof immunol, Sch Vet Med, La State Univ, Baton Rouge, 75-87; PROF IMMUNOL, DEPT PATHOBIOL, VA TECH, 87- *Concurrent Pos:* Res assoc, Ont Vet Col, Univ Guelph, 67-69. *Honors & Awards:* Fulbright sr lectr, 81; Small Animal Award, Ralston-Purina, 85. *Mem:* Am Asn Immunologists; Am Asn Vet Immunologists; World Asn Vet Microbiologists. *Res:* Immunology, research of the complement system; immunochemistry; clinical veterinary immunology; infectious diseases. *Mailing Add:* Va-Md Regional Col Vet Med Va Tech Dept Pathobiol Blacksburg VA 24061

BARTAL, ARIE H, b Iasi, Romania, Feb 9, 47; Israel citizen; m 69; c 2. MONOCLONAL ANTIBODIES & HYBRIDOMAS, CANCER INVESTIGATION. *Educ:* Hebrew Univ, Jerusalem, MD, 72; Technion Israel Inst Technol, DSc(immunol), 85. *Prof Exp:* Head, Hybridoma Lab, Ramborn Med Ctr, Inst Oncol, Haifa, Israel, 82-86; div chief, Biotherapeutics, Franklin, Tenn, 87-88; SCI DIR, IMMUNOSCI INC, 88- *Concurrent Pos:* Assoc researcher, Ramborn Med Ctr, Israel, 74-75, Sloan Kettering Cancer Ctr, NY, 79-82; vis scientist, Sloan Kettering Cancer Ctr, NY, 84, Nat Cancer Inst, 84; asst prof, Fac Med, Technion Inst Technol, Haifa, Israel, 86; co-ed, Method of Hybridoma Formation, 87. *Honors & Awards:* Oncol Award, Israel Med Asn, 79. *Mem:* Am Soc Clin Oncol; Am Asn Cancer Res; Am Fedn Clin Res; Am Asn Immunol; Fed Am Socs Exp Biol; Tissue Cult Asn. *Res:* Oncology; basic and applied research; immunology and biological response modifiers; hybridome and monoclonal antibody generation and characterization; immunopathology; tumor markers and prognostic factors; adoptive immunotherapy. *Mailing Add:* Immunosci 160 Community Dr Great Neck NY 11021

BARTEAU, MARK ALAN, b St Louis, Mo, Sept 8, 56; m 83; c 2. SURFACE SCIENCE, CATALYSIS. *Educ:* Wash Univ, BS, 76; Stanford Univ, MS, 77, PhD(chem eng), 81. *Prof Exp:* NSF postdoctoral fel, Physics Dept, Tech Univ München, 81-82; asst prof chem eng, 82-87, assoc prof chem eng & chem, 87-90, PROF CHEM ENG & CHEM, UNIV DEL, 82- *Concurrent Pos:* Assoc dir, Ctr Catalytic Sci & Technol, Univ Del, 82-; presidential young investr award, NSF, 85; vis prof, Dept Chem Eng, Univ Pa, 91-92. *Honors & Awards:* Victor K LaMer Award, Am Chem Soc, 82; Innovation Recognition Award, Union Carbide Corp, 88 & 89; Allan P Colburn Award, Am Inst Chem Engrs, 91. *Mem:* Am Inst Chem Engrs; Am Chem Soc; Catalysis Soc; Mat Res Soc; AAAS. *Res:* Synthesis of new catalytic materials; reactivity of metal oxides; surface science studies of intermediates and pathways in alcohol and hydrocarbon synthesis; halogen exchange catalysis. *Mailing Add:* Four Littlebrook Dr Wilmington DE 19807

BARTEL, ALLEN HAWLEY, b San Diego, Calif, July 26, 23; m 51; c 4. BIOPHYSICS, FLOW CYTOMETRY. *Educ:* Univ Calif, AB, 47, MS, 49, PhD, 54. *Prof Exp:* Instr biol sci, Univ Calif, 55-56; res fel chem, Calif Inst Technol, 56-58; from asst prof to assoc prof biol, Univ Houston, 58-67, prof biophys sci & chmn dept, 67-79, prof biochem & biophys sci & dir, Flow Cytometry Lab, 79-87, CHMN & PROF BIOCHEM & BIOPHYS SCI, UNIV HOUSTON, 87- *Mem:* Am Soc Biochem & Molecular Biol; Am Chem Soc; AAAS. *Res:* Computational biochemistry; comparative biochemistry. *Mailing Add:* Dept Biochem & Biophys Sci Univ Houston Houston TX 77204-5500

BARTEL, DONALD L, b Peoria, Ill, June 5, 39; m 62; c 4. MECHANICAL ENGINEERING. *Educ:* Univ Ill, BS, 61, MS, 63; Univ Iowa, PhD(mech), 69. *Prof Exp:* Asst prof eng, Black Hawk Col, 63-65; ASST PROF MECH ENG, CORNELL UNIV, 69- *Concurrent Pos:* NSF res grant, 70-71. *Mem:* Soc Indust & Appl Math; Am Soc Mech Engrs. *Res:* Optimum design of mechanical systems and components; application of mechanics principles to problems in orthopedic surgery. *Mailing Add:* Dept Mech & Aerospace Eng Cornell Univ Ithaca NY 14853

BARTEL, FRED F(RANK), b Milwaukee, Wis, Nov 4, 17; m 43; c 4. CIVIL ENGINEERING. *Educ:* Univ Wis, BS, 40; Univ Md, MS, 42. *Prof Exp:* Asst dir eng, Nat Sand & Gravel Asn, 46-49; chief engr, Tews Lime & Cement Co, 49-75, pres, 75-83; trustee in bankruptcy, 4x Corp, 85; RETIRED. *Concurrent Pos:* Chmn bd, Nat Ready Mixed Concrete Asn, 79-80; hon mem, Comt C-9, Am Soc Testing & Mat; hon dir, Nat Ready-Mixed Conrete Asn. *Mem:* Am Soc Civil Engrs; Am Soc Testing & Mat. *Res:* Portland cement concrete; aggregates for concrete and bituminous construction. *Mailing Add:* 5421 N Shoreland Ave Milwaukee WI 53217-5132

BARTEL, HERBERT H(ERMAN), JR, b Dallas, Tex, Mar 31, 24; m 50; c 2. CIVIL ENGINEERING. *Educ:* Southern Methodist Univ, BS, 44; Tex Univ, MS, 50; Tex A&M Univ, PhD(civil eng), 62. *Prof Exp:* Instr civil eng, Southern Methodist Univ, 46-53, from asst prof to prof, 53-72, chmn, Dept

Civil & Environ Eng, 69-72; chmn, Univ Tex, El Paso, 72-78, prof civil eng, 78-91; RETIRED. *Honors & Awards:* Excellence in eng teaching award, Gen Dynamics, 69. *Mem:* Am Soc Eng Educ; Am Soc Civil Engrs. *Res:* Highway illumination; highway traffic operation and pavement design. *Mailing Add:* 5801 Kingsfield Ave El Paso TX 79912-4815

BARTEL, LAVON L, b Salem, Ore, Nov 12, 51; m. LIPID METABOLISM, HUMAN NUTRITIONAL HEALTH. *Educ:* Ore State Univ, BS, 73, MS, 75; Univ Wis, Madison, PhD(nutrit), 79. *Prof Exp:* Res asst med sci, Med Sch, Univ Ore, 72; res asst, Dept Zool, Ore State Univ, 73, teaching asst, 73-75; res asst, Dept Nutrit Sci & Med, Univ Wis, Madison, 75-79; CHAIR ASST PROF, DEPT HOME ECON NUTRIT & FOOD SCI, WHITTIER COL, 79- *Concurrent Pos:* Teaching asst, Dept Nutrit Sci, Univ Wis, Madison, 77; dir dietetics & speaker, Speakers Bur, Whittier Col, 79-; consult, Pico Rivera Health Ctr, 79-; nutrit consult, Area Health Educ Coun, Univ Southern Calif & San Gabriel Health Educ Adv Bd, 81- *Mem:* AAAS; Am Dietetic Asn; Asn Women Sci; Inst Food Technologists. *Res:* Lipid metabolism and carnitine levels in hemodialysis patients and development of a rat model; management of obesity through nutrition education and behavior modification techniques. *Mailing Add:* Human Nutrit Foods Dept Univ Vt 310 Terrill Hall Burlington VT 05405

BARTEL, LEWIS CLARK, b Hillsboro, Kans, Dec 5, 34; m 56; c 2. GEOPHYSICS, THEORETICAL SOLID STATE PHYSICS. *Educ:* Univ Kans, BS, 58; Iowa State Univ, PhD(physics), 64. *Prof Exp:* Asst prof physics, Colo State Univ, 64-67; STAFF MEM, PHYSICS RES, SANDIA LABS, 67- *Concurrent Pos:* Consult, PEC Res Assocs, 65-67. *Mem:* Soc Explor Geophys. *Res:* Engineering and mining geophysics, electrical and electromagnetic geophysical methods; theoretical solid state physics, magnetism and ferroelectrics. *Mailing Add:* Sandia Labs Div 6258 P Box 5800 Albuquerque NM 87185

BARTEL, MONROE H, b Newton, Kans, Oct 3, 36; m 57; c 3. BIOLOGY, ENTOMOLOGY. *Educ:* Tabor Col, AB, 58; Kansas State Univ, MS, 60, PhD(parasitol), 63. *Prof Exp:* Instr zool, Kans State Univ, 62-63; from asst prof to assoc prof, 63-71, PROF BIOL, MOORHEAD STATE UNIV, 71- *Concurrent Pos:* Fel trop med, Sch Med, La State Univ, 66 & Histochem Inst, 67, Acarology Prog, Ohio State Univ, 78. *Mem:* Am Soc Parasitol; AAAS; Sigma Xi; Am Sci Affil. *Mailing Add:* Dept Biol Moorhead State Col Moorhead MN 56560

BARTELINK, DIRK JAN, b Heumen, Neth, Oct 28, 33; Can citizen; m 57; c 2. SOLID STATE ELECTRONICS. *Educ:* Univ Western Ont, BSc, 56; Stanford Univ, MS, 59, PhD(elec eng), 62. *Prof Exp:* Mem tech staff solid state electronics res, Bell Labs, NJ, 61-66; supvr semiconductor devices, 66-73; mgr device physics, Palo Alto Res Ctr, Xerox Corp, 73-80; MGR METAL-OXIDE-SEMICONDUCTOR TECHNOL, HEWLETT PACKARD LABS, 80- *Mem:* Inst Elec & Electronics Engrs. *Res:* Fundamental and applied phenomena in semiconductor and dielectrics and interfaces between these. *Mailing Add:* Hewlett Packard Lab 3500 Deer Creek Rd Palo Alto CA 94304

BARTELL, CLELMER KAY, b Kingstree, SC, Nov 1, 34. COMPARATIVE ANIMAL PHYSIOLOGY. *Educ:* Davidson Col, BS, 57; Univ Tenn, MS, 63; Duke Univ, PhD(zool), 66. *Prof Exp:* Instr zool, Duke Univ, 65-66; res assoc, Tulane Univ, 66-69; ASSOC PROF ZOOL, UNIV NEW ORLEANS, 69- *Mem:* Marine Biol Asn; Am Soc Zoologists. *Res:* Neuro-endocrine mechanisms in Crustacea; control of dermal chromatophores; Osmoregulation in animals. *Mailing Add:* Dept Biol Sci Univ New Orleans Lake Front New Orleans LA 70148

BARTELL, DANIEL P, b St Paul, Minn, Mar 30, 44. ENTOMOLOGY. *Educ:* Eastern Ill Univ, BSEd, 66; Purdue Univ, MS, 68, PhD(entom), 73. *Prof Exp:* Instr biol, parasitol, Davis & Elkins Col, 68-71; res assoc entom, Ill Nat Hist Surv, 73-75; asst prof, Ill Nat Hist Surv & Univ Ill, 75-76; asst prof, 76-78, assoc prof entom, Tex Tech Univ, 78-; AT COL AGR SCI, MONT STATE UNIV. *Mem:* Entom Soc Am; Entom Soc Can; Am Inst Biol Sci; Southwestern Entom Soc; Sigma Xi. *Res:* Pest management strategies for vegetable insects; host/parasite interactions; natural enemies of the red imported fire ant. *Mailing Add:* Dept Entom Okla State Univ 501 Life Sci W Stillwater OK 74078-0464

BARTELL, LAWRENCE SIMS, b Ann Arbor, Mich, Feb 23, 23; m 52; c 1. PHYSICAL CHEMISTRY, CHEMICAL DYNAMICS. *Educ:* Univ Mich, BS, 44, MS, 47, PhD(chem), 51. *Prof Exp:* Res asst, Manhattan Proj, Univ Chicago, 44-45; Rackham fel, Univ Mich, 51-52, res assoc, 52-53; from asst prof to prof chem, Iowa State Univ, 53-65; PROF CHEM, 65-, PHILIP J ELVING COL CHAIR, UNIV MICH, ANN ARBOR, 87- *Concurrent Pos:* Consult, Gillette Co, Ill, 56-62 & Mobil Oil Corp, NJ, 60-84; chemist, AEC, 56-65; assoc ed, J Chem Physics, 63-65; mem adv bd, Petrol Res Fund, 70-73; vis prof, Moscow State Univ, 72, Univ Paris XI, 73 & Univ Tex, Austin, 78 & 86; mem, Comm Electron Diffraction, Int Union Crystallog, 66-75. *Honors & Awards:* Creativity Award, NSF, 82. *Mem:* Am Chem Soc; Am Crystallog Asn; fel Am Phys Soc; fel AAAS. *Res:* Atomic and molecular structure by electron diffraction; surface and quantum chemistry; structure of molecular liquids and clusters. *Mailing Add:* Dept Chem Univ Mich Ann Arbor MI 48104

BARTELL, MARVIN H, b Wanatah, Ind, Sept 16, 38; m 63; c 3. VERTEBRATE ZOOLOGY, ENDOCRINOLOGY. *Educ:* Concordia Teachers Col, Ill, BS, 61; Univ Mich, MS, 62, PhD(zool), 69. *Prof Exp:* Instr biol, Concordia Teachers Col, Ill, 62-64; lectr endocrinol, Univ Mich, 67; asst prof, 68-72, assoc prof, 72-81, PROF BIOL, CONCORDIA COL, 81- *Mem:* AAAS; Am Soc Microbiol; Sigma Xi. *Res:* Developmental endocrinology; control of carbohydrate metabolism in lower vertebrates; medical microbiology. *Mailing Add:* Dept Biol Concordia Univ 7400 Augusta St River Forest IL 60305

BARTELL, STEVEN MICHAEL, b Beaver Dam, Wis, Oct 25, 48. SYSTEMS DESIGN, HYDROLOGY-WATER RESOURCES & ENVIRONMENTAL SCIENCE. *Educ:* Lawrence Univ, BA, 71; Univ Wis, MS, 73, PhD(limnol & oceanog), 78. *Prof Exp:* Res assoc, Savannah River Ecol Lab, 78-80; adj asst prof, Inst Ecol, Univ Ga, 78-80; RES STAFF, ENVIRON SCI DIV, OAK RIDGE NAT LAB, 80-; ADJ PROF ECOL, UNIV TENN, 84- *Concurrent Pos:* Panel mem, Fel Prog Rev, Nat Sci Found, NATO, 81 & Competitive Grant Proposal Rev Comt, US Environ Protection Agency, 82; co-ed, Savannah River Ecol Lab, 80- *Mem:* Ecol Soc Am; Int Soc Ecol Modelling. *Res:* Development and evaluation of simulation models that predict transport, accumulation and biological effects of toxicants in aquatic systems; ecological and human health risk analysis; global resource risk analysis. *Mailing Add:* Environ Sci Div Oak Ridge Nat Lab PO Box 2008 Oak Ridge TN 37831-6036

BARTELS, GEORGE WILLIAM, JR, b Hershey, Pa, June 23, 28; m 51; c 3. ORGANIC CHEMISTRY. *Educ:* Lebanon Valley Col, BS, 50; Univ Del, MS, 51, PhD(org chem), 53. *Prof Exp:* Asst chem, Univ Del, 50-51, Armstrong res fel, 51-53; res chemist org chem, E I du Pont de Nemours & Co, 53-58, from res supvr to sr res supvr, 58-65, tech supt, 65-67, process supt dacron, 67-69, indust prod supt, 69-72, asst to mfg dir, Dacron Div, Textile Fibers Dept, 72-79, mgr, specialty prod div, F&F Dept, 79-83, mgr, electronic prods, P&EP Dept, 83-85; RETIRED. *Mem:* AAAS; Am Chem Soc; Sigma Xi. *Res:* Organic research in linear high polymers; synthetic fibers; spun-bonded nonwoven fabrics. *Mailing Add:* 707 Severn Rd Wilmington DE 19803

BARTELS, PAUL GEORGE, b Yuma, Colo, Apr 9, 34; m 56; c 2. PLANT PHYSIOLOGY. *Educ:* Vanderbilt Univ, PhD(biol), 64. *Prof Exp:* NIH fel, 64-65; from asst prof to prof, 66-85, assoc biologist, Exp Sta, 70-78, PROF PLANT SCI, UNIV ARIZ, 78- *Concurrent Pos:* NSF grant, 67-69. *Mem:* Am Soc Plant Physiol. *Res:* Plant cell physiology; inhibition of chloroplast development by herbicides. *Mailing Add:* Dept Plant Sci Univ Ariz Tucson AZ 85721

BARTELS, PETER H, b Danzig, Ger, Jan 25, 29; US citizen; m 54; c 3. OPTICS, COMPUTER SCIENCE. *Educ:* Univ Gottingen, PhD(biophys), 54; Univ Giessen, Dr Habil, 58. *Prof Exp:* Asst prof, Univ Giessen, 54-58; dir res, E Leitz, Optical Co, Inc, 58-66; assoc prof, 66-70, comt comput sci, 70-76, PROF MICROBIOL & MED TECHNOL, UNIV ARIZ, 70-, OPTICAL SCI CTR, 70- *Concurrent Pos:* Assoc prof obstet & gynec, Univ Chicago, 67-69, prof, 69- *Mem:* Hon fel Inst Acad Cytol. *Res:* Image processing; pattern recognition; objectivation of diagnostic procedures through self learning computer programs; machine recognition of tumor cells. *Mailing Add:* Dept Optical Sci Univ Ariz Tucson AZ 85721

BARTELS, RICHARD ALFRED, b Saginaw, Mich, May 10, 38; m 59; c 3. SOLID STATE PHYSICS. *Educ:* Case Inst Technol, BS, 60, MS, 63, PhD(physics), 65. *Prof Exp:* Sloan fel, Princeton Univ, 64-66; from asst prof to assoc prof, 66-75, PROF PHYSICS, TRINITY UNIV, TEX, 75- *Honors & Awards:* Distinguished Serv Award, Am Asn Physics Teachers, 87. *Mem:* Am Asn Physics Teachers; Nat Sci Teachers Asn. *Res:* Effects of high pressures on the physical properties of solids; elastic and dielectric constants of solids; radiation and the atmosphere. *Mailing Add:* Dept Physics Trinity Univ San Antonio TX 78212

BARTELS, RICHARD HAROLD, b Ann Arbor, Mich, Jan 10, 39; m 68. COMPUTER SCIENCE. *Educ:* Univ Mich, BS, 61, MS, 63; Stanford Univ, PhD, 68. *Prof Exp:* Asst prof comput sci, Univ Tex, Austin, 68-74; asst prof math sci, Johns Hopkins Univ, 74-77, assoc prof, 77-80; assoc prof, 80-85, PROF COMPUT SCI, UNIV WATERLOO, 85- *Mem:* Soc Indust & Appl Math; Asn Comput Mat; Spec Interest Group Numerical Math; Spec Interest Group Comput Graphics & Interactive Tech. *Res:* Numerical data fitting; splines in computer graphics. *Mailing Add:* Dept Comput Sci Univ Waterloo Waterloo ON N2L 3G1 Can

BARTELS, ROBERT CHRISTIAN FRANK, b New York, NY, Oct 24, 11; m 38; c 1. MATHEMATICS. *Educ:* Univ Wis, PhB, 33, PhM, 36, PhD(math), 38. *Prof Exp:* Asst math, Univ Wis, 33-36, 37-38; instr, Univ Wis, 38-42; aeronaut engr, Bur Aeronaut, US Navy, 42-45; from asst prof to prof, 45-76, dir, Comput Ctr, 59-76, EMER PROF MATH, UNIV MICH, ANN ARBOR, 76- *Concurrent Pos:* Res partic, Oak Ridge Inst Nuclear Studies, 54; mem comput res study sect, NIH, 62- *Mem:* AAAS; Soc Indust & Appl Math; Asn Comput Mach; Am Math Soc; Math Asn Am; Sigma Xi. *Res:* Mathematical theory of elasticity; hydrodynamics; dynamics of compressible fluid; applied mathematics; Saint-Venant's flexure problem for the regular polygon; numerical analysis; magneto hydrodynamics. *Mailing Add:* Box 2240 Ann Arbor MI 48106

BARTELT, JOHN ERIC, b Milwaukee, Wis, Aug 11, 55; m 89. ELECTRON-POSITRON COLLISIONS, UNDERGROUND DETECTORS. *Educ:* Univ Wis, BS, 77; Univ Minn, PhD(physics), 84. *Prof Exp:* Teaching assoc astron, Univ Minn, 77-79, res asst, 79-83; res assoc, Stanford Linear Accelerator Ctr, Stanford Univ, 83-89; ASST PROF PHYSICS, VANDERBILT UNIV, 90- *Res:* High energy physics research using electron-positron colliding beams; detector construction and data analysis; proton decay and magnetic monopole searches; cosmic ray research. *Mailing Add:* Dept Physics & Astron Vanderbilt Univ PO Box 1807 Sta B Nashville TN 37235

BARTELT, MARTIN WILLIAM, b Brooklyn, NY, Apr 4, 41; m 65; c 2. NUMERICAL ANALYSIS, APPROXIMATION THEORY. *Educ:* Hofstra Univ, BA, 63; Univ Wis, MA, 65, PhD(math), 69. *Prof Exp:* Asst prof math, Rensselaer Polytech Inst, 69-75; PROF MATH, CHRISTOPHER NEWPORT COL, NEWPORT NEWS, VA, 75- *Concurrent Pos:* Vis asst prof, Univ RI, 73-74. *Mem:* Am Math Soc. *Res:* Approximation theory; strong unicity; Lipschitz conditions. *Mailing Add:* 465 Winterhaven Dr Newport News VA 23606

BARTER, JAMES T, b South Portland, Maine, May 31, 30; m 54; c 3. PSYCHIATRY. Educ: Antioch Col, BA, 52; Univ Ariz, MA, 55; Univ Rochester, MD, 61. Prof Exp: Asst prof psychiat, Sch Med, Univ Colo, 65-69, assoc chief, Colo Psychiat Hosp, 68-69; dep dir, 69-73, dir, Sacramento Ment Health Serv, 73-77; assoc clin prof psychiat, Univ Calif, Davis, 69-75; prof psychiat, Univ Cincinnati, 77-78; prof psychiat, Univ Calif, Davis, 78-85; DIR, ILL STATE PSYCHIAT INST, CHICAGO, 85- Concurrent Pos: Consult, USPHS Indian Health Serv, 68-71. Mem: Fel Am Psychiat Asn; AMA; fel Am Col Psychiat. Res: Cross cultural psychiatry; drugs and drug abuse; adolescents; suicide; administration. Mailing Add: Univ Ill Chicago Sch Med Ill State Psychol Inst 1601 W Taylor St Chicago IL 60612

BARTFELD, HARRY, b New York, NY, May 8, 13. IMMUNOLOGY, CELL BIOLOGY. Educ: NY Univ, AB, 36; NY Med Col, MD, 44. Prof Exp: Trainee tissue cult, Crocker Inst Cancer Res, Columbia Univ, 38-39; res asst animal oncol, Cancer Study Group, Post-grad Med Sch, 40, asst prof clin med, 50-56, ASSOC PROF CLIN MED, MED CTR, NY UNIV, 57-; HEAD IMMUNOL, CONNECTIVE TISSUE SECT, ST VINCENT'S HOSP & MED CTR, 64- Concurrent Pos: Fel med, Med Ctr, NY Univ, 48-49; dir lab connective tissue dis, Sch Med, NY Univ, 57-64; co-chmn conf rheumatoid factors & their biol significance, NY Acad Sci, 69; mem adv comt fundamental res related to multiple sclerosis, Nat Multiple Sclerosis Soc, 69-72; mem conf orgn comt, NY Acad Sci; dir, Anyotrophic Lateral Sclerosis Res Ctr, St Vincent's Hosp & Med Ctr, 74- Mem: Am Soc Cell Biol; Soc Exp Biol & Med; Am Asn Immunologists. Res: Cell biology studies of immunopathological phenomena associated with experimental and natural autoimmune disease. Mailing Add: St Vincent's Hosp & Med Ctr 153 W 11th St New York NY 10011

BARTH, CHARLES ADOLPH, b Philadelphia, Pa, July 12, 30; m 54; c 4. PHYSICS. Educ: Lehigh Univ, BS, 51; Univ Calif, Los Angeles, MA, 55, PhD(physics), 58. Prof Exp: Res geophysicist, Inst Geophys, Univ Calif, Los Angeles, 57-58; res physicist, Jet Propulsion Lab, Calif Inst Technol, 58-65; assoc prof, 65-67, PROF ASTRO-GEOPHYS, UNIV COLO, BOULDER, 67-, DIR LAB ATMOSPHERIC & SPACE PHYSICS, 65- Concurrent Pos: NSF fel, Bonn, Ger, 58-59. Mem: AAAS; Am Astron Soc; Am Geophys Union. Res: Aeronomy; planetary atmospheres. Mailing Add: Dept Astrophys Sci Univ Colo Boulder CO 80309

BARTH, HOWARD GORDON, b Boston, Mass, Nov 21, 46; m 72; c 3. ANALYTICAL CHEMISTRY. Educ: Northeastern Univ, BA, 69, PhD(anal chem), 73. Prof Exp: Res assoc clin chem, Hahnemann Med Col & Hosp, 73-74; res chemist, Hercules Res Ctr, 74-79, sr res chemist, 79-86, res scientist, 86-88, group leader, 83-88; asst prof chem, Northeastern Univ, 81; STAFF MEM, DUPONT CENT RES & DEVELOP, 88- Concurrent Pos: Ed, Del-Chem Bull, 78-80; assoc ed, J Appl Polymer Sci, 85-88; Instrumental Adv Panel, Anal Chem, 86-88; chmn & cofounder, Int Symp Polymer Anal & Characterization, 87-; chmn, Del Sect, Am Chem Soc, 88. Mem: Am Chem Soc; Soc Plastics Eng; AAAS; Am Inst Chemists. Res: Polymer characterization; solution properties of water soluble polymers; separation science; high performance liquid chromatography. Mailing Add: Du Pont Co Exp Sta Bldg 228 Wilmington DE 19880-0228

BARTH, KARL FREDERICK, b Houston, Tex, Sept 25, 38; div; c 2. MATHEMATICAL ANALYSIS. Educ: Rice Univ, BA, 60, MA, 62, PhD(math), 64. Prof Exp: Res mathematician, Ballistic Res Labs, 64-66; from asst prof to assoc prof, 66-77, PROF MATH, SYRACUSE UNIV, 77- Concurrent Pos: Sr vis fel, Brit Sci Res Coun, 75-76 & 87. Mem: Am Math Soc; Math Asn Am; London Math Soc. Res: Functions of a complex variable; potential theory. Mailing Add: Dept Math Syracuse Univ Syracuse NY 13244-1150

BARTH, KARL M, b Elizabeth, NJ, Dec 19, 27. ANIMAL NUTRITION. Educ: Del Valley Col, BS, 56; Rutgers Univ, MS, 58, PhD(nutrit), 64; WVa Univ, MS, 62. Prof Exp: Instr, Del Valley Col, 62; res assoc, Purdue Univ, 64-65; from asst prof to assoc prof, 65-79, PROF ANIMAL NUTRIT, UNIV TENN, KNOXVILLE, 79- Mem: Am Inst Nutrit; Am Soc Animal Sci. Res: Forage evaluation; non-protein nitrogen utilization. Mailing Add: Dept Animal Sci Univ Tenn Knoxville TN 37916

BARTH, ROBERT HOOD, JR, b Midland, Mich, Feb 10, 34; div. BEHAVIOR-ETHOLOGY. Educ: Princeton Univ, AB, 56; Harvard Univ, AM, 59, PhD(biol), 62. Prof Exp: Instr biol, Harvard Univ, 61-63, lectr, 63-64; NSF fel zool, Univ Sheffield, 64-65; asst prof, 65-70, ASSOC PROF ZOOL, UNIV TEX, AUSTIN, 70-, DIR FAC & STAFF, 73- Concurrent Pos: NSF res grants develop & regulative biol, 61-64, 66-80. Mem: AAAS; Am Soc Zoologists; Am Ornithologists Union; Animal Behav Soc. Res: Avain behavioral ecology; maturing systems; reproductive biology; neotropical ornithology; conservation biology. Mailing Add: Dept Zool Univ Tex Austin TX 78712

BARTH, ROLF FREDERICK, b New York, NY, Apr 4, 37; m 65; c 4. PATHOLOGY, IMMUNOLOGY. Educ: Cornell Univ, AB, 59; Columbia Univ, MD, 64; Am Bd Path, dipl, 70. Prof Exp: Intern surg, Columbia-Presby Hosp, New York, 64-65; fel tumor immunol, Dept Tumor Biol, Karolinska Inst, Sweden, 65-66; res assoc immunol, Nat Inst Allergy & Infectious Dis, NIH, 66-68, resident path, Path Br, Nat Cancer Inst, 68-70; from asst prof to prof path, Univ Kans Med Ctr, Kansas City, 70-77; clin prof path, Sch Med, Univ Wis-Madison & Med Col Wis, Milwaukee, 77-79; PROF PATH, COL MED, OHIO STATE UNIV, 79- Concurrent Pos: Clin prof path, Univ Wis Col Med, 78-; grantee, Nat Cancer Inst, Dept Energy. Mem: Am Asn Immunologists; Am Asn Pathologists; Am Asn Cancer Res; Soc Exp Biol Med; Asn Univ Path; Brit Soc Immunol; Am Asn Univ Profs. Res: Tumor immunology; monoclonal antibodies; neutron capture therapy of cancer; radiommunodetection of cancer. Mailing Add: Dept Path 175 Hamilton Hall 1645 Neil Ave Ohio State Univ Columbus OH 43210

BARTHA, RICHARD, b Budapest, Hungary, Nov 14, 34; US citizen; m 67; c 2. AGRICULTURAL ECOLOGY, MICROBIAL ECOLOGY. Educ: Eotvos Lorand Univ, Budapest, 52-56; Univ Gottingen, PhD(microbiol), 61. Prof Exp: USPHS fel microbiol, Univ Wash, 62-64; res assoc, 64-66, from asst prof to assoc prof, 66-73, PROF MICROBIOL, RUTGERS UNIV, 73- Mem: Am Soc Microbiol; AAAS. Res: Chemoautotrophic bacteria; microbial degradation of pesticides and oil pollutants; microbial ecology. Mailing Add: Dept Biochem & Microbiol Rutgers Univ New Brunswick NJ 08903

BARTHEL, HAROLD O(SCAR), b Milledgeville, Ill, Sept 17, 25; m 62; c 2. AERONAUTICAL ENGINEERING. Educ: Univ Ill, BS, 50, MS, 51, PhD(mech eng), 57. Prof Exp: Res assoc, 54-57, res asst prof, 57-61, ASSOC PROF AERONAUT & ASTRON ENG, UNIV ILL, URBANA-CHAMPAIGN, 61- Mem: Sigma Xi; Am Phys Soc. Res: Unsteady gas dynamics of weak and strong shock waves; shock tube flows; second shock initiation in spherical shock tubes. Mailing Add: 101 Arrow Lab B Univ Ill 102 S Burrill Urbana IL 61801

BARTHEL, ROMARD, b Evansville, Ind, Apr 8, 24. ULTRASOUND. Educ: Univ Notre Dame, BS, 47; Univ Tex, PhD(physics), 51. Prof Exp: From instr to assoc prof physics, 47-57, chmn div physics & natural sci, 54-68, PROF PHYSICS, ST EDWARD'S UNIV, 57- Mem: Am Asn Physics Teachers; Sigma Xi. Res: Ultrasonic interferometers; theory of solutions; philosophy of science; atomic and nuclear physics and electronics. Mailing Add: St Edward's Univ Austin TX 78704

BARTHEL, WILLIAM FREDERICK, b Arbutus, Md, Mar 12, 15; m 38; c 1. AGRICULTURAL CHEMISTRY. Prof Exp: Chemist, USDA, 40-45; chief chemist insecticide div, Victor Prods Corp, 46-48; chief chemist & vpres, Edco Corp, 48; chief chemist in-chg develop insecticide div, Innis Speiden & Co, 49-50; owner, W F Barthel Chem Co, 50-51; chemist pesticide chem res lab, USDA, 51-57, chemist-in-chg chem lab, Plant Pest Control, 58-62 & Methods Improv Labs, 62-67, supvry chemist, Pesticides Lab, Nat Commun Dis Ctr, 67-68, chief toxicol lab, 68; supvry res chemist, Pesticides Div, Atlanta Toxicol Br, US Food & Drug Admin, 68-71; toxicologist, Ctr Dis Control, 71-74; RETIRED. Honors & Awards: Serv Awards, USDA, 58 & 60. Mem: AAAS; Am Chem Soc; Entom Soc Am; Asn Official Agr Chem; Am Inst Chem. Res: Organic insecticides; structural studies of pyrethrins; cinerins; Barthel rearrangement; Barthrin and related insecticides of low mammalian toxicity. Mailing Add: Box 105 R 2 Mt Vernon IA 52314

BARTHOLD, LIONEL O, b Great Barrington, Mass, Mar 20, 26; m; c 2. ELECTRICAL ENGINEERING. Educ: Northwestern Univ, BA, 50. Prof Exp: Mgr eng, Gen Elec Co, 64-69; PRES & CHMN, PTI, 69- Mem: Nat Acad Eng; fel Inst Elec & Electronics Engrs; Soc Power Engrs. Mailing Add: PO Box 764 Schenectady NY 12301

BARTHOLD, STEPHEN WILLIAM, b San Francisco, Calif, Nov 10, 45; m 71; c 1. VETERINARY PATHOLOGY. Educ: Univ Calif, Davis, BS, 67, DVM, 69; Univ Wis-Madison, MS, 73, PhD(vet sci), 74. Prof Exp: Chief animal care, US Army Res Inst Environ Med, 69-71; res asst vet path, Univ Wis-Madison, 71-74; from asst prof to assoc prof, 74-89, PROF COMP MED, SCH MED, YALE UNIV, 89- Mem: Am Vet Med Asn; Am Col Vet Path; Am Asn Lab Animal Sci. Res: Comparative pathology; laboratory animal pathology; infectious disease. Mailing Add: Sect Comp Med Sch Med Yale Univ 333 Cedar St PO Box 3333 New Haven CT 06510

BARTHOLDI, MARTY FRANK, b Ballston Spa, NY, Apr 22, 52; m 79; c 2. CELL BIOLOGY, CANCER RESEARCH. Educ: Clarkson Univ, BS, 74, MS, 76, PhD (physics), 79. Prof Exp: Res assoc, Mem Sloan Kettering Cancer Ctr, 76-79; fel, Max Planck Inst Biophys Chem, 79-80; STAFF MEM, LOS ALAMOS NAT LAB, 80- Concurrent Pos: Prin investr, Nat Flow Cytometry Resource, 83-; consult, NIH, 85- Mem: Int Soc Anal Chem; AAAS; Am Phys Soc. Res: Analytical cytology; developed flow cytometry; laser microscopy and image processing for chromosome analysis and sorting; cell nuclear structure and functions; cancer research including cytogenetics and genome exposure. Mailing Add: 183 Piedra Loop Los Alamos NM 87544

BARTHOLMEY, SANDRA JEAN, b Chicago, Ill, Mar 22, 42; c 2. NUTRITION RESEARCH & INFANT NUTRITION, DIETARY SURVEY. Educ: Univ Ill, Urbana, BS, 64, PhD(food & nutrit), 86; Sangamon State Univ, MA, 76. Prof Exp: MGR NUTRIT, RES DIV, GERBER PRODS CO, 86- Mem: Am Inst Nutrit; Sigma Xi; Inst Food Technologists. Res: Iron bioavailability from cereals; dietary patterns of infants to 18 months of age-toddlers. Mailing Add: Nutrit Res Div Gerber Prods Co 445 State Rd Fremont MI 49412

BARTHOLOMEW, CALVIN HENRY, b Mt Pleasant, Utah, Mar 7, 43; m 65; c 5. CATALYSIS, AIR POLLUTION CONTROL. Educ: Brigham Young Univ, BES, 68; Stanford Univ, MS, 70, PhD(chem eng), 72. Prof Exp: Sr chem eng automotive emissions control, Corning Glass Works, 72-73; from asst prof to assoc prof, 73-82, PROF CHEM ENG, BRIGHAM YOUNG UNIV, 82- Concurrent Pos: Pvt consult, Bartholomew Consult Serv, Inc, 73-; founder & head, Catalysis Lab, Brigham Young Univ, 73-; assoc dir, Adv Combustion Eng Res Ctr, 86-; ed, Proc Fifth Int Catalysis Deactivation Meeting, 90-91. Honors & Awards: Utah Award, Am Chem Soc, 91. Mem: Am Chem Soc; Sigma Xi. Res: Research in catalysis, combustion and thermodynamics; investigator or co-investigator on over 40 grants and contracts; supervised more than 60 research students; author or co-author of over 175 technical publications, reports and proceedings of meetings. Mailing Add: 440 N 600 E Orem UT 84057

BARTHOLOMEW, DARRELL THOMAS, b Payson, Utah, Dec 24, 47; m 71; c 10. MEAT SCIENCE & MICROBIOLOGY, FOOD CHEMISTRY. Educ: Brigham Young Univ, BS, 73; NC State Univ, MS, 75, PhD(food sci), 78. Prof Exp: Asst prof meat indust sanitation, food processing & meat processing, Utah State Univ, 81-86; DIR RES & PROD DEVELOP, JEROME FOODS, INC, 86- Mem: Inst Food Technologists. Mailing Add: 561 N Mill St Barron WI 54812

BARTHOLOMEW, DUANE P, b Fargo, NDak, Sept 25, 34; m 56; c 3. AGRONOMY, CROP PHYSIOLOGY. *Educ:* Calif State Polytech Col, BS, 61; Iowa State Univ, PhD(plant physiol), 65. *Prof Exp:* Assoc plant physiologist, Pineapple Res Inst, 65-66, asst agronomist, 66-78, assoc agronomist, 78-83, AGRONOMIST, UNIV HAWAII, 83- *Mem:* Am Soc Hort Sci; Crop Sci Soc Am; Am Soc Agron. *Res:* Physiology of the pineapple; plant-water relations; plant-environment relations; computer modeling of development. *Mailing Add:* Dept Agron & Soil Sci Univ Hawaii Honolulu HI 96822

BARTHOLOMEW, GEORGE ADELBERT, b Independence, Mo, June 1, 19; m 42; c 2. ZOOLOGY, ECOLOGICAL PHYSIOLOGY. *Educ:* Univ Calif, AB, 40, MA, 41; Harvard Univ, PhD(zool), 47. *Hon Degrees:* DSci, Univ Chicago, 88. *Prof Exp:* Asst, Mus Vert Zool, Univ Calif, 40-41; asst zool & comp anat, Harvard Univ, 45-47; from instr to prof, 47-87, EMER PROF ZOOL, UNIV CALIF, LOS ANGELES, 87- *Mem:* Nat Acad Sci; Ecol Soc Am; Am Soc Mammal; Am Ornith Union; Cooper Ornith Soc; Am Soc Zool; Am Acad Arts & Sci. *Res:* Ecology and physiology of vertebrates and energetics of insects. *Mailing Add:* 85 Atherton Oaks Dr Novata CA 94945

BARTHOLOMEW, GILBERT ALFRED, b Nelson, BC, Apr 8, 22; m 52. NUCLEAR PHYSICS, NEUTRON SOURCES. *Educ:* Univ BC, BA, 43; McGill Univ, PhD(physics), 48. *Prof Exp:* Res officer nuclear physics, Atomic Energy Can Ltd, 48-62, head neutron & solid state physics br, 62-71, dir physics div, 71-83; RETIRED. *Concurrent Pos:* Mem, Int Union Pure & Appl Physics Comn Nuclear Physics, 78-87, secy, 81-87; chmn, Int Adv Comt Tokamak de Varennes, Hydro Québec, Varennes, Que, 82- *Mem:* Fel Am Phys Soc; Can Asn Physicists; Sigma Xi; fel Royal Soc Can; fel AAAS; Can Nuclear Soc. *Res:* Neutron capture gamma rays; low energy nuclear physics; nuclear structure; neutron sources; electronuclear breeding of fissile material. *Mailing Add:* Box 1258 Deep River ON K0J 1P0 Can

BARTHOLOMEW, JAMES COLLINS, b Dec 18, 42; m 65; c 2. CELL BIOLOGY. *Educ:* Hobart Col, BS, 65; Cornell Univ, PhD(biochem), 70. *Prof Exp:* Am Cancer Soc res fel, The Salk Inst, 70-72; SR STAFF CELL BIOLOGIST, LAWRENCE BERKELEY LAB, UNIV CALIF, 72- *Concurrent Pos:* Dep dir, Lab Chem Biodynamics, 81- *Mem:* AAAS; Am Soc Cell Biol. *Res:* The regulation of growth of mammalian cells in cultures and the factors that affect this regulation such as serum, viruses and chemical carcinogens. *Mailing Add:* Lab Chem Biodynamics Lawrence Berkeley Lab Univ Calif Berkeley CA 94720

BARTHOLOMEW, LLOYD GIBSON, b Whitehall, NY, Sept 15, 21; m 43; c 5. INTERNAL MEDICINE, GASTROENTEROLOGY. *Educ:* Union Col, BA, 41; Univ Vt, MD, 44; Univ Minn, MS, 52. *Hon Degrees:* LHD, Mountain Col, Vt, 84. *Prof Exp:* Asst, Div Med, Mayo Clin, 49-50, 52, staff, 52-53; from instr to assoc prof, 53-67, PROF MED, MAYO MED SCH, UNIV MINN & SR CONSULT GASTROENTEROL, MAYO CLIN, 78- *Concurrent Pos:* Attend physician, St Mary's Methodist & assoc hosps, 53-; head dept gastroenterol, Mayo Clinic, 67-77; deleg, Minn Med Asn, AMA, 78-84. *Honors & Awards:* Woodbury & Carbee Prizes, 44. *Mem:* AMA; Am Gastroenterol Asn; Am Bd Internal Med; Sigma Xi. *Res:* Pancreatic and liver diseases. *Mailing Add:* Mayo Clin 200 First St SW Rochester MN 55901

BARTHOLOMEW, MERVIN JEROME, b Altoona, Pa, Nov 22, 42. TECTONICS. *Educ:* Va Polytech State Univ, PhD(geol), 71. *Prof Exp:* CHIEF GEOL, MONT BUR MINES & GEOL, 83-; PROF GEOL, MONT TECH, 83- *Mem:* Fel Geol Soc Am. *Mailing Add:* Buxton Frontage Rd Silver Bow MT 59750

BARTHOLOMEW, ROGER FRANK, b Workington, Eng, May 21, 37; m 61; c 3. PHYSICAL CHEMISTRY. *Educ:* Univ London, BSc, 58, PhD(phys chem), 61. *Prof Exp:* Res fel, Nat Res Coun Can, 61-63; from res chemist to sr res chemist, 64-73, RES ASSOC CHEM, TECH STAFFS DIV, CORNING GLASS WORKS, 73- *Concurrent Pos:* Chmn, Coun Comt Technician Activ, Am Chem Soc, 84-86. *Honors & Awards:* Eugene C Sullivan Award, Am Chem Soc, 81. *Mem:* Am Chem Soc; fel Am Ceramic Soc; Royal Soc Chem. *Res:* Ion-exchange and diffusion studies in glasses and naturally occurring aluminosilicates; flow of gases through microporous activated carbon plugs; physical chemistry of molten salts in particular molten nitrates; low temperature glass forming systems; high energy battery systems; sodium-sulphur cells; application and development of ceramic materials for electronic devices; fluoride glasses for optical waveguides; erbium optical amplifiers. *Mailing Add:* Eight Overbrook Rd Painted Post NY 14870

BARTHOLOW, GEORGE WILLIAM, b Yale, Iowa, July 28, 30; c 3. PSYCHIATRY. *Educ:* Univ Iowa, BS, 51, MD, 55. *Prof Exp:* Intern, Wayne County Gen Hosp, Eloise, Mich, 55-56; resident psychiat, Univ Iowa, 56-59; assoc clin dir adult inpatient serv, Nebr Psychiat Inst, 61-74, from asst prof to assoc prof, 61-74, PROF PSYCHIAT, COL MED, UNIV NEBR, OMAHA, 74-,. *Concurrent Pos:* Chief psychiat serv, Vet Admin Hosp, Omaha, 62-; consult, 66-; prof psychiat, Creighton Univ Col Med, 78. *Mem:* Am Psychiat Asn. *Res:* Vocational rehabilitation of the mentally ill; suicidology; adult inpatient psychiatry. *Mailing Add:* Creighton Univ Vet Hosp 4101 Woolworth Ave Omaha NE 68105

BARTHOLOW, LESTER C, b San Diego, Calif, Sept 29, 36. NUTRITIONAL BIOCHEMISTRY. *Educ:* Stanford Univ, BS, 63; Bemidji State Univ, MA, 73; Mass Inst Technol, MS, 74; Harvard Univ, DSc(nutrit biochem), 81. *Prof Exp:* Mem staff, Int Bus Mach Corp, 66-69, Xerox Data Systs, 69-70, Youngstown State Univ, 74-77 & Kellogg Co, 81-83; EXEC VPRES & VPRES RES & DEVELOP, KABI PHARMACIA INC, 83- *Mem:* Am Inst Nutrit; Am Soc Oil Chemists. *Res:* Lipid biochemistry; parenteral and enteral nutrition; parenteral product development; physical chemistry; drug delivery. *Mailing Add:* Kabi Inc Hwy 70 Clayton NC 27520

BARTILUCCI, ANDREW J, b New York, NY, Nov 29, 22; m 50; c 3. PHARMACY. *Educ:* St John's Univ, NY, BS, 44; Rutgers Univ, MS, 49; Univ Md, PhD(pharm), 53. *Hon Degrees:* DSc, Union Univ, 90. *Prof Exp:* Anal chemist, US Armed Serv Med Procurement Off, 47-48; assoc res pharmacist, Merck & Co, 49-50; asst dean, Col Pharm, St John Univ, NY, 52-56, prof pharm, 52-88, dean, Col Pharm & Allied Health Professions, 56-88, vpres health professions, 80-91, EXEC VPRES HEALTH PROFESSIONS, ST JOHNS UNIV, NY, 91- *Concurrent Pos:* Mem, NY State Bd Pharm, 81- *Honors & Awards:* Distinguished Serv Profile Award, Am Found Pharmaceut Educ. *Mem:* AAAS; Am Col Apothecaries; Am Pharmaceut Asn; fel NY Acad Sci; Sigma Xi. *Res:* Pharmaceutical formulation and analysis. *Mailing Add:* Exec vpres St Johns Univ Jamaica NY 11439

BARTIMES, GEORGE F, b Aurora, Ill, Jan 29, 35; m 77; c 1. PLASTICS RECYCLING, ELECTROMECHANICAL SYSTEMS. *Educ:* Ill Inst Technol, BS, 57; Northwestern Univ, MBA, 62. *Prof Exp:* Engr, Automatic Elec, 56-58; mgr field serv, Am Mach & Foundry, 58-63; adv engr, Continental Plastic Containers, Inc, 63-67, mgr electromech develop, 67-75, mgr process anal, 75-88, dir res & develop, 88-89, mgr new venture technol, 89-91, SR ADV ENGR, CONTINENTAL PLASTIC CONTAINERS, INC, 91- *Mem:* Sr mem Inst Elec & Electronics Engrs; Soc Prof Engrs; Soc Mech Engrs. *Res:* In-mold labeling methods & appln; electronic/robotic control system; sortation of materials for recycling; unique bottle/spray system; numerous patents issued. *Mailing Add:* Continental Plastic Containers 2375 Touhy Ave Elk Grove Village IL 60007

BARTIS, JAMES THOMAS, b Pawtucket, RI, Oct 22, 45; m 84; c 2. ENERGY POLICY, RESEARCH PLANNING. *Educ:* Brown Univ, ScB, 67; Mass Inst Technol, PhD(chem physics), 72. *Prof Exp:* Asst prof chem, Cornell Univ, 74-75; Mem, Inst Defense Anal, 75-78; res chemist, Off Energy Res, US Dept Energy, 78-79, dir, Off Planning & Technol Assessment, 79-81, dir, fossil energy policy, 81-82; VPRES, EOS TECHNOL, INC, 82- *Concurrent Pos:* Fel, NATO, 72; vis scientist, Weizmann Inst, 72; fel, Cornell Univ, 72-74. *Mem:* Am Phys Soc; Am Chem Soc; AAAS. *Res:* Transport properties of fluids, theory of chemical reactions and properties of multicomponent solutions; defense systems evaluation and analysis; energy research and technology policy; coal, oil, gas and synthetic fuel policy; assessment of advanced coal technologies. *Mailing Add:* EOS Tech Inc 1601 N Kent St No 1102 Arlington VA 22209-2116

BARTISH, CHARLES MICHAEL CHRISTOPHER, b Easton, Pa, June 4, 47; m 69; c 2. INORGANIC CHEMISTRY, CATALYSIS. *Educ:* Villanova Univ, BS, 69; Lehigh Univ, PhD(chem), 73. *Prof Exp:* Res chemist, Corp Res Dept, 73-76, group leader, Corp Res Dept, 76-77, sect mgr carbon monoxide chem, 77-79, sect mgr amines res, 79-81, tech mgr develop, Indust Chem Div, 81-85, MGR RES, POLYURETHANE CHEM TECHNOL, AIR PROD & CHEM INC, 85- *Mem:* Am Chem Soc; Sigma Xi; Catalysis Soc. *Res:* Carbon monoxide chemistry; homogeneous and heterogeneous catalysis; organometallic chemistry; industrial chemistry; process development; research management; polyurethanes. *Mailing Add:* Air Prod & Chem Inc Allentown PA 18195

BARTKE, ANDRZEJ, b Krakow, Poland, May 23, 39; m 66. REPRODUCTIVE PHYSIOLOGY, ENDOCRINOLOGY. *Educ:* Jagiellonian Univ, MSc, 62; Univ Kans, PhD(zool), 65. *Prof Exp:* Vis scientist, Inst Cancer Res, Philadelphia, 65; asst prof dept animal genetics, Jagiellonian Univ, 65-67; fel, 67-69, staff scientist, 69-72, sr scientist, Worcester Found Exp Biol, 72-78; from assoc prof to prof obstet & gynec, Univ Tex Health Sci Ctr, San Antonio, 78-84; PROF & CHMN DEPT PHYSIOL, SCH MED, SOUTHERN ILL UNIV, CARBONDALE, 84- *Concurrent Pos:* Res career develop award, NIH, 72-77. *Mem:* AAAS; Soc Study Reproduction; Endocrine Soc; Brit Soc Endocrinol; Am Soc Andrology; Soc Study Fertility; Int Soc Neuroendocrinol. *Res:* Pituitary control of testicular steroidogenesis; role of prolactin in the male; effects of cannabinoids in the male; transgenic animals; seasonal breeding. *Mailing Add:* Dept Physiol Sch Med Southern Ill Univ Carbondale IL 62901-6512

BARTKO, JOHN, b New York, NY, Mar 11, 31; m 58. NUCLEAR PHYSICS. *Educ:* Columbia Univ, BA, 54; Fairleigh Dickinson Univ, BSEE, 60; Pa State Univ, PhD(nuclear physics), 66. *Prof Exp:* Res scientist, Nuclear Div, Martin Marietta Corp, 67-68; sr physicist, 68-73, fel scientist, 73-80, adv scientist, 80-81, mgr Radiation & Nucleonics Lab, Res & Develop Ctr, 81-84, ADV SCIENTIST, WESTINGHOUSE ELEC CORP, 84- *Mem:* Inst Elec & Electronics Engrs. *Res:* Proton induced reactions in potassium; radiation effects on materials and electronic devices; nondestructive testing via radiation techniques; beneficial applications of radiation in semiconductor devices; nuclear instrumentation (eg explosive detector). *Mailing Add:* Res Labs Westinghouse Elec Corp Pittsburgh PA 15235

BARTKO, JOHN JAROSLAV, b Massillon, Ohio, Nov 17, 37. MATHEMATICAL STATISTICS. *Educ:* Univ Fla, BA, 59; Va Polytech Inst, MS, 61, PhD(math statist), 62. *Prof Exp:* RES MATH STATISTICIAN, NIMH, 62-, INSTR, FOUND ADVAN EDUC IN SCI, INC, NIH GRAD SCH, 63-, CHMN, DEPT STATIST, 65- *Concurrent Pos:* Consult int pilot study schizophrenia, WHO, Geneva, Switz, 69- *Mem:* Fel Am Statist Asn; Biomet Soc. *Res:* Statistical theory and methodology; application of statistics in the life and behavioral sciences; research in reliability. *Mailing Add:* NIMH NIH Campus Bldg 10 Rm 3N-204 Bethesda MD 20892

BARTKOSKI, MICHAEL JOHN, JR, b Kansas City, Kans, July 1, 45; m 68; c 2. MICROBIOLOGY, BIOCHEMISTRY. *Educ:* Kans State Univ, BS, 67; Univ Mo, Kans City, 72; Univ Okla, PhD(microbiol), 74. *Prof Exp:* Postdoctoral, Univ Chicago, 75-77; asst prof microbiol, Uniformed Serv Univ, 77-81; proj leader, Tech Am Group Inc, 82-83, dir res & develop, 83-85, vpres oper, 85-87; V PRES BIOL, FERINENTA ANIMAL HEALTH, 87- *Mem:* Am Soc Microbiol; Am Soc Virol; AAAS; Soc Gen Microbiol. *Res:* Regulation of nuclei acid synthesis; animal health product vaccine and diagnosis. *Mailing Add:* Fermenta Animal Health 7410 NW Tiffany Springs Pkwy PO Box 901350 Kansas City MO 64153

BARTKUS, EDWARD PETER, b Thomas, WVa, Jan 29, 20; m 46; c 1. INFORMATION SCIENCE, CHEMICAL ENGINEERING. *Educ:* WVa Univ, BS, 41, BSChE, 46, MSChE, 47, PhD(chem eng), 50. *Prof Exp:* From instr to asst prof chem eng, WVa Univ, 47-51; eval engr chem processes, Eng Serv Div, E I du Pont de Nemours & Co Inc, 51-52, field res supvr pigment processes, 52-54, field sect mgr eng, 54-57, mgr eng comput & math applns, 57-61, mgr new eng prod, Eng Res Div, 61-66, spec asst mgr new info serv, Info Systs Div, Secy Dept, 66-74, mgr info resources, Info Systs Dept, 74-82; PRES/OWNER, EXEC INFO MGT, 82- *Concurrent Pos:* Mem, Nat Res Coun, Numerical Data Adv Bd, 72-78 & Nat Comt for Codata, 78-83, Del State Tech Serv Adv Comt, 70-80, Nat Res Coun-Nat Acad Sci-Nat Acad Eng Comt Energy Data, 80-82; consult, publ comt, Am Soc Testing & Mat, 71-89. *Mem:* Am Chem Soc; fel Am Inst Chem Engrs; Am Soc Eng Educ; Nat Soc Prof Engrs; Sigma Xi. *Res:* Engineering economics; systems engineering; operations research; information storage and retrieval, systems and processing; records management; futures issues research; office automation; competitor studies. *Mailing Add:* 507 Falkirk Rd Woodbrook Wilmington DE 19803-2445

BARTL, PAUL, b Prague, Czech, Apr 8, 28; m 53; c 1. BIOPHYSICS. *Educ:* Prague Tech Univ, MA, 51; Czech Acad Sci, PhD(phys chem), 54. *Prof Exp:* Scientist, Inst Org Chem & Biochem, Prague, 54-64; vis scientist, Inst Cancer Res, Villejuif, France, 65; fel biophys, Johns Hopkins Univ, 66-67, vis assoc prof, 67-68; sr scientist, Hoffmann-La Roche, Inc, 68-71; asst dir sci affairs, Roche Inst Molecular Biol, 71-78; assoc dir, Athymic Mouse Res Ctr, Univ Calif, San Diego, 78-80, acad adminr dept chem, 80-82; CONSULT, 88- *Concurrent Pos:* Vpres MeDiCa, Calif, 79- *Mem:* World Trade Asn. *Res:* High resolution electron microscopy of nucleic acids; water-miscible embedding media for electron microscopy; physical chemistry of nucleic acids and their constituents; immunodiagnostics; research administration. *Mailing Add:* 3305 Piragua Carlsbad CA 92009

BARTLE, ROBERT GARDNER, b Kansas City, Mo, Nov 20, 27; m 51, 82; c 2. ANALYSIS & FUNCTIONAL ANALYSIS. *Educ:* Swarthmore Col, BA, 47; Univ Chicago, SM, 48, PhD(math), 51. *Prof Exp:* AEC fel, Yale Univ, 51-52, instr math, 52-55; from asst prof to prof, 55-90, EMER PROF MATH, UNIV ILL, URBANA-CHAMPAIGN, 90-; PROF MATH, EASTERN MICH UNIV, UPSILANTI, 90- *Concurrent Pos:* Vis prof math, Ga Inst Technol, 75-76, 84; exec ed, Math Reviews, 76-78, 86-90. *Mem:* Am Math Soc; Math Asn Am; London Math Soc; Soc Indust & Appl Math. *Res:* Functional analysis; real analysis; spectral theory. *Mailing Add:* 3340 Alpine Dr Ann Arbor MI 48108-1704

BARTLES, JAMES RICHARD, CELL BIOLOGY. *Educ:* Wash Univ, St Louis, PhD(Biochem), 81. *Prof Exp:* Res assoc cell biol, John's Hopkins Univ, Sch Med, 81-87; AT NORTHWESTERN UNIV MED SCH, 87- *Mailing Add:* Dept Cell Molecular & Struct Biol Northwestern Univ Med Sch 303 E Chicago Ave Chicago IL 60611

BARTLESON, JOHN DAVID, b Detroit, Mich, Mar 17, 17; m 40; c 2. CHEMISTRY. *Educ:* Mich State Univ, BS, 38, MS, 39; Western Reserve Univ, PhD(org chem), 45. *Prof Exp:* Asst chem, Mich State Univ, 38-39; sect leader, Chem Res Div, Standard Oil Co, Ohio, 39-52; proj leader, Ethyl Int, 52-58, mgr antioxidant sale, 58-66, mgr new area develop, Petrol Chem Div, 81-82,; RETIRED. *Mem:* Am Chem Soc; Am Soc Testing & Mat; Chem Develop Asn; Chem Mkt Res Asn; Independent Oil Compounders Asn. *Res:* Lubricating oil additives; antioxidants and cetane improvers; engine evaluation of fuels and lubricants; refining, compounding and testing of lubricants; United States and international technical sales management; commercial development; market research; corporate planning; acquisition studies. *Mailing Add:* Box 137 Franklin MI 48025

BARTLETT, ALAN C, b Price, Utah, June 17, 34; m 56; c 4. GENETICS. *Educ:* Univ Utah, BA, 56, MS, 57; Purdue Univ, PhD(pop genetics), 62. *Prof Exp:* Instr pop genetics, Purdue Univ, 61-62; geneticist, Boll Weevil Res Lab, Entom Res Div, 62-67 & Western Cotton Insects Invests, 69-70, GENETICIST, WESTERN COTTON RES LAB, AGR RES SERV, USDA, 70- *Mem:* Entom Soc Am; Am Genetic Asn; Genetics Soc Am. *Res:* Radiation genetics of Drosophila, Tribolium castaneum and Anthonomus grandis; population genetics, selection and reproduction; genetic techniques for insect control and eradication; genetics and biology of pink bollworm; heliothis and boll weevil. *Mailing Add:* Western Cotton Res Lab 4135 E Broadway Phoenix AZ 85040

BARTLETT, ALBERT ALLEN, b Shanghai, China, Mar 21, 23; US citizen; m 46; c 4. PHYSICS. *Educ:* Colgate Univ, BA, 44; Harvard Univ, MA, 48, PhD(physics), 51. *Prof Exp:* Mem staff, Los Alamos Sci Lab, 44-46; teaching fel, Harvard Univ, 48-49; asst prof physics, 50-56, assoc prof, 56-62, 50-62, PROF PHYSICS, UNIV COLO, BOULDER, 62- *Concurrent Pos:* Res vis, Nobel Inst Physics, Sweden, 63-64. *Honors & Awards:* Distinguished Serv Citation, Am Asn Physics Teachers, 70, Robert A Millikan Award, 81, Melba Phillips Award, 90. *Mem:* Fel Am Phys Soc; Am Asn Physics Teachers (vpres, 76, pres-elect, 77, pres, 78); Sigma Xi; fel AAAS; Am Asn Univ Professors. *Res:* Nuclear physics; energy and resource problems. *Mailing Add:* Box 390 Dept Physics Univ Colo Boulder CO 80309

BARTLETT, CHARLES SAMUEL, JR, b Asheville, NC, Oct 20, 29; m 51; c 2. GEOLOGY, PETROLEUM EXPLORATION. *Educ:* Univ NC, BS, 51, MS, 67; Univ Tenn, PhD(geol), 73. *Prof Exp:* Field geologist, Gulf Oil Corp, 55-61; area geologist, J M Huber Corp, 61-65; instr geol, Pembroke State Univ, 66-67; from asst prof to assoc prof, Emory & Henry Col, 67-78; CHIEF GEOLOGIST, BARTLETT GEOL CONSULTS, ABINGDON, VA, 78- *Concurrent Pos:* Adj prof, King Col, 67-69; field geologist, Va Div Mineral Resources, 69-75; adj prof geol, Emory & Henry Col, 78-81; bd dirs, Archeol Soc, Va, 83-; sci adv bd, Va Mus Natural Hist, 86-88 & Va Cave Bd, 85-90. *Honors & Awards:* Sullivan Award, Univ NC, 51. *Mem:* Am Asn Petrol Geologists; Geol Soc Am; Nat Asn Geol Teachers; Asn Prof Geol Scientists; Nat Speleol Soc. *Res:* Natural gas exploration; space imagery applications in mineral exploration; antarctic meteorites recovery; environmental effects of coal extraction; geological archeology. *Mailing Add:* Bartlett Geol Consults 903 E Main St Abingdon VA 24210

BARTLETT, DAVID FARNHAM, b New York, NY, Dec 13, 38; m 60; c 4. GRAVITY, ELECTROMAGNETISM. *Educ:* Harvard Univ, AB, 59; Columbia Univ, AM, 61, PhD(physics), 65. *Prof Exp:* From instr to asst prof physics, Princeton Univ, 64-71; assoc prof, 71-82, PROF PHYSICS, UNIV COLO, BOULDER, 82- *Mem:* Fel Am Phys Soc; Am Asn Physics Teachers; AAAS. *Res:* Study of gravity and electromagnetism using cryogenic techniques. *Mailing Add:* Dept Physics Univ Colo Boulder CO 80309-0390

BARTLETT, DONALD, JR, b Hanover, NH, Aug 4, 37; m 65; c 3. PHYSIOLOGY. *Educ:* Dartmouth Col, AB, 59; Dartmouth Med Sch, BMS, 61; Harvard Med Sch, MD, 64. *Prof Exp:* Intern, Internal Med, Strong Mem Hosp, Rochester, NY, 64-65; asst res, 65-66; physician, Epidemiol Sect, Field Studies Br, Div Air Pollution, USPHS, 66-67; chief med res sect, Health Effects Prog, Nat Ctr Air Pollution Control, 67-68; from asst prof to assoc prof, 71-78, PROF PHYSIOL, DARTMOUTH MED SCH, 78-, ANDREW C VAIL PROF & CHMN, DEPT PHYSIOL, 89- *Concurrent Pos:* USPHS res fel physiol, Dartmouth Med Sch, 68-71. *Mem:* Am Physiol Soc. *Res:* Respiratory and comparative physiology. *Mailing Add:* Dept Physiol Dartmouth Med Sch Hanover NH 03756

BARTLETT, EDWIN S(OUTHWORTH), b Detroit, Mich, Nov 4, 28; m 51; c 7. METALLURGY. *Educ:* Mich Technol Univ, BS, 50; Ohio State Univ, MS, 59. *Prof Exp:* Res engr, Hoskins Mfg Co, 50-56; metallurgist, 56-61, res assoc, 61-70, tech adv, 70-81, SR RES SCIENTIST, BATTELLE MEM INST, 80- *Mem:* Am Inst Mining, Metall & Petrol Engrs; Am Soc Metals; Metall Soc. *Res:* Physical and mechanical metallurgy; special metalworking practices and techniques; refractory metals; nickel alloys; coatings; titanium. *Mailing Add:* 7868 Worthington Galena OH 43085

BARTLETT, FRANK DAVID, b Clarksburg, WVa, Nov 6, 28; m 50; c 4. SOILS, AGRICULTURE. *Educ:* Univ WVa, BS, 50, MS, 51; Univ Fla, PhD(soils), 55. *Prof Exp:* Conserv aid, USDA, 47-51, lab asst res, Fla Agr Exp Sta, 51-55, soil scientist, 55-58, supvry soil scientist, 58-62, soil scientist, Ark River Basin, 62-72, resource conservationist, 72-77, resource data specialist, River Basin Planning, Soil Conserv Serv, 77-83; RETIRED. *Concurrent Pos:* Cert herbicide applicator. *Res:* Soil hydrology and soil-water-plant relationships; environmental affairs watershed planning. *Mailing Add:* 2809 Hilltop Rd Alexander AR 72002

BARTLETT, GERALD LLOYD, b Portland, Ore, July 24, 39; m 65; c 3. PATHOLOGY. *Educ:* Seattle Pac Univ, BA, 61; Univ Wash, MD, 66; Univ Pa, PhD(path), 72. *Prof Exp:* Res assoc, Inst Cancer Res, Philadelphia, 66-70 & Nat Cancer Inst, Bethesda, Md, 70-72; from asst prof to prof path & immunol, Hershey Med Ctr, Pa State Univ, 72-89; PROF PATH & CHAIR DEPT, UNIV ILL COL MED, PEORIA, 89- *Concurrent Pos:* Ad hoc site vis mem, NIH, 71-91; mem, Construct Adv Group, NIH, 72-78. *Mem:* AAAS; Am Asn Pathologists; Am Asn Immunologists; Am Asn Cancer Res. *Res:* Pathogenesis of radiation fibrosis. *Mailing Add:* Dept Path Univ Ill Col Med Box 1649 Peoria IL 61656

BARTLETT, GRANT ROGERS, b Berkeley, Calif, June 16, 12; m 46; c 3. BIOCHEMISTRY. *Educ:* Stanford Univ, BS, 34; Univ Chicago, PhD(biochem), 42. *Prof Exp:* Res assoc, Dept Med, Univ Chicago, 42-45; Scripps Clin, 46-61; DIR LAB COMP BIOCHEM, 62- *Concurrent Pos:* With Off Sci Res & Develop; mem comt blood & related probs, Nat Acad Sci-Nat Res Coun. *Mem:* Am Chem Soc; Am Soc Biol Chem; Am Soc Hemat. *Res:* Carbohydrate metabolism; biochemistry of red blood cells. *Mailing Add:* 4620 Santa Fe St San Diego CA 92109

BARTLETT, J FREDERICK, b Ft Wayne, Ind, Aug 18, 36; m 84. SOFTWARE ENGINEERING. *Educ:* Yale Univ, SB, 58; Calif Inst Technol, MS, 61. *Prof Exp:* Staff astron, Calif Inst Technol, 68-69, comput, 69-70, physics, 70-78; APPL SCIENTIST, FERMI NAT ACCELERATOR LAB, 78- *Mem:* Asn Comput Mach. *Res:* Design and implementation of a large, distributed computer control system for the Fermilab particle accelerator; technical management of software engineering; neutrino structure functions. *Mailing Add:* 42 W540 Hidden Springs Dr St Charles IL 60175

BARTLETT, JAMES HOLLY, b Brooklyn, NY, Nov 2, 04. MECHANICS. *Educ:* Northeastern Univ, BCE, 24; Harvard Univ, AM, 26, PhD, 30. *Prof Exp:* From asst prof to prof theoret physics, Univ Ill, Urbana, 30-65; prof physics, Univ Ala, 65-75; EMER PROF THEORET PHYSICS, UNIV ILL, URBANA, 65-; EMER PROF PHYSICS, UNIV ALA, 75- *Concurrent Pos:* Fel, Rockefeller Found, 40-41; consult, Lockheed Aircraft Corp, 58-59; exchange prof to USSR, 61-63. *Mem:* Fel Am Phys Soc; Electrochem Soc; Biophys Soc; Am Astron Soc. *Res:* Ionization; quadrupole radiation; chemical valency; nuclear structure; properties of fast electrons; biophysics; dissolution of metals; anodic oxidation; passivity; stability of orbits; artificial intelligence. *Mailing Add:* Dept Physics & Astron Univ Ala Box 870324 Tuscaloosa AL 35487-0324

BARTLETT, JAMES KENNETH, b Lynden, Wash, Feb 2, 25; m 48; c 1. CHEMISTRY. *Educ:* Willamette Univ, BS, 49; Stanford Univ, PhD(chem), 55. *Prof Exp:* Instr chem, Univ Santa Clara, 53-54; asst prof, Long Beach State Col, 54-56; asst prof, 56-63, PROF CHEM, SOUTHERN ORE STATE COL, 63-, CHMN DEPT, 76- *Mem:* Am Chem Soc. *Res:* Colorimetric methods of analysis. *Mailing Add:* 1313 Woodland Dr Ashland OR 97520

BARTLETT, JAMES WILLIAMS, JR, b Baltimore, Md, Feb 2, 26; m 54; c 3. PSYCHIATRY. *Educ:* Harvard Univ, AB, 48; Johns Hopkins Univ, MD, 52; State Univ NY, 59-66. *Prof Exp:* From instr to asst prof psychiat, 57-68, asst dean, 58-65, assoc dean, 65-81, PROF PSYCHIAT, PROF HEALTH SERV & CHMN DEPT, MED SCH, UNIV ROCHESTER, 68-, SR ASSOC DEAN, 81- *Concurrent Pos:* Actg med dir, Strong Mem Hosp, 67-68, med dir, 68- *Mem:* AAAS; Am Psychiat Asn; Asn Am Med Cols. *Res:* Medical education; psychoanalysis. *Mailing Add:* Univ Rochester Sch Med Dent 300 Crittenden Blvd Rochester NY 14642

BARTLETT, JANETH MARIE, b Cooperstown, NY, Sept 10, 46. NUCLEAR PHARMACY, PHARMACEUTICAL SCIENCES. *Educ:* Temple Univ, BS, 69, MS, 71; Rutgers Univ, PhD(pharm sci), 81. *Prof Exp:* Res assoc, E R Squibb & Sons, Inc, 70-74, asst res investr, 74-81; asst prof nuclear pharm, Purdue Univ, 81-86; RES ASSOC, DOW CHEM CO, 86- *Mem:* AAAS; Soc Nuclear Med; Sigma Xi. *Res:* Radiopharmaceutical and novel drug delivery systems. *Mailing Add:* 209 Meadow Lane Midland MI 48640

BARTLETT, JOHN W(ESLEY), b Camden, NJ, Oct 18, 35; m 57; c 2. CHEMICAL & NUCLEAR ENGINEERING. *Educ:* Univ Rochester, BS, 57; Rensselaer Polytech Inst, MChE, 59, PhD(chem eng), 62. *Prof Exp:* Engr coolant technol, Knolls Atomic Power Lab, Gen Elec Co, 57-62; asst prof chem eng, Univ Rochester, 62-68; eng assoc, Pac Northwest Labs, Battelle Mem Inst, 68-69, mgr sodium coolant studies, 69-70, mgr chem systs technol, 70-73; pres interchange exec, Nat Bur Standards, 73-74; mgr process eval, Pac Northwest Labs, Battelle Mem Inst, 74-78; div dir, Anal Sci Corp, 78-89; DIR, OFF CIVILIAN RADIOACTIVE WASTE MGT, US DEPT ENERGY, 90- *Concurrent Pos:* Fulbright prof, Istanbul Univ, Turkey, 68-69; adj prof, Joint Ctr Grad Study, 71-78. *Mem:* Am Nuclear Soc. *Res:* Radioactive waste management; energy systems analysis; effluent control technology; transport processes. *Mailing Add:* 2151 Wolftrap Ct Vienna VA 22182

BARTLETT, NEIL, b Newcastle-upon-Tyne, Eng, Sept 15, 32; m 57; c 4. INORGANIC CHEMISTRY. *Educ:* Univ Durham, BSc, 54, PhD(inorg fluorine chem), 58. *Hon Degrees:* DSc, Univ Waterloo, 68, Colby Col, Maine, 72 & Univ Newcastle, Tyne, 81; Hon Dr, Univ Bordeaux, 76; Univ Ljubljana, 89, Univ Nantes, 90. *Prof Exp:* Sr chem master, Duke's Sch, Eng, 57-58; lectr chem, Univ BC, 58-59, from instr to prof chem, 59-66; prof chem, Princeton Univ, 66-69; PROF CHEM, UNIV CALIF, BERKELEY, 69- *Concurrent Pos:* Steacie Mem fel, 64-66; Sloan fel, 64-; Miller vis prof, Univ Calif, Berkeley, 67-68; vis Erskine Fel, Univ Canterbury, New Zealand, 83; vis fel, All Souls Col, Oxford, 84. *Honors & Awards:* Corday-Morgan Medal, The Chem Soc, 62; Noranda Award, Chem Inst Can, 63; Steacie Prize, 65; Res Corp Award, 65; William Lloyd Evans Mem Lectr, 66; Cresson Medal, Franklin Inst, Pa, 68; Kirkwood Medal & Award, 69; Inorg Chem Award, Am Chem Soc, 70; Dannie-Heineman Prize, Gottingen Acad, 71; 20th G N Lewis Mem Lectr, Univ Calif, Berkeley, 73; Robert A Welch Award, 76; Alexander von Humboldt US Sr Scientist Award, 77; Werner Lectr, Univ Kans, 77; R T Major Mem Lectr, Univ Conn, 85; W H Nichols Medal, Am Chem Soc, 83. *Mem:* Foreign assoc Nat Acad Sci; Gottingen Acad Sci; fel Chem Inst Can; Royal Soc Chem; fel Am Acad Arts & Sci; fel Royal Soc London; assoc mem Acad Sci France. *Res:* Fluorine inorganic chemistry; noble gas chemistry; high-energy oxidizers; x-ray crystallography; solid-state chemistry; nonaqueous solvent chemistry; thermochemistry. *Mailing Add:* Dept Chem Latimer Hall Univ Calif Berkeley CA 94720

BARTLETT, PAUL A, b Trenton, NJ, Jan 5, 48. BIO-ORGANIC CHEMISTRY, ORGANIC SYNTHESIS. *Educ:* Stanford Univ, PhD(org chem), 72. *Prof Exp:* PROF CHEM, UNIV CALIF, BERKELEY, 73- *Mailing Add:* 247 Stanford Ave Kensington CA 94708

BARTLETT, PAUL DOUGHTY, b Ann Arbor, Mich, Aug 14, 07; m 31; c 3. ORGANIC CHEMISTRY. *Educ:* Amherst Col, BA, 28; Harvard Univ, PhD(chem), 31. *Hon Degrees:* ScD, Amherst Col, 53 & Univ Chicago, 54; Dr, Univ Montpellier, 67, Univ Paris, 69 & Univ Munich, 77. *Prof Exp:* Nat Res fel chem, Rockefeller Inst & Columbia Univ, 31-32; instr org chem, Univ Minn, 32-34; from instr to prof chem, Harvard Univ, 34-48, Erving prof, 48-75, chmn dept, 51-54; Robert A Welch res prof, Tex Christian Univ, 74-85, Robert A Welch res emer prof chem, 85; Erving emer prof chem, Harvard Univ, 75-85; RETIRED. *Concurrent Pos:* Numerous lectureships US & abroad, 35-78; off invester, Nat Defense Res Comt, 41-45; vis prof, Univ Calif, Los Angeles, 50, Univ Chicago, 71, Univ Tex, Austin, 75 & Univ Munich, 77; Guggenheim fel, 57 & 71-72. *Honors & Awards:* Pure Chem Award, Am Chem Soc, 38, Willard Gibbs Medal & Roger Adams Medal, 63, James Flack Norris Award, 69 & 78; August Wilhelm von Hofmann Medal, German Soc Chem, 62; Theodore William Richards Medal, 66; Nat Medal Sci, 68; John Price Wetherill Medal, Franklin Inst, 70; Linus Pauling Medal, 75; Nichols Medal, 76; Alexander von Humboldt Senior Scientist Award, Freiburg, 76 & Munich, 77; Max Tishler Award, 81; Robert A Welch Award, 81. *Mem:* Nat Acad Sci; hon mem Royal Soc Chem; hon mem Swiss Chem Soc; hon mem Chem Soc Japan; Am Acad Arts & Sci; fel AAAS; Am Philos Soc; Franklin Inst. *Res:* Stereochemistry; kinetics and mechanism of organic reactions; polymerization; Walden inversion; molecular rearrangements; paraffin alkylation; highly branched compounds; free radicals; reactions of sulphur; cycloaddition; organdithium reactions; photo-oxidation; specially hindered olefins. *Mailing Add:* Brookhaven Lexington Amherest No A-311/1010 Waltham St Lexington MA 02173

BARTLETT, PAUL EUGENE, b Denver, Colo, Nov 21, 26; m 79; c 2. EDUCATIONAL ENGINEERING ADMINISTRATION. *Educ:* Univ Colo, BS(civil eng) & BS(bus mgt), 51, MS(civil eng), 56. *Prof Exp:* Instr appl math, Univ Colo, Denver, 51-56, from asst prof to assoc prof civil eng, 56-66, from asst dean to assoc dean, Col Eng, 62-71, vprovost, 70-71, asst vpres, 71-73, from assoc dean to dean, Col Eng, 73-89, EMER DEAN, COL ENG & EMER PROF CIVIL ENG, UNIV COLO, DENVER, 89-, ACTG VCHANCELLOR ACAD AFFAIRS, 90- *Concurrent Pos:* Consult, Martin-Marietta Corp, 58-62. *Mem:* Am Soc Civil Engrs; Am Soc Eng Educ; Nat Soc Prof Engrs. *Res:* Hydraulics; applied fluid mechanics; engineering education. *Mailing Add:* Col Eng Campus Box 104 Univ Colo PO Box 173364 Denver CO 80217-3364

BARTLETT, R(OBERT) W(ATKINS), b Salt Lake City, Utah, Jan 8, 33; m 54; c 4. METALLURGY, MINERAL ENGINEERING. *Educ:* Univ Utah, BS, 53, PhD(metall), 61. *Prof Exp:* Sr scientist, Ford Aeronutronic, 61-65; sr scientist, Stanford Res Inst, 65-67, assoc prof, dept mineral eng, Stanford Univ, 67-74; mgr hydrometall dept, MMD Res Ctr, Kennecott Copper Corp, 74-76; PRES, HYDROTEK MINING INC, 73-; DIR, MAT RES LAB, STANFORD UNIV, 76- *Honors & Awards:* Turner Award, Electrochem Soc; Extractive Metall Technol Award, Am Inst Mining, Metall & Petrol Engrs, 75. *Mem:* Am Inst Mining, Metall & Petrol Engrs; Electrochem Soc. *Res:* Extractive metallurgy; hydrometallurgy; diffusion; corrosion; high temperature materials; metallurgical process kinetics; process engineering research; geochemical kinetics. *Mailing Add:* Univ Idaho Col Eng Moscow ID 83843

BARTLETT, RICHMOND J, b Columbus, Ohio, Sept 23, 27; m 52; c 4. SOIL & WATER CHEMISTRY, PLANT NUTRITION. *Educ:* Ohio State Univ, BA, 49, PhD(soil chem), 58. *Prof Exp:* Newspaper reporter, 50-52; promotional writer, 52-55; asst soils, Ohio State Univ, 55-58; from asst prof to assoc prof, 58-67, PROF SOILS, UNIV VT, 67- *Mem:* Fel AAAS; fel Am Soc Agron; fel Soil Sci Soc Am; Int Soc Soil Sci. *Res:* Soil and water chemistry in relation to agriculture and environmental health; chemistry of root, soil, water and air interfaces; metal organic interactions; redox. *Mailing Add:* Dept Plant & Soil Sci Univ Vt Burlington VT 05405

BARTLETT, RODNEY JOSEPH, b Memphis, Tenn, Mar 31, 44; m 66; c 2. CHEMICAL PHYSICS, COMPUTER APPLICATIONS. *Educ:* Millsaps Col, BS, 66; Univ Fla, PhD(theoret chem), 71. *Prof Exp:* NSF fel theoret chem, Aarhus Univ, Denmark, 71-72; res assoc chem, Johns Hopkins Univ, 72-74; sr res scientist, Battelle Pac Northwest Lab, 74-77, sr res scientist & group leader chem physics, Battelle Columbus Lab, Battelle Mem Inst, 77-81; prof chem & physics, 81-88, GRAD RES PROF CHEM & PHYSICS, QUANTUM THEORY PROJ, UNIV FLA, 89- *Concurrent Pos:* Prin investr, Air Force Off Sci Res, 78-, Off Naval Res, 79- & Army Res Off, 79-; consult, Molecular Sci Res Ctr, Battelle, Pac Northwest Lab, 85-88, sr affil scientist, 89-; Guggenheim fel, Harvard Univ & Univ Calif, Berkeley, 87; chmn, Subdivision Theoret Chem, Am Chem Soc. *Mem:* Am Chem Soc; Sigma Xi; fel Am Phys Soc. *Res:* Development of the theory and applications of quantum mechanics in chemistry; correlation problem, particularly many-body, diagrammatic perturbation theory and coupled-cluster theory; studies of large molecules of biological interest; molecular structure and spectra; flame chemistry; analytical gradient and hessian methods; molecular hyperpolarizabilities and non-linear optics. *Mailing Add:* Quantum Theory Proj Williamson Hall Univ Fla Gainesville FL 32611

BARTLETT, ROGER JAMES, b Ft Madison, Iowa, Mar 19, 42; m 69; c 2. SOLID STATE PHYSICS. *Educ:* Iowa State Univ, BS, 64, MS, 68, PhD(physics), 70. *Prof Exp:* Fel optical physics, Ga Inst Technol, 70-72; staff mem low temperature physics, 72-79, STAFF MEM OPTICAL PROPERTIES OF MAT, LOS ALAMOS NAT LAB, 79- *Mem:* Am Phys Soc. *Res:* Optical properties of materials with emphasis on atomic & solid state effects in the soft x-ray energy range; superconductivity with emphasis on critical current and current and field profiles in superconductors. *Mailing Add:* Los Alamos Nat Lab P-14 MS D410 PO Box 1663 Los Alamos NM 87545

BARTLETT, WILLIAM ROSEBROUGH, b St Louis, Mo, July 25, 43; m 64; c 3. ORGANIC CHEMISTRY. *Educ:* Luther Col, BA, 65; Stanford Univ, PhD(org chem), 69. *Prof Exp:* NIH fel, Columbia Univ, 69-71; asst prof, Hamline Univ, 71-75 & St John's Univ, Minn, 75-78; from asst prof to assoc prof, 78-85, PROF CHEM, FT LEWIS COL, 85- *Concurrent Pos:* Res Corp Cottrell grant, Hamline Univ, 73-75 & St John's Univ, 76-77; Am Chem Soc-Petrol Res Fund grant, Ft Lewis Col, 79-82, 87-89 & 90-; NSF minigrant, 83; vis sr res assoc, Stanford Univ, 86-87; NSF-ROA, 86-87; chem sci eval panel, NRC, 86-89; ILIP grant, NSF, 89- *Mem:* Am Chem Soc; AAAS. *Res:* Synthesis of biologically-active compounds, insect pheromones. *Mailing Add:* Dept Chem Ft Lewis Col Durango CO 81301

BARTLEY, EDWARD FRANCIS, b Scranton, Pa, June 24, 16; m 40; c 2. MATHEMATICS. *Educ:* Univ Scranton, BA, 38; Columbia Univ, MA, 56. *Prof Exp:* Pub sch teacher, Pa, 39-43; assoc prof & chmn dept, Univ Scranton, 46-68, prof math, 68-; RETIRED. *Concurrent Pos:* Consult, Scranton Schs. *Mem:* Math Asn Am. *Res:* Algebra; secondary school curriculum. *Mailing Add:* Dept Math Univ Scranton Linden & Munroe Sts Scranton PA 18510

BARTLEY, JOHN C, b Pueblo, Colo, Apr 30, 32. METABOLISM. *Educ:* Colo State Univ, BS, 54, DVM, 56; Univ Calif, Davis, PhD(animal physiol), 63- *Prof Exp:* Assoc dir res, Children's Hosp Med Ctr N Calif, 68-78; res scientist, Peralha Cancer Res Inst, 78-82; STAFF SCIENTIST, LAWRENCE BERKELEY LAB, 82- *Mem:* Am Soc Biol Chemists; Biochem Soc; Am Soc Cell Biol; AAAS. *Mailing Add:* Biomed Div Bldg 934/6 Lawrence Berkeley Lab Berkeley CA 94708

BARTLEY, MURRAY HILL, JR, b Jamestown, NY, June 15, 33; m 56; c 3. ANATOMY, ORAL PATHOLOGY. *Educ:* Univ Ore, DMD, 58, cert, 65; Univ Utah, PhD(anat), 68. *Prof Exp:* Asst oral path, Univ Ore, 62-64; assoc dir res, D N Sharp Hosp, San Diego, 64-65; actg assoc prof oral biol, Sch Dent, Univ Calif, Los Angeles, 68-69; assoc prof, 69-77, chmn dept, 72-76, PROF PATH, DENT SCH, UNIV ORE, 77-, CHMN DEPT, 80- *Concurrent Pos:* Univ Ore & NIH teaching grants, 61-64; resident, Providence Hosp, 61-62; consult, Vet Admin, 70, Madigan Army Med Ctr, 76- & Coun Hosp Dent, Am Dent Asn, 77-79. *Mem:* AAAS; Am Dent Asn; Sigma Xi; Am Acad Oral; NY Acad Sci. *Res:* Cortisol effects on bone; immunopathology of oral lesions; aging in skeletal tissues determined by densitometric or histological methods; aging in skeletal tissues of anthropological material; heritable and metabolic diseases of the hard dental tissues. *Mailing Add:* Dept Path Univ Ore Dent Sch Portland OR 97201

BARTLEY, WILLIAM CALL, b Mason, Mich, Dec 4, 32; m 56; c 3. SCIENCE POLICY, MEDICAL & HEALTH SCIENCES. *Educ:* Mich State Univ, BS, 55, MS, 59. *Prof Exp:* Asst elec eng, Mich State Univ, 55, instr, 58-59; mgn engr, Apparatus Div, Tex Instruments, 59-60, mkt engr, Semiconductor Div, 60, design engr, 60-61, br mgr, 61-63; res scientist & mgr radiation exp develop, Univ Tex, Dallas, 63-67; exec secy, US Comt Solar

Terrestrial Res & Study, dir, Comt on Biol & Climate Effects of Emmissions in Stratosphere, Inst Med-Nat Acad Sci, 67-74; sr staff, White House Sci Adv, US Dept State, 74-77, asst dir energy res, Dept Energy, 77-80, assoc dir energy res, 80-82, minister counselor int health attache, Dept Health & Human Serv, US Dept State, US Mission, Geneva, Switz, 82-88, special asst to Off Comnr Food & Drugs, 88-89; SR POLICY ADV & DIR TRADE POLICY ENERGY & GLOBAL WARMING, OFF US TRADE REP, WHITE HOUSE, 89- *Concurrent Pos:* Sr investr NASA res contracts, 63-67; staff dir, study on photovoltaics & photosynthesis, 69 & study on space sci properties & life sci, Space Sci Bd, 70-72, dir study biol & med effects of nitrogen & oxide emmisions, Dept Transp, 72-75; sr policy analyst, Sci Technol & Int Affairs Directorate, NSF, 75-76; exec secy & mem, Fed Coord Coun for Sci, Eng & Technol Exec Off of the President, 76-78; prin staff person, Energy Res Adv Bd, Dept Energy, 78-80, US liaison, UN WHO, 81-88; vpresident's Competitiveness coun working groups on Orphan Drugs & Drug Approval, Biotechnol Working Group, 89- *Honors & Awards:* Award for Development of Satellite Aspect Comput, NASA, 74. *Mem:* Am Geophys Union; Inst Elec & Electronics Engrs; NY Acad Sci. *Res:* Experimental radiation research using detectors on satellites and deep space probes; analyzing public policy implications of advances in science and technology; identifying emerging national and international research and development problems related to environment, health, energy and industrial commercialization. *Mailing Add:* 10841 Stanmore Dr Potomac MD 20854

BARTLEY, WILLIAM J, b Richfield, Utah, Feb 27, 35; m 56; c 3. INDUSTRIAL CHEMISTRY. *Educ:* Brigham Young Univ, BS, 57, MS, 58; Univ Wash, PhD(org chem), 62. *Prof Exp:* SR RES SCIENTIST, UNION CARBIDE CHEM & PLASTICS CO, INC, 62- *Mem:* Am Chem Soc. *Res:* Agricultural chemicals; pesticide metabolism studies; organic synthesis; heterogeneous catalysis; new process development. *Mailing Add:* Union Carbide Chem & Plastics Co Inc PO Box 8361 South Charleston WV 25303

BARTLING, JUDD QUENTIN, b Muncie, Ind, July 24, 36; m 73; c 2. PHYSICS, ELECTRO-OPTICS & I R SIGNAL PROCESSING. *Educ:* Univ Calif, Berkeley, BA, 59; Purdue Univ, MS, 64; Univ Calif, Riverside, PhD(physics & math), 69. *Prof Exp:* Proj leader laser radar, US Army, 60-62; consult electro-magnetic, NAm Aviation, 65-66; res asst solid state, atomic & laser physics, Univ Calif, Riverside, 65-68; scientist electro-optics, Litton Guid & Control Div, Canoga Park, Calif, 69-71; PRES APPL, BASIC PHYSICS & BUS RES, AZAK CORP, 71- *Concurrent Pos:* Instr quantun chem, Gainesville, Fla, 69. *Res:* Solid state physics of insulators; electron band structure of solids; finite solids; optical properties of solids; laser physics; integrated optics; statistical properties of light; radar; signal processing; atomic scattering; laser radar; business research electromagnetics. *Mailing Add:* 21032 Devonshire St Chatsworth CA 91311

BARTLIT, JOHN R(AHN), b Chicago, Ill, June 1, 34; m 61; c 2. CHEMICAL ENGINEERING, CRYOGENICS. *Educ:* Purdue Univ, BSChE, 56; Princeton Univ, MSE, 57; Yale Univ, DEng, 63. *Prof Exp:* Instr, Yale Univ, 60-61; STAFF MEM, 62-, DEP PROJ MGR, TRITIUM SYSTS TEST ASSEMBLY, LOS ALAMOS NAT LAB, 82- *Concurrent Pos:* Consult, Georgetown Univ, Nat Coal Policy Proj, 76-78. *Honors & Awards:* Technol Utilization Award, 71. *Res:* Cryogenic engineering; heat transfer to cryogenic fluids; tritium handling; cryogenic distillation of hydrogen isotopes. *Mailing Add:* Los Alamos Nat Lab MS C348 Los Alamos NM 87545

BARTLOW, THOMAS L, b Johnson City, NY, May 14, 42. MATHEMATICS. *Educ:* State Univ NY Albany, BS, 63; State Univ NY Buffalo, MA, 66, PhD(math), 69. *Prof Exp:* Instr, 68-69, ASST PROF MATH, VILLANOVA UNIV, 69- *Mem:* Math Asn Am; Hist Sci Soc. *Res:* Algebra; power-associative quasigroups and loops; multiplicative systems; history of mathematics. *Mailing Add:* Dept Math Villanova Univ Villanova PA 19085

BARTNICKI-GARCIA, SALOMON, b Mexico City, Mex, May 18, 35; m 61; c 2. BIOCHEMISTRY. *Educ:* Nat Polytech Inst, Mex, prof degree, 57; Rutgers Univ, PhD(microbiol), 61. *Prof Exp:* Res assoc microbiol, Rutgers Univ, 61-62; asst microbiologist & lectr, Univ Calif, Riverside, 62-68, assoc prof & microbiologist, 68-71, PROF PLANT PATH & MICROBIOLOGIST, UNIV CALIF, RIVERSIDE, 71-, CHMN, DEPT PLANT PATH, 89- *Concurrent Pos:* Vis prof, Org Chem Inst, Univ Stockholm, 69-70; NIH grant, 63-; NSF grant 71-; lectr, Mycol Soc Am, 78; vis prof, French Ministry Educ, Univ Grenoble, France, 89; Fac res lectr, Univ Calif, Riverside, 89. *Honors & Awards:* New York Bot Garden Award, Bot Soc Am, 75; Ruth Allen Award, outstanding contributions plant path, Am Phytopath Soc, 83. *Mem:* Fel AAAS; Am Soc Microbiol; Am Phytopath Soc; hon mem Brit Mycol Soc; Am Soc Biol Chem. *Res:* Microbial biochemistry; biochemistry of morphogenesis; physiology of fungi; cell wall structure and biosynthesis; biochemistry of parasitism. *Mailing Add:* Dept Plant Path Univ Calif Riverside CA 92521

BARTNIKAS, RAY, b Kaunas, Lithuania, Jan 25, 36; Can citizen; m 67; c 2. MATERIALS SCIENCE ENGINEERING. *Educ:* Univ Toronto, BASc, 58; McGill Univ, MEng, 62, PhD(elec eng), 64. *Prof Exp:* Develop engr, Cable Labs, Northern Elec Co, Que, 58-63; mem sci staff, Phys Sci Dept, Bell-Northern Res Labs, Ont, 63-68; sr res scientist, Mat Res Dept, 68-70, dir mat res dept, 70-82, DISTINGUISHED SR SCIENTIST, INST RES, HYDRO QUEBEC, 82- *Concurrent Pos:* Auxiliary prof, McGill Univ, Can, 69-76; adj prof, 69-, Sandford-Fleming vis fel, Univ Waterloo, 81-; adj prof, Polytech, Univ Montreal, 80- *Honors & Awards:* Thomas Dakin Distinguished Sci Achievement Award, Inst Elec & Electronics Engrs, 80, Centennial Medal, 84, Morris Leeds Award, 89; Charles Dudley Medal, Am Soc Testing & Mat, 85, Arnold Scott Award, 85; Award of Merit, Am Soc Testing & Mat, 85; Award of Merit, Can Standards Asn, 86, John Jenkins Award, 89; Whitehead Mem Award, 87. *Mem:* Fel Inst Elec & Electronics Engrs; fel Am Soc Testing & Mat; Int Electrotech Comn; Can Standards Asn; Can Elec Asn; Order Engrs Can; fel Inst Physics, UK; fel Royal Soc Can. *Res:* Corona discharges and conduction and loss mechanisms in dielectric materials. *Mailing Add:* Inst de Recherche Hydro-Quebec 1800 Montee Ste-Julie Varennes PQ J3X 1S1 Can

BARTNOFF, SHEPARD, b Kobryn, Poland, Nov 6, 19; nat US; m 44 & 75; c 3. TECHNICAL MANAGEMENT, NUCLEAR ENGINEERING. *Educ:* Syracuse Univ, AB, 41, MA, 44; Mass Inst Technol, PhD(physics), 49. *Prof Exp:* Asst physics, Syracuse Univ, 41-43, instr, 43-45; from instr to asst prof, Tufts Univ, 48-53, assoc prof & exec secy, Physics Dept, 53-55; mem staff, Westinghouse Atomic Power Div, 55-68; nuclear fuels mgr, Gen Pub Utilities Corp, 68-69, mgr eng, 69-71, dir environ affairs, 71-72; PRES, JERSEY CENT POWER & LIGHT CO, 72- *Mem:* AAAS; Am Phys Soc; Am Nuclear Soc; Sigma Xi. *Res:* Compressible fluid flow; electromagnetic theory and properties of dielectrics; piezoelectric crystals; radar meteorology; nuclear power systems; ecology. *Mailing Add:* 280 Oakwood Rd Englewood NJ 07631

BARTOCHA, BODO, b Wroclaw, Poland, Dec 26, 28; m 57. SCIENCE POLICY. *Educ:* Philipps Univ, Marburg/Lahn, Ger, BS, 51, MS, 53, PhD(inorg chem), 56. *Prof Exp:* Milton fel postgrad res & teaching, Harvard Univ, 56-58; head, Propellant Br, Res US Naval Ord Lab, Corona, Calif, 58-61, dir res & actg dir develop, Res & Develop Mgt, Propellant Plant, Indian Head, Md, 61-64, dep chief scientist, Int Res & Develop Mgt, Off Naval Res, London, UK, 64-66, asst tech dir advan planning & develop, Naval Ord Sta, Indian Head, 66-67; staff assoc & dep head, Off Planning & Policy Studies, Policy Anal, 67-70, dep exec secy, Exec Coun, Mgt, 70-71, exec asst to asst dir, Nat & Int Progs, 71, dir, Div Int Progs, Int Res & Develop Mgt, NSF, 72-86; vis scholar & assoc dir, 86-88, dir off int prog, 88-89, PROF ARID LANDS, UNIV ARIZ, TUCSON, 89- *Mem:* Fel AAAS. *Res:* Synthetic organo-metal chemistry; reaction kinetics; solid propellants; weapons systems analysis; science policy; planning; resources allocation; technological forecasting; technology assessment; international research and development cooperation. *Mailing Add:* 1305 E Moonridge Rd Tucson AZ 85718-1142

BARTOK, WILLIAM, b Budapest, Hungary, May 1, 30; US citizen; m 57; c 3. PHYSICAL CHEMISTRY, CHEMICAL ENGINEERING. *Educ:* McGill Univ, BEng, 54, PhD(phys chem), 57. *Prof Exp:* From res chemist to sci adv, Exxon Res & Eng Co, 57-86, group head combustion sci, 72-81, group head high temperature chem, Corp Res Labs, 81-83, coordr, 83-86; SR VPRES, ENERGY & ENVIRON RES CORP, 86- *Concurrent Pos:* Mem, ad hoc Panel Nitrogen Oxide Emission Control, Nat Res Ctr, Nat Acad Sci, Nat Acad Eng, 71-72 & Adequacy of Sulfur Oxide Emission Control Technol Comt, 75-78, Peer Rev Comt, Environ Protection Agency, 83, Dept Energy, 84, Sandia Nat Lab, 86, Combustion Sci & Tech; bd dirs, Combustion Inst, 80- *Honors & Awards:* Award in Chem Contemporary Technol Problems, Am Chem Soc, 87. *Mem:* Am Chem Soc; Am Inst Chem Eng; Combustion Inst. *Res:* Combustion; air pollution control; kinetics and catalysis; free radical processes in oxidation and high temperature chemistry; nitrogen oxides formation and control; rheology of disperse systems. *Mailing Add:* 956 Wyandotte Trail Westfield NJ 07090

BARTOLINI, ROBERT ALFRED, b Waterbury, Conn, Apr 4, 42; m 64; c 3. ELECTROOPTICS. *Educ:* Villanova Univ, BS, 64; Case Inst of Technol, MS, 66; Univ Pa, PhD(elec eng), 72. *Prof Exp:* Mem tech staff, 66-80, res leader electrooptics, 80-83, group head optoelectroncis, 83-89, DIR, INTEGRATED CIRCUIT LAB, DAVID SARNOFF RES CTR, 89- *Concurrent Pos:* Lectr, LaSalle Univ, Philadelphia, 72-83, chmn dept, 83-90. *Honors & Awards:* Centennial Medal, Inst Elec & Electronics Engrs, 84. *Mem:* Fel Inst Elec & Electronics Engrs; fel Optical Soc Am; Am Inst Physics; Sigma Xi. *Res:* Development of materials and systems suitable for optical applications such as information storage and retrieval systems, holography and optical wave guides. *Mailing Add:* David Sarnoff Res Ctr Princeton NJ 08540

BARTON, ALEXANDER JAMES, b Mt Pleasant, Pa, May 9, 24; m 45; c 3. ECOLOGY, HERPETOLOGY. *Educ:* Franklin & Marshall Col, BS, 46; Univ Pittsburgh, MS, 56. *Prof Exp:* Park naturalist, Riverview Park, Pa, 45-46; herpetologist, Highland Park Zool Gardens, 46-52; instr biol, Stony Brook Sch, NY, 52-63, dir admis & financial aid, 57-63; dir, Savannah Natural Hist Mus, Ga, 57; from asst prof to assoc prof, Undergrad Educ Sci Div, Col Sci, NSF, 63-70, prog dir, Undergrad Instrnl Progs, 70-73, prog mgr, Student-Oriented Progs, 73-75, prog dir, Sci Educ Develop Prog, 76-84, prog dir, Instrumentation Prog, 84-88; RETIRED. *Concurrent Pos:* Pres, Asn Admis Officers Independent Sec Schs, 59-62; adj asst prof, C W Post Col, Long Island Univ, 61-63; consult, Doubleday & Co, 62-64; mem, inst serv subcomt, Col Entrance Exam Bd, 62-65, 62-65, col scholar serv comt, 65-68; chmn, Schs Scholar Serv, 62-63; mem comt endangered reptiles & amphibians, Int Union Conserv of Nature, 67-74; life-judge, Am Rose Soc, 70-, dir, 85-; mem environ educ subcomt, Fed Interagency Comt Educ, 74-, chmn, 81-82; chmn, Fed Interagency Subcomt, Int Environ Educ, 76-78; mem, US Deleg Intergovt Conf Environ Educ, Tbilisi, USSR, 78; naval mem on Secy Defense, ROPMA, steering group staff, 81-; serv to captain, US Naval Reserve. *Honors & Awards:* Silver Honor Medal, Am Rose Soc, 84. *Mem:* AAAS; Am Inst Biol Sci; Ecol Soc Am; Soc Study Amphibians & Reptiles. *Res:* Ecology and classification of amphibians and reptiles, especially turtles and salamanders; environmental education; design of programs for improving education in science. *Mailing Add:* PO Box 100 Monterey VA 24465

BARTON, BARBARA ANN, b Erie, Pa, May 20, 54. FORAGE INTAKE & DIGESTION BY RUMINANTS. *Educ:* Pa State Univ, BS, 76; Univ Wis-Madison, MS, 78, PhD(dairy sci), 81. *Prof Exp:* Res assoc dairy sci, Univ Wis-Madison, 76-81; asst prof animal sci, 82-87, ASSOC PROF ANIMAL SCI, UNIV MAINE, ORONO, 87- *Concurrent Pos:* Vis scientist, sustainable agr res & teaching, Ag Foras Taluntis, Ireland, 86; bd dir, Am Soc Animal Sci, 87-89. *Mem:* Am Dairy Sci Asn; Am Soc Animal Sci; Am Forage & Grassland Coun; Coun Agr Sci & Technol; Sigma Xi. *Res:* Nutritional physiology of early lactation dairy cows and gestating ewes as related to productivity and metabolic disorders; factors that influence the intake and digestibility of forages by ruminant animals. *Mailing Add:* Dept Animal & Vet Sci 24A Rogers Hall Univ Maine Orono ME 04469

BARTON, BEVERLY E, b New York, NY, May 6, 54; m 78; c 2. CYTOKINE IMMUNOLOGY, HEMATOPOIESIS OF OSTEOCLASTS. *Educ:* Johns Hopkins Univ, BA, 76, ScM, 79; Stanford Univ, PhD(med microbiol), 84. *Prof Exp:* Sr prof, Allergan Pharmaceut, Inc, 84-85, scientist, 85-87; sr scientist, 87-90, PRIN SCIENTIST, SCHERING-PLOUGH RES, SCHERING-PLOUGH CORP, 90- *Concurrent Pos:* Reviewer, Current Eye Res, 87- *Mem:* Am Asn Immunologists; NY Acad Sci; Soc Exp Biol & Med. *Res:* Roles of interleukin 6(IL-6) in septic shock, endotoxemia and multiple myeloma; discovering compounds which inhibit IL-6 synthesis in vivo and prevent symptoms of septic shock and endotoxemia. *Mailing Add:* Dept Immunol Schering-Plough Res 60 Orange St Bloomfield NJ 07003

BARTON, CHARLES JULIAN, SR, b Jellico, Tenn, Jan 16, 12; m 40; c 3. ENVIRONMENTAL HEALTH. *Educ:* Univ Tenn, BS, 33, MS, 34; Univ Va, PhD(anal chem), 39. *Prof Exp:* Jr phys sci aide, Tenn Valley Authority, 34-36; chemist, Chesapeake & Ohio Rwy Co, 39-40, Nat Aluminate Corp, 40 & Indust Rayon Corp, 40-47; res chemist & group leader, Int Minerals & Chem Corp, 47-48; res staff mem, Oak Ridge Nat Lab, 48-77; RETIRED. *Concurrent Pos:* Consult, Sci Applns, Inc, 77- & Evaluation Res Corp. *Mem:* Am Chem Soc; Am Nuclear Soc. *Res:* Photoelectric spectrophotometry applications in analytical chemistry; phase studies; plutonium chemistry; nuclear safety studies; radiological considerations in nuclear gas well stimulation; health effects of non-radioactive pollutants and radionuclides from uranium mill tailings; results of radiological surveys. *Mailing Add:* 237 Outer Dr Oak Ridge TN 37830

BARTON, CLIFF S, b Preston, Idaho, July 18, 19; m 42; c 5. ENGINEERING MECHANICS, EDUCATIONAL CONSULTANT. *Educ:* Utah State Univ, BSCE, 47; Rensselaer Polytech Inst, MCE, 53, PhD(eng mech), 59. *Prof Exp:* Design layout engr, Lockheed Aircraft Co, 41-45; instr appl mech, Rensselaer Polytech Inst, 47-53, asst prof mech, 53-59; assoc prof civil eng, 59-62, chmn dept, 63-69, PROF CIVIL ENG, BRIGHAM YOUNG UNIV, 62-, ASST DEAN, COL ENG SCI & TECHNOL, 77- *Concurrent Pos:* Consult, Watervliet Arsenal, 57-60; assoc ed, Am J Phys Anthropol, 83; contrib ed, Quar Rev Archaeol, 87. *Mem:* AAAS; Am Soc Civil Engrs; Soc Exp Stress Anal (pres, 69-70); Am Soc Eng Educ; Sigma Xi. *Res:* Wave motion in continous media; viscoelastic studies; development and utilization of residual stresses; effects of high pressure on material properties; experimental mechanics. *Mailing Add:* 1101 Elm Ave Provo UT 84604

BARTON, DAVID KNOX, b Greenwich, Conn, Sept 21, 27; m 49; c 8. ELECTRICAL ENGINEERING. *Educ:* Harvard Col, AB, 49. *Prof Exp:* Electronic engr, Signal Corps Eng Labs, 49-55 & RCA Corp, 55-63; consult scientist, Raytheon Co, 63-84; CONSULT, ANRO ENG, 84- *Honors & Awards:* David Sarnoff Award, RCA Corp, 58; M Barry Carlton Award, Inst Elec & Electronics Engrs, 62. *Mem:* Fel Inst Elec & Electronics Engrs. *Res:* Radar systems engineering and wave propagation effects on radar. *Mailing Add:* 180 Prospect Hill Rd Harvard MA 01451

BARTON, DEREK HAROLD RICHARD, b Eng, Sept 8, 18. CHEMISTRY. *Educ:* Univ London, BSc Hons, 40, PhD(org chem), 42. *Hon Degrees:* Numerous from US & foreign univs, 49-91. *Prof Exp:* dir, Inst Chem Natural Substances, Nat Ctr Sci Res, 77-86; prof, 57-78, EMER PROF ORG CHEM, IMP COL, UNIV LONDON, 78-; DISTINGUISHED PROF, TEX A&M UNIV, 86- *Concurrent Pos:* Vis prof & lectr, numerous US & foreign univs, 49-; hon dir, Inst Chem Natural Substances, Nat Ctr, Sci Res, 86- *Honors & Awards:* Nobel Prize in Chem, 69; Fritzsche Medal, Am Chem Soc, 56, Roger Adams Medal, 59, Second Centennial of the Priestly Chem Award, 74, Creativity in Org Synthesis Award, 89; Davy Medal, Royal Soc, 61, Royal Medal, 72, Copley Medal, 80; Order of the Rising Sun, Emperor of Japan, 72; Order of Chevalie de la Lègion d'Honneur, 74; First May & Baker Award, Royal Soc Chem, 87. *Mem:* Foreign assoc Nat Acad Sci; foreign hon mem Am Acad Arts & Sci; Int Union Pure & Appl Chem (pres, 69-72); hon mem Acad Pharmaceut Sci; hon fel Royal Soc Chem. *Mailing Add:* Dept Chem Texas A&M Univ College Station TX 77843-3255

BARTON, DONALD WILBER, b Fresno, Calif, June 12, 21; m 44; c 4. PLANT BREEDING. *Educ:* Univ Calif, BS, 46, PhD(genetics), 49. *Prof Exp:* Asst genetics, Univ Calif, 46-49; asst prof, Univ Mo, 50-51; assoc prof, 51-59, head dept, 59-60, prof veg crops, Cornell Univ, 59-, dir NY State Agr Exp Sta & assoc dir res, Col Agr, 60- EMER PROF. *Concurrent Pos:* AEC fel, Univ Mo, 49-50; res grant genetics, 50-51; res grant hort, Northwest Canners & Freezers Asn, 59-; vis prof, Ore State Col, 59-60. *Mem:* Genetics Soc Am; Am Soc Hort Sci; Sigma Xi. *Res:* Breeding and genetics of horticultural crops; genetic disease resistance. *Mailing Add:* Hort Sci NY State Agr Exp Sta Geneva NY 14456

BARTON, EVAN MANSFIELD, b Chicago, Ill, Nov 7, 03; wid; c 2. INTERNAL MEDICINE, RHEUMATOLOGY. *Educ:* Williams Col, AB, 24; Johns Hopkins Univ, MD, 29; Am Bd Internal Med, dipl, 42. *Prof Exp:* Intern, Presby Hosp, Chicago, 29-30; asst, Rush Med Col & Presby Hosp, 31-34; clin prof med, Med Col, Univ Ill, 51-71; PROF MED, RUSH MED COL, 71- *Concurrent Pos:* Fel path, Rush Med Col & Presby Hosp, 31-34, fel internal med, 36; vol asst path, Hamburg, Ger, 31; vol asst med res lab, Royal Infirmary, Aberdeen, Scotland, 35; attend physician, Presby-St Luke's Hosp, 36-; consult, rheumatol, Vet Admin Hosp, Hines, Ill, 46- *Mem:* Fel AMA; fel Am Col Physicians; Am Rheumatism Asn. *Res:* Rheumatic diseases. *Mailing Add:* 1725 W Harrison St Chicago IL 60612-3828

BARTON, FRANKLIN ELLWOOD, II, b Baltimore, Md, Aug 4, 42; m 65; c 2. PHYSICAL ORGANIC CHEMISTRY, PHYSICAL CHEMISTRY. *Educ:* NGa Col, BS 64; Univ Ga, PhD(chem), 69. *Prof Exp:* Res chemist, 71-77, SUPV RES CHEMIST, RUSSELL RES CTR, SCI EDUC ADMIN, USDA, 77- *Concurrent Pos:* Fel, Univ Ga, 70-71. *Mem:* Am Chem Soc; The Chem Soc; Am Oil Chemists Soc; Soc Appl Spectros. *Res:* Spectroscopic and chromatographic studies of the structure and association of plant polysaccharides and polyphenols and their utilization by ruminant animals. *Mailing Add:* Dept Agr RB Russell Agr Res Ctr PO Box 5677 Athens GA 30613

BARTON, FURMAN W(YCHE), b Greenville, SC, Sept 22, 32; m 63; c 2. ENGINEERING MECHANICS, STRUCTURAL ENGINEERING. *Educ:* Univ Va, BCE, 54; Univ Ill, MS, 59, PhD(struct mech), 62. *Prof Exp:* From res asst to res assoc struct, Univ Ill, 57-62, asst prof struct mech, 62-64; asst prof, Duke Univ, 64-67; assoc prof appl mech, 67-78, PROF CIVIL ENG, UNIV VA, 78- *Mem:* AAAS; Am Soc Civil Engrs; Am Soc Mech Engrs; Am Acad Mech. *Res:* Structural dynamics; dynamic stability; discrete field mechanics; computer methods of structural analysis; systems analysis; finite element applications; nonlinear analysis. *Mailing Add:* Dept Civil Eng Thornton Hall Univ Va Charlottesville VA 22903

BARTON, GERALD BLACKETT, b Arco, Idaho, Oct 16, 17; m 43; c 6. APPLIED CHEMISTRY. *Educ:* Brigham Young Univ, BS, 39, MS, 41; Ohio State Univ, PhD, 50. *Prof Exp:* Chemist, Am Smelting & Refining Co, Utah, 41-43, res investr, NJ, 43-45; asst chem, Ohio State Univ, 45-48; chemist, Gen Elec Co, 48-65; sr res scientist, Chem Dept, Pac Northwest Lab, Battelle Mem Inst, 65-70; res scientist, Westinghouse Hanford Co, 70-81; RETIRED. *Mem:* Am Chem Soc. *Res:* Chemical processing of nuclear reactor fuels; disposal of radioactive wastes; analysis for trace impurities in gases; stress corrosion cracking of stainless steel. *Mailing Add:* 12627 155th Ave SE Renton WA 08055

BARTON, HARVEY EUGENE, b Couch, Mo, Aug 26, 36; m 54; c 5. SYSTEMATIC ENTOMOLOY, ECONOMIC ENTOMOLOGY. *Educ:* Ark State Univ, BSE, 62, MSE, 63; Iowa State Univ, PhD(entom), 69. *Prof Exp:* From asst prof to assoc prof, 67-82, PROF ZOOL, ARK STATE UNIV, 82- *Mem:* Entom Soc Am; Sigma Xi; Lepidoptera Res Found; Lepidopterists Soc; Coleopterists Soc. *Res:* Systematics and morphology of heteroptera; dynamics of flying insect populations. *Mailing Add:* Dept Biol Ark State Univ Box 501 State University AR 72467

BARTON, HENRY RUWE, JR, b Geneva, Ill, Sept 7, 44; m 66; c 2. HIGH ENERGY PHYSICS. *Educ:* Univ Ill, Urbana, BS, 66; Univ Pa, MS, 67; Purdue Univ, PhD(physics), 72. *Prof Exp:* Res assoc physics, Univ Ill, Urbana, 71-73 & Ohio State Univ, 73-75; physicist, Fermi Nat Accelerator Lab, 75-83; physicist, DESY-F35, 83-90; PHYSICIST, SUPERCONDUCTING SUPER COLLIDER, 90- *Concurrent Pos:* Guest scientist, Argonne Nat Lab, 71-81. *Mem:* Am Phys Soc; Cryogenic Soc Am. *Res:* Dipion production at low and high momentum transfer; neutron charge exchange; charm production and applied super conductivity. *Mailing Add:* Superconducting Super Collider 2550 Beckleymeade Ave Dallas TX 75237

BARTON, JACQUELINE K, b New York, NY, May 7, 52; m 90; c 1. BIOPHYSICAL CHEMISTRY, BIOINORGANIC CHEMISTRY. *Educ:* Barnard Col, BA, 74; Columbia Univ, PhD(chem), 79. *Prof Exp:* Fel phys biochem, Dept Molecular Biol & Biochem, Yale Univ, 79-80; asst prof chem & biochem, Hunter Col, City Univ New York, 80-82; from asst prof to assoc prof, Columbia Univ, 83-85, prof chem & biol sci, 86-89; PROF CHEM, CALIF INST TECHNOL, 89- *Concurrent Pos:* Vis res assoc, Dept Biophys, Bell Labs, 79. *Honors & Awards:* Alan T Waterman Award, 85; Nat Fresenius Award, Phi Lambda Upsilon, 86; Eli Lilly Award in Biol Chem, Am Chem Soc, 87, Award in Pure Chem, 88. *Mem:* Am Chem Soc. *Res:* Metal complex interactions with nucleic acids; biophysical probes of DNA structure and conformation; design of metal complexes as site-specific DNA clearing-molecules. *Mailing Add:* Chem Div Calif Inst Tech 164-30 Pasadena CA 91125

BARTON, JAMES BROCKMAN, b Lubbock, Tex, July 12, 44. ELECTRONICS ENGINEERING. *Educ:* Univ Tex, Austin, BS & BA, 66; Mass Inst Technol, PhD(physics), 71. *Prof Exp:* Mem tech staff, Cent Res Labs, 72-76, sect mgr, Charge Coupled Device Mem Design, 76-79, BR MGR, COMPLIMENTARY METAL OXIDE SEMICONDUCTOR TECHNOL DEVELOP, TEX INSTRUMENTS, 79- *Mem:* Inst Elec & Electronics Engrs. *Res:* Complimentary metal oxide semiconductor manufacturing processes and design methodologies for very large-scale integrated circuits. *Mailing Add:* Tex Instruments MS 3669 PO Box 655303 Dallas TX 75265

BARTON, JAMES DON, JR, b Anna, Ill, Oct 25, 29. PLANT ECOLOGY. *Educ:* Northern Ill Univ, BS, 52, MS, 53; Purdue Univ, PhD(plant ecol), 56. *Prof Exp:* Instr sci, Boston Univ Jr Col, 56-58; asst prof ecol, Boston Univ Grad Sch, 58-63; assoc prof ecol & dir div nat sci, Southampton Col, Long Island Univ, 63-66, dean col, 66-68; prof bot, provost & vpres acad affairs, Alfred Univ, 68-75; prof spec environ studies, 75-76; dir inst res, 76-81, REGISTR, STATE UNIV NY AGR & TECH COL ALFRED, 82- *Concurrent Pos:* Consult, Appel & Assocs & Southampton Town Conserv Comn. *Mem:* Ecol Soc Am; Sigma Xi. *Res:* Forest ecology; preparation of computer data base of all trees for 20 acres of Donaldsons Woods, Mitchell, Indiana, 20 year follow up, 1974 and 30 year follow up, 1984; application of SPSS to ecological and environmental research problems. *Mailing Add:* RR 1 Box 2344 Newfane VT 05345

BARTON, JANICE SWEENY, b Trenton, NJ, Mar 22, 39; m 67. BIOPHYSICAL CHEMISTRY, PROTEINS & MICROTUBULES. *Educ:* Butler Univ, BS, 62; Fla State Univ, PhD(chem), 70. *Prof Exp:* Assoc res chemist, Eli Lilly & Co, 62-65; NIH fel, Johns Hopkins Univ, 70-72; asst prof chem, ETex State Univ, 72-78; asst prof chem, Tex Woman's Univ, 78-82; assoc prof, 82-87, PROF CHEM, WASHBURN UNIV, 88- *Concurrent Pos:* Res grants, NSF, 75 & 77-79, Petrol Res Fund, 84-86 & NIH, 85-; coordr, Meeting in Miniature, Dallas-Fort Worth, Am Chem Soc, 81; mem, Comt Educ Health Prof, Am Chem Soc, 84-89; coun, Kansas Acad Sci, 87-90; mem, NSF Rev Panel, Undergrad Fac Enhancement, 90; pres-elect, Kans Acad Sci, 91. *Mem:* Am Chem Soc; Biophys Soc; Neurochem Soc; AAAS; Sigma Xi. *Res:* The role and properties of proteins that undergo self-assembly into functional cellular macrostructures; the mechanism of tubulin assembly into microtubules; role of the nucleotide binding site of tubulin in assembly; characterization of tubulin's nucleotide site; nuclear magnetic resonance studies of nucleotide binding to tubulin. *Mailing Add:* 3401 SW Oak Pkwy Topeka KS 66614

BARTON, JOHN R, US citizen. INSTRUMENTATION. *Educ:* Ga Inst Technol, BS, 49. *Prof Exp:* Physicist flight res lab, Off Air Res, US Air Force, 51-53; physicist, Statham Labs, 53-54; physicist, Southwest Res Inst, 49-51 & 54-59, sr res engr dept electronics & elec eng, 59-61, mgr nondestructive eval, 61-68, dir, 69-72, tech vpres & dir, 72-74, vpres instrumentation res div, 75-85; CONSULT. *Mem:* Sigma Xi; fel Am Soc Nondestructive Testing. *Res:* New methods of nondestructive material evaluation; ultrasonics; eddy current; x-ray; gamma backscatter; new magnetic field perturbation methods. *Mailing Add:* HC 53 Box 3221 Bulverde TX 78163

BARTON, KENNETH RAY, b Elberton, Ga, July 12, 36; m 63; c 2. TEXTILE CHEMISTRY, SYNTHETIC FIBERS. *Educ:* Wofford Col, BS, 58; Univ Tenn, MS, 60, PhD(chem), 63. *Prof Exp:* Res chemist, Tenn Eastman Co, 64-66, sr res chemist, 66-75, res assoc, 76-87, tech assoc, Eastman Chem Div, 87-89, DEVELOP ASSOC, EASTEK INKS, EASTMAN KODAK CO, 90- *Mem:* Am Chem Soc; Am Asn Textile Chemists & Colorists; Am Asn Textile Technol. *Res:* Textile applications of water-dispersible polyesters; thermal bonding of non-woven fabrics; synthetic fiber lubricants, adhesives, coatings, textile sizes; aqueous inks. *Mailing Add:* 201 Claymore Dr Kingsport TN 37663

BARTON, LARRY LUMIR, b West Point, Nebr, May 13, 40; m 66; c 2. MICROBIOLOGY, BIOCHEMISTRY. *Educ:* Univ Nebr, BSc, 62, MS, 66, PhD(microbiol, biochem), 69. *Prof Exp:* Res assoc biochem, Univ Ga, 69-71; asst prof microbiol, Dept Pathobiol, Sch Health & Hyg, Johns Hopkins Univ, 71-72; asst prof, 72-76, ASSOC PROF MICROBIOL, DEPT BIOL, UNIV NMEX, 77- *Concurrent Pos:* Assoc biochemist, Leonard Wood Mem Leprosy Found, 71-72; prin investr, NMex Energy Res Inst, 74-77, US Dept Interior through NMex Water Resources Res Inst, 74-79, NSF, 77-78, NIH, 79-85 & Dept Energy through Waste Mgr Educ Res Consortium, 91-93. *Mem:* Am Soc Microbiol; Sigma Xi; AAAS. *Res:* Metabolism of sulfur by microorganisms, especially energetics of sulfide formations; biotechnology; interaction of soil organic acids and microorganisms; metal metabolism. *Mailing Add:* Dept Biol Albuquerque NM 87131

BARTON, LAWRENCE, b Preston, Eng, Aug 5, 38; m 64; c 3. INORGANIC CHEMISTRY. *Educ:* Univ Liverpool, BSc, 61, PhD(chem), 64. *Prof Exp:* Res assoc chem, Cornell Univ, 64-66; from asst prof to assoc prof, 66-86, PROF CHEM, UNIV MO-ST LOUIS, 86-, CHMN DEPT, 80- *Concurrent Pos:* Petrol Res Fund award, 67-69; NSF award, 69-72; sr res fel, Explosives Res & Develop Estab, Waltham, Abbey, Eng, 70-71; vis assoc prof, Wash Univ, 77 & Ohio State Univ, 77-78; chmn, St Louis Sect, Am Chem Soc, 80. *Mem:* Am Chem Soc; Royal Soc Chem; Am Asn Univ Prof. *Res:* Lower boranes and related compounds; boron oxygen ring systems, metalloboranes; organometallic chemistry; organoboranes and mass spectometry. *Mailing Add:* Dept Chem Univ Mo-St Louis St Louis MO 63121

BARTON, MARK Q, b Kansas City, Mo, June 5, 28; m 54; c 2. PHYSICS. *Educ:* Cent Methodist Col, AB, 50; Univ Ill, PhD(physics), 56. *Prof Exp:* Asst physics, Univ Ill, 50-55; physicist, Brookhaven Nat Lab, 56-74, chmn accelerator dept, 74-78, tech dir Isabelle proj, 78-79, dep chmn, Nat Synchrotron Light Source, 84-86; RES STAFF MEM, THOMAS J WATSON RES CTR, IBM, 89- *Mem:* Fel Am Phys Soc; fel Inst Elec & Electronics Engrs. *Res:* Intensity limitations of particle accelerators; accelerator design and development. *Mailing Add:* Thomas J Watson Res Ctr PO Box 218 Yorktown Heights NY 10598

BARTON, PAUL, b Heckscherville, Pa, July 9, 36; m 57; c 5. PETROLEUM. *Educ:* Pa State Univ, BS, 57, MS, 60, PhD(chem eng), 63. *Prof Exp:* From res asst to res assoc, Pa State Univ, 57-68, sr tech specialist, Penntap, 85-89, ASST PROF CHEM ENG, PA STATE UNIV, 68- *Concurrent Pos:* Consult, Am Aniline Prod, Inc, 66-67, Westvaco, 68, Union Carbide Corp, 70-71, Ventron Corp, 74-75, Argonne Nat Lab, 75, C-COR Electronics, Inc, 76-77, Baeuerle & Morris, Inc, 78-81, A M Todd Co, 80-81, 87-91, Leaf Protein Concentrate, Inc, 82-83, IGI Fermentation, Inc, 84, David Michael & Co, 84, Wm Wrigley Jr Co, 90, McCormick & Co, 90, Siemens Automotive, 90; vis scientist, Argonne Nat Lab, 71-72; sr principle engr, Air Products, 84-85. *Mem:* Am Inst Chem Engs; Am Chem Soc. *Res:* Phase equilibria; distillation; liquid extraction; dewaxing; supercritical extraction; pilot plant design; separational processes. *Mailing Add:* 132B Fenske Lab Pa State Univ University Park PA 16802

BARTON, PAUL BOOTH, JR, b New York, NY, Sept 30, 30; m 55; c 2. GEOLOGY, MINERALOGY-PETROLOGY. *Educ:* Pa State Univ, BS, 52; Columbia Univ, AM, 54, PhD(geol), 55. *Prof Exp:* GEOLOGIST, US GEOL SURV, 55- *Concurrent Pos:* Chmn, Geol Sect, Nat Acad Sci, 85-88. *Honors & Awards:* Roebling Medal, Mineral Soc Am, 84. *Mem:* Nat Acad Sci; fel Mineral Soc Am (pres, 86); fel Geol Soc Am; Geochem Soc; Soc Econ Geol (pres, 79); Mineral Asn Can. *Res:* Genesis of mineral deposits; chemical and physical nature of ore forming fluids; phase relations between minerals; thermodynamic properties of minerals; long range availability of resources; geochemistry. *Mailing Add:* US Geol Surv 959 National Ctr Reston VA 22092

BARTON, RANDOLPH, JR, b Wilmington, Del, Aug 15, 41; m 63; c 1. PHYSICAL CHEMISTRY, CRYSTALLOGRAPHY. *Educ:* Princeton Univ, AB, 63; Johns Hopkins Univ, MA, 65, PhD(phys chem), 68. *Prof Exp:* Res chemist, 67-80, sr res chemist, 80-88, RES ASSOC, TEXTILE FIBERS DEPT, E I DU PONT DE NEMOURS & CO, INC, 88- *Mem:* Am Crystallog Asn; Am Phys Soc; Am Chem Soc. *Res:* Polymer structure and morphology. *Mailing Add:* Exp Sta PO Box 80302 E I du Pont de Nemours & Co Inc Wilmington DE 19880-0302

BARTON, RICHARD DONALD, b Marlboro, Mass, Nov 20, 36; US & Can citizen; m 58; c 5. NUCLEAR PHYSICS. *Educ:* McGill Univ, BSc, 59, MSc, 61, PhD(nuclear physics), 63. *Prof Exp:* Asst prof physics, Loyola Col, Que, 63-64 & McGill Univ, 64-65; asst res officer, Nat Res Coun Can, 65-67; from asst prof to assoc prof physics, Carleton Univ, 67-83. *Concurrent Pos:* Nat Res

Coun Can grants in aid, 63-65 & 67-78. *Mem:* Am Phys Soc. *Res:* Delayed proton emission; muonic x-rays; positron and nuclear lifetimes; nuclear electronics; precision testing of speeds of photons under various conditions. *Mailing Add:* 151 Holland Ave Ottawa ON K1Y 0Y2 Can

BARTON, RICHARD J, b Painesville, Ohio, Aug 2, 28; m 55; c 3. PHYSICAL CHEMISTRY, PHYSICAL METALLURGY. *Educ:* Ohio Univ, BS, 50; Iowa State Univ, PhD(chem), 56. *Prof Exp:* Sr scientist, Mat Lab, Wright Air Develop Ctr, Wright Patterson Air Force Base, Ohio, 57-61; asst prof metall, Colo Sch Mines, 61-65; ASSOC PROF PHYSICS, UNIV REGINA, 65- *Mem:* Am Chem Soc; Electrochem Soc; Am Crystallogr Asn. *Res:* Structure and thermodynamic properties of non-stoichiometric compounds; electrochemical phenomena; structure of charge transfer compounds and organometallic compounds. *Mailing Add:* Dept Physics & Astron Univ Regina Regina SK S4S 0A2 Can

BARTON, STUART SAMUEL, b Toronto, Ont, Nov 16, 22; m 56. SURFACE CHEMISTRY. *Educ:* Univ Toronto, BA, 47, MA, 49; McGill Univ, PhD(phys chem), 56. *Prof Exp:* Res chemist, Courtaulds Can Ltd, Ont, 54-56; Defense Res Bd res assoc chem, 56-62, from lectr to prof chem, 62-90, CATARHQUI RES ASSOC, ROYAL MIL COL CAN, 90- *Concurrent Pos:* Contract res & develop grants, 70- *Mem:* Fel Royal Soc Chem; Sigma Xi. *Res:* Active carbon characterization; application of calorimetry to surface and solution chemistry. *Mailing Add:* 65 Gore St Kingston ON K7L 2L4 Can

BARTON, THOMAS J, b Dallas, Tex, Nov 5, 40; m 66. ORGANIC CHEMISTRY. *Educ:* Lamar State Col, BS, 62; Univ Fla, PhD(chem), 67. *Prof Exp:* NIH fel chem, Ohio State Univ, 67; instr, 67-69, from asst prof to prof, 69-83, DISTINGUISHED PROF ORG CHEM, IOWA STATE UNIV, 84-, DIR, MAT CHEM PROG, AMES LAB, 86- *Concurrent Pos:* Nat Acad Sci exchange scientist, Soviet Union, 75; NATO collab scientist, France 76; Japan Soc Prom Sci, 81; exchange prof, Nat Ctr Sci Res, Univ Montpellier, 83; res fel, Royal Chem Soc, 83. *Honors & Awards:* Frederic Stanley Kipping Award, Am Chem Soc, 82. *Mem:* Am Chem Soc; Mat Res Soc. *Res:* Synthetic and mechanistic organosilicon chemistry; high temperature gas-phase molecular rearrangements; generation of reactive intermediates; synthesis of novel precursors for chemical vapor deposition amorphous silicon and silicon-based ceramics; synthesis of conducting organic polymers. *Mailing Add:* Iowa State Univ Ames Lab 109 Off & Lab Bldg Ames IA 50011

BARTON, WALTER E, b Oak Park, Ill, July 29, 06; m 32; c 3. PSYCHIATRY. *Educ:* Univ Ill, BS, 28, MD, 31; Am Bd Neurol & Psychiat, dipl. *Prof Exp:* Intern, West Suburban Hosp, Ill, 30-31; resident psychiat, Worcester State Hosp, 31-34, sr psychiatrist, 34-38, asst supt, 38-42; supt, Boston State Hosp, 45-63; med dir, Am Psychiat Asn, 63-74; sr physician, Vet Admin Gen Hosp, White River Junction, 74-77; assoc prof, 52-63, CLIN PROF PSYCHIAT, SCH MED, BOSTON UNIV, 64-; EMER PROF PSYCHIAT, DARTMOUTH MED SCH, 74- *Concurrent Pos:* Lectr, Simmons Col, 36-37, Smith Col, 37-42, Clark Univ, 40-42, Columbia Univ & Yale Univ, 56-59, Med Sch, Tufts Univ, 58 & Med Sch, Georgetown Univ, 63-74; mem staff neurol, Nat Hosp, Eng, 38; assoc prof, 52-63, clin prof psychiat, Sch Med, Boston Univ, 64-; dir, Am Bd Neurol & Psychiat, 63-70; consult, NIMH & NIH, mem bd trustees, Joint Comn Ment Illness & Health & Joint Comn Ment Health of Children; mem spec res projs comt, NIMH & adv comt, Neurol-Psychiat Div, Vet Admin. *Honors & Awards:* Salmon Medal, NY Acad Med, 74; Distinguished Serv Award, Am Psychiat Asn, 73, Award in Admin Psychiat, 83. *Mem:* Fel AMA; fel Am Psychiat Asn (pres, 61-62); fel Am Col Physicians; fel Am Col Psychiatists; Am Col Ment Health Adminr (pres, 80-81). *Res:* Mental health administration; ethics; forensic psychiatry; history of psychiatry. *Mailing Add:* RFD 1 Box 188 Hartland VT 05048

BARTOO, JAMES BREESE, b Swanton, Vt, July 2, 21; m 43; c 5. MATHEMATICS. *Educ:* Pa State Teachers Col, Edinboro, BS, 47; Univ Iowa, MS, 49, PhD(math), 52. *Prof Exp:* Instr math, Univ Iowa, 47-52; from asst prof to prof, 52-84, head Dept Statist, 68-69, Dean, Grad Sch, 69-84, EMER PROF MATH STATIST, PA STATE UNIV, 84- *Concurrent Pos:* Actg vpres res, Pa State Univ, 71, exec vpres & provost, 83- 84. *Mem:* AAAS; Am Statist Asn. *Res:* Mathematical statistics. *Mailing Add:* 706 Windsor Ct State College PA 16801

BARTOS, DAGMAR, b Prague, Czech, Oct 18, 29; US citizen; m 44; c 2. CLINICAL BIOCHEMISTRY. *Educ:* Charles Univ, Prague, PhD(biochem), 53. *Prof Exp:* Dir biochem, Dept Nutrit Chem, Regional Off Hyg & Epidemiol, Hradec Kralove, 53-55; asst prof biochem, Charles Univ, Prague, 55-57, assoc prof, 57-67; res assoc med, 67-70, RES ASSOC PROF PEDIAT & SURG, ORE HEALTH SCI UNIV, PORTLAND, 73- *Mem:* Sigma Xi; Western Soc Pediat Res; Clin Ligand Assay Soc. *Res:* Study of the mechanisms of sepsis; separation, purification and characterization of the toxic substances from rat and human plasma responsible for the sepsis induced metabolic abnormalities. *Mailing Add:* Univ Ore Med Sch Health Sci Ctr 3181 SW Sam Jackson Pk Rd Portland OR 97201

BARTOS, FRANTISEK, b Prague, Czech, Dec 31, 26; US citizen; m 51; c 2. CLINICAL BIOCHEMISTRY. *Educ:* Charles Univ, Prague, PhD(biochem), 53. *Prof Exp:* Dir clin biochem, Med Sch Hosp, Czech, 53-55; dir biochem lab, Dept Plastic Surg, Charles Univ, Prague, 55-60; head dept radiotoxicol, Res Inst Radiation Hyg, 61-66; res assoc med, Sch Med, Univ Ore, 66-70; res scientist clin chem, United Med Lab Inc, Ore, 71-72; RES ASSOC PROF, HEALTH SCI CTR, UNIV ORE, 72- *Mem:* Sigma Xi; Western Soc Pediat Res; Clin Liganol Assay Soc. *Res:* Development of radioimmunoassays for determination of hormones, peptides, drugs and substances of biological importance; the biochemistry of polyamines and polyamine complexes; development of radioimmunoassays for determination of polyamines and their congeneres; isolation and characterization of polymaine peptide and protein complexes in biological fluids; study of the metabolic defects and toxic factors in sepsis; development of sensitive analytical assays for determination and characterization of the toxic substances in sepsis. *Mailing Add:* Epitope Inc 8505 SW Creekside Pl Beaverton OR 97005-7108

BARTOSHUK, LINDA MAY, TASTE PSYCHOPHYSICS, SENSORY PROCESSES. *Educ:* Brown Univ, PhD(psychol), 65. *Prof Exp:* PROF SENSORY PROCESSES & NUTRIT, YALE UNIV & FEL, PIERCE FOUND, 70- *Mailing Add:* Dept Surg Otolaryngol Yale Univ Sch Med 333 Cedar St PO Box 3333 New Haven CT 06510

BARTOSIK, ALEXANDER MICHAEL, b Warsaw, Poland, Sept 14, 24; US citizen; m 50. AGRICULTURAL ENGINEERING. *Educ:* Akademia Rolnicza, Poland, MS, 50, PhD(agr technol), 56. *Prof Exp:* Prof agr eng, Univ Mosul, Iraq, 71-74, Am Univ Beirut, 74-75; res assoc, Univ Hawaii, 75-77; prof, Univ Federal, Brazil, 77-79; farm coordr, Hawaiian Community, Waimanalo, 80; consult, Trop Agr Adv Serv, Honolulu, 80-81; prof agr eng, Am Univ, Beirut, 81-82; RETIRED. *Concurrent Pos:* Assoc researcher, Indust Bastifiber Inst, Poznan, Poland, 51-59; asst prof, Col Eng, Poznan, Poland, 59-60 & 61-71; vis prof, Univ Ill, Urbana-Champaign, 60-61. *Mem:* Am Soc Agr Engrs. *Res:* Development of international agriculture by promoting agricultural engineering or mechanization; author or coauthor of 100 publications. *Mailing Add:* 2440 Kuhio Ave No 1003 Honolulu HI 96815-3349

BARTOSIK, DELPHINE, b Chicago, Ill, Nov 22, 37. OBSTETRICS & GYNECOLOGY. *Educ:* Univ Chicago, BS, 57; Women's Med Col, MD, 61. *Prof Exp:* DIR, DIV REPRODUCTIVE ENDOCRINOL, DEPT OBSTET & GYNEC, HAHNEMANN UNIV, 78- *Mem:* Endocrine Soc; Am Physiol Soc; Soc Endocrinol Great Britain; Am Fertil Soc. *Mailing Add:* Dept Obstet & Gynec Hahnemann Univ 245 N Broad St Suite 300 Philadelphia PA 19107

BARTOVICS, ALBERT, b Roosevelt, NY, Dec 1, 16; m 44; c 2. POLYMER CHEMISTRY. *Educ:* Polytech Univ, BS, 37, MS, 39, PhD(chem), 43. *Prof Exp:* Asst chem, Polytech Univ, 37-39; analyst, Shellac Res Bur, 39-40; res chemist, Firestone Tire & Rubber Co, 43-50 & Armstrong Cork Co, 50-52; from develop chemist to prod develop mgr, Borg Fabric Div, Amphenol-Borg Electronics Corp, 52-64; sr res chemist, Textile Res Lab, E I du Pont de Nemours & Co, 64-79; RETIRED. *Mem:* Am Chem Soc; Am Inst Chemists. *Res:* Viscometric and osmotic molecular weight studies; chlorination of synthetic polymers; polyvinyl chloride and related copolymers for various applications; development of pile fabrics from natural and synthetic fibers; end-use research on synthetic fibers. *Mailing Add:* 707 Potter Dr Cedarcroft Kennett Square PA 19348

BARTRAM, JAMES F(RANKLIN), b Brooklyn, NY, Mar 18, 26; m 47; c 2. SIGNAL PROCESSING. *Educ:* Yale Univ, BE, 50, MEng, 51. *Prof Exp:* Mem tech staff tel commun, Bell Tel Labs, Inc, 51-56; sr assoc weapons systs studies, G C Dewey & Co, Inc, 56-59; fel engr, air arm div, Westinghouse Elec Corp, 59-61; prin engr, submarine signal div, Raytheon Co, 61-63; mem tech staff commun satellites, Int Tel & Tel Co, 63-64; CONSULT ENGR, SUBMARINE SIGNAL DIV, RAYTHEON CO, 64- *Concurrent Pos:* Assoc ed, Acoust Signal Processing, J Acoust Soc Am, 80-83. *Mem:* Sr mem Inst Elec & Electronics Engrs; fel Acoust Soc Am. *Res:* Space-time signal processing in communications, radar and sonar fields. *Mailing Add:* 94 Kane Ave Middletown RI 02840

BARTRAM, RALPH HERBERT, b New York, NY, Aug 16, 29; m 53; c 2. SOLID STATE PHYSICS. *Educ:* NY Univ, BA, 53, MS, 56, PhD(physics), 60. *Prof Exp:* Adv res physicist, Gen Tel & Electronics Labs, Inc, NY, 53-61; from asst prof to assoc prof, 61-71, PROF PHYSICS, 71-, PHYSICS DEPT HEAD, UNIV CONN, 86- *Concurrent Pos:* Guest assoc physicist, Brookhaven Nat Lab, 66-71, consult, 71-85; tech adv, Gen Telephone & Electronics 61-, US Army, 66-70, Am Optical Co, 66-78, Timex Corp, 81-82 & Polaroid Corp, 87-88; res assoc, Theoret Physics Div, Atomic Energy Res Estab, Harwell, Eng, 67-68; vis prof, Clarendon Lab, Oxford Univ, Eng, 78. *Mem:* Fel Am Phys Soc; Optical Soc Am. *Res:* Theory of color centers and radiation damage in solids; theory of radiationless transitions in solids. *Mailing Add:* Dept Physics Univ Conn Storrs CT 06268

BARTSCH, GLENN EMIL, b Mankato, Minn, Sept 11, 28. BIOSTATISTICS. *Educ:* Univ Minn, BS, 50, MA, 51; Johns Hopkins Univ, ScD, 57. *Prof Exp:* Asst, Univ Minn, 50-51 & Johns Hopkins Univ, 51-54; biostatistician, Army Chem Ctr, US Dept Army, 54-56; asst, Johns Hopkins Univ, 56-57, res assoc, 57-58; NIH fel, 58-59; asst prof biostatist, Western Reserve Univ, 60-65; ASSOC PROF BIOSTATIST, UNIV MINN, MINNEAPOLIS, 65- *Mem:* Am Statist Asn; Sigma Xi. *Res:* Effects of non-normality upon common statistical tests; statistical techniques as applied to biological research. *Mailing Add:* 2221 University Ave SE Apt 200 Minneapolis MN 55414-3080

BARTSCH, RICHARD ALLEN, b Portland, Ore, June 7, 40; m 66; c 2. ORGANIC CHEMISTRY. *Educ:* Ore State Univ, BA, 62, MS, 63; Brown Univ, PhD(org chem), 67. *Prof Exp:* Instr chem, Univ Calif, Santa Cruz, 66-67; NATO fel, Univ Wurzburg, 67-68; asst prof chem, Wash State Univ, 68-73; asst prog adminr, Petrol Res Fund, 73-74; from assoc prof to prof, 74-88, PAUL WHITFIELD HORN PROF CHEM, TEX TECH UNIV, 88- *Concurrent Pos:* Nat Acad Sci vis scientist, Czechoslovakia, 78. *Mem:* Am Chem Soc; Royal Soc Chem; Sigma Xi. *Res:* Organic reaction mechanisms and synthesis; elimination reactions; crown ethers and other neutral cation carriers; chromogenic and fluorogenic ligands; synthetic polymeric ionomer membranes; metal ion extraction and transport. *Mailing Add:* Dept Chem & Biochem Tex Tech Univ Lubbock TX 79409-1061

BARTSCHMID, BETTY RAINS, b Shreveport, La, Dec 27, 49. ANALYTICAL CHEMISTRY. *Educ:* Univ Tex, Austin, BS, 70; Univ Houston, PhD(anal chem), 74. *Prof Exp:* Fel anal chem, Colo State Univ, 74-75; ASST PROF ANALYTICAL CHEM, VA POLYTECH INST & STATE UNIV, 75- *Mem:* Am Chem Soc; Optical Soc Am; Soc Appl Spectros. *Res:* Atomic absorption, fluorescense and emission applied to practical problems of trace metal analysis; development of a versatile, automatic background correcting atomic fluorescence system; molecular emission techniques to determine non-metals. *Mailing Add:* 104 Maywood St Va Polytech Inst & State Univ Blacksburg VA 24060

BARTUS, RAYMOND T, b Chicago, Ill, May 19, 47; m 67; c 2. NEUROSCIENCE, GERONTOLOGY. *Educ:* Calif State Col, BA, 68; NC State Univ, MS, 70, PhD(physiol psychol), 72. *Prof Exp:* NIH res asst, NC State Univ, 68-69, NASA fel, 70-72; assoc, Naval Med Res Lab, 72-73; sr scientist, Dept Cent Nervous Syst, Parke-Davis Res Labs, Warner-Lambert, 73-78; GROUP LEADER, DEPT CENT NERVOUS SYST, LEDERLE LABS, DIV AM CYANAMID, 78-, DIR, GERIATRIC RES PROG, 80- *Concurrent Pos:* Lab & teaching asst, Calif State Col, 66-68; adj asst prof, Conn Col, 73; res affil, Tulane Univ, 78-; consult, Manhattan Vet Admin Hosp, 80-; ed-in-chief, Neurobiol Aging, 80-; adj prof, Sch Med, New York Univ, 80- *Mem:* Soc Neurosci; Am Aging Asn; Geront Soc Am. *Res:* Neurological and neurochemical factors involved with behavior, and the effects of drugs on these variables; deterioration that occurs during aging; development of animals models of neurodegenerative disorders; development of new drugs to treat neurological problems. *Mailing Add:* Dept Physics NY Univ Sch Med 550 First Ave New York NY 10016

BARTUSKA, DORIS G, b Nanticoke, Pa; c 6. ENDOCRINOLOGY, METABOLISM. *Educ:* Bucknell Univ, BS(biology), 49; Woman's Med Col Pa, MD, 54. *Prof Exp:* Asst prof med & clin path, Women's Med Col Pa, 66-71; assoc prof path, Med Col Pa, 71, dir div Endocrinol & Metab, 73-; assoc dean curriculum, 74-76, actg dir, Ctr Women Med, 76-77; prof of med, Med Col Pa, 77; CONSULT ENDOCRINOL, PA GERIAT CTR, 82. *Concurrent Pos:* Consult endocrinology, Vet Admin Hosp, 70- *Mem:* Am Thyroid Asn; Endocrine Soc; Asn Women Sci; fel Am Col Physicians; Am Fed Clin Res; AAAS; Am Col Physicians; Am Med Asn. *Res:* Behavioral and genetic aspects of the endocrine diseases; author or coauthor of over 25 publications; influence of progesterone on human thyroid function; estrogen assay of malignant tumor tissue of the breast, effects of testosterone and estrogen on the mammary glands of iodine deficient rats; endocrinology of Aging-Osteoporosis and altered thyroid function. *Mailing Add:* Hosp Med Col Penn 3300 Henry Ave Philadelphia PA 19129

BARTZ, JERRY A, b Fond du Lac, Wis, Mar 4, 42; m 68; c 2. PLANT PATHOLOGY. *Educ:* Univ Wis-Madison, BS, 64, MS, 66, PhD(plant path), 68. *Prof Exp:* Res fel, Environ Sci Prog, Univ Calif, Riverside, 68-69; asst prof, 69-75, ASSOC PROF PLANT PATH, UNIV FLA, 75- *Concurrent Pos:* vis prof, Univ Wisc, 82; sr ed, Plant Dis, 86- *Mem:* Am Phytopath Soc; Int Soc Plant Path. *Res:* Mode of action of fungicides; treatment and prevention of post-harvest diseases of vegetables. *Mailing Add:* Dept Plant Path Univ Fla Inst Food & Agr Sci 1453 Field Hall Gainesville FL 32611

BARTZ, WARREN F(REDERICK), b Sheldon, Iowa, Mar 18, 13; m 40; c 2. CHEMICAL ENGINEERING. *Educ:* Iowa State Col, BS, 35; Univ NC, PhD(chem), 39. *Prof Exp:* Asst, Univ NC, 35-39; chem engr, Barrett Div, Allied Chem Corp, Pa, 40-43; chemist, Rohm and Haas Co, 43-75; RETIRED. *Mem:* Am Chem Soc. *Res:* Resins; phenolics; urea formaldehyde; alkyds; ion exchange. *Mailing Add:* 646 Pine Tree Rd Jenkintown PA 19046

BARUCH, JORDAN J(AY), b New York, NY, Aug 21, 23; m 44; c 3. INSTRUMENTATION, COMPUTER SCIENCE. *Educ:* Mass Inst Technol, BS & MS, 48, ScD, 50. *Hon Degrees:* LLD, Franklin Pierce Law Ctr, 81. *Prof Exp:* Vpres, Bolt Beranek & Newman, Inc, 53-66; gen mgr medinet dept, Gen Elec Co, 66-68; pres, Educom, 68-70; independent consult, 70-71; lectr bus admin, Harvard Univ, 71-74; prof, Tuck Sch Bus & Thayer Sch Eng, Dartmouth Col, 74-77; asst sect sci & technol, US Dept Com, 77-81; CONSULT TECHNOL MGT, JORDAN BARUCH ASSOCS, 81- *Concurrent Pos:* Asst prof, Mass Inst Technol, 51-55, lectr, 55-71. *Mem:* Nat Acad Eng; fel NY Acad Sci; fel Acoust Soc Am; fel Inst Elec & Electronics Engrs; fel Am Acad Arts & Sci. *Res:* Computers in communication; acoustics; technology management. *Mailing Add:* Jordan Baruch Assocs 1200 18th St NW Washington DC 20036

BARUCH, MARJORY JEAN, b Boston, Mass, May 8, 51; m 79; c 1. ALGEBRA. *Educ:* Univ Chicago, BA, 72; Univ Pa, MA, 75, PhD(math), 79. *Prof Exp:* Res assoc math, Mich State Univ, 79-81; ASST PROF MATH, HAMILTON COL, 81- *Mem:* Sigma Xi. *Mailing Add:* 106 Cammot Lane Fayetteville NY 13066-1426

BARUCH, SULAMITA B, b Medellin, Colombia, Jan 6, 36. PHYSIOLOGY, MEDICINE. *Educ:* Univ Valle, Colombia, MD, 59; Cornell Univ, PhD(physiol), 63. *Prof Exp:* Intern, Univ Hosp, Sch Med, Univ Valle, Colombia, 57-58; instr physiol sci, Sch Med, 59-60, asst prof, 63-65; asst prof physiol, Med Col, Cornell Univ, 65-70, assoc prof, 70-80; ASSOC ATTEND EMERGENCY RM DIR, DEPT MED, LAGUARDIA HOSP. *Concurrent Pos:* Estab investr, Am Heart Asn, 67-72; adj assoc prof physiol, Med Col, Cornell Univ, 80- *Mem:* Am Soc Nephrology; Int Soc Nephrology; Am Physiol Soc. *Res:* Metabolic activity of the kidney as influenced by acid-base balance of the body; excretion of ammonia by the kidney; transport in isolated kidney membranes; renal handling of organic acids. *Mailing Add:* Emergency Dept LaGuardia Hosp 102-01 66th Rd Forest Hills NY 11375

BARUS, CARL, electrical engineering; deceased, see previous edition for last biography

BARUSCH, MAURICE R, b Yokohama, Japan, Sept 21, 19; m 42; c 2. CHEMISTRY. *Educ:* Stanford Univ, AB, 40, MA, 41, PhD(org chem), 44. *Prof Exp:* Res chemist, Calif Res Corp, 43-52, sr res chemist, 52-57, group supvr, 57-59, res assoc, 59-61, sr res assoc, 61-63, supvr, Fuel Additives Sect, 63-67, mgr, Fuel Additives Div, 67-69, mgr, Grease & Indust Oils Div, 69-73, mgr, Lubricating Oil Additives Div, 73-79, vpres, Lubricants Res Dept, Chevron Res Co, 79-82; RETIRED. *Honors & Awards:* Midgley Award, 77. *Mem:* Am Chem Soc; Combustion Inst; Soc Automotive Engrs. *Res:* Combustion of hydrocarbons; fuel and lubricating oil additives. *Mailing Add:* 651 Cypress Point Rd Richmond CA 94801

BARUT, ASIM ORHAN, b Turkey, June 24, 26; m 54; c 2. THEORETICAL PHYSICS, ELEMENTARY PARTICLES. *Educ:* Swiss Fed Inst Technol, dipl, 49, PhD, 52. *Hon Degrees:* Dr Sci hc, Turkey, 82, 87. *Prof Exp:* Asst prof, Swiss Fed Inst Technol, 51-53; fel physics, Univ Chicago, 53-54; asst prof, Reed Col, 54-55; Nat Res Coun Can fel, 55-56; from asst prof to assoc prof, Syracuse Univ, 56-61; mem staff, Lawrence Radiation Lab & Univ Calif, Berkeley, 61-62; PROF THEORET PHYSICS, UNIV COLO, BOULDER, 62- *Concurrent Pos:* Sr staff mem, Int Ctr Theoret Physics, Trieste, Italy, 64-65, 68-69 & 72-73 & 86-87; dir, NATO Advan Study Inst, 67, 70, 72, 77, 81, 83, 84 & 89; fac res fels, 69, 72, 78 & 84; Erskin fel, Univ Canterbury, 71 & Univ Frankfurt, 81; vis prof, Warsaw, 72, Dijon, 73, Stockholm, 73, Santiago, 74, Munich, 74-75, Mex, 75 & Caracas, 79 & Geneva, 84- 85; UN vis prof, Istanbul, 78; ed, Found Physics, Found Physics Letters, Hadronic J & Reports in Math Physics; fac res lectr, 83. *Honors & Awards:* Medal Sci, Turkey, 82; Alexander von Humboldt Found Award, Bonn, Germany, 74, 75 & 84. *Mem:* AAAS; fel Am Phys Soc; Europ Phys Soc; Swiss Phys Soc; Asn Math Physics; Turkish Phys Soc; NY Acad Sci. *Res:* Quantum theory of fields and particles; elementary particles; statistical mechanics; mathematical physics. *Mailing Add:* Dept Physics Univ Colo Boulder CO 80309

BARWICK, STEVEN WILLIAM, b Neenah, WI, Sept 16, 59; m 84. RADIOACTIVITY, HIGH ENERGY NUCLEAR PHYSICS. *Educ:* Mass Inst Technol, BA, 81; Univ Calif, Berkeley, MA, 83 & PhD(physics), 86. *Prof Exp:* ASSOC RES PHYSICIST, UNIV CALIF, BERKELEY, 86-88. *Mem:* Am Phys Soc. *Mailing Add:* Dept Physics Univ Calif PS 2 Irvine CA 92717

BAR-YAM, ZVI H, b Krakow, Poland, Apr 12, 28; US & Israeli citizen; m 51; c 3. ELEMENTARY PARTICLE PHYSICS, HIGH ENERGY PHYSICS. *Educ:* Mass Inst Technol, BS, 58, MS, 59, PhD(physics), 63. *Prof Exp:* Res asst high energy physics, Synchrotron Lab, Mass Inst Technol, 57-63, staff mem, Lab Nuclear Sci, 63-64; from assoc prof to prof physics, Southeastern Mass Univ, 64-70; assoc prof, Israel Inst Technol, 70-72 & 77-78; COMMONWEALTH PROF PHYSICS, SOUTHEASTERN MASS UNIV, 72- *Concurrent Pos:* Vis scientist, Div Sponsored Res, Mass Inst Technol, 64-65; prin investr, Cambridge Electron Accelerator, 65-71; vis scientist, Brookhaven Nat Lab, 73-, prin investr, 75- *Mem:* Am Phys Soc; Am Asn Physics Teachers. *Res:* Strong and electromagnetic interactions at high energies. *Mailing Add:* 19 Jordan Rd Brookline MA 02146

BAR-YISHAY, EPHRAIM, b Hertzlia, Israel, June 28, 48; m 71; c 3. RESPIRATORY PHYSIOLOGY. *Educ:* Newark Col Eng, BSc, 73; Univ Minn, PhD(biomed eng), 78. *Prof Exp:* Jr engr, Eng Develop Ctr, Johnson & Johnson Co, New Brunswick, NJ, 72; res asst, Dept Mech Eng, NJ Inst Technol, Newark, 72; res asst biomed eng, Univ Minn, 72-78; head, Pulmonary Function Lab & Pediat Pulmonary Res Lab, Hadassah Univ Hosp, Mt Scopus, Jerusalem, Israel, 78-90; Park B Francis fel pulmonary res, Dept Thoracic Dis & Physiol, Mayo Clin, Rochester, Minn, 82-83; asst prof physiol, Mayo Med Sch, Univ Minn, Rochester, 83; ASST PROF PEDIAT, FAC MED, OHIO STATE UNIV, COLUMBUS, 89-; RES SCHOLAR, DIV PULMONARY MED, CHILDREN'S HOSP, COLUMBUS, OHIO, 89- *Concurrent Pos:* Sect leader, Task Force Infant Pulmonary Function Testing, Am Thoracic Soc, 90-; dir, Pulmonary Function Labs, Hadassah Univ Hosps, Hebrew Univ, Jerusalem, Israel, 91. *Mem:* Am Thoracic Soc; Am Physiol Soc; Int Biomed Eng Fedn. *Res:* Respiratory physiology and mechanics; lung function testing in infants; whole body plethysmography; airway hyperreactivity and bronchial provocation tests; exercise-induced asthma; pulmonary Functuib Kabiratirt-testing and instrumentation; respiratory muscle fatigue and EMG. *Mailing Add:* Dept Pulmonary Med Children's Hosp 700 Children's Dr Columbus OH 43205

BARZEL, URIEL S, b Jerusalem, Israel, July 26, 29; US & Israeli citizen; m 52; c 3. ENDOCRINOLOGY & METABOLISM, GERIATRICS. *Educ:* Hebrew Univ, MA, 51; Columbia Univ, BSc, 54, MD, 58. *Prof Exp:* Intern, Montefiore Hosp, Bronx, NY, 58-59, trainee, Metab Ward, 59-61; sr resident & chief resident, Beilinson Hosp, Israel, 62-64; instr med, Tel Aviv Univ, 63-66; clin instr, 66-69, clin assoc, 69-71, from asst prof to assoc prof, 71-83, PROF MED, ALBERT EINSTEIN COL MED, 83- *Concurrent Pos:* Attend physician endocrinol & metab, Montefiore Med Ctr; physician & consult, Montefiore Med Group; book rev ed, J Am Geriat Soc. *Mem:* Fel Geront Soc Am; fel Am Geriat Soc; Endocrine Soc; Am Soc Bone & Mineral Res; NY Acad Sci. *Res:* Endocrinological and metabolic problems of aging; osteoporosis; disorders of calcium metabolisn in the elderly; abnormalities of calcium and thyroid metabolism. *Mailing Add:* Dept Med Montefiore Med Ctr 111 E 210th St Bronx NY 10467

BASAN, PAUL BRADLEY, sedimentology, paleoecology, for more information see previous edition

BASART, JOHN PHILIP, b Des Moines, Iowa, Feb 26, 38; m 60; c 2. RADIO ASTRONOMY, NONDESTRUCTIVE EVALUATION. *Educ:* Iowa State Univ, BS, 62, MS, 63, PhD(elec eng), 67. *Prof Exp:* Instr electronics technol, Iowa State Univ, 64-67; res assoc radio astron, Nat Radio Astron Observ, Va, 67-69; asst prof, 69-73, assoc prof radio astron & elec eng, 73-80, PROF ELEC ENG, IOWA STATE UNIV, 80- *Concurrent Pos:* Syst scientist, Nat Astron Observ, NMex, 79-81. *Mem:* Am Astron Soc; Inst Elec & Electronics Engrs; Am Geophys Union; Int Astron Union; Royal Astron Soc. *Res:* Radio astronomy observations of jets in radio galaxies and of planetay nebulas; improvement of radio astronomy maps by development of new signal processing procedures; signal processing; radio galaxies; processing x-ray images of materials. *Mailing Add:* Dept Elec & Comput Eng Iowa State Univ Ames IA 50011

BASAVAIAH, SURYADEVARA, b Munnalur, India, Dec 12, 38; m 69; c 1. SOLID STATE PHYSICS, ELECTRONICS. *Educ:* Madras Univ, BE, 60; Indian Inst Sci, Bangalore, ME, 61; Princeton Univ, MSE, 63; Univ Pa, PhD(elec eng), 68. *Prof Exp:* Sr res fel transistor sect, Nat Phys Lab, India, 61-62; instr elec eng, Drexel Univ, 63-66; RES STAFF MEM APPL RES, WATSON RES CTR, IBM CORP, NY, 68- *Mem:* Am Phys Soc; Inst Elec & Electronics Engrs. *Res:* Josephson tunnel junctions; superconductivity; thin films; semiconductor devices; active electronic circuits. *Mailing Add:* IBM Watson Res Ctr PO Box 218 Yorktown Heights NY 10598

BASBAUM, CAROL BETH, IMMUNOCHEMISTRY MUCINS. *Educ:* Univ Pa, PhD(anat), 72. *Prof Exp:* ASSOC PROF HISTOL, UNIV CALIF, SAN FRANCISCO, 86- *Res:* Tracheal secretion regulation. *Mailing Add:* Univ Calif 3rd St & Parnassus Ave San Francisco CA 94143

BASCH, JAY JUSTIN, b Philadelphia, Pa, May 23, 32; m 57; c 2. BIOLOGICAL CHEMISTRY. *Educ:* Univ Pa, BA, 56; Drexel Inst Technol, MS, 60; Temple Univ, PhD(chem), 68. *Prof Exp:* RES CHEMIST, EASTERN REGIONAL RES CTR, AGR RES, USDA, 56- *Mem:* Am Chem Soc; Am Dairy Sci Asn. *Res:* Preparation, isolation and characterization of the glycoproteins from milk fat globule membranes; detection of adulterated buttermilk; the factors which influence denaturation and reactivation of alkaline phosphatase of butter and cream; special technique for gel electrophoresis. *Mailing Add:* Eastern Regional Res Ctr USDA 600 E Mermaid Lane Philadelphia PA 19118

BASCH, PAUL FREDERICK, b Vienna, Austria, Nov 10, 33; nat US; m 66; c 2. PARASITOLOGY, PUBLIC HEALTH. *Educ:* City Col New York, 54; Univ Mich, MS, 56, PhD(zool), 58; Univ Calif, Berkeley, MPH, 67. *Prof Exp:* Asst prof biol, Kans State Teachers Col, 59-62; from asst res zoologist to assoc res zoologist, Hooper Found, Univ Calif, San Francisco, 62-70; assoc prof, 70-83, PROF INT HEALTH, STANFORD UNIV, 83- *Concurrent Pos:* Res zoologist, Inst Med Res, Malaysia, 63-65 & 69-70; consult parasitol, Calif State Dept Pub Health; mem panel parasitic dis, US-Japan Coop Med Sci Prog, NIH, 74-78 & Trop Med & Parasitol Study Sect, 77-81; dep chmn, US Schistosomiasis Deleg to People's Repub China, 75; mem biomed adv panel, Comt Scholarly Commun with People's Repub China, 76-79, distinguished scholar, 84. *Mem:* Am Soc Parasitol; Am Soc Trop Med & Hyg; Nat Coun Int Health; Royal Soc Trop Med & Hyg; Am Pub Health Asn. *Res:* Schistosomiasis; epidemiology and control of parasites; international health; developing countries. *Mailing Add:* Dept Health Res & Policy Stanford Univ Sch Med Stanford CA 94305-5092

BASCH, ROSS S, b Sept 6, 37; m 61; c 3. DEVELOPMENTAL IMMUNOLOGY, HEMATOLOGY. *Educ:* NY Univ, MD, 61. *Prof Exp:* From asst prof to assoc prof, 66-82, PROF PATH, SCH MED, NY UNIV, 82- *Concurrent Pos:* Lab med officer, Dept Immuno Chem, Walter Reed Army Inst Res, US Army, 62-64; guest investr, Biochem Genetics Lab, Rockefeller Univ, 64-66; consult, Ortho Pharmaceut Co, 75-, Lescarden, Inc, 81-83; vis fel, Dept Zool, Univ Col, London, 76-77; asst pathologist, Bellevue Hosp, New York, 77-; dir, Instrumentation Unit, Dept Path, NY Univ Sch Med, 80-, prog dir, Immunol Course, 81-; dir, Molecular Immunol Prog, Kaplan Cancer Ctr. *Mem:* Am Asn Immunologists; Am Asn Pathologists; Am Soc Flow Cytometry; Int Soc Develop & Comp Immunol; Soc Develop Biol; NY Acad Sci; Am Soc Microbiol; Soc Anal Cytol. *Res:* Developmental immunology; developmental hematology; maintenance of the differentiated state. *Mailing Add:* Dept Path Sch Med NY Univ 550 First Ave New York NY 10016

BASCO, N, b London, Eng, July 13, 29; m 56; c 2. PHOTOCHEMISTRY, ATMOSPHERIC CHEMISTRY. *Educ:* Univ Birmingham, BSc, 53, PhD(chem), 56; Univ Cambridge, PhD, 66. *Prof Exp:* Sr sci off, Civil Serv, UK, 56-58; Imp Chem Indust fel chem, Univ Cambridge, 58-61; lectr, Univ Sheffield, 61-64; from asst prof to assoc prof, 64-83, PROF CHEM, UNIV BC, 83- *Mem:* Royal Soc Chem; Inter-Am Photochem Soc. *Res:* Chemical kinetics studied by flash photolysis and kinetic spectroscopy; electronic absorption spectra of transient species; atmospheric photochemistry. *Mailing Add:* Dept Chem 2036 Main Mall Univ BC University Campus Vancouver BC V6T 1Y6 Can

BASCOM, WILLARD, b New York, NY, Nov 7, 16; m 47; c 1. OCEANOGRAPHY, ARCHAEOLOGY. *Prof Exp:* Mining engr, Idaho, Ariz, Colo & NY, 40-45; res engr, waves & beaches, Univ Calif, 45-51 & oceanog instruments, Scripps Inst Oceanog, 51-54; tech dir, Comt Civil Defense, 54-55; exec secy, Meteorol Comt, 56; sabbatical yr study Polynesian hist, Tahiti, 57; exec secy maritime res comt, Nat Res Coun, 58; dir, Mohole Proj, AMSOC Comt, Nat Acad Sci, 59-62; pres, Ocean Sci & Eng, Inc, 62-70; pres, Seafinders, Inc, 71-73; dir, Southern Calif Coastal Water Res Proj, 73-85; PROJ DIR, ANCIENT SHIPS SOC, 85- *Concurrent Pos:* Consult, Comt Amphibious Opers, 49, Rockefeller Bros spec study group, 56-57 & sci progs, Columbia Broadcasting Syst TV, 58-59; mem panel underwater swimmers, Nat Acad Sci-Nat Res Coun, 50-53, proj Nobska, 56; mem plowshare comn, AEC, 61-69; mem, Naval Res Adv Comt, 70-; mem Lab Bd for Undersea Warfare, 73-; mem ocean sci bd, Nat Acad Sci, 78-84; mem, Coastal Eng Res Bd, US Army Corps Engrs, 78-86. *Honors & Awards:* Compass Award for Outstanding Contrib to Oceanog, Marine Technol Soc, 70; Explorers Medal, Explorers Club, 80. *Mem:* Fel AAAS; Marine Technol Soc; Int Asn Water Pollution Res. *Res:* Oceanographic engineering and instrumentation; drilling in the deep sea; waves; beaches; science writing; undersea diamond mining; deep water archaeology; marine ecology and environmental studies. *Mailing Add:* 5137 Vista Hermosa Long Beach CA 90815

BASCOM, WILLARD D, b Watertown, Conn, Oct 27, 31; m 52; c 3. PHYSICAL CHEMISTRY. *Educ:* Worcester Polytech Inst, BS, 53; Georgetown Univ, MS, 60; Cath Univ Am, PhD, 71. *Prof Exp:* Res biochemist, Worcester Found Exp Biol, 53-56; head, Adhesion Sect, US Naval Res Lab, 66-80; MEM STAFF, COMPOSITES RES, HERCULES INC, MAGNA, UTAH, 81- *Mem:* Am Chem Soc; Sigma Xi; Soc Plastics Engrs. *Res:* Colloid and surface chemistry; nonaqueous systems; adhesion; polymer fracture; composite materials; graphite fiber. *Mailing Add:* 8769 Ida Lane Sandy UT 84093

BASCUNANA, JOSE LUIS, b Paris, France, Feb 10, 27; US citizen; m 57; c 5. MECHANICAL & AEROSPACE SCIENCES. *Educ:* Cath Inst Arts & Indust, Spain, BS, 47; Univ Rochester, MS, 61, PhD(mech & aerospace sci), 68; Eastern Mich Univ, MA, 78. *Prof Exp:* Design engr, Ctr Study Tech Automotive, Spain, 47-56, head car sect, 56-58; proj engr, Barreiros Diesel,

Spain, 61-63; tech assoc eng res, Univ Rochester, 63-64; sr res engr power train systs res dept, Ford Motor Co, 67-70; chief hwy vehicle sect, Div Emission Control Technol, Environ Protection Agency, 70-77; SR TECH ADV, TECHNOL ASSESSMENT DIV, NAT HWY TRAFFIC & SAFETY ADMIN, DEPT TRANSP, 77- *Concurrent Pos:* Prof engr. *Honors & Awards:* Honor, Res Soc Sigma Xi. *Mem:* Am Soc Mech Engrs; Soc Automotive Engrs; Air Pollution Control Asn; Combustion Inst; Am Soc Testing Mat; Tire Soc. *Res:* Engine design and research; energy conversion and power plants air pollution; tire research; automotive crash avoidance research; published 14 papers in professional journals. *Mailing Add:* 9104 Cranford Dr Potomac MD 20854

BASDEKAS, NICHOLAS LEONIDAS, b Kharkov, Russia, Mar 20, 24; US citizen; c 1. ENGINEERING MECHANICS. *Educ:* Columbia Univ, BS, 54, MS, 55, CE, 58; Cath Univ Am, PhD(solid mech), 69. *Prof Exp:* Instr, Columbia Univ, 55-58; structural res engr, Southwest Res Inst, 58-64; gen engr, Marine Eng Lab, 64-66; operations res analyst, David Taylor Naval Ship Res & Develop Ctr, 66-68; sci officer, Off Naval Res, 68-84; CONSULT, EG&G, 84- *Concurrent Pos:* Assoc prof lectr, George Washington Univ, 73-77, prof lectr, 78-; mem tech adv group, Shock & Vibration Info Ctr, 70-84; mem J Ship Res Comt, J Soc Naval Architects & Marine Engrs, 78- *Mem:* Am Soc Mech Engrs; Am Inst Aeronaut & Astronaut; Soc Naval Architects & Marine Engrs; Naval War Col Found; NY Acad Sci; AM Acad Mech. *Res:* Structural mechanics research related to the static and dynamic response of submersibles and their buckling strength. *Mailing Add:* 5133 Clovel Terr Rockville MD 20853

BASDEKIS, COSTAS H, b Manchester, NH, Feb 20, 21; m 45; c 4. POLYMER CHEMISTRY. *Educ:* Univ NH, BS, 42. *Prof Exp:* Polymer chemist, Plastics Div, Monsanto Plastics & Resins Co, 42-48, fibers chemist, Cent Res Dept, Dayton, Ohio, 48-50, group leader, 50-52, from group leader to sr group leader polymers, Plastics Div, 52-60, asst res dir polystyrene plastics, 60-64, mgr res new polymers, Plastics Div, 64-78, mgr res & develop explor polymers, New Prod, Develop Dept, 78-82; RETIRED. *Concurrent Pos:* Consult, 82- *Mem:* Am Chem Soc. *Res:* Polystyrene and acrylonitrile-butadiene-styrene copolymer plastics; high nitrile polymers for barrier applications; synthesis and application studies of new types of engineering thermoplastics; process and product development of reinforced thermoplastics; exploratory polymers. *Mailing Add:* 57 Warwick St Longmeadow MA 01106

BASEMAN, JOEL BARRY, b Boston, Mass, Apr 28, 42; m 68; c 3. MICROBIOLOGY. *Educ:* Tufts Univ, BS, 63; Univ Mass, MS, 65, PhD(microbiol), 68. *Prof Exp:* Asst prof, 71-76, assoc prof microbiol, Sch Med, Univ NC, Chapel Hill, 76-; AT DEPT MICROBIOL, UNIV TEX HEALTH SCI CTR. *Concurrent Pos:* NIH fel, Harvard Univ & Harvard Med Sch, 68-71; USPHS grant, Univ NC, Chapel Hill, 75-76 & 78-79; US Army Res & Develop Command grant, 75-76 & 78-79. *Mem:* AAAS; Am Soc Microbiol; Sigma Xi. *Res:* Biochemistry of pathogenesis; regulation of animal cell growth. *Mailing Add:* Dept Microbiol Health Sci Ctr Univ Tex 7703 Floyd Curl Dr San Antonio TX 78284

BASERGA, RENATO, b Milan, Italy, Apr 11, 25; nat US; m 53, 74; c 2. PATHOLOGY. *Educ:* Univ Milan, MD, 49. *Prof Exp:* Asst path, Univ Milan, 49-51; assoc oncol, Med Sch, Univ Chicago, 53-54; from instr to assoc prof path, Med Sch, Northwestern Univ, 58-65; res prof, Fels Res Inst, 65-68, chmn dept, 68-73, PROF, DEPT PATH, SCH MED, TEMPLE UNIV, 68-, CHMN DEPT, 80-, SR INVESTR, FELS RES INST, 65- *Concurrent Pos:* Consult, Argonne Nat Lab, 59; chmn path study sect, NIH, 71-73 & mem cancer spec prog adv comt, 75-; first ann fac res award, Temple Univ, 82; Wellcome vis prof cell biol, 83. *Honors & Awards:* Louis Gross Mem Lectr, 74; Searle Lectr, Brit Soc Cell Biol, 76. *Mem:* AAAS; Am Asn Cancer Res; Am Asn Path & Bact; Am Soc Exp Path; Am Soc Biol Chemists; Am Soc Microbiol. *Res:* Experimental pathology; control of cell division in normal and pathologic tissues; microinjection and hybrid cells to study gene expression during cell proliferation. *Mailing Add:* Dept Path Temple Univ Med Sch 3400 N Broad St Philadelphia PA 19140

BASFORD, ROBERT EUGENE, b Montpelier, NDak, Aug 21, 23; m; c 1. CELL BIOLOGY. *Educ:* Univ Wash, BS, 51, PhD(biochem), 54. *Prof Exp:* From asst prof to assoc prof, 58-70, PROF BIOCHEM, SCH MED, UNIV PITTSBURGH, 70- *Concurrent Pos:* Res fel, Inst Enzyme Res, Univ Wis, 54-58. *Mem:* AAAS; Am Soc Cell Biol; Am Thoracic Soc; Reticuloendothelial Soc; Oxygen Soc; Am Soc Biochem & Molecular Biol. *Res:* Terminal electron transport in the mitochondria of heart and brain; energy metabolism of brain; biochemical basis of phagocytosis; neutrophil-endothelial cell interaction. *Mailing Add:* Dept Molecular Genetics & Biochem Univ Pittsburgh Rm E1240 BST Pittsburgh PA 15261

BASH, FRANK NESS, b Medford, Ore, May 3, 37; m 60; c 2. RADIO ASTRONOMY. *Educ:* Willamette Univ, AB, 59; Harvard Univ, MA, 62; Univ Va, PhD(astron), 67. *Prof Exp:* Assoc astronomer, Nat Radio Astron Observ, 62-64, res assoc, 65-67; univ fel & fac assoc, 67-69, from asst prof to prof, 69-85, chmn dept, 82-86, EDMONDS REGENTS PROF ASTRON, UNIV TEX, AUSTIN, 85- *Concurrent Pos:* Blunk prof, Univ Tex, 82-83; prin investr, NSF, 72- *Mem:* Am Astron Soc; Int Sci Radio Union; Int Astron Union. *Res:* Observational radio astronomy; star formation in spiral galaxies and the dynamics of giant molecular clouds; interpretation of neutral hydrogen and carbon monoxide observations in the galaxy. *Mailing Add:* Dept Astron Univ Tex Austin TX 78712

BASH, PAUL ANTHONY, b Reno, Nev, Sept 27, 52. COMPUTATIONAL CHEMISTRY, ENZYMOLOGY. *Educ:* Univ Calif, Berkeley, AB, 74, PhD(biophys), 86. *Prof Exp:* Postdoctoral chem, Harvard Univ, 86-90; ASST PROF CHEM & BIOPHYS, FLA STATE UNIV, 90- *Mem:* Biophys Soc; Am Chem Soc; Am Phys Soc. *Res:* Computational methods to study structure-function relationships in biological macromolecules; theoretical investigations of enzyme reaction mechanisms and the process of protein folding. *Mailing Add:* Dept Chem B-164 Fla State Univ Tallahassee FL 32306-3006

BASHAM, CHARLES W, b Ponca City, Okla, July 25, 34; m 63. HORTICULTURE, PLANT PHYSIOLOGY. *Educ:* Okla State Univ, BS, 56; Univ Md, PhD(hort), 64. *Prof Exp:* Instr hort, Okla State Univ, Contract Imp Ethiopian Agr Col, 59-61; asst prof, Kans State Univ, 64-65; asst prof, 65-75, ASSOC PROF HORT, COLO STATE UNIV, 75- *Mem:* Am Soc Hort Sci; Nat Asn Col & Teachers Agr; Sigma Xi. *Res:* computer applications in agriculture; production of vegetable crops under irrigation. *Mailing Add:* Dept Comput Sci Colo State Univ Ft Collins CO 80523

BASHAM, JACK T, b Lachine, Que, Nov 2, 26; m 56; c 3. FOREST PATHOLOGY, STEM DECAY. *Educ:* Univ Toronto, BSc, 48, MA, 50; Queen's Univ, Ont, PhD(forest path), 58. *Prof Exp:* Res scientist, Can Dept Agr, 48-60; RES SCIENTIST, CAN FORESTRY SERV, 60- *Honors & Awards:* Can Forestry Sci Achievement Award, Can Inst Forestry, 83. *Mem:* Can Inst Forestry. *Res:* Ecology and succession patterns of fungi in their invasion and destruction of heartwood in living trees and of stem wood in recently killed trees, and their impact on forest management. *Mailing Add:* 2006 Queen St E Sault Ste Marie ON P6A 2H2 Can

BASHAM, JERALD F, b Alma, Ark, Apr 2, 42; m 80; c 3. SEMICONDUCTOR MANUFACTURING, OPTOELECTRONICS. *Educ:* Univ Ark, BSEE, 69; Stanford Univ, MSEE, 70. *Prof Exp:* Tech staff mem, Sandia Nat Labs, 69-75; dir opers, TRW Optoelectronics, 76-88; vpres, Alltest, Inc, 88-89; VPRES, OPTOSWITCH INC, 89- *Mem:* Inst Elec & Electronics Engrs. *Res:* Technical sales and marketing. *Mailing Add:* 3317 Bandolino Plano TX 75023

BASHAM, RAY S(COTT), b Selmer, Tenn, Nov 21, 21; m 45; c 3. ELECTRICAL ENGINEERING. *Educ:* US Mil Acad, BS, 45; Univ Ill, MS, 52, PhD(elec eng), 62. *Prof Exp:* Assoc prof elec eng, US Air Force Acad, US Air Force, 59-64, develop planning officer, Off of Dep Chief of Staff res & develop, Hqs, 64-65; ASSOC PROF ELEC ENG, UNIV MD, COLLEGE PARK, 65- *Mem:* Am Soc Eng Educ. *Res:* Network synthesis and radio communications. *Mailing Add:* Dept Elec Eng Univ Md College Park MD 20742

BASHAM, SAMUEL JEROME, JR, b Adairville, Ky, Feb 25, 27; m 49; c 2. NUCLEAR ENGINEERING. *Educ:* Univ Ky, BS, 53; Ohio State Univ, MS, 74. *Prof Exp:* Prin mech engr tribology, Battelle Mem Inst, 53-60, proj leader nuclear mat, 60-66, assoc div chief nuclear eng, Columbus Labs, 66-73 & assoc sect mgr, 73-78, dept mgr eng develop, 78-80, mgr, Waste Packaging Prog Off, 80-83 & Repository Proj Off, 83-85, asst prog mgr, Off Nuclear Waste Isolation, 85-89, PROG MGR, MICH LOW-LEVEL RADIATION WASTE PROJ, BATTELLE-LANSING, 89- *Mem:* Am Nuclear Soc; fel Am Soc Mech Engrs; Sigma Xi. *Res:* Nuclear fuel materials development; radiation effects on instrumentation and electromechanical devices; nuclear waste isolation. *Mailing Add:* 505 King Ave Columbus OH 43201

BASHAM, TERESA, b Roanoke, Va, Nov 3, 47; m 80; c 2. IMMUNOLOGY, CANCER RESEARCH. *Educ:* Va Commonwealth Univ, BS, 70; George Wash Univ, PhD(immunogenetics), 80. *Prof Exp:* Res assoc, 82-86, SR RES ASSOC, MED RES, STANFORD UNIV MED SCH, DEPT MED, DIV INFECTIOUS DIS, 86- *Mem:* Int Soc Interferon Res; Am Asn Immunologists. *Res:* Possible cures for non-hodgkins lymphoma using natural biological products alone and in combination. *Mailing Add:* Dept Med Stanford Univ Sch Med Stanford CA 94305

BASHARA, N(ICOLAS) M, b Bismarck, NDak, Mar 8, 17; m 41. ELECTRICAL ENGINEERING. *Educ:* Univ Nebr, BS, 47, MS, 58. *Prof Exp:* Physicist, Minn Mining & Mfg Co, 47-56; coordr optics res, Elec Mat Lab, 56-81, from assoc prof to prof elec eng, 61-73, res prof eng, 73-81, EMER RES PROF ENG, UNIV NEBR, LINCOLN, 81- *Concurrent Pos:* NSF grants, 58-81; Off Naval Res grants, 60-71; sr ed Proc 2nd & 3rd Int Conf Ellipsemetry, 69 & 76. *Mem:* Fel Optical Soc Am. *Res:* Co-author one book. *Mailing Add:* 5431 Cloudburst Lane Lincoln NE 68521

BASHAW, ELEXIS COOK, b Mt Juliet, Tenn, July 21, 23; m 45; c 2. CYTOGENETICS, PLANT BREEDING. *Educ:* Purdue Univ, BS, 47, MS, 48; Tex Agr & Mech Col, PhD(genetics), 54. *Prof Exp:* Asst agronomist, La Agr Exp Sta, 48-50; asst prof, Tex Agr Exp Sta, 52-55, GENETICIST, AGR RES SERV, USDA, TEX A&M UNIV, 55- *Honors & Awards:* Genetics & Plant Breeding Award, Nat Coun Com Plant Breeders, 81; Crop Sci Award, Crop Sci Soc Am, 82. *Mem:* Fel AAAS; fel Am Soc Agron; fel Crop Sci Soc Am. *Res:* Cytogenetics of grasses, especially genetics of apomixis, reproductive systems; radiation breeding and interspecific hybridization. *Mailing Add:* Dept Soil & Crop Sci Tex A&M Univ College Station TX 77843

BASHE, WINSLOW JEROME, JR, b Chicago, Ill, Mar 10, 20; m 53; c 3. PEDIATRICS, PREVENTIVE MEDICINE. *Educ:* Seton Hall Col, BS, 42; Loyola Univ, MD, 45; Columbia Univ, MPH, 59. *Prof Exp:* Intern gen med, Gorgas Hosp, CZ, 45-46, resident med, 46-47; resident pediat, Sea View Hosp, NY, 48-49; resident, Children's Hosp Philadelphia, 50, res assoc, 52-53; pvt pract, 53-57; epidemiologist commun dis, Ohio Dept Health, 57-58, chief div, 59-63; assoc clin prof pediat, Col Med, Univ Cincinnati, 63-70; assoc prof prev med & pediat, Ohio State Univ, 71-79; PROF COMMUNITY MED, WRIGHT STATE UNIV, 79- *Concurrent Pos:* Instr, Univ Pa, 51-53; instr & asst prof, Ohio State Univ, 57-63. *Mem:* AAAS; Am Pub Health Asn; Sigma Xi. *Res:* Virology; mumps and hepatitis; staphylococcal disease; bone maturation of children; influenza; trichinosis; sudden cardiac death; general epidemiology. *Mailing Add:* Sch Med Wright State Univ PO Box 927 Dayton OH 45401

BASHEY, REZA ISMAIL, b Bombay, India, Aug 28, 32; m 61; c 3. BIOCHEMISTRY. *Educ:* Univ Bombay, BSc, 52, MSc, 54; Rutgers Univ, PhD(biochem), 58. *Prof Exp:* Mem fac, Sch Med, Univ Miami, 62-64; res assoc biochem, Albert Einstein Col Med, 64-66, asst prof, 66-70; res asst prof, Hahnemann Med Col & Hosp, 70, res assoc prof med, 71-72, assoc prof med & biochem, 72-75, vis assoc prof biol & chem, 75-77; from asst prof to assoc prof med, 77-88, SR INVESTR, PHILADELPHIA GEN HOSP & ASSOC, RHEUMATOLOGY RES LAB, DEPT MED, UNIV PA, 75- *Concurrent Pos:* Fel biochem, Univ Southern Calif, 58-60; partic fel, Howard Hughes Med Inst, Fla, 60-62; pool officer med sci, Indian Coun Med Res, Ministry Health, India, 64; assoc prof, dept med, sect rheumatology, Thomas Jefferson Univ, 88, prof, 90. *Mem:* AAAS; Am Chem Soc. *Res:* Connective tissue; collagen and glycosammoglycans, biosynthesis and involvement in fibrillogenesis; heart valve biochemistry in normal and diseased conditions; scleroderma and other skin diseases. *Mailing Add:* 4055 Ctr Ave Lafayette Hill PA 19444

BASHIRELAHI, NASIR, b Tehran, Iran, Aug 26, 35; US citizen; m 78; c 1. BIOCHEMISTRY. *Educ:* Univ Tehran, BS, 60, Pharm D, 62, Univ Louisville, MS, 65, PhD(biochem), 68. *Prof Exp:* Assoc biochem, Harvard Med Sch, 70-71; asst dept surg, Mass Gen Hosp, 70-71; asst prof, 71-76, ASSOC PROF BIOCHEM, SCH DENT, UNIV MD, 76- & DEPT SURG, DIV UROL, SCH MED, 77- *Mem:* The Endocrine Soc; AAAS; Biophys Soc; Soc Exp Biol & Med; Am Asn Advan Aging Res. *Res:* Mechanism of action and metabolism of steroid hormones; hormones and cancer; aging and differentiation. *Mailing Add:* Dept Biochem Univ Md Sch Dent 666 W Baltimore St Baltimore MD 21201

BASHKIN, STANLEY, b Brooklyn, NY, June 20, 23; m 57; c 3. ATOMIC PHYSICS, NUCLEAR PHYSICS. *Educ:* Brooklyn Col, BA, 44; Univ Wis, PhD(physics), 50. *Prof Exp:* Asst physics, Manhattan Proj, 44-46; asst prof physics, La State Univ, 50-53; from res assoc to assoc prof, Univ Iowa, 53-62; dir, Van De Graaff Lab, 70-80, assoc dean, Col Liberal Arts, 80-82, PROF PHYSICS, UNIV ARIZ, 62- *Concurrent Pos:* Res fel, Calif Inst Technol, 59; Fulbright res scholar, Australian Nat Univ, 60, res fel, 69; mem comt atomic & molecular physics, Nat Acad Sci-Nat Res Coun; sr vis fel, Nuclear Physics Lab, Oxford Univ, 77-78; sr von Humboldt fel, Univ Bochum, 82. *Mem:* Optical Soc Am; Am Phys Soc; for mem Royal Soc Gothenburg. *Res:* Nuclear energy levels; atomic lifetimes; nuclear astrophysics; structures of highly ionized atoms. *Mailing Add:* Dept Physics Univ Ariz Tucson AZ 85721

BASHKOW, THEODORE R(OBERT), b St Louis, Mo, Nov 16, 21; m 60; c 1. DATA COMMUNICATION, COMPUTER SCIENCE. *Educ:* Wash Univ, BS, 43; Stanford Univ, MS, 47, PhD(elec eng), 50. *Prof Exp:* Mem tech staff, Radio Corp Am Labs, 50-52 & Bell Tel Labs, 52-58; assoc prof elec eng, Columbia Univ, 58-67, chmn dept, 68-71, prof elec eng, 67-79, prof comput sci, 79-91; RETIRED. *Concurrent Pos:* Consult, Dewey Electronics, Advan Comput Tech, World Health Orgn & Network Anal Corp, 59-; chmn, Sci Secy Comt, 63-65; partic, Int Fedn Info Processing Socs Cong, 65; corp secy, Sullivan Comput Corp, 79-87, CHOPP Computer Co, 87- *Mem:* Inst Elec & Electronic Engrs; Asn Comput Mach. *Res:* Digital computer architecture; distributed and parallel processing; data communications. *Mailing Add:* 92 Jay St Katonah NY 10536

BASHOUR, FOUAD A, b Tripoli, Lebanon, Jan 3, 24; US citizen. BIOLOGY, MEDICINE. *Educ:* Am Univ Beirut, BA, 44, MD, 49; Univ Minn, PhD(med), 57. *Prof Exp:* Intern, Am Univ Beirut, 49-50; med officer, UN Relief & Works Agency, 50-51; resident internal med, Hosps, Univ Minn, 51-54, instr med, 55-57; res assoc, Med Sch, Am Univ Beirut, 57, asst prof med & in chg cardiopulmonary lab sect, 57-59; from instr internal med to assoc prof med, 59-63, PROF MED, UNIV TEX HEALTH SCI CTR, DALLAS, 71-; PROF MED & PHYSIOL, METHODIST HOSP, DALLAS, 85- *Concurrent Pos:* Mem sr med staff, Parkland Mem Hosp, Dallas & Zale Lipshy Univ Hosp; med & cardiac consult, St Paul Hosp, Dallas, John Peter Smith Hosp, Ft Worth, Methodist Hosps, Dallas, Wilford Hall Hosp, USAF Lackland AFB & Tex State Dept Pub Welfare; cardiac consult, Wilford Hall Hosp & Lackland AFB; mem ad hoc comt coronary care unit & coun basic sci, Am Heart Asn; ad hoc proj site visit, Nat Heart Inst, 65; pres, Cardiol Fund; mem, Coun Circulation & Basic Sci, Am Heart Asn, Coun Circulation & Cent Res Rev Comt; Fouad Bashour distinguished chair, Cardiovasc Physiol, 84 & 90, Fouad Bashour prof, 81; dir, Cardiovasc Inst, Methodist Hosp, Dallas, 67-78; consult, Nat Heart & Lung Inst; fel, Circulation Group, Am Physiol Soc, 75; hon adv, Univ Kuwait, 74- *Honors & Awards:* Officer, Order of the Cedar of Lebanon, 71. *Mem:* Am Heart Asn; fel Am Col Chest Physicians; Am Fedn Clin Res; Am Physiol Soc; AMA; Cent Soc Clin Res; Sigma Xi; Asn Advan Med Instrumentation; Am Soc Internal Med. *Res:* Cardiovascular physiology and diseases; author of over 200 publications. *Mailing Add:* Univ Tex Southwestern Med Ctr Dallas TX 75235

BASILA, MICHAEL ROBERT, b Scranton, Pa, June 5, 30; m 54; c 6. PHYSICAL CHEMISTRY. *Educ:* Norwich Univ, BS, 52; Rensselaer Polytech Inst, PhD(phys chem), 58. *Prof Exp:* Res chemist, Gulf Res & Develop Co, Pa, 58-61, supvr chem physics sect, 61-64, supvr catalysis sect, 64-67, supvr catalytic mechanisms sect, 67-68; dir res & develop, Howe Baker Eng, Inc, Tex, 68-70; tech dir catalyst, Nalco Chem Co, Chicago, Ill, 70-76; head lab, Katalco Corp, Chicago, 76-80; CONSULT, 88- *Mem:* Catalysis Soc. *Res:* Infrared spectroscopy; molecular complexes; kinetics of catalytic reactions; process and catalyst development. *Mailing Add:* 807 Woodland Hills Tyler TX 75701

BASILE, DOMINICK V, b Yonkers, NY, Oct 15, 31; m 59; c 3. MORPHOLOGY, BRYOLOGY. *Educ:* Manhattan Col, BS, 58; Columbia Univ, MA, 62, PhD(bot), 64. *Prof Exp:* Teaching asst bot, Columbia Univ, 58-63, preceptor, 63-64; guest investr plant biol, Rockefeller Univ, 64-65, res assoc, 65-66; asst prof bot, Columbia Univ, 66-71; assoc prof, 71-77, PROF BOT, LEHMAN COL, 77- *Concurrent Pos:* NSF fel, 64-65; adj cur, NY Bot Garden, 71- *Mem:* Am Bryol & Lichenological Soc; Torrey Bot Club (vpres, 75, corresp secy, 80-); Am Inst Biol Sci; Sigma Xi; Am Soc Plant Physiol. *Res:* Chemical regulation of morphogenesis and phylogeny of plants, especially the Bryophyta. *Mailing Add:* Biol Sci Dept Lehman Col City Univ New York Bronx NY 10468

BASILE, LOUIS JOSEPH, b Chicago, Ill, Mar 20, 24; m 53; c 3. MOLECULAR SPECTROSCOPY, INORGANIC CHEMISTRY. *Educ:* Univ Chicago, BS, 48, MS, 49; St Louis Univ, PhD(chem), 54. *Prof Exp:* Assoc chemist, Argonne Nat Lab, 52-84; RETIRED. *Concurrent Pos:* Mem bd, sect comt N15, methods of nuclear mat control, Am Nat Standards Inst; assoc ed, Appl Spectros, 71-84. *Honors & Awards:* Meggers Award, Soc Appl Spectros, 75. *Mem:* Fel AAAS; Am Chem Soc; Sigma Xi; Soc Appl Spectros. *Res:* Energy transfer; plastics; hydrides of boron; rare earths; transplutonium chemistry; infrared spectroscopy; high pressure molecular spectroscopy. *Mailing Add:* 9544 S Springfield Evergreen Park IL 60642

BASILE, ROBERT MANLIUS, b Youngstown, Ohio, Mar 12, 16; m 45; c 3. PHYSICAL GEOGRAPHY, SOIL MORPHOLOGY. *Educ:* Washington & Lee Univ, BS, 38; Mich State Univ, MS, 40; Ohio State Univ, PhD(geog), 53. *Prof Exp:* Instr soils & geog, Northwestern State Col, Okla, 40-42; from instr to prof geog, Ohio State Univ, 50-69; prof, 69-81, EMER PROF GEOG, UNIV TOLEDO, 81- *Concurrent Pos:* Vis prof, Univ Winnipeg, 62, San Jose State, 66, Nat Defense Educ Act Inst, Univ SC, 67 & Univ Wyo, 81. *Mem:* Wilderness Soc. *Res:* Climatology and the morphology of soils. *Mailing Add:* Dept Geog Univ Toledo Toledo OH 43606

BASILEVSKY, ALEXANDER, b USSR, April 11, 43; Can citizen; m 66; c 2. ECONOMETRICS, ECONOMICS. *Educ:* Concordia Univ, Can, BA, 64; Univ Southampton, Eng, MA, 69, PhD(statist), 73. *Prof Exp:* Lectr statist, Univ Kent, Eng, 70-75; sr researcher econ, Manitoba Guaranteed Income Exp, 75-77; ASSOC PROF STATIST, UNIV WINNIPEG, CAN, 77- *Concurrent Pos:* Consult, Manitoba Guaranteed Income Exp, 77-78; pres-at-large, Statist Asn Manitoba; adj prof, Univ Manitoba, 79-80. *Mem:* Am Statist Asn; Can Statist Asn. *Res:* Linear statistical models and their application; estimation in presence of missing data; analysis of socio-economic experiments. *Mailing Add:* Dept Math Univ Winnipeg 515 Portage Ave Winnipeg NB R3B 2E9 Can

BASILI, VICTOR ROBERT, b Brooklyn, NY, Apr 13, 40; m 67; c 3. COMPUTER SCIENCE, SOFTWARE ENGINEERING. *Educ:* Fordham Univ, BS, 61; Syracuse Univ, MS, 63, Univ Tex, Austin, PhD(comput sci), 70. *Prof Exp:* Res asst comput ctr, Syracuse Univ, 61-63; from instr to asst prof math & comput sci, Providence Col, 63-67; teaching asst comput sci, Univ Tex, Austin, 67-69, curric coordr, Southwest Region Educ Comput Network, 69-70; from asst prof to prof, Univ Md, 70-88, chmn dept comput sci, 82-88; CONSULT, 88- *Concurrent Pos:* Prin investr grants, Off Naval Res, 72-85, NASA Goddard Space Flight Ctr 75-88 & Air Force Off Sci Res, 74-88; comput sci consult, Inst Comput Appln Sci & Eng, NASA-Langley Res, Hampton, Va, 74-; expert comput sci math, Naval Res Lab, Washington, DC, 75-; consult, Naval Surface Weapons Ctr, Dahlgren, Va, 75-77, Comput Sci Corp, Falls Church, Va, 77- & IBM Fed Systs Div, Gaithersburg, Md, 77-; assoc ed, J Systs & Software, 84- & ed in chief, Trans-Software Eng, 88- *Mem:* Fel Inst Elec & Electronic Engrs Comput Soc; Asn Comput Mach. *Res:* Development, analysis, and evaluation of software development techniques and the design, modeling, and implementation of programming languages for a variety of applications. *Mailing Add:* Dept Comput Sci Univ Md College Park MD 20742-3255

BASILICO, CLAUDIO, b Milan, Italy, Feb 7, 36; US citizen; m 61; c 3. GROWTH FACTORS, CELL CYCLE. *Educ:* Univ Milan, Italy, MD, 60. *Prof Exp:* Instr, Inst Clin Med, Univ Milan, 61-62; ricercatore qualificato, Int Lab Genetics & Biophys, Naples, Italy, 63-66; res assoc cell biol, Albert Einstein Col Med, 66-67; res scientist, 67-69, from assoc prof to prof path, 70-90, PROF MICROBIOL & CHMN DEPT, NY UNIV SCH MED, 90- *Concurrent Pos:* Res fel, Div Biol, Calif Inst Technol, 62; instr animal cell cult, Cold Spring Harbor Lab, 74, mem bd trustees, 81-86; mem, Adv Comt Virol & Cell Biol, Am Cancer Soc, 78 & Adv Comt Cellular & Develop Biol, 79-82; dir, Prog Viral & Molecular Oncol, NY Univ Cancer Ctr, 77-; chmn, Gordon Res Conf, Meriden, NH, 87; prof, Dept Genetics, Univ Milan, 87. *Mem:* Am Soc Microbiol; Am Soc Cell Biol; Am Asn Cancer Res; Am Soc Virol. *Res:* Molecular mechanisms of carcinogenesis; genes and gene-functions involved in regulating the proliferation of normal and cancer cells. *Mailing Add:* 550 First Ave New York NY 10016

BASIN, M(ICHAEL) A(BRAM), b Harbin, Manchuria, Apr 13, 29; US citizen; m 54; c 2. ELECTRICAL ENGINEERING. *Educ:* Calif Inst Technol, BS, 51, MS, 52, PhD(elec eng & physics), 54. *Prof Exp:* Elec engr, Gen Elec Co, 51; res engr, analog comput & aerodyn, Calif Inst Technol, 51-54, res engr, 54-55; mgr eng serv div, Comput Eng Assocs, 55-58; mgr sonar res & develop, Hughes Aircraft Co, 58-64; sr mem tech staff, Data Systs Div, Litton Industs, Inc, 64-66, tech mgr undersea warfare progs, 66-70, tech dir, Advan Navy Systs, 70-72; founder & pres, SDP Inc, 72-90; MGR, SHERMAN OAKS FIELD OFF, ORINCON-SDP, 90- *Concurrent Pos:* Instr eng exten, Univ Calif, Los Angeles, 58; consult, Comput Eng Assocs, 58-59; resident scientist, Tudor Hill Lab, Bermuda under subcontract to Litton Industs, Inc from Hudson Labs, Columbia Univ, 65-66. *Mem:* Inst Elec & Electronics Engrs; Am Inst Aeronaut & Astronaut; Acoustical Soc Am; Sigma Xi. *Res:* Sonar system design; marine technology; signal processing and display studies; direct analog computer technique development and utilization; aeroelastic dynamic and static analyses. *Mailing Add:* Orincon-SDP 13848 Ventura Blvd Suite D Sherman Oaks CA 91423

BASINGER, RICHARD CRAIG, b Kansas City, Kans, Oct 29, 38; m 61; c 2. MATHEMATICS, SYSTEMS ANALYSIS. *Educ:* Univ Mo-Rolla, BS, 60; Univ Kans, MA, 62, PhD(math), 65. *Prof Exp:* mem tech staff weapon systs anal, atomic energy comn, Sandia Corp, 65-79; ASSOC DIV LEADER, LAWRENCE LIVERMORE NAT LAB, 79- *Res:* Mathematical analysis; growth properties of meromorphic functions; weapons systems analysis; operations research; mathematical software. *Mailing Add:* 2712 Superior Dr Livermore CA 94550

BASKERVILLE, CHARLES ALEXANDER, b New York, NY, Aug 19, 28; m 53, 79; c 3. ENGINEERING GEOLOGY, LANDSLIDE RESEARCH. *Educ:* City Col New York, BS, 53; NY Univ, MS, 58, PhD(micropaleont, stratig), 65. *Prof Exp:* Asst civil engr, NY State Dept Transp, 53-65; from asst prof to prof eng geol, City Col New York, 65-79, dean, Sch Gen Studies, 70-79, emer prof dept earth & planetary sci, City Col, City Univ NY, 83; prof chief & res eng geol, US Geol Surv, 79-90; RETIRED. *Concurrent Pos:* Eng geol consult, Madigan-Hyland-Praeger Cavanaugh-Waterbury Engrs, 68-69, St Raymond's Cemetery, 70-73 & Consol Edison Co NY, 73; mem nat adv comt minority partic in geosci, Dept Interior, 72-75; panelist & chmn, NSF Grad Fel Prog, 79-80; US Nat Comt Tunnelling Technol, Nat Res Coun, chmn educ & training subcomt, 83-88; commonwealth vis prof geol, George Mason Univ, Fairfax, Va, 87-89; guest lectr, Syracuse Univ, Hofstra Univ, State Univ NY, Binghamton, Lafayette Col, Easton, Pa; prin investr planetary geol, NASA; mem, adv comt earth sci, NSF, 89- *Honors & Awards:* Award Excellence Eng Geol, Nat Consortium Black Prof Develop, 78. *Mem:* Fel Geol Soc Am; Am Inst Prof Geologists; Sigma Xi; NY Acad Sci; Asn Eng Geologists; Int Asn Eng Geol. *Res:* Landslide studies, State of Vermont; landslide process research and susceptibility mapping; engineering and geologic mapping of New York; ground failure hazards. *Mailing Add:* Dept Physics & Earth Sci Cent Conn State Univ 1615 Stanley St New Britian CT 06050

BASKERVILLE, GORDON LAWSON, b Emerson, Man, Can, Feb 20, 33; m 58; c 4. FORESTRY, ECOLOGY. *Educ:* Univ New Brunswick, BScF, 55; Yale Univ, MF, 57, PhD(forest ecol), 64. *Prof Exp:* Res scientist silvicult, Can Dept Forestry, 55-74; PROF FORESTRY, UNIV NEW BRUNSWICK, 74- *Concurrent Pos:* Mem Can subcomt terrestial productivity, Int Biol Prog, 65- *Mem:* Ecol Soc Am; Soc Am Foresters; Can Inst Forestry; Can Pulp & Paper Asn. *Res:* Physiological and ecological basis for silviculture; total biomass and productivity in forests. *Mailing Add:* Dept Forestry Univ New Brunswick Box 4400 Fredericton NB E3B 5A3 Can

BASKES, MICHAEL I, b Chicago, Ill, June 13, 43; m 68; c 2. ATOMISTIC CALCULATIONS, INTERATOMIC POTENTIALS. *Educ:* Calif Inst Technol, BSc, 65, PhD(mat sci), 70. *Prof Exp:* Staff mem, Sandia Nat Labs, 69-83, supvr sci comput, 83-84, supvr theoret, 84-89, SUPRVR MAT PROCESSES, SANDIA NAT LABS, 89- *Concurrent Pos:* Vis scientist, Univ Delft, Neth, 79; detailee, temp asst US Dept Energy, 90- *Honors & Awards:* Outstanding Res Award, US Dept Energy, 82 & 88. *Mem:* Mat Res Soc; Metals Soc. *Res:* Development of interatomic potentials; atomistic calculations of defects, surfaces, interfaces and dislocations; atomistic and phenomenological calculations of the interaction of hydrogen and helium with metals. *Mailing Add:* Sandia Nat Labs PO Box 969 Livermore CA 94551-0969

BASKETT, THOMAS SEBREE, b Liberty, Mo, Jan 23, 16; m. WILDLIFE BIOLOGY. *Educ:* Cent Methodist Col, Mo, AB, 37; Univ Okla, MS, 39; Iowa State Col, PhD(zool), 42. *Prof Exp:* Res asst, Iowa State Col, 38-41, exten wildlife specialist, USDA, 41, asst prof zool, 46-47; asst prof wildlife mgt, Univ Conn, 47-48; biologist, US Fish & Wildlife Serv, Mo, 48-68; chief, Div Wildlife REs, US Bur Sport Fisheries & Wildlife, Washington, DC, 68-73; biologist & leader, Mo Coop Wildlife Res Unit, US Fish & Wildlife Serv, 73-85; prof, 73-85, EMER PROF WILDLIFE & FISHERIES, UNIV MO, COLUMBIA, 85- *Concurrent Pos:* Ed, J Wildlife Mgt, 66-68. *Honors & Awards:* Aldo Leopold Award. *Mem:* Wildlife Soc (pres, 71); Am Ornithologists Union. *Res:* Wildlife management; ecology of pheasant, bobwhite, swamp rabbit and bullfrog; effects of timber harvest on deer foods; breeding biology and behavior of mourning dove; terrestrial habitat evaluation. *Mailing Add:* Sch Nat Resources Univ Mo Columbia MO 65211

BASKIN, DENIS GEORGE, b Minneapolis, Minn, Feb 7, 41; m 64; c 4. PEPTIDES, BRAIN. *Educ:* San Francisco State Col, BA, 62; Univ Calif, Berkeley, PhD(zool), 69. *Prof Exp:* NIH fel, Albert Einstein Col Med, 69-71; from asst prof to assoc prof zool, Pomona Col, 77-79; res asst prof, dept med & biol sci, Univ Wash, 79-85, res assoc prof, dept med & biol struct, 85-89, DIR CYTOHISTOCHEM LAB, DIABETES RES CTR, SCH MED & DIR CORE HISTOLOGY/EM LAB, UNIV WASH, 79-, RES PROF, DEPTS MED & BIOL STRUCT & VET ADMIN MED CTR, SEATTLE, 89- *Concurrent Pos:* NIH & NIMH res fel, Dept Anat, Sch Med, Univ Minn, Minneapolis, 77-78; ed bd, J Histochem-Cytochem, 80; mem coun, Histochem Soc 86-88. *Mem:* Am Asn Anatomists; Endocrine Soc; Am Diabetes Asn; Sigma Xi; Soc Neurosci; Histochem Soc (pres-elect & pres, 88-89). *Res:* Brain regulation of metabolism; insulin receptors in the brain; insulin-like growth receptors in the brain; localization of peptide and neurotransmitter receptors; neuropeptide regulation of body weight and obesity. *Mailing Add:* Div Metabolism Mail Stop 151 Vet Admin Med Ctr Seattle WA 98108-1532

BASKIN, JERRY MACK, b Covington, Tenn, July 27, 40; m 68. PLANT ECOLOGY. *Educ:* Union Univ, Tenn, BS, 63; Vanderbilt Univ, PhD(biol), 67. *Prof Exp:* Res assoc plant physiol, Univ Fla, 67-68; asst prof, 68-73, assoc prof, 73-81, PROF BOT, UNIV KY, 81- *Mem:* Ecol Soc Am; Bot Soc Am; Torrey Bot Club. *Res:* Autecology of herbaceous, vascular plants; biology and phytogeography of plant taxa endemic to rock outcrop communition of inglaciated eastern United States; biology of plants rare in Kentucky. *Mailing Add:* Dept Biol Sci Univ Ky Lexington KY 40506

BASKIN, RONALD J, b Joliet, Ill, Nov 25, 35; m 58; c 2. BIOPHYSICS. *Educ:* Univ Calif, Los Angeles, AB, 57, MA, 59, PhD(biophys), 60. *Prof Exp:* Fel biophysics, Univ Calif, Los Angeles, 60-61; asst prof biol, Rensselaer Polytech Inst, 61-64; from asst prof to assoc prof, 64-71, assoc dean col lett & sci, 67-71, chmn dept, 71-79, PROF ZOOL, UNIV CALIF, DAVIS, 71- *Mem:* AAAS; Biophys Soc Am; Soc Cell Biol; Am Physiol Soc. *Res:* Muscle biophysics; thermodynamics of biological systems; membrane structure and function; electron microscopy. *Mailing Add:* Dept Zool Univ Calif Davis CA 95616

BASKIN, STEVEN IVAN, b Los Angeles, Calif, Nov 14, 42; m 66; c 2. PHARMACOLOGY, TOXICOLOGY. *Educ:* Univ Southern Calif, PharmD, 66; Ohio State Univ, PhD(pharmacol), 71. *Prof Exp:* NIH trainee, Ohio State Univ Col Med, 67-71; res assoc, 71-73, res asst prof, 73-74, asst prof, 74-78, ASSOC PROF PHARMACOL, MED COL PA, 78-; TEAM LEADER, CARDIAC PATHOPHYSIOL, US ARMY MED RES INST CHEM DEFENSE. *Concurrent Pos:* Consult, Minn Mining & Mfg Co, 74-77, NSF, 75-, Syntex Pharmaceut Co, 77-, Nat Inst Aging, 77 & Dept Pub Welfare, Commonwealth Pa, 78- *Honors & Awards:* Pharmaceut Mfg Asn Award, 75-77; Am Heart Asn Southeastern Pa Award, 75-77. *Mem:* Am Heart Asn; Int Study Group Res Cardiac Metabolism; AAAS; Am Fedn Clin Res; Gerontol Soc. *Res:* Pharmacology of taurine and related amino acids; digitalis and other inotrophic agents; phenytoin and other antiepileptic drugs; neurotransmitters; mechanisms; kinetics and toxicity; aging and effects of drugs on aging. *Mailing Add:* USAMRICO ATTN-SGRD- UV-PB Pharm Div Aberdeen Proving Ground MD 21010-5425

BASKIR, EMANUEL, b New York, NY, July 13, 29; m 52; c 3. PHYSICS. *Educ:* Columbia Univ, BA, 51; Rochester Univ, PhD(physics), 57. *Prof Exp:* PHYSICIST, SHELL DEVELOP CO, 56- *Mem:* Am Phys Soc; Am Geophys Union; Sigma Xi; Soc Expl Geophys. *Res:* Nuclear physics and applications to geophysical problems; physics of rock magnetism; acoustic wave propagation; rock properties. *Mailing Add:* Shell Develop Co PO Box 481 Houston TX 77001

BASLER, EDDIE, JR, b Blanchard, Okla, Mar 25, 24; m 56; c 4. PLANT PHYSIOLOGY. *Educ:* Univ Okla, BS, 50, MS, 52; Washington Univ, PhD(bot), 54. *Prof Exp:* Res assoc, Washington Univ, 54-55, asst prof bot, 55-57; assoc prof, 57-67, PROF BOT, OKLA STATE UNIV, 67- *Mem:* Am Soc Plant Physiologists; Bot Soc Am; Weed Sci Soc Am. *Res:* Mechanisms of herbicidal action; translocation of herbicides and auxins; metabolism of herbicides. *Mailing Add:* Dept Bot & Microbiol Okla State Univ Stillwater OK 74074

BASLER, ROY PRENTICE, b Florence, Ala, Aug 12, 35; m 58; c 5. RADIOPHYSICS, IONOSPHERIC PHYSICS. *Educ:* Hamilton Col, AB, 56; Univ Alaska, MS, 61, PhD(physics), 64. *Prof Exp:* Res assoc, Geophys Inst, Univ Alaska, 58-64; staff physicist, ITT Electro-Physics Labs, 64-73; SR PHYSICIST, SRI INT, 73- *Concurrent Pos:* Mem comn III, Int Sci Radio Union. *Mem:* Am Geophys Union; AAAS; Fedn Am Scientists. *Res:* High frequency radio propagation in the ionosphere search for extraterrestrial intelligence; solar-terrestrial relations; solar power satellites; ionospheric radio wave absorption; scintillation of satellite radio signals. *Mailing Add:* 311 Claire Pl Menlo Park CA 94025

BASMADJIAN, DIRAN, b Dresden, Ger, Mar 4, 29; Can citizen. CHEMICAL ENGINEERING. *Educ:* Swiss Fed Inst Technol, Dipl Ing Chem, 52; Univ Toronto, MASc, 54, PhD(chem eng), 59. *Prof Exp:* Sci officer, Fuels Div, Dept Mines & Tech Survs, Can, 58-59; from asst prof to assoc prof chem eng, 59-72, PROF CHEM ENG, UNIV TORONTO, 72- *Concurrent Pos:* Consult, Chem Proj Ltd, Fibreglas Can Ltd, St Lawrence Cement Co, B F Goodrich Can Ltd, Spar Aerospace Ltd & Aerofall Mills Ltd. *Res:* Industrial gas adsorption and drying operations; mathematical analysis and modelling of heat and mass transport phenomena; biophysical studies of kidney function. *Mailing Add:* Dept Chem Eng Univ Toronto Toronto ON M5S 1A1 Can

BASMAJIAN, JOHN ARAM, b Binghamton, NY, Mar 10, 29; c 1. METALLURGY. *Educ:* Purdue Univ, BS, 56; Rensselaer Polytech Inst, MS, 61. *Prof Exp:* Process metallurgist, Knolls Atomic Power Lab, 58-59, irradiations engr, 59-62; staff mem, Los Alamos Sci Lab, Univ Calif, 62-67; sr res scientist, Pac Northwest Labs, Battelle Mem Inst, 67-70; prin engr, Westinghouse Hanford Co, 70-88; CONSULT, 88- *Mem:* Am Inst Mining, Metall & Petrol Engrs; Am Soc Metals; Am Nuclear Soc; NY Acad Sci. *Res:* Irradiation effects in materials; physical metallurgy of ferrous and nonferrous metals; nuclear metallurgy; refractory metal technology; liquid metal technology; plutonium and uranium metallurgy; nuclear ceramic materials. *Mailing Add:* Westinghouse Hanford Co PO Box 445 Richland WA 99352

BASMAJIAN, JOHN V, b Constantinople, Turkey, June 21, 21; Can citizen; m 47; c 3. NEUROSCIENCES. *Educ:* Univ Toronto, MD, 45; FRCP(C), FACRM. *Prof Exp:* Demonstr anat, Univ Toronto, 46-47, lectr, 49-51, from asst prof to prof, 51-57; prof & head dept, Queen's Univ, Ont, 57-69; prof anat & phys med & dir regional rehab ctr, Emory Univ, 69-77; prof med & dir rehab progs, 77-86, EMER PROF MED & ANAT, MCMASTER UNIV & CHEDOKE HOSPS, 86- *Concurrent Pos:* Asst resident surg, Hosp for Sick Children, 47-48, res assoc, 55-57; clin asst phys med, St Thomas Hosp, London, 53; hon secy, Banting Res Found, 55-57; Nat Res Coun exchange scientist, Soviet Acad Sci, 63; chief neurophysiol lab, Ga Ment Health Inst, 69-77; JCB grant award, 85. *Honors & Awards:* Starr Medal, 56; Kabakjian Award, 65; Gold Key Recipient, Am Cong Rehab Med, 77. *Mem:* AAAS; fel Am Col Angiol; Am Acad Neurol; Am Asn Anat; Can Asn Anat; fel Soc Behav Med; fel Royal Col Physicians. *Res:* Electromyography of normal anatomic and physiologic functions and of diseased muscle; nerve-muscle electrophysiology; studies of normal vascular patterns; biomechanics; bioengineering; psychophysiology. *Mailing Add:* 106 Forsyth N Hamilton ON L8S 4E4 Can

BASOLO, FRED, b Coello, Ill, Feb 11, 20; m 47; c 4. INORGANIC CHEMISTRY. *Educ:* Southern Ill Norm Univ, BEd, 40; Univ Ill, MS, 42, PhD(chem), 43. *Hon Degrees:* DSC, Southern Ill Univ, Carbondale, 84. *Prof Exp:* Res chemist, Rohm and Haas Chem Co, Pa, 43-46; from instr to assoc prof, 46-59, chmn dept, 69-72, PROF CHEM, NORTHWESTERN UNIV, EVANSTON, ILL, 59-, MORRISON PROF, 80- *Concurrent Pos:* Guggenheim fel, Tech Univ Denmark, 54-55; NSF sr fel, Inst Inorg Chem, Univ Rome, 61-62; NATO vis prof, WGermany, 69; bd trustees & bd chmn, Gordon Res Conf, 76; chmn chem sect, AAAS, 79; mem bd chem sci & technol, Nat Acad Sci-Nat Res Coun, 81-86. *Honors & Awards:* Award in

Inorg Chem, Am Chem Soc, 64, Award for Mechanisms of Inorg Reactions, 71 & Award for Distinguished Serv in Inorg Chem, 75; Bailar Medalist, 73; Dwyer Medalist, 77. *Mem:* Nat Acad Sci; Sigma Xi; Am Chem Soc (preselect, 82, pres, 83); fel Japanese Soc Prom Sci; AAAS; fel Am Acad Arts & Sci; Italian Acad Sci. *Res:* Coordination compounds; reaction mechanisms of inorganic complex compounds; metal nitrenes; synthetic oxygen-carriers. *Mailing Add:* Dept Chem Northwestern Univ Evanston IL 60201

BASORE, B(ENNETT) L(EE), b Oklahoma City, Okla, Aug 31, 22; div; c 3. COMMUNICATIONS THEORY, SCIENCE. *Educ:* Okla Agr & Mech Col, BS(math) & BS, 48; Mass Inst Technol, ScD(elec eng), 52. *Prof Exp:* Asst, Okla Agr & Mech Col, 46-48; asst, Mass Inst Technol, 48-49, res asst, res lab of electronics, 50-52, asst group leader, Lincoln Lab, 52; staff mem, Sandia Corp, 52-57; assoc tech dir, Dikewood Corp, 57-63; phys sci officer, US Arms Control & Disarmament Agency, 63-67; PROF ELEC ENG, OKLA STATE UNIV, 67-, HEAD, GEN ENG, 78- *Concurrent Pos:* Indust electronics fel, 51; interim assoc dean, Okla State Univ, 87-; consult, US Air Force, Jet Propulsion Lab, US Arms Control & Disarmament Agency, Sandia Nat Labs, Battelle-Columbus. *Mem:* Inst Elec & Electronics Engrs; Am Soc Eng Educ; Sigma Xi. *Res:* Communication theory and signal processing; radar systems; stochastic systems. *Mailing Add:* Dept Elec Eng Okla State Univ Stillwater OK 74078

BASRI, SAUL ABRAHAM, b Baghdad, Iraq, Feb 15, 26; nat US; m 50; c 2. ELEMENTARY PARTICLE PHYSICS. *Educ:* Mass Inst Technol, BS, 48; Columbia Univ, PhD(physics), 53. *Prof Exp:* Res asst physics, Columbia Univ, 52-53, res scientist, 53; from asst prof to assoc prof, 53-67, PROF PHYSICS, COLO STATE UNIV, 67- *Concurrent Pos:* Fulbright lectr, Univ Rangoon, Burma, 56 & Univ Ceylon, 65-66; vis prof, Israel Inst Technol, 73-74. *Mem:* Am Phys Soc; Am Asn Physics Teachers. *Res:* Elementary particle theory; algebraic s-matrix theory. *Mailing Add:* 500 Manhattan Dr B Four Boulder CO 80303

BASS, ALLAN DELMAGE, b Marcus, Iowa, Feb 12, 10; m 44; c 2. PHARMACOLOGY. *Educ:* Simpson Col, BS, 31; Vanderbilt Univ, MS, 32, MD, 39. *Prof Exp:* Intern, Vanderbilt Hosp, 39-40, asst resident physician, 40-41, resident & instr med, 43-44; instr pharmacol, Yale Univ, 42-43; prof pharmacol & chmn dept,Syracuse Univ, 45-55; prof & chmn dept, 52-73, dir neurosci prog, 71-73, assoc dean biomed sci, 73-75, actg dean med sch, 73-74, EMER PROF PHARMACOL, VANDERBILT UNIV, 75- *Concurrent Pos:* Mem Gen Med Res Prog Proj Comt, NIH, 64-67; mem, US Food & Drug Admin, 67-70; consult, Div Res Grants, NIH Res Anal & Eval, 68; mem Res Adv Comt, NIMH, 68-70; chmn bd, Tenn Neuropsychiat Inst, 70-75; Nat Bd Med examr & Ann Rev Pharmacol; mem ed bd, Int Quart Sci Rev J, 73-; mem Drug Res Bd, Nat Acad Sci, 74-76; assoc chief staff educ, Vet Admin Hosp, Nashville, Tenn, 75-81; interim dir, Tenn Neuropsychiat Inst, Nashville, 81-84; consult, Soc Security Disability Serv, 85- *Mem:* AAAS; Am Soc Pharmacol & Exp Therapeut (pres, 67-68); AMA; Am Col Physicians; Fedn Am Socs Exp Biol. *Res:* Autonomic and endocrine pharmacology; neuropsychopharmacology; steroid metabolism; cyanide poisoning; hypnotics; nucleoproteins. *Mailing Add:* Vanderbilt Univ Med Sch Nashville TN 37232

BASS, ARNOLD MARVIN, b New York, NY, Dec 22, 22; m 47; c 3. PHYSICS, PHYSICAL CHEMISTRY. *Educ:* City Col New York, BS, 42; Duke Univ, MA, 43, PhD(physics), 49. *Prof Exp:* Asst physics, Duke Univ, 42-43, instr, 43-44; staff mem lab insulation res, Mass Inst Technol, 49-50; physicist, Temperature Measurement Sect, 50-54, asst chief, 54-56, from asst chief to chief, Free Radicals Res Sect, 56-61, physicist, 61-66, chief, Molecular Energy Levels Sect, Heat Div, 66-69, asst chief, Phys Chem Div, 69-78, res physicist, Chem Kinetics Div, 78-79, RES PHYSICIST, GAS & PARTICULATE SCI DIV, NAT BUR STANDARDS, 79- *Honors & Awards:* Gold Medal, US Dept Com, 60. *Mem:* Fel Optical Soc Am; fel Am Phys Soc. *Res:* Fluorescence spectroscopy; molecular spectra and structure; spectra of liquids, solids, flames, hot gases and trapped radicals; fluorescence and vacuum ultraviolet spectra; infrared emissivities of hot gases; temperature measurements by radiation methods; low temperature spectra of solids; flash photolysis and pyrolysis; atmospheric spectroscopy and chemistry; environmental measurements for ozone. *Mailing Add:* 11920 Cold Stream Dr Potomac MD 20854

BASS, ARTHUR, b New York, NY, Apr 7, 41; m 66; c 2. ENVIRONMENTAL INFORMATION MANAGEMENT SYSTEMS, APPLIED REMOTE SENSING. *Educ:* Columbia Univ, BA, 61; Yale Univ, MS, 62; Mass Inst Technol, PhD(meteorol), 74. *Prof Exp:* Staff scientist, Mitre Corp, 63-68; sr scientist, Am Sci & Eng Inc, 68-69; res asst, Dept Meteorol, Mass Inst Technol, 69-74; res scientist, Flow Res Inc, 74-75; prin meteorologist, 75-81, dept mgr, Tech Develop Staff, Environ Res & Technol Inc, 81-85; MGR, ADVANCED PROG DEVELOP, PHYS RES, THE ANALYTICAL SCI CORP, 85- *Mem:* Am Meteorol Soc. *Res:* Meteorology; satellite remote sensing; air pollution meteorology. *Mailing Add:* 119 Parker St Newton MA 02159-2545

BASS, DANIEL MATERSON, JR, petroleum engineering, for more information see previous edition

BASS, DAVID A, b Oklahoma City, Okla, June 10, 41; m; c 2. MEDICINE. *Educ:* Yale Univ, BA, 64; Johns Hopkins Univ, MD, 68; Oxford Univ, Eng, DPhil, 73; Am Bd Internal Med, cert, 74. *Prof Exp:* Intern, Osler Med Serv, Johns Hopkins Hosp, 68-69, jr resident, 69-70, sr resident, 73-74; res fel, Nuffield Dept Med, Oxford Univ, Eng, 70-73, Commonwealth Fund fel, Radcliffe Infirmary, 70-73; from instr to asst prof, Dept Med, Univ Tex Southwestern Med Sch, 74-76; from asst prof to assoc prof, Dept Med, 76-83, ASSOC BIOCHEM, BOWMAN GRAY SCH MED, 81-, PROF, DEPT MED, 83-, SECT HEAD, SECT PULMONARY MED, 88- *Concurrent Pos:* NIH res fel infectious dis, Univ Tex Southwestern Med Sch, 74-76, grants, 79-; vis prof, Cambridge, Eng, 80; assoc dir, Oncol Res Ctr, Bowman Gray Sch Med, dir, Cell Biol Prog; affil, NC Bapt Hosp, Winston-Salem. *Mem:* Am Soc Clin Invest; fel Am Col Chest Physicians; fel Am Col Physicians; Am Soc Biol Chemists; Am Asn Immunologists; Am Thoracic Soc; Am Fedn Clin Res; Am Soc Microbiol; Reticuloendothelial Soc. *Res:* Elucidation of mechanisms of priming and stimulation of the bactericidal respiratory burst of human neutrophils. *Mailing Add:* Sect Pulmonary Med Dept Med Bowman Gray Sch Med 300 S Hawthorne Rd Winston-Salem NC 27103

BASS, DAVID ELI, ENVIRONMENTAL PHYSIOLOGY. *Educ:* Boston Univ, PhD(med physiol), 53. *Prof Exp:* Vis prof physiol, Univ Nev, Las Vegas, 79-81; RETIRED. *Res:* Human acclimatization to heat and cold. *Mailing Add:* 24 Gray Birch Terr Newtonville MA 02160

BASS, EUGENE L, b Brooklyn, NY, Sept 27, 42; m 68; c 1. ANIMAL PHYSIOLOGY, ZOOLOGY. *Educ:* City Univ NY, BS, 64; Univ Mass, PhD(zool), 70. *Prof Exp:* Asst prof biol, Salisbury State Col, 69-72; asst prof, 72-83, ASSOC PROF BIOL, UNIV MD EASTERN SHORE, 83- *Concurrent Pos:* Prin investr, USDA, 73-78, NIH, 81-85. *Mem:* Am Soc Zoologists; Am Inst Biol Sci; Sigma Xi. *Res:* Use of alternatives to vertebrate animals in toxicity testing and risk assessment; neuro- and neuromuscular toxicology of environmental contaminants. *Mailing Add:* PO Box 123 Fruitland MD 21826-0123

BASS, GEORGE F, b Columbia, SC, Dec 9, 32; m 60; c 2. NAUTICAL ARCHAEOLOGY, PRECLASSICAL ARCHAEOLOGY. *Educ:* Johns Hopkins Univ, MA, 55; Univ Pa, PhD(class archaeol), 64. *Hon Degrees:* Dr, Bogazici Univ, Istanbul, 87. *Prof Exp:* From asst prof to assoc prof class archaeol, Univ Pa, 64-73; prof, 76-80, DISTINGUISHED PROF ANTHROP, TEX A&M UNIV, 80- *Concurrent Pos:* Geddes-Harrower prof Greek art & archaeol, Univ Aberdeen, Scotland, 84. *Honors & Awards:* Nogi Award, Underwater Soc Am, 74; LaGorce Medal, Nat Geog Soc, 79; Gold Medal for Distinguished Achievement, Archaeol Inst Am, 86; Centennial Award, Nat Geog Soc, 88. *Mem:* Inst Nautical Archaeol (pres, 73-82); Archaeol Inst Am; Asn Field Archaeol; Am Orient Soc; Soc Hist Archaeol. *Res:* Excavations of ancient shipwrecks; preclassical eastern Mediterranean cultures. *Mailing Add:* 1600 Dominik Dr College Station TX 77840

BASS, HENRY ELLIS, b Tulsa, Okla, Aug 31, 43; m 78; c 4. PHYSICAL ACOUSTICS, CHEMICAL PHYSICS. *Educ:* Okla State Univ, BS, 65, PhD(physics), 71. *Prof Exp:* From assoc prof to prof, 70-88, DISTINGUISHED PROF PHYSICS, UNIV MISS, 88- *Concurrent Pos:* Staff officer, Asst Secy Army Res Develop & Acquisition, Washington, DC, 71-; mem working group S1-57, Am Nat Standards Inst, 72; pres, Arnold, Bass & Assoc, sci consults, 74-; mem res study group RSG-11, NATO, 82-; assoc ed, J Acoust Soc Am. *Honors & Awards:* Biennial Award, Acoust Soc Am, 78. *Mem:* Fel Acoust Soc Am; Sigma Xi. *Res:* Physical acoustics, especially molecular relaxation in the gas phase and long range sound propagation; chemical kinetics, especially molecular energy transfer. *Mailing Add:* Dept Physics & Astron Univ Miss University MS 38677

BASS, HYMAN, b Houston, Tex, Oct 5, 32; m 58; c 3. MATHEMATICS. *Educ:* Princeton Univ, BA, 55; Univ Chicago, MS, 56, PhD(math), 59. *Prof Exp:* Ritt instr math, Columbia Univ, 59-62; NSF fel, Col of France, 62-63; from asst prof to assoc prof, 63-65, chmn dept, 75-78, PROF MATH, COLUMBIA UNIV, 65- *Concurrent Pos:* Chmn dept, Barnard Col, Columbia Univ, 64-65; vis prof & lectr, numerous US & foreign univs, 62-; Sloan fel, 64-66; Guggenheim fel, Inst Higher Sci Studies, Paris, 68-69; ed, J Indian Math Soc, 68-70; Cambridge Tracts Pure & Appl Math, 68-; Bull Am Math Soc, 69-72 & 82-86; J Pure & Appl Algebra, 70-86; Am J Math, 71-77, Commun Algebra, 74-85 & Ann Math, 78-84. *Honors & Awards:* Hedrich Lectr, Math Asn Am, 68; Philips Lectr, Haverford Col, 70 & 75; Cole Prize Algebra, Am Math Soc, 75; Barrett Mem Lectr, Univ Tenn, 78; Karcher Lectr, Univ Okla, 79. *Mem:* Nat Acad Sci; Am Math Soc (vpres, 78-80); Math Soc France; London Math Soc; Math Asn Am; fel AAAS; Am Acad Arts & Sci. *Res:* Homological algebra; algebraic number theory; algebraic geometry; algebraic K-theory; group theory; author of 71 technical publications. *Mailing Add:* Dept Math Columbia Univ New York NY 10027

BASS, JACK, b New York, NY, Apr 1, 38; m 59; c 3. METAL PHYSICS, LOW TEMPERATURE PHYSICS. *Educ:* Calif Inst Technol, BS, 59; Univ Ill, MS, 61, PhD(physics), 64. *Prof Exp:* From asst prof to assoc prof, 64-73, chmn physics, 83-88, PROF PHYSICS, MICH STATE UNIV, 73- *Concurrent Pos:* Guest docent, Swiss Fed Inst Technol, 70-71; vis prof, Catholic Univ, Nijmegen, Netherlands, 78-79 & Bar-Ilan Univ, Israel, 79; vis prof, Max-Planck-Inst, Grenoble, France, 88-89. *Mem:* Am Phys Soc; Am Asn Physics Teachers; AAAS; Sigma Xi. *Res:* Electron transport properties of metals; low and ultra-low temperature physics; point defects in metals; artifical metallic superlattices; spin glasses. *Mailing Add:* Dept Physics Col Natural Sci Mich State Univ East Lansing MI 48824

BASS, JAMES W, b Shreveport, La, May 25, 30; m; c 2. PEDIATRICS, INFECTIOUS DISEASES. *Educ:* Tulane Univ, BS, 52, MPH, 68; La State Univ, New Orleans, MD, 57. *Prof Exp:* Researcher infectious dis, US Walter Reed Army Inst Res, Md, 60-63; asst chief pediat, Madigan Gen Hosp, Tacoma, Wash, 63-66; instr pediat, Sch Med, Tulane Univ & vis pediatrician, Charity Hosp La, New Orleans, 66-68; from asst dir to dir intern training, 68-71, asst chief pediat serv, 68, chief dept pediat, Tripler Army Med Ctr, Honolulu, 68-75; from asst clin prof to assoc clin prof pediat, Sch Med, Univ Hawaii, 68-73, clin prof pediat, Sch Med, Univ Hawaii, Manoa, 73-75 & trop med & med microbiol, 74-75; chief, Dept Pediat, Walter Reed Army Med Ctr, 75-81; prof & chmn, Dept Pediat, Sch Med, Uniformed Servs Univ Health Sci, 76-81; CHIEF, DEPT PEDIAT, TRIPLER ARMY MED CTR, HONOLULU, 81- *Concurrent Pos:* From asst clin prof to assoc clin prof pediat, Sch Med, Univ Hawaii, 68-73; consult pediat infectious dis, Kauikeolani Childrens Hosp, Honolulu, 70-75; assoc clin prof pediat, Univ Southern Calif, 72-; consult pediat, Surgeon General of US Army, 75-81; clin prof pediat, Sch Med, Georgetown Univ, Washington, DC, 75-78, prof, 78-81. *Mem:* AMA; Soc Pediat; Am Pediat Soc; Infectious Dis Soc Am; fel Am Col Physicians. *Res:* Microbiology; infectious diseases; fel Am Col Physicians. Res: Microbiology. *Mailing Add:* Dept Pediat Tripler Army Med Ctr Tripler AMC HI 96859-5000

BASS, JON DOLF, b Saginaw, Mich, Oct 31, 33; m 59; c 4. PHOTOGRAPHIC CHEMISTRY, PHOTOGRAPHIC PATENTS. *Educ:* Univ Mich, BS, 56; Univ Wis, PhD(org chem), 60. *Prof Exp:* Lab asst, Mellon Inst, 51-52; teaching asst, Univ Wis, 58-59; res chemist, Air Force Cambridge Res Labs, Mass, 60-62; res fel chem, Harvard Univ, 62-63; from res chemist to sr res chemist, 63-70, RES ASSOC, KODAK RES LABS, 70- *Mem:* Am Chem Soc; Soc Photog Sci & Eng. *Res:* organic semiconductors and photoconductors; photochemical organic synthesis; photographic chemistry; physical development; lithographic chemistry; spectral sensitization and image amplification processes; interrelation of patents, innovation, and technology transfer; technology forecasting. *Mailing Add:* Eastman Kodak Co Kodak Res Labs Bldg 82A Rochester NY 14650-2131

BASS, JONATHAN LANGER, b New York, NY, May 12, 38; m 71; c 2. PHYSICAL CHEMISTRY. *Educ:* Univ Rochester, BS, 59; Univ Minn, Minneapolis, PhD(phys chem), 65. *Prof Exp:* Chemist, maintenance lab, hq Ogden Air Mat Area, Hill AFB, Utah, 65-66 & weapons lab, Kirtland AFB, NMex, 66-68; chemist, Washington Res Ctr, W R Grace & Co, 68-77; supvr mat eval, 77-85, mgr anal characterization & testing, 85-90, PRIN SCIENTIST, PQ CORP, 90- *Mem:* Am Chem Soc; Soc Appl Spectros; Catalysis Soc; Am Soc Testing & Mat. *Res:* Characterization of inorganic oxides by infrared spectroscopy, porosimetry, scanning electron microscopy and x-ray analysis; degradation and tensile properties of polymers; atomic spectroscopy; thermal analysis; wet chemical analysis of inorganics; chemical modification of surfaces, separation and purification. *Mailing Add:* 39 Lee Rd Audubon PA 19403-2003

BASS, LEONARD JOEL, b Kalamazoo, Mich, Oct 30, 43; m 67; c 3. COMPUTER SCIENCES. *Educ:* Univ Calif, Riverside, BA, 64, MA, 66; Purdue Univ, PhD(comput sci), 70. *Prof Exp:* Prof computer sci, Univ RI, 70-86; SR MEM TECH STAFF, SOFTWARE ENG INST, CARNEGIE MELLON UNIV, 86- *Mem:* Asn Comput Mach; Inst Elec & Electronics Engrs. *Res:* Software engineering; data base management systems; user interfaces. *Mailing Add:* Software Eng Inst Carnegie Mellon Univ Pittsburgh PA 15213

BASS, MANUEL N, b Houston, Tex, July 20, 27; m 63; c 5. GEOLOGY, GEOCHEMISTRY. *Educ:* Calif Inst Technol, BS, 49, MS, 51; Princeton Univ, PhD(geol), 56. *Prof Exp:* Geologist, US Geol Surv, 51; asst prof geol, Northwestern Univ, 56-58; fel, Carnegie Inst, 58-60; asst prof geol, Calif Inst Technol, 60-62; asst prof, Univ Calif, San Diego, 62-69; mem staff, Geochem Br, NASA, 69-74; MEM STAFF, CHEVRON OIL FIELD RES CO, 74- *Mem:* Geol Soc Am; Mineral Soc Am; Am Geophys Union; Am Asn Petrol Geol. *Res:* Tectonics of age of Canadian Shield and buried Precambrian rocks of Central United States; sandstone petrology. *Mailing Add:* Chevron Oil Field Res Co PO Box 446 La Habra CA 90633-0446

BASS, MARY ANNA, b Clanton, Ala, June 1, 30; m 53; c 3. FOOD SCIENCE, NUTRITION. *Educ:* Ala Col Women, BS, 51; Walter Reed Army Hosp, dipl, 52; Univ Ky, MS, 56; Kans State Univ, PhD(food & nutrit), 72. *Prof Exp:* Instr food & nutrit, Univ Nebr, 59-60 & Univ Kans, 60-68; asst prof foods & nutrit, Col Home Econ, Univ Tenn, 71-77; DIR COMMUNITY FOOD & NUTRIT SERV, KNOXVILLE, 77- *Concurrent Pos:* Consult, Standing Rock Indian Reservation, 70-80, Human Resources Inst, Eastern Band Cherokee Indians, 75-; lectr, Dept Anthrop, Univ Tenn, 77- *Mem:* Am Dietetic Asn; Sigma Xi; Soc Nutrit Educ; Nutrit Today Soc; Am Anthrop Asn. *Res:* Determining food intake patterns and identifying those factors that influence them. *Mailing Add:* 8201 Bennington Dr Knoxville TN 37909

BASS, MAX H(ERMAN), b Troy, Ala, Dec 27, 34; m 56; c 4. ENTOMOLOGY. *Educ:* Troy State Col, BS, 57; Ala Polytech Inst, MS, 59; Auburn Univ, PhD(entom), 64. *Prof Exp:* Instr zool, Auburn Univ, 59-60, res entom, 60-80, from asst prof to prof entom, 63-80; PROF & HEAD, DEPT ENTOM, COASTAL PLAIN STA, UNIV GA, 80- *Mem:* Sigma Xi; Entom Soc Am. *Res:* Bionomics; control of insects affecting soybeans and peanuts. *Mailing Add:* PO Box 748 Tifton GA 31793-0748

BASS, MICHAEL, b New York, NY, Oct 24, 39; m 62; c 2. LINEAR OPTICS. *Educ:* Carnegie Mellon Univ, BS, 60; Univ Mich, MS, 62, PhD(physics), 64. *Prof Exp:* Actg asst prof physics, Univ Calif, Berkeley, 64-66; sr res scientist, res div, Raytheon Co, 66-73; from assoc dir to dir, Ctr Laser Studies, Univ Southern Calif, 73-84, prof elec eng, 81-84, chmn dept, 84-89; VPRES RES, UNIV CENT FLA, 89- *Concurrent Pos:* Consult, 73- *Mem:* Fel Inst Elec & Electronics Engrs; fel Optical Soc Am; Laser Inst Am; Sigma Xi; AAAS. *Res:* Light-matter interactions; laser damage; optical properties of matter; nonlinear optics; optically pumped solid state laser. *Mailing Add:* Vpres Res Univ Cent Fla PO Box 25000 Orlando FL 32816-0150

BASS, MICHAEL LAWRENCE, b Eskdale, WVa, Jan 10, 45; m 68; c 2. ANIMAL PHYSIOLOGY, ECOLOGY & TOXICOLOGY. *Educ:* Va Polytech Inst, BS, 66; Med Col Va, MS, 70; Va Polytech Inst & State Univ, PhD(zool), 75. *Prof Exp:* Instr biol, 68-73, asst prof, 74-77, ASSOC PROF BIOL, 77-, CHMN BIOL SCI, MARY WASHINGTON COL, 88- *Concurrent Pos:* Consult, Southeastern Inst Res, 70-; consult, Sharpley Labs Inc, 75-76, vpres, 76-78. *Mem:* AAAS; Am Soc Zoologists; Am Inst Biol Sci; Sigma Xi. *Res:* Environmental toxicology; aquatic ecology; physiological effects of pollutants on aquatic invertebrates and vertebrates with special interests on the effects of the interaction of temperature and toxicity. *Mailing Add:* Dept Biol Mary Washington Col Fredericksburg VA 22401

BASS, NORMAN HERBERT, b New York, NY, July 10, 36; m 61; c 3. NEUROCHEMISTRY, NEUROLOGY. *Educ:* Swarthmore Col, BA, 58; Yale Univ, MD, 62. *Prof Exp:* Intern med, Sch Med, Univ Wash, 62-63; resident neurol, Sch Med, Univ Va, 63-65; Nat Inst Neurol & Commun Disorders & Stroke fels neurochem, McLean Hosp Res Lab, Harvard Univ, 65-67; asst prof neurol, Sch Med, Univ Va, 67-70, assoc prof pharmacol, 70-73, prof neurol & dir, Clin Neurosci Res Ctr, 74-79; prof neurol & chmn dept,

Univ Ky Med Ctr, 79-85; PROF NEUROL & DEAN, MED COL GA, 85- *Concurrent Pos:* Markle scholar award, 69-74; Nat Inst Neurol & Commun Disorders & Stroke res career develop award, 71-76; vis prof, Inst Pharmacol, Univ Gothenburg, Sweden, 72-73; assoc ed, Neurochem Res, 78. *Honors & Awards:* S Weir Mitchell Res Award, Am Acad Neurol, 67. *Mem:* Am Soc Neurochem; Asn Univ Professors Neurol; fel Am Acad Neurol; Am Neurol Asn; Am Asn Anatomists. *Res:* Neurochemistry of developing brain; microchemistry of cerebral cortex maturation; malnutrition and hormonal imbalance during critical periods of brain development; experimental epilepsy; blood-brain-barrier; inborn errors of human brain metabolism; neuropharmacology of morphine abuse. *Mailing Add:* Dept Neurol Univ Ky Med Ctr Lexington KY 40536

BASS, PAUL, b Winnipeg, Man, Aug 12, 28; m 53; c 2. PHARMACOLOGY, PHYSIOLOGY. *Educ:* Univ BC, BSP, 53, MA, 55; McGill Univ, PhD(pharmacol), 57. *Prof Exp:* Asst, Ayerst, McKenna & Harrison, Can, 56; res pharmacologist, Parke Davis & Co, 60-70; PROF PHARMACOL, SCH PHARM & SCH MED, UNIV WIS-MADISON, 70- *Concurrent Pos:* Fel biochem, McGill Univ, 57-58; fel physiol, Mayo Found, 58-60. *Mem:* Am Soc Pharmacol; Soc Exp Biol & Med; Pharmacol Soc Can; Am Gastroenterol Asn. *Res:* Motility, secretion and absorption of the gastro-intestinal tract; cardiovascular and autonomic nervous system pharmacology. *Mailing Add:* Sch Pharm Univ Wis Madison WI 53706

BASS, ROBERT GERALD, b Durham, NC, May 2, 33; m 59; c 2. ORGANIC CHEMISTRY, POLYMER CHEMISTRY. *Educ:* Va Polytech Inst, BS, 54; Univ Va, PhD(chem), 61. *Prof Exp:* Res chemist, E I du Pont de Nemours & Co, Inc, 59-62; assoc prof, 62-72, PROF CHEM, VA COMMONWEALTH UNIV, 72- *Concurrent Pos:* Fel, Univ Va, 62. *Mem:* Am Chem Soc; Sigma Xi; SAMPE; AAAS. *Res:* Organic chemistry; thermally stable high performance polyers, synthesis and reactions of heterocyclic compounds; acetylenic compounds, carbenoid insertion reactions, conjugated systems, enol-keto and ring-chain tautomerism; mesoionic compounds. *Mailing Add:* Dept Chem Va Commonwealth Univ Richmond VA 23284-2006

BASS, STEVEN CRAIG, b Indianapolis, Ind, July 29, 43; m 65, 89; c 2. COMPUTER ENGINEERING. *Educ:* Purdue Univ, BSEE, 66, MSEE, PhD(elec eng), 71. *Prof Exp:* From asst prof to prof elec eng, Purdue Univ, West Lafayette, 71-88; PROF ELEC & COMPUTER ENG, GEORGE MASON UNIV, FAIRFAX, VA, 88- *Concurrent Pos:* Prin engr, Mitre Corp, 88. *Mem:* Inst Elec & Electronic Engrs; Audio Eng Soc. *Res:* Digital signal processing; computer-aided circuit design; very large scale integration of circuits; computer architecture. *Mailing Add:* Dept Elec & Computer Eng George Mason Univ 4400 University Dr Fairfax VA 22030

BASS, WILLIAM MARVIN, III, b Staunton, Va, Aug 30, 28; m 53; c 3. PHYSICAL ANTHROPOLOGY, ANTHROPOMETRICS. *Educ:* Univ Va, BA, 51; Univ Ky, MS, 56; Univ Pa, PhD(anthrop), 61; Am Bd Forensic Anthrop, dipl, 78. *Prof Exp:* Instr phys anthrop & anat, Grad Sch Med, Univ Pa, 56-60; instr anthrop, Univ Nebr, 60; from instr to prof, Univ Kans, 60-71; PROF ANTHROP & HEAD DEPT, UNIV TENN, KNOXVILLE, 71- *Concurrent Pos:* Res grant, Univ Kans, 61-68; NSF res grants, Smithsonian Inst, 62 & 63; Nat Park Serv grant, 63 & 66; NSF grants, 65, 67 & 69; Nat Geog grant, 68. *Mem:* Am Asn Phys Anthrop; Am Anthrop Asn; Soc Am Archaeol; Am Acad Forensic Sci. *Res:* Identification of human skeletal material. *Mailing Add:* Dept Anthrop 252 S Stadium Hall Univ Tenn Knoxville TN 37916

BASS, WILLIAM THOMAS, b Jefferson City, Tenn, Feb 18, 42; m 64; c 3. PHYSICS. *Educ:* Carson Newman Col, BS, 64; Univ Tenn, PhD(physics), 70. *Prof Exp:* ASSOC PROF PHYSICS, MACON JR COL, 70- *Concurrent Pos:* Instr, Governor's Hon Prog, Ga; adj prof, Wesleyan Col, 75 & 81. *Mem:* Am Asn Physics Teachers. *Res:* Nuclear-low-energy reactions; time-of-flight measurements; educational uses of computers & microcomputers. *Mailing Add:* 547 Billingswood Dr Macon GA 31210

BASSECHES, HAROLD, b Brooklyn, NY, Nov 27, 23; m 46; c 3. PHYSICAL CHEMISTRY. *Educ:* City Col New York, BS, 43; Ohio State Univ, PhD(phys chem), 51. *Prof Exp:* Mem tech staff, 52-59, supvr semiconductor mat group, Pa, 59-62, magnetic mat group, 62-67, head film technol dept, 67-70, HEAD THIN FILM MAT & TECHNOL DEPT, BELL TEL LABS, NJ, 70- *Mem:* Am Chem Soc; fel Am Inst Chem; Sigma Xi. *Res:* Preparation and processing of materials whose electric and magnetic properties are of use in modern solid state devices; development of thin film materials and processes for hybrid integrated circuits. *Mailing Add:* 618 N Glenwood St 555 Union Blvd Allentown PA 18104

BASSEL, ROBERT HAROLD, b Johnstown, Pa, Feb 16, 28; c 1. NUCLEAR PHYSICS. *Educ:* Univ Pittsburgh, BSc, 52, PhD(physics), 58. *Prof Exp:* Physicist, Oak Ridge Nat Lab, 58-66; assoc physicist, Brookhaven Nat Lab, 66-68; assoc prof physics, Univ Pittsburgh, 68-69; RES PHYSICIST, NAVAL RES LAB, 64- *Mem:* Fel Am Phys Soc; fel AAAS. *Res:* Nuclear and atomic reaction; evaluations of neutron induced reaction on cross sections. *Mailing Add:* Naval Res Lab Code 6672 Washington DC 20375

BASSETT, ALTON HERMAN, b Hartford, Conn, Nov 27, 30; m 56; c 2. TEXTILE CHEMISTRY. *Educ:* Middlebury Col, BA, 53. *Prof Exp:* Staff chemist, Am Viscose Corp, 55-58; proj dir sanit & surg prod, 58-60, dir prof develop, 60-69, asst dir res, 69-74, DIR RES, CHICOPEE MFG CO DIV, JOHNSON & JOHNSON, 74- *Mem:* AAAS; Am Asn Textile Technologists; Am Chem Soc. *Res:* Concentrated effort in development of new surgical dressings and applications; new textile constructions, nonwovens. *Mailing Add:* 73 Harriet Dr Princeton NJ 08540

BASSETT, ARTHUR L, b New York, NY, Feb 26, 35; m 60; c 2. CARDIOVASCULAR PHYSIOLOGY, PHARMACOLOGY. *Educ:* City Col New York, BS, 55; State Univ NY, PhD(physiol), 65. *Prof Exp:* Technician, Med Ctr, NY Univ, 56-58; res asst, State Univ NY Downstate Med Ctr, 58-60, asst physiol, 60-61 & 63- 65; fel pharmacol, Col Physicians & Surgeons, Columbia Univ, 65-67, assoc, 67-68, asst prof pharmacol, 68-73; assoc prof pharmacol & surg, 73-76, dir grad affairs comt, dept pharmacol, 75-82, PROF PHARMACOL, MED SCH, UNIV MIAMI, 76-; DIR, NIH PHARMACOL TRAINING GRANT, 76-82, 85- *Concurrent Pos:* Polachek Found fel med res, 67-; Nat Heart Inst res career develop award, 69-73; grants, AMA, 69-72, Heart Asn Gtr Miami, 73-, Fla Heart Asn, 74- & NIH, 76-; vpres, Heart Asn Greater Miami, 83-; NIH Pharmacol Study Sect, 84-88. *Mem:* Biophys Soc; Am Soc Clin Pharmacol & Therapeut; Am Soc Pharmacol & Exp Therapeut; Am Physiol Soc; Soc Gen Physiol. *Res:* Excitation and contraction cardiac muscle; cardiac arrhythmias and antiarrhythmic drugs; cardiac hypertrophy. *Mailing Add:* Dept Pharmacol Univ Miami Med Sch PO Box 016189 Miami FL 33101

BASSETT, CHARLES ANDREW LOOCKERMAN, b Crisfield, Md, Aug 4, 24; m 46; c 3. ORTHOPEDIC SURGERY, CELL PHYSIOLOGY. *Educ:* Columbia Univ, MD, 48, ScD(med), 55; Am Bd Orthop Surg, dipl. *Prof Exp:* From instr to assoc prof, 55-67, PROF ORTHOP SURG, COL PHYSICIANS & SURGEONS, COLUMBIA UNIV, 67- *Concurrent Pos:* Consult, Naval Med Res Inst, 53-54; assoc attend orthop surgeon, Presby Hosp, New York, 60-63, attend orthop surgeon, 63-; career scientist, New York City Health Res Coun, 61-71; consult, Food & Drug Admin; spec consult, NIH; mem comt skeletal syst, Nat Res Coun-Nat Acad Sci & NY State Rehab Hosp, West Haverstraw. *Honors & Awards:* Paralyzed Vet Am Award, 59; United Cerebral Palsy-Max Weinstein Award, 60; James Mather Smith Prize, 71. *Mem:* Orthop Res Soc (secy-treas, 61-64, pres, 68-69); Am Acad Orthop Surg; Am Orthop Asn; Am Soc Cell Biol; fel Am Col Surg. *Res:* Bioelectric mechanisms in bone growth and physiology; pulsing electromagnetic fields to control cell function in diseases of humans and animals; regenerating of central and peripheral nervous systems; tissue transplantations. *Mailing Add:* Dept Orthop Surg Bioelec Res Ctr Columbia Presby Terian Med Ctr 2600 Netherland Ave Riverdale NY 10463

BASSETT, DAVID R, b Winston-Salem, NC, Jan 23, 39; m 61; c 3. PHYSICAL CHEMISTRY, COLLOID CHEMISTRY. *Educ:* Lafayette Col, AB, 61; Lehigh Univ, MS, 63, PhD(phys chem), 68. *Prof Exp:* res chemist, 68-80, sr res chemist, 80-86, CORP FEL, UNION CARBIDE CORP, 86-. *Mem:* Am Chem Soc. *Res:* Surface chemistry; polymer colloids; polymers for electronics. *Mailing Add:* Tech Ctr Union Carbide Corp Charleston WV 25303

BASSETT, EMMETT W, b Martinsville, Va, Jan 23, 21; m 50; c 3. MICROBIOLOGY. *Educ:* Tuskegee Inst, BS, 42; Univ Mass, MS, 50; Ohio State Univ, PhD(dairy technol), 54. *Prof Exp:* Res assoc microbiol, Col Physicians & Surgeons, Columbia Univ, 55-59, asst prof, 59-67; sr scientist, Ortho Res Found, NJ, 67-69; asst prof, 69-73, ASSOC PROF MICROBIOL, COL MED & DENT NJ, 73- *Mem:* Am Soc Biol Chemists. *Res:* Aromatic synthesis in molds; studies on the combining region and biosynthesis of antibodies. *Mailing Add:* 157 Central Ave Greenport NY 11944

BASSETT, HENRY GORDON, b Newton, Mass, Nov 12, 24; m 49; c 3. STRATIGRAPHY. *Educ:* McGill Univ, BSc, 49, MSc, 50; Princeton Univ, AM & PhD(geol), 52. *Prof Exp:* Geologist, Shell Can Ltd, 52-72, head stratig serv group, Shell Develop Co, 72-76, sr staff geologist, Shell Oil Co, 76-80; sr staff geologist & mgr geol technol, Sohio Petrol Co, 80-85; RETIRED. *Mem:* Am Asn Petrol Geol; Can Asn Petrol Geol; fel Geol Asn Can; Asn Prof Engrs. *Res:* Geology of northern and western Canada. *Mailing Add:* 3320 Las Huertas Rd Lafayette CA 94549

BASSETT, JAMES WILBUR, b Greenville, Tex, July 22, 23; m 49; c 6. ANIMAL SCIENCE. *Educ:* Tex A&M Univ, BS, 48, PhD(animal breeding), 65; Mont State Univ, MS, 57. *Prof Exp:* Asst county agt, Ark Agr Exten Serv, 48-50; asst prof wool technol, Mont State Univ, 51-63; PROF ANIMAL SCI, TEX A&M UNIV, 63- *Mem:* Am Soc Animal Sci. *Res:* Wool and mohair technology, production and marketing. *Mailing Add:* 303 W Dexter Dr College Station TX 77840

BASSETT, JOSEPH YARNALL, JR, b Asheville, NC, Aug 2, 27; m 53; c 2. ORGANIC CHEMISTRY. *Educ:* Univ NC, BS, 51, PhD(chem), 58. *Prof Exp:* Raw mat qual control analyst, Am Enka Corp, 53; lab asst, Univ NC, 53-56; develop chemist, Union Carbide Chem Co, WVa, 57-64; asst prof chem, WVa State Col, 62-64; asst prof, Western Carolina Univ, 64-69, chmn div natural sci & math, 69-76, dept head, 84-87, PROF CHEM, MATH, WESTERN CAROLINA UNIV, 69-, FACIL & PLANNING COORDR, 77- *Mem:* AAAS; Am Chem Soc; Sigma Xi. *Res:* Esterification; plasticizers; telomerization; terpene chemistry. *Mailing Add:* Western Carolina Univ PO Box 1524 Cullowhee NC 28723

BASSETT, MARK JULIAN, b Washington, Ind, May 15, 40. PLANT BREEDING. *Educ:* Lake Forest Col, BA, 63; Univ Md, MS, 67, PhD(hort), 70. *Prof Exp:* from asst prof to assoc prof, 70-84, PROF VEG CROPS, UNIV FLA, 84- *Mem:* Am Soc Hort Sci; Am Soc Agron. *Res:* Genetic studies and varietal improvement of beans and carrots; special emphasis on induced mutations, linkage studies developing a primary drisomic series, and cytoplamic male sterility in bean. *Mailing Add:* Veg Crops Dept 1255 Fifield Hall Univ Fla Gainesville FL 32611

BASSETT, WILLIAM AKERS, b Brooklyn, NY, Aug 3, 31; m 62; c 3. MINERALOGY, GEOPHYSICS. *Educ:* Amherst Col, BA, 54; Columbia Univ, MA, 56, PhD(geol), 59. *Prof Exp:* Res assoc chem, Brookhaven Nat Lab, 60-61; from asst prof to prof geol, Univ Rochester, 60-78; PROF GEOL, CORNELL UNIV, 78- *Concurrent Pos:* Res collabr, Brookhaven Nat Lab, 58-62; res assoc, Columbia Univ, 59-62; Guggenheim fel, 85. *Mem:* AAAS; Geol Soc Am; Mineral Soc Am; Am Geophys Union; Geochem Soc. *Res:* Structure and chemistry of sheet silicates; radioactive age dating of volcanic rocks and Pre-Cambrian rocks; behavior of solids at pressures and temperatures comparable with the earth's interior. *Mailing Add:* Dept Geol Sci Snee Hall Cornell Univ Ithaca NY 14853

BASSHAM, JAMES ALAN, b Sacramento, Calif, Nov 26, 22; m 56; c 5. BIOCHEMISTRY. *Educ:* Univ Calif, BS, 45, PhD(chem), 49. *Prof Exp:* Dept head, Chem Bidynamics Div, Lawrence Berkeley Lab, Univ Calif, 76-81, chemist, Bio-org Chem Group, 49-77, sr staff chemist II, 78-85; RETIRED. *Concurrent Pos:* NSF sr fel, Oxford Univ, 56-57; lectr chem, Univ Calif, Berkeley, 57-59, adj prof biochem, 72-80; vis prof, dept biochem & biophys, Univ Hawaii, 68-69; consult, 86-88. *Mem:* Am Chem Soc; AAAS; Am Soc Plant Physiol. *Res:* Regulation of metabolism and gene expression in photosynthetic plant cells and plant cells in tissue culture. *Mailing Add:* 785 Balra Dr El Cerrito CA 94530

BASSI, SUKH D, b Kericho, Kenya; US citizen; m 71; c 1. DEVELOPMENTAL GENETICS. *Educ:* Knox Col, BS, 65; St Louis Univ, Mo, MS, 67, PhD(biol), 70. *Prof Exp:* Prog specialist biol, Clark Col, 70-71; from asst prof to assoc prof biol, Benedictine Col, 71-76; PROF & CHIEF MICROBIOLOGIST, MIDWEST SOLVENTS CO, INC, ATCHISON, KANSAS, 81- *Concurrent Pos:* Consult, Environ Coun Atchison, Kans, 71- *Mem:* Am Inst Biol Sci; Am Physiol Soc; AAAS; Am Chem Soc; Am Asn Cereal Chemists; Sigma Xi. *Res:* Interaction of juvenile hormone and molting hormone with carrier and receptor proteins in the large milkweed bug, Oncopeltus fasciatus; bioconversion of starch and cellulose into alcohol. *Mailing Add:* Midwest Grain Prods Inc 1300 Main PO Box 130 Atchison KS 66002

BASSICHIS, WILLIAM, b Cleveland, Ohio, Aug 9, 37. THEORETICAL NUCLEAR PHYSICS. *Educ:* Mass Inst Technol, BS, 59; Case Inst Technol, MS, 61, PhD(physics), 63. *Prof Exp:* Vis scientist, Weizmann Inst Sci, 63-64 & Saclay Nuclear Res Ctr, France, 64-65; asst prof physics, Mass Inst Technol, 66-70; ASSOC PROF PHYSICS, TEX A&M UNIV, 70- *Concurrent Pos:* Consult, AEC, Lawrence Radiation Lab, Univ Calif, Livermore, 66- *Res:* Nuclear structure and reactions. *Mailing Add:* Dept Physics Tex A&M Univ College Station TX 77843

BASSIM, MOHAMAD NABIL, b Cairo, Egypt, Oct 10, 44; Can citizen; m 75; c 2. MATERIALS SCIENCE, CHEMICAL ENGINEERING. *Educ:* Cairo Univ, BSc 65; Univ Va, MSc, 70, PhD(mat sci), 73. *Prof Exp:* Res engr metall, Atomic Energy Estab Egypt, 65-68; sr scientist mat sci, Univ Va, 73; sr res scientist metall, Ecole Polytechnique, Univ Montreal, 73-; PROF, DEPT MECH ENG, UNIV MAN. *Concurrent Pos:* Vpres, Tektrend Int, 78- *Mem:* Am Soc Metals; Can Inst Mining & Metall; Sigma Xi. *Res:* Physical metallurgy; dislocation theory, fracture and nondestructive testing; computer modeling. *Mailing Add:* Dept Mech Eng Univ Man 356 Eng Bldg Winnipeg MB R3T 2N2 Can

BASSIN, ROBERT HARRIS, b Washington, DC, May 2, 38; m 61; c 3. MICROBIOLOGY. *Educ:* Princeton Univ, AB, 59; Rutgers Univ, PhD(microbiol), 65. *Prof Exp:* From res asst to res assoc microbiol, Rutgers Univ, 61-66; Imp Cancer Res Fund fel, London, Eng, 66-68; staff fel, 68-73, STAFF MICROBIOLOGIST & HEAD VIRAL BIOCHEM SECT, NAT CANCER INST, 73- *Mem:* AAAS; Soc Gen Microbiol; Tissue Cult Asn; Am Soc Microbiol; Am Asn Cancer Res; Sigma Xi. *Res:* RNA tumor virology; replication and transfomation by murine sarcoma and leukemia viruses in cell culture. *Mailing Add:* Nat Cancer Inst Bldg 41 NIH Bethesda MD 20814

BASSINGTHWAIGHTE, JAMES B, b Toronto, Ont, Sept 10, 29; m 55; c 5. PHYSIOLOGY, BIOPHYSICS. *Educ:* Univ Toronto, BA, 51, MD, 55; Univ Minn, PhD(physiol), 64. *Prof Exp:* Intern, Toronto Gen Hosp, 55-56; gen practitioner, Ont, 56-57; house physician internal med, Hammersmith Hosp, London, 57-58; fel med & physiol, Mayo Grad Sch Med, Univ Minn, 58-61; teaching asst, Univ Minn, Minneapolis, 61-62; res asst, Mayo Grad Sch Med, Univ Minn, from asst prof to prof physiol, 67-75, assoc prof physiol, 72-75, prof med, 75; PROF BIOENG & DIR CTR BIOENG, COL ENG & SCH MED, UNIV WASH, 75-, MEM BIOMATH GROUP FAC, 75- *Concurrent Pos:* Minn Heart Asn fel, 59-62; NIH career develop award, 64-74; vis prof, Pharmacol Inst, Univ Berne, 70-71; mem study sect, Nat Heart & Lung Inst, 70-74; chmn biotechnol resources adv & rev comt, Div Res Resources, NIH, 77-79; mem, US Nat Comn, Int Union Physiol Sci, 78-86, chmn, 83-86 & chmn, Comn Bioeng in Physiol, 86-92. *Honors & Awards:* Alza Award, Biomed Eng Soc, 86. *Mem:* AAAS; Am Heart Asn; Am Physiol Soc; Soc Comput Simulation; Microcirculatory Soc (pres-elect, 89-90, pres, 90-91); Sigma Xi. *Res:* Cardiovascular and transport physiology; internal medicine; biomedical engineering; cardiology. *Mailing Add:* Ctr Bioeng Univ Wash WD-12 Seattle WA 98195

BASSIOUNY, MOHAMED ALI, b Dessouk, Egypt, Nov 21, 41. RESTORATIVE DENTISTRY, PERIODONTICS. *Educ:* Alexandria Univ, BDS, 65; Manchester Univ, MSc, 74, PhD(biomat), 77; Temple Univ Sch Dent, DMD, 84. *Prof Exp:* Lectr & res assoc appl dent mat, Victoria Univ, UK, 74-77, fel res assoc, 77-78; from asst prof to assoc prof, 78-86, PROF OPER DENT, SCH DENT, TEMPLE UNIV, 86- *Concurrent Pos:* Consult, Imp Chem Indust, 75-, Recket Toiletry Co, 77, Star Dent, Syntex Corp, 78-81, Premier Dent Prod, 81, Johnson & Johnson Dent Prod Co, 81- & Teledyne-Getz, 84-; pres collegial assembly, Sch Dent, Temple Univ, 86-87. *Mem:* Am Asn Dent Res; Int Asn Dent Res. *Res:* Anti-plaque agent; visible light cured composite concept; chemically activated veneer material; cosmetic restorative dentistry; dental adhesives; posterior composites; prosthodontics. *Mailing Add:* Dept Oper Dent Sch Dent Temple Univ 3223 N Broad St Philadelphia PA 19140

BASSLER, RICHARD ALBERT, information science, for more information see previous edition

BASSNER, SHERRI LYNN, b New York, NY, Sept 4, 62. MARKET DEVELOPMENT, APPLICATIONS DEVELOPMENT. *Educ:* Goucher Univ, BA, 84; Pa State Univ, PhD(chem), 88. *Prof Exp:* SR RES CHEMIST, AIR PROD & CHEM INC, 88- *Mem:* Am Chem Soc; Steel Struct Painting Coun. *Res:* Developed the synthesis and studied the reactions of several organometallic complexes; developed new compounds for aluminum vapor deposition; developing new resins for polyurethane coatings. *Mailing Add:* 7201 Hamilton Blvd Allentown PA 18195

BAST, ROBERT CLINTON, JR, TUMOR IMMUNOLOGY. *Educ:* Harvard Univ, MD, 71. *Prof Exp:* PROF MED, MICROBIOL & IMMUNOL & CO-DIR, DIV HEMAT & ONCOL, DUKE UNIV MED CTR & DIR CLIN RES PROG, DUKE COMPREHENSIVE CANCER CTR, 84- *Res:* Clinical application of monoclonol antibodies; autologous bone marrow transplantation. *Mailing Add:* Duke Univ Med Ctr Box 3843 Durham NC 27710

BASTIAANS, GLENN JOHN, b Oak Park, Ill, Oct 25, 47; m 74; c 1. ANALYTICAL CHEMISTRY. *Educ:* Univ Ill, Urbana, BS, 69; Ind Univ, Bloomington, PhD(chem), 73. *Prof Exp:* Fel chem, Colo State Univ, 73-74; asst prof chem, Georgetown Univ, 75-79; asst prof chem, Tex A&M Univ, 79-; SR VPRES & CHIEF TECH OFFICER, INTEGRATED CHEM SENSORS, 85- *Mem:* Am Chem Soc; Soc Appl Spectros. *Res:* Development of instrumentation for analytical chemistry; development of surface acoustic wave devices as biosensors, chemical sensors, surface coating monitors and surface reaction monitors; atomic and laser spectroscopy for analytical applications. *Mailing Add:* Dept Chem Tex A&M Univ College Station TX 77843

BASTIAN, JAMES W, b Indianapolis, Ind, Apr 17, 26; m 50; c 3. ENDOCRINOLOGY. *Educ:* Purdue Univ, PhD(biol sci), 54. *Prof Exp:* Pharmacologist, 54-67, head dept pharmacol, 67-77, mgr tech develop, 77-86, CONSULT, ARMOUR PHARMACEUT CO, 86- *Mem:* Am Soc Pharmacol & Exp Therapeut. *Res:* Small animal screening; cardiovascular and anti-inflammatory pharmacology; platelet aggregation; calcium metabolism; calcitonin. *Mailing Add:* 126 Chestnut St Park Forest IL 60466

BASTIAN, JOHN F, b Nov 12, 51. PEDIATRICS, IMMUNOLOGY ALLERGY. *Educ:* Univ Ill, BA, 76; Rush Med Col, MD, 77; Am Bd Pediat, cert, 83; Am Bd Allergy & Immunol, cert, 85. *Prof Exp:* Intern, Univ Calif, San Diego, 77-78, resident, 78-80, chief resident, 80-81, fel pediat immunol & allergy, 81-82; fel pediat immunol & allergy, Scripps Clin & Res Found, 82-83; DIR IMMUNOL, CHILDRENS HOSP, SAN DIEGO, 81-, DIR, PEDIAT IMMUNOL CLIN, 86-; ASSOC CLIN PROF, MED CTR UNIV CALIF, SAN DIEGO, 88- *Concurrent Pos:* Attend physician pediat, Childrens Hosp, San Diego, 84-, Pediat Intensive Care Unit, 84-; chmn, Physicians Emergency Ref Comt Qual Assurance; mem, Calif AIDS Leadership Subcomt, 88- *Mem:* Am Acad Pediat; Am Fedn Clin Res; AAAS; Am Asn Immunologists; Clin Immunol Soc. *Res:* Alternatives of cellular immunity and antibody responses after severe head injury in children; new uses of intravenous immune globulin therapy; post-traumatic meningitis in adolescents and children. *Mailing Add:* Dept Pediat Childrens Hosp 8001 Frost St San Diego CA 92123

BASTIAN, JOSEPH, b Mare Island, Calif, Feb 13, 44; m 66; c 3. NEUROBIOLOGY. *Educ:* Elmhurst Col, BS, 66; Univ Notre Dame, PhD(biol), 69. *Prof Exp:* Fel neurobiol, Univ Notre Dame, 69-70, Purdue Univ, 70-72; asst res neuroscientist, Sch Med, Univ Calif, San Diego, 72-74; GEORGE LYNDCROSS PROF, UNIV OKLA, 74- *Concurrent Pos:* Fel, Nat Inst Neurol Dis & Stroke, 70, res grant, 75. *Mem:* AAAS; Soc Neurosci. *Res:* Neurophysiology of the electrosensory system in weakly electric fish and the role of the cerebellum in processing sensory information; mechanisms of the excitation-coupling process in skeletal muscle. *Mailing Add:* Dept Zool 730 Van Vleet Oval Univ Okla Norman OK 73019

BASTON, JANET EVELYN, b Laramie, Wyo, Sept 13, 41; m 59; c 3. MANAGEMENT INFORMATION SYSTEMS, INDUSTRIAL & MANUFACTURING ENGINEERING. *Educ:* Univ Wyo, BS, 63; Idaho State Univ, MS, 72; Univ Va, MBA, 86. *Prof Exp:* Programmer, Aerojet Nuclear Corp, 72-75; sr programmer, EG&G Idaho Inc, 75-77, systs engr, 77-80, info sci specialist, 80-83; vpres admin, Phys Sci Inc, 83-89; mgr eng, Gibbs & Hill, 89-90; SR PROJ ENGR, APPL GEOSCI INC, 90- *Mem:* Soc Women Engrs; Asn Comput Mach. *Res:* Feasibility studies for remedial action at environmentally degraded subsurface sites. *Mailing Add:* 39120 Argonaut Way No 504 Fremont CA 94538-3000

BASTOS, MILTON LESSA, toxicology, for more information see previous edition

BASTRESS, E(RNEST) KARL, b Grand Rapids, Mich, Jan 20, 29; m 51, 68; c 6. ENGINEERING. *Educ:* Univ Rochester, BS, 50, MS 54; Princeton Univ, PhD(aeronaut eng), 61. *Prof Exp:* Instr eng, Univ Rochester, 53-54; jr engr, res labs, Air Reduction Co, 54-56; staff assoc chem thermodyn, Arthur D Little, Inc, 61-67; group dir, North Res & Eng Corp, 68-72; mgr environ res, IKOR Inc, 72-73; prog mgr, A D Little, Inc, 73-76; div dir, Energy Conservation Prog, US Dept Energy, 76-82; PHYS SCI ADMINR, OFFICE TECHNOL PLANNING & MGT, US ARMY, MAT COMMAND, 82- *Mem:* Am Soc Mech Engrs. *Res:* Combustion processes in propulsion systems. *Mailing Add:* 4620 DeRussey Pkwy Chevy Chase MD 20815

BASU, ASIT PRAKAS, b Mar 17, 37; US citizen; m 66; c 2. RELIABILITY THEORY. *Educ:* Univ Calcutta, BSc, 56, MSc, 58; Univ Minn, PhD(statist), 66. *Prof Exp:* Assoc lectr statist & math, Indian Inst Technol, Kharagpur, 60-62; asst statist, Rutgers Univ & Univ Minn, 62-66; asst prof, Univ Wis, 66-68; res staff mem, IBM Res Ctr, 68-70; asst prof indust eng, Northwestern Univ, 70-71; assoc prof math, Univ Pittsburgh, 71-74; chmn dept, 76-83, PROF STATIST, UNIV MO-COLUMBIA, 74- *Concurrent Pos:* Consult, Statist & Reliability Theory. *Mem:* fel Am Statist Asn; fel Inst Math Statist; fel Royal Statist Soc; Int Statist Inst; Biomet Soc; Am Soc Qual Control; fel AAAS. *Res:* Nonparametric statistics; life testing and reliability; biomedical applications; parametric and nonparametric inference in reliability; bayesian statistics. *Mailing Add:* Dept Statist Univ Mo Columbia MO 65211

BASU, DEBABRATA, b Dacca, Bangladesh, July 7, 24; Indian citizen; m; c 2. STATISTICS. *Educ:* Dacca Univ, BA, 45, MS, 46; Calcutta Univ, PhD(statist), 55. *Prof Exp:* From assoc prof to prof statist, Indian Statist Inst, Calcutta, 54-75; PROF STATIST, FLA STATE UNIV, 75- *Concurrent Pos:* Vis prof statist, Univ NC, Chapel Hill, 64-65 & Univ Chicago, 65-66; vis prof math, Am Univ Beirut, 64, Univ NMex, 68-69, Univ Manchester, Eng, 71-73 & Univ Western Australia, 75. *Mem:* Int Statist Inst. *Res:* Foundational questions in statistical inference. *Mailing Add:* Dept Statist Fla State Univ Tallahassee FL 32306

BASU, MANJU, b Calcutta, India, Aug 10, 42; US citizen; m 66; c 2. BIOCHEMISTRY, IMMUNOLOGY. *Educ:* Univ Calcutta, BSc, 63, MSc, 65, DSc(biochem), 74. *Prof Exp:* Res fel biochem, Bose Res Inst, 65-66; res asst biol, Johns Hopkins Univ, 67-70; res asst chem, 70-74, res assoc chem & biochem, 74-81, asst res prof, 81-85, ASSOC RES PROF CHEM & BIOCHEM, UNIV NOTRE DAME, 85- *Concurrent Pos:* New Investr Award, Nat Cancer Inst, NIH, Bethesda, Md, 83-86; vis scientist, Indian Inst Chem Biol, Calcutta, 83. *Mem:* Soc Complex Carbohydrates; Am Soc Biol Chemists. *Res:* Biosynthesis of tumor-specific fucose-containing glycosphigolipids; purification and properties of N-Acetylglucosaminyltransferases involved in the synthesis of I/i-antigens. *Mailing Add:* Dept Chem & Biochem Univ Notre Dame Stepan Chem Hall Rm 440 Notre Dame IN 46556

BASU, MITALI, b Calcutta, West Bengal, India, Nov 24, 51; m 82. RECEPTOR BIOCHEMISTRY, PROTEIN BIOCHEMISTRY. *Educ:* Calcutta Univ, India, BS, 73, MS, 75, PhD(biochem), 81. *Prof Exp:* Sr res fel biochem, Dept Biochem, Calcutta Univ, India, 80-82; postdoctoral res assoc biochem, Sloan-Kettering Inst Cancer Res, New York, NY, 82-83; postdoctoral res assoc biochem, Sch Med, Univ Pa, Philadelphia, Pa, 83-85; res asst prof biochem, Univ Fla, Gainesville, Fla, 85-88; res asst prof biochem, Univ Kans Med Ctr, Kansas City, Kans, 88-89; SR SCIENTIST BIOCHEM, HOFFMANN-LA ROCHE, NUTLEY, NJ, 89- *Mem:* Am Soc Cell Biol; Soc Leukocyte Biol. *Res:* Receptor-ligand interactions and signal transduction; growth-factors and immunomodulators expression, purification and characterization of recombinant proteins of pharmaceutical interest. *Mailing Add:* Hoffmann-La Roche 340 Kingsland St Nutley NJ 07110

BASU, PRASANTA KUMAR, b Mymensingh, Bangladesh, Apr 1, 22; Can citizen; m 48; c 1. PATHOLOGY, IMMUNOLOGY. *Educ:* Univ Calcutta, BSc, 41, MB, 46, DOMS, 51. *Hon Degrees:* FRCP(C). *Prof Exp:* Head ophthal, Ramakrishna Mission Hosp, Vrindaban, India, 51-59; Colombo Plan fel, Univ Toronto, 55-56, res assoc, 57-65, clin teacher, 58-59, assoc prof, 59-70, prof, 70-89, DIR OPHTHALMIC RES, UNIV TOROTO, 59-, EMER PROF OPHTHAL, 89-; SR OPHTHALMOLOGIST, TORONTO HOSP, 59- *Concurrent Pos:* res assoc, Univ Calif, San Francisco, 57; career investr, Med Res Coun Can, 65-; assoc med dir, Ont Div Eye Bank Can, 68- *Honors & Awards:* S Biswas Mem Medal, India, 82; S C Dutta Mem Medal, India, 86; Semi-Centennial Award, Can Ophthal Soc, 87 & 90. *Mem:* Asn Res Vision & Ophthal; Can Ophthal Soc; Can Implant Asn (secy, 81-86); Int Med Res Found (vpres, 81-86); Can Fedn Biol Socs; All India Ophthal Soc. *Res:* Corneal transplant immunology; preservation of cornea, sclera and fascia lata; ocular effects of air, water and electromagnetic pollutants; diabetic retinopathy; application of ocular cell and organ cultures in ophthalmic research; third world ophthalmology. *Mailing Add:* Univ Toronto Eye Bank One Spadina Crescent Toronto ON M5S 2J5 Can

BASU, SUBHASH CHANDRA, b Calcutta, India, May 28, 38; US citizen; m 66; c 2. GLYCOLIPID & DNA BIOSYNTHESIS. *Educ:* Univ Calcutta, BSc, 58, MSc, 60; Univ Mich, PhD(biochem), 66; Univ Calcutta, DSc(biochem), 76. *Prof Exp:* Lectr chem, Vidyasager Col, Univ Calcutta, 60-61; teaching asst biochem, Univ Mich, 61-64, teaching fel, 64-65, res asst, Rackham Arthritis Res Unit, 65-66; fel, Johns Hopkins Univ, 66-67, res assoc, 67-70; from asst prof to assoc prof, 70-82, PROF CHEM, UNIV NOTRE DAME, IND, 83-, CHMN, BIOCHEM, BIOPHYS & MOLECULAR BIOL PROG, 86- *Concurrent Pos:* Res fel, Helen Hay Whitney Found, NY, 67-70; vis scientist, Gothenburg Univ, 68; Kennedy Found travel fel, Sweden, 68. *Mem:* Am Soc Biol Chemists; Int Soc Neurochem; Soc Complex Carbohydrates; NY Acad Sci; Am Soc Neurochem; Am Soc Cell Biol; Sigma Xi; Soc Cryobiol. *Res:* Biosynthesis of gangliosides and blood group glycosphingolipids; differentiation of tumor cells; control of DNA polymerase activities of neural cells. *Mailing Add:* Dept Chem & Biochem Univ Notre Dame Stepan Chem Hall Rm 443 Notre Dame IN 46556

BASU, TAPAN KUMAR, b Burdwan, West Bengal, India; m; c 2. VITAMINOLOGY, DRUG-NUTRIENT INTERACTIONS. *Educ:* Calcutta Univ, India, BSc, 59, BVSc, 62; London Univ, Eng, MSc, 68; Surrey Univ, PhD(biochem), 71. *Prof Exp:* Head, Metab Unit Res, Marie Curie Mem Found, Eng, 71-75; lectr nutrit, dept biochem, Univ Surrey, Eng, 75-81; assoc prof, 81-83, PROF NUTRIT, DEPT FOOD & NUTRIT, UNIV ALTA, CAN, 83-, HON PROF MED, DEPT MED, 84- *Honors & Awards:* Borden Prize, Can Soc Nutrit Sci, 84. *Mem:* NY Acad Sci; fel Am Col Nutrit; Am Inst Nutrit; Can Soc Nutrit; Nat Inst Nutrit; Brit Soc Surg Oncol; fel Int Col Nutrit. *Res:* Bio and metabolic availability of vitamins in the presence of cancer, diabetes, and bone disorders; drugs, steroidal and non-steroidal, cytotoxic agents, and mega vitamins. *Mailing Add:* Dept Foods & Nutrit Univ Alta Rm 308 Edmonton AB T6G 2M8 Can

BATALDEN, PAUL B, b Minneapolis, Minn, Dec 4, 41; c 2. AMBULATORY CARE QUALITY ASSURANCE, PEDIATRICS. *Educ:* Augsburg, Col, Minneapolis, BA, 63; Univ Minn Med Sch, BS, MD, 67; Am Bd Ped, cert, 76. *Prof Exp:* Clin assoc, Nat Cancer Inst, NIH, 69; med dir, Job Corps Dept Labor, Off Admin, Health Serv & Mental Health Admin

Systs, Dept Health Educ Welfare, 69-72; asst surgeon gen & dir, Bur Community Health Serv, 72-75; dir, Quality Assurance Proj, Interstudy, Minneapolis, 75-76; pediatrician, Park Nicollet Med Ctr, 75-86, dir, Qual Assurance Prog & Health Serv Res Ctr, 76-84, exec vpres, 84-86; VPRES MED CARE, HOSP CORP AM, 86- *Concurrent Pos:* Pres, St Louis Park Med Ctr Found, 81-84; breech chmn, Dept Health Care Qual Improv, Henry Ford Health Sci Ctr, vpres, Henry Ford Health Syst. *Mem:* Ambulatory Pediatric Asn; Am Pub Health Asn; Inst Med Nat Acad Sci. *Res:* Author of numerous articles in various journals. *Mailing Add:* 308 Harpeth Ridge Dr Nashville TN 37221

BATAY-CSORBA, PETER ANDREW, b Budapest, Hungary, July 10, 45; Can; m 70. NUCLEAR PHYSICS. *Educ:* Mass Inst Technol, BS, 68; Calif Inst Technol, PhD(nuclear physics), 75; Univ Toronto, MD, 87. *Prof Exp:* Res asst nuclear physics, Kellogg Lab, Calif Inst Technol, 70-75; res assoc nuclear physics, Univ Colo, Boulder, 75-79; proj dir Nuclear Dept, Schlumberger-Doll Res, 79-83; ATMT SINAI HOSP, 87- *Mem:* Sigma Xi; Am Phys Soc; Can Med Soc; Ont Med Asn. *Res:* Oil well logging; medical physics. *Mailing Add:* 85 Thorncliff Park Dr Suite 4205 Toronto ON M4H 1L6 Can

BATCHELDER, ARTHUR ROLAND, b Haverhill, Mass, Apr 17, 32; m 55; c 3. CHEMICAL SAFETY. *Educ:* Univ Mass, BS, 54; Va Polytech Inst, MS, 66; Cornell Univ, PhD, 71. *Prof Exp:* Soil scientist agr res, Agr Res Serv, USDA, 57-87; PRES, HAZARD TRAINING ASSOC INC, 86- *Concurrent Pos:* Collateral duty safety officer, 79-87; mem, cert bd hazardous mat technicians & supervisors, World Safety Orgn, 89- *Mem:* Am Soc Agron; World Safety Orgn; Sigma Xi; Soil Sci Soc Am; Am Chem Soc; Nat Fire Protection Asn. *Res:* Chemical safety training. *Mailing Add:* Hazard Training Assoc Inc PO Box 8899 Ft Collins CO 80525

BATCHELDER, DAVID G(EORGE), b Hiawatha, Kans, Apr 30, 20; m 44; c 4. AGRICULTURAL ENGINEERING. *Educ:* Kans State Univ, BS, 55; Okla State Univ, MS, 62. *Prof Exp:* Farmer, 46-53; from instr to prof, 55-85, EMER PROF AGR ENG, OKLA STATE UNIV, 85- *Concurrent Pos:* Mem Coop in agr between US & USSR, SEA & State Dept, 77- *Mem:* Am Soc Agr Engrs; Sigma Xi. *Res:* Equipment design and testing of machinery for seed metering; seed bed preparation; planting; weed control; plant preparation for harvest; harvesting for cotton production in Oklahoma; field forage drying. *Mailing Add:* Dept Agr Eng Agr Hall 111 Okla State Univ Stillwater OK 74078

BATCHELDER, GERALD M(YLES), b Exeter, NH, June 4, 25; m 49; c 6. CIVIL ENGINEERING. *Educ:* Univ NH, BS, 50; Purdue Univ, MS, 52. *Prof Exp:* Asst, Purdue Univ, 51-52; res engr, Mass Inst Technol, 52; engr, NH Dept Pub Works & Hwys, 52-53; res assoc prof, Eng Exp Sta, Univ NH, 53-70, assoc prof math, Div Continuing Educ, 57-73, prof civil technol & adj prof civil eng, 70-90, prof, Sch Life Long Learning, Univ NH, 73-90; RETIRED. *Mem:* Am Soc Testing & Mat; Nat Soc Prof Engrs. *Res:* Physical properties of materials; nondestructive testing of concrete; strength, dynamic properties and durability studies of concrete; materials science; micro-hardness conversion. *Mailing Add:* PO Box 138 Stratham NH 03885

BATCHELOR, B(ARRINGTON) D(EVERE), b Lucea, Jamaica, WI, July 2, 28; m 60; c 3. CIVIL ENGINEERING. *Educ:* Univ Edinburgh, BSc, 56; Imp Col, dipl, 61, Univ London, PhD(civil eng), 63. *Prof Exp:* Asst engr, Sir William Halcrow & Partners, Consult Engrs, Eng, 56-58; exec engr, Ministry of Educ, WI, 58-60, sr exec engr, 63-64; dir, Caribbean Consult Engrs, Ltd, 64-66; from asst prof to assoc prof, 66-72, PROF CIVIL ENG, QUEEN'S UNIV, 72- *Concurrent Pos:* Partner, Franks & Batchelor, Consult Engrs, WI, 64-66; mem task force, Ont Hwy Bridge Design Code, 76-; mem adv comt sci & technol, Can Broadcasting Corp, 84-87; mem, Nat Bldg Code Can; mem task force, new draft Am Asn State Hwy & Transp Off, 89- *Mem:* Am Soc Civil Engrs; Am Concrete Inst; Asn Prof Engrs Ont; Can Soc Eng. *Res:* Engineering structures; concrete bridges. *Mailing Add:* Dept Civil Eng Queen's Univ Kingston ON K7L 3N6 Can

BATCHELOR, JOHN W, b Winchester, Ind, Nov 17, 14. ELECTRICAL ENGINEERING. *Educ:* Butler Univ, AB, 35; Purdue Univ, BSEE, 37. *Prof Exp:* Asst dept mgr, Westinghouse Elec Co, 37-38; RETIRED. *Mem:* Nat Acad Engrs; Inst Elec & Electronics Engrs; Am Soc Mech Eng. *Mailing Add:* 11709 Joan Dr Pittsburgh PA 15235

BATCHER, KENNETH EDWARD, b St Albans, NY, Dec 27, 35; m 67; c 2. ELECTRICAL ENGINEERING. *Educ:* Iowa State Univ, BS, 57; Univ Ill, MS, 62, PhD(elec eng), 64. *Prof Exp:* Trainee, Goodyear Atomic Corp, 57-58, develop engr, Goodyear Aircraft Corp, 58-60; res & teaching asst elec eng, Univ Ill, 60-64; staff engr comput, Goodyear Aerospace Corp, Akron, 64-77, prin engr, 77-87; PRIN ENGR, LORAL DEFENSE SYST DIV, 87- *Mem:* Asn Comput Mach; Sigma Xi. *Res:* Digital computers, especially design of parallel networks and parallel processors. *Mailing Add:* 2007 Kingsdale Dr Stow OH 44224

BATCHMAN, THEODORE E, b Great Bend, Kans, Mar 29, 40; m 61; c 3. ELECTRO-OPTICAL DEVICES, FIBER OPTICS. *Educ:* Univ Kans, BS, 62, MS, 63, PhD(elec eng), 66. *Prof Exp:* Engr sci specialist, LTV Missiles & Space Div, 66-70; sr lectr, elec engr, Univ Queensland, Autralia, 70-75; vis asst prof, 75-76, from asst prof to prof, elec eng, Univ Va, 76-88; PROF & DIR, SCH ELEC & COMPUT SCI, UNIV OKLA, 88- *Concurrent Pos:* Consult, Postmaster General's Dept, Australia, 72-74, Armott Morrow Proprietary, Australia, 75, Gen Lasers Proprietary, 75-76, Sperry Marine Systs, Va, 78-80, Commonwealth, Va, Consol Lab serv, 82-83, US Army Foreign Serv & Technol Ctr, 86-90. *Mem:* Instr Elec & Electronics Engr; Soc Photo-Optical Instrumentation Eng; Optical Soc Am. *Res:* Integrated optical devices for coherent communications, signal processing and optical computing; development of a miniature electric field probe for RF and biological measurement. *Mailing Add:* Sch Elec Eng & Comput Sci Univ Okla Norman OK 73019-0631

BATCHO, ANDREW DAVID, b Somerville, NJ, July 9, 34; m 70; c 1. ORGANIC CHEMISTRY. *Educ:* Rutgers Univ, BS, 55; Univ Calif, PhD(chem), 59. *Prof Exp:* Sr res chemist, Am Cyanamid Co, 58-61; sr res chemist, 61-76, res fel, 76-84, RES INVESTR, HOFFMANN-LA ROCHE, INC, 85- *Mem:* Am Chem Soc. *Res:* Synthetic organic chemistry; natural products. *Mailing Add:* Hoffmann-La Roche Inc Nutley NJ 07110-1199

BATDORF, ROBERT LUDWIG, b Reading, Pa, Sept 18, 26; m 53; c 2. SOLID STATE SCIENCE. *Educ:* Albright Col, BA, 50; Univ Minn, PhD(phys chem), 55. *Prof Exp:* Mem tech staff, Bell Labs, 55-65, supvr, Solid State Device Technol Group, 65-68, dept head semiconductor mat & appl chem, 68-73, semiconductor technol, 73-89; RETIRED. *Mem:* AAAS; Am Chem Soc. *Res:* Diode and field effect transistor development; integrated circuit process technology, especially ion implantation techniques, diffusion, and chemical processing; physical diagnostic analytical techniques such as scanning electron microscopy, electron beam and ion beam microprobes; micromachined silicon sensor technology. *Mailing Add:* 707 Warren St Reading PA 19601-1336

BATDORF, SAMUEL BURBRIDGE, b China, Mar 31, 14; m 40; c 2. COMPOSITES ENGINEERING MECHANICS, FRACTURE. *Educ:* Univ Calif, AB, 34, MA, 36, PhD(theoret physics), 38. *Prof Exp:* Instr physics, Univ Utah, 38; from asst prof to assoc prof, Univ Nev, 38-43; aeronaut res scientist, Langley Lab, Nat Adv Comt Aeronaut, 43-51; adv physicist, Westinghouse Res Labs, 51-52; mgr develop, Mat Eng Dept, Westinghouse Elec Corp, 52-55, dir develop, Eng Hq Staff, 55-56; asst dir res & mgr electronics div, Lockheed Missile Syst Div, Lockheed Aircraft Corp, 56-57, tech dir weapon syst div, 57-58; chmn man-in-space proj, Inst Defense Anal, 58-59, chmn commun satellite proj, 59-60; mgr prod planning, Aeronutronic Div, Ford Motor Co, 60-61, dir res physics & electronics, 61-62; dir off res, Aerospace Corp, 62-64, dir appl mech & physics subdiv, 64-70, group dir appl technol concepts, 70-71, prin staff scientist, 72-77; adj prof, Univ Calif, Los Angeles, 73-85; RETIRED. *Concurrent Pos:* Chmn, Langley Adv Study Comt, Nat Adv Comt Aeronaut, 45-51, Subcomt Aircraft Struct Mat, 45-51, mem Comt Aircraft Struct Mat, 47-51; mem Panel Physics of Solids, Res & Develop Bd, 49-51; adv Comt Aeromech, Air Force Off Sci Res, 67-74; Distinguished prof, Tsing Hua Univ, Taiwan, 69; adv Comt Solid Mech, NSF, 78-80. *Honors & Awards:* Silver Medallion, Am Soc Mech Engrs, 80. *Mem:* Fel Am Phys Soc; fel Am Inst Aeronaut & Astronaut; hon mem Am Soc Mech Engrs; fel Am Acad Mech (past pres). *Res:* Mechanics; elasticity; theory of plates and shells; plasticity; fracture; composites; geophysics. *Mailing Add:* 5536 B Via La Mesa Laguna Hills CA 92653

BATE, GEOFFREY, b Sheffield, Eng, Mar 30, 29; m 53; c 4. SOLID STATE PHYSICS. *Educ:* Univ Sheffield, BSc, 49, PhD(physics), 52. *Prof Exp:* Sci officer physics, Royal Naval Sci Serv, Eng, 52-56; res assoc, Univ BC, 56-57, asst prof, 57-59; staff physicist, Develop Lab, Data Systs Div, Int Bus Mach Corp, 59-60, adv physicist, 60-65, mgr advan rec technol, 63-65, sr physicist & mgr res physics, IBM Corp, 65-78; vpres, Advan Develop, 78-82, sr vpres eng, Verbatim Corp, 82-86; vis prof, 86-87, PROF & ASSOC DEAN, SCH ENG, SANTA CLARA UNIV, 87- *Concurrent Pos:* Adj prof, Syracuse Univ, 61-65, Colo State Univ, 72- *Honors & Awards:* Centennial Medal, Inst Elect & Electronics Engrs, 84. *Mem:* Am Phys Soc; Sigma Xi; fel Inst Elec & Electronics Engrs. *Res:* Ferromagnetism of high-coercivity materials; semiconductors and infrared photoconductors; physics of magnetic and optical recording; magnetic properties of recording materials; high-frequency properties of soft-magnetic materials. *Mailing Add:* 23344 Camino Hermoso Los Altos Hills CA 94024-6405

BATE, GEORGE LEE, b Newton Falls, Ohio, Feb 15, 24; m 46; c 3. PHYSICS, APPLIED MATHEMATICS. *Educ:* Princeton Univ, AB, 45; Calif Inst Technol, MS, 46; Columbia Univ, PhD, 55. *Prof Exp:* Instr physics, Wheaton Col, 47-51; res asst, Columbia Univ, 51-55; from asst prof to assoc prof physics, Wheaton Col, 55-65; sr scientist, Isotopes, Inc, 65-67; prof physics, 67-69, chmn dept physics & math, 70-75, PROF PHYSICS & APPL MATH, WESTMONT COL, 69- *Concurrent Pos:* Res assoc & consult, Argonne Nat Lab, 56-65. *Mem:* Am Asn Physics Teachers; Am Sci Affiliation. *Res:* Abundance of the elements by neutron activation analysis; induced fission of heavy elements. *Mailing Add:* Dept Chem & Physics Westmont Col 955 La Paz Rd Santa Barbara CA 93108

BATE, ROBERT THOMAS, b Denver, Colo, Apr 1, 31; m 51; c 5. SEMICONDUCTORS. *Educ:* Univ Colo, BS, 55; Ohio State Univ, MS, 57. *Prof Exp:* Physicist, Eng Lab, US Bur Standards, 55; res asst, Ohio State Univ, 56-57; physicist, Battelle Mem Inst, 57-64; physicist, 64-68, sr scientist, 68-72, BR MGR, CENT RES LABS, TEX INSTRUMENTS INC, DALLAS, 72- *Concurrent Pos:* Mem, Solid State Sci Comt, Nat Acad Sci, 81. *Honors & Awards:* Texas Instruments Fel Award, 85. *Mem:* Fel Am Phys Soc; sr mem Inst Elec & Electronics Engrs. *Res:* Semiconductor device research, nonvolatile semiconductor memory, physical limits on logic devices; electrical properties of semiconductors; quantum devices. *Mailing Add:* 3106 Kristin Ct Garland TX 75042

BATEMAN, ANDREW, b London, UK, Dec 12, 59; m 90. STRUCTURAL ANALYSIS, INFLAMMATION. *Educ:* Univ London, UK, BSc, 81, PhD(biochem), 85. *Prof Exp:* Postdoctoral peptide biol, 85-90, ASST PROF PEPTIDE BIOL, MCGILL UNIV, 90- *Mem:* Am Peptide Soc. *Res:* Isolation, structural analysis and biological activity of peptides involved in inflammation. *Mailing Add:* Endocrine Lab Rm L205 Royal Victoria Hosp 687 Pine Ave W Montreal PQ H3A 1A1

BATEMAN, BARRY LYNN, b Jacksonville, Tex, Sept 15, 43; m 63; c 2. COMPUTER SCIENCES. *Educ:* Tex A&M Univ, BA, 65, MS, 67, PhD(comput sci), 70. *Prof Exp:* From asst dir to actg dir comput ctr, Univ Southwestern La, 69-72, head dept comput sci, 70-72; chmn dept comput sci, Tex Tech Univ, 72-76; prof comput sci, bus admin & med educ, Sch Med, Southern Ill Univ, Springfield, 76-81; ASST VPRES, UNIV MD SYST, 81- *Concurrent Pos:* Consult, Sigma Syst, 72- & Potomac Inst, 74-; mem, comput servs adv coun, State Ill, 76-; exec dir comput affairs, Southern Ill Univ, Carbondale, 76- *Mem:* Asn Comput Mach; Soc Comput Simulation; Sigma Xi; Soc Indust & Appl Math. *Res:* Computer science education; computer center management; information storage and retrieval, data base management; simulation. *Mailing Add:* Univ Md Syst Adelphi MD 20783

BATEMAN, DURWARD F, b Tyner, NC, May 28, 34; m 53; c 3. PLANT PATHOLOGY. *Educ:* NC State Col, BS, 56; Cornell Univ, MS, 58, PhD(plant path), 60. *Prof Exp:* From asst prof to prof plant path, Cornell Univ, 60-79, chmn dept, 70-79; assoc dean, Sch Agr & Life Sci & Dir, NC Agr Res Serv, 79-86, DEAN, COL AGR & LIFE SCI, NC STATE UNIV, 86- *Concurrent Pos:* NIH spec fel, Dept Plant Path, Univ Calif, Davis, 67; vis prof plant path, NC State Univ, 74-75; chmn, Southern Agr Exp Sta Dirs Asn, 85; pres, Southern Asn Agr Scientists, 90; mem, bd trustees, Int Potato Ctr, Lima, Peru, 91-; trustee, Charles Valentine Riley Mem Found, Washington, DC, 85- *Mem:* AAAS; fel Am Phytopath Soc (vpres, 75, pres elect, 76, pres, 77, past pres, 78); Int Soc Plant Path; Sigma Xi. *Res:* Ecology of soil borne pathogens; physiology of parasitism and disease and pathogen physiology; biochemical basis of pathogenic mechanisms and host reactions to infection; enzymatic decomposition of plant cell walls by phytopathogenic organisms. *Mailing Add:* Col Agr & Life Sci NC State Univ Box 7601 Raleigh NC 27695-7601

BATEMAN, FELICE DAVIDSON, b Springfield, Mass, Sept 2, 22; m 48; c 1. MATHEMATICS. *Educ:* Smith Col, AB, 43; Univ Mich, AM, 44, PhD(math), 50. *Prof Exp:* Instr math, NJ Col Women, Rutgers Univ, 47-49; instr, 54-55 & 57-58, ASST PROF MATH, UNIV ILL, URBANA-CHAMPAIGN, 58- *Concurrent Pos:* Mem sch math, Inst Advan Study, 49-50; vis lectr, Swarthmore Col, 61-62 & Sarah Lawrence Col, 64-65; ed consult, Math Reviews, 80-81. *Mem:* Am Math Soc; Math Asn Am. *Res:* Linear algebras with radical. *Mailing Add:* 108 Meadows Ct Urbana IL 61801

BATEMAN, JOHN HUGH, b East St Louis, Ill, Dec 21, 41; m 64; c 3. POLYMER CHEMISTRY, MATERIALS OF ELECTRONICS. *Educ:* Ind Univ, Bloomington, AB, 64; Univ Kans, PhD(org chem), 68. *Prof Exp:* Sr res chemist, Ciba-Geigy Corp, 68-78, mem, res mgt staff, 78-85 & bus develop staff, 85-91, MEM BUS MGT STAFF, CIBA-GEIGY CORP, 91- *Mem:* Am Chem Soc; Inst Elec & Electronics Engrs; Semiconductor Equip & Mat Inst; Int Soc Hybrid Microcelectronics. *Res:* Coatings; thermoset plastics; weatherable materials; high temperature resistant polymers; materials for the electronic industry; photo-imagabable materials for electronics industry. *Mailing Add:* 12121 Viewoak Dr Saratoga CA 95070

BATEMAN, JOHN LAURENS, b Washington, DC, Mar 30, 26; m 50, 80; c 5. RADIOBIOLOGY, ONCOLOGY. *Educ:* Mass Inst Technol, SB, 46; Johns Hopkins Univ, MD, 56; Am Bd Nuclear Med, dipl, 74. *Prof Exp:* Intern, Stamford Hosp, 56-57; resident internal med, Greenwich Hosp Asn, 57-58; res assoc, 58-61, res collabr, 61-67, scientist, Med Res Ctr, Brookhaven Nat Lab, 67-71; ASSOC CHIEF NUCLEAR MED, VET ADMIN HOSP, 74-; ASST PROF MED, SCH MED, STATE UNIV NY, STONY BROOK, 74- *Mem:* Soc Nuclear Med; Radiation Res Soc; NY Acad Sci. *Res:* Group techniques in cancer patients; medical ethics and quality of terminal care; dynamic and metabolic aspects of nuclear medicine; radiation dose-rate effect. *Mailing Add:* Nuclear Med Serv (115) Vet Admin Med Ctr Northport NY 11768

BATEMAN, JOSEPH R, b Chicago, Ill, Sept 3, 22; m 48; c 2. HEMATOLOGY, ONCOLOGY. *Educ:* George Washington Univ, MD, 47; Am Bd Internal Med, dipl, 57. *Prof Exp:* Intern med, Gorgas Hosp, Balboa, CZ, 47-48; fel, Vet Admin Hosp, Grand Junction, Colo, 54-56; resident, Vet Admin Hosps, Albuquerque, NMex, 58-61, hematologist, Oakland, Calif, 56-60; from instr to assoc prof med, 60-75, prof, 75-82, EMER PROF MED, UNIV SOUTHERN CALIF, 82- *Concurrent Pos:* Fel hemat, Univ Southern Calif, 60-62; consult, Vet Admin Hosp, 64-; chmn, Western Cancer Study Group; chief med oncol div, Los Angeles County-Univ Southern Calif Med Ctr; dir, Home Care Serv Dept, Los Angeles County-Univ Southern Calif Med Ctr. *Mem:* Am Col Physicians; Am Soc Clin Oncol; Am Asn Cancer Res; Am Asn Cancer Educ; Am Fed Clin Res. *Res:* Clinical research; cancer chemotherapy. *Mailing Add:* 1491 Old CC Rd Colville WA 99114

BATEMAN, MILDRED MITCHELL, b Cordele, Ga, Mar 22, 22; m; c 2. PSYCHIATRY. *Educ:* J C Smith Univ, BS, 41; Woman's Med Col Pa, MD, 46. *Prof Exp:* Staff physician, Larkin State Hosp WVa, 47-48, clin dir, 51-52, 55-58, supt, 58-60; supvr, dir prof servs, WVa Dept Ment Health, 60-62, dir, State Capital, 62-77; prof & chmn, 77-82, PROF PSYCHIAT, MARSHALL UNIV SCH MED, 82-; STAFF PSYCHIAT, HUNTINGTON VET ADMIN MED CTR, 86- *Concurrent Pos:* Mem, Comn Ment Illness & Ment Retardation, Comn Aging, WVa Comn Ment Retardation, Gov WVa Comn Status of Women & Coop Health Statist Adv Comt, Nat Ctr Health Statist; trustee, Menninger Found; prof psychiat & chmn dept, Marshall Univ. *Honors & Awards:* Warren Williams Distinguished Award, Am Psychiat Asn, 91. *Mem:* Inst of Med of Nat Acad Sci; Am Psychiat Asn (vpres, 73); AMA. *Res:* published numerous articles in various journals. *Mailing Add:* Dept Psychiat Marshall Univ Sch Med Huntington WV 25701

BATEMAN, PAUL TREVIER, b Philadelphia, Pa, June 6, 19; m 48; c 1. MATHEMATICS. *Educ:* Univ Pa, AB, 39, AM, 40, PhD(math), 46. *Prof Exp:* Lectr statist, Bryn Mawr Col, 45-46; instr math, Yale Univ, 46-48; res, Off Naval Res Contract, Inst Advan Study, 48-50; from asst prof to assoc prof, 50-58, head dept, 65-80, PROF MATH, UNIV ILL, URBANA-CHAMPAIGN, 58- *Concurrent Pos:* NSF sr fel, Inst Advan Study, 56-57; researcher, Univ Pa, 61-62; mem math adv panel, NSF, 63-66; vis prof, City Univ New York, 64-65; CIC exchange prof, Univ Mich, 80-81. *Mem:* Am Math Soc; Math Asn Am; London Math Soc. *Res:* Number theory and related parts of algebra, analysis, and mathematics of computation. *Mailing Add:* Dept Math Univ Ill Urbana IL 61801

BATES, CARL H, b Rothwell, Yorkshire, Eng, Jan 1, 39. MATERIALS CHARACTERIZATION & FAILURE ANALYSIS, VITICULTURE & DENOLOGY. *Educ:* Sheffield Univ Yorkshire Eng, BS, 61; Pa State Univ, PhD(geochem), 65. *Prof Exp:* Sr sci officer ceramics, GEC Wembley Eng, 66-68; sr res scientist mat sci, Bell Aerospace Buffalo, 68-71; owner viticulture, Fourthriding Vineyards, 71-77; consult energy, Niagara Co Energy Comt, 77-80; sr process engr refractories, Carborundum Co N Falls, 80-86; sr res engr ceramics, Norton Co, Worcester, 86-89; ENVIRON SCIENTIST CONSULT, ESSEX THERMODYNAMICS CORP, 89- *Mem:* Am Ceramic Soc; Brit Ceramic Soc. *Res:* Casting of complex shapes of silicon carbide bonded silicon carbide ceramics for abrasion, high temperature and corrosion resistant applications; high purity ceramics; failure analysis and materials characterization; environmental implications of wood gasification for Essex hybonic heating system. *Mailing Add:* 116 Wildwood Ave Worcester MA 01603

BATES, CHARLES CARPENTER, b Harrison, Ill, Nov 4, 18; m 42; c 3. OCEANOGRAPHY. *Educ:* DePauw Univ, BS, 39; Univ Calif, Los Angeles, MA, 44; Tex A&M Univ, PhD, 53. *Prof Exp:* Geophys trainee, Carter Oil Co, Okla, 39-41; spec asst to pres, Am Meteorol Soc, Ill, 45-46; oceanog technician sci surv, Bikini Atomic Bomb Test, Woods Hole Oceanog Inst, 46; from oceanogr to dep div dir, Div Oceanog, Hydrog Off, Navy Dept, 46-57, tech coordr environ systs, Off Naval Res, 57-59 & Off Dept Chief Naval Opers, 59-60, chief, Vela Uniform Br, Advan Res Proj Agency, Off Secy Defense, 60-64, sci & tech dir, Naval Oceanog Off, 64-68; sci adv to Commandant, HQ, US Coast Guard, 68-79; PRIN ASSOC, SPECTRUM INT ASSOCS, 86- *Concurrent Pos:* Consult oceanogr & meteorologist, Bates & Glenn, Washington, DC, 47; consult meteorologist, A H Glenn & Assocs, La, 48-54; mem, Comt Meteorol Uses of Satellites, Space Sci Bd, Nat Acad Sci; State Dept alt observer, Austral Season Inspection by US under Antarctic Treaty, 63-64; mem, Earth Sci Div, Nat Res Coun, 63-66, Mem, Maritime Transp Res Bd, 68-71; mem, merchant marine coun, US Coast Guard, 68-71; co-chmn panel on marine facil, US-Japan Natural Resources Prog, 69-75. *Honors & Awards:* Silver Medal, US Dept Transp, 74, Gold Medal, 79; President's Award, Am Asn Petrol Geologists, 53. *Mem:* Am Meteorol Soc; Am Asn Petrol Geologists; Soc Explor Geophysicists (vpres, 65-66); fel AAAS; hon mem Soc Explor Geophysicists. *Res:* Forecasting wave and surf conditions; history of geophysics; delta formation; sea-ice; detection underground nuclear explosions; global budget of petroleum hydrocarbon influx into ocean from all sources. *Mailing Add:* 136 W La Pintura Green Valley AZ 85614

BATES, CHARLES JOHNSON, b Dayton, Ohio, May 4, 30; m 53; c 3. FOOD TECHNOLOGY, BIOLOGY. *Educ:* Calif Inst Technol, BS, 51; Mass Inst Technol, PhD(food technol), 57. *Prof Exp:* Develop engr food prod, Procter & Gamble Co, 57-62, technol serv mgr indust food prod, 62-67, mkt mgr indust detergents, 67-72; VPRES TECH, AM MAIZE PROD CO, INC, 72- *Mem:* Inst Food Technologists (pres, 85-86); Am Asn Cereal Chemists; Am Chem Soc; AAAS. *Mailing Add:* 760 Williams Dr Crown Point IN 46307

BATES, CLAYTON WILSON, JR, b New York, NY, Sept 5, 32; m 68; c 2. SOLID STATE PHYSICS. *Educ:* Manhattan Col, BS, 54; Polytech Inst Brooklyn, MEE, 56; Harvard Univ, ME, 60; Washington Univ, PhD(physics), 66. *Prof Exp:* Solid-state physicist, Sylvania Elec Prod, NY, 55-57; sr engr physics, Varian Assoc, 66-72; assoc prof mat sci, 72-77, PROF MAT SCI, STANFORD UNIV, 77- *Concurrent Pos:* Vis prof reader, Imperial Col Sci & Technol, 68; consult, Varian Assoc, 72-; mem panel optical physics, Nat Acad Sci, 74-; consult & part owner, Diagnostics Info Inc, 75-; consult, Spectra-Physics, Inc, 75-77 & Bell Tel Labs, Murray Hill, 78-79; vis prof, Princeton Univ, 78-79. *Mem:* Fel Am Phys Soc; Optical Soc Am; sr mem Inst Elec & Electronics Engrs; Soc Photo-Optical Instrumentation Engr; Royal Photog Soc Gt Brit. *Res:* Electrical and optical properties of crystalline and amorphous solids and surfaces; photoelectric emission; luminescence phosphors; electronically conducted glasses; interaction of intense radiation with matter. *Mailing Add:* Dept Mat Sci & Eng Stanford Univ Stanford CA 94305

BATES, DAVID JAMES, b Portland, Ore, Oct 22, 28; m 49; c 5. ELECTRONICS ENGINEERING. *Educ:* Ore State Univ, BS, 51, MS, 53; Stanford Univ, PhD(appl physics), 58. *Prof Exp:* Res assoc, W W Hansen Labs, Stanford Univ, 52-57; sect head power tube dept, Hughes Aircraft Co, 57-62; sect mgr, Phys Electronics Labs, 62-63; sr eng mgr, Wave Tube Dept, Varian Assocs, 63-65; sect head medium power res & develop, Tektronix, Inc, 65-73, mgr, Power Tube Eng Dept, 73-77, chief engr, Tube Div, 77-79, sr eng, 79-82, eng mgr, Display Devices & Components Group, 82-86, mgr technol & instrument group, 86-90; DIR, INDUST RELS, COL ENG, ORE STATE UNIV, 90- *Mem:* Inst Elec & Electronics Engrs; Sigma Xi; Sci Res Soc Asn. *Res:* Microwave amplifiers; electron linear accelerators; electron bombarded semiconductor amplifiers. *Mailing Add:* 6591 Failing St West Linn OR 97068

BATES, DAVID MARTIN, b Everett, Mass, May 31, 34; m 56; c 2. TAXONOMIC BOTANY. *Educ:* Cornell Univ, BS, 59; Univ Calif, Los Angeles, PhD(bot), 63. *Prof Exp:* From asst prof to assoc prof, 63-75, dir, L H Bailey Hortorium, 69-83, PROF BOT, CORNELL UNIV, 75- *Honors & Awards:* Gold Medal, Garden Club Am, 79. *Mem:* AAAS; Soc Econ Botanists; Am Soc Plant Taxonomists (pres, 83); Inst Asn Plant Taxonomists; Bot Soc Am. *Res:* Taxonomic and cytotaxonomic studies in the family Malvaceae; taxonomy of cultivated plants; ethnobotany. *Mailing Add:* L H Bailey Hortorium Cornell Univ 467 Mann Libr Ithaca NY 14853

BATES, DAVID VINCENT, b West Malling, Kent, Eng, May 20, 22; Can citizen; m 48; c 3. ATMOSPHERIC SCIENCES, SCIENCE POLICY. *Educ:* Cambridge Univ, MB Bchir, 45, MD, 54; Royal Col Physicians, London, MRCP, 48. *Prof Exp:* Physician med, Royal Victoria Hosp, 56-72; dean, 72-77, prof, 72-87, EMER PROF MED, UNIV BC, 87- *Concurrent Pos:* Chmn physiol, McGill Univ, Montreal, 67-72. *Honors & Awards:* Robert Cooke Medal, Am Acad Allergy, 66; Ramazzini Medal, Col Ramazzini, 85; Connaught Award, Can Lung Asn, 91. *Mem:* Can Thoracic Soc (pres, 64);

Am Thoracic Soc; Soc Occup & Environ Health; Am Physiol Soc; Air Pollution Control Asn. *Res:* Health effects due to air pollution; respiratory physiology; occupational lung disease. *Mailing Add:* 4891 College Highroad Vancouver BC V6T 1G6 Can

BATES, DONALD GEORGE, b Windsor, Ont, March 18, 33; m 57; c 2. HISTORY OF MEDICINE. *Educ:* Univ Western Ont, MD, 58, BA, 60; Johns Hopkins Univ, PhD, 75. *Prof Exp:* From instr to asst prof hist of med, Johns Hopkins Univ, 62-66; assoc prof hist med, 66-76, chmn dept, 66-82 & 87-88, assoc prof hist & social aspects med, 71-84, THOMAS F COTTON HIST MED, MCGILL UNIV, 76- *Concurrent Pos:* Vpres, Int Physicians for Prevention of Nuclear War. *Mem:* Am Asn Hist Med; Soc Social Hist Med; Int Physicians for Prevention of Nuclear War. *Res:* Medical dimensions of nuclear war; nature of scientific knowledge. *Mailing Add:* Dept Humanities & Social Studies Med McGill Univ 3655 Drummond St Montreal PQ H3G 1Y6 Can

BATES, DONALD W(ESLEY), b Frazee, Minn, Aug 14, 18; m 43; c 3. AGRICULTURAL ENGINEERING. *Educ:* NDak Agr Col, BS, 43; Cornell Univ, MS, 50. *Prof Exp:* Asst prof agr eng, Cornell Univ, 46-51; assoc prof, Univ Minn, St Paul, 51-61, prof agr, 61-87, adj prof large animal clin sci, Col Vet Med, 80-87; RETIRED. *Honors & Awards:* Hokkaido Medal, Japan, 77. *Mem:* Sr mem Am Soc Agr Engrs. *Res:* Farm building construction and environmental control, with particular emphasis on housing for dairy cattle; animal waste management; relationship between engineering design and bovine environmental health; interdisciplinary research of veterinary medicine. *Mailing Add:* Univ Minn 212 Agrieng St Paul MN 55108

BATES, DOUGLAS MARTIN, b Regina, Sask, July 5, 49; m 74; c 2. STATISTICAL COMPUTING, NONLINEAR REGRESSION. *Educ:* Queen's Univ, BSc, 71, PhD(statist), 78; Univ Calif, Los Angeles, MA, 73. *Prof Exp:* Asst prof statist, Dept Math, Univ Alta, 78-80; PROF STATIST, DEPT STATIST, UNIV WIS-MADISON, 80- *Concurrent Pos:* Consult, Statist Lab, Univ Wis, Madison, 80-82. *Mem:* Am Statist Asn; Royal Statist Soc; Asn Comput Mach; Statist Soc Can. *Res:* Statistical computing, particularly in the area of nonlinear models and applications of computational linear algebra. *Mailing Add:* Dept Statist Univ Wis Madison WI 53706

BATES, G WILLIAM, b Durham, NC, Feb 15, 40; m 69; c 3. REPRODUCTIVE ENDOCRINOLOGY. *Educ:* Univ NC, BS, 62, MD, 65; Mass Inst Technol, SM, 84. *Prof Exp:* Asst prof obstet & gynec, Univ Tenn, Knoxville, 72-76; res fel reprod endocrinol, Uinv Tex Southwestern Med Sch, 76-78; prof obstet & gynec, Univ Miss Med Ctr, 78-86; PROF OBSTET & GYNEC, MED UNIV SC, 86-, DEAN COL MED, 86- *Concurrent Pos:* Chmn indust comt, Am Fertility Soc, 87-; consult to surgeon gen, US Air Force, 87- *Mem:* Am Fertility Soc; Am Col Obstetricians & Gynecologists; Endocrine Soc; Soc Gynec Invest; Soc Gynec Surgeons. *Res:* Elucidated the role of the adipocyte in both thin and obese women on reproductive function; studies include the metabolism of estradiol 17-B and alternatives in the gonadotropin secretory pattern as affected by alterations in body weight. *Mailing Add:* Greenvill Hosp Syst 701 Grove Rd Greenville SC 29605

BATES, GEORGE WINSTON, NUTRITION, BIOAVAILABILITY OF IRON. *Educ:* Univ Southern Calif, PhD(biochem), 66. *Prof Exp:* PROF BIOCHEM & NUTRIT, FACIL SCI NUTRIT, TEX A&M UNIV, 69- *Mailing Add:* Dept Biochem & Biophys Texas A&M Univ College Station TX 77843

BATES, GRACE ELIZABETH, b Albany, NY, Aug 13, 14. MATHEMATICS. *Educ:* Middlebury Col, BS, 35; Brown Univ, ScM, 38; Univ Ill, PhD(math), 46. *Hon Degrees:* DSc, Middlebury Col, 72. *Prof Exp:* Teacher high sch, Vt, 35-36 & Pa, 38-43; instr math, Sweet Briar Col, 43-44; from instr to assoc prof, 46-56, prof, 56-79, EMER PROF MATH, MT HOLYOKE, 79- *Concurrent Pos:* Trustee, Teachers Ins & Annuity Asn, 65-69; mem, Comn Undergrad Prog Math, 70-74. *Mem:* Am Math Soc; Math Asn Am. *Res:* Modern algebra; free loops and nets and their generalizations; probability; mathematical statistics. *Mailing Add:* Pennswod Village No A109 Newtown PA 18940

BATES, HAROLD BRENNAN, JR, b Des Moines, Iowa, Feb 12, 35; m 64; c 4. ZOOLOGY. *Educ:* Drake Univ, Iowa, BA, 59, MA, 61; Iowa State Univ, PhD(zool), 70. *Prof Exp:* Instr biol, Burlington Col, 61-66; from asst prof to assoc prof, 70-80, PROF ZOOL, ALBANY STATE COL, 80- *Mem:* Am Asn Univ Profs. *Res:* Effects of prostaglandin E(1) on liver polyploidy; toxicology, use of nematode models; neuropharmacology, use of nematode models; quantitative cytochemistry. *Mailing Add:* 101 University Ave Albany GA 31707

BATES, HERBERT T(EMPLETON), b Ames, Iowa, Sept 9, 13; m 42; c 2. CHEMICAL ENGINEERING. *Educ:* Iowa State Univ, BS, 35, PhD(chem eng), 41; Va Polytech Inst, MS, 38. *Prof Exp:* Operator, Anaconda Lead Prods Co, East Chicago, Ind, 35-37; lab technician, Sherwin-Williams Co, Chicago, 37; instr chem & chem eng, Case Inst Technol, 41-46; from asst prof to assoc prof chem eng, Univ Nebr, 46-58; assoc prof, 58-59, prof, 59-78, EMER PROF CHEM ENG, KANS STATE UNIV, 78- *Concurrent Pos:* Chem engr, Glenn L Martin Co, Cleveland, 44-45, US Steel Co, Gary, Ind, 53, Argonne Nat Lab, Chicago, Ill, 59 & 60. *Mem:* Am Inst Chem Engrs; Am Chem Soc; Am Soc Eng Educ; Instrument Soc Am; Sigma Xi. *Res:* Thermodynamics; transport phenomena; separations; process economics; process dynamics; automatic controls; computers; education methods. *Mailing Add:* 1622 Fairview Ave Manhattan KS 66502

BATES, HOWARD FRANCIS, b Portland, Ore, Mar 27, 27; m 61. PHYSICS, ELECTRICAL ENGINEERING. *Educ:* Ore State Univ, BS, 50, MS, 56; Univ Alaska, PhD(geophys), 61. *Prof Exp:* Physicist, US Navy Electronics Lab, Calif, 51-52; physicist, Geophys Inst, Univ Alaska, 52-53, instr geophys, 57-61, assoc geophysicist, 61-62, assoc prof, 62-66; sr physicist, Stanford Res Inst, 66-70; prof geophys, Univ Alaska, Fairbanks, 70-80, prof elec eng, 70-85; RETIRED. *Concurrent Pos:* Elec engr, Eielson AFB, Alaska, 82- *Mem:* sr mem Inst Elec & Electronics Engrs. *Res:* Aeronomy; ice physics; seismological instrumentation. *Mailing Add:* PO Box 80463 Fairbanks AK 99708-0463

BATES, J(UNIOR) LAMBERT, b Ogden, Utah, Mar 21, 28; m 52; c 5. MATERIALS SCIENCE. *Educ:* Brigham Young Univ, BS, 53; Univ Utah, PhD(metall), 57. *Prof Exp:* Sr res scientist, Hanford Labs, Gen Elec Co, 57-64; res assoc reactor & mat technol, Battelle Mem Inst, 65-69, tech leader thermophys properties & mgr, Phys Ceramics Sect, 69-70, tech leader, Pac Northwest Labs, 70-86, sr staff scientist, 78-86, CHIEF SCIENTIST MAT DEPT, BATTELLE MEM INST, 86- *Concurrent Pos:* Bd trustees, Am Ceramic Soc, 82-86. *Mem:* Fel Am Ceramic Soc (vpres, 78-79, pres-elect, 82-83, pres, 83-84,). *Res:* High temperature thermal, electrical, physical and electrochemical properties of ceramics; MHD, fuel cells, thermoelectrics, nuclear, solar, fossil and other energy materials; corrosion and electrochemical interactions ceramics; electrolytes; electronically conducting high-temperature ceramics. *Mailing Add:* Pac Northwest Lab PO Box 999 K2-45 Richland WA 99352

BATES, JOHN BERTRAM, b Big Flats, NY, Sept 7, 14; m 44; c 2. COLLOID CHEMISTRY. *Educ:* Clarkson Col Technol, BS, 36; Rice Inst, MA, 39, PhD(colloid chem), 41. *Prof Exp:* Chemist, Corning Glass Works, NY, 36-37; res chemist, Sun Chem Corp, 41-42; Am Petrol Inst fel, Univ Mich, 42-44; leader, phys chem sect, Sun Chem Corp, 44-50, asst dir graphic arts labs, 50-56; vpres & tech dir, Ho-Par, Inc, 56-60; staff chemist, HRB-Singer, Inc, 60-62; tech dir, Flint Ink Corp, 62-82, vpres res & develop, 82-88; RETIRED. *Concurrent Pos:* Chmn steering comt, ink misting comt & newsink comt, Nat Asn Printing Ink Mfrs Tech Inst, 75-88. *Honors & Awards:* Ault Award, Nat Asn Printing Ink Mfrs, 82. *Mem:* AAAS; Am Chem Soc; Soc Rheol; Inter-Soc Color Coun; Soc Coating Technol. *Res:* Pigment and printing ink chemistry; x-ray diffraction; gas adsorption on pigments; rheology of pigment dispersions; iron cyanide chemistry; particle size methods; surface chemistry in pigment dispersions; spectrophotometry and spectrofluorometry. *Mailing Add:* 46701 Betty Hill Lane Plymouth MI 48170

BATES, JOHN BRYANT, b Harlan, Ky, Mar 11, 42; m 63; c 2. SOLID STATE PHYSICS, CHEMICAL PHYSICS. *Educ:* Univ Ky, AB, 64, PhD(chem), 68. *Prof Exp:* Res assoc chem, Univ Md, College Park, 68-69; staff scientist, 69-80, GROUP LEADER, SOLID STATE DIV, OAK RIDGE NAT LAB, 80- *Concurrent Pos:* Vis prof, Univ Paris, 84; adj prof, Univ Tenn, 84- *Mem:* Fel Am Physics Soc. *Res:* Solid state spectroscopy; vibrational spectroscopy, including experimental and theoretical aspects; structure and dynamics of inorganic and organic molecular crystals; radiation effects on insulating materials; hydrogen diffusion in solids; fast ion conduction in solid electrolytes; matrix isolated species; growth and characterization of thin films; RF sputtering, solid state microbatteries and other microionic devices; impedance spectroscopy; electrical response and microstructure of metal-electrolyte interfaces. *Mailing Add:* MS 6030 Solid State Div Oak Ridge Nat Lab PO Box 2008 Oak Ridge TN 37830-6030

BATES, JOSEPH H, b Little Rock, Ark, Sept 19, 33; m 55; c 4. MEDICINE, BACTERIOLOGY. *Educ:* Univ Ark, BS & MD, 57, MS, 63. *Prof Exp:* From instr to assoc prof med, 61-71, ASST PROF MICROBIOL, SCH MED, UNIV ARK, LITTLE ROCK, 63-, PROF MED, 71-, VCHMN DEPT, 78- *Concurrent Pos:* Fel infectious dis, 61-63; clin investr pulmonary dis, Vet Admin Hosp, Little Rock, 63-65, chief pulmonary & infectious dis serv, 65-67, chief med, 67- *Honors & Awards:* Abernathy Medal, Internal Med, Am Col Physicians, 81. *Mem:* Am Thoracic Soc; Am Fedn Clin Res; Infectious Dis Soc Am; Am Col Physicians. *Res:* Unclassified mycobacteria, mycobacteriophage and their interrelationships; genetics of atypical mycobacteria. *Mailing Add:* McClellan Med Ctr 4300 W Seventh St Little Rock AR 72205

BATES, LLOYD M, b Westville, NS, Dec 25, 24; m 46; c 2. RADIOLOGICAL PHYSICS. *Educ:* Univ NB, BSc, 50; Univ Sask, MSc, 52; Johns Hopkins Univ, ScD(radiol physics), 66. *Prof Exp:* Physicist, Prov NB, 52-57; asst prof radiol, Sch Med, Univ Md, 57-59; from instr to assoc prof radiol, Sch Pub Health, Johns Hopkins Univ, 59-75, radiation safety officer, Med Insts, 66-75; sr staff physicist, 74-77, DIR COORD, CRP COORD PROG, AM ASN PHYSICISTS IN MED, 77- *Mem:* Am Asn Physicists in Med; Sigma Xi; Am Col Radiol; Am Pub Health Asn; Health Physics Soc. *Res:* Evaluation of radiological imaging systems; application of radioactive materials to physiological studies; design and development of radiotherapy apparatus; radiation control at medical installations. *Mailing Add:* 7104 Sheffield Rd Baltimore MD 21212

BATES, LYNN SHANNON, b Salem, Ohio, May 7, 40; m 66; c 4. MICROSCOPY, GENETICS. *Educ:* Heidelberg Col, BS, 62; Purdue Univ, West Lafayette, MS, 66; Kans State Univ, PhD(grain sci), 72. *Prof Exp:* Biochemist, Rockefeller Found, Int Maize & Wheat Improv Ctr, Mex, 66-68; asst prof, Dept Grain Sci & Indust, Kans State Univ, 72-81; treas, Asima Corp, 85-88, vpres, 88-90; consult, LSB Prod, 81-86, PRES, ALTECA LTD, 86- *Concurrent Pos:* Rockefeller Found grant, 73-77; NSF travel grant, barley genetics III, 75. *Mem:* Am Asn Cereal Chemists; Am Chem Soc; Am Asn Feed Microscopists (pres, 87-88); Inst Food Technologists; Asn Off Anal Chem; Coun Agr Sci & Technol. *Res:* Biochemical basis of crossability barriers between field crop species particularly cereals and legumes; nutritional improvement of cereal grains and legumes; enzyme applications for rapid quality control testing kits; thermal processing (dry roasting) effects on food and feed quality. *Mailing Add:* 1814 Laramie Manhattan KS 66502

BATES, MARGARET WESTBROOK, b Boston, Mass, Oct 5, 26. BIOCHEMISTRY, NUTRITION. *Educ:* Wellesley Col, BA, 48; Cornell Univ, MS, 50; Harvard Univ, DSc(nutrit), 54. *Prof Exp:* Fel nutrit, Harvard Univ, 54-55; USPHS fel med res, Univ Toronto, 55-57; from res assoc to res asst prof biochem & nutrit, 57-67, from res asst prof to res assoc prof, Lab Clin

Sci, 67-73, res assoc prof psychiat, 73-77, RES ASSOC PROF EPIDEMIOL & HEAD NUTRIT CORE LAB, UNIV PITTSBURGH, 77- *Mem:* AAAS; Am Inst Nutrit; Am Chem Soc; Brit Biochem Soc. *Res:* Fat metabolism; ketone body metabolism; plasma transport of fatty acids. *Mailing Add:* 118 Mayflower Dr Pittsburgh PA 15238

BATES, PETER WILLIAM, b Manchester, Eng, Dec 27, 47; m 72; c 4. NONLINEAR PARTIAL DIFFERENTIAL EQUATIONS, DYNAMICAL SYSTEMS. *Educ:* Univ London, BSc, 69; Univ Utah, PhD(math), 76. *Prof Exp:* Asst prof math, Pan Am Univ, 76-79; from asst prof to assoc prof, Tex A&M Univ, 78-85; assoc prof, 84-88, PROF MATH, BRIGHAM YOUNG UNIV, 88- *Concurrent Pos:* Vis asst prof, Univ Tex, Austin, 78-79; res fel math, Heriot-Watt Univ, 82; prog dir appl math, NSF, 87-89; prog dir, Soc Indust & Appl Math, 90-; vis prof, Univ Utah, 91. *Mem:* Am Math Soc; Soc Indust & Appl Math. *Res:* Nonlinear partial differential equations and dynamical systems; qualitative behavior of solutions to models of phase transition. *Mailing Add:* 3343 Navajo Lane Provo UT 84604

BATES, RICHARD DOANE, JR, b Elizabeth, NJ, July 24, 44; m 71; c 2. PHYSICAL CHEMISTRY. *Educ:* Cornell Univ, BA, 66; Columbia Univ, MA, 67, PhD(chem), 71. *Prof Exp:* Preceptor chem, Columbia Univ, 67-68; asst prof, 73-79, ASSOC PROF CHEM, GEORGETOWN UNIV, 79- *Concurrent Pos:* Res fel, Northwestern Univ, 81. *Mem:* Sigma Xi; Am Chem Soc; Am Phys Soc; Am Asn Univ Profs; Laser Inst Am; Royal Soc Chem. *Res:* Molecular dynamics of reacting and non-reacting chemical systems; vibrational energy transfer and the role of vibrational energy in chemical reactions; solvent-solute interactions in solutions by magnetic resonance. *Mailing Add:* Dept Chem Georgetown Univ Washington DC 20057

BATES, RICHARD PIERCE, b Pennington, Tex, Jan 15, 26; m 52; c 1. AGRONOMY, PLANT BREEDING. *Educ:* Agr & Mech Col Tex, BS, 49, MS, 50; Univ Md, PhD(forage breeding), 53. *Prof Exp:* Technician, Seed Testing Lab, Agr & Mech Col Tex, 48-49, jr agronomist forage prod res, Substa 22, 50-51; asst agronomist, USDA, Md, 51-53; asst agronomist corn breeding & prod, Tex Substa 5, Agr & Mech Col Tex, 53-55; RES AGRONOMIST, SAMUEL ROBERS NOBLE FOUND, INC, 55- *Mem:* Am Soc Agron. *Res:* Forage crops, small grain and legume improvement and production. *Mailing Add:* Agr Div Samuel Roberts Noble Found Inc Ardmore OK 73401

BATES, ROBERT BROWN, b Huntington, NY, Dec 16, 33; m 55; c 3. ORGANIC CHEMISTRY. *Educ:* Rutgers Univ, BS, 54; Univ Wis, PhD(org chem), 57. *Prof Exp:* Res fel, Mass Inst Technol, 57-58; from instr to asst prof org chem, Univ Ill, 58-63; from asst prof to assoc prof , 63-69, PROF ORG CHEM, UNIV ARIZ, 69- *Concurrent Pos:* Sloan Found fel, 67-69. *Mem:* Am Chem Soc; Am Crystallog Asn. *Res:* Natural product structure, synthesis and biogenesis; carbanions. *Mailing Add:* Dept Chem Univ Ariz Tucson AZ 85721

BATES, ROBERT CLAIR, b Portland, Ore, Mar 8, 44; m 69; c 3. VIROLOGY, MICROBIOLOGY. *Educ:* Lewis & Clark Col, Ore, BA, 66; Wash State Univ, MS, 68; Colo State Univ, PhD(virol), 72. *Prof Exp:* From asst prof to assoc prof, 72-85, PROF VIROL, VA POLYTECH INST & STATE UNIV, 85- *Concurrent Pos:* Am Cancer Soc grant, Va Polytech Inst, 76-78; NIH grants, 76-79 & 80-83; consult, Carborundum Co, 76-78; NSF grants, Va Polytech Inst, 78-80 & 81-84, Am Cancer Soc grant, 84- 90. *Mem:* Sigma Xi; AAAS; Am Soc Microbiol; Am Soc Biol Chemists; Am Soc Virol. *Res:* Cellular and molecular aspects of parovirus replication, characterization of viral enzymes and viral replication intermediates, replication of viral DNA and viral morphogenesis. *Mailing Add:* Dept Biol Microbiol Sect Va Polytech Inst & State Univ Blacksburg VA 24061-0406

BATES, ROBERT LATIMER, b Brookings, SDak, June 17, 12; m 35; c 2. ECONOMIC GEOLOGY. *Educ:* Cornell Univ, AB, 34; Univ Iowa, MS, 36, PhD(stratig), 38. *Prof Exp:* Asst, Dept Geol, Univ Iowa, 34-38; jr geologist, The Texas Co, 38-40; geologist, NMex Bur Mines, 41-45, chief oil & gas div, 45-47; from asst prof to assoc prof geol, Newark Col, Rutgers Univ, 47-51; from assoc prof to prof, 51-77, EMER PROF GEOL, OHIO STATE UNIV, 77- *Concurrent Pos:* From instr to asst prof, NMex Sch Mines, 41-43; ed, J Nat Asn Geol Teachers, 60-64 & J Am Inst Prof Geologists, 69-70. *Honors & Awards:* Hardinge Award, Soc Mining Engrs, 78; Parker Mem Award, Am Inst Prof Geologists, 84. *Mem:* Distinguished mem Soc Mining Engrs; Asn Earth Sci Ed; Am Inst Prof Geol; Geol Soc Am. *Res:* Economic geology of the nonmetallics. *Mailing Add:* 180 Canyon Dr Columbus OH 43214-3106

BATES, ROBERT P(ARKER), b Cambridge, Mass, May 11, 32; m 56, 66; c 5. FOOD SCIENCE & TECHNOLOGY. *Educ:* Mass Inst Technol, BS, 59, PhD(food sci), 66; Univ Hawaii, MS, 63. *Prof Exp:* Res engr, Forte Eng Co, 59-60; fel, Inst Nutrit Cent Am & Panama, 66-67; assoc prof food sci, 67-81, PROF FOOD SCI & HUMAN NUTRIT, INST FOOD & AGR SCI, UNIV FLA, 81- *Mem:* Inst Food Technol. *Res:* Tropical food processing; appropriate food technology; food processing systems; research and teaching; product development; new food resources. *Mailing Add:* Dept Food Sci & Human Nutrit Univ Fla 329 FSB Bldg Gainesville FL 32601

BATES, ROBERT WESLEY, b Columbia, Iowa, Jan 31, 04; m 30; c 3. PHYSIOLOGICAL CHEMISTRY. *Educ:* Simpson Col, AB, 25; Univ Chicago, PhD(physiol chem), 31. *Hon Degrees:* DSc, Simpson Col, 87. *Prof Exp:* Asst prof chem, Simpson Col, 26-27; physiol chemist, E R Squibb & Sons, 27-29; res assoc, Sta for Exp Evolution, Carnegie Inst Technol, 31-32, investr, 33-41; biol chemist, Difco Labs, 41-42, foreman prod, 43-44; dept head, E R Squibb & Sons, 45-52; biochemist, NIH, 52-75; RETIRED. *Honors & Awards:* Koch Award, Endocrine Soc. *Mem:* AAAS; Am Soc Biol Chem; Endocrine Soc; fel NY Acad Sci. *Res:* Chemistry and assay of enzymes; isolation and assay of hormones of the anterior pituitary, especially prolactin and thyrotropin; chemical methods for tryptophane and natural estrogens determination; hormonal induction of diabetes. *Mailing Add:* 1522 Mission Way Nogales AZ 85621

BATES, ROGER GORDON, b Cummington, Mass, May 20, 12; m 41; c 1. ANALYTICAL CHEMISTRY. *Educ:* Univ Mass, BS, 34; Duke Univ, AM, 36, PhD(phys chem), 37. *Prof Exp:* Asst chem, Duke Univ, 34-37; Sterling fel, Yale Univ, 37-39; chemist, Nat Bur Standards, 39-57, chief electrochem anal sect, 57-69, asst chief anal chem div, 58-67; prof, 69-79, EMER PROF CHEM, UNIV FLA, 79- *Concurrent Pos:* Lectr, Trinity Col, DC, 47-49; USPHS fel, Univ Zurich, 53-54; chmn, Comn Electrochem Data, Int Union Pure & Appl Chem, 53-59, mem, US Nat Comt, 59-65, mem, Comn Electrochem, 59-67, mem, Comn Symbols, Terminology & Units, 63-71, chmn, Comn Electroanal Chem, 71-79, mem, Anal Div Comt, 79-83. *Honors & Awards:* Hillebrand Prize, 55; US Dept Com Except Serv Award, 57; Anal Chem Award, Am Chem Soc, 69; Anachem Award, 83. *Mem:* AAAS; Am Chem Soc. *Res:* Electrode potentials; homogeneous equilibria; thermodynamics of electrolytic solutions; nonaqueous solutions; seawater; standardization of pH measurements, dissociation of weak acids and bases. *Mailing Add:* Dept Chem Univ Fla Gainesville FL 32611

BATES, STEPHEN ROGER, b London, Eng, Aug 17, 44; US citizen; m 68; c 3. CHILD NEUROLOGY. *Educ:* Univ London, MD, 68; Royal Col Obstetricians & Gynecologists, dipl. *Prof Exp:* Rotating intern, St George's Hosp, London, 69-70; resident pediat, Children's Hosp Med Ctr, Cincinnati, Ohio, 71-72, pediat neurol, 72-73, asst to chief-of-staff, 75-82; resident neurol, Med Ctr, 73-74, chief resident, 74-75, instr pediat & neurol, 75-76, asst prof, 76-81, ASSOC PROF CLIN PEDIAT & CLIN NEUROL, DEPT PEDIAT & NEUROL, UNIV CINCINNATI, 81-; ASST CHIEF-OF-STAFF, CHILDREN'S HOSP MED CTR, 82-, ASSOC CHIEF-OF-STAFF, CONVALESCENT HOSP FOR CHILDREN, 84- *Concurrent Pos:* Attending pediatrician, Children's Hosp Med Ctr, 75-, dir, Electroencephalography Lab, 76-83 & Comprehensive Epilepsy Prog, 80-; attending neurologist, Ohio & Holmes Hosp Div, Univ Hosp, Cincinnati, 75-; consult child neurologist, Clermont County Health Serv & Diagnostic Ctr, 74-76, Univ Affil Cincinnati Ctr Develop Dis, 75-, Resident Home for Retarded, Cincinnati, 80-; mem, cerebral palsy sub-comt, Bur Crippled Children's Serv, Ohio State, 82-; staff consult pediat neurologist, The Christ Hosp, Our Lady of Mercy Hosp, 84-, Good Samaritan Hosp, 85-, Dearborn County Hosp, Ind, 85-, Bethesda Hosps, Ohio, 86-; vis prof child neurol, Sao Paulo, Brazil, 85. *Mem:* Brit Med Asn; Am Acad Neurol; Child Neurol Soc; Am Epilepsy Soc; Am Med EEG Asn; AAAS. *Res:* Clinical pediatric neurology, especially in fields of epilepsy and migraine. *Mailing Add:* Dept Pediat Univ Cincinnati Col Med 231 Bethesda Ave Cincinnati OH 45267

BATES, THOMAS EDWARD, b Irricana, Alta, July 13, 26; m 53; c 2. SOIL SCIENCE. *Educ:* Ont Agr Col, BSA, 55; NC State Col, MS, 54; Iowa State Univ, PhD(soil fertil), 61. *Prof Exp:* Res officer agron, Tobacco Res Bd, Rhodesia & Nyasaland, 51-58; asst soil fertil, Iowa State Univ, 58-61; asst prof soil fertil, Ont Agr Col, 61-65; assoc prof soil fertil, 65-68, PROF SOIL SCI, UNIV GUELPH, 68- *Concurrent Pos:* Assoc ed, Soil Sci Soc Am J, J Environ Qual, Fertilizer Res. *Mem:* Am Soc Agron; Agr Inst Can; Can Soil Sci Soc. *Res:* Soil fertility pertaining to fertilizer use and metal toxicities. *Mailing Add:* 303 Edinburgh Rd S Guelph ON N2G 2W1 Can

BATES, THOMAS FULCHER, b Evanston, Ill, Jan 2, 17; m 42, 76; c 3. MINERALOGY. *Educ:* Denison Univ, BA, 39; Columbia Univ, MS, 40, PhD(geol), 44. *Hon Degrees:* DSc, Denison Univ, 68. *Prof Exp:* From instr to assoc prof, Pa State Univ, 42-53, asst to vpres for res, 61-65, dir inst sci & eng, 63-65, asst dean grad sch, 64-65, vpres planning, 67-72, prof mineral, 53-77; phys scientist, US Geol Surv, 77-82; RETIRED. *Concurrent Pos:* Sci adv, Secy Interior, Washington, DC, 65-67. *Mem:* AAAS; fel Geol Soc Am. *Res:* X-ray and electron microscopic studies of clay and other minerals; origin of the Edwin clay, Ione, California; mineralogical investigation of fine-grained rocks; uraniferous black shales and lignites; clay minerals of Hawaii; application of earth sciences to land use and energy planning. *Mailing Add:* 903 Outer Dr State College PA 16801-8236

BATES, WILLIAM K, b Houston, Tex, Nov 16, 36; m 59; c 2. BIOCHEMISTRY, CELL BIOLOGY. *Educ:* Rice Univ, BA, 59, PhD(biochem), 63. *Prof Exp:* USPHS trainee biol sci, Stanford Univ, 63-66; from asst prof to assoc prof biol, 66-74, head dept, 79-88, PROF BIOL, UNIV NC, GREENSBORO, 88- *Mem:* AAAS; Am Soc Cell Biol; Am Electrophoresis Soc; Sigma Xi. *Res:* Intra-cellular distribution and kinetics of enzymes; cytoplasmic inheritance; radiotracer applications; computer applications in biological studies. *Mailing Add:* Dept Biol Univ NC Greensboro NC 27412

BATESON, ROBERT NEIL, b Hibbing, Minn, Apr 22, 31; m 54; c 3. ENGINEERING. *Educ:* Univ Minn, BME, 54, MSEE, 68. *Prof Exp:* Staff mem, Sandia Corp, 54-58; res engr, cent res lab, Gen Mills, Inc, 58-66; group leader, 66-68; INSTR ELECTRONIC ENG TECHNOL, ANOKA RAMSEY STATE JR COL, 68-, HEAD DEPT, 69- *Res:* New methods and equipment for processing food; new sensors for food processes; development of effective methods of teaching; development of engineering technology courses. *Mailing Add:* Dept Math & Eng Anoka-Ramsey Community 11200 Miss Blvd Coons Rapids MN 55433

BATEY, HARRY HALLSTED, JR, b Grand Island, Nebr, Feb 1, 23; m 45; c 4. INORGANIC CHEMISTRY. *Educ:* Cornell Col, AB, 46; Ohio State Univ, MSc, 48, PhD(chem), 51. *Prof Exp:* From instr to assoc prof, 51-70, PROF CHEM, WASH STATE UNIV, 70- *Concurrent Pos:* UNESCO sci teaching expert, Iran, 68-69. *Mem:* AAAS; Am Chem Soc; Sigma Xi. *Res:* Inorganic nitrogen compounds; oxyhalides; gas chromatography of inorganic substances; non-aqueous solvents; science education. *Mailing Add:* PO Box 464 Nordman ID 83848

BATEY, ROBERT WILLIAM, b Wilsall, Mont, Nov 16, 18; m 44. CHEMICAL ENGINEERING, FOOD ENGINEERING. *Educ:* Univ Minn, BS 47. *Prof Exp:* Proj engr, Gen Mills, Inc, 47-50 & Good Foods, Inc, 50-51; chief process engr, Pillsbury Mills, Inc, 51-55; plant tech dir, Dromedary Co, Nat Biscuit Co, 55-56; head process control sect, Tenco, Inc,

56-58; dir food technol, Foster D Snell, Inc, 58-63; sr res assoc, ITT Continental Baking Co, 63-84; PROP, B80 ENG ASSOCS, 85- *Mem:* Am Inst Chem Engrs; Am Chem Soc; Inst Food Technol; Sigma Xi. *Res:* Cereal; fats and oils processing; industrial waste disposal; microwave processing; development of new instruments for process analysis and control; uses of new sources of energy; evaluate and develop energy concepts; patents; rheology. *Mailing Add:* 38 Meadowlark Rd Rye Brook NY 10573-1220

BATH, DONALD ALAN, b Los Angeles, Calif, May 14, 47. ANALYTICAL CHEMISTRY, ENVIRONMENTAL CHEMISTRY. *Educ:* Rose Polytech Inst, BS, 69; Mont State Univ, PhD(chem), 75. *Prof Exp:* Res assoc environ chem, Univ Ill, 76-77; asst prof chem, Ill State Univ, 77-78; asst prof, 78-84, ASSOC PROF CHEM, WESTERN ILL UNIV, 84- *Concurrent Pos:* Vis asst prof, Inst Environ Studies, Univ Ill, 77-78; vis assoc prof chem, Univ Ill, 87-88. *Mem:* Soc Appl Spectros; Am Chem Soc. *Res:* Development of instrumentation and techniques for the analysis of organometallics in the environment and design and theoretical characterization of nonflame atomizers for atomic absorption spectrometry. *Mailing Add:* Dept Chem Western Ill Univ Macomb IL 61455

BATH, JAMES EDMOND, b Santa Ana, Calif, May 4, 38; m 58; c 3. ENTOMOLOGY, PLANT PATHOLOGY. *Educ:* Univ Calif, Davis, 60; Univ Wis, MS, 62, PhD(entom), 64. *Prof Exp:* From asst prof to assoc prof, 64-75, PROF ENTOM, MICH STATE UNIV, 75-, CHMN DEPT, 74- *Concurrent Pos:* Entomologist, Coop State Res Serv, USDA, 79; coordr integrated pest mgt prog, USDA, Sci Educ Admin, 79-80; chmn, North Cent Integrated Pest Mgt Steering Comn, 80-86. *Mem:* AAAS; Entom Soc Am; Am Phytopath Soc; Soc Nematol. *Res:* Mode of plant virus transmission by insects, especially circulative viruses and aphid vectors; development of interdisciplinary implementation of integrated pest management approach in agriculture. *Mailing Add:* Dept Entom Mich State Univ East Lansing MI 48823

BATHA, HOWARD DEAN, b Phillips, Wis, July 3, 25; m 48; c 4. PHYSICAL CHEMISTRY. *Educ:* Carroll Col, BA, 50; Univ Rochester, PhD, 54. *Prof Exp:* Group leader chem, Olin-Mathieson Chem Co, 54-56, sr res proj specialist, 56-58; res specialist, Carborundum Co, 58-63, mgr res br, 63-69, assoc dir res & develop, 69-74, dir develop, 74-81; tech asst to pres, Mat Int, Lexington, Mass, 80-83; MGR RES & ENG, BIDDEFORD, ME, 83- *Mem:* Am Chem Soc; AAAS; Sigma Xi; Am Ceramic Soc. *Res:* Free radical reactions; inorganic hydrides; high temperature reactions; mechanisms of crystal growth; silicon carbide; high performance composites; smart materials. *Mailing Add:* Eight Whippoorwill Circle Kennebunk ME 04043

BATHALA, MOHINDER S, b India, Dec 25, 40; US citizen; m 67; c 3. ANALYTICAL CHEMISTRY. *Educ:* Panjab Univ, India, BS, 64, MS, 66; Ohio State Univ, PhD(natural prod chem), 73. *Prof Exp:* Sr scientist, Cooper Labs, 76-77, A H Robins Co, 77-80; sr res investr, E R Squibb & Son Inc, 80-83; RES REL, BRISTOL-MYERS SQUIBB CO, 84- *Mem:* Am Pharmaceut Asn; Acad Pharmaceut Sci; Am Chem Soc. *Res:* Analyses of various drugs and their metabolites from biological fluid by gas chromatography, liquid chromatography and gas chromatography and mass spectrography methods; isolation and identification of drug metabolites from biological materials. *Mailing Add:* Drug Metab Dept Squibb Inst Med Res PO Box 4000 Princeton NJ 08540

BATHEN, KARL HANS, b New Haven, Conn, Nov 28, 34; c 3. PHYSICAL OCEANOGRAPHY. *Educ:* Univ Conn, BS, 56; Univ Hawaii, MS, 68, PhD(phys oceanog), 70. *Prof Exp:* Res & develop engr, Colorvision, Hycon Mfg Co, 57-62; res engr, Philco-Ford, Aeronutronic, 63-65; res asst dept oceanog, Univ Hawaii, 66-70; mgr environ studies, Dillingham Environ Co, 70-71; PROF & RESEARCHER, DEPT OCEAN ENG, UNIV HAWAII, 71-; PRES & DIR, SCI ENVIRON ANALYSES LTD, 78- *Mem:* Am Water Resources Asn; Marine Technol Soc; Sigma Xi. *Res:* Circulation systems around Hawaii; coastal environmental studies; numerical modeling of embayments; ocean thermal energy conversion; ocean heat storage; solar salinity gradient ponds; ocean instrumentations. *Mailing Add:* Dept Ocean Eng Univ Hawaii-Manoa Honolulu HI 96822

BATHER, ROY, virology, for more information see previous edition

BATHO, EDWARD HUBERT, mathematics; deceased, see previous edition for last biography

BATICH, CHRISTOPHER DAVID, b Jersey City, NJ, Dec 25, 43; c 2. PHYSICAL CHEMISTRY. *Educ:* Pa State Univ, BS, 65; Rutgers Univ, New Brunswick, PhD(org chem), 74. *Prof Exp:* Qual control chemist, White Labs, Kenilworth, NJ, 65-67; staff scientist, Cent Res & Develop Dept, Du Pont Exp Sta, 74-81; assoc prof, 81-88, PROF, MAT SCI & ENG DEPT, UNIV FLA, 88- *Mem:* Am Chem Soc; Sigma Xi. *Res:* Ultraviolet photoelectron and x-ray photoelectron studies of organic molecules; polymer surfaces, x-ray photoelectron, especially biological systems and synthetic polymers which are in contact with living cells; composite interface studies. *Mailing Add:* 217 Mat Sci & Eng Dept Univ Fla Gainesville FL 32611

BATKIN, STANLEY, b New York, NY, Nov 23, 12; m 52; c 3. NEUROSURGERY, GERIATRICS. *Educ:* NY Univ, BS, 33; Univ Edinburgh, MD, 44; Am Bd Neurol Surg, dipl. *Prof Exp:* Clin tutor surg neurol, Univ Edinburgh, 44-46; asst prof neurol, Med Sch, Univ Ark, 49-50; asst prof neurol & neurosurg, State Univ NY Upstate Med Ctr, 50-51, asst clin prof neurol & neurosurg & neurologist, Univ Hearing & Speech Ctr, 51-63; RETIRED. *Concurrent Pos:* Chief neurosurg div, Vet Admin Hosp, Little Rock, Ark, 49-50; neurosurgeon, Syracuse Mem Hosp, Syracuse Univ Hosp, Couse-Irving Hosp, Syracuse Gen Hosp, Midtown Hosp & Community Hosp of Syracuse, 51-63; neurol consult, NY State Educ Dept, Div Voc Rehab, 52-63; neurologist, Syracuse Cerebral Palsy Clin, Syracuse Univ Hearing & Speech Ctr, 52-63; lectr, Depts Spec Educ & Audiol, Syracuse Univ, 52-63; chief div neurol & neurosurg, Permanente Med Group, 63-79; neurol consult

& mem fac, Rehab Hosp Pac & Hyperbaric Treatment Facil, Univ Hawaii; lectr geront, Sch Public Health, Univ Hawaii; vis prof, Israel Naval Hyperbaric Inst, Haifa, 87; Vis neurologist, Dept Health, Scotland, 46-48. *Honors & Awards:* Res Award, Am Cancer Soc, 72. *Mem:* AAAS; Am Acad Neurol; Asn Res Nerv & Ment Dis; Am Asn Neurol Surgeons; Sigma Xi. *Res:* Neurophysiology; oncology; central nervous system regeneration. *Mailing Add:* Physiol Univ Hawaii Sch Med Honolulu HI 96822

BATLEY, FRANK, b Oldham, Eng, Dec 27, 20; m 46; c 3. RADIOLOGY. *Educ:* Univ Manchester, BSc, 41, MB, ChB, 44. *Prof Exp:* Registr radiation ther, Christie Hosp, Manchester, Eng, 45-51; asst, Ont Cancer Found, Hamilton, 52-56; assoc prof, Ohio State Univ Hosps, 56-57; dep dir, Ont Cancer Found, Kingston, 57-59; assoc prof radiol & chief radiation ther, State Univ NY Upstate Med Ctr, 59-67; PROF RADIOL & DIR RADIOTHERAPY, OHIO STATE UNIV HOSPS, 67- *Mem:* Am Radium Soc; Am Col Radiol; Radiol Soc NAm. *Res:* Cancer therapy and statistical methods. *Mailing Add:* 2154 Coach Rd N Columbus OH 43220

BATLOGG, BERTRAM, Austria citizen. SUPERCONDUCTIVITY, SOLID STATE PHYSICS. *Educ:* Swiss Fed Inst Technol, dipl, 74, PhD(nat sci), 79. *Prof Exp:* DEPT HEAD, SOLID STATE PHYSICS & MAT RES, AT&T BELL LABS, 78-, MEM STAFF, 79- *Mem:* Fel Am Phys Soc; Mat Res Soc. *Res:* Superconductivity. *Mailing Add:* AT&T Bell Labs Rm 1D369 600 Mountain Ave Murray Hill NJ 07974

BATOREWICZ, WADIM, b Poland, Nov 7, 34; US citizen; m 58. ORGANIC CHEMISTRY. *Educ:* Univ Minn, BCh, 60, PhD(org chem), 67. *Prof Exp:* Chemist, Minn Mining & Mfg Co, 60-62; SR RES CHEMIST, CHEM DIV, UNIROYAL, INC, 67- *Mem:* Am Chem Soc. *Res:* Chemistry of organophosphorus compounds; chemistry of polyurethanes; flame retardants for polymers; flammability of polymers; agriculture and food chemistry. *Mailing Add:* 999 Whalley Ave 1E New Haven CT 06515

BATRA, GOPAL KRISHAN, b Lahore, India, Oct 10, 43; m 60; c 2. BIOTECHNOLOGY. *Educ:* Panjab Univ, India, BSc, 62, MSc, 66; Univ Ga, PhD(virol), 72. *Prof Exp:* Lectr biol, Govt Col, Mandi, India, 65-66; sr sci asst microbiol, Indian Drugs & Pharmaceut, Virbhadra, 66-68; res assoc virol, Univ Ga, 72-73; res assoc cell biol, 73-74, sr res assoc virol, Emory Univ, 74-77; PRES, BIOSYSTS, INC, 76- *Concurrent Pos:* Tech serv mgr, FMC Corp, Atlanta, 78-80. *Mem:* AAAS; Am Soc Microbiol; Soc Gen Microbiol; Soc Indust Microbiol; Sigma Xi; Am Chem Soc; Am Soc Biotechnol; Am Soc Indust Hyg. *Res:* Vaccines; biological products; biological control of pests and pathogens; bioenergy. *Mailing Add:* Biosysts Inc 762 US Hwy 78 Loganville GA 30249-9522

BATRA, INDER PAUL, b India, June 25, 42; m 70; c 2. SOLID STATE PHYSICS. *Educ:* Univ Delhi, BSc, 62, MSc, 64; Simon Fraser Univ, PhD(physics), 68. *Prof Exp:* Nat Res Coun Can fel, Simon Fraser Univ, 68-69; tech staff, San Jose Res Lab, 79-80, vis scientist, Res Lab, Yorktown Heights, 80-81, PROF SCIENTIST, SURFACE SCI GROUP, RES LAB, IBM CORP, 69-, MGR, STRUCT MECH DEPT, 90- *Concurrent Pos:* vis scientist, IBM Zurich Res Lab, Switz, 85-86; dir tech staff, IBM Res Lab, San Jose, 79-80. *Mem:* Fel Am Phys Soc; Am Vacuum Soc. *Res:* Interband optical absorption; phonon assisted transitions in semiconductors; photoconductivity; stimulation scattering of laser light; theoretical studies of electronic transport; chemisorption and electronic structure of surfaces, metal semiconductor interfaces and polymers by molecular orbital cluster and band stucture techniques; linear combination of atomic orbital scheme and pseudopotential method for electronic structure of solids; theoretical studies of helium diffraction from surfaces. *Mailing Add:* IBM Corp Res Lab IBM Almaden Res Ctr San Jose CA 95120-6099

BATRA, KARAM VIR, b Jhelum, Panjab, India, Jan 3, 32; US citizen; m 63; c 4. PHARMACOLOGY, TOXICOLOGY. *Educ:* Panjab Univ, India, BPharm, 52; Univ Minn, MS, 57; Univ Chicago, PhD(pharmacol), 63. *Prof Exp:* Gen trainee res mfg, Glaxo Labs Ltd, India, 52-54; med rep, Alamagamated Chem & Dyestuff Co Ltd, India, 54-55; res fel, Inst Med Res, Chicago Med Sch, 63-64; res biochemist, Vet Admin Hosp, Hines, Ill & Northwestern Univ, 64-66; res pharmacologist, IIT Res Inst, 66-68; from instr to asst prof pharmacol, Chicago Col Osteop Med, 68-73; sr pharmacologist, Haskell Lab Toxicol & Indust Med, E I du Pont de Nemours & Co, Inc, 73-75; sr toxicologist, SRI Int, Arlington, 76-78; SR PHARMACOLOGIST, CTR DEVICES RADIOL HEALTH, FOOD & DRUG ADMIN, US DEPT HEALTH & HUMAN SERV, 78- *Concurrent Pos:* Consult tumor res, Vet Hosp, Hines, 70-73; adj asst prof psychopharmacol, Univ Del, 75-77. *Mem:* Am Soc Pharmacol & Exp Therapeut. *Res:* Metabolism and distribution of drugs; effect of drugs on cells in culture; radio-incorporation studies; biochemistry of carcinogenesis; chemical constitution and pharmacological-biological activity; lipoproteins; phytochemistry; cardiovascular pharmacology; preclinical investigations and screening; behavioral pharmacology and toxicology; factor VIII and prostacyclin in endothelial cells damage. *Mailing Add:* 2252 Senseney Lane Falls Church VA 22043-3105

BATRA, LEKH RAJ, b Panjab, India, Nov 26, 29; nat US; m 60; c 2. MYCOLOGY. *Educ:* Panjab Univ, India, BSc, 52, MSc, 54; Cornell Univ, PhD(mycol), 58. *Prof Exp:* Res scholar mycol, Govt India, 54-55; lectr biol, Desh Bandhu Col, India, 55; instr plant path, Cornell Univ, 57-58; lectr biol, Swarthmore Col, 58-60; res assoc, Univ Kans, 60-62, from asst prof to assoc prof bot, 62-68; res mycologist, Plant Protection Inst, 67-85, SCI ADV BELTSVILLE AREA, SCI & EDUC ADMIN, USDA, 86- *Concurrent Pos:* Directorate of Plant Quarantine, Govt India, 60; vis res fel, Fed Forest & Timber Mgt Res Ctr, Hamburg, Germany, 62; mem, US Fedn Cult Collection, 81-83, Comt Int Union, Nat Acad Sci & Am Type Cult Collection, 83-89, Int Sci Bd, 89-; adj prof, Towson State Univ. *Mem:* Mycol Soc Am; Bot Soc Am; NY Acad Sci; Am Soc Microbiol; Microbiol Soc (secy, 81-); Soc Conserv & Field Ecol (secy-treas, 80-84); Int Mycol Asn. *Res:* Morphology and taxonomy Hemiascomycetes and Discomycetes; ambrosia fungi, cereal, cotton and sugar cane diseases; plant embryo culture; fermented foods of the orient; insect-fungus mutualism and mimicry; author, editor and co-editor of several books. *Mailing Add:* Beltsville Dirs Off Agr Res Serv Beltsville MD 20705

BATRA, NARENDRA K, b India, Sept 15, 43. SOLID STATE PHYSICS, ULTRA SONICS NON-DESTRUCTIVE TESTING. *Educ:* Columbia Univ, MA, 67; Wayne State Univ, PhD(physics), 72. *Prof Exp:* Fac mem, Va Tech, 74-76; contractor, Wright Air Force Base, 77-82; RES PHYSICIST, NAVAL RES LAB, 82- *Mem:* Inst Elec Electronics Engrs; Am Soc Non-Destructive Testing; Am Phys Soc. *Mailing Add:* Naval Res Lab 5834 4555 Overlook Dr Washington DC 20375

BATRA, PREM PARKASH, b Jhang Maghiana, India, Apr 1, 36; US citizen; m 61; c 1. BIOCHEMISTRY. *Educ:* Panjab Univ, India, BS, 55, MS, 58; Univ Ariz, PhD(agr chem), 61. *Prof Exp:* Res assoc biochem, Univ Ariz, 61-63; fels, Univ Utah, 63-64 & Johns Hopkins Univ, 64-65; biochemist, aging res lab & geront br NIH, Vet Admin Hosp, 65; from asst prof to prof biol, 65-75, PROF BIOL CHEM, SCH MED, WRIGHT STATE UNIV, 75- *Concurrent Pos:* NIH spec fel, Univ Ky, 73-74. *Mem:* Am Chem Soc; fel Am Soc Biol Chemists. *Res:* Biosynthesis of carotenoids in plants and microorganisms; mechanism of photosynthetic phosphorylation; enzymology of nucleic acids. *Mailing Add:* Dept Biol Chem Sch Med Wright State Univ Dayton OH 45435

BATRA, ROMESH CHANDER, b Panjab, India, Aug 16, 47; m 72; c 2. MECHANICAL ENGINEERING. *Educ:* Panjabi Univ, India, BSc, 68; Univ Waterloo, MASc, 69; Johns Hopkins Univ, PhD(mech), 72. *Prof Exp:* Postdoc assoc mech, Johns Hopkins Univ, 72-73; res assoc eng mech, McMaster Univ, 73-74; asst prof, Univ Mo-Rolla, 74-76 & Univ Ala, 76-77; assoc prof, 77-81, PROF ENG MECH, UNIV MO-ROLLA, 81- *Concurrent Pos:* Univ Mo-Rolla grants, 76, 78, 79 & 81, 82-88; res engr, Athena Eng Co, Ala, 76-77; consult, Xerox Corp, 78-80, Dames & Moore, 80, Army Ballistic Res Lab, 85-88, Hewlett-Packard, 87 & Belloit Manhatten, 87; NSF grant, 80-82, 88-90; Army grant, 83-92; Varo grant, 85-91. *Mem:* Fel Am Soc Mech Engrs; Am Acad Mech; Soc Eng Sci; Soc Rheology; Am Soc Eng Educ. *Res:* Adiabatic shear banding; applications of finite element method to solving thermomechanical problems; penetration problem. *Mailing Add:* 125 Mech Eng Annex Univ Mo Rolla MO 65401

BATRA, SUBHASH KUMAR, b Bannu, Pakistan, Oct 4, 35; US citizen. FIBER-TEXTILE TECHNOLOGY & ANALYSIS. *Educ:* Univ Delhi, BS, 57; Mass Inst Technol, SM, 61, SM, 77; Rensselaer Polytech Inst, PhD(mech), 66. *Prof Exp:* Asst weaving dept, Ahmedabad New Cotton Mill, India, 57-58; res engr mech eng, Mass Inst Technol, 61-62; sr scientist lubrication technol & tribology, Battelle Mem Inst, 66-70; res assoc mech eng, Mass Inst Technol, 70-77; assoc prof textile mat & mgt, Sch Textiles, 77-84, PROF TEXTILE ENG & SCI, NC STATE UNIV, 84- *Concurrent Pos:* Vis fel, Sloan Sch Mgt, Mass Inst Technol, 75-76; consult, World Bank, Washington, DC, 76-77, Avtex Fibers Inc, Front Royal, Va, 78 & Monsanto Textile Co, NC, 79-; secy, Textile Div, Am Soc Mech Engrs, 78-83 & chmn, 83-84; vis scholar, Israel Fiber Inst, 82. *Mem:* Fel Am Soc Mech Engrs; Fiber Soc; Am Acad Mech; Sigma Xi; fel Textile Inst. *Res:* Fiber physics; mechanics of nonwovens and textile structures; textile technology; fiber fatigue. *Mailing Add:* Dept Textile Eng & Sci NC State Univ Box 8301 Raleigh NC 27695-8301

BATRA, SUZANNE WELLINGTON TUBBY, b New York, NY, Dec 15, 37; m 60; c 2. ENTOMOLOGY. *Educ:* Swarthmore Col, BA, 60; Univ Kans, PhD(entom), 64. *Prof Exp:* Res assoc entom, Univ Kans, 64-67; res, Utah State Univ, 68-69; consult, Apicult Res Br, 70-74, RES ENTOMOLOGIST, AGR RES SERV, USDA, 74- *Concurrent Pos:* Grants, Sigma Xi & Am Philos Soc, 64-65; sr res officer, Punjab Agr Univ, 65; ed, J Kans Entom Soc, 66-67; adj prof, Univ Md. *Mem:* AAAS; Sigma Xi. *Res:* Ecology, behavior of insects; evolution of social behavior; management of crop pollinators; insect-fungus symbioses; Apoidea; biological control of weeds. *Mailing Add:* Bee Res Lab Bldg 476 Beltsville Agr Res Ctr USDA Beltsville MD 20705

BATRA, TILAK RAJ, b Girotte, India, Dec 11, 36; Can citizen; m 62; c 2. ANIMAL BREEDING, QUANTITATIVE GENETICS. *Educ:* Jabalpur Univ, India, BVSc, 59; Univ Hawaii, MS, 66; Univ Ill, PhD(animal breeding), 68. *Prof Exp:* Lectr, Vet Col, India, 62-64; res asst, Univ Ill, 66-68, res assoc, 68-69; res programmer, Univ Guelph, 69-76; RES SCIENTIST, AGR CAN, OTTAWA, 76- *Mem:* Am Dairy Sci Asn; Can Soc Animal Sci; Nat Mastitis Coun; Int Soc Animal Blood Group Res (secy-treas); Can Soc Animal Sci. *Res:* Dairy cattle breeding. *Mailing Add:* Animal Res Ctr Res Br Agr Can Ottawa ON K1A 0C6 Can

BATRIN, GEORGE LESLIE, b Detroit, Mich, May 20, 57. ENGINEERING MATHEMATICS, ENGINEERING STATISTICS. *Educ:* Carnegie-Mellon Univ, BS, 85, MA, 86. *Prof Exp:* Safety engr, Carnegie-Mellon Univ, 84-85; APP ENGR, GATES RUBBER CO, 88- *Concurrent Pos:* Engr, US Tank Automotive Command, 87-88; process engr, Turnmatic, Inc, 87-88. *Mem:* Am Math Soc; Math Asn Am; Soc Indust & Appl Math; Soc Automotive Engrs; Soc Mfg Engrs; Indust Math Soc. *Res:* Mathematical investigation into modeling belt slippage during a rapid deceleration internal; reliability equations for machinery and belt mass reduction for power transmission; Department of Energy's testing and evaluation in the automotive industry. *Mailing Add:* 37965 Willowood Ct Mt Clemens MI 48045

BATSAKIS, JOHN G, b Petoskey, Mich, Aug 14, 29; m 57; c 3. MEDICINE, PATHOLOGY. *Educ:* Univ Mich, MD, 54. *Prof Exp:* Intern path, Univ Hosp, George Washington Univ, 54-55; resident, Univ Mich, 55-59; chief clin path, Walter Reed Gen Hosp, Washington, DC, 59-61; assoc pathologist, Bronson Methodist Hosp, Kalamazoo, Mich, 61-; from asst prof to prof path, Univ Mich, Ann Arbor, 62-79, from assoc dir to co-dir clin labs, Univ Hosp, 64-79; chief Path Dept, Maine Med Ctr, 79-81; CHMN & PROF, DEPT PATH, UNIV TEX M D ANDERSON HOSP, HOUSTON, 81-, RUTH LEGGET JONES PROF, 82- *Concurrent Pos:* Consult, Vet Admin Hosp, Ann Arbor, 62- & Armed Forces, 72- *Mem:* Fel Col Am Path; Asn Mil Surg US; fel Am Col Physicians; Am Asn Clin Chemists; fel Am Soc Clin Pathologists. *Res:* Clinical chemistry; head and neck, gastrointestinal and thyroid pathology; clinical enzymology. *Mailing Add:* Univ Tex M D Anderson Hosp 1515 Holcombe Blvd Houston TX 77030

BATSHAW, MARK LEVITT, b Montreal, Que, Sept 19, 45; m 69; c 3. DEVELOPMENTAL PEDIATRICS, METABOLISM. *Educ:* Univ Pa, BA, 67; Univ Chicago, MD, 71. *Prof Exp:* Intern pediat, Hosp Sick Children, Toronto, 71-72, resident, 72-73; fel, 73-75, instr, 75-76, from asst prof to assoc prof pediat, Johns Hopkins Med Inst, 76-88; W T GRANT PROF PEDIAT, UNIV PA SCH MED, 88- *Concurrent Pos:* Develop pediatrician, John F Kennedy Inst, Baltimore, 75-; physician-in-chief, Children's Seashore House, Philadelphia, 88- *Mem:* Soc Pediat Res; fel Royal Acad Physicians & Surg; Am Acad Pediat; Soc Inherited Metab Dis; Soc Develop Pediat. *Res:* Investigation of genetics, neurochemistry and therapy of inborn errors of urea cycle metabolism. *Mailing Add:* Children's Seashore House 3405 Civic Ctr Blvd Philadelphia PA 19104

BATSON, ALAN PERCY, b Birmingham, Eng, Sept 18, 32; m 57, 68; c 4. COMPUTER SCIENCE. *Educ:* Univ Birmingham, BSc, 53, PhD(physics), 56. *Prof Exp:* Fel, Univ Birmingham, 56-58; from instr to assoc prof physics, 58-67, dir, Comput-Sci Ctr, 62-72, PROF COMPUT SCI, UNIV VA, 67- *Mem:* Inst Elec & Electronics Engrs; Asn Comput Mach. *Res:* Computer systems. *Mailing Add:* Div Acad Comput Univ Va Charlottesville VA 22903

BATSON, BLAIR EVERETT, b Hattiesburg, Miss, Oct 24, 20; m 54. PEDIATRICS. *Educ:* Vanderbilt Univ, BS, 41, MD, 44; Johns Hopkins Univ, MPH, 54. *Prof Exp:* Intern pediat, Vanderbilt Univ Hosp, 44-45; asst resident, Johns Hopkins Hosp, 45-46; asst resident, Vanderbilt Univ Hosp, 48-49, instr pediat, Sch Med, Vanderbilt Univ, 49-52; instr pediat, Med Sch & pub health adminr, Div Maternal & Child Health, Sch Hyg & Pub Health, Johns Hopkins Univ, 52-54, asst prof, 54-55; chmn dept, 55-88, PROF PEDIAT, SCH MED, UNIV MISS, 55- *Concurrent Pos:* Resident pediatrician, Vanderbilt Univ Hosp, 49-50; consult pediat & US Air Force examr, Am Bd Pediat. *Mem:* AMA; Am Acad Pediat; Am Pub Health Asn; Am Pediat Soc. *Res:* Growth and development in children; immunization; handicapped children. *Mailing Add:* Univ Miss Sch Med Jackson MS 39216

BATSON, DAVID BANKS, b Tenn, April 14, 44. TOXICOLOGY. *Educ:* Austin Peay State Univ, BS, 66; Univ Tenn, MS, 68, PhD(animal sci), 72. *Prof Exp:* Regulatory scientist, Div Drugs Ruminant Species, 72-80, EXTRAMURAL RES COORDR & HEALTH SCIENTIST ADMIN, CTR VET MED, FOOD & DRUG ADMIN, 80- *Mem:* AAAS; Soc Study Reprod; Am Soc Animal Sci; Asn Gov Toxicologists. *Res:* Pharmacology and toxicology of animal drugs and feed additives; aquaculture research; analytical methods development. *Mailing Add:* 14608 Nadine Dr Rockville MD 20853

BATSON, GORDON B, b Addison, Maine, Oct 3, 32; m 61; c 3. CIVIL ENGINEERING, STRUCTURAL MECHANICS. *Educ:* Univ Maine, BS, 55, MS, 56; Carnegie-Mellon Univ, PhD (civil eng), 62. *Prof Exp:* From asst prof to assoc prof civil eng, 61-77, PROF CIVIL & ENVIRON ENG, CLARKSON UNIV, 77- *Mem:* Am Soc Civil Engrs; fel Am Concrete Inst; Am Soc Testing & Mat. *Res:* Fracture mechanics applied to concrete; crack arrest in concrete using fibers. *Mailing Add:* Dept Civil Eng Clarkson Univ Potsdam NY 13676

BATSON, MARGARET BAILLY, b New York, NY, July 13, 14; m 54. PSYCHIATRY, PEDIATRICS. *Educ:* Manhattanville Col, BA, 37; Columbia Univ, MA, 40, PhD(bact), 49; Univ Rochester, MD, 51. *Prof Exp:* Clin instr bact, Columbia Univ, 39-41; intern pediat, Johns Hopkins Univ Hosp, 51-52, asst resident, 53-54, instr, Sch Med, Univ & pediatrist-in-chg, Hosp, 54-55; asst resident pediat, Univ Minn, 52-53; from asst prof to assoc prof pediat, 55-80, assoc prof psychiat & chief, Div Human Behavior, 59-80, dir, Infant & Child Develop Clin, 65-80, asst prof microbiol & assoc mem grad fac, 67-80, EMER ASSOC PROF, UNIV MISS MED CTR, 80-; MED DIR, HUDSPETH RETARDATION CTR, 80- *Mem:* Sigma Xi. *Res:* Pediatric neurology; microbiology; immunology; human behavior. *Mailing Add:* 52 Eastbrooke Jackson MS 39216

BATSON, OSCAR RANDOLPH, b Hattiesburg, Miss, Oct 26, 16; m 50; c 4. PEDIATRICS. *Educ:* Vanderbilt Univ, BA, 38, MD, 42; Am Bd Pediat, dipl. *Prof Exp:* From intern to resident pediat, Vanderbilt Hosp, 42-44, fel, 46-47; from instr to assoc prof, 47-59, actg dean sch med, 62-63, dean & dir med affairs, 63-72, vchancellor med affairs & vchancellor alumni & develop, 72-74, PROF PEDIAT, SCH MED, VANDERBILT UNIV, 59- *Concurrent Pos:* Pres & physician in chief, Charles Henderson Child Health Ctr, 78- *Mem:* AMA; Am Pediat Soc; Am Col Physicians; Am Acad Pediat; Southern Med Asn. *Res:* Medical administration. *Mailing Add:* Vanderbilt Univ Box 928 Troy AL 36081

BATSON, WILLIAM EDWARD, JR, b Taylors, SC, Apr 3, 42; m 65; c 2. PLANT PATHOLOGY. *Educ:* Clemson Univ, BS, 65, MS, 66; Tex A&M Univ, PhD(plant path), 71. *Prof Exp:* Res assoc plant path, Tex A&M Univ, 67-71; from asst prof to assoc prof, 71-82, PROF PLANT PATH, MISS STATE UNIV, 82- *Mem:* Am Phytopath Soc. *Res:* Seedling diseases of cotton. *Mailing Add:* Dept Plant Path Miss State Univ Drawer PG Mississippi State MS 39762

BATT, ELLEN RAE, b New York, NY, Sept 24, 34. PHYSIOLOGY. *Educ:* Barnard Col, AB, 56, Columbia Univ, MA, 59, PhD(zool), 67. *Prof Exp:* NIH fel, 67-70, res assoc, 70-73, asst prof physiol, Columbia Univ, 73-81; FAC MEM, SARAH LAWRENCE COL, 87- *Concurrent Pos:* Consult, Harlem Hosp, 88- *Mem:* Am Soc Zoologists; AAAS; Sigma Xi; Am Physiol Soc; NY Acad Sci. *Res:* Transport through red blood cell, bacterial, and intestinal membranes; neonatal development of rodent intestine with respect to transport. *Mailing Add:* 5355 Henry Hudson Pkwy New York NY 10471

BATT, RUSSELL HOWARD, b Worcester, Mass, July 27, 38; m 60; c 2. PHYSICAL CHEMISTRY. *Educ:* Univ Rochester, BS, 60; Univ Calif, Berkeley, PhD(chem), 65. *Prof Exp:* Asst prof chem, Wesleyan Univ, 64-66; res instr, Dartmouth Col, 66-68; asst prof, 68-73, ASSOC PROF CHEM, KENYON COL, 73- *Mem:* Am Chem Soc; Sigma Xi. *Res:* Photoconductivity in molecular crystals; low temperature calorimetry; biology and chemistry of trace heavy metal environmental pollutants. *Mailing Add:* Dept Chem Kenyon Col Gambier OH 43022

BATT, WILLIAM, b Buffalo, NY, May 25, 31. BIOCHEMISTRY. *Educ:* Cath Univ Am, BA, 54, MS, 56; Georgetown Univ, PhD(biochem), 59. *Prof Exp:* Instr chem, La Salle Acad, RI, 54-55 & De La Salle Col, Washington, DC, 55-59; from inst to prof, 59-71, head dept, 64-72, dir admis, 72-81, PROF CHEM, MANHATTAN COL, 71- *Concurrent Pos:* NIH fel, Brookhaven Nat Lab, 63-64. *Mem:* AAAS; NY Acad Sci; Am Chem Soc; Am Inst Chemists. *Res:* Physical aspects of enzyme catalyzed reactions; kinetics and mechanisms of enzyme substrate interaction; thrombin. *Mailing Add:* Dept Comput Info Syst Manhattan Col Riverdale NY 10471

BATTAGLIA, FREDERICK CAMILLO, b Weehawken, NJ, Feb 15, 32; m; c 2. PEDIATRICS, PHYSIOLOGY. *Educ:* Cornell Univ, BA, 53; Yale Univ, MD, 57; Am Bd Pediat, cert, 64 & cert, neonatol-perinatol, 75. *Hon Degrees:* DSc, Univ Ind, 90. *Prof Exp:* Intern, Dept Pediat, Johns Hopkins Hosp, Md, 57-58, resident, 60-62; USPHS fel, Dept Biochem, Univ Cambridge, Eng, 58-59; Josiah Macy Found fel, Dept Physiol, Sch Med, Yale Univ, 59-60; asst prof pediat, Johns Hopkins Univ, 63-65; assoc prof, Dept Pediat & Obstet-Gynec, 65-69, dir, Div Perinatal Med, 70-74, chmn, dept pediat, 74-88, PROF, DEPT PEDIAT & OBSTET-GYNEC, SCH MED, UNIV COLO, 69- *Concurrent Pos:* Attend pediatrician, Johns Hopkins Hosp, 62, Union Mem Hosp, Md, 64-65, The Children's Hosp, Colo & Denver Gen Hosp, 67- & Fitzsimons Army Med Ctr, 70-; co-dir, Newborn Ctr, Colo Gen Hosp, 67-74; contrib ed, J Pediat, 66-74; assoc ed, Biol Neonate, 79-; mem sci adv bd, Joseph P Kennedy Jr Found, 83-; mem, Low Birthweight Comt, NIH, 84, Human Embryol & Develop Study Sect, Nat Inst Child Health & Human Develop, 86- *Honors & Awards:* E Mead Johnson Award Pediat Res, Am Acad Pediat, 69; Agnes Higgins Award, March of Dimes, 86. *Mem:* Inst Med-Nat Acad Sci; Soc Gynec Invest; Perinatal Res Soc (secy-treas, 70-73, pres, 74-75); Am Acad Pediat; Sigma Xi; AAAS; Am Pediat Soc; Soc Pediat Res (pres, 76-77); Asn Am Physicians. *Res:* Perinatal physiology; intrauterine growth retardation; biochemistry; author and co-author of several books on various aspects of fetal and maternal medicine. *Mailing Add:* Perinatal Pediat Univ Colo Med Ctr 4200 E Ninth Ave Denver CO 80262

BATTAILE, JULIAN, b Sept 26, 25; US citizen; m 58; c 2. BIOCHEMISTRY. *Educ:* La State Univ, BS, 47; Univ Ill, MS, 48; Ore State Univ, PhD(biochem), 60. *Prof Exp:* Instr chem, Southeast La Col, 47; asst prof, Southwestern La Inst, 48-50; chemist, Ethyl Corp, 51-54; instr chem, Ore State Univ, 57-58; NIH fel, Univ Calif, Davis, 60-62; from asst prof to prof chem, Southern Ore State Col, 62-88; RETIRED. *Mem:* Am Chem Soc; Sigma Xi. *Res:* Biosynthesis of monoterpenes. *Mailing Add:* Dept Chem Southern Ore State Col 1216 Tolman Creek Rd Ashland OR 97520

BATTAN, LOUIS JOSEPH, b New York, NY, Feb 9, 23; m 52; c 2. METEOROLOGY. *Educ:* NY Univ, BS, 46; Univ Chicago, MS, 49, PhD(meteorol), 53. *Prof Exp:* Radar meteorologist, US Weather Bur, 47-51; res meteorologist, Univ Chicago, 51-58; from assoc dir to dir, Inst Atmospheric Physics, 58-82, PROF ATMOSPHERIC SCI, UNIV ARIZ, 58- *Concurrent Pos:* Mem comt atmospheric sci, Nat Acad Sci, 66-76, chmn, 73-76, mem geophysics study comt, 75-82, vchmn, 78-82; mem US deleg, Cong World Meteorol Orgn, 67; secy, atmospheric & hydrospheric sci sect, AAAS, 68-74; mem US nat comt, Int Union Geod & Geophys, 74-76 & 78-84, chmn, 80-84; pres, meteorl sect, Am Geophys Union, 74-76; mem, Nat Comt Oceans & Atmospheres, 78-81. *Honors & Awards:* Meisinger Award, Am Meteorol Soc, 62; Brooks Award, 71. *Mem:* Fel AAAS; fel Am Meteorol Soc (pres, 66-67); fel Am Geophys Union. *Res:* Radar meteorology; cloud physics; mesometeorology; weather modification. *Mailing Add:* Inst Atmospheric Physics Univ Ariz Tucson AZ 85721

BATTARBEE, HAROLD DOUGLAS, b Highlands, Tex, July 25, 40; m 65, 82; c 4. ENDOCRINOLOGY, CARDIOVASCULAR PHYSIOLOGY. *Educ:* Univ Houston, BS, 66; Baylor Col Med, PhD(physiol), 71. *Prof Exp:* From instr to assoc prof, 71-85, PROF PHYSIOL, SCH MED, LA STATE UNIV, SHREVEPORT, 85- *Concurrent Pos:* Mem, med res & develop adv comt, US Army, 82-87; consult, US Dept Defense, US Army Aeromed Res Lab, 82-; assoc mem,Grad Fac, La State Univ, Shreveport, 73- *Mem:* AAAS; Am Physiol Soc; Am Heart Asn. *Res:* Cardiovascular physiology; stress physiology; role of dietary sodium and potassium in the pathogenesis of hypertension; baroreflex control of heart rate and peripheral resistances in hypertension; endocrine and sympathetic nervous system function changes induced by chemical warfare pretreatment and antidotal drugs; portal venous hypertension and visceral blood flow. *Mailing Add:* Dept Physiol PO Box 33932 La State Univ Med Ctr Shreveport LA 71130

BATTE, EDWARD G, b Ferris, Tex, Feb 4, 21; m 42; c 1. VETERINARY PARASITOLOGY. *Educ:* Agr & Mech Col Tex, BS, 42, MS, 49, DVM, 49. *Prof Exp:* Assoc parasitologist, Agr Exp Sta, Univ Fla, 49-51; res veterinarian, Calif Spray Chem Corp, 51-56; PROF VET PARASITOL, SCH VET MED, NC STATE UNIV, 56- *Mem:* Am Vet Med Asn; Am Col Vet Toxicologists. *Mailing Add:* Sch Vet Med NC State Univ Raleigh NC 27650

BATTE, WILLIAM GRANVILLE, b Jarratt, Va, Jan 26, 27; m 50; c 2. ELECTRICAL ENGINEERING. *Educ:* Va Polytech Inst, BS(elec eng) & BS(indust eng), 50; Case Inst Technol, PhD(elec eng), 65. *Prof Exp:* Scientist, NASA Langley Res Ctr, 54-67; assoc prof elec eng, Univ Va, 67-69; prof sci, Old Dominion Univ, 69-76, adj prof elec eng, 76-80; PRES, COMPUT RES INC, 80- *Mem:* Inst Elec & Electronics Engrs; Asn Comput Mach. *Res:* Digital computer and control engineering; computer time sharing and control; digital logic. *Mailing Add:* Comput Res Inc 1600 Pennwood Dr Hampton VA 23666

BATTEN, ALAN HENRY, b Tankerton, Eng, Jan 21, 33; Can citizen; m 60; c 2. ASTRONOMY. *Educ:* Univ St Andrews, BSc, 55, DSc, 74; Univ Manchester, PhD(close binary systs), 58. *Prof Exp:* Res assoc astron, Univ Manchester, 58-61; sci officer, Dom Astrophys Observ, 61-70, assoc res officer, 70-75, sr res officer, 76-91, GUEST WORKER, DOM ASTROPHYS OBSERV, 91- *Concurrent Pos:* Nat Res Coun Can fel, 59-61; lectr, Univ Victoria, 61-64; Can Nat Comt, 64-70 & 76-82, organizing comts & comns 30-42, 67-, pres, Comn 30, 76-79, mem, Organizing Comt 26, 73-79, pres comn 42, 82-85; on leave, Vatican Observ, Castel Gandolfo, Italy, spring & summer, 70; guest investr, Inst Astron & Space Physics, Buenos Aires, 72, chmn Can organizing comt for IAU, XVII Gen Assembly, 75-79; ed, J Royal Astron Soc Can, 80-88. *Mem:* Fel Royal Soc Can; Am Astron Soc; Royal Astron Soc Can (pres, 74-76); Int Astron Union (vpres, 85-91); fel Royal Astron Soc; Can Astron Soc (pres, 71-73). *Res:* Spectroscopic and photometric studies of close binary systems; radial velocities of stars; history of astronomy. *Mailing Add:* Dom Astrophys Observ 5071 W Saanich Rd Victoria BC V8X 4M6 Can

BATTEN, BRUCE EDGAR, CELLULAR ENDOCRINOLOGY. *Educ:* Med Col Va, PhD(anat), 77. *Prof Exp:* ASSOC PROF ANAT, SCH MED, TUFTS UNIV, 80- *Mailing Add:* Dept Anat Ohio State Univ 333 W Tenth St Columbus OH 43210

BATTEN, CHARLES FRANCIS, b Manchester, NH, Apr 15, 42. PHYSICAL CHEMISTRY. *Educ:* St Anselms Col, AB, 64; Univ Notre Dame, MS, 66, PhD(chem), 71. *Prof Exp:* Fel chem, Univ Houston, 71-72, vis asst prof, 72-73; res fel chem, Univ Nebr-Lincoln, 75-76; res assoc chem, 76-80, SR RES SCIENTIST, UNIV HOUSTON, 80- *Concurrent Pos:* Anal Lab mgr, NUS Corp. *Mem:* Sigma Xi; Am Chem Soc. *Res:* Atmospheric pressure ionization mass spectrometry; photoionization mass spectrometry; analytic detectors based on ionization and electron capture; kinetics and thermodynamics of negative ion reactions. *Mailing Add:* 7994 Locke Lane 29 Houston TX 77063

BATTEN, GEORGE L, JR, b Smithfield, NC, Oct 31, 52; m 77; c 4. HYDROGEN BONDING IN CELLULOSIC WEBS. *Educ:* Wake Forest Univ, BS, 75; Univ NC, Chapel Hill, PhD(chem physics), 78. *Prof Exp:* Res chemist, Westvaco Corp, 78-82, group leader, 82-89; res & develop mgr, 89-90, TECH SERV MGR, GA-PAC CORP, 90- *Mem:* Am Chem Soc; Tech Asn Pulp & Paper Indust. *Res:* Nature of hydrogen bonding in cellulosic webs; relations between hydrogen bond density and mechanical properties of cellulosic webs. *Mailing Add:* 1685 Planters Row Stone Mountain GA 30087

BATTEN, GEORGE WASHINGTON, JR, b Houston, Tex, Sept 4, 37; m 61; c 1. MATHEMATICS. *Educ:* Rice Univ, BA, 59, MA, 61, PhD(math), 63. *Prof Exp:* Res assoc math, Univ Ill, 63-64; asst biomathematician, Univ Tex M D Anderson Hosp & Tumor Inst, 64-65, asst prof biomath, 65-66; asst prof math, 66-81, ASSOC PROF ELEC ENG & MATH, UNIV HOUSTON, 81- *Concurrent Pos:* Vis lectr math, Rice Univ, 66. *Mem:* Soc Indust & Appl Math; Inst Elec & Electronics Engrs. *Res:* Probability theory; partial differential equations; numerical analysis. *Mailing Add:* 1656 Milford St Houston TX 77006

BATTEN, ROGER LYMAN, b Hammond, Ind, June 22, 23. INVERTEBRATE PALEONTOLOGY. *Educ:* Univ Wyo, BA, 48; Columbia Univ, PhD(geol), 55. *Prof Exp:* Geologist, US Geol Surv, 54-55; from asst prof to assoc prof geol, Univ Wis, 55-62; assoc prof, 62-68, PROF GEOL, COLUMBIA UNIV, 68-, CUR, AM MUS NATURAL HIST, 67- *Concurrent Pos:* Assoc cur, Am Mus Natural Hist, 62-67. *Mem:* Paleont Soc; Geol Soc Am. *Res:* Evolution and ecology of the primitive gastropods; Permian stratigraphy; recent marine ecology; population analysis. *Mailing Add:* Am Mus Natural Hist New York NY 10024

BATTENBURG, JOSEPH R, b Chicago, Ill, Jan 16, 34; m 54; c 4. MECHANICAL ENGINEERING. *Educ:* Andrews Univ, BS, 54; Mich Univ, BSE, 56; Univ Southern Calif, MSE, 60; Wis Univ, PhD(mech eng), 67. *Prof Exp:* Instr eng, Walla Walla Col, 56-58; mem tech staff, Hughes Aircraft Co, Inc, 58-60; sr design engr, NAm Aviation, Inc, 60-62; asst prof mech eng, Nev, 62-64; lectr & instr, Wis, 64-67; asst prof, San Jose State Col, 67-68; assoc prof mech eng & chmn eng, Purdue Univ, Calumet Campus, 68-76; exec engr advan technol, Eaton Corp, 76-81; PROF MECH ENG, CALIF STATE UNIV, FRESNO, 81- *Concurrent Pos:* NASA fel, Univ Houston, 67 & Stanford, 77; NSF fac res partic, Ford Motor Co, 75; US State Dept prof, Cairo Univ, Egypt, 75-76; tech consult, US State Dept, 81. *Mem:* Am Soc Eng Educ; Sigma Xi; Soc Automotive Engrs. *Res:* Fatigue testing and determination of fatigue properties of engineering materials; educational research in teaching techniques and metallurgical design. *Mailing Add:* Sch Eng Calif State Univ Fresno CA 93740

BATTER, JOHN F(REDERIC), JR, b Brooklyn, NY, Oct 23, 31; m 55; c 4. PRODUCT MANAGEMENT, VENTURE CAPITAL INVESTMENT. *Educ:* Mass Inst Technol, SB, 53, SM, 54; Reactor Engr, Oak Ridge Sch Reactor Technol, 56. *Prof Exp:* Proj engr, Pratt & Whitney Aircraft Co, 54-56; consult engr, Tech Opers, Inc, Mass, 57-63; mgr nuclear eng dept, Flow Corp, Mass, 63-67; dept mgr prod eng, Polaroid Corp, Cambridge, 67-72; chief engr, Raytheon Med Electronics, 72-74; dir Multinat Prof Off, 74-80, vpres, Reprographic Bus Group, Xerox Corp, 81-84; GROUP PROD MGR, EASTMAN KODAK CO, 84- *Mem:* Sigma Xi; Am Soc Mech Engrs; Asn Image & Info Mgt. *Res:* Commercial product development including xerography, digital based image communication storage and printing systems, and medical instrumentation. *Mailing Add:* Five Old Landmark Dr Rochester NY 14618

BATTERMAN, BORIS WILLIAM, b New York, NY, Aug 25, 30; c 3. X-RAY PHYSICS, EXPERIMENTAL SOLID STATE PHYSICS. *Educ:* Mass Inst Technol, BS, 52, PhD(physics). *Prof Exp:* Elec res electrostatic generators, Nat Bur Standards, 51; lab asst soil solidification, Mass Inst Technol, 51-52, asst physics, 52-53; mem tech staff, Bell Tel Labs, 56-65; assoc prof, 65-68, dir sch appl & eng physics, 74-78, PROF APPL ENG PHYSICS, CORNELL UNIV, 68-, DIR, CORNELL HIGH ENERGY SYNCHROTRON SOURCE LAB, 78-, WALTERS CARPENTER SR PROF ENG, 85- *Concurrent Pos:* Fulbright Scholar, Technish Hochschule, Stuttgart, Germany, 53; Guggenhein & Fulbright-Hays fels, 71; Alexander Von Humbolt, fel, Berlin, Germany, 83. *Mem:* AAAS; fel Am Phys Soc; Am Crystallog Asn. *Res:* X-ray, synchrotron radiation and neutron diffraction applied to solid state physics problems; studies of thermal vibrations in crystals. *Mailing Add:* Clark Hall Cornell Univ Ithaca NY 14853

BATTERMAN, ROBERT COLEMAN, b Brooklyn, NY, Apr 12, 11; m 47; c 3. PHARMACOLOGY. *Educ:* NY Univ, BS, 31, MD, 35. *Prof Exp:* Instr therapeut, NY Univ, 39-50, asst prof med, 49-50; assoc prof physiol, pharmacol & med & chief arthritis sect, New York Med Col, Flower & Fifth Ave Hosps, 50-59; dir & pres, Clin Pharmacol Res Inst, 59-81; DIR, CALIF HEALTH FACIL FINANCE AUTHORITY, 81- *Mem:* Am Soc Pharmacol & Exp Therapeut; Am Chem Soc; Am Soc Clin Invest; Am Col Physicians; Soc Exp Biol & Med. *Res:* Clinical pharmacology; arthritis; cardiovascular disease. *Mailing Add:* 2006 Dwight Way No 208 Berkeley CA 94704

BATTERMAN, STEVEN C(HARLES), b Brooklyn, NY Aug, 15, 37; m 59; c 3. APPLIED MECHANICS, BIOMECHANICS. *Educ:* Cooper Union, BCE, 59; Brown Univ, ScM, 61, PhD(eng), 64. *Hon Degrees:* MA, Univ Pa, 71. *Prof Exp:* Instr eng, Roger Williams Jr Col, 61-62 & Brown Univ, 63-64; asst prof to assoc prof appl mech, 64-74, assoc prof orthopedic surg res, SchMed, 72-75, prof appl mech, 74-79, prof biomech in vet med, Sch Vet Med, 75-84, PROF BIOENG, UNIV PA, 74-, PROF BIOENG IN ORTHO SURG, SCH MED, 75- *Concurrent Pos:* NSF res grant, 64-70, fel & vis assoc prof, Israel Inst Technol, 70-71; consult, Dept Defense, Advan Res Proj Agency, 68-69 & US Navy, Naval Air Develop Ctr, 69-70; Nat Inst Dental Res res grant, 73-77; vis prof, Tel-Aviv Univ, 84. *Honors & Awards:* Robert Ridgway Award, Am Soc Civil Engrs. *Mem:* Am Soc Civil Engrs; Am Soc Mech Engrs; Am Soc Eng Educ; Am Soc Safety Engrs; Biomed Eng Soc; Am Acad Mech; Soc Expt Mech; Sigma Xi; fel Am Acad Forensic Sci; Soc Automotive Engr; Asn Advan Automotive Med. *Res:* Solid and structural mechanics with applications to elasticity, plasticity and viscoelasticity; plates and shells; biomechanics in human and animal systems with applications to orthopedics, dentistry and veterinary medicine; biomechanics of injury. *Mailing Add:* Dept Bioeng Univ Pa Philadelphia PA 19104-6392

BATTERSHELL, ROBERT DEAN, b Dover, Ohio, Jan 23, 31; m 58; c 4. PESTICIDE CHEMISTRY. *Educ:* Bowling Green State Univ, BS, 52; Purdue Univ, PhD, 60. *Prof Exp:* Chemist, Callery Chem Co, 52, 54-55; sr res chemist, Diamond Alkali Co, 59-68; group leader, Diamond Shamrock Corp, 68-78 & chem Acquisitions, 78-83; mgr, licensing, SDS Biotech Corp, 83-85; MGR NEW BUS DEVELOP, FERMENTA PLANT PROTECTION CO, 85-; MGR NEW BUS DEVELOP, ISK BIOTECH CORP, 90- *Honors & Awards:* Outstanding Contrib to chem Award, Am Chem Soc, 83. *Mem:* Am Chem Soc; Chem Soc Japan; Am Phytopath Soc. *Res:* Synthesis of organic pesticides; chemistry of aliphatic and aromatic fluorine compounds; Hansch pi, rho, sigma analysis of chemically induced biological response. *Mailing Add:* Ten Bryn Mawr Dr Painesville OH 44077

BATTERSON, STEVEN L, b Newport News, Va, Nov 21, 50. DYNAMICAL SYSTEMS, NUMERICAL ANALYSIS. *Educ:* Col William & Mary, BA, 71; Northwestern Univ, MA, 72, PhD(math), 76. *Prof Exp:* Asst prof, 76-82, ASSOC PROF MATH, EMORY UNIV, 82- *Concurrent Pos:* Mem, Inst Advan Study, 80-81; vis scholar, Boston Univ, 85; res assoc, Univ Calif, Berkeley, 90. *Mem:* Am Math Soc; Asn Women Math. *Res:* Dynamics of eigenvalve computation. *Mailing Add:* Dept Math Emory Univ Atlanta GA 30322

BATTEY, JAMES F, b Boston, Mass, Nov 23, 52; m 81; c 2. GENETICS, NEUROCHEMISTRY. *Educ:* Stanford Univ, MD & PhD(biophys), 80. *Prof Exp:* Intern pediat, Stanford Univ, 80-81; postdoctoral fel, Dept Genetics, Harvard Univ, 81-83; sr investr, Med Oncol Br, Nat Cancer Inst, 83-88, SECT CHIEF, MOLECULAR NEUROSCI, LAB NEUROCHEM, NAT INST NEUROL DIS & STROKE, NIH, 88- *Concurrent Pos:* Adj prof genetics, George Washington Univ, 90- *Mem:* Am Soc Biol Chemists; Soc Neurosci. *Res:* Molecular genetic analysis of the structure, function and regulation of genes encoding neuropeptides and their receptors. *Mailing Add:* NIH Bldg 36 Rm 4D20 Bethesda MD 20892

BATTH, SURAT SINGH, toxicology, entomology, for more information see previous edition

BATTIFORA, HECTOR A, b Peru, Dec 11, 30; US citizen; m 59; c 3. PATHOLOGY. *Educ:* San Marcos Univ, Lima, BS, 50, MD, 57. *Prof Exp:* Assoc pathologist, Presby St Lukes Hosp, Chicago, 64-69; PROF PATH, MED SCH, NORTHWESTERN UNIV, CHICAGO, 69-; CHMN & DIR SURG PATH, CITY OF HOPE MED CTR, 87- *Mem:* Am Asn Path & Bact; Am Soc Exp Path; Soc Exp Biol & Med; Int Acad Path. *Res:* Experimental oncology; ultrastructure of human tumors; immunohistochemistry in tumor diagnosis. *Mailing Add:* 250 E Superior Chicago IL 60611

BATTIGELLI, MARIO C, b Florence, Italy, Dec 18, 27; US citizen; m 58; c 4. INDUSTRIAL MEDICINE. *Educ:* Univ Florence, MD, 51; Univ Pittsburgh, MPH, 57. *Prof Exp:* Resident instr med, Univ Milan, 52-53, instr, 54-55, assoc, 57-58; res assoc, Univ Pittsburgh, 59-61, asst prof occup med, 62-66; assoc prof, 66-69, assoc prof med, 69-74, PROF MED & ENVIRON SCI & ASSOC PROF FAMILY MED, SCH MED, UNIV NC, CHAPEL HILL, 74- *Mem:* AMA; Am Thoracic Soc; AAAS; Sigma Xi. *Res:* Environmental health; industrial toxicology; chest diseases; pulmonary physiology; inhalation experimental toxicology. *Mailing Add:* 2264 HSS OCC Med Morgantown WV 26506

BATTIN, RICHARD H(ORACE), b Atlantic City, NJ, Mar 3, 25; m 47; c 3. MATHEMATICS. *Educ:* Mass Inst Technol, SB, 45, PhD(appl math), 51. *Prof Exp:* Instr math, Mass Inst Technol, 46-52, res mathematician, 51-56; sr staff mem, Arthur D Little, Inc, 56-58; assoc dir, C S Draper Lab, 58-87, ADJ PROF AERONAUT & ASTRONAUT, MASS INST TECHNOL, 79- *Honors & Awards:* Louis W Hill Space Transp Award, Am Inst Aeronaut & Astronaut, 72, Mech & Control of Flight Award, 78. *Mem:* Nat Acad Eng; fel Am Inst Aeronaut & Astronaut; Int Astronaut Acad; fel Am Astronaut Soc. *Res:* General circulation of the atmosphere; random processes in automatic control; scientific computation; digital data processing; guidance and control of ballistic missiles and satellites; astrodynamics. *Mailing Add:* 15 Paul Revere Rd Lexington MA 02173

BATTIN, WILLIAM JAMES, b Philadelphia, Pa, Sept 2, 20; div. THERMODYNAMICS, AUTOMATIC CONTROL SYSTEMS. *Educ:* Swarthmore Col, BS, 50; Princeton Univ, MS, 55; Univ Ill, PhD(compressible fluid flow), 63. *Prof Exp:* Engr, Int Bus Mach Corp, 50-51 & Leeds & Northrup Co, 51-54; from asst prof to assoc prof eng, Clarkson Col Technol, 55-60; assoc prof, George Washington Univ, 63-64; PROF ENG, US NAVAL ACAD, 64- *Concurrent Pos:* Res assoc, Cambridge Univ, 68. *Mem:* Am Soc Mech Engrs; Soc Hist Technol; Am Soc Eng Educ. *Res:* Thermodynamics; compressible fluid flow; automatic control. *Mailing Add:* Dept Mech Eng 25 Wagner St Annapolis MD 21401

BATTIN, WILLIAM T, b Hackensack, NJ, Aug 21, 27; div; c 3. ZOOLOGY. *Educ:* Swarthmore Col, BA, 50; Univ Minn, PhD(zool), 56. *Prof Exp:* From instr to asst prof biol, Wesleyan Univ, 55-58; assoc prof, Simpson Col, 58-59; from asst prof to prof, 59-74, DISTINGUISHED TEACHING PROF BIOL, STATE UNIV NY BINGHAMTON, 74- *Mem:* AAAS; Am Soc Zool; Am Inst Biol Sci; Nat Asn Biol Teacher; Nat Sci Teachers Asn. *Res:* Physiology of nucleus and nuclear membrane; cytoplasmic DNA of amphibian oocytes; humanistic biology. *Mailing Add:* Dept Biol Sci State Univ NY Binghamton NY 13901

BATTINO, RUBIN, b New York, NY, June 22, 31; m 60. PHYSICAL CHEMISTRY, THERMODYNAMICS. *Educ:* City Col, NY, BA, 53; Duke Univ, MA, 54, PhD(chem), 57; Wright State Univ, MS, 78. *Prof Exp:* Res chemist, Leeds & Northrup Co, 56-57; from instr to asst prof chem, Ill Inst Technol, 57-66; assoc prof, 66-69, PROF CHEM, WRIGHT STATE UNIV, 69- *Concurrent Pos:* Sect ed, Chem Abstr, 63-73; vis prof, Hebrew Univ, 76 & 79, Ben Gurion Univ, 79, Univ New England, 79 & Univ Canterbury, 80 & 87-88, Rhodes Univ, 86, Okayama Univ Sci, 82 & 90; Sr Fulbright Scholar, Australia, 79. *Mem:* AAAS; Am Chem Soc; Int Union Pure & Appl Chem; Sigma Xi. *Res:* Thermodynamics of solutions of nonelectrolytes; gas solubilities; solubility of gases in liquids, particularly the high- precision solubility of gases in water; chemical education. *Mailing Add:* Dept Chem Wright State Univ Dayton OH 45435

BATTISTA, ARTHUR FRANCIS, b Can, Sept 7, 20; m 64. NEUROSURGERY, NEUROPHYSIOLOGY. *Educ:* McGill Univ, BSc, 43, MD, CM, 44; Univ Western Ont, MSc, 47; Am Bd Neurol Surg, dipl, 58; Hunter Col, MA, 69. *Prof Exp:* Intern, Royal Victoria Hosp, Montreal, 44-45; resident, NY Neurol Inst, 50-55; from instr to clin assoc prof, 56-64, assoc prof, 64-71, PROF NEUROSURG, MED SCH, NY UNIV, 71- *Concurrent Pos:* Am Physiol Soc Porter fel, Harvard Med Sch, 47-48, Milton fel, 48-49; from asst attend to assoc attend, NY Univ Hosp, 56-67, attend, 67-; asst vis surgeon, Bellevue Hosp, 57-; consult, Pub Health Marine Hosp, Staten Island, NY, 57- & Beekman-Downtown Hosp, 58-; attend, Vet Admin Hosp, New York, 58- *Mem:* Cong Neurol Surg; Am Asn Neurol Surg; Am Philos Asn; Asn Res Nervous & Ment Dis; NY Acad Sci. *Res:* Neurophysiology of dyskinesias, hypothermia and pain; clinical neurosurgical practice. *Mailing Add:* NY Univ Med Sch New York NY 10016

BATTISTA, ORLANDO ALOYSIUS, b Cornwall, Ont, June 20, 17; nat US; m 45; c 2. CHEMISTRY. *Educ:* McGill Univ, BSc, 40. *Hon Degrees:* ScD, St Vincent Col, 55 & Clarkson Univ, 85. *Prof Exp:* Res chemist, Am Viscose Corp, 40-51, sr res chemist, 52-53, leader pulping anal lab, 54, head anal group, 55-58, leader spec prod group, 58-59, spec prod sect, 59-60 & corp appl res sect, 60-61, mgr corp appl res dept, 61-63, asst dir res, Am Viscose Div, FMC Corp, 63, mgr interdisciplinary res, Chem Res & Develop Ctr, 63-65, asst dir cent res dept, 65-70; vpres sci & technol, Avicon, Inc, 70-74; CHMN & PRES, RES SERV CORP, 74-; PRES, O A BATTISTA RES INST, 81- *Concurrent Pos:* Spec consult, Avicon, Inc, 74-; adj prof chem, Univ Tex, Arlington, 75-77; chmn & pres, Knowledge Inc, 86, founder & pres, Knowledge Olympiads, 86. *Honors & Awards:* Chem Pioneer Award, Am Inst Chemists, 69; Capt of Achievement, Am Acad Achievement, 71; James T Grady Award, Am Chem Soc, 73, Anselme Payen Award, 85; Creative Invention Award, Am Chem Soc, 83; Appl Polymer Sci Award, Am Chem Soc, 87. *Mem:* Am Inst Chemists (pres-elect, 75-76, pres, 77-79); Am Chem Soc; Nat Asn Sci Writers; fel NY Acad Sci. *Res:* Molecular weight of celluose; fine structure of cellulose; oxidation, acid hydrolysis and fractionation of cellulose; heat extruding viscose; washable crepe fabrics; waterproofing cellulose articles; molding hydroplastics; characterization and evaluation of wood pulps; colloidal macromolecular phenomena; microcrystal polymer science. *Mailing Add:* 3863 SW Loop 820 Suite 100 Ft Worth TX 76133-2063

BATTISTA, SAM P, b Orange, NJ, June 22, 24; m 50; c 3. PHARMACOLOGY, PHYSIOLOGY. *Educ:* St John's Univ, NY, BS, 50; Univ Wis, MS, 57; Boston Univ, PhD(pharmacol), 68. *Prof Exp:* Res chemist, Merck & Co, Inc, 51-53; consult pharmacologist, Arthur D Little, Inc, 58-; RETIRED. *Mem:* Am Soc Pharmacol & Exp Therapeut. *Res:* Inhalation and effects of pharmacological agents on ciliary function. *Mailing Add:* 900 River Reach Dr Ft Lauderdale FL 33315

BATTISTE, DAVID RAY, b Mobile, Ala, Oct 15, 46; m 69; c 2. ORGANIC CHEMISTRY, POLYMER CHEMISTRY. *Educ:* Univ Fla, BS, 69; La State Univ, PhD(org chem), 74. *Prof Exp:* Sr develop chemist agr chemicals, Ciba-Geigy Corp, 74-75; res assoc alkaloid synthesis, Rice Univ, 75-77; res chemist new prod from terpenes, Glidden-Durkee, 77-78; res chemist molecular struct, Phillips Petrol Co, 78-80, corp res, 80-83, sr res chemist, transparent resins & stabilization, 83-91, TECH MGR, POLYMER STABILIZATION, PHILLIPS PETROL CO, 91- *Mem:* Am Chem Soc; Sigma Xi; Am Soc Mass Spectrometry; Soc Plastic Engrs. *Res:* Stereochemistry; heterocyclic chemistry; olefin reactions; Fourier transform infrared spectroscopy; reaction mechanisms; pillared clays; mass spectrometry; polymer additives. *Mailing Add:* 244 CPL-PRC Bartlesville OK 74004

BATTISTE, MERLE ANDREW, b Mobile, Ala, July 22, 33; m 60; c 2. ORGANIC CHEMISTRY. *Educ:* The Citadel, BS, 54; La State Univ, MS, 56; Columbia Univ, PhD(org chem), 59. *Prof Exp:* Res assoc chem, Univ Calif, Los Angeles, 59-60; from asst prof to assoc prof, 61-70, chmn org div, chem dept, 74-84, PROF CHEM, UNIV FLA, 70- *Concurrent Pos:* Alfred P Sloan res fel, 67-69; Fulbright-Hays sr res scholar, Ger, 73-74; Erskine fel, Univ Canterbury, NZ, 87. *Mem:* Am Chem Soc. *Res:* Synthetic and natural products chemistry; small-ring compounds; reaction mechanisms; stereochemistry and rearrangements of bridged polycyclic systems; organometallic chemistry; photochemical transformations. *Mailing Add:* Dept Chem Univ Fla Gainesville FL 32611

BATTISTI, ANGELO JAMES, b Little Falls, NY, July 26, 45; m 67; c 3. ORGANIC PHOTOCHEMISTRY, METAL FINISHING. *Educ:* State Univ NY at Albany, BS, 67, at Buffalo, PhD(chem), 71. *Prof Exp:* Vis res assoc, Ohio State Univ, 71-72; res chemist, GAF Corp, Binghamton, NY, 72-74, group leader, 74-76, mgr commercial develop, New York, NY, 76-79; tech mgr, Sel-Rex, Occidental Petrol Corp, 79-83; mgr coating technol, St Regis Corp, 83-84; tech dir, Olin Hunt Corp, 84-86; TECH DIR, J C DOLPH CO, 86- *Mem:* Am Chem Soc; Nat Elec Mfrs Asn; Am Soc Testing Mat; Int Electrotech Comn. *Res:* Organic photochemistry; photographic science; chemistry of semiconductor and microelectronic materials including photoresists, masking materials, metal finishing chemicals and insulating (dielectric) polymers; electrical and electronic dielectric compounds. *Mailing Add:* 20 Fairway Dr Cranbury NJ 08512

BATTISTO, JACK RICHARD, b Niagara Falls, NY, Sept 13, 22; m 50; c 2. IMMUNOLOGY, MICROBIOLOGY. *Educ:* Cornell Univ, BS, 49; Univ Mich, MS, 50, PhD(bact), 53. *Prof Exp:* Vis investr immunol, Rockefeller Inst Med Res, 53-55; asst prof microbiol, Sch Med, Univ Ark, 55-57; Pop Coun sr fel, Weizmann Inst Sci, Israel, 63-64; from asst prof to prof microbiol & immunol, Albert Einstein Col Med, 57-74; sci dir, Dept Immunol, 74-81, STAFF MEM, RES INST, CLEVELAND CLIN FOUND, 81- *Concurrent Pos:* NIH fel, 53-55; F G Novy res fel; pop coun res fel; adj prof, Case Western Reserve Univ, 76- *Mem:* AAAS; Am Asn Immunologists; Harvey Soc; NY Acad Sci; Transplantation Soc; Reticuloendothelial Soc. *Res:* Hypersensitivities to simple chemical compounds; immunological unresponsiveness to haptenes; naturally occurring delayed type iso-hypersensitivity; autologous mixed lymphocyte reaction; splenic influence on lymphoid tissue; regulation of immunological responses; cytolytic T cell reactions. *Mailing Add:* Sect Immunol Res Inst Cleveland Clin Found Cleveland OH 44195

BATTLE, HELEN IRENE, b London, Ont, Aug 31, 03. DEVELOPMENTAL BIOLOGY, HUMAN GENETICS. *Educ:* Univ Western Ont, BA, 23, MA, 24; Univ Toronto, PhD(marine biol), 28. *Hon Degrees:* LLD, Univ Western Ont & DSc, Carleton Univ, 71. *Prof Exp:* Demonstr, 23-24, from instr to prof, 24-72, EMER PROF ZOOL, UNIV WESTERN ONT, 72- *Honors & Awards:* Centennial Medal, Can, 67; F E J Fry Medal, Can Soc Zoologists, 77; J C B Grant Award of Merit, Can Asn Anatomists, 77. *Mem:* Hon mem Can Soc Zoologists; hon mem Nat Asn Biol Teachers; Can Asn Anatomists; Can Physiol Soc; Genetics Soc Can. *Res:* MacKinder's hereditary brachydactyly. *Mailing Add:* Dept Zool & Biol Univ Western Ont London ON N6A 3K7 Can

BATTLES, JAMES E(VERETT), b Alabama City, Ala, Apr 5, 30; m 51; c 4. METALLURGICAL ENGINEERING, CORROSION. *Educ:* Univ Ala, BS, 60; Ohio State Univ, MS, 61, PhD(metall eng), 64. *Prof Exp:* Helper, open hearth, Sheffield Steel Div, Armco Steel Corp, Tex, 48-49, receiving clerk, 50-58, res asst, 60; asst metallurgist, 64-70, metall engr, 70-85, leader, mat group, Chem Eng Div, 72-84, mgr, Electro Magnetic Casting Proj, 85-88, SR METALL ENGR, ARGONNE NAT LAB, 85-, MGR, IFR PYROMETALL PROCESS PROG, 88-, ASSOC DIV DIR, NUCLEAR PROGS, CHEM TECHNOL DIV, 90- *Mem:* Am Soc Metals; Am Inst Mining, Metall & Petrol Engrs; Electro Chem Soc; Metall Soc. *Res:* Reactions of tungsten and tungsten oxides with water vapor; mass spectrometric and thermodynamic studies of refractory metal oxides, nuclear metal oxides and nuclear metal carbides; corrosion nuclear fuel element materials; materials development; corrosion studies and failure analysis for high performance lithium/metal sulfide and sodium/sulfur secondary batteries for electric vehicle propulsion; electromagnetic casting process development for steel sheet; development of pyrometallurgical processing of metal and oxide nuclear fuels; nuclear waste treatment. *Mailing Add:* Chem Technol Div Argonne Nat Lab 9700 S Cass Ave Argonne IL 60439

BATTLES, WILLIS RALPH, b Erie, Pa, Nov 12, 14; m 41; c 3. PETROLEUM CHEMISTRY. *Educ:* Univ Calif, Los Angeles, AB, 36. *Prof Exp:* Inspector, Shell Oil Co, Calif, 36-42; chem operator, Trojan Powder Co, Ohio, 42-43 & Dow Chem Co, Calif, 43-44; res chemist, Union Oil Co, 44-46; bus mgr & partner, Calif Car Bed Co, 46-51; chemist, Fletcher Oil Co, 51-57, chief chemist, 57-59; chemist, Charles Martin & Co, San Pedro, 59-60; chemist, Atlantic Richfield Co, Wilmington, 60; sr anal chemist, Arco-Watson Refinery, 60-82; MGR, W B PROD CO, REDONDO BEACH, CALIF, 82- *Mem:* AAAS; Am Chem Soc; Am Inst Chemists. *Res:* Analytical methods research, especially for trace contaminants in petroleum products. *Mailing Add:* 560 S Helberta Ave Redondo Beach CA 90277-2598

BATTLEY, EDWIN HALL, b Detroit, Mich, Jan 24, 25. MICROBIOLOGY, BIOCHEMISTRY. *Educ:* Harvard Col, BA, 49; Fla State Univ, MS, 51; Stanford Univ, PhD(biol), 56. *Prof Exp:* Asst, Lab Microbiol, Technol Univ Delft, 55-56 & Sch Med, Wash Univ, 56-57; instr biol chem, Seton Hall Col Med & Dent, 57-60; asst prof biol, Dartmouth Col, 60-62; ASSOC PROF ECOL & EVOLUTION, STATE UNIV NY, STONY BROOK, 62- *Res:* Biochemistry and physiology of microorganisms; thermodynamics of biological processes; general biology. *Mailing Add:* Dept Ecol State Univ NY Stony Brook Main Stony Brook NY 11794

BATTOCLETTI, JOSEPH H(ENRY), b Bridgeport, Ohio, Mar 12, 25; m 51; c 7. ELECTRICAL ENGINEERING. *Educ:* Univ Detroit, BEE, 47; Northwestern Univ, MSEE, 49; Univ Calif, Los Angeles, PhD, 61. *Prof Exp:* Mem staff electronic eng develop, Motorola, Inc, Ill, 49-51; from instr to assoc prof elec eng, Loyola Univ, Los Angeles, 51-62, head dept, 62; Nat Acad Sci-Agency Int Develop electronics consult, Chilean Univs, 62-63; assoc prof elec eng, Marquette Univ, 63-66; res scientist, Badger Meter Mfg Co, 66-68, mgr, Dept Appl Res, 68-70; PROF BIOMED ENG, DEPT NEUROSURG, MED COL WIS, 70-; RESEARCHER BIOMED ENG, RES SERV, VET ADMIN CTR, WOOD, WIS, 78- *Honors & Awards:* Centennial Medal, Inst Elec & Electronics Engrs & Mem Award, Milwaukee Sect. *Mem:* Inst Elec & Electronics Engrs; Biomed Eng Soc; Soc Magnetic Res Med; Magnetics Soc. *Res:* Microwave interaction with plasmas; electronics; nuclear magnetic resonance flowmeters and imaging; biological effects of electric, magnetic and electromagnetic fields; permanent magnets; finite element analysis; nuclear magnetic resonance measurement of bone mineral content. *Mailing Add:* Dept Neurosurg 8700 W Wisconsin Ave Box 105 MCMC Milwaukee WI 53226

BATTS, BILLY STUART, b Raleigh, NC, July 14, 34; m 56; c 2. FISH BIOLOGY, ECOLOGY. *Educ:* NC State Univ, BS, 56, PhD, 70; Univ Wash, MS, 60. *Prof Exp:* Fisheries biologist, Univ Wash, 60-62; asst prof, 63-68, ASSOC PROF BIOL, LONGWOOD COL, 68- *Mem:* Am Fisheries Soc; Am Soc Ichthyologists & Herpetologists; fel Am Inst Fishery Res Biologists; Am Inst Biol Sci. *Res:* Lepidology of flounders; life history of skipjack tuna, especially age and growth, food habits, sexual maturity and reproduction. *Mailing Add:* 1004 Seventh Ave Farmville VA 23901

BATTS, HENRY LEWIS, JR, b Macon, Ga, May 24, 22; m 45; c 3. ECOLOGY. *Educ:* Kalamazoo Col, AB, 43; Univ Mich, MS, 47, PhD(zool), 55. *Hon Degrees:* ScD, Western Mich Univ, 76. *Prof Exp:* Asst biol & instr ornith, Kalamazoo Col, 40-43; asst zool, Univ Mich, 46-48, teaching fel, 48-50; from instr to assoc prof, 50-59, prof biol, 59-78, EMER PROF, KALAMAZOO COL, 78-; EXEC DIR, KALAMAZOO NATURE CTR, 61- *Concurrent Pos:* Instr, Exten Serv, Univ Mich, 53; ed, Bull Wilson Ornith Soc, 59-63; founding trustee, Environ Defense Fund, Inc, 67- *Mem:* fel AAAS; Wilson Ornith Soc (2nd vpres, 64-66, 1st vpres, 66-68, pres, 68-69); Am Ornithologists Union; Ecol Soc Am; Nat Audubon Soc. *Res:* Ecology of birds; nest activities of the American Goldfinch; bird nest identification; population and distribution of birds; pesticides; environmental degradation. *Mailing Add:* 2315 Angling Rd Kalamazoo MI 49008

BATTY, JOSEPH CLAIR, b Wallsburg, Utah, June 27, 39; m 59; c 3. THERMODYNAMICS, PHYSICAL CHEMISTRY. *Educ:* Utah State Univ, BS, 61, MS, 62; Mass Inst Technol, ScD(mech eng), 69. *Prof Exp:* Proj engr, Nielson & Maxwell Consult Engrs, 62-63; from asst prof to assoc prof, 63-77, PROF MECH ENG, UTAH STATE UNIV, 78- *Concurrent Pos:* NSF faculty fel, 67-69. *Mem:* Am Soc Mech Eng; Am Soc Engr Educ. *Res:* Thermodynamics of gas-solid reactions; ultrafiltration and reverse osmosis applications in food processing; utilization of animal wastes; energy use in irrigation, food systems, and at manmade water recreation areas; use of saline water in energy development, salt gradient solar panels. *Mailing Add:* Dept Mech Eng Utah State Univ Logan UT 84322-4130

BATY, RICHARD SAMUEL, b Lawrence, Kans, July 13, 37; m 63. OPTIMAL CONTROL, OPTIMAL FILTERING. *Educ:* Univ NMex, BS, 60; Air Force Inst Technol, MS, 66; Univ Calif Los Angeles, PhD(eng), 70. *Prof Exp:* Officer res & anal, USAF, 60-83; CHIEF SCIENTIST WEAPON SYSTS MODELING & SIMULATION, BDM INT, INC, 83- *Concurrent Pos:* Proj engr aerospace ground equip, USAF Manned Orbiting Lab, 67-68 & elec propulsion, Air Force Rocket Propulsion Lab, 71-73; study dir missile intercept effectiveness, Air Force Studies & Anal, 74-80; working group chmn, Mil Opers Res Soc, 78; prog mgr smart weapons, Defense Advan Res Projs Agency, 80-83; Nat Comt mem, Asst Secy Defense-C3I, 86-87. *Mem:* Sigma Xi. *Res:* Methods of implementing advanced missile guidance techniques; improving tracking of new generation penetrator systems. *Mailing Add:* 6620 Vista Del Monte NE Albuquerque NM 87109

BATZAR, KENNETH, b Brooklyn, NY, May 30, 38; m 62; c 2. FLUOROPOLYMER TECHNOLOGY. *Educ:* Brooklyn Col, BS, 59, MA, 62; City Univ New York, PhD(inorg chem), 66. *Prof Exp:* Sr res chemist, E I Du Pont De Nemours & Co, Newark, 66-77, res assoc, 77-86, MEM STAFF, DU PONT MARSHALL LAB, E I DU PONT DE NEMOURS CO, INC, 80-, SR RES ASSOC, 86- *Mem:* Am Chem Soc; Sigma Xi. *Res:* Pigment technology; fluoropolymer technology; high performance coatings; thermally stable polymers. *Mailing Add:* Du Pont Marshall Lab PO Box 3886 Philadelphia PA 19146

BATZEL, ROGER ELWOOD, b Idaho, Dec 1, 21; m 46; c 3. NUCLEAR CHEMISTRY. *Educ:* Univ Idaho, BS, 47; Univ Calif, PhD, 51. *Prof Exp:* Sr chemist, Calif Res & Develop Co, 51-53; assoc dir, 53-71, DIR, LAWRENCE LIVERMORE LAB, 71- *Mem:* Am Phys Soc. *Res:* High energy nuclear reactions; research and development. *Mailing Add:* 315 Bonanza Way Danville CA 94526

BATZER, HAROLD OTTO, b Gillett, Wis, Jan 22, 28; m 51; c 3. ENTOMOLOGY. *Educ:* Univ Minn, BS, 52, MS, 55, PhD, 65. *Prof Exp:* Entomologist, NCent Forest Exp Sta, US Forest Serv, 54-64, insect ecologist, 64-70, prin insect ecologist, 70-83; CONSULT, 88- *Concurrent Pos:* Leader subj group S2.06-8, Int Union Forest Res Orgns, 76-86; adj prof, Univ Minn, 76-; pest mgt consult, 83- *Mem:* Entom Soc Am; Soc Am Foresters; Int Soc Arboricult. *Res:* Forest insects, particularly the role of insects in natural forest ecosystems; completing emphasis on insects of aspen forests. *Mailing Add:* 791 Redwood Lane New Brighton MN 55112

BATZING, BARRY LEWIS, b Rochester, NY, May 6, 45; m 68; c 2. MICROBIOLOGY. *Educ:* Cornell Univ, BS, 67; Pa State Univ, MS, 69, PhD(microbiol), 71. *Prof Exp:* from asst prof to assoc prof, 76-84, PROF MICROBIOL, STATE UNIV NY COL CORTLAND, 84- *Concurrent Pos:* Investr, Immunol Carcinogenesis Sect, Biol Div, Oak Ridge Nat Lab-Grad Sch Biomed Sci, Univ Tenn, 71-73; guest scientist, Environ Sci Div, Oak Ridge Nat Lab, 81-82. *Mem:* Am Soc Microbiol; Sigma Xi; AAAS. *Res:* Microbial physiology, ultrastructure, and ecology. *Mailing Add:* Dept Biol Sci State Univ NY Col Cortland NY 13045

BATZLI, GEORGE OLIVER, b Minneapolis, Minn, Sept 23, 36; m 59; c 2. POPULATION & COMMUNITY ECOLOGY, NUTRITIONAL & ARCTIC ECOLOGY. *Educ:* Univ Minn, Minneapolis, BA, 59; San Francisco State Univ, MA, 65; Univ Calif, Berkeley, PhD(zool), 69. *Prof Exp:* Res assoc animal sci & nutrit, Univ Calif, Davis, 69-71; from asst prof to assoc prof, 71-80, head dept ecol, 83-88, PROF, UNIV ILL, URBANA-CHAMPAIGN, 81- *Concurrent Pos:* NIH fel, 69-71; sr scientist, NSF, 76-78; res scientist, ecol div, Dept Sci Indust Res, NZ, 79; res fel, Univ Oslo, 82; ecol panel, NSF, 84-87; vis scholar, Univ Calif, Berkeley, 89. *Mem:* Fel AAAS; Am Inst Biol Sci; Ecol Soc Am; Am Soc Mammal; Brit Ecol Soc; Am Soc Naturalists; Sigma Xi. *Res:* Dynamics of natural populations; community organization; interactions between herbivores and vegetation; arctic ecology; nutritional ecology. *Mailing Add:* Dept Ecol Ethology & Evolution Univ Ill 606 E Healy St Champaign IL 61820

BAU, HAIM HEINRICH, b Krakow, Poland, Jan 2, 47; US citizen; m 82; c 2. FLUID MECHANICS, HEAT TRANSFER. *Educ:* Technion, Israel, BS, 69, MS, 73; Cornell Univ, PhD(mech eng), 80. *Prof Exp:* Mem res & develop staff, Israeli Defense Force, 69-77; asst prof, 80-85, ASSOC PROF, UNIV PA, 85- *Concurrent Pos:* NSF presidential young investr, 84; consult, Panametrics Inc, 86-, Hewlett-Packard, Avondale Div, 91-; vis prof, Royal Inst Technol, Sweden, 89. *Honors & Awards:* R Teetor Award, Soc Automotive Engrs, 85. *Mem:* Am Soc Mech Engrs; Soc Indust & Appl Math; Am Phys Soc. *Res:* Active control of convective flows; buoyancy induced flows-transitions, bifurcations and chaos; transport in micro-structures; flow and boiling in porous media; interaction between stress waves transmitted in solid waveguides and adjecent media. *Mailing Add:* Univ Pa 111 Town Bldg-D3 Philadelphia PA 19104-6315

BAU, ROBERT, b Shanghai, China, Feb 10, 44; m 70; c 3. INORGANIC CHEMISTRY. *Educ:* Univ Hong Kong, BSc, 63; Univ Calif, PhD(chem), 68. *Prof Exp:* Res fel chem, Harvard Univ, 68-69; asst prof, 69-74, assoc prof, 74-77, PROF CHEM, UNIV SOUTHERN CALIF, 77- *Concurrent Pos:* Sloan Found res fel, 74-76; NIH res career develop award, 75-80. *Honors & Awards:* Alexander von Humboldt Sr US Scientist Prize, 85. *Mem:* Am Chem Soc; The Chem Soc; Am Crystallog Asn. *Res:* Structural investigations of transition metal hydride compounds and metal nucleotide complexes; neutron diffraction analysis of chiral organic compounds. *Mailing Add:* Dept Chem Univ Southern Calif Los Angeles CA 90089

BAUBLIS, JOSEPH V, pediatrics, virology, for more information see previous edition

BAUCHWITZ, PETER S, b Ger, 1920; US citizen; m 59. ORGANIC CHEMISTRY, PHYSICAL CHEMISTRY. *Educ:* Univ Chicago, PhD, 56. *Prof Exp:* Develop engr, Govt Labs, Akron, 47-51; res chemist, E I du Pont de Nemours Co, Inc, 56-85; RETIRED. *Mem:* Am Chem Soc. *Res:* High polymers; analytical chemistry. *Mailing Add:* 1420 Athens Rd Wilmington DE 19803

BAUDE, FREDERIC JOHN, b Milwaukee, Wis, Oct 13, 38; m 63; c 3. ANALYTICAL CHEMISTRY, ORGANIC CHEMISTRY. *Educ:* Univ Wis, BS, 60; Univ Minn, MS, 63, PhD(org chem), 65. *Prof Exp:* Sr res chemist anal chem, E I du Pont de Nemours & Co, 64-79; chemist anal chem, Iowa Testing Labs Inc, 79-80; SR ANALYTICAL CHEMIST, H B FULLER CO, 80- *Mem:* Am Chem Soc. *Res:* General agrichemical analysis, especially pesticide residue analysis, environmental chemistry of pesticides, and animal and plant metabolism of pesticides. *Mailing Add:* H B Fuller Co 1200 Wolters Blvd Vadnais Heights MN 55110

BAUDER, JAMES WARREN, b Gloversville, NY, Apr 22, 47; m 73; c 3. SOIL PHYSICS, AGRICULTURAL ENGINEERING. *Educ:* Univ Mass, BS, 69, MS, 71; Utah State Univ, PhD(soil physics), 74. *Prof Exp:* Asst prof soil sci, NDak State Univ, 74-78 & Univ Minn, St Paul, 78-80; ASSOC PROF SOIL SCI, MONT STATE UNIV, 80- *Concurrent Pos:* Consult, Int Joint Comn, 75-77 & NDak Irrig Task Force, 76-78. *Mem:* Soil Sci Soc Am; Am Soc Agron. *Res:* Soil physical properties as they relate to tillage, soil conservation erosion; irrigation return flow, nitrogen leaching and transformation. *Mailing Add:* Exten Serv Leon Johnson Hall Mont State Univ Bozeman MT 59717-0312

BAUER, ARMAND, b Zeeland, NDak, Nov 29, 24; m 49; c 4. SOIL SCIENCE, CHEMISTRY. *Educ:* NDak State Univ, BS, 50, MS, 55; Colo State Univ, PhD(soil sci), 63. *Prof Exp:* Soil scientist, US Dept Agr Soil Conserv Serv, 50-54, NDak State Univ, 55-76; SOIL SCIENTIST, AGR RES SERV, USDA, 76- *Mem:* Soil Sci Soc Am; fel Soil Conserv Soc Am; Am Soc Agron. *Res:* Soil management and conservation; soil water management for dryland and irrigated agriculture; soil fertility and plant nutrition. *Mailing Add:* 1814 N 20th Bismarck ND 58501

BAUER, BEVERLY ANN, b Cincinnati, Ohio, Aug 16, 49; m 71. ENZYME ASSISTED SYNTHESIS. *Educ:* Univ Cincinnati, BS, 71; Univ Georgia, MS, 73; La State Univ, PhD(biochem), 81. *Prof Exp:* Teaching asst food sci, Univ Georgia, 71-73; food researcher, Procter & Gamble Co, 73-76; res asst biochem, La State Univ, 77-80, res fel, 80-81; res chemist, 81-85, SR RES CHEMIST, ETHYL CORP, 85- *Mem:* Am Chem Soc; Sigma Xi; Inst Food Technologists. *Res:* Use of enzymes in organic synthesis; biochemical basis of diabetes; characterization of changes in enzyme activity in diabetes; isolation and characterization of enzymes; computers in chemistry. *Mailing Add:* Ethyl Corp R-D PO Box 14799 Baton Rouge LA 70898

BAUER, C(HARLES) L(LOYD), b Elizabeth, NJ, July 4, 33; m 59; c 2. THIN FILMS, SURFACES & INTERFACES. *Educ:* Rensselaer Polytech Inst, B Met Eng, 55; Yale Univ, MEng, 59, DEng(metall), 61. *Hon Degrees:* DSc, EPFL, Lausanne, Switz, 84. *Prof Exp:* Dir, Ctr Joining Mat, 73-79, PROF METALL & MAT SCI, CARNEGIE-MELLON UNIV, 61- *Concurrent Pos:* NSF res fel, FRG & Japan, 67-68; vis prof, EPFL, Lausanne, Switz, 70-71 & 84; surv visits USSR, Nat Acad Sci, 73 & 78, vis scientist EGer & France, 83-84. *Mem:* Am Inst Mining, Metall & Petrol Engrs. *Res:* Surfaces and interfaces; electron microscopy; thin films. *Mailing Add:* Dept Metall & Mat Sci Carnegie Mellon Univ Pittsburgh PA 15213-3890

BAUER, CARL AUGUST, b Marion Co, Kans, Nov 10, 16; m 41; c 3. ASTRONOMY. *Educ:* Univ Minn, BA, 42; Univ Chicago, MS, 44; Harvard Univ, PhD(astron), 49. *Prof Exp:* Instr, US Army Air Force, Univ Minn, 42; res asst, Yerkes Observ, Univ Chicago, 42-44; instr physics & math, reserve officers training corps, NDak Agr Col, 44; instr astron, Ind Univ, 45; instr & res assoc, observ, Univ Mich, 47-50 & Harvard Univ, 51; asst prof astron, Pa State Univ, 51-56, assoc prof, 56-80; RETIRED. *Mem:* AAAS; Am Astron Soc; Am Meteorol Soc; Int Astron Union; Sigma Xi. *Res:* Stellar and solar spectroscopy; meteoritics; production of helium by cosmic radiation in meteorites and other developments of the parent planet hypothesis; origin of comets. *Mailing Add:* 444 E McCormick Ave State College PA 16801

BAUER, CHARLES EDWARD, b Astoria, Ore; m; c 2. HIGH DENSITY HIGH PERFORMANCE ELECTRONICS PACKAGING, MANUFACTURING MATERIALS & PROCESSES. *Educ:* Stanford Univ, BS, 72; Ohio State Univ, MS, 75; Ore Grad Ctr, PhD(mat sci & eng), 80; Univ Portland, MBA, 88. *Prof Exp:* Res assoc, Stanford Ctr Mat Res, 74; mat sci engr III, Tektronix Inc, 78-80, eng mgr I, 80-82, eng mgr II, 82-85, mgr, Packaging Unit, 85-86, mgr, Packaging Opers, 86-88; dir, Res & Technol, MicroLithics Corp, 88-90; PRES, TECHLEAD CORP, 90- *Concurrent Pos:* Adj prof mech eng, Univ Portland, 81-82, adj asst prof bus admin, 85-89; vis prof mech eng, Fla Int Univ, 91. *Mem:* Int Soc Hybrid Microelectronics; Am Soc Metals Int; Int Electronics Packaging Soc (vpres, 89-90). *Res:* Electronics packaging and interconnection; manufacturing materials and processes; real time manufacturing; process and quality control; design for manufacturability; high speed, high performance electronics; high density surface mount manufacturing and assembly. *Mailing Add:* 31321 Island Dr Evergreen CO 80439

BAUER, CHRISTIAN SCHMID, JR, b Biloxi, Miss, Sept 19, 44. COMPUTER ENGINEERING, REAL-TIME SYSTEM SIMULATION. *Educ:* Univ Fla, BSIE, 66, MSE, 67, PhD(civil eng), 75. *Prof Exp:* From asst prof to assoc prof indust eng, 70-91, PROF & CHMN, DEPT COMPUTER ENG, UNIV CENT FLA, 91- *Mem:* Nat Soc Prof Engrs; Inst Elec & Electronics Engrs; Asn Comput Mach; AAAS; Inst Indust Engr. *Res:* Real-time systems simulation and modelling research; glass-cockpit displays for light aircraft. *Mailing Add:* Dept Computer Eng Univ Cent Fla Orlando FL 32816

BAUER, DANIEL A, b Vancouver, Wash, Nov 1, 52; m 81; c 2. PHYSICS, DETECTOR INSTRUMENTATION. *Educ:* Lewis & Clark Col, BA, 74; Mich State Univ, PhD(physics), 81. *Prof Exp:* Res assoc, Fermi Nat Accelerator Lab, 79-81; res asst prof, Univ Pa, 81-84; ASSOC RES PHYSICIST, UNIV CALIF, SANTA BARBARA, 84- *Mem:* Sigma Xi; Am Phys Soc. *Res:* colliding beam physics; two-photon interactions; particle detector instrumentation. *Mailing Add:* SLAC Bin 43 PO Box 4349 Stanford CA 94309

BAUER, DAVID FRANCIS, b Lehighton, Pa, Apr 13, 40; m 65; c 2. MATHEMATICAL STATISTICS. *Educ:* East Stroudsburg State Col, BS, 63; Ohio Univ, MS, 65; Univ Conn, PhD(math statist), 70. *Prof Exp:* Instr math, Denison Univ, 65-66; assoc mem tech staff, Traffic Studies Ctr, Bell Tel Labs, NJ, 66-67; asst prof statist & comput sci, Univ Del, 70-74; asst prof, 74-78, ASSOC PROF MATH SCI, VA COMMONWEALTH UNIV, 78-, ASST CHMN, 90- *Mem:* Sigma Xi; Math Asn Am; Am Statist Asn; Inst Math Statists. *Res:* Nonparametric methods and applications of statistics. *Mailing Add:* Dept Math Sci Va Commonwealth Univ Richmond VA 23284-2014

BAUER, DAVID ROBERT, b Portland, Ore, Aug 20, 49; m 71; c 2. POLYMER DEGRADATION & STABILITY. *Educ:* Calif Inst Tech, BS, 71; Stanford Univ, PhD(chem physics), 75. *Prof Exp:* Post doctorate, Univ Ill, 75-77; STAFF SCIENTIST, FORD MOTOR CO, 77- *Concurrent Pos:* ed, Polymer Degradation & Stabilization. *Mem:* Am Chem Soc; Sigma Xi; Fedn Soc Coatings Technol. *Res:* Chemical and physical characterization of coatings, including crosslinking, corrosion performance, rheology, adhesion, and weathering with emphasis on photodegradation and stabilization. *Mailing Add:* Ford Motor Co Res Staff E3198 PO Box 2053 Dearborn MI 48121

BAUER, DENNIS PAUL, b Pittsburgh, Pa, July 29, 48; m 71. ORGANIC CHEMISTRY, INORGANIC CHEMISTRY. *Educ:* Univ Cincinnati, BS, 71, PhD(org chem), 76; Univ Ga, MS, 73; La State Univ, MBA, 82. *Prof Exp:* Res assoc, Univ Calif, Berkeley, 76-77; from res chemist to sr res chemist, 77-83, sr com develop chemist, 83- 85, res supvr, 85-87, PROD MGR, PHARMACEUT INTERMEDIATES, ETHYL CORP, 88- *Concurrent Pos:* NSF grant, Univ Calif, Berkeley, 76-77. *Mem:* AAAS; Am Chem Soc; The Chem Soc; NY Acad Sci. *Res:* Organic synthesis; drug intermediate synthesis; synthesis of compounds of theoretical interest; transition and rare earth organometallic chemistry; alpha-kets carboxylic chemistry; amine and amino acid chemistry; aromatic alkylation chemistry; chemistry of thiophene compounds; surfactant chemistry. *Mailing Add:* Ethyl Corp 451 Florida St Baton Rouge LA 70801

BAUER, DIETRICH CHARLES, b Elgin, Ill, July 1, 31; m 54. IMMUNOLOGY. *Educ:* Univ Ill, BS, 54; Mich State Univ, MS, 57, PhD(microbiol), 59. *Prof Exp:* Res asst microbiol, Mich State Univ, 57-59; fel immunol, Case Western Reserve Univ, 59-61; from asst prof to assoc prof microbiol, 61-69, PROF MICROBIOL & IMMUNOL, IND UNIV, INDIANAPOLIS, 69-, CHMN DEPT, 81- *Mem:* AAAS; Am Soc Microbiol; Am Asn Immunol. *Res:* Virulence factors of Leptospirae; in vitro antibody synthesis; molecular forms of antibody; comparative reactivity of immunoglobulins; intracellular assembly of gamma globulin; suppression of immune response; immunological tolerance. *Mailing Add:* Dept Microbiol & Immunol Ind Univ Sch Med 635 Barnhill Dr MS 255 Indianapolis IN 46223

BAUER, DOUGLAS CLIFFORD, b Boston, Mass, May 11, 38; m 62; c 2. NUCLEAR ENGINEERING, MECHANICAL ENGINEERING. *Educ:* Cornell Univ, BME, 61, MS, 67; Carnegie-Mellon Univ, PhD(nuclear eng), 72. *Prof Exp:* Tech asst nuclear & coal plants, Am Elec Power Serv Corp, 65;

teaching & eng asst thermal power plants, Cornell Univ, 65-67; res assoc spatial nuclear reactor control, Carnegie-Mellon Univ, 71-72; sr engr nuclear thermal res, Bettis Atomic Power Lab, Westinghouse Elec Corp, 67-72; White House fel spec asst to Secy Dept Transp, 72-73; asst dir res & develop energy conserv, Dept Interior, Fed Energy Admin, 73-74; dir res planning, Fed Energy Admin, 74-75, assoc asst admin utility load mgt, 75-76; dir div nuclear & appln res, ERDA, 75-77; asst adminr utility systs regulation, Econ Regulatory Admin, Dept Energy, 77-80; CONSULT, 88- Mem: Sigma Xi. Res: Utility load management; rate structures; power generation techniques including co-generation; nuclear research including gas cooled reactors; advanced isotope separation; space nuclear systems; economic, environmental assessments; transmission; distribution. Mailing Add: 8310 Weller Ave McLean VA 22102

BAUER, ERNEST, b Vienna, Austria, Mar 24, 27; US citizen; m 59; c 2. ATMOSPHERIC PHYSICS. Educ: Cambridge Univ, BA, 47, PhD(theoret physics), 50. Prof Exp: Fel physics, Nat Res Coun Can, 50-52; res assoc, Inst Math Sci, NY Univ, 52-54, adj asst prof, Col Eng, 55-56, res assoc, 55-57, instr math, 53-54 & 56-57; asst prof physics, Univ NB, 54-55; prin scientist, Avco Res & Advan Develop, 57-59; staff scientist, Aeronutronic Div, Philco Corp, 59-65; RES STAFF MEM, SCI & TECHNOL DIV, INST DEFENSE ANALYSIS, 65- Concurrent Pos: Lectr, Cath Univ Am, 68-69 & City Col New York, 70-71; exhibitioner, scholar & prizeman, Selwyn Col, Cambridge Univ; consult, NASA, 79-81; mem, Air Pollution Indicators Task Force, Great Lakes Sci Adv Bd & US-Can Int Joint Comn, 85-86. Mem: Fel Am Phys Soc; Am Geophys Union. Res: Atmospheric sciences; stratospheric and tropospheric pollution; cloud and aerosol physics; remote sensing; ionospheric perturbations; nuclear weapons effects. Mailing Add: 8109 Fenway Rd Bethesda MD 20817-2762

BAUER, ERNST GEORG, b Schoenberg, Ger, Feb 27, 28; US citizen; m 55, 83; c 2. SURFACE PHYSICS. Educ: Univ Munich, MS, 53, PhD(physics), 55. Prof Exp: Asst, Phys Inst, Univ Munich, 54-58; br head, Naval Weapons Ctr, China Lake, Calif, 58-69; PROF, PHYS INST, CLAUSTHAL TECH UNIV, 69- Honors & Awards: Gaede Prize. Mem: Am Phys Soc; Am Vacuum Soc; Materials Res Soc. Res: Surface science; electron and crystal physics. Mailing Add: Phys Inst Clausthal Tech Univ Leibnizstr 4 Clausthal-Zellerfeld D3392 Germany

BAUER, EUGENE ANDREW, b Mattoon, Ill, Jun 17, 42; m 66; c 4. BIOLOGY, DERMATOLOGY. Educ: Northwestern Univ, BS & MD, 67, Nat Bd Med Examrs, dipl, 68, Am Bd Dermat, dipl, 73. Prof Exp: Instr dermat, Wash Univ Sch Med, 71-72, from asst prof to assoc prof, 74-88; PROF & CHMN DERMAT, STANFORD UNIV SCH MED, 88- Concurrent Pos: Dir, DEBRA Ctr Res & Ther Epidermolysis Bullosa, 82-88; assoc ed, J Invest ermat, 87; bd dirs, Soc Invest Dermat, 81-86; mem, NIH, 85-88; chmn study sect, Arth Musullosk Skin, 85-88; mem bd sci counselors, Div Cancer Biol, Nat Cancer Inst. Honors & Awards: Montagna Award, Soc Invest Dermat, 85; Sulzberger Award, Am Acad Dermat, 89. Mem: Am Dermat Asn; Am Acad Dermat; Am Fedn Clin Res; Am Soc Clin Invest; Asn Am Physicians. Res: Collagen biochemistry; degredation of extracellular matrix; collagenases and role in normal and pathologic remodelling in connective tissue; epidermolysis bullosa; tumor invasion. Mailing Add: Dept Dermat R-144 Stanford Univ Med Ctr Stanford CA 94305

BAUER, FRANCES BRAND, b New York, NY, July 5, 23; wid. APPLIED MATHEMATICS. Educ: Brooklyn Col, AB, 43; Brown Univ, MS, 45, PhD(appl math), 48. Prof Exp: Res assoc appl math, Brown Univ, 45-48; res assoc aero eng structures, Polytech Inst Brooklyn, 49-50; sr mathematician, Reeves Instrument Corp, 50-51 & 51-61; mathematician, Bur Stand, Am Univ, 51-52; res scientist, 61-80, SR RES SCIENTIST ELASTICITY, FLUID DYNAMICS & COMPUT, COURANT INST MATH SCI, NY UNIV, 80- Concurrent Pos: NASA-Ames Stanford fac fel, 89. Honors & Awards: Pub Serv Award, NASA, 76, Cert Recognition, 77 & 80. Mem: Am Math Soc; Asn Comput Mach. Res: Transonic flow; supercritical wing sections I; supercritical wing sections II & III; computational method in plasma physics; magnetohydrodynamic equilibrium and stability of stellarators; the beta equilibrium, stability, and transport codes. Mailing Add: 200 East End Ave New York NY 10128

BAUER, FREDERICK WILLIAM, b Wakefield, RI, Oct 22, 22; m 54; c 5. ORGANIC CHEMISTRY, RESEARCH ADMINISTRATION. Educ: Washington & Lee Univ, BS, 43; Princeton Univ, MA, 50, PhD(org chem), 52. Prof Exp: Jr asst res chemist, Tenn Eastman Corp, 43-44, asst res chemist, 46-47; technician, Textile Res Inst, 47, res chemist, 51-52; res chemist, Gen Labs, US Rubber Co, 52-54, org chemist, Textile Div, 54-56 & Res Ctr, 56-60; group leader, Cent Res Labs, 60-62, asst dir res admin, 62-64, dir res admin, 64-69, asst mgr employee rels, 69-80, MGR ENVIRON SERV, CORP TECHNOL, ALLIED CORP, 80- Mem: Am Chem Soc; Sigma Xi. Res: Structure of cellulose; rubber and polymers; textile finishing and chemical modifications; scientific personnel administration; polymer characterization; industrial hygiene, safety; product safety; hazardous waste disposal. Mailing Add: 19 Geneva Ct Packanack Lake Wayne NJ 07471

BAUER, GUSTAV ERIC, b New York, NY, Jan 26, 35; m 71; c 5. ANATOMY, CYTOCHEMISTRY. Educ: Queens Col, NY, BS, 57; Western Reserve Univ, MA, 59; Univ Minn, PhD(anat), 63. Prof Exp: Jr chemist, Sperry-Rand Corp, 57; from instr to assoc prof, 63-76, PROF ANAT, UNIV MINN, MINNEAPOLIS, 76- Concurrent Pos: USPHS fel, 65-66; mem corp, Marine Biol Lab, Woods Hole. Mem: Am Diabetes Asn; Am Soc Cell Biologists; Am Asn Anatomists. Res: Topographic organization, growth and regeneration of pancreatic islets of Langerhans; insulin metabolism and clearance by major salivary glands. Mailing Add: Dept Cell Biol & Neuroanat Univ Minn 321 Church St SE Minneapolis MN 55455

BAUER, HEINZ, b Vienna, Austria, Nov 28, 14; US citizen; m 39; c 2. PATHOLOGY. Educ: Univ Vienna, 33-38; Emory Univ, MD, 51. Prof Exp: Nat Cancer Inst trainee, Emory Univ, 54-56, from assoc to assoc prof path, 56-61; clin assoc prof, 61-63, assoc prof, 63-65, PROF PATH, SCH MED, GEORGETOWN UNIV, 65- Concurrent Pos: Res assoc, Mt Sinai Hosp, NY, 61-64. Mem: Fel Col Am Path; Am Asn Path & Bact; NY Acad Sci; Am Soc Exp Path; Int Acad Path. Res: Infectious diseases; immunopathology; germ-free animal research; rheumatology. Mailing Add: 4701 Willard Ave Apt 1510 Chevy Chase MD 20815

BAUER, HENRY, b Minneapolis, Minn, Nov 3, 14; m 38. BACTERIOLOGY. Educ: Univ Minn, PhD(bact), 49; Am Bd Microbiol, dipl. Prof Exp: Bacteriologist, State Dept Health, Minn, 38-41, supvr lab eval unit, 46-47, dep exec officer, 60-66, dir div med labs, 49-76; RETIRED. Concurrent Pos: Instr bact & immunol, Univ Minn, 47-48, lectr, Sch Pub Health, 49-73, adj prof, 74-76; consult bacteriologist, Off Surgeon Gen, 51-54 & NIH, 57-67; mem heart dis control adv comt, USPHS, 63-65; mem adv comt, Commun Dis Ctr, 63-69. Honors & Awards: Award, Minn Med Asn, 63; Dr Albert Justus Chesly Award, 77. Mem: Fel Am Acad Microbiol; Am Pub Health Asn; Asn State & Territorial Pub Health Lab Dirs. Res: Microbiology and virology; laboratory methodology. Mailing Add: 5127 34th Ave Minneapolis MN 55417

BAUER, HENRY HERMANN, b Vienna, Austria, Nov 16, 31; c 2. CHEMISTRY. Educ: Univ Sydney, BSc, 52, MSc, 53, PhD(chem), 56. Prof Exp: Res assoc chem, Univ Mich, 56-58; from lectr to sr lectr, Univ Sydney, 58-66; from assoc prof to prof, Univ Ky, 66-78, dean, Col Arts & Sci, 78-86, PROF CHEM & SCI STUDIES, VA POLYTECH INST & STATE UNIV, 86- Concurrent Pos: Fulbright travel award, 56-58; vis lectr chem & res scientist, Univ Mich, 65-66; vis prof, Univ Southampton, 72-73 & Japan Soc Promotion Sci, 74. Honors & Awards: Sydney S Negus Lectr, Va Acad Sci, 84. Mem: Am Chem Soc; Soc Sci Explor; Soc Social Studies Sci. Res: Case studies of Velikovsky controversy, Loch Ness monster controversy; distinguishing science and pseudo-science; interactions of science and society. Mailing Add: Chem Dept Va Polytech Inst & State Univ Blacksburg VA 24061-0212

BAUER, HENRY RAYMOND, III, b Meriden, Conn, Jan 16, 43; m 76. COMPUTER SCIENCE, COMPILER CONSTRUCTION. Educ: Brown Univ, ScB, 65; Stanford Univ, MS, 67, PhD(comput sci), 73. Prof Exp: From asst prof to assoc prof, 73-84, chmn, 78-84, PROF COMPUT SCI, UNIV WYO, 84-, chmn, 78-86. Concurrent Pos: Team chmn, CSAB, 85- Mem: Asn Comput Mach; Inst Elec & Electronics Engrs; Comput Soc. Res: Programming languages; compilers; parallel computation and compilers. Mailing Add: Dept Comput Sci Univ Wyo Box 3682 Laramie WY 82071

BAUER, JAMES H(ARRY), b Valley City, Ohio, June 5, 22; m 48; c 1. METALLURGICAL ENGINEERING. Educ: Ohio State Univ, BMetE, 51. Prof Exp: Metallurgist, Southwind Div, Stewart-Warner Corp, 51-53, chief metallurgist, 53-56; chief metallurgist, Ohio Injector Co, 56-61, mgr mat control, 61-63; chief metall chemist, Develop & Eng Dept, Am Standard Inc, 63-67, mgr standards & reliability, 67-71, mgr codes dept, 71-88; RETIRED. Honors & Awards: Award of Merit, Am Soc Testing & Mat. Mem: Am Soc Metals; Am Soc Testing & Mat; Am Welding Soc; Am Soc Plumbing Engrs; Am Soc Sanit Engrs. Res: Codes, standards and federal regulatory agency specifications in all phases of plumbing standards. Mailing Add: 3410 Country Club Dr Medina OH 44256-8713

BAUER, JERE MARKLEE, b Ft Worth, Tex, Apr 24, 15; m 61; c 2. MEDICINE. Educ: Univ Tex, MD, 41, BA, 44. Prof Exp: Intern, Med Br, Univ Tex, 41-42; from asst resident to instr internal med, 42-46, instr & res asst endocrinol & metab, 46-48, from asst prof to assoc prof, 48-72, PROF INTERNAL MED, UNIV MICH, ANN ARBOR, 72- Mem: AAAS; AMA; fel Am Col Physicians; Am Diabetes Asn; Am Heart Asn. Res: Endocrine and metabolic diseases; metabolic aspects of cancer. Mailing Add: 326 Juniper Lane Ann Arbor MI 48105

BAUER, JOHN HARRY, b Philadelphia, Pa, Apr 15, 43; m 66; c 2. NEPHROLOGY. Educ: State Univ NY Buffalo, BA, 65; Jefferson Med Col, MD, 69. Prof Exp: Intern, Med Ctr, Ind Univ, 69-70, resident, 70-71; med officer, Lemoore Naval Hosp, Calif, 71-73; fel renal med, Med Ctr, Ind Univ, 73-74 & hypertension, 74-75; from asst prof to assoc prof, 75-87, PROF MED NEPHROLOGY, MED CTR, UNIV MO-COLUMBIA, 87- Concurrent Pos: Staff physician, H S Truman Mem Vet Hosp, 75-85; res assoc, Vet Admin, 79-81. Mem: Am Fedn Clin Res; Am Soc Nephrology; Am Col Physicians; Am Heart Asn; Cent Soc Clin Res; Sigma Xi. Res: Hypertension; renal function; drug effects on renal function, body fluid composition and renin-aldosterone system; diabetes and kidney disease; Angiotensin II. Mailing Add: Dept Med N 403 Sch Med Univ Mo Columbia MO 65212

BAUER, KURT W, b Milwaukee, Wis, Aug 25, 29; m 55; c 3. SANITARY & ENVIRONMENTAL ENGINEERING. Educ: Marquette Univ, BS, 51; Univ Wis, MS, 55, PhD(civil eng), 61. Prof Exp: City planner, South Milwaukee, Wis, 53-55; instr civil eng, Univ Wis, 55-56 & 60-61; assoc engr, H C Webster & Sons Consult Engr, 56-59; chief current city planning, Madison, 59-60; asst dir, 61-62, EXEC DIR, SOUTHEASTERN WIS REGIONAL PLANNING COMN, 62- Concurrent Pos: Lectr civil eng, Marquette Univ, 61-62, adj prof, 68-; lectr, Univ Wis, 62-64; consult, Nat Water Comn, 70-73 & US Dept Interior, 89-90; mem exec comt, Transp Res Bd, 75-79. Honors & Awards: Harland Bartholomew Award, Am Soc Civil Engrs, 70; Pub Serv Award, US Dept Transp, Fed Hwy Admin, 77. Mem: Am Soc Civil Engrs; Am Inst Cert Planners; hon mem Soil Conserv Soc Am; Inst Munic Eng. Res: Municipal engineering; city and regional planning. Mailing Add: 2515 Broken Hill Ct Waukesha WI 53188

BAUER, LUDWIG, b Ger, July 27, 26; nat US; m 57; c 2. ORGANIC MEDICINAL CHEMISTRY. *Educ:* Univ Sydney, BSc, 49, MSc, 50; Northwestern Univ, PhD(chem), 52. *Prof Exp:* Res assoc, Harvard Univ, 52-53, Columbia Univ, 53 & Univ Sydney, 53-54; res chemist, Elkin Chem Co, 55; from asst prof to assoc prof, 55-65, PROF CHEM, UNIV ILL, CHICAGO, 65- *Mem:* Am Chem Soc; Int Soc Heterocyclic Chem. *Res:* Synthesis of potential medicinal agents; fundamental chemistry of aromatic heterocyclic compounds. *Mailing Add:* Med Chem M/C 781 Col Pharm Univ Ill Chicago IL 60680-6998

BAUER, MARVIN E, b Valparaiso, Ind, July 24, 43; m 69. REMOTE SENSING. *Educ:* Purdue Univ, BSA, 65, MS, 67; Univ Ill, PhD(agron), 70. *Prof Exp:* Res agronomist, dept agron & prog leader, Crop Inventory Res, Lab Appln Remote Sensing, Purdue Univ, 70-83; PROF & DIR, REMOTE SENSING LAB, UNIV MINN, 83- *Concurrent Pos:* Prin investr remote sensing agr, NASA, 79-83; ed, Remote Sensing Environ J, 80-; mem, Coun Agr Sci & Technol. *Mem:* Am Soc Agron; Crop Sci Soc Am; Am Soc Photogram & Remote Sensing; Inst Elec & Electronics Engrs; Sigma Xi. *Res:* Spectral properties of crops and soils; remote sensing of agricultural and natural resources. *Mailing Add:* Remote Sensing Lab Univ Minn 110 Green Hall 1530 N Cleveland Ave St Paul MN 55108

BAUER, MICHAEL ANTHONY, b Dayton, Ohio, Feb 18, 48; m 75; c 2. DISTRIBUTED COMPUTING, SOFTWARE ENGINEERING. *Educ:* Univ Dayton, BSc, 70; Univ Toronto, MSc, 71, PhD(computer sci), 78. *Prof Exp:* Researcher artificial intel, Edinburgh Univ, 74-75; ASSOC PROF COMPUTER SCI, UNIV WESTERN ONT, 75-, CHMN DEPT, 91- *Concurrent Pos:* Consult, Geac Computers Int, 84-88; mem bd, Can Info Processing Soc, 84-88 & Asn Comput Mach, 89-; adv, Int Bus Mach Ctr Advan Studies, 90-91. *Res:* Distributed computing, especially distributed algorithms, correctness, languages for distributed computing, verification; software engineering, including methodologies, formal specifications, development environments. *Mailing Add:* Dept Computer Sci Univ Western Ont London ON N6A 5B7 Can

BAUER, PENELOPE JANE HANCHEY, b Oberlin, La, Apr 12, 42; m. PHYTOPATHOLOGY, CYTOLOGY. *Educ:* McNeese State Col, BS, 64; La State Univ, MS, 66; Univ Ky, PhD(plant path), 68. *Prof Exp:* Res asst plant path, La State Univ, 64-67; res asst, Univ Ky, 67-68; asst prof, 69-74, assoc prof, 74-80, PROF BOT & PLANT PATH, COLO STATE UNIV, 81- *Concurrent Pos:* Vis scientist, Univ Calif, Berkeley, 82. *Mem:* Am Phytopath Soc; Bot Soc Am; Am Soc Plant Physiologists. *Res:* Physiology and ultrastructure of host-plant parasite relations. *Mailing Add:* Dept Biol Colo State Univ Ft Collins CO 80523

BAUER, PETER, b Reichenberg, Czech, Mar 2, 32; wid; c 2. ELECTRICAL & MECHANICAL ENGINEERING. *Educ:* Fed Tech Col, Vienna, MS(elec eng) & MS(mech eng), 53. *Prof Exp:* Develop engr, Laurence Scott & Electromotors Ltd, Eng, 53-54; sr engr, Pye Ltd, 54-56, dept mgr, 56-60; dept supvr, Univac Div, Remington Rand Group, Sperry Rand Corp, Pa, 60-62; sr eng proj mgr, 62-63, prin engr, 63-69, chief staff consult, 69-71, chief scientist, 71-76; CONSULT & PRES, ACT CO, MD, 76- *Mem:* Fel Brit Interplanetary Soc; Asn Austrian Eng & Archit. *Res:* Variable speed alternating current commutator machines; television transmission; digital information recording; fluidics and pure fluid systems. *Mailing Add:* 13921 Esworthy Rd Germantown MD 20874

BAUER, RICHARD G, b Kent, Ohio, Dec 9, 35; m 58; c 3. POLYMER CHEMISTRY, ORGANIC CHEMISTRY. *Educ:* Kent State Univ, BS, 56; Univ Akron, MS, 60, PhD(polymer sci), 66. *Prof Exp:* Res chemist, Gen Tire & Rubber Co, 56-60 & Air Reduction Chem & Carbide Co, 61-63; SECT HEAD, GOODYEAR TIRE & RUBBER CO, AKRON, 63- *Mem:* AAAS; Am Chem Soc; Soc Petrol Engrs; Soc Plastics Engrs. *Res:* Preparation and characterization of polymers and organic chemicals related to the polymer field. *Mailing Add:* 1624 Chadwick Dr Kent OH 44240-4410

BAUER, RICHARD M, b Appleton, Wis, Apr 30, 28; m 50; c 5. TECHNICAL MANAGEMENT. *Educ:* Lawrence Col, BS, 53. *Prof Exp:* From jr process engr to sr process engr, Marathon Corp, Am Can Co, Wis, 55-58, sr process engr, Marathon Div, 58-60, group leader, 60-65, mgr indust prod develop, 65-67, mgr converting paperboard & spec prod res & develop, 67-68, assoc dir, 68-70, lab dir consumer prod, 70-78, managing dir dry forming tech, 78-82; lab dir Consumer Prods Res & Develop, James River Corp, 82-84; vpres, Neenah Tech Ctr, 84-90; CONSULT, 90- *Concurrent Pos:* Marathon Div consult, Glamakote carton process to US & Europ countries, 61- *Res:* Research and development in the field of wood fiber treatments to manufacture soft, absorbent, buzky substrates from wet forming and dry forming processes. *Mailing Add:* 405 South Way Salem SC 29676

BAUER, ROBERT, b Grand Rapids, Mich, May 5, 26; m 51; c 3. CLINICAL BIOCHEMISTRY. *Educ:* Western Mich Univ, BS, 50, MS, 62. *Prof Exp:* Chemist, Socony Mobil Oil Co, Inc, 51-55; plastics chemist, Dow Chem Co, 55-58; chemist, Upjohn Co, 58-60; assoc res biochemist, 61-71, res scientist, 71-79, SR RES SCIENTIST, DIAG DIV, MILES, INC, 79-; STAFF SCIENTIST, 84- *Mem:* AAAS; Am Chem Soc. *Res:* Diagnostic aid and analytical aids for industry. *Mailing Add:* 56180 County Rd 21 Bristol IN 46507-9513

BAUER, ROBERT FOREST, b Declo, Idaho, Jan 10, 18; m 40; c 4. PETROLEUM ENGINEERING, OFFSHORE DRILLING. *Educ:* Southern Calif Univ, BS, 42. *Prof Exp:* Field res engr, Union Oil Co, Calif, 42-48, prod foreman, 48-50, dist prod foreman, 50-52, chief field engr, 52-54; mgr offshore group, Union, Continental, Shell & Superior Oil Co, 54-58; pres, Global Marine Explor Co, 58-66, chmn, Global Marine, Inc, 66-83; OWNER, BAUER LAND & CATTLE CO, 83- *Concurrent Pos:* Mem, UN Econ Comn Asia & Far East, 69; distinguished lectr marine affairs, Ore State Univ, 69. *Mem:* Nat Acad Eng; Am Inst Mining, Metall & Petrol Engrs; Am Petrol Inst. *Res:* Salt water based rotary mud; electromagnetic methods of geophysical exploration; design and fabrication of diving bell for gravimetric surveying of submerged land; extraction of bitumen from bituminous sands; stimulation of well production with salt solutions; use of surfactants to improve operating efficiency of producing wells; technology of drilling from a floating vessel. *Mailing Add:* 8570 Enramada Whittier CA 90605

BAUER, ROBERT OLIVER, b Chicago, Ill, Mar 2, 18; m 39; c 3. PHARMACOLOGY. *Educ:* Univ Mich, BS, 40; Wayne Univ, MS, 44, MD, 47; Am Bd Anesthesiol, dipl. *Prof Exp:* Intern, Wayne County Gen Hosp, Mich, 47-48; asst prof pharmacol, Sch Med, Boston Univ, 48-52; sr pharmacologist, Riker Lab, Inc, Los Angeles, 52-53; pharmacologist, Roswell Park Mem Inst, 55-58; from asst prof to assoc prof anesthesiol, 59-70, PROF ANESTHESIOL & PHARMACOL, CTR HEALTH SCI, UNIV CALIF, LOS ANGELES, 70- *Concurrent Pos:* Attend anesthesiologist, Ctr Health Sci, Univ Calif, Los Angeles, 58- *Mem:* AAAS; Am Soc Anesthesiol; Am Pharmaceut Asn; Am Soc Pharmacol & Exp Therapeut. *Res:* Analgesics; cardiovascular agents. *Mailing Add:* 12006 Chalon Rd Los Angeles CA 90049

BAUER, ROBERT STEVEN, b Brooklyn, NY, Dec 8, 44; m 67; c 4. EXPERIMENTAL SOLID STATE PHYSICS & CHEMISTRY. *Educ:* Rensselaer Polytech Inst, BEE, 66; Stanford Univ, MS, 67, PhD(elec eng), 71. *Prof Exp:* mem res staff surface & bulk electronic properties of solids, 70-79, SR SCIENTIST, XEROX PALO ALTO RES CTR, 79- *Concurrent Pos:* Mem tech comt, Annual Conf on Physics & Chem of Semiconductor Interfaces, 77-; vchmn, Users Orgn, Systs Simulation Res Lab, 80, chmn, 81, chmn, Users Meeting, 79 & 80; chmn, Sixth Annual Conf on Physics of Compound Semiconductor Interfaces, 79; mem tech comt, Int Conf on Physics of Metal-Oxide Semiconductor Insulators, 80; mem adv panel, Dept Defense Workshop on Submicron Devices, 81 & NSF Ctr for Res in Surface Sci, 81; dir, Second Int Union Pure & Appl Physics/UNESCO Semiconductor Symp, 82. *Mem:* Am Phys Soc; Am Vacuum Soc; Inst Elec & Electronics Engrs; Sigma Xi. *Res:* Photoelectron and modulated optical spectroscopy studies of surface, interface and bulk electronic states of crystalline and amorphous semiconductors, insulators and ionic conductors; use of synchrotron radiation for probing properties of solids; chemical bonding, semiconductor interface formation, (schottky barriers and heterostructures) oxidation, reaction, interdiffusion, impurities/defects, and electron/lattice effects; molecular beam epitaxy. *Mailing Add:* Xerox Park 3333 Coyote Hill Rd Palo Alto CA 94304

BAUER, ROGER DUANE, b Oxford, Nebr, Jan 17, 32; m 56; c 3. BIOCHEMISTRY. *Educ:* Beloit Col, BS, 53; Kans State Univ, MS, 57, PhD(biochem), 60. *Prof Exp:* From asst prof to assoc prof chem, 59-69, chmn dept, 66-75, dean, Sch Natural Sci, 75-88, PROF CHEM, CALIF STATE UNIV, LONG BEACH, 69- *Concurrent Pos:* Am Coun Educ fel, 70-71. *Mem:* Am Chem Soc; Radiation Res Soc; Sigma Xi. *Res:* Metabolism and structure of nucleoproteins; science education. *Mailing Add:* Sch Natural Sci Calif State Univ Long Beach CA 90840

BAUER, RONALD SHERMAN, b Huntington Park, Calif, Feb 24, 32; m 58; c 1. ORGANIC CHEMISTRY. *Educ:* Univ Calif, BS, 54, PhD(chem), 58. *Prof Exp:* CHEMIST, SHELL DEVELOP CO, 58- *Mem:* Am Chem Soc. *Res:* Synthesis and polymerization of halogenated allenes; polymerization of epoxides and related materials; process and product development on weatherable and high solids epoxy resins and coating systems. *Mailing Add:* Shell Develop Co Westhollow Res Ctr PO Box 1380 Houston TX 77251-1380

BAUER, RUDOLF WILHELM, b Rothenburg, Ger, Nov 28, 28; US citizen; m 58; c 3. ATOMIC & MOLECULAR PHYSICS. *Educ:* Amherst Col, BA, 52; Mass Inst Technol, PhD(physics), 59. *Prof Exp:* From instr to asst prof physics, Mass Inst Technol, 59-64; PHYSICIST, LAWRENCE LIVERMORE LAB, UNIV CALIF, 64- *Mem:* Am Phys Soc; Inst Elec & Electronics Engrs. *Res:* Nuclear physics; atomic and molecular physics; gamma and x-ray spectroscopy; neutron physics; plasma physics; diagnostics; accelerators; detectors. *Mailing Add:* L-296 Lawrence Livermore Lab Livermore CA 94550

BAUER, SIMON HARVEY, b Kovno, Lithuania, Oct 12, 11; nat US; m 38; c 3. CHEMICAL KINETICS, STRUCTURAL CHEMISTRY. *Educ:* Univ Chicago, BS, 31, PhD(chem), 35. *Prof Exp:* Fel, Calif Inst Technol, 35-37; instr fuel technol, Pa State Col, 37-39; from instr to prof, 39-77, EMER PROF CHEM, CORNELL UNIV, 77- *Concurrent Pos:* Consult, Atlantic-Richfield Oil Co, Los Alamos Nat Labs & Argonne Nat Lab; Guggenheim Mem Found fel, 49; NSF sr fel, 62-63; Alexander von Humboldt award, 79; adj prof, Inst Molecular Sci, Japan, 83. *Mem:* Fel AAAS; fel Am Phys Soc; fel Am Inst Chemists; Fedn Am Scientists; Am Chem Soc. *Res:* Electron diffraction; compounds of boron; molecular spectra; rates of energy transfer processes and of very rapid reactions as studied in shock tubes; chemical lasers; laser induced chemical conversions; structural studies with x-ray absorption fine structure; combustion mechanisms of hydrocarbons and boron hydrides. *Mailing Add:* Dept Chem Cornell Univ Ithaca NY 14853

BAUER, STEWART THOMAS, b Chicago, Ill, Apr 25, 09; m 38; c 2. ORGANIC CHEMISTRY. *Educ:* Univ Ill, BS, 32; Univ Minn, MS, 34. *Prof Exp:* Res chemist, Armour Co, 35-41; sect head, USDA, 41-46 & Drackett Co, 46-54; chief chemist, Crosby Chem, Inc, 54-66, from asst tech dir to assoc tech dir, 66-71; tech dir, 71-76; TECH CONSULT, 76- *Mem:* Am Chem Soc; Am Oil Chemists Soc. *Res:* Fatty acids and derivatives; terpenes and terpene polymers; tall oil products; rosins and derivatives. *Mailing Add:* 1306 N Beech St Picayune MS 39466

BAUER, WALTER, b Innsbruck, Austria, Mar 29, 35; US citizen; m 64. SOLID STATE PHYSICS. *Educ:* Univ Calif, Berkeley, AB, 57; Univ Ill, MS, 59, PhD(physics), 62. *Prof Exp:* Sr physicist, Atomics Int Div, NAm Aviation, Inc, 62-67, res specialist, 67-69; mem tech staff, 69-74, div supvr, 74-78, DEP MGR, SANDIA LABS, 78- *Concurrent Pos:* Exten instr, Univ Calif, Los Angeles, 63-66. *Mem:* AAAS; Am Phys Soc; Sigma Xi. *Res:* Defects in metals; electron radiation effects; inert gas migration in metals; ion implantation; magnetic fusion energy; science administration. *Mailing Add:* Dept 8340 Sandia Labs Livermore CA 94550

BAUER, WENDY HAGEN, b Aberdeen, Wash, May 11, 50; m 87; c 1. ASTRONOMY. *Educ:* Mt Holyoke Col, BA, 71; Univ Hawaii, MS, 74, PhD(astron), 77. *Prof Exp:* Fel astron, Harvard-Smithsonian Ctr Astrophys, 77-79; asst prof, 79-87, ASSOC PROF, WELLESLEY COL, 87- *Concurrent Pos:* Vis asst prof, Univ NMex, 84-85. *Mem:* Am Astron Soc; Sigma Xi; Astron Soc Pac; Int Astron Union. *Res:* Optical and infrared observation of mass loss from cool stars; mass loss from late-type stars. *Mailing Add:* Astron Dept Wellesley Col Wellesley MA 02181

BAUER, WILLIAM, JR, b Philadelphia, Pa, Aug 23, 36; m 65. ORGANIC CHEMISTRY, CHEMICAL ENGINEERING. *Educ:* Univ Pa, BS, 58, PhD(chem), 62. *Prof Exp:* Chemist org synthesis, 62-70, lab head plastics synthesis, 70-73, sr res assoc, 73-74, projs leader, 74-81, SECT MGR, ROHM AND HAAS CO, 81 - *Mem:* AAAS; Am Chem Soc. *Res:* Process research and development for commercial manufacture of vinyl and related monomers. *Mailing Add:* 2046 Winthrop Rd Huntingdon Valley PA 19006

BAUER, WILLIAM EUGENE, b Greensburg, Pa, Jan 21, 33; m 68; c 3. ANALYTICAL CHEMISTRY. *Educ:* Miami Univ, BA, 54; Pa State Univ, PhD(chem), 59. *Prof Exp:* INSTR CHEM, LUCKNOW CHRISTIAN COL, INDIA, 59-, HEAD DEPT, 70- *Concurrent Pos:* Vis asst prof & asst ed, Newslett, Adv Coun Col Chem, Wabash Col, 64-65. *Mem:* Am Chem Soc; Indian Chem Soc; Sigma Xi. *Res:* Polarography; emission and absorption spectroscopy; instrumentation. *Mailing Add:* c/o Lestor Curry 742 Bon Air Lansing MI 48917-2316

BAUER, WILLIAM R, b Friend, Nebr, July 10, 39. NUCLEIC ACID STRUCTURE, DNA & PROTEIN INTERACTIONS. *Educ:* Calif Inst Technol, PhD(phys chem), 69. *Prof Exp:* Asst prof chem, Univ Colo, Boulder, 69-73; assoc prof microbiol, 73-78, PROF MICROBIOL, STATE UNIV NY HEALTH SCI CTR, STONY BROOK, 78- *Concurrent Pos:* Guggenheim fel; NIH sr res fel. *Honors & Awards:* Am Can Soc Scholar. *Mem:* Am Soc Biol Chemists; Am Soc Virol; Am Soc Microbiol; Biophys Soc; Am Chem Soc. *Res:* Structure and properties of superhelical DNA; DNA-protein interactions, especially in vaccinia virus. *Mailing Add:* Dept Microbiol State Univ NY Health Sci Ctr Stony Brook NY 11794

BAUERLE, RONALD H, b Newport Ky, Dec 16, 37; m 61; c 2. MOLECULAR BIOLOGY. *Educ:* Thomas More Col, AB, 57, Univ Houston, MS, 59, Purdue Univ, PhD(microbiol), 63. *Prof Exp:* Instr microbiol, Purdue Univ, 61-62; res assoc molecular biol, Cold Spring Harbor Lab Quant Biol, 63-66; asst prof, Div Biol, Southwest Ctr Advan Studies, 66-69; assoc prof, 69-89, PROF BIOL, UNIV VA, 90- *Concurrent Pos:* Fel, NIH, Ctr Advan Studies, Univ Va, 63-66, NIH career develop award, 67-69, 71-76; vis prof, Univ Munster, 75; vis scholar, Stanford Univ, 84. *Mem:* AAAS; Genetics Soc Am; Am Soc Microbiol; Am Soc Biochem & Molecular Biol; Protein Soc. *Res:* Regulation of gene function in bacteria; allosteric mechanisms of enzyme control; genetics of bacteria; mechanism of protein synthesis; molecular genetics; structure, function, evolution and assembly of complex enzymes; molecular genetics of prokaryotes. *Mailing Add:* Dept Biol Gilmer Hall Univ Va Charlottesville VA 22901

BAUERMEISTER, HERMAN OTTO, b St Louis, Mo, Jan 18, 14; m 39. INDUSTRIAL ORGANIC CHEMISTRY. *Educ:* Armour Inst Technol, BS, 37; Ill Inst Technol, MS, 41; DePaul Univ, JD, 46. *Prof Exp:* Chemist, Commonwealth Edison Co, 37-40; patent examr insecticides & foods, US Patent Off, 40-42; res engr catalysis & synthetic rubber, Sinclair Refining Co, 42-46; patent att inorg chem & electronics, Monsanto Co, 46-82; RETIRED. *Mem:* Am Chem Soc. *Res:* Heat transfer and distillation; catalysis; organic phosphorus chemistry; detergents; electronic devices. *Mailing Add:* 7413 Truelight Ch Rd Charlotte NC 28227

BAUERNFEIND, JACOB (JACK) C(HRISTOPHER), b North Branch, NY, Apr 30, 14; m 39; c 3. FOOD TECHNOLOGY. *Educ:* Cornell Univ, BS, 36, MS, 39, PhD(nutrit, biochem & physiol), 40. *Prof Exp:* Asst nutritionist & instr, Cornell Univ, 36-40; nutritionist & res chemist, Hiram Walker & Son, Inc, Peoria, Ill, 40-44; chief appl nutrit, Hoffmann-LaRoche Inc, Nutley, NJ, 44-55, dir, Food & Agr Prod Develop Dept, 55-61, dir, Agr Res Dept, 61-68, dir agrochem & asst to vpres chem res, 68-71, coordr nutrit res, 72-79; RETIRED. *Concurrent Pos:* Consult food technol & nutrit, 79- *Mem:* Am Chem Soc; NY Acad Sci; Poultry Sci Asn; Am Soc Animal Sci; fel Inst Food Technologists; Fedn Am Soc Exp Biol; Am Inst Nutrit; Am Soc Clin Nutrit. *Res:* General nutrition; biochemistry; nutrients in human nutrition; nutrient delivery systems; nutrition of the elderly; prevention of blindness of dietary origin; experimental and practical feeding of animals; vitamins, antioxidants, drugs and carotenoids in human and agricultural applications; animal health and veterinary medicine; agricultural and food chemical research; food technology; food preservation. *Mailing Add:* 3664 NW 12th Ave Gainesville FL 32605

BAUGH, ANN LAWRENCE, b Freeport, Tex, Sept 4, 38; div. HAZARDOUS WASTES, PETROLEUM INDUSTRY. *Educ:* Southwestern Univ, Tex, BS, 60; Univ Tex, Austin, MA, 63, PhD(chem), 66; Pepperdine Univ, Malibu, Calif, MBA, 82; Univ Calif, Los Angeles, Hazardous Mat Mgt Cert, 87. *Prof Exp:* Asst prof chem, Southwestern Univ, Tex, 64-67; Welch fel, Univ Tex, Austin, 66-69; tech specialist & proj engr, Lockheed Propulsion Co, 69-75; sr res chemist & group leader, Occidental Res Corp, 75-82; prog develop mgr, Rockwell Int, 83-86; consult, Appl Geosci Corp, 87-89; ADMINR, ENVIRON PROGS, UNOCAL CORP, 89- *Honors & Awards:* Citation Merit Chem, Southwestern Univ, Tex, 72. *Mem:* Am Chem Soc; Am Inst Chem Engrs. *Res:* Microwave spectroscopy and molecular structure; aging studies of solid propellants and composites; instrument automation, computer programming; synthetic fuels; coal conversion, desulfurization, combustion kinetics and liquefaction; geothermal hydrogen sulfide abatement; phosphates research; hazardous materials. *Mailing Add:* 1615 Butternut Way Diamond Bar CA 91765

BAUGH, CHARLES M, b Fayetteville, NC, June 20, 31; m 52; c 4. BIOCHEMISTRY. *Educ:* Univ Chicago, SB, 58; Tulane Univ, PhD(biochem), 62. *Prof Exp:* Fel biochem, Tulane Univ, 62-63, from instr to asst prof, 63-65; asst prof pharmacol in med, Wash Univ, 65-67; assoc prof, 67-70, prof, Nutrit Div, 70-73, prof biochem & chmn dept, 73-81, assoc dean, 76-87, DEAN BASIC MED SCI, COL MED, UNIV S ALA, 87- *Concurrent Pos:* External assessor, Nat Res Coun Australia, 76-81; pres, Ala Acad Sci, 82; mem, bd trustees, Ala Acad Sci, 83-; pres, S Ala Med Sci Found, 82- *Mem:* Am Chem Soc; Am Inst Nutrit; Am Soc Biol Chem & Molecular Biol; NY Acad Sci; Soc Exp Biol Med. *Res:* Biosynthesis of the pteridine nucleus, as found in the vitamins, folic acid and riboflavin; antimetabolites in nucleic acids. *Mailing Add:* Dean Col Med Univ S Ala 170 CSAB Mobile AL 36688

BAUGHCUM, STEVEN LEE, b Atlanta, Ga, Dec 18, 50. KINETICS, MOLECULAR PHOTODISSOCIATION. *Educ:* Emory Univ, BS, 72; Harvard Univ, AM, 73, PhD(chem), 78. *Prof Exp:* Phys chemist, Air Force Cambridge Res Labs, 76; res assoc, Joint Inst Lab Astrophys, Univ Colo & Nat Bur Standards, 78-80; mem staff, Los Alamos Nat Lab, 80-87; prin res scientist & mgr chem physics prog, Spectra Technol Inc, 87-88; RES ANALYST, BOEING CO, 88- *Concurrent Pos:* Res asst, Emory Univ, 68-72; res asst, Harvard Univ, 72-78 & teaching fel, 73-76. *Mem:* Sigma Xi; Am Chem Soc; Am Phys Soc; Combustion Inst; Am Geophys Union; AAAS. *Res:* Modeling and analysis of combustion chemistry and atmospheric chemistry, with particular emphasis on the possible impact of supersonic aircraft on the stratospheric ozone layer. *Mailing Add:* 2215 185th Pl NE Redmond WA 98052

BAUGHMAN, GEORGE LARKINS, b Palatka, Fla, June 29, 38; m 58; c 2. ANALYTICAL CHEMISTRY, ENVIRONMENTAL CHEMISTRY. *Educ:* Fla State Univ, BS, 60. *Prof Exp:* Mfg engr anal chem, Martin-Marietta Corp, 60-61; chemist, Aero-Chem Res Lab, 61-63; aerospace engr anal chem, J F Kennedy Space Ctr, NASA, 63-70; RES CHEMIST, ENVIRON RES LAB, US ENVIRON PROTECTION AGENCY, 70- *Mem:* Am Chem Soc; Sigma Xi; Soc Environ Chem Toxicol. *Res:* Elucidation of transport and transformation mechanisms of aquatic pollutants; kinetics of environmental processes. *Mailing Add:* 165 Red Fox Run Athens GA 30605

BAUGHMAN, GLENN LAVERNE, b Dover, Pa, Apr 26, 31; m 56; c 2. ORGANIC POLYMER CHEMISTRY. *Educ:* Gettysburg Col, AB, 53; Pa State Univ, PhD(org chem), 61. *Prof Exp:* Chemist, Armstrong Cork Co, 55-57; res chemist, E I du Pont de Nemours & Co, 61-64; sr res chemist, 64-80, RES ASSOC, CHEM DIV, PPG INDUSTS, BARBERTON, 80- *Concurrent Pos:* Instr, Barberton Tech Sch, 64-65. *Mem:* Am Chem Soc. *Res:* Unsaturated polyesters for thermosetting castings; monomer structure-polymer property correlations; organometallics chemistry; photochemistry; synthesis of agricultural chemical intermediates. *Mailing Add:* 479 Dohner Dr Wadsworth OH 44281-2141

BAUGHMAN, RAY HENRY, b York, Pa, Jan 14, 43; m; c 4. CHEMICAL PHYSICS, POLYMER SCIENCE. *Educ:* Carnegie-Mellon Univ, BS, 64; Harvard Univ, MS, 66, PhD(mat sci), 71. *Prof Exp:* Staff scientist, Allied-Signal, Inc, 70-73, group leader polymer sci, Mat Res Ctr, 74-77, mgr org mat sci, 77-79, specialty polymers, 79-90, RES FEL CORP RES & DEVELOP, ALLIED-SIGNAL INC, 79- *Concurrent Pos:* Mem orgn comt, Int Conferences Synthetic Metals, 81-90, adv groups, NSF, NATO, Japan Found, Dept Energy & MRC. *Mem:* Fel Am Phys Soc; Am Chem Soc; Mat Res Soc. *Res:* Structures of disordered solids; molecular dynamics and phase transformations; electrical, optical and mechanical properties of polymers; solid state reactions; novel forms of carbon; new material sythesis, properties and applications; highly conducting organic polymers; high temperature superconductors. *Mailing Add:* Allied-Signal Inc 101 Columbia Rd PO Box 1021 Morristown NJ 07962-1021

BAUGHMAN, RUSSELL GEORGE, b Washington, DC; m 69; c 1. PHYSICAL CHEMISTRY. *Educ:* William Jewell Col, BA, 68; Iowa State Univ, PhD(phys chem), 77. *Prof Exp:* Grad asst, Iowa State Univ, 68-69, 73-77; asst prof, 77-84, ASSOC PROF PHYS CHEM, NORTHEAST MO STATE UNIV, 84- *Concurrent Pos:* Med technician, US Air Force, 69-73. *Mem:* Am Chem Soc; Am Crystallog Asn; Sigma Xi. *Res:* X-ray diffraction studies of pesticides and charge transfer complexes; quantum mechanics. *Mailing Add:* Div Sci Northeast Mo State Univ Kirksville MO 63501

BAUGHN, CHARLES (OTTO), JR, b Bicknell, Ind, Feb 25, 21; m 44; c 3. BIOLOGY. *Educ:* Ind State Teachers Col, BSc, 47; Univ NC, MS, 49, PhD(parasitol, bact), 52. *Prof Exp:* Instr parasitol, Univ NC, 50-52; parasitologist, Stamford Res Labs, Am Cyanamid Co, 52-55, group leader bact, Lederle Labs, 55-61, mgr microbial dis sect, Agr Res Ctr, 61-64; asst dir res, Hess & Clark Div, Richardson-Merrell, Inc, 64-69; sr investr, Squibb Inst Med Res, 69-84; RETIRED. *Mem:* AAAS; Am Soc Microbiol; NY Acad Sci; Sigma Xi. *Res:* Chemotherapy of microbial infections; resistance to infection. *Mailing Add:* 29 Maple Ave Flemington NJ 08822

BAUGHN, ROBERT ELROY, b Chanute, Kans, Jan 31, 40; m 65; c 2. SEXUALLY TRANSMITTED DISEASES. *Educ:* The Citadel, BS, 63; Univ Tenn, Memphis, MS, 66; Univ Cincinnati, PhD(microbiol & immunol), 75; Houston Baptist Univ, MBA, 80. *Prof Exp:* Chief bact br, Sixth US Army Med Lab, Ft Baker, Calif, 67-68, chief bact, serol & parasitol, 376th Med Detachment, US Army, Thailand, 68-69; microbiologist, John L Hutchenson-Tri-County Mem Hosp, Ft Oglethorpe, Ga, 69-71; from instr to asst prof, 75-83, ASSOC PROF MICROBIOL, IMMUNOL & DERMAT, BAYLOR MED COL, HOUSTON, 83-; DIR MICROBIOL, SYPHILIS RES LAB, VET AFFAIRS MED CTR, HOUSTON, 77-, ASSOC CAREER SCIENTIST, 90- *Concurrent Pos:* Clin assoc prof, med tech & cytogenetics, Sch Allied Health Sci, Univ Tex, Houston, 86- *Mem:* Sigma Xi; Am Soc Microbiol; Am Asn Immunologists; Reticuloendothelial Soc; Undersea Med Soc; Am Venereal Dis Asn; Am Soc Clin Pathologists. *Res:* Induction, regulation and control of immune responses in chronic infections; host-parasite relationships in infectious processes and mechanisms of pathogenesis; the role of immune complexes in pathogenesis and regulation of immune function. *Mailing Add:* Syphilis Res Lab Bldg 211 Rm 230 Vet Admin Med Ctr 2002 Holcome Blvd Houston TX 77030

BAUGUESS, CARL THOMAS, JR, b Lancaster Co, Pa, Sept 30, 28; m 65. PHARMACEUTICS. *Educ:* Univ NC, BS, 54, MS, 66; Univ Miss, PhD, 70. *Prof Exp:* Pharmacist, Jonesboro's Lee Drug Store, 54-57; instr & dir pharmaceut exten serv, Univ NC, 57-60, part-time instr, 60-64; pharm consult, NC Heart Asn, 64-67; asst prof pharmaceut, Northeast La State Univ, 67-68; assoc prof, 70-77, PROF PHARMACEUT, UNIV SC, 77-, DIR GRAD PROG, COL PHARM, 75- *Honors & Awards:* Lederle Award, 75. *Mem:* Am Pharmaceut Asn; Sigma Xi. *Res:* Investigations of new drugs for antitumor activity in mouse leukemia; pharmacokinetic studies of drug concentration changes after adminstration by various routes in larger animals; microencapsulation of basic drugs by spray dry techniques. *Mailing Add:* Col Pharm Univ SC Columbia SC 29208

BAUKNIGHT, CHARLES WILLIAM, JR, b Atlanta, Ga, June 29, 59. CHEMISTRY. *Educ:* Duke Univ, BA, 81; Clemson Univ, PhD(org chem), 87. *Prof Exp:* Postdoctoral, Chem Dept, Univ Iowa, 87-89; MEM STAFF, CHEM DEPT, TOWSON STATE UNIV, 89- *Mem:* Sigma Xi. *Res:* Synthesis of precursors to potentially crantioselective fluorinating agents; synthesis and chemistry of perfluorinated nitrites and nitrogen-bromine compounds. *Mailing Add:* Chem Dept Towson State Univ Baltimore MD 21204

BAULD, NATHAN LOUIS, b Clarksburg, WVa, Dec 12, 34; m 71; c 2. ORGANIC CHEMISTRY. *Educ:* WVa Univ, BS, 56; Univ Ill, PhD(org chem), 59. *Prof Exp:* NSF fel, Harvard Univ, 59-60; res chemist, Rohm and Haas Co, Pa, 60-61; from instr to assoc prof, 61-73, PROF CHEM, UNIV TEX, AUSTIN, 73- *Concurrent Pos:* Sloan Found fel, 66-68. *Mem:* Am Chem Soc. *Res:* Physical-organic and stereochemistry; anion radicals; molecular orbital theory; aromatic systems; cation radicals. *Mailing Add:* Dept Chem Univ Tex Austin TX 78712-1167

BAULD, NELSON ROBERT, JR, b Clarksburg, WVa, May 18, 31; m 51; c 2. ENGINEERING MECHANICS. *Educ:* WVa Univ, BSME, 58, MS, 60; Univ Ill, PhD(theoret & appl mech), 63. *Prof Exp:* Instr eng mech, WVa Univ, 58-60; instr, Va Polytech Inst, 60-61; assoc prof, 63-77, PROF ENG MECH, CLEMSON UNIV, 77- *Mem:* Am Soc Eng Educ. *Res:* Time-dependent studies of engineering members made from materials that creep, including columns, torsion members, plates and shells; applied mathematics. *Mailing Add:* Dept Mech Eng Clemson Univ Main Campus Clemson SC 29634

BAULE, GERHARD M, b Syracuse, NY, Jan 20, 34; m 55; c 4. ELECTRICAL ENGINEERING. *Educ:* Syracuse Univ, BEE, 56, MEE, 58, PhD(elec eng), 63. *Prof Exp:* Res asst elec eng, 56-58, from instr to asst prof, 58-69, ASSOC PROF ELEC ENG, SYRACUSE UNIV, 69- & ASSOC PROF COMPUT ENG, 77- *Honors & Awards:* Res Award, Sigma Xi, 64. *Res:* Application of electrical science to medicine. *Mailing Add:* 2-212 Ctr Sci & Technol Syracuse Univ Syracuse NY 13244-4100

BAULEKE, MAYNARD P(AUL), ceramics, for more information see previous edition

BAUM, BERNARD, b Boston, Mass, Sept 21, 24; m 53; c 1. POLYMER CHEMISTRY. *Educ:* Lowell Technol Inst, BS, 47; Clark Univ, MS, 49, PhD(chem), 50. *Prof Exp:* Chemist, Sherwin-Williams Co, 50-53; develop assoc, Union Carbide Plastics Co, 53-62; group leader, Allied Chem Co, 62-63 & Borden Chem Co, 63-64; mgr prod develop, Cast Nylon Dept, Budd Co, 64-67; mgr, Chem & Mat Develop Lab, Springborn Labs, Inc, 67-69, mgr, Chem & Mat Div, 69-72, vpres & mgr, Mat Res & Develop, 72-90; CONSULT, 90- *Mem:* Am Chem Soc; Air Pollution Control Asn; Soc Plastics Engr. *Res:* Polymer chemistry; synthesis; structure vs properties; coatings; elastomers; permanence properties of materials. *Mailing Add:* 44 Kirkwood Rd West Harvard CT 06117

BAUM, BERNARD R, b Paris, France, Feb 14, 37; m 61; c 1. TAXONOMY, BOTANY. *Educ:* Hebrew Univ, Jerusalem, MS, 63, PhD(tamarix taxon), 66. *Prof Exp:* Res scientist, Plant Res Inst, 67-73, sect chief, 73-87, PRIN RES SCI, BIOSYSTEMATICS RES INST, CAN DEPT AGR, 80- *Concurrent Pos:* Asst ed, Can J Bot, 74-, Euphytica, 87- *Honors & Awards:* Lawson Medal, Can Bot Asn. *Mem:* AAAS; Bot Soc France; Int Asn Plant Taxon; Can Bot Asn; fel Royal Soc Can; Bot Soc Am; Am Soc Plant Taxonomists. *Res:* Monograph of Tamarix, Avena and Hordeum; nomenclature; utilization of computers for taxonomy; taxometrics and statistics; evolution; international registries of cultivars. *Mailing Add:* Biosystematics Res Ctr Cent Exp Farm Ottawa ON K1A 0C6 Can

BAUM, BRUCE J, b Lynn, Mass, Oct 28, 45; m; c 2. ORAL MEDICINE. *Educ:* Univ Va, BA, 67; Tufts Univ, DMD, 71; Boston Univ, PhD(biochem), 74. *Prof Exp:* Sr investr, lab molecular aging, Nat Inst Aging, 78-82, CLIN DIR & CHIEF, CLIN INVEST & PATIENT CARE BR, NAT INST DENT RES, NIH, 82- *Concurrent Pos:* Wellcome vis prof, 88. *Honors & Awards:* Carl A Schlack Award, 87. *Mem:* AAAS; Am Soc Biol Chemists; Int Asn Dent Res; Geront Soc Am; Am Acad Oral Med. *Mailing Add:* 9402 Balfour Dr Bethesda MD 20814

BAUM, BURTON MURRY, b Brooklyn, NY, Dec 6, 34; m 65; c 2. ORGANIC CHEMISTRY, INORGANIC CHEMISTRY. *Educ:* Brooklyn Col, BS, 56; Univ Pittsburgh, PhD(org chem), 62. *Prof Exp:* Res chemist, 62-75, mgr, FMC, 75-83; MGR, ECOLAB INC, 83- *Mem:* Am Chem Soc. *Res:* Preparation and bleaching of textiles; flame retardants; detergents; organic synthesis; phosphorus chemistry; pulp and paper chemistry; odor counteractants; cleaners and disinfectants. *Mailing Add:* EcoLAB Inc 840 Sibley Hwy St Paul MN 55118

BAUM, CARL E(DWARD), b Binghamton, NY, Feb 6, 40. ELECTROMAGNETICS. *Educ:* Calif Inst Technol, BS, 62, MS, 63, PhD(elec eng), 69. *Prof Exp:* Proj officer, Air Force Weapons Lab, 63-67 & 68-71; SR SCIENTIST ELECTROMAGNETICS, PHILLIPS LAB, 71- *Concurrent Pos:* Mem, comn B US Nat Comt, Int Union Radio Sci, 75-, comn E, 82-, US deleg gen assembly, Peru 75, Finland, 78, Wash, DC, 81, Italy, 84, Israel, 87, Czech, 90, chmn, int comn E Working group, 81-; distinguished lectr, Inst Elec & Electronics Engrs, Antennas & Propagation Soc, 77-78, co-chmn, Joint Comt Nuclear Electromagnetic Pulse, 78-; adv, numerous US govt agencies. *Honors & Awards:* Richard R Stoddart Award, Inst Elec & Electronics Engrs, Electromagnetic Compatibility Soc, 84; Harry Diamond Mem Award, Inst Elec & Electronics Engrs, 87. *Mem:* Fel Inst Elec & Electronics Engrs; Int Union Radio Sci; Electromagnetics Soc (pres, 83-85); fel Int Biog Asn; fel Am Biog Asn. *Res:* Electromagnetic theory and applications; antennas, propagation, scattering, interaction, protection, and related pulse power for electromagnetic pulse, lightning, high power microwaves, and transient radar (high power electromagnetics); author of one book and numerous publications on electromagnetic theory. *Mailing Add:* 5116 Eastern SE Unit D Albuquerque NM 87108

BAUM, DAVID, b Saratoga Springs, NY, Aug 4, 27; m 59; c 4. PEDIATRIC CARDIOLOGY, PHYSIOLOGY. *Educ:* Dartmouth Col, AB, 51; Cornell Univ, MD, 55. *Prof Exp:* Intern med, Kings County Hosp, NY, 55-56; resident pediat, New York Hosp-Cornell Med Ctr, 56-59; asst pediat cardiol, Johns Hopkins Hosp, 59-60 & Mayo Clin, 60-61; from instr to assoc prof pediat, Sch Med, Univ Wash, 62-71, asst prog dir, Clin Res Ctr, 63-70; PROF PEDIAT & DIR DIV PEDIAT CARDIOL, SCH MED, STANFORD UNIV, 71- *Concurrent Pos:* Fel pediat cardiol, Johns Hopkins Hosp, 59-60 & Mayo Clin, 60-61; res trainee biochem, Sch Med, Univ Wash, 65-66; consult, Madigan Army Hosp, Ft Lewis & Rainier Sch, Buckley, Wash, 62-71 & Silas Hayes Army Hosp, Ft Ord, Calif. *Mem:* Soc Pediat Res; Am Pediat Soc; Soc Exp Biol & Med; Am Heart Asn; Am Physiol Soc; Perinatal Res Soc. *Res:* Metabolic aspects or hypoxia and heart failure as related to thermoregulation, execise and growth in children with cardiovascular disease; adipose tissue development and obesity. *Mailing Add:* Dept Pediat Stanford Univ Med Ctr 300 Pasteur Dr Stanford CA 94305

BAUM, DENNIS WILLARD, b Allentown, Pa, Dec 29, 40; m 64; c 2. HIGH ENERGY DENSITY SYSTEMS. *Educ:* Muhlenberg Col, BS, 63; Lehigh Univ, MS, 64, PhD(physics), 67. *Prof Exp:* Sr physicist gas dynamics, Physics Int Co, Inc, 67-72; mgr gas dynamics, Artec Assocs Inc, 72-76, vpres, 76-77, pres, 77-83; pres, Orinda Prod Corp, 83-88; PRES, LAWRENCE LIVERMORE, 88- *Res:* Development, design and utilization of compact chemical energy storage and rapid conversion into more useful kinetic and electrical forms, including explosive-metal acceleration, explosive electrical generators and high power, pulsed batteries. *Mailing Add:* 89 Longview Ct Danville CA 94526

BAUM, ELEANOR KUSHEL, b Poland; US citizen; m 62; c 2. ELECTRICAL ENGINEERING, ENGINEERING. *Educ:* City Col New York, BEE, 59; Polytech Inst NY, MEE, 61, PhD(elec eng), 64. *Prof Exp:* Engr, Sperry Corp, 61-63; chmn, 71-84, dean, Sch Eng, 84-87, PROF ELEC ENG, PRATT INST, 65-; SCH ENG, COOPER UNION, NY, 87-, DIR, RES FOUND. *Concurrent Pos:* Consult, 75-; prog visitor, Accreditation Bd Eng & Technol, 82-92, dir, 90-92; bd examiners, Grad Rec Exam, 85-90; lectr, var community groups; dir, Alleghany Power Systs Corp, 88-, Am Soc Eng Educ, 89-91. *Honors & Awards:* Emily Warren Roebling Award, 88; Upward Mobility Award, Soc Women Engrs, 90. *Mem:* Fel Inst Elec & Electronics Engrs; Am Soc Eng Educ; fel Soc Women Engrs. *Res:* Engineering education; engineering curricula; systems and control; career guidance for women in engineering. *Mailing Add:* Sch Eng Cooper Union 51 Astor Pl New York NY 10003

BAUM, GARY ALLEN, b New Richmond, Wis, Oct 2, 39; m 64; c 2. SOLID STATE PHYSICS. *Educ:* Wis State Univ, River Falls, BS, 61; Okla State Univ, MS, 64, PhD(physics), 69. *Prof Exp:* Res engr, Douglas Aircraft Co, Calif, 63-64; physicist, Dow Chem Co, Colo, 64-65, sr physicist, 65-66; from asst prof to assoc prof, 69-79, prof physics, Inst Paper Chem, 79-88, dir Paper Mat Div, 83-88; JVA fel, 88-89, DIR CORP RES & DEVELOP, JAMES RIVER CORP, 89- *Concurrent Pos:* Adj prof, Inst Paper Chem, 88; bd dirs, Tech Asn Pulp & Paper Indust. *Mem:* Am Asn Physics Teachers; Tech Asn Pulp & Paper Indust; Am Phys Soc. *Res:* Ferroelectric and antiferroelectric materials; vapor deposition, sputtering and accelerated ion techniques for thin film preparation; electrical and thermal properties of rutile; electrical properties of polymers; mechanical properties of paper and paperboard. *Mailing Add:* 426 Fidelis Appleton WI 54915

BAUM, GEORGE, b Hungary, Jan 11, 33; US citizen; m 54; c 3. IMMUNOLOGY. *Educ:* Case Inst Technol, BS, 54; Ohio State Univ, MS, 60, PhD, 66. *Prof Exp:* Engr solid fuel propellants, Wright Air Develop Command, 54-65; res engr synthetic hydraul fluids, Air Force Mat Lab, 55-65; res chemist, Corning Glass Works, 66-76; STAFF SCIENTIST, TECHNICON INSTRUMENT CORP, 76-; ADJ LECTR, PACE UNIV, 88- *Concurrent Pos:* Vis res assoc, Beth Israel Hosp, 75-76. *Mem:* NY Acad Sci; Am Chem Soc; AAAS. *Res:* Organometallic synthesis; reinforced plastics; peptide chemistry; fluoroaromatic chemistry; ion-selective electrodes; immobilized enzymes; affinity chromatography; hormone secretion by tumor cells; purification and derivatization of antibodies. *Mailing Add:* 3542 Sagamore Ave Mohegan Lake NY 10547-9666

BAUM, GERALD A(LLAN), b New York, NY, June 25, 29; m 54; c 2. POLYMER CHEMISTRY. *Educ:* City Col New York, BS, 51; Pa State Univ, PhD(chem), 55. *Prof Exp:* Engr insulation & chem develop sect, Westinghouse Elec Corp, 55-58; sr chemist res & develop, Resin Res Labs, Inc, 58-61; supvr plastics prod, M & T Chem, Inc, 61-67; res assoc, Ciba-Geigy Corp, 67-71, supvr new prod develop, Additives Dept, Plastics & Additives Div, 71-74; supvr tech serv, Tenneco Chem, Inc, 74-76; mgr additives res & develop, Dart Industs, Inc, 76-79, dir, compounded prod res & develop, Dart & Kraft, Inc, 79-85; mgr, plastic applns & develop, GAF Corp, 85-86; prog mgr, Core Technol, 86-89, SR DEVELOP CHEMIST, PROD TECHNOL, HOECHST CELANESE CORP, 89- *Concurrent Pos:* Guest lectr, NJ Inst Technol, 59-61; pres, Palisades Sect, Soc Plastic Engrs, 72-73. *Mem:* Am Chem Soc; Soc Plastics Engrs. *Res:* Plastics additives; stabilizers; specialty polymers. *Mailing Add:* 337 McKinley Blvd Paramus NJ 07652

BAUM, GERALD L, b Milwaukee, Wis, Dec 19, 24; m 51; c 3. MEDICINE. *Educ:* Univ Wis, BS, 45, MD, 47. *Prof Exp:* Intern, Jewish Hosp, Cincinnati, 47-48, resident med, 48 & 49-51; res chest med, Bellevue Hosp, New York, 51-52; fel bronchology, St Luke's Hosp, Chicago, 52; fel med mycol, Jewish Hosp, Cincinnati, 56-58; chief pulmonary sect, Vet Admin Hosp, Cincinnati, 58-65; prof med, Sch Med, Case Western Reserve Univ, 65-75; MED DIR, ISRAEL LUNG ASSOC, TEL AVIV, 75-, PROF MED, SCH MED, TEL AVIV UNIV, 75- *Concurrent Pos:* NIH fel, Tel Hashomer, Israel, 63-64 & Kupat Cholim award, Israel, 63-64; assoc prof, Univ Cincinnati, 64-65; chief pulmonary sect, Vet Admin Hosp, Cleveland, 65-73. *Honors & Awards:* Medalist, Am Col Chest Physicians, 89. *Mem:* Am Thoracic Soc; Am Col Physicians; Am Fedn Clin Res; Am Col Chest Physicians; Int Soc Human & Animal Mycol. *Res:* Pulmonary diseases, particularly fungus infections; rehabilitation of patients with chronic pulmonary insufficiency and tuberculosis; immunologic lung disease. *Mailing Add:* Rechov Hagolan 52 Givet Savyon Israel

BAUM, HARRY, b New York, NY, Oct 23, 15; m 44; c 2. ANALYTICAL CHEMISTRY. *Educ:* City Col New York, BS, 36; NY Univ, MS, 39. *Prof Exp:* Res chemist, Beth Israel Hosp, New York, 36-38; asst microchem, NY Univ, 38-39; res chemist, Gen Cigar Co, Pa, 40-41; supvr & plant chemist, US Rubber Co, Iowa, 41-43; supvr, Anal Res Lab, Publicker Indust, 46-47; supvr, Anal Develop Lab, Rohm and Haas Co, 48-69, supt qual control, 69-78, sr tech assoc, 78-80. *Concurrent Pos:* Referee, Collab Int Pesticide Anal Coun, 69-79; consult, 81- *Mem:* AAAS; Am Chem Soc; fel Asn Off Anal Chemists. *Res:* Analytical research and development. *Mailing Add:* 8210 Stockton Rd Elkins Park PA 19117

BAUM, HOWARD RICHARD, b New York, NY, April 3, 36; m 61; c 2. COMBUSTION FLUID MECHANICS, COMPUTATIONAL COMBUSTION. *Educ:* Polytech Inst Brooklyn, BS, 57, MS, 59; Harvard Univ, PhD(appl math), 64. *Prof Exp:* From lectr to asst prof mech eng, Harvard Univ, 64-71; sr scientist, Aerodyne Res Inc, 71-75; res physicist, Nat Bur Standards, 75-83; FEL, NAT INST STANDARDS & TECHNOL, 83- *Concurrent Pos:* Springer prof mech eng, Univ Calif, Berkeley, 85; panel nuclear winter res, Nat Acad Sci, 86-; Off Naval Res opportunities panel, Naval Studies Bd, Nat Res Coun, 86-87, 91; invited lectr, Second Int Symp Fire Safety Sci, 88, US/Japan Joint Sem, Computers in Heat Transfer Sci, 91. *Honors & Awards:* Silver Medal, US Dept Com, 81, Gold Medal, 83. *Mem:* Int Asn Fire Safety Sci; NY Acad Sci; Sigma Xi; Soc Indust & Appl Math; Combustion Inst. *Res:* Mathematical modeling of fire induced flow phenomena; turbulent combustion processes; ignition and flame spread in microgravity environment; development of computational techniques to study these phenomenas. *Mailing Add:* Nat Inst Standards & Technol Gaithersburg MD 20899

BAUM, J(AMES) CLAYTON, b Washington, DC, May 11, 46; m 77; c 2. MOLECULAR SPECTROSCOPY, H-BONDING. *Educ:* Williams Col, BA, 68; Princeton Univ, PhD(chem), 76. *Prof Exp:* Res assoc chem, Fla State Univ, 74-77; vis asst prof, Bowdoin Col, 77-79; asst prof, 79-82, ASSOC PROF CHEM, FLA INST TECHNOL, 82- *Concurrent Pos:* Lectr, Bowdoin Col, 80-82, vis assoc prof, 89-90. *Mem:* Am Chem Soc. *Res:* Molecular spectroscopy, including the study of photophysical and photochemical problems and energy transfer and relaxation processes; physical chemistry of molecular and aggregate phenomena that serve as prototypes for complex events in molecular biology. *Mailing Add:* Dept Chem Fla Inst Technol Melbourne FL 32901-6988

BAUM, JOHN, b New York, NY, June 2, 27; m 50; c 5. RHEUMATOLOGY, IMMUNOLOGY. *Educ:* NY Univ, BA, 49, MD, 54. *Prof Exp:* Instr med, Southwestern Med Sch, Univ Tex, 59-62, asst prof, 62-68; assoc prof med, 68-72, assoc prof prev med & community health, 69, assoc prof pediat, 70-72, PROF MED, PEDIAT, PREV MED & COMMUNITY HEALTH, SCH MED & DENT, UNIV ROCHESTER, 72-; DIR, ARTHRITIS & CLIN IMMUNOL UNIT & ASSOC MED DIR, MONROE COMMUNITY HOSP, ROCHESTER, 68- *Concurrent Pos:* Clin scholar, Arthritis Found, 64-69; dir arthritis clin, Parkland Mem Hosp, Dallas, Tex, 59-68, dir med clin, 65-67; consult rheumatol & co-dir arthritis clin, Scottish Rite Hosp Crippled Children, Dallas, 60-68; mem drug efficacy panel, Nat Acad Sci, 60-65 & DMSO panel, 72-74; sr assoc physician, Strong Mem Hosp, Rochester, NY, 68-72, dir pediat arthritis clin, 70-, physician & pediatrician, 72-; mem test comt rheumatol, Am Bd Internal Med, 71-76, consult, 78-79; mem merit rev bd immunol, Vet Admin, 70-76; coordr prob area, Eval Therapeut in Arthritis, US-USSR Coop Pub Health & Med Sci, 74-; mem adv panel clin immunol, US Pharmacopoeia, 75- *Mem:* Am Rheumatism Asn; Am Fedn Clin Res; Am Soc Human Genetics; Am Asn Immunologists; Heberden Soc; hon mem Polish Rheumatism Soc. *Res:* Clinical studies of delayed hypersensitivity and mechanisms of inflammation in humans; chemotaxis of polymorphonuclear leukocytes in human disease; immunology of rheumatoid arthritis and systemic lupus erythematosus; drug studies in juvenile arthritis; soft-tissue rheumatism. *Mailing Add:* Monroe Community Hosp 435 E Henrietta Rd Rochester NY 14620-4685

BAUM, JOHN DANIEL, mathematics; deceased, see previous edition for last biography

BAUM, JOHN W, b Clarion, Iowa, Mar 22, 31; m 55; c 6. HEALTH PHYSICS, BIOELECTROMAGNETICS. *Educ:* Univ Iowa, BS, 53; Univ Rochester, MS, 54; Univ Mich, PhD(environ health), 64. *Prof Exp:* Engr, Gen Elec Co, Richland, Wash, 54-56; chief health physics, Armour Res Found, Chicago, 56-58; supvr, Allis Chalmers Mfg Co, Milwaukee, 58-61; lectr radio health, Univ Mich, Ann Arbor, 61-65; assoc health physicist, 65-68, scientist, 68-75, SR SCIENTIST, BROOKHAVEN NAT LAB, 75- *Mem:* Health Physics Soc; Radiation Res Soc; Am Nuclear Soc. *Res:* Radiation protection; dosimetry; radiobiology; mutagenesis; carcinogenesis; bioelectromagnetics. *Mailing Add:* c/o Brookhaven Nat Lab Bldg 703M Upton NY 11973

BAUM, JOHN WILLIAM, b Highland Park, Ill, Mar 22, 40; m 62; c 2. ORGANIC CHEMISTRY. *Educ:* Univ Minn, BA & BChem, 62; Wash Univ, PhD(org chem), 67. *Prof Exp:* Res chemist, MacMillan Bloedel Res Ltd, BC, 67-70; dir chem develop, 71-81, sr res chemist, Zoecon Corp, 81-88, PRIN SCIENTIST, SANDOZ CROP PROTECTION CORP, 88- *Mem:* AAAS; Am Chem Soc. *Res:* Design and synthesis of biorational insecticides and herbicides; plant and insect biochemistry; synthesis of indoles and alicyclic compounds; photochemistry of alicyclic ketones. *Mailing Add:* Sandoz Crop Protection Res Div 975 California Ave Palo Alto CA 94304-1104

BAUM, JOSEPH HERMAN, b Chicago, Ill, Sept 9, 27; wid; c 2. PATHOLOGY. *Educ:* Roosevelt Univ, BS, 53; Northwestern Univ, PhD(path), 62. *Prof Exp:* Asst oncol, Chicago Med Sch, 53-54; fel path, Sch Med, Northwestern Univ, 62-63; instr, 63-66, asst prof path & dir student labs, 66-68; assoc prof, 68-81, asst dean, Grad Sch, Health Sci Ctr, 73-79, PROF PATH, SCH MED, TEMPLE UNIV, 81- *Mem:* Int Acad Path; Am Soc Cell Biologists; AAAS; Am Educ Res Asn; Am Asn Path. *Res:* Cellular and molecular pathology; tumor biology and cardiac pathology; biomedical education. *Mailing Add:* Temple Univ Sch Med 3420 N Broad St Philadelphia PA 19140

BAUM, JULES LEONARD, b New York, NY, Mar 13, 31; m; c 2. OPHTHALMOLOGY. *Educ:* Dartmouth Col, AB, 52; Tufts Univ, MD, 56; NY Univ, MS, 62; Am Bd Ophthal, dipl, 65. *Prof Exp:* Residency ophthal, Bellevue Hosp, NY, 62-64; NIH fel, Corneal Unit, Retina Found & Mass Eye & Ear Infirmary, Boston, 64-65; from asst prof to assoc prof, 65-74, PROF OPHTHAL, SCH MED, TUFTS UNIV, 74- *Concurrent Pos:* Consult, Vet Admin Hosp, Jamaica Plain, 68-, Educ Mat Proj Appraisal Panel, Asn Am Med Col, 79- & Adv Panel Ophthal, US Pharmacopeial Conv, 80-; ed, Invest Ophthal & Visual Sci, 78-82, Am J Ophthal, 85-, Ophthal Surg, 85- & Cornea, 89- *Honors & Awards:* William Warner Hoppin Award, NY Acad Med, 59; Sr Hon Award, Am Acad Ophthal, 90. *Mem:* AAAS; Asn Res Vision & Ophthalmol; fel Am Col Surgeons; Am Acad Ophthal; Sigma Xi; Am Ophthal Soc. *Res:* Tissue culture and biochemistry of cornea; ocular pharmacology of antibiotics; physiology of tearing; treatment of infectious diseases of the eye; corneal diseases. *Mailing Add:* Tufts-New Eng Med Ctr 750 Washington St Boston MA 02111

BAUM, LAWRENCE STEPHEN, b Scranton, Pa, Mar 3, 38; m 57; c 3. ONCOLOGY, CELL PHYSIOLOGY. *Educ:* Univ Ala, BS, 60, MS, 62, PhD(cell physiol), 65. *Prof Exp:* Asst prof biol, Exten Ctr, Univ Ala, 65-66; asst prof, 66-69, ASSOC PROF BIOL, NORTHEAST LA UNIV, 69-, DIR, CANCER RES CTR, 81- *Mem:* Am Physiol Soc; Sigma Xi. *Res:* Angiogenesis during tumor development; scanning electron microscope study of surface morphology of neoplastic urothelium; mode of action of various chemical carcinogens; incidence of cancer in Louisiana. *Mailing Add:* Dept Biol Northeast La Univ Monroe LA 71209

BAUM, LINDA LOUISE, b 1945. ACUTE PHASE REACTANTS, NATURAL KILLER CELLS. *Educ:* Mich State Univ, PhD(microbiol), 76. *Prof Exp:* ASSOC PROF IMMUNOL, CHICAGO MED SCH, 85- *Mem:* Am Soc Med; Am Asn Immunol; Reticuloendretielial Soc. *Res:* Antibody dependent cell-mediated cytotoxicity. *Mailing Add:* Dept Microbiol Chicago Med Sch 3333 Green Bay Rd North Chicago IL 60064

BAUM, MARTIN DAVID, b New York, NY, Jan 30, 41; m 64; c 2. PHOTOGRAPHIC CHEMISTRY. *Educ:* Mass Col Pharm, BS, 62, MS, 64; Univ Ill Med Ctr, PhD(org chem), 68. *Prof Exp:* Res chemist photochem, E I du Pont de Nemours & Co, Inc, 68-71, mkt rep, 72-73, sr res chem, 74, res supvr, 74-80, res mgr, photog syst, MKT TECH MGR, DIAGNOSTIC IMAGING, DUPONT, 87- *Mem:* Am Chem Soc. *Res:* Synthesis of dyes for use in photographic systems; development of products based on silver and non-silver photography. *Mailing Add:* 736 Tauton Rd Wilmington DE 19803

BAUM, O EUGENE, psychiatry, psychoanalysis; deceased, see previous edition for last biography

BAUM, PARKER BRYANT, b Memphis, Tenn, Dec 21, 23; m 54; c 2. PHYSICAL CHEMISTRY, INORGANIC CHEMISTRY. *Educ:* Col of William & Mary, BS, 47; Univ Tenn, MS, 52; Univ NC, PhD(phys chem), 62. *Prof Exp:* Instr chem, Old Dominion Col, 47-50, from asst prof to prof, 52-65; from assoc prof to prof chem, Skidmore Col, 67-87; RETIRED. *Mem:* AAAS; Am Chem Soc. *Res:* Thermodynamics; electromotive force measurements; applications of computers to science education. *Mailing Add:* Skimore Col Saratoga Springs NY 12866

BAUM, PAUL FRANK, b New York, NY, July 20, 36; m 61; c 2. MATHEMATICS. *Educ:* Harvard Univ, AB, 58; Princeton Univ, PhD(math), 63. *Prof Exp:* Instr math, Princeton Univ, 62-63; NSF fel, Oxford & Cambridge Univs, 63-64; fel, Inst Advan Study, 64-65; asst prof, Princeton Univ, 65-67; assoc prof, 67-72, PROF MATH, BROWN UNIV, 72- *Mem:* Am Math Soc. *Res:* Algebraic topology; lie groups. *Mailing Add:* 45 Boylston Ave Providence RI 02906

BAUM, PAUL M, b Brooklyn, NY, Feb 25, 35; m 62; c 2. EXPERIMENTAL PHYSICS. *Educ:* Columbia Univ, AB, 55; Univ Ill, MS, 57, PhD(physics), 62. *Prof Exp:* Res scientist, Grumman Aircraft Eng Corp, 62-63; ASSOC PROF PHYSICS, QUEENS COL, NY, 63- *Mem:* Am Asn Physics Teachers. *Res:* Study of properties of mesons using 300 mev betatron; energy levels of multiply ionized atoms. *Mailing Add:* Dept Physics City Univ New York Queens Col Flushing NY 11367

BAUM, PETER JOSEPH, b Lennox, Calif, June 4, 43; wid; c 1. PHYSICS. *Educ:* Univ Calif, Santa Barbara, BA, 65; Univ Nev, Reno, MS, 67; Univ Calif, Riverside, PhD(physics), 71. *Prof Exp:* Res engr microelectronics, Autonetics Div, NAm Rockwell Corp, 67-68; physicist comput anal, Corona Lab, Naval Weapons Ctr, 68-70; res assoc solar-plasma physics, Univ Calif, Riverside, 70-74, Air Force Off Sci Res grant, 72-74, res physicist solar-plasma physics,

74-85; MEM TECH STAFF, GEN RES CORP, SANTA BARBARA, CALIF, 85- *Concurrent Pos:* US deleg, Int Atomic Energy Agency Fourth Conf Plasma Physics & Controlled Nuclear Fusion Res, USAEC, 71; consult, Sandia Labs, Albuquerque, NMex, 73-75; NSF grant, instr astron, Riverside City Col, 75-80; mem tech staff, Energy Sci Prog, Univ Calif, 80-; Calif Space Inst grant, 81. *Mem:* Am Geophys Union; Am Phys Soc; Am Astron Soc. *Res:* Solar flares; laboratory study of magnetic field line reconnection; geomagnetic substorms; magnetic energy conversion; plasma radiation sources; controlled thermonuclear fusion; five-meter terrella experiment; wind energy potential in the San Gorgonio Pass region. *Mailing Add:* 1180 Garden Lane Santa Barbara CA 93108

BAUM, RICHARD T, b New York, NY, Oct 3, 19. MECHANICAL ENGINEERING. *Educ:* Columbia Univ, BA, 40, BS, 41, MS, 48. *Prof Exp:* Engr, Elec Boat Co, Groton, Conn, 41-43; assoc, 46-58, partner, 58-, PARTNER EMER & CONSULT TO FIRM, JAROS, BAUM & BOLLES, NYC, 86- *Honors & Awards:* Egleston Medalist Columbia Univ, 85. *Mem:* Nat Acad Eng; fel Am Soc Mech Engrs; fel Am Soc Heating, Refrigerating & Air Conditioning Engrs; fel Am Consult Engrs Coun; Nat Soc Prof Engrs; Nat Soc Energy Engrs. *Mailing Add:* Nine Ivy Hill Rd Chappaqua NY 10514

BAUM, ROBERT HAROLD, b New York, NY, June 15, 36; m 57; c 3. MICROBIOLOGY, BIOCHEMISTRY. *Educ:* Cornell Univ, BS, 57; Univ Ill, PhD(biochem), 62. *Prof Exp:* Res asst, Univ Ill, 61-62; Am Cancer Soc postdoctoral fel, Oak Ridge Nat Lab, 62-64; asst prof biochem, State Univ NY Col Forestry, Syracuse Univ, 64-71; asst prof microbiol, Sch Pharm, 71-74, assoc prof, Sch Dent, 74-87, ASSOC PROF MICROBIOL & IMMUNOL, SCH MED, TEMPLE UNIV, 87- *Mem:* AAAS; Sigma Xi; Am Soc Microbiol; Am Chem Soc; Am-Int Asn Dent Res. *Res:* Properties related to the adherence of cariogenic streptococci; analysis of nonspecific microbial inhibitors in saliva; metabolism of terpenoids by microorganisms; isolation and study of quinones of lactic acid bacteria; role of anaerobic microorganisms in endodontic infections. *Mailing Add:* Dept Microbiol Sch Med Temple Univ Philadelphia PA 19140-5101

BAUM, SANFORD, b San Francisco, Calif, Oct 22, 24; m 55; c 3. ENGINEERING ECONOMICS. *Educ:* Univ Calif, Berkeley, BS, 51; Stanford Univ, MS, 67, PhD(indust eng), 76. *Prof Exp:* Sr investr, US Naval Radiol Defense Lab, 51-60; design engr, Martin-Marietta Aerospace, 60-64; sr analyst, Stanford Res Inst, 64-75; prof indust eng, Univ Utah, 75-; RETIRED. *Mem:* Inst Indust Engrs; Inst Mgt Sci. *Res:* Strategic weapon analysis; civil defense; fall out; advanced space systems; energy conservation; multi-criterion optimization random search approaches to operations reseach; engineering economics. *Mailing Add:* Dept Mech & Indust Eng Univ Utah Salt Lake City UT 84112

BAUM, SIEGMUND JACOB, b Vienna, Austria, Nov 14, 20; nat US; m 47; c 5. PHYSIOLOGY, RADIOBIOLOGY. *Educ:* Univ Calif, Los Angeles, BA, 49, MA, 50; Univ Calif, Berkeley, PhD(physiol), 59. *Prof Exp:* Physiologist, US Naval Radiol Defense Lab, 50-60; group leader physiol & radiobiol, Douglas Missile & Space Systs, 60-62; head cellular radiobiol div, Armed Forces Radiobiol Res Inst, 62-64, chmn exp path dept, 64-76, chmn exp hemat dept, 76-82; RETIRED. *Concurrent Pos:* Adj prof physiol, Uniformed Serv Univ, Sch Med, 78-88, consult radiobiol, 82-; ed, Exp Hemat Today, 77-88. *Honors & Awards:* Sci Award, US Naval Radiol Defense Lab, 60. *Mem:* Am Physiol Soc; Radiation Res Soc; Int Soc Exp Hemat (treas, 81-89); Transplantation Soc. *Res:* Endocrinology; biological effects of radiation; erythrocyte precursor system; post-irradiation bone marrow therapy; experimental hematology; space radiobiology. *Mailing Add:* 6600 Greyswood Rd Bethesda MD 20817-1537

BAUM, STANLEY, b New York, NY, Dec 26, 29; m 58; c 3. RADIOLOGY. *Educ:* NY Univ, BA, 51; State Univ Utrecht, MD, 57; Univ Pa, cert med, 61. *Prof Exp:* Intern, Kings County Hosp Med Ctr, 57-58; Nat Cancer Inst resident radiol, Grad Hosp, Univ Pa, 58-61; fel cardiovasc radiol, Sch Med, Stanford Univ, 61, instr radiol, 62; from instr to prof, Univ Pa, 62-71; prof radiol, Harvard Med Sch, 71-75; PROF RADIOL & CHMN DEPT, SCH MED, UNIV PA, 75-, CHIEF RADIOL, HOSP UNIV PA, 75- *Concurrent Pos:* Asst radiologist, Grad Hosp, Univ Pa, 62-; consult, Vet Admin Hosp, Wilmington, Del, 63-66 & Vet Admin Hosp, Philadelphia, 65-; asst radiologist, Presby-Univ Pa Med Ctr, 66-71; radiologist, Mass Gen Hosp, 71-75. *Mem:* Asn Univ Radiol; Am Col Radiol; Radiol Soc NAm; Am Col Cardiol; Am Gastroenterol Asn. *Res:* Cardiovascular radiology; selective arteriography; pharmacologic control of portal hypertension. *Mailing Add:* Univ Hosp 3400 Spruce St Philadelphia PA 19104

BAUM, STEPHEN GRAHAM, b New York, NY, Apr 28, 37; wid; c 1. INTERNAL MEDICINE, INFECTIOUS DISEASES. *Educ:* Cornell Univ, AB, 58; NY Univ, MD, 62. *Prof Exp:* From intern to resident med, Harvard Div, Boston City Hosp, 62-68; assoc, 68-69, asst prof med, 69-73, assoc prof med & cell biol, 73-78, PROF MED, CELL BIOL, MICROBIOL & IMMUNOL, ALBERT EINSTEIN COL MED, 78-, DIR MD-PHD PROG, 73-, CO-DIR INFECTIOUS DIS DIV, 75-, SCI DIR ANALYSIS ULTRASTRUCTURE CTR & DIR OFF GRAD EDUC, 84- *Concurrent Pos:* Nat Inst Allergy & Infectious Dis res assoc virol, 64-66; Med Res Coun spec fel, Nat Inst Med Res, London, 66-67; NY Health Res Coun career scientist, 68-; Am Cancer Soc fac res assoc. *Mem:* Infectious Dis Soc Am; Am Soc Clin Invest; Harvey Soc; Am Soc Microbiol. *Res:* Viral oncology; electron microscopy. *Mailing Add:* Mt Sinai Sch Med Beth Israel Med Ctr First Ave 16th St New York NY 10003

BAUM, STUART J, b Brooklyn, NY, Mar 7, 39; m 64; c 2. ORGANIC CHEMISTRY, BIOORGANIC CHEMISTRY. *Educ:* Queens Col, BS, 60; Cornell Univ, MS, 63, PhD(molecular biol), 65. *Prof Exp:* Asst prof, 65-71, assoc prof, 71-76, PROF CHEM, STATE UNIV NY COL PLATTSBURGH, 65- *Mem:* Am Chem Soc. *Res:* Organic reaction mechanisms; organometallic chemistry; mechanisms of biochemical reactions; structure of coordination complexes; chemical education. *Mailing Add:* Dept Chem State Univ NY COL Plattsburgh NY 12901

BAUM, WERNER A, b Giessen, Ger, Apr 10, 23; nat US; m 45; c 2. METEOROLOGY, CLIMATOLOGY. *Educ:* Univ Chicago, BS, 43, MS, 44, PhD(meteorol), 48. *Hon Degrees:* ScD, Mt St Joseph Col, RI, 71, Univ RI, 74; DPA, Husson Col, 72. *Prof Exp:* From assoc prof to prof meteorol, Fla State Univ, 49-58, head dept, 49-58, dir univ res, 57-58, dean grad sch & dir res, 58-60, dean fac, 60-63, vpres acad affairs, 63; prof meteorol & vpres acad affairs, Univ Miami, 63-65; prof meteorol & vpres sci affairs, NY Univ, 65-67; dep adminr, Environ Sci Serv Admin, 67-68; prof physics & geog & pres, Univ RI, 68-73; prof geog & univ chancellor, Univ Wis-Milwaukee, 73-79; prof & dean, 79-90, EMER PROF METEOROL & DEAN, COL ARTS & SCI, FLA STATE UNIV, 90- *Concurrent Pos:* Ed, J Meteorol, Am Meteorol Soc, 49-61; mem, comt climat adv to US Weather Bur, Nat Acad Sci, 55-58, chmn, Panel on Educ, Comt Atmospheric Sci, 62-64; mem coun, Oak Ridge Inst Nuclear Studies, 58-62; mem exec reserve, US Weather Bur, 58-63, chmn adv comt educ & training, 64-66; trustee, Univ Corp Atmospheric Res, 59-63, 65-67 & 80-83, corp secy, 63-67; mem, Adv Panel on Atmospheric Sci Prog, NSF, 63-67, chmn, 65-67; dir, Fund for Overseas Res Grants & Educ, 65-74; mem sci adv coun, Tex Christian Univ Res Found, 67-73, 74-77; US rep, Panel of Experts on Meteorol Educ & Training, World Meteorol Orgn, UN, 71-80; mem, Nat Sea Grant Prog Adv Panel, 74-78; life trustee, Univ RI Found, 75. *Honors & Awards:* Spec Citation, Am Meteorol Soc, 62, Charles Franklin Brooks Award, 75, Cleveland Abbe Award, 88; Sci Freedom & Responsibility Award, AAAS, 85. *Mem:* Fel AAAS; fel Am Meteorol Soc (pres, 77-78); fel Am Geog Soc; fel Am Geophys Union. *Res:* Evaluation and analysis of foreign meteorological material; academic and scientific administration; climatology. *Mailing Add:* 2403 Perez Ave Tallahassee FL 32304

BAUM, WILLIAM ALVIN, b Toledo, Ohio, Jan 18, 24; m 61. GALAXIES, PLANETARY SCIENCE. *Educ:* Univ Rochester, BA, 43; Calif Inst Technol, MS, 45, PhD(physics), 50. *Prof Exp:* Physicist, Naval Res Lab, Washington, DC, 46-49; astronr, Mt Wilson & Palomar Observ, 50-65; dir, Planetary Res Ctr, Lowell Observ, 65-90; PROF, UNIV WASH, 90- *Concurrent Pos:* Guggenheim fel, 60-61; adj prof, Ohio State Univ, 69- & Northern Ariz State Univ, 73-; consult, Off Space Sci, NASA, 70-, mem bd dirs, Assoc Univ Res Astron, 76-79, Viking Orbiter Imaging Team, 70-79 & Hubble Space Telescope Camera Team, 77- *Honors & Awards:* Shapley lectr, Am Astron Soc, 61- *Mem:* Am Astron Soc; Int Astron Union; Royal Astron Soc Brit; Astron Soc Pac. *Res:* Photoelectric photometry; photoelectric image-receiving systems; optical instrument development; globular star clusters; stellar populations; magnitudes and redshifts of galaxies; planetary science; spacecraft instrumentation; observational cosmology; constancy of physical constants. *Mailing Add:* Univ Wash 2124 NE Park Rd Seattle WA 98105

BAUMAL, REUBEN, b Toronto, Ont, Jan 2, 39. RENAL PATHOLOGY. *Educ:* Univ Toronto, MD, 63, PhD(path), 69; FRCPath(C), 77. *Prof Exp:* Med res fel path, Med Res Coun Can, 64-71, med res scholar, 71-76; PROF, DEPT PATH, HOSP SICK CHILDREN, UNIV TORONTO, 76- *Mem:* Am Asn Pathologists; Am Soc Nephrology; Am Asn Cancer Res. *Mailing Add:* Dept Path Hosp Sick Children Univ Toronto 555 University Ave Toronto ON M5G 1X8 Can

BAUMAN, BERNARD D, b Rochester, NY, June 15, 46; c 2. ORGANIC CHEMISTRY. *Educ:* Eastern Nazarene Col, BA, 68; State Univ NY Albany, PhD(org chem), 73. *Prof Exp:* Scholar org chem, Pa State Univ, 73-74; sr chemist org chem, Rohm and Haas Co, 74-76; MGR, COMPOSITE PROD, AIR PROD & CHEM, INC, 76- *Mem:* Am Chem Soc; Soc Plastic Eng. *Res:* Surface modification of plastics and elastomers; development of novel composite systems. *Mailing Add:* RR 2-127-15 Acorn Dr Emmaus PA 18049-9802

BAUMAN, DALE E, b Dec 26, 42; m; c 3. NUTRITIONAL BIOCHEMISTRY. *Educ:* Univ Ill, PhD(nutrit biochem), 69. *Prof Exp:* PROF NUTRIT BIOCHEM, CORNELL UNIV, 79- *Concurrent Pos:* Mem, US Bd Agr & US Comt Biotechnol, 89- *Honors & Awards:* Nat Acad Sci, 88; Young Scientist Res Award, Am Dairy Sci Asn, 77; Nutrit Res Award, Am Soc Animal Sci, 82; Alexander von Humboldt Award, 85; USDA Super Serv Award, 86; Am Cyanamid Award, Am Dairy Sci Asn, 87; Liberty Hyde Bailey Professorship Cornell, 87. *Mem:* Nat Acad Sci; Am Inst Nutrit; Am Soc Animal Sci; Am Dairy Sci Asn; Am Asn Advan Sci. *Res:* Animal nutrition; physiology. *Mailing Add:* Cornell Univ 262 Morrison Ithaca NY 14853

BAUMAN, HOWARD EUGENE, b Woodworth, Wis, Mar 20, 25; m 48; c 2. FOOD SCIENCE. *Educ:* Univ Wis, BS, 49, MS, 51, PhD(bact), 53. *Prof Exp:* Head bact sect res & develop, Pillsbury Mills Inc, 53-55, head biol res br, Cent res, Pillsbury Co, 55-60, assoc dir res, 60-67, dir corp res, 67-69, vpres sci & technol, Minneapolis, 69-90; RETIRED. *Mem:* AAAS; Am Soc Microbiol; Soc Indust Microbiol; Am Asn Cereal Chemists; Inst Food Technol. *Res:* Nutrition; microbiology. *Mailing Add:* 4580 Greenwood Dr Minnetonka MN 55343

BAUMAN, JOHN E, JR, b Kalamazoo, Mich, Jan 18, 33; m 64; c 3. INORGANIC CHEMISTRY. *Educ:* Univ Mich, BS, 55, MS, 60, PhD(chem), 61. *Prof Exp:* Res chemist, Midwest Res Inst, 55-58; from asst prof to assoc prof, 61-75, PROF CHEM, UNIV MO-COLUMBIA, 75- *Concurrent Pos:* Mem staff, Scripps Inst Oceanog, Univ Calif, 70. *Mem:* Am Chem Soc; Sigma Xi; Calorimetry Conf; Nat Asn Corrosion Engrs. *Res:* Thermodynamics of inorganic coordination compounds; ionic reactions in solution; calorimetry; corrosion. *Mailing Add:* Dept Chem Univ Mo-Columbia Columbia MO 65211

BAUMAN, JOHN W, JR, b Stockton, Calif, Dec 17, 18; m 48; c 4. PHYSIOLOGY. *Educ:* Univ Southern Calif, AB, 48; Univ Calif, Berkeley, PhD(physiol), 55. *Prof Exp:* Instr physiol, Sch Med, NY Univ, 56-58, asst prof, 58-61; res scientist exp med, Bur Res, Princeton, NJ, 61-62, chief group med sect, 62-70; ASSOC PROF PHYSIOL, NJ COL MED & DENT, 70- *Concurrent Pos:* Fel biol, Princeton Univ, 68-70; adj asst prof, Med Sch, NY

Univ, 61-62; vis prof physiol, NJ Col Med & Dent, 68, 69. *Mem:* Am Physiol Soc; Am Soc Nephrol; Endocrine Soc; Brit Soc Endocrinol. *Res:* Endocrine effects on renal function; mechanisms of antidiuresis and water diuresis; proteinuria; sodium and hypertension. *Mailing Add:* 631 Mt Lucas Rd Princeton NJ 08540

BAUMAN, LOYAL FREDERICK, plant breeding; deceased, see previous edition for last biography

BAUMAN, NORMAN, b Brooklyn, NY, Aug 20, 32; m 57; c 3. IMMUNOLOGY, MEDICINE. *Educ:* Harvard Univ, AB, 53; NY Univ, MD, 57. *Prof Exp:* Intern, Bronx Munic Hosp, 57-58; fel rheumatology, Duke Univ, 60-61, asst res med, 61-62, fel rheumatology, 62-64, instr med, 62-64; SR SCIENTIST, LEDERLE LABS, AM CYANAMID CO, 64- *Concurrent Pos:* Res assoc, Nat Inst Arthritis & Metab Dis, 58-60. *Mem:* Math Asn Am; Am Rheumatism Asn. *Res:* Mechanism of drug action; complement; mediators of immune reactions; rheumatology. *Mailing Add:* Lederle Labs Am Cyanamid Co Pearl River NY 10965

BAUMAN, RICHARD GILBERT, b Warren, Ohio, Sept 14, 24; m 45; c 3. PHYSICAL CHEMISTRY. *Educ:* Case Inst Technol, BS, 44, MS, 48, PhD(phys chem), 51. *Prof Exp:* Instr phys chem, Fenn Col, 46-47 & Case Inst Technol, 47-50; res chemist, B F Goodrich Co, 50-58, mgr tire res, 58-64, dir prod res, 64-72, dir tire res, 73-88; CONSULT, R&D AUDITS, 89- *Concurrent Pos:* Chemist, Brookhaven Nat Lab, 53-54. *Mem:* Am Chem Soc. *Res:* Phase equilibria; polymerization reactions; physical properties of high polymers; polyelectrolytes; oxidation of high polymers; nuclear engineering; processing of reactor fuels; research administration; tire materials and construction. *Mailing Add:* 19 Glen Oaks Lane Berea OH 44017

BAUMAN, ROBERT ANDREW, b Rochester, NY, Apr 17, 23. ORGANIC CHEMISTRY. *Educ:* Univ Rochester, BS, 43; Univ Ill, PhD(chem), 46. *Prof Exp:* Spec res asst, Univ Ill, 43-46; res chemist, Gen Aniline & Film Corp, 46-48; asst, Univ Ill, 48-50; res chemist, Gen Aniline & Film Corp, 50-52; res chemist, Colgate-Palmolive-Peet Co, 52-53 & Colgate Palmolive Co, 53-63, res assoc, Colgate-Palmolive Co, 63-68, sr res assoc, 68-78, sr scientist, 78-83; RETIRED. *Mem:* Am Chem Soc. *Res:* Organic synthesis; surfactants; oral chemotherapy; fabric beneficiation. *Mailing Add:* Ten Landing Lane Apt 8C New Brunswick NJ 08901

BAUMAN, ROBERT POE, b Jackson, Mich, May 8, 28; m 49; c 4. SCIENCE EDUCATION, PHYSICAL CHEMISTRY. *Educ:* Purdue Univ, BS, 49, MS, 51; Univ Pittsburgh, PhD(physics), 54. *Prof Exp:* Asst, Purdue Univ, 49-51 & Mellon Inst Indust Res, 51-54; from instr to assoc prof physics, Polytech Inst Brooklyn, 54-67; chmn dept, 67-73, dir, Proj on Teaching & Learning in Univ Col, 75-78, PROF PHYSICS, UNIV ALA, BIRMINGHAM, 67- *Concurrent Pos:* Vis fel, Joint Inst Lab Astrophysics, 80-81. *Mem:* Optical Soc Am; Am Phys Soc; Am Asn Physics Teachers (vpres, 81, pres-elect, 82); Coblentz Soc (pres, 72-74). *Res:* Vibrational spectra and molecular structure; teaching-learning theory. *Mailing Add:* Dept Physics Univ Ala Birmingham University Sta Birmingham AL 35294

BAUMAN, THOMAS TROST, b Lafayette, Ind, Sept 11, 39; m 65; c 3. AGRONOMY. *Educ:* Purdue Univ, BS, 63, MS, 65, PhD(weed sci), 74. *Prof Exp:* PROF AGR EXTEN RES & TEACHING, PURDUE UNIV, 74- *Concurrent Pos:* Mem Task Force, Coun Agr Sci & Technol, 81-; guest lectr, Ill Spray Sch & Mich Pesticide Training Sch, 81. *Mem:* Weed Sci Soc Am; Am Agron Soc; Soil Sci Soc Am. *Res:* Soybean weed control; tillage systems on soybean weed control; weed competition. *Mailing Add:* Bot & Plant Path Dept 234 Connolly St West Lafayette IN 47906

BAUMANN, ARTHUR NICHOLAS, b Bogota, NJ, Dec 15, 22; m 47; c 2. INORGANIC CHEMISTRY, ANALYTICAL CHEMISTRY. *Educ:* Fla Southern Col, BS, 51. *Prof Exp:* Lab technician, phosphate rock div, Int Minerals & Chem Corp, Fla, 49-50; jr process engr, Davison Chem Co, 52; res chemist, Fla Exp Sta, 52-59, sr res chemist, tech div, 59-61, supvr tech serv, 61-64, chief chemist, 64-69, sr process eng, 69-74, mgr develop, Agr Prod Div, 74-80, mgr develop eng, New Wales Chem Opers, Int Minerals & Chem Corp, 80-86, PROCESS CONSULT, 86- *Concurrent Pos:* Counr, Am Chem Soc, 83; mem, Resource Mgt Comt, Fertilizer Inst; chmn, Tech Adv Comt, Fla Inst Phosphate Res, 81- *Mem:* Am Chem Soc; Cent Fla Sect of AIChE. *Res:* Production of alkali polyphosphates; defluorinated phosphate rock; anhydrous hydrogen fluoride; fluoride chemicals; potassium compounds; automation of wet chemical analytical methods for process control; solvent extraction; uranium recovery. *Mailing Add:* 2329 Rogers Rd Lakeland FL 33813-3139

BAUMANN, CARL AUGUST, b Milwaukee, Wis, Aug 10, 06. NUTRITIONAL BIOCHEMISTRY. *Educ:* Univ Wis, BS, 29, MS, 31, PhD(biochem), 33. *Prof Exp:* Asst biochem, Univ Wis-Madison, 29-34; Gen Educ Bd fel, Univ Heidelberg, Copenhagen Univ & Cambridge Univ, 34-36; res fel biochem, 36-38, from instr to prof, 38-76, EMER PROF BIOCHEM, UNIV WIS-MADISON, 76- *Concurrent Pos:* Consult, Off Sci Res & Develop, DC, 45-46; mem study sect nutrit, Am Cancer Soc; mem study sect nutrit & metab & nutrit, NIH, 48-60; mem nutrit surv, Interdept Comt Nutrit Nat Defense, NIH, Chile, 60; chief-of-party, Ford Found Basic Sci Proj, Agrarian Univ, Peru, 66-69; AID-Midwestern Univs Consortium Int Activities consult, Agr Univ, Indonesia, 74 & 75. *Mem:* AAAS; Am Chem Soc; Am Soc Biol Chem; fel Am Inst Nutrit. *Res:* Vitamins; nutrition; tumor development; skin and intestinal sterols; selenium metabolism. *Mailing Add:* 1840 N Prospect Ave No 821 Milwaukee WI 53202-1962

BAUMANN, DWIGHT MAYLON BILLY, b Ashley, NDak, May 12, 33; m 54; c 3. SYSTEMS DESIGN, SYSTEMS SCIENCE. *Educ:* NDak State Univ, BSME, 55; Mass Inst Technol, MSME, 57, ScD, 60. *Prof Exp:* Teaching asst physics, NDak State Univ, 54-55; instr eng design, Mass Inst Technol, 56-60, asst prof to assoc prof mech eng, 60-70; prof, 70-76, PROF

ENG DESIGN, CARNEGIE MELLON UNIV, 76- *Concurrent Pos:* Pres, Icon Corp, 64-69; dir, Graphic Sci Corp, 67-75; chmn, Eng Design Prog, 70-; pres & dir, Ctr Entrepreneurial Develop, 71-; co-dir, Transp Res Inst, Carnegie-Mellon Univ, 76- *Honors & Awards:* Chester Carlson Award, Am Soc Eng Educ, 82. *Mem:* Am Soc Mech Engrs; Am Soc Eng Educ; Sigma Xi. *Res:* Development of clinical approaches to engineering education. *Mailing Add:* 1235 Squirrel Hill Ave Pittsburgh PA 15217

BAUMANN, E(DWARD) ROBERT, b Rochester, NY, May 12, 21; m 46; c 2. SANITARY ENGINEERING. *Educ:* Univ Mich, BSE, 44; Univ Ill, BS, 45, MS, 47, PhD(sanit eng), 54. *Prof Exp:* Instr civil eng, Univ Ill, 47, res assoc sanit eng, 47-53; from assoc prof to prof, 53-72, Anson Morston Distinguished Prof civil eng, 72-91, EMER DISTINGUISHED PROF, CIVIL ENG, IOWA STATE UNIV, 91- *Concurrent Pos:* NSF fel, King's Col, Eng, 59-60. *Honors & Awards:* Gascoigne Medal, Water Pollution Control Fedn, 62, A S Bedell Award, 76, Philip F Morgan Award, 86; Pub Award, Am Waters Works Asn, 62 & 80, Purification Div Award, 65, A P Black Res Award, 78; Gold Medal Award, Filtration Soc, 70; Nalco Award, Asn Environ Eng Prof, 80. *Mem:* Am Soc Civil Engrs; Nat Soc Prof Engrs; Am Soc Eng Educ; Am Inst Chem Engrs; Sigma Xi; Asn Environ Eng Prof; Am Water Works Asn; Water Pollution Control Fedn; Am Acad Environ Engrs. *Res:* Water treatment; sewage treatment; hydraulics. *Mailing Add:* 1627 Crestwood Circle Ames IA 50010-5520

BAUMANN, ELIZABETH WILSON, b Pueblo, Colo; m 54; c 1. ANALYTICAL CHEMISTRY. *Educ:* Brigham Young Univ, BA, 45; Univ Kans, PhD(chem), 53. *Prof Exp:* Chemist, Tex Co, 45-49; asst instr, Univ Kans, 49-51; res chemist, Rayon Res Lab, E I du Pont de Nemours & Co, 53-54; res assoc, Univ Kans, 54-55; chemist, E I du Pont de Nemours & Co, 55-72, sr res chemist, 72-73, res staff chemist, Savannah River Lab, 73-90; SR FEL SCIENTIST, SAVANNAH RIVER SITE, WESTINGHOUSE, 90- *Mem:* Am Chem Soc; Sigma Xi. *Res:* Ion exchange; water purification; ionic equilibria; ion selective electrodes; nuclear reactor aqueous chemistry. *Mailing Add:* Savannah River Site Westinghouse Aiken SC 29808

BAUMANN, FREDERICK, b Los Angeles, Calif, Nov 26, 30; m 62; c 3. ANALYTICAL CHEMISTRY. *Educ:* Univ Calif, Los Angeles, BS, 52; Univ Wis, PhD(anal chem), 56. *Prof Exp:* Res chemist, Calif Res Corp, 56-65; mgr res, Varian Inst Div, 65-69, mgr res & eng, 70-79, managing dir, Varian Techtron, 80-81, tech dir, 81-82, mgr data syst oper, Varian Inst Group, 82-89, mgr strategic & tech planning, 89-90, TECH CONSULT, VARIAN ASSOCS, 91- *Concurrent Pos:* Instr gas chromatography, Exten, Univ Calif, 64-70; adv bd anal chem, Res & Eng Mgt, 72-74. *Mem:* Am Chem Soc; Sigma Xi. *Res:* Gas chromatography; liquid chromatography; laboratory automation; mass spectrometry; electrochemistry; expertise in development of scientific instruments. *Mailing Add:* Varian Chromatography Systs 2700 Mitchell Dr Walnut Creek CA 94598

BAUMANN, GEORGE P(ERSHING), chemical engineering; deceased, see previous edition for last biography

BAUMANN, GERHARD, b Basel, Switz, Sept 15, 41; m 72; c 1. ENDOCRINOLOGY. *Educ:* Univ Basel, MD, 67. *Prof Exp:* Vis scientist endocrinol, NIH, 74-77; from asst prof to assoc prof, 77-88, PROF MED, SCH MED, NORTHWESTERN UNIV, 88- *Concurrent Pos:* Fel, Peter Bent Brigham Hosp, 69-71. *Mem:* Am Fedn Clin Res; Endocrine Soc; Cent Soc Clin Res; Am Soc Clin Invest; Asn Am Physicians; Sigma Xi. *Res:* Pituitary gland; human growth hormone with its molecular variants; somatomedin; prolactin; insulin receptors; growth hormone receptors; growth hormone binding proteins. *Mailing Add:* Ctr Endocrinol Metab & Nutrit Northwestern Univ Sch Med Chicago IL 60611

BAUMANN, HANS D, m; c 2. FLUID MECHANICS, VALVE ACOUSTICS. *Educ:* Metallf Achschule Bielefelp, WGer, BS, 53; Columbia-Pac Univ, PhD(mech eng), 82. *Prof Exp:* chief engr, Welland S Texhorn Co, 54-58; sr develop engr, Masoneilan Co, 58-63; dir, Cashco, Inc, 63-66; mgr res & develop, Worthington S/A, 66-68; vpres, Masoneilan Int Inc, 68-74; consult mech eng, 74-77; pres, H D Baumann Assoc, 77-91; CONSULT, 91- *Concurrent Pos:* Dir, H B Serv Inc, 81-, dir StP, Instrument Soc Am, 89-; tech expert, IEC/SC 65B, 86-; guest prof, Kobe Univ, Japan & Korean Inst Advan Technol, 89. *Mem:* Fel Instrument Soc Am; Am Soc Mech Engrs. *Res:* author of 68 publications on valve sizing, cavitation, aerodynamic noise and co-author of four hand books on instrumentation, valves and acoustics; author of one handbook on control valves. *Mailing Add:* 32 Pine Rye NH 03870

BAUMANN, JACOB BRUCE, b Fremont, Ohio, Sept 11, 32; m 57; c 3. ORGANIC CHEMISTRY. *Educ:* Amherst Col, BA, 54; Univ Mich, MS, 57, PhD(chem), 61. *Prof Exp:* From instr to asst prof chem, Defiance Col, 60-65; asst prof, 65-71, ASSOC PROF CHEM, PA STATE UNIV, SCHUYLKILL CAMPUS, 71- *Mem:* Am Chem Soc. *Res:* Aromatic nucleophilic substitution. *Mailing Add:* Dept Chem Schuylkill Campus Pa State Univ State Hwy Schuylkill Haven PA 17972

BAUMANN, NORMAN PAUL, b Sylvan Grove, Kans, Nov 18, 27; m 54; c 1. NUCLEAR PHYSICS, ACOUSTIC MEASUREMENTS. *Educ:* Univ Kans, BS, 51, MS, 52, PhD(physics), 55. *Prof Exp:* PHYSICIST, SAVANNAH RIVER LAB, E I DU PONT DE NEMOURS & CO, 55- *Mem:* Am Phys Soc; Am Nuclear Soc. *Res:* Reactor physics and design; acoustic analysis of reactor systems; nondestructive assay of nuclear materials; criticality monitoring; radiation damage reactor materials. *Mailing Add:* Sci Comp Sect Savannah River Lab Westinghouse Savannah River Co Aiken SC 29801

BAUMANN, PAUL C, b Aurora, Ill, Oct 6, 46; m 84; c 1. ECOTOXICOLOGY, AQUATIC ECOLOGY. *Educ:* Beloit Col, Wis, BS, 68; Univ Wis-Madison, MS, 72, PhD(zool), 75. *Prof Exp:* Group leader, Ecol Serv Tex Instruments, 76-78; FISHERY BIOLOGIST AQUATIC TOXICOL, US FISH & WILDLIFE SERV, 78- *Concurrent Pos:* Adj asst prof, Dept Zool, Ohio State Univ, 78- & Sch Natural Resources, 80- *Mem:* Soc Environ Toxicol & Chem; Am Fisheries Soc. *Res:* Neoplasia in fish and potential carcinogen causes, particularly PAH; teratogenic and other reproductive effects of both organic and inorganic contaminants on aquatic vertebrates. *Mailing Add:* Nat Fisheries Contaminant Res Ctr Columbus Field Res Sta Ohio State Univ 473 Kottman Hall 2021 Coffey Rd Columbus OH 43210

BAUMANN, RICHARD WILLIAM, b Castle Dale, Utah, July 24, 40; m 64; c 6. SYSTEMATIC ENTOMOLOGY, AQUATIC BIOLOGY. *Educ:* Univ Utah, BA, 65, MS, 67, PhD(aquatic entom), 70. *Prof Exp:* Res guest aquatic entom, Max Planck Limnol Inst, 70-71; asst prof limnol, Southwest Mo Univ, 71-72; assoc cur entom, Smithsonian Inst, 72-75; asst prof, 75-80, ASSOC PROF ENTOM, BRIGHAM YOUNG UNIV, 80- *Concurrent Pos:* Supvr res Moraca River, Yugoslavia, Off Limnol & Oceanog, Smithsonian Inst, 73-75; investr ecol, Bur Land Mgt, 74-75; adj prof, NTex State Univ, 74- *Mem:* Sigma Xi; Entom Soc Am; Am Entom Soc; NAm Benthol Soc. *Res:* Studies on the systematics, phylogeny and ecology of the stoneflies, Insecta Plecoptera, of the northern hemisphere; ecological studies of aquatic insects in western North America. *Mailing Add:* 290 MLBM 574 WIDB Brigham Young Univ Provo UT 84602

BAUMANN, ROBERT JAY, b Chicago, Ill, Oct 22, 40; m 64; c 3. CHILD NEUROLOGY, NEUROEPIDEMIOLOGY. *Educ:* Tufts Univ, BS, 61; Western Reserve Univ, MD, 65. *Prof Exp:* Resident pediat, Univ Chicago, 65-67, resident neurol, 67-69; child neurologist, US Air Force, 69-71; resident neurol, Univ Chicago, 71-72; ASSOC PROF NEUROL, UNIV KY, 72-, HEAD, CHILD NEUROL SECT, 78- *Concurrent Pos:* Neuroepidemiologist, Nat Inst Neurol Commun Disorders & Stroke, Bethesda, Md, 78-79; mem res comt neuroepidemiol, World Fedn Neurol, 78- *Mem:* Child Neurol Soc; Am Acad Neurol; Soc Epidemiol Res; Am Acad Pediat; Int Child Neurol Soc. *Res:* Epidemiology of neurological disease; neonatal stroke and cerebral palsy. *Mailing Add:* Dept Neurol Univ Ky Lexington KY 40536-0084

BAUMANN, THIEMA WOLF, b St Louis, Mo, May 30, 21; m 50; c 1. NEUROHISTOCHEMISTRY, HISTOLOGY. *Educ:* Harris Col, AB, 42; Washington Univ, AM, 44; St Louis Univ, PhD(neuroanat), 66. *Prof Exp:* Res dir genetics, St Louis Univ, 60-77; prof histol-anat, Southern Ill Sch Dent Med, 80-81; PROF LIFE SCI, LOGAN COL, 77- *Concurrent Pos:* Prin investr, Found Orthod Educ & Res, 60-77; vis prof, Southern Ill Sch Dent Med, 80-81. *Mem:* Sigma Xi; AAAS. *Res:* Localization of neurotransmitters visually demonstrated by histochemistry; documentation of genetics and patterns of human skeletal growth; supra segmental inhibition of spinal reflex mechanisms; breathing reflexes; effects of therapeutic ionizing radiation on bone; phylogeny of the central nervous system; cell biology. *Mailing Add:* 6504 Devonshire St St Louis MO 63109

BAUMANN, WOLFGANG J, b Crailsheim, Ger, May 26, 36; m 67; c 2. MEMBRANE BIOPHYSICS, LIPID BIOCHEMISTRY. *Educ:* Univ Stuttgart, BS, 57, MS, 61, PhD(org chem), 63. *Prof Exp:* Asst chem, Univ Stuttgart, 61-63; res fel, 63-66, res assoc, 66-67, from asst prof to assoc prof, 67-74, asst to dir, 79-81, PROF, HORMEL INST, UNIV MINN, 74- *Concurrent Pos:* Assoc ed, Lipids, 74-84, ed, Communications, Lipids, 75-80, co-ed, Lipids, 85 & ed, Lipids, 86-; consult, NIH, NSF & USDA. *Mem:* Am Chem Soc; AAAS; Am Oil Chemists Soc; Am Soc Biochem Molecular Biol; Int Soc Magnetic Resonance; Am Heart Asn; Biophys Soc. *Res:* Nuclear magnetic resonance spectroscopy on biological and model membranes; membrane regulation by lysophospholipids; lysophospholipid/cholesterol interactions; endocytosis; phospholipid deacylation/reacylation; mechanism of demyelination; liposome/cell interactions. *Mailing Add:* Hormel Inst Austin MN 55912-3698

BAUMBER, JOHN SCOTT, b London, Eng, June 4, 37; m 61; c 4. PHYSIOLOGY. *Educ:* Univ Nottingham, BSc, 58; Queen's Univ, Ont, MSc, 60, PhD(physiol), 63; MD, 66. *Prof Exp:* Mo Heart Asn fel, 67-69; asst prof physiol, Univ Mo-Columbia, 69-70; asst prof, 70-72, ASSOC PROF PHYSIOL, UNIV CALGARY, 72-, ASST DEAN FAC MED, 74- *Mem:* Can Med Asn; Can Physiol Soc; Am Physiol Soc. *Res:* Cardiovascular, renal and endocrine physiology with emphasis on cardiodynamics, hemodynamics and heart failure; the renin-angiotensin-aldosterone system; hypoxia. *Mailing Add:* Dept Physiol Fac Med Univ Calgary 3330 Hospital Dr NW Calgary AB T2N 4N1 Can

BAUMEISTER, CARL FREDERICK, b Dolliver, Iowa, May 15, 07; m 30; c 1. INTERNAL MEDICINE. *Educ:* Univ Chicago, BS, 30; Univ Iowa, MD, 33. *Prof Exp:* Intern & resident, Univ Hosps, Ind Univ & Univ Louisville, 33-37; physician, Council Bluffs Clin, 37-43; asst prof clin med, Col Med, Univ Ill, 43-65; clin asst prof med, Stritch Sch Med, Loyola Univ, Chicago, 70-73; asst prof clin med, Col Med, Univ Ill, 73-81; RETIRED. *Concurrent Pos:* Head dept internal med, Suburban Med Ctr, Berwyn, 43; mem, Inst Med Chicago. *Mem:* AMA; Asn Am Med Cols; Am Heart Asn; Am Diabetes Asn; Am Med Writers Asn. *Res:* Role of the computer as a diagnostic aid in the surgical abdomen; mechanisms and prevention of type III sinus block headaches; recognition and prevention of variant vascular headaches. *Mailing Add:* 120 S Delaplaine Rd Riverside IL 60546

BAUMEISTER, PHILIP WERNER, b Troy, Ohio, Mar 17, 29; m 52; c 3. PHYSICS. *Educ:* Stanford Univ, BS, 50; Univ Calif, Berkeley, PhD(physics), 59. *Prof Exp:* From asst prof to prof optics, Univ Rochester, 59-78; chief scientist, Optical Coating Lab Inc, 79-85; RES & DEVELOP ENGR, COHERENT INC, 85- *Mem:* Optical Soc Am; Am Vacuum Soc. *Res:* Optical interference coatings; optical instrumentation; optical properties of thin, solid films; optical filter design. *Mailing Add:* 1588 Lilac Lane Auburn CA 95604

BAUMEL, JULIAN JOSEPH, b Sanford, Fla, July 26, 22; m 45; c 3. ANATOMY. *Educ:* Univ Fla, BS, 47, MS, 50, PhD, 53. *Prof Exp:* From instr to assoc prof, 53-64, PROF ANAT, CREIGHTON UNIV, 64- *Concurrent Pos:* Chmn, Int Comt Avian Anat Nomenclature; pres, Nebr State Anat Bd. *Mem:* Am Asn Anatomists; Am Soc Zoologists; Am Ornith Union; World Asn Vet Anat; Asn Avian Veterinarians. *Res:* Avian and human anatomy; vasculature; arthrology; peripheral nerves. *Mailing Add:* Dept Anat Creighton Univ Sch Med Omaha NE 68178

BAUMEL, PHILIP, b New York, NY, June 12, 32; m 56; c 4. HIGH ENERGY PHYSICS. *Educ:* City Col New York, BS, 53; Columbia Univ, PhD(physics), 60. *Prof Exp:* Asst prof physics, Fairleigh Dickinson Univ, 59-61; from instr to assoc prof, 61-72, from asst dean to assoc dean, 80-86, PROF PHYSICS, CITY COL NEW YORK, 72-, CHMN, 89- *Mem:* Am Asn Physics Teachers. *Res:* Particle physics. *Mailing Add:* Dept Physics City Col New York New York NY 10031

BAUMEYER, JOEL BERNARD, b St Louis, Mo, Nov, 30, 36. MATHEMATICS. *Educ:* St Mary's Col, Minn, BA, 58, MEd, 63; St Louis Univ, MA, 77, PhD(math), 79. *Prof Exp:* Instr math & sci, St George High Sch, Evanston, Bishop Carrol High Sch, Wichita, Kans & Christian Bros High Sch, Peoria, 58-77; asst prin, Bishop Caroll High Sch, Wichita, Kans, 65; ASST PROF MATH, CHRISTIAN BROS COL, MEMPHIS, 79- *Mem:* Nat Coun Teachers Math; Math Asn Am. *Mailing Add:* Dept Math & Sci Christian Bros Col 650 E Parkway S Memphis TN 38104

BAUMGARDNER, F WESLEY, ENVIRONMENTAL PHYSIOLOGY. *Educ:* Univ Buffalo, PhD(physiol), 71. *Prof Exp:* CHIEF CHEMICAL DEFENSE BR, USAF SCH AEROSPACE MED, TEX, 71- *Res:* Personal protective equipment. *Mailing Add:* USAF Sch Aerospace Med Crew Technol Div SAM/VN Brooks AFB TX 78235-5301

BAUMGARDNER, KANDY DIANE, b Peoria, Ill, Sept 16, 46. GENETICS. *Educ:* Bradley Univ, BS, 68; Utah State Univ, PhD(zool), 74. *Prof Exp:* From asst prof to assoc prof, 73-86, PROF ZOOL, EASTERN ILL UNIV, 86- *Mem:* AAAS; Sigma Xi; Am Genetic Asn. *Res:* The population biology of the lambdoid bacteriophages, particularly competition between lambda, phi-eighty and a hybrid type. *Mailing Add:* Dept Zool Eastern Ill Univ Charlestown IL 61920

BAUMGARDNER, MARION F, b Wellington, Tex, Feb 7, 26; m 55; c 3. AGRONOMY. *Educ:* Tex Tech Col, BS, 50; Purdue Univ, MS, 55, PhD(soil fertil, plant nutrit), 64. *Hon Degrees:* DSc, DePauw Univ, 80. *Prof Exp:* Lectr agron, Agr Inst, Allahabad, India, 50-53; admin secy int student conf, Nat Coun Churches, NY, 55-56; in-chg state soil testing opers, 58-61, from instr to assoc prof agron, 58-73, leader earth sci res progs, Lab Appln Remote Sensing, 72-80 PROF AGRON, PURDUE UNIV, WEST LAFAYETTE, 73-, ASSOC DIR, LAB APPL REMOTE SENSING, 80- *Concurrent Pos:* Purdue Univ rep, Ford Found Agr Develop Prog, Arg, 64-66; consult, Food & Agr Orgn, Arg, 70, Bulgaria, 71, 72 & Sudan, 74, Am Univ, Washington, DC & Lilly Endowment, Niger, 73 & Int Develop Res Centre, Ottawa, to Sudan, 75-77; Klinck lectr, Can, 72; mem agr panel, Nat Acad Eng, 74; vis scientist, Int Inst Aerial Surv & Earth Sci, Enschede, Neth, 74-75; mem, Int Comt Remote Sensing for Develop, Nat Acad Sci, 75-76; nat bd mem, United Methodist Comt on Relief, 76-; chmn study panel remote sensing, Agr Res Inst. *Mem:* Am Soc Agron; Int Soil Sci Soc; Soil Sci Soc Am; Am Soc Photogram; Int Asn Ecol; Sigma Xi. *Res:* Applications of remote sensing technology; digital analysis of multispectral data for identification, characterization and mapping of earth surface features; relationships between physicochemical and multispectral properties of soils. *Mailing Add:* Dept Agron Purdue Univ Lafayette IN 47907

BAUMGARDNER, RAY K, b Ridgway, Colo, Dec 26, 33; m 59; c 2. LIMNOLOGY. *Educ:* Ft Lewis Col, BS, 59; Adams State Col, BS, 61; Okla State Univ, MS, 62, PhD(zool), 66. *Prof Exp:* Asst prof aquatic biol, 65-69, assoc prof biol sci, 69-74, prof biol sci & head dept, 74-83, Univ Registrar, 83-87, DEAN & PROVOST, NORTHWESTERN STATE UNIV, FT POLK, LA, 87- *Mem:* Am Soc Limnol & Oceanog. *Res:* Water pollution; primary productivity and its relation to water quality; effect of impoundments on streams. *Mailing Add:* Dept Biol Northwestern State Col Natchitoches LA 71457

BAUMGARDT, BILLY RAY, b Lafayette, Ind, Jan 17, 33; m 52; c 3. ANIMAL NUTRITION. *Educ:* Purdue Univ, BS, 55, MS, 56; Rutgers Univ, PhD(agr biochem), 59. *Prof Exp:* Instr dairy sci, Rutgers Univ, 56-59; from asst prof to assoc prof, Univ Wis-Madison, 59-67; prof animal nutrit, Pa State Univ, Univ Park, 67-80, head dept, 70-79, assoc dir, Agr Exp Sta, 79-80; DIR AGR EXP STA & ASSOC DEAN AGR, PURDUE UNIV, 80- *Honors & Awards:* Am Dairy Sci Asn Nutrit Res Award, 66. *Mem:* Am Dairy Sci Asn; Am Soc Animal Sci; Am Inst Nutrit; fel AAAS; Sigma Xi. *Res:* Rumen physiology and biochemistry; food evaluation; regulation of food intake; relation of nutrition to body composition. *Mailing Add:* 116 Agr Ad Bldg Purdue Univ West Lafayette IN 47907

BAUMGART, RICHARD, b Princeton, Ind, Sept 2, 47; m 68; c 3. AUTOMOTIVE CHEMICALS. *Educ:* Murray State Univ, Ky, BS, 70. *Prof Exp:* Chemist, Champion Lab Inc, 72-86; SUPVR, MAC'S, ASHLAND OIL, KY, 86- *Mailing Add:* Ashland Oil 22nd & Front Sts Ashland KY 41101

BAUMGARTEN, ALEXANDER, b Warsaw, Poland, Nov 27, 35; Australian citizen; m 69; c 2. CLINICAL IMMUNOLOGY, LABORATORY MEDICINE. *Educ:* Sydney Univ, MB, BS, 59; Univ New South Wales, PhD(immunopath), 69. *Hon Degrees:* MA, Yale Univ. *Prof Exp:* Resident med, Royal Prince Alfred Hosp, Sydney, Australia, 59; resident, Royal Adelaide Hosp, 60; registr, Alfred Hosp, Melbourne, 61; res fel, Baker Med Res Inst, Melbourne, 61-64 & Med Sch, Univ New South Wales, Sydney, 65-69; from asst prof to assoc prof, 70-81, PROF LAB MED, MED SCH, YALE UNIV, 82- *Mem:* Am Asn Immunologists; Am Soc Microbiol; NY Acad Sci. *Res:* Clinical and general immunology; protein polymorphism; alpha-fetoprotin in pregnancy. *Mailing Add:* Dept Lab Med Yale New Haven Hosp New Haven CT 06504

BAUMGARTEN, HENRY ERNEST, b Texas City, Tex, Feb 27, 21; m 48; c 4. SYNTHETIC ORGANIC CHEMISTRY, COMPUTERS IN CHEMISTRY. *Educ:* Rice Univ, BA, 43, MA, 44, PhD(org chem), 48. *Prof Exp:* Lab asst, Rice Univ, 41-43; asst, Off Rubber Reserve, Univ Ill, 48-49; from instr to prof, 49-64, actg chmn dept, 70-71, chmn dept, 71-75, FOUND PROF ORG CHEM, UNIV NEBR-LINCOLN, 64- *Concurrent Pos:* Guggenheim fel & vis assoc, Calif Inst Technol, 62-63; consult, div instnl progs, NSF, 63-67; vis scholar & vis prof, Stanford Univ, 77-78; vis prof, Peking Univ, 85. *Mem:* Fel AAAS; Am Chem Soc. *Res:* Reactions of organic nitrogen compounds; alpha lactams; organic electrochemistry; organometallic chemistry; computers in chemical research. *Mailing Add:* Dept Chem Univ Nebr-Lincoln Lincoln NE 68588-0304

BAUMGARTEN, JOSEPH RUSSELL, b Louisville, Ky, July 6, 28; m 53; c 3. MECHANICAL ENGINEERING, MACHINE DESIGN. *Educ:* Univ Dayton, BME, 50; Purdue Univ, MSME, 55, PhD(mech eng), 58. *Prof Exp:* Assoc res engr, Res Labs, Whirlpool Corp, 58-60; eng assoc, Prod Eng Div, Nat Cash Register Co, 60-62, head consult, Serv Dept, 62-65; assoc prof mech eng, Ga Inst Technol, 65-70; prof mech eng, Univ Nebr, Lincoln, 70-78; PROF MECH ENG, IOWA STATE UNIV, 78- *Concurrent Pos:* Mem automation comt, Engrs Joint Coun, 64; vis prof, Polytecnica Warsaw, 75-76; vis prof, Technische Hogeschool Delft, 81-82, vis prof, Cath Univ, Leuven, 87-88; fac fel, NASA, 77, Sandia Nat Lab, 86; Fulbright fel, 81, NATO fel, 87. *Mem:* Am Soc Mech Eng. *Res:* Machinery vibrations and dynamics; dynamics of electromechanical design; design automation; vibration damping from viscoelastic coating; robotics; mechanics; elastic and plastic strain propagation; inflated structures; spacecraft nutation; path and position synthesis; kinematics; numerical methods. *Mailing Add:* Dept Mech Eng Iowa State Univ Ames IA 50011

BAUMGARTEN, PETER KARL, chemical process safety & scale-up, for more information see previous edition

BAUMGARTEN, REUBEN LAWRENCE, b New York, NY, Nov 19, 34; m 63; c 2. PHYSICAL ORGANIC CHEMISTRY. *Educ:* City Col New York, BS, 56; Univ Mich, MS, 58, PhD(chem), 62. *Prof Exp:* Instr chem, Hunter Col, 62-66, George N Shuster fel grant, 64-65; from asst prof to prof, 66-77, CHMN CHEM DEPT, LEHMAN COL, 78- *Mem:* Am Chem Soc; Sigma Xi. *Res:* Organic chemistry of nitrogen; reaction of hydroxylamines and derivatives; nuclear magnetic resonance; magnetic isomerism; aromatic ring current; prochirality in nuclear magnetic resonance; qualitative detection of amines, glycoproteins, Rimini-Simon test and mechanism. *Mailing Add:* Dept Chem Herbert H Lehman Col Bedford Park Blvd W Bronx NY 10468

BAUMGARTEN, RONALD J, b New York, NY, May 7, 35. ORGANIC CHEMISTRY, ENVIRONMENTAL CHEMISTRY. *Educ:* Brooklyn Col, BS, 56; Johns Hopkins Univ, MA, 58, PhD(org chem), 62. *Prof Exp:* Res assoc, Ind Univ, 61-62; fel, Brandeis Univ, 63-64; ASSOC PROF ORG CHEM, UNIV ILL, CHICAGO CIRCLE, 64- *Concurrent Pos:* Vis asst prof, Univ Ill, Urbana, 65-66; NSF fel, 70-72; vis scientist, State Univ Leiden, 71-72; vis prof, Univ Christchurch, NZ, 81. *Res:* New approaches to deamination; photochemical reductions; new synthetic procedures; aldehyde and ketone syntheses. *Mailing Add:* Dept Chem M/C 111 Univ Ill at Chicago Circle Chicago IL 60680

BAUMGARTEN, WERNER, b Berlin, Ger, Feb 14, 14; nat US; m 48; c 1. BIOCHEMISTRY. *Educ:* Univ Munich, Dipl, 37; Calif Inst Technol, PhD(org chem, biochem), 41. *Prof Exp:* Res chemist, Hiram Walker & Sons, 41-45, J T Baker Chem Co, NJ, 45-48 & Merck & Co, 48-49; res assoc, Sharp & Dohme, 49-56, res assoc, Merck Sharp & Dohme Res Labs, 56-65, sr res fel, 65-69, dir hemat, 69-71; sr info scientist, Res Labs, Hoffman-LaRoche, 71-73; asst dir clin res, Hoechst-Rossel Pharmaceut, 73-75; med writer, ICI United States Inc, 75-77, clin assoc, ICI Americas Inc, 77-84; CONSULT TO THE PHARMACEUT INDUST, 84- *Mem:* Am Chem Soc; fel NY Acad Sci; Am Soc Biol Chem; Am Soc Hematol; Int Soc Hematol; Sigma Xi; fel AAAS. *Res:* Sugar chemistry; heat of combustion of organic compounds; microbiological analysis of amino acids; determination of vitamins and carotenoids; isolation of natural products; fermentation; therapeutic enzymes; protein chemistry; blood coagulation; clinical research. *Mailing Add:* 2420 Oak Circle Huntingdon Valley PA 19006

BAUMGARTNER, GEORGE JULIUS, b Atchison, Kans, June 27, 24; m 54; c 6. ORGANIC CHEMISTRY. *Educ:* St Benedict's Col, BSc, 45; Univ Notre Dame, PhD, 53. *Prof Exp:* Prof chem, Mt St Scholastica Col, 50-71, dean acad affairs, Benedictine Col, 68-71, prof chem, 71-78, asst dean, 73-78, acad dean, 78-88; RETIRED. *Mem:* Am Chem Soc. *Res:* Reactions of tetrahydrofuryl compounds. *Mailing Add:* Benedictine Col Atchison KS 66002

BAUMGARTNER, LEONA, pediatrics, public health; deceased, see previous edition for last biography

BAUMGARTNER, WERNER ANDREAS, b Poertschach, Austria, May 6, 35; m 68; c 2. DRUG DETECTION BY HAIR ANALYSIS, BIOPHYSICAL CHEMISTRY. *Educ:* Univ New South Wales, BSc, 60, PhD(phys chem), 64. *Prof Exp:* Res asst phys chem, Commonwealth Sci & Indust Res Orgn, Australia, 55-56; res asst biophys chem, physics dept, St Vincent Hosp, Australia, 56-60; scholar, Univ New South Wales, 64-66; assoc scientist, Jet Propulsion Lab, Calif Inst Technol, 66; asst prof phys chem, Calif State Col, Long Beach, 67-70; dir biophys chem, Tritium & Dial Lab, Nuclear Dynamics, Inc, 70-71; FOUNDER & CHMN BD, PSYCHEMEDICS CORP, SANTA MONICA, 87- *Concurrent Pos:* Asst prof, Calif State Col, Long Beach, 67-68; assoc scientist, Scripps Inst Oceanog, 70 & Immunochem Lab, Vet Admin Hosp, Sepulveda, 70-71; res chemist, Vet Admin Hosp, Long Beach, 71-73 & Vet Admin Hosp, Wadsworth, 73- *Mem:* Am Chem Soc. *Res:* Theory of metal and enzyme catalysis; charge-transfer complexes in catalysis; immunochemistry and membrane processes; function of vitamin E and selenium; catalytic labeling of unstable biochemicals; analysis of drugs of abuse in hair. *Mailing Add:* PO Box 84573 Los Angeles CA 90073

BAUMHEFNER, DAVID PAUL, b Denver, Colo, Nov 18, 41; m 64; c 2. DYNAMICAL METEOROLOGY. *Educ:* Univ Calif, Los Angeles, BA, 65, MA, 66. *Prof Exp:* Res asst meteorol, Univ Calif, Los Angeles, 63-66; RES SCIENTIST METEOROL, NAT CTR ATMOSPHERIC RES, 66- *Mem:* Am Meteorol Soc; Can Meteorol Soc. *Res:* Numerical weather prediction; diagnostic analysis of atmospheric behavior. *Mailing Add:* 878 Pimlico Ct Boulder CO 80303

BAUMILLER, ROBERT CAHILL, b Baltimore, Md, Apr 15, 31. GENETICS. *Educ:* Loyola Col, Md, BS, 53; St Louis Univ, PhL & PhD(biol), 61; Woodstock Col, Md, STB, 65. *Prof Exp:* Nat Found fel, Univ Wis, 61-62; asst med, Sch Med, Johns Hopkins Univ, 62-67; assoc prof 67-80, PROF OBSTET, GYNEC & PEDIAT, SCH MED, GEORGETOWN UNIV, 80- *Concurrent Pos:* Prin investr, USPHS res grant, 64-67; AEC grant, 67-70; sr res fel, Kennedy Inst Study Reproduction & Bioethics; mem bd dir, Sex Info & Educ Coun US, 70-74; consult, Clin Pharmacol Assocs, 71-73; assoc ed, Linacre Quart, 71-78; mem med adv bd, Hemophilia Soc, Capitol Area, 74- & 76-; consult, Pope John XXIII Med-Moral Res Ctr, 75-; mem med adv bd, Childbirth Educ Asn, 77-; assoc ed, Downs Syndrome, 78-; consult, Nat Clearinghouse for Genetic Disease, Dept Health & Human Serv, 78-82; dir, Nat Ctr Educ Maternal & Child Health, 82-; mem bd trustees, Loyola Col, 90- *Mem:* AAAS; Genetics Soc Am; Am Soc Human Genetics; Environ Mutagen Soc; Sigma Xi. *Res:* Virus induced mutation both gene and chromosomal; oncogenetics; transformation of human leucocytes as affected by the menstrual cycle; human chromosome abnormalities. *Mailing Add:* Dept Obstet & Gynec Georgetown Univ Sch Med Washington DC 20007

BAUMMER, J CHARLES, JR, b Baltimore, MD, May 20, 45; m 68; c 4. AMMONIA CHEMISTRY & TOXICOLOGY, DREDGING. *Educ:* Loyola Col, BS, 67; Johns Hopkins Univ, MSE, 69, PhD(environ eng sci), 72. *Prof Exp:* Staff officer, Nat Acad Sci, 72-73; staff asst, Nat Comn Water Qual, 73-76; MGR, ENVIRON MGT, EA ENG SCI TECHNOL INC, 76- *Concurrent Pos:* Mem State Water Qual Adv Comt, 88. *Res:* Impact of dredging, hazardous waste and wastewater discharges on water quality. *Mailing Add:* EA Eng Sci & Technol Inc 15 Loveton Circle Sparks MD 21152

BAUMRUCKER, CRAIG RICHARD, b Jan 22, 44; m; c 2. ENDOCRINOLOGY, GROWTH FACTORS. *Educ:* Purdue Univ, PhD(animal sci), 74. *Prof Exp:* Asst prof dairy sci, Univ Ill-Urbana, 74-81; ASSOC PROF NUTRIT & PHYSIOL, PA STATE UNIV, 81- *Concurrent Pos:* Ed bd, J Dairy Sci; vis assoc prof, Inst Animal Breeding, Univ Bern, Switz, 89-90. *Mem:* Endocrine Soc; Am Soc Cell Biol; Am Dairy Sci Asn; Am Soc Animal Sci. *Res:* Effects of growth factors on bovine mammary and intestinal tissues; synthesis of IGF and IGF-binding proteins by mammary tissue; the assurance of IGF and IGF-binding proteins in mammary secretions; the effects of mammary secretions on the intestinal growth of the neonate. *Mailing Add:* Dept Dairy & Animal Sci 302 Henning Bldg Pa State Univ University Park PA 16802

BAUMSLAG, GILBERT, b Johannesburg, SAfrica, Apr 30, 33; m 59; c 2. MATHEMATICS. *Educ:* Univ Witwatersrand, BS, 53, Hons, 55, DSc, 76; Univ Manchester, PhD(math), 58. *Prof Exp:* Lectr math, Univ Manchester, 58-59; instr, Princeton Univ, 59-60; from asst prof to assoc prof, Courant Inst Math Sci, NY Univ, 62-64; prof, City Univ New York, 64-68; mem, Inst Advan Study, 68-69; prof, Rice Univ, 69-73; DISTINGUISHED PROF MATH, CITY COL CITY UNIV NEW YORK, 73- *Concurrent Pos:* Sloan fel, Rice Univ. *Mem:* Am Math Soc; London Math Soc. *Res:* Group theory. *Mailing Add:* Dept Math City Univ New York City Col Convent Ave at 138th St New York NY 10031

BAUMSTARK, JOHN SPANN, b Cape Girardeau, Mo, Dec 23, 27; m 52; c 3. BIOCHEMISTRY, MICROBIOLOGY. *Educ:* Southeast Mo State Col, BS, 51; Univ Mo, AM, 53, PhD(biochem), 57. *Prof Exp:* Instr agr chem, Univ Mo, 55-57; sr scientist biochem microbiol, Res Dept, Mech Div, Gen Mills, Inc, 57-60, prin scientist, 60-61; from instr to asst prof obstet & gynec, Med Sch, Tufts Univ, 63-72; assoc prof biol chem, 72-79, PROF & DIR RES OBSTET & GYNEC, SCH MED, CREIGHTON UNIV, 72-, PROF PATH, 74-, PROF BIOL CHEM, 79- *Concurrent Pos:* Sr biochemist, St Margaret's Hosp, Boston, Mass, 61-72. *Mem:* Am Asn Path; Am Soc Microbiol; Am Soc Biol Chem; Am Chem Soc; Soc Exp Biol Med; Electrophoresis Soc; Asn Prof Gynec Obstet. *Res:* Proteolytic enzymes; proteins; amino acids; chromatography; experimental pathology. *Mailing Add:* Dept Obstet & Gynec Creighton Univ Sch Med Omaha NE 68131

BAUR, FREDRIC JOHN, JR, b Toledo, Ohio, July 14, 18; m 43; c 3. PHYSIOLOGICAL CHEMISTRY, ORGANIC CHEMISTRY. *Educ:* Univ Toledo, BS, 39; Ohio State Univ, MS, 41, PhD(physiol chem), 43; cert com, Univ Cincinnati, 56. *Prof Exp:* Instr physiol chem, Ohio State Univ, 43-44; res chemist, Procter & Gamble Co, 46-56, in charge flavor chem, Food Div, 56-60, in charge anal labs, 60-65, in charge factory anal methods-sanit, Food Div, 65-74, in charge good mfg practices/qual assurance, Corp Compliance Rev, 74-78, spec assignment, 78-82, rep asn off anal chemist, 82-89; RETIRED. *Mem:* Am Chem Soc; Inst Food Technol; fel Asn Official Anal Chemists; Am Oil Chemists Soc; Am Asn Cereal Chemists. *Res:* Interesterification, composition and structure of fats; synthesis of compounds containing fatty groups; acetin fats; chemistry and utility of flavors; analysis of foods and food raw materials; sanitary design of buildings and equipment; measurement and control of infesting pests; good manufacturing practices for foods and drugs; regulatory compliance. *Mailing Add:* 1545 Larry Ave Cincinnati OH 45224

BAUR, JAMES FRANCIS, b McKeesport, Pa, Oct 22, 38; m 60; c 2. PLASMA PHYSICS, NUCLEAR RADIATION HARDENING. *Educ:* Univ Fla, BS, 60, MS, 61; Univ Colo, PhD(physics), 75. *Prof Exp:* Airborne, ranger, command & staff, US Army Corps Eng, 60-68, asst prof, lab dir, physics, US Military Acad, West Point, 65-68; res asst, Joint Inst Lab Astrophysics, 70-75; sr scientist, staff scientist, sr staff scientist, PHYSICS COORDR, FUSION DIV, GEN ATOMICS, 75-; PRIN INVESTR,

RADIATION HARDENING PROG, US DEPT ENERGY, 81- *Concurrent Pos:* Infrared consult, Atlantic Missile Range, Pan Am World Airways, 60-61; Fusion Core Exp, 83-86, Nat Design Team, Compact Ignition, Off Fusion Energy, 86-; sci fac fel, Nat Sci Found. *Mem:* Am Phys Soc; AAAS; Optical Soc Am; Sigma Xi; Inst Elec & Electronics Engrs. *Res:* Plasma Diagnostics; nuclear radiation hardening; fusion physics; spectroscopy. *Mailing Add:* Sci Solutions Inc 4011 Zenako St San Diego CA 92122-3431

BAUR, JOHN M, b Owensboro, Ky, Feb 2, 39; m 66; c 2. MICROBIOLOGY. *Educ:* Brescia Col, Owensboro, Ky, BA, 61; Kent State Univ, MA, 63. *Prof Exp:* Instr microbiol, Col St Teresa, Winona, Minn, 63-65, asst prof, 65-67; ASST PROF BIOL, BRESCIA COL, OWENSBORO, KY, 67-, CHMN, 70-80, 82- *Concurrent Pos:* Dir, Med-Tech Prog, Brescia Col, Owensboro, Ky, 70-80, 82-, chief, pre-prof adv, 82- *Res:* Determination of oxygen usage by certain lung parasites; transplantation techniques in mammals; thyroxine uptake by irradiated thyroid gland cells. *Mailing Add:* Brescia Col 717 Frederica St Owensboro KY 42301

BAUR, MARIO ELLIOTT, b Indianapolis, Ind, Aug 23, 34; m 60; c 5. PHYSICAL CHEMISTRY. *Educ:* Univ Chicago, AB, 53, MS, 55; Mass Inst Technol, PhD(chem), 59. *Prof Exp:* Asst chem, Mass Inst Technol, 55-56 & 58-59; NSF fel, State Univ Utrecht, 59-61; res assoc, Univ Calif, San Diego, 61-62; from asst prof to assoc prof, 62-79, PROF CHEM, UNIV CALIF, LOS ANGELES, 79- *Mem:* AAAS; Am Phys Soc; Royal Soc Chem; Netherlands Phys Soc; Am Geophys Union. *Res:* Physical geochemistry; paleochemistry; origin and evolution of the terrestrial atmosphere; chemical effects on evolutionary trends; statistical mechanics. *Mailing Add:* Dept Chem Univ Calif Los Angeles CA 90024

BAUR, THOMAS GEORGE, b Bad Axe, Mich, Mar 5, 44. ASTROPHYSICS, SOLAR PHYSICS. *Educ:* Univ Mich, BS, 66; Univ Colo, MS, 69. *Prof Exp:* Instr physics & math, St Dominic Col, 69-70; support scientist solar physics, Nat Ctr Atmospheric Res, 70-83; CHIEF EXEC OFFICER, MEADOWLAND OPTICS, 83- *Honors & Awards:* Technol Advan Award, Nat Ctr Atmospheric Res, 76. *Mem:* Soc Photo-Optical Instrumentation Engrs; Optical Soc Am (pres, Rocky Mt sect, 88). *Res:* Solar magnetic field measurement; polarimetry; optical design; stellar photometry. *Mailing Add:* 7460 Weld County Rd 1 Longmont CO 80504

BAUR, WERNER HEINZ, b Warsaw, Poland, Aug 2, 31; m 62; c 2. MINERALOGY, CRYSTALLOGRAPHY. *Educ:* Univ Gottingen, Dr rer nat, 56. *Prof Exp:* Sci officer, Univ Güttingen, 56-63, privat-dozent, 61; from asst prof to assoc prof mineral & crystallog, Univ Pittsburgh, 63-65; assoc prof, 65-68, head, Dept Geol Sci, 67-79, prof mineral & crystallog, 68-80, assoc dean lib arts & sci, 78-80, PROF GEOL SCI, UNIV ILL, CHICAGO, 80- *Concurrent Pos:* Fel, Univ Berne, 57; vis assoc chemist, Brookhaven Nat Lab, 62-63; vis res prof, Univ Karlsruhe, 71-72; res assoc, Field Mus Natural Hist, Chicago, 74-; assoc ed, Am Mineralogist, 75-78. *Mem:* Fel Mineral Soc Am; Am Crystallog Asn; Am Geophys Union. *Res:* Crystal chemistry of minerals and inorganic compounds; crystal structure determination by x-ray and neutron diffraction; hydrogen bonding; computer simulation of crystal structures; empirical theories of bonding in crystals; predictive crystal chemistry; zeolites; powder diffraction. *Mailing Add:* Dept Geol Sci Univ Ill Box 4348 Chicago IL 60680

BAURER, THEODORE, b New York, NY, June 20, 24; m 47; c 2. ENVIRONMENTAL SCIENCE, ENERGY RESOURCES. *Educ:* Queens Col, NY, BS, 45; Syracuse Univ, PhD(chem), 53; Temple Univ, JD, 80. *Prof Exp:* Asst, Syracuse Univ, 47-48 & Atomic Energy Comn, 51-53; res chemist, mineral beneficiation lab, Columbia Univ, 53-56; adv res engr, Sylvania Elec Prod, Inc & Sylvania Corning Nuclear Corp, 56-57; res phys chemist, Grumman Aircraft Eng Corp, 57-63; proj scientist, Gen Appl Sci Labs, Inc, 63-66; sr scientist, Gen Elec Co, Space Sci Lab, 66-80; RETIRED. *Concurrent Pos:* Guest scientist, free radicals res prog, Nat Bur Standards, 58-59; co-ed, Defense Nuclear Agency Reaction Rate Handbk; consult attorney & scientist, 80; attorney at Law, 80- *Mem:* AAAS; Am Chem Soc; Am Phys Soc; NY Acad Sci; Am Bar Asn. *Res:* Gas-phase kinetics; upper and lower atmosphere chemistry; propulsion and combustion chemistry; free radicals chemistry; spacecraft internal and external environmental quality; space experiments definition; chemical systems analyses; environmental science; law. *Mailing Add:* 751 Woodside Rd Rydal Jenkintown PA 19046-3331

BAURIEDEL, WALLACE ROBERT, biochemistry, for more information see previous edition

BAUS, BERNARD V(ILLARS), b Gramercy, La, Sept 27, 25; m 57; c 2. ENGINEERING, CHEMISTRY. *Educ:* Tulane Univ, BE, 45; Cornell Univ, PhD(chem eng), 50. *Prof Exp:* Chem engr plants technol sect, 50-55, asst tech supt, 55-56, tech supt, 56-58, sr supvr res & develop, 58-59, tech investr int dept, E I du Pont de Nemours & Co, 60-62; mgr spec proj, Commonwealth Oil Refining Co, Inc, 62-65, vpres chem develop, 65-70, pres, 70-71, vpres, Petrochem, Commonwealth Petrochem Co, 70-71; consult, Am-Rican Group, 71-73; PRES, INDUST CHEM CORP, 74- *Concurrent Pos:* Dir, Waste Mgt Asn, 84. *Mem:* AAAS; Am Inst Chem Engrs; Am Chem Soc. *Res:* Heat transfer; process development; plastics research; corporate planning and project development. *Mailing Add:* 17 Emajagua St San Juan PR 00913

BAUSHER, LARRY PAUL, b Reading, Pa, July 8, 39; m 78; c 1. ORGANIC CHEMISTRY. *Educ:* Franklin & Marshall Col, AB, 61; Univ Calif, Los Angeles, PhD(org chem), 67. *Prof Exp:* Res chemist, Hercules, Inc, 67-69; res fel chem, Wesleyan Univ, 69-71; res fel chmn, 71-83, assoc res scientist, 83-90, RES SCIENTIST, YALE UNIV, 90- *Mem:* AAAS; Am Chem Soc; Asn Res Vision & Ophthal; Sigma Xi; NY Acad Sci. *Res:* Physiology and pharmacology of aqueous humor dynamics; study of receptors and intracellular second messengers which regulate physiologic processes in the anterior segment of the eye. *Mailing Add:* 75 Forest Ct N Hamden CT 06518

BAUSHER, MICHAEL GEORGE, b Philadelphia, Pa, Mar 28, 45. PLANT PHYSIOLOGY, BIOCHEMISTRY. *Educ:* Del Valley Col, BS, 67; Rutgers Univ, MS, 69; Univ Fla, PhD(plant physiol), 74. *Prof Exp:* Res asst hort, Rutgers Univ, 67-69; res asst plant physiol, Univ Fla, 69-74, res assoc, 74-75; PLANT PHYSIOLOGIST, SCI & EDUC ADMIN-AGR RES, USDA, 75- *Concurrent Pos:* Adj asst prof, Univ Fla, 75- *Mem:* AAAS; Am Soc Hort Sci; Am Soc Plant Physiol. *Res:* Chemical plant growth regulators; chemistry of disease response in plants. *Mailing Add:* 2120 Camden Rd Orlando FL 32803

BAUSSUS VON LUETZOW, HANS GERHARD, b Nortorf, Ger, Feb 13, 21; US citizen; m 52; c 3. MATHEMATICS. *Educ:* Univ Kiel, BS & MS, 49, DSc(math), 63. *Prof Exp:* Sci asst opers res, State Energy Admin, Ger, 49-51; asst prof math & physics, Zimmermann Col & Com Col, 52-56; unit chief math anal, Aeroballistics Lab, US Army Ballistic Missile Agency, 56-57, dep sect & dep br chief anal invest & aerophys & geophys, 57-58, sci adv math statist, 59-60; sci adv earth & space sci, Aeroballistics Lab, Marshall Space Flight Ctr, 60-62; tech chief systs div, US Army Strategy & Tactics Anal Group, 62-63; prin scientist, Advan Progs, Gen Precision Co, 63-64; referent, Prog Coord Opers Res Budgeting, Off Fed Minister Defense, Ger, 64-66; sr scientist, US Army Topog Labs, 66-88; RETIRED. *Mem:* Am Geophys Union; assoc fel Am Inst Aeronaut & Astronaut; Acad Polit Sci. *Res:* Applied mathematics; dynamic meteorology; geodesy inertial navigation; operations research. *Mailing Add:* 8021 Garlot Dr Annandale VA 22003-1309

BAUST, JOHN G, b Flushing, NY, Oct 6, 42. CRYO BIOLOGY. *Educ:* Univ Alaska, PhD(physiol), 70. *Prof Exp:* DIR & LUYOT PROF CRYO BIOL, INST LOW TEMPERATURE BIOL, 84- *Concurrent Pos:* Vpres res & develop, Cryomed Sci Inc. *Mailing Add:* Dept Biol Sci State Univ NY Binghamton NY 13901

BAUSUM, HOWARD THOMAS, b Guilin, China, July 20, 33; US citizen; m 60; c 3. MICROBIOLOGY, ENVIRONMENTAL SCIENCES. *Educ:* Carson-Newman Col, BS, 54; Univ Tenn, MS, 56; Univ Tex, PhD(zool, genetics), 64. *Prof Exp:* Biologist, US Army Biol Labs, Ft Detrick, Md, 60-61; asst prof microbiol, Sch Med, Univ Kans, 64-65; asst prof genetics, Iowa State Univ, 65-68; microbiologist, US Army Biol Labs, 68-71; staff assoc, Comn Sci Educ, AAAS, 71-72; RES MICROBIOLOGIST, US ARMY BIOMED RES & DEVELOP LAB, 72- *Mem:* Sigma Xi; Am Soc Microbiol; Genetics Soc Am; NY Acad Sci. *Res:* Environmental microbiology; fate of pollutant chemicals. *Mailing Add:* US Army Biomed Res & Develop Lab Ft Detrick Frederick MD 21702-5010

BAUTZ, GORDON T, PHARMACOLOGY. *Educ:* Salem Col, WVa, BS, 64. *Prof Exp:* RES SCIENTIST, HOFFMAN LAROCHE, INC, 69- *Mem:* Am Soc Pharm & Exp Therapeut; Drug Info Asn; Pharmaceut Mfg Asn; NY Acad Sci. *Res:* Central nervous system. *Mailing Add:* Dept Preclin Develop Hoffman LaRoche Inc 340 Kingsland Ave Nutley NJ 07110-1150

BAUTZ, LAURA PATRICIA, b Washington, DC, Sept 3, 40. ASTRONOMY, SCIENCE ADMINISTRATION. *Educ:* Vanderbilt Univ, BA, 61; Univ Wis, PhD(astron), 67. *Prof Exp:* From instr to asst prof astron, Northwestern Univ, Evanston, 65-72; prog dir, NSF, 72-73; assoc prof astron, Northwestern Univ, Evanston, 73-75; sr staff assoc, 76-79, dep dir, Div Physics, 79-82, DIR, DIV ASTRON SCI, NSF, 82- *Concurrent Pos:* Trustee, Adler Planetarium, 73-76; vis scientist, Lawrence Berkeley Lab, 90-91. *Mem:* Am Astron Soc; AAAS; Int Astron Union; Am Phys Soc. *Res:* Clusters of galaxies. *Mailing Add:* No 506 1325 18th St Washington DC 20036

BAVISOTTO, VINCENT, b Buffalo, NY, Jan 21, 25; m 53; c 4. BIOCHEMISTRY. *Educ:* Univ Buffalo, BA, 48; Pa State Univ, MS, 50, PhD(biochem), 52. *Prof Exp:* Sr res chemist, Gen Biochem, Inc, 52-54; dir res, Paul Lewis Labs, Inc, 55-61, dir res & develop, Paul Lewis Labs Div, Chas Pfizer & Co, Inc, 61-67; tech dir, Theodore Hamm Co, 67-72; assoc dir res & develop, Heublein Inc, 72-74; VPRES BREWING & RES, MILLER BREWING CO, 74- *Mem:* Fel Am Chem Soc; Am Soc Brewing Chemists (pres, 72); Inst Food Technol; Master Brewers Asn Am (pres, 88-89); AAAS. *Res:* Basic and applied research on problems related to brewing, dairy, meat, enzyme, wine, spirited and chemical manufacturing industries. *Mailing Add:* 5780 Mary Lane Oconomowoc WI 53066

BAVISTER, BARRY DOUGLAS, b Cambridge, Eng, Mar 15, 43; m; c 3. FERTILIZATION, EMBRYOGENESIS. *Educ:* Univ Cambridge, Eng, BA, 67, PhD(reproductive physiol), 72. *Prof Exp:* Res fel, Med Sch, Univ Hawaii, 74-75; asst res pathologist, Med Sch, Univ Calif, Los Angeles, 76-78; lectr animal physiol, dept animal physiol, Univ Calif, Davis, 78-79; asst prof animal physiol, 79-83, assoc prof vet sci, Primate Res Ctr, 83-89, PROF VET SCI & SR SCIENTIST, PRIMATE RES CTR, UNIV WIS-MADISON, 89- *Mem:* Soc Study Fertil; Soc Study Reproduction; Tissue Culture Asn. *Res:* Physiology, morphology and biochemistry of mammalian gametes, fertilization and embryogenesis, and the development of culture media for study of these processes in vitro. *Mailing Add:* Dept Vet Sci Univ Wis 1655 Linden Dr Madison WI 53706

BAVLEY, ABRAHAM, b Boston, Mass, June 3, 15; m 46. CHEMISTRY. *Educ:* Tufts Col, BS, 36; Harvard Univ, MA & PhD(org chem), 40. *Prof Exp:* Chemist, Ansco Div, Gen Aniline & Film Co, Inc, NY, 40-43, leader colorformer group, 43-45; asst prof chem, Xavier Univ, Ohio, 45-46 & Gevaert Photo-Prod, Belg, 46-47; res chemist, Charles Pfizer & Co, Inc, NY, 47-51, res supvr, 51-58, mgr res, Greensborough Sect, 58-59; asst dir res, Gillette Safety Razor Co, Mass, 59-60; mgr res, Philip Morris, Inc, 60-65; vpres & tech dir, Marion Labs, Inc, Mo, 65-69; vpres & sci dir, Knoll Pharmaceut Co, 69-71; assoc, Res Corp, 74-83; RETIRED. *Concurrent Pos:* Consult, 71-74 & 83- *Res:* Organic synthesis; medicinal chemistry; chemistry of natural products. *Mailing Add:* 11 Black Watch Trail Morristown NJ 07960

BAW, PHILEMON S H, b Kobe, Japan, May 5, 39; m 63; c 2. HEAT TRANSFER, FLUID DYNAMICS. *Educ:* Cheng Kung Univ, Taiwan, BS, 59; Rutgers Univ, MS, 66, PhD(mech eng), 68. *Prof Exp:* Engr, Yue Loong Motor Co, China, 61-63; res asst, Rutgers Univ, 64-68; asst prof mech, aerospace & indust eng, La State Univ, Baton Rouge, 68-78; MECH ENGR, LEVY ASSOC, INC, 78- *Concurrent Pos:* Vis prof, Cheng Kung Univ, Taiwan, 72-73. *Honors & Awards:* Teetor Award, Soc Automotive Engrs, 70. *Mem:* Asn Energy Engrs; Am Soc Eng Educ. *Res:* Gas-solid suspension flow; unsteady state heat transfer; energy conservation and management. *Mailing Add:* 5420 Corporate Blvd Suite 301 Baton Rouge LA 70808

BAWA, KAMALJIT S, b Kapurthala, India, Apr 7, 39; m 69; c 2. POPULATION BIOLOGY, TROPICAL ECOLOGY. *Educ:* Panjab Univ, India, BSc, 60, MSc, 62, PhD(bot), 67. *Prof Exp:* Res assoc genetics, Univ Wash, Seattle, 67-72; Bullard & Cabot res fel, Harvard Univ, 72-73, res fel, 73-74; from asst prof to assoc prof, 74-81, PROF BIOL, UNIV MASS, BOSTON, 81-, CHMN, BIOL DEPT, 89- *Concurrent Pos:* Prin investr, NSF grants, 75-; mem, La Selva adv comt, Orgn Trop Studies, 78-85 & adv panel, NSF Conserv & Restoration Biol, 91; counr, Asn Trop Biol, 85-86; assoc ed, Conserv Biol, 87-; Guggenheim fel, 87-88. *Mem:* Soc Study Evolution; Bot Soc Am; Ecol Soc Am; Am Soc Naturalists; Asn Trop Biol. *Res:* Population biology of tropical rain forest trees; genetic variation in tropical tree populations; plant pollinator interactions; evolution of sexual systems; conservation and management of tropical rain forest resources. *Mailing Add:* Dept Biol Univ Mass Boston MA 02125

BAWA, MOHENDRA S, b Jagraon, Panjab, India, June 10, 31; m 63; c 2. CHEMICAL ENGINEERING, APPLIED CHEMISTRY. *Educ:* East Panjab Univ, India, BSc, 49; Calcutta Univ, MSc, 54; Univ Houston, PhD(chem eng), 65. *Prof Exp:* Asst plant mgr, Assoc Pigments Ltd, Calcutta, 55; sci asst surface coatings, Govt Test House, 55; chem eng asst process design, Kilburn & Co, 55-56; sr sci officer fuel chem, Cent Fuel Res Inst, 56-61; res fel energy conversion, Dept Chem Eng, Univ Houston, 61-65; mem tech staff, 65-80, SR MEM TECH STAFF RES & DEVELOP, TEX INSTRUMENTS, INC, DALLAS, 65- *Concurrent Pos:* Res engr, Retzloff Chem Co, Tex, 63-64 & Houston Res Inst, 64-65. *Honors & Awards:* Gold Medal, Directorate of Educ, Panjab, 45. *Mem:* Am Inst Chem Engrs; Sigma Xi; Am Asn Crystal Growth. *Res:* Process analysis and development; mass transfer processes; energy conversion processes; kinetics of electrode reactions; silicon and boron chemistry; coal science and fuel technology; new product development; surfactants; ablative heat shields; cermets; molten salt reaction media; single crystal silicon growth; hydrogen storage alloys. *Mailing Add:* 3313 Steven Dr Plano TX 75023

BAWDEN, JAMES WYATT, b St Louis, Mo, Apr 23, 30; m 51; c 5. PHYSIOLOGY, TOOTH ENAMEL FORMATION & FLUORIDES. *Educ:* Univ Iowa, DDS, 54, MS, 60, PhD(physiol), 61. *Prof Exp:* USN Dental Corps, 54-56, pvt pract, 56-58; fel physiol, Univ Iowa, 58-61; from asst prof to assoc prof pedodontics, 61-65, asst dean res, 63-66, dean, Sch Dent, 66-74, prof, 65-77, ALUMNI DISTINGUISHED PROF PEDIAT DENT, UNIV NC, CHAPEL HILL, 77- *Concurrent Pos:* NIH grant, 63-; vis prof, Sch Dent, Univ Lund, Sweden, 74-75; prin investr, NIH-Univ NC Dent Res Inst grant, 75-76 & W V Kellogg grant, 76-79. *Mem:* Inst Med-Nat Acad Sci; Int Asn Dent Res; Am Dent Asn; fel Int Col Dent; Am Acad Pediat Dent; fel Am Col Dent; Am Asn Dental Res (pres, 84-85). *Res:* Pediat dent; investigations in the placental transfer of calcium, fluoride, oxygen and carbon dioxide; the effect of maternal metabolic status on the metabolism of the fetus in utero; development of tooth enamel; trace elements; fluoride uptake in developing enamel; clinical studies on fluorides. *Mailing Add:* Sch Dent Univ NC Chapel Hill NC 27514

BAWDEN, MONTE PAUL, b Salt Lake City, Utah, June 3, 43; m 84; c 6. PARASITOLOGY, EDUCATIONAL ADMINISTRATION. *Educ:* Univ Calif, Riverside, BA, 65; Rutgers Univ, New Brunswick, PhD(zool), 70. *Prof Exp:* Res asst, Rutgers Univ, 65-66, predocfel, 68; fel, dept trop pub health, Sch Pub Health, Harvard Univ, 69-71, res assoc med parasitol, 71-73; scientist, Immuno-Parasitol Dept, Naval Med Res Inst, Bethesda, 73-76, head, Malaria Serol Br, 77-79, asst dir & admin officer Naval Med Res & Training Unit, Gorgas Mem Lab, Panama, 79-85, mil officer & dep head, Malaria Div, Bethesda, 85-87, adminr malaria prog, 87-88, head, Res Sci Dept, Cairo, Egypt, 88-90, actg exec officer, 88; PARASITOLOGIST, MED SERV CORPS, USN, 73-; DIR, MIL TRAINING NETWORK RESUSCITATIVE MED, MIL MED EDUC INST, UNIFORMED SERV UNIV HEALTH SCI, BETHESDA, 90-, ASST PROF, 90- *Concurrent Pos:* Lectr, Health Sci Educ & Training Command, Med Technologists Sch, Bethesda, 74-79, lectr med parasitol, Naval Med Res & Training Unit, Gorgas Lab Panama, 79-85, guest lectr, Naval Sch Health Care Admin, 73-76, Uniformed Serv Univ Health Sci, 78-79 & 90- & lectr clin med parasitol, Infectious Dis & Dermat Res Prog, Naval Hosp, Bethesda, 86-88; adj assoc & head, Photog Support Div, Naval Med Res & Training Unit, Gorgas Mem Lab, Panama, 79-85; jungle survival trainer, Jungle Oper Training Ctr, Fort Sherman, Panama, 80-85; clin asst prof trop med & med parasitol, La State Univ Med Ctr, 82-85; consult diagnosis parasitic dis, Santo Tomas Hosp & Gorgas Army Hosp, Panama, 79-85; prog adminr, US AID Proj, Egypt, 88-90. *Mem:* Am Soc Trop Med & Hyg; Royal Soc Trop Med & Hyg. *Res:* Malaria, trypanosomiasis, and schistosomiasis; vaccines and diagnostics; intestinal parasites. *Mailing Add:* Mil Training Network USUHS 4301 Jones Bridge Rd Bethesda MD 20814

BAWDON, ROGER EVERETT, b Highmore, SDak, Sept 6, 39; m 61; c 1. CLINICAL MICROBIOLOGY, MICROBIAL GENETICS. *Educ:* SDak State Univ, BS, 61, MS, 63; Kans State Univ, PhD(microbiol), 72. *Prof Exp:* Clin fel microbiol, St Joseph Mercy Hosp, 74-76; ASST PROF MICROBIOL, SCH MED, WAYNE STATE UNIV, 76-; DIR CLIN MICROBIOL LAB, DETROIT GEN HOSP, 76- *Concurrent Pos:* Fel, Univ Mich & St Joseph Mercy Hosp, 74-76. *Mem:* Sigma Xi; Am Soc Microbiol; Acad Clin Lab Physicians & Scientists. *Res:* Genetics of anaerobic bacteria. *Mailing Add:* Dept Obstet/Gynec Univ Tex Health Ctr 5323 Harry Hines Blvd Dallas TX 75235-9032

BAX, AD, b Zevenbergen, Neth, June 14, 56; m 86; c 1. NUCLEAR MAGNETIC RESONANCE. *Educ:* Delft Univ Technol, PhD(appl physics), 81. *Prof Exp:* Res fel, Delft Univ Technol, Neth & Oxford Univ, Eng, 78-81; assoc, dept chem, Colo State Univ, Ft Collins, 82-83; vis res assoc, 83-85, VIS RES SCIENTIST, NAT INST DIABETES, DIGESTIVE & KIDNEY DIS, NIH, BETHESDA, 85- *Concurrent Pos:* Assoc ed, J Biomolecular NMR, 90- *Res:* Two-dimensional and multiple quantum nuclear magnetic resonance and their applications in chemistry, biochemistry and biology. *Mailing Add:* Lab Chem Physics Nat Inst Diabetes Digestive & Kidney Dis NIH Bldg 2 Rm 109 Bethesda MD 20892

BAX, NICHOLAS JOHN, b Essex, Eng; US citizen. ECOSYSTEM ANALYSIS, SURVEY DESIGN & ANALYSIS. *Educ:* Cambridge Univ, Eng, MA, 79; Univ Wash, PhD(fisheries), 83. *Prof Exp:* Res assoc, Fisheries Res Inst, Univ Wash, 77-82; FISHERIES ANALYST, COMPASS SYSTS, INC, 82- *Concurrent Pos:* Researcher, Cambridge Coral Starfish Res Group, Port Sudan, 75-76. *Mem:* Pac Fisheries Biologists. *Res:* Fisheries ecosystems; migratory fish populations; sources of error influencing the description of ecosystems; statistical analysis and survey design. *Mailing Add:* Pentecntec 120 W Dayton Suite A-7 Edmonds WA 98020

BAXLEY, WILLIAM ALLISON, b Washington, DC, May 10, 33; m 56; c 3. INTERNAL MEDICINE. *Educ:* Duke Univ, BS, 55, MD, 62. *Prof Exp:* Intern med, Duke Hosp, Durham, NC, 62-63; med resident, Univ Wash, 63-64; asst med & res fel, Univ Wash & Vet Admin Hosp, 64-66; from instr to asst prof cardiol, Med Ctr, Univ Ala, 66-70; assoc prof med, Sch Med, Univ NMex, 70-72; assoc prof, 72-76, PROF MED, SCH MED, UNIV ALA, BIRMINGHAM, 76- *Concurrent Pos:* Mem coun clin cardiol, Am Heart Asn, 67. *Mem:* Am Fedn Clin Res; Am Heart Asn; fel Am Col Cardiol; fel Am Col Physicians. *Res:* Cardiovascular hemodynamics in patients with heart disease. *Mailing Add:* Dept Med Univ Ala Sch Med Birmingham AL 35294

BAXMAN, HORACE ROY, b Lansing, Ill, Jan 27, 21; m 44; c 4. CHEMISTRY. *Educ:* Ind Univ, BS, 43; Cornell Univ, PhD(inorg chem), 47. *Prof Exp:* Chemist, Monsanto Chem Co, Oak Ridge, 47-48; assoc chemist, Argonne Nat Lab, 48-52; mem staff, 52-73, SECT LEADER, LOS ALAMOS NAT LAB, 73- *Mem:* Fel Am Inst Chemists; Am Nuclear Soc; Am Chem Soc. *Res:* High Vacuum; boron compounds; spectrophotometry; unsymmetrical organoboron compounds; solvent extraction; fuel processing; uranium chemistry; process research on actinide elements; materials research on pyrolytic carbon. *Mailing Add:* Los Alamos Nat Lab Box 1663 Los Alamos NM 97544

BAXTER, ANN WEBSTER, b Evanston, Ill, July 25, 17; div; c 2. MICROBIAL PHYSIOLOGY. *Educ:* Rockford Col, BA, 38; Univ NC, Chapel Hill, PhD(bact immunol), 67. *Prof Exp:* Res technician physiol, sch med, Univ Louisville, 56-57; res technician bact, Am Sterilizer Co, Pa, 58-59; teacher, high sch, NC, 60-63; from asst prof to assoc prof microbiol, Clemson Univ, 67-82; RETIRED. *Honors & Awards:* 75th Anniversary Bronze Medallion, Am Soc Microbiol, 74; Serv Plaque, 75. *Mem:* AAAS; Am Soc Microbiol; Brit Soc Gen Microbiol; Sigma Xi. *Res:* Bacterial growth and metabolism; factors affecting bacterial membrane transport; mechanisms of active transport of amino acids in bacteria; effects of environmental pollutants on survival of bacteria. *Mailing Add:* 209 Wyatt Ave Clemson SC 29631

BAXTER, CHARLES RUFUS, b Paris, Tex, Nov 4, 29; c 3. SURGERY. *Educ:* Univ Tex, BA, 50; Univ Tex Southwest Med Sch Dallas, MD, 54. *Prof Exp:* From asst prof to assoc prof, 62-71, PROF SURG, UNIV TEX SOUTHWEST MED SCH, DALLAS, 71- *Concurrent Pos:* NIH res fel, 59-61; dir, Parkland Hosp Burn Ctr, Dallas, 61-, med dir, 72-; dir Student Health Servs, Univ Tex Southwest Med Sch Dallas, 62-72; chmn surg res adv bd burns & res, Shriners Hosps Crippled Children, 78-; mem surg A study sect, NIH, 78-; exec comt, Am Col Surgeons Comt Trauma, 78- *Honors & Awards:* H Steward Allen Award, Am Burn Asn, 81; G Whitdrer Int Burn Prize, 83; Curtiss P Artz Nat Trauma Award, 84. *Mem:* Am Surg Soc; Am Col Surgeons; Am Burn Asn (pres, 72-73); Am Asn Surg Trauma; Int Soc Burn Injuries. *Res:* Burns and trauma; electrolyte and fluid balance in burns and trauma; circulatory and fluid volumes changes in burns; leukocyte function in burns; preservation of transplantable human tissues and organs; experimental skin grafting. *Mailing Add:* Univ Tex Southwest Med Sch Parkland Mem Hosp 5323 Harry Hines Blvd Dallas TX 75235

BAXTER, CLAUDE FREDERICK, b Hamburg, Ger, July 24, 23; nat US; m 52; c 3. PHYSIOLOGY, BIOCHEMISTRY. *Educ:* Univ Calif, Davis, PhD, 54. *Prof Exp:* Res physiologist, Univ Calif, Davis, 54; fel, McCollum-Pratt Inst, Johns Hopkins Univ, 54-56; SR RES BIOCHEMIST, DIV NEUROBIOL, CITY OF HOPE MED CTR, DUARTE, CALIF, 56-; CHIEF NEUROCHEM LABS, VET ADMIN HOSP, SEPULVEDA, 63- *Concurrent Pos:* Assoc prof physiol, Univ Calif, Los Angeles, 64-74, res prof psychiat, 74- *Mem:* AAAS; Int Soc Neurochem; Am Soc Biol Chemists; Am Chem Soc; NY Acad Sci. *Res:* Milk synthesis; sulfur metabolism; neurochemistry; metabolism and functional aspects of nitrogenous compounds in nervous system; neurotransmitter substances; protein synthesis; osmotic regulation in nervous system; nerve regeneration; aging of nervous system and convulsive disorders. *Mailing Add:* Neurochem Labs 16111 Plummer St Vet Admin Med Ctr Univ Calif Los Angeles Sch Med Sepulveda CA 91343

BAXTER, DENVER O(LEN), b Crandall, Ga, June 26, 25. AGRICULTURAL ENGINEERING. *Educ:* Univ Ga, BS, 50; Univ Mo, MS, 51. *Prof Exp:* Exten agr engr, Univ Ky, 51-54; Univ Ga, 54-56; ASST PROF AGR ENG, UNIV TENN, KNOXVILLE, 56- *Concurrent Pos:* Reviewer, Fed Res Grants, USDA, collabr, functional struct & environ requirements for com animal prod struct; design engr, animal prod facil, Agr Exp Sta, Univ Tenn. *Mem:* Am Soc Agr Engrs; Am Soc Eng Educ. *Res:* Alternate energy systems; earth-tube heat exchangers; solar energy; environmental factors of plant and animal production; application of structural products; functional, structural and environmental design of structures for agriculture; rural housing. *Mailing Add:* Dept Agr Eng Univ Tenn Knoxville TN 37996

BAXTER, DONALD HENRY, b Schenectady, NY, Sept 6, 16; m 42; c 1. RADIOTHERAPY. *Educ:* Union Col, AB, 37; Albany Med Col, MD, 41. *Prof Exp:* Intern med, Albany Med Ctr Hosp, 41-42; resident radiol, Shreveport Charity Hosp, La, 46-47; radiologist, NLa Sanitarium, 47-49; from asst prof to assoc prof, 49-68, PROF RADIOL, ALBANY MED COL, 69- *Concurrent Pos:* Attend radiologist, Albany Med Ctr Hosp, 49- & Vet Admin Hosp, Albany, 51- *Mem:* AMA; Am Col Radiol; Sigma Xi. *Mailing Add:* Dept Radiation Ther Albany Med Col 47 New Scotland Ave Albany NY 12208

BAXTER, DONALD WILLIAM, b Brockville, Ont, Aug 24, 26. MEDICINE. *Educ:* Queen's Univ, Ont, MD, CM, 51; McGill Univ, MSc, 53; Am Bd Psychiat & Neurol, dipl, 58. *Prof Exp:* From instr to asst prof, Sch Med, Univ Sask, 57-62; PROF NEUROL, FAC MED, MCGILL UNIV, 63- *Concurrent Pos:* Dir, Montreal Neurol Inst, 84-91. *Mem:* Am Acad Neurol; Can Med Asn; Am Neurol Soc; Can Neurol Asn. *Mailing Add:* Montreal Neurol Inst 3801 University St Montreal PQ H3A 2B4 Can

BAXTER, GENE FRANCIS, b Sanish, NDak, July 25, 22; m 49, 70; c 3. POLYMER CHEMISTRY. *Educ:* Univ Wash, BS, 44. *Prof Exp:* Res chemist, Am Marietta Corp, 44-56, res group leader, 56-59, info coordr, 59-61, info coordr, Martin-Marietta Corp, 61-62; sr res scientist, Weyerhaeuser Co, 62-73; sr develop chemist, 73-77, chief develop chemist, 77-83, sr scientist, 83-85, CONSULT, GA-PAC CORP, 85- *Honors & Awards:* Distinguished Scientist Award, Georgia-Pacific Corp, 85. *Mem:* Am Chem Soc. *Res:* Polymers; adhesives; thermosetting resins; bonded wood products; plywood; laminated beams; construction materials; formaldehyde polymers; phenolic resins; aromatic amine reactions. *Mailing Add:* Chem Div Ga-Pac Corp 2883 Miller Rd Decatur GA 30035

BAXTER, GEORGE T, b Grover, Colo, Mar 19, 19; m 42; c 3. FISH BIOLOGY, LIMNOLOGY. *Educ:* Univ Wyo, BS & MS, 46; Univ Mich, PhD(zool), 52. *Prof Exp:* From instr to assoc prof, Univ Wyo, 47-59, actg head dept, 68-70, prof zool, 59-84; RETIRED. *Mem:* Am Fisheries Soc. *Res:* Limnology of alpine lakes; fishery management, primarily salmonids in the Rocky Mountains; ichthyology and herpetology of Wyoming. *Mailing Add:* 1614 Bonneville Laramie WY 82070

BAXTER, JAMES HUBERT, b Ashburn, Ga, Dec 19, 13; m 42; c 1. MEDICAL RESEARCH. *Educ:* Univ Ga, BS, 35; Vanderbilt Univ, MD, 41;. *Prof Exp:* intern med, Johns Hopkins Univ, 41-42; Nat Res Coun Welch fel, Dept Biochem, Med Sch, Cornell Univ, 46-47, Hosp, Rockefeller Inst, 47-48 & Dept Med, Johns Hopkins Univ Hosp, 48-50; sr staff mem, Lab Cellular Metab, Nat Heart & Lung Inst, 50-76; surgeon-med dir, USPHS, 52-76; RETIRED. *Mem:* Am Soc Clin Invest; Soc Exp Biol & Med; Am Soc Pharmacol & Exp Therapeut. *Res:* Metabolism. *Mailing Add:* 4511 Delmont Lane Bethesda MD 20814

BAXTER, JAMES WATSON, b Shamrock, Tex, Sept 9, 27. INDUSTRIAL MINERALS, FLUORSPAR DEPOSITS. *Educ:* Univ Ark, BS, 50, MS, 52; Univ Ill, PhD(geol), 58. *Prof Exp:* Asst mineral, Univ Ark, 50-51; res asst, Ark Inst Sci & Technol, 51-52; res asst, 52-54, asst geologist, 56-63, assoc geologist, 63-70, geologist, 70-82, HEAD, INDUST MINERALS & METALS SECT, ILL STATE GEOL SURV, 82- *Concurrent Pos:* Vis scientist, Université Catholique de Louvain, 74. *Mem:* Geol Soc Am; AAAS; Soc Econ Geologists; Soc Mining Engrs; Sigma Xi. *Res:* Stratigraphy and sedimentation; geology of industrial minerals; fluorspar deposits and MVT sulfide deposits; carbonate petrography; Mississippian biostratigraphy, foraminifera, calcareous algae. *Mailing Add:* 300A Natural Resource Bldg Ill State Geol Surv 615 E Peabody Dr Champaign IL 61820

BAXTER, JOHN DARLING, b Lexington, Ky, June 11, 40; m 63; c 2. BIOCHEMISTRY, ENDOCRINOLOGY. *Educ:* Univ Ky, BS, 62; Yale Univ, MD, 66. *Prof Exp:* Intern internal med, Yale-New Haven Hosp, 66-67, resident, 67-68; res assoc molecular biol, Nat Inst Arthritis & Metab Dis, 68-70; sr fel oncol, Dept Biochem & Biophys, Univ Calif, San Francisco, 70-72, from asst prof to prof med & biochem, 72-76, dir, endocrine res div, 76-81, DIR ENDOCRINOL DIV, UNIV CALIF, SAN FRANCISCO, 80-, DIR, METABOLIC RES UNIT, 81- *Concurrent Pos:* Nat Inst Arthritis & Metab Dis res career develop award, 72-75; prin investr, NIH grants, 74-, co-prin investr, 77-; prin investr, Howard Hughes Med Inst, 75-81. *Mem:* Endocrine Soc; Am Fedn Clin Res; Am Soc Clin Invest; Am Thyroid Asn; Asn Am Physicians; Am Chem Soc. *Res:* Molecular biology of the renin-angiotensim system; thyroid hormone action and growth hormone gene expression. *Mailing Add:* Univ Calif Parnassus & Third 671 HSE San Francisco CA 94143

BAXTER, JOHN EDWARDS, b Meridian, Miss, Aug 29, 37; m 68; c 3. PHYSICAL CHEMISTRY, BIOCHEMISTRY. *Educ:* Millsaps Col, BS, 58; Vanderbilt Univ, MS, 60; Duke Univ, PhD(phys chem), 66. *Prof Exp:* Res assoc phys chem, Duke Univ, 66; asst prof physics & chem, NC Wesleyan Col, 66-71; from asst to assoc prof, 71-81, dir, comput-based educ, 78-84, PROF BIOCHEM, CTR HEALTH SCI, UNIV TENN, MEMPHIS, 81-, ACTG DIR, HEALTH SCI COMPUT CTR, 81- *Mem:* Asn Develop Comput Based Instrnl Systs; Am Chem Soc; Sigma Xi. *Res:* Utilization of self instructional methods in biochemistry education. *Mailing Add:* 5292 N Clover Dr Memphis TN 38119

BAXTER, JOHN LEWIS, b San Diego, Calif, July 31, 25; m 47; c 2. FISHERIES MANAGEMENT. *Educ:* Univ Calif, Berkeley, AB, 51. *Prof Exp:* Marine biologist, fisheries lab, State Dept Fish & Game, Calif, 51-63, supvr, pelagic fish prog, 63-70, chief marine resources br, 70-71; spec asst to dir, Nat Marine Fisheries Serv, Washington, DC, 71-73; marine sport fish coordr, 73-74, supvr north ocean area, 74-81, regional mgr, Marine Resources Region, Calif Dept Fish & Game, 81-85; RETIRED. *Concurrent Pos:* Res fel, Scripps Inst Oceanog, 66-72; marine ed, Calif Fish & Game, 62-66; ed-in-chief, Fish Bull, 66-70; exec secy, Marine Fisheries Adv Comt, Dept of Commerce, 71-73. *Mem:* Am Inst Fishery Res Biol; Am Fisheries Soc. *Res:* Biology of sport and commercial fishes. *Mailing Add:* 5772 Garden Grove Blvd No 477 Westminster CA 92683-1858

BAXTER, LINCOLN, II, b Washington, DC, July 25, 24; m 50; c 2. APPLIED PHYSICS, OCEAN ACOUSTICS. *Educ:* Univ Richmond, BS, 46. *Prof Exp:* Teaching asst physics, Cornell Univ, 46-49; res asst physics, Polaroid Corp, 49-55; res assoc appl physics, 56-61, res specialist appl physics, Woods Hole Oceanog Inst, 61-86; RETIRED. *Mem:* Am Asn Physics Teachers; Acoust Soc Am; Optical Soc Am; Royal Astron Soc Can. *Res:* Ocean acoustics; underwater optics; computer data processing; navigation; optics of polarizers. *Mailing Add:* 53 Emmons Rd Falmouth MA 02540

BAXTER, LUTHER WILLIS, JR, b Lawrenceburg, Ky, Nov 25, 24; m 48; c 3. PLANT PATHOLOGY. *Educ:* Eastern Ky State Col, BS, 50; La State Univ, MS, 52, PhD, 54. *Prof Exp:* Plant pathologist, Kaiser Aluminum & Chem Corp, 54-55; assoc plant pathologist, Clemson Univ, 55-58; prof plant path & head dept agr, Western Ky State Col, 58-66; assoc prof, 66-70, PROF PLANT PATH & PHYSIOL, CLEMSON UNIV, 70- *Concurrent Pos:* Consult, Kaiser Aluminum & Chem Corp, 56. *Mem:* Am Phytopath Soc. *Res:* Diseases of camellias; physiology of reproduction in fungi. *Mailing Add:* Dept Plant Path Clemson Univ Main Campus Clemson SC 29634

BAXTER, NEAL EDWARD, b Bluffton, Ind, Sept 13, 08; m 52; c 5. INTERNAL MEDICINE, AEROSPACE MEDICINE. *Educ:* Ind Univ, AB, 32, MD, 35. *Prof Exp:* Intern, Indianapolis City Hosp, 35-36; physician gen pract, 36-54; MED DIR, WESTINGHOUSE ELEC CORP, 57- *Concurrent Pos:* Assoc, Woolery Clin, Bloomington, Ind, 36-37; specialist aerospace & internal med, pvt pract, 54-; dir found, Sch Bus, Ind Univ, 69-; aviation med examr, Fed Aviation Agency. *Honors & Awards:* John A Tamesia Award, Aerospace Med Asn, 65. *Mem:* Am Acad Gen Pract; Am Diabetes Asn; fel Am Acad Family Physicians; Aerospace Med Asn (pres, 65-66); Civil Aviation Med Asn (pres, 62). *Res:* Aviation physiology and oxygen equipment; syncopal reactions of anoxic subjects observed in the low pressure chamber; civil aviation medicine; medical aspects of business aviation. *Mailing Add:* 1624 Buffstone Ct Bloomington IN 47401

BAXTER, ROBERT MACCALLUM, b Summerside, PEI, Apr 25, 26; m 53; c 3. BIOCHEMISTRY, LIMNOLOGY. *Educ:* Mt Allison Univ, BSc, 47; McGill Univ, PhD(biochem), 52. *Prof Exp:* Res officer chem microbiol, Nat Res Coun, Ottawa, 52-61; prof chem, Haile Sellassie Univ, 61-73; RES SCIENTIST LIMNOL, NAT WATER RES INST, CAN, 73- *Concurrent Pos:* Fel, Tech Univ Norway, 59-60. *Mem:* Brit Biochem Soc; Can Biochem Soc; Can Soc Microbiol; Int Asn Theoret & Appl Limnol. *Res:* Physiology of microbial halophilism and psychrophilism; tropical limnology; environmental impacts of dams and impoundments; chemical limnology of humic substances; microbial metabolism of environmental contaminants. *Mailing Add:* Nat Water Res Inst PO Box 5050 Burlington ON L7R 4A6 Can

BAXTER, ROBERT WILSON, b CZ, June 7, 14; m 45; c 1. PALEOBOTANY. *Educ:* Wash Univ, AB, 37, MS, 47, PhD(paleobot), 49. *Prof Exp:* Asst instr bot, Univ Hawaii, 38-39; from asst prof to prof bot, Univ Kans, 49-83, chmn dept, 53-57, emer prof, 83; RETIRED. *Concurrent Pos:* Fulbright lectr, Univ Col WI, 55-56; NSF res award, 58, 61 & 63. *Mem:* Bot Soc Am; Int Soc Plant Morphol; Int Soc Plant Taxon. *Res:* Carboniferous flora of the central United States; cretaceous Dakota sandstone flora. *Mailing Add:* 1010 Wellington Rd Lawrence KS 66049

BAXTER, RONALD DALE, b Pittsburgh, Pa, Oct 17, 34; m 58; c 3. ELECTRICAL ENGINEERING, SOLID STATE ELECTRONICS. *Educ:* Ohio State Univ, BSEE, 60, MSc, 63. *Prof Exp:* From assoc to assoc mgr & scientist, Battelle Columbus Labs, 60-74; prin scientist, 74-83, corp scientist, 83-91, MGR, SOLID STATE TECHNOL, LEEDS & NORTHRUP CO, 91- *Mem:* Am Vacuum Soc. *Res:* Electrical properties of semiconductors and other solid state materials; solid state device development; thin film techniques for device fabrication including chemical and physical vapor deposition methods; solid state sensors development; micromachining-microfabrication. *Mailing Add:* Leeds & Northrup Tech Ctr Dickerson Rd North Wales PA 19454

BAXTER, ROSS M, b Erin, Ont, Oct 1, 18; m 46; c 3. PHARMACY. *Educ:* Univ Toronto, PhmB, 43; Univ Sask, BSc, 46; Univ Fla, MSc, 48, PhD, 51. *Prof Exp:* From asst prof to prof pharmacol, Univ Toronto, 51-79, dean pharm, 79-85; RETIRED. *Mem:* AAAS; NY Acad Sci; Can Pharmaceut Asn; Acad Pharmaceut Sci; Asn Fac Pharm Can. *Res:* Medicinal chemistry. *Mailing Add:* 3425 Harvester Rd Burlington ON L7N 3N1 Can

BAXTER, SAMUEL G, US citizen. ELECTRICAL ENGINEERING. *Educ:* Ga Inst Technol, BSEE, 60. *Prof Exp:* Asst design engr, Clark Bros Co, 60-61; asst design engr & proj engr, Int Controls Corp, 61-63; resident sr engr, George C Marshall Spaceflight Ctr, Brown Eng Co, 63-65; supvy electronic engr, NASA, 65-67; proj eng & mgr, Aerosci Inc, 67-68; prin engr, Sci Atlanta Inc, 68-; PRIN ENGR, ELECROMAGNETIC-SCI, INC. *Res:* VHF and UHF amateur radio equipment design; telemetry receiver design; satellite GCE equipment; Marisat GCE equipment; microwave components. *Mailing Add:* Electromagnetic-Sci Inc 125 Technology Park Norcross GA 30092

BAXTER, WILLARD ELLIS, b Chester, Pa, Dec 14, 29; m 54; c 3. MATHEMATICS. *Educ:* Ursinus Col, BS, 51; Univ Wis, MS, 52; Univ Pa, PhD, 56. *Prof Exp:* Asst prof math, Ohio Univ, 56-58; from asst prof to assoc prof, 58-67, chmn dept, 70-75, PROF MATH, UNIV DEL, 67- *Mem:* Am Math Soc; Math Asn Am; Sigma Xi. *Res:* Structure theory of rings. *Mailing Add:* Dept Math Sci Univ Del Newark DE 19716

BAXTER, WILLIAM D, b Larned, Kans, Sept 13, 36; m 59; c 3. IMMUNOBIOLOGY. *Educ:* Phillips Univ, AB, 60; Univ Kans, PhD(zool), 66. *Prof Exp:* Asst prof, 66-71, ASSOC PROF BIOL, BOWLING GREEN STATE UNIV, 71- *Concurrent Pos:* Adj prof, Med Col Ohio, Toledo, 68- *Mem:* AAAS; Electron Micros Soc Am; Sigma Xi. *Res:* Immunology, especially early cellular events associated with the establishment of immunity; electron microscopy of animal tissues. *Mailing Add:* Dept Biol Bowling Green State Univ Bowling Green OH 43403

BAXTER, WILLIAM JOHN, b Whinburgh, Eng, July 31, 35; US citizen; m 60; c 3. METAL PHYSICS. *Educ:* Oxford Univ, BA, 57, DPhil(physics), 61. *Prof Exp:* Res physicist, Fulmer Res Inst, Eng, 60-61; res assoc internal friction, Cornell Univ, 61-63; assoc sr res physicist, 64-66, sr res physicist, 66-78, staff res scientist, 78-82, SR STAFF RES SCIENTIST, RES LABS, GEN MOTORS CORP, 82- *Concurrent Pos:* Fel, Cornell Univ, 61-63. *Honors & Awards:* Second Prize Photog Exhib, Am Soc Testing & Mat, 73. *Mem:* Am Soc Testing & Mat; Am Phys Soc; Am Inst Mining, Metall & Petrol Engrs Metall Soc. *Res:* Exoelectron emission and photoelectron microscopy; fatigue and deformation of metals; crack detection; internal friction; metal composites. *Mailing Add:* Dept Physics Res Labs Gen Motors Corp Warren MI 48090-9055

BAXTER, WILLIAM LEROY, b San Diego, Calif, Dec 21, 29; m 58; c 2. VIROLOGY. *Educ:* Univ Calif, Los Angeles, AB, 56, PhD(microbiol), 61. *Prof Exp:* USPHS fel virol, Univ Calif, Los Angeles, 61-63; PROF MICROBIOL, SAN DIEGO STATE UNIV, 63- *Mem:* AAAS; Am Soc Microbiol; Sigma Xi. *Res:* Early steps of influenza virus infection; electron microscopy of virus-infected cell culture. *Mailing Add:* Dept of Biol San Diego State Univ San Diego CA 92182-0057

BAY, DARRELL EDWARD, b Hays, Kans, Dec 22, 42; m 75; c 1. MEDICAL ENTOMOLOGY, VETERINARY ENTOMOLOGY. *Educ:* Kans State Univ, BS, 64, MS, 67, PhD(entom), 74. *Prof Exp:* Entomologist, Walter Reed Army Inst Res, 68-71; PROF ENTOM, TEX A&M UNIV, 74- *Mem:* Entom Soc Am; Am Registry Prof Entomologists. *Res:* Biology, ecology and control of livestock insects. *Mailing Add:* Dept Entom Tex A&M Univ College Station TX 77843

BAY, ERNEST C, b Schenectady, NY, Aug 7, 29; m 63; c 2. MEDICAL ENTOMOLOGY, URBAN ENTOMOLOGY. *Educ:* Cornell Univ, BS, 53, PhD(entom), 60. *Prof Exp:* Specialist entom, Univ Calif, Riverside, 60-61, asst entomologist biol control, 61-69, head div, 69-71; head dept entom, Univ Md, College Park, 71-75; supt, 75-85, PROF ENTOM, WESTERN WASH RES & EXTEN CTR, WASH STATE UNIV, 85- *Concurrent Pos:* NIH grant, 61-65; WHO consult, Nicaraguan Ministry Health, 64, Far East, 67 & Nigeria, 72; secy gen, XV Int Cong Entom, 73-75. *Mem:* Entom Soc Am; Am Mosquito Control Asn. *Res:* Ecology and control of insects of medical and veterinary importance; ecology and control of chironomid midges; biological control of medically important insects. *Mailing Add:* Western Wash Res & Exten Ctr Wash State Univ Puyallup WA 98371

BAY, ROGER RUDOLPH, b La Crosse, Wis, Nov 27, 31; m 58; c 2. FORESTRY. *Educ:* Univ Idaho, BS, 53; Univ Minn, MF, 54, PhD(forestry), 67. *Prof Exp:* Forester, Forest Serv, USDA, 54, res forester, 56-61, proj leader, NCent Forest Exp Sta, 61-70, from asst chief to chief, Br Watershed & Aquatic Habitat Res, Div Forest Environ Res, 70-73, asst to dep chief res, 73-74, dir, Intermountain Forest & Range Exp Sta, 74-83, dir, Pac Southwest Forest & Range Exp Sta, 83-88; RETIRED. *Concurrent Pos:* Mem, Foreign Agr Orgn Wateshed Cong, Arg, 71; consult gen forestry, Arg, 81- *Mem:* AAAS; Am Forestry Asn; Soc Am Foresters; Am Geophys Union. *Res:* Watershed management research; forest hydrology. *Mailing Add:* 6931 Mogollon Dr Bozeman MT 59715

BAY, THEODOSIOS (TED), b Montreal, Que, Apr 22, 31; m 59; c 3. CHEMICAL ENGINEERING, SYSTEMS DESIGN. *Educ:* McGill Univ, BEng, 53; Princeton Univ, MSEng, 55. *Prof Exp:* Chem engr, Esso Res & Eng Co, 57-60; assoc scientist, Leeds & Northrup Co, 60-61, res scientist, 61-63, sr scientist systs anal, 63-75, prin scientist, 75-81; assoc engr, 81-88, SR ASSOC ENGR, MOBIL RES & DEVELOP CORP, 88- *Mem:* Am Inst Chem Engrs. *Res:* Automatic control of chemical processes; computer control; systems science. *Mailing Add:* 112 Knollwood Dr Lansdale PA 19446

BAY, ZOLTAN LAJOS, b Gyulavari, Hungary, July 24, 1900; m 47; c 3. QUANTUM PHYSICS. *Educ:* Univ Budapest, MS, 23, PhD(physics), 25. *Hon Degrees:* Dr, Univ Edinburgh, Eng, 78, Univ Budapest, Hungary, 86. *Prof Exp:* Prof theoret physics, Univ Szeged, 30-36; prof atomic physics, Tech Univ Budapest, 38-48; res prof, George Washington Univ, 48-55; physicist, Nat Bur Standards, 55-73; res prof, Am Univ, 86-; RETIRED. *Concurrent Pos:* Dir, Tungsram Res Lab, 36-48; tech eng mgr, United Incandescent Lamp & Elec Co, 44-48. *Honors & Awards:* Boyden Award, Franklin Inst, Philadelphia, 80; A Fono Award, Hungarian Astronaut Soc, Budapest, 81. *Mem:* Fel Am Phys Soc; hon mem Hungarian Acad Sci; hon mem Hungarian Phys Soc; sr mem Inst Elec & Electronics Engrs. *Res:* Atomic and molecular spectroscopy; nuclear excited states, techniques and theory of fast coincidence experiments; ionization of matter by high energy radiations; optical masers; unified standardization of time and length. *Mailing Add:* 151 Quincy St Chevy Chase MD 20815

BAYER, ARTHUR CRAIG, b Brooklyn, NY, Apr 5, 46; m 76; c 2. ORGANIC PROCESS RESEARCH, FLAME RETARDANT CHEMISTRY. *Educ:* Manhattan Col, BS, 67; State Univ NY, Buffalo, PhD(org chem), 72. *Prof Exp:* Res assoc, Univ Ore, 72-74; res chemist, Hooker Chem Corp, 74-78; sr res chemist, Stauffer Chem Co, 78-84; dir res, First Chem Corp, 84-88; SR GROUP LEADER, CIBA-GEIGY CORP, 88- *Mem:* AAAS; Am Chem Soc; Am Mgt Asn; Sigma Xi. *Res:* The invention and development of new syntheses of organic molecules as applied to agricultural and pharmaceutical chemical manufacturing processes. *Mailing Add:* 12335 Cardeza Ave Baton Rouge LA 70816-8906

BAYER, BARBARA MOORE, b Henderson, Tex, Dec 10, 48; m 72; c 1. PHARMACOLOGY. *Educ:* Univ Tex, El Paso, BS, 72; Ohio State Univ, PhD(pharmacol), 77. *Prof Exp:* Lab instr chem, Univ Tex, El Paso, 70-72; res assoc pharmacol, Toxicol Lab, Ohio State Univ, 72-76, pharmacol fac, 77; res fel, NIH, Bethesda, 77-79, staff fel, 79-81; spec asst sci, Food & Drug Admin, Rockville, 81-83; RES ASST PROF PHARMACOL, GEORGETOWN UNIV, 82- *Concurrent Pos:* Sci consult to comnr, Food & Drug Admin, 83-; prin investr, Nat Inst Arthritis & Infectious Disease, 83-85. *Mem:* AAAS; Found Advan Educ Sci; Am Soc Pharmacol & Exp Therapeut; Sigma Xi. *Res:* Cellular action of anti-inflammatory drugs, cell cycle analysis and the regulation and metabolism of histamine and polyamines in normal and rapidly growing tissues. *Mailing Add:* 9212 Farnsworth Ct Potomac MD 20854-4503

BAYER, CHARLENE WARRES, b Loredo, Tex, Aug 13, 50; m 75; c 2. INDOOR AIR RESEARCH, MASS SPECTROMETRY. *Educ:* Baylor Univ, BS, 72; Emory Univ, MS, 74, PhD(org chem), 81. *Prof Exp:* Chemist, Smith Kline Corp, 74-75, Emory Univ, 75-76 & US Geol Surv, 80-83; consult, Ga Inst Technol, 77-80; sr res scientist, 83-90, PRIN RES SCIENTIST, GA TECH RES INST, 90- *Concurrent Pos:* Consult, S C Johnson, Inc, 88- & DuPont, 91- *Mem:* Am Chem Soc; Am Soc Heating, Refrig & Air Conditioning Engrs; Am Soc Mass Spectrometry; Sigma Xi; Am Soc Testing & Mat. *Res:* Trace analysis and separations sciences and mass spectrometry of environmental samples, particularly of air quality analysis; air quality evaluation by chromatographic and mass spectrometric determination of volatile organic compounds; source characterization using environmental chambers; building policies; trace species detection and identification. *Mailing Add:* Emerson A112 Ga Tech Res Inst Atlanta GA 30332-0800

BAYER, DAVID E, b Grass Valley, Ore, Aug 1, 26. PLANT PHYSIOLOGY. *Educ:* Ore State Univ, BS, 51, MS, 53; Univ Wis, PhD(agron), 58. *Prof Exp:* Assoc agriculturalist, 59-62, asst botanist, 62-69, assoc botanist, 69-73, BOTANIST, UNIV CALIF, DAVIS, 73-, PROF, 78- *Mem:* Weed Soc Am. *Res:* Plant physiology, particularly pertaining to herbicides and their effects on plants. *Mailing Add:* Dept Bot Univ Calif Davis CA 95616

BAYER, DOUGLAS LESLIE, computer science, for more information see previous edition

BAYER, FREDERICK MERKLE, b Asbury Park, NJ, Oct 31, 21. ZOOLOGY. *Educ:* Univ Miami, BS, 48; George Washington Univ, MS, 54, PhD(zool), 58. *Prof Exp:* Asst dir, State Mus, Fla, 42-46; asst marine invertebrates, Marine Lab, Univ Miami, 46-47; from asst cur to assoc cur, US Nat Mus, 47-61; assoc prof, 62-63, prof, Sch Marine & Atmospheric Sci, Univ Miami, 64-75; vis cur, 71, CUR, NAT MUS NATURAL HIST, DEPT INVERT ZOOL, SMITHSONIAN INST, 75- *Concurrent Pos:* Mem, Bikini Sci Resurv Exped, USN, 47, Ifaluk Atoll Surv Team, Pac Sci Bd, Nat Res Coun-Nat Acad Sci, 53, Palau Island Exped, Vanderbilt Found, 55 & Gulf of Guinea Exped, Inst Marine Sci, Univ Miami, 64 & 65, Southwestern Caribbean Exped, 66, Hispaniola-Jamaica Exped, 70 & Caribbean Basin Exped, 72; ed, Bull Marine Sci of Gulf & Caribbean, 62-71; mem, Int Comn Zool Nomenclature, 72- *Mem:* Paleont Soc Am; Sigma Xi. *Res:* Taxonomy, zoogeography and ecology of Octocorallia. *Mailing Add:* Dept Invert Zool Smithsonian Inst Washington DC 20560

BAYER, GEORGE HERBERT, b Woodhaven, NY, Dec 7, 24; m 54. WEED SCIENCE. *Educ:* Cornell Univ, BS, 50, MS, 52, PhD(veg crops), 65. *Prof Exp:* Res asst, GLF Soil Bldg, 54-55, tech field rep, 55-61; res asst, Cornell Univ, 61-64; herbicide develop mgr, 64-65, Agway Inc, 64-65, sr staff scientist, 66-85; CONSULT, 88- *Concurrent Pos:* Courtesy prof, Veg Crops Dept, Cornell Univ, 80-; weed scientist consult, 85- *Mem:* Fel Weed Sci Soc Am (treas, 79-81); Coun Agr Sci & Technol; Sigma Xi. *Res:* Discover and evaluate the activity of the various classes of herbicides on vegetable, forage and tree fruit crops. *Mailing Add:* 216 Forest Home Dr Ithaca NY 14850

BAYER, HORST OTTO, b Stuttgart, Ger, Oct 21, 34; US citizen; m 55; c 2. ORGANIC CHEMISTRY. *Educ:* Rochester Inst Technol, BS, 57; Purdue Univ, PhD(org chem), 61. *Prof Exp:* Sr chemist, Pesticide Syntheses Group, 62-67, head res lab, 67-73, proj leader, 73-81, sect mgr, Res Lab, 81-85, DEPT MGR, HERBICIDES RES LAB, ROHM & HAAS CO, 85- *Concurrent Pos:* NSF-NATO fel, 61-62. *Mem:* Am Chem Soc. *Res:* Synthesis, structure and activity studies relating to herbicides, insecticides, fungicides and plant growth regulators. *Mailing Add:* 186 Red Rose Dr Levittown PA 19056

BAYER, MANFRED ERICH, b Goerlitz, Silesia, Sept 22, 28; WGer citizen; m 58; c 2. CELL BIOLOGY, VIROLOGY. *Educ:* Univ Hamburg, Dr med, 56; Inst Trop Med & Parasitol, dipl, 61. *Prof Exp:* Fel neurol, Univ Hosp, Hamburg, 56-58; fel virol, Inst Trop Dis, 58-59, res assoc, 59-60, asst mem, 60-62; asst mem molecular biol, 62-67, assoc mem, 67-78, mem, 78-86, SR MEM, INST CANCER RES, 86- *Concurrent Pos:* Vis scientist, Inst Cancer Res, France, 61; prin investr grants, NIH & NSF, 66-; adj prof, Dept Microbiol, Med Sch, Univ Pa, 71-, mem, Grad Group Microbiol, 76-; vis prof, Inst Virol, Kyoto Univ, Japan, 77; NSF panel mem, 82-84; biol instrumentation & study sections, Med Sci Spec Prog, NIH Microbiol Chem, 79-86; vis prof, Biozentrum Univ Basel, Switz, 84, Max Planck Inst for Immunobiol, Freiburg, WGer, 86; hon vis prof, Dept Microbiol, Dalhousie Univ Med Sch, Halifax, Can, 85- *Mem:* Am Soc Microbiol; Biophys Soc Am; Electron Microscope Soc Am; Electron Microscope Soc Can; Am Soc Cell Biol; Am Soc Virol. *Res:* Kinetics and mechanisms of virus infection; cell surface charge, electrophoresis; cell surface structure and function; isolation of receptor domains; metal interactions with membranes. *Mailing Add:* Inst Cancer Res 7701 Burholme Ave Fox Chase Cancer Ctr Philadelphia PA 19111-2497

BAYER, MARGRET HELENE JANSSEN, b 1931; m 58; c 2. MEMBRANE PHYSIOLOGY. *Educ:* Univ Hamburg, dipl, 58, Dr rer nat, 61, Dr habil(plant physiology), 76. *Prof Exp:* Res assoc biol, Univ Hamburg, 58-61; res assoc, Inst Cancer Res, 62-76, SR RES ASSOC BIOL & MICROBIOL, FOX CHASE CANCER CTR, 77- *Concurrent Pos:* Vis prof, Dept Biol, Univ Hamburg, 75, habilitation, 76; lectr, Univ Pa, Philadelphia, 79-80; NSF & NIH Fed grants. *Mem:* Am Soc Microbiol; Am Soc Plant Physiologists. *Res:* Host virus interactions; bacteriophage infection; membrane biochemistry; biology of plant tumors; plant hormone physiology; regulation and control of abnormal growth. *Mailing Add:* Inst for Cancer Res Fox Chase Cancer Ctr Philadelphia PA 19111-2497

BAYER, RAYMOND GEORGE, b New York, NY, June 9, 35; m 58; c 4. TRIBOLOGY, FAILURE ANALYSIS. *Educ:* St Johns Univ, NY, BS, 56; Brown Univ, RI, MS, 59. *Prof Exp:* SR ENG, IBM CORP, ENDICOTT, NY, 58- *Mem:* Am Soc Testing & Mat. *Res:* Tribology with particular emphasis on the development of approaches for predicting wear behavior; stress testing of electronic packages. *Mailing Add:* 4609 Marshall Dr W Vestal NY 13850

BAYER, RICHARD EUGENE, b Milwaukee, Wis, Jan 11, 32; m 57; c 4. ANALYTICAL CHEMISTRY, INORGANIC CHEMISTRY. *Educ:* Carroll Col, Wis, BS, 54; Ind Univ, PhD(chem), 59. *Prof Exp:* Assoc prof, 58-66, chmn dept, 67-76, PROF CHEM, CARROLL COL, WIS, 66-; PRES, BIONOMICS CORP, 71- *Honors & Awards:* Am Chem Soc Award. *Mem:* Am Chem Soc; Sigma Xi. *Res:* Rare earth chelates; atomic absorption; selective ion electrodes; chemical health and safety; solid waste disposal. *Mailing Add:* Dept Chem Carroll Col Waukesha WI 53186

BAYER, ROBERT CLARK, b New York, NY, July 4, 44; m 67; c 3. LOBSTER NUTRITION. *Educ:* Univ Vt, BS, 66, MS, 68; Mich State Univ, PhD(avian physiol), 72. *Prof Exp:* PROF ANIMAL VET & AQUATIC SCI, UNIV MAINE, ORONO, 72-, PROJ DIR, LOBSTER RES PROJ. *Mem:* AAAS; Nat Shellfisheries Asn; World Maricult Soc; Am Inst Nutrit. *Res:* Lobster nutrition, health and behavior. *Mailing Add:* 128 Hitchner Hall Univ Maine Orono ME 04473

BAYER, SHIRLEY ANN, b Evansville, Ind, Aug 20, 40; m 73. NEUROANATOMY. *Educ:* St Mary-of-the-Woods Col, BA, 63; Calif State Univ, Fullerton, MA, 69; Purdue Univ, PhD(biol), 74. *Prof Exp:* Teacher biol, Guerin High Sch, Ill, 63-65 & Our Lady of Providence High Sch, Ind, 65-66; teacher & chmn sci dept, Marywood Sch, Calif, 66-70; asst res biologist, 74-80, RES SCIENTIST, PURDUE UNIV, 80- *Mem:* Soc Neurosci. *Res:* Light microscopy studies on the development of the central and peripheral nervous systems in the rat; using x-irradiation and hydrogen-thymidine autoradiography. *Mailing Add:* Dept Biol Ind Univ Purdue Univ Indianapolis IN 46202

BAYER, THOMAS NORTON, b Elyria, Ohio, Dec 23, 34; m 74. GEOLOGY, PALEONTOLOGY. *Educ:* Macalester Col, BA, 57; Univ Minn, MS, 60, PhD(geol), 65. *Prof Exp:* Instr geol, Macalester Col, 57-64; asst prof, 64-68, PROF EARTH SCI, WINONA STATE UNIV, 68- *Concurrent Pos:* Lectr, Macalester Col, 61-64; consult classroom prod, KTCA-TV, 62-63. *Mem:* Sigma Xi; Geol Soc Am; Am Asn Petrol Geologists; Paleont Soc; Soc Econ Paleont & Mineral. *Res:* Paleoecology of lower Paleozoic invertebrate faunas. *Mailing Add:* Dept Geol & Earth Sci Winona State Univ Winona MN 55987

BAYES, KYLE D, b Colfax, Wash, Mar 3, 35; m 61; c 2. CHEMICAL DYNAMICS. *Educ:* Calif Inst Technol, BS, 56; Harvard Univ, PhD(chem), 59. *Prof Exp:* NSF fel, Univ Bonn, 59-60; chmn dept, 87-90, from asst prof to assoc prof, 60-71, PROF CHEM, UNIV CALIF, LOS ANGELES, 71- *Concurrent Pos:* Sloan vis lectr, Harvard Univ, 66; Erskine fel, Univ Canterbury, NZ, 88. *Mem:* AAAS; Am Phys Soc; Am Chem Soc. *Res:* Chemical kinetics and spectroscopy of gas phase reactions. *Mailing Add:* Dept Chem Univ Calif Los Angeles CA 90024

BAYEV, ALEXANDER A, b Chita, USSR, Jan 10, 04; m 44; c 2. BIOTECHNOLOGY. *Educ:* Kazan Univ, Physician, 27; USSR Acad Sci, Cand Biol, 47, Dr, 67. *Prof Exp:* Asst biochem & chair, Kazan Med Inst, 30-35; res scientist, Inst Biochem, 35-37 & 54-59, acad secy, Dept Biophys, Biochem, Physiol & Active Subst, 70-88, CHIEF LAB, INST MOLECULAR BIOL, USSR ACAD SCI, 59-, COUNR PRESIDIUM, 88- *Concurrent Pos:* Chmn, Human Genome Coun, State Comt on Sci & Technol & USSR Acad Sci, 89- *Mem:* Int Union Biochem; Am Soc Biochem & Molecular Biol; Humane Genome Orgn; Europ Molecular Biol Orgn. *Res:* Animal cell respiration and nitrogen metabolism; transfer RNA primary structure; primary structure of yeast ribosomal genes; genetic engineering and biotechnology human somatotropin and plasminogen tissue activator; human genome mapping and sequencing. *Mailing Add:* Inst Molecular Biol 117984 Moscow B-334 Vavilov Str 32 Moscow 7-095 USSR

BAYHURST, BARBARA P, b Vivian, La, Mar 31, 26; div; c 4. NUCLEAR CHEMISTRY, RADIO CHEMISTRY. *Educ:* La State Univ, BS, 46. *Prof Exp:* STAFF MEM RADIOCHEM, LOS ALAMOS SCI LAB, UNIV CALIF, 55- *Mem:* Am Chem Soc; AAAS. *Res:* Absolute counting; cross-section measurements; atomic devices as research tools; nuclear waste disposal. *Mailing Add:* 865 Camino Encantado Los Alamos NM 87544

BAYLESS, DAVID LEE, b Alliance, Ohio, June 12, 38; m 63; c 3. STATISTICS. *Educ:* Muskingum Col, AB, 60; Fla State Univ, MS, 63; Tex A&M Univ, PhD(statist), 68. *Prof Exp:* Asst, Fla State Univ, 61-63; assoc mem staff, Bell Tel Labs, NJ, 63-64; consult, Gen Food Tech Ctr, NY, 64-65; fel, Tex A&M Univ, 65-68; mgr prog develop & sci statist scientist, Res Triangle Inst, 68-83. *Concurrent Pos:* Adj assoc prof statist, NC State Univ, 68-; asst to dir res, Nat Assessment of Educ Proj, 70. *Mem:* Am Statist Asn. *Res:* Survey research methods and application in education, health and crime. *Mailing Add:* 660 Azalea Dr Rockville MD 20850-2003

BAYLESS, LAURENCE EMERY, b Richmond, Va, Aug 27, 38; m 62. BIOLOGY. *Educ:* Univ Tenn, AB, 61; Tulane Univ, MS, 62, PhD(biol), 66. *Prof Exp:* From asst prof to assoc prof, 66-77, PROF BIOL, CONCORD COL, 77- *Mem:* Am Soc Ichthyol & Herpet; Soc Study Amphibians & Reptiles; Ecol Soc Am; Sigma Xi. *Res:* Population ecology and life history of amphibians. *Mailing Add:* Dept Biol Concord Col Athens WV 24712

BAYLESS, PHILIP LEIGHTON, b Indianapolis, Ind, Feb 23, 28; m 49; c 3. ORGANIC CHEMISTRY. *Educ:* Oberlin Col, AB, 49; Duke Univ, PhD(chem), 54. *Prof Exp:* PROF CHEM & CHMN DEPT, WILMINGTON COL, 54- *Concurrent Pos:* Alexander von Humboldt fel, 63-64; fac fel, Nat Acad Sci, 70-71; assoc prof, Univ Petrol & Minerals, Dhahran, Saudi Arabia, 76-77. *Honors & Awards:* Gustav Ohaus-NSTA Award, Nat Sci Teacher's Asn, 75. *Mem:* Am Chem Soc. *Res:* Mechanisms of chemical reactions; metal-organic compounds. *Mailing Add:* 71 Faculty Pl Wilmington OH 45177

BAYLEY, HENRY SHAW, b Macclesfield, Eng, Aug 21, 38. BIOCHEMISTRY, NUTRITION. *Educ:* Univ Reading, BSc, 60; Univ Nottingham, PhD, 63; ARIC. *Prof Exp:* Asst lectr nutrit, Wye Col, Univ London, 63-65; from asst prof to assoc prof, 65-76, PROF NUTRIT, UNIV GUELPH, 76- *Honors & Awards:* Borden Award, Nutrit Soc Can, 77. *Mem:* Am Soc Animal Sci; Am Inst Nutrit; Can Soc Animal Sci; Nutrit Soc Can; Brit Nutrit Soc. *Res:* Energy metabolism: factors influencing digestion and absorption of major energy yielding nutrients, use of respiration calorimeter and substrate turnover rates as indicators of metabolic status in the developing animal. *Mailing Add:* Dept Nutrit Univ Guelph Guelph ON N1G 1E9 Can

BAYLEY, STANLEY THOMAS, b Grays, Essex, Eng, Nov 5, 26; Can citizen; m 50; c 2. BIOPHYSICS, BIOCHEMISTRY. *Educ:* Univ London, BSc, 46, PhD(physics), 50. *Prof Exp:* Nuffield res fel biophys, King's Col, Univ London, 50-52; from asst res officer to sr res officer, Nat Res Coun Can, 52-67; chmn dept, 68-74, PROF BIOL, MCMASTER UNIV, 67- *Honors & Awards:* Centennial Medal Can, 68, Silver Jubilee Medal, 78. *Mem:* Am Soc Biochem & Molecular Biol; Can Biochem Soc; Can Soc Cell Molecular Biol. *Res:* Molecular biology of DNA tumor viruses. *Mailing Add:* Dept Biol McMaster Univ 1280 Main St W Hamilton ON L8S 4K1 Can

BAYLIFF, WILLIAM HENRY, b Annapolis, Md, Aug 29, 28; m 69. FISHERIES. *Educ:* Western Md Col, BA, 49; Univ Wash, MS, 54, PhD, 65. *Prof Exp:* Biologist, State Dept Fisheries, Wash, 52-54, 57-58; scientist, 58-65, SR SCIENTIST, INTER-AM TROP TUNA COMN, SCRIPPS INST OCEANOG, 65- *Concurrent Pos:* Marine fishery biologist, Food & Agr Orgn, UN, 67-68, consult, Develop Prog Fishery Proj, Callao, Peru, 71, convener, working party tuna and billfish tagging, Panel Experts Facilitation Tuna Res, 70-87; mem ed comt, Fishery Bull, Nat Marine Fisheries Serv, 70-78, 87-90; consult, Direccion General de Recursos Marinos, Ministerio de Comercio e Industrias, Republic of Panama, 83. *Honors & Awards:* W F Thompson Award, Am Inst Fishery Res Biologists, 69. *Mem:* Am Inst Fishery Res Biologists; Am Fisheries Soc. *Res:* Biology and population dynamics of marine fishes. *Mailing Add:* Inter-Am Trop Tuna Comn Scripps Inst Oceanog 8604 La Jolla Shores Dr La Jolla CA 92037

BAYLIN, GEORGE JAY, radiology; deceased, see previous edition for last biography

BAYLINK, DAVID J, b Portland, Ore, May 24, 31. MINERAL METABOLISM. *Educ:* Loma Linda Univ, MD, 57. *Prof Exp:* CHIEF MINERAL METAB, PETTIS VET HOSP, 81-; PROF MED, LOMA LINDA UNIV, 81- *Mailing Add:* Vet Hosp 151 11201 Benton St Loma Linda CA 92357

BAYLIS, JEFFREY ROWE, b Jackson, Mich, Nov 1, 45. ETHOLOGY, ICHTHYOLOGY. *Educ:* Univ Calif, Santa Barbara, BA, 68; Univ Calif, Berkeley, MA, 72, PhD(zool), 75. *Prof Exp:* Fel zool, Rockefeller Univ, 74-76; asst prof zool, 76-81, ASSOC PROF ZOOL, UNIV WIS-MADISON, 81- *Mem:* Animal Behav Soc; Sigma Xi; Soc Study Evolution; Am Soc Ichthyologists & Herpetologists; AAAS. *Res:* Analysis of the social communicatory signals of animals, with special regard to the evolutionary origins of the signals and the environmental pressures that shaped them; evolution of sexual behavior and parental care; population biology of fishes. *Mailing Add:* Dept Zool Univ Wis Madison WI 53706

BAYLIS, JOHN ROBERT, JR, b Chicago, Ill, May 12, 27; m 54; c 2. GENETICS. *Educ:* Utah State Univ, BS, 51; Northwestern Univ, MS, 56; Fla State Univ, PhD(genetics), 66. *Prof Exp:* Biologist, Oak Ridge Nat Lab, 66-67; asst prof, 67-74, ASSOC PROF GENETICS, UNIV W FLA, 74- *Mem:* Genetics Soc Am; Am Inst Biol Sci; Environ Mutagen Soc; Sigma Xi. *Res:* Fungal genetics; environmental mutagenesis. *Mailing Add:* Dept Biol Univ WFla Pensacola FL 32504

BAYLIS, WILLIAM ERIC, b Providence, RI, Nov 28, 39; m 61; c 2. ATOMIC PHYSICS, THEORETICAL PHYSICS. *Educ:* Duke Univ, BSc, 61; Univ Ill, MSc, 63; Tech Univ Munich, DSc(physics), 67. *Prof Exp:* Res asst physics, Max Planck Inst Extraterrestrial Physics, Garching, WGer, 63-67; res assoc atomic physics, Joint Inst Lab Astro-Physics, Boulder, Colo, 67-69; asst prof to assoc prof, 69-78, PROF PHYSICS, UNIV WINDSOR, 78- *Concurrent Pos:* Vis scientist atomic physics, Max Planck Inst Fluid-dyn, Gottingen, WGer, 76-77; vchmn, Div Atomic & Molecular Physics, Can Asn Physicists, 84-85, chmn, 85-86; vis scientist atomic physics, Inst Fundamental Res, CEN-Saclay, France, 85. *Mem:* Fel Am Phys Soc; Can Asn Physicists. *Res:* Calculations of atomic and molecular structure; pseudopotential methods; pressure broadening of spectral lines; inelastic atomic collisions at thermal and keV energies; relativistic and correlation effects in electron/positron scattering of atoms; Pauli-algebra approach to relativistic physics. *Mailing Add:* Dept Physics Univ Windsor Windsor ON N9B 3P4 Can

BAYLISS, COLIN EDWARD, b Montreal, Que, Aug 13, 36; m 62; c 3. CARDIOVASCULAR SURGERY. *Educ:* Univ Toronto, MD, 61; FRCS(C), 69. *Prof Exp:* Resident surg, Univ Toronto, 61-68, Frenchay Hosp, Eng, 68; asst res surgeon, Univ Calif, Los Angeles, 69-71; staff surgeon cardiovasc surg, Toronto Western Hosp, 71-75; asst prof surg, 74-80, asst prof physiol, 76-80, ASSOC PROF SURG, UNIV TORONTO, 80-, ASSOC PROF PHYSIOL, SCH GRAD STUDIES, INST MED SCI, 80- *Concurrent Pos:* Res surgeon, Clin Sci Div, Univ Toronto, 71-89; Med Res Coun Can Scholar, 71-76; Starr Mem Grad Scholar, 76. *Honors & Awards:* Starr Medal, 78. *Mem:* Am Physiol Soc; Soc Thoracic Surgeons; Can Cardiovasc Soc; Am Fedn Clin Res; NY Acad Sci; Am Heart Asn; Can Asn Univ Surgeons; Int Soc Heart Res; Can Physiol Soc; Can Soc Cardiovasc & Thoracic Surgeons. *Res:* Myocardial function; development of isolated, perfused rabbit heart septum model (parabiotic); coronary microcirculation; cardiomyocyte culture; myocardial ischemia; no reflow phenomenon; myocardial edema; myocardial contracture; myocardial preservation. *Mailing Add:* Med Sci Bldg Rm 3238 Univ Toronto Toronto ON M5S 1A8 Can

BAYLISS, JOHN TEMPLE, b Richmond, Va, July 6, 39. PHYSICS. *Educ:* Bowdoin Col, BA, 61; Univ Va, PhD(physics), 67. *Prof Exp:* Asst prof physics, Va Commonwealth Univ, 67-73; conserv coordr, Va Energy Off, 75-78; prog develop officer, Va Dept Mines, Minerals & Energy, 78-79, energy div dir, Energy Div, 79-85. *Concurrent Pos:* Consult, energy & pub policy, 85- *Res:* Conservation techniques and state policy. *Mailing Add:* Tech Assoc Inc Eight N Harris Richmond VA 23225

BAYLISS, PETER, b Eng; m 66. X-RAY POWDER DIFFRACTION. *Educ:* Univ NSW, BE, 59, MSc, 62, PhD(geol), 67. *Prof Exp:* From asst prof to assoc prof, 67-78, asst head dept, 80-84, PROF GEOL, UNIV CALGARY, 78-; chmn, Finance Comt, Mineral Assoc, Canada, 88-92. *Concurrent Pos:* Abstractor, Mineral Abstr, 69-; chmn, Minerals Comt Powder Diffraction File, 77-; ed, Minerals, Powder Diffraction Standards, 89-; ed, Can Mineralogist, 87-92; adv bd, Powder Diffraction, 86-; chmn, Int Mineral Asn Comn Mineral Classification, 90-94. *Mem:* Mineral Asn Can; Clay Minerals Soc; Mineral Soc Gt Brit; fel Mineral Soc Am. *Res:* Mineral classification; powder x-ray diffraction data. *Mailing Add:* Dept Geol & Geophysics Univ Calgary Calgary AB T2N 1N4 Can

BAYLOR, CHARLES, JR, b Baltimore, Md, Dec 5, 40; c 2. ORGANIC CHEMISTRY, PHOTOGRAPHIC CHEMISTRY. *Educ:* Morgan State Col, BS, 62; Utah State Univ, PhD(chem), 67. *Prof Exp:* SR RES CHEMIST, E I DU PONT DE NEMOURS & CO, INC, 66- *Mem:* Am Chem Soc. *Res:* Synthesis, analytical methods and structure determination of organic compounds; surfactant applications, polymers and emulsion polymerization; gel technology and immunoassays. *Mailing Add:* Imaging Systs Dept Exp Sta E I du Pont de Nemours & Co Inc Wilmington DE 19898

BAYLOR, DENIS ARISTIDE, b Oskaloosa, Iowa, Jan 30, 40; m; c 2. NEUROPHYSIOLOGY. *Educ:* Knox Col, BA, 61; Yale Univ, MD, 65. *Prof Exp:* Fel neurophysiol, Sch Med, Yale Univ, 65-68; staff fel, Lab Neurophysiol, Nat Inst Neurol & Commun Disorders & Stroke, 68-70; USPHS spec fel, Physiol Lab, Cambridge Univ, 70-72; assoc prof physiol, Univ Colo Med Ctr, Denver, 72-74; assoc prof physiol, 74-75, assoc prof 75-78, PROF NEUROBIOL, SCH MED, STANFORD UNIV, 78- *Concurrent Pos:* Ed, J Physiology, 77-84, Neuron, 88. *Honors & Awards:* Sinsheimer Found Award Med Res, 75; Mathilde Solowey Award, Neurosci, 78; Rank Prize Optoelectronics, 80; Proctor Medal, Asn Res Vision & Ophthal, 86. *Mem:* Biophysical Soc; Physiol Soc; Soc Gen Physiologists; Asn Res Vis & Ophthal; Soc Neurosci. *Res:* Generation and transmission of neural signals in the vertebrate retina. *Mailing Add:* Dept Neurobiol Stanford Univ Sch Med Stanford CA 94305

BAYLOR, JOHN E, b Belvidere, NJ, Sept 16, 22; m 50; c 2. AGRONOMY. *Educ:* Rutgers Univ, BSc, 47, MSc, 48; Pa State Univ, PhD(agron), 58. *Prof Exp:* Asst exten specialist farm crops, Rutgers Univ, 48-49, assoc exten specialist, 49-55; assoc prof agron & assoc exten specialist, 57-65, prof agron & exten specialist, 65-83, EMER PROF, PA STATE UNIV, UNIVERSITY PARK, 83- *Concurrent Pos:* Consult, IRI Res Inst, Brazil, 64; chair gov bd XIV, Int Grassland Cong; pres, Forage & Grassland Found. *Honors & Awards:* Medallion Award, Am Forage & Grassland Coun, 71; Zur Crain Award, Nat Silo Asn, 71; Gamma Sigma Delta Exten Award, 75; Am Soc Agron Exten-Indust Award, 75; Pa Grassland Coun Bicentennial Award, 76; Agron Exten Educ Award, Am Soc Agron, 80. *Mem:* Fel Am Soc Agron; Am Forage & Grassland Coun (pres, 67-70, secy-treas, 73-); Coun Agr Sci & Technol; Am Soc Animal Sci. *Res:* Forage crop production and management. *Mailing Add:* 298 E McCormick Ave State College PA 16801

BAYLOUNY, RAYMOND ANTHONY, b Paterson, NJ, Aug 11, 32; m 59; c 4. ORGANIC CHEMISTRY. *Educ:* Seton Hall Univ, BS, 54; Univ Md, MS, 58, PhD(org chem), 60. *Prof Exp:* Instr org chem & biochem, Brooklyn Col, 60-63; asst prof org chem, 63-67, chmn dept, 67-69, assoc prof org chem, 67-77, PROF CHEM, FAIRLEIGH DICKINSON UNIV, 77- *Concurrent Pos:* Sr fel, Princeton Univ, 69-70; vis prof, Columbia Univ, 77 & 83-84; counr, Am Chem Soc, 86- *Mem:* Am Chem Soc. *Res:* Theoretical organic chemistry; biochemistry; thermal rearrangement of unsaturated esters; synthesis and structure analysis of natural products; newer methods of organic synthesis. *Mailing Add:* Fairleigh Dickinson Univ Madison Campus Madison NJ 07940

BAYLY, M BRIAN, b Northwood, Eng, Apr 16, 29; m 60; c 5. GEOLOGY. *Educ:* Cambridge Univ, BA, 52, MS & MSc, 62; Univ Chicago, PhD(geol), 62. *Prof Exp:* From asst prof to assoc prof, 62-74, PROF GEOL, RENSSELAER POLYTECH INST, 74- *Res:* Tectonics, structural geology, deformation processes, with emphasis on quantitative mechanical theories. *Mailing Add:* Dept Geol Rensselaer Polytech Inst Troy NY 12180-3590

BAYM, GORDON A, b New York, NY, July 1, 35; c 4. NUCLEAR, ASTROPHYSICS & CONDENSED MATTER PHYSICS. *Educ:* Cornell Univ, BA, 56; Harvard Univ, AM, 57, PhD(physics), 60. *Prof Exp:* NSF fel, Inst Theoret Physics, Denmark, 60-62; asst res physicist, Univ Calif, Berkeley, 62-63; from asst prof to assoc prof, 63-68, PROF PHYSICS, UNIV ILL, URBANA-CHAMPAIGN, 68- *Concurrent Pos:* A P Sloan res fel, 65-68; vis prof, Univ Tokyo, 68, Univ Kyoto, 68, Nordita, Copenhagen, 70, 76, Niels Bohr Inst, 76 & Nagoya Univ, 79; assoc ed, Nuclear Physics, 71-; mem, Inst Theoret Physics Adv Comt, Santa Barbara, 78-83, Adv Comt Physics, NSF, 82-85, Nuclear Sci Adv Comt, NSF-Dept Energy, 82-, adv panels, Brookhaven Nat Lab, 83, 84, 85-, adv comt, Quark Matter, Asilomar, Ca, 86, Muenster, WGer, 87 & Gatlinburg, Tenn, 91, Third Int Conf Nucleus-Nucleus Collisions, Fr, 88, Nat Adv Comt, Inst Nuclear Theory, Seattle, 89, adv comt, Nucleus-Nucleus Collisions, Fourth Int Conf, Kanazawa, Japan, 91; organizer, Quark Matter, 88, Lenox, Mass, 88 & Workshop Ultrarelativistic Heavy Ion Physics, Italy, 88; vis scientist, Acad Sinica, Beijing, 79; mem sub-comt, Theoretical Physics, NSF Adv Comt Physics, 81 & 84, 4 GeV Electron Accelerator Nuclear Physics, Nuclear Sci Adv Comt, 83-84; Alexander von Humboldt Found Fel, 83-89; NSF vis comt, Kellogg Lab, Caltech, 83-85; corresp, Comments Nuclear & Particle Physics, 86-; trustee, Associated Univs, Inc, 86-90; mem, Theoretical Physics, Nat Safety

Anal Ctr Subcomt, 86-88. *Honors & Awards:* Alexander von Humboldt Sr Scientist Award, 83. *Mem:* Nat Acad Sci; fel AAAS; fel Am Phys Soc; Am Astron Soc; fel Am Acad Arts & Sci; Int Astron Union; Inst Theoretical Physics. *Res:* Low temperature physics; astrophysics; theory of many body systems; nuclear physics. *Mailing Add:* Loomis Lab Physics Univ Ill 1110 Green St Urbana IL 61801

BAYMAN, BENJAMIN, b New York, NY, Dec 12, 30; m 57; c 2. NUCLEAR PHYSICS. *Educ:* Cooper Union, BChE, 51; Univ Edinburgh, PhD(physics), 55. *Prof Exp:* Asst prof, Princeton Univ, 60-65; assoc prof physics, 65-68, PROF PHYSICS, 68-, Fel, UNIV MINN, MINNEAPOLIS, 75- *Concurrent Pos:* Res fel, theoret physics, Univ Edinburg, 55-56; Ford Fuond fel, theoret nuclear physics, Inst Theoret Physics, Denmark, 56-60. *Mem:* Am Phys Soc. *Res:* Interpretation of experimental data on atomic nuclei obtained by nuclear reaction and radioactivity studies in terms of nuclear models. *Mailing Add:* Dept Physics Univ Minn Minneapolis 116 Church St SE Minneapolis MN 55455-0112

BAYNE, CHARLES KENNETH, b Pittsburgh, Pa, Aug 22, 44. STATISTICS. *Educ:* Blackburn Col, BA, 66; Wash Univ, MS, 68; NC State Univ, PhD(statist), 74. *Prof Exp:* Mathematician, US Naval Ordnance Lab, 68-70; RES ASSOC STATIST, NUCLEAR DIV, UNION CARBIDE CORP, 74- *Mem:* Am Statist Asn; Sigma Xi. *Res:* Experimental designs. *Mailing Add:* 7209 Stockton Dr Knoxville TN 37909

BAYNE, CHRISTOPHER JEFFREY, b Trinidad, WI, Aug 31, 41; m 63; c 3. COMPARATIVE IMMUNOBIOLOGY. *Educ:* Univ Wales, BSc, 63, PhD(zool), 67. *Prof Exp:* Fel, Marine Sci Labs, Menai Bridge, Wales, 66-67; lectr marine zool, Univ Wales, 67-68; res assoc molluscan physiol, Univ Mich, 68-71; from asst prof to assoc prof, 71-82, chair, 86-89, PROF ZOOL, ORE STATE UNIV, 82- *Concurrent Pos:* Investr, Univ Stockholm, 76-77, Univ Tromso, 84, Wageningen Univ, 85; chair, Div Comp Immunol, Am Soc Zool, 88-90. *Mem:* AAAS; Soc Invert Path; Am Soc Zool; Int Soc Develop Comp Immunol; Am Soc Parasitol. *Res:* Comparative immunology; molluscan and trematode cell culture; immuno-parasitology; mechanisms of immunity in deuterostome invertebrates; teleost immunology; mechanisms of natural immunity; phagocytic cells; stress hormones' influence on phagocytes; neuro-immunology, especially catecholamine regulation of defenses; fish and molluscan models incorporating helminth parasites. *Mailing Add:* Dept Zool Ore State Univ Corvallis OR 97331-2914

BAYNE, DAVID ROBERGE, b Selma, Ala, Jan 29, 41; m 64. LIMNOLOGY, AQUATIC FLORA. *Educ:* Tulane Univ, BA, 63; Auburn Univ, MS, 67, PhD(aquatic ecol), 70. *Prof Exp:* Asst prof biol, Ga Col, 70-71; ASSOC PROF FISHERIES & ALLIED AQUACULT, AUBURN UNIV, 72- *Concurrent Pos:* Fishery adv, El Salvador, Calif, 72-73. *Mem:* Am Fisheries Soc; Am Soc Limnol & Oceanog. *Res:* Limnological studies of large multipurpose reservoirs; fresh water plankton communities; effects of contamination on aquatic communities; limnology of water impoundments of southeast United States with emphasis on plankton communities and primary productivity; plankton management in aquaculture. *Mailing Add:* Dept Fish & Allied Aquacult Auburn Univ Auburn University AL 36849

BAYNE, ELLEN KAHN, b Oct 27, 1949; m 74; c 2. EXTRACELLULAR MATRIX, CYTOKINES. *Educ:* Northwestern Univ, PhD(cell biol), 76. *Prof Exp:* Sr res immunologist, 82-88, RES FEL, MERCK SHARP & DOHME RES LAB, 88- *Mem:* Am Soc Cell Biol. *Res:* Use of immunocytechnical techniques to examine metalloproteinases and inhibitors in cartilage and synorial tissue; role of inflammatory mediators in rheumatoid and osteoarthritis. *Mailing Add:* Dept Biochem & Molecular Biol Merck Sharp & Dohme Res Lab PO Box 2000 Rahway NJ 07065

BAYNE, GILBERT M, b Philadelphia, Pa, Mar 11, 21; m 45; c 4. PSYCHOPHARMACOLOGY, THERAPEUTICS. *Educ:* Ursinus Col, BS, 43; Univ Pa, MD, 47. *Prof Exp:* Intern, Hosp Univ Pa, Philadelphia, 47-48; asst med dir, 48-53, dir med res, 54-70, sr dir, 70, SR DIR LONG RANGE PLANNING MED RES, MERCK SHARP & DOHME RES LABS, 70- *Concurrent Pos:* Clin asst, Endocrinol Clin, Philadelphia Gen Hosp, 49-50, res asst, Med Serv, 50-52, asst vis physician, 52-53; res assoc, Dept Res Therapeut, Norristown State Hosp, Pa, 51-60; fel med, Dept Infectious Dis, Hosp Univ Pa, 51-52, asst instr, 52-53. *Mem:* NY Acad Sci; Am Heart Asn; Am Col Neuropsychopharmacol; Am Fedn Clin Res; AMA. *Res:* Medical research; mental health; infectious diseases. *Mailing Add:* Four Farrier Lane Blue Bell PA 19422

BAYNE, HENRY GODWIN, b New York, NY, Dec 2, 25; m 58; c 3. BACTERIOLOGY. *Educ:* Brooklyn Col, BA, 49, MA, 54. *Prof Exp:* Sr technician microbiol, Western Reserve Univ, 50-51; health inspector, NY Dept Health, 53-55, jr bacteriologist, 55; microbiologist, Western Utilization Res, USDA, 55-87; RETIRED. *Mem:* Am Soc Microbiol. *Res:* Investigation of biochemical and physiological processes of bacterial and other organisms found in agricultural products. *Mailing Add:* 246 Purdue Ave Berkeley CA 94708

BAYNES, JOHN WILLIAM, b Baltimore, Md, Dec 30, 40; m 75; c 2. BIOCHEMISTRY, CLINICAL CHEMISTRY. *Educ:* Loyola Col, Md, BS, 62; Marshall Univ, MS, 69; Johns Hopkins Univ, PhD(biochem), 73; Univ Minn, Minneapolis, MS, 76. *Prof Exp:* Fel lab med & biochem, Univ Minn, 73-76; ASST PROF BIOCHEM, UNIV SC, 76- *Mem:* AAAS; Am Chem Soc; Soc Complex Carbohydrates. *Res:* Chemistry-biochemistry of complex carbohydrates; glycoproteins; regulation of plasma protein catabolism; host-pathogen interactions. *Mailing Add:* 413 Leton Dr Columbia SC 29210

BAYNTON, HAROLD WILBERT, b Brandon, Man, Sept 16, 20; nat US; m 47; c 3. METEOROLOGY. *Educ:* Univ Mich, MS, 57, MA, 59, PhD(diffusion), 63. *Prof Exp:* Weather forecaster, Meteorol Serv Can, 42-52, res meteorologist, 52-55; res assoc meteorol, Univ Mich, 55-58; res meteorologist, Systs Div, Bendix Aviation Corp, 58-63; climatologist, Martin

Co, 63-64; METEOROLOGIST, NAT CTR FOR ATMOSPHERIC RES, BOULDER, 65- *Concurrent Pos:* Affil prof, Va Polytech Inst, 67-70. *Mem:* Am Meteorol Soc; Air Pollution Control Asn; fel Royal Meteorol Soc; Can Meteorol & Oceanog Soc. *Res:* Applications of probability and statistical methods to meteorology in the realm of applied pollution; radar. air pollution. *Mailing Add:* 415 Kiowa Pl Boulder CO 80303

BAYRD, EDWIN DORRANCE, b Chicago, Ill, Nov 12, 17; m 42; c 5. MEDICINE. *Educ:* Dartmouth Col, AB, 39; Harvard Med Sch, MD, 42; Univ Minn, MS, 47. *Prof Exp:* From instr to assoc prof, 47-67, PROF MED, MAYO MED SCH, 67- *Concurrent Pos:* Fel trop med & parasitol, Tulane Univ, 43-44; fel med, Mayo Found, Univ Minn, 47; consult, Mayo Clin, Methodist & St Mary's Hosps, 47-; ed-in-chief, Mayo Clin Proc, 62-; Sir Norman Paul vis prof, Sydney Hosp, Australia, 63; chmn div hemat, Mayo Clin, 67-, pres staff, 69- *Mem:* AAAS; AMA; Am Fedn Clin Res; Am Soc Hemat; Int Soc Hemat. *Res:* Clinical and protein aspects of plasma cell disease, especially multiple myeloma macroglobulinemia and systematized amyloidosis. *Mailing Add:* Mayo Med Ctr Rochester MN 55905

BAYS, JACKSON DARRELL, biochemical engineering, for more information see previous edition

BAYS, JAMES PHILIP, b West Frankfort, Ill, Feb 19, 41; m 63; c 2. ORGANIC CHEMISTRY. *Educ:* Northwestern Univ, BS, 63; Univ Wis, PhD(org chem), 68. *Prof Exp:* NIH fel chem, Yale Univ, 68-70; asst prof, Grinnell Col, 70-77; from asst prof to assoc prof, 77-87, PROF CHEM, ST MARYS COL, 87- *Concurrent Pos:* ACM fac fel, Argonne Nat Lab, 71-72; asst prof biochem, Med Sch, Rush Univ, 74-77; res assoc, Univ Notre Dame, 78-86; Lilly Found Open Fac Fel, 90-91. *Mem:* Am Chem Soc; Am Sci Affil; Sigma Xi. *Res:* Chemistry of dianions of phenylacetone and its derivatives; reactions of synthetic utility; organometallic chemistry; molecular modeling. *Mailing Add:* Dept Chem & Physics St Mary's Col Notre Dame IN 46556

BAYS, KARL D, medicine; deceased, see previous edition for last biography

BAYUZICK, ROBERT J(OHN), b Braddock, Pa, Sept 6, 37; m 61; c 2. PHYSICAL METALLURGY, SURFACE PHYSICS. *Educ:* Univ Pittsburgh, BS, 61; Univ Denver, MS, 63; Vanderbilt Univ, PhD(mat sci), 69. *Prof Exp:* Metall technician radiation damage, Bettis Atomic Power Lab, Westinghouse Elec Corp, 55-56 & 56-60; statist clerk, Fed Bur Invest, 56; metall engr, Bell Aerosysts Co, 61; res asst phase equilibria, Denver Res Inst, 61-63; res metallurgist, Battelle Mem Inst, 64-65; instr phys metall, 69, from asst prof to assoc prof mat sci, 69-77, dir mat sci, 74-75, PROF MAT SCI, VANDERBILT UNIV, 77- *Mem:* Am Soc Metals; Am Inst Mining, Metall & Petrol Eng; Am Asn Univ Profs; Sigma Xi. *Res:* Atomic structure and topography of grain boundaries; various aspects of problems in thin films; atomic mechanisms of crack initiation. *Mailing Add:* 7902 Highway 100 Nashville TN 37221

BAZ, AMR MAHMOUD SABRY, b Cairo, Egypt, Oct 12, 45; c 2. CONTROLS, ROBOTS. *Educ:* Cairo Univ, Egypt, BSc, 66; Univ Wis-Madison, MSc, 70, PhD(mech eng), 73. *Prof Exp:* Teaching asst, mech eng dept, Cairo Univ, 66-68, asst prof, 75-77, assoc prof, 79-83; res asst, Univ Wis, 68-73, teaching fel, 73-75, res assoc, 77-79; assoc prof, 79-86, PROF MECH ENG, CATHOLIC UNIV AM, 86- *Concurrent Pos:* Mem res staff, Western Elec Res Lab, 79; prin investr, Naval Sea Command Syst, 84-85; Res Triangle, NC, US Army, 85 & NASA- Goodard, 84. *Honors & Awards:* James Lincoln Found Award, 73; Nat Medal Excellence in Arts & Sci, Govt of Egypt, 80; Burno Dimiani Award. *Mem:* Am Soc Mech Eng; Robotics Int Soc Mfg Eng; Nat Bd Standards; Nat Oceanic Atomospheric. *Res:* Active control of structure; dynamics and vibration of bearings. *Mailing Add:* Dept Mech Eng Catholic Univ Am Washington DC 20064

BAZAN, NICOLAS GUILLERMO, b Los Sarmientos, Arg, May 22, 42; m 65; c 5. NEUROCHEMISTRY & NEUROLOGY, OPHTHALMOLOGY. *Educ:* Univ Tucuman, Arg, MD, 65, PhD(med sci), 71. *Prof Exp:* Teaching asst biol, Univ Tucuman, Arg, 60-65; fel enzym, Col Physicians & Surgeons, Columbia Univ, 65-66; fel, dept biol chem, Harvard Med Sch, 66-67, res assoc, 67-68; asst prof biochem, fac med, Univ Toronto, 68-70; asst dir, dept neurochem, Clarke Inst Psychiat, Toronto, 68-70; dir, Inst Biochem, UNS-CONICET, Blanca, Arg, 70-81; PROF OPHTHAL, NEUROL & BIOCHEM, MED SCH, LA STATE UNIV, NEW ORLEANS, 81-, ERNEST C & YVETTE C VILLERE CHAIR RES RETINAL DEGENERATION, 84-, DIR, NEUROSCI CTR, 88- *Concurrent Pos:* Fel enzym, Inst Biochem, Buenos Aires, 64; prin investr, CONICET, Arg, 76-81; vis prof ophthal, Baylor Col Med, 77; mem, Neurosci Steering Comt, La State Univ Med Ctr, 84-; pres, Brain Chem Tech Corp, 85-; sr ed, Molecular Neurobiol Rev, 86-; Jacob Javits Neurosci investr, 89- *Honors & Awards:* William & Mary Greve Int Award, Res to Prevent Blindness, Inc, 83. *Mem:* Int Soc Neurochem; Am Soc Neurochem; Am Soc Biol Chemists; Soc Neurosci; Am Epilepsy Soc; Brit Soc Nutrit Med. *Res:* Neurochemical research of excitable membranes; synapse; photoreceptors; retinal degeneration; nutrition; essential fatty acids; retinitis pigmentosa; cell biology of metabolism; visual cell renewal; retinal pigment epithelium; epilepsy; arachidonic acid; cell signaling and membrane lipids; inositol lipids and diacylglycerol calcium; eye inflammation; stroke cerebral ischemia and brain damage; prostaglandins and leukotrienes; biotechnology. *Mailing Add:* Eye Ctr La State Univ 2020 Gravier St Suite B New Orleans LA 70112-2234

BAZANT, ZDENEK P(AVEL), b Prague, Czech, Dec 10, 37; m 67; c 2. STRUCTURAL MECHANICS & ENGINEERING. *Educ:* Prague Tech Univ, civil engr, 60, Docent(concrete struct), 67; Czech Acad Sci, PhD(eng mech), 63; Charles Univ, Prague, dipl(theoret physics), 66. *Hon Degrees:* Dr, Czech Tech Univ Prague, CRUT, Czech, 91. *Prof Exp:* Bridge engr, Dopravoprojekt Bridge Consult Off, Czech, 61-63; adj prof, Bldg Res Inst, Prague Tech Univ, 64-67; sr res engr, 63-67; vis res fel, Centre d'Etude du Batiment et de Travaux Publics, Paris, France, 66-67; sr res fel civil eng, Univ Toronto, 67-68; vis assoc res engr, Univ Calif, Berkeley, 69; assoc prof civil

eng, 69-73, dir, Ctr Concrete & Geomat, 81-87, PROF CIVIL ENG, NORTHWESTERN UNIV, 73-, WALKER P MURPHY PROF, ENDOWED CHAIR CIVIL ENG, 90- *Concurrent Pos:* French Govt fel, 66; Ford Sci Found fel, 68; dir grants, NSF, Elec Power Res Inst, Nat Res Coun, Defense Nuclear Agency, Dept Energy, 70-; registered struct engr, State of Ill, 72-; dir, NSF grants, 71-, Air Force Off Sci Res grants, 75-, Dept Energy, 81-89, Off Naval Res, 90-; consult, Sargent & Lundy, Chicago, 73-, Argonne Nat Lab, 74-, Oak Ridge Nat Lab, 75-, Babcock & Wilcox, Ohio, 78-81, Sandia Labs, 79, Ont Hydro, 80, Teng & Assoc, 85- & M&M Engrs, 86-; chmn, Comt Properties Mat, Am Soc Civil Engrs, 75-77 & 81-83, comt Fracture Mech, Am Concrete Inst, 85-, Comt on Creep, Int Union Testing Labs Mat & Struct, 81-; vis prof, Royal Inst Technol, Stockholm, 77, Politevmico di Milano, 82, Swiss Fed Inst Technol, 83, Ecole Normale Supereur, Paris, 88, Tech Univ Munich, 90-91; dir, Ctr Concrete & Geomat, 81-; dir, Elec Power Res Inst contract, Los Alamos Sci Lab grant; Guggenheim fel, 78-79, Kajima Found Fel, Tokyo, 87 & NATO fel, sr guest scientist, France, 88; ed-in-chief J Eng Mech, 89- *Honors & Awards:* Gold Medal, Int Union Testing Labs in Mat & Struct, 75; W. L. Huber Civil Eng Res Prize, Am Soc Civil Engrs, 76; T. Y. Lin Restressed Concrete Award, Am Soc Civil Engrs, 77; Humboldt Award, 90; Gold Medal Bldg Res, Inst Spain, 90. *Mem:* Fel Am Soc Civil Engrs; fel Am Concrete Inst; fel Am Acad Mech; Int Union Testing Labs Mat & Struct; Am Soc Mech Engrs; Am Ceramic Soc; Int Asn Bridge & Structural Eng; Soc Eng Sci. *Res:* Creep; inelasticity and failure of concrete, rocks and soils; moisture and thermal effects in concrete; plasticity and viscoelasticity; structural stability; fracture and continuum mechanics; thin-wall structures; design of nuclear structures, bridges, tall buildings. *Mailing Add:* Dept Civil Eng Northwestern Univ Evanston IL 60201

BAZER, FULLER WARREN, b Shreveport, La, Sept 2, 38; m 62; c 2. ANIMAL SCIENCE, REPRODUCTIVE PHYSIOLOGY. *Educ:* Centenary Col, BS, 60; La State Univ, Baton Rouge, MS, 63; NC State Univ, PhD(physiol), 69. *Prof Exp:* From asst prof to assoc prof, 68-78, PROF ANIMAL SCI, UNIV FLA, 78- *Concurrent Pos:* Ed-in-chief, Biol of Reprod, 89-93. *Honors & Awards:* Physiol & Endocrinol Res Award, Am Soc Animal Sci; Res Award, Sigma Xi, 76; Res Award, Soc Study Reprod, 90. *Mem:* AAAS; Soc Study Reprod; Am Soc Animal Sci; Soc Theriogenology. *Res:* Uterine protein secretion of domestic animals as related to embryonic development and corpus luteum function; steroids and proteins secreted by the concepts. *Mailing Add:* Dept Animal Sci Univ Fla Gainesville FL 32611-0691

BAZER, JACK, b New York, NY, Dec 23, 24; m 51; c 2. MATHEMATICS, PHYSICS. *Educ:* Cornell Univ, BA, 47; Columbia Univ, MA, 49; NY Univ, PhD(math), 53. *Prof Exp:* Jr res scientist, 51-53, res assoc, 53-58, sr scientist, 58, from asst prof to assoc prof, 58-65, PROF MATH, COURANT INST MATH SCI, NY UNIV, 65- *Concurrent Pos:* Consult, Grumman Aircraft Eng Corp, Long Island, 60; mem comn IV, Int Sci Radio Union. *Mem:* Am Math Soc. *Res:* Diffraction theory of scalar and vector fields; magnetogasdynamics; plasma physics; probability theory; partial differential equations. *Mailing Add:* Dept Math NY Univ New York NY 10003

BAZETT-JONES, DAVID PAUL, b Toronto, Ont, Mar 26, 53; m 81. MOLECULAR BIOLOGY. *Educ:* Univ Waterloo, BSc, 75; Univ Toronto, MSc, 78, PhD(biophysics), 81. *Prof Exp:* FEL, SCRIPPS CLIN RES FOUND, 81- *Mem:* Can Soc Cell Biol; Micros Soc Can. *Res:* Relationship between structures and composition of euharyotic chromation with transcriptional regulation. *Mailing Add:* Dept Med Biochem Univ Calgary 3330 Hospital Dr NW Calgary AB T2N 4N1 Can

BAZINET, GEORGE FREDERICK, b Glens Falls, NY, July 16, 37; m 62; c 3. BIOCHEMICAL GENETICS. *Educ:* St Michaels Col, BA, 60; St Johns Univ, MS, 63; Hahnemann Med Col, PhD(genetics), 67. *Prof Exp:* Fel, Med Sch, Yale Univ, 69-70; PROF BIOL, SIENA COL, 70- *Res:* Biochemical genetics of Nicrassa; metabolic activites of halobacteria halobium. *Mailing Add:* Sci Div Siena Col Loudonville NY 12211

BAZINET, MAURICE L, b Haverhill, Mass, July 26, 18; m 53. CHEMISTRY. *Educ:* Univ Conn, BS, 49; Boston Univ, MS, 52. *Prof Exp:* Chemist, Nat Res Corp, 51-55 & Gulf Res & Develop Co, 55-56; CHEMIST, ARMY NATICK LABS, 56- *Mem:* AAAS; Am Soc Testing & Mat; Sigma Xi; Am Chem Soc; Am Soc Mass Spectrometry. *Res:* Analysis of food flavors and aromas; design and development of analytical methods and techniques in food research. *Mailing Add:* 43 Cypress Rd Natick MA 01760

BAZLEY, NORMAN WILLIAM, applied mathematics, for more information see previous edition

BAZZAZ, FAKHRI AL, b Baghdad, Iraq, June 16, 33; US citizen; m 58; c 2. PLANT ECOLOGY, ENVIRONMENTAL BIOLOGY. *Educ:* Univ Baghdad, BA, 53; Univ Ill, MS, 60, PhD(bot), 63. *Prof Exp:* Asst prof ecol, Univ Ill, 63-64; lectr bot, Univ Baghdad, 64-66; asst prof ecol, 66-72, assoc prof ecol & forestry, 72-77, PROF PLANT BIOL & FORESTRY, UNIV ILL, URBANA, 77-; PROF BIOL, HARVARD UNIV. *Concurrent Pos:* Head, dept plant biol, Univ Ill, 82-84 actg dir, Sch Life Sci, 83-84; Guggenheim fel, 88. *Mem:* Fel AAAS; Ecol Soc Am; Brit Ecol Soc. *Res:* Physiological ecology of successional plants; plant community organization; effects of environmental pollution. *Mailing Add:* Dept Organismic & Evolutionary Biol Harvard Univ 26 Oxford St Cambridge MA 02138

BAZZAZ, MAARIB BAKRI, b Baghdad, Iraq, Nov 27, 40; m 58; c 2. BIOPHYSICS, MICROBIOLOGY. *Educ:* Univ Ill, Urbana, BS, 61, MS, 63, PhD(biol), 72. *Prof Exp:* Instr plant physiol, Univ Baghdad, Col Sci, 64-66; res assoc photosynthesis, Inst Environ Studies, Univ Ill, 72-74; res assoc photosynthesis, Dept Hort, Univ Ill, Urbana, 74-78, res scientist, Dept Bot, 78-84; RES ASSOC CELLULAR & DEVELOP BIOL, HARVARD UNIV, CAMBRIDGE, MASS, 84- *Mem:* Am Soc Plant Physiologists; Sigma Xi. *Res:* Structural characterization and study of the biosynthetic and photosynthetic aspects of newly discovered chlorophylls isolated from a mutant of zea mays. *Mailing Add:* Biol Lab Harvard Univ 16 Divinity Ave Cambridge MA 02138

BEA, ROBERT G, CIVIL ENGINEERING. *Educ:* Univ Fla, BS, 59, MS, 60. *Prof Exp:* Sr staff civil engr, Shell Oil Co, 59-76; chief engr & vpres, Ocean Eng Div, Woodward-Clyde Consult, 76-81; vpres, PMB Syst Eng, Inc, 81-88; PROF, DEPT CIVIL ENG & NAVAL ARCH & OFFSHORE ENG, UNIV CALIF, BERKELEY, 88- *Concurrent Pos:* Consult prof civil eng, Stanford Univ, 85-89. *Honors & Awards:* J Hillis Miller Eng Award; Croes Medal, Am Soc Civil Engrs, 59. *Mem:* Nat Acad Eng. *Res:* Coastal, offshore and ocean engineering; methods to define design criteria for fixed and mobile offshore structures; development of guidelines for the requalifications and rehabilitation of marine structures and ships; evaluation of forces due to waves, currents, earthquakes, ice and sea floor slides; development of technology for evaluation of the dynamic response characteristics of marine foundations and structures. *Mailing Add:* Dept NAOE Rm 202 Naval Arch Bldg Univ Calif Berkeley CA 94720

BEACH, BETTY LAURA, b Falls City, Nebr. FOOD SCIENCE, NUTRITION. *Educ:* Univ Nebr, Lincoln, BS, 58; Univ Wis-Madison, MS, 67, PhD(food sci & admin), 74. *Prof Exp:* Dietetic intern, USPHS, Staten Island, NY, 58-59, staff dietitian, Detroit, 59-62, clin dietitian, Staten Island, 62-63, chief food procurement & prod, 66-68, asst dir dietetics, 68-69; from asst prof to prof dietetics, Univ Tenn, Knoxville, 69-82; dir, dietetic internship, Edward J Hines Jr Vet Admin Hosp, Ill, 82-87; CHIEF, DIETETIC SERV, VET ADMIN MED CTR, OMAHA, NEBR, 87- *Concurrent Pos:* Dir coordr diet prog, Nutrit Found Mary Swartz Rose fel, 72-82. *Mem:* Am Dietetic Asn; Soc Advan Food Serv Res; Food Systs Mgt Educ Coun; Sigma Xi; Inst Food Technologists. *Res:* Development of quantitative methods to control resources in food service systems; development and evaluation of clinical dietetic staffing guidelines; nutritional care protocols to shorten patient stay. *Mailing Add:* 1408 S 126th St Omaha NE 68144

BEACH, DAVID H, b Syracuse, NY, July 4, 39; m 61; c 3. LIPID BIOCHEMISTRY, ANALYTICAL CHEMISTRY. *Educ:* Syracuse Univ, BA, 61; State Univ NY Upstate Med Ctr, MS, 69, PhD(microbiol), 73. *Prof Exp:* from res asst to res assoc, 64-75, res asst prof, 75-81, UNIV INSTRNL SPECIALIST, STATE UNIV NY UPSTATE MED CTR, 81- *Mem:* AAAS; Am Soc Parasitologists; Soc Protozoologists; Am Oil Chemists Soc; Sigma Xi. *Res:* Lipid biochemistry and metabolic studies of protozoan blood flagellates of the genes Leishmania and Trypomasoma; gas chromatography; mass spectrometry; infrared and nuclear magnetic resonance spectrometry. *Mailing Add:* Dept Microbiol State Univ NY Health Sci Ctr Syracuse NY 13210

BEACH, ELIOT FREDERICK, biochemistry; deceased, see previous edition for last biography

BEACH, EUGENE HUFF, b Highland, Mich, Oct 9, 18; m 44; c 1. NUCLEAR PHYSICS, ENGINEERING MANAGEMENT. *Educ:* Univ Mich, BSE, 41, MS, 47, PhD(physics), 53. *Prof Exp:* Elec engr, Naval Ord Lab, 41-46, electronic scientist, Underwater Ord, 53-55, chief weapon mech div, 55-58, proj mgr, 58-59, chief underwater elec engr dept, 59-73, assoc head underwater weapons develop directorate, 73-75; head ord syst develop dept, Naval Surface Weapons Ctr, 75-77, dep tech dir, 77-78, head, Underwater Syst Dept, 78-79, assoc tech dir, 79-80; RETIRED. *Concurrent Pos:* Eng res assoc, Cyclotron Proj, Univ Mich, 50-52; instr, Mass Inst Technol & Univ Md, 80-85. *Mem:* Inst Elec & Electronics Engr; Am Phys Soc; Sigma Xi. *Res:* Angular distribution studies on phosphorous reactions. *Mailing Add:* 12201 Remington Dr Silver Spring MD 20902

BEACH, FRANK AMBROSE, neuropsychology, neuroendocrinology; deceased, see previous edition for last biography

BEACH, GEORGE WINCHESTER, chemistry; deceased, see previous edition for last biography

BEACH, HARRY LEE, JR, b Richmond, Va, Aug 16, 44; m 66; c 3. PROPULSION, COMBUSTION. *Educ:* NC State Univ, BS, 66, MS, 68, PhD(mech eng), 70. *Prof Exp:* Res engr, 70-75, head, Combustion Sect, 75-77, leader performance anal group, 77-80, asst, 80-81, HEAD HYPERSONIC PROPULSION BR, 81- *Concurrent Pos:* Asst prof lectr, Joint Inst Advan Flight Sci, George Washington Univ, 77-; adj prof, Christopher Newport Col, 80- *Mem:* Am Inst Aeronaut & Astronaut. *Res:* Direct and conduct research for supersonic combustion ramjet propulsion including combustion fundamentals, computational fluid dynamics, combustion diagnostics, inlet and combustor conceptual design and testing, and inlet-combustor component integration. *Mailing Add:* George Washington Univ Washington DC 98195

BEACH, LOUIS ANDREW, b Greenville, Ind, June 2, 25; m 56; c 4. EXPERIMENTAL NUCLEAR PHYSICS, NUCLEAR ENGINEERING. *Educ:* Ind Univ, BS, 44, MS, 47, PhD(physics), 49. *Prof Exp:* Asst physics, Ind Univ, 46-49; res assoc nuclear physics, Cornell Univ, 49-51; nuclear physicist, 51-55, head, Nuclear Reactions Br, 55-65, head, Physics I Sect, 65-71, head, Nuclear Physics Sect, 71-78, head, Mat Sect, 78-80, res physicist, US Naval Res Lab, 80-87; CONSULT PHYSICIST, SACHS & FREEMAN ASSOC, 87- *Concurrent Pos:* Lectr, Cath Univ Am, 60-66. *Mem:* AAAS; Am Phys Soc; Sigma Xi. *Res:* Disintegration studies of beta decay; interactions of Bremstrahlung radiation with matter; shielding of nuclear radiation; nuclear reactors; nuclear structure; studies with cyclotron beams; radiation damage to metallurgical systems; nuclear detectors and electronics. *Mailing Add:* 1200 Waynewood Blvd Alexandria VA 22308-1842

BEACH, NEIL WILLIAM, b Ann Arbor, Mich, Apr 11, 28; div; c 3. ZOOLOGY. *Educ:* Univ Mich, BS, 50, MS, 51, PhD(zool), 56. *Prof Exp:* Instr zool, Univ Mich, 56-57; asst prof biol, Lake Forest Col, 57-60; asst prof, 60-64, ASSOC PROF BIOL, GETTYSBURG COL, 64- *Concurrent Pos:* NSF grants, Mich Biol Sta, 55; Duke Marine Lab, 58 & 59; NSF sci fels, 60-62. *Mem:* Am Micros Soc; Sigma Xi; Am Soc Zool. *Res:* Invertebrate zoology and ecology; rotifera; distribution, life history and ecology of the oyster crab, Pinnotheres ostreum and pea crabs of Australia. *Mailing Add:* Dept Biol Gettysburg Col Gettysburg PA 17325

BEACH, R(UPERT) K(ENNETH), b Ringgold, Nebr, June 13, 18; m; c 4. ELECTRICAL ENGINEERING. *Educ:* Univ Wyo, BS, 40, EE, 58; Ill Inst Technol, MS, 41. *Prof Exp:* Instr mech eng, 46-47, PROF ELEC ENG, UNIV WYO, 47- *Mem:* Am Soc Eng Educ; Inst Elec & Electronics Engrs; Sigma Xi. *Res:* Automatic control systems; energy conversion. *Mailing Add:* 1005 Steele St Laramie WY 82070

BEACH, ROBERT L, b Kalamozoo, Mich, July 2, 38. OCEAN ENGINEERING. *Educ:* Mich Tech Univ, BS, 60; Univ RI, MS, 70. *Prof Exp:* Process engr, Union Carbide Corp, 60-68; proj engr, Ocean Systs, Inc, 70-74; VPRES ENG, SEAWARD INT, INC, 74- *Honors & Awards:* IR-100 Award, 73. *Mem:* Soc Naval Architects & Marine Engrs; Am Soc Testing Mat. *Res:* Oil spill control technology, including development of innovative oil skimmers, oil booms and disposal systems. *Mailing Add:* 3509 Queen Anne Dr Fairfax VA 22030

BEACH, SHARON SICKEL, b Indianapolis, Ind, May 5, 46; m; c 2. COMPUTER SCIENCE, INTELLIGENT SYSTEMS. *Educ:* Univ Wash, BA, 67, MS, 71, PhD(comput sci), 73. *Prof Exp:* Assoc engr comput, Boeing Co, 68-70; comput ctr dir, Univ Calif, Santa Cruz, 74-75, sr preceptor acad affairs, Crown Col, 75-77, asst prof comput sci, 73-79; PRES, BEACH ASN, INC, 78- *Concurrent Pos:* Off of Naval Res grant, 76- *Mem:* Asn Comput Mach; Inst Elec & Electronics Engrs; Asn Inst Admin Acct & Data Processing. *Res:* Automatic theorem proving expert systems. *Mailing Add:* 26 Moreno Dr Santa Cruz CA 95060

BEACHEM, CEDRIC D, b Beaufort, NC, May 2, 32. PHYSICAL METALLURGICAL ENGINEERING. *Educ:* NC State Univ, BS, 57. *Prof Exp:* METALL ENGR, NAVAL RES LAB, 57- *Honors & Awards:* Sam Tour Award, Am Soc Testing & Mat. *Mem:* Fel Am Soc Mat; Am Soc Testing & Mat; Nat Asn Corrosion Engrs. *Mailing Add:* Naval Res Lab Code 6327 Washington DC 20375

BEACHEY, EDWIN HENRY, infectious diseases; deceased, see previous edition for last biography

BEACHLEY, NORMAN HENRY, b Washington, DC, Jan 13, 33; m 59; c 3. MECHANICAL ENGINEERING. *Educ:* Cornell Univ, BME, 56, PhD(mech eng), 66. *Prof Exp:* Mem tech staff eng, Hughes Aircraft Co, 56-57; mem tech staff eng, Space Technol Labs, 59-63; PROF MECH ENG, UNIV WIS-MADISON, 66- *Concurrent Pos:* Consult, Lawrence Livermore Lab, 78-81; Nat Bur Standards, 78-; res fel, Sci & Engr Res Coun, 81-82. *Mem:* Am Soc Mech Engrs; Soc Automotive Engrs; Sigma Xi. *Res:* Energy storage powerplant systems for motor vehicles; mechanical design of internal combustion engines; fluid power systems; continously-variable transmissions. *Mailing Add:* Dept Mech Eng 1513 University Ave Madison WI 53706

BEACHLEY, ORVILLE THEODORE, JR, b East Orange, NJ, Nov 8, 37; m 62; c 2. INORGANIC CHEMISTRY. *Educ:* Franklin & Marshall Col, BSc, 59; Cornell Univ, PhD(inorg chem), 62. *Prof Exp:* NIH fel chem, Univ Durham, 63-64; asst prof, Cornell Univ, 64-66; from asst prof to assoc prof, 66-85, PROF CHEM, STATE UNIV NY BUFFALO, 85- *Mem:* Am Chem Soc; The Chem Soc; Sigma Xi. *Res:* Preparative and physical inorganic chemistry with emphasis on organometallic, hydride and heterocyclic derivatives of main group and transition elements. *Mailing Add:* Dept Chem State Univ NY Buffalo NY 14214

BEACHY, ROGER NEIL, b Plain City, Ohio, Oct 4, 44; m 67; c 2. PLANT VIROLOGY, DEVELOPMENTAL BIOLOGY. *Educ:* Goshen Col, BA, 66; Mich State Univ, PhD(plant path), 73. *Prof Exp:* Res assoc & NIH fel plant virol, Cornell Univ, 73-76; res assoc, US Plant, Soil & Nutrit Lab, Ithaca, NY, 76-78; from asst prof to assoc prof, 76-85, PROF BIOL, WASH UNIV, 85- *Honors & Awards:* Ruth Allen Award, Am Phytopath Soc, 90. *Mem:* Am Phytopath Soc; Am Soc Plant Physiologists; fel AAAS; Int Soc Plant Molecular Biol; Am Soc Virol; Am Soc Biol Chem. *Res:* Control of synthesis of soybean seed proteins; plant viral messenger RNAs; effects of virus gene products on infected host cells; genetic transformation of plants for virus resistance. *Mailing Add:* Dept Biol Wash Univ St Louis MO 63130

BEACOM, STANLEY ERNEST, b Ft Macleod, Alta, May 18, 27; m 57; c 2. RUMINANT NUTRITION. *Educ:* Univ Alta, BSc, 49, MSc, 51; McGill Univ, PhD(ruminant nutrit), 59. *Prof Exp:* Broiler super poultry prod & mgt, Swift Can, Calgary, Alta, 51-52; tech officer poultry & swine nutrit, Agr Can, Res Br, Melfort, 52-53, res officer swine & beef cattle nutrit, 53-61 & animal nutrit & pasture res, 61-66, res dir agr, 66-90; RETIRED. *Mem:* Agr Inst Can; Can Soc Animal Prod (pres, 68). *Res:* Forage crop harvesting and utilization with specialization on improved efficiency of use of forage crops in rations for growing beef heifers and finishing of beef steers. *Mailing Add:* Box 296 Melfort SK S0E 1A0 Can

BEADELL, DONALD ALBERT, chemical engineering, for more information see previous edition

BEADLE, BUELL WESLEY, b Port Barre, La, Sept 9, 11; m 34; c 2. CHEMISTRY. *Educ:* Kans State Col, BS, 35, MS, 38; Purdue Univ, PhD(agr biochem), 42. *Prof Exp:* Asst chemist, Exp Sta, Kans State Col, 35-42; res chemist, Am Meat Inst, Chicago, 42-47, chemist in charge phys & anal chem, Am Meat Inst Found, 47-48; dir & vpres, George W Gooch Labs, 48-50; head tech div & comndr staff, Naval Ord Test Sta, China Lake, 50-51, head staff, 52-54; chmn, Div Chem & Chem Eng, Southwest Res Inst, San Antonio, 55-57, dir chem & chem eng, Midwest Res Inst, Kansas City, Mo, 57-63; exec dir res & develop, Farmland Industs, Inc, 63-68, vpres res & develop, 68-76; RETIRED. *Concurrent Pos:* Asst, Purdue Univ, 40-42; from res assoc to asst prof pharmacol, Univ Chicago, 43-48. *Mem:* AAAS; Am Chem Soc; Am Soc Biol Chem; fel Am Inst Chemists; Am Oil Chemists Soc. *Res:* Physical and analytical methods; stability of vitamins; fat autoxidation, rancidity and antioxidants; spectrophotometric methods; high frequency electronic processing. *Mailing Add:* 7400 Robinson Overland Park KS 66204

BEADLE, CHARLES WILSON, b Beverly, Mass, Jan 24, 30; m 56; c 3. MECHANICAL ENGINEERING. *Educ:* Tufts Univ, BS, 51; Univ Mich, MSE, 54; Cornell Univ, PhD(eng mech), 61. *Prof Exp:* Res engr, Res Lab Div, Gen Motors Corp, 51-53 & RCA Res Lab, 54-57; PROF MECH ENG & CHMN DEPT, UNIV CALIF, DAVIS, 61- *Concurrent Pos:* Ford Found fel, Advan Testing Dept, Ford Motor Co, 65-66. *Mem:* Fel Am Soc Mech Engrs; Am Soc Eng Educ. *Res:* Machine design; dynamic structural analysis; computer-aided mechanical design. *Mailing Add:* Dept Mech Eng Univ Calif Davis CA 95616

BEADLE, GEORGE WELLS, plant genetics; deceased, see previous edition for last biography

BEADLE, RALPH EUGENE, b Plentywood, Mont, Apr 26, 43; m 68; c 2. VETERINARY PHYSIOLOGY. *Educ:* Colo State Univ, DVM, 67; Univ Ga, PhD(vet physiol), 73. *Prof Exp:* Instr clins, Sch Vet Med, Univ Ga, 67-68, res assoc physiol, 68-71, asst prof clins, 72; resident, Sch Vet Med, Mich State Univ, 72-73, asst prof anesthesiol, 73-74; asst prof physiol, 74-77, assoc prof, 77-81, ASSOC PROF CLIN SCI, SCH VET MED, LA STATE UNIV, 81- *Mem:* Am Soc Vet Anesthesiologists; assoc fel Am Col Vet Pharm & Therapeut; Am Soc Vet Physiologists & Pharmacologists. *Res:* Comparative aspects of chronic obstructive pulmonary disease and gas transport in the lungs. *Mailing Add:* Dept Clin Sci Sch Vet Med La State Univ Baton Rouge LA 70803

BEADLES, JOHN KENNETH, b Alva, Okla, Sept 22, 31; m 55; c 2. ICHTHYOLOGY, LIMNOLOGY. *Educ:* Northwestern State Col, BS, 57; Okla State Univ, MS, 63, PhD(ichthyol, limnol), 66. *Prof Exp:* High sch teacher, 57-62; res asst, Okla State Univ, 63-64; asst prof biol, Ark State Univ, 65-66; asst prof zool, Haile Selassie Univ, 66-68; PROF BIOL & CHMN, BIOL SCI DEPT, ARK STATE UNIV, 68- *Concurrent Pos:* Proj dir, Environ Inventory & Assessment, US Army Corps Engrs, 72-78 & Environ Inventory, US Soil Conserv Serv, 74-78. *Mem:* Am Soc Ichthyologists & Herpetologists; Sigma Xi; Am Fishery Soc. *Res:* Fishery biology; catfish farming; fish survey. *Mailing Add:* 1111 Thrush Rd Jonesboro AR 72401

BEADLING, LESLIE CRAIG, b Pittsburgh, Pa, Oct 15, 46; c 3. BIOCHEMISTRY. *Educ:* Col William & Mary, BS, 68; Yale Univ, PhD(biochem), 73. *Prof Exp:* Res asst biochem, Health Ctr, Univ Conn, 72-75; mgr tech servs, Pharmacia Fine Chems, 75-81, tech dir, 81-84, vpres, 84-87, SR VPRES, PROCESS SEPARATION DIV, PHARMACIA FINE CHEMS, 87- *Concurrent Pos:* NIH fel, Health Ctr, Univ Conn, 73-75. *Mem:* Am Chem Soc; AAAS; Am Soc Microbiol. *Res:* Biochemical separations techniques; chromatography, electrophoresis and centrifugation techniques; ultrafiltration, process control. *Mailing Add:* Pharmacia Inc 800 Centennial Ave Piscataway NJ 08854

BEAGRIE, GEORGE SIMPSON, b Peterhead, Scotland, Sept 14, 25; Can citizen; m 50; c 4. DENTISTRY, PERIODONTICS. *Educ:* Edinburgh Univ, LDS RCS, 47, FDS RCS, 54, DDS, 66 & 87; Royal Col Dent Surgeons, dipl periodont, 69, FRCD(C), 69, FICD(C), 78, FACD, 81, FDSRCS, 84. *Hon Degrees:* DSc, McGill Univ, 85; DDS, Edinburgh, 87. *Prof Exp:* Dent officer, Royal Air Force, 47-48; lectr dent surg, Edinburgh Univ, 51-58, sr lectr periodont, 61-63, prof restorative dent, 63-68; prof & chmn clin sci, Univ Toronto, Ont, 68-78, chmn grad dept, fac dent, 74-78; dean, 78-88, EMER DEAN FAC DENT, UNIV BC, 89- *Concurrent Pos:* Fel, Nuffield Found, UK, 57-58; consult, WHO, 75-; pres, Royal Col Dentists Can, 77-79; vchmn, Fedn Dentaire Int Dent Educ Comn, 76-80, chmn, 81-87; mem coun, Educ Can Dental Asn, 81-88; chmn exam comt, Nat Dental Exam Bd Can, 86-; chmn adv bd, World Health Orgn & Fedn Dent Int, 87- *Mem:* Int Asn Dent Res (pres, 77-78); Can Dent Asn; Brit Soc Periodontol; Brit Dent Asn; fel Int Col Dentists; hon mem Am Dent Asn. *Res:* Clinical trials, periodontal and pulp research in clinical science and practice; dental education. *Mailing Add:* Fac Dent Univ BC 350-2194 Health Sci Hall Vancouver BC V6T 1W5 Can

BEAHM, EDWARD CHARLES, b Philadelphia, Pa, July 9, 39; m 67; c 1. PHYSICAL CHEMISTRY, MATERIALS CHEMISTRY. *Educ:* Temple Univ, AB, 67; Pa State Univ, PhD(chem), 73. *Prof Exp:* Fel, Mat Res Lab, Pa State Univ, 73-74; MEM RES STAFF NUCLEAR FUEL CHEM, OAK RIDGE NAT LAB, 74- *Mem:* Am Nuclear Soc; Am Ceramic Soc; Am Soc Metals; Sigma Xi. *Res:* High temperature chemistry; thermodynamics; materials compatibility. *Mailing Add:* 106 Cooper Circle Oak Ridge TN 37830

BEAK, PETER, b Syracuse, NY, Jan 12, 36; m 59; c 2. ORGANIC CHEMISTRY. *Educ:* Harvard Univ, BA, 57; Iowa State Univ, PhD(org chem), 61. *Prof Exp:* From instr to assoc prof, 61-70, PROF ORG CHEM, UNIV ILL, URBANA-CHAMPAIGN, 70- *Concurrent Pos:* Sloan fel, 67-69; Guggenheim fel, 68-69. *Mem:* Fel AAAS. *Res:* New organic reactions; synthetic, structural and mechanistic organic chemistry. *Mailing Add:* 3618 Roger Adams Lab Univ Ill Urbana IL 61801

BEAKLEY, GEORGE CARROLL, JR, b Marble Falls, Tex, Feb 3, 22; m 44; c 4. MECHANICAL ENGINEERING. *Educ:* Tex Technol Univ, BSME, 47; Univ Tex, MSME, 52; Okla State Univ, PhD(mech eng), 56. *Prof Exp:* Assoc prof mech eng, Tarleton State Univ, 47-53; stress analyst, Bell Helicopter Corp, 53-54; prof eng, assoc dean eng & dir, Sch Eng, Ariz State Univ, 56-91, dean, Col Eng & Appl Sci, 87-89; RETIRED. *Concurrent Pos:* Consult, 52- *Honors & Awards:* Chester F Carlson Award, Am Soc Eng Educ, 73, Donald E Marlowe Award; Western Electric Award, Am Soc Eng Educ, 77. *Mem:* Fel Am Soc Eng Educ; fel Am Soc Mech Eng; Nat Soc Prof Engrs; Am Inst Indust Eng; Am Soc Heating Refrig & Air Conditioning; fel Accreditation Bd Engrs & Technologists. *Res:* Engineering design; engineering administration; author or co-author of 29 engineering textbooks. *Mailing Add:* Col Eng Ariz State Univ Tempe AZ 85282

BEAL, JACK LEWIS, b Harper, Kans, July 7, 23; m 48; c 3. PHARMACOGNOSY. *Educ:* Univ Kans, BS, 48, MS, 50; Ohio State Univ, PhD(pharmacog), 52. *Prof Exp:* Instr pharmacy & pharmacog, Univ Kans, 49-50; from asst prof to prof pharmacog, Ohio State Univ, 52-78, prof & asst dean, 78-81, prof & assoc dean, 81-86; RETIRED. *Concurrent Pos:* Ed, J Natural Prod, 77-84; NSF fac fel, 58-59; mem revision comt, US Pharmacopoeia, 75-80. *Honors & Awards:* Edwin L Newcomb Mem Award Pharmacog, 58; Res Achievement Award in Natural Prod Chem, Am Pharmaceut Asn & Acad Pharmaceut Sci, 82. *Mem:* Am Asn Cols Pharm; Am Soc Pharmacog (pres, 62-63); fel Acad Pharmaceut Sci; Am Pharmaceut Asn. *Res:* Drugs of biological origin; drug plant research and cultivation; isolation of plant constituents; plant biochemistry; chemical microscopy. *Mailing Add:* 5544 Rockwood Rd Columbus OH 43229

BEAL, JAMES BURTON, JR, b Galveston, Tex, Nov 22, 32; m 53; c 4. INORGANIC CHEMISTRY. *Educ:* Baylor Univ, BA, 53; Agr & Mech Col Tex, MS, 59, PhD(chem), 63. *Prof Exp:* Res chemist, Gen Chem Div, Allied Chem Corp, 62-63; assoc res dir, Chem Div, Ozark-Mahoning Co, 64-69; PROF CHEM, UNIV MONTEVALLO, 69- *Mem:* Am Chem Soc; Sigma Xi. *Mailing Add:* Dept Chem Univ Montevallo Montevallo AL 35115

BEAL, JIM C(AMPBELL), b London, Eng, July 15, 33; m 70. ELECTRICAL ENGINEERING, ELECTROMAGNETICS. *Educ:* Univ London, BSc, 58, PhD(elec eng, microwaves), 64. *Prof Exp:* Res engr, Redifon Ltd, Eng, 62-65; asst prof elec eng, Colo State Univ, 65-67; from asst prof to assoc prof, Queen's Univ, 67-74, prof elec eng, 74-, assoc dean res ser, 76-; ASST SUPV SUPPORT GROUPS, FLA POWER & LIGHT CO. *Mem:* Inst Elec & Electronics Engrs; Inst Elec Engrs, Eng. *Res:* Electromagnetic guided waves in communications systems and guided radar. *Mailing Add:* JSP Dept 700 Universe Blvd Juno Beach FL 33408

BEAL, JOHN ANTHONY, b Cleveland, Ohio, Mar 30, 45; m 69; c 3. ANATOMY, NEUROANATOMY. *Educ:* Xavier Univ, BS, 67; Univ Cincinnati, PhD(anat), 71. *Prof Exp:* Asst prof, Med Sch, Wayne State Univ, 71-78; ASSOC PROF ANAT, MED CTR, LA STATE UNIV, 78- *Res:* Fine structure of the nervous system; synaptology; spinal cord. *Mailing Add:* Dept Anat La State Univ Sch Med Shreveport PO Box 33932 Shreveport LA 71130

BEAL, KATHLEEN GRABASKAS, b Youngstown, Ohio, June 2, 51; m 76; c 1. ZOOLOGY. *Educ:* Ohio Dominican Col, BS, 73; Ohio State Univ, MS, 75, PhD(zool), 78. *Prof Exp:* Instr, Capital Univ, 78-81; asst prof zool, 81-85; INSTR MATH & STATIST, WRIGHT STATE UNIV, 88- *Mem:* AAAS; Am Ornith Soc; Animal Behav Soc; Ecol Soc Am; Am Arachnological Soc. *Res:* Behavioral ecology of birds, jumping spiders (salticidae) and terrestrial isopods. *Mailing Add:* 616 Xenia Ave Yellow Springs OH 45387

BEAL, MYRON CLARENCE, b New York, NY, Dec 4, 20; m 48; c 5. BIOMECHANICS. *Educ:* Univ Rochester, AB, 42, Chicago Col Osteopath Med, DO, 45; Univ Chicago, MS, 49. *Prof Exp:* Asst dir clinics osteopath med, Chicago Col Osteopath Med, 46-49; instr, London Col Osteopath, 49-51; pvt pract, Rochester NY, 51-74; prof biomech, 74-81, prof family med, 81-89, EMER PROF FAMILY MED, MICH STATE UNIV, 89- *Concurrent Pos:* Mem, Nat Bd Examiners Osteopath Physicians & Surgeons, 60-, NY State Bd Med, 61-73; trustee, Chicago Col Osteopath Med, 69-; actg chmn biomed, Mich State Univ, 75-77; chmn bd dir, Chicago Col Osteopath Med, 86- *Honors & Awards:* Gutensohn/Denslow Award, 88. *Mem:* Am Osteopath Asn; fel Am Acad Osteop; NAm Acad Mammal Med. *Res:* Biomechanics interrater reliability of palpatory examination procedures, and the use of physical tests to determine patient improvement; mechanics of joint motion; clinical management of cervical spine injury; influence of manipulation on static respiratory mechanics and neuroendocrine response. *Mailing Add:* 5873 Seneca Point Rd Naples NY 14512

BEAL, PHILIP FRANKLIN, III, b Brewster, NY, Oct 2, 22; m 47; c 4. ORGANIC CHEMISTRY. *Educ:* Williams Col, AB, 43; Ohio State Univ, PhD(org chem), 49. *Prof Exp:* Jr chemist, Winthrop Chem Co, 43-44; res assoc, UpJohn Co, 50-66, head chem process res & develop, 66-69, mgr, 69-70, group mgr chem process res & develop, 70-78, dir fine chem specialties prod & support servs, 78-79, dir fine chem prod, 79-80, dir fine chem res & develop, 80-84; RETIRED. *Mem:* Am Chem Soc. *Res:* Aliphatic diazo compounds; diazoketones; partial and total synthetic steroid work; synthetic prostaglandins. *Mailing Add:* 3411 Lorraine Ave Kalamazoo MI 49008

BEAL, RICHARD SIDNEY, JR, b Victor, Colo, May 7, 16; m 87; c 1. SYSTEMATICS. *Educ:* Univ Ariz, BS, 38; Univ Calif, PhD(entom), 52. *Prof Exp:* Asst prof biol, Westmont Col, 47-48; instr lib arts, San Francisco Baptist Col, 48-51; prof syst theol, Conservative Baptist Theol Sem, 51-56; entomologist, Agr Res Serv, USDA, 56-58; assoc prof entom, Ariz State Univ, 58-62; assoc dean arts & sci, 62-65, dean, Grad Sch, 65-81, prof zool, 81-83, emer prof entom, Northern Ariz Univ, 84-87; acad vpres, 87-89, PROF BIOL, COLO CHRISTIAN COL, 89- *Mem:* Coleopterists Soc; Am Entom Soc; Soc Syst Zool; Am Sci Affiliation. *Res:* Biology and systematics of coleopterous family Dermestidae; biology of Colorado insect. *Mailing Add:* Colo Christian Col 180 S Garrison St Lakewood CO 80226

BEAL, ROBERT CARL, b Boston, Mass, Dec 28, 40; m 62; c 1. GEOLOGY, PHYSICS. *Educ:* Mass Inst Technol, BS, 61; Univ Md, MS, 68. *Prof Exp:* Design engr, Johns Hopkins Appl Physics Lab, 61-68; syst engr, Itek Corp, 68-72; PRIN PHYSICIST, JOHNS HOPKINS APPL PHYSICS LAB, 73- *Concurrent Pos:* Consult, EG&G, Bedford, Mass, 72-73; vis scientist, Jet Propulsion Lab, 77-78; prin investr, Johns Hopkins Appl Physics Lab, 78-81; ed, Spaceborne Synthetic Aperture Radar Oceanog, 80-81. *Mem:* Am Geophys Union; Union Radio Scientists Int. *Res:* Interpretation and analysis of oceanographic data from satellite sensors, particularly microwave sensors such as the snythetic aperture radar; ocean wave generation and propagation; spatial evolution of ocean wave spectra. *Mailing Add:* 12222 Carrol Mill Rd Ellicot City MD 21043

BEAL, STUART KIRKHAM, b New York, NY, Nov 20, 32; m 54; c 2. MECHANICAL & CHEMICAL ENGINEERING. *Educ:* Va Polytech Inst, BS, 54; Univ Pittsburgh, MS, 63. *Prof Exp:* Jr engr nuclear systs design, Westinghouse Elec Corp, 54-55, assoc engr, 58-60, engr nuclear systs design & anal, 60-66, sr engr corrosion-prod res, 66-72, fel engr corrosion-prod res, environ modelling and anal, 72-75, adv engr math modelling corrosion-prof res, aerosol physics, Bettis Atomic Power Lab, 75-80; sr fel, Adv Comt Reactor Safeguards, 80-82, PRIN SCIENTIST, SC & A INC, 82- *Mem:* Am Nuclear Soc; Inst Elec & Electronics Engrs. *Res:* Mathematical modelling; environmental modelling; aerosol physics. *Mailing Add:* 3708 Camelot Dr An Annandale VA 22003

BEAL, THOMAS R, b Brookhaven, Miss, Apr 8, 29; m 56; c 3. ENGINEERING MECHANICS, ENGINEERING. *Educ:* Southwestern at Memphis, BS, 51; Wash Univ, MS, 57; Stanford Univ, PhD(eng mech), 63. *Prof Exp:* Tech analyst, McDonnell Aircraft Corp, 52-55; instr appl mech, Wash Univ, 57-58; advan study engr & scientist, Calif, 58-63, res specialist, Ala, 63-65, staff scientist, 65-67, mgr dynamics & guid, 67-72, SR STAFF ENGR, LOCKHEED MISSILES & SPACE CO, 73- *Concurrent Pos:* Assoc prof, Univ Ala, 65-72. *Mem:* Am Inst Aeronaut & Astronaut. *Res:* Aeroelastic stability of aircraft; structural dynamics of aerospace vehicles; orbital perturbations of earth satellites; inertial navigation systems; astronautical guidance for interplanetary flights; control; flight dynamics; trajectory optimization; astrodynamics; mission and operations analysis; performance evaluation; orbital mechanics. *Mailing Add:* Engr Dept Lockheed Missiles & Space 806 Flin Way Sunnyvale CA 94087

BEAL, VIRGINIA ASTA, b Hull, Mass, Oct 31, 18. NUTRITION. *Educ:* Simmons Col, BS, 39; Harvard Univ, MPH, 45. *Prof Exp:* Nutritionist, Maternal & Child Health, Sch Pub Health, Harvard Univ, 39-46; instr physiol growth, Sch Med, Univ Colo, 48-54, asst prof human growth, 54-59, asst clin prof pediat, 59-71, nutritionist, Child Res Coun, 46-71; assoc prof, 71-78, prof nutrit, 78-86, PROF EMER NUTRIT, UNIV MASS, AMHERST, 86- *Honors & Awards:* Medallion Award, Am Dietetic Asn. *Mem:* Fel Am Pub Health Asn; Am Inst Nutrit; Am Dietetic Asn; Soc Nutrit Educ; NY Acad Sci; Sigma Xi. *Res:* Nutritional intake during pregnancy and childhood and its relationship to physical and physiological growth and development; education. *Mailing Add:* Ten Highland Circle Hadley MA 01035

BEALE, GUY OTIS, b Cleveland, Ohio, June 16, 44; m 67; c 1. AUTOMATIC CONTROLS, DIGITAL SYSTEMS. *Educ:* Va Polytech Inst, BS, 67; Lynchburg Col, MS, 74; Univ Va, PhD(elec eng), 77. *Prof Exp:* Electronic engr, Babcock & Wilcox, 71-78, chief electronic engr, 78-81, eng prod mgr, 81; asst prof elec eng, Vanderbilt Univ, 81-86; ASSOC PROF ELEC & COMPUTER ENG, GEORGE MASON UNIV, 86- *Concurrent Pos:* Consult, Babcock & Wilcox & Sverdrup Technol, Inc, 82-, Merrick Eng, 84, Vanderbilt Univ Med Ctr, 84-85, David Taylor Res Ctr, 88-91. *Mem:* Inst Elec & Electronics Engrs; Sigma Xi. *Res:* Modern control and estimation theory; intelligent systems for design; optimal digital simulation; microprocessor-based industrial automation. *Mailing Add:* Elec & Computer Eng George Mason Univ 4400 University Dr Fairfax VA 22030-4444

BEALE, LUTHER A(LTON), b Jacksonville, Fla, July 30, 23; m 58; c 3. STRUCTURAL ANALYSIS. *Educ:* Ga Inst Technol, BS, 48, MS, 54; Univ Tex, PhD, 67. *Prof Exp:* Design engr, Robert & Co Assocs, 48-49; instr math, Ga Inst Tech, 49-50, asst prof eng mech, 54-55; expressway engr, McDougald Construct Co, 50-51; contractor, 51-52; opers engr, US Army Corps Engrs, 52-54; PROF CIVIL ENG & HEAD DEPT, LAMAR UNIV, 55- *Mem:* Am Soc Civil Engrs; Nat Soc Prof Engrs. *Res:* Structural analysis directed toward improvement of present design techniques and methods. *Mailing Add:* Lamar Univ PO Box 10024 Beaumont TX 77706

BEALE, PAUL DREW, b Wilmington, NC, Feb 20, 55; m 80. CRITICAL PHENOMENA, STATISTICAL MECHANICS. *Educ:* Univ NC, Chapel Hill, BS, 77; Cornell Univ, PhD(physics), 82. *Prof Exp:* Res fel physics, dept theoret physics, Oxford Univ, 82-84; ASST PROF PHYSICS, DEPT PHYSICS, UNIV COLO, BOULDER, 84- *Concurrent Pos:* Consult, Electromagnetic Applns, Denver, 85; vis researcher, Los Alamos Nat Lab, 84- *Mem:* Am Phys Soc. *Res:* Theoretical critical phenomena; critical and multicritical phenomena in pure and random systems; phenomenological finite-size scaling; statistical mechanics of model systems; theory of failure in random systems; ferroelectrics. *Mailing Add:* Dept Physics Univ Colo Box 390 Boulder CO 80309

BEALE, SAMUEL I, b Los Angeles, Calif, June 25, 42. CHLOROPLAST DEVELOPMENT, TETRAPYRROLE BIOSYNTHESIS. *Educ:* Univ Calif, Los Angeles, BS, 64, MS, 66, PhD(bot), 70. *Prof Exp:* Res Assoc, Brookhaven Nat Lab, 70-72; res botanist, Univ Calif, Davis, 72-73, res biochemist, 73-74; res fel, Rockefeller Univ, 74-76, asst prof, 76-77; lectr bot, Univ Calif, Davis, 77-78, asst res biochemist, Sch Med, 78-79; from asst prof to assoc prof, 79-86, PROF BIOL, BROWN UNIV, 86- *Concurrent Pos:* Jr investr, Summer Res Prog, Marine Biol Lab, Woods Hole, Mass, 74, assoc investr, 75, guest investr, 76-; mem adv panel, Cell Biol Prog, NSF, 83-84, Metab Biol Prog, 84-86 & cell biol Sci, 90- *Mem:* AAAS; Am Soc Biol Chemists; Am Soc Plant Physiologists; Int Soc Plant Molecular Biol; Sigma Xi; NY Acad Sci. *Res:* Biosynthesis of photosynthetic and respiratatory pigments and related products; expression of photosynthetic competence; regulation of chloroplast development. *Mailing Add:* Div Biol & Med Brown Univ Providence RI 02912

BEALER, STEVEN LEE, b Evansville, Ind, July 10, 49. CARDIOVASCULAR PHYSIOLOGY. *Educ:* Univ Wyo, BS, 71, MS, 73, PhD(psychol), 76. *Prof Exp:* Fel, Univ Iowa, 76-79; from asst prof to assoc prof, 79-89, PROF PHYSIOL, UNIV TENN, 89- *Mem:* Am Physiol Soc; Soc Neurosci. *Res:* Central nervous system control of cardiovascular regulation and fluid-electrolyte balance, including plasma volume regulation; hypertension, diabetes, vascular responses to hemorrhage and control of blood flow. *Mailing Add:* Dept Physiol Univ Tenn Memphis 894 Union Ave Memphis TN 38163

BEALES, FRANCIS WILLIAM, b Bristol, Eng, Feb 7, 19; Can citizen; m 54; c 5. GEOLOGY. *Educ:* Cambridge Univ, BA, 46; Univ Toronto, PhD(geol), 52. *Prof Exp:* Lectr geol, McMaster Univ, 47-51; assoc prof, 51-74, PROF GEOL, UNIV TORONTO, 74- *Mem:* Am Asn Petrol Geol; hon mem Soc Econ Paleont & Mineral; Geol Soc Am; Geol Asn Can; Int Asn Sedimentol; hon mem Can Asn Petrol Geol. *Res:* Stratigraphy with special interest in limestones as petroleum reservoirs or metallic mineral host rock. *Mailing Add:* Dept Geol Sci Univ Toronto Toronto ON M5S 1A1 Can

BEALL, ARTHUR CHARLES, JR, b Atlanta, Ga, Aug 17, 29; m 49; c 2. MEDICINE. *Educ:* Emory Univ, BS, 50, MD, 53. *Prof Exp:* Intern surg, Barnes Hosp, St Louis, Mo, 53-54; resident, Methodist Hosp, Houston, Tex, 54-55; resident surg affil hosps, 55-60, resident thoracic surg, 60-61, from instr to assoc prof, 59-71, PROF SURG, BAYLOR COL MED, 71- *Mem:* Am Asn Thoracic Surg; Am Col Cardiol; Am Col Chest Physicians; Am Col Surgeons; Soc Thoracic Surg. *Res:* Cardiovascular surgery, especially open heart surgery; techniques of extracorporeal circulation; pulmonary embolectomy; development of artificial heart valves; management cardiovascular trauma. *Mailing Add:* Dept Surg Baylor Col Med Houston TX 77030

BEALL, FRANK CARROLL, b Baltimore, Md, Oct 3, 33; m 63; c 3. WOOD SCIENCE. *Educ:* Pa State Univ, BS, 64; State Univ NY Col Forestry, Syracuse Univ, MS, 66, PhD(wood physics), 68. *Prof Exp:* Res technologist, US Forest Prod Lab, 66-68; asst prof wood sci & technol, Pa State Univ, University Park, 68-73, assoc prof, 73-75; assoc prof wood sci, Univ Toronto, 75-77; sr sci spec, Weyerhaeuser Co, 77-88; PROF & DIR, FOREST PROD LAB, 88- *Mem:* Soc Wood Sci & Technol (pres, 91-92); Forest Prod Res Soc; Am Soc Testing & Mat; Am Soc Non-Destructive Testing. *Res:* Non destructive evaluation; acoustic emission; ultrasonics; adhesion and adhesives; fire performance of wood. *Mailing Add:* Forest Prod Lab Univ Calif Berkeley 1301 S 46th St Richmond CA 94804

BEALL, GARY WAYNE, b Granbury, Tex, Aug 15, 50; m 76; c 1. PHYSICAL CHEMISTRY, INORGANIC CHEMISTRY. *Educ:* Tarleton State Univ, BS, 72; Baylor Univ, MS, 74, PhD(chem), 75. *Prof Exp:* Res assoc chem, Baylor Univ, 75-77; res staff chem, Oak Ridge Nat Lab, 77-80. *Mem:* Am Chem Soc; Sigma Xi. *Res:* Solution chemistry of the actinides; structural studies of rare earth and actinide compounds; transition metal cyanide complexes; environmental studies of the actinides. *Mailing Add:* PO Box 1 Fairfield KY 40020

BEALL, GEORGE HALSEY, b Montreal, Que, Oct 14, 35; m 62, 76; c 4. GEOCHEMISTRY, CERAMICS. *Educ:* McGill Univ, BSc, 56, MSc, 58; Mass Inst Technol, PhD(geol), 62. *Prof Exp:* From sr geologist to res geologist, 62-65, res assoc geol, 65-66, mgr glass-ceramic res dept, 66-77, RES FEL, CORNING GLASS WORKS, 77- *Mem:* AAAS; Mineral Soc Am; fel Am Inst Chemists; fel Am Ceramic Soc; Am Chem Soc. *Res:* Nucleation and growth of crystals in glass; glass-ceramics; synthetic mica; silica polymorph solid solutions; phase equilibria; petrology of basalt. *Mailing Add:* 106 Woodland Rd Big Flats NY 14814

BEALL, GILDON NOEL, b Raymond, Wash, Sept 11, 28; m 51; c 4. ALLERGY, IMMUNOLOGY. *Educ:* Univ Wash, BS, 50, MD, 53. *Prof Exp:* Asst med, Sch Med, Univ Wash, 57-60; instr, 60-61, from asst prof to assoc prof, 61-70, PROF MED, SCH MED, UNIV CALIF, LOS ANGELES, 70-, CHIEF ALLERGY & CLIN IMMUNOL & DIR MED CLIN, HARBOR-UCLA MED CTR, 68- *Mem:* Am Soc Clin Invest; Sigma Xi; fel Am Col Physicians; fel Am Acad Allergy. *Res:* Application of immunology to clinical medicine and allergy; mechanism of asthma pathogenesis; autoimmune thyroid disease. *Mailing Add:* 2716 Via Anita Palos Verdes Estates CA 90274

BEALL, HERBERT, b Chatham, Ont, Aug 26, 39; US citizen; m 77; c 3. INORGANIC CHEMISTRY, CHEMICAL EDUCATION. *Educ:* Univ Wis, BS, 61; Harvard Univ, PhD(chem), 67. *Prof Exp:* Sr res chemist, Olin Res Ctr, 67-68; asst prof, 68-73, assoc prof chem, 73-89, PROF CHEM, WORCESTER POLYTECH INST, 89- *Concurrent Pos:* Vis lectr, Univ Canterbury, 74-75. *Mem:* Am Chem Soc. *Res:* Inorganic chemistry of coal and carbon; chemical education. *Mailing Add:* Dept Chem Worcester Polytech Inst Worcester MA 01609

BEALL, JAMES HOWARD, b Grantsville, WVa, May 12, 45; m 66; c 1. ACTIVE GALACTIC NUCLEI, RELATIVISTIC PARTICLE BEAMS. *Educ:* Univ Colo, BA, 72; Univ Md, MS, 75, PhD(physics), 79. *Prof Exp:* Teaching asst physics, Univ Md, 72-75; res asst physics, Lab Astro & Solar Physics, Goddard Space Flight Ctr, 75-78; cong sci fel pub policy, Off Tech Assessment, US Cong, 78-79; proj scientist pub policy, Sci & Anal BKO, 79-81; resident assoc physics, Nat Acad Sci, 81-83; CONSULT PHYSICS, NAVAL RES LAB, 83-; PROF PHYSICS, ST JOHNS COL, ANNAPOLIS, MD, 88- *Concurrent Pos:* Bd dirs, The Word Works Inc, Washington, DC, 75-; prin investr, Astron Liberal Arts, NSF grant, 91-; adj prof, Space Sci Inst, George Mason Univ, 91-; sci & eng adv bd, High Frontier, Arlington, Va, 91- *Mem:* Am Phys Soc; Am Astron Soc; AAAS; Sci Res Soc NAm. *Res:* Theoretical and observational astrophysics and public policy; develop models for the energy loss mechanism of a beam of relativistic particles propagating through the ambient medium of an active galaxy; time variability of active galactic nuclei and quasars; emission mechanisms for the x-ray radiation observed in these objects; responsible for first report of concurrent radio and x-ray variability of an active galaxy; export control policy. *Mailing Add:* Code 4120 Naval Res Lab Washington DC 30375-5000

BEALL, JAMES ROBERT, b Stillwater, Okla, June 29, 40. TOXICOLOGY, PHYSIOLOGY. *Educ:* Okla State Univ, BS, 63; Univ Okla, MS, 65, PhD(physiol), 69. *Prof Exp:* Res asst physiol, Med Ctr, Univ Okla, 63-66; res asst reproductive physiol, Inst Health Sci, Brown Univ, 66-69; prin toxicologist, Dept Toxicol & Path, Schering Corp, 69-77; sr toxicologist & biol sci adminr toxic substances, Off Toxic Substance, Environ Protection Agency, 77-79; spec asst occup health, Directorate Tech Support, Occup

Safety & Health Admin, 79-81; TOXICOL PROG MGR, US DEPT ENERGY, 81- *Concurrent Pos:* Environ Protection Agency rep, Toxic Substances Control Act, Interagency Testing Comt, 77-78; chmn interagency regulatory liaison group, Work Group on Testing Guidelines, Environ Protection Agency, 77-79; US rep, Expert Work Group Orgn Econ Coop & Develop, 78-; consult, Sci Appl Inc, 79-, Litton Bionetics, Pharmacopathics, numerous legal associations; mem bd dir, Am Bd Toxicol, Inc, 81-85, Toxicol Lab Accreditation, Inc, 84-; mem bd dir, Asn Govt Toxicologists, 83-, pres, 83- *Mem:* Soc Toxicol; Am Col Toxicol; Am Teratology Soc; Environ Mutagen Soc; NY Acad Sci; Sigma Xi; Asn Govt Toxicologists. *Res:* Biochemistry of reproduction and teratology; indoor air pollution; toxicology. *Mailing Add:* 4804 Old Middletown Rd Jefferson MD 21755

BEALL, PAULA THORNTON, b Cameron, Tex, July 22, 46. BIOPHYSICS, CELL BIOLOGY. *Educ:* Rice Univ, BA, 68; Univ Tex, MS, 70, PhD(biophys), 76. *Prof Exp:* Spec chemist, Pitt Co Mem Hosp, 70-71; res assoc physiol, Med Sch, ECarolina Univ, 71-73; instr, 74-77, ASST PROF BIOPHYS, BAYLOR COL MED, 77- *Concurrent Pos:* Consult, Off Naval Res, 76-; adj asst prof, Univ Tex, 76-; lectr, Rice Univ, 77- *Mem:* Biophys Soc; Tissue Cult Asn; Asn Women Sci; Am Soc Cell Biol; Am Physiol Soc; Sigma Xi. *Res:* Role of water in physiological events; water-macromolecular interactions on the cellular level in growth, development and disease; cytoskeletal structure and cytoplasmic ground substance, nutritional requirements of cells. *Mailing Add:* 3007 Conway Houston TX 77025-2609

BEALL, ROBERT ALLAN, b Seattle, Wash, Mar 15, 20; m 42; c 3. PHYSICS, PHYSICAL METALLURGY. *Educ:* Univ Wash, BS, 41. *Prof Exp:* Res supvr phys metall, Bur Mines, 49-78; CONSULT REACTIVE METALS MELTING & CASTING, 78- *Concurrent Pos:* Mem subcomt, Mat Adv Bd Refractory Metals, Nat Acad Sci, 64 & Mat Adv Bd High Performance Steel & Ti Castings, 68; mem US working group electrometall, US-USSR, 73-77; consult, Union Carbide Nuclear Co, Oak Ridge, 74-76; mem expert melting zirconium, Int Atomic Energy Agency, Arg, 78-80. *Honors & Awards:* H R Russ Ogden Award, Am Soc Testing Mat, 86; Spec Melting Technol Award, Int Conf Vacuum Metall, 88. *Mem:* Sigma Xi. *Res:* Melting and casting reactive and refractory metals. *Mailing Add:* PO Box 446 Lincoln City OR 97367

BEALL, ROBERT JOSEPH, b Washington, DC, May 19, 43; m 67; c 2. BIOCHEMISTRY. *Educ:* Albright Col, BS, 65; State Univ NY Buffalo, MA, 70, PhD(biol), 70. *Prof Exp:* Asst biol, State Univ NY Buffalo, 65-70; from instr to asst prof physiol, Sch Med, Case Western Reserve Univ, 70-74, from instr to asst prof, Sch Dent, 71-74; grants assoc, NIH, 74-75, chief, Endocrinol & Metab Dis Br, 78-79; Dir, Metab Dis Prog, Extramural Progs, Nat Inst Arthritis, Metab & Digestive Dis, 75-78; med dir, 80-84, nat dir, 81-84, EXEC VPRES, CYSTIC FIBROSIS FOUND, BETHESDA, MD, 84- *Honors & Awards:* NIH Merit Award, 80. *Mem:* AAAS; Am Soc Human Genetics; NY Acad Sci; Am Soc Microbiol; Sigma Xi. *Res:* Molecular endocrinology, mechanism of action of adrenocorticotropic hormone; cyclic nucleotide synthesis and action, steriodogenesis, assay of adrenocorticotropic hormone in plasma, neoplastic endocrine tissue, genetic diseases, metabolic control; cystic fibrosis. *Mailing Add:* Cystic Fibrosis Found 6931 Arlington Rd Bethesda MD 20814

BEALL, S(AMUEL) E, JR, b Richland, Ga, May 16, 19; m 48; c 5. CHEMICAL ENGINEERING, NUCLEAR ENGINEERING. *Educ:* Univ Tenn, BS, 43. *Prof Exp:* Chem engr smokeless powder process develop, E I du Pont de Nemours & Co, 43-45; with Oak Ridge Nat Lab, 45-63, dir, Reactor Div, 63-75, dir, Energy Div, 75-77, dir, Planning-Energy Div, 77-79, CONSULT, ENERGY & ENVIRON PROGS, OAK RIDGE NAT LAB, 79-, DIR TENN ENERGY AUTHORITY, 79- *Mem:* Fel Am Nuclear Soc; AAAS. *Res:* Smokeless powders; nuclear reactor development and operation; energy systems; environmental effects; solar technology. *Mailing Add:* 1032 Craigland Ct Knoxville TN 37919

BEALMEAR, PATRICIA MARIA, b Dodge City, Kans, Oct 23, 29. EXPERIMENTAL PATHOLOGY. *Educ:* Mt St Scholastica Col, BS, 49; Univ Notre Dame, PhD(biol), 65. *Prof Exp:* Chmn sci dept, high schs, Kans, 52-62, Mo, 62-63; asst prof biol, Mt St Scholastica Col, 65-66; res scientist microbiol, Univ Notre Dame, 66-71; from asst prof to assoc prof exp biol, Baylor Col Med, 71-76; ASSOC CANCER RES SCIENTIST, DEPT DERMAT, ROSWELL PARK MEM INST, 76- *Concurrent Pos:* Bd trustees, Int Soc Exp Hemat, 75-77; exec secy, Int Asn Gnotobiology, 81- *Mem:* Am Asn Immunologists; Asn Gnotobiotics (exec secy, 75-); Int Soc Exp Hemat; Am Soc Exp Path; Soc Exp Biol & Med; AAAS; Am Asn Lab Animal Sci; Am Asn Pathologists; NY Acad Sci; Sigma Xi. *Res:* Bone marrow transplantation; cellular immunology; radiation pathology and treatment; application of germfree research to the clinical situation. *Mailing Add:* Dept Dermat Roswell Park Mem Inst Buffalo NY 14263

BEALOR, MARK DABNEY, b Shamokin, Pa, Sept 27, 21; m 51; c 1. ORGANIC CHEMISTRY. *Educ:* Univ Pa, BS, 49; Univ Ore, MA, 54, PhD(chem), 56. *Prof Exp:* Chemist, Sharp & Dohme Inc, 49-51; sr chemist, Niagara Div, Food Mach & Chem Co, Inc, 55-57; sr chemist, S C Johnson & Son, Inc, 57-68, res supvr, 68-77, technol invest & licensing, 78-85; RETIRED. *Mem:* Am Chem Soc; Soc Cosmetic Chem. *Res:* Chemical structure versus biological activity; carbonyl condensations; product development. *Mailing Add:* 408 Tar Landing New Bern NC 28562

BEALS, EDWARD WESLEY, b Wichita, Kans, July 1, 33; div; c 1. ECOLOGY. *Educ:* Earlham Col, BA, 56; Univ Wis-Madison, MS, 58, PhD(bot), 61. *Prof Exp:* Instr biol, Am Univ Beirut, 61-62; from asst prof to assoc prof, Haile Selassie Univ, 62-65; asst prof bot & zool, Univ Wis Ctr Syst, 65-66; asst prof bot, 66-67, lectr zool, 67-68, asst prof, 68-69, assoc prof zool & bot, 69-73, PROF ZOOL & BOT, UNIV WIS-MADISON, 73- *Mem:* Ecol Soc Am; Am Ornithologists Union; Am Soc Naturalists; Soc Study Evolution; Cooper Ornith Soc; Sigma Xi. *Res:* Mathematical analysis of ecological communities; interrelations between vegetation and animals, especially birds and mammals; factors in species diversity and niche specialization; allelopathy in plant communities. *Mailing Add:* Dept Zool 432 Birge Hall Univ Wis Madison WI 53706

BEALS, HAROLD OLIVER, b Mishawauka, Ind, July 4, 31; m 59, 82; c 1. WOOD TECHNOLOGY, PALEOBOTANY. *Educ:* Purdue Univ, BSF, 55, MS, 57, PhD(bot), 60. *Prof Exp:* Secy, Evans Prod, Inc, Kans, 58-59; asst prof, 60-69, ASSOC PROF FORESTRY, AUBURN UNIV, 69- *Concurrent Pos:* Consult; mem & housing rehab inspector, Southern Bldg Code Cong Int, Inc. *Mem:* AAAS; Forest Prod Res Soc (secy-treas, 62-64). *Res:* Plant anatomy. *Mailing Add:* Sch Forestry Auburn Univ Auburn AL 36849-5418

BEALS, R MICHAEL, b Erie, Pa, July 28, 54. MATHEMATICS. *Educ:* Univ Chicago, BA & MS, 76; Princeton Univ, PhD(math), 80. *Prof Exp:* ASSOC PROF MATH, RUTGERS UNIV, 81- *Mem:* Am Math Soc; Math Asn Am. *Mailing Add:* Dept Math Rutgers Univ New Brunswick NJ 08903

BEALS, RICHARD WILLIAM, b Erie, Pa, May 28, 38; m 62; c 3. PARTIAL DIFFERENTIAL EQUATIONS, FUNCTIONAL ANALYSIS. *Educ:* Yale Univ, BA, 60, MA, 62, PhD(math), 64. *Prof Exp:* Instr math, Yale Univ, 64-65; vis asst prof, Univ Chicago, 65-66, from asst prof to prof, 66-77; PROF MATH, YALE UNIV, 77- *Mem:* Am Math Soc; Math Asn Am. *Res:* Inverse scattering; pseudodifferential operators; complex analysis; global analysis; transport theory. *Mailing Add:* Dept Math Yale Univ PO Box 2155 New Haven CT 06520

BEALS, ROBERT J(ENNINGS), b Decatur, Ill, Nov 12, 23; m 52; c 3. CERAMIC ENGINEERING, ANALYTICAL CHEMISTRY. *Educ:* Univ Ill, BS, 47, MS, 50, PhD(ceramic eng), 55. *Prof Exp:* Asst ceramic eng, Univ Ill, 47-52, from instr to assoc prof, 52-62; assoc ceramic engr, Argonne Nat Lab, 62-69; asst tech dir, Charles Taylor Sons Co, 69-70, tech dir, 70-71; DIR RES & DEVELOP, THE HALL CHINA CO, 72- *Concurrent Pos:* Comnr, Eng Manpower Comn, Engrs Joint Coun, 70-; consult, E R Advan Ceramics; pres, Eng Acreditation Comn of Accreditation Bd, Eng & Technol. *Mem:* Fel Am Ceramic Soc (treas, 78-80, pres, 81-82); Nat Inst Ceramic Engrs. *Res:* Nuclear ceramic materials; ceramic chemistry; high temperature materials; high temperature x-ray analysis; manufacture of refractories; ceramic whitewares; electrical ceramics. *Mailing Add:* The Hall China Co PO Box 989 East Liverpool OH 43920

BEALS, RODNEY K, b Portland, Ore, Jan 4, 31; m 56; c 3. ORTHOPEDIC SURGERY. *Educ:* Willamette Univ, BA, 53; Univ Ore, MD, 56; Am Bd Orthop Surg, dipl, 64. *Prof Exp:* From instr to asst prof, 61-67, assoc prof, 67-80, PROF ORTHOP SURG, MED SCH, UNIV ORE, 80- *Mem:* Am Acad Orthop Surg. *Res:* Clinical orthopedics. *Mailing Add:* Div Orthop Rehab Univ Portland Med Sch Portland OR 97201

BEAM, CARL ADAMS, b Olympia, Wash, July 5, 20; m 52. MICROBIAL GENETICS, RADIOBIOLOGY. *Educ:* Brown Univ, BA, 42; Yale Univ, MS, 47, PhD(microbiol), 50. *Prof Exp:* Res assoc biophysics, Univ Calif, Berkeley, 50-56; asst prof, Yale Univ, 56-63; assoc prof biol, 63-75, PROF BIOL, BROOKLYN COL, 75- *Concurrent Pos:* NIH fel, 50-51, res grants & sr investr, 56-63. *Mem:* Phycol Soc Am; Radiation Res Soc; Biophys Soc; Genetics Soc Am; Soc Protozoologists. *Res:* Biological factors affecting radiation sensitivity; dinoflagellate genetics and diversification. *Mailing Add:* Dept Biol Brooklyn Col Brooklyn NY 11210

BEAM, CHARLES FITZHUGH, JR, b New York, NY, Sept 24, 40; m 83; c 2. ORGANIC CHEMISTRY, POLYMER CHEMISTRY. *Educ:* City Col New York, BS, 63; Univ Md, College Park, PhD(org chem), 70. *Prof Exp:* Asst gen & phys chem, Univ Md, 63-65, org polymer chem, 65-68; res assoc, P M Gross Chem Lab, Duke Univ, 68-71 & Lehigh Univ, 71-72; assoc prof chem, dir sci res & title III coordr, Newberry Col, 73-82; DEPT CHEM, COL CHARLESTON, 82- *Concurrent Pos:* Prin investr, numerous chem res grants, 72- *Mem:* Am Chem Soc; fel Am Inst Chemists; fel The Royal Chem Soc. *Res:* Preparation of new heterocyclic compounds with potential biological activity via strong-base multiple anion synthesis methods and techniques; organic synthesis with multiple anions; medicinal syntheses; pyrolysis studies; organic chemistry of high polymers. *Mailing Add:* Dept Chem Col Charleston Charleston SC 29401

BEAM, JOHN E, b Honesdale, Pa, Oct 27, 31; m 61; c 3. FOOD SCIENCE, DAIRY SCIENCE. *Educ:* Pa State Univ, BS, 57; Univ Mass, MS, 59, PhD(food sci), 62. *Prof Exp:* Instr dairy mfg, Univ Mass, 57-59, res, 59-62; proj leader res & develop, Ross Labs, Abbott Labs, 62-64; in charge spec projs food res & develop, Beech-Nut, Inc, 64-67, mgr processed food res, 67-68, food res, 68, mgr prod develop, 68-78; MEM STAFF, LIFESAVER INC, 78- *Concurrent Pos:* Mem lab comt, Wash Lab, Nat Canners Asn. *Mem:* AAAS; Inst Food Technol; Am Dairy Sci Asn; Sci Res Soc Am; Sigma Xi. *Res:* Dairy chemistry; gelling agents; infant nutrition; food product development; confections development; candy and chewing gum; cough and cold products; beverages. *Mailing Add:* 5140 River Chase Ridge Winston-Salem NC 27104

BEAM, KURT GEORGE, JR, b Chicago, Ill, Aug 13, 45; m 70; c 2. ELECTROPHYSIOLOGY, NEUROBIOLOGY. *Educ:* Pomona Col, BA, 67; Univ Wash, PhD(physiol), 74. *Prof Exp:* Fel, dept pharmacol, Yale Univ, 74-77; asst prof, dept physiol, Univ Iowa, 77-86; ASSOC PROF, COLO STATE UNIV, 86- *Concurrent Pos:* Res career develop award, NIH. *Mem:* Biophys Soc; Fedn Am Soc Exp Biol; Soc Neurosci; Soc Gen Physiologists; AAAS. *Res:* Voltage clamp techniques to characterize the electrical excitability of normal and denervated mammalian skeletal muscle, to understand developmental, hormonal, neurotrophic regulation of the physiology of electrically excitable cells; electrophysiological cell biological and molecular biological studies of nerve and muscle development; excitation-contraction coupling; calcium channels in nerve and muscle. *Mailing Add:* Dept Physiol Colo State Univ Ft Collins CO 80523

BEAM, THOMAS ROGER, b Elizabeth, NJ, July 12, 46; m 70; c 2. INFECTIOUS DISEASES, PHARMACOKINETICS. *Educ:* Univ Pa, AB, 68, Sch Med, MD, 72. *Prof Exp:* Intern resident, State Univ NY, Buffalo, 72-74, chief resident, 75, fel, 75-77, asst prof, 77-84, chief infectious dis, 77-87, ASSOC PROF MED, STATE UNIV NY, BUFFALO, 85-, ASSOC PROF

MICROBIOL, 89- *Concurrent Pos:* Consult, Erie County Med Ctr, 75-, W Seneca Develop Ctr, 83-, J N Adam Develop Ctr, Buffalo Gen Hosp & Roswell Park Mem Inst, 84- & Craig Develop Ctr, 86-; clin asst prof pharm & nursing, State Univ NY, Buffalo, 79-, assoc chief staff, 88-; chmn, adv comt antibiotics, Food & Drug Admin. *Mem:* AAAS; Am Soc Microbiol; Am Fedn Clin Res; NY Acad Sci; Infectious Dis Soc Am; Am Col Physicians. *Res:* Evaluation of treatment of bacterial meningitis by antibiotics; pharmacokinetics of drug penetration into the central nervous system; host defenses against bacterial infection of the central nervous system; adult immunizations; antibiotic usage patterns. *Mailing Add:* Buffalo VA Med Ctr 3495 Bailey Ave Buffalo NY 14215

BEAM, WALTER R(ALEIGH), b Richmond, Va, Aug 27, 28; c 2. ELECTRONICS ENGINEERING. *Educ:* Univ Md, BS, 47, MS, 50, PhD(elec eng), 53. *Prof Exp:* Instr elec eng, Univ Md, 47-51; res engr, RCA Labs, 52-56, mgr microwave appl res, 56-59; prof elec eng, Rensselaer Polytech Inst, 59-64, head dept, 59-62; mem res staff, Int Bus Mach Corp, 64-67, dir eng technol, IBM Syst Develop Div, 67-69; tech & mgt consult, 70-74; dep advan technol, Off Asst Secy US Air Force, 74-81; vpres res & develop, Sperry Div, 81-83, consult, 83-85; prof systs eng, George Mason Univ, 85-89; CONSULT, 89- *Concurrent Pos:* Mem, US Air Force Sci Adv Bd, 81-86 & USAF Studies Bd. *Honors & Awards:* Except Civilian Serv Decoration, USAF. *Mem:* Fel Inst Elec & Electronics Engrs; Asn Comput Mach; Am Arbit Asn; Am Defense Preparedness Asn; Asn Old Crows. *Res:* Computer technology and systems; electronic materials and devices; microwave electronics; application of computers in management; military applications of technology; author of 3 books. *Mailing Add:* 3824 Ft Worth Ave Alexandria VA 22304

BEAMAN, BLAINE LEE, b Portland, Ore, July 28, 42; m 65; c 1. MICROBIOLOGY. *Educ:* Utah State Univ, BS, 64; Univ Kans, PhD(microbiol), 68. *Prof Exp:* Asst prof med microbiol, Sch Med & Dent, Georgetown Univ, 70-75; from asst prof to assoc prof, 75-82, PROF MED MICROBIOL, SCH MED, UNIV CALIF, DAVIS, 82- *Concurrent Pos:* Fel, Sch Med, NY Univ, 68-70; Latin Am vis prof, Venezuela, 80. *Mem:* Am Soc Microbiol; AAAS; NY Acad Sci; fel, Am Acad Microbiol; Reticuloendothelial Soc. *Res:* Mechanisms of nocardial pathogenesis; induction, pathogenicity and biology of bacterial L-forms; the role of the alveolar macrophage in lung defense; infectious diseases of the lungs; bacterial cell wall biochemistry and bacterial ultrastructure. *Mailing Add:* Dept Med Microbiol Univ Calif Davis Sch Med Davis CA 95616

BEAMAN, DONALD ROBERT, b Chicago, Ill, Apr 2, 33; m 69; c 3. ELECTRON SPECTROSCOPY, METALLURGY. *Educ:* Univ Ill, BS, 58, MS, 61, PhD(metall), 63. *Prof Exp:* Metallurgist alloy steel develop, Inland Steel Co, 58-59; res asst physics, Univ Ill, 62-63, asst prof metall, 63-64; res metallurgist electron probe, Int Nickel Co, 64-65; SR RES SPECIALIST X-RAY & ELECTRON SPECTROS, DOW CHEM CO, 65- *Concurrent Pos:* US deleg water qual comt, Int Standards Orgn, 77-; mem water qual comt, Am Nat Standards Inst, 77- *Honors & Awards:* Environ Qual Award, US Environ Protection Agency, 76; 1978 Presidential Award, Microbeam Anal Soc, 78. *Mem:* Microbeam Anal Soc (treas, 70-71, 75-76, pres elect, 72, pres, 73); Am Soc Metals; Am Soc Testing & Mat. *Res:* X-ray and electron spectroscopy of materials using electron probe analysis, scanning and transmission electron microscopy, Auger electron spectroscopy and analytical transmission electron microscopy. *Mailing Add:* 930 Balfour Midland MI 48640-3328

BEAMAN, JOHN HOMER, b Marion, NC, June 20, 29; m 58; c 2. SYSTEMATIC BOTANY. *Educ:* NC State Col, BS, 51; State Col Wash, MS, 53; Harvard Univ, PhD, 57. *Prof Exp:* From asst prof to assoc prof, 56-68, PROF BOT, MICH STATE UNIV, 68-, CURATOR, BEAL-DARLINGTON HERBARIUM, 56- *Concurrent Pos:* Nat Acad Sci-Nat Res Coun sr vis res assoc, Smithsonian Inst, 65-66; mem ed comt, Flora NAm Prog, 66-73; mem bd dirs, Orgn Trop Studies, 72-79; prog dir syst biol, NSF, 79-81. *Mem:* Bot Soc Am; Am Soc Plant Taxon (pres, 80); Int Asn Plant Taxon; Soc Econ Bot (pres, 80-81); Ecol Soc Am; Sigma Xi. *Res:* Biosystematic studies in alpine flora of Mexico and Central America; monographic studies in Asteraceae; Compositae of Mexico; computer applications in systematic biology. *Mailing Add:* Dept Bot & Plant Path Mich State Univ Plant Biol Lab East Lansing MI 48824

BEAMER, PAUL DONALD, b Avonmore, Pa, Sept 21, 14; m 36; c 1. VETERINARY PATHOLOGY. *Educ:* Ohio State Univ, DVM, 41; Univ Ill, MS, 45, PhD, 51. *Prof Exp:* Asst bacteriologist & pathologist, State Dept Agr, Ohio, 41-42; asst vet path & hyg, State Dept Agr, Ill, 42-44; war food asst, Univ Ill, 44-46, from asst prof to prof vet path & hyg, 46-60; adv to dean col vet med, Uttar Pradesh Agr Univ, India, 60-64; PROF VET PATH & HYG, UNIV ILL, URBANA-CHAMPAIGN, 64-, ASST TO DEAN, COL VET MED, 77- *Mem:* Fel Am Vet Med Asn; Am Col Vet Path; Int Acad Path; Sigma Xi. *Res:* Infectious diseases of livestock; bovine tuberculosis; brucellosis; swine enteritis; Newcastle disease of poultry; lead poisoning in wild ducks. *Mailing Add:* Col Vet Med Univ Ill Urbana IL 61801

BEAMER, ROBERT LEWIS, b Pulaski, Va, June 9, 33; m 59; c 1. PHARMACY, BIOCHEMISTRY. *Educ:* Med Col Va, BS, 55, MS, 57, PhD(chem), 59. *Prof Exp:* Asst chem, Med Col Va, 55-58; from asst prof to assoc prof pharmaceut chem, 59-69, assoc dean col pharm, 72-75, PROF MED CHEM, UNIV SC, 69- *Concurrent Pos:* Mead Johnson res grant, 62-63; NIH res grant, 64-; vis prof, Cornell Univ Med Col, 77-78. *Mem:* Am Chem Soc; Am Pharmaceut Asn. *Res:* Synthetic estrogenic agents; antimetabolites in cancer chemotherapy; catalytic hydrogenation; asymmetric synthesis; enzyme inhibition; computer modeling of drugs. *Mailing Add:* Col Pharm Univ SC Columbia SC 29208

BEAMES, CALVIN G, JR, b Kingston, Okla, Oct 29, 30; m 52; c 3. PHYSIOLOGY. *Educ:* NMex Highlands Univ, AB, 55, MS, 56; Univ Okla, PhD(physiol, biochem), 61. *Prof Exp:* Asst physiol, Univ Okla, 55-56; teacher biol, Santa Fe High Sch, 56-57; asst physiol, Univ Okla, 57-60; trainee, Rice Univ, 60-61, NIH fel, 61-62; from asst prof to assoc prof, 62-70, PROF PHYSIOL, OKLA STATE UNIV, 70- *Mem:* Am Physiol Soc. *Res:* Carbohydrate and lipid metabolism and transport mechanisms. *Mailing Add:* Dept Zool Okla State Univ Stillwater OK 74075

BEAMES, R M, b Brisbane, Australia, Oct 12, 31; m 56; c 4. ANIMAL NUTRITION. *Educ:* Univ Queensland, BAgrSc, 54, MAgrSc, 62; McGill Univ, PhD(nutrit), 65. *Prof Exp:* Husb officer, Animal Res Inst Yeerongpilly, Queensland Dept Primary Industs, 54-65, sr husb officer, 65-68; from asst prof to assoc prof, 69-88, PROF ANIMAL SCI, FAC AGR SCI, UNIV BC, 88- *Mem:* Can Soc Animal Sci; Brit Nutrit Soc; Am Soc Animal Sci; Australian Soc Animal Prod. *Res:* Nutritive value of grains for pigs; use of amino acids in pig rations; value of grain screenings for pigs; the use of the rat for a model for pig nutrition; digestibility measurements with fish. *Mailing Add:* Dept Animal Sci Univ BC 2075 Westbrook Pl Vancouver BC V6T 1W5 Can

BEAMISH, FREDERICK WILLIAM HENRY, b Toronto, Ont, July 31, 35; m 58; c 3. ZOOLOGY. *Educ:* Univ Toronto, BA, 58, PhD(zool), 62. *Prof Exp:* Assoc scientist behav & physiol biol, Biol Sta, Fisheries Res Bd Can, 62-65; from asst prof to assoc prof, 65-72, dept chmn, 74-79, PROF ZOOL, UNIV GUELPH, 72- *Mem:* Am Fisheries Soc; Am Soc Ichthyol & Herpet. *Res:* Physiology and ecology of cyclostome and teleost fishes; fish energetics and nutrition; fish toxicology. *Mailing Add:* Dept Zool Univ Guelph Guelph ON N1G 2W1 Can

BEAMISH, ROBERT EARL, b Shoal Lake, Man, Sept 9, 16; m 43; c 3. CARDIOLOGY. *Educ:* Brandon Col, BA, 37; Univ Man, MD, 42, BSc, 44; FRCP(C), 50, Edinburgh, 61, London, 77, Brandon Univ, DSc, 88. *Hon Degrees:* DSc, Brandon Univ, 88, Univ Man, 89. *Prof Exp:* Demonstr med, Fac Med, Univ Man, 46-49, from lectr to prof, 49-81; med dir, Great-West Life Assurance Co, 70-81; EMER PROF, UNIV MAN, 81-; CONSULT, DIV CARDIOVASC SCI, ST BONIFACE RES CTR, 87- *Concurrent Pos:* Cardiologist, Manitoba Clin, 46-70; asst physician, Winnipeg Gen Hosp, 47-57, assoc physician, 57-; Nuffield Dom traveling fel, Gt Brit, 47-48; mem bd dir, Can Heart Found, 63-75; regional rep, Coun Clin Cardiol, Am Heart Asn, Can, 63-; ed-in-chief, Can J Cardiol, 84- *Mem:* Fel Am Col Cardiol; fel Am Col Physicians; Can Cardiovasc Soc (hon secy & treas, 57-, pres, 69-70); Inter-Am Soc Cardiol; Can Med Asn. *Res:* Clinical and laboratory research in ischemic heart disease with special reference to the therapeutic use of anticoagulants and stress-induced heart disease. *Mailing Add:* St Boniface Res Ctr 351 Tache Winnipeg MB R2H 2A6 Can

BEAMS, HAROLD WILLIAM, b Belle Plaine, Kans, Aug 3, 03; m 35; c 2. BIOLOGY. *Educ:* Fairmount Col, AB, 25; Northwestern Univ, AM, 26; Univ Wis, PhD(zool), 29. *Prof Exp:* DuPont fel histol & embryol, Dept Med, Univ Va, 29-30; from asst prof to prof, 30-72, EMER PROF ZOOL, UNIV IOWA, 72- *Concurrent Pos:* Res assoc, Argonne Nat Lab; Rockefeller traveling fel, 34-35; assoc ed, Microtomist's Vade-Mecum; mem, Corp Marine Biol Lab. *Mem:* AAAS; Am Soc Nat; Am Soc Zool (treas, 41-44); Am Micros Soc; Am Asn Anat; Soc Develop Biol; Am Soc Cell Biol. *Res:* Cytology; Golgi apparatus; mitochondria; polarity; sex chromosomes and spermatogenesis of man; salivary gland chromosomes; cytology of human ovary; colchocine studies upon cells; viscosity studies; effects of electric current upon polarity and growth of cells; studies with the electron microscopy on various cellular components. *Mailing Add:* Dept Biol Univ Iowa Iowa City IA 52240

BEAN, BARBARA LOUISE, b Doddington, Eng, Sept 20, 48; m 78; c 1. CARDIOVASCULAR PHARMACOLOGY, CARDIOVASCULAR PHYSIOLOGY. *Educ:* Univ Surrey, BSc, 70; Univ London, PhD(physiol), 76. *Prof Exp:* Res asst hypertension, London Hosp Med Col, 70-75; res fel, Med Res Coun Blood Pressure Unit, Western Infirmary, Glasgow, Scotland, 75-76 & NIH, Bethesda, Md, 77-78; ASST PROF CARDIOVASC PHARMACOL, MED CTR, GEORGE WASHINGTON UNIV, 78- *Concurrent Pos:* Res fel, Ciba-Geigy Corp, Gt Brit, 75-76. *Mem:* Med Res Soc; Fedn Am Soc Exp Biol. *Res:* Mechanisms in the development and maintenance of renal hypertension; antihypertensive factors in the renal medulla; central actions of angiotensins; mechanism of action of vasodilator drugs. *Mailing Add:* 1705 S Broadlee Trail Annapolis MD 21401

BEAN, BARRY, b Framingham, Mass, Nov 2, 42. CELLULAR MOTILITY, REPRODUCTIVE CELL BIOLOGY. *Educ:* Tufts Univ, BS, 64; Rockefeller Univ, PhD(life sci), 70. *Prof Exp:* Fel microbiol, Rockefeller Univ, 64-70; res assoc biochem, Indian Inst Sci, Bangalore, 70-71; res fel USPHS, Rockefeller Univ, NY, 71-73; asst prof, 73-79, chmn dept, 83-85, ASSOC PROF BIOL, LEHIGH UNIV, BETHLEHEM, PA, 79- *Honors & Awards:* Woodrow Wilson hon fel, 64. *Mem:* Am Soc Cell Biol; Am Fertil Soc; Am Soc Androl; NY Acad Sci; Am Soc Human Genetics; Europ Soc Human Reproduction & Embryol. *Res:* Sperm function and the events surrounding fertilization and gene transmission; cellular motility and behavior; fertility and diseases involving infertility. *Mailing Add:* Dept Biol & Cellular Molecular Biol Lehigh Univ Bldg 3 Bethlehem PA 18015

BEAN, BRENT LEROY, b Rexburg, Idaho, June 13, 41; m 65; c 6. LASERS, QUANTUM OPTICS. *Educ:* Brigham Young Univ, BS, 66, MS, 68; NMex State Univ, PhD(physics), 72. *Prof Exp:* Res asst laser physics, Univ Laval, Quebec, 72-74; vis asst prof physics, Emory Univ, 74-78; physicist, Sci Applications, Inc, 78-79; PHYSICIST, OPTIMETRICS INC, 79- *Concurrent Pos:* Consult physics, Eng Exp Sta, Ga Inst Technol, 75-78. *Mem:* Optical Soc Am. *Res:* Development of the far infrared laser; use of the far infrared laser for water vapor continuum measurements; atmospheric transmission measurements in natural and obscured environments. *Mailing Add:* OptiMetrics Inc 106 E Idaho Las Cruces NM 88005

BEAN, C THOMAS, JR, b Winchester, Ill, May 14, 20; m 44; c 2. POLYMER CHEMISTRY. *Educ:* Western Ill State Col, BEd, 42; Univ Kans, MA, 46, PhD(org chem), 49. *Prof Exp:* Anal control chemist, Hiram Walker & Sons, Inc, Ill, 42; asst instr chem, Univ Kans, 42-44; anal lab foreman, Tenn Eastman Corp & Clinton Eng Works, Tenn, 44-46; supvr asst instr qual anal, Univ Kans, 46-49; res chemist, J T Baker Co, 49-50; res chemist, Hooker Chem Corp, 50-56, res supvr, 56-62, mgr proj eval, 74-78, res sect mgr, plastics sect existing prod res, 62-; dir admin & tech serv, Ashland Oil Inc, Ashland Petrol Co, 81-85; RETIRED. *Concurrent Pos:* Mgr res planning & eng, Ashland Petrol Co, 79-81. *Mem:* Am Chem Soc; Am Inst Chem Engrs. *Res:* Vapor phase chlorination of hydrocarbons; insecticides; polyester resins; high temperature elastomers; corrosion control; epoxy resins; fire hazard reduction. *Mailing Add:* 2716 Auburn Ave Ashland KY 41102

BEAN, CHARLES PALMER, b Buffalo, NY, Nov 27, 23; m 47; c 5. PHYSICS, BIOPHYSICS. *Educ:* Univ Buffalo, BA, 47; Univ Ill, AM, 49, PhD(physics), 52. *Prof Exp:* Res assoc, Gen Elec Res & Develop Ctr, 51-85; INST PROF, RENSSELAER POLYTECH INST, 85- *Concurrent Pos:* Consult, US Dept State, 57-58 & Nat Acad Sci, 59; adj prof, Rensselaer Polytech Inst, 57-; vis scientist, Am Inst Physics, 58-; Coolidge fel, 71; guest investr, Rockefeller Univ, 73-78; assoc ed, Biophys J, 75-78; dir, Dudley Observ & Bellevue Res Found, 75-, pres, Dudley Observ, 83-90; mem exec comt, Assembly Math & Phys Soc, Nat Res Coun, 76-79; vis prof, Univ Ill, 78; adj prof, State Univ NY, Albany, 79- *Honors & Awards:* Indust Res Mag Award, 70. *Mem:* Nat Acad Sci; Am Acad Arts & Sci; Biophys Soc; Am Phys Soc; AAAS. *Res:* Solid state physics; membranes; neurophysiology; superconductivity. *Mailing Add:* Sch Sci Rensselaer Polytech Inst Troy NY 12180

BEAN, DANIEL JOSEPH, b Enosburg Falls, Vt, June 12, 34; m 58; c 3. LIMNOLOGY, PHYSIOLOGICAL ECOLOGY. *Educ:* Univ Vt, BA, 60, MS, 62; Univ RI, PhD(biol sci), 69. *Prof Exp:* Instr biol, Marist Col, 64-68; asst prof, St Michael's Col, 68-71, assoc prof, 71-80, chmn biol dept, 74-84, PROF BIOL, ST MICHAEL'S COL, VT, 80-, CHMN DEPT, 89- *Concurrent Pos:* Res assoc, Aquatec, Inc, Vt, 69-80. *Mem:* AAAS; Am Soc Limnol & Oceanog; Ecol Soc Am; Am Inst Biol Sci; Sigma Xi. *Res:* River zooplankton and botton fauna; lake zooplankton. *Mailing Add:* 163 Lyman Ave Burlington VT 05401

BEAN, GEORGE A, b Hempstead, NY, Apr 3, 33; m 57; c 2. PLANT PATHOLOGY. *Educ:* Cornell Univ, BS, 58; Univ Minn, MS, 60, PhD(plant path), 63. *Prof Exp:* Plant pathologist, Nat Capitol Region, US Dept Interior, 63-66; assoc prof bot, 66-72, assoc prof plant path, 72-76, assoc prof bot, 76-80, PROF BOT, UNIV MD, COLLEGE PARK, 80- *Mem:* Am Phytopath Soc. *Res:* Physiology and chemistry of plant disease causing fungi. *Mailing Add:* Dept Biol Univ Md College Park MD 20742

BEAN, GERRITT POST, b Amsterdam, NY, Apr 29, 29; m 56. ORGANIC CHEMISTRY. *Educ:* Northeastern Univ, BS, 52; Pa State Univ, PhD(org chem), 56. *Prof Exp:* Process develop chemist, Merck & Co, Inc, 56-57; res chemist, Althouse Chem Co, Inc, 57-60; asst prof chem, Douglass Col, Rutgers Univ, 60-66; vis prof, Univ East Anglia, 66-67; assoc prof, 67-71, PROF CHEM, WESTERN ILL UNIV, 71- *Mem:* AAAS; Am Chem Soc; The Chem Soc; fel Am Inst Chem. *Res:* Reactions and properties of 5-membered heterocycles; effect of substituents on reactions and properties of heterocycles; molecular orbital calculations. *Mailing Add:* 4415 Laquna Pl Apt 112 Boulder CO 80303

BEAN, JAMES J(OSEPH), b Salt Lake City, Utah, Mar 3, 03; m 29; c 2. MECHANICAL ENGINEERING. *Educ:* Univ Utah, BS, 24. *Prof Exp:* With, Westinghouse Elec & Mfg Co, 24-25; shop foreman, Utah Light & Traction Co, 25-26; concentration engr, asst mill supt, Int Smelting Co, 26-33; design & construct engr, Gen Eng Co, 33-35; field engr, Mineral Dressing, Am Cyanamid Co, 35-56; chief metallurgist, Miami Copper Co, 56-70; CONSULT METALLURGIST, 71- *Mem:* Am Inst Mining, Metall & Petrol Eng. *Res:* Flotation; cyanidation; heavy media separation; leaching. *Mailing Add:* Rte 1 Box 582B Miami AZ 85539

BEAN, JOHN CONDON, b Seattle, Wash, Sept 1, 50; c 2. SOLID STATE PHYSICS, ELECTRICAL ENGINEERING. *Educ:* Calif Inst Technol, BS, 72; Stanford Univ, MS, 74, PhD(appl physics), 76. *Prof Exp:* MEM TECH STAFF SOLID STATE ELECTRONICS, BELL TEL LABS, 76- *Mem:* Am Phys Soc; Inst Elect & Electronics Engrs. *Res:* Growth of semiconductor materials and devices by molecular beam epitaxy; interaction of ion and laser beams with semiconductors. *Mailing Add:* Bell Tel Labs 1C-326 600 Mountain Ave Murray Hill NJ 07974

BEAN, MICHAEL ARTHUR, b Alliance, Nebr, Sept 18, 40. PATHOLOGY, IMMUNOLOGY. *Educ:* Univ Colo, BA, 62, MD, 67. *Prof Exp:* Pathologist, Registry of Radiation Path, Armed Forces Inst Path, Washington, DC, 69; pathologist, Atomic Bomb Casualty Comn, Hiroshima, Japan, 69-70; assoc immunol, Sloan-Kettering Inst Cancer Res, 72-75, lab head tumor-host immunol, 74-75; sr investr, Va Mason Res Ctr, 75-77, assoc dir, 80-83, mem, 77-88; MEM, PAC NORTHWEST RES FOUND, 88- *Concurrent Pos:* Asst attend path, Mem Hosp, New York, 73-75; asst prof biol, Sloan-Kettering Div, Grad Sch, Cornell Univ, 74-75; affil investr, Fred Hutchinson Cancer Ctr, Seattle, 75-; affil assoc prof micro-immunology, Sch Med, Univ Wash, 76- *Mem:* AAAS; Am Asn Cancer Res; Am Asn Immunologists. *Res:* Tumor-host immunology; transplantation immunology. *Mailing Add:* Pac Northwest Res Found 720 Broadway Seattle WA 98122

BEAN, RALPH J, b Atlantic City, NJ, Aug 13, 33; m 59; c 3. MATHEMATICS. *Educ:* Univ Pittsburgh, BS, 57, MA, 60; Univ Md, PhD(math), 62. *Prof Exp:* Res scientist, Lewis Lab, NASA, 57-58; from instr to asst prof math, Univ Wis, 62-67; assoc prof, Univ Tenn, 67-71; PROF MATH, STOCKTON STATE COL, 71- *Concurrent Pos:* Math Asn Am vis lectr, 63-; NSF res grants, 65-71; vis prof math, Univ Houston, 81-82. *Mem:* Am Math Soc; Math Asn Am; Nat Coun Teachers Math; Am Fedn Teachers. *Res:* Point set topology; Euclidian spaces and manifolds. *Mailing Add:* Dept Math Stockton State Col Pomona NJ 08240

BEAN, ROBERT JAY, b Dallas, Tex, Aug 9, 24. GEOPHYSICS. *Educ:* Southern Methodist Univ, BS, 48; Harvard Univ, AM, 50, PhD(geophys), 51. *Prof Exp:* Geophysicist, Shell Oil Co, 51-89, geophys adv, 83-89; CONSULT, 90- *Mem:* Soc Explor Geophys; Europ Soc Explor Geophys. *Res:* Gravity and magnetic interpretation. *Mailing Add:* 1108 W Tri Oaks Lane No 147 Houston TX 77043

BEAN, ROBERT TAYLOR, b Cherokee, Iowa, Feb 8, 13; m 53; c 1. HYDROGEOLOGY, ENGINEERING GEOLOGY. *Educ:* Col Wooster, AB, 34; Ohio State Univ, MA, 42. *Prof Exp:* Asst geol, Stanford Univ, 46; geologist, US Geol Surv, 47; from jr to sr eng geologist, Calif Dept Water Resources, 47-56, supv eng geologist, 56-66; tech adv hydrogeol, UN, 66-71; CONSULT GEOLOGIST, 71- *Concurrent Pos:* Consult, Chile-Calif Prog, 64; lectr, Calif State Univ, Northridge, 72-78; lectr, Calif State Univ, Los Angeles, 74-89; vis prof, Univ Calif, Los Angeles, 89-90. *Mem:* Fedn Am Sci; Asn Eng Geologists (pres, 60-61); fel Geol Soc Am; Am Geophys Union; Asn Ground Water Scientists & Engrs. *Res:* Water resources evaluation; international peace. *Mailing Add:* 2729 Willowhaven Dr La Crescenta CA 91214

BEAN, ROSS COLEMAN, b Thatcher, Ariz, Apr 22, 24; m 49; c 3. BIOCHEMISTRY. *Educ:* Univ Calif, BS, 46, PhD(comp biochem), 53; Stanford Univ, MS, 48. *Prof Exp:* Chemist, Western Regional Res Lab, 48-50; asst, Univ Calif, Berkeley, 51-53, jr biochemist, 53-55; jr biochemist, Univ Calif, Riverside, 55-56, asst prof biochem, 56-63; supvr biochem & physiol, Res Staff, 63-74, PRIN SCIENTIST, AERONUTRONIC DIV, FORD AEROSPACE CORP, 74- *Concurrent Pos:* Consult, US Army Weapons Command, 67 & Off Saline Water, US Dept Interior, 69. *Mem:* AAAS; Biophys Soc; Am Soc Biol Chemists. *Res:* Photosynthesis; carbohydrate metabolism; membrane transport and electrophysiology; infrared detector materials. *Mailing Add:* Loral Aeronutronics Ford Rd Bldg 2 Rm 53 Newport Beach CA 92658

BEAN, VERN ELLIS, b La Grande, Ore, Jan 19, 37; m 61; c 4. HIGH PRESSURE PHYSICS, PRESSURE METROLOGY. *Educ:* Brigham Young Univ, BS, 62, MS, 64, PhD(physics), 73. *Prof Exp:* Physicist, Lawrence Radiation Lab, Univ Calif, Livermore, 64-67; physicist, Nat Bur Stand, 72-88, PHYSICIST, NAT INST STANDARDS & TECHNOL, 88- *Concurrent Pos:* Consult, Egyptian Orgn Standardization, 81 & 85; consult, Nat Phys Lab, New Delhi, India, 83; vis prof, Inst Metrol, "G Colonnetti", Italy, 88. *Res:* Development of techniques for the accurate measurement of static and dynamic pressure; development of techniques for precise measurements at high pressure. *Mailing Add:* Nat Inst Standards & Technol B126 PHY Gaithersburg MD 20899

BEAN, WENDELL CLEBERN, b Port Arthur, Tex, Apr 28, 34; m 61. ELECTRICAL ENGINEERING. *Educ:* Lamar State Col, BA & BS, 55; Univ Pittsburgh, MS, 58, PhD(elec eng), 61. *Prof Exp:* Jr engr, Westinghouse Elec Corp, 55, sr engr, Bettis Atomic Power Lab, 55-67; USPHS spec res fel, 67-68; PROF ELEC ENG, LAMAR UNIV, 68- *Mem:* Inst Elec & Electronics Engrs. *Res:* Signal processing; biological systems modeling. *Mailing Add:* Dept Elec Eng Lamar Univ Beaumont TX 77710

BEAN, WILLIAM BENNETT, medicine; deceased, see previous edition for last biography

BEAN, WILLIAM CLIFTON, b Paris, Tex, Sept 10, 38. MATHEMATICS, AEROSPACE ENGINEERING. *Educ:* Univ Tex, Austin, BA, 59, MA, 60, PhD(math), 71. *Prof Exp:* Spec instr math, Univ Tex, Austin, 60-63; aerospace technologist flight simulator software design, Flight Crew Support Div, 63-66, aerospace technologist, Mission Planning & Anal Div, 66-71, AEROSPACE ENG, MISSION PLANNING & ANALYSIS DIV, NASA-JOHNSON SPACECRAFT CTR, 71- *Concurrent Pos:* Asst instr, Univ Tex, Austin, 70-71. *Honors & Awards:* Apollo Achievement Award, NASA, 69. *Mem:* Am Inst Aeronaut & Astronaut; Math Asn Am; Tensor Soc; NY Acad Sci. *Res:* Optimum interplanetary trajectory design; studies on extensor structures in the calculus of variations; space shuttle communication and data handling. *Mailing Add:* 17345 Saturn Lane Houston TX 77058

BEAN, WILLIAM J, JR, b Mar 16, 45; US citizen; div; c 1. VIROLOGY. *Educ:* Univ Maine, BS, 67, MS, 69; Rutgers Univ, PhD(microbiol), 74. *Prof Exp:* Fel virol, Waksman Inst Microbiol, Rutgers Univ, 74-75; res assoc, 75-77, asst mem virol, 77-82, ASSOC MEM, VIROL & MOLECULAR BIOL DIV, ST JUDE CHILDREN'S RES HOSP, 82- *Concurrent Pos:* Asst prof microbiol, Ctr Health Sci, Univ Tenn, 81-85. *Mem:* Am Soc Microbiol; Int Soc Antiviral Res; Am Soc Virol. *Res:* Genetics and molecular biology of influenza virus. *Mailing Add:* St Jude Childrens Res Hosp 332 N Lauderdale Memphis TN 38101

BEANE, DONALD GENE, b Aurora, Ill, Aug 25, 29; m 52; c 3. MATHEMATICS, DISCRETE MATHEMATICS. *Educ:* Iowa Wesleyan Col, AB, 51; Univ Ill, MA, 58, PhD(math ed), 62; Ohio State Univ, MSc, 66. *Prof Exp:* Asst prof educ, 62-65; from asst prof to assoc prof, 66-75, PROF MATH, COL WOOSTER, 75- *Concurrent Pos:* Teaching fel, Great Lakes Cols Asn, 75-76. *Mem:* Math Asn Am; Nat Coun Teachers Math. *Res:* Mathematical education; discrete mathematics; applications of linear mathematical models. *Mailing Add:* Dept Math Col Wooster Wooster OH 44691

BEANS, ELROY WILLIAM, b Toledo, Ohio, Dec 25, 31; m 54. MECHANICAL ENGINEERING. *Educ:* Ohio State Univ, BME & MSc, 53; Pa State Univ, PhD(mech eng), 61. *Prof Exp:* Design engr, NAm Aviation, Inc, 54-56; instr mech eng, Pa State Univ, 56-58, res asst 58-59, instr aeronaut eng, 60-61; sr engr, Battelle Mem Inst, 61-64; group leader, NAm Aviation, Inc, 64-65, proj engr, 65-66; asst prof mech eng, Ohio State Univ, 66-70; tech specialist, Rockwell Int, 70-73; from asst prof to assoc prof, 73-88, PROF MECH ENG, UNIV TOLEDO, 88-, ASST CHMN, 89- *Mem:* Am Soc Mech Engrs; Combustion Inst. *Res:* Propulsion; IC engines; combustion; gas dynamics; therymodynamics. *Mailing Add:* 3719 Fairwood Dr Sylvania OH 43560

BEAR, DAVID GEORGE, b 1950. MECHANISMS OF GENE EXPRESSION. *Educ:* Univ Ariz, BS, 72; Univ Calif, Santa Cruz, PhD(chem), 78. *Prof Exp:* Res assoc, Univ Ore, 78-82; asst prof cell biol, 82-88, ASSOC PROF CELL BIOL, SCH MED, UNIV NMEX, 88- *Mem:* Elec Micros Soc Am. *Res:* Structure and function of transcriptional apparatus. *Mailing Add:* Dept Cell Biol Sch Med Univ NMex Albuquerque NM 87131

BEAR, HERBERT S, JR, b Philadelphia, Pa, Mar 13, 29; m 51, 84; c 2. MATHEMATICS. *Educ:* Univ Calif, Berkeley, BA, 50, PhD(math), 57. *Prof Exp:* Instr math, Univ Ore, 55-56; assoc, Univ Calif, Berkeley, 56-57; from instr to asst prof, Univ Wash, 57-62; assoc prof, Univ Calif, Santa Barbara, 62-67; prof, NMex State Univ, 67-69; chmn dept, 69-74, grad chmn, 80-83, PROF MATH, UNIV HAWAII, MANOA, 69- *Concurrent Pos:* Vis asst prof, Princeton Univ, 59-60; vis assoc prof, Univ Calif, San Diego, 65-66; trustee, Berkeley Math Sci Res Inst, 81- *Mem:* Am Math Soc; Math Asn Am. *Res:* Functional analysis; partial differential equations. *Mailing Add:* Dept Math Univ Hawaii at Manoa Honolulu HI 96822

BEAR, JOHN L, b Lampasas, Tex, Mar 6, 34; m 60; c 3. INORGANIC CHEMISTRY. *Educ:* Southwest Tex State Col, BS, 55, MA, 56; Tex Tech Col, PhD(inorg chem), 60. *Prof Exp:* Fel inorg chem, Fla State Univ, 60-62, asst prof, 62-63; assoc prof, 63-72, PROF CHEM, UNIV HOUSTON, 72-, CHMN CHEM, 75- *Mem:* Am Chem Soc. *Res:* Fast reactions in solution; thermodynamics of metal ion complex formation in solution; dimeric metal complexes. *Mailing Add:* Dept Chem Univ Houston Houston TX 77004-3995

BEAR, PHYLLIS DOROTHY, b New York, NY, Aug 7, 31. MICROBIAL GENETICS, MOLECULAR BIOLOGY. *Educ:* San Diego State Col, BS, 56; Univ Calif, Los Angeles, PhD(microbiol), 66. *Prof Exp:* Res technician marine genetics, Scripps Inst, Univ Calif, 56-62; assoc prof molecular biol, 68-74, PROF MOLECULAR BIOL, UNIV WYO, 74- *Concurrent Pos:* Carnegie fel, Carnegie Inst, 66-68. *Mem:* AAAS; Am Soc Microbiol. *Res:* Structure and function of microbial nucleic acid; regulation of gene expression in bacterial viruses; pathogenesis of Anaplasma marginale. *Mailing Add:* Dept Microbiol Univ Wyo Laramie WY 82071

BEAR, RICHARD SCOTT, b Miamisburg, Ohio, June 8, 08; m 31, 42, 72; c 2. MOLECULAR BIOLOGY, BIOPHYSICS. *Educ:* Princeton Univ, SB, 30; Univ Calif, Berkeley, PhD(chem), 33. *Prof Exp:* Nat Res Coun fel chem, Princeton Univ, 33-34; res assoc zool, Washington Univ, 34-38; asst prof chem, Iowa State Col, 38-41; from assoc prof to prof biol, Mass Inst Technol, 41-57; dean sci & humanities, Iowa State Univ, 57-61; dean grad sch, Boston Univ, 61-66, prof biol, 66-69; prof, 69-78, EMER PROF ANAT, UNIV NC, CHAPEL HILL, 78- *Concurrent Pos:* Mem panel cytochem, Comt Growth, Nat Res Coun, 47-50; mem study sect, Biophys & Biophys Chem, NIH, 60-64; mem sci adv comt, Helen Hay Whitney Found, 60-64. *Mem:* fel Am Acad Arts & Sci (vpres class II, 67-69); Am Crystallog Asn; Biophys Soc; Am Chem Soc. *Res:* X-ray diffraction and optical methods applied to the study of natural polymers and tissues. *Mailing Add:* 43 Winchester Ct Chapel Hill NC 27514

BEARCE, DENNY N, b Pittsburgh, Pa, Nov 1, 34; m 57; c 2. GEOLOGY. *Educ:* Brown Univ, BA, 56; Mo Sch Mines, BS, 62, MS, 63; Univ Tenn, PhD(geol), 66. *Prof Exp:* Geologist, Mobile Oil Co, 63-64; asst prof geol, Eastern Ky Univ, 66-67; from asst prof to assoc prof geol, Birmingham-Southern Col, 67-73, chmn dept, 70-73; MEM FAC, UNIV ALA, BIRMINGHAM, 77-, CHMN DEPT GEOL 77- & PROF GEOL, 81- *Mem:* Am Asn Petrol Geol; fel Geol Soc Am. *Res:* Mapping and describing structure and stratigraphy of lower Cambrian and Precambrian rocks in Blue Ridge province of eastern Tennessee; structure and stratigraphy of Talladega Metamorphic Belt of east Alabama. *Mailing Add:* Geology Dept Univ Alabama Birmingham Univ Sta Birmingham AL 35294

BEARCE, WINFIELD HUTCHINSON, b Lewiston, Maine, Oct 20, 37; m 61; c 2. ORGANIC CHEMISTRY. *Educ:* Bowdoin Col, AB, 59; Lawrence Col, MS, 61, PhD, 64. *Prof Exp:* Sr res chemist, St Regis Paper Co, 64-66; assoc prof chem, Mo Valley Col, 66-80, prof chem & acad dean, 80-87; High Point Col, 83-87; PROF, CENT COL, 87- *Mem:* Am Chem Soc. *Res:* Carbohydrate chemistry, particularly the stability, with respect to acid, of the glycosidic linkage in carbohydrate polymers; essential oils as a source of substrate for mechanistic studies. *Mailing Add:* Cent Col Pella IA 50219-1999

BEARD, BENJAMIN H, b Blair, Nebr, Jan 12, 18; m 45; c 1. PLANT BREEDING & GENETICS. *Educ:* Univ Mo, BS, 50, MS, 52; Univ Nebr, PhD(agron), 55. *Prof Exp:* Instr field crops, Univ Mo, 51-52; res geneticist, Sci & Educ Admin, Agr Res, USDA, Univ Minn, 55-60; res geneticist, Sci & Educ Admin, Agr Res, USDA, Univ Calif, 60-82; RETIRED. *Concurrent Pos:* Res assoc agron, Univ Nebr, 51-53; lectr, Univ Calif, 74-82. *Mem:* Am Genetic Asn; Crop Sci Soc; Sigma Xi; Am Soc Agron. *Res:* Radiation genetics; cytogenetics; breeding and genetics of oilseed crops; horticulture. *Mailing Add:* 6929 Radiance Circle Citrus Heights CA 95621

BEARD, CHARLES IRVIN, b Ambridge, Pa, Nov 30, 16; m 48; c 2. RADIOPHYSICS, REMOTE SENSING. *Educ:* Carnegie Inst Technol, BS, 38; Mass Inst Technol, PhD(physics), 48. *Prof Exp:* Asst engr res labs, Westinghouse Elec Corp, Pa, 38-39; sr physicist, Field Res Labs, Magnolia Petrol Co, Tex, 48-50 & Appl Physics Lab, Johns Hopkins Univ, 50-56; sr eng specialist physics, Sylvania Electronic Defense Lab, 56-62; staff mem, Boeing Sci Res Labs, Wash, 62-71; RES PHYSICIST, NAVAL RES LAB, 71- *Concurrent Pos:* Chmn, US Comt F, Int Sci Radio Union, 70-73, mem, US Nat Comt & assoc ed, Radio Sci, 73-75. *Honors & Awards:* Bolljahn Mem Award, Inst Elec & Electronics Engr, 62. *Mem:* AAAS; Am Phys Soc; Int Union Radio Sci; fel Inst Elec & Electronics Engrs. *Res:* Microwave spectroscopy; low frequency electromagnetic waves in conducting media; scattering of electromagnetic waves from random media, the ocean, atmospheric turbulence and internal waves. *Mailing Add:* 4115 S Capitol St Washington DC 20032

BEARD, CHARLES NOBLE, b India, Dec 27, 06; m 33; c 1. GEOLOGY. *Educ:* Ind Univ, BA, 35, MA, 36; Univ Ill, PhD(geol), 41. *Prof Exp:* EMER PROF GEOL, CALIF STATE UNIV, 70- *Concurrent Pos:* Cartographer, USAF, 42-46. *Mem:* Fel Geol Asn Am. *Res:* Columnar jointing of lava in California, Washington and Devil's Tower, Wyoming. *Mailing Add:* 1715 E Alluvial Ave Fresno CA 93720

BEARD, CHARLES WALTER, b Tifton, Ga, Nov 16, 32; m 60; c 4. VETERINARY VIROLOGY. *Educ:* Univ Ga, DVM, 55; Univ Wis, MS, 64, PhD(vet sci), 65. *Prof Exp:* Mem staff, 65-72, DIR, SOUTHEAST POULTRY RES LAB, AGR RES SERV, USDA, 72- *Mem:* Am Vet Med Asn; Am Asn Avian Path. *Res:* Aerobiology; respiratory diseases; viruses; techniques of aerosol exposure; influence of route of administration on host response. *Mailing Add:* USDA Agr Res Serv 934 College Station Rd Athens GA 30605

BEARD, DAVID BREED, b Needham, Mass, Feb 1, 22; div; c 4. THEORETICAL PHYSICS. *Educ:* Hamilton Col, BS, 43; Cornell Univ, PhD(physics), 51. *Prof Exp:* Instr nuclear physics, Cath Univ Am, 50-51; instr theoret physics, Univ Conn, 51-53; from asst prof to prof physics, Univ Calif, 53-64; prof physics & astron & chmn dept, 64-77, univ distinguished prof physics, 77-87, EMER PROF PHYSICS, UNIV KANS, 88- *Concurrent Pos:* Guggenheim fel & Fulbright scholar, Imp Col, Univ London, 65-66, Sci Res Coun Gt Brit sr res fel & vis prof, 78-79; consult, Naval Ord Lab, Cath Univ Am, 50-51, Pratt & Whitney Aircraft Corp, 52-53, Atomic Energy Comn Proj, Univ Conn, 53-54, Lawrence Radiation Lab, 54-56 & 63-65, Lockheed Spacecraft & Missiles Res, 56-64, Sandia Corp, 59-68, Goddard Space Flight Ctr, 61-69 & US Atomic Energy Hq, 65-72; NATO sr fel, 72. *Mem:* Fel AAAS; Am Geophys Union; fel Am Phys Soc; Fedn Am Scientists; Am Astron Soc. *Res:* Meson theory of nuclear forces; level densities in heavy nuclei; beta-ray spectrometry; plasma physics; astro-physics; space physics; magnetospheric physics; quantum mechanics. *Mailing Add:* 69 Springwood Rd South Portland ME 04106

BEARD, ELIZABETH L, b New Orleans, La, Apr 2, 32. PHYSIOLOGY. *Educ:* Tex Christian Univ, BA, 52, BS, 53, MS, 55; Tulane Univ, PhD(physiol), 61. *Prof Exp:* From instr to assoc prof, 55-68, PROF BIOL SCI, LOYOLA UNIV, LA, 68- *Concurrent Pos:* Prof biol sci, Med Reinforcement & Enrichment Prog, Sch Med, Tulane Univ, 68-; vis prof, dept physiol, Harvard Med Sch, 83-84. *Mem:* AAAS; NY Acad Sci; Am Asn Univ Professors; Am Physiol Soc; Soc Exp Biol & Med. *Res:* Blood proteolytic activity, particularly as related to stress in mammals and to atherosclerosis. *Mailing Add:* 6127 Garfield New Orleans LA 70118

BEARD, GEORGE B, b Marblehead, Mass, Feb 22, 24; m 55; c 5. EXPERIMENTAL NUCLEAR PHYSICS. *Educ:* Harvard Univ, AB, 47; Univ Mich, MS, 48, PhD(physics), 55. *Prof Exp:* Asst, Univ Mich, 50-51; from instr to asst prof physics, Mich State Univ, 54-60; assoc prof, 60-65, chmn dept physics, 73-77 & 80-81, actg chmn dept comput sci, 84-85, assoc dean, Col Lib Arts, 86, PROF PHYSICS, WAYNE STATE UNIV, 65- *Concurrent Pos:* Res prof, Dept Nuclear Eng, Univ Mich, 61; resident res assoc, Argonne Nat Lab, 63-65, resident assoc, 69-70; vis prof, Univ Melbourne, Australia, 81-82. *Mem:* Fel Am Phys Soc; Am Asn Physics Teachers. *Res:* Low energy nuclear spectroscopy; lifetimes of lowlying excited states; Mossbauer effect studies. *Mailing Add:* Dept Physics Wayne State Univ Detroit MI 48202

BEARD, JAMES B, b Piqua, Ohio, Sept 24, 35; m 55; c 2. STRESS PHYSIOLOGY, TURFGRASS. *Educ:* Ohio State Univ, BS, 57; Purdue Univ, MS, 59, PhD(environ plant physiol), 61. *Prof Exp:* From assoc prof to prof environ physiol grasses, Mich State Univ, 61-75; interim head soil & crop sci, 80, PROF ENVIRON PHYSIOL GRASSES, TEX A&M UNIV, 75- *Concurrent Pos:* NSF Post-doctoral fel, 69-70. *Honors & Awards:* Hon Serv Award, Int Turfgrass Soc, 73. *Mem:* Int Turfgrass Soc (pres, 69); Sigma Xi; fel Am Soc Agron; fel Crop Sci Soc Am (pres, 86); Am Soc Hort Sci; Coun Agr Sci & Technol. *Res:* Microenvironment and physiology of heat, cold, drought and wear stresses in perennial grasses; shade adaptation; rooting; turfgrass culture; sod production and utilization; sports field culture. *Mailing Add:* 1812 Shadowood Dr College Station TX 77840

BEARD, JAMES DAVID, b Dayton, Ohio, July 21, 37; m 59; c 2. MEDICAL PHYSIOLOGY. *Educ:* DePauw Univ, BA, 59; Univ Tenn, PhD(physiol), 63. *Prof Exp:* Instr & fel physiol, Sch Med, Marquette Univ, 63-65; from instr to asst prof physiol & biophys, 65-74, ASSOC PROF PSYCHIAT & DIR, ALCOHOL RES CTR, CTR HEALTH SCI, UNIV TENN, MEMPHIS, 74- *Concurrent Pos:* Physiologist, Res Serv, Vet Admin Hosp, Wood, Wis, 63-65; Nat Acad Sci-Nat Comn Int Unit Physiol Sci travel grant, Tokyo, 65; dir, Alcohol & Drug Res Ctr & Clin Labs, Memphis Mental Health Inst, 67-, dir res, 72-; mem bd, Nat Coun Alcoholism; mem res initial rev group, Study Sect, Nat Inst Alcohol Abuse & Alcoholism, 76-, chairperson biomed sect, 77-78, chairperson, 78- *Mem:* Sigma Xi; Res Soc Alcoholism; AAAS; Soc Neurosci; NY Acad Sci. *Res:* Pathophysiology of acute and chronic alcoholism and other drug ingestion; cardiovascular dynamics; fluid and electrolyte metabolism including trace metals; renal physiology; psychopharmacology, biological psychiatry. *Mailing Add:* Univ Tenn PO Box 40966 Memphis TN 38104-0966

BEARD, JAMES TAYLOR, b Birmingham, Ala, Oct 1, 39; m 65; c 2. HEAT TRANSFER, THERMAL SYSTEMS DESIGN & ANALYSIS. *Educ:* Auburn Univ, BME, 61; Okla State Univ, MS, 63, PhD(mech eng), 65. *Prof Exp:* Res asst eng, Okla State Univ, 64-65; asst prof, 65-69, asst provost, Acad Space Admin, 72-77, ASSOC PROF ENG, UNIV VA, 69- *Concurrent Pos:* Partner, Assoc Environ Consults, 71-; lectr solar energy & energy policy, Pakistan & Cyprus, US Info Angecy, 78; mem, Tech Adv Comt, Va Air Pollution Control Bd, 78-87. *Mem:* Am Soc Mech Engrs; Am Soc Heating, Refrig & Air-Conditioning Engrs; Air Pollution Control Asn. *Res:* Convective heat and mass transfer; solar energy; combustion; thermal sciences; radiation heat transfer. *Mailing Add:* Dept Mech & Aerospace Eng Univ Va Thornton Hall Charlottesville VA 22903

BEARD, JEAN, b Cedar Falls, Iowa, Apr 12, 34. SCIENCE EDUCATION. *Educ:* State Univ Iowa, BA, 56; Univ Northern Iowa, MA, 60; Ore State Univ, PhD(sci educ), 69. *Prof Exp:* Teacher jr high sch, Minn, 56-59 & 60-61; instr prof educ & sci supvr, Mankato State Col, Wilson Campus Sch, 61-65; part-time instr sci educ, Ore State Univ, 65-68; assoc prof, 69-80, PROF NATURAL SCI, SAN JOSE STATE UNIV, 80- *Mem:* Nat Asn Biol Teachers; Nat Sci Teachers Asn; Nat Asn Res Sci Teaching. *Res:* Assessment of science achievements by students in general education science, kindergarten-14th grade, including science processes, science knowledge and attitudes about natural phenomena and organisms; rational general education about energy alternatives. *Mailing Add:* Dept Biol Sci San Jose State Univ San Jose CA 95192

BEARD, KENNETH VAN KIRKE, b Cheyenne, Wyo, Apr 9, 37; m 63; c 3. ATMOSPHERIC SCIENCE, PHYSICS. *Educ:* Calif State Univ, Los Angeles, BSc, 66; Univ Calif, Los Angeles, MSc, 68, PhD(meteorol), 70. *Prof Exp:* Res meteorologist, Univ Calif, Los Angeles, 68-70 & 71-74; vis scientist, Nat Ctr Atmospheric Res, 70-71; PROF & PRIN SCIENTIST, DEPT ATMOSPHERIC SCI, ILL STATE WATER SURV, UNIV ILL, 74- *Concurrent Pos:* Environ Protection grant, 72-75; NSF grants, 75-91. *Mem:* Am Meteorol Soc; Am Geophys Union; AAAS; Am Asn Aerosol Res; Sigma Xi. *Res:* Cloud and precipitation physics; aerosol physics; atmospheric scavenging; radar meterology; drop dynamics and interactions. *Mailing Add:* Dept Atmospheric Sci Univ Ill 105 S Gregory Dr Urbana IL 61801

BEARD, LEO ROY, b West Baden, Ind, Apr 6, 17; m 39; c 3. WATER RESOURCES ENGINEERING, HYDROLOGY. *Educ:* Calif Inst Technol, BS, 39. *Prof Exp:* Engr, Corps Engrs, 39-49, Off, Chief of Engrs, 49-52 & Corps Engrs, Sacramento, 52-64, dir hydrol eng ctr, 64-72; prof civil eng & dir, Ctr Res Water Resources, Univ Tex, Austin, 72-80; SR STAFF ENGR, ESPEY HUSTON & ASSOC, 80- *Concurrent Pos:* Vis lectr, Univ Calif, 66-67, Utah State Univ, 67 & Univ Calif, Davis, 70-72; consult, H G Acres Ltd, Can, Tex Water Develop Bd, Calif Dept Water Resources, Sask-Nelson Basin Bd, Can, Aleyeska Pipeline Co, Electrobras of Brazil & others. *Honors & Awards:* Julian Hinds Award, Am Soc Civil Engrs. *Mem:* Nat Acad Eng; Am Geophys Union; hon mem Am Soc Civil Engrs; hon mem Am Water Resources Asn; fel Int Water Resources Asn; fel AAAS. *Res:* Analysis of water resource systems; simulation of stochastic processes in hydrology. *Mailing Add:* 606 Laurel Valley Rd Austin TX 78746

BEARD, LUTHER STANFORD, b Langley, SC, Feb 21, 29; m 52; c 2. SYSTEMATIC BOTANY. *Educ:* Furman Univ, BS, 50; Univ NC, MA, 59, PhD(bot), 64. *Prof Exp:* Teacher high sch, SC, 55-58; instr biol, Furman Univ, 58-59; assoc prof, 61-67, chmn, Biol Dept, 63-76, PROF BIOL, CAMPBELL UNIV, NC, 67- *Mem:* Am Soc Plant Taxon; Soc Study Evolution; Sigma Xi. *Res:* Taxonomy of the genus Schrankia and its relation to the genus Mimosa. *Mailing Add:* Box 366 Buies Creek NC 27506

BEARD, MARGARET ELZADA, b Washington, DC, Oct 17, 41. CELL BIOLOGY. *Educ:* Wellesley Col, BA, 63; Univ Mich, MS, 65, PhD(zool), 67. *Prof Exp:* Fel path, Albert Einstein Col Med, 67-69, asst prof, 69-70; asst prof res ophthal, Sch Med, Univ Ore, 72-75; asst prof biol, 75-81, sr res biologist & dir micros, Reed Col, 81-82; SR SCIENTIST & LECTR, DEPT BIOL, COLUMBIA UNIV, 82- *Concurrent Pos:* Horace H Rackham fel; NIH fel. *Mem:* Am Soc Cell Biol; NY Acad Sci; AAAS; Histochem Soc; Sigma Xi. *Res:* Investigations of the structure and function of peroxisomes; structure and secretory activity of bag cell neurons and of the atrial gland in Aplysia Californica. *Mailing Add:* Nathan Kline Inst Rockland City Psychiat Ctr Orangeburg NY 10962

BEARD, OWEN WAYNE, b Wattensaw, Ark, Nov 15, 16; m 44; c 3. INTERNAL MEDICINE, CARDIOLOGY. *Educ:* Okla State Univ, 37-39; Univ Ark, BSM & MD, 43; Am Bd Internal Med, dipl, 50. *Prof Exp:* Instr med, Med Br, Univ Tex, 45-47; from asst prof to prof med, Sch Med, Univ Ark, Little Rock, 47-84; chief geriat, Vet Admin Hosp, Little Rock, 75-84; CHIEF GERIAT, SCH MED SCI, UNIV ARK, LITTLE ROCK, 75-, EMER PROF MED, 84- *Concurrent Pos:* Asst chief med, Vet Admin Hosp, Little Rock, 55-75. *Mem:* Am Heart Asn; Am Col Physicians; fel Am Col Cardiol; Geront Soc; fel Am Geriat Soc. *Res:* Cardiovascular disease. *Mailing Add:* 7008 Rockwood Rd Little Rock AR 72207

BEARD, RICHARD B, b Boston, Mass, Dec 17, 22; m 48; c 3. ELECTRICAL & BIOMEDICAL ENGINEERING. *Educ:* Northeastern Univ, BS, 47; Lowell Technol Inst, cert, 48; Harvard Univ, SMEE, 50; Univ Pa, PhD(elec eng), 65. *Prof Exp:* Chemist, Weymouth Artificial Leather Co, 47-48; appln engr, Minneapolis-Honeywell Regulator Co, 50-54, res engr, 54-57; instr elec eng, Drexel Univ, 58-60; res assoc, Univ Pa, 60-65; assoc prof, 65-72, dir Biomed Eng & Sci Inst, 74-78, PROF ELEC ENG, DREXEL UNIV, 72- *Concurrent Pos:* Consult, Johnson & Johnson, 77- & Standard Pressed Steel, 78- *Mem:* Acoust Soc Am; Inst Elec & Electronics Engrs; Am Chem Soc; Soc Biomat; Electrochem Soc. *Res:* Biomaterials; bioelectrochemistry; biophysics; dielectric and ultrasonic measurements on macro molecules; polyelectrolytes and tissue; correlation of biocompatability and electrode impedance for smooth and porous stimulating electrodes. *Mailing Add:* 21 Willow Way Atco NJ 08004

BEARD, RODNEY RAU, b Guinda, Calif, Dec 27, 11; m 38; c 4. PREVENTIVE MEDICINE, ENVIRONMENTAL HEALTH. *Educ:* Stanford Univ, AB, 32, MD, 38; Harvard Univ, MPH, 40; Am Bd Prev Med, dipl, 49; Am Bd Indust Hyg, dipl, 62. *Prof Exp:* Clin lab technician, French Hosp, San Francisco, 33-37; intern, Gorgas Hosp, CZ, 37-38; asst resident pub health & prev med, Hosps & instr, Nursing Sch, Stanford Univ, 38-39; Rockefeller fel med sci, Sch Pub Health, Harvard Univ, 39-40; from instr to prof pub health & prev med, 40-69, exec, 49-69, dir rehab, 55-60, prof family, community & prev med, 69-77, EMER PROF PREV MED, SCH MED, STANFORD UNIV, 77- *Concurrent Pos:* Med dir, Pac Div, Pan Am Airways, 40-49; mem comn A, Comt Aviation Med, Nat Res Coun, 41-47; mem comt effects of atmospheric contaminants on human health & welfare,

68-70; clin prof, Sch Pub Health, Univ Calif, 52-64, lectr, 64-72; dept dir comn environ hyg, Armed Forces Epidemiol Bd, 54-56, dir, 56-66, mem, 66-73; consult indust med, US Vet Admin, 54-67; consult, Off Surgeon Gen, US Dept Army, 54-73; mem, Nat Adv Heart Coun, NIH, 57-61, mem, Nat Adv Coun Environ Health Sci, 71-74, consult to dir, Nat Inst Environ Health Sci, 74-75; vis prof, Univ Milan, 60-61; trustee, Am Bd Prev Med, 61-70; mem, Nat Adv Coun Pub Health Training, 65-69; consult, Surgeon Gen, US Air Force, 66-69; mem tech adv comt, Calif Air Resources Bd, 69-72; mem subcomt atherosclerosis, Intersoc Comn Heart Dis Resources, 69-71; consult, Calif Dept Health Servs, 52-; mem, Biomet & Epidemiol Contract Review Comt, Nat Cancer Inst, NIH, HEW, 82-85. *Mem:* Fel Am Pub Health Asn; fel Am Acad Occup Med; Soc Occup & Environ Health; Asn Teachers Prev Med (pres, 58-59); fel Am Col Prev Med; fel AAAS. *Res:* Occupational medicine; air pollution; epidemiology; behavior toxicology. *Mailing Add:* Stanford Med Ctr HRP8A Stanford CA 94305-5092

BEARD, WILLIAM CLARENCE, b Gallipolis, Ohio, June 9, 34; m 60. MINERALOGY. *Educ:* Ohio State Univ, BSc, 60, PhD(mineral), 65. *Prof Exp:* Res chemist, Linde Div, Union Carbide Corp, 65-67; asst prof, 67-69, ASSOC PROF GEOL, CLEVELAND STATE UNIV, 69- *Mem:* Mineral Soc Am; Nat Asn Geol Teachers. *Res:* High temperature phase equilibrium of minerals and ceramics; crystal chemistry; hydrothermal synthesis; crystal growth; thermal analysis of minerals and ceramics. *Mailing Add:* Rte 1 Box 260-101 Gallipolis OH 45631

BEARD, WILLIAM QUINBY, JR, b Beaufort, SC, Apr 10, 32; m 57; c 2. INDUSTRIAL CHEMISTRY. *Educ:* Univ NC, BS, 54; Duke Univ, PhD(org chem), 59. *Prof Exp:* Res fel chem, Duke Univ, 59-60; res chemist, Ethyl Corp, La, 60-68; SR RES CHEMIST, RES & DEVELOP LAB, CHEM DIV, VULCAN MAT CO, 68- *Mem:* Am Chem Soc. *Res:* Chlorinated hydrocarbons. *Mailing Add:* 9328 Briarwood Ct Wichita KS 67212

BEARDALL, JOHN SMITH, b Springville, Utah, Feb 4, 39; m 68; c 5. PHYSICS, ASTRONOMY. *Educ:* Brigham Young Univ, BA, 60, MS, 68. *Prof Exp:* Develop engr, Hercules Inc, 61-67; staff mem physics, Los Alamos Nat Lab, Univ Calif, 69-87; TECH STAFF, ANALYTICAL SCIS CORP, 87- *Mem:* Optical Soc Am. *Res:* Remote sensing, arms control verification, sensors. *Mailing Add:* 9802 Meadow Knoll Ct Vienna VA 22180

BEARDEN, ALAN JOYCE, b Baltimore, Md, Nov 23, 31. BIOPHYSICS. *Educ:* Johns Hopkins Univ, AB, 50, PhD(physics), 59. *Prof Exp:* Instr physics, Univ Wis-Madison, 59-60; asst prof, Cornell Univ, 60-64; NIH spec fel biophysics, Univ Calif, San Diego, 64-66, asst prof chem, 66-68; res biophysicist & lectr, 69-72, res biophysicist & adj assoc prof biophysics, Donner Lab & Div Med Physics, 73-76, chmn dept, 78-84, PROF BIOPHYSICS, DONNER LAB, UNIV CALIF, BERKELEY, 76- *Concurrent Pos:* USPHS res career develop award, 70-75; fac sr scientist, Lawrence Berkeley Lab. *Honors & Awards:* Teaching Apparatus Award, Am Asn Physics Teachers, 62. *Mem:* Fel Am Phys Soc; Biophys Soc; AAAS. *Res:* Auditory transduction, cochlear dynamics, bioenergetics and photosynthesis; physics of energy transduction processes in biology; applications of electron spin resonance and lasers to biophysics. *Mailing Add:* Dept Cell & Molecular Biol Univ Calif Berkeley CA 94720

BEARDEN, HENRY JOE, b Starkville, Miss, May 12, 26; m 46; c 3. REPRODUCTIVE PHYSIOLOGY. *Educ:* Miss State Univ, BS, 50; Univ Tenn, MS, 51; Cornell Univ, PhD(animal breeding), 54. *Prof Exp:* Instr dairy, Univ Tenn, 51; asst prof dairy exten & res, Cornell Univ, 54-60; PROF DAIRY SCI & HEAD DEPT, MISS STATE UNIV, 60- *Mem:* Am Dairy Sci Asn; Soc Study Reproduction; US Animal Health Asn; Sigma Xi. *Res:* Animal physiology; dairy cattle genetics. *Mailing Add:* Dept of Dairy Sci Miss State Univ Drawer DD Mississippi State MS 39762

BEARDEN, WILLIAM HARLIE, b Kileen, Tex, Aug 18, 49; m 70; c 3. PHYSICAL ORGANIC CHEMISTRY. *Educ:* Centenary Col, BS, 71; Univ Houston, PhD(chem), 75. *Prof Exp:* NSF res asst chem, Calif Inst Technol, 75-77; asst prof chem, Jackson State Univ, 77-84; app chemist, 84-87, NMR PROD MGR, JEOL, 87-, MGR, ANAL INSTRUMENTS DIV, JEOL, 90- *Mem:* Am Chem Soc; Am Soc Testing & Mat. *Res:* Structure determinations of organic compounds using nuclear magnetic resonance via nonclassical methods, solid state NMR. *Mailing Add:* JEOL 11 Dearborn Rd Peabody MA 01960

BEARDMORE, PETER, b Stoke-on-Trent, Eng, Aug 5, 35; m 58; c 1. PHYSICAL METALLURGY. *Educ:* Univ Sheffield, BMet, 58; Univ Liverpool, PhD(metall), 66. *Prof Exp:* Res metallurgist, English Elec Co, 58-60; mem res staff metall, Mass Inst Technol, 60-63; res fel, Univ Liverpool, 63-66; MGR, DEPT MAT SCI, SCI RES STAFF, FORD MOTOR CO, 66- *Mem:* Am Inst Mining, Metall & Petrol Eng; Am Soc Metals. *Res:* Deformation and fracture of materials; mechanical properties of body centered cubic metals and of polymeric materials, including fatigue behavior. *Mailing Add:* Dept Mat Sci Res Staff Ford Motor Co PO Box 2053 Dearborn MI 48121

BEARDMORE, WILLIAM BOONE, b Salem, Ohio, Aug 14, 25; m 48; c 2. MICROBIOLOGY. *Educ:* Ohio State Univ, BSc, 48, MSc, 50, PhD(bact), 53. *Prof Exp:* From asst to asst instr bact, Ohio State Univ, 48-53; assoc res microbiologist virol, Parke, Davis & Co, 53-56, mgr biol control dept, 56-63, from res virologist to sr res virologist, 63-67, lab dir, 67-81; MGR, CELL & TISSUE CULT RES DEPT, WARNER LAMBERT DIAG, 81- *Mem:* AAAS; Am Soc Microbiol; Sigma Xi. *Res:* Viruses and tissue cultures. *Mailing Add:* 213 Nesbit Lane Rochester MI 48309

BEARDSLEE, RONALD ALLEN, b Asheville, NC, Feb 1, 46; m 70. CLINICAL CHEMISTRY. *Educ:* Univ Calif, Santa Barbara, BS, 68, PhD(chem), 72. *Prof Exp:* Presidential intern protein res, Western Regional Res Ctr, USDA, 72-73; res scientist clin chem, 74-79, ASST DIR, RES DEPT, BIO-SCI LABS, 79- *Mem:* Am Chem Soc; Am Asn Clin Chem. *Res:* Development of new chemical and physical methods for the analysis of materials of clinical significance. *Mailing Add:* 27949 El Portal Dr Hayward CA 94542-2509

BEARDSLEY, GEORGE PETER, b New York, NY, Dec 29, 40; m 72; c 1. BIO-ORGANIC CHEMISTRY. *Educ:* Mass Inst Technol, BS, 67; Princeton Univ, PhD(chem), 71; Duke Univ, MD, 74. *Prof Exp:* Resident physician pediat, Yale New Haven Hosp, Yale Univ, 74-76; res fel pediat hemat, 76-79, instr pediat, 79-81, ASST PROF PEDIAT, MED SCH, HARVARD UNIV, 81-; ASSOC PROF PEDIAT & PHARMACOL, YALE UNIV, 85-, CHIEF, PEDIAT HEMAT/ONCOL. *Concurrent Pos:* Dir, Pediat Oncol Prog, Yale Comprehensive Cancer Ctr. *Mem:* Am Asn Cancer Res; Am Chem Soc; Am Soc Hemat; Soc Pediat Res. *Res:* Bio-organic chemistry of purines and pteridinos; mechanisms of drug action; pediatric hematology and oncology. *Mailing Add:* Dept Pediat/Pharmacol Yale Univ 333 Cedar St New Haven CT 06510

BEARDSLEY, IRENE ADELAIDE, b San Diego, Calif, Aug 18, 35; m 86; c 2. MAGNETIC RECORDING MAGNETISM, MAGNETIC MATERIALS. *Educ:* Stanford Univ, BS, 57, MS, 58, PhD(physics), 65. *Prof Exp:* Scientist, Space Physics Dept, Lockheed Missiles & Space Co, 58-62, res scientist, Mat Sci Lab, Calif, 66-69; consult, 69-76, RES STAFF MEM, ALMADEN RES CTR, IBM CORP, SAN JOSE, CA, 76- *Mem:* Am Phys Soc; Sigma Xi; sr mem Inst Elec & Electronics Engrs; Magnetics Soc. *Res:* Theoretical studies of magnetic recording. *Mailing Add:* 1013 Paradise Way Palo Alto CA 94306

BEARDSLEY, JOHN WYMAN, JR, b Los Angeles, Calif, Mar 25, 26; m 48, 83; c 4. ENTOMOLOGY. *Educ:* Univ Calif, BS, 50; Univ Hawaii, MS & PhD, 63. *Prof Exp:* Asst entomologist, US Trust Territory of Pac Islands, 52-54; from asst to assoc entomologist, Exp Sta, Hawaiian Sugar Planters Asn, 54-72; assoc prof entom, Univ Hawaii, 63-66; assoc res entomologist, Univ Calif, Berkeley, 66-68; assoc prof entom & assoc entomologist, 68-72, PROF ENTOM & ENTOMOLOGIST, UNIV HAWAII, 72- *Concurrent Pos:* Chmn, dept entom, Univ Hawaii, 81-; consult entomologist, ABA Int, Bangladesh, 84; Fulbright sr res scholar, La Trobe Univ, Melbourne Inst, 71-72. *Mem:* Am Entom Soc; Int Org Biol Control. *Res:* Taxonomy of scale insects and parasitic wasps; insect biology; biological control; sugar cane and pineapple entomology. *Mailing Add:* Dept Entom Univ Hawaii Manoa Honolulu HI 96822

BEARDSLEY, ROBERT CRUCE, b Jacksonville, Fla, Jan 28, 42; m 66; c 2. PHYSICAL OCEANOGRAPHY. *Educ:* Mass Inst Technol, BS, 64, PhD(oceanog), 68. *Prof Exp:* From asst prof to assoc prof oceanog, Mass Inst Technol, 67-75; assoc scientist, 75-81, SR SCIENTIST, WOODS HOLE OCEANOG INST, 81- *Mem:* AAAS; Am Geophys Union; Am Meteorol Soc. *Res:* Geophysical fluid dynamics; coastal oceanography. *Mailing Add:* Moses Pond Rd Teaticket MA 02536

BEARDSLEY, ROBERT EUGENE, b Walton, NY, June 11, 23; m 48; c 3. MICROBIOLOGY, GENETICS. *Educ:* Manhattan Col, BS, 50; Columbia Univ, AM, 51, PhD(zool), 60. *Prof Exp:* From instr to assoc prof, Manhattan Col, 51-68, dir lab plant morphogenesis, 62-69, prof biol, 68-77, head dept, 69-77, vis prof biol, 77-80; dean, Sch Arts & Sci, 77-83, PROF BIOL, IONA COL, 77- *Concurrent Pos:* Guggenheim fel, 66; vis investr, Inst Paris, 66-67 & Univ Paris, 75. *Mem:* Am Pub Health Asn; Sigma Xi; Am Soc Microbiol. *Res:* Mechanisms of tumor induction in crown gall. *Mailing Add:* Dept Biol Iona Col New Rochelle NY 10801

BEARE, STEVEN DOUGLAS, b Detroit, Mich, May 31, 44; m 65; c 2. ORGANIC CHEMISTRY, PHYSICAL CHEMISTRY. *Educ:* Oakland Univ, BA, 65; Univ Ill, Urbana-Champaign, MS, 67, PhD(org chem), 69. *Prof Exp:* Teaching asst chem, Univ Ill, 65-67, res asst org chem, 67-69; SR RES ASSOC, E I DU PONT DE NEMOURS & CO, 69- *Mem:* Am Chem Soc; Am Inst Aeronaut & Astronaut. *Res:* Determinations of optical purities and correlations of absolute configurations of alcohols, hydroxy acids, amino acids, amines, sulfoxides and phosphine oxides by nuclear magnetic resonance spectroscopy in chiral solvents. *Mailing Add:* Chestnut Run Bldg 701 E I du Pont de Nemours & Co Wilmington DE 19898

BEARER, ELAINE L, ANATOMIC PATHOLOGY. *Educ:* Univ Calif, San Francisco, PhD(exp biol), 82, MD, 83. *Prof Exp:* FEL BIOCHEM PATH, UNIV CALIF, SAN FRANCISCO, 86- *Mailing Add:* Dept Biochem Med Sch Univ Calif San Francisco CA 94143-0448

BEARE-ROGERS, JOYCE LOUISE, b Ont, Sept 8, 27; m 61; c 1. BIOCHEMISTRY, NUTRITION. *Educ:* Univ Toronto, BA, 51, MA, 52; Carleton Univ, Ont, PhD(biochem), 66. *Hon Degrees:* DSc, Univ Man, 85. *Prof Exp:* Instr physiol, Vassar Col, 54-56; res scientist, 56-75, CHIEF, NUTRIT RES, BUR NUTRIT SCI, FOOD DIRECTORATE, OTTAWA, 75- *Concurrent Pos:* Non-resident prof, Dept Biochem, Univ Ottawa. *Honors & Awards:* Borden Award, Nutrit Soc Can, 72; Queen's Silver Jubilee Medal, 77; Medaille Chevreul, 84; Normann Medal, 87. *Mem:* Am Inst Nutrit; Am Oil Chemists Soc; Can Inst Food Technol; Can Biochem Soc; Can Soc Nutrit Sci; Can Athero Soc; fel Royal Soc Can. *Res:* Effects of dietary components on tissue phospholipids; nutritional aspects of oils and fats. *Mailing Add:* Bur Nutrit Sci Food Directorate Ottawa ON K1A 0L2 Can

BEARMAN, RICHARD JOHN, b New York, NY, June 23, 29; m 61; c 2. PHYSICAL CHEMISTRY, CHEMICAL PHYSICS. *Educ:* Cornell Univ, AB, 51; Stanford Univ, PhD(chem), 56. *Prof Exp:* Asst, Yale Univ, 56-57; from asst prof to prof chem, Univ Kans, 57-71; PROF CHEM, UNIV NEW SOUTH WALES, 72- *Concurrent Pos:* Guggenheim fel, 62-63; vis reader, Univ New Eng, Australia, 68-69; hon fel, Res Sch Phys Sci, Australian Nat Univ, 68-69. *Mem:* Am Chem Soc; Am Phys Soc; fel Am Inst Chemists; fel Royal Australian Chem Inst. *Res:* Statistical mechanics and thermodynamics of irreversible processes; membrane permeation; thermal diffusion in liquids; equilibrium theory of fluids; P-V-T measurements. *Mailing Add:* Dept Chem Univ New South Wales ADFA Univ Col Northcutt Dr Campbell Act 2600 Australia

BEARN, ALEXANDER GORDON, b Cheam, Eng, Mar 29, 23; US citizen; m 52; c 2. PHARMACEUTICAL ADMINISTRATION, GENETICS. *Educ:* Univ London, MB & BS, 46, MD, 51; FRCP(E), 68; FRCP, 70. *Hon Degrees:* MD, Cath Univ, Korea, 68; Dr, Univ Rene Descartes, Paris, 75. *Prof Exp:* Intern, Guy's Hosp, London, 46-47, serv, 47-49; intern & resident, Post-Grad Med Sch, Univ London, 49-51; from asst to assoc, Rockefeller Univ, 51-57, from assoc prof to prof, 57-66, from asst physician to sr physician, 51-66; prof med & chem dept, Med Col, Cornell Univ, 66-77, Stanton Griffis distinguished med prof, 76-79; sr vpres, Med & Sci Affairs, Merck Sharp & Dohme Int, 79-88; ADJ PROF, ROCKEFELLER UNIV, 66- *Concurrent Pos:* Mem, Genetics Training Comt, USPHS, 61-65, Genetics Study Sect, 66-70, Comn Human Health Resources, Nat Acad Sci, 73-78, Adv Comt Radiation Effects, 75-81 & 82-87; physician-in-chief, NY Hosp, 66-77; vis prof, var US & foreign univs, 71-81. *Honors & Awards:* Lowell Lectr, Harvard Univ, 58; Murray Gordon Lectr, Downstate Med Ctr, 62; Augustus B Wadsworth Lectr, NY State Dept Health, 62; Stevenson Mem Lectr, Presby Hosp, 63; Louis Gross Mem Lectr, NY Univ, 69; Lilly Lectr, Royal Col Physicians, 73; Harvey Lectr, 75; Mary H Edens Lectr, Univ Tex, 75; Lettsomian Lectr, Med Soc London, 76; Abernethian Lectr, St Bartholomew's Hosp, 77; Rufus Cole Lectr, Rockefeller Univ, 80. *Mem:* Nat Acad Sci; fel AAAS; Am Philos Soc; Harvey Soc (pres, 72); Am Soc Human Genetics (pres, 71); Sigma Xi. *Res:* Human genetics; internal medicine. *Mailing Add:* Rockefeller Univ 1230 York Ave New York NY 10021-6399

BEARSE, GORDON EVERETT, b Cambridge, Mass, June 13, 07; m 31; c 2. POULTRY SCIENCE. *Educ:* Univ Mass, BS, 28. *Prof Exp:* Asst poultry genetics, Agr Exp Sta, Univ Mass, 29; asst poultryman, poultry husb, Western Wash Res & Exten Ctr, 29-38, poultryman, 38-43, res poultryman, poultry sci, 43-45; poultryman, 46-53, poultry scientist, 53-63, poultry scientist, Dept Animal Sci, 63-72, emer poultry scientist, Wash State Univ, 72-; RETIRED. *Mem:* Fel AAAS; fel Poultry Sci Asn (vpres, 57-59, pres, 59-60); World Poultry Sci Asn; Sigma Xi. *Res:* Effect of nutrition, genetic selection, managemental procedures and environment on performance of chickens and quality of their products. *Mailing Add:* 10712 62nd St Ct E Puyallup WA 98372-2798

BEARSE, ROBERT CARLETON, b Hartford, Conn, May 22, 38; m 63; c 2. APPLIED PHYSICS. *Educ:* Rice Univ, MA, 62, PhD(physics), 64. *Prof Exp:* Res assoc physics, Argonne Nat Lab, 64-68, asst physicist, 68-69; from asst prof to assoc prof physics, 69-75, assoc dean res admin, 75-78, PROF PHYSICS, UNIV KANS, 75-, ASSOC VCHANCELLOR RES GRAD STUDIES & PUB SERV, 78- *Concurrent Pos:* Vis staff mem, Los Alamos Sci Lab, NMex, 72- *Mem:* Sigma Xi; AAAS; Am Phys Soc. *Res:* Trace element analysis in biological materials; applications of nuclear physics in other disciplines. *Mailing Add:* 226 Stronge Ave Lawrence KS 66044

BEARY, DEXTER F, b Battle Creek, Mich, Nov 1, 24; m 47; c 2. ANATOMY, BIOLOGY. *Educ:* Andrews Univ, AB, 51; Western Mich Univ, MA, 59; Loma Linda Univ, PhD(anat), 67. *Prof Exp:* Res asst pharmaceut, Upjohn Co, 51-59; instr biol, Southwestern Union Col, 59-64, PROF & CHMN, DEPT BIOL, SOUTHWESTERN ADVENTIST COL, 64- *Mem:* Am Asn Anat; Am Inst Biol Sci; Sigma Xi. *Res:* Gross anatomy; histology; osteoporosis. *Mailing Add:* Dept Biol Sci Marymount Col Kans Salina KS 67401

BEASLEY, ANDREW BOWIE, b Upper Zion, Va, Sept 7, 31; div; c 2. ANATOMY, GENETICS. *Educ:* Va Polytech Inst, BS, 52, MS, 56; Med Col Pa, ScD(anat & genetics), 61. *Prof Exp:* Asst zool, Va Polytech Inst, 54-55; res asst genetics, Columbia Univ, 55-58; instr, 61-63, asst prof, 63-69, actg chmn dept, 71-73, asst dean student affairs, 75-77, ASSOC PROF ANAT, MED COL PA, 69-, ASSOC DEAN STUDENT AFFAIRS, 77- *Mem:* Am Asn Anatomists; Asn Am Med Cols. *Res:* Developmental genetics and anatomy of central nervous system. *Mailing Add:* Dept Anat Med Col Pa 3300 Henry Ave Philadelphia PA 19129

BEASLEY, CLARK WAYNE, b Pittsburg, Kans, June 6, 42; m 65; c 2. INVERTEBRATE ZOOLOGY. *Educ:* Kans State Col Pittsburg, BS, 64; Univ Okla, PhD(zool), 68. *Prof Exp:* Instr biol, Mo Southern State Col, 68-69; from asst prof to assoc prof, 69-77, PROF BIOL, MCMURRAY COL, 77-, CHMN DEPT, 73- *Mem:* AAAS; Am Inst Biol Sci; Am Micros Soc. *Res:* Invertebrate taxonomy, especially Tardigrada. *Mailing Add:* Dept Biol McMurray Univ Abilene TX 79697-0368

BEASLEY, CLOYD O, JR, b Florence, Ala, July 9, 33. PLASMA PHYSICS. *Educ:* Vanderbilt Univ, BA, 55, MA, 57; Univ Wis, PhD(physics), 62. *Prof Exp:* Appl sci rep physics, Int Bus Mach Corp, 57-58; SR RES SCIENTIST, FUSION ENERGY DIV, OAKRIDGE NAT LAB, 62- *Concurrent Pos:* Res assoc, Culham Lab, UK, 66-67 & Max-Planck Inst for Plasmaphysik, Fed Repub Ger, 77-78; consult, Div Magnetic Fusion Energy, US Dept Energy, 81-84. *Mem:* Am Phys Soc. *Res:* Plasma micro-instability theory; micro-instabilities in mirror-confined plasmas; combined numerical-analytical approach to investigation of equilibria and stability on Tokamaks; neoclassical transport in stellaratus and finite-aspect-ratio Tokamaks. *Mailing Add:* IAEA INIS Sect Wagramerstrasse 5 PO Box 200 A-1400 Vienna Austria

BEASLEY, DEBBIE SUE, SALT & WATER METABOLISM, RENAL BLOOD FLOW. *Educ:* Univ Mich, PhD(physiol), 82. *Prof Exp:* ASST RES PROF PHYSIOL & MED, UNIV HOSP, 84- *Mailing Add:* Dept Med Boston Univ Med Ctr 80 E Concord St Boston MA 02118

BEASLEY, EDWARD EVANS, b Oakland, Calif, Mar 19, 24; m 49; c 1. PHYSICS. *Educ:* US Naval Acad, BS, 44; Univ Md, MS, 57, PhD(physics), 63. *Prof Exp:* Asst physics, Univ Md, 56-60; chmn dept, 60-72, PROF PHYSICS, GALLAUDET UNIV, 65- *Concurrent Pos:* Dir, NSF Summer Sci Inst, Gallaudet Univ, 65-66. *Mem:* Am Phys Soc; Am Asn Physics Teachers. *Res:* Electron spin resonance in ultraviolet-irradiated aliphatic hydrocarbon glasses at liquid nitrogen temperature. *Mailing Add:* Dept Physics Gallaudet Univ Washington DC 20002

BEASLEY, JAMES GORDON, b Tela, Honduras, Nov 13, 28; US citizen; m 63; c 1. ORGANIC CHEMISTRY, MEDICINAL CHEMISTRY. *Educ:* Auburn Univ, BS, 51, MS, 55; Univ Va, PhD(org chem), 62. *Prof Exp:* Asst gen chem, Auburn Univ, 53-55; assoc prof chem, Memphis State Univ, 59-62; res assoc pharmaceut chem, Col Pharm, Univ Tenn, 62-63, asst prof pharmaceut & med chem, 63-65, assoc prof med chem, 65-71, dir chem enzym lab, 68-71; prof chem, Lambuth Col, 71-82; PROF MED CHEM, SAMFORD UNIV, 82- *Concurrent Pos:* Asst, Univ Va, 59; co-investr, NSF grant, 64-66, co-prin investr, NSF res grant, 66-72; res grants, Marion Labs, 66-72 & A H Robins, 70-73. *Mem:* Am Chem Soc; Am Asn Col Pharm; Sigma Xi. *Res:* Organic synthesis and biochemical evaluation of compounds with pharmacodynamic potential; correlation of molecular constitution and biochemical response. *Mailing Add:* Sch Pharm Samford Univ 800 Lakeshore Dr Birmingham AL 35229

BEASLEY, JOSEPH NOBLE, b Centerton, Ark, Mar 11, 24; m 56; c 1. VETERINARY PATHOLOGY. *Educ:* Tex A&M Univ, DVM, 49, MS, 56; Univ Okla, PhD(med sci), 64. *Prof Exp:* Instr vet sci, Univ Ark, 49-51; instr vet med & surg, Tex A&M Univ, 51-52; pathologist, Ark Livestock Sanit Bd, 52-55; from asst prof to assoc prof vet path, Tex A&M Univ, 56-65; PROF ANIMAL SCI, UNIV ARK, FAYETTEVILLE, 65- *Mem:* Am Vet Med Asn; Am Col Vet Path; Int Acad Path; Conf Res Workers Animal Dis; Am Asn Avian Path. *Res:* Comparative pathology of psittacosis ornithosis; canine dirofilariasis, Maren's disease. *Mailing Add:* Dept Animal Sci Univ Ark Fayetteville AR 72701

BEASLEY, MALCOLM ROY, b San Francisco, Calif, Jan 4, 40; m 62; c 3. LOW TEMPERATURE PHYSICS. *Educ:* Cornell Univ, BS, 62, PhD(physics), 68. *Prof Exp:* Res fel eng & appl physics, Harvard Univ, 67-69, from asst prof to assoc prof, 69-74; assoc prof, 74-80, PROF APPL PHYSICS & ELEC ENG, STANFORD UNIV, 80- *Honors & Awards:* Fel, Am Phys Soc. *Mem:* Am Phys Soc; AAAS. *Res:* Basic and applied superconductivity. *Mailing Add:* Dept Appl Physics Stanford Univ Stanford CA 94305

BEASLEY, PHILIP GENE, b Harrisburg, Ill, Dec 22, 27; m 52. ANIMAL PHYSIOLOGY, PLANT PHYSIOLOGY. *Educ:* Wash Univ, AB, 49; Auburn Univ, MS, 62, PhD(zool), 67. *Prof Exp:* With Fed Bur Invest, 50-59; instr fisheries, Auburn Univ, 60-66; from asst prof to assoc prof, 66-71, PROF BIOL & CHMN DEPT, UNIV MONTEVALLO, 71- *Concurrent Pos:* Pres, Ala Acad Sci, 86-87. *Mem:* AAAS; Sigma Xi. *Mailing Add:* 124 Shoshone Montevallo AL 35115

BEASLEY, THOMAS MILES, environmental & marine sciences, for more information see previous edition

BEASLEY, WAYNE M(ACHON), b Everett, Mass, May 23, 22; m 45; c 1. PHYSICAL CERAMICS, CRYSTALLOGRAPHY. *Educ:* Harvard Univ, SB, 46; Mass Inst Technol, SM, 65. *Prof Exp:* Physicist optics, R L Evans Assocs, 46-47; instr NH pub sch, 47-48; physicist optics, C P Shillaber, 48-51; engr corrosion films, Clarostat Mfg Co, 51-55; phys metallurgist nuclear fuel & instr physics, Metals & Controls Corp, 55-57; res asst prof x-ray diffraction, Dept Mech Eng, 57-66, res assoc prof & adj assoc prof mat sci, 66-72, assoc prof mat sci, Mat Sci Div, 72-84, grad prog cord, 82-84, CONSULT, MAT SCI DIV, 84- *Concurrent Pos:* Dir X-ray diffraction lab, Eng Exp Sta, Univ NH, 63-72; vis scientist, Mass Inst Technol, 78. *Mem:* Am Phys Soc; European Phys Soc; Int Soc Stereology. *Res:* X-ray diffraction; microstructures; structure determinations of sulfosalts; characterization of particle compacts of ceramic materials; characterization of ferromagnetic materials; x-ray diffraction studies of refractory, composite materials. *Mailing Add:* 22 Weeks Lane Rochester NH 03867-8017

BEASLEY, WILLIAM HAROLD, b Dallas, Tex, Oct 23, 44; m 67, 83; c 2. LIGHTNING, ATMOSPHERIC PHYSICS. *Educ:* Rice Univ, BA, 67, MS, 69; Univ Tex, Dallas, PhD(physics), 74. *Prof Exp:* Res & develop engr antennas, Collins Radio Co, 68-70; physicist, Environ Res Lab & Wave Propagation Lab, Nat Oceanic & Atmospheric Admin, 72-73; asst prof physics, WTex State Univ, 74-75; assoc, Univ Fla, 75-79, vis asst prof electromagnetics, 79-83; assoc prog dir meteorol, NSF, 83-86; sr staff officer, Nat Acad Sci/Nat Res Coun, 86-87; prog mgr, Ctr & Facil, Div Atmospheric Sci, NSF, 87-89; dep dir, Ctr Anal & Prediction Storms, 89-91, ASSOC PROF, SCH METEOROL, UNIV OKLA, 90-, DIR, 91- *Concurrent Pos:* Prin investr, NSF grants, W Tex State Univ, 75-76 Univ Fla, 79-81 & 81-83, Univ Okla, 91; consult, Lightning Location & Protection, Inc, 77-81; assoc ed, US Nat Report, Int Union Geodesy & Geophys, 87; secy, Sect W, AAAS, 87- & Atmos Sci Sect, Am Geophys Union, 88-90. *Mem:* Am Asn Physics Teachers; Am Geophys Union; Am Meteorol Soc; Sigma Xi; AAAS. *Res:* Observations of lightning electromagnetic fields and optical emissions; thunderstorms; infrasound from severe storms; antennas. *Mailing Add:* Sch Meteorol Univ Okla Norman OK 73019

BEASON, ROBERT CURTIS, b Ft Scott, Kans, May 12, 46; div; c 1. MAGNETIC SENSORY PERCEPTION, ANIMAL ORIENTATION. *Educ:* Bethanny Nazarene Col, AB, 68; Western Ill Univ, MS, 70; Clemson Univ, PhD(zool), 76. *Prof Exp:* Teaching asst, gen biol, Western Ill Univ, 68-70; res scientist, US Air Force, 70-74; res asst, Clemson Univ, 74-76; res scientist, US Forest Serv, 76; vis lectr behav ecol, Univ Calif, Irvine, 77; vis asst prof, Western Ill Univ, 77-78; from asst prof to assoc prof behav ecol, 78-90, PROF BIOL, STATE UNIV NY, GENESEO, 90- *Concurrent Pos:* Prin investr, Clemson Univ & US Dept Interior, 74-76, Res Found, State Univ NY, 79-80 & Nat Sci Found, 81, 87-88, NIH, 88-90, NSF, 90- *Mem:* Soc Neurosci; elective mem Am Ornithologists Union; Animal Behav Soc; Ecol Soc Am; Am Soc Naturalists. *Res:* Environmental and sensory cues used by migratory birds for migration, orientation and navigation and how each cue is utilized for migratory orientation; sensory basis of magnetic perception. *Mailing Add:* Biol Dept State Univ NY Geneseo NY 14454

BEATLEY, JANICE CARSON, b Columbus, Ohio, Mar 18, 19. PLANT ECOLOGY. *Educ:* Ohio State Univ, BA, 40, MSc, 48, PhD(bot), 53. *Prof Exp:* High sch teacher, Ohio, 43-45; asst instr & asst, Ohio State Univ, 45-54, instr, 55-56, Muellhaupt scholar, 57-58; asst prof, E Carolina Col, 54-55 & NC State Col, 56-57; res assoc, NMex Highlands Univ, 59; asst res ecologist, Lab Nuclear Med & Radiation Biol, Univ Calif, Los Angeles, 60-67, assoc res ecologist, 67-72; assoc prof biol, 73-77, PROF BIOL SCI, UNIV CINCINNATI, 77- *Concurrent Pos:* Instr & asst prof, Univ Tenn, 52-60. *Mem:* Ecol Soc Am; Am Soc Plant Taxon; Sigma Xi. *Res:* Desert and radiation ecology; ecology of deciduous forest tree species and wintergreen herbs. *Mailing Add:* 3469 Statewood Dr Cincinnati OH 45251

BEATON, ALBERT E, b Boston, Mass, Aug 9, 31; m 59; c 2. STATISTICS. *Educ:* State Teachers Col Boston, BS, 55; Harvard Univ, MEd, 56, EdD, 64. *Prof Exp:* Res asst, Harvard Univ, 56-57, managing dir, Littauer Statist Lab, 57-59, dir, Harvard Statist Lab, 59-62, actg mgr, Harvard Comput Ctr, 62, res assoc educ & IBM res fel, 62-64; adv statist & data anal, 64-70, dir, Off Data Anal Res, Educ Testing Serv, 70-81, DIR DATA ANALYSIS & RES, NAT ASSESSMENT EDUC PROGRESS, 83-; PROF, BOSTON COL, 90- *Concurrent Pos:* Res assoc, Educ Res Corp, 56-57; vis lectr, Sch Educ & Grad Sch, Boston Col, 60-61; pres, Albert E Beaton Assocs, Inc; consult, Cabot Corp, Ceir, Inc, Cides, Dominican Repub, Gen Elec Corp, Gillette Safety Razor Corp, Infrared Industs, Can Ministry of Youth, Nat Bur Econ Res, Prudential Life Ins Co, RCA Corp, S C Johnson & Co, Union Carbide Corp, US Res Corp, US Off Educ & US Off Econ Opportunity; vis res statistician & vis lectr statist, Princeton Univ, 66-78; vis res scientist, Patricks Col, Dublin, 79-80; vis lectr, Trinity Col, Dublin, 80. *Honors & Awards:* Wilcoxon Award, Am Soc Qual Control, 77; Award Tech Contrib, Nat Coun Measurement Educ. *Mem:* Am Educ Res Asn; Am Psychol Asn; Am Statist Asn; Asn Comput Mach; Psychomet Soc. *Res:* Design and analysis of national surveys of educational attainment and general statistical methods. *Mailing Add:* 171 Autumn Hill Rd Princeton NJ 08540

BEATON, DANIEL H(ARPER), b Paisley, Scotland, Aug 14, 10; US citizen; m 38; c 1. CHEMICAL ENGINEERING. *Educ:* Cooper Union, BS, 33. *Prof Exp:* Chem engr, Plastics Dept, E I du Pont de Nemours & Co, 33-36, control & develop engr, 36-43, process control supvr, 41-43, chem supvr, 43-45, develop engr, 45-48, tech sect engr, 48-58, res engr, 58-67, sr res engr, 67-74; RETIRED. *Mem:* Am Chem Soc. *Res:* Improved compositions for high polymers; reaction kinetics; rheology; degradation mechanisms. *Mailing Add:* 1505 14th St Parkersburg WV 26101

BEATON, GEORGE HECTOR, b Oshawa, Ont, Dec 20, 29; m 53; c 3. NUTRITION, BIOCHEMISTRY. *Educ:* Univ Toronto, BA, 52, MA, 53, PhD(nutrit), 55. *Prof Exp:* From asst prof to assoc prof nutrit, Sch Hyg, Univ Toronto, 55-63, prof & head dept, 63-75, actg dir, Sch Hyg, 74-75, actg dean fac food sci, 75-81, prof nutrit & food sci & chmn dept, 75-81, PROF, NUTRIT SCI, FAC MED, UNIV TORONTO, 81- *Concurrent Pos:* WHO fel, Inst Nutrit Cent Am & Panama, 61; mem expert adv comt nutrit in pregnancy & lactation & expert adv panel nutrit, WHO, 64, mem expert group in vitamin requirements, Food & Agr Orgn-WHO, 67, mem expert group vitamin & mineral requirements, 69, mem expert comt nutrit, 70, protein & energy requirements, 71 & 81; mem, diet & non-communicable dis comt, 89, trace element requirements comt, 90. *Honors & Awards:* Borden Award, Nutrit Soc Can, 68; E W McHenry Award, 84. *Mem:* Am Inst Nutrit; Nutrit Soc Can (secy, 57-60, vpres, 64-65, pres, 65-66). *Res:* Nutrition in pregnancy; biochemical assessment of nutritional status; intermediary metabolism and essential nutrients; public health applications of nutrition information; interpretation of nutrient requirements. *Mailing Add:* Dept Nutrit Sci Univ Toronto Fac Med Toronto ON M5S 1A1 Can

BEATON, JAMES DUNCAN, b Vancouver, BC, Aug 28, 30; m 52; c 3. SOIL FERTILITY. *Educ:* Univ BC, BSA, 51, MSA, 53; Utah State Univ, PhD(soils), 57. *Prof Exp:* Lab asst soil bact, Univ BC, 51-53, spec lectr, Dept Soil Sci, 56, instr, 57-59; phys chemist, Soil Sect, Exp Farm, Can Dept Agr, 59-61; soil scientist, Res & Develop Div, Consol Mining & Smelting Co Can, Ltd, 61-64, head soil sci res, Res & Corp Develop, 64-67; sr agronomist, Cominco Ltd, BC, 67-68; dir agr res, Sulphur Inst, 68-73; chief agronomist, Cominco Ltd, Alta, 73-78; western dir, Potash & Phosphate Inst, Alta, 78-86, vpres, 86-88, SR VPRES, INT PROGS, POTASH & PHOSPHATE INST, ATLANTA, GA, 88-, PRES, POTASH & PHOSPHATE INST CAN, SASKATOON, 88- *Mem:* Fel Soil Sci Soc Am; fel Can Soc Soil Sci; Brit Soc Soil Sci; Int Soc Soil Sci; fel Am Soc Agron; hon mem Nat Fertilizer Solutions Asn. *Res:* Soil fertility and reactions of fertilizers in soils; forest fertilization. *Mailing Add:* Potash & Phosphate Inst Can Suite 704 C N Tower Midtown Plaza Saskatoon SK S7K 1J5 Can

BEATON, JOHN MCCALL, b Huntly, Scotland, June 21, 44; m 68; c 2. PSYCHOPHARMACOLOGY. *Educ:* Univ Aberdeen, BSc, 66, MSc, 69; Univ Ala, Birmingham, PhD(physiol), 73. *Prof Exp:* Sr res assoc psychol, Addiction Res Found, Toronto, 69-71; from res assoc to instr, 71-74, from asst prof to assoc prof psychiat, 74-84, assoc prof psychol, 78-84, ASST PROF PHARMACOL, UNIV ALA, BIRMINGHAM, 75-, PROF PSYCHIAT & PSYCHOL 84- *Mem:* Brit Psychol Soc; Soc Neurosci; Am Soc Pharmacol & Exp Therapeut; Collegium Internationale Neuro-Psychopharmacologicum. *Res:* Psychopharmacological study of hallucinogenic agents using operant conditioning tecniques; animal models of psychoses especially with relationship to schizophrenia; the effects of amino acid loadings on behavior. *Mailing Add:* Dept Psychiat Univ Ala Birmingham Sta Birmingham AL 35294

BEATON, JOHN ROGERSON, b Oshawa, Ont, Sept 7, 25; m 48; c 4. NUTRITION. *Educ:* Univ Toronto, BA, 49, MA, 50, PhD(nutrit), 52. *Prof Exp:* Res fel nutrit, Univ Toronto, 49-51, res assoc, 51-52, asst prof, 52-55,; defense sci serv officer, Defence Res Bd, Can, 55-59, sect head, Defence Res Med Labs, 59-63; prof physiol, Univ Western Ont, 63-67; prof nutrit & chmn div, Univ Hawaii, 67-69; dean col human biol, Univ Wis-Green Bay, 69-70, dean cols, 70-75; dean col human ecol, Univ Md, College Park, 75-;

RETIRED. *Concurrent Pos:* Lectr food chem, Univ Toronto, 50-55. *Mem:* Soc Exp Biol & Med; Am Soc Biol Chemists; Am Inst Nutrit; Nutrit Soc Can; Soc Nutrit Educ; Can Physiol Soc. *Res:* Vitamin B-6; metabolic functions; amino acid metabolism in pregnancy and malignancy; physiology of cold exposure and hypothermia; metabolic effects of hormones; effects of exercise; fat mobilizing substances in urine; regulation of food intake; physiology and biochemistry of hyperphagia. *Mailing Add:* Univ Md 9211 Montpiler Dr Laurel MD 20708

BEATON, ROY HOWARD, b Boston, Mass, Sept 1, 16; m 39, 82; c 2. CHEMICAL ENGINEERING. *Educ:* Northeastern Univ, BS, 39; Yale Univ, DEng, 42. *Hon Degrees:* DSc, Northeastern Univ, 67. *Prof Exp:* Res chem engr, E I du Pont de Nemours & Co, Inc, Va, 42- 43; group leader, Metall Lab, Ill, 43 & Clinton Lab, Tenn, 44; sr supvr, Tech Div, Hanford Eng Works, Wash, 45-46; assoc prof chem eng, Univ Kans, 46; from chief supvr, Tech Div, to div engr, Design Div, Hanford Works, Gen Elec Co, 46-48, head separations, Tech Div, 48-51, asst mgr, Tech Div & chief process engr, 51, mgr design, 52-56, construct eng, 56-57, gen mgr, X-ray Dept Defense Prod, 57-62, spacecraft dept, 63-64 & Apollo Systs Dept, 64-67, vpres & gen mgr, Electronic Systs Div, 68-74, Energy Systs & Technol Div, 74-75 & Nuclear Energy Systs Div, 75-77, sr vpres & group exec, Nuclear Energy Group, 77-81; RETIRED. *Mem:* Nat Acad Eng; Am Inst Chem Engrs; fel Am Inst Chemists; Am Inst Aeronaut & Astronaut; sr mem Inst Elec & Electronics Engrs; Am Nuclear Soc. *Res:* Nuclear engineering; spacecraft design and manufacturing; manned space flight checkout; radar, sonar and weapon control systems design and manufacturing; nuclear reactors and fuel design and manufacturing. *Mailing Add:* PO Box 1018 Saratoga CA 95071

BEATTIE, ALAN GILBERT, b Oakland, Calif, Apr 2, 34; m 61; c 3. SOLID STATE PHYSICS. *Educ:* Univ Calif, Berkeley, BA, 59; Univ Wash, MS, 60, PhD(physics), 65. *Prof Exp:* Res engr, Algae Res Proj, Univ Calif, Berkeley, 59-60; MEM STAFF, SANDIA NAT LAB, 65- *Concurrent Pos:* Assoc ed, J of Acoustic Emissions, 81. *Mem:* Inst Elec & Electronics Engrs; Am Soc Testing & Mat; Acoustic Emission Working Group; Am Phys Soc; Sigma Xi. *Res:* Ultrasonics; acoustic emission. *Mailing Add:* Orgn 7552 Sandia Labs Albuquerque NM 87185

BEATTIE, CRAIG W, b Elizabeth, NJ, Oct 26, 43. ONCOLOGY, CELLULAR ENDOCRINOLOGY. *Educ:* Univ Del, PhD(biol), 71. *Prof Exp:* PROF PHARMACOL, SURG & PHARM, SCH MED, UNIV ILL, CHICAGO, 83- *Mailing Add:* Dept Surg Oncol & Pharmacol Univ Ill Sch Med 840 S Wood St Chicago IL 60612

BEATTIE, DIANA SCOTT, b Cranston, RI, Aug 11, 34; m 56; c 4. BIOCHEMISTRY. *Educ:* Swarthmore Col, BA, 56; Univ Pittsburgh, MS, 58, PhD(biochem), 61. *Prof Exp:* Res assoc biochem, Univ Pittsburgh, 61-68; from asst prof to prof biochem, Mt Sinai Sch Med, 68-85; PROF BIOCHEM & CHMN DEPT, SCH MED, WVA UNIV, 85- *Concurrent Pos:* Mem, Physical Biochem Study Sect, NIH, 79-85; mem, Grad Fac PhD Prog Biochem, Biol & Biomed Sci, City Univ NY, 69-85; vis prof, Univ Louvain, Lovain-la-Neuve, Belgium, 82; Fogarty Int fel, 82; mem, Grad Fac PhD Prog, Biochem, Biol & Biomed Sci, City Univ New York, 81- *Mem:* Am Soc Biol Chemists; Am Soc Cell Biologists; Biophys Soc. *Res:* Mitochondrial biogenesis, including various aspects of protein synthesis, enzymology and its regulation; bioenergetics of electron transport chain. *Mailing Add:* Dept Biochem WVa Univ Sch Med Morgantown WV 26506

BEATTIE, DONALD A, b New York, NY, Oct 30, 29; m 72; c 2. ENERGY, SPACE TECHNOLOGY. *Educ:* Columbia Univ, AB, 51; Colo Sch Mines, MS, 58. *Prof Exp:* Officer & aviator, US Navy, 51-56; regional geol, Mobil Oil Co, 58-63; aerospace tech, Nat Aeronaut & Space Admin, 63-82; div dir, Res & Develop, NSF, 73-75; asst admin, Energy Res & Develop Admin, 75-77; asst secy, Dept Energy, 77-78; div dir, res & develop, NASA, 78-82; vpres, BDM Corp, 83-84; CONSULT, 84- *Concurrent Pos:* Solar energy coordr, US & USSR Coop Sci & Tech, 74-76 & Vienna Inst Comp Econ Studies, 79; US rep, Vienna Inst Comp Econ Studies, 79; advisor, Montgomery Col, 78-81; vis lectr, Georgetown Univ, 82, Nat Space Club, Youth Educ, 84-; mem adv comt, NASA, 88- *Mem:* Geol Soc Am; Am Astronaut Soc; NY Acad Sci; AAAS; Naval Inst; Nat Space Club. *Res:* Study of lunar surface phenomena leading to the design and development of geological and geophysical experiments for the Apollo Program; renewable energy technologies. *Mailing Add:* 13831 Dowlais Dr Rockville MD 20853

BEATTIE, EDWARD J, b Philadelphia, Pa, June 30, 18; m 77; c 1. SURGERY. *Educ:* Princeton Univ, BA, 39; Harvard Univ, MD, 43; Am Bd Surg, dipl, 49; Am Bd Thoracic Surg, dipl, 51. *Hon Degrees:* LLD, Hampden-Sydney Col, 78. *Prof Exp:* Resident surg, Peter Bent Brigham Hosp, Boston, 42-43, from asst resident surgeon to resident surgeon, 44-45, jr assoc surg, 45-47; fel, George Washington Univ, 47-48, asst prof & Markle scholar, 48-52; from asst prof to prof, Univ Ill, 52-65; chief thoracic surgeon, Mem Hosp, New York, 65-75, chmn dept surg, 66-78, chief med officer, 66-83, gen dir & chief operating officer, 74-83; prof surg, Cornell Univ, 65-83, emer prof, 83; prof surg & oncol, Univ Miami, 83-85; CHIEF THORACIC SURGEON, DIR KRISER LUNG CANCER CTR & DIR CLIN CANCER PROGS, BETH ISRAEL MED CTR, 85- *Concurrent Pos:* Instr, Harvard Univ, 45-47; Moseley travel fel from Harvard Univ, Postgrad Med Sch, Univ London, 46-47; consult, Gallinger Munic Hosp, Washington, DC, Walter Reed Army Hosp, Newton D Baker & Mt Alto Vet Admin Hosps, 48-52; consult, Chicago State Tuberc Sanitorium, Univ Ill Hosps & Hines Vet Admin Hosp, 52-65 & Rockefeller Univ, 77; thoracic surgeon, Presby-St Luke's Hosp, Chicago, 54-65; mem bd, Am Bd Thoracic Surg, 60-65, from vchmn to chmn, 65-69; mem, Sloan Kettering Inst, 66-83. *Mem:* Am Asn Thoracic Surg; Am Col Surg; Soc Vascular Surg; Soc Clin Surg; Am Surg Asn. *Res:* Thoracic surgery; multidisciplinary approach, including early detection, surgery, internal and external radiotherapy and medical oncology, to decrease present high mortality from lung cancer. *Mailing Add:* Beth Israel Med Ctr First Ave at 16th St New York NY 10003

BEATTIE, HORACE S, b Utica, NY, July 19,09. HARDWARE SYSTEMS. *Educ:* Williams Col, BA, 31; Mass Inst Technol, BSME, 33. *Prof Exp:* Mem staff, IBM Corp, Washington, DC, 33-40, asst engr, East Orange Lab, 40-43, engr, Prod Develop Lab, Poughkeepsie, 43-44, advan mach develop, 52-56 & mgr, 56-57, mgr & engr, Elec Typewriter Div, 57-65, vpres eng, Off Prod Div, 65-72, vpres, 72-83; RETIRED. *Honors & Awards:* Am Soc Mech Engrs Medal, 71. *Mem:* Nat Acad Eng; Soc Mfg Engrs; Am Soc Mech Engrs. *Mailing Add:* 4000 Bryan Sta Pike Lexington KY 40511

BEATTIE, JAMES MONROE, b Washington, DC, Feb 14, 21; m 42; c 2. HORTICULTURE. *Educ:* Univ Md, BS, 41; Cornell Univ, PhD(pomol), 48. *Prof Exp:* From asst prof to prof hort, Ohio Agr Exp Sta, 48-63, from asst dir to assoc dir, Ohio Agr Res & Develop Ctr, 63-73; DEAN COL AGR, DIR AGR EXP STA & DIR COOP EXTEN SERV, PA STATE UNIV, UNIVERSITY PARK, 73- *Mem:* fel Am Soc Hort Sci (pres, 69-70); Am Inst Biol Sci. *Res:* Nitrogen and mineral nutrition of fruit crops; reclamation of strip mine spoils; culture and management of fruit plantings. *Mailing Add:* 1731 Princeton Dr State College PA 16801

BEATTIE, RANDALL CHESTER, b Orlando, Fla, June 12, 45; m 67; c 1. AUDIOLOGY. *Educ:* Northern Ill Univ, BS, 67; Univ Ill, MS, 69; Univ Southern Calif, PhD(audiol), 72. *Prof Exp:* Speech & hearing specialist, Georgetown, Ill, 67-69; audiologist, Los Angeles Outpatient Clin, Vet Admin, 69-72; educ audiologist, Alhambra Sch Dist, 72; asst prof, 72-77, ASSOC PROF AUDIOL, CALIF STATE UNIV, LONG BEACH, 77- *Mem:* Am Speech & Hearing Asn. *Res:* Speech audiometry; hearing aids. *Mailing Add:* Dept Commun Dis Calif State Univ 1250 Bellflower Blvd Long Beach CA 90840

BEATTIE, THOMAS ROBERT, b Philadelphia, Pa, Aug 31, 40; m 61; c 3. ORGANIC CHEMISTRY. *Educ:* Univ Pa, BS, 61; Univ Wis, PhD(org chem), 65. *Prof Exp:* Sr res chemist, 66-73, res fel, 73-77, SR RES FEL, MERCK & CO, INC, 77- *Concurrent Pos:* Mem, Atlanta Univ Ctr, Inc Chancellor's Circle, 81- *Mem:* Am Chem Soc. *Res:* Medicinal and synthetic organic chemistry; biomembranes; chemistry of phospholipids. *Mailing Add:* Four Heather Lane Scotch Plains NJ 07076-1299

BEATTIE, WILLARD HORATIO, b Oak Park, Ill, Mar 3, 27; m 54; c 3. PHYSICAL CHEMISTRY. *Educ:* Univ Chicago, BA, 51, MS, 54; Univ Minn, PhD(anal chem), 58. *Prof Exp:* Res chemist, Shell Chem Co, 58-62; sr res chemist, Beckman Instruments, Inc, 62; asst prof chem, Calif State Col, Long Beach, 62-67; mem staff, Los Alamos Sci Lab, 67-90; RETIRED. *Concurrent Pos:* Res prof, Univ Calif, Santa Barbara, 75-76. *Mem:* Am Chem Soc. *Res:* Light scattering; chemical laser development; laser isotope separation; radiation chemistry of gases; analytical chemical instrumentation; fuels from coal and natural gas; inorganic fluorine chemistry. *Mailing Add:* 4045 Pamela Pl Las Cruces NM 88005-5641

BEATTY, CHARLES LEE, b Kempton, Ind, Nov 7, 39; m 72; c 3. MATERIALS SCIENCE ENGINEERING. *Educ:* Purdue Univ, BS, 65; Case Western Reserve Univ, MS, 68; Univ Mass, MS, 71, PhD(polymer sci & eng), 72. *Prof Exp:* PROF, DEPT MAT SCI & ENG, UNIV FLA, 79- *Concurrent Pos:* Researcher, Xerox, 72-79; assoc prof polymer sci, Univ Rochester, 76-77, assoc prof mech & aerospace sci, 76-78; chmn-elect, Eng Properties & Struct Div, Soc Plastics Engrs; coun mem, Soc Plastics Engrs, 88-91. *Mem:* Am Inst Chem Engrs; Nat Asn Corrosion Engrs; Am Chem Soc; Am Phys Soc; Soc Plastics Engrs; Soc Biomaterials; Am Soc Testing & Mat. *Res:* Solid-state properties of polymers ranging from diffusion to calorimetry; mechanical properties in the linear and non-linear viscoelastic region. *Mailing Add:* Dept Mat Sci & Eng Univ Fla Gainesville FL 32611

BEATTY, CLARISSA HAGER, b Colorado Springs, Colo, June 3, 19; m 43; c 2. BIOCHEMISTRY. *Educ:* Sarah Lawrence Col, AB, 41; Columbia Univ, MA, 42, PhD(physiol), 45. *Prof Exp:* Asst physiol, Columbia Univ, 43-45, instr, 45-48; fel, Diabetic Res Found, Med Sch, Univ Ore, 48-53, from instr to asst prof, 53-62; assoc scientist, Ore Regional Primate Res Ctr, 61-67, scientist, 67-85; assoc prof biochem, Med Sch, Univ Ore, 62-85; RETIRED. *Mem:* Am Chem Soc; Am Physiol Soc; Soc Exp Biol & Med; Soc Study Reproduction. *Res:* Primate muscle metabolism (fetal to aged); smooth muscle metabolism. *Mailing Add:* 2958 Dosch Rd Portland OR 97201-1418

BEATTY, DAVID DELMAR, b Bellingham, Wash, Oct 10, 35; m 57; c 3. COMPARATIVE PHYSIOLOGY. *Educ:* Western Wash State Col, BA, 57; Univ Wyo, MSc, 59; Univ Ore, PhD(animal physiol), 64. *Prof Exp:* Instr zool, Univ Wyo, 59-60; from asst prof to assoc prof, 64-78, assoc dean sci, 76-83, PROF ZOOL, UNIV ALTA, 78- *Mem:* AAAS; Am Soc Zool; Can Soc Zool. *Res:* Biochemistry of fish visual pigment; physiology of fish. *Mailing Add:* Dept Zool Univ Alta Edmonton AB T6G 2M7 Can

BEATTY, JAMES WAYNE, JR, b Fargo, NDak, Sept 9, 34; wid; c 3. PHYSICAL CHEMISTRY. *Educ:* NDak State Univ, BSc, 56; Mass Inst Technol, PhD(phys chem), 60. *Prof Exp:* From instr to asst prof physics, Colby Col, 60-63; from asst prof to assoc prof chem, 63-74, chmn, dept chem, 81-87, PROF CHEM, RIPON COL, 74- *Concurrent Pos:* Consult-evaluator, NCent Asn Cols & Schs, 73-; deleg, Wis Gov Conf Libr & Info Serv, 78. *Mem:* Am Chem Soc; Am Phys Soc; Am Asn Univ Professors. *Res:* Measurement of transport properties of gases, particularly diffusion coefficients; determination of force law constants from this data; computer assisted instruction in chemistry. *Mailing Add:* Dept Chem Ripon Col Ripon WI 54971

BEATTY, JOHN C, b New York, NY, Nov 27, 47. COMPUTER SCIENCE, COMPUTER GRAPHICS. *Educ:* Princeton Univ, BS, 69; Univ Calif, Berkeley, PhD(computer sci), 78. *Prof Exp:* Asst prof, 78-84, ASSOC PROF COMPUTER SCI, UNIV WATERLOO, 84- *Concurrent Pos:* Vis scientist, Xerox Palo Alto Res Ctr, 85-86. *Mem:* Asn Comput Mach; Inst Elec & Electronics Engrs; Sigma Xi. *Mailing Add:* Computer Sci Dept Univ Waterloo Waterloo ON N2L 3G1 Can

BEATTY, JOSEPH JOHN, b Detroit, Mich, Feb 19, 47. ECOLOGY, BIOMETRICS. *Educ:* Univ Mo, Columbia, BS, 70, AM, 73; Ore State Univ, PhD(zool), 79. *Prof Exp:* Instr biol, Dept Zool, 79-90, SR INSTR & DIR, BIOL PROG, OREGON STATE UNIV, 90. *Mem:* Am Soc Ichthyologists & Herpetologists; Herpetologists League; Soc Study Amphibians & Reptiles. *Res:* Ecology and evolutionary biology of amphibians, primarily northwestern plethodontid salamanders. *Mailing Add:* Biol Prog Ore State Univ Corvallis OR 97331-2908

BEATTY, K(ENNETH) O(RION), JR, b East Lansdowne, Pa, Dec 18, 13; m 36; c 3. BIOMEDICAL ENGINEERING, FIRES & ARSON. *Educ:* Lehigh Univ, BS, 35, MS, 37; Univ Mich, PhD(chem eng), 46. *Prof Exp:* Chem engr, Dow Chem Co, 37-39; from instr to asst prof, Univ RI, 39-44; res assoc engr, Univ Mich, 44-46; prof chem eng, 46-61, actg head dept, 59-60, REYNOLDS PROF CHEM ENG, NC STATE UNIV, 61- *Concurrent Pos:* Eng consult, 49-; Proj engr, Pratt & Whitney, 57; consult, 61-; US del, Assembly Int Heat Transfer Conf, 67-72; mem sci coun, Int Ctr Heat & Mass Transfer, Yugoslavia, 68- 82; legal expert, 75- *Mem:* fel Am Inst Chem Engrs; Am Chem Soc; Int Asn Arson Investr; Sigma Xi. *Res:* Heat transfer; infrared thermography; fluid dynamics; devices for braille readers; biomedical engineering; legal expert in fires and arson; legal expert in slips, trips and falls. *Mailing Add:* 323 Shepherd St Raleigh NC 27607

BEATTY, KENNETH WILSON, b Oklahoma City, Okla, Feb 22, 29; m 65; c 3. ZOOLOGY, LIMNOLOGY. *Educ:* Univ Calif, Davis, BS, 52, MA, 60, PhD(limnol), 68. *Prof Exp:* Lab technician animal husb, Univ Calif, Davis, 54-60; INSTR BIOL, COL SISKIYOUS, 60- *Concurrent Pos:* Biologist, US Geol Surv, 71-73. *Mem:* Am Soc Limnol & Oceanog; Ecol Soc Am; Am Inst Biol Sci; Nat Asn Biol Teachers; Am Nature Study Soc. *Res:* Ecology of new reservoirs; benthic ecology of alpine lakes; behavior and reproductive cycles of woodpeckers; mastitis studies under controlled management practices. *Mailing Add:* 800 College Ave Weed CA 96094

BEATTY, MARVIN THEODORE, b Bozeman, Mont, Mar 13, 28; m 56; c 4. SOILS, RESOURCE MANAGEMENT. *Educ:* Mont State Col, BS, 50; Univ Wis, PhD(soils), 55. *Prof Exp:* Soil scientist, Soil Conserv Serv, 50-56; from asst prof to assoc prof soils, 56-66, chmn environ resources unit, 70-72, PROF SOILS, UNIV WIS-MADISON, 66-, DEAN COMMUN PROGS, 77- & CHMN, NATURAL & ENVIRON RESOURCES, UNIV EXTEN, 72- *Mem:* Am Soc Agron; Soil Conserv Soc Am; Nat Univs Exten Asn. *Res:* Soil surveys and the interpretation of physical and chemical properties of soils as they affect sustained use of soils for agriculture; engineering; urban development; soil conservation; natural resource and environmental education programs. *Mailing Add:* 4702 Waukesha St Madison WI 53705

BEATTY, MILLARD FILLMORE, JR, b Baltimore, Md, Nov 13, 30; m 51; c 3. NONLINEAR ELASTICITY, CONTINUUM MECHANICS. *Educ:* Johns Hopkins Univ, BES, 59, PhD(mech), 64. *Prof Exp:* Instr eng physics, Loyola Col, Md, 59-60; instr mech, Johns Hopkins Univ, 60-63; asst prof continuum mech, Univ Del, 63-67; assoc prof, 67-76, dir grad studies, 75-79, PROF THEORET MECH, UNIV KY, 76- *Mem:* Soc Natural Philos; Am Acad Mech; Am Asn Physics Teachers; Am Soc Mech Engrs; Soc Eng Sci; Sigma Xi. *Res:* Non-linear elasticity and elastic stability theory; rubber elasticity; biomechanics. *Mailing Add:* 866 Glendover Rd Lexington KY 40502

BEATTY, OREN ALEXANDER, b Nobob, Ky, July 3, 01; m 33; c 9. FOREST SCIENCE, TREE FARMER. *Educ:* Univ Louisville, MD, 30. *Prof Exp:* Med dir & supt, Richland Hosp, Mansfield, Ohio, 42-51, Hazelwood Sanatorium, Louisville, Ky, 51-62 & Richland Hosp, 62-72; RETIRED. *Concurrent Pos:* Asst prof, Sch Med, Univ Louisville, 52-62; tuberc controller, Lorain County Tuberc Clin, 65-72; consult, 72-81. *Mem:* AMA; fel Am Col Chest Physicians; Am Lung Asn. *Res:* Diseases of the chest; undulant fever in relation to chest involvement; tuberculosis cavities. *Mailing Add:* 3717 Hanover Rd Louisville KY 40207

BEATY, DAVID WAYNE, b St Louis, Mo, Nov 12, 53; m 78; c 2. ECONOMIC GEOLOGY, STABLE ISOTOPE GEOCHEMISTRY. *Educ:* Dartmouth Col, AB, 75; Calif Inst Technol, PhD(geol), 80. *Prof Exp:* Geologist, Noranda Explor, Inc, 80-86; GEOLOGIST, CHEVRON OIL FIELD RES CO, 86- *Mem:* Soc Econ Geologists; Am Geophys Union; Mineral Soc Am; Geochem Soc. *Res:* Mineral exploration; hydrothermal processes; igneous and ore petrology. *Mailing Add:* Chevron Oil Field Res Co PO Box 446 La Habra CA 90633-0446

BEATY, EARL CLAUDE, b Zeigler, Ill, Nov 6, 30; m 52; c 3. PHYSICS. *Educ:* Murray State Col, AB, 52; Washington Univ, PhD, 56. *Prof Exp:* PHYSICIST, NAT BUR STAND, 56-77 & 78-; FEL, JOINT INST LAB ASTROPHYS, UNIV COLO, BOULDER, 62- *Concurrent Pos:* Nat Res Coun fel, 56-57; consult, Int Atomic Energy Agency, 77-78. *Mem:* Am Phys Soc. *Res:* Atomic physics. *Mailing Add:* 3005 Stanford Ave Univ Colo Boulder CO 80303

BEATY, ORREN, III, b Las Cruces, NMex, Nov 13, 45; m 71; c 3. CARDIOVASCULAR PHYSIOLOGY, ADRENERGIC MECHANISMS. *Educ:* Richmond Col, BS, 68; Wake Forest Univ, PhD(physiol), 74; Kirksville Col Osteop Med, DO, 88. *Prof Exp:* Teaching asst chem, Richmond Col, 68-69; fel physiol, Sch Med, Mayo Grad Sch, Mayo Clin, 74-76, vis scientist, 77-79; asst prof, 79-84, vis asst prof physiol, Kirksville Col Osteop Med, 84-88; RESIDENT, MED/PEDIAT, E CAROLINA UNIV SCH MED, 88- *Concurrent Pos:* Vis asst prof physiol, Kirksville Col Osteop Med, 84- *Mem:* Sigma Xi; Am Physiol Soc. *Res:* Regulation of arterial blood pressure and peripheral circulation through both local and reflex mechanisms as well as the control of peripheral adrenergic neurotransmission. *Mailing Add:* 1504 Brownlea Dr Greenville NC 27834

BEAUBIEN, STEWART JAMES, b Vancouver, Can, Nov 15, 19; nat US; m 42; c 2. LUBRICATION, ENERGY ANALYSIS. *Educ:* Univ Calif, BS, 43. *Prof Exp:* Mech engr lubrication res, Shell Develop Co, 43-54, supvr res, 54-65, group leader, Shell Oil Co, 65-67, sr staff engr, 67-72, sr staff mem corp planning energy & forecasting, 73-77; RETIRED. *Mem:* Am Soc Mech Engrs; Sigma Xi. *Res:* Lubrication and wear of machine elements, particularly gears and bearings at high speeds and temperatures; influence of lubrication on processes involving metal cutting and working; automatic transmission and power transmission fluids; energy production, distribution and use. *Mailing Add:* 3009 N Mt Baker Circle Oak Harbor WA 98277

BEAUCHAMP, EDWIN KNIGHT, b Ely, Nev, July 16, 30; m 54; c 6. BRITTLE FRACTURE, HIGH TEMPERATURE DEFORMATION. *Educ:* Univ Nev, Reno, BS, 52; Northwestern Univ, Evanston, MS, 54; Univ Utah, Salt Lake City, PhD(physics), 61. *Prof Exp:* Res asst, Northwestern Univ, 52-53; jr engr develop, Boeing Aricraft Corp, 53; res engr, Lockheed Aircraft, 56-57; asst res & teaching, Univ Utah, 57-61; res scientist, Glass Res, Corning Glass Works, 61-64; DISTINGUISHED MEM TECH STAFF MAT & RES, SANDIA NAT LABS, 64- *Mem:* Am Ceramic Soc; Mat Res Soc. *Res:* Brittle fracture of ceramics and glasses; subcritical crack growth in glasses and glass-ceramics; internal friction in glasses; dielectric breakdown in ceramics; sintering of explosively shocked ceramics. *Mailing Add:* 7509 Harwood Ave NE Albuquerque NM 87110

BEAUCHAMP, ERIC G, b Grenville, Que, Jan 6, 36; m 58; c 3. SOIL FERTILITY & SOIL NITROGEN, PLANT NUTRITION. *Educ:* McGill Univ, BScAgr, 60, MSc, 62; Cornell Univ, PhD(soil sci), 65. *Prof Exp:* Res scientist, Res Br, Can Dept Agr, 65-67; asst prof soil sci, 67-70, assoc prof soil sci, 70-79, PROF SOIL SCI, UNIV GUELPH, 79- *Mem:* Am Soc Agron; Agr Inst Can; Can Soc Soil Sci; Int Soc Soil Sci. *Res:* Response of field crops to nitrogen fertilizers; soil nitrogen transformations. *Mailing Add:* Dept Land Resource Sci Univ Guelph Guelph ON N1G 2W2 Can

BEAUCHAMP, GARY KEITH, b Belvidere, Ill, Apr 5, 43; m 67; c 2. ANIMAL BEHAVIOR. *Educ:* Carleton Col, BA, 65; Univ Chicago, PhD(biopsychol), 71. *Prof Exp:* Fel chemosensation, Univ Pa, 71-73, asst prof psychol, Dept Otorhinolaryngol & Human Commun, Med Sch, 74-; ASSOC MEM, MONELL CHEM SENSES CTR, 73-, DIR, 90- *Mem:* AAAS; Asn Chemoreception Soc. *Res:* Investigation of the role of the chemical senses in regulating behavior and physiology in a variety of animal species, including humans. *Mailing Add:* Monell Chem Senses Ctr Philadelphia PA 19104

BEAUCHAMP, JESSE LEE, b Glendale, Calif, Nov 1, 42; m 64; c 3. PHYSICAL CHEMISTRY. *Educ:* Calif Inst Technol, BS, 64; Harvard Univ, PhD(chem), 67. *Prof Exp:* Noyes instr, 67-69, from asst prof to assoc prof, 69-74, PROF CHEM, CALIF INST TECHNOL, 74- *Concurrent Pos:* Alfred P Sloan Found fel, 68-70. *Honors & Awards:* Pure Chem Award, Am Chem Soc, 78. *Mem:* Nat Acad Sci; Am Chem Soc; Am Phys Soc. *Res:* Gas phase ion chemistry; structures, properties and reactions of organic ions; chemical applications of ion cyclotron resonance spectroscopy; surface science and catalysis; chemical applications of photoelectron spectroscopy. *Mailing Add:* A A Noyes Lab 127-72 Calif Inst Technol Pasadena CA 91125

BEAUCHAMP, JOHN J, b Nashville, Tenn, Sept 17, 37; m; c 2. APPLIED STATISTICS, DATA ANALYSIS. *Educ:* Vanderbilt Univ, BA, 59, MAT, 60; Fla State Univ, MS, 63, PhD(statist), 66. *Prof Exp:* Asst prof math, Birmingham-Southern Col, 60-61; biomet trainee statist, Vanderbilt Univ, 61-62; statistician, math div, Oak Ridge Nat Lab, 67-73; statistician, comput sci div, nuclear div, Union Carbide Corp, 73-84; RES STATISTICIAN, ENG, PHYSICS & MATH DIV, OAK RIDGE NAT LAB, 84- *Concurrent Pos:* Instr math & sci, Roane St Community Col, 76-79; instr biomed sci, Univ Tenn, 80, 86 & 88-90. *Mem:* Biomet Soc; Am Statist Asn. *Res:* Nonlinear estimation; model building; multivariate analysis; applied statistics; regression, design and analysis of experiments. *Mailing Add:* Oak Ridge Nat Lab PO Box 2008 Oak Ridge TN 37831-6367

BEAUCHAMP, LILIA MARIE, b New York, NY. MEDICINAL CHEMISTRY, ANTIVIRAL CHEMOTHERAPY. *Educ:* St John's Univ, BS, 53; Fordham Univ, MS, 55. *Prof Exp:* Anal chemist, Colgate Palmolive Co, 55-57; SR RES CHEMIST, BURROUGHS WELLCOME CO, 57- *Mem:* Am Chem Soc. *Res:* Synthesis of new drugs for biological testing; antiviral agents; acyclic nucleosides. *Mailing Add:* 3007 Wycliff Rd Raleigh NC 27607

BEAUCHENE, ROY E, b Sioux City, Iowa, Sept 4, 25; m 52; c 4. NUTRITION, GERIATRICS. *Educ:* Morningside Col, BS, 51; Kans State Univ, MS, 52, PhD, 56. *Prof Exp:* Asst chem, Kans State Univ, 51-56; from asst prof to assoc prof human nutrit res, Tex Woman's Univ, 56-58, prof nutrit & biochem res, 58-63; NIH res scientist, Geront Br, Baltimore City Hosps, 63-68; assoc prof nutrit, 68-72, head, Dept Food Sci, Nutrit & Food Syst Admin, 77-81, PROF NUTRIT, UNIV TENN, KNOXVILLE, 72- *Concurrent Pos:* Guest scholar, Kans State Univ, 73; pres, Univ Tenn, Knoxville Chap Sigma Xi, 90-91. *Mem:* Am Chem Soc; fel Geront Soc; Am Inst Nutrit; Sigma Xi; fel Am Col Nutrit. *Res:* Nutrition and aging; bone density. *Mailing Add:* Dept Nutrit Univ Tenn Knoxville TN 37996-1900

BEAUDET, ARTHUR L, b Woonsocket, RI, July 4, 42; m 67; c 2. HUMAN GENETICS,. *Educ:* Col Holy Cross, BS, 63; Yale Univ, MD, 67. *Prof Exp:* Intern & resident pediat, Johns Hopkins Hosp, 67-69; res assoc biochem, NIH, Bethesda, 69-71; from asst prof to assoc prof, 71-82, PROF PEDIAT, BAYLOR COL MED, HOUSTON, 82- *Concurrent Pos:* Investr, Howard Hughes Med Inst, 73-80, 85-; mem, Genetic Basis Dis Rev Comt, 80-84 & Biochem II Study Sect, NIH, 85-89. *Mem:* Soc Pediat Res; Fedn Am Soc Biol Chemists; Am Soc Human Genetics; Genetics Soc Am. *Res:* Molecular human genetics; human genetic diseases, inborn errors of metabolism; inborn errors of metabolism; cystic fibrosis. *Mailing Add:* Inst Molecular Genetics Baylor Col Med Howard Hughes Med Inst One Baylor Plaza Houston TX 77030

BEAUDET, ROBERT A, b Woonsocket, RI, Aug 18, 35; m 57; c 7. CHEMICAL PHYSICS, PHYSICAL CHEMISTRY. *Educ:* Worcester Polytech Inst, BS, 57; Harvard Univ, AM & PhD(phys chem), 62. *Prof Exp:* Res scientist, Jet Propulsion Lab, Calif Inst Technol, 61-63; from asst prof to assoc prof, 63-71, chmn dept, 74-76, PROF CHEM, UNIV SOUTHERN CALIF, 71- *Concurrent Pos:* Consult, Jet Propulsion Lab, Calif Inst Technol, 63-; NIH grant, Univ Southern Calif, 64; A P Sloan Found res fel, 67-71; consult, Army Sci Adv Panel, 68-75, mem, 75-77, assoc mem, Army Sci Bd, 78-79. *Honors & Awards:* Alexander von Humboldt Award, 74-75. *Mem:* Am Chem Soc; Am Phys Soc. *Res:* Microwave spectroscopy; rotational spectra; molecular structure; intra molecular interactions; laser spectroscopy; boron hydrides; carboranes. *Mailing Add:* 887 Vallombrosa Dr Pasadena CA 91107

BEAUDOIN, ADRIEN ROBERT, b Bury, Que, Sept 3, 40; m 64; c 3. CELL MORPHOLOGY, MEMBRANE FUNCTIONS. *Educ:* Univ Sherbrooke, BSc, 66; Univ Laval, DSc, 70. *Prof Exp:* Res assoc, Duke Univ, 70-71, Univ Laval, 71-72; from asst prof to prof, 72-81, FULL PROF, UNIV SHERBROOKE, 81-; dir, Ctr Res Mechanisms Secretion, 82-88. *Honors & Awards:* Diatome Prize, Am Soc Micros, 85. *Mem:* Am Soc Cell Biol; Micros Soc Can; Can Biochem Soc; Can-French Asn Advan Sci. *Res:* Cell secretory mechanisms; intracellular route and processing of the secretory proteins of pancreas; characterization of adenosine triphosphate-diphosphohydrolase in eukaryotic cells; membrane shedding mechanisms and cancer; microvisicular secretion. *Mailing Add:* Bept Biol Univ Sherbrooke 2500 Univ Blvd Sherbrooke PQ J1K 2R1 Can

BEAUDOIN, ALLAN ROGER, b New Britain, Conn, Aug 25, 27; m 50; c 4. TERATOLOGY, DEVELOPMENTAL TOXICOLOGY. *Educ:* Univ Conn, BS, MS, 51; Univ Iowa, PhD(zool), 54. *Prof Exp:* Instr zool, Univ Iowa, 53-54; instr zool, Vassar Col, 54-55, instr physiol, 55-56; from instr to asst prof anat, Col Med, Univ Fla, 56-61; from asst prof to assoc prof, 61-69, PROF ANAT, MED SCH, UNIV MICH, ANN ARBOR, 69- *Concurrent Pos:* Vis res assoc prof, Karolinska Inst, Sweden, 67-68. *Mem:* AAAS; Am Soc Zoologists; Am Asn Anatomists; Teratology Soc (pres, 78-79); Soc Develop Biol. *Res:* Teratology. *Mailing Add:* Dept Anat Univ Mich 4614 Med Sci II Ann Arbor MI 48109

BEAUDREAU, GEORGE STANLEY, b Ferndale, Wash, Dec 2, 25; m 46; c 4. BIOCHEMISTRY. *Educ:* Wash State Col, BS, 49; Ore State Col, MS, 52, PhD, 54. *Prof Exp:* Fel biol sci, Purdue Univ, 54-56; res assoc, Duke Univ, 56-63; assoc prof, 63-77, PROF CHEM & AGR CHEM, ORE STATE UNIV, 63- *Mem:* Am Chem Soc; Am Asn Cancer Res. *Res:* Intermediary metabolism and enzymology; antibiotics; tumor virus; tissue culture. *Mailing Add:* Dept Agr Chem Ore State Univ Corvallis OR 97330

BEAUDRY, BERNARD JOSEPH, b Albertville, Minn, Feb 26, 32; m 57; c 7. METALLURGY, MATERIAL SCIENCE. *Educ:* St John's Univ, BA, 54; Iowa State Univ, MS, 59. *Prof Exp:* Jr chemist, Ames Lab, Iowa State Univ, 54-59, res assoc metall, 59-61, asst metallurgist, 61-64, assoc metallurgist, 64-77, metallurgist, 77-89, SR METALLURGIST, AMES LAB, IOWA STATE UNIV, 89- *Mem:* Am Soc Metals; Metall Soc. *Res:* Rare earth metallurgy; preparation and properties of high purity rare earth metals; effect of impurities on their properties; preparation and characterization of rare earth alloys and compounds; thermoelectric properties of materials. *Mailing Add:* Ames Lab Iowa State Univ Ames IA 50011

BEAUFAIT, FREDERICK W, b Vicksburg, Miss, Nov 28, 36; m 64; c 2. ENGINEERING ADMINISTRATION. *Educ:* Miss State Univ, BS, 58; Univ Ky, MS, 61; Va Polytech Inst, PhD (structural eng), 65. *Prof Exp:* Structural eng, US Army Engrs, Vicksburg, 58-59; eng, L E Gregg & Assoc, Lexington, Ky, 59-60; vis lectr, Univ Liverpool, 60-61; instr, Va Polytech Inst, 61-65; prof civil eng, Vanderbilt Univ, 65-79; prof & chmn, Dept Civil Eng, WVa Univ, 83-86, assoc dean, Col Eng, 83-86; DEAN ENG, COL ENG, WAYNE STATE UNIV, 86. *Concurrent Pos:* Instr, Dept Civil Eng, Univ Ky, 59-60; structural eng, US Army Corps Engrs, 66 & 77; vis prof, Dept Civil & Structural Eng, Univ Wales, 75-76; vis prof, Div Basic Sci & Eng, Univ Autonoms Metropolitana, Mex, 83. *Mem:* Am Soc Civil Engrs; Am Soc Eng Educ; Am Concrete Inst; Nat Soc Prof Engrs. *Res:* The behavior of reinforced concrete and the nonlinear response and analysis of structural systems. *Mailing Add:* Col Eng Wayne State Univ Detroit MI 48202

BEAULIEU, J A, b Montreal, Que, Apr 17, 29; m 55; c 2. BACTERIOLOGY, IMMUNOLOGY. *Educ:* Univ Montreal, BSc, 51; McGill Univ, MSc, 53, PhD(bact, immunol), 55; Am Bd Microbiol, dipl. *Prof Exp:* Head dept bact, St Joan D'Arc Hosp, 53-56; lectr, Sch Med, Univ Ill, 56-57; from asst prof to assoc prof, Univ Ottawa, 57-65, assoc dean, 71-89, actg dean, 89-90, PROF BACT, FAC MED, UNIV OTTAWA, 65-, VDEAN, 90- *Concurrent Pos:* Head bact, Hosp Sacre Coeur, Hull, Que, 58-; Med Res Coun grants, 61-64; ed, Can Soc Microbiol News Bull, 64-67. *Mem:* Can Soc Microbiol. *Res:* Production of antibodies by monocytes in vitro; factors involved in monocytosis producing agent of Listeria; hospital infections. *Mailing Add:* Dept Bact Univ Ottawa Ottawa ON K1N 6N5 Can

BEAULIEU, JOHN DAVID, b Hanford, Wash, July 22, 44; m 67; c 3. STRATIGRAPHY, ENVIRONMENTAL GEOLOGY. *Educ:* Univ Wash, BS, 66; Stanford Univ, PhD(geol), 71. *Prof Exp:* Prof geol, Univ Ore, 69-70; geologist, 70-77, DEP STATE GEOLOGIST, ORE DEPT GEOL & MINERAL INDUSTS, 77- *Concurrent Pos:* fel, NSF, 67-69. *Res:* Cenozoic behavior of the San Andreas Fault in the Santa Cruz Mountains, California; structure and stratigraphy of Oregon; geologic history and hazards of Oregon; plate tectonics; geothermal potential and mineral potential of Oregon; reactor siting with respect to geology; science in natural resource management; resource management. *Mailing Add:* Dept Geol & Mineral Industs 910 State Off Bldg Portland OR 97201

BEAUMARIAGE, D(ONALD) C(URTIS), b Bridgeville, Pa, Aug 12, 25; m 47; c 1. ELECTRICAL ENGINEERING. *Educ:* Cornell Univ, BEE, 46; Carnegie Inst Tech, MSEE, 48, DSc, 50; Mass Inst Technol, MSIM, 60. *Prof Exp:* Res engr, Carnegie Inst Tech, 48-49, instr elec eng, 49-50; group leader systs anal, air armamant div, Sperry Gyroscope Co, 50-54; from mgr systs proj & electronic syst design to mgr data syst ctr, Radio Corp Am, 54-64; dir info sci, defense systs div, Bunker-Ramo Corp, Calif, 64-68, Eastern Tech Ctr, Md, 68-69; pres, Universal Info Technologies, Inc, 69-71; dir, Man Undersea Sci & Technol Prog, 71-80, CHIEF, OCEAN SYSTS DIV, NAT OCEANIC & ATMOSPHERIC ADMIN, 81- *Concurrent Pos:* Sloan fel, Sch Indust Mgt, Mass Inst Technol, 59-60; mem, Prog Sr Govt Mgrs, Harvard Bus Sch, 81; regist prof engr, Mass. *Mem:* Inst Elec & Electronics Engrs. *Res:* Electronic control systems; information sciences and systems; technical direction; management; ocean engineering. *Mailing Add:* Nat Oceanic Atmospheric Adm 6001 Executive Blvd Bldg One Rockville MD 20852

BEAUMARIAGE, TERRENCE GILBERT, b Washington, Pa, July 10, 61; m 86. COMPUTER SIMULATION, OBJECT ORIENTED DESIGN. *Educ:* Rochester Inst Technol, BS, 84; Okla State Univ, MS, 87, PhD(indust eng), 90. *Prof Exp:* Teaching asst indust eng, Dept Indust Eng & Mech, Okla State Univ, 84-85; mat handler engr, Telex Computer Prod, 87; res asst, Okla State Univ Ctr Computer Int Mfg, 85-89; ASST PROF INDUST ENG, DEPT INDUST & MGT SYSTS ENG, ARIZ STATE UNIV, 89-, ASSOC DIR RES, SYSTS SIMULATION LAB, 90- *Concurrent Pos:* Asst engr, Rochester Prod Div Gen Motors, 81-82, O'Donnell & Assocs, Inc, 82; cost engr, IBM Fed Systs Div, 82-83, mfg eng, 83-84; mfg engr, IBM Entry Systs Div, 85. *Mem:* Inst Indust Engrs; Am Soc Eng Educ; Am Soc Qual Control; Soc Computer Simulation. *Res:* Implementation of a comprehensive manufacturing systems engineering workbench within an object oriented programming environment; workbench will support object oriented simulation modeling; manufacturing system data management; shop floor control and manufacturing data analysis. *Mailing Add:* Dept Indust & Mgt Systs Eng Ariz State Univ Tempe AZ 84287-5906

BEAUMONT, RALPH HARRISON, JR, b Roxbury, NY, Nov 10, 23; m 50; c 6. CHEMISTRY. *Educ:* Colgate Univ, BA, 43; Rutgers Univ, PhD(chem), 53. *Prof Exp:* Asst, Rutgers Univ, 43-44, 46-59 & 51; res chemist spec prob sect, NB Lab, AEC, 49-51 & Allied Chem, 51-52; asst dir res, Huyck Corp, 52-57, dir chem res, 57-63; tech dir, W F Fancourt & Co, 64-66, vpres res & develop, 66-70; pres & dir res, 68-91, SR SCIENTIST, BRIN-MONT CORP, 91- *Mem:* Am Chem Soc; Sigma Xi. *Res:* Analytical chemistry; inorganic chemistry; physical chemistry; kinetics and mechanism of reactions; textile chemistry; general textile auxiliaries; surfactants and rubber chemicals; polymer. *Mailing Add:* Brin-Mont Corp 3921 Spring Garden St Greensboro NC 27407

BEAUMONT, RANDOLPH CAMPBELL, b Los Angeles, Calif, Oct 15, 41; m 66; c 3. INORGANIC CHEMISTRY. *Educ:* Whitman Col, BA, 63; Univ Idaho, PhD(inorg chem), 67. *Prof Exp:* Asst prof chem, Alma Col, Mich, 67-83; AT BEAUMONT CHEM CO, 89- *Concurrent Pos:* Postdoctoral work, McMaster Univ, Ont, 73-74. *Mem:* Am Chem Soc; Sigma Xi. *Res:* Inorganic, physical and analytical chemistry; nitriles of period 2 elements; peroxo and superoxo complexes of transition metals; carbonoanions (synthesis and properties). *Mailing Add:* 3085 Brenda Loop Flagstaff AZ 86001

BEAUMONT, ROSS ALLEN, b Dallas, Tex, July 23, 14; m 40; c 2. MATHEMATICS. *Educ:* Univ Mich, AB, 36, MS, 37; Univ Ill, PhD(math), 40. *Prof Exp:* From instr to assoc prof, 40-54, PROF MATH, UNIV WASH, 54- *Concurrent Pos:* Advan of Educ Fund fel, Inst Advan Study, 54-55; ed, Pac J Math, 56-59 & 73-; vis prof, Univ Ariz, 73. *Mem:* Am Math Soc; Math Asn Am. *Res:* Theory of groups; ring theory; linear algebra. *Mailing Add:* Dept Math Univ Wash Seattle WA 98195

BEAUREGARD, RAYMOND A, b New Bedford, Mass, Feb 10, 43; m 64; c 3. ALGEBRA. *Educ:* Providence Col, AB, 64; Univ NH, MS, 66, PhD(algebra), 68. *Prof Exp:* From asst prof to assoc prof, 68-82, PROF MATH, UNIV RI, 82- *Mem:* Am Math Soc; Math Asn Am. *Res:* Noncommutative rings; noncommutative integral domains, unique factorization; right LCM domains; principal right ideal domains; ring theory; author of linear algebra text book. *Mailing Add:* Dept Math Univ RI Kingston RI 02881

BEAVEN, MICHAEL ANTHONY, b London, Eng, Dec 4, 36; US citizen; m 64. PHARMACOLOGY & IMMUNOPHARMACOLOGY, BIOCHEMISTRY. *Educ:* Chelsea Col Sci & Technol, Univ London, London, BPharm, 59, PhD(med), 62. *Prof Exp:* Demonstr, Chelsea Col Sci & Technol, London, 59-62; vis fel, Lab Chem Pharmacol, Nat Heart Lung & Blood Inst, NIH, 62-66, res pharmacologist, 66-68, sr investr, Exp Therapeut Br, 68-74, sr investr, Pulmonary Br, 74-77, chief, Sect Cellular Pharmacol, Lab Cellular Metab, 77-82, DEP CHIEF, LAB CHEM PHARMACOL, NAT HEART LUNG & BLOOD INST, NIH, 82- *Concurrent Pos:* USPHS vis fel, Nat Heart Inst, 62-63, vis assoc, 64; assoc mem, Darwin Col, Cambridge Univ, Eng, 82-83; fel, Japanese Soc Prom Sci, 89. *Honors & Awards:* Dirs Award, NIH, 89. *Mem:* Am Asn Immunologist; Pharmaceut Soc Gt Brit; Am Soc Pharmacol & Exp Therapeut. *Res:* Mechanisms of mast cell degranulation; mechanisms of cell activation; role of histamine in pathological and physiological reactions; author of over 150 scientific articles. *Mailing Add:* PO Box 26 Garrett Park MD 20896-0026

BEAVEN, VIDA HELMS, b Bluefield, WVa, June 2, 39; m 64. HEALTH SCIENCE ADMINISTRATION. *Educ:* Ind Univ, BA, 61, MA, 62; George Wash Univ, PhD (biochem),68. *Prof Exp:* Staff fel, Nat Heart Inst, NIH, 68-71, grants assoc, Div Res Grants, 71-72, spec asst to assoc dir, Extramural Progs, Nat Inst Arthritis, Metab & Digestive Dis, 72-73; Biomed Health Progs adv, Off Asst Secy Health, Dept Health & Human Serv, 73-76; spec asst to dir, Nat Heart, Lung & Blood Inst, 76-77, actg dir, Fogarty Int Ctr, 80-81, spec asst to dep dir, 77-86, ASST DIR PROG COORD, NIH, 86- *Mailing Add:* NIH Blgd 1 Rm 114 Bethesda MD 20892

BEAVER, BONNIE VERYLE, b Minneapolis, Minn; m. VETERINARY ANIMAL BEHAVIOR. *Educ:* Univ Minn, BS, 66, DVM, 68; Tex A&M Univ, MS, 72. *Prof Exp:* Instr vet surg & radiol, Univ Minn, 68-69; instr vet anat, 69-72, from asst prof to prof vet anat, 72-86, PROF SMALL ANIMAL MED & SURG, TEX A&M UNIV, 86- *Concurrent Pos:* Mem, Vet Med Adv Comt, Dept Health, Educ & Welfare, 72-74; mem, Nat Adv Food & Drug Comt, Dept Health, Educ & Welfare, 75; mem, Comt Animal Models & Genetic Stocks, Nat Acad Sci, 84-; mem, Panel on Microlivestock, Nat Res Coun, 86-87, Adv Comt Pew Nat Vet Educ Prog, Pew Charitable Trusts, 87- *Mem:* Am Vet Med Asn; Am Animal Hosp Asn; Am Soc Vet Animal Behav; Am Asn Equine Practicioners; Am Vet Neurol Asn; Animal Behav Soc; Am Asn Vet Clinicians; Am Vet Comput Soc; AAAS. *Res:* Domestic animal behavior; specific projects which evaluate behavior and examine problem behaviors and how they develop. *Mailing Add:* Dept Small Animal Med & Surg Tex A&M Univ College Station TX 77843-4474

BEAVER, DONALD LOYD, b Hayden, Colo, Sept 16, 43; m 66; c 1. VERTEBRATE ECOLOGY. *Educ:* Colo State Univ, BS, 65, MS, 67; Univ Calif, Berkeley, PhD(zool), 72. *Prof Exp:* asst prof zool, 72-77, ASSOC PROF ZOOL, MICH STATE UNIV, 77- *Mem:* Am Ornithologists Union; Wilson Ornith Soc; Sigma Xi. *Res:* Ecological and behavioral aspects of foraging in birds, including selection of foraging patch, optimal food choice and time and energy spent foraging. *Mailing Add:* Dept Zool Mich State Univ East Lansing MI 48824

BEAVER, EARL RICHARD, b Newburgh, NY, Mar 2, 45; m 66. PHYSICAL CHEMISTRY. *Educ:* McMurry Col, BA, 66; Tex Tech Univ, 66-69, PhD(phys chem), 70. *Prof Exp:* Sr chemist, Monsanto Co, 69-73, process specialist, 73-74, group supvr, 74-77, mgr results mgt, 77-79, mgr process tech, 79-83, DIR BUS DEVELOP & TECHNOL, MONSANTO CO, 83- *Concurrent Pos:* Welsh Found fel. *Mem:* Am Chem Soc; Am Inst Chem Engrs. *Res:* Application of differential thermal analysis, thermogravimetric analysis and adsorption measurements to the solution of process problems in heterogeneous catalysis. *Mailing Add:* Monsanto Co 11444 Lackland Rd St Louis MO 63146

BEAVER, H H, b Steubenville, Ohio, July 24, 25. PETROLEUM. *Educ:* Univ Wis-Madison, PhD(geol), 54. *Prof Exp:* Asst prof, 53-59, CHMN GEOL, BAYLOR UNIV, 76- *Mem:* Fel Geol Soc Am. *Mailing Add:* Dept Geol Baylor Univ Waco TX 76798

BEAVER, PAUL CHESTER, b Glenwood, Ind, Mar 10, 05; m 31; c 1. PARASITOLOGY, TROPICAL MEDICINE. *Educ:* Wabash Col, AB, 28; Univ Ill, MS, 29, PhD(zool), 35; Am Bd Microbiol, dipl. *Hon Degrees:* DSc, Wabash Col, 63. *Prof Exp:* Asst zool, Univ Ill, Urbana, 28-29; instr, Univ Wyo, 29-31; asst, Univ Ill, 31-34; instr biol, Oak Park Jr Col, 34-37; asst prof, Lawrence Col, 37-42; biologist malaria control, Ga Dept Pub Health, 42-45; from asst prof to prof, 45-52, head dept parasitol, 56-71, Wm Vincent prof trop dis & hyg, 58-76, EMER WM VINCENT PROF TROP DIS & HYG, SCH MED & SCH PUB HEALTH & TROP MED, TULANE UNIV, 76- *Concurrent Pos:* Mem, Armed Forces Epidemiol Bd Comn Parasitic Dis, 53-73, dir, 67-73; mem microbiol fels rev panel, NIH, 60-63; ed, Am J Trop Med & Hyg, 60-66 & 72-84; mem, WHO Expert Panel on Parasitic Dis, 63-77; mem, NIH Parasitic Dis Panel, US-Japan Coop Med Sci Prog, 65-69; mem bd sci counr, Nat Inst Allergy & Infectious Dis, 66-68; dir, Int Ctr Med Res, Tulane Univ, 67-76; mem, Gorgas Mem Inst Trop & Prev Med Adv Sci Bd, 70-91. *Honors & Awards:* Walter Reed Medal, Am Soc Trop Med & Hyg, 90. *Mem:* Am Soc Trop Med & Hyg (vpres, 58, pres, 69); Am Soc Parasitol (pres, 68); fel Am Acad Microbiol; Am Micros Soc (vpres, 53); Am Pub Health Asn. *Res:* Occult and zoonotic helminthic infections; amebiasis; epidemiology of soil transmitted helminths. *Mailing Add:* Dept Trop Med Tulane Univ Med Ctr 1501 Canal St New Orleans LA 70112-2699

BEAVER, ROBERT JOHN, b Mt Carmel, Pa, Mar 27, 37; m 70. STATISTICS. *Educ:* Bloomsburg State Col, BS, 59; Bucknell Univ, MS, 64; Univ Fla, MStat, 66, PhD(statist), 70. *Prof Exp:* Statist consult, Teacher Eval Proj, Univ Fla, 66, dir comput-assisted instr proj, 66-67, instr statist, 67-68; med statistician, Med Div, Oak Ridge Assoc Univs, 68; PROF STATIST & STATISTICIAN, UNIV CALIF, RIVERSIDE, 70- *Res:* Model building as applied to problems of choice; paired comparisons. *Mailing Add:* Dept Statist Univ Calif Riverside CA 92521-0138

BEAVER, W DON, b Elkhart, Kans, May 7, 24; m 46; c 2. CHEMISTRY. *Educ:* Bethany Nazarene Col, AB, 46; Okla State Univ, MS, 53, PhD(chem), 55. *Prof Exp:* Teacher pub schs, Okla, 47-48; instr math & chem, Bethany Nazarene Col, 48-51; asst chem, Okla State Univ, 52-55; PROF CHEM, SOUTHERN NAZARENE UNIV, 55-, CHMN DIV NATURAL SCI, 68-, VPRES ACAD AFFAIRS, 82- *Concurrent Pos:* Assoc dean Inst Res, Southern Nazarene Univ, 76-, acad dean, 78- *Mem:* Am Chem Soc; Sigma Xi. *Res:* Use of borohydrides as reducing agents and coordination complexes of the transition metals in aqueous and non-aqueous solvents. *Mailing Add:* Dept Chem Southern Nazarene Univ 4509 N Peniel Bethany OK 73008

BEAVER, W(ILLIAM) L(AWRENCE), b Yucaipa, Calif, Oct 9, 20; m 44; c 2. EXERCISE PHYSIOLOGY, MEDICAL COMPUTER SYSTEMS. *Educ:* Univ Calif, BS, 44, PhD(elec eng), 51. *Prof Exp:* Res engr, Univ Calif, 47-51; sr scientist, Varian Assoc, 51-74; adj prof, Stanford Univ, 74-78; CONSULT, 78- *Mem:* Sigma Xi; Am Physiol Soc; Inst Elec & Electronics Engrs. *Res:* Computer analysis in physiology and medicine; exercise and respiratory physiology; on-line computer systems for gas exchange and cardio-respiratory analysis; ultra-sonic imaging. *Mailing Add:* PO Box 987 St Helena CA 94574

BEAVER, WILLIAM THOMAS, b Albany, NY, Jan 27, 33; m 61; c 3. CLINICAL PHARMACOLOGY, ANALGESIOLOGY. *Educ:* Princeton Univ, AB, 54; Cornell Univ, MD, 58. *Prof Exp:* Intern surg-med, Roosevelt Hosp, New York, 58-59; USPHS fel pharmacol, Med Col, Cornell Univ, 59-61, from instr to asst prof, 61-68; assoc prof, 68-79, PROF PHARMACOL & ANESTHESIOL, SCHS MED & DENT, GEORGETOWN UNIV, 79-, MEM GEN STAFF, DEPT ANESTHESIOL, UNIV HOSP, 69- *Concurrent Pos:* Clin asst, Mem Hosp, NY, 61-68; asst vis physician, Dept Med, James Ewing Hosp, 62-68; res assoc, Sloan-Kettering Inst Cancer Res, 63-68; attend physician, Calvary Hosp, 64-68; consult, Food & Drug Admin, 69-76, Fed Trade Comn, 70-75 & AMA Drug Eval, 71-; mem, Adv Panel Analgesics, Sedatives & Anti-inflammatory Agts, US Pharmacopoeia, 71-75, chmn, 76-80, mem, comt revision, 75-80. *Mem:* AAAS; Am Soc Clin Pharmacol & Therapeut; Int Asn Study Pain; Am Soc Pharmacol & Exp Therapeut. *Res:* Clinical pharmacology of analgesic drugs. *Mailing Add:* Dept Pharmacol Georgetown Univ Sch Med Washington DC 20007

BEAVERS, DOROTHY (ANNE) JOHNSON, b Worcester, Mass, July 11, 27; m 52; c 1. ORGANIC CHEMISTRY. *Educ:* Clark Univ, BA, 49; Duke Univ, PhD(org chem), 55. *Prof Exp:* Res chemist, Am Cyanamid Co, 49-51; asst gen & org chem, Duke Univ, 51-53; sr res chemist, 54-61, RES ASSOC, EASTMAN KODAK CO, 61- *Concurrent Pos:* USPHS grant, 53-54. *Mem:* AAAS; Am Chem Soc; Soc Photog Scientists & Engrs. *Res:* Synthetic organic chemistry; surface active agents; aromatic cyclodehydration; chemical transfer; rapid processing of photographic emulsions; photographic paper products; biochemistry of mental illness and retardation; emulsion addenda; biochemistry and clinical chemistry; sensitization of emulsions; color reversal systems. *Mailing Add:* 70 Rainbow Dr Rochester NY 14622

BEAVERS, ELLINGTON MCHENRY, b Atlanta, Ga; m 57; c 1. CHEMISTRY. *Educ:* Emory Univ, BS, 38, MS, 39; Univ NC, PhD(chem), 41. *Prof Exp:* Res chem, Rohm & Haas Co, 41-47, lab head, 47-55, res supvr, 55-59, assoc dir res, 60-69, dir, Res & Develop, 69-73, vpres, 70-73, mem, Bd Dirs, 71-81, sr vpres, 73-75, group vpres, 75-81; PRES, BEACON RES, INC, 81- *Mem:* Am Chem Soc (treas); Am Inst Chemists; Am Inst Chem Engrs. *Res:* Research and discovery with the objective of improving the biocompatibility of implantable medical devices. *Mailing Add:* 931 Coates Rd Meadowbrook PA 19046

BEAVERS, GORDON STANLEY, b Doncaster, Eng, Oct 16, 36; nat US; m 66; c 1. FLUID MECHANICS, ENGINEERING MECHANICS. *Educ:* Cambridge Univ, BA, 59; Harvard Univ, SM, 60; Cambridge Univ, PhD(mech eng), 63. *Prof Exp:* From asst prof to assoc prof, 63-74, assoc head dept, 72-83, PROF AEROSPACE ENG, UNIV MINN, MINNEAPOLIS, 74-, ASSOC DEAN, 83- *Mem:* Am Inst Aeronaut & Astronaut; Am Soc Mech Engrs; Soc Rheology; Am Soc Eng Educ. *Res:* Flows through porous media; fluid mechanics; gas dynamics; rheological fluid mechanics; mechanical properties of porous media. *Mailing Add:* Dept Aerospace Eng & Mech Univ Minn 110 Union St Minneapolis MN 55455

BEAVERS, LEO EARICE, b Miller's Grove, Tex, May 7, 20; m 58; c 1. ORGANIC CHEMISTRY. *Educ:* Harvard Univ, AB, 50; Duke Univ, PhD, 55. *Prof Exp:* Res chemist, Am Cyanamid Co, 50-51; asst, Duke Univ, 51-52; res chemist, 54-60, head color chem lab, 60-64, head NP systs lab, 64-74, mkt res & sr res assoc, Photomat Div, Eastman Kodak Co, 74-78; RETIRED. *Mem:* AAAS; Am Chem Soc; Soc Photog Scientists & Engrs; Soc Motion Picture & TV Engr. *Res:* Color photography; marketing research; electrophotography. *Mailing Add:* 70 Rainbow Dr Rochester NY 14622

BEAVERS, WILLET I, b Billings, Mont, Nov 13, 33; m 59; c 2. ASTROPHYSICS, ASTRONOMY. *Educ:* Univ Mo, BS, 55, MS, 59; Ind Univ, PhD(astrophys), 66. *Prof Exp:* Instr astron, Univ Mo, 63-65; asst prof, 65-69, ASSOC PROF PHYSICS, IOWA STATE UNIV, 69- *Mem:* Am Astron Soc; Optical Soc Am; Int Astron Union. *Res:* Observational and experimental astrophysics, design and construction of astronomical instruments. *Mailing Add:* 512 Lynn Ave Ames IA 50010

BEAVO, JOSEPH A, CYCLINUCLEOTIDES. *Educ:* Vanderbilt Col, PhD(physiol), 70. *Prof Exp:* PROF PHARMACOL, UNIV WASH, 77- *Mailing Add:* Dept Pharmacol SJ-30 Univ Wash Seattle WA 98195

BEBB, HERBERT BARRINGTON, b Wichita Falls, Tex, June 22, 35; m 58; c 2. SOLID STATE PHYSICS. *Educ:* Univ Okla, BS, 59; Syracuse Univ, MS, 64; Univ Rochester, PhD(optics), 65. *Prof Exp:* Assoc physicist, Fed Systs Div, Int Bus Mach Corp, 59-61, mem res staff, Res Lab, 61-62; res assoc, Inst Optics, Univ Rochester, 65-66; theoret physicist, Tex Instruments Inc, 66-69, mgr appl optics br, 69, dir advan technol lab, 69-74; mgr, Prod Develop Dept, 74-80, MGR ADVAN PROD & TECHNOL, XEROX CORP, 80- *Concurrent Pos:* Consult, Int Bus Mach Corp, 65. *Mem:* Am Phys Soc; Optical Soc Am; Math Asn Am. *Res:* Theoretical study of optical properties of solids; ultrasonic surface waves; liquid crystal phenomena; infrared sensors; semiconductor technology; systems engineering. *Mailing Add:* 1899 Clark Rd Rochester NY 14625

BEBBINGTON, W(ILLIAM) P(EARSON), b Painted Post, NY, Sept 10, 15; m 40; c 3. CHEMICAL ENGINEERING. *Educ:* Cornell Univ, BChem, 36, PhD(chem eng), 40. *Prof Exp:* Chem engr, Ammonia & Polychem Dept, E I du Pont de Nemours & Co, 40-50, chem engr explosives dept, Atomic Energy Div, 50-52, supt heavy water tech sect, Savannah River Plant, 52-55, asst sect dir, Savannah River Lab, 55-59, supt separations tech sect, Savannah River Plant, 59-62, gen supt, Works Tech Dept, 62-74; RETIRED. *Honors & Awards:* Robert E Wilson Award, Am Inst Chem Engrs. *Mem:* Fel Am Inst Chem Engrs. *Res:* Heavy water technology; chemical separations of nuclear fuels; environmental effects of nuclear industry. *Mailing Add:* 4275 Owens Rd Apt No 2307 Evans GA 30809

BEBERNES, JERROLD WILLIAM, b Cotesfield, Nebr, Apr 7, 35; m 56; c 3. APPLIED MATHEMATICS. *Educ:* Univ Nebr, BS, 57, MA, 59, PhD(math), 62. *Prof Exp:* Mathematician, US Naval Ord Lab, Calif, 57; instr math, Univ Nebr, 59-60; from asst prof to assoc prof, 62-71, PROF MATH, UNIV COLO, BOULDER, 71- *Mem:* Am Math Soc; Math Asn Am; Soc Indust & Appl Math. *Res:* Functional analysis, differential equations, reaction diffusion equations and combustion problems. *Mailing Add:* Dept Math Univ Colo Boulder Box 426 Boulder CO 80309-0426

BEBOUT, DON GRAY, b Moneson, Pa, Jan 23, 31; m 52; c 3. SEDIMENTOLOGY. *Educ:* Mt Union Col, BS, 52; Univ Wis, MS, 54; Univ Kans, PhD(geol), 61. *Prof Exp:* Asst micropaleont, Univ Kans, 56-57 & 59-60; sr res specialist geol, Exxon Prod Res Co, Tex, 60-72; res scientist geol, Bur Econ Geol, Univ Tex, Austin, 72-79; prof dept geol & dir, La Geol Surv, La State Univ, 79-81; SR RES SCIENTIST, BUR ECON GEOL, UNIV TEX, 81- *Concurrent Pos:* Lectr, Dept Geol Sci, Univ Tex, Austin, 75-79; distinguished lectr, Am Asn Petrol Geologists. *Mem:* Fel Geol Soc Am; Am Asn Petrol Geologists; hon mem, Soc Econ Paleontologists & Mineralogists. *Res:* Subsurface distribution of sedimentary facies along the Texas gulf coast; evaluating the potential of producing geothermal energy from Texas gulf coast geopressured reservoirs; characterization of carbonate reservoir. *Mailing Add:* Bur Econ Geol Univ Tex Austin TX 78713-7508

BECCHETTI, FREDERICK DANIEL, b Minneapolis, Minn, Mar 3, 43; m; c 2. PHYSICS. *Educ:* Univ Minn, BS, 65, MS, 68, PhD(physics), 69. *Prof Exp:* Vis asst prof physics, Niels Bohr Inst, 69-71; res assoc, Lawrence Berkeley Lab, 71-73; from asst prof to assoc prof, 73-82, PROF PHYSICS, UNIV MICH, ANN ARBOR, 83- *Concurrent Pos:* NSF fel, Niels Bohr Inst, 70-71. *Mem:* Am Phys Soc; Am Asn Physics Teachers. *Res:* Experimental nuclear physics; nuclear radiation detectors. *Mailing Add:* Dept Physics Univ Mich Ann Arbor MI 48109

BECHARA, IBRAHIM, b Bazzak, Syria, Jan 16, 43; US citizen; m 71; c 3. POLYMER CHEMISTRY. *Educ:* Hobart Col, BS, 63; Univ Del, PhD(chem), 67. *Prof Exp:* Res chemist, Air Prod & Chemicals, 67-80, sr prin res chemist, 80-84, mgr, urethane res, Polyvinyl Chemicals, 85-86; MGR POLYMER RES, WITCO CORP, 88- *Mem:* Am Chem Soc. *Res:* Synthesis of organic chemicals via liquid and gas phase methods; homogenous catalysis of organic reactions, mainly isocyanate reactions and structures and properties of polyurethanes; reaction injection molding and thermal and spectroscopic analysis of polymeric materials; water borne polyurethanes; chemistry and technology. *Mailing Add:* 241 Covington Ct Naperville IL 60565

BECHER, PAUL, b Brooklyn, NY, Mar 24, 18; m 45; c 2. PHYSICAL CHEMISTRY. *Educ:* Polytech Inst Brooklyn, BS, 40, MS, 42, PhD(chem), 49. *Prof Exp:* Assoc prof chem, NGa Col, 48-51; sr proj chemist, Colgate-Palmolive Co, 51-57; res assoc, Atlas Chem Industs, Inc, 57-67, prin scientist, ICI Americas Inc, 67-70, mgr, Phys Chem Sect, Chem Res Dept, 70, dir, Specialty Chem Res Dept, 75-79, assoc dir, Corp Res Dept, 80-81; PRES, PAUL BECHER ASSOCS, LTD, 81- *Honors & Awards:* Langmuir Lectr, Div Colloid & Surface Chem, Am Chem Soc, 83. *Mem:* fel Soc Cosmetic Chemists; Sigma Xi; Am Chem Soc; fel Am Inst Chemists. *Res:* Emulsions; surface chemistry. *Mailing Add:* 520 Windley Rd Wilmington DE 19803

BECHER, WILLIAM D(ON), b Bolivar, Ohio, Nov 26, 29; m 50; c 2. COMPUTER ENGINEERING, COMPUTER EDUCATION. *Educ:* Tri-State Col, BS, 50; Univ Mich, MS, 61, PhD(elec eng), 68. *Prof Exp:* Proj engr, Bogue Elec Co, 50-53; sr develop engr, Goodyear Aircraft Corp, 53 & 55-57; sr systs engr, Beckman Instruments, Inc, 57-58; staff engr, Bendix Systs Div, Bendix Corp, 58-60, sect supvr, 60-63; res engr, Willow Run Lab, 63-68; from asst prof to prof elec eng, Univ Mich, Dearborn, 68-78, chmn dept, 71-76; dept mgr, Environ Res Inst Mich, 77-79; dean, Col Eng, NJ Inst Technol, 79-81; assoc dir, Infrared & Optical Div, Environ Res Inst Mich, 81-88; CHAIR & PROF ELEC ENG, CALIF STATE UNIV, 88- *Concurrent Pos:* Lectr elec eng, Univ Mich, Dearborn, 64-66; adj prof, Univ Mich, Ann Arbor, 78-79 & 81-; consult, Widbec Eng, 76- *Mem:* Sr mem Inst Elec & Electronics Engrs; Am Soc Eng Educ; Soc Mech Engrs. *Res:* Computer aided design and instruction; digital computer design; switching theory; image processing; radar systems design; electronic circuits and systems; digital instrumentation systems. *Mailing Add:* 691 Spring Valley Dr Ann Arbor MI 48105

BECHHOFER, ROBERT E(RIC), b New York, NY, Mar 11, 19; m 52; c 4. MATHEMATICAL & ENGINEERING STATISTICS. *Educ:* Columbia Univ, AB, 41, PhD(math statist), 51. *Prof Exp:* Asst chief anal sect, Aberdeen Proving Ground, Md, 41-45; tech engr, Process Anal Sect, Carbide & Carbon Chem Corp, Tenn, 45; asst, Dept Indust Eng, Columbia Univ, 47-48, instr, 48-50, instr indust eng & asst dir statist consult serv, Dept Math Statist, 50-51, asst prof & dir, 51-52; res assoc & dir math statist proj, Air Res & Develop Command, Cornell Univ, 52-64, from assoc prof to prof indust eng & opers res, 53-89, chmn dept opers res, 67-75, EMER PROF INDUST ENG & OPERS RES, CORNELL UNIV, 89- *Concurrent Pos:* Vis assoc prof, Col Agr, Cornell Univ, 52-53, Stanford Univ, 58-59; NSF sci faculty fel, 62-63; vis prof, statist lab, Cambridge Univ, 66-67, Dept Math & Dept Mgt Sci, Imp Col, London, 73-74. *Mem:* Biomet Soc; fel Am Soc Qual Control; fel Inst Math Statist; fel Am Statist Asn; fel AAAS; Sigma Xi; Int Statist Inst; fel Royal Stat Soc; fel Am Soc Qual Control. *Res:* Statistical multiple decision procedures; statistical design, analysis and interpretation of experiments. *Mailing Add:* Opers Res Cornell Univ ETC Bldg Ithaca NY 14853-3801

BECHIS, DENNIS JOHN, b Boston, Mass, Aug 10, 51; m; c 1. CRT DESIGN, COMPUTER SIMULATION. *Educ:* Harvard Col, BA, 73; Univ Md, MS, 76, PhD(physics), 78. *Prof Exp:* Res asst high energy physics, Univ Md, 75-78; res assoc, Rutgers Univ, 78-80; MEM TECH STAFF, DAVID SARNOFF RES CTR, RCA LABS, PRINCETON, 80- *Honors & Awards:* David Sarnoff Award, 83. *Mem:* Am Phys Soc; Soc Info Display. *Res:* Computer experiments and theoretical analyses in electron optics and heat transfer to aid in the understanding and design of cathode ray tube display devices; development of computer models of electron beams and heat transfer. *Mailing Add:* David Sarnoff Res Ctr CN5300 Princeton NJ 08543-5300

BECHIS, KENNETH PAUL, b Boston, Mass, July 22, 49; m 76; c 2. ELECTRO-OPTIC SENSORS, SPACE TECHNOLOGY. *Educ:* Harvard Col, AB, 70; Mass Inst Technol, SM, 73; Univ Mass, Amherst, PhD(astrophys), 76. *Prof Exp:* Res asst, Res Lab Electronics, Mass Inst Technol, 71-74; res physicist, Radiometric Technol, Inc, 74-75; res asst, Dept Physics & Astron, Univ Mass, Amherst, 74-76, res assoc & lectr astron, 76-79; site mgr, Five Col Radio Astron Observ, 76-79; mem tech staff, 79-80, sect mgr, 80-82, DEPT MGR, TASC, 82- *Concurrent Pos:* Consult, Tasc, 78-79; payload specialist, Starlab Space Shuttle Mission, 87-90. *Mem:* Am Astron Soc; Inst Elec & Electronics Engrs; Sigma Xi; Soc Photo-Optical Instrumentation Engrs; Int Union Radio Sci. *Res:* New space technology; energy systems including lasers, particle beams, and microwaves; advanced optics and detectors for phenomenology, surveillance, reconnaissance, and guidance systems. *Mailing Add:* Tasc 55 Walkers Brook Dr Reading MA 01867

BECHT, CHARLES, IV, b Newark, NJ, Oct 5, 54; m 81; c 5. PROCESS EQUIPMENT, ELEVATED TEMPERATURE DESIGN. *Educ:* Union Col, Schenectady, BS, 76; Stanford Univ, MS, 77. *Prof Exp:* Lead engr, Energy Systs Group, Rockwell Int, 77-80; staff engr, Exxon Res & Engr Co, 80-86; PRIN, BECHT ENG CO, 86- *Concurrent Pos:* Mem, Design & Anal Comt Pressure Vessel & Piping & B31.3 Comt Chem Plant & Petrol Refinery Piping, Am Soc Mech Engrs, 88- *Mem:* Am Soc Mech Engrs; Am Soc Testing & Mat. *Res:* Design rules and analysis methods for high temperature equipment, piping and expansion joints; new equipment for synfuels and other severe service processes. *Mailing Add:* Shalebrook Dr Harding NJ 07960-6638

BECHTEL, DONALD BRUCE, b Paterson, NJ, Aug 1, 49; m 71; c 2. ELECTRON MICROSCOPY, NUCLEAR MAGNETIC RESONANCE SPECTROSCOPY. *Educ:* Iowa State Univ, BS, 71, MS, 74; Kans State Univ, PhD(microbiol), 82. *Prof Exp:* Chemist, 74-77, RES CHEMIST, US GRAIN MKT RES LAB, 77- *Concurrent Pos:* Adj asst prof, Kans State Univ, 83-; assoc ed, Am Asn Cereal Chemists, 85- *Mem:* Am Asn Cereal Chemists; Bot Soc Am; Sigma Xi. *Res:* Mechanisms of storage protein synthesis and deposition in wheat; post-translational modifications in developing kernels; improvement of methods for the identification and characterization of specific wheat storage proteins. *Mailing Add:* US Grain Mkt Res Lab 1515 Col Ave Manhattan KS 66502

BECHTEL, JAMES HARVEY, b Bellefontaine, Ohio, Mar 25, 45. PHYSICS. *Educ:* Miami Univ, BS, 67; Univ Mich, Ann Arbor, MS, 68, PhD(physics), 73. *Prof Exp:* Teaching asst physics, Univ Mich, 68-69; physicist, Night Vision Lab, US Army, 69-71; res asst physics, Univ Mich, 71-73; res fel, Harvard Univ, 73-76; RES SCIENTIST PHYSICS, GEN MOTORS RES LABS, 76- *Mem:* AAAS; Am Phys Soc; Optical Soc Am; Am Chem Soc. *Res:* Quantum electronics; nonlinear optics; Raman spectroscopy; damage in laser materials; combustion diagnostics; flame structure; flame quenching; ignition. *Mailing Add:* 2670 Del Mar Heights Rd Suite 207 Del Mar CA 92014

BECHTEL, MARLEY E(LDEN), electrical engineering, applied mathematics, for more information see previous edition

BECHTEL, PETER JOHN, b Minneapolis, Minn, Apr 9, 43; m 63; c 2. AGRICULTURAL BIOCHEMISTRY. *Educ:* Parsons Col, BS, 66; Mich State Univ, PhD(food sci), 71. *Prof Exp:* Chemist prod develop, Res Group, Quaker Oats Co, 66-67; biologist human nutrit, Human Nutrit Div, USDA, 70-71; scholar biochem, Univ Calif, Davis, 71-76; expert molecular biol, Nat Cancer Inst, NIH, 76-77; asst prof animal sci, Biochem & Food Technol, Iowa State Univ, 77-80; assoc prof, 80-86, PROF, DEPT ANIMAL SCI, FOOD SCI & PHYSIOL, UNIV ILL, 86- *Concurrent Pos:* Fel, Muscular Dystrophy Asn Am, 72-75; vis scientist, Sabbatical, Dept Cell Genetics, Nobel Inst & Karolinska Inst, Stockholm Sweden, 88-89. *Mem:* Inst Food Technologists; Am Soc Animal Sci; Am Inst Nutrit. *Res:* Biochemical regulatory mechanisms; muscle growth and development. *Mailing Add:* Univ Ill 1503 S Maryland Urbana IL 61801

BECHTEL, ROBERT D, b Chicago, Ill, Apr 2, 31; m 52; c 3. MATHEMATICS. *Educ:* McPherson Col, BS, 53; Kans State Univ, MS, 59; Purdue Univ, PhD(math educ), 63. *Prof Exp:* Teacher high schs, Kans, 55-58; asst prof math, Kans State Univ, 63-66; assoc prof, 66-71, PROF MATH EDUC, PURDUE UNIV, 71- *Mem:* Math Asn Am; Am Math Soc; Sigma Xi. *Res:* Mathematics education. *Mailing Add:* Div Math Sci Purdue Univ Calumet Campus Hammond IN 46323

BECHTEL, STEPHEN D, JR, b Oakland, Calif, May 10, 25. CIVIL ENGINEERING. *Educ:* Purdue Univ, BS, 46; Stanford Univ, MBA, 48. *Prof Exp:* Pres, 60-73, chmn, 73-90, ENGR, BECHTEL CORP, 41-, EMER CHMN, 90- *Honors & Awards:* Nat Medal of Technol, 91. *Mem:* Nat Acad Eng; fel Am Soc Civil Engrs; Am Inst Metall Eng. *Mailing Add:* PO Box 193965 San Francisco CA 94119

BECHTLE, DANIEL WAYNE, b Myrtle Creek, Ore, May 18, 49; m 74. OPTICS, ELECTRONICS. *Educ:* Univ Ore, BA(physics) & BA(math), 71; Univ Colo, PhD(physics), 77. *Prof Exp:* MEM TECH STAFF OPTICS, DAVID SARNOFF RES CTR, 78- *Concurrent Pos:* Consult, Burleigh Instruments, 75- *Mem:* Inst Elec & Electronic Engrs. *Res:* Optics, electronics and microwaves; Brilloun spectroscopy; fiber-optics communication. *Mailing Add:* David Sarnoff Res Ctr CN5300 RCA Labs Princeton NJ 08543-5300

BECHTLE, GERALD FRANCIS, b Ottawa, Kans, Nov 4, 21. ORGANIC CHEMISTRY. *Educ:* Ottawa Univ (Kans), BS, 43; Univ Kans, MS, 50. *Prof Exp:* Sr chemist, Sherwin-Williams Co, 47-54; assoc chemist, Midwest Res Inst, 55-67; CHEMIST, COOK PAINT & VARNISH CO, 67- *Res:* Resin; paint; oils and fats. *Mailing Add:* Cook Paint & Varnish Co PO Box 389 Kansas City MO 64141

BECHTOL, KATHLEEN B, b Oakland, Calif, July 13, 45. IMMUNOLOGY, DEVELOPMENTAL BIOLOGY. *Educ:* Stanford Univ, PhD(biol), 73. *Prof Exp:* Assoc prof develop biol, Wistar Inst, 83-85; sr scientist, Hyperdoma, Genentech Inc, 85-89; RETIRED. *Mailing Add:* 3012 Monterey St San Mateo CA 94403

BECHTOLD, MAX FREDERICK, b North Manchester, Ind, Jan 3, 15; m 37; c 4. PHYSICAL CHEMISTRY. *Educ:* Manchester Col, BS, 35; Purdue Univ, PhD(phys chem), 39. *Prof Exp:* Chemist, E I Du Pont De Nemours & Co, Inc, 39-67, res supvr, Cent Res Dept, Exp Sta, 67-80; RETIRED. *Mem:* AAAS; Sigma Xi; Am Chem Soc. *Res:* Inorganic polymerization kinetics; colloid chemistry of organic polymers; thin film techniques; solid state reactions in ceramics and metallic high temperature materials; refractory and magnetic powders; batteries, Rankine cycle engines and their fluids; solar energy conversion; scratch-resistant coatings; polysilicic acid chemistry; films and fibers from polymer dispersions. *Mailing Add:* 491 Bayard Rd Kennett Square PA 19348

BECHTOLD, WILLIAM ERIC, b New Orleans, La, Oct 10, 52; m 85. METABOLISM, HUMAN DOSIMETRY. *Educ:* Emory Univ, BS, 75; PhD(phys chem), 81. *Prof Exp:* Postdoctoral fel, 81-83, STAFF SCIENTIST, INHALATION TOXICOL RES INST, 83- *Mem:* Am Chem Soc. *Res:* Metabolism and disposition of inhalted xenobiotics; trace analysis of organic compounds in biological and environmental matricies; determination of human dosimetry to organic toxicants using biological monitoring. *Mailing Add:* PO Box 5890 Albuquerque NM 87185

BECK, AARON TEMKIN, b Providence, RI, July 18, 21; m 50; c 4. PSYCHIATRY. *Educ:* Brown Univ, BA, 42; Yale Univ, MD, 46. *Hon Degrees:* MA, Univ Pa, 70; DMS, Brown Univ, 82. *Prof Exp:* Asst chief dept neuro-psychiat, Valley Forge Army Hosp, Va, 52-54; instr, 54-57, assoc, 57-58, from asst prof to assoc prof, 58-71, PROF PSYCHIAT, UNIV PA, 71-, UNIV PROF, 83- *Concurrent Pos:* Sect chief, Philadelphia Gen Hosp, 58-; consult, Vet Admin Hosp, Philadelphia, 67-; spec consult, Ctr Studies Suicide Prev, NIMH, 69-72, chmn, Task Force Suicide Prev in 70's, 69-70; trustee, Am Acad Psychoanal, 70-74. *Honors & Awards:* Found fund prize res psychiat, Am Psychiat Asn, 79; Paul Hoch Award, Am Psychopath Asn, 83; Distinguished Sci Award, Am Psychol Asn, 89. *Mem:* Psychiat Res Soc; Am Psychopath Asn; Am Psychiat Soc; Asn Advan Behav Ther; Am Psychol Asn. *Res:* Depression; suicide; cognitive aspects of psychopathology; cognitive therapy; anxiety and panic disorders. *Mailing Add:* 133 S 36th St Philadelphia PA 19104

BECK, ALAN EDWARD, Can citizen; m 55; c 3. GEOPHYSICS. *Educ:* Univ London, BSc & BSc(physics), 51; Australian Nat Univ, PhD(geophys), 57. *Prof Exp:* Physicist, instruments develop sect, Brit Oxygen Co, Ltd, 51-52; scientist II, Nat Coal Bd Mining Res Estab, 56-57; Nat Res Coun Can fel, 57-58; from asst prof to assoc prof geophys, 58-65, actg head dept, 61-63, PROF GEOPHYS, UNIV WESTERN ONT, 65-, HEAD DEPT, 63- *Concurrent Pos:* Chmn, Can Nat Comt, Int Union Geod & Geophys, 80-84; mem, Comt on Data for Sci & Technol Task Group on Transp Properties, 78-84; chmn, Earth Sci Grants Selection Comt, Nat Sci & Eng Res Coun, 83-84, coun chmn Can Earth Sci Dept, 85-87; pres, Int Heat Flow Comn, 83-87; mem, exec comt IASPEI, 87- *Mem:* Am Geophys Union; Soc Explor Geophys; Geol Asn Can; Can Geophys Union; Can Soc Explor Geophysicists. *Res:* Terrestrial heat flow; energy balance of the earth; exploration methods. *Mailing Add:* Dept Geophys Univ Western Ont London ON N6A 5B8 Can

BECK, ALBERT J, b Nyack, NY, Aug 13, 35; m 59. ZOOLOGY. *Educ:* Univ Calif, Davis, AB, 57, MA, 61, PhD(zool), 66. *Prof Exp:* Assoc zool, Univ Calif, Davis, 63-64; asst res parasitologist, G W Hooper Found, Univ Calif, ICMRT, Malaysia, 65-69; asst res zoologist, Sch Pub Health, Univ Calif, Berkeley, 69-73; PRIN & SR ANALYST, ECO-ANALYSTS, 73- *Concurrent Pos:* Med zoologist, Inst Med Res, Malaysia, 66-68; lectr geog, Calif State Univ, Chico, 75-; trustee, Butte County Mosquito Abatement Dist, 73-; mem, Planning Comn, City of Chico, 75-79; pres, Calif Mosq Vector Control Asn Bd trustees, 86-87. *Mem:* AAAS; Am Soc Mammal; Wildlife Dis Asn; Soc Vector Ecologists; Asn Environ Prof; Am Planning Asn; Am Water Works Asn; Sigma Xi; Acoust Soc Am. *Res:* Ecology of ectoparasites; epidemiology of zoonotic diseases; biological indicators; host-parasite relationships. *Mailing Add:* PO Box 1187 Chico CA 95927

BECK, ANATOLE, b New York, NY, Mar 19, 30; div; c 2. MATHEMATICS. *Educ:* Brooklyn Col, BA, 51; Yale Univ, MA, 53, PhD(math), 56. *Prof Exp:* Ford instr math, Williams Col, 55-56; res assoc, Tulane Univ, 56-57; traveling fel, Yale Univ, 57-58; from asst prof to assoc prof, 58-66, PROF MATH, UNIV WIS-MADISON, 66- *Concurrent Pos:* Vis scholar, Hebrew Univ, Israel & Univ Gottingen, Germany, 64-65, London Sch Econ, 85; vis prof, Univ Md, 71 & Technische Univ Munich, 73; NSF sr fels, Univ Warwick, Univ London, Univ Erlangen & Hebrew Univ, 68-69; chair math, London Sch Econ, Univ London, 73-75. *Mem:* Sigma Xi; Math Asn Am. *Res:* Probability of Banach spaces; topological dynamics; ergodic theory; measure theory; mathematics for the Social Sciences. *Mailing Add:* Dept Math Univ Wis Madison WI 53706

BECK, BARBARA NORTH, IMMUNOGENETICS, T-CELL ACTIVATION. *Educ:* Univ Wash, Seattle, PhD(genetics), 78. *Prof Exp:* ASST PROF IMMUNOL, MAYO FOUND, 80- *Mailing Add:* Dept Immunol Mayo Clin 200 SW 1st St Rochester MN 55905

BECK, BENNY LEE, b Burlington, Ind, June 13, 32; m 56; c 3. ENVIRONMENTAL SCIENCES. *Educ:* Univ Wis, BS, 53, MS, 56, PhD(anal chem), 57. *Prof Exp:* From res chemist to sr res chemist, Humble Oil & Refining Co, 57-64; supv chemist, Anvil Points Oil Shale Res Ctr, Colo, 64-65; staff chemist, Chem Plant Lab, Exxon Chem Co, 65-77, sr sect supvr, 77-81, environ coordr, 81-84, lab head, Baytown Olefins Plant, 84-88, environ mgr, 88-90, SR STAFF ENVIRON ENGR, EXXON CHEM CO, 90- *Mem:* Am Soc Testing & Mat; Air & Waste Mgt Asn. *Res:* General petrochemical analysis. *Mailing Add:* Exxon Chem Co PO Box 400 Baytown TX 77522-0400

BECK, BETTY ANNE, b Indianapolis, Ind, Feb 12, 22. ENGINEERING. *Educ:* Univ Colo, BS, 47, MS, 63; Univ Denver, MA, 60. *Prof Exp:* Eng aid, Traffic Eng Sect, City & County of Denver, 48-51, eng draftsman, Civil Aeronaut Admin, 51-52; eng draftsman, Gen Res Corp, Calif, 52; jr engr, Douglas Aircraft Co, 53; supv eng draftsman, Naval Ordnance Test Sta, 53-54; res engr, Denver, 54-56; sr instr eng graphics & mach design, 56-63, asst prof eng design & econ eval, 63-78, ASST PROF MECH ENG, UNIV COLO, 78- *Concurrent Pos:* On leave, NSF Sci Fac fel for doctoral study & res, Mat Eng, 69-71. *Honors & Awards:* EDEE Fac Appreciation Award, 77. *Mem:* Am Inst Mining, Metall & Petrol Eng; Am Inst Mining & Met Engrs; Sigma Xi. *Res:* Mechanical metallurgy; human factors engineering; engineering design; history of engineering. *Mailing Add:* 8417 N Foothills Hwy Jamestown Star Rte Boulder CO 80302

BECK, CHARLES BEVERLEY, b Richmond, Va, Mar 26, 27; m 61; c 2. PLANT MORPHOLOGY. *Educ:* Univ Richmond, BA, 60; Cornell Univ, MS, 52, PhD(bot), 55. *Prof Exp:* Instr bot, Cornell Univ, 54-55; Cornell-Glasgow exchange fel, Glasgow Univ, 55-56; from instr to assoc prof, 56-65, chmn dept, 71-75 & 77-79, PROF BOT, UNIV MICH, ANN ARBOR, 65-, CUR MUS PALEONT, 79- *Concurrent Pos:* NSF sr fel, Univ Reading, 64; mem systs biol panel, NSF, 70-73. *Mem:* AAAS; Bot Soc Am; Am Inst Biol Sci; Int Orgn Paleobot. *Res:* Plant anatomy; Paleozoic paleobotany. *Mailing Add:* Dept Natural Sci Biol Univ Mich Main Campus 4042A Ann Arbor MI 48109-1048

BECK, CURT B(UXTON), b Dallas, Tex, Aug 6, 24; m 57; c 3. CHEMICAL ENGINEERING. *Educ:* Mass Inst Technol, BS, 45, MS, 52; Environ Engrs Intersoc, dipl. *Prof Exp:* Admin asst dir res, 45-66, corp energy utilization officer, 70-84, ASSOC DIR RES, CABOT CORP, 66- *Concurrent Pos:* Consult engr, 84-; ed, Environ Newsletter. *Mem:* Am Chem Soc; Am Inst Chem Engrs; Air Pollution Control Asn; Am Inst Chem Engrs; Am Acad Environ Engrs. *Res:* Combustion; heat transfer; fluid flow; air and water ecology. *Mailing Add:* 1940 Fir St Pampa TX 79065

BECK, CURT WERNER, b Halle, Ger, Sept 10, 27; nat US; m 53; c 2. ORGANIC CHEMISTRY, ARCHAEOLOGICAL CHEMISTRY. *Educ:* Tufts Univ, BS, 51; Mass Inst Technol, PhD(org chem), 55. *Prof Exp:* Instr, Franklin Technol Inst, 51-56; asst prof, Robert Col, Istanbul, 56-57; from lectr to assoc prof, 57-66, PROF CHEM, VASSAR COL, 66-, MATTHEW VASSAR JR PROF, 70- *Concurrent Pos:* Ed, Art & Archaeol Tech Abstracts; sect ed, Chem Abstracts. *Honors & Awards:* Res Award, Mid-Hudson Sect, Am Chem Soc. *Mem:* Am Chem Soc; Archaeol Inst Am; Royal Soc Chem; fel Royal Soc Arts; fel Int Inst Conserv Hist & Artistic Works. *Res:* Application of chemistry to archaeology; provenience analysis of amber artifacts. *Mailing Add:* Dept Chem Vassar Col Poughkeepsie NY 12601

BECK, DAVID PAUL, b Wilmington, Del, Aug 3, 44; m 66; c 2. BIOCHEMISTRY. *Educ:* Princeton Univ, AB, 66; Johns Hopkins Univ, PhD(biochem), 71. *Prof Exp:* Helen Hay Whitney fel, 71; fel biochem, Harvard Univ, 71-74; res assoc neurochem, Md Psychiat Res Ctr, Baltimore, 74-77; grants assoc, Nat Inst Gen Med Sci, Bethesda, Md, 77-78, adminr Genetics Prog, 78-81, chief, Molecular & Med Genetics Sect, Genetics Prog, 81-84; ASSOC DIR ADMIN, PUB HEALTH RES INST, NY, 84- *Mem:* Am Soc Cell Biol; AAAS. *Res:* Membrane structure and function; role of membranes in regulating cellular metabolism; substrate translocation processes; oxidative phosphorylation. *Mailing Add:* Pub Health Res Inst 455 First Ave Rm 1203 New York NY 10022

BECK, DONALD EDWARD, b Logan, Iowa, June 12, 34. SURFACE PHYSICS. *Educ:* Univ Calif, Berkeley, BS, 56, MA, 58, PhD(physics), 65. *Prof Exp:* Vis lectr physics, St Andrews Univ, 66-67; asst prof, Univ Va, 67-72; asst prof, 72-75, ASSOC PROF PHYSICS, UNIV WIS-MILWAUKEE, 75- *Mem:* AAAS; Am Phys Soc; Sigma Xi. *Res:* Theoretical studies of the electronic properties of metal and semiconductor surfaces. *Mailing Add:* Dept Physics Univ Wis Milwaukee WI 53201

BECK, DONALD RICHARDSON, b Paterson, NJ, Mar 31, 40; m 68; c 1. SOLID STATE THEORY, ATOMIC PHYSICS. *Educ:* Dickinson Col, BS, 62; Lehigh Univ, MS, 64, PhD(physics), 68. *Prof Exp:* Jr vis scientist physics, Joint Inst Lab Astrophys, Univ Colo, 68-69; res assoc chem, Yale Univ, 69-73, res assoc eng & appl sci, 73-74; asst prof chem, Belfer Grad Sch Sci, Yeshiva Univ, 74-76; PROF PHYSICS, MICH TECHNOL UNIV, 80- *Concurrent Pos:* Vis prof, Nat Hellenic Res Found, Athens, Greece, 76-78; vis res assoc prof, physics dept, Univ Ill, Urbana, 78-80. *Mem:* Am Phys Soc; Sigma Xi. *Res:* Quantum theory of atoms, molecules and solids, including relativistic and correlation effects, with application to transition metal atoms, ionic and energetic solids and alkane potential energy surfaces; properties; term, fine and hyperfine structure; transition probabilities; binding and Auger energies; autoionizing states; electron affinities. *Mailing Add:* Physics Dept Mich Technol Univ Houghton MI 49931

BECK, DORIS JEAN, b Blissfield, Mich. MICROBIAL GENETICS. *Educ:* Bowling Green State Univ, BS, 60; Mich State Univ, MS, 71, PhD(microbiol), 74. *Prof Exp:* Teacher sec educ chem, DeWitt Pub Schs, 61-63 & Holt Pub Schs, 63-68; asst microbiol, Mich State Univ, 68-71; asst prof microbiol, Bowling Green State Univ, 74-81. *Concurrent Pos:* Adj prof, Med Col Ohio, Toledo, 80-; res assoc, Yale Univ Sch Med, 81-82. *Mem:* AAAS; Am Soc Microbiol; Sigma Xi. *Res:* Studies of platinum antitumor agents which cause mutations and induce filament formation in cells of Escherichia coli; mutagenesis and DNA repair. *Mailing Add:* 1301 Bourgogne Bowling Green OH 43402

BECK, EDWARD C, b Spanish Fork, Utah, Feb 20, 18; m 44; c 4. PHYSIOLOGICAL PSYCHOLOGY. *Educ:* Brigham Young Univ, BA, 50; Univ Utah, PhD(psychol, neurophysiol), 54. *Prof Exp:* Res assoc physiol, Col Med, Univ Utah, 52-54; clin psychologist, Vet Admin Hosp, Salt Lake City, Utah, 54-56, res psychologist & dir, Neurophysiol & Psychophys Lab, 56-; AT DEPT PSYCHIAT & PSYCHOL, UNIV UTAH. *Concurrent Pos:* Res instr psychiat, Col Med, Univ Utah, 55-58, asst res prof, 58-, neurol, 59-64, assoc res prof, 64-, asst res prof pharmacol, 60-, lectr psychol, 62-67, res prof psychol & chmn div med psychol, 67-; USPHS sr scientist, Neurobiol Res Unit, Nat Inst Hyg, France, 60-61. *Mem:* Am Psychol Asn; Interam Soc Psychol; Am Physiol Soc; Asn Res Nerv & Ment Dis; NY Acad Sci. *Res:* Electrophysiological correlates of behavior. *Mailing Add:* 3156 E 4430th St Salt Lake City UT 84117

BECK, FRANKLIN H(ORNE), b Bethlehem, Pa, Apr 5, 20; m 44; c 3. METALLURGICAL ENGINEERING. *Educ:* Pa State Univ, BS, 43; Ohio State Univ, MS, 47, PhD(metall eng), 49. *Prof Exp:* Metall engr, E I du Pont de Nemours & Co, 43-46; res assoc metall eng, 46-49, res asst prof, 49-54, supvr, 54-59, from asst prof to assoc prof, 52-59, PROF METALL ENG, OHIO STATE UNIV, 59-, DIR METALS RES ENG EXP STA, 58- *Concurrent Pos:* Co-ed, Corrosion Jour. *Mem:* Am Soc Metals; Nat Asn Corrosion Eng. *Res:* Determination of corrosion mechanisms; effects of metallurgy of materials on corrosion. *Mailing Add:* 116 W 19th Ave Columbus OH 43210

BECK, GAIL EDWIN, b Dunn Co, Wis, July 25, 23; m 50; c 3. FLORICULTURE, PLANT PHYSIOLOGY. *Educ:* Mich State Univ, BS, 48, MS, 49; Univ Wis, PhD(floricult, plant physiol), 56. *Prof Exp:* Instr & exten specialist com floricult, Univ Wis-Madison, 49-52, from instr to prof, 52-85, EMER PROF FLORICULT, UNIV WIS- MADISON, 85- *Honors & Awards:* Alex Laurie Award for Floricult Educ & Res, Soc Am Florists, 83. *Mem:* Am Soc Hort Sci; Am Hort Soc. *Res:* Physiology of floriculture crops; florist crop production. *Mailing Add:* Dept Hort Univ Wis Madison WI 53706-1570

BECK, HAROLD LAWRENCE, b Detroit, Mich, Jan 7, 38. RADIATION PHYSICS, ENVIRONMENTAL HEALTH & SAFETY. *Educ:* Univ Miami, BS, 60. *Prof Exp:* Tech intern, AEC, NY, 62-63; physicist, Environ Measurements Lab, 63-88, DIR INSTRUMENTATION, ENVIRON MEASUREMENTS LAB, US DEPT ENERGY, 88- *Concurrent Pos:* Tech rep, Div Biomed & Environ Res, Washington, DC, US Energy Res & Develop Admin, 77-78. *Mem:* AAAS; Am Nuclear Soc; Health Physics Soc; Inst Elec & Electronic Engrs. *Res:* Environmental pollutant instrumentation; environmental radioactivity; radiation measurement techniques; radiation transport; naturally occurring radioactivity. *Mailing Add:* Environ Measurements Lab 376 Hudson St New York NY 10014

BECK, HENRY NELSON, b Troy, Ohio, July 14, 27; m 54; c 2. POLYMER CHEMISTRY. *Educ:* Univ Mich, BS, 49, MS, 50, PhD(phys org chem), 57. *Prof Exp:* Chemist, Dow Corning Corp, Mich, 56-60, proj leader chem, 60-62, group leader, 62-63; res chemist, 63-64, sr res chemist, 64-72, res specialist, Western Div, 72-82, RES ASSOC, CENT RES-WALNUT CREEK, DOW CHEM CO, 83- *Mem:* AAAS; Am Chem Soc; Sigma Xi. *Res:* Diazo and organosilicon compounds; silicones; polymer crystallization; nucleation; inorganic polymers; polymers; inorganic crystals; phase equilibria; polymer solubility fibers. *Mailing Add:* 2800 Mitchell Dr Walnut Creek CA 94598

BECK, HENRY V, b Colby, Kans, July 5, 20; m 42; c 2. QUATERNARY GEOLOGY, HYDROGEOLOGY. *Educ:* Kans State Univ, BS, 46, MS, 49; Univ Kans, PhD(geol), 55. *Prof Exp:* Instr geol, Kans State Univ, 47-50; asst instr, Univ Kans, 50-52; from asst prof to assoc prof, 52-60, PROF GEOL, KANS STATE UNIV, 60- *Mem:* Am Asn Petrol Geol; Geol Soc Am; Sigma Xi; Nat Water Well Asn. *Res:* Geomorphology. *Mailing Add:* 1863 Elaine Dr Manhattan KS 66502

BECK, IVAN THOMAS, b Budapest, Hungary, May 22, 24; nat Can; m 49; c 1. INTERNAL MEDICINE, GASTROENTEROLOGY. *Educ:* Univ Geneva, MD, 49; McGill Univ, dipl, 55, PhD(invest med), 63; FRCP(C). *Prof Exp:* Lectr pharmacol, McGill Univ, 49-52; actg head pharmacol, Univ Montreal, 54-56; res assoc, Royal Victoria Hosp, McGill Univ, 58-66, lectr, Dept Invest Med, 58-66, dir gastrointestinal lab, 60-66; assoc prof, 66-73, PROF MED, QUEENS UNIV, ONT, 73-, PROF PHYSIOL, 77- *Concurrent Pos:* Assoc physician & gastroenterologist in chg, St Mary's Hosp, 60-66; chmn, Med Adv Bd, Can Found Ileitis & Colitis. *Honors & Awards:* Gastroenterol Unit, Hotel Dieu Hosp, Queen's Univ named in Honor, 89. *Mem:* Fel Am Col Physicians; fel Am Col Gastroenterol (vpres, 79-80); Am Soc Pharmacol; Pharmacol Soc Can; Can Physiol Soc; Can Asn Gastroenterol (secy, 60, pres, 67). *Res:* Pharmacology; physiopathology and therapy of the gastrointestinal tract. *Mailing Add:* Div Gastroenterol Hotel Dieu Hosp 166 Brock St Kingston ON K7L 5G2 Can

BECK, JAMES DONALD, b York Co, Pa, Dec 24, 40; m 62; c 2. INORGANIC CHEMISTRY. *Educ:* Kutztown State Col, BS, 62; Univ Del, PhD(chem), 69. *Prof Exp:* Teacher chem, Piscataway Twp High Sch, NJ, 62-64; asst, Univ Del, 64-68; from asst prof to assoc prof, 68-78, head dept, 71-72, PROF CHEM, VA STATE UNIV, 78- *Mem:* Sigma Xi; Am Chem Soc; Nat Sci Teachers Asn; Am Asn Univ Prof. *Res:* Calorimetry of Group IIIA halide complexes; antioxidants in fats and oils; computers in chemical education. *Mailing Add:* Dept Chem Va State Univ Petersburg VA 23803

BECK, JAMES RICHARD, b Paris, Ill, Aug 17, 31. ORGANIC CHEMISTRY. *Educ:* Univ Calif, Berkeley, BS, 57; Univ Ill, PhD(org chem), 61. *Prof Exp:* NSF res fel, Swiss Fed Inst, Zurich, 61-62; res chemist, Eli Lilly & Co, 62-88; RETIRED. *Mem:* Am Chem Soc; Swiss Chem Soc; Sigma Xi; Int Soc Heterocyclic Chem. *Res:* Synthetic organic chemistry; herbicides; plant growth regulators; chemical hybridizing agents; heterocyclic chemistry; nonaqueous diazotization; aromatic nitro displacement. *Mailing Add:* 10023 Penrith Dr Indianapolis IN 46229

BECK, JAMES S, b Dallas, Tex, Dec 8, 31; wid; c 3. CELL BIOLOGY, SIMULATION. *Educ:* Wash Univ, MD, 57; Univ Calif, Berkeley, PhD(biophys), 62. *Prof Exp:* Intern, Mt Zion Hosp & Med Ctr, San Francisco, 57-58; Nat Cancer Inst fel, Univ Calif, Berkeley, 58-62, assoc res biophysicist, Lawrence Radiation Lab, 62; asst prof physiol & lectr physics, Univ Minn, Minneapolis, 62-69; assoc prof, 69-76, PROF MED BIOPHYS, UNIV CALGARY, 76- *Concurrent Pos:* Dir, Bragg Creek Inst Nat Philos, 74- *Mem:* Soc Math Biol; Biophys Soc; AAAS. *Res:* Mechanisms of cellular responses to liquid binding by receptors; physical properties of erythrocytes. *Mailing Add:* Div Med Biophys Univ Calgary Fac Med Calgary AB T2N 4N1 Can

BECK, JAMES V(ERE), b Cambridge, Mass, May 18, 30; m 60; c 2. MECHANICAL ENGINEERING. *Educ:* Tufts Univ, BS, 56; Mass Inst Technol, SM, 57; Mich State Univ, PhD(mech eng), 64. *Prof Exp:* Jr design engr, Pratt & Whitney Aircraft, Mass, 56-57; sr scientist, Res & Adv Develop Div, Avco Corp, 57-60, sr staff scientist, 60-62; from instr to assoc prof, 62-71, PROF MECH ENG, MICH STATE UNIV, 71- *Concurrent Pos:* Consult, Res & Adv Develop Div, Avco Corp, Mass, 62-64, Instrument Div, Lear Siegler, Inc, 64-68, Sandia Nat Labs, 77-79 & Oak Ridge Nat Labs, 66, 82-90. *Mem:* Am Soc Mech Eng; fel Am Soc Metall & Mining Engrs. *Res:* Numerical methods in transient heat conduction; transient determination of thermal properties and parameters; nonlinear estimation of parameters; inverse heat conduction. *Mailing Add:* Dept Mech Eng Mich State Univ East Lansing MI 48824

BECK, JAY VERN, b American Fork, Utah, Jan 15, 12; m 31; c 6. BIOCHEMISTRY, MICROBIOLOGY. *Educ:* Brigham Young Univ, AB, 33, AM, 36; Univ Calif, PhD(soil microbiol), 40. *Prof Exp:* Instr chem & math, High Sch, Utah, 35-36 & Dixie Jr Col, 36; technician plant nutrit, Univ Calif, 36-39; from jr chemist to assoc chemist, Food & Drug Admin, Fed Security Agency, 39-44; asst prof chem, Univ Idaho, 44-46; microbiologist, Pa Grade Crude Oil Asn, 46-47; from asst prof to assoc prof bact, Pa State Col, 47-51; prof, 51-77, EMER PROF BACT, BRIGHAM YOUNG UNIV, 77- *Concurrent Pos:* Spec instr, Univ San Francisco, 38; Guggenheim Found fel, Univ Sheffield, 57-58; spec fel, NIH, 65-; mem, Nat Sci Bd, 82-86. *Mem:* AAAS; Am Chem Soc; Am Soc Microbiol. *Res:* Uric acid as an index of filth in foods; minor element deficiency in plants; Utah sorgo syrup; purine fermentation by bacteria; purine determinations; microbial oxidation of sulfide minerals. *Mailing Add:* 4914 W Mountain View Highland UT 84003-9557

BECK, JEANNE CRAWFORD, b Mt Pleasant, Pa, Mar 18, 43; m 66; c 2. BIOCHEMISTRY. *Educ:* Wilson Col, AB, 65; Johns Hopkins Univ, PhD(biochem), 69. *Prof Exp:* USPHS-Nat Cancer Inst fel, Sch Med, Johns Hopkins Univ, 69-71 & Biol Labs, Harvard Univ, 71-74; sr staff fel, Geront Res Ctr, Nat Inst Aging, 74-80; mem fac, Dept Obstet & Gynec, Sch Med, Johns Hopkins Univ, 80-84; MEM FAC, DEPT MED, MT SINAI SCH MED, NY, 85- *Mem:* Am Chem Soc; Genetics Soc Am; Am Soc Cell Biol; Am Soc Nephrol. *Res:* Role of membrane components in membrane structure and function; mechanism of transport of inorganic ions, amino acids, and sugars in kidney brush border; alteration of membrane function in aging tissues. *Mailing Add:* 43 Montrose Rd Scarsdale NY 10583

BECK, JOHN CHRISTIAN, b Audubon, Iowa, Jan 4, 24; Can citizen; c 1. INTERNAL MEDICINE, GERIATRICS. *Educ:* McGill Univ, BSc, 44, MD, CM, 47, MSc, 51, dipl, 52; FRCP(L); FRCP(C), 64. *Hon Degrees:* PhD, Ben-Gurion Univ Negev. *Prof Exp:* Res asst, Royal Victoria Hosp, 49-53; res assoc & lectr med & clin med, 54-55, res fel Univ Clin, 55-57, from asst prof to assoc prof med, 55-64, chmn dept med & dir Univ Clin, 64-74; prof med, Univ Calif, San Francisco, 74-79, dir, Univ Calif, Los Angeles & Univ Southern Calif, Long Term Care Geront Ctr, 80-85; PROF MED, 79-, DIR, ACAD GERIAT RESOURCES CTR, UNIV CALIF, LOS ANGELES, 84- *Concurrent Pos:* Res asst, Cleveland Clin, Ohio, NEng Med Ctr, Boston, Righosp, Copenhagen & Univ Col Hosp, Eng, 49-53; clin asst, Royal Victoria Hosp, 52-54, chief, Endocrine-Metab Unit 54-64, dir, DIv Endocrine-Metab, 54-64, asst physician, 55-57, assoc physician, 57-60, physician-in-chief, 64-74, sr physician, dept med, 74-81; consult, various hosps, 64-88; Markel Scholar, 54-59; dir, Que Camp for Diabetic Children; dir & chmn, Am Bd Int Med; pres, Am Bd Med Spec, Inst Med; dir, Robert Wood Johnson Clin Scholar Orog & Sr consult, Robert Wood Johnson Found,73-78; attend staff, Univ Calif Med Ctr, 79-, Jewish Homes Aging, Los Angeles, 82-; dir, Acad Geriat Resource Ctr, Univ Calif, Los Angeles, 84-, Calif Geriat Educ Ctr, 87-, Long-Term Care, Nat Resource Ctr, Univ Calif Los Angeles USC, 88-; consult, Rand Corp, Koret Found, Delta Airlines, Bur Health Prof; mem, sub-comt aging, Am Col Physicians, 87-, Public Policy Comt, Asn Geront in Higher Educ, 90-91. *Honors & Awards:* Simeone Lectr, Brown Univ, 77; Ronald V Christie Award, Can Asn Prof Med; Milo D Leavitt Mem Award, Am Geriatrics Soc, 88; Duncan Graham Award, Royal Col Physicians & Surgeons, 90; Joseph T Freeman Award, Geront Soc Am 90; Irving S Wright Award, Am Fedn Aging Res, 90. *Mem:* Inst Med-Nat Acad Sci; Can Soc Clin Invest (pres); Am Fedn Clin Res; fel Am Col Physicians; Int Soc Endocrinol (secy-gen); fel Royal Soc Can; hon fel Royal Soc Physicians; Endocrine Soc; Am Diabetes Asn; AAAS. *Res:* Cinical and health services research in geriatrics. *Mailing Add:* Multicampus Div Geriat Med & Geronf Univ Calif 10833 Le Conte Ave (CHS) Los Angeles CA 90024

BECK, JOHN LOUIS, b Newark, NJ, Aug 29, 31; m 73; c 3. TECHNICAL MANAGEMENT, ANALYTICAL CHEMISTRY. *Educ:* Seton Hall Univ, BS, 53, MSc, 67, PhD(anal chem), 69. *Prof Exp:* From chemist to sr res chemist, Merck & Co, Inc, 55-70, res fel & sect leader, 70-72; dir qual control, Am Hoechst Corp, Somerville, NJ, 72-74; vpres qual control, J B Williams Co, Inc, 74-80, vpres, Tech Serv, 80-82; DIR, QUAL AFFAIRS, BERLEX LABS, INC, WAYNE, NJ, 83- *Mem:* Am Pharmaceut Asn; AAAS; Am Soc Qual Control. *Res:* Mass spectroscopy, metabolism and structure determination with particular emphasis on the use of spectroscopy; photochemical applications to analytical chemistry. *Mailing Add:* Berlex Labs Inc 300 Fairfield Rd Wayne NJ 07470

BECK, JOHN R, b Las Vegas, NMex, Feb 26, 29; m 51; c 4. VERTEBRATE ECOLOGY, ECONOMIC BIOLOGY. *Educ:* Okla A&M Col, BS, 50; Okla State Univ, MS, 57. *Prof Exp:* Wildlife asst, King Ranch, Tex, 50-51; asst instr zool, Okla State Univ, 51-53; control agt, US Fish & Wildlife Serv, 53-54, res biologist, 55-57, control biologist, 57-65; instr physiol, Univ Tenn, 54-55; dir,

Treasure Lake Job Corps Conserv Ctr, US Dept Interior, 65-67, state supvr, Div Wildlife Serv, Bur Sport Fisheries & Wildlife, 67-69; dir qual control, Bio-Serv Corp, 69-78, vpres, 70-78; PRIN CONSULT, BIOL ENVIRON CONSULT SERV, PHOENIX, 78-; spec asst to dep admn, Animal Plant Health Inspection Serv, USDA,86-87. *Concurrent Pos:* Lectr, Ohio State Univ, 59-65; grain sanit consult, 59-; guest lectr, Bowling Green State Univ, 60-79; adv bd, Sch Allied Health, Ferris State Col, 73-79; fac, Agr Div, Ariz State Univ, 80-90; chmn pesticides, Am Soc testing & Mat, 80-82; spec asst dept adminr, Animal Plant Health Inspection Serv, USDA, 86-87. *Honors & Awards:* Fel, Nat Explorers Club. *Mem:* Wildlife Dis Asn; Wildlife Soc; fel Royal Soc Health; Am Soc Testing & Mat; Nat Pest Control Asn (dir, 73-75); Sci Writers Soc; Sigma Xi. *Res:* Vertebrate pest control; development of avicides and control methods; zoonoses control; grain sanitation and quality control; desert crop research. *Mailing Add:* 3631 W Pasadena Phoenix AZ 85019

BECK, JOHN ROBERT, b Cleveland, Ohio, Sept 8, 53; m 75; c 3. MEDICAL DECISION MAKING, BLOOD BANKING. *Educ:* Dartmouth Col, AB, 74; Johns Hopkins Univ, MD, 78. *Prof Exp:* Resident path, Dartmouth-Hitchcock Med Ctr, 78-80; clin fel med, Div Clin Decision Making, New Eng Med Ctr, 81; instr path, Sch Med, Tufts Univ, 81; asst prof path, Dartmouth Med Sch, 82-87, dir, prog med info sci, 84-89, asst prof community & family med, 85-87, assoc prof path & community med, 87-89; PROF PATH MED & PUB HEALTH, ORE HEALTH SCI UNIV, 89- *Concurrent Pos:* Consult, Nat Comt Clin Lab Standards, 83-, Nat Libr Med, 85-86, WHO, 85-87; med dir, Blood Bank, Mary Hitchcock Mem Hosp, 84-89; ed-in-chief, Med Decision Making, 89-; res career develop award, Nat Libr Med, 86-91, biomed libr rev comt, 88-92; mem exec coun, Acad Clin Lab Physicians & Scientists, 88-91. *Honors & Awards:* Young Investr Awards, Acad Clin Lab Physicians & Scientists, 79, 80 & 81. *Mem:* Am Med Informatics Asn; Soc Med Decision Making (secy-treas, 85-87, vpres, 87-88); Am Fedn Clin Res; Col Am Pathologists; Acad Clin Lab Physicians & Scientists; Am Asn Blood Banks. *Res:* Theory of medical decision making; mathematical analysis of diagnostic tests; application of stochastic models to decision analysis; artificial intelligence and general microcomputer software in medical education. *Mailing Add:* BICC-OHSU 3181 SW Sam Jackson Park Rd Portland OR 97201-3098

BECK, JONATHON MOCK, mathematics, for more information see previous edition

BECK, KARL MAURICE, b Belleville, Ill, June 16, 22; m 45; c 3. ORGANIC CHEMISTRY. *Educ:* Monmouth Col, BS, 43; Univ Ill, PhD(org chem), 48. *Prof Exp:* Asst chem, Univ Ill, 43, mem, Nat Defense Res Comt, Off Sci Res & Develop, 44-45, mem, Comt Med Res, 45-46; res chemist, Abbott Labs, 48-53, tech serv rep, 53-56, head tech serv, Chem Mkt Div, 57-62, sci employ, 62-64; vpres, Cyclamate Corp Am, 64-66; mgr prod planning & serv, Abbott Labs, 67-70, mgr chem div, new prod res, 70-71, mgr mkt res & develop, chem & agr prod, 71-84; CONSULT, 84- *Concurrent Pos:* Tech ed, Food Prod Develop, 67-79. *Mem:* Am Chem Soc; Inst Food Technologists; Com Develop Asn. *Res:* Synthetic organic chemistry; chemical warfare with poison gases; reduction of nitro carboxylic esters; piperazines; diuretics; phenolic Mannich bases; sweetening agents; food and beverage technology; food chemicals; plastics additives; chemical antimicrobial agents; paint technology. *Mailing Add:* 224 E Sheridan Rd Lake Bluff IL 60044-2635

BECK, KEITH RUSSELL, b Hudson, Mich, Apr 25, 44; m 68; c 3. TEXTILE CHEMISTRY. *Educ:* Adrian Col, BS, 65; Purdue Univ, PhD(chem), 70. *Prof Exp:* Vis asst prof & res assoc, Purdue Univ, 69-70; asst prof chem, Elmhurst Col, 70-77; assoc prof textile sci, Purdue Univ, 77-86; PROF TEXTILE CHEM, NC STATE UNIV, 86- *Mem:* Am Asn Textile Chemists & Colorists; Am Chem Soc. *Res:* Synthesis and high-performance liquid chromatographic analysis of textile finishes; near infrared analysis of textile substrates; non-formaldehyde crosslinkers for cellulose; real-time data acquisition in batch dyeing. *Mailing Add:* 5909 Countryview Lane Raleigh NC 27606-9464

BECK, KENNETH CHARLES, RESPIRATORY PHYSIOLOGY, PULMONARY VASCULAR MECHANICS. *Educ:* Univ Wash, Seattle, PhD(physiol & biophys), 80. *Prof Exp:* ASST PROF PHYSIOL, MAYO CLIN & FOUND, 80- *Res:* Pulmonary function. *Mailing Add:* 821 10 1-2 St SW Rochester MN 55902

BECK, LLOYD, pharmacology, for more information see previous edition

BECK, LLOYD WILLARD, b Batesville, Ind, Aug 10, 19; m 45; c 3. CHEMISTRY. *Educ:* DePauw Univ, AB, 41; Univ Wis, PhD(org chem), 44. *Prof Exp:* Res chemist, Procter & Gamble Co, 44-55, assoc dir, Res Div, 55-71, assoc dir, Prof & Regulatory Serv Div, 71-81; exec vpres, Hilltop Res, Inc, 81-84; RETIRED. *Mem:* AAAS; Am Chem Soc; Soc Invest Dermat; Am Inst Nutrit; Soc Toxicol. *Res:* Synthesis of compounds related to female sex hormones; synthetic lubricants; food chemistry; nutrition; metabolism; hazard assessments; toxicology; clinical research. *Mailing Add:* 401 Wells Lane Versailles KY 40383

BECK, LUCILLE BLUSO, b Baltimore, Md, July 21, 49; m 73. AUDIOLOGY. *Educ:* Adelphi Univ, BA, 71; Univ Md, MA, 73, PhD(hearing & speech sci), 78. *Prof Exp:* supvry audiologist, Cent Audiol & Speech Path & Nat Hearing Aid Prog, 77-81, ASSOC CHIEF, AUDIOL & SPEECH PATH SERV, VET ADMIN MED CTR, 81-; INSTR AUDIOL, BIOCOMMUN LAB, DEPT HEARING & SPEECH SCI, UNIV MD, 73- *Concurrent Pos:* Prof lectr, George Wash Univ; mem, Ear, Nose & Throat Devices Panel, FDA & Working Group on Hearing Aids, Am Nat Standards Inst S3 48. *Mem:* Sigma Xi; Acoust Soc Am; Am Speech & Hearing Asn; Am Auditory Soc. *Res:* Auditory habilitation; auditory prosthetic devices; hearing aids. *Mailing Add:* Vet Admin Med Ctr 1509 Trinida Ave NE Washington DC 20002

BECK, MAE LUCILLE, b Buffalo, NY, Mar 27, 30. ORGANIC CHEMISTRY. *Educ:* Mich State Univ, BS, 51; Smith Col, AM, 55; Univ Pa, PhD(org chem), 60. *Prof Exp:* Polymer chemist, E I du Pont de Nemours & Co, 51-53; instr chem, Smith Col, 55-56; from asst prof to assoc prof, 60-83, EMER PROF CHEM, SIMMONS COL, 84- *Concurrent Pos:* NIH grant, 62-64; res fel, Harvard Univ, 66-67; artist, Boston Sch. *Mem:* Am Chem Soc; Sigma Xi. *Res:* Organic sulfur chemistry; antimetabolites; organic natural products and drugs; environmental health. *Mailing Add:* RFD Marlborough NH 03455

BECK, MYRL EMIL, JR, b Redlands, Calif, May 13, 33; m 83; c 3. PALEOMAGNETISM, TECTONICS. *Educ:* Stanford Univ, BA, 55, MS, 60; Univ Calif, Riverside, PhD(geol), 68. *Prof Exp:* Asst paleomagnetism, Dept Geophys, Stanford Univ, 61-62; geologist, Standard Oil Co Calif, 62; geologist, US Geol Surv, 62-66; assoc prof geol, 69-73, PROF GEOL, WESTERN WASH UNIV, 73- *Concurrent Pos:* Res assoc, Univ Puget Sound, 72-; vis prof, Geophys Inst, Swiss Fed Inst Technol, Zurich, 73, Northwestern Univ & Univ Ariz, 84-85. *Mem:* Fel Am Geophys Union; Geol Soc Am; Geol Soc Chile. *Res:* Application of paleomagnetism and other techniques to study of tectonic processes. *Mailing Add:* Dept Geol Western Wash Univ Bellingham WA 98225

BECK, PAUL ADAMS, b Budapest, Hungary, Feb 5, 08; nat US; m 51; c 2. PHYSICAL METALLURGY, MAGNETISM IN ALLOYS. *Educ:* Univ Budapest, ME, 31; Mich Technol Univ, MS, 29. *Hon Degrees:* Dr Min, Tech Univ Leoben. *Prof Exp:* Res engr, Vatea Elec Co, Hungary, 31-32 & Inst Phys Chem, Paris, 33-35; patent engr, Bernauer & Tavy Patent Attorneys, Budapest, 35-36; res metallurgist, Am Smelting & Ref Co, NJ, 37-41; chief metallurgist, Beryllium Corp, Pa, 41-42; head metall lab, Cleveland Graphite Bronze Co, 42- 45; from assoc prof to prof metall, Univ Notre Dame, 45-51; prof metall, 51-76, EMER PROF METALL, UNIV ILL, URBANA-CHAMPAIGN, 76- *Concurrent Pos:* Consult, US Steel Corp, Argonne Nat Lab. *Honors & Awards:* Sr Scientist Award, Humbolt Found, Bonn; Mathewson Gold Medal, Am Inst Min, Metall & Petrol Eng; Albert Sauveur Achievement Award, Am Soc Metals; Hume-Rothery Award, Am Inst Mining, Metall & Petrol Engrs Metall Soc; Heyn Mem Award, German Metall Soc. *Mem:* Nat Acad Eng; fel Am Soc Metals; fel Metall Soc; fel Am Phys Soc; hon mem Hungarian Phys Soc, Budapest; Am Inst Min, Metall & Petrol Eng. *Res:* Magnetism in alloys; effect of short range atomic order on magnetic properties. *Mailing Add:* Dept Mat Sci & Eng Univ Ill 1304 W Green St Urbana IL 61801

BECK, PAUL EDWARD, b Lancaster, Pa, Jan 14, 37. ORGANIC CHEMISTRY. *Educ:* Franklin & Marshall Col, BS, 58; Duquesne Univ, PhD(org chem), 63. *Prof Exp:* Res chemist, E I du Pont de Nemours & Co, Inc, 63-66; assoc prof, 66-74, chmn dept, 75-84, PROF CHEM, CLARION UNIV, 74- *Mem:* AAAS; Am Chem Soc; Am Inst Chemists; NY Acad Sci; Controlled Release Soc. *Res:* New synthetic methods; new polymers and polymer reactions; polymeric drug delivery systems. *Mailing Add:* Dept Chem Clarion Univ Clarion PA 16214

BECK, PAUL W, b Crosby, Minn, Mar 28, 16; m 42; c 2. PHYSICAL CHEMISTRY. *Educ:* Univ Minn, BA, 40, MS, 42; Ill Inst Technol, PhD(chem), 52. *Prof Exp:* Res engr, Solar Mfg Corp, 43-46; res chemist, Sinclair Res Labs, 46-52; staff physicist, Philips Labs, Inc, 53-68; assoc prof, 68-74, prof chem, Western Conn State Col, 74-86; RETIRED. *Mem:* Am Chem Soc; Sigma Xi. *Res:* Magnetics; dielectrics; photochemistry. *Mailing Add:* 25 Stormy View Rd Ithaca NY 14850

BECK, RAYMOND WARREN, b New Smyrna, Fla, Oct 20, 25; m 58; c 2. MICROBIOLOGY. *Educ:* Univ Fla, BS, 49, MS, 52; Univ Wis, PhD(bact), 56. *Prof Exp:* Asst bact, Univ Fla, 50-52 & Univ Wis, 52-55; from asst prof to assoc prof, 55-71, PROF BACT, UNIV TENN, 71-, DIR, BIOL CONSORTIUM, 87- *Mem:* Am Soc Microbiol. *Res:* Microbial physiology; energy metabolism in bacteria. *Mailing Add:* Biol Consortium Univ Tenn Knoxville TN 37996

BECK, ROBERT EDWARD, b Denver, Colo, June 7, 41; m 65; c 3. COMPUTATIONAL ALGEBRA. *Educ:* Harvey Mudd Col, BS, 63; Univ Pa, MA, 65, PhD(math), 69. *Prof Exp:* From instr to assoc prof, 66-78, PROF MATH, VILLANOVA UNIV, 78- *Concurrent Pos:* Chmn, Dept Math, Villanova Univ, 78-81 & dir comput sci prog, 85-; Fulbright exchange, 81-82. *Mem:* Asn Comput Mach; Am Math Soc; Sigma Xi. *Res:* Computational methods in nonassociative algebras; algorithms for operations research models; user/system interface design. *Mailing Add:* Dept Math Sci Villanova Univ Villanova PA 19085

BECK, ROBERT FREDERICK, b Flushing, NY, June 11, 43; m 71; c 2. HYDRODYNAMICS, NAVAL ARCHITECTURE. *Educ:* Univ Mich, BSE, 65; Mass Inst Technol, MS, 67, Naval Arch, 68, PhD(naval archit), 70. *Prof Exp:* Res fel appl math, Univ Adelaide, 71-72; from asst prof to assoc prof, 72-82, PROF NAVAL ARCHIT, UNIV MICH, ANN ARBOR, 82-, DIR SHIP HYDRODYN LAB, 87- *Concurrent Pos:* Mem, long range ship res adv comt, Nat Res Coun, 78-80, comn liquid slosh loading in cargo tanks, 83-85, loads adv group, 83-; chmn, anal ship wave rel panel H-5, Soc Naval Archit & Marine Engrs, 78- *Mem:* Soc Naval Archit & Marine Engrs. *Res:* Ship hydrodynamics, particularly ship wave resistance, ship maneuvering, ship motions and seakeeping. *Mailing Add:* Dept Naval Archit & Marine Eng N Campus Ann Arbor MI 48107

BECK, ROBERT NASON, b San Angelo, Tex, Mar 26, 28; m 58. NUCLEAR MEDICINE. *Educ:* Univ Chicago, AB, 54, BS, 55. *Prof Exp:* Chief scientist, Argonne Cancer Res Hosp, 57-67; assoc prof, 67-76, PROF RADIOL SCI, UNIV CHICAGO, 76-, DIR, FRANKLIN MCLEAN INST, 77- *Concurrent Pos:* Consult, Int Atomic Energy Agency, 66-68; mem, Int Comn on Radiation Units, 68-; mem, Nat Coun on Radiation, Protection & Measurements, 70- *Mem:* Soc Nuclear Med; Am Asn Physicists in Med. *Res:* Development of a theory of the process by which images can be formed of the distribution of radioactive material in a patient, in order to diagnose his disease. *Mailing Add:* Dept Radiol Univ Chicago Pritzker Sch Med 5841 Maryland Ave Chicago IL 60637

BECK, ROLAND ARTHUR, b Mountain Iron, Minn, Apr 16, 13; m 34; c 2. INORGANIC CHEMISTRY, PHYSICAL CHEMISTRY. *Educ:* Maryville Col, BA, 34; Univ Minn, Minneapolis, MS, 39. *Prof Exp:* Instr, Pub high schs, Minn, 34-41; res chemist, Texaco Res Labs, Texaco Inc, NY, 41-46, proj leader synthetic fuels, 46-48, supvr res, Texaco Res Lab, Calif, 49-60, dir, 60-68, mgr, Texaco Res Ctr, 68-75; dep dir coal conversion, Energy Res & Develop Admin, 75-76; asst dir, US Dept Energy, Los Angeles, 76-78, adminr, 78-81; RETIRED. *Concurrent Pos:* Teaching asst, Univ Minn, Minneapolis, 38-39; conf leader, Mgt Develop Div, Calif Inst Technol, 56-68; vis lectr, Univ Edinburgh, Scotland, 81; energy consult, Govt Indonesia, 80; consult, energy technol, 81- *Mem:* AAAS; fel, Am Inst Chem Engrs; emer mem Am Chem Soc; Sigma Xi. *Res:* Coal gasification; synthetic fuels. *Mailing Add:* 7374 S Forest Ave Whittier CA 90602-1922

BECK, RONALD RICHARD, b Tiltonsville, Ohio, Oct 8, 34; m 61; c 3. PHYSIOLOGY, ENDOCRINOLOGY. *Educ:* Univ Ohio, BS, 61; Ohio State Univ, MS, 67, PhD(physiol), 68. *Prof Exp:* From asst prof to assoc prof physiol, Sch Med, Ind Univ, Indianapolis, 68-75; ASSOC PROF PHYSIOL, SCH MED, UNIV SC, 75-, ASST DEAN GRAD SCH, 81- *Concurrent Pos:* NIH grant, 72-74, 78-81. *Mem:* Am Physiol Soc; Endocrine Soc. *Res:* Effects of endotoxin on metabolism and metabolic regulation. *Mailing Add:* 22217 Mylls Ct St Clair Shores MI 48081

BECK, SIDNEY L, b New York, NY, Mar 28, 35; m 55; c 2. DEVELOPMENTAL TOXICOLOGY, DEVELOPMENTAL GENETICS. *Educ:* City Col New York, BS, 55; Univ Kans, MA, 57; Brown Univ, PhD(biol), 60. *Prof Exp:* Instr zool, Univ Mich, 60-61, USPHS fel, 61-64; NIH spec fel, Univ Col, London, 64-65, res assoc genetics, 65; from asst prof to assoc prof biol, Univ Toledo, 65-69; prof biol, Wheaton Col, Mass, 69-85, chmn dept, 69-83; prof biol, Ball State Univ, 85-87, chmn dept, 85-86; PROF BIOL & CHMN, DEPAUL UNIV, 87- *Concurrent Pos:* Mem, Inst Lab Animal Resources Comt Genetic Standards, Nat Acad Sci, 66-69; biologist, US Environ Protection Agency, 75-76, prin investr, 77-80; sr fel path, Univ Wash, Seattle, 83-84; consult teratology, Mass Gen Hosp, 85-87; vis prof, Harvard Univ, 85-87. *Mem:* AAAS; Am Soc Zoologists; Teratology Soc; Genetics Soc Am; Soc Develop Biol; Am Genetic Asn; Sigma Xi; Am Asn Univ Profs. *Res:* Developmental toxicology; mammalian genetics and development; postnatal teratology. *Mailing Add:* Dept Biol DePaul Univ Chicago IL 60614

BECK, SIDNEY M, b Pleasant Grove, Utah, Mar 10, 19; m 48; c 4. BACTERIOLOGY. *Educ:* Brigham Young Univ, AB, 41, MA, 48; Pa State Univ, PhD(bact), 51. *Prof Exp:* Chemist, US Bur Mines, 42-44; from asst prof to prof bact, Univ Idaho, 51-91; RETIRED. *Mem:* Am Chem Soc. *Res:* Soil organic matter turnover; plant root exudates; bacterial nitrogen fixation; bacterial purine metabolism and nutrition. *Mailing Add:* 1205 Kouse St Moscow ID 83843-3827

BECK, STANLEY DWIGHT, b Portland, Ore, Oct 17, 19; m 43; c 4. ZOOLOGY. *Educ:* State Col Wash, BS, 42; Univ Wis, MS, 47, PhD(zool), 50. *Hon Degrees:* DSc, Luther Col Iowa, 72. *Prof Exp:* From instr to prof, 48-69, W A HENRY PROF, 69-88, EMER PROF ENTOM, UNIV WIS-MADISON, 89- *Honors & Awards:* Founders Mem Award, Entomological Soc Am. *Mem:* Nat Acad Sci; AAAS; Entom Soc Am (pres, 82). *Res:* Insect nutrition; metabolism and endocrinology; photoperiodism; development. *Mailing Add:* 6100 Gateway Green Monord WI 53716

BECK, STEPHEN D, b Broklyn, NY, Mar 29, 30. NUMERICAL. *Educ:* Tufts Univ, BS, 51; Iowa State Univ, MS, 55; Rensselaer Polytech Inst, PhD(mech), 69. *Prof Exp:* PROF MATH, BLOOMSBURG UNIV PA, 71- *Mem:* Math Asn Am; Soc Indust & Appl Math. *Mailing Add:* 220 W First St Bloomsburg PA 17815

BECK, STEVEN R, b Manhattan, Kans, Jan 26, 47; m 86; c 3. GASIFICATION & PYROLYSIS, ENZYME IMMOBILIZATION. *Educ:* Kans State Univ, BSChe, 69; Univ Tex, Austin, PhD(chem eng), 72. *Prof Exp:* Sr res engr, Atlantic Richfield Co, 72-77; from asst prof to prof chem eng, Tex Tech Univ, 77-88, chmn dept, 81-88; SR BIOCHEM ENGR, SYNTEX CHEMICALS, 88- *Mem:* Am Inst Chem Engrs; Am Chem Soc; Am Soc Eng Educ; Sigma Xi. *Res:* Utilization of biomass as a source of energy and chemicals; development of best method for production of ethanol from cellulosic residues; bioreactor design. *Mailing Add:* Syntex Group Technol Ctr 2075 N 55th St Boulder CO 80301-2880

BECK, THEODORE R(ICHARD), b Seattle, Wash, Apr 11, 26; m 51; c 2. CHEMICAL ENGINEERING, ELECTROCHEMISTRY. *Educ:* Univ Wash, BS, 49, MS, 50, PhD(chem eng), 52. *Prof Exp:* Res engr process develop, E I du Pont de Nemours & Co, 52-54; group leader reduction res, Kaiser Aluminum & Chem Corp, 54-59; sect head electrochem, Am Potash & Chem Corp, Nev, 59-61; res specialist electrochem & energy conversion, Aero-space Div, Boeing Co, 61-65, sr basic res scientist, Boeing Sci Res Labs, 65-72; div mgr, Flow Res, Inc, 72-75; PRES, ELECTROCHEM TECHNOL CORP, 75- *Concurrent Pos:* Res prof, Dept Chem Eng, Univ Wash, 72-80; mem, Panel Electrochem Corrosion, Nat Mat Adv Bd, 85-86. *Honors & Awards:* Outstanding Achievement Award, Electrochem Soc, 81. *Mem:* AAAS; Am Chem Soc; hon mem Electrochem Soc (pres, 75-76); Nat Asn Corrosion Eng; fel Am Inst Chem Engrs. *Res:* Fused salt and aqueous electrolysis; process development; batteries; fuel cells; energy conversion; corrosion; wear; electrokinetic phenomena. *Mailing Add:* 10035 31st Ave NE Seattle WA 98125

BECK, THOMAS W, b Spokane, Wash, Apr 20, 48. MUSCLE PHYSIOLOGY, ENDOCRINOLOGY. *Educ:* Wash State Univ, BS, 70, MS, 72; Mich State Univ, PhD(dairy sci), 76. *Prof Exp:* Res assoc, Okla State Univ, 76-78; asst prof reproduction, Ohio State Univ, 78-84; prin res scientist cellular & molecular biol, Battelle Mem Inst, Columbus, Ohio, 84-86; res technologist ophthal, 87-89, RES TECHNOLOGIST RADIOL, UNIV WASH, SEATTLE, 89- *Concurrent Pos:* Consult, Cellular & Molecular Biol Sect, Battelle Mem Inst, Columbus, Ohio, 83-84. *Res:* Understanding the process of muscle contraction at the level of the interaction between the molecules of actin and myosin. *Mailing Add:* 15841 197th Place NE Woodville WA 98072-9431

BECK, WARREN R(ANDALL), b Bethlehem, Pa, Feb 14, 18; m 39, 70; c 4. CERAMICS SCIENCE & TECHNOLOGY, GLASS SCIENCE. *Educ:* Pa State Col, BS, 42; Univ Minn, MS, 47. *Prof Exp:* Res asst ceramics, Pa State Col, 42-43; ceramist, Cent Res Lab, 43-47, sect leader inorg res, 48-55, mgr res & develop, Inorg Dept, 55-67 & Glass Bubble Proj, 67-74; CORP SCIENTIST, MINN MINING & MFG CO, 74- *Concurrent Pos:* Adj prof, Univ Wis, Menomonie, 87- *Honors & Awards:* IR-100 Award. *Mem:* Fel Am Ceramic Soc; Brit Soc Glass Technol. *Res:* Glass technology, unusual optical glasses; crystallographic inversions; polymorphism; solid state ceramics; ion exchange; glass microbubbles; foam composites. *Mailing Add:* 1567 Atlantic St St Paul MN 55106

BECK, WILLIAM CARL, b Chicago, Ill, Aug 24, 07; m 48; c 3. SURGERY. *Educ:* Northwestern Univ, BA, 28, MD, 32. *Prof Exp:* Intern, Robert Packer Hosp, Sayre, Pa, 32-33; resident surg, Univ Frankfurt, Ger, 33-35; assoc surg, Med Sch, Univ Ill, 36-42; from attend to chief, Surg Dept, Guthrie Clin & Robert Packer Hosp, Sayre, 46-72, pres, 68-84, EMER PRES, GUTHRIE FOUND MED RES, 85- *Concurrent Pos:* Attend surgeon, St Joseph Hosp, Chicago, 35-42; assoc surgeon, Cook County Hosp, Chicago, 36-42; lectr surg, Sch Grad Med, Univ Pa, 49-70; clin prof surg, Hahnemann Med Col, 49-66. *Mem:* Am Surg Asn; Int Soc Surg; fel Am Col Surg; Royal Soc Med; Illum Eng Soc; fel Illum Eng Soc NAm. *Res:* Hospital and surgical environment including lighting, ultraclean ventilation, noise control and design of hospitals; cancer control; delivery of health service. *Mailing Add:* Guthrie Found Guthrie Square Sayre PA 18840-1692

BECK, WILLIAM F(RANK), b Lansing, Mich, Aug 4, 38; m 60; c 3. CHEMICAL ENGINEERING. *Educ:* Univ Mich, BS(chem eng) & BS(mat eng), 60; Mass Inst Technol, ScD(chem eng), 64. *Prof Exp:* Res engr, 64-66, res supvr process eng, 66-68, asst dir prod res, 68-, VPRES CHEM PRODS GROUP, FMC CORP. *Mem:* Am Inst Chem Engrs; Am Inst Chem. *Res:* Crystallization from solutions; applied surface chemistry; engineering kinetics; chlorinated hydrocarbons; soda ash. *Mailing Add:* FMC Corp 2000 Market St Philadelphia PA 19103

BECK, WILLIAM J, b Fredericktown, Mo, Aug 2, 21; m 47; c 2. ENVIRONMENTAL HEALTH, BACTERIOLOGY. *Educ:* Southeast Mo State Col, BS, 47; Univ Mo, Columbia, MA, 50; Univ Mich, MPH, 59, PhD(environ health), 70. *Prof Exp:* Chief lab technician bacteriol, Vet Admin Hosp, 47-49; environ bacteriologist, Mo Div Health Labs, 50-58; food technologist, Washington, DC Health Dept, 60-61; chief, Northwest Water Hyg Lab, USPHS, 61-69, chief, Health Res Ctr, 69-71; asst regional health admin, Boston, Mass, 71-74; USPH Serv, 75-82; dep dir, Environ Control & Opidemol, 83-84; CONSULT, ENVIRON CONTROL, 84- *Concurrent Pos:* Lectr, Sch Fisheries, Univ Wash, 65-69. *Mem:* Am Pub Health Asn; Nat Shellfisheries Asn; Conf State & Prov Pub Health Lab Dirs; Sigma Xi. *Res:* Public health aspects of milk, food, fresh and estuarine water; environmental effects of man's health and well being in the Arctic and sub-Arctic. *Mailing Add:* 1610 Green Berry Rd Jefferson City MO 65101

BECK, WILLIAM SAMSON, b Reading, Pa, Nov 7, 23; m; c 4. BIOCHEMISTRY, HEMATOLOGY. *Educ:* Univ Mich, BS, 43, MD, 46. *Hon Degrees:* AM, Harvard Univ, 70. *Prof Exp:* Clin instr, Med Sch, Univ Calif, Los Angeles, 50-53, instr, 53-55, asst prof med, 55-57; from asst prof to assoc prof, 57-79, PROF MED, HARVARD UNIV, 79-; PROF HEALTH SCI & TECHNOL, MASS INST TECHNOL, 79- *Concurrent Pos:* Fel biochem, NY Univ, 55-57; chief hemat & med sects, Atomic Energy Proj, Univ Calif, Los Angeles, 51-57; estab investr, Am Heart Asn, 55-60; tutor biochem sci, Harvard Univ, 57-; chief hemat unit, 57-76, dir hemat res lab, Mass Gen Hosp, Boston, 57-; mem hemat study sect, NIH, 67-71; mem adv coun, Nat Inst Arthritis, Metab & Digestive Dis, 72-75. *Honors & Awards:* Wenner-Gren Prize, 55. *Mem:* AAAS; Asn Am Physicians; Am Soc Biol Chem; Am Soc Clin Invest; Am Asn Cancer Res. *Res:* Vitamin B12 folic acid; nucleic acid and bacterial metabolism; enzymology; blood cell biochemistry. *Mailing Add:* Dept Med Harvard Univ & Mass Gen Hosp Fruit St Boston MA 02114

BECKEL, CHARLES LEROY, b Philadelphia, Pa, Feb 7, 28; m 58; c 4. THEORETICAL PHYSICS. *Educ:* Univ Scranton, BS, 48; Johns Hopkins Univ, PhD(physics), 54. *Prof Exp:* Asst, Johns Hopkins Univ, 49; asst, Sch Pharm, Univ Md, 49-53; from asst prof to assoc prof physics, Georgetown Univ, 53-64; mem res staff, Inst Defense Anal, 64-66; assoc prof, 66-69, asst dean grad sch, 71-72, PROF PHYSICS, UNIV NMEX, 69- *Concurrent Pos:* Consult, Ballistics Res Lab, Aberdeen Proving Ground, Md, 55-57, Inst Defense Anal, 62-64 & 66-69, Dikewood Corp, 67-72 & 74-80, Albuquerque Urban Observ, 69-71, Los Alamos Nat Lab, 79-80, US Army Control & Disarm Agency, 81-84; Fulbright lectr, Univ Peshawar, 57-58 & Cheng Kung Univ, Taiwan, 63-64; actg dir, Inst social Res & Develop, Univ NMex, 72, actg vpres res, 72-73; vis prof theoret chem, Univ Oxford, 73; phys sci officer, US Arms Control & Disarm Agency, 80-81; vis prof chem & molecular sci, Univ Sussex, UK, 87. *Honors & Awards:* Award in Solid State Physics & Mat Sci, US Dept Energy, 88. *Mem:* Am Phys Soc; Bioelectromagnetics Soc; Sigma Xi. *Res:* Electronic and vibrational struct of solids (theory), molecules (theory); quantum mechanics; operations research; study of biomolecule conformation; electric field effects on biological cells. *Mailing Add:* Dept Physics & Astron Univ NMex Albuquerque NM 87131-1156

BECKEN, BRADFORD ALBERT, b Providence, RI, Oct 5, 24; m 46; c 4. ELECTRONICS ENGINEERING. *Educ:* US Naval Acad, BS, 46; US Naval Postgrad Sch, BS, 52; Univ Calif, Los Angeles, MS, 53, PhD(physics), 61. *Prof Exp:* Sonar prof officer, Oper Test & Eval Force, US Navy, 53-56, head surface ship sonar design br, Bur Ships, 56-58, sonar prog officer, US Navy Electronics Lab, 60-63, mgr ship sonar prog, Bur Ships, 63-65, head sonar adv develop br, Submarine Warfare Proj Off, 65-67; mgr syst eng lab, 67-68, prog mgr, 68-69, mgr eng, Submarine Signal Div, 70-83, DIR TECHNOL, RAYTHEON CO, 83- *Concurrent Pos:* Consult, Airtronics Inc, Dulles Int Airport, DC, 67- *Honors & Awards:* Silver Medal, Am Defense Preparedness Asn, 87. *Mem:* Fel Acoust Soc Am. *Res:* Directional distribution of ambient noise in the ocean; all aspects of the design, development, test, evaluation and operation of sonar systems for the detection of submarines. *Mailing Add:* Submarine Signal Div Raytheon Co 1847 W Main Rd Portsmouth RI 02871

BECKEN, EUGENE D, b Thief River Falls, Minn, Apr 29, 11; m 38; c 2. COMMUNICATIONS. *Educ:* Univ NDak, BSEE, 32; Univ Minn, MSEE, 33; Mass Inst Technol, MS, 52. *Prof Exp:* RCA Commun, Inc, 35, transmitting shift engr, 35-36, eng asst, 36-42, asst to station plant supt, 42-48, supt tech opers & maintenance & plant opers engr, 53-59, vpres opers eng, 59-60, vpres & chief engr, 60-68, vpres syst oper, 68-70, exec vpres opers, RCA Global Commun, Inc, 70-72, pres, 72-75, chmn exec comt bd dir, 75-76; RETIRED. *Concurrent Pos:* Mem, Nat Indust Adv Comt, Fed Commun Comn, 59-; Independent res, astro & energy physics, 76- *Mem:* Fel Inst Elec & Electronics Engrs; AAAS; NY Acad Sci; Sigma Xi. *Res:* International radio cable and satellite communications; energy transfers in atmospheric sciences and in astrophysics. *Mailing Add:* 52 Rutland Rd Glen Rock NJ 07452

BECKENDORF, STEVEN K, b Portland, Ore, Jan 17, 44; m 68. DEVELOPMENTAL BIOLOGY, GENETICS. *Educ:* Univ Calif, Los Angeles, AB, 66; Calif Inst Technol, PhD(biochem), 72. *Prof Exp:* Asst prof, 76-84, ASSOC PROF MOLECULAR BIOL, UNIV CALIF, BERKELEY, 84- *Mem:* Genetics Soc Am; Soc Develop Biol. *Res:* Eukaryotic gene expression; glue protein genes of drosophila melanogaster; mechanisms of pattern formation in early development. *Mailing Add:* Dept Molecular & Cell Biol Univ Calif 401 Barker Hall Berkeley CA 94720

BECKENSTEIN, EDWARD, b New York, NY, Oct 21, 40; m; c 1. MATHEMATICAL ANALYSIS. *Educ:* Polytech Inst Brooklyn, BS, 62, MS, 64, PhD(math), 66. *Prof Exp:* From instr to asst prof math, Polytech Inst Brooklyn, 65-67; asst prof, St John's Univ, NY, 67-68; from asst prof to assoc prof, Polytech Inst Brooklyn, 68-72; assoc prof, 72-77, chmn dept, 77-81, PROF MATH, ST JOHN'S UNIV, NY, 81- *Mem:* Am Math Soc. *Res:* Abstract algebra; topology; functional analysis; theory of commutative Banach algebras. *Mailing Add:* Five Flora Dr Holmdel NJ 07733

BECKER, AARON JAY, b Brooklyn, NY, Apr 28, 40; div; c 2. HIGH TEMPERATURE CHEMISTRY, MOLTEN SALT CHEMISTRY. *Educ:* Brooklyn Col, BS, 61; Univ Wash, MS, 64; Ill Inst Technol, PhD(chem), 71. *Prof Exp:* Scientist chem, Ill Inst Technol Res Inst, 65-68; resident assoc, Argonne Nat Labs, 68-71; fel, McMaster Univ, 71-73; scientist, 73-76, sr scientist, 76-78, staff scientist, 78-83, TECH SPECIALIST CHEM, ALCOA LABS, 83- *Concurrent Pos:* Vis scientist & lectr, Tech Univ, Norway, 88. *Mem:* Am Chem Soc; Sigma Xi; Am Inst Mining Engrs. *Res:* Production of metals from ores; fused salt chemistry; utilization of natural resources; production of chemicals from coal; electrochemistry; aluminum smelting; ceramic powder synthesis; plasmas. *Mailing Add:* 1378 Hillsdale Dr Monroeville PA 15146

BECKER, ALEX, b Bialystok, Poland, July 27, 35; US citizen; m 60; c 3. MINERAL EXPLORATION, INSTRUMENTATION. *Educ:* McGill Univ, BEng, 58, MSc, 61, PhD(physics), 64. *Prof Exp:* Res scientist geophys, Geol Surv, Can, 65-69; prof appl geophy, Ecole Polytech Montreal, 69-80; dir res, Questor Surv, 80-81; PROF APPL GEOPHYSICS, UNIV CALIF, 81- *Mem:* Soc Exp Geophys. *Res:* Airborne geophysical instrumentation; data acquisition and interpretation. *Mailing Add:* Dept Mat Sci & Mineral Eng Univ Calif Berkeley CA 84720

BECKER, ANTHONY J, JR, b Wilkes-Barre, Pa, June 17, 48; m 71; c 1. PHYSIOLOGY, IMMUNOLOGY. *Educ:* Mt St Mary's Col, BS, 70; WVa Univ, MS, 77, PhD(biol), 81. *Prof Exp:* Asst prof physiol, Pa State Univ, 81-84; asst prof, 84-90, ASSOC PROF PHYSIOL, MANSFIELD UNIV, 90- *Concurrent Pos:* Adj asst prof, WVa Univ, 82-85. *Mem:* AAAS; Am Soc Zoologists; NY Acad Sci. *Res:* Physiological and immunological responses to environmental stressors in freshwater and marine fishes and invertebrates. *Mailing Add:* Dept Biol Mansfield Univ Mansfield PA 16933

BECKER, BARBARA, b Chicago, Ill, Jan 10, 32. BIOCHEMISTRY. *Educ:* Marymount Col, NY, BA, 54; Cath Univ, MA, 55; Georgetown Univ, PhD(chem), 66. *Prof Exp:* Instr chem, Marymount Jr Col, Va, 55-62; instr, 65-66, ASST PROF CHEM, MARYMOUNT COL, NY, 66-, DEAN STUDENTS, 68- *Mem:* AAAS; Am Chem Soc. *Res:* Fatty acid synthesis; enzyme purification. *Mailing Add:* Dept Chem Marymount Col Tarrytown NY 10591

BECKER, BENJAMIN, b New York, NY, Apr 22, 16; m 51; c 2. MICROBIOLOGY, BIOCHEMISTRY. *Educ:* Rutgers Univ, BS, 37, MS, 62, PhD(microbiol), 65. *Prof Exp:* Asst prof microbiol, biochem & gen biol, Hamilton Col, 65-69; assoc prof cell biol, 69-75, prof cell biol, 75-77, prof, 77-81, EMER PROF BIOL SCI, PURDUE UNIV, 81- *Concurrent Pos:* Consult, indoor environ microbiol. *Mem:* Am Soc Microbiol; Am Chem Soc; Am Inst Biol Sci. *Res:* Actinomycetes; cell wall analyses; microbial transformations; asparaginase in leukemia; leprosy immunogens; new antifungal agents; antibiotics; nonspecific immunostimulaters; lures and traps for cat and dog fleas; alternatives to pesticides; soil screening for actinomycetes and drugs. *Mailing Add:* 5126 Fairfield Ave Ft Wayne IN 46807

BECKER, BERNARD ABRAHAM, toxicology, for more information see previous edition

BECKER, BRUCE CLARE, b Seattle, Wash, Dec 9, 29; m 55; c 7. CLINICAL PSYCHOLOGY, CLINICAL NEUROPSYCHOLOGY. *Educ:* St Ambrose Col, BA, 50; St Louis Univ, MA, 54; Loyola Univ, PhD(psychol), 62. *Prof Exp:* Clin psychologist, US Naval Training Ctr, Great Falls, Ill, 55-58, clin psychologist, Naval Hosp, 58-65; clin neuropsychologist, VA Hosp, Downey Ill, 65-68, chief, Psychol Serv, 68-69; dir, Psychol Training & Res, Naval Hosp Bethesda, Md, 69-84; PRES, NEUROPSYCHOL ASSOCS LTD, 84- *Concurrent Pos:* Assoc neurology & psychiat, Med Sch, North Western Univ, 67-69; clin assoc prof psychiat, Georgetown Univ Med Sch, 74-, clin assoc prof neurol, 78-; assoc clin prof psychol, George Wash Univ, 70-75; clin assoc psychol, Catholic Univ Am, 73-75; adj prof med psychol, Uniformed Servs Univ Health Sci, 78- *Mem:* Am Psychol Assoc; Int Neuropsychol Soc; Undersea Med Soc. *Res:* Interest in brain, behavior relationships with an emphasis on brain damage due to head injury, toxic exposure, and prolonged or repeated undersea diving. *Mailing Add:* 9508 Newbold Pl Bethesda MD 20817

BECKER, CARL GEORGE, b Philadelphia, Pa, Mar 18, 36; m 61; c 3. PATHOLOGY. *Educ:* Yale Univ, BS, 57; Cornell Univ, MD, 61. *Prof Exp:* From intern to resident path, New York Hosp-Cornell Med Ctr, 61-66, asst prof, 66-68; pathologist, Naval Hosp, St Albans, 68-70; assoc prof path & assoc pathologist, 70-76, PROF PATH & ATTENDING PATHOLOGIST, NEW YORK HOSP-CORNELL MED CTR, 76- *Concurrent Pos:* USPHS trainee path, Med Col, Cornell Univ, 62-66; mem, Thrombosis Coun, Am Heart Asn. *Mem:* Am Asn Path (pres, 85-86); Harvey Soc; Int Acad Path; Sigma Xi. *Res:* Arteriosclerosis; rheumatic heart disease; renal disease; immunopathology; hemostasis and thrombosis; microbiology. *Mailing Add:* Dept Path Cornell Univ Med Col 1300 York Ave New York NY 10021

BECKER, CARTER MILES, b Allentown, Pa, Mar 6, 40; c 2. PATHOLOGY, BIOPHYSICS. *Educ:* Pa State Univ, BA, 62; Jefferson Med Col, MD, 66. *Prof Exp:* Intern path, Yale New Haven Med Ctr, 66-67; postdoc fel biophys, Mass Inst Technol, 67-70, res assoc, 70-77; res fel hemat, Mass Gen Hosp, 75-78; resident path, Worcester Mem Hosp, 78-81, attending pathologist, 81-82; ASST PROF PATH & LAB MED & DIR, DIV LAB HEMAT, UNIV CINNCINNATI MED CTR, 82- *Concurrent Pos:* Mem, Coun Thrombosis, Am Heart Asn; res fel med, Harvard Med Sch, 75-78. *Mem:* Am Asn Pathologists; Col Am Pathologists; Int Acad Path; Am Heart Asn; Int Soc Thrombosis & Haemostasis; AAAS. *Res:* Investigations of molecular mechanisms involved in blood coagulation, hemostasis and thrombosis, especially the structure and function of fibrinogen association products and the effect of soluble fibrinogen-fibrin complexes on platelet function. *Mailing Add:* Dept Anatomic Pathol Christ Hosp 2139 Auburn Ave Cincinnati OH 45219

BECKER, CLARENCE DALE, b Albany, Ore, Aug 17, 30; m 62; c 4. FISHERIES. *Educ:* Ore State Univ, BS, 53, MS, 55; Univ Wash, PhD(fisheries), 64. *Prof Exp:* Fisheries biologist, Fisheries Res Inst, Univ Wash, 55-59, res asst, 59-63, res assoc Col Fisheries, 63-67; res scientist, 67-70, SR RES SCIENTIST, PAC NORTHWEST LABS, BATTELLE MEM INST, 70- *Concurrent Pos:* Fel, Am Inst Fisheries Res Biologists, 85. *Honors & Awards:* Dis Serv Award, Am Fisheries Soc, 80. *Mem:* Pac Fisheries Biologists; Am Fisheries Soc; fel Am Inst Fisheries Res Biologists. *Res:* Fisheries biology; aquatic toxicology; freshwater ecology; environmental assessments. *Mailing Add:* Earth Environ Sci Ctr K6-78 Pac Northwest Lab Richland WA 99352

BECKER, CLARENCE F(REDERICK), b Dazey, NDak, Mar 27, 20; m 44; c 2. AGRICULTURAL ENGINEERING. *Educ:* NDak Agr Col, BS, 43, MS, 49; Mich State Univ, PhD(agr eng), 56. *Prof Exp:* Dist rep, Food Mach & Chem Co, 47-49; AGR ENGR & HEAD DIV, UNIV WYO, 49- *Concurrent Pos:* Consult, Food & Agr Orgn UN, 64, 66; chief of party, Univ Wyo team, Kabul Univ, Afghanistan, 68-70. *Mem:* Am Soc Agr Eng. *Res:* Agricultural mechanization. *Mailing Add:* 1600 Joyce St Apt 1510C Arlington VA 22202

BECKER, DAVID MORRIS, neurochemistry, biochemistry, for more information see previous edition

BECKER, DAVID STEWART, b Chicago, Ill, Aug 28, 45. BIOPHYSICS, ARTIFICIAL INTELLIGENCE. *Educ:* Stanford Univ, BS, 67; Univ Wis, Madison, MS, 69; Ill Inst Tech, PhD(physics), 77. *Prof Exp:* Scientist, Hallicrafters, Northrop Corp, 69-71; lectr, North Park Col, 71; RES SCIENTIST, ILL INST TECHNOL RES INST, 72- *Res:* Toxic chemical risk analysis; artificial intelligence; natural language processing; decision theory; mathematical modeling. *Mailing Add:* 18 Augustus Rd Lexington MA 02173

BECKER, DAVID VICTOR, b New York, NY, May 24, 23; m 49; c 2. MEDICINE. *Educ:* Columbia Univ, AB, 43, MA, 44; NY Univ, MD, 48. *Prof Exp:* Rotating intern, Sinai Hosp, Baltimore, Md, 48-49; med intern, Maimonides Hosp, Brooklyn, 49-50; Runyan fel, Dept Clin Invest & Biophys, Sloan-Kettering Inst, New York, 50-52; chief radioisotope unit, Surg Res Unit, Brooke Army Med Ctr, Ft Sam Houston, Tex, 52-54; asst resident med serv, New York Hosp-Cornell Med Ctr, 54-55, radiologist, 57-62; from instr med & radiol to asst prof med, 57-61; assoc prof med, 61-75, prof radiol, Col Med, Cornell Univ, 61-77, prof Med, 75-77, DIR DIV NUCLEAR MED, NEW YORK HOSP-CORNELL MED CTR, 55- *Concurrent Pos:* Asst attend physician, New York Hosp-Cornell Med Ctr, 57-62, assoc attend physician, 62-74, attend radiologist, 71-, attend physician, 75-; mem, Mayor's Tech Adv Comt Radiation, New York, 73-; mem med adv comt, NY State Dept Health, 73-76; mem task force short lived radionuclides, Bur Radiol Health, Food & Drug Admin, 75-, consult, Bur Radiol Health Food & Drug Admin, 76; mem, Nat Coun Radiation Protection & Measurement Sci Comt, Nat Coop Thyroid Can Treat Study Group, 78; tech adv Comt Radiation to the Health Comnr NY City; NY Health Comnr Ad Study Comt for Disposal of low-level radioactive wastes, NY, 78-84; mem Med Radiation Adv Comt FDA, 82-86; consult Nat Ctr Devices & Radiol Health FDA; Nat Can Inst Thyroid & Iodine 131 Assessment Comt; mem NY State Dept Health Radiol

Adv Comt, 87- *Honors & Awards:* Hon mem Med Acad Sr Coratia, 88- *Mem:* Fel Am Col Physicians; Am Thyroid Asn (1st vpres, 76, pres-elect, 81); Soc Nuclear Med (trustee); Endocrine Soc; Am Fedn Clin Res; mem Publ & Radiation Effects Comt. *Res:* Endocrinology and radiation effects; clinical thyroid physiology and disease; nuclear medicine. *Mailing Add:* Div Nuclear Med 525 E 68th St New York NY 10021

BECKER, DONALD A, b Valley City, NDak, July 27, 38; m 60; c 3. BOTANY, PLANT ECOLOGY. *Educ:* Valley City State Col, BS, 60; Univ NDak, MS, 65, PhD(ecol), 68. *Prof Exp:* High sch teacher, Wyo, 60-62; instr biol, Univ NDak, 67-68; from asst prof to assoc prof biol, Midland Lutheran Col, 68-75; environ specialist, Mo Basin States Asn, Omaha, 75-81; ECOLOGIST, US ARMY CORPS ENGRS, OMAHA, 85- *Concurrent Pos:* NSF res fel, Okla State Univ, 70-71; Cottrell res grant, 75; study mgr, Mo River Floodplain Study, 80-81; veg res grant, US Nat Park Serv, 82-85. *Mem:* Ecol Soc Am. *Res:* Dispersal ecology; plant autecology; senescence abscission in plants; nitrogen fixation in legumes; environmental planning and assessment; ecology of great plains rivers; floodplain management studies, prairie management planning. *Mailing Add:* 21706 Edgevale Pl Elkhorn NE 68022

BECKER, DONALD ARTHUR, b Detroit, Mich, Feb 14, 38; m 60; c 2. NEUTRON ACTIVATION ANALYSIS, TRACE ELEMENT ANALYSIS. *Educ:* Valparaiso Univ, BS, 59; Fla State Univ, MS, 61. *Prof Exp:* Res chemist textile sect, Org & Fibrous Mat Div, 61-62, res chemist photochem sect, Phys Chem Div, 62-64, res chemist, Anal Chem Div, 64-73, chief activation anal sect, 73-75, asst chief, Anal Chem Div, 75-76, sci asst to dir, Inst Mat Res, 76, mgr recycled oil prog, 76-82, spec asst to dir, 82-84, SR RES CHEMIST, CTR ANALYTICAL CHEM, NAT BUR STANDARDS, 85- *Concurrent Pos:* Chmn, Nuclear Methods Task Group, Am Soc Testing & Mat. *Honors & Awards:* Silver Medal Award, US Dept Com, 82. *Mem:* Am Chem Soc; Soc Environ Geochem & Health; Am Soc Testing & Mat; Am Nuclear Soc. *Res:* Trace elemental analysis using neutron activation, including radiochemical separations; improvement of accuracy and precision, and evaluation of reactor neutron irradiation facilities; characterization of recycled oils by research on test procedures. *Mailing Add:* Ctr Analytical Chem Nat Bur Standards Gaithersburg MD 20899

BECKER, DONALD EUGENE, b Delavan, Ill, Feb 2, 23; m 49; c 5. ANIMAL NUTRITION, ANIMAL PHYSIOLOGY. *Educ:* Univ Ill, BS, 45, MS, 47; Cornell Univ, PhD(nutrit), 49. *Prof Exp:* Asst animal sci, Univ Ill, 45-47; asst animal husb, Cornell Univ, 47-49; assoc prof animal nutrit, Univ Tenn, 49-50; from asst prof to prof animal nutrit, Univ Ill, Urbana-Champaign, 58-84, head dept animal sci, 67-84; chmn, Dept Animal Sci, Ohio State Univ, Columbus, 84-87; PVT IND CONSULT, 87-; PRES, OXFORD FEEDERS, INC, 89- *Concurrent Pos:* Mem subcomt nutrient requirements of swine, Nat Res Coun; ed, J Animal Sci, 66-69. *Honors & Awards:* Am Feed Mfrs Award, 57; Morrison Award, 77. *Mem:* Fel AAAS; Am Soc Animal Sci (pres, 70-71); Am Soc Animal Nutrit Res Coun; Poultry Sci Asn; Am Inst Nutrit; Sigma Xi. *Res:* Metabolic function of cobalt in ruminant nutrition; chemistry and morphology of bovine blood; protein and amino acid nutrition for pregnancy, lactation, growth and fattening; antibiotics and nonruminant nutrition; comparative value of carbohydrates and available energy values of feeds. *Mailing Add:* 2209 Combes St Urbana IL 61801

BECKER, EDWARD BROOKS, b Emporia, Kans, Aug 12, 31; m 59; c 3. GENERAL MANAGEMENT, ENVIRONMENTAL MANAGEMENT. *Educ:* Kans State Teachers Col, AB, 53; Univ Kans, PhD(inorg chem), 59. *Prof Exp:* Sr res chemist, Chem Div, PPG Industs, 60-61; proj mgr, Alexandria Div, AMF, 61-63; res chemist, Gulf Res & Develop Co, 64-70; dir bur air pollution control & solid & hazardous waste mgt, State of Wis, 70-78; pres, RMT, Inc, 78-90; MGT & ENVIRON CONSULT, 90- *Concurrent Pos:* Chmn, State & Territorial Air Pollution Prog Admin, 74. *Mem:* AAAS; Am Chem Soc; Air Pollution Control Asn; Nat Solid Waste Mgt Asn; fel Am Inst Chemists. *Res:* High temperature plasma chemistry; air pollution control; solid waste management; nitrogen-phosphorus-potassium fertilizer chemistry; management of consulting firms. *Mailing Add:* 1132 University Bay Dr Madison WI 53705

BECKER, EDWARD SAMUEL, b Bisbee, Ariz, Sept 8, 29; m 51; c 2. PULP TECHNOLOGY, POLLUTION CONTROL. *Educ:* Ore State Col, BS, 51, MS, 53, PhD(forest prod chem), 57. *Prof Exp:* Sect leader develop pulping, Rayonier, Inc, 57-64 & G L Pulp & Papermaking Union Camp, 64-65; dept mgr, Columbia Cellulose Co, Ltd, 66-68; mgr tech develop, 69-71, dir tech & environ control, 72; PRES, ECONOTECH SERV LTD, 72- *Mem:* Tech Asn Pulp & Paper Indust; Can Pulp & Paper Asn; Chem Inst Can. *Res:* Technical consulting; research and development management; technical service; pulp and paper; pilot plant operation; management consulting. *Mailing Add:* Econotech Serv Ltd 852 Derwent Way Annacis Island New Westminster BC V3M 5R1 Can

BECKER, EDWIN DEMUTH, b Columbia, Pa, May 3, 30; m 53; c 2. PHYSICAL CHEMISTRY. *Educ:* Univ Rochester, BS, 52; Univ Calif, PhD(chem), 55. *Prof Exp:* Asst chem, Univ Calif, 53-54, instr, 55; phys chemist, 55-68, chief sect molecular biophys, 62-80, mem fac grad prog, 63, asst chief lab phys biol, 64-72, chief lab chem physics, Nat Inst Arthritis, Metab & Digestive Dis, 72-81, actg dir, Fogarty Int Ctr, 79-80, assoc dir, NIH, 80-88, CHIEF NMR SECT, NAT INST DIABETES & DIGESTIVE & KIDNEY DIS, 88- *Concurrent Pos:* Prof lectr, Georgetown Univ, 58-; sr asst scientist, USPHS, 55-58; mem adv bd, Off Critical Tables, 67-69; mem, Joint Comt Atomic & Molecular Data, 67-89, chmn comt, 78-81; mem prog comt, Exp NMR Conf, 68-70, chmn, 69; mem comn molecular structure & spectros, Int Union Pure & Appl Chem, 73-81, chmn comn, 75-81; mem bd dirs, Found Advan Educ Sci, 76-79; secy, Phys Chem Div, Int Union Pure & Appl Chem, 81-85; mem US Nat Comt, 79-86, chmn, Publications Comt, 88-; mem, Int Activ Comn, Am Chem Soc, 87-; mem coun Int Soc Magnetic Resonance, 90- *Honors & Awards:* Coblentz Mem Prize Chem Spectros, 66; Dept Health, Educ & Welfare Superior Serv Award, 74; NIH Dir Award, 89.

Mem: AAAS; Coblentz Soc; Am Soc Appl Spectros; Am Chem Soc; fel AAAS. *Res:* Molecular structure; infrared spectroscopy; nuclear magnetic resonance; hydrogen bonding. *Mailing Add:* Bldg 2 Rm 120 NIH Bethesda MD 20892

BECKER, EDWIN NORBERT, b Ossian, Iowa, Aug 6, 22; m 50; c 5. PHYSICAL CHEMISTRY. *Educ:* Iowa State Univ, BS, 47; Univ Wis, PhD(phys chem), 53. *Prof Exp:* Asst prof chem, Col of St Thomas, 53-55; from asst prof to prof, 55-83, EMER PROF CHEM, CALIF STATE UNIV, LONG BEACH, 83- *Mem:* AAAS; Am Chem Soc. *Res:* Reaction mechanisms. *Mailing Add:* Dept Chem Calif State Univ Long Beach CA 90840

BECKER, ELIZABETH ANN (WHITE), b Cincinnati, Ohio, Sept 2, 50; div. ENVIRONMENTAL ANALYSIS, PHARMACEUTICAL ANALYSIS. *Educ:* Miami Univ, BA, 72; Univ Cincinnati, MS, 74, PhD(chem), 80. *Prof Exp:* Scientist, Mead Johnson & Co, Div Bristol Myers, 80-87; GROUP LEADER, BRISTOL-MYERS US PHARMACEUT & NUTRIT GROUP EVANSVILLE, 87- *Mem:* Am Chem Soc. *Res:* Synthesis of analogs of natural products; analysis and stability determination of pharmaceutical compounds. *Mailing Add:* 1917 E Gum St Evansville IN 47714

BECKER, ELMER LEWIS, b Chicago, Ill, Feb 17, 18; m 46; c 3. IMMUNOLOGY. *Educ:* Univ Ill, MD, 45, PhD(biochem), 47. *Prof Exp:* Asst prof biochem, Col Med, Univ Ill, 47-49, asst prof bact, 50-51; chief dept immunochem, Walter Reed Army Inst Res, Washington, DC, 52-69; prof path, Health Ctr, 70-80, Boehringer Ingelheim, prof immunol, 84-89, EMER PROF, UNIV CONN, FARMINGTON, 89- *Concurrent Pos:* John Simon Guggenheim Mem Found fel, 69-70. *Honors & Awards:* Rouse-Whipple Award, Am Asn Pathol. *Mem:* AAAS; hon fel Am Acad Allergy; Am Asn Immunol. *Res:* Antigen-activated enzyme systems; chemotaxis; lysosomal enzyme release. *Mailing Add:* Dept Path Univ Conn Health Ctr Farmington CT 06034

BECKER, ERNEST I, b Cleveland, Ohio, Aug 18, 18; m 47; c 5. CHEMISTRY. *Educ:* Case Western Reserve Univ, BS, 41, MS, 43, PhD(chem), 46. *Prof Exp:* Instr, Ploytech Univ, 46, from asst prof to prof chem, 47-56; prof chem, Univ Mass, 65-88, chmn dept, 65-67 & 68-72; RETIRED. *Concurrent Pos:* Chmn, Div Nat Sci, Univ Mass, Boston, 65-70; indust consult, acad & govt; Fulbright prof, Ankara Univ, Turkey, 87-88. *Mem:* fel AAAS; Am Chem Soc; fel NY Acad Sci; Royal Soc Chem; Am Inst Chemists. *Res:* Synthesis reactions and spectra of tetracyclones; synthesis and properties of monomeric and polymeric fluors; organotin compounds; mechanism of organomagnesium reactions; oxazoles. *Mailing Add:* 32 Oxford Rd Newton MA 02159-2405

BECKER, F(LOYD) K(ENNETH), b Denver, Colo, June 10, 24; m 45; c 3. ELECTRONICS ENGINEERING. *Educ:* Univ Colo, BS, 45; Calif Inst Technol, MS, 47. *Prof Exp:* Dial equip engr, Mt States Tel & Tel Co, 47-51; mem tech staff, Bell Tel Labs, 51-66, head dept, Customer Switching Lab, 66-80, dir, 80-90, EMER DIR, TELECOMMUN SYSTS LAB, UNIV COLO, 90- *Concurrent Pos:* Spec probs engr, Off Naval Res, 52-53. *Mem:* Sr mem Inst Elec & Electronics Engrs; Sigma Xi. *Res:* Data communication; discrete systems theory; switching theory. *Mailing Add:* 5118 N 109th St Longmont CO 80501

BECKER, FREDERICK F, b New York, NY, July 23, 31; m 71. PATHOLOGY, ONCOLOGY. *Educ:* Columbia Univ, BA, 52; NY Univ, MD, 56. *Prof Exp:* Intern, Boston City Hosp, 56-57; asst & actg dir exp path, US Naval Med Res Inst, 60-61; from asst prof to prof path, Sch Med, NY Univ, 62-70; dir path, Bellevue Hosp, 70-76; chmn, Dept Anat & Res Path, 76-79, VPRES RES, M D ANDERSON HOSP, HOUSTON, 79- *Concurrent Pos:* Univ fel path, Sch Med, NY Univ, 57-60; career investr, NY Health Res Coun, 62-; consult, US Naval Hosp, St Albans, NY, 62- *Res:* Carcinogenesis. *Mailing Add:* Sci Dir Tumor Inst Univ Tex M D Anderson Cancer Ctr 1515 Holcombe Blvd Houston TX 77030

BECKER, GEORGE CHARLES, b Stevens Point, Wis, July 21, 35; m 64; c 3. FOREST ENTOMOLOGY, URBAN ECOLOGY. *Educ:* Univ Minn, BS, 60; Univ Wis, MS, 62, PhD(entomol), 65. *Prof Exp:* Asst prof, Univ Wis-Whitewater, 64-66; assoc prof, 66-69, PROF BIOL, 69- & CHMN DEPT BIOL, METRO STATE COL, 71- *Mem:* Sigma Xi. *Res:* Diprionidae. *Mailing Add:* Dept Biol Metro State Col Denver Campus Box 53 PO Box 173362 Denver CO 80217-3362

BECKER, GERALD ANTHONY, b Ossian, Iowa, May 30, 24; m 48; c 3. MATHEMATICS. *Educ:* Univ Iowa, BA, 50, MS, 51, PhD(math), 59. *Prof Exp:* High sch teacher, Iowa, 51-56; asst math, State Univ Iowa, 56-58; from asst prof to assoc prof, 58-65, EMER PROF MATH, SAN DIEGO STATE UNIV, 86- *Mem:* Math Asn Am; Sigma Xi. *Res:* Mathematics curriculum; mathematical models in the social sciences. *Mailing Add:* Dept Math San Diego State Univ San Diego CA 92182

BECKER, GERALD LEONARD, b San Diego, Calif, Oct 20, 40; m 65; c 2. BIOCHEMISTRY. *Educ:* Mass Inst Technol, BS, 62; Univ Chicago, MD, 66. *Prof Exp:* From intern to res assoc med, Beth Israel Hosp, Boston, 66-68; clin assoc Geront Br, Nat Inst Child Health & Human Develop, 68-70; fel instr physiol chem, Johns Hopkins Sch Med, 70-75; ASST PROF BIOCHEM, UNIV ALA, BIRMINGHAM, 75- *Mem:* AAAS. *Res:* Biochemistry of mitochondria and calcification. *Mailing Add:* 222 Byrd Ml Rd New Brockton AL 36351

BECKER, GORDON EDWARD, b St Louis, Mo, Sept 27, 20; m 56; c 2. SURFACE PHYSICS. *Educ:* Columbia Univ, BA, 42, AM, 43, PhD(physics), 46. *Prof Exp:* Asst physics, Columbia Univ, 42-44, res assoc, Radiation Lab, 43-46, instr physics, 46-47 & Stanford Univ, 47-51; sci staff, Hudson Labs, Columbia Univ, 51-53; MEM TECH STAFF, BELL TEL LABS, 53- *Mem:* Am Phys Soc; Inst Elec & Electronics Engrs. *Res:* Molecular beams. *Mailing Add:* 2761 Quail Hollow Rd Clearwater FL 33519

BECKER, HARRY CARROLL, b Bisbee, Ariz, Oct 31, 13; m 40; c 1. ANALYTICAL CHEMISTRY. *Educ:* Univ Ill, BS, 36, MS, 37, PhD(anal chem), 40. *Prof Exp:* Anal res, Texaco Inc, 40-43, supvr, 44-62, res assoc, 62-78; RETIRED. *Mem:* AAAS; Am Chem Soc; Sigma Xi. *Res:* Radiation chemistry; analysis of petroleum products; analytical studies with neutrons and energetic charged particles; scanning electron microscopy and electron microprobe. *Mailing Add:* 15 Walnut Hill Rd Poughkeepsie NY 12603

BECKER, HENRY A, b Castor, Alta, Apr 15, 29; m 57; c 3. CHEMICAL ENGINEERING. *Educ:* Univ Sask, BChE, 52, MSc, 55; Mass Inst Technol, ScD(chem eng), 61. *Prof Exp:* Jr res officer, Div Appl Biol, Nat Res Coun Can, 52-57, asst res officer, 57-61; asst prof chem eng, Mass Inst Technol, 61-63; from asst prof to assoc prof, 63-68, head dept, 76-87, PROF CHEM ENG, QUEEN'S UNIV, 68- *Concurrent Pos:* Ford Found fel, Mass Inst Technol, 61-63; SRC sr res fel, Univ Sheffield, 73. *Honors & Awards:* Eng Medal, Asn Prof Engrs, Ont, 90. *Mem:* Am Inst Chem Engrs; Combustion Inst; fel Chem Inst Can; Can Inst Energy; Am Soc Eng Educ; AAAS. *Res:* Diffusion in polymers; sedimentation; fluidization; drying; turbulence; jets and boundary layers; combustion; radiation; biochemical engineering; forest fire research. *Mailing Add:* Dept Chem Eng Queen's Univ Kingston ON K7L 3N6 Can

BECKER, JEFFREY MARVIN, b Baltimore, Md, July 13, 43; m 66; c 3. MICROBIOLOGY. *Educ:* Emory Univ, BA, 65; Ga State Univ, MS, 67; Univ Cincinnati, PhD(microbiol), 70. *Prof Exp:* Weizmann fel biophys, Weizmann Inst, Rehovot, Israel, 70-71, NATO-NSF fel, 71-72; from asst prof to assoc prof, 72-78, PROF MICROBIOL, UNIV TENN, KNOXVILLE, 78- *Concurrent Pos:* Consult, Oak Ridge Nat Lab, 74-; NIH res career develop award, 75-80; Fogarty sr int fel, 85-86. *Mem:* Am Soc Microbiol; AAAS; Am Asn Univ Prof; Am Soc Biol Chem. *Res:* Membrane transport; peptide transport and utilization in microorganisms and tissue culture; development of antifungal drugs. *Mailing Add:* Dept Microbiol Univ Tenn Knoxville TN 37916

BECKER, JERRY PAGE, b North Redwood, Minn, Mar 1, 37; m 59; c 3. MATHEMATICS. *Educ:* Univ Minn, Minneapolis, BS, 59; Univ Notre Dame, MS, 61; Stanford Univ, PhD(math educ), 67. *Prof Exp:* Pub sch teacher, Minn, 59-60 & 61-63; res asst math educ, Sch Educ, Stanford Univ, 63-64, res asst, Sch Math Study Group, 64-67; from asst prof to assoc prof, Grad Sch Educ, Rutgers Univ, 67-75; vis assoc prof math sci, Northern Ill Univ, 75-76; math teacher, Clinton Rosette Mid Sch, DeKalb, Ill, 77-78; assoc prof, 79-83, PROF, COL EDUC, SOUTHERN ILL UNIV, 83- *Concurrent Pos:* Instr, NSF Inst, Univ San Francisco, 65-67 & Rutgers Univ, 67-68; consult, NSF Sci Educ Improv Prog, India, 67 & 68, supvr, 69 & 70, staff scientist math, Sci Liaison Staff, New Delhi, 71-72; mem US comn math instr, Nat Acad Sci, 73-77, co-chmn panel US arrangements third in cong math educ, 74-76; consult, UN Develop Prog, 75; head deleg, Am Math Educ, Peoples Repub China, 77, 80, 84 & 87, Japan, 77, 88; co-prin investr, integrating prob solving into math teaching, 83-84, 87-89, prin investr, cross-cultural study prob solving behav maths 85-86, 89-90, co-prin investr intensive study prob solving, 88-89, prin investr cross-nat prob solving res, 90-91, prin investr develop middle sch math profs, 90-92. *Honors & Awards:* Serv Award, Sch Sci & Math Asns, 90. *Mem:* Math Asn Am; Nat Coun Teachers Math; Int Org Women & Maths Educ; Sch Sci & Math Asns; Nat Coun Supvrs Math. *Res:* Mathematics curriculum development; teacher training; psychology of learning; achievement testing in mathematics; evaluation of mathematics education programs; applications of microcomputers in mathematics teaching; international mathematics education; open-ended problem solving in maths; research, psychology, and women in math education. *Mailing Add:* Curr Inst Southern Ill Univ Carbondale IL 62901-4610

BECKER, JOHN ANGUS, b New York, NY, Nov 25, 36; m 79. NUCLEAR PHYSICS. *Educ:* Queens Col (NY), BS, 57; Fla State Univ, PhD(physics), 62. *Prof Exp:* Asst physics, Fla State Univ, 58-62, res assoc, 62; res assoc, Brookhaven Nat Lab, 62-64; res scientist, Lockheed Palo Alto Res Labs, 64-78; SR PHYSICIST, LAWRENCE LIVERMORE NAT LAB, LIVERMORE, CALIF, 78- *Concurrent Pos:* Sr fel, Oxford Univ, 71-72; assoc prof, Univ Minn, 74-75. *Mem:* Fel Am Phys Soc; AAAS. *Res:* Experimental, low-energy nuclear physics; nuclear reactions initiated by electrostatic generator beams; particle detectors and instrumentation. *Mailing Add:* Lawrence Livermore Nat Lab Livermore CA 94550

BECKER, JOHN HENRY, b Denver, Colo, Mar 29, 48; m 72; c 1. ION TRANSPORT, HIBERNATION. *Educ:* Loyola Univ, BS, 70; Univ Ill, MS, 73, PhD(physiol), 78. *Prof Exp:* Investr cell biol, Oak Ridge Nat Lab, 77-79; asst prof biol, Knox Col, 79-86; ASSOC PROF PHYSIOL, DR WILLIAM M SCHOLL COL PODIATRIC MED, 86- *Concurrent Pos:* Vis prof, Univ Ill, Urbana, 81; actg assoc dean, Dr William M Scholl Col Podiatric Med, 90- *Mem:* Sigma Xi; Am Physiol Soc. *Res:* Measurement of adaptations of membrane transport systems to various stresses: temperature, carcinogens, cell culture conditions. *Mailing Add:* Dr William M Scholl Col Podiatric Med 101 N Dearborn St Chicago IL 60610

BECKER, JOSEPH F, b Mt Vernon, NY, Feb 26, 27; m 64. CHEMISTRY, BIOCHEMISTRY. *Educ:* Harvard Univ, AB, 50; Univ Del, MEd, 54; Columbia Univ, MS, 58, DEd, 62; Seton Hall Univ, JD, 73. *Prof Exp:* Teacher, high sch, 51-58; assoc prof chem & biochem, 58-68, PROF CHEM, MONTCLAIR STATE COL, 68- *Mem:* AAAS; Am Chem Soc. *Res:* Biochemistry for non-science majors. *Mailing Add:* Dept Physics San Jose State Univ San Jose CA 95192

BECKER, JOSEPH WHITNEY, b New London, Conn, Apr 17, 43; m 76; c 1. BIOCHEMISTRY, STRUCTURAL CHEMISTRY. *Educ:* Yale Univ, BA, 64; Stanford Univ, PhD(phys chem), 70. *Prof Exp:* Postdoctoral fel develop & molecular biol, Rockefeller Univ, 70-71, res assoc, 71-72, from asst prof to assoc prof, 72-90; SR RES FEL, MERCK SHARP & DOHME RES LABS, MERCK & CO, INC, 90- *Mem:* Am Soc Biochem & Molecular Biol; Am Crystallog Asn; AAAS; Sigma Xi. *Res:* Macromolecular crystallography. *Mailing Add:* R80M-203 Merck & Co Inc PO Box 2000 Rahway NJ 07065-0900

BECKER, JOSHUA A, b Philadelphia, Pa, Nov 28, 32; m 59; c 2. MEDICINE, RADIOLOGY. *Educ:* Temple Univ, AB, 53, MD, 57, MS, 61. *Prof Exp:* Intern, Philadelphia Gen Hosp, 58; resident radiol, Univ Hosp, Temple Univ, 61, instr, Sch Med, 63-65; from asst prof to assoc prof, Col Physicians & Surgeons, Columbia Univ, 65-70; PROF RADIOL, CHMN DEPT & DIR, STATE UNIV NY DOWNSTATE MED CTR, 70- *Concurrent Pos:* James Picker Found scholar radiol res, 64-65; from asst attend physician to assoc attend physician, Presby Hosp, 65-70; consult, Bronx Vet Admin Hosp, 67-70 & Brooklyn Vet Admin Hosp, 70- *Mem:* Am Col Radiol; Asn Univ Radiol. *Mailing Add:* Dept Biol SUNY Health Sci Ctr Brooklyn Box 1198 Brooklyn NY 11203

BECKER, KENNETH LOUIS, b New York, NY, Mar 11, 31; m 54; c 2. INTERNAL MEDICINE, ENDOCRINOLOGY. *Educ:* Univ Mich, BA, 52; NY Med Col, MD, 56; Univ Minn, PhD(med, physiol), 64; Am Bd Internal Med, dipl, 66 & cert endocrinol & metab, 73. *Prof Exp:* Intern, Mt Sinai Hosp, NY, 56-57; resident internal med, Mayo Clin, 59-64; from asst prof to assoc prof med, George Washington Univ, 64-74; clin investr med & endocrinol, 64-65, dir endocrinol, 74-77, CHIEF SECT METAB, VET ADMIN HOSP, WASHINGTON, DC, 65-; PROF MED, GEORGE WASHINGTON UNIV, 74- *Concurrent Pos:* Consult, Wash Hosp Ctr, 64- *Honors & Awards:* Meritorious Res Award, Mayo Clin, 63. *Mem:* Fel Am Col Physicians; Am Fedn Clin Res; Am Soc Clin Pharmacol & Therapeut; Endocrine Soc; Am Diabeties Asn. *Res:* Calcium metabolism; calcitonin; gynecomastia; gonadal-pituitary inter-relationships. *Mailing Add:* Vet Admin Hosp 50 Irving St NW 151-J Washington DC 20422

BECKER, KURT H, b Zweibrücken, WGermany, Mar 24, 53. ATOMIC & MOLECULAR PHYSICS. *Educ:* Univ Saarland, Saarbrücken, WGermany, dipl, 78, DrRerNat(physics), 81. *Prof Exp:* Teaching & res assoc, Univ Saarland, Saarbrücken, WGermany, 78-81; res assoc, Univ Windsor, Ont, Can, 82-84; ASST PROF PHYSICS, LEHIGH UNIV, 85- *Mem:* Can Asn Physicists; Am Phys Soc; Sigma Xi; German Phys Soc. *Res:* Electron collisions with atoms and molecules; excitation, ionization and dissociation; polarization measurements in the vacuum ultraviolet region; spin effects in electron-atom collisions. *Mailing Add:* Dept Physics Lehigh Univ Bethlehem PA 18015

BECKER, KURT HEINRICH, b Zweibrucken, Ger, Mar 24, 53; m. PHYSICS. *Educ:* Univ Saarbrücken, dipl physics, 78, Dr rer nat, 81. *Prof Exp:* Postdoctoral physics res, Univ Windsor, Can, 82-84; vis asst prof physics, Lehigh Univ, Bethlehem, Pa, 84-86, asst prof, 86-88; assoc prof, 88-92, PROF PHYSICS, CITY COL NEW YORK, 92- *Mem:* Am Phys Soc; Can Asn Physicists; Geo Phys Soc; NY Acad Sci; AAAS. *Res:* Electron collision processes with atoms and molecules from fundamental to applied aspects, three-body problems, precision measurements of fundamental constants. *Mailing Add:* Physics Dept City Col NY New York NY 10031

BECKER, LAWRENCE CHARLES, b Schenectady, NY, Aug 1, 34; m 57; c 2. COMPUTER INTERFACING, DATA ACQUISITION. *Educ:* Carleton Col, BA, 56; Yale Univ, BD, 59, MS, 60, PhD(physics), 64. *Prof Exp:* From asst prof to assoc prof, 63-81, PROF PHYSICS, HIRAM COL, 81- *Concurrent Pos:* Res physicist, NASA Lewis Res Ctr, 69-70; tech writer, Ohio Scientific, 80-83; res scientist, 84-85, syst engr, Geocomp Corp, 87- *Mem:* AAAS; Am Phys Soc; Am Asn Physics Teachers. *Res:* Hardware & software development for a remote data acquisition system. *Mailing Add:* Dept Physics Box 1778 Hiram Col Hiram OH 44234-1778

BECKER, LEWIS CHARLES, HEART IMAGING, MYOCARDIAL PROTECTION. *Educ:* Johns Hopkins Univ, MD, 66. *Prof Exp:* PROF MED, DIV CARDIOL, SCH MED, JOHNS HOPKINS UNIV, 72- *Res:* Nuclear cardiology. *Mailing Add:* Halsted Bldg Rm 500 Sch Med Johns Hopkins Univ 600 N Wolfe St Baltimore MD 21205

BECKER, MARSHALL HILFORD, b New York, NY, Jan 18, 40; m 59; c 2. PUBLIC HEALTH, MEDICAL SOCIOLOGY. *Educ:* City Col NY, BS, 62; Univ Mich, MPH, 64, PhD(med sociol), 68. *Prof Exp:* From asst prof to assoc prof med sociol, Johns Hopkins Univ, 69-77; PROF HEALTH BEHAV, UNIV MICH, 77-, ASSOC DEAN SCH PUB HEALTH, 87- *Concurrent Pos:* Consult, NIH, Nat Acad Sci, various univs & founds, 69-; prin investr res grants, NIH, Nat Ctr Health Serv Res, 71-; ed, Health Educ Quart, 79-86. *Honors & Awards:* Mayhew Derryberry Award, Am Pub Health Asn, 82. *Mem:* Inst Med-Nat Acad Sci; Am Sociol Asn; Am Pub Health Asn; Asn Teachers Prev Med; Asn Social Sci Health; Asn Health Serv Res; Soc Pub Health Educ. *Res:* Determinants of, and strategies to alter, the health-related beliefs and behaviors of lay persons and health professionals in health and medical care settings; health promotion and disease prevention. *Mailing Add:* Assoc Dean Sch Public Health Univ Mich 1420 Washington Heights Ann Arbor MI 48109-2029

BECKER, MARTIN, b New York, NY, May 11, 40; m 61; c 6. NUCLEAR ENGINEERING. *Educ:* NY Univ, BS, 60; Mass Inst Technol, SM, 62, PhD(nuclear eng), 64. *Prof Exp:* Theoret physicist advan develop activity, Knolls Atomic Power Lab, Gen Elec Co, 64-65, supv physicist, 65-66; from assoc prof to prof nuclear eng, 66-90, assoc dean, 86-90, ADJ PROF, NUCLEAR ENG & ENGR PHYSICS, RENSSELAER POLYTECH INST, 90-; VICTOR P CLARKE PROF ENG, UNIV MIAMI, CORAL GABLES, 90- *Concurrent Pos:* Consult, Savannah River Lab, E I du Pont de Nemours & Co, Inc, 68, United Nuclear Corp, Gen Elec, Los Alamoso Nat Lab, NY Pub Serv Comn and others. *Honors & Awards:* Glenn Murphy Award, Am Soc Elec Eng. *Mem:* Fel Am Nuclear Soc; sr mem Inst Elec & Electronic Engrs; Am Soc Eng Educ. *Res:* Nuclear reactor physics; reactor dynamics and control; mathematical methods; calculus of variations; nuclear rocket propulsion; nuclear fuel management; optimization; radiation effects in electronics; energy modeling. *Mailing Add:* Eng Dept Univ Miami PO Box 248294 Coral Gables FL 33124-0620

BECKER, MARTIN JOSEPH, CELL MARKER ANALYSIS. *Educ:* Weizmann Inst Sci, PhD(chem immunol), 73. *Prof Exp:* DIR BIOL & COLLAB RES, SYVA CO, 80- *Res:* Infectious diseases. *Mailing Add:* Syva Co 900 Arastradero Rd Palo Alto CA 94303

BECKER, MICHAEL ALLEN, b New York, NY, Oct 3, 40; m 80; c 4. RHEUMATOLOGY. *Educ:* Univ Pa, BA, 61, MD, 65. *Prof Exp:* From intern & resident int med, Barnes Hosp, Wash Univ, 65-70; res assoc biochem, NIH, 66-69; 312-702-6899; fel rheumatol, 70-72, Univ Calif, San Diego, 70-72, from asst prof to assoc prof med, 77-79; PROF, MED, UNIV CHICAGO, 80- *Concurrent Pos:* Clin investr, US Vet Admin, 74-77; vis investr, Imperial Cancer Res Fund, 79-80; Guggenheim Found Fel, 79; sr res investr, Univ Mich, Ann Arbor, 87-88; prin investr, NIH res grants, 75- *Mem:* Asn Am Physicians; Am Rheumatism Asn; Am Col Physicians; Am Fedn Clin Res; AAAS; Am Soc Clin Invest. *Res:* Regulation of human biosynthetic process; investigation of inherited defects in purine metabolism. *Mailing Add:* Dept Med Rheumatol Sect Univ Chicago Pritzker Sch Med 5841 S Maryland Ave Box 74 Chicago IL 60637

BECKER, MICHAEL FRANKLIN, b Cheverly, Md, Nov 3, 47; m 70. ELECTRICAL ENGINEERING, LASER APPLICATIONS. *Educ:* Johns Hopkins Univ, BES, 69; Stanford Univ, MS, 70, PhD(elec eng), 73. *Prof Exp:* Res engr microwave develop, US Naval Res Lab, 66-69, res engr solid state develop, 69-71; res asst laser appln, Microwave Lab, Stanford Univ, 71-73; ASST PROF ELEC ENG, UNIV TEX, AUSTIN, 73- *Mem:* Inst Elec & Electronics Engrs; Optical Soc Am; Am Inst Physics; AAAS; Nat Soc Prof Engrs; Sigma Xi. *Res:* Nonlinear optics; multiphoton excitation; picosecond laser applications to semiconductor materials; optical processing. *Mailing Add:* Dept Elec Eng Univ Tex Austin TX 78712

BECKER, MILTON, b Chicago, Ill, July 13, 20; m 53; c 1. SOLID STATE PHYSICS. *Educ:* Univ Chicago, BS, 41; Purdue Univ, PhD(physics), 51. *Prof Exp:* Physicist, Electronics, Naval Res Labs, 42-47; physicist, semiconductors, Hughes Semiconductors, 51-58; PHYSICIST, SEMICONDUCTORS, CONTINENTAL DEVICE CORP, 58-, VPRES RELIABILITY, 61- *Concurrent Pos:* Consult, AEC Proj, 56-57; tech dir, Epidyne, Inc, 73-74. *Mem:* Am Phys Soc; Sigma Xi. *Res:* Microwave propagation in ionized media; optical properties of silicon and germanium in the infrared; electrical properties of semiconductor junctions. *Mailing Add:* 3100 Corda Dr Los Angeles CA 90049

BECKER, NORWIN HOWARD, b New York, NY, Apr 23, 30; m 51; c 3. PATHOLOGY, HISTOCHEMISTRY. *Educ:* Cornell Univ, AB, 50, MNS, 52; State Univ NY, MD, 55. *Prof Exp:* Intern med, Montefiore Hosp, NY, 55-56; asst resident path, Long Island Jewish Hosp, 56-57; asst resident, Montefiore Hosp, 57-58, Nat Cancer Inst trainee, 58-59, from asst pathologist to assoc pathologist, 59-67; from asst prof to assoc prof, 64-74, PROF PATH, ALBERT EINSTEIN COL MED, 74-; ATTEND PATHOLOGIST, MONTEFIORE HOSP, 67- *Mem:* Am Asn Path & Bact; Am Soc Exp Path; Histochem Soc; Am Soc Clin Pathologists. *Res:* Histochemistry and electron microscopy of the central nervous system; choroid plexus function; surgical pathology; hematopathology. *Mailing Add:* Dept Path Montefiore Med Ctr 111 E 210th St Bronx NY 10467

BECKER, RALPH SHERMAN, b Benton Harbor, Mich, Mar 14, 25; c 3. MOLECULAR SPECTROSCOPY, PHOTOPHYSICS. *Educ:* Univ Vt, BS, 49; Univ NH, MS, 50; Fla State Univ, PhD(chem), 55. *Prof Exp:* Dir phys & inorg res, Frederick S Bacon Labs, 52; from asst prof to assoc prof, 54-63, PROF CHEM, UNIV HOUSTON, 63- *Concurrent Pos:* Fulbright vis prof, Univ Barcelona, 62-63; vis prof, Weizmann Inst, 63 & 79, Univ Tubingen, Ger, 73 & 75, Univ Kyushu, 71, Univ Zurich, 71, Univ Paris, 76 & 78, Univ Wyo, 75, Univ Tokyo, 80, Inst Solid State Physics, 80, Univ Quebec, 89, Univ Coimbra, 90, Univ Perugia, Italy, 90, Univ Dublin, Ireland, 91; electronic transition in molecules, NSF, 64; consult, Phillips Petrol Co & others; sr fel, Sci Res Coun Eng, 81-83; vis, Soviet Union, Byelorussian Acad Sci, 85; fulbright vis prof, Lisbon, 88. *Honors & Awards:* Japanese Soc Promotion Sci Award, 71; Southwest Regional Award, Am Chem Soc, 81; Southeast Tex Sect Award, Am Chem Soc, 80, Esther Farfel Res Award, 79; Fulbright vis prof, Lisbon, 88. *Mem:* Int Soc Photobiol; Am Chem Soc; Interamerican Photochem Soc; Europ Photochem Asn. *Res:* Intercombinations of electronic states; photochemistry; visual pigments; photoisomerization; laser flash photolysis; photophysics; photochromism; liquid crystals; photosensitizers; non-linear optics. *Mailing Add:* Dept Chem Univ Houston Houston TX 77204-5641

BECKER, RANDOLPH ARMIN, b Bowler, Wis, Dec 29, 24; m 56; c 3. OPTICS, INSTRUMENTATION. *Educ:* Tex Col Arts & Indust, BS, 49, MS, 53. *Prof Exp:* Observing asst astron, Univ Chicago MacDonald Observ, Tex, 50; physicist, White Sands Missile Range, 50-57, chief optical res sect, Range Instrumentation Develop Div, 57-61; sr res engr space optics, Jet Propulsion Lab, Calif Inst Technol, 61-63, res specialist, 63-70, staff scientist, Space Photog Sect, 70-73; sr scientist, Xerox Electro-Optical Systs, 73-84; RETIRED. *Res:* Optical instrumentation; materials suitable for optical systems in interplanetary space; design and execution of television experiments on planetary space missions; deep-space mission design. *Mailing Add:* 616 E Mendocino Altadena CA 91001

BECKER, RICHARD ALAN, b Cincinnati, Ohio, Nov 20, 49; m 79. STATISTICAL COMPUTING, DATA ANALYSIS. *Educ:* Harvard Univ, AB, 72. *Prof Exp:* Engr, Procter & Gamble, 72-73; programmer, Nat Bur Econ Res, 73-74; MEM TECH STAFF, AT&T BELL LABS, 74- *Concurrent Pos:* Consult, New Eng Med Ctr, 73; adv bd mem, Comput Res Lab, NMex State Univ, 86-; mem, Mgt Comt, J Computational & Graphical Statist, 90-; prog chair graphics, Am Statist Asn, 90-91. *Honors & Awards:* Youden Prize, Technometrics, 87. *Mem:* Fel Am Statist Asn; Asn Comput Mach; Inst Math Statist; Inst Elec & Electronics Engrs Computer Soc. *Res:* Computer graphics; statistical computing and data analysis; created S language; auditing of data analyses; brushing scatterplots; dynamic graphics for statistics. *Mailing Add:* AT&T Bell Labs Rm 2C-259 600 Mountain Ave Murray Hill NJ 07974-2070

BECKER, RICHARD LOGAN, b Dayton, Ohio, Dec 29, 29; m 54; c 2. THEORETICAL NUCLEAR PHYSICS, ATOMIC PHYSICS. *Educ:* Harvard Univ, BA, 52; Yale Univ, MS, 53, PhD(physics), 57. *Prof Exp:* PHYSICIST, OAK RIDGE NAT LAB, 57- *Concurrent Pos:* Vis fel, Princeton Univ, 59-60; lectr & prof, Univ Tenn, 61- *Mem:* Fel Am Phys Soc. *Res:* Theoretical many-body physics; nuclear self-consistent field theory; nuclear structure theory; nuclear scattering theory; ion atom collision. *Mailing Add:* 2780 Oak Ridge Turnpike Oak Ridge TN 37830

BECKER, ROBERT ADOLPH, b Tacoma, Wash, Feb 10, 13; m 44; c 5. PLANETARY SCIENCE. *Educ:* Col Puget Sound, BS, 35; Calif Inst Technol, MS, 37, PhD(physics), 41. *Prof Exp:* Asst physics, Col Puget Sound, 33-35; asst fel & asst nuclear physics, Calif Inst Technol, 35-41; asst physicist, Dept Terrestrial Magnetism, Carnegie Inst, 41; from asst physicist to physicist, Nat Bur Standards, 41-43; physicist, Radiation Lab, Univ Calif, 43; mem staff appl physics lab, Univ Wash, 43-45, sr physicist, 45; from asst prof to prof physics, Univ Ill, 46-60; dir space physics lab, 60-68, assoc gen mgr lab opers, 68-73, consult physicist, Aerospace Corp, 73-75; RETIRED. *Concurrent Pos:* Guggenheim fel, 58-59. *Mem:* Fel AAAS; NY Acad Sci; fel Am Phys Soc; Am Astron Soc; fel Explorers Club. *Res:* Nuclear, space and atmospheric physics; astrophysics; proton and deuteron induced nuclear reactions; radioactive decay schemes from betatron produced radioactivities; element synthesis in the stars; structure of the upper atmosphere; planetary science. *Mailing Add:* PO Box 4609 Carmel CA 93921

BECKER, ROBERT HUGH, b Greenville, Pa, July 19, 34; m 63; c 7. ANALYTICAL CHEMISTRY, CATALYSIS. *Educ:* St Bonaventure Univ, BS, 56; Univ Notre Dame, PhD(org chem), 60. *Prof Exp:* NIH res assoc org chem, NMex Highlands Univ, 59-61; instr chem, Lawrence Col, 61-63; from asst prof to assoc prof, Gannon Col, 63-75, dir, Sci Res Prog, 69-75; SR RES CHEMIST, CALSICAT DIV, MALLINCKRODT INC, 75- *Concurrent Pos:* Petrol Res Fund res grant, 63-65; consult, Calsicat Div, Mallinckrodt Inc, 67-75. *Mem:* Am Chem Soc; Am Oil Chem Soc. *Res:* Instrumental and qualitative organic analysis; spectroscopy; chromatography; catalysis. *Mailing Add:* Calsicat Div Mallinckrodt Inc 1707 Gaskell Ave Erie PA 16503

BECKER, ROBERT O, b Rivers Edge, NJ, May 31, 23; m 46; c 3. ORTHOPEDIC SURGERY, BIOPHYSICS. *Educ:* Gettysburg Col, BA, 46; NY Univ, MD, 48. *Prof Exp:* Chief, Orthop Sect, Vet Admin Hosp, Syracuse, 56-80, med investr, 72-80; from instr to assoc prof, 56-66, PROF ORTHOP SURG, STATE UNIV NY UPSTATE MED CTR, 66- *Concurrent Pos:* Mem, Spec Study Sect Biomed Eng, NIH; mem, Sanguine Study Comt, US Navy; assoc chief of staff for res, Vet Admin Hosp, Syracuse, 65-72; consult biomed effects electromagnetic environ factors, 80-; dir, res, Becker Biomagnectics, Lowville, NY, 86- *Honors & Awards:* William S Middleton Award, 64; Nicholas Andre Award, 78. *Mem:* Inst Elec & Electronics Engrs; NY Acad Sci; Int Soc Bioelec; Bioelectromagnetics Soc. *Res:* Biological solid state; organic semiconductors; biological control systems as applied to growth control; regenerative mechanisms and relationship to central nervous system to physical factors of environment; bioeffects of electromagnetic fields; therapeutic effects electromagnetic fields. *Mailing Add:* Star Route Becker Biomagnetics Inc Lowville NY 13367

BECKER, ROBERT PAUL, b Murphysboro, Ill, Oct 15, 42; m 64; c 2. ENDOTHELIAL CELLS, CYTOCHEMISTRY. *Educ:* Univ Chicago, PhD(anat), 73. *Prof Exp:* Res assoc cell biol, Univ Chicago, 73-77; instr anat, 77-78; asst prof anat, 78-83, ASSOC PROF ANAT & CELL BIOL, COL MED, UNIV ILL, CHICAGO, 83- *Mem:* Am Soc Cell Biol; Histochem Soc; Am Asn Anatomists; Sigma Xi. *Res:* Cell surface structure and function; endocytosis; exocytosis; cell recognition; endothelial cell biology. *Mailing Add:* Dept Anat & Cell Biol MC 512 Univ Ill Col Med 808 S Wood St Chicago IL 60612

BECKER, ROBERT RICHARD, b Aitkin, Minn, Feb 16, 23; m 56; c 2. BIOCHEMISTRY. *Educ:* Univ NDak, BS, 48; Univ Wis, MS, 51, PhD(biochem), 52. *Prof Exp:* From instr to asst prof chem, Columbia Univ, 52-60; biochemist, Oak Ridge Nat Labs, 60-62; assoc prof chem, Ore State Univ, 62-67, prof biochem, 67-90, chmn biol prog, 78-86; RETIRED. *Concurrent Pos:* Vis scientist, Dept Biol, Brookhaven Nat Labs, 68-69; mem, NIH Pathobiol Chem Study Sect, 75-77, Biomed Sci Study Sect, 81-83, chmn, 83; vis scholar oceanog, Univ Wash, 86. *Mem:* AAAS; Am Soc Biol Chemists; Am Chem Soc. *Res:* Protein chemistry; polypeptides; thermophilic enzymes. *Mailing Add:* 3406 NW Polk Ave Corvallis OR 97330

BECKER, ROGER JACKSON, b Findlay, Ohio, June 9, 44; c 1. SOLID STATE PHYSICS, RAMAN SPECTROSCOPY. *Educ:* Lake Forest Col, BA, 67; Mass Inst Technol, MS, 72; Thomas Jefferson Univ, PhD(solid state physics), 77. *Prof Exp:* Fel, Univ Guelph, 76-77; RES PHYSICIST, UNIV DAYTON, 77- *Mem:* Am Inst Physics; Am Phys Soc. *Res:* Raman spectroscopy, surface chemistry and surface Raman, layered compound, turbulent flow and laser diagnostics in combustion; lattice dynamics; fluid dynamics. *Mailing Add:* Dept Physics Univ Dayton Dayton OH 45469

BECKER, ROLAND FREDERICK, b Methuen, Mass, Aug 12, 12; m 36; c 2. ANATOMY. *Educ:* Mass State Col, BS, 35, MS, 37; Northwestern Univ, PhD(anat), 40. *Prof Exp:* Asst psychol, Mass State Col, 35-37; asst anat, Med Sch, Northwestern Univ, 38-40, instr, 40-44, assoc, 44-45, asst prof, 45-46; assoc prof, Sch Med, Univ Wash, 47, chmn dept, 47-48; dir neurol div, Daniel Baugh Inst Anat, Jefferson Med Col, 49-51; assoc prof anat & dir lab neonatal sci, Duke Univ, 52-69; prof anat, Col Osteop Med, Mich State Univ, 69-71, prof biomechanics & coordr res prog, 71-75, coordr educ prog, 75-76; prof anat, Sch Med, E Carolina Univ, 76-78, emer prof, 78-87; RETIRED. *Concurrent Pos:* Consult, Eng Corps, US Army, 52-54, NIH, 57-59 & Greenbrier Col Osteop Med, 74; AID med educr, Thailand, 61-63; deans consult, Sch Med, East Carolina Univ, 78-87. *Mem:* Am Asn Anatomists; Soc Reproduction. *Res:* Neuroanatomy and fetal physiology; behavioral development; neuro-musculo-skeletal system; high altitude research; olfactory discrimination. *Mailing Add:* 102 Heartwood Dr Grimesland NC 27837

BECKER, SHELDON THEODORE, b Chicago, Ill, Sept 26, 38; m 74; c 1. ELECTRICAL ENGINEERING. *Educ:* Univ Ill, BS, 60, MS, 61; Ill Inst Technol, PhD(elec eng), 67. *Prof Exp:* Teaching asst, Ill Inst Technol, 62-63, instr elec eng, 63-67; mem tech staff, Bell Labs, 67-83, MEM TECH STAFF, BELL COMMUN RES, 83- *Mem:* Inst Elec & Electronics Engrs; Asn Comput Mach. *Res:* Computer systems, performance and evaluation, expert systems; Operations research. *Mailing Add:* 142 Flanders Dr Somerville NJ 08876

BECKER, STANLEY J, b New York, NY, Oct 21, 21; m 53; c 3. STRUCTUAL DYNAMICS, APPLIED MATHEMATICS. *Educ:* Cooper Union, BME, 48; Polytech Inst Brooklyn, MS, 53; Univ Pittsburgh, PhD(math), 64. *Prof Exp:* Jr engr, NYC Dept Pub Works, 48-51; stress analyst, heavy press prog, Loewy-Hydropress Div, Baldwin-Lima-Hamilton Corp, 51-57, consult stress anal, 68-69; sr engr, Bettis Atomic Power Lab, Westinghouse Elec Corp, 57-62; engr, Re-entry Systs Dept, Gen Elec Co, 62-68, specialist comput sci, Space Systs Div, 69; res engr, Gilbert Assocs, Inc, 69-76; sr engr, Nuclear Components Div, Allis-Chalmers Corp, 77; RETIRED. *Concurrent Pos:* Instr elec eng, Mass Inst Technol & King of Prussia Grad Ctr, Pa State Univ; sr eng fel, Villanova Univ, 90- *Mem:* Am Soc Mech Eng; Am Math Soc; Am Inst Physics; Am Acad Mech. *Res:* Pressure vessels; power piping. *Mailing Add:* 474 Regimental Rd King of Prussia PA 19406

BECKER, STANLEY LEONARD, b Syracuse, NY, June 19, 29; m 52; c 2. SCIENCE EDUCATION, HISTORY OF SCIENCE. *Educ:* NY State Col Environ Sci & Forestry, BS, 53; Univ Wis-Madison, PhD(hist sci), 68. *Prof Exp:* Chemist, Bristol Labs, Bristol-Myers Co, 52-60; asst prof gen sci, Univ Hawaii, 67-68; asst prof physics, 68-71, ASSOC PROF SCI, BETHANY COL, WVA, 71-, PROF PHILOS & COMPUT SCI, 83- *Concurrent Pos:* Vis prof hist sci, Conn Agr Exp Sta, 75-76; vis prof geog, Westminster Col, 75, 77, 79, 81 & 83. *Mem:* Hist Sci Soc; AAAS; NY Acad Sci. *Res:* History of agriculture, biochemistry and vitamins; hybrid corn; agriculture and patents. *Mailing Add:* Dept Math Bethany Col Bethany WV 26032

BECKER, STEPHEN FRALEY, b Toledo, Ohio, Aug 13, 42; m 64. THEORETICAL PHYSICS, COMPUTER SCIENCE. *Educ:* Miami Univ, AB, 65; Rutgers Univ, MS, 66, PhD(physics), 69. *Prof Exp:* Asst prof, 69-80, ASSOC PROF PHYSICS & HEAD DEPT, BUCKNELL UNIV, 80- *Concurrent Pos:* Instr, Lewisburg Fed Penitentiary, 71. *Mem:* Am Inst Physics; Am Phys Soc; Am Math Soc. *Res:* Three body problem; scattering theory; theoretical particle physics; complexity theory. *Mailing Add:* Dept Physics Bucknell Univ Lewisburg PA 17837

BECKER, STEVEN ALLAN, b Anamosa, Iowa, Nov 23, 38; m 61; c 2. PLANT ANATOMY, MORPHOLOGY. *Educ:* Cornell Col, BA, 60; Univ Iowa, MS, 66, PhD(bot), 68. *Prof Exp:* Asst prof, 68-74, assoc prof bot, Eastern Ill Univ, 74-, PROF BOT. *Mem:* Bot Soc Am; Am Inst Biol Sci. *Res:* Developmental plant anatomy-morphology. *Mailing Add:* Dept Bot Eastern Ill Univ Charleston IL 61920

BECKER, ULRICH J, b Dortmund, Ger, Dec 17, 38; m 66; c 3. HIGH ENERGY PHYSICS. *Educ:* Univ Marburg, Vordiplom, 60, Univ Hamburg, Diplom, 64, PhD(physics), 68, Dr habil, 76. *Prof Exp:* Sci employee exp physics, Deutsches Elektronen-Synchrotron, Hamburg, 64-68; res assoc high energy physics, 68-70, from vis asst prof to assoc prof, 77-77, PROF PHYSICS, MASS INST TECHNOL, 77- *Concurrent Pos:* Res coun mem, Deutsches Elektronen-Synchrotron, Hamburg, 70-71 & 80-82; mem, LEP Sci Coun, 85- & Nuclear Res Coun Europ, Geneva, Switz. *Mem:* Am Phys Soc. *Res:* High energy particle physics; co-discoverer of the new meson J (3.1) and study of the variety of these new particles, particularly vector mesons. *Mailing Add:* Dept Physics Mass Inst Technol 77 Massachusetts Ave Cambridge MA 02139

BECKER, WALTER ALVIN, b Pittsburgh, Pa, Dec 1, 20; m 52; c 2. GENETICS. *Educ:* Stanford Univ, AB, 52; Univ Calif, Berkeley, MS, 55, PhD(genetics), 56. *Prof Exp:* Res asst genetics, Univ Calif, 53-56; asst poultry scientist, Western Wash Exp Sta, Puyallup, 56-60; asst prof poultry sci & asst poultry scientist, 60-62, assoc poultry scientist, 62-68, from assoc prof to prof genetics, 62-82, actg chmn prog genetics, 67-68, poultry scientist, 68-82, chmn, prog genetics & cell biol, 79-82, EMER PROF GENETICS, WASH STATE UNIV, 83-; PRES, ACAD ENTERPRISES, 83- *Concurrent Pos:* Consult, Wash State Univ & US Forest Serv; vis researcher, Inst Animal Genetics, Univ Edinburgh, 64-65. *Mem:* Fel AAAS; Poultry Sci Asn. *Res:* Quantitative genetics. *Mailing Add:* SE 320 Nebraska St Pullman WA 99163-2240

BECKER, WAYNE MARVIN, b Waukesha, Wis, May 29, 40; m 63; c 2. PLANT PHYSIOLOGY, BIOCHEMISTRY & MOLECULAR BIOLOGY. *Educ:* Univ Wis-Madison, BS, 63, MS, 65, PhD(biochem), 67. *Prof Exp:* NATO fel molecular biol, Beatson Inst Cancer Res, Scotland, 67-68, NIH fel, 68-69; from asst prof to assoc prof bot, 69-75, PROF BOT, UNIV WIS-MADISON, 75- *Concurrent Pos:* Guggenheim res fel, Univ Edinburgh, 75-76; Midwest Univs Consortium Int Activ consult, chem, Univ Indonesia, 86. *Mem:* AAAS; Am Soc Plant Physiol; Int Soc Plant Molecular Biol. *Res:* Regulation of eukaryotic gene expression; developmental biochemistry of seed germination; nucleic acid metabolism in plants. *Mailing Add:* Dept Bot Univ Wis Madison WI 53706

BECKER, WILHELM, b Konigsberg, Ger, Feb 17, 42. FREE ELECTRON DEVICES, MULTIPHOTON PROCESSES. *Educ:* Tech Univ, Munich, dipl, 67, Drrernat, 69, Drhabil, 76. *Prof Exp:* Asst prof, Univ Tubingen, 70-79; vis scientist, Max-Planck Instit Quantenoptik, 80; vis scientist, physics Inst Mod Optics, 81-84, res assoc prof, 84-86, ASSOC PROF, DEPT PHYSICS & ASTRON, UNIV NMEX, 86- *Concurrent Pos:* Privatdozent, Univ Tubingen, 76-86. *Mem:* Am Phys Soc; Sigma Xi. *Res:* Interaction of matter with intense electromagnetic fields; multiphoton ionization, free-electron lasers, laser induced damage to thin films, quantum optics, quantum electrodynamics. *Mailing Add:* Dept Physics & Astron Univ NMex Albuquerque NM 87131

BECKERLE, JOHN C, b Mt Vernon, NY, Feb 14, 23; div; c 2. PHYSICAL OCEANOGRAPHY. *Educ:* Manhattan Col, BS, 48; Cath Univ Am, PhD(physics), 53. *Prof Exp:* Asst physics, Cath Univ Am, 48-51; physicist, Electronic Comput, Appl Physics Labs, Johns Hopkins Univ, 50; res assoc chem, Naval Ord Contract, Cath Univ Am, 51-53; mem tech staff, Bell Tel Labs, 53-60; mgr sonics res, Schlumberger Well Surv Corp, 60-64; assoc scientist, Woods Hole Oceanog Inst, 64-80; MEM STAFF, SCI APPLN RES ASSOC U SCI APPLN RES CORP AM, 80- *Honors & Awards:* Radford Gold Medal Physics. *Mem:* Acoust Soc Am. *Res:* Mathematical physics; electromagnetic and acoustic wave propagation; underwater sound; heat conduction in solids; sonic well logging; optical signal processing; internal gravity waves in the ocean; Rossby waves in the Sargasso Sea; geophysics; atmospheric sciences; earth sciences (seismology). *Mailing Add:* Sci Appln Res Assoc U Sci Appln Res Corp Am 38 Two Ponds Rd Falmouth MA 02540

BECKERMAN, BARRY LEE, b New York, NY, Jan 12, 41; m 65; c 2. OPHTHALMOLOGY. *Educ:* Cornell Univ, AB, 61; NY Univ, MD, 65. *Prof Exp:* Intern surgery, Tufts New Eng Med Ctr, 65-66; Lt Comdr surgery, USPHS, 66-68; from resident to chief resident ophthal, NY Univ-Bellevue Med Ctr, 68-72; instr, Albert Einstein Col Med, 72-73; dir electrophysiol, Retina Serv, Montefiore Hosp, Bronx, NY, 73-86; ASST CLIN PROF OPHTHAL, ALBERT EINSTEIN COL MED, BRONX, NY, 73- *Concurrent Pos:* Adj attend ophthal, Northern Westchester Hosp, Mt Kisco, NY, 73-81, chief ophthal, 81-; assoc attend, Westchester County Med Ctr, 79- *Mem:* Asn Res Vision & Ophthal; NY Acad Sci. *Res:* Clinical research in the diagnosis of ocular and especially retinal disorders, particularly development of diagnostic techniques and application to clinical disease; clinical practice. *Mailing Add:* 344 Main St Mt Kisco NY 10549

BECKERMANN, CHRISTOPH, b Hildesheim, Ger, July 24, 60; m 86; c 2. THERMAL SCIENCES, MATERIALS PROCESSING. *Educ:* Purdue Univ, MS, 84, PhD(mech eng), 87. *Prof Exp:* ASST PROF MECH ENG, UNIV IOWA, 87- *Concurrent Pos:* NSF presidential young investr award, 89- *Mem:* Am Soc Mech Engrs; Am Soc Eng Educ; Minerals Metals Mining Soc. *Res:* Heat transfer and fluid mechanics; transport phenomena in materials processing; solidification and welding; porous media; natural and double-diffusive convection; electronic cooling. *Mailing Add:* Dept Mech Eng Univ Iowa Iowa City IA 52242

BECKERS, JACQUES MAURICE, b Arnhem, Neth, Feb 14, 34; nat US; m 59; c 2. ASTROPHYSICS, OPTICS. *Educ:* Univ Utrecht, Drs, 59, DrAstron, 64. *Prof Exp:* Res physicist, High Altitude Observ, 62-64, Sacramento Peak Observ, 64-69 & High Altitude Observ, 69-70; res physicist, Sacramento Peak Observ, 70-79; dir, Mult Mirror Telescope Observ, 79-84; dir advan develop prog, Nat Optical Astron Observ, 84-88; EXP ASTROPHYSICIST, EUROP SOUTHERN OBSERV, 88- *Honors & Awards:* Henryk Arctowski Medal, Nat Acad Sci, 75. *Mem:* Am Astron Soc; Netherlands Astron Soc; Int Astron Union; Norweg Acad Sci. *Res:* Solar spectra; fine structures in the solar atmosphere; instrumentation for solar and stellar research; sunspots; interferometry; large telescopes; adaptive optics. *Mailing Add:* Europ Southern Observ Karl-Schwarzschild-Strasse Two D-8046 Garching bei Meunchen Germany

BECKERT, WILLIAM HENRY, b New York, NY, Sept 10, 20; m 48; c 2. CYTOLOGY, HISTOLOGY. *Educ:* St John's Univ, NY, BS, 49, MS, 51; NY Univ, PhD(cytol), 61. *Prof Exp:* From instr to assoc prof, 51-66, PROF BIOL, ST JOHN'S UNIV, NY, 66- *Mem:* AAAS; Am Soc Cell Biol; Am Inst Biol Sci; NY Acad Sci; Tissue Cult Asn; Sigma Xi. *Res:* Chromosome banding and sister chromatid exchange; karyotypes of nonmammalian vertebrates; cell cultures; sister chromatid exchange in cells from normal animal and those treated with carcinogenic agents or in pre-cancerous conditions. *Mailing Add:* Dept Biol St Johns Hosp Jamaica NY 11432

BECKETT, JACK BROWN, b Hutsonville, Ill, Aug 10, 25; m 70; c 1. PLANT GENETICS. *Educ:* Univ Wash, BS, 50; Univ Wis, MS, 52; PhD(hort genetics), 54. *Prof Exp:* Asst hort, Univ Wis, 50-54; plant geneticist, USDA, Univ Ill, 54-63, plant geneticist, USDA, Univ Mo, Columbia, 63-90, asst prof agron, 70-91; CONSULT, 91- *Concurrent Pos:* Asst prof genetics, Univ Mo-Columbia, 67-70. *Mem:* AAAS; Am Soc Agron; Genetics Soc Am; Am Genetic Asn. *Res:* Genetics and cytogenetics of maize, with particular emphasis on development and use of B-A translocations. *Mailing Add:* 607 Longfellow Lane Columbia MO 65203

BECKETT, JAMES REID, b Los Angeles, Calif, Apr 27, 46; m 70; c 1. ANALYTICAL CHEMISTRY, ORGANIC CHEMISTRY. *Educ:* Ft Hays State Univ, BS, 72; Univ Wyo, PhD(chem), 78. *Prof Exp:* Sr lab technician chem, Chemagro Agr Div, Mobay Chem Corp Kansas City, 67-71; consult chem, Wyo Anal Labs Inc, Laramie, 77-78; res chemist, Cities Serv Co, 78-81, region petrol engr, 81-86; EOR supvr, 87-89, MGR SPEC PROJ & FINANCIAL CONSULT, OX4 USA INC, 90- *Concurrent Pos:* Res assoc, Univ Wyo, 78. *Mem:* Am Chem Soc; Soc Petrol Engrs. *Mailing Add:* OX4 USA Inc Box 26100 Oklahoma City OK 73126

BECKETT, RALPH LAWRENCE, b Salt Lake City, Utah, June 6, 23; m 46; c 3. SPEECH PATHOLOGY. *Educ:* Univ Southern Calif, BA, 50, MA, 53, PhD(commun disorders), 68. *Prof Exp:* Asst prof speech, Los Angeles Harbor Col, 57-67; res fel, Univ Mo, Columbia, 68-69, asst prof speech path, 68-70; assoc prof, Calif State Univ, Fullerton, 70-78, prof commun disorders, 78-89; CONSULT, 89- *Concurrent Pos:* Vis lectr, Lincoln Univ, 68-69; consult, Speech & Lang Inst, Orange County, Calif, 72- *Mem:* Am Speech & Hearing Asn. *Res:* Laryngeal physiology and pathology; physiological phonetics. *Mailing Add:* 1238 Venice Ave Placentia CA 92670

BECKETT, ROYCE E(LLIOTT), b Hinch, Mo, May 31, 23; m 54; c 3. ENGINEERING MECHANICS. *Educ:* Univ Ill, BS, 44, MS, 49; Wash Univ, DSc(appl mech), 53. *Prof Exp:* From asst prof to prof mech, Univ Iowa, 53-68; dir, Res & Eng Lab, Army Weapons Command, 68-77; PROF MECH ENG & HEAD DEPT, AUBURN UNIV, 77- *Mem:* Am Soc Mech Engrs; Soc Indust & Appl Math; Sigma Xi. *Res:* Response of structures to dynamic loads; thermal stresses; numerical solutions. *Mailing Add:* 1235 Hickory Lane Auburn AL 36830

BECKETT, SIDNEY D, b Bruce, Mich, Feb 11, 32. CARDIOVASCULAR PHYSIOLOGY. *Educ:* Univ Mo, PhD(physiol), 66. *Prof Exp:* ASSOC DEAN RES & GRAD STUDIES, COL VET MED, AUBURN UNIV, 81- *Mem:* Am Physiol Soc; Am Vet Asn. *Mailing Add:* Rt 3 Box 101 Auburn AL 36830

BECKFIELD, WILLIAM JOHN, b Pittsburgh, Pa, Aug 25, 20; m 45; c 8. PATHOLOGY. *Educ:* Allegheny Col, AB, 41; Univ Pa, MD, 44. *Prof Exp:* Asst prof path, Grad Sch Med, Univ Pa, 55-61, assoc prof, 61-62; PATHOLOGIST, LUTHER HOSP, 62- *Concurrent Pos:* Consult pathologist, Wyeth Labs, Am Home Prod Corp, 52-70; pathologist, Grad Hosp, Univ Pa, 54-56, vis lectr, Grad Sch Med, 63-67; pathologist, Philadelphia Gen Hosp, Pa, 56-62. *Mem:* Am Asn Pathologists; Reticuloendothelial Soc; Am Soc Clin Path; Col Am Path; AMA. *Res:* Medical and surgical pathology; experimental pathology, toxicology and medicine. *Mailing Add:* 1221 Whipple St Eau Claire WI 54701

BECKHAM, ROBERT R(OUND), b Toledo, Ohio, May 8, 15; m 45; c 3. CHEMICAL ENGINEERING. *Educ:* Univ Mich, BS, 42. *Prof Exp:* Lab asst, Libbey-Owens-Ford Glass Co, 34-37, 38-40 & Univ Mich, 40-41; asst prod supt, Dow Chem Co, 42-47; res chem engr, Libbey-Owens-Ford Glass Co, 47-59, res specialist, 59-64, asst dir res, 64-67, sr scientist-consult, 67-80; CONSULT MECH ENG, 80- *Res:* Organic glass substitutes; laminated glass; plastic dyeing; films on glass; vacuum equipment; special machines. *Mailing Add:* 2553 Charlestown Ave Toledo OH 43612-4327

BECKHOFF, GERHARD FRANZ, b Ger, Oct 13, 29; m 63; c 2. ELECTRICAL ENGINEERING, COMPUTER SCIENCE. *Educ:* Inst Technol Ger, Dipl Ing, 56; Southern Methodist Univ, MS, 60; Laval Univ, DSc(elec eng), 66. *Prof Exp:* Engr, Continental Electronics Mfg Co, Tex, 57-59; instr elec eng & res asst, Laval Univ, 60-65; lectr, 65-66, asst prof, 66-69, actg chmn dept, 70-77, ASSOC PROF COMPUT SCI, UNIV WESTERN ONT, 69-; VIS PROF, DEPT ELEC ENG, UNIV PETROL & MINERALS. *Concurrent Pos:* Fel, Inst Prog, Fac Sci, Univ Paris, 69-70. *Mem:* Inst Elec & Electronics Engrs. *Res:* Network synthesis; computer-assisted circuit design of analog and digital filters; sequential machine theory; theory of algebraic languages and codes. *Mailing Add:* Dept Elec Eng Univ Petrol & Minerals Dhahran Saudi Arabia

BECKHORN, EDWARD JOHN, b Ithaca, NY; c 1. INDUSTRIAL ENZYMES, FERMENTATION. *Educ:* Cornell Univ, BS, 47, MS, 48, PhD(microbiol, genetics), 50. *Prof Exp:* Res assoc, Dept Genetics, Carnegie Inst Wash, 50-52; dir microbiol res, Wallerstein Co Div, Baxter Labs, 52-57, from asst dir res to dir res, 57-70; mgr sci serv, Lehn & Fink Prod Co Div, Sterling Drug Co, 71-73; dir sci serv, Cunningham & Walsh Inc, 74-, vpres, 80-86; CONSULT, 86- *Mem:* Am Soc Microbiol; Soc Indust Microbiol. *Res:* Production and industrial or pharmaceutical applications of fermentation products, enzymes, antibiotics, organic acids; formulation, efficacy and toxicology of disinfectants, antimicrobials and consumer household chemical specialties; food processing with enzymes; regulatory affairs; advertising claim substantiation. *Mailing Add:* 435 Nassau Court Marco Island FL 33937-4014

BECKING, GEORGE C, biochemistry, for more information see previous edition

BECKING, RUDOLF (WILLEM), b Blora, Java, Indonesia, Oct 19, 22; m 52; c 3. ENVIRONMENTAL SCIENCES, NATURAL RESOURCES. *Educ:* Univ Wageningen, MF temperate forestry & MF trop forestry, 52; Univ Wash, PhD(forest mgt), 54. *Prof Exp:* Forest officer, Dutch Forest Serv, 54-56; asst prof, Univ NH, 56-57 & Pa State Univ, 57-58; assoc prof forestry, Auburn Univ, 58-60; from assoc prof to prof forestry, 60-70, prof natural resources, 70-83, EMER PROF NATURAL RESOURCES, HUMBOLDT STATE UNIV, 83- *Concurrent Pos:* Mem, Sta Int Geobot Mediterranean & Alpine, France, 47-; forest inventory expert, Foreign Agr Orgn, Indonesia, 80-81. *Mem:* AAAS; Am Inst Biol Sci; Ecol Soc Am; Am Soc Photogram; Asn Trop Biol. *Res:* Phytosociology; forest ecology; photo interpretation; tropical rain forest dynamics; forest growth and yield. *Mailing Add:* Sch Natural Resources Humboldt State Univ Arcata CA 95521

BECKINGHAM, KATHLEEN MARY, b Sheffield, UK, May 8, 46; m 86. DEVELOPMENTAL MOLECULAR BIOLOGY. *Educ:* Cambridge Univ, BA, 67, MA, 68, PhD(biochem), 72. *Prof Exp:* Post doc, Inst Molecular Biol, Denmark, 70-72; Nat Inst Med Res, Mill Hill, London, Eng, 72-74, staff mem, 74-76; res assoc biochem, dept microbiol, Med Sch, Univ Mass, Worcester, 76-78, instr, 78-80; asst prof, 80-85, ASSOC PROF BIOCHEM, RICE UNIV, HOUSTON, TEX, 85- *Mem:* Am Soc Biol Chemists; Am Soc Cell Biol; Soc Develop Biol; Genetic Soc Am. *Res:* Characterization of genes uniquely active in oogenesis; function of calmodulin by a gene modification approach; RNA gene studies. *Mailing Add:* Dept Biochem Rice Univ PO Box 1892 Houston TX 77251

BECKLAKE, MARGARET RIGSBY, b London, Eng, May 27, 22; Can citizen; m 48; c 2. OCCUPATIONAL LUNG DISEASE, RESPIRATORY PHYSIOLOGY. *Educ:* Univ Witwatersrand, SAfrica, MB, BCh, 44, MD, 51; MRCP, 46; FRCP, 72. *Hon Degrees:* MD, Univ Witwatersrand, 74. *Prof Exp:* Lectr med, Univ Witwatersrand, 50-54; physiologist, Med Bur Occup Dis, SAfrica, 54-57; asst physician, Royal Victoria Hosp, Montreal, 64-66, assoc physician, 66-73; from asst prof to assoc prof exp med, 61-71, assoc prof epidemiol, 67-72, PROF MED & EPIDEMIOL, MCGILL UNIV, 72-; SR PHYSICIAN, ROYAL VICTORIA HOSP, MONTREAL, 74- *Concurrent Pos:* Career investr, Med Res Coun Can, 68-; Med Res Coun Can, 75-80; hon prof community health, Univ Witwatersrand, 84-85. *Honors & Awards:* Louis Mark Mem Lectr, Am Col Chest Physicians, 82. *Mem:* Am Thoracic Soc; Can Thoracic Soc (pres, 78-79); Med Res Coun Can; fel Royal Soc Can. *Res:* Cardiac output responses to exercise in athletes; the effects of pneumocomiosis on lung function; the natural history of asbestos related pulmonary fibrosis and other asbestos related disease; lung disease in occupationally exposed workers; environmental lung disease; respiratory disease; epidemiology. *Mailing Add:* Dept Epidemiol & Biostat McGill Univ Montreal PQ H3A 1A3 Can

BECKLER, DAVID (ZANDER), science administration, for more information see previous edition

BECKLEY, RONALD SCOTT, b Lebanon, Pa, Feb 23, 44:; m 67; c 2. ORGANIC CHEMISTRY. *Educ:* Lebanon Valley Col, BS, 66; Univ Colo, PhD(chem), 70. *Prof Exp:* Fel chem, Ohio State Univ, 70-72; vis lectr, Univ Colo, 72-73; RES CHEMIST, ROHM & HAAS CO INC, 73- *Mem:* Am Chem Soc; Sigma Xi. *Res:* Polymer chemistry including coating, emulsion polymerization and polymer synthesis. *Mailing Add:* 513 Kleman Rd Gilbertsville PA 19525

BECKMAN, ALEXANDER LYNN, b Los Angeles, Calif, June 9, 41; m 62, 83; c 2. NEUROPHYSIOLOGY. *Educ:* Univ Calif, Los Angeles, BA, 64; Univ Calif, Santa Barbara, PhD(psychol), 68. *Prof Exp:* USPHS fel, Inst Neurol Sci, 68-70, asst prof physiol, Sch Med, Univ Pa, 71-78; mem staff, Alfred I DuPont Inst, 78-86, sr res scientist, 78-86; PROF PSYCHOL, CALIF STATE UNIV, LONG BEACH, 86- *Concurrent Pos:* Univ Pa Plan to Develop Scientists in Med Res fel, 70-73. *Mem:* Soc Neurosci; Am Physiol Soc; Int Hibernation Soc. *Res:* Central nervous system mechanisms controlling body temperature; changes in central nervous system function during hibernation; actions of endogenous opioid peptides in central nervous system. *Mailing Add:* Dept Psychol Calif State Univ 120 Bellflower Blvd Long Beach CA 90840

BECKMAN, ARNOLD ORVILLE, b Cullom, Ill, Apr 10, 00; m 25; c 2. CHEMISTRY. *Educ:* Univ Ill, BS, 22, MS, 23; Calif Inst Technol, PhD(photochem), 28. *Hon Degrees:* LLD, Univ Calif, Riverside, Loyola Univ, Los Angeles & Pepperdine Univ; DSc, Chapman Col & Whittier Col. *Prof Exp:* Res engr, Bell Tel Labs, 24-26; instr chem, 26-29, asst prof, 29-40, Chairman Emer, Calif; vpres, Nat Tech Labs, 37-39, pres, 39-50, Arnold O Beckman, Inc, 42-57, pres, 50-65, FOUNDER & CHMN, BECKMAN INSTRUMENTS, INC, 50- *Concurrent Pos:* Pres, Helipot Corp, 44-58; trustee, Calif Inst Res Found & Calif Mus Found; chmn bd trustees, Syst Develop Found; dir, Scripps Clin & Res Found & Ear Res Found; chmn emer bd, Calif Inst Technol. *Honors & Awards:* Vermilye Medal, 87; Millikan Award, Calif Inst Technol, 85; Gold Medal, Am Inst Chemists, 87; Nat Medal Sci,89. *Mem:* Nat Acad Eng; hon fel Am Acad Arts & Sci; Instrument Soc Am (pres, 52); fel AAAS; hon fel Am Inst Chem; fel Asn Clin Scientists; fel Royal Soc Arts. *Res:* Applied chemistry; development of scientific instruments; photochemistry. *Mailing Add:* Beckman Ctr Nat Acad 100 Academy Dr Irvine CA 92715

BECKMAN, BARBARA STUCKEY, ERYTHROLEUKEMIA, HEMATOPHARMACOLOGY. *Educ:* Johns Hopkins Univ, PhD(pharmacol), 78. *Prof Exp:* ASSOC PROF PHARMACOL, SCH MED, TULANE UNIV, 78- *Res:* Arachidonic acid metabolism; cyclic nucleotides. *Mailing Add:* Dept Pharmacol Sch Med Tulane Univ 1430 Tulane Ave New Orleans LA 70112

BECKMAN, CARL HARRY, b Cranston, RI, May 9, 23; m 45, 67, 78; c 4. PLANT PATHOLOGY, HISTOPATHOLOGY. *Educ:* Univ RI, BS, 47; Univ Wis, PhD(plant path), 53. *Prof Exp:* Proj assoc, Univ Wis, 50-53; asst prof plant path, Univ RI, 53-58; plant pathologist, United Fruit Co, 58-62; from assoc prof to prof, 63-88, EMER PROF PLANT PATH, UNIV RI, 88- *Mem:* Am Phytopath Soc. *Res:* Wilt diseases; physiology of parasitism; disease resistance; histochemistry. *Mailing Add:* Dept Plant Sci Univ RI Kingston RI 02881

BECKMAN, DAVID ALLEN, b 1950; m 82; c 2. TERATOLOGY, FETAL PHYSIOLOGY. *Educ:* Univ Calif, Davis, BS, 72, MS, 74, PhD(physiol), 79. *Prof Exp:* Asst prof pediat, Dept Post-Grad Develop Biol, Jefferson Med Col, Thomas Jefferson Univ, 80-; RES DEPT DIV DEVELOP BIOL, ALFRED I DUPONT INST. *Mem:* Sigma Xi; Am Physiol Soc; Teratology Soc; Radiation Res Soc. *Res:* Developmental biochemistry; placental transport. *Mailing Add:* Res Dept Div Develop Biol Alfred I Dupont Inst 1600 Rockland Rd Wilmington DE 19803

BECKMAN, DAVID LEE, b Dayton, Ohio, May 11, 39; m 69. PULMONARY PHYSIOLOGY. *Educ:* Ohio State Univ, BS, 62, MS, 64, PhD(physiol), 67. *Prof Exp:* Res assoc physiol, Ohio State Univ, 64-65; res assoc biosci, Univ Mich, 67-69, res physiologist, 70-71; asst prof anesthesiol & physiol, Wayne State Univ, 71-73; Hill prof physiol, Univ NDak, 73-76; PROF PHYSIOL, EAST CAROLINA UNIV, 76- *Concurrent Pos:* Partic, Biospace Training Prog, NASA & Univ Va, 67. *Mem:* AAAS; Am Physiol Soc; Soc Exp Biol & Med; Undersea Med Soc; Aerospace Med Asn. *Res:* Respiratory physiology; biomechanics; environmental physiology; lung mechanics; hyperbaric medicine; cardiovascular physiology; autonomic pharmacology. *Mailing Add:* Dept Physiol East Carolina Univ Sch Med Brody Med Sci Bldg Greenville NC 27858-4354

BECKMAN, FRANK SAMUEL, b New York, NY, Apr 10, 21; m 51; c 3. MATHEMATICS, COMPUTER SCIENCE. *Educ:* City Col New York, BS, 40; Columbia Univ, AM, 47, PhD(math), 65. *Prof Exp:* Civilian instr aircraft eng, US Army Air Corps, 41-44; asst prof math, Pratt Inst, 47-51; asst to mgr, Sci Comput Ctr, IBM Corp, 51-56, prod planning rep, 56-57, mgr appl prog systs, 57-59, mgr educ & data processing, Watson Sci Comput Lab, 59-60, from asst to assoc dir comput sci & educ systs res inst, 60-66, res mgr spec projs, 66-71; prof comput & info sci & chmn dept, Brooklyn Col, 71-85; exec officer, PhD Computer Sci Prog, 85-88, PROF GRAD CTR, CITY UNIV, NY, 89- *Concurrent Pos:* Ed, SIAM Rev, 60-64; adj assoc prof, Columbia Univ, 65-71. *Mem:* Am Math Soc; Asn Comput Mach; Math Asn Am; Soc Indust & Appl Math. *Res:* Elliptic partial differential equations; education in computer-allied subjects; automata theory. *Mailing Add:* 16 Garwood Rd Fair Lawn NJ 07410

BECKMAN, JEAN CATHERINE, b Detroit, Mich, Sept 25, 51. ORGANIC CHEMISTRY. *Educ:* Colby Col, AB, 73; Ind Univ, PhD(chem), 77. *Prof Exp:* Res asst, Purdue Univ, 77-78; asst prof, 78-84, ASSOC PROF CHEM, UNIV EVANSVILLE, 84- *Mem:* Am Chem Soc; AAAS; Sigma Xi. *Mailing Add:* 208 S Taft Ave Evansville IN 47714

BECKMAN, JOSEPH ALFRED, b Macomb, Ill, Oct 30, 37; m 59; c 2. POLYMER CHEMISTRY. *Educ:* Western Ill Univ, AB, 60; Iowa State Univ, PhD(org chem), 65. *Prof Exp:* Res org chemist, 64-69, mgr org chem res, 69-71, asst dir res, 80-87, MGR ELASTOMER SYNTHESIS RES, CENT RES LABS, FIRESTONE TIRE & RUBBER CO, 71-; VPRES RES & DEVELOP, COPOLYMER, 87- *Concurrent Pos:* Mem, Polymer, Org & Rubber Divisions, Am Chem Soc, Akron Rubber Group. *Mem:* Am Chem Soc. *Res:* Synthesis, characterization, evaluation and application of organic and inorganic polymers. *Mailing Add:* Copolymer Rubber & Chem Corp Res & Develop PO Box 2591 Baton Rouge LA 70821

BECKMAN, WILLIAM A, b Detroit, Mich, Nov 11, 35; m 57; c 3. MECHANICAL ENGINEERING. *Educ:* Univ Mich, BSE, 58, MSE, 60, PhD(mech eng), 64. *Prof Exp:* Engr missile div, Chrysler Corp, 58-59; from asst prof to assoc prof, 63-77, PROF MECH ENG, UNIV WIS-MADISON, 77- *Honors & Awards:* Centennial Medal, Am Soc Mech Engrs, 81; Abbot Award, Am Sect Int Solar Energy Soc, 79. *Mem:* Am Soc Mech Engrs; Solar Energy Soc; Sigma Xi. *Res:* Radiation heat transfer; boiling heat transfer; solar energy utilization. *Mailing Add:* 4406 Fox Bluff Rd Middleton WI 53562

BECKMANN, PETR, b Prague, Czech, Nov 13, 24; m 65. RADIOPHYSICS. *Educ:* Tech Univ Prague, MSc, 49, PhD(elec eng), 55; Czech Acad Sci, DSc(elec eng), 61. *Prof Exp:* Res engr, Res Inst Telecommun, Czech, 51-54; scientist, Geophys Inst, Czech Acad Sci, 55 & Inst Radio Eng & Electronics, 55-63; PROF ELEC ENG, UNIV COLO, BOULDER, 63- *Mem:* Fel Inst Elec & Electronics Engrs; Am Nuclear Soc; Health Physics Soc; Sigma Xi. *Res:* Electromagnetism; probability theory; scattering by rough surfaces; computational linguistics; energy and public health. *Mailing Add:* 347 E Kelly Rd Sugarloaf Star Rte Boulder CO 80302

BECKMANN, ROBERT B(ADER), b St Louis, Mo, Sept 15, 18; m 42, 57; c 2. CHEMICAL ENGINEERING. *Educ:* Univ Ill, BS, 40; Univ Wis, PhD(chem eng), 44. *Prof Exp:* Engr, State Hwy Dept, Colo, 36; lab analyst, Merchants Exchange Lab, Mo, 40; instr chem eng, Univ Wis, 42-44; res chemist, Humble Oil & Ref Co, Texas, 44-46; from asst prof to prof chem eng, Carnegie Inst Tech, 46-61; head dept, 61-66, dean eng, 66-77, PROF CHEM ENG, UNIV MD, COLLEGE PARK, 61- *Concurrent Pos:* Consult chem engr, 46- *Mem:* AAAS; Am Soc Eng Educ; Am Chem Soc; Am Inst Chem Engrs; Sigma Xi. *Res:* Solvent extraction; kinetics and catalysis; process engineering and design; heat and mass transfer in agitated systems. *Mailing Add:* 10218 Democracy Lane Potomac MD 20854

BECKNER, EVERET HESS, b Clayton, NMex, Feb 24, 35; m 55; c 3. PLASMA PHYSICS. *Educ:* Baylor Univ, BS, 56; Rice Univ, MA, 59, PhD(physics), 61. *Prof Exp:* Staff scientist, Lockheed Missiles & Space Co, 56-57; staff mem, 61-66, div supvr, 66-69, dept mgr, 69-74, DIR, SANDIA NAT LABS, 74- *Mem:* Fel Am Phys Soc; AAAS. *Res:* Plasma physics and magnetohydrodynamics; plasma spectroscopy; pulsed electron accelerators; beam-plasma interactions; electron beam fusion; laser fusion; nuclear waste management. *Mailing Add:* 809 Warm Sands Trail SE Albuquerque NM 87123

BECKNER, SUZANNE K, b Hagertown, Md, July 30, 50; m 71. BIOCHEMISTRY, CELLULAR IMMUNOLOGY. *Educ:* Georgetown Univ, PhD(biochem), 80. *Prof Exp:* Staff fel, Nat Inst Arthritis, Diabetes & Digestive & Kidney Dis, NIH, 80-85, head lab cell, Prog Resources Inc, Nat Cancer Inst, 85-88; DIR, BIOL THER FACIL, BIOTHERAPEUTICS, INC, 88- *Mem:* AAAS; Tissue Cult Asn; Am Soc Bone & Mineral Asn. *Res:* Direct laboratory operations of biological therapy facility which supports experimental cancer clinical therapy trials; propose, execute and apply basic research projects to evaluate and define potential clinical therapy, involving biological response modifiers; contract programs. *Mailing Add:* Dept Cell Biol Life Technol Inc 8717 Grovemont Circle Gaithersburg MD 20877

BECKSTEAD, JAY H, b Pocatello, Idaho, Mar 4, 48. HEMATOPATHOLOGY. *Educ:* Univ Utah, BA, 70, MD, 74. *Prof Exp:* Prof, Dept Path, Univ Calif, 78-89; PROF, DEPT PATH, VET ADMIN ORE HEALTH SCI UNIV, 89- *Mailing Add:* Dept Path L113 Ore Health Sci Univ Portland OR 97201

BECKSTEAD, LEO WILLIAM, b Salt Lake City, Utah, Feb 9, 49; m 72; c 4. ENGINEERING, EXTRACTIVE METALLURGY. *Educ:* Univ Utah, BS, 71, MS, 73, PhD(metall), 75. *Prof Exp:* Res engr extractive metall, Occidental Res Corp, 75-77; res metallurgist, 77-80, SR METALL ENGR, AMAX, 80- *Honors & Awards:* Marcus A Grossmann Award, Am Soc Metals, 74. *Mem:* Am Soc Mining Engrs; Am Inst Mining, Metall & Petrol Engrs; Am Inst Chem Engrs. *Res:* Metallurgical treatment of ores and concentrates; unit processes such as leaching, solution purification, precipitation and roasting are optimized and integrated to develop an overall method of treatment; methods applied on a bench, pilot or full size plant scale. *Mailing Add:* Amax 5950 McIntyre St Golden CO 80403

BECKWITH, JOHN BRUCE, b Spokane, Wash, Sept 18, 33; m 54; c 3. PATHOLOGY. *Educ:* Whitman Col, BA, 54; Univ Wash, MD, 58. *Hon Degrees:* DSc, Whitman Col, 80. *Prof Exp:* Resident pediat path, Children's Hosp, Los Angeles, Calif, 59-60, resident path, Cedars of Lebanon Hosp, Los Angeles, 60-62; chief resident, Children's Hosp, Los Angeles, Calif, 62-64; pathologist, Children's Orthop Hosp & Med Ctr, 64-84; from instr to assoc prof path, 64-74, PROF PATH & PEDIAT, SCH MED, UNIV WASH, 74-; CHMN DEPT PATH, DENVER CHILDREN'S HOSP, 85- *Concurrent Pos:* NIH res grant, 66- *Mem:* AMA; Col Am Pathologists; Int Soc Pediat Oncol; Int Pediat Asn; Soc Pediat Path. *Res:* Sudden death in infancy; pathogenesis of tumors in children. *Mailing Add:* Denver Children's Hosp 1056 E 19th Ave Denver CO 80218

BECKWITH, JONATHAN ROGER, b Cambridge, Mass, Dec 25, 35; m 60; c 2. GENETICS, BACTERIAL GENETICS. *Educ:* Harvard Univ, BA, 57, PhD, 61. *Prof Exp:* NIH postdoctoral fel, 61-65; assoc, 65-66, from asst prof to prof, Dept Bact & Immunol, 66-69, PROF GENETICS, MED SCH, HARVARD UNIV, 69-, AM CANCER SOC RES PROF, 80- *Concurrent Pos:* Guggenheim fel, 70; chmn, Dept Microbiol & Molecular Genetics, Med Sch, Harvard Univ, 71-73, dir, Genetics Training Grant, 75-; vis prof, Univ Calif, Berkeley, 85; mem, adv bd, Coun Responsible Genetics & Eritrean Relief Comt, 85-, sci adv bd, Int Inst Genetics & Biophys, 86- & Ethics Working Group Off Human Genome Res, NIH, 89- *Honors & Awards:* Eli Lilly Award, 70; Jacques Monod Mem Lectr, Inst Pasteur, 88. *Mem:* Nat Acad Sci; Am Soc Exp Biologists; Am Soc Microbiol; AAAS; fel Am Acad Arts & Sci; assoc mem Europ Molecular Biol Orgn. *Res:* Use of bacterial genetics to study various problems of membrane biology, including protein secretion across membranes, the mechanism of cell division and membrane protein topology. *Mailing Add:* Dept Microbiol Med Sch Harvard Univ 25 Shattuck St Boston MA 02115

BECKWITH, RICHARD EDWARD, b San Jose, Calif, July 21, 27; m 82; c 3. STATISTICS, OPERATIONS RESEARCH. *Educ:* Stanford Univ, BS, 49, MS, 54; Purdue Univ, PhD, 59. *Prof Exp:* Statistician, US Fish & Wildlife Serv, 51-52; asst statist, Stanford Univ, 52-53; opers res, Case Inst Technol, 54-55; instr math, Purdue Univ, 55-58; sr res engr, Jet Propulsion Lab, Calif Inst Technol, 58-59; supvr opers res sect, Aeronutronic Div, Ford Motor Co, 59-62; assoc prof, Grad Sch Bus Admin, Univ Southern Calif, 62-68; prof quant methods & assoc dean, Sch Bus Admin, Ga State Univ, 68-69; prof mgt sci, Grad Sch Bus Admin, Tulane Univ, 69-81; PROF MGT INFO SCI, SCH BUS & PUB ADMIN, CALIF STATE UNIV, SACRAMENTO, 81- *Concurrent Pos:* Lectr univ exten, Univ Calif, 59-63. *Mem:* Am Statist Asn; Opers Res Soc Am; Inst Mgt Sci; Decision Sci Inst. *Res:* Industrial operations research; applied statistics; mathematical programming; Monte Carlo method; queuing; reliability; management control systems; weapon systems analysis; scientific programming in statistics, probability, management science and applied mathematics. *Mailing Add:* 3798 Sheridan Rd Cameron Park CA 95682-8930

BECKWITH, STERLING, b Carthage, Mo, Oct 29, 05; m 27; c 3. ELECTRICAL ENGINEERING, PHYSICS. *Educ:* Stanford Univ, BS, 27; Univ Pittsburgh, MS, 31; Calif Inst Technol, PhD(elec eng), 33. *Prof Exp:* Elec engr, Westinghouse Elec Co, Pittsburgh, 27-31; asst elec eng, Calif Inst Technol, 31-33; elec engr, Metrop Water Dist of Southern Calif, 33-35; elec engr in charge alternating current design, Allis-Chalmers Mfg Co, 35-52; CONSULT ENGR, 52- *Concurrent Pos:* Dir res, Chicago Stock Yards Res Div, 59-68; vpres eng & res, Dual Jet Refrig, 62-68; consult engr, Warren/ Dual Jet, 68-76. *Honors & Awards:* Lamme Medal, Inst Elec & Electronics Engrs, 59. *Mem:* Fel Inst Elec & Electronics Engrs; Int Conf Large Elec Systs. *Res:* Power factor bridge; rotating electrical machines; magneto plasma dynamics; liquid methane; helicopters; super-saturated air; air curtains. *Mailing Add:* 1824 Doris Dr Menlo Park CA 94025

BECKWITH, STEVEN VAN WALTER, b Madison, Wis, Nov 20, 51; m; c 2. ASTRONOMY. *Educ:* Cornell Univ, BS, 73; Calif Inst Technol, PhD(physics), 78. *Prof Exp:* assoc prof, 84-88, PROF ASTRON, CORNELL UNIV, 88- *Concurrent Pos:* Alfred P Sloan fel. *Honors & Awards:* Fullam Award, Dudley Observ. *Mem:* Am Astron Soc; Int Astron Union; AAAS; APS. *Res:* Observational programs using both ground-based and airborne astronomical telescopes for studies of star formation, stellar evolution and the interstellar medium; instrumentation for infrared astronomy; infrared imaging techniques including speckle interferometry. *Mailing Add:* Dept Astron Space Sci Lab Cornell Univ Ithaca NY 14850

BECKWITH, WILLIAM FREDERICK, b Marshalltown, Iowa, Oct 4, 34; m 59; c 4. CHEMICAL ENGINEERING. *Educ:* Iowa State Univ, BS, 57, MS, 61, PhD(chem eng), 63. *Prof Exp:* From asst prof to prof chem eng, 63-89, DIR FRESH ENG & ENG GRAPHICS, CLEMSON UNIV, 89- *Concurrent Pos:* Sabbatical leave, Univ Maine, 79. *Mem:* Am Inst Chem Eng; Am Soc Eng Educ; Sigma Xi; Tech Asn Pulp & Paper Indust. *Res:* Chemical engineering transport phenomena and heat transfer; nonwoven fabric formation by wet-lay process; wet end chemistry of paper making. *Mailing Add:* Dept Chem Eng Earle Hall Clemson Univ Clemson SC 29634-0909

BECKWITT, RICHARD DAVID, b Detroit, Mich, Apr 25, 49; m 74; c 2. MOLECULAR EVOLUTION, POPULATION GENETICS. *Educ:* Univ Calif, Berkeley, BA, 70; Univ Southern Calif, PhD(biol), 79. *Prof Exp:* Res assoc, Vantuna Res Group, Occidental Col, 80-85; postdoctoral molecular biol, Univ Southern Calif, 81-83; asst prof, 85-90, ASSOC PROF BIOL, FRAMINGHAM STATE COL, 90- *Concurrent Pos:* Consult, Biotechnol Group, US Army Natick Res, Develop & Eng Ctr, 89- *Mem:* AAAS; Am Soc Zoologists. *Res:* Molecular evolution and population genetics of invertebrates; genetic engineering of spider silk. *Mailing Add:* Dept Biol Framingham State Col Framingham MA 01701

BECRAFT, LLOYD G(RAINGER), b Glasgow, Mont, Apr 5, 36; m 58; c 2. CHEMICAL ENGINEERING. *Educ:* Mont State Col, BS, 58, PhD(chem eng), 62. *Prof Exp:* Reservoir engr, Shell Oil Co, 58; SR RES ENGR, CONOCO, INC, 62- *Res:* Petroleum processing, especially coking; petroleum products, especially petroleum cokes; crude oil assaying; statistical process and quality control. *Mailing Add:* 2013 Wildwood Ave Ponca City OK 74604

BECVAR, JAMES EDGAR, BIOCHEMISTRY. *Educ:* Univ Mich, PhD(biochem & biophys), 73. *Prof Exp:* ASSOC PROF BIOCHEM, UNIV TEX, EL PASO. *Mailing Add:* Dept Chem Univ Tex El Paso TX 79968-0513

BEDARD, DONNA LEE, b Adams, Mass, May 5, 47; m 77; c 2. BIODEGRADATION OF HALOGENATED AROMATICS, MOLECULAR GENETICS. *Educ:* Tufts Univ, BS, 69; Univ Chicago, PhD(biol), 73. *Prof Exp:* Res assoc, Univ Rochester, 74-75; fel res assoc, Johns Hopkins Univ, 75-78; prof biol, Bennington Col, 78-80; res scientist,

NY State Dept Health, 80-81; molecular biologist, 81-90, MICROBIOLOGIST, GROUP LEADER, GEN ELEC RES & DEVELOP CTR, 90- *Mem:* Am Soc Microbiol. *Res:* Biochemical and genetic basis for bacterial oxidation of polychlorinated biphenyls; congener specificity; degradative pathways; reductive dechlorination of polychlorinated biphenyls by anaerobic bacteria; microbial transformation of halogenated aromatic compounds. *Mailing Add:* Gen Elec Res & Develop Ctr Bldg K-1 PO Box 8 Schenectady NY 12301

BEDDOES, M(ICHAEL) P(ETER), b London, Eng, Feb 22, 24; m 48; c 3. ELECTRICAL ENGINEERING. *Educ:* Glasgow Univ, BSc, 45; Univ London, PhD(elec eng), 58. *Prof Exp:* Apprentice, Gen Elec Corp, Coventry, Eng, 45-48; sr engr, Plessey Corp, Ilford, 48-52; asst chief engr, McMichael Radio, Slough, 52-53; from asst prof to assoc prof, 56-77, PROF ELEC ENG, UNIV BC, 77- *Mem:* Assoc mem Brit Inst Elec Eng; sr mem Inst Elec & Electronics Engrs. *Res:* Television band compression using variable velocity scanning; speech band compression using analogue methods; aids for blind people; holography; epilepsy diagnoses through electron yield signals. *Mailing Add:* Dept Elec Eng Univ BC 2075 Westbrook Pl Vancouver BC V6T 1W5 Can

BEDELL, GEORGE NOBLE, b Harrisburg, Pa, May 1, 22; m 50, 70; c 9. INTERNAL MEDICINE. *Educ:* DePauw Univ, BA, 44; Univ Cincinnati, MD, 46. *Prof Exp:* Nat Heart Inst fel, 52-54; from asst prof to assoc prof, 54-68, PROF INTERNAL MED & DIR PULMONARY DIS DIV, UNIV IOWA, 68- *Concurrent Pos:* Nat Heart Inst spec fel, Univ Pa, 54-55; USPHS career develop award, 62; vis prof, Post-grad Sch Med, Univ London, 65-66. *Mem:* AAAS; Am Soc Clin Invest; Am Thoracic Soc; Am Col Chest Physicians; Am Fed Clin Res; Fedn Am Soc Exp Biol. *Res:* Pulmonary disease; clinical, cardiovascular and respiratory physiology. *Mailing Add:* Dept Internal Med Univ Iowa Iowa City IA 52240

BEDELL, LOUIS ROBERT, b Niagara Falls, NY, Nov 14, 39; m 64; c 5. SURFACE PHYSICS, ELECTRONICS. *Educ:* Auburn Univ, BS, 63; Brown Univ, MS, 68, PhD(physics), 71. *Prof Exp:* Res asst surface physics, Field Res Lab, Mobil Oil Co, 63-65; ASST PROF PHYSICS, NORTHEAST LA UNIV, 70- *Mem:* Am Phys Soc; Am Vacuum Soc. *Res:* Gas adsorption and absorption on clean metal surfaces using low energy electron diffraction and Auger electron spectroscopy. *Mailing Add:* Dept Physics Northeast La Univ Monroe LA 71209-0580

BEDELL, THOMAS ERWIN, b Santa Cruz, Calif, Dec 21, 31; m 83; c 3. ANIMAL SCIENCE & NUTRITION, RESOURCE MANAGEMENT. *Educ:* Calif State Polytech Col, BS, 53; Univ Calif, MS, 57; Ore State Univ, PhD, 66. *Prof Exp:* Jr specialist range mgt, Univ Calif, 57, asst agriculturist, Agr Exten, 57-63; from instr to asst prof range mgt, Ore State Univ, 66-70; exten range mgt specialist & assoc prof range mgt, Univ Wyo, 70-73; assoc prof agr, Exten & Area Livestock Exten Agt, 73-76, PROF, EXTEN RANGE LAND RESOURCES SPEC, ORE STATE UNIV, 76- *Mem:* Soc Range Mgt (pres, 89). *Res:* Range management and improvement. *Mailing Add:* Ore State Univ Corvallis OR 97331

BEDENBAUGH, ANGELA LEA OWEN, b Seguin, Tex, Oct 6, 39; m 61; c 1. CHEMISTRY. *Educ:* Univ Tex, BS, 61; Univ SC, PhD(chem), 67. *Prof Exp:* Geol mapping asst, 58-59, lab instr, Univ Tex, 60-61; res assoc, 66-80, RES ASSOC PROF CHEM, UNIV SOUTHERN MISS, 80- *Concurrent Pos:* Statewide Workshops Cross-over High Sch Chem Teachers, NSF. *Mem:* Am Chem Soc; Nat Sci Teachers Asn; Sigma Xi. *Res:* Dissolving metal reductions; development of new syntheses; developing new general chemistry experiments. *Mailing Add:* 63 Suggs Rd Hattiesburg MS 39402-9642

BEDENBAUGH, JOHN HOLCOMBE, b Newberry, SC, Apr 9, 31; m 61. ORGANIC CHEMISTRY. *Educ:* Newberry Col, BS, 53; Univ NC, MA, 57; Univ Tex, PhD(chem), 62. *Prof Exp:* Instr chem, Randolph-Macon Woman's Col, 56-58; assoc prof, Columbia Col, SC, 61-68; assoc prof, 68-74, PROF CHEM, UNIV SOUTHERN MISS, 74-, CHMN DEPT, 70- *Mem:* AAAS; Am Chem Soc; Nat Sci Teachers Asn; Sigma Xi. *Res:* Synthesis of heterocyclic compounds; chemical education; dissolving metal reductions; organic reagents for inorganic analysis. *Mailing Add:* 63 Suggs Rd Hattiesburg MS 39402

BEDERSON, BENJAMIN, b New York, NY, Nov 15, 21; m 56; c 4. EXPERIMENTAL ATOMIC MOLECULAR & OPTICAL PHYSICS. *Educ:* City Col New York, BSc, 46; Columbia Univ, MA, 48; NY Univ, PhD(physics), 50. *Prof Exp:* Res worker, Los Alamos Sci Lab, 44-45; mem staff, Res Lab Electronics, Mass Inst Technol, 50-51; from asst prof to assoc prof, NY Univ, 52-59, chmn dept, 73-76, dean grad sch Arts & Sci, 86-89, PROF PHYSICS, NY UNIV, 59- *Concurrent Pos:* Chmn Int Conf Physics of Electronic & Atomic Collisions, 58, 61 & 85; vis fel, Joint Inst Lab Astrophys, Nat Bur Stand, Univ Colo, 68-69; sci consult, Acad Press, NY; mem organizing comt, Int Conf Atomic Physics, 68; chmn atomic & molecular physics comt, Nat Acad Sci-Nat Res Coun, 70-; mem adv comt, Physics Div, Nat Sci Found, 72-75; ed, Phys Rev A, 78-; co-ed Advan in Atomic, Molecular & Optical Physics, 78. *Mem:* Fel Am Phys Soc; fel AAAS. *Res:* Atomic collisions and structure; polarized and laser excited beams; atomic and molecular clusters. *Mailing Add:* Dept Physics NY Univ Four Washington Pl New York NY 10003

BEDFORD, ANTHONY, b Baton Rouge, La, Oct 7, 38; m 61; c 2. CONTINUUM MECHANICS. *Educ:* Univ Tex, Austin, BS, 61; Calif Inst Technol, MS, 62; Rice Univ, PhD(mechanics), 67. *Prof Exp:* Engr res & develop, Douglas Aircraft Co, 62-64; mem tech staff res & develop, TRW Systs, 66-68; from asst prof to assoc prof, 75-81, PROF MECH, UNIV TEX, AUSTIN, 81- *Concurrent Pos:* NSF grants, 69-70, 72-74 & 75-77, ONR grants, 86, 87 & 88; consult, Tracor Inc, 73-74 & Sandia Labs, 73-83; tech ed, Appl Mech Rev, 74-83. *Mem:* Soc Natural Philos; Soc Eng Sci; Am Acad Mech. *Res:* Constitutive theories for mixtures; wave propagation; ocean sediment acoustics. *Mailing Add:* ASE-EM Dept Univ Tex Austin TX 78712

BEDFORD, BARBARA LYNN, b Odessa, Tex, Mar 24, 46; m 82; c 1. ECOSYSTEM SCIENCE, WETLAND ECOLOGY. *Educ:* Marquette Univ, AB, 68; Univ Wis-Madison, MS, 77, PhD(environ sci land resources), 80. *Prof Exp:* asst prof ecol & systs, 80-89, assoc dir, 80-90, ASST PROF, DEPT NATURAL RESOURCES, ECOSYSTEMS RES CTR, 89-, DIR, 91- *Concurrent Pos:* Consult, Am Soc Planning Officials, 74, Ga Dept Natural Resources, 76, Environ Law Inst, 76-78, Ill Natural Areas Inventory, 77, Wis Dept Justice, 77-80 & Nature Conservancy, Wis, 80; guest lectr, Sch Pub & Environ Affairs, Ind Univ, 77, 80 & Rocky Mtn Biol Lab, 79; pres-elect, Asn Ecosystem Res Ctrs, 90-91. *Mem:* Sigma Xi; Ecol Soc Am; Am Inst Biol Sci. *Res:* Responses of wetland ecosystems to natural and human induced stress; attempts to identify the appropriate and minimum set of system components and interactions necessary to predicting those responses. *Mailing Add:* Ecosysts Res Ctr 105 Wing Hall Cornell Univ Ithaca NY 14853-8101

BEDFORD, JAMES WILLIAM, b Colfax, Wash, Dec 3, 42; m 65; c 2. WATER CHEMISTRY, RISK ASSESSMENT & COMMUNICATION. *Educ:* Mich State Univ, BS, 65, MS, 67, PhD(entom), 70. *Prof Exp:* chemist, Dept Natural Resources, Mich Environ Lab, 70-84, asst dir, 73-84; sr toxicologist, Toxic Substance Control Comn, 84-89; OMBUDSMAN, ENVIRON HEALTH, MICH DEPT PUB HEALTH, 90- *Concurrent Pos:* Instr, Lansing Community Col, 80- *Mem:* Am Fisheries Soc; Sigma Xi; Soc Toxicol; Air Pollution Control Asn. *Res:* Biological monitoring of pesticides; pesticide residue analysis; pesticide and heavy metal pollution; lamellibranch physiology; environmental toxicology; risk communication. *Mailing Add:* Mich Dept Pub Health PO Box 30195 Lansing MI 48909

BEDFORD, JOEL S, b Denver, Colo, Feb 14, 38; m 62; c 3. RADIOBIOLOGY, CYTOGENETICS. *Educ:* Univ Colo, BA, 61, MS, 64; Oxford Univ, DPhil, 66. *Prof Exp:* Instr, 66-67, James Picker Found res fel, 67-68, from asst prof to assoc prof radiobiol, Col Med, Vanderbilt Univ, 68-75; PROF RADIOL & RADIATION BIOL, COLO STATE UNIV, 75- *Concurrent Pos:* NIH radiation study sect, 84-88, chmn, 86-88; assoc ed radiation res, 87- *Mem:* Radiation Res Soc. *Res:* Genetics control of cellular radiosensitivity; radiation genetics and cytogenetics; radiation effects on the cell cycle; cellular responses to hyperthermia; mechanisms of thermotolerance development in mammalian cells. *Mailing Add:* Dept Radiol Technol Colo State Univ Ft Collins CO 80523

BEDFORD, JOHN MICHAEL, b Sheffield, Eng, May 21, 32. ANATOMY, PHYSIOLOGY. *Educ:* Cambridge Univ, BA, 55, MA & VetMB, 58; Univ London, PhD, 65. *Prof Exp:* Jr fel surg, Vet Sch, Bristol Univ, 58-59; res assoc physiol, Worcester Found, Mass, 59-61; asst prof, Royal Vet Col, Univ London, 61-66; scientist, Worcester Found, Mass, 66-67; from asst prof to assoc prof anat, Columbia Univ, 70-72; PROF REPRODUCTIVE BIOL & ANAT, MED COL, CORNELL UNIV, 72- *Mem:* Soc Study Reproduction; Am Asn Anatomists; Endocrine Soc; Brit Soc Study Fertil. *Res:* Reproductive physiology; sperm maturation in the male; capacitation of sperm in the female; fertilization; physiology of the ovum. *Mailing Add:* Dept Obstet & Gynec Cornell Univ Med Col 1300 York Ave New York NY 10021

BEDFORD, RONALD ERNEST, b Manitoba, Can, June 26, 30; m 55; c 3. PHYSICS, TEMPERATURE STANDARDS. *Educ:* Univ Man BSc Hons, 52; Univ BC, MA, 53, PhD(physics), 55. *Prof Exp:* Asst res officer, Nat Res Coun Can, 55-61, assoc res officer, 61-70, sr res officer, Div Physics, 70-89, head, Heat Thermometry Sect, 80-87, HEAD, PHOTOM RADIOMETRY SEC, NAT RES COUN CAN, 87-, PRIN RES OFFICER, DIV PHYSICS, 90- *Concurrent Pos:* Can deleg, Consultative Comt Thermometry, 71-, mem working group 2, 71-, chmn, 78-; adv prof physics Inst of Technol, Harbin, China. *Mem:* Optical Soc Am. *Res:* Primary temperature standards; thermocouple and optical thermometry; radiometry. *Mailing Add:* Inst Nat Measurements Standards Nat Res Coun Can Ottawa ON K1A 0R6 Can

BEDFORD, WILLIAM BRIAN, b Omaha, Nebr, June 12, 47. ECOLOGY, MARINE BIOLOGY. *Educ:* Bemidji State Univ, BA(biol) & BA(chem), 69; Tex A&M Univ, PhD(marine Biol), 72. *Prof Exp:* Asst sci dir, Nature Conservancy, 72-73, dir natural area progs, 73-74; proj mgr ecol reserves, Inst Ecol, 74-76; asst prof marine biol, Tex A&M Univ, 75-76; marine studies coordr, Environ Studies Prog, Ministry Conserv, Victoria, Australia, 76-79; CONSULT, 80- *Concurrent Pos:* Rep, AAAS comt atmospheric & hydrospheric sci, 73-76; mem, stand comt, Am Nuclear Soc, 73-76; res assoc, Nature Conservancy, 74; rep Fed Comt Ecol Reserves, 74-76; bd mem, Am Coastal Soc, 74-76; vis asst prof, Kellogg Biol Sta, Mich State Univ, 75. *Mem:* The Nature Conservancy; Australian Marine Sci Asn; Ecol Soc Am. *Res:* The application of ecological concepts and approaches to resource management and utilization, particularly problems of the coastal zone and aquatic systems; the responses of organisms to environmental stress factors. *Mailing Add:* RR 1 Box 250 Nevis MN 56467

BEDGOOD, DALE RAY, b Saltillo, Tex, Aug 10, 32; m 59; c 2. MATHEMATICS, STATISTICS. *Educ:* ETex State Univ, BS, 54; Univ Ark, MA, 59; Okla State Univ, EdD(math), 64. *Prof Exp:* Asst math, Univ Ark, 58-59; instr, Northeast La State Col, 59-62, asst prof, 64-65, assoc prof & head dept, 65-67; PROF MATH & HEAD DEPT, ETEX STATE UNIV, 67- *Concurrent Pos:* Asst, Okla State Univ, 62-64; consult, NASA, Ala, 64-65, Cambridge Res Assoc, Mass, 66-67. *Mem:* Am Math Soc; Math Asn Am; Nat Coun Teachers Math. *Res:* Point-set topology; general analysis; teacher education and training at the elementary and secondary levels. *Mailing Add:* Dept Math ETex State Univ ETex Sta Commerce TX 75428

BEDIENT, JACK DEWITT, b Ithaca, NY, Jan 30, 26; m 49; c 4. MATHEMATICS. *Educ:* Albion Col, BA, 48; Univ Colo, MS, 60, EdD(math educ), 66. *Prof Exp:* High sch teacher, Wash, 50-59; asst prof math, 63-71, ASSOC PROF MATH, ARIZ STATE UNIV, 71- *Mem:* Math Asn Am. *Res:* Mathematics education; training secondary school mathematics teachers. *Mailing Add:* Dept Math Ariz State Univ Tempe AZ 85287

BEDIENT, PHILLIP E, b Foochow, China, Oct 15, 22; US citizen; m 43; c 2. MATHEMATICS. *Educ:* Park Col, AB, 43; Univ Mich, MA, 46, PhD(math), 59. *Prof Exp:* From instr to asst prof math, Juniata Col, 50-55; instr, Univ Mich, 56-59; from asst prof to assoc prof, 59-68, chmn, Dept Math & Astron, 72-81, prof, 68-87, EMER PROF MATH, FRANKLIN & MARSHALL COL, 87- *Mem:* Am Math Soc; Math Asn Am. *Res:* Special functions; partial differential equations. *Mailing Add:* Dept Math & Astron Franklin & Marshall Col Lancaster PA 17604

BEDINGER, CHARLES ARTHUR, JR, b Highlands, Tex, Apr 26, 42; m 70; c 2. ZOOLOGY. *Educ:* Tex A&M Univ, BS, 64, PhD(zool), 74; Sam Houston State Univ, MA, 67. *Prof Exp:* Res biologist, Res Found, Tex A&M Univ, 72-73; sr res biologist, 73-74, mgr biol res, 76-80, MGR PROG DEVELOP-HOUSTON, SOUTHWEST RES INST, 80- *Mem:* Am Soc Parasitol; Estuarine Res Fedn; Sigma Xi. *Res:* Aquatic ecosystems research; aquatic toxicology; nuclear waste handling; power plant environmental problems; fish parasitology; molluscan ecology. *Mailing Add:* PO Box 457 Aspen CO 81612-0457

BEDINGFIELD, CHARLES H(OSMER), b Kansas City, Mo, Aug 3, 16; m 41; c 1. CHEMICAL ENGINEERING. *Educ:* Univ Kans, BS, 36, MS, 37; Columbia Univ, PhD(chem eng), 48. *Prof Exp:* Develop engr, Columbia Chem Div, Pittsburgh Plate Glass Co, Ohio, 37-40; asst chem eng, Columbia Univ, 40-41; res engr, Armstrong Cork Co, Pa, 48-50; engr, Eng Dept, E I Du Pont de Nemours & Co, 50-62, sr consult, 63-72, prin consult, 72-81; RETIRED. *Mem:* Sigma Xi. *Res:* Process development; engineering and economic evaluation; research and development in research guidance. *Mailing Add:* 212 Sypherd Dr Newark DE 19711

BEDNAR, JONNIE BEE, b Shiner, Tex, Aug 15, 41; m 65; c 2. MATHEMATICS, COMPUTER SCIENCE. *Educ:* Southwest Tex State Col, BS, 62; Univ Tex, Austin, MA, 64, PhD(math), 68. *Prof Exp:* Teaching asst math, Univ Tex, Austin, 62-65; engr sci II, Tracor, Inc, Tex, 65-68; asst prof math, Drexel Inst Technol, 68-69; grad coordr, Univ Tulsa, 74-76, from asst prof to assoc prof math, 69-76, prof, 72-, chmn dept, 75-; MGR GEOPHYS SERV, AMERADA HESS, INC. *Mem:* Soc Indust & Appl Math; Inst Elec & Electronics Engrs; Soc Explor Geophysicists; European Asn Explor Geophysicists. *Res:* Ordered topological vector spaces: Choquet theory; numerical analysis. *Mailing Add:* Amerada Hess Inc PO Box 2040 Tulsa OK 74102

BEDNAR, RODNEY ALLAN, b Norristown, Pa, Nov 6, 55. ENZYMOLOGY, REACTION MECHANISMS. *Educ:* Juniata Col, BS, 77; Univ Del, PhD(chem), 81. *Prof Exp:* Chemist, E I Du Pont de Nemours & Co, 77; NIH fel, Brandeis Univ, 82-85; ASST PROF PHARMACOL SCI & CHEM, STATE UNIV NY, STONY BROOK, 85- *Mem:* Am Chem Soc; Sigma Xi; AAAS; Biophys Soc. *Res:* Mechanisms of enzyme action; general acid-base catalysis; chemical modification and site-directed mutagenesis of chalcane isomerase; mechanisms of proton transfer; structure-function relationship in enzyme catalysis. *Mailing Add:* Dept Pharmacol Sci State Univ NY Stony Brook NY 11794-8651

BEDNARCYK, NORMAN EARLE, b Buffalo, NY, Oct 12, 38; m 63; c 3. FOOD SCIENCE. *Educ:* Mass Inst Technol, BS, 60, MS, 62; Rutgers Univ, PhD(food sci), 67. *Prof Exp:* Res chemist, Lever Bros Co, 62-64; sr res chemist mix develop, Gen Mills, Inc, 67; head lipid chem, 67-72, head nutrit serv, 72-74, mgr sci serv, 75-81, mgr tech reg affairs, 82-89, MGR, NUTRIENT & FORMULA DATA RESOURCES, NABISCO BRANDS INC, 89- *Concurrent Pos:* Mem, Antioxidant Tech Comn, Int Life Sci Inst, 81- *Mem:* Inst Food Technologists; Am Asn Cereal Chemists. *Res:* Nutritional composition of foods; food regulations. *Mailing Add:* Nabisco Brands Inc Technol Ctr 200 De Forest Ave POBox 1944 East Hanover NJ 07936-1944

BEDNAREK, ALEXANDER R, b Buffalo, NY, July 15, 33; m 54; c 4. MATHEMATICS. *Educ:* State Univ NY Albany, BS, 57; Univ Buffalo, MA, 59, PhD(math), 61. *Prof Exp:* Instr math, Univ Buffalo, 58-60; mathematician, Goodyear Aerospace Corp, Ohio, 61-62; asst prof math, Univ Akron, 62-63; from asst prof to assoc prof, 63-69, PROF MATH & CHMN DEPT, UNIV FLA, 69-, CO-DIR, CTR APPL MATH, 76- *Concurrent Pos:* Vis staff mem, Los Alamos Sci Lab, 76- *Mem:* Am Math Soc; Math Asn Am; Polish Math Soc. *Res:* Topological algebra and the theory of relations, their applications to theoretical computer science. *Mailing Add:* Dept Math Univ Fla Gainesville FL 32611

BEDNAREK, JANA MARIE, b Bratislava, Czech; US citizen; m 66; c 1. MOLECULAR BIOLOGY, PROTEIN CHEMISTRY. *Educ:* Charles Univ, Master's, 59; Univ Va, Charlottesville, PhD(biochem), 73. *Prof Exp:* NIH postdoctoral fel, Dept Biochem, Med Sch, Univ Va, Charlottesville, 73-75; postdoctoral fel phys chem, Dept Chem, Univ SC, Columbia, 75-76, res fel & res asst prof biochem, 76-79; res assoc, Dept Biochem, Med Univ SC, Charleston, 79-80; RES CHEMIST, HEMAT-ONCOL SERV, DEPT HEMAT, WALTER REED ARMY MED CTR, 81- *Concurrent Pos:* Guest researcher, Dept Biochem, Nat Cancer Inst, NIH, Bethesda, Md, 81-90; prin investr, grant, Univ SC, Columbia, 78-79, ten approved protocols, Walter Reed Army Med Ctr, 81- *Mem:* Sigma Xi; NY Acad Sci; assoc mem Am Soc Biochem & Molecular Biol. *Res:* Protein chemistry; enzymology; biochemistry; molecular biology of human neutrophils and their precursors in human bone marrow, their maturation in normal and cancer state; isolation and purification and characterization of factor(s) regulating maturation of human neutrophils; using cultures of normal cells as well as cancer cell lines. *Mailing Add:* Dept Hemat Walter Reed Army Inst Res Georgia Ave Washington DC 20307

BEDNEKOFF, ALEXANDER G, b Seattle, Wash, June 13, 32; m 58; c 3. BIOCHEMISTRY. *Educ:* Univ Wash, BS, 54; Univ Wis, MS, 61, PhD(biochem), 62. *Prof Exp:* Asst prof biochem, SDak State Univ, 62-66; from asst prof to assoc prof, 66-73, PROF BIOCHEM, PITTSBURG STATE UNIV, 73- *Mem:* Am Chem Soc; Sigma Xi. *Res:* Blood group substances of cattle and sheep; urinary calculi. *Mailing Add:* Dept Chem Pittsburg State Univ Pittsburg KS 66762

BEDNORZ, JOHANNES GEORG, May 16, 50. CERAMIC PHYSICS. *Educ:* Univ Munster, BS, 76; Swiss Fed Inst Technol, PhD, 82. *Prof Exp:* RES STAFF MEM, IBM ZURICH RES LAB, SWITZ, 82- *Concurrent Pos:* Lectr, Swiss Fed Inst Technol, Zurich & Univ Zurich, 87- *Honors & Awards:* Nobel Prize in Physics, 87; Marcel-Benoist Prize, 86; Victor Moritz Goldschmidt Prize, 87; Robert Wichard Pohl Prize, 87. *Res:* Preparation, crystal growth and characterization of high refractive oxide materials (phase transition, quantum ferroelectricity); oxides with metallic conductivity and superconductivity. *Mailing Add:* IBM Zurich Res Lab Saemerstrasse Four 8803 Rueschlikon Zurich Switzerland

BEDNOWITZ, ALLAN LLOYD, b New York, NY, Oct 7, 39; m 84; c 3. X-RAY CRYSTALLOGRAPHY. *Educ:* Cooper Union, BChE, 60; Polytech Inst Brooklyn, MS, 63, PhD(chem physics), 66. *Prof Exp:* Fel x-ray crystallog, Brookhaven Nat Lab, 65-67; res staff mem x-ray crystallog, 67-70, mgr lab automation appln, 71-77, mgr consult, educ & comput serv anal, 78-80, WITH COMPUT GRAPHICS SUBSYST, THOMAS J WATSON RES CTR, IBM CORP, 80- *Concurrent Pos:* Mem tech planning staff, Thomas J Watson Res Ctr, IBM Corp, 83-84. *Mem:* Am Crystallog Asn; Am Phys Soc; Sigma Xi; Asn Comput Machines; NY Acad Sci. *Res:* Crystal structure analysis by direct methods; computer control of experiments in research laboratories; computer programming for problems in x-ray crystallography; laboratory automation; computer graphics. *Mailing Add:* Thomas J Watson Res Ctr IBM Corp PO Box 218 Yorktown Heights NY 10598

BEDO, DONALD ELRO, b Erie, Pa, Dec 25, 29. PHYSICS. *Educ:* Pa State Univ, BS, 51; Cornell Univ, PhD, 57. *Prof Exp:* Corning Glass Found fel, Cornell Univ, 56-58, res assoc physics, 58-60; RES PHYSICIST, US AIR FORCE CAMBRIDGE RES LABS, CRUU, 62- *Mem:* Am Phys Soc. *Res:* Soft x-ray and extreme ultraviolet spectroscopy; photoemission; electromagnetic theory. *Mailing Add:* AFGL/OPA Hanscom AFB Bedford MA 02174

BEDOIT, WILLIAM CLARENCE, JR, b Chattanooga, Tenn, Apr 20, 22; m 43; c 2. ORGANIC CHEMISTRY, PHYSICAL ORGANIC CHEMISTRY. *Educ:* Univ Tenn, BS, 47, MS, 48, PhD(phys org chem), 50. *Prof Exp:* Res chemist, Carbide & Carbon Chem Corp, 50-52 & Mallinckrodt Chem Works, 52-54; sr res chemist, Jefferson Chem Co, 54-57, mgr mkt develop, 57-70, mgr urethane mkt develop, 70-71; mkt mgr, Martin Sweets Co, 71-73, vpres, 73; pres, UCT Inc, 73-84; consult, Union Carbide Corp, 84-85; RETIRED. *Concurrent Pos:* Instr, Morris Harvey Col, 51-52. *Mem:* Am Chem Soc; Soc Plastics Indust; fel Am Inst Chemists; Com Develop Asn. *Res:* Catalytic hydrogenation; chemical kinetics; distillation; organic preparations; market development; organic chemistry; petrochemicals; catalysis; all aspects of new urethane technology. *Mailing Add:* 8618 Brow Lake Rd Soddy Daisy TN 37379-4500

BEDROSIAN, E(DWARD), b Chicago, Ill, May 22, 22; m 45, 71; c 5. ELECTRICAL ENGINEERING. *Educ:* Northwestern Univ, BS, 49, MS, 50, PhD(elec eng), 53. *Prof Exp:* Sr engr, Motorola, Inc, 53-57; MEM STAFF, RAND CORP, 57- *Concurrent Pos:* Adj prof elec eng, Univ Southern Calif, 68-71. *Mem:* Inst Elec & Electronics Engrs; Am Inst Aeronaut & Astronaut. *Res:* Radio and radar systems analysis; communication theory. *Mailing Add:* Rand Corp 1700 Main St Santa Monica CA 90406

BEDROSIAN, KARAKIAN, b Milford, Mass, June 29, 33; m 54; c 3. FOOD TECHNOLOGY. *Educ:* Univ Mass, BS, 54; Cornell Univ, MS, 56; Univ Ill, PhD(food technol), 58; Mich State Univ, MBA, 62. *Prof Exp:* Assoc res food technologist, Whirlpool Corp, 58-61, mgr food sci & technol, 61-62, dir res & develop, Tectrol Div, 62-63, dir res & develop, St Joseph Opers, 63-65; mgr new prod lab, Lever Bros Co, 65-66, tech mgr new prod & opers anal, 66-67; vpres res, DCA Food Indust, Inc, 67-71; PRES, BEDROSIAN & ASSOCS, 71-; PRES, NAT PAK SYSTS, 78- *Concurrent Pos:* Chmn bd, Natural Pak Produce. *Mem:* AAAS; Inst Food Technol; Am Soc Heating, Refrig & Air-Conditioning Engrs; Asn Res Dirs. *Res:* New product development; microwave processing; extruded foods; fruit and vegetable storage and distribution. *Mailing Add:* Sherwood Ct Alpine NJ 07620

BEDROSIAN, SAMUEL D, b Marash, Turkey, Mar 24, 21; m 51; c 2. SYSTEMS ENGINEERING, COMMUNICATIONS. *Educ:* State Univ NY, AB, 42; Polytech Inst Brooklyn, MEE, 51; Univ Pa, PhD(elec eng), 61. *Prof Exp:* Asst br chief, US Signal Corps Eng Labs, 46-55; systs engr, Burroughs Res Ctr, 55-60; prof & chmn, Systs Eng Dept, 75-80, FAC MEM, MOORE SCH, UNIV PA, 60- *Concurrent Pos:* Consult, RCA Corp, 68-87, Frankford Arsenal, 69-71, Jerrold Electronics Labs, 72-73, Gen Elec Space Div, 74, Aydin-Monitor, 76-77, Norden Systs, Div United Technol, 77-79, Gen Rad, 78-79, Siemens Corp, 80 & Sperry Gyroscope, 81-83; assoc ed, J Franklin Inst, 66-; Naval Electronics Syst Command res chair prof, Naval Postgrad Sch, Monterey, Calif, 80-81. *Mem:* Fel Inst Elec & Electronics Engrs; Sigma Xi. *Res:* Computer-communication networks; application of fuzzy concepts to image processing, fractal and graph theoretic modeling applications; automatic checkout for fault isolation and diagnosis; systems engineering studies; information storage and retrieval; knowledge representation for robotics. *Mailing Add:* Dept Elec Eng Univ Pa Philadelphia PA 19104-6390

BEDWELL, THOMAS HOWARD, b Forest, Idaho, Feb 17, 15; m 56. PHYSICS. *Educ:* Univ SDak, AB, 43, AM, 47, ME, 58; Univ Nebr, PhD(physics, educ), 66; Univ Okla, MS, 74, MNSc, 78. *Prof Exp:* Instr physics, Univ SDak, 44-48; asst prof, Fla State Univ, 48-50; plant engr, Freeman Co, SDak, 50-53; chmn elec div, Southern State Col (SDak), 53-57; assoc prof physics, Univ SDak, 58-64; from assoc prof to prof physics, Northern Ariz Univ, 64-81; RETIRED. *Concurrent Pos:* Fel, Univ Okla, 70-71; chmn, Am Physics Teachers Lect Series, 63-64; judge, Future Scientists of Am Found, Region VI, 58; environ monitoring consult, East SDak, Northern States Power, 63-64. *Mem:* Fel AAAS; Am Phys Soc; Inst Elec & Electronics Engrs; Sigma Xi; Am Meteorol Soc; Am Geophys Union. *Res:* Environmental science; radio physics; atmospheric science. *Mailing Add:* HC 50-Box 1075 Camp Verde AZ 86322

BEDWORTH, DAVID D, b Manchester, Eng, Nov 29, 32; US citizen; m 61; c 2. PRODUCTION CONTROL, COMPUTER-AIDED MANUFACTURING. *Educ:* Lamar State Col, BSIE, 55; Purdue Univ, MSIE, 59, PhD(indust eng), 61. *Prof Exp:* Methods engr indust eng, Boeing Airplane Co, Kans, 55; instr, Purdue Univ, 58-60, res asst, 60-61; sr systs analyst, Process Comput Sect, Gen Elec Co, Ariz, 61-62, consult analyst, 62-63; fac assoc, 62-63, from asst prof to assoc prof, 63-69, chair, dept indust engrs, 76-80, PROF INDUST ENG, ARIZ STATE UNIV, 74- *Honors & Awards:* Wickenden Award Am, Soc Eng Educ. *Mem:* Am Inst Indust Engrs; Am Soc Eng Educ; Soc Mfg Engrs. *Res:* Time-series analysis and forecasting; industrial engineering computer application; scheduling activities; computer aided manufacturing. *Mailing Add:* Dept Indust & Mgt Syst Eng Ariz State Univ Tempe AZ 85287-5906

BEEBE, GEORGE WARREN, b Eau Claire, Wis, Sept 9, 36; m 59; c 3. PHOTOGRAPHIC CHEMISTRY. *Educ:* Wis State Univ, Eau Claire, BS, 58; Mich State Univ, PhD(tetrazoles), 64. *Prof Exp:* SR RES SPECIALIST, PHOTO PROD LAB, 3M CO, 63- *Mem:* Am Chem Soc; Soc Photog Scientists & Engrs. *Res:* Nitrogen heterocycles; condensed ring hydrocarbons; fluorescence, tetrazole synthesis; chemiluminescence; organic dyes; polyurethanes; aliphatic epoxides; resin curing systems; color photographic chemistry; photographic emulsion making, digestion and coating. *Mailing Add:* 3M Co 3M Ctr Bldg 209-2C-08 St Paul MN 55144

BEEBE, GILBERT WHEELER, b Mahwah, NJ, Apr 3, 12; m 33; c 4. RADIATION BIOLOGY. *Educ:* Dartmouth Col, AB, 33; Columbia Univ, MA, 38, PhD(sociol, statist), 42. *Prof Exp:* Statistician, Nat Comt Maternal Health, 34-41; mem tech staff, Milbank Mem Fund, 41-46; statistician, Div Med Sci, Nat Acad Sci-Nat Res Coun, 46-77, dir, Follow-Up Agency, 57-77; EPIDEMIOLOGIST, CLIN EPIDEMIOL BR, NAT CANCER INST, NIH, 77- *Concurrent Pos:* Res assoc, Milbank Mem Fund, 39-41; chief, Reports & Anal Br, Control Div, Off Surgeon Gen, US Army, 43-46, consult, 46-50; consult, Comn Reorgn Exec Br, Hoover Comn, 48; chief epidemiol & statist dept, Atomic Bomb Casualty Comn, Japan, 58-60, 66-68 & 73-75, mem bd dirs & chief scientist, Radiation Effects Res Found, Japan, 75; mem subcomt somatic effects, Nat Acad Sci Adv Comt Biol Effects of Ionizing Radiation, 69-72 & 77-80; mem res training comt, Nat Inst Environ Sci, 70-73; mem, Comt Biol Effects Ionizing Radiations, Nat Acad Sci, 77-80; mem sci panel, Comt Interagency Radiation Res & Policy coord, Off Sci & Technol Policy, Exec Off Pres, 84- *Honors & Awards:* Cutter lectr prev med, Harvard Sch Pub Health, 79; Wade Hampton Frost lectr, Am Pub Health Asn, 80; Director's Award, NIH, 85; Special Recognition Award, PHS, 83. *Mem:* AAAS; Am Epidemiol Soc (vpres, 78-79); Soc Epidemiol Res; fel Am Statist Asn; Int Epidemiol Asn; Radiation Res Soc; Health Physics Soc. *Res:* Application of statistical methods to research in clinical medicine and epidemiology; estimation of the risks associated with exposure to ionizing radiation. *Mailing Add:* Clin Epidemiol Br Exec Plaza N Suite 400 Nat Cancer Inst Bethesda MD 20892

BEEBE, RICHARD TOWNSEND, b Great Barrington, Mass, Jan 22, 02; m 32; c 3. INTERNAL MEDICINE. *Educ:* Princeton Univ, BS, 24; Johns Hopkins Univ, MD, 28; Am Bd Internal Med, dipl. *Hon Degrees:* ScD, Albany Med Col, 82. *Prof Exp:* Res fel, Thorndike Mem Lab, 29-30; asst resident med, Johns Hopkins Univ, 30-32; assoc, 32-37, from assoc prof to prof, 37-67, head dept, 48, DISTINGUISHED PROF MED, ALBANY MED COL, 67- *Concurrent Pos:* Physician-in-chief, Albany Hosp, 48-67, sr physician, 67- *Mem:* Assoc AMA; Am Soc Clin Invest; Am Clin & Climat Asn; fel & master Am Col Physicians. *Mailing Add:* Albany Med Col Albany NY 12208

BEEBE, ROBERT RAY, b Butte, Mont, Apr 21, 28; m 56; c 1. EXTRACTIVE METALLURGY OF GOLD & BASE METALS. *Educ:* Mont Sch Mines, BS, 53, MS, 54. *Prof Exp:* Dir res, Carpco Res & Eng, 65-67, pres, 72-73; gen mgr proj develop, Marcona Corp, 67-73, vpres, 73-76; sr metall engr, Newmont Mining Corp, 76-80, vpres, 80-86; vpres, 86-88, SR VPRES, HOMESTAKE MINING CO, 88- *Concurrent Pos:* Vchmn, Comt Competitiveness, US Minerals & Metals Indust, 87-89; mem, Nat Mat Adv Bd, 90-92. *Honors & Awards:* Krumb Lectr, Am Inst Mining Metall & Petrol Engrs, 90. *Mem:* Nat Acad Eng; Soc Mining Minerals & Explor; Mining & Metall Soc Am (pres, 87-89). *Res:* Mining and mineral project development and administration; mineral processing and extractive metallurgical research; mineral economics and public policy. *Mailing Add:* Homestake Mining Co 650 California St Ninth Floor San Francisco CA 94108-2788

BEEBEE, JOHN CHRISTOPHER, b Glendale, Calif, Mar 7, 41; m 69; c 2. ENGINEERING PHYSICS. *Educ:* Pomona Col, BA, 63; Univ Calif, Riverside, MA, 65; Univ Wash, PhD(math), 72. *Prof Exp:* Asst prof math, Alaska Methodist Univ, 71-74; researcher, Alaska Res Assoc, 74-75; teacher math, East High Sch, 75-78; researcher, Energy Syst, Inc, 78-84; prof math, Calif State Col, Bakersfield, 87; PROF MATH, UNIV ALASKA, ANCHORAGE, 84-86 & 87- *Mem:* Math Asn Am. *Res:* Cost and scientific feasibility of various energy supply or conversion projects or concepts; properties of cyclic sequences. *Mailing Add:* 9571 Midden Way Anchorage AK 99507

BEECHAM, CURTIS MICHAEL, b Loma Linda, Calif, Sept 19, 47. TEACHING. *Educ:* Univ Calif, Riverside, BS, 70; Stanford Univ, PhD(org chem), 78. *Prof Exp:* Res chemist, Univ Calif, Riverside, 78-79, San Diego, 82-84; org chemist & supvr, Rachelle Labs, Inc, 79-80; org chemist, Ultrasystems, Inc, 81-82; asst prof chem, Mt St Mary's Col, 84-86; ASST PROF CHEM, STERLING COL, 86- *Mem:* Am Chem Soc; Nat Sci Teachers Asn. *Res:* Marine natural products chemistry. *Mailing Add:* 714 N Broadway Sterling KS 67579

BEECHER, CHRISTOPHER W W, b New York, NY, Aug 20, 48; m 81; c 2. BIOSYNTHESIS, PLANT TISSUE CULTURE. *Educ:* NY Univ, BA, 74, MS, 80; Univ Conn, PhD(pharmacog), 86. *Prof Exp:* Res asst, Univ Conn, 78-84; vis asst prof, 85-89, ASST PROF, COL PHARM, UNIV ILL, CHICAGO, 89- *Concurrent Pos:* Dir, Metmap Database, 87-; mgr & assoc dir, Napralert Database, 90- *Mem:* Am Soc Pharmacol; Am Soc Plant Physiol; Sigma Xi. *Res:* Information management techniques in biological studies; computer assisted drug discovery methodology. *Mailing Add:* 2131 W Bowler St Chicago IL 60612

BEECHER, GARY RICHARD, b Wilton, Wis, May 25, 39; m 62; c 3. NUTRITIONAL BIOCHEMISTRY. *Educ:* Univ Wis-Madison, BS, 61, MS, 63, PhD(biochem), 66. *Prof Exp:* Biochemist, US Army Med Res & Nutrit Lab, 66-68; asst prof biochem, Kans State Univ, 68-71; biochemist, Protein Nutrit Lab, Nutrit Inst, 71-82, RES LEADER, NUTRIENT COMPOS LAB, BELTSVILLE HUMAN NUTRIENT RES CTR, AGR RES SERV, USDA, 82- *Concurrent Pos:* Collab scientist, Med Sch Carolina Univ, 74-82; consult, Am Res Prod Corp, 80-82, Anal Proc Inc, 82-88. *Mem:* AAAS; Am Inst Nutrit; Sigma Xi; fel Am Inst Chem. *Res:* Analytical chemistry in biological and food systems. *Mailing Add:* 16132 Kenny Rd Laurel MD 20707

BEECHER, WILLIAM JOHN, b Chicago, Ill, May 23, 14. ORNITHOLOGY. *Educ:* Univ Chicago, BS, 47, MS, 49, PhD, 54. *Prof Exp:* Asst zool, Chicago Natural Hist Mus, 37-42, asst zool, Educ Dept, 46-54; asst zool, Conserv Dept, Cook County Forest Preserve Dist, 54-57; dir, Chicago Acad Sci, 58-83; PRES, BEECHER RES CO, 83-; RES ASSOC, BIRD DIV, FIELD MUS NAT HIST, 87- *Concurrent Pos:* Mem biol comt, Ill Bd Higher Educ; vchmn, Ill Chap Nature Conserv; mem environ comt, Northeastern Ill Plan Comn; mem open lands proj; mem, Wetlands Comt, Biol Comn, Univ Chicago, Adventurers Club, 67-, Explorers Club, 88- *Honors & Awards:* Nat Asn Biol Teachers Award; US Environ Protection Agency Environ Qual Award, 76; Cyrus Mark Award. *Mem:* AAAS; Am Ornith Union; Wilson Ornith Soc; Cooper Ornith Soc; Soc Study Evolution; fel Royal Soc Arts. *Res:* Birds, anatomy and classification, ecology and migration; other vertebrates, comparative anatomy of the ear, migration and homing; inventor of high-tech spectacle binoculars; convertible to stereomicroscope and benefic to victims of macular degeneration. *Mailing Add:* 1960 Lincoln Park W Chicago IL 60614

BEECHLER, BARBARA JEAN, b Rockford, Ill, Dec 13, 28. MATHEMATICS. *Educ:* Univ Iowa, BA, 49, MS, 51, PhD(math), 55. *Prof Exp:* Instr math, Smith Col, 52-54; asst prof, Wilson Col, 55-58, assoc prof & chmn dept, 58-60; assoc prof, Wheaton Col (Mass), 60-67; assoc prof, 67-68, PROF MATH, PITZER COL, 68- *Mem:* Math Asn Am; Am Math Soc. *Res:* Arthmetic properties of integral domains; local rings. *Mailing Add:* Dept Math Pitzer Col 1050 NMills Ave Claremont CA 91711

BEECKMANS, JAN MARIA, b Antwerp, Belg, June 17, 30; Can citizen; m 57, 70; c 3. CHEMICAL ENGINEERING. *Educ:* Univ London, BSc, 53; Univ Toronto, MASc, 54, PhD(phys chem), 58. *Prof Exp:* Res assoc chem eng, Imp Oil, Ltd, 58-59; chemist, Ont Res Found, 59-61; asst prof indust hyg, Univ Toronto, 61-66; PROF CHEM ENG, UNIV WESTERN ONT, 66- *Mem:* Chem Inst Can. *Res:* Fluidization; solids separation. *Mailing Add:* Dept Chem Eng Univ Western Ont London ON N6A 3K7 Can

BEEDE, CHARLES HERBERT, b Chelsea, Mass, Dec 4, 34; m 57; c 1. ORGANIC CHEMISTRY. *Educ:* Northeastern Univ, BS, 57; Mass Inst Technol, PhD(org chem), 62. *Prof Exp:* Res chemist, Org & Polymer Chem, Hercules Powder Co, 61-63; res chemist, 64-80, res assoc, 80-84, SR SCIENTIST, JOHNSON & JOHNSON, 85- *Mem:* NY Acad Sci; AAAS; Am Chem Soc; Sigma Xi. *Res:* Polymer synthesis, particularly vinyl, emulsion, stereoregular and condensation; organic synthesis; chemistry of small-ring compounds; pressure-sensitive adhesives; controlled release-transdermal delivery; absorbable polymers; hydrophilic polymers. *Mailing Add:* 6102 Tiffany Park Ct Arlington TX 76016-2037

BEEDLE, LYNN SIMPSON, b Orland, Calif, Dec 7, 17; m 46; c 5. CIVIL ENGINEERING. *Educ:* Univ Calif, BS, 41; Lehigh Univ, MS, 49, PhD(civil eng), 52. *Prof Exp:* Prog eng, Todd Calif Shipbldg Corp, 41; dep officer in charge inst group, Oper Crossroads, US Navy Bur, 46-47; officer in charge underwater exp, US Navy, Norfolk Naval Shipyard, 42-47; from asst prof to distinguished prof, Lehigh Univ, 52-60, dir, High-Rise Inst, 83-88; DIR, COUN TALL BUILDINGS & URBAN HABITATION, 78- *Concurrent Pos:* Student Naval archit, US Navy Postgrad Sch, Naval Acad, 41-42, instr, 42; consult, Indian Standards Inst, 57-58; ed, Costruzioni Metalliche Mag, 80; chmn, Structural Stability Res Coun, 66-70, dir, 70-; chmn, Coun Tall Bldg & Urban Habitat, 70-76, dir, 76-; consult, Freeman, Fox & Partners, 81; dir, Fritz Eng Lab, 60-84. *Honors & Awards:* Robinson Award, 52, Hillman Award, Lehigh Univ, 73; A F Davis Silver Medal, Am Welding Soc, 57; E E Howard Award, Am Soc Civil Engrs, 63. *Mem:* Nat Acad Eng; hon mem Am Soc Civil Engrs; Am Inst Steel Construct; hon mem Int Asn Bridge & Struct Eng. *Res:* Research in plastic design and steel, large bolted joints, column stability, planning and design of tall buildings. *Mailing Add:* 102 Cedar Rd Hellertown PA 18055

BEEHLER, JERRY R, b Lekhart, Ind, Jan 16, 43. MATHEMATICS. *Educ:* Univ Toronto, PhD(math), 73. *Prof Exp:* PROF MATH & DEAN ARTS & SCI, TRI-STATE UNIV, 69- *Mem:* Math Asn Am. *Mailing Add:* Dean Arts & Sci Tri State Univ Angola IN 46703

BEEHLER, ROGER EARL, b Rochester, Ind, May 8, 34; m 56; c 3. PHYSICS. *Educ:* Purdue Univ, BS, 56. *Prof Exp:* Physicist, Allison Div, Gen Motors Corp, Ind, 56-57; physicist, Nat Bur Standards, 57-60, proj leader atomic frequency standards, 60-65, asst sect chief, Atomic Time & Frequency Standards Sect, 65-66; physicist, Frequency & Time Div, Hewlett-Packard Co, 66-68; from asst chief to assoc chief time & frequency div, 68-75, CHIEF TIME & FREQUENCY SERV, NAT BUR STANDARDS, 75- *Honors & Awards:* US Dept Com Silver Medal, 64. *Res:* Atomic beam frequency standards; time; frequency. *Mailing Add:* 1505 Moss Rock Pl Boulder CO 80304

BEEK, JOHN, b Seattle, Wash, Nov 11, 06; m 30; c 2. CHEMICAL ENGINEERING. *Educ:* Univ Wash, BS, 27; George Washington Univ, AM, 32; PhD(phys chem), 44. *Prof Exp:* From jr chemist to chemist, Nat Bur Standards, 27-45; from chemist to supv, Shell Develop Corp, 46-70; RETIRED. *Concurrent Pos:* Res assoc chem, Univ Amsterdam, 71-72. *Mem:* Am Chem Soc; AAAS. *Res:* Effect of sulfuric acid on leather; constitution of collagen; effect of radiation on the combustion of rocket propellants; design and performance of chemical reactors. *Mailing Add:* 5620 Harbord Dr Oakland CA 94618

BEEKMAN, BRUCE EDWARD, b Upland, Calif, Apr 18, 30; c 4. COMPARATIVE PHYSIOLOGY. *Educ:* San Diego State Col, BA, 52; Ind Univ, PhD, 65. *Prof Exp:* From asst prof to assoc prof physiol, 58-71, chmn dept biol, 72-75, PROF PHYSIOL, CALIF STATE UNIV, LONG BEACH, 71- *Mem:* Am Soc Zoologists. *Res:* Comparative physiology and biochemistry of oxygen carriers. *Mailing Add:* Dept Anat & Physiol Calif State Univ Long Beach CA 90840

BEEKMAN, JOHN ALFRED, b LaCrosse, Wis, July 14, 31; m 55; c 2. MATHEMATICS. *Educ:* Univ Iowa, BA, 53, MS, 57; Univ Minn, Minneapolis, PhD(math), 63. *Prof Exp:* From asst prof to assoc prof, 63-69, PROF MATH, BALL STATE UNIV, 69- *Concurrent Pos:* Vis prof statist, Univ Iowa, 69-70. *Mem:* Am Math Soc; Soc Actuaries; Am Acad Actuaries; Math Asn Am; Sigma Xi. *Res:* Probability theory; Gaussian Markov and collective risk stochastic processes; mathematical demography including population projection techniques; actuarial applications of mathematical risk theory. *Mailing Add:* Dept Math Sci Ball State Univ Muncie IN 47306-0490

BEEL, JOHN ADDIS, b Butte, Mont, Sept 20, 21; m 44; c 2. ORGANIC CHEMISTRY. *Educ:* Mont State Col, BSc, 42; Iowa State Univ, PhD(org chem), 49. *Prof Exp:* Instr chem, Iowa State Univ, 45-49; from asst prof to prof chem, 49-84, head dept, 49-52, actg chmn sci div, 57, 64 & 68, actg assoc dean, 69, assoc dean, Arts & Sci, 71-, EMER PROF CHEM, UNIV NORTHERN COLO, 84- *Honors & Awards:* Meritorious Serv Award, Am Chem Soc, 69. *Mem:* Sigma Xi; Am Chem Soc. *Res:* Organometallic compounds; chemotherapy. *Mailing Add:* Dept Chem Univ Northern Colo Greeley CO 80631

BEELER, DONALD A, b Elmwood, NS, Aug 16, 31; m 53; c 4. BIOCHEMISTRY. *Educ:* McGill Univ, BSc, 53; Purdue Univ, MS, 59, PhD(biochem), 62. *Prof Exp:* Fel biochem, Univ Wis, 61-63; instr, 63-64, from asst prof to assoc prof, 64-75, PROF BIOCHEM, ALBANY MED COL, 75- *Mem:* Fedn Am Soc Exp Biol. *Res:* Hepatic, biliary and lung phospholipid metabolism. *Mailing Add:* Dept Biochem Albany Med Col Albany NY 12208

BEELER, GEORGE W, JR, b West Point, NY, Oct 5, 38; m 65; c 1. BIOMEDICAL ENGINEERING. *Educ:* Princeton Univ, BSE, 60; Calif Inst Technol, MS, 61, PhD(elec eng), 65. *Prof Exp:* Assoc consult, Mayo Clin & instr, Mayo Med Sch, 67-68, head sect info processing & systs, 80-86, head sect info systs technol, 87-90, CONSULT PHYSIOL & BIOPHYS, MAYO CLIN & ASST PROF, MAYO MED SCH, 69-, CHMN DIV INFO ARCH & TECHNOL, 90- *Mem:* AAAS; Inst Elec & Electronics Engr. *Res:* Analysis of biological systems; excitation-contraction coupling in cardiac muscle; application of computers to biomedical research and practice. *Mailing Add:* 200 SW First St Rochester MN 55901

BEELER, JOE R, JR, b Beloit, Kans, Aug 13, 24; m 47; c 2. SOLID STATE PHYSICS. *Educ:* Univ Kans, PhD(physics), 55. *Prof Exp:* Staff mem, Sandia Corp, 52-55; sr physicist, Gen Elec Co, 55-57, prin physicist, 57-61, supvr solid state physics, 61-64, consult physicist, 64-67; PROF MAT SCI ENG, NC STATE UNIV, 67- *Concurrent Pos:* Solid state physics ed, J Comput Physics Commun, 69-7; consult, Hanford Eng Develop Lab, 69-75 & Lawrence Livermore Nat Lab, 76-8; Battelle vis prof metall, Ohio State Univ, 70-71; vis staff mem, Los Alamos Nat Lab, 75- *Mem:* Am Phys Soc; Am Asn Eng Educ; Metall Soc of Am Inst Mining, Metall & Petrol Engrs; Am Soc Metals; fel Am Soc Metals. *Res:* Computer simulation in materials science and engineering; radiation effects computer simulation; active in research; computer aided education research; defects in solids. *Mailing Add:* Mat Sci Eng Dept NC State Univ PO Box 7907 Raleigh NC 27695-7907

BEELER, MYRTON FREEMAN, b Winthrop, Mass, Apr 27, 22. CLINICAL PATHOLOGY, CLINICAL CHEMISTRY. *Educ:* Harvard Univ, AB, 45; NY Med Col, MD, 49. *Prof Exp:* Resident path, Worcester City Hosp, Mass, 52-54; resident path, New Eng Deaconess Hosp, Boston, 54-56, asst pathologist, 56-58; dir labs, Ochsner Clin & Ochsner Found Hosp, 58-67; assoc prof, 67-71, prof, 71-87, dir grad prog clin chem, 70-87 EMER PROF PATH, LA STATE UNIV MED CTR, NEW ORLEANS, 87- *Concurrent Pos:* Assoc pathologist, Charity Hosp of La, New Orleans, 67-, dir clin chem, 75-81; consult clin path, Vet Admin Hosp, 71- & USPHS Hosp, 74-79; ed-in-chief, Am Jour Clin Path, 80- *Mem:* Am Soc Clin Path; Col Am Path Physicians & Scientists; AOA. *Res:* Medical decision making. *Mailing Add:* Dept Path La State Univ Med Ctr New Orleans LA 70112

BEELER, TROY JAMES, MUSCLE & MEMBRANE BIOCHEMISTRY. *Educ:* Univ Tenn, PhD(biochem), 78. *Prof Exp:* ASSOC PROF MED BIOCHEM, UNIFORMED SERV UNIV HEALTH SCI, 80- *Mailing Add:* 66 Timber Rock Rd Gaithersburg MD 20878-2229

BEELIK, ANDREW, b Nizne Valice, Czech, Dec 12, 24; US citizen; m 51. SYNTHETIC ORGANIC & NATURAL PRODUCTS CHEMISTRY. *Educ:* Univ Agrarian Sci, Hungary, dipl, 47; Univ Toronto, MSA, 52; McGill Univ, PhD(chem), 54. *Prof Exp:* Lectr org chem, Royal Mil Col, Can, 54-55; from res chemist to sr res chemist, ITT Rayonier Inc, 55-62, res group leader, 62-72, res supvr, 72-78, mgr lab serv, Rayonier Res Ctr, 78-82; RETIRED. *Mem:* Am Chem Soc; Sigma Xi. *Res:* Chemistry of wood constituents; cellulose acetate and viscose processes; instrumental analysis. *Mailing Add:* 2012 Walker Park Rd Shelton WA 98584

BEELMAN, ROBERT B, b Elyria, Ohio, May 16, 44; m 68. FOOD SCIENCE, ENOLOGY. *Educ:* Capital Univ, BS, 66; Ohio State Univ, MS, 67, PhD(food sci), 70. *Prof Exp:* Asst prof, 70-75, ASSOC PROF FOOD SCI, PA STATE UNIV, UNIVERSITY PARK, 75- *Mem:* Am Soc Microbiol; Inst Food Technologists; Am Soc Enol. *Res:* Effects of processing on the quality of fruits, vegetables and mushrooms; evaluation of vinification practices on eastern United States table wines; stimulation of malo-lactic fermentation in table wines; food microbiology. *Mailing Add:* 111 Borland Lab Pa State Univ University Park PA 16802

BEEM, JOHN KELLY, b Detroit, Mich, Jan 24, 42; m 64; c 1. GEOMETRY, APPLIED MATH. *Educ:* Univ Southern Calif, AB, 63, MA, 65, PhD(math), 68. *Prof Exp:* from asst prof to assoc prof, 68-79, PROF MATH, UNIV MO, COLUMBIA, 79- *Concurrent Pos:* Investr, NSF grant, 69-71, NSF grant, 84-88. *Mem:* Am Math Soc; Math Asn Am; Gen Relativity & Gravitation Soc; Sigma Xi. *Res:* Geometry of Lorentzian manifolds and general relativity. *Mailing Add:* Dept Math Univ Mo Columbia MO 65211

BEEM, MARC O, b Chicago, Ill, June 25, 23; m 46; c 4. MEDICINE. *Educ:* Williams Col, BA, 45; Univ Chicago, MD, 48. *Prof Exp:* From asst prof to assoc prof, 54-64, PROF PEDIAT, UNIV CHICAGO, 64- *Res:* Infectious diseases; virology. *Mailing Add:* Dept Pediat Univ Chicago Pritzker Sch Med Wyler Children's Hosp 5841 S Maryland Chicago IL 60637

BEEMAN, CURT PLETCHER, b Mt Vernon, Ohio, May 30, 44; m 66; c 2. ANALYTICAL CHEMISTRY. *Educ:* Univ Fla, BS, 67; Auburn Univ, MS, 69, PhD(chem), 71. *Prof Exp:* Res asst chem, Univ Alta, 71-73; res assoc, Mt Holyoke Col, 73-77; res assoc, Drinking Water Qual Res Ctr, Fla Int Univ, 77-79, RES ASSOC, INT PAPER CO, 79-; RES ASSOC, INT PAPER CO, 79- *Mem:* Am Chem Soc; Sigma Xi. *Res:* Gas chromatography and gas chromatography/mass spectrometry; instrumental analysis of wood, paper products and process components. *Mailing Add:* 6658 Hounds Run N Mobile AL 36608

BEEMAN, DAVID EDMUND, JR, b Sacramento, Calif, Dec 12, 38; m 65; c 2. THEORETICAL PHYSICS. *Educ:* Stanford Univ, BS, 61; Univ Calif, Los Angeles, MA, 63, PhD(physics), 67. *Prof Exp:* Res assoc theoret physics, Atomic Energy Res Estab, Eng, 67-69; asst prof, 69-74, ASSOC PROF PHYSICS, HARVEY MUDD COL, 69- *Mem:* AAAS; Am Phys Soc. *Res:* Theoretical solid state physics; amorphous materials. *Mailing Add:* 8217 Hygiene Rd Longmont CO 80503

BEEMAN, ROBERT D, b Los Angeles, Calif, Mar 23, 32; m 68; c 3. MARINE ZOOLOGY, INVERTEBRATE ZOOLOGY. *Educ:* Humboldt State Col, BS, 54; Univ Idaho, MS, 56; Stanford Univ, PhD(marine biol), 66. *Prof Exp:* Instr zool, Humboldt State Col, 52-54; res fel, Univ Idaho, 54-56; instr zool, Humboldt State Col, 56-57; instr zool & dir, Mt San Antonio Col, 58-64; sci fac fel marine biol, Stanford Univ, 62-66; chmn dept marine biol, 67-69, PROF MARINE ZOOL, SAN FRANCISCO STATE UNIV, 66- *Concurrent Pos:* NSF res grants, 66-72. *Mem:* AAAS; Inst Malacol; Soc Syst Zool; NY Acad Sci. *Res:* Migration ecology of elk; functional morphology of reproductive systems in anaspidean opisthobranch mollusks, including ultrastructure and capacitation of spermatozoa. *Mailing Add:* Alumni Assoc 1600 Hollway NAD 467 San Francisco CA 94132

BEEMAN, WILLIAM WALDRON, biophysics; deceased, see previous edition for last biography

BEEMON, KAREN LOUISE, b Jackson, Mich, Apr 11, 47; m 74; c 2. MOLECULAR BIOLOGY, VIROLOGY. *Educ:* Univ Mich, BS, 69; Univ Calif, Berkeley, MA, 72, PhD(molecular biol), 74. *Prof Exp:* NIH fel molecular biol, Univ Calif, Berkeley, 74-76, lectr, 75; NIH fel tumor virol, Salk Inst, 76-78, sr res assoc tumor virol, 78-81; asst prof, 81-87, ASSOC PROF BIOL, JOHNS HOPKINS UNIV, 87- *Honors & Awards:* Fac Res Award, Am Cancer Soc. *Mem:* Am Soc Microbiol; Am Soc Virol; Asn Women Sci. *Res:* Animal virology; retrovirus genome structure and gene expression; function of methylated nucleotides in messenger RNA; mechanism of cell transformation by tumor viruses. *Mailing Add:* Biol Dept Johns Hopkins Univ Baltimore MD 21218

BEENE, JAMES ROBERT, b Chattanooga, Tenn, Sept 29, 47; m 70; c 3. HEAVY ION NUCLEAR PHYSICS & REACTIONS, HYPERFINE INTERACTIONS PHYSICS. *Educ:* Univ of South, BA, 69; Oxford Univ, PhD(physics), 74. *Prof Exp:* Nat Res Coun postdoctoral fel, Chalk River Nuclear Lab, Atomic Energy Can Ltd, 74-76; res staff mem, 76-87, SR RES STAFF MEM & GROUP LEADER, OAK RIDGE NAT LAB, 87- *Mem:* Am Phys Soc; Nat Asn Advan Sci. *Res:* The study of the spectroscopic properties of highly excited collective nuclear states and the dynamics of intermediate and high energy heavy ion nuclear reactions. *Mailing Add:* Oak Ridge Nat Lab MS-6368 PO Box 2008 Oak Ridge TN 37831-6368

BEER, ALAN E, b Milford, Ind, Apr 14, 37; m 59; c 4. OBSTETRICS & GYNECOLOGY, TRANSPLANTATION BIOLOGY. *Educ:* Ind Univ, BS, 59, MD, 62. *Prof Exp:* Intern, Methodist Grad Med Ctr, Ind, 62-63; surgeon, USPHS Div Indian Health, USPHS Indian Hosp, Tuba City, Ariz, 63-65; from resident to chief resident obstet & gynec, Hosp Univ Pa, 65-69; fel med genetics & obstet & gynec, Sch Med, Univ Pa, 68-70, asst prof, 70-71; from asst prof to assoc prof cell biol & obstet & gynec, Univ Tex Health Sci Ctr, Dallas, 71-77, prof, 77-79; prof gynec, Women's Hosp, Univ Mich, 79-87, chmn dept, 79-87; PROF OBSTET & GYNEC, CHICAGO MED SCH, 87-, PROF MICROBIOL & IMMUNOL, 87- *Concurrent Pos:* Smith Kline & French foreign fel, Nigeria, 62-63. *Honors & Awards:* Carl Hartman Award, Am Fertil Soc, 70. *Mem:* AAAS; Transplantation Soc; Soc Study Reprod; Am Col Obstet & Gynec; Soc Gynec Invest. *Res:* Immunobiology of mammalian reproduction; immunological significance of the mammary gland; elicitation and expression of immunity in the uterus; transplantation immunobiology. *Mailing Add:* Reprod Immunol Clin Univ Health Sci Chicago Med Sch 3333 Greenbay Rd N Chicago IL 60064

BEER, ALBERT CARL, b Mansfield, Ohio, Mar 7, 20; m 49; c 3. PHYSICS. *Educ:* Oberlin Col, AB, 41; Cornell Univ, PhD(physics), 44. *Prof Exp:* Asst, Nat Defense Res Comt proj, Cornell Univ, 42-44, res assoc, 44-45; physicist appl physics lab, Johns Hopkins Univ, 45-51; prin physicist, Battelle Mem Inst, 51-52, asst chief div, 52-54, consult, 54-56, asst tech dir, 56-67, sr fel, 67-84, consult, 84-90; RETIRED. *Concurrent Pos:* Chmn panel defects, ad hoc comt Characterization of Mat, Mat Adv Bd, Nat Acad Sci, 65-67; adj prof, dept elec eng, Ohio State Univ, 69- *Mem:* Fel Am Phys Soc; Inst Elec & Electronics Engrs. *Res:* Transport theory; semiconductors; semimetals; metals. *Mailing Add:* 2788 Brandon Rd Columbus OH 43221-3347

BEER, BERNARD, BEHAVIORAL PHARMACOLOGY. *Educ:* George Washington Univ, PhD(physiol psychol), 66. *Prof Exp:* DIR CENTRAL NERVOUS SYST RES, LEDERLE LABS, 77-; PROF PSYCHIAT, CITY UNIV NEW YORK, 80-; PROF PSYCHIAT, NY UNIV, 80- *Res:* Neurochemistry. *Mailing Add:* CNS Res Am Cyanamid Lederle Labs Pearl River NY 10965

BEER, CHARLES, b Le Sueur Co, Minn, Dec 18, 23. AGRICULTURAL ECONOMICS. *Educ:* Univ Minn, St Paul, BS, 48; Mich State Univ, MS, 55, PhD(agr econ), 57. *Prof Exp:* Exten agent, Univ Minn, Anoka County, 49-53; asst prof agr econ, Mich State Univ, 57-62; prof, Univ Mo, Columbia, 62-64; specialist farm mgt, Exten Serv, USDA, 64-66, asst dir, 66-68, dir agr prod, Exten Serv, 68-74, dir independent family owned farm prog, 74-78, chief, Current & Future Priorities Staff, Sci & Educ Admin, 78-80, dep adminr, 81-85; RETIRED. *Mem:* Am Soc Farm Mgr & Rural Appraisers; Am Agr Econ Asn; Am Soc Agr Consult; World Futures Soc. *Mailing Add:* 490 M St SW Washington DC 20024

BEER, COLIN GORDON, b Waipukurau, NZ, Mar 13, 33; m 59; c 2. HISTORY & PHILOSOPHY OF ETHOLOGY. *Educ:* Univ Otago, NZ, BSc, 54, MSc, 56; Oxford Univ, DPhil, 60. *Hon Degrees:* MA, Oxford Univ, 70. *Prof Exp:* Jr lectr zool, Univ Otago, 56; sci officer, Ministry of Aviation, UK, 59-60; lectr, Univ Otago, 62-64; assoc prof, 64-69, prof, 69-80, PROF II PSYCHOL, RUTGERS UNIV, 80- *Concurrent Pos:* Fel, Inst Animal Behav, Rutgers Univ, 60-61; co-ed, Animal Behav Monographs, 67-73; mem, Psychobiol Adv Panel, NSF, 68-69 & 70-72; lectr & tutorial fel zool, Oxford Univ, 69-70; cosult ed, J Comp & Physiol Psychol, 69-80; assoc ed, Advan Study Behav, 73-, Animal Learning & Behav, 76-77; mem, Int Ethological Comt, 74-81. *Mem:* Am Ornithologists Union; Animal Behav Soc. *Res:* Breeding biology and social behavior of birds, concentrating on aspects of communication and social development in gulls; conceptual and historical aspects of ethology. *Mailing Add:* Dept Psychol Rutgers Univ Newark Campus Newark NJ 07102

BEER, CRAIG, b Keosauqua, Iowa, Feb 26, 27; m 49; c 4. AGRICULTURAL ENGINEERING. *Educ:* Iowa State Univ, BS, 50, MS, 57, PhD(agr & civil eng), 62. *Prof Exp:* From instr to assoc prof, 55-71, PROF AGR ENG, IOWA STATE UNIV, 71-, EMER PROF. *Mem:* Am Soc Agr Eng. *Res:* Derivation of quantitative expressions to express the various facets of watershed erosion; transport of sediment and watershed hydrology; on-site waste treatment systems. *Mailing Add:* Dept Agr Eng Iowa State Univ Ames IA 50012

BEER, FERDINAND P(IERRE), b Binic, France, Aug 8, 15; nat US; m 40; c 2. MECHANICS. *Educ:* Col Geneva, BachD, 33; Univ Geneva, licence, 35, PhD(potential theory), 37; Univ Paris, licence, 38. *Prof Exp:* Instr math, Goddard Col, 41-42 & Univ Kansas City, 42-43; asst prof physics, Williams Col, Mass, 43-46, asst prof math, 46-47; from asst prof to assoc prof, 47-51, head dept mech, Lehigh, Univ, 57-68, prof mech, 51-68, chmn dept mech & mech eng, 68-84; EMER UNIV DISTINGUISHED PROF, LEHIGH UNIV, 84- *Concurrent Pos:* Consult, Sprauge Elec Co, 45-47, Fed Civil Defense Admin, 51-52 & Boeing Airplane Co, 56-57. *Mem:* Am Math Soc; Am Soc Eng Educ; Am Soc Mech Engrs; Sigma Xi. *Res:* Potential theory; mechanics; vibrations; stochastic processes. *Mailing Add:* Dept Mech Eng Lehigh Univ Bethlehem PA 18015

BEER, GEORGE ATHERLEY, b Salmon Arm, BC, Jan 3, 35; m; c 1. NUCLEAR PHYSICS. *Educ:* Univ BC, BASc, 57, MSc, 59; Univ Sask, PhD(nuclear physics), 66. *Prof Exp:* Res officer reactor physics, Atomic Energy of Can Ltd, 59-61; Nat Res Coun fel nuclear physics, Inst Nuclear Physics, Darmstadt Tech Univ, 67-69; asst prof, 69-75, ASSOC PROF NUCLEAR PHYSICS, UNIV VICTORIA, 75- *Concurrent Pos:* Vis scientist, European Orgn Nuclear Res, 75-76. *Mem:* Can Asn Physicists. *Res:* Neutron gas scintillation spectroscopy; low-power reactor; nuclear structure investigations using inelastic electron scattering; intermediate energy physics and accelerator design. *Mailing Add:* Dept Physics Univ Victoria Box 1700 Victoria BC V8W 2Y2 Can

BEER, JOHN JOSEPH, b Saarbrücken, Germany, July 17, 27;US citizen; m 51; c 4. HISTORY OF CHEMISTRY. *Educ:* Earlham Col, AB, 50; Univ Ill, MS, 52, MA, 54, PhD(hist), 56. *Prof Exp:* Asst prof hist, Hanover Col, 56-57; asst prof chem, E Ill Univ, 57-58; asst prof hist, Okla State Univ, 58-61; asst prof, 61-63, ASSOC PROF HIST, UNIV DEL, 63- *Concurrent Pos:* Fulbright fel, 54- *Mem:* AAAS; Am Hist Asn; Hist Sci Soc; Soc Hist Technol; Soc Hist Alchemy & Chem. *Res:* History of chemistry and the chemical industry; history of appropriate technology. *Mailing Add:* Dept Hist Univ Del Newark DE 19716

BEER, MICHAEL, b Budapest, Hungary, Feb 20, 26; m 54; c 3. MOLECULAR BIOLOGY. *Educ:* Univ Toronto, BA, 49, MA, 50; Univ Manchester, PhD(phys chem), 53. *Prof Exp:* Instr physics, Univ Mich, 53-56; fel biophys, Nat Res Coun Can, 56-58; from asst prof to assoc prof, 58-63, chmn dept, 74-79, PROF BIOPHYS, JOHNS HOPKINS UNIV, 63- *Concurrent Pos:* NIH spec fel & chmn biophys training comt & sr fel, 67-68; mem, US Nat Comt of Int Union Pure & Appl Biophys, 74-; NATO fel, 75. *Mem:* Biophys Soc (pres, 75-76); Electron Micros Soc Am (pres, 80); AAAS; Sigma Xi. *Res:* Structure of biological macromolecules using optical methods; fine structure of cells; viruses and macromolecules as revealed by electron microscope. *Mailing Add:* Dept Biophys Johns Hopkins Univ Baltimore MD 21218

BEER, REINHARD, b Berlin, Ger, Nov 5, 35; m 60. IR SPECTROSCOPY. *Educ:* Univ Manchester, BSc, 56, PhD(physics), 60. *Prof Exp:* Res asst physics, Univ Manchester, 56-60, sr res asst astron, 60-63; sr scientist, 63-67, group leader, 67-68, mem tech staff, 68-71, GROUP SUPVR TROPOSPHERIC SCI, JET PROPULSION LAB, CALIF INST TECHNOL, 71-, SR RES SCIENTIST, 85-, MGR, ATMOSPHERIC & OCEANOG SCI, 90- *Concurrent Pos:* Vis assoc prof astron, Univ Tex, Austin, 74. *Honors & Awards:* Except Sci Achievement Medal, NASA, 74, Group Achievement Award, 80 & 86. *Mem:* Optical Soc Am; Int Astron Union; Am Astron Soc; AAAS. *Res:* Application of infrared interference spectroscopy to physics and astronomy; Fourier spectroscopy of atmospheres. *Mailing Add:* Space Sci Div 183/301 4800 Oak Grove Dr Pasadena CA 91109

BEER, ROBERT EDWARD, b Los Angeles, Calif, Apr 28, 18; m 41; c 1. ENTOMOLOGY. *Educ:* Univ Calif, BS, 47, MS, 48, PhD(entom), 50. *Prof Exp:* Res asst entom, Univ Calif, 47-49; from instr to prof, 50-88, chmn dept, 61-68 & 75-84, EMER PROF ENTON, UNIV KANS, 88- *Concurrent Pos:* From asst state entomologist to state entomologist, Kans, 53-63; from assoc prof to prof, Biol Sta, Univ Mich, 57-69; prof, Biol Sta, Univ Minn, 71-73 & Eastern Mich Univ, 81. *Mem:* AAAS; Entom Soc Am; Soc Study Evolution (treas); Sigma Xi. *Res:* Biology and taxonomy of mites and sawflies; agricultural entomology. *Mailing Add:* Dept Entom Univ Kans Lawrence KS 66045

BEER, STEVEN VINCENT, b Boston, Mass, July 19, 41; m 63; c 3. PLANT PATHOLOGY. *Educ:* Cornell Univ, BS, 65; Univ Calif, Davis, PhD(plant path), 69. *Prof Exp:* Asst prof, 69-77, ASSOC PROF PLANT PATH, CORNELL UNIV, 77- *Mem:* Am Phytopath Soc; AAAS; Am Soc Microbiol. *Res:* Epidemiology, physiology and control of plant diseases caused by bacteria with emphasis on fire blight; molecular genetics, physiology and ecology of phytopathogenic bacteria; biological control of phytopathogenic bacteria; biotechnology. *Mailing Add:* Dept Plant Path Plant Sci Bldg Cornell Univ Ithaca NY 14853

BEER, SYLVAN ZAVI, b New York, NY, Feb 5, 29; m 52; c 2. PHYSICAL CHEMISTRY. *Educ:* Brooklyn Col, BS, 51; Polytech Inst Brooklyn, PhD(phys chem), 58. *Prof Exp:* Chemist, Hexagon Labs, 51-52; instr chem, New York Community Col, 55-57; Crucible Steel Co fel high temperature chem, Mellon Inst, 57-60; staff scientist, Crucible Steel Co Am, 60-62; sr engr, Dept Chem, Westinghouse Elec Corp, 62-64 & Solid State Phenomena Dept, 64-67; mgr res & develop, Special Metals Corp, 67-70; vpres technol, Technostruct Corp, 70-72; consult, 72-76; vpres, Niagara Scientific, 76-88; CONSULT, 88- *Concurrent Pos:* Dir, Mfr Assistance Ctr, Syracuse Univ. *Mem:* AAAS; Am Chem Soc; Electrochem Soc; Am Soc Metals; Fedn Am Scientists; Sigma Xi. *Res:* High temperature chemistry; electrochemistry; air and water pollution; sensors/monitors development, instrumentation, superionic conductors, energy storage. *Mailing Add:* 315 Scott Ave Syracuse NY 13224

BEERBOWER, JAMES RICHARD, b Ft Wayne, Ind, Apr 5, 27; m; c 4. PALEONTOLOGY. *Educ:* Univ Colo, BA, 49; Univ Chicago, PhD(paleozool), 54. *Prof Exp:* From instr to assoc prof geol, Lafayette Col, 53-64; prof, McMaster Univ, 64-69; chmn, 69-86, PROF GEOL, STATE UNIV NY, BINGHAMTON 69-, CHMN, 84-86. *Mem:* Soc Econ Paleont & Mineral; Soc Vert Paleont; fel Geol Soc Am. *Res:* Paleoecology and evolution of vertebrate and invertebrate animals; vertebrate anatomy; sedimentology of alluvial deposits and marine shales. *Mailing Add:* Dept Geol State Univ NY Binghamton NY 13902-6000

BEERING, STEVEN CLAUS, b Berlin, Germany, Aug 20, 32; US citizen; m 56; c 3. EDUCATION ADMINISTRATION, MEDICINE. *Educ:* Univ Pittsburgh, BS, 54, MD, 58. *Hon Degrees:* DSc, Ind Cent Univ, 83, Univ Evansville, 84, Hanover Col, 86, Ramapo Col, 86, Anderson Col, 87 & Ind Univ, 88. *Prof Exp:* asst dean, Sch Med, Ind Univ, 69-70, assoc dean, 70-74, div med ctr, 74-83, dean, Sch Med, 74-83; PRES, PURDUE UNIV, 83- *Concurrent Pos:* Consult, Indianapolis Vet Admin Hosp, St Vincent Hosp, Indianapolis; chmn, Ind Med Educ Bd, 74-83, Coun Deans, Asn Am Med Cols, 80-81, Asn Am Med Cols, 82-83. *Honors & Awards:* Convocation Medal, Am Col Cardiol, 83. *Mem:* Nat Acad Sci Inst Med. *Mailing Add:* Off Pres Purdue Univ West Lafayette IN 47907

BEERMAN, HERMAN, b Johnstown, Pa, Oct 13, 01; m 24. DERMATOLOGY. *Educ:* Univ Pa, AB, 23, MD, 27, ScD(med), 35; Am Bd Dermat, dipl, 35. *Prof Exp:* Field asst, Bur Entom, USDA, 25-26; intern, Mt Sinai Hosp, Philadelphia, 27-28; asst instr dermat & syphil, Sch Med, 29-30, instr, 30-36, assoc, 36-37, from asst prof to prof, 37-70, from asst prof to prof, Grad Sch Med, 40-70, chmn dept, 49-67, EMER PROF DERMAT, SCH MED, UNIV PA, 70- *Concurrent Pos:* Dermatologist, Hosp Univ Pa, 29-70, asst chief dermat clin, 38-70, asst dir inst study venereal dis, 38-54, assoc serol, Pepper Lab, 50, head dept dermat, Grad Hosp, 53-67; Abbott fel, Univ Pa, 32-46; consult dermatologist, USPHS, 37-; asst dermatologist, Radium Clin, Philadelphia Gen Hosp, 38-40, dermatologist, 40-53, consult, 53-68, hon consult, 68-; dermatopathologist, Skin & Cancer Hosp, 48-54; treas & trustee, Inst Dermat Commun & Educ, 62-; consult, Grad Hosp, Univ Pa, 67-, Pa Hosp, 67- & Vet Admin Hosp, Coatesville, 67-; past assoc ed, Quart Rev Dermat & Quart Rev Syphilis; contrib ed, Am J Med Sci. *Honors & Awards:* Thomas Parran Award, Am Venereal Dis Asn, 74; Stephen Rothman Gold Medal Award, Soc Investigative Dermat, 85. *Mem:* Hon mem Am Dermat Asn (pres, 67-68); hon mem Am Acad Dermat (pres, 65-66); hon mem Soc Invest Dermat (pres, 47-58, secy-treas, 50-65); hon mem Am Soc Dermatopath (pres, 65-66); Asn Prof Dermat (pres, 67-68). *Res:* Dermatopathology; study of pathologic changes in patients; literary research. *Mailing Add:* Logan Sq E Apt 1412 Two Franklin Town Blvd Philadelphia PA 19103

BEERMANN, DONALD HAROLD, b Denison, Iowa, Oct 4, 49; m 75. MEAT SCIENCE, PHYSIOLOGY. *Educ:* Iowa State Univ, BS, 71; Univ Wis-Madison, MS, 74, PhD(muscle biol & human physiol), 76. *Prof Exp:* Asst prof, 78-84, ASSOC PROF ANIMAL SCI & FOOD SCI, CORNELL UNIV, 85- *Concurrent Pos:* Wis Heart Asn fel, Muscle Biol Lab, Univ Wis, 76-77, fel neurophysiol, Sch Med, 77-78. *Mem:* Am Soc Animal Sci; Am Meat Sci Asn; Inst Food Technologists; AAAS; Am Inst Nutrit. *Res:* Meats processing; muscle physiology; muscle biochemistry; muscle growth and development; animal physiology and metabolism. *Mailing Add:* Dept Animal Sci 53 Morrison Hall Cornell Univ Ithaca NY 14853-4801

BEERNTSEN, DONALD J(OSEPH), material science, physical chemistry, for more information see previous edition

BEERS, JOHN R, b Bridgeport, Conn, July 4, 33. BIOLOGICAL OCEANOGRAPHY. *Educ:* Bates Col, BS, 55; Univ NH, MS, 58; Harvard Univ, PhD(biol), 62. *Prof Exp:* Part-time instr biol, Univ NH, 57-58; marine biologist, Bermuda Biol Sta Res, Inc, 61-65, asst dir, 63-65; asst res zoologist, 65-71, lectr, 69-77, assoc res zoologist, 71-84, RES ZOOLOGIST, INST MARINE RESOURCES, UNIV CALIF, SAN DIEGO, 84- *Mem:* AAAS; Am Soc Limnol & Oceanog. *Res:* Biological oceanography; plankton ecology; protozoans in pelagic food webs. *Mailing Add:* Box 1299 La Jolla CA 92038

BEERS, ROLAND FRANK, JR, biomedical research; deceased, see previous edition for last biography

BEERS, THOMAS WESLEY, b Greensburg, Pa, Oct 23, 30; m 53; c 4. FOREST MENSURATION. *Educ:* Pa State Univ, BS, 55, MS, 56; Purdue Univ, PhD(forest mgt), 60. *Prof Exp:* From instr to assoc prof, 56-69, PROF FORESTRY, PURDUE UNIV, WEST LAFAYETTE, 69- *Mem:* Soc Am Foresters. *Res:* Forest mensuration techniques and instruments; inventory procedures and data processing methods, especially timber stand estimates by point sampling; forest management site productivity studies; computer programming of forestry problems. *Mailing Add:* Dept Forestry Purdue Univ West Lafayette IN 47907

BEERS, WILLIAM HOWARD, b Panama City, Panama, June 19, 43; US citizen; div; c 1. REPRODUCTIVE BIOLOGY, CELL BIOLOGY. *Educ:* Harvard Univ, AB, 65; Rockefeller Univ, PhD(life sci), 71. *Prof Exp:* Rockefeller Found fel, reproductive biol, Univ Ill, 71-73; from asst prof to assoc prof cell biol, Rockefeller Univ, 73-78; chmn dept, 78-80, PROF, DEPT BIOL, NY UNIV, 78- *Concurrent Pos:* Res career develop award, USPHS, Nat Inst Child Health & Human Develop, 77-78. *Res:* Mechanisms of ovarian function. *Mailing Add:* Scripps Res Clin 10666 N Torrey Pines La Jolla CA 92037

BEERS, YARDLEY, b Philadelphia, Pa, Apr 2, 13; m 45; c 2. MOLECULAR SPECTROSCOPY, ATOMIC & MOLECULAR PHYSICS. *Educ:* Yale Univ, BS, 34; Princeton Univ, MA, 37, PhD(physics), 41; Univ Colo, BA, 86. *Prof Exp:* Asst physics, Princeton Univ, 40; instr, NY Univ, 40-41 & Smith Col, 41-42; staff mem radiation lab, Mass Inst Technol, 42-45, res assoc, 46; from asst prof to prof physics, NY Univ, 46-61; chief millimeter wave res sect, Nat Bur Standards, 61-62, chief radio standards physics div, 62-68, consult, 68-79; RETIRED. *Concurrent Pos:* Consult, Brookhaven Nat Lab, 47-58; Fulbright grant, Nat Standards Lab, Australia & vis lectr, Univ Sydney, 56; vis lectr, Univ Colo, Boulder, 63-64 & 75; adj prof, Univ Denver, 65-67; vis prof, Colo State Univ, 67-68 & Univ Colo, Denver, 69; archeol vol, Peel Castle, Isle of Man, 85, 87. *Mem:* Inst Elec & Electronics Engrs; Sigma Xi; Am Asn Physics Teachers; fel Am Phys Soc. *Res:* Direct determination of the charge of the beta particle; microwave and millimeter wave spectra of gases; stark effect; laser power and energy measurements; theory of radiation. *Mailing Add:* 740 Willowbrook Rd Boulder CO 80302-7437

BEERSTECHER, ERNEST, JR, b Detroit, Mich, May 4, 19; m 63; c 4. BIOCHEMISTRY. *Educ:* Wayne Univ, BS, 40; Univ Tex, MA, 46, PhD(biochem), 48. *Prof Exp:* Asst exp med, Harper Hosp, Mich, 38-39; biochemist, Robison Labs, Inc, 40; toxicologist, Edgewood Arsenal, US Army, 40-42, res biochemist, 44-45; res biochemist, Biochem Inst, Univ Tex, 46-50; from asst prof to assoc prof, 50-55, PROF BIOCHEM, UNIV TEX DENT BR HOUSTON, 55- *Concurrent Pos:* Consult bacteriologist, Magnolia Petrol Co, Tex, 48-49. *Mem:* Am Chem Soc; Am Soc Biol Chemists. *Res:* Bioassay of drugs and metabolites; nutrition; endocrinology; bacterial metabolism; human metabolism; biosynthesis of amino acids; antibiotics; immunochemistry; dental biochemistry. *Mailing Add:* 5500 N Braeswood Blvd No 182 Houston TX 77096

BEERY, DWIGHT BEECHER, b Troy, Ohio, Dec 5, 37; m 62; c 3. PHYSICS. *Educ:* Manchester Col, BA, 59; Ind Univ, Bloomington, MS, 62; Mich State Univ, PhD(physics), 69. *Prof Exp:* Teaching asst physics, Ind Univ, Bloomington, 60-62; instr, Manchester Col, 62-64; teaching asst, Mich State Univ, 64-65, res asst, 65-69; from asst prof to assoc prof, 69-79, PROF PHYSICS, MANCHESTER COL, 79-, CHAIR, 86- *Mem:* Am Asn Physics Teachers; Am Phys Soc. *Res:* Nuclear physics gamma ray spectroscopy in the neutron deficient region. *Mailing Add:* Dept Physics Manchester Col North Manchester IN 46962

BEESACK, PAUL RICHARD, mathematical analysis; deceased, see previous edition for last biography

BEESE, RONALD ELROY, b Milwaukee, Wis, May 19, 29; m 54; c 2. PHYSICAL CHEMISTRY, INORGANIC CHEMISTRY. *Educ:* Lawrence Col, BS, 51. *Prof Exp:* From res chemist to sr res chemist, 51-67, res assoc, 67-78, SR RES ASSOC, BARRINGTON TECH CTR, AM CAN CO, 78- *Mem:* Am Chem Soc; Nat Asn Corrosion Eng; Electrochem Soc. *Res:* Electrochemical corrosion research involving the evaluation of corrosion mechanisms related to products packaged in metallic containers. *Mailing Add:* 618 Concord Pl Barrington IL 60010

BEESLEY, EDWARD MAURICE, b Belvidere, NJ, Jan 11, 15; m 40; c 3. MATHEMATICS. *Educ:* Lafayette Col, AB, 36; Brown Univ, ScM, 38, PhD(math), 43. *Prof Exp:* From instr to assoc prof, 40-55, actg head dept, 44-47, chmn dept, 47-79, prof math, 55-80, EMER PROF MATH, UNIV NEV, RENO, 80- *Mem:* Am Math Soc; Math Asn Am; AAAS. *Res:* Theory of functions of a real variable. *Mailing Add:* Dept Math Univ Nev Reno NV 89557

BEESLEY, ROBERT CHARLES, INTESTINAL TRANSPORT, MEMBRANE TRANSPORT. *Educ:* Univ Calif, Berkeley, PhD(physiol), 74. *Prof Exp:* ASSOC PROF GASTROINTESTINAL PHYSIOL, DEPT PHYSIOL & BIOPHYS, HEALTH SCI CTR, UNIV OKLA, 81- *Mailing Add:* Dept Physiol Univ Okla PO Box 26901 Oklahoma City OK 73190

BEESON, EDWARD LEE, JR, b Bartow, Fla, Mar 9, 28; m 59; c 3. MOLECULAR SPECTROSCOPY. *Educ:* Emory Univ, BA, 49, MS, 50; Ga Inst Technol, PhD(physics), 60. *Prof Exp:* Instr physics, Chattanooga Univ, 50-51 & Ga Inst Technol, 54-59; from asst prof to prof physics, Univ New Orleans, 60-83; OCEANOGR, NAVAL OCEANOG OFF, 84- *Concurrent Pos:* NIH res grant, 61-66. *Mem:* Am Phys Soc. *Res:* Microwave spectroscopy. *Mailing Add:* 7222 Claridge Ct New Orleans LA 70127

BEESON, JUSTIN LEO, b Des Moines, Iowa, Sept 6, 30. FUEL CHEMISTRY, SYNTHETIC FUELS. *Educ:* Iowa State Univ, BS, 54, MS, 69, PhD(chem eng), 73. *Prof Exp:* Develop chemist, Goodyear Tire & Rubber Co, 59-66; res assoc, Iowa State Univ, 73-75; eng supvr, Inst Gas Technol, 75-81; HEAD ENERGY SYSTS, MIDWEST RES INST, 81- *Mem:* Am Chem Soc; Am Inst Chem Engrs; Sigma Xi; Nat Soc Prof Engrs. *Res:* Synthetic fuels from coal, oil shale and biomass; hydrocarbon processing; coal beneficiation. *Mailing Add:* 2107 Peach St Morgantown WV 26505

BEESON, PAUL BRUCE, b Livingston, Mont, Oct 18, 08; m 42; c 3. MEDICINE, INFECTIOUS DISEASES. *Educ:* McGill Univ, MDCM, 33. *Hon Degrees:* DSc, Emory Univ, 68, McGill Univ, 71, Albany Med Col, 75 & Yale Univ, 75. *Prof Exp:* Intern, Univ Pa Hosp, 33-35; from asst resident to resident, Rockefeller Inst Hosp, 37-39; resident med, Peter Bent Brigham Hosp, Boston, 39-40; chief physician, Harvard Univ Field Hosp, Unit, Eng, 41-42; from asst prof to prof med, Med Sch, Emory Univ, 42-52, chmn dept, 46-52; prof & chmn dept, Med Sch, Yale Univ, 52-65; Nuffield prof clin med, Oxford Univ, 65-74; distinguished physician, US Vet Admin, 74-81; prof, 74-81, EMER PROF MED, UNIV WASH, 81- *Mem:* Nat Acad Sci; Am Soc Clin Invest; Soc Exp Biol & Med; Asn Am Physicians. *Res:* Infectious diseases; immunology; clinical medicine. *Mailing Add:* 21013 NE 122nd St Redmond WA 98053

BEESON, W MALCOLM, ANIMAL NUTRITION. *Educ:* Univ Wis, PhD(biochem), 35. *Prof Exp:* Prof animal sci, Purdue Univ, 52-77; RETIRED. *Mailing Add:* 1510 N Grant St West Lafayette IN 47906

BEESTMAN, GEORGE BERNARD, b Hammond, Wis, July 17, 39; m 65; c 2. AGRICULTURAL CHEMISTRY, WEED SCIENCE. *Educ:* Wis State Univ, River Falls, BS, 61; Univ Wis, MS, 67, PhD(soil chem), 69. *Prof Exp:* Sr res chemist, Monsanto Co, 68-73, res specialist, 73-78, SR RES SPECIALIST, MONSANTO AGR PROD CO, 78- *Mem:* Am Soc Agron; Soil Sci Soc Am; Sigma Xi; AAAS; Weed Sci Soc Am; Am Chem Soc; Am Soc Testing & Mat. *Res:* Fertilizer nutrient imbalance effects on crop quality; influence of soil properties on insecticide uptake and translocation by plants; herbicide dissipation modes in soils; pesticide formulation; pesticides suspension formulations; dry flowable pesticide formulations; microencapsulation. *Mailing Add:* DuPont Exp Sta Bldg 402 Rm 2112 Wilmington DE 19880

BEETCH, ELLSWORTH BENJAMIN, b Mankato, Minn, Jan 23; m 51; c 3. PHYSICAL CHEMISTRY. *Educ:* Mankato State Col, BS, 49; Kans State Univ, MS, 51, PhD(phys chem), 57. *Prof Exp:* Sr res chemist, Rahr Malting Co, 55-58; from asst prof to assoc prof, 58-66, PROF CHEM, MANKATO STATE COL, 66- *Mem:* Am Chem Soc. *Res:* Electrophoresis; interaction of proteins with small molecules; colorimetric determination of sulfur dioxide, radioisotopes; application of instruments to beer and wort color. *Mailing Add:* Dept Chem Mankato State Univ Mankato MN 56001

BEETHAM, KAREN LORRAINE, radiation biology, cell kinetics, for more information see previous edition

BEETON, ALFRED MERLE, b Denver, Colo, Aug 15, 27; m 45, 66; c 5. LIMNOLOGY. *Educ:* Univ Mich, BS, 52, MS, 54, PhD(zool), 58. *Prof Exp:* Instr biol, Wayne State Univ, 56-57; chief environ res prog, US Bur Com Fisheries, 57-66; prof zool & asst dir, Ctr Great Lakes Studies, Univ Wis-Milwaukee, 66-76, assoc dean grad sch, 73-76; dir, Great Lakes & Marine Waters Ctr, 76-86, PROF ATMOSPHERIC & OCEANIC SCI, COL ENG & PROF NATURAL RESOURCES, UNIV MICH, ANN ARBOR, 76-; DIR, GREAT LAKES ENVIRON RES LAB, 86- *Concurrent Pos:* Lectr civil eng, Univ Mich, 62-66; mem planning comt, Int Symp on Eutrophication, Nat Acad Sci, 65-68 & chmn panel on freshwater life & wildlife, Comt Water Qual Criteria, 70-72, chmn, Int Environ Prog Comm, 80-83; consult, US Army Corps Engrs & Metrop Sanit Dist of Greater Chicago, 68-79; Greater Cleveland Growth Asn, 70 & US Environ Protection Agency, 72-76; chmn res adv coun, Wis Dept Natural Resources, 69-71; US chmn, Sci Adv Bd, Int Joint Comn, 86-; comnr, Toxic Substances Control Comn, Mich, 87-89. *Honors & Awards:* James W Moffett Publ Award, US Bur Com Fisheries Lab, Mich, 67. *Mem:* Am Soc Limnol & Oceanog (treas, 62-81); Am Soc Zool; Int Soc Limnol; Int Asn Great Lakes Res. *Res:* Limnology of the Great Lakes; vertical migration and related behavior of planktonic Crustacea; limnology of man-made tropical lakes. *Mailing Add:* Great Lakes Environ Res Lab/NOAA 2205 Commonwealth Blvd Ann Arbor MI 48105

BEEVERS, HARRY, b Durham, Eng, Jan 10, 24; nat US; m 49; c 1. BIOCHEMISTRY. *Educ:* Univ Durham, BSc, 44, PhD, 47. *Hon Degrees:* DSc, Purdue Univ, 71, Univ Newcastle, Eng, 74 & Nagoya Univ, Japan, 86. *Prof Exp:* Asst plant physiol, Oxford Univ, 46-48, chief res asst, 48-50; vis prof, Purdue Univ, 50, from asst prof to prof, 51-69; PROF BIOL, UNIV CALIF, SANTA CRUZ, 69- *Concurrent Pos:* Demonstr & tutor, Oxford Univ, 46-50; Fulbright lectr, Australia, 62; NSF sr fel, 63-64. *Honors & Awards:* Stephen Hales Award, Am Soc Plant Physiol, 70; Von Humboldt Preistrager, 87. *Mem:* Nat Acad Sci; Am Soc Biol Chemists; Am Soc Plant Physiol (pres, 62); hon mem Ger Bot Soc; fel Am Acad Arts & Sci. *Res:* Plant metabolism. *Mailing Add:* Biol Dept Univ Calif Santa Cruz CA 95060

BEEVERS, LEONARD, b Co Durham, Eng, Aug 7, 34; m 60; c 2. PLANT PHYSIOLOGY. *Educ:* Univ Durham, BSc, 58; Univ Wales, PhD(agr bot), 61. *Prof Exp:* Res assoc agron, Univ Ill, Urbana-Champaign, 61-63, from asst prof to assoc prof hort, 63-71; PROF BOT, UNIV OKLA, 71-, CHMN DEPT, 81- *Mem:* Am Soc Plant Physiol (secy, 77-81, pres-elect, 81, pres, 82). *Res:* Intermediary nitrogen metabolism with specific reference to reserve protein accumulation; glycoprotein biosynthesis and post translational modification of proteins, general plant physiology; nitrate assimilation. *Mailing Add:* Dept Bot & Microbiol Univ Okla Norman OK 73019

BEEVIS, DAVID, b Surrey, UK, July 19, 40; Can citizen; m 61; c 2. HUMAN FACTORS. *Educ:* Royal Aircraft Estab, Tech Col Farnborough, Eng, cert, 61, 62; Col Aeronaut Cranfield, Eng, MSc, 64. *Prof Exp:* Sci asst ergonomics, Royal Aircraft Establishment, UK, 73; ergonomist, Ergonomics Lab, EMI Electronics Ltd, 74-76; human eng specialist, Canadair Ltd, 77-78; ergonomist, Ergonomics Lab, EMI Electronics Ltd, 78-81; tech officer, 81-83, DEFENCE SCIENTIST, DEFENCE & CIVIL INST ENVIRON MED, CAN, 83- *Concurrent Pos:* Vis scientist, TNO Inst fur Zintuigsfhysiologie, The Netherlands, 80-81; consult, Ergonomics Soc, UK, 79, soc rep, Comt Conf City in the year 2000, 69-70, asst ed, Newslett, 69-71; Can rep, NATO AC243, Workshop on Applications Systs, Ergonomics to Weapons Systs Dept, & RMC Shrivenham, 84; mem, Defense & Civil Inst Environ Med, univ grants comt human eng, 73-76. *Mem:* Human Factors Soc; Ergonomics Soc UK; Human Factors Asn Can (secy, 79-80). *Res:* Study of human performance as it affects the design of modern systems, and the study of the most effective means of incorporating human performance data into systems design. *Mailing Add:* 140 Newton Dr Willowdale ON M2M 2N3 Can

BEFELER, BENJAMIN, b San Jose, Costa Rica, Dec 8, 39; US citizen; m 65; c 3. CARDIOVASCULAR DISEASE. *Educ:* Nat Univ Mex, MD, 63; Am Bd Internal Med, dipl, 74, cert cardiovasc dis, 75. *Prof Exp:* Intern, Span Hosp & Nat Univ Mex, 62; intern med, Mt Sinai Hosp, Cleveland, Ohio, 64, asst resident, 65; resident med, Vet Admin Hosp & Case Western Reserve Univ Hosp, 66-67, resident cardiol, 67-68; USPHS fel med, Med Sch, Univ Calif, San Francisco, 68-69; chief cardiocasc lab, Vet Admin Hosp, Miami, 71-78; from instr to asst prof, 69-74, ASSOC PROF MED, SCH MED, UNIV MIAMI, 74- *Concurrent Pos:* Preceptor clin diag, Med Sch, Case Western Reserve Univ, 66-68; staff cardiologist, Mt Sinai Hosp, Miami Beach, 69-71; staff physician, Cardiopulmonary Lab, Div Cardiol, Dept Internal Med, Mt Sinai Hosp Greater Miami, 69-71; attend physician, Univ Miami Hosps & Clins, 71- & Univ Miami Jackson Mem Hosp, 71-; civilian consult cardiol, US Air Force Hosp, Homestead, Fla, 72-; ed consult, Chest, 74; mem res & educ comt, Vet Admin Hosp, Miami, 74-76; fel clin coun, Am Heart Asn; chief cardiol, Cedars Med Ctr, Miami, 86-88, chief staff, 89-90, mem, bd dirs, 89- *Mem:* Fel Am Col Physicians; fel Am Col Chest Physicians; fel Am Col Cardiol; fel Am Col Angiol; Am Heart Asn. *Res:* Mechanism of antiarrhythmias in patients with coronary artery disease; Wolff-Parkinson-White syndrome and the role of arrhythmias in determining patient prognosis. *Mailing Add:* 1321 NW 14th St No 202 Miami FL 33125

BEFUS, A(LBERT) DEAN, b Edmonton, Alta, Oct 8, 48; m 70; c 2. MUCOSAL IMMUNOPARASITOLOGY, MAST CELL BIOLOGY. *Educ:* Univ Alta, Can, BSc, 70; Univ Toronto, MSc, 72; Univ Glasgow, PhD(immunoparasitol), 75. *Prof Exp:* Asst prof mucosal immunol, Dept Path, McMaster Univ, 78-82, assoc prof immunol & parasitol, dept path, 82-85; assoc prof, 85-86, PROF MICROBIOL & INFECTIOUS DIS, UNIV CALGARY, ALTA, 86- *Concurrent Pos:* Univ Toronto, Open fel, 70-71; Ontario Grad Scholar, 71-72, Commonwealth Scholar, 74-75; Med Res Coun Can, fel, 75-78; Rockefeller Fel Trop & Geog Med, 80-84; vis scientist, Med Res Coun Lab, The Gambia, WAfrica, 83; Alta Heritage Found, Med Res Scholar, 85- *Honors & Awards:* Young Canadian Res Award, Int Develop Res Ctr, 83; Henry Baldwin Ward Medal, Am Soc Parasitol, 88. *Mem:* Am Asn Immunologists; Can Soc Immunol; Brit Soc Parasitol; Am Soc Trop Med Hyg; Am Soc Parasitol; Soc Muscosal Immunol. *Res:* Mechanisms of host resistance at lung and intestine; ontogeny, molecular differentiation and functional heterogeneity of mast cells during neuroimmunology. *Mailing Add:* Dept Microbiol & Infectious Dis Univ Calgary Calgary AB T2N 4N1 Can

BEG, MIRZA ABDUL BAQI, theoretical physics; deceased, see previous edition for last biography

BEGA, ROBERT V, b Milford, Mass, Aug 17, 28; m; c 1. PLANT PATHOLOGY. *Educ:* Univ Calif, BS, 53, PhD(path), 57. *Prof Exp:* Asst plant path, Univ Calif, Berkeley, 55-56, lectr, 56-57, assoc, Exp Sta, 57-85; RETIRED. *Concurrent Pos:* Plant pathologist, Pac Southwest Forest & Range Exp Sta, USDA, 56-88. *Mem:* Am Phytopath Soc; Am Forestry Asn; Sigma Xi. *Res:* Biology of rust fungi; root disease fungi; mycorrhiza; physiology of host-parasite relationship; aerobiology, microclimatology; biological control; air pollution. *Mailing Add:* 21275 Clydesdale Rd Grass Valley CA 95949

BEGALA, ARTHUR JAMES, b Newark, NJ, Sept 10, 40; m 66; c 2. PHYSICAL CHEMISTRY. *Educ:* Lehigh Univ, BS, 62; Rutgers Univ, PhD(chem), 71. *Prof Exp:* Chemist, E I du Pont de Nemours & Co, 62-63; res asst chem, Rutgers Univ, 64-68; sr res chemist, Am Cyanamid Co, 68-84; CHEMIST, NALCO CHEM CO, 84- *Mem:* Am Chem Soc; Sigma Xi. *Res:* Synthesis and characterization of polyelectrolytes and investigation of various polymer interactions with specific metallic and oxide substrates. *Mailing Add:* 833 Thornapple Dr Naperville IL 60540

BEGAY, FREDERICK, b Towaco, Colo, July 2, 32; m 52; c 7. PHYSICS. *Educ:* Univ NMex, BS, 61, MS, 63, PhD(physics), 72. *Prof Exp:* Res physicist, Air Force Weapons Lab, Kirtland AFB, NMex, 63-65; res assoc high energy neutron physics, Univ NMex, 65-71; staff mem x-ray spectros, 71-72, STAFF MEM LASER FUSION, LOS ALAMOS SCI LAB, UNIV CALIF, 72- *Concurrent Pos:* Mem, Nat Res Coun, 72- *Mem:* Am Phys Soc. *Res:* High energy gamma ray physics; solar wind physics; high energy neutron physics; plasma physics in exploding wire systems and other high energy pulsed power systems; satellite instrumentation; design and physics of high energy vacuum systems. *Mailing Add:* Los Alamos Nat Lab Univ Calif PO Box 1663 Los Alamos NM 87545

BEGELL, WILLIAM, b Wilno, Poland, May 18, 28; US citizen; m 48; c 2. TWO PHASE HEAT TRANSFER, THERMODYNAMIC PROPERTIES. *Educ:* City Col New York, BChE, 53; Polytech Inst Brooklyn, MChE, 58. *Hon Degrees:* DSc, Acad Sci BSSR, Minsk, 84. *Prof Exp:* Dir, Heat Transfer Res Facil, Columbia Univ, 55-59; exec pres, Scripta technica, 59-73; PRES, HEMISPHERE PUBL, 74- *Concurrent Pos:* Ed & publ, Hemisphere Publ Corp, 74-; vis prof, George Washington Univ, 76-77; mem, Bd Commun, Am Soc Mech Engrs, 80-, bd Asn Am Publ, 80-88 & Eng Libr, 88-; chmn, Publ Develop Bd, NY Acad Sci, 90- *Mem:* Am Soc Mech Engrs; Am Inst Chem Engrs; Am Chem Soc; Am Astronaut Soc. *Res:* Heat transfer in two phase systems, specifically in nuclear reactors; safety in thermal hydraulic systems; measurements in high temperature and high pressure power and chemical systems. *Mailing Add:* 46 E 91st St Apt 6A New York NY 10128

BEGG, DAVID A, b Lawrence, Mass, Dec 16, 43; m 68; c 2. CELL BIOLOGY, CELL MOTILITY. *Educ:* Colby Col, AB, 65; Univ Pa, MS, 68, PhD(cell biol), 75. *Prof Exp:* Fel, Univ Va, 76-79; asst prof anat, 79-85, ASSOC PROF ANAT & CELL BIOL, MED SCH, HARVARD UNIV, 85- *Mem:* Develop Biol Soc; Am Soc Cell Biol; Am Asn Anatomists; AAAS; NY Acad Sci. *Res:* Control of the organization of actin in the egg cortex; mechanism of fertilization; mechanism of cell division; actomyosin-based aspects of cell locomotion. *Mailing Add:* Lab Human Reproduction & Reproductive Biol Harvard Univ Med Sch 45 Shattuck St Boston MA 02115

BEGGS, WILLIAM H, b Ft Dodge, Iowa, Feb 19, 35; m 57; c 2. MICROBIAL PHYSIOLOGY. *Educ:* Univ Minn, BA, 56; Kans State Univ, MS, 59; Univ Cincinnati, PhD(microbiol), 64. *Prof Exp:* Jr scientist, Univ Minn, 59-61; RES MICROBIOLOGIST, GEN MED RES SERV, VET ADMIN MED CTR, 65- *Mem:* Fel Am Inst Chemists; Am Soc Microbiol. *Res:* Biological activities; modes of action; chemical properties and chemotherapeutic potentials of antifungal azole-containing drugs. *Mailing Add:* Gen Med Res Serv 151 Vet Admin Med Ctr Minneapolis MN 55417-2300

BEGGS, WILLIAM JOHN, b Pittsburgh, Pa, Apr 9, 42; m 65; c 3. DESIGN OF EXPERIMENTS, REGRESSION ANALYSIS. *Educ:* Allegheny Col, BS, 64; Univ Wis, MS, 66, PhD(statist), 69. *Prof Exp:* Res statistician, Chem Div, Am Cyanamid Co, 69-70; sr mathematician, 70-82, FEL MATHEMATICIAN, BETTIS ATOMIC POWER LAB, WESTINGHOUSE ELEC CORP, 82- *Concurrent Pos:* Lectr, Bettis Reactor Eng Sch. *Honors & Awards:* George Westinghouse Signature Award of Excellence, 89. *Mem:* Am Statist Asn. *Res:* Application of Kalman filter to engineering problems; design and analysis of accelerated test data. *Mailing Add:* 933 Bridgewater Dr Pittsburgh PA 15216

BEGIN-HEICK, NICOLE, b Que, Can, Oct, 30, 37; m 65; c 2. CELL BIOLOGY, METABOLISM. *Educ:* Laval Univ, BScD, 57; Cornell Univ, MS, 60; McGill Univ, PhD(biochem), 65. *Prof Exp:* Fel biochem, Univ Toronto, 64-65; fel physiol, Duke Univ, 65-67; from asst prof to assoc prof biochem, Laval Univ, 67-72; assoc prof, 72-80, vdean res, Sch Med, 87-89, PROF BIOCHEM, UNIV OTTAWA, 80-, DEAN, SCH GRAD STUDIES RES, 89- *Concurrent Pos:* Mem, Med Res Coun Can, 77-84; coordr nutrit prog, dept biochem, Univ Ottawa, 77-84; Nat Sci Eng Res Coun Can, 87- *Mem:* Am Inst Nutrit; Can Biochem Soc (secy, 72-75); Can Soc Nutrit Sci. *Res:* Metabolic aberations in experimental obesity; mechanisms of regulation of Adipocyte Adenylate Cyclase; mechanisms of insulin secretion. *Mailing Add:* Sch Grad Studies Res Univ Ottawa 115 Seraphin Marion Ottawa ON K1N 6N5 Can

BEGLAU, DAVID ALAN, b Detroit, Mich, Jan 11, 58; m 84; c 2. AMORPHOUS SILICON THIN FILM DEVICES, MATERIALS SCIENCE & DEPOSITION TECHNIQUES. *Educ:* Mich Technol Univ, BS, 81. *Prof Exp:* Scientist, 81-85, SR SCIENTIST, ENERGY CONVERSION DEVICES, INC, 85- *Res:* Solid-state physics; novel solid-state devices; silicon photovoltaics; chalcogenide thermoelectric devices; chalcogenide threshold switching devices; memory devices; liquid crystal displays; silicon transistors and diodes. *Mailing Add:* 47 Mechanic Oxford MI 48371

BEGLEITER, HENRI, b Nimes, France, Sept 11, 35; US citizen; m 63; c 2. PSYCHOPHYSIOLOGY, NEUROPHYSIOLOGY. *Educ:* New Sch Social Res, PhD(psychophysiol), 67. *Prof Exp:* Res assoc neurophysiol, 64-66, asst prof psychiat & psychophysiol, 67-72, assoc prof psychiat, 72-75, DIR NEURODYNAMICS LAB, SCH MED, STATE UNIV NY DOWNSTATE MED CTR, 66-, PROF PSYCHIAT, 75- *Honors & Awards:* Thorp Award; Isaacson Prize. *Mem:* Fel AAAS; Am Psychol Asn; Soc Psychophysiol Res; NY Acad Sci; fel Am Electroencephalogram Soc; Am Col Neurosychopharmacol. *Res:* Neurophysiological correlates of brain dysfunction. *Mailing Add:* Dept Psychiat State Univ NY Health Sci Ctr Brooklyn NY 11203

BEGLEY, JAMES ANDREW, b Trenton, NJ, Feb 18, 42; m 64; c 5. METALLURGICAL ENGINEERING, MECHANICAL ENGINEERING. *Educ:* Lehigh Univ, BS, 63, PhD(metall eng), 69. *Prof Exp:* Engr metall, Astronuclear Lab, Westinghouse Elec Corp, 63-65, sr engr mech, Res Lab, 68-75; assoc prof metall eng, Ohio State Univ, 75-80. *Mem:* Am Soc Testing & Mat; Am Soc Metals. *Res:* Elastic plastic fracture; fracture mechanics; fatigue; creep-fatigue; stress corrosion; corrosion fatigue; ductile to britile transitions. *Mailing Add:* 1373 Kings Rd Schenectady NY 12303

BEGOVAC, PAUL C, b Marquette, Mich, Mar 10, 56. CELLULAR BIOLOGY, CELLULAR INTERACTIONS. *Educ:* Northern Ariz Univ, BSc, 78; Univ Fla, PhD(anat & cell biol), 88. *Prof Exp:* Proj mgr med devices, W L Gore & Assoc, Inc, 78-82; postdoctoral fel cellular biochem, M D Anderson Cancer Ctr, Univ Tex, 88-90; ASST PROF CELL BIOL, UNIV OKLA, 90- *Concurrent Pos:* Prin investr, Okla Ctr Advan Sci & Technol, 91- *Mem:* Am Soc Cell Biol; AAAS. *Res:* Cellular and molecular basis of cell-cell and cell-extracellular matrix interactions; cell migration of neurite processes and capillary endothelial cells; cell surface receptor function. *Mailing Add:* Dept Zool Univ Okla 730 Van Vleet Oval Norman OK 73019

BEGOVICH, NICHOLAS A(NTHONY), b Oakland, Calif, Nov 29, 21; m 44. ELECTRONICS. *Educ:* Calif Inst Technol, BS, 43, MS, 44, PhD(elec eng), 48. *Prof Exp:* Asst & res engr, Calif Inst Technol, 44-46, instr elec eng, 47-48; res physicist, Hughes Aircraft Co, 48-51, staff consult, 51-52, dir eng, Ground Systs Group, 51-61, vpres, 61-69, asst group exec, 61-66, group exec, 66-69, vpres corp hq, 70; corp vpres, Litton Industs & pres, Data Systs Div, 70-74; MGT & TECH CONSULT, 75- *Concurrent Pos:* Consult, Off Secy Defense, 51-57 & 76, Appl Physics Lab, Johns Hopkins Univ, 77-, Off Secy Army, 83-; mem, Army Sci Bd, 83- *Mem:* Am Phys Soc; Opers Res Soc Am; fel Inst Elec & Electronics Engrs. *Res:* Theory of high frequency vacuum tubes; electro-magnetic field; radar detection theory and air defense system analysis. *Mailing Add:* 136 Miramonte Dr Fullerton CA 92635

BEGUE, WILLIAM JOHN, b Chicago, Ill, June 15, 31; m 52; c 3. MICROBIAL PHYSIOLOGY, ANALYTICAL BIOCHEMISTRY. *Educ:* Col St Thomas, BS, 53; Univ Minn, MS, 60, PhD(microbiol), 63. *Prof Exp:* From instr to asst prof microbiol, Univ Ky Med Ctr, 63-67; sr scientist, 67-74, RES SCIENTIST ANALYTICAL MICROBIOL, LILLY RES LABS, DIV ELI LILLY & CO, 74- *Mem:* Am Soc Microbiol; AAAS; Am Inst Biol Sci; Sigma Xi. *Res:* Development of analytical methods for new compounds, mostly antibiotics as found in animal feed and tissues; microbial nutrition, physiology and metabolism; biochemistry; enzymology; animal nutrition; antibiotics. *Mailing Add:* 2739 Saturn Dr Indianapolis IN 46229

BEGUIN , FRED P, b Brussels, Belg, Oct 13, 09; US citizen; m 34; c 1. ACOUSTICS, PHYSICS. *Educ:* State Tech Col Brussels, BS, 31; State Sch Cinematography, Paris, BSc, 33; Univ of City of Paris, prof radio commun, 33; Nat Radio Electronics Inst, Brussels, lic prof physics, 44; Colby Col, cert engr noise control, 71. *Prof Exp:* Patent analyst, French Thomson-CSF Co, Paris, France, 31-34; high tech proj engr, Philips Res Labs, Holland, 34-44; tech dir recording & mfg, Decca Records Belg & France, 44-46; Europ dir, Motorola Corp Chicago, 46-50; audio consult, Electronics Div, Gen Elec Co, Syracuse, NY, 50-59; chief scientist electroacoust res & develop, Am Optical Corp, 59-72; dir Hearing & Noise Control Ctr, Harrington Mem Hosp, Southbridge, 72-77; consult environ acoust physicist, Port Charlotte Fla, 83-86; RETIRED. *Concurrent Pos:* Chmn, TV Standardization Comn Belgian Eng Soc & Ministry Commun, Brussels, 44-46; tech consult to Belg embassy in US, Belg Ministry Foreign Trade Econ Affairs, 51-53; safety equip indust rep & mem Noise & Bioacoust Comts, Accoust Soc Am, Am Standards Inst, 63-72; acoustical consult, Worcester Mem Auditorium, Environ Protect Agency, Occupational Safety Health Admin & Pub Health Serv, Mass, 65-67. *Honors & Awards:* Nat Sci Achievement Award, Audio Eng Soc, 70; Nat Safety Coun Award, 77; Silver Cert Award, Acoust Soc Am, 85. *Mem:* Inst Elec & Electronics Engrs; fel Audio Eng Soc; emer mem Acoust Soc Am; emer mem NY Acad Sci; emer mem Am Inst Physics; hon mem Indust Safety Equip Asn. *Res:* Engineering; architectural acoustics; sound recording and reproduction; psychological and physiological acoustics; bioacoustics; noise; shock; vibration; hearing; audiology; audiometry; noise-abatement; speech communications in noise; environmental acoustics; measurement instrumentation. *Mailing Add:* Acoust Physicist PO Box 2750 Port Charlotte FL 33949-2750

BEGUN, GEORGE MURRAY, b Bedford, Mass, Aug 20, 21; m 48; c 4. PHYSICAL INORGANIC CHEMISTRY, CHEMICAL PHYSICS. *Educ:* Colo Col, BA, 43; Columbia Univ, MA, 44; Ohio State Univ, PhD(chem), 50. *Prof Exp:* Chemist, Electromagnetic Separation Plant, Tenn Eastman Corp, 44-46; res assoc photosurface proj, Ohio State Res Found, 50-51; chemist, Chem Div, 51-90, CONSULT, OAK RIDGE NAT LAB, 90- *Mem:* Am Chem Soc. *Res:* Electron impact studies; uranium chemistry; isotope separation and exchange; infrared, Raman and mass spectral studies; xenon and interhalogen compounds; fluorine and tritium chemistry; actinide chemistry, nuclear waste studies; applications of laser Raman spectroscopy and Raman microprobe to the study of advanced materials. *Mailing Add:* 106 Colby Rd Oak Ridge TN 37830

BEHAL, FRANCIS JOSEPH, b Yoakum, Tex, Oct 10, 31; m 55; c 2. BIOCHEMISTRY. *Educ:* St Edward's Univ, BS, 53; Univ Tex, MA, 56, PhD(chem), 58. *Prof Exp:* From asst prof to assoc prof biochem, Med Col Ga, 58-68, dir grad div, 64-66, dir sch grad studies, 66-68, prof biochem & microbiol & dean sch grad studies, 68-71; PROF BIOCHEM & CHMN DEPT, SCH MED, TEX TECH UNIV, 71- *Concurrent Pos:* Mem, Tenth Int Cancer Cong. *Mem:* Am Soc Biol Chemists; fel Am Acad Microbiol; fel Am Inst Chem; Soc Exp Biol & Med. *Res:* Comparative enzymology of mammalian and microbial peptidases. *Mailing Add:* Dept Biochem Tex Tech Univ Sch Med Lubbock TX 79430

BEHAN, MARK JOSEPH, b Denver, Colo, Jan 17, 31; c 3. PHYSIOLOGICAL ECOLOGY. *Educ:* Univ Denver, BA, 53; Univ Wyo, MS, 58; Univ Wash, PhD(plant physiol), 63. *Prof Exp:* Chmn wildlife biol degree progs, 70-74, PROF BOT, UNIV MONT, 61- *Concurrent Pos:* Fulbright lectr, Nepal, 82-83; consult, USAID, Nepal, 84-85 & 86-87 & US Fish & Wildlife Serv, India, 90. *Mem:* Ecol Soc Am; Int Mountain Soc. *Res:* Forest tree physiology; ecology. *Mailing Add:* Dept Bot Univ Mont Missoula MT 59812

BEHANNON, KENNETH WAYNE, b Houston, Tex, Jan 1, 34; m 55; c 1. SPACE PLASMA PHYSICS, MAGNETOSPHERIC PHYSICS. *Educ:* Univ Tex, BS, 63; Va Polytech Inst, MS, 67; Cath Univ Am, PhD(space sci), 76. *Prof Exp:* ASTROPHYSICIST, GODDARD SPACE FLIGHT CTR, NASA, 64- *Concurrent Pos:* Alexander von Humboldt Sr Scientist Award, Humboldt Found, Germany, 78. *Mem:* Am Geophys Union; Sigma Xi. *Res:* Large and small scale characteristics of magnetic fields in interplanetary space and their solar origins; magneto-plasma environments of the earth, moon and planets. *Mailing Add:* Goddard Space Flight Ctr NASA Code 692 Greenbelt MD 20770

BEHAN-PELLETIER, VALERIE MARY, b Drogheda, Ireland, Feb 15, 48; Can citizen; m 77. SOIL ZOOLOGY. *Educ:* Univ Col, Dublin, Ireland, BSc, 68; McGill Univ, Montreal, Can, MSc, 72, PhD(soil zool), 78. *Prof Exp:* Res asst soil zool, Univ Col, Dublin, Ireland, 68-69, Trinity Col, Dublin, 72-73 & Univ Alaska, Fairbanks, 75; lectr biol, John Abbott Col, Ste Anne de Bellevue, 78-81; RES SCIENTIST, BIOSYSTEMATICS RES CTR, AGR CAN, 81- *Concurrent Pos:* Chair, Can Nat Collection Curatorial Comt, Biosystematics Res Ctr, Ottawa, 82-86; mem, Sci Comt, Biol Surv Can, 83-; assoc ed, Can Entomologist, 86-91; adj prof, McGill Univ, Montreal, 88-; sci ed, Mem Entom Soc Can, 91- *Mem:* Entom Soc Can; Soil Ecol Soc. *Res:* Systematics and ecology of oribatida especially of North America; biodiversity of oribitada in arctic or tropical ecosystems. *Mailing Add:* Biosystematics Res Ctr Agr Canada K W Neatby Bldg C E F Ottawa ON K1A 0C6 Can

BEHAR, MARJAM GOJCHLERNER, b Luck, Poland, Dec 8, 25; US citizen; m 51; c 3. ANALYTICAL BIOCHEMISTRY. *Educ:* Univ Havana, DSc, 50. *Prof Exp:* Res chemist, Va Smelting Co, Cuba, 50-53; control res chemist, Lab Geol, 53-54; instr high sch, Cuba, 54-55; chemist, Garden State Tanning Corp, Pa, 55-57; jr prod engr, Lansdale Div, Philco Corp, 58-61; res assoc anesthesiol, 62-70, res specialist anesthesiol, Univ Hosp, Univ Pa, 70-79; health scientist admin & exec secy metallobiochem study sect, 80-84, HEALTH SCIENTIST ADMINR & EXEC SECY SPEC REV SECT, DIV RES GRANTS, NIH, MD, 84- *Mem:* Am Soc Anesthesiologists; Am Chem Soc; NY Acad Sci; Soc Appl Spectros; fel Am Inst Chemists. *Res:* Anesthesia research; effect of anesthetic agents on blood flows, metabolism and blood viscosity; radioimmuno assay; health science administration. *Mailing Add:* NIH Rm 2A11 Westwood Bldg 5333 Westbard Ave Bethesda MD 20892

BEHARA, MINAKETAN, b Jamshedpur, India, June 15, 37; Can citizen; m 66; c 2. MATHEMATICS, MATHEMATICAL STATISTICS. *Educ:* Univ Saarland, DSc(math statist), 63. *Prof Exp:* Sci asst decision theory & sequential anal, Inst Europ Statist, Univ Saarland, 61-63; asst prof math, Univ Waterloo, 63-65, adj prof, 65-67; assoc prof math, McMaster Univ, 67-82; PROF MATH, & STAT, MCMASTER UNIV, 82- *Concurrent Pos:* Adj asst prof, Univ Western Ont, 63-64; vis res prof, Inst Europ Statist, Univ Saarland, 64; vis prof, Univ Heidelberg, 67 & 72, actg chmn, 73; vis prof, Univ Karlsruhe, 68 & Univ Regensburg, 75; prof math & stat Univ Sao Paulo, 76-77; prof math, Kuwait Univ, 82-84; vis prof, Math & stat, Mu Newfoundland, 84; Nat Res Ctr (can) Grant Referee, 77-78; Nat Sci & Eng Res Coun (Can) Grant Referee, 82-83; Can Coun Killam Fel Referee, 86-87; mem Res Grant Comn, Kuwait Univ, 85-86. *Mem:* Am Math Soc; Inst Math Statist; Can Math Cong; fel Royal Statist Soc. *Res:* Information-theoretic contribution to statistical decision theory; entropy in coding and ergodic theories; probability theory; algebraic and geometric entrophies. *Mailing Add:* Dept Math & Statist McMaster Univ Hamilton ON L8S 4L8 Can

BEHBEHANI, ABBAS M, b Iran, July 27, 25; US citizen; m 58; c 3. CLINICAL VIROLOGY. *Educ:* Ind Univ, AB, 49; Univ Chicago, MS, 51; Southwestern Med Sch Univ Tex, PhD(virol), 55. *Prof Exp:* Assoc prof microbiol, Univ Tehran, 56-60; virologist, Med Ctr, Baylor Univ, 60-63, res asst prof virol, Col Med, 63-64; asst prof, 64-66, sr res assoc pediat, Sect Virus Res, 66-67, assoc prof path, Sch Med, 67-72, PROF PATH, SCH MED, UNIV KANS, 72- *Concurrent Pos:* Dir cent state health labs, Ministry Health, Tehran, 56-59; clin asst prof, Univ Tex Southwestern Med Sch Dallas, 61-64. *Mem:* AAAS; Am Soc Microbiol; Soc Exp Biol & Med; fel Am Acad Microbiol. *Res:* Properties and characteristics of enteroviruses; use of cell and organ culture for pathological studies of human disease; development of more rapid procedures in diagnostic virology, evaluation of anti-viral drugs in vitro; history of smallpox; Persian founders of Islamic medicine; over 65 articles and 3 books. *Mailing Add:* Dept Path & Oncol Univ Kans Sch Med Kansas City KS 66103

BEHER, WILLIAM TYERS, b Aurora, Ill, Dec 14, 22; m; c 5. BIOCHEMISTRY. *Educ:* NCent Col, BA, 44; Wayne State Univ, PhD(biochem), 50. *Prof Exp:* Res asst, 50-52, sr res assoc biochem, Edsel B Ford Inst Med Res, 52-77, sr staff investr biochem, 77-80, DIR, GASTROENTEROL RES LAB, HENRY FORD HOSP, 80- *Concurrent Pos:* Instr, Wayne State Univ, 52-60. *Mem:* AAAS; Am Chem Soc; Am Soc Biol Chem; Soc Exp Biol & Med; Am Heart Asn. *Res:* Steroid and bile acid metabolism and nutrition; steroid x-ray diffraction. *Mailing Add:* 2025 Laudrome Dr Royal Oak MI 48073-3957

BEHFOROOZ, ALI, b May 24, 42; Iranian citizen; m 71; c 1. COMPUTER SCIENCE, EXPERIMENTAL STATISTICS. *Educ:* Univ Tehran, BS, 65, MS, 66; Mich State Univ, MS, 72, MS, 73, PhD(comput sci), 75. *Prof Exp:* Instr math, Univ Tehran & Nat Univ Iran, 65-70; res asst comput, Mich State Univ, 72-74; dir, Inst Res & asst prof comput, Moorhead State Univ, 74-76, from assoc prof to prof comput sci, 76-86, chmn, 84-86; PROF, TOWSON

STATE UNIV, 87- *Concurrent Pos:* Grant, Educ Radio & TV of Iran, 75; NSF grant, 78-80; vis prof, Univ Calif, Santa Barbara, 81-82. *Mem:* Asn Comput Mach; Inst Elec & Electronics Engrs Comput Soc. *Res:* Cluster analysis and stochastic pattern recognition; formal languages, computer application in education and programming languages; publications: six text books in computer science and over 20 papers in computer science education and related topics. *Mailing Add:* Dept Comput & Info Sci Towson State Univ Towson MD 21204-7097

BEHKI, RAM M, b Nawashahr, India, Mar 31, 32; m 60; c 3. MOLECULAR BIOLOGY. *Educ:* Panjab Univ, India, BSc, 52; Univ Nagpur, MSc, 54, PhD(biochem), 59. *Prof Exp:* Sr sci asst biochem, Nat Chem Lab, Coun Sci & Indust Res, India, 57-59; res fel, Stritch Med Sch, Loyola Univ, Ill, 59-60, instr, 60-61; Jane Coffin Childs Mem Fund fel, Nat Cancer Inst, Md, 61-62; res fel, Mass Gen Hosp, Boston, 62-63; res fel, Western Reserve Univ, 63-65; res officer, Microbiol Res Inst, 65-67, res scientist, Cell Biol Res Inst, 67-72, SR RES SCIENTIST, CHEM & BIOL RES INST, CAN DEPT AGR, 73- *Concurrent Pos:* Vis scientist, Ctr Study Nuclear Energy, Mol, Belg, 73-74. *Mem:* Am Chem Soc; Am Soc Plant Physiologists. *Res:* Nucleic acids and protein metabolism in normal and abnormal growth; plant cell culture; protoplast technology; genetic modification of plant cells by molecular biology techniques; nucleic acid isolation and characterization; molecular biology nitrogen fixation; rhizobium genetics. *Mailing Add:* 31 Trimble Circle Ottawa ON K2H 7M9 Can

BEHL, WISHVENDER K, b Dhariwal, India, Dec 26, 35. PHYSICAL CHEMISTRY, ELECTROCHEMISTRY. *Educ:* Univ Delhi, BSc Hons, 55, MSc, 57, PhD(phys chem), 62. *Prof Exp:* Jr res fel phys chem, Univ Delhi, 58-62; res assoc, NY Univ, 62-64 & Brookhaven Nat Lab, NY, 64-67; RES CHEMIST, US ARMY ELECTRONICS TECHNOL & DEVICES LAB, FT MONMOUTH, NJ, 67- *Honors & Awards:* Jr Sci Exhib Prize, Univ Delhi, 55. *Mem:* Fel Am Inst Chemists; Am Chem Soc; Electrochem Soc. *Res:* Chemistry of molten salts, potential measurements polarography, chronopotentiometry, linear sweep voltammetry, thermodynamics, transference numbers and ionic mobilities measurements; diffusion coefficients; molten salt batteries; solid state galvanic cells; electroanalytical chemistry; lithium-inorganic electrolyte cells. *Mailing Add:* Power Sources Div US Army Electronics Technol & Devices Lab ATTN: SLCET-PR Ft Monmouth NJ 07703

BEHLE, WILLIAM HARROUN, b Salt Lake City, Utah, May 13, 09; m 34; c 2. ORNITHOLOGY. *Educ:* Univ Utah, AB, 32, AM, 33; Univ Calif, PhD(zool), 37. *Prof Exp:* Asst mus vert zool, Univ Calif, 33-37; from instr to assoc prof biol, 37-50, head dept gen biol, 48-54, dir biol gen educ, 54-63, prof, 50-77, EMER PROF BIOL, UNIV UTAH, 77- *Mem:* Am Ornithologists Union; Cooper Ornith Soc; Wilson Ornith Soc. *Res:* Geographic variation and distribution of birds; birds of Utah; systematic ornithology. *Mailing Add:* Dept Biol Univ Utah Salt Lake City UT 84112

BEHLING, ROBERT EDWARD, b Milwaukee, Wis, Sept 11, 41; m 70; c 1. GEOLOGY. *Educ:* Univ Wis-Milwaukee, BS, 63; Miami Univ, Oxford, Ohio, MS, 65; Ohio State Univ, PhD(geol), 71. *Prof Exp:* Res assoc geol, Inst Polar Studies, Ohio State Univ, 65-70; asst prof, 71-76, ASSOC PROF GEOL, WVA UNIV, 76-, ASST VPRES, ACAD AFFAIRS ADMIN, 78- *Concurrent Pos:* Co-prin investr, NSF grant, Off Polar Prog, 66-75; consult, WVa Geol & Econ Surv, 72-73; geol coordr, Argonne Nat Lab grant, 76-78; co-prin investr, Nat Park Serv grant, Nat Landmark Prog, 73-; consult, Quaternary Mapping Proj, US Geol Surv, 77-79; co-prin investr, Wetlands Study, US Army Corps Engrs, 81- *Mem:* Geol Soc Am; Glaciol Soc; Am Quaternary Asn; Clay Minerals Soc; Nat Asn Geol Teachers; Sigma Xi. *Res:* Physical and chemical weathering in ice-free areas of Antarctica; physical-chemical weathering and landform development in humid, temperate climates. *Mailing Add:* Dept Geol & Geog WVa Univ Morgantown WV 26506

BEHLOW, ROBERT FRANK, b Cleveland, Ohio, Mar 16, 26; m 54; c 3. ANIMAL PARASITOLOGY. *Educ:* Ohio State Univ, DVM, 53. *Prof Exp:* Prof animal path & head state diag lab, Univ Ky, 54-64; gen mgr, Blue Chip Mills, Ill, 64; prof animal sci & exten vet, NC State Univ, 64-86; RETIRED. *Concurrent Pos:* Pres, NC Qual Enterprises Inc. *Mem:* Am Vet Med Asn. *Res:* Isolation of Histoplasma capsulatum from a calf and a pig; swine parasite control. *Mailing Add:* 603 Macon Pl Raleigh NC 27609

BEHM, ROY, b Topeka, Kans, Feb 23, 30; m 54; c 2. ANALYTICAL CHEMISTRY. *Educ:* Univ Wash, BA, 55, PhD(chem), 62. *Prof Exp:* Teacher pub schs, Wash, 55-58; res chemist, Eastman Kodak Co, 62-63; assoc dean grad studies, 66-74, asst prof, 63-70, PROF CHEM, EASTERN WASH STATE UNIV, 70- *Mem:* AAAS; Am Chem Soc; Am Inst Chem. *Res:* Spectrophotometric studies of metal chelates; nonaqueous solvents; voltammetry. *Mailing Add:* Dept Chem Eastern Wash Univ Cheney WA 99004

BEHME, RONALD JOHN, b Evansville, Ind, Apr 12, 38; m 63; c 2. GENETICS, MICROBIOLOGY. *Educ:* Univ Evansville, AB, 60; Ind Univ, PhD(genetics), 69. *Prof Exp:* Lectr zool & genetics, Ind Univ, 65-67; res assoc genetics, Univ Waterloo, 67-69; Can Med Res Coun fel, Univ Western Ont, 69-72; vis prof genetics, Univ Waterloo, 72-73; asst prof, 73-79, ASSOC PROF MICROBIOL & IMMUNOL, UNIV WESTERN ONT, 79- *Concurrent Pos:* Owner, Microsystems Res. *Mem:* Sigma Xi; Am Soc Microbiol; Can Soc Microbiol; Can Col Microbiologists. *Res:* Genetics of antibiotic resistance in bacteria; plasmids; anaerobic bacteriology; automation in clinical microbiology. *Mailing Add:* 51 Brentwood Crescent London ON N6G 1X4 Can

BEHMER, DAVID J, b Milwaukee, Wis, May 29, 41; m 63; c 3. FISHERIES. *Educ:* Wis State Col, Stevens Point, BS, 63; Iowa State Univ, MS, 65, PhD(fisheries biol), 66. *Prof Exp:* Lectr fisheries, Humboldt State Col, 66-67; from asst prof to assoc prof, 67-81, PROF BIOL, LAKE SUPERIOR STATE UNIV, 81- *Mem:* Am Fisheries Soc. *Res:* Statistics; fishes or fisheries; genetics. *Mailing Add:* Dept Biol Lake Superior State Univ Sault Ste Marie MI 49783

BEHNKE, JAMES RALPH, b Milwaukee, Wis, May 2, 43; m 66; c 2. FOOD SCIENCE. *Educ:* Univ Wis-Madison, BS, 66, MS, 68, PhD(food sci), 71. *Prof Exp:* From group leader to sr group leader prod develop, Quaker Oats Co, 71-74; sect mgr, 73-74, mgr prod develop, 74-77, dir frozen foods, 77-78, refrig & frozen foods dir, 78-79, corp vpres, 79-80, sr vpres, opers & technol, 81-86, corp sr vpres, growth , 86-88, CORP SR VPRES GROWTH & TECHNOL, PILLSBURY CO, 86- *Mem:* Inst Food Technologists; Am Asn Cereal Chemists. *Res:* Identification, development and process engineering of consumer marketed and packaged food products. *Mailing Add:* Pillsbury Co 311 Second St S E Minneapolis MN 55414

BEHNKE, ROY HERBERT, b Chicago, Ill, Feb 24, 21; m 44; c 4. INTERNAL MEDICINE. *Educ:* Hanover Col, AB, 43; Ind Univ, MD, 46; Am Bd Internal Med, dipl, 56. *Hon Degrees:* ScD, Hanover Col, 72. *Prof Exp:* Markle scholar, Sch Med, Ind Univ, Indianapolis, 52-57, from instr to prof med, 52-72; PROF INTERNAL MED & CHMN DEPT, COL MED, UNIV SFLA, TAMPA, 72- *Concurrent Pos:* Chief med, Indianapolis Vet Admin Hosp, 57-72; chmn, Vet Admin Pulmonary Dis Res Comt, 62-68, mem, Vet Admin Coop Studies Res Comt, 66-70; chmn, Inter-Soc Comn Heart Dis Res, 69-77; mem residency rev comt, Internal Med, 70-75, chmn, 73-75; mem, Vet Admin Res Develop Career Comt, 79-, chmn, 81-83; mem res coord comt, Am Lung Asn, 83-85, bd dirs, 83-87, chmn, 85-87. *Mem:* Am Fedn Clin Res; Am Thoracic Soc/Am Lung Asn; Am Thoracic Soc; AMA; fel Am Col Physicians; Am Soc Clin Res; Am Col Chest Physicians. *Res:* Cardiopulmonary disease. *Mailing Add:* Dept Internal Med Univ SFla Col Med Tampa FL 33612

BEHNKE, WALLACE B, JR, b Evanston, Ill, Feb 5, 26; m; c 3. GENERAL ENGINEERING, NUCLEAR ENGINEERING. *Educ:* Northwestern Univ, BS, 45, BSEE, 47. *Prof Exp:* Mem staff, Commonwealth Edison, Chicago, 47-69, vpres, 69-73, exec vpres, 73-80, vchmn, 80-89, mem, bd dirs & trustee, 77-91; RETIRED. *Concurrent Pos:* Vchmn, Tech Activ Bd, Inst Elec & Electronics Engrs, chmn, Periodicals Coun; mem, Adv Comt, Idaho Nat Eng Lab; dir & chmn, Atomic Indust Forum; mem bd dirs, Arch Develop Corp, Calumet Indust, Ill Inst Technol, LaSalle Bank Lake View, Northwestern Mem Hosp, Ravenswood Hosp, Tuthill Pump Co, Paxall Group Inc, Standard Am Life Ins Co & United Way Chicago. *Honors & Awards:* James N Landis Medal, Am Soc Mech Engrs, 89. *Mem:* Nat Acad Eng; fel Inst Elec & Electronics Engrs; Am Nuclear Soc; Power Eng Soc (pres, 88-). *Mailing Add:* Commonwealth Edison PO Box 767 Chicago IL 60690

BEHNKE, WILLIAM DAVID, b Pasadena, Calif, Jan 15, 41; m 62; c 3. BIOLOGICAL CHEMISTRY, BIOPHYSICS. *Educ:* Univ Calif, Berkeley, AB, 63; Univ Wash, PhD(biochem), 68. *Prof Exp:* Fel, Harvard Med Sch, 68-72; asst prof chem, Univ SC, 72-74; ASSOC PROF BIOCHEM, COL MED, UNIV CINCINNATI, 74- *Mem:* Biophys Soc; Am Soc Biol Chemists; NY Acad Sci. *Res:* Molecular interaction of colipase with an artificial interface; metalloprotein structure and function. *Mailing Add:* Dept Molecular Genetics Biochem & Microbiol Univ Cincinnati Col Med 231 Bethesda Ave Cincinnati OH 45267

BEHOF, ANTHONY F, JR, b Chicago, Ill, Apr 30, 37. PHYSICS. *Educ:* DePaul Univ, BS, 59; Univ Notre Dame, PhD(nuclear physics), 65. *Prof Exp:* Instr physics, Univ Notre Dame, 65; res physicist & Nat Res Coun-Naval Res Lab fel, US Naval Res Lab, 65-67; ASST PROF PHYSICS, DEPAUL UNIV, 67- *Mem:* AAAS; Am Phys Soc; Am Asn Physics Teachers; Sigma Xi. *Res:* Experimental nuclear physics, especially low energy nuclear reactions; nuclear structure physics. *Mailing Add:* Dept Physics DePaul Univ 2323 N Seminary Ave Chicago IL 60614

BEHR, ELDON AUGUST, b Minneapolis, Minn, July 29, 18; m 50; c 2. WOOD SCIENCE, WOOD TECHNOLOGY. *Educ:* Univ Minn, BS, 40, PhD(agr biochem), 48. *Prof Exp:* Asst chemist, Am Creosoting Co, Ky, 40-42; wood technologist, Air Forces, US Army, NJ, 42-43; tech dir res, Chapman Chem Co, 47-50, vpres & mgr tech dept, 50-59; from assoc prof to prof forestry, Mich State Univ, 59-80; RETIRED. *Mem:* Am Wood Preserver's Asn; Railway Tie Asn; Railway & Locomotive Hist Soc. *Res:* Evaluation of wood preservatives; fibre preservatives; natural durability of wood; termite resistance of wood; test methods for decay of wood products. *Mailing Add:* 7767 Fawn Ridge Cove Cordova TN 38018-2904

BEHR, INGA, b Plovdiv, Bulgaria, Sept 8, 23; US citizen; m 51; c 2. CHEMISTRY. *Educ:* Univ Geneva, MS, 47, PhD(anal chem), 50. *Prof Exp:* Asst, Univ Geneva, 48-50; chemist, Weizmann Inst, 50-51; res chemist, Israel AEC, 52-58; res chemist, Bio-Sci Lab, Calif, 59; NIH grant, City of Hope, 60-62; from instr to assoc prof, Pasedena City Col, 62-76, prof gen chem & qual anal, 76-86; RETIRED. *Concurrent Pos:* Asst prof, Los Angeles State Col, 63-64. *Mem:* Sr mem Am Chem Soc. *Res:* Clinical biochemistry; nature of urochrome pigments; marine biology. *Mailing Add:* 395 S Oakland Ave Pasadena CA 91101

BEHR, LYELL CHRISTIAN, b Minneapolis, Minn, May 4, 16; m 54; c 4. ORGANIC CHEMISTRY. *Educ:* Univ Minn, BChem, 37; Univ Ill, PhD(org chem), 41. *Prof Exp:* Lab asst chem, Univ Ill, 37-40; Hormel fel, Univ Minn, 41-42; chemist, Chem Warfare Serv, Columbia Univ, 42-43; res chemist, E I du Pont de Nemours & Co, Del, 43-47; from asst prof to assoc prof chem, 47-54, head dept, 63-64, dean, Col Arts & Sci, 64-80, PROF CHEM, MISS STATE UNIV, 54-, EMER DEAN COL ARTS & SCI, 80- *Mem:* AAAS; Am Chem Soc; The Chem Soc. *Res:* Heterocyclic and synthetic organic chemistry; azoxy compounds. *Mailing Add:* Dept Chem Miss State Univ Mississippi State MS 39762

BEHR, STEPHEN RICHARD, b Portland, Ore, Jan 1, 52; c 2. NUTRITION, PHYSIOLOGY. *Educ:* Whitman Col, BA, 74; Columbia Univ, MS, 77; Cornell Univ, PhD(nutrit biochem), 83. *Prof Exp:* Postdoctoral fel, Cornell Univ, 83-84; postdoctoral fel, Sch Med, Stanford Univ, 84-85, Am Heart Asn Calif Affil res fel, 85-87; clin res assoc, Abbott Int, 87-90, SR CLIN RES

ASSOC, ROSS LABS, 90- *Concurrent Pos:* Am Inst Nutrit; Am Oil Chemists Soc. *Res:* Nutrition and cardiovascular disease: effects of dietary fats and fiber on blood lipids, lipoproteins and platelet composition and function; effect of antioxidants on lipoprotein metabolism; lipids and energy metabolism. *Mailing Add:* Ross Labs Med Dept 625 Cleveland Ave Columbus OH 43216

BEHRAVESH, MOHAMAD MARTIN, b Tehran, Iran, Mar 21, 45; m 77. PHYSICS, ULTRASONICS. *Educ:* Univ Kans, BS, 68; Iowa State Univ, MS, 70; Georgetown Univ, PhD(physics), 74. *Prof Exp:* Asst prof physics, Abadan Inst Technol, Iran, 74-78; group leader ultrasonil nondestructive eval, Battelle Mem Inst, 78-80; dep div mgr, Elec Power Res Inst, NDE Ctr, Charlotte, NC, 80-84; mgr, Nondestructive Eval Prog, 84-88, Corrosion Control Prog, 88-90, MGR, QUAL ASSURANCE PROG, ELEC POWER RES INST, 91- *Concurrent Pos:* Vis scholar, Georgetown Univ, 76-77; guest scientist, Nat Bur Standards, 77. *Mem:* Acoust Soc Am; Am Soc Mech Eng; Am Soc Qual Control. *Res:* Physical acoustics and nondestructive evaluation and testing; corrosion. *Mailing Add:* Nuclear Power Div Elec Power Res Inst 3412 Hillview Ave Palo Alto CA 94303

BEHREND, DONALD FRASER, b Manchester, Conn, Aug 30, 31; m 57; c 3. WILDLIFE MANAGEMENT, ENVIRONMENTAL SCIENCES. *Educ:* Univ Conn, BS, 58, MS, 60; State Univ NY Col Forestry, Syracuse Univ, PhD(forest zool), 66. *Prof Exp:* Asst forestry & wildlife mgt, Univ Conn, 58-60; forest game mgt specialist, Ohio Div Wildlife, 60; from res asst to res assoc forest zool, State Univ NY Col Forestry, Syracuse, 60-67; asst prof wildlife resources, Univ Maine, 67-68; dir wildlife res, Archer & Anna Huntington Wildlife Forest Sta, 68-73, actg dean grad studies, 73, exec dir, Inst Environ Prog Affairs & asst vpres res, 73-79, vpres acad affairs & prof, State Univ NY Col Environ Sci & Forestry, Syracuse, 79-85; provost, Univ Alaska, Anchorage, 85-88, vpres acad affairs, 85-87, exec vpres, 87-88, CHANCELLOR, UNIV ALASKA, ANCHORAGE, 88- *Concurrent Pos:* Leader deer res, Maine Dept Inland Fisheries & Game, 67-68. *Mem:* Soc Am Foresters; Wildlife Soc; Sigma Xi. *Res:* Ecology, behavior and management of forest wildlife with emphasis on snowshoe hares, white-tailed deer and songbirds; environmental research administration. *Mailing Add:* 1530 Crescent Dr Anchorage AK 99508

BEHRENDS, RALPH EUGENE, b Chicago, Ill, May 20, 26; div; c 2. ELEMENTARY PARTICLE PHYSICS. *Educ:* US Naval Acad, BS, 47; Univ Calif, Los Angeles, PhD(physics), 56. *Prof Exp:* Instr physics, Univ Calif, Los Angeles, 56-57; asst physicist, Brookhaven Nat Lab, 57-59; NSF fel physics, Inst Advan Study, 59-60; res assoc, Univ Pa, 60-61; from asst prof to assoc prof, 61-66, PROF PHYSICS, BELFER GRAD SCH SCI, YESHIVA UNIV, 66- *Mem:* Am Phys Soc. *Res:* Electromagnetic corrections to decay processes; symmetries of strong and weak interactions. *Mailing Add:* Dept Physics Grad Sch Sci Yeshiva Univ New York NY 10033

BEHRENDT, JOHN CHARLES, b Stevens Point, Wis, May 18, 32; div; c 2. MARINE GEOPHYSICS. *Educ:* Univ Wis, BS, 54, MS, 56, PhD(geophys), 61. *Prof Exp:* Asst seismologist, Ellsworth Station Antarctica, 56-58; proj assoc, Univ Wis, 58-64; geophysicist, Colo, 64-68 & 70-72, Geol Liberia, 68-70, geologist in charge, Off Marine Geol, Woods Hole Off, 72-74, chief br Atlantic-Gulf of Mex Geol, Off Marine Geol, US Geol Surv, Woods Hole, Mass 74-77; RES GEOPHYSICIST, US GEOL SURV, COLO, 77- *Concurrent Pos:* Coord, Antartic Res, 77; mem US deleg, Antartic Treaty Meeting, 77-; coord, Charleston Earthwake Res, 78-83. *Honors & Awards:* Explorers Club; Antarctia Serv Medal, NSF. *Mem:* Fel Royal Astron Soc; Am Geophys Union; Soc Explor Geophys; fel Geol Soc Am; AAAS. *Res:* Aeromagnetic aeroradioactivity, gravity and seismic investigations on West African continental margin, Antarctica and western United States; aeromagnetic, multichannel seismic and gravity research of the transition North America to the Atlantic Ocean; midcontinent rift system; West Antarctica rift system. *Mailing Add:* US Geol Surv MS964 Fed Ctr Denver CO 80225

BEHRENS, EARL WILLIAM, b Albany, NY, Nov 4, 35; m 66; c 2. MARINE GEOLOGY. *Educ:* Cornell Univ, BA, 56; Univ Mich, MS, 58; Rice Univ, PhD(geol), 63. *Prof Exp:* Res scientist assoc, 61- 74, Inst Marine Sci, Univ Tex, from asst prof to assoc prof geol, 65-74; RES SCIENTIST, UNIV TEX, 74-, ASSOC PROF MARINE SCI, 74- *Concurrent Pos:* Panelist, Shoreline Erosion Adv Panel, US Army Corps Engrs, 74-80. *Mem:* AAAS; Am Geophys Union; fel Geol Soc Am; Soc Econ Paleont & Mineral; Int Asn Sedimentologists; Sigma Xi. *Res:* Continental slope sedimentation and seismic stratigraphy; high resolution seismic profiling and data processing; quaternary stratigraphy and sedimentation; sea level fluctuations; sedimentary textures. *Mailing Add:* Inst Geophys 8701 N Mopac Austin TX 78759-8345

BEHRENS, ERNST WILHELM, b Piraeus, Greece, Nov 10, 31; div; c 2. POLYMER PHYSICS. *Educ:* Univ Gottingen, BS, 54, teacher's dipl & MS, 57, PhD(solid state physics), 61. *Prof Exp:* Physicist, Siemens Co, Ger, 61-66; scientist, Lockheed-Ga Co, 66-69; PHYSICIST & RES FEL, ARMSTRONG WORLD INDUSTS, 69- *Concurrent Pos:* Europ AEC res fel magnetic resonance br, Grenoble Nuclear Res Ctr, France, 61. *Mem:* Am Phys Soc. *Res:* Physical properties of high polymers and composite materials; materials science engineering; mathematical modeling. *Mailing Add:* 446 Haverhill Rd Lancaster PA 17601

BEHRENS, H WILHELM, b Zeven, Ger, July 16, 35; m 64; c 2. AERONAUTICS. *Educ:* Munich Tech Univ, Dipl Ing, 60; Von Karman Inst Fluid Mech, Belg, Dipl, 61; Calif Inst Technol, PhD(aeronaut), 66. *Prof Exp:* Res fel aeronaut, Calif Inst Technol, 66-67, asst prof, 67-73; sr staff scientist, 73-76, mgr appl physics dept, 76-85, MGR FLUID & THERMO PHYSICS DEPT, TRW SPACE & TECHNOL GROUP, TRW SYSTS INC, 85- *Mem:* Am Inst Aeronaut & Astronaut; Sigma Xi. *Res:* Supersonic and hypersonic aerodynamics; hypersonic wakes; flow separation; viscous and turbulent flows; chemical laser fluid mechanics; pulsed laser acoustics; aero-optics; dusty flows. *Mailing Add:* 4821 Falcon Rock Pl Palos Verdes Peninsula CA 90274

BEHRENS, HERBERT CHARLES, b Cedarburg, Wis, Dec 16, 04; m 30; c 3. OPHTHALMOLOGY. *Educ:* Univ Wis, BA, 27; Northwestern Univ, MD, 31; Univ Pa, MSc, 35. *Prof Exp:* Asst clin prof ophthal, Loma Linda Univ, 41, 48-52, assoc prof, 52-57; from assoc prof to prof, 57-71, EMER PROF OPHTHAL, UNIV SOUTHERN CALIF, 71- *Mem:* Am Acad Ophthal & Otolaryngol. *Mailing Add:* 5810 S Friends Ave Whittier CA 90601

BEHRENS, HERBERT ERNEST, b Milwaukee, Wis, Nov 9, 15; m 41; c 5. PHYSICS, PHOTOGRAPHY. *Educ:* Ill State Norm Univ, BEd, 37; Univ Wis, MS, 47. *Prof Exp:* Teacher high sch, Ill, 37-41; res physicist, Photo-Prod Dept, E I du Pont de Nemours & Co, Parlin, NJ, 47-80; RETIRED. *Mem:* Am Phys Soc; Soc Photog Sci & Eng. *Res:* Photographic materials and processes. *Mailing Add:* PO Box 128 Metuchen NJ 08840-0128

BEHRENS, MILDRED ESTHER, b Geneva, NY, Apr 26, 22. ZOOLOGY. *Educ:* Univ Rochester, BS, 50; Syracuse Univ, PhD(zool), 63. *Prof Exp:* Res scientist, Masonic Med Res Lab, 63-84; RETIRED. *Res:* Electrophysiology of the planarian photoreceptor and the Limulus lateral eye. *Mailing Add:* Five Terr Hill Dr New Hartford NY 13413

BEHRENS, OTTO KARL, biochemistry; deceased, see previous edition for last biography

BEHRENS, RICHARD, b Zenda, Wis, Nov 14, 21; m 50; c 2. WEED SCIENCE. *Educ:* Univ Wis, BS, 49, MS, 50, PhD(agron), 52. *Prof Exp:* Plant physiologist agr res serv, USDA, 52-58; assoc prof agron, 58-63, chmn plant physiol fac, 68-71, PROF AGRON, UNIV MINN, MINNEAPOLIS, 63- *Honors & Awards:* Publ Award, Weed Sci Soc Am, 62. *Mem:* Am Soc Plant Physiol; fel Weed Sci Soc Am (pres, 67). *Res:* Basic and applied aspects of weed control research; absorption, translocation and mode of action of herbicides; effect of environment on herbicidal action; effect of environment on weed competition. *Mailing Add:* Dept Agron Univ Minn St Paul MN 55108

BEHRENS, ROBERT GEORGE, b Mineola, NY, Mar 6, 43; m 70; c 1. PHYSICAL CHEMISTRY. *Educ:* Hartwick Col, BA, 65; Pa State Univ, PhD(chem), 71. *Prof Exp:* Fel, Dept Chem, Univ Kans, 71-74; adj asst prof & assoc prof dir, dept chem, Col of City Univ New York, 74-76; staff mem, 76-86, assoc group leader, Mat Chem Group, 86-87, PROJ LEADER, ADV RECOVERY PROCESSES, LOS ALAMOS NAT LAB, 88- *Concurrent Pos:* Mem, Tech Comt Space Mat Processing, Am Inst Aeronaut & Astronaut. *Mem:* Am Inst Aeronaut & Astronaut; Am Ceramic Soc. *Res:* High temperature chemistry and thermodynamics of nuclear and weapon related materials; plutonium and tritium technology; space materials processing; novel high temperature techniques for the synthesis of ceramics and alloys. *Mailing Add:* 1049 Cedro Ct Los Alamos NM 87544

BEHRENS, WILLY A, b Santiago, Chile, Aug 19, 38. VITAMINS. *Educ:* Univ Santiago, BA; Ottawa Univ, PhD(biochem), 73. *Prof Exp:* Postdoctoral fel, Ottawa Univ, 73-77; RES SCIENTIST, NUTRIT RES DIV, NAT HEALTH & WELFARE CAN, 77- *Mem:* Am Inst Nutrit. *Res:* Metabolic role of vitamin E and C. *Mailing Add:* Nutrit Res Div Nat Health & Welfare Can Banting Bldg Tunney Past Ottawa ON K1A 0L2 Can

BEHRENTS, ROLF GORDON, b Galesburg, Ill, July 21, 47; m 78; c 2. ORTHODONTICS, CRANIOFACIAL BIOLOGY. *Educ:* St Olaf Col, BA, 69; Meharry Med Col, DDS, 73; Case Western Reserve Univ, MS, 75; Univ Mich, PHD(human growth), 84. *Prof Exp:* Fel human growth, Univ Mich, 75-78; from asst prof to assoc prof & dir orthod, 78-84, ASSOC PROF ORTHOD & CHMN, UNIV TENN, 84- *Concurrent Pos:* Res assoc, Bolton-Bush Growth Study Ctr, 84. *Honors & Awards:* Hellman Res Award, Am Asn Orthodontists, 76 & 85. *Mem:* Am Asn Orthodontists. *Res:* Human craniofacial growth in childhood, adolescence and adulthood. *Mailing Add:* 875 Union Ave Memphis TN 38163

BEHRINGER, ROBERT ERNEST, b Springfield, Mass, May 18, 31; m 56; c 5. LASERS. *Educ:* Worcester Polytech Inst, BS, 53; Univ Calif, MA, 55, PhD(physics), 58. *Prof Exp:* Physicist, Gen Elec Co, 53; asst physics, Univ Calif, 53-58; assoc physicist, Int Bus Mach Corp, 58-60; consult, Atomics Int Div, NAm Aviation, Inc, 60-63; physicist, US Off Naval Res, 63-88; vpres, Ballena Systs Corp, 88-91; CONSULT, 91- *Concurrent Pos:* From asst prof to assoc prof physics, San Fernando Valley State Col, 60-65; consult, Marquardt Corp, 61-63. *Mem:* Am Phys Soc; Am Asn Physics Teachers; Sigma Xi. *Res:* Theoretical investigation in the field of ferro-magnetism; nuclear resonance; metallic conductivity; electrooptics; laser physics; free electron lasers. *Mailing Add:* 325 Tipperary Lane Alameda CA 94501

BEHRISCH, HANS WERNER, b Vienna, Austria, Nov 26, 41; Can citizen; m 67; c 3. BIOCHEMISTRY, COMPARATIVE PHYSIOLOGY. *Educ:* Univ BC, BSc, 64, PhD(comp biochem), 69; Ore State Univ, MA, 66. *Prof Exp:* Asst prof, 69-73, ASSOC PROF ZOOCHEM, INST ARCTIC BIOL, UNIV ALASKA, 73- *Concurrent Pos:* Vis prof, Inst Zoophysiol, Univ Innsbruck, 74-75; vis prof, Wenner-Aren Inst, Univ Stockholm, 80-81. *Mem:* AAAS; Can Physiol Soc; Am Soc Biol Chemists; Am Physiol Soc. *Res:* Molecular mechanisms of adaptation to extreme environments and environmental change; regulation of metabolism on three levels of organization: whole organism, organ tissue and enzymic. *Mailing Add:* Inst Arctic Biol Univ Alaska Fairbanks AK 99701

BEHRLE, FRANKLIN C, b Ansonia, Conn, June 4, 22; m 45; c 5. PEDIATRICS, NEONATAL & PERINATAL MEDICINE. *Educ:* Dartmouth Col, AB, 44; Yale Univ, MD, 46. *Prof Exp:* From instr to assoc prof pediat, Sch Med, Univ Kans, 51-61; chmn dept, 64-85, asst dean, 67-73, PROF PEDIAT, NJ MED SCH, 61- *Mem:* Soc Pediat Res; fel Am Acad Pediat. *Res:* Respiratory problems of newborn. *Mailing Add:* Dept Pediat Univ Hosp 100 Bergen St Newark NJ 07107

BEHRMAN, A(BRAHAM) SIDNEY, water chemistry, industrial chemistry; deceased, see previous edition for last biography

BEHRMAN, EDWARD JOSEPH, b New York, NY, Dec 13, 30; m 53; c 3. BIOCHEMISTRY. *Educ:* Yale Univ, BS, 52; Univ Calif, PhD(biochem), 57. *Prof Exp:* Fel biochem, Cancer Res Inst, Nat Cancer Inst, 57-60, res assoc, 60-64; asst prof res, Brown Univ, 64-65; from asst prof to assoc prof, 65-69, PROF BIOCHEM, OHIO STATE UNIV, 69- *Concurrent Pos:* Res fel biochem, Harvard Med Sch, 59-61, res assoc, 61-64, tutor, Harvard Col, 61-64. *Mem:* Royal Soc Chem; Am Chem Soc; Am Soc Biol Chem. *Res:* Peroxydisulfate oxidations; osmium chemistry; nucleic acid chemistry; sugar phosphates. *Mailing Add:* Dept Biochem Ohio State Univ Columbus OH 43210

BEHRMAN, HAROLD R, b Vidora, Sask, Nov 26, 39; m 81; c 3. PHYSIOLOGY, BIOCHEMISTRY. *Educ:* Univ Man, BSc, 62, MSc, 65; NC State Univ, PhD(physiol), 67. *Prof Exp:* Res asst sensory & digestive physiol, NC State Univ, 64-67; Can Med Res Coun res fel reproductive endocrinol, Harvard Med Sch, 67-70, assoc, 70-71, asst prof physiol, 71-72; dir dept reproductive biol, Merck Inst Therapeut Res, NJ, 72-75; assoc prof, 76-81, PROF OBSTET, GYNEC & PHARMACOL, SCH MED, YALE UNIV, 81-, DIR, REPRODUCTIVE BIOL SECT, 76- *Concurrent Pos:* Lalor fel, Harvard Med Sch, 71-73; Alta Heritage vis prof, 83. *Honors & Awards:* Lalor Award, 71. *Mem:* Endocrine Soc; Can Physiol Soc. *Res:* Polypeptide and peptide hormone interrelationships; prostaglandins; purines; oxygen radicals. *Mailing Add:* Dept Obstet & Gynec Sch Med Yale Univ 333 Cedar St New Haven CT 06510

BEHRMAN, RICHARD ELLIOT, b Philadelphia, Pa, Dec 13, 31; m 54; c 4. PEDIATRICS, NEONATOLOGY. *Educ:* Amherst Col, BA, 53; Harvard Univ, JD, 56; Univ Rochester, MD, 60. *Prof Exp:* Intern pediat, Johns Hopkins Hosp, 60-61, resident pediat & mem staff, Lab Obstet Physiol, 63-65; scientist, Lab Perinatal Physiol, Nat Inst Neurol Dis & Blindness, 61-63, sect chief physiol & biochem & actg lab chief, 62-63; from asst prof to assoc prof pediat, Med Sch, Univ Ore & from assoc scientist to scientist, Ore Regional Primate Res Ctr, 65-68, physician in-chg nursery serv, Med Sch & chmn dept perinatal physiol, Res Ctr, 65-68; prof pediat, Col Med, Univ Ill, 68-71; prof, 71-74, Carpentier prof pediat, Col Physicians & Surgeons, Columbia Univ, 74-76, chmn dept, 71-76; dir, Pediat Serv, Babies Hosp, 71-76 & Dept Pediat, Rainbow Babies & Children's Hosp, Univ Hosps, Cleveland, 76-82; Gertrude C Tucker prof & chmn, dept pediat, Sch Med, 76-82, DEAN, SCH MED, CASE WESTERN RESERVE UNIV, 80-, VPRES MED AFFAIRS, 87- *Concurrent Pos:* Whipple scholar, Univ Rochester & Johns Hopkins Hosp, 60-61; Wyeth fel pediat, 63-65; grants, Med Res Found, Ore, 65-66, United Health Found, 66-67, Nat Inst Child Health & Human Develop, 66-69, Pharmaceut Mfrs Asn Found, 70-72 & Robert Wood Johnson Found, 73-76; examr, Am Bd Pediat, 74-77, mem sub-bd neonatal-perinatal med, 75-77; mem bd maternal, child & family health res, Nat Res Coun, 74- *Mem:* Inst Med-Nat Acad Sci; fel Am Acad Pediat; Am Pediat Soc; Perinatal Res Soc; Sigma Xi; Soc Gynec Invest; Soc Pediat Res(vpres, 76-77). *Res:* Fetal, newborn and placental physiology; transfers of gases and solutes across the placenta; reproductive physiology; water and electrolyte balance; acid-base adjustments; membrane transport; bilirubin metabolism; protein binding. *Mailing Add:* David & Lucille Packard FD Ctr for Future of Children 300 Second St Los Altos CA 94022

BEHRMAN, RICHARD H, b New York, NY, May 2, 44; m 78; c 2. MEDICAL IMAGING. *Educ:* Lehigh Univ, BA, 66; McGill Univ, MSc, 68, PhD(physics), 71. *Prof Exp:* Fel, McGill Univ, 72-73; prof physics, Dawson Col, Montreal, 73-85; assoc prof physics, Simmons Col, Boston, 85-87; ASST PROF RADIATION ONCOL & RADIOL, SCH MED, TUFTS UNIV, BOSTON, 88- *Concurrent Pos:* Vis res assoc, McGill Univ & Univ Paris, 76-78; vis assoc prof, Swarthmore Col, 81-82. *Mem:* Am Phys Soc; Am Asn Physicists Med; Am Col Med Physics. *Res:* Diagnostic x-ray and nuclear medicine imaging. *Mailing Add:* 15 St Luke's Rd Apt 3 Allston MA 02134

BEHRMAN, SAMUEL J, b Worcester, SAfrica, Sept 10, 20; US citizen; m 56; c 2. OBSTETRICS & GYNECOLOGY. *Educ:* Univ Cape Town, MB, ChB, 44; Univ Mich, MS, 49; FRCOG, 60. *Prof Exp:* From instr to assoc prof obstet & gynec, 48-56, prof obstet & gynec, coordr postgrad med & res lectr pub health, Univ Mich, Ann Arbor, 56-77, lectr family planning & dir Ctr Res Reproductive Biol, 66-77; PROF OBSTET & GYNEC, WAYNE STATE UNIV, 77-; DIR & CHMN OBSTET & GYNEC, WILLIAM BEAUMONT HOSP, 77- *Concurrent Pos:* Res grants, Macy Found, 48-62 & NIH, 64-67; consult, Ypsilanti State Hosp, Wayne County Gen Hosp & Sinai Hosp, 56-; ed, Int J Fertil, 58-76 & J Int Fertil Asn, 62; guest prof, Univ London, 64; consult maternal health comt, AMA, 64-; mem sci adv bd, Human Life Found; consult, US AID; pres, Int Fedn Fertil Socs, 71-74; mem task force immunol, WHO, 72-; mem adv bd, Food & Drug Admin, 73- *Mem:* Fel Am Asn Obstet & Gynec; Am Fertil Soc (pres, 78-79); fel Am Col Surgeons; Am Col Obstet & Gynec; Royal Soc Med. *Res:* Reproductive physiology. *Mailing Add:* 122 E Brown Birmingham MI 48025

BEHRMANN, ELEANOR MITTS, b Williamstown, Ky, May 24, 17; m 46; c 2. CHEMISTRY, ORGANIC BIOCHEMISTRY. *Educ:* Univ Ky, BS, 38, MS, 39; Iowa State Univ, PhD, 43. *Prof Exp:* Chemist, Shell Develop Co, 43-46; instr chem, Harvard Univ, 46-48; from instr to assoc prof chem, Univ Cincinnati, 57-79; RETIRED. *Mem:* Am Chem Soc. *Res:* Glucose amine chemistry; phenolic resins; organic synthesis. *Mailing Add:* 5897 Crittenden Dr Cincinnati OH 45244-3827

BEHROOZI, FEREDOON, b Tehran, Iran, Jan 8, 41, US citizen; m; c 4. SUPERCONDUCTIVITY, LB FILMS. *Educ:* Univ Wash, BS, 64; Univ Pittsburgh, PhD(physics), 69. *Prof Exp:* Asst prof physics, Pahlavi Univ, Shiraz, Iran, 69-74; res assoc optics, Naval Res Lab, 74-75; from asst prof to assoc prof, 75-87, PROF PHYSICS, UNIV WIS-KENOSHA, 87-, CHMN DEPT, 83- *Concurrent Pos:* Resident scientist, Argonne Nat Lab, 82-83; vis scientist, Naval Res Lab, 87-88; vis scholar, Northwestern Univ, 86- *Mem:* Am Phys Soc; Am Asn Physics. *Res:* Electromagnetic properties of superconductors; anisostropy and critical phenomena in magnetic systems; 2D phase transitions; mathematical modeling of many-body systems; light scattering; lattice dynamics of rare gas solids; monolayers and LB films. *Mailing Add:* Dept Physics Univ Wis-Parkside Box 2000 Kenosha WI 53141-2000

BEICHL, GEORGE JOHN, b Philadelphia, Pa, Aug 20, 18; m 51; c 4. PHYSICAL CHEMISTRY, INORGANIC CHEMISTRY. *Educ:* St Joseph's Col, BS, 39; Univ Pa, MS, 42, PhD, 53. *Hon Degrees:* ScD, St Josephs Univ, 89. *Prof Exp:* Instr chem, St Joseph's Col, 40-44; asst, AEC, Los Alamos, 46; from asst prof to assoc prof, 47-56, chmn dept, 66-89, PROF CHEM, ST JOSEPH'S COL, 56- *Concurrent Pos:* Sr res investr, Univ Pa, 53-54; NSF fel, Munich, Ger, 58-59. *Mem:* Am Chem Soc. *Res:* Boron and copper compounds. *Mailing Add:* 6387 Drexel Rd Philadelphia PA 19151-2511

BEIDLEMAN, JAMES C, b Wilkes-Barre, Pa, Nov 13, 36; m; c 2. ALGEBRA. *Educ:* Bucknell Univ, BA, 58, MS, 59; Pa State Univ, PhD(math), 64. *Prof Exp:* From asst prof to assoc prof, 64-83, PROF MATH, UNIV KY, 83- *Concurrent Pos:* NSF grant on near-rings, 66-68, grant on groups & near-rings, 68-70; Ger acad exchange grant, 82. *Mem:* Am Math Soc; London Math Soc. *Res:* Fitting funetors of finite solvable groups; infinite solvable groups. *Mailing Add:* Dept Math Univ Ky Lexington KY 40506

BEIDLEMAN, RICHARD GOOCH, b Grand Forks, NDak, June 3, 23; m 46; c 3. ECOLOGY, HISTORY OF SCIENCE. *Educ:* Univ Colo, BA, 47, MA, 48, PhD(zool), 54. *Prof Exp:* Asst univ mus, Univ Colo, 46-48; asst prof zool, Colo State Univ, 48-56; asst prof biol, Univ Colo, 56-57; from asst prof to assoc prof, 57-64, PROF BIOL, COLO COL, 64- *Concurrent Pos:* Ford Found Fund Adv Ed grant, 54-55; Am Inst Biol Sci vis biologist prog comnr, Comn Undergrad Educ Biol Sci, NSF; mem consult bur steering comt, Biol Sci Curric Study. *Honors & Awards:* Romco Environ Award, 71. *Mem:* Fel AAAS; Am Soc Zool; Ecol Soc Am; Am Soc Mammal; Am Soc Ichthyol & Herpet; Sigma Xi. *Res:* Species association groups among birds; vertebrate ecology of Western biotic communities; winter bird population studies; small mammal population studies; significance of the American frontier on natural science; history of American biology. *Mailing Add:* 766 Bayview Ave Pacific Grove CA 43950

BEIDLER, LLOYD M, b Allentown Pa, Jan 17, 22; m 46; c 6. PHYSIOLOGY, BIOPHYSICS. *Educ:* Muhlenberg Col, BS, 43; Johns Hopkins Univ, PhD(biophysics), 51. *Hon Degrees:* LLD, Muhlenberg Col, 69. *Prof Exp:* Physicist, radiation lab, Johns Hopkins Univ, 44-45; from asst prof to prof physiol, 50-56, PROF BIOL SCI, FLA STATE UNIV, 56- *Concurrent Pos:* Bowditch lectr, 59; Nat lectr, Sigma Xi, 61; Tanner lectr, Inst Food Technol Sci Lectr, 68. *Mem:* Nat Acad Sci; Am Physiol Soc; AAAS; Am Acad Arts & Sci. *Res:* biophysical properties of chemoreceptors; mechanisms of taste stimulation; olfaction. *Mailing Add:* Dept Biol Sci Fla State Univ Tallahassee FL 32306

BEIER, EUGENE WILLIAM, b Harvey, Ill, Jan 30, 40; m 74; c 1. ELEMENTARY PARTICLE PHYSICS. *Educ:* Stanford Univ, BS, 61; Univ Ill, MS, 63, PhD(physics), 66. *Prof Exp:* Res assoc physics, Univ Ill, 66-67; from asst prof to assoc prof, 67-79, PROF PHYSICS, UNIV PA, 79- *Concurrent Pos:* Mem comt, Brookhaven Nat Lab High Energy Physics, 80-82, Prog Adv Comt, 89-92; prin investr res contract, Dept Energy, 81-90; consult, Univ Chicago, Argonne Nat Labs, 88- *Honors & Awards:* Ross Prize, Am Astron Soc, 89. *Mem:* Fel Am Phys Soc; Sigma Xi; Am Asn Univ Professors. *Res:* Experimental research in elementary particle physics; nuclear physics. *Mailing Add:* Dept Physics Univ Pa Philadelphia PA 19104-6396

BEIER, ROSS CARLTON, b Portage, Wis, Dec 27, 46; m 74; c 2. MASS SPECTROSCOPY, CHROMATOGRAPHY. *Educ:* Univ Wis-Stevens Pt, BS, 69; Mont State Univ, PhD(org chem), 80. *Prof Exp:* Res chem, Nat Cotton Path Res Lab, 79-80, RES CHEM, VET TOXICOL & ENTOM RES LAB, USDA, COLLEGE STATION, TEX, 80-, FOOD ANIMAL PROTECTION RES LAB. *Concurrent Pos:* Vis mem, Grad Fac Tex A&M Univ, 82- *Mem:* Am Chem Soc; NY Acad Sci; Am Soc Mass Spectrometry; fel Am Inst Chemists. *Res:* Chemistry and structural determination of natural products, pesticides and their metabolites in living systems. *Mailing Add:* RR 5 Box 810 College Station TX 77845

BEIERWAITES, WILLIAM HENRY, b Saginaw, Mich, Nov 23, 16; m 42; c 3. MEDICINE. *Educ:* Univ Mich, AB, 36, MD, 41; Am Bd Internal Med, dipl. *Prof Exp:* From intern to asst resident med, Cleveland City Hosp, 41-43; resident, 44, from instr to assoc prof, 45-59, PROF MED, UNIV HOSP, UNIV MICH, ANN ARBOR, 59-, DIR NUCLEAR MED, 52-; AT DEPT MED, HYPERTENSION RES LAB, HENRY FORD HOSP, DETROIT. *Concurrent Pos:* Guggenheim Found fel & Commonwealth Found fel, 66-67; mem adv comt, IAE Comn, 74- *Mem:* Fel Am Col Physicians; Am Thyroid Asn (vpres, 64); Am Fedn Clin Res (pres, 54); Soc Nuclear Med (pres-elect, 64, pres, 65); Am Cancer Soc. *Res:* Internal nuclear medicine; diagnosis and treatment of cancer, hypertension, thyroid and heart disease using radionuclide labeled compounds. *Mailing Add:* Hypertension Res Henry Ford Hosp 2799 W Grand Blvd Detroit MI 48202

BEIGEL, ALLAN, b Hamilton, Ohio, Apr 4, 40. PSYCHIATRY. *Educ:* Harvard Univ, BA, 61; Albert Einstein Col Med, MD, 65. *Prof Exp:* Dir, S Ariz Ment Health Ctr, 70-83; vpres, Univ Res & Develop, 83-89, PROF PSYCHIAT, UNIV ARIZ, 79-, PROF PSYCHOL, 85-, VPRES, UNIV AFFIL, 90- *Mem:* Inst Med-Nat Acad Sci; fel Am Psychiat Asn; fel Am Col Psychiatrists; AMA. *Mailing Add:* Univ Ariz Admin Bldg Rm 702 Tucson AZ 85721

BEIGELMAN, PAUL MAURICE, b Los Angeles, Calif, July 21, 24; m 53; c 1. MEDICINE. *Educ:* Univ Southern Calif, MD, 48; Am Bd Internal Med, dipl, 55. *Prof Exp:* Intern med, Los Angeles County Hosp, 47-48; asst resident, Stanford Univ, Hosps, 48-49; resident path, Cedars of Lebanon Hosp, 49-50; asst resident med, Peter Bent Brigham Hosp, 50-51, asst, 53-55; res asst anat & clin asst prof med, Sch Med, Univ Calif, Los Angeles, 55-56; sr physician, Sepulveda Vet Admin Hosp, 56; from asst prof to assoc prof med, 56-73, assoc prof physiol, 63-64, assoc prof pharmacol, 69-73, PROF MED & PHARMACOL, SCH MED, UNIV SOUTHERN CALIF, 73- *Concurrent Pos:* Res fel, Harvard Med Sch, 53-55 & Dazian Found, 54-55; attend physician, Los Angeles County Hosp, 56-; consult physician, Wadsworth Vet Admin Hosp, 62-66 & San Fernando Vet Admin Hosp, 66-72; attend physician, Rancho Los Amigos Hosp, Downey, 74- *Mem:* Endocrine Soc; Am Physiol Soc; Am Fedn Clin Res; fel Am Col Physicians; Soc Exp Biol & Med. *Res:* Bacteriology; antibiotics; epidemiology; virology; collagen diseases; endocrinology; metabolism; diabetes mellitus; electrophysiology. *Mailing Add:* Univ Southern Calif Med Sch 2025 Zonal Ave Los Angeles CA 90033

BEIGHLEY, CLAIR M(YRON), b Youngstown, Ohio, Dec 17, 24; m 51; c 6. MECHANICAL ENGINEERING, AEROSPACE ENGINEERING. *Educ:* Ohio State Univ, BS, 47; Purdue Univ, MS, 49, PhD(mech eng), 53. *Prof Exp:* Res engr, Ohio State Univ Res Found, 47-48; proj engr, Rocket Motor Lab, Sch Mech Eng, Purdue Univ, 48-52; group leader, Rocket Res Sect, Rocket Engine Div, Bell Aircraft Corp, 52-54; dir, Rocket Lab, Eng Res Inst, Univ Mich, 54-55; prin engr, Aerojet Gen Corp, Sacramento, 55-58, mgr, Res Dept, 58-60, mgr, Propulsion Systs Dept, 60-74, from staff to vpres res & develop, 74-76; consult, 76-77; dep dir, Stanford Linear Accelerator Ctr Site Off, 77-80, PROF MGR, SAN DIEGO OFF, US DEPT ENERGY, 80- *Concurrent Pos:* Consult, Proj SQUID, Princeton Univ, 49-50 & McGraw-Hill Info Systs Co, 60-77; lectr, Sch Mech Eng, Univ Mich, 54-55; teacher, exten, Univ Calif, Berkeley, 57-60. *Mem:* Am Nat Metric Coun; Am Phys Soc. *Res:* Energy research and development; project management of advanced energy facilities and construction projects. *Mailing Add:* 18655 W Bernardo Dr No 227 San Diego CA 92127

BEIGHTLER, CHARLES SPRAGUE, b Cincinnati, Ohio, Mar 18, 24; m 57; c 4. INDUSTRIAL ENGINEERING, OPERATIONS RESEARCH. *Educ:* Univ Mich, BS, 50, MS, 54; Northwestern Univ, PhD(indust eng), 61. *Prof Exp:* Aeronaut eng designer, Aeronca Mfg Co, 50-51; engr res labs, Gen Motors Corp, 54-55; opers res analyst, Arthur Anderson & Co, 55-56, Caywood-Schiller, Assocs, 56-57 & Ernst & Ernst, 57-58; from asst prof to assoc prof, 61-68, PROF MECH ENG, UNIV TEX, AUSTIN, 68- *Concurrent Pos:* Fulbright lectr, Univ Freiburg, 71-72. *Honors & Awards:* Lanchester Prize, Opers Res Soc Am. 67. *Mem:* Opers Res Soc Am; Am Inst Indust Engrs; Inst Mgt Sci; Sigma Xi; Am Statist Asn. *Res:* Optimization theory; optimal design. *Mailing Add:* 7007 Edgefield Dr Austin TX 78731

BEIL, GARY MILTON, b Clinton, Iowa, Dec 27, 38; m 60; c 2. PLANT BREEDING, STATISTICS. *Educ:* Iowa State Univ, BS, 60, MS, 63, PhD(plant breeding), 65. *Prof Exp:* Res assoc agron, Iowa State Univ, 60-65; plant breeder, Caladino Farm Seed, Inc, 65-67; dir, Northeastern Exp Sta, 67-79, mgr planning & develop, Int Seed Oper, De Kalb Agres, Inc, 79-85; RESEARCHER, SEED DIV, CONTINENTAL GRAIN CO, 85- *Mem:* Am Soc Agron; Crop Sci Soc Am; Am Entom Asn. *Res:* Genetics of maize. *Mailing Add:* Seed Div Continental Grain Co 277 Park Ave New York NY 10172

BEIL, ROBERT J(UNIOR), b Lima, Ohio, Jan 15, 24; m 44; c 1. ENGINEERING SCIENCE, APPLIED MATHEMATICS. *Educ:* ETenn State Univ, BS, 47; Okla Agr & Mech Col, MS, 49; Vanderbilt Univ, BE, 60; Purdue Univ, PhD(eng sci), 66. *Prof Exp:* Assoc prof eng mech, Vanderbilt Univ, 48-88, chmn dept, 70-88; RETIRED. *Mem:* Soc Eng Sci; Math Asn Am; Am Soc Eng Educ. *Res:* Continuum mechanics; structural changes in ceramic materials; electric and magnetic effects in solids. *Mailing Add:* Vanderbilt Univ 125 Westgrill Dr Palm Coast FL 32137

BEILBY, ALVIN LESTER, b Watsonville, Calif, Sept 17, 32; m 58; c 2. ELECTROANALYTICAL CHEMISTRY, CHEMICAL EDUCATION. *Educ:* San Jose State Col, BA, 54; Univ Wash, PhD(chem), 58. *Prof Exp:* Instr chem, 58-60, from asst prof to assoc prof, 60-72, chmn dept, 72-85, PROF CHEM, POMONA COL, 72- *Concurrent Pos:* Petrol Res Fund fac award, Univ Ill, 64-65; guest worker & res chemist, Nat Bur Standards, 71-72; NSF sci fac prof develop award, Lockheed Palo Alto Res Lab, 79-80; guest prof, Uppsala Univ, Sweden, 86-87. *Mem:* Am Chem Soc; Sigma Xi; AAAS. *Res:* Use and nature of carbon as an electrode material for electro-analytical methods; chemical education. *Mailing Add:* Seaver Chem Lab Pomona Col 645 N College Ave Claremont CA 91711-6338

BEILER, ADAM CLARKE, physics, for more information see previous edition

BEILER, THEODORE WISEMAN, b Meadville, Pa, Apr 29, 24; m 51; c 1. ORGANIC CHEMISTRY. *Educ:* Allegheny Col, BS, 48; Harvard Univ, MA, 50, PhD(chem), 52. *Prof Exp:* Sr asst scientist, USPHS, 51-53; from asst prof to assoc prof chem, Stetson Univ, 53-62; Fulbright lectr, Univ Panjam, WPakistan, 62-63; chmn dept chem, 63-89, prof, 63-68, WILLIAM KENAN PROF CHEM, STETSON UNIV, 68- *Concurrent Pos:* Vis scientist, NIH, 58; vis prof, Duke Univ, 79-80. *Mem:* Am Chem Soc; Sigma Xi; Am Asn Univ Professors. *Res:* Syntheses of alicyclic systems; transformations of amino acids. *Mailing Add:* 813 Oak Tree Terr DeLand FL 32724-3614

BEILSTEIN, HENRY RICHARD, b Philadelphia, Pa, Dec 2, 20; m 46; c 2. MICROBIOLOGY, PUBLIC HEALTH. *Educ:* Philadelphia Col Pharm & Sci, BSc, 43, MSc, 61, PhD(microbiol), 70. *Prof Exp:* Anal chemist immunochem & org chem, Merck, Sharpe & Dohme, 43-45; sr microbiologist, Dept Pub Health, City Philadelphia, 45-61, asst dir, Pub Health Labs, 61-70, dir, Pub Health Labs, 70-78; educ coordr instr microbiol & med lab technol, 78-79, dir & chmn, Dept Med Lab Technol, Manor Col, 79-86; INSTR, OGONTZ CAMPUS, PA STATE UNIV, 90- *Concurrent Pos:* Bacteriologist & consult, N Broad & Cherry Hill Clin Labs, 47-73; prof microbiol & med technol, Franklin Sch Sci & Arts, Philadelphia, 56-71; adj asst prof microbiol & immunol, Sch Med, Temple Univ, 69-82; vis assoc prof, dept microbiol & immunol, Hahnemann Med Col & Hosp, 70-; consult, Smith Kline Labs, Pa, 71-78; assoc prof microbiol, Pa Col Podiatric Med, 74-82 & 86-; adj prof biol, Beaver Col, Glenside, Pa, 86- *Mem:* Am Soc Microbiol; fel Am Pub Health Asn; Conf Pub Health Lab Dirs; Am Sci Affil; fel Am Acad Microbiol. *Res:* Clinical and laboratory aspects of sexually transmitted disease; immunofluorescence. *Mailing Add:* 1032 E Mt Pleasant Ave Philadelphia PA 19150

BEIN, DONALD, b New York, NY, Dec 13, 34; m 60; c 2. MATHEMATICS. *Educ:* Brooklyn Col, BS, 57; NY Univ, PhD(math), 64. *Prof Exp:* From asst prof to assoc prof, 64-74, PROF MATH, FAIRLEIGH DICKINSON UNIV, 74- *Mem:* Am Math Soc; Math Asn Am; Soc Indust & Appl Math. *Res:* Parabolic partial differential equations. *Mailing Add:* Dept Math & Comput Sci Fairleigh Dickinson Univ Teaneck Campus Teaneck NJ 07666

BEINDORFF, ARTHUR BAKER, b Omaha, Nebr, Apr 16, 25; m 46; c 3. POLYMER CHEMISTRY. *Educ:* Univ Nebr, BS, 47, MA, 49, PhD(biochem), 52. *Prof Exp:* From chemist to sr chemist, Chemstrand Corp, Monsanto Co, Ala, 52-58, group leader nylon polymer res, 58-59, head personnel sect Res Ctr, NC, 59-62, mgr patent liaison, 62-66, assoc dir patent liaison, Textiles Div, NY, 66-67, tech dir polyester, Ala, 67-70, dir tire technol, 70-74, dir apparel fibers technol, Monsanto Textiles Co, Fla, 74-75, dir, Acrylic Fiber Technol, 75-84, Decatur Ala Technol Ctr, 75-84; RETIRED. *Concurrent Pos:* Chmn dept chem, Athens Col, 52-55; prof & chmn, St Bernard Col, 54-56. *Mem:* Am Chem Soc; Soc Automotive Eng; Am Asn Textiles Technol. *Res:* Exploratory organic polymer chemistry; nylon; polyester; acrylic. *Mailing Add:* 2812 Burningtree Mt Rd S E Decatur AL 35603

BEINEKE, LOWELL WAYNE, b Decatur, Ind, Nov 20, 39; m 67; c 2. MATHEMATICS, GRAPH THEORY. *Educ:* Purdue Univ, BS, 61; Univ Mich, MA, 62, PhD(math), 65. *Prof Exp:* Res asst math inst soc res, Univ Mich, 62-63; from asst prof to assoc prof, 65-71, prof math, 71-86, JACK W SCHREY PROF MATH PURDUE UNIV, FT WAYNE, 86- *Concurrent Pos:* Res asst, Univ Col, Univ London, 66-67; consult inst soc res, Univ Mich, 65; tutor, Oxford Univ, 73-74; vis lectr, Polytechnic, North London, 80-81; assoc ed, J Graph Theory; Bd Gov, Math Asn Am, 90- *Mem:* Am Math Soc; Math Asn Am; Am Asn Univ Professors; Sigma Xi; London Math Soc. *Res:* Graph theory; network theory; mathematical models in the social sciences; combinatorial analysis; research and writing the theory and application of graphs; embedding and drawings of graph on surfaces, round-robin and bipartite tournaments, line graphs and line digraphs, multi- dimensional trees, and vulnerability of graphs. *Mailing Add:* Indiana Univ Purdue Univ Ft Wayne IN 46805

BEINEKE, WALTER FRANK, b Indianapolis, Ind, Mar 7, 38; m 61; c 2. FOREST GENETICS. *Educ:* Purdue Univ, BS, 60; Duke Univ, MF, 61; NC State Univ, PhD(forest genetics), 66. *Prof Exp:* ASSOC PROF FORESTRY, PURDUE UNIV, 64- *Mem:* Soc Am Foresters; AAAS. *Res:* Genetic improvement of black walnut, Juglans nigra, including selection, vegetative propagation, progeny testing, flowering and breeding; clones; tissue culture of hardwood trees. *Mailing Add:* Dept Forestry Purdue Univ West Lafayette IN 47906

BEINERT, HELMUT, b Lahr, Ger, Nov 17, 13; nat US; m 44; c 4. BIOINORGANIC CHEMISTRY, ENZYMOLOGY. *Educ:* Univ Leipzig, Dr rer nat(chem), 43. *Hon Degrees:* DSc, Univ Wis Milwaukee, 87. *Prof Exp:* Res assoc biol chem, Kaiser Wilhelm Inst Med Res, Ger, 43-45; biochemist, Air Force Aeromed Ctr, Ger, 46 & USAF Sch Aviation Med, 47-50; res assoc, Inst Enzyme Res, 51-52, from asst prof to prof enzyme chem, 52-84, chmn sect III, 58-84, prof biochem, 67-84, EMER PROF ENZYME CHEM, INST ENZYME RES & EMER PROF BIOCHEM, UNIV WIS-MADISON, 84-; PROF BIOCHEM & DISTINGUISHED SCHOLAR IN RESIDENCE, MED COL WIS, 85- *Concurrent Pos:* Permanent guest prof, Univ Konstanz, Ger, 67. *Honors & Awards:* Sr Scientist Award, Alexander von Humboldt Found, 81; Keilin Medal, 85; Sir Haus Krebs Medal, 89. *Mem:* Nat Acad Sci; Am Chem Soc; Am Soc Biol Chem; fel Am Acad Arts & Sci. *Res:* Structure and function of iron-sulfur proteins. *Mailing Add:* Dept Biochem Med Col Wis 8701 Watertown Plank Rd Milwaukee WI 53226

BEINFEST, SIDNEY, b Brooklyn, NY, Oct 29, 17; m 42; c 2. ORGANIC CHEMISTRY, CHEMICAL ENGINEERING. *Educ:* Brooklyn Col, BA, 38; Univ Ark, MS, 39; Polytech Inst Brooklyn, PhD(chem), 49. *Prof Exp:* Dir res & develop, Premo Pharmaceut Labs, Inc, 41-45; vpres in charge opers, Berkeley Chem Corp, 46-63; tech vpres, Millmaster Chem Corp, 63-64, vpres technol, Millmaster Onyx Chem Corp, Jersey City, 64-74, vpres corp tech, Millmaster Onyx Corp, NY, 74-86; RETIRED. *Mem:* AAAS; Am Inst Chemists; Am Chem Soc; Sci Res Soc Am; Am Inst Chem Eng; Sigma Xi. *Res:* Cyanine dyes; pyrazine derivatives; barbiturates; anti-malarial drugs; muscle relaxant; tranquillizing drugs. *Mailing Add:* 248 Emerson Lane Berkeley Heights NJ 07922

BEINFIELD, WILLIAM HARVEY, b St Louis, Mo, Apr 8, 18; m 50; c 2. INTERNAL MEDICINE, PHARMACOLOGY. *Educ:* Univ Wis, BA, 40; Columbia Univ, MD, 43; Am Bd Internal Med, dipl, 52. *Prof Exp:* From asst instr to instr med, 50-54, instr physiol, 50-52, assoc, 52-54, from assoc prof to prof physiol, 56-64, assoc prof pharmacol, 56-60, ASST PROF MED, NY MED COL, 56-, PROF PHARMACOL, 60- *Concurrent Pos:* Adv drug & formulary comt, Dept Hosp, NY; pres, Corlette Glorney Found Med Res, 81- *Mem:* Fel Am Col Physicians; Am Heart Asn; Am Thoracic Soc; fel NY Acad Med; fel NY Acad Sci. *Res:* Cardiopulmonary physiology and pharmacology; studies of spontaneous mechanical activity of airway smooth muscle in vivo and its modulation by endogenous mediator substances and drugs. *Mailing Add:* 150 E 69th Apt 16G New York NY 10021

BEINING, PAUL R, b Pittsburgh, Pa, Feb 2, 23. MICROBIOLOGY, BIOCHEMISTRY & IMMUNOLOGY. *Educ:* Spring Hill Col, BS, 49; Cath Univ Am, MS, 52, PhD(microbiol), 62. *Prof Exp:* Instr biol, Univ Scranton, 49-51; instr, St Joseph's Col, Pa, 62-63, asst prof, 63-66; assoc prof & chmn dept, Wheeling Col, 66-67; assoc prof, Univ Scranton, 67-75, prof biol, 75-91; RETIRED. *Concurrent Pos:* Researcher, Georgetown Univ, 69-71; guest researcher, NIH, 71-80, Food & Drug Admin, 81-85. *Mem:* AAAS; Am Soc Microbiol. *Res:* Biochemical and serological characteristics of in-vivo and in-vitro cultivated Staphyloccus aureus; increased protection with active immunization with influenza virus; polyphasic taxonomy of marine and non-marine Micrococci; ultrastructure and biochemical analysis of staphylococcal membrane systems; immune response in mice to staphylococcal lipo teichoilacid; enhanced immune response in mice to type VI capsular polysaccharide; antibody induced immunosuppression; immune response to conserved areas of the human immunodeficiency virus. *Mailing Add:* Dept Biol Univ Scranton Scranton PA 18510

BEIQUE, RENE ALEXANDRE, b Cornerbrook, Nfld, Nov 29, 25; m 51; c 3. RADIATION PHYSICS. *Educ:* Univ Montreal, BA, 47, BSc, 50, MSc, 51; Mass Inst Technol, PhD(physics), 58. *Prof Exp:* Jr physicist, High Voltage Lab, Mass Inst Technol, 52-53; physicist, Montreal Gen Hosp, 58-68; assoc prof, 69-71, PROF RADIOL, UNIV MONTREAL, 71-; PHYSICIST, NOTRE DAME HOSP, 68- *Concurrent Pos:* Lectr, McGill Univ, 60-69, asst prof, 69-; consult, Montreal Children's Hosp, 60- *Mem:* Soc Nuclear Med; Can Asn Physicists; Can Asn Radiol. *Res:* Physics of radiology, especially diagnostic radiology. *Mailing Add:* 405 Bienville Longueuil PQ J4H 2E8 Can

BEIRNE, BRYAN PATRICK, b Wexford, Ireland, Jan 22, 18; m 48; c 2. ECONOMIC ENTOMOLOGY. *Educ:* Univ Dublin, PhD(entom), 40, MSc, 41, MA, 42. *Prof Exp:* Asst lectr zool, Univ Dublin, 42-43, asst dir mus zool & comp anat, 42-49, lectr entom, 43-49; sr entomologist res br, Can Dept Agr, 49-55, dir, Res Inst, Belleville, Ont, 55-67; dir, Pestology Ctr, Simon Fraser Univ, 67-79, prof, 67-93, dean grad studies, 79-83, EMER PROF PEST MGT, SIMON FRASER UNIV, 83- *Concurrent Pos:* Consult/Adv various overseas aid projs. *Honors & Awards:* Gold Medal, Entom Soc Can. *Mem:* Am Entom Soc; Entom Soc Can; Royal Irish Acad. *Res:* Biological control; pest management; ecology of agricultural insects; insect taxonomy. *Mailing Add:* Dept Biol Sci Simon Fraser Univ Burnaby BC V5A 1S6 Can

BEISEL, WILLIAM R, b Philadelphia, Pa, Apr 8, 23; m 49; c 5. INFECTIOUS DISEASES, METABOLISM. *Educ:* Muhlenberg Col, BA, 46; Ind Univ, Indianapolis, MD, 48; Am Bd Internal Med, dipl, 55. *Prof Exp:* Intern, Fitsimmons Army Hosp, Denver, Colo, US Army, 48-49, resident med, Letterman Gen Hosp, San Francisco, 49-52, asst chief dept med, 21st Army Sta Hosp, Korea, 53-54, chief dept med, US Army Hosp, Ft Leonard Wood, Mo, 54-56, basic sci yr, Walter Reed Army Inst Res, DC, 56-57, chief dept metab, 57-60, chief phys sci div, Walter Reed Med Unit, Ft Detrick, Md, 62-68; sci adv, 69-80, dep scientist, US Army Med Res Inst Infectious Dis, 80-84, spec asst surgeon gen biotechnol, 84-85; ADJ PROF, JOHNS HOPKINS SCH HYG & PUB HEALTH, 85- *Concurrent Pos:* Fel metab, Sch Med, Univ Calif, San Francisco, 61-62; chief med sect 4, Walter Reed Army Hosp, Washington, DC & asst prof, Sch Med, Georgetown Univ, 57-60; mem, study sect pharmacol & exp therapeut, NIH, 59-60, study sect endocrinol, 64-68; consult metab dis to Surgeon Gen, US Army, 66-76; assoc prof, Sch Med, Univ Md, 67-77; mem infective agents res eval comt, Vet Admin, 69-71; mem, Grad Sch Adv Coun, Hood Col, 75-; chmn subcomt nutrit & infection, Nat Res Coun, 76-77; consult, Div Polar Progs, NSF, 76-80; Dept Defense rep, NIH Interagency Comt Recombinant DNA Res, 76-85; chmn, Nutrition Adv Group, AMA, 77-80; chmn, Life Sci Res Off Adv Coun, Fed Am Soc Exp Biol, 77-; chmn, Subcomt Nutrit & Infection, Int Union Nutrit Sci, 78- *Honors & Awards:* Hoff Gold Medal, Walter Reed Army Inst Res, 57; Award & Commendation, Army Sci Conf, 66; Stitt Award, Asn Mil Surgeons, 68; B L Cohen Award, Am Soc Microbiol, 82. *Mem:* Fel Am Col Physicians; Am Inst Nutrit; Infectious Dis Soc Am; Endocrine Soc; Am Soc Microbiol. *Res:* Broad aspects of endocrinology, metabolism, nutrition and immunology, especially as they pertain to and help regulate the responses of a host to the stress of infectious illness and immunonutrition. *Mailing Add:* 8210 Ridgelea Ct Frederick MD 21701

BEISER, CARL A(DOLPH), b Cleveland, Ohio, Apr 28, 29; m 50; c 2. PHYSICAL METALLURGY. *Educ:* Case Inst Technol, BS, 51, MS, 53, PhD(phys metall), 57. *Prof Exp:* Res asst metall, Case Inst Technol, 51-57; metall engr, Union Carbide Metals Co, 57-60; supvr phys metall, 60-64, asst div chief, Res & Develop Dept, 64-67, div chief mat res, 67-78, DIR RES, NAT STEEL CORP, 78- *Mem:* Am Soc Metals; Sigma Xi. *Res:* Ferrous materials. *Mailing Add:* 6108 G Gary Gate Lane Charlotte NC 28210-4050

BEISER, HELEN R, b Chicago, Ill, Nov 15, 14. PSYCHOANALYSIS, CHILD PSYCHIATRY. *Educ:* Univ Ariz, BSEd, 35; Univ Ill, MS & MD, 41. *Prof Exp:* From instr to assoc prof psychiat, 52-76, chmn child anal training, Inst Psychoanal, 69-78, clin prof, 76-79, EMER PROF PSYCHIAT, UNIV ILL COL MED, 79- *Concurrent Pos:* Fel staff & consult, Inst Juv Res, 48-80; med dir, North Shore Ment Health Clin, 53-54. *Honors & Awards:* Friend of Children, Ill Child Care Asn, 86. *Mem:* Am Psychoanal Asn; Am Acad Child Psychiat; Am Psychiat Asn. *Res:* Evaluation and measurement in child diagnosis and treatment; personality and achievement in medical students; process of supervision in psychotherapy; psychological meaning of games. *Mailing Add:* 5333 N Sheridan Rd 18C Chicago IL 60640

BEISER, MORTON, b Regina, Sask, Nov 16, 36; m 61; c 3. PSYCHIATRY. *Educ:* Univ BC, MD, 60. *Prof Exp:* Intern, Montreal Gen Hosp, 60-61; resident psychiat med ctr, Duke Univ, 61-64; from lectr to assoc prof behav sci, Harvard Sch Public Health, 67-76; PROF PSYCHIAT, UNIV BC, 76- *Concurrent Pos:* NIMH res fel med ctr, Duke Univ, 64; res fel med ctr, Cornell Univ, 65-66; asst psychiatrist, Mass Gen Hosp, 70-76; consult, Ment Health Br, Indian Health Serv, 72-75; prin investr grants, Dept Ment Health, Commonwealth of Mass, 72-75, Indian Health Serv, Dept Health, Educ Health & Welfare, Can, 72 & 81- & US NIMH, 82-; staff psychiatrist, Tufts New Eng Med Ctr, 72-76 & Children's Hosp Med Ctr, Boston, 74-76; Can Health & Welfare Nat Res Scholar, 81-83; res scientist, Can Health & Welfare Nat Res, 81-83 & 83-; chmn, epidemiol & serv eval rev comt, US NIMH, 84-87, chmn, Nat Task Force Ment Health of Migrants, Govt Can Ministry Health & Welfare & Ministry Multiculturalism, 86-88. *Honors & Awards:* Master of Res of First Degree, Nat Inst Health & Med Res, France, 72; Josiah Macy Jr Found Fac Scholar Award, 74-75. *Mem:* AAAS; Am Psychiat Asn; Am Psychopath Asn; hon mem Royal Soc Med, Belg; World Fedn Ment Health (vpres, 81-83); Am Pub Health Asn; Can Psychiat Asn. *Res:* Psychiatric epidemiology; cross-cultural psychiatry; epidemiology of childhood psychiatric disorders; refugee resettlement; course of mental disorders. *Mailing Add:* Dept Psychiat Univ BC Health Sci Ctr Hosp 2255 Wesbrook Mall Vancouver BC V6T 2A1 Can

BEISHLINE, ROBERT RAYMOND, b Ogden, Utah, Nov 28, 30; m 59; c 4. PHYSICAL ORGANIC CHEMISTRY. *Educ:* Brigham Young Univ, BS, 55, MS, 57; Pa State Univ, PhD(chem), 62. *Prof Exp:* Res chemist, Am Cyanamid Co, 62-63; asst prof chem, State Univ NY Albany, 63, Res Found fac res fel, 64; assoc prof, 66-74, PROF CHEM, WEBER STATE UNIV, 74- *Mem:* Am Chem Soc; Sigma Xi. *Res:* Chemical kinetics of organic reactions; reaction mechanisms; mechanism of coal hydrogenation by hydrogen donor solvents; pyrolysis of coal-related model compounds. *Mailing Add:* Dept Chem Weber State Univ Box 2503 Ogden UT 84408

BEISLER, JOHN ALBERT, b Hackensack, NJ, May 18, 37; m 68; c 2. MEDICINAL CHEMISTRY. *Educ:* Fairleigh Dickinson Univ, BS, 60; Rutgers Univ, PhD(org chem), 64. *Prof Exp:* Fel, Cambridge Univ, Eng, 64-66; staff fel steroid chem, Nat Inst Arthritis & Metab Dis, 66-71; sr investr med chem, Microbiol Assoc Whittaker Corp, 71-74; res chemist med chem, Nat Cancer Inst, 74-81; HEALTH SCIENTIST ADMIN, DIV RES GRANTS, NIH, 81- *Concurrent Pos:* Lectr, Am Univ, Washington, DC, 73. *Mem:* Am Chem Soc; Royal Soc Chem; AAAS; Am Crystallog Asn. *Res:* Rational design and synthesis of drug molecules having a potential chemotherapeutic value in the treatment of cancer in humans. *Mailing Add:* 8707 Garfield Bethesda MD 20817

BEISPIEL, MYRON, b New York, NY, Nov 21, 31; m 54; c 3. ORGANIC CHEMISTRY. *Educ:* Brooklyn Col, BS, 53; NY Univ, PhD(chem), 59. *Prof Exp:* Instr chem, NY Univ, 57-59; with Nuodex Prod Co, Div Heyden Newport Chem Co, 59-61; with Sci Design Co, Inc, 61-63; CHMN SCI DEPT, RANNEY SCH, 63- *Concurrent Pos:* Dir environ health lab, Woodbridge Twp, NJ; consult, Cent Jersey Regional Air Pollution Control Agency, NJ, 73-; asst dir, Middlesex County Pub Health Lab, 78- *Mem:* Am Chem Soc; Am Anal Chem Soc; Sigma Xi. *Res:* Organic and inorganic synthesis; analysis techniques, especially gas chromatography; photo chemistry; auto and catalytic oxidation; air and water pollution problems; analytical microscopy; environmental health and impact studies; public health. *Mailing Add:* 11 Gayle St New Monmouth NJ 07748-1419

BEISSER, ARNOLD RAY, b Santa Ana, Calif, Oct 5, 25; m 53. PSYCHIATRY. *Educ:* Stanford Univ, AB, 48, MD, 50. *Prof Exp:* Intern, Charity Hosp La, 49-50; resident, Metrop State Hosp, 53-55; resident, Los Angeles County Gen Hosp, 53-54; staff psychiatrist, Pasadena Guid Clin, 56; chief psychiatrist outpatient & aftercare dept, Metrop State Hosp, Dept Ment Hyg, State of Calif, 56-57, coordr residency training, 57-58, chief prof educ, 58-65, DIR CTR TRAINING IN COMMUNITY PSYCHIAT, DEPT MENT HYG, STATE OF CALIF, 65-; CLIN PROF, SCH MED, UNIV CALIF, LOS ANGELES, 74- *Concurrent Pos:* Consult, Metrop State Hosp, 65-, Los Angeles County Ment Health Dept, 66-, Ment Health Develop Comn Los Angeles County, 66-, Family Serv Los Angeles, 70- & Psychiat Residency Training Prog, Olive View Hosp, 70-; mem fac eng mgt sem, Grad Sch Bus Admin, Univ Calif, Los Angeles, 66-73, assoc clin prof psychiat, Sch Med, 67-73, lectr community ment health, Sch Pub Health, 67-73; mem training fac, Gestalt Ther Inst Los Angeles, 69-; teaching consult, Sch Med, Univ Southern Calif, 70- *Honors & Awards:* Physician of the Year Award, Calif Governor's Comt Employ of Handicapped, 71. *Mem:* AAAS; fel Am Psychiat Asn. *Res:* Development of human potential; community service systems; consultation; sports and leisure; psychotherapy. *Mailing Add:* 12301 First Helena Dr Brentwood CA 90049

BEISSNER, ROBERT EDWARD, b San Antonio, Tex, Oct 27, 33; m 54; c 6. THEORETICAL SOLID STATE PHYSICS. *Educ:* St Mary's Univ, BS, 55; Tex Christian Univ, MA, 60, PhD(physics), 65. *Prof Exp:* Nuclear engr, Westinghouse Elec Corp, 55-56 & Gen Dynamics Corp, 56-69; PHYSICIST, SOUTHWEST RES INST, 69- *Mem:* Am Phys Soc; AAAS. *Res:* Theory of defects in metals; applications of scattering theory in nondestructive evaluation. *Mailing Add:* Southwest Res Inst PO Drawer 28510 San Antonio TX 78228

BEISTEL, DONALD W, b Sunbury, Pa, Feb 29, 36; m 63; c 1. PHYSICAL CHEMISTRY. *Educ:* Bucknell Univ, BS, 58; Univ Del, PhD(chem), 63. *Prof Exp:* Mem tech staff, Directorate Res & Develop, US Army Missile Command, Redstone Arsenal, Ala, 63-64; asst prof chem, Franklin & Marshall Col, 64-66 & Marshall Univ, 66-68; asst prof, 68-69, ASSOC PROF CHEM, UNIV MO-ROLLA, 69- *Mem:* AAAS; Am Chem Soc; Soc Appl Spectros. *Res:* Elucidation of molecular structure and electronic configurations of polynuclear aromatic compounds by application of nuclear magnetic resonance; applications of computer methods to chemistry; molecular orbital calculations on chemical carcinogens. *Mailing Add:* Dept Chem Univ Mo Rolla MO 65401

BEITCH, IRWIN, b Brooklyn, NY, Nov 28, 37; m 63; c 2. EMBRYOLOGY, CYTOLOGY. *Educ:* Univ Richmond, BS, 60, MS, 62; Univ Va, PhD(biol), 68. *Prof Exp:* NIH trainee ophthal res, Col Physicians & Surgeons, Columbia Univ, 67-69; from asst prof to assoc prof, 69-76, PROF BIOL, QUINNIPIAC COL, 76- *Mem:* AAAS; Am Soc Zool; Am Inst Biol Sci; NY Acad Sci. *Res:* Electron microscopy of keratinization and feather development; electron microscopy of eye tissues; histology; histochemistry and electron microscopy of endorphin secretion in earthworms. *Mailing Add:* Dept Biol Quinnipiac Col Hamden CT 06518-9987

BEITCHMAN, BURTON DAVID, b Philadelphia, Pa, May 1, 26; m 56; c 2. PROCESS DEVELOPMENT, ANALYTICAL DEVELOPMENT. *Educ:* Temple Univ, BA, 48; Rutgers Univ, MS, 52, PhD(org chem), 54; Widener Col, MBA, 73. *Prof Exp:* Chemist, Publicker Industs, Inc, 48-49; asst res specialist, Rutgers Univ, 54-56; chemist, Nat Bur Standards, 57-60; chemist, Air Prod & Chem Inc, 60-66, sr res chemist, 66-80, sr prin res chemist, 66-82; SUPVR CHEM DEVELOP LAB, ALLIED CORP, 84- *Mem:* Am Chem Soc. *Res:* Organic synthesis; isocyanates; polyurethanes; chemical additives; amine catalysts and curing agents; peroxide decomposition accelerators; new products and process research and development; fluorochemicals; oxidation reactions, homogeneous and heterogeneous catalysis; process safety, process development, analytical methods development. *Mailing Add:* 134 Parkview Dr Springfield PA 19064-1735

BEITINGER, THOMAS LEE, b Prairie du Chein, Wis, Mar 4, 45; m 67; c 2. PHYSIOLOGICAL ECOLOGY. *Educ:* Hamline Univ, BS, 67; Univ RI, MS, 69; Univ Wis-Madison, PhD(zool), 74. *Prof Exp:* Ecologist, Great Lakes Thermal Effects Prog, Argonne Nat Lab, 74-76; asst prof, 76-82, ASSOC PROF, DEPT BIOL SCI, UNIV NTEX, 82- *Mem:* Am Fisheries Soc; Sigma Xi. *Res:* Behavioral and physiological responses of animals to environmental variables; temperature preference, avoidance and regulation of ectothermic organisms; aquatic toxicology. *Mailing Add:* Dept Biol Sci Univ NTex Denton TX 76203

BEITINS, INESE ZINTA, b Riga, Latvia. PEDIATRIC ENDOCRINOLOGY. *Educ:* Univ Toronto, MD, 62; FRCP(C), 67. *Prof Exp:* Rotating intern med & surg, Toronto Western Hosp, 62-63; resident pediat, Hosp Sick Children, Toronto, 63-65; fel pediat, Johns Hopkins Hosp, 65-66; resident path & med, Hosp Sick Children & Toronto Gen Hosp, 66-67; staff physician pediat, Hosp Sick Children, 67-68; fel pediat endocrinol, Johns Hopkins Hosp, 68-71; Busswell fel & res asst prof pediat, State Univ NY Buffalo, 72-73; asst prof pediat, Harvard Med Sch, 73-79, assoc prof, 79-; asst gynec, Mass Gen Hosp, 77-; AT DEPT PEDIAT, UNIV MICH, ANN ARBOR. *Mem:* Lawson Wilkins Pediat Endocrine Soc; Am Fedn Clin Res; Endocrine Soc; NY Acad Sci; Royal Col Med London. *Res:* Adrenal steroid measurement and their role in hypertension, maternal fetal interrelationships and disorders such as congenital virilizing adrenal hyperplasia; Gonadotropin heterogeneity, subunits and biological versus immunological activity; effects of nutrition and exercise on reproduction. *Mailing Add:* Dept Pediat Mass Gen Hosp 32 Fruit St Boston MA 02114

BEITZ, ALVIN JAMES, b Meadville, PA, Feb 16, 49; US citizen; m 71; c 7. NEUROCYTOLOGY, NEUROCHEMISTRY. *Educ:* Gannon Univ, BS, 71; Univ Minn, PhD(anat), 76. *Prof Exp:* Fel neuroanat, Harvard Med Sch, 76-78; asst prof anat, Univ SC, 78-82; from asst prof to assoc prof, 82-88, PROF VET BIOL & NEUROSCI, UNIV MINN, 88- *Concurrent Pos:* Review ed, J Synapse, 87-; mem NIH study sect, 87. *Mem:* Soc Neurosci; Int Brain Res Orgn; AAAS; Am Asn Anatomists; Am Pain Soc. *Res:* Elucidate chemical framework of key brain regions involved in pain, pain modulation, motor coordination, learning and memory; aging and alzheimers disease; mechanics underlying epilepsy. *Mailing Add:* Dept Vet Biol Univ Minn 1988 Fitch Ave St Paul MN 55108

BEITZ, DONALD CLARENCE, b Stewardson, Ill, Mar 30, 40; m 63; c 2. NUTRITIONAL BIOCHEMISTRY. *Educ:* Univ Ill, BS, 62, MS, 63; Mich State Univ, PhD(biochem, dairy sci), 67. *Prof Exp:* Ralston Purina fel, 62-65; NIH fel, 65-67; from asst prof to assoc prof, 67-77, PROF ANIMAL SCI & BIOCHEM, IOWA STATE UNIV, 77-, DISTINGUISHED PROF AGR, 89- *Honors & Awards:* Nutrit Res Award, Am Feed Indust Asn & Am Dairy Sci Asn. *Mem:* AAAS; Am Dairy Sci Asn; Am Soc Animal Sci; Am Inst Nutrit; Sigma Xi; Am Soc Biochem & Molecular Biol. *Res:* Lipid metabolism in farm animals; ketosis in dairy cattle; catecholamine metabolism in pigs; etiology of milk fever in dairy cattle; mitochondrial DNA and milk production. *Mailing Add:* Dept Animal Sci Iowa State Univ Ames IA 50011

BEIZER, BORIS, b Brussels, Belg, June 25, 34; US citizen; m 55; c 2. COMPUTER & INFORMATION SCIENCE. *Educ:* City Col NY, BS, 56; Univ Pa, MS, 62, PhD(comput sci), 66. *Prof Exp:* Technician med instrumentation, Columbia-Presby Med Ctr, 54-56; sr engr, Repub Aviation Corp, NY, 56-58; prin engr, Airborne Instrument Labs, 58-59; proj engr, comput div, Philco Corp, Pa, 59-61; chief logic design, Navig Comput Corp, 61-63; sr staff scientist, Pa Res Assocs, 63-66; chief scientist, Data Systs Analysts, Inc, Pennsauken, 66-87; CONSULT, SOFTWARE TESTING & QUAL ASSURANCE, 83- *Mem:* Inst Elec & Electronics Engrs; Asn Comput Mach; Asn Software Test Eng. *Res:* Computer system architecture; performance analysis and modeling; computer controlled communications; testing and quality assurance. *Mailing Add:* 1232 Glenbrook Rd Huntingdon Valley PA 19006

BEIZER, LAWRENCE H, b Scranton, Pa, Feb 17, 09; m 51; c 3. INTERNAL MEDICINE, HEMATOLOGY. *Educ:* Univ Pa, BS, 30; Harvard Med Sch, MD, 34; Univ Minn, MS, 42. *Prof Exp:* Intern, Philadelphia Gen Hosp, 34-36; resident clin path, Hosp Univ Pa, 36-37; fel, Mayo Clin, 37-43; instr med, Sch Med, Univ Pa, 46-49, assoc, 49-50, from asst prof to prof, 50-77, dir, Div Hemat, 50-70 & Div Oncol, 70-75, co-dir, Div Hemat & Oncol, Grad Hosp, 75-79, EMER PROF CLIN MED, SCH MED, UNIV PA, 77-, CONSULT, DIV HEMAT & ONCOL, GRAD HOSP, 79- *Concurrent Pos:* Assoc med, Woman's Med Col, 46-48, from clin asst prof to clin assoc prof, 48-56, consult, Hosp, 47-56; consult, Valley Forge Army Hosp, Phoenixville, 47-74, Vet Hosp, Philadelphia, 53-56, Walson Army Hosp, Ft Dix, NJ, 60-63 & Vet Hosp, Wilmington, Del, 62-66; physician, Presby-Univ Pa Med Ctr, 70- *Mem:* Fel Am Col Physicians; Am Soc Hemat; Am Soc Clin Oncol; Sigma Xi. *Res:* Clinical hematology, especially bone marrow aspiration biopsy and mutiple myeloma; oncology. *Mailing Add:* 1840 South St 2nd Floor Philadelphia PA 19146

BEJNAR, WALDEMERE, b Hamtramck, Mich, Feb 7, 20; m 46; c 5. GEOLOGY. *Educ:* Univ Mich, BS, 43, MA, 47; Univ Ariz, PhD(geol), 50; Univ NMex, MS, 65. *Prof Exp:* Geologist, US Dept Reclamation, 50 & Foreign Sect, US Geol Surv, Nigeria, 50-52; asst prof geol, NMex Inst Mining & Technol, 52-55; consult geologist, 55-58; assoc sponsor, Dale Carnegie & Assocs, 58-63; prof geol & geog, NMex Highlands Univ, 65-80. *Concurrent Pos:* Pres, Waldemere Bejnar & Assocs, Inc, 59-; consult geologist, 58- *Mem:* Am Inst Prof Geol; Sigma Xi. *Res:* Ore deposition; ground water; landslides. *Mailing Add:* Box 94 Star Rte 2 Socorro NM 87801

BEKEFI, GEORGE, b Prague, Czech, Mar 14, 25; m 61, 69; c 2. PHYSICS. *Educ:* London Univ, BSc, 48; McGill Univ, MSc, 50, PhD(physics), 52. *Prof Exp:* Res assoc, McGill Univ, 52-55, asst prof physics, 55-57; res assoc plasma physics, Res Lab Electronics, 57-61, from asst prof to assoc prof, 61-68, PROF PHYSICS, MASS INST TECHNOL, 68- *Concurrent Pos:* Chmn, Plasma Physics Div, Am Physical Soc, 78-79; Guggenheim fel, 72-73. *Honors & Awards:* Plasma Sci & Appln Prize, Inst Elec & Electronics Engrs, 89. *Mem:* Fel Am Phys Soc. *Res:* Microwave antennae and propagation; plasma physics; free electron lasers. *Mailing Add:* Dept Physics Mass Inst Technol Rm 36-213 Mass Ave Cambridge MA 02139

BEKENSTEIN, JACOB DAVID, b Mexico City, Mex, May 1, 47; US & Israeli citizen; m 78; c 3. RELATIVITY, ASTROPHYSICS. *Educ:* Polytech Inst Brooklyn, BS & MS, 69; Princeton Univ, PhD(physics), 72. *Prof Exp:* Res assoc, Ctr Relativity Theory, Dept Physics, Univ Tex, Austin, 72-74; sr lectr, Dept Physics, Ben-Gurion Univ Negev, Israel, 74-76, from assoc prof to prof, 76-91, arnow chair astrophys, 83-91; PROF, RACAH INST PHYSICS, HEBREW UNIV, ISRAEL, 91- *Concurrent Pos:* Vis prof, Univ Calif, Santa Barbara, 80, Princeton Univ, 80, Univ Tex, Austin, 86 & Univ Toronto, 87; steering comt, Jerusalem Winter Sch Theoret Physics; mem, Int Comt Gen Relativity. *Honors & Awards:* Bergmann Prize Sci, Israel, 77; Landau Prize Res, Israel, 81; Gravity Res Found Prize, 81; Rotschild Prize Physics, Israel, 88. *Mem:* Israeli Phys Soc; Int Astron Union; Soc Gen Relativity. *Res:* Physical properties of collapsed stars; thermodynamics of black holes; variability of particle properties; relativistic magnetohydrodynamics; gravitational effects on quantum processes; formation of galaxies, cosmology information theory. *Mailing Add:* Racah Inst Physics Dept Physics Hebrew Univ Givat Rarn Jerusalem 91904 Israel

BEKERSKY, IHOR, b USSR, Oct 25, 40; US citizen; m 65; c 3. BIOCHEMISTRY. *Educ:* Hunter Col, BA, 61; NY Univ, MS, 71, PhD(biochem), 78. *Prof Exp:* Res technician, Dept Environ Med, NY Univ, 61-64; res assoc, Pub Health Res Inst, 68-71; asst scientist drug metab, 71-72, assoc scientist, 72-75, scientist, 75-78, sr scientist biopharmaceuts, 78-81, ASST RES GROUP CHIEF BIOPHARAMACEUT, DEPT PHARMACOKINETICS, HOFFMAN LA ROCHE, INC, 81- *Mem:* Acad Pharmaceut Sci; Am Soc Pharmacol & Exp Therapeut; Sigma Xi. *Res:* Renal drug disposition and factors affecting the mechanisms of drug elimination; preclinical animal models to study drug disposition and pharmacokinetics. *Mailing Add:* 5895 Greenclover Lane Madison WI 53711

BEKEY, GEORGE A(LBERT), b Czech, June 19, 28; US citizen; m 51; c 2. COMPUTER SCIENCE. *Educ:* Univ Calif, BS, 50; Univ Calif, Los Angeles, MS, 52, PhD(eng), 62. *Prof Exp:* Tech asst & res engr, Sch Eng, Univ Calif, Los Angeles, 50-54; mgr, Computer Ctr, Beckman Instruments, Los Angeles, 56-58; group leader, sect head & sr staff engr, TRW Systs, Los Angeles, 58-62; chmn, Elec Eng Systs Dept, Univ Southern Calif, 78-82, dir, Robotics Inst, 83-86, chmn, Computer Sci Dept, 84-89, from asst prof to assoc prof, 62-68, PROF COMPUTER SCI, ELEC ENG & BIOMED ENG, UNIV SOUTHERN CALIF, 68- *Concurrent Pos:* Nat lectr, Sigma Xi, 76-77; ed, Trans Robotics & Automation, Inst Elec & Electronics Engrs, 85-; assoc ed, Math & Computers Simulation & Trans Soc Computer Simulation; grants, NSF, 87-89, 87- & 90-, NASA-Jet Propulsion Lab, 83-89, Kaprielian Res Innovation Fund, 89-; dir, Ctr Mfg Automation Res, Univ Southern Calif. *Mem:* Nat Acad Eng; fel Inst Elec & Electronics Engrs; fel AAAS; Asn Comput Mach; Am Asn Artificial Intel; Soc Computer Simulation; Int Neural Network Soc. *Res:* Intelligent robotic systems; artificial intelligence and robotics in medicine; biological signal processing; autonomous, intelligent robots equipped with sensory feedback which are capable of adaption to environmental change. *Mailing Add:* Dept Comput Sci Univ Southern Calif 941 W 37th Pl Los Angeles CA 90089-0782

BEKHOR, ISAAC, b Israel; US citizen. IN SITU HYBRIDIZATION, DNA CLONING. *Educ:* Univ Calif, Los Angeles, BS, 61; Univ Southern Calif, PhD(biochem), 66. *Prof Exp:* PROF MOLECULAR BIOL & BIOCHEM, UNIV SOUTHERN CALIF, LOS ANGELES, 70- *Concurrent Pos:* NIH career develop award, 70-75; consult, Scripps Labs, San Diego, 84- *Mem:* Asn Res Vision & Ophthal; Int Soc Eye Res; Fedn Biol Soc. *Res:* Expression of eye lens specific genes in both normal and cataractous lens; enzymology; clinical diagnostics. *Mailing Add:* Sch Dent Univ Calif Los Angeles CA 90089-0641

BEKOFF, ANNE C, b Denver, Colo, May 19, 47; m 70. DEVELOPMENTAL NEUROBIOLOGY, NEUROETHOLOGY. *Educ:* Smith Col, BA, 69; Wash Univ, St Louis, Mo, PhD(neurobiol & develop), 74. *Prof Exp:* Res assoc neurophysiol, 75-76, asst prof, 76-82, assoc prof, 82-89, PROF, DEPT ENVIRON POP & ORGANISMIC BIOL, UNIV COLO, 89- *Concurrent Pos:* Alfred P Sloan Found res fel, 79-81; John Simon Guggenheim res fel, 83-84; mem coun, Int Soc Neuroethol, 90-95. *Mem:* Soc Neurosci; fel AAAS; Int Brain Res Orgn; Animal Behav Soc; Int Soc Neuroethol. *Res:* Studies of the ontogeny of the motor control mechanisms underlying the production of coordinated behaviors in embryonic and posthatching chicks using neurophysiological, kinematic and behavioral techniques. *Mailing Add:* Dept of Environ Pop & Organismic Biol Univ Colo Boulder CO 80309-0334

BEKOFF, MARC, b Brooklyn, NY, Sept 6, 45; m 70. ANIMAL BEHAVIOR. *Educ:* Washington Univ, AB, 67, PhD(animal behav), 72; Hofstra Univ, MA, 68. *Prof Exp:* Asst prof biol, Univ Mo, St Louis, 73-74; asst prof, 74-80, PROF BIOL, UNIV COLO, BOULDER, 80- *Concurrent Pos:* Guggenheim Mem Found fel, 81. *Mem:* Animal Behav Soc; Am Soc Zoologists; Am Soc Mammalogists; US Antarctic Res Prog; Sigma Xi. *Res:* Social ecology of mammals, especially canids; mathematical models of social interaction; animal social development; communication processes. *Mailing Add:* Dept Environ Pop & Org Biol Univ Colo Boulder Box 334 Boulder CO 80309-0334

BELADY, LASZLO ANTAL, b Budapest, Hungary, Apr 29, 28; m 59; c 2. SOFTWARE SYSTEMS, COMPUTER SCIENCES. *Educ:* Tech Univ, Budapest, BS, 49, MS, 50. *Prof Exp:* Mem res staff comput sci, IBM, T J Watson Res Ctr, Yorktown, NY, 61-84; vpres software eng, Microelectronics & Comput Technol Corp, Austin, Tex, 84-91; CHMN & DIR, MITSUBISHI ELEC RES LABS, INC, 91- *Concurrent Pos:* Vis prof, Univ Calif, Berkeley, 71-72, Imperial Col, London Univ, 74; adj prof, NY Polytech Univ, 81-83. *Honors & Awards:* Warnier Prize, 90. *Mem:* Fel Inst Elec & Electronics Engrs; Asn Comput Mach. *Res:* Design of complex systems, particularly software systems which are controlling distributed, real time processes. *Mailing Add:* Mitsubishi Elec Res Labs Inc 201 Broadway Cambridge MA 02139

BELAND, GARY LAVERN, b Los Angeles, Calif, May 18, 42; m 85; c 3. ECONOMIC ENTOMOLOGY. *Educ:* Kearney State Col, BA, 65; Univ Nebr, Lincoln, MS, 68, PhD(entom), 72. *Prof Exp:* Instr entom, Univ Nebr, Lincoln, 69-72; res entomologist, USDA Sci & Educ Admin-Agr Res, 72-79; res entomologist, Funk Seeds Int, 79-84, mgr entom-path res, 84-87, mgr res support technol, 88; RES ENTOMOLOGIST, CIBA GEIGY SEED DIV, 89- *Mem:* Entom Soc Am; Am Registry Prof Entomologists; Sigma Xi. *Res:* Host plant resistance; biology and ecology of insect pests of corn, soybeans, sorghum and cotton. *Mailing Add:* Ciba Geigy Seed Div Box 2911 Bloomington IL 61701-2911

BELAND, JACQUES (ROBERT), geology, for more information see previous edition

BELAND, PIERRE, b Quebec, Que, Oct 4, 47; div; c 2. ECOLOGY, MARINE SCIENCES. *Educ:* Laval Univ, BA, 67, BSc, 71; Dalhousie Univ, PhD(pop biol), 74. *Prof Exp:* Consult ecologist, Nat Mus Natural Sci, Ottawa, 76-77, paleocologist, 77-80; ecologist, Can Dept Fisheries & Oceans, Rimo01ski, Que, 81-82, chief, Fisheries Ecol Res Ctr, 82-87; SCI DIR, ST LAWRENCE INST ECOTOXICOL, MONTREAL, 87- *Concurrent Pos:* Fel, Ctr Overseas Off Sci & Tech Res, New Caledonia, 74-75 & Dept Zool, Univ Queensland, 75; assoc prof, Univ Que Rimouski; invited prof, Nat Inst Sci Res-Oceanol, Rimouski. *Res:* Marine ecosystems: population dynamics, predation, food chains; contamination of marine mammals and effects on stocks; organic contaminants flow. *Mailing Add:* St Lawrence Inst Ecotoxicol 3872 Barc Lefontaine Montreal PQ H2L 3M6 Can3

BELANGER, ALAIN, b L'Islet, Que, June 30, 47; m 70; c 3. ENDOCRINOLOGY, CHEMISTRY. *Educ:* Laval Univ, BSc, 70, MSc, 71, PhD(org chem), 73. *Prof Exp:* SR RESEARCHER, LAB MOLECULAR ENDOCRINOL, LAVAL UNIV, 77-, ASSOC PROF. *Mem:* Fel Soc Can de Rech Clinique; Am Soc Andrology. *Res:* Steroid metabolism and ria. *Mailing Add:* Ecole de Readaptation Citee Univ Quebec PQ G1K 7P4 Can

BELANGER, BRIAN CHARLES, b Eveleth, Minn, Feb 2, 41; m 62; c 2. INSTRUMENTATION, ELECTRICAL ENGINEERING. *Educ:* Calif Inst Technol, BS, 63; Univ Southern Calif, PhD(elec eng), 68. *Prof Exp:* Res staff mem, Gen Elec Res & Develop Ctr, 68-72; chief, Advan Energy Delivery Systs Br, US Energy Res Develop Admin, 72-76; chief, Off Measurement Serv, US Nat Bur Standards, 77-83, liaison, Dept Defense, 83-86, assoc dir prog develop, Ctr Electronics & Elec Eng, 86-90,; DEP DIR, ADVAN TECHNOL PROG, NAT INST STANDARDS & TECHNOL, 90- *Concurrent Pos:* Fel, US Com Dept Sci & Technol, 83-84. *Honors & Awards:* Bronze Medal, US Dept Com, 80. *Mem:* Inst Elec & Electronics Engrs. *Res:* Measurement accuracy; metrology; standards. *Mailing Add:* 616 Nelson St Rockville MD 20850

BELANGER, DAVID GERALD, b Rockville, Conn, Dec 8, 44; m 71; c 2. SYSTEMS THEORY. *Educ:* Union Col, BS, 66; Case Western Reserve Univ, MS, 68, PhD(math), 71. *Prof Exp:* From asst prof to assoc prof, Univ S Ala, 71-79; DEPT HEAD, AT&T BELL LABS, 79- *Concurrent Pos:* Mathematician, US Army CEngr, 73-79. *Mem:* Soc Indust & Appl Math; Asn Comput Mach; Inst Elec & Electronics Engrs Comput Soc. *Res:* Automatic program generation; database management; software design and development; system simulation. *Mailing Add:* 20 Royce Brook Rd Belle Mead NJ 08502

BÉLANGER, JACQUELINE M R, b Mont-Joli, Que, Jan 8, 54. RESEARCH MANAGEMENT. *Educ:* Univ Moncton, BSc, 76, BEd, 77, MSc, 78; Carleton Univ, PhD(chem), 84. *Prof Exp:* Res scientist, Drug Id Div, Health & Welfare Can, 84-85; sci adv mgt, Indust, Sci & Technol CDN, 89-90; res scientist, Food Res Ctr, 85-89, ASST DEP DIR, LAND RESOURCE RES CTR, AGR CAN, 90- *Concurrent Pos:* Managing ed, Spectros: An Int J, 81-91; teaching master, Algonquin Col, Ottawa, 85, sci prog adv, 85; mem, Nat Adv Comt Agr & Natural Resources, Can Broadcasting Corp, 88-90; res assoc, Fac Grad Studies, Univ Moncton, 89-92; mem, Food Conserv & Innocuity Eval Comt, Corpaq, 90. *Honors & Awards:* Gold Medal, Soc Chem Indust, 76. *Mem:* Chem Inst Can; Can Soc Chem; Am Chem Soc; Am Soc Mass Spectrometry. *Res:* Extraction, isolation, purification and structure elucidation of natural compounds of biological interest from various sources; spectroscopy; chromatography. *Mailing Add:* 102 de Tadoussac Gatineau PQ J8T 7K1 Can

BÉLANGER, LUC, b Montreal, Que, July 2, 48. ONCOLOGY, DEVELOPMENTAL BIOLOGY. *Educ:* Petit Séminaire Chicoutimi, BA, 67; Laval Univ, MD, 71, PhD(biochem), 75. *Prof Exp:* Postdoctoral biochem & molecular biol, Med Res Coun-Univ Calif, San Diego, Salk Inst, Vanderbilt Univ, 75-78; from asst prof to assoc prof, 78-88, PROF BIOCHEM & MOLECULAR BIOL, LAVAL UNIV, 88-, DIR ONCOL, CANCER RES CTR, 84-; DIR ONCOL, L'HÔTEL-DIEU QUE RES CTR, 81- *Concurrent Pos:* Scholar molecular biol, Med Res Coun Can, 78-83; mem bd, Int Soc Oncodevelop Biol & Med, 80-90; mem, cancer grants panel, Med Res Coun, 80-83, scholar panel, Nat Cancer Inst, 80-82, fac med coun Laval Univ, 81-87, adv comt cancer, Que Ministry Health, 90-91; mem coun, Med Res Coun Can, 88-, Nat Cancer Inst Can, 88-; pres, scholar bd, Fonds Recherche Santé Que (FRSQ), 88-91, Med Res Coun Bd Eval Grants Progs Performances, 90- *Mem:* Can Soc Cellular & Molecular Biol; Int Soc Oncodevelop Biol & Med; Am Soc Cell Biol. *Res:* Cellular and molecular mechanisms underlying differentiation and transformation of hepatocytes; hereditary tyrosinemia; breast cancer. *Mailing Add:* L'Hôtel-Dieu de Que Res Ctr Quebec PQ G1R 2J6 Can

BELANGER, PATRICE CHARLES, b Montreal, Que, Dec 31, 40; m 64; c 3. ANALYTICAL CHEMISTRY, MEDICINAL CHEMISTRY. *Educ:* Univ Montreal, BSc, 62, MSc, 63, PhD(org chem), 66. *Prof Exp:* Sr res chemist, 66-77, res fel, 77-81, sr res fel, DIR QUAL ASSURANCE, MERCK FROSST LABS, PQ, CAN, 87. *Mem:* Chem Inst Can; Am Chem Soc; Ordre des Chimistes du Quebec. *Res:* Allergic asthma; muscle relaxants and analgesics; anti-inflammatories. *Mailing Add:* Merck Frosst Labs PO Box 1005 Pointe Claire-Dorval PQ H9R 4P8 Can

BELANGER, PIERRE ANDRE, b Pointe-au-Pic, Que, Nov 1, 41; m 66; c 1. LASERS. *Educ:* Univ Laval, BSc, 66, DSc, 71. *Prof Exp:* Res & develop specialist lasers, Gen-Tec, Inc, 71-72; res asst, 70-71, ASST PROF PHYSICS, LAVAL UNIV, 72- *Mem:* Optical Soc Am; Can Asn Physicists; Inst Elec & Electronics Engrs; Soc Photo-Optical Instrumentation Engrs. *Res:* Generation of very high peak power laser pulse; optics of high peak power laser beam. *Mailing Add:* Lab Res Optics & Laser Dept Physics Laval Univ Quebec PQ G1K 7P4 Can

BELANGER, PIERRE ROLLAND, b Montreal, Que, Aug 18, 37; m 63; c 3. ELECTRICAL ENGINEERING. *Educ:* McGill Univ, BEng, 59; Mass Inst Technol, SM, 61, EE, 62, PhD(elec eng), 64. *Prof Exp:* From instr to asst prof elec eng, Mass Inst Technol, 63-65; systs analyst, Foxboro Co, 65-67; assoc prof, 67-77, chmn, 78-84, PROF ELEC ENG, MCGILL UNIV, 77-, DEAN FAC, 84- *Concurrent Pos:* Mem, adv comt eng, Natural Sci & Eng Res Coun Can, 83-87; mem, Nat Adv Bd Sci & Technol, 87- *Honors & Awards:* Centennial Medal, Inst Elec & Electronics Engrs, 84. *Mem:* Inst Elec & Electronics Engrs. *Res:* Control engineering. *Mailing Add:* Dept Elec Eng McGill Univ Sherbrooke St W Montreal PQ H3A 2M5 Can

BELANGER, THOMAS V, b Carrington, NDak, May 25, 48; m. LIMNOLOGY. *Educ:* Ill State Univ, BS, 71; Univ Fla, MS, 74, PhD(environ eng sci), 79. *Prof Exp:* Aquatic biologist, Environ Sci & Eng, Inc, 71-74; asst prof, 78-79, ASSOC PROF ENVIRON SCI, FLA INST TECHNOL, 82- *Concurrent Pos:* Prin investr, Lake Washington Water Qual Study, Dept Environ Regulation, 78- & Upper St John's Diag Study, Environ Protection Agency, 81-82; consult groundwater seepage, E Lake Tohopekaliga & oxygen budgets, Everglades Water Conserv Areas, SFla Water Mgt Dist. *Mem:* Am Water Resources Asn; Am Soc Limnol & Oceanog; Sigma Xi. *Res:* Limnology; aquatic ecology; groundwater/surface water interaction; gas transfer at water surfaces. *Mailing Add:* Marine & Environ Studies Fla Inst Technol Melbourne FL 32901

BELARDINELLI, LUIZ, CARDIOLOGY, ELECTROPHYSIOLOGY. *Educ:* Med Cath Fac Found, Brazil, MD, 75. *Prof Exp:* ASSOC PROF MED & PHYSIOL, MED CTR, UNIV VA, 76- *Mailing Add:* Div Cardiol Univ Va Med Ctr Box 456 Charlottesville VA 22908

BELBECK, L W, b London, Ont, Aug 24, 43; c 2. VETERINARY MEDICINE. *Educ:* Ont Vet Col, DVM, 66. *Prof Exp:* Vet, Oakridge Animal Clin, 66-67; VET, HEALTH ANIMALS BR, CAN DEPT AGR, 66- *Mem:* Can Coun Animal Care; Can Asn Lab Animal Sci; Can Asn Lab Sci; Can Fedn Humane Soc; Can Vet Med Asn. *Mailing Add:* Seven Delbrook Ct Hamilton ON L8S 2B9 Can

BEL BRUNO, JOSEPH JAMES, b Passaic, NJ, June 30, 52; m 80; c 1. REACTION DYNAMICS, LASER PHOTOCHEMISTRY. *Educ:* Seton Hall Univ, BS, 74; Rutgers Univ, PhD(phys chem), 80. *Prof Exp:* Res chemist, Am Cyanamid Co, 74-76; res assoc, Princeton Univ, 80-82; asst prof, 82-88, ASSOC PROF CHEM, DARTMOUTH COL, 88- *Concurrent Pos:* Alexander Von Humboldt Fel, 88. *Mem:* Sigma Xi; Am Chem Soc; Am Phys Soc. *Res:* Laser photochemistry and spectroscopy of small organic molecules; reactions in intense, non-resonant laser fields; chemical dynamics of gas phase reactions; time-of-flight mass spectrometry. *Mailing Add:* Dept Chem Dartmouth Col Hanover NH 03755

BELCASTRO, PATRICK FRANK, b 1920; US citizen; m 63; c 2. PHARMACY. *Educ:* Duquesne Univ, BS, 42; Purdue Univ, MS, 51, PhD(pharm), 53. *Prof Exp:* Instr pharm & chem, Duquesne Univ, 46-49; asst prof pharm, Ohio State Univ, 53-54; asst prof pharm, 54-76, prof indust pharm, 76-90, EMER PROF INDUST PHARM, PURDUE UNIV, WEST LAFAYETTE, 90- *Concurrent Pos:* Contrib ed, Int Pharmaceut Abstr, Pharmaceut Technol. *Mem:* Am Pharmaceut Asn; Am Inst Hist Pharm; Am Asn Univ Professors. *Res:* Effects of x-irradiation on pharmacological action of certain drugs. *Mailing Add:* Sch Pharm & Pharmacal Sci Purdue Univ West Lafayette IN 47907

BELCHER, BASCOM ANTHONY, b Cairo, Ga, Sept 28, 02; m 66. SUGAR SORGHUM RESEARCH, SUGAR CANE BREEDING. *Educ:* Emory Univ, BS, 28, MS, 29. *Prof Exp:* Asst agronomist sugar plant invest, USDA, 29-32, assoc agronomist, 32-40, agronomist, 40-42, sr agronomist, 46-69, consult sugarcane res, 69-70; RETIRED. *Mem:* Hon mem Am Soc Sugar Cane Technologists; AAAS; Int Soc Sugar Cane Technologists; Am Soc Sugarcane Technologists. *Res:* Sugarcane breeding and variety development. *Mailing Add:* 19 West S Temple Suite 700 Salt Lake City UT 84101

BELCHER, MELVIN B, b Shawnee, Okla, Sept 4, 25; m 48; c 3. ELECTRICAL ENGINEERING. *Educ:* Univ Calif, Berkeley, BS, 51; Univ Southern Calif, MS, 69. *Prof Exp:* Test engr, Gen Elec Co, NY, 51-52, anal engr, 52-54, engr, 54-55, sales engr, Calif, 55-58; asst prof elec eng, Calif State Polytech Univ, Pomona, 58-59; mgr sales, Gen Elec Co, Nev, 59-61; from asst prof to assoc prof Calif State Polytech Univ, Pomona, 61-69, assoc dean eng, 75-80, prof elec eng, 69-85, dir, int prog, 80-85, emer prof, 85-; RETIRED. *Concurrent Pos:* Mem Calif State Polytech Univ, Pomona, Tech Col Prog, Dar es Salaam, Tanzania, Africa, 65-67 and 70-72. *Mem:* Am Soc Eng Educ; Inst Elec & Electronics Engrs; Inst Advan Eng; Nat Soc Prof Engrs. *Res:* Network analysis and synthesis; electrical machines; precision measurements, power system analysis. *Mailing Add:* 7475 Brydon Rd LaVerne CA 91750

BELCHER, ROBERT ORANGE, b Williamsburg, Ky, June 29, 18; m 38; c 2. BOTANY. *Educ:* Berea Col, AB, 38; Univ Mich, MS, 47, PhD(bot), 55; Mich Tech Univ, MS, 82. *Prof Exp:* Asst biol, Purdue Univ, 38-40; fel bot, Univ Mich, 46; from asst prof to prof biol, Eastern Mich Univ, 46-80, head dept, 58-66; RETIRED. *Concurrent Pos:* Staff scientist, Legis Office Sci Adv, Lansing, Mich, 81; curric consult, Fac Sci & Educ, Sana'a Univ, North Yemen, 80. *Honors & Awards:* US Typhus Comn medal, 46. *Mem:* AAAS; Int Asn Plant Taxon; Am Soc Plant Taxon. *Res:* Systematic botany of pantropical composite weeds; all species of Senecio in Australasia; utilization of fuel wood. *Mailing Add:* PO Box 242 Ypsilanti MI 48197-0242

BELDEN, DON ALEXANDER, JR, b Akron, Ohio, July 13, 26; m 52; c 2. ZOOPHYSIOLOGY. *Educ:* Middlebury Col, BA, 50; Williams Col, MA, 52; Wash State Univ, PhD(zool), 58. *Prof Exp:* Asst zool, Middlebury Col, 49-50; asst biol, Williams Col, 50-52; res fel zool, Wash State Univ, 52-56; asst prof, 56-65, ASSOC PROF ZOOL, UNIV DENVER, 65- *Mem:* Sigma Xi. *Res:* Cellular biology. *Mailing Add:* Dept Biol Sci Univ Denver Denver CO 80208

BELDEN, EVERETT LEE, b Bridgeport, Nebr, Aug 14, 38; m 61, 81; c 3. VETERINARY IMMUNOLOGY, VETERINARY MICROBIOLOGY. *Educ:* Univ Wyo, BS, 60, MS, 62; Univ Calif, Davis, PhD(microbiol), 71. *Prof Exp:* Instr microbiol, 62-65, from asst prof to assoc prof, 70-82, PROF IMMUNOL, UNIV WYO, 82- *Mem:* Am Soc Microbiol; Am Asn Vet Immunologists. *Res:* Humoral and cellular immune response in the bovine; infectious reproductive diseases in cattle and sheep; chlamydial diseases in domestic animals; monoclonal antibodies; reproductive immunology. *Mailing Add:* Dept Vet Sci Univ Wyo Laramie WY 82071-3354

BELDING, RALPH CEDRIC, poultry pathology; deceased, see previous edition for last biography

BELEW, JOHN SEYMOUR, b Waco, Tex, Nov 3, 20; m 44; c 2. ORGANIC CHEMISTRY. *Educ:* Baylor Univ, BS, 41; Univ Wichita, MS, 47; Univ Wis, PhD(chem), 51. *Prof Exp:* Res assoc org chem, Brown Univ, 51-53; actg asst prof chem, Univ Va, 53-56; from asst prof to assoc prof, Baylor Univ, 56-63, dean, Col Arts & Sci, 74-79, asst vpres undergrad affairs, 78-79, vpres acad affairs, 79-83, provost, 83-87, exec vpres acad affairs, 87-89, PROF CHEM, BAYLOR UNIV, 63-, VPRES & PROVOST, 90- *Mem:* AAAS; Am Chem Soc; Royal Soc Chem. *Res:* Reactions of ozone; heterocycle and o-quinone syntheses; oxidations with transition metal oxides. *Mailing Add:* VPres & Provost Baylor Univ Waco TX 76703

BELEW-NOAH, PATRICIA W, b Memphis, Tenn, June 12, 40; m 59, 84; c 2. PATHOLOGY, MICROBIOLOGY. *Educ:* Univ Tenn, BS, 61, PhD(path), 78. *Prof Exp:* Med technologist microbiol, City Memphis Hosp, Tenn, 61-69; res technician, Div Infectious Dis, 69-74 & Div Dermat, 74-78, INSTR MICROBIOL, DIV DERMAT, UNIV TENN, MEMPHIS, 78- *Concurrent Pos:* Dermat Found grant, 78-79. *Mem:* Soc Invest Dermat. *Res:* Activation of alternative complement pathway; normal cutaneous microflora; psoriasis and streptococcal associated psoriasis, seborrheic dermatitis. *Mailing Add:* 2749 Scarlet Rd Memphis TN 38138

BELFORD, GENEVA GROSZ, b Washington, DC, May 18, 32; m 54. COMPUTER SCIENCE. *Educ:* Univ Pa, BA, 53; Univ Calif, Berkeley, MA, 54; Univ Ill, PhD(comput theory), 60. *Prof Exp:* NIH fel, 61-64; res assoc chem, Univ Ill, Urbana-Champaign, 59-61, asst prof math, 64-72, res asst prof, Ctr Advan Comput, 72-77, assoc prof computer sci, 77-82, PROF COMPUTER SCI, UNIV ILL, URBANA-CHAMPAIGN, 82- *Mem:* Inst Elec & Electronics Engrs; Asn Comput Mach. *Res:* Analysis of computer systems, especially data management systems and distributed systems. *Mailing Add:* 2318 DCL Dept Comput Sci Univ Ill 1304 W Springfield Ave Urbana IL 61801

BELFORD, JULIUS, b Brooklyn, NY, Feb 3, 20; m 55; c 3. PHARMACOLOGY. *Educ:* Brooklyn Col, BA, 40; Long Island Univ, BS, 43; Yale Univ, PhD(pharm), 48. *Prof Exp:* From asst prof to assoc prof pharm, 49-69, asst to dean, 58-61, actg chmn dept pharmacol, 71-72, PROF PHARM, STATE UNIV NY DOWNSTATE MED CTR, 69, ASSOC DEAN OPERS, 77- *Concurrent Pos:* Exchange grant, St Mary's Hosp Med Sch, London, 52-53; USPHS spec fel, Mario Negri Inst Pharmacol Res, Milan, Italy, 65. *Mem:* Fel AAAS; Am Soc Pharmacol; Harvey Soc; Asn Am Med Cols. *Res:* Pharmacology of drugs affecting the cardiovascular and respiratory systems; experimental epilepsy and anticonvulsants; medical education. *Mailing Add:* RD 1 Box 332 Chenango Forks NY 13746

BELFORD, R LINN, b St Louis, Mo, Dec 13, 31; m 54. PHYSICAL CHEMISTRY, INORGANIC CHEMISTRY. *Educ:* Univ Ill, BS, 53; Univ Calif, Berkeley, PhD(chem), 55. *Prof Exp:* Chemist, Univ Ill, 52; asst, Univ Calif, 53, chemist radiation lab, 53-54; from instr to assoc prof, 55-81, PROF PHYS CHEM, UNIV ILL, URBANA, 81- *Concurrent Pos:* Sloan fel, 61-63; mem biophys & biophys chem fel comt, NIH, 65-69, sr fel, 66-67; prin investr, NSF, 58- *Mem:* AAAS; Am Phys Soc; Am Chem Soc; Soc Appl Spectros; Sigma Xi. *Res:* Physical and inorganic chemistry; spectra and bonding of transition-metal chelate compounds; electron paramagnetic resonance and nuclear quadrupole coupling; molecules excited by hypersonic shock waves. *Mailing Add:* 366 C Noyes Lab Univ Ill Urbana IL 61803

BELFORT, GEORGES, b Johannesburg, SAfrica, May 8, 40; US citizen; m 67; c 3. SEPARATIONS TECHNOLOGY. *Educ:* Univ Cape Town, BSc, 63; Univ Calif, Irvine, MS, 69, PhD(eng), 72. *Prof Exp:* Res engr chem engr sparations, Astropower Labs, McDouglas Corp, 64-70; vis sr lectr separations & desalination, Univ Cape Town, 72; sr lectr environ & separations eng, Sch Appl Sci & Technol, Hebrew Univ Jerusalem, 73-78; assoc prof chem eng, 78-81, PROF CHEM ENG, RENSSELAER POLYTECH INST, 81- *Concurrent Pos:* Mem bd adv, Nat Coun Res & Develop, Israel, 76-81; assoc prof civil eng, Northwestern Univ, 77-78; consult membranes, Amicon Corp, Mass, 78-; reviewer proposals, NSF, Dept Energy & NIH; consult, Danish Sugar Co, Denmark, 80-; water commr, SAfrica, 82-; W R Grace & Co & Lyonnaise de Eaux, France, Air Prod & Chem Inc, 88-, Dow Chem Co, 90- *Honors & Awards:* Plenary Lectr, NAm Membrane Soc, 88. *Mem:* Am Inst Chem Engrs; Am Chem Soc; NAm Membrane Soc. *Res:* Separations technology including cross-flow membrane processes (ultrafiltration and reverse osmosis); biochemical engineering; animal cell culture; thermodynamics and measurement of activated carbon absorption from aqueous phase; molecular surface force measurements; continuous bioreactor development; fluid mechanics. *Mailing Add:* Dept Chem Eng Rensselaer Polytech Inst Troy NY 12180-3590

BELFORT, MARLENE, b Cape Town, SAfrica, July 26, 45; US citizen; m 67; c 3. RECOMBINANT DNA. *Educ:* Univ Cape Town, BS, 66; Univ Calif, Irvine, PhD(microbiol genetics), 72. *Prof Exp:* Res assoc & teaching fel microbiol & chem, Hebrew Univ, Hadassah Med Sch, Israel, 73-77; res assoc molecular biol, Biochem Dept, Northwestern Univ, Evanston, Ill, 77-78; RES SCIENTIST, WADSWORTH LABS, NY STATE DEPT HEALTH, ALBANY, 78- *Concurrent Pos:* Assoc prof, Sch Pub Health Sci, State Univ, NY, 85-; adj assoc prof microbiol, Albany Med Col, NY, 84-; chairperson, Microbial Genetics Study Sect, NIH; ed, J Gene. *Mem:* Am Soc Microbiol; Am Soc Biochemists. *Res:* Regulation of gene expression in pro and eukaryotes; mobile DNA (DNA rearrangements); RNA splicing. *Mailing Add:* Wadsworth Ctr Lab Res NY State Dept Health Empire State Plaza Albany NY 12201

BELIAN, RICHARD DUANE, b Santa Fe, NMex, Feb 23, 38; m 62; c 3. PHYSICS. *Educ:* Univ NMex, BS, 65, MS, 67. *Prof Exp:* STAFF MEM SPACE PHYSICS, LOS ALAMOS NAT LAB, UNIV CALIF, 67- *Mem:* Am Astron Soc; Am Geophys Union. *Res:* Magnetospheric physics, substorm related particle events; solar energetic particle events. *Mailing Add:* Los Alamos Nat Lab Group SST-8 MS 438 Los Alamos NM 87545

BELILES, ROBERT PRYOR, b Louisville, Ky, Dec 21, 32; m 58; c 2. TOXICOLOGY. *Educ:* Univ Louisville, BA, 54, MS, 58; Iowa State Univ, PhD(pharmacol), 62. *Prof Exp:* Instr pharmacol, Iowa State Univ, 61-62; instr radiation biol, Univ Rochester, 62-64; pharmacologist, Woodard Res Corp, 64; chief environ toxicol div, 64-67; dir toxicol, Lakeside Labs, 67-75; assoc dir pharm toxicol, Litton Bionetics, Litton Indust, Inc, 75-77; dir toxicol, 77-80; MEM STAFF, OCCUP HEALTH ADMIN, US DEPT LABOR, 80- *Concurrent Pos:* Mem, Nat Agr Chem Comt Wildlife Pesticides, 65-67; asst clin prof pharmacol, Med Col Wis, 68-75. *Mem:* Teratology Soc; Am Indust Hyg Asn; Soc Toxicol; NY Acad Sci; Am Asn Lab Animal Sci. *Res:* Toxocology of environmental contaminants; iron compounds; behavioral and reproductive toxicology. *Mailing Add:* OSHA US Dept Labor Rm N3817 200 Constitution Ave NW Washington DC 20210

BELINFANTE, FREDERIK J, b The Hague, Holland, Jan 6, 13; nat US; m 37; c 3. PHYSICS. *Educ:* State Univ Leiden, BA, 33, MA, 36, PhD(theoret physics), 39. *Prof Exp:* Asst theoret physics, State Univ Leiden, 36-46, lectr, 45-46; assoc prof, Univ BC, 46-48; from assoc prof to prof, 48-79, EMER PROF THEORET PHYSICS, PURDUE UNIV, WEST LAFAYETTE, 80- *Concurrent Pos:* Adj Prof, Reed Col, Portland, Ore, 81- *Mem:* Fel Am Phys Soc; Neth Phys Soc. *Res:* General-relativistic quantum field theory; hidden-variables theories; foundations of quantum theory. *Mailing Add:* PO Box 901 Gresham OR 97030-0201

BELINFANTE, JOHAN G F, b Leiden, Neth, Feb 2, 40; US citizen; m 69. COMPUTER LANGUAGES, COMBINATORS. *Educ:* Purdue Univ, BS, 58; Princeton Univ, PhD(physics), 61. *Prof Exp:* NSF fel physics, Calif Inst Technol, 61-62; res asst, Univ Pa, 62-64; from asst prof to assoc prof, Carnegie-Mellon Univ, 64-73; assoc prof, 73-79, PROF MATH, GA INST TECHNOL, 79- *Mem:* Am Math Soc; Am Phys Soc; Math Asn Am; Soc Natural Philosophy. *Res:* Lie algebras; computers in algebra. *Mailing Add:* Dept Math Ga Inst Technol Atlanta GA 30332

BELINSKY, STEVEN ALAN, b High Point, NC, Mar 22, 56; m 89. DNA METHYLATION, ONCOGENE & SUPPRESSOR GENES. *Educ:* Univ NC, Chapel Hill, BS, 78, PhD(toxicol), 84. *Prof Exp:* Sr staff fel, Nat Inst Environ Health Sci, 84-90; MOLECULAR BIOLOGIST, INHALATION TOXICOL RES INST, 90- *Concurrent Pos:* Postdoctoral fel, Nat Res Serv Award, Nat Inst Environ Health Sci, 84-86; clin assoc prof, Col Pharm, Univ NMex, 91- *Mem:* Am Asn Cancer Res. *Res:* Identification of critical target genes involved in the development and progression of lung cancer. *Mailing Add:* Inhalation Toxicol Res Inst PO Box 5890 Albuquerque NM 87185

BELISLE, BARBARA WOLFANGER, b North Hornell, NY, Dec 14, 51; m 74; c 2. CELL BIOLOGY, INSECT CELL CULTURE. *Educ:* Keuka Col, BA, 73; La State Univ, MS, 77, PhD(physiol), 81. *Prof Exp:* Microbial pathogenesis, monoclonal antibodies, Dept Microbiol, Uniformed Serv Univ Health Sci, 82-85; dept head cell biol & immunol, res & develop human therapeut, IGB Prod, Ltd, 87-90; GROUP LEADER CELL BIOL, AM CYANAMID, SAN LEANDRO, CALIF, 91- *Mem:* Am Soc Cell Biol; Am Soc Metals; Sigma Xi; Tissue Cult Asn. *Res:* Production of recombinant proteins and biological pesticides using cultured insect cells; large scale production of recombinant proteins in mammalian and insect cell culture. *Mailing Add:* Am Cyanamid 14310 Catalina St San Leandro CA 94577

BELITSKUS, DAVID, b Cuddy, Pa, Dec 24, 38; m 68; c 4. PHYSICAL INORGANIC CHEMISTRY. *Educ:* Duquesne Univ, BS, 61; Univ Pittsburgh, PhD(phys chem), 64. *Prof Exp:* Res engr, 64-67, sr res scientist, 67-75, sci assoc, 75-78, sect head, 78-81, FEL, ALCOA LABS, ALUMINUM CO AM, 81- *Mem:* Am Ceramic Soc; Sigma Xi; Metall Soc. *Res:* X-ray crystal structure analyses; oxidation of light metal alloys; batteries and fuel cells; properties of lubricants and corrosion inhibitors; metallothermic reductions; metal matrix composites; carbon properties; ceramic matrix composites. *Mailing Add:* Alcoa Tech Ctr Alcoa Center PA 15069

BELIVEAU, GALE PATRICE, neurochemistry, metabolism, for more information see previous edition

BELJAN, JOHN RICHARD, b Detroit, Mich, May 26, 30; m 52; c 3. AEROSPACE MEDICINE, BIOMEDICAL ENGINEERING. *Educ:* Univ Mich, BS, 51, MD, 54. *Prof Exp:* Clin instr surg, Sch Med, Univ Mich, 56-59; dir med serv, Stuart Co, Calif, 65; from asst prof to assoc prof surg, Sch Med, Univ Calif, Davis, 66-74, asst to dean, 70-71, asst dean, 71-74, from asst prof to assoc prof eng, Col Eng, 68-74; dean, Sch Med, Wright State Univ, 74-80, prof surg & biomed eng, 74-83, vprovost, 74-78, vpres, Health Affairs, 78-80, provost & sr vpres, 81-83; provost & vpres acad affairs, Hahnemann Univ, 83-

85, prof surg & biomed eng, 83-86, spec adv to pres, 85-86; vpres acad affairs, 86-89, PROF ANAT, PHYSIOL & BIOMED ENG, CALIF STATE UNIV, LONG BEACH, 86-; PRES, NORTHROP UNIV, LOS ANGELES, CALIF, 89- Concurrent Pos: Mem ground support team, Dept Defense Task Force for Proj Gemini, 64-65; consult, Atlas Chem Industs, Inc, Del, 69-73, Dept Health, Educ & Welfare, 72-, Vet Admin, 75-, Booz-Allen & Hamilton, 79-85, AT Kearney, 85- Honors & Awards: Bauer Lectr, Aerospace Med Asn, 79. Mem: Aerospace Med Asn; Am Col Surgeons; Instrument Soc Am; Inst Elec & Electronics Engrs; AMA. Res: General surgery; aerospace medicine; biomedical engineering; telemetry. Mailing Add: 6490 Saddle Dr Long Beach CA 90815-4740

BELK, GENE DENTON, b Dallas, Tex, Dec 27, 38; m 63. ANIMAL HUSBANDRY. Educ: Univ Tex, Austin, BSEd, 66, MA, 69; Ariz State Univ, PhD(zool), 74. Prof Exp: PRES, BRACKENRIDGE STABLES, INC, 75-; ADJ PROF, OUR LADY OF THE LAKE UNIV, SAN ANTONIO, 78- Concurrent Pos: Vis asst prof invert zool, Ariz State Univ, 74-75. Honors & Awards: Wilks Award, 68; R M Harris Award, 77. Mem: Crustacean Soc (treas, 86-); Am Soc Zoologists; Sigma Xi; AAAS; Southwestern Asn Naturalists. Res: Systematics and distribution of the crustacean orders Anostraca, Spinicaudata and Laevicaudata; study of all aspects of their natural history and behavior. Mailing Add: TCS Bus Off 840 E Mulberry San Antonio TX 78212-3194

BELKIN, BARRY, b Philadelphia, Pa, May 4, 40; m 64; c 2. OPERATIONS RESEARCH. Educ: Mass Inst Technol, BS, 62; Univ Pa, MA, 64; Cornell Univ, PhD(math), 68. Prof Exp: V PRES OPERS RES & MATH, DANIEL H WAGNER, ASSOCS, 67- Honors & Awards: Rist Prize, Mil Opers Res Soc, 75. Mem: Inst Math Statist; Soc Indust & Appl Math. Res: First passage problems; tracking algorithms; theory of random walk. Mailing Add: Daniel H Wagner Assocs Sta Sq 2 Paoli PA 19301

BELKNAP, JOHN KENNETH, b Wilmington, NC, Jan 14, 43. PSYCHOPHARMACOLOGY, DRUG ABUSE. Educ: Univ Colo, Boulder, BA, 66, PhD(psychol), 71. Prof Exp: Asst prof psychol, Univ Tex, Austin, 72-78; assoc prof, 78-85, PROF PHARMACOL, SCH MED, UNIV NDAK, 85- Concurrent Pos: Fel, Univ Tex, Austin, 71-72; prin investr, Nat Inst Drug Abuse grant, 75- Mem: Behav Genetics Asn; Am Soc Pharmacol & Exp Therapeut; Soc Neuroscience. Res: Drug dependence; alcohol and barbiturates; pharmacogenetics of opioid analgesia. Mailing Add: Dept Pharmacol Sch Med Univ NDak Grand Forks ND 58202

BELKNAP, ROBERT WAYNE, b Omaha, Nebr, Jan 13, 24; m 53; c 2. PHYSIOLOGY. Educ: Creighton Univ, BS, 49, MS, 51; Univ Calif, PhD, 58. Prof Exp: Jr res physiologist, Univ Calif, 58-60; from asst prof to assoc prof, 60-76, PROF BIOL, CREIGHTON UNIV, 76- Concurrent Pos: Instr, Vet Admin Hosp, Omaha, 62- Mem: AAAS; NY Acad Sci; Am Soc Zool; Soc Nuclear Med; Am Statist Asn; Sigma Xi. Res: Mammalian physiology and environmental physiology; tissue metabolism; heat and cold exposure; hypothermia. Mailing Add: 3366 S 112th St Omaha NE 68144

BELL, A(UDRA) EARL, b Providence, Ky, May 9, 18; m 41; c 3. GENETICS. Educ: Univ Ky, BS, 39; La State Univ, MS, 41; Iowa State Col, PhD(genetics), 48. Prof Exp: Asst genetics, Iowa State Col, 41-42, 46-48; from asst prof to prof, 48-88, EMER PROF ANIMAL SCI, PURDUE UNIV, WEST LAFAYETTE, 88- Concurrent Pos: Guggenheim fel, 64-65; Japan SPS fel, Kyoto Univ, 87-88. Mem: Fel AAAS; Am Genetics Asn; Am Soc Animal Sci; Genetics Soc Am; Am Soc Naturalists. Res: Population genetics in poultry, drosophilia and tribolium; genetics of disease resistance. Mailing Add: 3913 Calle Real San Clemente CA 92672

BELL, ALAN EDWARD, b London, Eng, July 1, 48; m 72. PHYSICS. Educ: Univ London, BSc, 69, PhD(metal physics) & DIC, 73; Royal Col Sci, ARCS, 69. Prof Exp: Res asst metal physics, Imp Col, Univ London, 72-74; mem tech staff optical rec, David Sarnoff Res Ctr, RCA Corp, 74-; AT IBM RES LABS. Concurrent Pos: Vis res fel, David Sarnoff Res Ctr, RCA Corp, 73-74. Mem: Optical Soc Am; Inst Elec & Electronics Engrs. Res: High density optical recording of video and digital signals; and digital optics of thin films; thermal analysis applied to silicon crystal growth and laser annealing. Mailing Add: MS K47/282 IBM Res Labs 5600 Cottle Rd San Jose CA 95193

BELL, ALEXIS T, b New York, NY, Oct 16, 42. CHEMICAL ENGINEERING. Educ: Mass Inst Technol, BS, 64, ScD(chem eng), 67. Prof Exp: From asst prof to assoc prof, 67-76, PROF CHEM ENG, UNIV CALIF, BERKELEY, 76-; PRIN INVESTR, MAT & MOLECULAR RES DIV, LAWRENCE BERKELEY LAB, 75-, CHMN, DEPT CHEM ENG, 81- Concurrent Pos: Consult, Tracer Labs, Calif, 67-69; Int Plasma Corp, 69-, Tegal Corp & Lockheed Space & Missile Co. Honors & Awards: Curtis W McGraw Res Award, Am Soc Eng Educ, 81. Mem: Am Chem Soc; Am Inst Chem Engrs; Electrochem Soc; Sigma Xi. Res: Catalysis. Mailing Add: Dept Chem Eng Univ Calif 201 Gilman Hall Berkeley CA 94720-9989

BELL, ALFRED LEE LOOMIS, JR, b Englewood, NJ, Jan 12, 23; m 47; c 4. MEDICINE. Educ: Harvard Univ, MD, 47; Am Bd Internal Med, dipl, 58; Am Bd Cardiovasc Dis, dipl, 64. Prof Exp: From intern to asst resident, St Luke's Hosp, 47-50; NIH fel, USPHS. 50-51; mem fac internal med, Sch Aviation Med, Randolph AFB, 51-53; physician-in-charge, 53-64, from asst attend physician to assoc attend physician, 55-65, from asst cardiologist to assoc cardiologist, 55-65, assoc attend cardiologist, 65-69, CHIEF, CARDIOPULMONARY LAB, ST LUKE'S HOSP, 64-, ATTEND PHYSICIAN, 65-, ATTEND CARDIOLOGIST, 69-, CHIEF, DIV PULMONARY DIS, 70- Concurrent Pos: Instr, Col Physicians & Surgeons, Columbia Univ, 53-58, assoc, 58-64, asst prof, 64-73, assoc prof, 73-76, prof clin med, 77-; fel, Coun Clin Cardiol, Am Heart Asn. Mem: Fel Am Thoracic Soc; fel Am Col Physicians; fel Am Col Chest Physicians; fel NY Acad Med; Int Cardiovasc Soc; Am Soc Laser Med & Surg; NAm Soc Pacing & Electrophysiol. Res: Cardiopulmonary physiology. Mailing Add: St Lukes Roosevelt Hosp Ctr Amsterdam Ave & 114th St New York NY 10025

BELL, ALLEN L, b Bangor, Maine, Apr 13, 37; m 60; c 2. ELECTRON MICROSCOPY, HISTOLOGY. Educ: Univ Mich, BA, 59; State Univ NY, PhD(anat), 66. Prof Exp: Instr anat, Upstate Med Ctr, State Univ NY, 66-67; asst prof, Univ Colo Med Ctr, 67-72; ASSOC PROF ANAT, NEW ENG COL OSTEOP MED, 78- Concurrent Pos: Dir, Electron Micros Lab, Marine Biol Lab, Mass, 63-72. Res: Fine structure of photoreceptors of vertebrate and invertebrates; mammalian kidney disorders; fine structure of virus; spider silk spinning gland. Mailing Add: Anat Dept New Eng Col Osteop Med 605 Pool Rd Biddeford ME 04005

BELL, ALOIS ADRIAN, b Bloomfield, Nebr, Jan 25, 34; m 58; c 2. BIOCHEMISTRY, MICROBIOLOGY. Educ: Univ Nebr, BSc, 55, MSc, 58, PhD(bot), 61. Prof Exp: Asst prof bot, Univ Md, 61-65; res plant pathologist plant indust sta, USDA, 65-70, dir nat cotton path res lab, 70-77, res leader, supvr res plant path, Crops Res Div, 77-87, RES PLANT PATHOLOGIST, AGR RES SERV, USDA, COLLEGE STATION, TX, 87- Concurrent Pos: Grad Sch Res Coun grant, Univ Md, 62; USPHS res grant, 63-66; Competitive res grant, USDA, 72-75. Mem: Fel Am Phytopath Soc. Res: Biochemistry of virulence and natural products in phytopathogenic fungi; biochemistry and physiology of disease resistance and fungal pathogenesis of cotton; development of biocontrol agents and beneficial microorganisms for cotton; synthetic organic and natural products chemistry. Mailing Add: 3503 Spring Lane Bryan TX 77802

BELL, ANTHONY E, b Ramsgate, Eng, Aug 23, 37; m 63; c 1. SURFACE PHYSICS. Educ: Univ London, BS, 59, PhD(phys chem), 62. Prof Exp: Res assoc, Univ Chicago, 63-65; physicist, Field Emission Corp, 65-69; assoc prof chem, Linfield Col, 69-76; staff scientist, 76-85, ASSOC RES PROF PHYSICS, ORE GRAD CTR, 85- Concurrent Pos: NSF grants, 70-71, 80-82, 82-85 & 86-89. Mem: Am Chem Soc. Res: Field emission of electrons for application in scanning electron microscopes; electron beam lithography liquid metal field ion source development and characterization. Mailing Add: Ore Grad Ctr 19600 NW Walker Rd Beaverton OR 97005

BELL, BARBARA, b Evanston, Ill, Apr 1, 22. SOLAR PHYSICS, CLIMATOLOGY. Educ: Radcliffe Col, AB, 44, MA, 49, PhD(astron), 51. Prof Exp: Res asst, Harvard Observ, 48-57, sci asst to dir, 57-71, astronr, Harvard Col Observ, 71-73, ASTRONR, CTR ASTROPHYS, 73- Mem: Am Astron Soc; Am Res Ctr Egypt; Int Astron Union; Archaeol Inst Am. Res: Spectroscopy and the solar spectrum; solar-geomagnetic correlations; modern climatology, post-glacial climate fluctuations and their possible role in ancient history, especially of Egypt and the Nile. Mailing Add: Ctr Astrophys 60 Garden St Cambridge MA 02138

BELL, BRUCE ARNOLD, b Bronx, NY, Apr 4, 44; m 74; c 3. WASTEWATER TREATMENT. Educ: New York Univ, BCE, 68, MSCE, 69, PhD(environ eng), 74. Prof Exp: Asst civil engr, New York City Transit Authority, 64-68; instr civil eng, NY Univ, 69-73; systs mgr, Envirotech Corp, 73-75; vpres, Flood & Assocs Inc, 75-78; assoc prof environ eng, 78-82, PROF ENG, GEORGE WASHINGTON UNIV, 82- Concurrent Pos: Consult, 69-; sr sci adv, Natural Resources Defense Coun, 71-73; adj prof, Univ NFla, 76-78; scientist & consult, US Army Mgt Br, Eng Res & Develop Br, 80-81; vpres, Carpenter Environ Assoc, 79- Mem: Water Pollution Control Fedn; Am Soc Civil Eng; Int Asn Water Pollution Res; Am Acad Environ Engrs; Asn Environ Eng Professionals. Res: Wastewater treatment processes including biological and physical/chemical processes, sludge treatment and disposal and industrial wastewater treatment; hazardous waste treatment and disposal. Mailing Add: Sch Eng & Appl Sci George Washington Univ Washington DC 20052

BELL, C GORDON, b Kirksville, Mo, Aug 19, 34; m 59; c 2. COMPUTER SCIENCE, ELECTRICAL ENGINEERING. Educ: Mass Inst Technol, BS, 56, MS, 57. Prof Exp: Res engr, Res Lab Electronics, Mass Inst Technol, 59-61; mgr comput design, Digital Equip Corp, 61-66; from assoc prof to prof comput sci, 66-78, PROF ELEC ENG & COMPUT SCI, CARNEGIE-MELLON UNIV, 78- Concurrent Pos: Consult, Digital Equip Corp, 66-, vpres, 72-83; vchmn technol, Encore, 83-85; vpres, Ardent Comput, 85-; consult, Kubota Pac, Santa Clara. Honors & Awards: Nat Medal of Technol, 91; McDowell Award; Eckert-Mauchly Award. Mem: Nat Acad Eng; fel Inst Elec & Electronics Engrs; Asn Comput Mach. Res: Automatic design of digital computers; computer and information processing; systems structure analysis and design. Mailing Add: 450 Old Oak Ct Los Altos CA 94022

BELL, CARL F, b Otsego, Ohio, Nov 24, 33; m 61. PLANT PATHOLOGY. Educ: Muskingum Col, BS, 55; Miami Univ, MS, 58; Ohio State Univ, PhD(plant path), 61. Prof Exp: Assoc prof, 61-69, PROF BIOL, SHEPHERD COL, 69- Mem: Am Phytopath Soc. Res: Physiology of fungus parasitism. Mailing Add: Dept Sci & Math Shepherd Col Shepherdstown WV 25443

BELL, CARLOS G(OODWIN), JR, b Graham, Tex, Dec 18, 22; m 46; c 2. ENVIRONMENTAL & NUCLEAR ENGINEERING. Educ: Agr & Mech Col, Tex, BS, 48; Harvard Univ, SM, 49, SD, 55. Prof Exp: Sanit engr, Inst Inter-Am Affairs, Brazil, 49-51; AEC Proj asst, Harvard Univ, 52-54; assoc prof civil eng, Northwestern Univ, 54-59; lectr, Oak Ridge Sch Reactor Technol, 59-67, res staff mem reactor div, Oak Ridge Nat Lab, 67-70; CELANESE PROF CIVIL ENG, UNIV NC, CHARLOTTE, 70- Res: Sanitary engineering aspects of long range fall-out from nuclear detonations; environmental engineering. Mailing Add: Sch Eng Smith Bldg Rm 346 Univ NC Charlotte NC 28223

BELL, CHARLES BERNARD, JR, b New Orleans, La, Aug 20, 28; m 53; c 4. MATHEMATICS, STATISTICS. Educ: Xavier Univ, La, BS, 47; Univ Notre Dame, MS, 48, PhD, 53. Prof Exp: Asst math, Univ Notre Dame, 49-51; res engr, Douglas Aircraft Co, 51-55; asst prof math & physics, Xavier Univ, La, 55-57; instr math & res assoc statist, Stanford Univ, 57-58; from asst prof to prof math, San Diego State Col, 58-66; prof, Case Western Reserve Univ, 66-68; prof math & statist, Univ Mich, Ann Arbor, 68-71; prof math, Tulane Univ, 74-78; prof biostatist, Univ Wash, 78-81; PROF BIOSTATIST,

SAN DIEGO STATE UNIV, 81- *Mem:* Fel Inst Math Statist; fel Am Statist Asn; Am Math Soc; Soc Espanola Inv Oper. *Res:* Nonparametric statistics and applications of stochastic processes; biostatistics and signal detection. *Mailing Add:* Math Dept San Diego State Univ San Diego CA 92182

BELL, CHARLES E, JR, b Norwood, Mass, Aug 8, 31; m 55; c 4. PHYSICAL ORGANIC CHEMISTRY. *Educ:* Univ Va, BS, 54, MS, 55, PhD(chem), 61. *Prof Exp:* Chemist, Prod Res Div, Esso Res & Eng Co, Standard Oil, NJ, 60-62; from asst prof to assoc prof chem, 62-67, asst dean grad studies, 72-75, PROF CHEM, OLD DOMINION UNIV, 67-, ASSOC DEAN GRAD STUDIES, 75. *Mem:* Sigma Xi; Am Chem Soc. *Res:* Kinetics and mechanisms of organic reactions; analysis of chemical composition of jellyfish; organic pollutants in drinking water. *Mailing Add:* 6150 Eastwood Terr Norfolk VA 23508

BELL, CHARLES EUGENE, JR, b New York, NY, Dec 13, 32; m 67; c 1. FACILITIES PLANNING. *Educ:* Johns Hopkins Univ, BE, 54, MSE, 59. *Prof Exp:* Indust engr, Signode Industs, 57-61, asst plant mgr, 61-64, div indust engr, 64-69, asst div mgr, 69-76, ENG MGR, SIGNODE INDUSTS, 76- *Concurrent Pos:* Mem, Int Indust Eng Conf Comt, 84 & 92. *Mem:* Nat Soc Prof Engrs; Inst Indust Engrs; Soc Plastics Engrs. *Mailing Add:* 1021 W Old Mill Rd Lake Forest IL 60045

BELL, CHARLES VESTER, b Starkville, Miss, Mar 19, 34; m 54; c 3. ELECTRICAL ENGINEERING. *Educ:* Miss State Univ, BS, 56; Stanford Univ, MS, 57, PhD(elec eng), 60. *Prof Exp:* Asst prof physics, Walla Walla Col, 60-62; sect head, Hughes Aircraft Co, 62-72; assoc prof physics, 72-74, dean eng, Walla Walla Col, 74-84; ACAD VPRES, PAC UNION COL, 84- *Mem:* Sigma Xi; Inst Elec & Electronics Engrs; Am Soc Eng Educ. *Res:* Microwave electronics; biological effects of microwave radiation. *Mailing Add:* 175 Edgewood Pl Angwin CA 94508-9707

BELL, CHARLES W, b Cheyenne, Wyo, Sept 30, 31; m 51; c 6. PLANT PHYSIOLOGY. *Educ:* Univ Wyo, BA, 53; Univ Wash, MS, 57; Wash State Univ, PhD(bot), 62. *Prof Exp:* PROF NATURAL SCI & BIOL, SAN JOSE STATE UNIV, 62- *Concurrent Pos:* Partic, Biol Sci Curric Study Inst Prep Col Teachers, Univ Colo, 63. *Mem:* AAAS; Am Inst Biol Sci. *Res:* Plant mineral nutrition; computer modeling and simulation of biological systems. *Mailing Add:* Dept Biol Sci San Jose State Univ San Jose CA 95192

BELL, CLARA G, b Roumania, Oct 16, 37; US citizen; m 64; c 3. IMMUNOLOGY, IMMUNOGENETICS. *Educ:* Hebrew Univ Jerusalem, BSc, 59, MSc, 60, PhD(immunol), 64. *Prof Exp:* Fel immunol, Univ Sydney, 64, res fel immunochem, 66-68; USPHS res assoc immunol, 69-70, asst prof, 70-73, ASSOC PROF IMMUNOL, UNIV ILL MED CTR, CHICAGO, 73- *Concurrent Pos:* On leave, NIH career develop award & assoc prof, Karolinska Inst Tumor Biol, Stockholm, Sweden, 75-77. *Honors & Awards:* Boris Pragel Gold Medal, NY Acad Sci, 72. *Mem:* Am Asn Immunologists; AAAS; NY Acad Sci; Sigma Xi; Brit Biochem Soc. *Res:* Functional analysis of lymphocyte sub-populations in terms of membrane markers and specificities, DNA, immunoglobulin, antibody and lymphokine synthesis, secretion and cellular responsiveness in immunity and surveillance; immune repertoire to self-non-self, autoimmune mechanisms vaccination. *Mailing Add:* 5455 N Calif Unit 1A Chicago IL 60625

BELL, CLYDE RITCHIE, b Cincinnati, Ohio, Apr 10, 21; m 43. BOTANY. *Educ:* Univ NC, AB, 47, MA, 49; Univ Calif, Berkeley, PhD(bot), 53. *Prof Exp:* Instr bot, Univ Ill, 53-55; from asst prof to assoc prof, 55-66, PROF BOT, UNIV NC, CHAPEL HILL, 66- *Concurrent Pos:* Dir, NC Bot Garden, 66- *Mem:* Bot Soc Am (treas, 72-76, vpres, 78); Am Inst Biol Sci; Am Soc Plant Taxonomists (secy, 59-62, pres, 76); Int Asn Plant Taxonomists; fel AAAS. *Res:* Systematic botany; plant evolution; pollination biology; flora of the Southeastern United States. *Mailing Add:* 5000 Walnut Cove Rd No R16 Chapel Hill NC 27516

BELL, CURTIS CALVIN, SENSORY PHYSIOLOGY, SENSORY MOTOR. *Educ:* Univ Calif, Los Angeles, PhD(anat), 64. *Prof Exp:* SR SCIENTIST, NEUROL SCI INST, GOOD SAMARITAN HOSP & MED CTR, 65- *Mailing Add:* Neurol Sci Inst Good Samaritan Hosp 1120 NW 20th St Portland OR 97209

BELL, CURTIS PORTER, b Augusta, Ga, July 26, 34; m 61; c 1. MATHEMATICS. *Educ:* Wofford Col, BS, 55; Univ Ga, MA, 60, PhD(math), 63. *Prof Exp:* Instr math & physics, Limestone Col, 55-56; asst math, Univ Ga, 62-63; asst prof, 63-67, ASSOC PROF MATH, WOFFORD COL, 67- *Mem:* Am Math Soc; Math Asn Am. *Res:* Point set topology. *Mailing Add:* Dept Math Wofford Col Spartanburg SC 29301

BELL, DUNCAN HADLEY, b Stamford, Conn, June 23, 52; m 78; c 3. INSTRUMENTATION, AUTOMATION. *Educ:* Col Wooster, BA, 74; Mich State Univ, PhD(biochem), 79. *Prof Exp:* Res asst, Dept Bot & Plant Path, Mich State Univ, 74-79; res assoc, Radiation Lab, Univ Notre Dame, 79-82; res fel, Univ Glasgow, 82-85; sr res biochemist, 85-91, BIOMED ENGR, LEDERLE LABS, AM CYANAMID CO, 91- *Mem:* Sigma Xi; AAAS; Am Chem Soc; NY Acad Sci. *Res:* Instrumentation; computers, robotics and equipment design; radiation chemistry, enzyme inactivation, kinetics and pH gradients; fluorescent pH indicating probes. *Mailing Add:* Dept Mech Res & Develop Lederle Labs Am Cyanamid Co Pearl River NY 10965

BELL, EDWIN LEWIS, II, b Danville, Pa, May 13, 26; m 50; c 3. HERPETOLOGY. *Educ:* Bucknell Univ, BS, 48; Pa State Univ, MS, 50; Univ Ill, PhD(zool), Univ Ill, 54. *Prof Exp:* Asst prof biol, Moravian Col, 52-54; from asst prof to prof biol, Albright Col, 54-91, preprof adv health sci, 61-87, chmn dept biol, 65-84, emer prof, 91-; RETIRED. *Concurrent Pos:* Dir biol sci curric study teachers inserv inst, NSF, 66-67; spec collections asst archivist, Libr, Albright Col. *Mem:* AAAS; Am Soc Ichthyol & Herpet; Herpetologists League; Am Inst Biol Sci; Am Soc Human Genetics. *Res:* Herpetology of Huntingdon County, Pennsylvania; study of lizard Sceloporus occidentalis; sexing of Drosophila larvae; dorman panther legend. *Mailing Add:* 1454 Oak Lane Reading PA 19604

BELL, EUGENE, b New York, NY, Oct 28, 18. CELLMATIC INTERACTION. *Educ:* Brown Univ, PhD(biol), 54. *Prof Exp:* PROF BIOL, MASS INST TECHNOL, 57- *Mailing Add:* Dept Corp & Admin Organogenesis Inc 83 Rogers St Cambridge MA 02142

BELL, FRANK F, b Savannah, Tenn, Nov 6, 15; m 40; c 3. AGRONOMY. *Educ:* Univ Tenn, BS, 37, MS, 49; Iowa State Univ, PhD(agron), 56. *Prof Exp:* From instr to prof agron, Univ Tenn, Knoxville, 46-81; RETIRED. *Concurrent Pos:* Vis prof, Iowa State Univ, 61, NC State Univ, 67 & Univ Ky, 71; US rep, Int Soil Cong, Adelaide, Australia, 68; agron consult, US AID Prog, India, 70. *Mem:* AAAS; Am Soc Agron; Soil Sci Soc Am; Int Soc Soil Sci. *Res:* Soil productivity; management research. *Mailing Add:* 135 Judith Dr Knoxville TN 37920

BELL, FRED E, b Knoxville, Tenn, Feb 12, 25; m 50; c 1. BIOCHEMISTRY. *Educ:* Emory Univ, AB, 50, PhD(biochem), 57. *Prof Exp:* Instr biochem, Sch Med, Univ Va, 56-57; USPHS fel, Harvard Univ & Mass Gen Hosp, 57-58; from instr to asst prof, 58-72, ASSOC PROF BIOCHEM, SCH MED, UNIV NC, CHAPEL HILL, 72- *Mem:* AAAS; Am Chem Soc; Brit Biochem Soc; Sigma Xi. *Res:* Protein biosynthesis; nucleic acids; enzymes. *Mailing Add:* Dept Biochem Univ NC Sch Med Chapel Hill NC 27514

BELL, GEORGE IRVING, b Ill, Aug 4, 26; m 56; c 2. THEORETICAL BIOLOGY. *Educ:* Harvard Univ, AB, 47; Cornell Univ, PhD(physics), 51. *Prof Exp:* Asst, Cornell Univ, 47-49; mem staff, Los Alamos Nat Lab, 61-69, alt div leader, Theoret Div, 70-80, group leader Theoret Biol & Biophys, 74-90, div leader, 80-89, act dir, ctr Human Genome studies, 88-89, SR FEL, LOS ALAMOS NAT LAB, 89- *Concurrent Pos:* Vis lectr, Harvard Univ, 62-63; vis prof, Med Ctr, Univ Colo, 70-84; scholar human biol, Eleanor Roosevelt Cancer Res Ctr, Denver, 77-88, Nat Cancer Inst, Div Cancer Biol & Diag; mem, Basel Inst Immunol, Switz, 79-80, bd Sci Counr 85-89; chmn, oversight Group Supercomput, 87- *Honors & Awards:* Sowles Medal, 80. *Mem:* Fel AAAS; Am Phys Soc; Biophys Soc. *Res:* Theoretical physics and immunology; nuclear reactor theory; mathematical models in biophysics; sequencing the human genome. *Mailing Add:* T Div MS K710 Los Alamos Nat Lab Los Alamos NM 87545

BELL, GORDON M(ORETON), b Liverpool, Eng, Oct 17, 20; m 47; c 1. MINERAL ENGINEERING, EXTRACTIVE METALLURGY. *Educ:* Univ BC, BASc, 42; Mass Inst Technol, DSc, 53. *Prof Exp:* Plant asst, Falconbridge Nickel Mines, Ltd, 42-44; metallurgist, Aluminum Co Can, Ltd, 44-47; asst prof metall, Univ BC, 47-49; instr, Mass Inst Technol, 49-53; res engr, Extractive Metall Div, Alcoa Labs, Aluminum Co Am, 53-57, sr res engr, 57-65, sci assoc, 65-71, sect head, Geol & Raw Mat Res, 71-73, eng assoc, 73-85; RETIRED. *Mem:* Am Inst Mining, Metall & Petrol Eng; Sigma Xi. *Res:* Beneficiation of coal, clay, fluorspar, copper and aluminum ores; extraction of aluminum from ores; ore sampling, phases and properties of aluminas. *Mailing Add:* 825 Morewood Ave Condo G-8 Pittsburgh PA 15213-2911

BELL, GORDON RUSSELL, microbiology, for more information see previous edition

BELL, GRAHAM ARTHUR CHARLTON, b Leicester, Eng, Mar 3, 49; m 71; c 3. EVOLUTIONARY BIOLOGY, ECOLOGY. *Educ:* Oxford Univ, BA, 70, DPhil(ecol), 74. *Prof Exp:* Fel ecol, Univ York, 74-75; ASST PROF ECOL, MCGILL UNIV, 76- *Mem:* Soc Study Evolution; Ecol Soc Am; Brit Ecol Soc; Am Soc Naturalists. *Res:* Evolution of genetic systems and sexuality; life histories; sexual selection; population dynamics. *Mailing Add:* Dept Biol McGill Univ Sherbrooke St W Montreal PQ H3A 2M5 Can

BELL, GRAYDON DEE, b Paducah, Ky, May 5, 23; m 54; c 3. PHYSICS, ASTROPHYSICS. *Educ:* Univ Ky, BS, 49; Calif Inst Technol, MS, 51, PhD(physics), 57. *Prof Exp:* Asst prof physics, Robert Col, Istanbul, 51-54; res asst, Calif Inst Technol, 54-56, res fel, 56-57; from asst prof to assoc prof, 57-65, chmn dept, 71-81, PROF PHYSICS, HARVEY MUDD COL, 65- *Concurrent Pos:* NSF res grants, 60-70, fac fel, Nat Res Coun Can, 70-71; physicist, Nat Bur Standards, 63-64; vis scientist, Australia, 78-79; consult, GTE Sylvania, 79, 80 & 81. *Mem:* AAAS; Am Phys Soc; Am Astron Soc; Am Asn Physics Teachers. *Res:* Experimental measurement of f-values for spectral lines of heavy elements. *Mailing Add:* Dept Physics Harvey Mudd Col Claremont CA 91711

BELL, H(ARRY) R(ICH), b Vancouver, BC, May 15, 20. CIVIL ENGINEERING, SURVEYING ENGINEERING. *Educ:* Univ BC, BASc, 42; Univ Col London, Eng, dipl, 52, MSc(Eng), 56. *Prof Exp:* From instr to asst prof civil eng, 45-67, ASSOC PROF CIVIL ENG, SURV & PHOTOGRAM, UNIV BC, 67- *Concurrent Pos:* Engr, Topograph Surv, Dept Mines & Resources, 48, tech officer, 58; mem subcomt surv educ & res, Nat Adv Comt Control Surv & Mapping, 65-68. *Honors & Awards:* Subvention Award, NSF, 66; Ford Bartlett Award, Am Soc Photogram, 70. *Mem:* Am Soc Photogram; Eng Inst Can; Can Inst Surv; Brit Photogram Soc; fel Am Cong Surv & Mapping. *Res:* Surveying; photogrammetry; theory of error, least squares adjustment; photointerpretation; remote sensing. *Mailing Add:* 5151 45th Ave Delta BC V4K 1K6 Can

BELL, HAROLD, b New York, NY, June 25, 32; m 56; c 4. MATHEMATICS. *Educ:* Univ Miami, BS, 58, MS, 59; Tulane Univ, PhD(math), 64. *Prof Exp:* From asst prof to assoc prof, 64-74, PROF MATH, UNIV CINCINNATI, 74- *Mailing Add:* Dept Math Univ Cincinnati Main Campus Cincinnati OH 45221

BELL, HAROLD E, b Taber, Alta, Aug 7, 26; m 51; c 3. CLINICAL PATHOLOGY. *Educ:* Univ Alta, BSc, 47, MD, 49. *Prof Exp:* Lectr clin path, Univ Alta, 59-60, asst dir lab serv, Univ Hosp, 60; res asst lab med, Univ Minn, 61-62; asst dir lab serv, Univ Hosp Fac Med, 62-66, assoc prof, 66-71, chmn dept lab med, Hosp, 73-87, PROF CLIN PATH, UNIV ALTA, 71-; DIR, PROFICIENCY TESTING, ALTA, 80- *Mem:* Col Am Path; Can Asn

Med Biochem; Am Soc Clin Path; Can Med Asn; Can Asn Path. *Res:* Clinical biochemistry; lipoproteins; high-density lipoproteins; structure of immune globulins. *Mailing Add:* 8409 112th St Univ Alta Edmonton AB T6G 1K6 Can

BELL, HAROLD MORTON, b Monticello, Ky, May 7, 40; m 60; c 2. ORGANIC CHEMISTRY. *Educ:* Eastern Ky Univ, BBS, 60; Purdue Univ, PhD(chem), 64. *Prof Exp:* Asst prof, 66-75, ASSOC PROF CHEM, VA POLYTECH INST & STATE UNIV, 75- *Mem:* Am Chem Soc. *Res:* Metal hydride reductions; nuclear magnetic resonance; computers in chemistry. *Mailing Add:* Dept Chem Va Polytech Inst & State Univ Blacksburg VA 24061

BELL, HOWARD E, b Albany, NY, Jan 20, 37; m 61; c 2. RINGS, NEAR-RINGS. *Educ:* Union Univ, NY, BS, 58; Univ Wis, MS, 59, PhD(math), 61. *Prof Exp:* Asst prof math, Union Univ, NY, 61-63 & State Univ NY Binghamton, 63-66; assoc prof, Union Univ, NY, 66-67; assoc prof, 67-76, chmn dept, 75-78, PROF MATH, BROCK UNIV, 76- *Concurrent Pos:* Vis prof, Mount Holyoke Col, 79; bd gov, Math Asn Am, 88-91. *Mem:* Am Math Soc; Math Asn Am; Can Math Soc. *Res:* Algebra, especially rings and near-rings. *Mailing Add:* Dept Math Brock Univ St Catharines ON L2S 3A1 Can

BELL, IAN, b Chorley, Eng, Sept 29, 32; m 56; c 3. INDUSTRIAL ORGANIC CHEMISTRY. *Educ:* Univ Manchester, BSc, 53, PhD(org chem), 56; Penn State, MBA, 79. *Prof Exp:* Tech officer org chem, Imp Chem Indust, Ltd, Eng, 58-60; from res chemist to dir res, Nease Chem Co, Inc, 60-67; chemist, Winthrop Labs, Eng, 67-68; VPRES RES & DEVELOP, RUETGERS-NEASE CHEM CO, INC, 68- *Mem:* Am Chem Soc; The Chem Soc. *Res:* Acetylenic chemistry; steroid structure determination; oxo process; aromatic and aliphatic intermediates; hydrotropes. *Mailing Add:* Ruetgers-Nease Chem Co Inc 201 Struble Rd State College PA 16801-7499

BELL, J(OHN) M, b Tappan, NY, Oct 8, 35; m 60; c 4. CIVIL & SANITARY ENGINEERING. *Educ:* Clarkson Col Technol, BSCE, 57; Purdue Univ, MSCE, 59, PhD(refuse anal), 63. *Prof Exp:* Civil engr, Civil Aeronaut Admin, Idlewild Airport, NY, 57; NIH res proj asst, 58-62, asst prof sanit eng, 62-68, ASSOC PROF ENVIRON ENG, PURDUE UNIV, WEST LAFAYETTE, 69- *Mem:* Am Soc Civil Engrs; Water Pollution Control Fedn; Am Water Works Asn; Sigma Xi. *Res:* Collection and disposal of municipal refuse; statistical design and analysis of field and laboratory experiments; industrial wastes treatment; water supply and treatment; sludge concentration; waste treatment in lagoons; stream sampling; stream biology. *Mailing Add:* Three Concord Pl Lafayette IN 47905

BELL, JACK PERKINS, b San Francisco, Calif, Jan 24, 40; m 65; c 2. BIOCHEMISTRY, BIOLOGY. *Educ:* Univ Wash, BS, 62; Univ Wis-Madison, PhD(org chem), 67. *Prof Exp:* Biochemist, Stanford Res Inst, 67-69 & Syntex Res Div, 69-78; mem staff, Waters Assocs, 78-80; regional sales mgr, Appl Anal Indust, 80-81; prod line mgr, High Performance Liquid Chromatography, Varian Assocs, 81-89; WESTERN REGIONAL SALES, SUPREX CORP, 90- *Mem:* AAAS; Am Chem Soc; Sigma Xi. *Res:* RNA and DNA polymerases; nucleotides as probes of biochemical mechanisms and as drugs; synthesis and biological properties of polynucleotides; development of new drugs and immunoassays; identification of drug metabolites; development and marketing of high performance liquid chromatography; automated sample handling products; supercritical fluid extraction and chromatography. *Mailing Add:* 2950 Windtree Ct Lafayette CA 94549

BELL, JAMES F(REDERICK), b Melrose, Mass, Apr 21, 14; m 40; c 2. SOLID MECHANICS, CONTINUUM PHYSICS. *Educ:* NY Univ, BA, 40. *Prof Exp:* Gyrodynamics, Arma Corp, 40-45; prof, 45-79, EMER PROF SOLID MECH, JOHNS HOPKINS UNIV, 79-, PROF PHYSICS OF MUSIC, PEABODY CONSERV, 85- *Concurrent Pos:* Lectr at over 70 univs in Europe, Asia, Africa & USA, 51-; sr vis, Dept Appl Math & Theoret Physics, Cambridge Univ, Eng, 62-63; res assoc, Math Inst, Univ Bologna, 70-71; sr vis, Math Inst, Romanian Acad Sci; vis prof, Univ Ferrara, Italy; sr ed, Int J Plasticity, 85- *Honors & Awards:* B J Lazan Award, Soc Exp Stress Anal, 74; William M Murray Medal, Soc Exp Mech, 89. *Mem:* Soc Natural Philos; fel Am Acad Mech; Sigma Xi; fel Soc Exp Mech. *Res:* Dynamics of solid continuum; physics of large deformation in crystalline solids; finite amplitude and shock waves; musical acoustics; diffraction gratings; dialysis; history of the experimental foundations of solid mechanics; biomedical engineering; heart and lung research; theories of finite deformation in continuum physics; author of over 90 scientific papers and three books. *Mailing Add:* Dept Mech Johns Hopkins Univ Baltimore MD 21218

BELL, JAMES HENRY, b London, Eng, July 10, 17; m 41. ALGEBRA. *Educ:* Univ Western Ont, BA, 37; Univ Wis, AM, 38, PhD(math), 41. *Prof Exp:* Asst, Univ Wis, 39-41; sr res assoc, Nat Res Coun Can, 41-42, jr radio res engr, 42-45; asst prof math, Univ Man, 45-46; from asst prof to assoc prof, Mich State Col, 46-53; tech dir mace proj & dir guid & navig, AC Spark Plug Div, Gen Motors Corp, 58-63, dir reliability, 63-67, dir reliability, AC Electronics Div, 67-70, dir customer serv, Delco Electronics Div, 70-73, mgr eng & customer serv, Milwaukee Opers, Delco Electronics Div, 73-77; RETIRED. *Res:* Mathematics; matrix theory; radar, ultra-high frequency antennae and propagation; systems analysis; radar, missiles and inertial guidance; inertial navigation of aircraft. *Mailing Add:* 5620 S Kurtz Rd Hales Corners WI 53130

BELL, JAMES MILTON, b Portsmouth, Va, Nov 5, 21. PSYCHIATRY. *Educ:* NC Col, Durham, BS, 43; Meharry Med Col, MD, 47; Am Bd Psychiat & Neurol, dipl, cert psychiat, 56, child psychiat, 60. *Prof Exp:* Intern, Harlem Hosp, NY, 47-48; from asst physician to clin dir, Lakin State Hosp, WVa, 48-51; fel psychiat, Menninger Sch Psychiat, 53-56; asst sect chief, Children's Unit, Topeka State Hosp, Kans, 56-58; clin dir & psychiatrist, Bershire Farm Ctr & Serv for Youth, 59-86; from clin asst prof to clin assoc prof, 59-80, CLIN PROF PSYCHIAT, ALBANY MED COL, 80-; CONSULT, 85- *Concurrent Pos:* Resident psychiat, Winter Vet Admin Hosp, 53-56; civilian consult, Irvin Army Hosp, Kans, 57-58; fel child psychiat, Menninger Found, 57-58; chief consult psychiatrist, Parsons Child & Family Ctr, NY, 59-77 & staff psychiatrist, 77-; asst to dispensary psychiatrist, Albany Med Ctr Clin, 60-; trainee consult, Albany Child Guid Ctr, 60-; mem instrnl staff, Frederick A Moran Inst Delinq & Crime, St Lawrence Univ, 65-70; mem & vchmn subcomt returning vet, NY State Post Vietnam Planning Comt, 68-; consult, Astor Home for Children, NY; mem, Gov State Wide Comt for 1970 White House Conf Children & Youth; mem bd dirs & exec comt, Guild Farm, Mass; deleg, White House Conf Children & Youth; consult on adolescence & mem med adv bd, NY State Div for Youth, 65-76; consult, West Point Mil Acad Hosp, NY, 77; chmn, Comt & Adolescent on Psychiat Facilities for Children & Adolescents, Am Acad Child Psychiat, 73-75; chmn, Coun Nat Affairs & Comt Psychiat Facilities for Affairs, Am Psychiat Asn, 73-75; sr child adolescent psychiatrist, Bershire Farm Ctr & Serv for youth, 86- *Honors & Awards:* President's Award, Meharry Med Col, 72. *Mem:* Group Advan Psychiat; fel Am Psychopath Asn; fel Am Psychiat Asn; Am Acad Child & Adolescent Psychiat; fel Am Col Psychiat; fel AAAS. *Res:* Child psychiatry; delinquency. *Mailing Add:* Hudsonview Old Post Rd North-Croton-on-Hudson NY 10520

BELL, JAMES PAUL, b Pulaski, Tenn, Nov 4, 34; m 55; c 4. CHEMICAL ENGINEERING, POLYMER SCIENCE. *Educ:* Lehigh Univ, BSc, 56; Mass Inst Technol, DSc(chem eng), 66. *Prof Exp:* Sr res engr polymer res, Res & Develop Div, plastics dept, E I du Pont de Nemours, 56-62; res specialist, Textile Fibers Div & New Enterprise Div, Monsanto Co, 66-69; assoc prof, 69-75, PROF CHEM ENG, INST MAT SCI, UNIV CONN, 75- *Concurrent Pos:* Fulbright-Hays sr lectr, Coun Int Exchange Scholars, Univ Freiburg, WGer, 75-76; vis prof & Lady Davis fel, Israel Inst Technol & vis prof, Univ Naples, 83. *Mem:* Am Inst Chem Engrs; Am Phys Soc; Am Chem Soc. *Res:* mechanical properties of polymers; structure and properties of epoxy resins; adhesion of polymers to metals; electropolymerization; graphite fiber/polmer composites. *Mailing Add:* Dept Chem Eng Box U-139 Univ Conn Storrs CT 06269

BELL, JEFFREY, b Sept 10, 52. IMMUNOLOGY. *Educ:* Fairleigh Dickinson Univ, BS, 75; Temple Univ, PhD, 89. *Prof Exp:* Res asst, Health Res Inst, Madison, NJ, 75-76; path asst. Health Res Inst, Madison, NJ, 75-76; res assoc, Div Allergy/Immunol, 89-90, RES ASSOC, DIV INFECTIOUS DIS, WASH UNIV SCH MED, 90- *Mem:* Am Soc Pharmacol & Exp Therapeut; Int Soc Immunopharmacol; Am Soc Microbiol. *Res:* Development of several antigen delivery systems in attenuated Salmonella typhimurium strains that carry Haemophilus influenzae outer membrane protein P2 to interact with the secretory immune system via the Peyer's Patches; investigate the interaction with and regulation of the secretory immune response by recombinent P2 proteins presented by these antigen delivery systems and their role as possible vaccine candidates; modification of these antigen delivery systems by introduction of recombinent lymphokine proteins in order to enhance specific immune responses. *Mailing Add:* Div Infectious Dis Wash Univ Sch Med 400 S Kingshighway Blvd St Louis MO 63110

BELL, JERRY ALAN, b Davenport, Iowa, June 28, 36; m 61, 84; c 4. PHYSICAL CHEMISTRY. *Educ:* Harvard Univ, AB, 58, PhD(chem), 62. *Prof Exp:* Res assoc chem, Brandeis Univ, 61-62; asst prof, Univ Calif, Riverside, 62-67; assoc prof, 67-72, PROF CHEM, SIMMONS COL, 72- *Concurrent Pos:* Vis res prof, Brandeis Univ, 80-81; vis prof, Harvard Univ, 87-88; div dir, directorate Sci & Engr Educ, NSF, 84-86. *Honors & Awards:* Catalyst Award, Mfg Chemists Asn, 77. *Mem:* Am Chem Soc. *Res:* Kinetics of fast reactions; methylene reactions; flash photolysis; photochemistry; enzyme kinetics and mechanism; chemical education. *Mailing Add:* Dept Chem Simmons Col 300 The Fenway Boston MA 02115

BELL, JIMMY HOLT, b Burlington, NC, Jan 26, 35; m 61; c 3. ANALYTICAL CHEMISTRY. *Educ:* Elon Col, BS, 56. *Prof Exp:* Chemist, 57-60, res chemist, 60-72, assoc scientist, 72-80, SCIENTIST, LORILLARD, INC, 80- *Mem:* Am Chem Soc. *Res:* Analysis of the chemical composition of tobacco from harvest to curing; chemical changes during manufacturing processes; isolation and identification of the components of tobacco combustion. *Mailing Add:* 1826 New Garden Rd Greensboro NC 27410-2006

BELL, JIMMY TODD, b Hazlehurst, Ga, Dec 17, 38; m 61; c 2. PHYSICAL CHEMISTRY. *Educ:* Berry Col, AB, 60; Univ Miss, PhD(phys chem), 63. *Prof Exp:* Res chemist chem technol div, 63-74, group leader, 74-80, MGR FUEL CYCLE CHEM, OAK RIDGE NAT LAB, 80- *Mem:* AAAS; Am Chem Soc; Am Nuclear Soc. *Res:* High temperature and pressure spectroscopy of transuranium element aqueous solutions; photochemistry of actinides; tritium equilibrias; tritium permeation; high temperature materials chemistry; fuel cycle chemistry. *Mailing Add:* Rte 4 Box 344 Kingston TN 37763-9648

BELL, JOHN FREDERICK, b Ashland, Ore, Jan 7, 24; m 50; c 4. FOREST MENSURATION, FOREST MANAGEMENT. *Educ:* Ore State Univ, BS, 49; Duke Univ, MF, 51; Univ Mich, PhD, 70. *Prof Exp:* Tech asst, Ore State Forestry Dept, 49-50, inventory forester, 51-53, unit forester, 53-55, supvr forest inventory, 55-59; from asst prof to assoc prof, Col Forestry, Ore State Univ, 59-72, prof forest mensutation, 72-85; RETIRED. *Concurrent Pos:* Chmn, Forest Inventory Working Group, Soc Am Foresters. *Mem:* Soc Am Foresters. *Res:* Forest measurements and sampling. *Mailing Add:* Sch Forestry Ore State Univ Corvallis OR 97331

BELL, JOHN MILTON, b Islay, Alta, Jan 16, 22; m 44; c 5. ANIMAL NUTRITION. *Educ:* Univ Alta, BScA, 43; McGill Univ, MSc, 45; Cornell Univ, PhD, 48. *Hon Degrees:* DSc, McGill, 86. *Prof Exp:* From asst prof to assoc prof animal husb, Univ Sask, 48-54, prof & head dept, 54-75, assoc dean, 75-80, Burford Hooke Prof, 80-84, prof animal sci, Res Col Agr, 84-89, EMER PROF ANIMAL SCI, UNIV SASK, 89- *Concurrent Pos:* Chmn comt animal nutrit, Can Dept Agr; chmn, Sask Adv Coun Animal Prod; mem bd gov, Univ Sask, 77-83. *Honors & Awards:* Borden Award, 62; Canola Res Award, 81; Jas McAnsh Award, 88. *Mem:* Am Soc Animal Sci; Am Inst Nutrit; Can Soc Animal Sci (pres, 52); fel Agr Inst Can; fel Royal Soc Can. *Res:* Swine nutrition; dairy calf nutrition; toxic factors in rapeseed; environmental temperature effects on animals. *Mailing Add:* Animal Sci Bldg Rm 112 Univ Sask Saskatoon SK S7N 0W0 Can

BELL, JOHN URWIN, b Elgin, Morayshire, Scotland, Apr 4, 48; Can citizen; div; c 2. TOXICOLOGY, PHARMACOLOGY. *Educ:* Dalhousie Univ, BSc, 70, PhD(pharmacol), 74. *Prof Exp:* Davies Mem Res Fel, Univ Calgary, 74-76; asst prof pharmacol & toxicol, WVa Univ, 76-80; ASSOC PROF TOXICOL, UNIV FLA, GAINESVILLE, 80- *Mem:* Soc Toxicol; Soc Toxicol Can; Pharmacol Soc Can; Am Soc Pharmacol & Exp Therapeut. *Res:* The influence of xenobiotics on mammalian development; perinatal toxicology; toxicity of heavy metals on mammalian reproduction; fetal drug metabolism; heavy metal toxicology. *Mailing Add:* Dept Physiol Sci Univ Fla Box J-137 JHMHC Gainesville FL 32610-0137

BELL, JONATHAN GEORGE, b Spokane, Wash. MATHEMATICS, BIOMATHEMATICS. *Educ:* San Diego State Univ, BA, 69, MS, 71; Univ Calif, Los Angeles, PhD(appl math), 77. *Prof Exp:* Mathematician acoust, Naval Undersea Ctr, 74; programmer, Logicon, Inc, 75; asst prof appl math, Tex A&M Univ, 77-; AT DEPT MATH, HEALTH SCI CTR, STATE UNIV NY. *Mem:* Soc Indust & Appl Math; Am Math Soc; Sigma Xi. *Res:* Mathematical modeling of the nerve conduction process, analyzing the models for solution behavior; models of nerve bundles and flow of nutrients in trees. *Mailing Add:* Dept Math Diefendorf Hall State Univ NY Buffalo NY 14214

BELL, JULIETTE B, b Apr 15, 55. MOLECULAR GENETICS. *Educ:* Talladega Col, BA, 77; Atlanta Univ, PhD(chem, biochem), 87. *Prof Exp:* Asst chemist, Southern Res Inst, Birmingham, Ala, 77-78; Carolina minority postdoctoral scholar, Univ NC, Chapel Hill, 87-87, postdoctoral fel, 89-90; STAFF FEL, NAT INST ENVIRON HEALTH SCI, RESEARCH TRIANGLE PARK, NC, 90- *Concurrent Pos:* Sub teacher, SC Pub Sch, 78-79; teaching asst, Clark Col, Atlanta, 82. *Mem:* Asn Women in Sci (co-pres, 91-92); AAAS; Am Soc Biochem & Molecular Biol. *Res:* Molecular genetic and biochemical studies of protein structure/function relationships in nucleotide biosynthesis and metabolism. *Mailing Add:* Lab Molecular Genetics E3-01 Nat Inst Environ Health Sci PO Box 12233 Research Triangle Park NC 27709

BELL, KENNETH J(OHN), b Cleveland, Ohio, Mar 1, 30; m 56; c 4. CHEMICAL ENGINEERING. *Educ:* Case Inst Technol, BSchE, 51; Univ Del, MChE, 53, PhD (chem eng), 55. *Prof Exp:* Engr heat transfer, Hanford oper, Gen Elec Co, 55-56; asst prof chem eng, Case Inst Technol, 56-61; from assoc prof to prof, 61-76, REGENTS PROF CHEM ENG, OKLA STATE UNIV, 76-, KERR-MCGEE CHAIR CHEM ENG, 90- *Concurrent Pos:* Mem vis fac, Oak Ridge Sch Reactor Technol, 58; consult to numerous indust & govt agencies; sr res engr, Heat Transfer Res, Inc, 68-69; sr chem engr, Argonne Nat Lab, 85- *Honors & Awards:* Don Q Kern Award, Am Inst Chem Engrs, 78. *Mem:* Fel Am Inst Chem Engrs; Am Soc Eng Educ; Nat Soc Prof Engrs. *Res:* Design of heat transfer equipment. *Mailing Add:* Sch Chem Eng Okla State Univ Stillwater OK 74078

BELL, M(ARION) W(ETHERBEE) JACK, b Wichita Falls, Tex, Aug 17, 27; m 50; c 6. NAVAL SCIENCE. *Educ:* US Merchant Marine Acad, BS, 49. *Prof Exp:* Propulsion engr, Columbus Div, NAm Aviation, Inc, 54-57, proj engr missile preliminary design, 57-61, chief sci space systs, Space & Info Systs Div, 62-63, dir lunar & planetary systs, 63-66, mgr advan systs, 66-69, chief prog eng, Space Div, NAm Rockwell Corp, 69-71, proj mgr, Advan Shuttle Proj, 72-77, dir advan syst, Shuttle Orbiter Div, 77-81, ASST CHIEF ENGR PRELIMINARY DESIGN & MISSION ANALYSIS, SPACE GROUP, ROCKWELL INT CORP, 81- *Mem:* Am Inst Aeronaut & Astronaut; Brit Interplanetary Soc. *Res:* Design and development of advanced space vehicle systems; synthesis and analysis of advanced spacecraft and launch vehicle system concepts; synthesis, analysis, design and test of rocket propulsion systems. *Mailing Add:* 3332 Corinna Dr Palos Verdes Peninsula CA 90274

BELL, MALCOLM RICE, b New Britain, Conn, Aug 22, 28; m 60; c 3. ORGANIC CHEMISTRY. *Educ:* Rensselaer Polytech Inst, BS, 50; Cornell Univ, PhD(chem), 55. *Prof Exp:* Develop chemist, Allied Chem & Dye Corp, 50-51; asst, Cornell Univ, 51-55; res assoc, 55-61, res chemist-group leader, 61-65, SECT HEAD, STERLING-WINTHROP RES INST, 65- *Mem:* Am Chem Soc. *Res:* Gliotoxin; synthetic analgesic agents; synthetic therapeutic agents. *Mailing Add:* Sterling Winthrop Res Inst Rensselaer NY 12144-3493

BELL, MARCUS ARTHUR MONEY, b Victoria, BC, Mar 1, 35; m 59; c 3. PLANT ECOLOGY, ETHNOBOTANY. *Educ:* Univ BC, BSF, 57, PhD(plant ecol), 64; Yale Univ, MF, 58. *Prof Exp:* Lectr plant & human ecology, Bobot, morphol & taxon, 61-63, instr, 63-64, asst prof bot, 64-70, ASSOC PROF BIOL, UNIV VICTORIA, 70-, CUR, HERBARIUM BOT COLLECTIONS, 63- *Concurrent Pos:* Nat Res Coun Can res grants, 65-, travel grant, Van Dusen fel & Ger Acad Exchange Serv grant, 65; Neth Educ Dept grant & Brit Coun grant, 66. *Mem:* AAAS; Ecol Soc Am; Can Bot Asn; Int Asn Plant Taxon; Int Asn Plant Ecol. *Res:* Forest ecology; classification and analysis of plant communities and environments; forest productivity, environmental information; ecological aspects of waste disposal; ethnobotany of British Columbia native groups. *Mailing Add:* Dept Biol Sci Univ Victoria Box 1700 Victoria BC V8W 2Y2 Can

BELL, MARVIN CARL, b Centertown, Ky, Nov 24, 21; m 48; c 2. ANIMAL NUTRITION. *Educ:* Univ Ky, BS, 47, MS, 49; Okla Agr & Mech Col, PhD(animal nutrit), 52. *Prof Exp:* Res asst animal husb, Univ Ky, 47-48; asst, Okla Agr & Mech Col, 48-51; assoc prof animal husb & assoc animal husbandman, 51-65, from assoc scientist to scientist, AEC, 58-74, PROF ANIMAL SCI, UNIV TENN, KNOXVILLE, 65- *Concurrent Pos:* Mem

staff, Travel Lect Prog, Oak Ridge Inst Nuclear Studies, 59-; mem protection & measures sci comt, Nat Coun Radiation, 63. *Mem:* Fel AAAS; Am Inst Nutrit; fel Am Soc Animal Sci; Nutrit Today Soc; fel Nat Feed Ingredients Asn. *Res:* Metabolism studies with cattle and sheep; urea and stilbestrol feeding to ruminants; fescue poisoning investigations; feeding value of wood molasses; radioisotope tracers in nutrition studies; irradiation effects on metabolism in large animals; mineral metabolism; hypomagnesemia tetany; vitamin injections. *Mailing Add:* 4608 Hill Top Rd Knoxville TN 37920

BELL, MARVIN DRAKE, b Tulsa, Okla, Jan 14, 29; m 50; c 4. SOLID STATE PHYSICS. *Educ:* Okla State Univ, BS, 55, MS, 56, PhD(physics), 64. *Prof Exp:* Asst chemist, Phillips Petrol Co, 52-54; res physicist, 56-58; res assoc physics, Okla State Univ, 64-65; assoc prof, Western Ky Univ, 65-67; chmn dept, 67-81, PROF PHYSICS, CENT MO STATE UNIV, 67- *Concurrent Pos:* Fac fel, Solar Energy Res Inst, 80. *Mem:* Am Phys Soc; Optical Soc Am; Am Asn Physics Teachers; Solar Energy Soc; Laser Inst Am. *Res:* Electron spin resonance in semiconducting diamonds; photoconductivity; photovoltaic effect; electrical properties of diamonds; holography; coherent optics; solar energy materials research. *Mailing Add:* Dept Physics Cent Mo State Univ Warrensburg MO 64093

BELL, MARY, b Providence, RI, May 21, 37. CELL BIOLOGY, HUMAN ANATOMY. *Educ:* Brown Univ, AB, 58; Yale Univ, PhD(anat), 65. *Prof Exp:* Asst scientist, Dept Cutaneous Biol & Electron Micros, Ore Regional Primate Res Ctr, 64-71, assoc scientist, 71-74; asst prof, Dept Environ Health & Dept Anat, Col Med, Univ Cincinnati, 74-81, assoc prof, Dept Anat, Cell Biol & Dermat, 81- 84; prin scientist, Gillette Res Inst, 85-89; HEALTH SCIENTIST ADMIN, NIH, 89- *Concurrent Pos:* Assoc prof, Div Dermat, Sch Med, Univ Ore, 70-74. *Mem:* Am Soc Cell Biol; Am Asn Anatomists; Electron Micros Soc Am; Soc Investigative Dermat. *Res:* Ultrastructure of skin; effects of topical agents in epithelia. *Mailing Add:* c/o NIH-NCI-DEA-GRB 5333 Westbard Ave Westward Bldg Rm 838 Bethesda MD 20892

BELL, MARY ALLISON, b Halifax, NS, Aug 31, 36; Can citizen. NEUROLOGICAL SCIENCES, MICROVASCULAR RADIOLOGY. *Educ:* Univ Kings Col, Can, BSc, 57; Dalhousie Univ, MSc, 69; Oxford Univ, Eng, DPhil(anat), 77. *Prof Exp:* Lectr & asst prof anat, Dalhousie Univ, Halifax, Can, 69-73; res fel path, Univ Western Ont, London, 77-84; RES ASST PROF RADIOL, BOWMAN GRAY SCH MED, WINSTON-SALEM, NC, 84- *Mem:* Am Asn Neuropathologists; Royal Micros Soc; Can Soc Anatomist; Soc Neurosci; Brit Neuropath Soc; Sigma Xi. *Res:* X-ray, light and electron microscopy of blood vessels in brain and peripheral nerves; comparisons of the human cerebrovasculature in youth, age, Alzheimer's dementia and hypertension. *Mailing Add:* Radiol Dept Bowman Gray Sch Med Winston-Salem NC 27103

BELL, MAURICE EVAN, b New Castle, Ohio, Sept 10, 10; wid; c 1. OPERATIONS RESEARCH. *Educ:* Kenyon Col, ScB, 32; Mass Inst Technol, PhD(physics), 37. *Prof Exp:* Res engr, Westinghouse Elec & Mfg Co, 37-41; res assoc underwater sound, Harvard Univ, 41-42; res assoc opers res, Nat Defense Res Comt, Off Sci Res & Develop & US Navy, 42-47; sci liaison officer, Off Naval Res, London, 48, dep sci dir, 49-50, sci dir, 51-53; eng mgr, Sylvania Missile Systs Lab, Sylvania Elec Prod, Inc, 54, mgr solid state br, Physics Lab, 54-56; asst dean res, Col Earth & Mineral Sci, 56-72, prof, 56-76, assoc dean, 72-76, EMER PROF GEOPHYS & ASSOC DEAN, PA STATE UNIV, UNIVERSITY PARK, 76- *Res:* Ionization probabilities; dielectric properties of solids; high vacuum lubricants; operations research; missile systems; semiconductor physics. *Mailing Add:* 1008 Glenn Circle N State College PA 16803-3464

BELL, MAX EWART, b Milton, Iowa, Nov 19, 27; m 52; c 3. PLANT MORPHOLOGY. *Educ:* Parsons Col, BS, 50; Iowa State Col, MS, 52, PhD, 54. *Prof Exp:* Asst bot, Iowa State Col, 50-54; PROF BOT, NORTHEAST MO STATE UNIV, 54- *Mem:* Bot Soc Am; Sigma Xi. *Res:* Electron microscopy; corn embryology and anatomy. *Mailing Add:* Sci Div Northeast Mo State Univ Kirksville MO 63501

BELL, MICHAEL ALLEN, b Brooklyn, NY, Sept 17, 47; m 73; c 3. EVOLUTION, PALEOBIOLOGY. *Educ:* Univ Calif, Los Angeles, BA, 70, MA, 75, PhD(biol), 76. *Prof Exp:* Lectr biol, Univ Calif, Los Angeles, 78; asst prof, 78-84, ASSOC PROF ECOL & EVOLUTION, STATE UNIV NY, STONY BROOK, 84- *Concurrent Pos:* Res assoc ichthyol, Natural Hist Mus Los Angeles County, 76- *Mem:* AAAS; Am Soc Ichthyologists & Herpetologists; Soc Study Evolution (secy, 88-90); Soc Syst Zool; Soc Vert Paleont. *Res:* Evolutionary biology of stickleback fish; roles of natural selection and gene flow in formation of patterns of spatial differentiation and variation within populations; patterns and rates of evolution in fossils; mechanisms of morphological change in evolution. *Mailing Add:* Dept Ecol & Evolution State Univ NY Stony Brook NY 11794-5245

BELL, MICHAEL J(OSEPH), chemical & nuclear engineering, for more information see previous edition

BELL, MILO C, b 1905; US citizen. FISHERIES, ENGINEERING. *Educ:* Univ Wash, BS, 23. *Prof Exp:* Mem staff, Wash Dept Fisheries, 30-33, 35-43; from instr to prof, 40-75, EMER PROF FISHERIES, UNIV WASH, 75- *Concurrent Pos:* Mem, Int Pac Salmon Fisheries Comn, 43-51; consult, Wash Dept Fisheries, 51-57, var US power co, 57- & Wash Water Res Ctr, 68-69. *Mem:* Nat Acad Eng. *Mailing Add:* PO Box 23 Mukilteo WA 98275

BELL, NORMAN H, b Gainesville, Ga, Feb 11, 31; m 59, 68, 72; c 3. ENDOCRINOLOGY, METABOLISM. *Educ:* Emory Univ, AB, 51; Duke Univ, MD, 55. *Prof Exp:* From intern to asst resident med, Duke Univ Med Ctr, 55-57; clin assoc, Nat Inst Allergy & Infectious Dis, 57-59; staff investr endocrinol, Nat Heart Inst, 59-63; assoc med, Med Sch, Northwestern Univ, 63-65; asst prof, 65-68; assoc prof, Sch Med, Ind Univ, Indianapolis, 68-71; prof med pharmacol & dir, Clin Res Ctr, 71-79; PROF MED PHARMACOL,

DIR, DIV BONE & MINERAL METAB & DIR, GEN CLIN RES CTR, MED UNIV SC, CHARLESTON, 79- *Concurrent Pos:* Attend physician, Res Vet Admin Hosp, 64-68; chief, Sect Metab, Indianapolis Vet Admin Hosp, 68-74, assoc chief staff, 71-78; med investr, Vet Admin Hosp, Charleston, 81-; mem, Gen Med B Study Sect, NIH, 82-86, chmn, 85-86. *Mem:* AAAS; Asn Am Physicians; Am Soc Bone & Mineral Res (secy-treas, 77-85, pres, 86-87); Am Soc Nephrol; Am Soc Clin Invest; Am Soc Pharm Exp Therapeut. *Res:* Bone and mineral metabolism. *Mailing Add:* Vet Admin Med Ctr 109 Bee St Charleston SC 29403

BELL, NORMAN R(OBERT), b Syracuse, NY, May 5, 18; m 45; c 2. ENGINEERING CONSULTING. *Educ:* Lehigh Univ, BS, 39; Cornell Univ, MS, 45. *Prof Exp:* Control engr, Gen Elec Co, 39-41; instr elec eng, Cornell Univ, 41-46; asst prof, Bucknell Univ, 46-50; assoc prof, Brooklyn Polytech, 50-58; assoc prof elec eng, NC State Univ, 58-82; RETIRED. *Concurrent Pos:* Sr Engr, Bakelite, 53; consult, Proj Matterhorn, AEC, 54-58; consult engr, NC & NY. *Mem:* Inst Elec & Electronics Engrs; Am Soc Eng Educ. *Res:* Digital computers; automata; bio-engineering. *Mailing Add:* 2312 Woodrow Dr Raleigh NC 27609-7627

BELL, PAUL BURTON, JR, b Memphis, Tenn, June 24, 46. CELL BIOLOGY, ELECTRON MICROSCOPY. *Educ:* Washington Univ, St Louis, AB, 68; Yale Univ, PhD(biol), 75. *Prof Exp:* Fel cell biol, Wallenberg Lab, Uppsala Univ, Sweden, 74; fel biol, div biol, Calif Inst Tech, 74-76, res fel, 76-79; asst prof biol, 79-86, ASSOC PROF BIOL, DEPT ZOOL, UNIV OKLA, NORMAN, 86- *Concurrent Pos:* Actg asst prof biol, dept biol, Univ Calif, Los Angeles, 76-78, adj asst prof, 78-79; vis researcher, dept path, Linkoping Univ, Sweden, 84-88. *Mem:* AAAS; Am Soc Cell Biol; Am Soc Zool; Soc Develop Biol; Electron Micros Soc Am; Scand Soc Electron Micros. *Res:* Mechanism of cell motility and contact inhibition of movement; tumor invasion and metastasis; structure and function of the cytoskeleton, especially intermediate filaments; electron microscopic and immunoelectron microscopic methods; quantitative immunofluorescence light microscopy. *Mailing Add:* Dept Zool Univ Okla Norman OK 73019

BELL, PAUL HADLEY, b Cornerville, Ohio, Aug 3, 14; m 41; c 4. BIOCHEMISTRY. *Educ:* Marietta Col, AB, 36; Pa State Col, MS, 38, PhD(phys chem), 40. *Hon Degrees:* DSc, Marietta Col, 60. *Prof Exp:* Asst, Pa State Col, 37-40; res chemist, Lederle Labs, Am Cyanamid Co, 40-56, head biochem res dept, 56-69, res fel biochem, 69-76; RETIRED. *Mem:* Am Chem Soc; fel NY Acad Sci. *Res:* Mechanism of drug action; purification and chemistry of antibiotics and pituitary hormones; enzyme kinetics; metabolism of drugs; intermediate metabolism; steroid action; fibrinolytic enzymes; immunological control systems; transfer factor. *Mailing Add:* 1208 N Lee St No 130 Leesburg FL 32748

BELL, PERSA RAYMOND, b Ft Wayne, Ind, Apr 24, 13; m 41, 67; c 1. NUCLEAR MEDICINE, CLIMATOLOGY. *Educ:* Howard Col, BSc, 36, DSc, 54. *Prof Exp:* Mem staff, Nat Defense Res Comt Proj, Univ Chicago, 40-41; mem staff radiation lab, Mass Inst Technol, 41-46; physicist, Oak Ridge Nat Lab, 46-67 & 70-78; chief, Lunar & Earth Si Div, Manned Spacecraft Ctr, Houston, NASA, 67-70; staff mem, Inst Energy Anal, Oak Ridge Assoc Univs, 79-86; RETIRED. *Mem:* Am Phys Soc; Am Geophys Union; Sigma Xi. *Res:* Electronic and physics instrumentation; nuclear physics; nuclear medical instrumentation; planetary research. *Mailing Add:* 132 Westlook Circle Oak Ridge TN 37830

BELL, PETER M, b New York, NY, Jan 3, 34; m 59; c 4. GEOCHEMISTRY, GEOPHYSICS. *Educ:* St Lawrence Univ, BS, 56; Univ Cincinnati, MS, 59; Harvard Univ, PhD(geophys sci), 63. *Prof Exp:* Fel, Carnegie Inst, 63-64; corp technologist, Solid State Sci, 83-84, vpres sci & technol, Norton Christenson, 84-86, Norton Co, 86-90, VPRES CORP RES & CHIEF SCIENTIST, NORTON CO,90-; GEOPHYSICIST, CARNEGIE INST, 64- *Concurrent Pos:* Mat Adv Bd, Nat Res Coun, Mass Inst Technol, bd gov, Assoc Inst Physics. *Mem:* Geol Soc Am; fel Mineral Soc Am; fel Am Geophys Union. *Res:* High pressure-temperature experimental geochemistry and geophysics. *Mailing Add:* Norton Co 120 Front St Suite 800 Worcester MA 01608-1446

BELL, RAYMOND MARTIN, b Weatherly, Pa, Mar 21, 07; m 42; c 3. ASTRONOMY. *Educ:* Dickinson Col, AB, 28; Syracuse Univ, AM, 30; Pa State Univ, PhD(physics), 37. *Hon Degrees:* ScD, Washington & Jefferson Col, 76. *Prof Exp:* Asst physics, Syracuse Univ, 28-30 & Pa State Univ, 30-37; from instr to assoc prof, 37-46, prof & chmn dept, 46-75, EMER PROF PHYSICS, WASHINGTON & JEFFERSON COL, 75- *Mem:* Am Asn Physics Teachers; Am Inst Physics. *Res:* Raman spectra; propagation of radio waves; Allison magnetooptic effect; dielectric measurements of porous materials; barium titanate. *Mailing Add:* 413 Burton Ave Washington PA 15301-3303

BELL, RICHARD DENNIS, b Prague, Okla, Apr 21, 37; m 61. PHYSIOLOGY. *Educ:* Univ Okla, BS, 60, MS, 65, PhD(physiol), 68; Okla State Univ, BS, 61. *Prof Exp:* Res asst prof urol & asst prof physiol, biophys & allied health educ, Health Sci Ctr, Univ Okla, 68-80; MEM STAFF, CHICAGO COL OSTEOP MED, 80- *Concurrent Pos:* USPHS res grant, 69-71. *Mem:* NY Acad Sci; Am Physiol Soc; Sigma Xi. *Res:* Renal physiology; body fluids and electrolytes; blood and lymph vascular systems of the kidney and their relationships to renal function. *Mailing Add:* 1018 Meadow Crest La Grange Park IL 60525

BELL, RICHARD OMAN, b Havre, Mont, Feb 16, 33; m 57; c 4. SOLID STATE PHYSICS, PHOTOVOLTAICS. *Educ:* Mont State Univ, BS, 55; Univ Calif, Los Angeles, MS, 58; Boston Univ, PhD(physics), 68. *Prof Exp:* Mem tech staff, Hughes Res Labs, Calif, 56-58; staff scientist, Raytheon Co, Mass, 60-62; sr scientist, Tyco Labs, 62-66; group leader electronic mat, Tyco Labs, Waltham, 67-74, RES ASSOC, MOBIL SOLAR ENERGY CORP, 75- *Mem:* Am Phys Soc; Inst Elec & Electronics Engrs. *Res:* Electronic materials and devices for solar energy applications; evaluation of physical and electronic properties of semiconductor, ferroelectric and ferromagnetic materials. *Mailing Add:* 24 Austin Rd Sudbury MA 01776

BELL, RICHARD THOMAS, b Haileybury, Ont, May 1, 37; m 62; c 3. URANIUM RESOURCE ASSESSMENT, URANIUM METALLURGY. *Educ:* Univ Toronto, BA, 60, MA, 62, Princeton Univ, PhD(geol), 66. *Prof Exp:* Tech officer, Geol Surv Can, 62-63, res scientist, 66-69; from asst prof to assoc prof sedimentology & struct, Brock Univ, 69-75; RES SCIENTIST, GEOL SURV CAN, 75- *Mem:* Fel Geol Soc Am. *Res:* Metallurgy research & regional assessment of uranium in Western Canada & northern Quebec; regional geological studies in precambrian rocks in western Canada, Central Keewatin. *Mailing Add:* Geol Surv Can 601 Booth St Ottawa ON K1A 0E8 Can

BELL, ROBERT ALAN, b Detroit, Mich, Apr 13, 34; m 62; c 2. MECHANICAL ENGINEERING. *Educ:* Univ Calif, Berkeley, BS, 58, MS, 60. *Prof Exp:* Proj engr mech eng, William M Brobeck & Assoc, 60-65; mech engr, Stanford Linear Accelerator Ctr, 65-69; vis scientist mech eng, Europ Coun Nuclear Res, Geneva, Switz, 69-72; MECH ENGR, STANFORD LINEAR ACCELERATOR CTR, 72- *Mem:* Soc Automotive Engrs. *Res:* Particle accelerator design and development. *Mailing Add:* Stanford Linear Accelerator Ctr PO Box 4349 Stanford CA 94305

BELL, ROBERT EDWARD, b Eng, Nov 29, 18; m 47; c 1. NUCLEAR PHYSICS. *Educ:* Univ BC, BA, 39, MA, 41; McGill Univ, PhD(physics), 48. *Hon Degrees:* DSc, Univ NB, Univ Laval, Univ Montreal, Univ BC, McMaster Univ, Carlton Univ & Mc Gill Univ, LLD, Univ Toronto & Concordia Univ, Bishop Univ. *Prof Exp:* Physicist, Nat Res Coun Can, 41-45; res physicist, Atomic Energy Proj, Chalk River, Ont, 45-56; assoc prof physics, 56-60, dir, Foster radiation Lab, 60-69, vdean phys sci, 64-67, dean grad studies & res, 69-70, prin & vchancellor, 70-79, Rutherford prof physics, 60-83, EMER PROF PHYSICS, MCGILL UNIV, 83- *Concurrent Pos:* Dir, Arts, Sci & Technol Ctr, Vancouver, 83-85. *Honors & Awards:* Physics Medal, 68; Centennial Medal, Royal Soc Can, 82. *Mem:* Fel Am Phys Soc; Can Asn Physicists (pres, 65-66); fel Royal Soc Can (pres, 78-81); fel Royal Soc. *Res:* Beta and gamma ray spectroscopy; high speed coincidence techniques; proton induced nuclear reactions; positron annihilation and positronium; delayed proton radioactivity; teaching and administration. *Mailing Add:* 1120 Berry Circle Norman OK 73072

BELL, ROBERT EUGENE, b Marion, Ohio, June 16, 14; m 38; c 2. ARCHAEOLOGY, PHYSICAL ANTHROPOLOGY. *Educ:* Univ NMex, BA, 40; Univ Chicago, MA, 43, PhD(anthrop), 47. *Prof Exp:* From asst prof to prof anthrop, Univ Okla, 47-67, chmn dept, 47-55, head cur, Social Sci Div, 47-80, George Lynn Cross Res prof anthrop, 67-80; RETIRED. *Concurrent Pos:* Wenner-Gren Found res grant, 55-56; ed elect, Soc Am Archaeol, 65-66. *Mem:* AAAS; Soc Am Archaeol; fel Am Anthrop Asn; Am Asn Phys Anthrop; Sigma Xi. *Res:* Archaeology of Southern Plains; early man in the new world. *Mailing Add:* 1120 Berry Circle Norman OK 73072

BELL, ROBERT GALE, biochemistry, for more information see previous edition

BELL, ROBERT JOHN, b Lewisburg, WVa, Dec 18, 34; m 57; c 2. SPECTROSCOPY, SOLID STATE PHYSICS. *Educ:* Va Polytech State Univ Inst, BS, 56, PhD(solid state physics), 63; Rice Inst, MA, 58. *Prof Exp:* Asst instr physics, Va Polytech Inst, 62-63; sr res mem, Southwest Res Inst, 63-65; from asst prof to assoc prof, 65-72, actg chmn dept, 70-71, PROF PHYSICS, UNIV MO-ROLLA, 72- *Concurrent Pos:* Asst prof, Trinity Univ, Tex, 64-65; vis scientist, Max Planck Inst Solid State Physics, Stuttgart, Ger, 71-72. *Honors & Awards:* Res Award, Univ Mo-Rolla, 71. *Mem:* Fel Am Phys Soc. *Res:* Submillimeter and millimeter spectroscopy studies of semiconductors, metals, dielectrics and superconductors at room and cryogenic temperatures; surface physics and chemistry; small particle studies. *Mailing Add:* Dept Physics Univ Mo Rolla MO 65401

BELL, ROBERT LLOYD, b McKeesport, Pa, Sept 3, 23; m 51; c 2. MEDICINE, NEUROSURGERY. *Educ:* Washington & Jefferson Col, BS, 43; Univ Pittsburgh, MD, 47. *Prof Exp:* Asst instr neurosurg, State Univ NY Downstate Med Ctr, 54-55, from instr to assoc prof, 55-61; chief neurosurg, US Vet Admin Ctr, Wadsworth, Kans, 61-67; chief neurosurg, 64-69, CHIEF NUCLEAR MED, VET ADMIN HOSP, 69-; CHIEF NEUROSURG, CHESTER COUNTY HOSP, 70- *Mem:* Am Col Nuclear Med (pres, 81-); Am Bd Nuclear Med; Am Bd Neurol Surg Soc Nuclear Med; fel Am Col Surg; Cong Neurol Surg. *Res:* Nuclear instrumentation; brain tumor localization. *Mailing Add:* Coatesville Vet Admin Med Ctr 51 S 12th Ave Coatesville PA 19320

BELL, ROBERT MAURICE, b Lincoln, Nebr, Mar 24, 44; m 66; c 2. BIOCHEMISTRY, MOLECULAR BIOLOGY. *Educ:* Univ Nebr, Lincoln, BS, 66; Univ Calif, Berkeley, PhD(biochem), 70. *Prof Exp:* NSF fel, Wash Univ, 70-72; from asst prof to prof, 72-87, JAMES B DUKE PROF BIOCHEM, MED CTR, DUKE UNIV, 87-, DEP DIR, DUKE COMPREHENSIVE CANCER CTR, 87-, HEAD, SECT CELL GROWTH, REGULATION & ONCOGENESIS, 87- *Concurrent Pos:* NSF grant, Med Ctr, Duke Univ, 73-74, NIH grant, 74-, Nat Found-March of Dimes grant, 74-76; estab investr, Am Heart Asn, 74-79, grant, 76-78; Macy fac scholar, Med Res Coun Lab Molecular Biol, Cambridge, Eng; assoc ed, J Biol Chem, Duke Univ Med Ctr, 87-; chmn, Basic Sci Planning Comt, Duke Univ Comprehensive Cancer Ctr, 88-, mem, numerous comts, 84-; chmn, Sphinx Biotechnol Adv Comt, Med Col Va, 87-; Sci adv comt, Gallo Res Clin, Univ Calif, San Diego, 88- *Mem:* Am Soc Biol Chemists; Am Soc Microbiol. *Res:* Regulation of enzymes of phospholipid and triacylglycerol synthesis; membrane biogenesis; diacylglycerols function as intracellular regulators, second messengers, of protein kinases; phospholipid bilayer assembly; transbilayer movement of phospholipids; mechanism of oncogene action in cellular transformation; signal transduction mechanisms; regulation of diacylglycerol second messenger generation and attenuation; diacylglycerol metabolism; diacylglycerols as intracellular regulators of protein kinase C; diacylglycerol protein kinase C structure-function relationships; sphingosine/lysosphingolipid second messengers and signal transduction; sphingosineprotein kinase C structure activity relationships; biological activities and function of sphingoglycolipids, sphingolipid cycles; mechanism of negative growth factors action; regulation of sphingosine/lysosphingosine production & metabolism; structure and function of protein kinase C; structure and function of enzymes of lipid metabolism; role of sphingolipid breakdown production cellular regulation. *Mailing Add:* Dept Biochem Duke Univ Med Ctr Durham NC 27710

BELL, ROBIN GRAHAM, b Canberra, Australia, July 12, 42; div; c 3. RESISTANCE TO HELMINTHS, IMMUNOGENETICS. *Educ:* Australian Nat Univ, BSc, 68, PhD(immunol), 71. *Prof Exp:* Med res fel microbiol, Univ Western Australia, 71-73, res officer immunol & microbiol, Princess Margaret Hosp, 73-76; res officer, 76-79, from asst prof to assoc prof, 79-90, PROF IMMUNOL, J A BAKER INST ANIMAL HEALTH, NY STATE COL VET MED, CORNELL UNIV, 90- *Concurrent Pos:* Mem, Trop Med & Parasitol Study Sect, Nat Inst Allergy & Infectious Dis, 87-91. *Mem:* Am Asn Immunologists; Am Soc Trop Med & Hyg. *Res:* Immunity at mucosal surfaces, mechanisms of anti-parasite immunity, genetics of resistance and susceptibility to parasites in mammalian species; effects of nutritional factors on resistance to infection. *Mailing Add:* J A Baker Inst NY State Col Vet Med Ithaca NY 14853

BELL, ROGER ALISTAIR, b Walton-on-Thames, Eng, Sept 16, 35; m 60; c 2. ASTROPHYSICS. *Educ:* Univ Melbourne, BSc, 57; Australian Nat Univ, PhD(astron), 62. *Hon Degrees:* Dr, Uppsala Univ, Sweden, 82. *Prof Exp:* Lectr physics, Univ Adelaide, 62-63; from asst prof to prof, 63-87, PROF ASTRON & DIR, UNIV MD, 87- *Concurrent Pos:* Prog dir, Div Astron Sci, NSF, 81-84. *Mem:* Am Astron Soc; Royal Astron Soc; Int Astron Union. *Res:* Stellar atmospheres, spectroscopy and photometry. *Mailing Add:* Astron Prog Univ Md College Park MD 20742

BELL, ROMA RAINES, b Clark, SDak, July 21, 44; m 71. NUTRITION. *Educ:* SDak State Univ, BS, 66; Univ Calif, PhD(nutrit), 70. *Prof Exp:* Res assoc nutrit, Univ Ill, Urbana-Champaign, 70-74, asst prof nutrit, 74-77; lectr nutrit, 78-80, HEAD, DEPT NUTRIT & DIETETICS, CURTIN UNIV TECHNOL, 81- *Mem:* Am Inst Nutrit; Nutrit Soc Australia. *Res:* Nutritional surveys; mineral metabolism; calcium, phosphorus and selenium; energy utilization and energy balance. *Mailing Add:* Sch Pub Health Curtin Univ Technol Perth Western Australia 6001 Australia

BELL, RONDAL E, b Kennett, Mo, Dec 29, 33; m 54; c 2. CELL PHYSIOLOGY. *Educ:* William Jewell Col, BA, 55; Univ NMex, MS, 60; Univ Miss, PhD(immunol, path), 71. *Prof Exp:* Asst biol, Univ NMex, 57-60; prof biol, Millsaps Col, 60-78; DEAN COL, CENT METHODIST COL, 78- *Concurrent Pos:* Dir, NSF grants, Univ Colo, 64-66. *Mem:* AAAS; Am Soc Microbiol; Sigma Xi; NY Acad Sci. *Res:* Biochemical and immunological aspects of hormonally induced polyarteritis nodosa. *Mailing Add:* Off of the Dean Cent Methodist Col Fayette MO 65248

BELL, ROSS TAYLOR, b Apr 23, 29; m 57. SYSTEMATIC ZOOLOGY. *Educ:* Univ Ill, BA, 49, MS, 50, PhD(entom), 53. *Prof Exp:* Instr, 55-58, from asst prof to assoc prof, 58-72, PROF ZOOL, UNIV VT, 73- *Mem:* Soc Study Evolution. *Res:* Systematics morphology and zoogeography of the carabid beetles, especially rhysodine beetles. *Mailing Add:* Dept Zool Univ Vt Burlington VT 05405

BELL, RUSSELL A, b Christchurch, NZ, Feb 3, 35; m 62; c 2. ORGANIC CHEMISTRY. *Educ:* Univ Victoria, NZ, MSc, 58; Univ Wis, MS, 60; Stanford Univ, PhD(chem), 63. *Prof Exp:* Res assoc org chem, Univ Mich, 62-63; fel, Nat Res Coun Can, 63-64; from asst prof to assoc prof, 64-73, PROF ORG CHEM, MCMASTER UNIV, 73- *Mem:* Am Chem Soc; Royal Soc Chem. *Res:* Total synthesis of organic natural products related to macrolide antibiotics; nuclear magnetic resonance; nuclear overhauser effect. *Mailing Add:* Dept Chem McMaster Univ Hamilton ON L8S 4M1 Can

BELL, SANDRA LUCILLE, b Dupo, Ill, Dec 20, 35. BOTANY, CYTOGENETICS. *Educ:* Eastern Ill Univ, BS, 57; Univ Chicago, PhD(bot), 60. *Prof Exp:* From instr to assoc prof bot, Univ Tenn, 60-80. *Concurrent Pos:* Consult biol div, Oak Ridge Nat Lab, 62-69; res assoc, Nat Inst Agron, Paris, France, 66-67. *Mem:* AAAS; Am Soc Plant Physiol; Am Soc Cell Biol. *Res:* Chromosome cytology of higher plants; chemical effects on chromosomes. *Mailing Add:* 9408 Sarasota Dr Knoxville TN 37923

BELL, STANLEY C, b Philadelphia, Pa, June 23, 31; m 67; c 2. ORGANIC CHEMISTRY, MEDICINAL CHEMISTRY. *Educ:* Univ Pa, BA, 52; Temple Univ, MA, 54, PhD(org chem), 59. *Prof Exp:* Res assoc medicinal chem, Merck Sharp & Dohme, 54-59; sr res chemist, Wyeth Labs Div, Am Home Prod Corp, 59-63; group leader, 63-68, mgr Med Chem Sect, 68-82; dir, 82-90, SR DIR, MED CHEM, WORLDWIDE, R W JOHNSON PHARMACEUT RES INST, RARITAN, NJ, 90- *Concurrent Pos:* Former secy, chmn-div med chem, Am Chem Soc; vis lectr, P M A; book reviewer, J Med Chem; med chem rep, Int Union Pure & Appl Chem; consult ed, Bioorg & Med Chem Lett. *Mem:* Am Chem Soc; NY Acad Sci; Franklin Inst; Sigma Xi. *Res:* Synthesis of new organic compounds as potential medicinal products; central nervous system; allergy; gastrointestinal immunomodulators; cardiovascular; B-lactam antibiotics; 200 patents and publications. *Mailing Add:* 732 Braeburn Lane Narberth PA 19072

BELL, STOUGHTON, b Waltham, Mass, Dec 20, 23; m 49; c 4. MATHEMATICS, COMPUTER SCIENCE. *Educ:* Univ Calif, AB, 50, MA, 53, PhD, 55. *Prof Exp:* Researcher math, Univ Calif, 52-55; staff mem math res dept, Sandia Corp, 55-64, supvr systs anal div, 64-66; assoc prof math, 67-71, dir comput ctr, 67-79, PROF MATH & COMPUT SCI, UNIV NMEX, 71- *Concurrent Pos:* Vis lectr, Univ NMex, 56-66; nat lectr, Asn Comput Mach, Inc, 73-74. *Mem:* AAAS; Soc Indust & Appl Math; Am Statist Asn; Am Math Soc; Math Asn Am. *Res:* Rarefied gas dynamics; ordinary differential equations; operations research; weapons systems analysis; computing systems. *Mailing Add:* Comput Sci Dept Univ NMex Main Campus Albuquerque NM 87131

BELL, THADDEUS GIBSON, b Leominster, Mass, Feb 6, 23; m 50; c 2. ELECTRONICS ENGINEERING. *Educ:* Yale Univ, BS, 45. *Prof Exp:* Physicist acoust, US Naval Underwater Systs Ctr, 47-85; CONSULT, 85- *Honors & Awards:* Solberg Award, Am Soc Naval Engrs, 68; Bushnell Award, Am Defense Preparedness Asn, 86. *Mem:* Fel Acoust Soc Am; Am Soc Naval Engrs. *Res:* Sonar design and analysis; underwater sound propagation; operations research; military oceanography. *Mailing Add:* Nine Colonial Dr Waterford CT 06385

BELL, THOMAS NORMAN, b Runcorn, Eng, Apr 18, 32. PHYSICAL CHEMISTRY. *Educ:* Univ Durham, BSc, 53, PhD(kinetics), 56. *Prof Exp:* Fel, Cambridge Univ, 56-57; fel, Univ Manchester, 57-58; sr lectr chem, Univ Adelaide, 58-67; assoc prof, 67-75, chmn dept, 70-75, PROF CHEM, SIMON FRASER UNIV, 75- *Concurrent Pos:* Nat Res Coun Can fel, 63-64. *Mem:* Fel Chem Inst Can. *Res:* Free radical reactions in gas phase; reaction mechanisms. *Mailing Add:* Dept Chem Simon Fraser Univ Vancouver BC V5A 1S6 Can

BELL, THOMAS WAYNE, b Indianapolis, Ind, Jan 12, 51. MOLECULAR RECOGNITION, MOLECULAR ARCHITECTURE. *Educ:* Calif Inst Technol, BS, 74; Univ Col London, PhD(organic chem), 80. *Prof Exp:* Nat Inst Health postdoctoral, Cornell Univ, 80-82; asst prof chem, State Univ NY, Stony Brook, 82-87; ASSOC PROF CHEM, STATE UNIV NY, STONY BROOK, 87- *Concurrent Pos:* Vis prof, Univ Louis Pasteur de Strasbourg, France, 90. *Mem:* Am Chem Soc; AAAS; Sigma Xi. *Mailing Add:* Dept Chem State Univ NY Stony Brook NY 11794

BELL, VERNON LEE, JR, b Omaha, Nebr, June 2, 27; m 49; c 2. POLYMER CHEMISTRY, ORGANIC CHEMISTRY. *Educ:* Doane Col, AB, 50; Ala Polytech Inst, MS, 53; Univ Nebr, Lincoln, PhD(chem), 58. *Prof Exp:* Water anal chemist, US Geol Surv, 51; fiber res chemist, Chemstrand Corp, Ala, 53-55; water anal chemist, US Geol Surv, 55-56; res chemist, Film Dept, Exp Sta, E I du Pont de Nemours & Co, Inc, Del, 58-61 & Mylar Res Lab, Ohio, 62-63; polymer chemist, Langley Res Ctr, NASA, 63-68, head polymeric mat sect, 68-70, head nonmetallic mat br, 70-74, staff chemist, 74-80, sr scientist, 80-85; RETIRED. *Concurrent Pos:* Liaison mem comt aerospace struct adhesives, Nat Mat Adv Bd, 72-74; chief scientist, Graphite Fiber Risk Anal Prog, 78-80; adv coun, Col Sci & Math, Auburn Univ, 86-90. *Mem:* Am Chem Soc. *Res:* Aromatic organophosphorus chemistry; benzacridines; Zeigler catalysis; polyhydrocarbons; polytrienes and Diels-Alder adducts; aromatic polyesters, synthesis and radiation effects; synthesis and properties of thermostable aromatic-heterocyclic polymers; polyimides; polyimidazopyrrolones; adhesives; composites; carbon fibers. *Mailing Add:* 701 Dockside Village Rte 3 Hayes VA 23072-9342

BELL, WARREN NAPIER, b Winnipeg, Man, May 8, 21; m 50; c 3. HEMATOLOGY. *Educ:* Univ Man, MD, LM, CC, 44; Univ Pa, MSc, 52, DSc, 55; FRCP(C), 53. *Prof Exp:* Assoc med, Univ Pa, 41-53, Damon Runyan scholar, 52-53, mem hemat staff, Hosp, 51-53; Storey Fund fel, Cambridge Univ, 53-54; assoc prof med & prof clin lab sci, Univ Miss & dir labs, Univ Hosp, 54-88; RETIRED. *Honors & Awards:* Chown Prize, Univ Man, 44. *Mem:* AMA; Sigma Xi. *Res:* Mechanisms of coagulation and hemolysis. *Mailing Add:* 3928 Eastwood Dr Jackson MS 39211

BELL, WILLIAM AVERILL, b Havre, Mont, Nov 29, 34; m 61; c 3. PATHOLOGY, ORAL PATHOLOGY. *Educ:* Mont State Univ, BS, 57; Wash Univ, DDS, 61. *Prof Exp:* Res path, Nat Inst Dent Res, 67-69; res assoc embryol, Nat Inst Child Health & Human Develop, 69-70; HEAD, PATH DIV, SCH DENT, MARQUETTE UNIV, 74- *Concurrent Pos:* Consult oral path, Vet Admin, San Francisco, 69-70; NIH Spec fel, Scripps Inst Oceanog, Univ Calif, San Diego, 70-72; consult oral path, Vet Admin, Wood, Wis, 74-; consult duel speciality, Am Dent Asn, 75-76. *Mem:* Am Acad Oral Path. *Res:* Dentistry; pulp pathology; embryology; the effect of current on sea urchin embryos. *Mailing Add:* 120 Lakeview Ave NE Atlanta GA 30305

BELL, WILLIAM E, b Fairmont, WVa, July 31, 29; m 58; c 2. PEDIATRIC NEUROLOGY. *Educ:* Univ WVa, AB, 51, BS & MS, 53; Med Col Va, MD, 55; Am Bd Psychiat & Neurol, dipl neurol, 63; Am Bd Pediat, dipl, 69. *Prof Exp:* Instr neurol, Univ Iowa, 62-64, asst prof pediat & neurol, 64; asst prof pediat, Univ Tex, 64-65; asst from asst prof to assoc prof, 65-72, PROF PEDIAT & NEUROL, UNIV IOWA, 72- *Mem:* Fel Am Acad Neurol; fel Am Acad Pediat. *Res:* Infections and metabolic disorders of the nervous system in childhood. *Mailing Add:* Dept Pediat & Neurol Univ Iowa Col Med Iowa City IA 52242

BELL, WILLIAM HARRISON, b St Louis, Mo, Mar 28, 27; m 65; c 4. ORAL SURGERY. *Educ:* St Louis Univ, BS, 50, DDS, 54; Am Bd Oral Surg, dipl. *Prof Exp:* Intern oral surg, Metrop Hosp, New York, 54-55; resident, Univ Tex M D Anderson Hosp & Tumor Inst, Houston, 55-56; teaching fel, Univ Tex Dent Br Houston, 56-57, asst prof, Dent Sci Inst, 57-66, asst mem inst, 66-72; assoc prof, 72-79, PROF ORAL SURG, SOUTHWESTERN MED SCH, UNIV TEX HEALTH SCI CTR DALLAS, 79- *Concurrent Pos:* Consult, Vet Admin Hosp, Dallas, 61-; pvt pract, 61- *Honors & Awards:* Res Recognition Award, Am Soc Oral Surgeons, 77; John J Geis Oral Surg Award, 84. *Mem:* Am Dent Asn; Am Soc Oral Surg; Am Equilibration Soc; Int Asn Dent Res. *Res:* Bone graft and bone graft substitutes; developmental and acquired jaw deformities; surgical-orthodontics and bone healing; bone physiology; neuro-musculo-skeletal adaptation to surgical repositioning of the jaws; orthognathic surgery. *Mailing Add:* Coastal Surg Specialists 800 Hospital Dr New Bern NC 28560

BELL, WILLIAM ROBERT, JR, b Greece, NY; m 66; c 2. INTERNAL MEDICINE, HEMATOLOGY. *Educ:* Univ Notre Dame, BS, 57; George Washington Univ, MS, 60, PhD(pharmacol), 61; Harvard Med Sch, MD, 63; Univ London, HRA, 68. *Hon Degrees:* HRA, Univ London. *Prof Exp:* Consult, Lab Physiol, Nat Cancer Inst, 59-61; intern med, Johns Hopkins Hosp, 63-64, asst resident, 66-67; clin assoc, NIH, 64-66; vis scientist, Royal Postgrad Med Sch London, Hammersmith Hosp, 67-68; sr resident med,

Harvard Med Ctr-Peter Bent Brigham Hosp, 68-69; from asst prof to assoc prof, 70-70, PROF MED & RADIOL, JOHNS HOPKINS UNIV, 81- *Concurrent Pos:* Admin officer, Nat Cancer Inst, 65-66, exec co-dir lymphoma task force, 64-66; consult, Twyford Labs, London, 67-68, Walter Reed Army Hosp & Nat Inst Health. *Honors & Awards:* C U Mosby Award. *Mem:* Am Fedn Clin Res. *Res:* Blood coagulation; thrombotic disorders. *Mailing Add:* Dept Hemat Johns Hopkins Hosp Baltimore MD 21205-2101

BELL, ZANE W, b New York, NY, Aug 15, 52; m; c 2. IMAGE PROCESSING, PHYSICS. *Educ:* Rensselaer Polytech, BS, MS, 73; Univ Ill, PhD(nuclear physics) 79; Univ Tenn, MS, 86. *Prof Exp:* AT MARTIN-MARIETTA ENERGY SYSTS, 79- *Concurrent Pos:* Pvt consult. *Mem:* Am Phys Soc; Inst Elec & Electronics Engrs. *Res:* Image processing; ultrasonics; radiography. *Mailing Add:* Y12 Plant Bldg 9103 MS 8141 PO Box 2009 Oak Ridge TN 37831-8141

BELLA, IMRE E, b Hungary; Can citizen; c 1. FOREST MENSURATION & MANAGEMENT. *Educ:* Univ BC, BSF, 58, PhD(forest mgt), 70; Univ Wash, MF, 64. *Prof Exp:* Res forester, 64-70, RES SCIENTIST GROWTH & YIELD, CAN FORESTRY SERV, 71-, PROJ LEADER, 78- *Mem:* Can Inst Forestry. *Res:* Forest stand growth and yield simulation modeling (individual tree distance dependent); growth and yield research on thinning, spacing, general stocking-density-growth relationships and yield prediction and modeling. *Mailing Add:* 4803 154th St Edmonton AB T6H 5K6 Can

BELLABARBA, DIEGO, b Rome, Italy, Aug 13, 35; Can citizen; m 63; c 3. ENDOCRINOLOGY. *Educ:* Univ Rome, MD, 59. *Prof Exp:* Asst physician, Univ Hosp, Rome, Italy, 60-64; clin res fel med, Mass Gen Hosp, Boston, 64-65; res assoc, Bronx Vet Admin Hosp, NY, 66-68; asst to assoc prof med, 68-79, PROF MED, MED SCH, UNIV SHERBROOKE, 79- *Concurrent Pos:* Fulbright scholar, 64-65; Med Res Coun Can scholar, 69-74; Que Health Res Coun res scholar, 75-81; mem endocrin comt, Med Res Coun Can, 79-81. *Mem:* AAAS; Am Fedn Clin Res; Endocrine Soc; Am Thyroid Asn; Can Soc Clin Invest. *Res:* Ontogeny of hormone receptors; metabolism of triiodothyronine in health and disease. *Mailing Add:* Dept Endocrinol Univ Sherbrooke 2500 Univ Blvd Sherbrooke PQ J1K 2R1 Can

BELLACK, JACK H, b Newberry, Mich, Mar 1, 26; m 48; c 3. SUBSTATION DESIGN & APPLICATION, BATTER STANDARDS. *Educ:* Mich Technol Univ, BSEE, 49; Case Western Reserve Univ, MSEA, 63. *Prof Exp:* Engr, US Bur Reclamation, 49-53 & AEC, 53-54; instr, Fenn Col Eng, 56-66; engr, Cleveland Elec Illum Co, 54-70, gen supv engr, 70-87; CONSULT BATTERY STANDARDS, INST ELEC & ELECTRONICS ENGRS, 87- *Concurrent Pos:* Co-chmn, Working Group Auxiliaries, Inst Elec & Electronics Engrs, 67-70, chmn, Working Group Batteries, 69-87, mem, Substas Comt, 70-78. *Mem:* Fel Inst Elec & Electronics Engrs. *Res:* Battery standards for use in the industry and for nuclear plants; battery standards for all photovoltaic, UPS systems and new battery types. *Mailing Add:* 126 Deer Run Denver NC 28037

BELLAH, ROBERT GLENN, b Nocona, Tex, June 21, 27; m 57; c 4. ENVIRONMENTAL BIOLOGY. *Educ:* McPherson Col, BS, 54; Kans State Univ, MS, 57, PhD(bot), 69. *Prof Exp:* Teacher pub sch, Kans, 55; from instr to prof, Bethany Col, Kans, 57-90, chmn, Div Natural Sci & Math, 71-85, Milfred Riddle McKeown distinguished prof sci, 80-90, prof & head dept biol, 80-90, EMER PROF, BETHANY COL, KANS, 90- *Mem:* Sigma Xi; Am Inst Biol Sci. *Res:* Social behavior of vertebrates, especially domestic chickens; plant ecology; forest successions on flood plains of rivers; floristics. *Mailing Add:* 319 S Third Lindsborg KS 67456

BELLAK, LEOPOLD, b Vienna, Austria, June 22, 16; nat US; div; c 2. PSYCHIATRY, PSYCHOANALYSIS. *Educ:* Hamerling Gym, Vienna, Matura, 35; Boston Univ, MA, 39; Harvard Univ, MA, 41; NY Med Col, MD, 44. *Prof Exp:* Instr psychol, Boston Ctr Adult Educ, 39-42; instr, City Col New York, 42-44; from intern to med officer, St Elizabeths Hosp, DC, 44-46; instr psychol, George Washington Univ & USDA Grad Sch, 45-46; assoc psychiat, NY Med Col, 46-50, clin asst prof, 50-56; vis clin prof, 65-75, CLIN PROF PSYCHIAT, ALBERT EINSTEIN COL MED, NY UNIV, 75- *Concurrent Pos:* Dir, Dept Psychiat, City Hosp, Elmhurst, 58-64; NIMH grant proj dir trouble shooting clin, Elmhurst Hosp & prin investr, Schizophrenia Res Proj; chief psychiat consult, Altro Health & Rehab Serv, Inc, 47-58; vis prof psychiat & behav sci, Sch Med, George Wash Univ, 75-79; clin prof psychol, Fel Prog Psychother, NY Univ, 75- *Honors & Awards:* Frieda Fromm Reichmann Award, Am Acad Psychoanal, 81. *Mem:* Fel Am Psychiat Asn; Am Psychoanal Asn; fel Am Psychol Asn; Am Psychol Asn. *Res:* Projective techniques; manic-depressive psychosis; schizophrenia; ego function assessment; geriatric psychiatry; author 35 books, 300 articles. *Mailing Add:* 22 Rockwood Dr Larchmont NY 10538

BELLAMY, DAVID, b Rochester, NY, July 17, 26; m 58; c 4. PLASTICS. *Educ:* Yale Univ, BS, 49; Univ Chicago, MBA, 69. *Prof Exp:* Chief engr, Fenwal Labs, Inc; biomed engr, Baxter Health Care Corp, 60-65, dir biomed eng, 65-75; vpres tech admin, 75-84, VPRES VENTORE TECHNOL, BAXTER HEALTH CARE CORP, 84. *Mem:* Soc Cryobiol; NY Acad Sci; Am Asn Blood Banks. *Res:* Blood processing equipment; hospital devices. *Mailing Add:* Baxter Healthcare Corp One Baxter Pkwy Deerfield IL 60015

BELLAMY, DAVID P, b Bristol, Va, Apr 25, 44; div; c 1. TOPOLOGY. *Educ:* King Col, AB, 64; Mich State Univ, MS, PhD(math), 68. *Prof Exp:* From asst prof to assoc prof math, 68-79, PROF MATH SCI, UNIV DEL, 79- *Mem:* Am Math Soc. *Res:* Point set topology; continua theory. *Mailing Add:* Dept Math Univ Del Newark DE 19716

BELLAMY, JOHN C, b Cheyenne, Wyo, Apr 18, 15; m 40; c 5. CIVIL ENGINEERING. *Educ:* Univ Wyo, BS, 36; Univ Wis, PhM, 38; Univ Chicago, PhD(meteorol), 46. *Prof Exp:* Partner, Bellamy & Sons, Engrs, Wyo, 38-42; asst prof meteorol, Univ Chicago, 42-47; assoc dir, Cook Res Labs, Ill, 47-60; prof res engr, 60; prof civil eng, dir res & dir, Nat Resources Res Inst, Univ Wyo, 60-81; RETIRED. *Concurrent Pos:* Dir inst trop meteorol, Univ PR, 43-44; mem, World Meteorol Orgn, 46-47, US Data Processing Comt, Int Geophys Yr, 55-57, US Meteorol Panel, 56-58 & Gov Adv Comt Indust Develop, 66-67; mem, Western Interstate Nuclear Bd, 67-80. *Honors & Awards:* Losey Award, Inst Aeronaut Sci, 44; Thurlow Award, Am Inst Navig, 46; Medal of Freedom, Pres US, 46. *Mem:* Am Meteorol Soc; Am Inst Navig (pres, 62-63); Geophys Union; Am Soc Eng Educ; Sigma Xi. *Res:* Operational engineering; navigation; meteorology. *Mailing Add:* 2308 Holliday Dr Laramie WY 82071

BELLAMY, RONALD FRANK, CORONARY PHYSIOLOGY. *Educ:* State Univ NY, Buffalo, MD, 61. *Prof Exp:* Vchmn dept mil med, 85-88, STAFF THORACIC SURG SERV, UNIFORMED SERV UNIV HEALTH SCI, 88- *Res:* Thoracic surgery. *Mailing Add:* Dept Thoracic Surg Walter Reed Army Med Ctr Washington DC 20307

BELLAMY, WINTHROP DEXTER, b Philadelphia, Pa, Oct 7, 15; m 47; c 2. MICROBIOLOGY, WASTE RECYCLING. *Educ:* Cornell Univ, BS, 38, PhD(biochem & microbiol), 45. *Prof Exp:* Teaching asst bact, Cornell Univ, 41-43, instr, 43-45; res assoc microbiol, Sterling Winthrop Chem Co; res biochemist, Res & Develop Ctr, Gen Elec Co, 49-78; CONSULT WASTE RECYCLING, 78- *Concurrent Pos:* Adj prof food sci dept, Cornell Univ, Ithaca, NY, 78-84. *Mem:* Fel AAAS; fel NY Acad Sci; Am Soc Microbiol; fel Am Acad Microbiol; Am Chem Soc; Sigma Xi. *Res:* Microbial degradation of organic wastes; single cell protein production from cellulosic wastes; use of thermophilic microorganisms for waste conversion. *Mailing Add:* 5548 Hamlet Lane Ft Myers FL 33919

BELLAN, PAUL MURRAY, b Winnepeg, Can, Apr 18, 48. PLASMA PHYSICS. *Educ:* Univ Man, BS, 70; Princeton Univ, PhD(physics), 76. *Prof Exp:* Assoc prof, 77-90, PROF PHYSICS, CALIF INST TECHNOL, 90- *Mem:* Am Phys Soc. *Res:* Current drive laser induced-flourescent physics; anomalous current diffusion; spheromak injection into tokamales. *Mailing Add:* 128-95 Calif Inst Technol-Watson Pasadena CA 91125

BELLAND, RENÉ JEAN, b Edmonton, Alta, June 23, 54; m 81; c 1. MOLECULAR BIOLOGY. *Educ:* Univ Alta, BSc, 78; Mem Univ Nfld, MSc, 81, PhD(biol), 85. *Prof Exp:* Postdoctoral fel bryol, 86-88, RES SCIENTIST, UNIV BC, 88- *Mem:* Can Bot Asn; Am Bryological & Lichenological Soc. *Res:* Evolution of bryophyte distributions using molecular biological techniques; bryophytes of the national parks of eastern Canada. *Mailing Add:* Dept Bot Univ BC Vancouver BC V6T 1Z4 Can

BELLANGER, MAURICE G, b France, June 21, 41; m 65; c 2. COMMUNICATION SYSTEMS, SIGNAL PROCESSING. *Educ:* ENST, Paris, Engr, 65; Univ Paris, DSc, 81. *Prof Exp:* Prof automatic control & signal processing, Univ Paris, 84-87; res & develop engr, 67-74, dept head, 75-84, SCI DIR TELECOMMUN, TRT-PHILIPS CO, FRANCE, 88- *Concurrent Pos:* Vis prof, Univ Calif, Santa Barbara, 87; adj prof, Ecole Superieure D'Electricite, France, 87-; chmn, Tech Prog, Inst Elec & Electronics Engrs, 88. *Honors & Awards:* Centennial Medal, Inst Elec & Electronics Engrs, 84. *Mem:* Inst Elec & Electronics Engrs; Europ Asn Signal Processing (pres, 88-92). *Res:* Signal processing mainly for application in communication; digital signal processing; adaptive digital filter and signal analysis; theory, algorithms, structures for implementation and integrated circuits for realization. *Mailing Add:* 54 Rue de Rennes Paris 75006 France

BELLANTI, JOSEPH A, b Buffalo, NY, Nov 21, 34; m 58; c 7. PEDIATRICS, MICROBIOLOGY. *Educ:* Univ Buffalo, MD, 58; Am Bd Pediat, cert, 64; Am Bd Allergy & Immunol, dipl, 74. *Prof Exp:* Intern, Millard Fillmore Hosp, Buffalo, 58-59; resident pediat, Children's Hosp Buffalo, 59-61; NIH spec trainee immunol, J Hillis Miller Health Ctr, Univ Fla, 61-62; res virologist, Dept Virus Dis, Walter Reed Army Inst Res, Washington, DC, 62-64; from asst prof to assoc prof, 63-70, PROF PEDIAT & MICROBIOL, SCH MED, GEORGETOWN UNIV, 70-, DIR CTR INTERDISCIPLINARY STUDIES IMMUNOL, 75- *Concurrent Pos:* Mead Johnson grant pediat res, 64; mem growth & develop comt, NIH, 70-75; mem, med adv comt, Nat Kidney Found, 71-; chmn, Infectious Dis Comt, 72; dir, Am Bd Allergy & Immunol, 75- *Honors & Awards:* William Peck Sci Res Award, 66; Sci Exhibit Award, Am Acad Clin Pathologists & Col Am Path, 66; E Mead Johnson Award, Am Acad Pediat, 70. *Mem:* AMA; fel Am Acad Pediat; fel Am Acad Allergy; fel Am Acad Allergists; fel Am Asn Clin Immunol & Allergy. *Res:* Immunologic aspects of facultatively slow virus infections; biochemical changes in human polymorphonuclear leukocytes during maturation. *Mailing Add:* Dept Pediat & Microbiol Sch Med Immunol Ctr Georgetown Univ 3800 Reservoir Rd Washington DC 20007

BELLAVANCE, DAVID WALTER, b Attleboro, Mass, Aug 30, 43; m 69; c 1. SOLID STATE CHEMISTRY, INORGANIC CHEMISTRY. *Educ:* Tufts Univ, BS, 65; Brown Univ, PhD(chem), 70. *Prof Exp:* Mem tech staff mat growth, Allied Chem Corp, 70-73; mem tech staff mat growth, 73-86, SEMICONDUCTOR PROCESS ENG MGR, TEX INSTRUMENTS INC, 87- *Mem:* Electrochem Soc; Mat Res Soc. *Res:* Crystal growth and thin film epitaxy of silicon, III-V compounds, magnetic garnets, infrared materials, laser materials, and opto-electronic materials. *Mailing Add:* 7410 La Cosa Dr Dallas TX 75248

BELLEAU, BERNARD ROLAND, organic chemistry, biochemical pharmacology; deceased, see previous edition for last biography

BELLENOT, STEVEN F, b Glendale, Calif, Aug 4, 48; m 75; c 2. MATHEMATICAL ANALYSIS, COMPUTER SCIENCE. *Educ:* Harvey Mudd Col, BS, 70; Claremont Grad Sch, PhD(math), 74. *Prof Exp:* From asst prof to assoc prof, 74-84, PROF MATH, FLA STATE UNIV, 84- *Concurrent Pos:* Vis assoc prof math & comput sci, Clarkson Col, 80-81; vis prof math, Univ Tex, Austin, 87-88. *Mem:* Am Math Soc; Math Asn Am; AAAS; Asn Comput Mach. *Res:* Functional analysis, especially banach spaces; parallel and distributed simulations, especially time warp. *Mailing Add:* Dept Math Fla State Univ Tallahassee FL 32306

BELLER, BARRY M, b New York, NY, Dec 12, 35; m 58; c 2. CARDIOLOGY. *Educ:* Columbia Univ, AB, 56, MD, 60. *Prof Exp:* Intern, Univ Chicago Hosps, 60-61, from asst resident to sr resident internal med, 61-64, USPHS res trainee cardiol, 63-64; assoc med, Albert Einstein Col Med, 66-67, asst prof, 67-68; assoc prof med & head sect cardiovasc dis, Med Sch, Univ Tex, San Antonio, 68-72; MEM STAFF, CARDIOVASC ASSOCS, 72- *Concurrent Pos:* Consult, USPHS Surv, 62; consult tech rev comt artificial heart prog, Nat Heart Inst-NIH, 69; mem coun clin cardiol, Am Heart Asn, 69. *Mem:* AAAS; Am Fedn Clin Res; Fel Am Heart Asn; fel Am Col Cardiol; fel Am Col Physicians. *Res:* Myocardial protein synthesis; hemodynamics; electrocardiography. *Mailing Add:* Humana Hosp Metro GL 72 Metro Prof Bldg San Antonio TX 78212

BELLER, FRITZ K, b Munich, Ger, May 17, 24; US citizen; m 48; c 2. OBSTETRICS & GYNECOLOGY. *Educ:* Univ Berlin, Dr med, 48; Univ Giessen, MedSciD, 55. *Hon Degrees:* Dr, Univ Akita, Japan. *Prof Exp:* Docent obstet & gynec, Univ Giessen, 52-56; docent, Univ Tubingen, 56-58, head dept, 58-61, prof, 61; vis assoc prof, Sch Med, NY Univ, 61-63, from assoc prof to prof, 63-72; prof obstet & gynec & chmn dept, Univ Munster, 72-88; PROF OBSTET & GYNEC, UNIV IOWA, 88- *Concurrent Pos:* Vis fac, Ben Gurion Univ. *Mem:* Fel Am Col Obstet & Gynec; Am Col Surg; Am Soc Exp Path; Soc Gynec Invest; fel Royal Soc Med; hon fel Polish Soc Obstet Gynec; fel Am Gynec Soc; fel Am Asn Obstet Gynec; fel Am Gynec Obstet Soc. *Res:* Hematological problems in obstetrics; intravascular coagulation and proteolysis; gynecological oncology. *Mailing Add:* 852 Cypress Ct Iowa City IA 52242

BELLER, GEORGE ALLAN, b New York, NY, Dec 23, 40. NUCLEAR CARDIOLOGY. *Educ:* Dartmouth Col, AB, 62; Univ Va, MD, 66. *Prof Exp:* Med intern, Univ Wis Hosp, 66-67, asst resident, 67-68; sr asst resident, Boston City Hosp, 68-69; clin & res fel cardiol, Mass Gen Hosp, 73-74; from instr to asst prof med, Harvard Med Sch, 74-77; PROF MED, UNIV VA SCH MED, 77-, HEAD DIV CARDIOL, 77- *Concurrent Pos:* Mem cardiovasc & pulmonary study sect, Nat Heart, Lung & Blood Inst, 87-; estab investr, Am Heart Asn, 77. *Honors & Awards:* William S Middleton Lectr, Wis Med Soc, 68. *Mem:* Am Heart Asn; Am Fedn Clin Res; Am Soc Clin Investr; Am Physiol Soc; fel Am Col Physicians; fel Am Col Cardiol. *Res:* Development of nuclear cardiology techniques for assessing regional myocardial blood flow and function; evaluating the effects of coronary reperfusion on myocardial viability; investigation of the role of inotropic agents for treatment of congestive heart failure; the assessment of the roles of neutrophils on mediating myocardial ischemis injury. *Mailing Add:* Div Cardiol Univ Va Med Sch Box 158 Charlottesville VA 22908

BELLER, MARTIN LEONARD, b New York, NY, Apr 30, 24; m 47; c 3. ORTHOPEDIC SURGERY. *Educ:* Columbia Univ, AB, 44, MD, 46; Am Bd Orthop Surg, dipl, 55. *Prof Exp:* Intern, Mt Sinai Hosp, New York, 46-47; resident orthop surg, Hosp Joint Dis, 49-52; assoc prof, 72-77, CLIN PROF ORTHOP SURG, SCH MED, UNIV PA, 77- *Concurrent Pos:* Attend orthop surgeon, Albert Einstein Med Ctr, 60-70, chmn dept orthop surg, Daroff Div, 70-79, emer chmn, 79-; attend orthop surgeon, Hosp Univ Pa, 63- *Mem:* Fel Am Acad Orthop Surgeons; fel Int Soc Orthop Surg & Traumatol; AMA; fel Am Col Surgeons; Am Orthop Asn. *Res:* Radioactive phosphorus and other isotopes in evaluating circulation of bone; x-ray densitometry of bone with aluminum step wedge; bone metabolism; osteoporosis; Paget's disease; total joints; spine survey. *Mailing Add:* RD 1 Box 256B Gaines PA 16921

BELLES, FRANK EDWARD, b Cleveland, Ohio, Feb 28, 23; m 46; c 3. FUEL SCIENCE. *Educ:* Western Reserve Univ, BS, 47; Case Inst Technol, MS, 52. *Prof Exp:* Aeronaut res scientist, NASA Lewis Res Ctr, 47-57, head gas dynamics sect, 57-64, head kinetics sect, 64-71, chief propulsion chem br, 71-72, dir aerospace safety res & data inst, 72-74; dir, Labs Res & Develop, 74-82, CONSULT COMBUSTION & ENERGY UTILIZATION TECHNOL, AM GAS ASN, 82- *Mem:* Am Chem Soc; Combustion Inst; Am Soc Gas Engrs; AAAS; Sigma Xi. *Res:* Flame quenching and fire extinguishing; high-temperature reactions; gaseous detonations; rocket propellant safety; combustion research and development. *Mailing Add:* 29308 Wolf Rd Cleveland OH 44140

BELLET, EUGENE MARSHALL, b Hollywood, Calif, July 3, 40; m 68; c 5. PESTICIDE CHEMISTRY. *Educ:* Univ Calif, Riverside, BS, 63, MS, 69, PhD(chem & entom), 71. *Prof Exp:* Staff physicist residue chem, Univ Calif, Riverside, 65-68, environ sci fel, 69-71; res fel, Univ Calif, Berkeley, 71-73; chemist, Off Pesticide Prog, Environ Protection Agency, 73-74; mgr prod regist div, Weslaco Tech Ctr, Ansul Co, 74-76; DIR RES & DEVELOP, KALO LAB, 76- *Mem:* Am Chem Soc; Entom Soc Am; Sigma Xi. *Res:* Synthesis, chemistry and toxicology of organophosphorus compounds and bicyclic phosphates; chemistry of organoarsenicals; synthesis and development of new pesticides. *Mailing Add:* Chem Consults Int Inc 7270 W 98th Terr Suite 100 Overland Park KS 66212

BELLET, RICHARD JOSEPH, b East Orange, NJ, Oct 26, 27; m 55; c 4. POLYMER CHEMISTRY. *Educ:* Princeton Univ, AB, 49; Columbia Univ, AM, 51, PhD(org chem), 56. *Prof Exp:* Chemist, Berkeley Chem Corp, 49-51; sr res scientist, Hooker Chem Co, 56-57; sr res chemist, Tex-US Chem Co, Texaco, Inc & US Rubber Co, 57-61, group leader, 61-63; sr res chemist, Plastics Div, Allied Chem Corp, 63-65, tech supvr, 65-66, mgr, 66-70, corp res coordr, 70-76, RES ASSOC, ALLIED CORP, 76- *Mem:* Am Chem Soc. *Res:* Fine organics; pharmaceuticals; ultraviolet stabilizers; butadiene chemistry; antioxidants; emulsion polymerization; synthetic rubber; polymer characterization, development and commercialization; impact and condensation polymers; polyamides; applications research. *Mailing Add:* 14 Tolchester Heights RR 2 Chestertown MD 21620-9438

BELLETIRE, JOHN LEWIS, b Chicago, Ill, Aug 23, 43; m 75. SYNTHETIC ORGANIC & NATURAL PRODUCTS CHEMISTRY. *Educ:* Univ Chicago, SB, 65; Northwestern Univ, PhD(org chem), 72. *Prof Exp:* Res scientist med chem, Pfizer Inc, 74-78; res fel, Dept Chem, Univ Wis, 78-81; asst prof chem, 81-86, ASSOC PROF CHEM, UNIV CINCINNATI, 86- *Mem:* Am Chem Soc; The Chem Soc; Sigma Xi. *Res:* Medicinal chemistry of diabetes; organic synthesis of novel heterocycles; anthracyclinones; heterolytic fragmentation reactions; photochemistry; oxidative coupling of electron-rich intermediates. *Mailing Add:* Dept Chem Univ Cincinnati Cincinnati OH 45221

BELLHORN, MARGARET BURNS, b Sharon, Pa, Nov 26, 39. BIOCHEMISTRY. *Educ:* Allegheny Col, BS, 61; Yale Univ, MS, 62; Albert Einstein Col Med, PhD(biochem), 71. *Prof Exp:* From instr ophthal to instr biochem, 72-75, ASST PROF OPHTHAL, ALBERT EINSTEIN COL MED, 73-, ASST PROF BIOCHEM, 75- *Mem:* Am Chem Soc; Asn Res Vision Ophthal. *Res:* Biochemical and morphological analysis of spontaneous and induced retinal degenerations and transport of substances through the retinal vessels; secondary ion mass spectrometry as a tool for chemical analysis of biological systems. *Mailing Add:* 33 Prescott Ave Montclair NJ 07042

BELLIN, JUDITH SCHRYVER, biochemistry, environmental chemistry, for more information see previous edition

BELLINA, JOSEPH HENRY, b New Orleans, La, Jan 30, 42; m; c 2. OBSTETRICS & GYNECOLOGY, BIOENGINEERING. *Educ:* La State Univ, MD, 65; Am Bd Obstet & Gynec, dipl, 73; Century Univ, Calif, PhD(Engr); Am Bd Laser Surg, 85. *Prof Exp:* Lectr, Univ Southwestern La, 68-69; clin instr, 69-70, clin asst prof, 70-78, clin assoc prof, 78-80, CLIN PROF, SCH MED, LA STATE UNIV, 80-; RES ASSOC, ELECTROSCI & BIOPHYS RES LAB, TULANE UNIV, 78- *Concurrent Pos:* Vis surgeon, Charity Hosp, 69-; co-dir, Ultra-sound Diagnostic Sect & Ctr Reproductive Biol, Jo Ellen Smith-F Edward Hebert Mem Hosp, dir, Laser Sect; chmn, Laser Res Found; laser consult, Sch Med, Johns Hopkins Univ & Baylor Col Med; carbon dioxide laser res award, Cavitron Corp, 76-, Cooper Corp, 79, Laser Indust Israel, 79, Hamilton Pierce Co, 79 & Advan Surg Techol, Inc, 80; laser res award, Coherent Radiation, 76-; A Ward Ford Mem Inst grant, 79; Nat Cancer Inst grant, 80; nat adv, Nat Inst Child Health & Human Develop, NIH. *Honors & Awards:* William B Mark Award, 85; Sigma Xi, 88. *Mem:* Fel Am Col Obstet & Gynec; Am Asn Gynec Laparoscopists; Am Soc Colposcopy & Colpomicros; Int Soc Study Vulva Dis; Am Inst Ultrasound Med; Fallopine Int Soc. *Res:* Cervical intraepithelia neoplasia biomolecular structure responses to carbon dioxide laser, and other neoplastic processes and their relationship to molecular engineering; reproductive microsurgery. *Mailing Add:* 833 S Carrollton Ave New Orleans LA 70118

BELLINA, JOSEPH JAMES, JR, b Orange, NJ, Mar 27, 40; m 62; c 5. SURFACE PHYSICS, INTERFACE PHYSICS. *Educ:* Univ Notre Dame, BS, 61, PhD(physics), 66. *Prof Exp:* Res assoc physics, Barus Surface Physics Lab, Brown Univ, 66-68; res scientist, McDonnell Douglas Corp, 68-74; vis asst prof, Southern Ill Univ-Edwardsville, 74-75; asst prof, 75-76, ASSOC PROF PHYSICS, ST MARY'S COL, 76-; NSF col prof, dept metall eng & mat sci, Notre Dame Univ, Ind, 76-; AT DEPT CHEM & PHYSICS, ST MARY'S COL. *Mem:* Am Phys Soc; Am Asn Physics Teachers; Sigma Xi. *Res:* Low energy electron diffraction; ellipsometry; corrosion; stress-corrosion cracking; vacuum technology; auger spectroscopy; surface chemistry; autoradiography; affect of hydrogen on bonding at surfaces and interfaces of high strength isothermal remanent magnetization and nickel based alloys. *Mailing Add:* Dept Chem & Physics St Mary's Col Notre Dame IN 46556

BELLINA, RUSSELL FRANK, b Newark, NJ, June 24, 42; m 66; c 2. ORGANIC CHEMISTRY. *Educ:* Fairleigh Dickinson Univ, BS, 64; Seton Hall Univ, MS, 66, PhD(org chem), 68. *Prof Exp:* Res chemist, 68-73, res supvr synthesis chem, 73-74, lab adminr, 74-76, area supvr-tech, 76, chief chemist, Control Lab, Belle WVa Plant, 76-78, res mgr synthesis, Biochem Dept, 78-80, prin consult life sci res planning, Cent Res & Develop Dept, 80-81 & Corp Plans Dept, 81, MGR, AGRICHEMICALS DEVELOP, AGR PRODS DEPT, E I DU PONT DE NEMOURS & CO, INC, 81- *Mem:* Am Chem Soc. *Res:* Agricultural chemistry, synthesis and market development of biologically active chemicals designed for use as insecticides, fungicides, bactericides, herbicides, and plant growth modifiers. *Mailing Add:* 708 Westcliff Rd Edenridge III Wilmington DE 19803-1712

BELLINGER, LARRY LEE, b Oakland, Calif, May 12, 47; m 73; c 3. PHYSIOLOGY. *Educ:* Univ Calif, Davis, BS, 69, PhD(physiol), 74. *Prof Exp:* NIH fel physiol, State Univ NY, Buffalo, 74-76; from asst prof to assoc prof, 76-85, PROF PHYSIOL, BAYLOR COL DENT, DALLAS, GRAD SCH, BAYLOR UNIV, WACO TEX, 85- *Mem:* Endocrine Soc; Am Physiol Soc; Soc Neurosci; Soc Exp Biol Med; Am Inst Nutrit; Soc Study Ingestive Behav. *Res:* The study of neurophysiological and neuroendocrine parameters involved in food intake and body weight control. *Mailing Add:* Dept Physiol 3302 Gaston Ave Dallas TX 75246

BELLINGER, PETER F, b New Haven, Conn, June 15, 21; m 53. ENTOMOLOGY. *Educ:* Yale Univ, BA, 42, PhD, 52. *Prof Exp:* Lectr zool, Univ Col WI, 52-56; instr, Yale Univ, 56-58; from asst prof to assoc prof, 58-65, PROF BIOL, CALIF STATE UNIV, NORTHRIDGE, 65- *Mem:* Entom Soc Am; Lepidopterists Soc. *Res:* Collembola. *Mailing Add:* Dept Biol Calif State Univ 18111 Nordoff State Northridge CA 91330

BELLINO, FRANCIS LEONARD, b Elizabeth, NJ, Dec 7, 38; m 63; c 11. BIOCHEMICAL ENDOCRINOLOGY, MOLECULAR BIOLOGY. *Educ:* St Francis Col, Pa, BA, 60; Univ Notre Dame, BS, 61; Univ Pittsburgh, MS, 65; State Univ NY Buffalo, PhD(biophys sci), 71. *Prof Exp:* Assoc engr, Westinghouse Elec Corp, 61-65; instr physics, St Francis Col, Pa, 65-67; assoc res scientist, Med Found Buffalo, 71-81; from res asst prof to res assoc prof, State Univ NY, Buffalo, 81-90; GRANTS ASSOC, NIH, 90- *Concurrent Pos:* Prin investr, NIH grant, 77-90. *Mem:* Inst Elec & Electronics Engrs; Biophys Soc; Endocrine Soc; AAAS; Am Soc Biochem & Molecular Biol. *Res:* Purification and characterization of estrogen synthetase from human placenta; mechanism of physiological regulation of estrogen synthetase in normal and malignant human trophoblast in culture. *Mailing Add:* NIH Bldg 31 Rm 5B35 Bethesda MD 20892

BELLION, EDWARD, b Liverpool, Eng, Sept 6, 44; US citizen; m 84; c 3. MEMBRANES, PROTEINS. *Educ:* Univ Leeds, Eng, BSc, 65, PhD(biochem), 68. *Prof Exp:* Res fel chem, Univ Minn, 68-70; from asst prof to assoc prof, 70-90, PROF CHEM, UNIV TEX, ARLINGTON, 90- *Concurrent Pos:* Adj assoc prof, dept biochem, Health Sci Ctr Dallas, Univ Tex, 70-86, vis assoc prof, dept pharmacol, 84-85. *Mem:* Am Chem Soc; Am Soc Microbiol; Am Soc Biol Chem & Molecular Biol. *Res:* Biochemistry of microbial growth on one-carbon compounds, including enzymology, transport, membrane studies and assembly of penoxisomes. *Mailing Add:* Dept Chem Univ Tex Box 19065 Arlington TX 76019-0065

BELLIS, EDWARD DAVID, b Ridley Park, Pa, June 28, 27. ZOOLOGY. *Educ:* Pa State Univ, BS, 51, Univ Okla, MS, 53, Univ Minn, PhD(zool), 57. *Prof Exp:* Asst zool, Univ Okla, 51-53; teaching asst, Univ Minn, 53-57; instr, Univ Ga, 57; from asst prof to assoc prof, 58-63, PROF ZOOL, PA STATE UNIV, 71- *Mem:* AAAS; Am Soc Ichthyol & Herpet; Ecol Soc Am; Am Soc Nat. *Res:* Herpetology; speciation; general ecology. *Mailing Add:* Pa State Univ Biol 208 Life Sci Bldg University Park PA 16802

BELLIS, HAROLD E, b Middletown, Conn, Sept 18, 30; m 54, 81; c 2. PHYSICAL CHEMISTRY, MATHEMATICS. *Educ:* Cent Conn State Col, BSc, 51; Univ Conn, PhD(chem), 57. *Prof Exp:* Res chemist, Electrochem Dept, 57-69, staff scientist, 69-76, sr staff scientist, Indust Chem Dept, 76-78, RES ASSOC CHEMICALS & PIGMENTS, 78-86, RES FEL, E I DU PONT DE NEMOURS & CO, INC, 87- *Mem:* Am Chem Soc; Am Soc Metals; Am Electroplaters Soc. *Res:* High temperature chemistry and metallurgy of molten metal/salt systems; molecular spectroscpy; homogeneous catalysis; electroless plating; reactions of carbon monoxide; oxidation mechanisms. *Mailing Add:* Jackson Lab C&P Dept Chambers Works E I du Pont de Nemours & Co Inc Deepwater NJ 08023

BELLIS, VINCENT J, JR, b Penn Yan, NY, Jan 18, 38; m 58; c 2. AQUATIC ECOLOGY. *Educ:* NC State Univ, BS, 60, MS, 63; Univ Western Ont, PhD(bot), 66. *Prof Exp:* PROF BIOL, E CAROLINA UNIV, 66- *Mem:* AAAS; Phycol Soc Am; Am Inst Biol Sci; Brit Phycol Soc; Int Phycol Soc. *Res:* General estuarine and coastal ecology; effects of development on natural ecosystems of eastern North Carolina. *Mailing Add:* Dept Biol E Carolina Univ Greenville NC 27834

BELLIVEAU, LOUIS J, b Boston, Mass, Nov 18, 26; m 52; c 6. PHYSICS, APPLIED MATHEMATICS. *Educ:* Boston Col, BS, 51, MS, 52. *Prof Exp:* Physicist explosions, Naval Ship Res & Develop Ctr, Md, 52-54, naval architect submarines, 54-57; physicist explosions, Naval Surface Weapons Ctr, Naval Ord Lab White Oak, 57-63; br chief explosives, Navy Dept, 63-67; physicist fluids, Defense Nuclear Agency, 67-72; SUPVRY PHYSICIST, HARRY DIAMOND LABS, US ARMY LAB COMMAND, 72- *Mem:* Am Phys Soc. *Res:* Explosives; explosions, chemical and nuclear; vulnerability of military equipment to blast and shock. *Mailing Add:* Harry Diamond Labs US Army Lab Command 2800 Powder Mill Rd Adelphi MD 20783-1197

BELLMER, ELIZABETH HENRY, b New York, Sept 30, 27. BOTANY, ZOOLOGY. *Educ:* Trinity Col (DC), AB, 59; Cath Univ Am, MS, 62, PhD(bot), 68. *Prof Exp:* From instr to assoc prof, 59-84, PROF BIOL, TRINITY COL, DC, 84- *Mem:* AAAS; Am Genetic Asn; Am Inst Biol Sci; Sigma Xi. *Res:* Time of embryonic fusion of the malleus and incus of the guinea pig; distribution, variation and chromosome number in the Appalachian shale barren endemic Eriogonum Allenii Watson. *Mailing Add:* Dept Biol Trinity Col Washington DC 20017

BELLMORE, MANDELL, b Washington, DC, May 22, 35; m 57; c 2. ELECTRICAL ENGINEERING, OPERATIONS RESEARCH. *Educ:* Univ Md, BS, 57; Cath Univ Am, MEE, 61; Johns Hopkins Univ, PhD(opers res), 65. *Prof Exp:* Sr design engr, Electronics Div, Am Car Foundry, Md, 60-62; eng specialist, Martin-Marietta Corp, 62-64; mem tech staff, Mitre Corp, Va, 64-65; from asst prof to assoc prof opers res, Johns Hopkins Univ, 65-72; MEM STAFF, BLOCK MCGIBONY & ASSOCS, 72- *Concurrent Pos:* Consult, Task Force Sci & Technol, President's Comn Law Enforcement & Criminal Justice, 66; adv, Law Enforcement Assistance Agency, US Dept Justice, 68; mem adv comt, Secy Health & Ment Hyg, State of Md, 70. *Mem:* Opers Res Soc Am; Inst Elec & Electronics Engrs; Inst Mgt Sci. *Res:* Graph and network theory; theory of optimization; mathematical programming; computers. *Mailing Add:* Block McGibony Bellmore Assoc 5525 Twin Knolls Rd-327 Columbia MD 21045

BELLO, JAKE, b Detroit, Mich, Feb 22, 28; m 57. PHYSICAL BIOCHEMISTRY. *Educ:* Wayne Univ, BS, 48, PhD(chem), 52. *Prof Exp:* Fel org chem, Purdue Univ, 52-53; fel protein chem, Calif Inst Technol, 53-56; chemist, Eastman Kodak Co, NY, 56-58; sr res assoc protein chem, Polytech Inst Brooklyn, 58-59; assoc scientist, 60-67, dir grad studies chem, Roswell Park Div, 73-80, PRIN SCIENTIST, ROSWELL PARK MEM INST, 67-, CHMN, CHEM DEPT, 80- *Concurrent Pos:* Mem comt biophys training, NIH. *Mem:* AAAS; Am Chem Soc; NY Acad Sci. *Res:* Protein and nucleic acid chemistry and structure; mechanism of gelation of gelatin; polymers; interferon inducers. *Mailing Add:* Roswell Park Mem Inst Buffalo NY 14263

BELLO, P(HILLIP), b Lynn, Mass, Oct 22, 29; m 52; c 2. ELECTRICAL ENGINEERING. *Educ:* Northeastern Univ, BS, 53; Mass Inst Technol, SM, 55, ScD(elec eng), 59. *Prof Exp:* Asst, Mass Inst Technol, 53-55; res assoc, Northeastern Univ, 55-56, asst prof commun, 56-57; eng specialist, Sylvania Appl Res Lab, 59-61; sr scientist, Adcom, Inc, 61-65; dir commun res & vpres, Signatron, Inc, 65-72; pres, CNR, Inc, 72-82; Milcom, Inc, 82-88; CONSULT ENGR, MITRE, 88- *Mem:* Fel Inst Elec & Electronics Engrs. *Res:* Communications; administrations research and development. *Mailing Add:* Mitre Corp Burlington Rd Bedford MA 01730

BELLOLI, ROBERT CHARLES, b St Louis, Mo, Nov 3, 42; m 68; c 2. ORGANIC CHEMISTRY. *Educ:* St Louis Univ, BS in Chem, 64; Univ Calif, Berkeley, PhD(org chem), 68. *Prof Exp:* Teaching asst chem, Univ Calif, Berkeley, 64-65; from asst prof to assoc prof, 68, dept chair, 83-89, PROF CHEM, CALIF STATE UNIV, FULLERTON, 76- *Concurrent Pos:* Res Corp grant, 69-70; Petrol Res Fund grant, Am Chem Soc, 72-74; NSF-SOS grant, 78; Instrumentation grant HPLC, NSF, 84. *Mem:* Am Chem Soc; AAAS. *Res:* Physical-organic chemistry; organic analytical problems; chemical education. *Mailing Add:* Dept Chem & Biochem Calif State Univ Fullerton CA 92634

BELLONCIK, SERGE, b Aleppo, Syria, Feb 19, 44; Can citizen. AGRICULTURE, VIROLOGY. *Educ:* Univ Montpellier, France, Agr Eng, 67, DEA, 68, Dr(biol sci), 69. *Prof Exp:* PROF MICROBIOL, UNIV QUE, TROIS RIVIERES, 72-; PROF VIROL, INST ARMAND FRAPPIER, UNIV QUE, LAVAL, 78- *Concurrent Pos:* Fel, Univ Montreal, 69-72; examr grant, Dept Educ, Que, 77-; head grad students comt, Univ Que, 77-78. *Mem:* Am Mosquito Control Asn; Can Micros Soc; Can Soc Microbiologists; Am Soc Microbiologists. *Res:* Invertebrate virology; biological control of insects; virus transmitted by arthropods (arbovirus); epidemiological survey by virus isolation and serology; insect tissue culture. *Mailing Add:* Inst Armand Frappier PO Box 100 Station LDR Laval PQ H7N 4Z3

BELLONE, CLIFFORD JOHN, b San Francisco, Calif, Feb 9, 41; m 63; c 5. IMMUNOBIOLOGY. *Educ:* Univ Notre Dame, BS, 63, PhD(microbiol), 71. *Prof Exp:* Fel immunol, Univ Calif, San Francisco, 71-74; asst prof immunol, Sch Med, St Louis Univ, 74-80 AT DEPT MICROBIOL-PATH. *Res:* Use of synthetic antigens to study the nature of the immune recognition unit at both the T and B cell level. *Mailing Add:* Dept Microbiol-Path Sch Med St Louis Univ 1402 S Grand Blvd St Louis MO 63104

BELLOW, ALEXANDRA, b Bucharest, Rumania, Aug 30, 35. MATHEMATICS. *Educ:* Univ Bucharest, MS, 57; Yale Univ, PhD(math), 59. *Prof Exp:* Res assoc math, Yale Univ, 59-61; res assoc, Univ Pa, 61-62, asst prof, 62-64; assoc prof, Univ Ill, Urbana, 64-67; PROF MATH, NORTHWESTERN UNIV, 67- *Concurrent Pos:* Ed, Transactions Am Math Soc, 74-77. *Mem:* NY Acad Sci; Am Math Soc; Sigma Xi. *Mailing Add:* Dept Math Northwestern Univ Evanston IL 60201

BELLOW, DONALD GRANT, b Winnipeg, Man, Aug 5, 31; m 56; c 2. MECHANICAL ENGINEERING, APPLIED MECHANICS. *Educ:* Univ BC, BASc, 56; Univ Alta, MSc, 60, PhD(mech eng), 63. *Prof Exp:* Design engr, Can Industs Ltd, 56-57; proj engr, Gen Motors Diesel Ltd, 57-58; sessional lectr mech eng, 58-63, from asst prof to assoc prof, 63-70, chmn dept, 75-84, PROF MECH ENG, UNIV ALTA, 70-, ASSOC VPRES FACIL, 89- *Honors & Awards:* L C Charlesworth Award, Asn Prof Eng Geologists & Geophys Alta. *Mem:* Am Soc Mech Engrs; Soc Exp Mech; fel Can Soc Mech Engrs; Asn Prof Eng Geologists & Geophysicists Alta. *Res:* Static and dynamic analyses of machine elements; vibration problems; problems related to metal fatigue and brittle fracture as applied to the petroleum industry and steel industry; wear of metals and non-metallics. *Mailing Add:* VPres Admin 3-16 University Hall Univ Alta Edmonton AB T6G 2J9 Can

BELLPORT, BERNARD PHILIP, water resources; deceased, see previous edition for last biography

BELLUCE, LAWRENCE P, b Chester, Pa, June 15, 32; div; c 2. MATHEMATICS. *Educ:* Univ Calif, Berkeley, BA, 58; Univ Calif, Los Angeles, MA, 61, PhD(math), 64. *Prof Exp:* Asst prof, math, Univ Calif, Riverside, 63-67; EMER ASSOC PROF MATH, UNIV BC, 88- *Res:* Functional analysis; mathematical logic; theory of rings. *Mailing Add:* 3880 W 17th Ave Univ BC 2075 Wesbrook Pl Vancouver BC V6S 1A4 Can

BELLUM, JOHN CURTIS, b Lakeland, Fla, Aug 2, 45; m 77; c 3. ADAPTIVE OPTICS, NON-LINEAR OPTICS. *Educ:* Ga Inst Technol, BS, 68; Univ Fla, PhD(physics), 76. *Prof Exp:* Fel, Dept Chem, Univ Rochester, 76-78; sci assoc, Dept Physics, Univ Kaiserslautern, Fed Repub Ger, 78-80; vis asst prof, Dept Physics, Univ SFla, 80-81; res assoc, Quantum Theory Proj, Dept Chem, Univ Fla, 81; asst prof, Dept Physics & Astron, Inst Modern Optics, Univ NMex, 81-85; prin scientist, Litton Laser Systs (Albuquerque Opers), 85-87; SR STAFF ENGR, HUGHES AIRCRAFT CO, 87- *Concurrent Pos:* Lectr physics, Univ Md, Europ Div, Fed Repub Ger, 80; guest scientist, Max-Planck-Inst Quantum Optics, Fed Repub Germany, 82. *Mem:* Am Phys Soc; Sigma Xi. *Res:* Molecular collision theory; laser-induced molecular rate processes; electronic-vibrational-rotational energy transfer in molecular collisions; nonlinear optical wavefront control; optical phase conjugation; laser radar; nonconventional adaptive imaging. *Mailing Add:* Hughes Aircraft Co Albuquerque Eng Lab 1600 Randolph Ct SE Albuquerque NM 87106

BELLVÉ, ANTHONY REX, b Wanganui, NZ, July 16, 40; m 61, 83; c 5. MAMMALIAN REPRODUCTION, SPERMATOGENESIS. *Educ:* Massey Col, BAgSc, 64, MAgSc, 67; NC State Univ, PhD(biochem & reproduction), 70. *Prof Exp:* Agr adv officer, NZ Dept Agr, 64-66; staff scientist reprod, Ruakura Agr Res Ctr, 67-68; res fel, NC State Univ, 68-70 & Johns Hopkins Univ, 70-71; res assoc, Harvard Med Sch, 71-72, from asst prof to assoc prof physiol, 72-85; PROF, COL PHYSICIANS & SURGEONS, COLUMBIA UNIV, 85- *Concurrent Pos:* Dep chmn, Div Med Sci, Harvard Med Sch, 79-81, assoc chmn, Div Med Sci, 81-82, mem, Lab Human Reprod & Biol, 71-85; consult, Ctr Pop Res, Nat Inst Child Health & Human Develop, 77-88. *Honors & Awards:* Am Soc Animal Sci Award; Goding Res Award. *Mem:* Am Soc Cell Biol; Am Soc Develop Biol; Am Soc Andrology; Soc Study Reproduction; Soc Study Fertil; Endocrine Soc. *Res:* Regulation of mammalian spermatogenesis, including biochemical processes controlling the proliferation of stem cells, purification and characterization of a spermatogenic growth factor, and the molecular events involved in biogenesis, assembly and reorganization of the developing germ cell; endocrinology. *Mailing Add:* Prof Col Physicians & Surgeons Columbia Univ 650 W 168th St New York NY 10032

BELLWARD, GAIL DIANNE, b Brock, Sask, May 27, 39; div. BIOCHEMICAL PHARMACOLOGY. *Educ:* Univ BC, BSP, 60, MSP, 63, PhD(pharmacol), 66. *Prof Exp:* Instr med chem & pharmacol, Univ BC, 60-61, lectr pharmacol, 66-67, from asst prof to prof, 67-85, chmn dept, 81-85, asst dean, Grad Studies & Res, Fac Pharmaceut Sci, 85-89. *Concurrent Pos:* Fel med, Sch Med, Emory Univ, 68-69; mem, pharm sci grants comt, Med Res Coun Can, 70-74; vis prof, Univ Toronto, 71-72, Univ Sask & Dalhousie Univ, 75-76 & Royal Postgrad Med Sch, London, 75; mem, Pharm Exam Bd Can, 76-79; mem subcomt women pharmacol, Am Soc Pharmacol & Exp Therapeut, 80-85; mem expert adv comt dioxins, Dept Nat Health, Welfare Environ Can, 82-83; mem expert adv comt, Can Drug Scheduling Study Group, 85-86; Killam sr fel, 89-90. *Mem:* Soc Toxicol Can; Am Soc Pharmacol & Exp Therapeut; Pharmacol Soc Can (secy, 77-80, vpres, 85-87, pres, 87-89, past-pres, 89-91). *Res:* Pharmacology; drug metabolism and interactions; toxicology; enzymology; cytochromes P-450. *Mailing Add:* Fac Pharm Sci Univ BC Vancouver BC V6T 1W5 Can

BELLY, ROBERT T, b Trenton, NJ, Feb 19, 45; m; c 1. MICROBIOLOGY, CELL BIOLOGY. *Educ:* Rutgers Univ, AB, 66; Pa State Univ, PhD(microbiol), 70. *Prof Exp:* Res assoc microbiol, Ind Univ, 70-71 & Univ Wis, 71-73; SR RES BIOLOGIST, EASTMAN KODAK CO, 73- *Concurrent Pos:* NIH fel, Univ Wis, 72-73. *Mem:* Am Soc Microbiol; Soc Indust Microbiol (pres, 88); AAAS; Sigma Xi. *Res:* Microbial degradation; cell biology; cellular diagnostics; digital microscopy. *Mailing Add:* 736 Blue Creek Dr Webster NY 14580

BELMAKER, ROBERT HENRY, b Los Angeles, Calif, July 8, 47; m 67; c 6. BIOLOGICAL PSYCHIATRY. *Educ:* Harvard Univ, BA, 67; Duke Med School, MD, 71. *Prof Exp:* Clin assoc, NIMH, 72-74; dir res, Jerusalem Mental Health Ctr, 74-85; PROF PSYCHIAT, BEN GURION UNIV, 85- *Concurrent Pos:* Chmn, Int Col Neuro, 82. *Honors & Awards:* Bennett Award, Soc Biological Psychiatry, 80; Anna Monika Prize, 83. *Mem:* Corresp fel Am Psychiat Asn. *Res:* Manic-depressive illness. *Mailing Add:* PO Box 4600 Beersheba Israel

BELMAN, SIDNEY, b New York, NY, 1926; m 51; c 2. BIOCHEMISTRY. *Educ:* City Col New York, BS, 48; Polytech Inst Brooklyn, MS, 52; NY Univ, PhD(biochem), 58. *Prof Exp:* Instr indust med, 58-61, asst prof environ med, 61-67, assoc prof, 67-76, PROF ENVIRON MED, NY UNIV, 76- *Mem:* AAAS; NY Acad Sci; Am Chem Soc; Am Asn Cancer Res. *Res:* Cancer. *Mailing Add:* NY Univ Dept of Environ Med 550 First Ave New York NY 10016

BELMONT, ARTHUR DAVID, meteorology, for more information see previous edition

BELMONT, DANIEL THOMAS, b Ft Thomas, Ky, Feb 2, 57; m 80; c 1. CHEMICAL PROCESS DEVELOPMENT. *Educ:* Northern Ky Univ, BS, 80; Ohio State Univ, PhD(org chem), 85. *Prof Exp:* RES ASSOC, WARNER-LAMBERT/PARKE-DAVIS, 85- *Mem:* Am Chem Soc. *Res:* Investigate and develop chemical processes for the synthesis of bulk pharmaceuticals. *Mailing Add:* Parke-Davis Pharmaceut Res & Develop 188 Howard Ave Holland MI 49424-6596

BELMONT, HERMAN S, b Philadelphia, Pa, Mar 13, 20; m 46; c 3. PSYCHIATRY, PSYCHOANALYSIS. *Educ:* Univ Pa, AB, 40, MD, 43. *Prof Exp:* Rockefeller fel psychiat, Pa Hosp Ment & Nerv Dis, Inst Pa Hosp, 44, 46 & 47; instr sch med, Univ Pa, 47-48, assoc, 48-52; dir child anal training, Inst Philadelphia Asn Psychoanal, 57-63; assoc prof & sr attend child psychiat, Hahnemann Med Col & Hosp, 52-60, clin prof, 60-63, prof ment health sci & dir div child psychiat, 63-87, dep chmn, 72-87, EMER PROF MENT HEALTH SCI, HAHNEMANN MED COL & HOSP, 88-; SUPVR & TRAINING ANALYST CHILD & ADULT PSYCHOANAL, INST PHILADELPHIA ASN PSYCHOANALYSIS, 56- *Concurrent Pos:* Mem, Govt Comprehensive Ment Health Planning Adv Comt, 64-66; mem adv comt child psychiat planning, Comnr Ment Health, State of Pa, 64; mem task force II, Joint Comn Ment Health Children, 67. *Mem:* Am Psychoanal Asn; Int Psychoanal Asn; Am Orthopsychiat Asn; Am Psychiat Asn; Am Acad Child Psychiat. *Res:* Undergraduate and graduate training; administration of a program in child psychiatry integrating individual psychodynamics and social applications; application of individual psychodynamic knowledge to preventive work with children. *Mailing Add:* 3472 NE Causeway Blvd No 3-104 Jensen Beach FL 34957-4251

BELMONTE, ALBERT ANTHONY, b Lynn, Mass, July 17, 44; m 67; c 3. PHARMACEUTICS. *Educ:* Northeastern Univ, BS, 67, MS, 69; Univ Conn, PhD(pharmaceut), 72. *Prof Exp:* Res asst pharm, Univ Conn, 70-72; asst prof pharmaceut, Sch Pharm, Auburn Univ, 72-78, assoc prof, 78-81; assoc dean, 81-88, DEAN & PROF, COL PHARM, ST JOHN'S UNIV, 88- *Concurrent Pos:* Consult, Natchaug Hosp, 70-72 & Bur Drugs, Food & Drug Admin, 77-; Auburn Univ Res Found fel, 73-74. *Honors & Awards:* Award, McKesson & Robbins, 67; Lederle Pharm Fac Award, 76. *Mem:* AAAS; Am Pharmaceut Asn; Am Asn Pharmaceut Sci; Am Asn Cols Pharm; Am Soc Hosp Pharmacists; Nat Asn Retail Druggists. *Res:* Membrane phenomena and model systems; surface chemistry of carcinogens; electrical phenomena at interfaces; dosage form design and their parameters. *Mailing Add:* Col Pharm & Allied Health Professions St John's Univ Jamaica NY 11439

BELMONTE, ROCCO GEORGE, b New York, NY, July 6, 15. BIOLOGY. *Educ:* Georgetown Univ, AB, 40; Woodstock Col, Md, PhL, 41, STL, 48; Fordham Univ, MS, 44; Cath Univ Am, PhD(biol), 54. *Prof Exp:* Instr biol, Loyola Col, Md, 41-43 & Fordham Univ, 49-51; asst prof, Canisius Col, 54-61 & Fordham Univ, 61-62; asst prof, 62-64, ASSOC PROF BIOL, ST PETER'S COL, NJ, 64- *Mem:* AAAS; NY Acad Sci; Am Jesuit Sci; Am Inst Biol Sci. *Res:* Effects of adrenal extracts on blood-forming organs in amphibians; effects of inhibiting agents on mitosis in Allium cepa. *Mailing Add:* Dept Biol St Peter's Col Jersey City NJ 07306

BELNAP, JAYNE, MICROBIOLOGY, BOTANY-PHYTOPATHOLOGY. *Educ:* Univ Calif, Santa Cruz, BA(biol) & BA(environ studies), 80; Stanford Univ, MS, 84; Brigham Young Univ, PhD(bot), 90. *Prof Exp:* BIOLOGIST RES, NAT PARK SERV, 86- *Mem:* Ecol Soc Am; Am Soc Biologists & Lichenologists. *Res:* Ecology of cold-desert organisms, specifically cyanobacterial soil crusts; role of soil crusts in community dynamics, plant nutrition and soil stability; use of cyanobactrim and soil lichens as biomonitors for air and soil pollution. *Mailing Add:* Box 50 Moab UT 84532

BELOHOUBEK, ERWIN F, b Vienna, Austria, May 7, 29; US citizen; m 54; c 2. MICROWAVES, ELECTRONICS. *Educ:* Vienna Tech Univ, dipl ing, 53, Dr Tech(microwaves), 55. *Prof Exp:* Proj engr, RCA Tube Div, 56-57, David Sarnoff Res Ctr, mem tech staff, Microwave Advan Develop Lab, 57-65, group leader, 65-69, head Microwave Circuits Technol, Microwave Technol Ctr, 69-87, head microwave circuits res, Microwave & Design Automation Lab, 87-89, CONSULT, DAVID SARNOFF RES CTR, 89- *Honors & Awards:* Microwave Appln Award, Inst Elec & Electronics Engrs, 80. *Mem:* Fel Inst Elec & Electronics Engrs. *Res:* Slow wave circuits for traveling wave tubes and magnetrons; crossed-field microwave delay tubes; solid state amplifiers, oscillators and multipliers; hybrid integrated microwave circuits; solid state radar systems; microwave supconductor applications. *Mailing Add:* David Sarnoff Res Ctr CN5300 Princeton NJ 08543

BELON, ALBERT EDWARD, b Gap, Hautes Alpes, France, May 2, 30; US citizen; m 55; c 3. AERONOMY, REMOTE SENSING. *Educ:* Univ Alaska, BS, 52; Univ Calif, Los Angeles, MA, 54. *Prof Exp:* Physicist geophys, Int Geophys Inc, 53-54; res geophysicist, Univ Calif, Los Angeles, 54-56; instr, Geophys Inst, Univ Alaska, 56-61, assoc prof physics, 61-69; prog dir solar-terrestrial res, NSF, 68-70; prof physics, Geophys Inst, Univ Alaska, 69-86, assoc dir, 76-86; RETIRED. *Concurrent Pos:* Consult, Los Alamos Sci Lab, 67-77; mem, Inter-Union Comn Solar-Terrestrial Physics, 68-70 & Comt Polar Res, Panel on Upper Atmospheric Physics, Nat 73-76. Sci, 73- *Honors & Awards:* Meritorious Serv Award, NSF, 70; Except Sci Achievement Medal, NASA, 74. *Mem:* Am Geophys Union; Int Union Geod & Geophys. *Res:* Spectrophotometry of upper atmospheric phenomena; auroral morphology and magnetospheric convection; applications of satellite and aircraft imagery to Alaskan resources surveys. *Mailing Add:* Geophysics Dept Univ Alaska Fairbanks AK 99775-0800

BELOVE, CHARLES, electrical engineering; deceased, see previous edition for last biography

BELSER, WILLIAM LUTHER, JR, b Hershey, Pa, May 28, 25; m 53. GENETICS. *Educ:* Lafayette Col, BA, 48; Yale Univ, PhD(microbiol), 55. *Prof Exp:* USPHS fel marine genetics, Scripps Inst, 55-57, asst res biologist, 57-61, asst prof microbiol, 61-66, assoc prof, 66-72, PROF MICROBIOL, UNIV CALIF, RIVERSIDE, 72- *Mem:* Am Soc Microbiol; Brit Soc Gen Microbiol; Sigma Xi. *Res:* Regulation of metabolic flow in bacteria with emphasis on tryptophan and pyrimidine biosynthesis; comparative biochemistry in the Enterobacteriaceae; enzyme formation and function. *Mailing Add:* Dept Biol Univ Calif Riverside CA 92521

BELSERENE, EMILIA PISANI, b New Rochelle, NY, Dec 12, 22; wid; c 2. ASTRONOMY. *Educ:* Smith Col, AB, 43; Columbia Univ, AM, 47, PhD(astron), 52. *Prof Exp:* Asst, Lick Observ, 43-44; assoc in astron, Columbia Univ, 54-73; mem fac physics & astron, Herbert H Lehman Col, City Univ New York, 52, 56-78; DIR, MARIA MITCHELL OBSERV, 78- *Concurrent Pos:* Vis lectr, Pa State Univ, 78-79; res assoc, Harvard Col Observ, 79-83; lectr, Univ Mass, Boston, 85, adj assoc prof, 87. *Mem:* Am Astron Soc. *Res:* Photographic photometry; variable stars. *Mailing Add:* Maria Mitchell Observ Three Vestal St Nantucket MA 02554

BELSHE, JOHN FRANCIS, b Marshall, Mo, Feb 6, 35; m 57; c 2. ZOOLOGY, ECOLOGY. *Educ:* Cent Mo State Col, BS, 57; Univ Miami, MS, 61, PhD(zool), 67. *Prof Exp:* Instr biol, Miami-Dade Jr Col, 61-63, asst prof, 63-64; from asst prof to assoc prof, 64-74, PROF BIOL, CENT MO STATE UNIV, 74- *Concurrent Pos:* Lic Designated Qualified Person (DQP), Horse Protection Act. *Mem:* AAAS; Am Fisheries Soc; Am Soc Zool; Nat Asn Biol Teachers; Sigma Xi. *Res:* Aquatic ecology; aquatic entomology; population studies of Tribolium; environmental education; odonata of Missouri and the Ozark Plateau. *Mailing Add:* Dept Biol Cent Mo State Univ Warrensburg MO 64093

BELSHEIM, ROBERT OSCAR, b Leland, Iowa, Aug 26, 24. RESEARCH ADMINISTRATION. *Educ:* Iowa State Col, BS, 44; Purdue Univ, MS, 48, PhD(eng mech), 53. *Prof Exp:* Mech engr, Naval Res Lab, 44-47, chief specialist, 44-45, res engr, 50-58, head, Structures Br, 58-70, consult, 70-77; res asst, Purdue Univ, 47-48, instr, 48-50; consult, NKF Eng Assoc, Inc, 77-85; RETIRED. *Mem:* Am Soc Mech Engrs; Sigma Xi. *Res:* Structural dynamics; shock and vibration; mechanical properties of materials; underwater technology; stress analysis; engineering mechanics. *Mailing Add:* 2475 Virginia Ave NW Apt 514 Washington DC 20037

BELSKY, MELVIN MYRON, b Brooklyn, NY, Apr 26, 26; m 52. BIOLOGY. *Educ:* Brooklyn Col, BS, 49, MA, 51; Univ Pa, PhD(bot), 55. *Prof Exp:* From asst prof to assoc prof, 62-68, PROF BIOL, BROOKLYN COL, 68- *Concurrent Pos:* Res assoc, Queens Univ, Can, 55, Univ Calif, Berkeley, 63-64. *Mem:* NY Acad Sci; Am Soc Plant Physiologists; Am Soc Microbiol; AAAS; Sigma Xi. *Res:* Intermediary metabolism; marine mycology; hormone action; enzymology; membrane transport. *Mailing Add:* Dept Biol Brooklyn Col Brooklyn NY 11210

BELSKY, THEODORE, b New Brunswick, NJ, Dec 24, 30; m 52; c 1. HAZARDOUS WASTE, INDUSTRIAL HYGIENE. *Educ:* Univ Calif, Berkeley, BS, 60, PhD(biophys), 66. *Prof Exp:* Biophysicist, Space Sci Lab, Univ Calif, Berkeley, 63-66; asst res geochemist, Space Sci Lab, Univ Calif, Los Angeles, 66-68; asst chemist, Western Regional Lab, USDA, Calif, 69; asst pub health chemist, 69-74, pub health chemist, 74-85, ENVIRON

BIOCHEMIST, CALIF STATE DEPT HEALTH SERV, 86- *Concurrent Pos:* Intersoc comt mem, 71-78. *Mem:* AAAS; Am Chem Soc. *Res:* Organic microanalysis of air pollutants and occupational and environmental contaminants, especially dioxins. *Mailing Add:* Calif State Dept Health Serv 2151 Berkeley Way Berkeley CA 94704

BELT, CHARLES BANKS, JR, b New York, NY, Dec 11, 31; m 57; c 5. ECONOMIC GEOLOGY. *Educ:* Williams Col, BS, 53; Columbia Univ, MA, 55, PhD(econ geol), 59. *Prof Exp:* Asst explor geologist, Mineracao Hannaco, Brazil, 58-59; geologist, Bear Creek Coord Unit, 59-60; res assoc eng exp sta, Univ Utah, 61; asst prof geol & geol eng inst technol, 61-66, ASSOC PROF GEOL, ST LOUIS UNIV, 66- *Mem:* Geol Soc Am; Inst Mining, Metall & Petrol Engrs. *Res:* Origin of hydrothermal ore deposits; relation between igneous intrusions and ore deposition; atomic absorption spectrophotometry and its use in the analysis of rocks and minerals for trace and major chemical elements; environmental geology; geohydrology; statistical applications to the study of changes on the lower Missouri River and the Osage River, particularly natural and man-induced fluvial constriction and erosion. *Mailing Add:* 2559 Oak Springs Lane St Louis MO 63131

BELT, EDWARD SCUDDER, b New York, NY, Aug 4, 33; m 61; c 4. SEDIMENTOLOGY. *Educ:* Williams Col, BA, 55; Harvard Univ, AM, 57; Yale Univ, MA, 59, PhD(geol), 63, Amherst Col, MA, 78. *Prof Exp:* Asst prof geol, Villanova Univ, 62-66; from asst prof to assoc prof, 66-78, chmn dept, 71-76, PROF GEOL, AMHERST COL, 78-; DIR, PRATT GEOL MUSEUM, 87- *Concurrent Pos:* NSF grant, 64-67 & 85-88; Am Philos Soc & Nat Geog Soc grant, 72-73; Quebec grant, 70-79; NRC grant, UK, 77; vis prof, Colo State Univ, 79-80; geologist, Coal Br, US Geol Surv, 80-83, 86-88, NDak Geol Surv, 83 & Earth Sci & Res Inst, Univ SC, 83-84; ROA NSF grant, 84-86, 88-90. *Mem:* Am Inst Mining, Metall & Petrol Eng; Am Asn Petrol Geol; Soc Econ Paleont & Mineral; Geol Soc Am; Int Asn Sedimentology. *Res:* Stratigraphic relationships of facies, depositional environments and regional tectonics of Carboniferous strata in eastern Canada, Scotland and Appalachian Basin USA; Ordovician strata in Quebec; Pleistocene sediments in North Carolina; Paleocene sediments in North Dakota; Karoo and younger strata in Madagascar; sedimentology of fluvial, lacustrine, estuarine, deltaic, and deep-sea paleoenvironments. *Mailing Add:* Dept Geol Amherst Col Amherst MA 01002

BELT, ROGER FRANCIS, b Springfield, Ohio, Mar 20, 29; m 56; c 3. CRYSTALLOGRAPHY. *Educ:* Ohio State Univ, BS, 50; Duquesne Univ, MS, 52; Univ Iowa, PhD(chem), 56. *Prof Exp:* Chemist, E I du Pont de Nemours & Co, 53; sr chemist, B F Goodrich Co, 56-61; sr scientist, Harshaw Chem Co, 61-64; staff scientist, 64-71, RES DIR, AIRTRON DIV, LITTON INDUSTS, INC, 71- *Mem:* AAAS; Am Chem Soc; Am Crystallog Asn; fel Am Inst Chem; Am Inst Physics; Am Phys Soc. *Res:* Solid state physics; x-ray crystallography, crystal growth; morphology; physical properties; perfection. *Mailing Add:* Airtron 200 E Hanover Ave Morris Plains NJ 07950-2193

BELT, WARNER DUANE, b Bellaire, Ohio, Dec 29, 25; m 50; c 4. ANATOMY, CYTOLOGY. *Educ:* Bethany Col, BS, 48; Ohio State Univ, MSc, 50, PhD(anat), 55. *Prof Exp:* Asst instr anat, Ohio State Univ, 51-53, instr, 53-55; res assoc, Univ Calif, Los Angeles, 55-56; asst prof, Emory Univ, 56-59; from asst prof to assoc prof, Med Col Va, 59-64; assoc prof, 64-67, chmn, 73-77, PROF ANAT, SCH MED, TUFTS UNIV, 67- *Mem:* Am Asn Anat. *Res:* Fine structural cytology; structure of endocrine and exocrine secretory cells; fine structure of steroids secretory organs; cytoplasmic interrelationships. *Mailing Add:* Dept Anat & Cellular Biol Tufts Univ Sch Med Boston MA 02111

BELTON, GEOFFREY RICHARD, b Rotherham, Eng, Feb 22, 34; m 60; c 3. METALLURGY, PHYSICAL CHEMISTRY. *Educ:* Univ London, BSc, 57, PhD(metall), 60. *Prof Exp:* Metallurgist, Robert Jenkins & Co, Ltd, Eng, 51-57; asst prof metall, Mass Inst Technol, 60-61; from asst prof to assoc prof, 61-77, PROF METALL & MAT SCI, UNIV PA, 77- *Concurrent Pos:* Vis lectr, Sir John Cass Col, Eng, 58-60. *Mem:* Am Inst Mining, Metall & Petrol Engrs. *Res:* High temperature physical chemistry of metallic and nonmetallic solutions; volatile hydrated compounds; mass spectroscopy of high temperature systems; interaction of ions and monatomic gases with solids. *Mailing Add:* Central Res Lab Broken Hill Co Pty Ltd Shortland NSW 2307 Australia

BELTON, MICHAEL J S, b Bognor Regis, Eng, Sept 29, 34; m 61; c 2. ASTRONOMY. *Educ:* Univ St Andrews, BSc, 59; Univ Calif, Berkeley, PhD(astron), 64. *Prof Exp:* ASTRONR, KITT PEAK NAT OBSERV, 64- *Mem:* AAAS; Am Meteorol Soc; Am Astron Soc. *Res:* Planetary atmospheres and spectroscopy; space photography. *Mailing Add:* Kitt Peak Nat Observ 950 N Cherry Ave Tucson AZ 85719

BELTON, PETER, Driffield, UK, Sept 6, 30; Can citizen; m 57; c 3. BIOLOGY, PEST-MANAGEMENT. *Educ:* Univ London, BSc, 55; Univ Glasgow, PhD(insect physiol), 70. *Prof Exp:* Asst zool, Univ Glasgow, 57-60; res assoc, Col Physicians & Surgeons, Columbia Univ, 60-61; res officer, Res Inst, Can Agr, 61-67; ASSOC PROF BIOL & PEST MGT, SIMON FRASER UNIV, 67- *Concurrent Pos:* Mem, Armed Forces Comt Pest Control, Dept Nat Defense, Can, 75-; expert comt, Vert Pests, Can Agr, 81-; resource person, Can Biting Fly Ctr, Winnipeg, Can, 84- *Mem:* Fel Royal Entom Soc; Can Entom Soc; Am Mosquito Control Asn. *Res:* Analysis of the sensory physiology and behavior of pests, particularly insects; mosquito biology and behavior; bioacoustics. *Mailing Add:* Dept Biol Sci Simon Fraser Univ Burnaby BC V5A 1S6 Can

BELTRAMI, EDWARD J, b New York, NY, Apr 21, 34; m 62. MATHEMATICS. *Educ:* Polytech Inst Brooklyn, BS, 56; NY Univ, MS, 59; Adelphi Univ, PhD(math), 62. *Prof Exp:* Res mathematician, Grumman Aircraft Eng Corp, 56-62; res mathematician, Saclant Anti-Submarine Warfare Res Ctr, 62-63 & Grumman Aircraft Eng Corp, 63-66; prof appl math & urban sci, State Univ NY Stony Brook, 66-80. *Mem:* Opers Res Soc; Soc Indust & Appl Math; Math Asn Am. *Res:* Mathematical optimization techniques and abstract analysis, with emphasis on optical resource allocation. *Mailing Add:* Dept Math & Statist State Univ NY Stony Brook Main Campus Stony Brook NY 11794

BELTZ, CHARLES R(OBERT), b Pittsburgh, Pa, Feb 23, 13; m 35; c 6. AERONAUTICS, REFRIGERATION. *Educ:* Cornell Univ, ME, 35; Univ Pittsburgh, AE, 37. *Prof Exp:* Proj engr, Skycar, Stout Eng Labs, 39-42; aeronaut engr cycle-weld aeronaut, Chrysler Corp, 42-46; proj mgr exp flight controls, Fairchild Eng & Airplane Corp, 46-48; PRES AERONAUT & REFRIG BELTZ ENG LABS, 50- *Concurrent Pos:* Designer, Beltemp Ice Skating Rinks. *Mem:* Am Soc Heating, Refrig & Air Conditioning Engrs; Am Inst Aeronaut & Astronaut. *Res:* Aeronautical experimental research and design of simplified control systems; refrigeration and low temperature development and designs. *Mailing Add:* 500 Lakeland Grosse Point MI 48230

BELTZ, RICHARD EDWARD, b Loma Linda, Calif, June 20, 29; m 51; c 2. BIOCHEMISTRY. *Educ:* Walla Walla Col, BS, 50; Univ Southern Calif, MA, 51, PhD(biochem, nutrit), 56. *Prof Exp:* From instr to assoc prof, 56-70, PROF BIOCHEM, SCH MED, LOMA LINDA UNIV, 70- *Concurrent Pos:* USPHS sr res fel & career develop award, 60-64. *Mem:* AAAS; Am Chem Soc; Sigma Xi. *Res:* Nucleic acids; control of DNA biosynthesis; metabolism of deoxynucleotides and deoxynucleosides; liver regeneration. *Mailing Add:* Dept Biochem Loma Linda Univ Sch Med Loma Linda CA 92350

BELYEA, GLENN YOUNG, b Portland, Maine, Apr 3, 43; m 65; c 2. WILDLIFE ECOLOGY. *Educ:* Univ Maine, BA, 65, MS, 67; Mich State Univ, PhD(wildlife ecol), 76. *Prof Exp:* Entomologist vector control, US Army 67-69; WILDLIFE RES BIOLOGIST, MICH DEPT NAT RESOURCES, EAST LANSING, 72- *Mem:* Wildlife Soc; Raptor Res Found; Lepidopterist's Soc. *Res:* Management and population status of raptors and sandhill cranes in Michigan; hunting and non-hunting recreational uses of public and private lands in Southern Michigan. *Mailing Add:* 8051 Clark Rd Bath MI 48808

BELZER, FOLKERT O, b Soerabaja, Indonesia, Oct 5, 30; US citizen; m; c 3. SURGERY. *Educ:* Colby Col, MA, 53; Boston Univ, MS, 54, MD, 58. *Prof Exp:* Instr surg, Univ Ore Med Sch, 63-64, asst resident surgeon, 64; sr lectr surg, Guy's Hosp, Eng, 64-66; asst prof surg & chief surg outpatient clin, Med Ctr, Univ Calif, San Francisco, 66-69, asst chief transplantation, 69, assoc prof surg, 69-73, prof surg, 73-77, co-dir transplant serv, 69-77; PROF & CHMN DEPT SURG, UNIV WIS CLIN SCI CTR, MADISON, 77- *Concurrent Pos:* Attend surgeon, Moffitt Hosp & San Francisco Gen Hosp, 69-77. *Res:* General surgery; transplantation. *Mailing Add:* Dept Surg Univ Wis Hosp-Clins Univ Wis Med Sch 600 Highland Ave Madison WI 53792

BELZILE, RENE, b Kapuskasing, Ont, Mar 6, 30; m 57; c 3. ANIMAL NUTRITION. *Educ:* Univ Ottawa, BA, 51; McGill Univ, BScAgr, 55; Purdue Univ, MSc, 57; Univ Sask, PhD(animal nutrit), 63. *Prof Exp:* Lectr biochem, Ont Agr Col, 57-60; from asst prof to assoc prof animal nutrit, 63-72, chmn dept animal sci, 74-77, PROF ANIMAL NUTRIT, LAVAL UNIV, 72- *Concurrent Pos:* Feedstuffs Act Can res grant, 65-67; asst ed, Can J Animal Sci, 75-77; chmn, Fur Bearing Animal Comt, Quebec Coun Animal Prod, 77-84; prof & chmn dept animal soc, Nat Univ Rwanda, 88-90. *Mem:* Nutrit Soc Can; Can Soc Animal Sci. *Res:* Plant proteins for mink; evaluation of new feeding programs for foxes and mink; energy & protein requirements of mink; mycotoxins and swine nutrition. *Mailing Add:* Fac Agr & Foods Laval Univ Quebec PQ G1K 7P4 Can

BEMENT, A(RDEN) L(EE), JR, b Pittsburgh, Pa, May 22, 32; m 52; c 8. MATERIALS SCIENCE, NUCLEAR ENGINEERING. *Educ:* Colo Sch Mines, EMet, 54; Univ Idaho, MS, 59; Univ Mich, PhD(metall eng), 63. *Prof Exp:* Metall engr, Hanford Labs, Gen Elec Co, 54-59, sr engr, 59-65; mgr reactor metals res, Pac Northwest Lab, Battelle Mem Inst, 65-66, mgr metall res, 66-68, mgr fuels & mat dept, 68-70; prof nuclear mat, Mass Inst Technol, 70-76; dir, Off Mat Sci, Defense Advan Res Projs Agency, Dept Defense, 76-79, dep undersecy defense, Res & Eng, 79-81; vpres tech resources, 81-89, VPRES SCI & TECHNOL, TRW INC, 89- *Concurrent Pos:* Lectr, Ctr Grad Studies, Univ Wash, Richland, 63-70; adj assoc prof, metall dept, Ore State Univ, 68-70; prof, dept nuclear eng & dept metall & mat sci, Mass Inst Technol, 70-76; adj prof, metall & mat sci dept, Case Western Reserve Univ, 83-; numerous comts, govt, nat & int. *Honors & Awards:* Distinguished Civilian Serv Medal, US Dept Defense, 80. *Mem:* Nat Acad Eng; fel Am Soc Metals; fel Am Nuclear Soc; fel Am Inst Chem; Sigma Xi. *Res:* Irradiation effects; deformation and fracture; energy materials technology; defense materials technology; materials science administration. *Mailing Add:* TRW Inc 1900 Richmond Rd Cleveland OH 44124-3760

BEMENT, ROBERT EARL, b Denver, Colo, Jan 24, 18; m 39; c 2. RANGE SCIENCE. *Educ:* Colo State Univ, BS, 40, MF, 47, PhD, 68; US Army Command & Gen Staff Col, dipl, 62. *Prof Exp:* Instr range forage plants, Colo State Univ, 47-48; range conservationist soil conserv serv, USDA, 48, soil conservationist, 48-55, res range conservationist cent plains exp range, 55-61, res range conservationist in charge, 61-65, res range scientist in charge, 65-73, range scientist agr res serv, 66-69, range scientist in charge forage & range res br crops res lab, Colo State Univ, 69-73; CONSULT, 73- *Concurrent Pos:* Spec lectr col forestry, Colo State Univ, 59-; agr res serv rep, Tech Comt Seedling Estab Range Plants, Great Plains Coun, 60-, comt chmn, 62-63, agr res serv rep, Range & Livestock Mgt Comt, 70-, consult, Food & Agr Orgn, UN, Iceland, 73-78; consult, Int Prog Ecol Mgt of Arid and Semi-arid Rangelands in Africa and Near East, 74-75 & USAID on grazing management in eleven african countries, 77-; expert witness, Colorado Water Courts, 77-78; sr range adv to Niger Range & Livestock Proj. *Honors & Awards:* Trail Boss Award, Soc Range Mgt, 84. *Mem:* Am Soc Range Mgt; Sigma Xi; Am Soc Animal Sci. *Res:* Relationship of season and degree of grazing to vegetation, gains per animal and animal gain per acre; artificial range revegetation, including role of temperature and moisture as modified by mulches. *Mailing Add:* PO Box 524 Mancos CO 81328

BE MENT, SPENCER L, b Detroit, Mich, Apr 1, 37; m 62; c 3. BIOENGINEERING, ROBOTICS. *Educ:* Univ Mich, BSE, 60, MSE, 62, PhD(bioeng), 67. *Prof Exp:* From instr to assoc prof, 63-77, PROF ELEC & COMP ENG, UNIV MICH, 77- *Concurrent Pos:* Vis assoc prof, Med Sch, Univ Vt, 76. *Mem:* AAAS; Inst Elec & Electronics Engrs; Soc Neurosci; Biomed Eng Soc; Sigma Xi. *Res:* Communications and systems theory applied to neural information coding and transmission; passive and active electrical properties of biological neural tissue; digital signal processing; bioinstrumentation. *Mailing Add:* 5500 Warren Rd Ann Arbor MI 48105

BEMILLER, JAMES NOBLE, b Evansville, Ind, Apr 7, 33; m 60; c 2. CARBOHYDRATE CHEMISTRY, ORGANIC & NATURAL PRODUCTS CHEMISTRY. *Educ:* Purdue Univ, BS, 54, MS, 56, PhD(biochem), 59. *Prof Exp:* Asst biochem, Purdue Univ, 54-59, asst prof, 59-61; from asst prof to prof chem & biochem, Southern Ill Univ, Carbondale, 61-85, actg chmn dept, 66-67, prof, Sch Med, 71-85, actg dean, Col Sci, 76-77, asst dean curric, 77-79, chmn, dept Med Biochem, 80-83; PROF FOOD SCI & DIR, WHISTLER CTR FOR CARBOHYDRATE RES, PURDUE UNIV, WEST LAFAYETTE, IND, 86- *Concurrent Pos:* Chmn, div carbohydrate chem, Am Chem Soc, 65-66. *Mem:* Fel AAAS; Am Chem Soc; Am Asn Cereal Chem; fel Am Inst Chem (secy, 85, pres, 88); Am Soc Biochem & Molecular Biol. *Res:* Carbohydrate chemistry; poly saccharides chemistry; industrial uses of carbohydrates. *Mailing Add:* Purdue Univ Ctr Carbohydrate Res Smith Hall West Lafayette IN 47907

BEMILLER, PARASKEVI MAVRIDIS, US citizen. HUMAN DEVELOPMENT, CELLULAR AGING. *Educ:* Am Univ Beirut, BA, 53, MS, 55; Purdue Univ, PhD(physiol biochem), 61. *Prof Exp:* Instr biol, Am Univ Beirut, 55-58; researcher, Marine Biol Lab, Woods Hole, Mass, 59; res assoc physiol, Purdue Univ, 61; res assoc, Sch Med, Southern Ill Univ, 66-68, teaching assoc, 68-71, asst prof biol, 72-79, assoc prof embryol, 79-86; ASSOC PROF ANAT, PURDUE UNIV, 86- *Mem:* Am Soc Cell Biol; Am Asn Anatomists; Am Aging Asn; Sigma Xi; Tissue Cult Asn. *Res:* Study of changes which occur in cells during aging; use of human diploid fibroblasts to study morphological and biochemical changes which occur as cells age in vitro; nucleolar changes and ribosomal RNA processing. *Mailing Add:* 2829 Bentbrook Lane West Lafayette IN 47906-5274

BEMIS, CURTIS ELLIOT, JR, b Boston, Mass, Feb 18, 40; m 62; c 3. NUCLEAR PHYSICS, NUCLEAR CHEMISTRY. *Educ:* Univ NH, BS, 61; Mass Inst Technol, PhD(nuclear chem), 65. *Prof Exp:* Res asst nuclear chem, Mass Inst Technol, 61-64; Sweden-Am Found & Fulbright-Hayes fels nuclear spectros, Nobel Inst Physics, Stockholm, Sweden, 64-65; researcher, 65-66, res mem staff, Nuclear Spectros, 66-79, SR STAFF, PHYSICS DIV, OAK RIDGE NAT LAB, 79- *Mem:* AAAS; fel Am Phys Soc; Sigma Xi. *Res:* Nuclear spectroscopy and low energy nuclear physics; heavy ion physics; transuranium element nuclear and atomic structure physics and health physics; production and identification of new transuranium elements and isotopes; laser spectroscopy and lasers in nuclear physics. *Mailing Add:* Physics Div Bldg 6000 Oak Ridge Nat Lab Oak Ridge TN 37831-6371

BEMIS, WILLIAM PUTNAM, genetics; deceased, see previous edition for last biography

BEMPONG, MAXWELL ALEXANDER, b Akim-Oda, Ghana, Sept 14, 38; m 68; c 2. GENETIC TOXICOLOGY. *Educ:* Lincoln Univ, Mo, BSc, 64; Tenn State Univ, MS, 65; Mich State Univ, PhD(cytogenetics), 67. *Prof Exp:* Teaching fel, Univ Nev, 69-70; fel, Johns Hopkins Univ, 75-77; from asst prof to assoc prof, 67-74, PROF BIOL, NORFOLK STATE UNIV, 74- *Concurrent Pos:* Ed, J Basic Appl Sci, 81-; dir, Biomed Res Norfolk State Univ, 75-; ed, J Toxicol & Biomed Sci & Trans Nat Inst Sci. *Mem:* Environ Mutagen Soc; Nat Inst Sci; Tissue Cult Asn; Am Col Toxicol; NY Acad Sci; Sigma Xi; Europ Acad Sci; Europ Soc Arts & Lett. *Res:* Genetic toxicological analysis of anti-neoplastic drugs and industrial toxicants; in vivo and in vitro estimation of mutagenesis and carcinogenesis in laboratory rodents; chemical-induced morphological modulation in malignant cells in vitro and examination of the comparative in vivo tumorigenicity of the revertant and parental cell types. *Mailing Add:* Dept Biol Norfolk State Univ 2401 Corprew Ave Norfolk VA 23504

BEMRICK, WILLIAM JOSEPH, b Superior, Wis, Jan 19, 27; m 67; c 6. PARASITOLOGY. *Educ:* Superior State Col, BS, 50; Univ Wis, MS, 55, PhD(zool), 57. *Prof Exp:* Res asst zool & parasitol, Univ Wis, 53-57, teaching asst, 57; from asst prof to prof vet med parasitol, Univ Minn, St Paul, 57-90; RETIRED. *Concurrent Pos:* Presiding officer, Annual Midwestern Conf Parasitologists, 74. *Mem:* Am Soc Parasitol. *Res:* Host parasite relationships; parasitic immunology; electron microscopy of parasites; Coccidia and coccidiosis morphology and epidemiology of Giardia species. *Mailing Add:* Dept Pathobiol Col Vet Med Univ Minn St Paul MN 55101

BEN, MANUEL, b Syracuse, NY, July 30, 16; wid; c 2. TECHNICAL MANAGEMENT, WATER POLLUTION. *Educ:* Univ Mich, BS, 39. *Prof Exp:* Asst bearing metallurgist, AC Spark Plug Div, Gen Motors Corp, Flint, Mich, 40-41, bearing chemist supvr, 41-45, sr res chemist & supvr, 45-53, sr res chemist, Gen Motors Res Labs, Warren, Mich, 53-54, supvr, 54-66, supvr mfg staff, Mfg Develop, 66-72, sr design engr, Chevrolet Motor Co Div, 72-74, eng supvr & staff develop engr, Gen Motors Mfg Develop, 74-76, Plant Eng Develop, 76-81; RETIRED. *Concurrent Pos:* Consult & teacher, 81- *Honors & Awards:* C H Proctor Mem Award, 64. *Mem:* Am Electroplaters Soc (vpres, 59-62, pres, 62-63, past pres, 63-64); Electrochem Soc; Brit Inst Metal Finishing. *Res:* Decorative, corrosion protection, and engineering electroplating (electrochemistry); anodizing and electroless plating (electrochemistry); strategies and equipment technology and systems to control water pollution and air emissions. *Mailing Add:* 23605 Sutton Dr No 1421 Southfield MI 48034-3354

BEN, MAX, b Utica, NY, Oct 12, 26; m 52; c 4. PHARMACOLOGY, PHYSIOLOGY. *Educ:* Syracuse Univ, BA, 51; Princeton Univ, MA, 53, PhD(physiol), 54. *Prof Exp:* Lab asst gen biol, Utica Col, 50-51; asst biol & biochem, Princeton Univ, 51-53, asst endocrinol to Prof Swingel, 53-54; assoc surg res, Harrison, Dept Surg Res, Med Sch, Univ Pa, 54-56 & Merck, Sharp & Dohme, 56-58; sect head, Endocrine Res & Eval, Squibb Inst, 58; sr scientist, Warner-Lambert Res Inst, 58-62; sci dir pharmacol & co-dir mkt, Hazelton Labs, Inc, 62-64, sr pharmacologist & sci dir develop, 64-65; UN tech assistance expert pharmacol & adv to Nat Coun Res & Develop, Prime Minister's Off, Israel, 65-68; exec secy, Nat Biol Cong, 69-70; exec vpres & dir, Health Sci Assocs, Bethesda, Md, 70-74; pres, Mediserv Int, 74-77; dir corp res, Dept Toxicol & Biochem Pharmacol, Miles Labs, Inc, 77-79; med dir, Med Prod Div, US Packaging Corp, 79-81; vpres & sci dir, Life Sci, Inc, 81-83; vpres & med dir, Belmac Corp, 83-86; pres & dir, Valcor Sci Ltd, 86-89; CONSULT CLIN STUDIES, FOOD & DRUG ADMIN, 89- *Concurrent Pos:* Consult pharmacol & toxicol, NIH contract, 58. *Mem:* Soc Toxicol; Am Physiol Soc; fel NY Acad Sci; Am Soc Clin Pharmacol & Therapeut; Am Soc Pharmacol & Exp Therapeut; AAAS; Sigma Xi; Am Inst Biol Sci; Am Acad Clin Toxicol; Am Asn Poison Control Ctrs; Genetic Toxicol Asn. *Res:* Over 90 publications in endocrinology, shock, pharmacology and toxicology; investigation of cancer chemotherapeutic agents; cardiovascular and endocrine pharmacology and physiology; studies of drug abuse in man. *Mailing Add:* 812 Amelia Court NE St Petersburg FL 33702

BENACERRAF, BARUJ, b Caracas, Venezuela, Oct 29, 20; US citizen; m 43; c 1. IMMUNOLOGY, EXPERIMENTAL PATHOLOGY. *Educ:* Univ Paris, Lic es lett, 40; Columbia Univ, BS, 42; Med Col Va, MD, 45; Univ Geneva, MD, 80. *Hon Degrees:* MD, Univ Geneva, 80; DSc, NY Univ & Va Commonwealth Univ, 81, Yeshiva Univ & Univ Aix-Marseille, 82, Columbia Univ, 85, Adelphi Univ, 88; PhD, Weizmann Inst Sci, 89. *Prof Exp:* Intern, Queens Gen Hosp, NY, 45-46 & US Army, 46-48; res fel, dept microbiol, Col Physicians & Surgeons, Columbia Univ, 48-50; sr researcher, Nat Ctr Sci Res, Hosp Broussais, Paris, 50-56; from asst prof to prof path, Sch Med, NY Univ, 56-68; chief, Lab Immunol, Nat Inst Allergy & Infectious Dis, NIH, 68-70; Fabyan prof comp path & chmn, Dept Path, 70-91, FABYAN EMER PROF COMP PATH, MED SCH, HARVARD UNIV, 91-; PRES & CHIEF EXEC OFFICER, DANA FARBER CANCER INST, BOSTON, 80- *Concurrent Pos:* Mem, immunol study sect, NIH, 65-69; sci adv immunol, WHO, 65-; trustee & mem, sci adv bd, Trudeau Found, 70-76; mem, sci adv bd, Mass Gen Hosp, 71-74; assoc ed, Am J Path, Lab Invest, J Exp Med, J Immunol, J Exp Path & Immunol Commun; mem bd govs, Weizmann Inst Sci, 76; mem & chmn, sci adv comt, Ctr Immunol Marseille, Nat Ctr Sci Res-Nat Inst Health & Med Res, France, 78-84; mem, adv coun, Nat Inst Allergy & Infectious Dis & adv comt, Basel Inst Immunol, 85. *Honors & Awards:* Nobel Prize Physiol or Med, 80; R E Dyer Lectr, NIH, 69; Harvey Lectr, 71 & 72; Rabbi Shai Shacknai Lectr & Prize Immunol & Cancer Res, Hebrew Univ Jerusalem, 74; T Duckett Jones Mem Award, Helen Hay Whitney Found, 76; J S Blumenthal Lectr Allergy & Immunol, 80; Waterford Biomed Sci Award, 80; Rous-Whipple Award, Am Asn Pathologists, 85; Margaret Byrd Rawson Award, Am Dyslexia Soc, 88; Nat Medal of Sci, 90. *Mem:* Nat Acad Sci; Inst Med-Nat Acad Sci; Am Asn Immunologists (pres, 73-74); Am Asn Pathologists & Bacteriologists; fel Am Acad Arts & Sci; Soc Exp Biol & Med; NY Acad Sci; Am Soc Exp Path; Fedn Am Soc Exp Biol (pres, 74-75); Int Immunol Soc (pres, 80-83). *Res:* Antibody synthesis and structure; hypersensitivity; immunopathology; immunogenetics. *Mailing Add:* Dana Farber Cancer Inst 44 Biney St Boston MA 02115

BENACH, JORGE L, b Havana, Cuba, June 27, 45; US citizen; m 68; c 2. MEDICAL ENTOMOLOGY. *Educ:* Upsala Col, BA, 66; Seton Hall Univ, MS, 68; Rutgers Univ, PhD(entom), 71. *Prof Exp:* Med entomologist, Bur Acute Commun Dis Control, Albany, 71-72, SR MED ENTOMOLOGIST, WHITE PLAINS REGIONAL OFF, NY STATE DEPT HEALTH, 72- *Mem:* AAAS; Am Asn Trop Med & Hyg; NY Acad Sci; Am Mosquito Control Asn. *Res:* Characterization and identification of Rickettsial pathogens and the epidemiology of Rickettsial diseases in New York. *Mailing Add:* Eight Patriot Ct Stony Brook NY 11790

BEN-AKIVA, MOSHE E, b Tel Aviv, Israel, June 11, 44; m 68; c 3. ECONOMETRICS. *Educ:* Technion-Israel Inst Technol, BSc, 68; Mass Inst Technol, SM 71, PhD(transp), 73. *Prof Exp:* Res asst, 68-73, PROF TRANSP, MASS INST TECHNOL, 73- *Concurrent Pos:* Prin, Cambridge Syst, Inc, 72-; vis prof, Technion-Israel Inst Technol, 78-79 & 81-82; dir, Hague Consult Group, 85- *Res:* Transportation systems analysis; behavioral models of travel and mobility; transportation policy analysis; transportation economics; urban and regional models; econometric models. *Mailing Add:* Dept Civil Eng Mass Inst Technol Cambridge MA 02139

BENARD, BERNARD, b Dorval, Que, Oct 16, 33; Can citizen; m 62; c 4. ENDOCRINOLOGY, MEDICINE. *Educ:* Univ Montreal, BA, 55, MD, 60; McGill Univ, MSc, 65. *Prof Exp:* Resident med, Notre-Dame Hosp, Montreal, 60-63; res fel, Dept Anat, McGill Univ, Montreal, 63-65, Thyroid Unit, Mass Gen Hosp, Boston, 65-66 & Food & Nutrit Sci, Mass Inst Technol, 66-67; vdean res, 79-81, vice rector res, 81-89, PROF MED, MED SCH, UNIV SHERBROOKE, 67- *Concurrent Pos:* Res fel, Med Res Coun Can, 63-67; coun mem, Can Soc Clin Invest, 68-71; consult & mem, MRC Comt Clin Res, 71-75; consult, Que Ministry Educ, 73-77 & Que Ministry Social Affairs, 76- *Mem:* Endocrine Soc; Am Thyroid Asn; Can Soc Clin Res; Royal Col Physicians, Can; Can Soc Endocrinol. *Res:* Studies on thyroid tumors; studies on viral transformation of thyroid cells; studies of steroid receptors. *Mailing Add:* Dept Med Sch Med Univ Sherbrooke 3001 12th Ave N Fleurimont PQ J1H 5N4 Can

BENARD, MARK, b Pittsburgh, Pa, Jan 4, 44; m 73. ANALYSIS OF ALGORITHMS, SOFTWARE ENGINEERING. *Educ:* Tulane Univ, BA, 65; Yale Univ, MPhil, 67, PhD(math), 69. *Prof Exp:* Instr math, Yale Univ, 69-70; asst prof, 70-75, assoc prof math, 75-80, ASSOC PROF COMPUTER SCI, TULANE UNIV, 80- *Mem:* Am Math Soc; Computer Soc; Inst Elec & Electronics Engrs; Sigma Xi. *Res:* Design and analysis of algorithms; distributed processing; software engineering. *Mailing Add:* Computer Sci Dept Tulane Univ New Orleans LA 70118

BENARDE, MELVIN ALBERT, b Brooklyn, NY, June 15, 23; m 51; c 3. EPIDEMIOLOGY, PUBLIC HEALTH. *Educ:* St John's Univ, NY, BS, 48; Univ Mo, MS, 50; Mich State Univ, PhD(bact), 54. *Prof Exp:* Res asst virus res, Pub Health Res Inst, NY, 50-51; food specialist, Off Naval Res, Naval Supply Depot, Bayonne, 54-55; asst prof, Seafood Processing Lab, Univ Md, 55-61; assoc prof, Bioeng Lab, Dept Civil Eng, Rutgers Univ, 61-67; assoc prof, 67-74, PROF COMMUNITY MED, HAHNEMANN MED COL, 74-, ACTG CHMN, COMMUNITY MED & ENVIRON HEALTH, 81-; AT ENVIRON STUDIES INST, DREXEL UNIV. *Concurrent Pos:* WHO fel, Sch Hyg & Trop Med, Univ London, 63; mem bd dirs, Am Coun Sci & Health. *Mem:* Fel Am Pub Health Asn; Sigma Xi; Asn Teachers Prev Med; fel Royal Soc Health. *Res:* Nosocomial infections; microbial metabolism of synthetic detergents; effects of synthetic chemicals on tissue cultures; germicidal effects of chlorine compounds; evaluation of medical care delivery; environmental factors in pancreatic cancer. *Mailing Add:* Environ Studies Inst Drexel Univ Philadelphia PA 19104

BENAROYA, HAYM, b 1954; US citizen; m 83; c 2. SPACE BUSINESS. *Educ:* Cooper Union, BECE, 76; Univ Pa, MS, 77, PhD(eng), 81. *Prof Exp:* Sr res engr, Weidlinger Assocs, 81-89; FAC MECH & AEROSPACE ENG, RUTGERS UNIV, 89- *Concurrent Pos:* Adj prof eng, Cooper Union, 81-89; founder & dir, Lab Extraterrestrial Struct Res, Rutgers Univ; tech ed, Appl Mech Reviews. *Mem:* Am Soc Civil Engrs; NY Acad Sci; Am Acad Mech; Am Inst Aeronaut & Astronaut; Am Soc Mech Eng; Soc Indust & Appl Math. *Res:* Probabilistic methods in mechanics; inverse vibration problems; space and lunar structures; mechanics of aging aircraft. *Mailing Add:* Dept Mech & Aerospace Eng Rutgers Univ PO Box 909 Piscataway NJ 08855-0909

BENATAR, SOLOMON ROBERT, b Selukwe, South Rhodesia, Feb 6, 42; m 65; c 3. HOSPITAL ADMINISTRATION. *Educ:* Univ Capetown, MBChB, 65; Col Med South Africa, FFA(SA), 71, MRCP, 72, FRCP, 81. *Prof Exp:* Intern med & obstet gynecol, Groote Schuur Hosp, Cape Town, 66, res anesthesiol, 68-70, res internal med, 70-71, physician internal med & chest dis, 74-80; family dr, primary health care, pvt pract, 67-68; res fel anesthesiol, Northwick Park Hosp, London, 72; res fel chest dis, Brompton Hosp, London, 72, res, 73; PROF & CHMN, DEPT MED, UNIV CAPETOWN, 80- *Concurrent Pos:* Res fel, Imp Chem Industs, 72; chmn Fac Med, Col Med SAfrica, 80-, vpres, 89-92; physician-in-chief, Dept Med, Groote Schuur Hosp, 80-; int adv respiratory med, 87-; vis prof, several univs, US & UK; vpres, Col Med SAfrica, 89- *Honors & Awards:* African Oxygen Gold Medal, Col Med SAfrica, 71. *Mem:* Foreign assoc mem Nat Acad Sci Inst Med; Med Asn SAfrica. *Res:* Clinical and epidemiological aspects of asthma, tuberculosis, venous thromboembolism and sarcoidosis; health economics and health services research; medical ethics and philosophy of medicine; national health service; author of over 200 publications. *Mailing Add:* Dept Med Univ Cape Town Pvt Bag Rondebosch Cape Town 7700 South Africa

BENBOW, ROBERT MICHAEL, b San Pedro, Calif, Nov 10, 43; m 75; c 7. DNA REPLICATION, GENETIC TRANSFORMATION. *Educ:* Yale Univ, BS, 67; Calif Inst Technol, PhD(biophysics), 72. *Prof Exp:* Helen Hay Whitney res fel, Med Res Coun, Lab Molecular Biol, 72-75; from asst prof to assoc prof biol, Johns Hopkins Univ, 75-85; PROF BIOL, IOWA STATE UNIV, 85-, DIR, NUCLEIC ACID FACILITY, 86- *Concurrent Pos:* Mem cell biol panel, NSF, 80-84; biotechnol Coun, Iowa State Univ, 87-90. *Mem:* Am Soc Biochem & Molecular Biol. *Res:* Molecular mechanisms of DNA replication and genetic transformation in early embryos of the frog, Xenopus laevis; soybean Glycine max. *Mailing Add:* Dept Zool & Genetics Iowa State Univ Ames IA 50011-3223

BENBROOK, CHARLES M, b Los Angeles, Calif, Nov 26, 49; m 73; c 3. AGRICULTURAL TECHNOLOGY RISK ASSESSMENT. *Educ:* Harvard Unv, BA, 71; Univ Wis-Madison, MS, 80, PhD(agr econ), 81. *Prof Exp:* Policy analyst, Coun Environ Qual, 80-82; staff dir, Subcomt on Dept Opers Res & Foreign Agr, Comt Agr, US House Representatives, 82-84; exec dir, Bd Agr, Nat Res Coun, 84-90; BENBROOK CONSULT SERV, 91- *Mem:* Am Agr Econ Asn; Soil Conserv Soc Am. *Res:* Agricultural policy analysis, particularly food safety and natural resource use issues; international competitiveness. *Mailing Add:* 24222 White's Ferry Rd Dickerson MD 20842

BENCALA, KENNETH EDWARD, b Detroit, Mich, 1951. GEOCHEMISTRY, HYDROBIOLOGY. *Educ:* Harvey Mudd Col, BS, 73; Calif Inst Technol, MS, 75, PhD(chem eng), 79. *Prof Exp:* CHEM ENGR, WATER RESOURCES DIV, US GEOL SURV, 78- *Mem:* Am Geophys Union; Sigma Xi; Am Chem Soc. *Res:* Develop quantitative formulations of the processes determining the rate and extent of the reactions of stream solutes with sediments. *Mailing Add:* US Geol Surv MS 496 345 Middlefield Rd Menlo Park CA 94025

BENCE, ALFRED EDWARD, b Saskatoon, Sask, Aug 30, 40. ELECTRON MICROPROBE THEORY & APPLICATIONS. *Educ:* Univ Sask, BEng, 62; Univ Tex, Austin, MA, 64; Mass Inst Technol, PhD(geochem), 66. *Prof Exp:* Res fel, Calif Inst Technol, 66-68; asst prof petrol, State Univ NY, Stony Brook, 68-71, assoc prof, 71-74, prof, 74-80; res assoc, Exxon Prod Res, 80-82, sr res assoc, Exxon Mat, Exxon Prod Res Co, 82-89, EXXON VALDEZ, NRDA LITIGATION SUPPORT, EXXON USA, 89- *Concurrent Pos:* Vis fel, Australian Nat Univ, 74-75 & 79; expert, Facilities Subcomt of Lunar & Planetary Sample Planning Team, Johnson Space Ctr, NASA, 76-79, prin investr, 76-80, prin investr, Spec Recognition Lunar Prog, 79; assoc ed, J Geophys Res, 78-81; counr, Geochem Soc, 83-86. *Mem:* Am Geophys Union; Geochem Soc; Microbeam Anal Soc; Am Asn Petrol Geologists. *Res:* Chemical and mineralogical study of terrestrial and extraterrestrial basaltic systems; the genesis of volcanogenic massive sulfide deposits with an emphasis on rock-water interactions; geochemistry of petroleum source rocks; hydrocarbon geochemistry of subtidal sediments; degradation of petroleum hydrocarbons in the marine environment. *Mailing Add:* NRDA Litigation Support Exxon USA PO Box 2080 Houston TX 77252

BENCOMO, JOSÉ A, b Caracas, Venezuela, Nov 27, 46; US citizen; m 76; c 2. RADIOLOGICAL PHYSICS, PHYSICS OF IMAGING. *Educ:* Cent Univ, Caracas, Venezuela, BS, 74; Univ Tex Houston, MS, 78, PhD(biophysics), 82. *Prof Exp:* Instr physics, Univ Simon, Bolivar, Venezuela, 74-75; res asst diag imaging, MD Anderson Cancer Ctr, Grad Sch Biomed Sci, Univ Tex, 77-82, asst physicist diag imaging, 83-85, ASST PROF PHYSICS DIAG IMAGING, M D ANDERSON CANCER CTR, GRAD SCH BIOMED SCI, UNIV TEX, 86-, ASSOC PHYSICIST RADIATION PROTECTION, 90- *Concurrent Pos:* Prin investr, Advan Technol & Advan Res Prog, Tex Higher Educ Coord Board Grant, 88-90; qual assurance physicist, MD Anderson Cancer Ctr, Houston, 90-; chmn, Specialized Master Comt, Grad Sch Biomed Sci & Exec Comt, 90- *Honors & Awards:* Itek Award, Soc Photog Scientists & Engrs, 84. *Mem:* Am Asn Physicist Med; Am Col Med Physics; Am Col Radiol; Mat Res Soc. *Res:* Quantitative medical imaging; digital radiography; associated image processing for application in diagnostic radiology; development of x-ray receptors and image processing techniques related to specific diagnostic tools; evaluation of display and viewing requirements. *Mailing Add:* M D Anderson Cancer Ctr Univ Tex Box 57 Houston TX 77030

BENCOSME, SERGIO ARTURO, b Montecristi, Dominican Repub, Apr 27, 20; Can citizen; m; c 5. PATHOLOGY. *Educ:* Univ Montreal, MD, 47; McGill Univ, MSc, 48, PhD(path), 50. *Prof Exp:* Asst prosector, Royal Victoria Hosp, Montreal, 50-51; asst prof path, Univ Ottawa, 51-53; assoc prof, Queen's Univ, Ont, 53-57; asst res pathologist, Univ Calif, Los Angeles, 57-59; assoc prof, 59-65, PROF PATH, QUEEN'S UNIV, ONT, 65- *Concurrent Pos:* Can Life Ins Officer's Asn fel, 53-56; Can Cancer Soc traveling fel, US, 54; Ont Cancer Treatment & Res Found traveling fel, Univ Calif, Los Angeles, 56, lectr, 57; actg dir, Hotel Dieu Hosp, Ont, 57; pres, Burton Soc Electron Micros, 59; mem coun basic sci, Am Heart Asn, 62-; dir inst biomed, Univ Pedro Henriquez Urena, Santo Domingo, 73. *Mem:* Am Soc Cell Biol; Am Asn Anat; Am Asn Path & Bact; Am Soc Exp Path; Soc Exp Biol & Med. *Res:* Pancreatic islets; kidney pathology; myocardium pathology; cancer of the breast. *Mailing Add:* Dept of Path Queen's Univ Kingston ON K7L 3N6 Can

BENCSATH, KATALIN A, b Szeged, Hungary, 1945; US citizen; m 70. COMBINATORIAL GROUP THEORY, DISCRETE MATHEMATICS. *Educ:* Eötvös Loránd Univ, Budapest, Cert(math), 68; Queens Col City Univ NY, MA, 76; Grad Sch City Univ NY, PhD(math), 83. *Prof Exp:* Mathematician, Közti, Inst Archit, Budapest, 68-69; mathematician & appln programmer, opers res, Infelor Systs Eng Inst, Budapest, 69-72; instr, 81-83, asst prof, 83-88, ASSOC PROF MATH & COMPUT SCI, DEPT MATH & COMPUT SCI, MANHATTAN COL, 88- *Concurrent Pos:* Consult, Füti Data Processing Inst, Budapest, 68-69; adj lectr math, Hunter Col City Univ NY, 76-81; Univ assoc, NY Univ, 86- *Mem:* Am Math Soc; Math Asn Am; NY Acad Sci; Sigma Xi. *Res:* The structural and residual properties of one-relator groups and their generalization; problems concerning the existence of certain algorithms in various classes of infinite discrete groups. *Mailing Add:* Dept Math & Comput Sci Manhattan Col Riverdale NY 10471

BEND, JOHN RICHARD, b Teulon, Can, Aug 10, 42; m 67; c 2. COMPARATIVE PHARMACOLOGY, BIOCHEMICAL TOXICOLOGY. *Educ:* Univ Manitoba, BSc, 64, MSc, 67; Univ Sydney, PhD(pharm & chem), 71. *Prof Exp:* Asst lectr pharm, Univ Manitoba, 66-67; vis assoc, Nat Inst Environ Health Sci, NIH, 70-76, sect head, 76-78, actg chief, 78-79, res chemist, Lab Pharmacol, 80-, chief, 80-; CHMN DEPT PHARM & TOXICOL, UNIV WESTERN ONT. *Concurrent Pos:* Vis scientist, C V Whitney Marine Lab, Univ Fla, 76-; mem, Comt Environ Pharmacol, Am Soc Pharmacol Exp Therapeut, 76-; adj assoc prof, NC State Univ, 78-; assoc ed, Metab Basis Detoxification, 80-81, Reviews Biochem Toxicol, 77-; ed, Chemico-Biol Interactions, 78-; adj assoc prof, Univ NC, 80- *Mem:* Am Soc Pharmacol & Exp Therapeut; AAAS. *Res:* Relationships between xenobiotic metabolism and target organ and cell toxicity in mammalian and aquatic species; characteristics of the enzyme systems responsible for the metabolic activation and detoxication of chemicals. *Mailing Add:* Dept Pharmacol & Toxicol Univ Western Ont Med Sci Bldg London ON N6A 5C1 Can

BENDA, GERD THOMAS ALFRED, b Berlin, Ger, Nov 19, 27; nat US; m 56; c 2. PLANT PHYSIOLOGY. *Educ:* Princeton Univ, AB, 46; Yale Univ, MS, 49, PhD(bot), 51. *Prof Exp:* Instr plant physiol, Yale Univ, 52-53; vol worker dept plant path, Rothamsted Exp Sta, 53-54; USPHS res fel virus lab, Univ Calif, 54-56, jr res biologist, 56-57; from asst prof to assoc prof biol, Univ Notre Dame, 57-64; Rockefeller Found grant virol sect, Inst Agron, Brazil, 64-65; plant physiologist, Agr Res Serv, USDA, 66-87; RETIRED. *Mem:* Bot Soc Am; Am Soc Sugar Cane Technol. *Res:* Physiology of plant virus diseases. *Mailing Add:* 118 Fahey St Houma LA 70360-7598

BENDA, MIROSLAV, b Czech, Oct 28, 44; US citizen. MATHEMATICS, COMPUTER SCIENCE. *Educ:* Univ Warsaw, MSc, 68; Univ Wis, MA, 69, PhD(math), 70; Univ Wash, MSc, 81. *Prof Exp:* Lectr math, Univ Calif, Berkeley, 70-72; asst prof math, Univ Wash, 72-80; MODELING ANALYST, BOEING COMPUT SERV, 80- *Concurrent Pos:* Prof, Inst Math, Sao Paulo, Brazil, 76. *Mem:* Am Math Soc; Asn Comput Mach; Soc Comput Simulation. *Res:* Theory of models; optimization; computer simulation. *Mailing Add:* Boeing Comput Serv PO Box 24346 M/S9C-01 Seattle WA 98124

BENDALL, VICTOR IVOR, b London, Eng, Dec 14, 35; m 66; c 2. ORGANIC CHEMISTRY. *Educ:* Univ London, BSc, 57; Bucknell Univ, MS, 60; Brown Univ, PhD(chem), 64. *Prof Exp:* Res chemist, Brit-Thompson-Houston Co, Eng, 57-58; res assoc, Univ Chicago, 63-64; asst prof chem, George Peabody Col, 64-66; assoc prof, 66-69, PROF CHEM, EASTERN KY UNIV, 69- *Concurrent Pos:* Res grant, Sch Pharm, Univ London, 77-78. *Mem:* Am Chem Soc; Royal Soc Chem. *Res:* Small and medium alicyclic compounds; carbenes and reactive intermediates. *Mailing Add:* 105 Hammons Richmond KY 40475-2307

BENDAPUDI, KASI VISWESWARARAO, b Eluru, India, July 28, 46; US citizen; m 77; c 2. STRUCTURAL ENGINEERING, ANALYSIS & DESIGN OF STRUCTURES. *Educ:* Andhra Univ, India, BE, 68; Univ Mo-Rolla, MS, 71. *Prof Exp:* Struct engr, bridge design, Wilbur Smith & Assocs, 71-73; Struct engr, 73-85, asst mgr struct eng, 85-88, MGR STRUCT ENG, LOCKWOOD GREENE ENGRS, INC, 88- *Concurrent Pos:* Mem, Standards Subcomt Perm Railing, Am Soc Civil Engrs, 87-, Task Comt Design of D L Grids, 88-, Comt Spec Structures, 88- & Standards Subcomt Seismic Rehab of Buildings, 90-; off evaluator, Accreditation Bd Eng & Technol, 90- *Mem:* Am Soc Civil Engrs; Am Inst Steel Construct; Am Concrete Inst; Am Asn Artificial Intel. *Res:* Design of industrial structures; vibration control; seismic design of structures; industrial floors; artificial intelligence in structural engineering; analysis and design of special structures. *Mailing Add:* 14034 Prestwick Dr Dallas TX 75234

BENDAT, JULIUS SAMUEL, b Chicago, Ill, Oct 26, 23; m 47; c 2. RANDOM DATA ANALYSIS, APPLIED MATHEMATICS. *Educ:* Univ Calif, AB, 44; Calif Inst Technol, MS, 48; Univ Southern Calif, PhD(math), 53. *Prof Exp:* Res asst physics radiation lab, Univ Calif, 42-45; asst appl mech, Calif Inst Technol, 46-48; asst prof aeronaut eng col aeronaut, Univ Southern Calif, 48-49, lectr math, 49-58; res engr, Northrop Aircraft, Inc, 53-55; mem sr staff, Thompson-Ramo-Wooldridge, Inc, 55-62; pres, Measurement Anal Corp, 62-70; CONSULT, 70- *Concurrent Pos:* Asst, Univ Calif, 44-46; res physicist, Calif Inst Technol, 47-48. *Mem:* Am Math Soc; Math Asn Am; Soc Indust & Appl Math; Sigma Xi. *Res:* Abstract analysis; prediction and filter theory; system engineering; mathematical physics; information theory and circuit theory; random noise theory; vibration and acoustics research. *Mailing Add:* J S Bendat Co 833 Moraga Dr-10 Los Angeles CA 90049

BENDAYAN, MOISE, b Tanger, Morocco, June 21, 49; m 76; c 2. HISTOLOGY, ELECTRON MICROSCOPY. *Educ:* Univ Bordeaux, BS, 68; Univ Montreal, BSc, 72, MSc, 74, PhD(anat), 76. *Prof Exp:* Fel histol, Univ Geneva, 76-79; from asst prof to assoc prof, 79-88, PROF ANAT, UNIV MONTREAL, 88-, CHMN ANAT, 90- *Concurrent Pos:* Vis prof, Univ Western Ont, 84, Weizmann Inst Sci, Israel, 86, Mich State Univ, 87, Univ Minn, 88, Inst Sci Res Cancer, France, 81 & 82, Hebrew Univ, Jerusalem, 87, Univ Ill, 90, Univ Reims, France, 91; scholar, Med Res Coun Can, 80-85, scientist, 85-90. *Honors & Awards:* Vector Award, Am Soc Histochem, 83; Robert Feulgen Award, Soc Histochem, 84; Murray L Barr Award, Can Asn Anat, 82. *Mem:* Am Soc Histochem; Can Soc Electron Microscopy; Can Asn Anatomists; Am Soc Histochem (pres, 89-90). *Res:* Perform cytochemistry at the electron microscope level; introduced the colloidal gold techniques in cytochemistry; studies of diabetes complications, such as pancreas alterations in diabetes and modifications of the vascular walls during microangiopathy. *Mailing Add:* Dept Anat Univ Montreal CP 6128 Succursale A Montreal PQ H3C 3J7 Can

BENDEL, WARREN LEE, b Ravenna, Ohio, Jan 19, 25; m 56; c 1. NUCLEAR PHYSICS. *Educ:* Kent State Univ, BS, 46; Univ Ill, MS, 47, PhD(physics), 53. *Prof Exp:* Lab asst physics, Kent State Univ, 46; asst, Univ Ill, 46-50; res asst, Los Alamos Sci Lab, 49; res asst, Univ Ill, 50-53; physicist, US Naval Res Lab, 53-84; RETIRED. *Concurrent Pos:* Vis fac mem, Univ Md, 71-72. *Mem:* Am Phys Soc. *Res:* Nuclear spectroscopy; photonuclear reactions; electron accelerators; electron scattering; radiation effects on spacecraft. *Mailing Add:* 2714 Stone Edge San Antonio TX 78232-4218

BENDELOW, VICTOR MARTIN, cereal chemistry, for more information see previous edition

BENDER, CARL MARTIN, b Brooklyn, NY, Jan 18, 43; m 66; c 2. HIGH ENERGY PHYSICS, MATHEMATICAL PHYSICS. *Educ:* Cornell Univ, AB, 64; Harvard Univ, AM, 65, PhD(theoret physics), 69. *Prof Exp:* Mem physics res, Inst Advan Study, 69-70; from asst prof to assoc prof math, Mass Inst Technol, 70-77; PROF PHYSICS, WASH UNIV, 77- *Concurrent Pos:* Vis scientist, Imperial Col, London, 74 & 86-87; sci consult, Los Alamos Nat Lab, 79-; Woodrow Wilson fel, 64-65, NSF fel, 64-69, Sloan Found fel, 72-77. *Mem:* Fel Am Physics Soc. *Res:* Quantum field theory; perturbation theory; group theory; applied mathematics. *Mailing Add:* Dept Physics Wash Univ St Louis MO 63130

BENDER, DANIEL FRANK, b Cincinnati, Ohio, Mar 22, 36; m 59; c 7. PHYSICAL ORGANIC CHEMISTRY, ANALYTICAL CHEMISTRY. *Educ:* Xavier Univ, Ohio, BS, 59, MS, 63; Univ Cincinnati, PhD(phys org chem), 67. *Prof Exp:* Chemist, Nat Air Pollution Control Admin, USPHS, US Dept Health, Educ & Welfare, 60-61, res chemist, 61-67, res chemist environ control admin, Bur Solid Waste Mgt, 67-69, supvr res chemist, Res Serv, Chem Studies Group, 69-73; RES CHEMIST WATER POLLUTION, ENVIRON PROTECTION AGENCY, 73- *Concurrent Pos:* Pyrolysis session moderator, Nat Indust Solid Wastes Mgt Conf, 70, pyrolysis session chmn, Am Chem Soc Solid Waste Chem Symp, 70; adj asst prof chem, Raymond Walters Br, Univ Cincinnati, 73-77; part-time fac, dept phys sci, N Ky Univ, 83-86; adj asst prof chem, Univ Col, Univ Cincinnati, 87- *Mem:* Am Chem Soc; The Chem Soc; Sigma Xi. *Res:* Structure-activity relationships; reaction mechanisms; organic molecular orbital theory; absorption and fluorescence spectroscopy; column and thin-layer chromatography; ultramicroorganic analysis of environmental pollutants; pyrolytic synthesis; water pollution; analytical methodology research; environmental research; quality assurance management. *Mailing Add:* 9536 Lansford Dr Cincinnati OH 45242

BENDER, DAVID BOWMAN, b Boston, Mass, Aug 7, 42. NEUROPHYSIOLOGY, NEUROBIOLOGY. *Educ:* Harvard Col, BA, 64; Princeton Univ, MA, 72, PhD(psychol), 74. *Prof Exp:* Res assoc, Psychol Dept, Princeton Univ, 74-76; asst prof, 76-82, ASSOC PROF, PHYSIOL DEPT, STATE UNIV NY, 82- *Concurrent Pos:* Proj dir, Nat Eye Inst grant, 78-89; guest res, NIMH, 90-91. *Mem:* Asn Res Vision & Opthalmol; AAAS; NY Acad Sci. *Res:* Neurobiol of the central nervous system; neurophysiology of vision. *Mailing Add:* 4234 Ridge Lea Rd Buffalo NY 14226

BENDER, DONALD LEE, b Powell, Wyo, Sept 7, 31; m 54; c 5. CIVIL ENGINEERING, HYDROLOGY. *Educ:* Univ Wyo, BS, 53; Colo State Univ, MS, 55; Univ Wis, PhD(hydrol), 63. *Prof Exp:* Hydraul res engr, Colo State Univ, 55-56; from instr to assoc prof civil eng, 56-71, prof civil eng, Wash State Univ, 71-, assoc dean eng, 73-; AT DEPT CIVIL ENG, GONZAGA UNIV. *Concurrent Pos:* Prog intern, Am Coun Educ Acad Admin, 70-71. *Mem:* Am Soc Civil Engrs; Am Geophys Union; Am Soc Eng Educ; Nat Soc Prof Engrs. *Res:* Water resources development, particularly surface hydrology; dimensionless unit hydrograph and the prediction of flood peaks and flood flows in rivers; mechanics of infiltration. *Mailing Add:* E 1532 Pinecrest Spokane WA 99163

BENDER, EDWARD ANTON, b Brooklyn, NY, Feb 7, 42; m; c 3. MATHEMATICS. *Educ:* Calif Inst Technol, BS, 63, PhD(math), 66. *Prof Exp:* Instr math, Harvard Univ, 66-68; res mathematician, Inst Defense Anal, 68-74; assoc prof, 74-77, PROF MATH, UNIV CALIF, SAN DIEGO, 78- *Res:* Combinatorics. *Mailing Add:* Dept Math Univ Calif San Diego La Jolla CA 92093-0112

BENDER, HARVEY ALAN, b Cleveland, Ohio, June 5, 33; m 56; c 3. GENETICS, BIOMEDICAL LEGAL ETHICS. *Educ:* Case Western Reserve Univ, AB, 54; Northwestern Univ, MS, 57, PhD(biol sci), 59, Am Bd Med Genetics, dipl, 82. *Prof Exp:* USPHS fel, Univ Calif, 59-60; from asst prof to assoc prof, 60-69, PROF BIOL, UNIV NOTRE DAME, 69-, MEM SR STAFF RADIATION LAB, 64-; DIR, N CENT IND REGIONAL GENETICS CTR, 79- *Concurrent Pos:* Vis prof in-serv inst, Purdue Univ, 62; Gosney fel & USPHS spec fel, Calif Inst Technol, 65-66; vis prof human genetics, Yale Univ Sch Med & res assoc, Sch Law, 73-74; exec dir, Nat Ctr Law & Handicapped; at-large mem, Yale Univ Task Force Genetics & Reproduction, 73-; consult ed, Mental Retardation; consult, President's Comt Ment Retardation, Nat Comn Protection Human Subjects of Biomed & Behav Res; adj prof med genetics, Sch Med, Ind Univ, 78- *Mem:* Fel AAAS; Am Genetic Asn; Am Soc Zoologists; Am Soc Human Genetics; Radiation Res Soc; Soc Develop Biol. *Res:* Developmental genetics; human genetics; biomedical legal ethics; histochemistry. *Mailing Add:* Dept Biol Sci Univ Notre Dame Notre Dame IN 46556-0369

BENDER, HOWARD L, b Williamstown, WVa, Sept 14, 1893; m 18; c 2. PLASTICS CHEMISTRY. *Educ:* Marietta Col, AB, 17, AM, 18; Case Inst, BS, 19; Columbia Univ, PhD(chem), 25. *Hon Degrees:* DSc, Marietta Col, 53. *Prof Exp:* Res chemist, Dow Chem Co, 19-21; instr chem, Marietta Col, 21; res chemist, Union Carbide Plastics Co, 22-51, asst dir res, 51-58, consult, 58-66; INDEPENDENT CONSULT PLASTICS, 66- *Honors & Awards:* Hyatt Award, 53. *Mem:* Am Chem Soc. *Res:* Thermosetting resins and plastics; synthetic resins; acetylene gas reactions; polymers; phenolformaldehyde resins. *Mailing Add:* 32 Sylavan Dr Morris Plains NJ 07950-2193

BENDER, HOWARD SANFORD, b Brooklyn, NY, Aug 19, 35; m 61; c 2. POLYMER CHEMISTRY, ORGANIC CHEMISTRY. *Educ:* State Univ NY Buffalo, BS, 57; Bucknell Univ, MS, 60; Univ Del, PhD(org chem), 62. *Prof Exp:* Res chemist cent res labs, Interchem Corp, 62-65; from assoc sr res chemist to sr res chemist, Gen Motors Corp, 65-71, supvr coatings, Res Lab, 71-75, dept res scientist, 75-84; dept head coatings res, Masonite Corp, 84-87; DIR RES & DEVELOP, SANNCOR INDUSTS INC, 87- *Honors & Awards:* Roon Award, 70. *Mem:* Am Chem Soc; Fedn Socs Paint Technol; Sigma Xi. *Res:* Surface coatings; paint; organic synthesis; adhesives; urethanes. *Mailing Add:* Sanncor Indust Inc 300 Whitney St PO Box 703 Leominster MA 01453

BENDER, JAMES ARTHUR, applied physics, for more information see previous edition

BENDER, LEONARD FRANKLIN, b Philadelphia, Pa, Oct 2, 25; m 48; c 4. MEDICINE. *Educ:* Jefferson Med Col, MD, 48; Univ Minn, MS, 52; Am Bd Phys Med & Rehab, dipl. *Prof Exp:* From asst prof to prof phys med & rehab, Med Sch, Univ Mich, Ann Arbor, 54-75, vpres & med dir, Rehab Inst, Detroit, 75-79; PROF PHYS MED & CHMN DEPT, SCH MED, WAYNE STATE UNIV, 75-; PRES & CHIEF EXEC OFF, REHAB INST, DETROIT, 80- *Concurrent Pos:* Consult, Wayne County Gen Hosp, Saginaw Community Hosp & Detroit Cerebral Palsy Ctr, Allen Park Vet Adm in Hosp & Detroit Inst Children; chief, Phys Med & Rehab Serv, Detroit Receiving Hosp. *Honors & Awards:* Resolution of Appreciation, Nat Easter Seal Soc. *Mem:* AMA; Am Asn Electromyog & Electrodiag; Am Acad Cerebral Palsy & Develop Med (pres, 80); Am Cong Rehab Med; Am Acad Phys Med & Rehab (pres, 74); Am Col Physician Exec. *Res:* Ultrasound; rheumatoid arthritis; rehabilitation; electromyography; prosthetics; orthotics. *Mailing Add:* CEO Rehab Inst Wayne State Univ 261 Mack Blvd Detroit MI 48201

BENDER, MARGARET MCLEAN, b Easthampton, Mass, May 14, 16; m 43. ORGANIC CHEMISTRY. *Educ:* Mt Holyoke Col, BA, 37, MA, 39; Yale Univ, PhD(org chem), 41. *Prof Exp:* Instr chem, Conn Col, 41-42; proj assoc, Yale Univ, 42-43; proj assoc, Chem Dept, Univ Wis-Madison, 43-45, lectr, Exten Div, 51-69, proj assoc radiocarbon dating, Ctr for Climatic Res, 63-73, assoc scientist, 73-77, sr scientist, 77-81; RETIRED. *Res:* Radiocarbon dating; carbon isotope ratios of plants. *Mailing Add:* 3305 Kingston Dr Madison WI 53713

BENDER, MAX, b Boston, Mass, Oct 23, 14; m 47; c 2. PHYSICAL CHEMISTRY, COLLOID CHEMISTRY. *Educ:* Northeastern Univ, SB, 36; Mass Inst Technol, SM, 37; NY Univ, PhD(chem), 50. *Prof Exp:* Charge of lab, Manton-Gaulin Mfg Co, Inc, 38-40; sr res chemist & group leader colloid & phys chem, Interchem Corp, 41-48; sr res chemist, Am Cyanamid Co, 50-61; from asst prof to prof, 61-85, EMER PROF CHEM, FAIRLEIGH DICKINSON UNIV, 85- *Concurrent Pos:* Consult. *Mem:* AAAS; fel NY Acad Sci; Am Chem Soc; Am Inst Chem; Sigma Xi. *Res:* Electrokinetics; dispersions; emulsions; rheology; interfactial chemistry; sorption; diffusion; gelation; membranes; biophysics. *Mailing Add:* 16 S Woodland Ave East Brunswick NJ 08816

BENDER, MICHAEL A, b New York, NY, July 25, 29; m 50, 69, 79; c 4. CYTOGENETICS, ENVIRONMENTAL HEALTH. *Educ:* Univ Wash, BS, 52; Johns Hopkins Univ, PhD(biol), 56. *Prof Exp:* Nat Cancer Inst fel biol, Johns Hopkins Univ, 56-58; biologist, Oak Ridge Nat Lab, 58-69; assoc prof radiol, Sch Med, Vanderbilt Univ, 69-71; geneticist, US AEC, 71-73; vis prof radiol, Sch Med, Johns Hopkins Univ, 73-75; SR SCIENTIST, BROOKHAVEN NAT LAB, 75- *Mem:* Fel AAAS; Am Soc Cell Biol; Radiation Res Soc; Am Soc Photobiol; Environ Mutagen Soc. *Res:* Eukaryotic chromosome structure; mechanisms of mutagenesis; environmental mutagenesis and oncogenesis. *Mailing Add:* Med Dept Brookhaven Nat Lab Upton NY 11973

BENDER, MICHAEL E, b Spring Valley, Ill, Feb 18, 39; m 61; c 2. POLLUTION BIOLOGY. *Educ:* Southern Ill Univ, BA, 61; Mich State Univ, MS, 62; Rutgers Univ, PhD(environ sci), 68. *Prof Exp:* Biologist, USPHS, 62-64; asst prof environ health, Univ Mich, 67-70; sr marine biologist & chmn, Dept Ecol & Pollution, 70-73, asst dir & head, Div Environ Sci & Eng, Va Inst Marine Sci, 73-82; assoc prof, 70-79, PROF, SCH MARINE SCI, COL WILLIAM & MARY, 79- *Concurrent Pos:* Vpres, Environ Control Technol Corp, Mich. *Mem:* Soc Environ Chem & Toxicol; Sigma Xi; Am Fisheries Soc; Water Pollution Control Fedn. *Res:* Water quality; bioassays; coastal zone management; wetlands; information transfer to management organizations. *Mailing Add:* Marine Sci Col William & Mary Williamsburg VA 23185

BENDER, PATRICK KEVIN, b Cleveland, Ohio, Apr 8, 53; m 85; c 2. CALCIUM CALMADULIN REGULATION, GENETICS OF PHASPHORYLASE KINASE. *Educ:* Kent State Univ, BS, 75; Univ Va, PhD(biol), 82. *Prof Exp:* Postdoctoral res, Univ Tex Med Sch, Houston, 81-83; postdoctoral res, Univ Va, 83-84, res asst prof biol, 84-89; ASST PROF BIOCHEM, VA POLYTECH INST, 89- *Mem:* Am Soc Cell Biologists; AAAS. *Res:* Calcium/calmodulin regulation of metabolic processes with particular efforts in the calmodulin regulation of the phosphorylase kinase enzyme; additional investigations of the regulation of calmodulin gene(s) expression in cultured muscle cells. *Mailing Add:* Dept Biochem Va Polytech Inst Rm 124 Engel Hall Blacksburg VA 24061

BENDER, PAUL A, solid state physics; deceased, see previous edition for last biography

BENDER, PAUL ELLIOT, b Long Beach, NY, Dec 11, 42; m 65; c 2. SYNTHETIC ORGANIC CHEMISTRY. *Educ:* State Univ NY Stony Brook, BS, 63; Univ Ill, Urbana, MS, 66, PhD(org chem), 69. *Prof Exp:* SR INVESTR MED CHEM, SMITH KLINE & FRENCH LABS, 69- *Mem:* Am Chem Soc; NY Acad Sci. *Res:* Design and synthesis of leukotrienes, cannabinoids, peptides, conjugated proteins and various heterocyclic compounds to investigate their potential for medicinal use. *Mailing Add:* Smith Kline Beecham PO Box 1539 King of Prussia PA 19406

BENDER, PAUL J, b Mansfield, Ohio, Nov 20, 17; m 43. PHYSICAL CHEMISTRY. *Educ:* Yale Univ, BS, 39, PhD(phys chem), 42. *Prof Exp:* Phys chemist, Off Sci Res & Develop proj, Yale Univ, 42; instr phys chem, 42-45, from asst prof to prof, 45-79, EMER PROF CHEM, UNIV WIS-MADISON, 79- *Res:* Thermodynamic properties of solutions; Raman spectroscopy; vapor pressures for liquids; nuclear magnetic resonance spectroscopy. *Mailing Add:* 3305 Kingston Dr Madison WI 53713

BENDER, PETER LEOPOLD, b New York, NY, Oct 18, 30; m 53; c 3. PHYSICS. *Educ:* Rutgers Univ, BS, 51; Princeton Univ, PhD(physics), 56, MA, 57. *Prof Exp:* PHYSICIST, NAT BUR STANDARD, 56- *Concurrent Pos:* Fulbright grant, Leiden Univ, 51-52; mem, Joint Inst Lab Astrophys, Nat Bur Stand & Univ Colo, 62-, chmn, 69-70; adj prof, Univ Colo, 64-; NSF adv panel for physics, 72-75; NASA lunar adv comn, 73-75, sub-panel on Relativity & Gravity, 74-76; Nat Acad Sci comn on planet explor, 76-77, comn on geod, 76-79, chmn comn earth sci in space, 77-, space sci bd, 77- & comn gravitational physics, 78- *Honors & Awards:* Gold Medal, US Dept Com, 59; Samuel Wesley Stratton award, Nat Bur Stand, 62. *Mem:* Fel AAAS; Am Astron Soc; Am Geophys Union; fel Optical Soc Am; fel Am Phys Soc. *Res:* Atomic physics; geophysics; astronomy. *Mailing Add:* Joint Inst Astrophys Univ Colo Boulder CO 80309

BENDER, PHILLIP R, b Milwaukee, Wis, Sept 20, 27; m 56; c 5. MATHEMATICS. *Educ:* Purdue Univ, BS, 51; Marquette Univ, MS, 59; Iowa State Univ, PhD(math), 66. *Prof Exp:* Qual control engr, A-P Controls Corp, 51-55; instr math, Milwaukee Sch Eng, 50-60; instr, 60-63, asst prof, 66-74, ASSOC PROF & ASST CHMN DEPT, MARQUETTE UNIV, 74- *Mem:* Math Asn Am; Soc Indust & Appl Math. *Res:* Differential equations; asymptotic behavior of solutions; numberical solutions. *Mailing Add:* Dept Math & Statist Marquette Univ Milwaukee WI 53233

BENDER, ROBERT ALGERD, b Bethlehem, Pa, Apr 17, 46. MICROBIOLOGY, MOLECULAR BIOLOGY. *Educ:* Mass Inst Technol, BS, 68, PhD(microbiol), 76; Univ Southern Maine, MS, 72. *Prof Exp:* Fel molecular biol, Albert Einstein Col Med, 76-78; ASST PROF MICROBIOL, DIV BIOL SCI, UNIV MICH, 78- *Concurrent Pos:* Jr fac res award, Am Cancer Soc, 78-81. *Mem:* Am Soc Microbiol; Genetics Soc Am. *Res:* Microbiol regulation: nitrogen metabolism in the bacterium Klebsiella and regulation of the cell cycle in the bacterium Caulobacter. *Mailing Add:* Dept Biol Univ Mich 4042a Nat Sci Ann Arbor MI 48109-1048

BENDER, WELCOME W(ILLIAM), b Elizabeth, NJ, Nov 30, 15; m 46; c 7. AERONAUTICAL ENGINEERING. *Educ:* Mass Inst Technol, BS, 38, MS, 39. *Prof Exp:* Tech dir pilotless aircraft sect, Martin Co Div, Martin Marietta Corp, 39-48, mgr electromech dept, 49-51, chief electronics engr, 52-55, dir res inst advan studies, 55-62, co dir res, 62-66, dir space explor group, 64-66, Voyager prog scientist, 66-67, prog scientist planetary systs, 67-69, proj scientist, Viking Proj, 69-; AT DEPT BIOL CHEM, HARVARD MED SCH. *Concurrent Pos:* Vpres, J R Nelson & Assoc, Inc, Consult Engrs, 72-; pres, Custom Eng, Inc, 75-81. *Mem:* AAAS; assoc fel Am Inst Aeronaut & Astronaut; Am Astronaut Soc; fel Inst Elec & Electronics Engrs. *Res:* Research management in the aerospace industry. *Mailing Add:* Dept Biochem Harvard Med Sch 25 Shattuck St Boston MA 02115

BENDERSKY, MARTIN, b New York, NY, Apr 15, 45; m 71; c 1. TOPOLOGY. *Educ:* City Col New York, BS, 66; Univ Calif, Berkeley, PhD(math), 71. *Prof Exp:* Asst math, Aarhus, Denmark, 71-72; vis lectr, 72-73, asst prof math, Univ Wash, 73-; AT DEPT MATH, RIDER COL. *Res:* Study of the Adams-Novikov spectral sequence, particularly the description of an unstable version and an appropriate Curtis-Kan algebra. *Mailing Add:* Dept Math City Univ NY-Hunter College New York NY 10021

BENDICH, ADRIANNE, b New York, NY; m 84; c 2. VITAMINS RESEARCH, ANTIOXIDANTS. *Educ:* City Col NY, BS, 65; Iowa State Univ, MS, 66; Rutgers Univ, PhD(immunol), 81. *Prof Exp:* Sr scientist, 81-88, SR CLIN RES COORDR, HOFFMANN-LA ROCHE INC, 88- *Concurrent Pos:* Ed, J Nutrit Immunol, 90-; mem adv bd, Fedn Am Soc Exp Biol. *Mem:* Am Asn Immunologists; NY Acad Sci; Am Inst Nutrit; Am Soc Clin Nutrit; Inst Food Technologists; Sigma Xi. *Res:* Beneficial effects of supplemental vitamins on immune function; over 30 publications. *Mailing Add:* Hoffmann-La Roche Inc 340 Kingsland St Bldg 76/4 Nutley NJ 07110-1199

BENDICH, ARNOLD JAY, b New York, NY, May 5, 42. MOLECULAR BIOLOGY, BIOCHEMISTRY. *Educ:* Univ Vt, BA, 62; Univ Wash, PhD(microbiol), 69. *Prof Exp:* Res assoc molecular biol, Dept Terrestrial Magnetism, Carnegie Inst Wash, 64-65; ASSOC PROF BOT, UNIV WASH, 70- *Concurrent Pos:* Mem, Genetic Biol Panel, Nat Sci Found. *Mem:* Int Soc Plant Molecular Biol; fel AAAS; Genetics Soc Am. *Res:* Nucleic acids in plants; C-value paradox. *Mailing Add:* Dept Bot Univ Wash Seattle WA 98195

BENDISZ, KAZIMIERZ, b Warsaw, Poland, Feb 11, 14; US citizen; m 50; c 2. STRESS ANALYSIS, NUMERICAL METHODS. *Educ:* Univ Warsaw, Poland, BS, 39; Univ Munich, Ger, MS, 47. *Prof Exp:* Supvr mach design, Power Press Design Bur, Warsaw, Poland, 48-53; consult engr mach design, diverse institutions, 54-63; sr asst numerical methods, Dept Mach Design, Warsaw Univ, 63-65; researcher seals, Univ Stuttgart, Ger, 65-68; design engr valves, Anchor Valve Co, Hayward, Calif, 68-69 & piping, Pac Gas & Elec Co, Calif, 70-79; WRITER, TECH BOOKS, 80- *Concurrent Pos:* Teacher mech, Radio a Telecom Technicum, Warsaw, Poland, 53-63; design engr, Inst Heat Engines, Univ Warsaw, Poland, 57-60, supvr & design engr, Inst Astron, 60-63; instr mach design & mech, Heald Eng Col, Calif, 70-79. *Mem:* Am Soc Mech Engrs. *Res:* Separation of liquids of different viscosities by mechanical means; rotary shaft seals for chemical industry; stress analysis in mechanical elements. *Mailing Add:* 1442 Sanchez St San Francisco CA 94131-2024

BENDITT, EARL PHILIP, b Philadelphia, Pa, Apr 15, 16; m 45; c 4. PATHOLOGY. *Educ:* Swarthmore Col, BA, 37; Harvard Univ, MD, 41. *Prof Exp:* From instr to assoc prof path, Sch Med, Univ Chicago, 44-57; chmn dept, 57-81, prof path, 57-87, EMER PROF PATH, SCH MED, UNIV WASH, 87- *Concurrent Pos:* Asst dir res, La Rabida Sanitarium, Ill, 50-56; vis scientist & Commonwealth Fund fel, Sir William Dunn Sch Path, Oxford, 65; coun mem, Nat Inst Environ Health Sci, 70-73 & 83-87; consult, USPHS; vis prof, Sir William Dunn Sch Path, Oxford, 79-80. *Honors & Awards:* Rous-Whipple Award, Am Asn Path, 80 & Gold Headed Cane, 84. *Mem:* Nat Acad Sci; Am Soc Exp Path (vpres, 74-75, pres, 75-76); Soc Exp Biol & Med; Histochem Soc (vpres, 62-63, pres, 63-64); Am Asn Path & Bact; Am Soc Biol Chemists; Am Soc Cell Biol. *Res:* Cell injury; inflammation; wound healing; atherosclerosis and heart diseases; amyloidosis and Alzheimers disease. *Mailing Add:* Dept Path Univ Wash Sch Med Seattle WA 98195

BENDIX, SELINA (WEINBAUM), b Pasadena, Calif, Feb 16, 30; m 53; c 3. ENVIRONMENTAL MANAGEMENT. *Educ:* Univ Calif, Los Angeles, BS, 50; Univ Calif, PhD(zool), 57. *Prof Exp:* Asst plant physiol, Earhart Lab, Calif Inst Technol, 50-51; asst res physicochem biol, Dept Zool, Univ Calif, 51-54; jr res biologist, Lab Comp Biol, Kaiser Found Res Inst, 57-59, asst res biologist, 59-64; asst prof biol, San Francisco State Col, 64-65; lectr, Mills Col, 65-69; sci ed, Freedom News, 70-72; res dir, Bendix Res-Environ Consult, 72-80; environ rev officer, City & Co of San Francisco, Dept City Planning, 74-80; PRES, BENDIX ENVIRON RES, INC, 80- *Concurrent Pos:* Consult, Berkwood Sch, 59-70; mem, Calif Atty Gen Task Force Environ Probs of San Francisco Bay Area; mem, Environ Protection Agency Adminr Toxic Substances Adv Comt, 77-82. *Mem:* AAAS; Asn Environ Prof; Scientists' Inst Pub Info; Am Chem Soc; Asn Women in Sci; Sigma Xi; Environ Mutagen Soc; Int Soc Environ Toxicol & Cancer. *Res:* Algology; cellular physiology; genetics; effect of climatic factors on growth; human ecology; heavy metal pollution; environmental impact assessment; environmental policy and decision-making process; combustion toxicology; occupational toxicology. *Mailing Add:* 1390 Market 418 San Francisco CA 94102

BENDIXEN, HENRIK H, b Fredriksberg, Denmark, Dec 2, 23. ANESTHESIOLOGY. *Educ:* Univ Copenhagen, MD, 51. *Hon Degrees:* Dr, Jagiellonian Univ, Krakow, Poland, 85, Univ Copenhagen, 87. *Prof Exp:* Med dir, Univ Hosp, San Diego, 71-72; actg provost & vpres health sci, Columbia Univ, 80-81, alumni prof, 84, E M Papper prof anesthesiol, 85-86, vpres health sci & dean, Fac Med, 84-89, PROF ANESTHESIOL, DEPT ANESTHESIOL, COL PHYSICIANS & SURGEONS, COLUMBIA UNIV, 73-, SR ASSOC VPRES HEALTH SCI & SR ASSOC DEAN, FAC MED, 89- *Concurrent Pos:* Chmn, comt prof safety & liability, Presbyn Hosp, NY, 82-83; chmn, Postgrad Assembly in Anesthesiology; consult, Dept Anesthesia, Vassar Brothers Hosp, Poughkeepsie, NY, 77-; chmn policy comt sci & technol, Columbia Univ, NY, 82-; chmn Orthop Search Comt, Columbia Univ, NY, 82-; dir anesthesiol serv, Presby Hosp, 73-85. *Honors & Awards:* Wesley Bourne Lectr, McGill Univ, 69; Husfeldt Lectr, Danish Soc Anesthesiol, 74; Benjamin Howard Robbins Mem Lectr, Vanderbilt Univ, 75; John J Bonica Lectr, Univ Wash, 82; Ralph Waters Award, 85. *Mem:* Inst Med-Nat Acad Sci; fel AAAS; fel Am Col Anesthesiol; AMA; NY Acad Sci;

Am Soc Pharmacol & Exp Therapeut; Am Soc Anesthesiologists; Int Anesthesia Res Soc; Asn Am Med Col; Am Heart Asn. *Res:* Critical, intensive and respiratory care; respirators and resuscitation; hypoxia; catecholamines; monitoring; cost effectiveness of care; epidemiology and statistics in clinical medicine; published numerous articles in various journals. *Mailing Add:* Dept Anesthesiol Col Physicians & Surgeons Columbia Univ 630 W 168th St New York NY 10032

BENDIXEN, LEO E, b Mills, Utah, Oct 21, 23; m 49; c 4. PLANT PHYSIOLOGY. *Educ:* Utah State Univ, BS, 53; Univ Calif, Davis, MS, 56, PhD(plant physiol), 60. *Prof Exp:* Sr lab technician agron, Univ Calif, Davis, 55-61; asst prof, 61-66, assoc prof, 66-70, PROF AGRON, OHIO STATE UNIV, 70- *Concurrent Pos:* Sr fel, East-West Ctr, 75-76. *Mem:* Int Weed Sci Soc; Weed Sci Soc Am; Coun Agr Sci & Technol. *Res:* Physiological and biochemical aspects of herbicides; physiological and ecological aspects of plant reproduction and growth; physiological and anatomical aspects of plagiotropic growth; weed hosts as pest reservoirs of organisms affecting crops. *Mailing Add:* Dept Agron Ohio State Univ Main Campus 2021 Kottman Hall Columbus OH 43210-1086

BENDLER, JOHN THOMAS, b Derby, Conn, Oct 5, 44; m. POLYMER PHYSICS, STATISTICAL MECHANICS. *Educ:* Holy Cross Univ, AB, 66; Yale Univ, PhD, 74. *Prof Exp:* Instr chem, Northern Ill Univ, 72-75; res assoc, Midland Macro Molecular Inst, 75-77; POLYMER PHYSICIST, GEN ELEC CO, 78- *Concurrent Pos:* Postdoctoral res, Wash Univ, 77; vis prof, Colo State Univ, 90; GE Coolidge fel. *Honors & Awards:* Woodrow Wilson fel, Yale Univ, 67. *Mem:* Fel Am Phys Soc; Am Chem Soc. *Mailing Add:* Corp Res & Develop Ctr K1/4B7 Gen Elec Co PO Box 8 Schenectady NY 12301

BENDT, PHILIP JOSEPH, b Syracuse, NY, Dec 21, 19. CRYOGENICS, EXPERIMENTAL NUCLEAR PHYSICS. *Educ:* Mass Inst Technol, BS, 42; Columbia Univ, PhD(physics), 51. *Prof Exp:* Staff mem, Los Alamos Nat Lab, 51-80; CONSULT, 80- *Mem:* Am Phys Soc. *Res:* Superfluid hydrodynamics; electron-electron paramagnetic resonance; neutron capture gamma spectroscopy. *Mailing Add:* 154 W Zia Rd Sante Fe NM 87505

BENDURE, RAYMOND LEE, b Middletown, Ohio, July 1, 43; m 65; c 2. PRODUCT DEVELOPMENT, PROCESS DEVELOPMENT. *Educ:* Purdue Univ, BS, 65; Iowa State Univ, PhD(phys chem), 68. *Prof Exp:* Group leader basic res, Miami Valley Labs, Procter & Gamble, 68-72, process develop, Paper Div, 72-74, sect head mfg, Charmin Paper Prods, 74-75, sect head paper & personal care, Paper Prod Develop, 75-78, dir, W Ger, 78-82, dir, Japan, 82-84, assoc dir food prod develop, 84-86; corp dir prod develop, Helene Curtis Indust Inc, 86-88; WORLDWIDE DIR RES & DEVELOP, COLGATE-PALMOLIVE CO, 88- *Mem:* Am Chem Soc; Am Oil Chem Soc. *Res:* Surface physical chemistry; detergents, foams; consumer product development for sanitary paper, personal care, household products and food products; process development and commercialization. *Mailing Add:* Colgate-Palmolive Co 909 River Rd Piscataway NJ 08854-5503

BENDURE, ROBERT J, b Steubenville, Ohio, May 2, 20; m 42; c 2. INORGANIC CHEMISTRY, POLLUTION CHEMISTRY. *Educ:* Miami Univ, BA, 42. *Prof Exp:* Chem analyst, Armco Inc, 42-44 & 46-47, sr analyst, 47-49, from chemist to sr chemist, 49-61, supv chemist, 61-68, mgr chem labs, 68-70, dir chem res, 70-80, corp dir res & technol, 80-83; RETIRED. *Honors & Awards:* Lundell Bright Mem Award, Am Soc Testing & Mat, 72. *Mem:* AAAS; Am Chem Soc; fel Am Soc Testing & Mat. *Res:* Inorganic analytical chemistry using both classical and instrumental methods; ferrous and nonferrous metals analysis; inclusions; intermetallic compounds and dissolved gases in steel; air and water pollution control; industrial hygiene; corrosion. *Mailing Add:* 825 W Lane Lebanon OH 45036

BENEDEK, ANDREW, b Budapest, Hungary, May 29, 43; Can citizen; m 66. CHEMICAL & ENVIRONMENTAL ENGINEERING. *Educ:* McGill Univ, BEng, 66; Univ Wash, PhD(chem eng), 70. *Prof Exp:* Asst prof, 70-76, assoc prof chem eng, McMaster Univ, 76-86; CHMN & CHIEF EXEC OFFICER, ZERON ENVIRONMENTAL INC, 87- *Concurrent Pos:* Consult govt & chem indust, 75-; coordr water res group, 76- *Honors & Awards:* L K Cecil Award, Am Inst Chem Eng, 82. *Mem:* Am Inst Chem Engrs; Chem Inst Can; Can Soc Chem Engrs. *Res:* Mass transfer in gas-liquid dispersions; polarographic oxygen probe response; water treatment processes such as coagulation; activitated carbon adsorption and ozonation. *Mailing Add:* Zenon Environmental Inc 845 Harrington Court Burlington ON L7N 3P3 Can

BENEDEK, GEORGE BERNARD, b New York, NY, Dec 1, 28; m 55; c 2. PHYSICS. *Educ:* Rensselaer Polytech Inst, BS, 49; Harvard Univ, MA, 52, PhD(physics), 53. *Prof Exp:* Mem staff, Lincoln Lab, Mass Inst Technol, 53-55; res fel, Harvard Univ, 55-57, lectr solid state physics, 57-58, asst prof appl physics, 58-61; from assoc prof to prof physics, 61-79, ALFRED H CASPARY PROF PHYSICS & BIOL PHYSICS, MASS INST TECHNOL, 79- *Concurrent Pos:* Guggenheim fel, Stanford Univ, 60; prof fel, Atomic Energy Res Estab, Harwell, Eng, 67; consult physics, Retina Found, Boston, 68-; lectr physics, Harvard Univ, 75-85; mem adv comt physics, NSF, 83-86; mem vision res prog comt, Nat Eye Inst, 83-87; Walker-Ames prof, Univ Wash, Seattle, 84. *Honors & Awards:* Debye Lectr chem, Cornell Univ, 72. *Mem:* Nat Acad Sci; Inst Med-Nat Acad Sci; fel Am Phys Soc; fel Am Acad Arts & Sci. *Res:* Magnetic resonance; high pressure physics; solid state physics; critical phenomena and light scattering with lasers; biological physics. *Mailing Add:* Dept Physics Mass Inst Technol Cambridge MA 02139

BENEDEK, ROY, b New York, NY, July 2, 45. THEORETICAL PHYSICS. *Educ:* Cornell Univ, BS, 66, MS, 67; Brown Univ, PhD(physics), 72. *Prof Exp:* Res assoc, physics, Cornell Univ, 71-74; PHYSICIST, ARGONNE NAT LAB, 74- *Mem:* Am Phys Soc; AAAS. *Mailing Add:* Mat Sci Div Argonne Nat Lab 9700 S Cass Ave Argonne IL 60439

BENEDETTI, JACQUELINE KAY, b Chicago, Ill, July 22, 48. BIOSTATISTICS. *Educ:* Univ Calif, Los Angeles, BA, 70; Univ Wash, PhD(biomath), 74. *Prof Exp:* NIH gen med sci fel, Sch Med, Univ Calif, Los Angeles, 74-76, asst prof in residence biomath, 76-80; AT DEPT BIOMATH, UNIV WASH. *Mem:* Inst Math Statist; Am Statist Asn; Biomet Soc. *Res:* Analysis of categorical data, including multidimensional contingency table analysis, and measures of association for two-way tables; clinical trials statistics. *Mailing Add:* Dept Biostatist Univ Wash Seattle WA 98195

BENEDETTO, JOHN, b Melrose, Mass, July 16, 39; m 68; c 1. ANALYTICAL MATHEMATICS. *Educ:* Boston Col, BA, 60; Harvard Univ, MA, 62; Univ Toronto, PhD(math), 64. *Prof Exp:* Assoc mem tech staff, Radio Corp Am, Burlington, 61-62, mem tech staff, 62-64; teaching fel math, Univ Toronto, 62-64; asst prof, NY Univ, 64-65; res assoc, Inst Fluid Dynamics & Appl Math, 65-66, asst prof, from asst prof to assoc prof, 66-76, PROF MATH, UNIV MD, COLLEGE PARK, 77- *Concurrent Pos:* Assoc mathematician, Int Bus Mach Corp, Mass, 64-; prof, Scuola Normale Superiore, Pisa, 70-71. *Mem:* Am Math Soc; Math Asn Am. *Res:* Theory of generalized function; Laplace and Fourier type transforms and topological vector spaces. *Mailing Add:* Dept Math Univ Md College Park MD 20742-4015

BENEDICT, ALBERT ALFRED, b Pasadena, Calif, Nov 26, 21; m 47; c 1. IMMUNOLOGY. *Educ:* Univ Calif, AB, 48, MA, 50, PhD(bact), 52. *Prof Exp:* Res assoc bact, Univ Calif, 49-52; res assoc virol med br, Univ Tex, 52-53, asst prof prev med & pub health, 53-57; assoc prof bact, Univ Kans, 57-63; PROF MICROBIOL & CHMN DEPT, UNIV HAWAII, 63- *Concurrent Pos:* WHO consult, 71. *Mem:* AAAS; Am Soc Microbiol; Soc Exp Biol & Med; Am Asn Immunol. *Res:* Avian immunoglobulins; structure, genetics and synthesis of immunoglobulins. *Mailing Add:* Dept Microbiol Univ Hawaii Honolulu HI 96822

BENEDICT, BARRY ARDEN, b Wauchula, Fla, Feb 7, 42; m 75; c 3. HYDRAULIC ENGINEERING. *Educ:* Univ Fla, BCE, 65, MSE, 67, PhD(civil eng), 68. *Prof Exp:* Res assoc coastal & oceanog eng dept, Univ Fla, 68-69; from asst prof to assoc prof hydraulic & water resources eng & prog dir, Vanderbilt Univ, 69-75; assoc prof civil eng, Tulane Univ, 75-77; assoc prof civil eng, Univ SC, 78-80; PROF, DEPT CIVIL ENG, UNIV FLA, GAINESVILLE, 80- *Mem:* Am Soc Eng Educ; Int Asn Hydraul Res; Am Soc Civil Engrs. *Res:* Sediment transport and hydrodynamics of flow over a rough bed; mechanics of heated discharges into water bodies; modeling thermal stratification in reservoirs; water quality modeling and spill modeling; groundwater. *Mailing Add:* Dept Civil Eng La Tech Univ Ruston LA 71272

BENEDICT, C R, b Lake Placid, NY, June 26, 30; m 54; c 2. PLANT BIOCHEMISTRY, PLANT PHYSIOLOGY. *Educ:* Cornell Univ, BS, 54, MS, 56; Purdue Univ, PhD(plant biochem), 60. *Prof Exp:* Asst prof biochem, Wayne State Univ, 62-66; sr plant physiologist, USDA, 66-69; prof plant biochem, Tex A&M Univ, 69-79, interim head dept, 79-81, asst head, Dept Plant Biochem, 81-83, PROF BIOCHEM, TEX A&M UNIV, 83- *Mem:* Am Chem Soc; Am Soc Biochem & Molecular Biol; Am Soc Plant Physiologists; Am Soc Agron. *Res:* Enzymology and biochemistry of natural products including the regulation of carotenes, polyisoprenoid and the anti-ovarian cancer drug taxol. *Mailing Add:* Dept Biochem & Biophys Tex A&M Univ College Station TX 77843

BENEDICT, CHAUNCEY, b Lake Placid, NY, June 10, 30; m 54; c 1. PLANT BIOCHEMISTRY. *Educ:* Cornell Univ, BS, 54, MS, 56; Purdue Univ, PhD(plant biochem), 60. *Prof Exp:* Res assoc biochem, Brookhaven Nat Lab, 60-61; fel & asst prof biol, Dartmouth Med Sch, 61-62; asst prof biochem, Wayne State Univ, 62-66; sr scientist, USDA, 66-69, PROF PLANT PHYSIOL, TEX A&M UNIV, 69- *Concurrent Pos:* Mem subcomt biol chem, Nat Res Coun, 64-67. *Mem:* Am Soc Plant Physiol. *Res:* Plant metabolism and carotene synthesis; stable carbon isotopes; enzymes; terpene synthesis; crop science. *Mailing Add:* Dept Plant Sci Tex A&M Univ College Station TX 77843

BENEDICT, ELLEN MARING, b Eugene, Ore, Aug 28, 31; m 51; c 5. BIOLOGY, SPELEOLOGY. *Educ:* Portland State Univ, BS, 65, MS, 69, PhD(environ sci biol), 78. *Prof Exp:* Instr gen sci biol, Portland State Univ, 69-72; chief investr, Nat Speleological Soc Res Adv grant, 74-75; RES & WRITING, 78- *Concurrent Pos:* Res assoc biol, Portland State Univ, 78; mem, Int Ctr Arachnology Doc; instr, Malheur Field Sta, 74-, adj prof biol, 88-; proj dir, Diamond Craters Geol Area, Bur Land Mgt, 79; adj prof biol, Pac Univ, 79-88. *Mem:* Am Arachnological Soc; Brit Arachnological Soc; fel Nat Speleological Soc; Sigma Xi. *Res:* Ecology and systematics of pseudoscorpions; soil arthropod population densities and monitoring techniques; biota of caves; cave microclimates. *Mailing Add:* 8106 SE Carlton Portland OR 97206

BENEDICT, GEORGE FREDERICK, b Los Angeles, Calif, Mar 17, 45; m 67; c 2. ASTROMETRY, IMAGE PROCESSING. *Educ:* Univ Mich, BS & BS, 67; Northwestern Univ, MS, 69, PhD(astron), 72. *Prof Exp:* ASTRONOMER, UNIV TEX, AUSTIN, 72-, FAC MGR, ASTRON DATA REDUCTION & ANALYTICAL SYST, 81- *Concurrent Pos:* Consult, Boller & Chivens Div, Perkin Elmer Corp, 73-76 & Monsanto Mound Res Facil, 78-81; mem, NASA Space Telescope Astrometry Team, 77- *Mem:* Am Astron Soc; Int Astron Union. *Res:* Astrometry; space astronomy; galaxy photometry; digital image processing and analysis. *Mailing Add:* McDonald Observ Univ Tex Austin TX 78712

BENEDICT, IRVIN J, biology, public health, for more information see previous edition

BENEDICT, JAMES HAROLD, b Floral Park, NY, Mar 9, 22; m 48; c 4. BIOCHEMISTRY. *Educ:* NY Univ, AB, 44; Univ Pittsburgh, PhD(chem), 50. *Prof Exp:* Asst chem, Univ Pittsburgh, 46-47; chemist, Procter & Gamble Co, 49-73, toxicologist, 73-83; RETIRED. *Mem:* Fel AAAS; Am Chem Soc; Am Acad Clin Toxicol; fel Am Soc Testing & Mat; Am Oil Chem Soc. *Res:* Fat metabolism; trace analyses; toxicology. *Mailing Add:* 5583 Jessup Rd Cincinnati OH 45247

BENEDICT, JOHN HOWARD, JR, b San Antonio, Tex, Mar 31, 44; m 65; c 2. BEHAVIOR, ECOLOGY. *Educ:* Calif State Univ, Los Angeles, BA, 69; Univ Calif, Davis, PhD(entom), 75. *Prof Exp:* Res entomologist Vi, Rockefeller Found-Univ Calif, Davis, 75-76; asst prof, 77-81, assoc prof, 82-88; PROF ENTOM, AGR EXP STA, TEX A & M UNIV, CORPUS CHRISTI, 88- *Concurrent Pos:* Consult Hopkins Agr Serv & Agr Chem Indust; vis prof, dept biol Tex A&I Univ, 88-, teacher, 88-; consult, Monsanto Co, 90- *Mem:* AAAS; Entom Soc Am; Am Chem Soc; Int Soc Chem Ecol; Crop Sci Soc Am. *Res:* Pest management and evaluation genetically engineered plants and host plant resistance throughout the US; to identify and utilize sources of resistance in cotton to plant bugs, bollweevil and heliothis; behavioral and semichemical mechanisms of insect-plant interactions; mechanistic modeling insect-plant-environment crop systems. *Mailing Add:* Tex A&M Univ Dept Entomology Rte 2 Box 589 Corpus Christi TX 78410

BENEDICT, JOSEPH T, b Chicago, Ill, May 21, 20; m 57. INORGANIC CHEMISTRY, ACADEMIC ADMINISTRATION. *Educ:* Univ Chicago, BS, 43; Mass Inst Technol, PhD(inorg chem), 50. *Prof Exp:* Asst, Mass Inst Technol, 44-50; res assoc chem, Columbia Univ, 50-53; res chemist, Nickel Processing Corp, Div Int Nickel Co, 53-57; control mgr electronic chem, Merck & Co, Inc, 57-58; mgr tech develop, Int Div, Stauffer Chem Co, 58-63; from asst prof to assoc prof, 63-71, PROF CHEM, FAIRLEIGH DICKINSON UNIV, 71-, DIR MBA PROG CHEM MKT & ECON, PHARMACEUT MKT & ECON, 74- *Mem:* Am Chem Soc. *Res:* Fused salt technology; chemical-pharmaceutical marketing and economics. *Mailing Add:* Dept Chem Fairleigh Dickinson Univ Teaneck NJ 07666

BENEDICT, MANSON, b Lake Linden, Mich, Oct 9, 07; m 35; c 2. CHEMICAL & NUCLEAR ENGINEERING. *Educ:* Cornell Univ, BChem, 28; Mass Inst Technol, MS, 32, PhD (phys chem), 35. *Prof Exp:* Res chemist, Nat Aniline & Chem Co, 29-30 & 37-38 & Nat Res Coun fel, 35-36; res assoc geophys, Harvard Univ, 36-37; res chemist, M W Kellogg Co, NJ, 38-42; head process develop, Kellex Corp, NY, 42-46; dir process develop, Hydrocarbon Res, Inc, 46-51; chief opers anal staff, Atomic Energy Comn, 51-52; sci dir, Nat Res Corp, 51-57; from prof to inst prof nuclear eng, 51-73, head dept, 58-71, EMER INST PROF NUCLEAR ENG, MASS INST TECHNOL, 73- *Concurrent Pos:* Mem gen adv comt, Atomic Energy Comn, 58-68, chmn, 62-64; dir, Burns & Roe, Inc, 78-87. *Honors & Awards:* Nat Medal of Sci; Walker Award, Wilson Award, Fritz Medal & Founders Award, Am Inst Chem Engrs; Compton Award, Am Nuclear Soc, Seaborg Medal; Perkin Medal, Soc Chem Indust; Founders Award, Nat Acad Engrs; Fermi Award, Atomic Energy Comn, 72. *Mem:* Nat Acad Sci; Nat Acad Eng; Am Inst Chem Engrs; Am Nuclear Soc (vpres, 61-62, pres, 62-63); Am Chem Soc; Am Philos Soc. *Res:* Isotope separation; nuclear chemical technology. *Mailing Add:* 108 Moorings Park Dr Naples FL 33942-2155

BENEDICT, PETER CARL, geology, for more information see previous edition

BENEDICT, ROBERT CURTIS, b Bridgeport, Conn, May 17, 32; m 58; c 3. BIOCHEMISTRY. *Educ:* Univ Conn, BS, 53; Cornell Univ, MS, 55; Univ Pa, PhD(biochem), 65. *Prof Exp:* Pa Plan Med Res fel, 64-66; RES CHEMIST, EASTERN REGIONAL RES CTR, USDA, WYNDMOOR, 66- *Mem:* AAAS; Am Chem Soc; Am Soc Microbiol; Inst Food Technologists; Am Meat Sci Assoc. *Res:* Biochemistry of meat and muscle; molecular biology of pathogenic bacteria. *Mailing Add:* 1915 Huntingdon Rd Huntingdon Valley PA 19006

BENEDICT, ROBERT GLENN, b Ionia Co, Mich, July 16, 11; m 45; c 5. MICROBIOLOGY. *Educ:* Mich State Univ, BS, 36; Va Polytech Inst, MS, 38; Univ Wis, PhD(agr bact), 42. *Prof Exp:* Asst gen bact, Va Polytech Inst, 36-38; res asst agr bact, Univ Wis, 38-41, instr, 41-42; cur, Northern Regional Res Lab, USDA, 42-46, investr, 45-53; investr classified res, Chem Corps, US Army, 52-56; in charge microbial technol polymers unit, Northern Regional Lab, USDA, 56-58 & new prod explor & res unit, 58-60; res asst prof pharmacog, Col Pharm, Univ Wash, 60-63, actg asst prof, 63-64, res assoc prof civil eng, Univ, 73, res assoc prof pharmacog, 64-73 & 74-75. *Concurrent Pos:* Consult & expert witness, Antibiotic Litigations, Fed Trade Comn, 59-60 & Patent Litigations, Am Cyanamid Co, Can, Europe & Mid East, 63-73; gen med res grants, NIH, 60-68, 69-72; vis assoc prof microbiol, Univ Puget Sound, 73-74. *Mem:* Am Soc Pharmacog; Am Soc Microbiol. *Res:* Mushroom toxins and secondary metabolites of possible interest in medicine; taxonomy of microorganisms in disposal of sewage and solid wastes. *Mailing Add:* 4212 NE 203rd St Seattle WA 98155

BENEDICT, WINFRED GERALD, b Wallaceburg, Ont, Mar 18, 19; m 43; c 4. PLANT PATHOLOGY. *Educ:* Ont Agr Col, BSA, 49; Univ Toronto, PhD(plant path & mycol), 52; Univ Leeds, MPhil, 71; Univ Windsor, BA, 81. *Prof Exp:* Asst-assoc plant pathologist, Can Dept Agr, 52-57; from assoc prof to prof, 57-84, EMER PROF BIOL, UNIV WINDSOR, 86- *Concurrent Pos:* Vis prof, Univ Leeds, 69-70. *Mem:* Am Phytopath Soc; Can Phytopath Soc; fel Linnean Soc London. *Res:* Diseases of forage and field crops; physiology of plant infection; mycology, myxomycetes; effect of light intensity and wavelength on early stages of leaf spot infections. *Mailing Add:* Dept Biol Sci Univ Windsor Windsor ON N9B 3P4 Can

BENEDICTY, MARIO, b Trieste, Italy, July 16, 22; m 47; c 2. GEOMETRY, MATH EDUCATION. *Educ:* Univ Rome, MathD, 46, Libera Docenza, 52. *Prof Exp:* Asst prof anal geom, Univ Rome, 48-51, assoc prof algebraic geom, 51-58; assoc prof, 57-60, chmn dept, 63-69, actg chmn, 78-79, PROF MATH, UNIV PITTSBURGH, 60- *Concurrent Pos:* Co-worker math, Dizionario Enciclopedico dell Enciclopedia Italiana, 54-56, ed, 56-57, contribr, 60; vis prof, Univ BC, 60-62 & Colo State Univ, 76. *Mem:* AAAS; Am Math Soc; Math Asn Am; NY Acad Sci. *Res:* Algebraic geometry and related fields; projective geometries; math education; co-authored a book in discrete mathematics. *Mailing Add:* Dept Math & Statist Univ Pittsburgh Pittsburgh PA 15260

BENEKE, EVERETT SMITH, b Greensboro, NC, July 6, 18. BOTANY, MICROBIOLOGY. *Educ:* Miami Univ, AB, 40; Ohio State Univ, MS, 41; Univ Ill, PhD(bot), 48. *Prof Exp:* From asst prof to prof, Dept Bot & Microbiol, 58-67, prof, 67-88, EMER PROF BOT & MICROBIOL, DEPT MICROBIOL & PUB HEALTH, MICH STATE UNIV, 88- *Concurrent Pos:* Tech consult, Pan Am Union, San Paulo, Brazil, 60; WHO fel, Indonesia, 69; chmn, Mycol Sect, Int Union Microbiol Socs, 71-78. *Honors & Awards:* Sigma Xi Jr Award , 58; William A Western Award, Mycol Soc Am, 84; Rhods Benham Award, Med Mycol Soc Am, 86; Plaque Award, Int Union Microbiol Soc, 87. *Mem:* Am Soc Microbiol; Med Mycol Soc Am (secy-treas, 76-79, pres, 81); Int Union Microbiol Socs; hon mem Mycol Soc Am (actg secy-treas, 54, secy-treas, 56-59, vpres, 60, pres, 62); fel AAAS. *Res:* Saprolegniales and other related water molds; mycology, medical mycology; industrial mycology. *Mailing Add:* 1664 Forest Hills Dr Okemos MI 48864

BENEMANN, JOHN RUDIGER, b Germany, Oct 18, 43; US citizen; m 65; c 2. BIOCHEMISTRY, SOLAR ENERGY. *Educ:* Univ Calif, Berkeley, BS, 65, PhD(biochem), 70. *Prof Exp:* Res chemist biochem, Univ Calif, San Diego, 70-74; res biochem solar energy, Lawrence Berkeley Lab, Univ Calif, 75-76, assoc res biochem waste water treat, Sanit Eng Res Lab, 76-80; mem staff, Ecoenergetics Inc, 80-84; assoc prof, Ga Inst Technol, 84-88; CONSULT, 88- *Concurrent Pos:* Consult, Off Technol Assessment, US Congress, 78-, State Water Resources Control Bd, Calif, 78-, Off Appropriate Technol, Calif, 78-, Nat Bur Standards, 80-88, Dupont Co, 3MG, Kimberly Clark. *Mem:* AAAS; Am Soc Microbiol; Sigma Xi; World Agr Soc. *Res:* Biological solar energy; nitrogen fixation; microalgae mass culture; blue green algae; aquatic plants; waste water aquaculture; toxicology; biofuels; biophotolysis. *Mailing Add:* 1212 Kelly Ct Pinole CA 94564

BENENATI, R(OBERT) F(RANCIS), b Brooklyn, NY, Jan 28, 21; m 44; c 5. NUCLEAR ENGINEERING, COMPUTER SCIENCE. *Educ:* Polytech Inst Brooklyn, BChE, 42, MChE, 47, DChE, 59. *Prof Exp:* Head operating dept, Kellex Corp, 44-46; design engr, Hydrocarbon Res Co, 46-49; from instr to prof chem eng, Polytech Inst Brooklyn, 49-91, dir comput ctr, 65-70, head chem eng dept, 72-74 & 83-85, head nuclear eng dept, 83-85; RETIRED. *Concurrent Pos:* Consult, H K Ferguson Co, 51-52, Walter Kidde Nuclear Labs, 52-54, Sanderson & Porter, 54-67 & Brookhaven Nat Labs, 71- *Mem:* Am Chem Soc; Am Nuclear Soc; Am Inst Chem Engrs; Sigma Xi. *Res:* Nuclear engineering; nuclear reactor design; heat transfer operations; mass transfer operations; water desalination; computer simulation systems. *Mailing Add:* Dept Chem Eng 333 Jay St Brooklyn NY 11201

BENENSON, ABRAM SALMON, b Napanoch, NY, Jan 22, 14; m 39; c 4. MEDICAL RESEARCH, MICROBIOLOGY. *Educ:* Cornell Univ, AB, 33, MD, 37; Am Bd Path, Am Bd Prev Med & Am Bd Med Microbiol, dipl. *Prof Exp:* Rotating intern, Queens Gen Hosp, New York, 37-39; psychiat intern, Bellevue Hosp, 39; asst & chief lab serv, Tripler Gen Hosp, Honolulu, 41-42; asst virus lab, Army Med Dept Res & Grad Sch, US Dept Army, Washington, DC, 46-47, instr path & bact, Med Sch Tufts Col, 47-49, prof mil sci & tactic, 47-49, dir, Second Army Area Med Lab, Ft George G Meade, Md, 49-52, dir, Trop Res Med Lab, San Juan, PR, 52-54, dir exp med, Camp Detrick, Md, 54-55, dir div immunol, Walter Reed Army Inst Res, 56-60, dir div commun dis & immunol, 60-62; dir, Pakistan-SEATO Cholera Res Lab, 62-66; prof prev med & microbiol, Jefferson Med Col, 66-69; prof community med & chmn dept, Col Med, Univ Ky, 69-77; dir, Gorgas Mem Lab, 77-81; PROF & HEAD DIV EPIDEMIOL & BIOSTATISTICS RES, GRAD SCH PUB HEALTH, SAN DIEGO STATE UNIV, 82- *Concurrent Pos:* From assoc mem to mem comn immunization, Armed Forces Epidemiol Bd, 56-72, dir, 67-72, mem, 77-83, 85-, dir, Subcomt Dis Control, 77-; mem study group on typhoid vaccines, WHO & expert comt on cholera, 62-73, mem expert adv panel comt bact dis, 73-84, consult, Expert Comt Biol Stand, 58 & Expert Comt on Smallpox, 64, 67; mem bd sci counsr, Div Biol Stand, NIH, 67-71, chmn, 69-71, mem cholera adv comt, 68-72; ed, Control Commun Dis of Man, Am Pub Health Asn, 70, 75, 80, 85 & 90; mem, Cholera Panel, US-Japan Coop Sci Prog, 73-77; res fel, Sch Microbiol, Univ Melbourne, 76-77. *Honors & Awards:* First Award, Army Sci Conf, 57. *Mem:* AAAS; Am Pub Health Asn; Am Epidemiol Soc (secy-treas, 73-77); AMA; Am Asn Immunol; Am Soc Trop Med & Hyg. *Res:* Immunology; virology; epidemiology; clinical pathology; infectious diseases; diseases of the tropics. *Mailing Add:* Grad Sch Pub Health San Diego State Univ San Diego CA 92182

BENENSON, DAVID MAURICE, b Brooklyn, NY, Jan 22, 27; m 55; c 2. PLASMA PHYSICS. *Educ:* Mass Inst Technol, BS, 50; Calif Inst Technol, MS, 53, PhD(aeronaut), 57. *Prof Exp:* Proj engr, Calif Inst Technol, 50-53; res engr, Res Labs, Westinghouse Elec Corp, 57-63; assoc prof, State Univ NY, 63-67, dir, Lab Power & Environ Studies, 77-78, assoc chmn dept elec & comput eng, 78-83, chmn, 83-89, PROF, PLASMA & GAS DYNAMICS, STATE UNIV NY, BUFFALO, 67- *Concurrent Pos:* Instr, Carnegie Inst Technol, 58-62; investr, NSF, 64-; consult, Gen Elec Co, 74- *Mem:* Inst Elec & Electronics Engrs; Am Soc Mech Engrs. *Res:* Plasma and gas dynamics; turbomachinery; boundary layer analysis; flow problem; heat transfer; aerodynamic noise; energy conversion; power distribution. *Mailing Add:* 53 Andover Lane Williamsville NY 14221

BENENSON, RAYMOND ELLIOTT, b New York, NY, Dec 12, 25; m 56; c 3. PARTICLE-SOLID INTERACTIONS, PHYSICS EDUCATION. *Educ:* Mass Inst Technol, BS, 46; Univ Chicago, MS, 49; Univ Wis, PhD(physics), 55. *Prof Exp:* Assoc physicist, Brookhaven Nat Lab, 54-56; physicist, Wright Air Develop Ctr, Brookhaven, 56; instr physics, City Col New York, 56-60, from asst prof to assoc prof, 60-67; assoc prof, 67-70, PROF PHYSICS, STATE UNIV NY, ALBANY, 70- *Concurrent Pos:* Res assoc, Pegram Lab, Columbia Univ, 56-; co-worker, Phys Inst Basel, 62-63; resident visitor, Bell Labs, Murray Hill, NJ, 78-79; collabr, Chem Eng, Grenoble, France, 85-86. *Mem:* Am Phys Soc; Am Asn Physics Teachers; Mat Res Soc. *Res:* Fast neutron and gamma ray spectroscopy in conjunction with Van de Graaff accelerators; hydrogen profiling in solids, nuclear reaction and back scattering analysis; channeling; teaching experiments. *Mailing Add:* Dept Physics 1400 Washington Ave Albany NY 12222

BENENSON, WALTER, b New York, NY, Apr 27, 36; m 69; c 2. NUCLEAR PHYSICS. *Educ:* Yale Univ, BS, 57; Univ Wis, MS, 59, PhD(physics), 62. *Prof Exp:* Res assoc nuclear physics, Nuclear Res Inst, Strasbourg, 62-63; from asst prof to assoc prof nuclear physics, Mich State Univ, 63-72; assoc dir, 80-82, ASSOC DIR, NAT SUPERCONDUCTING CYCLOTRON LAB, 90-; PROF NUCLEAR PHYSICS, MICH STATE UNIV, 72- *Concurrent Pos:* Vis fel, Australian Nat Univ, 68; vis prof, Univ Grenoble, 70; Nat Acad Sci fel, Poland & Czech, 74; mem, Int Union Pure & Appl Physics on Atomic Masses & Fundamental Constants, 76-; vis consult, Lawrence Berkeley Lab, 78-79. *Honors & Awards:* A V Humboldt Sr Scientist Award, 89. *Mem:* Fel Am Phys Soc. *Res:* Nuclear temperature measurements; heavy ion gamma production; mass measurements of nuclei far from stability; direct nuclear reactions; heavy ion pion production. *Mailing Add:* Cyclotron Lab Mich State Univ East Lansing MI 48824

BENEPAL, PARSHOTAM S, b Bhagomajra-Punjab, India, Feb 25, 33; m 54; c 3. PLANT GENETICS, PLANT PHYSIOLOGY. *Educ:* Punjab, India, BS, 53, MS, 58; Kans State Univ, PhD(plant genetics, hort), 67. *Prof Exp:* Res asst, Regional Res Lab, USDA, Kans State Univ, 62-65; PROF LIFE SCI & CHMN DEPT, VA STATE COL, 67- *Concurrent Pos:* USDA Agr Res Serv grant, 68-75. *Mem:* AAAS; Am Genetic Asn; fel Royal Hort Soc. *Res:* Plant sciences; genetics and biochemistry of insect and disease resistance; breeding for quality and agronomic and physiological research on crop plants. *Mailing Add:* Dept Life Sci Va State Col Petersburg VA 23803

BENERITO, RUTH ROGAN, b New Orleans, La, Jan 12, 16; m 50. PHYSICAL CHEMISTRY. *Educ:* Sophie Newcomb Mem Col, BS, 35; Tulane Univ, MS, 38; Univ Chicago, PhD(chem), 48. *Hon Degrees:* DSc, Tulane Univ, 81. *Prof Exp:* Instr chem, Randolph-Macon Women's Col, 40-43; asst prof, Tulane Univ, 43-53; phys chemist, Southern Regional Res Ctr, USDA, 53-58, head colloidal chem invest, Cotton Chem Lab, 58-61, head Phys Chem Group, Natural Polymers Lab, 61-86; RETIRED. *Concurrent Pos:* Adj prof, Med Sch & Grad Sch, Tulane Univ, 53-; lectr, Univ New Orleans, 81- *Honors & Awards:* Fed Woman Award, 68, Southern Chemist Award, 68; Garvan Medalist, 70; Southwest Regional Award, Am Chem Soc, 72. *Mem:* AAAS; Am Chem Soc; Sigma Xi; fel Am Inst Chemists; Am Asn Textile Chem Colorists. *Res:* Thermodynamics of electrolytic solutions; kinetics and reaction mechanisms; cellulose reactions; surface phenomena; equilibrium; x-ray diffraction. *Mailing Add:* 4733 Marigny St New Orleans LA 70122

BENES, ELINOR SIMSON, b Yreka, Calif, Oct 5, 24; m 45, 76; c 4. MICROSCOPIC ANATOMY, HERPETOLOGY. *Educ:* San Jose State Col, AB, 45; Ariz State Univ, MS, 60; Univ Calif, Davis, PhD(zool), 66. *Prof Exp:* NIH fel, Univ Calif, Davis, 66-67; from instr to assoc prof, 67-80, PROF BIOL SCI, CALIF STATE UNIV, SACRAMENTO, 80- *Concurrent Pos:* Res assoc, dept herpet, Calif Acad Sci & Conserv Agency. *Mem:* Soc Study Amphibians & Reptiles; Am Inst Biol Sci; Am Soc Ichthyol & Herpet; Am Soc Zool; Orgn Trop Studies. *Res:* Behavioral response to color; microanatomy; growth and natural history of lizards. *Mailing Add:* 4604 Ravenwood Ave Sacramento CA 95821

BENES, NORMAN STANLEY, b Detroit, Mich, July 1, 21; m 45, 72; c 4. QUANTITATIVE PRECIPITATION FORECASTING. *Educ:* Univ Wash, BS, 49. *Prof Exp:* Asst flight supvr, aircraft dispatch, Southwest Airways, 50-51; chief meteorologist, weather forecasting, Hawthorne Sch Aeronaut, 51-55; meteorologist, US Weather Bur, Dept Com, 55-57; meteorologist in charge weather res, NSF-Int Geophys Yr, 57-59; meteorologist, fire weather forecasting, US Weather Bur, Dept Com, 59-60; sta sci leader weather res, US Antarctic Res Prog-NSF, 60-62; meteorologist agr weather forecasting, Nat Weather Serv, Nat Oceanic & Atmospheric Admin, 62-67, supv meteorologist, 67- 84; RETIRED. *Honors & Awards:* Benes Peak, US Bd Geog Names & NSF, 63; Antarctic Serv Medal, US Dept Defense, 65. *Mem:* Nat Weather Asn; Am Meteorol Soc; Am Geophys Union; AAAS. *Res:* Measurement of Antarctic energy balance in short and long wavelengths plus below grade, in soil, temperatures; evaluation and construction of minimum temperature prediction formulae for springtime orchard use. *Mailing Add:* PO Box Box 2184 Fair Oaks CA 95628

BENES, VACLAV EDVARD, b Brussels, Belgium, July 24, 30; nat US; m 51; c 2. MATHEMATICS. *Educ:* Harvard Univ, BA, 50; Princeton Univ, PhD(logic), 53. *Prof Exp:* Instr logic & philos, Princeton Univ, 52-53; MEM TECH STAFF, BELL TEL LABS, INC, 53- *Mem:* Am Math Soc; Math Asn Am; Soc Indust & Appl Math; Inst Math Statist. *Res:* Probability; stochastic processes; logic; set theory; linear systems; circuit theory; functional equations; information theory; theory of queues and congestion. *Mailing Add:* 275 Ashland Rd Summit NJ 07901

BENESCH, RUTH ERICA, b Paris, France, Feb 25, 25; US citizen; m 46; c 2. BIOCHEMISTRY. *Educ:* Univ London, BS, 46; Northwestern Univ, PhD(biochem), 51. *Prof Exp:* Demonstr chem, Univ Reading, 46-47; res assoc biochem, Johns Hopkins Univ, 47-48; fel, State Univ Iowa, 52; fel, Enzyme Inst, Univ Wis, 55; independent investr, Marine Biol Lab, Woods Hole Oceanog Inst, 56-60; res assoc, 60-64, from asst prof to assoc prof, 64-80, PROF BIOCHEM, COL PHYSICIANS & SURGEONS, COLUMBIA UNIV, 80- *Mem:* Am Chem Soc; Am Soc Biol Chemists; Biophys Soc; Am Soc Hemat. *Res:* Physical biochemistry; protein chemistry; enzymology; structure-function relationships in hemoglobin; sickle cell anemia; hematology. *Mailing Add:* Dept Biochem Col Physicians & Surgeons Columbia Univ 630 W 168th St New York NY 10032

BENESCH, SAMUEL ELI, b Baltimore, Md, Aug 19, 24; m 54; c 2. SYSTEMS THEORY. *Educ:* Johns Hopkins Univ, BA, 50; Univ Ill, MA, 51, PhD(math), 53. *Prof Exp:* Res supvr exterior ballistics, Calif Inst Technol, 53-62; staff lab dept mgr space systs, TRW Systs Group, 62-70; exec dir psychol, Los Angeles Ctr Group Psychother, 70-71; dir training crisis intervention, Childrens Hosp, Los Angeles, 71-72; staff scientist defense systs, 72-76, DEPT MGR, TRW SYSTS GROUP, REDONDO BEACH, 77- *Mem:* AAAS; Am Math Soc; Am Soc Psychical Res; Sigma Xi. *Res:* Systems models of consciousness. *Mailing Add:* 5155 1/2 Village Green Los Angeles CA 90016

BENESCH, WILLIAM MILTON, b Baltimore, Md, Apr 22, 22; m 46; c 3. MOLECULAR PHYSICS, MOLECULAR SPECTROSCOPY. *Educ:* Lehigh Univ, BA, 42; Johns Hopkins Univ, MA, 49, PhD(physics), 52. *Prof Exp:* Jr instr physics, Johns Hopkins Univ, 42-44; mem, Comn Relief Belg Fel, Univ Liege, 52-53; asst prof, Univ Pittsburgh, 53-60; res fel, Weizmann Inst Sci, 60; assoc prof, 62-66, PROF PHYSICS, UNIV MD, COLLEGE PARK, 66- *Concurrent Pos:* Dir inst molecular physics, Univ Md, College Park, 73-76; consult, Argonne Nat Lab, 74-77; physics dept fel, Johns Hopkins Univ, 77-, assoc ed, J Optical Soc Am, 77-84. *Mem:* Fel Am Phys Soc; fel Optical Soc Am; Soc Appl Spectros; Sigma Xi. *Res:* Infrared spectroscopy; solar and atmospheric spectroscopy; molecular interactions; atomic and molecular structure; recent publications have been on the spectrum of the aurora borealis and the underlying atom and molecular processes which give rise to its excitation and development. *Mailing Add:* Inst Molecular Physics Univ Md College Park MD 20742

BENET, LESLIE Z, b Cincinnati, Ohio, May 17, 37; m 60; c 2. PHARMACOKINETICS. *Educ:* Univ Mich, AB, 59, BS, 60, MS, 62; Univ Calif, PhD(pharmaceut chem), 65. *Hon Degrees:* PharmD, Uppsala Univ, Sweden, 87. *Prof Exp:* Asst prof pharm, Wash State Univ, 65-69; from asst prof to assoc prof, 69-76, vchmn dept, 73-78, PROF PHARM & PHARMACEUT CHEM, MED CTR, UNIV CALIF, SAN FRANCISCO, 76-, CHMN DEPT PHARM, SCH PHARM, 78- *Concurrent Pos:* Assoc ed, J Pharmacokinetics & Biopharmaceut, 72-75, ed, 76-; co-dir, Drug Studies Unit, Univ Calif, 77-, dir, Drug Kinetics & Dynamics Ctr, 79-; chmn, Pharmacol Study Sect, NIH, 79-81; mem bd dirs, Am Found Pharmaceut Educ, 87-; mem sci adv bd, Smithkline Beecham Pharmaceuticals, 90; mem, Generic Drugs Adv Comt, Food & Drug Admin, 91. *Honors & Awards:* Res Achievement Award Pharmaceut, Acad Pharmaceut Sci, 82; Pfizer Lectr, La State Univ Med Ctr, 85; Edward R Garrett Lectr, Univ Fla, 87; Varro E Tyler Distinguished Lectr, Purdue Univ, 88; Distinguished Serv Award, Am Col Clin Pharmacol, 88; Volwiler Distinguished Res Achievement Award, Am Asn Cols Pharm, 91. *Mem:* Inst Med-Nat Acad Sci; fel Acad Pharmaceut Res & Sci; fel AAAS; fel Am Asn Pharmaceut Scientists (pres, 86, treas, 87); Am Asn Cols Pharm; Am Asn Univ Professors; Am Col Clin Pharmacol; Am Pharmaceut Asn; Am Soc Clin Pharmacol & Therapeut; Am Soc Pharmacol & Exp Therapeut. *Res:* Development of the correlation of pharmacokinetics and pharmacodynamics for drugs in various patient populations with specific study areas of immunosuppressive agents; nitroglycerin and its dinatrate metabolites; nonsteroidal anti-inflammatory agents and anti-hypertensive agents. *Mailing Add:* Dept Pharm Univ Calif San Francisco CA 94143-0446

BENEVENGA, NORLIN JAY, b San Francisco, Calif, Feb 16, 34; m 56; c 3. BIOCHEMISTRY, NUTRITION. *Educ:* Univ Calif, Davis, BS, 59, EE & MS, 60, PhD(nutrit), 65. *Prof Exp:* Res assoc amino acid metab, Mass Inst Technol, 65; proj assoc, 65-66, from asst prof to assoc prof, 66-74, PROF MEAT & ANIMAL SCI & NUTRIT SCI, UNIV WIS-MADISON, 74- *Concurrent Pos:* Mem, Nat Res Coun Sub-Comt Lab Animal Nutrit, 74. *Mem:* AAAS; Am Nutrit Soc; Am Inst Nutrit; Am Soc Animal Sci; Brit Biochem Soc. *Res:* Nutritional and metabolic effects of amino acid excesses and imbalances and digestion and metabolism of lactose. *Mailing Add:* Dept Meat & Animal Sci 1675 Observatory Dr Madison WI 53706

BENEVENTO, LOUIS ANTHONY, b Waterbury, Conn, Nov 17, 40. NEUROPHYSIOLOGY, NEUROANATOMY. *Educ:* Rensselaer Polytech Inst, BS, 62, MS, 64; Univ Md, PhD(physiol), 68; Brown Univ, MD, 70. *Prof Exp:* Nat Inst Neurol Dis & Stroke res fel neurosci, Brown Univ, 68-70; spec fel neurophysiol, Max Planck Inst Psychiat, Ger, 70-71; asst prof physiol, Sch Med, Univ Va, 71-72; from assoc prof to prof anat & ophthal, Col Med, Univ Ill Med Ctr, 76-89; PROF & HEAD ANAT, UNIV MD, 89- *Concurrent Pos:* Vis scientist, John Curtin Sch Med, Australia, 75-76; vis prof, Johns Hopkins Univ Sch Med, 87- *Mem:* AAAS; NY Acad Sci; Am Asn Univ Prof; Am Acad Arts & Sci; Am Anat Asn; Soc Neurosci. *Res:* Anatomical and physiological correlates of the non-primary visual pathways-visuomotor, multimodal and associational systems. *Mailing Add:* Col Dent Surg Anat Dept Univ Md 666 W Baltimore St Baltimore MD 21201-1586

BENFEY, BRUNO GEORG, b Dusseldorf, Ger, Oct 9, 17; m 49; c 3. PHARMACOLOGY, THERAPEUTICS. *Educ:* Univ Hamburg, Md, 48; Univ Gottingen, dipl chem, 50. *Prof Exp:* Asst org chem, Univ Gottingen, 48-50; asst, Second Clin Internal Med, Univ Hamburg, 50-51; sci asst, Inst Physiol Chem, Univ Gottingen, 51; USPHS fel physiol chem, Yale Univ, 51-52; lectr, 52-54, asst prof, 54-60, assoc prof pharmacol, 60-75, PROF PHARMACOL & THERAPEUT, MCGILL UNIV, 75- *Mem:* Am Soc Pharmacol & Exp Therapeut; Pharmacol Soc Can; Can Physiol Soc. *Res:* Molecular weight of heparin; isolation of pikromycin from actinomycetes; enzymatic studies with amino acids and peptides; chemistry of pituitary hormones; pharmacology and biochemistry of sympathomimetic and adrenergic blocking agents. *Mailing Add:* Dept Pharmacol & Therapeut McGill Univ Montreal PQ H3G 1Y6 Can

BENFEY, OTTO THEODOR, b Berlin, Ger, Oct 31, 25; nat US; m 49; c 3. CHEMICAL EDUCTION. *Educ:* Univ London, BSc, 45, PhD(org chem), 47. *Prof Exp:* From asst prof to assoc prof chem, Haverford Col, 48-56; from assoc prof to prof, Earlham Col, 56-73; Dana prof chem, 73-88, EMER PROF CHEM & HISTORY SCI, GUILFORD COL, 89-; ADJ PROF, DEPT HIST & SOCIOL SCI, UNIV PA, 90- *Concurrent Pos:* Res fel, Columbia Univ, 47-48 & Harvard Univ, 55-56; lectr, Univ Dublin, 61, Univ Sao Paulo, 63 & Univ Santiago, Chile, 66; mem Adv Coun Col Chem, 65-68; Danforth Found E Harris Harbison fel, 67-68; Fulbright-Hays res & study award, Kwansei-Gakuin Univ, Japan, 70-71; vis prof, Intl Christian Univ Tokyo, Japan, 85-86; chmn, Div Hist Chem, Am Soc Chem, 66; ed, Nat Found Hist Chem, 89- *Mem:* Am Chem Soc; Hist Sci Soc; Sigma Xi; Soc Social Responsibility Sci (pres, 51-53). *Res:* Mechanism of organic reactions; solvent and salt effects in unimolecular solvolysis; history of chemistry; Oriental science; chemical education. *Mailing Add:* Nat Found Hist Chem 3401 Walnut St Philadelphia PA 19104-6228

BENFIELD, CHARLES W(ILLIAM), b Crossnore, NC, Aug 2, 21; m 53. ELECTRONICS, NAVIGATION. *Educ:* Univ Fla, BEE, 48. *Hon Degrees:* PhD, Fla Res Inst, 72. *Prof Exp:* Electronics technician, Flagler Radio Corp & Miami Army Air Field, 45; electronic scientist, US Naval Res Lab, Washington, DC, 48-50; radio engr, Isle of Dreams Broadcasting Corp, Fla, 50-51; electronics engr develop lab, Bendix Pac Div, Calif, 51-52; electronic scientist, USN Underwater Sound Ref Lab, 53-59; sr develop engr aeronaut div, Honeywell, Inc, St Petersburg, 59-65, staff engr advan planning, 65-66; sr engr marine systs anal, RCA Missile Test Proj, 66-74; staff engr, Electronics Br, Tech & Logistics Serv Div, US Army Aeromed Res Lab, Ala, 74-76, chief engr, 76-78, chief engr, Data Systs & Instrumentation Br, Res Systs Div, US Army Aviation Develop Test Activ, Ala, 78-79, chief, Instrumentation Lab, 79-82; RETIRED. *Concurrent Pos:* Prin instr, Orange County Adult & Vet Inst, Orlando, 56-59; guest lectr, Univ SFla, 61-62 & Long Beach Col Reserve Unit, 62; lectr, Fla Vis Sci Prog, 63-66; consult, Nicoa Corp, Neo-Dyne Res Corp, Aim Mfg Co, 82- *Honors & Awards:* Burka Award, Inst Navig, 66. *Mem:* Am Inst Navig; sr mem Inst Elec & Electronics Engrs; assoc fel Am Inst Aeronaut & Astronaut. *Res:* Astronautics; gravity; parapsychology; oceanography; geodetic survey of undersea transponders; biomedical instrumentation; data acquisition systems. *Mailing Add:* 1000 W Bonanza Rd Las Vegas NV 89106

BENFIELD, ERNEST FREDERICK, b Bangor, Maine, Feb 1, 42; m 63; c 3. STREAM ECOLOGY. *Educ:* Appalachian State Univ, BS, 64, MA, 65; Va Polytech Inst & State Univ, PhD(zool), 71. *Prof Exp:* Instr biol, Gordon Mil Col, 65-67; res assoc, 70-71, from asst prof to assoc prof, 71-83, PROF, VA POLYTECH INST & STATE UNIV, 83- *Concurrent Pos:* Am Elec Power Serv Corp grant, 70-77; FMC Corp grant, 73-76; USERDA grants, 74-77; USDA Forest Serv grant, 76-81; Va Comn Game Inland Fisheries grant, 77-80; Va State Water Control Bd, 78-80; consult, Merck & Co, E I du Pont de Nemours & Co, Inc, Tenn Valley Authority, US Army Corps Engrs, 70-78; NSF grants, 79-88 & 91-; US Environ Protection Agency grant, 80-82 & 89-91; assoc ed, Am Midland Naturalist, 82-85, J NAm Benthological Soc, 85-88; interim managing ed, J NAm Benthological Soc, 89-91. *Mem:* Ecol Soc Am; NAm Benthological Soc; Sigma Xi. *Res:* Organic matter dynamics in streams; ecology of aquatic macroinvertebrates; pollution ecology of lakes and streams. *Mailing Add:* Dept Biol Va Polytech Inst & State Univ Blacksburg VA 24061

BENFIELD, JOHN R, b Vienna, Austria, June 24, 31; US citizen; m 63; c 3. SURGERY. *Educ:* Columbia Univ, AB, 52; Univ Chicago, MD, 55; Am Bd Surg, dipl, 63; Am Bd Thoracic Surg, dipl, 65. *Prof Exp:* Instr surg, Sch Med, Univ Chicago, 61-64; asst prof, Univ Wis, 64-67; assoc prof surg, Sch Med, Univ Calif-Harbor Gen Hosp, 67-71; prof surg, Univ Calif, Los Angeles, 72-76; Jame Utley prof & chmn surg, Univ Hosp, Boston, 77; chmn surg, City of Hope Med Ctr, 78-87; PROF & CHIEF CARDIOTHORACIC SURG, UNIV CALIF, DAVIS, 87- *Concurrent Pos:* Consult, US Naval Hosp, San Diego & US Naval Hosp, Camp Pendelton, Calif; dir, Am Bd Thoracic Surg; gov, Am Col Surgeons. *Mem:* Am Surg Asn; fel Am Col Surg; Soc Univ Surg; Am Asn Thoracic Surg. *Res:* Lung cancer; lung transplantation; obesity. *Mailing Add:* Univ Calif Davis 4301 K St Rm 2310 Sacramento CA 95817

BENFORADO, JOSEPH MARK, b New York, NY, June 20, 21; m 48; c 6. CLINICAL PHARMACOLOGY. *Educ:* City Col New York, BS, 41; Columbia Univ, MA, 42; State Univ NY, MD, 51. *Prof Exp:* Intern, Boston City Hosp, 51-52; res fel pharmacol, Harvard Med Sch, 52-53, instr, 53-57, assoc, 57-58; Life Ins med res fel, Oxford Univ, 57-58; assoc prof pharmacol, Sch Med, State Univ NY Buffalo, 58-67; EMER PROF MED, SCH MED, UNIV WIS-MADISON, 67- *Mem:* Am Soc Pharmacol; Sigma Xi. *Res:* Clinical medicine; drug abuse. *Mailing Add:* 4034 Mandan Circle Madison WI 53711-3005

BENFORD, ARTHUR E, b Benton Harbor, Mich, July 21, 31; m 52; c 3. PLASTICS CHEMISTRY & ENGINEERING. *Educ:* Western Mich Univ, BS, 57. *Prof Exp:* Assoc chemist res labs, 57-58, chemist plastics, 58-60, assoc res chemist, 60-64, mgr mat res, 64-68, Res & Eng Ctr, 68-71, Corp Mgr Plastic Mfg Systs, 71-73, mgr plastics res & eng, refrig group, 73-75, LEAD PRODUCT ENGR PLASTICS, REFRIG GROUP, WHIRLPOOL CORP, 75- *Mem:* Am Chem Soc; Sigma Xi; Soc Plastics Indust; Soc Plastics Engrs. *Res:* Polyurethane chemistry and application; thermoplastic and thermoset; plastics chemistry, process, equipment and engineering; four patents. *Mailing Add:* 6624 Whispering Hills Dr Evansville IN 47720

BENFORD, GREGORY A, b Mobile, Ala, Jan 30, 41. PLASMA PHYSICS, ASTROPHYSICS. *Educ:* Univ Okla, BS, 63; Univ Calif, San Diego, MS, 65, PhD(physics), 67. *Prof Exp:* res physicist, Lawrence Radiation Lab, 67-71; PROF PHYSICS, UNIV CALIF, IRVINE, 71- *Concurrent Pos:* Woodrow Wilson fel; fel, Cambridge Univ. *Mem:* Am Phys Soc; Royal Astron Soc; Soc Scientific Exploration. *Res:* Extremeley strong plasma turbulence; high energy astrophysics. *Mailing Add:* Dept Phys Univ Calif Irvine CA 92717

BENFORD, HARRY (BELL), b Schenectady, NY, Aug 7, 17; m 41; c 3. NAVAL ARCHITECTURE, MARINE ENGINEERING. *Educ:* Univ Mich, BSE, 40. *Prof Exp:* Staff supvr, ship repair cost estimator & other positions, Newport News Shipbldg & Dry Dock Co, 40-48; from asst prof to prof naval archit & marine eng, 48-82, chmn dept, 67-72, EMER PROF NAVAL ARCHIT & MARINE ENG, UNIV MICH, 82- *Concurrent Pos:* Exec dir maritime res adv comt, Nat Acad Sci-Nat Res Coun, 59-60; mem at large maritime res, Nat Res Coun, 64-69. *Honors & Awards:* Linnard Prize, Soc Naval Architects & Marine Engrs, 63, David W Taylor Medal, 76. *Mem:* Am Soc Eng Educ; Am Asn Cost Engrs; fel Soc Naval Archit & Marine Engrs; fel Royal Inst Naval Archit. *Res:* Preliminary design of ships; shipbuilding cost and weight analysis; engineering economy in ship design and operation. *Mailing Add:* Six Westbury Ct Ann Arbor MI 48105

BENGELE, HOWARD HENRY, b Sewickley, Pa, Sept 17, 37. VOLUME REGULATION, GRAVITATIONAL PHYSIOLOGY. *Educ:* Blackburn Col, Carlinville, Ill, AB, 63; Univ Iowa, PhD(physiol & biophys), 69. *Prof Exp:* Fel physiol, Univ Toronto, 69-71; res assoc, Univ NMex, Albuquerque, 71-73; ASSOC PROF MED & PHYSIOL, BOSTON UNIV, 73- *Mem:* Am Physiol Soc; Am Soc Nephrology; Can Physiol Soc; Sigma Xi; Am Fedn Clin Res; NY Acad Sci. *Res:* The role of the inner medullary collecting duct of the kidney in acid-base regulation. *Mailing Add:* Thorndike Mem Lab Boston City Hosp 818 Harrison Ave Boston MA 02118

BENGELSDORF, IRVING SWEM, b Chicago, Ill, Oct 23, 22; m 49; c 3. SCIENCE WRITING. *Educ:* Univ Ill, BS, 43; Univ Chicago, MS, 48, PhD(chem), 51. *Prof Exp:* Res fel chem, Calif Inst Technol, 51-52; instr, Univ Calif, Los Angeles, 52-54 & Gen Elec Res Lab, 54-59; group leader org chem res, Texus Res Ctr, 59-60; sr scientist, US Borax Res Corp, 60-63; sci ed, Los Angeles Times, 63-70; lectr sci commun & dir, Inst, 71-80, TECH WRITER SPECIALIST, JET PROPULSION LAB, CALIF INST TECHNOL, 81- *Concurrent Pos:* Sr lectr, Sch Jour, Univ Southern Calif, 66, 68 & Dept Chem, 71, 73, 74; sr lectr, Dept Chem, Univ Calif, Los Angeles, 67, 68, 71; sci columnist & contrib ed, Enterprise Sci Serv, Newspaper Enterprise Asn, 71-74; distinguished vis prof sci, Whittier Col, 74; vis distinguished prof chem, Univ Southern Calif, 77-; contrib sci columnist, Los Angeles Herald Examr, 78-; mem tech staff, Aerospace Corp, 81. *Honors & Awards:* Jean M Kline Mem Award, Am Cancer Soc, 65; James T Grady Award, Am Chem Soc, 67; Westinghouse Writing Award, AAAS, 67, 69; Claude Bernard Sci Jour Award, Nat Soc Med Res, 68. *Mem:* AAAS; Am Chem Soc; Chem Soc London; NY Acad Sci; hon fel Soc Tech Commun. *Res:* Explanation of principles and philosophy of science and technology to nonscientists. *Mailing Add:* 3778 Via Las Villas Oceanside CA 92056-5182

BENGTSON, GEORGE WESLEY, b Sealy, Tex, Nov 3, 30; m 59; c 3. FOREST BIOLOGY, FOREST ECOLOGY. *Educ:* La State Univ, BS, 52; Duke Univ, MF, 55; Yale Univ, PhD(plant physiol, soils), 61. *Prof Exp:* Res forester physiol, soils, Forest Serv, USDA, 55-65 & Nat Fertilizer Develop Ctr, US Tenn Valley Authority, 65-79; assoc dean, Col Forestry, Ore State Univ, 79-89; dir, Ctr Forested Wetlands Res, Forest Serv, USDA, 89-91; CONSULT, INT FERTILIZER DEVELOP CTR, 91- *Honors & Awards:* John Beale Mem Award, Soc Am Foresters, 91. *Mem:* Fel Soc Am Foresters; Soil Sci Soc Am; Int Union Forest Res Orgn. *Res:* Forest tree nutrition with emphasis on fertilizer; utilization of wastes as soil amendments, ecological implications of intensive silviculture; forest wetland ecology. *Mailing Add:* Sch Forestry Auburn Univ Auburn AL 36849-5418

BENGTSON, HARLAN HOLGER, b Maquokata, Iowa, Oct 4, 41; m 65; c 4. CHEMICAL ENGINEERING. *Educ:* Iowa State Univ, BS, 63, MS, 65; Univ Colo, PhD(chem eng), 71. *Prof Exp:* Process develop engr, E I du Pont de Nemours & Co, 63; prod develop engr, Minn Mining & Mfg, 64-65; asst prof eng sci, Rockhurst Col, 71-75; dir, Environ Resources Training Ctr, 84-88, PROF CIVIL ENG, SOUTHERN ILL UNIV, EDWARDSVILLE, 75-, ASST DEAN, SCH ENG, 88- *Concurrent Pos:* Consult, Panhandle Eastern Pipeline Co, 71 & Am Steel Foundries, 77. *Mem:* Am Inst Chem Engrs; Am Soc Eng Educ; Am Water Works Asn. *Res:* Energy production from wastes by anaerobic digestion; mathematical modeling of biological wastes treatment. *Mailing Add:* Dept Civil Eng Southern Ill Univ Edwardsville IL 62026-1800

BENGTSON, KERMIT (BERNARD), b Boston, Mass, Aug 21, 22; m 51; c 4. CHEMICAL ENGINEERING. *Educ:* Univ Wash, BS, 47, MS, 55, PhD(chem eng), 57. *Prof Exp:* Instr inorg & anal chem, geol & gen sci, Western Wash Col, 48-51; asst, Univ Wash, 53-55; group leader chlorination res, Kaiser Aluminum & Chem Corp, 57-58; dir ctr grad study, Univ Wash, Richland, 58-68; sect head, Extractive Metall, Chem & Specialty Metals Div, Ctr Technol, Kaiser Aluminum & Chem Corp, 68-85; RETIRED. *Mem:* Am Chem Soc; Metal Soc; Sigma Xi. *Res:* Application of chlorine and hydrochloric acid technology to extractive metallurgy of metals, magnesium and aluminum; glacial geology and glaciology. *Mailing Add:* 19 Hampton Ct Alameda CA 94501

BENGTSON, ROGER D, b Wausa, Nebr, Apr 29, 41; m 63; c 2. PLASMA PHYSICS, ATOMIC PHYSICS. *Educ:* Univ Nebr, BS, 62; Va Polytech Inst, MS, 64; Univ Md, PhD, 68. *Prof Exp:* Fac assoc, 68-70, from asst prof to assoc prof, 70-81, PROF PHYSICS, UNIV TEXAS, AUSTIN, 81-, CHMN DEPT, 83- *Mem:* Am Phys Soc; Sigma Xi. *Res:* Plasma spectroscopy; experimental plasma physics. *Mailing Add:* Dept Physics Univ Tex Austin TX 78712

BENHAM, CRAIG JOHN, b Chicago, Ill, Sept 1, 46; m 75; c 2. PHYSICAL CHEMISTRY, MOLECULAR BIOLOGY. *Educ:* Swarthmore Col, AB, 68; Princeton Univ, MA, 71, PhD(math), 72. *Prof Exp:* Asst prof math, Univ Notre Dame, 72-76, Lawrence Univ, 77-78, Univ Ky, 78-81; res fel biol, Calif Inst Technol, 76-77; assoc prof math, Univ Ky, 81-88; PROF BIOMATH SCI, MT SINAI SCH MED, 88- *Concurrent Pos:* Vis res scientist, NY State Dept Health, 78, Cancer Res Inst, 79; fel, Alfred P Sloan Found, 82. *Mem:* AAAS. *Res:* Theoretical calculations of the equilibria and kinetics of DNA conformations; superhelically stressed molocules, including mechanical equilibria, conformational transitions and superhelical regulation of DNA activities. *Mailing Add:* Dept Biomath Sci Mt Sinai Sch Med New York NY 10029-6574

BENHAM, JUDITH LAUREEN, POLYMER CHEMISTRY. *Educ:* Univ Rochester, BA, 69; Univ Wis-Madison, PhD(org chem), 78. *Prof Exp:* Res specialist, 81-82, res supvr, 82-83, RES MGR, SPECIALTY CHEM DIV, 3M, 83- *Mem:* Am Chem Soc; Sigma Xi. *Res:* Production of chemicals; high performance polymers; high temperature polymers; conductive polymers; radiation curable materials and electronically active materials. *Mailing Add:* 1628 McKusick Lane Stillwater MN 55082

BENHAM, ROSS STEPHEN, b Calgary, Alta, Feb 13, 11; m 40; c 2. MICROBIOLOGY. *Educ:* Univ Chicago, SM & PhD(microbiol), 57; Am Bd Microbiol, dipl. *Prof Exp:* Dir clin lab servs, Muskoka Hosp, 39-48; dir, Clin Microbiol Labs, Hosps & Clins, Univ Chicago, 48-69, asst prof med, 57-69, res assoc, Dept Microbiol, 59-69, assoc prof med & clin microbiol, 67-69; chief microbiologist, Prof Serv Corp, 69-77; CONSULT MICROBIOLOGIST, 77- *Concurrent Pos:* Eli Lilly fel, Univ Chicago, 66; hon mem staff, St Joseph Mercy Hosp, Pontiac, 69- *Mem:* AAAS; fel Am Pub Health Asn; Am Soc Microbiol; Am Soc Med Technol; NY Acad Sci; Sigma Xi. *Res:* Clinical microbiology; immunology; bacterial and protozoan physiology; epidemiology. *Mailing Add:* 4611 Burnley Dr Bloomfield Hills MI 48013

BENI, GERARDO, b Florence, Italy, Feb 21, 46; US citizen; m; c 1. THEORETICAL ROBOTICS. *Educ:* Univ Florence, Laurea in physics, 70; Univ Calif, Los Angeles, PhD(physics), 74. *Prof Exp:* Res physicist, Scattering & Low Energy Physics Dept, AT&T Bell Labs, Murray Hill, 74-76, Theoret Physics Dept, 76, mem tech staff, Solid State Physics Res Dept, Holmdel, 77-81, Robotic Systs & Robotic Technol Res Dept, 82-84; PROF ELEC & COMPUTER ENG, UNIV CALIF, SANTA BARBARA, 84-, DIR, CTR ROBOTIC SYSTS IN MICROELECTRONICS, 84- *Concurrent Pos:* Mem, Comn Eng & Tech Systs, Mfg Studies Bd, Nat Res Coun; ed, J Robotic Systs. *Mem:* Fel Am Phys Soc; Electrochem Soc; NY Acad Sci; AAAS; Inst Elec & Electronics Engrs; Soc Mfg Engrs. *Res:* Electrochromism - displays, optical switches, electro-wetting, ionic devices; electrocatalysis; electron-hole liquids; extended x-ray absorption fine structure; organic semiconductors; one-dimensional systems; thermoelectricity; avalanche phenomena in semiconductors. *Mailing Add:* Dept Elec/Comput Eng Univ Calif Santa Barbara CA 93106

BENICEWICZ, BRIAN CHESTER, b Danbury, Conn, June 1, 54; m; c 2. POLYMER CHEMISTRY, ORGANIC CHEMISTRY. *Educ:* Fla Inst Technol, BS, 76; Univ Conn, MS, 78, PhD(polymer chem), 80. *Prof Exp:* Res chemist, Celanese Res Co, 80-82; sr scientist, Ethicon, Inc, 82-85; STAFF MEM, LOS ALAMOS NAT LAB, 85- *Mem:* Am Chem Soc; Mat Res Soc; Soc Advan Mat & Plastics Eng. *Res:* Chemistry of high polymers; high temperature polymers; liquid crystal polymers; high performance polymers; piezoelectric and conducting polymers; biomedical polymers. *Mailing Add:* PO Box 1663 MS E549 Los Alamos NM 87545

BENIN, DAVID B, b Chicago, Ill, May 19, 41; m 83; c 1. CONDENSED MATTER PHYSICS, PHOTOPHYSICS OF ORGANIC MOLECULES. *Educ:* Cornell Univ, BA, 63; Univ Rochester, MS, 66, PhD(physics), 68. *Prof Exp:* Res assoc physics, Cornell Univ, 67-69 & Univ Ore, 69-70; asst prof, 70-76, ASSOC PROF PHYSICS, ARIZ STATE UNIV, 76- *Concurrent Pos:* Vis assoc prof, Brown Univ, 76-77. *Mem:* Am Phys Soc. *Res:* Condensed matter physics theory; lattice dynamics and transport properties of phonon systems; theory and transport properties of superfluid helium; photophysical properties of organic molecules. *Mailing Add:* Dept Physics & Astron Ariz State Univ Tempe AZ 85287-1504

BENINGER, RICHARD J, b Walkerton, Ont, Feb 11, 50; m 89; c 1. BIOLOGICAL PSYCHOLOGY, PSYCHOPHARMACOLOGY. *Educ:* Univ Western Ont, BA, 73; McGill Univ, MA, 74, PhD(psychol), 77. *Prof Exp:* Fel psychopharmacol, Univ BC, 77-80; from asst prof to assoc prof, 80-88, PROF PSYCHOL, QUEENS UNIV, CAN, 88- *Concurrent Pos:* Career scientist res award, Ministry Health, Ont, 82-92; asst prof psychiat, Queens Univ, 83-; vis scientist, Dept Pharmacol Pitié Salpétriè re, Paris, 87-88. *Mem:* AAAS; Can Col Neuropsychopharmacol; Can Psychol Asn; Europ Behav Pharmacol Soc; Soc Neurosci. *Res:* Central nervous system transmitter substances in behavior; the role of dopamine in the control of locomotor activity and learning; the role of acetylcholine in memory. *Mailing Add:* Dept Psychol Queens Univ Kingston ON K7L 3N6 Can

BENINGTON, FREDERICK, b Chelsea, Mass; m 40; c 2. ORGANIC CHEMISTRY, MATHEMATICS. *Educ:* Tufts Univ, BSc, 39. *Hon Degrees:* PhD, Univ Wis. *Prof Exp:* Res chemist, Innis Speiden Co, 39-41 & Niacet Chem Corp, 41-45; physicist, Corning Glass Works, 45-47; res engr, Battelle Mem Inst, 47-49, sr res assoc, 49-52; assoc prof, 62-71, prof med chem, 71-79, PROF PSYCHIAT, MED CTR, UNIV ALA, 78- *Concurrent Pos:* Instr, Niagara Univ, 42-44; consult chemist, 49- *Mem:* Am Chem Soc; Nat Speleological Soc; Sigma Xi; Cave Res Found; Explorer's Club. *Res:* Thermodynamics and kinetics of gas reactions; free radical reactions; organometallic reactions; syntehsis of psychotomimetic compounds; author or co-author of numerous publications. *Mailing Add:* Neurosci Prog CDLD Bldg 9th Fl Univ Ala Med Ctr Birmingham AL 35233

BENIOFF, PAUL, b Pasadena, Calif, May 1, 30; m 59; c 3. ENVIRONMENTAL EARTH & MARINE SCIENCES. *Educ:* Univ Calif, Berkeley, PhD(chem), 59. *Prof Exp:* Nuclear chemist, Lab Electronics, Tracerlab, Inc, 52-54; Weizmann fel, Weizmann Inst, Israel, 60-61; Ford grant, Inst Theoret Physics, Copenhagen Univ, 61; nuclear theoret chemist, 61-78, ENVIRON SCIENTIST, ARGONNE NAT LAB, 78- *Concurrent Pos:* Vis prof, Dept Hist & Philos Sci, Tel Aviv Univ, 80, CNRS, Univ Marseille, Luminy France, 80. *Mem:* AAAS. *Res:* Molecular structure; relationship between foundations of physics and mathematics. *Mailing Add:* Argonne Nat Lab 9700 S Cass Ave Argonne IL 60439

BENIRSCHKE, KURT, b Gluckstadt, Ger, May 26, 24; nat US; m 52; c 3. PATHOLOGY. *Educ:* Univ Hamburg, MD, 48. *Prof Exp:* Assoc path, Harvard Med Sch, 57-60; prof & chmn dept, Dartmouth Med Sch, 60-70; dir res, San Diego Zoo, 75-86; PROF REPRODUCTIVE MED & PATH, UNIV CALIF, SAN DIEGO, 70- *Honors & Awards:* Presidential Medal, NY Acad Sci, 84. *Mem:* Col Am Path; Am Asn Path & Bact; Sigma Xi. *Res:* Placental pathology; developmental anomalies and twinning processes; comparative mammalian cytogenetics; cause of sterility in hybrids; pathology of human fetus and placenta; gemellology; mammalian hybrids. *Mailing Add:* Dept Path Univ Calif at San Diego La Jolla CA 92093

BENISEK, WILLIAM FRANK, b Los Angeles, Calif, Oct 12, 38; m 60; c 2. BIOCHEMISTRY. *Educ:* Calif Inst Technol, BS, 60; Columbia Univ, MA, 61; Univ Calif, Berkeley, PhD(biochem), 66. *Prof Exp:* USPHS fel biophys, Yale Univ, 66-68; biochemist, Sch Med, Stanford Univ, 68-69; asst prof, 69-75, ASSOC PROF BIOCHEM, SCH MED, UNIV CALIF, DAVIS, 75- *Mem:* Am Soc Biol Chem; Am Chem Soc. *Res:* Protein chemistry; structures and conformations of protein molecules; chemical mechanisms of enzyme-catalyzed reactions. *Mailing Add:* Dept Biol Chem Univ Calif Davis CA 95616

BENISON, BETTY BRYANT, b Irvine, Ky, Aug 25, 39; wid; c 1. HEALTH EDUCATION. *Educ:* La State Univ, Baton Rouge, BS, 55; Univ Mich, Ann Arbor, MA, 58; Univ NMex, PhD(health, phys educ), 68; Tex Woman's Univ, MS, 90. *Prof Exp:* Instr phys educ, C E Byrd High Sch, Shreveport, La, 55-56 & Miss State Col Women, 56-57; asst, Univ Mich, Ann Arbor, 57-58; asst prof, Stephen F Austin Univ, 58-59, NTex State Univ, 59-64 & Univ NMex, 64-69; PROF BIOMECH & PHYSIOL, TEX CHRISTIAN UNIV, 69- *Concurrent Pos:* Mem, Nat Coun Outdoor Educ, Am Alliance Health, Phys Educ & Recreation, 73-; consult, Col Educ Adv Bd, Am Heart Asn, 74-75 & Adapted & Corrective Ther, Standford Convalescent Ctrs, 74-; grant, Tex Christian Univ, 74-75, Am Heart Asn, 75 & NIH-NASA, Southwest Med Sch, Dallas, 76; Tex Christian Univ fel, Cooper Aerobics Ctr, Dallas, 75-76; sabbatical mem staff, Aerobics Ctr, Dallas, 76; Tex Christian Univ fel, Cooper Aerobics Ctr, Dallas, Tex, 75-76; NIH-NASA grant, Southwest Med Sch, Dallas, 76; sabbatical mem staff, Aerobics Ctr, Dallas, 76. *Mem:* Am Pub Health Asn; Am Col Sports Med; Int Soc Sports Psychol; Soc Behav Kinesiology; Am Heart Asn; Am Asn Sex Educrs, Counrs & Therapists. *Res:* Proper electrode placement in stress testing; comparative studies of cardiac impairments; development and evaluation of high-level physical fitness programs. *Mailing Add:* 3501 Bellaire Dr N No 11 Ft Worth TX 76109

BEN-ISRAEL, ADI, b Rio de Janeiro, Brazil, Nov 6, 33; m 64; c 2. OPTIMIZATION, LINEAR ALGEBRA. *Educ:* Israel Inst Technol, BSc, 55, Dipl Ing, 56, MSc, 59; Northwestern Univ, PhD(eng sci), 62. *Prof Exp:* Vis asst prof statist, Carnegie Inst Technol, 62-63; sr lectr math, Israel Inst Technol, 63-65; assoc prof systs eng, Univ Ill, Chicago Circle, 65-66; from assoc prof to prof eng sci, Northwestern Univ, 66-70; prof appl math & chmn dept, Technion-Israel Inst Technol, 70-75; prof indust eng dept, Northwestern Univ, 75-76; chmn opers res prog, 76-80, H FLETCHER BROWN PROF MATH, UNIV DEL, 76- *Mem:* Oper Res Soc Am; Am Math Soc; Soc Indust Appl Math; Math Prog Soc. *Res:* Matrix theory; mathematical programming; operations research; fishery management; natural resource and environmental management; optimal engineering design; numerical analysis. *Mailing Add:* Rutgers Univ 249 University Ave Newark NJ 07102

BENITEZ, ALLEN, organic chemistry, for more information see previous edition

BENITEZ, FRANCISCO MANUEL, b Cuba, Apr 12, 50; US citizen; m 77. ORGANIC SYNTHETIC CHEMISTRY, POLYMER CHEMISTRY. *Educ:* Wilmington Col, BA, 72; Miami Univ, PhD(chem), 77. *Prof Exp:* RES CHEMIST, EXXON CHEM CO 77- *Mem:* Am Chem Soc; Sigma Xi. *Res:* Synthetic and mechanistic aspects of sulfur chemistry; synthetic organic acids; hydrocarbon resins. *Mailing Add:* Exxon Chem Co PO Box 241 Baton Rouge LA 70821-0241

BENJAMIN, BEN MONTE, b St Petersburg, Fla, Feb 16, 23; m 48; c 3. ORGANIC CHEMISTRY, FUEL CHEMISTRY. *Educ:* Univ Fla, BS, 48, MS, 49, PhD(org chem), 52. *Prof Exp:* Org chemist, 52-80, GROUP LEADER CHEM & SR SCIENTIST, COAL CONVERSION, OAK RIDGE NAT LAB, 80- *Mem:* Am Chem Soc. *Res:* Organic reaction mechanisms; isotopic tracer studies; nuclear magnetic resonance; coal chemistry. *Mailing Add:* 848 W Outer Dr Oak Ridge TN 37830

BENJAMIN, CHESTER RAY, b Alliance, Ohio, Jan 23, 23; m 47; c 1. MYCOLOGY. *Educ:* Mt Union Col, BS, 48; Univ Iowa, MS, 54, PhD(bot), 55. *Prof Exp:* Chemist res & develop div, Babcock-Wilcox Co, 48-51; asst bot, Univ Iowa, 51-53, asst mycol, 53-55; mycologist cult collection invest, Fermentation Lab, Northern Utilization Res & Develop Div, Agr Res Serv, USDA, 55-60, prin mycologist, cur nat fungus collections & leader mycol invests, Crops Protection Res Br, Plant Sci Res Div, 60-71, liaison officer assigned to Dept of State, Int Progs Div, 71-74, asst dir int progs staff, Sci & Educ Admin, 75-78, actg dir, 79, assoc coordr, Int Orgn Affairs, Off Int Coop & Develop, 80-84; RETIRED. *Concurrent Pos:* Mem, Nat Res Coun, 61-67, hon res assoc, Smithsonian Inst, 61-68; chmn toxic micro-organisms panel, Joint US-Japan Coop Develop Natural Resources Prog, 64-69, collabr, 69-75; mem, US Nat Comt, Int Union Biol Sci, 65-71, secy, 69-71; mem spec comt fungi & lichens, Tenth & 11th Int Bot Cong, 64-72; chmn US deleg, US-Japan Joint Panel Conf Toxic Micro-Organisms, 66-68; alt US deleg, 17th Gen Assembly, Int Union Biol Sci, 70; mem, US deleg to 16th-21st sessions conf & numerous sessions coun, Food & Agr Orgn UN, 71-81, alt US deleg regional conf Latin Am, 72, conf mem, Staff Pension comt, 71-73; US coordr, Joint US-Japan Coop Develop Natural Resources Prog, 76-80. *Honors & Awards:* Plaque Award, Japanese Toxic Micro-Organisms Panel, US-Japan Coop Natural Resources Prog, 69. *Mem:* Fel AAAS; Mycol Soc Am (pres, 66-67). *Res:* Classification of the ascomycetes and mucorales; nomenclature of fungi; international agriculture and organization affairs related to agriculture. *Mailing Add:* 315 Timberwood Ave Silver Spring MD 20901

BENJAMIN, DAVID CHARLES, b Aug 19, 36. PROTEIN FOLDING, MICROBIOLOGY. *Educ:* Univ Calif, Los Angeles, BS, 65, PhD(biol), 69. *Prof Exp:* From asst prof to assoc prof, 71-80, PROF MICROBIOL, SCH MED, UNIV VA, 80- *Concurrent Pos:* Consult, Regulatory Biol Panel & Cell Biol Panel, NSF, 74-77, ad hoc reviewer, 77-, Immunol Sci Study Sect, NIH, 81-85; vis prof, Transplantation Biol Unit, Clin Res Ctr, Harrow, Eng, 77-78, Univ Sydney, Australia, 87-88; assoc provost res, Univ Va, 84-87, dir, Commonwealth Protein & Nucleic Acid Sequencing Ctr, actg dir, Beirne B

Carter Ctr Immunol Res, Provosts Acad Adv Comt. *Mem:* Sigma Xi. *Res:* Cellular immunology; protein folding; structural immunology; numerous technical publications. *Mailing Add:* Dept Microbiol Health Sci Ctr Univ Va Box 441 Charlottesville VA 22908

BENJAMIN, DAVID MARSHALL, b Boston, Mass, July 16, 46. CLINICAL PHARMACOLOGY. *Educ:* Boston Univ, AB, 68; Univ Vt, MS, 70, PhD(pharmacol), 73. *Prof Exp:* Asst dir sci affairs, Pfizer Pharmaceut, 73-74; clin res scientist Clin Pharmacol, Hoffman La Roche, 74-86; CONSULT, CLIN PHARMACOL & TOXICOL, 86- *Concurrent Pos:* Asst adj prof pharmacol, Col Med, Cornell Univ, 76- *Mem:* Am Soc Clin Pharmacol & Therapeut; Am Col Clin Pharmacol; Acad Pharmaceut Sci; Am Acad Clin Toxicol; Sigma Xi. *Res:* Clinical pharmacology and pharmacokinetics of drugs; correlations between pharmacokinetics and pharmacodynamics of drugs; early development of new drug entities; drug interactions; clinical toxicology. *Mailing Add:* Two Hammond Pond Pkwy Suite 605 Chestnut Hill MA 02167

BENJAMIN, FRED BERTHOLD, b Darmstadt, Ger, Oct 24, 12; nat US. PHYSIOLOGY. *Educ:* Bonn Univ, DMD, 35; Univ Ill, MS, 49; Loyola Univ, PhD(physiol), 53. *Prof Exp:* Pvt pract dent, Kashmir, India, 36-46; res assoc physiol, Univ Ill, 49-53; asst prof, Univ Pa, 53-60; sr res coordr, Life Sci Lab, Repub Aviation Corp, 60-64; chief eval br, Off Space Med, NASA, 64-66, staff asst, Apollo Med Support, 66-70; sr res physiologist, Res Inst, Nat Hwy Traffic Safety Admin, 70-79; RETIRED. *Concurrent Pos:* Prof lectr, Sch Med, George Washington Univ, 64-66. *Mem:* AAAS; Soc Exp Biol & Med; Am Physiol Soc; Aerospace Med Asn. *Res:* Pain sensation; reaction to heat; effects of alcohol and narcotic drugs. *Mailing Add:* 15300 Beaverbrook Ct Silver Spring MD 20906

BENJAMIN, HIRAM BERNARD, b Austria, July 4, 01; nat US; m 27. PHYSIOLOGY, ANATOMY. *Educ:* Marquette Univ, MD, 30, MSc, 49, PhD, 79. *Prof Exp:* Intern, Milwaukee County Hosp, Wis, 31, resident, 31-38; pathologist, St Anthony Hosp, 39; assoc prof, 49-74, ADJ PROF ANAT, MED COL WIS, 74-; CHIEF SURG, EVANGEL DEACONESS HOSP, 54- *Mem:* Am Col Angiol; Am Col Chest Physicians; AMA; Am Chem Soc; Asn Mil Surg US. *Res:* Gastrointestinal, blood expanders. *Mailing Add:* Med Col Wis 561 N 15th St Milwaukee WI 53233

BENJAMIN, PHILIP PALAMOOTTIL, b Eraviperur, India, Sept 5, 32; m 60; c 4. NUCLEAR CHEMISTRY, NUCLEAR MEDICINE. *Educ:* Univ Madras, BSc, 52; St John's Col, India, MSc, 55; McGill Univ, PhD(radiochem), 65. *Prof Exp:* Lectr chem, Ewing Christian Col, Univ Allahabad, 55-60, asst prof, 65-66; asst prof radiol & radiochem, Case Western Reserve Univ, 66-75; dir, Medi-Nuclear Inst, 75-89; PROF CHEM, MO BAPTIST COL, 90- *Concurrent Pos:* Res grants, Squibb Inst Med Sci, 68-69; Am Cancer Soc res grant, 69-; consult, Squibb Inst Med Res & Abbott Labs. *Mem:* AAAS; Soc Nuclear Med; fel Am Inst Chem; NY Acad Sci; Chem Inst Can. *Res:* Radiochemical investigations of nuclear fission products; distribution of nuclear charge in fission; applications of radiochemistry to nuclear medicine. *Mailing Add:* 1418 Old Farm Dr Creve Coeur MO 63141

BENJAMIN, RICHARD KEITH, b Argenta, Ill, Apr 9, 22; m 46; c 2. MYCOLOGY. *Educ:* Univ Ill, BS, 47, MS, 49, PhD(bot), 51. *Prof Exp:* Nat Res Coun fel bot, Harvard Univ, 51-52; from asst prof to assoc prof, 52-62, PROF BOT, CLAREMONT GRAD SCH, 62-; MYCOLOGIST, RANCHO SANTA ANA BOT GARDEN, 52- *Concurrent Pos:* Ed-in-chief, Mycol Soc Am, 71-75, Aliso, 58- *Mem:* Bot Soc Am; Mycol Soc Am (secy-treas, 60-62, vpres, 63, pres, 65); Brit Mycol Soc. *Res:* Laboulbeniales; Mucorales. *Mailing Add:* 1542 Bates Pl Claremont CA 91711

BENJAMIN, RICHARD WALTER, b Albany, NY, Dec 8, 35; m 61; c 1. NUCLEAR PHYSICS, NUCLEAR ENGINEERING. *Educ:* Lamar State Col, BS, 58; Southern Methodist Univ, MS, 61; Univ Tex, Austin, PhD(physics), 65. *Prof Exp:* Design engr, Los Alamos Sci Lab, 58-59; res scientist physics, Tex Nuclear Corp, 61-66; res assoc, Swiss Fed Inst Technol, 66-68; staff physicist, Savannah River Lab, E I Du Pont, 68-83; tech adv, Off Nuclear Mat Prod, US Dept Energy, 83-84; res supvr, Environ Transp, Savannah River Lab, E I du Pont de Nemours & Co, 84-88; MGR, OPER PLANNING, WESTINGHOUSE SAVANNAH RIVER CO, 88- *Concurrent Pos:* Lectr, Paine Col, 69-73. *Mem:* Am Phys Soc; Am Nuclear Soc; AAAS; Swiss Phys Soc; Europ Phys Soc. *Res:* Nuclear fuel cycle studies and planning; neutron cross section measurement, evaluation and testing, particularly of the actinide nuclides; environmental transport, aqueous and atmospheric; technical planning. *Mailing Add:* Westinghouse Savannah River Co Savannah River Site Aiken SC 29808

BENJAMIN, ROBERT FREDRIC, b Washington, DC, Jan 19, 45; m 68; c 2. EXPERIMENTAL PHYSICS. *Educ:* Cornell Univ, BS, 67; Mass Inst Technol, PhD(physics), 73. *Prof Exp:* STAFF PHYSICIST, LOS ALAMOS NAT LAB, 73- *Concurrent Pos:* Vis scientist, Maxwell Labs, 80, 86. *Res:* Laser-plasma interactions, x-ray and optical diagnostics and short pulse phenomena; high energy density phenomena; fluid instabilities. *Mailing Add:* 315 Rover Blvd Los Alamos NM 87544

BENJAMIN, ROBERT MYLES, b Bronxville, NY, Aug 13, 27; m 60; c 4. NEUROPHYSIOLOGY. *Educ:* St Lawrence Univ, BS, 49; Brown Univ, MS, 51, PhD(psychol), 53. *Prof Exp:* USPHS fel, 54-56, from instr to assoc prof physiol, 56-65, PROF NEUROPHYSIOL & PHYSIOL, SCH MED, UNIV WIS-MADISON, 65- *Concurrent Pos:* Mem neurol study sect, NIH, 67-71. *Mem:* Am Physiol Soc; Am Asn Anatomists; Soc Neurosci. *Res:* Sensory neurophysiology. *Mailing Add:* Dept Neurophysiol 283 Med Sci Bldg Madison WI 53706

BENJAMIN, ROBERT STEPHEN, b Brooklyn, NY, Apr 20, 43; m 65; c 2. MEDICAL ONCOLOGY, CLINICAL PHARMACOLOGY. *Educ:* Williams Col, AB, 64; NY Univ, MD, 68. *Prof Exp:* Sr clin assoc, Baltimore Cancer Res Ctr, Nat Cancer Inst, 72-73; from asst prof med to physician specialist, Los Angeles County-Univ Southern Calif Med Ctr, 73-74; asst internist & asst prof clin pharmacol, 74-77, assoc internist, 77-81, PROF CLIN PHARMACOL, CHIEF, SECT CLIN PHARMACOL & ASSOC PROF MED, DEPT DEVELOP THERAPEUT, UNIV TEX SYST CANCER CTR, M D ANDERSON HOSP & TUMOR INST, 77-, INTERNIST, 81- *Concurrent Pos:* Clin assoc, Nat Cancer Inst, Baltimore Cancer Res Ctr, 70-72; mem, NIH, Nat Cancer Inst Clin Trials Comt, 75-, Am Cancer Soc jr fac clin fel, 75. *Honors & Awards:* J D Lane Award, USPHS, 73. *Mem:* Am Fedn Clin Res; Am Soc Clin Oncol; Am Asn Cancer Res. *Res:* Clinical trials in medical oncology; clinical pharmacology of chemotherapeutic agents; particular interest in clinical and pharmacologic studies of anthracycline antibiotics. *Mailing Add:* Univ Tex M D Anderson Cancer Ctr 1515 Holcombe Blvd Med Clin Houston TX 77030

BENJAMIN, ROLAND JOHN, b Williamsfield, Ill, May 18, 28; m 50; c 1. APPLIED MECHANICS, OPTICS. *Educ:* Univ Ill, BS, 50, MS, 51, PhD(theoret & appl mech), 55. *Prof Exp:* Sr struct engr, Downey Div, NAm Aviation, Inc, 52-56; sr engr, Technol Ctr Div, Cook Elec Co, 56-58, staff engr, 58-62, dir eng, Aerospace Sect, 62-64, mgr, 64-67; dir optical eng, Bell & Howell Co, 67-70, dir eng, 70-74, asst vpres optics, 74-77, chief scientist, Optical Div, 77-88; mem staff, Hughes Optic Prod, Inc; RETIRED. *Res:* Aircraft and missile structural analysis; development of supersonic and hypersonic recovery systems; photographic optics, cine and still cameras and projectors; plastic optics development; diamond machining of aspheric optical surfaces; optical systems analysis. *Mailing Add:* Dept 6701 Hughes Optic Prod Inc 2000 S Wolf Rd Des Plaines IL 60018

BENJAMIN, STEPHEN ALFRED, b New York, NY, Mar 27, 39; c 4. RADIATION BIOLOGY, TOXICOLOGY. *Educ:* Brandeis Univ, AB, 60; Cornell Univ, DVM, 64, PhD(path), 68; Am Col Vet Path, dipl, 69. *Prof Exp:* Instr comp path & pathologist, Col Med, Pa State Univ, 67-68, asst prof, 68-70; exp path, Inhalation Toxicol Res Inst, Lovelace Found, 70-77; assoc prof, 77-81, PROF PATH & RADIATION BIOL, COL VET MED & BIOMED SCI, 81-, DIR, RADIOL HEALTH LAB, 77-, ASSOC DEAN, GRAD SCH, COLO STATE UNIV, 86- *Concurrent Pos:* Vis fel comp path, Dept Path, Sch Med, Johns Hopkins Univ, 66-67; clin assoc path, Dept Path, Univ NMex, Sch Med, 71-77. *Mem:* Am Col Vet Pathologists; Am Asn Pathologists; Int Acad Pathologists; Am Asn Lab Animal Sci; Radiation Res Soc. *Res:* Effects of radiation and other toxic agents on mammalian systems with emphasis on immunologic effects and carcinogenesis; effects of injurious agents on the developing prenatal and early postnatal mammal. *Mailing Add:* Dept Path Col Vet Med & Biomed Sci Colo State Univ Ft Collins CO 80523

BENJAMIN, WILLIAM B, b Brooklyn, NY, Jan 28, 34. MOLECULAR BIOLOGY, ENDOCRINOLOGY. *Educ:* Columbia Col, BA, 51; Col Physicians & Surgeons, MD, 59. *Prof Exp:* PROF, DEPT PHYSIOL & BIOPHYS, SCH MED, STATE UNIV NY, 82- *Mem:* Am Soc Biochem & Molecular Biol; Fedn Am Socs Exp Biol. *Mailing Add:* Dept Physiol & Biophys Sch Med State Univ NY Stony Brook NY 11794

BENJAMINI, ELIEZER, b Tel-Aviv, Israel, Feb 8, 29; US citizen; m 53; c 2. IMMUNOLOGY. *Educ:* Univ Calif, Berkeley, BS, 52, MS, 54, PhD(insect toxicol), 58. *Prof Exp:* Toxicologist, Nat Canners Res Asn Lab, Calif, 54-57; jr specialist, Citrus Exp Sta, Univ Calif, Riverside, 57-58; from asst res scientist to assoc res scientist, Kaiser Found Res Inst, San Francisco, 59-66, res scientist & asst dir lab med entom, 66-70; PROF IMMUNOL, SCH MED, UNIV CALIF, DAVIS, 70- *Concurrent Pos:* Prin investr, var res grants, 61-; mem, Calif Cancer Res Coord Comt, 76-; mem adv panel, NSF, 77-80. *Honors & Awards:* Fac Res Award, Sch Med, Univ Calif Davis, 77. *Mem:* Am Chem Soc; Am Asn Immunol; Sigma Xi. *Res:* Immune response to antigenic determinants and its manipulation; characterization and synthesis of antigenic determinants of proteins with particular emphasis on the relationship between structure and immunological activity; regulation of the immune response; immunochemical and immunobiologic studies of retroviruses. *Mailing Add:* Dept Med Microbiol & Immunol Univ Calif Sch Med Davis CA 95616

BENJAMINOV, BENJAMIN S, b Sofia, Bulgaria, Mar 21, 23; US citizen; m 47. ORGANIC CHEMISTRY. *Educ:* Univ Kans, BA, 52; Allegheny Col, MS, 58; Weizmann Inst, PhD(org chem), 64. *Prof Exp:* Instr chem, Univ Mass, 53-54 & Rockford Col, 54-56; assoc prof & head dept, Alliance Col, 56-59; assoc prof org chem, 59-66, PROF CHEM, ROSE-HULMAN INST TECHNOL, 66- *Concurrent Pos:* Nat Cancer Inst fel, 62-64; Fulbright grant & vis prof, Univ Strasbourg, 70-71. *Mem:* AAAS; Am Chem Soc; The Chem Soc; Sigma Xi; Israel Chem Soc. *Res:* Chemistry of natural products, especially terpenes and steroids; reaction mechanisms; stereochemical, elucidative and structural problems; coordination compounds; bicyclic and small ring compounds; biosynthetic and synthetic investigations; lipid chemistry. *Mailing Add:* Dept Chem Rose-Hulman Inst of Technol 5500 Wabash Ave Terre Haute IN 47803

BENJAMINS, JOYCE ANN, b Bay City, Mich, June 1, 41; m 65; c 2. NEUROCHEMISTRY. *Educ:* Albion Col, BA, 63; Univ Mich, PhD(biochem), 67. *Prof Exp:* Fel pediat & genetics, Sch Med, Stanford Univ, 67-68; res assoc neurol, Sch Med, Johns Hopkins Univ, 68-69, from instr to asst prof, 69-73; asst prof biochem & res scientist, Biol Sci Res Ctr, Sch Med, Univ NC, Chapel Hill, 73-75; from asst prof to assoc prof, 75-85, PROF NEUROL, SCH MED, WAYNE STATE UNIV, 85- *Concurrent Pos:* Mem, Neurol B Study Sect, Nat Inst Neurol & Commun Disorders & Stroke, 78-82; mem, Sci Review Comt, Amyotrophic Lateral Sclerosis Soc Am, 81-84; mem Adv Comt Res, Nat Multiple Sclerosis Soc, 85-88; Javitz neuroscience res award, 87-94. *Mem:* Soc Neurosci; Am Soc Neurochem. *Res:* Developmental neurochemistry; lipid and protein synthesis assembly of membranes in developing nervous system especially myelination; glial cultures and cytoskeleton; glycolipid antibodies. *Mailing Add:* 1310 S Oxford Grosse Pointe Woods MI 48236

BENJAMINSON, MORRIS AARON, b Bronx, NY, Aug 6, 30; div; c 2. MICROBIOLOGY, DEVELOPMENTAL GENETICS. *Educ:* Long Island Univ, BS, 51; NY Univ, MS, 61, PhD(biol), 67. *Prof Exp:* Med technician, First Med Field Lab, US Army, 52-54; sr biol technician, Sloan-Kettering Inst Cancer Res, 54-55; med technician microbiol, Vet Admin Hosp, New York, 56-59; res asst, Margaret M Caspary Inst Vet Res, 59-61; res assoc, Bronx-Lebanon Hosp Ctr, 61-64; sr task leader, Naval Appl Sci Lab, 64-69; asst prof, Dent Sch, NY Univ, 69-74; assoc prof allied health & coordr med technol, York Col, NY, 74-75; assoc prof, microbiol, NY Col Osteop Med, 84-90; DIR N STAR RES INC, 75- *Concurrent Pos:* Grant, Am Cancer Soc, NY Univ, 65, Navy Independent Explor Res, 66, Lenk & Fink Prod Group of Sterling Drug Co, 80, Sci & Technol Found, State NY, 88; US Navy contract, NY Univ, 69-71, NSF grant, 74-75; adj asst prof, City Col New York, 69-75; consult, Dept Air Resources, City of New York, 71-; mem, Environ Rev Comt, Continuing Educ Units, Am Soc Microbiol, 77; USDA coop agreement, 78-; vis res scientist, Dental Res Inst, NY Univ, 78-; treas/exec vpres & dir res & develop, BioDor Chem Prods Ltd, Bridgeport, Conn, 78-80; assoc prof pub health, Arnold & Marie Schwartz Col Pharm Health Sci, Long Island Univ, 79-83; expert witness, Microbiol Drift Panel, Pub Serv Comn, State NY, 81; consult, biogenetic toxicity testing, Electro-Optics Devices Corp; reviewer, J Histochem & Cytochem, 90. *Mem:* AAAS; Am Soc Microbiol; NY Acad Sci; Am Asn Textile Chemists & Colorists; Sigma Xi; Histochem Soc; Am Inst Biol Sci; Int Asn Comp Res Leukemia & Related Dis. *Res:* Clinical and industrial microbiology; cytochemistry; automation; aerobiology; public health; developmental genetics; gravitational biology. *Mailing Add:* Box 1212 Church St Sta New York NY 10007

BEN-JONATHAN, NIRA, b Holon, Israel, Nov 23, 40. NEUROENDOCRINOLOGY, REPRODUCTIVE PHYSIOLOGY. *Educ:* Univ Tel-Aviv, BSc, 67; Univ Ill, Urbana, MSc, 69, PhD(physiol), 72. *Prof Exp:* Teacher chem, Alpha High Sch, Tel-Aviv, 66-67; from teaching asst to res asst physiol, Univ Ill, Urbana, 68-71; fel endocrinol, Univ Tex Southwestern Med Sch, 72-74, asst prof physiol, 74-75; from asst prof to assoc prof, 76-83, PRO PHYSIOL, SCH MED, IND UNIV, 83- *Concurrent Pos:* Fel, Pop Coun, 72-73; res career develop award, NIH, 78-82, Endocrinol Study Sect, 84-87, chmn, 87-88. *Mem:* Am Endocrine Soc; Am Physiol Soc; Soc Neurosci; Int Soc Neuroendocrinol; Soc Study Reproduction. *Res:* Neuronal-endocrine interrelations; hypothalamic catecholamines and releasing hormones and the control of tropic hormones secretion by the anterior pituitary gland; catecholamines and ovarian function; catecholamines during fetal development. *Mailing Add:* Dept Physiol & Biophysics Ind Univ Sch Med 635 Barnhill Dr Indianapolis IN 46223

BENKENDORF, CAROL ANN, b Indianapolis, Ind, Feb 8, 40; m 60; c 9. INDUSTRIAL TOXICOLOGY, CHEMICAL CARCINOGENESIS. *Educ:* St Mary's Col, BS, 61; Univ Mich, MS, 75, PhD(toxicol), 78; Am Bd Toxicol, dipl, 81. *Prof Exp:* Lab asst, Northern Regional Lab, USDA, 61-62; consult, Clement Assocs, 77-78; asst toxicologist, Ford Motor Co, 78-82; PRES, TOXICOL DATA SERV INC, 83-, TOXDATA SYSTS INC, 86- *Concurrent Pos:* Mem, Am Conf Chem Labeling; Secy-treas, Hazard Commun Resources, Inc, 86- *Mem:* Am Chem Soc; Am Indust Hyg Asn; Genetic Toxicol Asn; AAAS; Nat Safety Coun. *Res:* Consulting. *Mailing Add:* 380 Sutton Rd Barrington IL 60010

BENKESER, ROBERT ANTHONY, b Cincinnati, Ohio, Feb 16, 20; m 46; c 5. ORGANIC CHEMISTRY. *Educ:* Xavier Univ, Ohio, BS, 42; Univ Detroit, MS, 44; Iowa State Col, PhD(org chem), 47. *Prof Exp:* From asst prof to assoc prof, 46-54, head dept, 74-78, PROF CHEM, PURDUE UNIV, WEST LAFAYETTE, 54- *Honors & Awards:* F S Kipping Award, Am Chem Soc, 69. *Mem:* Am Chem Soc. *Res:* Organometallics and synthetic organic chemistry; reactions of organosilicon compounds; reductions in amine solvents; metallations and reactions of benzylic Grignard systems. *Mailing Add:* Dept Chem Purdue Univ West Lafayette IN 47907

BENKOVIC, STEPHEN J, b Orange, NJ, Apr 20, 38; m. ORGANIC CHEMISTRY, PHYSICAL CHEMISTRY. *Educ:* Lehigh Univ, AB & BS, 60; Cornell Univ, PhD(org chem), 63. *Prof Exp:* Res assoc, Univ Calif, Santa Barbara, 64-65; from asst prof to assoc prof, 65-70, PROF CHEM, PA STATE UNIV, 70-, EVAN PUGH PROF, 77- *Concurrent Pos:* Aldred P Sloan fel, Pa State Univ, 68-70, NIH career develop award, 69-74, Guggenheim fel, 76; mem NIH Med Chem Panel, 72-76; co-chmn, Third Biannual Conf Biochem Mech, Los Angeles, 72, NATO Conf Catalysis, Sardinia, 72 & Gordon Conf, 73 & 76; mem med chem, NIH Physiol Chem Study Sect, 78-; mem, NJ Comm Molecular Biol, 84, externa adv group, Geisinger Hosp, 84, Sci Adv Bd Pa State Biotechnol Inst, 85 & Searle Scholars Prog, Chicago Community Trust, 85; univ chair biol sci, Pa State Univ, 84; Eberly chmn chem, Pa State Univ, 87- *Honors & Awards:* Pfizer Enzyme Award, Am Chem Soc, 77. *Mem:* Nat Acad Sci; Am Chem Soc; Fedn Am Biologists; Sigma Xi; Royal Soc Chem; Am Acad Arts & Sci. *Res:* Mechanisms of enzymatic reactions. *Mailing Add:* Dept Chem Pa State Univ University Park PA 16802

BENMAMAN, JOSEPH DAVID, b Tetuan, Morocco, Dec 12, 24; US citizen; m 60; c 2. PHARMACOKINETICS. *Educ:* Univ Madrid, Lic pharm, 54, PhD(pharmaceut chem), 60. *Prof Exp:* Asst prof phys chem, Univ Madrid, 54-60; prof phys chem & chmn dept, Univ Oriente, Venezuela, 60-63; res chemist, Sch Pharm, Univ Calif, San Francisco, 63-65, res chemist, Sch Med, 65-68; from asst prof to assoc prof phys chem, 68-74, PROF PHARMACOKINETICS, COL PHARM, MED UNIV SC, 74- *Mem:* Am Chem Soc; Am Pharmaceut Asn; Acad Pharmaceut Sci; Pan-Am Fedn Pharm & Biochem (secy, Sci Div, 72-75). *Res:* Methods of separation and analysis of biological materials; kinetics and mechanisms of reactions; drug stability; kinetics of absorption, distribution, metabolism and excretion of drugs; clinical pharmacokinetics; drug design. *Mailing Add:* Col Pharm Med Univ SC Charleston SC 29403

BENNER, BLAIR RICHARD, b Braddock, Pa, Mar 29, 47; m 74; c 3. CHEMICAL METALLURGY, MINERAL BENEFICIATION. *Educ:* Pa State Univ, BS, 69; Stanford Univ, MS, 71. *Prof Exp:* Jr metallurgist, NMex State Bur Mines & Mineral Resources, 71-73; res metallurgist extractive metall, Deepsea Ventures Inc, Tenneco Co, 73-76; res engr, US Steel Res Lab, 76-80, sr res engr, 80-86; SCIENTIST, NATURAL RESOURCES RES INST, UNIV MINN, DULUTH, 86- *Mem:* Am Inst Mining, Metall & Petrol Engr. *Res:* Extraction of uranium from phosphoric acid; hydrometallurgical processing of deep ocean nodules; restoration of in-situ uranium mines; gravity concentration of iron ore; by product recovery from in-situ uranium leaching; computer modeling of grinding circuits; iron ore flotation; process control. *Mailing Add:* Coleraine Res Lab Coleraine MN 55722

BENNER, GERELD STOKES, b Waukeegan, Ill, July 27, 33; m 76; c 1. CATALYSIS, INORGANIC CHEMISTRY. *Educ:* Univ NC, Chapel Hill, BS, 61; Ohio State Univ, MS, 63, PhD(inorg chem), 66. *Prof Exp:* Sr res chemist, Goodyear Tire & Rubber Co, 66-74; res chemist catalysis, MW Kellogg, Houston, 74-87; CONSULT, 87- *Mem:* Am Chem Soc; Sigma Xi; Catalysis Soc Am; NY Acad Sci. *Res:* Development and improvement of catalysis and catalytic processes in the fields of petroleum refineries, petrochemicals and ammonia synthesis. *Mailing Add:* 22707 Goldstone Katy TX 77450-1617

BENNER, ROBERT E, b Lock Haven, Pa. OPTICS, MATERIALS SCIENCE. *Educ:* Lehigh Univ, BS, 71; Univ Rochester, PhD(mat sci), 78. *Prof Exp:* Vis instr physics, Univ Toledo, 77; mem res staff, Yale Univ, 77-78, asst prof appl sci, 78-80; mem staff, Appl Physics Div, Sandia Labs, Livermore, Calif, 80-; AT DEPT ELEC ENG, UNIV UTAH. *Mem:* Am Phys Soc. *Res:* Brillouin and Raman scattering; nonlinear optics; Raman monitoring of aerosols; evanescent wave spectroscopy. *Mailing Add:* Dept Elec Eng Univ Utah 3280 Merrill Eng Salt Lake City UT 84112

BENNER, RUSSELL EDWARD, b Quakertown, Pa, Dec 19, 25; m 54; c 2. MECHANICAL ENGINEERING. *Educ:* Cornell Univ, BME, 47; Lehigh Univ, MSME, 51, PhD(mech eng), 59. *Prof Exp:* Designer, Yale & Towne Mfg Co, Pa, 47-49; instr mech eng, Lehigh Univ, 49-59; res engr, E I du Pont de Nemours & Co, 59-62; PROF MECH ENG, LEHIGH UNIV, 62- *Concurrent Pos:* Consult, Fuller Co Div, Gen Am Transp Corp, 62-67. *Mem:* Am Soc Mech Engrs; Soc Exp Stress Anal; Am Soc Eng Educ. *Res:* Machine design; stress analysis; system dynamics; structural and systems reliability. *Mailing Add:* Dept Mech Eng Lehigh Univ Packard Lab 19 Bethlehem PA 18015

BENNET, ARCHIE WAYNE, b Rocky Mount, Va, May 5, 37; m 58; c 2. EDUCATIONAL ADMINISTRATION. *Educ:* Va Polytech Inst & State Univ, BS, 60, MS, 63; Univ Fla, PhD(elec eng), 66. *Prof Exp:* Systs engr, Gen Elec, 60-61; from instr to assoc prof elec eng, Va Tech, 61-74, chmn, Computer Eng Group, 72-81, prof elec eng & computer sci, 74-81; head elec & computer eng, 81-88, ASSOC DEAN & SPEC ASST TO PRES, CLEMSON UNIV, 88- *Concurrent Pos:* Bd dirs, Am Soc Eng Educ, 76-78; eng consult, Litton Industs, 68-81, Babcock & Wilcox, 77; computer systs consult, IBM, 85; educ consult, State of Fla, 88; mgt consult, Pryor Resources, 89-; vchmn & chmn, Nat Asn Elec Eng Dept Heads. *Honors & Awards:* Outstanding Serv Award, Am Soc Eng Educ, 75. *Mem:* Am Soc Eng Educ; Inst Elec & Electronics Engrs; Int Asn Math & Computers in Simulation (vpres, 79-85); Nat Asn Elec Eng Dept Heads; Asn Comput Mach. *Res:* Continuous and discrete control theory; telecommunication and systems networking; system simulation; optimal and adaptive control. *Mailing Add:* 608 Ridgecrest Dr Clemson SC 29631

BENNET, GEORGE KEMBLE, JR, b Jacksonville, Fla, Apr 2, 40; m 66; c 2. OPERATIONS RESEARCH, INDUSTRIAL ENGINEERING. *Educ:* Fla State Univ, BS, 62; San Jose State Col, MS, 68; Tex Tech Univ, PhD(indust eng), 70. *Prof Exp:* Assoc engr, Martin Marietta Co, 62-63; math analyst, Lockheed Missiles & Space Co, 63-66; asst dir comput ctr, Tex Tech Univ, 66-69; asst prof indust eng & opers res, Va Polytech Inst & State Univ, 70-74; PROF & HEAD INDUST SYST ENG, UNIV SFLA, 74- *Concurrent Pos:* Ed, Logistics Sprectrum, 84- *Mem:* Am Inst Indust Engrs; Soc Logistics Engrs; Am Soc Eng Educ. *Res:* Development of smooth empirical Bayes estimation techniques; design and development of consumer commodity regulatory systems; design of cost based quality control systems. *Mailing Add:* 2616 E 19th Ave Tampa FL 33605

BENNETT, A(RTHUR) D(AVID), b Brooklyn, NY, Oct 31, 09; m 33; c 1. MECHANICAL ENGINEERING. *Educ:* Stevens Inst Technol, ME, 31. *Prof Exp:* Rock drill designer, Ingersoll-Rand Co, 35-42; dep chief Appl Electronics Br, Eng Res & Develop Labs, Ft Belvoir, 46-48; sr res engr, geophys res, Pan Am Petrol Corp, 48-52; res group supvr, Amoco Prod Co, 52-72; RETIRED. *Res:* Methods for determining points of fluid entry in oil wells; method for logging densities of subsurface geological formations; special instrumentation problems; seismic model studies; digital processing of seismic data; experimental seismic field investigations. *Mailing Add:* 4107 E 46th Pl Tulsa OK 74135

BENNETT, ALAN JEROME, b Philadelphia, Pa, June 13, 41; m 63; c 3. SEMICONDUCTOR PHYSICS, DEVICE PHYSICS. *Educ:* Univ Pa, BA, 62; Univ Chicago, MS, 63, PhD(physics), 65. *Prof Exp:* Physicist, Phys Sci Br, Res & Develop Ctr, Gen Elec, 66-74, mgr, Solid State Common Br, 75-77, mgr rel ultrasonic imager prog, 77-78, mgr planning & resources, electronics sci & eng, 78-79; dir, Gould Electronics Lab, 79-84; VPRES RES, VARIAN ASSOCS, 84- *Concurrent Pos:* Fel, NSF, Cambridge Univ, 66; vis assoc prof, Cornell Univ, 70, Technion,72. *Mem:* Am Phys Soc; Inst Elec & Electronics Engrs; Indust Res Inst; AAAS; Am Mgt Asn. *Res:* Solid state communications; ultrasonic imaging; semiconductor fabrication equipment; computer systems; laboratory instrumentation; medical therapeutics. *Mailing Add:* Varian Assocs 611 Hansen Way Palo Alto CA 94303

BENNETT, ALBERT FARRELL, b Whittier, Calif, July 18, 44; m 77; c 4. COMPARATIVE PHYSIOLOGY, PHYSIOLOGICAL ECOLOGY. *Educ:* Univ Calif, Riverside, AB, 66; Univ Mich, PhD(zool), 71. *Prof Exp:* Lectr zool, Univ Mich, 70-71; Miller fel, Univ Calif, Berkeley, 71-73, actg asst prof, 73-74; from asst prof to assoc prof, 74-83, actg dean, 86-87, PROF BIOL SCI, UNIV CALIF, IRVINE, 83-, CHAIR, DEVELOP & CELL BIOL, 84- *Concurrent Pos:* Prin investr, NSF grants, 75-; NIH Career Develop Award, 78-83; vis res assoc, Univ Chicago, 81-82; vis prof fel, Univ Adelaide, SAustralia, 83-84. *Mem:* Am Physiol Soc; Am Soc Naturalists; Am Soc Zoologists; Ecol Soc Am; Soc Exp Biol. *Res:* Activity and exercise physiology of vertebrates; comparative respiratory and muscle physiology; physiological ecology; ecological energetics. *Mailing Add:* Sch Biol Sci Univ Calif Irvine CA 92717

BENNETT, ALBERT GEORGE, JR, b Booneville, Miss, Aug 8, 37; m 60; c 2. COMPOSITE AIRCRAFT DEVELOPMENT, AUTOMATED COMPOSITE MANUFACTURING. *Educ:* Miss State Univ, BS, 59; Univ Ill, MS, 64, PhD(aeronaut & astronaut eng), 70. *Prof Exp:* Aerodynamicist, Douglas Aircraft Co, Santa Monica, Calif, 59-62; asst prof aeronaut eng, Miss State Univ, Miss, 63-66; instr aeronaut & astronaut eng, Univ Ill, 66-76; assoc prof, 69-76, PROF AERONAUT ENG, MISS STATE UNIV, MISS, 76-, DIR, RASPET FLIGHT RES LAB, 79- *Concurrent Pos:* Fac fel, NASA-Am Soc Eng Educ, summer fac fel res, Langley, 70-75; mem, Commun & Instrumentation Support Serv, Miss State Univ, 81-, Strategic Defens Initiative Orgn Tech Appl Adv Comt, Dept Defense, 86-89; consult, Lockheed-Ga, co-chmn, Safety Rev Bd, 84-87, Aero Corp, 85, Aerovironment Inc, 85; chmn, Power Generation Subcomt, 86-89. *Mem:* Sigma Xi; Soc Aeronaut Eng; Am Inst Aeronaut & Astronaut. *Res:* Prototype composite aircraft development; parameter identification of aircraft performance; automation of composite fabrication; ultrasonic imaging of composite. *Mailing Add:* Dept Aerospace Eng Raspet Flight Res Lab Drawer A Mississippi State MS 39762

BENNETT, ARCHIE WAYNE, b Rocky Mount, Va, May 5, 37; m 58; c 2. ELECTRICAL ENGINEERING, COMPUTER SCIENCE. *Educ:* Va Polytech Inst, BSEE, 60, MSEE, 63; Univ Fla, PhD(elec eng), 66. *Prof Exp:* Systs engr, Gen Elec Co, 60-61; from instr to assoc prof elec eng, Va Polytech Inst & State Univ, 61-72, prof elec eng & comput sci, 72-81; PROF ELEC ENG & COMPUT ENG & HEAD DEPT, CLEMSON UNIV, 81- *Mem:* Inst Elec & Electronics Engrs; Am Soc Eng Educ; Asn Comput Mach; Int Asn Math & Comput Simulation. *Res:* Digital computer design; computer graphics; digital simulation. *Mailing Add:* Elec & Comput Eng Dept Clemson Univ 102 Riggs Hall Clemson SC 29631

BENNETT, BASIL TAYLOR, b Durham, NC, July 20, 44; m 64; c 2. LABORATORY ANIMAL MEDICINE, EXPERIMENTAL PATHOLOGY. *Educ:* Auburn Univ, DVM, 69; Univ Ill, PhD(path), 74. *Prof Exp:* Resident investr, Hines Vet Admin Hosp, Ill, 69-74; dir prof serv, 74-76, asst admin, 76-78, DIR, BIOL RESOURCES LAB, UNIV ILL, 78- *Concurrent Pos:* Chief, Res Lab Animal Med Sci & Technol, 72-79; bd dir, Am Col Lab Animal Md, 87-89, Am Soc Lab Animal Practrs, 89-91. *Honors & Awards:* Flynn Award, 79. *Mem:* Am Vet Med Asn; Am Soc Lab Animal Practr; Am Col Lab Animal Med; Am Asn Lab Animal Sci (pres, 85); Am Soc Primatology; Asn Primate Vet. *Res:* Laboratory animal medicine and technology as they apply to improving the experimental models that are available. *Mailing Add:* Biol Resources Lab 1840 W Taylor St Chicago IL 60612

BENNETT, BURTON GEORGE, environmental sciences, for more information see previous edition

BENNETT, BYRON J(IRDEN), b Loraine, Tex, Oct 4, 20; m 47; c 7. ELECTRICAL ENGINEERING. *Educ:* Tex Technol Col, BS, 43; Stanford Univ, MS, 49, PhD(elec eng), 52. *Prof Exp:* Test engr, Gen Elec Co, 40-41; radio engr, Philco Co, 42-43; instr elec eng, Tex Technol Col, 43-44, assoc prof, 46-52; mgr comput lab, Stanford Res Inst, 52-58; mgr advan technol, Int Bus Mach Corp, 58-60; prof elec eng & head dept, 60-63, DEAN COL ENG & DIR ENG EXP STA, MONT STATE UNIV, 63- *Concurrent Pos:* Instr, Temple Univ, 42-43; lectr, Stanford Univ, 56-60; asst prof, San Jose State Col, 59-60; consult, Electronics Res Lab, Endowment & Res Found, 60-; Westinghouse fel; tech prog dir, Navy Res & Develop Clin, 63-64; mem bd dirs, Montronics, Inc, Fluke Mfg Co, Inc, 63-; mem bd dirs, Develop Technol, Inc, 70- *Mem:* Sigma Xi; Inst Elec & Electronics Engrs. *Res:* Transistors; magnetics; electroluminescence; photoconductors. *Mailing Add:* Elec Eng Mont State Univ Bozeman MT 59715

BENNETT, C FRANK, b Farmington, NMex, Nov 1, 56. PHARMACOLOGY. *Educ:* Univ NMex, BS, 80; Baylor Col Med, PhD(pharmacol), 85. *Prof Exp:* From postdoctoral scientist to sr scientist, Smith Kline & French Labs, Philadelphia, Pa, 85-89; sr scientist, GROUP LEADER, ISIS PHARMACEUT INC, 89- *Mem:* AAAS; Am Soc Cell Biol. *Mailing Add:* Dept Biol Sci Isis Pharmaceut Inc 2280 Faraday Ave Carlsbad CA 92008

BENNETT, C LEONARD, b Lowell, Mass, Oct 5, 39; m 66; c 2. SATELLITE COMMUNICATIONS, ANTENNAS & PROPAGATION. *Educ:* Lowell Technol Inst, BS, 61; NC State Univ, MS, 64; Purdue Univ, PhD(elec eng), 68. *Prof Exp:* Mem tech staff, Sperry Res Ctr, 68-73, mgr syst appln, Radar Target Class, 73-83; prin engr commun, 83-86, CONSULT ENGR COMMUN SYST, RAYTHEON CO, 86- *Concurrent Pos:* Lectr, Advan Study Inst, N Atlantic Treaty Orgn, 79; Short Courses Comput Methods Electromagnetics, 80-84; adj prof, Univ Man, 76-77 & Northeastern Univ, 82-83; mem Union Radio Sci Int Comt. *Mem:* Fel Inst Elec & Electronics Engrs; Inst Elec & Electronics Engrs Antenna Propagation Soc. *Res:* Development and applications of computational methods in electromagnetics, transient electromagnetic and radar target classification; design and development of satellite communication systems. *Mailing Add:* 304 Reedy Meadow Rd Groton MA 01450

BENNETT, CARL ALLEN, b Winfield, Pa, Nov 22, 21; m 44; c 2. STATISTICS. *Educ:* Bucknell Univ, AB, 40, MA, 41; Univ Mich, AM, 42, PhD(math), 52. *Hon Degrees:* DSc, Bucknell Univ, 72. *Prof Exp:* Jr chemist, Chem & Metall Lab, Univ Chicago, 44; sr chemist & qual control supvr, Tenn Eastman Corp, 44-46; from chief statist & head appl math to mgr appl math dept, Hanford Labs, Gen Elec Co, 47-65; mgr, Appl Math Dept, 65-68, mgr, Systs & Electronics Div, 68-70, sr staff scientist, 70-71, STAFF SCIENTIST, HUMAN AFFAIRS RES CENTERS, BATTELLE MEM INST, 71- *Concurrent Pos:* Res assoc math, Princeton Univ, 50; vis prof statist, Stanford Univ, 64. *Mem:* Fel AAAS; fel Am Statist Asn; fel Am Soc Qual Control; fel Inst Nuclear Mat Mgt; Biomet Soc. *Res:* Application of statistical and mathematical techniques to industrial problems; order statistics; nonparametric methods; variance component analysis. *Mailing Add:* 11121 SE 59th St Bellevue WA 98006

BENNETT, CARL LEROY, b Stambaugh, Mich, May 22, 35; m 56; c 4. WILDLIFE RESEARCH, BIOMETRY. *Educ:* Mich State Univ, BS, 60, MS, 65. *Prof Exp:* Res biologist waterfowl, Erie Fishing & Shooting Club, 61; biometrician, Wildlife Div & Res & Develop, 61-70, supvr forest wildlife res, 70-81, SECT HEAD WILDLIFE RES, WILDLIFE DIV, MICH DEPT NATURAL RESOURCES, LANSING, 81- *Concurrent Pos:* Instr wildlife biomet, Mich State Univ, 77 & 81. *Mem:* Wildlife Soc; Sigma Xi. *Res:* Interactive dynamics of animals and their habitats, especially white-tailed deer in its northern environment. *Mailing Add:* 6467 Shoeman Rd Haslett MI 48840

BENNETT, CARROLL G, b Richmond, Va, Dec 8, 33; m 56; c 2. DENTISTRY. *Educ:* Randolph-Macon Col, BS, 55; Med Col Va, DDS, 59, MS, 62; Am Bd Pedodont, dipl, 67. *Prof Exp:* Instr dent anat & pedodontics, Sch Dent, WVa Univ, 62-63, from asst prof to prof pedodontics & chmn dept, 63-73, asst dean student affairs, 72-73; prof pedodontics & chmn div, Univ Fla, 73-74, prof pediat dent & chmn dept, 74-84, dir admissions & student financial aid, 85; CONSULT, 84- *Concurrent Pos:* Chmn, Am Bd Pedodontics, 78. *Mem:* Am Dent Asn; Am Soc Dent for Children; Int Asn Dent Res; Am Acad Pediat Dent. *Res:* Pedodontics; growth, development and clinical research in children's dentistry. *Mailing Add:* Col Dent Univ Fla Box J445 Gainesville FL 32610

BENNETT, CARROLL O(SBORN), b New Britain, Conn, Apr 1, 21; m 49; c 3. CHEMICAL ENGINEERING. *Educ:* Worcester Polytech Inst, BS, 43; Yale Univ, DEng(chem eng), 50. *Prof Exp:* From asst prof to prof chem eng, Purdue Univ, 49-59; mgr process res & develop, Lummus Co, 59-64; PROF CHEM ENG, UNIV CONN, STORRS, 64- *Concurrent Pos:* Fulbright lectr, Nancy Univ, 52-53; Nat Ctr Sci Res, Ministry Educ, France, 57-58; AID prof, Univ Santa Maria, Chile, 64; vis lectr, Yale Univ, 66-70; vis prof, Univ Lyon, France, 77-78, Univ Paris, 84- *Honors & Awards:* Warren K Lewis Award, Am Inst Chem Engrs. *Mem:* Am Chem Soc; Am Inst Chem Engrs. *Res:* High pressure; thermodynamics; heat and mass transfer; heterogeneous catalysis. *Mailing Add:* Sch Eng U 139 Univ Conn Storrs CT 06269-3139

BENNETT, CECIL JACKSON, b Eau Claire, Wis, Oct 4, 27; m 51; c 2. ANIMAL GENETICS, BEHAVIORAL GENETICS. *Educ:* Univ Wis, BS, 49, PhD(zool & genetics), 59; Wash Univ, MA, 53. *Prof Exp:* Lab maintenance man zool, Univ Okla, 53-55; res asst genetics, Univ Wis, 55-57; from asst prof to prof, 57-90, EMER PROF BIOL, NORTHERN ILL UNIV, 90- *Mem:* AAAS; Soc Study Evolution; Genetics Soc Am; Am Genetics Asn; Am Inst Biol Sci. *Res:* Population genetics; gene action; behavior in Drosophila. *Mailing Add:* Dept Biol Northern Ill Univ De Kalb IL 60115-2861

BENNETT, CHARLES FRANKLIN, b Oakland, Calif, Apr 10, 26; m 47; c 1. BIOGEOGRAPHY, ECOLOGY. *Educ:* Univ Calif, Los Angeles, BA, 55, PhD(geog), 60. *Prof Exp:* From instr to assoc prof geog, 59-69, chmn dept, 83-87, PROF BIOGEOG, UNIV CALIF, LOS ANGELES, 69- *Concurrent Pos:* Consult, Tex Instruments Corp, 63-64 & AEC, 65-67; Off Naval Res fel res in Panama, Univ Calif, Los Angeles, 63; res assoc, Smithsonian Trop Res Inst, 66-; directorate 1 trop forests, Man & the Biosphere, Dept State-UNESCO, 78-; Guggenheim fel, 70-71. *Mem:* Fel AAAS; Ecol Soc Am; Brit Ecol Soc; Am Geog Soc; Asn Trop Biol; Soc Conserv Biol. *Res:* Ecology of human modified ecosystems; ecology of the humid tropics; ecology of agricultural systems; conservation of resources in underdeveloped countries. *Mailing Add:* Dept Geog Univ Calif Los Angeles CA 90024

BENNETT, CHARLES L, b New Brunswick, NJ, Nov 16, 56; m 84; c 1. RADIO ASTRONOMY, COSMOLOGY. *Educ:* Univ Md, BS, 78; Mass Inst Technol, PhD(physics), 84. *Prof Exp:* Res asst radio astron, res lab electronics, Mass Inst Technol, 78-84; ASTROPHYSICIST, GODDARD SPACE FLIGHT CTR, NASA, 84- *Mem:* Am Astron Soc; Am Inst Physics; Int Astron Union; Am Phys Soc; Sigma Xi. *Res:* Radio astronomy; submillimeter wave astronomy; cosmology; instrumentation for astronomy. *Mailing Add:* NASA/Goddard Space Flight Ctr Code 685 Greenbelt MD 20771

BENNETT, CHARLES LOUGHEED, b Duluth Minn, Dec 27, 49. NUCLEAR PHYSICS. *Educ:* Univ Minn, BA, 72; Univ Rochester, MA, 75, PhD(physics), 77. *Prof Exp:* Res asst to res assoc physics, Univ Rochester, 72-78; asst prof physics, Princeton Univ, 78-84; PHYSICIST, LAWRENCE LIVERMORE NAT LABS, 84- *Mem:* AAAS; Am Phys Soc. *Res:* Nuclear reactions and nuclear structure, especially multi-nucleon transfer spectroscopy; radiocarbon and other radioisotope dating; solar neutrino physics. *Mailing Add:* Lawrence Livermore Nat Labs L-45 Livermore CA 94551

BENNETT, CLARENCE EDWIN, physics; deceased, see previous edition for last biography

BENNETT, CLEAVES M, b Champaign, Ill, July 9, 34; m 85; c 3. PREVENTIVE MEDICINE, HYPERTENSION. *Educ:* Carleton Col, BA, 55;Univ Rochester, MD, 60; Am Bd Internal Med, dipl, 70;dipl, nephrology, 74. *Prof Exp:* Fel, dept biochem, Univ Rochester, 57-58, 79-; intern, dept med, UCLA Med Ctr, 60-61, jr resident, 61-62; sr resident, dept med, Albert Einstein Col Med, 62-63, chief resident, 63-64; fel, Lab Kidney & Electrolyte Metab, Nat Heart Inst, NIH, 64-65, investr, USPHS, 65-67; clin investr, Vet Admin Hosp, Durham, NC, 67-69; assoc chief, div nephrology, Harbor-UCLA Med Ctr, 70-76, assoc clin prof med, 77-83, CLIN PROF MED, 83-, DIR, RENAL HYPERTENSION CLIN, 78- *Concurrent Pos:* Dir, Renal Hypertension Clin,Harbor-UCLA Med Ctr, 70-76,from asst prof to assoc prof med, 69-76; med dir, Biomed Applns Dialysis Ctr, 72-75; chief, div nephrology, King Faisal Specialist Hosp, 76-77; dir,Hypertension Clin, Vet Admin Wadsworth Hosp Ctr, 77-78; med dir, Pritkin Longevity Ctr, 78-80; med dir, Innerhealth Progs, 80- *Mem:* Am Physiol Soc; Am Soc Nephrology; Int Soc Nephrology; Am Fedn Clin Res; fel Am Col Physicians. *Res:* Kidney physiology; hypertension; renal disease and dialysis; nutrition and exercise physiology. *Mailing Add:* Inner Health 1000 W Ninth St Los Angeles CA 90015

BENNETT, CLIFTON FRANCIS, b Tillamook, Ore, July 27, 25; m 56; c 3. WOOD CHEMISTRY, SPECIALTY PRODUCTS CHEMISTRY. *Educ:* Lewis & Clark Col, 49; Ore State Univ, MS, 52; McGill Univ, PhD(wood chem), 56. *Prof Exp:* Res chemist, Weyerhaeuser Co, 51-53; res chemist, Crown Zellerbach Corp, 55-81; prin chemist, Container Corp Am, 81-84; PRIN CHEMIST, PACE NAT CORP, 85- *Concurrent Pos:* Nat dir, US Jr Chamber of Com, 58-59; instr, Portland Continuation Ctr, Ore State Syst Higher Educ, 63-65; adv, Chem Explorer Post 404, 67-77; Nat Acad Sci exchange scientist, Czech & Romania, 72-73 & 79, Poland, 84 & Romania, 91. *Mem:* Am Chem Soc; Tech Asn Pulp & Paper Indust. *Res:* Cellulose reactions; substitutions and fractionations; organic sulfur compounds; lignin; organic chemical synthesis; pulp and paper chemistry; specialty chemicals. *Mailing Add:* 9202 Northeast 126th Pl Kirkland WA 98034-2773

BENNETT, COLIN, b Newcastle upon Tyne, Eng, Jan 16, 46; m 67; c 2. MATHEMATICS. *Educ:* Univ Newcastle, BSc, 67, PhD(math), 71. *Prof Exp:* Harry Bateman res instr math, Calif Inst Technol, 71-73, asst prof, 73-76; assoc prof math, McMaster Univ, 76-79; assoc prof, 79-82, PROF MATH, UNIV SC, 82-, CHMN, DEPT MATH, 85- *Concurrent Pos:* NSF res grants, 74-76 & 80-82; Nat Sci & Eng Res Coun Can res grant, 77-80. *Mem:* Am Math Soc; Math Asn Am; Can Math Soc. *Res:* Harmonic analysis; approximation theory; interpolation of operators. *Mailing Add:* Dept Math Univ SC Columbia SC 29208

BENNETT, DAVID ARTHUR, b Cleveland, Ohio, Dec 9, 42; c 2. BIOINORGANIC CHEMISTRY. *Educ:* Muskingum Col, BS, 64; Northwestern Univ, MAT, 65; Cornell Univ, PhD(chem), 73. *Prof Exp:* Instr sci, Am Sch, Switz, 65-67; res assoc chem, Purdue Univ, 72-74; asst prof chem, Middlebury Col, 74-80. *Mem:* AAAS; Am Chem Soc. *Res:* Chemical and biochemical studies of porphyrins and porphyrin containing biological molecules. *Mailing Add:* 7820 Overhill Rd Bethesda MD 20814-1115

BENNETT, DEBRA A, b Boundbrook, NJ. ANXIOLYTICS. *Educ:* Drew Univ, BA, 76; Fairleigh-Dickinson Univ, MA, 78; Univ RI, PhD, 81. *Prof Exp:* ASSOC DIR CLIN DEVELOP, CIBA-GEIGY CORP, 81- *Mem:* Am Psychol Asn; Neurosci Soc; Am Soc Pharmacol & Exp Therapeut. *Mailing Add:* Ciba-Geigy Corp 556 Morris Ave Summit NJ 07901

BENNETT, DONALD RAYMOND, b Mishawaka, Ind, Feb 16, 26; m 47; c 2. PHARMACOLOGY. *Educ:* Univ Mich, BS, 49, MS, 51, MD, 55, PhD, 58; Am Bd Family Pract, cert, 76, 83 & 90. *Prof Exp:* From asst prof to assoc prof pharmacol, Univ Mich, 57-65; mgr, Biosci Res Lab, Dow Corning Corp, 65-74; sr scientist, Dept Drugs, 76-80, assoc dir, 80-85, DIR, DIV DRUGS & TOXICOL, AMA, 85- *Mem:* Am Soc Pharmacol & Exp Therapeut; Soc Toxicol; Am Acad Family Physicians; Am Soc Clin Pharmacol & Therapeut. *Mailing Add:* Div Drugs & Toxicol 515 N State St Chicago IL 60610

BENNETT, DOROTHEA, molecular biology; deceased, see previous edition for last biography

BENNETT, DWIGHT G, JR, b Pittsburgh, Pa, Aug 27, 35; m 59; c 2. VETERINARY MEDICINE. *Educ:* Univ Ill, BS, 57, DVM, 59; Univ Wis, MS, 63, PhD(vet sci), 64. *Prof Exp:* Private practice, 59; asst vet sci, Univ Wis, 62-63, USPHS fel, 63-64; from asst prof to assoc prof vet parasitol, Purdue Univ, 64-73, prof vet parasitol & large animal clin, 73-75; PROF CLIN SCI, COLO STATE UNIV, 75- *Concurrent Pos:* Mem, Nat Bd Vet Med Examrs, 71- *Mem:* Am Vet Med Asn; Am Asn Vet Parasitol; World Asn Advan Vet Parasitol; Am Asn Equine Practitioners; Sigma Xi. *Res:* Chemotherapy and immunology of nematode parasites of ruminants; chemotherapy of parasites of horses. *Mailing Add:* 2307 Tanglewood Dr Ft Collins CO 80525

BENNETT, E MAXINE, otolaryngology, for more information see previous edition

BENNETT, EDGAR F, b Colebrook, NH, July 29, 29; m 70. REACTOR PHYSICS. *Educ:* Univ NH, BS, 51, MS, 53; Princeton Univ, PhD(physics), 57. *Prof Exp:* PHYSICIST, DIV REACTOR PHYSICS, ARGONNE NAT LAB, 57- *Mem:* Am Nuclear Soc; Sigma Xi. *Res:* Radiation detection; radiation instrumentation. *Mailing Add:* 913 Clyde Dr Downers Grove IL 60516

BENNETT, EDWARD LEIGH, b Hood River, Ore, Nov 20, 21; m 54; c 3. BIOCHEMISTRY, NEUROBIOLOGY. *Educ:* Reed Col, BA, 43; Calif Inst Technol, PhD, 49. *Prof Exp:* Asst chem, Calif Inst Technol, 42-49; Am Cancer Inst fel, Inst Cytophysol, Copenhagen Univ, 51-52; res chemist, Radiation Lab, 49-51 & Lawrence Berkeley Lab, 52-86, RES PSYCHOLOGIST, UNIV CALIF, BERKELEY, 86- *Mem:* Am Chem Soc; Am Soc Biol Chem; Am Soc Neurochem: Soc Neurosci. *Res:* Nucleic acid metabolism; biochemical psychology; biochemistry and memory. *Mailing Add:* 2719 Marin Berkeley CA 94708-1529

BENNETT, EDWARD OWEN, b St Louis, Mo, Mar 16, 26; m 47; c 2. BACTERIOLOGY. *Educ:* Univ Houston, BS, 49; Univ Iowa, MS, 51; Baylor Univ, PhD, 58. *Prof Exp:* From asst prof to assoc prof, 51-63, chmn dept bact, 64-67, 79-81, assoc dean col arts & sci, 67-73, PROF BACT, UNIV HOUSTON, 63- *Concurrent Pos:* Consult, various co. *Mem:* Am Soc Microbiol; Soc Indust Microbiol; fel Am Acad Microbiol; fel Soc Tribology & Lubrication Engrs. *Res:* Antimicrobial agents. *Mailing Add:* Dept Biol Univ Houston Houston TX 77204-5513

BENNETT, ELBERT WHITE, b Texarkana, Tex, Jan 24, 29; m 62; c 3. EXPERIMENTAL NUCLEAR PHYSICS. *Educ:* Univ Tex, BS, 51, MA, 52, PhD(physics), 57. *Prof Exp:* Staff mem, 56-63, assoc group leader, 63-67, MEM STAFF, LOS ALAMOS SCI LAB, 67- *Mem:* Am Phys Soc. *Res:* Neutron scattering; diagnostic and effects measurements of nuclear explosions. *Mailing Add:* 263 Dos Brazos Los Alamos NM 87544

BENNETT, F LAWRENCE, b Troy, NY, Apr 4, 39; m 62; c 2. COLD REGIONS CONSTRUCTION, PROJECT MANAGEMENT. *Educ:* Rensselaer Polytechnic Inst, BcE, 61; Cornell Univ, MS, 63, PhD(civil eng), 66. *Prof Exp:* Planning engr, United Engrs & Constructors Inc, 65-68; vis assoc prof civil eng, NC State Univ, 74-75; actg vpres acad affairs, Alaska Pac Univ, 82-83; from assoc prof to prof eng mgt, Univ Alaska, Fairbanks, 68-74, asst to chancellor, 77-79, vice chancellor for acad affairs, 79-82, DEPT HEAD ENG MGT, UNIV ALASKA, FAIRBANKS, 69-78, 83-, PROF, 83- *Concurrent Pos:* Eng & mgt consult, 69-; prin investr, Sch Eng, Univ Alaska, 70- *Mem:* Am Soc Civil Engrs; Nat Soc Prof Engrs; Am Soc Eng Educ; Sigma Xi. *Res:* Cold regions construction techniques and management; construction productivity; project management; network scheduling; computer applications; engineering economy. *Mailing Add:* Dept Eng Mgt Univ Alaska Fairbanks AK 99775

BENNETT, FOSTER CLYDE, b Wilmette, Ill, Oct 14, 14; m 40; c 7. ENGINEERING PHYSICS. *Educ:* Univ Ill, BS, 36; Calif Inst Technol, MS, 37. *Prof Exp:* Engr res & develop, Dow Chem Co, 37-40, supt electrothermal magnesium plant, 40-43, supt sodium prod plant, 44-49, supt pressure die casting, 49-61, supvr, Metals Lab, 61-70, assoc scientist, Inorganic Chem Dept, 70-72; sr res scientist, Battelle Columbus Labs, 72-82; pres, 72-90, VPRES, DIE CASTING CONSULT, INC, 90- *Honors & Awards:* Doehler Award, Am Die Casting Inst, 64. *Mem:* Soc Die Casting Eng. *Res:* Pressure and gravity casting techniques, nonferrous alloys; molten metal handling; heat transfer and process analysis. *Mailing Add:* 3700 Waldo Pl Columbus OH 43220

BENNETT, FREDERICK DEWEY, b Miles City, Mont, June 2, 17; m 39, 80; c 6. FLUID MECHANICS. *Educ:* Oberlin Col, AB, 37; Pa State Col, MS, 39, PhD(physics), 41. *Prof Exp:* Asst physics, Pa State Col, 38-41; from instr to asst prof, Univ NH, 41-43; physicist, Spec Projs Lab, AMC, Wright Field, Ohio, 43-46; from asst prof to assoc prof elec eng, Univ Ill, 46-48; physicist, Ballistics Res Lab, Aberdeen Proving Ground, 48-63, chief exterior ballistics div, 62-70; CONSULT, 70- *Concurrent Pos:* Fel, Dept Mech & Mat Sci, Johns Hopkins Univ, 71-80, Dept Chem Eng, 81- *Mem:* AAAS; fel Am Phys Soc; Am Geophys Union. *Res:* Physics of hydrogen palladium system; aircraft antennas; diffusion of hydrogen through metals; optical methods in analysis of airflow; exploding wires; volcanic ash formation. *Mailing Add:* Chem Engr Dept Johns Hopkins Univ Baltimore MD 21218

BENNETT, G(ARY) F, b Windsor, Ont, July 22, 35; m 63; c 2. BIOCHEMICAL & CHEMICAL ENGINEERING. *Educ:* Queen's Univ, Ont, BSc, 57; Univ Mich, MSE, 60, PhD(chem eng), 63. *Prof Exp:* From asst prof to assoc prof, 63-72, PROF BIOCHEM ENG, UNIV TOLEDO, 72- *Concurrent Pos:* Consult, Great Lakes Container Corp, 76-84 & Assoc Chem & Environ Serv, 79-84, Envirosafe, 84-; ed, J Hazardous Mat, 80-, Environmental Progress, 82- *Honors & Awards:* Environ Award Chem Eng, Am Inst Chem Engrs, 75, Serv to Soc Award, 82. *Mem:* Am Inst Chem Engrs; Am Acad Environ Engrs; Water Pollution Control Fedn. *Res:* Oxygen transfer; industrial waste treatment; hazardous chemical spills; air flotation. *Mailing Add:* Col Eng Univ Toledo Toledo OH 43606

BENNETT, GARY COLIN, b Montreal, Que, Oct 18, 39; m 67. CELL BIOLOGY. *Educ:* Sir George Williams Univ, BA & BSc, 64; McGill Univ, MSc, 67, PhD(anat & cell biol), 71. *Prof Exp:* From lectr to assoc prof, 70-82, PROF ANAT, McGILL UNIV, 82- *Concurrent Pos:* Fel, Ctr Nuclear Study, Saclay, France, 71-72. *Mem:* Am Asn Anatomists; Am Asn Cell Biol. *Res:* Synthesis and fate of glycoproteins in various cell types as studied by autoradiography; membrane biogenesis. *Mailing Add:* Dept Anat McGill Univ 3640 University St Montreal PQ H3A 2B2 Can

BENNETT, GARY LEE, b Sioux City, Iowa, Mar 5, 51; m 75; c 2. PRODUCTION SYSTEMS. *Educ:* Iowa State Univ, BS, 73; Ohio State Univ, MS, 75, PhD, 77. *Prof Exp:* Fel, Univ Nebr, 77-81; scientist, NZ Min Agr Fish, 81-85; RES GENETICIST, MEAT ANIMAL RES CTR, AGR RES SERV, USDA, 85- *Mem:* Am Soc Animal Sci; Sigma Xi. *Res:* Modeling and simulation of livestock systems; livestock genetics; statistical methods. *Mailing Add:* US Meat Animal Res Ctr PO Box 166 Clay Center NE 68933-0166

BENNETT, GEORGE KEMBLE, b Jacksonville, Fla, Apr 2, 40; m 82; c 4. QUALITY ENGINEERING, RELIABILITY & MAINTAINABILITY. *Educ:* Fla State Univ, BS, 62; San Jose State Univ, MS, 68; Texas Tech Univ, PhD(indust eng), 70. *Prof Exp:* Assoc engr, Martin Co, 62-63; sr engr, Lockheed Missiles & Space Co, 63-66; vis scientist, NASA, 69-70; asst prof indust eng, Va Tech, 70-73; staff engr, Honeywell Avionics, 79-86; prof indust eng, 73-79, chmn dept, Univ SFla, 79-86; HEAD INDUST ENG, TEX A&M UNIV, 86-, HALLIBURTON PROF, 86- *Mem:* Fel Inst Indust Engrs; fel Soc Logistics Engrs; Am Soc Eng Educ. *Res:* Development of cost based sampling plans and procedures; the effect of inspection error in sampling plan design; empirical Bayes parameter estimation procedures for reliability engineering. *Mailing Add:* 5918 Blue Ridge Dr Col Station TX 77845

BENNETT, GEORGE NELSON, REGULATION, TRANSCRIPTION. *Educ:* Purdue Univ, PhD(biol sci), 74. *Prof Exp:* ASSOC PROF MOLECULAR BIOL, RICE UNIV, 78- *Res:* Prokaryotic molecular biology. *Mailing Add:* Dept Biochem Rice Univ 6100 Main St Houston TX 77251

BENNETT, GERALD WILLIAM, b Hempstead, NY, June 15, 33; m 55; c 6. RADIATION PHYSICS, MEDICAL PHYSICS. *Educ:* Brooklyn Polytech Inst, BME, 55; Hofstra Univ, MA, 62; State Univ NY, Stony Brook, PhD(physics), 68; Am Bd Nuclear Med, dipl. *Prof Exp:* Jr engr, NY Naval Shipyard, 56; mech engr, Fairchild Engine Div, NY, 56-59; physicist, Ger Electron Synchrotron, Hamburg, 68-69; oper engr, Cosmotron Div, Brookhaven Nat Lab, 59-64, develop engr, 64-68, physicist, Accelerator Dept, 69-76, physicist, Med Dept, 76-88, PHYSICIST, AGS DEPT, BROOKHAVEN NAT LAB, 88-; PROF, CLIN CAMPUS, STATE UNIV NY, STONY BROOK, 75- *Concurrent Pos:* prin invest, Nat Cancer Inst Res Grant, 75-78; consult nuclear med, Northport Vet Admin Hosp, 76- *Mem:* Inst Elec & Electronics Engrs; Am Col Nuclear Med; Soc Nuclear Med; Am Col Med Physics. *Res:* Radiation physics; instrumentation; medical applications of protons, anti-protons and heavy ions. *Mailing Add:* Brookhaven Nat Lab Upton NY 11973-5000

BENNETT, GLENN ALLEN, b La Follette, Tenn, Nov 18, 38; m 71; c 2. ORGANIC CHEMISTRY, BIOCHEMISTRY. *Educ:* Bradley Univ, BA, 62; Iowa State Univ, MS, 66. *Prof Exp:* Chemist biol insecticides, 62-72, RES CHEMIST MYCOTOXINS, SCI & EDUC ADMIN-AGR RES, USDA, PEORIA, ILL, 72- *Mem:* Am Chem Soc; Sigma Xi; Am Oil Chemists Soc; Asn Off Anal Chemists. *Res:* Analysis of agricultural commodities for fungal metabolites, specifically mycotoxins, by high-pressure liquid chromatography, gas chromatography, gas chromatography-mass spectrometry and thin layer chromatography; development of methods to detoxify contaminated grains. *Mailing Add:* 117 High St RR 4 Metamora IL 61548

BENNETT, GLENN TAYLOR, b Allentown, Pa, Feb 1, 56; m 79. NOVEL LASER & NONLINEAR OPTIC SYSTEMS, ULTRAFAST DIAGNOSTICS. *Educ:* Rensselaer Polytech Inst, BS, 78; Univ Tex, Austin, PhD(physics), 84. *Prof Exp:* Vis asst prof physics, Tex A&M Univ, 84-85; postdoctoral fel, 85-87; MEM TECH STAFF NONLINEAR OPTICS, ROCKETDYNE DIV, ROCKWELL INT, 87- *Mem:* Am Phys Soc; Optical Soc Am; Inst Elec & Electronics Engrs; Soc Photo-Optical Instrumentation Engrs. *Res:* Experimental research includes high brightness photoelectron sources for accelerators and free electron lasers; novel solid state and FEL laser systems, ultrafast phenona, laser ignition for rocket engines and nonequilibrium materials for nonlinear optics. *Mailing Add:* 4115 Gadshill Lane Agoura CA 91301

BENNETT, GORDON DANIEL, b Elmira, NY, Oct 5, 31; m 58; c 3. WELL HYDRAULICS, REGIONAL FLOW SYSTEM ANALYSIS. *Educ:* Univ Notre Dame, BS, 56; Pa State Univ, MS, 61. *Prof Exp:* Geophysicist, US Geol Surv, 56-61, tech adv, USAID, 62-66, hydrologist, US Geol Surv, 66-81, chief ground water br, 81-82, asst chief, water resources dir, 82-85, res hydrologist, 85-86; mgr, Kearney Inc, 86-87; SR ASSOC, S S PAPADOPULOS & ASSOC, INC, 87- *Honors & Awards:* O E Meinzer Award, Geol Soc Am, 81; Distinguished Serv Award, US Dept Interior, 86. *Mem:* Fel Geol Soc Am; Am Geophys Union; Asn Ground Water Scientists & Engrs. *Res:* Effects of wellbores on regional flow; well hydraulics; movement of contaminants; freshwater-saltwater relationships; characterization of regional flow systems. *Mailing Add:* 5402 Water Crest Ct Fairfax VA 22032

BENNETT, GORDON FRASER, b Ootacamund, India, Aug 20, 30; Can citizen; m 59. ENTOMOLOGY, PROTOZOOLOGY. *Educ:* Univ Toronto, BA, 53, MA, 54, PhD(entom), 57. *Prof Exp:* Res fel parasitol, Ont Res Found, 57-68; assoc prof, 68-74, PROF BIOL & HEAD INT REF CTR, AVIAN HAEMATOZOA, MEM UNIV NFLD, 74- *Concurrent Pos:* Vis scientist, NIH, 63-65. *Mem:* Am Soc Parasitol; Am Soc Protozool; Wildlife Dis Asn; Can Soc Zool; Royal Soc Trop Med & Hyg. *Res:* Biology and taxonomy of myiasis-producing Diptera; biology of vectors of hematozoa; biology and taxonomy of hematozoa. *Mailing Add:* Dept Biol Mem Univ Nfld St Johns NF A1B 3X9 Can

BENNETT, GUDRUN STAUB, b New York, NY, Nov 13, 40; m 62; c 2. CELL BIOLOGY, NEUROBIOLOGY. *Educ:* Vassar Col, AB, 61; Rockefeller Univ, PhD(life sci), 68. *Prof Exp:* Instr, Dept Path, Univ Ky, 71-72; res assoc, Fla State Univ, 72-76; res assoc, Dept Anat, Univ Pa, 76-79, from res asst prof to res assoc prof, 79-86; RES SCIENTIST, DEPT ZOOL & ASSOC PROF, DEPT ANAT & CELL BIOL & DEPT NEUROSCI, UNIV FLA, 86- *Mem:* AAAS; Am Soc Cell Biol; Soc Neurosci. *Res:* Neuronal cytoskeletons; function and metabolism of neurofilaments; neuronal differentiation; cell differentiation. *Mailing Add:* Dept Anat & Cell Biol Univ Fla Box J-235 Gainesville FL 32610-0235

BENNETT, HAROLD EARL, b Missoula, Mont, Feb 25, 29; m 52, 84; c 2. OPTICAL PHYSICS, OPTICAL TESTING. *Educ:* Univ Mont, BA, 51; Pa State Univ, MS, 53, PhD(physics), 55. *Prof Exp:* Physicist, Nat Bur Standards, 53 & Wright Air Develop Ctr, 55-56; head phys optics br, 60-78, PHYSICIST, NAVAL WEAPONS CTR, 56-, ASSOC HEAD PHYS DIV, 71- *Concurrent Pos:* Fel Ord Sci, Naval Weapons Ctr, 72. *Honors & Awards:* L T E Thompson Award, Naval Weapons Ctr, 74; Captain Robert Dexter Conrad Award, US Navy, 79; Technol Achievement Award, Int Soc Optical Eng, 83. *Mem:* Fel Optical Soc Am; Am Phys Soc; fel Int Soc Optical Eng (pres, 88). *Res:* Optical properties of solids; solid state physics; optical instrumentation; thin films; laser components; large optics. *Mailing Add:* Code 38101 Naval Weapons Ctr China Lake CA 93555

BENNETT, HARRY, industrial chemistry; deceased, see previous edition for last biography

BENNETT, HARRY JACKSON, zoology; deceased, see previous edition for last biography

BENNETT, HENRY STANLEY, b Tottori, Japan, Dec 22, 10; US citizen; m 35; c 4. ANATOMY, CELL BIOLOGY. *Educ:* Oberlin Col, AB, 32; Harvard Univ, MD, 36. *Hon Degrees:* DSc, Monmouth Col, 62. *Prof Exp:* Nat Res Coun fel & univ res fel anat, Harvard Med Sch, 37-39, instr anat & pharmacol, 39-41, assoc anat, 41-48; prof & head dept, Sch Med, Univ Wash, 48-60; prof biophys & dean div biol sci, Sch Med, Univ Chicago, 61-65, prof anat, 61-69, Robert R Bensley prof biol & med sci, 66-69; chmn, Dept Anat, 69-77, dir, Lab Reproductive Biol, 69-78, prof, 69-81, EMER PROF ANAT & SARAH GRAHAM KENAN PROF BIOL SCI, UNIV NC, CHAPEL HILL, 81- *Concurrent Pos:* Asst prof cytol, Mass Inst Technol, 45-48; mem bd trustees, Salk Inst Biol Studies, 63-76; mem, US-Japan Comt Sci Coop, 64-76, US co-chmn, 69-76. *Honors & Awards:* Eastman Mem lectr, Univ Rochester, 58; Phillips lectr, Haverford Col, 59. *Mem:* Am Asn Anat (pres, 60-61); Asn Am Physicians; Am Soc Cell Biol; Am Acad Arts & Sci; Am Physiol Soc. *Res:* Histochemistry and cytochemistry of sulfhydryl groups; cell ultrastructure, molecular structure and dynamic behavior of membranes; development of methods for structural analysis of cells and tissues; general functions of actin and myoin. *Mailing Add:* 3300 Darby Rd No 204 Haverfod PA 19041-1016

BENNETT, HOLLY VANDER LAAN, b Alexandria, Va, Feb 13, 57; m 88. CELL BIOLOGY, ACTIN-MEMBRANE INTERACTION. *Educ:* Wheaton Col, BA, 78; Albert Einstein Col Med, PhD(anat & struct biol), 86. *Prof Exp:* FEL, WHITEHEAD INST, 86- *Mem:* NY Acad Sci; Am Soc Cell Biol; Electron Micros Soc; Biophys Soc. *Res:* Intestinal epithelial ankyrin. *Mailing Add:* Whitehead Inst Biomed Res Nine Cambridge Ctr Cambridge MA 02142

BENNETT, HUGH DEEVEREAUX, b Brooklyn, NY, Jan 19, 18; m 41. MEDICINE. *Educ:* Univ Chicago, BS, 40, MD, 42. *Prof Exp:* Instr med, Med Sch, Northwestern Univ, 51-53; from asst prof to assoc prof, Col Med, Baylor Univ, 53-62; PROF MED & ASSOC DEAN, HAHNEMANN MED COL, 62- *Concurrent Pos:* Asst chief med serv, Vet Admin Hosp, 48-62; chief med serv, Houston, 53-62; attend physician, Philadelphia Gen Hosp, 62-; vis prof, Free Univ Lille, France, 75- *Mem:* Fel Am Col Physicians; AMA. *Res:* Hepatic and gastrointestinal diseases. *Mailing Add:* Hahnemann Univ Sch Med Univ Hosp 235 N 15th St Philadelphia PA 19102

BENNETT, IAN CECIL, b Bedington, Eng, Aug 2, 31; US citizen; div; c 2. DENTISTRY. *Educ:* Univ Liverpool, BDS, 56; Univ Toronto, DDS, 59; Univ Wash, MSD, 64. *Prof Exp:* Assoc prof pedodontics, Dalhousie Univ, 63-65; asst prof, Univ Ky, 65-68, dir med ctr commun, 67-68; from assoc prof to prof pedodontics & from assoc dean to dean, Col Med & Dent, NJ, 69-76; prof pedodontics & dean fac dent, Dalhousie Univ, 76-86; vis prof, Univ Calif, San Francisco, 87-88; PROF PEDODONTICS, DALHOUSIE UNIV, 86- *Mem:* AAAS; Am Asn Dent Schs; Am Acad Pedodontics; Can Acad Pedodontics (pres elect, 78-80, pres, 80-83); Can Dent Asn; fel Am Col Dentists; fel Int Col Dentists. *Res:* Calcification of bone and tooth substance; effect of tetracycline on calcification and on organ culture of bone. *Mailing Add:* Fac Dent 5981 University Ave Suite 5160 Halifax NS B3H 3J5 Can

BENNETT, IVAN FRANK, b Hartford, Conn, Sept 6, 19; m 44; c 2. PSYCHIATRY. *Educ:* Trinity Col, BS, 41; Jefferson Med Col, Thomas Jefferson Univ, MD, 44. *Prof Exp:* Instr psychiat, Sch Med, Univ Pa, 54-56, guest lectr, Sch Dent, 54-56; clin asst prof, Sch Med, Georgetown Univ, 56-58; from asst prof to assoc prof, Sch Med, IND UNIV, INDIANAPOLIS, 72- *Concurrent Pos:* Asst physician, State Hosp, Harrisburg, Pa, 48-50; asst chief acute intensive treatment serv & chief physiol treatment sect, Vet Admin Hosp, Coatesville, Pa, 50-56; chief psychiat res, Psychiat & Neurol Serv, Dept Med & Surg, Vet Admin, Washington, DC, 56-58; physician, Lilly Lab Clin Res, Eli Lilly & Co, Indianapolis, 58-81, from physician to sr physician, Clin Invest Div, Lilly Res Div, 63-76, clin investr, 76-; dir, Lilly Psychiat Clin, Wishard Mem Hosp, Indianapolis, 59-84. *Mem:* AMA; fel Am Col Physicians; fel Am Psychiat Asn; charter fel Am Col Neuropsychopharmacol. *Res:* Physiological and pharmacological therapies in psychiatry. *Mailing Add:* 8452 Green Braes N Dr Indianapolis IN 46234

BENNETT, IVAN LOVERIDGE, JR, pathology; deceased, see previous edition for last biography

BENNETT, JAMES ANTHONY, b Buffalo, NY, Jan 9, 48; m 75. CANCER. *Educ:* Univ Notre Dame, BA, 69; State Univ NY, Buffalo, MS, 72, PhD(pharmacol), 76. *Prof Exp:* From res asst to res assoc, Yale Univ, 76-78, res fel, 78-79; ASST PROF IMMUNOL, ALBANY MED COL UNION UNIV, 79- *Mem:* Am Asn Cancer Res; Am Asn Immunologists. *Res:* Investigation of immunological changes during preclinical and clinical chemoimmunotheraphy. *Mailing Add:* Dept Surg Albany Med Col New Scotland Ave Albany NY 12208

BENNETT, JAMES AUSTIN, b Taber, Alta, Jan 29, 15; nat US; m 40; c 5. ANIMAL BREEDING. *Educ:* Utah State Univ, BS, 40, MS, 41; Univ Minn, PhD(animal breeding), 57. *Prof Exp:* Livestock asst, Dom Dept Agr Can, 41-45; asst prof, 45-50, head dept, 50-76, PROF ANIMAL SCI, UTAH STATE UNIV, 50- *Mem:* AAAS; Am Soc Animal Sci; Sigma Xi; Am Genetics Asn. *Res:* Breeding phases of beef cattle and sheep. *Mailing Add:* Dept Animal Sci Utah State Univ Logan UT 84322-4815

BENNETT, JAMES GORDY, JR, b Washington, DC, Aug 29, 32; m 56; c 2. POLYMER CHEMISTRY. *Educ:* NY State Col Teachers, Albany, BS, 54; Rensselaer Polytech Inst, PhD(chem), 59. *Prof Exp:* Assoc res chemist, Parke Davis & Co, Mich, 59-60; sr res chemist, Huyck Felt Co, NY, 60-63; res chemist, 63-72, plant chemist, 72-79, MGR, PROCESS TECHNOL PLASTICS DIV, GEN ELEC CO, 79- *Mem:* Am Chem Soc. *Res:* Organic and polymer synthesis; oxidative coupling chemistry; homogeneous and heterogeneous catalysis. *Mailing Add:* 610 Cortland St Albany NY 12208

BENNETT, JAMES PETER, b Chicago, Ill, Aug 25, 44. AIR POLLUTION ECOLOGY. *Educ:* Washington Univ, BA, 66; Univ Mich, Ann Arbor, MA, 69; Univ BC, PhD(plant sci), 75. *Prof Exp:* Sr scientist ecol, Hudson River Valley Comn, Tarrytown, NY, 69-70; instr, Pratt Inst, 70-71; asst prof & olericult crop ecol, Dept Veg Crops, Univ Calif, Davis, 75-80; RES ECOLOGIST, AIR QUAL DIV, NAT PARK SERV, US DEPT INTERIOR, 80- *Mem:* AAAS; Am Inst Biol Sci; Am Soc Hort Sci; Crop Sci Soc Am. *Res:* Plant competition, density, spacing, productivity; whole plant development and growth analysis; allometry; yield component compensation; experimental design; air pollution effects on native vegetation; plant ecology. *Mailing Add:* 5066 Blackhawk Dr Danville CA 94526-4556

BENNETT, JEAN MCPHERSON, b Kensington, Md, May 9, 30; div. OPTICAL PHYSICS. *Educ:* Mt Holyoke Col, BA, 51; Pa State Univ, MS, 53, PhD(physics), 55. *Prof Exp:* Physicist, Nat Bur Standards, 51, 53 & Wright Air Develop Ctr, 55-56; PHYSICIST, NAVAL WEAPONS CTR, 56-, RES SCIENTIST, 60- *Concurrent Pos:* Vis prof, Univ Ala, Huntsville, 86-87; vis scientist, Uppsala Univ Uppsala, Inst Optical Res, Stockholm, Sweden, 88; sr fel, Naval Weapons Ctr, 89. *Honors & Awards:* Woman Scientist of the Year, Naval Weapons Ctr, 79, LTE Thompson Award, 88; David Richardson Medal, Optical Soc Am, 90. *Mem:* Fel Optical Soc Am (vpres, 84, pres elect, 85, pres, 86-87). *Res:* Optical properties of solids; solid state physics; interferometry; thin films; characterization of optical surfaces. *Mailing Add:* Code 38103 Michelson Lab Naval Weapons Ctr China Lake CA 93555

BENNETT, JESSE HARLAND, b Lehi, Utah, June 21, 36; m 58; c 3. AGRICULTURE, FORESTRY. *Educ:* Utah State Univ, BS, 61; Univ Utah, PhD(bot) & cert environ toxicol, 69. *Prof Exp:* USPHS trainee environ toxicol, Ctr Environ Biol, Univ Utah, 65-68; NIH fel & res assoc biophysics, Ctr Biol Natural Syst, Wash Univ, 69-70; res assoc, Dept Biol, Univ Utah, 70-74; plant physiologist, Plant Stress Lab, Plant Physiol Inst, Agr Res Serv Sci & Educ, 74-84, PLANT PHYSIOLOGIST, AGR RES STA, USDA, LOGAN, UTAH, 84- *Concurrent Pos:* Air pollution consult & researcher, Ajax Presses, Utah Power & Light Co, 71-73; sci reviewer, J Environ Qual Sci, Nat Acad Sci, Coun Environ Qual, NSF, Energy Res & Develop Admin, Dept Energy, Environ Protection Agency & USDA, 74-81. *Honors & Awards:* Res Excellence Award, Environ Protection Agency, 87. *Mem:* AAAS; Am Soc Plant Physiologists; Ezra Taft Benson Agr & Food Inst; Bot Soc Am. *Res:* Physiology, metabolism and control of plant growth; environmental pollution and stress. *Mailing Add:* Ars Res Sta FRRL USDA UMC 6300 Utah State Univ Logan UT 84322

BENNETT, JOAN WENNSTROM, b Brooklyn, NY, Sept 15, 42; m 66; c 3. GENETICS, MYCOLOGY. *Educ:* Upsala Col, BS, 63; Univ Chicago, MS, 64, PhD(bot), 67. *Hon Degrees:* DLitt, Upsala Col, 90. *Prof Exp:* Res assoc, NSF fel, Univ Chicago, 67-68; Nat Res Coun fel, 68-70; from asst to assoc prof biol, 70-81, PROF BIOL, TULANE UNIV, 81- *Concurrent Pos:* NSF fel, 70-71; lectr, Am Soc Microbiol Found, 81; chmn, Fermentation & Biotech Div, Am Soc Microbiol, 85-86; bd, Soc Indust Microbiol, 87; vpres, Brit Mycol Soc, 88. *Mem:* AAAS; Soc Indust Microbiol; Am Soc Human Genetics; Genetics Soc Am; Am Soc Microbiol (pres, 90-91); Mycol Soc Am. *Res:* Aflatoxin biosynthesis; genetics of secondary metabolism; Down's Syndrome. *Mailing Add:* Dept Cell & Molecular Biol Tulane Univ New Orleans LA 70118

BENNETT, JOE CLAUDE, b Birmingham, Ala, Dec 12, 33; m 58; c 3. RHEUMATOLOGY. *Educ:* Howard Col, AB, 54; Harvard Med Sch, MD, 58; Am Bd Internal Med, dipl, 68, rheumatology, cert, 72. *Prof Exp:* Intern med, Med Ctr, Univ Ala, 58-59, asst resident, 59-60; Arthritis & Rheumatism Found fel, 60-62; res assoc molecular biol, NIH, 62-64; sr res fel biol chem, Calif Inst Technol, 64-65; assoc prof med & microbiol & asst dir div clin immunol & rheumatology, 66-70, prof & chmn microbiol, 70-82, prof dept med & dir div clin immunol & rheumatology, Med Ctr, 70-83, PROF & CHMN DEPT MED, UNIV ALA, BIRMINGHAM, 82- *Concurrent Pos:* Clin fel rheumatology, Mass Gen Hosp, Harvard Med Sch, 60-61, res fel, 61-62; Markle scholar acad med, 65-70; Nat Inst Gen Med Sci res career develop award, 65-75; mem study sect for training in rheumatology, NIH, 69-72, chmn, 70-72, mem & chmn study sect for allergy & immunol, 75-78; mem, Comt Eval Clin Competance, Am Bd Internal Med, 87-, chmn, Subcomt Methods, 90, Comt Res & Develop, 87-, mem gov bd, 87-; mem sci adv comt, Biocryst, Inc, Warren Alpert Found, Harvard Med Sch, Charles E Culpeper Found, Inc, 89-; mem prog comt, Inst Med-Nat Acad Sci, 90-; numerous lectrs & vis professorships. *Mem:* Inst Med-Nat Acad Sci; Genetics Soc Am; Am Rheumatism Asn (secy-treas, 74-76); Am Asn Immunologists; Am Soc Clin Invest; Am Asn Physicians; fel AAAS; master Am Col Physicians; Am Acad Allergy & Immunol; Am Acad Microbiol. *Res:* Genetic determinants of protein structure; structural aspects of immunoglobulins; molecular basis of disease states; immunochemistry; cell surface structure. *Mailing Add:* Dept Med Univ Ala Sch Med Univ Sta Birmingham AL 35294

BENNETT, JOHN E, b El Centro, Calif, Mar 6, 33; m 58; c 2. MYCOLOGY. *Educ:* Stanford Univ, Calif, BS, 55; Johns Hopkins Univ, MD, 59. *Prof Exp:* Intern, Johns Hopkins, 59-60; asst resident med, Univ Wash, Seattle,, 60-61 & Wash Univ, St Louis, 64-65; clin assoc, 61-64, actg head, infectious dis sect, 65-66, head, 66-71, HEAD, CLIN MYCOL SECT, LCi, NAT INST ALLERGY & INFECTIOUS DIS, NIH, 71- *Concurrent Pos:* Prof med, Dept Med, Uniformed Serv Univ Health Sci, 76-; lectr, Dept Micros & Med, Sch Med, Johns Hopkins Univ, 76- *Honors & Awards:* R R Hawkins Award, 79. *Mem:* Am Col Physicians; Asn Am Physicians; Am Soc Clin Invest; Infectious Dis Soc. *Res:* Host defense, chemotherapy and rapid diagnostic tests for systemic mycoses; infectious diseases. *Mailing Add:* Clin Mycol Sect Bldg 10 Rm 11N107 LCI, NIAID, NIH 9000 Rockville Pike Bethesda MD 20892

BENNETT, JOHN FRANCIS, b Palo Alto, Calif, Jan 13, 25; m 64; c 1. HISTORY OF SCIENCE, PHILOSOPHY OF SCIENCE. *Educ:* Stanford Univ, AB, 46, PhD(biol), 62. *Prof Exp:* Physicist, US Navy Electronics Lab, 50-51; asst biol, Hopkins Marine Sta, Calif, 53-54; asst, Stanford Univ, 54-56, actg instr, 57-60; NIH fel microbiol, 61-63; asst prof hist & philos sci, Univ Pa, 63-69; vis scholar biol sci, Stanford Univ, 69-71; proj dir natural sci, Sullivan Assocs, 71-74; staff mem, Encycl Britannica, 75; lectr embryol, human ecol & grad biol, San Jose State Univ, 76-78; lectr interdisciplinary sci, San Francisco State Univ, 78-82, instr physics, Castilleja Sch, 83-84; RES ASSOC, UNIV CALIF, BERKELEY, 84- *Concurrent Pos:* Fulbright guest prof, Inst Statist, Univ Vienna, 67-68. *Mem:* AAAS; Sigma Xi. *Res:* Macroevolution theory; energetics. *Mailing Add:* 3654 Oxford Common Fremont CA 94536

BENNETT, JOHN M, b Boston, Mass, Apr 24, 33; m 57; c 3. INTERNAL MEDICINE, HEMATOLOGY. *Educ:* Harvard Univ, AB, 55; Boston Univ, MD, 59. *Prof Exp:* Instr med, Harvard Med Sch, 65-66; head morphol & histochem sect, Clin Path Dept, NIH, 66-68; asst prof med, Sch Med, Tufts Univ, 68-69; asst prof, 69-71, assoc prof med, 71-76, PROF ONCOL MED, SCH MED, UNIV ROCHESTER, 76- ASSOC DIR CLIN ONCOL, UNIV ROCHESTER CANCER CTR, 74- *Concurrent Pos:* Dir outpatient labs, Boston City Hosp, 68-69; mem lymphoma task force, Nat Cancer Inst, 68-74; dir hemat & med oncol, Highland Hosp, Rochester, 69-74; head med oncol unit, Strong Mem Hosp, 74- *Mem:* AAAS; Am Soc Hemat; Am Fedn Clin Res; Am Soc Clin Oncol; Am Col Physicians. *Res:* Diagnosis by cyctochemical techniques and treatment of malignant disorders of the hematopoietic system. *Mailing Add:* Univ Rochester Cancer Ctr 601 Elmwood Ave Rochester NY 14642

BENNETT, KENNETH A, b Butler, Okla, Oct 3, 35; m 59; c 2. BIOLOGICAL & FORENSIC ANTHROPOLOGY, HUMAN GENETICS. *Educ:* Univ Tex, BA, 61; Univ Ariz, MA, 66, PhD(anthrop), 67. *Prof Exp:* Asst prof anthrop, Univ Ore, 67-70; assoc prof, 70-75, PROF ANTHROP, UNIV WIS-MADISON, 75- *Concurrent Pos:* Ed, Yrbk Phys Anthrop, 76-81; rev article ed, Human Biol, 81-87; burial sites preserv bd, State Wis, 87-89. *Mem:* Human Biol Coun; Am Asn Phys Anthrop; Soc Study Evolution; Soc Study Human Biol; Soc Syst Zool; Am Soc Nat; Am Acad Forensic Sci. *Res:* Human skeletal anatomy; human genetics; forensic anthrop. *Mailing Add:* Dept Anthrop Univ Wis Madison WI 53706

BENNETT, LARRY E, b San Diego, Calif, Feb 29, 40. INORGANIC CHEMISTRY. *Educ:* San Diego State Univ, BS, 62; Stanford Univ, PhD(chem), 65. *Prof Exp:* NSF fel chem, Columbia Univ, 65-66; asst prof, Univ Fla, 66-70; assoc prof, 70-74, PROF CHEM, SAN DIEGO STATE UNIV, 74- *Concurrent Pos:* Vis prof, Stanford Univ, 77. *Res:* Coordination complexes of transition metals; mechanisms of electron transfer and substitution processes; design and synthesis of complexes for investigation of mechanism and possible new modes of reaction; bioinorganic redox processes. *Mailing Add:* Dept Chem San Diego State Univ San Diego CA 92182

BENNETT, LAWRENCE HERMAN, b Brooklyn, NY, Oct 17, 30; m 53; c 3. PHYSICS, MATERIALS SCIENCE. *Educ:* Brooklyn Col, BA, 51; Univ Md, MS, 55; Rutgers Univ, PhD(physics), 58. *Prof Exp:* Physicist, US Naval Ord Lab, 51-53; physicist, Metal Physics Sect, Nat Bur Standards, 58-63, chief, Alloy Physics Sect, 63-81, div scientist, Metall Div, 81-83; LEADER MAGNETIC MAT GROUP, NAT INST STANDARDS & TECHNOL, 83- *Concurrent Pos:* Letr, Grad Sch, Univ Md, 58-61, assoc prof, 61-76, prof, 76- *Honors & Awards:* Gold Medal, Dept of Commerce, 71; Burgess Mem Award, Am Soc Metals, 75. *Mem:* Fel Am Phys Soc; fel Am Soc Metals; Am Inst Mining, Metall & Petrol Eng; Am Soc Test & Mat; Mat Res Soc. *Res:* Magnetic properties of materials; nuclear magnetic resonance; Mossbauer effect; alloy theory; superconductivity. *Mailing Add:* 6524 E Halbert Rd Bethesda MD 20817

BENNETT, LEE COTTON, JR, b Philadelphia, Pa, Mar 14, 33; m 53; c 3. MARINE GEOPHYSICS. *Educ:* Haverford Col, BA, 55; Temple Univ, MS, 58; Bryn Mawr Col, PhD, 66. *Prof Exp:* Res asst solid state physics, Franklin Inst Labs, Pa, 55-58, res physicist, 58-59; res asst geophys, Woods Hole Oceanog Inst, 63-66; asst prof oceanog, Univ Wash, 66-73, asst prof geophys, 68-73; geophysicist, L C B Consults, 73-82; SR INSTR, 82-, SR MUS EDUCR, FRANKLIN INST SCI MUS, 86- *Concurrent Pos:* Actg asst prof, Univ Wash, 64; instr, Shoreline Community Col, 75-76; vol instr, Franklin Inst Mus, 79-82. *Mem:* Am Geophys Union; Soc Explor Geophys. *Res:* Continuous seismic profiling; seismic absorption in unconsolidated sediments; shoreline processes. *Mailing Add:* 224 Haverford Ave Swarthmore PA 19081

BENNETT, LEON, b New York, NY, Sept 10, 27; m 62; c 1. INSTRUMENTATION, FLUID MECHANICS. *Educ:* New York Univ, BAE, 45, MAE, 48. *Prof Exp:* Sr res scientist, New York Univ, 45-76; CHIEF, TISSUE STRESS LAB, VET ADMIN, 76- *Concurrent Pos:* Adj asst prof bioeng, Rensselaer Polytech Inst, 77- *Honors & Awards:* Jobst Award, Am Cong Rehab Med, 84. *Res:* Bioengineering; stress within flesh; pressure sore etiology; blood flow modeling; devices to monitor cutaneous arteriolar blood flow; oscillating aerodynamics; insect flight. *Mailing Add:* 6 Rivercrest Rd Bronx NY 10471

BENNETT, LEONARD LEE, JR, b Savannah, Ga, Nov 10, 20; m 49; c 3. BIOCHEMICAL PHARMACOLOGY. *Educ:* Vanderbilt Univ, AB, 42, MS, 43; Univ NC, PhD(org chem), 49. *Prof Exp:* Asst, Vanderbilt Univ, 42-43; instr, Univ Ga, 43-44; asst & fel, Univ NC, 45-47; res assoc, Southern Res Inst, 45, res chemist, 48-56, head biochem div, 56-64, dir biochem res, 64-85; RETIRED. *Concurrent Pos:* Mem adv comt res ther cancer, Am Cancer Soc, 56-58 & 60-63; mem chemother study sect, NIH, 64-68; assoc ed, Cancer Res, 77-86. *Mem:* AAAS; Am Asn Cancer Res; Am Soc Biol Chemists; Am Soc Pharmacol & Exp Therapeut. *Res:* Biochemistry of cancer; mechanisms of drug action; biochemistry of purines and pyrimidines. *Mailing Add:* Southern Res Inst PO Box 55305 Birmingham AL 35255-5305

BENNETT, LESLIE R, b Denver, Colo, Feb 13, 18; m 48; c 4. RADIOLOGY. *Educ:* Univ Calif, AB, 40; Univ Rochester, MD, 43. *Prof Exp:* Intern, Strong Mem Hosp, Rochester, NY, 44; asst radiation biol & pediat, Univ Rochester, 46-49; asst clin prof radiol, 49-52, actg chief div radiobiol, Atomic Energy Proj, 50-52, from asst prof to assoc prof radiol, 52-60, PROF RADIOL, SCH MED, UNIV CALIF, LOS ANGELES, 60-, CHIEF CLIN RADIOISOTOPE SERV, 52- *Mem:* Am Roentgen Ray Soc; Am Soc Exp Path; Radiation Res Soc; Soc Nuclear Med. *Res:* Radiobiology. *Mailing Add:* Div Nuclear Med Univ Calif Ctr Health Sci Los Angeles CA 90024

BENNETT, LLOYD M, b Columbus, Ind, Nov 22, 28; m 56; c 2. SCIENCE EDUCATION. *Educ:* Ball State Teachers Col, AB, 50; Butler Univ, MS, 52; Univ Tex, MA, 58; Fla State Univ, PhD(marine biol, sci educ), 63. *Prof Exp:* Elem sch teacher, Ind, 50-51; res chemist, Firestone Tire & Rubber Co, Ind, 51-53; high sch teacher, Ind, 53-55; instr & asst biol, Univ Tex, 55-57; teacher & chmn dept sci, Tex, 57-59; assoc prof biol, Amarillo Col, 59-60; prof biol & chem & dir eve div, Indian River Jr Col, 60-61; asst physiol, Fla State Univ, 61-63; asst prof educ & sci educ, 63-66, assoc prof sci educ, 66-71, univ grants, 63-70, PROF SCI EDUC, TEX WOMAN'S UNIV, 71- *Concurrent Pos:* Hogg Found ment health grant, Univ Tex & Hogg Found, 64; co-researcher, Hogg Found grant, 65-66; UNESCO consult elem sci & math educ, Govt Philippines, 74. *Mem:* AAAS; Nat Asn Res Sci Teaching; Nat Asn Biol Teachers; Nat Sci Teachers Asn; Sch Sci & Math Asn. *Res:* Devising, writing, standardizing and implementing the new Twupps curricular program in science for early childhood, including hands-on experiences; research in science for preschool level; preschool curricular developmental research in science. *Mailing Add:* Tex Woman's Univ Sta PO Box 23029 Curriculum Denton TX 76204

BENNETT, LONNIE TRUMAN, b Gorum, La, June 9, 33; m 55; c 4. MATHEMATICS, STATISTICS. *Educ:* Northwest State Col (La), BS, 55; La State Univ, MA, 60; Okla State Univ, PhD(statist), 66. *Prof Exp:* Mathematician, Western Geophys Co Am, 55; teacher pub sch, 57-59; instr math, La State Univ, 60-62; asst prof, 66-69, ASSOC PROF MATH, NORTHEAST LA STATE UNIV, 69- *Concurrent Pos:* Consult, Marshall Space Flight Ctr, NASA, Ala, 66-67. *Mem:* Math Asn Am; Am Statist Asn. *Res:* Experimental design. *Mailing Add:* Dir Acad Comput Northeast La Univ Monroe LA 71209

BENNETT, MARVIN HERBERT, neuroanatomy, neurophysiology, for more information see previous edition

BENNETT, MARY KATHERINE, b Waterbury, Conn, Jan 30, 40. MATHEMATICS. *Educ:* Albertus Magnus Col, BS, 61; Univ Mass, MA, 65, PhD(math), 66. *Prof Exp:* Teacher high sch, Conn, 61-63; asst prof math, Univ Mass, Amherst, 66-68; John Wesley Young res instr, Dartmouth Col, 68-70; from asst to assoc prof math, 70-72, PROF MATH, UNIV MASS, AMHERST, 79- *Mem:* Am Math Soc; Math Asn Am. *Res:* Lattice theory, particularly lattices of convex geometries and geometric lattices. *Mailing Add:* Dept Math Univ Mass Amherst Campus Amherst MA 01003

BENNETT, MICHAEL, b Goose Creek, Tex, Feb 11, 36; m 60; c 4. EXPERIMENTAL PATHOLOGY, HEMATOLOGY. *Educ:* Baylor Univ, MD, 61. *Prof Exp:* Intern, Cleveland Metrop Gen Hosp, Ohio, 62; resident path, Philadelphia Gen Hosp, Pa, 63; biologist, Oak Ridge Nat Labs, 63-65; mem staff, Roswell Park Mem Inst, 65-68, assoc cancer res scientist, 68-72; assoc prof path, Sch Med, Boston Univ, 72-78, prof path & micro, 78-81; PROF PATH, SOUTHWESTERN MED SCH, 81- *Concurrent Pos:* Resident path, Buffalo Gen Hosp, 69-71. *Mem:* Am Soc Exp Path; Am Asn Immunol; Int Asn Hemat; Reticuloendo Soc. *Res:* Function and morphology of hemopoietic stem cells; cellular immunology; bone marrow transplantation. *Mailing Add:* Dept Path Univ Tex Southwestern Med Sch 5323 Harry Hines Blvd Dallas TX 75235

BENNETT, MICHAEL VANDER LAAN, b Madison, Wis, Jan 7, 31; m 63; c 2. NEUROPHYSIOLOGY. *Educ:* Yale Univ, BS, 52; Oxford Univ, PhD(zool), 57. *Prof Exp:* Res worker, dept neurol, Col Physician & Surgeons, Columbia Univ, 57-58, res assoc, 58-59, from asst prof to assoc prof neurol, 59-66; prof anat, 67-74, PROF NEUROSCI & DIR DIV CELLULAR NEUROBIOL, ALBERT EINSTEIN COL MED, 74-, CHMN NEUROSCI, 82-, SYLVIA & ROBERT S OLNICK PROF, 86- *Concurrent Pos:* James Ali Ben Hagen Lounsberry Scholarship, Yale Univ, 50-51; Rhodes scholar, 52; pub health pre doc fel, Nat Inst Neurol Dis & Blindness, 55-56; Grass fel, 58; sr res fel, NIH, 60-62; co dir neurobiol course, Marine Biol Lab, 70-74; chmn, Biol Sci Sect, NY Acad Sci, 78 & bd gov, 80-82; mem coun, Soc Neurosci, 76-80 & prog comt, 80-83; ed, J Cell Biol, 83-85, J Neurobiol, 69- & assoc ed, 79-85, J Neurocytol, 80-82 & J Neurosci, 81-85; adv group, comt teaching & res basic neurol & commun sci, NAS, 73-74, proj review & comt, Neurol Disorders Prog, 77-81, sci adv comt, Irma T Hirschl Trust, 77-81, comt models biomed res, NAS, 83-85, US Nat Comt, Int Union Physiol Sci, 83- *Mem:* Nat Acad Sci; Sigma Xi; fel Nat Neurol Res Found; fel AAAS; Am Asn Anatomists; Am Physiol Soc; Am Soc Cell Biol; Am Soc Zoologists; Biophys Soc; fel NY Acad Sci. *Res:* Author of over 180 publications in morphology and physiology. *Mailing Add:* Albert Einstein Col Med Yeshiva Univ Bronx NY 10461

BENNETT, MIRIAM FRANCES, b Milwaukee, Wis, May 17, 28. BIOLOGICAL RHYTHMS, COMPARATIVE PHYSIOLOGY. *Educ:* Carleton Col, BA, 50; Mt Holyoke Col, MA, 52; Northwestern Univ, PhD(biol), 54. *Hon Degrees:* MA, Colby Col, 73. *Prof Exp:* Asst zool, Mt Holyoke Col, 50-52; asst biol, Northwestern Univ, 52-53; from instr to prof, Sweet Briar Col, 54-73, chmn dept, 64-73; Dana prof, 73-80, CHMN DEPT BIOL, COLBY COL, 73-, KENAN PROF, 80- *Concurrent Pos:* Guest investr, Inst Zool, Univ Munich, 60-61 & 68; NSF fel, 61; mem, Marine Biol Lab, Woods Hole; trustee, Kents Hill Sch, 75- *Mem:* Fel AAAS; Animal Behav Soc; Am Zool Soc; NY Acad Sci; Am Micros Soc; Int Soc Chronobiol. *Res:* Biological rhythmicity; invertebrate and amphibian hormones; regeneration. *Mailing Add:* Dept Biol Colby Col Waterville ME 04901

BENNETT, OVELL FRANCIS, b Middleboro, Mass, Nov 5, 29; m 55; c 2. ORGANIC CHEMISTRY. *Educ:* State Teachers Col, Mass, BS, 53; Boston Col, MS, 55; Pa State Univ, PhD(chem), 58. *Prof Exp:* Asst, Boston Col, 53-55; asst, Pa State Univ, 55-56; res chemist, E I du Pont de Nemours & Co, NJ, 58-61; asst prof chem, 61-64, ASSOC PROF CHEM, BOSTON COL, 64- *Mem:* AAAS; Am Chem Soc. *Res:* Organic mechanisms; organosulfur chemistry. *Mailing Add:* Dept Chem Boston Col Chestnut Hill MA 02167

BENNETT, PETER BRIAN, b Portsmouth, Eng, June 12, 31; m 56; c 2. ANESTHESIOLOGY, PHYSIOLOGY. *Educ:* Univ London, BSc, 51; Univ Southampton, PhD(physiol, biochem), 64, DSc(physiol biochem), 84. *Prof Exp:* Asst head surg sect, Royal Navy Physiol Lab, Alverstoke, Eng, 53-56, head inert gas narcosis sect, 56-66; head pressure physiol, Can Defence & Civil Inst Environ Res, 66-68; head pressure physiol, Royal Navy Physiol Lab, 68-72; prof biomed eng, 72-75, dir res anesthesiol, 73-84, co-dir, F G Hall Environ Res Lab, 74-77, dir, F G Hall Environ Res Lab, 77-88, PROF ANESTHESIOL, MED CTR, 72-, ASSOC PROF PHYSIOL, 75-, SR DIR HYPERBARIC CTR, DUKE UNIV, 88- *Honors & Awards:* Oceaneering Int Award, Undersea Med Soc, 75; Sci Award, Underwater Soc Am, 80; Albert Behnke Award, Undersea Med Soc, 83; Greenstone Award, Nat Asn Underwater Instrs, 85. *Mem:* AAAS; NY Acad Sci; Undersea Med Soc (pres, 75-76); Soc Neurosci; Aerospace Med Asn. *Res:* Inert gas narcosis; mechanisms of anesthesia; high pressure nervous syndrome; decompression sickness; oxygen toxicity; hyperbaric physiology and medicine. *Mailing Add:* F G Hall Lab Duke Univ Med Ctr Box 3823 Durham NC 27710

BENNETT, PETER HOWARD, b Farnworth, Eng, June 21, 37; m 63; c 1. EPIDEMIOLOGY, MEDICINE. *Educ:* Univ Manchester, BSc, 58, MB, ChB, 61, FFCM, 73, FRCP, 79. *Prof Exp:* House physician, Royal Infirmary, Manchester, Eng, 62; house surgeon, 62-63; house phsyician, Postgrad Med Sch, London, Eng, 63-64; house physician, Nat Hosp Nerv Dis, Queen Square, 64; vis assoc arthritis, NIH, 64-68, assoc chief epidemiol, Clin Field Studies Unit, 68-70, chief epidemiol & field studies br, 70-85, CHIEF PHOENIX EPIDEMIOL & CLIN RES BR, NAT INST DIABETES, DIGESTIVE & KIDNEY DIS, PHOENIX, ARIZ, 85- *Concurrent Pos:* Vis fel arthritis, NIH, 63; res fel, Postgrad Med Sch, London, 64; assoc, Col Med, Univ Ariz, 68-; epidemiologist, SPac Comn, 80-81; consult, WHO. *Honors & Awards:* Lilly Award, Am Diabetes Asn, 77; Luther Terry Award, USPHS, 79. *Mem:* Am Rheumatism Asn; Am Diabetes Asn; Am Epidemiol Soc; Am Rheumatism Asn; Am Diabetes Asn. *Res:* Clinical rheumatology; epidemiologic studies of arthritis and diabetes. *Mailing Add:* Nat Inst Diabetes Digestive and Kidney Dis 1550 E Indian School Rd Phoenix AZ 85014

BENNETT, RALPH DECKER, physics, for more information see previous edition

BENNETT, RAYMOND DUDLEY, b Meriden, Conn, May 9, 31; m 61; c 2. AGRICULTURAL BIOCHEMISTRY, NATURAL PRODUCTS CHEMISTRY. *Educ:* Univ Conn, BS, 53; Purdue Univ, MS, 55, PhD(pharm chem), 58. *Prof Exp:* Fel org chem, Univ Va, 58-60; res chemist, Nat Insts Health, 60-63; res chemist, Calif Inst Technol, 64-69, RES CHEMIST, FRUIT & VEG CHEM LAB, AGR RES SERV, US DEPT AGR, 69- *Mem:* Am Chem Soc; AAAS; Phytochem Soc NAm. *Res:* Structure determination of citrus constituents; nuclear magnetic resonance; biochemistry of citrus fruits. *Mailing Add:* 827 N Altadena Dr Pasadena CA 91107-1850

BENNETT, RICHARD BOND, b Grove City, Pa, Nov 28, 32; m 61; c 2. ORGANIC CHEMISTRY. *Educ:* Grove City Col, BS, 54; Ohio State Univ, MSc, 55, PhD(carbohydrate chem), 64. *Prof Exp:* Res chemist, Gulf Res & Develop Co, 58; asst prof chem, Millikin Univ, 61-63; from asst prof to assoc prof, 67-74, chmn dept, 69-71 & 80-86, PROF CHEM, THIEL COL, 74- *Mem:* Am Chem Soc. *Res:* Aldonamide hydrolysis; ketose synthesis. *Mailing Add:* Dept Chem Thiel Col Greenville PA 16125-2182

BENNETT, RICHARD HAROLD, b Washington, DC, July 27, 39; c 3. MARINE GEOLOGY, GEOTECHNIQUE. *Educ:* Am Univ, BS, 67; Tex A&M Univ, PhD(oceanog), 76. *Prof Exp:* Geol asst, US Geol Surv, Dept Interior, 58-64; field crew chief surv, Gravity Meter Explor Co, 64-65; cartographer, Environ Sci Serv Admin, Dept Com, 65-66, res oceanographer, Nat Oceanic & Atmospheric Admin-Atlantic Oceanog & Meteorol Labs-Dept Com, 66-, MGR, MARINE GEOTECH-RATIONAL USE OF SEA FLOOR PROG, NAT OCEANIC & ATMOSPHERIC ADMIN, 76-, CHIEF SCI, SEAFLOOR DIV, NORDA, US NAVY. *Concurrent Pos:* Adv, Deep Sea Drilling Prog, Int Phase Ocean Drilling; USSAC adj prof, Oceanic Eng Div, Univ Miami & Texas A&M Univ. *Mem:* Soc Econ Paleontologists & Mineralogists; Asn Int L'Etude Argiles; Am Soc Civil Engrs. *Res:* Marine geotechnique; sedimentology; clay fabric of submarine sediments; seafloor stability studies of continental margins; in situ measurement of submarine sediment properties; pore pressures in surficial sediments; geophysical and hydrographic surveys, mapping, sediment & transport processes, seafloor stability. *Mailing Add:* PO Box 1756 Picayune MS 39466

BENNETT, RICHARD HENRY, b Hart Co, Ky, Mar 4, 44; m 65; c 3. ORGANIC CHEMISTRY. *Educ:* Western Ky Univ, BS, 65; Vanderbilt Univ, PhD(org chem), 69. *Prof Exp:* Res chemist petrol chem, Texaco Res Lab, 69-75; res chemist liquid chromatography, Jefferson Chem Co, 75-80; SR STAFF CHEMIST, TEXACO CHEM CO, 80- *Mem:* Am Chem Soc; Sigma Xi. *Res:* Improvement of refinery processes; isolation and identification of trace organics present in refinery waste water streams; high pressure liquid chromatography. *Mailing Add:* 3013 Creekview Ct Missouri TX 77459

BENNETT, RICHARD THOMAS, plastics chemistry, for more information see previous edition

BENNETT, ROBERT BOWEN, b Tillamook, Ore, Mar 28, 27; m 49; c 4. PHYSICS. *Educ:* Willamette Univ, BA, 50, Univ Ore, MA, 54, PhD(physics), 58. *Prof Exp:* From asst prof to assoc prof physics, Whiteman Col, 57-65; UNESCO sr lectr, Univ Col Rhodesia & Nyasaland, 65-66, UNESCO lectr, Univ Zambia, 66-67; assoc prof, 67-80, chmn dept, 84-88 PROF PHYSICS, CENT WASH UNIV, 80- *Concurrent Pos:* Staff physicist, Comn Col Physics, Univ Md, College Park, 68-69. *Mem:* Am Asn Physics Teachers; Am Phys Soc; Am Geophys Union. *Res:* Collision process in atomic systems; pressure broadening of spectral lines. *Mailing Add:* Dept Physics Cent Wash State Univ Ellensburg WA 98926

BENNETT, ROBERT M, b Berkhamsted, Eng, Nov 30, 40; US citizen; m 65; c 3. RHEUMATOLOGY, IMMUNOLOGY. *Educ:* Univ London, MB & BS, 64, MD, 78; FRCP, 85. *Prof Exp:* House surgeon, Middlesex Hosp, London, 64; house physician, Taplow Hosp, Maidenhead, UK, 65; med officer, Royal Air Force, 65-70; registr rheumatology, Hammersmith Hosp, London, 70-72; asst prof med, Univ Chicago, 72-76; assoc prof, 76-81, PROF MED, ORE HEALTH SCI UNIV, 81-, HEAD DIV ARTHRITIS & RHEUMATIC DIS, 85- *Concurrent Pos:* Consult, Vet Admin Hosp, Portland, Ore, 76- *Mem:* Am Rheumatism Asn; Am Asn Immunologists; Brit Soc Rheumatology; Med Res Soc Gt Brit; Am Fedn Clin Res; AAAS; fel Am Col Physicians. *Res:* Arthritis and rheumatic diseases; biological role of lactoferrin; DNA receptors, basic pathophysiology and abnormalities of receptor function in systemic lupus erythematosus and allied disorders; fibromyacgia, clinical and basic studies. *Mailing Add:* Dept Med L329A Ore Health Sci Univ Portland OR 97201

BENNETT, ROBERT PUTNAM, b Hartford, Conn, Dec 11, 32; m 57; c 4. ORGANIC CHEMISTRY. *Educ:* Trinity Col, Conn, BS, 55, MS, 57; Case Inst Technol, PhD(org chem), 60. *Prof Exp:* Fel chem, Pa State Univ, 60-61; res chemist Am Cyanamid Co, 61-66, sr res chemist, 66-69; lab dir, 69-74, dir chem res & develop, 74, asst vpres res & develop, 75-78, vpres & tech dir, Apollo Technologies Inc, 78-83; vpres, Gus Inc, 83-88; DIR PROCESS DEVELOP & ENG, NATEC RES INC, 88- *Mem:* Am Chem Soc. *Res:* Catalytic reactions; combustion reactions; catalysis petroleum chemistry; air pollution chemistry and control; coal chemistry. *Mailing Add:* 5915 Penrose Ave Dallas TX 75206

BENNETT, SARA NEVILLE, b Bulloch Co, Ga, Mar 8, 31; m 48; c 3. FUNGAL GENETICS. *Educ:* Ga Southern Col, BS & BSEd, 64, MS, 67; Univ Ga, PhD(microbiol), 75. *Prof Exp:* From instr to prof biol, Ga Southern Col, 66-90, PROF BIOL, GA SOUTHERN UNIV, 90- *Mem:* Genetics Soc Am; AAAS; Sigma Xi; Am Inst Biol Studies. *Res:* Genetics of osmotic-sensitive and morphological mutants of the fungus Neuyospora. *Mailing Add:* Dept Biol Ga Southern Univ Statesboro GA 30460-8042

BENNETT, STEPHEN LAWRENCE, b Winnipeg, Man, Oct 14, 38; US citizen; m 70; c 2. CEMENTED CARBIDES. *Educ:* Univ Man, BSc, 60, MSc, 61; Queen's Univ, Ont, PhD(solid state chem), 66. *Prof Exp:* Res officer solid state chem, Div Appl Chem, Nat Res Coun, 61-62; res assoc high temperature chem, Univ Kans, 66-68; assoc physicist, Math & Physics Div, Midwest Res Inst, Kansas City, Mo, 68-70; res assoc, Rockefeller Univ, New York, 70-72 & Rice Univ, Houston, Tex, 72-74; res staff mem, Metals & Ceramics Div, Oak Ridge Nat Lab, 74-78; mat scientist, Kennametal Inc, Greensburg, Pa, 78-86; RES ENGR, GTE VALENITE CORP, TROY, MICH, 86- *Mem:* Sigma Xi; Am Chem Soc; Am Ceramic Soc. *Res:* Electrical and magnetic properties of solids; single crystal growth; high temperature mass spectrometry; thermodynamic properties of solids; vaporization phenomena; chemical ionization mass spectrometry; negative ionization mass spectrometry; electron affinities; directional solidification; carbon control in cemented carbides; advanced cutting tool materials; modifications to surface microstructure of cemented carbides. *Mailing Add:* GTE Valenite Corp 1711 Thunderbird St Troy MI 48084

BENNETT, STEWART, b New York, NY, Feb 14, 33; m 56; c 2. PHYSICS. *Educ:* Cornell Univ, AB, 53, PhD(nuclear physics), 61. *Prof Exp:* Res physicist anti-submarine warfare, Opers Res Inc, 60-61; staff scientist plasma physics, Res & Adv Develop Div, Avco Corp, 61-63, chief sect, 63-69; mgr spec proj, 69-79, GEN MGR, POLARIZER DIV, POLAROID CORP, 79- *Mem:* Am Phys Soc. *Res:* Plasma and discharge physics; cosmic rays. *Mailing Add:* 133 Annursnac Hill Rd Concord MA 01742

BENNETT, THOMAS EDWARD, b Norristown, Pa, Mar 14, 50; m 75; c 3. PHYSIOLOGY, BIOLOGY. *Educ:* Pa State Univ, BS, 72, MS, 75, PhD(physiol), 79. *Prof Exp:* Lab instr biol, Pa State Univ, 72-77, asst, 74-77; instr anat & physiol, Sch Nursing, Philipsburg State Gen Hosp, 73-80; MEM FAC, DEPT BIOL, BELLARMINE COL, 80-, CHMN, DEPT BIOL, 85- *Concurrent Pos:* Histologist, Depts Poultry Sci & Food Sci, Pa State Univ, 72-74. *Res:* Sensory, neural and endocrine responses to environmental stressors; auditory physiology; effects of gravity on cardiovascular system; space physiology. *Mailing Add:* Dept Biol Bellarmine Col 2001 Newburg Rd Louisville KY 40205-0671

BENNETT, THOMAS P(ETER), b Lakeland, Fla, Oct 8, 37. EXPERIMENTAL BIOLOGY. *Prof Exp:* DIR BIOL, FLA MUS NATURAL HIST, UNIV FLA, 86- *Mailing Add:* Fla Mus Natural Hist Univ Fla Museum Rd Gainesville FL 32611

BENNETT, W DONALD, b Sheffield, Eng, May 2, 24; Can citizen; m 57; c 4. PHYSICS, METALLURGY. *Educ:* Univ Sheffield, BSc, 43, PhD(physics), 49. *Prof Exp:* Nat Res Coun Can fel physics, 49-51; res scientist metall, Dept Mines & Technol Survs, Ont, 51-55, sr sci officer, 55; design specialist physics, Canadair, Ltd, Que, 55-57; adv engr metall, Can Westinghouse Co, Ltd, 57-62; head phys metall res, Falconbridge Nickel Mines, Ltd, 62-68; sci adv, Sci Coun Can, 68-72, Sci Coun Can Embassy, Washington, DC, 72-75; sr sci adv, 75-78, dir prog eval, 78-88, MINES & RESOURCES MGT CONSULT, 88- *Mem:* Am Soc Metals; Am Inst Mining, Metall & Petrol Engrs; fel Brit Inst Physics & Phys Soc. *Res:* Order-disorder in binary alloy systems; metallurgy of titanium, beryllium, zironium and uranium; nickel alloy development; high purity metals; science policy. *Mailing Add:* 17 Borduas Ct Kanata ON K2K 1K9 Can

BENNETT, W SCOTT, b Wash, DC, Nov 29, 29. ELECTROMAGNETIC COMPATIBILITY ENGINEERING. *Educ:* Syracuse Univ, BEE, 63, MSEE, 65, PhD(elec eng), 67. *Prof Exp:* Instr elec eng, Syracuse Univ, 65-67; asst prof, Va Polytech Inst & State Univ, 67-70; staff engr, Burroughs Corp, 70-74; mem tech staff, Hewlett-Packard Co, 74-90; CONSULT ELECTROMAGNETIC COMPATIBILITY ENG, 90- *Concurrent Pos:* Distinguished lectr, Inst Elec & Electronics Engrs, Electromagnetic Compatibility Soc, 88-90; assoc ed, Inst Elec & Electronics Engrs, Electromagnetic Compatibility Soc Trans, 89- *Mem:* Inst Elec & Electronics Engrs; Inst Elec & Electronics Engrs Electromagnetic Compatibility Soc. *Res:* Incorporation of electromagnetic compatibility into the initial design of electronic equipment; measurement of unintentional radiated electromagnetic emissions. *Mailing Add:* 1324 Meadow Ridge Ct Loveland CO 80537

BENNETT, WILLIAM EARL, b Eskridge, Kans, Dec 29, 23; m 52. INORGANIC CHEMISTRY. *Educ:* Sterling Col, BS, 47, Univ Kans, PhD(chem), 51. *Prof Exp:* Res assoc chem, Univ Chicago, 51-52 & Mass Inst Technol, 52-53; asst prof, 53-59, ASSOC PROF INORG CHEM, UNIV IOWA, 59- *Mem:* Am Chem Soc. *Res:* Coordination compounds; ion exchange. *Mailing Add:* Dept Chem Univ Iowa Iowa City IA 52242

BENNETT, WILLIAM ERNEST, b Salters, SC, Feb 2, 28; m 53; c 1. IMMUNOPATHOLOGY. *Educ:* Lincoln Univ, Pa, AB, 50; Temple Univ, MS, 55; Univ Pa, PhD(med microbiol), 60. *Hon Degrees:* ScD, Morehouse Col, 80. *Prof Exp:* Instr-asst prof microbiol, Meharry Med Col, 60-63; NIH fel investr cellular biol, Rockefeller Univ, 63-65; chief immunopath sect, Med Lab Div, US Biol Lab, 66-69; microbiologist sci admin, Div Res Grants, NIH, 69-70, health sci adminr, Div Physician & Health Professions Educ, 70-74; chief, Inst Resources Br Sci Admin, Div Med, Health Resources Admin, Dept Health, Educ & Welfare, 74-80; CHIEF RES MANPOWER DEVELOP, NAT INST ALLERGY INFECTIOUS DIS, NIH, 81- *Concurrent Pos:* Consult, Oak Ridge Nat Lab, 62-65; NIH grants assoc, 69-70. *Mem:* Am Asn Path; Soc Reticuloendothelial Syst; Sigma Xi; Am Soc Microbiol; Fedn Am Soc Exp Biol. *Res:* Mechanisms of cellular immunity; ultrastructural cytochemistry; biochemistry and physiology of macrophages; science administration. *Mailing Add:* Rm 7A-03 Westwood Bldg NIH Off Dir NIAID Bethesda MD 20205

BENNETT, WILLIAM FRANKLIN, PROTEIN BIOCHEMISTRY. *Educ:* Univ Tex Health Sci Ctr, PhD(biochem), 76. *Prof Exp:* SR SCIENTIST, GENENTECH, INC, 82- *Res:* Recovery development; protein structure. *Mailing Add:* Genentech Inc 460 Point San Bruno Blvd South San Francisco CA 94080

BENNETT, WILLIAM FREDERICK, b Plainview, Ark, Jan 23, 27; m 50; c 3. AGRONOMY, PLANT NUTRITION. *Educ:* Okla State Univ, BS, 50; Iowa State Univ, MS, 52, PhD, 58. *Prof Exp:* Exten area agronomist, Iowa State Univ, 51-54, exten agronomist, 54-57; exten soil chemist, Tex A&M Univ, 57-63; chief agronomist, Elcor Chem Corp, 63-68; PROF AGRON, TEX TECH UNIV, 68-, ASSOC DEAN AGR SCI, 70- *Concurrent Pos:* Consult, King Ranch, Arg. *Mem:* Am Soc Agron; Soil Sci Soc Am. *Res:* Soil fertility and fertilizer use on agronomic and vegetable crops; plant analysis and soil testing; developing diagnostic tools. *Mailing Add:* Agr Sci Tex Tech Univ Lubbock TX 79409-2123

BENNETT, WILLIAM M, b Chicago, Ill, May 6, 38; m 77; c 5. INTERNAL MEDICINE, NEPHROLOGY. *Educ:* Northwestern Univ, BS, 60, MA, 63. *Prof Exp:* Instr med, Harvard Med Sch, 69-70; asst prof, 70-74, assoc prof, 74-78, PROF MED, ORE HEALTH SCI UNIV, 78-, PROF PHARMACOL, 81- *Mem:* Am Soc Nephrol; Int Soc Nephrol; Am Soc Pharmacol & Exp Therapeut; Am Fedn Clin Res; fel Am Col Physicians. *Res:* Investigations into the mechanisms of drug nephrotoxicity and the proper prescribing of drugs to patients with renal failure. *Mailing Add:* Div Nephrol Dept Med Ore Health Sci Univ 3181 SW Sam Jackson Park Rd Portland OR 97201

BENNETT, WILLIAM RALPH, b Des Moines, Iowa, June 5, 04; m 28; c 3. ELECTRICAL ENGINEERING, PHYSICS. *Educ:* Ore State Univ, BS, 25; Columbia Univ, AM, 28, PhD(physics), 49. *Prof Exp:* Mem tech staff elec eng, Bell Tel Labs, Inc, 25-65, data commun consult, 60-62, head data theory dept, 62-65; prof, 65-68, Charles Batchelor prof, 68-72, EMER CHARLES BATCHELOR PROF ELEC ENG, COLUMBIA UNIV, 72- *Concurrent Pos:* James Mackay Lectr elec eng, Univ Calif, Berkeley, 64; prin investr, NSF grant, 67-71. *Honors & Awards:* Mervin J Kelly Award, Inst Elec & Electronics Engrs, 68. *Mem:* Fel Inst Elec & Electronics Engrs; fel Am Phys Soc; Sigma Xi. *Res:* Electrical communication. *Mailing Add:* 102 Dunham Lab Yale Univ New Haven CT 06520

BENNETT, WILLIAM RALPH, JR, b NJ, Jan 30, 30; m 52; c 3. PHYSICS. *Educ:* Princeton Univ, BA, 51; Columbia Univ, PhD(physics), 57; Yale Univ, MA, 65. *Hon Degrees:* DsC, Univ New Haven, 74. *Prof Exp:* Asst physics, Radiation Lab, Columbia Univ, 52-54, asst, Cyclotron Group, 54-57; instr, Yale Univ, 57-59, res assoc, 59; mem tech staff, Bell Tel Labs, 59-62; from assoc prof to prof physics & appl sci, 62-70, master, Stillman Col, 81-87, CHARLES BALDWIN SAWYER PROF ENG & APPL SCI & PROF PHYSICS, YALE UNIV, 70- *Concurrent Pos:* Consult, Tech Res Group, Inc, 62-67; vis scientist, Am Inst Physics Prog, 63-64; Sloan Found fel, 63-65; consult, Inst Defense Anal, 63-70; mem adv panel, Nat Acad Sci to Nat Bur Standards, 63-69, chmn, 66; consult, Army Res Off, Durham, 65- & CBS Labs, 66-68; John S Guggenheim Found fel, 67; guest of Soviet Acad Sci, 67, 69 & 79; consult & mem bd dirs, Laser Sci Corp, 68-69; mem lab adv bd for res, Naval Res Adv Comt, 68-76 & Avco Corp, 78. *Honors & Awards:* Morris N Liebmann Award, Inst Elec & Electronics Eng, 65; Western Elec Fund Award, Am Soc Eng Educr, 77. *Mem:* Fel Am Phys Soc; fel Optical Soc Am; fel Inst Elec & Electronics Engrs. *Res:* Atomic and molecular physics; positronium; optical pumping; inelastic collisions; radiative lifetimes; excitation processes; optical and rf spectroscopy; gas lasers. *Mailing Add:* Dunham Lab Yale Univ New Haven CT 06520

BENNETT, WORD BROWN, JR, b Nashville, Tenn, Oct 24, 15; m 39; c 4. NATURAL PRODUCTS CHEMISTRY. *Educ:* Vanderbilt Univ, MS, 40. *Prof Exp:* Chief chemist, US Tobacco Co, 40-56, res dir, 56-66, vpres res & develop, 66-80; RETIRED. *Mem:* Am Chem Soc. *Res:* Dark tobacco. *Mailing Add:* 2700 Overhill Dr Nashville TN 37214

BENNETTE, JERRY MAC, b Abilene, Tex, July 16, 52; m 74; c 1. IRRIGATION MANAGEMENT. *Educ:* Tex Tech Univ, BS, 74, MS, 76; Univ Nebr, PhD(agron), 79. *Prof Exp:* Res asst agron, Tex Tech Univ, 74-76 & Univ Nebr, 76-79; from asst prof to assoc prof, 79-90, PROF, UNIV FLA, 90- *Mem:* Crop Sci Soc Am; Soil Sci Soc Am; Am Soc Agron. *Res:* Physiological and morphological factors which are related to heat and drought resistance of agronomic crops; efficient irrigation scheduling techiques; environmental effects of crop growth, physiology and yield. *Mailing Add:* Bldg 164 Agron Physiol Lab Univ Fla Gainesville FL 32611

BENNICK, ANDERS, SALIVARY PROTEIN, PROTEIN STRUCTURE. *Educ:* Univ Toronto, PhD(biochem), 70. *Prof Exp:* PROF BIOCHEM & CHMN, GRAD DEPT DENT, UNIV TORONTO, 72- *Mailing Add:* Dept Biochem & Fac Dent Univ Toronto Toronto ON M5S 1A8 Can

BENNIGHOF, R(AYMOND) H(OWARD), b Westminster, Md, Apr 3, 28; m 55; c 4. ELECTRONICS ENGINEERING. *Educ:* Western Md Col, BS, 48; Johns Hopkins Univ, BE, 50; Mass Inst Technol, SMEE, 52. *Prof Exp:* Asst proj engr, Radio Div, Bendix Aviation Corp, 52-53; res asst, Johns Hopkins Univ, 53-55, res staff asst, 55-56, res assoc, 56; proj engr, Electronic Commun Inc, 56-59; sr design engr electronics, Aircraft Armaments, Inc, 59-64, prin develop engr, 64-69; prin staff scientist, AAI Corp, 69-88; RETIRED. *Res:* Product development in electronic circuits field. *Mailing Add:* 1910 Stockton Rd Phoenix MD 22131

BENNING, CALVIN JAMES, organic chemistry, for more information see previous edition

BENNING, CARL J, JR, b Springfield, Mo, Dec 4, 30; m 55; c 1. ELECTRICAL ENGINEERING. *Educ:* US Naval Acad, BS, 53; Univ NMex, MS, 64. *Prof Exp:* Staff mem systs eng, Sandia Corp, 57-62, reliability eng, 63-64; res asst, Univ NMex, 62-63; sr engr, Braddock-Dunn-McDonald, 64-65 & Tracor, Inc, 65-67; sr mem tech staff, Tex Instruments, Inc, 67-90; RETIRED. *Mem:* Inst Elec & Electronics Engrs. *Res:* Submarine antenna design; microwave breakdown in wave guides; reliability analysis; moving target indicator radar design; deceptive electronic counter measures systems; signal processing; IR image processing for target detection; system analysis. *Mailing Add:* 6522 Redpine Dallas TX 75248

BENNINGHOFF, WILLIAM SHIFFER, b Ft Wayne, Ind, Mar 23, 18; m 41, 69; c 2. PLANT ECOLOGY. *Educ:* Harvard Univ, SB, 40, MA, 42, PhD(plant geog), 48. *Prof Exp:* Asst bot mus, 38-40, lab instr, Univ exten, Harvard Univ, 46-48; botanist, 48-57, chief Alaska terrain & permafrost sect, US Geol Surv, 52-57; assoc prof, 57-60, assoc prof biol sta, 57, prof, 61, 63, 66, asst dir bot gardens, 65-66, dir matthaei bot gardens, 77-86, PROF BOT, UNIV MICH, 60-, EMER PROF, 88- *Concurrent Pos:* Mem panel biol & med sci, chmn, 66-75, Nat Acad Sci-Nat Res Coun Comt Polar Res, 62-; mem, Sci Comt Antarctic Res working group in biol, 68-80, chmn, 74-80; dir, US Aerobiol prog Int Biol Prog, 67-72, chmn, Int Aerobiol Working Group, 68-72; mem, secy-gen comn Aerobiol, Int Union Biol Sci, 74-82; adv comt mem, Smithsonian Inst Res Awards, 67-74; mem, Nat Acad Sci, Nat Res Coun, US Comt Int Union Biol Sci, 70-76; adv panel mem, Polar Prog, Nat Sci Found, 71-74; organizer, Int Asn Aerobiology, under Int Union Biol Sci, 73, pres, 74, Past pres, 75-83, hon mem, 82, hon mem, Int Asn Areobiol, 82. *Honors & Awards:* Antarctic Serv Medal, US Am, 73. *Mem:* AAAS; fel Geol Soc Am; Bot Soc Am; Am Soc Limnol & Oceanog; fel Arctic Inst NAm; hon mem Int Asn Aerobiol; Sigma Xi. *Res:* Spore dispersal; historical phytogeography; phytocoenology; Arctic and Antarctic ecology. *Mailing Add:* Dept Bot Univ Mich Ann Arbor MI 48109-1048

BENNINGTON, JAMES LYNNE, b Evanston, Ill, Apr 29, 35; m 59; c 2. PATHOLOGY. *Educ:* Univ Chicago, MD, 59, MS, 62. *Prof Exp:* Intern, Presby-St Lukes Hosp, Chicago, 59-60; trainee path, USPHS, 60-62; resident, Kaiser Found Hosp, Oakland, Calif, 62-64; assoc prof, Sch Med, Univ Wash, 64-69; CLIN PROF PATH, STANFORD MED SCH, CHMN DEPT PATH & CLIN LAB, CHILDRENS HOSP SAN FRANCISCO & ADULT MED CTR, SAN FRANCISCO, 70- *Concurrent Pos:* Chief & dir labs, King County Hosp, Wash, 66-69. *Mem:* Am Soc Clin Path; Col Am Path; Int Acad Path. *Res:* Oncology; cellular kinetics of normal and neoplastic tissue; cytomorphometry of neoplastic cells. *Mailing Add:* 3536 Jackson St San Francisco CA 94118

BENNINGTON, KENNETH OLIVER, b Forsyth, Mont, Oct 13, 16; m 69; c 2. GEOCHEMISTRY. *Educ:* Mont State Univ, BS, 47, BA, 49; State Col Wash, MS, 51; Univ Chicago, PhD(geol), 60. *Prof Exp:* Res assoc geol, Univ Chicago, 57-58; crystallographer, Univ Wash, 59-63, res assoc prof geophys, 64-68; RES CHEMIST, ALBANY RES CTR, BUR MINES, US DEPT INTERIOR, 68- *Mem:* Geochem Soc; Am Geophys Union. *Res:* Compositional changes and mineralogical reactions attributable to mechanical stress; thermodynamic properties of silicate minerals. *Mailing Add:* 2021 Ferry St SW-21 Albany OR 97321-3989

BENNINK, MAURICE RAY, b Coopersville, Mich, May 31, 44; m 68; c 2. HUMAN NUTRITION, NUTRITIONAL BIOCHEMISTRY. *Educ:* Mich State Univ, BS, 66; Colo State Univ, MS, 68; Univ Ill, PhD(nutrit sci), 73. *Prof Exp:* Lab technician clin biochem, Blodgett Mem Hosp, Grand Rapids, Mich, 68-70; asst prof, 73-78, ASSOC PROF FOOD SCI & HUMAN NUTRIT, MICH STATE UNIV, 78- *Mem:* Inst Food Technol; Am Inst Nutrit. *Res:* Interactions of diet and colon cancer; diet and lipid metabolism. *Mailing Add:* Dept Food Sci & Human Nutrit Mich State Univ East Lansing MI 48823

BENNION, DOUGLAS NOEL, b Ogden, Utah, Mar 10, 35; m 56; c 6. CHEMICAL ENGINEERING, ELECTROCHEMISTRY. *Educ:* Ore State Univ, BS, 57; Univ Calif, Berkeley, PhD(chem eng), 64. *Prof Exp:* Chem engr, Dow Chem Co, 57-60; from asst prof to assoc prof eng, Univ Calif, Los Angeles, 64-75, prof eng, 75-; CHEM ENG DEPT, BRIGHAM YOUNG UNIV. *Mem:* Electrochem Soc; Am Inst Chem Engrs; Am Chem Soc. *Res:* Mass transport in electrochemical systems and membranes, associated problems in electrode kinetics and hydrodynamics. *Mailing Add:* Dept Chem Eng Brigham Young Univ 350 CB Provo UT 84602

BENNION, DOUGLAS WILFORD, b Cardston, Alta, June 2, 31; m 57; c 3. PETROLEUM ENGINEERING, COMPUTER SCIENCE. *Educ:* Univ Okla, BS, 56; Pa State Univ, MS, 62, PhD(petrol eng), 65. *Prof Exp:* Trainee engr, Mobil Oil Can Ltd, 56, jr engr, 57-60, prod engr, 60-61; from asst prof to assoc prof chem eng, 65-72, PROF CHEM & PETROL ENG, UNIV CALGARY, 72- *Mem:* Can Inst Mining & Metall; Am Inst Mining, Metall & Petrol Engrs; Can Inst Chem Engrs. *Res:* Flow of fluid in heterogeneous porous media; use of numerical models to simulate fluid flow behavior. *Mailing Add:* Dept Chem Eng Univ Calgary Calgary AB T2N 1N4 Can

BENNISON, ALLAN P, b Stockton, Calif, Mar 8, 18; m 41; c 3. GEOLOGY. *Educ:* Univ Calif, Berkeley, AB, 40. *Prof Exp:* Fel geol, Antioch Col, 40-42; photogrammet engr, US Geol Surv, 42-45; asst chief geologist, Tri-Pet Corp, Columbia, 45-49; staff stratigrapher, Sinclair Oil & Gas Co, Okla, 49-68; GEOL CONSULT, 68- *Concurrent Pos:* Lectr, Univ Tulsa, 64; deleg, Int Geol Cong, Prague, Czech, 68-; compiler geol hwy map ser, Am Asn Petrol Geol, 72- *Honors & Awards:* Distinguished Serv Award, Am Asn Petrol Geol, 86; Distinguished Serv Award, Soc Econ Paleont & Mineral, 90. *Mem:* AAAS; Am Asn Petrol Geol; Soc Econ Paleont & Mineral; Am Asn Petrol Geologists; Geol Soc Am; Explorers Club. *Res:* Cretaceous stratigraphy of California; stratigraphy of Oklahoma, Kansas, Colombia, Somalia and Mexico; tectonics of Alaska, California, Wyoming and Utah. *Mailing Add:* Pratt Tower 125 W 15 St Suite 401 Tulsa OK 74119-3801

BENNISON, BERTRAND EARL, b Boston, Mass, Apr 18, 15; m 43; c 4. MEDICINE. *Educ:* Mass Inst Technol, SB, 37; Harvard Univ, MD, 41; Univ Pittsburgh, MPH, 54. *Prof Exp:* Med intern, US Marine Hosp, NY, 41-42; med officer, USPHS, 42-54; asst dir med res div, Esso Res & Eng Co, 54-59; asst dir res, Ortho Pharmaceut Corp, NJ, 60-66; prof biol sci & head dept, Drexel Univ, 66-71; dir, Leon County Health Dept, Fla, 71-76; health prog supvr, Fla Dept Health & Rehab Serv, 77-78; RETIRED. *Concurrent Pos:* Adj prof, Div Reproductive Biol, Hahnemann Med Col; vis prof, Dept Urban & Regional Planning, Fla State Univ, 72-75; clinician, 78-81. *Mem:* AAAS; Am Pub Health Asn. *Res:* Preventive medicine. *Mailing Add:* PO Box 224 Eastham MA 02642-0224

BENNUN, ALFRED, b Buenos Aires, Arg, July 9, 34; m 61; c 1. BIOCHEMISTRY, BIOPHYSICS. *Educ:* Nat Univ Cordoba, BS, 54, MS, 57, PhD(biochem, pharm), 63. *Prof Exp:* Lab instr, Dept Psychiat, Sch Med, Nat Univ Cordoba, 56-57, instr histol & embryol, 57-59; instr biol chem, Univ Buenos Aires, 62-63; fel biochem, Weizmann Inst, 63-64; res assoc, Duke Univ, 64-65; NIH fel, New York, 65-66 & spec fel, Cornell Univ, 66-67; lectr prof, Univ PR, 67-68, asst prof, 68-69; ASSOC PROF BIOCHEM, NEWARK COL ARTS & SCI, RUTGERS UNIV, 69- *Concurrent Pos:* Resident, Clin Path Lab, Ment Health Hosp of State Inst Health, Cordoba, Arg, 56-57; fel, Arg Coun Sci Res, 61 & 63; res grants, NIH, 68-70, Res Corp, 69-72, Rutgers Res Fund, 69-73 & Charles & Johanna Busch Mem Fund, 74-76; mem comt bioenergetics, Int Union Pure & Appl Biophys. *Mem:* Am Chem Soc; Am Soc Plant Physiol; Fedn Am Soc Exp Biol; Biophys Soc; NY Acad Sci. *Res:* Phosphorus metabolism; enzymology; cellular respiration; oxidative and photophosphorylation; hormone receptors and enzymatic regulation; theoretical bioenergetics. *Mailing Add:* Dept Biol Sci Rutgers State Univ 195 University Ave Newark NJ 07102

BENOIT, GUY J C, b Montreal, Que, Oct 6, 26; m 54; c 3. BIOLOGY, MICROBIOLOGY. *Educ:* Univ Montreal, BS, 49, MS, 51, PhD(bact), 55. *Prof Exp:* Res asst bact, Inst Microbiol & Hyg, 53-70, ASSOC PROF MICROBIOL, FAC MED, UNIV MONTREAL, 70- *Mem:* Can Soc Microbiol. *Res:* Biology of mycobacteria. *Mailing Add:* Dept Microbiol Fac Med Univ Montreal Montreal PQ H3C 3J7 Can

BENOIT, PETER WELLS, b Boston, Mass, Dec 9, 39; m 65; c 2. DENTISTRY, ANATOMY. *Educ:* Tufts Univ, BS, 61, DMD, 65, MS, 67, PhD(anat), 71. *Prof Exp:* Consult pharmacol, Astra Pharmaceut Prod Inc, Mass, 71-72; assoc prof oral biol, Sch Dent, Emory Univ, 72-88; PVT PRACT, ORAL/FACIAL PAIN, 89- *Concurrent Pos:* USPHS fel oral path, Tufts Univ, Boston, 65-67; instnl training grant anat, 67-71; residency, Facial Pain Ctr, Univ Fla, 88-89. *Mem:* Int Asn Den Res; Am Acad Oral Path; Sigma Xi. *Res:* Dental research. *Mailing Add:* 3364 Thornwood Dr Atlanta GA 30340

BENOIT, RICHARD J, b Chicopee Falls, Mass, July 10, 22. LIMNOLOGY. *Educ:* Yale Univ, BS, 48, PhD(zool), 57. *Prof Exp:* Supvr biochem processes & chief marine sci, Gen Dynamics, 59-70; prof, 72-84, EMER PROF NATURAL SCI, MOHEGAN COMMUNITY COL, 84-, DIR, ECOSCI LAB, NORWICH, CONN, 70- *Concurrent Pos:* Adj assoc prof bot, Conn Col, New London, 84- *Mem:* Sigma Xi; Int Soc Limnol. *Res:* Ecological value scales; aquatic geochemistry; natural purification in aquatic ecosystems. *Mailing Add:* 162 Beach St Westerly RI 02891

BENOITON, NORMAND LEO, b Somerset, Man, Sept 30, 32; m 73. BIOCHEMISTRY, ORGANIC CHEMISTRY. *Educ:* Loyola Col, Can, BSc, 53; Univ Montreal, MSc, 55, PhD(biochem), 56. *Prof Exp:* Vis scientist, Nat Insts Health, 56-58; Imp Chem Industs res fel, Univ Exeter, 58-60; assoc res officer, Nat Res Coun Can, 60-61; from asst prof to assoc prof, 61-70, PROF BIOCHEM, UNIV OTTAWA, 70- *Concurrent Pos:* Career investr, Med Res Coun Can, 61- *Mem:* Royal Soc Chem; Am Chem Soc; Can Biochem Soc; Biochem Soc. *Res:* Mechanistic, analytical, and synthetic aspects of amino acid, N-methylamino acid, and peptide chemistry; peptide chemistry. *Mailing Add:* Dept Biochem Univ Ottawa Ottawa ON K1H 8M5 Can

BENOKRAITIS, VITALIUS, b Keturkaimis, Lithuania, July 25, 41; m 67; c 1. NUMERICAL ANALYSIS, GRAPHICS APPLICATIONS. *Educ:* Kent State Univ, BS, 64; Univ Ill, MS, 66; Univ Tex, PhD(comput sci), 74. *Prof Exp:* Asst prof comput sci, Va Commonwealth Univ, 74-75; mathematician, Ballistic Res Labs, US Army, 75-81; DESIGN ANALYST, AAI CORP, 81- *Mem:* Soc Indust & Appl Math; Asn Comput Mach; Am Math Soc. *Res:* Iterative solution of large linear systems; numerical solution of partial differential equations; mathematical modeling; real time simulation; computer natural language processing. *Mailing Add:* 4404 Starview Ct Glen Arm MD 21057

BENOLKEN, ROBERT MARSHALL, b St Paul, Minn, May 11, 32; m 57; c 6. CELL PHYSIOLOGY. *Educ:* Marquette Univ, BS, 54; Johns Hopkins Univ, PhD(biophys), 59. *Prof Exp:* Asst physics, Marquette Univ, 53-54; asst biophys, Johns Hopkins Univ, 54-56, res assoc, 56-59; from asst prof to assoc prof, Univ Minn, Minneapolis, 59-68; prof neural sci, Grad Sch Biomed Sci, Univ Tex, 68-80; INSTR PHYSICS, CHEMEKETA COMMUNITY COL, 80- *Mem:* Soc Gen Physiologists; Biophys Soc. *Res:* Cellular physiology; physiology of visual sense cells; electrophysiology. *Mailing Add:* Sci Dept Chemeketa Community Col Salem OR 97309

BENOS, DALE JOHN, b Cleveland, Ohio, Sept 30, 50. PHYSIOLOGY, BIOPHYSICS. *Educ:* Case Western Reserve Univ, BA, 72; Duke Univ, PhD(physiol & pharmacol), 76. *Prof Exp:* Postdoctoral physiol, Med Ctr, Duke Univ, 76-78; from asst prof to assoc prof physiol-biophys, Harvard Med Sch, 78-85; assoc prof, 85-87, RES SCIENTIST, CF RES CTR, UNIV ALA, BIRMINGHAM, 85-, PROF PHYSIOL-BIOPHYS, 87- *Concurrent Pos:* Nat Res Serv Award, Nat Inst Arthritis, Metabolism & Digestive Dis, NIH, 76-78; prin investr, NIH, 79-, Off Naval Res; mem, Cellular Biol & Physiol Study Sect, NIH, 85-89. *Mem:* Am Physiol Soc; Am Soc Cell Biol; Biophys Soc; Soc Gen Physiologists; Soc Study Reproduction; NY Acad Sci. *Res:* Mechanisms of cation transport across epithelial and cellular membranes; development aspects of ionic transport and metabolic function in preimplantation mammalian embryos and cultured neural and epithelial cell lines. *Mailing Add:* Dept Physiol & Biophys Univ Ala UAB Sta Birmingham AL 35294

BEN-PORAT, TAMAR, b Worms, Ger, Sept 4, 29; US citizen; m 59; c 2. VIROLOGY, BIOCHEMISTRY. *Educ:* Hebrew Univ Jerusalem, MS, 54; Univ Ill, PhD(microbiol), 59. *Prof Exp:* Assoc mem staff virol, Albert Einstein Med Ctr, 58-70; res assoc prof microbiol, Sch Med, Temple Univ, 70-72; assoc prof, 72-78, PROF MICROBIOL, SCH MED, VANDERBILT UNIV, 78- *Concurrent Pos:* Mem rev comt, Cancer Res Ctr, Nat Cancer Inst; assoc ed, Virology, 78-81, 81-84. *Mem:* Am Soc Microbiol; AAAS; Int Soc Veneral Dis. *Res:* Analysis of virus growth at the molecular level. *Mailing Add:* Dept Microbiol Vanderbilt Univ Sch Med 21st Ave S & Garland Nashville TN 37232

BENRUD, CHARLES HARRIS, b Goodhue, Minn, Apr 30, 21; m 49; c 3. STATISTICS, AGRICULTURAL ECONOMICS. *Educ:* Univ Minn, BS, 48, MS, 49, PhD(agr econ), 63. *Prof Exp:* Asst agr agent, Agr Exten Serv, Faribault County, Minn, 49-50, 4 H Club agent, Hennepin County, 50-53 & agr agent, Ramsey County, 53-55; asst prof econ & statist & asst economist, SDak State Col, 55-58, assoc prof & assoc economist, 58-63; sr statistician, 63-83, RES STATISTICIAN, RES TRIANGLE INST, 83- *Res:* Sampling; variances of estimates; population projection; farm management. *Mailing Add:* 929 Warren Ave Cary NC 27511

BENS, FREDERICK PETER, b Detroit, Mich, Oct 20, 13. MARKET DEVELOPMENT REFRACTORY METALS, MOLTEN SALT BATH TECHNOLOGY. *Educ:* Wayne State Univ, BS, 36. *Prof Exp:* Anal chemist to res supvr, Climax Molybdenum Co, 36-57, asst mgr, 57-59, mgr, Automotive Dept, 74; mgr mkt develop, 74-89, MGR, CORP DEVELOP & CONSULT, KOLENE CORP, 89- *Honors & Awards:* Cert of Appreciation, Soc Automotive Engrs, 75; Allan Ray Putnam Serv Award, Am Soc Metals Int, 91. *Mem:* Am Foundrymen's Soc; Am Soc Metals Int; Am Inst Mining & Metall Engrs; fel Soc Automotive Engrs; Am Soc Testing & Mat; Soc Mfg Engrs; Am Ord Asn; Nat Asn Corrosion Engrs. *Res:* Chromium-base alloys for high velocity machine gun liners; super-charger buckets for gas turbines; Nine United States patents and one British patent. *Mailing Add:* 51 Claireview Rd Grosse Pointe Shores MI 48236

BENSADOUN, ANDRE, b Fes, Morocco, Nov 10, 31; US citizen; m 59; c 3. NUTRITION, PHYSIOLOGY. *Educ:* Univ Bordeaux, BS, 51; Nat Sch Advan Agron, France, BSEng, 56; Univ Toulouse, MS, 56; Cornell Univ, PhD(nutrit), 60. *Prof Exp:* Res nutritionist, Cornell Univ, 59-65; asst prof nutrit, Univ Ill, Urbana, 65-67; assoc prof physiol, 67-76, PROF NUTRIT BIOCHEM, CORNELL UNIV, 76- *Mem:* Am Inst Nutrit; Soc Exp Biol & Med; Am Soc Biochem & Molecular Biol; Am Soc Cell Biol. *Res:* Lipid metabolism; lipid transport; lipoproteins; lipases; molecular biology of lipases. *Mailing Add:* Div Nutrit Sci 242 Savage Hall Cornell Univ Ithaca NY 14853

BENSCH, KLAUS GEORGE, b Miedar, Ger, Sept 1, 28; nat US; m 55; c 3. HUMAN PATHOLOGY. *Educ:* Univ Erlangen, MD, 53; Am Bd Path, dipl, 59. *Prof Exp:* From instr to assoc prof path, Sch Med, Yale Univ, 58-68; prof & actg chmn path, 68-84, PROF & CHMN PATH, SCH MED, STANFORD UNIV, 84- *Mem:* Am Asn Path & Bact. *Res:* Cellular pathology and its underlying molecular mechanisms. *Mailing Add:* Dept Path Stanford Univ Med Ctr L-235 Stanford CA 94305

BENSCHOTER, REBA ANN, b Smithland, Iowa, June 14, 30; m 56; c 4. BIOMEDICAL COMMUNICATIONS, ALLIED HEALTH PROFESSIONS ADMINISTRATION. *Educ:* Briar Cliff Col, BA, 52; Iowa State Univ, MS, 56; Univ Nebr, Lincoln, PhD(adult educ), 78. *Prof Exp:* Dir AV Aids Sect, Nebr Psychiat Inst, 60-62, from instr to asst prof med teaching aids, 62-69; assoc prof, 69-70, ASSOC PROF BIOMED EDUC, SCH ALLIED HEALTH, MED CTR, UNIV NEBR, 78-, ASSOC PROF COL NURSING, 79-, ASSOC PROF DEPT ADULT & CONTINUING EDUC, 84-; DIR BIOMED COMMUN, 70-, ASSOC DEAN, ALLIED HEALTH, UNIV NEBR MED CTR, 85- *Concurrent Pos:* Dir Commun Sect, Nebr Psychiat Inst, 62-88; Consult, Nat Libr Med, 72-86, Vet Admin, 74-76, Nat Heart, Lung & Blood Inst, 76-; reader-consult, Pub Health Serv Health Manpower Progs, Dallas region, 74-76; mem, Nat Res & Demonstration Rev Comn, 74-76, consult for rev, Nat Heart Lung & Blood Inst, NIH, 76-; consult, Proj Hope, Zhejiang Med Univ, China, 87, Univ Mich, 87 & George Washington Univ, 90- *Honors & Awards:* Golden Raster Award, Health Sci Commun Asn, 81. *Mem:* Health Sci Commun Asn (pres, 76-77); Asn Biomed Commun Dirs; Asn Allied Health Professions; Asn Med Illustrators; Asn Educ & Commun Technol. *Res:* Use of two-way television systems in education and psychiatric care; training biomedical communications specialists; computer applications in health sciences administration; education of allied health personnel for rural practice; distance learning; communication systems. *Mailing Add:* 2528 S 40th St Omaha NE 68105

BENSEL, JOHN PHILLIP, b Glen Ridge, NJ, Oct 30, 45; m 70. SOLID STATE PHYSICS. *Educ:* Stevens Inst Technol, BS, 67; Univ Pa, PhD(physics), 73. *Prof Exp:* Res assoc physics, Univ Ill, Urbana, 73-76; mem staff, 76-80, RES ENGR, DEPT PHYSICS, COL WILLIAM & MARY, 80- *Mem:* Am Phys Soc. *Res:* Instrumentation and fabrication techniques; production of custom made instruments with special expertise in scintilants and light guides, vacuum systems, crygenics; general mechanical fabrication as applied to all fields of physics. *Mailing Add:* Dept Physics Col William & Mary Williamsburg VA 23185

BENSELER, ROLF WILHELM, b San Jose, Calif, Sept 24, 32; m 61; c 2. DENDROLOGY, FOREST BIOLOGY. *Educ:* Univ Calif, Berkeley, BS, 57, PhD(bot), 68; Yale Univ, MF, 58. *Prof Exp:* Jr specialist, Sch Forestry, Univ Calif, 58-60; instr biol, Modesto Jr Col, 61-63; lectr, San Jose State Univ, 67; from asst prof to assoc prof, 68-75, PROF BIOL SCI, CALIF STATE UNIV, HAYWARD, 75- *Mem:* Am Inst Biol Scientists; Bot Soc Am; Am Soc Plant Taxonomists. *Res:* Reproductive biology and biosystematics of woody plants. *Mailing Add:* Dept Biol Sci Calif State Univ Hayward CA 94542

BENSEN, DAVID WARREN, b Paterson, NJ, Feb 6, 28; m 54; c 4. SOIL CHEMISTRY. *Educ:* Rutgers Univ, BSA, 54, MS, 55, PhD(soil chem), 58. *Prof Exp:* Chemist, Hanford Atomic Prod Oper, 58-63; sr res scientist, Isotopes, Inc, 63-66; tech mgr gov contracts, US Naval Radiol Defense Lab, 66-67; res analyst, Res & Eng Off, Civil Defense, Off Secy Army, 67-73, dep dir hazard eval & vulnerability reduction div, Defense Civil Preparedness Agency, 73-78, prog mgr nuclear hazards, Off Res, 78-81, sr scientist, Fed Emergency Mgt Agency, 81-86; SR STAFF OFFICER & PROG ADMINR, NAT ACAD SCI/NAT RES COUN, 86- *Mem:* Sigma Xi; Health Physics Soc; Am Nuclear Soc. *Res:* Soil chemistry; soil-plant relationships; soil fertility; plant nutrition; radioactive waste disposal; radionuclide reactions with soils and minerals; fallout contamination in the food chain; fallout phenomenology; radiological hazard assessment; radiological countermeasures and protection; civil defense. *Mailing Add:* 4624 Sunflower Dr Rockville MD 20853

BENSEN, JACK F, b Chicago, Ill, Jan 4, 23; c 3. SPEECH & HEARING SCIENCES. *Educ:* Univ Miami, BA, 49; Univ WVa, MA, 51; Univ Fla, PhD, 61. *Prof Exp:* Speech therapist, Univ Fla, 51-53; dean men, Dana Col, 54-55; prof speech, Univ Miami, 55-88, dir, Speech & Hearing Clin, 68-88; RETIRED. *Concurrent Pos:* Mem staff, S Fla Cleft Palate Clin, 56-58; consult, United Cerebral Palsy, 65-68; mem bd dirs, Miami Hearing & Speech Ctr, 65-68. *Mem:* Am Speech & Hearing Asn; Am Cleft Palate Asn. *Res:* Cleft palate; speech analysis of the cerebral palsied. *Mailing Add:* Theater Arts Dept Univ Miami PO Box 248273 Coral Gables FL 33124

BENSINGER, DAVID AUGUST, b St Louis, Mo, May 14, 26; div; c 2. DENTISTRY, PERIODONTOLOGY. *Educ:* Wash Univ, BA, 44; St Louis Univ, DDS, 48. *Prof Exp:* From instr to asst prof, 49-55, Wash Univ, 49-55, asst dean, 69-71, assoc prof dent med, Sch Dent, 55-, assoc dean, 71-; RETIRED. *Concurrent Pos:* Consult, US Air Force, 59- *Mem:* Fel Am Col Dent; Am Acad Periodont; Am Soc Periodont; Int Asn Dent Res. *Res:* Oral diagnosis; time-lapse alterations in periodontal disease. *Mailing Add:* Sch Dent Wash Univ St Louis MO 63110

BENSINGER, JAMES ROBERT, b Washington, DC, Aug 20, 41; m 66; c 2. ELEMENTARY PARTICLE PHYSICS. *Educ:* Bucknell Univ, BS, 63; Univ Wis, PhD(physics), 70. *Prof Exp:* Asst prof physics, Univ Pa, 73-74; from asst prof to assoc prof, 74-89, PROF PHYSICS, BRANDEIS UNIV, 89- *Mem:* Am Phys Soc. *Res:* Superconducting super collider physics. *Mailing Add:* Dept Physics Brandeis Univ Waltham MA 02154-9110

BENSLEY, EDWARD HORTON, history of medicine, for more information see previous edition

BENSON, ANDREW ALM, b Modesto, Calif, Sept 24, 17; m 42, 71; c 2. BIOCHEMISTRY, PLANT PHYSIOLOGY. *Educ:* Univ Calif, Berkeley, BS, 39; Calif Inst Technol, PhD(org chem), 42. *Hon Degrees:* DPhil, Univ Oslo, 65; Dr, Univ Pierre et Marie Curie, Paris, 86. *Prof Exp:* Instr chem, Univ Calif, Berkeley, 42-43, asst dir, Bio-Org Group, Radiation Lab, 46-55; res assoc, Stanford Univ, 44-45; asst, Calif Inst Technol, 45-46; from assoc prof to prof agr & biol chem, Pa State Univ, 55-61; prof in residence physiol chem & biophys & res biochemist, Lab Nuclear Med, Univ Calif, Los Angeles, 61-62; chmn, Dept Marine Biol, Scripps Inst Oceanog, 65-70, assoc dir inst, 66-70, chmn, Marine Biol Res Div, 67-70, actg chmn, 70-71, dir, Physiol Res Lab, 70-77, prof, 62-88, EMER PROF BIOL, SCRIPPS INST OCEANOG, UNIV CALIF, SAN DIEGO, 88- *Concurrent Pos:* Fulbright lectr, Agr Col Norway, 51-52; consult, Molecular Biol Prog, NSF, 61-63 & Space Biol Panel, Am Inst Biol Sci/NASA, 81-87; adv coun, Cousteau Soc, 77-; mem, adv bd, Australian Inst Marine Sci, 75 & Marine Biotechnol Inst, Co, Tokyo, 90-; sr Queens fel, Australia, 79. *Honors & Awards:* Sugar Res Found Award, 50; Lawrence Mem Award, US Atomic Energy Comn, 62; Stephen Hales Award, Am Soc Plant Physiol, 72; Supelco Sci Award, Am Oil Chemists Soc, 87.

Mem: Nat Acad Sci; Am Soc Biol Chem; Am Chem Soc; Am Soc Plant Physiol; Royal Norweg Soc Sci & Lett. *Res:* Thyroxine analog synthesis; path of carbon in photosynthesis; tracer methodology; neutron activation analysis; radiochromatography; hot-atom chemistry; lipid biochemistry; biological membrane surfactants; phosphonic and sulfonic acid metabolism; arsenic metabolism; coral metabolism; wax ester metabolism. *Mailing Add:* Scripps Inst Oceanog Univ Calif San Diego CA 92093-0202

BENSON, ANN MARIE, DETOXIFICATION ENZYMES, CANCER CHEMOTHERAPY. *Educ:* Univ Hawaii, PhD(biochem), 69. *Prof Exp:* ASST PROF BIOCHEM, UNIV ARK MED SCI, 84- *Res:* Biochemical mechanisms of protection against chemical carcinogenesis and toxicity. *Mailing Add:* PO Box 24104 Little Rock AR 72221-4104

BENSON, BARRETT WENDELL, b Brattleboro, Vt, Sept 4, 39; m 57; c 4. ORGANIC CHEMISTRY. *Educ:* Middlebury Col, AB, 61; Univ Vt, PhD(org chem), 65. *Prof Exp:* Asst prof chem, Fresno State Col, 65-67; assoc prof, 67-72, chem dept, 72-77, PROF CHEM, BLOOMSBURG STATE COL, 72- *Concurrent Pos:* NSF res grants, 68-; vis prof, Dartmouth Col, 74. *Mem:* Am Chem Soc. *Res:* Organic structure determination; organic reaction mechanisms. *Mailing Add:* Dept Chem Bloomsburg Univ Bloomsburg PA 17815

BENSON, BRENT W, b Chicago, Ill, Apr 10, 41; m 62; c 2. BIOPHYSICS, RADIATION CHEMISTRY. *Educ:* Knox Col, Ill, BA, 63; Pa State Univ, MS, 65, PhD(biophys), 69. *Prof Exp:* Asst prof physics, Southern Ill Univ, 69-72; ASSOC PROF PHYSICS, LEHIGH UNIV, 72- *Concurrent Pos:* Consult, Bell Tel Labs, 80-83; res scientist, Max Planck Inst, Stuttgart, 87. *Mem:* AAAS; Biophys Soc; Am Phys Soc; Radiation Res Soc. *Res:* Electron spin resonance studies of radiation induced free radicals in important biological molecules; biological and molecular effects due to decay of incorporated radioisotopes; reactions of microcircuit materials with atomic oxygen and other species produced by electrical discharge of gas mixtures; point defects in semiconductors. *Mailing Add:* Physics Bldg 16 Lehigh Univ Bethleham PA 18015

BENSON, BRUCE BUZZELL, chemical physics, oceanography; deceased, see previous edition for last biography

BENSON, CARL SIDNEY, b Minneapolis, Minn, June 23, 27; m 55; c 3. GLACIOLOGY. *Educ:* Univ Minn, BA, 50, MS, 56; Calif Inst Technol, PhD(geol), 60. *Prof Exp:* Physicist, Snow, Ice & Permafrost Res Estab, 52-56; asst prof geophys res, 60-61, assoc res geophysicist, 61-62, assoc prof, 62-69, chmn dept geol, 69-73, prof, 69-87, EMER PROF GEOPHYS INST, UNIV ALASKA, 87-; HYDROLOGIST, US GEOL SURV, 90- *Concurrent Pos:* Glaciol panel, Comt Polar Res, Nat Res Coun-Nat Acad Sci, 64-67; mem, Gov Environ Adv Bd Alaska, 75- *Mem:* Fel AAAS; Am Geophys Union; Meteorol Soc; fel Arctic Inst NAm; Int Glaciol Soc; Geol Soc Am. *Res:* Glacier-volcano interactions; freezing processes; physical processes in seasonal snow cover; low temperature air pollution. *Mailing Add:* Geophys Inst Univ Alaska Fairbanks AK 99775-0800

BENSON, CHARLES EVERETT, b Dayton, Ohio, Dec 15, 37; m 60; c 2. BIOCHEMICAL GENETICS, INFECTIOUS DISEASES. *Educ:* Franklin Col, AB, 60; Miami Univ, MS, 64; Wake Forest Univ, PhD(microbiol), 69; Univ Pa, MA. *Hon Degrees:* MA, Univ Pa. *Prof Exp:* Teacher gen sci, Cornell Heights Elem Sch, 60-61; res asst cardiac physiol, Dept Res, Miami Valley Hosp, 63-65; trainee, 69-71, scholar, pa Plan Develop Scientists Med Res, 71-74, res assoc biochem genetics, 73-75, asst prof, Sch Allied Med Prof & Sch Med, 75-79, assoc prof, 79-80, assoc prof microbiol & dir clin microbiol, Large Animals Sch Vet Med, 80-88, PROF MICROBIOL, LARGE ANIMALS SCH VET MED, UNIV PA, 88-, CHMN CLIN STUDIES, NEW BOLTON CTR, 89- *Honors & Awards:* President's Award, Am Soc Metals, 69. *Mem:* Am Soc Microbiol; NY Acad Sci; Soc Gen Microbiol; Sigma Xi; Am Asn Vet Lab Diagnosticians. *Res:* Interrelationship of gene and enzyme function in purine metabolism of enteric bacteria the mechanism of salmonella pathogenesis; mechanism of host bacteria interaction in the disease process; Potomac Horse Fever-isolation of the etiological agent and characterization of the disease; salmonella enteritidis in poultry. *Mailing Add:* 123 Hawthorne Ave Haddonfield NJ 08033

BENSON, DALE B(ULEN), b Madison, Wis, Aug 21, 30; m 59; c 3. CHEMICAL ENGINEERING. *Educ:* Mont State Univ, BA, 52, MS, 55, PhD(chem eng), 59. *Prof Exp:* Res engr chem, Calif Res Corp, 58-59; sr res engr, Chevron Res Co, Standard Oil Co Calif, 59-85; RETIRED. *Mem:* Am Inst Indust Engrs; Am Inst Chem Engrs. *Res:* Catalytic hydrotreating of shale oil; petroleum process development; process planning. *Mailing Add:* 2808 Appaloosa Ct Pinole CA 94564

BENSON, DAVID MICHAEL, b Dayton, Ohio, Aug 28, 45; m 67; c 3. PLANT PATHOLOGY. *Educ:* Earlham Col, BS, 67; Colo State Univ, MS, 68, PhD(plant path), 73. *Prof Exp:* Postdoctoral fel plant path, Univ Calif, Berkeley, 73-74; from asst prof to assoc prof, 74-84, PROF PLANT PATH, NC STATE UNIV, 84- *Mem:* Am Phytopath Soc; Sigma Xi. *Res:* Ecology of soil-borne plant pathogens; phytophthora diseases; ornamental diseases. *Mailing Add:* NC State Univ Box 7616 Raleigh NC 27695-7616

BENSON, DEAN CLIFTON, b Hazelton, NDak, Oct 25, 18; m 43; c 2. MATHEMATICS. *Educ:* Sioux Falls Col, BA, 41; Iowa State Col, MS, 47, PhD(math), 54. *Prof Exp:* Aeronaut res photographer, Nat Adv Comt Aeronaut, Langley Field, 44-45; asst prof math, Sioux Falls Col, 47-50; instr, Iowa State col, 50-54; asst prof, SDak Sch Mines & Tech, 54-56; asst prof, Chico State Col, 56-60; prof math, SDak Sch Mines & Technol, 60-81, chmn dept, 65-81; RETIRED. *Mem:* Am Math Soc; Math Asn Am. *Res:* Analysis; differential equations; complex variables; vector analysis. *Mailing Add:* No 5125 Fir Lane Fir Grove No 1 Rathdrum ID 83858

BENSON, DENNIS ALAN, b New York, NY, Apr 8, 44; m 69; c 1. NEUROPHYSIOLOGY, PHYSIOLOGICAL PSYCHOLOGY. *Educ:* Univ Fla, BS, 69, PhD(physiol psychol), 73. *Prof Exp:* Fel, Sch Med, Johns Hopkins Univ, 74-77, res scientist neurosci, Dept Biomed Eng, 77-81. *Concurrent Pos:* NIH fel, 74-76; NATO travel grant, Conf Animal Behav, Ulm, Ger, 77; Computer Res Specialist, Nat Libr Med, 80-81. *Mem:* Acoust Soc Am; Soc Neurosci; AAAS. *Res:* Neurophysiology of auditory system; electrophysiological correlates of behavior; lateralization of function. *Mailing Add:* Dept Biomed Eng Johns Hopkins Univ Sch Med 720 Rutledge Ave Baltimore MD 21205

BENSON, DONALD CHARLES, b Modesto, Calif, June 6, 27; m 54; c 2. DIFFERENTIAL EQUATIONS, SOFTWARE SYSTEMS. *Educ:* Pomona Col, BA, 50; Stanford Univ, MS, 52, PhD(math), 54. *Prof Exp:* Instr math, Princeton Univ, 54-55; asst prof, Carnegie Inst Technol, 55-57; from asst prof to assoc prof, 57-68, prof, 68-83, EMER PROF MATH, UNIV CALIF, DAVIS, 83- *Concurrent Pos:* Freelance, Software Develop, 83- *Mem:* Am Math Soc; Math Asn Am; Soc Indust Appl Math. *Res:* Analysis; inequalities; nonlinear ordinary differential equations. *Mailing Add:* 3505 Ridgeview Circle Ct Palm Springs CA 92264-5042

BENSON, DONALD WARREN, b Jamestown, NY, Aug 17, 21; m 46; c 3. ANESTHESIOLOGY. *Educ:* Univ Chicago, BS, 48, MD, 50, PhD(pharmacol), 57. *Prof Exp:* From instr to asst prof anesthesiol, Univ Chicago, 53-56; from assoc prof to prof, Johns Hopkins Univ, 56-75; PROF ANESTHESIOL & CHMN DEPT, UNIV CHICAGO, 75- *Mem:* Am Soc Anesthesiol; AMA; Int Anesthesia Res Soc. *Res:* Respiratory physiology; mechanical respirators; hypothermia. *Mailing Add:* Dept Anesthesiol Univ Chicago Box 428 Chicago IL 60637

BENSON, EDMUND WALTER, b Woburn, Mass, July 27, 38; m 73; c 2. INORGANIC CHEMISTRY, ANALYTICAL CHEMISTRY. *Educ:* Univ NH, BS, 61; Univ Tenn, MS, 63, PhD(chem), 67. *Prof Exp:* Res assoc chem, Univ Ill, 66-67; asst prof, 67-72, ASSOC PROF CHEM, CENT MICH UNIV, 72- *Mem:* Am Chem Soc; Sigma Xi. *Res:* Water analysis and quality. *Mailing Add:* Dept Chem Cent Mich Univ Mt Pleasant MI 48859

BENSON, ELLIS STARBRANCH, b Xuchang, China, Oct 28, 19; m 47; c 3. PATHOLOGY. *Educ:* Augustana Col, BA, 41; Univ Minn, MD, 45. *Hon Degrees:* DSc, Midland Lutheran Col, Fremont, NE, 86. *Prof Exp:* From instr to asst prof lab med, 50-53, asst prof, lab med, 53-57, from asst dir to dir clin labs, 53-66, assoc prof path, lab med, 57-61, head lab med, 66-73, PROF LAB MED & BIOCHEM, UNIV MINN, 61-, HEAD LAB MED & PATH, 73- *Concurrent Pos:* USPHS sr res fel, 57-63; trustee, Am Bd Path, 71-83. *Honors & Awards:* G T Evans Award, Acad Clin Lab Physicians & Scientists, 72; George H Whipple lectr, Univ Rochester, 77; Ward Burdick Award, Am Soc Clin Path, 78. *Mem:* Am Soc Clin Path; Am Soc Biol Chemists; Am Asn Path & Bact; Biophys Soc; Am Soc Cell Biol; Am Soc Hemat. *Res:* Muscle structure and function; hemoglobin and red blood cell structure and function. *Mailing Add:* Dept Lab Med Univ Minn 0242 Mayo Minneapolis MN 55455-0309

BENSON, ERNEST PHILLIP, JR, b New Castle, Pa, Feb 21, 36; m 60. INORGANIC CHEMISTRY. *Educ:* Geneva Col, BS, 58; Mich State Univ, PhD(inorg chem), 63. *Prof Exp:* Res fel, Pa State Univ, 63-64; asst prof chem, Fairleigh Dickinson, Florham-Madison Campus, 64-66; asst prof, 66-74, ASSOC PROF CHEM, IDAHO STATE UNIV, 74- *Mem:* Am Chem Soc; Sigma Xi. *Res:* Electron transfer reactions; preparation of coordination compounds; reactions of oxyanions. *Mailing Add:* Dept Chem Box 8160 Idaho State Univ Pocatello ID 83209-8160

BENSON, FRED J(ACOB), b Grainfield, Kans, Sept 27, 14; m 37; c 3. CIVIL ENGINEERING. *Educ:* Kans State Univ, BS, 35; Agr & Mech Col Tex, MS, 36. *Hon Degrees:* Dr Eng, Kans State Univ, 80. *Prof Exp:* Draftsman, Kans Hwy Comn, 35; instr, Purdue Univ, 36-37; draftsman, Int Boundary Comn, Tex, 37; from instr to assoc prof civil eng, Tex A&M Univ, 37-47, prof, 47-80, dean eng, 57-78, vpres & dir res found, 63-80, vpres eng & non-renewable resources, 78-79, vchancellor, 79-80, PRES, TEX A&M RES FOUND, TEX A&M UNIV, 80- *Concurrent Pos:* Asst city engr, Wichita Falls, 38; city engr, College Station, 47-63; exec officer, Tex Transp Inst, 55-62; vdir, Tex Eng Exp Sta, 56-57, dir, 62-78; eng consult. *Honors & Awards:* Edmund Friedman Prof Recognition Award, Am Soc Civil Engrs, 73. *Mem:* Soc Civil Engrs; Am Road Builders Asn; Sigma Xi; hom mem Am Soc Civil Engrs. *Res:* Highway and airfield pavements; aggregates; bituminous materials; concrete. *Mailing Add:* 817 Rosemary Dr Bryan TX 77802

BENSON, FREDERIC RUPERT, b Cape Girardeau, Mo, Sept 21, 15; m 39; c 1. ORGANIC CHEMISTRY, INFORMATION SCIENCE. *Educ:* Wesleyan Univ, AB, 36; NY Univ, MS, 38, PhD(org chem), 47. *Prof Exp:* Res chemist, Hambrock Chem Corp, NJ, 37; instr chem, Bergen Jr Col, 38-41; res chemist, Picatinny Arsenal, US War Ord Dept, NJ, 41-45; chief res chemist, Res Lab, Remington Rand, Inc, 46-54; sr res chemist, Metalectro Corp, 54-56; supvr info br, Chem Res Dept, Atlas Chem Industs, Inc, 54-56, supvr info br, Chem Res Dept, 56-59, mgr info sect, 59-70, mgr info serv, ICI Americas, Inc, 70-81; RETIRED. *Mem:* AAAS; Am Chem Soc; Drug Info Asn. *Res:* Organic nitrogen chemistry; tetrazoles; triazoles; hydrazine chemistry; polyols; surfactants; automation of chemical information; information research. *Mailing Add:* 410 Northwood Rd Wilmington DE 19803

BENSON, GEORGE CAMPBELL, b Toronto, Ont, July 25, 19; m 46; c 2. THERMODYNAMICS. *Educ:* Univ Toronto, BA, 42, MA, 43, PhD(chem), 45. *Prof Exp:* Res chemist, Chem Div, Nat Res Coun Can, 45-49; mem solid state physics group, Bristol Univ, 49-50; res chemist, chem div, Nat Res Coun Can, 51-84; HON SR SCIENTIST, DEPT CHEM ENG, UNIV OTTAWA, 84- *Honors & Awards:* Bronze Medal, Brit Asn, 42. *Mem:* Am Phys Soc; Am Chem Soc; Royal Soc Can; Chem Inst Can. *Res:* Thermodynamic studies of mixtures of nonelectrolytes. *Mailing Add:* Dept Chem Eng Univ Ottawa 770 King Edward Ave Ottawa ON K1N 9B4 Can

BENSON, GILBERT THOMAS, b Los Angeles, Calif, Oct 3, 29; m 57; c 1. GEOLOGY. *Educ:* Stanford Univ, BS, 52, MS, 53; Yale Univ, PhD(geol), 63. *Prof Exp:* Geologist, Texaco, Inc, 53-58; asst prof geol, Univ Ore, 62-68; asst prof geol, 68-74, ASSOC PROF EARTH SCI, PORTLAND STATE UNIV, 74- *Mem:* Geol Soc Am; Am Asn Petrol Geologists; Sigma Xi. *Res:* Structural, Areal, and petroleum geology. *Mailing Add:* 2705 SW Glen Eagles Rd Lake Oswego OR 97034

BENSON, H(OMER) E(DWIN), chemical engineering, for more information see previous edition

BENSON, HARRIET, b Kansas City, Mo, May 17, 41. ORGANIC CHEMISTRY, SCIENTIFIC BIBLIOGRAPHY. *Educ:* Wellesley Col, BA, 63; Univ Kans, PhD(org chem), 67. *Prof Exp:* Lit specialist, Shell Develop Co, 63-64; fel, State Univ Groningen, 68-69; MEM STAFF, TECH INFO & REGULATORY AFFAIRS, ALZA CORP, 69- *Mem:* AAAS; Am Chem Soc; fel Am Med Writers Asn; Drug Info Asn. *Res:* Conformational factors in free radical and photochemical reactions; scientific literature related to pharmaceutical sciences including medicinal, pharmaceutical, organic, physical and polymer chemistry, clinical and veterinary medicine and toxicology; technical writing and editing. *Mailing Add:* 2825 Ramona Palo Alto CA 94306

BENSON, HERBERT, b Yonkers, NY, Apr 24, 35; m 62; c 2. CARDIOLOGY. *Educ:* Wesleyan Univ, BA, 57; Harvard Med Sch, MD, 61. *Prof Exp:* Assoc physician med & dir Hypertension Sect, Beth Israel Hosp, 74-87, dir, Div Behav Med, 78-87; CHIEF, DIV BEHAV MED, NEW ENG DEACONESS HOSP, 87-, PRES, MIND/BODY INST, 88- *Concurrent Pos:* USPHS fel cardiol, Thorndike Mem Lab, Harvard Med Sch, 65-67, Med Found grant behav physiol, 67-69; prog dir, Gen Clin Res Ctr, Boston City Hosp, 72-74 & Gen Clin Res Ctr, Beth Israel Hosp, 74-78; expert consult, Spec Action Off Drug Abuse Prev, Exec Off of Pres, Washington, DC, 74; mem, Med Adv Bd, Coun High Blood Pressure Res, 76-; mem, Acad Behav Med Res, 78- *Mem:* AAAS; Am Fedn Clin Res; Am Physiol Soc; Am Psychosom Soc; fel Am Col Cardiol. *Res:* Behavioral aspects of cardiovascular disease and other diseases related to so-called stress; the counterpart of the fight or flight response-the relaxation response. *Mailing Add:* Div Behav Med Deaconess Hosp 185 Pilgrim Rd Boston MA 02215

BENSON, HERBERT LINNE, JR, b Kansas City, Mo, Apr 26. 34; m 57; c 3. PHYSICAL CHEMISTRY. *Educ:* Bethany Col, Kans, BS, 56; Univ Wis, PhD(radiation chem), 61. *Prof Exp:* Res chemist, 61-66, SR RES CHEMIST, SHELL OIL CO, DEER PARK, 66- *Mem:* Am Chem Soc; Sigma Xi. *Res:* Radiation chemistry of organic compounds; petroleum chemistry. *Mailing Add:* 13106 King Circle Dr Cypress TX 77429

BENSON, JAMES R, b NJ, 42; m 84; c 1. LIQUID CHROMATOGRAPHY, BIOCHEMISTRY. *Educ:* Stanford Univ, PhD(med microbiol), 74. *Prof Exp:* PRES, INTERACTION CHEMS, INC, 79- *Mem:* Am Soc Biochem & Molecular Biol; Am Soc Microbiol. *Mailing Add:* Interaction Chemicals Inc 1615 Plymouth St Mountain View CA 94043

BENSON, JOHN ALEXANDER, JR, b Manchester, Conn, July 23, 21; m 47; c 4. GASTROENTEROLOGY. *Educ:* Wesleyan Univ, BA, 43; Harvard Med Sch, MD, 46. *Prof Exp:* Instr med, Harvard Med Sch, 56-59; assoc prof med, 59-65, head div gastroenterol, 59-75, PROF MED, MED SCH, UNIV ORE, 65- *Concurrent Pos:* Attend in med, Vet Admin Hosps, Boston, 58-59 & Portland, Ore, 60-75; consult gastroenterol, Vancouver, Wash, 60-; mem, sub specialty gastroenterol bd, Am Bd Internal Med, 61-66, chmn, 61-66, mem, 69-, secy-treas, 72-75, pres, 75-; consult, Madigan Gen Army Hosp. Hosp. *Mem:* Inst Med-Nat Acad Sci; Am Col Physicians; AMA; Am Asn Study Liver Diseases; Am Soc Internal Med; Am Clin & Climat Asn; Am Fedn Clin Res; Am Gastroenterol Asn (secy, 70-73, vpres, 75-76, pres-elect, 76-77, pres, 77-78). *Res:* Normal and abnormal intestinal absorption. *Mailing Add:* Ore Health Sci Univ 200 SW Market St No 1770 Portland OR 97201

BENSON, KATHERINE ALICE, b Roanoke, Va, June 29, 38; m 62; c 2. DEVELOPMENTAL PHYSIOLOGY. *Educ:* Col William & Mary, BS, 60; Univ Va, PhD(biol), 65. *Prof Exp:* Asst prof zool, La State Univ, 65-67; assoc prof biol, Radford Col, 67-75; CONSULT, 75- *Mem:* Am Soc Zool. *Res:* Mechanisms of histolysis of insect tissues during metamorphosis; reorientation of physiological mechanisms in developing cells of insects and the lower vertebrates. *Mailing Add:* 612 W Fifth St Morris MN 56267

BENSON, KEITH RODNEY, b Portland, Ore, July 22, 48. HISTORY OF BIOLOGY, HISTORY OF MEDICINE. *Educ:* Whitworth Col, BA, 70; Ore State Univ, MA, 73, PhD(hist sci & gen sci), 79. *Prof Exp:* asst prof biol/hist sci, Pac Lutheran Univ, 79-81; asst prof, 81-87, ASSOC PROF HIST BIOL, UNIV WASH, 87- *Concurrent Pos:* Adj prof, hist dept, Univ Wash, 85-; book rev ed, J Hist Biol, 85-; mem, Forum Hist Sci Am; NSF Scholar's Award, 89. *Mem:* Sigma Xi; Hist Sci Soc; Am Soc Zoologists; Int Soc Hist Philos & Soc Social Studies Biol. *Res:* History of American biology; role of biological institutions in growth of American biology; nineteenth century morphology, European and American; history of ethics in science. *Mailing Add:* Dept Med Hist & Ethics SB-20 Univ Wash Seattle WA 98195

BENSON, LOREN ALLEN, b St Louis, Mo, Mar 19, 32; m 58; c 3. PHYSICAL CHEMISTRY. *Educ:* Wash Univ, AB, 52; St Louis Univ, PhD, 59. *Prof Exp:* Oper analyst, Inst Defense Analysis, 58-65; mem staff, Lambda Corp, 65-67; chief scientist, Keystone Comput Assocs, 67-70; CHIEF SCIENTIST, DATRAN CORP, 70- *Mem:* AAAS; Asn Comput Mach. *Res:* Computer science; operations research; communications analysis. *Mailing Add:* 204 Richard Burbydge Williamsburg VA 23185

BENSON, NORMAN G, b Berlin, Conn, May 7, 23; m 51; c 5. BIOLOGY, LIMNOLOGY. *Educ:* Univ Maine, BS, 48; Univ Mich, MA, 51, PhD(fisheries), 53. *Prof Exp:* Fishery res biologist, US Fish & Wildlife Serv, 57-61, chief, N Cent Res Invest, US Bur Sport Fisheries & Wildlife, 61-75, fish & wildlife adminr & team leader nat stream alteration team, US Fish & Wildlife Serv, 75-78, fishery biologist, nat coastal ecosyst team, 78-84. *Concurrent Pos:* Consult, James Bay Energy Co, Montreal, 76-85. *Mem:* Am Fisheries Soc; Ecol Soc Am; fel Am Inst Fishery Res Biol. *Res:* Trout stream ecology; fish population dynamics; large reservoir limnology and fish populations; water management. *Mailing Add:* 1303 Northface Ct Colorado Springs CO 80919

BENSON, RALPH C, JR, b Birmingham, Ala, Jan 16, 42; c 2. SURGERY. *Educ:* Stanford Univ, BS, 63; Johns Hopkins Med Sch, MD, 67. *Prof Exp:* Intern, Univ Ore, 67-68; resident, Mayo Grad Sch, 70-75; asst prof urol, Univ Wis Med Sch, 76-79; from instr to asst prof urol, Mayo Med Sch, 80-81; consult, dept urol, Mayo Clin, 80-86; PROF UROL, MAYO MED SCH & CHMN, DEPT UROL, MAYO CLIN, JACKSONVILLE, FLA, 86- *Concurrent Pos:* Consult, div urol, Univ Wis Hosps & Clins, 76-79. *Mem:* Sigma Xi; Am Med Asn; Am Urol Asn; Am Col Surgeons; Am Soc Laser Med & Surgery; Int Soc Urol. *Res:* Phenochromocytoma of the bladder. *Mailing Add:* 2201 Murphy Ave Suite 115 Nashville TN 37203

BENSON, RICHARD C, b Kalamazoo, Mich, Jan 23, 44; m 72; c 2. MICROELECTRONIC MATERIALS, SURFACE SCIENCE. *Educ:* Mich State Univ, BS, 66; Univ Ill, PhD(phys chem), 72. *Prof Exp:* Res assoc, Univ Ill, 71-72; sr chemist, 72-80, PRIN PROF STAFF, APPL PHYSICS LAB, JOHNS HOPKINS UNIV, 80-, GROUP SUPVR, 90- *Mem:* Am Phys Soc; Am Vacuum Soc; Inst Elec & Electronic Engrs; Mat Res Soc. *Res:* Surface science; properties of microelectronic materials; stability of spacecraft materials; molecular spectroscopy; mass spectrometry; gas chromatography; application of optical techniques to surface science. *Mailing Add:* Appl Physics Lab Johns Hopkins Univ Laurel MD 20723

BENSON, RICHARD CARTER, b Newport News, Va, July 29, 51; m 76; c 1. OPTIMAL DESIGN. *Educ:* Princeton Univ, BSE, 73; Univ Va, MS, 74; Univ Calif, Berkeley, PhD(mech eng), 77. *Prof Exp:* Tech specialist & proj mgr, Xerox Corp, 77-80; asst prof, 80-83, ASSOC PROF MECH ENG, UNIV ROCHESTER, 83- *Concurrent Pos:* Consult, 80-; assoc ed, Appl Mech Rev, 85- *Honors & Awards:* Henry Hess Award, Am Soc Mech Engrs, 84. *Mem:* Am Soc Mech Engrs; Am Soc Eng Educ; Soc Tribiologists & Lubrication Engrs; Am Acad Mech. *Res:* Nonlinear plate and shell theory, structural mechanics, and machine dynamics; applications to very flexible structures such as paper sheets, computer floppy disks, and collapsible shells; stability and control of mechanical structures. *Mailing Add:* Dept Mech Eng Univ Rochester Rochester NY 14627

BENSON, RICHARD EDWARD, b Racine, Wis, May 8, 20; m 42; c 2. ANALYTICAL CHEMISTRY. *Educ:* Ariz State Teachers Col, BA, 42; Univ Nebr, MA, 44, PhD(org chem), 46. *Prof Exp:* Lab asst, Univ Nebr, 42-46; res chemist, Chem Dept, E I Du Pont de Nemours & Co, Inc, 46-56, res supvr, 57-67, assoc res dir, 67-80, dir analytical sci, Cent Res Dept, Exp Sta, 80-85; RETIRED. *Mem:* Am Chem Soc. *Res:* Synthetic organic chemistry; allene; ferrocene. *Mailing Add:* 112 E Pembrey Dr Wilmington DE 19803

BENSON, RICHARD HALL, b Huntington, WVa, May 19, 29; m 57. PALEONTOLOGY, OCEANOGRAPHY. *Educ:* Marshall Col, BS, 51; Univ Ill, MS, 53, PhD(geol), 55; Univ Leicester, MSc, 71, DSc, 83. *Prof Exp:* Asst, State Geol Surv, Ill, 51-53, asst geologist, 55-56; from asst prof to prof geol, Univ Kans, 55-82; res paleobiologist & cur, 64-84, SR SCIENTIST, SMITHSONIAN INST, 84- *Concurrent Pos:* Vis scientist, Soc Nat Elf-Aquitain, Pau, France, 76-77. *Honors & Awards:* Nat Mus Nat Hist Dirs Award, Smithsonian Inst, 74. *Mem:* Paleont Soc; Soc Syst Zool; Paleont Asn Gt Brit; fel AAAS; Int Paleont Asn (treas). *Res:* Micropaleontology; Recent and Cenozoic Ostracoda; deep-sea environmental studies; paleoecology; Cenozoic stratigraphy; analysis of morphological change through computer simulation; studies in North African geology and the origin of the Mediterranean. *Mailing Add:* Dept Paleobiol Smithsonian Inst Mus Natural Hist Washington DC 20560

BENSON, RICHARD NORMAN, b Sioux City, Iowa, Nov 11, 35; m 59; c 1. PETROLEUM GEOLOGY. *Educ:* Augustana Col, Ill, BA, 58; Univ Minn, Minneapolis, PhD(geol), 66. *Prof Exp:* Geologist, Exxon, 66-69; asst prof geol, Augustana Col, Ill, 69-75; scientist, 75-81, SR SCIENTIST, DEL GEOL SURV, 81-, JOINT APPOINTMENT ASSOC PROF, DEPT GEOL, UNIV DEL, 85- *Concurrent Pos:* Shipboard scientist, Deep Sea Drilling Proj, Legs 12, 65; geologist, Water Resources Div, US Geol Surv, 74-79; alt Del rep, Mid-Atlantic Regional Tech Working Group, & OCS Policy Com, Outer Continental Shelf Adv Bd, US Dept Interior, 79-; ed, Del Geol Surv publ; prin investr, res contracts, Continental Margins Prog Minerals Mgt Serv, Asn Am State Geologists, 83- *Mem:* AAAS; Am Asn Petrol Geol; Sigma Xi; NAm Microplant Sect, Soc Econ Paleontologists & Mineralogist; Geol Soc Am. *Res:* Geological history stratigraphy and hydrocarbon resources of the Atlantic continental margin of North America; radiolarian and foraminiferal biostratigraphy and paleoecology; paleoenvironmental interpretation of marine sedimentary rocks; petroleum geology; interpretation of seismic reflection profiles; seismic stratigraphy. *Mailing Add:* Del Geol Surv Univ Del Newark DE 19716

BENSON, ROBERT FRANKLIN, b Cumberland, Md, Jan 22, 41; m 63; c 1. EXPERIMENTAL DESIGN, LIQUID CHROMATOGRAPHY. *Educ:* Wva Univ, BS, 63; Univ SC, MS, 65; Rensselaer Polytech Inst, PhD(org chem), 73. *Prof Exp:* Asst res chemist, Sterling-Winthrop Res Inst, 65-73; sr res chemist, GAF Res Lab, 75-77; sr scientist, Waters Assocs, 77-91; MGR ENG MGR, BIO-RAD, 91- *Concurrent Pos:* Assoc, Rensselaer Polytech Inst, 73-75. *Mem:* Am Chem Soc. *Res:* Development and production of azo dyestuffs. *Mailing Add:* 2523 Tulare Ave El Cerrito CA 94530

BENSON, ROBERT FREDERICK, b Minneapolis, Minn, Mar 16, 35; m 58; c 3. SPACE PHYSICS, IONOSPHERIC PHYSICS. *Educ:* Univ Minn, BS, 56, MS, 59; Univ Alaska, PhD(geophys), 63. *Prof Exp:* Int Geophys Year scientist, Arctic Inst NAm, 56-58; asst geophysicist, Geophys Inst, Univ

Alaska, 59-63; asst prof astron, Univ Minn, 63-64; GEOPHYSICIST, GODDARD SPACE FLIGHT CTR, NASA, 65- *Concurrent Pos:* Nat Acad Sci-Nat Res Coun resident res assoc, 64-65. *Mem:* Am Polar Soc; Am Geophys Union; Int Union Radio Sci; Antarctican Soc. *Res:* Radio star scintillations; ionospheric cross modulation; plasma waves, stimulated instabilities and non-linear phenomena in the topside ionosphere; ionospheric electron temperature; equatorial ionospheric plasma bubbles; planetary radio emissions. *Mailing Add:* Code 692 NASA/Goddard Space Flight Ctr Greenbelt MD 20771

BENSON, ROBERT HAYNES, b Hanover, NH, 24; m 50; c 3. ANIMAL BREEDING. *Educ:* Univ NH, BS, 49; WVa Univ, MS, 51; Univ Wis, PhD, 55. *Prof Exp:* Dairy husbandman, Univ Wis, 53-55; exten dairyman, Univ Conn, 55-86; RETIRED. *Concurrent Pos:* Mem coord group, Nat Dairy Herd Improv Asn Prog, 68-71, policy bd, 83-85 & rules comt, 83-; chmn, New Eng Green Pastures Dairymen of Year Comt, 83- & Conn Mastitis Coun, 82- *Mem:* Am Dairy Sci Asn. *Res:* Type classification of dairy cattle; environmental influences and their effect on milk production; dairy cattle breeding. *Mailing Add:* 1440 Ashley Dr Virginia Beach VA 23454

BENSON, ROBERT LELAND, b Tucson, Ariz, Oct 27, 41; m 66. ENTOMOLOGY, ENZYMOLOGY. *Educ:* Pomona Col, BA, 63; Univ Ill, Urbana-Champaign, PhD(entom), 69. *Prof Exp:* Staff fel, Geront Res Ctr, NIH, Md, 68-70; asst prof entom, Wash State Univ, 70-76; mem staff biochem & biophys, Univ Calif, 76-80. *Mem:* AAAS; Sigma Xi. *Res:* Synthesis of aminosugars and chitin in insects; properties of enzymes involved in aminosugar metabolism; hormonal control metabolic processes; ultrastructural localization of disaccharidases in mammalian intestine. *Mailing Add:* 5120 Cowell Blvd Davis CA 95616

BENSON, ROBERT WILMER, b Grand Island, Nebr, Jan 21, 24; m 52; c 3. ELECTRICAL ENGINEERING, PHYSICS. *Educ:* Wash Univ, BSEE, 48, MSEE, 49, PhD(elec eng), 51. *Prof Exp:* Res assoc physics, Cent Inst Deaf, St Louis, 49-54; asst dir, Armour Res Found, Chicago, 54-60; prof elec eng, Vanderbilt Univ, 60-62; PRES, BONITRON, INC, 62- *Concurrent Pos:* Consult, Am Acad Ophthal & Otolaryngol, 52-; mem hearing & bioacoust comt, Nat Res Coun, 56-63, chmn helmets comt, 68-73; mem exec bd dirs, Wilkerson Hearing & Speech Ctr, 68- & Am Red Cross, Nashville, 74- *Mem:* Fel Acoust Soc Am; Inst Elec & Electronics Engrs. *Res:* Acoustics and electronic instrumentation. *Mailing Add:* Bonitron Inc 2970 Sidco Dr Nashville TN 37204

BENSON, ROYAL H, b Galveston, Tex, Oct 25, 25; m 54; c 4. ORGANIC CHEMISTRY, ANALYTICAL CHEMISTRY. *Educ:* Univ Houston, BS, 48, MS, 56. *Prof Exp:* Lab instr chem, Univ Houston, 47-48; res chemist, M D Anderson Hosp & Tumor Inst, 48-50; res chemist, Radioisotope Unit, Vet Admin, 50-56; res specialist, 56-70, SR RES SPECIALIST, MONSANTO CO, 70-, SR FEL, 77- *Mem:* Am Chem Soc; Am Nuclear Soc. *Res:* Chemical and analytical research using radioisotopes; radioactivity assay techniques, liquid scintillation and gamma spectrometry. *Mailing Add:* 1522 19th Ave N Texas City TX 77590-8294

BENSON, SIDNEY WILLIAM, b New York, NY, Sept 26, 18; m 86; c 2. PHYSICAL CHEMISTRY. *Educ:* Columbia Univ, BA, 38; Harvard Univ, MA & PhD(phys chem), 41. *Hon Degrees:* Dr, Univ Nancy, France, 89. *Prof Exp:* Res assoc, Harvard Univ, 41; instr chem, City Col New York, 42-43; res chemist, Manhattan Proj, Kellex Corp, 43; from asst prof to prof chem, Univ Southern Calif, 43-63, dir chem physics prog, 62-63; chmn, Dept Kinetics & Thermochem, Stanford Res Inst, 63-76; PROF CHEM, UNIV SOUTHERN CALIF, 76-, SCI DIR, HYDROCARBON RES INST, 77- *Concurrent Pos:* Consult, Goodyear Tire & Rubber Co, 57-62 & 70-75, Douglas Aircraft Co, 58-70, Jet Propulsion Labs, 61-65 & Aerospace Labs, 61-; ed-in-chief, Int J Chem Kinetics, 67-83; mem adv coun, Gordon Res Conf, 68-71, 78-81; mem comt selection postdoctoral res fels, NSF, 69; mem org solvents adv comt, Nat Air Pollution Control Admin, 69-72; chem eval panel phys chem, Nat Bur Stand, 69-72; mem Comt on Data for Sci & Technol & chmn task group chem kinetics, Int Coun Sci Unions, 69-74; mem, Comn Motor Vehicle Exhausts, Nat Acad Sci, 71-74 & Sci Adv Panel, State of Calif Air Resources Bd, 72-76, chmn, 75-76; vis prof, Univ Utah, 71, Univ Paris, 50, 71-72 & 79, Tex A&M Univ, 78 & Univ Lausanne, 79; hon vis prof, Univ St Andrews, Scotland, 73, Univ Guelph & Waterloo, Can, 81 & hon Brotherton prof, Univ Leeds, Eng, 84; adv panel, Biol & Climate Effects, US Environ Protection Agency, 77; consult, Occidential Res Corp, 78, TRW, 78-, Seoul Nat Univ, 79-80 & Gas Res Inst, 79-80, Panel Fossil Energy Res, Dept Energy, 81-82; mem bd dirs, Pyrotech, KTI, 80-84; assoc mem, comm chem kinetics, Int Union Pure & Appl Chem. *Honors & Awards:* Glidden lectr, Purdue Univ, 61; Petrol Chem Award, Am Chem Soc, 77, Tolman Award, 78 & Langmuir Award, 86; Frank Gucker Mem Lectr, Indiana Univ, 84; Polanyi Medal, Royal Soc Eng, 86. *Mem:* Nat Acad Sci; fel AAAS; fel Am Phys Soc; Am Chem Soc; fel Japan Soc Advan Sci; assoc mem Int Union Pure & Appl Chem. *Res:* Chemical lasers; photochemistry; kinetics; theory of liquid structure; catalysis adsorption; statistical mechanics; free radicals; laser chemistry; thermochemistry; quantum mechanics. *Mailing Add:* Dept Chem Univ Southern Calif Los Angeles CA 90089-1661

BENSON, WALTER RODERICK, b Chicago, Ill, Oct 16, 29; m 57; c 3. PHARMACEUTICAL CHEMISTRY, ANALYTICAL CHEMISTRY. *Educ:* Univ Ill, BS, 51; Univ Colo, PhD(chem), 58. *Prof Exp:* Chemist, Griffith Labs, Ill, 53-54; asst, Univ Colo, 54-57; fel, Purdue Univ, 58-59; asst prof chem, Colo State Univ, 59-63; sect head, Pesticides Br, Div Food Chem, Food & Drug Admin, 63-69, br chief, Instrumental Appln Res Br, 69-77, dir, Div Drug Chem, Bur Drugs, 77-85; World Health Org/Stc, Zimbabwe, 87-91; UNICEF-TANZANIA, 91- *Concurrent Pos:* Prof lectr, USDA Grad Sch, 65-71 & Am Univ, 67; consult chemist, 86- *Mem:* Asn Off Analytical Chemists; Am Chem Soc; Sigma Xi; assoc mem Am Pharmaceut Asn. *Res:* Electron spin immunoassay; radioactive drug analysis; synthetic organic chemistry; bridgehead rearrangements; organic mechanisms; carbamate pesticides; oximes; agricultural and pharmaceutical chemistry; photochemistry; drug analysis. *Mailing Add:* 6209 Crathie Lane Bethesda MD 20816

BENSON, WALTER RUSSELL, b Tamaqua, Pa, July 27, 20; m 50; c 4. PATHOLOGY, MEDICINE. *Educ:* Duke Univ, MD, 44. *Prof Exp:* Instr path, Sch Med, Duke Univ, 52-54, assoc, 54-55; asst prof, Sch Med, Univ Louisville, 55-56; from asst prof to assoc prof, 56-67, PROF PATH, MED SCH, UNIV NC, CHAPEL HILL, 67- *Mem:* Am Asn Path; Am Soc Exp Path; Int Acad Path. *Res:* Cardiovascular pulmonary diseases; protein metabolism; neoplasia. *Mailing Add:* 310 Cedar St Mt Bolus NC 27514

BENSON, WILLIAM EDWARD BARNES, b West Haven, Conn, May 15, 19; m 44; c 3. GEOLOGY. *Educ:* Yale Univ, BA, 40, MS, 42, PhD(geol), 52. *Prof Exp:* Geologist, Conn State Geol & Nat Hist Surv, 40-41; geologist, US Geol Surv, 42-54, actg chief, Gen Geol Br, 53-54; exec secy, Div Eearth Sci, Nat Res Coun, 54-55; chief geologist, Manidon Mining Inc, 55-56; mem staff, NSF, 56-79, from prog dir to head earth sci sect, 57-75, chief scientist, Earth Sci Div, 75-79; SR STAFF OFFICER, NAT ACAD SCI, 80- *Concurrent Pos:* Adv, Pres Off Sci & Technol Policy, 76-77; vis prof, Univ Hawaii, 80; consult, Environ Geol & Nuclear Waste Mgt, 80- *Honors & Awards:* Penfield Prize; Hurlbut-Runk Prize. *Mem:* AAAS; Am Geophys Union; Geol Soc Am. *Res:* Marine geology; geomorphology; glacial geology; tertiary stratigraphy; economic geology. *Mailing Add:* Off Earth Sci US Nat Acad Sci Washington DC 20418

BENSON, WILLIAM HAZLEHURST, b Philadelphia Pa, Nov 21, 54; Brit citizen; m 77; c 2. AQUATIC TOXICOLOGY, ECOTOXICOLOGY. *Educ:* Fla Inst Technol, BS, 76; Univ Ky, MS, 80, PhD(toxicol), 84. *Prof Exp:* Sr scientist toxicol, Cannon Labs, Inc, 76-78; from asst prof to assoc prof toxicol, Northeast La Univ, 84-88; ASSOC PROF TOXICOL, UNIV MISS, 88- *Concurrent Pos:* Dir, Toxicol Prog, Northeast La Univ, 87-88; coordr, Environ Toxicol Res Prog, Univ Miss, 89- *Honors & Awards:* Cert Merit, Environ Chem Div, Am Chem Soc, 88. *Mem:* Am Chem Soc; Soc Environ Toxicol & Chem; Soc Toxicol. *Res:* Assessment of the acute and chronic health effects of environmental contaminants; research areas also include reproduction studies in aquatic animals; use of biomarkers in environmental monitoring; residue health effects. *Mailing Add:* Sch Pharm Univ Miss University MS 38677

BENSTON, MARGARET LOWE, b Longview, Wash, Oct 1, 37; div. THEORETICAL PHYSICAL CHEMISTRY. *Educ:* Willamette, BA, 59; Univ Wash, PhD(phys chem), 64. *Prof Exp:* Proj assoc theoret chem, Univ Wis, 64-66; asst prof chem, 66-76, asst prof chem & comput sci, 76-81, ASST PROF COMPUT SCI & WOMEN'S STUDIES, SIMON FRASER UNIV, 81- *Res:* Social implications of computing; women and science. *Mailing Add:* Dept Comput Sci Simon Fraser Univ Burnaby BC V5A 1S6 Can

BENT, BRIAN E, b Minneapolis, Minn, Oct 18, 60; m 91. PHYSICAL CHEMISTRY, SURFACE CHEMISTRY. *Educ:* Carleton Col, BA, 82; Univ Calif, Berkeley, PhD(chem), 86. *Prof Exp:* Postdoctoral, AT&T Bell, 86-88; PROF CHEM, COLUMBIA UNIV, 88- *Concurrent Pos:* Nat Sci found fel; presidential young investr, 89- *Mem:* AAAS; Am Vacuum Soc; Am Chem Soc; Am Phys Soc; Mat Res Soc. *Res:* Surface chemistry, heterogeneous catalysis, materials deposition and growth. *Mailing Add:* Dept Chem Columbia Univ Box 252 New York NY 10027

BENT, DONALD FREDERICK, b Clinton, Mass, Nov, 16, 25; m 51; c 3. MICROBIOLOGY. *Educ:* Univ NH, BS, 48, MS, 53; Univ Md, PhD(microbiol), 57. *Prof Exp:* Spec staff asst, Am Coun Ed, 56; instr bact & immunol, Harvard Med Sch, 56-59; owner & dir, Health Serv Lab, 60-78; vpres, Biospheric Consult Int, Inc, 74-77; DIR, N H WATER SUPPLY & POLLUTION CONTROL COMN LAB, CONCORD, 77- *Concurrent Pos:* Div dir, Biol Labs, Mass Dept Pub Health, 56-59; mem fac, Colby Jr Col Women, 56-69; vpres, Bio-tronics Res, Inc, 61-66; Fulbright lectr, Med Fac, Univ Malaya, 66-67; dir, Clin Microbiol Div, Metpath Lab, Inc, NJ, 69-71; consult, Town Health Officer; bd, State Health Officer Biosphenic; consult, Corp BCI; vpres, Water Quality Pub Health; dir, Libratary State NH Environ Serv; found partner, Water Test Corp; tech dir, 88- *Mem:* Am Soc Microbiol; Nat Environ Health Asn. *Res:* Clinical and environmental microbiology; bacterial toxins; immunology; diagnostic and automated microbiology; water quality; environmental health laboratory management. *Mailing Add:* Lamson Lane Box 32 New London NH 03257

BENT, HENRY ALBERT, b Cambridge, Mass, Dec 21, 26; m 59; c 2. PHYSICAL CHEMISTRY. *Educ:* Oberlin Col, AB, 49; Univ Calif, PhD(chem), 52. *Prof Exp:* Asst, Univ Calif, 49-51; instr phys chem, Univ Conn, 52-55; res fel, Univ Minn, Minneapolis, 55-57, from asst prof to prof inorg chem, 57-69; prof chem, NC State Univ, 69-89; PROF CHEM, UNIV PITTSBURGH, 89- *Concurrent Pos:* Lectr, Gulbenkian Inst Advan Study, Portugal, 70, 71 & 72; Chautaugua course dir, AAAS, 72-74, 77, 82, 84 & 85; chmn, Div Chem Educ, Am Chem Soc, 78; job lectr, Mem Univ Nfld, 81; vis prof, Willimas Col, 81. *Honors & Awards:* Award, Am Chem Soc, 80; James Flack Norris Award Chem Educ, 84. *Mem:* Fel AAAS; Am Chem Soc; Am Phys Soc. *Res:* Thermodynamics; molecular structure; intermolecular interactions; localized molecular orbitals; history of chemistry; philosophy of science; chemical education. *Mailing Add:* Dept Chem Univ Pittsburgh Pittsburgh PA 15260

BENT, RICHARD LINCOLN, b Rochester, NY, Oct 25, 17. ORGANIC CHEMISTRY. *Educ:* Cambridge Univ, BA, 39. *Prof Exp:* Chemist, Eastman Kodak Co, 41-48, res chemist, 48-71, res assoc, Color Photog Div, 71-79; RETIRED. *Mem:* Am Chem Soc; Soc Photog Sci & Eng. *Res:* Synthesis of photographic developing agents. *Mailing Add:* PO Box 25506 Rochester NY 14625-0506

BENT, ROBERT DEMO, b Cambridge, Mass, Dec 22, 28; m 56; c 3. PHYSICS. *Educ:* Oberlin Col, AB, 50; Rice Inst, AM, 52, PhD(physics), 54. *Prof Exp:* Res assoc physics, Rice Inst, 54-55; res assoc, Columbia Univ, 55-58; from asst prof to assoc prof, 58-66, PROF PHYSICS, IND UNIV, BLOOMINGTON, 66- *Concurrent Pos:* Guggenheim fel, Oxford & Atomic Energy Authority Res Estab, Harwell, Eng, 62-63. *Mem:* Fel Am Phys Soc. *Res:* Intermediate energy nuclear physics. *Mailing Add:* Dept Physics Ind Univ Bloomington IN 47401

BENT, SAMUEL W, b Gardner, Mass, May 11, 55; m; c 2. ANALYSIS OF ALGORITHMS. *Educ:* Cornell Univ, AB, 75; Stanford Univ, PhD(comput sci), 82. *Prof Exp:* Asst prof comput sci, Univ Wis-Madison, 81-85; ASST PROF COMPUT SCI, DARTMOUTH COL, 85- *Mem:* Asn Comput Mach; Soc Indust & Appl Math; Math Asn Am. *Res:* Mathematical analysis of algorithms; design of data structures; theory of computing. *Mailing Add:* Dept Math & Comput Sci Dartmouth Col Bradley Hall Hanover NH 03755

BENTALL, RAY, b Grand Rapids, Mich, June 28, 17; m 38, 81; c 2. GEOLOGY. *Educ:* Univ Mich, BS, 38; Univ Tenn, MS, 40. *Prof Exp:* Geologist, State Div Geol, Tenn, 40-43; geologist, US Geol Surv, Tenn, 43-45 & Nebr, 45-67, hydrologist, Nebr, 67-73; HYDROLOGIST, CONSERV & SURV DIV, INST AGR & NATURAL RESOURCES, UNIV NEBR, LINCOLN, 74- *Mem:* Sigma Xi. *Res:* Hydrology. *Mailing Add:* 4410 N 14th St Lincoln NE 68521

BENTLEY, BARBARA LEE, b Los Angeles, Calif, Dec 14, 42; m 80; c 2. ECOLOGY. *Educ:* Willamette Univ, BA, 64; Univ Calif, Los Angeles, MA, 66; Univ Kans, PhD(ecol), 74. *Prof Exp:* Asst prof, 73-79, dean grad sch, 83-87, ASSOC PROF ECOL, STATE UNIV NY, STONY BROOK, 79- *Concurrent Pos:* Instr, Biol Sta, Univ Mich, 74, Johnson Co Community Col, 69-70 & Cuttington Col, West Africa, 66-68; vis scientist, Ins Nac de Pesquises da Amazonia, Brazil, 74; coordr, Orgn Trop Studies, 72-, vpres, 79-82; consult, Brookhaven Nat Lab, 78-85; Nat Sci Found vis prof, 88-89; Fulbright fel, 87; bd dir, Am Inst Biol Sci, 91-94. *Mem:* Am Soc Naturalists (secy, 88-91); Ecol Soc Am (vpres, 89-90); Brit Ecol Soc; Asn Trop Biol; Int Soc Study Social Insects; Bot Soc Am; Soc Study Evolution; NY Acad Sci; fel AAAS. *Res:* Nitrogen fixation in tropical environments; plant and insect interactions; geographic variation in termite populations; plant/herbivore interactions in Lupinus. *Mailing Add:* Dept Ecol & Evolution State Univ NY Stony Brook NY 11794

BENTLEY, CHARLES RAYMOND, b Rochester, NY, Dec 23, 29; m 64; c 2. ANTARCTIC GEOPHYSICS, ANTARCTIC GLACIOLOGY. *Educ:* Yale Univ, BS, 50; Columbia Univ, PhD(geophys), 59. *Prof Exp:* Asst, Columbia Univ, 52-55, geophysicist, 55-56; traverse co-leader seismologist, Int Geophys Year, Antarctic Prog, Arctic Inst NAm, 56-58, traverse leader, 58-59; proj assoc geophys, 59-61, from asst prof to assoc prof, 61-68, prof geophys, 67-87, A P CRARY PROF GEOPHYS, UNIV WIS, MADISON, 87- *Concurrent Pos:* NSF sr fel, Mass Inst Technol, 68-69; alt US deleg, Sci Comt Antarctic Res, 81-, vpres, 90-; chmn, Bd Assoc Ed, Antartic Res Series, Am Geophys Union, 74-90; secy chmn, Working Group Solid Earth Geophysics, Sci Comn Antartic Res, 76-82; mem, Polar Res Bd, 78-, chmn, 82-86, chmn, Comt Glaciol, 78-81; vpres, Int Glaciological Soc, 78-81; mem, Bd of Arctic Inst NAm, 87-; vpres, Int Comn on Snow Ice, 87- *Honors & Awards:* Bellingshausen-Lazarev Medal, Acad Sci, USSR, 71; Seligman Crystal, Int Glaciol Soc, 90. *Mem:* Fel AAAS; Am Quaternary Asn; fel Am Geophys Union; Geol Soc Am; Int Glaciological Soc; Seismol Soc Am; Sigma Xi; Soc Explor Geophys; Artic Inst NAm. *Res:* Geophysical studies on polar ice sheets and glaciological applications of geophysical methods. *Mailing Add:* Geophys & Polar Res Ctr 1215 W Dayton St Madison WI 53706

BENTLEY, CLEO L, vertebrate anatomy, parasitology, for more information see previous edition

BENTLEY, DAVID R, b Lansing, Mich, June 18, 40; c 2. NEUROBIOLOGY, DEVELOPMENTAL BIOLOGY. *Educ:* Williams Col, AB, 62; Univ Mich, MS, 64, PhD(zool), 66. *Prof Exp:* Fel, Inst fur Vergleichende Tierphysiologie, 66-67 & Stanford Univ, 67-68; asst prof, 68-73, assoc prof, 73-78, PROF NEUROBIOL, ZOOL DEPT, UNIV CALIF, BERKELEY, 78-, CHAIRPERSON DEPT, 82- *Concurrent Pos:* Prin investr, NIH, 69-; Alexander von Humboldt Sr Scientist Award, WGer Govt, 74. *Mem:* Soc Neurosci. *Res:* Long-distance pathfinding by nerve cell growth coneso, the establishment of nerve trunks and neural differentiation during embryonic development of the grasshopper. *Mailing Add:* Dept Zool Univ Calif Berkeley CA 94720

BENTLEY, DONALD LYON, b Los Angeles, Calif, Apr 25, 35; m 57; c 3. MATHEMATICAL STATISTICS. *Educ:* Stanford Univ, BS, 57, MS, 58, PhD(statist), 61. *Prof Exp:* Asst prof math statist, Colo State Univ, 61-64; from asst prof to assoc prof, 64-74, PROF MATH, POMONA COL, 74-, LINGURN H BURKHEAD CHAIR MATH, 82- *Concurrent Pos:* Prin investr, NSF grant, 62-64; NIH grant, 68-70; NSF fac fel, 70-71; consult, 65- *Mem:* Inst Math Statist; fel Am Statist Asn; Biomet Soc; Royal Statist Soc; Soc Epidemiol Res. *Res:* Mathematical statistics and stochastic processes and their application to biomedical phenomena. *Mailing Add:* Dept Math Pomona Col Claremont CA 91711

BENTLEY, GLENN E, b Los Alamos, NMex, Oct 25, 46; m 73; c 2. ANALYTICAL CHEMISTRY, RADIOCHEMISTRY. *Educ:* NMex Inst Mining & Technol, BS, 69; Ariz State Univ, PhD(analytical chem), 76. *Prof Exp:* Res assoc analytical chem, Univ Colo, 75-76; MEM STAFF RADIO-ANALYTICAL CHEM, LOS ALAMOS SCI LAB, 76- *Mem:* Am Chem Soc; Soc Appl Spectros; Sigma Xi. *Res:* Development of separation methods for novel radioisotopes; hot cell process development; trace element determinations and methods development for many elements from many materials. *Mailing Add:* 412 Cheryl Ave Los Alamos NM 87544

BENTLEY, HARRY THOMAS, III, b Oak Park, Ill, Jan 30, 42; m 77; c 4. PHYSICS, ELECTROOPTICS. *Educ:* NC State Univ, BS, 64, MS, 66; Univ Tenn Space Inst, MS, 87. *Prof Exp:* Engr physics, Sverdrup AEDC, Inc, 66-74, sr res engr physics & sr proj mgr, 74-81, mgr, Adv Propulsion diag group, 81-85, ENG SPECIALIST, SVERDRUP AEDC, INC, 85-; SYSTS ENGR, LEWIS SPACE SHUTTLE MAIN HEALTH MONITORING PROG, NASA. *Concurrent Pos:* Consult, Indust Nucleonics Corp, 64-66, Sci Appl, Inc, 74-76 & Sci Metrics, Inc, 74-76. *Mem:* Soc Photo-Optical Instrument Eng; Nat Mgt Asn; Sr mem Am Inst Aeronauts & Astronauts; Sigma Xi. *Res:* Electro-optical instrumentation; laser velocimetry, particle sizing, combustion gas analysis and plasma physics; the measurement and modeling of the radiative transfer process associated with turbine engine propulsion systems; advanced electro-optical techniques for gas species identification; developed techniques for assessing propulsion related observables and countermeasures; over 30 papers published. *Mailing Add:* Sverdrup MS 900/ Arnold Eng Develop Ctr Arnold Air Force Base TN 37389

BENTLEY, HERSCHEL LAMAR, b Sylacauga, Ala, July 14, 39; m 59; c 4. MATHEMATICS. *Educ:* Univ Ariz, BS, 61, MS, 63; Rensselaer Polytech Inst, PhD(math), 65. *Prof Exp:* Eng trainee, Hughes Aircraft Co, 59-60; grad asst, Univ Ariz, 61-63; mathematician, Gulf Res Develop, 63; asst prof math, Rensselaer Polytech Inst, 65-66, Univ NMex, 66-69 & Bucknell Univ, 69-71; assoc prof, 71-75, PROF MATH, UNIV TOLEDO, 75- *Concurrent Pos:* Doctoral fel, NDEA, 63-65. *Res:* General topology; categorical topology. *Mailing Add:* Dept Math Univ Toledo Toledo OH 43606

BENTLEY, J PETER, b Oldham, Eng, Sept 15, 31; US citizen; m 57; c 3. BIOCHEMISTRY. *Educ:* Univ Ore, MS, 61, PhD(biochem, physiol), 63. *Prof Exp:* Res asst, Harvard Med Sch, 57-59; res asst, 59-62, res assoc, 62-64, instr exp biol, 64-65, asst prof exp biol & biochem, 65-68, assoc prof biochem, 68-76, PROF BIOCHEM, SCH MED, ORE HEALTH SCI UNIV, 76- *Mem:* Am Soc Biol Chem; Soc Complex Carbohydrates. *Res:* Biosynthesis of connective tissue components; collagen and proteoglycans. *Mailing Add:* Ore Health Sci Univ Sch Med 3181 SW Sam Jackson Park Rd Portland OR 97201

BENTLEY, JOHN JOSEPH, b Washington, DC, Nov 13,46; m 52; c 1. PHOTOACOUSTIC SPECTROSCOPY, INTERMOLECULAR POTENTIAL CALCULATIONS. *Educ:* Col Holy Cross, BS, 67; Carnegie-Mellon Univ, MS, 70, PhD(phys chem), 75. *Prof Exp:* Res assoc, Chem Dept, Carnegie-Mellon Univ, 74; res assoc, Radiation Lab, 75-78, ASST SPECIALIST, UNIV NOTRE DAME, 78- *Mem:* Am Chem Soc. *Res:* Study of energy disposal in photoinitiated reactions by photoacoustic and infrared spectroscopy; theoretical determination of mechanisms for penning ionization processes. *Mailing Add:* 1342 Longfellow Ave South Bend IN 46615

BENTLEY, KENNETH CHESSAR, b Montreal, Que, Sept 22, 35; m 61; c 2. ORAL SURGERY. *Educ:* McGill Univ, DDS, 58, MD, CM, 62. *Prof Exp:* Jr rotating intern, Montreal Gen Hosp, 62-63, jr asst resident, 63-64; partic prog oral surg, Bellevue Hosp, New York, 64-66, chief resident, 65-66; from asst prof to assoc prof, 66-75, PROF ORAL SURG, McGILL UNIV, 75-, CHMN DEPT ORAL SURG & DIR DIV SURG & ORAL MED, 68-; DENT SURGEON-IN-CHIEF, MONTREAL GEN HOSP, 70- *Concurrent Pos:* Dean fac dent, McGill Univ, 77-87. *Mailing Add:* Montreal Gen Hosp 1650 Cedar Ave Montreal PQ H3G 1A4 Can

BENTLEY, KENTON EARL, b Detroit, Mich, June 1, 27; m 53. ANALYTICAL CHEMISTRY, REMOTE SENSING. *Educ:* Univ Mich, BS, 50; Univ NMex, PhD(analytical chem), 59. *Prof Exp:* Lab asst electrochem, Univ Mich, 49-50; phys chemist, Res Dept, Consol Electrodynamics Corp, 56-57; consult chemist, 57-59; asst prof chem, Am Univ Beirut, 59-61; res scientist, Lockheed-Calif Co, Lockheed Aircraft Corp, 62-63; res specialist space sci, Jet Propulsion Lab, Calif Inst Technol, 63-65; res scientist, Electrochem Lab, Hughes Aircraft Co, 65-67; asst dept mgr, Space Environ Dept, Manned Spacecraft Ctr, 67, mgr, Space Physics Dept, 67-70, mgr, Earth Resources Dept, 68-70, mgr, Iran Earth Resources Prog, 75-77, DIR, SCI & APPLN BR, SYSTS & SERV DIV, LOCKHEED ENG & MGT SERV CO, INC, 70-, PROG MGR, LAS VEGAS PROGS, ENVIRON PROTECTION AGENCY, 80- *Concurrent Pos:* Rockefeller Found res grant, 59-61. *Mem:* AAAS; Am Chem Soc; Combustion Inst; Am Astronaut Soc; Sigma Xi. *Res:* Instrumental methods of chemical analysis; chemistry of the lower valence states of transition metals; electrochemical studies; earth resources and remote sensing program operations and management. *Mailing Add:* PO Box 339 Moffett Field CA 94035-0339

BENTLEY, MICHAEL DAVID, b Jacksonville, Fla, Feb 7, 39; m 63; c 2. ORGANIC CHEMISTRY. *Educ:* Auburn Univ, BSChem, 63, MS, 65; Univ Tex, Austin, PhD(org chem), 69. *Prof Exp:* NIH fel, Univ Calif, Berkeley, 68-69; asst prof, 69-74, assoc prof, 74-81, PROF CHEM, UNIV MAINE, 81- *Mem:* Entom Soc Am; AAAS; Sigma Xi; Am Chem Soc. *Res:* Insect chemistry; fungal metabolites; organosulfur chemistry. *Mailing Add:* Dept Chem Univ Maine Orono ME 04473

BENTLEY, ORVILLE GEORGE, b Midland, SDak, Mar 6, 18; m 42; c 2. ANIMAL SCIENCE & NUTRITION, NUTRITION. *Educ:* SDak State Col, BS, 42; Univ Wis, MS, 47, PhD(biochem), 50. *Hon Degrees:* DSc, SDak State Univ, 74 & Univ Wis, 84. *Prof Exp:* From asst prof to prof animal sci, Ohio Agr Exp Sta & Dept Animal Sci, Ohio State Univ, 50-58; dean & dir agr, Col Agr & Agr Exp Sta, SDak State Univ, 58-65; dean agr, Col Agr, Univ Ill, Urbana-Champaign, 65-82; asst secy agr, sci & educ, USDA, Washington, DC, 82-89; EMER DEAN AGR, COL AGR, UNIV ILL, URBANA-CHAMPAIGN, 89- *Concurrent Pos:* Res award, Animal Nutrit, Am Soc Animal Sci, 58; mem, Bd Trustees, Univ Beirut, 73-, Bd Int Food & Agr, USAID, 76-79, Sci Forum, Helsinki Accord, Hamburg, 80, Nat Res Councils Govt-Univ-Indust Roundtable, 84-89 & President's Adv Panel for US-Japan Joint High-Level Sci & Tech Agreement, 89-; mem & leader, Indo-Am Sub-comn on Agr, 80, 84 & 87; co-chair, Venezuela-US Agr Comn, 90- *Mem:* AAAS; Am Inst Nutrit; Am Soc Animal Sci. *Res:* Ruminant nutrition, especially on microbial breakdown of cellulose; academic administration coordinating teaching, research and extension education in agriculture, home economics and agricultural economics. *Mailing Add:* Eight Concord Lane Urbana IL 61801

BENTLEY, PETER JOHN, b Perth, Australia, Jan 13, 30; m 54; c 2. PHYSIOLOGY, PHARMACOLOGY. *Educ:* Univ Western Australia, BSc, 52, PhD(physiol), 60. *Prof Exp:* Res asst pharmacol, Univ Col, Univ London, 53-54; asst physiol, Univ Western Australia, 55-58, lectr, 58-60; res fel pharmacol, Bristol Univ, 61-62, lectr, 62-65; res assoc zool, Duke Univ, 65-66, assoc prof pharmacol, 66-68; PROF PHARMACOL & OPHTHAL, MT SINAI SCH MED, 68-; PROF PHARMACOL, DEPT ANAT, PHYSIOL, SCI & RADIOL, SCH VET MED, BURROUGHS WELLCOME DISTINGUISHED PROF, NC STATE UNIV, 88- *Mem:* Fel AAAS; Am Zool Soc; Am Physiol Soc; Brit Soc Endocrinol. *Res:* Endocrinology; zoology; pharmacology. *Mailing Add:* Dept Anat, Physiol, Sci & Radiol Sch Vet NC State Univ 4700 Hillsborough St Raleigh NC 27606

BENTLEY, RONALD, b Derby, Eng, Mar 10, 22; US citizen; m 48; c 2. BIOCHEMISTRY. *Educ:* Univ London, BSc, 43, PhD(org chem), 45. *Hon Degrees:* DSc, Univ London, 65. *Prof Exp:* Commonwealth Fund fel, Col Physicians & Surgeons, Columbia Univ, 46-47, res assoc biochem, 51-52; mem sci staff, Nat Inst Med Res, London, 48-51; from asst prof to assoc prof biochem, 53-60, chmn dept, 72-76, PROF BIOCHEM, UNIV PITTSBURGH, 60- *Concurrent Pos:* Guggenheim fel, Inst Biochem, Univ Lund, 64; mem microbial chem study sect, NIH, 72-76. *Honors & Awards:* Pittsburgh Award, Am Chem Soc, 79. *Mem:* AAAS; Am Soc Microbiol; Am Soc Biol Chemists; Am Soc Chem; Royal Soc Chem. *Res:* Biochemistry of microorganisms; carbohydrate biochemistry; biosynthesis of aromatic compounds; stereochemical implications in biology. *Mailing Add:* Dept Biol Sci Univ Pittsburgh Pittsburgh PA 15260

BENTON, ALLEN HAYDON, b Ira, NY, Sept 4, 21; m 47; c 3. ZOOLOGY, ENTOMOLOGY. *Educ:* Cornell Univ, BS, 48, MS, 49, PhD(vert zool), 52. *Prof Exp:* From instr to assoc prof biol, State Univ NY Albany, 49-62; prof, 62-73, chmn dept, 66-69, DISTINGUISHED TEACHING PROF BIOL, 73-, FAC EXHANG PROF, 78-, EMER PROF BIOL, STATE UNIV NY COL FREDONIA, 84- *Concurrent Pos:* Vis prof biol, Concord Col, 69-70; vis prof, Univ Minn, 70; freelance writer, Sci & Nature, columnist, Dunkirk NY Evening Observer, 86- *Honors & Awards:* Kasling Lectr, State Univ NY Col Fredonia, 79. *Mem:* Am Ornith Union; Am Soc Mammal. *Res:* Bird distribution in New York State; life histories of small mammals; taxonomy, distribution and host relationships of American Siphonaptera. *Mailing Add:* 292 Water St Fredonia NY 14063

BENTON, ALLEN WILLIAM, b Greenwich, Conn, June 8, 31; m 54; c 3. BIOCHEMISTRY, PHYSIOLOGY. *Educ:* Univ Conn, BS, 58, MS, 62; Cornell Univ, PhD(entom), 65. *Prof Exp:* Soil scientist, Soil Conserv Serv, USDA, 54-62; asst prof entom & apicult, Rutgers Univ, 65-66; asst prof, 66-69, assoc prof, 69-77, PROF ENTOM, PA STATE UNIV, 77- *Concurrent Pos:* Consult, Center Labs, 65-, Pharmacia, 75-, Good Samaritan Hosp, Johns Hopkins Univ, 75- & Buffalo Gen Hosp, 74-; owner, Vespid Labs, mfr Vespid venoms. *Mem:* Entom Soc Am; Bee Res Asn. *Res:* Venoms, biochemistry and physiological effects upon animals; biochemistry of allergens. *Mailing Add:* 27329 Rest Circle Easton MD 21601

BENTON, ARTHUR LESTER, b New York, NY, Oct 16, 09; m 39; c 3. NEUROPSYCHOLOGY. *Educ:* Oberlin Col, AB, 31, AM, 33; Columbia Univ, PhD, 35. *Hon Degrees:* DSc, Cornell Col, 78; PscholD, Univ Rome, 90. *Prof Exp:* Assoc prof med psychol, Sch Med, Univ Louisville, 46-48; PROF NEUROL & PSYCHOL, UNIV IOWA, 48- *Concurrent Pos:* USPHS res grant, Univ Iowa, 54-82, USPHS spec fel neuropsychol, 58-59; consult, NIH, 61-; vis prof, Fac Med, Univ Milan, 64, Free Univ, Amsterdam, 71, Dept Neurol, Univ Helsinki, 74 & Tokyo Metrop Inst Geront, 74; dir studies, Sch Higher Studies, Paris, 79; Lansdowne Scholar, Univ Victoria (Can), 80. *Mem:* Am Psychol Asn; Am Neurol Asn; Am Acad Neurol; Int Neuropsychol Soc (pres, 70-72); Am Orthopsychiat Asn (pres, 64-65). *Res:* Behavioral disabilities associated with brain disease. *Mailing Add:* 504 Manor Dr Iowa City IA 52246-2918

BENTON, BYRL E, b Armstrong, Iowa, Sept 4, 12; m 39; c 2. PHARMACOLOGY, PHARMACY. *Educ:* State Univ, BSPharm, 35, MSPharm, 39; Univ Ill, PhD(pharmacol), 47. *Prof Exp:* Pharmacist, Hermanson Drug Co, 35-37; asst pharm, SDak State Univ, 37-40; from instr to assoc prof, Univ Ill, 40-49; dean, 49-77, EMER PROF PHARM & DEAN, COL PHARM, DRAKE UNIV, 77- *Concurrent Pos:* Prescription ed, Nat Asn Retail Druggists J, 45-48; sci ed, La Farmacia Mod, 46-51; mem, Am Bd Dipl Pharm. *Mem:* Am Pharmaceut Asn. *Res:* Cardiac research involving digitalis; compressed tablet research, disintegration. *Mailing Add:* 7510 College Dr Des Moines IA 50322-5737

BENTON, CHARLES HERBERT, b Kansas City, Mo, Nov 16, 25; m 50; c 2. RESEARCH MANAGEMENT, PRODUCT DEVELOPMENT. *Educ:* Univ Kans, BS, 48; Univ Ill, PhD(org chem), 51. *Prof Exp:* Res chemist, 51, sr res chemist, 52-59, sales supvr, 60-64, head prod develop, 65-69, res assoc, 70, sr res assoc, 71, mgr sales develop & tech serv, 72, dir Health & Nutrit Res Div, Eastman Chem Div, 73-87; RETIRED. *Concurrent Pos:* Consult, 88- *Mem:* Am Chem Soc. *Res:* Consultation on research and development related to products for human and animal health and nutrition. *Mailing Add:* 4308 Gray Fox Dr Kingsport TN 37664-4417

BENTON, DUANE ALLEN, b Waterloo, Iowa, June 1, 31; m 55; c 3. BIOCHEMISTRY. *Educ:* Mich State Univ, BS, 52; Univ Wis, MS, 54, PhD(biochem), 56. *Prof Exp:* Asst prof biochem, Okla State Univ, 56-59; cancer res scientist, Roswell Park Mem Inst, 59-63; assoc prof animal nutrit, Cornell Univ, 63-66; head biochem invest, USDA, 66-70; DIR NUTRIT RES, ROSS LABS, 70- *Mem:* AAAS; Fedn Am Soc Exp Biol; Am Inst Nutrit; Am Chem Soc. *Res:* Amino acid interrelationships in the nutrition of animals; dietary interrelations and their effects on metabolism; infant nutrition. *Mailing Add:* Dept Med Ross Labs 625 Cleveland Ave Columbus OH 43216

BENTON, DUANE MARSHALL, b Savannah, Ohio, Nov 1, 33; m 60; c 3. ENVIRONMENTAL CHEMISTRY. *Educ:* Ashland Col, AB, 55; Univ Utah, BS, 58; Ohio State Univ, MS, 64; George Washington Univ, MS, 67. *Prof Exp:* Asst chem, Ohio State Univ, 55-57; asst anal res chemist, Hess & Clark Div, Richardson-Merrell Inc, 58-59; instr physics, meteorol, math & chem, Ashland Col, 59-60 & 62-64, from asst prof to assoc prof phys sci, 64-73; environ phys scientist, Dept Army Hq, Environ Off, Washington, DC, 73-78, ENVIRON PROTECTION SPECIALIST, ARMY MAT COMMAND ENVIRON DIV & CHIEF, POLLUTION ABATEMENT OPERS CTR, DEPT ARMY HQ, ALEXANDRIA, VA, 78- *Concurrent Pos:* AEC res grant, 64-67; comdr, 164 Weaflt, 71-76 & 204 Weaflt, 78-80; weather & civil eng staff officer, State Hq, NJ Air Nat Guard, 76-78, dir plans & progs, 82-85; lectr chem, physics & meteorol, Northern Va Community Col, 80-; consult & planner, Army Environ Noise & Audit; planner radio activ, hazardous & multihazardous mats, Fed Emer Mgt Agency. *Honors & Awards:* Nat Guard Minuteman Award, 83. *Mem:* Nat Asn Environ Prof; Am Meterol Soc; Environ Auditing Roundtable. *Res:* Photosynthesis; C-14 uptake in algae of Lake Erie. *Mailing Add:* 5000 King Richard Dr Annandale VA 22003-4356

BENTON, EDWARD ROWELL, b Milwaukee, Wis, Jan 20, 34; m 87; c 3. GEOPHYSICS. *Educ:* Harvard Univ, AB, 56, AM, 57, PhD(appl math), 61. *Prof Exp:* Asst sr engr, appl mech group, Arthur D Little, Inc, 60-62; lectr math, Univ Manchester, 62-63; staff scientist, Nat Ctr Atmospheric Res, 63-65, asst dir adv study prog, 67-69; asst prof, Univ Colo, Boulder, 65-67, prof & chmn astro-geophys, 69-74; consult, Aberdeen Proving Grounds, Ballistic Res Labs, US Army, 74-76; PROF ASTRO-GEOPHYS, UNIV COLO, BOULDER, 76- *Concurrent Pos:* Consult, Arthur D Little, Inc, 62-63; Environ Sci Serv Admin res grant, 66-67; NSF res grant, 70-, NASA res grants, 79-; spec asst univ rels, Univ Corp Atmospheric Res, 75-77; consult, NASA, 78-80 & US Geol Surv, 79-; Green Found scholar, 91. *Honors & Awards:* Group Achievement Award, NASA, 84. *Mem:* AAAS; Am Geophys Union; Am Phys Soc; Sigma Xi; Royal Astron Soc. *Res:* Magnetohydrodynamics; rotating fluids; applied mathematics; boundary layer theory; turbulence theory; dynamo theory of geomagnetism. *Mailing Add:* APAS Dept Campus Box 391 Univ Colo Boulder CO 80309-0391

BENTON, EUGENE VLADIMIR, b Russia, July 23, 37; US citizen; m 61; c 2. RADIOLOGICAL PHYSICS. *Educ:* San Jose State Col, BA, 58, MA, 60; Stanford Univ, PhD(physics), 68. *Prof Exp:* Instr physics, San Jose State Col, 58-60; from jr investr to investr, US Naval Radiol Defense Lab, 61-69; res prof, 69-73, PROF PHYSICS, UNIV SAN FRANCISCO, 73- *Concurrent Pos:* Consult, Lawrence Berkeley Lab, Univ Calif, 69-; mem comt space res, 85-; mem adv sci comt 75, Nat Coun Radiation Protection & Measurements, 84- *Honors & Awards:* Gold Medal Sci Achievement, Naval Radiol Defense Lab, 69; Alexander von Humboldt Prize, 79; Komos Achievement Award, NASA, 75, 78 & 81. *Mem:* AAAS; Am Phys Soc. *Res:* Medical physics; nuclear photographic emulsions; radiation dosimetry; materials science; solid state physics; radiation effects in solids; nuclear particle track detectors. *Mailing Add:* Dept Physics & Eng Univ San Francisco San Francisco CA 94117

BENTON, FRANCIS LEE, b Moxahala, Ohio, Feb 12, 12; m 40; c 3. ORGANIC CHEMISTRY. *Educ:* Ohio State Univ, BChE, 35, MS, 36, PhD(chem), 40. *Prof Exp:* Instr chem, Univ Notre Dame, 40-42, from asst prof to assoc prof, 42-52; res chemist, Armour Lab, 52-59; assoc prof chem, 59-62, prof, 62-77, EMER PROF CHEM, ST MARY'S COL, IND, 77- *Mem:* Am Chem Soc. *Res:* General organic synthesis; isolation of naturally occurring substances of biological origin and investigation of their molecular structure. *Mailing Add:* 1714 E Cedar St South Bend IN 46617

BENTON, GEORGE STOCK, b Oak Park, Ill, Sept 24, 17; m 45; c 4. METEOROLOGY, SCIENCE POLICY. *Educ:* Univ Chicago, SB, 42, PhD(meteorol), 47. *Prof Exp:* From asst to asst prof meteorol, Univ Chicago, 42-48; civil engr, 48-52, assoc prof, 52-57, actg chmn dept civil eng, 58-60, chmn dept mech, 60-66, chmn dept earth & planetary sci,69-70, dean fac arts & sci, 70-72, PROF METEOROL, JOHNS HOPKINS UNIV, 57- *Concurrent Pos:* Dir, Insts Environ Res, Environ Sci Serv Admin, Boulder, Colo, 66-69; vpres, Homewood Divisions, Johns Hopkins Univ, 72-77; assoc admr, Nat Oceanic & Atmospheric Admin, 78-81; US rep, World Met Org, 77-81. *Mem:* Fel Am Meteorol Soc (pres, 69-70); fel Am Geophys Union; fel Am Acad Arts & Sci. *Res:* Flow of stratified and rotating fluid; general circulation of the atmosphere; mesoscale meteorology. *Mailing Add:* Dept Earth & Planetary Sci Johns Hopkins Univ 34th & Charles Sts Baltimore MD 21218

BENTON, JOHN WILLIAM, JR, b Enterprise, Ala, July 3, 30; m 55; c 3. PEDIATRICS, NEUROLOGY. *Educ:* Univ Ala, BS, 51; Med Col Ala, MD, 55; Am Bd Pediat, dipl, 62; Am Bd Psychiat & Neurol, dipl, 69. *Prof Exp:* Intern pediat, Univ Ala Hosp, Birmingham, 55-56; resident, Univ Utah, 56-57 & Univ Minn, Minneapolis, 59-60; from instr to asst prof pediat & neurol, 60-66, dir, Ctr Develop & Learning Disorders, 62-68, assoc prof pediat & assoc prof Div Neurol, 66-69, interim chmn dept, & dir, Children's Hosp, 69-83, PROF PEDIAT & NEUROL, SCH MED, UNIV ALA, BIRMINGHAM, 69- *Concurrent Pos:* NIH trainee neurol, Med Col, Univ Ala, 60-62; fel, Mass Gen Hosp, Harvard Med Sch, 64-66. *Mem:* Am Pediat Soc; Am Acad Neurol; Am Acad Pediat. *Res:* Acute ataxia; acute encephalopathy of undetermined origin; convulsive disorders. *Mailing Add:* Univ Ala Children's Hosp 1601 Sixth Ave S Birmingham AL 35233

BENTON, KENNETH CURTIS, b Whitinsville, Mass, Sept 15, 41. POLYMER CHEMISTRY. *Educ:* Worcester Polytech Inst, BS, 63; Univ Akron, PhD(polymer chem), 69. *Prof Exp:* Sr res chemist, Copolymer Rubber & Chem Corp, 67-70, group leader polymers, CPC Int, Inc, 70-74; SR RES CHEMIST, STANDARD OIL CO, OHIO, 74- *Mem:* Am Chem Soc. *Res:* Synthesis of acrylonitrile barrier resins for packaging applications. *Mailing Add:* 8472 Bobolink Dr Macedonia OH 44056

BENTON, STEPHEN ANTHONY, b San Francisco, Calif, Dec 1, 41; m 64; c 2. IMAGING PHYSICS. *Educ:* Mass Inst Technol, BS, 63; Harvard Univ, MS, 65, PhD(appl physics), 68. *Prof Exp:* Asst prof appl physics, Div Eng & Appl Physics, Harvard Univ, 68-73; sr scientist, res labs, Polaroid Corp, 61-85; PROF MEDIA TECHNOL, MASS INST TECHNOL, 82- *Concurrent Pos:* Vis prof physics, Univ Reading, Eng, 72-73; vis scientist, Harrison Spectros Lab, Mass Inst Technol, 81-82; chmn, US Nat Comn, Int

Comn for Optics, 82-84; gov, Soc Photo-optical Instrumentation Engrs; trustee, Mus Holography. *Mem:* Fel Optical Soc Am; fel Soc Photographic Scientists & Engrs; Inst Elec & Electronic Engrs; Soc Photo-optical Instrumentation Engrs. *Res:* Image communication systems, holography. *Mailing Add:* 319 S Great Rd RFD No 3 Lincoln MA 01773-4304

BENTON, WILLIAM J, b Franklin, Va, June 28, 33; m 55; c 2. VETERINARY VIROLOGY. *Educ:* Univ Ga, DVM, 56; Univ Del, MS, 59, PhD(biol sci), 66. *Prof Exp:* Res assoc poultry path & virol, 56-66, asst prof, 66-67, actg chmn dept animal sci & agr biochem, 67-68, prof & chmn dept, 68-71, asst dean col agr sci & asst dir agr exp sta, 71-74, PROF ANIMAL SCI, UNIV DEL, 71-, ASSOC DEAN COL AGR SCI & ASSOC DIR AGR EXP STA, 74- *Mem:* Am Vet Med Asn; Am Asn Avian Path. *Res:* Application of pathology, serology, immunology and virology in the study of avian diseases such as infectious laryngotracheitis, infectious synovitus, avian leukosis, mycoplasma gallisepticum infection, Newcastle disease, infectious bronchitis and infectious bursal disease. *Mailing Add:* Col Agr Sci Univ Del Newark DE 19711

BENTRUDE, WESLEY GEORGE, b Waterloo, Iowa, Mar 13, 35; m 55; c 2. ORGANIC CHEMISTRY, ORGANOPHOSPHORUS CHEMISTRY. *Educ:* Iowa State Univ, BS, 57; Univ Ill, PhD(org chem), 61. *Prof Exp:* Res chemist, Celanese Chem Co, 61-63; res assoc, Univ Pittsburgh, 63-64; from asst prof to assoc prof, 64-67, PROF CHEM, UNIV UTAH, 72- *Concurrent Pos:* Vis scholar, Univ Calif, Los Angeles, 75-76; Humboldt fel, 77 & 81; assoc prof, Univ Marseille, France, 83. *Mem:* Am Chem Soc; Am Sci Affiliation. *Res:* Organophosphorus chemistry; mechanisms of free-radical reactions; physical organic chemistry; bio-organic chemistry; photochemistry; exploring photorearrangements and free radical chemistry of organophosphorus compound; use NMR methods to probe conformational properties of molecules structurally analogous to the anticancer agent cyclophospharnide and the bioregulator camp. *Mailing Add:* Dept Chem Univ Utah Salt Lake City UT 84112

BENTZ, ALAN P(AUL), b Minneapolis, Minn, Apr 1, 27; m 47; c 3. ENVIRONMENTAL CHEMISTRY. *Educ:* Pa State Univ, BS, 47; Univ Minn, MS, 49; Rutgers Univ, PhD, 58. *Prof Exp:* Asst, Univ Minn, 47-49; asst prof, US Naval Acad, 49-51; pharmaceut develop chemist, Am Cyanamid Co, 51-56, polymer res chemist, 57-63; flavor chemist, Gen Foods Corp, 63-73; sr res chemist, US Coast Guard Res & Develop Ctr, 73-80, chief, Chem Br, 80-81, proj mgr chem protective clothing, 82-90, MAT SCI MARINE ENG BR, US COAST GUARD RES & DEVELOP CTR, 91- *Concurrent Pos:* Adj prof org chem, Univ Conn, 77-; consult & expert witness, oil spill identification, 81- *Mem:* Am Chem Soc; Am Soc for Testing & Mat; Coblentz Soc. *Res:* Management; chemical permeation of protective clothing materials, forensic oil spill identification; hazardous chemical pollution; Fourier transform infrared spectroscopy; gas chromatography; calorimetry; patents in pharmaceutical processing and food flavor compositions. *Mailing Add:* US Coast Guard Res & Develop Ctr Avery Point Groton CT 06340

BENTZ, GREGORY DEAN, b Cleveland, Ohio, Nov 6, 45; m 88; c 3. BIOLOGY. *Educ:* Allegheny Col, BS, 67; Kent State Univ, MA, 69; Univ Pittsburgh, PhD(biol), 76; Autonomous Univ City Juarez, Mexico, MD, 88. *Prof Exp:* Postdoctoral fel, Smithsonian Inst, 76-77, res assoc, 77-85; from asst prof to assoc prof biol, Mt Vernon Col, 78-85, chmn, Dept Math & Sci, 84-85; CHIEF RESIDENT, DEPT FAMILY MED, EASTERN VA GRAD SCH MED, 88- *Mem:* Am Acad Family Physicians. *Res:* Processes of evolution of structural adaptations in vertebrates; anatomical descriptions of structural variations and determining how such structures may have evolved; medical essays; vertebrate morphology; family medicine; biosystematics. *Mailing Add:* 3031 S Reese Dr Portsmouth VA 23703

BENTZ, RALPH WAGNER, b Pennsylvania, Mar 9, 19; m 50; c 2. PHYSICAL CHEMISTRY, ORGANIC CHEMISTRY. *Educ:* Albright Col, BS, 43; Lehigh Univ, MS, 44, PhD(chem), 48. *Prof Exp:* Lab technician, A Wilhelm Co, 36-43; dept supvr, Manhattan Proj, 44-45; sr chemist, Tenn Eastman Co, 48-53; mkt analyst, Eastman Chem Prod, Inc, 53-66, SR MKT ANALYST, EASTMAN CHEM PROD, INC, 66- *Mem:* Am Chem Soc; Chem Mkt Res Asn. *Res:* Petrochemicals; plastics; packaging. *Mailing Add:* 2344 Inglewood Dr Kingsport TN 37664-2873

BENTZEL, CARL JOHAN, b Vicksburg, Miss, June 29, 34; m 62; c 4. EPITHELIAL PHYSIOLOGY, NEPHROLOGY. *Educ:* Univ Ala, BS, 54, MD, 58. *Prof Exp:* Intern Med, Presbyterian Hosp Col Physicians & Surgeons, 58-59, resident, 59-61; clin assoc, Nat Cancer Inst, NIH, 61-63; res fel, biophysics, Biophys Lab, Med Sch, Harvard Univ, 64-66; from asst prof to prof med, State Univ NY, Buffalo, 66-84; PROF, DEPT MED, E CAROLINA UNIV, 84- *Concurrent Pos:* Buswell fel, State Univ NY, Buffalo, 68-71; clin investr, Vet Admin Hosp, 71-73; estab investr, Am Heart Asn, 74-79. *Mem:* Am Soc Clin Invest; Am Physiol Soc; Am Soc Nephrology; Am Bd Internal Med & Nephrology. *Res:* Epithelial transport with particular reference to the role of the paracellular pathway and cytoplasmic mechanisms for control of tight junction permeability. *Mailing Add:* Dept Med East Carolina Univ Sch Med Brody Bldg Greenville NC 27858

BENUCK, MYRON, b Chicago, Ill, July 7, 34; m 68; c 2. NEUROCHEMISTRY. *Educ:* Univ Chicago, BA & BS, 57, MS, 62; Univ Ill, PhD(physiol), 66. *Prof Exp:* Fel neurochem, Albert Einstein Col Med, 65-67; SR RES SCIENTIST BIOCHEM, CTR NEUROCHEM, N S KLINE INST, 67- *Mem:* Am Soc Neurochem; Am Soc Biol Chemists; Int Soc Neurochemistry; Soc Neurosci. *Res:* Brain metabolism. *Mailing Add:* Ctr Neurochem N S Kline Inst Orangeburg NY 10962

BENUMOF, REUBEN, b New York, NY, Nov 30, 12; m 36, 52; c 2. ENGINEERING PHYSICS. *Educ:* City Col New York, BS, 33, MS, 37; NY Univ, PhD(physics), 45. *Prof Exp:* Elec engr, Fed Power Comn, 37-38; lectr physics, City Col New York, 45-56; assoc prof, Stevens Inst Technol, 51-56; prof physics & head dept, 56-77, PROF APPL SCI, COL STATEN ISLAND, 77-; CONSULT, JET PROPULSION LAB, 85- *Concurrent Pos:* Res partic, Oak Ridge Nat Lab, 56-58; fel, NSF, 59-61, dir optical pumping equip proj, 63-65; dir nuclear reactor kinetics res proj, 65-70, dir physics eng, 71-; fel, State Univ NY, 67, 69, 72; vis prof, Univ Zambia, 73; fel, NASA-Am Soc Eng Educ, 77; mem staff hazardous waste mgt, Argonne Nat Lab, 81; fel, Jet Propulsion Lab, 83 & 85; fel, Naval Ocean Systs Ctr, 84; fel, Naval Res Lab, 86-87, Air Force Geophysics Lab, 88-90. *Mem:* AAAS; Am Phys Soc; Am Asn Physics Teachers; Am Soc Eng Educ; Inst Elec & Electronic Engrs. *Res:* Astronomical theory of climate; nuclear reactor physics; mathematical modeling of microelectronic devices. *Mailing Add:* 21-15 34th Ave Long Island City NY 11106

BENVENISTE, JACOB, b Portland, Ore, Dec 21, 21; m 44; c 3. NUCLEAR PHYSICS. *Educ:* Reed Col, BA, 43; Univ Calif, PhD(physics), 52. *Prof Exp:* Radio engr, Naval Res Lab, 43-46; asst physics, Reed Col, 46; asst physics, Univ Calif, 47-50, physicist, Lawrence Radiation Lab, 50-63; physicist, Aerospace Corp, 63-65, sr staff scientist advan concepts, 65, dir nuclear effects subdiv, 65-68; dir res, Physics Int Co, 68-72, vpres, 69-72; SR STAFF SCIENTIST, AEROSPACE CORP, 72-; CHIEF SCIENTIST, NORTHROP RES & TECH CTR. *Concurrent Pos:* Mem, Nuclear Cross Sect Adv Group, 56-63; deleg, Geneva Conf, 58; mem, Defense Nuclear Agency Adv Panels, 65-68. *Mem:* AAAS; Am Phys Soc; Sigma Xi. *Res:* Solid state physics; radiation damage and effects; instrumentation; energy sources; laser gyroscopes. *Mailing Add:* 1364 Via Coronel 1325 Via Gabriel Palos Verdes Peninsula CA 90274

BENYAJATI, SIRIBHINYA, RENAL PHYSIOLOGY, COMPARATIVE PHYSIOLOGY. *Educ:* Brown Univ, PhD(physiol), 77. *Prof Exp:* Res assoc physiol, Col Med, Univ Ariz, 78-87; ASST PROF PHYSIOL, COL MED, UNIV OKLA, 87- *Mailing Add:* Dept Physiol Col Med Univ Okla Health Sci Ctr Oklahoma City OK 73190

BEN-YOSEPH, YOAV, b Petah-Tikva, Israel, Jan 8, 41; m 74; c 2. BIOCHEMICAL GENETICS, IMMUNOCHEMISTRY. *Educ:* Hebrew Univ, Israel, BS, 65, MS, 68, PhD(immunol), 73. *Prof Exp:* Lectr immunochemistry, Weizmann Inst Sci, 73-75; res assoc genetics, Children's Mem Hosp, 75-81; asst prof pediat, Sch Med, Northwestern Univ, Chicago, 75-79, assoc prof, 79-81; assoc prof 81-88, PROF PEDIAT, BIOCHEM & OBSTET/GYNEC, SCH MED, WAYNE STATE UNIV, 88- *Concurrent Pos:* Res fel, Med-Chem Inst, Berne Univ, Switz, 74. *Mem:* Am Soc Human Genetics; Soc Pediat Res; Soc Inherited Metabolic Dis. *Res:* Lysosomal enzymes deficiencies; cystic fibrosis; post-translational process in glycoprotein synthesis; autoimmune diseases. *Mailing Add:* 5200 Anthony Wayne Dr Detroit MI 48202

BENYSHEK, LARRY L, b Concordia, Kans, Feb 26, 47; m 66; c 2. ANIMAL BREEDING. *Educ:* Kans State Univ, BS, 69; Va Polytech Inst & State Univ, MS, 71, PhD(animal breeding), 73. *Prof Exp:* Dir res & educ beef cattle performance prog, NAm Limousin Found, 73-74; asst prof agr, Ft Hays Kans State Col, 74-76; ASST PROF BEEF CATTLE BREEDING, UNIV GA, 76- *Concurrent Pos:* Consult, NAm Limousin Found, 73- *Mem:* Sigma Xi; Am Soc Animal Sci; Am Genetic Asn. *Res:* Estimation of genetic parameters in new breeds of cattle; crossing of newly introduced breeds of cattle for increased commercial production; selection response using National Sire Evaluation in beef cattle. *Mailing Add:* Dept Animal & Dairy Sci Livestock-Poultry Bldg Univ Ga Athens GA 30602

BENZ, EDMUND WOODWARD, b Nashville, Tenn, May 8, 11; m 45; c 4. SURGERY. *Educ:* Vanderbilt Univ, AB, 37, MD, 40. *Prof Exp:* Res asst physiol, Sch Med, 38-40, surg training, Hosp, 40-45, resident surgeon, 44-45, from asst prof to assoc prof, 52-83, EMER CLIN PROF SURG, VANDERBILT UNIV, 83- *Concurrent Pos:* Pvt pract, 45-; mem surg staff, Hosps; chmn cancer study group, Tenn/Mid-South Regional Med Prog, 68-72; consult, Disability Determination Serv Social Security Agency, 83- *Mem:* AAAS; AMA; Am Col Surg. *Res:* Traumatic shock; wound healing and temperature; control of respiration; action potentials in peripheral nerve, splenectomy in hemophilia. *Mailing Add:* 3704 M Estes Rd Nashville TN 37215-1732

BENZ, EDWARD JOHN, b Pittsburgh, Pa, June 11, 23; m 45; c 4. CLINICAL PATHOLOGY, MICROBIOLOGY. *Educ:* Univ Pittsburgh, MD, 46; Univ Minn, MS, 52. *Prof Exp:* Dir labs & pathologist, 53-84, VPRES, MED AFFAIRS, ST LUKE'S HOSP, 84- *Concurrent Pos:* Consult, Palmerton Hosp; adj prof microbiol, Lehigh Univ, 65-; chmn med adv comt labs, Pa Dept Health, 74-76. *Mem:* Am Asn Path & Bact; Int Acad Path; Col Am Pathologists; Am Soc Clin Pathologists; Am Acad Med Dirs. *Mailing Add:* 57 Cindy Lane Guilford CT 06437

BENZ, FREDERICK W, b White Plains, NY, June 5, 44; m 69; c 2. PHARMACOLOGY, BIOCHEMISTRY. *Educ:* Manhattan Col, BS, 66; Univ Iowa, MS, 68, PhD(pharmacol), 70. *Prof Exp:* NIH fel, Molecular Pharmacol Unit, Med Res Coun, Cambridge, Eng, 70-72; sr sci staff, Nat Inst Med Res, Mill Hill, London, Eng, 72-73; asst prof pharmacol, Univ Wis-Madison, 73-78; ASSOC PROF PHARMACOL, COL MED, UNIV LOUISVILLE, 78- *Mem:* AAAS; Am Chem Soc; Protein Soc; Sigma Xi. *Res:* Biological applications of magnetic resonance; protein structure and folding; drug-protein interactions; sulfhydryl-disulfide biochemistry and pharmacology. *Mailing Add:* Sch Med Health Sci Ctr Univ Louisville Louisville KY 40292

BENZ, GEORGE WILLIAM, b Ulm, Ger, June 26, 22; US citizen; m 47; c 5. ORGANIC CHEMISTRY. *Educ:* Dartmouth Col, AB, 43; Univ Tex, MA, 46, PhD(org chem), 50. *Prof Exp:* Chemist, Winthrop Chem Co, 43-44, Heyden Chem Co, 49-52, Callery Chem Co, 52 & Polak's Frutal Works, 52-55; PRES, ULBECO, INC, 55- *Concurrent Pos:* Prof, Ulster County Community Col, Stone Ridge, NY, 69-, chmn dept phys sci, 71- *Mem:* Am Chem Soc. *Res:* Organic synthesis; organometallic compounds; perfume and flavor chemicals. *Mailing Add:* 159 Sheryl St Kingston NY 12401

BENZ, M G, b New York, NY, June 23, 35; m 59; c 4. METALLURGY. *Educ:* Middlebury Col, BA, 56; Mass Inst Technol, MS, 59, ScD(metall), 61. *Prof Exp:* Metallurgist, 61-80, MGR, PHYS METALL BRANCH, GEN ELEC RES & DEVELOP CTR, 80- *Honors & Awards:* Geisler Award, Am Soc Metals, 70. *Mem:* Am Inst Mining, Metall & Petrol Engrs; Am Soc Metals; Inst Elec & Electronics Engrs. *Res:* Process and physical metallurgy; iron carbon phase diagram; high temperature calorimetry; continuous casting of metals; process thermodynamics and kinetics; superconducting materials; permanent magnet materials; superalloys; oxidation and hot corrosion. *Mailing Add:* 11 Parkwood Dr Burnt Hills NY 12027

BENZ, WOLFGANG, b Heilbronn, Ger, Mar 16, 32; m 60; c 1. MASS SPECTROMETRY. *Educ:* Univ Heidelberg, dipl chem, 60, Dr rer nat (chem), 62. *Prof Exp:* Res assoc chem, Mass Inst Technol, 62-63; asst, Inst Org Chem, Univ Heidelberg, 63-64; chemist, Badische Anilin & Sodafabrik, 64-68; SR CHEMIST, HOFFMANN-LA ROCHE INC, 68- *Mem:* Am Soc Mass Spectrometry; Am Chem Soc; Ger Chem Soc. *Res:* Organic mass spectrometry; mechanisms of mass spectral fragmentations; analytical use of mass spectrometry; use of computers for data processing in mass spectrometry. *Mailing Add:* Hoffmann-La Roche Inc 340 Kingsland St Nutley NJ 07110

BEN-ZE'EV, AVRI, b Tirgu-Mures, Romania, Feb 11, 47; Israeli citizen; m 75; c 2. MOLECULAR BIOLOGY OF THE CYTOSKELETON. *Educ:* Hebrew Univ, Jerusalem, Israel, BSc, 70, MSc, 72, PhD(virol), 77. *Prof Exp:* Postdoctoral cell biol, Mass Inst Technol, 77-79; res assoc, 79-81, sr scientist, 81-84, PROF CELL BIOL, WEIZMANN INST SCI, 85- *Concurrent Pos:* Vis prof, Dept Surg Children's Hosp, Harvard Med Sch, Boston, Mass & Dept Biochem, Med Sch, Boston Univ, Mass, 86-87. *Mem:* Am Soc Cell Biol; Israeli Soc Cell Biol (treas, 91-); Israeli Soc Biochem; Israeli Soc Electronmicros. *Res:* Cytoskeletal proteins at areas of cell-cell and cell-matrix contact in the regulation of cell growth, differentiation and tumor invasion. *Mailing Add:* Dept Molecular Genetics & Virol Weizmann Inst Sci Rehovot 76100 Israel

BENZER, SEYMOUR, b New York, NY, Oct 15, 21; m 42, 80; c 3. MOLECULAR NEUROGENETICS. *Educ:* Brooklyn Col, BA, 42; Purdue Univ, MS, 43, PhD(physics), 47. *Hon Degrees:* DSc, Purdue Univ, 68, Columbia Univ, 74, Yale Univ, 77, Brandeis Univ & City Univ New York, 78, Univ Paris, 83. *Prof Exp:* From instr to asst prof physics, Purdue Univ, 45-48; biophysicist, Oak Ridge Nat Lab, 48-49; res fel biophys, Calif Inst Technol, 49-51; Fulbright res scholar, Pasteur Inst, Paris, 51-52; from asst prof to Stuart distinguished prof biophys, Purdue Univ, 53-67; prof biol, 67-75, BOSWELL PROF NEUROSCI, CALIF INST TECHNOL, 75- *Concurrent Pos:* NSF sr res fel, Cambridge Univ, 57-58. *Honors & Awards:* Sigma Xi Award; Howard Taylor Ricketts Award; Gairdner Award, Can, 64; McCoy Award, 65; Lasker Award, 71; T Duckett Jones Award & Prix Charles-Leopold Mayer, French Acad Sci, 75; Louisa Gross Horwitz Prize, Columbia Univ, 76; Harvey Prize, Technion, Israel & Warren Triennial Prize, 77; Dickson Award, 78; Nat Medal Sci, 83; Rosentiel Award, 86; Thomas Hunt Morgan Medal, Genetics Soc Am, 86; Karl Spencer Lashley Award, Am Philos Soc, 88. *Mem:* Nat Acad Sci; Am Acad Arts & Sci; Am Philos Soc; Biophys Soc Am; Soc Neurosci; Indian Acad Sci; Genetics Soc Am. *Res:* Behavioral and developmental genetics and neurophysiology of drosophila; molecular genetics; genetics and physiology of the nervous system and behavior; behavioral and developmental genetics and neurophysiology of drosophila; author of numerous publications and articles. *Mailing Add:* Div Biol 156-29 Calif Inst Technol Pasadena CA 91125

BENZIGER, CHARLES P(HILLIP), engineering geology, rock mechanics, for more information see previous edition

BENZIGER, THEODORE MICHELL, b Waterloo, Iowa, Nov 14, 22; m 43; c 3. CHEMICAL ENGINEERING. *Educ:* Iowa State Univ, BS, 44. *Prof Exp:* Res engr chem eng, Union Carbide Corp, 46-52; asst group leader, Explosives Res & Dev, Los Alamos Nat Lab, Univ Calif, 52-82; CONSULT, CHEM PROCESS DEVELOP, 82- *Honors & Awards:* Crozier Prize, Am Defense Preparedness Asn, 78. *Res:* Explosives-organic synthesis, process development and design, formulation, fabrication and materials evaluation; manufacture of insensitive, thermally-stable explosives; explosives manufacturing processes; plastic-bonded explosives. *Mailing Add:* 1122 Taos Hwy Santa Fe NM 87501

BENZING, DAVID H, b Chicago, Ill, Oct 13, 37. BIOLOGY. *Educ:* Miami Univ, BA, 59; Univ Mich, MS, 62, PhD(bot), 65. *Prof Exp:* Assoc prof, 65-77, chmn dept biol, 81-82 & 83-88, PROF BIOL, OBERLIN COL, 77- *Concurrent Pos:* Robert S Danforth prof biol, Oberlin Col. *Mem:* NSF sci fac fel, Univ SFla, 71-72; Bot Soc Am; Sigma Xi; Asn Trop Biol. *Res:* Adaptive biology of vascular epiphytes; demography, nutrition, water balance, and interaction with host vegetation, with major emphasis on Bromeliaceae and Orchidaceae. *Mailing Add:* Dept Biol Oberlin Col Oberlin OH 44074

BENZING, GEORGE, III, b Hamilton, Ohio, Nov 18, 26. PEDIATRIC CARDIOLOGY, PHYSIOLOGY. *Educ:* Univ Cincinnati, BS, 51, MD, 58; Am Bd Pediat, dipl, 63, cert pediat cardiol, 64. *Prof Exp:* From asst to assoc prof, 64-72, PROF PEDIAT, COL MED, UNIV CINCINNATI, 72-; ATTEND CARDIOLOGIST, CHILDREN'S HOSP, 70- *Concurrent Pos:* Assoc attend cardiologist, Children's Hosp, 66-70. *Mem:* Am Heart Asn; Sigma Xi. *Res:* Myocardial performance as evaluated by ventricular function curves and cardiopulmonary bypass. *Mailing Add:* Div Cardiol Children's Hosp Cincinnati OH 45229

BENZINGER, HAROLD EDWARD, JR, mathematics; deceased, see previous edition for last biography

BENZINGER, JAMES ROBERT, b Buffalo, NY, Nov 26, 22; m 48; c 3. ELECTRICAL INSULATION, SUPERCONDUCTIVITY. *Educ:* Canisius Col, BS, 48, MS, 54. *Prof Exp:* Plant chemist, Spaulding Fibre Co, Inc, 48-53, res chemist, 53-56, group leader, 56-58, mgr plastics res & develop, 58-67, mgr advan develop, 67-76, mgr corp prod develop, 76-86, tech dir, 86-90; CONSULT, 90- *Mem:* Am Chem Soc; Soc Plastics Eng. *Res:* Product and process development of thermosetting composites and filament wound structures for electrical insulation; industrial thermosetting laminates; chemistry of high polymers and adhesives, including phenolics, epoxies, melamines, polymides and polyesters. *Mailing Add:* 135 Hillside Dr Orchard Park NY 14127

BENZINGER, ROLF HANS, b Rostock, Ger, Dec 4, 35; div; c 2. BIOCHEMICAL GENETICS. *Educ:* Johns Hopkins Univ, BA, 56, PhD(biochem genetics), 61. *Prof Exp:* Ger Res Asn fel, Max Planck Inst Biochem, 61-65; State of Geneva assistantship, Lab Biochem Genetics, Geneva, 65-67; asst prof biol, 67-70, ASSOC PROF BIOL, UNIV VA, 70- *Concurrent Pos:* Ctr Advan Studies res grant, Univ Va, 67-68; NIH res grants, 68-76; NIH res career develop award, 71-76; prog assoc prokaryotic genetics, NSF, 84-85; mem, Ctr Bioprocess Develop, Univ Va, 90- *Mem:* AAAS; Am Soc Microbiol. *Res:* Plasmid instability in B subtilis; injection of T5 phage DNA and control of membrane structure. *Mailing Add:* Dept Biol Gilmer Hall Univ Va Charlottesville VA 22903

BENZINGER, WILLIAM DONALD, b Pittsburgh, Pa, Feb 6, 40; m 62; c 2. INDUSTRIAL CHEMISTRY, MEMBRANE TECHNOLOGY. *Educ:* Univ Notre Dame, BS, 61; Pa State Univ, PhD(inorg chem), 67. *Prof Exp:* Res chemist, Res Labs, US Army Edgewood Arsenal, Md, 67-68; sr res chemist, Penwalt Corp, 69-75, proj leader, 76-83; MGR MEMBRANE TECHNOL, CULLIGAN INT CO, 83- *Mem:* Am Chem Soc; NAm Membrane Soc. *Res:* Coordination chemistry; inorganic polymers; fiber reinforced composites; coatings; electrochemistry; industrial processes; membrane technology; ultrafiltration; waste treatment; polymer properties; reverse osmosis; water treatment. *Mailing Add:* 150 Kings Row Barrington IL 60010-4827

BENZO, CAMILLO ANTHONY, b Utica, NY, Dec 31, 42. DEVELOPMENTAL ENDOCRINOLOGY, EXPERIMENTAL DIABETES. *Educ:* Syracuse Univ, BA, 64; Univ Pa, PhD(anat), 69. *Prof Exp:* Fel gross anat, Med Sch, Univ Pa, 69-70; asst prof, 70-75, ASSOC PROF GROSS ANAT, STATE UNIV NY UPSTATE MED CTR, 75- *Concurrent Pos:* Dir Cent New York Cadaver Donor Prog, State Univ NY Upstate Med Ctr, 71-; adj assoc prof biol, Sch Environ Sci & Forestry, State Univ NY, 75- *Mem:* Am Asn Anatomists; Am Soc Zoologists. *Res:* Problems in embryonic development and in biochemical events during cellular differentiation; problems in the development and regulation of glycogen metabolism in avian and in diabetic mammalian systems. *Mailing Add:* Dept Anat State Univ NY Health Sci Col Med 750 E Adams St Syracuse NY 13210

BEOUGHER, ELTON EARL, b Gove, Kans, Mar 22, 40; m 60; c 2. MATHEMATICS. *Educ:* Ft Hays State Univ, BS, 61, MA, 64; Univ Mich, PhD(math educ), 68. *Prof Exp:* Teacher, Winona Consol Schs, Kans, 61-64 & Garden City Schs, 64-65; from asst prof to assoc prof, 68-76, chmn dept, 73-82, PROF MATH, FT HAYS STATE UNIV, 76- *Mem:* Math Asn Am; Nat Coun Teachers Math. *Res:* Mathematical education and curriculum; psychology of learning mathematics. *Mailing Add:* Dept Math Ft Hays State Univ Hays KS 67601

BEQUETTE, B WAYNE, b Fayetteville, Ark, July 13, 57. PROCESS CONTROL, PROCESS DESIGN. *Educ:* Univ Ark, BS, 80; Univ Tex, MSE, 85, PhD(chem eng), 86. *Prof Exp:* Fel chem eng, Univ Tex, 86-87; vis lectr chem eng, Univ Calif-Davis, 87-88; ASST PROF CHEM ENG, RENSSELAER POLYTECH INST, 88- *Mem:* Am Inst Chem Engrs. *Res:* Nonlinear control of chemical processes; nonlinear dynamics; data reconciliation; multi-rate sampled data control; control of drug delivery systems. *Mailing Add:* Dept Chem Eng Rensselaer Polytech Inst Troy NY 12180-3590

BERAN, DONALD WILMER, b Wheatland, Wyo, Aug 15, 35; m 56; c 1. METEOROLOGY. *Educ:* Utah State Univ, BSc, 58; Colo State Univ, MSc, 66; Univ Melbourne, PhD(meteorol), 70. *Prof Exp:* Engr-meteorologist, Martin Co, 62-64; res asst meteorol, Colo State Univ, 64-66; sr meteorologist, Allied Res Assocs, 66-67; res meteorologist, 70-75, supvry res meteorologist, 75-78, chief remote sensor appln, Wave Propagation Lab, 78-81, DIR, PROFILER PROF OFF, NAT OCEANIC & ATMOSPHERIC ADMIN, 81- *Honors & Awards:* Gold Medal Award, Dept Com, 85. *Mem:* Am Meteorol Soc; Am Geophys Union. *Res:* Satellite meteorology; development of satellite based lidar for global wind measurements; development of Prototype Regional Observing and Forecasting Service Program for improving local weather services; lee waves and mountain winds; clear air turbulence and its detection; development of acoustic sounding methods for remote sensing of boundary layer phenomena. *Mailing Add:* Environ Res Lab /FSL/R/E/FS3 Nat Oceanic & Atmospheric Admin 325 Broadway Boulder CO 80303-3328

BERAN, GEORGE WESLEY, b Riceville, Iowa, May 22, 28; m 54; c 3. VETERINARY PUBLIC HEALTH. *Educ:* Iowa State Univ, DVM, 54; Kans Univ, PhD(med microbiol), 59; Am Col Vet Prev Med, dipl, Am Col Epidemiol, dipl. *Hon Degrees:* LHD, Silliman Univ, Philippines, 73. *Prof Exp:* Epidemic intel serv officer, USPHS, 54-56; prof microbiol, Silliman Univ, Philippines, 60-73, dir, Van Houweling Lab, 61-73; PROF VET MICROBIOL, COL VET MED, IOWA STATE UNIV, 73- *Concurrent Pos:* Consult, WHO and/or Pan Am Health Orgn, India, 71, Malaysia, 71, Laos, 72, Philippines, 73, 74, 76 & 78, Jamaica, 79, Ecuador, 84, 85, 86 & 89, Mexico, 85, Surinam, 85; Fulbright prof, Ahmadu Bello Univ, Zaria, Nigeria, 80; vis lectr, Chuinan & Taipei, Taiwan, 83, Loja & Guayaquil, Ecuador, 83, Beijing & Wuhan, China, 88, Budapest, Hungary, 88 & 90; exchange deleg, USSR, 89 & 90, Costa Rica, 90. *Honors & Awards:* James H Steele Award,

World Vet Epidemiol Soc, 79. *Mem:* Am Vet Med Asn; Conf Pub Health Veterinarians; Asn Teachers Vet Pub Health; Am Col Vet Prev Med; Am Asn Food Hyg Veterinarians; US Animal Health Asn; fel Royal Soc Health; Sigma Xi. *Res:* Epidemiology and immunology of rabies; epidemiology of pseudorabies; food virology. *Mailing Add:* Vet Microbiol & Prev Med Iowa State Univ Col Vet Med Ames IA 50011

BERAN, JO ALLAN, b Odell, Nebr, Aug 24, 42; m 64; c 2. ENVIRONMENTAL & INORGANIC CHEMISTRY. *Educ:* Hastings Col, BA, 64; Univ Kans, PhD(chem), 68. *Prof Exp:* From asst prof to assoc prof, 68-76, chmn dept, 75-81, PROF CHEM, TEX A&I UNIV, 76- *Mem:* Am Chem Soc. *Res:* Gas phase studies of ions produced by photoionization; investigations in chemical education. *Mailing Add:* Dept Chem Texas A&I Univ Box 161 Kingsville TX 78363

BERAN, MARK JAY, b New York, NY, Aug 19, 30; m 53; c 3. PHYSICS, ENGINEERING SCIENCE. *Educ:* Mass Inst Technol, SB, 52; Harvard Univ, SM, 53, PhD, 55. *Prof Exp:* Hydrodynamicist, Hydrodyn Lab, Mass Inst Technol, 51-55; group leader math, Waltham Airborne Systs Lab, Radio Corp Am, 55-56; res physicist, Air Force Cambridge Res Ctr, 56-59 & Tech Opers, 59-61; from assoc prof to prof mech eng, Univ Pa, 61-74; PROF ENG, TEL-AVIV UNIV, 74- *Concurrent Pos:* Fulbright scholar, Dept Electronics, Weizmann Inst Sci, Israel, 67-68. *Mem:* Fel Optical Soc Am; fel Acoust Soc Am; Am Phys Soc. *Res:* Underwater acoustics; applied physics; fluid mechanics; electromagnetic theory; plasma physics; statistical continuum theory; coherence theory. *Mailing Add:* Sch Eng Tel-Aviv Univ Tel-Aviv Israel

BERAN, ROBERT LYNN, b Eau Claire, Wis, Oct 1, 43; m 71; c 2. PROCESS & PRODUCT OPTIMIZATION. *Educ:* Univ Wis-Madison, BS, 67, PhD(mech eng), 72. *Prof Exp:* Design engr, Monitor Data Systs, Inc, 71-72; res engr, 72-77, sr res engr, 77-79, group leader, 79-81, RES DIR, WESTVACO CORP, 81- *Honors & Awards:* George Olmsted Award, Am Paper Inst, 78. *Mem:* Tech Asn Pulp & Paper Indust; AAAS. *Res:* Theoretical and experimental investigations of papermaking processes satisfying technical needs and business goals of manufacturing, converting, and product development groups; physics of paper and papermaking. *Mailing Add:* Covington Res Lab Westvaco Covington VA 24426

BERANBAUM, SAMUEL LOUIS, b Toronto, Ont, Can, Apr 20, 15; US citizen; m 49; c 3. RADIOLOGY. *Educ:* Univ Toronto, BA, 37, MD, 40. *Prof Exp:* Intern, Mt Sinai Hosp, Cleveland, 40-42; resident radiol, Postgrad Hosp, New York, 42-45; instr, Columbia Univ, 45-48; from asst clin prof to assoc clin prof, 48-60, CLIN PROF RADIOL, NY UNIV, 60-, ATTEND RADIOLOGIST, UNIV HOSP, BELLEVUE MED CTR, 55- *Concurrent Pos:* Assoc attend radiologist, Postgrad Hosp, New York, 45-48; dir radiol, St Barnabas Hosp, 45-70; consult, St Lukes Hosp, 70-; consult radiology, Vet Admin Hosp & Lenox Hill Hosp, 75- *Mem:* Fel Am Col Gastroenterol; fel Am Col Radiol; Radiol Soc NAm; Am Roentgen Ray Soc; Asn Gastro-Intestinal Radiologists. *Res:* Special procedures in roentgen diagnostics; fluoroscopy and radiography of the gastro-intestinal tract; right side diverticular disease; gastric volvulus; elusive abdominal tumors. *Mailing Add:* 121 E 60th St New York NY 10022

BERANEK, DAVID T, b LaCrosse, Wis, Jan 31, 44; m 68; c 4. MUTAGENESIS, CARCINOGENESIS. *Educ:* Univ Wis, LaCrosse, BA, 68, MA, 76; Univ Ark, PhD, 89. *Prof Exp:* Med res technician, Med Sch, Wash Univ, 76-77; RES CHEMIST, NAT CTR TOXICOL RES, 77- *Mem:* Am Chem Soc; AAAS; Environ Mutagen Soc. *Res:* Biochemical mechanisms of mutagenesis and carcinogenesis caused by alkylating agents and aromatic amines and amides; site specific mutagenesis of aromatic amines in eucaryotic organisms. *Mailing Add:* Nat Ctr Toxicol Res HFT-321 Jefferson AR 72079

BERANEK, LEO LEROY, b Solon, Iowa, Sept 15, 14; m 41, 85; c 2. ACOUSTICS, COMMUNICATIONS. *Educ:* Cornell Col, AB,36; Harvard Univ, MS, 37, ScD(physics), 40. *Hon Degrees:* ScD, Cornell Col, 46; DEng, Worcester Polytech Inst, 71; DCS, Suffolk Univ, 79; LLd, Emerson Col, 82; DPS, Northeastern Univ, 84. *Prof Exp:* From instr to asst prof physics & communication eng, Harvard Univ, 40-43, dir electro-acoustic lab, 43-46 & systs res lab, 45-46; Guggenheim Mem Found fel, Harvard Univ & Mass Inst Technol, 46-47; assoc prof commun eng, Mass Inst Technol, 47-58, tech dir, Acoustics Lab, 47-53, lectr, 58-80; pres, elec off, Bolt, Beranek & Newman Inc, 53-69, dir, 53-83, chief scientist, 69-71; pres, elec off, 71-79, chmn bd, Boston Broadcasters Inc, 83-87; CONSULT, ACOUSTICS, ELECTRONICS, 87- *Concurrent Pos:* Pres, elec off, Bolt, Beranek & Newman, Inc, 53-69, dir, 53-83, chief scientist, 69-71; trustee, Cornell Col, 56-71 & Emerson Col, 74-79; chmn bd, Mueller-BBN GmbH, Munich, 62-86; mem, Aeronaut & Space Eng Bd, Nat Acad Eng, 65-76, comn pub eng policy, 66-71, marine bd, 67-71, coun, 69-71; vchmn, Mass Comn Ocean Mgt, 68-71; mem vis comt, Dept Soc Rels & Psychol, Harvard Univ, 64-70, dept biol, 71-77, 86-, mem adv comt mgt develop progs, Grad Sch Bus Admin, 65-71, mem advan mgt prog, 65-71, vis comt, 87-90; bd overseers, Harvard Univ, 84-90. *Honors & Awards:* Biennial Award, Acoust Soc Am, 44; Wallace Clement Sabine Award, Acoust Soc Am, 61; Gold Medal, Audio Eng Soc, 71; Gold Medal, Acoust Soc Am, 75; Abe Lincoln TV Award, 76. *Mem:* Nat Acad Eng; fel AAAS; fel Acoust Soc Am (vpres, 49-50, pres, 54-55); fel Audio Eng Soc (pres, 67-68); fel Am Acad Arts & Sci (pres, 89-92); fel Inst Elec & Electronic Engrs; fel Am Inst Phys. *Res:* Architectural acoustics; electromechanico-acoustical devices; acoustic measurements; noise control techniques; technology assessment. *Mailing Add:* 975 Memorial Dr Suite 804 Cambridge MA 02138-5717

BERANEK, WILLIAM, JR, b Chicago, Ill, Jan 8, 46. BIOCHEMISTRY. *Educ:* Univ Wis, BS, 67; Calif Inst Technol, PhD(chem), 73. *Prof Exp:* Res assoc biochem, Med Ctr, Duke Univ, 73-75, res fel, 75-77; asst dir, Holcombe Res Inst, Butler Univ, 77-78, assoc dir, 78-90; PRES, IND ENVIRON INST, 90- *Concurrent Pos:* Consult prog planning, WNET-TV, New York, 76-; adj assoc prof chem, Butler Univ, 77- *Mem:* AAAS; Am Chem Soc; NY Acad Sci. *Res:* Protein-sugar interactions; biochemical aspects of ecological and environmental issues; public policy issues involving chemistry and biochemistry. *Mailing Add:* 150 W Market St Suite 816 Indianapolis IN 46204

BERARD, ANTHONY D, JR, b Lynn, Mass, Sept 23, 42. TOPOLOGY. *Educ:* The Citadel, BS, 64; Case Western Reserve Univ, MA, 67, PhD(math), 68. *Prof Exp:* Asst prof math, US Air Force Inst Technol, 68-72; from assoc prof to prof math, 72-87, PROF MATH & COMPUT SCI, KING'S COL, 87- *Mem:* Am Math Soc; Math Asn Am. *Res:* Characterization of metric spaces by the use of their midsets. *Mailing Add:* Dept Math King's Col Wilkes-Barre PA 18711

BERARD, COSTAN WILLIAM, b Cranford, NJ, Dec 23, 32; m 58; c 2. MEDICINE, PATHOLOGY. *Educ:* Princeton Univ, AB, 55; Harvard Univ, MD, 59. *Prof Exp:* Intern surg, Univ Rochester, 59-60; res assoc, Walter Reed Inst Res, 60-62, dept chief, 62-63; resident, Nat Cancer Inst, 63-66, staff pathologist, 66-80, head, Hematopath Sect, 70-80; PATHOLOGIST, ST JUDE CHILDREN'S RES HOSP, MEMPHIS, 80- *Mem:* AMA; Fedn Am Soc Exp Biol; Am Soc Exp Path. *Res:* Metabolic response to cancer with particular reference to immunology; metabolic response to trauma with particular reference to wound healing. *Mailing Add:* 4576 Park Ave Memphis TN 38117

BERARD, MICHAEL F, b Des Moines, Iowa, Aug 9, 38. CERAMICS ENGINEERING. *Educ:* Iowa State Univ, BS, 60, MS, 62, PhD(ceramic eng), 68. *Prof Exp:* Develop engr, Aerojet-Gen Corp, 62-64; instr, dept ceramic eng & jr engr, Ames Lab, Atomic Energy Comn, 64-67, asst prof dept ceramic eng & assoc engr, 67-71, assoc prof dept ceramic eng & engr, 71-77, PROF DEPT MAT SCI & ENG & SR ENGR, AMES LAB, IOWA STATE UNIV, 77- *Mem:* Am Ceramic Soc; Nat Inst Ceramic Engrs; Am Soc Eng Educ. *Res:* Diffusion, defect structure and electrical conductivity in ceramic materials; properties of rare earth oxides. *Mailing Add:* Dept Mat Sci Iowa State Univ 110 Eng Annex Ames IA 50011

BERARDINELLI, FRANK MICHAEL, b Newark, NJ, June 6, 20; m 57; c 2. POLYMER CHEMISTRY. *Educ:* Seton Hall Univ, BS, 43. *Prof Exp:* Chemist, 43-47, res chemist, 47-49, sr res chemist, 49-60, group leader polyacetal res & develop, 60-74, RES SUPVR, CELANESE RES CO, 74- *Mem:* Am Chem Soc; Sigma Xi. *Res:* Synthesis and processing of cellulosic plastics; textile finishes and plasticizer synthesis and evaluation; polyethylene film studies; emulsion, condensation and ionic polymerization of polymers; polyacetals; composites; high temperature polymers. *Mailing Add:* Four Garland Dr Brick NJ 08723

BERARDO, PETER ANTONIO, b Los Angeles, Calif, Apr 25, 39; m 63; c 2. PARTICLE PHYSICS, COMPUTER MODELS. *Educ:* Univ Calif, Los Angeles, BS, 64, MS, 65, PhD(physics), 71. *Prof Exp:* Physicist dosimetry, Armed Forces Radiobiol Res DNA Inst, 70-73; physicist strategic planning, Joint Strategic Target Planning Staff, Joint Chiefs Staff, 73-75; PHYSICIST TREAT PLANNING DEVELOP, LOS ALAMOS NAT LAB, 75- *Honors & Awards:* Joint Serv Commendation Medal, Defense Nuclear Agency, 73. *Mem:* Am Phys Soc; Am Asn Physicists Med. *Res:* Utility of pions in cancer radiotherapy; practical applications of particle physics; improved treatment-planning with CT data. *Mailing Add:* MS H809 Box 1663 MS 809 Los Alamos NM 87545

BERBARI, EDWARD J, b Massillon, Ohio, May 6, 49; m 77; c 4. MEDICAL INSTRUMENTATION, CARDIAC ELECTROPHYSIOLOGY. *Educ:* Carnegie-Mellon Univ, BS, 71; Univ Miami, MS, 73; Univ Iowa, PhD(elec eng), 80. *Prof Exp:* ASSOC PROF, UNIV OKLA HEALTH SCIS CTR, 80- *Mem:* Inst Elec Electronic Engrs; Int Soc Comput Electrocardiography. *Res:* High resolution computer techniques for recording electrical activity of the heart to identify individuals at risk for sudden cardiac death due to arrhythmias. *Mailing Add:* Vet Admin Med Ctr 151F 921 NE 13th St Oklahoma City OK 73104

BERBEE, JOHN GERARD, b Hamilton, Can, Oct 12, 25; m 50; c 3. PLANT PATHOLOGY, FORESTRY. *Educ:* Univ Toronto, BScF, 49; Yale Univ, MF, 50; Univ Wis, PhD(plant path), 54. *Prof Exp:* Asst forest biologist, Can Dept Agr Sci Serv, 54-57; assoc prof plant path, 57-69, PROF PLANT PATH & FORESTRY, UNIV WIS-MADISON, 69- *Concurrent Pos:* Sr lectr plant sci, Univ Ife, Nigeria, 64-67. *Mem:* Am Phytopath Soc; Soc Am Foresters. *Res:* Disease of forest trees and breeding for disease resistance; forest tree virology; tropical plant pathology. *Mailing Add:* Dept Plant Path Univ Wis Madison WI 53706

BERBERIAN, PAUL ANTHONY, b Boston, Mass, Sept 10, 45; m 79. ARTERIAL WALL METABOLISM, PATHOBIOLOGY OF ARTERIOSCLEROSIS. *Educ:* Boston Univ, BA, 68; Univ Miami, Fla, PhD(physiol-biophysics), 76. *Prof Exp:* Res assoc biochem & dermat, S L Hsia Lab, Univ Miami, Fla, 76; fel biochem cytol, C deDuve Lab, Rockefeller Univ, 76-79; ASSOC PROF MICROANAT & CELL BIOL, DEPT ANAT, BOWMAN GRAY SCH MED, WAKE FOREST UNIV, 79-, ASSOC MED CARDIOL, 88- *Concurrent Pos:* Prin investr, NIH, 80- & NC Heart Asn, 81-82; mem teaching fac, Vis Scientists for Minority Insts, Fedn Am Socs Exp Biol, 85-; Sigma Xi Nat lectr comt, 88- *Mem:* Am Heart Asn; Am Soc Cell Biol; Soc Exp Biol & Med; Am Asn Anatomists; Am Asn Pathologists; Sigma Xi. *Res:* Arterial prostaglandins stress proteins and lysosomal function during atherogenesis and aging of the vasculature, especially with respect to intracellular lipid accumulation and cell death. *Mailing Add:* Dept Neurobiol & Anat Bowman Gray Sch Med Wake Forest Univ Winston-Salem NC 27103

BERBERIAN, STERLING KHAZAG, b Waukegan, Ill, Jan 15, 26; m 61; c 2. MATHEMATICAL ANALYSIS, ALGEBRA. *Educ:* Mich State Univ, BS, 48, MS, 50; Univ Chicago, PhD(math), 55. *Prof Exp:* Instr math, Fisk Univ, 50, Southern Ill Univ, 51-52 & Univ Ill, 52-53; from instr to asst prof, Mich State Univ, 55-57; from asst prof to prof, Univ Iowa, 57-66; ed, Math

Rev, Ann Arbor, Mich, 66-68; PROF MATH, UNIV TEX, AUSTIN, 68- *Concurrent Pos:* Vis prof, Ind Univ, Bloomington, 70-71, Univ Reading & Univ Poitiers; lectr, Int Ctr Pure & Appl Math, France, 86. *Mem:* French Math Soc; Math Asn Am. *Res:* Hilbert space; integration theory; operator algebras; regular rings; Baer rings. *Mailing Add:* Dept Math Univ Tex Austin TX 78712

BERCAW, JAMES ROBERT, b Canton, Ohio, Aug 10, 23; m 44; c 4. TEXTILE PRODUCTS. *Educ:* William Jewell Col, AB, 48; Ohio State Univ, PhD(chem), 54. *Prof Exp:* Res chemist, 54-57, from res supvr to sr supvr, 57-63, tech supt, 63-65, indust tech mgr, 65-66, asst nylon prod mgr, 66-68, asst tech serv mgr, 68-74, mgr prod serv & technol, Textile Fibers dept, 74-80, mgr int trade & regulatory affairs, E I Du Pont de Nemours & Co, Inc, 80-85; RETIRED. *Mem:* Am Chem Soc; Am Burn Asn. *Res:* Man-made fiber research, development, evaluation and production; international trade and regulatory affairs. *Mailing Add:* 13 Sorrel Dr Surrey Park Wilmington DE 19803-1928

BERCAW, JOHN EDWARD, b Cincinnati, Ohio, Dec 3, 44; m 65; c 2. ORGANOMETALLIC CHEMISTRY. *Educ:* NC State Univ, BS, 67; Univ Mich, Ann Arbor, PhD(chem), 71. *Prof Exp:* Arthur Amos Noyes res fel chem, 72-74, from asst prof to assoc prof, 77-79, PROF CHEM, CALIF INST TECHNOL, 79- *Concurrent Pos:* Alfred P Sloan res fel, Calif Inst Technol, 76-78, Dreyfus teacher scholar, 77-82; consult, Exxon Res & Eng Co, 79, Lawrence Berkeley Lab, Mat & Chem Sci Div Rev Panel, 84; Shell distinguished prof, 85-90; mem, adv bd, Inorganica Chemica Acta, 88-, sci rev comt, OER/Basic Chem Sci, Solar Energy Res Inst, 90-; chmn elect, Div Inorg Chem, 87, chmn, 88; vchmn, Gordon Res Conf on Organometallic Chem, 90. *Honors & Awards:* Am Chem Soc Award in Organometallic Chem, 90. *Mem:* Nat Acad Sci; Am Chem Soc; fel AAAS. *Res:* Synthetic and mechanistic organotransition metal chemistry with special emphasis on activation of small molecules at early transition metal centers. *Mailing Add:* Div Chem & Chem Eng 127-72 Calif Inst Technol Pasadena CA 91125

BERCH, JULIAN, b Winnipeg, Man, Oct 13, 16; nat US; m 44; c 4. CHEMISTRY. *Educ:* Univ Wash, BS, 38. *Prof Exp:* Chemist, Fish & Wildlife Serv, 39-40; chemist, Nat Bur Standards, 42-43; res assoc, Textile Res Found, 43-44 & Harris Res Labs, 45-67; RES SUPVR, GILLETTE RES INST, 67- *Mem:* AAAS; Am Chem Soc; Am Asn Textile Chem & Colorists. *Res:* Surface active agents and detergents; technology of non-woven fabrics; mechanism of wool felting and shrinkage control; wool chemistry; textile application of rubber latices and permanent-press finishes; water pollution. *Mailing Add:* 2100 Washington Ave Silver Spring MD 20910

BERCHTOLD, GLENN ALLEN, b Pekin, Ill, July 1, 32; m 59; c 3. ORGANIC CHEMISTRY. *Educ:* Univ Ill, BS, 54; Univ Ind, PhD(org chem), 59. *Prof Exp:* From instr to assoc prof, 60-69, PROF CHEM, MASS INST TECHNOL, 69- *Mem:* Am Chem Soc. *Res:* Organic synthesis. *Mailing Add:* Dept Chem Mass Inst Technol Cambridge MA 02139-4307

BERCOV, RONALD DAVID, b Edmonton, Alta, Dec 14, 37; m 65; c 2. ALGEBRA. *Educ:* Univ Alta, BSc, 59; Calif Inst Technol, PhD(math), 62. *Prof Exp:* Res assoc math, Cornell Univ, 62-63; from asst prof to assoc prof, 63-74, PROF MATH, UNIV ALTA, 74- *Concurrent Pos:* Vis assoc prof, Univ Wash, 69; ed, Can Math Bull, 70. *Mem:* Am Math Soc; Math Asn Am; Can Math Cong. *Res:* Group theory, primarily in the field of permutation groups. *Mailing Add:* Dept Math Univ Alta Edmonton AB T6G 2G1 Can

BERCOVITZ, ARDEN BRYAN, b Los Angeles, Calif, Oct 18, 45; c 1. REPRODUCTIVE ENDOCRINOLOGY. *Educ:* Calif State Polytech Univ, Pomona, BS, 68; Wash State Univ, MS, 70; Univ Mo, Columbia, PhD(physiol), 76. *Prof Exp:* Fel environ physiol, Univ Mo, 70-73; res assoc animal sci, La State Univ, 73-75; res fel, San Diego Zoo, 77-78; instr human physiol, San Diego City Col, 77; avian reproductive physiologist, res & ornith depts, San Diego Zoo, 78-88; RES ASSOC, HUBB/SEA WORLD RES INST, 77-; REPRODUCTIVE ENDOCRINOLOGIST, CTR REPRODUCTION ENDANGERED SPECIES, ZOOL SOC SAN DIEGO, 88- *Concurrent Pos:* NSF grant, 78-80; res fel, Scripps Found, 77; fel, USPHS, 70-73; spec educ consult, San Diego Unified Sch Dist, 83-; adj fac, San Diego State Univ, 84-; Inst Mus Serv grant, 85-86; spec educ consult, San Diego Univ Sch Dist, 85-; lectr, animal metaphores, wild side res & intuitive prob solving. *Mem:* Am Asn Zool Parks & Aquaria; Sigma Xi; Am Soc Zool. *Res:* Reproductive endocrinology of endangered species; non invasive methodologies, neonatal sex steroids and paleo- endocrinology. *Mailing Add:* 10565 Caminito Banyon San Diego CA 92131

BERCZI, ISTVAN, b Bekes, Hungary, Nov 12, 38; Can citizen; m 67; c 3. IMMUNOLOGY. *Educ:* Budapest Vet Sch, Hungary, DVM, 62; Univ Man, PhD(immunol), 72. *Prof Exp:* Res scientist microbiol, Vet Med Res Inst, Hungarian Acad Sci, Budapest, 62-67; asst prof, 74-82, ASSOC PROF IMMUNOL, UNIV MAN, 82- *Mem:* Am Asn Immunologists; Am Asn Cancer Res; Transplantation Soc; Can Soc Immunol; NY Acad Sci; AAAS. *Res:* Cancer immunology; regulation of the immune response; function and the pituitary gland; prolactins importance as an immunoregulatory hormone. *Mailing Add:* Dept Immunol Univ Man Fac Med Winnipeg MB R3E 0W3 Can

BERDAHL, DONALD RICHARD, b Newcastle, Wyo, July 19, 54; m 79; c 3. ORGANIC CHEMISTRY, POLYMER CHEMISTRY. *Educ:* Univ Wyo, BS, 76; Yale Univ, MS, 78, PhD(chem), 83. *Prof Exp:* STAFF CHEMIST, GEN ELEC CORP RES & DEVELOP, 82- *Res:* Design and development of monomer synthetic methods; polymer synthesis and process chemistry for high temperature; solvent resistant polymides; photostabilization of thermoplastics; organic and polymer sonochemistry. *Mailing Add:* RD 4 Johnson Rd Scotia NY 12302-9803

BERDAHL, PAUL HILLAND, b Washington, DC, Aug 1, 45; m; c 2. INFRARED RADIATION, SEMICONDUCTORS. *Educ:* Rice Univ, BA, 67; Stanford Univ, MS, 68, PhD (theoret physics), 72. *Prof Exp:* Res assoc, dept physics, Stanford Univ, 72-73 & Univ Wash, Seattle, 73-75; STAFF SCIENTIST, LAWRENCE BERKELEY LAB, UNIV CALIF, 76- *Concurrent Pos:* Chmn, Solar Radiation Div, Am Solar Energy Soc, 82. *Mem:* Am Phys Soc; Mat Res Soc. *Res:* Infrared radiative properties of narrow-band gap semiconductors and other materials; transport properties of superconductors and theory of superfluidity; radiation transfer in atmosphere. *Mailing Add:* Lawrence Berkeley Lab Bldg 2 Rm 318 Berkeley CA 94720

BERDAN, JEAN MILTON, b New Haven, Conn, May 9, 16. INVERTEBRATE PALEONTOLOGY. *Educ:* Vassar Col, AB, 37; Yale Univ, MS, 43, PhD(geol), 49. *Prof Exp:* Geologist, Ground Water, US Geol Surv, 42-46, geologist, paleont & stratig, 49-85; RETIRED. *Mem:* Geol Soc Am; Am Paleont Soc; Brit Paleont Asn; Am Paleont Res Inst; Int Paleont Asn. *Res:* Lower Paleozoic ostracode faunas. *Mailing Add:* 510 21st NW No 708 US Nat Mus Natural Hist Washington DC 20006

BERDANIER, CAROLYN DAWSON, b East Brunswick, NJ, Nov 14, 36; m 57; c 3. NUTRITIONAL BIOCHEMISTRY. *Educ:* Pa State Univ, BS, 58; Rutgers Univ, MS, 63, PhD(nutrit), 66. *Prof Exp:* Res asst nutrition, Douglass Col, 62-63; NIH fel nutrition, Rutgers Univ, 66-67; res nutritionist, Human Nutrit Res Div, Agr Res Serv, USDA, 68-75; from res asst prof to res assoc prof, Depts Med & Biochem, Col Med, Univ Nebr, 75-77; prof foods & nutrit & head dept, 77-88, PROF NUTRIT, UNIV GA, 77- *Concurrent Pos:* Asst prof nutrit, Univ Md, 70-75. *Honors & Awards:* Lamar Dodd Award, 84; Geo Nutrit Coun Award, 82. *Mem:* Am Inst Nutrit; Am Soc Clin Nutrit; Endocrine Soc; Am Home Econ Asn; Soc Exp Biol & Med; NAm Soc Study Obesity; Sigma Xi. *Res:* Hormones and metabolic control mechanisms; nutrient-genetic interactions; carbohydrate nutrition; diet-drug interactions; obesity; diabetes; lipemia. *Mailing Add:* Dept Nutrit Univ Ga Athens GA 30602

BERDICK, MURRAY, b New Rochelle, NY, June 27, 20; m 47; c 1. PHARMACEUTICAL PRODUCTS, CONSUMER PRODUCT INTEGRITY. *Educ:* George Washington Univ, BS, 42; Polytech Inst Brooklyn, MS, 49, PhD(polymer chem), 54. *Prof Exp:* Asst, Am Electroplaters Soc, Bur Stand, Washington, DC, 40-42; test engr, Gen Elec Co, NY, 42-43, chemist, 43-46; proj leader, Evans Res & Develop Corp, 46-51, coord res, 53-60, vpres & dir res, 60-61; res mgr, Chesebrough-Pond's Inc, 62-64, dir, Clinton Labs, 64-69, dir, Res Labs, 69-70, dir appl res, 71-75, dir regulatory affairs, 75-80; CONSULT, 80- *Concurrent Pos:* Chmn, Food & Drug Admin-Toilet Goods Asn Sci Liaison Comt, 69-78; chmn, Inter-Indust Color Comt, 71-76; lectr, Yale Univ, 80-82. *Honors & Awards:* Cosmetic Sci & Indust Award, Cosmetic, Toiletry & Fragrance Asn, 71; Soc Cosmetic Chemists Medal Award, 75. *Mem:* Am Chem Soc; Am Inst Chemists; NY Acad Sci; Asn Res Dirs; AAAS; Sigma Xi. *Res:* Interaction of proteins and polyelectrolytes; sustained release drugs; transepidermal moisture loss; food, drug and cosmetic color safety; polymerization; unsaturated polyesters; fiber physics; properties of high polymers; cosmetic and pharmaceutical products. *Mailing Add:* PO Box 836 Branford CT 06405

BERECEK, KATHLEEN HELEN, b Chicago, Ill. CARDIOVASCULAR PHYSIOLOGY, HYPERTENSION RESEARCH. *Educ:* St Xavier Col, BS, 68; Univ Mich, Ann Arbor, MS, 71, PhD(physiol), 76. *Prof Exp:* Scholar, Dept Physiol, Univ Mich, 76-77; res assoc, Dept Pharmacol, Univ Heidelberg, 78-79; res fel, Cardiovasc Ctr, Univ Iowa, 79-81; ASSOC PROF PHYSIOL & BIOPHYSICS, UNIV ALA, BIRMINGHAM, 81- *Concurrent Pos:* Assoc scientist, Hypertension Prog, Univ Ala, Birmingham, 81; estab investr, Am Heart Asn, 82, mem high blood pressure res coun, basic sci circulation. *Honors & Awards:* Marion Young Scholars Award, Am Soc Hypertension, 88. *Mem:* Am Physiol Soc; Soc Exp Biol & Med; Am Heart Asn; Sigma Xi; AAAS; Am Fed Clin Res; Soc Neurosci; Am Soc Hypertension. *Res:* Role of the central nervous system and angiotensin II in arterial pressure regulation and the pathogenesis of hypertension; exploration of the role of vasopressin in arterial pressure regulation and hypertension. *Mailing Add:* Hypertension Res Prog Univ Ala 1016 Zeigler Res Birmingham AL 35294

BEREK, JONATHAN S, b Fremont, Nebr, Apr 21, 48; m 70; c 3. ONCOLOGY. *Educ:* Brown Univ, BA, 70, MMSc, 73; Johns Hopkins Univ, MD, 75. *Prof Exp:* Intern & resident, obstet & gynecol, Brigham Women's Hosp, Harvard Med Sch, 75-79; oncol fel, Univ Calif Los Angeles, Sch Med, 79-81, from asst prof to assoc prof, gynecol & oncol, 84-86, PROF & DIR, DIV GYNECOL ONCOL, SCH MED, UNIV CALIF, LOS ANGELES, 86- *Mem:* Soc Gynecol Oncologists; Am Soc Clin Oncologists; Am Radium Soc; Soc Surgical Oncologists; Am Assoc Cancer Res. *Res:* Clinical gynecologic oncology, ovarian cancer, tumor immunology, gynecologic surgery. *Mailing Add:* Div Gynec Oncol Univ Calif Los Angeles Sch Med Rm 24-127 CHS Los Angeles CA 90024

BEREMAN, ROBERT DEANE, b Clinton, Ind, Oct 2, 43; m 65; c 2. INORGANIC CHEMISTRY, BIOINORGANIC CHEMISTRY. *Educ:* Butler Univ, BS, 65; Mich State Univ, PhD(inorg chem), 69. *Prof Exp:* Res asst chem, Butler Univ, 61-65; asst inorg chem, Mich State Univ, 65-69; NSF fel phys-inorg chem, Univ Ill, Urbana, 69-70; from asst prof to assoc prof inorg chem, State Univ NY, 70-79; PROF CHEM, NC STATE UNIV, 79-, ASSOC DEAN PHYS & MATH SCI, 82- *Concurrent Pos:* Dreyfus Found fel, 75- *Mem:* Am Chem Soc; Sigma Xi. *Res:* Preparation and characterization by magnetic resonance techniques--nuclear magnetic, electron spin and nuclear quadrupole resonance--of theoretically interesting transition metal complexes; magnetic resonance spectroscopy; bioinorganic model complexes; copper enzymes. *Mailing Add:* Dept Chem NC State Univ Box 8201 Cox Hall Rm 113 Raleigh NC 27695-8201

BEREN, SHELDON KUCIEL, b Marietta, Ohio, Oct 7, 22; m 46, 68; c 7. CHEMISTRY. *Educ:* Harvard Univ, BS, 44. *Prof Exp:* Chem engr, AEC, 44-46; dir res & prod, Marco Chem Co, Tex, 46-47; mem staff, Okmar Oil Co, 48-70; CHMN, BEREN CORP, 70-; PRES, BERENERGY CORP, 81- *Res:* Crude oil production. *Mailing Add:* 1635 Tennyson St Denver CO 80204

BERENBAUM, MAY ROBERTA, b Trenton, NJ, July 22, 53. CHEMICAL ECOLOGY. *Educ:* Yale Univ, BS, 75; Cornell Univ, PhD(ecol & syst), 80. *Prof Exp:* Asst prof, 80-85, ASSOC PROF ENTOM, UNIV ILL, URBANA-CHAMPAIGN, 85-, AFFIL, INST ENVIRON STUDIES, 83- *Concurrent Pos:* Assoc ed entom, Am Midland Naturalist, 82-85; Presidental Young Investr award, NSF, 84-; univ scholar, Univ Ill, Urbana-Champaign, 85- *Mem:* Sigma Xi; AAAS; Entom Soc Am; Ecol Soc Am; Phytochem Soc Am; Int Soc Chem Ecol; Am Genetic Asn. *Res:* Chemical aspects of insect-plant interaction; evolutionary ecology of insects; phototoxicity of plant products; host-plant resistance. *Mailing Add:* Dept Entom 320 Morrill Hall Univ Ill 505 S Goodwin Urbana IL 61801-3795

BERENBAUM, MORRIS BENJAMIN, b Chicago, Ill, Dec 19, 24; m 46; c 3. POLYMER CHEMISTRY, ORGANIC CHEMISTRY. *Educ:* City Col New York, BCheE, 44; Polytech Inst Brooklyn, PhD(org chem), 51. *Prof Exp:* Res engr, Stand Brands, Inc, 44, jr engr, Process Res on Food Prods, 44-45; chemist, Stauffer Chem Corp, 45-48, res chemist, 51-55; supvr res sect, Thiokol Chem Co, 55-58, mgr res dept, 58-60, dir res dept, 60-62, tech dir, 62-69; tech dir, 69-71, vpres res & develop, 71-79, dir res, Specialty Chem Div, 79-84, CHIEF SCIENTIST, CHEM SECT, ALLIED SIGNAL CORP, 84- *Concurrent Pos:* Mem bd dirs, Frontier Sci & Technol Res Found. *Mem:* Am Chem Soc; Sigma Xi; Soc Plastics Eng; NY Acad Sci. *Res:* fluorocarbons; fluorochemicals; specialty chemicals; fibers and fiber finishes. *Mailing Add:* 59 Crest Dr Summit NJ 07901-4115

BERENBERG, WILLIAM, b Haverhill, Mass, Oct 29, 15; wid; c 2. MEDICINE. *Educ:* Harvard Univ, AB, 36; Boston Univ, MD, 40. *Prof Exp:* Assoc prof, 53-70, PROF PEDIAT, HARVARD MED SCH, 70-; ASSOC PHYSICIAN IN CHIEF, CHILDRENS HOSP BOSTON, 70- *Concurrent Pos:* Consult, Mass Gen Hosp, 50-; chief med serv, Childrens Hosp Boston, 53-70; mem, Mass Cerebral Palsy Comn; chmn res coun, Nat United Cerebral Palsy. *Honors & Awards:* Jacobi Award, 49; Weinstein Award, United Cerebral Palsy, 69; Presidential Medal Merit, Ecuador, 70. *Mem:* Soc Pediat Res; Am Pediat Soc; Am Acad Pediat; Am Acad Cerebral Palsy; Am Acad Neurol. *Res:* Handicapped children; pulmonary disease. *Mailing Add:* Childrens Hosp 300 Longwood Boston MA 02115

BERENBOM, MAX, b Saskatoon, Sask, Sept 4, 19; nat US; m 47; c 2. BIOCHEMISTRY. *Educ:* Univ Sask, BA, 41, MA, 43; Univ Toronto, PhD(biochem), 47. *Prof Exp:* Sr res fel, Nat Cancer Inst, 47-49, spec res fel, 49-50, vis scientist, 50-51; assoc oncol, Med Ctr, Univ Kans, 51-54, asst prof oncol & biochem, 54-57; dir biochem, Menorah Med Ctr, 57-89; RETIRED. *Mem:* Am Chem Soc; fel AAAS; Am Asn Clin Chem. *Res:* Intermediary metabolism; enzymes and tissue components in normal and abnormal states; radiation effects; carcinogenesis. *Mailing Add:* 10750 Bridlespur Terr Kansas City MO 64114-5047

BERENDES, HEINZ WERNER, b Dortmund, Ger, May 1, 25; US citizen; m 84; c 3. PUBLIC HEALTH, EPIDEMIOLOGY. *Educ:* Univ Goettingen, Ger, MD, 49; Univ Munich, MD, 52; Johns Hopkins Univ, MHS, 72. *Prof Exp:* From instr to asst prof, dept pediat, Univ Minn Hosp, 54-60; asst dir, Collab Res, Nat Inst Neurol Dis & Blindness, NIH, 60, chief, Perinatal Res Br, 60-73, chief, Contraceptive Eval Br, 74-79, dir res prog, 79-86, DIR RES PROG, NAT INST CHILD HEALTH & HUMAN DEVELOP & DIR DIV PREV RES, 90- *Concurrent Pos:* Clin prof pediat, Howard Univ, 61-; temporary adv, expert comt prevention perinatal mortality & morbidity, WHO, 69 & Sci Group Steroid Contraception & Risk Neoplasia, 77; mem, Sci Adv Group, Special Prog Res & Training Human Reproduction, WHO, 79, chmn, 79; consult, fertility & maternal health drug adv comt, Food & Drug Admin, 79-, Environ Protection Agency, 80-; sr assoc, Dept Epidemiol, Johns Hopkins Univ, 81-; bd dirs, Better Babies Proj, Inc. *Mem:* Soc Epidemiol Res; Am Col Epidemiol; Am Pub Health Asn; Int Epidemiol Soc; Soc Pediat Epidemiol. *Res:* Perinatal factors in child development including developmental disorders of infancy and childhood, epidemiology of high-risk pregnancies, and adverse medical effects of contraceptive methods currently in use. *Mailing Add:* Nat Inst Child Health & Human Develop NIH EPN Rm 640 Bethesda MD 20892

BERENDSEN, PETER BARNEY, b Los Angeles, Calif, July 14, 37; m 62; c 3. ANATOMY, CELL BIOLOGY. *Educ:* St Marys Col, Calif, BS, 60; George Washington Univ, MS, 65, PhD(anat), 72. *Prof Exp:* Res asst hemat, Armed Forces Inst Path, 60-65; res assoc path, Univ Mich, 65-67; guest worker physiol, NIH, 70-72, res physiologist, 72; instr anat, 72-73, asst prof, 73-81, ASSOC PROF ANAT, UNIV MED & DENT NJ, 81-, ASST DEAN, 90- *Mem:* Electron Micros Soc Am; Am Asn Anat; Sigma Xi; Microcirculatory Soc; Am Soc Cell Biol. *Res:* Intestinal absorption; lipid metabolism; reticuloendothelial and endothelial function; pulmonary ultrastructure. *Mailing Add:* Univ Med & Dent NJ 185 S Orange Ave Newark NJ 07103

BERENDZEN, RICHARD, b Walters, Okla, Sept 6, 38; m 64; c 2. ASTRONOMY, HISTORY OF SCIENCE. *Educ:* Mass Inst Technol, BS, 61; Harvard Univ, MA, 67, PhD, 69. *Hon Degrees:* DSc, Univ Colombo; LHD, Bridgewater Col; LLD, Univ Charlestown, Kean Col NJ, Seton Hall Univ, WVa Weylayan Col. *Prof Exp:* Res scientist, Geophys Corp Am, 59-64; fel, Harvard Univ, 61-64; from lectr to prof astron, Boston Univ, 65-73, actg chmn dept, 71-72; dean, Col Arts & Sci, Am Univ, Washington, DC, 74-76, provost, 76-79, pres, 80-90, PROF PHYSICS, AM UNIV, WASHINGTON, DC, 74- *Concurrent Pos:* Res scientist, Astronaut Div, Ling-Temco-Vought Corp, 62-63; staff mem astron ed, Proj Physics, Harvard Univ, 65-66; mem comn teaching astron & comn hist sci of Int Astron Union; consult, Educ Coop, Natick, Mass; adv, NAm Fedn Planetarium Educators; ed, J Col Sci Teaching, Nat Sci Teachers Asn & Cosmic Search; mem Astron Surv Comt, Nat Acad Sci; astron adv bd, Ctr Hist & Philos Physics, Am Inst Physics; invited partic sem, Int Union Pure & Appl Physics, 70; arranger & chmn, Int Conf Educ & Hist Mod Astron, Am Astron Soc-NY Acad Sci; dir, Nat Sci Found prog, Case Studies Proj Develop Mod Astron; arranger & chmn, Conf Life Beyond Earth & Mind of Man, NASA, Boston, 72; consult, Off Acad Affairs, Am Coun Educ & Space Sci Bd, Nat Acad Sci, 73-74; mem energy task force, Wash Bd Trade, 74-75 & comts col sci teaching & publ of Nat Sci Teachers Asn, 74-; consult or adv, Am Inst Physics, NASA, US State Dept, US Info Agency, Smithsonian Inst, UNESCO & US House Rep; educ & astron expert, WUSA-TV & WTOP Newsradio, Wash, DC. *Mem:* Fel AAAS; Am Astron Soc; Am Hist Sci Soc; Am Asn Higher Educ; Int Astron Union. *Res:* Astronomy education; cosmology; search for extraterrestrial intelligence; sociology of science; science education and planning; American higher education; international education; academic administration. *Mailing Add:* Physics Dept AM Univ Washington DC 20016

BERENS, ALAN PAUL, b Cincinnati, Ohio, June 15, 34; m 57; c 5. MATHEMATICAL STATISTICS. *Educ:* Univ Dayton, BS, 55; Purdue Univ, MS, 57, PhD(math statist), 63. *Prof Exp:* Instr math, Purdue Univ, 58-62; head statist dept, Technol Inc, Ohio, 62-68; vpres, Beta Industs, 68-69; res statistician, 69-80, SR RES STATISTICIAN, UNIV DAYTON RES INST, 80- *Mem:* Am Statist Asn; Am Soc Testing & Mat. *Res:* Application of mathematical statistics and probability to physical science research; NDI reliability, structural integrity. *Mailing Add:* 4940 James Hill Rd Kettering OH 45429

BERENS, ALAN ROBERT, b Oak Park, Ill, Sept 28, 25; m 49; c 2. POLYVINYL CHLORIDE, DIFFUSION IN POLYMERS. *Educ:* Harvard Univ, AB, 47; Case Western Reserve Univ, MS, 49, PhD(chem), 51. *Prof Exp:* Res chemist, 50-60, from res assoc to sr res assoc, 60-71, res fel, 71-78, sr res fel polymerization, 78-87, CONSULT, B F GOODRICH CO, 88- *Mem:* Am Chem Soc; Am Inst Chem Engrs; fel Am Inst Chemists. *Res:* Vinyl polymerization; polyvinyl chloride structure, rheology, stability and morphology; diffusion of small molecules in polymers and polymer relaxations; appoximately fifty scientific publications. *Mailing Add:* RD 2 Box 3510 Middlebury VT 05753-8904

BERENS, RANDOLPH LEE, b Los Angeles, Calif, June 13, 43; m 68; c 1. CELL BIOLOGY, PROTOZOOLOGY. *Educ:* Univ Calif, Irvine, BS, 67, PhD(cell biol), 75. *Prof Exp:* Res asst cell biol, Univ Calif, Irvine, 71-75; fel, Sch Med, Washington Univ, 75-78; ASST PROF PHYSIOL, SCH MED, ST LOUIS UNIV, 78-; ASST PROF MED & MICROBIOL, UNIV COLO HEALTH SCI CTR. *Concurrent Pos:* Instr physiol, Cerritos Col, 74-75; prin investr grants, WHO, 79-82, NIH, 81- *Mem:* Am Soc Protozoologists; Am Soc Microbiologists; AAAS; NY Acad Sci; Am Soc Parasitologists. *Res:* Various intermediate metabolic pathways in pathogenic protozoans with the primary goal of finding metabolic differences between these organisms and their mammalian hosts. *Mailing Add:* Dept Infect Dis Univ Colo Health Sci Ctr Box B 4200 E Ninth Ave Denver CO 80262

BERENSON, GERALD SANDERS, b Bogalusa, La, Sept 19, 22; m 51; c 4. INTERNAL MEDICINE, CARDIOLOGY. *Educ:* Tulane Univ, BS, 43, MD, 45. *Prof Exp:* Asst med, Sch Med, Tulane Univ, 48-49, instr, 49-52; USPHS fel pediat, Univ Chicago, 52-54; from asst prof to assoc prof, Sch Med, La State Univ Med Ctr, New Orleans, 54-63, head sect cardiol, 75, dir, Specialized Ctr Res in Arteriosclerosis, 72-87, PROF MED, SCH MED, LA STATE UNIV MED CTR, NEW ORLEANS, 63- *Concurrent Pos:* Dir Nat Res & Demonstration Ctr Arteriolsclerosis, NIH, Nat Heart, Lung & Blood Inst, 85-87; Boyd prof, La State Univ. *Honors & Awards:* Searle Distinguished Award for Res on Hypertension in Blacks. *Mem:* Am Chem Soc; Soc Exp Biochem & Med Biol; Am Soc Bio Chem & Molecular Biol; fel Am Col Physicians; fel Am Col Cardiol. *Res:* Cardiology; atherosclerosis; biochemistry of connective tissues; cardio-vascular connective tissue in arteriosclerosis; epidemiology of cardio-vascular disease; studying a total community of children, greater than five-thousand, for the natural history of cardio-vascular disease and essential hypertension; cardio-vascular health education and health promotion for children from kindergarten through sixth-grade. *Mailing Add:* Dept Med La State Univ Med Ctr New Orleans LA 70112

BERENSON, LEWIS JAY, mathematics education, for more information see previous edition

BERENSON, MALCOLM MARK, GASTROENTEROLOGY, HEPATOLOGY. *Educ:* Univ Pittsburgh, MD, 63. *Prof Exp:* ASSOC PROF GASTROENTEROL, SCH MED, UNIV UTAH, 71- *Res:* Nutrition. *Mailing Add:* Dept Internal Med Univ Utah 50 N Medical Dr Salt Lake City UT 84132

BERENT, STANLEY, b Norfolk, Va, Mar 10, 41; m; c 3. CLINICAL PSYCHOLOGY, NEUROPSYCHOLOGY. *Educ:* Old Dom Univ, BS, 66; Va Commonwealth Univ, MS, 68; Rutgers Univ, PhD(clin psychol), 72; Am Bd Prof Psychol, dipl. *Prof Exp:* Instr psychol, Va Commonwealth Univ, 67-68; res asst psychiat, Rutgers Med Sch, 68-71; intern psychol, NIHM, St Elizabeths Hosp, 71-72; asst prof psychiat, Med Sch, Univ Va, 72-77, asst prof psychol, 75-77, assoc prof psychiat & psychol, 77-79; chief psychol, Vet Admin Med Ctr, Ann Arbor, 79-85; assoc prof & dir neuropsychol prog, 79-88, PROF, UNIV MICH, 88- *Concurrent Pos:* Consult, NIMH, St Elizabeths Hosp, 72-75 & Univ Va, 79-81, Vet Admin, NIH & var others. *Mem:* Fel Am Psychol Asn; Soc Neurosci; Int Soc Neuropsychol; Am Acad Neurol; Am Epilepsy Soc; AAAS. *Res:* Behavioral and electrophysiological studies of epilepsy and other neurological disorders, lateralization and other functional organization of brain in normal and pathological conditions and implications of disordered cognition for psycho-social functioning and behavior; pet and other imaging techniques for the study of brain-behavior relationships; effects of environmental toxins on the central nervous system and behavior. *Mailing Add:* Univ Mich Med Ctr Box 0840 480 Med Inn Ann Arbor MI 48109-0840

BERERA, GEETHA POONACHA, b Mercara, Karnataka, India. MATERIALS RESEARCH, ELECTRO-OPTICS. *Educ:* Mysore Univ, India, BSc, 77, MSc, 79; Tufts Univ, MS, 85. *Prof Exp:* Lectr physics, Kaveri Col, Coorg, Karnataka, India, 80-81; ADJ FAC PHYS SCI, SUFFOLK UNIV, 90- *Mem:* Am Phys Soc; Mat Res Soc. *Res:* Growing single xtals and characterization of electrochromic materials; fabricating and characterizing the optical, structural, electrical and chemical properties of mixed and transparent conductors; fabricating electro-chromic devices. *Mailing Add:* Dept Physics & Eng Suffolk Univ 41 Temple St Boston MA 02114

BERES, JOHN JOSEPH, b Beaver Falls, Pa, May 1, 47; m 75. PHYSICAL POLYMER CHEMISTRY. *Educ:* Geneva Col, BS, 69; Carnegie-Mellon Univ, MS, 73, PhD(phys chem), 75. *Prof Exp:* Fel polymers, Univ Mass, 75-77; sr res chemist, Gen Corp, 77, group leader polymer characterization, 80, sect head polymer structure, 81, mgr anal res & servs, 88, POLYMER CHARACTERIZATION, GEN CORP, 77- *Mem:* Am Crystallog Asn; Sigma Xi; Am Chem Soc; Soc Appl Spectroscopy. *Res:* X-ray structural investigations of polymers; structure-property relationships in polymers, especially polyurethanes and polyphosphazenes; solution characterization of elastomers, spectroscopy, thermal analysis, lims, microscopy. *Mailing Add:* 6014 Echodell NW North Canton OH 44720

BERES, WILLIAM PHILIP, b Peabody, Mass, Jan 8, 36; m 66; c 3. NUCLEAR PHYSICS. *Educ:* Mass Inst Technol, BS, 59, PhD(physics), 64. *Prof Exp:* Asst physics, Mass Inst Technol, 60-64; res physicist, GCA Corp, 64; res assoc physics, Univ Md, 64-66; asst prof, Duke Univ, 66-69; assoc prof, 69-75, PROF PHYSICS, WAYNE STATE UNIV, 75- *Concurrent Pos:* Fulbright lectr sci educ, Hebrew Univ, Jerusalem, 77-78. *Mem:* Am Phys Soc; Am Asn Physics Teachers. *Res:* Theory of nuclear structure and reactions; microscopic nuclear calculations; quasiparticles; local and non-local potentials; photonuclear reactions; heavy ion reactions; nuclear processes in stars. *Mailing Add:* Dept Physics Wayne State Univ Detroit MI 48202

BERESFORD, WILLIAM ANTHONY, b London, Eng, Aug 13, 36; m 57; c 3. HISTOLOGY, NEUROANATOMY. *Educ:* Oxford Univ, BA, 59, DPhil (histol), 64. *Prof Exp:* Lectr histol, Univ Liverpool, 63-65; asst prof anat, Am Univ Beirut, 65-68; from asst prof to assoc prof, 68-76, PROF ANAT, WVA UNIV, 76- *Mem:* Bone & Tooth Soc; Anat Soc Gt Brit & Ireland; Ger Anat Soc; Int Soc Differentiation; Am Soc Zoologists. *Res:* Metaplasia; ectopic bone and cartilage; bone induction; reptilian liver histology. *Mailing Add:* Dept Anat WVa Univ Med Ctr Morgantown WV 26506

BERESNIEWICZ, ALEKSANDER, b Lithuania, Apr 5, 27; nat US; m 54; c 5. PHYSICAL CHEMISTRY. *Educ:* Marquette Univ, BS, 51; Univ Ill, PhD(chem), 54. *Prof Exp:* Asst, Univ Ill, 51-53; res chemist, 54-68, RES ASSOC, E I DU PONT DE NEMOURS & CO, INC, 68-, RES FEL, 78- *Mem:* Am Chem Soc; Sigma Xi; NY Acad Sci. *Res:* Physical chemistry of high polymers; relation between the structure of polymers and their physical properties. *Mailing Add:* 2501 Elmdale Lane Wilmington DE 19810-3418

BERESTON, EUGENE SYDNEY, b Baltimore, Md, Feb 21, 14; m 42, 80; c 3. DERMATOLOGY, MYCOLOGY. *Educ:* Johns Hopkins Univ, AB, 33; Univ Md, MD, 37; Univ Pa, MSc, 45, DSc, 55. *Prof Exp:* Intern, Conemaugh Valley Mem Hosp, Johnstown, Pa, 37-38 & Mercy Hosp, Baltimore, 38-39, grad student Dermat Univ of Pa, 39-40; asst dermat, Skin & Cancer Unit, NY Postgrad Hosp, Columbia Univ, 40-41; resident, Montefiore Hosp, NY, 40-41; asst dermatologist, Outpatient Dept, Univ Hosp, 46-50, instr, Sch Med, 46-47, assoc, 47-52, from asst prof to assoc prof dermat, 52-72, PROF MED DERMAT, SCH MED, UNIV MD, BALTIMORE CITY, 72-, DERMATOLOGIST, UNIV HOSP, 50- *Concurrent Pos:* Asst, Sch Med, Johns Hopkins Univ, 46-50, instr, 50-60, chief fungus lab, Hosp, 46-51, dermatologist, 50-60; attend dermatologist, Sinai Hosp, 47-80, consult staff, 80-; dermatologist, Mercy Hosp, 50-67, chief dermatologist, 67-; asst investr, US Dept Army Fungus Res Proj, 51-57; consult dermatologist, US Vet Admin Hosp, Baltimore & Ft Howard, 46-77 & Spring Grove State Hosp, 47-82; mem, dermat staff, US Vet Admin Hosp, Washington, DC, 77-83. *Mem:* Fel AMA; fel Am Col Physicians; Am Acad Dermat; fel Royal Soc Health; Royal Soc Med Affil. *Res:* Aspergillus infection of nails; nutritional requirements of microsporum group of fungi and Trichophyton Tonsurans; fluorescent compound in fungus infected hairs; fungal antibiotics. *Mailing Add:* 22 E Eager St Baltimore MD 21202

BERETS, DONALD JOSEPH, b New York, NY, July 6, 26; m 56; c 2. PHYSICAL CHEMISTRY. *Educ:* Harvard Univ, AB, 46, MA, 47, PhD(chem), 49. *Prof Exp:* Res chemist, 49-54, group leader, 54-60, sect mgr, Energy Conversion Res, 60-67, mgr, Mat Sci, 67-73, MGR CATALYST RES, CHEM RES DIV, AM CYANAMID CO, 73- *Concurrent Pos:* Assoc, Mass Inst Technol, 49. *Mem:* Am Chem Soc; AAAS. *Res:* New product development; catalysis; surface chemistry; pigments; solid state physics and chemistry. *Mailing Add:* 33 Arrow Head Dr Stamford CT 06903-3814

BEREZIN, ALEXANDER A, b Sverdlovsk, USSR, Apr 26, 44; Can citizen; m 67; c 2. THEORETICAL SOLID STATE PHYSICS. *Educ:* Leningrad State Univ, MSc, 66, PhD(theor physics), 70. *Prof Exp:* Assoc prof physics, Higher Naval Eng Col, Leningrad, 74-77; res assoc solid state physics, Univ Alta, Edmonton, 78-80; ASSOC PROF ENG PHYSICS, MCMASTER UNIV, HAMILTON, ONT, 80- *Concurrent Pos:* Vis prof, Tokyo Denki Univ, Japan, 87. *Mem:* Am Phys Soc; Can Asn Physicists. *Res:* Hopping conductivity in semiconductors, color centers in ionic crystals, recombination, trapping, energy transfer, Auger processes; thermoelectricity, optoletelectronics, thin films, positron annihilation, isotopic effects, non-linear dynamics, philosophical problems of physics. *Mailing Add:* Dept Eng Physics McMaster Univ Hamilton ON L8S 4M1 Can

BEREZNEY, RONALD, b New York, NY, Dec 25, 43; c 2. CELL BIOLOGY, BIOCHEMISTRY. *Educ:* Fairleigh Dickinson Univ, BS, 66; Purdue Univ, PhD(membrane biochem), 71. *Prof Exp:* Res asst molecular biol, Inst Cancer Res, Col Physician & Surgeons, Columbia Univ, 65-66; fel cell biol, Univ Freiburg, Ger, 71-72; res assoc pharmacol, Sch Med, Johns Hopkins Univ, 72-75; from asst prof to assoc prof, 75-85, PROF CELL & MOLECULAR BIOL, STATE UNIV NY BUFFALO, 86- *Mem:* Am Soc Cell Biol; Am Chem Soc; AAAS; Am Soc Biochem & Molecular Biol. *Res:* Molecular biology of the eucaryotic nucleus; nuclear matrix organization and function in normal and cancer cells; DNA replication; molecular properties and topography of nuclear proteins. *Mailing Add:* Dept Biol Sci State Univ NY Buffalo NY 14260

BERG, ARTHUR R, b Clay Center, Kans, Feb 9, 37; m 61; c 3. PLANT MORPHOLOGY. *Educ:* Tex Tech Univ, BS, 60; Univ Calif, Davis, PhD(bot), 66. *Prof Exp:* Plant physiologist, Pac Southwest Forest & Range Exp Sta, US Forest Serv, 66-72; LECTR BOT, UNIV ABERDEEN, 72- *Mem:* Bot Soc Am. *Res:* Plant morphogenesis; growth regulator physiology; developmental anatomy and morphology. *Mailing Add:* Dept Plant & Soil Sci Univ Aberdeen St Machar Dr Aberdeen AB9 2UD Scotland

BERG, BENJAMIN NATHAN, b New York City, NY, Dec 8, 97; m 24; c 1. NUTRITION. *Educ:* Columbia Univ, MD, 20. *Prof Exp:* Dept path, Columbia Presby Hosp, New York; RETIRED. *Mem:* Am Soc Exp Path; Soc Exp Biol & Med; Harvey Soc; Teratology Soc; AAAS. *Res:* Relation of nutrition to reproduction and aging; endocrine factors in reproduction; teratology; nephrosis; muscular dystrophy; spontaneous diseases of lab animals; physiology of stomach, gall bladder, liver and pancreas; mechanisms of fetal resorptions and malformations. *Mailing Add:* 40 E 88th St New York NY 10128

BERG, CARL JOHN, JR, b Bridgeport, Conn, June 6, 44; m 85; c 2. MARINE BIOLOGY, AQUACULTURE & FISHERIES MANAGEMENT. *Educ:* Univ Conn, BA, 66; Univ Pac, MS, 68; Univ Hawaii, PhD(zool), 71. *Prof Exp:* Asst prof biol, City Univ New York, 71-76; res assoc mollusks, Harvard Univ, 76-80; assoc scientist marine biol, Marine Biol Lab, 79-86; biol scientist, Fla Marine Res Inst, 86-90; PVT CONSULT, 90- *Concurrent Pos:* Adj asst prof, Pace Univ & Univ Southern Calif, 72; assoc, Am Mus Natural Hist, 72-78; investr & instr, Marine Biol Lab, 74-79; res assoc, Columbia Univ, 78-81; intermittent fac, US Nat Marine Fisheries Serv, 80-82. *Mem:* Sigma Xi; Explorers Club; Ecol Soc Am; AAAS; World Aquacult Soc. *Res:* Aquaculture development and fisheries management; marine invertebrates; larval dispersion around islands. *Mailing Add:* PO Box 769 Kilauea HI 96754-0769

BERG, CLAIRE M, b Mt Vernon, NY, Apr 24, 37. BACTERIAL GENETICS, SEQUENCING STRATEGIES. *Educ:* Cornell Univ, BS, 59; Univ Chicago, MS, 62; Columbia Univ, PhD(genetics), 66. *Prof Exp:* NATO fel, Med Res Coun-Microbiol Gen Res Unit, Hammersmith Hosp, London, Eng, 66-67; Jane Coffin Childs fel, Molecular Biol Lab, Univ Geneva, Switz, 67-68; from asst prof to assoc prof, 68-82, PROF BIOL, UNIV CONN, 82- *Mem:* Genetics Soc Am; Am Soc Microbiol; Fedn Am Scientists; AAAS. *Res:* Uses of transposable elements in genetic analysis and DNA sequencing; molecular genetics; physiological genetics of amino acid biosynthesis in bacteria. *Mailing Add:* Box U-131 Molecular Cell Biol Dept Univ Conn Storrs CT 06269-2131

BERG, CLYDE C, b Meriden, Kans, Nov 2, 36; m 58; c 3. PLANT BREEDING, PLANT GENETICS. *Educ:* Kans State Univ, BS, 58; Okla State Univ, MS, 61; Wash State Univ, PhD(genetics), 65. *Prof Exp:* Asst prof agron & genetics & asst agronomist, Wash State Univ, 65-66; res geneticist, 66-79, RES AGRONOMIST, AGR RES SERV, USDA, 79- *Mem:* Am Genetic Asn; Am Forage & Grassland Coun; Crop Sci Soc Am; Genetics Soc Can; Am Soc Agron. *Res:* Genetics, cytogenetics and breeding Festuca arundinacea, Festuca pratnesis, Lolium perenne, Lolium multiforum, Dactylis glomerata, Bromus inermis, Phleum pratense (cool season perennial forage grasses) for improved forage quality and resistance to diseases. *Mailing Add:* US Regional Pasture Res Lab University Park PA 16802

BERG, CLYDE H O, b Minneapolis, Minn, Apr 12, 15. CHEMICAL ENGINEERING. *Educ:* Univ Minn, BChE, 36, PhD(chem eng), 40. *Prof Exp:* Develop engr, Union Oil Co, Calif, 40-42, supvr process develop, 42-49, asst to vpres in chg res & develop, 49-51, mgr develop dept, 51-55 & design div, 55-60; dir, Clyde Berg & Assocs, Long Beach, 60-68; PRES, EARTH ENERGY, INC, 68- *Concurrent Pos:* Rep, primary processes subcomt, Petrol Indust War Coun, 43-46; chmn, shale oil processing comt & comt synthetic liquid fuels prod costs, Nat Petrol Coun, 49-53; chmn energy, Oil Shale Comt, Nat Petrol Coun, 52-56; chmn petrochem symp & short course, Univ Calif, Los Angeles, 62-63; mem bd, Thermal Dynamics Corp, Utah, 62-; mem pres's coun, Chapman Col, 70- *Mem:* Am Inst Chem Engrs; Am Chem Soc; Am Soc Mech Engrs. *Res:* Chemical and petroleum processing; solids circulating techniques; reforming of gasoline solvent extraction; distillation; gas adsorption; alternate sources of energy. *Mailing Add:* 3655 E Ocean Blvd 2H Long Beach CA 90803

BERG, DANA B, b Chicago, Ill, July 8, 21; m 52; c 1. SCIENCE WRITING. *Educ:* Ill Inst Technol, BS, 43. *Prof Exp:* Res & develop engr, Linde Air Prods Lab, NY, 43-48; from asst ed to ed, Putman Pub Co, 48-66, chmn ed bd, 66-67, staff consult, 67-68; sr tech ed, Argonne Nat Lab, 68-85; RETIRED. *Res:* Technical writing; editing; staff administration. *Mailing Add:* 1629 77th Ave Elmwood Park IL 60635

BERG, DANIEL, b New York, NY, June 1, 29; m 56; c 3. SCIENCE POLICY, TECHNICAL MANAGEMENT. *Educ:* City Col New York, BS, 50; Yale Univ, MS, 51, PhD(phys chem), 53. *Prof Exp:* Res phys chemist, Insulation Dept, Westinghouse Res Labs, 53-60, mgr phys & inorg chem sect, 60-65, mgr inorg mat, Sci & Technol Res & Develop, 65-67, mgr insulation & chem technol, 67-69, dir, Energy Systs Res, 69-76; tech dir, Uranium Resources, Westinghouse Elec, 76-77; prof sci & technol & dean, Col of Sci, Carnegie-Mellon Univ, 77-81, provost, 81-83; vpres acad affairs & provost, Rensselaer Polytech Inst, 83-84, actg pres, 84-85, pres, 85-87, INST PROF SCI & TECHNOL, RENSSELAER POLYTECH INST, 83- *Concurrent Pos:* Mem, Nat Acad Sci-Nat Res Coun Conf Elec Insulation, 66, chmn, Conf Elec Insulation & Dielectric Phenomena; mem bd dirs, Duquesne Light, Hy-tech, Perfect-A-Tec, Inc; mem bd trustees, Argonne Univ Assoc, 78-; mem mfg bd, Dielectrics & Dielectric Phenomena, Nat Res Coun, Nat Acad Sci; mem rev comt, Nat Bur Standards Exp Technol Incentives Prog; fel, Carnegie-Mellon Robotics Inst; chmn, Educ Adv Bd, Nat Acad Eng. *Honors & Awards:* Belden Prize; Wilbur Cross Medal; Townsend Harris Medal. *Mem:* Nat Acad Eng; Am Chem Soc; fel Inst Elec & Electronics Engrs; fel Am Inst Chemists; fel AAAS; Am Phys Soc; Technol & Serv Soc. *Res:* Dielectrics; electrical breakdown; energy conversion. *Mailing Add:* Rensselaer Polytech Inst 110 Eighth St Troy NY 12180-3590

BERG, EDUARD, b Trier, Ger, Nov 9, 28. GEOPHYSICS. *Educ:* Univ Saarlandes, dipl phys, 53, Dr rer nat(physics), 55. *Prof Exp:* Mem, Inst Sci Res Cent Africa, Congo, 55-63; from assoc prof to prof geophys, Univ Alaska, 63-72, dir seismol prog, 63-72; chmn, Dept Geol & Geophys, Univ Hawaii, 81-84; GEOPHYSICIST, HAWAII INST GEOPHYS, 72- *Concurrent Pos:* Res mem fac, Univ Bonn, 57 & Univ Calif, Berkeley, 61; seismol consult, Mining Union Upper Katanga, 60-61; head seismic & volcanic depts & time serv, Inst Sci Res Cent Africa, 59-63; vis prof, Hawaii Inst Geophys, 71-72. *Honors & Awards:* Harry Wood Oscar Award, Carnegie Inst Wash, 63. *Mem:* Seismol Soc Am; Am Geophys Union; Soc Explor Geophysicists. *Res:* Seismology; geodesy; volcanology. *Mailing Add:* Hawaii Inst Geophys Univ Hawaii 2525 Correa Rd Honolulu HI 96822

BERG, EUGENE WALTER, b Dade City, Fla, Nov 10, 26; m 47; c 3. ANALYTICAL CHEMISTRY. *Educ:* Miss Col, BS, 49; Univ Tex, PhD(chem), 52. *Prof Exp:* From asst prof to prof chem, La State Univ, 52-83, chmn dept, 73-76; RETIRED. *Mem:* Am Chem Soc; assoc Int Union Pure & Appl Chem. *Res:* Ion exchange; chromatography; separation techniques; solvent extraction; beta-dikotane chelates. *Mailing Add:* 425 Nelson Dr La State Univ Baton Rouge LA 70808

BERG, GEORGE G, b Warsaw, Poland, May 27, 19; nat US; m 52; c 3. TOXICOLOGY, CYTOCHEMISTRY. *Educ:* Temple Univ, BA, 42; Columbia Univ, MS, 47, PhD(zool), 54. *Prof Exp:* Res assoc, Sch Med, Georgetown Univ, 49-51; res collabr, Dept Biol, Brookhaven Nat Lab, 54-55; lectr gen biol & embryol, Univ Sch, 55-63, instr pharmacol, 56-64, asst prof, 60-64, assoc prof radiation biol biophys, 64-88, EMER PROF TOXICOL, SCH MED, UNIV ROCHESTER, 88- *Concurrent Pos:* Vis prof environ sci, Zhejiang Univ, China, 89. *Honors & Awards:* Sol Feinstone Environ Award, 87. *Mem:* Fedn Am Sci; Sigma Xi; AAAS; Soc Toxicol; Soc Risk Anal. *Res:* Mercurials; environmental toxicity. *Mailing Add:* 1242 Wildcliff Pkwy Atlanta GA 30329

BERG, GERALD, b New York, NY, Nov 3, 28; m 55; c 3. VIROLOGY, MICROBIOLOGY. *Educ:* Syracuse Univ, AB, 51, MS, 52, PhD(bact), 55. *Prof Exp:* Asst, Upstate Med Ctr, State Univ NY, 53-54; res fel virol, Children's Hosp Res Found, Cincinnati, Ohio, 55; researcher, USPHS, 55-64, chief virus dis studies, 64-67, chief virol, Cincinnati Water Res Lab, Fed Water Qual Admin, 67-74; chief biol methods br, Environ Monitoring & Support Lab, US Environ Protection Agency, 74-81, tech adv, Dir Off, 81-83; RES PROF, DEPT CIVIL & ENVIRON ENG, UNIV CINCINNATI, 83- *Concurrent Pos:* Adj prof, Univ Cincinnati, 69-83. *Mem:* Fel Am Acad Microbiol; Am Soc Microbiol; Am Pub Health Asn; Am Water Works Asn. *Res:* Kinetic virucidal and bactericidal studies in disinfection; molecular basis, indicators of fecal pollution, viruses and drinking, renovated, waste and other waters; viruses in sludges and in solids associated with various waters; virus multiplication, isolation and identification; epidemiology of virus dissemination. *Mailing Add:* Dept Civil & Environ Eng Mail Stop 71 Univ Cincinnati Cincinnati OH 45221

BERG, GLEN V(IRGIL), civil engineering, for more information see previous edition

BERG, GUNNAR JOHANNES, b Ullsfjord, Norway, Mar 1, 30; m 62; c 4. ELECTRICAL ENGINEERING. *Educ:* Tech Univ Norway, Siv ing, 57; Univ BC, MASc, 60. *Prof Exp:* Instr elec eng, Tech Univ Norway, 57-58; res asst, Univ BC, 58-59, instr, 59-60, 62-64; staff engr, BC Energy Bd, 60-64; from asst prof to assoc prof elec eng, 64-74, prof elec eng, 74-88, EMER PROF ELEC ENG, UNIV CALGARY, 88- *Concurrent Pos:* Proj mgt, Univ Calgary, 89- *Mem:* Inst Elec & Electronics Engrs; Can Elec Asn; Norweg Elec Asn. *Res:* Induction motor modeling and performance analysis; power system load behavior and representation; power system planning. *Mailing Add:* Dept Elec Eng Univ Calgary Calgary AB T2N 1N4 Can

BERG, HENRY CLAY, b Brooklyn, NY, Apr 23, 29; m 82; c 1. GEOLOGY. *Educ:* Brooklyn Col, BA, 51; Harvard Univ, AM, 56. *Prof Exp:* Geologist US Geol Surv, 56-84, chief, Menlo Tech Report Unit, 65-67, mgr, Alaskan Mineral Resources Assessment Prog, 74-80, geologist in charge, Alaskan Geol Br, 80-82; CONSULT GEOLOGIST, 84- *Mem:* Fel Geol Soc Am. *Res:* Areal geology tectonics and mineral resources of southeastern Alaska. *Mailing Add:* 115 Malvern Ave Fullerton CA 92632

BERG, HOWARD CURTIS, b Iowa City, Iowa, Mar 16, 34; m 64; c 3. BIOPHYSICS. *Educ:* Calif Inst Technol, BS, 56; Harvard Univ, AM, 60, PhD(chem physics), 64. *Prof Exp:* Jr fel, Harvard Univ, 63-66, from asst prof biol to assoc prof biochem, 66-70, chmn, Bd Tutors Biochem Sci, 66-70; assoc prof to prof molecullar, cellular & develop biol, Univ Colo, Boulder, 74-79, chmn dept, 76-77; prof biol, Calif Inst Technol, 79-86; PROF BIOL, HARVARD UNIV, 86- *Concurrent Pos:* Mem, Rowland Inst Sci, 86- *Honors & Awards:* Biol Physics Prize, Am Phys Soc, 84. *Mem:* Nat Acad Sci; fel AAAS; Am Phys Soc; Am Soc Biol Chemists; NY Acad Sci; Am Soc Microbiologist; Am Acad Arts & Sci. *Res:* Chemical structure of cell membranes; motility and chemotaxis of bacteria. *Mailing Add:* Cellular & Develop Biol Harvard Univ 16 Divinity Ave Cambridge MA 02138

BERG, HOWARD MARTIN, b Chicago, Ill, Aug 31, 42; m; c 3. MATERIALS SCIENCE. *Educ:* Univ Ill, BS, 65; Northwestern Univ, PhD(mat sci), 71. *Prof Exp:* Metallurgist, Pratt & Whitney Aircraft, 65-67; STAFF SCIENTIST SEMICONDUCTOR GROUP, MOTOROLA, INC, 71- *Mem:* Am Soc Metals; Inst Elec & Electronics Engrs; Int Soc Hybrid Microelectronics. *Res:* Assembly and packaging of semiconductor devices; x-ray diffraction; field ion microscopy. *Mailing Add:* 12017 N 84th St Scottsdale AZ 85260

BERG, IRA DAVID, b New York, NY, Nov 27, 31; m 62; c 3. MATHEMATICS. *Educ:* Univ Pa, BS, 53; Lehigh Univ, MS, 59, PhD(math), 62. *Prof Exp:* Lectr math, Yale Univ, 62-64; from asst prof to assoc prof, Univ Ill, Urbana, 64-70; assoc prof, Queen's Univ, Ont, 70-71; assoc prof, 71-74, PROF MATH, UNIV ILL, URBANA, 74- *Mem:* Am Math Soc; Sigma Xi. *Res:* Functional and harmonic analysis; differential geometry. *Mailing Add:* 401 Eliot Dr Urbana IL 61801

BERG, J(OHN) ROBERT, b Chicago, Ill, Feb 18, 15; m 40; c 1. GEOLOGY, CHEMISTRY. *Educ:* Augustana Col, AB, 38; Univ Iowa, MS, 40, PhD(geol), 42. *Prof Exp:* Asst mineral, petrol & econ geol, Univ Iowa, 39-42; supvr mil explosives, E I Du Pont de Nemours & Co, Inc, 42-44; sub-surface geologist, Shell Oil Co, Kans, 44-46; from assoc prof to prof geol, Wichita State Univ, 46-81, head dept, 53-63, dean Univ col, 62-69, chmn dept, 76-81, EMER PROF GEOL, WICHITA STATE UNIV, 81- *Concurrent Pos:* Mem bd dirs & secy-treas, Petrol Resources Fund, Inc, 69-; ed, Trans, Kans Acad Sci, 85- *Mem:* Am Chem Soc; Geol Soc Am; Am Asn Petrol Geologists; Nat Asn Geol Teachers (pres, 56-58). *Res:* Petrology of alkaline igneous rocks; pre-Cambrian stratigraphy and petrology; petroleum exploratory statistics; geological education. *Mailing Add:* Dept Geol Wichita State Univ Wichita KS 67208

BERG, JAMES IRVING, b Minneapolis, Minn, May 5, 40; m 66; c 2. GLASS SCIENCE. *Educ:* Univ Minn, BS, 62; Ohio State Univ, PhD(physics), 69. *Prof Exp:* Prin res scientist physics, Honeywell Corp Res Ctr, 69-71; sr physicist, Graphics Res & Develop Ctr, Addressograph Multigraph Corp, 72-74; res assoc mat sci, Case Western Reserve Univ, 74-76; sr scientist, 76-81, RES ASSOC, TECH CTR, OWENS-CORNING FIBERGLAS CORP, 81- *Mem:* Am Phys Soc; Am Ceramic Soc; Sigma Xi. *Res:* Infrared physics; glass science; solid state electronics. *Mailing Add:* 65 Brynwood Circle Granville OH 43023

BERG, JEFFREY HOWARD, b New York, NY, Feb 6, 43; m 66; c 2. ORGANIC CHEMISTRY. *Educ:* Yeshiva Col, BA, 64; NY Univ, PhD(org chem), 69, MBA, 73. *Prof Exp:* Teaching asst chem, NY Univ, 64-69; sr chemist, Gen Foods Corp, 69-73; qual control supvr & chief chemist, I Rokeach & Sons, Inc, 73-74; sr group leader, Johnson & Johnson, 74-78; SR GROUP LEADER, ORTHO PHARMACEUT CORP, 78- *Mem:* Am Chem Soc; Inst Food Technologists; Am Asn Med Instrumentation. *Res:* Chlorinated anthraquinodimethanes and related compounds; crowded anthracenes; food chemistry with emphasis on flavors and carbohydrates; orthopaedic and textile new product development; medical electrodes; contraceptive devices. *Mailing Add:* Three Fairhill Rd Edison NJ 08817-2927

BERG, JOHN CALVIN, b Minneapolis, Minn, Nov 18, 37. CHEMICAL ENGINEERING. *Educ:* Carnegie Inst Technol, BS, 60; Univ Calif, Berkeley, PhD (chem eng), 64. *Prof Exp:* From asst prof to assoc prof, 64-73, PROF CHEM ENG, UNIV WASH, 73- *Mem:* Am Inst Chem Engrs; Am Chem Soc. *Res:* Interfacial phenomena; thermodynamics. *Mailing Add:* Dept Chem Eng BF-10 Univ Wash Seattle WA 98105

BERG, JOHN RICHARD, b Chippewa Falls, Wis, Apr 24, 32; m 56; c 4. PHYSICAL CHEMISTRY, INORGANIC CHEMISTRY. *Educ:* Col St Thomas, BS, 54; Iowa State Univ, PhD(phys chem, physics), 61. *Prof Exp:* Asst chem, Iowa State Univ, 54-61; sr chemist, Minn Mining & Mfg Co, 61-66, supvr, 66-69, mgr, 69-83, govt serv mgr, 83-86; prin dept asst sec conserv & renewable energy, US Dept Energy, 86-88, asst sec, 88-90; CONSULT, 90- *Mem:* Am Chem Soc; AAAS; Am Inst Chemists. *Res:* Photoconductivity; solid state chemistry; preparative inorganic chemistry; imaging chemistry; optics; data processing. *Mailing Add:* 3202 N Tacoma St Arlington VA 22213-1340

BERG, JONATHAN H, b Milaca, Minn, April 10, 47. IGNEOUS. *Educ:* Univ Minn, BA, 69; Franklin & Marshall Col, MS, 71; Univ Mass, PhD(geol), 76. *Prof Exp:* Res assoc, Univ Mass, 76-77; PROF GEOL, NORTHERN ILL UNIV, 77- *Concurrent Pos:* Vis fel, Australian Nat Univ, 90. *Mem:* Fel Geol Soc Am; Mineral Soc Am; Mineral Soc Can; Am Geophys Union. *Res:* Composition and evolution of granulites and the lower crust; geochemistry and origin of flood basalts; origin of anorthosites; nature of contact metamorphism; geology of Antarctica. *Mailing Add:* Dept Geol Northern Ill Univ De Kalb IL 60115

BERG, JOSEPH WILBUR, JR, b Essington, Pa, Oct 6, 20; m 50; c 3. GEOPHYSICS. *Educ:* Univ Ga, BS, 47; Pa State Univ, MS, 52, PhD(geophys), 54. *Prof Exp:* Instr, Armstrong Col (Ga), 47-48; asst prof, Univ Tulsa, 54-55; assoc prof geophys, Univ Utah, 55-60; prof, Ore State Univ, 60-66; exec secy, Off Earth Sci, Nat Acad Sci, 66-87, CONSULT, 87- *Concurrent Pos:* Geophysicist, Inst Defense Analysis, 60-61; vis prof, Cornell Univ, 69-70. *Mem:* Soc Explor Geophys (vpres, 75); Seismol Soc Am (pres, 68); Am Geophys Union; Earthquake Eng Res Inst; Geol Soc Am. *Res:* Earth structure as determined by seismic, gravitational and magnetic methods; heat flow. *Mailing Add:* 3319 Dauphine Dr Falls Church VA 22042

BERG, LLOYD, b Paterson, NJ, Aug 8, 14. CHEMICAL ENGINEERING. *Educ:* Lehigh Univ, BS, 36; Purdue Univ, PhD(chem eng), 42. *Prof Exp:* Lacquer formulator, Pittsburgh Plate Glass Co, NJ, 36; tech salesman, Sherwin-Williams Co, 36-39; instr powder & explosives, Purdue Univ, 41-42; res chem engr & group leader, Gulf Res & Develop Co, Pa, 42-46; assoc prof chem eng, Univ Kans, 46; prof chem eng & head dept, 46-79, PROF CHEM ENG, MONT STATE UNIV, 80- *Mem:* Fel Am Inst Chem Engrs; Am Chem Soc; Am Soc Eng Educ. *Res:* Desulfurization of petroleum; battery active manganese dioxide; fuels from shale oil; coal liquefaction; azeotropic and extractive distillation. *Mailing Add:* Dept Chem Eng Mont State Univ Bozeman MT 59717

BERG, M(ORRIS), b Columbus, Ohio, June 9, 21; m 56; c 2. CERAMICS. *Educ:* Ohio State Univ, BCerE, 42; Mass Inst Technol, ScD(ceramics), 53. *Prof Exp:* Process engr, AC Spark Plug Div, Gen Motors Corp, 42-44, ceramic res engr, 46-48; res asst, Mass Inst Technol, 49-53; ceramic proj engr, Bur Mines, 53-54; sr engr, Radio Corp Am, 54-63; staff res scientist, 63-76, STAFF ENGR INORG MAT, AC SPARK PLUG DIV, GEN MOTORS CORP, 76- *Honors & Awards:* Ferro Award, 52; David Sarnoff Award, 62. *Mem:* Am Ceramic Soc; Am Soc Metals; Soc Automotive Engrs. *Res:* Electronic ceramics; microelectronics; catalysts; metallurgy; physics. *Mailing Add:* 1076 Maple Grove Rd Lapeer MI 48446

BERG, MARIE HIRSCH, b Mannheim, Germany, Mar 20, 09; nat US; wid; c 1. BIOCHEMISTRY. *Educ:* Univ Heidelberg, PhD(chem), 34. *Prof Exp:* Prod chemist, Fat & Oil Indust, Germany, 34-36; asst to Prof G Bredig, Univ Karlsruhe, 36-37; Lambert Pharmacol Co Fel, Colgate-Palmolive-Peet Col fel & res assoc, Northwestern Univ, 41-47; chief chemist, Dept Path, St Luke's Hosp, Chicago, Ill, 47-48; biochemist, Dept Dermat, Univ Mich, 48-51; mem staff, Wayne County Gen Hosp, Eloise, Mich, 51-52; mem dept biol, Hamline Univ, 52-59 & dept chem, 55-59; res assoc, Sch Med, Univ Minn, 53-59 & dept agr biochem, 59-60; prof chem & chmn div natural sci & math, Northwestern Col, Minn, 60-67, lectr sci, 67-76, adv fire protection prog, Metrop Community Col, 76; RETIRED. *Concurrent Pos:* Fire protection instr, Suburban Hennepin Voc Tech Schs; adj prof, Fire Protection, Univ Minn, 75-87; lectr, 85-88. *Mem:* Fel AAAS; fel Am Sci Affil; Int Fire Serv Instr; Am Chem Soc; Am Asn Univ Women. *Res:* Science education; pyorhea dental research porphyrines and lead poisoning; fire science. *Mailing Add:* 121 Washington Ave South Apt 1917 Minneapolis MN 55401-2123

BERG, MARK ALAN, b Dec 1, 58. ULTRAFAST SPECTROSCOPY. *Educ:* Univ Minn, BS, 79; Univ Calif, Berkeley, PhD(chem), 85. *Prof Exp:* Postdoctoral assoc chem, Stanford Univ, 85-87; ASST PROF CHEM, UNIV TEX, 87- *Concurrent Pos:* Dreyfus distinguished young fac award, Dreyfus Found, 87; NSF presidential young investr, 90. *Mem:* Am Phys Soc. *Res:* Using time-resolved non-linear spectroscopies and ultrafast laser techniques to study the molecular motions in liquids and the resulting solvent effects on chemical reactions. *Mailing Add:* Chem Dept Univ Tex Austin TX 78720

BERG, MYLES RENVER, b Riverside, Calif, Sept 6, 32; m 59; c 4. ELECTRICAL ENGINEERING. *Educ:* Stanford Univ, BS, 55, MS, 56. *Prof Exp:* Res engr radio physics lab, SRI Int, 55-65; mem tech staff, Aerospace Corp, 66-72; group dir advan systs, Lincoln Lab, Mass Inst Technol, 72-73; mgr strategic systs, Technol Serv Corp, 73-74, sr res specialist probe systs, 75-77; SR STAFF ENGR SIGNAL PROCESSING SYST, SPACE SYST DIV, LOCKHEED MISSILES & SPACE CO, 77- *Concurrent Pos:* Mem reentry physics panel, Nat Acad Sci, 64-67; dir, Big Ten Oil Co, 76-88, vpres, 77-88; dir, Eagle Creek Mining & Drilling Co, 88-, Suprise Oil Co, 89- *Mem:* Inst Elec & Electronics Engrs; Sigma Xi. *Res:* Signal processing; satellite systems; communication systems; oil production; computer sciences; radio propagation; radar systems; counter measures; reentry physics. *Mailing Add:* 52 Marianna Lane Atherton CA 94027

BERG, NORMAN J, b Chicago, Ill. SOLID STATE PHYSICS. *Educ:* Ill Inst Technol, BSEE, 65, MSEE, 66; Univ Md, PhD(elec eng), 75. *Prof Exp:* Staff scientist mat res, 66-76, GROUP LEADER DEVICE DEVELOP, HARRY DIAMOND LABS, 76- *Concurrent Pos:* Guest researcher, Nat Bur Standards, 72-75. *Honors & Awards:* Wilbur S Hinman Award, Harry Diamond Labs, 77. *Mem:* Inst Elec & Electronics Engrs. *Res:* Linear and nonlinear interaction of high frequency sound waves with coherent laser light in piezoelectric crystals and the utilization of these phenomena of sophisticated signal processing devices. *Mailing Add:* Off Commander Harry Diamond Labs S3T0 Bldg 202 Rm 36120 2800 Powder Mill Rd Adelphi MD 20783

BERG, PAUL, b New York, NY, June 30, 26; m 47; c 1. BIOCHEMISTRY, MOLECULAR GENETICS OF EUKARYOTES. *Educ:* Pa State Univ, BS, 48; Western Reserve Univ, PhD, 52. *Hon Degrees:* DSc, Univ Rochester & Yale Univ, 78, Wash Univ, 86 & Ore State Univ, 89. *Prof Exp:* From asst to assoc prof microbiol, Sch Med, Wash Univ, 55-59; from assoc prof to prof, 59-69, chmn dept, 69-74, WILLSON PROF BIOCHEM, MED CTR, STANFORD UNIV, 70-, DIR, BECKMAN CTR MOLECULAR & GENETIC MED, 84- *Concurrent Pos:* Am Cancer Soc res fel, Inst Cytophysiol, Copenhagen & Sch Med, Wash Univ, 52-54 & scholar, 54-57; ed, Biochem & Biophys Res Commun, 59-68; mem study sect physiol chem, NIH, 62-66; chmn, Gordon Conf Nucleic Acids, 66-75; mem gen med basic sci comt & biochem comt, Nat Bd Med Examr, 68-69; mem adv bd, Jane Coffin Childs Found, 71; Salk Inst fel, 73-; mem adv bd, NIH, 75. *Honors & Awards:* Nobel Prize chem, 80; Sarasota Med Award, 79; NY Acad Sci Award, 80; Albert Lasker Award, 80; Sci Freedom & Responsibility Award, AAAS, 82; Nat Medal of Sci, 83; Nat Libr Med Metal, 86. *Mem:* Nat Acad Sci; Inst Med-Nat Acad Sci; Am Acad Arts & Sci; Am Soc Biol Chem (pres, 75); Am Chem Soc; fel AAAS. *Res:* Mechanism of gene expression in higher organisms, particularly the interplay of viral and cellular genes in regulating growth and division. *Mailing Add:* Beckman Ctr B062 Med Ctr Stanford Univ Stanford CA 94305

BERG, PAUL WALTER, b New York, NY, Mar 18, 25; wid; c 2. MATHEMATICS. *Educ:* NY Univ, BA, 47, PhD(math), 53. *Prof Exp:* Asst math, NY Univ, 47-51; instr, Rutgers Univ, 51-53; Nat Res Coun fel, 53-54; actg asst prof math, 54-55, from asst prof to assoc prof, 55-67, vchmn dept, 67-80, PROF MATH, STANFORD UNIV, 67- *Res:* Partial differential equations; calculus of variations; applied mathematics. *Mailing Add:* Dept Math Stanford Univ Stanford CA 94305

BERG, RICHARD A, b Spokane, Wash, Apr 27, 45. TISSUE ENGINEERING, BIOMATERIALS & WOUND HEALING. *Educ:* Univ Chicago, BS, 67; Univ Pa, PhD, 72. *Prof Exp:* From asst prof to assoc prof biochem, Col Med & Dent, NJ, 72-80; PROF BIOCHEM, UNIV MED & DENT NJ, 86- *Concurrent Pos:* Pulmonary Br, NIH, 78-80. *Honors & Awards:* Sinsheimer Award, 76. *Mem:* Am Soc Biochem & Molecular Biol; Am Chem Soc; Inst Elec & Electronics Engrs; Am Mgt Asn. *Res:* Connective tissue biochemistry; skin, bone and cartilage; regulation of gene expression of macromolecules; tissue engineering. *Mailing Add:* Dept Biochem Univ Med Dent NJ Robert Wood Johnson Med Sch 675 Hoes Lane Piscataway NJ 08854-5635

BERG, RICHARD ALLEN, b Chicago, Ill, June 12, 42; m 68. ASTRONOMY. *Educ:* Univ Ill, Urbana, BSc, 64; Univ Va, MA, 66, PhD(astron), 70. *Prof Exp:* Phys scientist astron & cartog, Aeronaut Chart & Info Ctr, St Louis, Mo, 64-70; fel, Dept Physics, Univ Del, 70-72, asst prof astron, 72-73; asst prof astron, Univ Rochester, 73-78; phys scientist, DMA Aerospace Ctr, St Louis, 78-80; CONSULT, RAND CORP, 71- *Mem:* Am Astron Soc; Int Astron Union; Sigma Xi. *Res:* Spectroscopy and photometry of spectrum variable stars; high-speed photometry of transient events in cataclysmic variables; computer applications in astronomical observing; image processing. *Mailing Add:* 4902 Ft Sumner Dr Bethesda MD 20816

BERG, RICHARD BLAKE, b Portland, Ore, Mar 7, 37; m 65; c 2. GEOLOGY, MINERALOGY-PETROLOGY. *Educ:* Beloit Col, BS, 59; Univ Mont, PhD(geol), 64. *Prof Exp:* Instr geol, State Univ NY Col Plattsburgh, 64-66; ECON GEOLOGIST & CURATOR MINERAL MUS, MONT BUR MINES & GEOL, MONT COL MINERAL SCI & TECHNOL, 66- *Mem:* Geol Soc Am; Mineral Soc Am; Am Inst Mining Metall & Petrol Engrs. *Res:* Mineralogy; geology of industrial minerals in Montana; Precambrian geology. *Mailing Add:* Mont Bur Mines & Geol Butte MT 59701

BERG, RICHARD HAROLD, b Seattle, Wash, Aug 6, 37; m 60; c 2. SANITARY ENGINEERING, ENVIRONMENTAL ENGINEERING. *Educ:* Univ Wash, BS, 59, MS, 61; Ore State Univ, PhD(civil eng), 66. *Prof Exp:* Instr eng, Skagit Valley Col, 61-62 & Univ Wash, 62-63; assoc prof civil eng, Seattle Univ, 68-71; ASSOC PROF CIVIL ENG, WESTERN WASH UNIV, 71- *Concurrent Pos:* Sr scientist, Dept Interior res grant, 69-70. *Mem:* Water Pollution Control Fedn; Asn Environ Eng Prof; Am Asn Univ Prof. *Res:* Liquid waste treatment; biological environmental control; environmental systems analysis; and mathematical modeling. *Mailing Add:* Western Wash Univ Huxley Col Bellingham WA 98225

BERG, ROBERT R, b St Paul, Minn, May 28, 24; m 46; c 3. GEOLOGY. *Educ:* Univ Minn, BA, 48, PhD(geol), 51. *Prof Exp:* Geologist, oil & gas explor, Calif Co, 51-56; div geologist, Cosden Petrol Corp, 57-58; consult geologist, Berg & Wasson, 59-66; dir res, 72-82, prof geol & head dept, 67-72, MICHEL T HALBOUTY CHAIR GEOL, TEX A&M UNIV, 82- *Mem:* Nat Acad Eng; fel Geol Soc Am; hon mem Am Asn Petrol Geol; Am Inst Prof Geol (pres, 71). *Res:* Stratigraphy; geology of petroleum; origin and properties of sandstones that are reservoirs for oil and gas; hydrostatic and hydrodynamic trapping of oil and gas; prediction of reservoir properties for enhanced recovery of oil. *Mailing Add:* Dept Geol Tex A&M Univ College Station TX 77843-3115

BERG, ROBERT W, b Welch, Minn, Apr 9, 17; m 44; c 3. POULTRY GENETICS. *Educ:* Univ Minn, BS, 41, MS, 50, PhD(poultry husb, animal breeding), 53. *Prof Exp:* Inspector, Minn Poultry Improve Bd, 45-48; asst geneticist, Western Coop Hatcheries, 53; geneticist & hatchery mgr, Jerome Turkey Hatchery, Inc, 53-58; POULTRY EXTEN SPECIALIST, INST AGR, UNIV MINN, ST PAUL, 58- *Mem:* Poultry Sci Asn; World Poultry Sci Asn; Sigma Xi. *Res:* Poultry husbandry and genetics; effects of light intensity on growth of turkeys in a total confinement program. *Mailing Add:* 643 Birch Lane S Shareview MN 55126

BERG, ROY TORGNY, b Millicent, Atla, Apr 8, 27; m 51; c 4. ANIMAL GENETICS. *Educ:* Univ Alta, BSc, 50; Univ Minn, MS, 54, PhD(animal breeding), 55. *Hon Degrees:* DSc, Univ Guelph, 91. *Prof Exp:* Lectr animal husb, Univ Alta, 50-52, from asst prof to prof, 55-89, dept chmn, 77-82, dean fac agr forestry, 83-88, EMER PROF ANIMAL SCI, UNIV ALTA, 89- *Concurrent Pos:* Nat Res Coun sr res fel, 64-65; Vis prof, Edinburgh, 72, Copenhagen, 74 & 77, Cairo, 80. *Honors & Awards:* Genetic & Physiol Medal, Can Soc Animal Sci, 81; Cert Merit, 82; Sir Frederick Haultain Prize in Sci, 89; Sci & Technol Award, Can Meat Coun, 89. *Mem:* Fel Agr Inst Can; Am Soc Animal Sci; Am Meat Sci Asn; Genetics Soc Can; Agr Inst Can; Am Genetics Asn; Brit Soc Animal Prod. *Res:* Animal breeding; genetics. *Mailing Add:* Fac Agr Forestry Univ Alta Edmonton AB T6G 2P5 Can

BERG, STEVEN PAUL, b Minneapolis, Minn, Dec 28, 48; m 70; c 2. BIOCHEMISTRY. *Educ:* Pac Lutheran Univ, BS, 70; Purdue Univ, PhD(biochem), 75. *Prof Exp:* Res assoc photosynthesis, Dept Biol, Wayne State Univ, 75-76, NSF fel, 76-77; ASST PROF BIOL, UNIV DENVER, 77- *Concurrent Pos:* Prin investr, USDA grant, 77- *Mem:* Am Soc Plant Physiologists; Am Soc Biol Chemists. *Res:* Photosynthetic electron transport and photophosphorylation; chloroplast membrane proteins and membrane structure. *Mailing Add:* Dept Biol Winona State Univ Winona MN 55987

BERG, VIRGINIA SEYMOUR, b Mass, July 30, 48. PLANT PHYSIOLOGY. *Educ:* Tufts Univ, BS, 72; Univ Wash, MS, 75, PhD(plant physiol), 80. *Prof Exp:* fel, Dept Land, Air & Water Resources, Univ Calif, Davis, 81-84; asst prof, 84-88, ASSOC PROF, DEPT BIOL, UNIV NORTHERN IOWA, CEDAR FALLS, 88- *Concurrent Pos:* Vis prof, Plant Ecol Group, Univ Andes, Venezuela, 82; vis prof, Fitotechnia Inst, Córdoba, Spain; vis scientist, Int Ctr Trop Agr, Cali, Columbia. *Mem:* AAAS; Am Soc Plant Physiologists; Sigma Xi. *Res:* Plant water relations; energy budgets; effects of acidity on plant surfaces. *Mailing Add:* Dept Biol Univ Northern Iowa Cedar Falls IA 50614-0421

BERG, WILLIAM ALBERT, b Sterling, Colo, Jan 12, 30; m 60; c 4. SOIL CONSERVATION. *Educ:* Colo State Univ, BS, 53, MS, 57; NC State Univ, PhD(soils), 60. *Prof Exp:* Soil scientist, US Forest Serv, 60-68; from asst prof to prof agron, Colo State Univ, 68-80; SOIL SCIENTIST, AGR RES SERV, USDA, 80- *Mem:* Am Soc Agron; Soc Range Mgt; Soil & Water Conserv Soc. *Res:* Application of soils information in range management; vegetative stabilization of disturbed lands and mine wastes. *Mailing Add:* Southern Plains Range Res Sta 2000 18th Woodward OK 73801

BERG, WILLIAM EUGENE, b Round Mountain, Nev, Dec 6, 18; m 47. BIOLOGY. *Educ:* Calif Inst Technol, BS, 39, MS, 40; Stanford Univ, PhD(biol), 46. *Prof Exp:* Asst biol, Stanford Univ, 41-43, res assoc, 43; res assoc, Univ Southern Calif, 43; from instr to prof biol, Univ Calif, Berkeley, 47-70, prof zool, 70-80, vchmn dept, 73-77, EMER PROF ZOOL, UNIV CALIF, BERKELEY, 80- *Mem:* Am Soc Zoologists; Sigma Xi. *Res:* Physiological embryology; aviation physiology; human exercise physiology. *Mailing Add:* PO Box 1810 Pollock Pines CA 95726-1810

BERG, WILLIAM KEITH, b Cloquet, Minn, July 20, 43; m 67; c 1. PSYCHOLOGY, DEVELOPMENTAL PSYCHOPHYSIOLOGY. *Educ:* Univ Minn, Duluth, BA, 65; Univ Wis, MA, 68, PhD(psychophys), 71. *Prof Exp:* Asst prof child behavior & develop, Univ Iowa, 70-73; asst prof, 73-77, PROF PSYCHOL, UNIV FLA, 77- *Concurrent Pos:* Grant reviewer, NSF, NIH, NIMH, March of Dimes. *Mem:* Soc Res in Child Develop; Soc Psychophysiol Res; Am Psychol Asn; fel AAAS; fel Am Psychol Asoc; Int Soc Develop Psychobiol. *Res:* Habituation, attention, and learning behavior in young infants assessed with physiological response systems; Physiological evaluation of visual stress during video display viewing. *Mailing Add:* Dept Psychol Univ Fla Gainesville FL 32611

BERGBREITER, DAVID EDWARD, b Chicago, Ill, Nov 10, 48; m 71; c 2. POLYMER SURFACE CHEMISTRY, HOMOGENEOUS CATALYSIS. *Educ:* Mich State Univ, BS, 70; Mass Inst Technol, PhD(chem), 74. *Prof Exp:* Chmn, Org Div, 83-87, PROF ORG CHEM, TEX A&M UNIV, 74- *Concurrent Pos:* Texaco fel, Mass Inst Technol, 71-73. *Mem:* Am Chem Soc. *Res:* Organic, organometallic and polymer chemistry; new catalysts, new methods for synthesis and mechanistic studies in organometallic chemistry and asymmetric synthesis. *Mailing Add:* Dept Chem Tex A&M Univ College Station TX 77843

BERGDOLL, MERLIN SCOTT, b Petersburg, WVa, Sept 23, 16; div; c 3. BIOCHEMISTRY. *Educ:* WVa Univ, BS, 40; Purdue Univ, PhD(agr biochem), 46. *Prof Exp:* Asst chemist, Purdue Univ, 42-46; instr, Biochem & Food Res Inst, Univ Chicago, 46-50, from asst prof to assoc prof, Food Res Inst, 50-66; prof, Dept Foods & Nutrit, 66-68, prof, Food Res Inst & Dept Food Sci, 68-88, EMER PROF, UNIV WIS-MADISON, 88- *Honors & Awards:* Ray A Kroc Prof Award for Res on Food Toxicol. *Mem:* Am Chem Soc; Am Soc Microbiol; Inst Food Technol; Am Soc Biol Chem. *Res:* Production, purification, physico-chemical properties, structure, immunology and analysis of the staphylococcal enterotoxins; toxic shock syndrome toxin-1 purification and characterization. *Mailing Add:* Food Res Inst Univ Wis Madison WI 53706

BERGDOLT, VOLLMAR EDGAR, b Evansville, Ind, Dec 29, 18; m 45; c 4. MECHANICAL ENGINEERING. *Educ:* Purdue Univ, BSME, 39, MSME, 46, PhD(heat transfer), 56. *Prof Exp:* Asst engr, Am Creosoting Co, Ky, 39-40; mech engr, Ballistic Res Labs, Aberdeen Proving Ground, Md, 46-53; res asst mech eng, 53-56, asst prof mech & nuclear eng, 56-58, from assoc prof to prof, 58-64, asst dean grad sch, 64-69, prof mech eng, 69-84, EMER PROF MECH ENG, PURDUE UNIV, 84- *Concurrent Pos:* Vis prof, Indian Inst Technol, Kanpur, 63-64; prof mech eng, Univ Zambia, Lusaka, 71-73. *Mem:* Am Soc Eng Educ; Am Soc Mech Engrs. *Res:* Energy conversion in mechanical systems; kinematics and mechanical design. *Mailing Add:* 1209 Sunset Lane West Lafayette IN 47906-2428

BERGE, DOUGLAS G, b Montevideo, Minn, Sept 19, 38; m 66; c 1. ANALYTICAL CHEMISTRY. *Educ:* Mankato State Col, BS & BA, 61; Univ Iowa, PhD(analytical chem), 65. *Prof Exp:* Asst prof chem, Adrian Col, 65-66; from asst prof to assoc prof, 66-82, PROF CHEM, UNIV WIS-OSHKOSH, 82- *Mem:* Am Chem Soc. *Res:* Equilibrium studies; separations. *Mailing Add:* Dept Chem Univ Wis Oshkosh WI 54901-8045

BERGE, JOHN WILLISTON, b Madison, Wis, July 29, 30; m 56; c 3. POLYMER CHEMISTRY. *Educ:* Univ Wis, BS, 51, PhD(chem), 59. *Prof Exp:* Chemist, E I du Pont de Nemours & Co, Inc, 51-52, res chemist, 58-63; sr res chemist, S C Johnson & Son, Inc, 63-83, sr scientist, 83-87; RETIRED. *Mem:* Am Chem Soc. *Res:* Scanning electron microscopy of skin, hair, household products; dynamical mechanical properties of high polymers; synthetic polymers and resins for polishes; gel permeation chromatography. *Mailing Add:* 1529 Crabapple Dr Racine WI 53405

BERGE, JON PETER, b Madison, Wis, Feb 4, 35; m 59; c 2. EXPERIMENTAL HIGH ENERGY PHYSICS. *Educ:* Univ Calif, BA, 56, MA, 58, PhD(physics), 63. *Prof Exp:* Res assoc physics, Lawrence Radiation Lab, Univ Calif, Berkeley, 56-67; res officer, Univ Oxford, 67-70; physicist, Ctr Europ Nuclear Res, Geneva, 70-74; PHYSICIST, FERMI NAT ACCELERATOR LAB, 74- *Mem:* Am Phys Soc; AAAS. *Res:* Bubble chamber data analysis computer programming. *Mailing Add:* Fermi Nat Accelerator Lab Batavia IL 60510

BERGE, KENNETH G, b Wahkon, Minn, Feb 9, 26; m 48; c 3. INTERNAL MEDICINE. *Educ:* Univ Minn, BA, 48, BS, 49, BMed, 51, DMed, 52, MS, 55. *Prof Exp:* Asst med, Mayo Clin, 55, consult, 55-70; NW area found prof community health, 77-84; from instr to assoc prof, Mayo Grad Sch Med, 57-74, PROF MED, MAYO SCH, UNIV MINN, 74- *Concurrent Pos:* Consult & vchmn steering comt coronary drug proj, Nat Heart & Lung Inst, 62-87, mem epidemiol & biomet adv comt, 70-72, mem policy adv bd, Hypertension Detection & Followup Prog, 71-87, mem policy bd, Aspirin Myocardial Infarction Study, 75-80; fel coun arteriosclerosis & epidemiol, Am Heart Asn; mem sect arteriosclerosis & ischaemic heart dis, Int Soc Cardiol; mem ad bd dir, Rochester Methodist Hosp, 70-84; mem, Olmsted County Bd Health, 75-81, chmn, 77; pres voting staff, Mayo Clin, 76, head sect med, 70-80; State Health Adv Coun, 80-82; chmn steering comt, systolic hypertension in the elderly prog, NIH, 84- *Honors & Awards:* Billings Silver Medal, AMA, 57. *Mem:* Fel Am Col Physicians; AMA; Am Fedn Clin Res. *Res:* Cardiovascular research, especially epidemiology, atherosclerosis and hypertension. *Mailing Add:* Mayo Clin Rochester MN 55901

BERGE, TRYGVE O, VIROLOGY, IMMUNOLOGY. *Educ:* Univ Calif, Berkeley, PhD(bacteriol), 51. *Prof Exp:* Bacteriologist, Am Type Culture Collection, 64-75; RETIRED. *Mailing Add:* 2305 Hillsdale Way Boulder CO 80303-5623

BERGELAND, MARTIN E, b Madison, Minn, Oct 14, 35; m 59; c 2. VETERINARY PATHOLOGY. *Educ:* Univ Minn, BS, 57, DVM, 59, PhD(vet path), 65. *Prof Exp:* From instr to assoc prof vet path, Univ Minn, St Paul, 59-70; prof vet path & dir vet diag med, Univ Ill, Urbana-Champaign, 70-73; PROF VET SCI, SDAK STATE UNIV, 73- *Mem:* Am Col Vet Path; Am Vet Med Asn; Am Asn Avian Path; Nat Conf Vet Diagnosticians. *Res:* Pathology of clostridial infections of animals. *Mailing Add:* Dept Vet Sci SDak State Univ Brookings SD 57007

BERGELIN, OLAF P(REYSZ), b Big Rapids, Mich, Nov 8, 11; m 42; c 3. CHEMICAL ENGINEERING. *Educ:* Univ Mich, BSE, 35, MS, 36, ScD(chem eng), 42. *Prof Exp:* Asst eng res, Univ Mich, 35-37; student engr, Union Oil Co, Calif, 39; chem engr, Hercules Powder Co, Wilmington, 41-42; from assoc prof to prof chem eng, Univ Del, 46-56; educ consult, Tex Agr & Mech Col Pakistan Proj, Dacca, E Pakistan, 56-62; proj dir, Kabul Afghan-Am Prog, Educ Serv, Inc, Afghanistan, 63-68; assoc dean col grad studies & coord res, 68-77, EMER PROF CHEM ENG, UNIV DEL, 77- *Concurrent Pos:* Res Corp grant, Fulbright res fel, NZ Geothermal Power Proj, 51; technol transfer consult, Univ Del, 77- *Honors & Awards:* Medal, Govt of Afghanistan, 68. *Mem:* Am Chem Soc; Am Inst Chem Engrs; Am Soc Eng Educ; Am Soc Mech Engrs; Am Numis Soc. *Res:* Heat transfer in tubular heat exchangers; two phase fluid flow; boiling. *Mailing Add:* 1410 SW 20th St Cape Coral FL 33914

BERGEMAN, THOMAS H, b Ft Dodge, Iowa, Nov 11, 33; m 80; c 2. ATOMS IN EXTERNAL FIELDS, DIATOMIC SPECTROSCOPY. *Educ:* Calif Inst Technol, BS, 56; Harvard Univ, MA, 58, PhD(chem physics), 71. *Prof Exp:* Res asst, radiation lab, Columbia Univ, 70-72; asst prof physics, Fordham Univ, 72-74 & 75-76, adj asst prof, 74-75 & 76-79; SR RES ASSOC PHYSICS, STATE UNIV NY, STONY BROOK, 79- *Concurrent Pos:* Vis assoc prof, Univ Paris-S, Orsay, 78; vis scientist, Res Inst Atomic Physics, Stockholm, Sweden, 80; consult & vis staff mem, Meson Physics Facil, Los Alamos Nat Lab, 82-84. *Mem:* Am Phys Soc. *Res:* Calculations in atomic physics on the effects of external electric fields, trapped neutral atoms and microwave ionization; optical spectra of diatomic molecules; stochastic laser fields and atomic interactions. *Mailing Add:* Dept Physics State Univ NY Stony Brook NY 11794-3800

BERGEN, CATHARINE MARY, b Garden City, NY, Jan 16, 12. SCIENCE EDUCATION. *Educ:* Wellesley Col, AB, 33; Columbia Univ, AM, 35, PhD(sci educ), 42. *Prof Exp:* Sci consult pub sch, Garden City, NY, 34-37; asst physics, Hofstra Col, 39-40; teacher elem sci, Lincoln Sch, Teachers Col, Columbia Univ, 40-43; from assoc prof to prof, 41-72, EMER PROF SCI, JERSEY CITY STATE COL, 72- *Mem:* AAAS; Am Phys Soc; Am Asn Physics Teachers. *Res:* Sources of science information used by children; analysis of science material published for laymen; use of mathematics in science. *Mailing Add:* 83 Bentley Ave Jersey City NJ 07304

BERGEN, DONNA CATHERINE, b Crawfordsville, Ind, Mar 17, 45. NEUROLOGY. *Educ:* Vassar Col, BA, 67; Univ Ill, MD, 71. *Prof Exp:* Intern med, Evanston Hosp, Ill, 71-72; resident neurol, 72-75; staff neurologist, 75-80, ASSOC PROF NEUROL, MED COL, RUSH UNIV, 80- *Mem:* Am Acad Neurol; Am Epilepsy Soc; Am EEG Soc; Am Bd Qualification EEG. *Res:* Clinical electroencephalography. *Mailing Add:* Rush Med Col Rush Presby St Lukes Hosp 1753 W Congress Pkwy Chicago IL 60612

BERGEN, JAMES DAVID, b Spokane, Wash, May 29, 32; m 74; c 1. METEOROLOGY, AIR POLLUTION. *Educ:* Univ Wash, BS, 55; Colo State Univ, MS, 64, PhD(atmospheric sci), 67. *Prof Exp:* Res aeronaut engr, Space Task Group, NASA, 59-61; climatologist, US Navy Hydrographic Off, 61-62; res meteorologist, Rocky Mountain Sta, Subalpine Hydrol Res Unit, 62-74, Mountain & Forest Meteorol Proj, 74-78; res meteorologist, Snow Zone Hydrol Proj, Pac Southwest Forest & Range Exp Sta, 78-82, meteorologist, Forest Fire Meteorol Proj, 82-86; AIR QUAL RES SPECIALIST, PIMA CO DEPT ENVIRON QUAL, 86- *Concurrent Pos:* Consult, Task Force Aerial Applications, 78; mem spec grad affil fac, Dept Watershed Mgt, Colo State Univ, 67-70; vis scientist, Dept Watershed Mgt, Univ Ariz, 81-82; mem, Canopy & Diffusion Comts, Coop Field Exp, US Army-US Forest Serv, 83-86; prin investr, mesometeorol study, DPG, US Army-US Forest Serv, 85-86; vol scientist, Forest Fire Meteorol Proj, Pac Southwest Forest Range Exp Sta, 86- *Mem:* Am Meteorol Soc; Sigma Xi. *Res:* Forest meteorology; boundary layer in mountain terrain; snow physics; analysis of airflow and turbulence in and near forest stands; modeling and measurement of airflow in complex terrain for input into pollution transport models. *Mailing Add:* Pima Co Dept Environ Qual PCAQCD 151 W Congress Tucson AZ 85701

BERGEN, LAWRENCE, CELL CYCLE CONTROL, DICTYOSTELIUM DEVELOPMENTS. *Educ:* Univ Wis, PhD(molecular biol), 80. *Prof Exp:* RES ASSOC CELL BIOL, UNIV WIS, 86- *Mailing Add:* Dept Bact Univ Wis Madison WI 53709

BERGEN, ROBERT LUDLUM, JR, b Islip, NY, Oct 29, 29; m 51; c 4. POLYMER CHEMISTRY. *Educ:* Williams Col, AB, 51; Cornell Univ, MS, 53, PhD(phys chem), 55. *Prof Exp:* Sr res chemist chem div, US Rubber Co, 55-60, sr res specialist, 60-69; mgr textile & plastic res, Res Ctr, Uniroyal Inc, 69-72, mgr Paracril res & develop, 72-74, mgr synthetic rubber res & develop, Chem Div, 74-77, group mgr chem & polymers res & develop, 77-79, mgr corp res & dev 79-81, dir res & develop, Uniroyal Eng Prod Co, 81-85, dir corp eng, Uniroyal Inc, 85; CONSULT, 86- *Concurrent Pos:* Instr, New Haven Col, 64-67; adj prof, 67-69; adj prof math, Univ New Haven, 86- *Mem:* Fel AAAS; Am Chem Soc; Sigma Xi. *Res:* Physical chemistry and physics of polymers; rheology; statistical experimentation; physics of polymer blends; environmental stress cracking of plastics; fiber spinning; fibers for pollution control; polymer flammability; rubber chemistry; synthetic elastomers for specialty applications; chemicals for specialty applications; thermoplastic elastomers; synthetic lubricants. *Mailing Add:* 79 Lebanon Rd Bethany CT 06525

BERGEN, STANLEY S, JR, b Princeton, NJ, May 2, 29; m 65; c 5. INTERNAL MEDICINE. Educ: Princeton Univ, AB, 51; Columbia Univ, MD, 55. Prof Exp: Med dir, St Luke's Hosp, Greenwich, Conn, 62-64; actg chief med serv, Ft Jay Army Hosp, New York, NY, 60-62; dir med, Cumberland Hosp, 64-68; chief commun med, Brooklyn-Cumberland Med Ctr, 68-70; sr vpres, New York City Health & Hosps Corp, 70-71; PROF INTERNAL MED, RUTGERS MED SCH & NJ MED SCH & PRES, COL MED & DENT NJ, 71- Concurrent Pos: Zabriskie fel, St Luke's Hosp, New York, 58-59; Hartford Found grant, 63-65; assoc prof med, State Univ NY Downstate Med Ctr, 64-71. Mem: NY Acad Sci; fel Am Col Physicians; fel Am Soc Clin Nutrit; Asn Am Med Cols; Sigma Xi. Res: Carbohydrate and hepatic metabolism; endocrinology; development of new health care delivery system. Mailing Add: Univ Med-Dent NJ 30 Bergen St Newark NJ 07107

BERGEN, WERNER GERHARD, b Warstade, Ger, Apr 23, 43; US citizen; m 66; c 1. ANIMAL NUTRITION, GROWTH BIOLOGY. Educ: Ohio State Univ, BSc, 64, MSc, 66, PhD(protein nutrit), 67. Prof Exp: PROF GROWTH BIOL & NUTRIT, MICH STATE UNIV, 67- Honors & Awards: Nutrit Res Award, Am Soc Animal Sci, 79. Mem: Am Soc Animal Sci; Am Inst Nutrit; Fedn Am Soc Exp Biol; Soc Exp Biol & Med. Res: Animal nutrition; plasma amino acid and protein metabolism; genome regulation of animal growth; protein turnover in animals; protein digestion; muscle biology. Mailing Add: Dept Animal Sci Col Agr Mich State Univ East Lansing MI 48824-1225

BERGENBACK, RICHARD EDWARD, b Allentown, Pa, Oct 23, 26; m 47; c 3. GEOLOGY. Educ: Lafayette Col, AB, 48; Lehigh Univ, MS, 50; Pa State Univ, PhD, 64. Prof Exp: Asst, Lehigh Univ, 48-50; geologist, Cities Serv Res & Develop Co, 57-58; geologist-in-chg fuels br, Tenn Valley Authority, 58-60; teacher, Baltimore County, Md, 60-61; asst, Pa State Univ, 61-64, res assoc, 64-65; asst prof geol, Southern Conn State Col, 65-66; assoc prof geol & geog & chmn dept, Howard Univ, 66-68; assoc prof, 68-80, PROF GEOL, UNIV TENN, CHATTANOOGA, 80- Concurrent Pos: Instr eve col, Univ Chattanooga, 58-59. Mem: Geol Soc Am; Am Asn Petrol Geologists. Res: Petrology and geochemistry of sedimentary rocks, especially carbonate rocks; regional stratigraphy, especially coal-bearing rocks. Mailing Add: Dept Geoscis Univ Tenn Chattanooga 615 McCallie Chattanooga TN 37403

BERGENDAHL, MAXIMILIAN HILMAR, b Reading, Pa, Apr 24, 21; m 45, 78; c 1. ECONOMIC GEOLOGY. Educ: Brown Univ, AB, 50. Prof Exp: Geologist, US Geol Surv, 51-67; sr geologist, Amax Explor Inc, 67-81; CONSULT. 82- Concurrent Pos: Explor supvr, Phillips Petrol Co, 82. Mem: Fel Explorers Club; Am Inst Prof Geologists. Res: Geology and stratigraphy of Florida phosphate deposits; late Cenozoic stratigraphy of south-central Florida; uranium deposits of Wyoming; stratigraphy of northeastern Wyoming; pre-Cambrian geology; ore deposits of central Colorado; gold deposits of the United States; metals exploration of western United States, Middle Eastern countries and South America. Mailing Add: PO Box 723 Poteau OK 74953

BERGER, ALAN ERIC, b Plainfield, NJ, Sept 4, 46. NUMERICAL ANALYSIS. Educ: Rutgers Univ, AB, 68; Mass Inst Technol, MS, 69, PhD(math), 72. Prof Exp: RES MATHEMATICIAN, NAVAL SURFACE WARFARE CTR, 72- Mem: Am Math Soc; Soc Indust & Appl Math; Math Asn Am; Am Meteorol Soc; AAAS. Res: Numerical solution of partial differential equations; numerical analysis of physical models. Mailing Add: Appl Math Br Code R44 Naval Surface Warfare Ctr Silver Spring MD 20903-5000

BERGER, ALBERT JEFFREY, b New York, NY, Jan 7, 43; m 65; c 1. NEUROPHYSIOLOGY, RESPIRATORY PHYSIOLOGY. Educ: Cornell Univ, BChE, 64; Princeton Univ, MA, 65, PhD(chem eng), 67; Univ Calif, San Francisco, PhD(physiol), 76. Prof Exp: Res engr, Shell Develop Co, 67-70; asst prof chem eng, Rensselaer Polytech Inst, 70-72; adj asst prof physiol, Univ Calif, San Francisco, 76-78; from asst prof to assoc prof, 78-85, PROF PHYSIOL, UNIV WASH, 85- Concurrent Pos: Spec fel, NIH, USPHS, 72-76, Res Career Develop Award, 77-82, Jacob Javits Neurosci Invest Award, 85-; assoc staff mem, Cardiovasc Res Inst, Univ Calif, San Francisco, 77-78; Guggenheim fel, 81-82, Fogarty Sr Int fel, NIH, 88-89; vis scientist, Inst Neurol, London, Eng, 81-82; vis prof, Dept Physiol, Kyoto Univ, Kyoto, Japan. Mem: Am Physiol Soc; Soc Neurosci; AAAS. Res: Neural regulation of respiration; neuronal connectivity; neuroanatomy. Mailing Add: Dept Physiol & Biophys Univ Wash Med Sch SJ 40 Seattle WA 98195

BERGER, ANDREW JOHN, b Warren, Ohio, Aug 30, 15; div; c 2. ANATOMY, ORNITHOLOGY. Educ: Oberlin Col, AB, 39; Univ Mich, AM, 47, PhD, 50. Prof Exp: From instr to assoc prof anat, Med Sch, Univ Mich, 50-64; chmn dept zool, 65-71, prof zool, Univ Hawaii, 64-81, EMER PROF, 81- Concurrent Pos: Asst ed, Wilson Bull, Wilson Ornith Soc, 50-51 & The Auk, Am Ornithologists' Union, 53-54; Guggenheim fel, 63; Fulbright lectr, Univ Baroda, 64-65; Univ Hawaii Res Admin grant, 65; NSF grants, 66-69, 70-75; ecol consult, 71- Mem: Am Asn Anatomists; fel Am Ornithologists' Union; Wilson Ornith Soc (second vpres, 71-73, first vpres, 73-75, pres, 75-77); Cooper Ornith Soc; fel AAAS. Res: Human anatomy; avian anatomy and life history. Mailing Add: Dept Zool Univ Hawaii Honolulu HI 96822

BERGER, ANN ELIZABETH, LYMPHOCYTE ACTIVATION, HYBRIDOMAS. Educ: Duke Univ, PhD(immunol), 78. Prof Exp: RES SCIENTIST, UPJOHN CO, 81- Mailing Add: 6537 Rothbury Portage MI 49081

BERGER, BERNARD BEN, b New York, NY, Aug 21, 12. WATER SUPPLY, WASTEWATER TREATMENT. Educ: Mass Inst Technol, BS, 35; Harvard Univ, MS, 48. Prof Exp: Sanit engr, Kent County, Mich, 39-41; prof, 66-78, EMER PROF CIVIL ENG, UNIV MASS, 78- Mem: Nat Acad Eng; Am Water Works Asn; Am Acad Environ Eng; Water Pollution Control Fedn; hon fel Inst Water Pollution Control UK; Int Water Resources Asn. Mailing Add: Dept Civil Eng Univ Mass Amherst MA 01003

BERGER, BEVERLY JANE, b Morristown, NJ, Apr 28, 39; M. RISK ANALYSIS. Educ: Univ NMex, BA, 61, MEdSci, 65, MS, 67; Univ Calif, Davis, PhD(genetics), 71. Prof Exp: NIH fel biostatist, Sch Pub Health, Univ Calif, Berkeley, 71-73; math biologist, Lawrence Livermore Lab, Univ Calif, 73-76; prog mgr chem hydrogen storage systs, US Dept Energy, 76-79, prog mgr biomass energy systs, 79, dir, Biomass Energy Technol Div, 79-86; asst dir life sci, Off Sci & Technol Policy, Exec Off Pres, 87-89; REP, FED LAB CONSORTIUM TECHNOL TRANSFER, WASHINGTON, DC, 90- Concurrent Pos: Mem, Domestic Policy Coun Working Groups on Health Policy, Agent Orange, Energy, Environ & Natural Resources, 87-89; chmn, Fed Coord Coun Sci, Eng & Technol, Comt Life Sci, 87-89; mem, Fed Coord Coun Sci, Eng & Technol Comts, Biotechnol Sci Coord Comt, Comt Interagency Radiation Res & Policy Coord & Comt Earth Sci, 87-89. Mem: Genetics Soc Am; AAAS; Sigma Xi; Soc Risk Anal. Res: Liaison between consortium and Washington based organizations including federal government; policy formulation related to health, biological, physical sciences. Mailing Add: 1601 N Randolph St Arlington VA 22207

BERGER, BEVERLY KOBRE, b Passaic, NJ, May 22, 46; m 70. GENERAL RELATIVITY, QUANTUM GRAVITY. Educ: Univ Rochester, BS, 67; Univ Md, PhD(physics), 72. Prof Exp: Res assoc, Joint Inst Lab Astrophys, Univ Colo & Nat Bur Standards, 72-74; res staff physicist & lectr, Yale Univ, 74-77; PROF PHYSICS, OAKLAND UNIV, ROCHESTER, MICH, 77- Concurrent Pos: Res assoc, Enrico Fermi Inst, Univ Chicago, 80; vis asst physicist, Inst Theoret Physics, Univ Calif, Santa Barbara, 81; vis asst prof, dept physics & astron, Univ Md, 82; vis assoc prof, astron dept, Univ Mich, 83-84; physicist, Lawrence Livermore Nat Lab, 86. Mem: Am Phys Soc; Am Math Soc; Sigma Xi; Int Soc Gen Relativity & Gravitation. Res: Classical general relativity; cosmological models; quantum gravity; quantum field theory in curved spacetime; astrophysics; plasma physics. Mailing Add: Physics Dept Oakland Univ Rochester MI 48309

BERGER, BRUCE S, b Philadelphia, Pa, May 23, 32; m 58; c 2. ENGINEERING MECHANICS. Educ: Univ Pa, BSc, 54, MSc, 58, PhD(eng mech), 62. Prof Exp: Math analyst, Gen Elec Co, 55-56; instr eng, Univ Pa, 56-60; asst prof appl math, Vanderbilt Univ, 61-63; assoc prof mech eng, 63-69, PROF MECH ENG, UNIV MD, COLLEGE PARK, 69- Concurrent Pos: Naval Res & Develop Ctr grant, 68-70; Off Naval Res grant, 71-76; Dept Energy res grant, 81-85; vis prof, Imp Col Sci & Tech, London, 83-84, 91. Mem: Am Soc Mech Engrs; Soc Indust & Appl Math; Sigma Xi. Res: Continuum mechanics; fluid solid interaction and the analysis of related numerical problems; dynamical systems and chaos. Mailing Add: Dept Mech Eng Univ Md College Park MD 20742

BERGER, BYRON ROLAND, b Seattle, Wash, Mar 27, 44; m 65; c 2. ECONOMIC GEOLOGY, GEOCHEMISTRY. Educ: Occidental Col, BA, 66; Univ Calif, Los Angeles, MS, 75. Prof Exp: Geologist, Continental Oil Co, 71-77, res scientist geol, 77; GEOLOGIST, US GEOL SURV, 77- Mem: Geol Soc Am; Am Geophys Union. Res: Geochemistry of mineral deposits; exploration geochemistry; geochemistry of supergene processes. Mailing Add: 209 Wright St No 1-306 Denver CO 80228

BERGER, DANIEL RICHARD, b Oakland, Calif, Nov 3, 33; m 58. ORGANIC CHEMISTRY. Educ: Stanford Univ, BS, 54; Northwestern Univ, PhD(chem), 58. Prof Exp: Instr chem, Northwestern Univ, 57-58; chemist, Am Cyanamid Co, 58-59; chemist, Richards Co, 59-64, tech serv mgr, Org Chem Div, 64-74, prod develop mgr, 74-78, tech dir, Org Chem Div, 78-82; chief chemist, Far Best Co, 82-84; sect head, 84-90, TECH SERV MGR, AKZO CHEM INC, 90- Concurrent Pos: Lectr eve div, Northwestern Univ, 59-68. Mem: Am Oil Chem Soc; Am Chem Soc. Res: Rearrangements of alpha-halogenated ethers; mechanism; ethers; surfactant chemistry. Mailing Add: 1324 Allison Lane Schaumburg IL 60194

BERGER, DANIEL S, b Brooklyn, NY, Aug 11, 23; m 50; c 2. ELECTRICAL ENGINEERING, BIOPHYSICS. Educ: Cooper Union, BEE, 48; Ill Inst Technol, MSEE, 54. Prof Exp: Jr engr cent sta develop, Western Union, 48-50; sr engr flow meter develop, Mittelman Electronics, Chicago, 50-52; chief engr component res, Sci Specialties, Inc, 52-56; proj engr med electronics, Bionics, Inc, Philadelphia, 56-60; asst prof med physics, Sch Med, Temple Univ, 60-85, sr assoc prof, 85-; RETIRED. Concurrent Pos: Instr electronics, Pa State Univ, 59-61. Mem: Inst Elec & Electronics Engrs; Optical Soc Am; Illum Eng Soc; AAAS. Res: Developed diagnostic and therapeutic light sources; meters of effective light intensity for various biological systems; tissue oxygen electrodes and pulsing circuits; blind guidance devices; phonocardiography apparatus. Mailing Add: 6655 Lawnton Ave Philadelphia PA 19126

BERGER, EDMOND LOUIS, b Salem, Mass, Dec 5, 39; m 64; c 3. ELEMENTARY PARTICLE, THEORETICAL PHYSICS. Educ: Mass Inst Technol, BS, 61; Princeton Univ, PhD(physics), 65. Prof Exp: Asst prof physics, Dartmouth Col, 65-68; res physicist, Lawrence Radiation Lab, 68-69; asst physicist, 69-70; assoc physicist, 70-76, HIGH ENERGY THEORY GROUP LEADER, ARGONNE NAT LAB, 74-, SR PHYSICIST, 76- Concurrent Pos: Vis scientist theory div, CERN, Geneva, Switz, 72-74 & 83-84; vis scientist, Theory Group, Stanford Linear Accelerator Ctr, 78-79; mem, Physics Adv Comt, Fermi Nat Accelerator Lab, 80-83, Dept Energy Revs High Energy Physics Progs, Brookhaven Nat Lab, 85, Stanford Linear Accelerator Ctr, 85, Lawrence Berkeley Lab, 86 & 89, Sci Prog Comt, Rencontres de Moriond, Les Arcs, France, 88-91, Nat Comt Particle Theory, 89-90 & High Energy Physics Adv Comt Subpanel on US High Energy Physics Res Prog, 90; fel comt chmn, Div Particles & Fields, Am Phys Soc, 90 & 91; chmn, Organizing Comt, 1990 Summer Study High Energy Physics, Res Directions for the Decade, Snowmass, Colo, 90. Mem: Fel Am Phys Soc. Res: Quantum chromodynamics; gauge theories; strong interactions of elementary particles; phenomenological models of quasi-two-body and multiparticle production; nuclear dependence of short distance processes. Mailing Add: High Energy Physics Bldg 362 Argonne Nat Lab Argonne IL 60439

BERGER, EDWARD ALAN, b Bronx, NY, Mar 25, 48. BIOCHEMISTRY, DEVELOPMENTAL BIOLOGY. *Educ:* City Col New York, BS, 68; Cornell Univ, PhD(biochem), 72. *Prof Exp:* Fel biochem, Cornell Univ, 72-73, neurobiol, Sch Med, Stanford Univ, 73-76 & cell & develop biol, Scripps Clin & Res Found & Univ Calif, San Diego, 76-77; staff scientist cell, Worcester Found Exp Biol, 77-; SCIENTIST, NIH. *Concurrent Pos:* Lt Joseph P Kennedy Jr fel neurosci, Sch Med, Stanford Univ, 73-75; staff scientist res grants, Am Cancer Soc, NIH, March Dimes Birth Defects Found, Worcester Found, 78-; mem, biomed sci study sect, NIH. *Honors & Awards:* Merck Award, NSF. *Mem:* Am Soc Cell Biol; Nat Soc Autistic Children; Am Soc Biol Chemists; AAAS; Soc Develop Biol. *Res:* Structure and function of biological membranes; cell surface interactions during development; molecular basis of cell-cell recognition; control of cell differentiation by cell-cell contact. *Mailing Add:* Berger Lab NIAID NIH Bldg 4 Rm 210 Bethesda MD 20842

BERGER, EDWARD MICHAEL, b New York, NY, May 2, 44; m 67; c 4. GENETICS. *Educ:* Hunter Col, BA, 65; Syracuse Univ, MS, 67, PhD(genetics), 69. *Prof Exp:* NIH fel develop biol, Harvard Univ, 69-70, Univ Chicago, 70-71 & State Univ NY Albany, 71-75; from asst prof to assoc prof, 75-84, PROF BIOL, DARTMOUTH COL, 84- *Concurrent Pos:* Prog dir cellbiol, NSF, 86-87. *Mem:* AAAS; Genetics Soc Am; Soc Zool; Am Soc Naturalists. *Res:* Molecular genetics. *Mailing Add:* Dept Biol Dartmouth Col Hanover NH 03755

BERGER, EUGENE Y, medicine, for more information see previous edition

BERGER, FRANK MILAN, b Pilsen, Czech, June 25, 13; nat US; c 2. MEDICAL RESEARCH, PHARMACOLOGY. *Educ:* Prague Univ, MD, 37. *Hon Degrees:* DSc, Philadelphia Col Pharm & Sci, 66. *Prof Exp:* Chief res, Monsall Hosp, Manchester, 41-43; chief bacteriologist, W Riding of Yorkshire, 43-45; head pharm & bact, British Drug House, 45-47; asst prof pediat, Univ Rochester, 47-49; dir res, Wallace Labs, 49-58, pres, 58-73; PROF PSYCHIAT & CO-DIR PSYCHOPHARMACOL DOCTORAL PROG, UNIV LOUISVILLE SCH MED, 74- *Concurrent Pos:* Mem adv coun biol, Princeton Univ, prof lectr, 69-76; chmn ad hoc study group clin pharmacol, Walter Reed Army Med Ctr, Washington, DC, 74-80; hon prof microbiol, Waksman Inst microbiol, Rutgers Univ, 82- *Mem:* Fel AAAS; fel NY Acad Sci; Am Soc Pharmacol & Exp Therapeut; Am Soc Microbiol; fel Royal Soc Med. *Res:* Muscle relaxants mephenesin, carisoprodol; tranquilizers; meprobamate; non-specific immunity, protodyne, adjuvants. *Mailing Add:* 190 E 72nd St New York NY 10021

BERGER, FRANKLIN GORDON, b Providence, RI, Sept 26, 47. BIOCHEMISTRY, MOLECULAR BIOLOGY. *Educ:* State Univ NY Buffalo, BA, 69; Purdue Univ, PhD(biochem), 74. *Prof Exp:* Res asst biochem, Purdue Univ, 69-74; res assoc, Cornell Univ, 74-76; res fel biochem, Roswell Park Mem Inst, 76-79, cancer res scientist, 79-; DEPT BIOL, UNIV SC. *Concurrent Pos:* NIH fel, Cornell Univ, 75-76. *Res:* Biochemical genetics; hormonal and developmental regulation of mammalian gene expression. *Mailing Add:* Dept Biol Univ SC Columbia SC 29208

BERGER, HAROLD, b New York, NY, Aug 9, 27; m 51; c 3. PARASITOLOGY. *Educ:* NY Univ, BA, 52, MS, 57, PhD(biol), 62. *Prof Exp:* Technician cancer res, Montefiore Hosp, NY, 52-53; biologist physiol, Pearl River, NY, 53-57, res parasitologist, Princeton, NJ, 57-75, SR RES PARASITOLOGIST, AM CYANAMID CO, 75- *Concurrent Pos:* NIH fel parasitol, Univ Mex, 62-63. *Mem:* Am Soc Parasitologists; Am Asn Vet Parasitologists. *Res:* Testing and development of compounds in domestic animals in order to discover and evaluate materials of possible utility in the treatment against parasitic helminth infection. *Mailing Add:* Am Cyanamid Co Agr Ctr PO Box 400 Princeton NJ 08540

BERGER, HAROLD, b Syracuse, NY, Oct 7, 26; m 52; c 5. NONDESTRUCTIVE TESTING, RADIOLOGICAL PHYSICS. *Educ:* Syracuse Univ, BS, 49, MS, 51. *Prof Exp:* Physicist x-ray dept advan develop lab, Gen Elec Co, 50-59; sr physicist solid state devices div, Battelle Mem Inst, 59-60; assoc physicist metall div, Argonne Nat Lab, 60-70, group leader, Nondestructive Testing Mat Sci Div, 65-73, sr physicist, 70-73; physicist, Reactor Radiation Div, Nat Bur Standards, 73-76, prog mgr nondestructive eval, 75-78, chief, Off Nondestructive Eval, 78-81; PRES, INDUST QUALITY, INC, 81- *Concurrent Pos:* Vis scientist, Ctr Nuclear Studies, Grenoble, 68-69; tech ed, Mat Eval, Am Soc Nondestructive Testing, 69-87, fel, 73, Mehl hon lectr, 75; chmn, Tech Comt 135, Subcomt 3, acoust methods, Int Standardization Orgn, 83-86. *Honors & Awards:* Achievement Award, Am Soc Nondestructive Test, 67, Gold Medal, 82; Radiation Indust Award, Am Nuclear Soc, 74; Silver Medal, US Dept Com, 79; Award of Merit, Am Soc Testing & Mat, 88. *Mem:* Am Phys Soc; hon mem Am Soc Nondestructive Testing; Sigma Xi; fel Am Nuclear Soc; fel Am Soc Testing & Mat; Am Soc Metals; Am Welding Soc. *Res:* X-ray and light detection techniques; ultrasonic imaging methods for nondestructive evaluation; x-radiography & proton radiography; neutron radiography; development of nondestructive techniques. *Mailing Add:* 9832 Canal Rd Gaithersburg MD 20879

BERGER, HARVEY J, b New York, NY, June 6, 50; m 76; c 2. NUCLEAR MEDICINE. *Educ:* Colgate Univ, AB, 72; Yale Univ, MD, 77. *Prof Exp:* Dir cardiovasc imaging, Yale-New Haven Hosp, 81-84; from asst prof to assoc prof radiol & med, Yale Univ, 83-84; prof radiol & med & dir, Div Nuclear Med, Emory Univ, 84-86; sr vpres, Med Affairs, 86-87, sr vpres, Res & Develop, 87-89, EXEC VPRES & PRES, RES & DEVELOP DIV, CENTOCOR, INC, 89- *Concurrent Pos:* Estab investr, Am Heart Asn, 81-84; mem, Nat Bd Trustees, Soc Nuclear Med, 83-85; mem task force nuclear cardiol, Int Soc & Fedn Cardiol & WHO, 83-84; NHLBI site visitor, NIH & Veteran's Admin Study Sects, 84-; US ed-in-chief, Nuclear Med Commun, 85-87; assoc ed, Am J Cardiac Imaging, 85-88; adj prof, Univ Pa, 86-; mem sci affairs & drug regulatory affairs comn, Indust Biotechnol Asn, 87-; mem bd dirs, Centocor Develop Corp, 89-; mem, US-Japan Dialogue on Working Environ for Res, Nat Res Coun, 90-, Panel Govt Role Civilian Technol, Comn on Sci, Eng & Public Policy, 90-; mem biotechnol adv comt, Univ Pa, 90- *Honors & Awards:* Mem Medal Award, Asn Univ Radiologists, 79; Marc Tetalman Award, Soc Nuclear Med, 81; Cline Fixott Mem Lectr, Am Acad Dent Radiol, 84. *Mem:* Fel Am Col Cardiol; fel Am Col Physicians; Soc Nuclear Med; Soc Exp Biol & Med; Fel Am Heart Asn Coun Cardiovasc Radiol & Circulation; Am Physiol Soc; NY Acad Sci; AAAS; Sigma Xi; Am Fedn Clin Res. *Res:* Cardiovascular biology; biotechnology; clinical trial design and clinical epidemiology; development and validation of cardiovascular imaging techniques; discovery and development of monoclonal antibodies, proteins and small molecules as pharmaceutical products; technology transfer between academia, government and the private sector. *Mailing Add:* Exec Vpres Centocor Inc 244 Great Valley Pkwy Malvern PA 19355

BERGER, JACQUES, b New York, NY, Apr 14, 34; m 54. PROTOZOOLOGY, HISTORY OF BIOLOGY & ECOTOXICOLOGY. *Educ:* Pa State Univ, BSc, 55; Univ Ill, MS, 58, PhD(zool), 64. *Prof Exp:* Instr zool, Univ Ill, 62-63 & Duke Univ, 63-64; asst prof, NC State Univ, 64-65; from asst prof to assoc prof, 65-78, PROF ZOOL, UNIV TORONTO, 78- *Concurrent Pos:* NSF fel marine biol, Duke Univ, 64-65; sr fel, Massey Col, Univ Toronto, 73-, assoc, Inst Hist Sci, Univ Col, res assoc, Royal Ontario Mus. *Mem:* Soc Protozool; Am Micros Soc; Asn Fr Speaking Protistologists. *Res:* Multivariate morphometrics of ciliate protozoa; behavioral ecology and ethology of ciliate protozoa; petroleum microbiology; effects of pollutants on protozoa; history of microscopy. *Mailing Add:* Dept Zool Univ Toronto Toronto ON M5S 1A1 Can

BERGER, JAMES DENNIS, b Spokane, Wash, July 16, 42; m 66; c 2. CELL BIOLOGY. *Educ:* Ind Univ, AB, 64, MA, 65, PhD(zool), 69. *Prof Exp:* ASSOC PROF ZOOL, UNIV BC, 70- *Mem:* Soc Protozoologists; Can Soc Cell Biol. *Res:* Genetic and physiological basis of regulation of the cell cycle and the sexual pathway in Paramecium. *Mailing Add:* Dept Zool Univ BC 2075 Wesbrook Pl Vancouver BC V6T 1W5 Can

BERGER, JAMES EDWARD, b Pinson Fork, Ky, Sept 10, 35; m 56; c 4. PHARMACOLOGY. *Educ:* Univ Cincinnati, BS, 59, MS, 61; Univ Fla, PhD(pharmacol), 67. *Prof Exp:* Res asst bact, Surg Bact Res Lab, Cincinnati Gen Hosp, Ohio, 60-62, res assoc, 62-64; from asst prof to assoc prof, 67-80, PROF PHARMACOL, COL PHARM, BUTLER UNIV, 80- *Concurrent Pos:* Mead-Johnson grant, 70-71; consult pharmacologist, Vet Admin Hosp, 69-73, Wishard Mem Hosp, 71-75 & St Vincent Hosp, 75-; consult several legal firms in Indiana. *Honors & Awards:* Meritorious Serv Medal, Am Heart Asn, 86. *Mem:* Sigma Xi; Am Heart Asn; AAAS; Am Pharmaceut Asn; Am Hosp Asn. *Res:* Staphylococcal toxin purification and production; action of quinidine on the heart as related to alpha and beta adrenergic receptors in the tissue; calcium antagonist in smooth and myocardial tissue; morphine addiction and alpha and beta adrenergic antagonist. *Mailing Add:* Dept Pharmacol Butler Univ Indianapolis IN 46208

BERGER, JAY MANTON, b New York, NY, Oct 29, 27; m 48; c 3. THEORETICAL PHYSICS, COMPUTER SCIENCE. *Educ:* Univ Mich, BS, 47; Columbia Univ, AM, 48; Case Inst Technol, PhD(physics), 53. *Prof Exp:* Instr physics, Fenn Col, 48-51; res assoc, Princeton Univ, 52-56; sr engr, Adv Systs Develop Div, 56-71, prod assurance adv, Data Processing Group, 71-72, adv litigation anal, 72-80, PROG MGR, DATA PROCESSING GROUP, IBM CORP, 81- *Concurrent Pos:* Consult, Radiation Lab, Univ Calif, 54 & Lockheed Missile Systems, 55; vis prof math, physics, & comput sci, Savannah State Col, 80-81. *Mem:* Am Phys Soc. *Res:* Noise and information theory; applied mathematics; theoretical nuclear physics; plasma physics; astrophysics; field theory; systems design. *Mailing Add:* 57 Fuller Rd Briarcliff Manor NY 10510

BERGER, JERRY EUGENE, b Jefferson City, Mo, Aug 8, 33; m 54; c 4. ATMOSPHERIC CHEMISTRY. *Educ:* Westminster Col, BA, 55; Univ Ky, MS, 57, PhD(phys chem), 59; Univ Houston, JD, 78. *Prof Exp:* Res chemist, 59-65, sr res chemist, 65-67, group leader, 66-67, sr chemist, 67-70, spec analyst, 70-75, sr staff chemist, 75-77, mgr health, safety & environ support, 77-80, WASHINGTON REP, SHELL OIL CO, 81- *Mem:* Air Pollution Control Asn; Am Chem Soc; Soc Automotive Engrs. *Res:* Environmental conservation. *Mailing Add:* RR 66 Box 73 Gamaliel AR 72537

BERGER, JOEL GILBERT, b Brooklyn, NY, June 2, 37; m 62; c 1. MEDICINAL CHEMISTRY. *Educ:* City Col New York, BS, 57; NY Univ, MS, 60, PhD(org chem), 62. *Prof Exp:* Chemist, Esso Res & Eng Co, 62-64 & Wallace & Tiernan, Inc, 64-65; res assoc, Princeton Univ, 65-66; sr med chemist, Endo Labs Inc, 66-73, group leader, 73-76; SECT HEAD CENT NERVOUS SYST, SCHERING CORP, 76- *Mem:* Am Chem Soc; NY Acad Sci. *Res:* Reaction mechanisms; chemistry of pyrroles and indoles; synthesis of drugs affecting the central nervous system. *Mailing Add:* Schering Corp 60 Orange St Bloomfield NJ 07003-4799

BERGER, JOSEF, b Hradec Kralove, Czech, Oct 30, 40; US citizen; m 63; c 2. PHYSICAL ELECTRONICS. *Educ:* Czech Inst Technol, Prague, MSc, 63; Czech Acad Sci, PhD(phys electronics), 68. *Prof Exp:* Proj leader phys electronics, Inst Radio Eng & Electronics, Czech Acad Sci, 63-68; res assoc silicon integrated circuits, Stanford Univ, 68-72; mem tech staff, Solid State Lab, 72-74, lab proj mgr, 74-75, RES & DEVELOP PROG MGR, SILICON INTEGRATED CIRCUIT LAB, HEWLETT-PACKARD CO, 77- *Mem:* Sr mem Inst Elec & Electronics Engrs; Sigma Xi. *Res:* Integrated circuits for biomedical applications; metal-oxide-silicon integrated circuits; charge coupled devices; semiconductor memories; silicon integrated circuit technology. *Mailing Add:* 1460 Topar Ave Los Altos CA 94022

BERGER, KENNETH WALTER, b Evansville, Ind, Mar 22, 24; m 46; c 4. AUDIOLOGY. *Educ:* Evansville Col, BA, 49; Ind State Univ, MA, 50; Southern Ill Univ, MS, 60, PhD, 62. *Prof Exp:* Speech & hearing therapist, Carmi Pub Schs, Ill, 55-61; asst audiol, Southern Ill Univ, 61-62; MEM FAC

& DIR AUDIOL, KENT STATE UNIV, 62- *Mem:* Am Acad Audiol; fel Am Speech & Hearing Asn. *Res:* Hearing aids; hearing loss in children; audiometric testing of discrimination for speech. *Mailing Add:* Dept Audiol Kent State Univ Kent OH 44242

BERGER, LAWRENCE, b New York, NY, Apr 6, 26; m 50; c 4. MEDICINE, NEPHROLOGY. *Educ:* NY Univ, BA, 47; Chicago Med Sch, MD & MB, 51; Am Bd Internal Med, dipl in Nephrology. *Prof Exp:* Intern, Mt Sinai Hosp, 51-52, resident med, 54-56; resident, Montefiore Hosp, 52-53; NIH res fel, Med Col, Cornell Univ, 53-54; asst, Goldwater Mem Hosp, 56-57, asst attend, 56-58; res fel, 57-69; clin asst physician, 57-63, asst attend physician, 63-72, res assoc med, 69-79, assoc attend, 72-79, assoc clin prof med, 72-79, ATTEND PHYSICIAN, MT SINAI HOSP, 79-, CLIN PROF MED, 79- *Mem:* AAAS; AMA; Am Fedn Clin Res; fel Am Col Physicians; fel NY Acad Med; Am Soc Nephrology. *Res:* Renal function; physiology; renal uric acid excretion. *Mailing Add:* 119 E 84th St New York NY 10028

BERGER, LESLIE RALPH, b London, Eng, Dec 18, 28; nat citizen; m 54; c 2. MICROBIAL PHYSIOLOGY, ENVIRONMENTAL SCIENCES. *Educ:* Univ Cincinnati, BS, 50; Univ Wash, MS, 53; Univ Calif, PhD(microbiol), 57. *Prof Exp:* Res microbiologist, Scripps Inst Oceanog, 57-59; asst biochemist, Univ Calif, Davis, 59-60; from asst prof to assoc prof microbiol, 60-69, chmn, 78-85, PROF MICROBIOL, UNIV HAWAII, 69- *Mem:* Am Chem Soc; Am Soc Microbiol; Am Soc Gen Microbiol. *Res:* Physiology of growth at increased hydrostatic pressure; cytology and physiology of photosynthetic organisms at increased hydrostatic pressure; kinetics of allosteric enzymes at extreme physical environments; corrosion of heat exchangers in nuclear reactors and OTEC systems. *Mailing Add:* Dept Microbiol 2538 Mall Univ Hawaii Honolulu HI 96822

BERGER, LUC, b Morges, Switz, May 2, 33; m. MAGNETISM. *Educ:* Univ Lausanne, BS, 55, PhD(physics), 60. *Prof Exp:* Swiss Nat Sci Found fel physics, 60-61, from instr to assoc prof, 61-74, PROF PHYSICS, CARNEGIE-MELLON UNIV, 74- *Mem:* Am Phys Soc; Inst Elec & Electronics Engrs. *Res:* Electronic structure of transition metals; ferromagnetism; transport processes in ferromagnets. *Mailing Add:* Dept Physics Carnegie-Mellon Univ 5000 Forbes Ave Pittsburgh PA 15213

BERGER, MARTIN, b New York, NY, May 23, 26; m 47; c 3. PHYSICS. *Educ:* Columbia Univ, BS, 49. *Prof Exp:* Physicist, Tire Div, US Rubber Co, 50-55; engr, Res Labs, Chrysler Corp, 55-56; physicist, Chem Res Div, Esso Res & Eng Co, 56-60, sr physicist, Ctr Basic Res, 60-62, res assoc polymers, 62-64, sect head elastomer res, Enjay Polymer Lab, Exxon Res & Eng Co, 64-68, dir phys & eng sci, Corp Res Lab, 68-75, dir, Govt Res Labs, 75-76; PRES, OCCIDENTAL RES CORP & SR VPRES RES & DEVELOP, OCCIDENTAL PETROL CORP, 77- *Res:* Polymer structure and properties, their relationship and methods of changing structure to improve properties; separations, electrochemistry and materials. *Mailing Add:* Invention Submission Corp 903 Liberty Ave Pittsburgh PA 15222

BERGER, MARTIN JACOB, b Vienna, Austria, July 12, 22; nat US; c 3. MATHEMATICAL PHYSICS. *Educ:* Univ Chicago, PhD, 51. *Prof Exp:* Fel statist, Univ Chicago, 51-52; PHYSICIST, NAT BUR STAND, 52- *Concurrent Pos:* Mem comt nuclear sci, Nat Acad Sci-Nat Res Coun. *Mem:* Am Phys Soc; Radiation Res Soc. *Res:* Radiation transport theory; Monte Carlo methods; atmospheric and auroral physics; nuclear medicine; shielding. *Mailing Add:* Nat Bur Stand Ctr Radiation Res Washington DC 20234

BERGER, MELVIN, b Philadelphia, Pa, Mar 7, 50; m 71; c 2. IMMUNOLOGY. *Educ:* Western Reserve Univ, BA, 70; Case Western Reserve Univ, MD & PhD(biochem), 76. *Prof Exp:* Resident pediat, Childrens Hosp Med Ctr, Boston, 76-78; clin assoc allergy & immunol, Nat Inst Allergy & Infectious Dis, NIH, 78-81; asst chief, Allergy-Immunol Serv, Walter Reed Army Med Ctr, Washington, DC, 81-84; chief, Allergy-Immunol Div, Dept Pediat, 84-87, CHIEF, SECT IMMUNOL & INFECTIOUS DIS, DEPT PEDIAT, SCH MED, CASE WESTERN RESERVE UNIV, 87- *Concurrent Pos:* Clin fel pediat, Harvard Med Sch, 76-78; asst prof pediat & med, Uniformed Serv Univ of Health Sci, 81-84; clin care consult, Nat Inst Health Clin Ctr, 81-84; assoc prof pediat & path, Sch Med, Case Western Reserve Univ, 84- *Mem:* Soc Pediat Res; Am Acad Allergy & Immunol; Am Asn Immunologists; Am Acad Pediat. *Res:* Host defense mechanisms and their defects; role of infection and the immune response in cystic fibrosis. *Mailing Add:* Dept Pediat Case Western Reserve Univ Sch Med Rainbow Babies & Children's Hosp 2101 Adelbert Rd Cleveland OH 44106

BERGER, MELVYN STUART, b Brooklyn, NY, Aug 23, 39; m 77; c 2. MATHEMATICS. *Educ:* Univ Toronto, BA, 61; Yale Univ, MA, 63, PhD(math), 64. *Prof Exp:* From asst prof to assoc prof math, Univ Minn, Minneapolis, 64-69; from assoc prof to prof math, Belfer Grad Sch, Yeshiva Univ, 69-78; dir, Ctr Appl Math & Math Sci, 80-82 & 85, PROF MATH, UNIV MASS, AMHERST, 78-, DIR CTR APPL MATH, NONLINEAR SCI & MATH COMPUTATION, 89- *Concurrent Pos:* Vis mem, Courant Inst, 66-68; vis mem, Inst Advan Study, Princeton, 72-73, mem, 77-79; invited mem, Univ Paris, 81, 83 & 84, Princeton Univ, 83, Mass Inst Technol, 85, Oxford Univ, Eng, 85. *Mem:* Am Math Soc; Soc Indust & Appl Math; AAAS. *Res:* Nonlinear analysis; partial differential equations, mathematical physics and differential geometry; functional analysis; applied mathematics; nonlinear aspects of high technology; computer graphics. *Mailing Add:* Dept Math Univ Mass Amherst MA 01002

BERGER, MITCHELL HARVEY, b Miami Beach, Fla, Mar 19, 49; m 72; c 1. ORGANIC CHEMISTRY, POLYMER CHEMISTRY. *Educ:* City Col New York, BS, 70; Univ Rochester, MS, 73, PhD(chem), 75. *Prof Exp:* Res assoc org chem, Cornell Univ, 74-76; chemist, Chem & Plastics Div, Union Carbide Corp, 76-80; proj scientist, Coatings Mat Div, 80-81, SR CHEMIST, COATED FILMS DIV, SEALED AIR CORP, 81- *Mem:* Am Chem Soc; Sigma Xi. *Mailing Add:* 24970 White Plains Rd Novi MI 48050-5654

BERGER, NEIL EVERETT, b New York, NY, Oct 8, 42; div; c 1. APPLIED MATHEMATICS. *Educ:* Columbia Univ, BS, 63; NY Univ, MS, 65, PhD(appl math), 68. *Prof Exp:* Asst prof, 68-75, ASSOC PROF MATH, UNIV ILL CHICAGO, 75- *Honors & Awards:* Monroe Martin Prize, Inst Fluid Dynamics & Appl Math, Univ Md, 74. *Mem:* Soc Indust & Appl Math; AAAS; Am Math Soc; Math Asn Am. *Res:* Partial differential equations of elasticity; fluid dynamics; asymptotic expansions. *Mailing Add:* Dept Math Univ Ill Chicago Box 4348 Chicago IL 60680

BERGER, PAUL RAYMOND, b Lafayette, Ind, May 8, 63; m 89. OPTOELECTRONICS. *Educ:* Univ Mich, Ann Arbor, BSE, 85, MSE, 87, PhD(elec eng), 90. *Prof Exp:* POST MEM TECH STAFF, AT&T BELL LABS, 90- *Mem:* Inst Elec & Electronics Engrs; Optical Soc Am. *Res:* III-IV semiconductor growth as well as heterostructure optical and electronic devices and their monolithic integration; author of various publications. *Mailing Add:* AT&T Bell Labs Rm 6E-401 600 Mountain Ave Murray Hill NJ 07974

BERGER, PHILIP JEFFREY, b Newark, NJ, June 28, 43; m 65; c 2. ANIMAL BREEDING, STATISTICAL ANALYSIS. *Educ:* Del Valley Col, BS, 65; Ohio State Univ, MS, 67, PhD(animal breeding), 70. *Prof Exp:* Res & teaching asst animal breeding, Ohio State Univ, 65-70, fel, 71; asst prof, 72-78, assoc mem grad fac, 74-79, assoc prof animal breeding, 78-81, MEM GRAD FAC, IOWA STATE UNIV, 79-, PROF ANIMAL SCI, 82- *Mem:* Am Soc Animal Sci; Am Dairy Sci Asn; Biomet Soc; Sigma Xi. *Res:* Development of computing procedures for best linear unbiased prediction of breeding values, predictability of correlated responses in breeding populations, economic evaluation of productive traits and recovery of selection response. *Mailing Add:* 2518 Kellogg Ames IA 50010

BERGER, PHYLLIS BELOUS, fluid dynamics, thermodynamics, for more information see previous edition

BERGER, RALPH JACOB, b Vienna, Austria, Sept 16, 37; m 63; c 2. SLEEP THERMOREGULATION. *Educ:* Cambridge Univ, BA, 60, MA, 64; Univ Edinburgh, PhD(psychol), 63. *Prof Exp:* Res asst, Dept Psychol Med, Univ Edinburgh, 60-63; res psychologist, Lab Paranatal Physiol, Nat Inst Neurol Dis & Blindness, NIH, San Juan, PR, 63-65; asst res anatomist, Univ Calif, Los Angeles, 65-68; asst prof psychol, 68-69, assoc prof, 69-74, PROF BIOL & PSYCHOL, UNIV CALIF, SANTA CRUZ, 74- *Concurrent Pos:* Vis asst prof, Univ PR, 63-64. *Mem:* AAAS; Soc Neurosci; Sleep Res Soc; Europ Sleep Res Soc. *Res:* Sleep; thermoregulation; metabolism; hibernation; epistemological implications of neurobiological research. *Mailing Add:* Thimann Labs Univ Calif Santa Cruz CA 95064

BERGER, RICHARD DONALD, b Macungie, Pa, Jan 15, 34; m 62; c 2. PLANT PATHOLOGY. *Educ:* Kutztown State Col, 55; Univ Wis, PhD(plant path), 62. *Prof Exp:* Res fel, Univ Wis, 62-63; plant pathologist, Pa State Univ, 63-66; PROF PLANT PATH, UNIV FLA, 66- *Concurrent Pos:* Assoc ed, J Am Phytopath Soc, 76-; res award, Fla Fruit & Veg Asn, 72. *Honors & Awards:* Campbell Award, Am Phytopath Soc & Campbell Soup Co, 74. *Mem:* Fel Am Phytopath Soc. *Res:* Epidemiology of vegetable and field crop diseases; Cercospora taxonomy; disease modeling; computer simulation of epidemics. *Mailing Add:* IFAS Univ Fla Gainesville FL 32611-0513

BERGER, RICHARD LEE, b Belleville, Ill, July 31, 35. CIVIL ENGINEERING, CERAMICS ENGINEERING. *Educ:* Univ Ill, BS, 58, MS, 60, PhD(geol), 65. *Prof Exp:* Res scientist, Am Cement Corp, 62-70; assoc prof, 71-74, PROF CIVIL & CERAMIC ENG, UNIV ILL, URBANA, 74- *Concurrent Pos:* vis prof, Tech Inst, Denmark, 76; vis scientist, Oak Ridge Nat Lab, 83-84. *Honors & Awards:* Bates Award, Am Soc Testing & Mat, 74. *Mem:* Fel Am Ceramic Soc; Sigma Xi; Am Concrete Inst. *Res:* Influence of microstructure on the engineering performance of construction materials. *Mailing Add:* Dept Civil Eng Univ Ill Urbana IL 61801

BERGER, RICHARD LEO, b Salem, Mass, May 13, 44; m 70. PLASMA PHYSICS. *Educ:* Princeton Univ, AB, 66; Univ Md, PhD(physics), 72. *Prof Exp:* Res assoc physics, Dept Astro-Geophys, Univ Colo, 72-74; res assoc plasma physics, Plasma Physics Lab, Princeton Univ, 74-77; RES SCIENTIST PLASMA PHYSICS, KMS FUSION INC, 77- *Mem:* Am Phys Soc. *Res:* Theoretical plasma physics: parametric instabilities; large-amplitude nonlinear waves; plasma kinetic theory. *Mailing Add:* KMS Fusion Inc PO Box 1567 Ann Arbor MI 48106

BERGER, RICHARD S, b Brooklyn, NY, Apr 28, 29; m 54; c 2. ORGANIC POLYMER CHEMISTRY. *Educ:* Stanford Univ, BS, 50; Univ Wis, PhD(org chem), 54. *Prof Exp:* Fel, Univ Minn, 54; chemist, Shell Develop Co, 56-65; mgr fiber chem br, Res Div, Phillips Petrol Co, 65-70, dir res projs, 70-73, DIR STRATEGIC PLANNING, PHILLIPS FIBERS CORP, 73- *Res:* Synthetic polymer chemistry; polymer and fiber stabilization; chemical modification of polyamide, polyester and polyolefin polymers and fibers. *Mailing Add:* 344 Pimlico Rd Greenville SC 29607

BERGER, ROBERT, b New York, NY, July 10, 38; m 67; c 3. ELECTRICAL ENGINEERING. *Educ:* Rensselaer Polytech Inst, BEE, 60; Harvard Univ, MS, 61, PhD(appl math), 65. *Prof Exp:* STAFF MEM, LINCOLN LAB, MASS INST TECHNOL, 64- *Mem:* Inst Elec & Electronics Engrs. *Res:* Wafer-scale integrated circuits; computer architecture. *Mailing Add:* Lincoln Lab Rm B-143 Mass Inst of Technol 244 Wood St PO Box 73 Lexington MA 02173

BERGER, ROBERT ELLIOTT, b Baltimore, Md, July 22, 47; m 69; c 2. TECHNOLOGICAL INNOVATION ASSESSMENT. *Educ:* Case Western Reserve Univ, BS, 68; Johns Hopkins Univ, PhD(biomech), 73. *Prof Exp:* Assoc, US Nat Bur Standards, 73-74; res assoc, Optical Mfrs Asn, 74-75; mech engr, Prod Safety Technol Div, 75-79, prog analyst off dir, 79-81, gen phys scientist, Ctr Chem Eng, 81-82, dir tech assessment, Off Naval Res, 82-

83, DIR INNOVATION & TECHNOL DIV, SMALL BUS ADMIN, 83- *Mem:* Am Soc Mech Engrs; Am Soc Testing & Mat; Am Nat Standards Inst. *Res:* Measurements and modelling of solid fluid flows with industrial applications; identification of injury mechanisms associated with products; development of human injury tolerance criteria and product safety test methods. *Mailing Add:* 7013 Woodscape Dr Clarksville MD 21029

BERGER, ROBERT LEWIS, b Omaha, Nebr, Sept 2, 25; m 50, 81; c 4. THERMOMETRY, INSTRUMENTATION. *Educ:* Colo Agr & Mech Col, BS, 50; Pa State Univ, MS, 53, PhD(physics), 56. *Prof Exp:* Instr physics, Park Col, 50-51; fel, Brit-Am Exchange, Am Cancer Soc, Cambridge Univ, 56; asst prof physics, Utah State Univ, 57-60, assoc prof, 60-62; mem staff, Lab Tech Develop, 67-77, chief, Sect Biophys Instrumentation, Lab Tech Develop, 77-89, MEM STAFF, LAB BIOPHYS CHEM, NAT HEART, LUNG & BLOOD INST, 89- *Concurrent Pos:* Vis scientist, Dept Chem, Univ Calif, San Diego, 69-71; ed, Nat Bur Stand publ, 70-72. *Mem:* Fel AAAS; fel Am Phys Soc; Soc Gen Physiol; Biophys Soc; Am Soc Biochem & Molecular Biol; Sigma Xi. *Res:* Enzyme kinetics and mechanisms; spectroscopy, fluid dynamics and computers; microcalorimetry; dye laser spectroscopy; Mossbauer spectroscopy; near ir and roman spectroscopy. *Mailing Add:* Nat Heart Lung & Blood Inst Bldg 3 Rm B1-03 Bethesda MD 20892

BERGER, ROBERT S, b Tours, Tex, Jan 2, 33; m 59; c 4. INSECT TOXICOLOGY. *Educ:* Tex A&M Univ, BS, 54, MS, 57; Cornell Univ, PhD(biochem), 61. *Prof Exp:* Entomologist agr res serv, USDA, 61-63; assoc prof insect toxicol, 63-70, PROF ZOOL & ENTOM, AUBURN UNIV, 70- *Mem:* AAAS; Entom Soc Am; Am Chem Soc. *Res:* Insecticide chemistry; insect attractants. *Mailing Add:* Dept Entom Auburn Univ Auburn AL 36849-5413

BERGER, S EDMUND, b Osijek, Yugoslavia, Nov 13, 22; nat US; m 47; c 2. ORGANIC CHEMISTRY. *Educ:* Univ Rome, ChD(org chem), 48. *Prof Exp:* Asst org chem, Univ Rome, 48-49; res chemist, Dept Legal Med, Harvard Med Sch, 49-51; res chemist, 51-63, res supvr, 63-80, group leader, 80-82, tech supvr, Allied Corp, 82-87; RETIRED. *Mem:* Am Chem Soc; Am Translr Asn. *Res:* Polyurethane chemistry; polyesters; diisocyanates; food acidulants; surfactants; liquid ion exchange; exploratory organic synthesis; catalysis; fluorocarbon solvents; precision cleaning. *Mailing Add:* 298 Grayton Rd Tonawanda NY 14150

BERGER, SELMAN A, b Brooklyn, NY, Aug 31, 42; m 67; c 2. ANALYTICAL CHEMISTRY, ENVIRONMENTAL CHEMISTRY. *Educ:* Brooklyn Col, BS, 64; Univ Conn, MS, 67, PhD(analytical chem), 69. *Prof Exp:* Teaching asst chem, Univ Conn, 64-69; postdoctoral fel, Dalhousie Univ, Halifax, NS, 69-70; sr chemist, Toxicol Lab, Bur Labs, NYC Dept Health, 70-71; from asst prof to assoc prof, 71-80, dep dept chmn, 72-82, PROF CHEM, JOHN JAY COL, CITY UNIV NEW YORK, 81-, DEPT CHMN, 82- *Concurrent Pos:* Doctoral fac chem, City Univ New York, Grad Ctr, 73- *Mem:* Am Chem Soc; Sigma Xi. *Res:* Spectroscopic methods for trace analysis; flame methods; complexation and solvent extraction methods; mathematical modeling of solvent extraction equilibria. *Mailing Add:* Dept Sci John Jay Col Crim Just City Univ NY 445 W 59th St New York NY 10019

BERGER, SHELBY LOUISE, b New York, NY, Jan 5, 41; m 73. BIOCHEMISTRY, CELL & MOLECULAR BIOLOGY. *Educ:* Bryn Mawr Col, AB, 62; Harvard Univ, PhD(biophys), 68. *Prof Exp:* Fel biochem, Nat Inst Arthritis & Metab Dis, 68-70, staff fel, 70-71, sr staff fel biochem & cell biol, Nat Inst Dent Res, 72-74 & Nat Cancer Inst, 74-76, RES CHEMIST BIOCHEM & CELL BIOL, NAT CANCER INST, NIH, 76- *Mem:* Am Soc Biol Chemists. *Res:* Control of the metabolism of messenger ribonucleic acid and heterogeneous nuclear ribonucleic acid in physiologically resting, growing and malignant human lymphocytes; interferon. *Mailing Add:* Nat Cancer Inst NIH Bldg 10 Rm B1B38 Bethesda MD 20892

BERGER, SHELDON, b Chicago, Ill, Nov 12, 28; m 50; c 4. MEDICINE, ENDOCRINOLOGY. *Educ:* Univ Ill, BA, 49, BS, 51, MD, 53. *Prof Exp:* Physician in charge, Radioisotope Lab, 59-63, asst dir dept metab & endocrine res, 60-63; assoc dir dept res & educ, Evanston Hosp, 63-65, coordr res labs, 65-69; from asst prof to assoc prof med, Northwestern Univ, Chicago, 65-76; ASSOC PROF MED, NORTHWESTERN UNIV, CHICAGO, 76- *Concurrent Pos:* NIH trainee metab & endocrine, Michael Reese Hosp, Chicago, 58-59; NIH trainee thyroid res methods, Beth Israel Hosp, Boston, 59. *Mem:* Am Col Physicians; Endocrine Soc. *Res:* Diabetes; general metabolism; thyroid physiology; internal medicine. *Mailing Add:* Northwestern Univ Med Sch Northwestern Mem Hosp 211 E Chicago Ave Suite 930 Chicago IL 60611

BERGER, STANLEY A(LLAN), b Brooklyn, NY, Aug 9, 34; c 2. FLUID MECHANICS. *Educ:* Brooklyn Col, BS, 55; Brown Univ, PhD(appl math), 59. *Prof Exp:* Res assoc aeronaut eng, Princeton Univ, 59-60; lectr aeronaut sci, 61, from asst prof to assoc prof, 61-70, PROF ENG SCI, UNIV CALIF, BERKELEY, 70- *Concurrent Pos:* Consult, Rand Corp, Lockheed Space & Missile Co, Sci Applications, Inc, Flow Res, Inc, IBM & Alcoa Res Lab. *Mem:* Fel Am Physics Soc; Soc Indust & Appl Math; fel Am Soc Mech Engrs; fel AAAS; assoc fel Am Inst Aeronaut & Astronaut. *Res:* Fluid mechanics; explosions; wake flows; viscous flows; physiological fluid mechanics; vortex breakdown; flow in curved tubes; numerical methods in fluid mechanics. *Mailing Add:* Dept Mech Eng Univ Calif Berkeley CA 94720

BERGER, STEVEN BARRY, b New York, NY, Dec 29, 46. APPLIED MATHEMATICS. *Educ:* Mass Inst Technol, SB, 67, PhD(physics), 73. *Prof Exp:* Res assoc physics, Boston Univ, 73-74; fel physics, Mass Inst Technol, 74-78; mem staff,TRW Systs Group, 78-80; mem staff, Dept Elec Eng & Computer Sci, George Washington Univ, 80-; DIR & PROF COMPUTER SCI & MATH DEPT, LEWIS UNIV. *Concurrent Pos:* Fight-for-Sight fel, NY, 74-76; asst ed, Am J Physics, 75-78; sr scientist, Itek Corp, 76-77. *Mem:* Sigma Xi; Am Phys Soc; Am Optical Soc; NY Acad Sci; Math Asn Am. *Mailing Add:* Computer Sci Dept Lewis Univ Rte 53 Romeoville IL 60441

BERGER, TOBY, b Sept 4, 40; US citizen; m 61; c 2. INFORMATION THEORY, COMMUNICATIONS. *Educ:* Yale Univ, BE, 62; Harvard Univ, MS, 64, PhD(appl math), 66. *Prof Exp:* Assoc scientist, Raytheon Co, 62-66, sr scientist, 66-68; from asst prof to assoc prof, 68-77, PROF ELEC ENG, CORNELL UNIV, 77- *Concurrent Pos:* Consult, Raytheon Co, 68-, IBM Fed Systs Div, 75-, Schlumberger, 81-, Teknekron Commun Systs, 86-, AT&T Bell Labs, 87-; Guggenheim fel, 75-76, fel Japan Soc Prom Sci, 80-81, fel Ministry Educ People's Republic of China, 81-; ed-in-chief, Inst Elec & Electronics Engrs Trans Info Theory, 87- *Honors & Awards:* Frederick E Torman Award, ASEE, 82. *Mem:* AAAS; Sigma Xi; Inst Elec & Electronics Engrs. *Res:* Information theory and communication theory, especially source encoding; processing of analog and digital radar data. *Mailing Add:* 422 Highland Rd Ithaca NY 14850

BERGER, WILLIAM J, b Arnold, Pa, Nov 10, 21; wid. MATHEMATICS, COMPUTER SCIENCE. *Educ:* Carnegie Inst Technol, BS, 50, MS, 51, PhD(math), 54. *Prof Exp:* Mathematician & engr, Radio Corp Am, Fla, 55; mathematician, Hastings-Raydist Corp, Va, 56-57; res engr & mathematician, Convair-Astronaut, Gen Dynamics Corp, Calif, 57-58; res mathematician, Lockheed Aircraft Corp, 58-60; res engr, Aeronutronic Div, Ford Motor Co, 60; consult mathematician, Gen Elec Co, Va, 60-65; assoc prof math, Howard Univ, 65-69; prof math, DC Teachers Col, 69-77; prof math, Univ DC, 77-88; RETIRED. *Mem:* Math Asn Am; Nat Coun Teachers Math; Sigma Xi. *Res:* Scientific computing; astronautics; functional analysis; geometry. *Mailing Add:* PO Box 1334 Rockville MD 20849-1334

BERGER, WOLFGANG HELMUT, b Erlangen, Ger, 1937; m 66; c 2. MICROPALEONTOLOGY, OCEAN HISTORY. *Educ:* Univ Erlangen, cand geol, 61; Univ Colo, Boulder, MSc, 63; Univ Calif, San Diego, PhD(oceanog), 68. *Prof Exp:* Asst res oceanogr, Scripps Inst Oceanog, 68-70; asst geol, Univ Kiel, 70-71; from asst prof to assoc prof, 71-81, PROF OCEANOG, SCRIPPS INST OCEANOG, 81- *Concurrent Pos:* Consult, Oceanog Div, NSF, 73-75 & Deep Sea Drilling Proj, Joint Oceanog Inst Deep Earth Sampling, 74-; guest prof, Univ Kiel, 77 & 80. *Honors & Awards:* Bigelow Medal, Woods Hole Oceanog Inst, 79; Huntsman Medal, Bedford, 84; Humboldt Award, Bonn, 86; Ewing Medal, Am Geophys Union, 89. *Mem:* AAAS; Geol Soc Am; Am Geophys Union. *Res:* Ecology and paleoecology of planktonic foraminifera; biogenous sedimentation in the deep ocean; paleo-oceanography; history of the carbon cycle and climate. *Mailing Add:* Scripps Inst Oceanog Univ Calif San Diego La Jolla CA 92093

BERGERON, CLIFTON GEORGE, b Los Angeles, Calif, Jan 5, 25; m 50; c 2. CERAMIC ENGINEERING. *Educ:* Univ Ill, BS, 50, MS, 59, PhD(ceramic eng), 61. *Prof Exp:* Ceramic res engr, A O Smith Corp, 50-53, sr res engr, 53-55; staff engr res & develop labs, Whirlpool Corp, 55-57; res assoc ceramic eng, Univ Ill, 57-61, from asst prof to prof ceramic eng, 61-88, head dept, 79-86, EMER PROF, UNIV ILL, 88- *Honors & Awards:* Friedberg Lectr, Nat Inst Ceramic Engrs, 85; Outstanding, Educr Award, Am Ceramic Soc, 87. *Mem:* AAAS; fel Am Ceramic Soc; Nat Inst Ceramic Engrs; Am Soc Testing & Mat; Am Soc Eng Educ; Am Asn Univ Prof. *Res:* Nucleation and crystal growth in glasses; dielectric properties of ceramic coating; high temperature chemical reactions and coatings for metals; electrical conduction in ceramic materials. *Mailing Add:* 208 W Michigan Urbana IL 61801

BERGERON, CLYDE J, JR, b New Orleans, La, July 2, 32; m 56; c 3. PHYSICS. *Educ:* Loyola Univ, BS, 55; La State Univ, PhD, 59. *Prof Exp:* Res asst, Loyola Univ, 54-55; asst, 55-60, from asst prof to assoc prof, 60-68, PROF PHYSICS, UNIV NEW ORLEANS, 68-, CHMN DEPT, 74- *Concurrent Pos:* Oak Ridge Nat Lab equip subcontract, 63-; Res Corp grant, 67-68; consult, Electronuclear Div, Oak Ridge Nat Lab, 61 & 62-66. *Mem:* Am Phys Soc; Sigma Xi; Am Asn Physics Teachers. *Res:* Solid state low temperature physics; physics of metals, especially superconductivity. *Mailing Add:* Dept Physics Univ New Orleans New Orleans LA 70148

BERGERON, GEORGES ALBERT, b Quebec, Que, Oct 11, 16; m 45; c 5. PHYSIOLOGY. *Educ:* Laval Univ, BA, 37, MD, 42; FRCP(C), 47. *Prof Exp:* Resident med, Hotel-Dieu de Quebec, 43; from lectr to asst prof, 43-51, PROF PHYSIOL, MED SCH, LAVAL UNIV, 51-, VDEAN, 64-, HEALTH SCI DIR, 70-, ASST TO DEAN, 75- *Concurrent Pos:* Fel, Western Reserve Univ, 44; Markle Found scholar, 50-55. *Mem:* AAAS; Am Physiol Soc; Can Physiol Soc. *Res:* Circulation; medical education. *Mailing Add:* 965 Ernest Gagnon Quebec City PQ G1S 3R5 Can

BERGERON, JOHN ALBERT, b Cumberland, RI, June 13, 29; m 52; c 3. CYTOLOGY. *Educ:* Brown Univ, BA, 51; Cornell Univ, PhD(zool), 56. *Prof Exp:* Asst hist embryol, Cornell Univ, 51-55; assoc physiologist, Brookhaven Nat Lab, 55-61, physiologist, 61-64; MEM STAFF, GEN ELEC RES LAB, 64-, RES SCIENTIST, RES & DEVELOP CTR, 64-, MGR DIAGNOSTIC PROJS, CORP RES & DEVELOP, 70- *Concurrent Pos:* Mem, Biol Stain Comn. *Mem:* Soc Gen Physiol; Am Soc Microbiol; Am Soc Cell Biol; Fedn Am Soc Exp Biol; Soc Cryobiol. *Res:* Electromicroscopy; sub-cellular fractions; physicochemical aspects of the behavior of dyes and photosynthetic pigments; hemostatic mechanisms. *Mailing Add:* Bldg K1-Knolls Gen Elec Co Res & Develop Ctr Schenectady NY 12301

BERGERON, JOHN JOSEPH MARCEL, b Belleville, Ont, Dec 22, 46; m 71; c 2. CELL BIOLOGY. *Educ:* McGill Univ, BSc, 66; Oxford Univ, DPhil(biochem), 69. *Prof Exp:* Res assoc cell biol, Rockefeller Univ, 69-71; scientist, Nat Inst Med Res, 71-74; from asst prof to assoc prof, 74-82, PROF ANAT, MCGILL UNIV, 82- *Honors & Awards:* Murray J Barr Award, Can Asn Anatomists. *Mem:* Biochem Soc; Am Soc Cell Biol; Am Soc Biol Chem; Endocrine Soc. *Res:* Membrane biogenesis; hormone action; glycosylation. *Mailing Add:* Dept Anat Strathcona Anat & Dent Bldg 3640 University St Montreal PQ H3A 2B2 Can

BERGERON, KENNETH DONALD, b Jacksonville, Fla, May 20, 46; div. NUCLEAR ENGINEERING, SOFTWARE SYSTEMS. *Educ:* Brown Univ, ScB, 68; Brandeis Univ, MS, 69, PhD(physics), 75. *Prof Exp:* Physicist, Plasma Theory Div, 74-78, syst analyst, Solar Energy Dept, 78-81, TECH SUPVR, REACTOR SAFETY ANALYSIS, SANDIA NAT LABS, 81- *Res:* Ion emission in diodes; magnetic insulation; insulator flashover; sheaths in plasma; solar industrial process heat; energy economics; nuclear reactor severe accident analysis; computer code development. *Mailing Add:* 1633 Monte Largo Dr NE Albuquerque NM 87185

BERGERON, MICHEL, b Alma, Que. PHYSIOLOGY, NEPHROLOGY. *Educ:* Laval Univ, BA, 53, MD, 59; McGill Univ, MSc, 64. *Prof Exp:* Intern internal med, Huntington Mem Hosp, 59-60; resident, Lahey Clin, 61-62; Med Res Coun Can res fel, Nuclear Res Ctr, France, 64-67, scholar, 67-72; from asst prof to assoc prof physiol, 67-75, PROF PHYSIOL, UNIV MONTREAL, 75-, CHMN DEPT, 86- *Concurrent Pos:* Ed-in-chief, Med Sci, Paris, Montreal. *Mem:* AAAS; Am Soc Nephrology; Can Soc Physiol (pres, 86-); Can Soc Nephrology; Asn Sci, Eng & Technol Community Can (pres, 78). *Res:* Amino acid transport; micropuncture techniques; electron microscopy; radioautography, renal physiology; membrane transport. *Mailing Add:* Dept Physiol Univ Montreal Montreal PQ H3C 3T8 Can

BERGERON, ROBERT F(RANCIS) (TERRY), JR, b Gloucester, Mass, Jan 23, 42; m 79; c 3. APPLIED MATHEMATICS, COMPUTER SCIENCE. *Educ:* Brown Univ, ScB, 64; Mass Inst Technol, PhD(appl math), 68. *Prof Exp:* Instr math, Mass Inst Technol, 68-69; mem tech staff appl math & fluid mech, 69-72, supvr, AMDF Dept, 72-76, supvr Advan Transaction Systs Dept, 78-83, supvr video systs, 83-85, SUPVR TRANS SYSTS DEPT, AT&T BELL LABS, 85- *Concurrent Pos:* Vis scientist, Lab Comput Sci, Mass Inst Technol, 76-77. *Mem:* Sigma Xi; Asn Comput Mach. *Res:* Nonlinear waves in fluids; aerodynamically generated noise; numerical fluid mechanics; programming languages; programming methodology; data base systems; software testing. *Mailing Add:* 3W-T01 AT&T Bell Labs Warren NJ 07060

BERGES, DAVID ALAN, b Evansville, Ind, Aug 5, 41; m 65; c 4. ORGANIC CHEMISTRY, MEDICINAL CHEMISTRY. *Educ:* Evansville Col, BA, 63; Ind Univ, PhD(chem), 67. *Prof Exp:* Assoc org chem, Nat Ctr Sci Res, Inst Chem Natural Substances, France, 67-68; SR INVESTR, SMITH KLINE BEECHAM PHARMACEUTICALS, 68- *Mem:* Am Chem Soc; AAAS. *Res:* Isolation, structure elucidation and synthesis of natural products; development of new organo-chemical synthetic methods; conformational analysis; biological properties, chemistry and synthesis of betalactam antibiotics; synthesis of enzyme inhibitors; synthesis of antiviral agents. *Mailing Add:* Smith Kline Beecham Pharmaceuticals PO Box 1539 King of Prussia PA 19406-0939

BERGESON, GLENN BERNARD, plant nematology, for more information see previous edition

BERGESON, HAVEN ELDRED, b Logan, Utah, Dec 22, 33; m 57; c 8. HIGH ENERGY PHYSICS. *Educ:* Univ Utah, BS, 58, PhD(physics), 62. *Prof Exp:* Physicist, Space Sci Lab, Gen Elec Co, 61-63; asst prof physics, 63-64, asst res prof, 64-68, assoc prof, 68-75, chmn, physics dept, 76-80, PROF PHYSICS, UNIV UTAH, 75- *Mem:* Am Phys Soc; Am Asn Physics Teachers. *Res:* Cosmic ray physics; cosmic ray anisotropies; high energy neutrino interactions; muon production; ultra high energy collisions; extensive air showers; air scintillation. *Mailing Add:* Dept Physics Univ Utah 201b J Fletcher Bldg Salt Lake City UT 84112

BERGEY, GREGORY KENT, b Bryn Mawr, Pa, Nov 9, 49; m 72; c 2. NEUROLOGY, NEUROPHYSIOLOGY. *Educ:* Princeton Univ, AB, 71; Univ Pa, MD, 75. *Prof Exp:* From intern to resident internal med, Yale-New Haven Hosp, 75-77; res assoc neurobiol, Lab Develop Neurobiol, NICHD, NIH, 77-81; resident neurol, Johns Hopkins Hosp, 79-83; asst prof neurol & physiol, 83-89, ASSOC PROF NEUROL, SCH MED, UNIV MD, 89-; DIR, MD EPILEPSY CTR, 87- *Mem:* Soc Neurosci; Am Acad Neurol; Am Epilepsy Soc; AAAS. *Res:* Mechanisms of action of convulsants and anticonvulsants; actions of clostridial neurotoxins; developmental neurobiology in dissociated tissue culture; intractable seizures. *Mailing Add:* Dept Neurol Sch Med Univ Md 22 S Greene St Baltimore MD 21201

BERGEY, JAMES L, b Allentown, Pa, May 22, 45; m 68; c 2. CARDIOVASCULAR PHARMACOLOGY. *Educ:* Temple Univ, BS, 68, PhD(cardiovasc pharmacol), 76. *Prof Exp:* Res assoc electrophysiol, Sch Med, Temple Univ, 75-76; sr res pharmacologist, Res Div, Wyeth Labs Inc, 76-81; sr pharmacologist, Pharm Div, Ciba-Geigy Corp, 81-82; SR RES FEL, SQUIBB INST MED RES, PRINCETON, NJ, 82- *Concurrent Pos:* mem, Am Heart Asn. *Mem:* Sigma Xi; NY Acad Sci. *Res:* Enzymatic regulation of ionic permeability in biological membranes; role of calcium in pathological processes, including hypertension and cardiac arrhythmias; cardiac cellular electrophysiology. *Mailing Add:* 117 Shady Lane Lansdale PA 19446

BERGGREN, GERARD THOMAS, JR, b New Orleans, La, Dec 8, 46; m 67; c 2. PLANT PATHOLOGY. *Educ:* Southeastern La Univ, BS, 69; La State Univ, MS, 71, PhD(plant path), 74. *Prof Exp:* Res plant pathologist, Ansul Co, 74-75; mgr prod develop, Kalo Lab, Marion Labs, 75-76, mgr field res, 76-77; ASST SPECIALIST PLANT PATH, COOP EXTEN, LA STATE UNIV, BATON ROUGE, 77- *Mem:* Am Phytopath Soc. *Res:* Applied plant pathology in the areas of cotton, soybean, sugarcane, fruits, nuts and small grain. *Mailing Add:* Dept Plant Path La State Univ Baton Rouge LA 70803-1720

BERGGREN, MICHAEL J, b Menominee, Mich, Feb 22, 39. MEDICAL PHYSICS. *Educ:* Univ Mich, BS, 61; Stanford Univ, PhD(physics), 69. *Prof Exp:* Teaching asst physics, Stanford Univ, 61-63, res asst, 63-69; asst prof, Mankato State Col, 69-73; vis scientist, Mayo Found, Mayo Clin, 74-75, res fel, Biophys Sci Unit, 75-78; res asst prof bioeng, 78-84, RES ASSOC PROF BIOENG, UNIV UTAH, 85- *Concurrent Pos:* Res asst, Kaman Nuclear, Colo, 62. *Mem:* Am Asn Physicists in Med; Sigma Xi; Am Phys Soc; Inst Elec & Electronics Engrs. *Res:* Medical imaging techniques with emphasis on computerized tomography from ultrasonic, x-ray and radioisotope sources. *Mailing Add:* Dept Bioeng Univ Utah Merrill Eng Bldg Salt Lake City UT 84112

BERGGREN, RONALD B, b Staten Island, NY, June 13, 31; m 54; c 2. SURGERY. *Educ:* Johns Hopkins Univ, AB, 53; Univ Pa, MD, 57; Am Bd Surg, dipl; Am Bd Plastic Surg, dipl. *Prof Exp:* From asst instr to instr, Univ Pa, 58-65; from asst prof to prof, 65-86, dir plastic surg, Col Med, Univ Hosps & Childrens Hosp, Columbus, 65-85, EMER PROF SURG, OHIO STATE UNIV, 86- *Concurrent Pos:* Pres, Plastic Surg Educ Found, 86-87, chmn, Am Bd Plastic Surg, 87-88, vpres, Am Asn Plastic Surgeons, 88-89; chmn designate, Accreditation Coun Grad Med Educ. *Mem:* AAAS; fel Am Col Surg; Am Soc Plastic & Reconstruct Surg; Am Soc Surg Trauma; Am Asn Plastic Surg. *Res:* Low temperature preservation of tissues and organs; clinical problems in reconstruction of traumatic deformities of the face. *Mailing Add:* 3732-E Olentangy River Rd Columbus OH 43214

BERGGREN, WILLIAM ALFRED, b New York, NY, Jan 15, 31; div; c 4. GEOLOGY, MICROPALEONTOLOGY. *Educ:* Dickinson Col, BA, 52; Univ Houston, MSc, 57; Univ Stockholm, PhD(geol), 60, DSc(geol), 62. *Prof Exp:* Micropaleontologist, Shell Oil Co, 55-57; res assoc micropaleont, Univ Stockholm, 57-62, Doktorand stipendiat, 60-62; res paleontologist, Oasis Oil Co, Libya, 62-65; SR SCIENTIST, WOODS HOLE OCEANOG INST, 65- *Concurrent Pos:* Fel, Princeton Univ, 60-61; Am Asn Petrol Geol grant-in-aid, 61-62; Sigma Xi grant-in-aid, 62; exchange student, Soviet-Am Cult Exchange Prog, 62; Rumanian-Am Acad Sci Exchange Prog, 69; mem adv panel, Joint Oceanog Deep Earth Sampling Prog Atlantic, Inst Comn Deep Sea Stratig, 67-68; working group on biostratig zonation of Cretaceous & Cenozoic for correlation in marine geol, Comn Stratig, Int Union Geol Sci, 68-; co-ed, J Paleont. *Honors & Awards:* Mary Clark Thompson Medal Paleont, Nat Acad Sci, 82. *Mem:* Nat Acad Sci; Am Asn Petrol Geol; Soc Explor Paleont & Mineral; Swiss Geol Soc; Sigma Xi; fel Geol Soc Am; Paleont Soc. *Res:* Studies in Mesozoic and Cenozoic planktonic Foraminifera, including their evolutionary development and world-wide stratigraphic correlation based on their occurrence. *Mailing Add:* Dept Geol & Geophys Woods Hole Oceanog Inst Woods Hole MA 02543

BERGH, ARPAD A, b Hungary, Apr 26, 30; US citizen; m 56; c 4. PHYSICAL CHEMISTRY. *Educ:* Univ Szeged, MS, 52; Univ Pa, PhD(phys chem), 59. *Prof Exp:* Res assoc, Geophys Inst, Hungary, 52-56; mem tech staff, Bell Tel Labs, 59-63, supvr planar transistor tech, 64-69, head, lightwave devices dept, 69-86, DIV MGR, BELL COMMUN RES, BELL LABS, 86- *Mem:* Fel Inst Elec & Electronics Engrs. *Res:* Reaction kinetics; homogeneous reactions and catalysis in gases and aqueous solutions; physical chemistry of electron devices; processing, operation and failure mechanisms of semiconductor devices; opto-electronic devices including III-V semiconductor materials and the physics and fabrication of LED displays and opto-isolators. *Mailing Add:* Bellcore 445 South St Morristown NJ 07960-1910

BERGH, BERTHOLD ORPHIE, b Sask, Jan 30, 25; m 48; c 2. AVOCADO, EVOLUTION. *Educ:* Univ Sask, BS, 50; Ohio State Univ, MSc, 51; Univ Calif, PhD(genetics), 56. *Prof Exp:* Res officer, Fruit Breeding, Can Dept Agr, 51-52; asst, Genetics Dept, 52-56; asst geneticist, Citrus Exp Sta, 56-67, GENETICS SPECIALIST, CITRUS EXP STA, UNIV CALIF, RIVERSIDE, 67- *Concurrent Pos:* Consult, UN Foreign Agr Orgn, Kenya, 80; avocado consult various countries. *Mem:* AAAS; Bot Soc Am; Am Genetic Asn; Am Soc Hort Sci; Sigma Xi; Am Inst Biol Sci. *Res:* Basic genetic studies of the tomato and the pepper; commercial improvement of the avocado. *Mailing Add:* Bot Sci Univ Calif Riverside CA 92521-0124

BERGHOEFER, FRED G, b Chicago, Ill, June 7, 21; m 50; c 3. OPERATIONS RESEARCH. *Educ:* Rose Polytech, BS, 43; Univ Chicago, SM, 49. *Prof Exp:* Radar engr, Naval Res Lab, 43-45, analyst, Opers Eval Group, 50-66, STUDY DIR, CTR NAVAL ANALYSES, 66- *Mem:* AAAS; Opers Res Soc Am; Inst Elec & Electronics Engrs. *Res:* Operations analysis of naval operations, especially antisubmarine warfare, air warfare, and guided missiles. *Mailing Add:* 2720 N Fillmore St Arlington VA 22207

BERGIN, MARION JOSEPH, b Lavoye, Wyo, May 2, 27; m 58; c 3. GEOLOGY. *Educ:* Univ Wyo, BS, 51. *Prof Exp:* GEOLOGIST, US GEOL SURV, 51- *Res:* Geology of mineral fuels. *Mailing Add:* 100 Yeonas Dr SW Vienna VA 22180

BERGLES, ARTHUR E(DWARD), b New York, NY, Aug 9, 35; m 60; c 2. MECHANICAL ENGINEERING. *Educ:* Mass Inst Technol, SB & SM, 58, PhD(mech eng), 62. *Prof Exp:* Engr, US Steel Corp, 55, Stone & Webster Eng Corp, 57 & Joseph Kaye & Co, 58; res asst mech eng, Mass Inst Technol, 59-62, fel & asst prof, 63-68, chmn eng proj lab, 67-68, assoc dir heat transfer lab, 66-69, assoc prof, 68-69, vis prof, 70, lectr, 71; prof mech eng, Ga Inst Technol, 70-72; prof & chmn dept, Iowa State Univ, 72-83, distinguished prof eng & prof mech eng, 83-86; CLARKE & CROSSAN PROF ENG & DIR HEAT TRANSFER LAB, RENSSELAER POLYTECH INST, 86-, DEAN ENG, 89- *Concurrent Pos:* Consult numerous indust. *Honors & Awards:* Heat Transfer Mem Award, Am Soc Mech Engrs, Dedicated Serv Award; US Sr Scientist Award, Alexander von Humboldt Found, Ger; Fel Award, Int Centre Heat & Mass Transfer; Lamme Award, Am Soc Eng Educ; Teetor Award, Soc Automotive Engrs; Kern Award, Am Inst Chem Engrs. *Mem:* Am Soc Heating, Refrig & Air Conditioning Engrs; fel Am Soc Eng Educ; fel Am Soc Mech Engrs (vpres, 81-85, bd gov, 85-89, pres, 90-91); Am Inst Chem Engrs; fel AAAS; Soc Automotive Engrs. *Res:* Augmentation of convective heat transfer; two-phase gas-liquid flow and heat transfer; heat transfer to laminar internal flows; cooling of electronic equipment; heat recovery systems. *Mailing Add:* Sch Eng Rensselaer Polytech Inst Troy NY 12180-3590

BERGLUND, CARL NEIL, b Ft William, Ont, July 21, 38; m 61; c 3. ELECTRICAL ENGINEERING. *Educ:* Queen's Univ, Ont, BSc, 60; Mass Inst Technol, MS, 61; Stanford Univ, PhD(elec eng), 64. *Prof Exp:* Mem tech staff semiconductor device lab, Bell Labs, Inc, 64-72, supvr, 65-72, mgr, Bell Northern Res, 72-73; vpres, Microsysts Int, 73-74; dir silicon technol lab, Bell Northern Res, 74-78; mgr, Intel Corp, 78-; PRES, ATEQ CORP. *Mem:* Am Phys Soc; Inst Elec & Electronics Engrs; Electrochem Soc. *Res:* Photoemission from solids; semiconductor device research; silicon integrated circuits; silicon device and process technology. *Mailing Add:* Northwest Technol Group Consult Inc 6700 SW 105th Ave Suite 207 Beaverton OR 97005

BERGLUND, DONNA LOU, inorganic chemistry, for more information see previous edition

BERGLUND, LARRY GLENN, b Minneapolis, Minn, Oct 17, 38; m 62; c 2. MECHANICAL ENGINEERING, BIOENGINEERING. *Educ:* Univ Minn, Minneapolis, BME, 62, MSME, 65; Kans State Univ, PhD(mech eng), 71. *Prof Exp:* Develop engr absorption refrig, The Trane Co, Wis, 65-68; USPHS bioenviron eng trainee mech eng, Kans State Univ, 68-71; asst prof mech eng, Mich Technol Univ, 72-75; asst fel, 75-81, ASSOC FEL, JOHN B PIERCE FOUND LAB, 81- *Concurrent Pos:* Lectr environ technol, Col Archit, Yale Univ, 75- *Honors & Awards:* Teetor Award, Soc Automotive Engrs, 74; Nevins Award, Am Soc Heating, Refrig & Air Conditioning Engrs, 79. *Mem:* Am Soc Heating, Refrig & Air Conditioning Engrs; Am Soc Mech Engrs. *Res:* Biothermal engineering; thermal physiology; physiological modeling; ventilation; thermal comfort; clothing; energy utilization in buildings; heating and air conditioning systems; air quality. *Mailing Add:* John B Pierce Found Lab 290 Congress Ave New Haven CT 06519

BERGMAN, ABRAHAM, b Seattle, Wash, May 11, 32; c 5. MEDICINE, PEDIATRICS. *Educ:* Reed Col, BA, 54; Western Reserve Univ, MD, 58; Am Bd Pediat, dipl, 63. *Prof Exp:* Intern, Children's Hosp Med Ctr, Boston, 58-59; jr asst resident, 59-60; exchange registr, Pediat Unit, St Mary's Hosp, London, Eng, 60-61; res fel pediat, State Univ NY Upstate Med Ctr, 61-63; PROF PEDIAT, SCH MED & PROF HEALTH SERV, SCH PUB HEALTH & COMMUNITY MED, UNIV WASH, 64-; DIR DEPT PEDIAT, HARBORVIEW MED CTR, 83- *Mem:* Fel Am Acad Pediat; Ambulatory Pediat Asn; Am Pediat Soc. *Res:* Epidemiology; public policy in health; medical care; injury prevention. *Mailing Add:* Univ Wash Harborview Med Ctr 325 9th Ave Seattle WA 98104

BERGMAN, ELLIOT, b Brooklyn, NY, Feb 2, 30; m 52; c 3. ORGANIC CHEMISTRY. *Educ:* Brooklyn Col, BS, 51; Cornell Univ, PhD(org chem), 55. *Prof Exp:* Instr org chem, Univ Calif, 55-56; chemist, Shell Develop Co, Emeryville, Calif, 56-64, group leader, 64, supvr org chem, Agr Res Labs, Modesto, Calif, 65-67, head org chem div, Woodstock Agr Res Ctr, Shell Res Ltd, Eng, 67-70, asst to mgr res & develop, Agr Div, Shell Chem Co, 70-71, proj mgr consumer prod, 71-74; vpres res & develop, Liquid Crystal Technol Inc, 74-76; vpres original equip mfr & int mkt, 75-76; mgr, crop chem res & develop, Mobil Chem Co, 76-80, planning mgr, 81; asst dir, Stauffer Chem Co, Western Res Ctr, Richmond, Calif, 81-86; DIR RES, GRIFFIN CORP, VALDOSTA, GA, 86- *Concurrent Pos:* Off Naval Res asst, Cornell Univ, 51-53; Proctor & Gamble Fel; mem, Dirs Comt Nat Agr Chems Asn; chmn-elect, Gordon Res Conf, 81-83, chmn, 84-85; mem res dirs comt, Nat Agr Chemicals Asn. *Mem:* Am Chem Soc; Plant Growth Regulator Soc Am. *Res:* Fluorine, organometallic, organophosphorus, polymer and agricultural chemistry; consumer products; cholesteric liquid crystal materials for consumer, industrial and medical products (visual displays); biotechnology and genetic engineering. *Mailing Add:* Griffin Corp PO Box 1847 Valdosta GA 31603-1847

BERGMAN, EMMETT NORLIN, b Slayton, Minn, May 6, 29; m 53; c 4. NUTRITION. *Educ:* Univ Minn, BS, 50, DVM & MS, 53, PhD(physiol), 59. *Prof Exp:* Instr physiol, Univ Minn, 50-53; vet lab off, Walter Reed Army Inst Res, 53-55; from instr to asst prof physiol, Univ Minn, 55-61; assoc prof, 61-66, PROF PHYSIOL, CORNELL UNIV, 66- *Concurrent Pos:* NSF sr fel, 63-64; mem metab study sect, Div Res Grants, NIH, 66-70; vis prof physiol, Med Ctr, Univ Calif, San Francisco, 69-70; consult, Coun Grad Schs US; vis scientist, Inst Res Animal Diseases, Compton, Berkshire, UK, 77-78. *Mem:* AAAS; Am Physiol Soc; Am Vet Med Asn; Conf Res Workers Animal Dis; NY Acad Sci; Am Inst Nutrit. *Res:* Ketone body, carbohydrate and amino acid metabolism; electrolyte metabolism; ruminant metabolism; visceral blood flow; liver metabolism. *Mailing Add:* Dept Physiol Cornell Univ NY State Col Ithaca NY 14853

BERGMAN, ERNEST L, b Munich, Ger, July 12, 22; nat US; m 48. HORTICULTURE, PLANT NUTRITION. *Educ:* Kanton Landwirtsch, Schwand, Switz, dipl landwirt, 41; Ore State Col, BS, 55; Mich State Univ, MS, 56, PhD(hort), 58. *Prof Exp:* Res asst veg crops, Kanton Gartenbau Schule, Oeschberg, Switz, 45; pomologist, Schweizerisch Obstbauzentrale, 46; pvt enterprise, 47-53; from asst prof to prof, 58-87, EMER PROF PLANT NUTRIT, PA STATE UNIV, 87- *Concurrent Pos:* Var int assignments Arg, Uruguay, People's Repub China & WAfrica; vpres int affairs, Am Soc Hort Sci, 86-87. *Honors & Awards:* Kenneth Post Award, Am Soc Hort Sci, 68. *Mem:* Fel AAAS; fel Am Soc Hort Sci; Am Soc Plant Physiol; Am Soc Agron; Int Soc Hort Sci. *Res:* Mineral nutrition of horticultural plants, vegetables and fruits; physiological, pathological, and virus disorders in connection with nutrient deficiency or toxicity problems; plant nutrition and agriculture on international level. *Mailing Add:* 1421 Harris St State College PA 16803-3024

BERGMAN, HAROLD LEE, b Sault St Marie, Mich, July 8, 41; m 68; c 2. PHYSIOLOGICAL ECOLOGY. *Educ:* Eastern Mich Univ, BA, 68, MS, 71; Mich State Univ, PhD(fisheries biol), 73. *Prof Exp:* Fishery biologist, Great Lakes Fishery Lab, US Fish & Wildlife Serv, Ann Arbor, Mich, 68-71; res asst fishery biol, Dept Fisheries & Wildlife, Mich State Univ, East Lansing, 71-73, res assoc environ physiol, Dept Physiol, 74; res assoc environ impact, Environ

Sci Div, Oak Ridge Nat Lab, Oak Ridge, Tenn, 74-75; asst prof, 75-80, ASSOC PROF ZOOL & PHYSIOL, UNIV WYO, 80- *Concurrent Pos:* Ed, Black Thunder Study, Wyo Environ Inst, 75. *Mem:* Sigma Xi; Am Fisheries Soc; AAAS; NAm Benthological Soc; Am Soc Testing & Mat. *Res:* Physiological ecology of fishes and the effects of environmental contaminants on aquatic animals. *Mailing Add:* Dept Zool & Physiol Univ Wyo Univ Sta Box 3166 Laramie WY 82071

BERGMAN, HYMAN CHAIM, b Latvia, Oct 10, 05; nat US; m 37; c 2. PHARMACOLOGY. *Educ:* Univ Calif, Los Angeles, AB, 28; Univ Southern Calif, PhD(physiol), 37. *Prof Exp:* Jr chemist, Bur Animal Indust, USDA, 28-29; biochemist, Scripps Metab Clin, 29-33 & Hormone Assay Lab, 37-38; pharmacologist, Wilson Labs, 38-43; res assoc, Cedars Lebanon Hosp, Los Angeles, 43-48; dir, Joffe Labs, 48-51, Primorganics, Inc, 51-57 & Bergman Labs, 57-62; sr chemist, NAm Aviation, Inc, 62-69; SR CHEMIST, BERGMAN LABS, 69- *Mem:* AAAS; Am Chem Soc; Soc Exp Biol & Med; NY Acad Sci. *Res:* Carbohydrate metabolism; hormones; anterior pituitary; adrenal cortex; pharmacology of tissue extracts; mechanism of shock; cardiovascular physiology. *Mailing Add:* 2006 Chariton St Los Angeles CA 90034-1504

BERGMAN, KENNETH DAVID, b Shafter, Calif, Mar 24, 49; m 78; c 2. BIOLOGY. *Educ:* Univ Calif, Berkeley, AB, 71; Harvard Univ, PhD(cell biol), 77. *Prof Exp:* Fel, Dept Biol, Harvard Univ, 77-78; ASSOC PROF BIOL, CELL BIOL & GENETICS, KEENE STATE COL, NH, 79- *Mem:* AAAS; Am Pub Health Asn. *Mailing Add:* Dept Sci Univ NH Keene State Col 249 Pako Ave Keene NH 03431

BERGMAN, KENNETH HARRIS, b Plainfield, NJ, Dec 11, 35; m 78. ATMOSPHERIC SCIENCES & CLIMATE. *Educ:* Rutgers Univ, BA, 58; Univ Wash, PhD(atmospheric sci), 69. *Prof Exp:* Design engr, Semiconductor Div, Radio Corp Am, 58-59; asst, dept atmospheric sci, Univ Wash, 60-65; asst prof meteorol, San Jose State Col, 65-70; fel, Tex A&M Univ, 70-71; res meteorologist, Nat Hurricane Res Lab, Nat Weather Serv, Nat Oceanic & Atmospheric Agency, 71-73 & Nat Meteorol Ctr, 73-79; assoc prog dir, Climate Dynamics Prog, NSF, 79-82, actg prog dir, 82-83; meteorologist, Climate Anal Ctr, Nat Weather Serv, Nat Oceanic & Atmospheric Admin, 83-87; SR PROG OFFICER, BD ATMOSPHERIC SCI & CLIMATE, NAT RES COUN, NAT ACAD SCI, 87- *Concurrent Pos:* Ed, J Climate Appl Meteorol, Am Meteorol Soc, 82-84. *Mem:* Am Meteorol Soc; AAAS; Sigma Xi. *Res:* Mass exchange between stratosphere and troposphere; dynamics and dynamic stability of convective atmospheric vortices; atmospheric transport and diffusion in the 100 meter boundary layer; dynamic meteorology, numerical modeling and weather prediction; numerical objective analysis of synoptic data; climate change and prediction; climate monitoring and information services. *Mailing Add:* 600 Independence Ave SW Washington DC 20546

BERGMAN, MOE, b Brooklyn, NY, Mar 28, 16; m 38; c 1. EDUCATIONAL ADMINISTRATION, MEDICAL & HEALTH SCIENCES. *Educ:* Univ Ill, AB, 37; Columbia Univ, MA, 39, EdD(audiol), 49. *Prof Exp:* Dir spec educ, Peekskill Union Free Schs, 38-43; chief audiologist, NY Regional Off, Vet Admin, 46-53; prof lect audiol, Col Physicians & Surgeons & Teachers Col, Columbia Univ, 48-58; dir, Speech & Hearing Ctr, Hunter Col, NY, 53-69; prof 76-85, EMER PROF AUDIOL, SCH MED, TEL-AVIV UNIV, ISRAEL, 85-; EMER PROF, HUNTER COL, NY, 75- *Concurrent Pos:* consult, Indust Home for Blind, 59-64 & New York Dept Health, 60-75; exec officer, PhD Prog Speech, City Univ New York, 65-69; vis prof, Med Sch, Tel-Aviv Univ, 68-69 & 71. *Mem:* AAAS; fel & hon mem Am Speech & Hearing Asn; Acoust Soc Am; Israel Speech & Hearing Asn (hon pres); hon mem Israeli Med Asn Otolaryngolog; Am Inst Physics; Acad Rehab Audiol. *Res:* Binaural and stereophonic hearing; audiology in audiosurgery; auditory perception in blind persons; development of audiology services; hearing in primitive peoples; presbycucis; aging speech perception; hearing and aging; central auditory dysfunction in brain pathology; noise-induced changes in cochlear function. *Mailing Add:* Ten Wissotzky St Tel-Aviv Israel

BERGMAN, NORMAN, b Seattle, Wash, Oct 14, 26; m 52; c 2. ANESTHESIOLOGY. *Educ:* Reed Col, BA, 49; Univ Ore, MD, 51. *Prof Exp:* Assoc anesthesiol, Columbia Univ, 54-58; asst attend, Presby Hosp, New York, 54-58; from asst prof to prof, Col Med, Univ Utah, 58-70; chief anesthesiol serv, Vet Admin Hosp, 58-70; PROF ANESTHESIOL, MED SCH, UNIV ORE & DIR ANESTHESIA SERV, MED SCH HOSPS, 70- *Concurrent Pos:* Vis res assoc, Royal Col Surgeons & Post-Grad Med Sch, Univ London & Hammersmith Hosp, 63-64; Nat Heart Inst spec fel, 63-64. *Mem:* Am Soc Anesthesiol; AMA; Fedn Am Soc Exp Biol. *Res:* Respiratory changes during anesthesia; physiology of artificial respiration. *Mailing Add:* Ore Health Sci Univ Univ Hosp 3181 SW Sam Jackson Park Rd Portland OR 97201

BERGMAN, RAY E(LDON), b Drumright, Okla, Nov 6, 28; m 59; c 2. APPLIED MATHEMATICS, OPERATIONS RESEARCH. *Educ:* Okla State Univ, BS, 50; Univ Denver, MA, 58. *Prof Exp:* Res engr, Physics Div, Denver Res Inst, 53-54; petrol engr, reservoir anal, Creole Petrol Corp, 54-56; instr math, Univ Denver, 58-59; opers res analyst, develop dept, Syst Develop Corp, 59-61; mathematician, Test Data Div, Pac Missile Range, 61-68; OPERS RES ANALYST, NAVAL CIVIL ENG LAB, 68- *Concurrent Pos:* Pac Missile Range scholar, Univ Calif, Los Angeles, 65-66. *Mem:* Sigma Xi. *Res:* Management and operational systems analysis; instrumentation analysis; inversion of large matrices. *Mailing Add:* 1572 E Prima Ct Camarillo CA 93010

BERGMAN, RICHARD N, b Chicago, Ill, Nov 7, 44; m; c 2. METABOLIC RESEARCH. *Educ:* Univ Pittsburgh, PhD(physiol), 71. *Prof Exp:* PROF PHYSIOL, SCH MED, UNIV SOUTHERN CALIF, 80-, HEAD, METABOLIC UNIT. *Mem:* Am Physiol Soc; Am Fedn Clin Res; Endocrine Soc; AAAS. *Mailing Add:* Dept Physiol & Biophys Sch Med Univ Southern Calif 2025 Zonal Ave Los Angeles CA 90033

BERGMAN, ROBERT, b New York, NY, Sept 2, 31; c 2. MECHANICAL ENGINEERING. *Educ:* NY Univ, BME, 54; City Univ New York, MME, 62. *Prof Exp:* Engr aircraft radar, Gen Precision Lab, 58-63; lead engr satellites, Radio Corp Am, 63-68; bus planning reentry vehicles, Gen Elec Co, 68-74; asst prog mgr, 74-81, head proj safety, Fusion Energy, 81-83, head safety training, Princeton Univ, 83-86; syst safety mgr for satellites, Sci Appln Int Corp, 86-87; SYST ENG & PROJ MGR, AT&T BELL LAB, 87- *Mailing Add:* 300 Dodds Lane Princeton NJ 08540

BERGMAN, ROBERT GEORGE, b Chicago, Ill, May 23, 42; m 65; c 2. ORGANOMETALLIC CHEMISTRY, INORGANIC CHEMISTRY. *Educ:* Carleton Col, BA, 63; Univ Wis, PhD(chem), 66. *Prof Exp:* Arthur Amos Noyes instr chem, Calif Inst Technol, 67-69, from asst prof to prof chem, 69-77; PROF CHEM, UNIV CALIF, BERKELEY, 77- *Concurrent Pos:* NATO fel chem, Columbia Univ, 66-67; Alfred P Sloan Found fel, 70-72; NIH & NSF grants; Dreyfus Found Teacher-Scholar award, 70-75. *Honors & Awards:* Award Organometallic Chem, Am Chem Soc, 86; Cope Scholar Award, Am Chem Soc, 87. *Mem:* Nat Acad Sci; Am Chem Soc; Am Acad Arts & Sci. *Res:* Synthesis and reaction mechanisms of organic and transition metal organometallic compounds; catalysis. *Mailing Add:* Dept Chem Univ Calif Berkeley CA 94720

BERGMAN, ROBERT K, b Great Falls, Mont, Apr 5, 34; m 57; c 6. MEDICAL RESEARCH. *Educ:* Mont State Univ, Bozeman, BS, 56, MS, 57; Univ Mo, Columbia, PhD, 63. *Prof Exp:* Res asst, Dairy Indust Dept, Mont State Univ, Bozeman, 56-57, instr, 58-60; res asst, Dairy Husbandry Dept, Univ Mo, Columbia, 60-63; staff fel, Lab Microbial Struct & Function, Rocky Mountain Labs, Hamilton, Mont, Nat Inst Allergy & Infectious Dis, NIH, 63-66, sr scientist, 66-74, scientist dir, 74-81, asst to chief, 81-85, CHIEF, OPERS BR, ROCKY MOUNTAIN LABS, HAMILTON MONT, NAT INST ALLERGY & INFECTIOUS DIS, NIH, 85- *Concurrent Pos:* Ralston Purina fel, 60-61, NSF fel, 61-62 & 62-63. *Honors & Awards:* Commendation Medal, USPHS, 88. *Res:* Phenomena related to hypersensitivity in mice; experimental allergic encephalomyelitis in rats. *Mailing Add:* NIH Nat Inst Allergy & Infectious Dis Rocky Mountain Opers Br Rocky Mountain Labs 903 S Fourth St Hamilton MT 59840

BERGMAN, RONALD ARLY, b Chicago, Ill, June 21, 27; m 54; c 4. ANATOMY, PHYSIOLOGY. *Educ:* Univ Ill, BS, 50, MS, 53, PhD(physiol, zool & chem), 55. *Prof Exp:* Nat Found Infantile Paralysis fel, Karolinska Inst, Sweden, 55-56; instr epidemiol, Sch Hyg & Pub Health, Johns Hopkins Univ, 56-57, res assoc anat, Sch Med, 57-59, from asst prof to assoc prof, 68-74; prof human morphol, Sch Med, Am Univ Beirut, 74-80; PROF ANAT, COL MED, UNIV IOWA, 80- *Concurrent Pos:* Vis assoc prof anat, Am Univ Beirut, 69. *Mem:* Am Asn Anatomists; Am Soc Cell Biol; Electron Micros Soc Am. *Res:* Comparative and cell biology; structure, function and pathology of striated and smooth muscle and myogenesis. *Mailing Add:* Dept Anat Univ Iowa Col Med Iowa City IA 52242

BERGMAN, THEODORE L, b Seneca, Kans, Oct 12, 56. HEAT TRANSFER, FLUID DYNAMICS. *Educ:* Univ Kans, BSME, 78; Purdue Univ, MSME, 81, PhD(mech eng), 85. *Prof Exp:* Mech engr, Black & Veatch Consult Engrs, 78-80; ASST PROF MECH ENG, UNIV TEX, AUSTIN, 85- *Mem:* Am Soc Mech Engrs; Sigma Xi. *Res:* Basic and applied heat transfer and fluid dynamics; natural convection; double-diffusive convection. *Mailing Add:* Dept Mech Eng Univ Tex Austin TX 78712

BERGMANN, DIETRICH R(UDOLF), b Detroit, Mich, Aug 12, 38; m 77. URBAN TRANSPORTATION SYSTEMS ENGINEERING. *Educ:* Univ Mich, BS, 61; Stanford Univ, MS, 64, PhD(civil eng), 69. *Prof Exp:* Res civil engr, Syst Anal Div, Battelle Mem Inst, 64-65; asst prof civil eng, Wayne State Univ, 69-70, asst prof indust eng & opers res, 70-72; sr res engr, Gen Motors Res Labs, 72-74; eng projs mgr, Gen Motors Transp Systs Div Ctr, 74-78, mgr technol assessment, 78-80; PRES & DIR ENG, RAILWAY SYSTS ENG CORP, 81- *Concurrent Pos:* Consult, West Bay Rapid Transit Authority, Calif, 66, DeLeuw, Cather & Co, Calif, 66-67, Automotive Safety Found, Washington, DC, 67 & Transp Technol Inc, Colo, 70; mem, Transp Res Bd, Nat Res Coun, Nat Acad Sci & Nat Acad Eng, 63, chmn, Comt Comput Graphics & Interactive Techniques, 79-; grant, NATO, 72. *Mem:* Am Inst Indust Engrs; Am Rwy Eng Asn; Am Soc Civil Engrs; Opers Res Soc Am. *Res:* Analysis, design, and management of urban public transportation system operations; economic evaluation and cost-effectiveness analysis of governmental investments in transportation systems and other types of public works; freight railroad systems engineering analysis and design. *Mailing Add:* PO Box 351 St Clair Shores MI 48080

BERGMANN, ERNEST EISENHARDT, b New York, Nov 2, 42; m 67; c 3. LASERS, FIBER OPTICS. *Educ:* Columbia Univ, AB, 64; Princeton Univ, MA, 66, PhD(physics), 69. *Prof Exp:* From asst prof to prof physics, Lehigh Univ, 69-85; mem tech staff, 84-88, DISTINGUISHED MEM TECH STAFF, AT&T BELL LABS, 88- *Concurrent Pos:* Vis staff mem, Los Alamos Sci Lab, 77-78; Consult, Bell Labs, Allentown, 81-84. *Mem:* Am Phys Soc; Optical Soc Am; Am Asn Univ Professors. *Res:* Fiber-optic communication; theory and experiment in gas lasers and optical resonators; quantum mechanics. *Mailing Add:* AT&T Bell Labs Rte 222 Breinigsville PA 18031

BERGMANN, FRED HEINZ, b Feuchtwagen, Ger, Jan 26, 28; nat US; m 66; c 2. GENETICS. *Educ:* Mass Inst Technol, BS, 50, MS, 51; Univ Wis, PhD(biochem), 57. *Prof Exp:* Jr res chemist, Ethicon Suture Labs, 51-53; USPHS fel, Microbiol Dept, Med Sch, Washington Univ, 57-59; fel biol, Brandeis Univ, 59-61; res biochemist, NIH, 61-66; sci adminr, Res Grants Br, 66-72; dir genetics prog, Nat Inst Gen Med Sci, 72-88. *Honors & Awards:* USPHS Superior Serv Award, 74. *Mem:* Am Soc Human Genetics; Genetics Soc Am. *Res:* Carboxyl activation of amino acids; protein synthesis; science policy. *Mailing Add:* 5430 41 Pl NW Washington DC 20015

BERGMANN, LOUIS LAWRENCE, anatomy, for more information see previous edition

BERGMANN, OTTO, b Vienna, Austria, Feb 7, 25; m 57; c 2. THEORETICAL PHYSICS. *Educ:* Univ Vienna, PhD(physics), 49. *Prof Exp:* Scholar, Dublin Inst Adv Studies, 51-52; sr res fel, Dept Math Physics, Univ Adelaide, 52-55; sr res fel physics, Univ New Eng, Australia, 56-58; res physicist, Res Inst Adv Study, Univ Baltimore, 58-60; assoc prof physics, Univ Ala, 60-62; assoc prof, 62-67, PROF PHYSICS, GEORGE WASHINGTON UNIV, 67- *Concurrent Pos:* Vis prof, Univ Ala, 58-59; vis prof, Inst Theoret Physics, Univ Vienna, Austria, 73-74. *Mem:* Am Phys Soc; Sigma Xi. *Res:* Classical and quantized field theories; theory of relativity. *Mailing Add:* Samson Hall Rm 209 2036 H St NW Washington DC 20052

BERGMANN, PETER GABRIEL, b Berlin, Ger, Mar 24, 15; US citizen; m 36; c 2. THEORETICAL PHYSICS. *Educ:* Prague Univ, PhD(theoret physics), 36. *Hon Degrees:* Tech Univ, Dresden, 79. *Prof Exp:* Mem & asst, Sch Math, Inst Adv Study, 36-41; asst prof physics, Black Mountain Col, 41-42 & Lehigh Univ, 42-44; staff mem & asst dir sonar anal group, Div War Res, Columbia Oceanog Inst, Woods Hole, 44-47; from assoc prof to prof, 47-82, EMER PROF PHYSICS, SYRACUSE UNIV, 82-, RES PROF, 82-; RES PROF, NYU, 83- *Concurrent Pos:* Adj prof, Polytech Inst Brooklyn, 47-57; vis prof, Brandeis Univ, 57 & King's Col, Univ London, 58, Queens Col, 87; vis prof, Yeshiva Univ, 59-63, prof & chmn dept, 63-64 & vis prof, Belfer Grad Sch Sci, 70-78; lectr, Marburg Univ, 55, Univ Stockholm, 58 & Int Sch Cosmology & Gravitation, Int Ctr Sci Culture, Erice, Italy, 75 & 77, dir, 79 & 85; mem, Int Comt Gen Relativity & Gravitation, pres, 77-80,. *Honors & Awards:* Boris Pregel Award, NY Acad Sci, 70. *Mem:* AAAS; Fedn Am Sci (vchmn, 61, chmn, 64); Am Phys Soc; Am Math Soc; Europ Phys Soc; Sigma Xi. *Res:* Relativistic field theories, including quantization; wave propagation and scattering; electron optics; tactosols; stochastic problems; irreversible processes. *Mailing Add:* 640 Riverside Dr New York NY 10031

BERGMANN, STEVEN R, b New York, NY, Feb 4, 51. CARDIOVASCULAR DISEASE, NUCLEAR CARDIOLOGY. *Educ:* George Wash Univ, BA, 72; Wash Univ, St Louis, Mo, MD, 85; Hahnemann Med Col, PhD(physiol & biophys), 78. *Prof Exp:* Fel cardiovasc dis, 77-80, res instr, 79-80, asst prof, 80-89, ASSOC PROF, CARDIOVASC DIV, WASH UNIV, ST LOUIS, 89- *Concurrent Pos:* Mem, Basic Sci Coun, Am Heart Asn. *Mem:* Am Physiol Soc; Am Heart Asn; Soc Nuclear Med; AAAS; fel Am Col Cardiol; Am Soc Clin Invest. *Res:* Positron emission tomography for measurement of myocardial blood flow and metabolism; thrombolytic therapy for salvage of ischemic myocardium. *Mailing Add:* Cardiovasc Div Box 8086 Wash Univ 660 S Euclid Ave St Louis MO 63110

BERGMARK, WILLIAM R, b Mankato, Minn, Oct 28, 40; m 63. ORGANIC CHEMISTRY. *Educ:* St Olaf Col, BA, 62; Mass Inst Technol, PhD(org chem), 66. *Prof Exp:* Res chemist, Am Cyanamid Co, 66-67; res assoc org chem, State Univ NY Buffalo, 67-68; asst prof, 68-73, PROF ORG CHEM, ITHACA COL, 73- *Concurrent Pos:* Res grants, Petrol Res Fund & Res Corp, 69- *Mem:* Am Chem Soc. *Res:* Photochemistry; small ring compounds; electron transfer reactions. *Mailing Add:* Dept Chem Ithaca Col Ithaca NY 14850

BERGNA, HORACIO ENRIQUE, b La Plata, Arg, Feb 11, 24; nat US; m 51; c 1. COLLOID CHEMISTRY, SURFACE CHEMISTRY. *Educ:* La Plata Nat Univ, Lic, 48, Dr Chem, 50. *Prof Exp:* Chemist colloid sci, Lab Testing Mat & Tech Res, 44-50; Fr Govt fel, Univ Paris, 51; US Govt & Orgn Am States fels, Mass Inst Technol, 52, mem staff surface chem, 52-56; sr res chemist, 56-74, staff chemist, 74-78, res assoc, 78-88, SR RES ASSOC, E I DU PONT DE NEMOURS & CO, 88- *Concurrent Pos:* Asst anal chem, La Plata Nat Univ, 44-46. *Honors & Awards:* Croix de Chevalier dans l'ordre des Palmes Academiques, French Govt, 70. *Mem:* AAAS; Am Chem Soc; Sigma Xi; Kolloid Gesellschaft; Arg Chem Asn. *Res:* Surface and colloid chemistry of silica, other oxides and aluminosilicates. *Mailing Add:* 34 Vining Lane Westhaven Wilmington DE 19807

BERGO, CONRAD HUNTER, b Evanston, Ill, Jan 5, 43; m 77; c 2. ORGANIC CHEMISTRY, PHYSICAL CHEMISTRY. *Educ:* St Olaf Col, BA, 65; Univ Minn, Minneapolis, PhD(org chem), 72. *Prof Exp:* Lectr org chem, Chiengmai Univ, US Peace Corps, 72-75; res assoc pharmacol, Col Med, Univ Ky, 75-77; asst prof chem, Alliance Col, 77-80; ASSOC PROF, E STROUDSBURG UNIV, 80- *Mem:* Am Chem Soc; Sigma Xi. *Res:* Nonabusive analgesics; infrared spectroscopy; study of carbonyl frequencies in aromatic systems; flash points of mixtures. *Mailing Add:* 204 Analomink St East Stroudsburg PA 18301-2604

BERGOFSKY, EDWARD HAROLD, b Baltimore, Md, June 18, 27. PHYSIOLOGY. *Educ:* Univ Md, BS, 48, MD, 52. *Prof Exp:* Intern med, Mt Sinai Hosp, NY, 52-53; asst resident, Bellevue Hosp, 53-54; from asst resident to resident, Mt Sinai Hosp, 54-57; Polachek Found fel, NY, 57-59; NY Heart Asn sr res fel, 59-62; from asst prof to prof physiol, Sch Med, NY Univ, 62-74; PROF MED & HEAD PULMONARY DIS SECT, STATE UNIV NY STONY BROOK, 74- *Concurrent Pos:* Res fel physiol, Presby Hosp, Columbia Univ, 55-56; instr, Columbia Univ, 61-62. *Mem:* AAAS; Am Soc Clin Invest; Am Physiol Soc; Am Fedn Clin Res; NY Acad Sci. *Res:* Respiratory and circulatory physiology; regulation of pulmonary and systemic ciruclations; distribution of oxygen and carbon dioxide in tissues; mechanical performance and biophysics of muscles of respiration. *Mailing Add:* Health Sci Ctr TI7/040 State Univ NY Stony Brook NY 11794-8172

BERGOMI, ANGELO, b Milan, Italy, Oct 22, 33; US citizen. ORGANIC CHEMISTRY. *Educ:* Univ Milan, Dr(org chem), 60; Univ Birmingham, MSc, 65. *Prof Exp:* Chemist org chem, Montecatini-Edison, 61-69; SR RES CHEMIST ORG CHEM, GOODYEAR TIRE & RUBBER CO, 69- *Mem:* Am Chem Soc. *Res:* Catalytic isomerization of hydrocarbons; preparation of hydrocarbon resins; synthesis of rubber chemicals; fluorocarbon chemistry. *Mailing Add:* 2015 White Pond Dr Chem Res & Develop Akron OH 44313-7242

BERGOUGNOU, MAURICE A(MEDEE), b Bach, France, Sept 7, 28; m 55; c 2. CHEMICAL ENGINEERING. *Educ:* Univ Nancy, MS, 53; Univ Minn, PhD(chem eng), 58. *Prof Exp:* Asst chem eng, Univ Minn, 54-58; sr engr petrol develop, Exxon Res & Eng Co, Standard Oil Co, 58-62, eng res, 62-67; PROF CHEM ENG, UNIV WESTERN ONT, CAN, 67- *Concurrent Pos:* Pvt consult fluidized reactors. *Mem:* AAAS; Am Chem Soc; Am Inst Chem Engrs; fel Chem Inst Can; Can Soc Chem Eng. *Res:* Chemical reactor technology; fluidization; chemical process development; fluidized reactor design; application of fluidization to chemical and energy processes; design and scale-up of conventional industrial fluidized reactors; development of the ultra-rapid fluidized reactor and of ultra rapid fluidized crackers and pyrolyzers for all kinds of hydrocarbon and carbonaceous feedstocks. *Mailing Add:* Dept Chem & Biochem Eng Univ Western Ont London ON N6A 5B9 Can

BERGQUIST, HARLAN RICHARD, geology; deceased, see previous edition for last biography

BERGQUIST, JAMES WILLIAM, b Ottumwa, Iowa, Apr 23, 28; m 51. COMPUTER SCIENCES, APPLIED MATHEMATICS. *Educ:* Iowa State Univ, BS, 50; Univ Southern Calif, MS, 55, PhD(math), 63. *Prof Exp:* Math analyst, Lockheed Aircraft Corp, 51-54; design engr, Gilfillan Bros Electronics, 55-57; mathematician, 58-68, SCI REP, IBM CORP, 68- *Concurrent Pos:* IBM consult, Calif Inst Technol, vis assoc, 63-68, 83-86. *Mem:* Am Math Soc; Soc Indust & Appl Math; Math Asn Am; Asn Comput Mach. *Res:* Combinatorial analysis; mathematical theory of vision; computers and computation; symbolic manipulation. *Mailing Add:* 4705 Daleridge Rd La Canada CA 91011

BERGS, VICTOR VISVALDIS, b Kuldiga, Latvia, Dec 20, 23; US citizen; m 60; c 1. VIROLOGY. *Educ:* Boston Univ, AM, 55; Univ Pa, PhD(med microbiol), 58. *Prof Exp:* Res asst virol, Children's Med Ctr, Boston, 54-55; res assoc, Children's Hosp, Philadelphia, 58; asst prof, Inst Microbiol, Rutgers Univ, 58-62; res virologist, Stanford Res Inst, 63-65; from asst prof to assoc prof microbiol, Sch Med, Univ Miami, 65-73; ASSOC VIROLOGIST & MGR, LIFE SCI, INC, 73- *Concurrent Pos:* Asst dir, Variety Children's Res Found, 66-70. *Mem:* Am Asn Cancer Res; Am Asn Immunol; Soc Exp Biol & Med; fel Am Acad Microbiol. *Res:* Cocarcinogens and oncogenic viruses; immuno prevention and chemotherapy of virus-induced diseases; tissue culture. *Mailing Add:* 14401 Tanglewood Dr N Largo FL 34644

BERGSAGEL, DANIEL EGIL, b Outlook, Sask, Apr 25, 25; m 50; c 4. INTERNAL MEDICINE, ONCOLOGY. *Educ:* Univ Man, MD, 49; Oxford Univ, DPhil, 55; Am Bd Internal Med, dipl, 59; FRCP(C), 68. *Prof Exp:* Asst resident, Winnipeg Gen Hosp, 50-51 & Salt Lake County Gen Hosp, Utah, 51-52; assoc internist, Univ Tex M D Anderson Hosp, 55-65; PROF MED, UNIV TORONTO, 65-; CHIEF MED, PRINCESS MARGARET HOSP, 65- *Mem:* AMA; Am Fedn Clin Res; Int Soc Hemat. *Res:* Tumor growth; cancer chemotherapy. *Mailing Add:* 20 Highland Cir Toronto ON M4W 2S7 Can

BERGSTEIN, JERRY MICHAEL, PEDIATRICS, NEPHROLOGY. *Educ:* Univ Minn, MD, 65. *Prof Exp:* PROF PEDIAT & HEAD PEDIAT NEPHROLOGY, SCH MED, IND UNIV, 77- *Mailing Add:* James Whitcomb Riley Hosp Children Univ Ind Sch Med Indianapolis IN 46223

BERGSTEN, JANE WILLIAMS, b Detroit, Mich, June 25, 27; m 56; c 3. SURVEY SAMPLE DESIGN. *Educ:* Univ Mich, BA, 49, MA, 51; Univ Iowa, PhD(appl statist), 72. *Prof Exp:* Sampling statistician, Surv Res Ctr, Univ Mich, 49-56; statistician, Nat Ctr Health Statist, 56-60; SR STATISTICIAN, RES TRIANGLE INST, 73- *Concurrent Pos:* Res psychologist, Am Col Testing, 73. *Mem:* Am Statist Asn; Int Asn Surv Statisticians; Am Educ Res Asn. *Res:* Survey design and implementation; sample design and selection; data analysis and statistical report writing; project management and direction; methodological research. *Mailing Add:* Ctr Res & Statist Res Triangle Inst Research Triangle Park NC 27709

BERGSTRALH, JAY THOR, b Minneapolis, Minn, Aug 23, 43; m 66; c 3. PLANETARY SCIENCE. *Educ:* Carleton Col, BA, 65; Univ Tex, Austin, MA, 68, PhD(astron), 72. *Prof Exp:* Nat Res Coun fel, 72-74, sr scientist planetary sci, 74-79, MEM TECH STAFF, JET PROPULSION LAB, 79- *Mem:* Am Astron Soc; Int Astron Union; Sigma Xi. *Res:* Radiative transfer, especially spectroscopic line formation in the atmospheres of the giant planets. *Mailing Add:* Jet Propulsion Lab 4800 Oak Grove Dr Pasadena CA 91109

BERGSTRESSER, KENNETH A, b Lower Saucon Township, Pa, May 25, 12; m 39; c 3. BIOLOGY, ZOOLOGY. *Educ:* Albright Col, BS, 34; Univ Pittsburgh, MS, 37; Lehigh Univ, PhD(cell biol), 74. *Prof Exp:* Asst prof biol, Beaver Col, 37-41; field dir, Am Red Cross, 43-45; assoc prof biol, Moravian Col, 46-52; chief bacteriologist, R K Laros Co, 52-58; assoc prof biol, 58-71, prof, 71-77, EMER PROF BIOL, MORAVIAN COL, 77- *Concurrent Pos:* Northhampton County Comm Col, 77-82. *Mem:* AAAS; Am Soc Microbiol; Sigma Xi; Am Soc Zoologists. *Res:* Bacteriology; industrial fermentations; cytology; cell biology. *Mailing Add:* 2066 Chester Ave Apt 4B Bethlehem PA 18017

BERGSTRESSER, PAUL RICHARD, b Ottawa, Kans, Aug 24, 41; m 69; c 2. DERMATOLOGY. *Educ:* Sch Med, Stanford Univ, MD, 68; Am Bd Dermat, cert, 76. *Prof Exp:* Intern med, Sch Med, Univ NMex, 68-69; resident dermatol, Sch Med, Stanford Univ, 69-70; resident, Sch Med, Univ Miami, 72-74, res assoc, 74-75, asst prof dermatol, 75-76; asst prof, 76-80, ASSOC PROF INTERNAL MED, HEALTH SCI CTR, UNIV TEX, DALLAS, 80- *Mem:* Soc Invest Dermatol; fel Am Col Physicians; Am Acad Dermatol; Am Fedn Clin Res; AAAS. *Res:* Immunology in dermatology, control mechanisms in epidermal proliferation. *Mailing Add:* Dept Dermat Univ Tex Southwestern Med Ctr 5323 Harry Hines Blvd Dallas TX 75235

BERGSTROM, CLARENCE GEORGE, b Chicago, Ill, Oct 23, 25; m 48; c 3. ORGANIC CHEMISTRY. *Educ:* Ill Inst Technol, PhD(chem), 51. *Prof Exp:* res chemist, G D Searle & Co, 50-77, head, Stability Lab, 80-86, consult, 86-89; RETIRED. *Mem:* Am Chem Soc. *Res:* Steroids; synthetic organic chemistry; medicinal chemistry; stability testing; pharmaceutical products. *Mailing Add:* 101 Summit Ave Apt 614 Park Ridge IL 60068

BERGSTROM, DONALD EUGENE, b Tacoma, Wash, Oct 9, 43; m 70; c 3. NUCLEIC ACID CHEMISTRY. *Educ:* Univ Wash, BS, 65; Univ Calif, Berkeley, PhD(chem), 70. *Prof Exp:* Res assoc, Univ Ill, 70-72; asst prof org chem, Rockefeller Univ, 72-74 & Univ Calif, Davis, 74-80; assoc prof, 80-85, PROF ORG CHEM, UNIV NDAK, 86- *Concurrent Pos:* Consult, Riker Labs, 3M, St Paul, Minn, 82-; mem, bio-organic & natural prods study sect, Nat Inst Health, 88-91. *Honors & Awards:* Sci Res Soc Fac Award Outstanding Res, Sigma Xi, 85. *Mem:* Am Chem Soc; AAAS; NY Acad Sci; Sigma Xi. *Res:* Synthetic, organic and bioorganic chemistry; synthesis and reactions of nucleosides; nucleotides and nucleic acids; organopalladium chemistry; synthesis of fluorinated bioactive compounds; design and synthesis of antiviral agents. *Mailing Add:* Dept Chem Box 7185 Univ Sta Univ NDak Grand Forks ND 58202

BERGSTROM, GARY CARLTON, b Chicago, Ill, May 12, 53; m 83; c 1. DISEASES OF FIELD CROPS, PHYSIOLOGY OF PARASITISM. *Educ:* Purdue Univ, BS, 75, MS, 78; Univ Ky, PhD(plant path), 81. *Prof Exp:* Teaching asst biol, Purdue Univ, 75-76, res asst plant path, 76-77; res asst plant path, Univ Ky, 78-81; asst prof, 81-87, ASSOC PROF PLANT PATH, CORNELL UNIV, 87- *Concurrent Pos:* Co-chmn NY State Pest Control Conf, 84; pres, Northeast Corn Improv Conf, 87-; team leader, Field Crops, Fungicide Assessment Proj, Nat Agr Pesticide Impact Assessment Prog-USDA, 89-91, chair, NCent Regional Corn & Sorghum Dis, 91. *Mem:* Am Phytopath Soc; Am Soc Agron; Crop Sci Soc Am; Int Soc Plant Path; Can Soc Plant Path. *Res:* Biology and management of leaf and stem diseases of corn, small grain cereals, and forage legumes of importance in New York State. *Mailing Add:* Dept Plant Path 316 Plant Sci Bldg Cornell Univ Ithaca NY 14853-5908

BERGSTROM, K SUNE D, b Stockholm, Sweden, Jan 10, 16; m 43. CHEMISTRY. *Educ:* Karolinska Inst, MD & DMedSci, 44. *Hon Degrees:* Dr, Univ Basel, 60, Univ Chicago, 60, Harvard Univ, 76, Mt Sinai Med Sch, 76, Med Acad Wroclaw, 76, DSc, McMaster Univ, 88. *Prof Exp:* Asst, Biochem Dept, Med Nobel Inst, Karolinska Inst, Stockholm, 44-47; prof physiol chem, Univ Lund, 47-58; prof chem, Karolinska Inst, 58-80, dean med fac, 63-66, rector, 69-77, PROF, DEPT BIOCHEM, KAROLINSKA INST, SWEDEN, 90- *Concurrent Pos:* Res fel, Basel Univ, 46-47; chmn, bd dirs, Nobel Found, 75-87; Albert Lasker Basic Med Res Award, 77; chmn adv comt med res, WHO, 77-82; chmn Sci & Tech Adv Group Human Reproduction, WHO, Geneva, 85-90. *Honors & Awards:* Dunham Lectr, Harvard Univ, 72; Dohme Lectr, Johns Hopkins Univ, 73; Merrimon Lectr, Univ NC, 73; V D Mattia Lectr, Roche Inst, 74; Harvey Lectr, Harvey Soc, 74; Francis Amory Prize, Am Acad Arts & Sci, 75; Benjamin Franklin Medal, Am Philos Soc, 88. *Mem:* Inst Med-Nat Acad Sci; Royal Swed Acad Sci (pres, 65); Am Acad Arts & Sci; Am Soc Biol Chemists; Am Philos Soc. *Res:* Biochemistry; physiology. *Mailing Add:* Karolinska Inst Solnavagen 1 POB 60250 Stockholm 104 01 Sweden

BERGSTROM, ROBERT CHARLES, b Newcastle, Wyo, Mar 23, 22; m 48; c 2. VETERINARY PARASITOLOGY. *Educ:* Colo State Univ, BS, 50; Univ Wyo, MS, 56, PhD(zool), 64. *Prof Exp:* Surv supvr, Bur Entom & Plant Quarantine, USDA, 50-51; instr high sch, Wyo & Mont, 51-57; technologist, Dept Entom & Parasitol, Univ Wyo, 57-58; instr high sch, Wyo, 58-61; teaching asst zool, 61-62, PARASITOLOGIST, DIV MICROBIOL & VET MED, UNIV WYO, 62- *Concurrent Pos:* NIH fel, Med Sch, Mex, 66; res vet micro & pathol, Wash State Univ, 81. *Mem:* Am Soc Parasitologists; World Asn Advan Vet Parasitol; Am Asn Vet Parasitol; NY Acad Sci; Sigma Xi. *Res:* Helminths of domestic animals, especially research physiological responses of the host animal to nematode parasite infections; dual infections; domestic ruminants. *Mailing Add:* Dept Vet Sci Univ Wyo Box 950 Laramie WY 82070

BERGSTROM, ROBERT EDWARD, b Rock Island, Ill, Mar 27, 23; m 46; c 3. GEOLOGY. *Educ:* Augustana Col (Ill), AB, 47; Univ Wis, MS, 50, PhD(geol), 53. *Prof Exp:* Instr geol, Beloit Col, 50-51; integrated lib studies, Univ Wis, 52-53; asst geologist, Ill Geol Surv, 53-55, assoc geologist, 55-59, geologist, 59-61; proj geologist, Kuwait, Parsons Corp, 61-63; geologist & head groundwater geol & geophys explor sect, 63-74, prin geologist, 74-81, chief, 81-83, EMER CHIEF, ILL GEOL SURV, 83- *Concurrent Pos:* Mem comt geol sci, Nat Acad Sci, 69-71; mem comt surface mining & reclamation, Nat Res Coun, 78-80. *Mem:* Geol Soc Am; Am Inst Prof Geol; Asn Groundwater Scientists & Engrs. *Res:* Ground water geology; geology of Kuwait; geology of waste disposal. *Mailing Add:* 2004 Galen Dr Champaign IL 61821

BERGSTROM, STIG MAGNUS, b Skovde, Sweden, June 12, 35; m 82. GEOLOGY, INVERTEBRATE PALEONTOLOGY. *Educ:* Lund Univ, Sweden, Fil Kand, 59, PhD(geol), 61. *Hon Degrees:* Dr Lund Univ, 87. *Prof Exp:* Amanuensis geol, Lund Univ, 58-60; res asst, Ohio State Univ, 60-61; lectr, Lund Univ, 62-68; from asst prof & cur to assoc prof & cur, 68-72, PROF GEOL, OHIO STATE UNIV, 72- *Concurrent Pos:* Scand-Am Found fel, 64. *Mem:* Geol Soc Am; Paleont Soc; Paleont Asn; Am Asn Petrol Geologists. *Res:* Stratigraphy, fossils and geologic history of the lower Paleozoic of North America and Northwestern Europe. *Mailing Add:* Dept Geol Sci Ohio State Univ 125 S Oval Mall Columbus OH 43210

BERGSTROM, WILLIAM H, b Bay City, Mich, Jan 1, 21; m 44; c 4. PEDIATRICS. *Educ:* Amherst Col, BA, 42; Univ Rochester, MD, 45. *Prof Exp:* From assoc prof to prof, 55-77, EMER PROF PEDIAT, STATE UNIV NY MED CTR, 85- *Concurrent Pos:* Adj prof biochem, 76-85. *Mem:* AAAS; Am Pediat Soc; Soc Pediat Res; Soc Exp Biol & Med; Am Soc Bone & Mineral Res. *Res:* Electrolyte metabolism; blood-bone equilibria. *Mailing Add:* Dept Pediat State Univ NY Upstate Med Ctr Syracuse NY 13210

BERGTROM, GERALD, b New York, NY, Feb 14, 45. MOLECULAR BIOLOGY, INSECT PHYSIOLOGY. *Educ:* City Col New York, BSc, 67; Brandeis Univ, PhD(biol), 74. *Prof Exp:* Fel insect develop, Univ Conn, 73-75; asst prof biol, Vanderbilt Univ, 75-78; asst prof, 78-81, ASSOC PROF BIOL SCI, UNIV WIS-MILWAUKEE, 81- *Mem:* AAAS; Am Soc Cell Biol; Am Soc Zoologists; Am Soc Microbiol. *Res:* Hemoglobin synthesis during insect development and in response to insect growth homones; molecular biology of hemoglobin synthesis, assembly and secretion; insect globin gene structure and regulation; molecular biology of fibronectin synthesis by cultured chick hepatocytes; molecular biology of plant drought tolerance. *Mailing Add:* Dept Biol Sci Univ Milwaukee Milwaukee WI 53201

BERI, AVINASH CHANDRA, b Jullunder, India, Oct 28, 49. HYPERFINE INTERACTIONS, SURFACE CHEMICAL PHYSICS. *Educ:* Univ Delhi, BSc Hons, 69; State Univ NY, Albany, MS, 74, PhD(physics), 79. *Prof Exp:* Fel, 79-80, RES ASSOC CHEM, UNIV ROCHESTER, 80- *Mem:* Am Phys Soc; Am Asn Physics Teachers; Sigma Xi. *Res:* Optical magnetic and hyperfine properties of ionic solids; phenomenological and ab initio study of laser stimulated processes at a gas-solid interface. *Mailing Add:* 108 York St Rochester NY 14611

BERING, CHARLES LAWRENCE, b Houston, Tex, July 29, 47. BIOCHEMISTRY, MICROBIOLOGY. *Educ:* Univ Houston, BS, 70; Purdue Univ, PhD(biochem), 75. *Prof Exp:* Res assoc & lectr biochem, Northwestern Univ, 75-78; chmn dept nat sci, 80-84, assoc prof chem, St ate Univ NY, Col Technol, 78- 88; ASSOC PROF CHEM, CLARION UNIV PA, 88- *Concurrent Pos:* Workshop dir, Monroe County Community Col, 84; sci adv, Cobleskill Agr & Technol Col, 84-87; vis prof, Imperial Col, London, 85; vis res assoc, Univ Col-Cardiff. *Mem:* AAAS; Am Chem Soc; Am Soc Biochem & Molecular Biol; Am Soc Microbiol. *Res:* Microbial degradation of aromatic compounds; microbial synthesis of rare intermediates in path of degradation. *Mailing Add:* Dept Chem Clarion Univ Pa Clarion PA 16214

BERING, EDGAR ANDREW, JR, b Salt Lake City, Utah, Feb 18, 17; m 44; c 3. MEDICINE, NEUROSURGERY. *Educ:* Univ Utah, AB, 37; Harvard Univ, MD, 41. *Prof Exp:* Surg house officer, Boston City Hosp, 41-42; spec res assoc phys chem, Med Sch, Harvard Univ, 42; asst neurosurg, spec res assoc & demonstr anat, NY Med Col & Flower & Fifth Ave Hosps, 46-48; Moseley traveling fel from Harvard Med Sch, Nat Hosp Queens Sq, London, 48-49; resident neurosurg, Children's Hosp, Boston, 49-50, asst neurosurgeon, 50-54; dir neurosurg res lab, Children's Med Ctr, 52-63, assoc neurosurgeon, 54-64; actg chief prog anal, Nat Inst Neurol Dis & Stroke, 64-65, spec asst to dir, 65-71, chief spec proj br, 71-74; assoc clin prof neurosurg, Sch Med, Georgetown Univ, 69-83; RETIRED. *Concurrent Pos:* Resident, Peter Bent Brigham Hosp, Boston, 49-50, jr assoc, 50; res fel surg, Harvard Med Sch, 50-52; Cushing fel, 50-51; Nat Res Coun sr fel, 50-52; from asst to asst clin prof, Harvard Med Sch, 52-64; mem, Am Bd Neurol Surg, 52; from consult to sr consult, Lemuel Shattuck Hosp, Boston, 54-64; attend neurosurgeon, West Roxbury Vet Admin Hosp, 55-64; vis lectr, Univ Calif, Los Angeles, 58; vis scientist, Nat Inst Neurol Dis & Stroke, 63-64; chief of staff, Mem Hosp, Easton, Md, 83-85. *Mem:* AAAS; Am Asn Neurol Surg; Neurosurg Soc Am; Am Acad Neurol; Soc Neurosci. *Res:* Metabolism and physiology of the nervous system, cerebrospinal fluid; problems relating to the surgery of the nervous system; biophysics. *Mailing Add:* 4211 Windrush Rd Oxford MD 21654

BERING, EDGAR ANDREW, III, b New York, NY, Jan 9, 46; m 85; c 1. MAGNETOSPHERIC PHYSICS, SPACE PLASMA PHYSICS. *Educ:* Harvard Col, BA, 67; Univ Calif, Berkeley, PhD(physics), 74. *Prof Exp:* Res sci, 74, from vis asst prof to assoc prof, 75-89, PROF PHYSICS, UNIV HOUSTON, 89- *Concurrent Pos:* Partner, 84-87, vpres, IF&G Tech Consults Inc, 87- *Mem:* Am Geophys Union; Am Astron Soc; Sigma Xi; AAAS; NY Acad Sci. *Res:* Study of the earth's magnetosphere and ionosphere; electrodynamics of the polar cusp; development of an x-ray camera for auroral studies. *Mailing Add:* Dept Physics Univ Houston Houston TX 77204-5504

BERINGER, ROBERT, b Pittsburgh, Pa, Oct 14, 17; m 42; c 3. PHYSICS. *Educ:* Washington & Jefferson Col, BS, 39; Yale Univ, PhD(physics), 42. *Prof Exp:* Asst radiation lab, Mass Inst Technol, 42-46; asst prof physics, 46-50, from assoc prof to prof physics, Yale Univ, 50-, dir heavy ion accelerator lab, 88; RETIRED. *Mem:* Am Phys Soc. *Res:* Counting techniques; microwave physics; electronics; nuclear particle accelerators. *Mailing Add:* 3997 Whitney Ave New Haven CT 06517

BERINGER, THEODORE MICHAEL, May1, 44. CELL BIOLOGY, CELLULAR ULTRASTRUCTURE. *Educ:* Loyola Univ, PhD(anat), 72. *Prof Exp:* ASSOC PROF HISTOL & CELL BIOL, UNIV MO, 80- *Mem:* Am Soc Cell Biol; Am Asn Anatomists. *Res:* Liposome mediated chemotherapy; renal transport mechanisms. *Mailing Add:* Univ Mo 2411 Holmes Kansas City MO 64108

BERK, ARISTID D, b Istanbul, Apr 28, 25; m 48; c 2. ELECTRICAL ENGINEERING. *Educ:* Robert Col, Istanbul, BS, 47; Rensselaer Polytech Inst, MS, 51; Mass Inst Technol, ScD(elec eng), 54. *Prof Exp:* Mem tech staff, Hughes Aircraft Co, 54-55, group head, 55-56, asst sect head & sr mem tech staff, 56-58, sect head solid state sect, 58-60; exec vpres, Micromega Corp, 60-75; vpres eng, Electronic Systs Div, Sams Bunker Ramo Corp, 76-77, pres, Amphenol Div, 77-82; RETIRED. *Concurrent Pos:* Lectr, Univ Calif, Los Angeles, 55-58. *Mem:* Inst Elec & Electronics Engrs; Am Phys Soc; Sigma Xi. *Res:* Electromagnetic theory; ferromagnetism at microwave frequencies. *Mailing Add:* 24603 Skyline View Dr Malibu CA 90265

BERK, ARNOLD J, b Los Angeles, Calif, Dec 23, 49; m 71; c 2. MICROBIOLOGY. *Educ:* Univ Calif, Berkeley, BA, 71; Stanford Univ, MD, 76. *Prof Exp:* Postdoctoral molecular biol, Mass Inst Technol, 76-79; from asst prof to assoc prof microbiol, 79-86, PROF MICROBIOL & MOLECULAR GENETICS, UNIV CALIF, LOS ANGELES, 86-, CHAIR DEPT, 88- *Concurrent Pos:* Mem, Sci Adv Bd, Amgen, Inc, Thousand Oaks, Calif, 80- & Genetics Panel, NSF, 84-88; ed, Virol, 85-; chmn, Gordon Conf Animal Cells & Viruses, 86. *Honors & Awards:* Trygve Tuve Lectr, NIH, 79, Merit Award, 87. *Mem:* Am Soc Microbiol; Fedn Am Socs Exp Biol. *Res:* Biochemistry of transcription initiation by RNA polymerases II and III; transformation by adenovirus EIB proteins. *Mailing Add:* Molecular Biol Inst Univ Calif Los Angeles CA 90024-1570

BERK, BERNARD, b Brooklyn, NY, Dec 23, 15; m 40; c 1. CHEMISTRY. *Educ:* Ind Univ, BA, 37, MA, 38, PhD(org chem), 41; NY Univ, MChE, 47. *Prof Exp:* Jr chemist, Ind Ord Works, 41-42; res chemist, Lawrence Richard Bruce, Conn, 42-46; res & develop chemist & head dept, Tech Serv Lab, 46-66, mgr mfg, Overseas Plants, 53-55, assoc tech dir, Overseas Div, 55-58, dir, Steroid Prep Lab, Squibb Inst Med Res, 58-66, dir, Res Prod Lab, 66-70, MGR, ANTIBIOTIC MFG PLANNING, E R SQUIBB & SONS, 70-, DIR CHEM DEVELOP, 79- *Mem:* Assoc Am Chem Soc; Am Inst Chem Eng; assoc NY Acad Sci. *Res:* Organic synthesis; antibiotics; wool dyeing processes; vapor phase catalytic reduction over intermetallic catalysts; crystallization of penicillin salts. *Mailing Add:* 45 Ocean Ave No 5H Monmouth Beach NJ 07750

BERK, JACK EDWARD, b Philadelphia, Pa, Nov 24, 11; m 37; c 2. MEDICINE. *Educ:* Univ Pa, BA, 32, MSc, 39, DSc(med), 43; Jefferson Med Col, MD, 36; Am Bd Internal Med & Am Bd Gastroenterol, dipl, 43. *Prof Exp:* Instr gastroenterol, Grad Sch Med, Univ Pa, 41-46; asst prof, Med Sch, Temple Univ, 46-54, res assoc & asst dir, Fels Res Inst, 46-54; assoc prof clin med, Wayne State Univ, 54-62, prof clin med, 62-63; prof med & chmn dept, Univ Calif, Irvine, 63-73, head, Div Gastroenterol, 63-79, asst dean, 79-90, DISTINGUISHED PROF, UNIV CALIF, IRVINE, 79- *Concurrent Pos:* Consult, Surgeon Gen, US Army, 47-, Cedars Sinai & White Mem Hosps, Los Angeles, Mem Hosp, Long Beach, 63-, Long Beach Vet Admin Hosp, 63- & Sinai Hosp Detroit, 63- dir dept med, Sinai Hosp, Detroit, Mich, 54-63. *Honors & Awards:* Schindler Award, Am Soc Gastrointestinal Endoscopy, 66; Rorer Award, Am Col Gastroenterol, 70, 74, 78 & 79; Distinguished Physician Award, Nat Found for Ileitis & Colitis, 80; Distinguished Sci Achievement Award, Am Col Gastroenterol, 82; Clin Achievement Award, Am Col Gastroenterol, 88; Laureate Award, Am Col Physicians, 90- *Mem:* Am Soc Gastrointestinal Endoscopy (pres, 58); Bockus Int Soc Gastroenterol (pres, 67-71); master Am Col Physicians; Am Gastroenterol Asn; master Am Col Gastroenterol (pres, 75-76). *Res:* Characterization of serum amylase; acidity of the first part of duodenum in ulcer; tetracycline fluorescence in gastric lesions. *Mailing Add:* Univ Calif Irvine Med Ctr 101 The City Dr Irvine CA 92668

BERK, KENNETH N, b Takoma Park, Md, Sept 24, 38; m 68; c 2. MATHEMATICS. *Educ:* Carnegie Inst Technol, BS, 60; Univ Minn, PhD(math), 65. *Prof Exp:* Asst prof math, Northwestern Univ, 65-68; asst prof, Northeastern Ill State Col, 68-69; assoc prof, 69-78, PROF MATH, ILL STATE UNIV, 78- *Concurrent Pos:* Chair, Am Statist Asn, Statist Comput Sect, 89. *Honors & Awards:* Youden Prize, Am Soc Qual Control, 78. *Mem:* Inst Math Statist; fel Am Statist Asn; Biomet Soc; Int Asn Statist Comput. *Res:* Statistical computing; regression analysis; time series. *Mailing Add:* Dept Math Ill State Univ Normal IL 61761

BERK, LAWRENCE B, ophthalmology, electrochemistry & air electrodes, for more information see previous edition

BERK, RICHARD SAMUEL, b Chicago, Ill, Oct 7, 28. MICROBIOLOGY, BIOCHEMISTRY. *Educ:* Roosevelt Univ, BS, 50; Univ Minn, MS, 55; Univ Chicago, PhD(microbiol), 58. *Prof Exp:* Researcher, Univ Calif, Los Angeles, 58-61; sr res biochemist, Magna Prods, Calif, 61-62; assoc prof microbiol, 62-69, PROF MICROBIOL, COL MED, WAYNE STATE UNIV, 69- *Mem:* Am Chem Soc; Am Soc Microbiol; Soc Exp Biol & Med. *Res:* Bacteriocins; endotoxins; sulfur metabolism; bacterial cytology; electron microscopy; pathological study of exotoxins. *Mailing Add:* Col Med Wayne State Univ Detroit MI 48201

BERK, ROBERT NORTON, b Pittsburgh, Pa, Sept 3, 30; m 56; c 3. RADIOLOGY. *Educ:* Univ Pittsburgh, BS, 51, MD, 55. *Prof Exp:* From instr to asst prof radiol, Sch Med, Univ Pittsburgh, 61-65; chief radiol, Passavant Hosp, Pittsburgh, 66-68; assoc prof, Sch Med, Univ Calif, San Diego, 68-74; prof radiol & chmn dept, Univ Tex Health Sci Ctr, Dallas, 74-77; PROF RADIOL & CHMN DEPT, UNIV CALIF, SAN DIEGO, 77- *Concurrent Pos:* Consult radiol, Balboa Naval Hosp, San Diego & Vet Admin Hosp, La Jolla, 70-74; dir radiol, Parkland Mem Hosp, Dallas, 74-77; consult radiol, Vet Admin Hosp, Dallas, 74-77 & Baylor Univ Med Ctr, 74-77; chief radiol serv, Children's Med Ctr, Dallas, 74-77. *Mem:* Asn Univ Radiologists; Am Col Radiol; Radiol Soc NAm; Am Roentgen Ray Soc; Soc Gastrointestinal Radiologists. *Res:* Pharmacokinetics of radiographic contrast materials for cholangiography and cholecystography. *Mailing Add:* Dept Radiol H-756 Univ Calif-San Diego Box 109 La Jolla CA 92093

BERK, TOBY STEVEN, b Chicago, Ill, Jan 15, 44; m 65. COMPUTER SCIENCES. *Educ:* Univ Mich, BSE, 65; Purdue Univ, MS, 68, PhD(comput sci), 72. *Prof Exp:* Instr comput sci, Ind Univ-Purdue Univ, Indianapolis, 70-72; from asst prof to assoc prof, 72-86, chmn, math sci dept, 81-85, prof comput sci, Fla Int Univ, 87; ASSOC DIR, SCH COMPUT SCI, 87- *Concurrent Pos:* Vis assoc prof, Ohio State Univ, 80-81; vis prof, Univ Colo, 89-90. *Mem:* Asn Comput Mach; Inst Elec & Electronics Engrs. *Res:* Computer graphics, graphic languages, distributed processing. *Mailing Add:* Sch Comput Sci Fla Int Univ Miami FL 33199

BERKA, LADISLAV HENRY, b Bay Shore, NY, July 16, 36; m 59; c 2. INORGANIC CHEMISTRY. *Educ:* Union Col, NY, BS, 57; Univ Calif, Berkeley, MS, 60; Univ Conn, Storrs, PhD(chem), 65. *Prof Exp:* Teacher chem, Mineola High Sch, 59-61; instr, Univ Conn, 64-65; asst prof, 65-69, assoc prof, 69-79, PROF CHEM, WORCESTER POLYTECH INST, 79- *Concurrent Pos:* Consult, Sherman Assocs, 76-; prog dir, Cent Maine Metric Consortium, 77- *Honors & Awards:* Gustav Ohaus Award, Nat Sci Teachers Asn, 77. *Mem:* Am Chem Soc; Sigma Xi. *Res:* Synthesis and characterization of coordination compounds; protonated organic bases; anomalous liquids. *Mailing Add:* 14 Walbridge Rd Paxton MA 01612

BERKEBILE, CHARLES ALAN, b Jamaica, NY, Mar 4, 38; m 79; c 1. GROUNDWATER, COMPUTER MODELING. *Educ:* Allegheny Col, BS, 60; Boston Univ, AM, 61, PhD(geol), 64. *Prof Exp:* Mem staff crystal growth, Lab for Insulation Res, Crystal Physics Sect, Mass Inst Technol, 63-64; asst prof earth sci, Southampton Col, Long Island, 64-67; mem res staff, Tech Staffs Div, Corning Glass Works, 67-69; assoc prof geol & chmn dept, Southampton Col, Long Island Univ, 69-75, prof geol & marine sci, 75-79 & 80-81; PROF GEOL, DIV SCI, COL SCI & TECHNOL, CORPUS CHRISTI STATE UNIV, 81-, CHAIRPERSON, 90- *Concurrent Pos:* Vis assoc scientist & res collab, Crystal Struct Anal Group, Chem Dept, Brookhaven Nat Lab, 66-67; environ consult; vis sr res geologist, Dept Geol & Geophys Sci, Princeton Univ, 79-80. *Mem:* Fel Geol Soc Am; Nat Water Well Asn; Nat Asn Geol Teachers; Asn Ground Water Scientists & Engrs. *Res:* Ground-water resources; field geology. *Mailing Add:* Col Sci & Technol Corpus Christi State Univ Corpus Christi TX 78412

BERKELHAMER, LOUIS H(ARRY), b Chicago, Ill, Apr 23, 12; m 38; c 2. CERAMICS, MINERAL TECHNOLOGY. *Educ:* Univ Ill, BS, 33; Univ Wash, MS, 36. *Prof Exp:* Res ceramic eng, Ill State Geol Surv, 34; res engr ceramic div, Battelle Mem Inst, 36-38; res engr & petrographer, Saranac Lab, Trudeau Found, NY, 38-41; mineral technologist, US Bur Mines, 41-45; ceramic engr, Ohmite Mfg Co, 45-50, dir res & chief ceramics engr, 50-64, dir res & develop eng, 64-68, vpres res & develop eng, 68-71, vpres eng, 71-78, consult eng, 78-81; RETIRED. *Concurrent Pos:* Tech ed, Ceramic Catalogs & Cyclopedia, 34. *Mem:* Am Ceramic Soc; Nat Inst Ceramic Eng; Sigma Xi. *Res:* Properties of clays and non-metallic minerals; mineralogical aspects of silicosis and pneumoconiosis problems; differential thermal analysis studies of non-metallic minerals; properties and composition of molding sands; study of domestic talcs; coatings and composition for electrical resistors and rheostats; new electronic components. *Mailing Add:* 1051 Hohfelder Rd Glencoe IL 60022

BERKELHAMMER, GERALD, b Newark, NJ, Feb 3, 31; m 54; c 3. ORGANIC CHEMISTRY. *Educ:* Brown Univ, AB, 52; Univ Wash, PhD(org chem), 57. *Prof Exp:* Res chemist, 57-60, group leader, 60-70, mgr org synthesis, 70-85, DIR CHEM DISCOVERY, AGR RES DIV, AM CYANAMID CO, 85- *Mem:* Am Chem Soc; AAAS. *Res:* Agricultural and medicinal chemistry. *Mailing Add:* Agr Res Div Am Cyanamid Co Box 400 Princeton NJ 08543-0400

BERKELHAMMER, JANE, TUMOR IMMUNOLOGY. *Educ:* Purdue Univ, PhD(immunol), 72. *Prof Exp:* SCIENTIST, AMC CANCER RES CTR, LAKEWOOD, COLO, 83- *Res:* Melanoma; metastasis. *Mailing Add:* Dept Cell Biol AMC Cancer Res Ctr 1600 Pierce St Lakewood CO 80214

BERKELMAN, KARL, b Lewiston, Maine, June 7, 33; m 59; c 3. EXPERIMENTAL HIGH ENERGY PHYSICS. *Educ:* Univ Rochester, BS, 55; Cornell Univ, PhD(physics), 59. *Prof Exp:* Instr & res assoc physics, 59-60, from asst prof, 61-67, PROF PHYSICS, CORNELL UNIV, 67-, DIR LAB NUCLEAR STUDIES, 85- *Concurrent Pos:* NSF fels, Ital Nat Synchrotron Lab, Frascati, 60-61; European Orgn Nuclear Res, Geneva, Switz, 60-61 & 67-68; sci staff mem, Ger Electron Synchrotron, Hamburg, 74-75. *Mem:* Am Phys Soc. *Res:* Meson photoproduction experiments; high energy electron scattering experiments; phenomenological theory of high energy particle reactions; electron-positron colliding beam experiments. *Mailing Add:* Lab Nuclear Studies Cornell Univ Ithaca NY 14853

BERKER, AHMET NIHAT, b Istanbul, Turkey, Sept 20, 49; div; c 1. PHASE TRANSITION THEORY, STATISTICAL MECHANICS. *Educ:* Mass Inst Technol, BS(chem) & BS(physics), 71; Univ Ill, Urbana, MS, 72, PhD(physics), 77. *Prof Exp:* Res fel physics, Harvard Univ, 77-79; from asst prof to assoc prof, 79-88, PROF PHYSICS, MASS INST TECHNOL, 88- *Mem:* Am Phys Soc; Am Asn Univ Professors. *Res:* Statistical mechanics; theory of phase transitions and critical phenomena in condensed matter systems. *Mailing Add:* Dept Physics 12-135 Mass Inst Technol Cambridge MA 02139

BERKES, JOHN STEPHAN, b Buzias, Rumania, Sept 5, 40; US citizen; m 69; c 4. XEROGRAPHIC MATERIALS & PROCESS. *Educ:* Mich Technol Univ, BS, 62; Pa State Univ, MS, 64, PhD(solid state sci), 68. *Prof Exp:* Scientist mat sci, 68-73, mgr, 73-82, SR MEM RES STAFF, XEROX CORP, 82- *Mem:* Soc Photog Scientists & Engrs; Am Vacuum Soc; Sigma Xi. *Res:* Structure, thermal properties, viscous flow, bulk and surface stability of amorphous chalcogenide photoconducting materials; thin films technology; xerographic materials and process. *Mailing Add:* 800 Phillips Rd Bldg 114 Webster NY 14580

BERKEY, DENNIS DALE, b Wooster, Ohio, May 27, 47. DIFFERENTIAL EQUATIONS, APPLIED ANALYSIS. *Educ:* Muskingum Col, BA, 69; Miami Univ, MA, 71; Univ Cincinnati, PhD(math), 74. *Prof Exp:* Instr math, Miami Univ, 73-74; asst prof, 74-79, chmn dept, 79-83, assoc vpres acad affairs, 83-85, vice provost, 85-87, ASSOC PROF MATH, BOSTON UNIV, 79-, PROVOST, 88- *Concurrent Pos:* Danforth Assoc, 79. *Mem:* Am Math Soc; Math Asn Am; Soc Indust & Appl Math; Asn Comput Mach. *Res:* Qualitative theory for linear systems; periodic and almost periodic solution of differential systems; perturbation methods; bifurcation problems; optimal control theory. *Mailing Add:* Provost's Off Boston Univ 145 Bay State Rd Boston MA 02215

BERKEY, DONALD C, ENGINEERING. *Prof Exp:* RETIRED. *Mem:* Nat Acad Eng. *Mailing Add:* Three Fiddlers Landing Harwichport MA 02646

BERKEY, GORDON BRUCE, b DuBois, Pa, June 1, 42; m 64; c 2. ASTROPHYSICS, PHYSICS. *Educ:* Cornell Univ, AB, 64; Purdue Univ, MS, 67, PhD(physics), 69. *Prof Exp:* From asst prof to assoc prof, 69-82, PROF PHYSICS, MINOT STATE COL, 82- *Mem:* Sigma Xi; Am Asn Phys Teachers. *Res:* Propagation of galactic cosmic-ray electrons; production of x-rays and gamma rays by cosmic-ray electrons. *Mailing Add:* Dept Physics Minot State Univ Minot ND 58701

BERKHEIMER, HENRY EDWARD, b Williamsport, Pa, Oct 13, 29; m 81; c 3. RUBBER CHEMISTRY. *Educ:* Dickinson Col, AB, 51; Bucknell Univ, MS, 53; Pa State Univ, PhD(org chem), 58. *Prof Exp:* Res fel, Harvard Univ, 58-59; res chemist, Jackson Lab, E I Du Pont de Nemours & Co, Inc, NJ, 59-62; sr res chemist, A E Staley Mfg Co, 62-63; res chemist, Elastomer Chem Dept, Exp Sta, E I du Pont de Nemours & Co Inc, Chestnut Run, 63-69, res chemist, 69-79, sr res chemist, 79-80, res assoc, 80-84, tech consult, Elastomer Lab, 84-88; ASST PROF, LYCOMING COL, WILLIAMSPORT, PA, 88- *Concurrent Pos:* Lectr chem, Glassboro State Col, 77 & Widener Univ, 77-81, Pa State Univ, Del Co Campus, 87. *Mem:* Am Chem Soc. *Res:* Polymer synthesis and mechanisms of polymerization; latex technology; formation and properties of films from polymer solutions and dispersions. *Mailing Add:* 2035 N Konkle Rd Montoursville PA 17754-9607

BERKHEISER, SAMUEL WILLIAM, b Ashland, Pa, July 29, 22; m 46; c 1. CLINICAL PATHOLOGY. *Educ:* Western Reserve Univ, BS, 43, MD, 46. *Prof Exp:* Intern, Geisinger Mem Hosp, Danville, Pa, 46-47; res path & bact, Albany Hosp & Med Sch, NY, 49-50; path & clin path, Youngstown Hosp Asn, Ohio, 50-52, asst pathologist, 52; from asst pathologist to assoc pathologist, Guthrie Clin, Robert Packer Hosp, Sayre, 52-55; assoc pathologist, Harrisburg Polyclin Hosp, Pa, 55-69, dir labs, 69-79; RETIRED. *Concurrent Pos:* Spec study, Armed Forces Inst Path, 52; consult, Vet Admin Hosp, Lebanon, Pa. *Mem:* Col Am Pathologists; Am Soc Clin Pathologists; AMA; Aerospace Med Asn. *Res:* Nevi and malignant melanomas of skin; tumors of thyroid and parotid gland; cysts of mediastinum and diaphragm; hypothyroidism; metabolic craniopathy; problems in diagnosis of skin lesions; epithelial hyperplasia of lung associated with cortisone and thromboembolism. *Mailing Add:* Dept Labs Harrisburg Polyclin Med Ctr Harrisburg PA 17110

BERKHOUT, AART W J, b Dinteloord, Netherlands, June 13, 39; m 66; c 2. GEOPHYSICS, GEOLOGY. *Educ:* Technol Univ Delft, MSc, 64; Queen's Univ (Ont), PhD(geol, geophys), 68. *Prof Exp:* Res scientist, Tulsa Res Ctr, Sinclair Oil Corp, 67-69; res geophysicist, Prod Res Lab, Atlantic Richfield Co, 69-71; assoc geophysicist, Mobile Oil Co, Dallas, Tex, 71-74, geophys specialist, 74-79, supvr gravity magnetic appln, 79-82, coordr seismic interpretation, 82-89, GEOPHYS INTERPRETATION COORDR, NEW EXPLOR VENTURES, MOBIL OIL, CO, DALLAS, TEX, 89-, ASSOC GEOPHYSICIST, EXPLOR SAM. *Concurrent Pos:* Polar Continental Shelf Proj, Can govt acquisition of gravity data in the Arctic. *Mem:* Soc Explor Geophysicists; Royal Netherlands Geol & Mining Soc; Netherlands Royal Inst Engrs. *Res:* Interpretation of potential field data, especially gravity and magnetic observations, in terms of geological prospects for oil or mineral accumulation; regional gravity field of the Canadian Arctic; special seismic studies; amplitude-hydrocarbon relationships; amplitude with offset related to lithology; seismic stratigraphy. *Mailing Add:* Mobil Oil Co MNEV PO Box 650232 Dallas TX 75265-0232

BERKLAND, JAMES OMER, b Glendale, CA, July 31, 30; m 66; c 2. EARTHQUAKE HISTORY & PREDICTION. *Educ:* Univ Calif, Berkeley, BS, 58; San Jose State Univ, MS, 69. *Prof Exp:* Geologic asst, US Geol Surv, 58-64, eng geologist, US Bur Reclamation, 64-69; consult, 69-72; asst prof geol, Appalachian State Univ, 72-73; COUNTY GEOLOGIST, SANTA CLARA COUNTY, 73- *Mem:* Fel Geol Soc Am; Asn Eng Geologists; AAAS; Ctr Study Early Man; Seismol Soc Am; Earthquake Eng Res Asn. *Res:* Coined the term 'seismic window'; extinct Miocene shrew named for him; over 50 scientific publications. *Mailing Add:* 14927 E Hills Dr San Jose CA 95127

BERKLAND, TERRILL RAYMOND, b Mason City, Iowa, Oct 17, 41; m 67. SCIENCE EDUCATION. *Educ:* Loras Col, BS, 64; Drake Univ, MA, 66; Univ Iowa, PhD(sci educ), 73. *Prof Exp:* Instr gen sci, 66-70, from asst prof to assoc prof, 72-90, PROF EARTH SCI & EDUC, CENT MO STATE UNIV, 90- *Res:* Understanding of science process and interest in science teaching among prospective elementary teachers. *Mailing Add:* Dept Earth Sci Cent Mo State Univ Warrensburg MO 64093

BERKLEY, DAVID A, b Brooklyn, NY, Apr 28, 40; m 73; c 3. ACOUSTICS. *Educ:* Cornell Univ, BEE, 61, PhD(appl physics), 66. *Prof Exp:* Res assoc electron & med physics, Chalmers Tech Sweden, 66-68; mem tech staff acoust res, Bell Tel Labs, 68-80; supvr, 80-90, HEAD, ACOUST RES DEPT, AT&T BELL LABS, 90- *Mem:* Acoust Soc Am; NY Acad Sci. *Res:* Acoustics signal processing and electroacoustics; speech and hearing. *Mailing Add:* AT&T Bell Labs Rm 2D537 Murray Hill NJ 07974

BERKLEY, JOHN LEE, b Lawrence, Kans, May 17, 48; m 77; c 2. GEOLOGY, IGNEOUS PETROLOGY. *Educ:* Univ Minn, Duluth, BA, 70; Univ Mo-Columbia, MA, 72; Univ NMex, PhD(geol), 77. *Prof Exp:* Engr aid defense contracts, M B Assocs, San Ramon, Calif, 73; meteorites & ocean basalts, Inst Meteoritics, Univ NMex, 77-79; res assoc, Lunar Planetary Lab, Univ Ariz, 79-82; ASSOC PROF IGNEOUS-METAMORPHIC PETROL, STATE UNIV NY, FREDONIA, 82-, ASST DEAN, 90- *Mem:* Geol Soc Am; Mineral Soc Am; Am Geophys Union; Meteoritical Soc; Sigma Xi. *Res:* Achondrite meteorites; petrology and geochemistry of Ureilite and HED achondrite meteorites (NASA grant NAG9-82). *Mailing Add:* Geosci Dept State Col Fredonia NY 14063

BERKMAN, JAMES ISRAEL, b Cambridge, Mass, Nov 14, 13; m 42; c 2. MEDICINE, PATHOLOGY. *Educ:* Harvard Univ, AB, 35, MA, 36; NY Univ, MD, 40. *Prof Exp:* From asst pathologist to assoc pathologist, Montefiore Hosp, NY, 48-53; CHMN DEPT LABS, LONG ISLAND JEWISH-HILLSIDE MED CTR, 53-; PROF PATH, STATE UNIV NY STONY BROOK, 71- *Concurrent Pos:* Attend pathologist, Montefiore Hosp, NY, 63-; pathologist in chief, Queens Hosp Ctr, 64-; prof lectr, State Univ NY Downstate Med Ctr; consult path, Creedmoor State Hosp, 75- *Mem:* Am Asn Path & Bact; Int Acad Path; Col Am Pathologists; Am Soc Clin Path; Am Soc Nephrology. *Res:* Vascular disease in diabetes; renal disease. *Mailing Add:* LI Jewish-Hillside Med Ctr 70 E Tenth St New York NY 10003

BERKMAN, MICHAEL G, b Poland, Apr 4, 17; nat US; m 41; c 2. CHEMISTRY. *Educ:* Univ Chicago, BS, 37, PhD(org chem), 41; De Paul Univ, JD, 58; John Marshall Law Sch, LLM, 62. *Prof Exp:* Res chemist, Deavitt Labs, Univ Chicago, 39-41, Am Can Co, Ill, 41-42 & Argonne Nat Lab, 46-51; chief chemist, Colburn Labs, Inc, 51-59; patent atty, Mann, Brown & McWilliams, 59-63; patent atty, Kegan, Kegan & Berkman, 63-84; patent atty, Trexler, Bushnell, Giangiorgi & Blackstone, 84-91; PATENT ATTY, MICHAEL G BERKMAN & ASSOC, 91- *Concurrent Pos:* Patent consult. *Mem:* Am Chem Soc; Sigma Xi; Am Bar Asn. *Res:* Industrial research and development; synthetic coatings; halogenation; atomic energy; heavy metal complexes; chlorination of hydrocarbons; food technology; federal food and drug laws. *Mailing Add:* 939 Glenview Rd Glenview IL 60025

BERKMAN, SAMUEL, b Bronx, NY, July 6, 35; m 64; c 2. CHEMICAL VAPOR DEPOSITION, SHAPED CRYSTAL GROWTH. *Educ:* Fairleigh Dickinson Univ, BS, 65. *Prof Exp:* Mem tech staff, RCA/GE/SRI/David Sarnoff Res Ctr, 66-90; VPRES, NU-TEC CORP, 90- *Concurrent Pos:* Consult, David Sarnoff Res Ctr. *Honors & Awards:* Recognition Award, NASA, 79 & 80. *Mem:* Am Asn Crystal Growth. *Res:* High temperature materials processing, including semiconductors, optical crystals, and ceramics; analyzed the dynamics of heat, fluid, and mass transport while designing various equipment and processes for crystal growth; fifteen US patents; eight publications including principal co-author of a chapter of one technical book. *Mailing Add:* 1818 Boat Point Dr Point Pleasant NJ 08742

BERKNER, KLAUS HANS, b Dessau, Ger, Mar 2, 38; US citizen; div; c 2. ACCELERATOR DEVELOPMENT, FUSION TECHNOLOGY. *Educ:* Mass Inst Technol, SB, 60; Univ Calif, Berkeley, PhD(physics), 64. *Prof Exp:* Physicist, 64-78, dep head, Accelerator & Fusion Res Div, 82-84, actg assoc dir, 84-85, STAFF SR SCIENTIST, LAWRENCE BERKELEY LAB, UNIV CALIF, BERKELEY, 78-, DIR, ACCELERATOR & FUSION RES DIV, 85- *Concurrent Pos:* NSF fel, UK Atomic Energy Authority, Culham Lab, Eng, 65-66; NATO res grantee, 79-82. *Mem:* Fel Am Phys Soc. *Res:* Accelerator development; high power neutral beams for heating fusion plasmas; measurement of atomic collision cross sections; electron capture and loss; inner shell excitation. *Mailing Add:* Lawrence Berkeley Lab Univ Calif Berkeley CA 94720

BERKO, STEPHAN, physics, solid state physics; deceased, see previous edition for last biography

BERKOF, RICHARD STANLEY, b Brooklyn, NY, Mar 2, 41; m 62; c 3. MECHANICAL ENGINEERING. *Educ:* City Col New York, BME, 62; Columbia Univ, MSME, 63; City Univ New York, PhD(eng mech), 69. *Prof Exp:* Engr, Syska & Hennessy, Inc, 62; design engr, Gibbs & Cox, Inc, 63-64; res asst, City Col New York, 64-65, lectr, 65-67, res asst, 67-68; res assoc, Princeton Lab, Am Can Co, 68-73; mgr advan technol, Wickes Gulf & Western Advan Develop & Eng Ctr, 73-79, mgr proposals & planning, 79-84, mgr planning & tech serv, 84-89; MGR ADVAN MFG, UNITED ENG & CONST, 89- *Concurrent Pos:* Consult, Bright Lithographing Corp, 63; ed dynamics, Mechanism & Mach Theory, 75-85; ed spec features, Transactions Am Soc Mech Engrs: J Mech Design, 77-79, tech ed pro tem, 79-80. *Mem:* Soc Mfg Engrs; Am Soc Eng Educ; Am Soc Mech Engrs; Am Nuclear Soc; US Comt Theory Mach & Mechanisms. *Res:* Dynamics and kinematics of machines. *Mailing Add:* United Eng & Const 30 S 17th St Philadelphia PA 19101

BERKOFF, CHARLES EDWARD, b London, Eng, Sept 29, 32; US citizen; m 61; c 3. PHARMACEUTICAL PRODUCT LICENSING, MEDICINAL CHEMISTRY. *Educ:* Univ London, BSc, 56, PhD(org chem); Imp Col, Univ London, dipl, 59. *Prof Exp:* Jr chemist, Brit Drug Houses, London, 49-54; Monsanto bursary, 56-59; Fulbright travel scholar & res assoc, Johns Hopkins Univ, 59-60; fel, Univ Southampton, 60-61; asst to dir res, Nicholas Res Inst, Eng, 61-62; asst dir chem res & develop, Biorex Labs, 62; group leader org chem, Wyeth Labs, Am Home Prod Co, Pa, 63-64; sr scientist, Sci Info Dept, 64-65, group leader, 65-70, asst dir chem support, 70-71, mgr technol assessment & long range planning, Planning & Opers Dept, 72-74, mgr, 74-75, dir, Org Chem Dept, 75-81, dir chem technol, Res & Develop Div, Smith Kline & Fr Labs, 81-83; exec vpres, Imutech, 83-84, pres & chief exec officer, Antigenics, Inc, 84-89; PRES & CHIEF EXEC OFFICER, CREATIVE LICENSING INT, INC & CEBRAL, 86- *Concurrent Pos:* Mem adv coun, Smithsonian Sci Info Exchange, 74-82. *Mem:* AAAS; Am Chem Soc; fel Royal Inst Chem; fel Am Inst Chemists; fel Royal Soc Chem; Entom Soc Am; NY Acad Sci; Am Inst Chem Engrs; Licensing Exec Soc. *Res:* Terpenoids, steroids and other natural products; small ring and heterocyclic compounds, especially those of medicinal interest; chemistry and biochemistry of insect hormones. *Mailing Add:* C L I Inc PO Box 537 Willow Grove PA 19090-0537

BERKOVITS, SHIMSHON, b Berlin, Germany, Oct 5, 36; US citizen; m 63; c 7. CRYPTOGRAPHY, SECURITY. *Educ:* Mass Inst Technol, SB, 57; Univ Chicago, MS, 60; Northeastern Univ, PhD(math), 73. *Prof Exp:* Mem tech staff, Mitre Corp, 63-70; asst prof math, Boston Univ, 68-74; asst prof math, 74-80, chmn dept comput sci, 79-82, assoc prof comput sci, 80-86, ASSOC PROF MATH, UNIV LOWELL, 86-; MEM TECH STAFF, MITRE CORP, 76- *Mem:* Asn Comput Mach. *Res:* Modern uses of cryptography in communications and computer security; design and implementation of public key systems; appropriate system parameter selection; potential attack avoidance. *Mailing Add:* Dept Math Univ Lowell Lowell MA 01854

BERKOVITZ, LEONARD DAVID, b Chicago, Ill, Jan 24, 24; m 53; c 2. MATHEMATICS. *Educ:* Univ Chicago, PhD(math), 51. *Prof Exp:* AEC fel, Stanford Univ, 51-52; res fel, Calif Inst Technol, 52-54; mathematician, Rand Corp, 54-62; head dept, 75-80, PROF MATH, PURDUE UNIV, 62- *Concurrent Pos:* Ed comt, Math Reviews; assoc ed, J Optimization Theory & Appln, J Math Anal & Appln. *Mem:* Am Math Soc; Math Asn Am; Soc Indust & Appl Math. *Res:* Calculus of variations; optimal control; game theory. *Mailing Add:* Dept Math Purdue Univ West Lafayette IN 47907

BERKOWER, IRA, b New York, NY, Aug 15, 48; m 72; c 3. VIRAL IMMUNITY, ANTIGEN PROCESSING. *Educ:* Yale Univ, BS & MS, 69; Albert Einstein Col Med, PhD(molecular biol), 74, MD, 75. *Prof Exp:* Intern & resident, Columbia Presby Med Ctr, 75-78; res fel rheumatology, Kunkel Lab, Rockefeller Univ, 78-79; res assoc immunol, Nat Cancer Inst, NIH, 79-82, cancer expert, 82-83; SR INVESTR, MOLECULAR IMMUNOL LAB, DIV BIOCHEM & BIOPHYS, CTR BIOLOGICS, FOOD & DRUG ADMIN, 83- *Concurrent Pos:* Lectr, Found Advan Educ Sci, NIH, 87-88, Univ Nebr & Mass Inst Technol, 89, Univ Tex & Downstate Med Ctr, 90; mem, Ad Hoc Study Sect Cancer AIDS, NIH, 89-90; guest lectr, Dept Biochem, George Washington Univ, 90-91. *Mem:* Am Asn Immunol; Am Fedn Clin Res; Am Col Physicians. *Res:* Immunity to viral protein antigens; mapping sites on viral proteins which are targets of neutralizing antibodies and soluble receptor proteins and the mechanism of inactivation; pathways of antigen processing leading to T cell recognition of viral antigens, including the signals for entry into each processing pathway; vaccine design to improve potency. *Mailing Add:* Ctr Biologics Food & Drug Admin 8800 Rockville Pike Bldg 29 Rm 523 Bethesda MD 20892

BERKOWITZ, BARRY ALAN, b Brookline, Mass, Dec 29, 42; m 63; c 2. PHARMACOLOGY. *Educ:* Northeastern Univ, BS, 64; Univ Calif, PhD(pharmacol), 68. *Prof Exp:* Fel, Roche Inst Molecular Biol, 68-70, res assoc, 70-71, assoc mem pharmacol, 71-80; MEM STAFF, SMITH KLINE & FRENCH LABS, PHILADELPHIA, 80-, VPRES BIOL SCI & BIOPHARMACEUT RES & DEVELOP. *Concurrent Pos:* Vis assoc prof pharmacol, Med Col, Cornell Univ, 71- *Mem:* AAAS; NY Acad Sci; Am Soc Pharmacol & Exp Therapeut. *Res:* Neurochemistry and pharmacology of blood vessels; pharmacology of narcotics and narcotic antagonist analgesics; biogenic amines; cardiovascular disease. *Mailing Add:* Pres & CEO Magainin Sci Inc 550 Pinetown Rd Ft Washington PA 19034

BERKOWITZ, BARRY JAY, metallurgy, materials science, for more information see previous edition

BERKOWITZ, DAVID ANDREW, b Brooklyn, NY, Mar 21, 52; m 75; c 3. BIOMEDICAL ENGINEERING, MECHANICAL ENGINEERING. *Educ:* Rensselaer Polytech Inst, BS, 74, ME, 75, cert clin eng, 75. *Prof Exp:* Proj engr mech design & develop, Automatech Indust, 75-77; proj officer biomed eng, 77-79, sr proj engr & consult, 80-86, VPRES CONSULT SERV, EMERGENCY CARE RES INST, 86- *Concurrent Pos:* Instr biomed eng, Bridgeport Eng Inst, 77; consult biomed technol, Park City Hosp, 77. *Honors & Awards:* Eng Design Award, Am Soc Eng Educ, 74. *Mem:* Am Soc Mech Engrs; Asn Advan Med Instrumentation; Nat Soc Prof Engrs. *Res:* Medical devices; their impact on health care; product evaluation; accident investigation; teaching; consulting on medical equipment, systems and facilities; special interest in medical equipment assessment and Acquisition. *Mailing Add:* Emergency Care Res Inst 5200 Butler Pike Plymouth Meeting PA 19462

BERKOWITZ, HARRY LEO, b New York, NY, Mar 10, 37; m 60; c 2. RADIATION PHYSICS, THEORETICAL PHYSICS. *Educ:* Adelphi Col, AB, 59; Columbia Univ, MA, 61; Stevens Inst Technol, PhD(physics), 72. *Prof Exp:* Lectr physics lab, Columbia Univ, 60-61; res physicist, US Army Electronics Technol Develop Lab, 61-87, res physicist, solid state develop, 87-90. *Mem:* Am Phys Soc. *Res:* Use of nuclear reactions for determining dopant profiles in silicon. *Mailing Add:* 2635 Woodmont Lane Wexford PA 15050

BERKOWITZ, JEROME, b Brooklyn, NY, Oct 2, 28; m 54; c 2. MATHEMATICS. *Educ:* NY Univ, PhD(math), 51. *Prof Exp:* Mathematician, Reeves Instrument Corp, 48; res asst, 50-56, from asst prof to assoc prof math, 56-65, PROF, COURANT INST MATH SCI, NY UNIV, 65- *Mem:* Am Math Soc; Math Asn Am; Sigma Xi. *Res:* Functional analysis; magnetohydrodynamics. *Mailing Add:* 170 Second Ave New York NY 10003

BERKOWITZ, JESSE M, b New York, NY, Dec 10, 28; m; c 3. GASTROENTEROLOGY, MEDICAL SCIENCE. *Educ:* Wash Univ, BA, 50; Cornell Univ, MS, 52; Chicago Med Sch, MD, 56. *Prof Exp:* Intern, Montefiore Hosp, NY, 56-57, jr asst res, Dept Med, 57-58; asst res, Dept Med, Mt Sinai Hosp, NY, 58-59, res Gastroenterology, 61-62, postdoctoral fel, 62-63, res asst, 63- 67, res asst, 68-78; res assoc, Dept Chem, Polytech Inst, Brooklyn, NY, 65-67; assoc prof med, 71-80, prof, 80-82, PROF CLIN MED, SCH MED, HEALTH SCI CTR, STATE UNIV NY, STONY BROOK, 82- *Concurrent Pos:* Vis res asst, Dept Biol, Brookhaven Nat Lab, 52, res collabr, 73-79; consult, Gastroenterology, Vet Admin Hosp, Northport, NY, 71-81; chief, div Gastroenterology, Dept Med, Nassau County Med Ctr, 67-82, consult, 82-; attend physician med, S Nassau Community Hosp, Oceanside, NY, 74-, Mercy Hosp, Rockville Centre, NY, 75-, Winthrop Univ Hosp, Mineola, NY, 80- *Res:* Animal physiology; endocrinology; biochemistry; pharmacology. *Mailing Add:* Ryan Med Arts Bldg Sch Med State Univ NY Stony Brook 2000 N Village Ave Rockville Center NY 11516

BERKOWITZ, JOAN B, b Brooklyn, NY, Mar 13, 31; m 59; c 1. PHYSICAL CHEMISTRY. *Educ:* Swarthmore Col, BA, 52; Univ Ill, PhD(phys chem), 55. *Prof Exp:* NSF fel, Yale Univ, 55-57; phys chemist, 57-80, VPRES, ARTHUR D LITTLE, INC, 80-, SECT MGR, 81- *Mem:* Am Chem Soc; Electrochem Soc; Am Phys Soc. *Res:* Thermodynamics of inorganic systems at high temperatures; oxidation of refractory metals and alloys; electrochemistry in flames; mass spectrometry; solid-gas interactions; inorganic coating technology; heterogeneous catalytic recombination; high temperature vaporization; hazardous waste management. *Mailing Add:* 1940 35th St NW Washington DC 20007

BERKOWITZ, JOSEPH, b Czech, Apr 22, 30; US citizen; m 58; c 2. CHEMICAL PHYSICS. *Educ:* NY Univ, BE, 51; Harvard Univ, PhD(phys chem), 55. *Prof Exp:* Jr chem engr, Brookhaven Nat Lab, 51-52; res assoc physics, Univ Chicago, 55-57; physicist, 57-73, SR PHYSICIST, ARGONNE NAT LAB, 73- *Concurrent Pos:* Vis asst prof, Univ Ill, 59-60; Guggenheim Found fel phys inst, Univ Freiburg, 65-66; vis prof, Northwestern Univ, 73. *Honors & Awards:* Alexander von Humboldt Award, 88. *Mem:* Am Chem Soc; Am Phys Soc; Am Soc Mass Spectrometry. *Res:* Molecular structure; chemical kinetics; high temperature thermodynamics; mass spectrometry; photoionization; photoelectron spectroscopy. *Mailing Add:* Chem Div Bldg 203 Argonne Nat Lab 9700 S Cass Ave Argonne IL 60439

BERKOWITZ, LEWIS MAURICE, b New York, NY, Apr 12, 31; m 68; c 1. ORGANIC CHEMISTRY. *Educ:* City Col New York, BS, 52; Columbia Univ, AM, 53, PhD(chem), 57. *Prof Exp:* Chemist, Engelhard Indust Inc, 57-58, US Vitamin & Pharmaceut Indust Corp, 59-60 & Gen Foods Corp, 60-63; CHEMIST, US ARMY, EDGEWOOD ARSENAL, 63- *Mem:* Am Chem Soc; Sigma Xi. *Res:* Structure-activity relationship of chemotherapeutic and physiologically active compounds; organic synthesis; enzyme immobilization. *Mailing Add:* 627 Leafydale Terr Baltimore MD 21208

BERKOWITZ, RAYMOND S, b Philadelphia, Pa, Feb 21, 23; m 46; c 3. RADAR SYSTEMS. *Educ:* Univ Pa, BSEE, 43, MSEE, 48, PhD, 51. *Prof Exp:* Engr, Radio Corp Am, 43-44; from res asst to assoc prof, 47-77, PROF ELEC ENG, UNIV PA, 77- *Concurrent Pos:* Consult, Radio Corp Am, 59- *Mem:* AAAS; Am Soc Eng Educ; Inst Elec & Electronics Engrs; Soc Indust & Appl Math; Sigma Xi. *Res:* Radar; general systems engineering; statistical communication theory. *Mailing Add:* 511 Woodbrook Lane Philadelphia PA 19119

BERKOWITZ, SIDNEY, b Perth Amboy, NJ, Aug 1, 21; m 45; c 2. INDUSTRIAL ORGANIC CHEMISTRY. *Educ:* Rutgers Univ, BA, 65. *Prof Exp:* Res chemist, 58-70, sr res chemist, 70-76, res assoc, 76-80, SR RES ASSOC, FMC CORP, PRINCETON, 80- *Mem:* Catalyst Soc; fel Am Inst Chemists. *Res:* Heterocyclic chemistry with emphasis on the reactions of isocyanic acid and many of its polymeric derivatives, both linear and cyclic; high temperature catalytic and hot tube oxidations. *Mailing Add:* 6101 Wirt Ave Baltimore MD 21215

BERKSON, DAVID M, b Chicago, Ill, Oct 16, 28; m 73; c 3. MEDICINE. *Educ:* Univ Ill, BS, 49; Northwestern Univ, Chicago, MD, 53. *Prof Exp:* Assoc prof community health & prev med, Sch Med, Northwestern Univ, Chicago, 72-82; heart dis control officer, Chicago Bd Health, 73-90; PROF COMMUNITY HEALTH, PREV MED & MED, SCH MED, NORTHWESTERN UNIV, 82- *Concurrent Pos:* Assoc attend physician, Michael Reese Hosp, Chicago, 61-80; head sect cardiovasc dis, St Joseph Hosp, Chicago, 63- *Mem:* AAAS; Am Heart Asn; AMA; fel Am Col Physicians; fel Am Col Cardiol. *Res:* Cardiovascular diseases; epidemiology of hypertensive and atherosclerotic coronary heart disease. *Mailing Add:* 2800 N Sheridan Rd Chicago IL 60657

BERKSON, EARL ROBERT, b Chicago, Ill, June 6, 34; m 60; c 3. MATHEMATICS. *Educ:* Univ Calif, Los Angeles, BS, 56, MA, 57; Univ Chicago, PhD(math), 60. *Prof Exp:* From instr to asst prof math, Univ Calif, Los Angeles, 60-64; assoc prof, 64-67, PROF MATH, UNIV ILL, URBANA, 73- *Concurrent Pos:* Vis asst prof, Univ Calif, Berkeley, 64-65. *Mem:* Am Math Soc. *Res:* Operator theory; spectral theory; complex analysis. *Mailing Add:* Dept Math Univ Ill 1409 W Green St Urbana IL 61801

BERKSON, HAROLD, b Easton, Pa, Oct 30, 29; m 58; c 2. ENVIRONMENTAL BIOLOGY. *Educ:* Rutgers Univ, BA, 51; Amherst Col, MA, 53; Univ Calif, San Diego, PhD(environ physiol), 63. *Prof Exp:* Asst comp anat, Amherst Col, 51-53; asst eng scientist, NY Univ, 54-55; biochem technician, M D Anderson Hosp & Tumor Inst, Univ Tex, 55-56; asst biol oceanogr, Narragansett Marine Lab, Univ R I, 56-58; asst physiol & ecol, Scripps Inst, Univ Calif, 58-63; marine biologist, Fed Water Pollution Control Admin, 63-68; chief biol & ecol sci sect, Estuarine & Oceanog Progs Br, 68-70; specialist environ policy, Cong Res Serv, Libr of Congress, 70-72; sr environ specialist, 72-80, sr prog & planning analyst, 80-83, chief, Planning & Resources Anal Br, 83-84, spec asst environ policy, 84-85, SR PROJ MGR, NUCLEAR REGULATORY COMN, 85- *Mem:* Sigma Xi; Am Soc Limnol & Oceanog; Ecol Soc Am; Am Fisheries Soc; Am Inst Fishery Res Biologists. *Res:* Estuarine ecology; adaptation to the environment; prolonged and deep-diving by air-breathing vertebrates; environmental management; environmental impact assessment. *Mailing Add:* Marine Biol 12001 Whippoorwill Lane Rockville MD 20852

BERKSON, JONATHAN MILTON, b Chicago, Ill, Feb 23, 41; m 75; c 3. MARINE GEOPHYSICS, UNDERWATER ACOUSTICS. *Educ:* Univ Ill, BS, 63; Univ Wis, MS, 69, PhD(geophysics), 72. *Prof Exp:* Res asst, Geophys & Polar Res Ctr, Univ Wis, 66-72, fel, 72-73; geophysicist, US Naval Oceanog Off, 73-76; geophysicist, Naval Oceanog & Atmospheric Res Lab, 76-83; sr scientist, Saclant Undersea Res Ctr, 83-86; GEOPHYSICIST, NAVAL RES LAB, 86- *Mem:* Acoustical Soc Am; Soc Explor Geophysicists; Am Geophys Union; Sigma Xi. *Res:* Marine geophysics; underwater acoustics. *Mailing Add:* Code 5160 Naval Res Lab Washington DC 20375-5000

BERKUT, MICHAEL KALEN, b NY, June 30, 15; m 38; c 2. BIOLOGICAL CHEMISTRY. *Educ:* NC State Col, BS, 41; Univ NC, PhD(biol chem), 53. *Prof Exp:* Chemist, NC State Lab Hyg, 41; instr chem, NC State Col, 42; from instr to assoc prof biol chem, Univ NC, Chapel Hill, 47-74, assoc prof biochem, nutrit & physiol, 74-77, assoc prof physiol & prof biochem, 77-80; RETIRED. *Concurrent Pos:* Res biochemist, US AEC, 62-64. *Mem:* AAAS; Am Chem Soc; Am Asn Clin Chemists. *Res:* Mineral metabolism; mechanism of biological calcification; neural and humoral influences on hematopoiesis. *Mailing Add:* 807 Old Mill Rd Chapel Hill NC 27514

BERKY, JOHN JAMES, b Billings, Mont, Nov 19, 24; m 50; c 3. LABORATORY SAFETY, BIOLOGY PROGRAM MANAGEMENT. *Educ:* NDak State Univ, BS, 49; Pa State Univ, MS, 51, PhD(microbial physiol), 54; Nat Registry Microbiol, registered, 63; Am Acad Microbiol, cert specialist microbiol, Pub Health & Medical Lab Microbiol, 70, Food, Dairy & Sanit Microbiol, 75. *Prof Exp:* Food res microbiologist, Armour Res Div, Chicago, 54-56, asst sect chief bacteriol, 56-58; asst chief, Lab Div, US Army Biol Opers, Pine Bluff, Ark, 58-59, chief, Biol Lab Div, 59-71; sr investr, Virol & Cell Cult Lab, 71-77, dir, Div Anal Serv, Microbiol & Immunol Div, 77-79, dir, qual assurance & safety staff, Nat Ctr for Toxicol Res, 79-86; RETIRED. *Honors & Awards:* Award of Merit, Food & Drug Admin, 78. *Mem:* NY Acad Sci; Sigma Xi. *Res:* Cell culture systems in toxicology, toxicity testing, carcinogenesis; virus-chemical interactions; microbial metabolism, food research microbiology, microbial processes fermentations by-products purification and assay; contract administration and laboratory management. *Mailing Add:* 201 Raven Dr P-4 Hot Springs AR 71913

BERL, SOLL, b New York, NY, June 12, 18; m 56; c 4. BIOCHEMISTRY, NEUROCHEMISTRY. *Educ:* St John's Univ, BS, 40; Univ Wis, MS, 43; Western Reserve Univ, MD, 50. *Prof Exp:* Intern, Long Island Col Hosp, 50-51; resident psychiat, Bellevue Med Ctr, NY Univ, 51-54; NIH fel, 56; sr res scientist neurochem, NY Psychiat Inst, 57-61, assoc res scientist, 61-62; res assoc, 62-64, from asst prof to assoc prof neurochem, Col Physicians & Surgeons, Columbia Univ, 64-73; res prof, 73-88, EMER RES PROF NEUROL, MT SINAI SCH MED, 88- *Concurrent Pos:* Vet Admin training fel, 51-54; asst attend, Vanderbilt Clin, Columbia-Presby Med Ctr, 57-64; NIH res career develop award, 62. *Honors & Awards:* Lucy G Moses Prize, 70. *Mem:* AAAS; Am Psychiat Asn; Am Soc Neurochem; Am Soc Biol Chemists; Am Chem Soc; Int Soc Neurochem. *Res:* Amino acid and protein metabolism of the central nervous system in relation to structure and function; psychiatry. *Mailing Add:* SR23-RD1 Box 329 Carryville NY 12521-9774

BERL, WALTER G(EORGE), b Vienna, Austria, July 3, 17; nat US; m 52, 89; c 1. COMBUSTION, FIRE RESEARCH. *Educ:* Carnegie Inst Technol, BS, 37, PhD(phys chem), 41; Harvard Univ, MS, 39. *Prof Exp:* Res fel & instr, Carnegie Inst Technol, 42-45; group supvr, Fire Probs Group, 77-80, PRIN STAFF CHEMIST, APPL PHYSICS LAB, JOHNS HOPKINS UNIV, 45- *Concurrent Pos:* Foster Mem Lectr, State Univ NY, Buffalo, 58; ed, Fire Res Abstr & Rev, 58-64; assoc ed, Am Rocket Soc J, 60-62 & Am Inst Aeronaut & Astronaut J, 63-64; mem comt fire res, Nat Acad Sci-Nat Res Coun, 65-68; US ed, Combustion & Flame, 66-71; mem adv comt, NSF-Res Appl Nat Needs, 72; consult, Nuclear Regulatory Comn, 76-80; ed-in-chief, Johns Hopkins Appl Tech Digest, 80-88, ed, Appl Tech Rev, 88-90. *Mem:* Sigma Xi; fel AAAS; fel NY Acad Sci; Am Chem Soc; Am Inst Aeronaut & Astronaut. *Res:* Combustion; propulsion; fuels; propellants; instrumental analysis; origin of petroleum; fuel cells. *Mailing Add:* Appl Physics Lab Johns Hopkins Univ Laurel MD 20707

BERLAD, ABRAHAM LEON, b New York, NY, Sept 20, 21; m 49; c 3. CHEMICAL PHYSICS, MECHANICAL ENGINEERING. *Educ:* Brooklyn Col, BA, 43; Ohio State Univ, PhD(physics), 50. *Prof Exp:* Aeronaut res scientist, Nat Adv Comt Aeronaut, 51-54, head, Combustion Fundamentals Sect, 54-56; sr staff scientist, Convair Sci Res Lab, 56-63; vis prof, dept aeronaut sci, Univ Calif, Berkeley, 63-64; mem staff, Physics Dept, Gen Res Corp, 64-66; prof eng, EMER PROF ENG, STATE UNIV NY, STONY BROOK, 84- *Concurrent Pos:* Consult, AEC, 73-75, Energy Res & Develop Agency, US Dept Energy, NASA & Nat Res Coun; vis prof, Univ Calif, San Diego & Hebrew Univ Jerusalem, 74; vis sr scientist, Brookhaven Nat Lab, 75; fel, E I Du Pont Co; adj prof combustion sci, Energy Ct, Univ Calif, San Diego, 84- *Mem:* Am Phys Soc; Combustion Inst. *Res:* Irreversible processes; unstable media; crystallization and combustion theory; radiation phenomena; stability of reaction processes; atmospheric and environmental rate processes; thermokinetic systems; nuclear containment safety. *Mailing Add:* Energy Ctr B-010 Univ Calif San Diego La Jolla CA 92093

BERLANDI, FRANCIS JOSEPH, b Boston, Mass, Mar 23, 41. INDUSTRIAL HYGIENE, CHEMISTRY. *Educ:* Mass Inst Technol, BS, 62; Univ Mich, MS, 64, PhD(anal-nuclear chem), 66. *Prof Exp:* Mgr, Geonuclear Dept, Teledyne Isotopes, 68-73; mgr anal serv indust hyg & chem, Environ Res & Technol, Inc, 73-76; mgr environ serv indust hyg, Environ Sci Assocs, Inc, 76-80; MEM STAFF, TOUCHSTONE ENVIRON CONSULTS, 80- *Mem:* Am Indust Hyg Asn; Sigma Xi. *Res:* Physical-chemical in situ measurements; design, management and interpretation of large scale environmental field studies. *Mailing Add:* Touchstone Environ Consult 33 Thompson St Winchester MA 01890

BERLEKAMP, ELWYN R(ALPH), b Dover, Ohio, Sept 6, 40; m 66; c 3. MATHEMATICS, ELECTRICAL ENGINEERING. *Educ:* Mass Inst Technol, BS & MS, 62, PhD(elec eng), 64. *Prof Exp:* Asst prof elec eng, Univ Calif, Berkeley, 64-67; consult, Bell Tel Labs, 67-71; pres, Cyclotomics, Eastman Kodak, 83-89; PROF MATH, ELEC ENG & COMPUTER SCI, UNIV CALIF, BERKELEY, 71-; PRES, HAZELNUT FUTURES, 90- *Honors & Awards:* Centennial Medal, Inst Elec & Electronics Engrs, 84, Koji Kobayashi Computers & Commun Award, 90, Richard W Hamming Medal, 91. *Mem:* Nat Acad Eng; Am Math Soc; fel Inst Elec & Electronics Engrs. *Res:* Information theory, coding and related problems in combinatorial mathematics and computer science; recent extensions of game theory; use of statistical information theory to forecast stock and commodity prices. *Mailing Add:* 2140 Shattuck Ave Suite 907 Berkeley CA 94704

BERLEY, DAVID, b Brooklyn, NY, Mar 25, 30; m 62; c 2. PHYSICS. *Educ:* Union Col, NY, BS, 51; Cornell Univ, PhD(physics), 57. *Prof Exp:* Res assoc, Columbia Univ, 57-59 & 62; Louis de Broglie fel, Col de France, 59-61; physicist, Brookhaven Nat Lab, 62-74, head exp div, Accelerator Dept, 69-74; prog dir, NSF, 74-77; actg head accelerator opers br, Div High Energy Physics, US Dept Energy, 77-80; PROG DIR ELEM PARTICLE PHYSICS, NSF, 80- *Concurrent Pos:* Fulbright fel, 59 & 60; vis prof physics, Univ Md, 88-89. *Mem:* AAAS; fel Am Phys Soc; Sigma Xi. *Res:* Elementary particle physics; weak interactions; decay of muons, hyperons and kaons; strong interactions; cosmic rays; administration of elementary particle physics programs. *Mailing Add:* 3535 Chesapeake St NW Washington DC 20008

BERLIN, BYRON SANFORD, b Detroit, Mich, Mar 19, 21; m 46; c 6. VIROLOGY. *Educ:* Wayne State Univ, BA, 42, MD, 45; Am Bd Med Microbiol, dipl, 68. *Prof Exp:* Coman fel, Univ Chicago, 50-52, resident med, 52-54; res assoc, Sch Pub Health, Univ Mich, 54-59, asst prof, 59-67; ASSOC PROF MED & PATH, MED SCH, NORTHWESTERN UNIV, CHICAGO, 67-; COORDR PHYS, BUR LAB & EPIDEMIOL SERV, MICH DEPT PUB HEALTH. *Concurrent Pos:* Mem, Am Bd Med Microbiol. *Mem:* Am Asn Immunologists; Soc Exp Biol & Med. *Res:* Immunology; infection; vaccinology. *Mailing Add:* Bur Dis Control & Lab Serv Mich Dept Pub Health 3500 N Logan Lansing MI 48909

BERLIN, CHARLES I, b New York, NY, Dec 26, 33; m 58; c 4. AUDIOLOGY. *Educ:* NY Univ, BS, 53; Univ Wis, MA, 54; Univ Pittsburgh, PhD(hearing & speech), 58. *Prof Exp:* Audiologist & speech pathologist, US Vet Admin Hosp, Calif, 59-61; asst prof otolaryngol, Johns Hopkins Hosp, 63-67; assoc prof, 67-70, PROF OTOLARYNGOL, MED SCH, LA STATE UNIV MED CTR, NEW ORLEANS, 70-, DIR, KRESGE HEARING RES LAB OF THE SOUTH, 68-. *Concurrent Pos:* Fel med audiol, Johns Hopkins Hosp, 62-63. *Honors & Awards:* Nat Inst Neurol Dis & Blindness res career develop award, 63-67. *Mem:* AAAS; fel Acoust Soc Am; fel Am Speech & Hearing Asn; Sigma Xi; Am Acad Audiol. *Res:* Hearing and speech sciences; evoked potentials; speech perception; communication in pathological states in humans and animals; otoacoustic emissions; hearing disorders and hearing aids. *Mailing Add:* 6001 Pratt Dr New Orleans LA 70122

BERLIN, CHESTON MILTON, b Pittsburgh, Pa, Mar 28, 36; m 60; c 4. PEDIATRICS. *Educ:* Haverford Col, BA, 58, Harvard Med Sch, MD, 62. *Prof Exp:* Asst prof pediat, Sch Med, Univ Ala, 67-68 & George Washington Univ, 68-71; assoc prof pediat & pharmacol, 71-75, asst dean student affairs, 72-87, PROF PEDIAT & PHARMACOL, 75-, UNIV PROF, COL MED, PA STATE UNIV, 86-. *Concurrent Pos:* Markle scholar acad med, 69-74. *Mem:* Am Soc Exp Pharmacol & Therapeut; Am Soc Clin Pharmacol & Therapeut; Am Pediat Soc; Am Acad Pediat. *Res:* The excretion of drugs and chemicals in human milk; antibody formation in lupus; salt excretion in hypertension; pharmacology of movement disorders in children. *Mailing Add:* Dept Pediat Hershey Med Ctr PO Box 850 Hershey PA 17033

BERLIN, ELLIOTT, b Baltimore, Md, Jan 4, 34; m 56; c 5. BIOCHEMISTRY, NUTRITION. *Educ:* Johns Hopkins Univ, BS, 57, MA, 59; Univ Md, College Park, PhD(biochem), 66. *Prof Exp:* RES CHEMIST, AGR RES SERV, USDA, 62-. *Concurrent Pos:* Lectr, Catonsville Community Col, 65-68; lectr, Univ Md, College Park, 69-70 & 80-; instr, Bowie State Col, 71-72; adj prof, Southeastern Univ, 73-74. *Mem:* AAAS; Biophys Soc; Am Chem Soc; Am Inst Nutrit; Am Oil Chemists Soc; NY Acad Sci. *Res:* Nutritional factors affecting physical chemical properties and biological function of biomembranes; membrane fluidity; lipoprotein chemistry; water binding by proteins and other membranes. *Mailing Add:* Beltsville Human Nutrit Res Ctr Agr Res Serv USDA Beltsville MD 20705

BERLIN, FRED S, b Pittsburgh, Pa, July 27, 41; m 70; c 4. PSYCHIATRY. *Educ:* Univ Pittsburgh, BA, 64; Fordham Univ, MA, 66; Dalhousie Univ, MD & PhD(psychol), 74. *Prof Exp:* ASSOC PROF, SCH MED, JOHNS HOPKINS UNIV, 78-. *Res:* Biological determinants of sexual phenomology relations between biological factors and states of mind. *Mailing Add:* 101 Meyer Bldg Johns Hopkins Hosp Baltimore MD 21205

BERLIN, GRAYDON LENNIS, b St Petersburg, Pa, May 21, 43; m 64; c 1. PHYSICAL GEOGRAPHY. *Educ:* Clarion State Col, BS, 65; Ariz State Univ, MA, 67; Univ Tenn, PhD(geog), 70. *Prof Exp:* Asst prof & res assoc geog, Fla Atlantic Univ, 68-69; asst prof, 69-80, ASSOC PROF GEOG, NORTHERN ARIZ UNIV, 80-. *Concurrent Pos:* Res assoc, US Geol Surv, 70-. *Mem:* Am Soc Photogram; Asn Am Geogr; Nat Res Soc NAm; Sigma Xi. *Res:* Environmental impacts and modifications; remote sensing of natural resources; fossil landscapes; urban climatology; land use. *Mailing Add:* 1180 Rockridge Rd Flagstaff AZ 86001

BERLIN, IRVING NORMAN, b Chicago, Ill, May 31, 17; m 43; c 3. CHILD & ADOLESCENT PSYCHIATRY. *Educ:* Univ Calif, Los Angeles, BA, 39; Univ Calif, San Francisco, MD, 43. *Prof Exp:* Lectr psychiat, Sch Med, Univ Calif, San Francisco, 50-52, clin instr, 52-54, asst res psychiatrist, 55-56, asst clin prof psychiat, 56-60, assoc clin prof, 60-65; prof psychiat & pediat & head div child psychiat, Sch Med, Univ Wash, 65-75; prof psychiat & pediat & head child psychiat, Univ Calif, Davis, 75-78; PROF PSYCHIAT & PEDIAT, SCH MED, UNIV NMEX, 78-. *Concurrent Pos:* Fel child psychiat, Langley Porter Inst, 46-50; consult, McAuley Neuropsychiat Inst, Calif, 60-, Ctr Training Community Psychiat, 61- & Training Br, NIMH, 74-; Grant Found grant, 73-75; Robert Wood Johnson grant, 87-89; Sr consult, Ment Health Indian Health Serv. *Honors & Awards:* McGrath Award, Asn Suicidology. *Mem:* Fel Am Orthopsychiat Asn; fel Am Psychiat Asn; fel Am Pub Health Asn; fel Am Acad Child Psychiat (pres, 75). *Res:* Childhood psychosis; mental health consultation and community psychiatry; training in mental health specialties; early intervention and prevention child mental illness; prevention adolescent Indian suicide and Indian pregnancy and alcoholism. *Mailing Add:* Dept Psychiat Sch Med Univ NMex Med Ctr Albuquerque NM 87131

BERLIN, JERRY D, b Trenton, Mo, Aug 28, 34; m 58; c 3. CELL BIOLOGY. *Educ:* Univ Mo, BS, 60, MA, 61; Iowa State Univ, PhD(cell biol), 64. *Prof Exp:* Fel, Iowa State Univ, 64; biol scientist, Gen Elec Co, 64-65; sr res scientist, Pac Northwest Lab, Battelle Mem Inst, 65-68; assoc prof, 68-73, prof biol & chmn dept biol sci, Tex Tech Univ, 73-87, adj prof anat, Sch Med, 74-87; PROF BIOL & DEAN SCI & MATH, SOUTHWEST MO STATE UNIV, 88-. *Concurrent Pos:* Grants, AEC, Cotton Producers Inst & Cotton Inc, 73-. *Mem:* AAAS; Am Micros Soc; Am Soc Cell Biol; Bot Soc Am; Electron Micros Soc Am; NSF. *Res:* Cotton fiber initiation and development; effect of water stress on plants, especially at the cellular and subcellular level. *Mailing Add:* Col Sci & Math Southwest Mo State Univ Springfield MO 65804-0094

BERLIN, KENNETH DARRELL, b Quincy, Ill, June 12, 33; m 58; c 2. ORGANIC CHEMISTRY, MEDICINAL CHEMISTRY. *Educ:* NCent Col, Ill, BA, 55; Univ Ill, PhD(org chem), 58. *Prof Exp:* Fel chem, Univ Fla, 58-60; from asst prof to prof, 60-71, REGENTS PROF CHEM, OKLA STATE UNIV, 71-. *Concurrent Pos:* App mem, Res Rev Panel, NSF postdoctoral fac fels, 69 & predoctoral fac fels, 70; mem, Med Chem Study Sect B, NIH, 69-73; develop therapeut rev comt, Nat Cancer Inst, 83-87. *Honors & Awards:* Award in Res, Sigma Xi, 69. *Mem:* Sr mem Am Chem Soc; The Chem Soc. *Res:* Organophosphorus chemistry; heterocyclic medicinal compounds; nuclear magnetic resonance spectroscopy; organometallics; synthetic organic chemistry. *Mailing Add:* Dept Chem Okla State Univ Stillwater OK 74078-0447

BERLIN, NATHANIEL ISAAC, b New York, NY, July 4, 20; m 53; c 2. MEDICAL RESEARCH. *Educ:* Western Reserve Univ, BS, 42; Long Island Col Med, MD, 45; Univ Calif, PhD(med physics), 49. *Prof Exp:* Intern, Kings County Hosp, Brooklyn, 45-46, res pathologist, 46-47; fel, Univ Calif, 49-50, res assoc, 50-51, instr, 51, lectr & res assoc, 51-52, lectr & assoc res med physicist, 52-53; Nat Heart Inst spec res fel, Nat Inst Med Res, London, 53-54; med officer, Anal Br, Effects Div, Hqs Armed Forces Spec Weapons Proj, 54-56; head metab serv, Gen Med Br, NIH, 56-59, chief, 59-61, clin dir, 61-71, sci dir gen labs & clin, Nat Cancer Inst, 69-72, dir div cancer biol & diag, 72-75; Genevieve B Teuton Prof Med & Dir, Cancer Ctr, Northwestern Univ, Chicago, 75-87; PROF ONCOL, DEP DIR, SYLVESTER COMPREHENSIVE CANCER CTR, MIAMI, 87-. *Concurrent Pos:* Mem med staff, Highland-Alameda County Hosp, Calif, 49-54; consult, US Naval Hosp, 55-65, Radiation Lab, Univ Calif, 55-57 & Defense Atomic Support Agency, Dept Defense, 57-59; assoc ed, Cancer Res; mem adv comt blood dis & resources, Nat Heart & Lung Inst, 76-80, rev comt, Cancer Pre-Clin Prog Proj, Nat Cancer Inst, 81-85; alumni lectr, State Univ NY. *Honors & Awards:* Superior Serv Award, Dept Health, Educ & Welfare. *Mem:* AAAS; Am Soc Clin Invest; Soc Exp Biol & Med; Am Physiol Soc; Am Asn Cancer Res; Asn Am Physicians; Am Clin & Climat Asn. *Res:* Erythropoiesis; use of radioactive isotopes in biological and medical research; cancer research. *Mailing Add:* Sylvester Comprehensive Cancer Ctr 1475 NW 12 Ave Rm 4037L Miami FL 33136

BERLINCOURT, TED GIBBS, b Fremont, Ohio, Oct 29, 25; m 53; c 1. PHYSICS. *Educ:* Case Inst Technol, BS, 49; Yale Univ, MS, 50, PhD(physics), 53. *Prof Exp:* Lab asst, Yale Univ, 49-50, res asst, 50-51; proj physicist, Naval Res Lab, 52-55; supvr electronic properties unit, Atomics Int Div, NAm Aviation, Inc, 55-64, group leader electronic properties, NAm Rockwell Sci Ctr, 64-65, assoc dir, 65-70, chief scientist, NAm Rockwell Microelectronics Co, 69-70; prof physics & chmn dept, Colo State Univ, 70-72; dir, Phys Sci Div, 72-81, dir math & phys sci, Off Naval Res, 81-86; DIR, RES & LAB MGT, OFF DEFENSE DIR RES & ENG, 86-. *Concurrent Pos:* Mem Instrumentation Panel, Nat Acad Sci-Nat Res Coun Physics Surv, 71 & Interface & Applns Panel, 84; Navy liaison rep, Div Phys Sci, Nat Res Coun, 72-76; orgn comt, Appl Superconductivity Confs, 66-72. *Mem:* Fel AAAS; fel Am Phys Soc; Sigma Xi; Inst Elec & Electronics Engrs. *Res:* Superconductivity; Fermi surfaces; de Haas-van Alphen effect; galvanomagnetic effects; magnetism; cryogenics; pulsed magnetic fields; metal-oxide-semiconductor large-scale-integration technology; scintillation counters; x-ray diffraction; research and development management and policy. *Mailing Add:* Off Defense Dir Res & Eng The Pentagon Rm 3E114 Washington DC 20301-3080

BERLIND, ALLAN, b New York, NY, Dec 24, 42; m 68; c 2. NEUROSCIENCES. *Educ:* Swarthmore Col, BA, 64; Harvard Univ, MA, 65, PhD(biol), 69. *Prof Exp:* NIH fel, Univ Calif, Berkeley, 69-71; from asst prof to assoc prof, 71-83, PROF BIOL, WESLEYAN UNIV, 83-. *Concurrent Pos:* Vis assoc prof zool, Univ Hawaii, Manoa, 80-81. *Mem:* Soc Neurosci. *Res:* Neurophysiology; neuroendocrinology. *Mailing Add:* Dept Biol Wesleyan Univ Middletown CT 06457

BERLINER, ERNST, b Katowice, Ger, Feb 18, 15; nat US; m 47; c 1. ORGANIC CHEMISTRY. *Educ:* Harvard Univ, MA, 40, PhD(org chem), 43. *Prof Exp:* Pvt asst antimalarials, Harvard Univ, 43-44; lectr, Bryn Mawr Col, 44-45, from asst prof to prof org chem, 45-85; RETIRED. *Concurrent Pos:* Guggenheim fel, 62. *Honors & Awards:* Award, Am Chem Soc, 71. *Mem:* AAAS; Am Chem Soc; Royal Soc Chem. *Res:* Physical-organic aspects of aromatic chemistry; hyperconjugation; aromatic substitution; isotope effects; polynuclear aromatic reactivities; halogenation of acetylenes. *Mailing Add:* Dept Chem Bryn Mawr Col Bryn Mawr PA 19010

BERLINER, FRANCES (BONDHUS), b Oskaloosa, Iowa, Nov 21, 21; m 47; c 1. ORGANIC CHEMISTRY. *Educ:* William Penn Col, BA, 43; Bryn Mawr Col, MA, 44, PhD(org chem), 47. *Prof Exp:* Demonstr, Bryn Mawr Col, 46-50, from instr to asst prof chem, 50-57, lectr, 57-71, assoc prof, 71-85; RETIRED. *Mem:* Am Chem Soc. *Res:* Aromatic chemistry. *Mailing Add:* Dept Chem Bryn Mawr Col Bryn Mawr PA 19010

BERLINER, HANS JACK, b Berlin, Ger, Jan 27, 29; US citizen; m 69. COMPUTER SCIENCE. *Educ:* George Washington Univ, BA, 54; Carnegie-Mellon Univ, PhD(comput sci), 75. *Prof Exp:* Develop syst analyst, IBM, 61-69; SR PRIN SCIENTIST COMPUT SCI, CARNEGIE-MELLON UNIV, 69-. *Mem:* Fel Am Asn Artificial Intel; Asn Comput Mach; Int Joint Conf Artificial Intel. *Res:* Construction of intelligent systems, specializing in game playing and search; developed the B* tree search algorithm. *Mailing Add:* Computer Sci Dept Carnegie-Mellon Univ Pittsburgh PA 15213

BERLINER, JUDITH A, b Cincinnati, Ohio, Sept 9, 39. BLOOD VESSEL DISEASES. *Educ:* Brown Univ, PhD(biol), 68. *Prof Exp:* ASSOC PROF PATH, SCH MED, UNIV CALIF, LOS ANGELES, 81-. *Mem:* Am Soc Cell Biol; Am Heart Asn; Am Diabetes Asn. *Mailing Add:* Dept Path Univ Calif Los Angeles Sch Med Ctr Health Sci Los Angeles CA 90024-1732

BERLINER, LAWRENCE J, b Los Angeles, Calif, Sept 18, 41; m 79; c 3. BIOPHYSICAL CHEMISTRY. *Educ:* Univ Calif, Los Angeles, BS, 63; Stanford Univ, PhD(phys chem), 67. *Prof Exp:* Res chemist, Sun Oil Co, 62 & Chevron Res Corp, Stand Oil Co, Calif, 63; Brit Heart Found-Am Heart Asn fel molecular biophysics, Oxford Univ, 67-68; Am Heart Asn fel phys chem, Stanford Univ, 68-69; from asst prof to assoc prof, 69-82, PROF CHEM, OHIO STATE UNIV, 82- *Concurrent Pos:* Estab investr, Am Heart Asn, 75-80; exchange vis to Bulgaria & Hungary, Nat Acad Sci, 79; UNESCO-Orgn Am States lectr, Chile, 79 & NSF-Nat Coun Sci & Technol lectr, US-Romania, 81; Lady Davis vis prof, Technion, 81; mem, thrombosis coun, Am Heart Asn; vis prof, Ministry Sci, Japan, 89; assoc ed, J Protein Chem. *Mem:* Am Chem Soc; Am Soc Biol & Molecular Biol; Soc Magnetic Resonance Med; Int Soc Magnetic Resonance; NY Acad Sci; AAAS. *Res:* Structure-function relations in biomolecules by magnetic resonance methods; small molecule binding; conformational changes; x-ray structural investigations; applications of magnetic resonance to biology; electron magnetic resonance imaging; in-vivo ESR. *Mailing Add:* Dept Chem Ohio State Univ 120 W 18th Ave Columbus OH 43210-1173

BERLINER, MARTHA D, b Antwerp, Belg, Nov 18, 28; US citizen; m 52; c 2. MICROBIOLOGY, MYCOLOGY. *Educ:* Hunter Col, BA, 49; Univ Mich, MS, 50; Columbia Univ, PhD(mycol), 53. *Prof Exp:* Jr scientist cytol, Sloan Kettering Inst, 50-51; sr scientist microbiol, Res & Advan Develop Div, Med Sci Dept, Avco Corp, 60-62, sr staff scientist space microbiol, 62-65; PROF BIOL, VA COMMONWEALTH UNIV, 78-87; from instr to assoc prof, 65-78, PROF BIOL, SIMMONS COL, 78-; ADJ PROF MICROBIOL, EASTERN VA MED SCH, NORFOLK, 87- *Concurrent Pos:* Instr, Lynn Hosp, Mass, 53-62; sr res assoc, Dept Microbiol, Sch Pub Health, Harvard Univ, 65-74; ed consult, Charles River Breeding Labs, 75-; prog assoc, NSF, 81-82. *Mem:* Am Mycol Soc; Am Soc Microbiol; NY Acad Sci; Int Soc Human & Animal Mycol; Med Mycol Soc Americas. *Res:* Medical mycology; microscopy; biological rhythms; radiation and environmental effects; protoplasts and wall deficient variants; biological parameter of pathogenicity in fungi; protoplasts of fungi and algae. *Mailing Add:* 2224 Bayville Rd Richmond VA 23455

BERLINER, ROBERT WILLIAM, b New York, NY, Mar 10, 15; m 41; c 4. PHYSIOLOGY, INTERNAL MEDICINE. *Educ:* Yale Univ, BS, 36; Columbia Univ, MD, 39. *Hon Degrees:* DSc, Med Col Wis & Yale Univ, 73. *Prof Exp:* Intern, Presby Hosp, NY, 39-41; resident physician, Goldwater Mem Hosp, 42-43; asst med, Col Med, NY Univ, 43-44, instr, 44-47; asst prof med, Columbia Univ, 47-50; chief, Lab Kidney & Electrolyte Metab, Nat Heart Inst, 50-62, dir intramural res, 54-68; dir labs & clins, NIH, 68-69, dep dir for sci, 69-73; prof, 73-85, dean, 73-84, EMER PROF & EMER DEAN, SCH MED, YALE UNIV, 85- *Concurrent Pos:* Res fel, Goldwater Mem Hosp, NY, 43-44, res asst, 44-47; res assoc, Dept Hosps, NY, 47-50; lectr, Sch Med, George Washington Univ, 51-73; Am Soc Clin Invest rep, Nat Res Coun, 57-60; prof lectr, Sch Med & Dent, Georgetown Univ, 64-73; coun, Nat Acad Sci, 78-81; mem, bd dirs, AAAS, 83- *Honors & Awards:* Homer W Smith Award, 65; Res Achievement Award, Am Heart Asn, 70. *Mem:* Nat Acad Sci; Am Acad Arts & Sci; AAAS (vpres, 72); Am Physiol Soc (pres, 67); Soc Exp Biol & Med (pres, 81-82); Am Asn Physicians; Sigma Xi. *Res:* Physiology of kidney; electrolyte transport. *Mailing Add:* Sch Med Yale Univ 333 Cedar St New Haven CT 06510

BERLINGHIERI, JOEL CARL, b Boston, Mass, Dec 18, 42; m 71; c 1. SPECTROSCOPIC INSTRUMENTATION DESIGN. *Educ:* Boston Col, BS, 64; Univ Rochester, MS, 66, PhD(physics), 70. *Prof Exp:* Asst prof physics, Colgate Univ, 69-71; from asst prof to assoc prof, 71-81, PROF PHYSICS, THE CITADEL, 81- *Concurrent Pos:* Res assoc, Univ Houston, 73-75. *Mem:* Inst Elec & Electronics Engrs; Am Asn Physics Teachers; Acoust Soc Am; Optical Soc Am. *Res:* Design and development of novel optical instruments including interferometer based spectrometers. *Mailing Add:* Dept Physics The Citadel Charleston SC 29409-0270

BERLINROOD, MARTIN, b New York, NY, Aug 20, 43; m 86; c 2. DEVELOPMENTAL BIOLOGY. *Educ:* City Col NY, BS, 64; Univ Tex, Austin, PhD(zool), 69. *Prof Exp:* Res assoc biol sci, State Univ NY Albany, 68-70; from asst prof to assoc prof biol sci, Goucher Col, 70-88; ASST DEAN, SCHOOL HUMAN & SCI, ITHACA COL, ITHACA, 88- *Concurrent Pos:* Registr, Goucher Col, 81-84, asst dean, 82-84, actg dean, 84-85, assoc dean, 85-88; asst dean, Sch Humanities, Ithaca Col, 88- *Res:* Axoplasmic transport in embryonic nerve fibers; axonal outgrowth and guidance in amphibian embryos. *Mailing Add:* Muller 206 Sch Humanities & Sci Ithaca Col Ithaca NY 14850

BERLMAN, ISADORE B, b St Louis, Mo, Jan 13, 22; m 53; c 2. PHYSICS. *Educ:* Wash Univ, PhD(physics), 50. *Prof Exp:* Assoc physicist radiol physics, Argonne Nat Lab, 50-71; assoc prof, 71-79, PROF PHYSICS, HEBREW UNIV, JERUSALEM, 79- *Mem:* Am Phys Soc; Radiation Res Soc. *Res:* Energy levels of the light nuclei; neutron detection; nuclear physics; luminescence; radiation damage. *Mailing Add:* Dept Physics Hebrew Univ Jerusalem Israel

BERLOW, STANLEY, b New York, NY, June 16, 21; m 47; c 6. PEDIATRICS. *Educ:* Univ Mich, BA, 40; Harvard Univ, MA, 41, MD, 50. *Prof Exp:* Intern pediat, Mass Gen Hosp, 50-51, chief resident, 53-54; resident, Babies Hosp, NY, 51-53; from instr to asst prof, Med Sch, Marquette Univ, 54-65; assoc prof pediat, Chicago Med Sch & med dir dysfunctioning child prog, Michael Reese Hosp & Med Ctr, Chicago, 65-69; ASSOC PROF PEDIAT, MED SCH, UNIV WIS-MADISON, 71- *Res:* Biochemistry of mental retardation. *Mailing Add:* 3100 Lake Mendota Dr Madison WI 53705

BERLOWITZ TARRANT, LAURENCE, b New York, NY, Oct 20, 34. DEVELOPMENTAL BIOLOGY, BIOMATERIALS. *Educ:* Univ Calif, Berkeley, AB, 54, PhD(genetics), 65; Univ Calif, Los Angeles, MA, 58. *Prof Exp:* Mem tech staff, Thompson-Ramo-Wooldridge Corp, 58-60; human factors scientist, Western Develop Lab, Philco Corp, 60-61; instr biol sci,

Chabot Col, 61-64; res fel med res coun epigenetics res group, Inst Animal Genetics, Edinburgh, Scotland, 65-66; from asst prof to assoc prof biol, State Univ NY Buffalo, 66-75, co-chmn dept, 68-69; prog dir genetic biol, NSF, 75-76, Directorate Biol, Behav, Soc Sci, 76-77; prof biol & asst vpres acad affairs, NY Univ, 77-81; provost, vpres Acad Affairs & prof biochem, Clark Univ, 81-83; dir, Mass Biotechnol Res Inst, 83-85; CONSULT BIOTECHNOL & BIOMAT, 83- *Concurrent Pos:* NIH spec fel, Univ Nijmegen, Neth, 72-73, grants assoc, 74-75; sr res fel, Ctr Sci & Technol Policy, NY Univ, 81- *Honors & Awards:* John Belling Prize in Genetics, 70. *Mem:* AAAS; Am Soc Cell Biol; Genetics Soc Am; Soc Indust Microbiol; Soc Develop Biol; Am Chem Soc. *Res:* Molecular and ultrastructural changes in the nucleus during differentiation; the chemical interface between materials and cells and tissues. *Mailing Add:* 73 Oak Hill Rd Harvard MA 01451

BERLYN, GRAEME PIERCE, b Chicago, Ill, Sept 6, 33; m 58; c 2. PLANT ANATOMY, TREE PHYSIOLOGY. *Educ:* Iowa State Univ, BS, 56, PhD(anat), 60. *Hon Degrees:* MA, Yale Univ, 78. *Prof Exp:* From instr to assoc prof wood anat, 60-78, PROF ANAT & PHYSIOL, SCH FORESTRY & ENVIRON STUDIES, YALE UNIV, 78- *Concurrent Pos:* Res collabr, Brookhaven Nat Lab, 63-66; ed, J Sustainable Forestry. *Mem:* Soc Wood Sci & Technol; Am Soc Plant Physiol; Int Asn Wood Anatomists; Bot Soc Am; fel Int Acad Wood Sci. *Res:* Quantitative cytochemistry; electron microscopy; reaction wood; developmental and experimental anatomy of pine embryos and seedlings; cell wall structure; microtechnical methods; effect of enzymatic hydrolysis on cell wall structure and free space; variation in DNA content of nuclei from populations of Pinus rigida; biotechnology of temperate and tropical trees; alpine ecophysiology. *Mailing Add:* Sch Forestry & Environ Studies Yale Univ 370 Prospect New Haven CT 06511

BERLYN, MARY BERRY, b Iowa Co, Iowa, Dec 28, 38; m 58; c 2. GENETICS. *Educ:* Iowa State Univ, BS, 59; Yale Univ, PhD(genetics), 66. *Prof Exp:* Res staff biologist, Yale Univ, 67-71, res assoc, 71-73; staff scientist, Conn Agr Exp Sta, 74-82; assoc prof, Lehman Col & Grad Ctr, City Univ New York, 82-88; RES SCI SCH FORESTRY, YALE UNIV, 88- *Mem:* AAAS; Genetics Soc Am. *Res:* Somatic cell genetics of plants; RNA polymerase mutants of E coli; histidine and polyaromatic biosynthetic pathways in prokaryotes and eukaryotes. *Mailing Add:* Dept Biol Sch Forestry Yale Univ PO Box 6666 New Haven CT 06511

BERLYN, ROBIN WILFRID, b Montreal, Que, Dec 3, 34; m 59; c 3. MECHANICAL ENGINEERING. *Educ:* McGill Univ, BEng, 58; Univ London, DIC, 61. *Prof Exp:* Prod develop engr, Can Ingersoll-Rand Co Ltd, Que, 58-60; sr res asst woodlands res, 60, assoc engr, 61-66, mech engr, 66-75, sr scientist, Forest Eng Res Inst Can, 75-79, sr scientist, 79-85, SR RES ENGR, PULP & PAPER RES INST CAN, 85- *Mem:* Tech Asn Pulp & Paper Indust; Can Pulp & Paper Asn. *Res:* Removal of bark from wood; off-road mobility of logging vehicles; measurement of forest environmental factors and their effect on the productivity of logging machines; upgrading wood chips for use in pulp and paper manufacture. *Mailing Add:* 381 Elm Ave Westmount PQ H3Z 1Z4 Can

BERLYNE, GEOFFREY MERTON, b Britain, May 11, 31; US citizen; m 59; c 3. NEPHROLOGY, RENAL PHYSIOLOGY. *Educ:* Univ Manchester, MBChB, 54, MD, 66; MRCP, 56; FRCP, 69. *Prof Exp:* Reader med, Univ Manchester, 68-70; prof life sci & med, Univ Negev, Israel, 70-80; PROF MED, STATE UNIV NY, BROOKLYN, 76- *Concurrent Pos:* Ed, Nephron, 71-, Contributions to Physiology, 75- & Renal Physiol, 78-; vis prof, Univ Otago, 80. *Mem:* Am Soc Nephrol; Am Soc Artificial Internal Organs; Am Col Nutrit; NY Acad Sci; Am Col Physicians; Am Soc Trace Element Res. *Res:* Trace elements in renal disease; electrolyte physiology; nutrition in renal disease. *Mailing Add:* Dept Med-Nephrol State Univ NY Health Sci Col Med 450 Clarkson Ave Brooklyn NY 11203

BERMAN, ABRAHAM S, fluid mechanics; deceased, see previous edition for last biography

BERMAN, ALAN, b Brooklyn, NY, Nov 2, 25; m 62; c 5. PHYSICS, PHYSICAL OCEANOGRAPHY. *Educ:* Columbia Univ, AB, 47, PhD(physics), 52. *Prof Exp:* Res physicist, Hudson Labs, Columbia Univ, 52-57, assoc dir, 57-63, dir, 63-67; dir res, US Naval Res Lab, 67-; AT ROSENSTIEL SCH MARINE & ATMOSPHERIC SCI. *Mem:* Fel Am Phys Soc; fel Acoust Soc Am. *Res:* Atomic beams. *Mailing Add:* 5300 Holmes Run Pkwy Apt 804 Alexandria VA 22304

BERMAN, ALEX, b New York, NY, Feb 7, 14; m 43. HISTORY OF PHARMACY. *Educ:* Fordham Univ, BS, 47; Univ Wis, MS, 51, PhD(hist of pharm & sci), 54. *Prof Exp:* Pharmacist, Vet Admin, 48-49; asst hist of pharm, Alumni Res Found, Univ Wis, 50-53; staff assoc, Am Pharmaceut Asn, 53-54; pharmacist, Univ Mich Hosp, 54-55; asst prof hist of pharm, Univ Wis, 55-56; asst prof pharm, Univ Mich, 56-61; assoc prof, Univ Tex, 61-68; prof hist & social studies in pharm, 68-74, EMER PROF HIST & HIST STUDIES IN PHARM, COL PHARM, UNIV CINCINNATI, 75- *Concurrent Pos:* Guggenheim Mem Found fel, 58-59; NSF grant, 62-64; hon pres, Am Inst Hist Pharm, 85-87. *Honors & Awards:* Kremers Award, Am Inst Hist Pharm, 63. *Mem:* Am Pharmaceut Asn; hon mem Am Soc Hosp Pharmacists; Am Asn Hist Med; Am Inst Hist Pharm (pres, 67-69); Int Acad Hist Pharm. *Res:* Nineteenth century pharmaceutical and medical Americana and French pharmacy; history of hospital pharmacy. *Mailing Add:* 310 Bryant Cincinnati OH 45220

BERMAN, ALVIN LEONARD, b Baltimore, Md, July 19, 24; m 49; c 3. NEUROANATOMY, NEUROPHYSIOLOGY. *Educ:* Johns Hopkins Univ, AB, 45, PhD(physiol), 57. *Prof Exp:* USPHS fel neuroanat, Johns Hopkins Hosp, 57-59, fel psychiat, 59-60; asst prof physiol, Univ Md, 60-61; from asst prof to assoc prof anat & neurophysiol, Univ Wis-Madison, 61-69, dir, Multi-Disciplinary Labs & coordr interdisciplinary curric, 69-78, prof neurophysiol & neurol, 69-90, EMER PROF NEUROPHYSIOL & NEUROL, UNIV WIS-MADISON, 90- *Mem:* Am Asn Anat; Soc Neurosci. *Res:* Structure and function of mammalian central nervous system; cytoarchitecture of the the brain stem, thalamus and basal telencephalon of the cat and monkey; connections of interpeduncular complex in mammals; electrophysiology of somatic and auditory areas of cat cerebral cortex. *Mailing Add:* Dept Neurophysiol Med Sch Univ Wis 623 Waisman Ctr Madison WI 53705

BERMAN, BARRY L, b Chicago, Ill, Mar 8, 36; m 63; c 2. PHOTONUCLEAR REACTIONS, ELECTROMAGNETIC INTERACTIONS. *Educ:* Harvard Univ, BA, 57; Univ Ill, MS, 59, PhD(physics), 63. *Prof Exp:* Nuclear physicist, Lawrence Livermore Lab, 63-85; PROF PHYSICS, GEORGE WASHINGTON UNIV, 85- *Concurrent Pos:* Vis prof, Yale Univ, 69-70; vis prof, Univ Toronto, 70 & 75; guest prof, Univ Frankfurt, 74; vis prof, Univ Sao Paulo, 77; Sir Thomas Lyle fel, Univ Melbourne, 78; mem subcomt photonuclear reactions, US Nuclear Data Comt, 73; mem subcomt basic sci, 73-75; vis scientist, Ctr d'Etudes Nucleaires, Saclay, France, 80-81; vis prof, Mass Inst Technol, 82; guest scientist, Lawrence Berkeley Lab, 86, Los Alamos Nat Lab, 85, 88 & 89. *Mem:* Fel Am Phys Soc. *Res:* Photonuclear reactions, electron and proton scattering, pion physics; neutron physics; nuclear astrophysics; applications; channeling and transition radiation. *Mailing Add:* Dept Physics George Washington Univ Washington DC 20052

BERMAN, DAVID ALBERT, b Rochester, NY, Nov 4, 17; m 45; c 2. PHARMACOLOGY. *Educ:* Univ Southern Calif, BS, 40, MS, 48, PhD(pharmacol), 51. *Prof Exp:* Life Ins Med res fel, 51-52; from instr to assoc prof, 52-63, PROF PHARMACOL & TOXICOL, UNIV SOUTHERN CALIF, 63- *Concurrent Pos:* USPHS spec fel, 60-61. *Mem:* Am Soc Pharmacol & Exp Therapeut. *Res:* Cardiac metabolism in relation to drug action. *Mailing Add:* Dept Pharmacol Sch Med Univ Southern Calif Los Angeles CA 90033

BERMAN, DAVID ALVIN, b Milwaukee, Wis, Apr 13, 24; m 70. INORGANIC CHEMISTRY, CORROSION. *Educ:* Univ Wis, BS, 48; Univ Mich, MS, 50, PhD(inorg chem), 54. *Prof Exp:* Instr, Wis State Teachers Col, Oshkosh, 50; asst prof chem, Exten Div, Univ Wis, 55-63 & Carroll Col, Wis, 64-65; chemist, Naval Air Eng Ctr, Naval Air Develop Ctr, 67-86; RETIRED. *Mem:* AAAS; Am Chem Soc; Sigma Xi. *Res:* Corrosion; electrochemistry; hydrogen embrittlement. *Mailing Add:* 173 Deerfield Dr Souderton PA 18964

BERMAN, DAVID MICHAEL, b Boston, Mass, May 14, 46. MATHEMATICS. *Educ:* Dartmouth Col, AB, 67; Univ Pa, PhD(appl math), 73. *Prof Exp:* Vis instr math, Franklin & Marshall Col, 73-74; from instr to assoc prof, 74-90, PROF MATH, UNIV NEW ORLEANS, 90- *Concurrent Pos:* Vis scholar, Smith Col, 87. *Mem:* Soc Indust Appl & Math. *Res:* Graph theory. *Mailing Add:* Dept Math Univ New Orleans New Orleans LA 70148

BERMAN, DAVID S, b Los Angeles, Calif, Jan 10, 40; div. VERTEBRATE PALEONTOLOGY. *Educ:* Univ Calif, Los Angeles, BA, 62, MA, 65, PhD(zool), 69. *Prof Exp:* Asst cur vert paleont, 70-77, assoc cur, 77-89, CUR, CARNEGIE MUS NATURAL HIST, 89- *Mem:* Paleont Soc; Am Soc Ichthyologists & Herpetologists; Soc Study Evolution. *Res:* Paleontology of late Paleozoic vertebrates. *Mailing Add:* Carnegie Mus Natural Hist 4400 Forbes Ave Pittsburgh PA 15213

BERMAN, DAVID THEODORE, b Brooklyn, NY, June 14, 20; m 44; c 2. VETERINARY SCIENCE. *Educ:* Brooklyn Col, BA, 41; Cornell Univ, DVM, 44; Univ Wis, MS, 46, PhD, 49. *Prof Exp:* Res asst, 44-46, from instr to assoc prof, 46-57, chmn dept, 64-68, assoc dean grad sch, 69-75, PROF VET SCI & BACT, UNIV WIS-MADISON, 57- *Concurrent Pos:* Coop agent, USDA, 46-57; mem, Conf Res Workers Animal Dis NAm, 47-; consult, Nat Brucellosis Comt, 58-; grad sch G I Haight traveling fel, State Serum Inst, Copenhagen, Denmark & Vet Lab, Weybridge, Eng, 63; expert comt brucellosis, WHO, 63-70; mem, 70-85; mem bd sci counr, Nat Inst Allergy & Infectious Dis, 74-78, chmn, 78, chmn, Nat Brucellosis Tech Comn, 76-79; chmn, Div Microbial Pathogenesis, Am Soc Microbiol, 85. *Honors & Awards:* Animal Health Award, USDA, 80. *Mem:* AAAS; Am Soc Microbiol; Am Vet Med Asn; US Animal Health Asn; Am Asn Immunol. *Res:* Immunology and bacteriology of brucellosis; microbial variation; staphylococcal infection; hypersensitivity in experimental arthritis; antigenic analysis mycobacteria; chlamydia; immune response mucosae. *Mailing Add:* Dept Vet Sci & Bacteriol Univ Wis 1655 Linden Dr Madison WI 53706

BERMAN, DONALD, virology, water pollution, for more information see previous edition

BERMAN, ELEANOR, b Duluth, Minn, Oct 5, 21. PHARMACOLOGY, BIOCHEMISTRY. *Educ:* Wis State Col, Superior, BS, 42; Univ Minn, Minneapolis, MS, 50, PhD(pharmacol), 57. *Prof Exp:* Teacher pub schs, Mich, 42-45 & Wis, 45-46; med technologist, St Mary's Hosp, Duluth, Minn, 46-48; clin biochemist, Ill Masonic Hosp, Chicago, 53-66; toxicologist, Hektoen Inst, 66-71 & Div Biochem, Cook County Hosp, 71-91; RETIRED. *Mem:* Am Asn Clin Chem; fel Am Acad Forensic Sci; Am Chem Soc; Soc Appl Spectros. *Res:* Clinical toxicology; lead poisoning and trace metals in clinical chemistry. *Mailing Add:* 2500 Lakeview Chicago IL 60614

BERMAN, ELLEN MYRA, b Pittsburgh, Pa, July 4, 57. PHARMACEUTICAL CHEMISTRY, NATURAL PRODUCTS CHEMISTRY. *Educ:* Univ Pittsburgh, BS & MS, 79; Yale Univ, MS & PhD(org chem), 83. *Prof Exp:* Scientist, 83-85, SR SCIENTIST, WARNER-LAMBERT PARKE-DAVIS PHARM RES, 85- *Mem:* Am Chem Soc. *Res:* Total synthesis of organic compounds of potential use as cancer chemotherapeutic agents; medicinal chemistry; synthetic methods development. *Mailing Add:* 2205 Winchell Dr Ann Arbor MI 48104-4774

BERMAN, ELLIOT, b Quincy, Mass, Jan 13, 30; m 53; c 3. ORGANIC CHEMISTRY. *Educ:* Brown Univ, ScB, 51; Boston Univ, PhD(org chem), 56. *Prof Exp:* Sr chemist, Fundamental Res Dept, Nat Cash Register Co, 55-57, head org sect, 57-59; head chem sect, Res Div, Itek Corp, 59-61, mgr chem dept, 61-64, tech dir res, 64-65, dir, Lexington Res Labs, 65-67, dir photosensitive mat, 67-69; consult, 69-73; pres & chmn, Solar Power Corp, 73-75; dir, Ctr Energy Studies, Boston Univ, 75-78; res prof appl sci, 78-79; chief scientist, Arco Solar Inc, 79-89; PRES, PROJ SUNRISE, INC, 90- *Concurrent Pos:* Instr, Univ Dayton, 56-57; asst prof, Univ Cincinnati, 56-58; pres, Sun Systs, Inc, 78- *Mem:* AAAS; Am Chem Soc; Solar Energy Soc; Royal Soc Chem; Inst Elec & Electronics Engrs. *Res:* Reversible photochemical systems; kinetics and mechanisms of reactions at solid interfaces; photochemistry; photophysics; solar energy utilization; photography; pollution control. *Mailing Add:* Proj Sunrise Inc 6377 San Como Lane Camarillo CA 93012

BERMAN, GERALD, b Can, Nov 12, 24; US citizen; m 48; c 4. MATHEMATICS. *Educ:* Univ Toronto, BA, 47, MA, 48, PhD(math), 50. *Prof Exp:* From instr to assoc prof math, Ill Inst Technol, 50-59; chmn, Dept Combinatorics & Optimization, Univ Waterloo, 66-78, prof math, 59-89; RETIRED. *Concurrent Pos:* Consult, Inst Air Weapons Res, Chicago, 57-59, Martin-Marietta Corp, Colo, 59-63 & Defense Res Bd Can, 64-68. *Mem:* Am Math Soc; Math Asn Am; Soc Indust & Appl Math; Opers Res Soc Am; Can Math Cong. *Res:* Applied mathematics; combinatorics; optimization. *Mailing Add:* Four Atlantis Unionville ON L3P 7B5 Can

BERMAN, GERALD ADRIAN, b Bridgeport, Conn, Dec 26, 34; m 60; c 2. ELECTRICAL ENGINEERING. *Educ:* Univ Miami, BS, 55; La State Univ, MS, 57; Univ Detroit, DrEng(elec eng), 68. *Prof Exp:* Instr elec eng, La State Univ, 56-57; from instr to asst prof, Univ Miami, 57-66; from instr to assoc prof & chmn elec eng, Univ Detroit, 66-77; proj mgr & group leader, Nat Inst Standards & Technol, 77-83; CHIEF, TECHNOL ASSESSMENT & PLANNING, VOICE AM, 83- *Concurrent Pos:* Res consult, Detroit Edison Power Co, 70-; res consult, Ford Motor Co, 73-74; res consult, Bendix Res Labs, 76; res consult, US Army, 75; NASA fac fel, 72. *Mem:* Am Soc Testing & Mat; Inst Elec & Electronics Engrs. *Res:* Oceanographic engineering; underwater acoustics and instrumentation; information display systems; environmental engineering; electrostatic precipitors; biological effects of magnetic fields; automotive instrumentation; radio propogation; broadcast system engineering; computer graphics. *Mailing Add:* Voice Am 330 Independence Ave SW Washington DC 20547

BERMAN, HELEN MIRIAM, b Chicago, Ill, May 19, 43. BIOLOGICAL STRUCTURE. *Educ:* Columbia Univ, AB, 64; Univ Pittsburgh, PhD(crystallog), 67. *Prof Exp:* NIH traineeship biochem crystallog, Univ Pittsburgh, 67-69; res assoc molecular struct, 69-73, asst mem, 73-78, MEM, INST CANCER RES, 78- *Concurrent Pos:* Dir, Res Comput Facil; mem, NIH Study Sect, 80-84; US Nat Comt Crystallog, 80- *Mem:* AAAS; Am Crystallog Asn; Biophys Soc; Am Chem Soc; Am Soc Biol Chemists. *Res:* Crystal structures of nucleic acid components and proteins; structural interactions between nucleic acids and proteins. *Mailing Add:* Dept Chem Rutgers Univ PO Box 939 Piscataway NJ 08855-0939

BERMAN, HERBERT JOSHUA, b Boston, Mass, Oct 20, 24. PHYSIOLOGY. *Educ:* Univ RI, BS, 45; Boston Univ, AM, 48, PhD(physiol), 53. *Prof Exp:* Instr zool, Univ Mass, 48-49; asst, 52-53, res assoc biol, 54-61, from asst res prof to assoc res prof, 61-69, PROF BIOL, BOSTON UNIV, 69- *Mem:* Radiation Res Soc; Am Soc Hemat; Microcirc Soc (secy-treas, 59-); Am Soc Zoologists; Am Physiol Soc; Sigma Xi. *Res:* Physiology of circulation; small blood vessels; hypertension; thromboembolism and blood coagulation; ecaluation of plasma expanders; physiology of aging. *Mailing Add:* 142 Colidge St Brookline MA 02146

BERMAN, HORACE AARON, b Brooklyn, NY, Nov 21, 15; m 39; c 3. APPLIED CHEMISTRY. *Educ:* Columbia Univ, BS, 35; ChE, 36. *Prof Exp:* Chemist, Union Chem Corp, NJ, 36; chem engr, Herstein Labs, Inc, NY, 37; draftsman, Westminster Tire Corp, 38-40; chemist San Francisco field off, Nat Bur Stand, 40-57, head paint lab, 45-57, DC, 57-68; res chemist, Fed Hwy Admin, 68-76; transl, sci arts, Bay Area Transl, Inc, 77-91; RETIRED. *Concurrent Pos:* Teacher, Drew Sch, San Francisco, 56; Emerson Inst, 59-61; abstractor, Chem Abstracts, 58-72; auditor, Calorimetry Conf, 63; consult, 77-91. *Mem:* Am Chem Soc; Sigma Xi. *Res:* Thermochemistry and phase equilibria of portland cement constituents; analytical chemistry; paint and varnish technology; synthetic resins; colorimetry of paint surfaces. *Mailing Add:* 234 Elm St Apt 301 San Mateo CA 94401-2625

BERMAN, HOWARD MITCHELL, b Pittsburgh, Pa, May 1, 36; m 58; c 2. BIOCHEMISTRY, IMMUNOLOGY. *Educ:* Univ Pittsburgh, BS, 57. *Prof Exp:* Res chemist, Vet Admin Med Ctr, Pittsburgh, 60-72; health sci adminr, Vet Admin Cent Off, Washington, DC, 72-79, chief prog rev div, 79-87; health sci adminr, Nat Inst Allergy & Infectious Dis, 87-89, HEALTH SCI ADMINR, DIV RES GRANTS, NIH, 89- *Concurrent Pos:* Vet Admin rep, Fed Interagency Recombinant DNA Comt, 79-87. *Mem:* Am Inst Chemists; AAAS. *Res:* Biochemical pharmacology and use of gas chromatography and mass spectrometry for analysis; enzyme purification; cell membrane ATPase and structure-function relationships; cell membrane aging. *Mailing Add:* 3225 Birchtree Lane Silver Spring MD 20906

BERMAN, IRWIN, b New York, NY, Oct 16, 25; div; c 2. SOLID MECHANICS, MECHANICAL ENGINEERING. *Educ:* City Col New York, mech eng, 48; Stevens Inst Technol, MS, 53; Polytech Inst NY, PhD(appl mech), 59. *Prof Exp:* Head anal sect, Wright Aero Div, Curtiss Wright Corp, 48-54; res assoc appl mech, Polytech Inst NY, 54-56; mgr solid mech, Res Div, 56-76, tech dir res & develop, 76-78, CHMN, TECH DIRECTORATE, FOSTER WHEELER DEVELOP CORP, 78- *Concurrent Pos:* Adj prof, Col Eng, NY Univ, 60-71; consult, Govt Arg, 74; sr tech ed, J Pressure Vessel Technol, 73-77; reviewer, Appl Mech Reviews, 73-; mem bd dirs, Am Soc Mech Engrs, 85-87. *Honors & Awards:* Centennial Medallion

& Pressure Vessels & Piping Medal, Am Soc Mech Engrs. *Mem:* Fel Am Soc Mech Engrs (vpres, 77-81, sr vpres, 81-85); Soc Exp Mech. *Res:* Theoretical and experimental analysis of structures; methods of fabrication and use of analysis in design codes; plasticity; creep; thermal stress; stability strain measurement; high pressure vessels; explosive forming; solar applications. *Mailing Add:* Foster Wheeler Develop Corp 12 Peach Tree Hill Rd Livingston NJ 07039

BERMAN, IRWIN, b New York, NY, Dec 4, 24; m 51; c 2. ANATOMY, HEMATOLOGY. *Educ:* Seton Hall Univ, BS, 49; NY Univ, MS, 51, PhD(physiol), 55. *Prof Exp:* Asst radiobiol, Brookhaven Nat Lab, 55-56; res assoc, Sch Med, Stanford Univ, 56-59; asst prof, 59-62, ASSOC PROF ANAT, SCH MED, UNIV MIAMI, 62- *Concurrent Pos:* Am Cancer Soc fel, Brookhaven Nat Lab, 55-56 & Sch Med, Stanford Univ, 56-57; NIH res fel, 57-58; USPHS spec res fel, Harvard Med Sch, 63-64. *Mem:* AAAS; Am Soc Cell Biologists; Am Asn Anat; Radiation Res Soc; Sigma Xi. *Res:* Electron microscopy; radiation biology; oncology. *Mailing Add:* Dept Anat & Cell Biol R124 Univ Miami Sch Med PO Box 016960 Miami FL 33101

BERMAN, JEROME RICHARD, b New York, NY, Aug 7, 20; m 46; c 1. INTERNAL MEDICINE. *Educ:* Univ Cincinnati, BA, 41, MD, 44. *Prof Exp:* Fel, 47-50, from instr to assoc prof med, 48-67, clin prof, 67-74, PROF MED, UNIV CINCINNATI, 74-; ATTEND PHYSICIAN, HOLMES HOSP, UNIV CINCINNATI HOSP, 74- *Concurrent Pos:* Med consult, Integrated Acad Info Mgt Syst, 87-89, Med Info, 87-89, consult, Time Proj, NLM-NIH. *Mem:* AAAS; Am Fedn Clin Res; Am Asn Study Liver Dis; fel Am Col Physicians; fel Am Col Gastroenterol; Sigma Xi; Am Gastroenterol Asn; Am Soc Gastrointestinal Endoscopy. *Res:* Gastroenterology. *Mailing Add:* 3904 Wess Park Dr Cincinnati OH 45217

BERMAN, JOEL DAVID, b Minneapolis, Minn, Jan 1, 43. MATHEMATICS. *Educ:* Univ Minn, Minneapolis, BA, 65; Univ Wash, PhD(math), 70. *Prof Exp:* From asst prof to assoc prof, 70-82, PROF MATH, UNIV ILL, CHICAGO, 82- *Mem:* Am Math Soc; Math Asn Am. *Res:* Lattice theory and universal algebra. *Mailing Add:* Dept Math Univ Ill Chicago Circle Chicago IL 60680

BERMAN, JULIAN, b New York, NY, Oct 4, 24; m 48; c 3. AERONAUTICAL ENGINEERING. *Educ:* City Col New York, BS, 44; NY Univ, MS, 48. *Prof Exp:* Mathematician aerial survs, Fairchild Aerial Surv Inc, 47-48, engr, Fairchild Aircraft Co, 48-50; res scientist, Nat Adv Comt Aeronaut, 50-56; engr, Repub Aviation Corp, 56-64; engr, Grumman Aircraft & Eng Corp, Bethpage, 64-73; CHIEF FLUTTER & VIBRATION ENGR, FAIRCHILD REPUB CO, 73- *Concurrent Pos:* Exten instr, George Washington Univ, 54-55. *Mem:* Assoc fel Am Inst Aeronaut & Astronaut. *Res:* Aeroelasticity; vibrations; flutter; structural mechanics; oscillatory aerodynamics. *Mailing Add:* Six Nutley Ct Plainview NY 11803

BERMAN, LAWRENCE URETZ, b Chicago, Ill, Aug 4, 19; m 50; c 1. POLYMER CHEMISTRY, ENVIRONMENTAL SCIENCE. *Educ:* Univ Ill, Urbana, BS, 41; Northwestern Univ, MS, 47. *Prof Exp:* Asst chem org chem dept, IIT Res Inst, 48-56, assoc chemist, 56, res chemist, 56-65; res chemist bus equip group, Explor Res Dept, Bell & Howell Co, 65-68; res chemist, Res & Develop Div, Kraft Inc, 68-82; CONSULT, 82- *Mem:* AAAS; Am Chem Soc; Am Inst Chem. *Res:* Synthesis of new organic compounds, including monomers and polymers; development of organic reaction processes by new techniques and commercial scale-up. *Mailing Add:* 8025 N Hamlin Ave Skokie IL 60076-3403

BERMAN, LOUIS, b London, Eng, Mar 21, 03; nat US; m 34; c 1. ASTRONOMY. *Educ:* Univ Minn, AB, 25, AM, 27; Univ Calif, PhD(astrophys), 29. *Prof Exp:* Asst, Univ Minn Observ, 25-27; instr astron, Carleton Col, 29-31; instr astron & math, San Mateo Col, 31-35 & City Col San Francisco, 35-68; lectr, Univ San Francisco, 68-75; adj prof astron, 75-81; RETIRED. *Mem:* AAAS; Am Astron Soc; Astron Soc Pac. *Res:* Comet positions and orbits; spectroscopic and visual binaries; variable stars; nebular and stellar spectroscopy; galactic rotation. *Mailing Add:* 1020 Laguna Ave Burlingame CA 94010

BERMAN, M LAWRENCE, b Stanford, Conn, July 13, 29; m ; m ; c 2. ANESTHESIOLOGY, CLINICAL PHARMACOLOGY. *Educ:* Univ Conn, BS, 51; Univ Wash, MS, 54, PhD, 56; Univ NC, MD, 64. *Prof Exp:* Toxicologist & asst chief environ health lab, Air Materiel Command, Wright-Patterson AFB, 56-57, prin investr pharmacol-biochem, Aerospace Med Lab, Wright Air Develop Ctr, 57-60; resident anesthesia, Sch Med, Univ NC, Chapel Hill, 65-67; from asst prof to assoc prof, Sch Med, Northwestern Univ, 67-74; PROF ANESTHESIA, SCH MED, VANDERBILT UNIV, 74-, ASSOC PROF PHARMACOL, 76- *Mem:* Am Soc Clin & Exp Pharmacol; Int Anesthesia Res Soc; Am Soc Clin Res; Am Soc Anesthesiol; Asn Univ Anesthetists. *Res:* Man in space program; humoral response to stress; catabolism of catechol amines; psychotropic drugs; drug metabolism. *Mailing Add:* Sch Med Vanderbilt Univ Nashville TN 37205

BERMAN, MARLENE OSCAR, b Philadelphia, Pa, Nov 21, 39; div; c 1. NEUROPSYCHOLOGY, PHYSIOLOGICAL PSYCHOLOGY. *Educ:* Univ Pa, BA, 61; Bryn Mawr Col, MA, 64; Univ Conn, PhD(psychol), 68. *Prof Exp:* Instr psychol, Univ Conn, 66-68; USPHS fel, Harvard Univ, 68-70, instr, 70; res assoc, 71-72, asst prof, 72-75, assoc prof, 76-81, PROF NEUROL & PROF PSYCHIAT, SCH MED, BOSTON UNIV, 82-; RES PSYCHOLOGIST, VET MED CTR HOSP, 73- *Concurrent Pos:* Res assoc psychol, Vet Admin Hosp, Boston, 70-73; clin investr, 73-76; instr psychol, Exten Sch, Harvard Univ, 70-76; Nat Inst Neurol & Commun Dis & Stroke res grant, 72-76; res career develop award, 76-81; affil prof, Clark Univ, 73-; mem ment health small grant comt, NIMH, 74-78; Sigma Xi Nat lectr, 80-81; res sci develop award, Nat Inst Alcohol Abuse & Alcoholism, 81-86, mem biomed res rev comt, 87-91, chmn, 90-91; Fulbright sr scholar award, 91- *Honors & Awards:* Vet Admin Med Clin Invest Award, 73. *Mem:* Sigma Xi; fel Am Psychol Asn; Soc Neurosci; Psychonomic Soc; Int Neuropsychol Soc.

Res: Normal and pathological brain function, including mechanisms of cerebral specialization, Korsakoff's syndrome, aphasia, Huntington's disease, Parkinsonism and dementia; emphasis on understanding mechanisms of disordered memory, perception and cognition. *Mailing Add:* Psychol Serv Vet Admin Med Ctr 150 Huntington Ave Boston MA 02130

BERMAN, MARTIN, physics, engineering physics; deceased, see previous edition for last biography

BERMAN, MICHAEL ROY, b Brooklyn, NY, Jan 22, 47; m 72; c 2. CROSSBRIDGE KINETICS. *Educ:* City Col, City Univ New York, BS, 69; Ohio State Univ, MS, 77, PhD(physiol), 79. *Prof Exp:* Assoc, Univ Vt, 79-81, fel, 81-82, res assoc physiol & biophysics, 82-85; ASST PROF, DEPT ANAT, JOHNS HOPKINS UNIV, 85- *Res:* Chemical and geometrical factors which modulate crossbridge kinetics in striated muscle. *Mailing Add:* 9076 Meadow Heights Randallstown MD 21133

BERMAN, NANCY ELIZABETH JOHNSON, b Paducah, Ky, Aug 14, 46; m 70; c 3. NEUROPHYSIOLOGY, NEUROANATOMY. *Educ:* Lawrence Univ, BA, 68; Mass Inst Technol, PhD(psychol & brain sci), 72. *Prof Exp:* Res asst brain sci, Mass Inst Technol, 68-72; fel anat, Sch Med, Univ Pa, 72-74; res assoc neuroanat & neurophysiol, Sch Med, Wash Univ, 74-76; asst prof physiol, Med Col Pa, 76-80, asst prof anat, 77-80, assoc prof physiol & anat, 81-87; PROF ANAT, UNIV KANS MED CTR, 87- *Concurrent Pos:* Mem neurobehav sect, NIMH, 83-87. *Mem:* Soc Neurosci; Am Asn Anatomists; Asn Res Vision & Ophthal; Cajal Club. *Res:* Anatomy, physiology and development of the visual system; effects of experience on brain development. *Mailing Add:* Dept Anat Univ Kans Med Ctr 39th & Rainbow Blvd Kansas City KS 66103

BERMAN, NEIL SHELDON, b Milwaukee, Wis, Sept 21, 33; m 62; c 2. CHEMICAL ENGINEERING. *Educ:* Univ Wis, BS, 55; Univ Tex, MS, 61, MA & PhD(chem eng), 62. *Prof Exp:* Designs engr, Standard Oil Co Calif, 55-56; res scientist, E I du Pont de Nemours & Co, 62-64; from asst to assoc prof eng, 64-70, PROF ENG, ARIZ STATE UNIV, 70- *Concurrent Pos:* Prin investr grants, NSF, Petrol Res Fund & US Dept Transp; vis prof, Pa State Univ, 72-73 & Univ Tex, Austin, 80 & 83; grad col distinguished res prof, 84-85. *Mem:* Fel AAAS; fel Am Inst Chem Engrs; Am Chem Soc; Am Soc Mech Engrs; Am Phys Soc. *Res:* Thermodynamics; applied mathematics and numerical analysis; fluid dynamics and transport processes related to small particles and polymers; air pollution modeling; laser doppler anemometry; particle counting methods in fluid mechanics. *Mailing Add:* Dept Chem Biol & MHS Eng Ariz State Univ Tempe AZ 85287-6006

BERMAN, PAUL RONALD, b New York, NY, Feb 5, 45; m 87; c 1. ATOMIC PHYSICS, LASERS. *Educ:* Rensselaer Polytech Inst, BS, 65; Yale Univ, MS, 66, PhD(physics), 69. *Prof Exp:* Instr physics, Yale Univ, 69-71; from asst prof to assoc prof, 71-82, PROF PHYSICS, NY UNIV, 82- *Mem:* Fel Am Phys Soc. *Res:* Theoretical atomic and laser physics with special emphasis on collision effects. *Mailing Add:* Dept Physics NY Univ Four Wash Pl New York NY 10003

BERMAN, REUBEN, b Minneapolis, Minn, Feb 8, 08; m 31; c 6. CARDIOLOGY. *Educ:* Univ Minn, Minneapolis, BA, 29, MD, 33. *Prof Exp:* Clin prof med, Univ Minn, Minneapolis, 63-77, emer prof, 77-85; RETIRED. *Concurrent Pos:* Prin investr, Coronary Drug Proj, Mt Sinai Hosp, Minneapolis; pvt pract med, 37- *Mem:* Am Col Physicians; Am Col Cardiol; Am Heart Asn. *Mailing Add:* 2809 Drew Ave Minneapolis MN 55416

BERMAN, ROBERT HIRAM, b New York, NY, Apr 1, 48; m 86. APPLIED MATHEMATICS, PLASMA PHYSICS. *Educ:* Mass Inst Technol, BS, 70, PhD(appl math), 75. *Prof Exp:* Res fel, Univ Reading, 75-78; res assoc, Mass Inst Technol, 78-86. *Mem:* Soc Indust & Appl Math; Royal Astron Soc; Am Astron Soc; Am Phys Soc; Sigma Xi. *Res:* Computational physics; large scale numerical simulations of plasma turbulence, using supercomputers for scientific computing applications and using computer algebra for applications in artificial intelligence. *Mailing Add:* 50 Richfield Rd Arlington MA 02174

BERMAN, SAM MORRIS, b Worcester, Mass, Feb 4, 33; m 57. THEORETICAL PHYSICS. *Educ:* Univ Miami, Fla, BS, 54, MS, 55; Calif Inst Technol, PhD(physics), 59. *Prof Exp:* Instr, Calif Inst Technol, 55-59; NSF fel, Univ Copenhagen & Bohr Inst, Denmark, 59-64; assoc prof, Linear Accelerator Ctr, 64-69, PROF, LINEAR ACCELERATOR CTR, STANFORD UNIV, 69- *Concurrent Pos:* Consult, IBM, 58-59; consult, Space Tech Labs, 59- *Res:* Quantum theory of fields; quantum electrodynamics and related high energy physics; plasma physics and hydromagnetics. *Mailing Add:* Lawrence Berkeley Labs Univ Calif Mail Stop 46-125 Berkeley CA 94720

BERMAN, SANFORD, b Syracuse, NY, Apr 22, 27; m 50; c 2. MICROBIOLOGY. *Educ:* Syracuse Univ, BS, 49, MS, 51, PhD(microbiol), 54. *Prof Exp:* Bacteriologist, Ralph M Parsons Co, Md, 53-55; bacteriologist, 55-60, res bacteriologist, 60-62, chief microbiol sect, 62-81, CHIEF, DEPT BIOL RES, WALTER REED ARMY INST RES, 81- *Mem:* AAAS; Am Soc Microbiol; Soc Cryobiol. *Res:* Methods of production and evaluation of biologics, particularly in the dry state, such as bacterial enzymes, bacterial, viral and rickettsial vaccines. *Mailing Add:* Berman & Aisenberg 1730 Rhode Island NW Suite 809 Washington DC 20036-3186

BERMAN, SHIER, b Toronto, Ont, Oct 4, 29; m 51; c 3. ANALYTICAL CHEMISTRY. *Educ:* Univ Toronto, BA, 52, MA, 55, PhD(analytical chem), 57. *Prof Exp:* From asst res officer to sr res officer, 57-83, head analytical chem, 81-90, PRIN RES OFFICER ANALYTICAL CHEM, CHEM DIV, NAT RES COUN CAN, 83-, HEAD MEASUREMENT SCI, INST ENVIRON CHEM, 90- *Concurrent Pos:* Vis scientist, Hebrew Univ Jerusalem, Israel, 65-66 & vis lectr, 70; vis lectr, Queen's Univ, Ont, 75; vis fel, Western Australia Inst Technol, Perth, 85; Can deleg, Int Coun Explor Sea Working Groups, Marine Chem & Marine Sediments, 81-; chmn, XXV

Colloquium Spectroscopicum Int, 87. *Honors & Awards:* Fisher Sci Lectr Award, 85. *Mem:* Fel Chem Inst Can; Spectros Soc Can (pres, 62-63). *Res:* Inorganic analytical chemistry; trace analysis; optical emission spectroscopy; xray fluorescence spectroscopy; atomic absorption; inductively coupled plasma; photon activation analysis; trace metals in seawater and marine samples; marine reference materials for trace metals. *Mailing Add:* Inst Environ Chem Nat Res Coun Montreal Rd Ottawa ON K1A 0R6 Can

BERMAN, SIMEON MOSES, b Rochester, NY, Mar 28, 35; m 55; c 6. STOCHASTIC PROCESSES. *Educ:* City Col New York, BA, 56; Columbia Univ, MA, 58, PhD(math statist), 61. *Prof Exp:* Lectr math, City Col New York, 57-60; res assoc math statist, Columbia Univ, 60-61, asst prof, 61-65; assoc prof, 65-77, PROF MATH, COURANT INST MATH SCI, NY UNIV, 77- *Concurrent Pos:* Assoc ed, Annal Probability, 79-85; Prin investr, NSF res grant, stochastic processes, NY Univ, 66-73, 76-; stat consult, EPA acid rain res grant, NY Univ, 79-83; prin investr, short term res, 85 & contract, US Army Res Off, NY Univ, 86-; prin investr, NIH res contract, HIV math res, NY Univ, 89- *Mem:* Fel Inst Math Statist; Sigma Xi. *Res:* Probability theory, especially stochastic processes. *Mailing Add:* Courant Inst Math Sci NY Univ 251 Mercer St New York NY 10012

BERMES, BORIS JOHN, b Merchantville, NJ, Mar 2, 26; m; c 3. GEOHYDROLOGY. *Educ:* Colo Sch Mines, Geol Eng, 50; Univ Utah, MS, 54. *Prof Exp:* Geologist, US Geol Surv, 50-54; exploitation engr, Shell Oil Co, 54-56; hydraulic engr, US Dept Interior, 56-73; sr scientist water resources, Geraghty & Miller Inc, West Palm Beach, Fla, 73-85; chief hydrogeologist, Cayman Islands Govt, 85-87; RETIRED. *Concurrent Pos:* Model specialist and drainage adv, UN Develop Prog, 69 & 77; tech officer, Food & Agr Orgn UN, 69-71; consult, World Bank, 78. *Mem:* Int Asn Hydrogeologists; Am Inst Hydrol. *Res:* Application of electronic technology to resource appraisal and management for purposes of exploitation and/or environmental protection. *Mailing Add:* PO Box 302 Dunnellon FL 32630

BERMES, EDWARD WILLIAM, JR, b Chicago, Ill, Aug 20, 32; m 57; c 5. CLINICAL CHEMISTRY. *Educ:* St Mary's Col, Minn, BS, 54; Loyola Univ Ill, MS, 56, PhD(biochem), 59. *Prof Exp:* Asst biochem, Loyola Univ, Ill, 54-56; chief biochemist, Hektoen Inst Med Res, Cook County Hosp, Chicago, 58-60; clin instr, 59-65, from asst prof to assoc prof, 66-74, assoc dir, Clin Labs, 78-81, PROF BIOCHEM & PATH, STRITCH SCH MED, LOYOLA UNIV, CHICAGO, 74-, DIR CLIN CHEM, HOSP, 69-, ASSOC DIR, SCH MED TECHNOL, 70-, DIR, CLIN LABS, 81- *Concurrent Pos:* Chief biochemist, W Suburban Hosp, Chicago, 60-62; dir biochem, St Francis Hosp, 62-68; ed, Chicago Clin Chemist, 69- *Mem:* AAAS; Am Chem Soc; Am Asn Clin Chem (secy, 78-80); Am Asn Clin Path; Asn Clin Scientists; Sigma Xi. *Res:* Method development in clinical chemistry; test interference in clinical laboratory testing. *Mailing Add:* 1907 Sunnyside Circle Northbrook IL 60062

BERMON, STUART, b Philadelphia, Pa, Dec 2, 36; m 61; c 1. SOLID STATE PHYSICS. *Educ:* Univ Pa, BA, 58; Univ Ill, MS, 59, PhD(physics), 64. *Prof Exp:* Staff mem physics, Lincoln Lab, Mass Inst Technol, 64-66; asst prof eng, Brown Univ, 66-73; assoc prof physics, City Col New York, 73-78; RES STAFF MEM, IBM WATSON RES CTR, 78- *Mem:* Am Phys Soc. *Res:* Electron tunneling; superconductivity; properties of thin films; transport in amorphous semiconductors; cyclotron resonance in semiconductors and metals. *Mailing Add:* IBM Watson Res Ctr PO Box 218 Yorktown Heights NY 10598

BERMUDEZ, VICTOR MANUEL, b New York, NY, Jan 24, 47. SOLID STATE SCIENCE. *Educ:* Mass Inst Technol, BS, 67; Princeton Univ, MA, 69, PhD(phys chem), 76. *Prof Exp:* RES PHYSICIST, NAVAL RES LAB, 72- *Mem:* Sigma Xi; Am Phys Soc; Am Vacuum Soc. *Res:* Surface chemistry and surface physics of materials. *Mailing Add:* Code 6833 Naval Res Lab 4555 Overlook Ave SW Washington DC 20375

BERN, HOWARD ALAN, b Montreal, Que, Jan 30, 20; US citizen; m 46; c 2. ENDOCRINOLOGY, TUMOR BIOLOGY. *Educ:* Univ Calif, Los Angeles, AB, 41, MA, 42, PhD(zool), 48. *Prof Exp:* Nat Res Coun fel, 46-48; from instr to prof zool, 48-90, chair, Group Endocrinol, 62-90, RES ENDOCRINOLOGIST, CANCER RES LAB, UNIV CALIF, BERKELEY, 60-, EMER PROF INTEGRATIVE BIOL, 90-; PROF ZOOL & RES ENDOCRINOLOGIST, CANCER RES LAB, UNIV CALIF, BERKELEY, 61-, CHAIR, GROUP ENDOCRINOL, 62- *Concurrent Pos:* Guggenheim fel, 51-52; NSF sr fel, 58-59, 65-66; fel, Ctr Adv Study in Behav Sci, Stanford Univ, 60; res prof, Miller Inst Basic Res Sci, 61; vis prof, Bristol Univ, 65-66, Univ Kerala, Trivandrum, India, 67, Univ Tokyo, 71, 86, Univ PR, 73 & 74, Nat Mus Natural Hist, Paris, 81, Toho Univ, Funabashi, Japan, 82-84, 86-89, Watkins vis prof, Wichita State Univ, 84, James vis prof, Xavier Univ, Antigonish, NS, 86, Yokohama City Univ, Japan, 88; mem, US Deleg to Indo-Am Scholars Conf, Delhi, 74; mem coun, Int Soc Neuroendocrinol, 77-80; Walker-Ames prof, Univ Wash, 77; memAdv Comn, NIH Eval Res Needs Endocrin Metab, 78-79; mem, Comt Basic Res, US-France Agreement Cancer Res, Nat Cancer Inst, 79-; transatlantic lectr, British Soc Endocrinol, 80; distinguished vis, Univ Alta, 81; mem Comt Int Human Res, Nat Res coun, 81-82; plenary lectr, Am Soc Zoologists, 84, Centennial plenary lectr, Boston, 89; mem Sloan Med Selection Comt, Gen Med Cancer Res Fedn, 84-85; mem adv comt, Neurobiol Lab, Univ PR, 85-88; mem sci adv bd, Staz Zool, Naples, 87-; mem sci adv bd, Staz Zoologica, Naples. *Honors & Awards:* Nieuwland Lectr, Univ Notre Dame, 72; Eli Lilly Lectr, Endocrine Soc, 75; Geschwind Mem Lectr, Univ Calif, Davis, 84; Cowper Distinguished Vis Lectr, State Univ NY, Buffalo, 84; Huang-Chan Mem Lectr, Univ Hong Kong, 85; David Tyler Mem Lectr, Univ SFla, 88; Scharrer-Bargmann Lectr, Malaga, Spain, 89; Distinguished Serv Award, Soc Advan Chicanos & Native Americans in Sci, 90. *Mem:* Nat Acad Sci; Am Soc Zool (pres, 67); Am Physiol Soc; Am Asn Cancer Res; fel AAAS; foreign fel Indian Nat Sci Acad; Am Acad Arts & Sci. *Res:* Comparative endocrinology, especially of prolactin; perinatal hormonal carcinogenesis; hormones and normal and abnormal genital tract and mammary gland growth; neurosecretion and comparative neuroendocrinology; caudal neurosecretory system of fishes; endocrinology of salmon smoltification, hormones and early development of fish; cell culture of genital epithelia; striped bass growth and osmoregulation. *Mailing Add:* Dept Integrative Biol Univ Calif Berkeley CA 94720

BERNABEI, AUSTIN M, b New Rochelle, NY, July 5, 28. NUCLEAR PHYSICS. *Educ:* Manhattan Col, BCE, 50; Cath Univ Am, MS, 53; NY Univ, MNucE, 59, PhD(physics), 64. *Prof Exp:* ASSOC PROF PHYSICS, MANHATTAN COL, 56- *Mem:* Am Phys Soc; Am Asn Physics Teachers. *Res:* Neutron resonance analysis, particularly the Doppler distortion effect and Mossbauer effect. *Mailing Add:* Dept Physics Manhattan Col Bronx NY 10471

BERNABEI, STEFANO, b San Martino in Rio, Italy, Feb 29, 44. PLASMA PHYSICS. *Educ:* Univ Milan, PhD(physics), 69. *Prof Exp:* Res asst physics, Univ Milan, 69-71; res assoc, 71-74, res staff mem, 74-85, PRIN RES PHYSICIST, PLASMA PHYSICS LAB, PRINCETON UNIV, 85- *Concurrent Pos:* Nat Coun Res Italy fel, Plasma Physics Lab, Princeton Univ, 70-71. *Honors & Awards:* Award, Plasma Physics Res, Am Phys Soc, 84. *Res:* Plasma heating with radio frequency waves; particularly heating at the lower hybrid frequency in toroidal machines. *Mailing Add:* Plasma Physics Lab Forrestal Campus Princeton Univ Princeton NJ 08544

BERNACKI, RALPH J, b Buffalo, NY, Oct 9, 46; m 69; c 2. CANCER RESEARCH, CELL BIOLOGY. *Educ:* Rensselear Polytech Inst, BS, 68; Univ Rochester, PhD(pharmacol), 72. *Prof Exp:* Cancer res specialist I-IV, 72-81, CANCER RES SPECIALIST V, DEPT EXP THERAPEUT, ROSWELL PARK CANCER INST, 81- *Concurrent Pos:* Prin investr, USPHS-Nat Cancer Inst Grants, 73-; res prof pharmacol, State Univ NY, Buffalo, 82-, dir, Pharmol Prog, Roswell Park Grad Div, 82-85; consult, Nat Cancer Inst-DIC, 85- & Am Cancer Soc, 90- *Mem:* Am Asn Cancer Res; Am Soc Pharmacol & Exp Therapeut; NY Acad Sci; Sigma Xi; Metastasis Soc. *Res:* Design and development of new antitumor agents; tumor cell growth and metastasis; tumor cell surface glycoproteins as a target for chemotherapy. *Mailing Add:* Roswell Park Cancer Inst Buffalo NY 14263

BERNADY, KAREL FRANCIS, b Baltimore, Md, Feb 1, 41; m 64; c 2. PHARMACEUTICAL CHEMISTRY. *Educ:* Loyola Col, Md, BS, 63; Fordham Univ, PhD(org chem), 68. *Prof Exp:* Res chemist org chem, Lederle Labs, 67-75, group leader, 75-81, mgr pharmaceut bulk mfg, 81-83, prod supt, 83-85, mgr chem prod, Lederle Labs, 85-86, PLANT MGR, BOUND BROOK PLANT, AM CYANAMID CO, 86- *Mem:* Am Chem Soc; NY Acad Sci; AAAS. *Res:* Chemistry of natural products; process research on pharmaceuticals and fine chemicals. *Mailing Add:* 48 Appert Tr Mahwah NJ 07430-3001

BERNAL, IVAN, b Barranquilla, Colombia, Mar 28, 31; m 57; c 1. CHEMISTRY. *Educ:* Clarkson Col, BS, 54; Univ Va, MS, 56; Columbia Univ, PhD(chem), 63. *Prof Exp:* Staff mem chem, RCA Labs, NJ, 56-59; Harvard Corp fel, 63-64; asst prof chem, State Univ NY Stony Brook, 64-67; chemist, Brookhaven Nat Lab, 67-73; assoc prof, 73-75, PROF CHEM, UNIV HOUSTON, 75- *Concurrent Pos:* vis fel, Australian Nat Univ, 80, 86; ed, Elsevier Series, Stereochemistry of Organometallic & Inorganic Compounds. *Honors & Awards:* US Sr Scientist Award, Alexander von Humboldt Found, 75 & 80; Am Chem Soc Award, 85. *Mem:* Am Chem Soc; fel NY Acad Sci; Am Crystallog Asn. *Res:* Problems of valence states and of stereochemistry of transition metal ions; spectroscopic and x-ray crystallographic work; absolute configurations of organometallics having metals as chiral centers; structures and absolute configurations of organosulfur compounds, particularly those containing disulfide linkages; the molecular nature of the phenomenon of conglomerate crystallization. *Mailing Add:* Dept Chem Univ Houston Houston TX 77024

BERNAL G, ENRIQUE, b Barranquilla, Colombia, July 30, 38; US citizen; m 61; c 2. OPTICAL PHYSICS, LASERS. *Educ:* Col St Thomas, BS, 60; Univ Minn, MS, 63. *Prof Exp:* Res asst & res scientist, Honeywell Inc, 63-66, prin res scientist, 66-71, sr prin res scientist, 71-75, staff scientist, 75-76, prog mgr, 76, dept mgr, Electro-optics Dept, 77-80, assoc dir, 80-81, dir, Phys Sci Ctr, 81-85, dir eng, Electro-optics Div, 85-89; VPRES CVD MAT, MORTON INT, 90- *Mem:* Inst Elec & Electronics Engrs; Optical Soc Am; Soc Photo-Optical & Instrument Engrs; Am Soc Testing & Mat. *Res:* Properties of transparent infrared materials and the effect of laser irradiation on optical properties; optical materials, interferometry and effects of laser irradiation of opaque solids. *Mailing Add:* Morton Int Advan Mat 185 New Boston St Woburn MA 01801-6203

BERNAL-LLANAS, ENRIQUE, pharmacology, microbiology, for more information see previous edition

BERNARD, BERNIE BOYD, b Bryan, Tex, Sept 1, 52; m 74; c 2. ORGANIC GEOCHEMISTRY, EXPLORATION GEOCHEMISTRY. *Educ:* Abilene Christian Univ, BS, 74; Tex A&M Univ, PhD(oceanog), 78. *Prof Exp:* Asst prof geol, Univ Okla, 78-80; VPRES, RES & DEVELOP, O I CORP. *Concurrent Pos:* Design consult, Oceanog Inst Corp, 78- *Mem:* AAAS; Am Chem Soc; Am Geophys Union; Sigma Xi. *Res:* Stable carbon isotope geochemistry; organic diagenesis; geochemical hydrocarbon exploration; microbial methane generation. *Mailing Add:* O I Corp PO Box 2980 College Station TX 77841

BERNARD, DANE THOMAS, b Evansville, Ind, Dec 29, 48; m 78; c 2. FOOD MICROBIOLOGY. *Educ:* Purdue Univ, BS, 70; Univ Md, MS, 79. *Prof Exp:* Specialist, food inspection, US Army, 70-73; microbiologist, 73-77, head microbiol, 77-84, dir, microbiol packaging & processing, 84-, VPRES, EASTERN RES, NAT FOOD PROCESSORS ASN. *Mem:* Inst Food Technologists; Am Soc Microbiol; Int Asn Milk Food & Environ Sanitarians. *Res:* Food safety research, mostly dealing with Clostridium botulinum; heat resistance; growth characteristics in various classes of food products; effects of inhibitory substances on growth or toxin production; process establishment for thermally processed foods. *Mailing Add:* Nat Food Processors Asn 1401 New York Ave NW Suite 400 Washington DC 20005

BERNARD, DAVY LEE, b Vermillion Parish, La, July 16, 36; m 60; c 3. NUCLEAR PHYSICS. *Educ:* Univ Southwestern La, BS, 58; Rice Univ, MA, 60, PhD(physics), 65. *Prof Exp:* Asst prof physics, Univ Va, 65-68; assoc prof, 68-74, PROF PHYSICS, UNIV SOUTHWESTERN LA, 74- *Concurrent Pos:* Res Corp grant, 70; pres, La Acad Sci, 82-83. *Mem:* Am Phys Soc. *Res:* Atomic physics, especially energy loss and charge exchange of ions passing through matter; neutron physics, particularly measurement of neutron scattering cross sections. *Mailing Add:* 218 Sunny Lane Lafayette LA 70506

BERNARD, DOUGLAS ALAN, b Barrie, Ont, Feb 1, 53. COMPUTER SIMULATION, POLYMER PHYSICS. *Educ:* Univ Sask, BSc, 75; Princeton Univ, PhD(physics), 80. *Prof Exp:* Mem res staff, Xerox Res Ctr Can, Xerox Corp, 80-83; sr mem tech staff, Philips Res Labs Sunnyvale, Signetics Corp, 83-90; SR MEM TECH STAFF, TECHNOL MODELING ASSOCS, INC, PALO ALTO, CALIF, 90- *Concurrent Pos:* Instr, DeAnza Col, 89- *Mem:* Am Phys Soc. *Res:* Process simulation-microelectronics. *Mailing Add:* 5714 McKellar Dr San Jose CA 95129-4125

BERNARD, EDDIE NOLAN, b Houston, Tex, Nov 23, 46; m 70; c 1. TSUNAMIS. *Educ:* Lamar Univ, BS, 69; Tex A&M Univ, MS, 70, PhD(phys oceanog), 76. *Prof Exp:* Geophysicist, Pan Am Petrol Co, 69; res oceanogr, Tex A&M Univ, 69-70; mem staff, Geophys Off, 70-72, res oceanogr, Joint Tsunami Res Effort, 72-77, dir, Nat Tsunami Warning Ctr, 77-80, DIR, PAC MARINE ENVIRON LAB, NAT OCEANIC & ATMOSPHERIC ADMIN, 80- *Concurrent Pos:* Mem, Int Union Geodesy & Geophysics Tsunami Comn, 79-, secy, 83, chmn, 87, US/Japan Coop Prog Natural Resources, 81-; proj leader, US/USSR Integration Tsunami Warning Systs, 80-; sr fel, Joint Inst Marine & Atmospheric Res, 80-; mem admin bd, Joint Inst Study Atmosphere & Oceans, 82- *Mem:* Am Geophys Union; Am Meterol Soc; Marine Technol Soc. *Res:* Application of linear hydrodynamics for mesoscale tsunami processes; science administration. *Mailing Add:* Pac Marine Environ Lab Nat Oceanog & Atomspheric Admin 7600 Sand Pt Way NE Seattle WA 98115

BERNARD, ERNEST CHARLES, b Detroit, Mich, Aug 20, 50; m 74; c 2. NEMATOLOGY, SOIL ZOOLOGY. *Educ:* Mich State Univ, BS, 72, MS, 74; Univ Ga, PhD(nematol), 77. *Prof Exp:* From asst prof to assoc prof, 77-86, PROF NEMATOL, UNIV TENN, KNOXVILLE, 86- *Mem:* Soc Nematologists; Am Micros Soc; Sigma Xi. *Res:* Population dynamics and pathogenicity of plant-parasitic nematodes; taxonomy and biology of protura. *Mailing Add:* Entomol & Plant Path Dept Univ Tenn Knoxville TN 37901

BERNARD, GARY DALE, b Everett, Wash, Feb 19, 38; m 64; c 4. INSECT VISION, OPTICAL PHYSIOLOGY. *Educ:* Univ Wash, BSEE, 59, MSEE, 60, PhD(elec eng), 64. *Prof Exp:* From instr to asst prof elec eng, Mass Inst Technol, 65-68; asst prof ophthal eng & appl sci, Sch Med, Yale Univ, 68-71, assoc prof ophthal & visual sci & eng & appl sci, 71-76, assoc prof ophthal & visual sci, 71-79, sr res scientist, 79-88; LEAD ENGR, ADVAN SENSORS GROUP, MFG RES & DEVELOP, BOEING COM AIRPLANES, 88- *Concurrent Pos:* Ford Found fel, Mass Inst Technol, 65-66, consult, Lincoln Lab, 66-68; res career develop award, 71-75, prin investr grant, Nat Eye Inst, 73-86; vis prof, Dept Zool, Univ Zurich, 78, 83 & 89; consult, Gen Tel & Electronics Corp, 81-86; vis prof, Dept Biophysics, Univ Groningen, 82; Prin investr, Army Res Off, 86-88, NSF, 88-90, Nat Geog Soc, 88. *Mem:* AAAS; Asn Res Vision & Ophthal; Optical Soc Am; Lepidopterists Soc; Inst Elec & Electronics Engrs. *Res:* Visual information processing by photoreceptor arrays and associated optics; physiology, photochemistry, and waveguide optics of insect photoreceptors; invertebrate vision; visual ecology of butterflies; advanced sensor systems for control of manufacturing processes. *Mailing Add:* 30714 19th Ave S Federal Way WA 98003

BERNARD, GEORGE W, b New York, NY, Aug 22, 25; c 2. ELECTRON MICROSCOPY, CELL-TISSUE CULTURE. *Educ:* Washington Univ, St Louis, DDS, 47; Univ Calif, Los Angeles, BS, 63, PhD(anat), 67. *Prof Exp:* PROF ANAT & ORAL BIOL, UNIV CALIF, LOS ANGELES, 67- *Concurrent Pos:* Vis prof, Inst Sci Res on Cancer, France, 70-71, Cordeliers Inst Sch Med, Paris, 80-81; site vis mem, NIH, 72-, NSF, 83-; mem, oral med study sect, NIH, 77-80; consult, Vet Admin, Long Beach, 79-; NIH fel oral biol. *Mem:* Sigma Xi; fel AAAS; Am Asn Anatomists; Int Asn Dent Res; Electron Micros Soc Am; Int Soc Develop Biol. *Res:* Bone and bone marrow seeking elements of the calcification mechanism; the origin of various bone cells and osteogenic enhancing factors. *Mailing Add:* Dept Anat Sch Med Univ Calif Los Angeles CA 90024

BERNARD, JOHN MILFORD, b Duluth, Minn, May 22, 33; m 63; c 1. PLANT ECOLOGY. *Educ:* Univ Minn, Duluth, BS, 58; Rutgers Univ, MS, 60, PhD(plant ecol), 63. *Prof Exp:* Tech asst biol, Rutgers Univ, 58-62; instr, Franklin & Marshall Col, 62-64; from asst prof to assoc prof biol, 64-72, chmn dept, 73-82, PROF BIOL, ITHACA COL, 72- *Concurrent Pos:* Fel, Univ London, 84-85. *Mem:* Ecol Soc Am; Torrey Bot Club; Sigma Xi. *Res:* Structure and function of wetland ecosystems. *Mailing Add:* Dept Biol Ithaca Col Ithaca NY 14850

BERNARD, JOHN WILFRID, b Canton, Ohio, Jan 26, 28; m 51; c 3. CHEMICAL ENGINEERING, INSTRUMENTATION. *Educ:* Case Inst Technol, BS, 52, MS, 61. *Prof Exp:* Instr engr, Dow Chem Co, Midland, Mich, 52-60; systs engr, Indust Nucleonics, Midland, Mich & Ohio State Univ Systs Res Group, Columbus, Ohio, 60; res engr, Foxboro Co, 61-66, mgr syst technol & simulation, 66-70, mgr systs res, 70-71, dir corp res, 71-80, dir Europ technol, 81-85; PRES, BERNARD AUTOMAT CONSULT, 86- *Concurrent Pos:* Found Instrumentation Educ & Res fel, 60-61; co-rep, Indust Res Inst, Inc, 71-85 & chmn rules comt, 77-79; treas, Am Automatic Control Coun, 74-80; mem, Am Automatic Control Coun Technol Comt, Automation Res Coun & NSF Indust Panel Sci & Technol. *Mem:* Instrument Soc Am; Nat Soc Prof Engrs; Sigma Xi; Soc Mech Engrs; Soc Mach Intel. *Res:* Systems research in the application of process management and control systems for fluid process industries. *Mailing Add:* 3520 Rocky Point Rd Springfield OH 45502-9745

BERNARD, PETER SIMON, b New York, NY, Jan 14, 50; m 79; c 2. FLUID DYNAMICS, TURBULENCE. *Educ:* City Univ NY, BE, 72; Univ Calif, Berkeley, MS, 73, PhD(mech eng), 77. *Prof Exp:* Fel, Lawrence Berkeley Lab, 77; vis asst prof, Inst Phys Sci Technol, 77-79, ASST PROF MECH ENG, UNIV MD, 79- *Mem:* Soc Indust & Appl Math; Am Soc Mech Engrs. *Res:* Methods of predicting the mean properties of turbulent flows. *Mailing Add:* Dept Mech Eng Univ Md College Park MD 20742

BERNARD, RICHARD FERNAND, b Lewiston, Maine, Aug 19, 34; m 59. ZOOLOGY. *Educ:* Univ Maine, BA, 56; Mich State Univ, MS, 59, PhD(zool), 62. *Prof Exp:* Assoc prof zool, Wis State Univ, Superior, 62-69; assoc prof biol, 69-73, PROF BIOL, QUINNIPIAC COL, 73- *Concurrent Pos:* Bd Regents Instnl study grant, 63-64. *Mem:* Am Ornith Union; Am Soc Mammalogists; Nat Audubon Soc. *Res:* Effects of insecticides on wildlife. *Mailing Add:* Dept Biol Quinnipiac Col Mt Carmel Ave Hamden CT 06518

BERNARD, RICHARD LAWSON, b Detroit, Mich, Aug 12, 26; m 52; c 4. PLANT BREEDING. *Educ:* Ohio State Univ, BS, 49, MS, 50; NC State Col, PhD(field crops), 60. *Prof Exp:* Res instr agron, NC State Col, 54; res agronomist, 54-61, RES GENETICIST, USDA, DEPT AGRON, UNIV ILL, URBANA-CHAMPAIGN, 61- *Mem:* Am Soc Agron; Am Genetic Asn; Crop Sci Soc Am; Sigma Xi. *Res:* Qualitative genetics, germplasm and breeding of soybeans. *Mailing Add:* Dept Agron Turner Hall Univ Ill 1102 S Goodwin Urbana IL 61801

BERNARD, RICHARD RYERSON, b Jacksonville, Fla, Dec 24, 17; m 43; c 3. MATHEMATICS. *Educ:* Univ Va, PhD(math), 49. *Prof Exp:* Instr math, Univ Va, 48-49; from instr to asst prof, Yale Univ, 49-55; assoc prof, 55-59, PROF MATH, DAVIDSON COL, 59- *Mem:* Math Asn Am. *Mailing Add:* Davidson Col Box 656 Davidson NC 28036

BERNARD, ROBERT SCALES, b Conway, SC, Mar 5, 46. SOLID MECHANICS, FLUID MECHANICS. *Educ:* Miss State Univ, BS(aero eng) & BS(appl physics), 69, PhD(gen eng), 81; Stanford Univ, MS, 71. *Prof Exp:* Tech staff mem mechanics, Sandia Labs, 69-73; RES PHYSICIST MECH, US ARMY ENGR WATERWAYS EXP STA, 73- *Mem:* Am Phys Soc. *Res:* Terradynamics; impact phenomena; computational fluid dynamics. *Mailing Add:* 115 Bellwood Dr Vicksburg MS 39180

BERNARD, RUDY ANDREW, b New York, NY, May 31, 30; m 60; c 3. NEUROPHYSIOLOGY, BEHAVIOR. *Educ:* Univ Montreal, BA, 53; Cornell Univ, MNS, 60, PhD(animal physiol), 62. *Prof Exp:* Asst nutrit, Cornell Univ, 58-60; fel neurophysiol, Univ Wis, 62-64; asst prof physiol, State Univ NY Downstate Med Ctr, 64-66 & Rockefeller Univ, 66-69; assoc prof, 69-75, PROF PHYSIOL, MICH STATE UNIV, 75- *Mem:* AAAS; Am Physiol Soc; Soc Neurosci. *Res:* Anatomy of central taste pathways; electrophysiology of peripheral and central taste mechanisms; behavioral studies of taste preferences; stress hormones, taste, salt intake and hypertension. *Mailing Add:* 111 Giltner Hall Mich State Univ East Lansing MI 48864

BERNARD, SELDEN ROBERT, b Scobey, Mont, Nov 10, 25; c 3. MATHEMATICAL BIOLOGY. *Educ:* Univ Denver, BS, 48; Univ Chicago, PhD(math biol), 60. *Prof Exp:* Asst chemist, Kaiser Aluminum & Chem Co, 49; fel radiol physics, Oak Ridge Nat Lab, 49-50; health physicist, Union Carbide Nuclear Co, 50-54; res health physicist, 54-57, MEM RES STAFF, OAK RIDGE NAT LAB, 60- *Mem:* AAAS; Health Physics Soc. *Res:* Mathematical biology and health physics; mathematical studies of biological phenomena, especially blood-gas equilibria and compartmental models for metabolism of inorganic ions and molecules; estimation of radiation exposure to organs and tissues from internally contained radionuclides. *Mailing Add:* Health Physics Div Oak Ridge Nat Lab 109 Packer Rd Oak Ridge TN 37830

BERNARD, WALTER JOSEPH, b Manchester, NH, Dec 25, 23; m 48; c 4. PHYSICAL CHEMISTRY. *Educ:* Univ NH, BS, 50, MS, 51; Mass Inst Technol, PhD(chem), 54. *Prof Exp:* Assoc dir eng, 64-68, dir res & develop, 68-71, SR SCIENTIST, SPRAGUE ELEC CO, 71- *Concurrent Pos:* Pvt consult, 87- *Mem:* Electrochem Soc; Am Chem Soc. *Res:* Inorganic fluorides; silicon chemistry; oxide films on metals; dielectrics. *Mailing Add:* Bee Hill Rd Williamstown MA 01267

BERNARD, WILLIAM, b Philadelphia, Pa, Dec 22, 27; m 53; c 3. PHYSICS. *Educ:* Univ Pa, AB, 49, MS, 53, PhD(physics), 59; George Wash Univ, MA, 88. *Prof Exp:* From sr scientist to prin scientist, Raytheon Res Div, Waltham, Ma, 59-65; from sr scientist to chief, Advan Res Br, Electronic Components Lab, Electronics Res Ctr, NASA, 65-70; staff assoc, Mat Res Labs Sect, Div Mat Res, 72-78, sr staff assoc, Directorate Math & Phys Sci, 78-90, SR EDUC POLICY ADV, DIRECTORATE EDUC & HUMAN RESOURCES, NSF, 90- *Concurrent Pos:* Actg dep asst dir, Math & Phys Sci, NSF, 80-81 & 82-83. *Mem:* AAAS. *Res:* Condensed matter theory; irreversible thermodynamics; nonequilibrium statistical mechanics. *Mailing Add:* Directorate Educ & Human Resources NSF Washington DC 20550

BERNARD, WILLIAM HICKMAN, b New Orleans, La, July 7, 32; m 71; c 4. PHYSICS. *Educ:* Tulane Univ, BS, 54, PhD(physics), 63. *Prof Exp:* Asst prof physics, 62-69, PROF PHYSICS, LA TECH UNIV, 69- *Mem:* Am Asn Physics Teachers. *Res:* X-ray crystallography. *Mailing Add:* Dept Physics La Tech Univ Ruston LA 71271

BERNARDEZ, TERESA, b Buenos Aires, Argentina, June 11, 31; US citizen; c 1. PSYCHIATRY. *Educ:* Liceo N 1 de Senoritas, Argentina, BA, 48; Univ Argentina, MD, 56; Am Bd Psychiat & Neurol, cert, 63. *Prof Exp:* Staff psychiatrist, C F Menninger Mem Hosp, 62-65, Menninger Found, 65-71; from asst prof to assoc prof, Mich State Univ, 71-79, prof, Dept Psychiat, 79-89; CONSULT, 89- *Concurrent Pos:* Fel, Menninger Found, 60-62, Bunting Inst, Harvard Univ, 84- 85; clin assoc, Tavistock Clin, London, 77-78; vis prof, dept psychol, Univ Bergen, Norway, 77-78; consult, NIMH, 81, Am Psychol Asn, 79. *Honors & Awards:* Peace Award, Pawlowski Found,

74. *Mem:* Fel Am Psychiat Asn; Am Group Psychother Asn; Am Orthopsychiat Asn. *Res:* Sex-role socialization of women and men and its impact on health; conflicts of contemporary women; gynecological disorders of women and their psychological import. *Mailing Add:* 835 Westlawn East Lansing MI 48823

BERNARDIN, JOHN EMILE, b Santa Monica, Calif, Feb 28, 37; m 58; c 2. PHYSICAL BIOCHEMISTRY. *Educ:* Univ Calif, Riverside, BA, 63; Univ Ore, PhD(phys biochem), 70. *Prof Exp:* Chemist, 63-70, asst dir, 82-84, RES CHEMIST, WESTERN REGIONAL RES CTR, AGR RES SERV, USDA, 70-82, 84- *Mem:* Int Soc Plant Molecular Biol; Am Asn Cereal Chemists. *Res:* Chromosomal nonhistone proteins. *Mailing Add:* Western Regional Res Ctr USDA 800 Buchanan St Albany CA 94710

BERNARDIN, LEO J, b Lawrence, Mass, June 4, 30; m 56; c 4. PAPER CHEMISTRY. *Educ:* Tufts Univ, BS, 52; Inst Paper Chem, MS, 54, PhD(wood chem), 58. *Prof Exp:* Res scientist, Marathon Div, Am Can Co, 57-59; res scientist, 59-66, sr res scientist, 66-73, res mgr, Res & Eng, 73-81, res assoc, 81-86, SR RES FEL, KIMBERLY-CLARK CORP, 86- *Mem:* Am Chem Soc. *Res:* Wood and cellulose chemistry; prehydrolysis of wood; hemicellulose chemistry; synthetic fiber paper; chemical modification of cellulose; absorbent product research and development. *Mailing Add:* Res & Eng Kimberly-Clark Corp Neenah WI 54956

BERNARDIS, LEE L, b Graz, Austria, Sept 18, 26; US citizen; m 58; c 1. NEUROVISCERAL PHYSIOLOGY, NEUROENDOCRINOLOGY. *Educ:* Graz Univ, DrPhil, 49; Univ Western Ont, PhD(physiol, biochem), 61. *Prof Exp:* Res tech liver dis, Hosp for Sick Children, Toronto, Ont, 53-56; res mem meat processing & diet res, Can Packers, Ltd, 56-57; res asst physiol, Univ Western Ont, 57-61; res assoc exp path, State Univ NY Buffalo, 61-63, from res assoc prof to res prof surg, 63-80; RES PHYSIOLOGIST & DIR NEUROSCI, VET ADMIN MED CTR, BUFFALO, 78-, RES PROF MED, 80-, ASSOC RES CANCER SCIENTIST, DEPT VET AFFAIRS, 89- *Concurrent Pos:* NIH, NSF, & VA grants, 68-; Proj & dir neuroendocrine-neurovisceral res unit, Dept Surg, State Univ NY Buffalo, 73-78, res prof med, 81. *Mem:* Am Asn Univ Prof; Fedn Am Scientists; Am Inst Nutrit; Int Brain Res Orgn; Int Soc Neuroendocrinol; Am Physiol Soc; Int Soc Psychoneuroendocrinol; Endocrine Soc; Can Physiol Soc. *Res:* Hypothalamic regulation of growth, endocrines and metabolism; regulation of body weight and fat; set point regulation; hypothalmic regulation of food intake. *Mailing Add:* Div Endocrinol Vet Admin Med Ctr 3495 Bailey Ave Buffalo NY 14215

BERNASCONI, CLAUDE FRANCOIS, b Zurich, Switz, Feb 17, 39; m 63; c 2. PHYSICAL ORGANIC CHEMISTRY. *Educ:* Swiss Fed Inst Technol, dipl, 63, PhD(chem), 65. *Prof Exp:* Res asst chem, Swiss Fed Inst Technol, 65-66 & Max Planck Inst Phys Chem, 66-67; from asst prof to assoc prof, 67-77, PROF CHEM, UNIV CALIF, SANTA CRUZ, 77- *Concurrent Pos:* Res grants, Petrol Res Fund, 69-71, 72-74 & 75-77, 86-88, 89-90, Res Corp, 70, Alfred P Sloan Found, 71-73 & NSF, 71-91. *Mem:* Fel AAAS; Am Chem Soc; The Chem Soc. *Res:* Mechanisms of nucleophilic aromatic and vinylic substitution reactions; electron transfer; acid-base catalysis fast reaction kinetics; proton transfer; nucleophilic addition to olefins. *Mailing Add:* Dept Chem Univ Calif Santa Cruz CA 95064

BERNASEK, STEVEN LYNN, b Holton, Kans, Dec 14, 49; m 71; c 2. SURFACE CHEMICAL PHYSICS. *Educ:* Kans State Univ, BS, 71; Univ Calif, Berkeley, PhD(phys chem), 75. *Prof Exp:* Asst prof, 75-81, assoc prof, 81-86, PROF CHEM, PRINCETON UNIV, 86- *Honors & Awards:* Exxon Solid State Chem Prize, Am Chem Soc, 81; Alexander von Humboldt Award, 86. *Mem:* Am Chem Soc; AAAS; Am Vacuum Soc. *Res:* Gas-solid interactions; dynamics of reactions occurring on well characterized transition metal and transition metal compound surfaces. *Mailing Add:* Dept Chem Princeton Univ Princeton NJ 08544

BERNATH, L(OUIS), b New York, NY, Dec 31, 21; m 44; c 3. CHEMICAL ENGINEERING. *Educ:* City Col New York, BChE, 43; Univ Del, PhD(chem eng), 50. *Prof Exp:* Asst, City Col New York, 42-43; develop engr, Schenley Res Inst, Inc, 43-45; staff engr, Schenley Labs, Inc, 46-47; Off Naval Res fel, Proj Squid, Univ Del, 47-50; sr engr, E I du Pont de Nemours & Co, Inc, 50-61; chief proj engr, Atomics Int, NAm Aviation, Inc, 61-66, tech adv, 66-69, proj mgr, NAm Rockwell Corp, 69-73; mgr breeder reactor design, Gen Atomic, 73-74; chief nuclear eng, San Diego Gas & Elec, 85-86, mgr generation eng & res & develop, 78-82, mgr nuclear, 82-85, prin scientist energy technols, 85-865-86; RETIRED. *Mem:* Am Inst Chem Engrs; Am Soc Mech Engrs; Am Nuclear Soc; Nat Soc Prof Engrs; Sigma Xi; fel Am Instan Advan Eng. *Res:* Boiling heat transfer; reactor technology; energy technology. *Mailing Add:* San Diego Gas & Elec 8043 Hemingway Ave San Diego CA 92120

BERNATH, PETER FRANCIS, b Ottawa, Can, Dec 7, 53; Can citizen; m 78; c 2. PHYSICAL CHEMISTRY, MOLECULAR ASTRONOMY. *Educ:* Univ Waterloo, BS, 76; Mass Inst Technol, PhD(chem), 80. *Prof Exp:* Res assoc, Herzberg Inst Astrophysics, Nat Res Coun Can, 80-82; from asst prof to assoc prof chem, Univ Ariz, 82-90; PROF CHEM, UNIV WATERLOO, 91- *Mem:* Am Phys Soc; Am Chem Soc; Optical Soc Am. *Res:* The laser and Fouries transform spectroscopy of transient molecules such as free radicals, ions and high temperature molecules.; infrared and microwave molecular astronomy. *Mailing Add:* Dept Chem Univ Waterloo Waterloo ON N2L 3G1 Can

BERNATH, TIBOR, b Bucharest, Romania, Feb 26, 34; US citizen; m 79; c 1. ENVIRONMENT & TOXICOLOGY. *Educ:* Hebrew Univ, BSc, 63, MSc, 65; Univ Paris, France, DSc, 70; Harvard Sch Pub Health, Boston, MSc, 80. *Prof Exp:* Res assoc, Dept Chem, Ind Univ, Bloomington, 70-71; sr res assoc, Brandeis Univ, 71-75; sr lectr environ sci, Ctr Occup Health, Tel Aviv Univ, Israel, 75-77; res assoc, Harvard Sch Pub Health, Boston, 77-79; sr indust hygienist petrochem, Med Dept, Exxon Corp, NJ, 79-85; CONSULT, 86-

Concurrent Pos: Mem staff, Develop Toxicol Sect, Smith-Kline Leary Lab, Boston, 72; consult environ health serv, ABT Assoc, Cambridge, 77; lectr, Harvard Col, Cambridge, 78; consult prog toxicol, Mass Col Pharm, Boston, 78. *Mem:* Am Chem Soc; Am Bd Indust Hyg; Brit Occup Health Soc; Europ Soc Toxicol; Am Soc Toxicol. *Res:* Aerometric (aerosols and vapors) and biological methods development in environmental health; health hazard assessments; risk assessments. *Mailing Add:* 220 W Jersey St PO Box 1108 Elizabeth NJ 07207-1108

BERNAU, SIMON J, b Wanganui, NZ, June 12, 37; m 59; c 2. MATHEMATICS. *Educ:* Univ Canterbury, BSc, 58, MS, 59; Cambridge Univ, BA, 61, PhD(math), 64. *Prof Exp:* Lectr math, Univ Canterbury, 64-65, sr lectr, 65-66; prof, Univ Otago, NZ, 66-69; prof math, Univ Tex, Austin, 69-85; prof math & head dept, Southwest Mo State Univ, Springfield, 85-88; PROF MATH & CHMN DEPT, UNIV TEX, EL PASO, 88- *Mem:* Am Math Soc. *Res:* Lattice ordered algebraic systems; operator theory; functional analysis. *Mailing Add:* Dept Math & Sci Univ Texas El Paso El Paso TX 79968-0514

BERNAUER, EDMUND MICHAEL, b Chicago, Ill, Sept 27, 26. EXERCISE PHYSIOLOGY. *Educ:* Univ Ill, PhD(physiol), 62. *Prof Exp:* PROF EXERCISE PHYSIOL, UNIV CALIF, DAVIS, 62- *Mem:* Am Col Sports Med; Am Physiol Soc. *Mailing Add:* 508 Scripps Dr Davis CA 95616-1062

BERNAYS, ELIZABETH ANNA, b Chihchilla, Queensland, Australia, Dec 31, 40; Brit citizen; m 83. PHYSIOLOGICAL BASIS OF BEHAVIOR, PHYSIOLOGICAL ECOLOGY. *Educ:* Univ Queensland, BSc, 62; Univ London, MSc, 67, PhD(entom), 70. *Prof Exp:* From sr scientist to prin scientist, Ctr Overseas Pest Res, UK, 70-83; prof entom, Univ Calif, Berkeley, 83-90, prof zool, 86-89; PROF ENTOM & PROF ECOL & EVOLUTIONARY BIOL & HEAD DEPT ENTOM, UNIV ARIZ, 89- *Concurrent Pos:* Ed, Exp & Appl Entom, 78-; CRC Series Plant-Insect Interactions, 87- *Honors & Awards:* Gold Medal, Pontifical Acad Sci, 86. *Mem:* Int Soc Chem Ecol; Am Soc Zoologists; Entom Soc Am. *Res:* Physiological basis of behavior and ecology of plant-feeding insects; evolutionary and adaptive relationships of plants and insects. *Mailing Add:* Dept Entom Univ Ariz Tucson AZ 85721

BERNAYS, PETER MICHAEL, b New York, NY, July 19, 18; m 47; c 3. PHYSICAL INORGANIC CHEMISTRY, INFORMATION SCIENCE. *Educ:* Mass Inst Technol, BS, 39; Univ Ill, MS, 40, PhD(anal chem), 42. *Prof Exp:* From instr to asst prof chem, Ill Inst Technol, 46-50; assoc prof, Southwestern La Inst, 50-54; asst ed, Chem Abstr, 54-57, assoc ed, 58, head assignment dept, 59-61, asst to dir res & develop, 61-65, personnel asst, 65-66, sr assoc ed, 66-69, sr ed, 69-71, asst mgr, Phys, Analytical & Inorg Dept, 71-88; RETIRED. *Concurrent Pos:* Res Corp Cotrell grant, 48-50. *Mem:* AAAS; Am Chem Soc. *Res:* X-ray diffraction; inorganic compounds; chemistry of scandium; structural properties of lime; chemical documentation. *Mailing Add:* 2391 Eastcleft Dr Columbus OH 43221

BERNDT, ALAN FREDRIC, physical chemistry, dental research; deceased, see previous edition for last biography

BERNDT, BRUCE CARL, b St Joseph, Mich, Mar 13, 39; m 63; c 3. MATHEMATICS. *Educ:* Albion Col, AB, 61; Univ Wis-Madison, MS, 63, PhD(math), 66. *Prof Exp:* Asst lectr math, Glasgow Univ, 66-67; from asst prof to assoc prof, 67-75, PROF MATH, UNIV ILL, URBANA-CHAMPAIGN, 75- *Concurrent Pos:* Mem, Inst Advan Study, Princeton, NJ, 73-74. *Honors & Awards:* Allendorfer Award, 79; Ford Award, 89. *Mem:* Am Math Soc; Math Asn Am. *Res:* Analytic number theory; classical analysis. *Mailing Add:* Dept Math Univ Ill 1409 W Green St Urbana IL 61801

BERNDT, DONALD CARL, b Toledo, Ohio, Apr 11, 35; m 60; c 2. PHYSICAL ORGANIC CHEMISTRY. *Educ:* Ohio State Univ, BSc, 57, PhD(org chem), 61. *Prof Exp:* NSF fel org chem, Ohio State Univ, 62; from asst prof to assoc prof, 62-74, PROF CHEM, WESTERN MICH UNIV, 74- *Mem:* Am Chem Soc. *Res:* Elucidation of reaction mechanisms in organic chemistry, including the study of medium, catalytic, micellar and structural influences. *Mailing Add:* Dept Chem Western Mich Univ Kalamazoo MI 49008

BERNDT, WILLIAM O, b St Joseph, Mo, May 33; m 54; c 5. RENAL TOXICOLOGY & PHARMACOLOGY, GENERAL TOXICOLOGY. *Educ:* Creighton Univ, BS, 54; Univ Buffalo, PhD(pharmacol), 59. *Prof Exp:* Fel pharmacol, Sch Med, Univ Buffalo, 59; from instr to prof, Dartmouth Med Sch, 59-74; prof pharmacol & toxicol & chmn dept, Med Ctr, Univ Miss, 74-81; PROF PHARMACOL & DEAN GRAD STUDIES & RES, UNIV NEBR MED CTR, 82-, VCHANCELLOR ACAD AFFAIRS, 85- *Concurrent Pos:* Estab investr, Am Heart Asn, 64-69; consult, US Environ Protection Agency, Miss State Univ Epilemiol Studies, 78-81; mem, Pharmacol Study Sect, NIH, 80-84; mem, Adv Panel Electrolytes, large vol parenterals & renal drugs, US Pharmacopeial Conv, 80-85 & Cosmetic Ingredient Rev Expert Panel, Cosmetic Toiletry & Fragrance Asn, 83-86, chmn, 86-91; bd sci counrs, DHHS Agency for Toxic Substance & Dis, 88-91; mem, Am Bd Toxicol, 88-92. *Mem:* Am Chem Soc; Soc Toxicol; Am Soc Pharmacol & Exp Therapeut; Am Soc Nephrol; AAAS. *Res:* Electrolyte and organic acid transport; renal physiology, pharmacology and toxicology; metabolic aspects of transport processes; diffusion processes; nephrotoxicity studies; effects of mycotoxins; teratogenicity studies. *Mailing Add:* Univ Nebr Med Ctr 600 S 42nd St Omaha NE 68198-6810

BERNDTSON, WILLIAM EVERETT, b Middletown, Conn, Oct 1, 44; m 69; c 3. REPRODUCTIVE PHYSIOLOGY. *Educ:* Univ Conn, BS, 66; Cornell Univ, PhD(reprod physiol), 71. *Prof Exp:* Trainee reprod, Okla State Univ, 71-72; res assoc, Colo State Univ, 72-75, asst prof reprod, 75- *Mem:* Soc Study Reprod; Am Soc Animal Sci; Sigma Xi. *Res:* Cryopreservation of mammalian spermatozoa for artificial insemination; influence of hormones, drugs and exteroceptive stimuli on quantitative aspects of spermatogenesis. *Mailing Add:* RFD 2 Dover Rd Durham NH 03824

BERNE, BERNARD H, b New York, NY, Sept 26, 38. IMMUNOLOGY, RHEUMATOLOGY. *Educ:* Johns Hopkins Univ, AB, 60; NY Univ Sch Med, MD, 64; Univ Ill, PhD(immunol & microbiol), 72. *Prof Exp:* Intern surg, NY Hosp, Cornel Med Ctr, 64-65; fel, Mayo Grad Sch Med, Univ Minn, 65-66; sr asst surg resident, Univ Chicago Hosp & Clin, 66-67; NIH trainee microbiol & immunol, Univ Ill Med Ctr, 67-68, Schweppe Found fel, 68-71; asst prof path, Med Sci Prog, Ind Univ, 71-74; prin investr immunol, Meloy Labs, Inc, Springfield, Va, 74-75; res med officer immunol & rheumatol, Rheumatol Serv, Walter Reed Army Med Ctr, Washington, DC, 76-84, microbiologist & immunologist, Allergy & Immunol Serv, 84-87; MICROBIOLOGIST & MED OFFICER (REVIEWER), OFF DEVICE EVAL, CTR DEVICES & RADIOL HEALTH, FOOD & DRUG ADMIN, SILVER SPRING, MD, 87- *Mem:* Am Asn Immunologists; Am Asn Clin Chemists; Fedn Am Soc Exp Biol. *Res:* Immunochemistry and genetics of rabbit allotypes; immunological relationships of serum macroglobulins; radioimmunoassay and immunodiffusion techniques; characterization and clinical significance of pregnancy proteins; assays for immune complexes and antibodies in autoimmune diseases and relationships of antibodies to disease activity; characterization of allergen extracts. *Mailing Add:* Div Surg & Rehab Devices Off Device Eval Food & Drug Admin 1390 Piccard Dr Rockville MD 20850

BERNE, BRUCE J, b New York, NY, Mar 8, 40; m 61. STATISTICAL MECHANICS, COMPUTER SIMULATION. *Educ:* Brooklyn Col, BS, 61; Univ Chicago, PhD(chem physics), 64. *Prof Exp:* NATO fel, 66; mem chem fac, 66-69, assoc prof chem, 69-72, PROF CHEM, COLUMBIA UNIV, 72- *Concurrent Pos:* Alfred P Sloan Found fel, 67-70; Guggenheim fel, 71-72; chmn adv panel, div chem, NSF; vis prof, Univ Tel Aviv, 78-80; mem policy bd, Nat Resource Comput Chem, Lawrence Berkeley Lab, 80-82; comt Chem Sci, NAS-NRC, 76-78; chmn, Subdiv of Theoret Chem of Am Chem Soc, 88-89. *Honors & Awards:* Sackler Distinguished Lectr, Univ Tel Aviv. *Mem:* Fel Am Phys Soc; AAAS. *Res:* Equilibrium and non-equilibrium statistical mechanics, structure and dynamics of the condensed phases of matter, quantum Monte Carlo, computer simulation of chemical process in liquids. *Mailing Add:* Dept Chem Havermeyer Hall Columbia Univ Box 755 New York NY 10027

BERNE, ROBERT MATTHEW, b Yonkers, NY, Apr 22, 18; m 44; c 4. MEDICAL PHYSIOLOGY. *Educ:* Univ NC, AB, 39; Harvard Med Sch, MD, 43. *Hon Degrees:* DSc, Med Col Ohio, 73. *Prof Exp:* From asst prof to prof physiol, Case Western Reserve Univ, 52-66, asst prof med, 57-66; Charles Slaughter prof & chmn dept, 66-88, ALUMNI PROF PHYSIOL, UNIV VA MED CTR, 88- *Concurrent Pos:* Ed, Circulation Res, 70-75; mem panel, Heart & Blood Vessel Dis Task Force, Nat Heart & Lung Inst, 72-, mem proj comt, Heart & Lung Prog, 75-77; mem panel, Heart & Vascular Dis Nat Res & Demonstration Rev Comt, NIH, 73-74; mem selection comt, Ciba Found Award Hypertension, 75-77; mem, Med Adv Bd, Coun High Blood Pressure Res, Am Heart Asn, 76-80, Coun Subcomt Pub Policy, 77-78; chmn-elect, Coun Acad Socs, Am Asn Med Cols, 76-77 & 77-78; mem, Sci Adv Bd, Alfred I Dupont Inst, 78-82, bd dir, Am Heart Asn, 79-80 & 83-85, bd, Nat Med Exam Study Comt, 83-84, Inst Med Mem Comt, 84-85, bd, Inst Med Mem Comt, 84-85, bd, Inst Med Health Sci Policy, 84-85, award comt, Lita A Hazen Awards, 84- & Nat Adv Coun, Pew Scholars prog, 84-89; chmn, Long Range Planning Comt, Am Physiol Soc, 80-84; Pub Comt, Am Heart Asn, 81-85; ed, Ann Rev Physiol, 82-88, field ed, Pflugers Arch, 81-84; pres, Va Affil, Am Heart Asn, 86. *Honors & Awards:* Carl J Wiggers Award, Am Physiol Soc, 75; Res Achievement Award & Award of Merit, Am Heart Asn, 79, Gold Heart Award, 85; Jacobi Medallion Award, 87; Ray G Daggs Award, Am Physiol Soc, 90- *Mem:* Inst Med-Nat Acad Sci; Am Physiol Soc (pres, 72-73); Asn Chmn Depts Physiol (pres, 70-71); Am Soc Clin Invest; Am Heart Asn; fel AAAS; hon fel Am Col Cardiol. *Res:* Cardiovascular physiology, particularly the chemical factors involved in the local regulation of organ blood flow. *Mailing Add:* Dept Physiol Jordan Hall Univ Va Med Ctr Charlottesville VA 22908

BERNECKER, RICHARD RUDOLPH, b Allentown, Pa, Feb 23, 36; m 59; c 4. PHYSICAL CHEMISTRY. *Educ:* Muhlenberg Col, BS, 57; Cornell Univ, PhD(phys chem), 62. *Prof Exp:* Res chemist, Standard Oil, Ind, 61-64, Los Alamos Sci Lab, Univ Calif, 64-69; SR SCIENTIST, ENERGETIC MAT DIV, NAVAL SURFACE WARFARE CTR, WHITE OAK LAB, MD, 69- *Mem:* Combustion Inst; Am Phys Soc; Am Chem Soc; Sigma Xi. *Res:* Sensitivity and detonability of explosives and propellants; mechanism of transition from deflagration to detonation in explosives and propellants. *Mailing Add:* 204 Pewter Lane Silver Spring MD 20905

BERNER, LEO DEWITTE, JR, b Pasadena, Calif, Feb 11, 22; m 47; c 2. BIOLOGICAL OCEANOGRAPHY. *Educ:* Pomona Col, BA, 43; Scripps Inst Oceanog, Univ Calif, MS, 52; Univ Calif, PhD(oceanog), 57. *Prof Exp:* Asst, Pomona Col, 48-49; res biologist, Scripps Inst Oceanog, Univ Calif, 52-56, jr res biologist, 58-61; fisheries res biologist, US Fish & Wildlife Serv, 57-58; assoc prog dir div sci personnel & educ, NSF, 61-65; res scientist, 65-66, from assoc prof to prof, 66-87, from asst dean to assoc dean, 67-84, DEAN GRAD COL, TEX A&M UNIV, 84-, EMER PROF OCEANOG, 87- *Concurrent Pos:* Biologist, US Navy Oper Wigwam, 55. *Mem:* Fel AAAS; Am Soc Limnol & Oceanog; Sigma Xi. *Res:* Plankton zeography and ecology; pelagic tunicates; taxonomy of tunicata. *Mailing Add:* Dept Oceanog Tex A&M Univ College Station TX 77840

BERNER, LEWIS, b Savannah, Ga, Sept 30, 15; m 45; c 2. ZOOLOGY. *Educ:* Univ Fla, BS, 37, MS, 39, PhD(zool), 41. *Prof Exp:* From asst prof to assoc prof biol, 46-54, actg dir div biol sci, 70-73, dir, 73-75, head biol sci, 59-75, PROF BIOL, UNIV FLA, 54- *Concurrent Pos:* NIH grant, 53-60; consult entomologist, Gold Coast, Brit WAfrica, 50 & Nyasaland, Brit Cent Africa, 52; ed, Fla Entomologist, 50-63; prof entom, Lake Itasca Biol Sta, Univ Minn, 58-62, 68-69, 73, 75 & 77; USPHS grant, 63-66. *Mem:* Fel Entom Soc; Sigma Xi. *Res:* Ecology and taxonomy of Ephemeroptera; biology; parasitology; mayflies of Southeast. *Mailing Add:* 309 Bartram Hall Univ Fla Gainesville FL 32611

BERNER, ROBERT A, b Erie, Pa, Nov 25, 35; m 59; c 3. SEDIMENTARY GEOCHEMISTRY. *Educ:* Univ Mich, BS, 57, MS, 58; Harvard Univ, PhD(geol), 62. *Hon Degrees:* Dr Hon Causa, Univ Aix-Marseille III, France, 91. *Prof Exp:* Sverdup fel oceanog, Scripps Inst, 62-63; from asst prof to assoc prof geophys sci, Univ Chicago, 63-65; from assoc prof to prof, 65-87, ALAN M BATEMAN PROF GEOL & GEOPHYS, YALE UNIV, 87- *Concurrent Pos:* Alfred P Sloan Found fel phys sci, 68-72; Guggenheim Found fel, Swiss Fed Inst Technol, 72; assoc ed, Am J Sci, 68-80, 90-, ed, 80-90; mem, US Nat Comt Geochem, 78-81; NATO vis scientist, Univ Paris, 81 & 84; distinguished vis scientist, State Univ New York, Stony Brook, 85, Nat Ctr for Atmospheric Res, 87; chmn, Gordon Conf Chem Oceanog, 86, Nat Res Coun Panel on Geochem Cycles, Nat Acad Sci, 88-90; mem, Adv Comt on Earth Sci, NSF, 90-93. *Honors & Awards:* Award, Mineral Soc Am, 71. *Mem:* Nat Acad Sci; Geochem Soc (vpres, 82, pres, 83); fel Mineral Soc Am; fel Geol Soc Am. *Res:* Kinetics of geochemical processes, chemical diagenesis and weathering; biogeochemistry; computer modeling of geochemical cycles; evolution of atmospheric and oceanic composition. *Mailing Add:* Dept Geol & Geophys Yale Univ New Haven CT 06511

BERNERS, EDGAR DAVIS, b Milwaukee, Wis, Aug 22, 27; m 60; c 4. ACCELERATOR PHYSICS. *Educ:* Col Holy Cross, BS, 49; Univ Wis, PhD(physics), 57. *Prof Exp:* From instr to asst prof physics, Marquette Univ, 57-67; from asst fac fel to assoc fac fel, 67-77, FAC FEL, UNIV NOTRE DAME, 77- *Res:* Electrostatic accelerators. *Mailing Add:* Dept Physics Univ Notre Dame Notre Dame IN 46556

BERNETTI, RAFFAELE, b Florence, Italy, Aug 24, 32; m 60; c 4. ORGANIC CHEMISTRY, ANALYTICAL CHEMISTRY. *Educ:* Univ Pisa, MS, 55; Univ Pa, PhD(org chem), 59; Univ Chicago, MBA, 70. *Prof Exp:* Lab dir, Milan Lab, 60-65, sect head org chem, Moffett Tech Ctr, Argo, 65-67, asst dir prod methodology, 67-72, dir com develop, Corp Hq, Englewood Cliffs, NJ, 72-74, dir res & develop, Analysis Dept, Moffet Tech Ctr, CPC Int, 75-87; DIR, ANALYSIS & QUAL SUPPORT, CORN PROD TECH CTR, CPC INT, 87- *Concurrent Pos:* Instr, DePaul Univ, 70-72; chmn & ed chief, J Asn Analytical chemist, 87-89. *Mem:* Am Chem Soc; AAAS; Am Asn Cereal Chemists; fel Asn Off Analysis Chemists. *Res:* Heterocyclic chemistry; carbohydrate chemistry; starch and dextrose production; dextrose, levulose sweeteners; analytical methods for carbohydrate and food products; project management. *Mailing Add:* 12325 S 83rd Ave Palos Park IL 60464-1919

BERNEY, CHARLES V, b Walla Walla, Wash, July 11, 31; m 56, 76; c 5. NEUTRON SCATTERING, MOLECULAR SPECTROSCOPY. *Educ:* Whitman Col, BA, 53; Univ Wash, PhD(phys chem), 62. *Prof Exp:* Res fel, Mellon Inst, 62-65; asst prof chem, Univ NH, 65-72; sr res assoc, Air Force Rocket Propulsion Lab, 72-73; sr res assoc, Nuclear Eng Dept, 73-80, SR RES ASSOC, CHEM ENG DEPT, MASS INST TECHNOL, 80- *Concurrent Pos:* Prin investr, NSF grants, 73, 76, 80 & 83; supvr, Polymer Cent Fac, Mass Inst Technol, 83- *Mem:* AAAS; Am Phys Soc; Sigma Xi. *Res:* Molecular structure; spectroscopy; application of neutron scattering to polymer structure. *Mailing Add:* 13-1018 Chem Eng Dept Mass Inst Technol Cambridge MA 02139

BERNEY, REX LEROY, b St Paul, Nebr, Sept 25, 50; m 74, 89. SOLID STATE PHYSICS. *Educ:* Univ Mo-Columbia, BS, 71, MS, 73, PhD(physics), 78. *Prof Exp:* Asst prof, 78-85, ASSOC PROF PHYSICS, UNIV DAYTON, 85- *Mem:* Am Phys Soc; Sigma Xi. *Res:* Photochromic materials; microcomputers in the laboratory. *Mailing Add:* Dept Physics Univ Dayton Dayton OH 45469

BERNEY, STEVEN, b Germany, Dec 31, 35; US citizen; m 61; c 3. RHEUMATOLOGY, IMMUNOLOGY. *Educ:* Rutgers Univ, BA, 58; State Univ NY Upstate Med Ctr, MD, 62. *Prof Exp:* Intern & resident internal med, Ohio State Univ Hosp, 62-63 & 65-67; USPHS fel rheumatic dis & immunol, NY Univ, 67-71, Arthritis Found fel, 70-71; asst prof med, 71-76, ASSOC PROF MED & MICROBIOL, HEALTH SCI CTR, TEMPLE UNIV, 76- *Mem:* Am Rheumatism Asn; Reticuloendothelial Soc; Am Fedn Clin Res. *Res:* Lymphocyte physiology and immunology; isolation and purification of membrane antigen receptors; circulatory behavior of lymphocytes; mechanism of immunologic depression in aging. *Mailing Add:* Dept Med Temple Univ Health Sci Ctr Philadelphia PA 19140

BERNEY, STUART ALAN, b Albany, NY, Aug 28, 45; div; c 2. BIOCHEMICAL PSYCHOPHARMACOLOGY, PHARMACOKINETICS. *Educ:* Union Univ, BS, 69; Vanderbilt Univ, PhD(pharmacol), 74. *Prof Exp:* Lectr pharmacol, Univ Toronto, 75-76; res assoc pharmacol, Vanderbilt Univ, 77, res asst prof psychiat, 78-82. *Concurrent Pos:* Grant award, Dept Health & Human Serv, Nat Inst Drug Abuse fel, 75. *Mem:* Soc Neurosci; AAAS; Sigma Xi; Am Pub Health Asn. *Res:* Biochemical pharmacology of the central nervous system; clinical psychopharmacology and pharmacokinetics. *Mailing Add:* 301 Fairfax Ave No B Nashville TN 37212-4006

BERNFELD, PETER, b Leipzig, Ger, June 1, 12; nat US; m 40; c 2. ENZYMOLOGY, BIOCHEMISTRY. *Educ:* Univ Leipzig, MS, 35; Univ Geneva, PhD(chem), 37. *Prof Exp:* From res fel to chief chemist org chem, Univ Geneva, 37-49, privat docent, 47-49; from asst prof to assoc prof biochem & nutrit & biochemist, Cancer Res Control Unit, Tufts Univ, 49-57; SR VPRES & DIR RES, BIO-RES INST & BIO-RES CONSULTS, 57- *Honors & Awards:* Werner Medal, Swiss Chem Soc, 48. *Mem:* AAAS; Am Chem Soc; Am Soc Biol Chem; Soc Exp Biol & Med; Am Inst Chemists; Am Asn Cancer Res; Am Col Toxicol. *Res:* Proteins; enzymes; polysaccharides; immobilized enzymes and antigens; studies on serum glycoproteins; humoral antigens prognosticating susceptibility to mammary cancer in women and mice; inhalation toxicity of smoke including tobacco smoke. *Mailing Add:* 247 Farm Lanes Westwood MA 02090-1111

BERNFIELD, MERTON RONALD, b Chicago, Ill, Apr 9, 38; m 59; c 3. CELL BIOLOGY, PEDIATRICS. *Educ:* Univ Ill, Urbana, BS, 59, Chicago, MS & MD, 61. *Prof Exp:* Intern, Res & Educ Hosp, Univ Ill, Chicago, 61-62; asst resident pediat, New York Hosp, Cornell Med Ctr, 62-63; res assoc, Lab Biochem Genetics, Nat Heart Inst, NIH, 63-65; res investr, Nat Inst Child Health & Human Develop, Univ Calif, San Diego, 65-66; chief resident pediat, Med Ctr, from asst prof to prof pediat, Sch Med, Stanford Univ, 67-89, Josephine K Knowles prof human biol, 77-89; DIR, JOINT PROG NEONATOL, HARVARD MED SCH, 89-, CLEMENT A SMITH PROF PEDIAT, 89-, PROF ANAT & CELLULAR BIOL, 89- *Concurrent Pos:* Dir, Med Scientist Training Prog, Stanford Univ, 74-77, chmn, Prog Human Biol, 77-80, dir, Fel Prog Molecular Path of Cell Membranes, 76-85; chmn, Gordon Res Conf on Basement Membranes, 86-; assoc ed, Develop Biol; dir, Neonatal Develop Biol Training Prog, Stanford Univ. *Mem:* Am Soc Biol Chemists; Am Soc Cell Biol; Perinatal Res Soc; Soc Develop Biol; Soc Pediat Res; Sigma Xi. *Res:* Determine the chemical events which control cellular behavior during their formation of specific organs and structures during development; how cells interact with each other and with extracellular matrix molecules. *Mailing Add:* Joint Prog Neonatol Harvard Med Sch Enders Nine 300 Longwood Ave Boston MA 02115

BERNHAGEN, RALPH JOHN, b Toledo, Ohio, Aug 2, 10; m 40; c 2. GEOLOGY. *Educ:* Ohio State Univ, BA, 37, MA, 39. *Prof Exp:* Jr paleontologist, Shell Oil Co, Tex, 39-41; eng geologist, Lockwood, Andrews & Duller Eng Co, 41; chief geologist, Ohio Water Resources Bd, 41-52; asst chief, Ohio Div, Geol Surv, 52-57; div chief & state geologist, Ohio Geol Surv, 57-68; chief water planning sect, 68-72, adminr shoreland mgt sect, 72-74, GEOL PROG SUPVR, DIV WATER, OHIO DEPT NATURAL RESOURCES, 74- *Concurrent Pos:* Bd Counr, Dept Geol, Wright State Univ; mem adv comt, Dept Geol & Minerol, Ohio State Univ. *Mem:* Fel Geol Soc Am; Am Asn Petrol Geologists; Asn Am State Geologists; Am Inst Prof Geologists; Sigma Xi. *Res:* Stratigraphy of Ohio and the Appalachian Basin; ground water and petroleum geology. *Mailing Add:* 5916 Linworth Rd Worthington OH 43085

BERNHARD, JEFFREY DAVID, b Buffalo, NY, Oct 31, 51. PSORIASIS, PRURITUS. *Educ:* Harvard Col, BA, 73; Harvard Med Sch, MD, 78. *Prof Exp:* Intern med, Beth Israel Hosp, Boston, 78-79; res dermat, Mass Gen Hosp, 79-81, chief resident, 81-82; fel, 82-83, asst prof, 83-87, ASSOC PROF MED, UNIV MASS MED SCH, 87- *Concurrent Pos:* Ed bd, Yearbook Dermat & J Am Acad Dermat, 82; dir, Phototherapy Ctr, Univ Mass Med, 83, Div Dermat, 86; assoc ed, Dermat Capsule & Comment, 84. *Mem:* Soc Investigative Dermat; Am Acad Dermat; Sigma Xi. *Res:* Itching; psoriasis; photosensitivity; phototherapy; dermatology. *Mailing Add:* Div Dermat Univ Mass Med Ctr Worcester MA 01655

BERNHARD, RICHARD ALLAN, b Pittsburgh, Pa, Oct 29, 23; m 53; c 2. ORGANIC CHEMISTRY. *Educ:* Stanford Univ, BS, 50; Calif Inst Technol, PhD(chem), 55. *Prof Exp:* Res chemist, Lemon Prods, Adv Bd, Calif, 55-57; from asst prof to assoc prof food sci & technol, 57-70, PROF FOOD SCI & TECHNOL, UNIV CALIF, DAVIS, 70- *Mem:* Am Chem Soc; Inst Food Technol. *Res:* Analytical methods for determination of micro-constituents in materials of biological origin; flavor chemistry; theory of chromatographic processes; mode and mechanism of pyrazine formation in model systems. *Mailing Add:* Dept Food Sci & Technol Univ Calif Davis CA 95616

BERNHARD, RICHARD HAROLD, b New York, NY, Dec 11, 33; m 69; c 3. ENGINEERING ECONOMIC ANALYSIS, DECISION ANALYSIS. *Educ:* Cornell Univ, BME, 56, PhD(opers res), 61; Mass Inst Technol, SM, 58. *Prof Exp:* Asst prof indust eng, Cornell Univ, 61-69; assoc prof, 69-80, PROF INDUST ENG, NC STATE UNIV, 80- *Concurrent Pos:* Consult, Arthur D Little Inc, 57-59; NSF sci fac fel, Univ NC, Chapel Hill, 71-72; vis prof, Norweg Sch Econ & Bus Admin, 77-78 & Norweg Inst Technol, 88-89; vis lectr, Univ Canterbury, NZ, 81-82; Fulbright scholar, Fed Univ Santa Catarina, Brazil, 86. *Honors & Awards:* Eugene L Grant Award, Am Soc Eng Educ, 84. *Mem:* Inst Indust Engrs; Am Soc Eng Educ; Financial Mgt Asn. *Res:* Engineering and managerial economis and decision analysis; mathematical modeling, especially in the presence of constraints and uncertainty, for capital investment planning, finance, production management, cost analysis and cost control. *Mailing Add:* Dept Indust Eng NC State Univ Box 7906 Raleigh NC 27695-7906

BERNHARD, VICTOR MONTWID, b Milwaukee, Wis, June 29, 27; m 60; c 3. MEDICINE. *Educ:* Northwestern Univ, BS, 47, MD, 51. *Prof Exp:* From clin instr to clin asst prof surg, Med Col Wis, 59-64; NIH fel cardiovasc surg, Col Med, Baylor Univ, 65; from assoc prof to prof surg, Med Col Wis, 74-84; PROF SURG & CHIEF SECT VASCULAR SURG, ARIZ HEALTH SCI CTR, 84- *Concurrent Pos:* Mem, Coun Cardiovasc Surg, Am Heart Asn. *Mem:* Soc Vascular Surg; fel Am Heart Asn; AMA; Am Col Surg; Int Cardiovasc Soc. *Res:* Vascular diseases; vascular surgery. *Mailing Add:* Ariz Health Sci Ctr 1501 N Campbell Tucson AZ 85719

BERNHARD, WILLIAM ALLEN, b Philadelphia, Pa, Oct 9, 42; m 65; c 2. RADIATION CHEMISTRY. *Educ:* Union Col, BS, 64; Pa State Univ, MS, 66, PhD(biophysics), 68. *Prof Exp:* Asst biophysics, Pa State Univ, 64-68; res fel biol & med, Argonne Nat Lab, 68-70; asst prof, 70-76, assoc prof radiation biol & biophysics, 76-85, PROF BIOPHYSICS, UNIV ROCHESTER, 85- *Concurrent Pos:* Career develop award, NIH, 72, res grant awards, 75 & 78; vis prof biol, Univ Regensburg, 75-76; assoc ed, Radiation Res, 75-78; NSF res grant awards, 78 & 80. *Honors & Awards:* Merit Award, NIH, 91. *Mem:* Biophys Soc; Radiation Res Soc; Am Phys Soc. *Res:* Effects of ionizing radiation on the primary structure of nucleic acids; electron spin resonance spectroscopy. *Mailing Add:* Dept Biophysics Univ Rochester Rochester NY 14642

BERNHARDT, ANTHONY F, b Los Angeles, Calif, Apr 16, 45. APPLIED PHYSICS. *Educ:* Univ Calif, Los Angeles, AB, 67; Univ Calif, Davis, MEng, 73, PhD(appl sci), 75. *Prof Exp:* Staff scientist physics, Lawrence Livermore Lab, Univ Calif, 68-72; fel appl sci, Fannie & John Hertz Found, 73-75; staff scientist physics, Lawrence Livermore Nat Lab, Univ Calif, 76-80; QUANTA-RAY, 80- *Concurrent Pos:* Consult, W G Reynolds Sr, 75-76. *Mem:* Am Phys Soc. *Res:* Laser isotope separation; atomic and molecular structure and spectroscopy; photochemistry; atomic interactions in strong resonant optical fields. *Mailing Add:* Four Uplands Berkeley CA 94705

BERNHARDT, ERNEST C(ARL), b Ger, Feb 20, 23; nat US; m 51; c 2. ENGINEERING. *Educ:* Purdue Univ, BS, 43; Univ Del, MChE, 48; Tech Hochschule, Ger, DIng, 52. *Prof Exp:* Chem engr, Calco Chem Div, Am Cyanamid Co, 43-46 & Monsanto Chem Co, 48-50; sales serv engr, Polychem Dept, E I Du Pont De Nemours & Co, 52-56, supvr process develop, 56-64, sales mgr plastics, Int Dept, Switz, 64-68, mkt mgr, Int Dept, Switz, 68-74, mgr admin, Europe, 74-81; pres, Plastics & Comput, Inc, 82-90; CONSULT, 90- *Concurrent Pos:* Consult, Europe, 50-52; mem, bd trustees, Polymer Processing Inst. *Mem:* Soc Plastics Indust; Soc Plastic Engrs. *Res:* Computer aided engineering for injection molding; processing of thermoplastic materials. *Mailing Add:* 17 Vera Place Montclair NJ 07042

BERNHARDT, RANDAL JAY, b Quincy, Ill, Nov 25, 56; m 81; c 4. SURFACTANT CHEMISTRY. *Educ:* Quincy Col, BS, 78; Univ Iowa, PhD(chem), 84. *Prof Exp:* Teaching asst chem, Univ Iowa, 78-84; res chemist, 84-87, group leader, 87-88, MGR RES, STEPAN CO, 88- *Mem:* Am Chem Soc; Am Oil Chemists Soc. *Res:* Synthesis, characterization and application of organic and organometallic chemicals for the surfactant and specialty chemical industries. *Mailing Add:* Stepan Co Edens & Winnetka Northfield IL 60096

BERNHARDT, ROBERT L, III, b Salisbury, NC, Apr 28, 39; m 75; c 1. MODULE & RING THEORY, TORSION THEORY. *Educ:* Univ NC, Chapel Hill, BS, 61, MA, 64; Univ Ore, PhD(math), 68. *Prof Exp:* Asst prof math, Univ NC, Greensboro, 68-74; assoc prof math, Chicago State Univ, 74-83, chairperson dept 77-83; CHMN MATH DEPT, E CAROLINA UNIV, 83- *Mem:* Am Math Soc; Math Asn Am; Sigma Xi. *Res:* Torsion theory of modules; splitting properties of torsion theories. *Mailing Add:* Dept Math E Carolina Univ E Fifth St Greenville NC 27858

BERNHEIM, FREDERICK, b West End, NJ, Aug 18, 05; m 28; c 1. PHARMACOLOGY. *Educ:* Harvard Univ, AB, 25; Cambridge Univ, PhD(biochem), 28. *Prof Exp:* Nat Res Coun fel, 29-30; from asst prof to assoc prof physiol, 30-45, PROF PHARMACOL, SCH MED, DUKE UNIV, 45- *Mem:* Am Soc Biol Chem; Am Soc Pharmacol & Exp Therapeut; NY Acad Sci. *Res:* Oxidative processes in the animal body; biochemical aspects of pharmacology. *Mailing Add:* Duke Univ Med Ctr Durham NC 27710

BERNHEIM, ROBERT A, b Hackensack, NJ, June 8, 33; c 2. PHYSICAL CHEMISTRY, CHEMICAL PHYSICS. *Educ:* Brown Univ, BS, 55; Harvard Univ, MA, 57; Univ Ill, PhD(chem), 59. *Prof Exp:* Res fel chem, Columbia Univ, 59-61; from asst prof to assoc prof, 61-69, PROF CHEM, PA STATE UNIV, UNIVERSITY PARK, 69- *Concurrent Pos:* NSF sr fel, 67-68; Guggenheim fel, 74-75; vis fel, J Inst for Lab Astrophys, 82-83; Fulbright fac scholar, 88-89. *Honors & Awards:* Alexander von Humboldt Sr Scientist Award, 89. *Mem:* Fel Optical Soc Am; Am Chem Soc; fel Am Phys Soc. *Res:* Optical pumping; laser spectroscopy; molecular structure and dynamics. *Mailing Add:* 152 Davey Lab Pa State Univ University Park PA 16802

BERNHEIMER, ALAN WEYL, b Philadelphia, Pa, Dec 9, 13; m 42; c 1. MICROBIOLOGY, TOXINOLOGY. *Educ:* Temple Univ, BS, 35, AM, 37; Univ Pa, PhD(med sci), 42. *Prof Exp:* Asst biol, Temple Univ, 35-37; instr bact, Pa State Col Optom, 37-39; chmn basic med sci, 69-74, from instr to prof bact, 41-58, prof, 58-84, EMER PROF MICROBIOL, SCH MED, NY UNIV, 84- *Concurrent Pos:* Mem, Marine Biol Lab; consult, Surgeon Gen, 57-; NIH res career award, 62-84; trustee, Cold Spring Harbor Lab Quant Biol, 63-68. *Honors & Awards:* Lily Award, 48. *Mem:* Am Soc Microbiol; fel Am Acad Microbiol; Am Micros Soc; Mineral Soc Am; Am Asn Immunol. *Res:* Extracellular toxins and enzymes of bacteria; streptococci and staphylococci; cell membranes; toxins of various invertebrates. *Mailing Add:* Dept Microbiol NY Univ Sch Med New York NY 10016

BERNHEIMER, HARRIET P, b New York, NY, May 27, 19; m 42; c 1. MICROBIAL GENETICS. *Educ:* Hunter Col, BA, 39; NY Univ, MS, 41, PhD(microbiol), 50. *Prof Exp:* Asst biochem, Col Physicians & Surgeons, Columbia Univ, 50-53; from instr to assoc prof 58-79, EMER PROF MED, STATE UNIV NY DOWNSTATE MED CTR, 85- *Concurrent Pos:* NIH grant, 63-70; Health Res Coun NY career scientist award, 64-70; Irma T Hirschl trust career scientist award, 73-77; dir, Med PhD prog, 78- *Mem:* AAAS; Genetics Soc Am; Am Soc Microbiol; Harvey Soc; Sigma Xi. *Res:* Genetic control of capsule production, studied through the bacterial transformation system; biosynthesis of capsular polysaccharide in pneumococcus; pneumococcus bacteriophages. *Mailing Add:* 51 Fifth Ave New York NY 10003

BERNHOLC, JERZY, b Szczecin, Poland, Feb 12, 52; US citizen; m 82; c 1. INTERFACES, PHYSICS. *Educ:* Univ Lund, Sweden, BSc, 73, PhD(physics), 77. *Prof Exp:* Postdoc fel, IBM JT Watson Res Ctr, 78-80; sr physicist, Exxon Res & Engr Co, 80-86; assoc prof, 86-90, PROF PHYSICS, NC STATE UNIV, 90- *Concurrent Pos:* Vis scientist, Dept Theoret Physics, Univ Lund, Sweden, 78, Calif Inst Technol, Pasadena, 82. *Mem:* Am Phys Soc. *Res:* Theoretical condensed matter physics, materials science and chemistry; defects and impurities in semiconductors; clusters; interfaces; surface reactions. *Mailing Add:* Phys Dept NC State Univ Box 8202 Raleigh NC 27695

BERNHOLZ, WILLIAM FRANCIS, b New York, NY, Jan 26, 24; m 49; c 3. LIPID CHEMISTRY, SURFACE CHEMISTRY. *Educ:* Manhattan Col, BS, 45; Fordham Univ, MS, 48. *Prof Exp:* Assoc technologist, Cent Labs, Gen Foods Corp, 48-57; develop chemist, Avon Prod, Inc, 57-62; res chemist, Colgate-Palmolive Co, Inc, 62; chief chemist textile lab, Drew Chem Corp, 63-66, mgr indust res lab, Drew Div, Pac Veg Oil Corp, 66-75; dir, Int Tech Serv, PVO Int, Inc, 75-77, dir res, 77-80; PRES, BBK INC, 81- *Mem:* Am Chem Soc; Am Oil Chemists Soc. *Res:* Chemistry of coffee, gelatin; gel forming proteins and carbohydrates; caramelization of sugars; browning reaction; enzyme isolation; colorimetric, chromatographic and high vacuum techniques; product development; convenience foods and cosmetics; aerosol formulation and packaging; textile lubricant and antistat specialties; military-approved, hindered polyol type-turbo jet engine lubricants; hazardous material control; air quality monitoring of potential environmental hazards; water quality. *Mailing Add:* 11 Ledge Rd Pines Lake Wayne NJ 07470

BERNI, RALPH JOHN, b New Orleans, La, Nov 1, 31; m 57, 87; c 6. ANALYTICAL CHEMISTRY. *Educ:* La State Univ, BS, 54; Tulane Univ, MS, 61, PhD(inorg chem), 66. *Prof Exp:* Res assoc chemist, Southern Regional Res Ctr, Agr Res Serv, USDA, 55-65, res chemist, 65-74, res leader, 75-77, lab chief, 77-85, res leader, Compos & Properties Res Unit, 85-88; CHEMIST & CONSULT, ENVIRON ANALYSIS INC, 89- *Concurrent Pos:* Instr chem, Univ New Orleans, 87- *Mem:* Am Chem Soc; Am Inst Chemists; Sigma Xi. *Res:* Etherification and esterification of cellulose, including reaction mechanisms; metal ion complexes of urea derivatives; transition elements; instrumental analysis. *Mailing Add:* SRRC Agr Res Serv USDA PO Box 19687 New Orleans LA 70179

BERNICK, SOL, anatomy, oral pathology; deceased, see previous edition for last biography

BERNIER, CHARLES L(EWELLYN), cooperative intelligence, ecosystems; deceased, see previous edition for last biography

BERNIER, CLAUDE, b St Boniface, Man, June 18, 31; m 58; c 3. PHYTOPATHOLOGY. *Educ:* Univ Man, BA, 53, BSA, 57, MSc, 61; Univ Minn, PhD(plant path), 64. *Prof Exp:* Res officer veg crops, Can Dept Agr, 57-61; from asst prof to assoc prof phytopath, 65-74, PROF PHYTOPATH, UNIV MAN, 74- *Mem:* Am Phytopath Soc; Can Phytopath Soc. *Res:* Resistance to tan spot of wheat; biology and genetics of the tan spot fungus; diseases of lentil; biological control of take-all. *Mailing Add:* Dept Plant Sci Univ Man Winnipeg MB R3T 2N2 Can

BERNIER, JOSEPH LEROY, b Chicago, Ill, Apr 5, 09; m 36; c 2. ORAL PATHOLOGY. *Educ:* Univ Ill, DDS, 32, MS, 34; Am Bd Oral Path, dipl, 48; Am Bd Periodont, dipl; Am Bd Oral Med, dipl, 57;. *Prof Exp:* Instr oral path, Univ Ill, 32-34; mem staff, Walter Reed Gen Hosp, US Army, 34-38, chief oral path br, Armed Forces Inst Path, 38-39, chief dept dent lab, CZ, 39-41, chief oral path br, Armed Forces Inst Path, 41-42, dent coordr, 15th Hosp Ctr, 42-43, dent surgeon, Camp Polk, La, 43, dir dent div, McCloskey Gen Hosp, Temple, Tex, 43-44, instr, Ft Lewis, Wash, 44, dent surgeon, 254th Gen Hosp, 44-45, chief oral path div, Armed Forces Inst Path, 45-60, asst surgeon gen & chief, Army Dent Corps, 60-67; prof oral path & chmn dept, Georgetown Univ, 45-77; CONSULT, 77- *Concurrent Pos:* Mem comt dent, Nat Res Coun & Nat Adv Dent Health Coun, 54; pres, Am Bd Oral Path, 59-60; prof lectr, Jefferson Med Col, 59-; spec lectr, Dent Sch, Fairleigh Dickinson Univ, 61-; chmn dent res adv comt, US Dept Army. *Honors & Awards:* Award, Hinman Clin, Ga, 55; Callahan Gold Medal Award, Ohio State Dent Asn, 55; Gold Medal & Cert, Pierre Fauchard Acad, 61; Miller Medal, Am Acad Oral Med, 67; William J Gies Award, Am Acad Periodont, 69. *Mem:* Fel Am Col Dent; Am Soc Oral Surg; Am Acad Oral Path (pres, 59-60); Am Acad Periodont (pres, 59-60); fel Int Acad Oral Path (secy). *Res:* Pathology of periodontitis; etiology of oral cancer; atomic effects on oral structures; pathology of dental pulp diseases. *Mailing Add:* 6905 Hillmead Rd Bethesda MD 20817

BERNIER, PAUL EMILE, b St Michel, Que, Oct 22, 11; US citizen; m 40. POULTRY GENETICS. *Educ:* Laval, BSA, 32; Univ Calif, PhD(genetics), 47. *Prof Exp:* Poultry husbandman, Ecole Superieure d'Agr, Que, 28-45, lectr poultry husb, 32-45, lectr genetics, 39-45; chief inspector nat poultry breeding prog, Can Dept Agr, 45-47; from assoc prof to prof poultry genetics, 47-77, EMER PROF POULTRY GENETICS, ORE STATE UNIV, 77- *Honors & Awards:* Res Prize, Poultry Sci Asn, 52. *Mem:* Fel AAAS; Biomet Soc; Poultry Sci Asn; Genetics Soc Am; Teratology Soc; World's Poultry Sci Asn. *Res:* Genetics of domestic fowl and teratology. *Mailing Add:* Dept Poultry Sci Ore State Univ Corvallis OR 97331-3402

BERNING, WARREN WALT, b Cincinnati, Ohio, July 29, 20; m 46; c 2. PHYSICS, METEOROLOGY. *Educ:* Univ Cincinnati, BA, 42; Calif Inst Technol, MS, 46. *Prof Exp:* Aerodynamicist, McDonnell Corp, St Louis, 46 & Curtiss-Wright Corp, Columbus, 46-47; meteorologist, Ballistic Res Lab, Aberdeen Proving Ground, 47-50, physicist, 50-67; dep chief radiation directorate, Defense Atomic Support Agency, 67-71, asst to dep dir theoret res, Defense Nuclear Agency, 71-76; sr oper analyst, SRI Int, 76-77; SR SCIENTIST, PHYS SCI LAB, NMEX STATE UNIV, 77- *Mem:* AAAS; assoc fel Am Inst Aeronaut & Astronaut; Am Meteorol Soc; Am Geophys Union; fel Am Phys Soc. *Res:* Electromagnetic propagation through atmosphere; lower level meteorology; upper atmosphere physics. *Mailing Add:* 11007 Candelight Lane Potomac MD 20854

BERNIUS, MARK THOMAS, b Brooklyn, NY, Oct 24, 57. ELECTRON & ION OPTICS, MICROSCOPY. *Educ:* Polytech Inst, BSc, 79; State Univ NY, MSc, 81; Cornell Univ, PhD(nuclear sci & appl physics), 87. *Prof Exp:* Postdoctoral physics, Jet Propulsion Lab, 87-89; sr res fel geol & planetary sci, Calif Inst Technol, 89-91; SR RES PHYSICIST PHYSICS, DOW CHEM CO, 92- *Concurrent Pos:* Consult space sci, Jet Propulsion Lab, 89- *Honors & Awards:* Achievement Awards, NASA, 89. *Mem:* Am Phys Soc; Europ Phys Soc; Am Chem Soc. *Res:* Experimental materials research utilizing electron, ion and light microscopies; physics of ion formation and electron-molecule reactions at threshold; theoretical and experimental charged-particle optics; high-precision mass spectrometry. *Mailing Add:* Cent Mat Res Bldg No 1702 Dow Chem Co Midland MI 48674

BERNKOPF, MICHAEL, b Boston, Mass, Jan 11, 27; m 65; c 3. MATHEMATICS. *Educ:* Dartmouth Col, BA, 49; Columbia Univ, MA, 58; NY Univ, PhD(hist math), 65. *Prof Exp:* Asst prof math, Fairleigh Dickinson Univ, 65-66; assoc prof, 66-70, PROF MATH, PACE UNIV, 70- *Mem:* Fel Nat Acad Sci; Sigma Xi; Math Asn Am. *Res:* History of modern mathematics, particularly analysis. *Mailing Add:* Pace Plaza Pace Univ New York NY 10038

BERNLOHR, ROBERT WILLIAM, b Columbus, Ohio, Apr 20, 33; m 55; c 4. BACTERIOLOGY, PHYSIOLOGY. *Educ:* Capital Univ, BS, 55; Ohio State Univ, PhD(biochem), 58. *Hon Degrees:* DSc, Capital Univ, 80. *Prof Exp:* Fel biol, Oak Ridge Nat Lab, 58-60; asst prof biochem, Ohio State Univ, 60-62; from asst prof to prof microbiol & biochem, Univ Minn, 63-74; prof microbiol & head dept, 74-75, prof biochem & biophys & head dept, 75-79, prof biochem & head dept microbiol, cell biol, biochem & biophys, 79-84, PROF BIOCHEM, DEPT MOLECULAR & CELL BIOL, PA STATE UNIV, 84- *Concurrent Pos:* USPHS res career develop award, 62-72; mem adv panel metab biol prog, NSF, 68-71. *Mem:* AAAS; Am Soc Biol Chem; Biochem Soc; Am Soc Microbiol. *Res:* Bacterial metabolism; enzymology; regulation of growth; spore formation; bacterial sensing. *Mailing Add:* 108 Althouse Pa State Univ University Park PA 16802

BERNOCO, DOMENICO, b Cherasco, Italy, Apr 6, 35; m 72. IMMUNOLOGY, VETERINARY GENETICS. *Educ:* Univ Torino, Italy, DVM, 59; Ministry Pub Inst, Rome, Italy, liberadocenza, 71. *Prof Exp:* Asst prof, Observ Animal Genetics, Torino, Italy, 61-62 & 64-67; ast prof, Inst Med Genetics, Univ Torino, 62-64 & 69-70; mem BII, Basel Inst Immunol, Basel, Switz, 70-76; assoc res immunologist, Tissue Typing Lab, Dept Surg, Univ Calif, Los Angeles, 77-81, res fel, Dept Vet Microbiol, Davis, 67-68, assoc res immunologist, 79-81, ASSOC PROF, DEPT VET MED REPRODUCTION, UNIV CALIF, DAVIS, 79- *Concurrent Pos:* Coun-at-large, Am Asn Clin Histocompatibility Testing, 80-83; mem adv bd, J Animal Genetics, 86- *Mem:* Am Asn Immunologists; Int Soc Animal Genetics; Am Soc Histocompatibility & Immunogenetics. *Res:* Histocompatibility complex and blood groups; use of genetic markers for individual identification; pedigree analysis; matching for organ transplantation and identification of animals with superior genetic resistance to disease to improve quantity and quality of food animal products. *Mailing Add:* Univ Calif Davis CA 95616

BERNOFSKY, CARL, b Brooklyn, NY, Nov 22, 33; m 57; c 2. IMMUNOLOGY, TOXICOLOGY. *Educ:* Brooklyn Col, BS, 55; Univ Kans, PhD(biochem), 63. *Prof Exp:* Res asst biochem, Am Meat Inst Found, 56-58; res fel, Case Western Reserve Univ, 62-67; asst prof biochem, Mayo Grad Sch Med, Univ Minn, 67-73, assoc prof biochem, Mayo Med Sch, 73-75; vis assoc prof, 75-80, res assoc prof, 80-83, RES PROF BIOCHEM, SCH MED, TULANE UNIV, 83- *Concurrent Pos:* Consult, Mayo Clin & Found, 67-74; NSF grants, 71-73, 74-80, 83-85 & 90-93; NIH grants, 73-82 & 86-90. *Mem:* AAAS; Am Chem Soc; NY Acad Sci; Am Soc Biol Chem Molecular Biol; Oxygen Soc. *Res:* Pyridine nucleotide chemistry and metabolism; free radical mechanisms of inflammatory tissue damage; molecular toxicology. *Mailing Add:* Dept Biochem Tulane Univ Sch Med New Orleans LA 70112

BERNOR, RAYMOND LOUIS, b Los Angeles, Calif, Feb 15, 49. PALEOBIOLOGY. *Educ:* Univ Calif, Los Angeles, BA, 71, MA, 73, PhD(anthrop), 78. *Prof Exp:* Res assoc, Dept Earth Sci, Univ Calif, Riverside, 78-80, curator vertebrate paleontol, 79-81; res fel, 80-81, ASSOC PROF ANAT, COL MED, HOWARD UNIV, 81- *Concurrent Pos:* Lectr anthrop, Univ Calif, Los Angeles, 71-79, ed, Anthrop, 78-80; chief vertebrate paleontologist, Archaeol Resource Mgt Corp, Garden Grove, Calif, 78-80; prin proj dir, Paleontol & Multidisciplinary Studies, Environ Res Archaeologists Corp, Los Angeles, Calif, 80; NSF fel; Alexander von Humboldt fel. *Mem:* AAAS; Soc Vertebrate Paleontol; Am Asn Phys Anthropologists; Soc Syst Zool. *Res:* Evolution of Miocene Age mammalian zoogeographic provinces of Eurasia and Africa, and its pertinence to early hominid and pongid evolution and environments; complete mammal faunas and specialization in hipparionine horses and primates. *Mailing Add:* Dept Anat Col Med Howard Univ 520 West St NW Washington DC 20059

BERNS, DONALD SHELDON, b Bronx, NY, June 27, 34; m 56; c 3. PHYSICAL CHEMISTRY, BIOPHYSICS. *Educ:* Wilkes Col, BS, 55; Univ Pa, PhD(phys chem), 59. *Prof Exp:* Calif Res Corp grant chem, Yale Univ, 59-60 & Rockefeller Found grant, 60-61; resident res assoc chem, Argonne Nat Lab, 61-62; sr res scientist phys chem, Div Labs & Res, NY State Dept Health, Albany, 62-67, assoc res scientist, 67-71; from asst prof to assoc to assoc prof biochem, Albany Med Col, 62-76; dir phys chem lab, 68-74, prin res scientist phys chem, Wadsworth Ctr labs & res, 71-77, lab chief molecular biol, 82-83, lab chief biophys, 83-86, sci admin, New Scotland Ave Infectious Dis Labs, 83-86, DIR DIV CLIN SCI, NY STATE DEPT HEALTH, 86- *Concurrent Pos:* NSF grant, Div Labs & Res, NY State Dept Health, 62-75; adj prof chem, Rensselaer Polytech Inst, 71-; Environ Protection Agency spec sr fel & vis prof physiol & microbiol chem, Hadassah Med Sch, Hebrew Univ, Israel, 72-73; NIH grant, 75-; Lady Davis vis prof, Hebrew Univ, Israel, 85. *Mem:* Am Chem Soc; Biophys Soc; Am Soc Biol Chem; Am Soc Photobiol. *Res:* Energy transduction and transfer in model membrane systems; protein structure and function, particularly algal biliproteins, protein-lipid interaction and membrane assembly; physical chemical properties of biological macromolecules; monolayer studies of lipids and proteins; serology of HIV infection; staging of HIV infection. *Mailing Add:* 47 Hizkiyahu Hamelech Jerusalem Israel

BERNS, KENNETH, b Cleveland, Ohio, June 14, 38; m 64; c 2. MOLECULAR BIOLOGY. *Educ:* Johns Hopkins Univ, AB, 60, PhD(biol), 64, MD, 66. *Prof Exp:* Mem house staff, Johns Hopkins Univ Hosp, 66-67; staff assoc, Nat Inst Arthritis & Metab Dis, 67-68, staff assoc, Nat Inst Allergy & Infectious Dis, 68-70; from asst prof to assoc prof microbiol, 70-76, asst prof pediat, 70-76, dir, Yr One Prog, Sch Med, Johns Hopkins Univ, 73-76; prof & chmn dept immunol & med microbiol, Col Med, Univ Fla,

Gainesville, 76-84; R A REES PRITCHETT PROF & CHMN, DEPT MICROBIOL, MED COL, CORNELL UNIV, 84- *Concurrent Pos:* Shell Oil fel, 63-64; sr asst surgeon, USPHS, 66-67, surgeon, 67-70, sr surgeon, 70-; fel pediat, Johns Hopkins Univ, 66-67; Howard Hughes Med Investr, 70-75; fac res award, Am Cancer Soc, 75-80; consult, Vet Admin Hosp, Gainesville, Fla, 78-84; consult, Bd Army Sci & Technol, Nat Acad Sci-Nat Res Coun; chmn, Basic Sci Bd, 79-82, mem exec comt, Col Med, Univ Fla; mem exec comt, Grad Sch Med Sci, Med Col, Cornell Univ; mem, Recombinant DNA Adv Comt, NIH, 80-83, chmn, 83; mem, Genetic Biol Panel, NSF, 81-85; mem, Int Comt Taxon Viruses (Parvoviruses), 81-; Guggenheim hon fel, 82-83; Fogarty Sr Int fel, Weizmann Inst Sci, Israel, 82-83; sci adv, Showa Univ Res Inst Biomedicine in Fla, 83-85; prof pediat, Med Col, Cornell Univ, 84-; virol study sect, NIH, 85-; mem, Rev Panel Virol & Microbiol, Am Cancer Soc, 86-; mem exec bd, Nat Bd Med Examrs, 87-; chmn, Pub Sci Affil Bd, Am Soc Microbiol, 89-; chmn, Coun Acad Soc, Asn Am Med Col, 90-; pres, Asn Med Sch Microbiol & Immunol Chrs, 86. *Mem:* Fel Am Acad Microbiol; Am Soc Microbiol; Am Soc Virol (pres elect, 87, pres, 88); Am Soc Biol Chemists; Soc Gen Microbiol; Soc Pediat Res; Harvey Soc; NY Acad Sci (vpres, 90-); Asn Am Med Col; Sigma Xi. *Res:* DNA structure; structure and replication of animal virus chromosomes; latent infection. *Mailing Add:* Dept Microbiol Med Col Cornell Univ 1300 York Ave New York NY 10021

BERNS, MICHAEL W, b Burlington, Vt, Dec 1, 42; m 63; c 2. DEVELOPMENTAL BIOLOGY, CELL BIOLOGY. *Educ:* Cornell Univ, BS, 64, MS, 66, PhD(biol), 68. *Prof Exp:* Assoc dir laser biol, Pasadena Found Med Res Calif, 69-70; asst prof zool, Univ Mich, 70-72; assoc prof develop biol & cell biol, 72-75, PROF DEVELOP BIOL & CELL BIOL, UNIV CALIF, IRVINE, 75- CHMN DEPT, 75- Concurrent NIH GMS res grants grant laser microbeam effects on chromosomes & mitosis, 70, 72 & 80; NIH-Heart & Lung Inst grant, 71-80; co-prin investr laser microbeam studies on mitosis, NSF, 73-75; co-dir carcinogenesis training grant, NIH-Nat Cancer Inst, 75-80; NSF equip grant, 75; US Air Force grant effects of pollutants on cells, 77-79; Nat Acad Sci ad hoc comt mem, lasers in biomed res, 77; Int Union Against Cancer ICRETT award, US-USSR Joint Prog Estab, 77-80; chmn, Gordon Conf, lasers in med & biol, 78. *Mem:* AAAS; Tissue Cult Asn; Am Soc Cell Biol; Am Soc Photobiol; Soc Develop Biol. *Res:* Laser microbeam studies on chromosomes; nucleoli and mitochondria of tissue culture cells; studies on mitosis of cells; laser instrumentation for biomedical research; cellular and embryonic development. *Mailing Add:* Dept Develop Biol & Cell Biol Univ Calif Irvine CA 92717

BERNSEN, SIDNEY A, b Chicago, Ill, Apr 20, 28; m 51; c 3. NUCLEAR & MECHANICAL ENGINEERING. *Educ:* Purdue Univ, BS, 50, MS, 51, PhD(mech eng), 53. *Prof Exp:* Assoc mech engr, Argonne Nat Lab, 52-57; mem res staff, Gen Atomics Div, Gen Dynamics Corp, 57-59; mgr nuclear develop lab, Advan Technol Labs Div, Am Standard Corp, 59-62; chief engr, Sci Develop Dept, Bechtel Corp, 62-68, eng mgr & mgr qual assurance, Power & Indust Div, 68-73; mgr qual assurance, 73-75, exec asst, 75-76, chief nuclear engr, 76-79, mgr nuclear eng, 79-82, asst proj mgr, 82-84, MGR QUAL ASSURANCE, BECHTEL POWER CORP, 84- *Concurrent Pos:* Vchmn, Am Nat Standards Comt N-45, 73-75; chmn comt, Nuclear Qual Assurance, Am Soc Mech Engrs, 75-78, mem, Nuclear Codes & Standards Bd, 75-, Nuclear Qual Assurance, 79-, Exec Standards Coun, 86-; chmn, Nuclear Technol Adv Group, Am Nat Standards Inst, 81-; sr examr, Bd Examiners. *Honors & Awards:* Malcolm Baldrige Nat Qual Award, 88. *Mem:* Am Nuclear Soc; Am Soc Mech Engrs; Am Soc Qual Control. *Res:* Heat transfer; reactor power plant and reactor design; space environmental facilities; rocket engine support and test facility design. *Mailing Add:* 50 Beale St San Francisco CA 94105

BERNSTEIN, ABRAM BERNARD, b New York, NY, July 11, 35; div; c 2. ENVIRONMENTAL MANAGEMENT, NATURAL & TECHNOLOGICAL HAZARDS. *Educ:* City Col NY, BS, 56; Pa State Univ, MS, 59; Univ Wash, PhD(atmos sci), 73. *Prof Exp:* Res meteorologist, Nat Oceanic & Atmospheric Admin, 58-73; staff scientist, Nat Adv Comt Oceans & Atmosphere, 73-78; sr prog officer, Nat Res Coun, Nat Acad Sci, 78-87; vis assoc prof, Dept Technol & Soc, State Univ NY, Stony Brook, 87-90; ADMIN DIR, RESOURCE MGT PROG, DEPT ENVIRON STUDIES, ANTIOCH NEW ENG GRAD SCH, 90- *Mem:* Am Meteorol Soc; AAAS; Sigma Xi; Am Geophys Union; Soc Risk Anal. *Res:* Atmospheric and oceanic science; natural, technological and environmental hazards; application of new technology to buildings and cities; science, technology and public policy; interdisciplinary communication. *Mailing Add:* Dept Environ Studies Antioch New Eng Grad Sch Rexbury St Keene NH 03431

BERNSTEIN, ALAN, b Brooklyn, NY, Nov 21, 26; m 48; c 3. VIROLOGY. *Educ:* Philadelphia Col Pharm, BS, 51; Univ Pa, PhD(med microbiol), 54. *Prof Exp:* Bacteriologist, Philadelphia Childrens Hosp, 48-51, res asst, 51-53; bacteriologist, virus & rickettsia sect, Communicable Dis Ctr, USPHS, 53-58; bacteriologist, Wyeth Labs, Marietta, 58-68, managing dir, 68-87; RETIRED. *Concurrent Pos:* Consult, 87- *Res:* Production of bacterial and viral vaccines; control testing of biologicals; serology; epidemiology; production of parenteral pharmaceutical products; diagnostic virology. *Mailing Add:* 24 Bentley Summit Lancaster PA 17603

BERNSTEIN, ALAN D, b New York, NY, Jan 31, 38; wid. CARDIAC PACING, CARDIAC ELECTROPHYSIOLOGY. *Educ:* NJ Inst Technol, MS, 75, PhD(elec eng), 79. *Prof Exp:* Instr physics & dir gen physics labs, Rutgers Univ, Newark, 70-78, res specialist physics, 78-79; res assoc surg, 79-80, sr res assoc, 80-83, DIR TECH RES SURG & TECH DIR PACEMAKER CTR, NEWARK BETH ISRAEL MED CTR, 83- *Concurrent Pos:* Adj assoc prof surg, Univ Med & Dent NJ, 90-; chmn, Mode Code Comt, NAm Soc Pacing & Electrophysiol, 91- *Mem:* Am Col Cardiol; Biomed Eng Soc; Inst Elec & Electronics Engrs; Asn Advan Med Instrumentation. *Res:* Cardiac pacing and electrophysiology; cardiac-vibration analysis; biomedical signal processing; computer applications in medicine. *Mailing Add:* Newark Beth Israel Med Ctr 201 Lyons Ave Newark NJ 07112

BERNSTEIN, ALLEN RICHARD, b Chicago, Ill, Feb 14, 41; m 72; c 2. MATHEMATICS. *Educ:* Calif Inst Technol, BS, 62; Univ Calif, Los Angeles, MA, 64, PhD(math), 65. *Prof Exp:* Asst prof math, Univ Wis-Madison, 65-68; from assoc prof to prof math, Univ Md, College Park, 68-83; PRES, ACAD INFO SERV, INC, 83- *Mem:* Am Math Soc; Asn Symbolic Logic. *Res:* Mathematical logic and its applications; analysis. *Mailing Add:* 7809 Chestnut Ave Bowie MD 20715

BERNSTEIN, ALVIN STANLEY, b New York, NY, Nov 2, 29; m 68; c 1. PSYCHIATRY. *Educ:* NY Univ, BA, 50; Univ Buffalo, PhD(psychol), 58. *Prof Exp:* Ward psychologist, Vet Admin Hosp, Montrose, NY, 58-61, res psychologist, 61-66; ASSOC PROF PSYCHIAT, COL MED, STATE UNIV NY DOWNSTATE MED CTR, 66- , ASSOC PROF PROGS PSYCHIAT & BIOL PSYCHOL, GRAD SCH, 70- *Concurrent Pos:* Lectr psychol, Queens Col, City Univ NY, 60-65; res asst to asst prof psychiat, Sch Med, Cornell Univ, 65-66; spec res fel, State Univ NY, Downstate Med Ctr, NIMH, 66-68; vis prof, dept psychol, Univ York, Eng, 76; mem, Res Rev Comt Psychopath & Clin Biol, NIMH, 79-81; consult, Natural Sci & Rec Coun Can, 82; assoc ed, Psychophysiol, 87- *Mem:* Soc Psychophysiol Res; fel Int Orgn Psychophysiol; Am Psychol Asn; AAAS. *Res:* Psychophysiological studies of attentional processes in the normal population; specific dysfunctions in schizophrenia and in depression. *Mailing Add:* Dept Psychiat Box 32 State Univ NY Downstate Med Ctr 450 Clarkson Ave Brooklyn NY 11203

BERNSTEIN, ARON, b New York, NY, Apr 6, 31. NUCLEAR PHYSICS. *Educ:* Union Col, BA, 52; Univ Pa, PhD(physics), 57. *Prof Exp:* Res fel & asst prof, Princeton Univ, 57-61; PROF PHYSICS, MASS INST TECHNOL, 61- *Concurrent Pos:* Guggenheim Fel, 68. *Mem:* Fel Am Phys Soc. *Res:* Working on nuclear arms control; exploring the basic structure of the nuclei and nucleus using electro magnetic interaction. *Mailing Add:* Mass Inst Technol Bldg 26 419 Cambridge MA 02139

BERNSTEIN, BARBARA ELAINE, b Washington, DC, Sept 26, 48; m 72. MATHEMATICS, EDUCATIONAL PSYCHOLOGY. *Educ:* Univ Chicago, BA, 70; Univ Md, MEd, 71, PhD(human develop), 73. *Prof Exp:* Instr math & statist, Bowie State Col, 73-77; RES & WRITING, 77-; MGR, ACAD INFO SERV, INC, 81- *Res:* Analysis and empirical validation of effective methods for communicating mathematical concepts; psychiatric and physiological factors which influence learning; psychoanalytic interpretations of classroom interactions and other behaviors relevant to learning. *Mailing Add:* 12761 Midwood Lane Bowie MD 20715

BERNSTEIN, BARRY, b New York, NY, Nov 20, 30; m 54; c 2. APPLIED MATHEMATICS. *Educ:* City Col, BS, 51; Ind Univ, MA, 54, PhD(math), 56. *Prof Exp:* Mathematician, US Naval Res Lab, 51-53, 56-61 & Nat Bur Stand, 61-66; PROF MATH, ILL INST TECHNOL, 66-, ACTG DIR CTR APPL MATH, 70- *Concurrent Pos:* Lectr, Univ Md, 56-57, 64- & Catholic Univ, 58-64; consult, Nat Bur Stand, 68- *Mem:* Am Math Soc; Soc Rheology; Soc Indust & Appl Math. *Res:* Rheology; continuum mechanics; thermodynamics; mathematical analysis; gas dynamics; biomechanics. *Mailing Add:* Dept Math Ill Inst Technol Chicago IL 60616

BERNSTEIN, BRADLEY ALAN, b Madison, Wis, May 18, 51. FOOD SCIENCE, BIOCHEMISTRY. *Educ:* Univ Wis, BS, 73, MS, 74, PhD(food sci), 77. *Prof Exp:* Staff mem, 77-80, dir res & develop food chem, Amber Labs Div, Milbrew, Inc, 80-83; PRES, SCHULTZ BERNSTEIN & ASSOC INC, 83- *Honors & Awards:* Steinbock-Borden Award, Univ Wis & S-B Awards Comt, 73. *Mem:* Sigma Xi; Am Chem Soc; Inst Food Technologists; Am Dairy Sci Asn. *Res:* Investigation of the inhibition of cholesterol biosynthesis by orotic acid; isolation and characterization of cytochromes P-450 and P-420 in mammary tissues; development of methods for the utilization of dairy by-products via fermentation. *Mailing Add:* 270 Nob Hill E Colgate WI 53017

BERNSTEIN, BURTON, b Brooklyn, NY, Dec 11, 20; m 61; c 2. LASERS, OPTICAL PHYSICS. *Educ:* Brooklyn Col, BA, 42; Columbia Univ, MA, 49; Mass Inst Technol, PhD(physics), 54. *Prof Exp:* Res scientist res lab, Philco Corp, 54-57; staff scientist, United Nuclear Corp, 57-58; res physicist, Gen Precision Labs, 58-61; staff physicist, Loral Electronics Corp, 61-62; sr staff scientist, Repub Aviation Corp, 62-64; mgr crystal physics & laser develop lab, Airtron Div Litton Industs, 64-66; assoc prof physics, 66-69, PROF PHYSICS, STATE UNIV NY COL NEW PALTZ, 69- *Mem:* Am Phys Soc; Sigma Xi. *Res:* Semiconductor surface physics; solid state laser materials. *Mailing Add:* Physics Dept State Univ Col New Paltz NY 12561

BERNSTEIN, CAROL, b Paterson, NJ, Mar 20, 41; m 62; c 3. MOLECULAR BIOLOGY, GENETICS. *Educ:* Univ Chicago, BS, 61; Yale Univ, MS, 64; Univ Calif, Davis, PhD(genetics), 67. *Prof Exp:* NIH fel zool, Univ Calif, Davis, 67-68; res assoc microbiol, 68-77, adj asst prof, 77-81, RES ASSOC PROF, COL MED, UNIV ARIZ, 81- *Concurrent Pos:* Mem, panel grad fels, NSF, 84-86; mem, Nat Coun, Am Asn Univ Prof, 86-89; treas, Nat Assembly State Conferences, Am Asn Univ Prof, 86-89; mem, panel grad fels, NSF, 91-93. *Mem:* Genetics Soc Am; Am Soc Microbiol; Am Asn Univ Prof. *Res:* DNA repair; structure and replication of DNA of microorganisms; mechanisms of chemical mutagenesis; molecular basis of aging; the sexual cycle; etiology of colon cancer. *Mailing Add:* Dept Microbiol & Immunol Col Med Univ Ariz Tucson AZ 85724

BERNSTEIN, DAVID, otorhinolaryngology; deceased, see previous edition for last biography

BERNSTEIN, DAVID MAIER, b Hartford, Conn, Jan 27, 47. INHALATION TOXICOLOGY, AEROSOL PHYSICS. *Educ:* Yeshiva Univ, BA, 68; City Univ New York, MA, 72; NY Univ, PhD(environ health sci), 77. *Prof Exp:* Instr, Queens Col, City Univ New York, 70-72; asst res scientist, Inst Environ Med, NY Univ Med Ctr, 73-77; res assoc, dept med, Brookhaven Nat Lab, 77-81; sect head, Inhalation Toxicol, Battelle Geneva

Res Ctr, 81-87; DIR, RES & CONSULT CO, 87- *Concurrent Pos:* Instr, Sch Basic Health Sci, State Univ NY Stony Brook, 78-81. *Mem:* AAAS; Am Indust Hyg Asn; Am Assoc Aerosol Res; Soc Toxicol; Soc Aerosol Invest; Assoc Inhalation Toxicol; Swiss Assoc Hyg (secy). *Res:* Inhalation toxicology and pathology of fibers, dusts and vapors; the design, conduct and interpretation of chronic and sub-chronic inhalation toxicity studies; Aerosol physics; generation and inhalation exposure system design; data management and acquisition systems for toxicology labs. *Mailing Add:* Res & Consult Co AG 1 Rt de Troinex Carouge 1227 Switzerland

BERNSTEIN, DOROTHY LEWIS, mathematics; deceased, see previous edition for last biography

BERNSTEIN, ELAINE KATZ, b Baltimore, Md, May 14, 22; m 55; c 2. BIOCHEMISTRY, SCIENCE WRITING. *Educ:* Goucher Col, BA, 41; Oberlin Col, MA, 42. *Prof Exp:* Asst scientist, Metall Lab, Univ Chicago, 43-46; assoc scientist, Argonne Nat Lab, 46-58, consult clin chem, 58-60; sci info specialist, John Crerar Libr, Chicago, 63-70; writer & ed med & clin chem, 70-74; SCI WRITER & ED, 74- *Mem:* AAAS; Sigma Xi; Fedn Am Scientists; Scientists in Pub Interest; Radiation Res Soc. *Res:* Clinical chemistry; liver function; toxicology; radiation-induced biochemical changes; leukemia. *Mailing Add:* 21 Court of Island Pt Northbrook IL 60062

BERNSTEIN, ELLIOT R, b New York, NY, Apr 14, 41; m 65; c 2. CHEMICAL PHYSICS, SPECTROSCOPY. *Educ:* Princeton Univ, BA, 63; Calif Inst Technol, PhD(chem physics), 67. *Prof Exp:* Res assoc paramagnetic resonance, Univ Chicago, 67-69; asst prof chem, Princeton Univ, 69-75; assoc prof, 75-79, PROF CHEM, COLO STATE UNIV, 79- *Concurrent Pos:* Consult, Los Alamos Nat Lab, 75-, Philip Morris, 84-, Du Pont, 85- *Mem:* Am Chem Soc; Am Phys Soc. *Res:* Spectroscopy of inorganic and organic molecular crystals; exchange interactions, magnetic ordering and exciton structure in molecular solids; light scattering studies of phase transition in molecular solids and liquid crystals; spectroscopy of cryogenic liquid solutions and molecular clusters; structure of non rigid molecules; chemical reactions in clusters. *Mailing Add:* Dept Chem Colo State Univ Ft Collins CO 80523

BERNSTEIN, EMIL OSCAR, b New York, NY, Dec 16, 29; m 51; c 2. CELL BIOLOGY. *Educ:* Syracuse Univ, AB, 51, MS, 53; Univ Calif, Los Angeles, PhD(zool), 56. *Prof Exp:* Am Cancer Soc fel, Univ Calif, Los Angeles, 56-57; from instr to assoc prof zool, Univ Conn, 57-65; assoc prof, Univ Md, 65-68; sr res assoc, Gillette Res Inst, 68-73, prin res assoc, 73-80; MEM STAFF, DEPT BIOL, ALLEGHENY COL, 80- *Concurrent Pos:* NSF grant, 59-, fel biol inst, Carsberg Found, Copenhagen, 63-64. *Mem:* AAAS; Am Soc Cell Biol; Soc Gen Physiol; Electron Micros Soc Am; Sigma Xi. *Res:* Cell physiology; growth and division of cells; metabolic problems; problem of obligate autotrophy; scanning electron microscopy; skin morphology and physiology; ultrastructure; cell biology and cell physiology. *Mailing Add:* 23 Waterglen Circle Sacramento CA 95826

BERNSTEIN, EUGENE F, b New York, NY, Oct 9, 30; m 54; c 3. SURGERY. *Educ:* State Univ NY Downstate Med Ctr, MD, 54; Univ Minn, MS, 61, PhD(surg), 64; Am Bd Thoracic Surg, dipl, 68. *Prof Exp:* From instr to assoc prof surg, Univ Minn, 63-69; PROF SURG, UNIV CALIF, SAN DIEGO, 69- *Concurrent Pos:* Markle scholar acad med, 63-68; consult, Vet Admin Hosp, San Diego, Naval Hosp, San Diego & Medtronic, Inc, Minn. *Mem:* Fel Am Col Surg; Am Soc Artificial Internal Organs; Soc Vascular Surg; Am Surg Asn; Int Cardiovasc Soc. *Res:* Prolonged mechanical circulatory support; vascular surgery; revascularization of the lower extremity; noninvasive diagnostic techniques in vascular disease. *Mailing Add:* 10666 N Torrey Pines Rd La Jolla CA 92037

BERNSTEIN, EUGENE H, b New York, NY; m 52; c 3. BIOCHEMISTRY, VIROLOGY. *Educ:* US Merchant Marine Acad, BS, 47; Rutgers Univ, BS, 51, MS, 52, PhD(biochem, microbiol), 55. *Prof Exp:* USPHS fel, Sloan-Kettering Inst Cancer Res, New York, 55-56; biochemist, Colgate-Palmolive Co Dent Res, NJ, 56-58, sect head, 58-59; OWNER & DIR, UNIV LABS, INC, 59- *Concurrent Pos:* Vis investr, Inst Microbiol, Rutgers Univ, 59-68, vis prof, Waksman Inst Microbiol, 82- *Mem:* AAAS; Am Chem Soc; NY Acad Sci; Am Soc Microbiologists; Int Leukemia Soc. *Res:* Enzymology; virology, especially avian and mammalian tumor and leukemia viruses; toxicology; microbiology; nutrition; carcinogenesis; cancer chemotherapy. *Mailing Add:* Univ Labs Inc 810 N Second Ave Highland Park NJ 08904

BERNSTEIN, EUGENE MERLE, b Baltimore, Md, Feb 13, 31; m 60; c 2. ATOMIC & NUCLEAR PHYSICS. *Educ:* Duke Univ, BS, 53, MA, 54, PhD(physics), 56. *Prof Exp:* Instr & res assoc physics, Duke Univ, 56-57; instr & res assoc physics, Univ Wis, 57-59, lectr & res assoc, 60-61; NSF fel, Inst Theoret Physics, Copenhagen, 59-60; from asst prof to assoc prof physics, Univ Tex, 61-65; vis staff mem, Los Alamos Sci Lab, 65-67; prof physics, Univ Tex, 67-68; chairperson dept, 80-89, PROF PHYSICS, WESTERN MICH UNIV, 68- *Concurrent Pos:* Vis prof, Univ Ariz, 75-76; vis scientist, Lawrence Berkeley Lab, 86 & GSI-Darmstadt. *Mem:* Fel Am Phys Soc. *Res:* Atomic and low energy nuclear physics. *Mailing Add:* Dept Physics Western Mich Univ Kalamazoo MI 49008

BERNSTEIN, GERALD SANFORD, b Trenton, NJ, July 4, 28; m 52; c 3. REPRODUCTIVE BIOLOGY, OBSTETRICS & GYNECOLOGY. *Educ:* Temple Univ, BA, 50; Univ Mass, MA, 52; Univ Del, PhD(chem), 55; Univ Southern Calif, MD, 62. *Prof Exp:* Pop Coun fel, 55-56, USPHS fel, 56-58; res fel pharmacol, 59-62, intern, Los Angeles County-Univ Southern Calif Med Ctr, 62-63, resident physician obstet & gynec, 63-67, from instr to assoc prof, 67-82, PROF OBSTET & GYNEC, UNIV SOUTHERN CALIF, 82- *Concurrent Pos:* Dir family planning serv, Los Angeles County-Univ Southern Calif Med Ctr, 68-; chmn bd dir, Los Angeles Regional Family Planning Coun, 68-71; med dir, Los Angeles Affil Planned Parenthood, 68-80. *Mem:* AAAS; NY Acad Sci; Am Fertil Soc; Am Soc Study Reproduction; Am Soc Andrology. *Res:* Andrology; family planning; physiology of sperm. *Mailing Add:* Woman's Hosp Los Angeles County Univ Southern Calif Med Ctr 1200 N State St Los Angeles CA 90033

BERNSTEIN, HARRIS, b Brooklyn, NY, Dec 12, 34; m 62; c 3. GENETICS. *Educ:* Purdue Univ, BS, 56; Calif Inst Technol, PhD(genetics), 61. *Prof Exp:* USPHS res fel genetics, Yale Univ, 61-63; asst prof, Univ Calif, Davis, 63-68; assoc prof, 68-74, PROF MICROBIOL, COL MED, UNIV ARIZ, 74- *Concurrent Pos:* NSF res grant, 64-79; NIH grant, 79-87. *Mem:* Genetics Soc Am; Am Soc Microbiol; Am Soc Biol Chemists; Fedn Am Scientists; fel Am Acad Microbiol. *Res:* Bacteriophage genetics; mechanisms of recombination; evolution of sexual reproduction; DNA replication; DNA repair and mutation; DNA damage as the basis of aging. *Mailing Add:* Dept Microbiol & Immunol Univ Ariz Med Col Tucson AZ 85724

BERNSTEIN, HERBERT J, b Washington, DC, Apr 21, 43; m 71; c 2. RECONSTRUCTIVE KNOWLEDGE, SCIENCE POLICY. *Educ:* Columbia Univ, BS, 63; Univ Calif, MS, 65, PhD(theoret physics), 67. *Prof Exp:* Mem, Inst Advan Studies, Princeton, 67-69; PROF PHYSICS, HAMPSHIRE COL, 71-, MASS INST TECHNOL, 80-; DIR, INST SCI & INTERDISCIPLINARY STUDIES, AMHERST, 89- *Concurrent Pos:* Tech Dir, Econ Develop Agency; consult World Bank, Dept of Energy; Pres Sci Adv, AAAS. *Honors & Awards:* Proctor Prize, Sigma Xi, 87. *Mem:* Am Phys Soc. *Res:* Matter wave interferometry and foundations of physical epistemology; particle interferometry and foundations of physical epistemology; science policy, especially energy and international science; reconstructive science for social benefit by including its moral and cultural concommittants, assumptions, and effects. *Mailing Add:* 266 Shays St Amherst MA 01002

BERNSTEIN, HERBERT JACOB, b Brooklyn, NY, Sept 26, 44; m 68; c 2. COMPUTING. *Educ:* NY Univ, BA, 64, MS, 65, PhD(math), 68. *Prof Exp:* Adj assoc res prof math, NY Univ, 68-69, assoc res scientist, AEC Comput & Appl Math Ctr, Courant Inst Math Sci, 68-70; sci programmer analyst, appl math dept, Brookhaven Nat Lab, 70-71, sr comput sci analyst, 71-74, assoc scientist, chem dept, 74-78, scientist, 78-83, chmn, Comput Users Orgn, 81-82; vis mem, NY Univ, 82-83, sr res scientist, Courant Inst Math Sci, 83-91, adminr comput projs & technol eval & assessment, 85-89, dir adminr comput, 89-90; CONSULT, 90- *Concurrent Pos:* Comput consult, Bernstein & Sons, 70 -; adj assoc prof, comput sci dept, Courant Inst Math Sci, NY Univ, 82-87, adj prof, 87-91; res collabr, dept appl sci & dept chem, Brookhaven Nat Lab, 83 -; mem bd trustees, Consortium Sci Comput, 84-91; mem bd dirs, NY State Educ & Res Network, Inc, 85-89. *Mem:* Am Math Soc; Am Crystallog Asn; Soc Indust & Appl Math; Math Asn Am. *Res:* Applied computer science; high speed parallel processing; robotics; data communications; numerical linear algebra; evaluation of computer related technology. *Mailing Add:* PO Box V Five Brewster Lane Bellport NY 11713

BERNSTEIN, I LEONARD, b Jersey City, NJ, Feb 17, 24; m 48; c 4. CLINICAL IMMUNOLOGY ALLERGY, PULMONARY DISEASES. *Educ:* Univ Cincinnati, MD, 49; Am Bd Internal Med, dipl, 58; Am Bd Allergy, dipl, 77; Am Bd Diag Immunol, dipl, 89. *Prof Exp:* Resident, Jewish Hosp, Cincinnati, 50-51; med officer, USAF, 51-53; resident, Bellevue Hosp, 53-55; fel, Northwestern Univ, Chicago, 55-56; intern, 49-50, CLIN PROF MED & ATTEND PHYSICIAN, CINCINNATI GEN HOSP, 56- *Concurrent Pos:* Consult, VA Hosp, Cincinnati, 64-; chmn educ coun & past pres, Am Acad Allergy, 78-; chmn, Antihistamine Panel Drug Interaction Proj, Am Pharmaceut Asn, 75-; co-dir, Allergy Res Lab & Allergy Training Prog, Univ Cincinnati Med Ctr, 58-; dir, Am Bd Allery & Immunol, 84-90; gov, Am Bd Internal Med, 86-89. *Mem:* Fel Am Acad Allergy; fel Am Col Physicians; Am Asn Immunologists; Am Thoracic Soc; AAAS; Cent Soc Clin Res. *Res:* Immediate hypersensitivity; regulation of the IgE type of immune response; industrial allergens; platelet function in allergic individuals; clinical investigation of various new drugs and procedures; reproductive immunology. *Mailing Add:* 8464 Winton Rd Cincinnati OH 45231

BERNSTEIN, I(RVING) MELVIN, b New York, NY, Oct 14, 38; m 64; c 1. METALLURGY, MATERIALS SCIENCE. *Educ:* Columbia Univ, BS, 60, MS, 62, PhD(metall), 65. *Prof Exp:* Res asst metall, Henry Krumb Sch Mines, Columbia Univ, 60-65; Cent Elec Generating Bd res fel, Berkeley Nuclear Lab, Eng, 66-67; scientist, E C Bain Fundamental Res Lab, US Steel Corp, 67-72; from asst prof to prof metall eng & mat sci, Carnegie-Mellon Univ, 72-87, assoc dean eng, 79-82, head dept, 82-87; provost, 87-90, CHANCELLOR & SR VPRES, ILL INST TECHNOL, 90- *Concurrent Pos:* Chief consult, Mat Consults & Labs, 72-; liaison scientist, Off Naval Res, London, 77-78; assoc ed, Metall Trans. *Honors & Awards:* Outstanding Young Metallurgist Award, Am Soc Metals, 74; C S Barrett Medal, Am Soc Metals, 83; Jules Garnier Prize, 86. *Mem:* AAAS; Am Soc Eng Educ; Am Inst Mining, Metall & Petrol Engrs; fel Am Soc Metals. *Res:* Microstructure-mechanical properties relationships, with emphasis on environmental effects; structure of interfaces and grain boundaries; engineering education and administration. *Mailing Add:* Chancellor Ill Inst Technol Chicago IL 15213

BERNSTEIN, IRA BORAH, b Bronx, NY, Nov 8, 24; m 55; c 2. THEORETICAL PHYSICS. *Educ:* City Col NY, BChE, 44; NY Univ, PhD(physics), 50. *Prof Exp:* Res physicist, Westinghouse Res Labs, 50-54; sr res physicist, Plasma Physics Lab, Princeton Univ, 54-64; PROF APPL PHYSICS, YALE UNIV, 64- *Concurrent Pos:* Fulbright scholar, 62-63; Guggenheim fel, 69. *Honors & Awards:* Maxwell Prize, Am Phys Soc. *Mem:* Nat Acad Sci; fel Am Phys Soc. *Res:* Theoretical plasma physics. *Mailing Add:* Dept Appl Physics Yale Univ New Haven CT 06520-2159

BERNSTEIN, IRWIN S, b New York, NY, Mar 9, 34; m 65. MATHEMATICS. *Educ:* City Col New York, BS, 55; Mass Inst Technol, SM, 56, PhD(math), 59. *Prof Exp:* Sr staff scientist, Avco Res & Advan Develop, 59-63 & Aerospace Res Ctr, Gen Precision, Inc, 63-68; asst prof math, 68-74, ASSOC PROF MATH, CITY COL NEW YORK, 74- *Mem:* Am Math Soc. *Res:* Uniqueness of solutions of elliptic partial differential equations; numerical solutions of partial differential equations; analytical celestial mechanics. *Mailing Add:* Dept Math City Col New York 185 West End Ave New York NY 10023

BERNSTEIN, IRWIN SAMUEL, b Brooklyn, NY, July 11, 33; div; c 4. PRIMATOLOGY. *Educ:* Cornell Univ, BA, 54; Univ Chicago, MA, 55, PhD(psychol), 59. *Prof Exp:* Res fel & res assoc psychobiol, 60-67, SOCIOBIOLOGIST & RES PROF, YERKES REGIONAL PRIMATE RES CTR, 67-; PROF PSYCHOL & ZOOL, UNIV GA, 68- *Concurrent Pos:* NSF, NIH, NIMH, Walter Reed Army Inst Res & Wenner Gren Found Anthrop Res grants & contracts. *Mem:* Am Soc Primatologists (pres, 80-82); Animal Behav Soc; Int Primatol Soc. *Res:* Primate social behavior; principles of group organization; communication and comparative patterns of activities in both field and captive group studies; control of aggression; endocrine correlates of sex, stress and aggression; adolescent socialization; dominance. *Mailing Add:* Dept Psychol Univ Ga Athens GA 30602

BERNSTEIN, ISADORE A, b Clarksburg, WVa, Dec 23, 19; m 42; c 2. BIOCHEMISTRY, TOXICOLOGY. *Educ:* Johns Hopkins Univ, AB, 41; Western Reserve Univ, PhD(biochem), 52. *Prof Exp:* Res assoc biochem, Western Reserve Univ, 51-52, sr instr, 52-53; res assoc, Inst Indust Health, Univ Mich, 53-56 & 59-70, from instr to assoc prof biol chem, 54-61, from assoc prof to prof indust health, 61-67, prof biol chem, 68-90, prof environ & indust health, 70-90, EMER PROF BIOL CHEM & ENVIRON & INDUST HEALTH, INST ENVIRON & INDUST HEALTH, UNIV MICH, ANN ARBOR, 90- *Concurrent Pos:* USPHS spec fel & vis prof, Inst Protein Res, Osaka Univ, 63-64; WHO travel fel, Western Europe, 72 & China, 86; consult, Vet Admin Hosp, Allen Park, Mich, 61-70 & 72-; vis prof, Rockefeller Univ, 77-78; vis scientist, Hebrew Univ, 78; assoc dir res, Inst Environ & Indust Health, Univ Mich, Ann Arbor, 78-; dir, toxicol prog, Univ Mich, Ann Arbor, 83-87; vis scientist, NIH, 88; hon prof, Kunming Med Col, 88- *Honors & Awards:* Taub Int Mem Award Res Psoriasis, Baylor Col Med, 59; Stephan Rothman Mem Award, Soc Invest Dermatol, 81. *Mem:* Fel AAAS; Am Soc Biochem & Molecular Biol; Am Soc Microbiol; Am Soc Cell Biol; hon Soc Invest Dermat (vpres, 73-74); Am Chem Soc; Soc Toxicol; Am Col Toxicol. *Res:* Cutaneous metabolism, keratinization, neoplasia and differentiation; biochemical mechanisms of accommodation. *Mailing Add:* Sch Pub Health Rm 1528 Univ Mich Ann Arbor MI 48109-2029

BERNSTEIN, JAY, b New York, NY, May 14, 27; m 57; c 2. PEDIATRIC PATHOLOGY, RENAL PATHOLOGY. *Educ:* Columbia Univ, BA, 48; State Univ NY, MD, 52. *Prof Exp:* Instr path, Sch Med, Wayne State Univ, 58-60, asst prof pediat, 60-62; from asst prof to assoc prof path, Albert Einstein Col Med, 62-69; dir anat path, William Beaumont Hosp, 69-90, DIR, WILLIAM BEAUMONT HOSP RES INST, 84-, ASSOC MED DIR RES, 90- *Concurrent Pos:* Attend pathologist, Children's Hosp Mich, Detroit, 57-62; sr res assoc, Child Res Ctr, Wayne State Univ, 60-62; contrib ed path, J Pediat, 68-84; vis assoc prof path, Albert Einstein Col Med, 69-74, vis prof, 74-; partic pathologist, Int Study Kidney Dis in Childhood, 68-86; co-ed, Perspective in Pediat Path, 78-; mem sci adv bd, Mich Kidney Dis Found, 72-, chmn, 86-88; consult pathologist, Children's Hosp of Mich, 75-; mem sci adv bd, Nat Kidney Found, 76-82; clin prof path, Sch Med, Wayne State Univ, 77-; clin prof health sci, Ctr Health Sci, Oakland Univ, Rochester, Mich, 77-; mem, Comt Classification & Nomenclature Renal Diseases, WHO, 75-77 & 81-; chmn, Renal Path Club, 90-92. *Honors & Awards:* Farber Lectr, 82. *Mem:* Am Asn Pathologists; Int Acad Path; Am Soc Nephrology; Am Pediat Soc; Soc Pediat Path (pres, 69); Am Soc Clin Pathologists. *Res:* Congenital renal abnormalities; renal disease-pathogenesis and ultrastructure; publications on pediatric renal pathology. *Mailing Add:* William Beaumont Hosp Res Inst 3601 W 13 Mile Rd Royal Oak MI 48073-6769

BERNSTEIN, JERALD JACK, b Brooklyn, NY, Mar 30, 34; m 57, 86; c 2. NEUROANATOMY, NEUROSCIENCE. *Educ:* Hunter Col, BA, 55; Univ Mich, MS, 57, PhD, 59. *Prof Exp:* Asst fisheries, Univ Mich, 56-58, asst zool, Mus Zool, 58-59, asst psychol, 58-59; assoc prof biol, Clarion State Col, 59-60; res biologist, Lab Neuroanat Sci, NIH, 60-65; asst prof anat, Ctr Neurobiol Sci, Col Med, Univ Fla, 65-73, prof neurosci & ophthal, Ctr Neurobiol Sci & Ctr Human Prosthetic Res, 69-81; RES PROF NEUROSURG & PHYSIOL, COL MED, GEORGE WASHINGTON UNIV, 80-; CHIEF & CAREER RES SCIENTIST, LAB CENT NERVOUS SYST INJURY & REGENERATION, VET AFFAIRS MED CTR, 80- *Concurrent Pos:* Res prof anat, Hadassah Med Sch, Jerusalem, Israel, 74 - *Mem:* AAAS; Soc Neurosci; Am Asn Pathologists; Am Asn Anat; Am Physiol Soc. *Res:* Morphological and neurochemical basis of spinal cord regeneration; neuro-oncology. *Mailing Add:* Vet Admin Med Ctr 151Q 50 Irving St NW Washington DC 20422

BERNSTEIN, JEREMY, b Rochester, NY, Dec 31, 29. THEORETICAL PHYSICS. *Educ:* Harvard Univ, BA, 51, MA, 53, PhD(physics), 55. *Prof Exp:* Res assoc physics, Cyclotron Lab, Harvard Univ, 55-57; mem, Inst Advan Study, 57-59; assoc, Brookhaven Nat Lab, 60-62; assoc prof, NY Univ, 62-67; PROF PHYSICS, STEVENS INST TECHNOL, 67- *Concurrent Pos:* NSF fel, 59-61; Ferris prof, Princeton Univ, 80-82; Rabi vis prof, Columbia Univ, 83-84. *Honors & Awards:* Westinghouse-AAAS Writing Prize, 64; Am Inst Physics-US Steel Found Sci Writing Prize, 70; Brandeis Award, 79; Brittanica Award, 87. *Mem:* Fel Am Phys Soc; Franklin Fel Royal Soc Arts. *Res:* Elementary particles weak interactions, and cosmology. *Mailing Add:* Dept Physics Stevens Inst Technol Castle Pt Sta Hoboken NJ 07030

BERNSTEIN, JOEL EDWARD, b Chicago, Ill, Apr 8, 43; m 64; c 3. DERMATOPHARMACOLOGY, NEUROPHARMACOLOGY. *Educ:* Carleton Col, BA, 64; Univ Chicago, MD, 69. *Prof Exp:* Res medicine, Univ Chicago, 70-73, clin pharm fel, 76-78, sr med fel, 77-79, asst prof, 79-82; vpres med affiliate, Schering-Plough Corp, 73-75; assoc dir clin res , Abbott Lab, 75-76; assoc prof, Northwestern Univ, 82-83; PRES, GENDERM CORP, 83- *Concurrent Pos:* Clin instr, Univ Tenn, 74-75. *Mem:* Am Acad Dermat; Am Soc Clin Pharmacol & Ther; Soc Investigative Dermat. *Res:* Dermato-pharmacology with a special interest in the role of neuropeptides and the skin; published over 100 research papers and textbook chapters. *Mailing Add:* Unit Ten 425 Huehl Rd Northbrook IL 60062

BERNSTEIN, JOSEPH N, b Moscow, USSR, Apr 18, 45; US citizen; m 71, 89; c 2. REPRESENTATION THEORY, DIFFERENTIAL EQUATIONS. *Educ:* Moscow State Univ, MA, 68, PhD(math), 72. *Prof Exp:* Jr researcher, Lab Math Methods in Biol, Moscow State Univ, 71-78; vis prof, dept math, Univ Md, College Park, 81-83; DEPT MATH, HARVARD UNIV, 83- *Mem:* Am Math Soc. *Res:* Representations of lie groups (mostly reductive) over finite p-adic and real fields; representations of lie algebras; modules over differential operators; applications of the theory of categories. *Mailing Add:* Harvard Univ Cambridge MA 02138

BERNSTEIN, LAWRENCE, b New York, NY, Jan 20, 40; m 75; c 3. SOFTWARE PROJECT MANAGEMENT, SOFTWARE TESTING. *Educ:* Rensselaer Polytech Inst, BEE, 61; NY Univ, MEE, 63. *Prof Exp:* Asst vpres res & develop, Bellcore, 84-85; dir res & develop, AT&T, Bell Labs, 79-84 & 85-86, exec dir systs anal, 86-89, vpres prod mgt, Computer Systs, 89-90, VPRES PROD MGT, AT&T, NETWORK SYSTS, 90- *Mem:* Fel Inst Elec & Electronics Engrs; Asn Comput Mach. *Res:* Software management including design development and deployment with a special emphasis of software for telecommunication management. *Mailing Add:* AT&T Bell Labs 480 Red Hill Rd Middletown NJ 07748

BERNSTEIN, LESLIE, b Poland, Apr 26, 24; US citizen; m 57. OTOLARYNGOLOGY, MAXILLOFACIAL SURGERY. *Educ:* Univ Witwatersrand, BDS, 47, MB, BCh, 54; Royal Col Physicians & Surgeons, Eng, dipl laryngol & otol, 59; Am Bd Otolaryngol, dipl, 63. *Prof Exp:* From asst prof to assoc prof otolaryngol, Col Med, Univ Iowa, 60-69, chmn sect maxillofacial-plastic surg, 69, prof otolaryngol & maxillofacial surg, 69-72; prof otorhinolaryngol & chmn dept, Med Sch, Univ Calif, Davis, 72-86. *Honors & Awards:* Harris P Mosher Award, 68; Award of Merit, Am Acad Ophthal & Otolaryngol, 70; Ira Tresley Res Award, 72. *Mem:* Am Broncho-Esophagol Asn; fel Am Acad Ophthal & Otolaryngol; fel Am Acad Facial Plastic & Reconstruct Surg; fel Am Laryngol, Rhinol & Otol Soc; Int Asn Maxillofacial Surg (vpres). *Res:* Ear; Larynx; broncho-esophagology; cleft palate. *Mailing Add:* 77 Scripts Dr Suite 105 Sacramento CA 95825

BERNSTEIN, LESLIE, b Los Angeles, Calif, Oct 9, 39; m 58; c 3. EPIDEMIOLOGY OF HORMONE DEPENDENT TUMORS. *Educ:* Univ Calif, Los Angeles, BA, 64; Univ Southern Calif, MS, 78, PhD(biometry), 81. *Prof Exp:* Res assoc biometry & epidemiol, 81-82, res asst prof, 82-85, asst prof 85-88, ASSOC PROF BIOSTATIST & EPIDEMIOL, UNIV SOUTHERN CALIF, 88- *Concurrent Pos:* Consult, div geriatric med, Univ Calif Los Angeles, 81-84, Res Triangle Inst, 85-86, Nat Inst Child Abuse & Human Develop, 86, Lawrence Berkely Labs, 80- & Nat Inst Justice, 88; ad hoc reviewer, NIH, 84; prin investr epidemiol studies, Nat Cancer Inst, 85- *Mem:* Am Col Epidemiol; Soc Epidemiol Res; Am Statist Asn; Biometrics Soc. *Res:* Evaluation of risk factors for hormone dependent tumors (breast, testis, prostate & thyroid cancer); evaluation of risk factors for multiple primary cancers, particularly the potential of cancer treatment induced second primary tumors; the development of methods for and analysis of animal carcinogenesis bioassays to generate data for human risk evaluation. *Mailing Add:* Dept Prev Med Pub Health Univ Southern Calif Sch Med 2025 Zonal Ave Los Angeles CA 90033

BERNSTEIN, LIONEL M, b Chicago, Ill, Sept 10, 23; m 52; c 3. INFORMATION SCIENCE, GASTROENTEROLOGY. *Educ:* Univ Ill, BS, 44, MD, 45, MS, 51, PhD(physiol), 54; Am Bd Internal Med, cert, 54. *Prof Exp:* Res assoc med & clin sci, Univ Ill, 52-53, instr med & physiol, 53-54; chief metab res div, Med Nutrit Lab, Fitzsimons Army Hosp, Denver, Colo, 54-55; physician sr grade, Vet Admin Hosp, Sepulveda, Calif, 55-56; chief gastroenterol sect, Vet Admin Hosp, Hines, Ill, 56-57, assoc chief staff, 57-62; chief med serv, Vet Admin West Side Hosp, Chicago, 62-67; dir res serv, Vet Admin Cent Off, 67-70; assoc dir extramural progs, Nat Inst Arthritis & Metab Dis, 70-73; dir off prog opers, Dept HEW, 73-74, spec asst, Off Asst Secy Health, 75-77; asst dep dir res & educ, Lister Hill Nat Ctr Biomed Commun, Nat Libr Med, 77-78, dir, 78-83; prof, Health Professions Educ, 85-88, actg head, Dept Med Educ, 88-90, PROF MED EDUC, COL MED, UNIV ILL, 88- *Concurrent Pos:* Vet admin mem, Gen Med Study Sect, NIH, 62-67; clin prof med, George Washington Univ, 82-; pres, Knowledge Systs, Inc, 83- *Mem:* Inst Med-Nat Acad Sci; AMA; Am Gastroenterol Asn; Am Fedn Clin Res; fel Am Col Physicians; AAAS. *Res:* Renal physiology of clinical disease states; gastroenterology; author or co-author of over 70 publications. *Mailing Add:* Ctr Educ Develop Univ Ill 808 S Wood St Chicago IL 60612

BERNSTEIN, MARVIN HARRY, b Los Angeles, Calif, Dec 30, 43; m 66; c 2. COMPARATIVE PHYSIOLOGY. *Educ:* Univ Calif, Los Angeles, BA, 65, MA, 66, PhD(zool), 70. *Prof Exp:* NIH trainee neurophysiol, Univ Calif, Los Angeles, 70-71; from temp instr to temp asst prof zool, Duke Univ, 71-73; from asst prof to assoc prof, 73-82, PROF BIOL, NMEX STATE UNIV, 82-, HEAD, 90- *Concurrent Pos:* Fulbright-Hayes Int Travel Award, Israel, 80. *Mem:* Am Physiol Soc; Am Soc Zoologists; Sigma Xi; fel AAAS; NY Acad Sci. *Res:* Respiration, gas transport, cardiovascular function and temperature compensation in vertebrates during rest and locomotion, growth and adulthood, at sea level and high altitude. *Mailing Add:* Dept Biol NMex State Univ Las Cruces NM 88003

BERNSTEIN, MAURICE HARRY, b St Louis, Mo, Apr 3, 23; m 67; c 1. ANATOMY. *Educ:* Wash Univ, AB, 47, PhD(zool), 50. *Prof Exp:* Am Cancer Soc-Nat Res Coun fels, Univ Mo-Columbia, 50-51 & Univ Calif, Berkeley, 51-52; jr staff mem genetics, Carnegie Inst Wash, 52-54; asst res biologist, Virus Lab, Univ Calif, Berkeley, 54-56; res physiologist, Res Inst, Mt Sinai Hosp, Los Angeles, Calif, 56-57; asst res anatomist, Sch Med, Univ Calif, Los Angeles, 57-58; from asst prof to prof, 58-88, EMER PROF ANAT, SCH MED, WAYNE STATE UNIV, 88- *Concurrent Pos:* Lalor Found fel, Marine Biol Lab, Woods Hole, Mass, 59; consult, Vet Admin, 65-70 & NIH, 67; vis prof anat & reprod biol, Univ Hawaii, 71-72; consult, cell biol; fel, AAAS, 78. *Mem:* Fel AAAS; Am Asn Anat; Am Soc Cell Biol; Asn Res Vision & Ophthal; Soc Study Reprod. *Res:* Role of retinal pigment epithelium organelles in visual function; lysosomes and receptor-mediated endocytosis in photoreceptor disk shedding and phagocytosis; regulatory cell interactions, lysosomes, and receptor-mediated endocytosis in testis. *Mailing Add:* Dept Anat & Cell Biol Wayne State Univ Sch Med Detroit MI 48201

BERNSTEIN, RALPH, b Zweibrucken, Germany, Feb 20, 33; US citizen; m 59; c 3. DIGITAL IMAGE PROCESSING, COMPUTER SCIENCE. *Educ:* Univ Conn, BSEE, 56; Syracuse Univ, MSEE, 60. *Prof Exp:* Engr & scientist, 56-60, assoc engr, 60-62, staff engr, 62-68, sr engr & mgr, Syst Design Dept, 68-71, sr engr, scientist & mgr, Adv Image Processing Anal Develop Dept, NY, 71-79, staff mem, 79-85, SR TECH STAFF MEM & MGR, IMAGE SCI & APPLNS DEPT, PALO ALTO SCI CTR, INT BUS MACH CORP, CALIF, 85- *Concurrent Pos:* Mem space sci bd, Nat Res Coun-Nat Acad Sci, 77-82, chmn, Comt Data Mgt & Computation, 78-81, mem space appln bd, 82-; consult, NASA Space Appln Adv Comt, 82-85; adj prof, Univ Calif, San Francisco, 83- *Honors & Awards:* Except Sci Achievement Medal, NASA, 74; Cert of Appreciation-Outstanding Serv, Nat Res Coun, 87. *Mem:* Fel Inst Elec & Electronics Engrs; Am Soc Photogram. *Res:* Digital image science and research studies for Earth observation sensor data and medical image science; image system research and development; earth resources technology Landsat satellite studies; navigation systems and positioning devices; oceanographic and geophysical computer system development; automatic control systems analysis and development. *Mailing Add:* IBM Sci Ctr 1530 Page Mill Rd Palo Alto CA 94304

BERNSTEIN, RICHARD BARRY, physical chemistry; deceased, see previous edition for last biography

BERNSTEIN, ROBERT LEE, b Rochester, NY, Nov 6, 44; m 68; c 2. PHYSICAL OCEANOGRAPHY, REMOTE SENSING. *Educ:* Calif Inst Technol, BS, 66; Columbia Univ, MS, 69, PhD(oceanog), 71. *Prof Exp:* ASST RES OCEANOGR, SCRIPPS INST OCEANOG, UNIV CALIF, SAN DIEGO, 72- *Concurrent Pos:* Mem, NPac Exp Exec Comt, NSF, 75-; consult, World Meteorol Orgn, 76-77; mem, Sea Satellite Altimeter Exp Team, NASA, 77- *Mem:* Am Geophys Union; AAAS. *Res:* Ocean circulation; mesoscale eddy dynamics; large scale air-sea interaction; methods of satellite oceanography. *Mailing Add:* 5360 Bothe Ave San Diego CA 92122

BERNSTEIN, ROBERT STEVEN, b New York, NY, Aug 11, 43; m; c 2. EPIDEMIOLOGY. *Educ:* Univ Pa, AB, 65; Pa State Univ, PhD(biochem), 71; Univ Conn, MD, 77; Johns Hopkins Univ, MPH, 79. *Prof Exp:* Fel neuropharmacol, Sch Med, Yale Univ, 71-73; intern med & surg, USPHS Hosp, San Francisco, 77-78; epidemic intel serv officer, 79-81, MED EPIDEMIOLOGIST, CTR ENVIRON HEALTH, CTR DIS CONTROL, 81- *Concurrent Pos:* Adj asst prof, Sch Med, WVa Univ, 79-81; vis prof, Dept Community Med, Emory Univ, 87-91. *Honors & Awards:* Presidential Citation, 80. *Mem:* AAAS; NY Acad Sci; Am Pub Health Asn; Am Thoracic Soc; Soc Epidemiol Res. *Res:* Epidemiology of acute and chronic diseases in environmental and occupational health; basic science research in enzyme structure and function and in neurochemistry. *Mailing Add:* Ctr Dis Control Int Health Prog Office FSD F-03 Atlanta GA 30333

BERNSTEIN, SANFORD IRWIN, b Brooklyn, NY, June 10, 53; m 83. DEVELOPMENTAL GENETICS. *Educ:* State Univ, NY, Stony Brook, BS, 74; Wesleyan Univ, PhD(biol), 79. *Prof Exp:* Fel molecular biol, Univ Va, 79-82; from asst prof to assoc prof, 83-88, PROF BIOL, SAN DIEGO STATE UNIV, 88-, ASSOC DIR MOLECULAR BIOL INST, 87- *Concurrent Pos:* Co-dir, Recombinant DNA Cert Prog, 83-; estab investr, Am Heart Asn, 89-94. *Mem:* AAAS; Genetics & Soc Am; Sigma Xi; Am Soc Cell Biol; Am Soc Biochem Molecular Biol; Soc Develop Biol. *Res:* Molecular structure of Drosophila muscle genes; analysis of genetic defects in muscle genes; mechanism of alternative RNA splicing. *Mailing Add:* Dept Biol Molecular Biol Inst San Diego State Univ San Diego CA 92182-0057

BERNSTEIN, SELDON EDWIN, b Bangor, Maine, Jan 3, 26; m 50, 59; c 4. MEDICAL GENETICS, HEMATOLOGY. *Educ:* Univ Maine, BA, 49, MA, 52; Brown Univ, PhD(biol), 56. *Prof Exp:* Asst, Univ Maine, 49-52; asst, Brown Univ, 52-54; assoc staff scientist, 56-58, staff scientist, 58-67, asst dir, 64-72, sr staff scientist, 67-87, EMER SR STAFF SCIENTIST, JACKSON LAB, 87- *Concurrent Pos:* Mem bd dirs, Col of the Atlantic, 69-79, chmn, 72-79; lectr zool, Univ Maine, 71-86. *Mem:* Int Soc Exp Hemat; Am Genetic Asn; Am Soc Hemat; NY Acad Sci. *Res:* Normal and pathological gene-environment interactions in the development and regulation of mammalian red cell function; experimental hematology. *Mailing Add:* 69 Cotlage St Bar Harbor ME 04609

BERNSTEIN, SHELDON, b Milwaukee, Wis, Mar 23, 27; m 48; c 3. PHYSIOLOGICAL CHEMISTRY. *Educ:* Univ Wis, BS, 49, PhD(physiol chem), 52. *Prof Exp:* Asst physiol chem, Univ Wis, 50-52; res biochemist, Upjohn Co, 52-53; vpres & dir res, 53-65, pres & dir res, Milbrew Inc, 65-83; DIR, TECH DEVELOP, UNIV FOODS CORP, 83- *Concurrent Pos:* Pres, Bader & Bernstein, Inc, 61-83. *Honors & Awards:* Van Lanen Award, Am Chem Soc. *Mem:* AAAS; Am Chem Soc; Am Soc Microbiol; fel Am Inst Chemists; Inst Food Technologists; Sigma Xi. *Res:* Yeast and yeast products; enzymes; microbiological nutrients; nucleic acids; amino acids; vitamins; custom fermentations and fermentation research; food supplements; dairy products; whey fractionation; alcohol fermentation and production. *Mailing Add:* 1600 W Green Tree Rd No 127 Milwaukee WI 53209

BERNSTEIN, SOL, b West New York, NJ, Feb 3, 27; m 63; c 1. MEDICINE. *Educ:* Univ Southern Calif, AB, 52, MD, 56. *Prof Exp:* Resident internal med, Los Angeles County Hosp, Calif, 57-60; asst prof, 62-67, assoc prof med, 67-74, MED, DIR LOS ANGELES COUNTY-UNIV SOUTHERN CALIF MED CTR, 74-; ASSOC DEAN, UNIV SOUTHERN CALIF SCH MED, 87- *Concurrent Pos:* Fel cardiol, St Vincents Hosp, Los Angeles, 58-59; asst dir dept med, Los Angeles County-Univ Southern Calif Med Ctr, 66-72, chief prof serv, Gen Hosp, 72-74. *Mem:* Fel Am Col Cardiol; Am Heart Asn; fel Am Col Physicians; Am Diabetes Asn; Am Fedn Clin Res; Sigma Xi. *Res:* Cardiology; endocarditis, myocarditis and pulmonary embolus; infectious diseases; peripheral vascular disease; metabolism; lactic acidosis; diabetes; health care; health education. *Mailing Add:* 4966 Ambrose Ave Los Angeles CA 90027

BERNSTEIN, STANLEY CARL, b New York, NY, May 19, 37; m 60; c 2. PHYSICAL ORGANIC CHEMISTRY. *Educ:* Queens Col, NY, BS, 58; Univ Mich, MS, 61, PhD(chem), 63. *Prof Exp:* Res assoc chem, Ohio State Univ, 62-64, instr, 63-64; asst prof, Wright State Univ, 64-70; from asst prof to assoc prof, 70-84, PROF CHEM, ANTIOCH COL, 84-, CHMN DEPT, 79- *Mem:* Am Chem Soc; AAAS; Nat Sci Teachers Asn. *Res:* The synthesis of potential fluorogenic pre-column HPLC derivatizing reagents for primary and secondary ammines; science and society. *Mailing Add:* Dept Chem Antioch Col Yellow Springs OH 45387-1697

BERNSTEIN, STANLEY H, b Brooklyn, NY, Sept 10, 24. INTERNAL MEDICINE, CARDIOLOGY. *Educ:* Col William & Mary, BS, 45; NY Univ, MD, 48. *Prof Exp:* Vis prof, All India Med Inst, 64; assoc prof med, Univ Conn, 68-74; ASSOC PROF CLIN MED, SCH MED, UNIV MIAMI, 74- *Concurrent Pos:* Dir med serv, Mt Sinai Hosp, 68-74. *Mem:* Am Fedn Clin Res; fel Am Col Physicians; Am Heart Asn; AMA; fel NY Acad Med; fel Am Col Cardiol. *Res:* Pathogenesis of rheumatic fever and acute glomerulonephritis following streptococcal infection; the relation of streptococci to pharyngeal infections; myoglobin-antimyoglobin reactions in cardiac muscle after various injuries and its determination by immunofluorescent techniques. *Mailing Add:* 3800 S Ocean Dr Hollywood FL 33019

BERNSTEIN, STEPHEN, b Rochester, NY, Nov 14, 33; m 60. NEUROPSYCHOLOGY. *Educ:* Princeton Univ, AB, 55; Univ Wis, MA, 59, PhD(psychol), 62. *Prof Exp:* NIMH fels, Zurich, Switz, 62-64 & Paris, France, 64-65; trainee, Ment Health Training Prog, Brain Res Inst, Univ Calif, Los Angeles, 65-66, asst prof psychiat, 66-76; RES ASSOC, E P O BIOL, UNIV COLO, BOULDER, 76- *Concurrent Pos:* NIMH career develop award, 66-76. *Mem:* AAAS; Sigma Xi. *Res:* Brain research; emotional development and complex learning in rhesus monkeys; neural mechanisms of behavior in ants, bees and wasps. *Mailing Add:* Dept E P O Biol Univ Colo Boulder CO 80309-0334

BERNSTEIN, THEODORE, b Milwaukee, Wis, Dec 1, 26; m 61; c 2. ELECTRICAL ENGINEERING. *Educ:* Univ Wis, BS, 49, MS, 55, PhD(elec eng), 59. *Prof Exp:* Eng designer, Boeing Airplane Co, Wash, 49-52; sr proj engr, AC Electronics, Wis, 52-56; instr elec eng, Univ Wis, 56-59; mem tech staff, TRW Systs Group, Calif, 59-62; PROF ELEC & COMPUTER ENG, UNIV WIS, MADISON, 62- *Mem:* Inst Elec & Electronics Engrs; Sigma Xi. *Res:* Magnetic materials and devices; electrical and lightning safety. *Mailing Add:* Dept Elec & Computer Eng Univ Wis 1415 Johnson Dr Madison WI 53706-1691

BERNSTORF, EARL CRANSTON, b Judsonia, Ark, June 17, 21; m 44; c 4. ANATOMY. *Educ:* Taylor Univ, BS, 44; Ind Univ, MA, 48, PhD(human anat), 50. *Prof Exp:* Teacher high schs, Ind, 44-46; from instr to asst prof anat, Hahnemann Med Col, 50-58; Ind Univ-AID contract prof, Postgrad Med Ctr, Karachi, Pakistan, 58-63; vis assoc prof, Med Ctr, Ind Univ, 63-64; assoc prof, 64-80, PROF ANAT, MEHARRY MED COL, 80- *Mem:* Endocrine Soc; Am Asn Anat; Pakistan Asn Advan Sci. *Res:* Histology; neuroanatomy; endocrinology. *Mailing Add:* Dept Anat Meharry Med Col Nashville TN 37208

BERNTHAL, FREDERICK MICHAEL, b Sheridan, Wyo, Jan 10, 43; c 1. NUCLEAR CHEMISTRY, NUCLEAR PHYSICS. *Educ:* Valparaiso Univ, BS, 64; Univ Calif, Berkeley, PhD(chem), 69. *Prof Exp:* Res staff chemist, Heavy Ion Accelerator Lab, Yale Univ, 69-70; asst prof chem & physics, Mich State Univ, 70-75, assoc prof, 75-80; legis asst, 78-80, chief legis asst to US Sen Howard Baker, 80-83; comnr, US Nuclear Regulatory Comn, 83-88; asst secy state, Oceans & Int Environ & Sci Affairs, 88-90; DEPT DIR, NSF, 90- *Concurrent Pos:* Vis scientist, Niels Bohr Inst, Univ Copenhagen, 76-77; NATO sr scientist fel, 77; Cong sci fel, Am Phys Soc, 78-79. *Mem:* AAAS; Am Chem Soc; Am Phys Soc; Sigma Xi. *Res:* Nuclear spectroscopy and structure; behavior of nuclei at very high spin; nuclear decay; applications of nuclear technology. *Mailing Add:* NSF 1800 G St Washington DC 20550

BERNTHAL, JOHN E(RWIN), b Sheridan, Wyo, July 18, 40; m 66; c 2. SPEECH PATHOLOGY. *Educ:* Wayne State Col, BA, 62; Univ Kans, MA, 64; Univ Wis, PhD(commun disorders), 71. *Prof Exp:* Assoc prof speech path, Mankato State Univ, 70-73; asst prof speech path & dir, Speech Clin, Univ Md, 73-79; HEAD & PROF SPEECH PATH, UNIV NORTHERN IOWA, 79-; DIR BARKLEY MEM CTR, UNIV NEBR, LINCOLN. *Mem:* Fel Am Speech Hearing Lang Asn, 79. *Res:* Child phonology and management of individuals with articulation disorders. *Mailing Add:* Spec Ed Dept East Campus Univ Nebr Lincoln NE 68583

BERNTSEN, ROBERT ANDYV, b Northfield, Minn, Feb 22, 17; m 43; c 2. INORGANIC CHEMISTRY. *Educ:* St Olaf Col, BA, 39; NY Univ, MS, 41; Purdue Univ, PhD(chem), 49. *Prof Exp:* Instr chem, St Olaf Col, 41-43; head dept, Minot State Teachers Col, 43-46; from asst prof to assoc prof, 48-57, head dept, 68-80, PROF CHEM, AUGUSTANA COL, 57- *Mem:* Am Chem Soc. *Res:* Oxidation of carbohydrates with oxides of nitrogen. *Mailing Add:* 4536 25th Ave Rock Island IL 61201-5710

BERON, PATRICK, b France, Mar 8, 52; Can citizen; m 76; c 2. URBAN HYDROLOGY, WATER & WASTE TREATMENT. *Educ:* Nat Inst Appl Sci, Toulouse, France, BEng, 74; Montreal Polytech, MASc, 76, PhD(civil eng), 83. *Prof Exp:* Teacher phys sci, Nat Ministry Educ, Tunisia, 76-78; res assoc, Montreal Polytech, 78-79 & 82, lectr, 80-83, fel, 83; ASST PROF WATER RESOURCES, UNIV QUE, MONTREAL, 83- *Mem:* Am Water Resources Asn; Int Asn Water Pollution Res & Control; Int Ozone Asn. *Res:* Improvement of urban drainage techniques; use of ozone as an alternative wastewater treatment; water management. *Mailing Add:* 1580 Filion Montreal PQ H4L 4E8 Can

BEROZA, MORTON, b New Haven, Conn, Mar 7, 17; m 46; c 2. ANALYTICAL & ENVIRONMENTAL CHEMISTRY. *Educ:* George Washington Univ, BS, 43; Georgetown Univ, MS, 46, PhD(org chem, biochem), 50. *Prof Exp:* Sci asst drugs, Food & Drug Admin, 39-42; chemist & chem engr plastics, US Naval Ord Lab, 46-48; res chemist, USDA, Beltsville, 48-74; CONSULT, 75- *Honors & Awards:* Hillebrand Prize, 63; Award in Chromatography & Electrophoresis, Am Chem Soc, 69; Harvey W Wiley Award, 70; Gold Medal Award, Synthetic Org Chem Mfrs Asn, 73; Int Award in Pesticide Chem, Am Chem Soc, 77. *Mem:* AAAS; Am Chem Soc; Entom Soc Am; Soc Exp Biol & Med; Am Inst Chemists. *Res:* Insecticidal alkaloids of plants, isolation and chemical structure; insecticidal synergists of natural origin; synthesis of insect repellents, attractants and insecticides; spectroscopy; chromatography; pesticide residue analysis; analytical micromethodology; pheromones. *Mailing Add:* 821 Malta Lane Silver Spring MD 20901

BERRA, TIM MARTIN, b St Louis, Mo, Aug 31, 43; m 67. ICHTHYOLOGY, ZOOGEOGRAPHY. *Educ:* St Louis Univ, BS, 65; Tulane Univ, MS, 67, PhD(biol), 69. *Prof Exp:* NIH trainee environ biol, Tulane Univ, 66-69; Fulbright postdoctoral scholar zool, Australian Nat Univ, 69-70, demonstr, 70; sr tutor biol, Univ Papua New Guinea, 70-71; from asst prof to assoc prof, 72-85, PROF ZOOL, OHIO STATE UNIV, MANSFIELD, 85- *Concurrent Pos:* Fulbright sr res scholar zool, Monash Univ, Melbourne, Australia, 79; ed, Ohio J Sci, 81-85; res assoc, Western Australian Mus, Perth, 86 & 88-89. *Mem:* Am Soc Ichthyologists & Herpetologists; Am Fisheries Soc; Soc Syst Zool; Australian Soc Limnol; Australian Soc Fish Biol; AAAS. *Res:* Ecology, taxonomy evolution and zoogeography of freshwater fishes; bibliography of naturalists, history of biology and natural history of Australia and New Guinea; human evolution. *Mailing Add:* Dept Zool Ohio State Univ Mansfield OH 44906

BERREMAN, DWIGHT WINTON, physics, for more information see previous edition

BERREND, ROBERT E, b Wausau, Wis, July 12, 25; m 57; c 2. INVERTEBRATE ZOOLOGY. *Educ:* Univ Wis, BA, 49, MA, 52, PhD(protozool), 58. *Prof Exp:* Instr zool, Mont State Univ, 58-60; asst prof biol, 60-64, ASSOC PROF BIOL, SAN FRANCISCO STATE UNIV, 64- *Mem:* AAAS; Sigma Xi. *Res:* Distribution of and locomotion in testaceans; estuarine ecology. *Mailing Add:* Dept Biol San Francisco State Col San Francisco CA 94132

BERRESFORD, GEOFFREY CASE, b New York, NY, May 25, 44; m 70; c 2. STOCHASTIC PROCESSES. *Educ:* Lawrence Univ, BA, 67; New York Univ, MS, 71, PhD(math), 74. *Prof Exp:* Lectr natural sci, State Univ NY, Purchase, 73-74, asst prof math, 74-75; from asst prof to assoc prof, 75-85, PROF MATH, CW POST CTR, LONG ISLAND UNIV, 85- *Concurrent Pos:* Chmn speakers bur, Math Asn Am, 85- *Mem:* Math Asn Am; Am Math Soc; Soc Indust & Appl Math. *Res:* Probability and partial differential equation; author of technical articles on algorithms, ordinary differential equations, logic, complex analysis; author of one textbook. *Mailing Add:* Dept Math C W Post Ctr-Long Island Univ Greenvale NY 11548

BERRETH, JULIUS R, b Artas, SDak, Aug 15, 29; m 58; c 3. NUCLEAR CHEMISTRY, NUCLEAR WASTE MANAGEMENT. *Educ:* Cent Wash Col, BA, 52; Wash State Univ, MS, 59. *Prof Exp:* Chemist, Scott Paper Co, 56-57; chemist, Phillips Petrol Co, 57-59, res chemist, Atomic Energy Div, 59-68; res chemist & group leader nuclear chem, Aerojet Nuclear Co, 68-73; group leader, Allied Chem Corp, 73-80; mgr advan process develop, Exxon Nuclear Idaho Co, 80-85; MGR, WASTE MGT RES, WESTINGHOUSE IDAHO NUCLEAR. *Mem:* Am Chem Soc; Am Nuclear Soc. *Res:* Preparations and separations of radioactive elements and determination of integral and differential cross-sections; neutron activation analysis on biological and inorganic samples; nuclear waste management; high-level waste immobilization, waste form and process development. *Mailing Add:* 780 S Bellin Idaho Falls ID 83402

BERRETT, DELWYN GREEN, b Menan, Idaho, July 27, 35; m 63; c 5. ORNITHOLOGY. *Educ:* Brigham Young Univ, BS, 57, MS, 58; La State Univ, PhD(ornith), 62. *Prof Exp:* Mem staff zool, Ricks Col, 62-64; assoc prof zool, Brigham Young Univ, Hawaii Campus, 64-78; RICKS COL, 82- *Mem:* Am Ornith Union; Cooper Ornith Soc; Wilson Ornith Soc. *Res:* Taxonomy, classification and distribution of birds. *Mailing Add:* Div Natural Sci Ricks Col Rexburg ID 83440

BERRETT, PAUL O(RIN), b Riverton, Utah, Mar 1, 28; m 52; c 8. ELECTRICAL ENGINEERING, PHYSICS. *Educ:* Univ Utah, BS, 53, PhD(elec eng, physics), 65; Univ Southern Calif, MS, 55. *Prof Exp:* Res engr, Hughes Aircraft Co, 53-56; instr elec eng, Univ Utah, 57-62, res fel, microwave devices lab, 62-64; from asst prof to assoc prof elec eng, 64-71, PROF ELEC ENG, BRIGHAM YOUNG UNIV, 71-; CONSULT EYRING RES INST, PROVO, UT, 82- *Concurrent Pos:* Res engr, Upper Air Res Labs, Univ Utah, 59-61; res engr, Battelle Northwest Labs, 65; workshop, Goddard Space Flight Ctr, NASA, 67; sabbatical leave, Naval Weapons Ctr, China Lake, Calif, 70-71. *Mem:* Sr mem Inst Elec & Electronics Engrs; Am Inst Physics. *Res:* Microwaves; field theory; antennas; upper atmosphere measurements; electromagnetic waves in plasmas and plasma diagnostics; rocket plume-electromagnetic wave interactions; microwave systems; active antennas; ground interactive antennas. *Mailing Add:* Dept Elec & Computer Eng 459 CB Brigham Young Univ Provo UT 84602

BERRETTINI, WADE HAYHURST, b Wilkes-Barre, Pa, July 22, 51; m 86; c 1. PSYCHIATRY, MOLECULAR BIOLOGY. *Educ:* Dickinson Col, BS, 73; Jefferson Med Col, MD, 77, PhD (pharm), 79. *Prof Exp:* STAFF PSYCHIATRIST, CLIN NEUROGENETICS BR, NIMH, 81-; ADJ ASST PROF DEPTS PHARMACOL & PSYCHIAT, JEFFERSON MED COL, 82- *Mem:* Am Psychopath Asn; Am Psychiat Asn; Soc Biol Psychiat. *Mailing Add:* Bldg 10 Rm 3N-218 NIH Bethesda MD 20892-1000

BERRIE, DAVID WILLIAM, b Mason City, Iowa, Feb 15, 37; m 58; c 5. COUNTERMEASURES TO HOSTILE AIR DEFENSE SYSTEMS. *Educ:* Iowa State Col, BS, 58; Ohio State Univ, MS, 64, PhD(elec eng), 74. *Prof Exp:* Electronics engr, Aeronaut Systs Div, Wright Patterson AFB, Ohio, 59-87; ELECTRONICS ENGR, 6585 TEST GROUP, HOLLOMAN AFB, NMEX, 87- *Mem:* Inst Elec & Electronics Engrs. *Res:* Development of countermeasures to hostile air defense systems. *Mailing Add:* 6585 Test Group 6585 TG-CA Holloman AFB NM 88330-5000

BERRIER, HARRY HILBOURN, b Norborne, Mo, July 6, 17; m 50. VETERINARY PATHOLOGY. *Educ:* Univ Mo, BSc, 41, MSc, 60; Kans State Univ, DVM, 45. *Prof Exp:* Instr pub schs, Mo, 41-42; practicing veterinarian, Mo, 45-46; from assoc prof to emer prof, Col Vet Med, Univ Mo-Columbia, 85-88; RETIRED. *Concurrent Pos:* NSF fel, Vet Path Div, US Armed Forces Inst Path, 63-64. *Mem:* Am Vet Med Asn; Am Soc Vet Clin Path; fel Am Col Vet Toxicol. *Res:* Ovine pregnancy diseases; veterinary medicine diagnostic aids; canine surgical mouth speculum. *Mailing Add:* 1250 Cedar Grove Blvd R2 Columbia MO 65201

BERRILL, MICHAEL, b Montreal, Que, Apr 18, 44; m 69. ANIMAL BEHAVIOR. *Educ:* McGill Univ, BSc, 64; Univ Hawaii, MS, 65; Princeton Univ, PhD(animal behav), 68. *Prof Exp:* Asst prof biol, 68-74, ASSOC PROF BIOL, TRENT UNIV, 74-, CHMN, DEPT BIOL, 77- *Res:* Behavioral ecology and development of behavior of marine and freshwater crustaceans, and of lower vertebrates. *Mailing Add:* Dept Biol Trent Univ Peterborough ON K9J 7B8 Can

BERRIMAN, LESTER P, b Berkeley, Calif, Feb 5, 25; m 56; c 1. CHEMICAL ENGINEERING. *Educ:* Univ Calif, Berkeley, BS, 50. *Prof Exp:* Jr chem engr, Stanford Res Inst, 50-52, assoc chem engr, 52-56, sr chem engr, 56-62, asst mgr agr sci res, 62-64, mgr process eng res, 64-68, mgr chem & mech eng labs, 68-70; dir, Eng Advan Technol Ctr Dresser Industs, 70-87; CONSULT ENGR TECHNOL, 87- *Mem:* Am Chem Soc; Am Inst Chem Engrs; Sigma Xi. *Res:* Automotive air pollution and control; water pollution control; reverse osmosis; solid waste collection, transportation and reclamation; energy conversion and recovery processes; fluid dynamic systems; fluid machinery; materials. *Mailing Add:* 18871 Portofino Dr Irvine CA 92715

BERRY, B(RIAN) S(HEPHERD), b Manchester, Eng, Feb 27, 29; nat US; m 52; c 2. MATERIALS SCIENCE, SOLID-STATE PHYSICS. *Educ:* Univ Manchester, BSc, 49, MSc, 51, PhD(metall), 54. *Prof Exp:* Asst lectr metall, Univ Manchester, 50-54; asst, Yale Univ, 54-56; investr phys metall, Fulmer Res Inst, Eng, 56-58; MAT SCIENTIST, THOMAS J WATSON RES CTR, IBM CORP, 58- *Mem:* Minerals Metals & Mat Soc; Am Phys Soc. *Res:* Anelastic behavior of solids; imperfections in crystals; diffusion; magnetic properties; thin films; amorphous materials; hydrogen in metals. *Mailing Add:* Thomas J Watson Res Ctr IBM Corp PO Box 218 Yorktown Heights NY 10598

BERRY, BRADFORD WILLIAM, b Washington, DC, May 2, 41; m 63; c 3. SENSORY EVALUATION. *Educ:* Wash State Univ, BS, 63; NMex State Univ, MS, 65; Texas A&M Univ, PhD(meat sci). *Prof Exp:* Inst & scientist meat sci, Wash State Univ, 65-70; assoc prof meat sci, Colo State Univ, 72-77; RES FOOD TECHNOLOGIST MEAT SCI, US DEPT AGR, 77- *Mem:* Am Meat Sci Asn; Inst Food Technologists; Am Soc Animal Sci; Food Distrib Res Soc; Am Soc Testing & mats. *Res:* The effects of production, processing and cooking on sensory, instrumental texture, chemical, storage life and microbiological properties of meat products. *Mailing Add:* Meat Sci Res Lab Bldg 201 BARC-E 10300 Baltimore Ave Beltsville MD 20705-2350

BERRY, CHARLES ARTHUR, b Ketchum, Idaho, Aug 19, 29; m 54; c 1. PHARMACOLOGY. *Educ:* Univ Idaho, BS, 51; Idaho State Univ, BS, 54; Univ Iowa, MS, 59, PhD(pharmacol), 61. *Prof Exp:* Instr pharmacol, Univ Iowa, 61-62; fel, Stanford Univ, 62-64; asst prof, 64-68, ASSOC PROF PHARMACOL, NORTHWESTERN UNIV, 68- *Mem:* AAAS; Am Soc Pharmacol & Exp Therapeut; Soc Neurosci. *Res:* Problems of attention, learning, memory, motivation and animal behavior; investigation of drugs and neural systems interactions; definition of cortical-subcortical relationships. *Mailing Add:* Dept Pharmacol 2107 E Wing Ave Evanston IL 60201

BERRY, CHARLES DENNIS, b Jacksboro, Tex, May 31, 40; m 66; c 2. RESEARCH MANAGEMENT, PLANT PHYSIOLOGY. *Educ:* Tex A&M Univ, BS, 62; Purdue Univ, MS, 65, PhD(plant breeding), 69. *Prof Exp:* Asst prof agron, Auburn Univ, 68-74; mgr sorghum res, Cargill Inc, 74-89, mgr prod res/sorghum & sunflowers, 89-91; DIR RES, STONEVILLE PEDIGREE SEEDS, 91- *Concurrent Pos:* Int res direction & liaison, Cargill Sorghum, 78-89. *Mem:* Am Soc Agron; Crop Sci Soc Am. *Res:* Variety development in cotton, sorghum, tall fescue and canary grass. *Mailing Add:* Rte 1 Box 39 Leland MS 38756

BERRY, CHARLES RICHARD, b Morgantown, WVa, Mar 8, 27; m 48; c 2. PLANT PATHOLOGY. *Educ:* Glenville State Col, AB, 49; WVa Univ, MS, 55, PhD(plant path), 58. *Prof Exp:* Plant pathologist, Div Forest Dis Res, 57-62, PLANT PATHOLOGIST, DIV FOREST PROTECTION RES, SOUTHEASTERN FOREST EXP STA, USDA, 66-, PROJ LEADER AIR POLLUTION, 62- *Mem:* Soil Sci Soc Am; Am Phytopath Soc. *Res:* Reclamation; air pollution; forest disease. *Mailing Add:* 179 Tara Way Athens GA 30601

BERRY, CHARLES RICHARD, JR, b Salisbury, Md, May 11, 45; m 72; c 2. FISH & WILDLIFE SCIENCES. *Educ:* Randolph-Macon Col, BS, 67; Fordham Univ, MS, 70; Va Polytech Inst & State Univ, PhD(fisheries), 76. *Prof Exp:* Biologist, Va Water Control Bd, 68-72; res asst, Va Comn Game & Inland Fisheries, 72-75; asst leader pollution & fish biol, US Fish & Wildlife Serv, Utah State Univ, 76-85; LEADER AQUATIC ECOL & FISH BIOL, US FISH & WILDLIFE SERV, SDAK STATE UNIV, 85-, ASSOC PROF, 85- *Concurrent Pos:* Asst prof, Interdept Prog, Utah State Univ, 79-85; assoc ed, Am Fisheries Soc, 84-87. *Mem:* Am Fisheries Soc; Am Inst Fishery Res Biologists. *Res:* Effects of water pollution and habitat alteration on fish; fishery management in wetlands; endangered fish biology; fish health. *Mailing Add:* SDak Coop Unit Dept Wildlife/Fish Sci SDak State Univ Brookings SD 57007

BERRY, CHRISTINE ALBACHTEN, b San Diego, Calif, Nov 27, 46; m 69. PHYSIOLOGY, BIOPHYSICS. *Educ:* Stanford Univ, BA, 68; Yale Univ, PhD(physiol), 74. *Prof Exp:* Fel renal physiol, 76, lectr physiol, 76-77, ADJ ASST PROF PHYSIOL, UNIV CALIF, SAN FRANCISCO, 77- *Concurrent Pos:* Asst res physiologist, Cardiovasc Res Inst, San Francisco, 76-77. *Mem:* Am Soc Nephrology; Int Soc Nephrology; AAAS; Sigma Xi; Am Physiol Soc. *Res:* Salt and water transport across proximal convoluted tubule of mammalian nephron. *Mailing Add:* Dept Physiol 1065 HSE Med Sch Univ Calif San Francisco CA 94143

BERRY, CLARK GREEN, b Ilion, NY, Sept 28, 08; m 36; c 2. CHEMISTRY. *Educ:* Clarkson Col Technol, BS, 30. *Prof Exp:* Chemist, Atmospheric Nitrogen Corp, 30-31 & Skenandoa Rayon Corp, 33-37, control supvr, 37-38; res chemist, Inst Paper Chem, 38-42 & Skenandoa Rayon Corp, 42-46; chief chemist, Del Rayon Co, 46-52; lab dir, New Bedford Rayon Co, 52-53, chief chemist & dir res, 53-66; supvry chemist & br head, Bur Engraving & Printing, US Treas Dept, 66-77; RETIRED. *Mem:* Am Chem Soc; Tech Asn Pulp & Paper Indust. *Res:* Viscose rayon; cellulose; synthetic fibers. *Mailing Add:* 4927 25th St S Arlington VA 22206-1050

BERRY, CLYDE MARVIN, b Posey, Ill, June 18, 13; m 40; c 3. INDUSTRIAL HYGIENE, CHEMICAL ENGINEERING. *Educ:* McKendree Col, BS, 33; Univ Ill, MS, 36; Univ Iowa, MSChE, 40, PhD(indust hyg), 41. *Prof Exp:* Indust hyg engr, USPHS, 41-48 & Esso Standard Oil Co, 48-55; assoc dir, Inst Agr Med, Univ Iowa, 55-85; RETIRED. *Concurrent Pos:* Mem adv comt, Div Occup Health, USPHS, 57-60, consult, 60-, mem adv comt, Div Accident Prev, 60-63, consult, 63-; consult, Nat Comn Community Health Serv, 64-66. *Mem:* Fel Am Pub Health Asn; Am Indust Hyg Asn (pres, 67); Am Conf Govt Indust Hygienists; Am Soc Safety Engrs. *Res:* Bacterial quality of air; industrial carcinogens; charged aerosols. *Mailing Add:* 201 First Ave N No 109 Iowa City IA 52245

BERRY, DAISILEE H, b Honolulu, Hawaii. PEDIATRICS, HEMATOLOGY. *Educ:* Univ Ark, BS, 49, MD, 59; Western Reserve Univ, MN, 52. *Prof Exp:* Intern resident pediat, Univ Ark, 59-61; resident fel, Washington Univ, 61-63; fel hemat oncol, Univ Ark, 63-64, asst prof pediat hemat oncol, 64-67; fel hemat, Duke Univ, 67-69; from asst prof to assoc prof, 69-76, PROF PEDIAT HEMAT ONCOL, UNIV ARK, LITTLE ROCK, 76- *Mem:* Am Soc Hemat; Am Soc Clin Oncol; Acad Pediat; Am Asn Cancer Educ. *Res:* Pediatric oncology; clinical manifestations of sickle cell anemia. *Mailing Add:* Ark Childrens Hosp 800 Marshall St Little Rock AR 72202

BERRY, DONALD S(TILWELL), b Vale, SDak, Jan 1, 11; m 37; c 2. CIVIL ENGINEERING. *Educ:* SDak Sch Mines & Technol, BS, 31; Iowa State Col, MS, 33; Univ Mich, PhD(civil eng), 36. *Hon Degrees:* DEng, SDak Sch Mines & Technol, 64. *Prof Exp:* Traffic engr, Nat Safety Coun, 36-43, dir traffic & transp div, 43-48; prof transp eng, Univ Calif, 48-55 & Purdue Univ, 55-56; prof civil eng, 56-77, chmn dept, 62-69, Walter P Murphy Prof, 67-78, EMER PROF CIVIL ENG, NORTHWESTERN UNIV, EVANSTON, 78- *Concurrent Pos:* With Fed Bur Invest & Off Civilian Defense, 44; chmn hwy res bd, Nat Acad Sci-Nat Res Coun, 65- *Mem:* Nat Acad Eng; Am Soc Civil Engrs; Inst Traffic Eng. *Res:* Skidding of vehicles on road surfaces; causes of traffic accidents; methods for evaluating the stability of macroscopic materials; vehicle speeds, travel times and intersection delays; capacity of arterial streets. *Mailing Add:* 2520 Park Pl Evanston IL 60201

BERRY, E JANET, b Wheatland, Ind, May 28, 17; m 69. ORGANIC CHEMISTRY, CHEMICAL ENGINEERING. *Educ:* Purdue Univ, BS, 42, PhD, 46; NY Univ, JD, 52. *Prof Exp:* Res chemist & patent asst, Am Cyanamid Co, 46-48; patent atty, Esso Res & Eng, 48-53; patent atty & dept mgr, Nat Distillers & Chem Corp, 53-63; PATENT ATTY, 63- *Concurrent Pos:* Trustee, Carnegie-Mellon Univ, 86- & Animal Med Ctr, 80-; nat dir, Am Inst Chemists, 86- *Mem:* Am Inst Chemists (pres, 80-82); Asn Consult Chemist & Chem Engrs (pres, 76-78); Am Chem Soc; Chem Indust Asn; Soc Chem Indust. *Res:* Chemistry; pharmaceutical chemistry; inventions; patents; innovation; research administration; technology transfer licensing. *Mailing Add:* Two Horatio St Apt 15G New York NY 10014

BERRY, EDWARD ALAN, b Ft Carson, Colo, July 14, 52; m; c 1. HEME PROTEIN BIOCHEMISTRY, MEMBRANE PROTEIN PURIFICATION. *Educ:* Cornell Univ, PhD(biochem), 81. *Prof Exp:* Res assoc, Biochem Dept, Dartmouth Med Sch, 81-84 & Physiol & Biophysics Dept, Univ Ill, Urbana-Champaign, 84-87; STAFF SCIENTIST, LAWRENCE BERKELEY LAB, 87- *Mem:* Biophys Soc. *Res:* Function and structure of membrane proteins involved in energy conservation; techniques of purification, enzymology, reconstitution, crystallization. *Mailing Add:* Donner Lab Lawrence Berkeley Lab Univ Calif Berkeley CA 94720

BERRY, EDWIN X, b San Francisco, Calif, June 20, 35; m 57, 73; c 3. ATMOSPHERIC PHYSICS. *Educ:* Calif Inst Technol, BS, 57; Dartmouth Col, MS, 60; Univ Nev, PhD(physics), 65. *Prof Exp:* Res asst ionospheric physics, Thayer Sch Eng, Dartmouth Col, 59-60; radio propagation engr, Advan Commun Eng Div, Cook Elec Co, 60-61; from res asst to res assoc atmospheric physics, Desert Res Inst, Univ Nev, Reno, 62-72; prog mgr weather modification, NSF, 73-74, phys scientist, 74-76; pres, Atmospheric Res & Technol Inc, 76-87; PRES, EDWIN X BERRY & ASSOC, 87- *Concurrent Pos:* Mem comt on land use planning, 72; chmn comt weather modification, Am Meterol Soc, 74. *Mem:* AAAS; Am Meteorol Soc; Am Wind Energy Asn. *Res:* Physics; wind energy. *Mailing Add:* Edwin X Berry & Assocs 6040 Verner Ave Sacramento CA 95841

BERRY, GAIL W(RUBLE), b Kalamazoo, Mich, Nov 7, 39. PSYCHIATRY, PSYCHOANALYSIS. *Educ:* Kalamazoo Col, AB, 60; NY Univ Sch Med, MD, 64. *Prof Exp:* Intern pediat, Kings County Hosp, Brooklyn, NY, 64-65; resident psychiat, NY Med Col, Metrop Hosp, NY, 65-68; staff psychiatrist, Beth Israel Med Ctr, NY, 68-74; from clin instr psychiat to clin assoc, Mt Sinai Sch Med, NY, 68-75; ASST CLIN PROF PSYCHIAT, MT SINAI SCH MED, NY, 75-, ASSOC ATTEND PSYCHOANALYST, MT SINAI HOSP, 83- *Concurrent Pos:* Assoc ed, Academy, Am Acad Psychoanalysis, 76-83; training & supvry psychoanalyst, NY Med Col, 80-; asst ed, J Am Acad Psychoanalysis, 84- *Mem:* Am Psychiat Asn; Am Acad Psychoanalysis. *Res:* Psychoanalytic research on incest. *Mailing Add:* 1474 Third Ave New York NY 10028

BERRY, GEORGE WILLARD, b Poolville, NY, Feb 22, 15; m 46; c 2. GEOLOGY. *Educ:* Colgate Univ, AB, 36; Cornell Univ, MS, 38, PhD(struct geol), 41. *Prof Exp:* From asst to sr asst geol, Cornell Univ, 36-38, instr, 38-41; geol scout, Texas Co, Tex, 41-42; geologist, Fla, 46 & Wyo, 46-48; dist geologist, Sun Oil Co, Wyo, 48-52, asst div geologist, Colo, 52-61; sr geologist, Tex, 61-64 & Colo, 64-70; sr res geologist, Cordero Mining Co, 70-72; CONSULT GEOLOGIST, 72- *Mem:* Am Asn Petrol Geologists; Am Inst Mining, Metall & Petrol Engrs; Geol Soc Am. *Res:* Rocky Mountain geology; geothermal energy. *Mailing Add:* 600 Spruce St Boulder CO 80302

BERRY, GEORGE WILLIAM, b Parker, Colo, Feb 6, 07; m 30; c 5. CHEMISTRY. *Educ:* Colo State Col, BSc, 30; Univ Nebr, MSc, 32, PhD(phys chem), 34. *Prof Exp:* Instr org chem, Univ Nebr, 34-35; res chemist, Socony-Vacuum Oil Co, Kans, 35-45; res chemist, Flintkote Co, 45-56, mgr built-up roofing dept, 56-63, tech mgr, Archit Prod Div, 63-66; sr res assoc, Johns-Manville Corp, 66-72; SR ASSOC, ROBT M STAFFORD INC, 72- *Concurrent Pos:* Fac mem, Roofing Indust Educ Inst, 80-86. *Mem:* Am Chem Soc; fel Am Inst Chem. *Res:* Utilization of asphalt in building materials; petroleum refining; surface tension of dilute solutions of sodium palmitate. *Mailing Add:* 4363 Woodlark Lane Charlotte NC 28211

BERRY, GREGORY FRANKLIN, b Brooklyn, NY, June 2, 43; m 66; c 2. MAGNETOHYDRODYNAMICS. *Educ:* Worcester Polytech Inst, BS, 65; Univ Conn, MS, 67, PhD(mech eng), 70. *Prof Exp:* Sr engr, Gen Elec Co, 69-72; MECH ENGR, ARGONNE NAT LAB, 72- *Concurrent Pos:* Pres, Symp Eng Aspects & Magnetohydrodynamics, 90-91. *Mem:* Am Soc Mech Engrs; Am Inst Aeronaut & Astronaut. *Res:* Magnetohydrodynamic power generation; erosion in fluidized bed combusters; systems analyses of power conversion systems. *Mailing Add:* 9700 S Cass Ave Argonne IL 60439

BERRY, GUY C, b Greene Co, Ill, May 11, 35; m 57; c 3. POLYMER CHEMISTRY. *Educ:* Univ Mich, BS, 57, MS, 58, PhD(chem eng), 60. *Prof Exp:* Fel polymer sci, Mellon Inst, 60-67; sr fel & assoc prof chem & polymer sci, Carnegie Mellon Univ, 67-73, actg dean, Col Sci, 81-82, actg head chem, 83-84, PROF CHEM & POLYMER SCI, CARNEGIE MELLON UNIV, 73-, HEAD CHEM, 90- *Concurrent Pos:* Vis prof, Univ Tokyo, 73, Colo State Univ, 79 & Univ Kyoto, 84; adj prof metall & mat eng, Univ Pittsburgh, 77- *Honors & Awards:* Bingham Medal, Soc Rheology, 90. *Mem:* Am Chem Soc; Soc Rheology; Japan Soc Rheology. *Res:* Physical chemistry of polymers; solution viscosity; static and dynamic light scattering; rheology; liquid crystalline polymers, glassy polymers. *Mailing Add:* Dept Chem Carnegie Mellon Univ Pittsburgh PA 15213

BERRY, HENRY GORDON, b Huddersfield, Eng, July 25, 40; m 68; c 4. ATOMIC PHYSICS, SCIENCE EDUCATION. *Educ:* Oxford Univ, BA, 62; Univ Wis-Madison, MS, 63, PhD(physics), 67. *Prof Exp:* Instr physics, Univ Wis-Madison, 67-68; fel physics, Univ Ariz, 68-69; guest researcher, Res Inst, Stockholm, Sweden, 69-70; maitre de conf, Univ Lyon, France, 70-72; asst prof physics, Univ Chicago, 72-79; PHYSICIST, ARGONNE NAT LAB, 76- *Concurrent Pos:* Founder, developer & dir, Acad Math & Sci Teachers Chicago, 90-91. *Mem:* Fel Am Phys Soc. *Res:* Beam foil spectroscopy; laser-fast beam spectroscopy, studies of radiative lifetimes, relativistic atomic structure, spectral analysis and hyperfine structures of heavy ions; heavy ion collisions in solids and gases. *Mailing Add:* Physics Div Argonne Nat Lab Argonne IL 60439

BERRY, HERBERT WEAVER, b Syracuse, NY, Dec 17, 13; m 42; c 2. PHYSICS. *Educ:* Syracuse Univ, AB, 37, AM, 39; Washington Univ, PhD(physics), 42. *Prof Exp:* Instr, Case Inst Technol, 42-43; res physicist, Radiation Lab, Univ Calif, 43-45; asst prof physics, Univ Okla, 46; from asst prof to assoc prof physics, Syracuse Univ, 46-54, prof, 54-80; RETIRED. *Mem:* Am Phys Soc; Am Asn Physics Teachers. *Res:* Plasma physics; ionization of gases; scattering in gases. *Mailing Add:* 2649 E Genese St Syracuse NY 13224

BERRY, IVAN LEROY, b Mt Vernon, Mo, Sept 9, 37; m 59; c 3. AGRICULTURAL ENGINEERING. *Educ:* Univ Mo, BS, 60, MS, 61; Tex A&M Univ, PhD(agr eng), 69. *Prof Exp:* Agr engr, Agr Res Serv, USDA, 60-65, supvry agr engr, 65-73, res leader, 73-85; PROF, UNIV ARK, 85- *Mem:* Am Soc Agr Engrs; Entom Soc Am; Int Soc Biometeorol. *Res:* Environmental physiology of livestock; livestock shelter design; livestock sprayer design; insect attractants; livestock insect physiology and ecology; integrated pest management. *Mailing Add:* Dept Agr Eng Univ Ark Fayetteville AR 72701

BERRY, JAMES FREDERICK, b Washington, DC, Dec 22, 47; m 73; c 2. HERPETOLOGY, EVOLUTIONARY BIOLOGY. *Educ:* Fla State Univ, BS, 70, MS, 73; Univ Utah PhD(biol), 78. *Prof Exp:* Chemist biochem, Fla Dept Agr, 73; asst prof, 78-84, ASSOC PROF BIOL, ELMHURST COL, 84- *Concurrent Pos:* Res assoc, Carnegie Mus Nat Hist, Pittsburgh, Pa. *Mem:* Am Soc Zoologists; Soc Study Evolution; Soc Syst Zool; Am Soc Ichthyologists & Herpetologists. *Res:* Evolutionary biology, ecology, and systematics of North and Central American reptiles; ecomorphology in modern turtles. *Mailing Add:* Dept Biol Elmhurst Col Elmhurst IL 60126

BERRY, JAMES FREDERICK, b Baltimore, Md, Nov 11, 27; m 52; c 4. BIOCHEMISTRY, NEUROSCIENCE. *Educ:* Johns Hopkins Univ, BA, 49; Univ Rochester, PhD(biochem), 53. *Prof Exp:* USPHS fel biochem, Univ Western Ont, 53-55, sr res fel, 53-56, lectr, 55-56; mult sclerosis fel, Agr Res Coun Inst Animal Physiol, Babraham, Eng, 56-57; fel physiol chem, Johns Hopkins Univ, 58-59, instr, 59-61; assoc prof, 61-66, PROF NEUROL, MED SCH, UNIV MINN, MINNEAPOLIS, 66- *Concurrent Pos:* Assoc dir biochem res, Sinai Hosp, Baltimore, 57-61. *Mem:* Fel AAAS; Am Soc Biol Chemists; Am Oil Chemists' Soc; Int Soc Neurochem; Am Soc Neurochem. *Res:* Biochemistry of degenerating and regenerating nerve; lipid and fatty acid composition and biosynthesis in nervous tissue; biosynthesis of choline esters; alcohol and acetaldehyde metabolism. *Mailing Add:* 1059 Woodhill Dr Roseville MN 55113

BERRY, JAMES G(ILBERT), b Brooklyn, NY, Mar 16, 25; m 47; c 4. INFORMATION RESOURCE MANAGEMENT POLICY. *Educ:* Univ Mich, BS, 49, MS, 50, PhD(eng mech), 54. *Prof Exp:* Instr eng mech, Univ Mich, 50-55; mem sr tech staff, Space Technol Labs, Calif, 55-63, dir Titan Prog Off, 63-64; dir Washington opers, TRW Systs, 64-71; SR STAFF SPECIALIST, OFF SECY DEFENSE, 71- *Mem:* Am Soc Mech Engrs; Am Inst Aeronaut & Astronaut. *Res:* Elasticity; shell theory; structural dynamics; stochastic process theory. *Mailing Add:* 1407 Cola Dr McLean VA 22101

BERRY, JAMES WESLEY, b Rankin, Ill, Mar 23, 26; m 47; c 5. ORGANIC CHEMISTRY. *Educ:* Augustana Col, AB, 49; Univ Ill, PhD(org chem), 53. *Prof Exp:* Res chemist, Rayonier, Inc, 53-55; asst prof chem, Whitworth Col, Wash, 55-56; prof nutrit & food sci & dept Head, Univ Ariz, 56-90; RETIRED. *Mem:* Am Chem Soc; Inst Food Technologists. *Res:* Polysaccharides; arid land plants as food sources. *Mailing Add:* 8871 Driftwood Trail Tucson AZ 85749-9656

BERRY, JAMES WILLIAM, b Bristol, Va, Dec 7, 35; m 70; c 1. SPIDER ECOLOGY. *Educ:* E Tenn State Univ, BS, 57; Va Polytech Inst, MS, 58; Duke Univ, PhD(zool), 66. *Prof Exp:* From instr to assoc prof, 65-77, PROF ZOOL, BUTLER UNIV, 77- *Concurrent Pos:* USPHS fel, 66-67; ed, J Arachnology, 91- *Mem:* Am Arachnological Soc (secy, 87-91); Brit Arachnological Soc. *Res:* Spider ecology in North Carolina piedmont; spiders of the Florida Everglades; spider distribution on Pacific Islands; ecology of intertidal spiders. *Mailing Add:* Dept Biol Sci Butler Univ Indianapolis IN 46208

BERRY, JEWEL EDWARD, parasitology; deceased, see previous edition for last biography

BERRY, JOE GENE, b Sayre, Okla, Feb 19, 44; m 65; c 3. FOOD SCIENCE, POULTRY SCIENCE. *Educ:* Okla State Univ, BS, 65, MS, 67; Kans State Univ, PhD(food sci), 70. *Prof Exp:* ASSOC PROF ANIMAL SCI, PURDUE UNIV, 70-; DEPT ANIMAL SCI, OKLA STATE UNIV, STILLWATER. *Mem:* Am Registry Prof Animal Scientists; Poultry Sci Asn. *Res:* Prevention of egg shell damage in production, processing and retailing; poultry products technology. *Mailing Add:* Dept Animal Sci Okla State Univ Stillwater OK 74078

BERRY, KEITH O, b Ft Collins, Colo, Aug 6, 38; m 60; c 2. INORGANIC CHEMISTRY, CHEMICAL EDUCATION. *Educ:* Colo State Col, BA, 60; Iowa State Univ, PhD(inorg chem), 66. *Prof Exp:* From asst prof to assoc prof, 65-76, PROF CHEM, UNIV PUGET SOUND, 77- *Mem:* AAAS; Am Chem Soc; Forensic Sci Soc; Am Acad Forensic Sci. *Res:* Transition metal complexes with oxygen-donor ligands; microcrystalline detection of trace elements. *Mailing Add:* Dept Chem Univ Puget Sound Tacoma WA 98416

BERRY, LEE ALLEN, b La Junta, Colo, Mar 18, 45; m 66; c 2. PHYSICS, PLASMA PHYSICS. *Educ:* Univ Calif, Riverside, BA, 66, MA, 69, PhD(physics), 70. *Prof Exp:* Res staff mem, 70-73, group leader, Tokamak Opers Group, 73-74, sect head, Tokamak Exp, 74-77, prog dir fusion develop, 77-80, mgr, EBT Prog, 80-84, ASSOC DIR, DIV DEVELOP & TECHNOL, OAK RIDGE NAT LAB, 84- *Mem:* Fel Am Phys Soc. *Res:* Experimental plasma physics; emphasis has been on large scale fusion experiments. *Mailing Add:* Oak Ridge Nat Lab PO Box 2009 Oak Ridge TN 37831-8072

BERRY, LEONARD, b Malmesbury, Eng, June 5, 30; m 66; c 4. PHYSICAL GEOGRAPHY, RESOURCE MANAGEMENT. *Educ:* Bristol Univ, BSc, 51, MSc, 56, PhD(geog), 69. *Prof Exp:* Asst lectr geog, Univ Hong Kong, 54-57; from lectr to sr lectr, Univ Khartoum, 57-65; prof, Univ Col, Dar es Salaam, 65-69, dean fac arts & soc sci, 67-69; dir resource mgt, Univ Dar es Salaam, 69-71; actg dir, Grad Sch Geog, 71-72 & 78-79, dir, Int Develop Prog, 74-78, dean, Grad Sch, 75-78, dir, ID Prog, 76-83, dir, Grad Sch Geog, 79-83, PROF GEOG, CLARK UNIV, 69-, PROVOST, 83- *Concurrent Pos:* Consult, Int Develop Res Ctr, Ottawa, 72; mem, comt to Indonesia on natural resources planning, Nat Acad Sci, 72; comt environ implications of US mat policy, 72-73; comt on remote sensing for develop, 75, chmn, OTA comt trop forests, 82-83 & comt Sahel, 82-84. *Mem:* Royal Geog Soc; Inst Brit Geogr; Asn Am Geogr. *Res:* Tropical geomorphology; rural water development; regional planning; natural resource planning; environmental problems and development; rural development Africa. *Mailing Add:* Acad Dean Fla Atlantic Univ Boca Raton FL 33431

BERRY, LEONIDAS HARRIS, b Woodsdale, NC, July 20, 02; m 37, 59; c 2. MEDICINE. *Educ:* Wilberforce Univ, BS, 24; Univ Chicago, SB, 25, MD, 29; Univ Ill, MS, 33; Am Bd Internal Med, dipl, 46; Am Bd Gastroenterol, dipl, 46. *Hon Degrees:* ScD, Wilberforce Univ, 45; LLD, Lincoln Univ, 83. *Prof Exp:* From jr attend physician to assoc attend physician, Provident Hosp, 33-43, chmn div digestive dis, 34-60, from vchmn to chmn dept med, 43-48; PROF MED, COOK COUNTY GRAD SCH MED, 46- *Concurrent Pos:* Sr attend physician, 43-70, emer sr attend physician, Provident Hosp, 70-; mem dept med, Michael Reese Hosp, 46-63, sr attend Physician, 64; from clin asst prof to clin assoc prof, 52-74; emer clin assoc prof, Sch Med, Univ Ill, 74-; consult gastroenterologist, Women's & Children's Hosp, Chicago, 56- &

Alexian Bros Hosp, 61-71; int lectr, Cult Affairs Div, US Dept State, Africa, Asia & Europe, 65, 66 & 70; mem bd trustees, Cook County Grad Sch Med; mem nat adv coun, Fed Regional Med Progs Versus Heart Dis, Cancer & Stroke; mem nat tech adv comts, Medicare Prog; spec dep, Prof Community Affairs, Cook County Hosp Governing Comn. *Honors & Awards:* Rudolf Schindler Award, Am Soc Gastrointestinal Endoscopy, 77; First Clin Achievement Award, Am Col Gastroenterol, 87. *Mem:* Nat Med Asn (1st vpres, 59, pres elect, 64, pres, 65); fel Am Col Gastroenterol; AMA; fel Am Col Physicians. *Res:* Sociological and pathological aspects of tuberculosis and drug addiction; techniques of gastroscopy; gastro-biopsy instrument; gastroscopic pathology and therapy of chronic gastritis and peptic ulcer; gastric cancer; medical history; narcotic rehabilitation; textbook endoscopy biography. *Mailing Add:* 5142 S Ellis Ave Chicago IL 60615

BERRY, MAXWELL (RUFUS), b Atlanta, Ga, June 7, 10; m 34; c 4. INTERNAL MEDICINE. *Educ:* Cornell Univ, AB, 31, MD, 35; Univ Minn, PhD(med), 42. *Prof Exp:* Bursar, Knickerbocker Fund, Cornell Univ, 33-35; intern med, Bellevue Hosp, NY, 35-37; res physicians, Mayo Clinic, 37-42; assoc med, Hosp Med Col Va, 42, asst prof, 43-44; assoc, Sch Med, Emory Univ, 44-49; dir cancer clin, St Joseph's Infirmary, 47-88, dir Berry Clin, 48-88; RETIRED. *Concurrent Pos:* Consult, Grady Hosp, 44-49; pres, Northwest Hosp Corp, 66-70; chmn bd dirs, West Paces Ferry Hosp, 66-70; chmn bd trustees, Annandale at Suwanee, Inc, 66-73, pres, 73 - *Mem:* Fel Am Col Physicians; fel Am Col Gastroenterol (pres, 65-66); Sigma Xi. *Res:* Tomac oxygen nebulizer; plethysmograph; physiology of circulation and respiration; studies on etiology of duodenal ulcer; clinical gastroenterology. *Mailing Add:* One Magnolia Point Panama City Beach FL 32408

BERRY, MICHAEL JAMES, b Chicago, Ill, July 17, 47; m 67, 84; c 2. CHEMICAL PHYSICS, LASER APPLICATIONS. *Educ:* Univ Mich, BS, 67; Univ Calif, Berkeley, PhD(chem), 70. *Prof Exp:* From asst prof to assoc prof chem, Univ Wis-Madison, 70-76; mgr photon chem, Allied Chem Corp, 76-79; dir, Rice Quantum Inst, Rice Univ, 79-86; dir, Laser Applns Res Ctr, HARC, 84-90; PRES, ANTROPIX CORP, 82-; ROBERT A WELSH PROF CHEM, RICE QUANTUM INST, RICE UNIV, 79- *Concurrent Pos:* Alfred P Sloan res fel, 75-76; Camille & Henry Dreyfus found teacher-scholar, 75-79; John Simon Guggenheim Mem Found fel, 81-82; consult, Dow Chem, USA, Mar Chem, Tex Eastern, Dresser, Houston Area Res Ctr, AMOCO. *Honors & Awards:* Award Pure Chem, Am Chem Soc, 83; Robert G Denkewalter Lectr, 84. *Mem:* Am Phys Soc; Am Chem Soc; Am Soc Photobiol; fel AAAS; Am Soc Laser Med & Surg; Optical Soc Am. *Res:* Applications of lasers in chemistry, medicine and materials science; use of lasers in chemistry to produce and analyze highly vibrationally excited molecules, and to explore rate processes (energy redistribution and chemical reaction) involving these excited species; diagnostic and therapeutical medical applications of lasers; laser-materials interactions. *Mailing Add:* Dept Chem Rice Univ PO Box 1892 Houston TX 77251

BERRY, MICHAEL JOHN, b Southport, Eng, Oct 5, 40; Can citizen; m; c 2. SEISMOLOGY. *Educ:* Univ Toronto, BSc, 61, MA, 62, PhD(seismol), 65. *Prof Exp:* Res asst inst geophys & planetary sci, Univ Calif, Los Angeles, 65-66; res scientist, Seismol Div, Earth Physics Br, Gov Can, 67-74, dir, Div Seismol & Geothermal Studies, 74-82 & dir, Div Seismol & Geomagnatism, 82-86; DIR GEOPHYSICS DIV, GEOL SURV CAN, 86- *Concurrent Pos:* Assoc ed, Geoscience Can, 74-79, Can Earth Sci, 74-84; mem, Can Nat Comt Earthquake Eng; chmn, Gov Coun Int Seismol Ctr, 83-87; chmn, Coord Comt, Nat Rep, Int Union Comn Lithosphere, 85-89, secy-gen Comn, 90- *Mem:* Seismol Soc Am; Am Geophys Union; Can Geophys Union; Int Union Geodesy & Geophysics; Fedn Global Digital Seismograph Networks (pres, 86-89). *Res:* Studies of the Earth's crust and upper mantle using reflection and refraction seismology; seismicity and seismic risk and hazard. *Mailing Add:* Geophys Div One Observ Cres Ottawa ON K1A 0Y3 Can

BERRY, MYRON GARLAND, b Franklin, NH, May 24, 19; m 48; c 3. PHYSICAL CHEMISTRY. *Educ:* Colby Col, BA, 40; Harvard Univ, MA, 42; Syracuse Univ, PhD(phys chem), 51. *Prof Exp:* Teacher, Dept Phys Sci, Urbana Jr Col, 46-49; asst prof chem, Ohio Wesleyan, 51-55; from asst prof to assoc prof, 55-62, admin asst, 62-69, prof chem, 62-84, EMER PROF CHEM, MICH TECHNOL UNIV, 84- *Mem:* Am Chem Soc; Sigma Xi. *Res:* Surface chemistry; photochemistry. *Mailing Add:* PO Box 25 Copper Harbor MI 49918-0025

BERRY, RALPH EUGENE, b Gering, Nebr, June 14, 40; m 66; c 4. ENTOMOLOGY, INSECT PEST MANAGEMENT. *Educ:* Colo State Univ, BS, 63, MS, 65; Kans State Univ, PhD(entom), 68. *Prof Exp:* Entomologist, Pesticide Regulation Div, Agr Res Serv, USDA, 65; from asst prof to assoc prof entom, 68-81, PROF & CHMN ENTOM, ORE STATE UNIV, 81- *Concurrent Pos:* Actg assoc dir, Agr Exp Sta, Ore State Univ, 85-86. *Mem:* Entom Soc Am. *Res:* Applied entomology; biology and control of soil arthropods; ecology of soil insects; investigations of non-insecticidal control of insect pests; pest population ecology; insect-plant interactions; insect pest management. *Mailing Add:* Dept Entom Ore State Univ Corvallis OR 97331

BERRY, RICHARD C(HISHOLM), b Salem, Mass, July 14, 28; m 51; c 5. CHEMICAL ENGINEERING, POLYMER CHEMISTRY. *Educ:* Mass Inst Technol, SB, 48, SM, 49. *Prof Exp:* Develop engr fibrous prod, 49-53, supvr fibrous prod develop, 53-58, mgr prod develop, 58-63, tech dir, 64-66, VPRES RES, DEVELOP & ENG, ROGERS CORP, 66- *Mem:* AAAS; Am Inst Chem Engrs; Inst Elec & Electronics Engrs; Soc Plastics Engrs; Am Chem Soc. *Res:* Engineering materials; fibers, elastomers, plastics and combinations. *Mailing Add:* Mashentuck Rd RFD 4 Box 363 Danielson CT 06239-9804

BERRY, RICHARD EMERSON, b Washington, NJ, Nov 11, 33; m 54; c 5. OCEANOGRAPHY. *Educ:* Lafayette Col, BS, 54; Princeton Univ, MA, 56, PhD(physics), 58. *Prof Exp:* Researcher, Gen Elec Co, 57-58; asst prof physics, Lafayette Col, 58-62; assoc prof, Tex Technol Col, 62-65; chmn dept,

65-74, PROF PHYSICS, INDIANA UNIV PA, 74-; CONSULT, LASER APPLN, NASA, 81- *Mem:* Am Phys Soc; Sigma Xi; Am Asn Physics Teachers. *Res:* Electron spin resonance in solid state; theory of electricity and magnetism; electromagnetic theory and unidentified flying objects effects; oceanography; LIDAR. *Mailing Add:* 340 College Lodge Rd Indiana PA 15701

BERRY, RICHARD G, b Bethel, Conn, Jan 29, 16; m 42; c 5. NEUROPATHOLOGY. *Educ:* Wesleyan Univ, BA, 37; Albany Med Col, MD, 42. *Prof Exp:* Asst psychol, Wesleyan Univ, 38; intern, US Naval Hosp, Newport, RI, 42-43; res neurol, Jefferson Hosp, Philadelphia, 46-47; staff neurologist, US Naval Hosp, 47-50; instr, US Naval Med Sch, 50-53; from assoc prof to prof, 54-84, dir neuropath lab, 54-84, EMER PROF NEUROL, JEFFERSON MED COL, 84- *Concurrent Pos:* Fel neuropath, US Armed Forces Inst Path, 53-54; clin asst prof, Sch Med & staff neurologist, Univ Hosp, Georgetown Univ, 50-54; consult, Vet Admin Hosps, Coatesville, 58- & Germantown Hosp, Philadelphia, 87- *Mem:* AAAS; Am Neurol Asn; Am Acad Neurol; Am Asn Neuropath; Asn Res Nerv & Ment Dis; Sigma Xi. *Res:* Cerebral vascular disease. *Mailing Add:* 108 N Rolling Rd Springfield PA 19064

BERRY, RICHARD LEE, b Shelby, Ohio, July 28, 42; m 64; c 2. MEDICAL ENTOMOLOGY, INSECT TAXONOMY. *Educ:* Univ Notre Dame, BS, 64; Tulane Univ, MS, 67; Ohio State Univ, PhD(entom), 70. *Prof Exp:* MED ENTOMOLOGIST, OHIO DEPT HEALTH, VECTOR BORNE DIS UNIT, 70- *Mem:* Entom Soc Am; Am Mosquito Control Asn; Am Soc Trop Med & Hyg; Sigma Xi. *Res:* Epidemiology of mosquito borne encephalitis viruses; arbovirus vector potential of mosquitoes; biology and ecology of mosquitoes; biological control of mosquitoes; taxonomy of Tenebrionidae. *Mailing Add:* 899 Bmicker Blvd Columbus OH 43221

BERRY, RICHARD STEPHEN, b Denver, Colo, Apr 9, 31; m 55; c 3. PHYSICAL CHEMISTRY. *Educ:* Harvard Univ, AB, 52, AM, 54, PhD(chem), 56. *Prof Exp:* Instr chem, Harvard Univ, 56-57 & Univ Mich, 57-60; asst prof, Yale Univ, 60-64; assoc prof, 64-67, PROF, DEPT CHEM & JAMES FRANCK INST, UNIV CHICAGO, 67-, PROF, COMT PUB POLICY STUDIES, 75-, JAMES FRANCK DISTINGUISHED SERV PROF, 89- *Concurrent Pos:* Alfred P Sloan fel, 62-66; guest prof, Copenhagen Univ, 67, 79; Arthur D Little prof, Mass Inst Technol, 68; lectr, Int Sch Physics, Varenna, 68 & 88 & Int Union Pure & Appl Chem Cong, Sydney, 69; chmn, Subpanel Physics Chem, Physics Surv comt, Nat Acad Sci, 70-73; assoc ed, J Chem Physics, 72-74 & Reviews Mod Physics, 83-; Guggenheim fel, 73; dir, Bull Atomic Scientists, 75-81; trustee, Aspen Ctr Physics, 76-82; mem, numerous sci comts, var socs & govt agencies, 72-; mem, Nat Res Coun, Chem Sci Comt, 80-, chmn, Numerical Data Adv Bd, 81-86; Newton Abraham Prof, 86-87; sr adv & consult, Defense Sci Study Group, Inst Defense Anal, 85-; MacArthur fel, 83. *Honors & Awards:* Seydel-Woolley Lectr, Ga Inst Technol, 71-73; Snider Lectr, Univ Toronto, 79; Hinshelwood Lectr, Oxford Univ, 80; Walter Kaskan Mem Lectr, State Univ NY, 82; L06wdin Lectr, Uppsala Univ, 89; Phillips Lectr, Haverford Col, 90. *Mem:* Nat Acad Sci; fel Am Acad Arts & Sci; Am Chem Soc; Royal Danish Acad Sci; fel Am Phys Soc; fel Japan Soc Prom Sci; fel AAAS. *Res:* Atomic-molecular processes; spectroscopy; thermodynamics. *Mailing Add:* Dept Chem Univ Chicago 5735 S Ellis Ave Chicago IL 606371

BERRY, RICHARD WARREN, b Quincy, Mass, June 21, 33; m; c 3. MARINE GEOLOGY, MARINE GEOCHEMISTRY. *Educ:* Lafayette Col, BS, 55; Wash Univ, MA, 57, PhD(geochem), 63. *Prof Exp:* Instr geol, Trinity Col, Conn, 59-61; from asst prof to assoc prof, 61-71, assoc dir, Ctr Marine Studies, 75-77, chmn, Dept Geol Sci, 76-80, PROF GEOL, SAN DIEGO STATE UNIV, 71-, DIR, CLAY MINERALS ANALYSIS LAB, 80- *Concurrent Pos:* Fulbright prof, Univ Baghdad, 65-66; Royal Norweg Coun Indust & Sci Res fel, Univ Oslo, 68-69; consult, Nat Coun Educ Geol Sci, various mining & petrol indust co. *Mem:* AAAS; Mineral Soc Am; fel Geol Soc Am; Soc Explor Paleontologists & Mineralogists; Clay Minerals Soc; fel Am Inst Chemists. *Res:* Mineralogy and geochemistry of clay sized ocean bottom sediments and their relationship to cationic concentrations in sea water; distribution of unconsolidated Quaternary sediments on the continental shelf; diagenesis and clay mineralogy. *Mailing Add:* Dept Geol Sci San Diego State Univ San Diego CA 92182-0337

BERRY, ROBERT EDDY, b East Prairie, Mo, Jan 23, 30; m 51; c 3. SUBTROPICAL FOODS. *Educ:* Vanderbilt Univ, AB, 51; Univ Mo, MS, 57, PhD(agr biochem), 59. *Prof Exp:* From asst to instr agr biochem, Univ Mo, 55-59; appl res chemist, Nestle Co, Ohio, 59-63; invests head, US Fruit & Veg Prod Lab, 63-72, DIR, US SUBTROP PROD LAB, 72- *Mem:* AAAS; Am Chem Soc; Inst Food Technol. *Res:* New citrus products; pollution abatement; space foods; analysis of natural products; development of food processing, especially methods of dehydration; chemical changes in foods processing and storage; chemistry of citrus; nucleotides and nucleic acids. *Mailing Add:* Subtrop Prod Res Lab USDA 3000 W Lake Hartridge Dr No 312 Winter Haven FL 33881

BERRY, ROBERT JOHN, b Belleville, Ont, Dec 11, 29; m 55; c 2. SOLID STATE PHYSICS. *Educ:* Queen's Univ, Ont, BSc, 51, MSc, 52; Ottawa Univ, PhD(solid state physics), 72. *Prof Exp:* Assoc res officer, 52-69, SR RES OFFICER, HEAT & SOLID STATE PHYSICS, NAT RES COUN CAN, 69- *Mem:* Can Asn Physicists. *Res:* Temperature standards and scales; resistance thermometry; electrical restivity of metals. *Mailing Add:* Physics Div Nat Res Coun Ottawa ON K1A 0R6 Can

BERRY, ROBERT WADE, b Granbury, Tex, July 21, 30; m 55; c 3. PLANT PATHOLOGY. *Educ:* Tex A&M Univ, 52, MS, 60; Univ Wis, PhD(plant path), 63. *Prof Exp:* County agr agent, Tex Agr Exten Serv 54-58, plant pathologist, 63-86; RETIRED. *Mem:* Am Phytopath Soc. *Res:* Disease control in plants. *Mailing Add:* 513 E Kent Lubbock TX 79403

BERRY, ROBERT WALTER, b Atlanta, Ga, Oct 27, 28; m 54; c 2. INORGANIC CHEMISTRY, PHYSICAL CHEMISTRY. *Educ:* Clemson Agr Col, BS, 50; Mich State Univ, PhD(chem), 56. *Prof Exp:* mem tech staff, Bell Tel Labs, 56-67, head, thin film technol dept, 67-77; head, components & mat dept, AT&T Bell Labs, 77-87; RETIRED. *Mem:* Am Vacuum Soc; Inst Elec & Electronics Engrs; Sigma Xi; Optical Soc Am. *Res:* Electro-optic and dielectric materials and thin metallic and dielectric films as applied to electrical and optical components. *Mailing Add:* 425 Rockhill Circle Bethlehem PA 18017

BERRY, ROBERT WAYNE, b Gilmer, Tex, Sept 14, 44; m 75; c 2. NEUROBIOLOGY. *Educ:* Calif Tech Inst, BS, 67; Univ Ore, MS, 68, PhD(biol), 70. *Prof Exp:* Asst prof, 73-78, ASSOC PROF CELL BIOL & ANAT, SCH MED, NORTHWESTERN UNIV, CHICAGO, 78- *Concurrent Pos:* USPHS fel neuropath, Albert Einstein Col Med, 70-71; Nat Inst Neurol Dis & Stroke fel, Calif Inst Technol, 71-72; vis scientist, Alberta Heritage Found, Univ Calgary, 81. *Mem:* Am Soc Neurochem; Soc Neurosci; AAAS. *Res:* Neuronal protein and RNA synthesis. *Mailing Add:* Dept Cell Biol & Anat Northwestern Univ Med Sch 303 E Chicago Ave Chicago IL 60611

BERRY, ROY ALFRED, JR, b New Hebron, Miss, Dec 11, 33; m 57; c 2. ORGANIC CHEMISTRY. *Educ:* Miss Col, BS, 56; Univ NC, PhD(org chem), 62. *Prof Exp:* NSF res fel org chem, Univ Fla, 61-62; asst prof, 62-69, prof, 69-77, J B PRICE PROF CHEM, MILLSAPS COL, 77- *Concurrent Pos:* R J Reynolds res fel, 59-60; Petrol Res Fund asst, 60-61. *Mem:* Am Chem Soc. *Res:* The basicities of the ferrocenylazobenzenes halogenation decarboxylation. *Mailing Add:* Dept Chem Millsaps Col Jackson MS 39210

BERRY, SPENCER JULIAN, b Quincy, Mass, May 25, 33; m 57; c 3. INSECT PHYSIOLOGY. *Educ:* Williams Col, BA, 55; Wesleyan Univ, MA, 57; Western Reserve Univ, PhD(biol), 65. *Prof Exp:* Technician biochem, Harvard Univ Huntington Labs, Mass Gen Hosp, 57-58; from asst to assoc prof biol, 64-77, PROF BIOL, WESLEYAN UNIV, 77- *Mem:* Soc Develop Biol; AAAS; Int Soc Develop Biol. *Res:* Physiology of insect development, particularly at the cellular and molecular level. *Mailing Add:* Dept Biol Atwater Sci Ctr Wesleyan Univ Middletown CT 06457

BERRY, STANLEY Z, b NY, May 10, 30; m 59; c 2. PLANT BREEDING. *Educ:* Cornell Univ, BS, 52; Univ NH, MS, 53; Univ Calif, PhD(plant path), 56. *Prof Exp:* Plant scientist, USDA, 56-60; plant breeder, Campbell Soup Co, 60-67; assoc prof, 67-77, PROF HORT, OHIO STATE UNIV & OHIO AGR RES & DEVELOP CTR, 77- *Mem:* Am Phytopath Soc; Am Soc Hort Sci. *Res:* Plant breeding and the utilization of disease resistance; tomato and vegetable breeding; processing tomato quality. *Mailing Add:* Dept Hort Ohio State Univ Agr Res & Develop Ctr Wooster OH 44691

BERRY, VERN VINCENT, b Dayton, Ohio, March 25, 41. CHROMOTOGRAPHY, SEPARATIONS. *Educ:* Mass Inst Technol, BS, 64; Northeastern Univ, MS, 70, PhD(chem), 72, MBA, 77. *Prof Exp:* Protein chemist, Found Res Nervous Syst, 64-67; synthetic chemist, Collab Res, 67-69; teaching asst chem, Northeastern Univ, 69-70, res fel chromatography, 70-72; res fel, Univ Saarbrucken, Ger, 72-73; group leader, Gillette Corp, 73-77; res group leader, Polaroid Corp, 77-83; asst prof, 83-88, ASSOC PROF, SALEM STATE COL, 88-; PRES, SEPCON SEPARATIONS CONSULTS, BOSTON, 83- *Concurrent Pos:* Lectr liaison, Gillette Corp, 74-77; adj prof continuing educ, Northeastern Univ, 79-83; vis researcher, Knaver Corp, Berlin, 83, Univ Nagoya, 84, Univ Gent, Belgium, 85, Shinalzre Corp, Kyoto, Japan, 85; vis researcher & prof, Inst Chromatography, Univ Pretoria Repub SAfrica, 86, 87 & 88; pres, Jour Liquid Chromatography, 86- *Mem:* Am Chem Soc; Electrophoresis Soc. *Res:* Develop universal gas and liquid chromatography methods that allow all components to be quantitated in a first run to eliminate method development; automate sample prep for chromatography and metals analysis. *Mailing Add:* 326 Reservoir Rd Boston MA 02167-1451

BERRY, VINOD K, b Lahore, India, Nov 30, 36; m 69; c 2. MATERIALS SCIENCE, ELECTRON MICROSCOPY. *Educ:* Agra Univ, BSc, 57; Eastern NMex Univ, MS, 76; NMex Inst Mining & Technol, PhD(metall), 77. *Prof Exp:* Sci asst electron micros, Electron Micros Div, Defence Sci Lab, Delhi, India, 60-72; res assoc metall, NMex Inst Mining & Technol, 77; asst prof, Health Sci Ctr, Univ Tex, San Antonio, 78-80; res leader, Dow Chem, 82-89; TECH MGR, ELEC DEPT, GEN ELEC PLASTICS, 89- *Mem:* Mat Res Soc; Am Inst Mining, Metall & Petrol Engrs; Am Soc Cell Biol; Electron Micros Soc Am; Microbeam Analytical Soc. *Res:* Bulk and surface studies of materials using transmission and scanning electron microscopy and various analytical electron optical techniques; morphological and ultrastructural studies of biological tissues and microbiological specimens. *Mailing Add:* GE Plastics Tech Ctr Gen Elec Plastics Washington DC 26181

BERRY, WILLIAM B(ERNARD), b Shelby, Ohio, July 23, 31; m 55; c 4. ELECTRICAL ENGINEERING, ELECTRONICS. *Educ:* Univ Notre Dame, BS, 53, MS, 57; Purdue Univ, PhD(elec eng), 64. *Prof Exp:* From instr to asst prof elec eng, Marquette Univ, 57-61; from asst prof to assoc prof, 63-70, asst dean res, 74-85, prof elec eng, 70-, PROG MGR, COLD WEATHER TRANSIT TECHNOL, 79-, ACTG DEPT CHMN, UNIV NOTRE DAME, 88- *Concurrent Pos:* Sr res assoc, Nat Res Coun, 72-73; vis prof, SERI, 86-87; Eng Res Coun. *Mem:* Sigma Xi; Electrochem Soc; Am Soc Eng Educ; Inst Elec & Electronics Engrs. *Res:* Solid state energy conversion including thermoelectric phenomena; photovoltaic and thin film electronic phenomena. *Mailing Add:* Dept Elec & Computer Eng 402 Pokagon South Bend IN 46617

BERRY, WILLIAM BENJAMIN NEWELL, b Boston, Mass, Sept 1, 31; m 61; c 1. PALEOOCEANOGRAPHY. *Educ:* Harvard Univ, AB, 55; Yale Univ, PhD(geol), 57. *Prof Exp:* Asst prof geol, Univ Houston, 57-58; vis asst prof paleont, Univ Calif, Berkeley, 58-60, from asst prof to assoc prof, 60-68, assoc dir mus paleont, 63-69, vchmn dept paleont, 67-69, dir mus paleont & chmn dept, 75-87, prof paleont, 68-91, CUR PALEOZOIC INVERT, MUS

PALEONT, UNIV CALIF, BERKELEY, 60-, PROF GEOL, 91- *Concurrent Pos:* Guggenheim fel, 66-67. *Mem:* Paleont Soc; Norweg Geol Soc; Explorers Club. *Res:* Graptolites; paleozoic biostratigraphy; paleo-oceanography; environmental science. *Mailing Add:* Dept Geol Univ Calif Berkeley CA 94720

BERRY, WILLIAM LEE, b Auburn, NY, Sept 1, 27; m 50; c 2. ORGANIC CHEMISTRY. *Educ:* Cornell Univ, BA, 51; Univ Mich, PhD(chem), 56. *Prof Exp:* Res chemist, 55-57, group leader, 57-60, dir org pigments res & develop, 60-63, prod mgr, Pigments Div, 63-67, from asst tech dir to tech dir, 67-71, dept mgr color pigments, 71-73, dept mgr, Inorg Chem Div, 73-78, vpres, 78-80, pres, Formica Div, 80-81, PRES, ORG CHEM DIV, AM CYANAMID CO, 81- *Mem:* Am Chem Soc; Sigma Xi. *Res:* Synthetic organic chemistry; colored hetero and carbocyclic systems and their utilization; high performance thermoplastic and thermoset resins. *Mailing Add:* 545 Steel Gap Rd Bridgewater NJ 08807

BERRYHILL, DAVID LEE, b Council Bluffs, Iowa, Mar 9, 44; m 68. BACTERIOLOGY. *Educ:* Simpson Col, BA, 66; Iowa State Univ, MS, 69, PhD(bact), 71. *Prof Exp:* Asst prof, 71-75, ASSOC PROF BACT, NDAK STATE UNIV, 75- *Mem:* Am Soc Microbiol; Int Soc Plant Molecular Biol; Genetics Soc Am; Sigma Xi. *Res:* Bacterial genetics. *Mailing Add:* Dept Bact NDak State Univ 195 Van ES Fargo ND 58105

BERRYMAN, ALAN ANDREW, b Tanzania, Africa, Jan 5, 37; m; c 3. ECOLOGY, ENTOMOLOGY. *Educ:* Univ London, BSc, 59; Univ Calif, Berkeley, MS, 61, PhD(entom), 64. *Prof Exp:* From asst prof to assoc prof, 64-74, PROF ENTOM, WASH STATE UNIV, 74- *Mem:* Entom Soc Am; Ecol Soc Am; Entom Soc Can. *Res:* Population dynamics of forest insects; mathematical and computer models of population dynamics; resistance of conifers to insect and fungus invasion; theoretical ecology. *Mailing Add:* Dept Entom Wash State Univ Pullman WA 99164

BERRYMAN, GEORGE HUGH, b South Shields, Eng, Apr 3, 14; nat US; m 39; c 3. NUTRITION, ALLERGY & IMMUNOLOGY. *Educ:* Univ Scranton, BS, 35; Pa State Col, MS, 36; Univ Minn, PhD(biochem, human nutrit), 41; Univ Chicago, MD, 50. *Prof Exp:* Asst animal nutrit, Univ Ill, 37-38; chief, Div Food & Nutrit, Army Med Sch, Washington, DC, 41-44; commanding officer & dir res, Med Nutrit Lab, Chicago, 44-46; head, Nutrit Br, Qm Food & Container Inst, Chicago, 48-49; rotating intern, USPHS Hosp, Staten Island, 50-51; assoc, 51-54, head clin invest, 54-58, dir med sci proj, 58-60, dir clin develop, 60-64, med dir, 65-70, vpres med affairs, Dept Med, Abbott Labs, 70-73; assoc prof med, 53-62, EMER PROF MED, STANFORD UNIV, 85- *Concurrent Pos:* From clin asst prof to clin assoc prof med, Univ Ill, 53-77; assoc prof med, 71-85, emer prof med, Stanford Univ, 85- *Mem:* AAAS; Am Med Asn; Am Soc Clin Nutrit; Am Acad Allergy; fel Am Col Physicians; Am Asn Cert Allergists. *Res:* Nutritional status; appraisal; relation of food; food composition; military nutrition; metabolic disease; allergy; drug reactions. *Mailing Add:* One Gem Ave Los Gatos CA 95032

BERRYMAN, JACK HOLMES, b Salt Lake City, Utah, July 28, 21; m 41, 82; c 2. RESOURCE MANAGEMENT, FISH & WILDLIFE MANAGEMENT. *Educ:* Westminster Col, AA, 40; Univ Utah, BS, 47, MS, 48. *Prof Exp:* Proj leader big game, Utah State Dept Fish & Game, 47-48, actg coordr wildlife, 48-50; asst regional supvr, US Fish & Wildlife Serv, NMex, 50-53, Minn, 53-59; assoc prof wildlife, Utah State Univ, 59-65; chief div wildlife serv, Bur Sport Fisheries & Wildlife, US Dept Interior, 65-74; actg dep assoc dir, US Fish & Wildlife Serv, 74, chief, Div Tech Assistance, 74-76, chief, Off Exten Educ, 76-78; exec vpres, 78-88, EMER COUNR, INT ASN FISH & WILDLIFE AGENCIES, 88- *Concurrent Pos:* Consult, Off Secy Interior, 62-64 & Gov Comt State Recreation Plan, 64-; deleg, White House Conf Conserv, 62; secy, Int Task Force Wetlands, 82-84; chmn, conserv comt, Boy Scouts Am, 84-; secy, Agr Adv Comt Animal Damage, 88-; consult, 90- *Honors & Awards:* Am Motors Conserv Prof Award, 76; Distinguished Serv Award, US Dept Interior, 79; Seth Gordon Award, 85. *Mem:* Wildlife Soc (vpres, 60-62, pres, 64); NY Acad Sci; Am Fisheries Soc; Sigma Xi; Am Forestry Asn; hon mem Can Wildlife Fedn; hon mem Int Asn Fish & Wildlife Agencies. *Res:* Relationships of land and resource use; energy developments; population requirements to fish and wildlife resource management and use. *Mailing Add:* 10503 Linfield St Fairfax VA 22032

BERRYMAN, JAMES GARLAND, b Hutchinson, Kans, June 26, 47; m 73; c 2. MATHEMATICAL PHYSICS. *Educ:* Univ Kans, BA & BS, 69; Univ Wis, Madison, MS, 70, PhD(physics), 75. *Prof Exp:* Asst scientist, Math Res Ctr, Univ Wis, 75-76; res physicist, Explor Res Div, Conoco, 76-77; vis mem, Courant Inst, NY Univ, 77-78 & 87-88; mem tech staff, Ocean Syst, Bell Labs, 78-81; physicist, Earth Sci Dept, 81-85, physicist & group leader electronics eng, 85-90, PHYSICIST & GROUP LEADER, EARTH SCI DEPT, LAWRENCE LIVERMORE NAT LAB, 90- *Concurrent Pos:* Woodrow Wilson fel, 69; NSF Postdoctoral fel, 77-78; assoc ed, J Math Physics, 79-81. *Mem:* Am Phys Soc; Am Geophys Union; Soc Explor Geophysics. *Res:* Applied mathematical physics; wave propagation in porous materials and composites; fluid flow through porous media; nonlinear waves; nonlinear diffusion; tomography and inverse problems in geophysics. *Mailing Add:* Lawrence Livermore Nat Lab PO Box 808 L-202 Livermore CA 94550

BERS, ABRAHAM, b Cernauti, Romania, May 28, 30; US citizen; m 66; c 2. ELECTRICAL ENGINEERING, PLASMA PHYSICS. *Educ:* Univ Calif, Berkeley, BS, 53; Mass Inst Technol, SM, 55, ScD(elec eng), 59. *Prof Exp:* Res asst elec eng, 53-58, from instr to asst prof, 58-63, assoc prof elec commun, 63-71, PROF ELEC ENG, MASS INST TECHNOL, 71- *Concurrent Pos:* Consult sci & eng, various orgns; Ford Found fac exchange fel, Tech Univ, Berlin, 66; John Simon Guggenheim Mem fel, 68-69; assoc prof, Univ Paris, 68-69 & 79-80; dir res, Physics Lab, Polytech Sch, Paris, 79-80, vis prof, Univ Paris, Orsay, 81-88; prin investr grants, Nat Sci Found & Dept Energy, 71- *Mem:* AAAS; Am Phys Soc; Inst Elec & Electronics Engrs. *Res:* Microwave electron beam devices; network and field theory; acoustic surface waves; plasmas; fusion; plasma electrodynamics for fusion energy generation; plasma heating with electromagnetic power sources; surface acoustoelectronic devices; surface wave convolvers; memory correlators. *Mailing Add:* Dept Elec Eng 38-260 Cambridge MA 02139

BERS, LIPMAN, b Riga, Latvia, May 22, 14; nat US; m 38; c 2. MATHEMATICS. *Educ:* Univ Prague, Dr rer nat(math), 38. *Hon Degrees:* DSc, State Univ NY, Stony Brook, 85. *Prof Exp:* Asst dynamics & mech, Brown Univ, 42-43, res instr, 43-44, sr res mathematician, 44-45; from asst prof to assoc prof math, Syracuse Univ, 45-51; mem staff, Inst Advan Study, 49-51; vis prof, NY Univ, 51-53, prof, 53-64, chmn dept, Grad Sch, 59-64; prof, Columbia Univ, 64-73, chmn dept, 72-75, Davies prof math, 73-82, special prof, 82-84; vis prof, Grad Ctr, City Univ NY, 84-88; RETIRED. *Concurrent Pos:* Guggenheim fel, 59-60 & 78; Fulbright award, 59-60; ed, Trans, Am Math Soc, 59-64; chmn math sect, Nat Acad Sci, 67-70; chmn math sci, Nat Res Coun Div, 69-71; at Nat Bur Standards & Nat Adv Comt Aeronaut; chmn, US Nat Comt Math, 77-81. *Honors & Awards:* Steele Prize, Am Math Soc, 74. *Mem:* Nat Acad Sci; fel Am Acad Arts & Sci; fel AAAS; Am Math Soc (vpres, 64-65, pres, 75-77); fel Am Philos Soc; foreign mem Finnish Acad Sci & Lett; hon mem NY Acad Sci; 050757860inland Asn Arts & Sci. *Res:* Complex function theory and its generalizations; partial differential equations; gas dynamics. *Mailing Add:* 111 Hunter Ave New Rochelle NY 10801

BERSCH, CHARLES FRANK, b Sheboygan, Wis, Jan 23, 27; m 50, 72; c 2. MATERIALS ENGINEERING, AERONAUTICAL & ASTRONAUTICAL ENGINEERING. *Educ:* St Norbert Col, BS, 51. *Prof Exp:* Chemist polymers, Nat Bur Standards, 51-58, phys chemist, 58-59; mat engr nonmetals, Bur Aeronaut, US Navy, 59 & Bur Naval Weapons, 59-66, mat engr, 66-73, supvy mat engr nonmetals, Naval Air Systs Command, 73-79, head, Mat Br, Naval Air Systs Command, 79-80; mgr composites prog, NASA, 80-84, actg mgr, Mat & Structures Off, 82-84; CONSULT, INST DEFENSE ANALYSIS, 84- *Concurrent Pos:* Navy liaison mem, Nat Mat Adv Bd, Nat Acad Sci-Nat Acad Eng, 65-79; US Nat Leader & Navy rep, Tech Coop Prog Polymeric Mat, 73-80; chmn Naval Adv Coun Mat, 76-78; consult, Dept Energy, 79-83. *Mem:* Fel Am Ceramic Soc; Am Chem Soc; AAAS. *Res:* Composite,particularly property relationships of polymers and ceramics; interface phenomena of composites and adhesives; ceramic bearings; properties and design of transparents; processing and application of advanced composites. *Mailing Add:* 4986 Sentinel Dr Bethesda MD 20816-3518

BERSHAD, NEIL JEREMY, b Brooklyn, NY, Oct 20, 37; c 2. COMMUNICATION THEORY, SIGNAL PROCESSING. *Educ:* Rensselaer Polytech Inst, BEE, 58, PhD(elec eng), 62; Univ Southern Calif, MS, 60. *Prof Exp:* Staff engr radar signal processing, Hughes Aircraft Co, 58-62; lectr network theory, Northeastern Univ, 62-65; staff engr radar signal processing, Hughes Aircraft Co, 65-69; PROF COMMUN INFO THEORY, UNIV CALIF, IRVINE, 66- *Concurrent Pos:* consult underwater acoustic signal processing, 66-; assoc ed, Transactions Communs, Inst Elec & Electronics Engrs, 74-80, Transaction Acoust, Speech & Signal Processing, 88-90. *Mem:* Fel Inst Elec & Electronics Engrs; Sigma Xi. *Res:* Communication theory; stochastic signal processing; acoustic array processing. *Mailing Add:* 1621 Santiago Dr Newport Beach CA 92660

BERSHADER, DANIEL, b New York, NY, Mar 14, 23; c 2. AEROPHYSICS. *Educ:* Brooklyn Col, AB, 42; Princeton Univ, MA, 46, PhD(physics), 48. *Prof Exp:* Instr physics, Princeton Univ, 43-44; flight res electronic engr, Bell Aircraft Corp, NY, 44; physicist, Naval Ord Lab, Washington, DC, 44-45; instr physics, Palmer Phys Lab, Princeton Univ, 48-49; res assoc, Univ Md, 49-51, assoc res prof, 51-52; res assoc & assoc prof, Princeton Univ, 52-56; assoc prof, 56-64, PROF AEROPHYSICS, STANFORD UNIV, 64-, ASSOC CHMN DEPT, 74- *Concurrent Pos:* Mgr gas dynamics, Lockheed Res Lab, 56-64; NATO lectr, France & Italy, 67; distinguished vis prof, Syracuse Univ, 69, vis prof, Imperial Col, London, 67, Technion, Israel, 76 & Univ Sydney, Australia, 83; consult, res & develop. *Honors & Awards:* Vielle Lectr, Japan, 75. *Mem:* Fel Am Phys Soc; fel Am Inst Aeronaut & Astronaut; Am Soc Eng Educ; Am Asn Physics Teachers. *Res:* Kinetic processes in high speed air flow; optical methods in fluid dynamics; plasma dynamics; physics of planetary entry. *Mailing Add:* Dept Aeronaut & Astronaut 937 Wing Pl Stanford CA 94305

BERSOHN, MALCOLM, b New York, NY, May 13, 25; m 64. COMPUTATIONAL LINGUISTICS, ARTIFICIAL INTELLIGENCE. *Educ:* Harvard Univ, BS, 43; Columbia Univ, MA, 57, PhD(chem), 60. *Prof Exp:* Res scientist, Calif Res Corp, 60-62; asst prof, 62-66, PROF CHEM, UNIV TORONTO, 66- *Concurrent Pos:* Guggenheim fel, Comput Sci Dept, Stanford Univ, 69-71; vis prof, Univ Tokyo, 78. *Mem:* Am Chem Soc; Am Asn Artificial Intel. *Res:* Natural language processing of technical documents. *Mailing Add:* Dept Chem Univ Toronto Toronto ON M5S 1A1 Can

BERSOHN, RICHARD, b New York, NY, May 13, 25; c 4. CHEMICAL DYNAMICS. *Educ:* Mass Inst Technol, BS, 44; Harvard Univ, MA, 47, PhD(chem physics), 50. *Prof Exp:* From asst prof to assoc prof chem, Cornell Univ, 51-59; assoc prof, 59-62, HIGGINS PROF CHEM, COLUMBIA UNIV, 62-, CHMN DEPT, 90- *Concurrent Pos:* NSF fel, Univ Paris, 58; head, Div Chem Physics, Am Phys Soc, 71; Guggenheim fel, Univ Tel Aviv, 72; chmn, Adv Comt Chem Dept, Brookhaven Nat Lab, 81-84; mem, Comt Atomic & Molecular Sci, Nat Res Coun, 84-87. *Honors & Awards:* Herbert Broida Prize, Am Phys Soc, 85. *Mem:* Nat Acad Sci; Am Acad Arts & Sci. *Res:* Dynamics of chemical reactions and photodissociation; state distributions of reaction products. *Mailing Add:* 959 Havemeyer Hall Columbia Univ Box 959 New York NY 10027

BERSON, ALAN, b New York, NY. CARDIOVASCULAR INSTRUMENTATION. *Educ:* City Col New York, BEE, 50; George Washington Univ, MSE, 65, PhD(physiol & biophysics), 72. *Prof Exp:* Sect chief comput, Litton, College Park, Md, 54-58; proj engr, Melpar, Arlington,

Va, 58-60; chief data processing & instrument cardiovasc res, Vet Admin Med Ctr, Washington, 60-78; DEP CHIEF, DEVICES & TECHNOL BR, NAT HEART, LUNG & BLOOD INST, 78- *Concurrent Pos:* Mem, stroke coun, Am Heart Asn, electrocardiograph comt, 62-78 & comput subcomt, 85-; res assoc, Georgetown Univ, 66-71; asst res prof, George Washington Univ, 71-78; chmn, electrocardiographs, cardiac monitors & sphygmomanometers subcomts & bd dirs mem, Asn Advan Med Instrumentation; mem, Bd Dirs, Int Soc Computerized Electrocardiology. *Mem:* Inst Elec & Electronics Engrs; AMA; Am Advan Med Instrumentation; Am Heart Asn. *Res:* Cardiovascular instrumentation including electrocardiography, phonocardiography, ultrasound and computers in biological and medical research. *Mailing Add:* NIH Devices & Tech Br Fed Bldg Rm 312 Bethesda MD 20892

BERSON, JEROME ABRAHAM, b Sanford, Fla, May 10, 24; m 46; c 3. ORGANIC CHEMISTRY. *Educ:* City Col New York, BS, 44; Columbia Univ, AM, 47, PhD(chem), 49. *Prof Exp:* Asst chemist, Hoffmann-La Roche, Inc, 44; lab asst, Columbia Univ, 46-49; Nat Res Coun fel, Harvard Univ, 49-50; from asst prof to prof chem, Univ Southern Calif, 50-63; prof, Univ Wis, 63-69; chmn dept, Yale Univ, 71-74, prof chem, 69-79, div, Dir Phys Sci & Eng, 83-90, IRENE E DU PONT PROF, YALE UNIV, 79- *Concurrent Pos:* Sloan fel, 57-61; NSF sr fel, 59-60; vis prof, Univ Calif, Los Angeles, 62, Univ Cologne, 65 & Univ Western Ont, 67; mem adv panel chem, NSF, 64-68; consult, Goodyear Tire & Rubber Co, 65-74; mem med chem study sect, NIH, 69-73; chmn div org chem, Am Chem Soc, 71; Sherman Fairchild distinguished scholar, Calif Inst Technol, 74-75; mem adv comt, Assembly Math & Phys Sci, Nat Res Coun, 75-77; Alexander von Humboldt sr US scientist award, 80. *Honors & Awards:* Sect Award, Am Chem Soc, Calif, 63 & Nichols Medal, NY sect, 85; James Flack Norris Award, Phys Org Chem, 78; Roger Adams Award, Am Chem Soc, 87. *Mem:* Nat Acad Sci; Am Chem Soc; Royal Soc Chem; Am Acad Arts & Sci. *Res:* Synthetic and theoretical organic chemistry; mechanisms of organic reactions. *Mailing Add:* Dept Chem Box 6660 Yale Univ New Haven CT 06511

BERSTED, BRUCE HOWARD, b Chicago, Ill, Sept 25, 40; m 68; c 2. POLYMER PHYSICS. *Educ:* Beloit Col, BA, 63; Univ Ill, MS, 65, PhD(phys chem), 69. *Prof Exp:* Sr res chemist, 69-81, RES ASSOC, AMOCO CHEM CORP, 81- *Concurrent Pos:* App to write test procedure vapor pressure osmometry, Am Soc Testing & Mat, 75- *Mem:* Am Phys Soc; Soc Rheology. *Res:* Methods of molecular weight determination; polymer rheology and its relation to molecular weight distribution; dynamic mechanical properties of polymers and composites; dependence of polymer toughness and impact resistance on molecular structure. *Mailing Add:* Amoco Chem Corp Res Ctr 916 Peidmont Circle Naperville IL 60565

BERSTEIN, ISRAEL, b Brichany, USSR, June 23, 26; div. MATHEMATICS. *Educ:* Univ Bucharest, MS, 54; Bucharest Inst Math, PhD(math), 58. *Prof Exp:* Res assoc math, Bucharest Inst Math, 54-58, sr res assoc, 58-59, calculator, 58-61; lectr, Israel Inst Technol, 61-62; from asst prof to assoc prof, 62-67, PROF MATH, CORNELL UNIV, 67- *Mem:* Am Math Soc. *Res:* Algebraic topology, especially homotopy theory. *Mailing Add:* Dept Math Cornell Univ White Hall Ithaca NY 14853

BERSU, EDWARD THORWALD, b Duluth, Minn, May 6, 46. ANATOMY, CYTOGENETICS. *Educ:* Univ Minn, Duluth, BA, 68; Univ Wis-Madison, PhD(anat), 76. *Prof Exp:* Asst prof, 76-82, ASSOC PROF ANAT, UNIV OF WIS-MADISON, 82- *Mem:* AAAS; Teratology Soc; Am Soc Human Genetics; Am Asn Cin Anatomists. *Res:* Detailed anatomical studies of human, genetically determined multiple congenital malformation syndromes; development analysis of trisomy syndromes in the mouse. *Mailing Add:* Rm 163 Med Sci Ctr Dept Anat Univ Wis 1300 University Ave Madison WI 53706

BERT, CHARLES WESLEY, b Chambersburg, Pa, Nov 11, 29; m 57; c 2. ENGINEERING MECHANICS, MECHANICAL ENGINEERING. *Educ:* Pa State Univ, BS, 51, MS, 56; Ohio State Univ, PhD(eng mech), 61. *Prof Exp:* Aeronaut design engr, Fairchild Aircraft Div, Fairchild Engine & Airplane Corp, Md, 54-56; prin mech engr, Appl Mech Div, Battelle Mem Inst, 56-59, solid & struct mech group, 61, sr engr, 61-62, prog dir, 62-63; instr eng mech, Ohio State Univ, 59-61; from assoc prof to prof, 63-78, BENJAMIN H PERKINSON PROF ENG, UNIV OKLA, 78-, GEORGE L CROSS RES PROF, 81-, DIR, SCH AEROSPACE, MECH & NUCLEAR ENG, 72-77 & 90- *Concurrent Pos:* Prin mech engr, Battelle Mem Inst, 59-61, consult, 63-65; consult, Inst Defense Anal & Univ Engrs, Inc, 64-66, Sandia Labs, 69-76, Firestone Tire & Rubber Co, 76-80 & Univ Technologists, Inc, 77-80; assoc ed, Exp Mech, 82-87 & Appl Mech Rev, 84-87; mem ed bd, Composite Structures, 82- & J Sound & Vibration, 88- *Honors & Awards:* Pressure Vessel & Piping Lit Award for 1975, Am Soc Mech Engrs, 76. *Mem:* Fel Am Soc Mech Engrs; fel Am Acad Mech; Am Soc Eng Educ; Nat Soc Prof Engrs; fel AAAS; Soc Eng Sci; fel Soc Exp Mech. *Res:* Composite-material structures; mechanical behavior of materials; structural dynamics; shell structures; thermal stress analysis. *Mailing Add:* Rm 212 Sch Aerospace & Mech Eng Univ Okla 865 Asp Ave Norman OK 73019

BERT, MARK HENRY, b Lima, Peru, May 1, 16; US citizen; m 60; c 2. NUTRITIONAL BIOCHEMISTRY, STATISTICS. *Educ:* Nat Col Agr & Vet Sci, Lima, BS, 39; Univ Ill, Urbana, MS, 48, PhD(nutrit biochem), 55. *Prof Exp:* Specialist in flax fiber processing, Dept Agr, Peru, 40-41; tech dir, Desfibradora de Lino, SA, 41-43; Inter-Am Trade scholar flax fiber processing firms, Ore, 44-46; consult var indust firms & Dept Agr, Peru, 46-47; res biochemist lab, Corn Prod Co, Mass, 55-58, dir res lab, 58-63, biochemist in-chg biochem sect, NJ, 63-65; asst prof nutrit biochem, 65-68, asst prof nutrit & food & head dept, 68-70, assoc prof, 70-80, PROF NUTRIT & FOOD, UNIV MASS, AMHERST, 80- *Concurrent Pos:* Univ Mass res grant, 66-68. *Mem:* Animal Nutrit Res Coun; Am Chem Soc; Am Statist Asn; Inst Food Technologists; fel Am Inst Chemists. *Res:* Amino acid metabolism; food irradiation effects on protein quality, vitamins, enzyme activity, lipids; nutritive value of algae; vitamin metabolism; statistical design and analysis; computer applications to nutrition; atherosclerosis; radioisotopic techniques. *Mailing Add:* 63 Tracy Circle Amherst MA 01002

BERTALANFFY, FELIX D, b Vienna, Austria, Feb 20, 26; nat Can; m 54. ANATOMY, HISTOLOGY. *Educ:* McGill Univ, MSc, 51, PhD(anat), 54. *Prof Exp:* Res asst, McGill Univ, 53-54; from asst prof to assoc prof, 55-64, PROF ANAT, UNIV MAN, 64- *Mem:* Am Asn Cancer Res; Am Asn Anatomists; Can Asn Anatomists; Can Soc Cytol; Int Soc Stereology; fel Royal Micros Soc; Int Soc Chronobiol. *Res:* Cancer; histochemistry and cytochemistry; exfoliative cytology; cytodynamics of normal and cancerous tissues; chemotherapeutic agents; respiratory system; fluorescence microscopy. *Mailing Add:* Dept Anat Univ Man Fac Med & Dent Winnipeg MB R3E 0W3 Can

BERTANI, GIUSEPPE, b Como, Italy, Oct 23, 23; nat US; m 54; c 2. MOLECULAR GENETICS, BIOTECHNOLOGY. *Educ:* Univ Milan, DrNatSc, 45. *Hon Degrees:* Dr, Univ Uppsala, 83. *Prof Exp:* Res fel biol, Zool Sta, Naples, 45-46; asst zool, Univ Milan, 46-47 & Univ Zurich, 47-48; res fel genetics, Carnegie Inst, Cold Spring Harbor, NY, 48-49; res assoc bact, Ind Univ, 49-50 & Univ Ill, 50-54; sr res fel biol, Calif Inst Technol, 54-57; assoc prof med microbiol, Med Sch, Univ Southern Calif, 57-60; vis prof microbial genetics, Karolinska Inst, Sweden, 60-64, prof, 64-83; SR SCIENTIST, JET PROPULSION LAB, CALIF INST TECHNOL, PASADENA, 81- *Mem:* AAAS; Genetics Soc Am; Am Soc Microbiol; Brit Soc Gen Microbiol; Europ Molecular Biol Orgn. *Res:* Genetics of bacteria and bacterial viruses; applied genetics; methanogenic bacteria. *Mailing Add:* Jet Propulsion Lab MS 125-112 4800 Oak Grove Dr Pasadena CA 91109

BERTANI, LILLIAN ELIZABETH, b Ind, July 9, 31; m 54; c 2. GENETICS, MOLECULAR BIOLOGY. *Educ:* Univ Mich, BS, 53; Calif Inst Technol, PhD(virol), 57. *Prof Exp:* Res assoc med microbiol, Univ Southern Calif, 57-60; res assoc microbial genetics, Karolinska Inst, Sweden, 61-65; Med Res Coun fel, Swedish Med Res Coun, 66-75; docent med microbiol, Karolinska Inst, 75-84; SR RES BIOLOGIST, CALIF INST TECHNOL, PASADENA, 85- *Concurrent Pos:* USPHS fel, 60-61; docent microbiol, Univ Stockholm, 65; vis assoc biol, Calif Inst Technol, Pasadena, 81-84. *Mem:* Am Soc Microbiol. *Res:* Temperate bacteriophages; plasmids; regulation. *Mailing Add:* Biol Div Calif Inst Technol Pasadena CA 91125

BERTAUT, EDGARD FRANCIS, b Chicago, Ill, May 23, 31; m 57; c 5. COMPUTER SCIENCE. *Educ:* Loyola Univ, Ill, BSc, 53; Carnegie Inst Technol, MS, 56, PhD(inorg chem), 58. *Prof Exp:* Instr chem, Carnegie Inst Technol, 57-58; asst prof, Pa State Univ, 58-60, fel mineral sci, 60-62; asst prof chem, Univ Detroit, 62-68; assoc prof natural sci, Univ DC, 68-70, prof & chmn div, 70-71, prof comput sci & chmn prog, 74-80. *Mem:* Am Chem Soc; Nat Sci Teachers Asn; Asn Comput Mach; AAAS. *Res:* High temperature spectra; televised instruction; application of computers to chemistry; computer graphics; operating systems. *Mailing Add:* 6635 Western Ave NW Washington DC 20015-2335

BERTEAU, PETER EDMUND, b Maidstone, Eng, Nov 21, 29; US citizen; m 86; c 2. TOXICOLOGY, ORGANIC CHEMISTRY. *Educ:* Queen Mary Col, Univ London, BSc, 52; Johns Hopkins Univ, MA, 57; Univ Calif Med Ctr, San Francisco, PhD(pharm chem), 64. *Prof Exp:* Sci officer, Dept Govt Chem, London, 52-57; res chemist, Chevron Chem Co, Richmond, Calif, 63-66; scientist, World Health Org, Switz, 68-72; asst res toxicologist, Univ Calif Naval Biosci Lab, Oakland, 73-77; sr toxicol specialist, Monsanto Co, 77-81; STAFF TOXICOLOGIST, CALIF DEPT HEALTH SERV, BERKELEY, 81- *Concurrent Pos:* Asst specialist, Dept Entom Univ Calif, 66-68, fel, 73. *Mem:* Am Chem Soc; Soc Toxicol. *Res:* Structural analogs of thyroxine; synthesis and insecticidal activity of pyrethroids; comparative toxicology of insecticide aerosols; inhalation toxicology of industrial chemicals. *Mailing Add:* Calif Dept Health Serv Rm 619 2151 Berkeley Way Berkeley CA 94704-1011

BERTELL, ROSALIE, b Buffalo, NY, Apr 4, 29. BIOMATHEMATICS. *Educ:* D'Youville Col, BA, 51; Cath Univ Am, MA, 59, PhD(math), 66. *Hon Degrees:* Dr, Mt St Vincent Univ, 85. *Prof Exp:* Teacher Bishop O'Hern High Sch, NY, 56-57; asst math, Cath Univ Am, 57-58; instr, Sacred Heart Jr Col, Pa, 58-62, assoc prof, 65-68; coordr & teacher, D'Youville Acad, Diocese Atlanta, 68-69; assoc prof math, D'Youville Col, 69-73; asst res prof, Grad Sch, State Univ NY Buffalo, 74-80; CONSULT, 80- *Concurrent Pos:* Sr cancer res scientist, Roswell Park Mem Inst, 70-78; cancer res grant, Cancer Res scientist & cancer res consult, 75-; staff expert for Pub Health & Energy Concern, Global Educ Assocs; found & pres bd dir, Ministry of Concern Pub Health, Buffalo & Int Inst of Concern Pub Health, Toronto; ed, Int Perspectives Pub Health. *Honors & Awards:* Hans Adalbert Schweigart Medal, 83. *Mem:* Sigma Xi; Health Phys Soc; Am Acad Polit & Soc Sci; Am Pub Health Asn; Int Biometric Soc; NY Acad Soc. *Res:* Mathematical statistics; analysis; measure theory; aging effect in humans associated with exposure to ionizing radiation; updating relative risk methodology for biomedical applications; life-style and chronic diseases. *Mailing Add:* Int Inst Concern Health 830 Bathurst St Toronto ON M5R 3G1 Can

BERTELO, CHRISTOPHER ANTHONY, b Orange, NJ, Apr 22, 47; m 73. ORGANOMETALLIC CHEMISTRY, PROCESS RESEARCH. *Educ:* Bates Col, BS, 69; Princeton Univ, MS, 73, PhD(chem), 75. *Prof Exp:* Res assoc chem, Univ Iowa, 75-77; vis asst prof, Grinnell Col, 77; sr chemist, Tenneco Chem, Inc, 78-82, Tenneco Polymers, 82-85; SR PROJ CHEMIST, M&T CHEMICALS, RAHWAY, NJ, 85- *Mem:* Am Chem Soc. *Res:* Plastic additives; polymer stabilizers; catalysts and impact modifiers. *Mailing Add:* 2694 Crest Lane Scotch Plains NJ 07076-1299

BERTELSEN, BRUCE I(RVING), b Portland, Maine, Sept 11, 26; m 51; c 3. PHYSICS, SURFACE & TRACE ANALYSIS. *Educ:* Univ Maine, BS, 54. *Prof Exp:* From physicist to adv physicist vacuum deposition, xerography & magnetic thin films, Gen Prod Div, Int Bus Mach Corp, NY, 54-60, res staff mem thin magnetic film memory device process res, Res Lab, Switz, 60-61, from group mgr to sr physicist thin magnetic film process develop, Components Div, 61-68, dept mgr magnetic film device & process develop, Systs Develop Div, 65-68, mgr, Device Thin Film Prod Eng, 68-70, mgr,

Tooling & Process Control, Integrated Circuit Prod Eng, 70-71, proj mgr process & mat develop, 71-73, dept mgr int circuit process line automation, 73-76, dept mgr surface & trace analysis, 76-81; sr physicist, IBM Corp, 80-83, consult tech writer, Patent Dept, 83-91; RETIRED. *Mem:* Sr mem Inst Elec & Electronics Engrs; Am Vacuum Soc. *Res:* Vacuum techniques; electrical and magnetic properties as function of deposition process parameters; computer applications for thin film components; magnetic thin film memory devices; surface and trace analysis, Auger, ESCA, SIMS, ISS, EDX, WDX, radioisotopes; materials and technology for very large scale integrated circuits. *Mailing Add:* Thayers Beach Rd Colchester VT 05446

BERTELSON, ROBERT CALVIN, b Milwaukee, Wis, Nov 5, 31; m 60; c 2. ORGANIC PHOTOCHEMISTRY, PHOTOCHROMIC DYESTUFFS. *Educ:* Univ Wis, BS, 52; Mass Inst Technol, PhD(org chem), 57. *Prof Exp:* Sr res chemist, Nat Cash Register Co, 57-61, group leader, 61-74; CHMN & PRES, CHROMA CHEM INC, 74- *Concurrent Pos:* Sr res assoc, Nat Acad Sci-Nat Res Coun, 72-74. *Mem:* NY Acad Sci; Am Chem Soc; Soc Photog Scientists & Engrs; Sigma Xi; Am Inst Chemists; Interamerican Photochem Asn. *Res:* Photochromic and photosensitive dyestuffs for optical information storage, exposure indicators and selectively absorbing filters. *Mailing Add:* 5312 Bliss Pl Dayton OH 45440

BERTERA, JAMES H, b Springfield, Mass, March 18, 48. VISION RESEARCH. *Educ:* Fairfield Univ, Conn, BA, 70; Hollins Col, Va, MA, 72, Univ Okla Med Ctr, PhD(biopsychol), 77. *Prof Exp:* Res assoc comput learning, NC State Univ, 71-72; mgr, Neuropsychol Lab, Univ Okla Health Sci, 72-77; res scientist vision, Systs Res Labs, Inc, 77-78; RES ASSOC VISION, PSYCHOL DEPT, UNIV MASS, 78-, STAFF ASSOC, VISION, EYE RES INST, 86-, ASSOC SCIENTIST, 90- *Res:* Human visual information processing investigated using eye-controlled display techniques; visual disease simulation with artificial scotoma on computer display; eye controlled typewriter. *Mailing Add:* 20 Staniford Boston MA 02114

BERTHOLD, JOHN WILLIAM, III, b York, Pa, May 24, 45; m 74; c 2. FIBER OPTIC SENSORS, INDUSTRIAL APPLICATIONS OF LASERS. *Educ:* Gettysburg Col, BA, 67; Univ Ariz, MS, 74, PhD(optical sci), 76. *Prof Exp:* Sr tech aide, Bell Tel Labs Inc, 67-69; from res asst to res assoc, Optical Sci Ctr, Univ Ariz, 69-76; physicist, US Dept Defense, 76-79; sr res physicist, Babcock & Wilcox Co, 79-83, group supvr, 83-84, sect mgr, 84-85, tech adv, 85-90, SCIENTIST, BABCOCK & WILCOX CO, 90- *Concurrent Pos:* Consult, Iota Eng Inc, 73-76; lectr optics & fiber optics. *Honors & Awards:* Tech Brief Award, NASA, 77. *Mem:* Optical Soc Am; Soc Photo-Optical Instrumentation Engrs; Inst Elec & Electronics Engrs; Instrument Soc Am. *Res:* Fiber optic intensity sensors for measurement of pressure, acceleration, temperature, strain, flow rate, liquid level, gas concentration, and process chemistry. *Mailing Add:* Babcock & Wilcox 1562 Beeson St Alliance OH 44601

BERTHOLD, JOSEPH ERNEST, b Paterson, NJ, Sept 30, 47; m 69; c 2. PHYSICS, ELECTRONICS ENGINEERING. *Educ:* Seton Hall Univ, BS, 69; Brown Univ, MS, 73, PhD(physics), 76. *Prof Exp:* Res assoc physics, Lab Atomic & Solid State Physics, Cornell Univ, 75-77; mem tech staff integrated circuits, 77-80, SUPVR, HIGH VOLTAGE INTEGRATED CIRCUITS, BELL LABS, 80- *Mem:* Electrochem Soc; Inst Elec & Electronics Engrs. *Res:* Semiconductor devices, integrated circuits; development of high voltage integrated circuit technology. *Mailing Add:* Div Mar Bell Commun Res 331 Newman Springs Rm 32-331 Redbank NJ 07701

BERTHOLD, ROBERT, JR, b Paterson, NJ, Aug 2, 41; m 66; c 2. APICULTURE, ENTOMOLOGY. *Educ:* Juniata Col, BS, 63; Rutgers Univ, MS, 65; Pa State Univ, PhD(entom), 68. *Prof Exp:* Res asst entom, Rutgers Univ, 63-66 & Pa State Univ, 66-68; asst prof, 68-76, asst chmn biol, 79-80, ASSOC DEAN SCI, 87- *Concurrent Pos:* Apiary inspector, Pa Dept Agr. *Mem:* Entom Soc Am; Am Entom Soc; Sigma Xi. *Res:* Insect behavior; German cockroach; honey bee; development of new honey products including marketing studies. *Mailing Add:* Dept Biol Delaware Valley Col Doylestown PA 18901

BERTHOLF, DENNIS E, b Harper, Kans, Aug 19, 41; m 62; c 3. MATHEMATICS. *Educ:* Univ Kans, BS, 63; NMex State Univ, MA, 65, PhD(math), 68. *Prof Exp:* Asst prof math, 68-74, ASSOC PROF MATH, OKLA STATE UNIV, 74- *Mem:* Math Asn Am. *Res:* Abelian group theory; homological algebra. *Mailing Add:* Dept Math & Statist Okla State Univ Stillwater OK 74078

BERTHOUEX, PAUL MAC, b Delwein, Iowa, Aug 15, 40. WATER SUPPLY & POLLUTION CONTROL, ENVIRONMENTAL STATISTICS. *Educ:* Univ Iowa, BS, 63, MS, 64; Univ Wis, Madison, PhD(civil eng), 69. *Prof Exp:* Instr prev med, Univ Iowa, 64-65; asst prof civil eng, Univ Conn, 65-67; chief res engr, GKW Consult, Manheim, WGer, 69-71; from asst prof to assoc prof, 71-76, PROF ENVIRON ENG, UNIV WIS, MADISON, 76- *Concurrent Pos:* Consult, Tenn Valley Authority, 72-77, Asian Develop Bank, 75-81, 85 & 89-90, WHO, 78-; mem, Comt Nat Statist, Nat Res Coun, 75-76. *Honors & Awards:* Eddy Medal, Water Pollution Control Fedn, 71, Radebaugh Award; Rudolph Hering Medal, Am Soc Civil Engrs, 75. *Mem:* Am Soc Civil Engrs; Int Asn Water Pollution Res & Control; Water Pollution Control Fedn; AAAS; Am Water Works Asn. *Res:* Drinking water supply and water pollution control, including toxic and hazardous waste treatment, with special interests in optimal process design and environmental statistics. *Mailing Add:* 3204 Eng Univ Wis Madison WI 53706

BERTHRONG, MORGAN, b Aurora Hills, Va, July 17, 18; m 43, 72; c 9. MEDICINE. *Educ:* Harvard Med Sch, MD, 43. *Prof Exp:* Resident path, Johns Hopkins Univ & Hosp, 46-50, from instr to assoc prof, 49-53; prof & head dept, 59-61, dir labs, Penrose Hosp, 67-79, CLIN PROF PATH, SCH MED, UNIV COLO, 61-, PATHOLOGIST, PENROSE HOSP, COLORADO SPRINGS, 54- *Concurrent Pos:* Vis prof path & actg head dept, Sch Med, Stanford Univ, 66-67; adj prof path, Univ NMex, 73-; sci adv bd, Armed Forces Inst Path, 82-88. *Mem:* Am Asn Path; Am Soc Clin Path; Int Acad Path; Col Am Path. *Res:* General anatomic pathology, especially diseases of lungs, liver, kidneys and congenital heart; experiments with tuberculosis, hypersensitivity and so-called collagen vascular diseases. *Mailing Add:* Dept Path Penrose Hosp Colorado Springs CO 80907

BERTIE, JOHN E, b London, Eng, Mar 24, 36; Can citizen. PHYSICAL CHEMISTRY, CHEMICAL PHYSICS. *Educ:* Univ London, PhD(phys chem), 60. *Prof Exp:* Fel chem, Div Appl Chem, Nat Res Coun Can, 60-62, asst res off, 62-65, assoc res off, 65-67; assoc prof, 67-75, PROF CHEM, UNIV ALTA, 75- *Concurrent Pos:* Pres, Asn Acad Staff, Univ Alta, 78-79, gov, 89-92. *Mem:* Am Phys Soc; Optical Soc Am; Chem Inst Can; Can Asn Physicists; Spectros Soc Can. *Res:* Infrared and Raman spectroscopy of molecules and crystals; infrared optical and dielectric constant of liquids; emphasis on hydrogen-bonded systems; spectroscopic and diffraction properties of molecular crystals. *Mailing Add:* Dept Chem Univ Alta Edmonton AB T6G 2G2 Can

BERTIN, JOHN JOSEPH, b Milwaukee, Wis, Oct 13, 38; m 62; c 4. AERODYNAMICS. *Educ:* Rice Inst, BA, 60; Rice Univ, MS, 62, PhD(mechand aerospace eng), 66. *Prof Exp:* Aerospace technologist, NASA Manned Spacecraft Ctr, Tex, 62-66; from asst prof to assoc prof aerospace eng, Univ Tex, Austin, 66-77, Beattie Margaret Smith prof eng, Aerospace Eng & Eng Mech & Res Scientist, Appl Res Labs, 77-89. *Concurrent Pos:* Consult, Tracor, 79-83, Difesa C Spazio, Italy, 80-81 & Sandia Nat Lab, 81- *Mem:* NY Acad Sci; fel Am Inst Aeronaut & Astronaut. *Res:* Reentry aerothermodynamics; aerodynamics; boundary-layer transition; reentry convective heat transfer; viscid/inviscid interactions; architectural aerodynamics; computational fluid dynamics; rocket exhaust flows. *Mailing Add:* 3313 LaSala Del Oenta NE Albuquerque NM 87111

BERTIN, MICHAEL C, b Bronx, NY, Sept 25, 42; m 64; c 2. DESIGN & CONSTRUCTION OF ON-LINE MEASUREMENT SYSTEMS, QUALITY CONTROL SYSTEMS. *Educ:* Mass Inst Technol, BS, 63; Rutgers Univ, MS, 65, PhD(physics), 68. *Prof Exp:* Res assoc, Physics Dept, Stanford Univ, 68-71; sr scientist, Nucleonic Data Systs, Div Sentrol Systs Inc, 71-74, dir eng, 74-79, dir res & develop, 79-81; pres, Gamma Instruments Inc, Subsid Sensor Control Corp, 81-88, consult, 88-89; PRES, QUALITY PROFILES, INC, 90- *Concurrent Pos:* Res assoc, Physics Dept, Rutgers Univ, 68; lectr physics, San Jose Univ, 71; assoc prof, Calif State Univ, 81-82. *Mem:* Am Phys Soc. *Res:* Application of measurement and analysis technology to on-line industrial processing; industrial gauging of thickness, width, length, density and other physical properties, and analysis and quality control for these processes; developed gauges and on-line computer systems for use in the metals, rubber, plastics and paper industries; on-line quality control and SPC data gathering and reporting systems. *Mailing Add:* Qual Profiles Inc PO Box 5032 Irvine CA 92716

BERTIN, ROBERT IAN, b Shreveport, La, Aug 4, 52. PLANT REPRODUCTIVE ECOLOGY. *Educ:* Hobart Col, BS, 73; Univ Conn, MS, 76; Univ Ill, PhD(biol), 80. *Prof Exp:* Vis asst prof, Biol, Bucknell Univ, 80-81, 83-84, Miami Univ, 81-83; asst prof 84-87, ASSOC PROF, COL HOLY CROSS, 87. *Mem:* Ecol Soc Am; Bot Soc Am; Soc Study Evolution. *Res:* Plant reproductive ecology; plant-animal interactions; campsis radicans; effects of pollination on fruit production and seed number and quality. *Mailing Add:* Biol Dept Holy Cross Col Worcester MA 01610

BERTINI, HUGO W, b Chicago, Ill, Dec 12, 26; m 54; c 3. PHYSICS. *Educ:* Northwestern Univ, MS, 51; Univ Tenn, PhD(physics), 62. *Prof Exp:* PHYSICIST, OAK RIDGE NAT LAB, 53- *Mem:* Am Phys Soc; Sigma Xi. *Res:* Nuclear reactor physics; high energy nuclear reactions. *Mailing Add:* PO Box 2003 Bldg K1546C Oak Ridge TN 37831

BERTINO, JOSEPH R, b Port Chester, NY, Aug 16, 30; m 56; c 4. PHARMACOLOGY, MEDICINE. *Educ:* State Univ NY Downstate Med Ctr, MD, 54. *Hon Degrees:* MA, Yale Univ, 69. *Prof Exp:* Intern, Grad Hosp, Univ Pa, 54-55; resident internal med, Vet Admin Hosp, Philadelphia, 55-56; asst prof pharmacol, 61-64, assoc prof, 64-69, prof med & pharmacol, Sch Med, Yale Univ, 69-86; HEAD, PROG MOLECULAR PHARMACOL & THERAPEUT, SLOAN-KETTERING CANCER CTR, 87- *Concurrent Pos:* USPHS res fel hemat & biochem, Sch Med, Univ Wash, 58-61, career develop award, Nat Cancer Inst, 64-74; consult, Nat Serv Ctr, 64-; Am Cancer Soc prof, 76- *Honors & Awards:* Exp Therapeut Award, Am Soc Pharmacol, 70; Richard & Hilda Rosenthal Award, Am Asn Cancer Res, 78. *Mem:* Am Soc Hemat; Am Asn Cancer Res; Am Fedn Clin Res; Am Soc Clin Invest; Am Soc Biol Chemists. *Res:* Leukocyte enzymes; folic acid and vitamin B12 metabolism; cancer chemotherapy; drug resistance; gene amplification; gene transfer. *Mailing Add:* Mem Sloan-Kettering Cancer Ctr 1275 York Ave New York NY 10021

BERTKE, ELDRIDGE MELVIN, histology, pathology; deceased, see previous edition for last biography

BERTLAND, ALEXANDER U, b Budapest, Hungary, 31; US citizen; m 65; c 2. HEPATITIS RESEARCH, BIOCHEMISTRY. *Educ:* Lafayette Col, AB, 57; Univ Del, MA, 60; Univ Ill, PhD(biochem & microbiol), 65. *Prof Exp:* SR RES FEL, MERCK INST THERAPEUT RES, 71- *Concurrent Pos:* Harvard Med Sch, Mass Gen Hosp. *Mem:* Sigma Xi. *Res:* Biochemistry; protein conformation, fluorescence, bacteriophage isolation HIV; microbiol. *Mailing Add:* Dept Virol Merck Inst Therapeut Res West Point PA 19486

BERTLES, JOHN F, b Spokane, Wash, June 8, 25; m 48, 81; c 3. INTERNAL MEDICINE, HEMATOLOGY. *Educ:* Yale Univ, BS, 45; Harvard Med Sch, MD, 52; Am Bd Internal Med, dipl, 61. *Prof Exp:* Intern & asst resident med, Presby Hosp, New York, 52-55; instr, Harvard Med Sch, 59-61; CHIEF, HEMAT ONCOL DIV, ST LUKE'S-ROOSEVELTS HOSP CTR, 62-, DIR, TRANSFUSION SERV; PROF MED, COL PHYSICIANS & SURGEONS,

COLUMBIA UNIV, 74- *Concurrent Pos:* USPHS res fel, Sch Med & Dent, Univ Rochester, 55-56; USPHS res fel, 56-57 & res fel, Harvard Med Sch, 56-59; from asst attend physician to attend physician, St Luke's Hosp Ctr, 62- *Mem:* Am Soc Clin Invest; Am Physiol Soc; fel Am Col Physicians; Am Soc Hemat; Am Fedn Clin Res. *Res:* Human hemoglobinopathies. *Mailing Add:* Hemat-Oncol Div St Luke's Hosp Ctr New York NY 10025

BERTNOLLI, EDWARD CLARENCE, b Kansas City, Mo, Aug 8, 35; m 56; c 4. ELECTRICAL ENGINEERING. *Educ:* Kansas State Univ, BS, 58, MS, 61, PhD(elec eng), 66. *Prof Exp:* Instr elec eng, Kansas State Univ, 58-65; assoc prof, 65-68, PROF ELEC ENG, UNIV MO-ROLLA, 68- *Concurrent Pos:* NSF res grant, 66-67; eng prog evaluator, Accreditation Bd Eng & Technol, 88. *Mem:* Inst Elec & Electronics Engrs; Am Soc Eng Educ. *Res:* System theory and design. *Mailing Add:* UMR Eng Ctr 8001 Natural Bridge Rd St Louis MO 63121

BERTOCCI, UGO, Derna, Libya, Jan 31, 26; US citizen. ELECTROCHEMISTRY, CORROSION. *Educ:* Univ Milan, Italy, Dr (chem), 50. *Prof Exp:* Asst prof phys chem, Milan Polytech Inst, 50-57; group leader, Italian Comt Nuclear Res, 58-61; chemist, Oak Ridge Nat Lab, 61-71; RES CHEMIST, NAT INST STANDARDS & TECHNOL, 71- *Honors & Awards:* Silver Medal, Dept of Com. *Mem:* Int Soc Electrochem; Electrochem Soc; Am Chem Soc; AAAS. *Res:* Electrodeposition; electrochemical properties of metal single crystals; computer simulation of crystal growth; effects of alternating current on corrosion; electrochemical methods for corrosion studies; electrochemical noise measurements; stress corrosion cracking. *Mailing Add:* Nat Inst Standards & Technol Gaithersburg MD 20899

BERTOLACINI, RALPH JAMES, b Pawtucket, RI, Aug 8, 25; m 53; c 3. INORGANIC CHEMISTRY. *Educ:* Univ RI, BS, 49; Mich State Univ, MS, 51. *Prof Exp:* Asst, Univ RI, 48-49 & Mich State Univ, 49-51; sr res scientist, Am Oil Co, 51-68, proj mgr & res assoc, Amoco Oil, 68-76, proj mgr & sr res assoc, 76-79, dir catalysis res, Res & Develop Div, 79-, mgr explor & catalysis res & develop, 82-; RETIRED. *Concurrent Pos:* Adj prof chem eng, Catalysis Tech Ctr, Univ Delaware, 91- *Honors & Awards:* Eugene Houdry Award, NAm Catalysis Soc, 87. *Mem:* Am Chem Soc; Sigma Xi; Am Inst Chem Engrs; Am Soc Testing & Mat; Catalysis Soc. *Res:* Catalysis; analytical chemistry. *Mailing Add:* 6S572 Millcreek Lane Naperville IL 60540

BERTOLAMI, CHARLES NICHOLAS, b Lorain, Ohio, Dec 31, 49; m 77; c 2. ORAL BIOLOGY, ORAL & MAXILLOFACIAL SURGERY. *Educ:* Ohio State Univ, DDS, 74; Harvard Univ, DMedSc, 79. *Prof Exp:* Intern oral surg, Mass Gen Hosp, 75-76, asst resident, 78-79, chief resident, 79-80; asst prof oral & maxillofacial surg, Univ Conn, 80-89, asst prof surg, 80-89; PROF, DEPT DENT, UNIV CALIF, LOS ANGELES, 89- *Concurrent Pos:* Clin fel, Harvard Sch Dent Med, 80- *Mem:* Int Asn Dent Res; Am Asn Dent Res; Am Dent Asn; Sigma Xi. *Res:* Wound healing as a developmental phenomenon; role of glycosaminoglycans in tissue repair; experimental animal model systems for hypertrophic scarring; phenomenon of wound contraction. *Mailing Add:* Dept Dent Univ Calif 10833 Le Conte Ave Los Angeles CA 90024

BERTOLDI, GILBERT LEROY, b Fresno, Calif, Mar 10, 38; m 58; c 3. HYDROLOGY, HYDROGEOLOGY. *Educ:* Calif State Univ, Fresno, BS, 60. *Prof Exp:* Hydrologist/geologist, US Geol Surv, Calif, 65-75, chief proj sect, 75-79, chief, Cent Valley Aquifer Proj, 79-81, regional groundwater specialist, Western Region, 81-85, CHIEF, CALIF DIST, US GEOL SURV, 85- *Concurrent Pos:* Instr, Calif State Univ, Fresno, 64-65, Sacramento, 70-71 & Cosumnes River Col, 79-80; appointee, Cosumnes River Col Curriculum Comt, 76- *Mem:* Geol Soc Am; Am Soc Advan Sci; Asn Eng Geologists. *Res:* Methodology and evaluation of hydraulic properties of alluvial systems, flow modeling, land subsidence, and water-mineral phase equilibrium; developed open-channel hydraulic analysis models; author or coauthor of 31 publications. *Mailing Add:* 8490 Berry Rd Wilton CA 95693

BERTOLETTE, W(ILLIAM) DEB(ENNEVILLE), b Wilmington, Del, June 4, 14; m 41; c 3. CHEMICAL ENGINEERING. *Educ:* Pa State Univ, BS, 36. *Prof Exp:* Chemist, E I du Pont de Nemours & Co, 36-38, supvr, 38-42, supvr tetra ethyl lead mfr, 42-46, asst prod mgr, Petrol Chem, 47-50, asst tech mgr, 51-55, supvr fuel oil group, Petrol Lab, 56-68, supt lab serv, Dept Org Chem, 68-71, opers coordr, Petrol Chem Div, 71-74; RETIRED. *Mem:* Am Chem Soc; Am Inst Chem Engrs. *Res:* Additives for petroleum products; antiknocks, stabilizers, antioxidants, metal deactivators, dyes, or corrosion inhibitors for gasoline or fuel oils. *Mailing Add:* 4621 Drusilla Dr Baton Rouge LA 70809-6948

BERTOLINI, DONALD R, b New York, NY, Nov 28, 49; m 73; c 2. CELL BIOLOGY, BONE BIOLOGY. *Educ:* Ithaca Col, BA, 71; C W Post Col, MS, 73; Wesleyan Univ, PhD(biol), 83. *Prof Exp:* Res assoc, Health Sci Ctr, Univ Tex, 82-86; res scientist, Otsuka Pharmaceut Co Ltd, 86-90; SR INVESTR, SMITH KLINE BEECHAM, 90- *Honors & Awards:* Norwich Eaton Res Award, 86. *Mem:* AAAS; Am Soc Cell Biol; Am Soc Bone & Mineral Res. *Res:* Cellular communications through cell contact or specfic molecules; bone cell biology. *Mailing Add:* Smith Kline Beecham Pharmaceut Co L-109 PO Box 1539 King of Prussia PA 19406

BERTON, JOHN ANDREW, b Villa Park, Ill, June 22, 30; m 52; c 5. GEOMETRY. *Educ:* Univ Ill, AB, 55, MA, 57, PhD(geom), 64. *Prof Exp:* From instr to asst prof math, Ind State Col, 59-64; assoc prof, Ripon Col, 64-67; chmn dept, 67-75, PROF MATH, OHIO NORTHERN UNIV, 67- *Mem:* AAAS; Am Math Soc; Math Asn Am; Soc Indust & Appl Math; Asn Comput Mach. *Res:* Differential geometry. *Mailing Add:* Dept Math Ohio Northern Univ Ada OH 45810

BERTON, WILLIAM MORRIS, b Fresno, Calif, Feb 8, 24; m 49; c 3. PATHOLOGY. *Educ:* Univ Calif, MD, 49. *Prof Exp:* Intern, US Naval Hosp, Mass, 49-50; res path, Methodist Hosp, Ind, 50-51; from intern to resident, Duke Univ Hosp, 51-54, instr, Sch Med, Duke Univ, 52-54; head lab serv & chief res, US Naval Hosp, Camp Pendleton, 55-56; from asst prof to prof path, Univ Tenn, 56-68; prof path, 69-86, EMER PROF PATH & MICROBIOL, UNIV NEBR MED CTR, OMAHA, 86- *Mem:* Asn Am Med Cols; Col Am Path; Am Soc Clin Path; Am Asn Pathologists; NY Acad Sci. *Res:* Chromatographic methods as applied to pathology and bacteriology. *Mailing Add:* 13136 Mason St Omaha NE 68154

BERTONCINI, PETER JOSEPH, b Chicago, Ill, Nov 29, 39; m 66; c 2. COMPUTER SCIENCE. *Educ:* Univ Ill, BS, 61; Univ Wis, PhD(phys chem), 68. *Prof Exp:* Fel quantum chem, Argonne Nat Lab, 68-71; fel quantum chem, Fac for Physics, Univ Freiburg, 71-73; fel quantum chem, 73-74, asst computer scientist reactor physics, 74-80, ASST COMPUTER SCIENTIST, COMPUTER SERV, ARGONNE NAT LAB, 80- *Mem:* Am Phys Soc. *Mailing Add:* 335 S Lombard Oak Park IL 60302

BERTONI, HENRY LOUIS, b Chicago, Ill, Nov 15, 38; c 2. ELECTROPHYSICS. *Educ:* Northwestern Univ, BS, 60; Polytech Inst Brooklyn, MS, 62, PhD(electrophys), 67. *Prof Exp:* From instr to assoc prof, 66-75, PROF ELECTROPHYS, POLYTECH UNIV 75-, HEAD DEPT ELEC ENG, 90- *Concurrent Pos:* Consult, Magi, Inc, NY, Anderson Labs, Conn Brockwell Int, Calif Loral, Yonkers, NY & Parametrics, Waltham, MA; guest res fel, Univ Col, London, 82-83; summer res fel, Hanscom Air Force Base, Mass, 83. *Mem:* Fel Inst Elec & Electronics Engrs; Int Sci Radio Union; Acoust Soc Am; Sigma Xi. *Res:* Electromagnetic, elastic and acoustic wave propagation and scattering, as applied to ultrasonic nondestructive evaluation, acoustic microscopy and UHF radio propagation. *Mailing Add:* Polytech Univ 333 Jay St Brooklyn NY 11201

BERTONIERE, NOELIE RITA, b New Orleans, La, Oct 17, 36. ORGANIC CHEMISTRY. *Educ:* St Mary's Dominican Col, BS, 59; Univ New Orleans, PhD(org chem), 71. *Prof Exp:* RES CHEMIST TEXTILES & FOOD, SOUTHERN REGIONAL RES CTR, AGR RES SERV, USDA, 60- *Mem:* Am Chem Soc; Am Asn Textile Chemists & Colorists; Sigma Xi; Fiber Soc. *Res:* Cellulose chemistry; durable press cotton fabric; permeation chromatography; supramolecular structure of cellulose; photochemistry of small nitrg heterocycles and carbenes. *Mailing Add:* Southern Regional Res Ctr PO Box 19687 New Orleans LA 70179

BERTOSSA, ROBERT C, b Chicago, Ill, May 2, 16. METALLURGY. *Educ:* Univ Ill, BS, 49. *Prof Exp:* Asst mgr, Mat Qual Serv Dept, Bechtel Corp, 70-80; RETIRED. *Mem:* Am Soc Metals; Am Welding Soc. *Mailing Add:* PO Box 1257 Aptos CA 95001

BERTOZZI, EUGENE R, b Lowell, Mass, Sept 26, 15; m 43; c 3. POLYMER CHEMISTRY. *Educ:* Worcester Polytech Inst, BS, 38; Brooklyn Polytech Inst, MS, 50. *Prof Exp:* Chemist res & develop, Benzol Prod Co, 38-42; mgr develop, Dept Res & Develop, Thiokol Corp, 42-64, asst tech dir res & develop, 64-68, dir advan technol, 68-72, prin scientist, Chem Div, 72-82; CONSULT, 82- *Mem:* Am Chem Soc; Sigma Xi. *Res:* Process development of organic polysulfide polymers; new polymer products development in condensation and hydrocarbon polymers. *Mailing Add:* 2115 Stackhouse Dr Yardley PA 19067

BERTOZZI, WILLIAM, b Framingham, Mass, June 9, 31; m 56; c 3. PHYSICS. *Educ:* Mass Inst Technol, SB, 53, PhD(physics), 58. *Prof Exp:* Staff mem, Div Sponsored Res, Lab Nuclear Sci, 57-58, from instr to assoc prof, 58-68, PROF PHYSICS, MASS INST TECHNOL, 68- *Mem:* Am Phys Soc; Am Asn Physics Teachers; AAAS. *Res:* Nuclear physics. *Mailing Add:* Eight Castle Rd Lexington MA 02173

BERTRAM, J(OHN) E(LWOOD), computer science; deceased, see previous edition for last biography

BERTRAM, LEON LEROY, b Corry, Pa, Jan 8, 17; m 44; c 2. CELLULOSE CHEMISTRY. *Educ:* Pa State Univ, BS, 41. *Prof Exp:* Chemist, Hercules, Inc, 41-45, develop supvr, 46-50, chief chemist, 50, oper supvr, 50-51, asst mgr, 51-54, supt, Carboxymethyl Cellulose Dept, 55-69, asst plant mgr, Hopewell Plant, 69-77, mgr tech interchange, Hercules France SA, 77-79; RETIRED. *Mem:* Am Chem Soc. *Res:* Cellulose chemistry. *Mailing Add:* 507 Central Terr Hopewell VA 23860-1910

BERTRAM, ROBERT WILLIAM, b Toronto, Ont, Nov 19, 42; m 89; c 5. THIN FILM PHYSICS, OPTICAL COATINGS. *Educ:* Univ Toronto, BS, 65; Univ Waterloo, MS, 67. *Prof Exp:* Res scientist, Ont Res Found, 67-81; TECH DIR, CAMETOID LTD, 81- *Concurrent Pos:* Chmn, Mat & Processes Comt, Aerospace Industs Asn Can, 87-, vchmn, Res & Develop Comt, 89- *Mem:* Am Vacuum Soc; Optical Soc Am; Soc Vacuum Coaters; Am Inst Physics. *Res:* Physical, electrical and optical properties of vacuum-deposited thin films; development and optimization of composite coatings based on graded composition inhomogeneous films; creation of novel solid state materials by inhomogeneous thin film modeling; low temperature superconductivity; solar energy; granted one patent; author of various publications. *Mailing Add:* Cametoid Ltd 1449 Hopkins St Whitby ON L1N 2C2 Can

BERTRAM, SIDNEY, b Winnipeg, Man, July 7, 13; US nat; m 44; c 4. SYSTEMS ENGINEERING, ELECTROMAGNETIC THEORY. *Educ:* Calif Inst Technol, BS, 38; Ohio State Univ, MS, 41, PhD(physics), 51. *Prof Exp:* Instr radio & math, Radio Inst Calif, 34-36; res engr, Int Geophys Co, 38-39, Res Found, Ohio State Univ, 41-42, Div War Res, Univ Calif, 42-45 & Boeing Aircraft Co, Wash, 45-46; asst prof commun eng, Ohio State Univ, 46-51; engr, Rand Corp, 51-53 & Syst Develop Corp, 53-57; sr staff scientist, Bunker Ramo Corp, 57-71; sr staff engr, Missile Systs Div, Hughes Aircraft Co, 71-80; lectr, Calif Polytech State Univ, 80-86; RETIRED. *Honors &*

Awards: Fairchild Photogram Award, Am Soc Photogram, 69. *Mem:* Fel Inst Elec & Electronics Engrs. *Res:* Automation of photogrammetric instruments; sonar; radar; simulators; field theory; applied mathematics; missile image matching and infrared seekers. *Mailing Add:* 1210 Oceanaire Dr San Luis Obispo CA 93401

BERTRAM, TIMOTHY ALLYN, b Le Mars, Iowa, Aug 26, 55; m 78; c 3. IMMUNOPATHOLOGY. *Educ:* Iowa State Univ, DVM, 79, PhD(path), 83. *Prof Exp:* Resident toxicol, Iowa Vet Diag Lab, 79-80; res assoc path, Nat Animal Dis Ctr, 80-83, postdoctoral assoc cellular path, 83-85; assoc prof path, Univ Ill, 85-89; SECT HEAD DIGESTIVE & NUTRIT TOXICOL, PROCTER & GAMBLE CO, 89- *Concurrent Pos:* Prin investr, Am Lung Asn, 86-88, USDA, 87-89, NSF, Nat Supercomputer Ctr, 88-89; vis scientist biochem path, Nat Inst Environ Health Sci, 87-88; scholar chmn, Am Col Vet Pathologists. *Honors & Awards:* Cert Merit, USDA, 84. *Mem:* Am Vet Med Asn; Am Col Vet Pathologists; Int Acad Path. *Res:* Cellular and biochemical mechanisms of injury induced by infectious/or toxic agents; organ systems include reproductive, respiratory and gastrointestinal. *Mailing Add:* Procter & Gamble Co Box 398707 Cincinnati OH 45239

BERTRAMSON, BERTRAM RODNEY, b Potter, Nebr, Jan 25, 14; m 38; c 3. AGRONOMY. *Educ:* Univ Nebr, BS, 37, MS, 38; Ore State Univ, PhD(soils), 41. *Hon Degrees:* DAgr, Univ Nebr, 78. *Prof Exp:* From asst prof physics to asst instr soils, Univ Nebr, 37-38; asst soil fertil, Ore State Col, 38-41; instr soils, Soil Lab, Colo State Col, 41-42; asst prof soil chem, Univ Wis, 46; assoc soil chemist, Dept Agron, Purdue Univ, 46-49; chmn, Dept Agron, Col Agr, Wash State Univ, 49-67, dir resident instr, 67-79; party chief, Morocco Proj, Univ Nebr, 81-82; CONSULT AGRONOMIST, 82- *Mem:* Fel AAAS; fel Am Soc Agron (vpres, 60, pres, 61); Crop Sci Soc Am; Soil Sci Soc Am; Soil Conserv Soc Am. *Res:* Soil physics; physical properties of the soil as affected by manuring and fertilizers; irrigation efficiency; phosphorus analyses; soil fertility; comparative efficiency of organic phosphorus and of superphosphate; research on potash, manganese, boron, sulfur and magnesium. *Mailing Add:* SE 510 Crestview Pullman WA 99163

BERTRAN, CARLOS ENRIQUE, cardiology, for more information see previous edition

BERTRAND, FOREST, b St Pie-de-Guire, Que, May 31, 18; m 45; c 4. AGRICULTURE. *Educ:* Laval Univ, BA, 39, BSc, 43; McGill Univ, MSc, 45; Cornell Univ, PhD, 56. *Hon Degrees:* Dsc, McGill Univ, 77. *Prof Exp:* Res officer, Can Dept Agr, 45-60; tech adv, Que Dept Agr, 60-62, head res div, 62-67, dir res & educ, 67-73, dir gen res & educ, 73-79; CONSULT AGROLOGIST, 79- *Concurrent Pos:* Secy, Que Agr Res Coun, 60-63, chmn, 63-72, mem, 72- *Honors & Awards:* Silver Merit Medal, Diocesan of Pocatiere, 61; Govt Can Centenary Medal, 67; Comdr Agronomic Merit Order, 76. *Mem:* Hon life mem Potato Asn Am. *Mailing Add:* 1570 Chemin Ste Foy Quebec PQ G1S 2P3 Can

BERTRAND, FRED EDMOND, b New Orleans, La, Sept 27, 38; m 65; c 2. NUCLEAR REACTION MECHANISM. *Educ:* Southwestern Memphis, BS, 60; La State Univ, MS, 62, PhD(physics), 68. *Prof Exp:* Res assoc, Univ Southern Calif, 68-70; staff mem, 70-78, group leader, 78-86, SECT HEAD, PHYSICS DIV, OAK RIDGE NAT LAB, 86. *Mem:* Am Phys Soc. *Res:* Experimental studies of the excitation and decay of nuclear giant resonances following excitation in medium energy hadronic reactions; studies of direct reactions with medium energy nuclear probes; nuclear spectroscopy using direct reactions. *Mailing Add:* Physics Div Oak Ridge Nat Lab PO Box 2008 Oak Ridge TN 37831-6368

BERTRAND, GARY LANE, b Lake Charles, La, Sept 25, 35; m 64; c 2. PHYSICAL CHEMISTRY. *Educ:* McNeese State Col, BS, 57; Tulane Univ, PhD(phys chem), 64. *Prof Exp:* Fel chem, Carnegie Inst Technol, 64-65, asst prof, 65-66; From asst prof to assoc prof, 66-75, PROF CHEM, UNIV MO-ROLLA, 75- *Mem:* Calorimetry Conf (chmn-elect, 82); Am Chem Soc; Am Soc Enologists. *Res:* Investigations of solvent effects and deuterium isotope effects on the thermodynamic properties of dissolved species; solution calorimetry and dilatometry; physical chemistry of wine. *Mailing Add:* Dept Chem Univ Mo-Rolla Rolla MO 65401

BERTRAND, HELEN ANNE, b Wilmington, Del, Nov 24, 39. GERONTOLOGY. *Educ:* Univ Del, BA, 61; NC State Univ, PhD(biochem), 70. *Prof Exp:* Res assoc biol, Univ Del, 61-63; res technician plant biol, Rockefeller Univ, 63-64; from instr to asst prof biochem, Med Col Pa, 71-75; Asst prof physiol, 75-81, ASSOC PROF PHYSIOL, UNIV TEX HEALTH SCI CTR, SAN ANTONIO, 81- *Concurrent Pos:* NIH training grant, Med Col Pa, 70-71. *Mem:* Am Phys Soc; Am Phys Soc; Geront Soc; Sigma Xi. *Res:* aging; mechanism of life-prolonging action of restricted food intake; adipocyte lipolysis. *Mailing Add:* Dept Physiol Univ Tex Health Sci Ctr 7703 Floyd Curl Dr San Antonio TX 78284

BERTRAND, HELMUT, b Asuncion, Paraguay, Dec 29, 37; US citizen; m; c 2. MOLECULAR GENETICS. *Educ:* Bethel Col, Kans, BA, 63; Kans State Univ, MSc, 67, PhD(molecular genetics), 69. *Prof Exp:* NIH res biochem, Albert Einstein Col Med, 69-70; asst prof, 70-74, assoc prof, 74-78, prof biol, Univ Regina, 78-88, PROF & CHMN MICROBIOL, UNIV GUELPH, 88- *Concurrent Pos:* Vis scientist physiol chem, Univ Munich, 79; vis scientist biochem, Sch Med, St Louis Univ, 80. *Mem:* Genetics Soc Can; Can Soc Cell Biol; Am Soc Microbiol; Can Soc Microbiol. *Res:* Genetic control of the formation and function of the mitochondrial electron-transport system in neurospora crassa; cyanide-sensitive and cyanide-insensitive respiratory systems; senescence-inducing plasmids in Neurospora; indigenous plasmids and transposons of brassica. *Mailing Add:* Dept Biol Univ Regina Regina SK S4S 0A2 Can

BERTRAND, JOSEPH AARON, b Lake Charles, La, Mar 20, 33; m 57; c 4. INORGANIC CHEMISTRY. *Educ:* McNeese State Col, BS, 55; Tulane Univ, MS, 56, PhD(inorg chem), 61. *Prof Exp:* Instr chem, McNeese State Col, 56-59; res assoc, Mass Inst Technol, 61-62; from asst prof to assoc prof, 62-70, dir chem, 74-81, PROF CHEM, GA INST TECHNOL, 70- *Concurrent Pos:* Alfred P Sloan res fel, 66-68; vis prof, Univ Kans, 68. *Mem:* Am Chem Soc; Am Crystallog Asn. *Res:* Structure of transition metal complexes. *Mailing Add:* 1671 Homestead Ave NE Atlanta GA 30332

BERTRAND, JOSEPH E, b Kaplan, La, June 20, 24; m 61; c 4. ANIMAL NUTRITION. *Educ:* Southwestern La Inst, BS, 48; La State Univ, MS, 52, PhD(ruminant nutrition), 60. *Prof Exp:* Res assoc animal indust, La State Univ, 52-56; asst prof, Ark State Col, 56-58; ruminant nutritionist, Com Solvents Corp, New York, 60-66; assoc prof, 66-77, PROF ANIMAL SCI, AGR RES CTR, UNIV FLA, 77- *Mem:* Am Soc Animal Sci; Poultry Sci Asn; Am Dairy Sci Asn. *Res:* Ruminant nutrition; beef cattle production, especially growing and developing light-weight calves to desired feedlot weights and finishing cattle for market, emphasizing maximum use of locally grown roughages and concentrates. *Mailing Add:* Agr Res & Educ Ctr Rte 3 Box 575 Jay FL 32565-9524

BERTRAND, RENE ROBERT, b Manchester, NH, Feb 5, 36; m 62. PHYSICAL CHEMISTRY, CHEMICAL ENGINEERING. *Educ:* Worcester Polytech Inst, BS, 57; Mass Inst Technol, PhD(phys chem), 62. *Prof Exp:* Chemist, Esso Res & Eng Co, Stand Oil Co, 61-66; prog mgr res & develop, Am Cryogenics, Inc, 66-68; sr res chemist, Govt Lab, Exxon Res & Eng Co, 68-72, analyst, Gas Dept, Exxon Corp, 72, eng assoc, Govt Lab, Exxon Res & Eng Co, 73-77, sr eng assoc, 77-80. *Mem:* Am Chem Soc. *Res:* Fluidized bed combustion of coal; enviromental aspects of coal conversion processes. *Mailing Add:* 1403 Jewett Dr Zanesville OH 43701

BERTSCH, CHARLES RUDOLPH, b Long Island, NY, June 13, 31; m 66; c 2. PUBLISHING CHEMICAL JOURNALS. *Educ:* Syracuse Univ, BS, 52; Pa State Univ, PhD(chem), 55. *Prof Exp:* Res chemist, US Naval Propellants Plant, 55-56, Olin Mathieson Chem Corp, 57-58 & Pennsalt Chem Corp, 58-61; sr prod ed, 61-69, mgr ed prod, 69-73, HEAD, JOURNALS DEPT, AM CHEM SOC, 73- *Mem:* Am Chem Soc; Coun Biol Ed; Soc Scholarly Publ. *Mailing Add:* Am Chem Soc 1155 16th St NW Washington DC 20036

BERTSCH, GEORGE FREDERICK, b Oswego, NY, Nov 5, 42; m 64; c 2. NUCLEAR PHYSICS. *Educ:* Swarthmore Col, BA, 62; Princeton Univ, PhD(physics), 65. *Prof Exp:* Fel physics, Niels Bohr Inst, Copenhagen, Denmark, 65-66; from instr to asst prof, Princeton Univ, 66-69; asst prof, Mass Inst Technol, 69-70; asst prof, 70-74, PROF PHYSICS, MICH STATE UNIV, 74- *Concurrent Pos:* Alfred P Sloan Found fel, 69-71; Humbolt fel, 86. *Mem:* AAAS; Am Phys Soc. *Res:* Theoretical studies of nuclear structure based on shell model; qualitative and quantitative understanding of experimental results. *Mailing Add:* Cyclotron Lab Mich State Univ East Lansing MI 48824

BERTSCH, HANS, b St Louis, Mo, Nov 2, 44; div. INVERTEBRATE ZOOLOGY, MALACOLOGY. *Educ:* San Luis Rey Col, BA, 67; Franciscan Sch Theol, BTh, 71; Univ Calif, Berkeley, PhD(zool), 76. *Prof Exp:* Asst prof biol, Chaminade Univ Honolulu, 76-78; cur marine invert, San Diego Nat Hist Mus, 78-80; prof marine sci, Univ Autonoma de Baja Calif, Ensenada, 81-83; ADJ FAC, NAT UNIV, 83- *Concurrent Pos:* Prin investr, Ctr Field Res, 78-80; Nat Teaching fel, Chaminade Univ Honolulu, 76-78; res & tech consult, Mex Nat Fisheries, 84-85; res assoc, Los Angeles Co Mus Nat Hist, 77- & Calif Acad Sci, 72- *Mem:* AAAS; Soc Syst Zool; Sigma Xi; Malacol Soc Japan; Western Soc Malacologists (pres, 89-). *Res:* Opisthobranch taxonomy; functional morphology; intertidal and subtidal ecology of benthic marine invertebrates; marine zoogeography. *Mailing Add:* 640 The Village No 203 Redondo Beach CA 90277

BERTSCH, ROBERT JOSEPH, b Philadelphia, Pa, May 6, 48; m 70; c 2. FREE RADICAL POLYMERIZATION, THERMOSET RESINS. *Educ:* Drexel Univ, BS, 70; Ohio State Univ, PhD(chem), 75. *Prof Exp:* RES CHEMIST, POLYMER SYNTHESIS, B F GOODRICH CO, 75- *Concurrent Pos:* Undergrad trainee/researcher, Eastern Regional Lab, USDA, 66-71. *Mem:* Am Chem Soc. *Res:* Free radical polymerization, thermoset resin chemistry, synthesis of new organic monomers and polymers and the characterization of polymeric species. *Mailing Add:* B F Goodrich Co 9921 Brecksville Rd Brecksville OH 44141

BERTSCH, WOLFGANG, b Ludwigsburg, Ger, Mar 17, 40; c 1. ANALYTICAL CHEMISTRY, ENVIRONMENTAL CHEMISTRY. *Educ:* Tech Acad, Isny, Ger, BS, 68; Univ Houston, PhD(chem), 73. *Prof Exp:* Asst prof, 74-78, ASSOC PROF ANALYTICAL CHEM, DEPT CHEM, UNIV ALA, 78- *Concurrent Pos:* Environ Protection Agency grant, 76-78; consult, Ionics Res, Houston, 77-78; Off Water Resources Res & Develop grant, 77-78; mem organizing comt, Int Symp on Glass Capillary Columns, Hindelang, Ger, 77-78; ed, J of High Resolution Chromatography & Chromatography Commun, Ger, 78- *Mem:* Am Chem Soc. *Res:* High performance gas chromatography and gas chromatography/mass spectrometry; development of glass capillary columns, environmental and medicinal applications. *Mailing Add:* Dept Chem Univ Ala Box 870336 Tuscaloosa AL 35487-0336

BERTY, JOZSEF M, b Budapest, Hungary, Oct 25, 22; US citizen; m 45; c 6. CHEMICAL ENGINEERING. *Educ:* Budapest Tech Univ, dipl chem eng, 44, DSc, 50. *Hon Degrees:* Dr, Veszprem Univ, 88. *Prof Exp:* Engr, Nitrochem Works, Ltd, Hungary, 44-45 & Agr & Chem Works, Ltd, 45-46; asst prof, Budapest Tech Univ, 46-50; head dept petrochem, Hungarian Oil & Gas Res Inst, 50-56; vis prof, Inst Chem Technol, Ger, 56-57; engr chem div, Union Carbide Corp, 57-60, group leader res & develop, 60-68, tech mgr, 68-69, asst dir, 69-77; vpres, Autoclave Engrs, Inc, 77-78; PRES, BERTY REACTION ENGRS, LTD, 78- *Concurrent Pos:* Assoc prof, Inst Chem Technol, Hungary, 52-54, head dept, 54-56; sr Fulbright scholar, Tech Univ,

Munich, Ger, 73-74; adj prof, Univ Akron, 79-82, prof chem eng, 82-89. *Honors & Awards:* Sr Fulbright Lectr, T U Munchen, 73-74. *Mem:* Am Chem Soc; fel Am Inst Chem Engrs. *Res:* Chemical reaction engineering; chemical reactor design and stability problems; development of basic petrochemical processes; computer simulation experiments. *Mailing Add:* Berty Reaction Engrs Ltd Four Bent Pine Hill Fogelsville PA 18051-9712

BERTZ, STEVEN HOWARD, b Fond du Lac, Wis, Dec 24, 50; m 74; c 2. MECHANISTIC CHEMISTRY, THEORETICAL CHEMISTRY. *Educ:* Univ Wis-Madison, BS, 73; Harvard Univ, AM, 75, PhD(chem), 78. *Prof Exp:* NSF fel, 73-76; mem tech staff, 78-85, DISTINGUISHED MEM TECH STAFF, BELL LABS, 85- *Concurrent Pos:* Vis prof, Univ Wis-Madison, 85. *Mem:* Am Chem Soc. *Res:* Chemistry of organocopper reagents; arenesulfonylhydrazones; metal-carbene complexes; glyoxal and malondialdehyde; organotitanium reagents; study of chemical topology, molecular complexity and chemical applications of graph theory and information theory. *Mailing Add:* Bell Labs 600 Mountain Ave Rm 1T206 Murray Hill NJ 07974

BERU, NEGA, b Maichew, Ethiopia, Aug 22, 51. MOLECULAR BIOLOGY, BIOCHEMISTRY. *Educ:* Univ Chicago, BA, 74, PhD(biochem), 83. *Prof Exp:* Writer, Sci Curric Develop Ctr, Ministry Educ, Addis Ababa, Ethiopia, 74-76; lectr, Dept Biol, Addis Ababa Univ, Ethiopia, 76-78; postdoctoral res assoc & instr, Dept Biochem & Molecular Biol, 84-88, instr, 88-90, ASST PROF, DEPT MED, SECT HEMAT/ONCOL, UNIV CHICAGO, 90- *Mem:* Am Soc Biochem & Molecular Biol; Int Soc Exp Hemat; Sigma Xi. *Res:* Regulation of growth factor gene expression both in normal as well as in leukemic cells at the molecular level; hematopoietic differentiation including signal transduction; study of the second messengers and transcription factors involved in the process. *Mailing Add:* Dept Med Hemat Oncol Univ Chicago 5841 S Maryland Box 420 Chicago IL 60637

BERUBE, GENE ROLAND, physical biochemistry, for more information see previous edition

BERUBE, ROBERT, b Boston, Mass, Apr 23, 35; m 65; c 2. STERILIZATION, MICROBIAL CONTROL. *Educ:* Boston Col, BS, 57; Univ Mass, MS, 59; Kans State Univ, PhD(bact), 67. *Prof Exp:* Asst prof, McNeese State Univ, Lake Charles, La, 67-71; actg br chief, microbiol qual assurance, Becton, Dickinson & Co, 71-73, br chief, 73-75, mgr, 75-77; specialist, Prod Develop, Infection Control, 3M, 77-83, microbiol qual assurance, Med Div, 83-86, Med-Surg Div, 86-88, SPECIALIST, HEALTH CARE & STERILIZATION, MICROBIOL SERV, 3M, 88- *Concurrent Pos:* Instr, Kans State Univ, Manhattan, Kans, 65-67. *Mem:* Soc Indust Microbiol; Am Soc Qual Control. *Res:* Develop and validate sterilization cycles (E&O, steam, radiation, dry heat) for medical-surgical devices; develop and validate microbial control systems (sanitization, disinfection); bioburden assessment and validation; microbiological control of water. *Mailing Add:* 3M Ctr Bldg 270-SN-01 St Paul MN 55144-1000

BERY, MAHENDERA K, b Bombay, India; US citizen; m 74; c 2. BIOCHEMICAL ENGINEERING, FERMENTATION ENGINEERING. *Educ:* Birkenheed Technol, Eng, dipl, 68; Univ Birmingham, Eng, MSc, 69; Univ Aston, Eng, PhD(chem eng), 73. *Prof Exp:* Lectr chem eng, Univ Aston, Eng, 70-72; chem engr process develop, Tech Ctr, Gen Foods, 74-77; SR RES ENGR BIOENERGY, GEORGIAN TECHNOL, ATLANTA, 77- *Concurrent Pos:* Develop engr prod develop, Barrier Chem Inc, 74-77; process engr pet food, Ralston Purina, St Louis, 77. *Mem:* Am Inst Chem Engrs; Am Chem Soc. *Res:* Production of substitute natural gas from poultry and dairy cow manure and agricultural residues; production of ethanol from cellulosic biomass; extraction and fractionation of chemicals from wood; process design and economic feasibility study of various fermentation plants; biological treatment of waste. *Mailing Add:* ENJ Gallo-Winery PO Box 1130 Modesto CA 95353

BERZOFSKY, JAY ARTHUR, b Baltimore, Md, Apr 13, 46; m 69; c 2. IMMUNOGENETICS, ANTIGENIC STRUCTURE. *Educ:* Harvard Univ, BA, 67; Albert Einstein Col Med, PhD(molecular biol), 71, MD, 73. *Prof Exp:* Resident internal med, Mass Gen Hosp, 73-74; res assoc biochem, Nat Inst Arthritis, Metab & Digestive Dis, 74-76, investr immunol, Nat Cancer Inst, 76-79, sr investr immunol, 79-87, CHIEF MOLECULAR IMMUNOGENETICS & VACCINE RES SECT, NAT CANCER INST, NIH, 87- *Concurrent Pos:* Assoc ed, J Immunol, 80-84; fac mem, Found Advan Educ in Sci, 84-; adv ed, Molecular Immunol J, 85- *Honors & Awards:* Sophia Freund Prize, Harvard Univ, 67. *Mem:* NY Acad Sci; Am Asn Immunologists; Am Fedn Clin Res; Am Soc Biol Chemists; Am Soc Clin Investr (secy-treas, 89-92). *Res:* Antigen-specific genetic, cellular and biochemical regulation of the immune response; mechanism of T-lymphocyte activation by antigen and antigen presentation; antigenic structure of proteins; vaccine development; idiotype interactions. *Mailing Add:* Bldg 10 Rm 6B12 NIH Nat Cancer Inst Bethesda MD 20892

BESANCON, ROBERT MARTIN, optical physics, cosmic ray physics; deceased, see previous edition for last biography

BESCH, EMERSON LOUIS, b Hammond, Ind, June 9, 28; m 55; c 4. ENVIRONMENTAL PHYSIOLOGY. *Educ:* Southwest Tex State Col, BS, 52, MA, 55; Univ Calif, Davis, PhD(physiol), 64. *Prof Exp:* Instr biol, Southwest Tex State Col, 54-55; res asst animal physiol, Univ Calif, Davis, 60-64, res physiologist & lectr animal physiol, 64-67; from assoc prof to prof physiol & mech eng, Kans State Univ, 67-74; assoc dean, Univ Fla, 74-80 & 81-87, actg dean, 80-81, exec assoc dean, 87-88, PROF, COL VET MED, UNIV FLA, 74- *Concurrent Pos:* NIH trainee, 61-64; chmn comt lab animal housing, Inst Lab Animal Resources, Nat Acad Sci, 75-77; assoc ed, Lab Animal Sci, 77-82; mem, Ilar Coun, Nat Res Coun, 78-83; chmn comt environ conditions in lab animal rooms, Inst Lab Animal Resources, Nat Acad Sci/ Nat Res Coun, 82-84, comt lab animal housing, 75-78, environ health comt, Am Soc Heating, Refrig & Air Conditioning Eng, 86- *Honors & Awards:* Res Award, Am Asn Lab Animal Sci, 75; Cert Aerospace Physiologist, Aerospace Physiologist Soc, 79; Prof Excellence Award, Life Sci & Biomed Eng Br, Aerospace Med Asn, 87. *Mem:* Fel Aerospace Med Asn; Am Physiol Soc; Am Asn Lab Animal Sci; Am Soc Heating Refrigerating & Air-Conditioning Engrs; Soc Exp Biol Med. *Res:* Aviation physiology; effects of mechanical forces on biological systems; definition of environmental requirements for lab animals; adaptation to high altitude; comparative physiology of environmental stresses. *Mailing Add:* Col Vet Med Box J-144 JHMHC Univ Fla Gainesville FL 32610-0144

BESCH, GORDON OTTO CARL, b Lacrosse, Wis, March 16, 22; m 49; c 4. PHYSICS EDUCATION, PUBLIC PHYSICS. *Educ:* Univ Wis, Lacrosse, BS, 51; US Merchant Marine Acad, BS, 47; Harvard Univ, Ed Med, 54; Ohio State Univ, PhD(sci ed & physics), 69. *Prof Exp:* Radio officer, marine radiotelegraphy, Am South-African Co, 43-44; third officer, marine sciences, Lykes Steamship Lines, 47-78; div chmn, sci & math, Concordia Teachers Col, 58-68; PROF PHYSICS, UNIV WIS-SUPERIOR, 69- *Concurrent Pos:* Consult, Dept Agr, State Wis, 72-77; lectr physics, N Park Col, 61-68; univ dir, Nat Sci Found Grants Physics Educ, 72-77. *Mem:* Am Asn Physics Teachers; Am Radio Relay League. *Res:* Public lecture demonstrations on theories in physics such as special theory of relativity and neutron activation analysis of crude protein in grain. *Mailing Add:* Dept Chem & Physics Univ Wis Superior Superior WI 54880

BESCH, HENRY ROLAND, JR, b San Antonio, Tex, Sept 12, 42; m 86; c 1. PHARMACOLOGY, CARDIOLOGY. *Educ:* Ohio State Univ, BSc, 64, PhD(pharmacol, biophys), 67. *Prof Exp:* Instr steroid biochem, Ohio State Univ, 67-68; asst prof pharmacol, 71-73, asst prof med biophys, 72-73, assoc prof pharmacol & med, 73-77, SHOWALTER PROF PHARMACOL & CHMN DEPT, SCH MED, IND UNIV, INDIANAPOLIS, 77- *Concurrent Pos:* USPHS fel, Baylor Col Med, 68-70; Med Res Coun fel, Royal Free Hosp Sch Med, London, Eng, 71-72, vis sr lectr, 73-74; sr res assoc, Krannert Inst Cardiol, 74-; Showalter awardee, 75- *Mem:* Am Soc Pharmacol & Exp Therapeut; Brit Biochem Soc; NY Acad Sci; Int Study Group Res Cardiac Metab; Am Heart Asn. *Res:* Cardiac pharmacology; subcellular mechanisms of inotropic and chronotropic drugs. *Mailing Add:* Dept Pharmacol & Toxicol Ind Univ Sch Med Indianapolis IN 46202-5120

BESCH, PAIGE KEITH, b San Antonio, Tex, June 23, 31; m 57; c 2. BIOCHEMICAL PHARMACOLOGY. *Educ:* Trinity Univ, BS, 54, MS, 55; Ohio State Univ, PhD(physiol, pharmacol), 60. *Prof Exp:* Biochemist-in-chg metab sect, Surg Res Unit, Brooke Army Med Ctr, Ft Sam Houston, Tex, 53-54; sr biochemist, Clin Lab, Robert B Green Mem Hosp, San Antonio, 54-55; sr res asst to chmn dept endocrinol, Southwest Found Res & Educ, 55-58; asst, Dept Physiol, Col Med, Ohio State Univ, 58-60, res assoc, Steroid Res Lab, 60, instr pharmacol, Univ, 60-62, co-dir med sch & dir hosp, 60-66, asst prof physiol, obstet & gynec, Med Sch, 62-66, asst prof pharmacol, 63-66, assoc prof obstet & gynec, Univ Hosp, 66-68; assoc prof to prof, 68-83, DISTINGUISHED PROF OBSTET & GYNEC, BAYLOR COL MED, 83- *Concurrent Pos:* Res assoc, Columbus Psychiat Inst & Hosp, Ohio, 58-59; dir steroid res lab, Grant Hosp Res & Educ Found, 61-; dir labs, Midwest Found Res & Educ, 64-68; sr consult, Med Res Consult, Inc, 62-68; mem, med lab sci rev comt, USPHS, NIH, 74-77. *Honors & Awards:* Smith Kline Award, Am Asn Clin Chem, 79. *Mem:* AAAS; Am Asn Clin Chemists (pres, 81); Sigma Xi; sr mem Am Chem Soc; Soc Exp Biol & Med; fel Nat Acad Clin Biochem. *Res:* Steroid metabolism; pharmacological effects of drugs; ability to alter steroids; fertility and sterility; chemical contraception. *Mailing Add:* Dept Obstet & Gynec Baylor Col Med Houston TX 77030

BESCHORNER, W E, b Aurora, Ill, June 4, 47. TRANSPLANT IMMUNOLOGY, AIDS. *Educ:* Augustana Col, AB, 68; St Louis Univ, MS, 72; Univ Ill, MD, 76. *Prof Exp:* Resident path, 76-80, asst prof, 80-91, ASSOC PROF, DEPT PATH, JOHNS HOPKINS HOSP, 91- *Mem:* Am Asn Pathologists; Int Asn Pathologists. *Res:* Mechanisms of tolerance to transplant allografps-proposed a novel approach to inducing specific immune tolerance; immunopathology of AIDS myocarditis and enteropathy. *Mailing Add:* Dept Path Johns Hopkins Med Insts 600 N Wolfe St Baltimore MD 21205

BESHARSE, JOSEPH CULP, b Hickman, Ky, Jan 21, 44; m 66; c 2. CELL BIOLOGY. *Educ:* Hendrix Col, BA, 66; Southern Ill Univ, MA, 69, PhD(zool & physiol), 73. *Prof Exp:* Asst prof biol, Old Dominion Univ, 72-75; NIH fel, Col Physicians & Surgeons, Columbia Univ, 75-77; from asst prof to assoc prof cell biol, Sch Med, Emory Univ, 77-84, prof anat, cell biol & ophthal, 84-89; PROF & CHMN ANAT & CELL BIOL, SCH MED, UNIV KANS, KANSAS CITY, 89- *Concurrent Pos:* NIH Res Career Develop Award, 79-84; mem, Vision Res Review Comt, Nat Eye Inst, 82-86; prin investr, NIH, 78- *Mem:* AAAS; Am Soc Cell Biol; Am Soc Zoologists; Asn Res in Vision & Opthal; Soc Neurosci. *Res:* Cell biology with emphasis on cell-cell interaction and membrane turnover in retinal tissue; photosensitive membrane turnover in retinal photoreceptors; interactions of photoreceptors with support cells; role of neurons in control circadian rhythmicity. *Mailing Add:* Dept Anat & Cell Biol Univ Kans Med Ctr 39th & Rainbow Blvd Kansas City KS 66103

BESHERS, DANIEL N(EWSON), b Chicago, Ill, Aug 13, 28; m 53; c 4. PHYSICAL METALLURGY, SOLID STATE PHYSICS. *Educ:* Swarthmore Col, BA, 49; Univ Ill, MS, 51, PhD(physics), 56. *Prof Exp:* Physicist, Metall Div, US Naval Res Lab, Washington, DC, 55-57; from asst prof to assoc prof, 57-69, PROF METALL, HENRY KRUMB SCH MINES, COLUMBIA UNIV, 68- *Concurrent Pos:* Vchmn, Gordon Conf Phys Metall, 64, chmn, 65; vis prof, Univ Poitiers, 79-80. *Mem:* Am Phys Soc; Minerals Metals & Mat Soc; Am Soc Metals Int. *Res:* Elastic constants, anelasticity, nonlinear internal friction, hysteresis, and acoustic harmonic generation; acoustic emission; internal stress; motion of dislocations, domain walls, and point defects. *Mailing Add:* 1105 SW Mudd Bldg Columbia Univ New York NY 10027

BESIK, FERDINAND, environmental & chemical engineering, for more information see previous edition

BESKID, GEORGE, b Erie, Pa, Mar 20, 29; m 54; c 3. BACTERIOLOGY. *Educ:* Univ Buffalo, BA, 50; Syracuse Univ, MS, 54, PhD(bact), 59. *Prof Exp:* Asst bact, Syracuse Univ, 57-59; res assoc microbiol, Hahnemann Med Col & Hosp, 59-61, instr microbiol & surg, 61-63; sr bacteriologist, 63-71, res group chief, 71-78, sr res group chief, 78-84, RES INVESTR, HOFFMANN-LA ROCHE, INC, 84- *Mem:* AAAS; Am Soc Microbiol; Sigma Xi. *Res:* Bacterial sporulation; endotoxin shock; surgical infections; amylase producing bacteria; chemotherapy. *Mailing Add:* 93 Rolling Hills Rd Clifton NJ 07013

BESLEY, HARRY E(LMER), agricultural engineering, for more information see previous edition

BESMANN, THEODORE MARTIN, b New York, NY, Feb 18, 49. NUCLEAR ENGINEERING, MATERIALS SCIENCE. *Educ:* NY Univ, BE, 70; Iowa State Univ, MS, 71; Pa State Univ, PhD(nuclear eng), 76. *Prof Exp:* MEM RES STAFF HIGH TEMPERATURE CHEM NUCLEAR MAT, OAK RIDGE NAT LAB, 75-; ASSOC PROF, NUCLEAR ENG DEPT, UNIV TENN, 80- *Concurrent Pos:* Instr, Roane State Commun Col, 77- *Mem:* Am Nuclear Soc; Sigma Xi; Am Ceramic Soc; AAAS. *Res:* High temperature chemistry of materials for energy systems; national energy policy issues. *Mailing Add:* Bldg 4501 PO Box X Oak Ridge TN 37830

BESNER, MICHEL, b Verdun, Que, March 16, 47; m; c 2. AQUACULTURE, FISH PHYSIOLOGY. *Educ:* Univ Montreal, BA, 67, BSc, 70, MSc, 75; Univ Wash, PhD(fisheries), 80. *Prof Exp:* coordr aquacult, Univ Montreal, 80-81, researcher fisheries, 82-83; RES ASSOC, INRS OCEANOLOGIE, RIMOUSKI, QUE, 83- *Mem:* Can Asn Fr Advan Sci. *Res:* Effect of natatory exercise on growth, metabolism and diseases resistance of salmonids; adaptation to seawater of salmonids; adaptation of brook trout (Salvelinus forminalis) and salmon to seawater; metabolism fluctuations of the mussel Mytilus educis related to the carrying capacity of the environment. *Mailing Add:* Univ Quebec Rimouski 300 Ave de Ursulines Rimouski PQ G5L 3A1 Can

BESOZZI, ALFIO JOSEPH, b New York, NY, Dec 2, 21; m; c 4. ORGANIC CHEMISTRY. *Educ:* NY Univ, AB, 43, PhD(org chem), 51. *Prof Exp:* Jr pharmaceut chemist, Winthrop Chem Co Div, Sterling Drug, Inc, 43-44; proj leader & consult chem, Evans Res & Develop Corp, 51-56; sr res chemist, 56-67, res group head, 67-74, INDUST HYG CORRDR, PETRO-TEX CHEM CORP, 74- *Mem:* Am Chem Soc; Catalysis Soc. *Res:* Petrochemical research including heterogeneous catalysis, polymerization and synthesis; pharmaceutical chemistry. *Mailing Add:* 5534 Dumfries Houston TX 77096-4004

BESS, JOHN CLIFFORD, audiology, speech pathology; deceased, see previous edition for last biography

BESS, ROBERT CARL, b Houston, Tex, Sept 24, 44; m 70; c 2. ANALYTICAL CHEMISTRY, ELECTROCHEMISTRY. *Educ:* Tex A&M Univ, BS, 73, PhD(analytical chem), 76; Sussex (England), MBA, 88. *Prof Exp:* Lab technician eng, Spencer Buchanan & Assoc, 67; lab technician chem, Nalco Chem Co, 68-72; res chem & res fel, Tex A&M Res Found, 77; sr chemist, Texaco, Inc, 77-79; supvr, Arabian Am Oil Co, 79-90; CHIEF CHEM ENGR, CALTEX INDONESIA, 91- *Concurrent Pos:* Tech reviewer, corrosion journals. *Mem:* Am Chem Soc; Electrochem Soc; Nat Asn Chem Engrs; Res Soc NAm; Soc Petrol Engrs; Soc Electroanalytical Chem. *Res:* Inorganic and organic electrochemistry; enhanced oil recovery; corrosion science; reverse osmosis; water treatment; oilfield chemical treatment; scaling in the petroleum industry; geothermal oil recovery; oilfield facilities engineering. *Mailing Add:* Rumbai Camp c/o Amoseas PO Box 237 Orchard Point Post Off Singapore 9123 Singapore

BESSE, JOHN C, b Rayne, La, June 23, 33; m 62; c 3. PHARMACOLOGY, PHYSIOLOGY. *Educ:* Univ Southwestern La, BSc, 56; Vanderbilt Univ, PhD(pharmacol), 66. *Prof Exp:* Fel pharmacol, State Univ NY Downstate Med Ctr, 65-67; asst prof, 67-76, ASSOC PROF PHARMACOL, MED CTR, UNIV ALA, BIRMINGHAM, 76- *Mem:* Fedn Am Soc Exp Biol. *Res:* Autonomics and cardiovascular pharmacology; mechanism of action of drugs and hormones in smooth muscle at the level of receptors and membranes; cardiovascular physiology; biophysics of circulation and its correlation to drug activity on the circulation. *Mailing Add:* Dept Pharmacol Univ Ala Med Ctr Birmingham AL 35294

BESSER, JOHN EDWIN, b Louisville, Ky, Jan 25, 47; US citizen; m 69; c 2. THERMODYNAMICS & MATERIAL PROPERTIES, TECHNICAL MANAGEMENT. *Educ:* Univ Louisville, BS, 69; Mid Tenn State Univ, MBA, 90. *Prof Exp:* Develop chemist, Eastman Chem Prods Inc, 69-72; qual control supvr, Celotex Corp, Div Jim Walter Corp, 72-74; mgr qual, 74-76, mgr tech plant, 76-78; lab mgr, Aladdin Synergetics Inc, 78-81; assoc dir chem res, 81-82, TECH DIR, ALADDIN INDUSTS INC, 82- *Mem:* Am Chem Soc; Am Soc Qual Control; Soc Plastic Engrs. *Res:* Polymeric failure modes in products; relationships between neat polymer characteristics and performance in molded products; balance of economic factors and product performance for cost/performance relationships; polyurethane rigid foam chemistry and application; insulation materials and vacuum technology. *Mailing Add:* 703 Murfreesboro Rd Franklin TN 37064

BESSETTE, FRANCE MARIE, b Magog, Que, Can, June 26, 44. MOLECULAR BIOPHYSICS. *Educ:* Univ Sherbrooke, BSc, 66, PhD(phys chem), 69. *Prof Exp:* Phys sci specialist, Ministry Health, Que, 70; prof phys chem, Col Gen & Voc Educ, Prov Que, Can, Lionel-Groulx, 71; ASSOC PROF BIOPHYS, UNIV SHERBROOKE, 71- *Mem:* Biophys Soc; Spectros Soc Can. *Res:* Mathematical characterization of the ECG using spline functions and correlation with numerical elements symptomatic for certain heart diseases in view of diagnostic support; fluorescence properties of grape anthoryanins and of their polymerization with catechins in wine; analytical method of distinguishing between mono and diglucorides; stability-coloration relationship of dimus and trimus for their use as food dyes. *Mailing Add:* Dept Biophys Fac Med Univ Sherbrooke Sherbrooke PQ J1H 5N4 Can

BESSETTE, RUSSELL ROMULUS, b New Bedford, Mass, July 21, 40; m 64; c 2. ANALYTICAL ELECTROCHEMISTRY, CHEMICAL INSTRUMENTATION. *Educ:* Univ Rhode Island, BS, 62; Univ Mass, Amherst, MS, 65, PhD, 67. *Prof Exp:* Air Force Off Sci Res res asst, Univ Mass, Amherst, 65-67; res chemist, Res Ctr, Olin Corp, 67-68; from asst prof to assoc prof, 68-76, chmn, Dept Chem, 86-90, PROF CHEM, SOUTHEASTERN MASS UNIV, 76- *Concurrent Pos:* Chem consult; reviewer, Analytical Grants Div, NSF. *Mem:* Am Chem Soc; Sigma Xi. *Res:* Electrochemistry of copper lithium and zirconium in non aqueous media; anode-cathode studies for high energy density battery systems; use of new materials as indicating electrodes; determination of disolved oxygen utilizing a pulsed technique with an electrochemical oxygen sensor. *Mailing Add:* 86R Aucoot Rd Mattapoisett MA 02739-2429

BESSEY, PAUL MACK, b Cudahy, Wis, Mar 23, 25; m 47; c 5. HORTICULTURE. *Educ:* Univ Wis, BS, 49, MS, 51; Mich State Univ, PhD(hort), 57. *Prof Exp:* Asst veg crops & hort, Univ Wis, 49-51; instr & asst hort, Univ Maine, 51-54; asst hort, Mich State Univ, 54-57; asst hort, 57-65, assoc prof hort, 65-78, assoc prof plant sci & assoc res spec hort, Exten Spec Hort, Univ Ariz, 78-86; Pima County Urban Hort Agt, 86-91; RETIRED. *Concurrent Pos:* Landscape consult, 91- *Mem:* Am Soc Hort Sci; Inst Food Technologists. *Res:* Postharvest physiology of fruits and vegetables, potato breeding, vegetable flavor in relation to market quality standards, controlled environments, vegetable adaptation. *Mailing Add:* 2602 E Arroyo Chico Tucson AZ 85716-5606

BESSEY, ROBERT JOHN, physics; deceased, see previous edition for last biography

BESSEY, WILLIAM HIGGINS, b East Lansing, Mich, Mar 18, 13; m 45; c 2. SOLID STATE PHYSICS. *Educ:* Univ Chicago, BS, 34; Carnegie Inst Technol, MS, 35, DSc, 40. *Prof Exp:* Asst, Carnegie Inst Technol, 35-39; instr physics, SDak Sch Mines, 39-40 & NC State Col, 40-42; from instr to asst prof, Carnegie Inst Technol, 42-52; assoc prof, Sch Mines & Metal, Univ Mo, 52-56; prof physics & head dept, Butler Univ, 56-80; RETIRED. *Mem:* Am Phys Soc; Am Asn Physics Teachers; Sigma Xi. *Res:* Diffraction of molecular rays; high speed testing of metals; shaped charges; electrical properties of evaporated metal films; thermal conductivity of ionic crystals. *Mailing Add:* 6421 Peace Pl Indianapolis IN 46268

BESSMAN, ALICE NEUMAN, b Washington, DC, Nov 7, 22; m 45; c 2. INTERNAL MEDICINE. *Educ:* Smith Col, BA, 43; George Washington Univ, MD, 49; Am Bd Internal Med, dipl, 59. *Prof Exp:* Fel med, DC Gen Hosp, George Washington Univ, 51-52; Nat Found Infantile Paralysis fel pediat, Mass Gen Hosp, 52-53; fel med, DC Gen Hosp, George Washington Univ, 53-54; staff physician, Group Health Asn, Inc, Washington, DC, 54-56; vis physician, Dept Med, Baltimore City Hosps, Johns Hopkins Univ, 56-62, hosp physician, 62-67, asst chief chronic & community med, 67-68; assoc prof, 68-78, PROF, MED SCH, UNIV SOUTHERN CALIF, 78-; CHIEF DIABETES SERV, RANCHO LOS AMIGOS HOSP, 68- *Concurrent Pos:* Dept Health, Educ & Welfare Social Rehab Serv grant, 65- *Mem:* Am Fedn Clin Res; AMA; Am Col Physicians; Am Diabetes Asn. *Res:* Significance of elevated serum ammonia levels in gastrointestinal bleeding and hepatic failure; ketone and lactate metabolism in diabetes mellitus; delivery of outpatient care via nurse practitioners; microbiology of infected diabetic gangrene. *Mailing Add:* Dept Med Sch Med Univ Southern Calif Los Angeles CA 90033

BESSMAN, JOEL DAVID, b Norfolk, Va. HEMATOLOGY, ANEMIA. *Educ:* Columbia Univ, BA, 67, MD, 72. *Prof Exp:* Intern internal med, Univ Southern Calif, 72-73; res fel drug develop, Nat Cancer Inst, 73-75; resident internal med, Univ Southern Calif, 75-77; fel internal med, Johns Hopkins Med Inst, 77-79; asst prof med, 79-84, ASSOC PROF MED & PATH, UNIV TEX MED BR, 84- *Concurrent Pos:* Mem adv learning res comt, Am Soc Hemat, 87-90; pres, Island Anal, Inc, 86- *Mem:* Am Soc Hemat; Am Soc Clin Path; Soc Anal Cytol; Am Col Physicians. *Res:* Biologic significance of non-discrete hererogenity; new methods for well subject cancer screening; physiology of platelet function; interactive computer education. *Mailing Add:* John Sealy Hosp Univ Tex Med Br Galveston TX 77550

BESSMAN, MAURICE JULES, b Newark, NJ, July 31, 28; m 52; c 4. BIOCHEMISTRY. *Educ:* Harvard Univ, AB, 49; Tufts Univ, MS, 52, PhD(biochem), 56. *Prof Exp:* Fel, Nat Cancer Inst, USPHS, 56-58; instr, Sch Med, Wash Univ, 58; from asst prof to assoc prof, 58-66, PROF BIOL, McCOLLUM-PRATT INST, JOHNS HOPKINS UNIV, 66- *Mem:* AAAS; Am Soc Biol Chem; Am Chem Soc. *Res:* Enzymology of nucleic acid metabolism. *Mailing Add:* Dept Biol Johns Hopkins Univ Baltimore MD 21218-2680

BESSMAN, SAMUEL PAUL, b NJ, Feb 3, 21; m 45; c 2. PEDIATRICS, BIOCHEMISTRY. *Educ:* Wash Univ, MD, 44; Am Bd Pediat, dipl. *Prof Exp:* Intern, St Louis Children's Hosp, 44-46; pathologist, USPHS, US Marine Hosp, Va, 46-47; fel biochem, Children's Hosp, Washington, DC, 47-48, dir biochem res, 48-54; from assoc prof to prof pediat, Sch Med, Univ Md, 54-68, assoc prof biochem, 58-68; PROF PHARMACOL & CHMN DEPT, SCH MED, UNIV SOUTHERN CALIF, 68-, PROF PEDIAT, 69- *Concurrent Pos:* Fel, Psychiat Inst, NY, 48; assoc prof clin pediat, Sch Med, George Washington Univ, 52; fel, Harvard Univ, 52-53; ed, Biochem Med; assoc ed, Anal Biochem; sr attend physician, Cedars-Sinai Med Ctr, Los Angeles, 72-79, hon attend physician, 80-; ed, Biochem Med. *Honors & Awards:* Aaron Brown Lectr, Baylor Col Med, 79. *Mem:* Soc Pediat Res; Am Chem Soc; Am Acad Pediat; Am Inst Nutrit; Am Soc Biol Chemists. *Res:* Biochemistry of disease; mechanism of mental symptoms; carbohydrate diseases; artificial pancreas. *Mailing Add:* Dept Pharmacol Univ Southern Calif Los Angeles CA 90033

BESSO, MICHAEL M, b Brooklyn, NY, Mar 30, 30; m 55; c 2. POLYMER STABILIZATION, SPECIALTY CHEMICALS. *Educ:* Lowell Technol Inst, BSc, 50; Lehigh Univ, MS, 57, PhD(org chem), 59. *Prof Exp:* Chemist, Pinatel Piece Dye Works, Can, 50-51 & Nat Lead Co, 53-55; res chemist, Chas Pfizer & Co, 59-65; sr res chemist, Celanese Corp, 65-68, group leader fibers, plastics & coatings, 69-74, mgr polymer chem, 74-78, mgr explor res, 78-84; vpres res & develop, M&T Chemicals, 84-89; VPRES, ATOCHEM NAM, 89- *Mem:* Am Chem Soc; NY Acad Sci; Am Inst Chemists. *Res:* Polymer flammability; plastics, and polyester fiber technology; polymer synthesis; powder coating technology; increasing productivity of research; polymer stabilization; fermentation chemicals, membrane fabrication and applications; electrochemistry; materials science. *Mailing Add:* 74 Crestmont Rd W Orange NJ 07052

BEST, CECIL H(AMILTON), b Tillsonburg, Ont, Oct 3, 28; US citizen; m 53; c 2. ENGINEERING MATERIALS. *Educ:* Univ Calif, Berkeley, BS, 55, MS, 56, PhD(eng sci), 60. *Prof Exp:* From res engr to asst specialist, Forest Prod Lab, Univ Calif, Berkeley, 55-60; from assoc prof to prof appl mech, 61-75, assoc dean col eng, 68-74, PROF CIVIL ENG, KANS STATE UNIV, 75- *Concurrent Pos:* Royal Norweg Coun Sci & Indust res fel, Norweg Inst Technol, 60-61; vis scholar, Univ Calif, Berkeley, 80; res civ engr, US Army Engr, Waterwasy Exp Sta, 85. *Mem:* Fel Am Concrete Inst; fel Am Soc Civil Engrs; Am Soc Testing & Mat; Nat Soc Prof Engrs. *Res:* Structure and properties of engineering materials, primarily cement and concrete. *Mailing Add:* Dept Civil Eng Seaton Hall Kans State Univ Manhattan KS 66506

BEST, DAVID M, b Cardiff, Wales, UK, July 12, 45; m. GEOPHYSICS, MATHEMATICS. *Educ:* Univ NC, BS, 67, MS, 70, PhD(geol), 77. *Prof Exp:* asst prof, 78-84, ASSOC PROF GEOL, NORTHERN ARIZ UNIV, 84- *Concurrent Pos:* Interim chair, 84-86. *Mem:* Geol Soc Am; Sigma Xi; Am Geophys Union. *Res:* Geophysics of Proterozoic boundaries in Arizona; modelling Cenozoic basins in central Arizona. *Mailing Add:* Dept Geol Northern Ariz Univ Box 6030 Flagstaff AZ 86011-6030

BEST, EDGAR ALLAN, b Enumclaw, Wash, Aug 17, 25; m 50; c 1. MARINE BIOLOGY. *Educ:* Univ Wash, BS, 53. *Prof Exp:* Fishery biologist, Ore Fish Comn, 53; marine biologist, Calif Dept Fish & Game, 54-57, proj leader, 57-64, supvr Point Arguello surv, 64-65; supvr small fish invest, 65-78, SUPVR AGE INVEST, INT PAC HALIBUT COMN, UNIV WASH, 78- *Concurrent Pos:* Mem int trawl fisheries comt, US State Dept, 59-65; res assoc ichthyol, Santa Barbara Mus Natural Hist, 65-68. *Mem:* Am Inst Fishery Res Biologists (secy, 75-81); Pac Fish Biologists. *Res:* Marine biology and ecology; ecology; life history of marine fishes. *Mailing Add:* 7316 50th Ave NE Seattle WA 98115

BEST, GARY KEITH, b Weatherford, Okla, Oct 8, 38; m 62; c 2. MICROBIOLOGY. *Educ:* Southwestern State Col, Okla, BS, 60; Okla State Univ, PhD(microbiol), 65. *Prof Exp:* Res microbiologist, Agr Res Serv, USDA, 65-68; from asst prof to assoc prof, 68-80, PROF MICROBIOL, MED COL GA, 80- *Mem:* Am Soc Microbiol. *Res:* Bacterial cell wall synthesis; effect of antibiotics on microorganisms; bacterial virulence mechanisms. *Mailing Add:* Dept Cell & Molecular Biol Med Col Sch Med 1120 15th St Augusta GA 30912

BEST, GEORGE E(DWARD), b South Portland, Maine, Nov 15, 12; m 40; c 1. ELECTROCHEMICAL ENGINEERING. *Educ:* Mass Inst Technol, 34. *Prof Exp:* Res investr process metal, NJ Zinc Co of Pa, 34-43; prod supvr, Pa Salt Mfg Co, 43-47; res investr fused refractories, Carborundum Co, 47; develop chemist, Martin Dennis Co, 47-48; mgr tech serv, Mutual Chem Co Am, 48-58; tech asst to mgr tech serv, Solvay Process Div, Allied Chem Corp, 58-59; chem engr, Mfg Chemists Asn Inc, 59-68, tech dir, 68-73, vpres & secy-treas, 73-78, sr vpres, 78; RETIRED. *Mem:* Am Chem Soc; Am Electroplaters Soc; Am Inst Chem Engrs; Electrochem Soc; Nat Asn Corrosion Eng (pres, 60-61). *Res:* Trade association technical programs in environmental management, chemical packaging and transportation, precautionary labeling, safety, international trade, occupational health, plastics, reactive metals; zinc smelting and refining; powdered metals production; chromium chemicals applications. *Mailing Add:* 4416 Wickford Rd Baltimore MD 21210-2810

BEST, GEORGE HAROLD, b Chicago, Ill, Aug 4, 20; m 44; c 3. PHYSICS. *Educ:* Purdue Univ, BS, 42; Northwestern Univ, MS, 48, PhD(physics), 49. *Prof Exp:* mem staff, Los Alamos Nat Lab, Univ Calif, 49-83; CONSULT, 83- *Mem:* Am Phys Soc. *Res:* Nuclear physics; nuclear weapons effects; nuclear reactors; directed energy. *Mailing Add:* 332 Manhattan Loop Los Alamos NM 87544

BEST, JAY BOYD, US Citizen; m 83; c 1. PHYSIOLOGY, BIOPHYSICS. *Educ:* Univ Tex, AB, 47; Univ Chicago, PhD(physiol), 53. *Prof Exp:* Res assoc psychol, Univ Chicago, 53-55; biomathematics, Walter Reed Army Inst Res, 55-60; assoc prof physiol, Col Med, Univ Ill, 60-63; prof physiol & biophys, 63-87, prof environ health & physiol, 87-88, EMER PROF ENVIRON HEALTH, COLO STATE UNIV, 88- *Concurrent Pos:* Res fel, Univ Chicago, 53-55; NIH grant, 61-72, NASA grants, 64-69, USDA grants, 79-81 & NSF grant, 85-90; vis scientist, Argonne Lab, 77-78. *Honors & Awards:* Sigma Xi Award, 70. *Mem:* Fel AAAS; Am Soc Zoologists; Soc Math Biol; Sigma Xi; Soc Neurosci. *Res:* Transphyletic animal similarities and predictive toxicology; invertebrate metazoa as models for assessing toxic hazards for higher animals and humans; comparative regulatory biology between flatworms and higher animals; developmental neuroscience. *Mailing Add:* Dept Environ Health Colo State Univ Ft Collins CO 80523

BEST, LAVAR, b Salt Lake City, Utah, Mar 28, 31; m 54; c 4. BIOACOUSTICS. *Educ:* Univ Utah, BS, 53, MS, 57, PhD(audiol), 64. *Prof Exp:* Instr res, Med Ctr, Univ Colo, 64-66; asst prof speech & hearing sci, 67-71, assoc prof speech & hearing sci, Univ Denver, 71-78; dir, Neurosensory Diag Ctr, Swed Med Ctr, Colo, 78-85; DIR, NEUROSCI LAB, CRAIG HOSP, ENGLEWOOD, COLO, 85- *Mem:* Fel Am Soc Evoked Potential Monitoring. *Res:* Cortical processing of auditory stimuli. *Mailing Add:* Craig Hosp 3425 S Clarkson Englewood CO 80110

BEST, LOUIS BROWN, b Ogden, Utah, May 7, 44; m 68; c 4. AVIAN ECOLOGY. *Educ:* Weber State Col, BS, 68; Mont State Univ, MS, 70; Univ Ill, Urbana-Champaign, PhD(zool), 74. *Prof Exp:* From asst to assoc prof, 74-79, PROF ECOL, DEPT ANIMAL ECOL, IOWA STATE UNIV, 79- *Mem:* Ecol Soc Am; Am Ornithologists Union; Cooper Ornith Soc; Wilson Ornith Soc. *Res:* Vertebrate ecology, with particular emphasis on various aspects of avian ecology, including breeding biology, feeding behavior, social organization and habitat selection. *Mailing Add:* Dept Animal Ecol Iowa State Univ Ames IA 50011

BEST, PHILIP ERNEST, b Perth, Australia, July 28, 38; m 60; c 3. SECONDARY ELECTRON EMISSION, SURFACE PHYSICS. *Educ:* Univ Western Australia, BSc, 59, PhD(physics), 63. *Prof Exp:* Instr & res assoc physics, Lab Atomic & Solid State Physics, Cornell Univ, 63-65; exp physicist, Res Labs, United Aircraft Corp, 65-70; fel, Inst Mat Sci, 70-71, assoc prof, 71-79, PROF PHYSICS, UNIV CONN, 79- *Mem:* Am Phys Soc; Inst Elec & Electronics Engrs; Brit Inst Physics & Phys Soc. *Res:* Interaction of slow electrons and x-rays with matter; electronic excitations at surfaces and in the bulk of solids; electronic properties of small molecules. *Mailing Add:* Dept Physics & Inst Mat Sci Univ Conn Storrs CT 06269-3046

BEST, STANLEY GORDON, b Chicago, Ill, July 10, 18; m 41; c 4. MECHANICAL ENGINEERING, ELECTRONICS ENGINEERING. *Educ:* Purdue Univ, BSME, 40. *Prof Exp:* Anal engr, Hamilton Standard Div, United Aircraft Corp, 40-45, proj engr, 45-55, chief anal, 55-59, chief tech staff, 59-62, chief res, 62-63; tech consult, Hamilton Stand Div, United Technol, 63-81; eng consult, Best Eng, Inc, 81-86; RETIRED. *Res:* Vibration and structural analysis; aerodynamics; stability; control dynamics; thermodynamics; heat transfer; electronic circuitry; computer programming; finite element method development for structures; vibrations; heat transfer; fluid flow; circuit analysis; controls. *Mailing Add:* 53 Coburn Rd Manchester CT 06040

BEST, TROY LEE, b Fort Sumner, NMex, Aug 30, 45; c 2. VERTEBRATE ECOLOGY, ECOSYSTEM DYNAMICS. *Educ:* Eastern NMex Univ, BS, 67; Univ Okla, MS, 71, PhD(zool), 76. *Prof Exp:* Preparator zool, Stovall Mus Sci & Hist, Univ Okla, 68-71, res & teaching asst, dept zool, 71-74; asst prof biol, Dept Biol, Northeastern Univ, 74-76; res prof, Llano Estacado Ctr for Advan Prof Studies Res, 76-81; asst prof, Dept Biol, Univ NMex, 83-88; ASSOC PROF, DEPT ZOOL & WILDLIFE SCI, AUBURN UNIV, 88- *Concurrent Pos:* Index ed, Syst Zool, 72-; prin consult, US Bur Reclamation, US Fish & Wildlife Serv, USAF, US Army Corp Eng, Sandia Nat Labs, Westinghouse Elec Corp, 81- *Mem:* Am Soc Mammalogists; Ecol Soc Am; Soc Syst Zool; Am Ornithologists Union; AAAS; Sigma Xi. *Res:* Assessment of interrelationships between variation in ecological and morphological characteristics of vertebrates; mammalian systematics utilizing univariate and multivariate statistical techniques; effect of character selection on data analyses; vertebrate feeding ecology. *Mailing Add:* Dept Zool & Wildlife Sci Auburn Univ 331 Funchess Hall Auburn University AL 36849-5414

BEST, WILLIAM ROBERT, b Chicago, Ill, July 14, 22; m 44; c 2. HEMATOLOGY, BIOSTATISTICS. *Educ:* Univ Ill, BS, 45, MD, 47, MS, 51; Am Bd Internal Med, dipl, 57, cert hemat, 72. *Prof Exp:* From clin asst to assoc prof med, Univ Ill Med Ctr, 48-70, asst dir, Aeromed Lab, 64-67, prof, 70-81, assoc dean, Abraham Lincoln Sch Med, 72-81, chief staff, Univ Ill Hosp, 75-81; PROF MED & ASSOC DEAN, STRITCH SCH MED, LOYOLA UNIV CHICAGO MED CTR, HINES, 81-; CHIEF STAFF, HINES VET ADMIN HOSP, 81- *Concurrent Pos:* Attend physician, Ill Res & Educ Hosp, 53-81; consult, Vet Admin Hosp, Hines, 53-81, chief midwest res support ctr, 67-72; consult, Grant Hosp, 54-67. *Mem:* AAAS; Am Fedn Clin Res; Am Soc Hemat; Int Soc Hemat; Am Statist Asn; Sigma Xi. *Res:* Statistical methods; cooperative clinical trials; computers in medicine; leukemia and pernicious anemia; Crohn's disease; sickle-cell trait. *Mailing Add:* Blythe Rd Riverside IL 60546

BESTE, CHARLES EDWARD, b Evansville, Ind, Sept 3, 39; div; c 2. WEED SCIENCE, HORTICULTURE SCIENCE. *Educ:* Purdue Univ, BS, 61, MS, 69, PhD(weed sci), 71. *Prof Exp:* Develop engr, Gen Elec Co, Ind, 61-67; asst prof, 71-76, ASSOC PROF HORT, UNIV MD, 76- *Concurrent Pos:* Mem, Integrated Pest Mgt Report Comt, Coun Agr Sci & Technol, 78-81; Sabbatical, Univ Calif, Davis, 86. *Honors & Awards:* Past-pres award, Northeastern Weed Sci Soc, 78; IR-4 Meritorious Serv Award, 90. *Mem:* Weed Sci Soc Am; Am Soc Hort Sci; Sigma Xi; Int Weed Sci Soc. *Res:* Weed control studies and sustainable agriculture. *Mailing Add:* Salisbury Facil LESREC Univ Md Rte 5 Box 246 Salisbury MD 21801

BESTUL, ALDEN BEECHER, b Forestburg, SDak, Sept 1, 21; m 47; c 4. PHYSICAL CHEMISTRY. *Educ:* St Olaf Col, BA, 42; Pa State Col, MS, 44; Univ Md, PhD(chem), 53. *Prof Exp:* Res asst cryogen, Pa State Col, 44-45; res thermochemist, Stamford Res Labs, Am Cyanamid Co, 46-48; res physicist, Nat Bur Standards, 48-70; gen phys scientist, Off Spec Studies, Nat Oceanic & Atmospheric Admin, 70-85; RETIRED. *Concurrent Pos:* Com Sci & Technol fel, 69-70. *Mem:* AAAS; Am Chem Soc; Am Phys Soc. *Res:* Molecular structure in the vitreous state; rheology of polymers; thermochemistry of organic compounds; research and development evaluation; environmental policy analysis; vitreous state; polymer rheology; thermodynamics and thermochemistry. *Mailing Add:* 9400 Overlea Dr Rockville MD 20850

BETH, ERIC WALTER, b Vienna, Austria, June 7, 12; nat US. THEORETICAL PHYSICS. *Educ:* Univ Vienna, PhD(physics), 34. *Prof Exp:* Asst physics, Univ Vienna, 36-37; res fel, Univ Calif, 37-38, asst, 38; asst physics, Reed Col, 39-40; vis instr physics & math, Mills Col, 40-41; instr physics, Ill Inst Technol, 41-42; spec consult, Electronic Res Labs, Mass, 46-53; mem sci staff, Melpar, Inc, 53-54; physicist, Union Switch & Signal Div, Westinghouse Air Brake Co, 54-62; asst prof, 62-82, EMER ASST PROF PHYSICS, STATE UNIV NY BUFFALO, 82- *Mem:* NY Acad Sci; Sigma Xi. *Res:* Wave mechanics; quantum theory of the non-idea gas; upper air properties. *Mailing Add:* Dept Physics & Astron State Univ NY Buffalo NY 14260

BETHE, HANS ALBRECHT, b Strasbourg, Ger, July 2, 06; nat US; m 39; c 2. NUCLEAR PHYSICS, ASTROPHYSICS. *Educ:* Univ Munich, PhD(physics), 28. *Hon Degrees:* DSc, Polytech Inst Brooklyn, 50, Univ Denver, 52, Univ Chicago, 53, Univ Birmingham, England, 56, Harvard Univ, 58, Univ Munich & Tech Univ Munich, 68, Univ Seoul & Univ Delhi, 69, Univ Tübingen & Weizmann Inst, 78. *Prof Exp:* Instr physics, Univ Frankfurt, 28-29 & Univ Stuttgart, 29; lectr, Univ Munich & Univ Tübingen, 30-33 & Univ Manchester, Eng, 33-34; fel, Univ Bristol, 34-35; from asst prof to John Wendell Anderson prof, 35-75, EMER PROF PHYSICS, CORNELL UNIV, 75- *Concurrent Pos:* Rockefeller Found Int Educ Bd fel, Cambridge Univ & Univ Rome, 30-32; staff mem radiation lab, Mass Inst Technol, 42-43; chief theoret physics div, Los Alamos Sci Lab, NMex, 43-46; vis prof, Columbia Univ, 41 & 48; consult, Los Alamos Sci Lab, 47-, Atomic Power Develop Assocs, 53-63 & Avco Everett Res Lab, 55-87; vis prof, Univ Wash, 50, 60, 67, 73, 76, & Cambridge Univ, 55-56; mem, President's Sci Adv Comt, 56-59; mem, US Delegation to Discussions on Discontinuance of Nuclear Weapons Tests, Geneva, 58-59; vis prof, Calif Inst Technol, 64, 82, 85, 87, Nordita, Copenhagen, Denmark, 64, 70, 74, 78, 81, 84 & Mass Inst Technol, 75, 77, Harvard Univ, 82. *Honors & Awards:* Nobel Prize Physics, 67; Morrison Prize, NY Acad Sci, 38 & 40; US Medal Merit, 46; Draper Medal, Nat Acad Sci, 48; Planck Medal, 55; Enrico Fermi Prize, US AEC, 61; Nat Medal Sci, 76. *Mem:* Nat Acad Sci; fel Am Phys Soc (pres, 54); Am Astron Soc; Am Philos Soc. *Res:* Quantum theory of atoms; theory of metals; quantum theory of collisions; theory of atomic nuclei; meson theory; energy production in stars; neutron stars and their formation; quantum electrodynamics; shock wave theory; theory of supernovae. *Mailing Add:* 320 Newman Lab Cornell Univ Ithaca NY 14853

BETHEA, CYNTHIA LOUISE, b Mobile, Ala, Nov 12, 52. ENDOCRINOLOGY, REPRODUCTION. *Educ:* Winthrop Col, BA, 72; Clemson Univ, MS, 74; Emory Univ, PhD(physiol), 78. *Prof Exp:* ASST SCIENTIST, ORE REGIONAL PRIMATE RES CTR, 81- *Concurrent Pos:* Fel endocrinol, Univ Calif, San Francisco, 81. *Mem:* Soc Neurosci; Fedn Am Soc Exp Biol. *Res:* Examination of the cellular regulation of prolactin and luteinizing hormone secretion and the hypothalamic factors which modify the production of these hormones. *Mailing Add:* Dept Reprod Biol & Behav Ore Regional Primate Res Ctr 505 NW 185th Ave Beaverton OR 97006

BETHEA, ROBERT MORRISON, b Raleigh, NC, Dec 12, 35; m 61. CHEMICAL ENGINEERING, AIR POLLUTION. *Educ:* Va Polytech Inst, BS, 57; Iowa State Univ, MS, 59, PhD(chem eng), 64. *Prof Exp:* Instr chem eng, Iowa State Univ, 59-64; aerospace technologist environ monitoring & control, Langley Res Ctr, NASA, 64-66; from asst prof to assoc prof chem eng, 66-74, PROF CHEM ENG, TEX TECH UNIV, 75- *Concurrent Pos:* Res assoc, Iowa Eng Res Inst, Iowa State Univ, 59-64; consult atmosphere control systs, Collins Radio Co, 63-64; chem engr, Indust Environ Res Lab, US Environ Protection Agency, NC, 72; consult, numerous co & agencies; dipl, Am Acad Environ Engrs, Cert Air Pollution Control. *Mem:* Am Inst Chem Engrs; Am Chem Soc; Am Indust Hyg Asn; Air Pollution Control Asn. *Res:* Air pollution control system evaluation and design; instrumental methods for air quality control and measurement; evaluation of solar collector mirror survivability in erosive environments; diffusion mechanisms; dust storm prediction and modeling. *Mailing Add:* Dept Chem Eng 5206 17th St Lubbock TX 79416

BETHEA, TRISTRAM WALKER, III, b Lancaster, SC, July 29, 38; m 70; c 4. ORGANIC CHEMISTRY, POLYMER CHEMISTRY. *Educ:* Davidson Col, BS, 60; Clemson Univ, MS, 62, PhD(org chem), 65. *Prof Exp:* Res scientist elastomer synthesis, Firestone Tire & Rubber Co, 67-69, sr res scientist elastomer synthesis & polymer physics, 69-78, assoc scientist mat sci, 79-85, RES ASSOC POLYMER SCI, BRIDGESTONE/FIRESTONE RES INC, 85- *Concurrent Pos:* Res assoc, Case Inst Technol, 64-66. *Mem:* Am Chem Soc. *Res:* Polymer chemistry; polymer synthesis; elastomers; structure-property relationships. *Mailing Add:* 950 Sandin Dr Akron OH 44333-1348

BETHEIL, JOSEPH JAY, b New York, NY, Apr 16, 24; wid; c 2. BIOCHEMISTRY. *Educ:* City Col New York, BS, 44; Univ Wis, MS, 47, PhD(biochem), 49. *Prof Exp:* Res asst, Dept Chem, Off Sci Res & Develop Proj, Columbia Univ, 44-45; USPHS fel, Inst Enzyme Res, Univ Wis, 49-50; from instr to asst prof biochem, Sch Med & Dent, Univ Rochester, 50-55; asst prof, 55-58, ASSOC PROF BIOCHEM, ALBERT EINSTEIN COL MED, YESHIVA UNIV, 58- *Concurrent Pos:* USPHS spec fel, Nat Inst Med Res, London, 64-65. *Mem:* AAAS; Am Soc Biol Chemists; Am Inst Nutrit; Harvey Soc; Am Chem Soc. *Res:* Protein chemistry and structure; intermediary metabolism; study of cellular aging; regulation of protein biosynthesis; nutrition. *Mailing Add:* 92 Shoreview Dr Apt 2 Yonkers NY 10710-1351

BETHEL, EDWARD LEE, b Strawn, Tex, Feb 18, 26. MATHEMATICS. *Educ:* Southern Methodist Univ, BA, 50; NTex State Univ, MA, 57; Univ Tex, PhD(math), 67. *Prof Exp:* Radar instr, US Civil Serv, Keesler AFB, Miss, 50-51; systs engr, Chance Vought Aircraft Corp, Tex, 51-53; aerophys engr, Gen Dynamics/Convair, 57; instr math, Univ Tex, 57-63 & Knox Col, Ill, 63-64; assoc prof, Clemson Univ, 64-67; asst prof, 67-74, assoc prof math, Kent State Univ, 74-85; RETIRED. *Mem:* Am Math Soc; Math Asn Am. *Res:* General topology and convex sets in metric spaces. *Mailing Add:* HCR 5 Box 1063 Burnet TX 78611

BETHEL, JAMES SAMUEL, b New Westminster, BC, Aug 13, 15; m 41; c 3. WOOD TECHNOLOGY. *Educ:* Univ Wash, BSF, 37; Duke Univ, MF, 39, DF, 47. *Prof Exp:* Instr forestry, Pa State Col, 39-41; asst prof, Va Polytech Inst, 41-42; asst mgr, Tidewater Plywood Co, Ga, 46-48, prod mgr, 48-49; assoc prof wood technol, NC State Col, 49-50, prof & dir wood prod lab, 50-58, head wood prod dept & actg dean, Grad Sch, 58-59; head spec proj, Sci Educ Sect, NSF, 59-62; assoc dean, Grad Sch & Col Forest Res, Univ Wash, 62-64, prof forestry, 62-85, dean, 64-81, emer dean, Col Forest Res, 81-85, emer prof forestry, 85; dir, Ctr Int Forestry, 81-85; RETIRED. *Concurrent Pos:* UN consult, Yugoslavia, 52; adv, Econ Develop Admin, PR, 56; consult, US Dept Health, Educ & Welfare, 56-, Gov Res Triangle Comt, NC, 57-59 & NSF, 62-; dir, Orgn Trop Studies, 63-; vpres, 64-65, pres, 66-67; ed, J Soc Am Foresters; consult, President's Coun Environ Qual; treas & bd mem, Int Bot Cong, 69; mem comt effects herbicides Vietnam, Nat Acad Sci; consult, Comt Eng Policy, Nat Acad Eng, 72-73; mem bd agr & renewable res, Nat Acad Sci, 73-75, chmn, Task Force on Educ Agr & Renewable Res & Comt on Renewable for Indust Mat, 74-75, mem, Firewood Crop Panel Task Force & Panel Rev US/USSR Agreement Coop Fields Sci & Technol, Bd Int Sci Exchange, 77; mem prog comt, Int Mat Cong, 77-78; mem, Nat Forest Systs Adv Comt, US Forest Serv, USDA, 78-79; pres, Univ Int Forestry, 79-80, bd mem, 79-81; consult, UN Food & Agr Organ, 82-87, US Agency Int Develop, 84-88. *Mem:* Fel AAAS; fel Soc Am Foresters; fel Int Acad Wood Sci; Soc Wood Sci & Technol (pres, 58-59, 60-61); fel Inst Wood Sci; Forest Prod Res Soc. *Res:* Wood science, especially anatomy and morphology; relationship of environment to structure of wood in forest trees; wood moisture relationships; tropical forest utilization. *Mailing Add:* 3816 E Mercer Way Mercer Island WA 98040

BETHEL, WILLIAM MACK, parasitology, ecology, for more information see previous edition

BETHKE, PHILIP MARTIN, b Chicago, Ill, Mar 22, 30; m 55; c 7. GEOLOGY. *Educ:* Amherst Col, BA, 52; Columbia Univ, MA, 54, PhD(geol), 57. *Prof Exp:* Asst geol, Columbia Univ, 52-53, res asst, 53-54; asst prof, Mo Sch Mines, 55-59; chief, Br Exp Geochem & Minerol,80-83, GEOLOGIST, US GEOL SURV, 57- *Mem:* Geol Soc Am; Geochem Soc; Mineral Soc Am (treas, 73-74); Soc Econ Geol (vpres, 92); Am Geophys Union. *Res:* Chemistry of ore deposition; mineralogy of ore minerals; time-space relations in ore deposits; phase equilibria of ore minerals. *Mailing Add:* US Geol Surv Nat Ctr MS 959 Reston VA 22092

BETHLAHMY, NEDAVIA, b Tel-Aviv, Palestine, July 23, 18; US citizen; m 43; c 2. FOREST HYDROLOGY. *Educ:* Pa State Col, BS, 39; Yale Univ, MF, 40; Cornell Univ, PhD, 56. *Prof Exp:* Forester Northeastern Forest Exp Sta, US Forest Serv, 46-55, res forester, Pac Northwest Forest & Range Exp Sta, 56-63, proj leader, Intermountain Forest & Range Exp Sta, 63-66; watershed mgr expert, Food & Agr Orgn, UN, Taiwan, 66-68; res forester, 68-69, prin forest hydrologist, Intermountain Forest & Range Exp Sta, US Forest Serv, 69-76; CONSULT, 80- *Concurrent Pos:* Adj prof forest hydrol, Univ Idaho, 69-76; UN Develop Prog, Food & Agr Orgn, Tehran, Iran, 78-79. *Honors & Awards:* Nat Res, Sigma Xi. *Mem:* Soc Am Foresters; Am Geophys Union. *Res:* Hydrology of wildlands, with special emphasis on effects of major wild fires; watershed management. *Mailing Add:* 4043 First Ave NE Seattle WA 98105

BETHLENDY, GEORGE, nuclear engineering, for more information see previous edition

BETHUNE, DONALD STIMSON, b Philadelphia, Pa, July 2, 48; m 76; c 5. NONLINEAR OPTICS, LASER PHYSICS. *Educ:* Stanford Univ, BS, 70; Univ Calif, Berkeley, PhD(physics), 77. *Prof Exp:* Scientist, 77-82, RES STAFF MEM, WATSON RES CTR, IBM RES LAB, IBM CORP, SAN JOSE, CALIF, 82- *Mem:* Am Phys Soc; Optical Soc Am. *Res:* Nonlinear optics; dipole forbidden nonlinear susceptibilities and processes; nonlinear optical method for capturing broadband infared spectra on a nansecond time scale; time resolved studies of chemical processes. *Mailing Add:* IBM Almaden Res Ctr K33/801 650 Harry Rd San Jose CA 95120-6099

BETHUNE, JOHN EDMUND, b Mar 23, 27; Can citizen; m 48; c 2. ENDOCRINOLOGY. *Educ:* Acadia Univ, BA, 47, BSc, 48; Dalhousie Univ, MD, 53; FRCPS(C), 57. *Prof Exp:* Asst prof med, Dalhousie Univ, 58-61; from asst prof to assoc prof, 61-70, head sect endocrinol & dir clin res ctr, 67-72, PROF MED, UNIV SOUTHERN CALIF, 70-, CHMN DEPT, 72- *Concurrent Pos:* Consult, Los Angeles Vet Admin Hosp, 68 & Sepulveda Vet Admin Hosp, 70. *Mem:* Fel Am Col Physicians; Endocrine Soc; Am Fedn Clin Res; Am Soc Nephrology; Am Physiol Soc. *Res:* Adrenal hormone activity, with special reference to calcium and bone metabolism. *Mailing Add:* Sch Med Dept Med Univ Southern Calif Los Angeles CA 90033

BETHUNE, JOHN LEMUEL, b Baddeck, NS, July 22, 25; m 58. PHYSICAL CHEMISTRY, BIOPHYSICS. *Educ:* Acadia Univ, BSc, 47; Clark Univ, PhD(chem), 61. *Prof Exp:* Prof chemist, Can Breweries, Ltd, 48-58; asst prof chem, Clark Univ, 61; res assoc, 61-64, assoc, 64-66, asst prof, 66-69, assoc prof biochem, 69-75, PROF RADIOL, HARVARD MED SCH, 75- *Mem:* Fedn Am Soc Exp Biol. *Res:* Effect of chemical reactions upon the interpretation of transport experiments. *Mailing Add:* Brigham & Women's Hosp 75 Francis St Boston MA 02115

BETSCHART, ANTOINETTE, b Manteca, Calif, Apr 12, 39. NUTRITION QUALITY FOOD. *Educ:* Cornell Univ, PhD(food sci & nutrit), 71. *Prof Exp:* RES FOOD TECHNOLOGIST, WESTERN REGIONAL RES CTR, AGR RES SERV, USDA, 71. *Mailing Add:* Western Regional Res Ctr ARS USDA 800 Buchanan St Albany CA 94710

BETSO, STEPHEN RICHARD, b Brooklyn, NY, May 28, 45; m 68. ANALYTICAL CHEMISTRY, POLYMER CHEMISTRY. *Educ:* St Johns Univ, NY, BS, 66; Ohio State Univ, MS, 70, PhD(chem), 71. *Prof Exp:* Res fel analytical chem, Univ Ga, 71-73; analytical chemist, 73-77, plastics engr, 77-80, chelation chemist, 80-85, POLYMER CHEMIST, DOW CHEM CO, 85- *Mem:* Am Chem Soc; Sigma Xi. *Res:* New polymer development. *Mailing Add:* 1019 Oleander Lake Jackson TX 77566

BETTELHEIM, FREDERICK A, b Gyor, Hungary, June 3, 23; nat US; m 47; c 1. PHYSICAL CHEMISTRY. *Educ:* Cornell Univ, BS, 53; Univ Calif, MS, 54, PhD(phys chem), 56. *Prof Exp:* Asst chemist, Agr Exp Sta, Israel, 47-51; analyst, NY Agr Exp Sta, 52-53; asst, Univ Calif, Davis, 53-56; instr, Univ Mass, 56-57; chmn dept, 85-91, from asst prof to assoc prof, 57-64, PROF PHYS CHEM, ADELPHI UNIV, 65- *Concurrent Pos:* Lalor Found fel, 58; vis prof, Uppsala Univ, Sweden, 65, Technion, Israel, 65, Weizmann Inst,

Israel, 73 & Univ Fla, 81; mem adv bd mil personnel supplies, Nat Acad Sci, 67-71; mem study sects, Nat Eye Inst, NIH, 78-, vis scientist, 85; Fulbright prof, Weizman Inst, Israel, 84; ed, Exp Eye Res, 84- *Mem:* Am Chem Soc; NY Acad Sci; Fedn Am Socs Exp Biol; Asn Res Vision & Ophthal; Int Soc Eye Res (treas, 84-89). *Res:* Physical chemistry of high polymers; mucopolysaccharides; glycoproteins; lens proteins; x-ray diffraction, birefringence; light scattering high vacuum vapor sorption; infra-red dichroism; molecular, supermolecular structures in cornea, lens, vitreous of the eye; textbook writer: general, organic, and biochemistry. *Mailing Add:* Dept Chem Adelphi Univ Garden City NY 11530

BETTEN, J ROBERT, b Omaha, Nebr, Nov 10, 32; m; c 2. COMMUNICATIONS. *Educ:* Iowa State Univ, BS, 55, MS, 59, PhD(elec eng), 62. *Prof Exp:* Res asst elec eng, Iowa State Univ, 54-55, asst & instr, 58-62; sr elec engr, Gen Dynamics/Convair, 57-58; assoc prof, 62-64, PROF ELEC ENG, UNIV MO-ROLLA, 64-, CHMN DEPT, 67- *Concurrent Pos:* Res & teaching consult, Continental Oil Co, 66- *Mem:* Inst Elec & Electronics Engrs; Am Soc Eng Educ; Sigma Xi. *Res:* Electrical communications, including statistical detection, statistical filtering, information theory, signal analysis and modulation systems. *Mailing Add:* Rte 4 Box 161 Rolla MO 65401

BETTENCOURT, JOSEPH S, JR, b Cambridge, Mass, Mar 5, 40; m 63; c 3. PARASITOLOGY. *Educ:* Suffolk Univ, AB, 62; Univ NH, MS, 65, PhD(zool), 76. *Prof Exp:* Instr, 65-67, asst prof, 69-77, ASSOC PROF BIOL, MARIST COL, 78- *Concurrent Pos:* Dir, Physically Handicapped Sci Prog, NSF, 79-81. *Mem:* Am Soc Parasitologists; AAAS; Sigma Xi. *Res:* Marine fish and fresh water fish hematozoa; life cycles of Trypanoplasmea, host specificity and vector transmission. *Mailing Add:* Dept Biol Marist Col Poughkeepsie NY 12601

BETTERTON, JESSE O, JR, physical metallurgy, solid state physics; deceased, see previous edition for last biography

BETTGER, WILLIAM JOSEPH, b St Louis, Mo, Sept 10, 53. TRACE MINERALS, MEMBRANE BIOCHEMISTRY. *Educ:* Univ Mo, BSA, 75, PhD(biochem), 79. *Prof Exp:* Res assoc, dept molecular, cellular & developing biol, Univ Colo, Boulder, 79-80 & dept biochem, Sch Med, St Louis Univ, 80-82; asst prof, 82-88, ASSOC PROF, DEPT NUTRIT SCI, UNIV GUELPH, 88- *Mem:* Am Inst Nutrit; Can Soc Nutrit Sci; Tissue Cult Asn. *Res:* Effect of trace minerals on membrane structure and function; nutritional requirements of cells in culture. *Mailing Add:* 12 Lambert Crescent Guelph ON N1G 2R5 Can

BETTI, JOHN A, b Ottawa, Ill, Jan 6, 31. AUTOMOTIVE ENGINEERING. *Educ:* Ill Inst Technol, BS, 52; Chrysler Inst Eng, MAutomotive Eng, 54. *Prof Exp:* Chassis engr, Chrysler Corp, 52-57, asst managing engr chassis prod planning, 57, exec asst to elec eng chassis & truck design, 57-58, managing engr, Valiant Prog, 58-59, asst chief engr chassis design, 59-60 & elec design, 60-62; exec engr body eng, Ford Motor Co, 62-67, chief elec systs eng, 68, light truck eng, 68-72, chief engine engr, 72-74, chief car planning mgr, 74-75, chief car eng NAm, 75, vpres truck & recreation vehicles, Ford NAm, 75-76, vpres prod develop, Ford Europe, 76-79, vpres power train & chassis, 79-83, vpres mfg & bus develop, Ford NAm, 83-84, exec vpres tech affairs & operating staffs, 84-88 & diversified prods, 88-89; undersecy acquisitions, Dept Defense, 89-91; PVT INVESTOR & CONSULT, 91- *Concurrent Pos:* Chmn bd trustees, Gen Motors Eng & Mgt Inst; mem bd dirs, Ford Motor Co, 85-90. *Mem:* Nat Acad Eng. *Mailing Add:* 1231 Stuart Robeson Dr McLean VA 22101

BETTICE, JOHN ALLEN, b Columbus, Ind, Nov 27, 42; m 67. MAMMALIAN PHYSIOLOGY. *Educ:* Univ Dayton, BS, 65; Johns Hopkins Univ, PhD(physiol), 72. *Prof Exp:* ASST PROF PHYSIOL, SCH MED, CASE WESTERN RESERVE UNIV, 74- *Concurrent Pos:* fel Sch Med, Johns Hopkins Univ, 72-73; fel, Univ Kans Med Ctr, 73-74. *Mem:* AAAS; Am Physiol Soc. *Res:* Mechanisms of the physiological reactions which buffer changes in acid-base balance of the body and the effects of such inbalances on body functions and cellular metabolism. *Mailing Add:* Curric Admin Off Med Educ Case Western Reserve Univ Sch Med Cleveland OH 44106

BETTINGER, DONALD JOHN, b Cincinnati, Ohio, Aug 20, 27; m 51; c 4. INORGANIC CHEMISTRY. *Educ:* Miami Univ, BS, 47; Univ Cincinnati, MS, 50; Univ NC, PhD(chem), 55. *Prof Exp:* Asst prof chem, Davidson Col, 50-51 & WVa Univ, 54-58; from asst prof to assoc prof, Denison Univ, 58-63; prof chem, Ohio Northern Univ, 63-79, chmn dept, 63-74, head, Div Math & Natural Sci, 67-74; prof chem, 79-89, ASST TO DEAN COL, ALICE LLOYD COL, 89- *Mem:* Am Chem Soc. *Res:* Chemical education; preparative inorganic chemistry; inorganic aggregation and polymerization phenomena; structure and stereochemistry of coordination compounds; chemical education. *Mailing Add:* Asst to Dean Col Alice Lloyd Col Pippa Passes KY 41844

BETTINGER, RICHARD THOMAS, b Dayton, Ohio, Aug 3, 31; m 52; c 7. SPACE PHYSICS. *Educ:* Syracuse Univ, BS, 55; Univ Md, PhD(physics), 64. *Prof Exp:* Asst prof physics, Univ Md, College Park, 64-74; mem staff, Betco Electronics, 74-85; PROG MGR, WEBS SYSTS PROJ STAFF, US TREASURY, 86- *Res:* In situ probe systems for the measurement of ionospheric parameters; vertical ozone distribution measured from satellites. *Mailing Add:* 15000 Donna Dr Silver Spring MD 20905

BETTMAN, JEROME WOLF, b San Francisco, Calif, June 22, 09; m 35; c 2. MEDICINE, OPHTHALMOLOGY. *Educ:* Univ Calif, AB, 31, MD, 35. *Prof Exp:* From asst resident to resident ophthal, Stanford Univ, 35-37, from assoc instr to assoc clin prof, 37-57, clin prof surg, Med Sch, 57-74; clin prof ophthal, Med Sch, Univ Calif, San Francisco, 64-76; EXEC DIR BASIC SCI COURSE IN OPHTHAL & UPDATE COURSE, MED SCH, STANFORD UNIV, 70-, EMER CLIN PROF SURG, 74-; EMER CLIN PROF OPHTHAL, MED SCH, UNIV CALIF, SAN FRANCISCO, 76- *Concurrent Pos:* consult, Marine Hosp, USPHS & San Francisco Hosp, 57-59; chief ophthal, Presby Med Ctr, 59-66; mem comt sci assemblies, NIH-Calif Med Asn Sci Bd; chmn ethics comt, Am Acad Ophthal. *Honors & Awards:* Award of Honor, Am Acad Ophthal & Otolaryngol, 58; Sr Honor Award, Am Acad Opthal, 85; Edward Jackson Lectr, Univ Colo 52. *Mem:* Asn Res Vision & Ophthal; Am Acad Ophthal (first vpres); Am Col Surgeons. *Res:* Factors influencing the blood volume of the choroid and the retina; toxic cataracts; malpractice in ophthalmology. *Mailing Add:* 3910 Sand Hill Rd Woodside CA 94062

BETTMAN, MAX, b Ger, Nov 15, 25; nat US; m 57; c 1. PHYSICAL CHEMISTRY. *Educ:* Reed Col, BA, 49; Calif Inst Technol, PhD(chem), 52. *Prof Exp:* Res chemist, Calif Res & Develop Co, 52-53 & Union Carbide Corp, 53-64; res chemist, Sci Lab, Ford Motor Co, 64-87; RETIRED. *Mem:* Am Chem Soc; Sigma Xi. *Res:* Materials research for solid state devices; automobile exhaust catalysis; x-ray crystallography. *Mailing Add:* 28328 E Kalong Circle Southfield MI 48034

BETTONVILLE, PAUL JOHN, b St Louis, Mo, July 19, 18; m 50; c 2. PHARMACOLOGY. *Educ:* St Louis Univ, MD, 42. *Prof Exp:* Resident internal med, Univ Hosps, 46-48, instr pharmacol & internal med, Sch Med, 48-59, ASST PROF CLIN MED, SCH MED, ST LOUIS UNIV, 59- *Mem:* AAAS; AMA. *Res:* Chemical carcinogenesis. *Mailing Add:* 135 W Adams St Louis MO 63122

BETTS, ATTIE L(ESTER), b Fairy, Tex, July 30, 16; m 40; c 2. ELECTRICAL ENGINEERING. *Educ:* Agr & Mech Col Tex, BS, 38, MS, 39; Univ Tex, PhD(elec eng), 52. *Prof Exp:* Asst elec eng, Agr & Mech Col Tex, 38-39; engr, Gulf States Utilities, 39-41; instr elec eng, Okla State Univ, 41-42 & 46, from asst prof to prof, 46-55; prof, 55-78, head dept, 58-78, EMER PROF ELEC ENG, WASH STATE UNIV, 78- *Mem:* Sr mem Inst Elec & Electronics Engrs; Am Soc Eng Educ. *Res:* Electrical power needs and the alternatives. *Mailing Add:* 1927 Jungo Ct Gardnerville NV 89410

BETTS, AUSTIN WORTHAM, b Westwood, NJ, Nov 22, 12; m 34; c 3. CIVIL ENGINEERING, NUCLEAR ENERGY. *Educ:* US Mil Acad, BS, 34; Mass Inst Technol, MS, 38. *Prof Exp:* Assoc dir nuclear weapons, Los Alamos Sci Lab, 46-48; chief, Atomic Energy Br, US Army Gen Staff, 49-54; dep dir guided missiles, Off Secy Defense, 57-60; dir, Advan Res Proj Agency, Off Defense, 60-61; dir mil appln nuclear weapons, Atomic Energy Chmn Staff, 61-63; chief res & develop multi-discipline, US Army Gen Staff, 66-70; sr vpres multi-discipline, Southwest Res Inst, 71-83; RETIRED. *Concurrent Pos:* Mem, Army Sci Adv Panel, 71-74, consult, 74-79; chmn, Sci Adv Coun, Ballistic Res Lab, 72-77; consult, Energy Res & Develop Admin, 76-77; mem, Army Sci Bd, 78-83; consult, 83-; exec secy, Nat Conf Advan Res, 87- *Mem:* Am Inst Aeronaut & Astronaut; fel Soc Am Mil Engrs; fel Inst Environ Sci; AAAS. *Res:* Competitive interrelationship among available sources of energy. *Mailing Add:* Southwest Res Inst PO Drawer 28510 San Antonio TX 78228

BETTS, BURR JOSEPH, b Denver, Colo, May 5, 45; m 68; c 2. ANIMAL BEHAVIOR, ECOLOGY. *Educ:* Purdue Univ, BS, 67; Univ Mont, PhD(zool), 73. *Prof Exp:* Res assoc, Univ Mont, 73-75; from asst prof to assoc prof, 75-87, PROF BIOL, EASTERN ORE STATE COL, 87- *Concurrent Pos:* Dir, George Ott Wildlife Res Area, 83- *Mem:* Animal Behav Soc; Am Soc Mammalogists; Sigma Xi. *Res:* Social behavior and organizations; time and energy budgets; feeding strategies of birds and mammals. *Mailing Add:* Sch Arts & Sci Eastern Ore State Col La Grande OR 97850-2899

BETTS, DONALD DRYSDALE, b Montreal, Que, May 16, 29; m 86; c 6. PHYSICS. *Educ:* Dalhousie Univ, BSc, 50, MSc, 52; McGill Univ, PhD(math physics), 55. *Prof Exp:* Nat Res Coun Can fel, Univ Alta, 55-56, from asst prof to assoc prof physics, 56-66, prof physics, 66-80, dir, Theoret Physics Inst, 72-78; dean arts & sci, 80-88, dean sci, 88-90, PROF PHYSICS, DALHOUSIE UNIV, 90- *Concurrent Pos:* NATO sci fel, King's Col, London, 63-64; Nuffield fel, 70-71; mem, Int Comn Thermodyn & Statist Mech; chmn comn thermodyn & statist mech, Int Union Pure & Appl Physics, 72-75; vis prof chem & physics, Cornell Univ, 75; res fel, Japan Soc Prom Sci, 82; vis prof, Univ New South Wales, 91. *Mem:* Fel Royal Soc Can; Can Asn Physicists (vpres, 68-69, pres, 69-70). *Res:* Theoretical solid state and statistical physics. *Mailing Add:* Dept Physics Dalhousie Univ Halifax NS B3H 3J5 Can

BETTS, HENRY BROGNARD, b New Rochelle, NY, May 25, 28; m 70. REHABILITATION MEDICINE. *Educ:* Princeton Univ, BA, 50; Univ Va, MD, 54. *Prof Exp:* Staff psychiatrist phys med & rehab, 63-64, assoc med dir, 64-65, MED DIR, REHAB INST CHICAGO, 65-, EXEC VPRES, 69-; PROF REHAB MED & CHMN DEPT, MED SCH, NORTHWESTERN UNIV, CHICAGO, 69- *Concurrent Pos:* Consult staff mem, Northwestern Mem Hosp, 67-; dir, Res & Training Ctr 20; mem gov comt, Employ of Handicapped. *Honors & Awards:* Meritorious Serv Citation, US Pres Comt Employ of Handicapped, 65. *Mem:* Am Cong Rehab Med (vpres, 71-, pres, 75-76); Am Acad Phys Med & Rehab. *Res:* Research of prosthetics and orthotics; neuromuscular studies; sociological research. *Mailing Add:* Rehab Inst Chicago 345 E Superior Chicago IL 60611

BETZ, A LORRIS, b La Crosse, Wis, Feb 9, 47. PEDIATRICS. *Educ:* Univ Wis, BS, 69, MD & PhD(biochem & physiol), 75. *Prof Exp:* Intern pediat, Univ Calif, San Francisco, 75-76, resident, 76-77, postdoctoral fel neurol & pediat, 77-79; from asst prof to assoc prof neurol & pediat, 79-87, PROF PEDIAT, SURG & NEUROL, UNIV MICH, 87-, DIR, NEUROSURG RES, 87- *Concurrent Pos:* Estab investr award, Am Heart Asn, 81; assoc chair, Dept Pediat, Univ Mich, 89-; mem bd dirs, Int Soc Cerebral Blood Flow & Metab, 91- *Mem:* Soc Pediat Res; Am Physiol Soc; Int Soc Cerebral Blood Flow & Metab. *Mailing Add:* Dept Pediat Surg & Neurol Univ Mich D3227 Med Prof Bldg Box 0718 Ann Arbor MI 48109-0718

BETZ, EBON ELBERT, b Springport, Mich, Sept 3, 14; m 43; c 4. MATHEMATICS. *Educ:* Albion Col, AB, 34; Univ Mich, AM, 35; Univ Pa, PhD(math), 39. *Prof Exp:* Asst instr math, Univ Pa, 35-38; instr, Haverford Col, 39-41; from instr to assoc prof, 41-57, PROF MATH, US NAVAL ACAD, 57- *Mem:* Am Math Soc. *Res:* Accessibility and separation by simple closed curves; topology. *Mailing Add:* Dept Math US Naval Acad Annapolis MD 21402

BETZ, GEORGE, b Asherville, Kans, Apr 27, 34; m 57; c 3. OBSTETRICS & GYNECOLOGY, ENDOCRINOLOGY. *Educ:* Kans State Col, BS, 56; Univ Kans, MD, 60, PhD(biochem), 68. *Prof Exp:* Resident gen surg, Denver Gen Hosp, 61-62; resident obstet & gynec, Univ Kans, 62-65, instr, 65-68; from asst prof to assoc prof, 68-77, from res assoc to assoc prof biochem, 71-77, PROF OBSTET & GYNEC, UNIV COLO, DENVER, 77-; HEAD REPRODUCTIVE ENDOCRINOL, KAISER FOUND, 88- *Concurrent Pos:* Sr fel, NIH, 81-82. *Mem:* Am Soc Biol Chemists; Endocrine Soc; Soc Gynec Invest. *Res:* Biochemistry of reproduction with primary emphasis on steroid converting enzymes; endocrinology of reproduction. *Mailing Add:* 2045 Franklin Denver CO 80205

BETZ, JOHN VIANNEY, b Philadelphia, Pa, Apr 5, 37; m 62; c 2. BACTERIOLOGY, VIROLOGY. *Educ:* St Joseph's Col, Pa, BSc, 58; St Bonaventure Univ, PhD(microbiol), 63. *Prof Exp:* Fel microbiol, Ind Univ, 62-63, instr bact, 63; asst prof, 63-69, ASSOC PROF BOT & BACT, UNIV SFLA, 69- *Mem:* Am Soc Microbiol. *Res:* Interactions of host bacteria and bacteriophages; nature of lysogeny; genetic factors affecting lysogenesis; mechanism of flagellar movement. *Mailing Add:* Dept Biol Univ SFla Tampa FL 33620

BETZ, MATHEW J(OSEPH), b Chicago, Ill, Jan 27, 32; m 63; c 1. CIVIL ENGINEERING. *Educ:* Northwestern Univ, BS, 55, MS, 56, PhD(civil eng), 61. *Prof Exp:* Asst prof civil eng, Univ Khartoum, 56-57; instr, Northwestern Univ, 58-59; from asst prof to prof civil eng, Ariz State Univ, 61-76, asst dean, 69-70, assoc dean grad cp;. 70-76, assoc dean vpres, 76-83, dir, Ctr Advan Res in Transp, 83-88, VPROVOST PLANNING, ARIZ STATE UNIV, 83- *Concurrent Pos:* Civil engr, Brookings Inst, DC, 63; consult, Valley Area Transp Study, Ariz, 66-80 & Off Educ, 70-71; mem nat coop hwy res adv panel, Nat Acad Eng-Nat Acad Sci, 70-74; NSF SEED grant, Kenya, 73; mem hwy res bd, Nat Res Coun, Nat Acad Sci; consult, World Bank, 77-, US Dept Com, 84. *Mem:* Am Soc Civil Engrs; Inst Transp Engrs; Am Planning Asn; African Studies Asn. *Res:* Transportation engineering and planning; city planning; urban transport planning; transport in developing areas. *Mailing Add:* Ariz State Univ ADM 211 Tempe AZ 85287-2803

BETZ, NORMAN LEO, b Baton Rouge, La, Jan 23, 38; m 60; c 5. BEHAVIORAL PHYSIOLOGY, FOOD SCIENCE. *Educ:* La State Univ, BA, 61, MS, 63, PhD(food sci), 66. *Prof Exp:* Res assoc, Agr Res Serv, Entom Res Div, USDA, 62-66; chief field messing br/instr, US Army Quartermaster Sch, 66-68; assoc sr scientist, Mallinckrodt Chem Works, 68-72; scientist, Ralston Purina, 72-79, sr scientist, 79-81, dir res & develop planning, 81-82 & dir palatability res, Corp Res & Develop, 82-85; PRES, TECHNOL RESOURCES, LTD, 85-; CHAIR EXCELLENCE, UNIV TENN, MARTIN. *Concurrent Pos:* Exec ed, Cereal Foods World, Am Asn Cereal Chemists, 74-80; chmn, Am Asn Cereal Chemists, 83. *Honors & Awards:* Geddes Memorial Award, Am Asn Cereal Chemists, 80. *Mem:* Inst Food Technologists; Am Asn Cereal Chemists (pres-elect, 81, pres, 82); European Chemoreception Res Orgn; Asn Chemoreception Sci. *Res:* Short and long range plans for corporate research and development; identification of appropriate technologies necessary for successful implementation; food technology. *Mailing Add:* Technol Resources Ltd Rte 1 Box 607C Dresden TN 38225

BETZ, ROBERT F, b Chicago, Ill, Jan 25, 23; m 51; c 3. BIOCHEMISTRY, ECOLOGY. *Educ:* Ill Inst Technol, BS, 48, MS, 52, PhD(biochem), 55. *Prof Exp:* Instr biochem & microbiol, Ill Inst Technol, 50-52; assoc prof biol, 55-68, PROF BIOL SCI, NORTHEASTERN ILL STATE COL, 68- *Res:* Biochemistry and physiology of Myxomycetes, especially their enzymes; regeneration of prairies in abandoned cemeteries; autecology and conservation of prairie plants. *Mailing Add:* Dept Biol Northeastern Ill Univ 5500 N St Louis Ave Chicago IL 60625

BETZ, WILLIAM J, b Berwyn, Ill, Feb 3, 43. NEUROPHYSIOLOGY. *Educ:* Wash Univ, BS, 65; Yale Univ, PhD(physiol), 69. *Prof Exp:* Postdoctoral fel, Dept Biophys, Univ Col, London, 69-71; from asst prof to assoc prof, 71-84, PROF, DEPT PHYSIOL, SCH MED, UNIV COLO, 84- *Mem:* AAAS; Soc Neurosci. *Mailing Add:* Dept Physiol Box C-240 Sch Med Univ Colo 4200 E Ninth Denver CO 80262

BETZER, PETER ROBIN, b Delavan, Wis, May 14, 42; m 65; c 2. GEOCHEMISTRY, ANALYTICAL CHEMISTRY. *Educ:* Lawrence Col, BA, 64; Univ RI, PhD(oceanog), 71. *Prof Exp:* From asst prof to prof teaching res, 71-82, PROF ADMIN RES & CHMN DEPT, UNIV SFLA, 82- *Concurrent Pos:* Vis prof oceanog, Dept Oceanog, Univ Hawaii & Hawaii Inst Geophysics, 81-82; mem, Ocean Chem Panel, NSF, 87. *Honors & Awards:* Distinguished Authorship Award, Nat Oceanic & Atmospheric Admin, 85; Plenary Lectr, Australian Marine Sci Meetings, 87. *Mem:* AAAS; Am Geophys Union; Am Inst Chemist. *Res:* Chemical, physical and biological processes influencing the delivery and flux of inorganic and organic materials in the ocean; application of chemical tracers and computer-controlled analytical systems to oceanographic problems; biomass production and conversion. *Mailing Add:* Dept Marine Sci 1830 Seventh St N St Petersburg FL 33704-3322

BEUCHAT, CAROL ANN, b Calif, June 16, 55. ECOLOGICAL PHYSIOLOGY, COMPARATIVE PHYSIOLOGY. *Educ:* Occidental Col, AB, 76, MA, 77; Cornell Univ, PhD(physiol), 82. *Prof Exp:* NIH postdoctoral fel, Univ Ariz, 82-85, res asst prof, 89-90; asst res physiol, Colgate Univ, 85-88; ASST PROF PHYSIOL, SAN DIEGO STATE UNIV, 90-; RES

ASSOC, HUBBS MARINE RES CTR, SEA WORLD RES INST, 90- *Mem:* AAAS; Am Physiol Soc; Am Soc Zoologists; Ecol Soc Am; Am Soc Ichthyologists & Herpetologists; Am Ornith Union. *Res:* Comparative integrative and ecological physiology of vertebrates, especially osmoregulatory physiology, allometry of form and function, animal energetics and thermal biology. *Mailing Add:* Dept Biol San Diego State Univ San Diego CA 92182

BEUCHAT, LARRY RAY, b Meadville, Pa, July 23, 43. FOOD MICROBIOLOGY. *Educ:* Pa State Univ, BS, 65; Mich State Univ, MS, 67, PhD(food sci), 70. *Prof Exp:* Group leader res & develop, Quaker Oats Co, 70-72; asst prof, 72-76, assoc prof, 76-81, PROF FOOD MICROBIOL, UNIV GA, 81- *Honors & Awards:* Samuel Cate Prescott Award, Inst Food Technol, 77. *Mem:* Inst Food Technologists; Am Soc Microbiol; Soc Appl Bact; Int Asn Milk, Food & Environ Sanitarians. *Res:* Environmental factors affecting growth and death of food bore pathogens; chemical changes occurring in peanuts fermented or infected with fungi; conditions for growth of microbial pathogens on nuts and fruits. *Mailing Add:* Dept Food Sci & Technol Univ Ga Griffin GA 30223

BEUERMAN, ROGER WILMER, b Washington, DC, Oct 3, 43; m 65; c 2. NEUROSCIENCES, NEUROPHYSIOLOGY. *Educ:* Hood Col, BS, 65; Fla State Univ, PhD(sensory physiol), 73. *Prof Exp:* Sr res fel neurophysiol, Univ Wash, 73-75; res assoc, Div Ophthal, Stanford Univ Med Ctr, 75-81; ASSOC PROF OPHTHAL, LA STATE UNIV MED SCH, 81- *Mem:* Int Asn Study Pain; Soc Neurosci; Asn Res Vision Ophthal; Sigma Xi. *Res:* Sensory function and pain in the trigeminal system. *Mailing Add:* La State Univ Eye Ctr 2020 Gravier St-B New Orleans LA 70112-2234

BEUG, MICHAEL WILLIAM, b Austin, Tex, May 18, 44; m 68; c 2. ORGANIC CHEMISTRY, ENVIRONMENTAL CHEMISTRY. *Educ:* Harvey Mudd Col, BS, 66; Univ Wash, PhD(chem), 71. *Prof Exp:* Asst prof chem, Harvey Mudd Col, 71-72; MEM FAC CHEM, EVERGREEN STATE COL, 72-, SR ACAD DEAN, 86- *Mem:* AAAS. *Res:* The effects of pesticides, polychlorinated biphenyls, heavy metals and petroleum on terrestrial and aquatic ecosystems; the detection of pollutants by chromatography and gas chromatography; fungal natural products; agricultural ecology. *Mailing Add:* Acad Dean Evergreen State Col Olympia WA 98505

BEUHLER, ROBERT JAMES, JR, b Lake Charles, La, Feb 21, 42; m 65; c 3. PHYSICAL CHEMISTRY. *Educ:* Univ Mich, BS, 63; Univ Wis, Madison, PhD(phys chem), 68. *Prof Exp:* Res assoc, Argonne Nat Lab, 68-70; res assoc, 70-72, asst chemist, 72-75, CHEMIST, BROOKHAVEN NAT LAB, 75- *Mem:* Am Soc Mass Spectros; Am Chem Soc. *Res:* focusing and orientation of molecules in molecular beams; ion-molecule reactions; cluster ion impact phenomena; biological mass spectrometry; cluster ion mass spectroscopy. *Mailing Add:* Chem Dept Brookhaven Nat Lab Upton NY 11973

BEUKENS, ROELF PIETER, b Venlo, Neth, May 4, 43. NUCLEAR PHYSICS, MASS SPECTROMETRY. *Educ:* State Univ Groningen, Drs, 71; Univ Toronto, PhD(nuclear physics), 76. *Prof Exp:* Fel, 76; res assoc 76-80, ASSOC PROF PHYSICS, UNIV TORONTO, 80- *Concurrent Pos:* Consult, Watts, Griffis & McOuat Ltd, 77-78 & Interex Comput Systs Ltd, 77-; Killam fel, 78-79. *Mem:* Can Archaeol Asn. *Res:* Development of ultrasensitive radio isotope dating techniques and ultrasensitive mass spectroscopy using electrostatic accelerators. *Mailing Add:* 50 Walmer Rd Toronto ON M5R 2X4 Can

BEUS, STANLEY S, b Salt Lake City, Utah, July 31, 30; m 53; c 5. INVERTEBRATE PALEONTOLOGY, STRATIGRAPHY. *Educ:* Utah State Univ, BS, 57, MS, 58; Univ Calif, Los Angeles, PhD(geol), 63. *Prof Exp:* From asst prof to prof geol, 62-89, chmn, Dept Geol & Geog, 66-69, REGENTS PROF, NORTHERN ARIZ UNIV, 89- *Concurrent Pos:* Paleontologist, Mus Northern Ariz, 73-; co-ed, J Paleont, 75-80; consult, Bendix Corp, May Oil Co & US Forest Serv, Grand Canyon Nat Park; secy, Rocky Mountain Sect, Geol Soc Am, 76-84. *Mem:* Geol Soc Am; Paleont Soc; Soc Econ Paleont & Mineral; Am Asn Petrol Geol. *Res:* Paleozoic stratigraphy and paleontology with emphasis on the Grand Canyon area of northern Arizona; invertebrate paleontology; paleoecology. *Mailing Add:* Dept Geol Box 6030 Northern Ariz Univ Flagstaff AZ 86011

BEUSCH, JOHN U, b Erie, Pa, Apr 22, 38; m 61; c 2. ELECTRICAL ENGINEERING, APPLIED MATHEMATICS. *Educ:* Rochester Inst Technol, BS(mech eng) & BS(elec eng), 61; Mass Inst Technol, SM, 62, PhD(elec eng), 65. *Prof Exp:* Coop student employee, Gen Elec Co, Pa, 57 & Gen Signal Co, NY, 57-61; staff mem, Lincoln Lab, 64, consult, Lincoln Lab & res asst, Opers Res Ctr, 64-65, staff mem, 65-71, GROUP LEADER, LINCOLN LAB, MASS INST TECHNOL, 71- *Mem:* Inst Elec & Electronics Engrs; Am Inst Aeronaut & Astronaut; Opers Res Soc Am. *Res:* Control systems; control of complex systems; communications theory; operations research; queuing theory; communications satellite planning and preliminary design; air traffic control systems design; tactical communications systems; radar systems. *Mailing Add:* 416 Taylor Rd Stow MA 01775

BEUTE, MARVIN KENNETH, b Jenison, Mich, Mar 3, 35; m 55; c 2. PLANT PATHOLOGY. *Educ:* Calvin Col, BA, 63; Mich State Univ, PhD(plant path), 68. *Prof Exp:* PROF PLANT PATH, NC STATE UNIV, 68- *Mem:* AAAS; Am Phytopath Soc. *Res:* Ecology of soil borne pathogens; disease complexes; diseases of the peanut. *Mailing Add:* Box 7616 Plant Pathol NC State Univ Raleigh Main Campus Raleigh NC 27695

BEUTER, JOHN H, b Chicago, Ill, Dec 24, 35; m 60; c 2. FOREST ECONOMICS. *Educ:* Mich State Univ, BS, 57, MS, 58; Iowa State Univ, PhD(forestry econ), 66. *Prof Exp:* Resource analyst, US Forest Serv, 61-63, economist & proj leader, 65-68; economist, Mason, Bruce & Girard Consult Foresters, 68-70; from assoc prof to prof forest mgt, Ore State Univ, 76-87, head dept, 76-84, assoc dean, 84-91; DEP ASST SECY, NAT RES & ENVIRON FOREST SERV, 91- *Mem:* Soc Am Foresters; Forest Prod Res Soc; Am Forestry Asn. *Res:* Economics of forest management and forest products marketing and utilization; general resource economics; operations research. *Mailing Add:* US Forest Serv 3100 Connecticut Ave NW No 205 Washington DC 20008

BEUTHER, HAROLD, b Pittsburgh, Pa, Sept 3, 17; m 41; c 3. CHEMICAL ENGINEERING. *Educ:* Carnegie Inst Technol, BS, 39. *Prof Exp:* Chem engr, 39-50, asst head process develop, 50-54, from asst head process res to head process res, 54-61, staff engr process div, 61-67, dir process res, 67-76, mgr catalyst & chem res, 76-78, mgr synthetic fuels, 78-81, MGR SCIENTIFIC STAFF, GULF RES & DEVELOP CO, 81- *Mem:* Am Chem Soc; Am Inst Chem Engrs. *Res:* Petroleum processing and catalysis. *Mailing Add:* 1331 White Heron Ln Vero Beach FL 32963-1399

BEUTLER, ERNEST, b Berlin, Ger, Sept 30, 28; US citizen; m 50; c 4. MEDICINE. *Educ:* Univ Chicago, PhB, 46, BS, 48, MD, 50. *Prof Exp:* From instr to asst prof med, Univ Chicago, 55-59, attend physician, Univ Clins, 55-59; chmn, Div Med, City of Hope Med Ctr, 59-78; head, Div Hemat-Oncol, 78-87, chmn, Dept Clin Res, 79-82, CHMN, DEPT MOLECULAR & EXP MED, SCRIPPS CLIN & RES FOUND, 82- *Concurrent Pos:* Assoc clin prof med, Univ Southern Calif, 60-64, clin prof, 64-65 & 67-79; clin prof, Univ Calif-Calif Col Med, 65-67; mem hemat study sect, NIH, 70-74; clin prof, Univ Calif, San Diego, 80- *Honors & Awards:* Gairdner Found Award, 75; James Blundell Prize, 85; Nat Heart, Lung & Blood Inst Merit Award, NIH, 87; Acad Clin Biochem Lectureship Award, Kodak Instruments, 90. *Mem:* Nat Acad Sci; Am Acad Arts & Sci; Asn Am Physicians; Am Soc Human Genetics; Am Soc Hemat (pres, 78-79); Am Soc Clin Invest. *Res:* Red cell biochemistry and physiology; biochemical genetics; hematology. *Mailing Add:* Scripps Clin & Res Found 10666 N Torrey Pines Rd La Jolla CA 92037

BEUTLER, FREDERICK J(OSEPH), b Berlin, Ger, Oct 3, 26; nat US; m 50; c 3. ENGINEERING, PROBABILITY. *Educ:* Mass Inst Technol, SB, 49, SM, 51; Calif Inst Technol, PhD(eng sci & math), 57. *Prof Exp:* Jr engr, AC Spark Plug Div, Gen Motors Corp, 49-50; staff mem, Instrumentation Lab, Mass Inst Technol, 50-51, asst, 55-56; res engr, Autonetics Div, NAm Aviation, Inc, 51-54; tech engr, Ramo-Woolridge Corp, 56-57; from asst prof to assoc prof aeronaut eng, Univ Mich, 57-63, prof instrumentation, 63-67, chmn, Comput, Info & Control Eng Prog, 70 & 76-89, chmn, Elec Eng Systs Grad Prog, 84-89, PROF INFO & CONTROL ENG, UNIV MICH, 67- *Concurrent Pos:* Vis scholar, Univ Calif, Berkeley, 64-65; vis prof, Calif Inst Technol, 67-68; managing ed, Soc Indust & Appl Math J Appl Math, 70-75, ed, 84-90. *Mem:* AAAS; Am Math Soc; fel Inst Elec & Electronics Engrs; Am Soc Eng Educ; Soc Indust & Appl Math; Opers Res Soc Am. *Res:* Stochastic optimization; applications of functional analysis to systems theory; probability theory and functional analysis applied to communication and large scale systems; queueing networks; operations research; analysis and functional analysis. *Mailing Add:* EECS Dept Univ Mich EECS Bldg Ann Arbor MI 48109-2122

BEUTNER, EDWARD C, b Tucson, Ariz, May 22, 39. STRUCTURAL GEOLOGY. *Educ:* Penn State Univ, PhD(geol), 68. *Prof Exp:* Res structural geol, Shell Oil Co, 68-70; PROF GEOL, FRANKLIN & MARSHALL COL, 70- *Mem:* Fel Geol Soc Am. *Mailing Add:* Dept Geol Franklin & Marshall Col Box 3003 Lancaster PA 17604

BEUTNER, ERNST HERMAN, b Berlin, Ger, Aug 27, 23; nat US; m 49; c 3. MICROBIOLOGY. *Educ:* Univ Pa, PhD, 51, Am Bd Med Microbiol, dipl; Am Bd Med Lab Immunol, dipl. *Prof Exp:* Res supvr, Sias Res Labs, Brooks Hosp, 51-55; res assoc, Harvard Sch Dent Med, 55-56; asst prof bact & immunol, 62-68, PROF MICROBIOL & DERMAT, STATE UNIV NY BUFFALO, 68- *Concurrent Pos:* Dir diag; pres, Int Serv for Immunol Labs; fel, Philadelphia Col Physicians. *Honors & Awards:* Rocha Lima Award, 68; Vincenzo Chiarugi Award, 89. *Mem:* Am Acad Dermat; Soc Investigative Dermat; Am Soc Immunol; NY Acad Sci. *Res:* Quantitation of immunofluorescence staining techniques; tissue immunology of pituitary and salivary glands; thyroiditis; systemic lupus erythematosus; myasthenia gravis; pemphigus bullous pemphigold, psoriasis, dermatitis herpetiformis, celiac disease; immunopathology of the skin. *Mailing Add:* 22 Brantwood Rd Buffalo NY 14226

BEVAK, JOSEPH PERRY, b Detroit, Mich, Mar 8, 29. PHYSICAL CHEMISTRY. *Educ:* Wayne State Univ, BS, 50; Mass Inst Technol, PhD(chem), 55. *Prof Exp:* From instr to assoc prof, 56-68, from actg head dept to head dept, 58-69, PROF CHEM, SIENA COL, 68- *Mem:* Am Chem Soc. *Res:* Thermodynamics of solutions. *Mailing Add:* Dept Chem Siena Col Loudonville NY 12211

BEVAN, DONALD EDWARD, b Seattle, Wash, Feb 23, 21. FISH BIOLOGY. *Educ:* Univ Wash, BS, 48, PhD(fisheries), 59. *Prof Exp:* Sci asst, Fisheries Res Inst, Univ Wash, 47-48, res assoc, 48, proj leader, Kodiak Island Res, 48-58, proj supvr pink salmon res, 58-60, res assoc prof, Col Fisheries, 60-64, assoc prof, 64-66, assoc dean, Col Fisheries, 65-69 & 76-80, dir comput ctr, 68-69, asst vpres res, 69-76, dean, 80-85, prof, Col Fisheries, 66-86; adj prof, Inst Marine Studies, 73-86; EMER PROF, COL OCEAN & FISHERY SCI, 86- *Concurrent Pos:* Consult fisheries indust, 48-, pulp & paper indust, 62-; fel, Moscow State Univ, 59-60. *Mem:* AAAS; Am Inst Fishery Res Biologists; Sigma Xi. *Res:* Population dynamics; biometrics; computer science; Soviet fisheries. *Mailing Add:* Univ Wash HN-15 Seattle WA 98195

BEVAN, JOHN ACTON, b London, Eng, Apr 24, 30; m 56; c 4. PHARMACOLOGY. *Educ:* Univ London, BSc, 50, MB & BS, 53. *Prof Exp:* Demonstr pharmacol, St Bartholemew's Med Col, 50-52; actg chmn, Dept Pharmacol, Univ Calif, Los Angeles, 65-66, from asst prof to assoc prof, 57-67, prof, 67-83; PROF & CHMN, COL MED, UNIV VT, 83- *Concurrent Pos:* Mem cardiovasc & renal ad hoc study sect, Nat Heart & Lung Inst, 78-; mem exp cardiovasc sci study sect, NIH, 81-84. *Mem:* Am Soc Pharmacol & Exp Therapeut; Am Physiol Soc. *Res:* Pharmacology of the cardiovascular system. *Mailing Add:* Dept Pharmacol Col Med Univ VT Burlington VT 05405

BEVAN, ROSEMARY D, VASCULAR DEVELOPMENT. *Educ:* Univ London, MBBS, 53. *Prof Exp:* ASSOC PROF PHARMACOL, COL MED, UNIV VT, 84- *Res:* Neural control of vasculature. *Mailing Add:* Dept Pharmacol Given Bldg Col Med Univ Vt Burlington VT 05405

BEVAN, WILLIAM, b Plains, Pa, May 16, 22; m 45; c 3. BEHAVIOR. *Educ:* Franklin & Marshall Col, AB, 42; Duke Univ, MA, 43, PhD(exp psychol), 48. *Hon Degrees:* ScD, Fla Atlantic Univ, 68, Emory Univ, 74, Franklin & Marshall Col, 79, Univ Md, 81, Kansas State Univ, 87; LLD, Duke Univ, 72; DHL, Southern Ill Univ, 89. *Prof Exp:* Asst, Duke Univ, 42-43, res asst, Frangible Bullet Proj, 43-44, instr psychol, 47; from instr to asst prof, Heidelberg Col, 46-48; from asst prof to prof, Emory Univ, 48-59; prof & chmn dept, Kans State Univ, 59-62, dean, Sch Arts & Sci, 62-63, vpres acad affairs, 63-66; vpres & provost, Johns Hopkins Univ, 66-70, prof, 66-74; dir sci & pub affairs, 74-83, chmn dept, 77-78, provost, 79-82, WILLIAM PRESTON FEW PROF PSYCHOL, DUKE UNIV, 74-; VPRES, JOHN D & CATHERINE T MACARTHUR FOUND, 82- *Concurrent Pos:* Res psychologist, Equipment Acceptability Proj, US Army, 52; Fulbright res scholar & guest prof, Univ Oslo, Norway, 52-53; opers res specialist, Lockheed Aircraft Corp, 55, staff scientist, 56 & 57; consult, Lockheed Aircraft Corp, 55-59, Baylor Univ, 57-62, Emory Univ, 58-59, Midwest Res Inst, 59-61, Menninger Found, 60-66 & Univ Md, 77-; NSF sr fel & fel, Ctr Advan Study Behav Sci, Stanford, Calif, 65-66; mem sci info coun, Nat Acad Sci, 66-70; exec officer, AAAS, 70-74; publ, Sci, 70-74; assoc ed, Am Psychologist, 76-85; univ lectr, Brown Univ, 84. *Honors & Awards:* Sandia Lectr, Univ NMex, 81. *Mem:* Inst Med-Nat Acad Sci; Soc Exp Psychologists; Am Psychol Asn (pres-elect, 81, pres, 82); assoc Am Ecol Soc; fel AAAS; Psychonomic Soc; Soc Social Study Sci; Acad Behav Med Res; Sigma Xi; Hist Sci Soc. *Res:* Perception and other cognitive processes; vision and other sensory processes; human engineering and human factors; public policy; experimental psychology. *Mailing Add:* John D & Catherine T MacArthur Found 140 S Dearborn St Suite 700 Chicago IL 60603

BEVANS, ROWLAND S(COTT), b Elizabeth, NJ, Apr 7, 19; m 50; c 4. CHEMICAL ENGINEERING. *Educ:* Yale Univ, BE, 40; Carnegie Inst Technol, MS, 41. *Hon Degrees:* ScD, Mass Inst Technol,46. *Prof Exp:* Staff mem, Div Indust Coop, Mass Inst Technol, 42-49, indust liaison officer, 49-52; mem res planning staff, Ethyl Corp, 52-54, prod develop staff, 54-56; mgr appl mech, Res & Develop Lab, Am Stardard, Inc, 57-75; sr process engr, Chem-Pro Equip Corp, 75-84; RETIRED. *Concurrent Pos:* Chem engr, Standard Oil Develop Co, NJ, 46. *Mem:* AAAS; Am Chem Soc; Am Inst Chem Engrs; Am Soc Heating, Air Conditioning & Refrig Engrs. *Res:* High output combustion of volatile liquid fuels. *Mailing Add:* Glen Alpin Rd Morristown NJ 07960

BEVC, VLADISLAV, b Ljubljana, Yugoslavia, Apr 9, 32; US citizen; m 61; c 2. ELECTROMAGNETICS THEORY, PLASMA PHYSICS. *Educ:* Univ Calif, Berkeley, BSc, 57, MSc, 58, PhD(microwave electronics), 61. *Prof Exp:* Nat Acad Sci-Nat Res Coun & Air Force Off Sci Res res fel, Univ Eng Lab & St Catherine's Col, Oxford Univ, 62-63; assoc prof elec eng, Naval Postgrad Sch, 63-66; mem tech staff, Aerospace Corp, 65-69; physicist, Lawrence Livermore Lab, Univ Calif, 69-73; PRES, BEVC ENG INC, 73- *Concurrent Pos:* Lectr & sr lectr, Univ Col, Univ Exten & Community Serv Div, Univ Southern Calif, 67-70 & San Francisco State Univ, 77-78. *Mem:* Am Phys Soc; sr mem Inst Elec & Electronics Engrs; Sigma Xi. *Res:* High energy particle beam transport and dynamics, controlled thermonuclear fusion; electromagnetic theory; wave propagation in plasmas; electron optics. *Mailing Add:* Bevc Eng Inc 51 Hardester Ct Danville CA 94526

BEVELACQUA, JOSEPH JOHN, b Waynesburg, Pa, Mar 17, 49; m 71; c 6. NUCLEAR THEORY, HEALTH PHYSICS. *Educ:* Calif State Col, BS, 70; Fla State Univ, MS, 74, PhD, 76; Am Bd Health Physics, cert comprehensive, 85, cert power reactor, 87. *Prof Exp:* Teaching asst physics, Univ Maine, Orono, 70-72, res asst, 72; radiol engr, Bettis Atomic Power Lab, 73, sr radiol engr, 76-78; teaching asst, Fla State Univ, 73-74, from res asst to res assoc, 74-76; nuclear physicist, US Dept Energy, 78-83; sr nuclear engr, Three Mile Island, GPU Nuclear Corp, 83-84, mgr emergency preparedness, 84-86, mgr safety RE group, 86-89, radiol controls dir, 89; SUPT HEALTH PHYSICS, POINT BEACH NUCLEAR PLANT, WIS ELEC POWER CO, 89- *Concurrent Pos:* Consult, Advan Isotope Separation Selection Bd, US Dept Energy, 81-82; mem nuclear utility coord group, KMC, Inc, 84-88; mem, Babcock & Wilcox Owners Group, 84-86; von Humboldt fel, Univ Hamburg, 76-79. *Mem:* Am Phys Soc; Am Nuclear Soc; Health Physics Soc; NY Acad Sci; Am Acad Health Physics; Soc Nuclear Med; Am Acad Health Physics; Am Bd Health Physics. *Res:* Low energy nuclear theory involving structure and reaction properties in light nuclei; decontamination and decommissioning of nuclear facilities; emergency preparedness; health physics; nuclear reactor safety; author of 62 publications in nuclear physics including physical review letters covering low energy nuclear theory with a focus upon the structure and reaction properties of light nuclei. *Mailing Add:* 934 Regent Lane Green Bay WI 54311-5949

BEVELANDER, GERRIT, b West Sayville, NY, Apr 6, 05; m 35; c 1. HISTOLOGY, ANATOMY. *Educ:* Hope Col, AB, 26; Univ Mich, AM, 28; Johns Hopkins Univ, PhD(zool), 32. *Prof Exp:* Assoc, US Bur Fisheries, 28-29; instr biol, Union Univ, NY, 31-33; res assoc, Col Dent, NY Univ, 33-34, from instr to assoc prof anat, Univ, 34-47, prof histol, 47-62; prof, 62-72, EMER PROF HISTOL, UNIV TEX DENT BR HOUSTON, 72-; CONSULT, MONTEREY ABALONE FARMS, 73- *Concurrent Pos:* Consult, USPHS, 57-61 & 63-67; mem staff, Marine Biol Lab, Woods Hole & Bermuda Biol Sta. *Honors & Awards:* Res Basic Biomineralization Award, Int Asn Dent Res, 88. *Mem:* Am Soc Zoologists; Soc Exp Biol & Med; Am Asn Anatomists; fel NY Acad Sci; Int Asn Dent Res (pres). *Res:* Comparative histology of fishes; experimental approach to problems in calcification; histochemistry; integumentary derivatives; dynamic aspects of calcification; marine biology; electron microscope studies relating to mineralization. *Mailing Add:* PO Box 2656 Carmel CA 93921

BEVER, CHRISTOPHER THEODORE, b Munich, Ger, Mar 12, 19; nat US; m 44; c 4. PSYCHIATRY. *Educ:* Harvard Univ, BA, 40, MD, 43. *Prof Exp:* Intern, Hartford Hosp, Conn, 44; resident psychiat, St Elizabeths Hosp, Washington, DC, 47-48, psychiatrist, 48-50; psychiatrist, Washington Inst Ment Hyg, 50-51; dir, Montgomery County Ment Hyg Clin, 51-54; assoc prof psychiat, Sch Med, Univ NC, 54-56; mem bd dirs, 74-77, MEM FAC, WASHINGTON SCH PSYCHIAT, 56-, CHMN PSYCHOTHER PRECEPTORSHIP & SEM PROG, 66- *Concurrent Pos:* Clin instr, Sch Med, Georgetown Univ, 49-51; instr, Washington Psychoanalysis Inst, 54-61, teaching analyst, 61-; dir psychiat outpatient ctr, NC Mem Hosp, 54-56; consult, Family Serv Agency, 56-58, St Elizabeths Hosp, Washington, DC, 59-61 & 73-74 & Walter Reed Army Hosp, 71-75; assoc psychiat, Med Sch, George Washington Univ, 57-60, from asst clin prof to assoc clin prof, 60-74, clin prof, 74-; pres, DC Inst Ment Hyg, 66-68, consult, 66-72; pres, Community Psychiat Clin, 73-75; trustee, William Alanson White Psychiat Found, 75- *Mem:* AMA; fel Am Psychiat Asn; Am Psychoanalysis Asn; Am Orthopsychiat Asn; fel Acad Psychoanalysis. *Res:* Continuing education in psychotherapy. *Mailing Add:* Dept Advan Psychol Washington Sch Psychol 1610 New Hamp NW Washington DC 20009

BEVER, JAMES EDWARD, b Bellingham, Wash, July 7, 20; m 46; c 1. GEOLOGY. *Educ:* Wash State Univ, BS, 42; Univ Mich, MS, 49, PhD(mineral), 54. *Prof Exp:* Asst purchasing agent, Traub Mfg Co, 47-48; instr mineral, Univ Mich, 50-54, instr conserv & acad counr, 52-54; from asst prof to prof, 54-85, EMER PROF GEOL, MIAMI UNIV, 85- *Concurrent Pos:* Consult, Oliver Mining Co, 53-54, Encyclop Americana, 56-57, Nat Res Coun, 57-58 & Gen Elec Co, 65-; vpres & dir, Timberline Minerals, Inc, 71-76. *Mem:* AAAS; Geol Soc Am; Mineral Soc Am; Mineral Asn Can. *Res:* Field geology; petrography and petrology; mineral resources; gems. *Mailing Add:* Dept Geol Miami Univ Oxford OH 45056

BEVER, MICHAEL B(ERLINER), b Schmargendorf, Germany, Aug 7, 11; nat US; m 36; c 3. MATERIALS POLICY, RESOURCE MANAGEMENT. *Educ:* Univ Heidelberg, Dr iur, 34; Harvard Univ, MBA, 37; Mass Inst Technol, SM, 42, ScD 44. *Prof Exp:* Asst metall, 40-44, from instr to prof, 44-76, EMER PROF MAT SCI & ENG, MASS INST TECHNOL, 77-, SR LECTR, 77- *Concurrent Pos:* Hon res assoc, Harvard Univ, 66-67; mem, Adv Res Proj Agency Mat Res Council, 67-74; sr res assoc, Inst Econ Anal, NY Univ, 78-88; consult ed, Environ Impact Assessment Rev, 79-85, co-chmn ed policy comt; ed, Conserv & Recycling, 76-87; corresp mem, Berliner Wissenschaftliche Gesellschaft, 79-; ed-in-chief, Encycl Mat Sci & Eng, 77-86; sr adv ed, Advanes Mat Sci & Eng, 86- *Honors & Awards:* Mathewson Gold Medal, Metall Soc, 65; Recycling Award, Nat Asn Sec Mat Indust, 72. *Mem:* Am Inst Mining, Metall & Petrol Engrs; Mat Res Soc; fel Am Acad Arts & Sci; fel AAAS; fel Am Soc Metals. *Res:* Physical metallurgy; application of thermodynamics to metallurgy; calorimetry; deformation of metals and intermetallic compounds; order-disorder phenomena; surface hardening of metals; characterization of structures; materials engineering; conservation and recycling of materials; environmental aspects of materials production and consumption. *Mailing Add:* Dept Mat Sci & Eng Mass Inst Technol Rm 13-5026 Cambridge MA 02139

BEVERIDGE, DAVID L, b Coshocton, Ohio, Jan 29, 38; m 64; c 3. BIOPHYSICS. *Educ:* Col Wooster, BA, 59; Univ Cincinnati, PhD(phys chem), 65. *Prof Exp:* Asst operating chemist, Monsanto Res Corp, Ohio, 60-62; USPHS fel, Ctr Appl Wave Mech, Paris, 65-66 & Carnegie-Mellon Univ, 66-68; asst prof pharmacol, Mt Sinai Sch Med, 68-70; from assoc prof to prof chem, Hunter Col, 74-86; PROF CHEM, WESLEYAN UNIV, 86- *Concurrent Pos:* NIH res career develop award, 72-77; Study sect, NIH, 78-82; Thomas Hunter distinguished prof, 84-86; Merit Award, NIH, 89- *Mem:* AAAS; Am Chem Soc; Biophys Soc. *Res:* Theoretical chemistry and computational molecular biophysics; statistical thermodynamics and computer simulation of biological molecules; hydration and ion atmosphere of nucleic acids; organization of water in nucleic acid hydrates; solvent effects on conformational stability; dynamical structure of proteins and DNA. *Mailing Add:* Dept Chem Wesleyan Univ Middletown CT 06457

BEVERIDGE, JAMES MACDONALD RICHARDSON, b Dunfermline, Scotland, Aug 17, 12; nat Can; m 40; c 7. BIOCHEMISTRY. *Educ:* Acadia Univ, BSc, 37; Univ Toronto, PhD(biochem), 40; Univ Western Ont, MD, 50. *Hon Degrees:* DSc, Acadia Univ, 62; LLD, Mt Allison Univ, 66 & Queen's Univ, 78. *Prof Exp:* Asst, Dept Med Res, Banting Inst, 40; assoc biochemist, Pac Fisheries Exp Sta, 44-46; lectr path chem, Univ Western Ont, 46-50; Craine prof biochem & head dept, Queen's Univ, Ont, 50-64, dean sch grad studies, 60-64; pres, Acadia Univ, 64-78; RETIRED. *Concurrent Pos:* Officer, Order Can, 77. *Mem:* Royal Soc Can. *Res:* Analysis of protein hydrolysates; choline and fatty liver problem; dietary liver necrosis and lipid and bile acid metabolism; effect of dietary changes on plasma cholesterol levels in human subjects using homogenized rations. *Mailing Add:* RR-1 Canning NS B0P 1H0 Can

BEVERLEY-BURTON, MARY, b Abergavenny, Wales, June 10, 30; m 57, 68; c 3. ZOOLOGY. *Educ:* Univ Wales, BSc, 53; Univ London, PhD(parasitol), & dipl, Imp Col, 58. *Prof Exp:* Exp officer pest control, Scotland Dept Agr, Glasgow, 53-54, sci officer seed potato cert, Seed Testing Sta, 54-55; asst lectr parasitol, Imp Col, Univ London, 55-58; Nuffield res fel, Univ Col Rhodesia & Nyasaland, 58-60; asst ed helminth, Commonwealth Agr Bur, Eng, 67-68; asst prof zool, 68-72, assoc prof, 73-85, PROF ZOOL, UNIV GUELPH, 86- *Mem:* Am Soc Parasitol; Can Soc Zool; Brit Soc Parasitol. *Res:* Helminth taxonomy; biology of parasitic helminths of fishes. *Mailing Add:* Dept Zool Univ Guelph Guelph ON N1G 2W1 Can

BEVERLY, ROBERT EDWARD, III, b Atlanta, Ga, Mar 13, 48. GAS LASER DEVELOPMENT. *Educ:* Ga Inst Technol, BS, 69; Univ Cincinnati, PhD(nuc eng), 73. *Prof Exp:* Prin physicist, Phys Sci Sect, Battelle Columbus Labs, 73-78; SR SCIENTIST, R E BEVERLY III & ASSOCS, 78- *Mem:* Am Phys Soc; Asn Comput Mach. *Res:* Development and applications of high-energy and high-power lasers including the aspects of gas-discharge physics; kinetic modeling, optics design, laser propagation and laser-target interactions; laser interaction phenomena. *Mailing Add:* 854 Pipeston Dr Worthington OH 43235

BEVERUNG, WARREN NEIL, JR, b New Orleans, La, Sept 3, 41; m 64; c 2. MEDICINAL CHEMISTRY, CHEMICAL SEPARATIONS. *Educ:* La State Univ, New Orleans, BS, 64, PhD(org chem), 68. *Prof Exp:* Asst org chem, La State Univ, New Orleans, 64-68; instr & fel, Univ Ill, Chicago Circle, 68-69; sr res chemist, Med Chem Div, Bristol Labs, 69-77; tech rep org synthesis div, 77-80, SE REGIONAL MGR, WATERS ASSOCS, 80- *Mem:* Am Chem Soc. *Res:* Total synthesis of 9-azasteroids; general photochemical processes; synthesis of new biologically active heterocyclic systems. *Mailing Add:* 104 Amesbury Lane Cary NC 27511-5505

BEVIER, MARY LOU, b Chicago, Ill, Mar 3, 53; m 85. RADIOGENIC ISOTOPIC STUDIES, URANIUM-LEAD GEOCHRONOLOGY. *Educ:* Univ Calif Santa Cruz, BS, 75; Univ BC, MSc, 78; Univ Calif Santa Barbara, PhD(geol sci), 82. *Prof Exp:* Res asst geol earth sci, Los Alamos Nat Lab, 77 & 78; teaching assoc geol, Univ Calif, Santa Barbara, 78-82; asst prof geol, Western State Col, Colo, 82-85; postdoctoral geochronology, 85-87, CONTRACT RES GEOLOGIST RADIOGENIC ISOTOPES & GEOCHRONOLOGY, GEOL SURV CAN, 87- *Concurrent Pos:* Vis prof, Dept Geol Sci, Univ Ore, 84. *Mem:* Geol Soc Am; fel Geol Asn Can; Am Geophys Union; Sigma Xi. *Res:* Radiogenic isotope geology; uranium-lead geochronology; lead-neodymium-strontium tracer studies. *Mailing Add:* Geol Surv Can 601 Booth St Ottawa ON K1A 0E8 Can

BEVILL, RICHARD F, JR, b Christopher, Ill, May 18, 34; m 53; c 2. VETERINARY PHARMACOLOGY. *Educ:* Univ Ill, DVM, 64, PhD(pharmacol), 72. *Prof Exp:* Asst prof, 72-79, PROF PHARMACOL, COL VET MED, UNIV ILL, URBANA, 79- *Mem:* Am Soc Pharmacol & Exp Therapeut; Am Vet Med Asn. *Res:* Pharmacokinetics of antibacterial drugs following their administration to domestic animals; relationships between plasma, urine and tissue residues are stressed. *Mailing Add:* Col Vet Med 3519 VMBSB Univ Ill 2001 S Lincoln Ave Urbana IL 61801

BEVILL, VINCENT (DARELL), b San Diego, Calif, Mar 31, 28; m 48; c 3. MECHANICAL ENGINEERING. *Educ:* Fresno State Col, BS, 52; Univ Calif, MSc, 65. *Prof Exp:* Chief engr, Griffith-Dyer Co, 52-58; from assoc prof to prof mech eng, Fresno State Col, 64-80; RETIRED. *Concurrent Pos:* Consult mech engr, 58- *Mem:* Nat Soc Prof Engrs; Solar Energy Soc; Am Soc Heat, Refrig & Air-Conditioning Eng. *Res:* Solar energy, environmental control for poultry houses and computer programs for total energy and air conditioning load calculations. *Mailing Add:* 7265 N Doolittle Dr Fresno CA 93705

BEVIN, A GRISWOLD, b New Haven, Conn, Dec 7, 35; m 58; c 5. PLASTIC SURGERY, RECONSTRUCTIVE SURGERY. *Educ:* Wesleyan Univ, AB, 56; Yale Univ, MD, 60; Am Bd Surg, dipl, 67; Am Bd Plastic Surg, dipl, 71. *Prof Exp:* Instr surg, Sch Med, Yale Univ, 66-68; From asst prof to assoc prof plastic surg, 69-79, chief, Div Plastic & Reconstruct Surg, 73-79, PROF PLASTIC SURG, SCH MED, UNIV NC, CHAPEL HILL, 79- *Concurrent Pos:* Consult, Watts Hosp, Durham, NC, 69-78. *Mem:* Am Asn Univ Prof; Am Col Surgeons; Am Asn Surg Trauma; Am Soc Surg Hand; Am Soc Plastic & Reconstruct Surgeons. *Res:* Wound healing; surgery of the hand; maxillofacial trauma. *Mailing Add:* Div Plastic & Reconstruct Surg Univ NC Med Sch CB No 7195 Chapel Hill NC 27599-7195

BEVIS, HERBERT A(NDERSON), b Perry, Fla, Sept 28, 29; m 52; c 2. ENVIRONMENTAL ENGINEERING, RADIOLOGICAL HEALTH. *Educ:* Univ Fla, BCE, 51, MSE, 52, PhD(sanit eng), 63; US Naval Postgrad Sch, MS, 58. *Prof Exp:* Sanit engr, USPHS, NY, 52-53, DC, 53-54, sanit engr radiol health prog, Wash, 54-55, sanit engr, R A Taft Sanit Eng Ctr, Ohio, 55-56; chief ionizing radiation prog, Tex State Dept Health, 58-61; teaching assoc environ eng, 61-63, assoc prof, 64-74, vchmn dept, 74-77, asst dean, Col Eng, 78-79, PROF ENVIRON ENG, UNIV FLA, 74-, ASSOC DEAN, COL ENG, 81- *Concurrent Pos:* Mem, Conf Fed Environ Engrs Dipl, Am Bd Health Physics & Am Environ Eng Intersoc Bd; dir, Water & Air Res, Inc, 71- *Mem:* Nat Soc Prof Engrs; Am Soc Civil Engrs; Am Soc Eng Educ. *Res:* Radiological health; environmental radiation surveillance; application of radiological techniques in environmental engineering. *Mailing Add:* Col Eng Univ Fla Gainesville FL 32601

BEVIS, JEAN HARWELL, b Miami, Fla, Dec 6, 39; m 60; c 2. ALGEBRA, SYSTEMS THEORY. *Educ:* Univ Fla, BS, 61, MS, 62, PhD(math), 65. *Prof Exp:* Asst prof math, Va Polytech Inst, 65-69; assoc prof, 69-73, PROF MATH, GA STATE UNIV, 73- *Mem:* Am Math Soc; Math Asn Am. *Res:* Lattices; adjacency matrices; graphs; algebraic representation of properties of graphs. *Mailing Add:* Dept Math & Computer Sci Ga State Univ Atlanta GA 30303

BEVK, JOZE, b Ljubljana, Yugoslavia, April 1, 43; US citizen; m 77; c 1. MICRO-COMPOSITE MATERIALS, SOLID STATE PHYSICS. *Educ:* Univ Ljubljana, Yugoslavia, BS, 65; Univ Wis-Madison, MS, 68, PhD(metall eng), 70. *Prof Exp:* Res fel metal physics, Mellon Inst Sci, Carnegie-Mellon Univ, Pittsburgh, 70-72, asst prof, 72-73; from asst prof to assoc prof appld physics, Harvard Univ, 73-80; MEM TECH STAFF, BELL LABS, MURRAY HILL, NJ, 81- *Concurrent Pos:* Prin Investgr, Prep & Characterization Improv Superconducting Mats, NSF, 74-80; vis prof appl physics, Stanford Univ, 80; assoc ed, Mats Letters, 81- *Mem:* Am Physi Soc; Am Soc Metals; Am Inst Mining, Metall & Petrol Engrs; Mat Res Soc. *Res:* Superconductivity; composite materials; effect of microstructure on physical properties of materials; electronic structure and stability of metals and alloys; artificially layered materials; metallic glasses; physical properties of micro-composite materials. *Mailing Add:* 82 Sycamore Ave 600 Mountain Ave Murray Hill NJ 07974

BEVOLO, ALBERT JOSEPH, b St Louis, Mo, Oct 20, 40; m 66; c 2. SOLID STATE PHYSICS. *Educ:* St Louis Univ, BS, 62; MS, 64, PhD(physics), 70. *Prof Exp:* Fel physics, St Louis Univ, 70-72; Presidential intern, US AEC, 72-73, ASST PHYSICIST, AMES LAB, ENERGY RESOURCE & DEVELOP ADMIN, 73- *Mem:* Am Phys Soc. *Res:* Low temperature specific heat of tungsten bronzes; IV-VI semicomputers and soft mode superconductors; electrochemical energy conversion. *Mailing Add:* Ames Lab A205 Physics Iowa State Univ Ames IA 50011

BEWLEY, GLENN CARL, b Middletown, Ohio, July 19, 42; m 65; c 4. DEVELOPMENTAL & MOLECULAR GENETICS. *Educ:* Miami Univ, BSEd, 65, MA, 67; Univ NC, Chapel Hill, PhD(genetics), 74. *Prof Exp:* Aerospace physiologist, US Naval Reserve, 67-70; NIH fel dept biol, Univ Va, 74-75; asst prof, 75-81, assoc prof, 81-85, PROF GENETICS, BIOTECHNOL FAC, NC STATE UNIV, 85- *Concurrent Pos:* Mem, NIH study sect, comt genetics & molecular biol aging, 85-; ed, Isozyme Bull, 81-84. *Mem:* Genetics Soc Am; AAAS. *Res:* Developmental and molecular genetics of higher eukaryotes; regulation of tissue-specific gene expression; antioxidant enzymes and oxygen free radical metabolism. *Mailing Add:* Box 7614 Dept Genetics NC State Univ Raleigh Main Campus Raleigh NC 27695-7614

BEWLEY, JOHN DEREK, b Preston, Eng, Dec 11, 43; m 66; c 2. PLANT PHYSIOLOGY, BIOCHEMISTRY. *Educ:* Queen Elizabeth Col, Univ London, BSc, 65, PhD(plant physiol), 68, DSc, 83. *Prof Exp:* Fel plant biochem, Inst Cancer Res, 68-70; from asst prof to prof biol, Univ Calgary, 70-85; chmn, Dept Bot, 85-90, PROF BOT, UNIV GUELPH, 85- *Concurrent Pos:* E W R Steacie Mem fel, Nat Sci & Eng Res Coun, Can, 79-81; vis prof, McGill Univ, 81; mem, Can Strategic Grants Panel Environ & Toxicol, Nat Sci & Eng Res Coun, 81-82, mem & chmn, Plant Biol Grants Selection Panel, 86-90; mem seed storage adv panel, Int Bd Genetic Resources, 81-87; rapporteur & convener, Plant Biol Div, Sci Acad, Royal Soc Can, 84-87. *Honors & Awards:* C D Nelson Award, Can Soc Plant Physiol, 78. *Mem:* Am Soc Plant Physiol; Can Soc Plant Physiol (secy, 83-85, vpres, 87-88, pres, 88-89, past pres, 89-90); Brit Soc Exp Biol; fel Royal Soc Can. *Res:* Dormancy and survival mechanisms in plants, molecular, biochemical and physiological aspects; desiccation tolerance in plants, mosses and ferns; weed root biology; development, germination and growth physiology of seeds. *Mailing Add:* Dept Bot Univ Guelph Guelph ON N1G 2W1 Can

BEWLEY, LOYAL V, b Republic, Wash, Dec 19, 1898; m 23; c 2. ENGINEERING. *Educ:* Univ Wash, BS, 23; Union Col, MS, 28. *Prof Exp:* Res engr, General Elec Co, 23-40; dean eng, admin & teaching, Lehigh Univ, 40-62; consult, analysis, General Elec Co, 62-70; RETIRED. *Honors & Awards:* Coffin Award, Gen Elec Co, 34; Lamme Medal, Inst Elec & Electronics Engrs, 64. *Mem:* Fel Inst Elec & Electronics Engrs. *Res:* Author of six books; alternating current mach; flux linkages and electromagnetic induction; two-dimensional fields in electronical engineering; tensor analysis of circuits and mechines; protection of transmission systems against lightning; Traveling wares on transmissions systems; author of six books. *Mailing Add:* 2145 Orchard Park Dr Schenectady NY 12309

BEWTRA, JATINDER KUMAR, b Nawabshah, India, May 1, 35; m 63; c 2. SANITARY & BIOCHEMICAL ENGINEERING. *Educ:* Univ Roorkee, BE, 55; Univ Iowa, MS, 60, PhD(sanit eng), 62. *Prof Exp:* Asst engr, Local Self Govt Eng Dept, India, 55-58; scientist in charge, Nat Environ Eng Res Inst, Delhi, India, 62-66; vis assoc prof sanit eng, Shiraz Univ, Iran, 66-68; PROF CIVIL ENG, UNIV WINDSOR, 68- *Mem:* Am Soc Civil Engrs; Am Water Works Asn; Water Pollution Control Fedn; Int Asn Water Pollution Res; Can Soc Civil Eng. *Res:* Physical, chemical and biological aspects of water purification and waste-water treatment. *Mailing Add:* Dept Civil Eng Univ Windsor Windsor ON N9B 3P4 Can

BEX, FREDERICK JAMES, b Syracuse, NY, Nov 1, 47; m 72; c 3. BONE METABOLISM, REPRODUCTIVE PHYSIOLOGY. *Educ:* Cornell Univ, BA, 69; Univ Conn, MS, 74, PhD(endocrinol), 76. *Prof Exp:* res asst, Univ Conn, 71-75; res assoc, Worcester Found for Exp Biol, 75-77; group leader, 77-85, mgr Endocrinol Sect, 85-88, ASSOC DIR ENDOCRINOL, WYETH-AYERST RES, WYETH LABS, INC, 88- *Mem:* Endocrine Soc; Soc Study Reprod; Am Soc Andrology; Am Soc Bone Mineral Res. *Res:* Neuroendocrinology of mammalian reproduction; specifically the development of hypothalamic releasing hormone derived and gonadal peptides as fertility regulating agents and as therapy for reproductive pathologies; contraceptive research and general endocrine evaluation of drugs; bone metabolism and calciotropic hormone regulation. *Mailing Add:* Wyeth-Ayerst Res CN 8000 Princeton NJ 08543

BEYAD, MOHAMMED HOSSAIN, b Iran, Nov 28, 50; m 76. CHEMISTRY. *Educ:* Indian Inst Technol, BS, 71, MS, 73; Kans State Univ, PhD(chem), 78. *Prof Exp:* Asst prof chem, Kans State Univ, 79-80; res chemist, Mobil Chem Co, 80-; ALBRIGHT & WILSON, INC. *Concurrent Pos:* Consult, DOM Assocs Int, 79- *Mem:* Am Chem Soc. *Res:* Analytical methods to support programs in polymer res and develop industrial chemicals and chemical coatings; specialize in areas of spectroscopy: fourier transform nuclear magnetic resonance, fourier transform infrared and gas chromatography-mass spectrometry. *Mailing Add:* Albright & Wilson Inc PO Box 26229 Richmond VA 23260-6229

BEYEA, JAN EDGAR, b Englewood, NJ, Dec 16, 39; div; c 2. NUCLEAR ACCIDENT ASSESSMENT. *Educ:* Amherst Col, BA, 62; Columbia Univ, PhD(physics), 68. *Prof Exp:* Res assoc physics, Columbia Univ, 68-70; asst prof physics, Holy Cross Col, 70-76; mem res staff, Princeton Univ, 76-80; SR STAFF SCIENTIST, NAT AUDUBON SOC, 80- *Concurrent Pos:* Consult, US Coun Environ Qual, 80, Mass Atty Gen, 88 & Nat Res Coun, 89. *Mem:* Am Phys Soc; AAAS; Health Physics Soc. *Res:* Risk assessment of energy sources; solid waste management; United States energy policy. *Mailing Add:* Nat Audubon Soc 950 Third Ave New York NY 10022

BEYEN, WERNER J, b Sidney, NY, Dec 24, 29; m 59; c 2. SEMICONDUCTOR PHYSICS. *Educ:* Syracuse Univ, BS, 52, MS, 54, PhD(physics), 62. *Prof Exp:* Res asst, Dept Physics, Syracuse Univ, 52-58; RETIRED. *Mem:* AAAS; Am Phys Soc; Inst Elec & Electronics Engrs. *Res:* Semiconductor devices, materials, and technology. *Mailing Add:* 19641 Marble Dr Sun City West AZ 85375

BEYENBACH, KLAUS WERNER, b Mainz, WGer, Mar 19, 43; US citizen; m; c 1. PHYSIOLOGY. *Educ:* St Mary's Univ, Tex, BA, 68; Southwest Tex State Univ, MA, 70; Wash State Univ, PhD(physiol), 74. *Prof Exp:* Teaching asst, Southwest Tex State Univ, 68-70; NIH trainee, Wash State Univ, 70-74; NIH res fel, Col Med, Univ Ariz, 75-78; from asst prof to assoc prof, 78-88, PROF PHYSIOL, CORNELL UNIV, 88- *Concurrent Pos:* Nat Kidney Found fel, Dept Physiol, Col Med, Univ Ariz, 74-75; Fogarty sr int fel, 85. *Mem:* Am Soc Physiologists; Am Soc Nephrology; Biophys Soc; Am Soc Zoologists; AAAS. *Res:* Renal electrolyte transport and electrophysiology; epithelial transport mechanism; comparative epithelial transport physiology. *Mailing Add:* Sect Physiol VRT 826 Cornell Univ Ithaca NY 14853

BEYER, ANN L, b Louisville, Ky, Feb 12, 51; m; c 2. RNA PROCESSING. *Educ:* Ctr ColKy, BS, 73; Vanderbilt Univ, PhD(molecular biol), 77. *Prof Exp:* Asst Prof,82-88, ASSOC PROF DEPT MICROBIOL, UNIV VA, 88- *Mem:* Am Soc Cell Biol; Sigma Xi; Am Soc Microbiol. *Res:* Ultrastructural analysis of RNA splicing and ribonucleo protein formation. *Mailing Add:* Dept Microbiol Box 441 Univ Va Sch Med Charlottesville VA 22908

BEYER, ARTHUR FREDERICK, b Toledo, Ohio. PALEOBOTANY. *Educ:* Ohio Univ, BSc, 43; Ohio State Univ, MSc, 45; Univ Cincinnati, PhD(bot), 50. *Prof Exp:* Instr bot, Western Reserve Univ, 47-50; assoc prof, 50, chmn, Dept Biol, 60-81, PROF BOT, MIDWESTERN STATE UNIV, 51- *Concurrent Pos:* Bd mem, Tex Syst Natural Labs. *Mem:* Bot Soc Am; Sigma Xi. *Res:* Tamarix gallica morphology; ecology; paleoxylotomy. *Mailing Add:* Dept Sci Midwestern State Univ 3400 Taft Blvd Wichita Falls TX 76308

BEYER, CARL FREDRICK, b Louisville, Ky, July 22, 47; m 80; c 1. CELL BIOLOGY, NEUROBIOLOGY. *Educ:* Bellarmine Col, BA, 69; Rockefeller Univ, PhD(biochem), 74. *Prof Exp:* USPHS postdoctoral fel cell biol & immunol, Rockefeller Univ, 74-77, asst prof, 77-85; SR RES SCIENTIST, MED RES DIV, AM CYANAMID CO, 85- *Concurrent Pos:* Vis fel, Int Inst Cellular & Molecular Path, Brussels, Belg, 78-79. *Mem:* AAAS; Am Chem Soc; Am Soc Cell Biol; NY Acad Sci; Sigma Xi. *Res:* New approaches to anti-cancer therapy, especially monoclonal antibody-directed drug delivery. *Mailing Add:* Lederle Labs Pearl River NY 10965

BEYER, EDGAR HERMAN, b Melrose Park, Ill, Apr 27, 31; m 54; c 3. PLANT BREEDING, PLANT GENETICS. *Educ:* Univ Ill, BS, 58; Purdue Univ, MS, 62, PhD(plant breeding & genetics), 64. *Prof Exp:* Asst prof forage breeding, Univ Md, College Park, 63-66; res dir, Farm Seed Res Corp, 66-74; veg & flower seed prod mgr, Ferry-Morse Seed Co, 74-80; res dir, Growers Seed Asn, Lubbock, Tex, 80-81; PROF CROP SCI, CALIF POLYTECH STATE UNIV, SAN LUIS OBISPO, 81- *Concurrent Pos:* Mem, Nat Cert Alfalfa Variety Rev Bd, 69-72. *Mem:* Am Soc Agron; Crop Sci Soc Am; Sigma Xi; Nat Asn Col Teachers Agr. *Res:* Testing performance of forage crop varieties; combining ability studies with alfalfa single crosses; improvement of Trifolium pratense by interspecific hybridization; alfalfa breeding and development; cauliflower improvement through selection. *Mailing Add:* Crop Sci Dept Calif Polytech State Univ San Luis Obispo CA 93407

BEYER, ELMO MONROE, JR, b Corpus Christi, Tex, Mar 22, 41; m 64; c 2. PLANT BIOCHEMISTRY, CHEMISTRY. *Educ:* Tex Technol Col, BS, 63; Tex A&M Univ, PhD(plant physiol), 69. *Prof Exp:* Res biologist, Ctr Res Dept, E I du Pont de Nemours & Co, Inc, 69-74, res supvr, 74-79, res mgr, Agr Prod Dept, 79-88, DIR DISCOVERY RES, AGR PROD DEPT, E I DU PONT DE NEMOURS & CO, INC, 88- *Mem:* Am Soc Plant Physiologists; Plant Growth Regulator Soc; Coun Agr Sci & Technol. *Res:* Phytohormones, especially mechanism of ethylene action; effect of ethylene on translocation; plant and cell biology; plant growth regulators, utility and mechanism of action; herbicides; insecticides and fungicides; agrichemicals. *Mailing Add:* 4005 Valley Green RD Hillside Farms Wilmington DE 19807

BEYER, GEORGE LEIDY, b Philadelphia, Pa, Jan 12, 19; m 46; c 3. POLYMER CHEMISTRY. *Educ:* Juniata Col, BS, 41; Rutgers Univ, PhD(anal chem), 45. *Prof Exp:* Lab asst, Rutgers Univ, 41-43, instr chem, 43-44; phys chemist, 45-55, res assoc res labs, Eastman Kodak Co, 55-83; RETIRED. *Mem:* Am Chem Soc. *Res:* Polymer molecular characterization; light-scattering; exclusion chromatography. *Mailing Add:* 674 Lake Rd Webster NY 14580

BEYER, GERHARD H(AROLD), b Fowler, Mich, July 28, 23; m 47; c 4. CHEMICAL ENGINEERING. *Educ:* Univ Wis, BS, 44, MS, 47, PhD(chem eng), 49. *Prof Exp:* Prof chem eng, Iowa State Col, 49-55; prof & chmn dept, Univ Mo, 56-64; prof chem eng & chmn dept, 64-77, PROF CHEM & NUCLEAR ENG, VA POLYTECH INST & STATE UNIV, 77- *Concurrent Pos:* Assoc engr, Ames Lab, AEC, 50-55. *Mem:* Sigma Xi. *Res:* Solvent extraction; zirconium process metallurgy; vacuum technology. *Mailing Add:* Dept Chem Eng Va Polytech Inst Blacksburg VA 24061-0211

BEYER, JAMES B, b Horicon, Wis, July 7, 31; m 55; c 4. ELECTRICAL ENGINEERING. *Educ:* Univ Wis, BS, 57, MS, 59, PhD(elec eng), 61. *Prof Exp:* From asst prof to assoc prof, 61-69, PROF ELEC ENG, UNIV WIS-MADISON, 69-, ASSOC CHMN DEPT, 78- *Concurrent Pos:* Vis prof, Tech Univ Braunschweig, 68-69. *Mem:* Inst Elec & Electronics Engrs; Sigma Xi. *Res:* Microwaves, especially interaction of microwaves with plasmas and semiconductors. *Mailing Add:* Dept Elec Eng Univ Wis Madison WI 53706

BEYER, KARL HENRY, JR, b Henderson, Ky, June 19, 14; m; c 2. CLINICAL PHARMACOLOGY. *Educ:* Western Ky State Col, BS, 35; Univ Wis, PhD(physiol), 40, MD, 43. *Hon Degrees:* DSc, Univ Wis, 72. *Prof Exp:* Sr vpres res, Merck, Sharp & Dohme, 43-73; VIS PROF PHARMACOL, SCH MED, PA STATE UNIV, 73-; VIS PROF, CTR BIOCHEM & BIOPHYS SCI & MED, HARVARD UNIV, 85- *Concurrent Pos:* Chmn, Expert Panel - Cosmetic Ingredient Review 76-86. *Honors & Awards:* Alvert Lasker Award, 75. *Mem:* Nat Acad Sci; Am Soc Pharmacol & Exp Therapeut; Am Physiol Soc; Am Col Physicians. *Res:* Renal pharmacology, primarily hypertension. *Mailing Add:* Box 7387 Penllyn PA 19422

BEYER, LOUIS MARTIN, b Paducah, Ky, Nov 7, 39; m 59; c 2. NUCLEAR PHYSICS. *Educ:* Murray State Univ, BS, 62; Mich State Univ, MS, 63, PhD(nuclear physics), 67. *Prof Exp:* From asst prof to assoc prof, 67-75, PROF PHYSICS, MURRY STATE COL, 75-; SPARTA INC, HUNTSVILLE. *Concurrent Pos:* Vis staff mem, Los Alamos Sci Lab, 77- *Mem:* Am Asn Physics Teachers. *Res:* Low energy nuclear structure by methods of beta and gamma ray spectroscopy; atomic structure and lifetime by beam-foil spectroscopic means; x-ray produced by heavy charged particle reactions; two-dimensional image processing; small computer applications in education. *Mailing Add:* Dept Physics & Astron Murray State Univ Murray KY 42071

BEYER, ROBERT EDWARD, b Englewood, NJ, Feb 20, 28; div; c 3. BIOCHEMISTRY. *Educ:* Univ Conn, AB, 50, MSc, 52; Brown Univ, PhD(biol), 54. *Prof Exp:* Asst biol, Brown Univ, 51-53; USPHS fel, Wenner-Gren Inst, Stockholm, 54-56; instr physiol, Sch Med, Tufts Univ, 56-58, asst prof, 58-62; asst prof enzyme chem, Enzyme Inst, Univ Wis-Madison, 62-65; from assoc prof to prof zool, 65-76, PROF BIOL SCI, UNIV MICH, ANN ARBOR, 76- *Concurrent Pos:* USPHS sr res fel, Sch Med, Tufts Univ, 58-59, vis prof biochem, 75-76, NIH Fogarty sr fel, Univ Stockholm, 85-86; NIH res career develop award, Tufts Univ, 59-62 & Univ Wis, 62-65. *Mem:* AAAS; Am Soc Biol Chemists; Biophys Soc; Geront Soc Am. *Res:* Hormonal control of metabolic systems; biochemistry of aging; anticancer enzymes and loenzyme Q function. *Mailing Add:* Dept Biol Univ Mich Ann Arbor MI 48109-1048

BEYER, ROBERT THOMAS, b Harrisburg, Pa, Jan 27, 20; m 44; c 4. ACOUSTICS. *Educ:* Hofstra Univ, AB, 42; Cornell Univ, PhD, 45. *Hon Degrees:* DSc, Hofstra Univ, 85. *Prof Exp:* Asst physics, Cornell Univ, 42-45; from instr to assoc prof physics, Brown Univ, 45-58, exec officer, 66-68, 81-85, chmn dept, 58-74, prof, 58-85, EMER PROF PHYSICS, BROWN UNIV, 85- *Concurrent Pos:* Advan Educ Fund fel, Univ Calif, Los Angeles, 53-54; consult Russian trans, Am Inst Physics, 55-, Raytheon, 61-71 & Off Naval Res, 74-75; mem, Int Comn Acoust, 75-78, chmn, 78-84. *Honors & Awards:* Gold Medal, Acoust Soc Am, 84. *Mem:* Fel Am Phys Soc; fel Acoust Soc Am (vpres, 61-62, pres, 68-69); fel Inst Elec & Electronics Engrs; fel AAAS. *Res:* Underwater sound; acoustic relaxation times; ultrasonic absorption in liquids and solids; nonlinear acoustics; liquid state; physics in the Soviet Union. *Mailing Add:* 132 Cushman Ave East Providence RI 02914

BEYER, TERRY, b Van Nuys, Calif, Nov 26, 39; div. COMPUTER SCIENCE. *Educ:* Univ Ore, BA, 62, MA, 64; Mass Inst Technol, PhD(math), 69. *Prof Exp:* Mem tech staff, Bell Tel Labs, 64-65; asst prof comput sci, Univ Ore, 69-74 & 80-83, sr systs programmer, 74-80; software engr, Methus-Computervision, 83-86; prin engr, Rosetta Technols, 86-89; SOFTWARE ENGR, MENTOR GRAPHICS CORP, 89- *Concurrent Pos:* Independent software contractor; adj asst prof fine arts, Univ Ore, 82-83. *Mem:* Asn Comput Mach. *Res:* Programming language design. *Mailing Add:* 7930 SE 30th Portland OR 97202

BEYER, W NELSON, b New York, NY, Jan 14, 50; m 72; c 2. ENVIRONMENTAL TOXICOLOGY, SOIL ORGANISMS. *Educ:* Columbia Univ, BA, 71; Cornell Univ, PhD(ecol), 75. *Prof Exp:* RES ZOOLOGIST, PATUXENT WILDLIFE RES CTR, 77- *Mem:* Soc Environ Toxicol & Chem. *Res:* Movement of environmental contaminants into wildlife food chains; soil ingestion by wildlife; concentrations of environmental; contaminants in earthworms; problems associated with disposing of contaminated dredged material. *Mailing Add:* Patuxent Wildlife Res Ctr Laurel MD 20708

BEYER, WILLIAM A, b Tyrone, Pa, Nov 9, 24; m 55; c 2. MATHEMATICS. *Educ:* Pa State Univ, BS, 49, PhD(math), 59; Univ Ill, Urbana, MS, 50. *Prof Exp:* Teaching asst math, Pa State Univ, 52-53, instr, 54-55 & 57-59; teaching asst, Queen's Univ, Ont, 53-54; mathematician, Gen Elec Co, Ohio, 55-57; STAFF MEM, LOS ALAMOS NAT LAB, 60- *Concurrent Pos:* Adj prof, Univ NMex, 59- *Mem:* AAAS; Am Math Soc; Math Asn Am; Soc Indust & Appl Math; Sigma Xi. *Res:* Mathematical biology; probability; dynamic systems; fractional dimension theory; geometry; optimization. *Mailing Add:* 343 Rim Rd Los Alamos NM 87544

BEYER, WILLIAM HYMAN, b Akron, Ohio, Mar 8, 30; m 59; c 3. MATHEMATICAL STATISTICS, MATHEMATICS. *Educ:* Univ Akron, BS, 52; Va Polytech Inst, MS, 54, PhD(statist), 61. *Prof Exp:* Group leader reliability & qual control, Goodyear Aerospace Corp, 53-57; corp staff statistician, Gen Tire & Rubber Co, 57-58; asst prof math, Va Polytech Inst, 58-61; from asst prof to assoc prof, 61-66, head dept math sci, 69-90, PROF MATH & STATIST, UNIV AKRON, 66-, ASSOC DEAN, COL ARTS & SCI, 90- *Concurrent Pos:* Instr, Univ Akron, 54-56; partic, NSF In-Serv Inst Sec Sch Teachers Math & lectr, Univ Akron, 63-66; ed statist, CRC Press, Inc, Ohio, 66-, ed math, 75- *Mem:* Am Statist Asn; Math Asn Am; Sigma Xi. *Res:* Experimental design; analysis of variance. *Mailing Add:* Col Arts & Sci Univ Akron 302 E Buchtel Ave Akron OH 44325

BEYERLEIN, FLOYD HILBERT, b Frankenmuth, Mich, Apr 15, 42; m 65; c 6. ANALYTICAL CHEMISTRY. *Educ:* Mich State Univ, BS, 64, MS, 67, PhD(analytical chem), 70. *Prof Exp:* RESEARCH ASSOC ANALYTICAL CHEM, S C JOHNSON & SON, INC, 70- *Mem:* Am Chem Soc. *Res:* Methods development in support of product research; update analysis techniques already available in the areas of gas and liquid chromatography; performing research employing super critical fluid chromatography and ion chromatography. *Mailing Add:* S C Johnson & Son Inc 1525 Howe St Racine WI 53403-5011

BEYERS, ROBERT JOHN, b Long Beach, Calif, Sept 13, 33; m 54; c 4. ECOLOGY, AQUATIC BIOLOGY. *Educ:* Univ Miami, BS, 54; Univ Tex, PhD(zool), 62. *Prof Exp:* NSF fel ecol, Inst Marine Sci, Univ Tex, 61-62, res assoc scientist, 62-64, head ecol prog, 63-64; from asst prof to assoc prof zool, Lab Radiation Ecol, Univ Ga, 64-74; dir, Savannah River Ecol Lab, 67-74;

PROF BIOL SCI & CHMN DEPT, UNIV SALA, 74- *Concurrent Pos:* Co-prin investr, AEC contract, 67-74; Fed Water Qual Admin grant, 70-74; vpres, Echo Environ Consults, Inc, 75-77; chmn, Environ Biol Rev Panel, Environ Protection Agency, 81-85; co-prin investr, TVA-AUTRC Contract, 89-91. *Mem:* AAAS; Am Soc Limnol & Oceanog; Ecol Soc Am; Inst Soc Limnol. *Res:* Limnology; marine science; microcosm techniques; measurement of carbon dioxide metabolism in aquatic organisms and ecosystems; effects of radiation and pollution on aquatic ecosystems; fish behavior; microcosms. *Mailing Add:* Dept Biol Sci Univ SAla Mobile AL 36688

BEYLER, ROGER ELDON, b Nappanee, Ind, May 20, 22; m 44; c 3. ORGANIC CHEMISTRY. *Educ:* NCent Col, BA, 44; Univ Ill, MS, 47, PhD(chem), 49. *Prof Exp:* Res chemist, Merck & Co, Inc, 49-59; dean, Col Lib Arts & Sci, Southern Ill Univ, Carbondale, 66-74, prof chem, 59-87; RETIRED. *Concurrent Pos:* Orgn Econ Coop & Develop sci fel, Univ Strasbourg, 64. *Mem:* Am Chem Soc. *Res:* Synthesis of the alkaloid sparteine; adrenal steroid total synthesis; steroid synthesis; teratogens. *Mailing Add:* 32 Pinewood RR Nine Carbondale IL 62901-5219

BEYLERIAN, NUREL, b Istanbul, Turkey, May 8, 37. STRUCTURAL ENGINEERING, COMPUTER SCIENCE. *Educ:* Robert Col, Istanbul, BS, 59, MS, 60; Mich State Univ, PhD(civil eng), 65. *Prof Exp:* Asst prof civil eng, Univ PR, 65-68; instr computer sci, 68-69, educ mgr, 69-76, DIR, INST ADVAN TECHNOL, 77- *Res:* Computer sciences, especially software; peaceful uses of nuclear devices; structural reliability and dynamics. *Mailing Add:* 91 Edmonton Dr Willowdale ON M2J 3W9 Can

BEYSTER, J ROBERT, SYSTEMS SCIENCE. *Prof Exp:* CHIEF EXEC OFFICER & CHMN BD, SCI APPLICATIONS INT CORP, 69- *Mem:* Nat Acad Eng. *Mailing Add:* Sci Applications Int Corp 10260 Campus Point Dr San Diego CA 92121

BEZDEK, JAMES CHRISTIAN, b Harrisburg, Pa, Oct 22, 39; m 63; c 4. APPLIED MATHEMATICS. *Educ:* Univ Nev, Reno, BSCE, 69; Cornell Univ, PhD(appl math), 73. *Prof Exp:* Instr eng mech, Cornell Univ, 72; asst prof math, State Univ NY Col Oneonta, 73-74; asst prof, Marquette Univ, 74-76; assoc prof, Utah State Univ, 76-83; prof & chmn computer sci, Univ SC, 83-87; MGR, INFO PROC LAB, BOEING, 87- *Concurrent Pos:* Pres, Int Fuzzy Systs Asn; ed, IJAR, 87-; eminent scholar, Div Computer Sci, Univ WFla. *Mem:* Classification Soc; Soc Indust & Appl Math; Pattern Recognition Soc; Inst Elec & Electronics Engrs; Computer Soc. *Res:* Pattern recognition, cluster analysis and unsupervised learning using fuzzy sets and graphs; applied probability and statistics; numerical fluid mechanics. *Mailing Add:* Div Computer Sci Univ WFla Bldg 79 Rm 113 Pensacola FL 32514

BEZDICEK, DAVID FRED, b Jackson, Minn, Sept 18, 38; m 62; c 2. SOIL MICROBIOLOGY. *Educ:* SDak State Univ, BS, 60; Univ Minn, MS, 64, PhD(soil sci), 67. *Prof Exp:* Field supvr, Calif Packing Corp, 60-61; asst prof agron, Univ Md, College Park, 67-74; assoc prof, 74-79, PROF SOILS, WASH STATE UNIV, 79- *Concurrent Pos:* Vis scientist, Battelle Pacific Northwest Lab, Richland, Wash, 85-86. *Mem:* Fel Am Soc Agron; Am Soc Microbiol; fel Soil Sci Soc Am. *Res:* Microbial ecology; symbiotic nitrogen fixation; molecular biology. *Mailing Add:* Dept Crop & Soil Sci Wash State Univ Pullman WA 99163

BEZELLA, WINFRED AUGUST, b Milwaukee, Wis, Mar 1, 35; m 68; c 2. NUCLEAR ENGINEERING, CHEMICAL ENGINEERING. *Educ:* Univ Wis, BS, 57; Univ Mich, MS, 59; Purdue Univ, PhD(nuclear eng), 72. *Prof Exp:* Engr nuclear eng, Allis Chalmers Mfg Co, 59-64; sr engr nuclear eng, Westinghouse Elec Corp, 64-68; NUCLEAR ENGR, ARGONNE NAT LAB, 71- *Mem:* Am Nuclear Soc. *Res:* Thermal and hydraulic design of nuclear reactors; breeder reactor safety analysis and design of in-reactor safety experiments; reactor engineering, reactor physics and reactor kinetic problems associated with reactor development. *Mailing Add:* Argonne Nat Lab 9700 S Cass Ave Argonne IL 60439

BEZKOROVAINY, ANATOLY, b Riga, Latvia, Feb 11, 35; US citizen; m 64; c 2. BIOCHEMISTRY. *Educ:* Univ Chicago, BS, 56; Univ Ill, MS, 58, PhD(biochem), 60; Ill Inst Technol, JD, 77. *Prof Exp:* Res assoc biochem, Oak Ridge Nat Lab, 60-61; res chemist, Nat Animal Dis Lab, USDA, Iowa, 61-62; from asst biochemist to assoc biochemist, 62-70, assoc prof, 70-73, PROF BIOCHEM & SR BIOCHEMIST, RUSH-PRESBY-ST LUKE'S MED CTR, 73-, DIR EDUC PROGS, 80- & ASSOC CHMN, BIOCHEM DEPT, 83- *Concurrent Pos:* From asst prof to assoc prof biochem, Univ Ill Col Med, 62-70, res prof, Dept Path, 74-80; mem bd dirs, Nat Acad Clin Biochemists, 84-86. *Mem:* Am Soc Biol Chemists; Am Chem Soc; Am Inst Nutrit; Am Dairy Sci Asn; Nat Acad Clin Biochem. *Res:* Iron metabolism; biochemistry of milk; microbiological iron metabolism. *Mailing Add:* Dept Biochem Rush-Presby-St Luke's Med Ctr Chicago IL 60612

BEZMAN, RICHARD DAVID, b Pittsburgh, Pa, Oct 2, 46; m 68. PHYSICAL CHEMISTRY. *Educ:* Pa State Univ, BSc, 67; Harvard Univ, PhD(chem), 72. *Prof Exp:* Mem tech staff, GTE Lab, Inc, 73-75; sr staff chemist, Linde Div, Union Carbide Corp, 75-80; MEM STAFF, CHEVRON RES CO, 80- *Mem:* Am Chem Soc. *Res:* Heterogeneous catalysis; electrochemistry; solid state chemistry. *Mailing Add:* Chevron Res Co 576 Standard Ave Richmond CA 94802

BEZUSZKA, STANLEY JOHN, b Wilna, Poland, Jan 26, 14; nat US. MATHEMATICAL PHYSICS. *Educ:* Boston Col, AB, 39, AM, 40, MS, 42; Weston Col, STL, 47; Brown Univ, PhD(physics), 53. *Prof Exp:* From instr to asst prof physics, Boston Col, 41-43; instr math, Weston Col, 43-45; from asst prof to assoc prof, 53-69, chmn dept, 53-67, PROF MATH, BOSTON COL, 69-, DIR MATH INST, 57- *Mem:* AAAS; Am Math Soc; Math Asn Am; Acoust Soc Am; Am Phys Soc; Sigma Xi; Nat Coun Teachers Math. *Res:* Theoretical physics; scattering of ultrasonic waves; matrix theory; vector analysis and mathematical physics; mathematics education. *Mailing Add:* Mat Inst Boston Col Chestnut Hill MA 02167

BEZWADA, RAO SRINIVASA, b Jammulapalem, India, July 5, 45; US citizen; m 61; c 2. POLYMER CHEMISTRY. *Educ:* Univ Madras, India, BTech, 68; Stevens Inst Technol, MS, 73, PhD(chem), 81. *Prof Exp:* Chemist, A J & J O Pilar Inc, Newark, NJ, 69-73; supvr Rubber Testing Lab, 74-76, scientist, 77-79, res scientist, chem res div, Am Cyanamid Co, 81-; AT ETHICON, INC. *Mem:* Am Chem Soc; Sigma Xi. *Res:* Synthesis and solution properties of polyurethanes; polyurethane castable elastomers and thermoplastic elastomers; polyurethane coatings; structure-property relations in polyurethanes; adhesion promoters for bonding steel cords to rubber; solution properties of model comb-branched polyurethanes. *Mailing Add:* Ethicon Inc Rte 22 Somerville NJ 08876

BHADA, ROHINTON(RON) K, b Bombay, India, Mar 23, 35; US citizen; m 59; c 5. WASTE MANAGEMENT, ENVIRONMENTAL RESTORATION. *Educ:* Univ Mich, BS, 55, MS, 57, PhD(chem eng), 68; Univ Akron, MBA, 74. *Prof Exp:* Res engr, Babcock & Wilcox Co, 59-72, sect mgr, 72-78, dept mgr, 78-88; DEPT HEAD & PROF CHEM ENGR, NMEX STATE UNIV, 88- *Concurrent Pos:* Adj prof, Youngstown State Univ, 78-88; dir, Waste Mgt Educ & Res Consortium, 89- *Mem:* Am Inst Chem Engrs; Nat Soc Prof Engrs; Am Soc Eng Educ; Coun Chem Res. *Res:* Waste-management and environmental restoration. *Mailing Add:* Dept 3805 Col Eng NMex State Univ PO Box 30001 Las Cruces NM 88003-0001

BHADURI, SAUMYA, b Calcutta, India, Sept 9, 42; US citizen; m 74; c 1. MOLECULAR VIROLOGY, RECOMBINANT DNA TECHNOLOGY. *Educ:* Calcutta Univ, India, BSc, 63, MSc, 65, PhD(biochem), 70. *Prof Exp:* Fel molecular biol & biochem, Chem Dept, Nebr Univ, 70; res assoc molecular virol, Inst Molecular Virol, 71-74, instr, 74-76; trainee molecular biol, Dept Path, Med Sch, Washington Univ, 76-77, res instr molecular biol & virol, 77-81; MICROBIOLOGIST MOLECULAR BIOL, USDA, 81- *Mem:* Am Soc Microbiol; Genetic Toxicol Asn; Soc Indust Microbiol. *Res:* Initiation of protein synthesis and coding properties of transfer RNA; transcription mapping of tumor viruses; mapping of phage DNA; characterization of the clustridium botalinum by microbial finger printing; identification and characterization of virulence gene in yesinia enderocolitica to establish key determinants for virulence in Y enderocolitica. *Mailing Add:* 600 E Mermaid Lane Philadelphia PA 19118

BHAGAT, HITESH RAMESHCHANDRA, Indian citizen; m 88; c 1. PREPARATION & EVALUATION OF DRUG DELIVERY SYSTEMS, BIODEGRADABLE VACCINE DELIVERY SYSTEMS. *Educ:* Bombay Univ, BS, 81; Mass Col Pharm, MS, 84, PhD(pharmaceut), 88. *Prof Exp:* Res pharmacist, Hoechst Pharmaceut Ltd, 81-82, Pfeiffer Pharmaceut Sci Lab, 83-88; res affil, Mass Inst Technol, 87-88; ASST PROF, UNIV MD, BALTIMORE, 88- *Mem:* Am Asn Pharmaceut Scientists; Controlled Release Soc. *Res:* Pharmaceutical research and development related to oral and parenteral controlled drug delivery systems of biodegradable polymeric microspheres and implants containing antigens, biotechnology products and conventional drugs. *Mailing Add:* Univ Md Sch Pharm 20 N Pine St Rm 654 Baltimore MD 21201

BHAGAT, PHIROZ MANECK, b Poona, India, Oct 28, 48; m 79; c 2. THERMAL SCIENCES, COMBUSTION. *Educ:* Indian Inst Tech, BTech, 70; Univ Mich, MSE, 71, PhD(mech eng), 75. *Prof Exp:* Res fel, Harvard Univ, 75-77; asst prof, Columbia Univ, 77-81; staff engr, 81-83, SR STAFF ENGR, EXXON RES & ENG CO, 83- *Concurrent Pos:* Adj asst prof, Columbia Univ, 81-84. *Mem:* NY Acad Sci; Am Soc Mech Engrs; Am Inst Chem Engrs; Sigma Xi; Int Neural Networks Soc. *Res:* Neural nets and scientific computing; thermal sciences; combustion and petroleum and fuel systems. *Mailing Add:* 519 Alden Ave Westfield NJ 07090

BHAGAT, PRAMODE KUMAR, b Ranchi, India, Oct 7, 44; US citizen; m 73; c 1. BIOMEDICAL ENGINEERING, CONTROL ENGINEERING. *Educ:* Indian Inst Technol, Madras, BTech, 65; Univ Cincinnati, MS, 69; Ohio State Univ, Columbus, PhD(elec eng), 72. *Prof Exp:* Proj eng elec eng, Systs Res Labs, Dayton, Ohio, 66-67; res assoc ultrasonics, Ohio State Univ, Columbus, 72-73; res assoc, Univ Ky, 73-74, from asst prof to assoc prof mech eng, 74-86, assoc prof biomed eng, 86-89; MAT RES ENGR, WRIGHT-PATTERSON AFB, 89- *Concurrent Pos:* Consult elec eng, Vet Admin Hosp, Lexington, Ky, 75- *Mem:* Inst Elec & Electronics Engrs. *Res:* Biomedical engineering; ultrasonics; microwave spectroscopy; control theory; instrumentation; digital signal processing. *Mailing Add:* WL/MLLP Bldg 655 Area B Wright Patterson AFB OH 45433-6533

BHAGAT, SATINDAR M, b Jammu, India, July 19, 33; m 64. PHYSICS. *Educ:* Univ Jammu & Kashmir, India, BS, 50; Univ Delhi, MSc, 53, PhD(physics), 56. *Prof Exp:* Lectr physics, Univ Delhi, 55-57; Govt of India res fel, Clarendon Lab, Oxford Univ, 57-60; res assoc, Carnegie Inst Technol, 60-62, instr, 61-62; from asst prof to assoc prof, 62-73, PROF PHYSICS, UNIV MD, COLLEGE PARK, 73- *Concurrent Pos:* Vis prof, Indian Inst Technol, Kanpur, India, 68-69 & Uppsala Univ, Sweden, 75. *Res:* Thermodynamic and hydrodynamic properties of liquid helium; properties of magnetic systems at low temperatures and other problems in solid state physics; high temperature superconductivity. *Mailing Add:* Dept Physics Univ Md College Park MD 20740

BHAGAT, SURINDER KUMAR, b Sargodha, Pakistan, May 14, 35; m 63; c 1. ACADEMIC ADMINISTRATION, ENGINEERING EDUCATION. *Educ:* Univ Panjab, India, BSc, 54; Univ Tex, BS, 61, MS, 62, PhD(environ health eng), 66. *Prof Exp:* Asst prof sanit eng, 65-69, asst prof civil eng & asst sanit engr, Col Eng Res Div, 69-71, asst head sanit eng, 71-72, assoc prof, 71-77, head environ eng, 72-80, actg dir, Water Res Ctr, 77-78, PROF CIVIL & ENVIRON ENG, WASH STATE UNIV, 77-, CHMN DEPT, 80- *Concurrent Pos:* Chmn, Environ Eng Div, Am Soc Eng Educ. *Honors & Awards:* Arthur Sidney Bedell Award, Water Pollution Control Fedn, 80- *Mem:* Water Pollution Control Fedn; Am Soc Civil Engrs; Am Soc Eng Educ; Asn Environ Eng Prof; Sigma Xi. *Res:* Transport and behavior of radionuclides and other materials in an aquatic environment; industrial waste treatment; water quality problems of lakes and rivers. *Mailing Add:* Dept Civil Environ Eng Wash State Univ 146 Sloan Hall Pullman WA 99164

BHAGAVAN, HEMMIGE, b Hemmige, India, Oct 23, 34. NUTRITION, BIOCHEMISTRY. *Educ:* Univ Mysore, BS, 53; Indian Inst Sci, Bangalore, MS, 58; Univ Ill, Urbana-Champaign, PhD(nutrit, biochem), 63. *Prof Exp:* Res asst biochem & nutrit, Cent Food Tech Res Inst, Univ Mysore, India, 57-58, jr sci asst biochem, 58-59; res assoc biol sci, Purdue Univ, 63-64; chief biochem lab, Res Inst, St Joseph Hosp, 64-76; chief, Biochem Lab & dir, Div Nutrit Sci, North Nassau Ment Health Ctr, 76-79; asst dir, Clin Pharmacol & res assoc prof, Dept Med, Thomas Jefferson Univ Med Col, 79-80; NUTRIT COORDR, DEPT CLIN NUTRIT, HOFFMANN-LA ROCHE INC, 80- *Mem:* Am Inst Nutrit; Am Col Nutrit; NY Acad Sci; Brit Biochem Soc; Soc Exp Biol Med; Am Asn Clin Chem; Soc Biol Psychiat. *Res:* Metabolism and function of vitamins; biochemical effects of vitamin deficiencies; central nervous system function. *Mailing Add:* Dept Clin Nutrit Bldg 76 Rm 412 Hoffmann-La Roche Inc Nutley NJ 07110

BHAGAVAN, NADHIPURAM V, b Mysore, India, Oct 5, 31; m 62; c 2. CLINICAL BIOCHEMISTRY. *Educ:* Univ Mysore, BSc, 51; Univ Bombay, MSc, 55; Univ Calif, PhD(pharmaceut chem), 60. *Prof Exp:* Asst res biochemist, Med Ctr, Univ Calif, San Francisco, 61-65; asst biochemist, 65-70, asst prof anat, 66-70, from assoc prof to prof biochem & med technol, 70-82, PROF & CHMN BIOCHEM & BIOPHYSICS, MED SCH, UNIV HAWAII, MANOA, 82- *Concurrent Pos:* Assoc prof biol & chem, Hawaii Loa Col, 69-70; consult biochemist, Kaiser Found Hosp, Honolulu, 72- *Mem:* Am Chem Soc. *Res:* Immunochemistry; biochemical and immunochemical studies on normal and malignant cell nuclei. *Mailing Add:* Dept Biochem & Biophys Univ Hawaii at Manoa Honolulu HI 96822

BHAGAVATULA, VIJAYAKUMAR, b Porumamilla, India, Aug 15, 53; m 82; c 3. OPTICAL PROCESSING & PATTERN RECOGNITION, NEURAL NETWORKS. *Educ:* Indian Inst Technol, BTech, 75 & MTech, 77; Carnegie Mellon Univ, PhD(elec eng), 80. *Prof Exp:* From res asst from res assoc, 77-82, asst prof, 82-87, ASSOC PROF ELEC ENG, CARNEGIE MELLON UNIV, 87- *Concurrent Pos:* Consult, Westinghouse Elec, 85-88 & Two-Six, Inc, Pittsburgh, 87-88, Teledyne Brown Eng, Huntsville, Ala, 89-; prin investr pattern recognition algorithms, Sandia Labs, Albuquerque, 87-90, correlators automatic target recognition, Litton Data Syst, 90-; chmn, Max Born Award Comt, Optical Soc Am, 87-89, conf session, Soc Photo-Optical Instrumentation Engrs. *Mem:* Optical Soc Am; Soc Photo-Optical Instrumentation Engrs; sr mem Inst Elec & Electronics Engrs; Int Neural Networks Soc; Sigma Xi. *Res:* Published over 120 research papers in areas of optical processing, pattern recognition, signal and image processing, coding theory and magnetic recording. *Mailing Add:* Dept Elec & Computer Eng Carnegie Mellon Univ Pittsburgh PA 15213

BHAGIA, GOBIND SHEWAKRAM, b Karachi, Pakistan, Dec 7, 35; US citizen; m 72; c 1. AGRICULTURAL ECONOMICS, STATISTICS. *Educ:* Univ Rajasthan, BA, 62; Tex A&M Univ, MS, 67; Ore State Univ, PhD(agr econ), 71. *Prof Exp:* Res assoc agr econ, Ore State Univ, 71-75 & Wash State Univ, 75-76; asst prof bus & coordr finance, Prairie View A&M Univ, 76-81; PROF & CHMN, DIV BUS & ECON, VOORHEES COL, 81- *Concurrent Pos:* Consult local econ develop community, Howard Univ, 77- *Mem:* Am Agr Econ Asn. *Res:* Health economics; environmental economics; microeconomic theory; financial management; investment securities. *Mailing Add:* Dept Bus Paine Col 1235 15th St Augusta GA 30910

BHAKAR, BALRAM SINGH, b Wardha, India, Jan 1, 37; m 68; c 1. THEORETICAL NUCLEAR PHYSICS. *Educ:* Agra Univ, BSc, 57; Aligarh Muslim Univ, India, MSc, 60; Univ Delhi, PhD(physics), 65. *Prof Exp:* Res assoc physics, Bonner Nuclear Lab, Rice Univ, 66-68 & Univ Sussex, 68-69; ASSOC PROF PHYSICS, UNIV MANITOBA, 69- *Mem:* Am Phys Soc. *Res:* 3-nucleon system using separable potential and Faddeev theory and study of nuclear matter; 4-nucleon system. *Mailing Add:* Phys Dept Univ Man Winnipeg MB R3T 2N2 Can

BHALLA, AMAR S, b Panjab, India. MATERIALS SCIENCE. *Educ:* Rajasthan Univ, India, BS, 61, MS, 63; Pa State Univ, PhD(solid state sci), 71. *Prof Exp:* Lectr physics, Rajasthan Univ, 63-64; res scholar, Banaras Univ, India, 65; res scientist, Nat Phys Lab, India, 66-67; res fel, Pa State Univ, 67-71, fel, 72; res assoc, NASA, 73-75; fac res assoc, 75-81, sr fac res assoc, Mat Res Lab & assoc prof Solid State Sci, 81-86, SR SCIENTIST & PROF SOLID STATE SCI, PA STATE UNIV, 86- *Concurrent Pos:* Vis scientist, Univ Turku, 71 & Nat Phys Lab, 71; vis scientist, Nat Phys Lab, India, 71-72. *Mem:* Am Ceramic Soc; Indian Crystallog Soc; Electron Micros Soc; Optical Soc Am; Mat Res Soc. *Res:* Ferroelectric, pyroelectric and piezoelectric properties of materials; x-ray and electron microscopic studies of materials; crystal growth and characterization; glass ceramics and electronic devices; macro, micro and nano composites; superconductivity. *Mailing Add:* Mat Res Lab Pa State Univ University Park PA 16802

BHALLA, CHANDER P, b Hariana, India, Sept 15, 32; US citizen; m 62; c 3. ATOMIC PHYSICS. *Educ:* Punjab Univ, India, BS, 52, Hons, 54, MS, 55; Univ Tenn, PhD(physics), 60. *Prof Exp:* Asst physics, Univ Tenn, 56-60; sr scientist, Westinghouse Elec Corp, Pa, 60-64; from asst prof to assoc prof physics, Univ Ala, 64-66; assoc prof, 66-72, PROF PHYSICS, KANS STATE UNIV, 72- *Concurrent Pos:* Consult, Oak Ridge Nat Lab, 58-60, Nat Bur Standards, DC, 62-64, space div, Northrop Corp, Ala, 65-69 & Argonne Nat Lab, 66-69; vis prof, FOM Inst Atomic & Molecular Physics, Amsterdam, Netherlands, 73-74. *Mem:* Am Nuclear Soc; fel Am Phys Soc. *Res:* Heavy ion interactions; Auger effect; radiative transitions; internal conversion processes; inelastic and elastic energy loss of heavy ions; atomic structure. *Mailing Add:* Dept Physics Cardwell Hall Kans State Univ Manhattan KS 66506

BHALLA, RAMESH C, b Jullundur, India, June 7, 35; US citizen; m 62; c 2. ANATOMY, CELL STRUCTURE. *Educ:* Panjab Univ, India, DVM, 57; Univ Wis-Madison, PhD(physiol), 70. *Prof Exp:* Asst prof vet med, U P Agr Univ, 60-66; asst prof, 73-76, assoc prof, 76-82, PROF ANAT & CELL STRUCTURE, UNIV IOWA MED CTR, 82- *Mem:* Am Physiol Soc; Am

Soc Biol Chemists; Am Asn Anatomists. *Res:* Subcellular mechanisms of cardiovascular dysfunction in the hypertensive state; role of cyclic adenylic acid and calcium 2 in the regulation of cardiac and vascular smooth muscle function. *Mailing Add:* Dept Anat Col Med 1-611 Bowen Sci Bldg Univ Iowa Iowa City IA 52242

BHALLA, RANBIR J R SINGH, b Anandpur, India, July 24, 43; m 68; c 2. SOLID STATE SCIENCE. *Educ:* Univ Jabalpur, BS, 61, MS, 63, PhD(appl physics), 69; Pa State Univ, PhD(solid state sci), 70; Fairleigh Dickonson Univ, MBA, 76. *Prof Exp:* Lectr physics, Govt Sci Col, Univ Jabalpur, 64-66; res physicist inorg mats sci, Cent Res Lab, Am Cyanamid Co, 70-72; sr res engr, inorg mat sci, 72-78, fel res engr, High Intensity Discharge, 78-79, mgr, High Pressure Sodium Discharge Develop, Lamp Div, Westinghouse Elec Corp, 79-83; PRES, BHALLA LIGHTNING INC, 83- *Mem:* Illum Eng Soc. *Res:* Preparation, properties and characterization of inorganic materials; display materials, including cathodo and photochromics and phosphors; high intensity discharge. *Mailing Add:* 29 Sylvan Dr Pine Brook NJ 07058

BHALLA, SUSHIL K, b Ghaziabad, India, Jan 7, 39; US citizen. CHEMICAL ENGINEERING. *Educ:* Kanpur Univ, India, BChe, 61; Indian Inst Technol, Kharagpur, MS, 63; Ohio State Univ, PhD(chem eng), 68. *Prof Exp:* Res engr, 69-71, prin res engr, 71-74, mgr process eng, 74-76, mgr eng develop, 76-78, dir eng develop, 78-83, MGR TECHNOL EVAL, FMC CORP, 83- *Mem:* Am Inst Chem Engrs. *Res:* Peroxide and peroxide chemicals; development of new methods for manufacturing cyanuric acid; emerging technology evaluation and development. *Mailing Add:* 907 Revolution St Havre De Grace MD 21078

BHALLA, VINOD KUMAR, b Lahore, India, Aug 4, 40; m 66; c 3. ENDOCRINOLOGY. *Educ:* Agra Univ, BS, 62, MS, 64, PhD(natural prod), 68. *Prof Exp:* Fel org chem, Univ Ga, 68-69, res assoc biochem & reproductive physiol, 69-72; res assoc biochem & endocrinol, Emory Univ, 72-74; from asst prof to assoc prof, 74-82, PROF ENDOCRINOL, MED COL GA, 82- *Concurrent Pos:* Mem, Endocrine Study Sect, 84-87. *Mem:* Endocrine Soc; Soc Study Reproduction; Am Chem Soc; Soc Complex Carbohydrates; NY Acad Sci. *Res:* Mechanism of action of gonadotropin in testicular functions. *Mailing Add:* Dept Endocrinol Med Col Ga Augusta GA 30912

BHANDARKAR, DILEEP PANDURANG, b Bombay, India, July 16, 49; m 73; c 2. COMPUTER SCIENCE. *Educ:* Indian Inst Technol, Bombay, BTechnol, 70; Carnegie-Mellon Univ, MS, 71, PhD(elec eng), 73. *Prof Exp:* Mem tech staff comput sci, Tex Instruments Inc, 73-77; TECH DIR, COMPUT SYSTS ARCHIT, DIGITAL EQUIP CORP, 78- *Mem:* Inst Elec & Electronics Engrs. *Res:* Computer performance evaluation; computer architecture. *Mailing Add:* Digital Equip Corp BXB1-1E11 85 Swanson Rd Boxborough MA 01719

BHANDARKAR, MANGALORE DILIP, b Mangalore, India, Mar 12, 46; US citizen; m 76; c 2. FAILURE ANALYSIS, ALLOY DESIGN. *Educ:* Indian Inst Technol, BTech, 68; Univ Calif, Berkeley, MS, 70, DEng, 73. *Prof Exp:* Res metallurgist, Lawrence Berkeley Lab, 73-74; Nat Res Coun res assoc failure anal, Langley Res Ctr, NASA, 74-76; assoc develop engr res & proj mgt, Dept Mat Sci & Eng, Univ Calif, Berkeley, 76-77; PRIN METALLURGIST & MGR SPEC PROJ, ANAMET LABS, 77- *Mem:* Am Soc Metals; Am Soc Testing & Mat; Metall Soc Am; Inst Mining & Metall Engrs. *Res:* Strain-induced transformations in austenitic steels; microstructure and elevated temperature mechanical properties of ferritic alloys containing Laves phases; microstructure and fracture behavior of aluminum alloys and metal-matrix composites; design of abrasion resistant steels; development of improved high-speed tool steels; failure analysis. *Mailing Add:* Anamet Labs Inc 3400 Investment Blvd Hayward CA 94545-3811

BHANDARKAR, SUHAS D, b India, Oct 8, 64. MATERIALS SCIENCE, CHEMICAL ENGINEERING. *Educ:* Mangalore Univ, India, BS, 86; Univ RI, MS, 88, PhD(chem eng), 90. *Prof Exp:* POSTDOCTORAL FEL MAT, AT&T BELL LABS, MURRAY HILL, NJ, 90- *Mem:* Am Inst Chem Engrs. *Res:* Topics of ultrafine materials; investigation has ranged from the synthesis, using novel techniques the characterization and their application; application has involved optical and electronic materials and films. *Mailing Add:* AT&T Bell Labs Rm 6C-312 600 Mountain Ave Murray Hill NJ 07974-2070

BHANSALI, PRAFUL V, b Bombay, India, Oct 20, 49; m 82. HARDWARE SYSTEMS, SOFTWARE SYSTEMS. *Educ:* Indian Inst Technol, BS, 74; Univ Wis-Madison, MS, 77, PhD(elec eng), 80. *Prof Exp:* Res assoc biomed eng, Univ Wis-Madison, 75-80; asst prof, Tex Tech Univ, 80-81; proj specialist, Univ Wis-Madison, 81; DESIGN ENGR SOFTWARE, HEWLETT PACKARD, 82- *Res:* Respiratory physiology; developing hardware and software for signal processing of biological signals; programming and microcomputer applications in medicine. *Mailing Add:* Box 75018 Seattle WA 98125

BHANU, BIR, b Etah, UP, India, Jan 8, 51; US citizen; m 82; c 2. MACHINE INTELLIGENCE, SIGNAL & IMAGE PROCESSING. *Educ:* Banaras Hindu Univ, India, BS, 72; Birla Inst Technol & Sci,Pilani, India, ME, 74; Mass Inst Technol, Cambridge, SM & EE, 77; Univ Southern Calif, Los Angeles, Phd(elec eng), 81; Univ Calif, Irvine, MBA, 84. *Prof Exp:* Lectr electronics eng, Birla Inst Technol & Sci, Pilani, India, 74-75; acad assoc res, comput sci dept, IBM Res Lab, San Jose, Calif, 78; res fel, Nat Inst Res Informatics & Automation, France, 80-81; eng specialist res & develop, Ford Aerospace & Communications Corp, 81-84; asst prof comput sci, Univ Utah, 84-86, dir, Grad Admis, 85-86; sr Honeywell fel, Honeywell Systs & Res Ctr, 86-91; CONSULT, 91- *Concurrent Pos:* Consult, Evolving Technol Inst, San Diego, 83-85; Sch Med, Univ Calif, Irvine, 83-84 & Bonneville Sci, Salt Lake City, 85-86; prin investr, Dept Defense, NASA, NSF, Defense Advan Res Projs, Ford & var other agencies, 85- *Mem:* Sr mem Inst Elec & Electronics Engrs; Sigma Xi; Am Asn Artificial Intel; Asn Comput Mach; Soc Photo-Optical & Instrumentation Engrs; Int Asn Pattern Recognition. *Res:*

Scientific visualization; navigation; outdoor robotics; computer vision; machine learning; signal and image processing; computer graphics; obstacle detection and avoidance; multisensor integration; photointerpretation; automatic target recognition and technology transfer. *Mailing Add:* Col Eng Univ Calif Riverside CA 92521-0425

BHAPKAR, VASANT PRABHAKAR, b India, Apr 8, 31; m 61; c 3. MATHEMATICAL STATISTICS. *Educ:* Univ Bombay, BSc, 51, MSc, 53; Univ NC, PhD(math statist), 59. *Prof Exp:* Lectr statist, Univ Poona, 54-60, reader, 60-68, prof, 68-72; assoc prof, 69-73, interim chmn dept, 79-81, PROF STATIST, UNIV KY, 73- *Concurrent Pos:* Vis assoc prof statist, Univ NC, 64-66; vis sr prin res sci, Commonwealth Sci & Indust Res Orgn, Australia, 78; vis prof, Dept Biostatist, Univ Mich, 84-85. *Mem:* Inst Math Statist; fel Am Statist Asn; Int Statist Inst. *Res:* Categorical data, non-parametric methods in statistics; multivariate analysis; statistical inference. *Mailing Add:* Dept Statist Univ Ky Lexington KY 40506-0027

BHAPPU, ROSHAN B, b Karachi, Pakistan, Sept 14, 26; nat US; m 52; c 3. METALLURGICAL ENGINEERING. *Educ:* Colo Sch Mines, ME & MS, 50, DSc, 53. *Prof Exp:* Proj engr, Colo Sch Mines Res Found, 53-55; metallurgist, Miami Copper Co, Ariz, 55-59; sr metallurgist, Bur Mines & Mineral Resources, NMex Inst Mining & Technol, 59-72, res prof metall & mat eng, 67-72, actg chmn dept, 71-72; gen mgr & vpres, 72-, PRES, MOUNTAIN STATES RES & DEVELOP. *Concurrent Pos:* Consult, Pakistan Industs, Ltd, 50 & 54; tech expert, UNESCO, Turkey, 69-70; vpres res, NMex Inst Mining & Technol Res Found, 71- *Mem:* Am Chem Soc; Am Inst Mining, Metall & Petrol Engrs. *Res:* Extractive metallurgy and mineral beneficiation plant design and construction. *Mailing Add:* PO Box 310 Vail AZ 85641-0310

BHARADVAJ, BALA KRISHNAN, b Cuddalore, Tamil Nadu, Dec 22, 52; m 78; c 2. UNSTEADY AERODYNAMICS, AEROLASTICITY,. *Educ:* Indian Inst Technol, Madras, BTech, 74; Ga Inst Technol, Atlanta, MS, 75, PhD(aerospace eng), 79. *Prof Exp:* Res asst fluid mech, Sch Aerospace Eng, Ga Inst Technol, Atlanta, 74-79, post-doctoral res fel fluid mech, 79-80; asst prof mech & aero eng, Fluid Mech, Instrumentation, Propulsion, Rotor Aerodyn, Col Eng, Boston Univ, 80-87; prin eng scientist, 87-91, SR PRIN ENG SCIENTIST, DOUGLAS AIRCRAFT CO, LONG BEACH, CALIF, 91- *Concurrent Pos:* Co-prin investr grant, NIH, 81-83; vis prof, Univ Rome, Italy, 83; consult, Inst Comput Appln Res & Utilization Sci, Inc, Lincoln, Mass, 83-87; adj, prof, Calif Polytechnic, Pomona, 90- *Honors & Awards:* Monie A Ferst Award, Sigma Xi, 80. *Mem:* Fel Am Inst Aeronaut & Astronaut; Sigma Xi; Am Helicopter Soc. *Res:* Computational fluid dynamics applied to the study of viscous flows past airplanes and helicopter rotors; flow through the human arteries and the human lungs; unsteady aerodynamics of airplane wings and complete airplanes; aeroelastic behaviour of airplane wings and complete airplanes; finite difference and integral equation methods. *Mailing Add:* 5535 Via Verano Yorba Linda CA 92687-4931

BHARADWAJ, PREM DATTA, b Gorakhpur, India, May 20, 31; m 49; c 4. NUCLEAR PHYSICS. *Educ:* NREC Col, India, BSc, 50; Agra Univ, MSc, 52; State Univ NY Buffalo, PhD, 64. *Prof Exp:* Asst prof physics, BR Col, India, 52-54; lectr physics, GPI Col, 54-56 & Govt Col, Meerut, 56-59; asst prof physics, BR Col, 59-60; from asst prof to assoc prof, 62-66, chmn dept, 76-86, PROF PHYSICS, NIAGARA UNIV, 66- *Res:* High energy physics; theoretical physics. *Mailing Add:* Dept Physics Niagara Univ Niagara Falls NY 14109

BHARATI, SAROJA, Nat US. CARDIOLOGY. *Educ:* Univ Madras, MB & BS, 66. *Prof Exp:* DIR, CONGENITAL HEART & CONDUCTION SYST CTR, HEART INST, CHILDREN CHRIST HOSP, 88-; PROF PATH, RUSH MED COL, CHICAGO, ILL, 89- *Concurrent Pos:* Assoc dir, Congenital Heart Dis Res & Training Ctr, Hektoen Inst Med Res, 76-82; res prof med, Abraham Lincoln Sch Med, Univ Ill, 82-87; consult, JACC, chest circulation & Am J Cardiol, 77-; clin prof path, Milton S Hershey Med Ctr, Penn State Univ Med Sch, 83-88, Temple Univ, 83-88, Univ Med & Dent NJ-Robert Wood Johnson Med Sch, 86-88; chair, Dept Path, Deborah Heart & Lung Ctr, Browns Mills NJ, 82-88. *Mem:* Am Heart Asn; AMA; fel Am Col Cardiol; fel Am Col Chest Physicians. *Res:* Pathology of congenital heart disease; conduction system of the heart. *Mailing Add:* Congenital Heart & Conduction Syst Ctr 11745 Southwest Hwy Palos Heights IL 60463

BHARGAVA, HEMENDRA NATH, b Delhi, India, Sept 30, 42. PHARMACOLOGY. *Educ:* Banaras Hindu Univ, India, BPharm, 63, MPharm, 65; Univ Calif, San Francisco, PhD(pharm chem), 69. *Prof Exp:* Fel pharmacol, Univ Calif, San Francisco, 69-72, res pharmacologist, 72-75, lectr, 74-75; from asst prof to assoc prof, 75-81, PROF PHARMACOL, UNIV ILL MED CTR, 81-, AFFIL MED CHEM, 80- *Concurrent Pos:* Res grants, Ill Dept Ment Health & Develop Disabilities & Nat Inst Drug Abuse, Am Heart Asn; univ grants comn fel, Govt India, 63-65. *Honors & Awards:* Madan Mohan Malviya Prize, Banaras Hindu Univ, 63, Univ Gold Medal, 65; Travel Award, Am Soc Pharmacol & Exp Therapeut, 78. *Mem:* Am Soc Pharmacol & Exp Therapeut. *Res:* Mechanisms in narcotic tolerance and physical dependence, endorphins and other opiate peptides, drug metabolism, drug effects in disease states, neuropharmacology; treatment of parkinsonism, hypertension and tardive dyskinesias; pharmacology of hypothalamic peptide hormones; aging; receptors. *Mailing Add:* Dept Pharmacodynamics Health Sci Ctr Univ Ill 833 S Wood St Rm 310 Chicago IL 60612

BHARGAVA, HRIDAYA NATH, b Jodhpur, India, Dec 24, 35; US citizen; m 65; c 2. PHARMACEUTICAL & FOOD PROCESSING. *Educ:* Saugor Univ, India, BS, 58; NDak State Univ, PhD(med chem), 65. *Prof Exp:* Group leader prod & process develop, Revlon Inc, NY, 70-72; res fel prod develop, Merck & Co, Inc, Rahway, NJ, 72-79; PROF INDUST PHARM, MASS COL PHARM, 79- *Concurrent Pos:* Consult, Pfeiffer Pharmacuet Sci Labs, 79-, Stiefel Res Labs, 80-, Repub Surinam & World Health Orgn, 81-& Agency Int Develop, 85; prin investr, res grants, NIH, 81- *Mem:* Am Pharmaceut Asn; Acad Pharmaceut Sci; Soc Cosmetic Chemists. *Res:*

Pharmaceutical dosage form design and development; application of surfactants in solubilization of drugs; pharmaceutical process evaluation; hydrocolloids in emulsions and suspensions; evaluation of excipeints in solid pharmaceuticals. *Mailing Add:* Dept Pharm Mass Col Pharm 179 Longwood Ave Boston MA 02115

BHARGAVA, MADHU MITTRA, PROTEIN CHEMISTRY. *Educ:* Bombay Univ, India, PhD(biochem), 68. *Prof Exp:* ASSOC PROF BIOCHEM, ALBERT EINSTEIN COL MED, 80- *Mailing Add:* Dept Biochem Albert Einstein Col Med 1300 Morris Park Ave Bronx NY 10461

BHARGAVA, RAMESHWAR NATH, b Allahabad, India, Dec 25, 39; m 65; c 2. PHYSICS, SOLID STATE PHYSICS. *Educ:* Univ Allahabad, BS, 57, MS, 59; Columbia Univ, PhD(physics), 66. *Prof Exp:* Res asst physics, Watson Labs, Int Bus Mach Corp, 62-66, consult, Watson Res Ctr, 66-67; mem tech staff, Bell Tel Labs, 67-70; mem tech staff, Philips Labs Div, NAm Philips Corp, 70-74, sr prog leader, 74-78, dept head, 78-88, ASSOC DIR, PHILIPS LABS DIV, NAM PHILIPS CORP, 88- *Mem:* Fel Inst Elec & Electronics Engrs; fel Am Phys Soc. *Res:* Galvanomagnetic properties of semimetals and semiconductors; optical properties of semiconductors primarily in gallium arsenide, gallium phosphide and II-VI semiconductors; device work in gallium phosphide and zinc selenide diodes; deep states and nonradiative processes in semiconductors; non-linear optics; superconductivity; nanocomposites. *Mailing Add:* Philips Lab 345 Scarborough Rd Briarcliff Manor NY 10510

BHARGAVA, TRILOKI NATH, b Lucknow, India, Aug 21, 33; US citizen; c 5. APPLIED STATISTICS. *Educ:* Lucknow Univ, BSc, 52, MSc, 54; Mich State Univ, PhD(math & statist), 62. *Prof Exp:* Lectr math & statist, Khalsa Col, India, 54-55; lectr, Gujerat Univ, India, 55-57; asst prof, Agra Univ, 57-58; res asst statist, Mich State Univ, 59-61; from asst prof to assoc prof, Kent State Univ, 62-67, PROF MATH & STATIST, KENT STATE UNIV, 67- *Concurrent Pos:* NASA grant, Kent State Univ, 63-67, NSF grant, 64, Environ Protection Agency res grant, 71-; res assoc, Ctr Urban Regionalism, Kent State Univ, 70- *Mem:* Fel Royal Statist Soc; Am Math Soc; Am Statist Soc; Inst Math Statist; Math Asn Am. *Res:* Applied probability and statistics; graph theory; binary systems; ecological systems. *Mailing Add:* Dept Math Kent State Univ Kent OH 44242

BHARJ, SARJIT SINGH, b Kitale, Kenya, Mar 31, 50; US citizen; m; c 4. MONOLITHIC MICROWAVE INTEGRATED CIRCUITS, MICROWAVE ENGINEERING. *Educ:* King's Col, London, BSc, 73; Univ Col, London, MSc, 74, dipl, 74. *Prof Exp:* Appln engr components, Microwave Assocs, 73-76; chief engr systs, Vega Cantley Instruments, 76-78; group leader res, Marconi Space & Defense Ltd, 78-82; proj mgr amplifiers, Microwave Semiconductor Corp, 82-85; mgr MMIC's Anadigics Inc, 85-89; MEM TECH STAFF RES, DAVID SARNOFF RES CTR, 89- *Concurrent Pos:* Prin investr, Small Sus Innovation Res Prog contracts, 85-89; proj mgr, govt contracts, 85-; vchmn, Packaging Panel, Microwave Theory & Technique, 88, chmn, Spread Spectrum Panel, 89, Spread Spectrum Workshop, 90; reviewer, Microwave Theory & Technique Publ. *Mem:* Inst Elec Eng; sr mem Inst Elec & Electronics Engrs. *Res:* Monolithic microwave integrated circuit design; development of novel planar components; advancement of evanescent mode waveguide components. *Mailing Add:* 118 Robin Dr Mercerville NJ 08619

BHARTENDU, SRIVASTAVA, b Banda, India, Dec 15, 35; Can citizen; m 64; c 2. METEOROLOGY, ATMOSPHERIC PHYSICS. *Educ:* Univ Allahabad, BSc, 55, MSc, 58; Univ Sask, PhD(atmospheric physics), 64. *Prof Exp:* Res asst physics, Univ Sask, 60-64; fel, NMex Inst Mining & Technol, 65-66; res scientist, 66-82, METEOROLOGIST, ATMOSPHERIC ENVIRON SERV, DEPT ENVIRON, CAN, 82- *Concurrent Pos:* Fel, Univ Sask, 66; mem working group joint comt atmospheric elec, Int Asn Meteorol & Atmospheric Physics & Int Asn Geomagnetism & Aeronomy, 67-71; mem subcomn II, Int Comn Atmospheric Elec, Int Asn Meteorol & Atmospheric Physics, 71-; mem, Can Comt Atmospheric Elec, 72-75, chmn, 75-77; mem, Can Nat Comt, Int Union Radio Sci, 74-80; vis scientist, Nat Phys Lab, New Delhi, India, 77-78; vis prof, Merut Univ, Meerut, India, 80; co-chmn, Worging Group Air, Phase Two, Environ Audit East Bayfront/Port Indust Area, Royal Comn Future Toronto Waterfront, 90-91. *Mem:* Am Geophys Union; Am Meteorol Soc; Can Meteorol Soc. *Res:* Atmospheric electricity including its relationship with meteorology, fair weather atmospheric electricity, biological effects of atmospheric electricity; atmospheric acoustics, including thunder, gravity and pressure waves from nuclear and chemical explosions; applied climatology, including socioeconomic impacts; carbon dioxide induced climate change impacts; agrometeorology; air pollution meteorology, biometerology, long range transport of air pollutant (acid rain). *Mailing Add:* Atmospheric Environ Serv Dept Environ 25 St Clair Ave E 3rd Floor Toronto ON M4T 1M2 Can

BHARTIA, PRAKASH, b Calcutta, India, Jan 6, 44; Can citizen; m 71; c 2. MICROSTRIP ANTENNAS, MICROWAVE & MILLIMETER WAVE CIRCUITS. *Educ:* Indian Inst Technol, Bombay, BTech Hons, 66; Univ Man, Winnipeg, MSc, 68, PhD(elec eng), 71. *Prof Exp:* Spec lectr & asst prof elec eng, Univ Regina, 73-75, asst dean & assoc prof, 75-77; scientist, Defence Res Estab Ottawa, 77-82, head, Electromagnetics Sect, 82-85, dir, Aerospace Res & Develop, Nat Defence Hq, 85-86, dir, Res & Develop Commun & Space, Ottawa, 86- 89, DIR, SONAR DIV, DEFENCE RES ESTAB ATLANTIC, 89- *Concurrent Pos:* Consult, Winter Haven Indust, Gerling Moore Inc & Lockheed Aircraft & Missiles, 71-; mem, bd gov, Int Microwave Power Inst, 75-77 & Can-US Defense Develop & Prod Sharing, 86-89; adj prof elec eng, Univ Ottawa, 77-; Can deleg, various comts, NATO, 78-; nat leader, Panel 3 Physics & Electronics, NATO, 86- *Mem:* Fel Inst Elec & Electronics Engrs; fel Inst Elec & Telecommun Engrs; Electromagnetics Acad; Int Union Radio Scientists; Int Microwave Power Inst; Inst Elec & Electronics Engrs Microwave Theory & Tech Soc; Inst Elec & Electronics Engrs Antennas & Propagation Soc. *Res:* Millimeter wave technology and microstrip antennas; millimeter wave and microwave circuits; microwave, millimeter wave monolithic integrated circuits, devices, components and systems; author of over 100 publications and co-author of five books. *Mailing Add:* Sonar Div Defence Res Estab Atlantic Dartmouth NS B2Y 3Z7 Can

BHARUCHA, KEKI RUSTOMJI, b Bombay, India, Feb 4, 28. ORGANIC CHEMISTRY. *Educ:* Univ Bombay, BSc, 46, MSc, 49; Univ London, PhD(org chem) & DIC, 52. *Prof Exp:* Res assoc, Univ Toronto, 52-53; SR RES CHEMIST, SR SCIENTIST, GROUP LEADER & RES MGR, RES LABS, CAN PACKERS, LTD, 53- *Mem:* Am Chem Soc. *Res:* Synthetic organic chemistry. *Mailing Add:* 236 Lonsmount Dr Toronto ON M5P 2Z1 Can

BHARUCHA, NANA R, b Bombay, India, Oct 20, 26; m 76; c 3. ELECTROCHEMISTRY, SURFACE CHEMISTRY. *Educ:* Univ Bombay, BSc, 46, MSc, 48, PhD(chem), 50; Univ Manchester, PhD(chem technol), 53. *Prof Exp:* Res chemist, Paint Res Sta, Eng, 54-60; sr investr chem, Brit Non-Ferrous Metals Res Asn, 60-66, head appl chem div, 66-68; group leader, Noranda Res Ctr, 68-69, head Electrochem & Corrosion Dept, 69-70, mgr, Res Div, 70-85; res assoc, McGill Univ, 81-84; CONSULT, 85- *Honors & Awards:* Gold Medal Award, Am Electroplaters Soc, 73; Dr D J I Evans Award, Am Inst Mining, Metall & Petrol Eng, 83. *Res:* Corrosion; chemical metallurgy; metal finishing. *Mailing Add:* 376 Princess Louise Dr Orleans ON K4A 1W7

BHASIN, MADAN M, b Lahore, India, June 23, 38; US citizen; m 61; c 2. PHYSICAL CHEMISTRY, SURFACE CHEMISTRY. *Educ:* Univ Delhi, BSc, 58; Notre Dame Univ, PhD(phys chem), 64. *Prof Exp:* Teaching asst phys chem, Ind Univ, 59-60; res fel, Univ Notre Dame, 60-63; proj chemist, Union Carbide Corp, 63-69, proj scientist, 69-77, res scientist, 77-81, sr res sci & group supvr, 81-88, CORP FEL, UNION CARBIDE CORP, 88- *Concurrent Pos:* Atomic energy estab of India, 58-59. *Honors & Awards:* Tech Achievement Award, Union Carbide Corp, 85. *Mem:* Am Chem Soc; NAm Catalysis Soc. *Res:* Heterogenous catalysis; selective oxidation of olefins and paraffins; selective hydrogenations & hydrocracking; spectroscopic investigation of catalyst surfaces; AUGER; electron spectroscopy for chemical analysis; ion-scattering spectrometry; secondary ion mass spectrometry; diffusion in porous catalysts; membrane diffusion. *Mailing Add:* Indust Chem Div Union Carbide C & P Inc PO Box 25303 South Charleston WV 25303

BHASKAR, SURINDAR NATH, b Rasul, India, Jan 7, 23; nat US; m 50; c 3. PATHOLOGY, PERIODONTICS. *Educ:* Punjab Univ, BDS, 42; Northwestern Univ, DDS, 46; Univ Ill, MS, 48, PhD(anat), 51; Am Bd Oral Path & Am Bd Oral Med, dipl. *Prof Exp:* Instr oral path & histol, Col Dent, Univ Ill, 51-52, assoc prof path, 52-55; chief oral tumors br, Armed Forces Inst Path, US Army, 55-60, chief, Dept Oral Path, 60-70, dir, Army Inst Dent Res, Walter Reed Army Med Ctr, 70-73, dir personnel, Off Surgeon Gen, 73-75, asst surgeon gen & chief dent corps, 75-78; PVT PRACT PERIDONT, 78- *Concurrent Pos:* Prof, Sch Dent & Med, Georgetown Univ. *Mem:* Fel AAAS; fel Am Acad Oral Path; Am Acad Oral Med; Am Dent Asn; Int Acad Oral Path; Am Acad Periodont. *Res:* General and oral pathology; human and experimental tumors of salivary glands; oral tumors and oral diseases; author of four text books on histology, pathology, radiology. *Mailing Add:* 333 El Dorado Monterey CA 93940

BHASKARAN, GOVINDAN, b Mavelikara, India, Feb 12, 35; m 62; c 2. INSECT ENDOCRINOLOGY, DEVELOPMENTAL BIOLOGY. *Educ:* Univ Kerala, BSc, 55, MSc, 57; Univ Bombay, PhD(zool), 62. *Prof Exp:* Asst res officer, Indian Coun Med Res, Haffkine Inst, Bombay, 61; sci officer, Bhabha Atomic Res Ctr, Trombay, 61-68; res assoc entom, Univ Ill, Urbana-Champaign, 68-70; sr scientist, Biol Dept, Zoecon Corp, 70-73; sr scientist, Inst Develop Biol, 73-85, assoc prof biol, 75-81, interim dir, 83-85, PROF BIOL, TEX A&M UNIV, 82- *Mem:* AAAS; Soc Develop Biol; Am Soc Zoologists; Entom Soc Am. *Res:* Insect hormones and development; regulation of juvenile hormone biosynthesis. *Mailing Add:* Dept Biol Sci Tex A&M Univ College Station TX 77843

BHAT, GOPAL KRISHNA, b India, Dec 31, 25; US citizen; m 50; c 3. METALLURGICAL ENGINEERING, MATERIALS SCIENCE. *Educ:* Banaras Hindu Univ, India, BSc, 48; Lehigh Univ, MS, 51, PhD(metall eng), 55. *Prof Exp:* Staff metallurgist high strength steels, Crucible Steel Res Div, Colt Industs, 55-58; fel & head Scaife fellowship metals technol, 58-65, sr fel & head missile mfg eng & technol, 65-72, sr fel & mgr mat process eng, 72-76, DIR METALL & MAT TECHNOL, CARNEGIE-MELLON INST RES, 76- *Concurrent Pos:* Spec adv, US Navy Spec Proj Off, 60-62; tech consult, Titan Proj, Lockheed Aircraft Corp, 62-66 & Fast Deployment Logistics Ships Proj, 66-68; mfg tech consult, Air Force Mat Lab, Dayton, 66-74; mat consult, Space Shuttle & Lunar Explorer Projs, NASA, 68-71; vpres & tech dir, Electroslag Inst, 68-71; tech dir, Nutek Mat Inc, 71-76; mem ad hoc comt, Nat Mat Adv Bd, 73-74. *Mem:* Fel Am Soc Metals; Am Vacuum Soc; The Metals Soc; Am Welding Soc; Sigma Xi. *Res:* Plasma heat application technology; electroslag melting, welding, casting technology; materials fabrication and manufacturing technology. *Mailing Add:* 172 Boxfield Rd Pittsburgh PA 15241

BHAT, MULKI RADHAKRISHNA, b Mulki, India, May 7, 30; m 67; c 1. NUCLEAR PHYSICS. *Educ:* Univ Bombay, BSc, 51; Univ Poona, MSc, 54; Ohio State Univ, PhD(nuclear physics), 61. *Prof Exp:* Nat Res Coun Can fel, 61-63; res assoc neutron physics, 64-66, from asst physicist to assoc physicist, 66-73, physicist, 73-78, SR PHYSICIST, BROOKHAVEN NAT LAB, 78- *Mem:* AAAS; Am Phys Soc. *Res:* Neutron physics; measurement of resonance parameters and neutron capture gamma rays; nuclear spectroscopy; neutron cross-section evaluation; nuclear structure and decay data. *Mailing Add:* Brookhaven Nat Lab Upton NY 11973

BHAT, RAMA B, b Badur, Kerala, India, Oct 2, 43; Can citizen; m 74; c 2. ROTOR DYNAMICS, RANDOM VIBRATIONS. *Educ:* Mysore Univ, BE, 66; Indian Inst Technol, MTech, 68, PhD(mech eng), 73. *Prof Exp:* Scientist, Indian Space Res Orgn, 72-76; res assoc, NASA Langley Res Ctr, Hampton, Va, 76-79; from asst prof to assoc prof, 79-88, PROF MECH ENG, CONCORDIA UNIV, 88- *Concurrent Pos:* Alt mem, Mach Dynamics Subcomt, Nat Res Coun Can, 84- *Mem:* Am Soc Mech Engrs; Inst Engrs

India; Can Soc Mech Eng. *Res:* Dynamic behavior of rotors supported on fluid film bearings; geared rotors, vibration of plates and beams using beam characteristic orthogonal polynomials; response of structures to random excitations; model testing and analysis; structural acoustics. *Mailing Add:* Mech Eng Concordia Univ 1455 De Maisonneuve Blvd W Montreal PQ H3G 1M5 Can

BHAT, UGGAPPAKODI NARAYAN, b Vittal, India, Nov 17, 33; m 59; c 2. APPLIED PROBABILITY, STATISTICS. *Educ:* Univ Madras, BA, 53, BT, 54; Karnatak Univ, India, MA, 58; Univ Western Australia, PhD(math statist), 64. *Prof Exp:* Teacher high sch, India, 54-56; lectr statist, Karnatak Univ, India, 58-61; asst math, Univ Western Australia, 61-64, temp lectr, 64-65; asst prof statist, Mich State Univ, 65-66; from asst prof to assoc prof opers res, Case Western Reserve Univ, 66-69; from assoc prof to prof opers res & statist, Southern Methodist Univ, 69-79, head dept, 75-79, assoc dean, 76-79, chmn, Div Math Sci, 79-80, vprovost & dean grad studies, 80-82, chmn, dept statist sci, 87-89, PROF STATIST, SOUTHERN METHODIST UNIV, 79-; ASSOC DEAN ACAD AFFAIRS, DEDMAN COL, 89- *Concurrent Pos:* Assoc ed, Opsearch, 68-, Opers Res, 68-75 & Mgt Sci, 69-74, queueing systems, 86- *Mem:* Fel Am Statist Asn; Inst Math Statist; Opers Res Soc Am; Inst Mgt Sci; Sigma Xi. *Res:* Queueing theory; probabilistic models for computer and information systems; applied probability; stochastic processes. *Mailing Add:* Dept Statist Sci Southern Methodist Univ Dallas TX 75275

BHAT, VENKATRAMANA KAKEKOCHI, b Padre, India, July 26, 33; US citizen; m 67; c 2. PHARMACEUTICAL CHEMISTRY, MEDICINAL CHEMISTRY. *Educ:* Univ Madras, BSc, 55, BPharm, 58; Univ Wash, PhD(pharmaceut chem), 63. *Prof Exp:* Res assoc carbohydrate chem, Georgetown Univ, 63-67; pharmaceut chemist, Gulf South Res Inst, 67-71; CONSULT CHEMIST, 71- *Mem:* AAAS; Am Chem Soc; Am Inst Chemists; NY Acad Sci; Sigma Xi. *Res:* Synthesis of sulfur containing compounds related to Ephedrine; synthesis of nucleosides, carbohydrates and antimalarials; analytical biochemistry, especially analysis of drugs, herbicides and pesticides using gas chromatography. *Mailing Add:* 8002 53rd Ave W Mukilteo WA 98275-2630

BHATHENA, SAM JEHANGIRJI, b Bombay, India, Sept 18, 36; US citizen; m 75. ENDOCRINOLOGY, HUMAN NUTRITION. *Educ:* Univ Bombay, BSc, 61, MSc, 64, PhD(biochem), 70. *Prof Exp:* Vis fel biochem, NIH, 71-73, vis assoc, 74; res biochemist, Diabetes Res Lab, Vet Admin Med Ctr, Washington, DC, 75-82; RES CHEMIST, CARBOHYDRATE NUTRIT LAB, BELTSVILLE HUMAN NUTRIT RES CTR, MD, 83- *Concurrent Pos:* Res instr, dept med, Georgetown Univ, Washington, DC, 79-80, res asst prof, 81-83, adj asst prof, 83- *Mem:* Endocrine Soc; Am Diabetes Asn; NY Acad Sci; AAAS; Soc Exp Biol & Med; Am Fedn Clin Res. *Res:* Study of dietary regulation of peptide hormone receptors; endocrine aspects of nutritional disorders; role of opiates, hormones and neuropeptides in the control of food intake and satiety. *Mailing Add:* Carbohydrate Nutrition Lab Beltsville Human Nutrit Res Ctr Bldg 307 Rm 315 Beltsville MD 20705

BHATIA, ANAND K, b W__Pakistan, Jan 26, 34; m 63; c 1. ATOMIC PHYSICS. *Educ:* Univ Delhi, BSc, 53, MSc, 55; Univ Md, PhD(physics), 62. *Prof Exp:* Asst prof physics, Wesleyan Univ, 62-63; res assoc, Nat Acad Sci, 63-65; AEROSPACE TECHNOLOGIST, GODDARD SPACE FLIGHT CTR, NASA, 65- *Res:* High energy physics; molecular physics. *Mailing Add:* Goddard Space Flight Ctr NASA Code 681 Greenbelt MD 20771

BHATIA, DARSHAN SINGH, b Lahore, India, Apr 12, 23; m 45; c 3. FOOD TECHNOLOGY, NUTRITION. *Educ:* Punjab Univ, BSc, 43, MSc, 45; Mass Inst Technol, ScD, 50. *Prof Exp:* Asst dir, Cent Food Technol Res Inst, Mysore, India, 50-64; chmn, Training Prog, FAO Ctr, Mysore, 64-65; tech mgr, Coca-Cola Export Corp, New Delhi, 65-72; DIR RES & DEVELOP, COCA-COLA CO, 72- *Concurrent Pos:* Mem panel experts, Adv Comt Appl Sci & Technol to Develop, UN, 67; mem, vis comt, Dept Nutrit & Food Sci, Mass Inst Technol, 75-78. *Mem:* Inst Food Technologists; Am Inst Nutrit; NY Acad Sci. *Res:* Food science and technology; process development, nutrition, management of research and development in food and allied fields. *Mailing Add:* 2290 Bohler Rd NW Atlanta GA 30327

BHATIA, KISHAN, b Poona, India, Mar 23, 36; c 2. ORGANIC CHEMISTRY, ANALYTICAL CHEMISTRY. *Educ:* Univ Poona, BSc, 58, MSc, 60; Univ Ark, PhD(org chem), 65. *Prof Exp:* Demonstr chem, Fergusson Col, Univ Poona, 59-60; jr res fel org chem, Nat Chem Lab, India, 60-61; asst, Univ Ark, 61-65; sr res chemist, US Steel Corp, 65-70; fel chem, Mellon Inst, Carnegie-Mellon Univ, 71-75; chromatography sect leader, Allied Chem Corp, 75-76; sr scientist, Pullman Kellogg, Pullman Inc, 76-80; DEVELOP SCIENTIST, N L TREATING CHEM, EXXON CHEM CO, 81- *Mem:* Am Chem Soc; Sigma Xi; Nat Asn Corrosion Engrs; Soc Petrol Engrs. *Res:* Feedstock characterization technology; advanced process instrumentation technology; petroleum and solid fuel chemistry, radiation chemistry, organic chemistry; design and development of chrometographic instrumentations, sectrometry, isotope techniques; computerization and automation technology; corrosion engineering; gas processing technology; corrosion inhibitors and other chemicals for treatment of produced gas and fluids; refinery and petro chemicals; synthetic fuels. *Mailing Add:* 20222 Lavereton Katy TX 77450

BHATIA, NAM PARSHAD, b Lahore, India, Aug 24, 32; m 62; c 3. MATHEMATICS. *Educ:* Agra Univ, BSc, 52, MSc, 54 & 56; Dresden Tech Univ, Dr rer nat(math), 61. *Prof Exp:* Lectr math, REI Degree Col, Agra Univ, 55-56; asst prof math, Birla Col, India, 56-58; aspirant, Dresden Tech Univ, 58-61; asst prof math, Birla Col, India, 61-62; vis mathematician, Res Inst Advan Studies, div Martin Co, Md, 62-63; from asst prof to assoc prof math, Case Western Reserve Univ, 63-68; vis assoc prof, dept math & inst fluid dynamics & appl math, Univ Md, College Park, 68-69, PROF MATH, UNIV MD, BALTIMORE COUNTY, 69- *Concurrent Pos:* NSF res grant, 64-73; ed, Math Systs Theory, 68-72. *Res:* Dynamical and semi-dynamical systems; theory and application of ordinary differential equations; control theory; stability theory; chaos theory. *Mailing Add:* Dept Math Univ Md Baltimore Co 5401 Wilkens Ave Catonsville MD 21228

BHATIA, SHYAM SUNDER, b Rawalpindi, India, July 7, 24; US citizen; m 50; c 2. POPULATION GEOGRAPHY,. *Educ:* Univ Panjab, BSc, 43, MA, 47; Univ Kans, PhD(geog), 59. *Prof Exp:* Lectr geog, Univ Col Panjab, New Delhi, 48-56; reader, Univ Delhi, 59-66; assoc prof, Univ Wis-Oshkosh, 66-70, PROF GEOG, UNIV WIS-OSHKOSH, 70- *Concurrent Pos:* Reader, Indian Sch Int Studies, 60-64; exchange vis prof, San Diego State Univ, 75-76. *Mem:* Asn Am Geogr; Sigma Xi. *Res:* Population; agricultural geography; geography of South Asia. *Mailing Add:* Dept Geol Univ Wis Oshkosh 800 Algoma Blvd Oshkosh WI 54901

BHATLA, MANMOHAN N (BART), b India. ENGINEERING. *Educ:* Punjab Eng Col, India, BSc, 60, Univ Roorkee, India, ME, 61, Okla State Univ, PhD(sanitary & pub health), 64. *Prof Exp:* Proj mgr, Proj Eng Lab Dir, 64-70, tech mgr, Weston-Europe, 70-78, VPRES, INDUST WASTE MGT, WESTON, 78- *Honors & Awards:* George Bradley Gascoigne Medal, Water Pollution Control Fedn, 76. *Mem:* Am Acad Environ Engrs. *Res:* Management and supervision of environmental projects in solid waste and hazardous waste management and water pollution control field; author or coauthor of over 25 publications. *Mailing Add:* 706 Apricot Lane West Chester PA 19380

BHATNAGAR, AJAY SAHAI, b Muree, India, Sept 26, 42; m 70. ORGANIC CHEMISTRY, REPRODUCTIVE ENDOCRINOLOGY. *Educ:* Cambridge Univ, BA, 63, MA, 67; Univ Basel, PhD(org chem), 67. *Prof Exp:* Res assoc physiol, 68-70, ASST PROF OBSTET & GYNEC, MED COL VA, 70-, ASST PROF BIOCHEM, 73- *Concurrent Pos:* Fel org chem, Univ Basel, 67. *Mem:* AAAS; NY Acad Sci; Am Chem Soc; Swiss Chem Soc; Endocrine Soc. *Res:* Analytical methodology for the estimation of steroids; steroid metabolism in the female; endocrinology of pregnancy. *Mailing Add:* Dept Obstet & Gynec Va Commonwealth Richmond VA 23298

BHATNAGAR, DINECH C, b Lahore, WPakistan, Apr 14, 34; m 62; c 2. INORGANIC CHEMISTRY, ANALYTICAL CHEMISTRY. *Educ:* Univ Delhi, BSc, 53, MSc, 55; Wayne State Univ, PhD(chem), 63. *Prof Exp:* Res asst chem, Nat Chem Lab, India, 56-59; instr, Detroit Inst Technol, 62-63; fel, Univ Ariz, 63-65; asst prof, La Verne Col, 65-67; master, Sch Mines, Haileybury, Ont, Can, 67-68; MASTER CHEM, ALGONQUIN COL APPL ARTS & TECHNOL, 68- *Concurrent Pos:* Consult, Garett Res Corp, Calif, 65. *Mem:* Am Chem Soc; Chem Inst Can; Sigma Xi. *Res:* Optical rotary dispersion of inorganic complex ions; luminescence of metal complexes. *Mailing Add:* Dept Chem Algonquin Con 200 Lees Ave Ottawa ON K1S 0C5 Can

BHATNAGAR, GOPAL MOHAN, b Lucknow, India, July 15, 37; US citizen. MOLECULAR BIOLOGY. *Educ:* Lucknow Univ, India, BSc, 55, MSc, 57, PhD(biochem), 61. *Prof Exp:* Res assoc biochem, Chicago Med Sch, 63-65; res scientist protein chem, Commonwealth Sci & Indust Res Orgn, Melbourne, Australia, 65-69; staff scientist muscle proteins, Boston Biomed Res Inst, 69-72; assoc prof, Lucknow Univ, India, 73-74; prin assoc epidermal proteins, Sch Med, Harvard Univ, 74-77; assoc prof dermat, Sch Med, Johns Hopkins Univ, 77-87; SR CHEMIST, FOOD & DRUG ADMIN, ROCKVILLE, MD, 87- *Concurrent Pos:* vis scientist, Univ Montreal, 73-; prin investr, res grant, NIH, 76-86; consult, Food & Drug admin, 80-85; guest res scientist, Nat Inst Aging, NIH, 80-88. *Mem:* Am Soc Biol Chemists; AAAS; Soc Invest Dermat. *Res:* Molecular changes involved in cellular aging; epithelial differentiation and keratinization; biochemistry and physiology of cardiac muscle; molecular biology of epidermal proteins; regulation of wound healing devices. *Mailing Add:* Food & Drug Admin 1390 Piccard Dr Rockville MD 20850

BHATNAGAR, KUNWAR PRASAD, b Gwalior, India, Mar 21, 34; m 61; c 2. NEUROANATOMY, BIOLOGY OF BATS. *Educ:* Agra Univ, BSc, 56; Vikram Univ, India, MSc, 58; State Univ NY Buffalo, PhD(anat), 72. *Prof Exp:* From lectr to asst prof zool, Madhya Pradesh Educ Serv, India, 58-67; teaching asst anat, State Univ NY Buffalo, 68-72; from asst prof to assoc prof, 72-85, PROF ANAT, SCH MED, UNIV LOUISVILLE, 85- *Concurrent Pos:* Sabbatical leave, Max Planck Inst Brain Res, Frankfurt & Inst Anat, Essen, WGer, 78-79, 86-87; ed, Bat Research News, 82-86; guest ed, J Electron Micros Tech, 91. *Mem:* Am Soc Mammal; Am Asn Anatomists; Sigma Xi. *Res:* Mammalian olfaction; rhinencephalon in Chiroptera; biology of bats; pineal ultrastructure; human development. *Mailing Add:* Health Sci Ctr Dept Anat Scis & Neurobiol Univ Louisville Sch Med Louisville KY 40292-0001

BHATNAGAR, RAJENDRA SAHAI, b Lucknow, India, Mar 10, 36; m 66; c 2. BIOCHEMISTRY, BIOPHYSICS. *Educ:* Agra Univ, BS, 54, MS, 56; Duke Univ, MS, 63, PhD(biochem), 64. *Prof Exp:* Lectr, Govt Col, Rupar, 57-58; tech asst, Tech Develop Dept, Ministry Indust, 58-60; asst instr med & biochem, Sch Med, Univ Pa & Philadelphia Gen Hosp, 65-67; sr res assoc biochem, Med Sch, Northwestern Univ, 67-68; assoc prof, 68-74, PROF BIOCHEM, SCH DENT, UNIV CALIF, SAN FRANCISCO, 74- *Concurrent Pos:* Fel, Vet Admin Hosp, Hines, Ill, 64-65 & Sch Med, Univ Pa & Philadelphia Gen Hosp, 65-67; res career develop award, 69; vis prof biophys & biochem, Indian Inst Sci, Bangalore, India, 72-73. *Mem:* AAAS; Am Chem Soc; Am Soc Biol Chemists; NY Acad Sci; fel Am Inst Chemists. *Res:* Biology of connective tissue, especially the biosynthesis and regulation of collagen; structure and interactions of collagen. *Mailing Add:* Sch Dent 661 HSW Univ Calif San Francisco CA 94143-0515

BHATNAGAR, RANBIR KRISHNA, b India; US citizen. PHARMACOLOGY. *Educ:* Univ Lucknow, BSc, 54; Agra Univ, BVSc, 58; Mich State Univ, MS, 63, PhD(pharmacol), 71. *Prof Exp:* Veterinarian, Govt Uttar Pradesh, India, 58-59; res assoc physiol, Col Vet Sci, Mathura, India, 59-61; res assoc pharmacol, Univ Chicago, 65-67; res assoc, 71-72, from asst prof to assoc prof, 72-81, PROF PHARMACOL, UNIV IOWA, 81- *Concurrent Pos:* Assoc ed, J Pharmacol Methods. *Mem:* Am Soc Pharmacol & Exp Therapeut; Am Soc Neurochem; Soc Neurosci; Int Soc Neurochem. *Res:* Neurochemistry and neuropharmacology with particular emphasis on central nervous system neurotransmitters; factors which influence the growth and differentiation of neurons in central nervous system. *Mailing Add:* 2310 Bowen Sci Bldg Dept Pharmacol Univ Iowa Iowa City IA 52242

BHATNAGAR, YOGENDRA MOHAN, b Gorakhpur, UP, India, Feb 5, 45; m 74; c 1. MOLECULAR BIOLOGY, EMBRYOLOGY. *Educ:* Univ Lucknow, India, BSc, 62, MSc, 64; Boston Col, PhD(biol), 74. *Prof Exp:* Instr biol, Boston Col, 73-74; res fel physiol, Med Sch, Harvard Univ, 74-76, res assoc, 76-77; sr scientist, EIC Corp, Newton, Mass, 77-79; asst prof pediat, 79-80, asst prof anat, 80-84, ASSOC PROF ANAT, UNIV SALA, MOBILE, 84- *Concurrent Pos:* Res fel, Pop Coun, NY, 74-76; adj lectr, Northeastern Univ, Boston, Mass, 77-79; prin investr, NIH, 80-, res career develop award, 85-90. *Mem:* AAAS; Am Soc Biol Chemists; Soc Study Reprod. *Res:* Biology and biochemistry of mammalian reproduction. *Mailing Add:* Depts Struct & Cell Biol Col Med 2042 MSB Univ SAla Mobile AL 36688

BHATNAGER, DEEPAK, b India, Aug 5, 49; US citizen; m; c 1. PLANT PHYSIOLOGY. *Educ:* Univ Udaipur, India, BSc Hons, 72; Indian Agr Res Inst, New Delhi, MSc, 74, PhD(agr physics), 77. *Prof Exp:* ICAR jr res fel, Indian Agr Res Inst, New Delhi, 72-74, sr res fel, 74-77; officer, Dept Biophys, All India Inst Med Sci, New Delhi, 77-78; postdoctoral res fel, Dept Biol, Purdue Univ, West Lafayette, Ind, 79-81; sr res assoc, Dept Biochem, Sch Med, La State Univ, 81-85; GENETICIST, SOUTHERN REGIONAL RES CTR, USDA AGR RES SERV, NEW ORLEANS, 85- *Concurrent Pos:* Grants, Dr George S Bell Mem Res Award, Am Heart Asn, 82-83 & 83-84, US Dept Agr, Agr Res Serv Res Assoc Prog, 89-91; tech reviewer, J Exp Mycol. *Mem:* Am Soc Biochem & Molecular Biol; Am Chem Soc; Am Soc Plant Physiologists; Am Soc Microbiol; AAAS; Sigma Xi; NY Acad Sci. *Res:* Genetic manipulation of the pertinent fungal genome for the development of the long term procedures for control of the aflatoxin contamination process; utilization of natural products in control of aflatoxin biosynthesis by the fungus. *Mailing Add:* USDA Agr Res Ctr Southern Reg Res Ctr 1100 Robert E Lee Blvd New Orleans LA 70124

BHATT, GIRISH M, b Coimbatore, India, June 21, 46; m 75; c 2. STRUCTURE PROPERTY RELATIONS, PROCESSING-STRUCTURE RELATIONSHIPS. *Educ:* Univ Madras, India, BS, 68; Univ Conn, Storrs, MS, 70, PhD(chem eng & polymer sci), 74. *Prof Exp:* Res engr, Mill Spinning Dept, Am Enka, 70-71; asst prof man made fibers, Textile Tech Dept, Grad Sch, PSG Col Technol, India, 75-76; asst prof, Chem Eng Dept, Caimbatore Inst Technol, India, 77; vis prof, Dept Chem Eng, Univ Conn, 78; asst prof, Home Econ Dept, Univ Vt, 79-80; TECH DIR, JOHNSON FILAMENTS, INC, 79- *Mem:* Am Inst Chem Engrs; Soc Plastic Engrs. *Res:* Processing techniques and formulating polymer blends and alloys to yield high performance fibers and monofilaments. *Mailing Add:* Johnson Filaments Inc 36 Industrial Ave Williston VT 05495

BHATT, JAGDISH J, b Umreth, India, Feb 17, 39; US citizen; m 71; c 2. MARINE GEOLOGY. *Educ:* Univ Baroda, BSc Hons, 61; Univ Wis-Madison, MS, 63; Univ Wales, PhD(marine geol), 72. *Prof Exp:* Instr chem & phys sci, Jackson Community Col, Mich, 64-65; instr phys sci & geol, Okla Panhandle State Univ, 65-66; res scientist geol, Stanford Univ, 71-72; asst prof environ sci & oceanog, Rachel Carson Col, Univ Buffalo, 72-74; assoc prof, 74-84, PROF GEOL & OCEANOG, COMMUNITY COL RI, WARWICK, 84- *Concurrent Pos:* Mem exec comt, Gov Ocean Tech Task Force, RI, 75-76; co-chmn comt on joint offshore tech prog, Univ RI & RI Jr Col, 75-76; chmn comt ocean training prog, RI Jr Col, 77-78; mem sci adv comt, Pac Marine Tech, 90 & 92. *Res:* Geological and geochemical studies of marine sediments, notably carbonates; education techniques and programs for improving undergraduate training in marine, earth and environmental sciences; author of fifty articles including five books. *Mailing Add:* Community Col RI 400 East Ave Warwick RI 02886

BHATT, PADMAMABH P, b Bulsar, India, Aug 1, 57; m 83. TRANSDERMAL DRUG DELIVERY, PHYSICAL PHARMACY. *Educ:* Univ Bombay, BPharm, 80, MPharm, 84; Univ Kans, MS, 87, PhD(pharmaceut chem), 89. *Prof Exp:* Mfg pharmacist, Unique Pharmaceut Lab, 82-83; res pharmacist, Sandoz Ltd, India, 83; SR RES SCIENTIST, HERCON LABS, 89- *Mem:* Am Asn Pharmaceut Scientists; Controlled Release Soc. *Res:* Transdermal drug delivery. *Mailing Add:* 144 Cedar Village Dr York PA 17402

BHATT, PRAVIN NANABHAI, BIOLOGY, HOST RESPONSES. *Educ:* Banaras Hindu Univ, India, AMS, 48; Tulane Univ, MD, 64. *Prof Exp:* SR RES SCIENTIST, VIROL & MICROBIOL, SCH MED, YALE UNIV, 68- *Mailing Add:* Sect-Lab Sci-Med Sch Yale Univ 333 Cedar St New Haven CT 06510

BHATT, RAVINDRA NAUTAM, b New Delhi, India, Jan 20, 52; m 80; c 1. THEORETICAL PHYSICS, ENGINEERING PHYSICS. *Educ:* Univ Delhi, BSc, 71; Univ Ill, MS, 74, PhD(physics), 76. *Prof Exp:* Mem tech staff, AT&T Bell Labs, 76-89, dept head, Theoret Physics Res Dept, 89-90; PROF ELEC ENG & PHYSICS, PRINCETON UNIV, 90- *Concurrent Pos:* Fel, CNRS, Ecole Normale Superieure, Paris, 83; sr vis fel, Imp Col, London, Sci & Eng Res Coun, 84; vis, Inst Theoret Physics, Santa Barbara, 86; trustee, Aspen Ctr Physics; mem, Supercomputer Allocations Comt, Cornell Theory Ctr. *Mem:* Fel Am Phys Soc. *Res:* Metal insulator transitions; disordered systems; spin glasses; quantum fluids and solids; superfluid helium; high temperature superconductivity; charge density waves; structural phase transitions; doped semiconductors; expanded metals; correlated and magnetic systems. *Mailing Add:* Dept Elec Eng Princeton Univ Princeton NJ 08544

BHATTACHARJEE, HIMANGSHU RANJAN, b Comilla, Bangladesh, Mar 21, 42; US citizen; m 72; c 2. MATERIALS RESEARCH, OVERALL RESEARCH & DEVELOPMENT. *Educ:* Univ Dacca, BSc, 64; Univ Col Sci & Technol, Calcutta, MSc, 67; Wayne State Univ, PhD(chem), 73. *Prof Exp:* Postdoctoral res fel laser chem, Radiation Lab, Univ Notre Dame, 73-74; postdoctoral res fel laser spectros, Univ NB, Can, 74-75; res assoc & chem lectr, Univ Toronto, Can, 75-77; res assoc laser spectros, Univ Calif, Santa Cruz, 77-78; sr res chemist, Allied Signal, Inc, Morristown, NJ, 78-90; SR RES SCIENTIST, TEXWIPE CO, UPPER SADDLE RIVER, NJ, 90- *Concurrent Pos:* Prin investr several indust res & develop progs. *Mem:* Am Chem Soc; AAAS; Chem Soc Can. *Res:* Laser spectroscopy; printing plates; coatings; photochemistry; clean room products; indicators-time-temperature; lithium dosage; radiation monitoring; polymer modifications; polypeptides; biopolymers; 18 publications and 14 US patents. *Mailing Add:* 56 Randolph Ave Randolph NJ 07869

BHATTACHARJEE, JNANENDRA K, b Sylhet, Bangladesh, Feb 1, 36; m; c 2. MICROBIAL GENETICS. *Educ:* MC Col, Sylhet, BS, 57; Univ Dacca, MS, 59; Southern Ill Univ, PhD(microbiol), 66. *Prof Exp:* Res assoc microbiol, Albert Einstein Med Ctr, Pa, 65-66, asst mem, 66-68; assoc prof, 68-73, PROF MICROBIOL, MIAMI UNIV, 73- *Concurrent Pos:* NSF res grants, 69, 71, 75, 78 & 79; Lilly Res Found grant, 70, 72; S & H Found res grant, 74; NIH res grant, 85, NSF, 88. *Honors & Awards:* President's Award, President of Pakistan, 60. *Mem:* AAAS; Genetics Soc Am; Am Soc Microbiol; fel Am Acad Microbiol. *Res:* Genetics of yeast; biosynthetic mechanism of lysine and other related amino acids in yeast; regulation of gene action in eucaryotic organisms; cloning molecular studies and evolutionary diversity of lysine genes of yeast. *Mailing Add:* Dept Microbiol Miami Univ Oxford OH 45056

BHATTACHARJI, SOMDEV, b Calcutta, India, Apr 23, 32; US citizen; m 62; c 3. STRUCTURAL GEOLOGY. *Educ:* Univ Calcutta, BS, 50, Hons, 51; Indian Sch Mines, MS, 54; Univ Chicago, MS, 57, PhD(geol), 59. *Prof Exp:* Res assoc geol & instr ling, Univ Chicago, 59-60, res assoc & lectr S Asian lang, 60-61, asst prof, 61-62; fel geol, Nat Res Coun Can, 62-64; from instr to assoc prof geol, 64-72, dep chmn dept, 69-72 & 79-82, PROF GEOL, BROOKLYN COL, CITY UNIV NEW YORK, 72- *Concurrent Pos:* NSF grants, 66-68, 70-72, 74-75 & 80-83; City Univ New York fac res grant, 73-78; mem, Int Geodynamic Comt, 72; chmn, Geol Sci Sect, NY Acad Sci, 76-77; sr Fulbright fel to India & fel of Indo-US subcomn on Educ & Culture, 77-78; vis res fel, Princeton Univ, 77; City Univ New York grant, 79-81; Am Chem Soc grant, 80-82; fel, Indian Inst Penninsular Geol; adv, Comt Sci & Technol in Develop Countries Prog, Asia & India; fac res grant, City Univ New York, 79-82, 83-86 & 86-89; proj dir, Indo-US Coop Prog Crumansonata, 82-86; NSF grants, 83-86; UN Develop Prog consult, Nat Geophys Res Inst, India; selection comt mem, Indo-US Subcomn Educ & Culture, 87-89. *Mem:* Fel Geol Soc Am; fel Geol Soc India; Am Geophys Union; Geol Soc Edinburgh, Scotland; Int Soc Planetology; fel India Soc Earth Sci; fel India Soc Pennin Geol. *Res:* Tectonophysics; structural geology; petrogenesis; languages (Indic). *Mailing Add:* 685 Kelly Rd Arkville NY 12406

BHATTACHARYA, AMAR NATH, b Calcutta, India, Oct 1, 34; m 66; c 1. PHARMACOLOGY, PHYSIOLOGY. *Educ:* Bengal Vet Col, India, BVetS, 57; Ohio State Univ, MS, 63, PhD(pharmacol), 67. *Prof Exp:* State vet, Directorate Vet Serv, Govt WBengal, India, 57-59; demonstr, Dept Clin Vet Med & Pharmacol, Bengal Vet Col, 59-61; res asst pharmacol, Col Med, Ohio State Univ, 62-67, res assoc pharmacol & med, 68; asst prof, 70-73, ASSOC PROF PHARMACOL, OHIO NORTHERN UNIV, 74- *Concurrent Pos:* Ford Found fel physiol, Sch Med, Univ Pittsburgh, 68-70. *Mem:* Soc Study Reprod; Brit Soc Study Fertil; Indian Sci Cong Asn. *Res:* Neuroendocrine control mechanisms of corticotropin and gonadotropins, role of central catecholamines and serotonin; gonad-anterior pituitary feedback interrelationship in subhuman primates, studies with castration, steroids and psychotropic drugs; prostaglandins. *Mailing Add:* Pharmacol Dept Raabe Col PharmACOL Ohio Northern Univ Ada OH 45810

BHATTACHARYA, ASHOK KUMAR, b Allahabad, India, Oct 14, 35; m 64. NUCLEAR ENGINEERING, PHYSICS. *Educ:* Univ Lucknow, BSc, 54, MSc, 56; Univ Ill, Urbana-Champaign, PhD(nuclear eng), 66. *Prof Exp:* Res assoc gas electronics, Univ Ill, Urbana-Champaign, 66-67; RES PHYSICIST, GEN ELEC CO, 67- *Res:* Gaseous discharge; electronic atomic and molecular processes in a gaseous plasma. *Mailing Add:* Gen Elec Lighting Res Lab Nela Park Cleveland OH 44112

BHATTACHARYA, DEBANSHU, b Uttarpara, India, June 26, 47. METALLURGY, MATERIALS SCIENCE. *Educ:* Regional Eng Col, Durgapur, India, BE, 69; Wash State Univ, MS, 71, PhD(eng sci), 75. *Prof Exp:* Res assoc geosci, Pa State Univ, 75-76; res engr, 76-79, SR RES ENGR, INLAND STEEL RES LABS, 79- *Concurrent Pos:* Res & teaching asst, Wash State Univ, 60-75. *Honors & Awards:* Chipman Award, Iron & Steel Soc, 88. *Mem:* Am Soc Metals; Am Inst Metall Engrs. *Res:* Thermodynamics of alloys; machinability of materials; high temperature mechanical properties of metals; phase transformations in metals. *Mailing Add:* Inland Steel Res Labs 3001 E Columbus Dr East Chicago IN 46312

BHATTACHARYA, JAHAR, b Calcutta, India, Mar 12, 46; m 72; c 2. RESPIRATORY PHYSIOLOGY, MICROCIRCULATORY PHYSIOLOGY. *Educ:* All-India Inst Med Sci, MBBS, 69; Oxford Univ, MSc, 72, DPhil, 76. *Prof Exp:* Schorstein lectr, Regius Dept Med, Oxford Univ, 73-76; Neizer fel, Div Nephrology, Stanford Univ, 76-77; fel physiol, 77-80, ASST PROF, DEPT PHYSIOL & MED, UNIV CALIF, SAN FRANCISCO, 85- *Honors & Awards:* Res Career Develop Award, 85. *Mem:* Am Physiol Soc; Am Microcirculatory Soc. *Res:* Lung microcirculation to determine mechanisms of microvascular fluid flux by a new micropunctive technique. *Mailing Add:* Col P & S Columbia Univ Roosevelt Hosp 428 W 59th St New York NY 10019

BHATTACHARYA, MALAYA, b Cooch-behar, India, Jan 16, 46; m 72; c 2. IMMUNOLOGY, BIOCHEMISTRY. *Educ:* Presidency Col, Calcutta, India, BS, 63; Univ Col Sci, Calcutta, MS, 65, PhD(biochem), 69. *Prof Exp:* Fel, Immunol & Immunochem, 69-71, sr cancer res scientist, 71-78, CANCER RES SCIENTIST IV, ROSWELL PARK MEM INST, 79- *Concurrent Pos:* Nat Cancer Inst res grants, 76-79; NSF res grant, 80-81; Am Cancer Soc res grants, 83-86; NIH res grant, 89-92. *Mem:* Am Asn Cancer Res; Am Asn Immunologists. *Res:* Immunobiology of human tumors; immunodiagnosis and immunotherapy of cancer; biological response modifiers. *Mailing Add:* Roswell Park Mem Inst 666 Elm St Buffalo NY 14263

BHATTACHARYA, PALLAB KUMAR, b Calcutta, India, Dec 6, 49; m 75; c 2. COMPOUND SEMICONDUCTOR SYNTHESIS, OPTOELECTRONICS & INTEGRATED OPTICS. *Educ:* Univ Calcutta, BS, 68, BTech, 70, MTech, 71, MEng, 76, PhD(elec eng), 78. *Prof Exp:* Sr res asst microwaves, Radar & Communication Ctr, Ind Inst Technol, 72-73, com exec, Hindustan Steel Ltd, Rourkela, India, 73-75; from asst prof to assoc prof elec & computer eng, Ore State Univ, 78-83; assoc prof, 84-87, PROF, UNIV MICH, ANN ARBOR, 87- *Concurrent Pos:* Consult, Tektronix Inc, Beavertown, Ore, 80-83, prin investr, grants & contracts from NSFS, Dept Eng & NASA, 80-, invited prof microelectron, Swiss Federal Inst Technol, 81-83, consult, Int Tel & Tel, Roanoke, Va, 83-85; Army Res Orgn & Dow Chem, 85-87, Hughes Res Labs, 85-86, Ford Sci Labs, 86-, Rockwell Int, 87; vis prof, Aachen Tech Univ, WGer, 88. *Mem:* Am Phys Soc; sr mem Inst Elec & Electronics Engrs. *Res:* Molecular beam epitaxy of III-V compound semiconductor heterostructures and superlattices and strained-layer heterostructures, their electrical and optical properties and their use in high-speed electronic and opto-electronic devices; optical detectors and modulators, non-linear optical phenomena and integrated optics; over 150 contributed and invited papers. *Mailing Add:* Dept Elec Eng & Computer Sci Ore State Univ Corvallis OR 97331

BHATTACHARYA, PRABIR, b Gaya, Bihar, India, Jan 24, 48; US citizen; m 75; c 2. BIOCHEMISTRY, CELL BIOLOGY. *Educ:* Univ Calcutta, India, BS, 67, MS, 69, PhD(biochem), 74. *Prof Exp:* Scientist, Am Red Cross, Bethesda, Md, 84-87; sr prin scientist, 87-90, DIR RES & DEVELOP, ALPHA THERAPEUT CORP, 90- *Concurrent Pos:* Mem, Thrombosis Coun, Am Heart Asn. *Mem:* Am Heart Asn; Am Soc Biol & Molecular Biol. *Res:* Developing therapeutic agents from human plasma. *Mailing Add:* Alpha Therapeut Corp 5555 Valley Blvd Los Angeles CA 90032

BHATTACHARYA, PRADEEP KUMAR, b Dacca, Brit India, Jan 12, 40. PLANT PHYSIOLOGY, PLANT PATHOLOGY. *Educ:* Banaras Hindu Univ, BSc, 57, MSc, 59; Univ Sask, PhD(biol), 66. *Prof Exp:* Lectr bot, MLK Degree Col, Gorakhpur Univ, India, 59-61; res assoc biol & plant physiol, Univ Sask, 66-67; fel bot, Univ Western Ont, 67-69; fel plant path, Univ Wis-Madison, 69-70; asst prof biol, Rockford Col, 70-73; chmn dept, 73-83 ASSOC PROF BIOL, IND UNIV NORTHWEST, 73-,. *Concurrent Pos:* Nat Res Coun Can assoc, Dept Biol, Univ Sask, 66-67; Nat Res Coun Can fel, Dept Bot, Univ Western Ont, 67-69; NIH fel, Dept Plant Path, Univ Wis-Madison, 69-70. *Mem:* Am Inst Biol Scientists; Am Soc Cell Biol; Am Asn Univ Prof; Sigma Xi; AAAS. *Res:* Physiology of host-parasite relations with particular reference to obligate parasitism; cytochemical and related biochemical events in plant-parasite interaction studies in wheat rust, barley mildew and root rot of cabbage; senescence in plant and human tissue. *Mailing Add:* Dept Biol Ind Univ NW 3400 Broadway Gary IN 46408

BHATTACHARYA, RABI SANKAR, b Silchar, India, Feb 19, 48; m 79; c 2. ION IMPLANTATION, THIN FILM. *Educ:* Gauhati Univ, BS, 67; Indian Inst Technol, MS, 69; Saha Inst, Univ Calcutta, PhD(physics), 75. *Prof Exp:* Vis scientist physics, Fom Inst Atomic & Molecular Physics, Amsterdam, 75-77; vis prof, Univ Giessen, WGer, 77-78; vis scientist plasma physics, Max Planck Inst Plasma Physics, WGer, 78-79; sr res assoc eng physics McMaster Univ, Can, 79-80; sr res assoc mat res, Univ Fla, 80-81; sr scientist, 81-86, PRIN RES SCIENTIST MAT RES, UNIV ENERGY SYSTS, 86- *Mem:* Am Phys Soc; Am Soc Metals; Nat Asn Corrosion Engrs. *Res:* Applications of ion beams in materials; surface modification of structural ceramics to improve their strength, joinability, reliability, hardness and toughness using ion beams; synthesis of novel surface alloys and composites using vapor deposition and ion beams; high TC superconductor thin films; self-lubricating surfaces by ion beam processing; Am Phys Soc. *Mailing Add:* Univ Energy Systs Inc 4401 Dayton-Xenia Rd Dayton OH 45432

BHATTACHARYA, RABINDRA NATH, b Barisal, Bangladesh, Jan 11, 37; m 67; c 2. MATHEMATICAL STATISTICS. *Educ:* Univ Calcutta, MSc, 59; Univ Chicago, PhD(statist), 67. *Prof Exp:* Asst prof, Univ Calif, Berkeley, 67-72; from assoc prof to prof math, Univ Ariz, 72-82; PROF MATH, IND UNIV, 82- *Concurrent Pos:* Assoc ed, Ann Probability, 76-81, Econometric Theory, 84-, J Statist Planning & Info, 84-88 & J Multivariate Anal, 86- *Mem:* Fel Inst Math Statist; Am Math Soc. *Res:* Central limit theorems; markov processes; Edgeworth expansions in statistics. *Mailing Add:* Dept Math Ind Univ Bloomington IN 47405

BHATTACHARYA, SYAMAL KANTI, b Calcutta, India, Feb 13, 49; nat US; m 69; c 3. MINERAL METABOLISM, NEUROMUSCULAR DISEASES. *Educ:* Univ Calcutta, BSc Hons, 68, BA, 69; Murray State Univ, MS, 76; Washington Univ, AM, 78; Memphis State Univ, PhD(anal chem), 79; Am Bd Bioanal, dipl, 80. *Prof Exp:* Asst instr chem, Netaji Sikshyatan, Calcutta, 68-69; sr instr, Bhabanath Inst, Calcutta, 69-70; res & develop chemist, Swastik Household & Indust Prod, Bombay, 70-74; teaching asst, Murray State Univ, 74-76; sr res technician pediat, Med Ctr, Washington Univ, 76-77, teaching-res fel chem, 76-78; res assoc med, Med Ctr, 79-80, instr, 80-82, from instr to asst prof surg, 83-86, ASST PROF MED CHEM, UNIV TENN, MEMPHIS, 85-, ASSOC PROF SURG, 86-, DIR, SURG RES LABS & CHEM & NUTRIENT DATA OUTPUT LAB, 82- *Concurrent Pos:* Res fel, Memphis State Univ, 78-79; NIH Nat Res Serv Awardee, Univ Tenn, Ctr Health Sci, 79-81; prin investr, res grants, USPHS, 81-83, Muscular Dystrophy Asn Am, 83-85, Univ Physician Found, 85-86 & Am Heart Asn, 86-87; consult clin chem, William Bowld Hosp & Regional Med Ctr, Memphis, 82-, Baptist Mem & Affil Hosps, 83-, Le Bonheur Children's Med Ctr, 84-, Crittenden County Mem Hosp, Ark, 84- & St Joseph Hosp, Methodist Cent & Affil Hosps & Med Express Regional Labs, 85-; vis prof surg, Sch Med, Yale Univ, 85; vis prof pediat, Med Ctr, Univ Cincinnati & Cincinnati Children's Hosp, 85. *Mem:* Fel Am Inst Chemists; fel Royal Soc Chem; Am Chem Soc; NY Acad Sci; fel Indian Chem Soc; fel Am Col Nutrit; fel Am Heart Asn; fel Inst Chemists India; Am Fedn Clin Res; Soc Neurosci; Int Soc Brain Res; AAAS; Sigma Xi. *Res:* Mineral metabolism in muscular dystrophy and acute pancreatitis: applications of surgical procedures and calcium-antagonists to ameliorate these conditions by inhibiting intracellular calcium shift; prevention of myocardial ischemia and endotoxin-shock by calcium-antagonists and fructose 1, 6 diphosphate. *Mailing Add:* Surg Dept Univ Tenn Med Ctr 956 Court Ave Memphis TN 38163

BHATTACHARYA-CHATTERJEE, MALAYA, b Cooch-Behar, India, Jan 16, 46; US citizen; m 72; c 2. CLINICAL IMMUNOLOGY, IMMUNOTHERAPY. *Educ:* Presidency Col, Calcutta, India, BS, 63; Univ Col Sci, Calcutta, India, MS, 65, PhD(biochem), 69. *Prof Exp:* Asst cancer res sci, Dept Immunol & Immunochem, 69-71, cancer res scientist III, 72-78, CANCER RES SCIENTIST IV IMMUNOL, DEPT GYNEC ONCOL & CLIN IMMUNOL, ROSWELL PARK CANCER INST, 79- *Concurrent Pos:* Prin investr grants, NIH, Bethesda, Md, 76-79, NSF, 80-81 & Am Cancer Soc, 83-86; asst res prof, Dept Exp Path, State Univ NY, Buffalo, 89-; spec reviewer, Nat Cancer Inst, Bethesda, Md, 89- & NIH Study Sect, 91- *Mem:* Am Asn Cancer Res; Am Asn Immunologists. *Res:* Tumor immunology and immunochemistry; experimental immunotherapy; vaccines and biological response modifiers. *Mailing Add:* Div Clin Immunol Roswell Park Cancer Inst Buffalo NY 14263

BHATTACHARYYA, ASHIM KUMAR, b Kanpur, India, July 9, 36; m 66; c 2. PHYSIOLOGY, NUTRITION. *Educ:* Univ Calcutta, BSc, 57, MSc, 59, PhD(physiol), 65, DSc(physiol), 89. *Prof Exp:* Fel physiol, Univ Minn, 66-68; fel med, Clin Res Ctr, Univ Iowa, 69-70, assoc res scientist, 70-74, res scientist, 74-75; from asst prof to assoc prof, 75-89, PROF PATH & PHYSIOL, LA STATE UNIV MED CTR, 89- *Concurrent Pos:* Assoc mem, Sch Grad Studies, La State Univ Med Ctr, 75-, dir, Lipid Chem Lab, Dept Path, 76-78; fel, Coun on Atherosclerosis, Am Heart Asn, 71- *Mem:* Sigma Xi; Am Soc Clin Nutrit; Am Inst Nutrit; Soc Exp Biol & Med; Am Physiol Soc; NY Acad Sci. *Res:* Regulation and mechanism of sterol absorption in the gastrointestinal tract; sterol and bile acid metabolism; atherosclerosis; dietary fatty acids, their transport and tissue storage. *Mailing Add:* Dept Path La State Univ Med Ctr 1901 Perdido St New Orleans LA 70112-1393

BHATTACHARYYA, BIBHUTI BHUSAN, b Calcutta, India, Mar 22, 38; m 65; c 1. ELECTRICAL ENGINEERING. *Educ:* Indian Inst Technol, Kharagpur, BTech, 58, MTech, 59; NS Tech Col, PhD(elec eng), 68. *Prof Exp:* Trainee tech teaching, Govt of India, 59-62; lectr elec eng, Indian Inst Technol, Madras, 62-65; asst prof, Univ Calgary, 68-69, assoc prof, 70; assoc prof, 70-77, PROF ELEC ENG, SIR GEORGE WILLIAMS CAMPUS, CONCORDIA UNIV, 77- *Mem:* Inst Elec & Electronics Engrs. *Res:* Lumped and distributed, passive and active linear network theory. *Mailing Add:* Concordia Univ Concordia Univ 1455 de Maisonneuve Montreal PQ H3G 1M8 Can

BHATTACHARYYA, BIBHUTI BHUSHAN, b Bhatpara, India, Aug 1, 35; m 62. MATHEMATICAL STATISTICS. *Educ:* Univ Calcutta, BSc, 53, MSc, 55; Univ London, PhD(economet), 59. *Prof Exp:* Res scholar statist, Indian Statist Inst, 56; vis statistician, NC State Col, 59-61; res fel economet, Inst Econ Growth, 62; lectr math & statist, Univ Toronto, 62-63; asst prof, 63-69, ASSOC PROF STATIST, NC STATE UNIV, 69- *Mem:* Inst Math Statist. *Res:* Econometrics; mathematical programming and its application to national planning as well as production planning of individual firms; sequential estimation problems. *Mailing Add:* Box 8203-Statist NC State Univ Raleigh Main Campus Raleigh NC 27695

BHATTACHARYYA, GOURI KANTA, b Hooghly, WBengal, India; Jan 12, 40; m 62; c 2. STATISTICS, MATHEMATICAL STATISTICS. *Educ:* Univ Calcutta, BSc, 58, MSc, 60; Univ Calif, Berkeley, PhD(statist), 66. *Prof Exp:* Lectr statist, R K Mission, Narendrapur, India, 61; res officer, River Res Inst, Calcutta, 61-63; asst prof, Univ Calif, Berkeley, 66; from asst prof to assoc prof, 66-75, PROF STATIST, UNIV WIS-MADISON, 75- *Concurrent Pos:* Consult sch med, Univ Wis, 67-; prof, Indian Inst Mgt, Ahmedabad, 69-70; vis prof, Indian Statist Inst, Calcutta, 75-76, Sch Comput & Info Sci, Syracuse Univ, 79-80 & Calcutta Univ, 83. *Mem:* Fel Inst Math Statist; fel Am Statist Asn; fel Royal Statist Soc; fel Int Statist Inst; Indian Asn Qual, Productivity, Reliability & Technometrics. *Res:* Statistical inference, life testing and reliability studies; nonparametric methods. *Mailing Add:* Dept Statist Univ Wis 1210 W Dayton Madison WI 53706

BHATTACHARYYA, MARYKA HORSTING, b Glen Ridge, NJ, Sept 17, 43; m 71; c 1. BIOCHEMISTRY, TOXICOLOGY. *Educ:* Tufts Univ, BS, 65; Univ Wis, PhD(biochem), 70. *Prof Exp:* Fel biochem, Univ Wis, 70-71, res assoc, 71-74; asst biochemist, 74-79, BIOCHEMIST, ARGONNE NAT LAB, 79- *Concurrent Pos:* Adj prof, Northern Ill Univ, 79-; mem, Task Group Neptunium, NCRP, 82- *Honors & Awards:* R & D 100 Award, Blood Cadmium Assay Kit. *Mem:* Soc Toxicol; Sigma Xi; Radiation Res Soc. *Res:* Influence of pregnancy, lactation, and ovariectomy on cadmium-induced bone loss; gastrointestinal absorption of metals including cadmium, lead, plutonium, and other actinides. *Mailing Add:* Argonne Nat Lab 9700 S Cass Ave Argonne IL 60439-4833

BHATTACHARYYA, MOHIT LAL, b May 28, 44; m; c 1. ELECTROPHYSIOLOGY, MEMBRANE BIOPHYSICS. *Educ:* Univ Wyo, PhD(physics), 71. *Prof Exp:* ASST PROF CELL BIOL & CARDIOVASC PHYSIOL, MEHARRY MED COL, 82- *Mailing Add:* Dept Physiol Meharry Med Col 1005 18th Ave N Nashville TN 37208

BHATTACHARYYA, PRANAB K, b India, Aug 9, 38; m 69; c 2. PHYSICAL CHEMISTRY, ANALYTICAL CHEMISTRY. *Educ:* Calcutta Univ, BS, 59, MS, 62; Columbia Univ, PhD(phys chem), 71. *Prof Exp:* Res assoc nuclear magnetic resonance,Cchem dept, Columbia Univ, 71-74; sr scientist & supvr spectros technol, Analytical Res Sect, 74-79, tech fel, Res & Diag Prod Sect, 79-83, group leader, Pharmaceut Res Prod Sect, 83-87, MGR, PHARMACEUT RES PROD SECT, QUAL CONTROL DEPT, HOFFMANN-LA ROCHE, INC, 87- *Mem:* NY Acad Sci; Am Chem Soc; Am Phys Soc; Int Soc Magnetic Resonance. *Res:* Quantitative analysis and characterization of drug and biological molecules and complex organic compounds using nuclear magnetic resonance and molecular spectroscopy; development of analytical methods and validation of methodology for testing the purity and stability of bulk drugs and dosage forms; preparation of investigational new drug and new drug application submissions to the Food and Drug Administration on test procedures for new drug products. *Mailing Add:* 18 Valley Rd Nutley NJ 07110-2225

BHATTACHARYYA, RAMENDRA KUMAR, b India, Mar 29, 31; m. MATHEMATICS. *Educ:* Univ Calcutta, BSc, 51, MSc, 53; Stanford Univ, PhD(math), 64. *Prof Exp:* Lectr math, Taki Govt Col, India, 54-57 & Jadavpur Univ, 57-59; asst prof math, Univ Ariz, 64-65 & Southern Ill Univ, Carbondale, 65-69; ASSOC PROF MATH, PAC UNIV, 69- *Mem:* Am Math Soc; Math Asn Am; alcutta Math Soc. *Res:* Theory of affine connections; system analysis. *Mailing Add:* Dept Math Pac Univ Forest Grove OR 97116

BHATTACHARYYA, SAMIT KUMAR, b Calcutta, India, Mar 21, 47; US citizen; m 71; c 1. NUCLEAR REACTOR ENGINEERING. *Educ:* Indian Inst Technol, BTech Hons, 68; Univ Wis, MS, 70, PhD(nuclear eng), 73. *Prof Exp:* Asst nuclear eng physics, 74-78, nuclear eng, 78-79, group leader saref treat upgrade proj, reactor core design, 79-80, MGR REACTOR PHYSICS SECT, TREAT UPGRADE PROJ, ARGONNE NAT LAB, 80- *Mem:* Am Nuclear Soc; Sigma Xi. *Res:* Experimental and analytical studies on fast breeder reactor physics; safety related physics; reactor design. *Mailing Add:* Argonne Nat Lab 9700 S Cass Ave Argonne IL 60439

BHATTACHARYYA, SHANKAR P, b Rangoon, Burma, June 23, 46; US citizen; m 71, 85; c 3. CONTROL SYSTEMS. *Educ:* Indian Inst Technol, BTech, 67; Rice Univ, MS, 69, PhD(elec eng), 71. *Prof Exp:* From asst prof to prof elec eng, Fed Univ, Rio de Janeiro, Brazil, 71-80, chmn dept, 78-80; assoc prof, 80-84, PROF ELEC ENG, TEX A&M UNIV, 84- *Concurrent Pos:* Nat Res Coun res fel, Marshall Space Flight Ctr, NASA, 74-75; prin investr, NSF res contracts, 82-; assoc ed, Trans on Automatic Control, Inst Elec & Electronics Engrs, 85-, bd gov, Control Syst Soc, 86; Fulbright lectr, 89; fel Tex Eng Exp Sta, 89. *Mem:* Fel Inst Elec & Electronics Engrs; Soc Indust & Appl Math; Control Systs Soc Inst Elec & Electronics Engrs. *Res:* Design of automatic control systems with emphasis on robust linear and multivariable control systems; author of more than seventy papers. *Mailing Add:* Elec Eng Dept Tex A&M Univ College Station TX 77843

BHATTI, RASHID, b Karachi, July 7, 39; US citizen; c 1. TUMOR IMMUNOLOGY IMMUNOBIOLOGY. *Educ:* Univ Karachi, BS, 61, MS, 63; Eastern Ill Univ, Charleston, MS, 70; Harvard Med Sch, Boston, dipl tumor biol, 76. *Prof Exp:* Assoc pharmacologist drug eval, Bio-Test Labs, Northbrook, Ill, 70-71; res assoc prostate cancer, Rush Presby St Luke Med Ctr, Chicago, Ill, 72-74; DIR RES PROSTATE TUMORS, COOK COUNTY HOSP UROL RES LAB, CHICAGO, 74- *Concurrent Pos:* Lectr, Int Found Microbiol, 70-72; consult, UN, 85-; res asst prof surg, Col Med, Univ Ill, Chicago, 86- *Mem:* Am Asn Cancer Res; Europ Asn Cancer Res; NY Acad Sci; Asn Pakistani Scientists & Engrs NAm (secy, 87-88, vpres, 91-93); Pakistan Acad Med Sci (vpres, 90-92). *Res:* Early detection, monitoring and management of prostate cancer, tumor markers and a variety of therapeutic modalities; dissemination of metastasis of cancer. *Mailing Add:* Urol Res Lab Cook County Hosp 1835 W Harrison Chicago IL 60612

BHATTI, WAQAR HAMID, b Sohawa, WPakistan, Nov 22, 31; m 64. PHYTOCHEMISTRY, HEMATOLOGY. *Educ:* Univ Panjab, BSc, 55, MSc, 57; Philadelphia Col Pharm, PhD(pharmacog), 63. *Prof Exp:* Lectr pharmacog, Univ Panjab, Pakistan, 57; lectr bot, Univ Peshawar, 58-59; asst biol, Philadelphia Col Pharm, 59-63; asst prof pharmacog, Sch Pharm, NDak State Univ, 63-67; res fel internal med, Univ Iowa Hosp, 67-68; res assoc, Sch Pharm, Univ Pittsburgh, 68-69; asst prof, 69-74, ASSOC PROF PHARMACOG, BUTLER UNIV, 74- *Concurrent Pos:* Res assoc, Einstein Med Ctr, 62-63 & Vet Admin, 63-; teaching fel pharmacog, NSF Inst Prog High Sch Students & Teachers, 64, prin investr, NSF grant, 64- *Mem:* AAAS; Am Pharmaceut Asn; Am Soc Pharmacog; Am Chem Soc. *Res:* The chemical investigation of both higher and lower plants and the isolation of their active principles which may serve as new drugs useful in the prophylaxis or treatment of cancer or cardiac and hemolytic disorders. *Mailing Add:* Col Pharm Butler Univ Indianapolis IN 46208-3485

BHAUMIK, MANI LAL, b Calcutta, India, Jan 5, 32; US citizen. PHYSICS, QUANTUM ELECTRONICS. *Educ:* Univ Calcutta, BS, 51, MS, 53; Indian Inst Technol, PhD(physics), 58. *Prof Exp:* Fel chem physics, Univ Calif, Los Angeles, 59-61; group leader lasers, Xerox Corp, 61-67; sr physics specialist, Aerojet Gen Corp, 67-68; res mgr lasers, Northrop Corp, 78-86; CONSULT, 86- *Concurrent Pos:* Lectr, Calif State Univ, Long Beach, 67-68. *Mem:* Fel Am Phys Soc; Inst Elec & Electronics Engrs; Optical Soc Am; Sigma Xi. *Res:* Discovery and development of new lasers; investigation of laser applications to various fields. *Mailing Add:* PO Box 24050 Los Angeles CA 90024

BHAVANANDAN, VEER P, b Jaffna, Sri Lanka, Nov 1, 36; m 65; c 2. CANCER, CYSTIC FIBROSIS. *Educ:* Univ Edinburgh, PhD(chem), 62; DSc, 83. *Prof Exp:* Res assoc, Columbia Univ, NY, 65-67; agr chemist, Tea Res Inst, Sri Lanka, 67-71; staff scientist, Med Res Found, Okla City, 71-73; res assoc, 73-74, from asst prof to assoc prof, 75-85, PROF BIOCHEM, M S HERSHEY MED CTR, PA STATE UNIV, 86- *Concurrent Pos:* Mem, NIH Biochem Study sect, 87-; fel, Univ Edinburg, 60-62; NIH grant, 76-; Am Cancer Soc Eleanor Roosevelt Int Cancer fel, 82-83; NIH sr Fogarty Int fel, 82. *Mem:* Fel Royal Soc Chem; Soc Complex Carbohydrates; Am Chem Soc; Am Soc Biochemists; AAAS. *Res:* Investigation of cancer-associated mucin glycoproteins from human cancer cells, cancer diagnostic tests, host-bacterial interaction in pulmonary infection of cystic fibrosis patients, structure-function of mucins. *Mailing Add:* Milton Hershey Med Ctr Pa State Univ Hershey PA 16802

BHAVNANI, BHAGU R, b Hyderabad, India, Feb 16, 36; Can citizen; m 73; c 3. BIOCHEMISTRY, PHARMACEUTICAL CHEMISTRY. *Educ:* Univ Bombay, BS, 58 & 60, MS, 62; Mass Col Pharm, PhD(biochem), 66. *Prof Exp:* Res assoc steroid biochem, Worcester Found Exp Biol, 62-66; lectr, McGill Univ, 68-69, asst prof investigative med, 69-75; assoc scientist, Royal Victoria Hosp, Montreal, 68-75; asst prof obstet & gynec & clin biochem, Univ Toronto, 75-76, assoc prof obstet & gynec, 76-77; staff scientist, St Michael's Hosp, Toronto, 75-76; from assoc prof to prof obstet & gynec, McMaster Univ, Hamilton, 77-88, dir, reprod biol res prog, 86-88; PROF OBSTET & GYNEC, UNIV TORONTO, 88-; DIR RES, DEPT OBSTET & GYNEC, ST MICHAEL'S HOSP, 88- *Concurrent Pos:* Fel biochem, McGill Univ, 66-68; res oper grants, Med Res Coun Can, 68-, res grant, 70-75; res grants, Ontario Cancer Treatment & Res Found, 78-82; fel, Nat Cancer Inst, 81-86 & Ministry Health, 85-88. *Mem:* AAAS; Endocrine Soc; NY Acad Sci; Can Soc Endocrinol & Metab; Soc Obstet & Gynec Can; Soc Gynec Invest. *Res:* Genetic aspects of cortisol metabolism in the guinea pig; steroid formation and metabolism during pregnancy in humans; formation of ring B unsaturated estrogens in the pregnant mare; epidemiologic relation between use of exogenous estrogens and carcinoma of the endometrium; estrogen metabolism in postmenopausal women; diabetic pregnancy and fetal lung maturation; developmental aspects of fetal glycogen metabolism; infertility and luteal function; role of opiates in the regulation of the menstrual cycle; exercise induced secondary ammenhorea. *Mailing Add:* St Michael's Hosp 30 Bond St Toronto ON M5B 1W8 Can

BHORJEE, JASWANT S, b Ajmer, India, July 23, 35. CHROMATIN STRUCTURE FUNCTION. *Educ:* Univ Sask, Saskatoon, PhD(microbiol), 66. *Prof Exp:* Asst prof, Univ Ill, Chicago, 78-84; ASST PROF MICROBIOL, MEHARRY MED COL, NASHVILLE, 84- *Mem:* AAAS; Sigma Xi; Am Soc Cell Biol. *Mailing Add:* 6612 Clearbrook Dr Nashville TN 37205

BHOWN, AJIT SINGH, PROTEIN CHEMISTRY. *Educ:* Univ Rajasthan, PhD(biochem), 65. *Prof Exp:* ASSOC PROF, UNIV ALA, 74- *Res:* Structural studies of proteins; high-pressure liquid chromatography. *Mailing Add:* Dept Med Univ Ala 19th St Eighth Ave S Birmingham AL 35294

BHUSHAN, BHARAT, b Jhinjhana, India, Sept 30, 49; m 75; c 2. MECHANICAL ENGINEERING, MATERIAL SCIENCE. *Educ:* Birla Inst Technol, India, BE Hons, 70; Mass Inst Technol, SM, 71; Univ Colo, MS, 73, PhD(mech eng), 76; Rensselaer Polytechnol Inst, MBA, 80; Univ Trondheim, Norway, DSc, 90. *Prof Exp:* Res asst, Mass Inst Technol, 70-71, mem res staff mech eng, 71-72; instr & res asst, Univ Colo, 72-76; prof mgr, Advan Technol, Mech Technol Inc, 76-80; res scientist, Technol Serv Div, SKF Industs Inc, 80-81; engr & mgr, GPD Lab, Tucson, 81-86, SR ENGR & MGR, ALMADEN RES CTR, IBM CO, SAN JOSE, 86- *Concurrent Pos:* Expert investr, Automotive Specialists, Denver, 73-76; proprietor & mgr, Time Sharing Comput Serv, India, 77-, pvt consult, 76-, merit scholar, U P State Educ Bd, India, 61-65, Govt India, 65-70; res fel, Ford Found, 70-71; grad fel, Univ Colo, 73-74; res grants, US Navy, US Air Force, NASA, Dept Energy, E I Du Pont de Nemours & Co Inc, Chrysler & United Technol Corp, 76-81. *Honors & Awards:* Burt L Newkirk Award, Am Soc Mech Engrs, 83, Tribology Div Award, 89; Tech Excellence Award, Am Soc Engrs. *Mem:* Fel Am Soc Mech Eng; Soc Tribologists & Lubrication Engrs; Am Acad Mech; Soc Automotive Eng; Sigma Xi; fel NY Acad Sci. *Res:* Technical involvement in and management of research and development activities in the fields of friction, wear, lubrication, and interface temperature problems encountered in magnetic storage systems and high-speed computer printers. *Mailing Add:* 14527 Eastview Dr Los Gatos CA 95030-1705

BHUVA, ROHIT L, b Balwa, India, May 24, 54; m 80; c 2. PROGRAMMABLE INTEGRATED CIRCUIT DESIGNER, ANALOG INTEGRATED CIRCUIT DESIGNER. *Educ:* MS Univ, Baroda, India, BEng, 76; Ill Inst Technol, Chicago, MS, 77. *Prof Exp:* Design mgr, 78-89, tech liaison officer, 90, MGR, LINEAR DESIGN CTR, TEX INSTRUMENTS, 91- *Res:* Design and development of field programmable integrated circuits. *Mailing Add:* 8330 LBJ Freeway Dallas TX 75243

BHUVANESWARAN, CHIDAMBARAM, b India, Nov 10, 34. BIOCHEMISTRY. *Educ:* Univ Bombay, BSc, 55 & 57, PhD(biochem), 63. *Prof Exp:* Jr sci officer, Cent Food Tech Res Inst, Mysore, India, 60-64; sci officer, Atomic Energy Estab, Bombay, 64-65; res assoc biochem, Ore State Univ, 65-68; res assoc biochem, Fac Med, Univ Man, 68-72; asst prof, 77-80, ASSOC PROF BIOCHEM, UNIV ARK, MED SCI CAMPUS, 80- *Mem:* AAAS; Am Chem Soc; Asn Clin Scientists; Sigma Xi. *Res:* Electron transport and oxidative phosphorylation; lipid metabolism. *Mailing Add:* Dept Biochem Univ Ark Med Sci Campus 4301 W Markham Little Rock AR 72205

BHUYAN, BIJOY KUMAR, b Calcutta, India, Aug 30, 30; m 55; c 2. CELL BIOLOGY, EXPERIMENTAL CANCER CHEMOTHERAPY. *Educ:* Utkal Univ, India, BSc, 48; Univ Calcutta, MSc, 51; Univ Wis, MS, 54, PhD(biochem), 56. *Prof Exp:* Res assoc & fel biochem, Univ Wis, 56-57; fel, Nat Res Coun Can, 57-58; sr sci officer, Hindustan Antibiotics, India, 58-59; res assoc microbiol, 60-63, res assoc biochem, 63-73, SR RES SCIENTIST, CANCER RES, UPJOHN CO, 73- *Concurrent Pos:* Assoc ed, Cancer Res, Cell & Tissue Kinetics. *Mem:* Tissue Cult Asn; Am Asn Cancer Res; Cell Kinetic Soc. *Res:* Penicillin production; anti-tumor antibiotics; tissue culture; cell cycle; cell kinetics; human tumor cloning; developing conditions under which antitumor drugs will be more effective in killing cancer cells based on studies involving cell cycle, drug combinations, drug resistance, etc. *Mailing Add:* Upjohn Co Kalamazoo MI 49001

BIAGETTI, RICHARD VICTOR, b Woonsocket, RI, Jan 13, 40; m 64; c 2. BATTERIES, ELECTROCHEMISTRY & INORGANIC CHEMISTRY. *Educ:* Providence Col, BS, 60, MS, 62; Univ NH, PhD(inorg chem), 66. *Prof Exp:* Mem staff, 65-69, SUPVR ELECTROCHEM, BELL TEL LABS, 69- *Mem:* Electrochem Soc. *Res:* Coordination chemistry of nonaqueous transition metal nitrates; failure mechanism analyses and development of improved lead-acid battery systems. *Mailing Add:* Ten Rye St Piscataway NJ 08854

BIAGGIONI, ITALO, b Lima, Peru, Sept 4, 54; m 83. CLINICAL PHARMACOLOGY. *Educ:* Universidad Peruana Cayetano Heredia, MD, 80. *Prof Exp:* Intern, Cayetano Heredia Univ Hosp, 78-79, from resident to chief resident, 80-83; assoc investr med, Nat Inst Salud, Peru, 82-83; res fel pharmocol, 84-86, instr, 86-87, ASST PROF MED, VANDERBILT SCH MED, 87- *Concurrent Pos:* Physician, Peruvian Health Serv, 79-80; clin assoc physician, Elliot V Neuman Clin Res Ctr, Vanderbilt Univ, 86- *Mem:* Am Fedn Clin Res. *Res:* The autonomic nervous system and cardiovascular control; the role of adenoisine in the cardiovascular system. *Mailing Add:* Clin Res Ctr-AA3228 MCIV Vanderbilt Univ Nashville TN 37232

BIAGINI, RAYMOND E, b San Francisco, Calif, Oct 10, 49; m 76; c 2. IMMUNOTOXICOLOGY, BIO-MONITORING. *Educ:* Univ San Francisco, BS, 71; Univ Nev, Reno, MS, 76; Univ Cincinnati, PhD(toxicol), 84; Am Bd Toxicol, dipl, 86. *Prof Exp:* Qual control anal chemist & microbiol consult, Best Foods Div, CPC Int, San Francisco, Calif, 73-74; res biochemist, Dept Biochem, Sch Biomed Sci, Univ Nev, Reno, 76-77; criteria mgr, Criteria Develop Br, Nat Inst Occup Safety & Health, 77-78, res toxicologist, Chronic Toxicol Sect, Exp Toxicol Br, 78-82 & 84-89, grad student trainee, Off of Dir, 82-84, RES TOXICOLOGIST, IMMUNOCHEM RES SECT, APPL BIOL BR, NAT INST OCCUP SAFETY & HEALTH, CINCINNATI, OHIO, 89- *Concurrent Pos:* Sr scientist officer, USPHS; Nat Res Coun mentor, Robert A Taft Lab, Nat Inst Occup Safety & Health; expert adv, WHO. *Mem:* Am Asn Immunologists; Soc Toxicol. *Res:* Effects of xenobiotics on the immune system of animals and humans; pulmonary pharmacology; sub-cellular bioenergetics of cardiac muscle. *Mailing Add:* ABPB CDC NIOSH 4676 Columbia Pkwy MS-C26 Cincinnati OH 45226

BIAGLOW, JOHN E, b Cleveland, Ohio, Apr 1, 37; m 60; c 9. BIOCHEMISTRY, ONCOLOGY. *Educ:* John Carroll Univ, BS, 59; Loyola Univ, Chicago, MS, 61, PhD(biochem), 63. *Hon Degrees:* MA, Univ Pa, 87. *Prof Exp:* NIH fel biochem, 63-65,from asst prof to assoc prof, 65-84, PROF RADIATION BIOCHEHEM, CASE WESTERN RESERVE UNIV, 84-, DIR RADIATION BIOL, 84-; DIR, BIOCHEM DIV, DEPT RADIATION ONCOL, UNIV PA & DIR TUMOR METAB, DIV CANCER CTR. *Concurrent Pos:* Pres, Custom Biochem Co; asst prof radiol, Case Western Reserve Univ 67-76, assoc prof Biochem, 76-85, prof biochem & environ health, 85-, dir biochem oncol, 85-; mem, Federated Soc for Exp Biol, 76-; consult, Adamantech, Sun Oil, 85- *Mem:* Am Chem Soc; NY Acad Sci; Int Soc Oxygen Transp to Tissue; AAAS; Oxygen Soc. *Res:* Enzymology; kinetics; control and organization of enzyme systems; carcinogenesis; cell culture; oncology; physiological controls; oxygen utilization; artificial oxygen carriers. *Mailing Add:* Dept Radiation Oncol Univ Pa Philadelphia PA 19104-4283

BIALAS, WAYNE FRANCIS, b Middletown, NY, Aug 5, 49. OPERATIONS RESEARCH, STATISTICS. *Educ:* Clarkson Col Technol, BS, 71; Cornell Univ, MS, 74, PhD(oper res), 75. *Prof Exp:* Res assoc & instr, Col Eng, Cornell Univ, 75; ASSOC PROF INDUST ENG, STATE UNIV NY BUFFALO, 76- *Concurrent Pos:* Vis asst prof civil & environ eng, Cornell Univ, 81-82; vis assoc prof civil eng, Mass Inst Technol, 83-84. *Mem:* Am Statist Asn; Asn Comput Mach; Inst Indust Eng; Am Soc Civil Engrs; Inst Mgt Sci; Opers Res Soc Am. *Res:* Multilevel optimization; asymptotic properties of probability measures; economic models for flood prone regions. *Mailing Add:* Dept Indust Eng Bell Hall Rm 319 State Univ NY Buffalo Buffalo NY 14260

BIALECKE, EDWARD P, b Chicago, Ill, Dec 3, 34. ELECTRICAL ENGINEERING, PLASMA PHYSICS. *Educ:* Univ Ill, BSEE, 56, MS, 58, PhD(elec eng), 63. *Prof Exp:* Asst prof elec eng, Univ Ill, 65-69; sr engr, 69-80, TECH SPECIALIST, MCDONNELL DOUGLAS CORP, 80- *Mem:* Inst Elec & Electronics Engrs; Geophys Union; Sigma Xi. *Res:* Gaseous electronics; plasma physics; lasers; holography; ionospheric physics; reentry physics. *Mailing Add:* 11134 Cricket Hill St Louis MO 63141

BIANCANI, PIERO, b Rosignano, Italy, Sept 3, 1944. GASTROENTEROLOGY. *Educ:* Yale Univ, PhD(biomech), 71. *Prof Exp:* Assoc prof, 83-85, PROF MED, RI HOSP & BROWN UNIV, 85- *Res:* Neuro regulation gastrointestinal smooth muscle function. *Mailing Add:* Dept Med Brown Univ Providence RI 02902

BIANCHI, CARMINE PAUL, b Newark, NJ, Apr 9, 27; m 57; c 3. CELL PHYSIOLOGY. *Educ:* Columbia Univ, AB, 50; Rutgers Univ, MS, 53, PhD(biochem, physiol), 56. *Prof Exp:* Vis scientist & res assoc, NIH, 58-60; asst mem, Inst Muscle Dis, New York, 60-61; from assoc to assoc prof physiol, 61-68, prof physiol, Sch Med, Univ Pa, 68-76; prof pharmacol & chmn dept, 76-87, PROF PHARMACOL, THOMAS JEFFERSON UNIV, PHILADELPHIA, 87- *Concurrent Pos:* Res fel pub health, NIH, 56-58. *Mem:* Biophys Soc; Am Soc Zoologists; Soc Gen Physiol; Am Physiol Soc; Am Soc Pharmacol & Therapeut. *Res:* Role of mono- and divalent cations in controlling cell membrane function and metabolism. *Mailing Add:* Dept Pharmacol Thomas Jefferson Univ Philadelphia PA 19107

BIANCHI, DONALD ERNEST, b Santa Cruz, Calif, Nov 22, 33; m 56; c 3. PHYSIOLOGY, MYCOLOGY. *Educ:* Stanford Univ, AB, 55, AM, 56; Univ Mich, PhD, 59. *Prof Exp:* Asst prof bot, San Fernando Valley State Col, 59-63; assoc prof biol, 63-66, PROF BIOL, CALIF STATE UNIV, NORTHRIDGE, 66-, DEAN SCH SCI & MATH, 73- *Concurrent Pos:* NSF sci fac fel, 65-66; sr researcher, Univ Geneva, 70. *Mem:* Sigma Xi; Mycol Soc Am. *Res:* Physiology of fungi; cell biology. *Mailing Add:* Sch Sci & Math Calif State Univ Northridge CA 91324

BIANCHI, ROBERT GEORGE, b Chicago, Ill, Mar 20, 25; m 48; c 4. PHARMACOLOGY. *Educ:* Franklin & Marshall Col, BS, 45. *Prof Exp:* Jr investr biol res, 50-60, SR INVESTR PHARMACOL, SEARLE LABS, 60- *Mem:* Sigma Xi; AAAS. *Res:* Gastrointestinal pharmacology; discovering compounds useful in the treatment of gastrointestinal diseases. *Mailing Add:* 8336 Caldwell Rd Niles IL 60648

BIANCHINE, JOSEPH RAYMOND, b Albany, NY, Sept 7, 29; m 56. PHARMACOLOGY. *Educ:* Siena Col, BS, 51; Albany Med Col, PhD(pharmacol), 59; State Univ NY, Syracuse, MD, 60. *Prof Exp:* Physician, Johns Hopkins Univ Sch Med, 66-72; prof & chmn, Dept Pharmacol, Tex Tech Univ, 72-74; prof med, Dept Med & Prof & chmn, Dept Pharmacol, Ohio State Univ Sch Med, 74-83; vpres med res, Hoffman-LaRoche, 83-85 & DuPont de Nemours, 85-87; VPRES & DIR MED RES CTR, ADRIA LABS ERBAMONT, 87- *Concurrent Pos:* Chmn, Gen Clin Res Comt, Div Res Resources, NIH, 80-81. *Mem:* Am Soc Clin Pharmacol & Therapeut (pres, 87-88); fel Am Col Physicians; fel Am Col Clin Pharmacol; Am Soc Pharmacol & Exp Therapeut; Asn Med Sch Pharm; Soc Neurosci. *Res:* Clinical pharmacology; neurochemistry; internal medicine. *Mailing Add:* Adria Labs Erbamont Inc PO Box 16529 Columbus OH 43216-6529

BIARD, JAMES R, b Paris, Tex, May 20, 31; m 52; c 3. SEMICONDUCTOR DEVICE PHYSICS, INTEGRATED CIRCUIT DESIGN. *Educ:* Tex A&M Univ, BS, 54, MS, 56, PhD(elec eng), 57. *Prof Exp:* Sr engr, Semiconductor Res & Develop Lab, Tex Instruments, 57-69; vpres res & develop, Spectronics, Inc, 69-78; chief scientist, Optoelectronics Div, 78-88, CHIEF SCIENTIST, MICROSWITCH DIV, HONEYWELL, 88- *Concurrent Pos:* Adj prof, Dept Elec Eng, Tex A&M Univ, 80- *Mem:* Nat Acad Eng; fel Inst Elec & Electronics Engrs; Am Phys Soc. *Res:* Design and development of semiconductor devices and integrated circuits; holder of 30 US and 16 foreign patents. *Mailing Add:* Honeywell Microswitch Div 830 E Arapaho Rd Richardson TX 75081-2241

BIAS, WILMA B, b Muskogee, Okla, Dec 23, 28; m 47; c 3. HUMAN GENETICS, IMMUNOLOGY. *Educ:* Univ Okla, BS, 49; Johns Hopkins Univ, PhD(genetics), 63. *Prof Exp:* Fel med genetics, Sch Med, Johns Hopkins Univ, 63-64; from instr to prof med, 64-84, prof epidemiol, Sch Hyg, 67-78, SURG, SCH MED, 69-, PROF MED, JOHNS HOPKINS UNIV, 84- *Concurrent Pos:* Prof immunol & infectious dis, Sch Hyg, Johns Hopkins Univ, 73- *Mem:* AAAS; Am Soc Human Genetics; Int Transplantation Soc; Am Asn Immunol; Am Soc Histocompatibility & Immunogenetics; Clin Immunol Soc. *Res:* Immunogenetics. *Mailing Add:* Dept Med Sch Med Johns Hopkins Univ Baltimore MD 21205

BIASELL, LAVERNE R(OBERT), b Detroit, Mich, Oct 26, 15; m 40; c 4. AERONAUTICAL & ASTRONAUTICAL ENGINEERING. *Educ:* Univ Detroit, AB, 37. *Prof Exp:* Lab asst, Ethyl Gasoline Corp, 34-35; asst lab instr physics, Univ Detroit, 35-37; proj engr, Stinson Div, Consol Vultee Corp, 37-40; aircraft engr, Res Labs, Gen Motors Corp, 41-45; chief prod engr, Exp Eng Div, Bendix Aviation Corp, 45-46; assoc dir, Willow Run Res Ctr, Univ Mich, 46-54; chief engr, Missile Div, Chrysler Corp, 54-63, dir res & develop, Defense & Diversified Prod Divs, 65-81; RETIRED. *Honors & Awards:* Continental Aircraft Design Award, 37; Am Legion Award, 37; Space Pioneer Medal, NASA, 64. *Res:* Defense and space missles, vehicles and equipment; personal aircraft. *Mailing Add:* 86 Windsor Dr Englewood FL 34223

BIBB, HAROLD DAVID, b Centralia, Ill, Feb 12, 40; m 71; c 3. DEVELOPMENTAL BIOLOGY, NEURAL-DEVELOPMENT. *Educ:* Univ Iowa, PhD(zool), 69. *Prof Exp:* ASSOC PROF ZOOL, UNIV RI, 72- *Mailing Add:* Dept Zool Univ RI Kingston RI 02881

BIBB, WILLIAM ROBERT, b Salisbury, NC, May 28, 32; m 55; c 2. IMMUNOLOGY. *Educ:* Univ NC, BS, 57, MS, 59, PhD(bact). 62. *Prof Exp:* Res assoc virol, virus lab, Sch Med, Univ NC, Chapel Hill, 62-63, USPHS trainee, 63-64, asst prof bact, 64-65; mem med res br, Div Biol & Med, US Atomic Energy Comn, Washington, DC, 65-71, tech asst to comnr, 71-73, chief res & develop, 73-78, DIR RES DIV, US DEPT ENERGY, OAK RIDGE, 78- *Mem:* Am Soc Microbiol; Electron Micros Soc Am; Reticuloendothelial Soc; Am Nuclear Soc. *Res:* Metabolism; amino acid transport; enzyme synthesis; bacterial cell structure; amino acid decarboxylases; organ and tissue transplantation; effects of low-dose radiation; radiation protection. *Mailing Add:* PO Box 861 Oak Ridge TN 37830

BIBBO, MARLUCE, b Sao Carlos, Brazil, July 14, 39. CYTOPATHOLOGY. *Educ:* Dr Alvaro Guiao Inst Educ, Brazil, BS, 57; Univ Sao Paulo, MD, 63, DSc, 68. *Prof Exp:* Resident obstet & gynec, Univ Sao Paulo, 63-65, trainee exfoliative cytol, 65, trainee nucleic acids, 65-66, instr morphol, Fac Med, 65-68, asst prof morphol & obstet & gynec & chief cytol serv, 68-70; res assoc, 69-70, from asst prof to assoc prof, 70-77, PROF OBSTET & GYNEC & PATH, UNIV CHICAGO, 78-, DIR CYTOL LAB & SCHS CYTOTECHNOL & CYTO CYBERNET, 90- *Concurrent Pos:* Mem coun cytopath, Am Soc Clin Path. *Mem:* Fel Int Acad Cytol (pres elect, 90); Am Soc Cytol (pres, 83); fel Asn Clin Sci; World Asn Gynec Cancer Prev; Int Acad Path; Am Soc Clin Path. *Res:* Diagnostic and experimental cytology; computerized cell evaluations; pattern recognition; endocrinologic cell assessments; cancer detection techniques. *Mailing Add:* Cytopath Lab Mail Box 445 Univ Chicago Chicago IL 60637

BIBBY, MALCOLM, b Mountain Park, Alta, Aug 28, 39; m 66. PHYSICAL METALLURGY. *Educ:* Univ Alta, BSc, 61, MSc, 63, PhD(phys metall), 65. *Prof Exp:* Mem tech staff, Solid State Device Labs, Bell Tel Labs, 66-68; asst prof eng, 68-77, ASSOC PROF ENG, CARLETON UNIV, ONT, 77- *Mem:* Am Soc Metals. *Res:* Magnetic properties of materials. *Mailing Add:* Dept Eng Mech Carleton Univ Ottawa ON K1S 5B6 Can

BIBEAU, ARMAND A, zoology; deceased, see previous edition for last biography

BIBEAU, THOMAS CLIFFORD, b Palmer, Mass, Mar 15, 49; m 71; c 2. FOOD COLORIMETRY. *Educ:* Univ Mass, BS, 71, MS, 74 & 88, PhD(food sci), 77. *Prof Exp:* Res chemist, Heublein Inc, RJR Industs, 76-78, mgr, prod develop, 78-81, assoc mgr, 81-82, mgr, 82-84, dir, beverage prod, 86-88, brand develop, 88-90; GROUP DIR, RES & DEVELOP, GRAND MET, 90- *Concurrent Pos:* Inst Food Technologists scholar, 70 & 71. *Mem:* Inst Food Technologists; Flavor & Extract Mfrs Asn. *Res:* The study of organic acid changes on processing of vegetable products; methods of measuring specific product's color; the evaluation of stability of subsidiary coal tar dyes; organic acid profiles of vegetables. *Mailing Add:* 190 Oxford Lane Windsor CT 06095

BIBEL, DEBRA JAN, b San Francisco, Calif, Apr 6, 45; c 1. MEDICAL MICROBIOLOGY, MICROBIAL ECOLOGY. *Educ:* Univ Calif, Berkeley, AB, 67, CPhil, 71, PhD(immunol), 72. *Prof Exp:* Bacteriologist, Letterman Army Inst Res, 72-76, microbiologist, 76-78; res assoc, Kaiser Found, 81-83; prod mgr, Tago, Inc, 83-86; DIR, ELIE METCHNIKOFF MEM LIBR, 78-; RES ASSOC, KAISER FOUND, UNIV CALIF, SAN FRANCISCO, 78- *Concurrent Pos:* Lectr, Sch Pub Health, Univ Calif, Berkeley, 75, Ctr Health Studies, Antioch Col W, 75 & Univ Calif Exten, Berkeley, 77; USPHS traineeship, 67-72; tech writer, Hoefer Sci Instruments, 79; consult, 80, 86. *Mem:* Am Soc Microbiol; AAAS; Asn Women Sci; Fedn Am Scientists. *Res:* Microbiology and ecology of human skin and nose; topical antimicrobial agents; host-parasite relationships; history and philosophy of medical microbiology and immunology. *Mailing Add:* 230 Orange St No 6 Oakland CA 94610-4139

BIBER, MARGARET CLARE BOADLE, b Melbourne, Australia, Jan 18, 43; m 69; c 1. NEUROPHARMACOLOGY. *Educ:* Univ London, BSc, 64; Oxford Univ, DPhil(biochem, pharmacol), 67. *Prof Exp:* Res assoc pharmacol, Sch Med, Yale Univ, 68-69, from instr to asst prof, 69-75; assoc prof, 75-87, PROF PHYSIOL, MED COL VA, VA COMMONWEALTH UNIV, 87- *Concurrent Pos:* A B Coxe Mem fel pharmacol, Sch Med, Yale Univ, 67-68; estab investr, Am Heart Asn, 75-80; mem, Basic Psychopharmacol & Neuropsychopharmacol Res Rev Comt, Alcohol, Drug Abuse, and Mental Health Admin, HEW, 81-85; prin investr, grant support, NIH, NINCDS, 79- *Mem:* Am Soc Pharmacol & Exp Therapeut; Soc Neurosci; Am Soc Neurochem; AAAS. *Res:* Studies on the regulation of neurotransmitter formation, particularly of catecholamines and serotonin, by ongoing nervous activity, drugs and hormones. *Mailing Add:* Dept Physiol Med Col Va Va Commonwealth Univ Richmond VA 23298-0551

BIBER, THOMAS U L, MEMBRANE PHYSIOLOGY. *Educ:* Univ Bern, MD, 56. *Prof Exp:* PROF PHYSIOL & BIOPHYS, MED COL VA, 75- *Mailing Add:* Sanger Hall Rm 3009 Med Col Va 1101 E Marshall St Richmond VA 23298

BIBERMAN, LUCIEN MORTON, b Philadelphia, Pa, May 31, 19; m; c 3. PHYSICS. *Educ:* Rensselaer Polytech Inst, BS, 40. *Prof Exp:* Phys chemist, Congoleum-Nairn, Inc, NJ, 41-42; physicist, US Navy, SC, 42-44 & US Naval Ord Test Sta, 44-57; dir syst div, Midway Labs, Univ Chicago, 57-59, assoc dir, Labs Appl Sci, 59-63; MEM SR RES STAFF, INST DEFENSE ANALYSIS, 63- *Concurrent Pos:* Adj prof, Univ RI, 69-, vis prof, 71-72. *Mem:* AAAS; fel Optical Soc Am; fel Soc Photo-Optical Instrument Engrs; fel Inst Elec & Electronics Engrs; fel Soc Info Display. *Res:* Military applications of optics and electronics with emphasis on infrared and ultraviolet techniques. *Mailing Add:* Inst Defense Analysis 1801 Beauregard St Alexandria VA 22311

BIBERSTEIN, ERNST LUDWIG, b Breslau, Ger, Nov 11, 22; nat US; m 49; c 4. VETERINARY BACTERIOLOGY. *Educ:* Univ Ill, BS, 47; Cornell Univ, DVM, 51, MS, 54, PhD(vet bact), 55; Am Col Vet Microbiologists, dipl. *Prof Exp:* Actg asst prof path, NY State Vet Col, Cornell Univ, 55-56; asst prof vet med, Univ Calif, Davis, 56-60, assoc prof, 60- 66, chmn dept, 69-74, chief, Microbiol Serv, Vet Med Teaching Hosp, 66-75, prof, 66-90, EMER PROF MICROBIOL, UNIV CALIF, DAVIS, 90- *Concurrent Pos:* NIH spec fel, 63-64 & 68-69. *Mem:* Am Vet Med Asn; Am Soc Microbiol; fel Am Acad Microbiol; Path Soc Gt Brit & Ireland; Am Col Vet Microbiol. *Res:* Infectious diseases of animals; pasteurellosis; haemophilus infections; clinical microbiology. *Mailing Add:* Dept Vet Microbiol Univ Calif Davis CA 95616

BIBLER, NED EUGENE, b Bucyrus, Ohio, July 25, 37; m 63; c 3. RADIATION CHEMISTRY. *Educ:* Denison Univ, BSc, 59; Ohio State Univ, MSc, 62, PhD(chem), 65. *Prof Exp:* Res chemist, Savannah River Lab, E I du Pont de Nemours & Co Inc, 65-73, staff chemist, 73-79, radiation chem, 79-91, RADIATION CHEM, SAVANNAH RIVER LAB, WESTINGHOUSE SAVANNAH RIVER CO, 91- *Mem:* Am Chem Soc; Mat Res Soc; Am Soc Testing & Mat. *Res:* Radiation chemistry associated with radioactive isotope production, nuclear fuel reprocessing, and radioactive waste storage; nuclear waste vitrification and disposal. *Mailing Add:* Savannah River Lab Westinghouse Savannah River Co Aiken SC 29801

BIBLIS, EVANGELOS J, b Thessaloniki, Greece, Apr 8, 29; US citizen; m 61; c 2. WOOD SCIENCE & TECHNOLOGY. *Educ:* Univ Thessaloniki, BS, 53; Yale Univ, MF, 61, PhD(forestry), 65. *Prof Exp:* Qual control engr, Kaman Aircraft Corp, 61-63; PROF WOOD TECHNOL, AUBURN UNIV, 65- *Concurrent Pos:* Consult, Indust & Atty. *Mem:* Soc Wood Sci & Technol; Forest Prod Res Soc. *Res:* Mechanical behavior of wood and wood structures. *Mailing Add:* Dept Forestry Auburn Univ Main Campus Auburn AL 36849

BIBRING, THOMAS, GAMETES, MICRO TUBULES. *Educ:* Univ Calif, Berkeley, PhD(biophys), 62. *Prof Exp:* ASSOC PROF MOLECULAR BIOL, VANDERBILT UNIV, 70- *Mailing Add:* Dept Molecular Biol Vanderbilt Univ Nashville TN 37232

BIC, LUBOMIR, b Iglau, Czech, Nov 28, 51; German citizen; m 81; c 1. PARALLEL PROCESSING. *Educ:* Tech Univ, Darmstadt, Germany, MS, 76; Univ Calif, Irvine, PhD(computer sci), 79. *Prof Exp:* Computer scientist, Siemens, Munich, Germany, 79-80; asst prof, 80-85, ASSOC PROF COMPUTER SCI, UNIV CALIF, IRVINE, 86- *Mem:* Asn Comput Mach; Inst Elec & Electronics Eng. *Res:* Development of computational models suitable to highly parallel processing. *Mailing Add:* Dept ICS Univ Calif Irvine CA 92717

BICE, DAVID EARL, b Cornville, Ariz, Apr 8, 38; m 60; c 6. IMMUNOLOGY. *Educ:* Utah State Univ, BS, 62; Univ Ariz, MS, 64; La State Univ, PhD(trop med), 68. *Prof Exp:* La State Univ res assoc, Int Ctr Med Res & Training, San Jose, Costa Rica, 64-68, from instr to asst prof med & microbiol, La State Univ Med Ctr, New Orleans, 68-75; SR SCIENTIST, INHALATION TOXICOL RES INST, LOVELACE BIOMED & ENVIRON RES INST, 75- *Concurrent Pos:* Res fels immunol, La State Univ, 68-69 & Harvard Med Sch & Tufts Univ, 70-71; scientist, Dept Med, La State Univ Div Charity Hosp, 68-75; fac affil, Dept Path, Colo State Univ, 84-, Fort Collins, Colo, 84-; consult, Nat Acad Sci. *Mem:* Am Asn Immunol; Am Thoracic Soc; AAAS. *Res:* Lung immunology and the effects of inhaled pollutants on lung immune defenses. *Mailing Add:* Inhalation Toxicol Res Inst PO Box 5890 Cell Toxicol Albuquerque NM 87185

BICHARD, J W, solid state physics, for more information see previous edition

BICHSEL, HANS, b Basel, Switz, Sept 2, 24; nat US; m 59; c 2. RADIATION PHYSICS, NEUTRON DOSIMETRY. *Educ:* Univ Basel, MA & PhD(physics), 51. *Prof Exp:* Exchange fel physics, Princeton Univ, 51-52, res asst physics, 52-54; lectr optics, Univ Basel, 54-55; res assoc physics, Rice Inst, 55-57; asst prof physics, Univ Wash, 57-59; from asst prof to assoc prof, Univ Southern Calif, 59-68; assoc prof, Univ Calif, Berkeley, 68-69; from assoc prof to prof radiol, Univ Wash, 69-80; CONSULT, 80- *Concurrent Pos:* Mem subcomt penetration charged particles in matter, Nat Acad Sci-Nat Res Coun, 62-; NSF grant, 62-64; consult, Pac Northwest Labs, Battelle Mem Inst, 66-69 & Int Comn Radiation Units & Measurements, 70-; consult, 80-; lectr, Aarhus Univ, 82; vis scientist, GSF Muenchen, 84-85; consult, IAEA, Vienna, 90- *Mem:* Am Phys Soc; Swiss Soc Physics & Natural Hist. *Res:* Interaction of radiation with matter, use of radiation in medicine and biology; photon and neutron radiation dosimetry. *Mailing Add:* 1211 22nd E Seattle WA 98112-3534

BICHTELER, KLAUS RICHARD, b Leipzig, Ger, Mar 15, 38; m 69; c 1. MATHEMATICS. *Educ:* Univ Hamburg, dipl, 62, PhD(gen relativity), 65. *Prof Exp:* Asst math, Univ Heidelberg, 65-68; asst prof, Southwest Ctr Advan Studies, 66-69; asst prof, 69-71, ASSOC PROF MATH, UNIV TEX, AUSTIN, 71- *Mem:* Am Math Soc; Math Asn Am. *Res:* Stochastic analysis. *Mailing Add:* Dept Math Univ Tex Austin TX 78712

BICK, GEORGE HERMAN, b Neptune, La, Sept 23, 14; m 45; c 2. ENTOMOLOGY, ZOOLOGY. *Educ:* Tulane Univ, BS, 36, MS, 38; Cornell Univ, PhD(entom), 47. *Prof Exp:* Inspector, Food & Drug Admin, USDA, 38-41; field biologist, State Dept Conserv, La, 41-42; asst prof, Univ Miss, 47-48 & Tulane Univ, 48-56; assoc prof, Univ Southwestern La, 56-59; prof, Clarion State Col, 59-60; prof biol, St Mary's Col, Inc, 60-80, emer prof, 80-81; RETIRED. *Mem:* Fel AAAS; Entom Soc Am. *Res:* Biology of Odonata. *Mailing Add:* 1928 SW 48 Ave Gainesville FL 32608

BICK, KATHERINE LIVINGSTONE, b PEI, Can; US citizen; wid; c 2. NEUROBIOLOGY, DEMENTIA. *Educ:* Acadia Univ, BSc, 51, MSc, 52; Brown Univ, PhD(biol), 57. *Prof Exp:* Res assoc biochem, Collip Med Res Lab, Univ Western Ont, 52-54; asst res pathologist, Sch Med, Univ Calif, Los Angeles, 59-61; asst prof biol, Calif State Univ, Northridge, 61-66 & Georgetown Univ, 70-76; dep dir, neurol dis prog, 76-81, actg dep dir, 81-83, dep dir, Nat Inst Neurol & Commun Dis & Stroke, 83-87; dep dir extramural res, NIH, 87-90; SCI LIAISON, CENTRO SMID USA, 90- *Concurrent Pos:* Mem, Pew Neurosci Prog, Nat Adv Comt, 86-, Nat Mult Sclerosis Soc Res Progs Adv Comt, 91- & Comt Fed Role in Educ Res, Nat Res Coun - Nat Acad Sci, 91-92; chair, World Fedn Neurol Res Group Dementia, 86-93. *Honors & Awards:* Super Serv Award, USPHS. *Mem:* Am Acad Neurol; AAAS; Asn Res Nerv & Ment Dis; Soc Neurosci; Int Brain Res Orgn; Sigma Xi; World Fedn Neurol. *Res:* Alzheimer's disease, senile dementia and related degenerative disorders of the nervous system; demyelinating disorders; histochemistry of osmoregulation in fishes; histochemistry of calcification. *Mailing Add:* Centro Smid USA 1775 K St NW Suite 800 Washington DC 20006

BICK, KENNETH F, b Janesville, Wis, Feb 14, 32; m 58; c 3. GEOLOGY. *Educ:* Yale Univ, BS, 54, MS, 56, PhD(geol), 58. *Prof Exp:* Asst prof geol, Washington & Lee Univ, 58-61; assoc prof, 61-66, chmn dept geol, 62-68, PROF GEOL, COL WILLIAM & MARY, 66- *Mem:* AAAS; Geol Soc Am; Am Asn Petrol Geologists; Am Geophys Union. *Res:* Structural and stratigraphic geology; application of computers to geological problems. *Mailing Add:* Dept Geol Col William & Mary Williamsburg VA 23185

BICK, PETER HAMILTON, b Toledo, Ohio, Sept 30, 48; m 71; c 1. IMMUNOLOGY. *Educ:* Albion Col, AB, 70; Univ Mich, PhD(microbiol), 75. *Prof Exp:* Guest scientist immunol, Div Immunobiol, Karolinska Inst, Stockholm, Sweden, 75-77; fel genetics, Wash Univ, 77-78; ASST PROF IMMUNOL, DEPT MICROBIOL, MED COL VA, 78- *Concurrent Pos:* Arthritis Found fel, 75-78. *Res:* Activation and regulation of immunocompetent cells. *Mailing Add:* Dept Biochem Eli Lilly & Co Lilly Corp Ctr Indianapolis IN 46285

BICK, RODGER LEE, b San Francisco, Calif, May 21, 42. HEMATOLOGY, ONCOLOGY. *Educ:* Univ Calif, Irvine, MD, 70. *Prof Exp:* From intern to resident internal med, Kern Gen Hosp, Bakersfield, Calif, 70-72; fel, 72-73, MEM FAC HEMAT-ONCOL, UNIV CALIF CTR HEALTH SCI, LOS ANGELES, 73-; MED DIR, BAY AREA HEMAT ONCOL RES LABS, 73- *Concurrent Pos:* Dir med educ & hemat-oncol, Kern Gen Hosp, 73-74; attend hematologist, Santa Monica Med Ctr, Calif, 74-, dir oncol, 75-; attend hematologist, St John's Health Sci Ctr, Santa Monica & Univ Calif Ctr Health Sci, Los Angeles, 74-; mem thrombosis coun, Am Heart Asn; med dir, Calif Coagulation Labs; med dir, San Joaquin Hemat Oncol Med Group, Bakersfield, Calif; med dir, Oncol Unit, San Joaquin Community Hosp; chief staff, San Joaquin Community Hosp. *Mem:* Int Soc Thrombosis & Hemostasis; Am Soc Hemat; Am Cancer Soc; fel Am Col Physicians; fel Am Soc Clin Pathologists. *Res:* Hemostasis and blood coagulation; alterations of hemostasis associated with cardiopulmonary bypass; hypercoagulability and thrombosis, acquired antithrombin-III deficiency; alterations of hemostasis associated with malignancy, nature of disseminated intravascular coagulation. *Mailing Add:* San Joaquin Hemat Oncol Group 3550 Q St No 105 Bakersfield CA 93301

BICK, THEODORE A, b Brooklyn, NY, Dec 18, 30; m 56; c 4. MATHEMATICS. *Educ:* Union Col, BS, 58; Univ Rochester, MS, 60, PhD(ergodic theory), 64. *Prof Exp:* Instr math, Hobart & William Smith Cols, 61-64, from asst prof to assoc prof, 64-66; assoc prof, 66-81, PROF MATH, UNION COL, NY, 81- *Mem:* Math Asn Am. *Res:* Ergodic theory; functional analysis; real variable theory; measure and integration. *Mailing Add:* Dept Math Union Col Schenectady NY 12308

BICKART, THEODORE ALBERT, b New York, NY, Aug 25, 35; m 58, 82; c 3. ELECTRICAL ENGINEERING, COMPUTER ENGINEERING. *Educ:* Johns Hopkins Univ, BES, 57, MSE, 58, DEng, 60. *Prof Exp:* Instr elec eng, Johns Hopkins Univ, 58-59, lectr, 59-61; from asst prof to prof elec & computer eng, Syracuse Univ, 63-89, dean eng, 84-89; PROF ELEC ENG, MICH STATE UNIV, 84-, DEAN ENG, 89- *Concurrent Pos:* Vis scholar, Dept Elec Eng & Computer Sci, Univ Calif, Berkeley, 77; Vis lectr, Nanging Inst Technol, People's Repub China, 81; mem bd dirs eng, Deans Coun, Am Soc Eng Educ, 90-; comnr eng, Manpower Comn, Am Asn Eng Soc, 90- *Honors & Awards:* Fulbright Lectr, Kiev Polytech Inst, USSR, 81. *Mem:* Fel Inst Elec & Electronics Engrs; Am Mat Soc; Soc Indust & Appl Math; Asn Comput Mach; Nat Soc Prof Engrs; Am Soc Eng Educ. *Res:* Nonlinear systems analysis; linear systems analysis; system simulation; numerical analysis & functional analysis in systems analysis; stability theory; digital hardware design. *Mailing Add:* Eng Dean's Off 300 Eng Bldg Mich State Univ East Lansing MI 48824-1226

BICKEL, EDWIN DAVID, b Louisville, Ky, Nov 11, 41; m 67; c 2. ECONOMIC GEOLOGY, ENVIRONMENTAL GEOLOGY. *Educ:* Univ Louisville, AB, 63, MSc, 65; Ohio State Univ, PhD(geol), 70. *Prof Exp:* Asst prof geol, 70-73, dir coop educ, Minot State Col, 73-76; adminr geol, N Am Coal Corp, 77-83; geologist, Thames Bickel Assoc, 84-89; GEOLOGIST & HYDROLOGIST, RECLAMATION DIV, PUB SERV COMN, 89- *Concurrent Pos:* Work-group researcher, Northern Great Plains Resources Prog, 74-75; contracted malacologist, Off Endangered Species, US Fish & Wildlife Serv, 74-; consult, 75-77; proj prin investr, NDak Regional Environ Assessment Prog, 76-77. *Mem:* Am Asn Petrol Geologists; Soc Mining Engrs; Sigma Xi; Paleont Res Inst; Am Inst Prof Geologists; Am Inst Mining Engrs; Nat Water Well Asn. *Res:* Malacology; systematics and ecology of living and fossil non-marine Mollusca; Cretaceous-Tertiary biostratigraphy, coal stratigraphy; socio-economic and environmental impacts of energy development; petroleum geology and groundwater hydrology. *Mailing Add:* Reclamation Div State Capital NDak Pub Serv Comn Bismark ND 58505

BICKEL, JOHN HENRY, b Chicago, IL, June 23, 50; m 73; c 3. NUCLEAR POWER PLANT SAFETY ANALYSIS. *Educ:* Univ VT, BS, 72, MS, 74; Rensselaer Polytech Inst, MS, 76, PhD(eng sci), 80. *Prof Exp:* Nuclear Eng, Combustion Eng, 75-79; Adv comt Reactor Safegards Fel, US Nuclear Regulatory Comn, 79-80; supvr probabilistic risk assessment, Northeast Utilities Serv Co, 80-89; MGR RISK ANALYSIS, EG&G IDAHO, 89- *Concurrent Pos:* Consult & lectr, Int Atomic Energy Agency, Vienna, Soviet State Comt Supr Nuclear Energy, Moscow, Swedish Power Bd, Stockholm; lectr, Argonne Nat Lab, Nuclear power plant oper safety, 86- 87; mem peer rev, NRC, 87-88. *Honors & Awards:* Young Mem Eng Achievement Award, Am Nuclear Soc, 83. *Mem:* Am Nuclear Soc. *Res:* Development and application of probabilistic safety and reliability methods to nuclear power plants; operational decision making and reliability improvements; reliability engineering and risk assessment. *Mailing Add:* EG&G Idaho Inc PO Box 1625 Mail Stop 2405 Idaho Falls ID 83415-2405

BICKEL, PETER J, b Bucharest, Romania, Sept 21, 40; US citizen; m 64; c 2. PROBABILITY. *Educ:* Univ Calif, Berkeley, AB, 60, MA, 61, PhD(statist), 63; Hebrew Univ, Jerusalem, PhD(hc). *Prof Exp:* From asst prof to assoc prof statist, 63-70, chmn dept, 75-78, PROF STATIST, UNIV CALIF, BERKELEY, 70- *Concurrent Pos:* Lectr, Imp Col, Univ London, 65-66; J S Guggenheim Found fel, 70; NATO sr sci fel, 74-75; J D & Catherine T MacArthur Found fel, 84-88. *Honors & Awards:* Comt Presidents Statist Socs Award, 79. *Mem:* Nat Acad Sci; fel Inst Math Statist (pres, 80); Royal Statist Soc; fel Am Statist Asn; mem Int Statist Inst; Am Acad Arts & Sci; Am Statist Asn. *Res:* Nonparametric statistics; robustness; asymptotic methods. *Mailing Add:* Dept Statist Univ Calif Berkeley CA 94720

BICKEL, ROBERT JOHN, b Louisville, Ky, Nov 8, 16; m 42. MATHEMATICS. *Educ:* Univ Louisville, AB, 37; Northwestern Univ, MA, 41; Univ Pittsburgh, PhD(math), 60. *Prof Exp:* Teacher, pub schs, Ky, 37-41; from instr to assoc prof, 46-60, actg head dept, 68-69, 73-74, PROF MATH, DREXEL UNIV, 60-, ASSOC HEAD DEPT, 63- *Mem:* Math Asn Am; Soc Indust & Appl Math (treas, 55-62). *Res:* Analysis; summability; divergent series. *Mailing Add:* 415 Ellis Woods Rd Pottstown PA 19464

BICKEL, THOMAS FULCHER, b Detroit, Mich, Nov 20, 37; m 64. ALGEBRA. *Educ:* Univ Mich, BS, 59, MA, 60, PhD(math), 65. *Prof Exp:* Instr math, Mass Inst Technol, 65-67; from asst prof, 67-80, chmn dept, 85-87, PROF MATH, DARMOUTH COL, 80- *Mem:* Am Math Soc. *Res:* Finite group theory; theory of finite groups of Lie type; permutation groups. *Mailing Add:* Dept Math Dartmouth Col Hanover NH 03755

BICKEL, WILLIAM SAMUEL, b Ottsville, Pa, June 8, 37. EXPERIMENTAL PHYSICS, SPECTROSCOPY. *Educ:* Pa State Univ, BS, 59, PhD(physics), 65. *Prof Exp:* Res assoc physics, 65, from asst prof to assoc prof, 65-75, PROF PHYSICS, UNIV ARIZ, 75- *Mem:* Am Phys Soc; Biophys Soc; fel Am Soc Eng Educ. *Res:* Spectroscopic diagnostics; atomic physics; spectroscopy of fast excited ions; measurement of mean lives of excited ions; vacuum ultraviolet and visible spectroscopy; biophysics; time-resolved polarized light scattering from biological macromolecules, spheres, fibers and irregular particles; acoustics and physics of music; surface physics. *Mailing Add:* Dept Physics PAS Univ Ariz Tucson AZ 85721

BICKELHAUPT, R(OY) E(DWARD), b Waterloo, Ill, Jan 14, 28; m 48; c 1. CERAMICS ENGINEERING. *Educ:* Univ Ill, BS, 50, PhD(ceramic eng), 63; Rensselaer Polytech, MS, 59. *Prof Exp:* Instr ceramic eng, Clemson Col, 50-52; engr, Prod Div, Bendix Corp, 52-55, actg chief engr cerametallic eng, Marshall Eclipse Div, 55-59; res assoc ceramic eng, Univ Ill, 59-63; sr ceramic engr, Southern Res Inst, 63-83; PRES & TECH CONSULT, BICKELHAUPT ASSOCS INC, 83- *Concurrent Pos:* Int fel, Second Int Conf Electrostaic Precipitation. *Mem:* Am Chem Soc; Am Ceramic Soc; Air Pollution Control Asn. *Res:* Electrostatic precipitation; electrical resistivity of particulates; coal technology. *Mailing Add:* 1206 S New Wilkie Rd Arlington Heights IL 60005

BICKERMAN, HYLAN A, b New York, NY, Oct 26, 13; m 41; c 2. RESPIRATORY PHYSIOLOGY, CLINICAL PHARMACOLOGY. *Educ:* Columbia Univ, BA, 34, MA, 35; NY Univ, MD, 39. *Prof Exp:* CHIEF ASTHMA-EMPHYSEMA CLIN, COLUMBIA-PRESBY MED CTR, 61- *Concurrent Pos:* Res fel med & chest dis, Columbia Univ Res Serv, Goldwater Mem Hosp, New York, 42-47, vis physician, 57-68; chief med serv, Cushing Gen Hosp, Boston, 45-46; asst clin prof, Col Physicians & Surgeons, Columbia Univ, 57-58, assoc clin prof, 58-78, clin prof, 78-80, spec lectr, 80-; consult, St Barnabas Hosp, New York, 65-71, Brookhaven Mem Hosp, Suffolk County, NY & Arden Hill, Monroe, NY, 68-; dir respiratory lab & inhalation ther dept, Francis Delafield Hosp, New York, 68-75; panelist on emphysema, Int Cong Chest Dis, Lausanne, Switz, 70; asst attend physician, Presby Hosp, New York, 70-78, attend physician, 78-; mem adv panel, Food & Drug Admin, 72-76; dir pulmonary function labs, Doctors Hosp, New York. *Honors & Awards:* Golden Tree of Life Award, Am Asn Inhalation Therapists, 67. *Mem:* AAAS; fel AMA; fel Am Col Chest Physicians; fel Am Col Physicians; Am Thoracic Soc. *Res:* Respiratory physiology and management of patients with chronic obstructive lung disease, especially pharmacologic therapy; development of various modalities of inhalation therapy including aerosols and pressure breathing. *Mailing Add:* 215-30 28th Ave Bayside NY 11360

BICKERT, WILLIAM GEORGE, b Bismarck, NDak, Apr 9, 37; m 61; c 2. AGRICULTURAL ENGINEERING. *Educ:* NDak State Univ, BS, 59, MS, 60; Mich State Univ, PhD(agr eng), 64. *Prof Exp:* Res asst agr eng, NDak State Univ, 59-60; from res asst & instr to assoc prof, 61-72, PROF AGR ENG, MICH STATE UNIV, 72- *Mem:* Am Soc Agr Engrs. *Res:* Livestock production facilities, ventilation systems; milking systems, equipment, design, labor utilization; environmental control. *Mailing Add:* Dept Agr Eng Mich State Univ East Lansing MI 48824-1323

BICKERTON, ROBERT KEITH, b East Liverpool, Ohio, Oct 1, 34; m 57; c 2. PHARMACOLOGY. *Educ:* Univ Pittsburgh, BS, 56, MS, 58, PhD(pharmacol), 60. *Prof Exp:* Sr res pharmacologist, Norwich Pharmacol Co, 60-62, unit leader, 62-63, from asst chief to chief sect pharmacol, 63-65, dir pharmacometrics div, 65-68, dir res, 68-69, vpres res, 69-73, vpres sci affairs, 73-77, sr vpres sci affairs, 77-80; MENTOR GROUP, INC, 80-; DIR, CO-OPER RES, YALE UNIV, 87- *Mem:* Am Pharmaceut Asn; Am Soc Pharmacol & Exp Therapeut. *Res:* Mechanism of action and development of antihypertensive drugs. *Mailing Add:* Nine Old Still Rd Woodbridge CT 06525

BICKFORD, LAWRENCE RICHARDSON, b Elmira, NY, Nov 24, 21; m 43, 70; c 3. SOLID STATE PHYSICS. *Educ:* Alfred Univ, BS, 43; Mass Inst Technol, PhD(physics), 49. *Prof Exp:* Assoc prof physics, State Univ NY Col Ceramics, Alfred, 49-54; sr physicist res lab, IBM, 54-63, dir gen sci, Res Ctr, 62-63, dir res lab, Tokyo, Japan, 63-65, dir memory & mat res, Res Lab, NY, 65-70, mgr mat sci & technol, 70-73, res staff mem, IBM Res Lab, Calif, 73-77, prog dir tech commun, IBM White Plains, NY, 77-82; dep exec secy, Am Phys Soc, 82-84; RETIRED. *Mem:* Fel Am Phys Soc; Inst Elec & Electronics Engrs. *Res:* Ferrites; titanates; glass; organic conductors; contact effects on conductivity in anisotropic organic conductors; superconductivity. *Mailing Add:* 175 E 62nd St New York NY 10021

BICKFORD, MARION EUGENE, b Memphis, Tenn, Aug 30, 32; m 54; c 3. PETROLOGY, GEOCHEMISTRY. *Educ:* Carleton Col, BA, 54; Univ Ill, MS, 58, PhD(geol), 60. *Prof Exp:* Asst prof geol, San Fernando Valley State Col, 60-63; asst res geophysicist, Inst Geophys, Univ Calif, Los Angeles, 63-64; from asst prof to assoc prof, 64-73, PROF GEOL, UNIV KANS, 73- *Mem:* Fel Mineral Soc Am; fel Geol Soc Am. *Res:* Geochronology and petrology of the Precambrian of Saskatchewan, Colorado, and the buried crust of the midcontinent; geochronology of deformation in core complexes. *Mailing Add:* Dept Geol Univ Kans Lawrence KS 66045

BICKFORD, REGINALD G, b Brewood, Eng, Jan 20, 13; nat US; m 45; c 2. NEUROSCIENCES. *Educ:* Cambridge Univ, BA, MD & BCh, 36; FRCP, 71. *Prof Exp:* Mayo res assoc, Univ Minn, 46-48, from assoc prof to prof physiol, 53-69; head EEG Lab, 69-80, PROF NEUROSCI, SCH MED, UNIV HOSP, UNIV CALIF, SAN DIEGO, 69- *Concurrent Pos:* Med Res Coun fel, Univ London, 37-40; consult EEG, Mayo Clin, 46-69; mem, neurol study sect, USPHS, 60-64, mem, comput study sect, 64-68, mem adv comt epilepsies, 64-70, mem, clin pract comt deleg to Soviet Union, 68 & mem, sci info prog adv comt, Nat Inst Neurol Dis & Stroke, 70- *Mem:* Am EEG Soc (pres, 56); Am Physiol Soc; Int League Against Epilepsy; Am Neurol Asn. *Res:* Electrical activity of the brain in man and animals; experimentally produced changes and their relation to behavioral and psychological effects; computer analysis and automation of the electroencephalogram. *Mailing Add:* Dept Neurosci Univ Calif San Diego M-024-Mailcode La Jolla CA 92093

BICKHAM, JOHN W, b Chillicothe, Ohio, Mar 4, 49; m 81; c 2. CYTOGENETICS, SYSTEMATICS. *Educ:* Univ Dayton, BS, 71, MS, 73; Tex Tech Univ, PhD(zool), 76. *Prof Exp:* Teaching asst biol, Univ Dayton, 72-73; teacher zool, Texas Technol Univ, 73-76; from asst prof to assoc prof, 76-87, PROF WILDLIFE & FISHERIES SCI, TEX A&M UNIV, 87- *Mem:* Am Soc Mammalogists; Am Soc Ichthyologists & Herpetologists; Soc Syst Zool; AAAS. *Res:* Application of cytogenetics to the study of evolution, speciation and taxonomy of vertebrates, chiefly mammals and reptiles. *Mailing Add:* Dept Wildlife & Fisheries Sci Tex A&M Univ College Station TX 77843-2258

BICKLEY, WILLIAM ELBERT, b Knoxville, Tenn, Jan 20, 14; m 41; c 4. ENTOMOLOGY. *Educ:* Univ Tenn, BS, 34, MS, 36; Univ Md, PhD(entom), 40. *Prof Exp:* Teacher, pub schs, Tenn, 35-37; teaching fel entom, Univ Md, 37-40, instr, 40-42; from asst entomologist to sr asst sanitarian, malaria control in war areas, USPHS, 42-46; asst prof biol, Univ Richmond, 46-49; from assoc prof to prof, 48-78, head dept, 57-71, EMER PROF ENTOM, UNIV MD, 78-; RES ASSOC, WALTER REED BIOSYSTEMATICS UNIT, SMITHSONIAN INST, 78- *Concurrent Pos:* Entomologist, State

Dept Pub Health, Va, 47-49; ed, Mosquito News, 73-81, emer ed, 81-; assoc ed, J Am Mosquito. *Mem:* Entom Soc Am; Mosquito Control Asn (pres, 61); Am Soc Trop Med & Hyg; Sigma Xi. *Res:* Chrysopidae; insect morphology; Japanese beetle; mosquitoes; vegetable insects; alfalfa weevil; dog to dog transmission of heartworm by mosquitoes. *Mailing Add:* 6516 40th Ave University Park MD 20782

BICKLING, CHARLES ROBERT, b Wilmington, Del, May 29, 24; m 47. CHEMICAL ENGINEERING. *Educ:* Univ Del, BChE, 45; Univ Wis, MS, 46; PhD(chem eng), 50. *Prof Exp:* Res engr, Textile Fibers Dept, 50-53, group supvr & tech supvr Chatta nylon, 53-58, tech supvr Seaford nylon, 58-63, process supvr, 63-66, fibers planning mgr, Ducilo SAIC, Arg, 66-70, sr supvr process, Va, 70-77, SPECIALIST, TEXTILE FIBERS DEPT, E I DU PONT DE NEMOURS & CO, INC, 77- *Mem:* AAAS; Am Chem Soc; Am Inst Chem Engrs. *Mailing Add:* 916 Dundee Ct Martinsville VA 24112

BICKMORE, JOHN TARRY, b Logan, Utah, Jan 15, 28; m 53; c 3. XEROGRAPHIC DEVELOPMENT, XEROGRAPHIC DEVELOPERS. *Educ:* Idaho State Col, BS, 50; Univ Rochester, PhD(biophys), 56. *Prof Exp:* Res assoc, Univ Rochester, 52-55; res physicist, 55-75, SR SCIENTIST, XEROX CORP, 75- *Mem:* AAAS; fel Soc Photog Sci & Eng. *Res:* Chemical effects of ionizing radiation; xerographic process research, particularly the mechanisms, materials properties and failure modes in the development process. *Mailing Add:* Joseph C Wilson Ctr for Technol 800 Philips Rd Webster NY 14580

BICKNELL, EDWARD J, b Kansas City, Mo, Jan 23, 28; m 52. COMPARATIVE PATHOLOGY, CLINICAL PATHOLOGY. *Educ:* Univ Mo, BA, 48; Kans State Univ, MS, 51, DVM, 60; Mich State Univ, PhD(path), 65. *Prof Exp:* Instr path, Mich State Univ, 60-64; asst prof, Kans State Univ, 65-66; assoc prof, Iowa State Univ, 66-67; assoc prof vet sci, SDak State Univ, 68-72; animal pathologist & exten veterinarian, Agr Exp Sta, Univ Ariz, 72-; VET PATHOLOGIST, ANIMAL DIAG LAB, UNIV ARIZ. *Res:* Pathology of range cattle disease; pathology of swine disease. *Mailing Add:* Animal Diag Lab Univ Ariz Tucson AZ 85721

BICKNELL, WILLIAM EDMUND, b Kearney, Nebr, Nov 22, 35; m 57; c 3. ELECTRICAL ENGINEERING. *Educ:* Univ Ill, BS, 57; Mass Inst Technol, MS, 58; Stanford Univ, PhD(elec eng), 64. *Prof Exp:* Asst elec eng, Stanford Univ, 60-63; electronics engr, Res & Develop Lab, US Army, 63-65; specialist, Appl Res Lab, Sylvania Elec Co, Gen Tel & Electronics Corp, 66-69; STAFF MEM, LINCOLN LAB, MASS INST TECHNOL, 69- *Concurrent Pos:* Consult, Sylvania Elec Co, Gen Tel & Electronics Corp, Calif, 62 & 63; lectr, Monmouth Col, 63- *Mem:* Am Phys Soc; Inst Elec & Electronics Engrs. *Res:* Microwave and quantum electronics; maser and laser devices; electro-optic systems design and analysis. *Mailing Add:* Lincoln Lab MS KB 276 Mass Inst Technol PO Box 73 Lexington MA 02173

BICKNELL-BROWN, ELLEN, b Jackson, Miss, Nov 24, 44; c 1. VIBRATIONAL SPECTROSCOPY, BIOMOLECULAR CONFORMATION. *Educ:* Rice Univ, AB, 66; Brown Univ, PhD(phys chem), 71. *Prof Exp:* Res assoc, Univ Ore, 71-73, Univ Fla, 73-77; asst prof chem, Wayne State Univ, 77-87; spec expert chem, Nat Lib Med, 87-90, DR CHOU LAB, NIH, 90- *Mem:* Am Chem Soc; Sigma Xi; AAAS. *Res:* Raman and infrared spectroscopy used to probe structure and interactions in biological molecules and systems such as membranes, proteins, lipids, and nucleic acids; resonance-enchanced near ultraviolet raman spectroscopy to investigate electronic excited states of proteins. *Mailing Add:* NIH Dr Chou Lab 9000 Rockville Pike Bethesda MD 20894

BIDDINGTON, WILLIAM ROBERT, b Piedmont, WVa, Mar 30, 25; m 47; c 1. DENTISTRY. *Educ:* Univ Md, DDS, 48. *Prof Exp:* From instr to assoc prof, Dent Sch, Univ Md, 48-59; PROF ENDODONTICS & HEAD DEPT, SCH DENT, WVA UNIV MED CTR, 59-, DEAN SCH DENT, 68- *Concurrent Pos:* Vchmn, Joint Comn Nat Dent Exam; mem, Comn Dent Health, Coun Sports Med, US Olympic Comn, 80- *Mem:* Am Dent Asn; Int Asn Dent Res; Am Col Dent; Am Asn Endodontists. *Res:* Endodontics; periodontics. *Mailing Add:* Off Dean WVa Univ Sch Dent Morgantown WV 26505

BIDDLE, JOHN WILBUR, chemical engineering, for more information see previous edition

BIDDLE, RICHARD ALBERT, b Philadelphia, Pa, July 16, 30; m 55; c 4. INORGANIC CHEMISTRY. *Educ:* Pa State Univ, BS, 52; Iowa State Univ, MS, 55. *Prof Exp:* Instr & res assoc chem, Iowa State Univ, 52-56; res chemist, Elkton Div, 57-60, head inorg & propellant chem group, Res Dept, 60-66, scientist res dept, 66-85, HEAD INGREDIENTS SYNTHESIS, ELKTON DIV, MORTON THIOKOL INC, 85- *Mem:* AAAS; Am Chem Soc; Am Defense Preparedness Asn. *Res:* Inorganic chemical syntheses with emphasis on boron hydrides and high vacuum techniques; thermal decomposition of perchlorates and kinetics of solid phase reactions; space simulation studies; thermoanalytical techniques; hygrometry; solid rocket propellant chemistry and formulation; hazards studies in propellants and explosives. *Mailing Add:* Morton Thokol Inc 55 Thokol Rd Elkton MD 21921-0241

BIDDLECOM, WILLIAM GERARD, chiral organic synthesis; deceased, see previous edition for last biography

BIDE, MARTIN JOHN, b Portsmouth, UK, Nov 7, 51. TEXTILES, DYESTUFFS. *Educ:* Univ Bradford, UK, BTech Hons, 74, PhD(dyestuff chem), 79. *Prof Exp:* Develop chemist dyes, Yorkshire Chem, UK, 77-80; develop chemist herbicides, A H Marks, Bradford, UK, 80-81; ASSOC PROF TEXTILE CHEM, SOUTHEASTERN MASS UNIV, 81- *Concurrent Pos:* Consult, res proj, USDA. 83-85, dyeing indust Mass & RI, 81-; lectr, US Navy Clothing & Textile Res Facil, Natrick, Mass, 82-85 & AATCC, 85. *Mem:* Am Asn Textile Chemists & Colorists; Soc Dyes & Colorists; Royal Soc Chem. *Res:* Dyestuff chemistry; dyeing, especially wool; wool scouring; color. *Mailing Add:* Textile Sci Dept SE Mass Univ N Dartmouth MA 02747

BIDE, RICHARD W, b Calgary, Alta, Sept 5, 39; c 5. PATHOLOGICAL CHEMISTRY, TOXICOLOGY. *Educ:* Univ Alta, BSc, 59, MSc, 61; Aberdeen Univ, PhD(biochem), 64. *Prof Exp:* Nat Res Coun Can fel radiobiol, Atomic Energy Can, Ltd, 64-66; vet clin chem & biochem, Animal Res Inst, Lethbridge, Can Dept Agr, 66-81; toxicol & biochem residues, Animal Path Lab, Agr Can, Sask, 81-84; head, Biomed Sect, 84-87, DEFENSE SCIENTIST, DEFENSE RES ESTAB, SUFFIELD, RALSTON, ALTA, 87- *Mem:* NY Acad Sci. *Res:* Enzymology; radiobiology; veterinary pathological chemistry; biomedical profiling; toxicology. *Mailing Add:* Biomed Defense Sect Defense Res Estab Suffield Box 4000 Medicine Hat AB T1A 8K6 Can

BIDELMAN, WILLIAM PENDRY, b Los Angeles, Calif, Sept 25, 18; m 40; c 4. ASTRONOMY. *Educ:* Harvard Univ, SB, 40; Univ Chicago, PhD(astron), 43. *Prof Exp:* Asst, Yerkes Observ, Univ Chicago, 41-43; physicist, Ballistic Res Lab, Aberdeen Proving Ground, US Ord Dept, 43-45; from instr astron to asst prof astrophys, Univ Chicago, 45-53; from asst astronr to assoc astronr, Lick Observ, Univ Calif, 53-62; prof astron, Univ Mich, 62-69 & Univ Tex, 69-70; chmn dept astron, Case Western Reserve Univ & dir, Warner & Swasey Observ, 70-75, prof, 70-86, EMER PROF ASTRON, CASE WESTERN RESERVE UNIV, 86- *Mem:* Am Astron Soc; Int Astron Union; Astron Soc Pac. *Res:* Spectral classification; observational astrophysics; astronomical data. *Mailing Add:* Case Western Reserve Univ Cleveland OH 44106

BIDER, JOHN ROGER, b Lachine, Que, Nov 23, 32; m 56; c 6. ECOLOGY. *Educ:* Univ Montreal, BSc, 56, MSc, 59, PhD(ecol), 66. *Prof Exp:* PROF ECOL, FAC AGR, McGILL UNIV, 65- *Concurrent Pos:* Sci consult, Sci Coun Can, 69-70; consult, Environ Adv Coun, Gov Que, 73-77, Food & Agr Orgn, Rome & Can Int Develop Agency, Burundi, 74, Upper Volta, 74 & Mali, 88; consult, Environ Mgt Hydro-Que, 75-85, Coun Ecol Reserves, 75-77, Can Int Develop Agency, Haiti, 76-80 & Environ Coun Gov Que, 86-90. *Mem:* Ecol Soc Am; Am Soc Mammalogists; Wildlife Soc; Can Soc Zoologists; Sigma Xi. *Res:* Terrestrial vertebrate ecology with emphasis on factors which affect animal activity; the temporal and spatial utilization of the environment; the relationship between activity and numbers; life history strategies and population dynamics of turtles. *Mailing Add:* Macdonald Col 21111 Lakeshore Ste Anne de Bellevue PQ H9X 1C0 Can

BIDGOOD, BRYANT FREDERICK, b Port Colborne, Ont, Apr 29, 37; m 64; c 2. STATISTICS. *Educ:* Ont Col, BSA, 62; Univ Guelph, 65; Univ Alta, 72. *Prof Exp:* Dist biollogist fisheries, Alta Dept Nat Resources, 65-68; RES SCIENTIST FISHERIES, ALTA DEPT FORESTRY LANDS & WILDLIFE, 68- *Mem:* Am Fish Soc; fel Am Inst Fish Res Biol; Can Soc Zoologist. *Mailing Add:* N Tower Petrol Plaza 9945 108th St Edmonton AB T5K 2C9 Can

BIDINOSTI, DINO RONALD, b Winnipeg, Man, Mar 27, 33; m 57; c 4. PHYSICAL CHEMISTRY. *Educ:* Univ Man, BSc, 55, MSc, 56; McMaster Univ, PhD(chem), 59. *Prof Exp:* Defense Res Bd Can fel phys chem, Univ Ottawa, 59-60; res assoc & Air Res & Develop Command fel, Cornell Univ, 60-61; assoc prof, 61-66, ASSOC PROF PHYS CHEM, UNIV WESTERN ONT, 66- *Mem:* Chem Inst Can; Royal Soc Chem. *Res:* Application of mass spectrometry to chemical systems; thermochemistry. *Mailing Add:* Dept Chem Univ Western Ont London ON N6A 5B8 Can

BIDLACK, DONALD EUGENE, b Oakwood, Ohio, Apr 16, 32; m 53; c 3. ANIMAL HUSBANDRY, ANIMAL SCIENCE & NUTRITION. *Educ:* Ohio State Univ, BSc, 54, DVM, 57. *Prof Exp:* Pvt pract, Ind, 57-65; vet, Eaton Labs, Norwich Pharmacol Co, 65-73; res vet, 73-78, assoc pathologist, 78-85, RES PATHOLOGIST, SYNTEX RES, 85- *Mem:* Am Vet Med Asn. *Res:* Industrial pharmaceutical research and development. *Mailing Add:* Syntex Res Div 3401 Hillview Ave Palo Alto CA 94304

BIDLACK, VERNE CLAUDE, JR, b South Bend, Ind, Mar 5, 23; m 46; c 4. ORGANIC CHEMISTRY. *Educ:* Univ Mich, BS, 44; Pa State Univ, MS, 48, PhD(chem), 50. *Prof Exp:* Res chemist, E I Du Pont de Nemours & Co, 50-51; chief org chemist pharmaceuts, Henry K Wampole Co, Inc, 51-55; sr com develop engr, Archer-Daniels-Midland Co, 55-62; mgr mkt & prod develop, Atlantic Refining Co, 62, chem mkt res, 62-64; dir long planning, Chem Group, W R Grace & Co, 64-66; develop engr & purchasing agent, FMC Corp, 66-76; mem staff, Res Dept, C Lever Co, Inc, 77-80, dir res, 80-87; RETIRED. *Res:* Fatty acids and vegetable oils; natural and synthetic resins; protective coatings; adhesives; medicinal chemicals; plastics; fertilizers; plastic films, dyes and pigments. *Mailing Add:* 1599 W Rightstown Rd Newtown PA 18940

BIDLACK, WAYNE ROSS, b Waverly, NY, Aug 12, 44; m 68. TOXICOLOGY. *Educ:* Pa State Univ, Univ Park, BS, 66; Iowa State Univ, Ames, MS, 68; Univ Calif, Davis, PhD(biochem), 72. *Prof Exp:* Fel pharmacol, 72-74, asst prof, 74-80, ASSOC PROF PHARMACOL & NUTRIT, SCH MED, UNIV SOUTHERN CALIF, 80- *Concurrent Pos:* Chmn bd & bd dirs, Nat Coun Against Health Fraud, 85-86; USC postgrad, 87-88; chmn, Southern Calif Inst Food Technol, 88-89, Toxicol & Safety Eval Div, 89-90. *Mem:* Am Inst Nutrit; Inst Food Technologists; Soc Toxicol; Am Soc Pharmacol & Exp Therapeut. *Res:* Hepatic drug and toxicant metabolism; evaluation of nutritional status; nutrition and the elderly; vitamin and mineral metabolism; effects of nutritional parameters. *Mailing Add:* 900 N Alamansor St Alhambra CA 91801

BIDLEMAN, TERRY FRANK, b Chicago, Ill, Feb 17, 42. ANALYTICAL CHEMISTRY, ENVIRONMENTAL CHEMISTRY. *Educ:* Ohio Univ, BS, 64; Univ Minn, PhD(analytical chem), 70. *Prof Exp:* Fel chem, Dalhousie Univ, 70-72; res assoc food & resource chem, Univ RI, 72-75; asst prof chem, 75-81, ASSOC PROF CHEM & MARINE SCI, UNIV SC, 81- *Mem:* Am Chem Soc; Soc Environ Toxicol & Chem. *Res:* Pesticide residue analysis; transport of pesticides in the environment; chemical aspects of air, sea transfer; solvent extraction equilibria; equilibria in natural water systems. *Mailing Add:* Dept Chem Univ SC Columbia SC 29208

BIDLINGMAYER, WILLIAM LESTER, b Cleveland, Ohio, July 7, 20; m 50; c 2. ENTOMOLOGY. *Educ:* Univ Fla, BSA, 49, MS, 52. *Prof Exp:* Entomologist, Fla State Bd Health, 50-51; asst sanitarian, tech develop lab, Commun Dis Ctr, USPHS, 51-55; entomologist, Div Health, Fla Med Entom Lab, Univ Fla, 55-79, prof, 79-87; RETIRED. *Mem:* Am Mosquito Control Asn; Entom Soc Am. *Res:* Biology and ecology of mosquitoes and sandflies; sampling and population dynamics of mosquito populations. *Mailing Add:* Rte 3 Box 24 C Monticello FL 32344

BIDLINGMEYER, BRIAN ARTHUR, b Dallas, Tex, Aug 8, 44; m 72; c 1. ANALYTICAL CHEMISTRY. *Educ:* Kenyon Col, AB, 66; Purdue Univ, PhD(analytical chem), 71. *Prof Exp:* Res chemist, Standard Oil Co Ind, 71-74; MGR ORG BIOSCI RES, WATERS ASSOCS, 75- *Mem:* Am Chem Soc; Am Soc Testing & Mat. *Res:* Application of liquid chromatography to difficult separations in organic and biological research, specifically pesticide residues, long-chain fatty acids and lipids; data handling, manipulation and interpretation in chromatography. *Mailing Add:* 26 Teresa Rd Hopkinton MA 01748

BIDWELL, LAWRENCE ROMAINE, b Olean, NY, Jan 8, 31; m 52; c 4. PHYSICAL METALLURGY. *Educ:* Ohio State Univ, BMetE, 53, MSc, 54, PhD(thermodyn), 62. *Prof Exp:* Res assoc metall, Res Found, Ohio State Univ, 53-54; task scientist, Aerospace Res Labs, 54-60, res metallurgist, 60-67, res metallurgist, Mat Lab, 67-70, sr res scientist, 70-76, GROUP LEADER, MAT LAB, USAF, 76- *Mem:* AAAS; Am Soc Metals; Am Inst Mining, Metall & Petrol Engrs; Sigma Xi. *Res:* Development of corrosion resistant, high temperature, rapidly solidified and aerospace structural alloys; phase equilibria and thermodynamic properties of metal and non-metal systems; electrical properties of refractory oxides; solid electrolytes. *Mailing Add:* 5787 Redbird Ct Dayton OH 45431

BIDWELL, LEONARD NATHAN, b Camden, NJ, Dec 20, 34; m 63; c 2. MATHEMATICS. *Educ:* Univ Pa, BA, 56, MA, 57, PhD(math), 60. *Prof Exp:* Vis asst prof math, Haverford Col, 60-61; mathematician, Gen Elec Co, 61-62; asst prof math, 62-67, ASSOC PROF MATH, RUTGERS UNIV, CAMDEN, 67- *Mem:* Am Math Soc; Math Asn Am. *Res:* Topology; functional analysis. *Mailing Add:* Rutgers Univ Camden NJ 08102

BIDWELL, ORVILLE WILLARD, b Whitehouse, Ohio, Jan 14, 18; m 44; c 2. SOIL MORPHOLOGY, ALTERNATIVE AGRICULTURE. *Educ:* Oberlin Col, AB, 40; Ohio State Univ, BSc, 42, PhD(agron), 49. *Prof Exp:* Asst agron, Ohio State Univ, 46-49; from asst prof to prof, 50-84, actg head dept agron, 70-71, emer prof soils, Kans State Univ, 84; RETIRED. *Concurrent Pos:* Vis prof, Ahmadu Bello Univ, Nigeria, 73. *Honors & Awards:* Merit Award, Soil Conserv Soc Am, 63. *Mem:* Fel AAAS; fel Am Soc Agron; fel Soil Sci Soc Am; fel Soil Conserv Soc Am. *Res:* Soil classification and development. *Mailing Add:* 3815 Emerald Circle Manhattan KS 66502-7514

BIDWELL, ROGER GRAFTON SHELFORD, b Halifax, NS, June 8, 27; m 50; c 4. PLANT PHYSIOLOGY. *Educ:* Dalhousie Univ, BS, 47; Queen's Univ, Ont, BA, 50, MA, 51, PhD, 54. *Hon Degrees:* FRSC, 72. *Prof Exp:* Res officer, Defense Res Bd, 52-56; botanist, Atlantic Regional Lab, Nat Res Coun, 56-59; assoc prof bot, Univ Toronto, 59-65; prof biol, Case Western Reserve Univ, 65-69, chmn dept, 67-69; prof biol, Queen's Univ, Ont, 69-79; res prof biol, Dalhousie Univ, Halifax, NS, 81-85; dir, Atlantic Inst Biotechno, 85-88; dir, Atlantic Res Assocs, 82-90. *Concurrent Pos:* Secy, Biol Coun Can, 72-76; Nat Biotechnol Adv Comt, 83-89. *Honors & Awards:* Queen Elizabeth II Silver Jubilee Medal, 79; Gold Medal, Can Soc Plant Physiol, 80. *Mem:* Fel AAAS; Am Soc Plant Physiol; Can Soc Plant Physiol (secy-treas, 63-65, vpres, 71-72, pres, 72-73); fel Royal Soc Can. *Res:* Intermediary metabolism in plants; carbohydrates and amino acids; process of photosynthesis; metabolism of marine algae. *Mailing Add:* Atlantic Res Assocs RR 1 Wallace NS B0K 1Y0 Can

BIEBEL, PAUL JOSEPH, b Belleville, Ill, Feb 26, 28; m 51; c 6. PHYCOLOGY. *Educ:* Univ Notre Dame, BS, 49; St Louis Univ, MS, 55; Ind Univ, PhD(phycol), 63. *Prof Exp:* Instr biol, Spring Hill Col, 59-63; from asst prof to assoc prof, 63-74, PROF BIOL, DICKINSON COL, 74- *Honors & Awards:* DAAD Awardee. *Mem:* Bot Soc Am; Phycol Soc Am; Am Bryol & Lichenological Soc; Int Phycol Soc; Sigma Xi. *Res:* Life cycles, morphology, taxonomy, genetics and morphogenesis of algae, especially saccoderm desmids. *Mailing Add:* Dept Biol Dickinson Col Carlisle PA 17013

BIEBER, ALLAN LEROY, b Mott, NDak, Aug 14, 34; m 58; c 2. BIOCHEMISTRY. *Educ:* NDak State Univ, BS, 56, MS, 58; Ore State Univ, PhD(biochem), 62. *Prof Exp:* NIH training grant, Sch Med, Yale Univ, 61-63; from asst prof to assoc prof, 63-75, PROF CHEM, ARIZ STATE UNIV, 75- *Mem:* Am Soc Biol Chemists; Am Chem Soc; Sigma Xi. *Res:* Metabolism of purines; purine analogs and their respective nucleotides; biochemistry of snake venoms and toxins. *Mailing Add:* Dept Chem Ariz State Univ Tempe AZ 85287

BIEBER, HAROLD H, b New York, NY, Apr 4, 27; m 50; c 3. CHEMICAL ENGINEERING. *Educ:* City Col New York, BChE, 47; Polytech Inst Brooklyn, MChE, 50, DChE, 56. *Prof Exp:* Chem engr, Am Cyanamid Co, NJ, 47-54; sr res assoc, Polytech Inst Brooklyn, 54-56; sr chem engr, 56-64, chem eng fel, 64-68, mgr process design, 68-78, proj dir, 78-79, dir eng design, 79-81, DIR CHEM PROCESS PLANT DEPT, HOFFMANN-LA ROCHE, INC, 82- *Mem:* Fel Am Inst Chem Engrs. *Res:* Unit operations. *Mailing Add:* 68 E Brook Terr Livingston NJ 07039

BIEBER, HERMAN, b Berleburg, Ger, Jan 13, 30; US citizen; m 57; c 3. CHEMICAL ENGINEERING. *Educ:* Columbia Univ, BS, 52, MS, 53, DEngSci(chem eng), 62. *Prof Exp:* Instr chem eng, Columbia Univ, 53-54; engr, Esso Res & Eng Co, Linden, NJ, 55-57, group head, 57-58, sr engr, 58-61, res assoc, 61-62, sect head tech info, 62-68, sr res assoc, Corp Res Dept, 68-81, sr res assoc, Comput & Info Sci Div, 81-86; CONSULT, BIEBER

ENTERPRISES, 86- *Concurrent Pos:* Lectr, City Col New York, 54-55; adj prof, Stevens Inst Technol, 58-64; trustee, Midwest Res Inst, 67-90; mem, bd & exec comt, Eng Found, 81-; mem bd, Eng Soc Libr, 84- *Mem:* Am Inst Chem Engrs; fel Am Inst Chemists; Sigma Xi. *Res:* Ion exchange kinetics; residual fuels and heavy crudes refining; information research; technology transfer; creative problem solving. *Mailing Add:* Bieber Enterprises 14 Dorset Dr Kenilworth NJ 07033-1417

BIEBER, IRVING, medicine, psychiatry; deceased, see previous edition for last biography

BIEBER, LORAN LAMOINE, b Mott, NDak, Apr, 33; m 55; c 3. BIOCHEMISTRY, SCIENCE ADMINISTRATION. *Educ:* NDak State Univ, BS, 55, MS, 56; Ore State Univ, PhD(biochem), 63. *Prof Exp:* Asst agr chem, NDak State Univ, 56-57; lab technician, Hosp Lab, Ft Carson, Colo, 57-58 & Chem Ctr, Edgewood, Md, 58-59; PROF BIOCHEM, MICH STATE UNIV, 59-, ASSOC DEAN RES MED, 83- *Concurrent Pos:* NIH fel biochem, Univ Calif, Los Angeles, 63-65; NIH fel, Mich State Univ, 66-74 & NSF fel, 67-71; vis prof cell physiol, Wenner-Gren Inst, Stockholm, 74; distinguished vis scholar, Univ Adelaide, 83. *Mem:* AAAS; Am Soc Biol Chemists; Am Chem Soc; Sigma Xi; Am Inst Nutrit. *Res:* Functions of carnitine; metabolic control; control of fatty acid and branched-chain amino acid oxidation. *Mailing Add:* Dept Biochem Mich State Univ East Lansing MI 48823

BIEBER, MARK ALLAN, b Cleveland, Ohio, Sept 16, 46. BIOCHEMISTRY, NUTRITION. *Educ:* Univ Pittsburgh, BS, 68; Mich State Univ, PhD(biochem), 73. *Prof Exp:* Fel nutrit, Col Physicians & Surgeons, Columbia Univ, 73-77; sr nutritionist, 78-80, prin nutritionist, 80-82, NUTRIT RES ASSOC, BEST FOODS, UNIT CPC INT, INC, 82- *Concurrent Pos:* Matheson Found fel brain res, Col Physicians & Surgeons, Columbia Univ, 74-77; vis fel Pediat, 74-77. *Mem:* Am Chem Soc; Am Oil Chemists Soc; Soc Nutrit Educ; Am Heart Asn; Am Inst Nutrit; Am Col Nutrit. *Res:* Lipids, fats and oils; essential fatty acids and deficiency; lipoproteins; atherosclerosis diet and cancer; brain growth and development; insect hormones; mass spectrometry. *Mailing Add:* Best Foods Res & Eng Ctr 1120 Commerce Ave Union NJ 07083

BIEBER, RAYMOND W, CELL BIOLOGY. *Educ:* Univ Colo, PhD(cell biol), 72. *Prof Exp:* ASSOC PROF CELL BIOL, MED CTR, UNIV NEBR, 72- *Mailing Add:* RFD 1 Lincoln NE 68583

BIEBER, SAMUEL, b US, Feb 5, 26; m 49; c 2. DEVELOPMENTAL BIOLOGY. *Educ:* NY Univ, BA, 44, MS, 48, PhD(vert zool), 52. *Prof Exp:* Teaching fel, Washington Sq Col, NY Univ, 48-51, res fel, 51-52; sr biologist, res labs, Burroughs Wellcome & Co, 52-62; prof biol, assoc dean sci & assoc dean grad sch, Long Island Univ, 62-64, assoc dean grad fac & spec adv to provost for sci, 64-66, dean, Conolly Col, 66-69; campus dean, Fairleigh Dickinson Univ, 69-71, prof biol sci & provost, Teaneck-Hackensack Campus, 71-78, actg vpres acad affairs, 78-80; vpres acad affairs, 80-82, PROF, BIOL SCI DIV, OLD DOMINION UNIV, 80- *Concurrent Pos:* Sci collabr, New York Aquarium, 52-62; from adj asst prof to adj assoc prof grad sch, Long Island Univ, 57-62; vis fel, Cambridge Univ, 76. *Mem:* AAAS; Am Chem Soc; fel NY Acad Sci; Am Asn Cancer Res; Int Soc Develop Biol; Soc Develop Biol; Soc Exp Biol & Med; Sigma Xi. *Res:* Renotropic effects of steroids; antimetabolites and embryogenesis and regeneration; nucleic acid metabolism during gametogenesis and embryogenesis; experimental cancer chemotherapy; chemical suppression of the immune response. *Mailing Add:* Dept Biol Sci Old Dominion Univ Norfolk VA 23529-0266

BIEBER, THEODORE IMMANUEL, b Zurich, Switz, July 6, 25. ORGANIC CHEMISTRY, BIOCHEMISTRY. *Educ:* NY Univ, BA, 45, MS, 46, PhD(biochem), 51. *Prof Exp:* Asst chem, NY Univ, 47-51; asst prof, Coe Col, 51-52, Adelphi Col, 53-56 & Ga Inst Technol, 56-57; from assoc prof to prof, Univ Miss, 57-63; PROF CHEM, FLA ATLANTIC UNIV, 63- *Mem:* AAAS; Am Chem Soc; NY Acad Sci; The Chem Soc; Sigma Xi. *Res:* Organophosphorus and organosulfur chemistry; stereochemistry; amide synthesis; iodine-containing dyes; heterocycles; dehydrogenation; respiratory chain phosphorylation; metallocene chemistry. *Mailing Add:* Dept Chem Fla Atlantic Univ Boca Raton FL 33431

BIEBERMAN, ROBERT ARTHUR, b Rock Island, Ill, Apr 3, 23; m 44; c 2. PETROLEUM GEOLOGY. *Educ:* Ind Univ, AB, 48, AM, 50. *Prof Exp:* Petrol geologist, NMex Bur Mines, 50-88; RETIRED. *Mem:* Am Asn Petrol Geologists. *Res:* Subsurface geology. *Mailing Add:* 601 Fitch Ave Socorro NM 87801

BIEBUYCK, JULIEN FRANCOIS, b S Africa, Feb 2, 35; US citizen; m 61; c 3. NEUROCHEMISTRY, NATIONAL ACADEMIC ADMINISTRATION. *Educ:* Univ Cape Town, MB & ChB, 59; Univ Oxford, Eng, DPhil(biochem), 71. *Prof Exp:* Nuffield res fel biochem, Univ Oxford, Eng, 69-72; asst prof anesthesia, Med Sch, Harvard Univ & asst anesthetist, Mass Gen Hosp, 72-74, lectr, Med Sch, Harvard Univ, 74-76; prof anesthesia, 77-83, ERIC A WALKER PROF ANESTHESIA, COL MED, PA STATE UNIV, 84-, CHMN DEPT, 77- *Concurrent Pos:* Fel, Med Found, Boston, 72-74; mem sci adv bd, Asn Univ Anesthetists, 80-83; ed, Anesthesiol, 85-; fel fac Anesthetiats, Royal Australiasian Soc Surg, 87. *Mem:* Biochem Soc Eng; Am Soc Neurochem; Am Physiol Soc; Asn Univ Anesthetists; Am Soc Anesthesiologists; Soc Acad Anesthesia (pres, 85-87); Soc Neurosci. *Res:* Metabolic basis of altered consciousness including metabolic coma and anesthesia; non-respiratory functions of the lung; organ substate utilization and metabolic control. *Mailing Add:* Dept Anesthesia Col Med Pa State Univ Hershey PA 17033

BIEDEBACH, MARK CONRAD, b Pasadena, Calif, Apr 21, 32; m 64; c 2. BIOPHYSICS. *Educ:* Univ Southern Calif, BE, 56, MS, 58; Univ Calif, Los Angeles, PhD(biophys), 64. *Prof Exp:* NIH res fel biol systs, Calif Inst Technol, 64-65; Nat Inst Neurol Dis & Blindness res fel neurophysiol, Lab

Cellular Neurophysiol, Paris, France, 65-66; ASSOC PROF BIOL, CALIF STATE UNIV, LONG BEACH, 67- *Mem:* Soc Neurosci; Sigma Xi. *Res:* Biophysics of invertebrate nervous system-integration and photoresponsive mechanisms. *Mailing Add:* 612 Balboa Dr Seal Beach CA 90740

BIEDENBENDER, MICHAEL DAVID, b Wichita, Kans, Apr 28, 61. MICROWAVE DEVICES & CIRCUITS, MICROELECTRONIC PROCESS DEVELOPMENT. *Educ:* Univ Cincinnati, BS, 84, MS, 87, PhD(elec eng), 91. *Prof Exp:* Res asst, Northrop, 84; grad asst, 84-85, RES ASST ELEC ENG, UNIV CINCINNATI, 85- *Mem:* Inst Elec & Electronics Engrs; Electrochem Soc. *Res:* Designed, fabricated and characterized microwave power transistors on indium phosphide; ion implantation, encapsulation, and rapid thermal annealing of indium phosphide; indium phosphide surfaces using auger electron spectroscopy and x-ray photoelectron spectroscopy. *Mailing Add:* 2362 Stratford Apt 3 Cincinnati OH 45219

BIEDENHARN, LAWRENCE CHRISTIAN, JR, b Vicksburg, Miss, Nov 18, 22; m 50; c 2. NUCLEAR PHYSICS. *Educ:* Mass Inst Technol, BS, 44, PhD(physics), 49. *Prof Exp:* From res asst to res assoc, Mass Inst Technol, 48-50; physicist, Oak Ridge Nat Lab, 50-52; asst prof physics, Yale Univ, 52-54; assoc prof, Rice Univ, 54-61; PROF PHYSICS, DUKE UNIV, 61- *Concurrent Pos:* Sr Fulbright fel, 57-58; Guggenheim fel, 58; NSF sr fel, 64-65; Erskine fel, NZ, 73; consult, Los Alamos Sci Lab; vis prof, Calif Inst Technol, 68, Inst Theoret Phys, Univ Karlsruhe, Germany, 72 & Univ Tex, Austin, 83-84; vis scientist, Inst Higher Studies, Bures-sur-Yvette, France, 78 & 80 & theoret div, Stanford Linear Accelerator Lab, Stanford, Calif, 81; ed, J Math Physics, 85- *Honors & Awards:* Alexander von Humboldt Found Award, Ger, 76; Jesse Beams Prize, Am Phys Soc, 79. *Mem:* Sigma Xi; fel Inst Physics; fel Brit Phys Soc; Swiss Phys Soc; fel Am Phys Soc. *Res:* Theoretical physics; nuclear reactions. *Mailing Add:* Dept Physics Duke Univ Durham NC 27706

BIEDERMAN, BRIAN MAURICE, cytogenetics, genome organization; deceased, see previous edition for last biography

BIEDERMAN, EDWIN WILLIAMS, JR, b Stamford, Conn, June 30, 30; m 58; c 4. OIL PETROLOGY, GAS RESERVOIRS & MINEROLOGY. *Educ:* Cornell Univ, BA, 52; Pa State Univ, PhD(mineral), 58. *Prof Exp:* Res geologist & tech group leader, 58-68, res planner, Geochem & Sedimentology, Cities Serv Res & Develop, Co, 68-72; asst dir, 72-77, energy exten serv coordr & tech specialist, Pa Tech Assistance Prog, 77-80, TECH SPECIALIST, PA STATE UNIV, 80- *Concurrent Pos:* Mem bd adv, Micropaleont Press, Am Mus Natural Hist, 70-73. *Mem:* Am Asn Petrol Geologists; Soc Econ Paleontologists & Mineralogists; Geochem Soc; Am Inst Prof Geologists; AAAS; Int Platform Soc. *Res:* Sedimentary petrology of oil and gas reservoir rocks; recent sediments; geochemistry; origin of oil; geological statistics; economic geology; photomicrography; general and physical geology; electron microscopy; glacial deposits; precious metals; five United States patents awarded. *Mailing Add:* 232 Mineral Sci Bldg Pa State Univ University Park PA 16802

BIEDERMAN, RONALD R, b Hartford, Conn, Oct 19, 38; m 62; c 1. METALLURGY, MATERIALS SCIENCE. *Educ:* Univ Conn, BS, 60, MS, 62, PhD(mat sci), 68. *Prof Exp:* Instr metall, Univ Conn, 62-68; prof mech eng, 68-81, DIR ELECTRON MICROSCOPE FACIL, WORCESTER POLYTECH INST, 68-, GEORGE F FULLER PROF, 87- *Concurrent Pos:* Fel, Am Soc Metals, 86- *Honors & Awards:* Stanley P Rockwell Mem Lectr, Am Soc Metal, 81. *Mem:* Fel Am Soc Metals; Am Inst Mining, Metall & Petrol Engrs; Sigma Xi; Electron Micros Soc Am; Int Metallog Soc. *Res:* Strengthening mechanisms; x-ray and electron metallography; powder metallurgy; bearing materials; high temperature materials; titanium & Zirconium; ceramic membranes. *Mailing Add:* Dept Mech Eng Worcester Polytech Inst 100 Institute Rd Worcester MA 01609

BIEDERMAN-THORSON, MARGUERITE ANN, b Ft Leavenworth, Kans, Jan 9, 36; m 64. NEUROPHYSIOLOGY. *Educ:* Univ Wash, Seattle, BS, 59; Univ Calif, Los Angeles, PhD(neurophysiol), 65. *Prof Exp:* NATO fel, Max Planck Inst Biol, Tubingen, 65-66 & Inst Exp Psychol, Oxford Univ, 66-67; asst res scientist neurosci, Univ Calif, San Diego, 67-69; Humboldt Stipend, Max Planck Inst Physiol Behav, 70-72; CONSULT, 72- *Concurrent Pos:* Freelance translr, 72- *Res:* Neurophysiology and behavior of invertebrates and birds; editing papers and books on physiology, ecology, zoology and medicine. *Mailing Add:* Old Marlborough Arms Combe Oxford 0X7 2NQ England

BIEDLER, JUNE LEE, b New York, NY, June 24, 25. CELL GENETICS. *Educ:* Vassar Col, AB, 47; Columbia Univ, MA, 54; Cornell Univ, PhD(biol), 59. *Prof Exp:* Exchange investr, Inst Gustave-Roussy, France, 59-60; res fel, 59-60, res assoc, 60-62, assoc, 62-72, sect head, 66-72, assoc mem, 72-78, MEM, SLOAN-KETTERING INST CANCER RES, 78-, LAB HEAD, 75-; ASSOC PROF BIOL, SLOAN-KETTERING DIV, GRAD SCH MED SCI, CORNELL UNIV, 73- *Concurrent Pos:* From instr to asst prof, Sloan-Kettering Div, Grad Sch Med Sci, Cornell Univ, 62-73; USPHS res career develop award, 63. *Mem:* Fel AAAS; Am Asn Cancer Res; Am Soc Cell Biol; Genetics Soc Am; NY Acad Sci. *Res:* Cytogenetics and somatic cell genetics; tumor biology; chromosome structure-function relationships; drug resistance of mammalian cells. *Mailing Add:* Sloan-Kettering Mem Cancer Ctr 1275 York Ave New York NY 10021

BIEFELD, PAUL FRANKLIN, b Brownsville, Pa, Nov 5, 25; m 50; c 3. PHYSICAL CHEMISTRY. *Educ:* Denison Univ, BS, 48; Mich State Univ, MS, 51. *Prof Exp:* Asst, Mich State Univ, 48-51 & chem res infrared spectros, State Hwy Dept, Mich, 51-56; chemist, Characterization Lab, Owens-Corning Fiberglas Corp, Newark, 56, mgr, Analytical Lab, Granville, 56-71, sr scientist, 71-77, RES ASSOC, DEPT CHEM TECHNOL, TECH CTR, OWENS-CORNING FIBERGLAS CORP, 77- *Concurrent Pos:* Mem, Analytical & Phys Equip Comt, Owens-Corning Fiberglas Corp. *Mem:* Soc Appl Spectros; Coblentz Soc. *Res:* Application of instrumental analysis in the fields of glass and organic plastics and resins. *Mailing Add:* 345 Jefferson Newark OH 43055

BIEGEL, JOHN E, b Eau Claire, Wis, Nov 19, 25; m; c 3. INDUSTRIAL ENGINEERING, SYSTEMS DESIGN & SYSTEMS SCIENCE. *Educ:* Mont State Univ, BS, 48; Stanford Univ, MS, 50; Syracuse Univ, PhD(mat sci), 72. *Prof Exp:* Instr math, Mont State Col, 48-49; asst indust eng, Stanford Univ, 49-50; from instr to asst prof, Univ Ark, 50-52; engr, Ford Motor Co, 52-53 & Sandia Corp, 53-58; from asst prof to prof indust eng, Syracuse Univ, 58-78; prof indust eng, Kans State Univ, 78-82; PROF ENG, UNIV CENT FLA, 82- *Concurrent Pos:* Consult, Pass & Seymour, Inc, NY, 58-64; adj prof, Col Environ Sci & Forestry, State Univ NY, 78. *Mem:* Soc Mfg Engrs; Inst Indust Engrs; Am Asn Artificial Intel. *Res:* intelligent simulation training systems, intelligent simulations, object-oriented simulation. *Mailing Add:* Dept Indust Eng Univ Cent Fla Box 25000 Orlando FL 32816

BIEGELSEN, DAVID K, b St Louis, Mo, Oct 18, 43; m 66. EXPERIMENTAL SOLID STATE PHYSICS. *Educ:* Yale Univ, BA, 65; Wash Univ, MA & PhD(physics), 70. *Prof Exp:* PRIN SCIENTIST PHYSICS, XEROX PALO ALTO RES CTR, 70- *Mem:* Am Phys Soc; Inst Elec & Electronics Engrs; Mat Res Soc. *Res:* Using acousto-optic probe of photoelastic properties of solids; amorphous semiconductors; heteroepitaxy; thin film crystal growth. *Mailing Add:* Xerox Palo Alto Res Ctr 3333 Coyote Hill Rd Palo Alto CA 94304

BIEGER, DETLEF, b Schoenberg, Ger, Oct 1, 39; c 3. NEUROPHARMACOLOGY, NEUROPHYSIOLOGY. *Educ:* Univ Kiel, WGer, Dr, 67. *Prof Exp:* Res assoc & instr pharmacol, Univ Kiel, 65-69; res assoc, Univ Toronto, 69-72; asst prof, Col Med, Univ Ill, 72-78; assoc prof, 78-84, PROF PHARMACOL, MEM UNIV, NFLD, 84- *Mem:* Ger Soc Pharmacol & Toxicol; Soc Neurosci; Pharmacol Soc Can; Can Asn Neurosci; Res Defence Soc Brit. *Res:* Neural basis of deglutition; neuropharmacology of myenteric ganglia; neural control of smooth muscle contractility. *Mailing Add:* Div Basic Sci Mem Univ Nfld Fac Med St Johns NF A1B 3V6 Can

BIEGLER, LORENZ THEODOR, b Chicago, Ill, Sept 10, 56; m 87; c 1. PROCESS OPTIMIZATION, PROCESS DESIGN & CONTROL. *Educ:* Ill Inst Technol, BS, 77; Univ Wis, MS, 79, PhD(chem eng), 81. *Prof Exp:* From asst prof to assoc prof, 81-90, PROF CHEM ENG, CARNEGIE-MELLON UNIV, 90- *Concurrent Pos:* Presidential young investr, NSF, 85; vis scientist, Argonne Nat Lab, 90-91. *Mem:* Am Inst Chem Engrs; Am Chem Soc; Soc Indust & Appl Math; Opers Res Soc Am. *Res:* Process flowsheet optimization; nonlinear programming; parameter estimation; process simulation; process dynamics and control/optimization; synthesis of reactor networks; large-scale modelling. *Mailing Add:* Chem Eng Dept Carnegie-Mellon Univ Pittsburgh PA 15213

BIEHL, ARTHUR TREW, physics, for more information see previous edition

BIEHL, EDWARD ROBERT, b Pittsburgh, Pa, July 14, 32; m 55; c 4. ORGANIC CHEMISTRY. *Educ:* Univ Pittsburgh, BS, 58, PhD(chem), 61. *Prof Exp:* Sr res chemist, Monsanto Res Corp, 61-62; from asst prof to assoc prof, 62-66, PROF CHEM, SOUTHERN METHODIST UNIV, 72- *Mem:* Am Chem Soc; Sigma Xi. *Res:* Benzyne chemistry. *Mailing Add:* Dept Chem Southern Methodist Univ Dallas TX 75275

BIEHL, JOSEPH PARK, b Berkeley, Calif, June 14, 22; m 49; c 4. NEUROLOGY. *Educ:* Stanford Univ, AB, 43, MD, 46. *Prof Exp:* Intern internal med, Boston City Hosp, 45-46; sr asst res physician, Cincinnati Gen Hosp, 48-49, res neurologist, 49-50; from instr to prof, 50-80, EMER PROF NEUROL, COL MED, UNIV CINCINNATI, 80- *Concurrent Pos:* Fel med, Col Med, Univ Cincinnati, 50-52; consult, Vet Admin Hosp; attend neurologist, Cincinnati Gen Hosp, Drake Mem Hosp & Longview State Hosp. *Mem:* Asn Res Nerv & Ment Dis. *Res:* Clinical neurology; electroencephalography. *Mailing Add:* 5315 Hickory Trail Lane - Blue Ash Cincinnati OH 45242

BIEHLER, SHAWN, b Jersey City, NJ, May 20, 37; m 60; c 2. GEOPHYSICS. *Educ:* Princeton Univ, BSE, 58; Calif Inst Technol, MS, 61, PhD(geophys), 64. *Prof Exp:* Res fel geophys, Calif Inst Technol, 64-66; asst prof, Mass Inst Technol, 66-70; assoc prof, 70-, PROF GEOPHYS, UNIV CALIF, RIVERSIDE. *Mem:* AAAS; Am Geophys Union; Am Soc Explor Geophys; Seismol Soc Am; Europ Asn Explor Geophys. *Res:* Application of geophysics to geothermal areas; relationship of gravity anomalies to geologic structure. *Mailing Add:* Dept Earth Sci Univ Calif Riverside Riverside CA 92521

BIELAJEW, ALEXANDER FREDERICK, b Montreal, Que, June 16, 53; m 73; c 2. MONTE CARLO THEORY, IONIZING RADIATION THEORY. *Educ:* McGill Univ, BSc, 78; Stanford Univ, PhD(physics), 82. *Prof Exp:* RES OFFICER, INST NAT MEASUREMENT STANDARDS, NAT RES COUN CAN, 82- *Honors & Awards:* Farrington Daniels Award, Am Asn Physicists Med, 85; Sylvia Fedoruk Prize, Can Orgn Med Phys, 89. *Res:* Theoretical research applied to the development of Monte Carlo techniques for use in radiation dosimetry and radiotherapy; theory of interaction of radiation with matter; theory of response of radiation-measuring instruments. *Mailing Add:* Inst Nat Measurement Standards Nat Res Coun-CN Montreal Rd Bldg M35 Ottawa ON K1A 0R6

BIELAK, JACOBO, b Mexico City, Mex. CIVIL ENGINEERING. *Educ:* Nat Univ Mex, BS, 63; Rice Univ, MS, 66; Calif Inst Technol, PhD(civil eng), 71. *Prof Exp:* Asst prof civil eng, Nat Univ Mex, 72-74, assoc prof, 74-78; ASSOC PROF CIVIL ENG, CARNEGIE-MELLON UNIV, 78- *Concurrent Pos:* Sect head dynamics, Inst Eng, Nat Univ Mex, 74-78; vis assoc res engr, Univ Calif, San Diego, 77-78. *Mem:* Am Soc Chem Engrs; Am Soc Mech Engrs; Am Acad Mech; Seismol Soc Am; Sigma Xi. *Res:* Structural mechanics; analysis of structural systems; earthquake engineering with emphasis on dynamic soil-structure interaction. *Mailing Add:* Dept Civil Eng Carnegie-Mellon Univ Schenley Park Pittsburgh PA 15213

BIELAT, KENNETH L, b Yonkers, NY, Jan 4, 45; m 82. TUMOR ULTRASTRUCTURAL BIOLOGY & IMMUNOLOGY, IMPLANTOLOGY-BIOMATERIALS. *Educ:* Niagara Univ, BS, 66; Univ Vienna, Austria, Zeugnis, 65; State Univ NY, Buffalo, PhD(biol), 75. *Prof Exp:* CANCER RES SCIENTIST, LAB MGR & COMPUTER DATABASE MGR, PATH & ULTRASTRUCTURE, ROSWELL PARK MEM INST, 68- *Concurrent Pos:* Asst prof, Erie Community Col, 75-; immunohematologist, Buffalo Gen Hosp, 79-; tutor, Empire State Col, 85-; clin asst prof, Col Med, State Univ NY, Buffalo, 87-, res asst prof, Col Dent, 88- *Mem:* Am Asn Immunologists; Am Asn Cancer Res; Electron Micros Soc Am; Int Soc Prev Oncol; Soc Ultrastruct Path; AAAS; NY Acad Sci; Am Asn Dental Res; Int Asn Dental Res. *Res:* Ultrastructural evaluation and diagnosis of cancer; immunotherapy with monoclonal antibodies, mononuclear cells and biological response modifiers; immunoelectron microscopy and flow cytometry analyses and immunophenotyping; ultrastructural evaluation and microanalysis of tissue implants composed of various biomaterials; tumor markers; ion microscopy and x-ray microanalysis. *Mailing Add:* Dept Path Roswell Park Cancer Inst Elm & Carlton Sts Buffalo NY 14263

BIELECKI, EDWIN J(OSEPH), b North Attleborough, Mass, Mar 19, 24; m 48; c 3. INORGANIC CHEMISTRY. *Educ:* Mass Inst Tehcnol, SB, 46. *Prof Exp:* Res engr, Beryllium Corp, 46-48, asst dir res, 48-51; proj mgr, Nat Res Corp, 51-56; develop engr, Kawecki Chem Co, 56-58, res mgr, 59-68, mgr res & develop chem, Kawecki Berylco Industs, Inc, 68-78; mgr, KBI Div, Cabot Corp, 78-81, mgr, res & develop chem, Cabot Eng Prod Group, 81-85; RETIRED. *Concurrent Pos:* Consult, field of rare metals. *Mem:* Am Chem Soc; Am Inst Chem Engrs; Electrochem Soc. *Res:* Rare metal; beryllium, titanium, zirconium, tantalum, columbium, boron, selenium, rubidium, germanium, cesium and others. *Mailing Add:* 310 E Sixth St Boyertown PA 19512-1212

BIELER, BARRIE HILL, b Pasadena, Calif, June 17, 29; m 55; c 3. MINERALOGY & PETROLOGY. *Educ:* Calif Inst Technol, BS, 51, MS, 52; Pa State Univ, PhD(mineral), 55. *Prof Exp:* Field asst reconnaissance geol mapping, US Geol Surv, 52; res asst mineral, Pa State Univ, 52-55; geologist, US Geol Surv, 55-57; ceramist, Dow Chem Co, 57-67, res ceramist, Western Div, 67-70, sr res ceramist, 70-74, res specialist, 74-80, res leader, 81-86; RETIRED. *Concurrent Pos:* Assoc prof gen sci, John F Kennedy Univ, 68- *Mem:* Fel Photog Soc Am. *Res:* Geothermal exploration; ceramics and refractories; slags in basic oxygen surface steelmaking; melting, fining, sintering and grinding of glasses; develop porous ceramic membranes. *Mailing Add:* 737 Wiget Lane Walnut Creek CA 94598

BIELER, RÜDIGER, b Hamburg, Ger, Apr 9, 55. PHYLOGENETIC SYSTEMATICS, COMPARATIVE ANATOMY. *Educ:* Univ Hamburg, MSc, 82, PhD(zool), 85. *Prof Exp:* Lectr zool, Univ Hamburg, 82-85; postdoctoral res fel, Smithsonian Inst, Wash, 85-86, Smithsonian Marine Sta, Ft Pierce, Fla, 86-87, NATO postdoctoral fel, Marine Sta, 87-88; asst cur & div head, Dept Malacol, Del Mus Natural Hist, 88-90, actg exec dir, Mus, 90; ASST CUR & DIV HEAD, DEPT ZOOL, FIELD MUS NATURAL HIST, 90- *Concurrent Pos:* Res assoc, Div Mollusks, Nat Mus Natural Hist, Wash, 88- & Mus Com Zool, Harvard Univ, 88-; adj asst prof, Grad Col Marine Studies, Univ Delaware, 88-; vis scientist, Smithsonian Marine Sta, 88-; counr, Am Malacol Union, 90-91; mem, Coun Syst Malacologists. *Mem:* Am Malacol Union; Soc Syst Zool; Sigma Xi. *Res:* Evolution, biology, and systematics of mollusks, especially marine snails and clams. *Mailing Add:* Field Mus Natural Hist Roosevelt Rd at Lake Shore Dr Chicago IL 60605

BIELLIER, HAROLD VICTOR, b Bois D'Arc, Mo, Jan 22, 21; m 42; c 2. PHYSIOLOGY. *Educ:* Univ Mo, BS, 43, PhD, 55. *Prof Exp:* Asst prof mil sci & tactics, 50-53, assoc prof, 53-69, PROF POULTRY HUSB, UNIV MO-COLUMBIA, 69- *Concurrent Pos:* Assoc ed, Poultry Sci, 67-82. *Mem:* Fel AAAS; Poultry Sci Asn; Sigma Xi; Worlds Poultry Sci Asn. *Res:* Endocrine and reproductive physiology of domestic poultry. *Mailing Add:* 5149 Animal Sci Ctr Univ Mo Columbia MO 65211

BIELSKI, BENON H J, b Poland, June 12, 27; US citizen; m 69. RADIATION CHEMISTRY. *Educ:* Harvard Univ, BA, 51; Columbia Univ, MA, 56, PhD(chem), 57. *Prof Exp:* Asst prof chem, Univ Fla, 57-58; res assoc, Brookhaven Nat Lab, 58-60, assoc chemist, 60-63, chemist, 66-86, SR CHEMIST, BROOKHAVEN NAT LAB, 86- *Concurrent Pos:* Vis prof, State Univ NY Downstate Med Ctr, 71-83; consult, Procyte Corp, 87- *Mem:* Am Chem Soc; Am Soc Biol Chemists. *Res:* Radiation chemistry of inorganic and biological systems; chemistry of free radicals. *Mailing Add:* Dept Chem Brookhaven Nat Lab Upton NY 11973

BIEMANN, KLAUS, b Innsbruck, Austria, Nov 2, 26; m 56; c 2. ORGANIC CHEMISTRY, ANALYTICAL CHEMISTRY. *Educ:* Univ Innsbruck, PhD(chem), 51. *Prof Exp:* Instr chem, Univ Innsbruck, 51-55; res assoc, 55-57, from instr to assoc prof, 57-63, PROF CHEM, MASS INST TECHNOL, 63- *Honors & Awards:* Stas Medal, Belg Chem Soc, 62; Powers Award, Am Acad Pharmaceut Sci, 73; Except Sci Achievement Medal, NASA, 77; Fritz Pregl Medal, Austrian Microchem Soc, 77; Field & Franklin Award Mass Spectrometry, Am Chem Soc, 86. *Mem:* Am Chem Soc; Am Acad Arts & Sci; Am Soc Mass Spectrometry; hon mem Chem Soc Belg; Protein Soc. *Res:* Mass spectrometry of organic compounds. *Mailing Add:* Dept Chem Mass Inst Technol Cambridge MA 02139

BIEMPICA, LUIS, b Orense, Spain, Aug 14, 25; US citizen; m 53; c 2. GASTROENTEROLOGY, PATHOLOGY. *Educ:* Univ Buenos Aires, MD, 52, cert nutrit, 56, PhD(med), 62. *Prof Exp:* Resident med, Bronx Munic Hosp Ctr, 66-67 & resident path, 68-71, from asst attend to assoc attend pathologist, 71-78; from asst prof to assoc prof path, 64-78, assoc med, 64-70, ASST PROF MED, ALBERT EINSTEIN COL MED, YESHIVA UNIV, 70-, CLIN PROF PATHOL, 78-; ATTEND PATHOLOGIST, BRONX MUNIC HOSP CTR, 78- *Concurrent Pos:* Arg Res Coun fel, Albert Einstein Col Med, 61-62, USPHS res grant, 67-70. *Honors & Awards:* Riopedre Prize, Arg Asn Advan Sci, 69. *Mem:* Am Asn Pathologists; Am Soc Cell Biol; Am Asn Study Liver Dis; fel Col Am Pathologists; AMA. *Res:* Liver diseases; cytochemistry; collagen degradation; ultrastructure; collagen; enzymology. *Mailing Add:* Dept Path Albert Einstein Col Med Yeshiva Univ 300 Morris Park Ave Bronx NY 10461

BIENENSTOCK, ARTHUR IRWIN, b New York, NY, Mar 20, 35; m 57; c 3. AMORPHOUS MATERIALS, SYNCHROTRON RADIATION. *Educ:* Polytech Inst Brooklyn, BS, 55, MS, 57; Harvard Univ, PhD(appl physics), 62. *Prof Exp:* Asst prof appl physics, Harvard Univ, 62-67; assoc prof mat sci, 67-72, vprovost fac affairs, 72-77, PROF MAT SCI & ENG, STANFORD UNIV, 72-, DIR, STANFORD SYNCHROTRON RADIATION LAB, 78- *Concurrent Pos:* NSF fel, Atomic Energy Res Estab, Eng, 62-63; mem ad hoc comt, Nat Acad Sci-Nat Res Coun, 69-70. *Mem:* Am Crystallog Asn; fel Am Phys Soc; NY Acad Sci; fel AAAS; Sigma Xi. *Res:* Structure and properties of imperfectly crystallized and amorphous materials; amorphous semiconductor devices; determination of atomic arrangements in amorphous materials and of the relationships of these arrangements to electrical, optical and thermal properties; synchrotron radiation. *Mailing Add:* 967 Mears Ct Stanford CA 94305

BIENFANG, PAUL KENNETH, b Watertown, Wis, Apr 14, 48; m 73. BIOLOGICAL OCEANOGRAPHY. *Educ:* Univ Hawaii, BS, 71, MS, 74, PhD, 77. *Prof Exp:* Res asst oceanog, Hawaii Inst Marine Biol, 69-73; res asst, 73-74, SR SCIENTIST OCEANOG, OCEANIC INST, 73-, VPRES, 83- *Concurrent Pos:* Oceanog consult, Environ Consult Inc, 69-74, Oceanic Inst, 72, Sunn, Low, Tom & Hara Inc, 75 & Oceanic Eng Dept, Univ Hawaii, 76; affil fac, dept oceanog, Univ Hawaii. *Mem:* Am Soc Limnol & Oceanog; World Maricult Soc; Phycological Soc Am; AAAS. *Res:* Phytoplankton ecology, particularly nutritional control of sinking rates of phytoplankton and dynamics of substrate limited growth in phytoplankton; thermal pollution assessment, especially effects on phytoplankton biomass and productivity; ocean thermal energy conversion impacts and alternate uses. *Mailing Add:* The Oceanic Inst Makapuu Point Waimanalo HI 96795

BIENIAWSKI, ZDZISLAW TADEUSZ, b Cracow, Poland, Oct 1, 36; US citizen; m 64; c 3. ROCK MECHANICS, MINING & TUNNELING. *Educ:* Univ Witwatersrand, BS, 61, MS, 63; Univ Pretoria, PhD(rock mech), 68. *Prof Exp:* Sr res scientist rock mech, Coun Sci & Indust Res, 64-66, head dept geomech, 66-77; PROF MINERAL ENG, PA STATE UNIV, UNIV PARK, 78- *Concurrent Pos:* Chmn, US Nat Comn Tunneling Technol, 84-85; vis prof, Stanford Univ, 85, Harvard Univ, 90; mem, US Nat Comn Rock Mech, 80-84. *Honors & Awards:* Rock Mech Award, Am Inst Metall & Petrol Engrs, 84. *Mem:* Int Soc Rock Mech (vpres, 74-79); Am Soc Eng Educ; Am Soc Mech Engrs; Am Inst Metall & Petrol Engrs; Am Soc Civil Engrs; Sigma Xi. *Res:* Improved design of longwall coal mining; geomechanics classification of rock masses; design theory and methodology. *Mailing Add:* 122 Mineral Sci Bldg Pa State Univ University Park PA 16802

BIENIEK, MACIEJ P, b Wilno, Poland, Jan 5, 27; nat US; m 61; c 2. APPLIED MECHANICS, STRUCTURAL ENGINEERING. *Educ:* Danzig Tech Univ, BCivilEng & MS, 48, DSc(appl mech), 51. *Prof Exp:* Lectr mech, Danzig Tech Univ, 52-55, prof, 55-57; prof, Polish Acad Sci, 57-58; vis scholar, Columbia Univ, 58-59; vis lectr, Princeton Univ, 59-60; assoc prof civil eng & mech, Columbia Univ, 60-63; prof, Univ Southern Calif, 63-69; PROF CIVIL ENG, COLUMBIA UNIV, 69- *Concurrent Pos:* Consult, Chinese Acad Sci, 56-57. *Mem:* Am Inst Aeronaut & Astronaut; Am Soc Civil Engrs; Sigma Xi. *Res:* Dynamics; mechanics of inelastic solids; dynamics of structures; theory of plates and shells. *Mailing Add:* Dept Civil Eng Columbia Univ 624 SW Mudd Bldg New York NY 10027

BIENIEK, RONALD JAMES, b South Gate, Calif, Aug 30, 48; div; c 2. ATOMIC & MOLECULAR COLLISIONS. *Educ:* Univ Calif, Riverside, BS 70; Harvard Univ, AM, 73, PhD(physics), 75. *Prof Exp:* Physicist, Naval Undersea Res & Develop Ctr, 70; NSF fels, Mass Inst Technol & Harvard Univ, 70-73 & Univ Colo, Boulder, 75; asst prof phys sci, Univ Ill, Urbana-Champaign, 75-81, asst prof astron, 76-81, asst prof humanities, 78-79; asst prof, 81-84, ASSOC PROF PHYSICS, UNIV MO, ROLLA, 84- *Concurrent Pos:* Prin investr, NSF, 76-86, mem, rev panel, 81 & 85; vis scientist, Paris-Meudon Observ, 85; Fulbright fel, 88-89; vis scientist, Univ Kásers lautern, 87 & 90, Univ Paris-Sud, 89. *Honors & Awards:* Philips Distinguished Vis Lectr, Haverford Col, 84. *Mem:* Am Phys Soc; Hist Sci Soc; Am Asn Physics Teachers; Sigma Xi. *Res:* Quantum mechanical and semiclassical investigations of atomic and molecular processes, such as ionization and detachment, spectral line shapes, collisional redistribution of polarized radiation, laser-induced phenomena, vibrational-rotational excitations in molecular collisions. *Mailing Add:* Dept Physics Univ Mo Rolla MO 65401-0249

BIENIEWSKI, THOMAS M, b Posen, Poland, July 12, 36; US citizen; m 66; c 2. MATERIAL SYNTHESIS USING DENSE PLASMAS, PLASMA SPRAYING & DEPOSITION. *Educ:* Univ Detroit, BS, 58; Calif Inst Technol, PhD(physics), 65. *Prof Exp:* Staff mem, Aerospace Res Lab, USAF, 63-72; STAFF MEM, LOS ALAMOS NAT LAB, 72- *Res:* Atomic and molecular spectroscopy of actinide metals and compounds; development of plasma physics application for materials synthesis with plasma; handling and generating ultrapure tritium gas streams for thermonuclear energy applications. *Mailing Add:* Los Alamos Nat Lab MS-C348 Los Alamos NM 87545

BIENSTOCK, D(ANIEL), b Brooklyn, NY, Dec 1, 17; m 47; c 3. CHEMICAL ENGINEERING. *Educ:* Brooklyn Col, BA, 37; NY Univ, MS, 40; Carnegie Inst Technol, MS, 57. *Prof Exp:* Chemist, Edgewood Arsenal, US Dept War, 41-46; supvry chem engr, US Bur Mines, 46-60, asst proj coordr, 60-65, proj coordr, 65-77; mgr, 77-82, DEP ASSOC DIR, COMB DIV, PITTSBURGH ENERGY RES CTR, DEPT ENERGY, 82- *Honors & Awards:* Vermeil Medal, France, 79; McAfee Award, Am Inst Chem Engrs, 83. *Mem:* Am Inst Chem Engrs; Am Soc Mech Engrs; Am Chem Soc; Nat Soc Prof Engrs; Air Pollution Control Asn; NY Acad Sci. *Res:* Coal-slurry preparation and combustion; high-temperature combustion; air pollution; oxides of sulfur and nitrogen; coal-fired magnetohydrodynamics; alkalized alumina removal of sulfur oxide in flue gas; two-stage combustion in control of nitrogen oxides; Fischer-Tropsch synthesis; corrosion in hot carbonate systems. *Mailing Add:* 6611 Dalzell Pl Pittsburgh PA 15217

BIENSTOCK, DANIEL, b Feb 10, 60; US citizen. COMBINATORIAL OPTIMIZATION, DISCRETE STRUCTURES. *Educ:* Brandeis Univ, BA, 82; Mass Inst Technol, PhD(opers res), 85. *Prof Exp:* Asst prof opers res, GSIA, Carnegie Mellon Univ, 85-86; mem tech staff, Bell Commun Res, 86-89; ASSOC PROF INDUST ENG & OPERS RES, COLUMBIA UNIV, 89- *Concurrent Pos:* Consult, Bell Commun Res, 90-; NSF presidential young investr award, 90. *Mem:* Opers Res Soc Am; Math Prog Soc. *Res:* Area of applied mathematics known as combinatorial optimization; discrete mathematics; theoretical computer science. *Mailing Add:* Dept Indust Eng & Opers Res Columbia Univ New York NY 10027

BIENVENUE, GORDON RAYMOND, b Fall River, Mass, Oct 11, 46; m 68; c 2. AUDIOLOGY, PSYCHOACOUSTICS. *Educ:* Univ Mass, Amherst, BA, 68; Mich State Univ, MA, 69; Pa State Univ, PhD(audiol), 75; Am Speech & Hearing Asn, Clin Cert Audiol. *Prof Exp:* Chief audiol, Brooke Gen Hosp, 69-71; res asst audiol & psychoacoust, Pa State Univ, 71-83; ASST PROF AUDIOL & PSYCHOACOUST, STATE UNIV NY, NEW PALTZ, 83- *Concurrent Pos:* Instr audiol, US Army Med Field Serv Sch, San Antonio, Tex, 69-71; lectr audiol, Brooke Gen Hosp, 69-71; consult, Indust Audiol, 71- *Mem:* Am Speech & Hearing Asn; Sigma Xi; Acoust Soc Am; Am Audiol Soc; Mil Audiol & Speech Pathol Soc. *Res:* Investigation of psychoacoustic and physiological acoustic phenomena with particular emphasis on hearing conservation and hearing diagnostic procedures especially in the study of the effects of high level noise. *Mailing Add:* Speech SUNY Col New Paltz New Paltz NY 12561

BIENZ, DARREL RUDOLPH, b Bern, Idaho, Apr 1, 26; m 50; c 5. PLANT BREEDING. *Educ:* Univ Idaho, BS, 50; Cornell Univ, PhD(plant breeding), 55. *Prof Exp:* Asst prof hort, Univ Idaho, 54-58, horticulturist, Agr Res Serv, USDA & Univ Idaho, 58-59; from asst prof to prof, 59-91, EMER PROF HORT, WASH STATE UNIV, 91- *Concurrent Pos:* Fulbright exchange prof, Turkey, 67-68; partic, Survey Tropical Agr & World Hunger, SAsia & Africa, 75-76; sr fac adv, Lambung, Mangkurat Univ Banjarbaru, S Kalimantan, Indonesia, 81; vis prof, Univ Hawaii, 75; vis res scientist, Mayaguez Agr Res Sta, PR, 85 & China Asn Sci & Technol, 85; sr adv, Jordan Valley, Jordan, 87. *Mem:* Am Soc Hort Sci; Int Hort Soc; Am Genetic Asn; Am Hort Soc; Am Inst Biol Sci. *Res:* Vegetable breeding and incompatability; general vegetable research; international agriculture; breeding all male asparagus cultivars resistant to field decline and early disease resistant tomato cultivars suitable for processing. *Mailing Add:* Dept Hort Wash State Univ Pullman WA 99164-6414

BIER, DENNIS MARTIN, b Hoboken, NJ, July 24, 41; m 71; c 3. METABOLISM, MASS SPECTROMETRY. *Educ:* LeMoyne Col, BS, 62; NJ Col Med, MD, 66. *Prof Exp:* Intern pediat, Univ Calif, San Francisco, 66-67, resident, 67-68; Bay Area Heart Asn fel, 68-69; lieutenant comdr, Naval Regional Med Ctr, US Naval Reserves, Oakland, 69-71; NIH fel, Sch Med, Wash Univ, 71-73; asst prof, Univ Calif, San Francisco, 73-76; from asst prof to assoc prof, Med & Pediat, Sch Med, 76-86, PROF MED & PEDIAT, WASH UNIV, 86- *Concurrent Pos:* Vis staff scientist, Los Alamos Sci Lab, 75- *Honors & Awards:* Grace A Goldsmith Award, Am Col Nutrit; E V McCollun Award, Juv Diabetes Found; Mary Jane Keeger Award, Juv Diabetes Found. *Mem:* Soc Pediat Res; Am Diabetes Asn; Am Soc Mass Spectrometry; Fedn Am Soc Exp Biol; Am Soc Clin Invest; Am Soc Clin Nutrit. *Res:* Metabolic fuel transport; stable isotope tracers; diabetes; lipoprotein metabolism; inborn errors of metabolism. *Mailing Add:* Metab Div 630 Wohl Hosp Sch Med Wash Univ 660 S Euclid Ave St Louis MO 63110

BIER, MILAN, b Vukovar, Yugoslavia, Dec 7, 20; nat US; m 52; c 3. BIOPHYSICS. *Educ:* Univ Geneva, License Sci Chim, 46; Fordham Univ, PhD(biochem), 50. *Prof Exp:* Asst prof chem, Fordham Univ, 50-62; res biophysicist, Vet Admin Hosp, 62-77; vis res prof chem, 62-77, PROF ENG & MICROBIOL, UNIV ARIZ, 77- *Concurrent Pos:* Head chem res lab, Inst Appl Biol, 52-60; vis res prof chem, Univ Ariz, 62-77; consult, NASA, Abbott Sci Prod, Ionics Inc, Univs Space Res Asn & others, 73- *Mem:* Am Chem Soc; Am Soc Biol Chem. *Res:* Biomedical engineering, electrophoresis, membrane processes and artificial kidney technology; protein and enzyme biophysics, plasma fractionation, applied immunology. *Mailing Add:* Ctr Separation Sci Univ Ariz Bldg 20 Tucson AZ 85721

BIERBAUM, VERONICA MARIE, b Allentown, Pa, Aug 2, 48; m 82; c 2. GAS PHASE ION CHEMISTRY. *Educ:* Cath Univ Am, BA, 70; Univ Pittsburgh, PhD(phys chem), 74. *Prof Exp:* Vis asst prof & lectr, 74-76, res assoc, 76-78, SR RES ASSOC CHEM, UNIV COLO, BOULDER, 78-, SPEC MEM GRAD FAC, 81- *Concurrent Pos:* Co-prin investr, Air Force Off Sci Res, 78-91, US Army Res Off, 78-88 & NSF, 79-92; mem, Joint Inst Lab Astrophysics, 78- *Mem:* Am Chem Soc; Am Soc Mass Spectrometry (secy, 89-91); Sigma Xi. *Res:* Kinetics, mechanisms and thermochemistry of gas phase ion-molecule reactions; infrared chemiluminescence and laser induced fluorescence studies of product state distributions of gas phase ion-molecule reactions. *Mailing Add:* Dept Chem & Biochem Univ Colo Boulder CO 80309-0215

BIERENBAUM, MARVIN L, b Philadelphia, Pa, Aug 30, 26; m 51; c 2. CARDIOLOGY, NUTRITION. *Educ:* Rutgers Univ, BS, 47; Hahnemann Med Col, MD, 53. *Prof Exp:* Intern med, Beth Israel Hosp, Newark, NJ, 53-54; resident med & cardiol, Vet Admin Hosp, Brooklyn, 54-57; DIR ATHEROSCLEROSIS RES PROJ, ST VINCENT'S HOSP, 57- *Concurrent Pos:* Prog coordr heart dis control, NJ Dept Health, 57-; prin investr, NIH grant, 60-; consult, Fairleigh Dickinson Univ, 62-; mem coun arteriosclerosis, Am Heart Asn, 64; clin assoc prof med, NJ Col Med & Dent, 67- *Mem:* AMA; fel Am Col Physicians; Am Inst Nutrit; Am Soc Clin Invest; fel Am Col Cardiol. *Res:* Nutritional management in prevention and therapy of arteriosclerotic coronary heart disease. *Mailing Add:* Kenneth L Jordan Cardiac Ctr 48 Plymouth St Montclair NJ 07042

BIERI, JOHN GENTHER, b Norfolk, Va, May 24, 20; m 43; c 3. BIOCHEMISTRY, NUTRITION. *Educ:* Antioch Col, BA, 43; Pa State Col, MS, 44; Univ Minn, PhD(biochem), 49. *Prof Exp:* Nutritionist, Naval Med Res Inst, 44-46; instr biochem, Univ Minn, 48-49; assoc prof, Med Br, Univ Tex, 49-55; biochemist, Nat Inst Arthritis, Metab & Digestive Dis, 55-82; NUTRIT CONSULT, 82- *Honors & Awards:* Meade-Johnson Award, Am Inst Nutrit, 65. *Mem:* AAAS; Am Soc Biol Chemists; Am Inst Nutrit. *Res:* Nutritional biochemistry; metabolism of vitamin E and vitamin A; polyunsaturated fatty acids; selenium. *Mailing Add:* 3612 Pimlico Pl Silver Spring MD 20906

BIERI, ROBERT, b Washington, DC, Feb 7, 26; c 3. OCEANOGRAPHY, ECOLOGY. *Educ:* Antioch Col, BS, 49; Univ Calif, MS, 53, PhD(oceanog), 58. *Prof Exp:* Res biologist, Scripps Inst, Univ Calif, 54-55; oceanogr, Lamont Geol Observ, 55-57; assoc dir personnel, Antioch Col, 57-59, asst prof biol, 59-67, chmn dept biol, 59-63, chmn Environ Studies Ctr, 68-72, prof biol, 67-86; RETIRED. *Mem:* Am Soc Limnol & Oceanog; Marine Biol Asn UK; AAAS. *Res:* Zooplankton, Chaetognatha; Neuston; sea surface community life histories, food web and behavior; forest community succession. *Mailing Add:* 175 Brookside Dr Yellow Springs OH 45387

BIERKAMPER, GEORGE G, PHARMACOLOGY, TOXICOLOGY. *Educ:* WVa Univ, PhD(pharmacol), 76. *Prof Exp:* ASSOC PROF PHARMOCOL, SCH MED, UNIV NEV, 83- *Mailing Add:* Dept Pharmacol Univ Nev Sch Med Savitt Med Sci Bldg Reno NV 89557-0046

BIERLEIN, JAMES A(LLISON), b Dayton, Ohio, Oct 1, 21; m 50; c 3. CHEMICAL ENGINEERING, PHYSICAL CHEMISTRY. *Educ:* Purdue Univ, BS, 42; Ohio State Univ, MSc, 48, PhD(chem eng), 51. *Prof Exp:* Develop engr, Wright Field, Wright-Patterson AFB, 46-49, res scientist, Aerospace Res Labs, 50-56, chief chem br, 56-64, dir chem res, Aerospace Res Labs, 64-75; CONSULT, 76- *Concurrent Pos:* Fel phys chem, Cambridge Univ, 62; dir sci div, Off Naval Res, London, 66-67. *Mem:* Am Chem Soc; NY Acad Sci. *Res:* Thermodynamics and thermochemistry of rocket propellants; combustion technology; reaction kinetics; phase equilibria; molecular transport properties; electrochemical energy technology; statistical and mathematical modeling; stochastic phenomena; instrumental methods of analysis; technical management. *Mailing Add:* 658 Rockhill Ave Dayton OH 45429

BIERLEIN, JOHN CARL, b Cleveland, Ohio, Nov 13, 36; m 60; c 2. TRIBOLOGIST, MATERIALS ENGINEERING. *Educ:* Mich State Univ, BS, 59; Univ Wis, MS, 61, PhD(mat eng, sci), 65; Univ Detroit, MBA, 77. *Prof Exp:* Instr mat eng, Univ Wis, 59-65; sr res engr, Gen Motors Res Lab, Warren, 65-81, staff res engr mat, 81-85; PRIN ENGR, EATON ENG & RES CTR, SOUTHFIELD, MICH, 85- *Concurrent Pos:* Adj prof, Lawrence Technol Univ, 90- *Honors & Awards:* Arch T Colwell Merit Award, Soc Automotive Engrs, Ralph R Teetor Indust Lectureship Prog. *Mem:* Am Soc Metals; Am Inst Mining, Metall & Petrol Engrs; Soc Automotive Engrs; Am Soc Mech Engrs; Soc Tribological & Lubrication Engrs. *Res:* Design and material development of tribosystems for lubrication, friction, wear and scoring and embedability; failure analysis; study of materials in sliding contact under marginal lubrication conditions such as bearings, elastomeric seals, gears, polymeric products and hydraulic components. *Mailing Add:* 5481 Vincent Trail Washington MI 48094

BIERLEIN, JOHN DAVID, b May 22, 40; m 64; c 2. MATERIAL SCIENCE-NON LINEAR OPTICS, LASER PHYSICS & ENGINEERING. *Educ:* Univ Kans, BS, 63, MS, 66, PhD(physics), 68. *Prof Exp:* Res physicist, Cent Res Dept, 68-73, res assoc, Photo Prods Dept, 73-76, res supvr, 76-85, res physicist, Cent Res Dept, 85-86, Res Leader, 86-88, RES SURVR, CENT RES DEPT, E I DU PONT & CO, 88- *Honors & Awards:* Res & Develop 100 Award, 89. *Mem:* Am Phys Soc. *Res:* Discovery and development of new non linear optical and electro-optical materials. *Mailing Add:* Cent Res Dept Du Pont Exp Sta E356/284 Wilmington DE 19880

BIERLEIN, THEO KARL, b Ansbach, Ger, Feb 6, 24; nat US; m 49; c 6. PHYSICAL CHEMISTRY. *Educ:* Univ Wash, BS, 45, PhD(chem), 50. *Prof Exp:* Chemist, Gen Elec Co, 50-52, engr, 52-55, sr scientist, 55-63; mgr phys metall unit, Reactor & Mat Technol Dept, Pac Northwest Labs, Battelle Mem Inst, 63-70, mgr damage anal, Mat Dept, 70-77, mgr alloy microstruct, Mat Eng Dept, 77-79, MGR MAT PROPERTIES, WESTINGHOUSE-HANFORD, 79- *Mem:* Am Chem Soc; Am Soc Metals; Electron Micros Soc Am; Am Nuclear Soc. *Res:* Relationship between microstructure and properties of metals and alloys; deformation and fracture mechanism; effect of reactor irradiation on structure of fuel and its cladding. *Mailing Add:* 1005 Warren Ct Richland WA 99352

BIERLY, EUGENE WENDELL, b Pittston, Pa, Sept 11, 31; m 53; c 3. METEOROLOGY, SCIENCE ADMINISTRATION. *Educ:* Univ Pa, AB, 53; US Naval Postgrad Sch, cert, 54; Univ Mich, MS, 57, PhD, 68. *Prof Exp:* Asst, dept civil eng, meteorol labs, Univ Mich, 56-60, asst res meteorologist, dept eng mech, 60-63, lectr, 61-63; meteorologist, US Atomic Energy Comn, 63-66; prog dir meteorol, 66-71, coordr global atmospheric res prog, 71-74, head, Off Climate Dynamics, 74-75, head, Climate Dynamics Res Sect, 75-79, DIR, DIV ATMOSPHERIC SCI, NSF, 79- *Concurrent Pos:* Consult, Reactor Develop Co, 61-62 & Pac Missile Range, 62-63; Cong fel, 70-71. *Honors & Awards:* Charles Franklin Brooks Award, Am Meteorol Soc, 90. *Mem:* Fel AAAS; fel Am Meteorol Soc (pres, 84); Air Pollution Control Asn; Royal Meteorol Soc; Am Polit Sci Asn; Chinese Meteorol Soc. *Res:* Air pollution; diffusion; lake breezes; atmospheric tracers; science management; environmental problems and science policy. *Mailing Add:* NSF 1800 G St NW Washington DC 20550

BIERLY, JAMES N, JR, b Lewisberg, Pa, May 6, 22; m 47. PHYSICS. *Educ:* Kutztown State Teachers Col, BS, 43; Bucknell Univ, MS, 47; Temple Univ, PhD(physics), 61; Am Bd Radiol, dipl, 55. *Prof Exp:* Instr physics, Bucknell Univ, 47; physicist radiol dept, Jefferson Med Col, 51-55; physicist, Vet Admin Hosp, 55-56; res physicist, Franklin Inst, 56-62; from asst prof to assoc prof physics, 62-73, PROF PHYSICS, PHILADELPHIA COL PHARM & SCI, 73- *Concurrent Pos:* Consult, Misericordia Hosp; radiol physicist, Mercy Catholic Med Ctr. *Mem:* Am Phys Soc; Sigma Xi; Am Asn Physicists Med; Am Asn Physics Teachers. *Res:* Thermoelectricity in semiconductors; consulting work in medical radiation physics. *Mailing Add:* 110 Oak Ridge Ct Williamsburg VA 23188

BIERLY, MAHLON ZWINGLI, JR, b Philadelphia, Pa, Apr 24, 22; m 44; c 2. MEDICINE. *Educ:* Franklin & Marshall Col, BS, 43; Jefferson Med Col, MD, 46; Am Bd Pediat, dipl, 55. *Prof Exp:* Resident & asst chief, Philadelphia Hosp Contagious Dis, 49-50; resident pediat, Children's Hosp Philadelphia, 51-52; pvt pract, 52-53; MEM STAFF, MED DEPT, WYETH LABS, 53- *Mem:* Am Pub Health Asn; Am Acad Pediat. *Res:* Biological agents; vaccines, toxoids and antisera. *Mailing Add:* 1707 Thomas Rd Wayne PA 19087

BIERMAN, ARTHUR, b Vienna, Austria, Oct 14, 25; US citizen; m 83; c 2. THEORETICAL PHYSICS. *Educ:* Univ Chicago, PhD(math, biol), 54; Columbia Univ, MA, 57. *Prof Exp:* Nat Found Infantile Paralysis fel, 54-55; from instr to asst prof physics, City Col New York, 58-61; sr res scientist, Lockheed-Calif Co, 62-63; assoc prof physics, Los Angeles State Col, 63; from asst prof to assoc prof, 64-69, actg assoc provost, 72-73, PROF PHYSICS, CITY COL, CITY UNIV NEW YORK, 69- *Concurrent Pos:* Consult, Lockheed-Calif Co, 62 & Lockheed Ga Co, 65-66; US Atomic Energy Comn grant, 64-71. *Mem:* Am Phys Soc. *Res:* Theory of molecular excitons in large molecules. *Mailing Add:* 542 Arapahoe Ave Boulder CO 80302

BIERMAN, EDWIN LAWRENCE, b Far Rockaway, NY, Sept 17, 30; m 56; c 2. INTERNAL MEDICINE, METABOLISM. *Educ:* Brooklyn Col, AB, 51; Cornell Univ, MD, 55. *Prof Exp:* Intern, NY Hosp, 55-56, asst physician, Diabetes Study Group, 56-57; from asst chief to chief metab res div, US Army Med Res & Nutrit Lab, Fitzsimons Army Hosp, Denver, 57-59; asst resident, NY Hosp, 59-60, outpatient physician, 60-62; from asst prof to assoc prof, 62-68, PROF MED, SCH MED, UNIV WASH, 68-, HEAD, DIV METAB & ENDOCRINOL, 75- *Concurrent Pos:* Asst physician, Rockefeller Inst Hosp, NY, 56-57, assoc physician & asst prof, Rockefeller Inst, 60-62. *Honors & Awards:* Herman Award, Am Soc Clin Nutrit; Goldberger Award, Am Med Asn. *Mem:* Endocrine Soc; Am Diabetes Asn; Asn Am Physicians; Am Soc Clin Invest; Am Col Physicians. *Res:* Lipid and carbohydrate metabolism, with particular emphasis on diabetes and hyperlipemia. *Mailing Add:* Sch Med Univ Wash Seattle WA 98195

BIERMAN, HOWARD RICHARD, b Newark, NJ, Jan 27, 15; m; c 3. MEDICINE. *Educ:* Wash Univ, BSc & MD, 39; Am Bd Internal Med, dipl. *Prof Exp:* From jr to sr intern, Barnes Hosp, 40, resident house officer, 41; clin physiologist & prin clin investr, Nat Cancer Inst, 46-53; chief clin sect, Lab Exp Oncol & assoc clin prof med, Sch Med, Univ Calif, 47-53; dir, Hosp Tumors & Allied Dis, City of Hope Med Ctr, 53-56, med dir, 56-59, chmn, Dept Internal Med & sci dir, 53-59; DIR, INST CANCER & BLOOD RES, BEVERLY HILLS, 59-; CLIN PROF MED, SCH MED, LOMA LINDA UNIV, 59- *Concurrent Pos:* Chief resident, St Louis Isolation Hosp, Contagious Dis, 40; spec consult, Nat Cancer Inst, 53-; emer attend physician, Los Angeles County Hosp, USN. *Mem:* Fel Am Col Physicians; fel NY Acad Sci; fel Am Col Angiol; fel Int Soc Hemat; fel Am Soc Clin Oncol. *Res:* Hematology and oncology; chemotherapy and clinical pharmacology; bone marrow metabolism; biophysical characteristics of tumors; aviation medicine; barometrics and acceleration; nutrition and vitamins; leukocyte kinetics. *Mailing Add:* Inst Cancer & Blood Res 170 N Robertson Blvd No 350 Beverly Hills CA 90211

BIERMAN, SIDNEY ROY, b Comfort, Tex, Jan 24, 28; m 49; c 3. EXPERIMENTAL NUCLEAR CRITICALITY RESEARCH, EXPERIMENTAL NEUTRON KINETICS. *Educ:* Tex Technol Univ, BS, 56; Univ Wash, MS, 63. *Prof Exp:* Process engr, Standard Oil Co, NJ, 56-59; process engr, 59-63, res scientist, 63-65, sr res scientist, 65-78, STAFF SCIENTIST, BATTELLE MEM INST, 78- *Concurrent Pos:* Chmn, Nuclear Criticality Safety Div, Am Nuclear Soc, 79-80. *Mem:* Am Nuclear Soc. *Res:* Nuclear criticality; kinetics. *Mailing Add:* 2352 Enterprise Richland WA 99352

BIERMANN, ALAN WALES, b Newport News, Va, Feb 5, 39; m 68; c 2. COMPUTER SCIENCE. *Educ:* Ohio State Univ, BEE & MSc, 61; Univ Calif, Berkeley, PhD(computer sci), 68. *Prof Exp:* Asst prof computer sci, San Fernando Valley State Col, 62-64 & 68-69; res assoc, Stanford Univ, 69-71; asst prof, Ohio State Univ, 71-73; from asst prof to assoc prof, 74-86, PROF COMPUTER SCI, DUKE UNIV, 86- *Concurrent Pos:* Pres, Asn Computational Ling, 88; vis scientist, SRI Int, Menlo Park, Calif, 88. *Mem:* Sigma Xi; Inst Elec & Electronics Engrs; Asn Comput Mach; Asn Computational Ling. *Res:* Learning and inference theory; automatic program synthesis; natural language processing; author of one book. *Mailing Add:* Dept Computer Sci Duke Univ Durham NC 27706

BIERMANN, CHRISTOPHER JAMES, b Winnipeg, Man, Mar 16, 58; m 90. WOOD CHEMISTRY. *Educ:* Univ Maine, Orono, BS, 80; Miss State Univ, PhD(wood chem), 83. *Prof Exp:* Grad res asst, Miss Forest Prod Utilization Lab, 80-83; res asst, dept nutrit sci, Univ Wis-Madison, 84-85; res assoc, Lab Renewable Resources Eng, Purdue Univ, 85-87; ASST PROF, DEPT FOREST PROD, ORE STATE UNIV, 87- *Mem:* Am Chem Soc; Forest Prod Res Soc; Soc Wood Sci & Technol; Can Pulp & Paper Asn; Tech Asn Pulp & Paper Indust. *Res:* Pretreatment of lignocellulosic materials; carbohydrate analysis; grafting of polymers onto cellulose derivatives; pulp and paper chemistry. *Mailing Add:* Dept Forest Prod Oregon State Univ Corvallis OR 97331-5709

BIERMANN, JANET SYBIL, b Winnipeg, Man, Mar 14, 61; US citizen. TRAUMA, ONCOLOGY. *Educ:* Cornell Univ, BS, 83; Stanford Univ, MD, 87. *Prof Exp:* Internship, 87-88, ORTHOP RESIDENT, DEPT ORTHOP, UNIV IOWA HOSP, 88- *Concurrent Pos:* Teaching asst, Stanford Univ, 83-84, res asst, 83-86. *Mem:* Am Med Asn; Am Acad Orthop Surgeons. *Res:* Severe tibial injuries; examining the clinical utility of the unilateral bone transport system using distraction osteogenesis techniques. *Mailing Add:* Dept Orthop Surg Univ Iowa Hosp Iowa City IA 52242

BIERNBAUM, CHARLES KNOX, b Woodbury, NJ, Dec 7, 46. INVERTEBRATE ZOOLOGY, CRUSTACEAN BIOLOGY. *Educ:* Wake Forest Univ, BS, 68; Univ Conn, PhD(biol), 74. *Prof Exp:* From asst prof to assoc prof, 74-86, PROF BIOL, COL CHARLESTON, 86- *Mem:* Am Soc Zoologists; Crustacean Soc; Am Inst Biol Sci; Estuarine Res Fedn; Asn Southeastern Biologists. *Res:* Examination of the ecology and distribution of marine and freshwater crustaceans, especially amphipods. *Mailing Add:* Grice Marine Biol Lab 205 Ft Johnson Charleston SC 29412

BIERON, JOSEPH F, b Buffalo, NY, Oct 19, 37; m 59; c 5. ORGANIC CHEMISTRY. *Educ:* Canisius Col, BS, 59, MS, 61; State Univ NY Buffalo, PhD(chem), 65. *Prof Exp:* From asst prof to assoc prof, 66-77, dean Col Arts & Sci, 71-79, PROF CHEM, CANISIUS COL, 78-, CHMN DEPT, 84- *Mem:* Am Chem Soc. *Res:* Oxidation of polymeric materials and conformational analysis of organic molecules by nuclear magnetic resonance spectroscopy. *Mailing Add:* Canisius Col 2001 Main St Buffalo NY 14208

BIERSCHENK, WILLIAM HENRY, geological engineering; deceased, see previous edition for last biography

BIERSDORF, WILLIAM RICHARD, b Salem, Ore, Sept 27, 25; m 66; c 2. VISUAL PHYSIOLOGY, OPHTHALMOLOGY. *Educ:* Wash State Univ, BS, 50, MS, 51; Univ Wis, PhD(psychol), 54. *Prof Exp:* Exp psychologist vision, Walter Reed Army Inst Res, 54-67, asst chief, Dept Sensory Psychol, 62-66, actg chief, Dept Psychophysiol, 66-67; assoc prof ophthal, Ohio State Univ, 67-78, assoc prof biophys, 70-78, proj supvr vision res, Inst Res in Vision, 67-78; PROF OPHTHAL, UNIV SFLA, 78-, HEALTH SCI OFFICER, VET ADMIN MED CTR, 78- *Concurrent Pos:* US Army res grant vision, Ohio State Univ, 67-68, USPHS res grant, 68-80, 90-94; Am Psychol Asn deleg, Inter-Soc Color Coun, 66-75; adv, Comt Vision Nat Res Coun-Armed Force, 68-; mem Visual Sci Study Sect, Div Res Grants, NIH, 71-75; Vet Admin res grant, 80-88. *Mem:* Int Soc Clin Electrophysiol Vision; Asn Res Vision & Ophthal. *Res:* Human visual electrophysiology, electroretinograms and electro-oculograms from the eye, and visual evoked potentials from the brain; normal and abnormal vision, hereditary eye disease and visual brain dysfunction. *Mailing Add:* Vet Hosp 13000 N Bruce B Downs Blvd Tampa FL 33612

BIERSMITH, EDWARD L, III, b Kansas City, Mo, Feb 26, 42; m 68; c 1. ORGANIC CHEMISTRY, BIO-ORGANIC CHEMISTRY. *Educ:* Rockhurst Col, AB, 63; Univ Kans, PhD(org chem), 68. *Prof Exp:* Asst chem, Rockhurst Col, 62-63; asst org chem, Univ Kans, 63-68; res assoc & instr, Univ Okla, 69-70; assoc prof chem, 70-77, DIR UNIV RES, NORTHEAST LA UNIV, 77- *Mem:* Am Chem Soc; Royal Soc Chem; Sigma Xi. *Res:* Photolysis of 2-cyclopropyl cyclohexanone: a three-carbon photochemical ring expansion; total synthesis of bicyclo heptan-2-one; isolation and identification of sterols from marine life, especially Renilla, Acrapora palmata; reactions of cyclic amides. *Mailing Add:* 1913 Richard Circle Monroe LA 71201

BIERSNER, ROBERT JOHN, b Walla Walla, Wash, Sept 22, 41; m 63; c 2. PHYSIOLOGICAL PSYCHOLOGY, HEALTH PSYCHOLOGY. *Educ:* Cent Wash Univ, BA, 63; McGill Univ, MA, 64, PhD(physiol psychol), 66. *Prof Exp:* Asst biol, Calif Inst Technol, 66-67; researcher psychol, Naval Health Res Ctr, Naval Biodynamics Lab, 70-73, human factors specialist, training eval, Naval Educ & Training Support Command, 73-76, res projs officer, Naval Submarine Med Res Lab, 76-78, prog mgr, Naval Med Res & Develop Command, 78-82, assoc dir prog planning, 82-84, cmndg officer, Naval Biodynamics Lab, 84-87; at Nat Inst Occup Safety & Health, 87-91; ATTY, OCCUP SAFETY & HEALTH, 91- *Concurrent Pos:* Instr psychol, San Diego City Col & Mesa Col, 71-73; consult, NSF Antarctic Res Prog, 80-81 & Airline Pilots Rights Asn, 82-84; adv, Nat Res Coun Comt on Hearing, Bioacoust & Biomechs, 85- *Mem:* AAAS; Am Psychol Asn; US Naval Inst; Asn Mil Surgeons; Am Polar Soc; Sigma Xi. *Res:* Techniques to select individuals for stressful occupations; epidemiological designs and methods to assess effects of stressful occupations. *Mailing Add:* 9914 Capital View Ave Silver Spring MD 20910

BIERWAGEN, GORDON PAUL, b Wauwatosa, Wis, Mar 3, 43; m 64; c 3. COLLOID CHEMISTRY, LIQUID SURFACE CHEMISTRY. *Educ:* Valparaiso Univ, BS, 64; Iowa State Univ, PhD(phys chem), 68. *Prof Exp:* Actg asst prof chem eng, Univ Minn, 68-69; res chemist, Dept Electrochem Eng, Battelle Mem Inst, 69-70; sr scientist, Sherwin-Williams Res Ctr, 78-80, dir, Phys Anal Lab, Paint Res Dept, 80-90; PROF, POLYMERS DEPT, NDAK STATE UNIV, 90- *Concurrent Pos:* Consult, Foil Div, Gould, Inc, 74-75. *Honors & Awards:* Roon Award, Fedn Socs Coatings Technol, 72. *Mem:* Am Chem Soc; Am Inst Chem Engrs; AAAS; Soc Advan Mat & Process Eng; Fine Particle Soc. *Res:* Effects of particle packing in coatings and composites; fine particle characterization; emulsion and suspension polymerization; physical chemistry of coatings; surface dynamics in thin liquid films; film formation in coatings; computer modeling of coatings properties. *Mailing Add:* Polymers Dept NDak State Univ Fargo ND 58105

BIERZYCHUDEK, PAULETTE F, b Chicago, Ill, Aug 25, 51. PLANT POPULATION BIOLOGY. *Educ:* Univ Wash, BA & BS, 74; Cornell Univ, PhD(ecol & evolutionary biol), 81. *Prof Exp:* Asst prof, 80-86, ASSOC PROF, DEPT BIOL, POMONA COL, 86-, DEPT CHAIR BIOL, 90- *Concurrent Pos:* Vis asst prof, Dept Bot, Duke Univ, 83-84. *Mem:* Ecol Soc Am; Soc Study Evolution; Am Soc Naturalists. *Res:* Evolution of plant breeding systems, especially apomixis and sex-changing; ecological consequences of producing genetically variable progeny; plant demography; evolution of flower color polymorphisms. *Mailing Add:* Dept Biol Pomona Col Claremont CA 91711

BIES, DAVID ALAN, b Los Angeles, Calif, Aug 15, 25; m 54; c 1. ACOUSTICS. *Educ:* Univ Calif, Los Angeles, BA, 48, MA, 51, PhD, 53. *Prof Exp:* Sr consult, Bolt, Beranek & Newman, Inc, 53-72; sr res fel, Univ Adelaide, S Australia, 72-75, reader mech eng, 76-90; RETIRED. *Concurrent Pos:* Vis res fel, Univ Adelaide, S Australia, 90- *Mem:* AAAS; Acoust Soc Am; Australian Acoust Soc; Inst Noise Control Eng; Sigma Xi. *Res:* Sound propagation in ducts; uses of anechoic and reverberant rooms; uses of holography for study of vibration and sound radiation; general problems of noise control. *Mailing Add:* Dept Mech Eng Univ Adelaide Adelaide SA 5001 Australia

BIESELE, JOHN JULIUS, b Waco, Tex, Mar 24, 18; m 43; c 3. CYTOLOGY. *Educ:* Univ Tex, AB, 39, PhD(zool), 42. *Prof Exp:* Int Cancer Res Found fel, 42 & 43-44; res assoc genetics, Carnegie Inst Technol, 44-46; asst, Sloan-Kettering Inst Cancer Res, 46, res fel, 47, head cell growth sect, 47-58, assoc, 47-55, mem, 55-58; asst prof anat, Med Col, Cornell Univ, 50-52, assoc prof biol, Grad Sch Med Sci, Sloan-Kettering Div, 52-55, prof biol, 55-58; prof zool, 58-73, dir, Genetics Found, 69-77, prof zool & educ, 73-78, MEM GRAD FAC, UNIV TEX, AUSTIN, 58-, EMER PROF, 78- *Concurrent Pos:* Instr, Univ Pa, 43-44; res assoc, Mass Inst Technol, 46-47; mem cell biol study sect, NIH, 58-63; assoc scientist, Sloan-Kettering Inst Cancer Res, 59-78; dir, AAAS, 60-63; counr, Cancirco, 62-90; assoc chmn, Conf Advan Sci & Math Teaching, Tex, 65, chmn, 66; mem discussion group on chem carcinogenesis, Nat Cancer Inst, 67, cancer res training comt, 69-72, res career awards, 62-77. *Mem:* Fel AAAS; Am Soc Cell Biol; Am Asn Cancer Res. *Res:* Electron microscopy; cytochemistry of cancerous and normal mammalian tissue; effects of differentiation, aging, hormones and drugs on cell chemistry and morphology; tumor chemotherapy; tissue culture; antimetabolites and mitotic poisons. *Mailing Add:* 2500 Great Oaks Pkwy Austin TX 78756

BIESENBERGER, JOSEPH A, b Newark, NJ, Nov 23, 35; m 57. CHEMICAL ENGINEERING. *Educ:* Newark Col Eng, BS, 57; Princeton Univ, MS, 59, PhD(chem eng), 63. *Hon Degrees:* MEng, Stevens Inst Technol. *Prof Exp:* Montecatini res fel high polymers & consult, Milan Polytech Inst, 62-63; asst prof chem, Newark Col Eng, 63; from asst prof to assoc prof, 63-70, head, Dept Chem & Chem Eng, 71-78, PROF CHEM ENG, STEVENS INST TECHNOL, 71- *Concurrent Pos:* Dir res & educ progs, Polymer Processing Inst; ed, Polymer Process Eng res jour. *Mem:* Am Inst Chem Engrs; Soc Plastics Engrs. *Res:* Polymerization engineering, polymer processing and polymer devolatilization. *Mailing Add:* Dept Chem Eng Stevens Inst Technol Hoboken NJ 07030

BIESIOT, PATRICIA MARIE, b St Cloud, Minn, Mar 23, 50; m 88. MARINE CRUSTACEANS. *Educ:* Bowling Green State Univ, BS, 72, MS, 75; Mass Inst Technol & Woods Hole Oceanog Inst, PhD(biol oceanog), 86. *Prof Exp:* Postdoctoral fel, Oak Ridge Nat Lab, 86-88; ASST PROF ZOOL, UNIV SOUTHERN MISS, 89- *Mem:* Am Soc Zool; Crustacean Soc; Sigma Xi; Western Soc Naturalists; World Aquacult Soc. *Res:* Marine invertebrate zoology; physiological and biochemical adaptions of marine invertebrates; early life history stages of decapod crustaceans. *Mailing Add:* Dept Biol Sci Univ Southern Miss Southern Sta Box 5018 Hattiesburg MS 39406

BIESTER, JOHN LOUIS, b Aurora, Ill, Aug 29, 18; m 47; c 2. CHEMISTRY, ACADEMIC ADMINISTRATION. *Educ:* Beloit Col, BA, 41; Syracuse Univ, MS, 43. PhD, 59. *Prof Exp:* Asst, Syracuse Univ, 42-43; res chemist, Standard Oil Co, Ind, 43; asst Syracuse Univ, 46-47; from asst prof to assoc prof chem, 48-80, assoc dir field placement, 64-69, dir field placement & career coun, 69-83, EMER PROF CHEM & EMER DIR FIELD & CAREER SERV, BELOIT COL, 83- *Concurrent Pos:* Assoc prog dir, acad yr insts, NSF, 61-62, consult, sci personnel & educ div, 62-69. *Res:* Science education; career education; experience based education; preparation of science teachers. *Mailing Add:* 1737 Arrowhead Dr Beloit WI 53511-3807

BIESTERFELDT, HERMAN JOHN, JR, mathematics, for more information see previous edition

BIETZ, JEROLD ALLEN, b Mayville, NDak, Feb 22, 42; m 63; c 2. PROTEIN CHEMISTRY, CEREAL CHEMISTRY. *Educ:* Mayville State Col, BS, 63; Univ NDak, MS, 66. *Prof Exp:* RES CHEMIST, NAT CTR AGR UTILIZATION RES, AGR RES SERV, USDA, PEORIA, 66- *Mem:* AAAS; Am Chem Soc; Am Asn Cereal Chemists. *Res:* Concerning isolation, characterization and comparison of cereal endosperm proteins relates structure to functionality, quality and genetic background through peptide characterization, electrophoresis, HPLC and amino acid sequence analysis. *Mailing Add:* Nat Ctr Agr Utilization Res USDA-Agr Res Serv 1815 N University Peoria IL 61604

BIEVER, KENNETH DUANE, b Hot Springs, SDak, Jan 8, 40; m 60; c 2. BIOLOGICAL CONTROL. *Educ:* SDak State Univ, BS, 62; Univ Calif, Riverside, PhD(entom), 67. *Prof Exp:* RES ENTOMOLOGIST, YAKIMA AGR RES LAB, AGR RES SERV, USDA, 66- *Mem:* Entom Soc Am; Soc Invert Path; Int Orgn Biol Control. *Res:* Design and development of biocontrol pest management systems; biological control of insects; medical entomology; insect pathology. *Mailing Add:* Yakima Agr Res Labs 3706 W Nob Hill Blvd Yakima WA 98902

BIGELEISEN, JACOB, b Paterson, NJ, May 2, 19; m 45; c 3. NUCLEAR CHEMISTRY. *Educ:* NY Univ, AB, 39; State Col Wash, MS, 41; Univ Calif, PhD(chem), 43. *Prof Exp:* Asst chem, Univ Calif, 41-43; res scientist, SAM Labs, Manhattan Proj, Columbia Univ, 43-45; res assoc, Cryogenic Lab, Ohio State Univ, 45-46; fel, Inst Nuclear Studies, Univ Chicago, 46-48; from assoc chemist to sr chemist, Brookhaven Nat Lab, 48-68; prof chem, Univ

Rochester, River Campus, 68-73, chmn dept, 70-75, Tracy H Harris prof, 73-78; vpres res & dean grad studies, State Univ NY Stony Brook, 78-80, prof chem, 78-89, distinguished prof, 89, DISTINGUISHED EMER PROF, STATE UNIV NY STONY BROOK, 89- *Concurrent Pos:* Vis prof, Cornell Univ, 53-54; NSF fel & hon vis prof, Swiss Fed Inst Technol, 62-63; John Simon Guggenheim fel, 74-75; chmn, Assembly Math & Phys Sci, Nat Res Coun-Nat Acad Sci, 76-80. *Honors & Awards:* Am Chem Soc Award, 58; Gilbert N Lewis Lectr, 63; E O Lawrence Mem Award, 64. *Mem:* Nat Acad Sci; Am Acad Arts & Sci; Am Chem Soc; Am Phys Soc; AAAS. *Res:* Theoretical basis of isotope chemistry; utilization of isotope effects; kinetics of chemical reactions; semi-quinones; photochemistry in rigid media; color of organic compounds; low temperature; spectroscopy; use of isotopes as tracers; dissociation of strong electrolytes; thermodynamics of electrolytes; acids and bases; isotope separation; intra-and inter-molecular forces in liquids and solids. *Mailing Add:* Dept Chem State Univ NY Stony Brook NY 11794-3400

BIGELOW, CHARLES C, b Edmonton, Alta, Apr 25, 28; m 53, 77; c 2. CHEMISTRY. *Educ:* Univ Toronto, BASc, 53; McMaster Univ, MSc, 55, PhD(chem), 57. *Prof Exp:* Nat Res Coun Can fel protein chem, Carlsberg Lab, Copenhagen, 57-59; assoc, Sloan-Kettering Inst, 59-62; from asst prof to assoc prof chem, Univ Alta, 62-65; from assoc prof to prof biochem, Univ Western Ont, 65-74; prof biochem & head dept, Mem Univ Newfoundland, 74-76; prof chem & dean sci, St Mary's Univ, 77-79; dean sci, 79-89, PROF CHEM, UNIV MAN, WINNIPEG, 79-, EMER DEAN SCI, 90- *Concurrent Pos:* Res grants, Nat Res Coun Can, 63-69 & 77-; Med Res Coun Can, 65-77; vis prof biochem, Univ Toronto, 73-74; Fla State Univ, 65; vis scientist, Nat Inst Med Res, London, 84-85. *Mem:* Fel Chem Inst Can; Am Chem Soc; Am Soc Biol Chem; Can Biochem Soc; AAAS. *Res:* Physical chemistry of proteins. *Mailing Add:* Univ Man Winnipeg MB R3T 2N2 Can

BIGELOW, HOWARD ELSON, b Greenfield, Mass, June 28, 23; m 56. BOTANY. *Educ:* Oberlin Col, AB, 49, MA, 51; Univ Mich, PhD(bot), 56. *Prof Exp:* Attache de recherche, Univ Montreal, 56-57; from instr to assoc prof bot, 57-70, PROF BOT, UNIV MASS, AMHERST, 70- *Mem:* AAAS; Mycol Soc Am (vpres, 74-75, pres-elect, 75-76, pres, 76-77); Bot Soc Am; Am Inst Biol Sci; Ger Soc Mycol; Sigma Xi. *Res:* Mycology; taxonomy and ecology of fleshy fungi, especially the Agaricales. *Mailing Add:* 275 Shad Hole Rd Dennis Port MA 02639

BIGELOW, JOHN E(ALY), b Hammond, Ind, Jan 28, 29; m 54; c 3. CHEMICAL ENGINEERING, RADIOCHEMICAL PROCESSING. *Educ:* Purdue Univ, BS, 50; Mass Inst Technol, SM, 52, ScD(chem eng), 56. *Prof Exp:* Develop engr, Oak Ridge Nat Lab, 56-58; eng opers analyst, US Atomic Energy Comn, 58-61; DEVELOP ENGR, OAK RIDGE NAT LAB, 61- *Mem:* Am Chem Soc; Am Nuclear Soc; Sigma Xi. *Res:* Chemical processing of nuclear fuel, including nonaqueous methods; production and purification of transuranium elements; design and fabrication of californium-252 neutron sources. *Mailing Add:* 111 Concord Rd Oak Ridge TN 37830

BIGELOW, JOHN EDWARD, b Worcester, Mass, July 25, 22; m 48; c 5. ENGINEERING, PHYSICS. *Educ:* Worcester Polytech Inst, BS, 44 & 46. *Prof Exp:* Develop engr, Gen Eng Lab, Gen Elec Co, 46-54, sr engr, X-ray Dept, 54-59, mgr, Advan Eng Lab, 59-64, Info Eng Advan Tech Labs, 64-66, mgr display progs, 66-; RETIRED. *Res:* Scientific instruments; process instruments; diagnostic devices; optical and electronic information processors; detectors; computer driven displays; data terminals. *Mailing Add:* 834 Riverview Rexford NY 12148

BIGELOW, MARGARET ELIZABETH BARR, b Elkhorn, Man, Apr 16, 23; m 56. MYCOLOGY. *Educ:* Univ BC, BA, 50, MA, 52; Univ Mich, PhD(bot), 56. *Prof Exp:* Nat Res Coun fel, Univ Montreal, 56-57; from instr to prof, 57-89, Ray Ethan Torrey prof, 86-89, EMER PROF BOT, UNIV MASS, 89- *Mem:* Mycol Soc Am (vpres, 79, pres, 81-82); Am Inst Biol Sci; Int Asn Plant Taxonomists; Brit Mycol Soc. *Res:* Systematics and ecology of pyrenomycetous ascomycetes. *Mailing Add:* 9475 Inverness Ave Sidney BC V8L 3S1 Can

BIGELOW, SANFORD WALKER, b Winchester, Mass, Sept 28, 56; m 82; c 3. SCIENCE POLICY, MEDICAL & HEALTH SCIENCE. *Educ:* Am Univ, BS, 78, MS, 81, PhD(chem), 84; Am Bd Toxicol, cert. *Prof Exp:* Chem researcher, Am Univ, 78-79; chemist, NIH, 79-85; postdoctoral fel toxicol, Uniformed Serv Univ Health Sci, 85-86; toxicologist, Karch & Assocs, Inc, 86-87 & Environ Protection Agency, 87-89; PROJ DIR FOOD SAFETY, NAT ACAD SCI, 89- *Concurrent Pos:* Consult, MultiSci, Inc, 87- *Mem:* Soc Toxicol; Inst Food Technologists; Am Chem Soc; Am Soc Plant Physiologists; Am Col Toxicol; Toxicol Forum. *Res:* Food safety, specifically in the area of purity of food additives and natural constituents in foods. *Mailing Add:* Food & Nutrit Bd Nat Acad Sci 2101 Constitution Ave NW Washington DC 20418

BIGELOW, WILBUR CHARLES, b Wyoming Co, Pa, Mar 18, 23; m 50; c 2. MATERIALS SCIENCE ENGINEERING. *Educ:* Pa State Univ, BS, 44; Univ Mich, MS, 48, PhD(phys chem), 52. *Prof Exp:* Res chemist, US Naval Res Lab, Washington, DC, 44-46; res assoc res inst, 51-56, asst prof sci, 56-59, from assoc prof to prof chem & metall eng, 59-71, PROF MAT ENG, UNIV MICH, 71- *Mem:* Am Chem Soc; Electron Micros Soc Am; Microbeam Analytical Soc. *Res:* Electron microscopy, diffraction and microprobe analysis; microstructures of metals and ceramics systems and their relationships to composition and physical properties. *Mailing Add:* Dept Mat Sci & Eng Univ Mich Ann Arbor MI 48109

BIGGER, CYNTHIA ANITA HOPWOOD, b Sheffield, Ala, Mar 3, 42; m 81; c 1. GENETIC TOXICOLOGY, CHEMICAL CARCINOGENESIS. *Educ:* La State Univ, Baton Rouge, BS, 66, MS, 69, PhD(microbiol), 71. *Prof Exp:* NIH fel, Dept Molecular Biol, Univ Edinburgh, 71-72; vis instr res, Dept Chem, 73, res assoc, Dept Microbiol, 74, res assoc, Sch Vet Med, La State Univ, Baton Rouge, 74-76, scientist, Basic Res Prog, Nat Cancer Inst, Frederick Cancer Res Ctr, 76-91; STUDY DIR, GENETIC TOXICOL DIV,

MICROBIOL ASSOCS, INC, ROCKVILLE, MD, 91- *Mem:* Environ Mutagen Soc; Am Asn Cancer Res. *Res:* Genetic toxicology; molecular methods in environmental mutagenesis; molecular aspects of chemical carcinogenesis, interactions of polycyclic aromatic hydrocarbons with nucleic acids. *Mailing Add:* Microbiol Assocs Inc 9900 Blackwell Rd Rockville MD 20850

BIGGER, J THOMAS, JR, b Cambridge, Mass, Jan 17, 35. CARDIOLOGY, PHARMACOLOGY. *Educ:* Emory Univ, AB, 55; Med Col Ga, MD, 60. *Prof Exp:* Intern & asst resident, Columbia Univ Med Div, Bellevue Hosp, 60-62; asst resident & resident cardiol, Presby Hosp, New York, 63-65; instr, 66-67, assoc, 67-68, from asst prof to assoc prof med, 68-72, assoc prof, 72-75, PROF MED & PHARMACOL, COL PHYSICIANS & SURGEONS, COLUMBIA UNIV, 75- *Concurrent Pos:* Vis fel cardiol, Col Physicians & Surgeons, Columbia Univ, 65-66, fel, 66-67; NY Heart Asn fel, 65-68; Nat Heart & Lung Inst res career develop award, 72-77; asst physician, Presby Hosp, New York, 65-68, from asst attend physician to assoc attend physician, 68-75, attend physician, 75-; sr investr, NY Heart Asn, 68-72; assoc ed, Circulation, 70- *Mem:* Am Soc Clin Invest; Am Heart Asn; Am Physiol Soc; Am Soc Pharmacol & Exp Therapeut; fel NY Acad Sci. *Res:* Cardiac electrophysiology and arrhythmias; clinical and laboratory evaluation of cardioactive drugs, particularly cardiac antiarrhythmic drugs. *Mailing Add:* Col P & S Columbia Univ 630 W 168th St New York NY 10032

BIGGERS, CHARLES JAMES, b Gastonia, NC, Feb 7, 35; m 58; c 3. BIOLOGY, GENETICS. *Educ:* Wake Forest Col, BS, 57; Appalachian State Teachers Col, MA, 59; Univ SC, PhD(biol), 69. *Prof Exp:* Instr biol, Col of Orlando, 59-61, asst prof, 62-66, chmn dept, 61-66; exten instr, Univ SC, 66-69; assoc prof, 69-77, PROF BIOL, MEMPHIS STATE UNIV, 77- *Mem:* AAAS; Am Genetic Asn; Genetics Soc Am; Sigma Xi; Am Soc Mammalogists. *Res:* Genetic studies of protein polymorphisms in Peromyscus polionotus, Anthonomus grandis, Bufo woodhouseii, and Ictaluras punctatus, with primary interest in biochemical analysis and chromosomes. *Mailing Add:* 7250 Abercrombie Lane Memphis TN 38119

BIGGERS, JAMES VIRGIL, b Dallas, Tex, Nov 6, 28; m 56. CERAMICS SCIENCE & TECHNOLOGY. *Educ:* Purdue Univ, BS, 60; Pa State Univ, PhD(metall), 66. *Prof Exp:* Res engr, Union Carbide Corp, 60-62; dir res, Erie Technol Corp, 66-69; from asst prof to assoc prof ceramic sci, 69-76, SR RES ASSOC, MAT RES LAB, PA STATE UNIV, UNIVERSITY PARK, 76- *Mem:* Fel Am Ceramic Soc. *Res:* Processing and structure-property relationships in electronic ceramics. *Mailing Add:* 1321 Old Bogisburg Rd State Col PA 16801

BIGGERS, JOHN DENNIS, b Gt Brit, Aug 18, 23; div; c 3. PHYSIOLOGY. *Educ:* Univ London, BSc(vet sci) & BSc(spec), 46; PhD(physiol), 52, DSc, 65. *Prof Exp:* Student demonstr physiol, Royal Vet Col, Univ London, 44-45, demonstr, 45-47; asst lectr, Univ Sheffield, 47-48; lectr vet physiol, Univ Sydney, 48-53, sr lectr, 53-55; sr lectr physiol, Royal Vet Col, Univ London, 55-59; assoc mem, Wistar Inst & assoc prof, Sch Vet Med, Univ Pa, 59-61, King Ranch res prof reproductive physiol, 61-66; prof pop dynamics & assoc prof obstet & gynec, Johns Hopkins Univ, 66-72; PROF PHYSIOL & MEM LAB HUMAN REPRODUCTION & REPRODUCTIVE BIOL, HARVARD MED SCH, 72- *Concurrent Pos:* Vis scientist, Strangeways Res Lab, Eng, 54-55; Commonwealth fel, St John's Col, Cambridge Univ, 54-55; Damon Runyon Mem Fund grant, 60-62; NIH grant, 60-65 & 67-; Pop Coun grant, 60-; Lalor Found grant, 61; NSF grant, 61-64; adv, WHO, 64 & 71-; mem study sect reproductive biol, NIH, 67-71, adv, 69-; mem, ICRO Comt, 68-; Ford Found grant, 69-70; ed, Biol of Reproduction, 70-74; Ford Found adv, 74-; prog proj dir, NIH Grant, 74-; mem corp, Marine Biol Lab, Woods Hole. *Honors & Awards:* Hartman Award, Soc Study Reproduction, 87; Pioneer Award, Int Embryo Transfer Soc, 90; Marshall Medal, Soc Study Fertility, 90. *Mem:* Am Statist Asn; Soc Study Reproduction (pres, 68-69); fel Royal Col Vet Surg; fel Royal Statist Soc; Am Inst Biol Sci; Tissue Cult Asn; Physiol Soc; Soc Study Fertility; fel AAAS. *Res:* Physiology of fertilization, preimplantation development and implantation; growth and differentiation; chemically defined media; statistics and theory of experimental design; history of reproductive biology; ethics of reproductive biology. *Mailing Add:* Dept Physiol Harvard Med Sch Boston MA 02115

BIGGERSTAFF, JOHN A, b Berea, Ky, Sept 10, 31. PHYSICS. *Educ:* Univ Ky, BS & MS, 53, PhD(physics), 61. *Prof Exp:* Physicist, eng lab, US Army Signal Corps, 55; sr engr nucleonics, Martin Co, 55-57; res assoc nuclear physics, Univ Ky, 61-63; PHYSICIST, OAK RIDGE NAT LAB, 62- *Mem:* Am Phys Soc. *Res:* Low energy nuclear spectroscopy; instrumentation for nuclear spectroscopy. *Mailing Add:* Physics Div Oak Ridge Nat Lab B 6000 PO Box 2008 Oak Ridge TN 37831-6368

BIGGERSTAFF, ROBERT HUGGINS, b Richmond, Ky, June 1, 27; m 50; c 2. PHYSICAL ANTHROPOLOGY, ORTHODONTICS. *Educ:* Howard Univ, BS, 51, DDS, 55; Univ Pa, MS, 66, PhD(anthrop), 69; cert orthod, 73. *Prof Exp:* NIH trainee, Univ Pa, 64-66, res investr, 67-69; from asst prof to assoc prof orthod, Univ Ky, 73-82; PROF ORTHOD, UNIV TEX HEALTH SCI CTR, SAN ANTONIO, 82- *Mem:* Am Asn Phys Anthrop; Am Anthrop Asn; Am Dent Asn; Am Asn Orthod. *Res:* Analysis of morphological and mensurational variations in the post-canine dentition; the metric description of occlusion; analysis of arch form; environmental variables which can cause deviations in tooth form. *Mailing Add:* Univ Tex Health Sci Ctr 7703 Floyd Curl Dr San Antonio TX 78284

BIGGERSTAFF, WARREN RICHARD, b Folsom, NMex, May 2, 18; m 42; c 3. ORGANIC CHEMISTRY. *Educ:* Willamette Univ, BA, 40; Ore State Col, MS, 42; Univ Wis, PhD(org chem), 48. *Prof Exp:* Asst chem, Ore State Col, 40-42; asst, Univ Wis, 42-44, instr org chem, 47-48; prof, 58-81, EMER PROF CHEM, CALIF STATE UNIV, FRESNO, 81- *Concurrent Pos:* Chmn dept chem, Fresno State Col, 61-66, assoc vpres acad planning, 69-70; vis assoc, Sloan-Kettering Inst Cancer Res, 56-57; vis prof, Univ Lund, 67. *Mem:* Am Chem Soc; Sigma Xi. *Res:* Synthetic hormone substitutes; diethylstilbestrol and analogs related to progesterone and desoxycorticosterone; ring D metabolites of the estrogens; unsaturated thiolactones. *Mailing Add:* 2911 Juniper Ave Morro Bay CA 93442

BIGGINS, JOHN, b Sheffield, Eng, Mar 30, 36; m 63. PLANT PHYSIOLOGY, PLANT BIOCHEMISTRY. *Educ:* Univ London, BSc, 60; Univ Calif, Berkeley, PhD(plant physiol), 65. *Prof Exp:* Asst prof bot, Univ Pa, assoc prof biol, 70-78, PROF BIOL, BROWN UNIV, 78-, CHMN BIOCHEM, 85- *Mem:* Am Soc Photobiol; Am Soc Plant Physiol. *Res:* Correlation of structure and biochemical function of energy transducing systems; mechanism of photosynthesis and respiration in higher plants and algae; marine microbiology; phytoplankton physiology. *Mailing Add:* Dept of Biol & Med Sci Brown Univ Providence RI 02912

BIGGS, ALAN RICHARD, b Lewisburg, Pa, June 22, 53; m 81; c 1. HISTOPATHOLOGY, EPIDEMIOLOGY. *Educ:* Pa State Univ, BSc, 76, MSc, 78, PhD(plant path), 82. *Prof Exp:* Instr plant path, Pa State Univ, 82; res scientist, Agr Can, 83-89, head, Plant, Path Sect, 87-89; ASSOC PROF PLANT PATH, WVA UNIV, 89- *Concurrent Pos:* Adj prof, dept bot & plant path, Univ Toronto, 84-87. *Mem:* Can Phytopath Soc; Int Asn Wood Anatomists; Am Phytopath Soc. *Res:* Fungal and bacterial diseases of deciduous tree fruits; wound response and periderm regeneration in tree bark; breeding trees for resistance to trunk and limb diseases. *Mailing Add:* Univ Exp Farm WVa Univ PO Box 609 Kearneysville WV 25430

BIGGS, ALBERT WAYNE, b Toledo, Ohio, Oct 4, 26; m 59; c 2. ELECTRICAL ENGINEERING. *Educ:* Wash Univ, BSEE, 47; Stanford Univ, MBA, 49; Univ Wash, PhD(elec eng), 65. *Prof Exp:* Sales analyst, Surface Combustion Co, Ohio, 50-51; standards engr, Wright Air Develop Ctr, 51-52; design engr, Boeing Co, 52-55, res engr, Aerospace Div, 57-63, res specialist, Missile & Space Info Div, 65-67; instr elec eng, Univ Wash, 55-57, res instr, 63-65; assoc prof, 67-73, PROF ELEC ENG, UNIV KANS, 73- *Concurrent Pos:* Mem Comn II, Int Sci Radio Union, 65- *Mem:* Am Phys Soc; Inst Elec & Electronics Engrs; Sigma Xi; Optical Soc Am; Int Sci Radio Union. *Res:* Microwave systems, antennas and radio wave propagation; radar systems in guided missiles sources, high energy microwave and particle accelerators (neutrons). *Mailing Add:* 224 Shawnee Ct SE Albuquerque NM 87108

BIGGS, DAVID FREDERICK, b London, Eng, Jan 3, 39; m 62; c 2. PHARMACOLOGY. *Educ:* Univ Nottingham, BPharm, 61; King's Col, Univ London, 67. *Prof Exp:* Pharmacol dept head, Res Labs, May & Baker Ltd, Eng, 63-69; asst prof biopharm, 69-71, assoc prof, 71-78, McCalla prof, 83-84, PROF PHARMACOL, UNIV ALTA, 78- *Mem:* Brit Inst Biol; Brit Pharmacol Soc; Pharmaceut Soc Gt Brit; Pharmacol Soc Can. *Res:* Pharmacological properties of novel chemical compounds; relationship between structure and activity among neuromuscular blocking agents; pharmaceutical toxicology; respiratory physiology and pharmacology; toxicology of natural products; metabolism of heavy metals. *Mailing Add:* Fac Pharm & Pharmaceut Sci Univ Alta Edmonton AB T6G 2N8 Can

BIGGS, DONALD LEE, geology; deceased, see previous edition for last biography

BIGGS, DOUGLAS CRAIG, b Lancaster, Pa, Oct 17, 50; m; c 1. BIOLOGICAL OCEANOGRAPHY. *Educ:* Franklin & Marshall Col, Pa, AB, 72; Mass Inst Technol, PhD(biol oceanog), 76. *Prof Exp:* Fel biol oceanog, Joint Prog, Mass Inst Technol-Woods Hole Oceanog Inst, 72-76; res assoc, Marine Sci Res Ctr, State Univ NY Stony Brook, 76-77; asst prof, 77-83, ASSOC PROF OCEANOG, TEX A&M UNIV, 83- *Concurrent Pos:* Instr zooplankton ecol, Bermuda Biol Sta, 83, 86, 87, 89; mgr, Tech Support Group Tex A&M Univ, 86- *Mem:* Am Soc Limnol & Oceanog; AAAS; Am Geophys Union; Sigma Xi; Oceanog Soc; Marine Technol Soc. *Res:* Physiology and ecology of marine plankton; modeling of energy flow through intermediate trophic levels in open-ocean and estuarine systems; Antarctic marine ecosystems; nutrient cycling. *Mailing Add:* Dept Oceanog Tex A&M Univ College Station TX 77843-3146

BIGGS, FRANK, b Langdon, Mo, Dec 16, 27; m 64; c 7. MATHEMATICAL PHYSICS. *Educ:* Univ Ark, BS, 56, MS, 57, PhD(physics), 65. *Prof Exp:* STAFF MEM, SANDIA LAB, 56-, RES PHYSICIST, 65- *Concurrent Pos:* Physicist, Sandia Lab, 57-65, consult, 60-65; prof, Univ NMex, 75-76. *Mem:* Am Phys Soc; Am Soc Mech Engrs. *Res:* X-ray cross sections and instrumentation; analysis of experimental data; applied mathematics; energy studies. *Mailing Add:* 3515 Monte Vista NE Albuquerque NM 87106

BIGGS, HOMER GATES, b Greene, NY, Feb 16, 30; m 52; c 2. BIOCHEMISTRY. *Educ:* State Univ NY Binghamton, BA, 51; Univ Iowa, MS, 54, PhD(biochem), 56. *Prof Exp:* Instr & asst dir clin chem, Univ Tenn, 56-59, asst prof, Div Clin Chem, Dept Med Labs, 59-63; assoc prof path, Med Col & dir clin chem, Univ Hosp, Univ Ala, 63-70, dep dir educ, Dept Clin Path, 69-70; dir clin chem, Med Ctr, Ind Univ, Indianapolis, 70-74, prof clin path, 70-75; vpres & tech dir, Biozyme Labs, Inc, 75-78; dir, Prev Med Res Ctr of Cleveland & King James Med Labs, Inc, 79-86; pres & lab dir, BDI Labs, Inc, 86-88; CONSULT, 88-; DIR, WILLIAMS-LYNCH MED LAB & FRANKLIN PARK MED LAB, 89- *Concurrent Pos:* mem bd dirs & fel, Nat Acad Biochem. *Mem:* Am Chem Soc; Am Asn clin chemists; AAAS; Asn Clin Scientists; fel Nat Acad Clin Biochem; Sigma Xi; Fel Am Soc Lab Clin Scientists. *Res:* Serum proteins minerals and enzymes. *Mailing Add:* RR 2 Box 252 Spencer IN 47460-0252

BIGGS, MAURICE EARL, b Sebree, Ky, Oct 27, 21; m 46; c 4. SEISMOLOGY, GRAVITY. *Educ:* Ind Univ, AB, 48, MA, 50, PhD(geophysics), 74. *Prof Exp:* Head, Geophysics Sect, Int Geolog Surv, 51-87, asst state geologist, 60-87; RETIRED. *Mem:* Sigma Xi; Europ Asn Explor Geophysicists. *Res:* Seismic refraction and reflection, gravity and magnetic methods to study the configuration and type of basement rocks in Indiana; correlate physical properties of rock with rock type. *Mailing Add:* 1213 S Brooks Dr Bloomington IN 47401

BIGGS, PAUL, b Okmulgee, Okla, Nov 15, 13; m 38. PETROLEUM ENGINEERING. *Educ:* Okla Agr & Mech Col, BS, 37. *Prof Exp:* Petrol engr & prod supvr, Pan Am Petrol Corp, 37-49; prof supvr petrol eng, Petrol & Natural Gas Br, US Bur Mines, 49-60, proj coordr mineral resource develop, 60-74; LECTR PETROL ENG, COL PETROL ENG, UNIV WYO, 67- *Concurrent Pos:* Consult. *Honors & Awards:* Meritorious Serv Honor Award, US Dept Interior, 75, Unit Award, 75. *Mem:* Am Soc Petrol Engrs; Am Inst Mining, Metall & Petrol Engrs; hon mem Nat Soc Prof Engrs. *Res:* Oil and gas development and conservation in Rocky Mountain region, Oregon, Washington and Alaska; oil and gas studies in Missouri River basin; author over fifty government publications. *Mailing Add:* 1929 Thornburgh Dr Laramie WY 82070

BIGGS, R B, b Baltimore, Md, Feb 27, 37; m 60; c 2. MARINE GEOLOGY. *Educ:* Lehigh Univ, BA, 59, MS, 61, PhD(geol), 63. *Prof Exp:* Asst geol, Lehigh Univ, 60-62; res assoc, Chesapeake Biol Lab, 62-64; from asst prof to assoc prof geol,Col Marine Studies, Univ Del, 64-78,assoc dean, 77-80; VPRES, R F WESTON INC, 88- *Mem:* AAAS; Sigma Xi; Geol Soc Am; Am Soc Limnol Oceanog; Am Geophys Union. *Res:* Marine geology, especially the geochemistry of modern sediments; coastal pollution. *Mailing Add:* Geosci Dept R F Weston Inc West Chester PA 19380

BIGGS, R DALE, b Amarillo, Tex, Dec 14, 28; m 52; c 4. CHEMICAL ENGINEERING. *Educ:* Rice Univ, BSChE, 50; Univ Mich, MSChE, 51, PhD(chem eng), 54. *Prof Exp:* Chem engr, Ethyl Corp, La, 54-56; proj engr, Plastics Prod Dept, Dow Chem Co, 56-59, from res engr to sr res engr, Process Fundamentals Lab, 59-67, prod develop eng supvr, Plastics Prod Dept, 67-68, spec assignment, Technol Serv & Develop, 68-69, sr res engr, Phys Res Lab, 69-70, res mgr, Dow Interdisciplinary Group, 70-74, lab dir eng, Cent Res Labs, 74-79, assoc scientist, La Div, 79-86; RETIRED. *Concurrent Pos:* Lectr, Grad Exten Prog, Univ Mich, Midland, 57-68. *Mem:* Am Inst Chem Eng; Am Chem Soc. *Res:* Reaction kinetics; mass transfer; fluid mechanics, particularly polymer melts and solutions; chemical process development; coal gasification; process research. *Mailing Add:* 15903 Malvern Hill Baton Rouge LA 70817

BIGGS, ROBERT HILTON, b US, May 5, 31; m 53, 78; c 3. PLANT PHYSIOLOGY. *Educ:* ECarolina Univ, BS, 53; Purdue Univ, MS, 55, PhD(plant physiol), 58. *Prof Exp:* From asst prof biochem to assoc prof biochem, 58-69, prof & chmn, 75-77, PROF BIOCHEM, UNIV FLA, 69- *Concurrent Pos:* Sci & Educ Admin-Coop Res, USDA, Washington, DC, 78- *Mem:* Am Inst Biol Sci; AAAS; Sigma Xi; Am Soc Plant Physiol; Am Soc Hort Sci. *Res:* Abscission; ultraviolet light; chemical regulators; dormancy. *Mailing Add:* Five NW 28th Terrance Gainesville FL 32607

BIGGS, RONALD C(LARKE), b Ottawa, Ont, Feb 5, 38; m 61; c 2. BUILDING SERVICES, HEAT TRANSFER. *Educ:* Queens Univ, Ont, BSc, 61; Stanford Univ, MS, 62; McGill Univ, PhD(mech eng), 68. *Prof Exp:* Develop engr, Can Gen Elec, Ont, 63-65; assoc prof eng, Carleton Univ, Ont, 68-74; mgr energy res & develop, Pub Works Can, 74-78, dir, solar energy progs, 78-80; sr tech adv solar energy progs, 80-81, head bldg serv sect, Div Bldg Res, 81-86, DIR RES INST FOR RES IN CONST, NAT RES COUN CAN, 86- *Mem:* Am Soc Heating, Refrig & Air Conditioning Engrs. *Res:* Combined free and forced convective heat transfer for laminar flow of diathermanous and althermanous gases in a vertical tube; heat transfer characteristics of turbulent boundary layer flows with the wall and pressure gradients; solar energy; building research. *Mailing Add:* 22 Wren Rd Gloucester ON K1J 7H7 Can

BIGGS, WALTER CLARK, JR, b Wilmington, NC, June 17, 31; m 63; c 2. VERTEBRATE ZOOLOGY, ANIMAL BEHAVIOR. *Educ:* ECarolina Univ, BS, 53; Tex A&M Univ, MS, 60; NC State Univ, PhD(zool), 69. *Prof Exp:* Asst prof biol, 60-68, assoc prof, 69-78, PROF BIOL, UNIV NC, WILMINGTON, 78- *Mem:* Animal Behav Soc; Am Soc Mammalogists. *Res:* Social behavior and organization of animals, especially mother-young interactions among mammals; suckling behavior of domestic and feral swine compared in a variety of environmental regimes; population density, territoriality, and home range of small mammals in pine and hardwood forests of the coastal plains of North Carolina; mammalian septematics. *Mailing Add:* Dept Biol Univ NC Wilmington Wilmington NC 28403

BIGHAM, CLIFFORD BRUCE, b Kamloops, BC, June 12, 28; m 54; c 3. ACCELERATOR. *Educ:* Queen's Univ, Ont, BS, 51, MS, 52; Univ Liverpool, PhD(nuclear physics), 54. *Prof Exp:* From assoc res officer to sr res officer reactor physics, 54-66, SR RES OFFICER, ACCELERATOR PHYSICS BR, ATOMIC ENERGY CAN, LTD, 66- *Mem:* Can Asn Physicists; Can Nuclear Soc; Am Phys Soc. *Res:* Experimental reactor physics; nuclear physics data pertaining to and parameters of nuclear reactor lattices; RF power for linear accelerators; superconducting cyclotron design. *Mailing Add:* Accelerator Physics Br Physics Div Atomic Energy Can Ltd Chalk River ON K0J 1J0 Can

BIGHAM, ERIC CLEVELAND, b Kannapolis, NC, Mar 26, 47; m 70; c 1. MEDICINAL CHEMISTRY. *Educ:* NC State Univ, BS, 69; Princeton Univ, MA, 71, PhD(chem), 75. *Prof Exp:* res scientist med chem, Cent Res Div, Pfizer Inc, 73-78; SR RES SCIENTIST, BURROUGHS WELLCOME CO, 78-, PROJ COMT LEADER, 85- *Concurrent Pos:* Teaching asst, Princeton Univ, 69-73. *Mem:* Am Chem Soc; Sigma Xi; AAAS. *Res:* Design and synthesis of novel chemical compounds as potential drugs for the treatment of CNS and cardiovascular diseases; application of QSAR and computer methods to drug design. *Mailing Add:* Wellcome Res Labs 3030 Cornwallis Rd Research Triangle Park NC 27709

BIGHAM, ROBERT ERIC, b Lampasas, Tex, Feb 10, 40; m 70; c 1. EARTHWORK IN CIVIL ENGINEERING. *Educ:* A&M Col Tex, BS, 63; Tex A&M Univ, MS, 69. *Prof Exp:* Res asst, Tex Transp Inst, 63-64; US Army officer, US Army, 64-67; self employed, 67-68; ENGR, BUCHANAN SOIL MECHS INC, 68- *Concurrent Pos:* Vchmn, Drainage Syst Adv Bd, Bryan,

Tex, 86- *Mem:* Am Soc Civil Engrs; Int Soc Soil Mech & Found Eng. *Res:* Civil engineering earth-type materials and in computational methods; computational methods in civil engineering. *Mailing Add:* Buchanan Soil Mechanics Inc PO Box 672 Bryan TX 77806-0672

BIGLER, RODNEY ERROL, b Pocatello, Idaho, Mar 15, 41; m; c 4. NUCLEAR PHYSICS, BIOPHYSICS. *Educ:* Portland State Univ, BS, 66; Univ Tex, Austin, PhD(nuclear physics), 71. *Prof Exp:* Res assoc, Mem Sloan-Kettering Cancer Ctr, 71-73, assoc biophys, 73-82, asst mem, 82-84, asst lab mem, 84; asst prof biophys, Sloan-Kettering Div, 74-84, ASSOC PROF, MED COL, CORNELL UNIV, 84-, GRAD SCH MED SCI, 86- *Concurrent Pos:* Res collabr, Brookhaven Nat Lab, 73-79, 83-; attend physicist, New York Hosp, 84- *Mem:* Health Physics Soc; Am Asn Physics Teachers; Am Phys Soc; Am Asn Physicists in Med; Soc Nuclear Med; AAAS; NY Acad Sci; Radiation Res Soc; Soc Magnetic Resonance Med; Sigma Xi. *Res:* New methods to diagnose, treat and evaluate the results of therapy directed primarily toward malignant disease; positron emission tomography, nuclear magnetic resonance and monoclonal antibodies. *Mailing Add:* Dept Radiol Div Nuclear Med New York Hosp-Cornell Med Ctr 525 E 68th St New York NY 10021

BIGLER, STUART GRAZIER, b Johnstown, Pa, Oct 21, 27; m 50; c 5. METEOROLOGY. *Educ:* Pa State Univ, BS, 52; Agr & Mech Col Tex, MS, 57. *Prof Exp:* Assoc, Ill State Water Surv, 52-55; from res assoc to res scientist III, Tex A&M Univ, 55-59; meteorologist & head radar sferics unit, US Weather Bur, 59-64, supvr radiation systs sect, 64-70; dir, Alaska Region, Nat Weather Serv, 70-87; RETIRED. *Concurrent Pos:* Chmn, working group uses radar in meteorol, World Meteorol Orgn, 63-65; US Dept Com sci & tech training fel, 65-66. *Honors & Awards:* Spec Award, Am Meteorol Soc, 57; Dept Com Gold Medal, 70. *Mem:* Fel Am Meteorol Soc. *Res:* Radar meteorology and cloud physics. *Mailing Add:* Alaska Region Nat Weather Serv 701 C St Box 23 Anchorage AK 99513

BIGLER, WILLIAM NORMAN, b Oakland, Calif, Aug 29, 37; m 61; c 3. BIOCHEMISTRY. *Educ:* Univ Calif, Berkeley, AB, 60; San Jose State Col, MS, 63; Univ Colo, Boulder, PhD(biochem), 68. *Prof Exp:* Instr chem, Univ Colo, Boulder, 66-67; NIH fel, Univ Calif, Los Angeles, 68-70, actg asst prof biochem, 70; from asst prof to assoc prof chem, Rochester Inst Technol, 70-81, dir clin chem, 76-81; PROF CHEM & DIR, CTR ADVAN MED TECHNOL, SAN FRANCISCO STATE UNIV, 81- *Concurrent Pos:* Vis assoc prof biochem & biophys, Univ Hawaii, 77-78. *Mem:* NY Acad Sci; Am Asn Clin Chemists; AAAS; Am Chem Soc. *Res:* Biochemistry of cell division, regulatory enzymes, effects of ionizing radiation and slime molds; high pressure liquid chromatography, von Hipple-Lindau syndrome. *Mailing Add:* Ctr Advan Med Technol San Francisco State Univ 1600 Holloway Ave San Francisco CA 94132

BIGLEY, HARRY ANDREW, JR, b Wilkinsburg, Pa, July 18, 13; m; c 2. AERONAUTICAL & MECHANICAL ENGINEERING. *Educ:* Univ Pittsburgh, BS, 35. *Prof Exp:* Test engr, Gulf Res & Develop Co, 36-38, rd test engr, 38-47, fuels engr, 48-52, sect head automotive fuels, 53- 61, sr proj engr, Automotive Div, 61-65, staff engr, Petrol Prods Dept, 65-75; RETIRED. *Concurrent Pos:* Instr, Univ Pittsburgh, 36-37. *Mem:* Soc Automotive Engrs. *Res:* Fuels and lubricants for passenger cars and trucks and performance under various operating and climatic conditions. *Mailing Add:* 604 Nutmeg Pl Sun City Center FL 33573-5711

BIGLEY, NANCY JANE, b Sewickley, Pa, Feb 1, 32. MICROBIOLOGY, IMMUNOLOGY. *Educ:* Pa State Univ, BS, 53; Ohio State Univ, MSc, 55, PhD(bact, immunol), 57. *Prof Exp:* Res assoc immunol, Ohio State Univ, 57-65, asst prof microbiol, 65-68, assoc prof, Fac Microbiol & Cellular Biol, 68-69; from assoc prof to prof microbiol, Chicago Med Sch, 69-76; dept chmn & prog dir, Sch Med & Col Sci & Eng, 76-86, PROF MICROBIOL & IMMUNOL, WRIGHT STATE UNIV, 76- *Concurrent Pos:* Am Found Microbiol lectr, 75-76. *Mem:* AAAS; Am Soc Microbiol; Am Asn Immunologists; Reticuloendothelial Soc; Sigma Xi. *Res:* Viral alterations of tissue components; autoimmunization; Rh antigen structural aspects; nucleic acid antigens; RNA-protein subfractions of Salmonella typhimurium involved in protective immunity; tumor-specific DNA-protein antigens; virus impairment of immunity; role of viruses in insulin-dependent diabetes. *Mailing Add:* Dept Microbiol & Immunol Wright State Univ Dayton OH 45435

BIGLEY, ROBERT HARRY, b Auburn, Wash, Aug 11, 29; m 59; c 2. INTERNAL MEDICINE, GENETICS. *Educ:* Univ Wash, BS, 51; Univ Ore, MD, 53. *Prof Exp:* From instr to assoc prof med, 60-75, PROF MED & MED GENETICS, SCH MED, ORE HEALTH SCI UNIV, 75- *Concurrent Pos:* USPHS res fel coagulation, Med Sch, Univ Ore, 59-60. *Mem:* Am Soc Hemat; Int Soc Hemat. *Res:* Blood cell function; ascorbate metabolism. *Mailing Add:* 3181 SW Sam Jackson Park Rd L 103 Univ Ore Health Sci Ctr Portland OR 97201

BIGLIANO, ROBERT P(AUL), b New York, NY, Dec 6, 29; m 53; c 3. ELECTRICAL ENGINEERING. *Educ:* Rutgers Univ, BSEE, 51; Univ Del, MBA, 69. *Prof Exp:* Res engr, E I du Pont de Nemours Co Inc, 53-60, sr res engr, 60-63, res assoc, 64-72, res supvr, 72-74, eng mgr, 74-77, eng mgr, Du Pont Instrument Prod, 77-78, opers mgr, Far East Opers, 78-80, mgr Advan Instrument Systs, 86-89, MGR INST ENG, CLIN SYST DIV, E I DU PONT DE NEMOURS & CO, INC, 80- *Concurrent Pos:* Electronics res officer, Air Force Cambridge Res Ctr; pres, eng mgt & develop, Vicon Co. *Mem:* AAAS; Inst Elec & Electronics Engrs; Sigma Xi. *Res:* Instrument development for process control; adaptive systems for measurement and control; signal interpreting devices and systems; medical instrumentation; liquid chromatograph instrument development and marketing; computer displays for graphic arts color printing. *Mailing Add:* Three Berrywood Ct Wilmington DE 19810

BIGLIERI, EDWARD GEORGE, b San Francisco, Calif, Jan 17, 25; m 53; c 3. INTERNAL MEDICINE, ENDOCRINOLOGY. *Educ:* Univ San Francisco, BS, 48; Univ Calif, San Francisco, MD, 52. *Hon Degrees:* DSc, Univ S Fla, 85. *Prof Exp:* Intern, Med Ctr, Univ Calif, San Francisco, 52-53, resident, 53-54; resident, Vet Admin Hosp, San Francisco, 54-56; clin assoc, Nat Heart Inst, 56-58; asst res physician, Metab Unit, 58-61, from asst prof to assoc prof, 62-71, PROF MED, MED CTR, UNIV CALIF, SAN FRANCISCO, 71-; CHIEF, ENDOCRINE METAB DIV, SAN FRANCISCO GEN HOSP, 62-, PROG DIR, CLIN STUDY CTR, 64- *Concurrent Pos:* Vis scientist, Prince Henry's Hosp, Monash Univ, Australia, 67; consult, Clin Invest Ctr, Oak Knoll Naval Hosp, Calif, 67-76; mem coun high blood pressure res, Am Heart Asn; transatlantic lectr, British Endo Soc, 91. *Honors & Awards:* Tigerstadt Award, Am Soc Hypertension, 87. *Mem:* Asn Am Physicians; fel Am Col Physicians; Asn Am Physicians; Am Soc Clin Invest; Endocrine Soc; Am Soc Hypertension (pres). *Res:* Adrenal; mineralocorticoid hormones. *Mailing Add:* Clin Study Ctr San Francisco Gen Hosp San Francisco CA 94110

BIGNELL, RICHARD CARL, b Toronto, Ont, Oct 22, 43; m 68; c 2. ASTRONOMY. *Educ:* Univ Western Ont, BA, 66, MSc, 67; Univ Toronto, PhD(astron), 72. *Prof Exp:* Teaching asst astron, Univ Toronto, 67-72; Nat Res Coun Can fel & res assoc, Nat Radio Astron Observ, 72-74; Nat Res Coun Can fel, Dominion Radio Astrophys Observ, 74-75; syst scientist, 76-78, head, 78-81, DEP ASST DIR, VERY LARGE ARRAY TELESCOPE OPERS DIV, NAT RADIO ASTRON OBSERV, 81- *Mem:* Am Astron Soc; Can Astron Soc. *Res:* Radio properties of planetary nebulae; radio polarization properties of extragalactic radio sources; radio properties of normal galaxies with peculiar nuclei. *Mailing Add:* 600 Western Ave Socorro NM 87801

BIGRAS, FRANCINE JEANNE, b Montreal, Que, Mar 13, 48; m 71; c 2. STRESS PHYSIOLOGY, SEEDLING PRODUCTION. *Educ:* Univ Montreal, BS, 68; Laval Univ, MS, 72, PhD(plant sci), 87. *Prof Exp:* Teacher biol, Cegep de Sainte-Foy, 71-73; res asst biol, Laval Univ, 73-80, res asst plant sci, 87; postdoctoral forestry, 88, RES SCIENTIST FORESTRY, FORESTRY CAN, 88- *Concurrent Pos:* Vis scientist, Univ Minn, 84; adj prof, Laval Univ, 88- *Mem:* Can Soc Plant Physiol; Can Soc Hort Sci; Am Soc Hort Sci; Int Soc Hort Sci. *Res:* Conifer seedling production; stress physiology. *Mailing Add:* Forestry Can Que Region PO Box 3800 Sainte-Foy PQ G1V 4C7

BIHLER, IVAN, b Osijek, Yugoslavia, Aug 12, 24; Can citizen; m 48. BIOPHYSICS, ANIMAL PHYSIOLOGY. *Educ:* Hebrew Univ, Jerusalem, MSc, 54, PhD(biochem), 57. *Prof Exp:* Asst biochem, Hebrew Univ, Jerusalem, 53-58; from asst prof to assoc prof, 63-67, PROF PHARMACOL & THERAPEUT, UNIV MAN, 67- *Concurrent Pos:* Fel pharmacol, Univ Rochester, 58-59; fel biochem, Washington Univ, 59-61; Imp Chem Indust res fel, Univ Edinburgh, 61-63; career investr, Med Res Coun, Can, 64. *Honors & Awards:* Upjohn Award, Pharmacol Soc Can, 87- *Mem:* Am Soc Pharmacol & Exp Therapeut; Pharmacol Soc Can; NY Acad Sci; Can Biochem Soc; Int Soc Heart Res. *Res:* Mechanisms of cell membrane permeability and its regulation; membrane transport of sugars in intestine and muscle; effect of drugs and hormones. *Mailing Add:* Dept Pharmacol & Therapeut Univ Man Winnipeg MB R3E 0W3 Can

BIHOVSKY, RON, b Ithaca, NY, Nov 9, 48; m 85; c 2. SYNTHETIC ORGANIC CHEMISTRY. *Educ:* State Univ NY, Stony Brook, BS, 70; Univ Calif, Berkeley, PhD(chem), 77. *Prof Exp:* Chemist, Entomol Res Div, USDA, 71-72; NIH fel, Univ Wis-Madison, 78-80; prof chem, State Univ NY, Stony Brook, 80-88; SCIENTIST, BERLEX LABS, 87- *Mem:* Am Chem Soc. *Res:* Synthesis of medicinal compounds; enzyme inhibitors; heterocyclic compounds; peptide mimetics. *Mailing Add:* Berlex Labs 110 E Hanover Ave Cedar Knolls NJ 07927

BIJOU, SIDNEY WILLIAM, b Baltimore, Md, Nov 12, 08; m 34; c 2. DEVELOPMENTAL PSYCHOLOGY, CLINICAL PSYCHOLOGY. *Educ:* Univ Fla, BS, 33; Columbia Univ, MA, 36, Univ Iowa, PhD(psychol), 41. *Prof Exp:* Psychologist, Del State Hosp & Mental Health Clin, 37-39; res child psychologist, Wayne County Training Sch, 41-42 & 46-47; asst prof psychol, Ind Univ, 46-48; prof & dir, Inst Child Develop, Univ Wash, 48-65; prof & dir child behav lab, Univ Ill, 65-75; EMER PROF, UNIV ILL, 75-; EMER ADJ PROF PSYCHOL, SPEC EDUC & REHAB, UNIV ARIZ, 75- *Concurrent Pos:* Postdoctoral fel, Harvard Univ, 61-62; consult, Nat Asn Retarded Citizens, 65-88, Portage Proj Early Childhood Educ, 73-; pres, Div Child Develop, Am Psychol Asn, 65-66; bd mem, Intermountain Ctrs Human Develop, 77-, Desert Survivors, 82-90; nat adv comt mental retardation res, Univ Kans, 90-; career res sci award, Am Acad Mental Retardation, 80. *Honors & Awards:* G Stanley Hall Award Child Develop Am Psychol Asn, 80, Edgar A Doll Award Mental Retardation, 84. *Mem:* Am Psychol Asn; Asn Behav Analysts (pres, 77-78); Am Asn Univ Prof; Psychonomic Soc; Soc Res Child Develop; Am Psychol Soc; Am Asn Appl & Prev Pschol. *Res:* Theory and research relating to normal and deviant human development from a natural science perspective; study of language development. *Mailing Add:* 5131 N Soledad Primera Tucson AZ 85718-4822

BIKALES, NORBERT M, b Berlin, Ger, Jan 7, 29; nat US; m 51; c 2. MATERIALS SCIENCE, ORGANIC & POLYMER CHEMISTRY. *Educ:* City Col, BS, 51; Polytech Inst Brooklyn, MS, 56, PhD(chem), 61. *Prof Exp:* Res chemist, Am Cyanamid Co, 51-62; tech dir, Gaylord Assoc, Inc, 62-65; consult chem, 65-76; prof chem & dir continuing educ sci, Rutgers Univ, New Brunswick, 73-79; DIR POLYMERS PROG, NSF, 76- *Concurrent Pos:* Ed, Encycl Polymer Sci & Technol, 62-77; polymer workshop dir, Fairleigh Dickinson Univ, 66-68; adj prof, Upsala Col & consult to trustees, Columbia Univ & Strickman Found, 67-73; mem, comt lead paints, Nat Acad Sci, 72-74, adv bd, J Polymer Sci, 78- & bd dirs, EPS Div, Soc Plastics Engrs, 79-82; assoc mem comn macromolecules, Int Union Pure & Appl Chem, 74-77; titular mem & secy, 77-87; ed bd, Encyclopedia Polymer Sci & Eng, 82-; chmn, Polymer Div, Am Chem Soc, 83, counr, 87-90; vis sr scientist, Ctr Res

Macromolecules, Strasbourg, France, 84; chmn polymer conf, Gordon Res Conf, 85, counr, 85-88, Bd Trustees, 90-; distinguished lectr, Soc Polmer Sci, Japan, 86; adv bd, J Appl Polymer Sci, 89- *Honors & Awards:* Great Medal, City of Paris, France, 85. *Mem:* Fel AAAS; sr mem Soc Plastics Engrs; Am Chem Soc; fel NY Acad Sci; Am Phys Soc; Mat Res Soc. *Res:* Polymer synthesis, reactions, and properties; chemical modification of natural polymers; water-soluble polymers; materials research; polymer education. *Mailing Add:* NSF 1800 G St NW Washington DC 20550

BIKERMAN, MICHAEL, b Berlin, Ger, July 30, 34; US citizen; m 56; c 3. GEOCHRONOLOGY. *Educ:* Queens Col, NY, BS, 54; NMex Inst Mining & Technol, BS, 56; Univ Ariz, MS, 62, PhD(geol), 65. *Prof Exp:* Instr sci & eng, Ft Lewis Agr & Mech Col, 56-57; assayer, Holly Minerals, Cinnabar Mine, Idaho, 57; instr sci, Boise Jr Col, 58-60; asst geochem, geochronology lab, Univ Ariz, 63-65; asst prof geol, Wichita State Univ, 65-67; asst prof, 67-71, ASSOC PROF GEOL & PLANETARY SCI, UNIV PITTSBURGH, 71- *Concurrent Pos:* Res assoc, Carnegie Mus Natural Hist, 79- *Mem:* Fel Geol Soc Am; Geochem Soc; AAAS. *Res:* Isotope geology and geochronology of volcanic rocks; isotope geology and geochronology of metamorphic terranes; history of Appalachian epeirogeny. *Mailing Add:* Dept Geol & Planetary Sci Univ Pittsburgh Pittsburgh PA 15260

BIKIN, HENRY, b Chicago, Ill, Oct 7, 18; m 43; c 2. PHARMACEUTICAL CHEMISTRY. *Educ:* Purdue Univ, BS, 40, MS, 48, PhD(pharmacy), 50. *Prof Exp:* Hosp pharmacist, Univ Hosp, Univ Mich, 40-41; pharmaceut chemist, Burroughs Wellcome & Co, 41-43; lab asst, Purdue Univ, 47-48; pharmaceut chemist, Eli Lilly & Co, 49-57; mgr pharmaceut develop & prod, Corvel, Inc, 57-67; res scientist, Eli Lilly & Co, 67-79; RETIRED. *Mem:* AAAS; fel Am Inst Chem; Am Chem Soc; Am Pharmaceut Asn. *Res:* Incompatibilities; stabilities of liquid products; tablet products; formulations; stabilities; manufacturing trouble shooter. *Mailing Add:* 109 Longhorn Trail Kerrville TX 78028

BIKLE, DANIEL DAVID, b Harrisburg, Pa, Apr 25, 44; m 65; c 2. INTERNAL MEDICINE, BIOCHEMISTRY. *Educ:* Harvard Univ, AB, 65; Univ Pa, MD, 69, PhD(biochem), 74. *Prof Exp:* Clin fel internal med, Peter Bent Brigham Hosp, 69-71; res internist internal med & biochem, Letterman Army Inst Res, 74-80; dir, spec diagnostic & treatment unit, Univ Calif-Vet Admin Med Ctr & asst prof med, 80-86, ASSOC PROF MED, UNIV CALIF, SAN FRANCISCO, 86- *Honors & Awards:* Henderson Mem Prize in Biochem; Baldwin Luckie Mem Prize; Horatio C Wood Prize in Pharmacol; Spencer Morris Prize. *Mem:* AAAS; Am Soc Clin Invest; Endocrinol Soc; Am Soc Bone & Mineral Res; Am Inst Nutrit; Am Soc Clin Nutrit; Am Col Physicians. *Res:* Mechanism of action of Vitamin D; metabolic bone disease. *Mailing Add:* Dept Med Univ Calif-Vet Admin Med Ctr 4150 Clement San Francisco CA 94121

BILAN, M VICTOR, b Ukraine, June 6, 22; US citizen; m 57; c 4. FOREST PHYSIOLOGY, ECOLOGY. *Educ:* Univ Munich, BS, 49; Duke Univ, MF, 54, DF, 57. *Prof Exp:* Res asst forest biol, Duke Univ, 53-57; from asst prof to assoc prof forestry, 57-66, prof tree physiol & ecol, 67-73, distiguished prof, 73-84, REGENTS PROF TREE PHYSIOL & ECOL, STEPHEN F AUSTIN STATE UNIV, 84- *Concurrent Pos:* NSF travel grant, Austria, 61, res grant, 62-66; Nat Acad Sci/NSF exchange scientist, Bulgaria & Romania, 74. *Mem:* Soc Am Foresters; Ecol Soc Am; Sigma Xi; Am Inst Biol Sci. *Res:* Effect of environment on root growth; root-shoot growth correlation; seed production; growth and development of southern pines; moisture relations in trees, drought resistance. *Mailing Add:* Sch Forestry Stephen F Austin State Univ Nacogdoches TX 75962

BILANIUK, LARISSA TETIANA, b Ukraine, July 15, 41; US citizen; m 64; c 2. NEURORADIOLOGY, MAGNETIC RESONANCE IMAGING. *Educ:* Wayne State Univ, BA, 61, MD, 65; Am Bd Radiol, radiol specialty dipl, 71. *Prof Exp:* Intern Philadelphia Gen Hosp, 65-66; resident, 66-67 & 68-71, assoc, 72-74, from asst prof to assoc prof, 74-82, PROF RADIOL, HOSP & SCH MED, UNIV PA, 82- *Concurrent Pos:* Res fel, Cancer Res Ctr, Heidelberg, Ger, 67-68; NIH fel, 68-69 & 70-71; clin fel, Am Cancer Inst, 69-70; trainee, Armed Forces Inst Path, Washington, DC, 70; Rothschild Ophthal Found fel, Paris, 72 & 80; vis prof, Grosshadern Clin, Univ Munich, Ger, 88. *Mem:* Ukrainian Med Asn NAm; Radiol Soc NAm; fel Am Col Radiol; Sigma Xi; Am Soc Neuroradiol; Europ Soc Neuroradiol. *Res:* Computed tomography and magnetic resonance imaging in the diagnosis and treatment of brain tumors. *Mailing Add:* Dept Radiol Hosp Univ Pa Philadelphia PA 19104-4283

BILANIUK, OLEKSA-MYRON, b Ukraine, Dec 15, 26; US citizen; m 64; c 2. NUCLEAR & SUBNUCLEONIC STRUCTURE. *Educ:* Univ Louvain, Cand Eng, 49; Univ Mich, BSE, 52, MS, 53, AM, 54, PhD(physics), 57. *Prof Exp:* Instr, Univ Mich, 57-58; asst prof physics, Univ Rochester, 59-64; assoc prof, 64-70, prof physics, 70-82, SWARTHMORE CENTENNIAL PROF, SWARTHMORE COL, 82- *Concurrent Pos:* vis scientist, Arg Atomic Energy Comn, 62-63; NSF sci fac fel, Max Planck Inst, Heidelberg, 67-68 & Univ Paris, Orsay, 72; Nat Acad Sci exchange scholar, Kiev, Ukraine, 76; Vis scientist, Institut de Physique Nucleaire, Orsay, France, 80, Laboratori Nazionali di Frascati, Italy, 84 & Universität München, Garching, Germany, 88. *Mem:* Am Phys Soc; Europ Phys Soc; Sigma Xi; Ukrainian Acad Arts & Sci in US. *Res:* Nuclear spectroscopy; nuclear and hyper-nuclear reaction mechanisms; possibility of existence of supraluminal quanta. *Mailing Add:* Dept Physics Swarthmore Col Swarthmore PA 19081-1397

BILBAO, MARCIA KEPLER, b Rochester, Minn, Jan 14, 31; m 54; c 3. RADIOLOGY. *Educ:* Univ Minn, Minneapolis, BS, 52; Columbia Univ, MD, 57; Am Bd Radiol, dipl, 62. *Prof Exp:* From instr to prof radiol, Med Sch, Univ Ore, 61-82; CHIEF RADIOL, VET ADMIN MED CTR, SALT LAKE CITY, UTAH, 82- *Concurrent Pos:* Consult, Kaiser Permanaente Clin, 61-63. *Honors & Awards:* Janeway Prize, Col Physicians & Surgeons, Columbia Univ, 57; Award of Merit, Am Women's Med Asn, 57; Cert of Merit, AMA, 68. *Res:* Diagnostic radiology; gas contrast radiography; gastrointestinal radiography. *Mailing Add:* VA Med Ctr Dept Rad No 114 500 Foothill Blvd Salt Lake City UT 84148

BILD, RICHARD WAYNE, b Washington, DC, June 17, 46; m 74; c 1. ANALYTICAL NUCLEAR CHEMISTRY, METEORITICS. *Educ:* Calif Inst Technol, BS, 68; Univ Calif, Los Angeles, PhD(chem), 76. *Prof Exp:* Mem staff cosmochem, Max Planck Inst Nuclear Physics, 76-77; res assoc exp geochem, Lunar & Planetary Lab, Univ Ariz, 77-78; mem tech staff analytical chem, 78-87, SR MEM TECH STAFF RADIATION PHYSICS & HOT CELL APPLNS, SANDIA LABS, 87- *Honors & Awards:* R F Bunshah Award, Am Vacuum Soc, 85. *Mem:* Meteoritical Soc; Am Nuclear Soc; Am Geophys Union. *Res:* Radio and nuclear analytical chemistry; analytical geochemistry; hot cell operations and applications. *Mailing Add:* Div 6454 Sandia Nat Labs PO Box 5800 Albuquerque NM 87185

BILDERBACK, DAVID EARL, b Salem, Mass, Nov 27, 43; m 70; c 2. PLANT TISSUE CULTURE. *Educ:* Univ Ore, Eugene, BS, 65, PhD(biol), 68. *Prof Exp:* Vis asst prof, dept bot, Univ Mont, Missoula, 68-69; NIH fel, AEC Plant Res Lab, Mich State Univ, 69-70; vis asst prof, dept biol, Univ Ore, Eugene, 70-71 & dept bot sci, Univ Calif, Los Angeles, 71-72; from asst prof to assoc prof, 72-78, chmn dept, 83-88, PROF BOT, UNIV MONT, MISSOULA, 78-, ASSOC DEAN, COL ARTS & SCI, 88- *Mem:* Bot Soc Am; Am Soc Plant Physiologists; Sigma Xi. *Res:* Growth and development of vascular plants. *Mailing Add:* Col Arts & Sci Univ Mont Missoula MT 58812

BILDERBACK, DONALD HEYWOOD, b Usumbura, Africa, March 6, 47; US citizen; m 69; c 1. ENGINEERING PHYSICS. *Educ:* Seattle Pac Univ, BS, 69; Purdue Univ, MS, 72, PhD(physics), 75. *Prof Exp:* Res mgr, X-ray Facil, Mat Sci Ctr, 75-78; oper mgr, Cornell High Energy Synchrotron Source, 277 Wilson Lab, 78-83, staff scientist, 83-89, ASSOC DIR, CORNELL UNIV, 90- *Concurrent Pos:* Pres, Multiwire Lab Ltd, NY, 81- *Mem:* Am Crystallog Asn; Am Phys Soc; Soc Photooptical Instrumentation Engrs. *Res:* Field of x-ray diffraction; synchrotron radiation research and two dimensional area detectors; x-ray optics (wide bandpass monochromators, transmission and reflection x-ray mirrors). *Mailing Add:* Cornell High Energy Synchrotron Source Wilson Lab Cornell Univ Ithaca NY 14850

BILDSTEIN, KEITH LOUIS, b Hoboken, NJ, Jan 19, 50. BEHAVIORAL ECOLOGY, ORNITHOLOGY. *Educ:* Muhlenberg Col, BS, 72; Ohio State Univ, MS, 76, PhD(zool), 78. *Prof Exp:* Vis asst prof biol, Col William & Mary, 78; from instr to assoc prof, 78-88, PROF BIOL, WINTHROP COL, 88-; RES ASSOC, BARUCH INST, UNIV SC, 80- *Mem:* AAAS; Am Ornithologists Union; Wilson Ornith Soc; Animal Behav Soc; Soc Conserv Biologists. *Res:* Winter behavior of diurnal raptors; behavioral development of sex-specific hunting behavior in diurnal raptors; behavioral development of hunting behavior in White Ibis; coastal ecology. *Mailing Add:* Dept Biol Winthrop Col Rock Hill SC 29733

BILELLO, JOHN CHARLES, b Brooklyn, NY, Oct 15, 38; m 59; c 3. METALLURGY, MATERIALS SCIENCE. *Educ:* NY Univ, BMetE, 60, MS, 62; Univ Ill, PhD(metall), 65. *Prof Exp:* Res engr, Gen Tel & Electronics Labs, Inc, 65-68; asst prof eng, State Univ NY, Stony Brook, 68-71, from assoc dean to dean, Col Eng & Appl Sci, 76-77, prof eng, 71-86; dean, Col Eng & Computer Sci, Calif State Univ, Fullerton, 86-89; PROF MAT SCI ENG & APPL PHYSICS, UNIV MICH, ANN ARBOR, 89- *Concurrent Pos:* NATO sr fac fel, 73; vis prof, Enrico Fermi Ctr Nuclear Studies, Milan, 74; consult eng, NAm Philips Corp, 75-81, Brookhaven Nat Lab, 78-86, NATO Sr Fac Exchange, Oxford Univ, England, 85-90. *Mem:* Am Inst Mining, Metall & Petrol Engrs; Am Soc Metals; Am Phys Soc; Sigma Xi; Mat Res Soc. *Res:* Synchrotron x-ray diffraction topography imaging; dislocation mechanisms of yielding, work hardening and fracture of ionic, semiconductor and metallic crystals; internal friction studies of defect interactions with dislocations; precipitation and dispersion-hardened alloy systems. *Mailing Add:* 3646 W Huron River Dr Ann Arbor MI 48103

BILELLO, MICHAEL ANTHONY, b NY, Oct 24, 24; m 51; c 5. METEOROLOGY, CLIMATOLOGY. *Educ:* Univ Wash, BS, 50; McGill Univ, MSc, 72. *Prof Exp:* Meteorologist, hydrol sect, US Weather Bur, 51-54 & northern hemisphere map unit, 54-55, nat weather anal ctr, 55-56; meteorologist, snow, ice & permafrost res estab, US Army Corps Engrs 56-68, res metrorologist, Geophys Sci Br, Cold Regions Res & Eng Lab, 68-82; CONSULT, SCI & TECHNOL CORP, 85- *Concurrent Pos:* Vis instr, Dartmouth Col, 64, instr glaciol sem, 68; original cataloging, Baker Lib, Dartmouth Col, 83-84; researcher river ice mgt, Ice Eng Facil, Cold Regions Res & Eng Lab, 85. *Honors & Awards:* Ann Grad Award, Can Asn Geogr, 73; US-Latin Am Coop Sci Award, NSF, 75. *Mem:* Am Meteorol Soc; Can Geog Soc. *Res:* Meteorology and climatology in association with physical properties of snow cover; formation, growth, and decay of sea, lake and river ice; special projects on cold regions environmental research. *Mailing Add:* 12 Spencer Rd Hanover NH 03755

BILES, CHARLES MORGAN, b Yakima, Wash, Nov 29, 39; c 3. MATHEMATICS. *Educ:* St Martin's Col, BS, 61; Univ Ariz, MS, 64; Univ NH, PhD(math), 69. *Prof Exp:* Instr math & chem, St Martin's Col, 64-66; assoc prof, 69-77, PROF MATH, HUMBOLDT STATE UNIV, 77- *Concurrent Pos:* Mem, Comt Alliance for Progress, US Dept Com. *Mem:* Am Math Soc; Math Asn Am; Consortium Math & Appln Proj; Resource Modeling Asn (treas, 86-). *Res:* Mathematical modeling of microbial systems and renewable resource systems. *Mailing Add:* Dept Math Humboldt State Univ Arcata CA 95521

BILES, JOHN ALEXANDER, b Del Norte, Colo, May 4, 23; m 43; c 2. PHARMACEUTICAL CHEMISTRY. *Educ:* Univ Colo, BS, 44, PhD(chem), 49. *Prof Exp:* Asst chem, Univ Colo, 44-47, instr pharm, 47-48, asst chem, 48-49; prof pharmaceut chem, Midwestern Univ, 49-50; asst prof pharm, Ohio State Univ, 50-52; from asst prof to prof, 52-57, PROF PHARMACEUT CHEM, UNIV SOUTHERN CALIF, 57-, DEAN, SCH PHARM, 68-, JOHN STAUFFER DEAN'S CHAIR, 88- *Concurrent Pos:* mem bd dirs, Marion Merrell Dow, Kansas City, MO; mem, educ comt, Nat Assoc Retail Druggists; mem bd dirs, exec comt, Am Found Pharmaceut Educ. *Honors & Awards:* Lehn & Fink Gold Medal, 45. *Mem:* Fel Am Pharmaceut Asn; fel Acad Pharmaceut Sci; fel Am Asn Pharmaceut Sci; Am Asn Hosp Pharmacists; Am Asn Col Pharm (pres, 90-91). *Res:* Pharmaceutical economics; education, health care delivery systems. *Mailing Add:* Sch Pharm Univ Southern Calif 1985 Zonal Ave Los Angeles CA 90033-1086

BILES, ROBERT WAYNE, b West Palm Beach, Fla, Oct 9, 47; m 73; c 3. TOXICOLOGY. *Educ:* Univ Calif, Santa Barbara, BA, 70; San Diego State Univ, MS, 74; Univ Tex Med Br Galveston, PhD(environ toxicol), 78. *Prof Exp:* Toxicologist, Exxon Corp, 78-81, sr toxicologist, 81-83, staff toxicologist, 83-87, sr staff toxicologist, sect head chem toxicol, 84-, MKT TECH SERVS, EXXON CO USA, HOUSTON. *Concurrent Pos:* Lectr, Sch Pub Health, Columbia Univ, 79- & Rutgers Univ, 81- *Mem:* Soc Occup & Environ Health. *Res:* Biomedical, toxicological research and consultation. *Mailing Add:* Exxon USA 800 Bell St Houston TX 77002

BILES, W(ILLIAM) R(OY), b Cincinnati, Ohio, Dec 20, 23; m 45; c 3. CHEMICAL ENGINEERING, PROCESS CONTROL. *Educ:* Univ Cincinnati, BS, 49; Pa State Univ, MS, 52, PhD(chem eng), 54. *Prof Exp:* Supvr chem prod, Am Cyanamid Co, 49-51; res chem engr, Houston Res Lab, Shell Oil Co, 54-56, group leader eng res, 56-58, group leader math & comput, 58-60, exchange scientist, Royal Dutch/Shell Lab, Amsterdam, 60-61, asst to mgr mfg res, NY, 61-62, sr technologist, 62-63, coordr comput process control, 64-69; vpres, Davis Comput Systs, Inc, 69-70; PRES & CHIEF ENGR, W R BILES & ASSOCS, INC, 70- *Mem:* Am Inst Chem Engrs; Am Chem Soc; Instrument Soc Am; Inst Elec & Electronics Engrs; Am Oil Chem Soc. *Res:* Computer control of batch and continuous processes and factories.; economic evaluation and justification of applications; development of new techniques and application of standard techniques to new fields; distillation. *Mailing Add:* 6161 Savoy Suite 500 Houston TX 77036

BILETCH, HARRY, organic polymer chemistry; deceased, see previous edition for last biography

BILGER, HANS RUDOLF, b Singen, Germany, May 17, 35; m 58; c 5. PHYSICS, ELECTRICAL ENGINEERING. *Educ:* Univ Basel, PhD(solid state physics), 61. *Prof Exp:* Res assoc solid state physics, Univ Basel, 61-62; res assoc digital computer sic, Univ Ill, Urbana-Champaign, 62-63; asst prof elec eng, Okla State Univ, 63-65; vis scientist, Argonne Nat Lab, 65-67; assoc prof, 68-75, PROF ELEC ENG, OKLA STATE UNIV, 75- *Concurrent Pos:* Res fel, Nuclear Res Ctr, Ger, 69-70; res assoc, Calif Inst Technol, 73-74; vis prof, Univ Sci & Technol, Montpellier, France, 76-78; consult, 81-; Erskine fel, Canterbury Univ, New Zealand. *Mem:* Am Phys Soc; Swiss Phys Soc; Am Asn Univ Profs; Sigma Xi. *Res:* Noise in semiconductors; resolution in germanium nuclear radiation detectors; space charge limited devices; ring lasers; ion implantation and backscattering; very low frequency noise; precision thermostats. *Mailing Add:* Sch Elec & Computer Eng Okla State Univ Stillwater OK 74078-0321

BILGUTAY, NIHAT MUSTAFA, b Ankara, Turkey, Mar 31, 52; US citizen; m 77; c 3. DIGITAL SIGNAL PROCESSING, COMMUNICATIONS. *Educ:* Bradley Univ, BS, 73; Purdue Univ, MS, 75, PhD(elec eng), 81. *Prof Exp:* Drexel fel elec eng, Col Eng, Drexel Univ, 81-82, vis asst prof, 82-83, asst prof, 83-87, asst dept head, 89-90, ASSOC PROF ELEC ENG, COL ENG, DREXEL UNIV, 87-, ASSOC DEAN ENG, 90- *Concurrent Pos:* Mem, Exec Comt, Educ Soc, Inst Elec & Electronics Engrs, 84-88. *Mem:* Sr mem Inst Elec & Electronics Engrs; Am Soc Nondestructive Testing; Am Soc Eng Educ; Sigma Xi. *Res:* Industrial and medical ultrasonic testing; digital signal processing; radar detection; imaging. *Mailing Add:* Col Eng Drexel Univ Philadelphia PA 19104-9984

BILHORN, JOHN MERLYN, b Hillsdale, Mich, Apr 4, 26; m 48; c 5. ELECTROCHEMICAL ENGINEERING, BATTERY RESEARCH & ENGINEERING. *Educ:* Cornell Univ, BS 49, MS, 89; Northern Baptist Theol Sem, BD, 54, MDiv, 73. *Prof Exp:* Res engr, Pure Oil Co, 49-51; pastor, Mayfield Congregational Church, Sycamore, Ill, 54-62; Mat engr mgr, Rayovac Corp, 61-75; proj mgr, Yardney Elec Corp, 75-82; SR ENG MGR, ENERGY RES CORP, 82- *Concurrent Pos:* Pastor, Fulton Congregational Church, Edgerton, Wis, 62-75, Road Meetinghouse (First Congregational), Stonington, Conn, 75- *Honors & Awards:* Young Scientist Award, Bausch & Lomb, 43. *Mem:* Am Inst Chem Engrs; Am Sci Affil. *Mailing Add:* 888 Pequot Trail Stonington CT 06378

BILIMORIA, MINOO HORMASJI, b Barkhera, India, Aug 24, 30; m 67; c 2. MICROBIOLOGY, BIOCHEMISTRY. *Educ:* Univ Bombay, BSc, 52, MSc, 54; Indian Inst Sci, PhD(microbiol), 63. *Prof Exp:* Microbiologist & chemist, Geoffrey Manners & Co Ltd, India, 54-58; res asst microbiol, Indian Inst Sci, Bangalore, 60-63; fel biochem, Sch Med, Duke Univ, 64-67; biochemist, 68-72, SR RES SCIENTIST, IMP TOBACCO LTD, 72- *Concurrent Pos:* Vis scientist, Path Inst, McGill Univ, Montreal, Can, 75-88. *Mem:* Can Biochem Soc; Indian Soc Biol Chem; Can Soc Microbiologists; NY Acad Sci; Environ Mutagen Soc. *Res:* Thermophilic bacilli, pectolytic enzymes from a variety of microorganisms; microsomal electron transport; L-asparaginases; biological effects of tobacco smoke inhalation; aryl hydrocarbon hydroxylase; detection of mutagenic activity of chemicals and tobacco smoke in microbial systems, including the Ames system. *Mailing Add:* 5210 Westmore Montreal PQ H4V 1Z5 Can

BILLEN, DANIEL, b New York, NY, Nov 27, 24; m 51; c 3. MICROBIOLOGY. *Educ:* Cornell Univ, BS, 48; Univ Tenn, MS, 49, PhD(microbiol), 51. *Prof Exp:* Instr microbial metab, Univ Tenn, 50-51; res biologist radiobiol, Oak Ridge Nat Lab, 51-57; assoc prof, Post Grad Sch Med, Univ Tex, 57-60, biologist & prof biol & chief sect radiation biol, Univ Tex M D Anderson Hosp & Tumor Inst, 60-66; prof radiation biol & mem dept microbiol, biochem and radiol, Col Med, Univ Fla, 66-73; prof biol & dir, Oak Ridge Grad Sch Biomed Sci, Univ Tenn, 74-77, dir, Inst Radiation Biol, 77-80, prof, 74-88, EMER PROF BIOMED SCI, UNIV TENN,

KNOXVILLE, 88- *Concurrent Pos:* Prog dir metab biol prog, NSF, 60-61, consult, Career Develop Rev Br, NIH, 66-70, consult radiation study sect, 69-73; ed, Microbios J & Cytobios J; hon ed, Photochem & Photobiol J, 71-73; mem, Cancer Res Manpower Rev Comt, Nat Cancer Inst, 77-80; ed-in-chief, Radiation Res, 79-88. *Mem:* Am Soc Microbiol; Radiation Res Soc. *Res:* Microbial physiology, biochemistry and cytology; radiobiology; microbiological genetics; DNA repair and replication. *Mailing Add:* 1087 W Outer Dr Oak Ridge TN 37830

BILLENSTIEN, DOROTHY CORINNE, b Easton, Pa, Feb 27, 21. ANATOMY. *Educ:* Boston Univ, BS, 42; Univ Colo, MS, 51, PhD(anat), 56. *Prof Exp:* From instr to asst prof anat, Albert Einstein Col Med, 56-61; from asst prof to assoc prof anat, Colo State Univ, 61-80. *Mem:* Am Asn Anatomists; Am Soc Zoologists. *Res:* Neurosecretion; neuroendocrinology. *Mailing Add:* 4821 Skyline Dr Ft Collins CO 80521

BILLER, JOSE, b Montevideo, Uruguay, Jan 18, 48; US citizen; m 72; c 3. CEREBROVASCULAR DISEASES. *Educ:* AV Acevedo Inst, Uruguay, BS, 65; Univ Repub Sch Med, MD, 74; Am Bd Psychiat & Neurol, cert neurol, 82. *Prof Exp:* Resident, Maciel Hosp, Pub Health Ministry, 74-76, Henry Ford Hosp, 77-78, Loyola-Univ & Hines Vet Admin Hosp, 78-79, chief resident, Loyola Univ, 79-80; fel cerebrovasc, Bowman Gray Sch Med, Wake Forest Univ, 80-81, instr neurol, 81; asst prof neurol, Stritch Med Sch, Loyola Univ, 82-84; from asst prof to assoc prof, 84-89, PROF NEUROL, UNIV IOWA COL MED, 90- *Concurrent Pos:* Fel Stroke Coun, Am Heart Asn, 81; prin investr heparinoid study, NIH Grant, Univ Iowa, 85-89, magnetic resonance imaging, 87-89; lectr, Cash County Sch Med, 83-; reviewer, J Archives Neurol & Stroke, 83-, Cerebrovascular Dis, J Stroke & Cerebrovascular Dis, 83-; consult, Neurology Serv, Vet Admin Hosp, Iowa City, 84- *Mem:* Fel Am Heart Asn; fel Am Acad Neurol; Am Neurol Assoc; Am Med Asn; Am Soc Neurol Invest; Interam Col Physicians & Surgeons. *Res:* Cerebrovascular disease, its clinical aspects and therapeutic interventions. *Mailing Add:* Dept Neurol Univ Iowa Hosp & Clin Iowa City IA 52242

BILLERA, LOUIS JOSEPH, b New York, NY, Apr 12, 43; m 64; c 2. DISCRETE GEOMETRY, ALGEBRAIC COMBINATORICS. *Educ:* Rensselaer Polytech Inst, BS, 64; City Univ New York, MA, 67, PhD(math), 68. *Prof Exp:* Asst prof opers res, 68-73, assoc prof opers res & math, 73-79, PROF OPERS RES & MATH, CORNELL UNIV, 80- *Concurrent Pos:* NSF fel math, Hebrew Univ, Israel, 69; vis res assoc, Brandeis Univ, 74-75; vis prof, Univ Cath Louvain, Belg, 80; prof math, Rutgers Univ, 85-89. *Mem:* Am Math Soc; Soc Indust & Appl Math; Math Asn Am. *Res:* Algebraic methods for combinatorial problems arising in geometry; convex polytopes; multivariate splines. *Mailing Add:* Dept Math White Hall Cornell Univ Ithaca NY 14853-7901

BILLESBACH, DAVID P, b Hastings, Nebr, May 22, 56. BRILLOUIN & RAMAN SPECTROSCOPY, FERROELECTRICS. *Educ:* Univ Nebr, BS, 79, MS, 82, PhD(physics), 87. *Prof Exp:* Res assoc, 88-90, vis asst prof, Dept Physics & Astron, 90- 91, RES ASST PROF, CTR LASER ANALYTICAL STUDIES TRACE GAS DYNAMICS, UNIV NEBR, 91- *Mem:* Am Phys Soc; Sigma Xi. *Res:* Experimental studies of structural phase transitions in solids especially brillouin scattering and birefringence; use of tunable diode lasers for the detection of trace gases in the atmosphere especially methane. *Mailing Add:* Dept Elec Eng 201 N WSEC Univ Nebr Lincoln NE 68588-0111

BILLETTE, JACQUES, b Valleyfield, Que, Can, Aug 8, 40; c 5. CARDIAC PHYSIOLOGY, ELECTROPHYSIOLOGY. *Educ:* Univ Montreal, MD, 66, PhD(physiol), 72; Can Med Coun, LMCC, 67. *Prof Exp:* Intern med, Montreal Gen Hosp, 66-67; Med Res Coun fel physiol, Univ Montreal, 67-71, lectr, 71-72; Med Res Coun fel physiol, Univ Amsterdam, 72-74; asst prof, 74-78, ASSOC PROF PHYSIOL, UNIV MONTREAL, 79- *Concurrent Pos:* Med Res Coun scholar, Univ Montreal, 74-79. *Mem:* Am Physiol Soc; Am Heart Asn; Int Soc Heart Res; Can Cardiovasc Soc; Can Physiol Soc; Can Biomed Eng Soc. *Res:* Electrophysiology of the atrioventricular node in the heart; definition of the nodal functional properties and the establishment of their cellular counterparts; rhythmic and arrhythmic responses generated by the node. *Mailing Add:* Dept Physiol Fac Med Univ Montreal CP6128 Montreal PQ H3C 3J7 Can

BILLHEIMER, FOSTER E, b Drums, Pa, June 24, 36; m 59; c 3. CELL BIOLOGY, ELECTRON MICROSCOPY. *Educ:* Pa State Univ, BS, 59; Univ Tex, BA, 66; Rutgers Univ, PhD(cell biol), 69. *Prof Exp:* Teacher gen sci, Newburgh, NY, 62-65; teaching asst, Rutgers Univ, 66-69; PROF BIOL, CALIFORNIA UNIV, PA, 69- *Concurrent Pos:* NSF fel, Acad Yr Inst, Univ Tex, 65-66; co-dir, Electron Micros Inst, 78; sr staff mem, NSF Pre-Col Teacher Develop Prog, 78-79. *Mem:* Am Soc Cell Biol; Am Soc Microbiol; Sigma Xi. *Res:* Heterotrophic nutrition and ultrastructure of the acidophilic bacterium Thiobacillus ferrooxidans; the relationship of mitochondrial DNA to mitochondrial function; cloning of cyanobacterial genes. *Mailing Add:* Dept Biol & Environ Sci Calif Univ Pa California PA 15419

BILLHEIMER, JEFFREY THOMAS, b Huntington, WVa, Oct 31, 46; m 72; c 3. ENZYMOLOGY, LIPID METABOLISM. *Educ:* Purdue Univ, BS, 68; NC State Univ, Raleigh, PhD(biochem), 75. *Prof Exp:* Fel, Cornell Univ, 75-77 & Univ Mo-Columbia, 77-79; asst prof biochem, Drexel Univ, 79-84; prin investr, 84-89, SR RES BIOCHEMIST, E I DU PONT, INC, 89- *Mem:* Am Soc Biochem & Molecular Biol; Am Heart Asn. *Res:* Membrane-bound enzymes involved in cholesterol (sterol) metabolism in health and disease and the regulation of these enzymes: intracellular transport of lipids especially by cytosolic carrier proteins (z-protein and sterol carrier protein). *Mailing Add:* Cardiovasc Sci Dupont Merck Pharmaceut Exp Sta 400-3231 Wilmington DE 19880

BILLICA, HARRY ROBERT, b Spokane, Wash, July 14, 19; m 46; c 3. CHEMISTRY & PHYSICS OF MAN-MADE FIBERS. *Educ:* Univ NC, BS, 41; Univ Wis, PhD(org chem), 48. *Prof Exp:* Shift supvr, Exp Sta, E I du Pont de Nemours & Co, 41-42, chemist, 42-47, res chemist, 47-50, group leader, Tech Sect, Nylon Plant, Tenn, 50-51, tech supvr, Dacron Polyester Fiber Plant, NC, 51-59, res mgr, Fiber Surface Res Sect, Textile Fibers Dept, 59-80; RETIRED. *Concurrent Pos:* Fiber Soc lectr, 70-71; chmn fiber sci, Gordon Res Conf, 73; consult fiber sci, 80-; lectr fiber sci, prof develop courses, 80- *Mem:* Fiber Soc (pres, 76); Textured Yarn Asn Am. *Res:* Chemical warfare; protective clothing; high pressure catalysis; hydrogenation; polymerization; catalyst development; synthetic fibers; physics and chemistry of fiber and film surfaces; polymer morphology; consulting in areas of synthetic fiber production and processing; emphasis on fiber processing finishes; polymer, fiber and fabric character, uniformity and interrelationships. *Mailing Add:* Eight Sweet Gum Ct Hilton Head Island SC 29928

BILLIG, FRANKLIN A, b Los Angeles, Calif, Feb 11, 23; m 57; c 1. ORGANIC CHEMISTRY, SCIENCE EDUCATION. *Educ:* Univ Southern Calif, AB, 54; Kansai Gakuen Univ, Japan, MS, 67. *Prof Exp:* Res chemist, Am Potash & Chem Corp, 54-58, sr res chemist, 58-64; supvr chem labs, 64-75, MGR LABS, UNIV SOUTHERN CALIF, 75- SAFETY OFFICER, DEPT CHEM, 73-, MGR HAZARDOUS CHEM WASTE, 84- *Concurrent Pos:* Consult, Environ Health & Lab Safety. *Mem:* Fel Am Inst Chemists; NY Acad Sci; Am Chem Soc; fel AAAS; Sigma Xi; fel L Pasteur Inst Advan Med Studies. *Res:* Organometallics; science education techniques; theoretical organic chemistry; reduction and detoxification of hazardous waste. *Mailing Add:* Dept Chem Univ Southern Calif Los Angeles CA 90089-1062

BILLIG, FREDERICK S(TUCKY), b Pittsburgh, Pa, Feb 28, 33; m 55; c 4. MECHANICAL ENGINEERING. *Educ:* Johns Hopkins Univ, BE, 55; Univ Md, MS, 58, PhD(mech eng), 64. *Prof Exp:* Assoc engr, Appl Physics Lab, Johns Hopkins Univ, 55-58, sr engr, 58-63, proj supvr hypersonic propulsion, 64-73, prin prof staff, 65, supvr submarine physics, 77-84, asst supvr, Aeronaut Div, 77-78, CHIEF SCIENTIST, ASSOC SUPVR, APPL PHYSICS LAB, JOHNS HOPKINS UNIV, 84- *Concurrent Pos:* Mem comt aeroballistics, Naval Bur Ord, 57-59; lectr, Univ Md, 65, mem grad fac, 67-; US chmn, Int Airbreathing Propulsion Comt; mem sci adv bd, USAF, 88- *Honors & Awards:* Silver Medal, Combustion Inst, 70. *Mem:* Fel Am Inst Aeronaut & Astronaut; Combustion Inst. *Res:* Hypersonic aerodynamics; thermodynamics; boundary layers; heat transfer and applied research; design and development work on subsonic and supersonic combustion ramjet engines and propulsion test facilities. *Mailing Add:* Appl Physics Lab Johns Hopkins Univ Johns Hopkins Rd Laurel MD 20723-6099

BILLIG, OTTO, psychiatry; deceased, see previous edition for last biography

BILLIGHEIMER, CLAUDE ELIAS, b Breslau, Ger, Nov 25, 30; Can citizen; m 55. ORDINARY DIFFERENTIAL EQUATIONS. *Educ:* Univ Melbourne, BA, 52, BSc, 53, MA, 58; Univ Toronto, PhD(math), 66. *Prof Exp:* Lectr math, Royal Mil Col, Canberra, 55-57; res officer statist, Commonwealth Sci & Indust Res Orgn, Australia, 58; lectr math, Australian Nat Univ, 59-61 & Univ Toronto, 62-66; from asst prof to assoc prof, 66-79, PROF MATH, MCMASTER UNIV, 79- *Concurrent Pos:* Nat Res Coun Can grant, 69-; mem bd gov, Maimonides Col, Can, 75 & McMaster Univ, Can, 90. *Mem:* Sigma Xi; Am Math Soc; Can Math Soc; Australian Math Soc; Soc Indust & Appl Math; AAAS. *Res:* Mathematical analysis; ordinary differential equations; differential and difference operations; spectral theory; eigenfunction expansions; differential operators in Banach algebras; functional analysis; mathematical physics. *Mailing Add:* Dept Math McMaster Univ Hamilton ON L8S 4K1 Can

BILLINGHAM, EDWARD J, JR, b Lebanon, Pa, Dec 6, 34; m 58. ANALYTICAL CHEMISTRY. *Educ:* Lebanon Valley Col, BS, 56; Pa State Univ, PhD(chem), 61. *Prof Exp:* Assoc prof chem & chmn dept, Thiel Col, 60-65; from asst prof to prof, 65-87, EMER PROF CHEM, UNIV NEV, LAS VEGAS, 88- *Mem:* Am Chem Soc. *Res:* Thermometric precipitation processes in both aqueous and fused salt media, including determination of thermodynamic properties as well as analytical implications; trace metal determinations. *Mailing Add:* Dept Chem Univ Nev Las Vegas NV 89154

BILLINGHAM, JOHN, b Worcester, Eng, Mar 18, 30; US citizen; m 56; c 2. AEROSPACE MEDICINE, EXOBIOLOGY. *Educ:* Oxford Univ, BA, 51, MA, BM & BCh, 54. *Hon Degrees:* DHL, Hawaii Loa Col, 81. *Prof Exp:* Med res officer aviation physiol, Royal Air Force Inst Aviation Med, Farnborough, Eng, 56-63; fbr chief environ physiol & space med, NASA Johnson Spacecraft Ctr, Houston, Tex, 63-65, div chief biotechnol & aerospace med, Ames Res Ctr, 66-75, chief, Extraterrestrial Intel, 76-85, CHIEF, LIFE SCI DIV & ACTG CHIEF, SEARCH FOR EXTRATERRESTRIAL INTEL, AMES RES CTR, NASA, MOFFETT FIELD, 86- *Concurrent Pos:* Staff mem marine sci coun, Exec Off of the President, 69. *Mem:* Fel Aerospace Med Asn. *Res:* Exobiology and search for extraterrestrial intelligent life; aerospace medicine and physiology; biotechnology; bioengineering. *Mailing Add:* 33 Campbell Lane Menlo Park CA 94025

BILLINGHAM, RUPERT EVERETT, b Warminster, Eng, Oct 15, 21; m 51; c 3. CELL BIOLOGY, IMMUNOLOGY. *Educ:* Oxford Univ, BA, 43, MA, 47, DPhil(zool), 50, DSc(zool), 57. *Hon Degrees:* DSc, Trinity Col, Conn, 65. *Prof Exp:* Lectr zool, Univ Birmingham, 47-51; hon res assoc, Univ Col, Univ London, 51-57; prof zool, Univ Pa, 57-71, prof med genetics & chmn dept, Sch Med, 65-71, dir, Henry Phipps Inst Med Genetics, 65-71; prof & chmn dept, 71-86, EMER PROF CELL BIOL, UNIV TEX HEALTH SCI CTR, DALLAS, 90- *Honors & Awards:* Alvarenga Prize, Am Col Physicians, 63; Hon Award, Am Asn Plastic Surgeons, 64. *Mem:* Fel Royal Soc; fel Am Acad Arts & Sci; Am Asn Immunologists; Soc Develop Biol; Am Fertil Soc. *Res:* Biology and immunology of transplantation; immunobiology of mammalian reproduction; wound healing and regeneration. *Mailing Add:* Rte 2 102B Vineyard Haven MA 02568

BILLINGS, BRUCE HADLEY, b Chicago, Ill, July 6, 15; m 75; c 4. OPTICS. *Educ:* Harvard Univ, AB, 36, MA, 37; Johns Hopkins Univ, PhD(physics), 43. *Hon Degrees:* PhD, Chinese Cult Univ. *Prof Exp:* Mem radiol safety sect atomic bomb test, Bikini, 46; dir res, Baird-Atomic, Inc, 47-63; exec vpres, 55-58, asst dir defense res & eng, US Dept Defense, 59-60; dir, Baird-Atomic Holland, 60-62; vpres & gen mgr lab opers, Aerospace Corp, 63-68; spec asst to ambassador for sci & technol, Taipei, Taiwan, 68-72; vpres, Aerospace Corp, 73-76; pres, Thagard Res Corp, 76-80; BD CHMN, INT TECHNOL ASSOC, 77-, VPRES, 80- *Concurrent Pos:* Deleg, Marseille, Conf Thin Films, France, 49 & 63; assoc ed J Optical Soc Am, 50; mem bd, Ealing Corp & Diffraction, Ltd, 60; mem, Sci Adv Bd, USAF, 62-; Am comnr, Joint Comn Rural Reconstruct, Taipei, Taiwan; mem UN adv comt, Appln Sci & Technol, 72-; vpres, Int Comn Optics, 73-75; mem bd, Electro Thermal Corp, Phys Optics Corp. *Honors & Awards:* Brilliant Star, Govt Repub China, 72. *Mem:* Fel Am Acad Arts & Sci (secy, 55-59); Am Phys Soc; Acoust Soc Am; Optical Soc Am (pres, 71). *Res:* Infrared receivers and crystal optics; development of tunable and fixed narrowband filters; Fourier transform spectroscopy; energy conversion. *Mailing Add:* 7303 N Marina Pacifica Dr Long Beach CA 90803

BILLINGS, CHARLES EDGAR, JR, b Boston, Mass, June 15, 29; m 55; c 1. AEROSPACE MEDICINE, ENVIRONMENTAL PHYSIOLOGY. *Educ:* NY Univ, MD, 53; Ohio State Univ, MSc, 60. *Prof Exp:* Resident physician, Univ Vt, 57-58; res med officer, Aviation Safety Res Off, 73-76, chief, 76-80, asst chief res, Man-Vehicle Systs Res Div, 80-84, sr scientist, 84; from instr prev med to prof prev med, physiol & aviation, 60-73, MEM FAC, DEPT PREV MED, OHIO STATE UNIV, 80- *Concurrent Pos:* Consult, Webb Assocs, 60-73; med consult, Beckett Aviation Corp, 62-73 & US Army, 64-77; tech consult, Fed Aviation Agency, 63-66, mem med adv comt, 66-70; mem comt hearing, acoust & biomech, Nat Acad Sci-Nat Res Coun, 65-69 & 75-; clin prof prev med & aviation, Ohio State Univ, 73-83, emer prof, 83-; Ames fel, 90. *Honors & Awards:* Boothby Res Award, Aerospace Med Asn, 73; Tamisea Mem Award, Aerospace Med Asn, 80; Laura Taber Barbour Mem Medal Aviation Safety, 81; NASA Outstanding Leadership Medal, 81, 90; Jeffrey Aerospace Med Res Award, Am Inst Aeronaut & Astronaut, 86. *Mem:* Fel Aerospace Med Asn; fel Am Col Prev Med; fel Am Acad Occup Med; Int Acad Aviation & Space Med; fel Royal Aeronaut Soc. *Res:* Effects of environmental stress on man's ability to perform in the work environment, particularly the flight environment. *Mailing Add:* NASA Ames Res Ctr 262-4 Moffett Field CA 94035

BILLINGS, CHARLES EDGAR, b Boston, Mass, May 26, 25; m 56; c 2. OCCUPATIONAL HEALTH ENGINEERING, AEROSOL SCIENCE. *Educ:* Northeastern Univ, BSME, 50; Harvard Univ, ScM, 53; Calif Inst Technol, PhD(environ health eng), 66. *Prof Exp:* Asst prof occup hyg, Harvard Sch Pub Health, 53-61, assoc prof, 62-64; res assoc aerosol sci, Calif Inst Technol, 64-66; assoc prof occup hyg, Univ Mich, 66-67; dept mgr, GCA Corp, Mass, 68-70; pres, Environ Eng Sci, 70-77; assoc prof environ sci, Johns Hopkins Univ, 77-85. *Concurrent Pos:* Air pollution spec fel, US Pub Health Serv, US Dept Health, Educ & Welfare, 62-66. *Mem:* AAAS; Am Indust Hyg Asn; Brit Occup Hyg Soc; Health Physics Soc; Sigma Xi; Am Asn Aerosol Res; Air Pollution Control Asn. *Res:* Development of morphology of aerosol particle aggregates in accumulation processes; contaminant control technology; occupational hygiene. *Mailing Add:* Dept Poli Sci Fla State Univ Tallahassee FL 32306

BILLINGS, MARLAND PRATT, b Boston, Mass, Mar 11, 02; m 38; c 2. ENGINEERING GEOLOGY, PETROGRAPHY. *Educ:* Harvard Univ, AB, 23, AM, 25, PhD(geol), 27. *Hon Degrees:* DSc, Wash Univ, 60 & Univ NH, 66. *Prof Exp:* From asst to instr geol, Harvard Univ, 22-28; assoc geol, Bryn Mawr Col, 28-29, assoc prof, 29-30; from asst prof to prof, 30-72, chmn div geol sci, 46-51, EMER PROF GEOL, HARVARD UNIV, 72- *Concurrent Pos:* Lectr, Am Asn Petrol Geologists, 48; consult, Pa Turnpike, 38 & Metrop Dist Comn, 58- *Honors & Awards:* Penrose Medal, Geol Soc Am, 87. *Mem:* Nat Acad Sci; AAAS (vpres, 46-47); Am Acad Arts & Sci; fel Geol Soc Am (vpres, 51, pres, 59); fel Mineral Soc Am. *Res:* Structural geology; petrology; New England geology; engineering geology. *Mailing Add:* RFD W Side Rd North Conway NH 03860

BILLINGS, R GAIL, b Orem, Utah, Apr 12, 34; m 57; c 5. BIOPHYSICS, BIOMEDICAL ENGINEERING. *Educ:* Univ Utah, BS, 56, PhD(biophysics), 75; Utah State Univ, MS, 65. *Prof Exp:* Res engr telemetry, Lockheed Aircraft Corp, 56-58; electronics engr flight simulator, US Air Force, Hill AFB, 58-59; instrumentation engr, Marquardt Corp, 59-62; eng mgr, Thiokol Chem Corp, 62-69; res engr, Space Data Corp, 69-70; res investr heart res, Primary Children's Med Ctr, 74-76; res & develop mgr med comput, Tenet Info Systs, Inc, 76-90; RES & DEVELOP ENGR, UTAH MED PROD, 90- *Concurrent Pos:* Nat Inst Gen Med Sci fel, 70-74. *Mem:* Inst Elec & Electronics Engrs. *Res:* Medical data processing; computer diagnosis; waveform processing. *Mailing Add:* Utah Med Prod Inc 7043 S 300 W Midvale UT 84047

BILLINGS, WILLIAM DWIGHT, b Washington, DC, Dec 29, 10; m 58. BOTANY, ECOLOGY. *Educ:* Butler Univ, AB, 33; Duke Univ, AM, 35, PhD(bot), 36. *Hon Degrees:* DSc, Butler Univ, 55. *Prof Exp:* Instr bot, Univ Tenn, 36-37; from instr to prof biol, Univ Nev, 38-52, head dept, 50-52; from assoc prof to prof, 52-67, JAMES B DUKE PROF BOT, DUKE UNIV, 67- *Concurrent Pos:* Ed, Ecol, Ecol Soc Am, 51-56 & Ecol Monogr, 68, 69; Fulbright res scholar, Univ NZ, 59; res fel, Australian Nat Univ, 77. *Honors & Awards:* Cert of Merit, Bot Soc Am, 60; Mercer Distinguished Serv Award, Ecol Soc Am, 62. *Mem:* AAAS; Am Soc Naturalists; hon mem Brit Ecol Soc; Bot Soc Am; Ecol Soc Am (vpres, 60, pres, 78-79). *Res:* Effect of geologic substratum on plant growth and distribution; desert, arctic and alpine ecology; physiological ecology; ecology of ecological races; environment especially carbon dioxide in biosphere. *Mailing Add:* Dept Bot Duke Univ Durham NC 27706

BILLINGSLEY, PATRICK PAUL, b Sioux Falls, SDak, May 3, 25; m 53; c 5. MATHEMATICS. *Educ:* US Naval Acad, BS, 48; Princeton Univ, MA, 52, PhD(math), 55. *Prof Exp:* NSF fel, Princeton Univ, 57-58; asst prof statist, 58-62, assoc prof math & statist, 62-67, PROF MATH & STATIST, UNIV CHICAGO, 67- *Concurrent Pos:* Vis prof, Copenhagen Univ, 64-65; ed, Ann Probability, 76-79. *Mem:* Am Math Soc; fel Inst Math Statist. *Res:* Probability theory; stochastic process theory. *Mailing Add:* Dept Statist & Math Univ Chicago Chicago IL 60637

BILLINGTON, DAVID PERKINS, b Bryn Mawr, Pa, June 1, 27; m 51; c 6. STRUCTURAL ENGINEERING. *Educ:* Princeton Univ, BSE, 50. *Hon Degrees:* LHD, Union Col, 90. *Prof Exp:* Struct engr, Roberts & Schaefer Co, 52-60; vis lectr, 58-60, assoc prof civil eng, 60-64, PROF CIVIL ENG, PRINCETON UNIV, 64- *Concurrent Pos:* Mem del vis engrs, USSR, 58; NSF sci fac fel, Univ Delft, 66-67; visitor, Inst Advan Studies, 74-75 & 77-78; Andrew O White prof-at-large, Cornell Univ, 87; Fulbright Fel, Belg, 50-52. *Honors & Awards:* Dexter Prize, Soc His Technol, 79. *Mem:* Am Soc Civil Engrs; Am Concrete Inst; Soc Hist Technol; Nat Acad Eng. *Res:* Thin shell concrete structures; concrete cooling towers; relationship between structure and architecture; history and aesthetics of structures; life and works of Robert Maillart; design of bridges. *Mailing Add:* Dept Civil Eng Princeton Univ Princeton NJ 08544

BILLINGTON, DOUGLAS S(HELDON), b Spearfish, SDak, June 5, 12; m 36; c 1. MATERIALS SCIENCE, INORGANIC CHEMISTRY. *Educ:* Yankton Col, AB, 35; Univ Iowa, MS, 41, PhD(chem metall), 42. *Prof Exp:* Chemist, Utah & Idaho Sugar Co, SDak, 34; chemist, Beryllium Corp, Pa, 35-37, chief chemist, 37-39; asst chem, Univ Iowa, 39-42; res metallurgist, Beryllium Corp, Pa, 42-43 & Linde Aid Prod Co, NY, 43-46; metallurgist, Naval Res Lab, Washington, DC, 46-49; metallurgist, 47-49, dir solid state div, 49-72, sr staff adv, 73-77, CONSULT, OAK RIDGE NAT LAB, 77- *Concurrent Pos:* Lectr, Univ Tenn; mem solid state sci adv panel, Nat Res Coun; mem adv panel, Metall Div, Nat Bur Standards; mem ed adv bd, Int J Chem & Physics of Solids, 56-; consult, Inst Energy Anal, Oak Ridge, 78. *Mem:* Fel AAAS; fel Am Phys Soc; fel Am Nuclear Soc; Am Soc Metals; Am Chem Soc. *Res:* Process and physical metallurgy of beryllium; cobalt-iron alloys; physics of solids; magnetic alloys; radiation effects. *Mailing Add:* 35 Outer Dr Oak Ridge TN 37830

BILLINTON, ROY, b Leeds, Eng, Sept 14, 35; Can citizen; m 56; c 5. ELECTRICAL ENGINEERING. *Educ:* Univ Man, BScEE, 60, MSc, 63; Univ Sask, PhD(elec eng), 67. *Hon Degrees:* DSc, Univ Sask, 76. *Prof Exp:* Staff engr, Man Hydro, 60-62, cost control engr, 62-64; from asst prof to assoc prof, 64-71, head dept, 76-84, PROF ELEC ENG, UNIV SASK, 71-, ASSOC DEAN GRAD STUDIES & RES, 84- *Concurrent Pos:* Spec lectr, Univ Man, 63-64. *Mem:* Can Elec Asn; fel Inst Elec & Electronics Engrs; fel Royal Soc Can; fel Eng Inst Can. *Res:* Power system analysis pertaining to operation and planning; reliability engineering, particularly power system reliability. *Mailing Add:* Dept Elec Eng Univ Sask Saskatoon SK S7N 0W0 Can

BILLMAN, FRED RICHARD, b El Paso, Tex, Apr 20, 43; m 66; c 3. MECHANICAL METALLURGY, PHYSICAL METALLURGY. *Educ:* Univ Tex, El Paso, BS, 65; Ohio State Univ, MS, 67. *Prof Exp:* Metallurgist ingot plant, Alcoa-Rockdale Works, 67-68; res scientist oxidation, Battelle-Metal Sci Group, 68-69; qual assurance metallurgist, Ti Forgings, Alcoa-Cleveland Works, 69-70, sr res eng, Fabricating Metall Div, 71-76, sr res eng, Ingot Casting Div, 77-78, TECH SPECALIST RAPID SOLIDIFICATION & POWDER METALL FABRICATION PROCESSES, FABRICATING TECHNOL DIV, ALCOA LABS, 79- *Mem:* Am Soc Metals; Sigma Xi. *Res:* Rapid solidification; powder metallurgy; aluminum and titanium forging processes; ingot casting process development. *Mailing Add:* 52 Greenview Dr RD 2 Jeannette PA 15644

BILLMAN, GEORGE EDWARD, b Ft Worth, Tex, July 23, 54; m 75; c 2. CARDIOVASCULAR PHYSIOLOGY, AUTONOMIC NEUROPHYSIOLOGY. *Educ:* Xavier Univ, BS, 75; Univ Ky, PhD(physiol & biophys), 80. *Prof Exp:* Res assoc, Health Sci Ctr, Univ Okla, 80-82, asst prof res, 82-84; asst prof, 84-90, ASSOC PROF, OHIO STATE UNIV, 90- *Concurrent Pos:* Prin investr, Am Heart Asn grant, 82-84, new investr award, NIH, 83-86; consult, Glaxo Inc, 89-; prin investr, NIH ROI grant, 86-90, NIDA ROI grant, 90- *Mem:* Am Physiol Soc; Am Heart Asn; Sigma Xi. *Res:* Neural control of cardiovascular system especially control during exercise and behavioral stress; regulation of coronary blood flow; role of neural factors in the genesis of cardiac arrhythmias. *Mailing Add:* Dept Physiol Ohio State Univ 333 W Tenth Ave Columbus OH 43210-1239

BILLMAN, JOHN HENRY, b Brooklyn, NY, Feb 8, 12; m 37; c 2. ORGANIC CHEMISTRY. *Educ:* Univ Va, BS, 34; Princeton Univ, AM, 35, PhD(chem), 37. *Prof Exp:* Instr org & inorg chem, Univ Ill, 37-39; from instr to prof, 39-78, EMER PROF ORG CHEM, IND UNIV, BLOOMINGTON, 78- *Concurrent Pos:* Vis lectr, Yale Univ, 46-47; prof, Univ Del, 58; consult, US Dept Health, Educ & Welfare, Am Viscose Corp & New Castle Prod; mem Pharmacol & Endocrinol Fel Panel; civilian investr, Off Sci Res & Develop. *Mem:* Am Chem Soc. *Res:* Pharmaceutical chemistry; organic synthesis; organic analytical reagents; compounds for use as plant stimulants; antitumor and antiviral agents; catalysis; amino acids. *Mailing Add:* 10850 Main Rd RR 1 East Marion NY 11939

BILLMAN, KENNETH WILLIAM, b Covington, Ky, Jan 9, 33; m 54; c 4. FUSION. *Educ:* Thomas More Col, AB, 55; Univ Cincinnati, MS, 58, PhD(physics), 59. *Prof Exp:* Instr physics, Univ Cincinnati, 57-59; from instr to asst prof, Mass Inst Technol, 59-67; staff physicist, Electronics Res Ctr, NASA, 67-70, staff physicist, Ames Res Ctr, 70-72, asst chief phys gas-dynamics & lasers br, 72-77, advan proj group leader, 77-79; fusion proj mgr, Elec Power Res Inst, 79-83; SR SCIENTIST, TITAN SYSTS, INC, 84- *Concurrent Pos:* Adj prof physics, Colo State Univ, 75-82. *Mem:* Am Phys Soc. *Res:* Inertial confinement fusion; orbiting mirrors for terrestrial energy supply; power transmission via laser; laser energy conversion; laser induced chemistry; isotope separation and development; space power; laser development and application; strategic defense. *Mailing Add:* 1942 Limetree Lane Mountain View CA 94040

BILLMEYER, FRED WALLACE, JR, b Chattanooga, Tenn, Aug 24, 19; m 51; c 3. POLYMER CHEMISTRY, COLOR SCIENCE. *Educ:* Calif Inst Technol, BSc, 41; Cornell Univ, PhD(phys chem), 45. *Prof Exp:* Res chemist, Plastics Dept, E I Du Pont de Nemours & Co, Inc, 45-57, res assoc, 57-64; prof, 64-84, EMER PROF CHEM, RENSSELAER POLYTECH INST, 84- *Concurrent Pos:* lectr, Univ Del, 52-64; vis prof, Mass Inst Technol, 60-61. *Honors & Awards:* Bruning Award, Fed Soc Coatings Tech, 77; Macbeth Award, Inter-Soc Color Coun, 78, Nickerson Serv Award, 83; Award of Merit, Am Soc Testing & Mat, 90. *Mem:* Fel AAAS; Am Chem Soc; Inter-Soc Color Coun (pres, 69-70, secy, 70-82); fel Optical Soc Am; fel Am Phys Soc; fel Am Soc Testing Mat. *Res:* Molecular structure of polymers; color science and technology; optical properties of plastics. *Mailing Add:* 1294 Garner Ave Schenectady NY 12309-5716

BILLO, EDWARD JOSEPH, b Brantford, Ont, Aug 3, 38; m 61; c 3. INORGANIC CHEMISTRY, ANALYTICAL CHEMISTRY. *Educ:* McMaster Univ, BSc, 61, MSc, 63, PhD(chem), 67. *Prof Exp:* Res assoc, Purdue Univ, 67-69; asst prof, 69-74, ASSOC PROF CHEM, BOSTON COL, 75- *Mem:* Am Chem Soc. *Res:* Solution equilibria of transition metal chelates; kinetics of fast reactions of metal chelates. *Mailing Add:* Dept Chem Boston Col Chestnut Hill MA 02167-3809

BILLS, CHARLES WAYNE, b Meeker, Colo, Nov 6, 24; m 47; c 3. ORGANIC CHEMISTRY. *Educ:* Colo Agr & Mech Col, BS, 47; Univ Colo, MS, 52, PhD(chem), 54. *Prof Exp:* Analyst, Los Alamos Sci Lab, Univ Calif, 47, org chem & radiochem synthesis, 48-50; asst, Univ Colo, 50-52, qual anal, 53; res chemist corrosion, Prod Res Dept, Stanolind Oil & Gas Co, 54-55 & Mallinckrodt Chem Works, Mo, 55-56; chief geochem & geophys res & develop br, US AEC, Colo, 56-58, nuclear chemist, Chem Separations Br, Hanford Opers Off, Wash, 58-59, dep dir health & safety div, Nat Reactor Testing Sta, Idaho Opers, 59-62, dir nuclear technol div, 62-74, asst mgr prod & tech support, Idaho Opers, Energy Res & Develop Admin, 74-77; asst to gen mgr, EG&G Serv, Inc, 77-81, vpres & mgr bus admin, 81-86; CONSULT, 86- *Concurrent Pos:* Princeton Univ fel pub affairs, 62-63; ex-officio dir, Eastern Idaho Nuclear Indust Coun, 66- & Intermountain Sci Experience Ctr, Inc, 73- *Mem:* Am Chem Soc. *Res:* Aluminum stearates; geochemistry of uranium; processing nuclear power reactor fuels; transuranic and fission product recovery; nuclear reactor operations, safety and technology; geothermal, solar and low head hydroelectric technology; radioactive waste management; organizational development. *Mailing Add:* 1090 E 21st St Idaho Falls ID 83404

BILLS, DANIEL GRANVILLE, b Wenatchee, Wash, Sept 8, 24; m 45; c 3. VACUUM PHYSICS. *Educ:* Wash State Univ, BS, 49, MS, 51; Harvard Univ, PhD(physics), 57. *Prof Exp:* Instr physics, Wash State Univ, 55-57, asst prof, 57-60; pres & chmn bd dirs, 54-88, CHMN & CEO, GRANVILLE-PHILLIPS CO, 88- *Mem:* Am Phys Soc; hon mem Am Vacuum Soc (treas, 67-70, pres, 72). *Res:* Surface physics; physical electronics; ultrahigh vacuum. *Mailing Add:* Granville-Phillips Co 5675 E Arapahoe Boulder CO 80303

BILLS, DONALD DUANE, b Hillsboro, Ore, Dec 4, 32; m 56; c 4. FOOD SCIENCE. *Educ:* Ore State Univ, BS, 59, MS, 64, PhD(food sci), 66. *Prof Exp:* From asst prof to prof food sci & technol, Ore State Univ, 65-75; CHIEF, PLANT PROD LAB, USDA, 75- *Mem:* Am Chem Soc; Phytochem Soc; Am Potato Asn; Sigma Xi; Inst Food Technologists. *Res:* Trace components in foods, food safety, plant biochemistry. *Mailing Add:* ACS USDA ARS Nat Agr Lib Rm 414 Beltsville MD 20705

BILLS, JAMES LAVAR, b Murray, Utah, Aug 4, 35; m 58; c 3. INORGANIC CHEMISTRY. *Educ:* Univ Utah, BS, 58; Mass Inst Technol, PhD(inorg chem), 63. *Prof Exp:* From asst prof to assoc prof, 62-77, PROF CHEM, BRIGHAM YOUNG UNIV, 77- *Mem:* Am Chem Soc; Sigma Xi. *Res:* Theoretical interpretations of molecular properties. *Mailing Add:* 1720 W 1400 N Provo UT 84604

BILLS, JOHN LAWRENCE, b Moore, Idaho, July 11, 20; m 46; c 7. CHEMISTRY. *Educ:* Stanford Univ, AB, 42, PhD(chem), 47. *Prof Exp:* Res chemist, Union Oil Co, Calif, 44-46, 47-49, group leader, 49-52; instr chem, Stanford Univ, 46-47; mkt develop engr, Brea Chem, Inc, 52-57; mgr mkt res, Am Potash & Chem Corp, 57-64; mgr corp develop, 64-79, VPRES CORP STUDIES & PROG, KERR-MCGEE CORP, 79- *Mem:* Am Chem Soc. *Res:* Hydroforming and desulfurization catalysts; organic synthesis of isoquinoline derivatives; oxidation and combustion of petroleum products as a synthetic tool; acetylene chemistry, hydrogen cyanide; phthalic acids; polymers; fertilizers and related products. *Mailing Add:* 4940 NW 30th Pl Oklahoma City OK 73122

BILLUPS, NORMAN FREDERICK, b Portland, Ore, Oct 15, 34; m 57; c 2. PHARMACY, PHARMACEUTICAL CHEMISTRY. *Educ:* Ore State Univ, BS, 58, MS, 61, PhD(pharmaceut), 63. *Prof Exp:* Instr pharm, Ore State Univ, 58-59, asst, 59-62, NIH res fel phys pharm, 62-63; from assoc prof to prof pharm, Univ Ky, 63-77; DEAN & PROF, COL PHARM, UNIV TOLEDO, 77- *Concurrent Pos:* Fel, Am Found Pharmaceut Educ, 62-63; consult, Blue Cross-Blue Shield & Hipple Nat Res Ctr; auth, Ann Am Drug Index, 77- *Honors & Awards:* Lyman Award, Am Asn Cols Pharm, 71; Res Achievement Award, Am Soc Hosp Pharmacists, 75. *Mem:* Am Pharmaceut Asn; Am Soc Hosp Pharmacists; Acad Gen Pract Pharm; Am Asn Cols Pharm. *Res:* Various dosage forms of pharmaceutical products, especially disintegrating agents and ointment technology; drug release from selected ointment bases; kinetics of drug absorption and distribution; protein binding of drugs; complexation; percutaneous adsorption; socio-economics of pharmacy. *Mailing Add:* Col Pharm Univ Toledo Toledo OH 43606

BILLUPS, W EDWARD, b Huntington, WVa, Apr 7, 39; m 66; c 1. ORGANIC CHEMISTRY. *Educ:* Marshall Univ, BS, 61, MS, 65; Pa State Univ, PhD(chem), 70. *Prof Exp:* Res chemist, Union Carbide Corp, WVa, 61-68; from asst prof to assoc prof, 70-81, PROF CHEM, RICE UNIV, 81- *Concurrent Pos:* Alfred P Sloan fel, 73-75; consult, UNESCO, 80. *Mem:* Am Chem Soc; Royal Soc Chem. *Res:* Chemistry of small ring systems, reactive intermediates, synthetic methods; organo-transition metal chemistry. *Mailing Add:* Dept Chem Rice Univ PO Box 1892 Houston TX 77251

BILODEAU, GERALD GUSTAVE, b Waterville, Maine, Nov 2, 29; m 56; c 6. MATHEMATICAL ANALYSIS. *Educ:* Univ Maine, BA, 50; Harvard Univ, AM, 51, PhD(math), 59. *Prof Exp:* Sr mathematician, Westinghouse Atomic Power Div, Pa, 55-59; adv res engr, Sylvania Electronics Systs, 59-60; from asst prof to assoc prof, 60-71, PROF MATH, BOSTON COL, 71- *Mem:* Am Math Soc; Math Asn Am; Sigma Xi; Soc Indust & Appl Math. *Res:* Orthogonal functions; integral transforms; partial differential equations. *Mailing Add:* 200 Harvard Circle Newtonville MA 02160

BILOEN, PAUL, heterogeneous catalysis, surface phenomena; deceased, see previous edition for last biography

BILOW, NORMAN, b Chicago, Ill, Sept 9, 28; m 54; c 3. CHEMISTRY, ORGANIC CHEMISTRY. *Educ:* Roosevelt Univ, BS, 49; Univ Chicago, MS, 52, PhD(chem), 56. *Prof Exp:* Res chemist, Emulsol Corp, 49-52; asst, Univ Chicago, 55-56; res chemist, Dow Chem Co, 56-59; staff scientist, Hughes Aircraft Co, 59-65, sr staff chemist, 65-67, head, Chem Synthesis Group, 67-69, head, Polymer & Chem Technol Sect, Aerospace Group, 69-70, head, Polymer & Phys Chem Sect, 70-73, sr scientist, Advan Technol Lab, 73-80, chief scientist, Mat & Processes Lab, 80-84; RETIRED. *Concurrent Pos:* Dir, res & develop, Furane Prods Co, Subsid Rohm P Haas, 84-86, Furane Prod Co, Div Ciba Geigy, 86-89; mgr, qual assurance & consult, Ciba Geigy, 89-90. *Honors & Awards:* Indust Res Mag Award, 70 & 74; L A Hyland Award, Hughes Aircraft Co, 76; New Technol Awards (4), NASA. *Mem:* Am Chem Soc; Sigma Xi; fel Am Inst Chemists; NY Acad Sci; AAAS. *Res:* Stereospecific polyolefins; polyphenylenes; phenolics; ablative resins; free radicals; photochemistry; diazo compounds; surfactants; polyurethanes; intumescent paints; electrical insulation; polyferrocenes; poly-aromatic-heterocyclic polymers; thermosetting acetylene-substituted polymers; epoxy resins; microelectronic materials; high temperature polymers; polyimides; electrically conductive polyimines; conductive adhesives. *Mailing Add:* 16685 Calneva Dr Encino CA 91436

BILPUCH, EDWARD GEORGE, b Connellsville, Pa, Feb 10, 27; m 52. PHYSICS. *Educ:* Univ NC, BS, 50, MS, 52, PhD, 56. *Prof Exp:* Res assoc, Duke Univ, 56-62, from asst prof to assoc prof, 62-71, dep dir, 68-78, PROF PHYSICS, DUKE UNIV, 71-, DIR, TRIANGLE UNIV NUCLEAR LAB, 78- *Concurrent Pos:* Sr US scientist Humboldt awardee, 83-84; hon prof, Fudan Univ, Shanghai, 86; Henry W Newson prof physics, Duke Univ, 87. *Mem:* Fel Am Phys Soc; AAAS. *Res:* High resolution neutron and charged particle cross section measurements; fine structure of isobaric analogue states. *Mailing Add:* Dept Physics Duke Univ Durham NC 27706

BILS, ROBERT F, b Harvey, Ill, Jan 10, 31; m 54; c 3. CYTOLOGY, ELECTRON MICROSCOPY. *Educ:* Univ Ill, BS, 54, MS, 58, PhD(bot), 60. *Prof Exp:* Asst bot, Univ Ill, 56-57, Electron Micros Lab, 57-60; res assoc biol, Mass Inst Technol, 60-61; from asst prof to assoc prof, 61-70, PROF BIOL, UNIV SOUTHERN CALIF, 70-, DIR, CTR ELECTRON MICROSCOPY & MICROANALYSIS, 84- *Concurrent Pos:* NIH fel, Mass Inst Technol, 60-61; dir electron micros lab, Allan Hancock Found, 61-84; vis prof, Path Inst & Air Hyg Inst, Univ Dusseldorf, 68-69; vis prof, Res Unit Comp Animal Resp, Univ Bristol, 76-77. *Mem:* Fel Royal Micros Soc; AAAS; NY Acad Sci; Electron Micros Soc Am; Am Soc Cell Biol. *Res:* Cell ultrastructure, electron microscopy; cytologic effects of toxic air pollutants on lung cells; connective tissues of lung; development and aging of cells; author of book on electron microscopy. *Mailing Add:* Dept Biol Sci Univ Southern Calif Los Angeles CA 90007

BILTONEN, RODNEY LINCOLN, b Sudbury, Ont, Aug 24, 37; US citizen; m 60; c 2. BIOPHYSICAL CHEMISTRY, BIOCHEMISTRY. *Educ:* Harvard Univ, AB, 59; Univ Minn, PhD(phys chem), 65. *Prof Exp:* NIH fel, 65-66; asst prof phys chem, Johns Hopkins Univ, 66-72, assoc prof pharmacol & biochem, 72-77; assoc dean med, 79-81, assoc provost res, 81-84, PROF PHARMACOL & BIOCHEM, SCH MED, UNIV VA, 77- *Concurrent Pos:* Counr, Biophys Soc, 83-87. *Mem:* Am Chem Soc; AAAS; Am Soc Biol Chemists; Biophys Soc. *Res:* Structure and thermodynamics of macromolecules of biological significance; thermodynamics of ligand binding to macromolecules; application of calorimetric techniques to the study of biological systems; thermodynamic aspects structural transitions of membranes; mechanism of activation of phospholipase A2. *Mailing Add:* Dept Pharm Univ Va Jordan Med Educ Bldg Charlottesville VA 22908

BILYEU, RUSSELL GENE, b Krum, Tex, Jan 22, 30; m 53; c 4. MATHEMATICAL ANALYSIS. *Educ:* NTex State Univ, BS, 52, MS, 57; Univ Kans, PhD(math), 60. *Prof Exp:* Engr systs eng, Chance Vought Aircraft, Inc, 52-57; from asst prof to assoc prof, 60-70, PROF MATH, N TEX STATE UNIV, 70- *Mem:* Am Math Soc; Math Asn Am. *Res:* Normed linear spaces. *Mailing Add:* 1709 Westchester Denton TX 76201

BIMBER, RUSSELL MORROW, b Warren, Pa, Mar 26, 29; m 51; c 3. CHEMICAL PROCESS RESEARCH & DEVELOPMENT. *Educ:* Antioch Col, BS, 52; Western Reserve Univ, MS, 62. *Prof Exp:* From chemist to sr chemist, Diamond Shamrock Corp, 52-80, res assoc, 80-86; RES ASSOC, RICERCA INC, 86- *Concurrent Pos:* Consult, Penn Cent, 69, Pub Rel Comt Voorhees, 82. *Mem:* AAAS; Am Chem Soc. *Res:* Chemical process research and development, including vapor phase catalytic reactions such as ammoxidation, chlorination and ketone synthesis. *Mailing Add:* 10471 Prouty Rd Painesville OH 44077

BINA, MINOU, b Teheran, Iran, Oct 15, 45; US citizen. BIOPHYSICS, MOLECULAR BIOLOGY. *Educ:* Temple Univ, BA, 70; Yale Univ, MS, 72, PhD(chem), 74. *Prof Exp:* Fel biochem, Nat Cancer Inst, NIH, 75-77, sr fel virol, Nat Inst Allergy & Infectious Dis, 77-79; PROF, PURDUE UNIV, 79- *Concurrent Pos:* Vis scholar, Ctr Cancer Res, Mass Inst Technol, 87-88. *Honors & Awards:* Am Chem Soc Award, 68. *Mem:* AAAS; Biophys Soc; Am Soc Biol Chem; Am Soc Cell Biol; Am Soc Microbiol; NY Acad Sci; Sigma Xi. *Res:* Protein - DNA interactions; chromatin structure; regulation of gene expression; analysis of the human genome. *Mailing Add:* Dept Chem Purdue Univ Brown Bldg West Lafayette IN 47907-1393

BINCER, ADAM MARIAN, b Krakow, Poland, Apr 25, 30; US citizen; m; c 2. THEORETICAL PHYSICS. *Educ:* Mass Inst Technol, SB, 53, PhD(physics), 56. *Prof Exp:* Res assoc physics, Brookhaven Nat Lab, 56-58; asst res, Univ Calif, Berkeley, 58-60; from asst prof to assoc prof, 60-68, PROF PHYSICS, UNIV WIS, MADISON, 68- *Concurrent Pos:* Fulbright scholar, Univ Sao Paulo, Brazil, 65. *Mem:* Am Phys Soc. *Res:* Theory of fundamental particles; analyticity properties of scattering amplitudes; group theory. *Mailing Add:* Dept Physics Univ Wis Madison WI 53706

BINDER, BERND R, b Vienna, Austria, Jan 7, 45; m 74; c 4. HAEMATOLOGY, FIBRINOLYSIS. *Educ:* Univ Vienna, Dr, 69, Dozent, 73 & 89. *Prof Exp:* Dir res & develop, Serotherapeut, Inst Vienna, 81-87; HEAD, DIV CLIN EXP PHYSIOL, UNIV VIENNA, 78-, PROF PHYSIOL, DEPT MED PHYSIOL, 82-; DIR RES & DEVELOP, TECHNOCLONE INC, VIENNA, AUSTRIA, 87- *Concurrent Pos:* Vis lectr med, Harvard Med Sch, Boston, 77-78; actg chmn, Dept Med Physiol, Univ Vienna, 81-84; vis investr, Scripps Clin & Res Found, La Jolla, Calif, 85-86. *Mem:* Austrian Sci Found; Int Comt Thrombosis & Haemostasis; Int Comt Fibrinolysis; NY Acad Sci; Am Asn Immunologists. *Res:* Biochemistry; cell and molecular biology; physiology and pathophysiology of the fibrinolytic system involved in thrombolysis and tumor cell biology. *Mailing Add:* Clin Exp Physiol Univ Vienna Schwarzspanierstr 17 Vienna A-1090 Austria

BINDER, BERNHARD, b WGer, Jan 25, 36; US citizen; m 59; c 2. INORGANIC CHEMISTRY, ANALYTICAL CHEMISTRY. *Educ:* Western NMex Univ, BA, 64; Stanford Univ, MS, 66, PhD(inorg chem), 68. *Prof Exp:* Chemist, Kennecott Copper Corp, 60-64; from asst prof to assoc prof, 68-80, chmn dept, 85-89, PROF CHEM, SOUTHERN ORE STATE COL, 80-, DIR, SCH SCI, 89- *Concurrent Pos:* Vis prof, Univ Konstanz, WGer, 74-75. *Mem:* Am Chem Soc; Sigma Xi. *Res:* Mixed-metal, binuclear coordination compounds; computer-assisted chemistry instruction. *Mailing Add:* Dept Chem Southern Ore State Col Ashland OR 97520

BINDER, DANIEL, b New York, NY, Feb 20, 27; m 56. PHYSICS. *Educ:* City Col New York, BS, 47; Yale Univ, PhD(physics), 50. *Prof Exp:* Physicist, Oak Ridge Nat Lab, 50-60; sr staff, Hughes Aircraft Co, 60-84, sr scientist, 84-89; RETIRED. *Mem:* Am Phys Soc. *Res:* Radioactivity; radiation damage and dosimetry; transient radiation effects on electronics, single event upset, EMP. *Mailing Add:* 30004 Via Borica Rancho Palos Verdes CA 90274

BINDER, FRANKLIN LEWIS, b Bristol, Pa, Nov 14, 45; m 65; c 3. MICROBIAL PHYSIOLOGY, PHARMACEUTICAL MICROBIOLOGY. *Educ:* Ind Univ Pa, BA, 67; WVa Univ, MS, 69, PhD(microbiol), 71. *Prof Exp:* Res asst, WVa Univ, 67-71; prog dir med & cytotechnol, 77, PROF MICROBIOL, MARSHALL UNIV, 71- *Concurrent Pos:* Instr microbiol, Ohio Univ, Portsmouth, 73-75; fac res grants, Marshall Univ, 75, 76, 77 & 79. *Honors & Awards:* Res Award, Sigma Xi, 75. *Mem:* Am Soc Microbiol; AAAS; Sigma Xi. *Res:* Biochemical studies with the haustorial mycoparasite Tieghemiomyces Parasiticus in axenic culture; transport mechanism in fungi; microbial decomposition of metalworking fluids; pharmaceutical microbiology; sterilization/disinfection. *Mailing Add:* Dept Biol Sci Marshall Univ Huntington WV 25701

BINDER, HENRY J, b New York, NY, Dec 5, 36; m 61; c 2. INTERNAL MEDICINE, GASTROENTEROLOGY. *Educ:* Dartmouth Col, AB, 57; NY Univ, MD, 61; Yale Univ, MA, 78. *Hon Degrees:* MA, Yale Univ, 78. *Prof Exp:* Instr med, Yale Univ, 65-66; clin instr, Univ Calif, San Francisco, 67-68; asst prof, Univ Chicago, 68-69; from asst prof to assoc prof, 75-78, PROF MED, YALE UNIV, 78- *Mem:* Am Soc Clin Invest; Am Physiol Soc; Am Gastroenterol Asn; Asn Am Physicians. *Res:* Gastrointestinal physiology; intestinal transport. *Mailing Add:* Dept Internal Med Yale Univ 333 Cedar St New Haven CT 06510

BINDER, IRWIN, b Salzburg, Austria, Apr 9, 49; US citizen; m 76; c 2. RADIOCHEMISTRY. *Educ:* Univ Calif, Los Angeles, BS, 70; Univ Calif, Berkeley, PhD(chem), 77. *Prof Exp:* STAFF MEM RADIOCHEM, LOS ALAMOS NAT LAB, 77- *Mem:* Am Phys Soc; Am Chem Soc. *Res:* Analytical radiochemistry; neutron activation analysis; heavy-ion nuclear reactions; geochemistry. *Mailing Add:* 1960 Camino Mora Los Alamos NM 87544

BINDER, MARC DAVID, b Brookline, Mass, June 8, 49; m 81; c 1. NEUROPHYSIOLOGY. *Educ:* Columbia Univ, AB, 71; Univ Southern Calif, MS, 72, PHD(biol & bioeng), 74. *Prof Exp:* Res asst biol, Univ Southern Calif, 72-74; consult environ impact, Ocean Sci & Eng, Inc, 72-73; res assoc neurophysiol, Pac Med Ctr, 74-75; instr english lit, Univ Calif Davis, 75; res assoc physiol, Col Med, Univ Ariz, 75-78; from asst prof to assoc prof, 78-86, PROF PHYSIOL & BIOPHYSICS, SCH MED, UNIV WASH, 86- *Concurrent Pos:* Vis asst prof physiol, Col Med, Univ Ariz, 78-81; vis assoc prof physiol, Harvard Med Sch, 82-83. *Mem:* Soc Neurosci; AAAS; Sigma Xi. *Res:* Neural control of movement; neurophysiological experiments designed to reveal the role of proprioceptive feedback on the behavior of neurons controlling muscle contraction. *Mailing Add:* Dept Physiol & Biophysics Sch Med Univ Wash Seattle WA 98195

BINDER, MICHAEL, b Waterbury, Conn, Apr 18, 51. CHEMISTRY, PHYSICS. *Educ:* Brooklyn Col, BS, 68, MS, 72, PhD(chem-physics), 78. *Prof Exp:* Asst prof chem, Queens Univ, 78-79; scientist, Power Conservation Induct, Mt Vernon, NY, 79-81; RES SCIENTIST, US ARMY ELECTRONIC TECHNOL & DEVICES, FT MONMOUTH, NJ, 81- *Mem:* Electrochem Soc; Sigma Xi. *Res:* Spectroscopy and electrochemistry of nonaqueous solvents for battery applications; dielectrics. *Mailing Add:* US Army LABCOM Ft Monmouth NJ 07703-5000

BINDLOSS, WILLIAM, b Westerly, RI, Dec 16, 37; m 63; c 3. SOLID STATE PHYSICS, MATERIALS SCIENCE. *Educ:* Yale Univ, BE, 60; Univ Calif, Berkeley, PhD(physics), 67. *Prof Exp:* Res scientist physics, 67-70, res group leader, 70-73, res supvr, cent res dept, 73-79, sr res assoc, photo prod dept, 79-85, MEM TECH STAFF, CENT RES DEPT, E I DU PONT DE NEMOURS & CO, INC, 85- *Mem:* Am Phys Soc; AAAS. *Res:* Magnetic, electrical, and optical properties of inorganic and organic materials. *Mailing Add:* Cent Res & Develop Dept DuPont Co Exp Sta 352 Wilmington DE 19898

BINDRA, JASJIT SINGH, b Rawalpindi, India, Oct 20, 42; m 70; c 2. ORGANIC CHEMISTRY. *Educ:* Agra Univ, MSc, 64, PhD(chem), 68. *Prof Exp:* Scientist med chem, Cent Drug Res Inst, India, 68-69; res assoc chem, Ind Univ, Bloomington, 69-71; res chemist, Pfizer Inc, 71-74, sr res chemist, 74-76, sr res investr, 76-80, prin res investr, 80-83, asst dir, 83-89, SR SCI ADVISOR, PFIZER, INC, 89- *Mem:* Am Chem Soc. *Res:* Organic synthesis; medicinal agents. *Mailing Add:* Cent Res Pfizer Inc Eastern Point Rd Groton CT 06340

BINFORD, JESSE STONE, JR, b Freeport, Tex, Nov 1, 28; m 55; c 2. CALORIMETRY, LIPID MEMBRANE THERMODYNAMICS. *Educ:* Rice Univ, BA, 50, MA, 52; Univ Utah, PhD(chem), 55. *Prof Exp:* Instr chem, Univ Tex, 55-58; from asst prof to assoc prof, Univ Pac, 58-61; assoc prof, 61-71, PROF CHEM, UNIV SFLA, 72- *Concurrent Pos:* Fulbright prof chem & chmn dept, Nat Univ Honduras, 68-69; consult, Fla Consort, AID, Honduras, 69 & Exxon Prod Res Co, Houston, 74; vis scientist, Sweden Thermochem Lab, Univ Lund, 82-83. *Mem:* AAAS; Am Chem Soc; Calorimetry Conf; Sigma Xi. *Res:* Solution calorimetry; biochemical calorimetry. *Mailing Add:* 1905 E 111th Ave Tampa FL 33612

BINFORD, LAURENCE CHARLES, b Chicago, Ill, Jan 11, 35. VERTEBRATE ZOOLOGY, ORNITHOLOGY. *Educ:* Univ Mich, BS, 57; La State Univ, PhD(vert zool), 68. *Prof Exp:* Instr zool, La State Univ, 65-66; asst cur birds & mammals, Calif Acad Sci, 68-73, assoc cur & chmn dept, 73-80. *Concurrent Pos:* Vpres bd dirs, Point Reyes Bird Observ, 69-72, pres bd dirs, 72-76; fel, Calif Acad Sci, 72; chmn ed bd, Western Birds, 73-; mem coun, Cooper Ornith Soc, 76-78, secy bd, 78-81; mem bd dirs, Western Field Ornithologists, 76-, pres bd dirs, 84-; charter mem, Calif Bird Records Comt, 69-77, vice secy, 78-79 & 81- *Mem:* Am Ornith Union; Cooper Ornith Soc; Wilson Ornith Soc. *Res:* Ornithology, especially avian taxonomy; ornithogeography, particularly Mexico. *Mailing Add:* 330 Grove St Glencoe IL 60022

BINFORD, MICHAEL W, b Hutchinson, Kans, May 21, 51; m 85; c 2. PALEOLIMNOLOGY, ECOSYSTEM ECOLOGY. *Educ:* Kans State Univ, BS, 73; La State Univ, MS, 75; Ind Univ, PhD(zool & geol), 80. *Prof Exp:* Res assoc, Fla State Mus, Univ Fla, 80-86; asst prof, 86-90, ASSOC PROF, HARVARD UNIV, 90- *Concurrent Pos:* Assoc ed, J Paleolimnology, 88- *Mem:* Ecol Soc Am; Am Soc Limnol & Oceanog; Int Asn Theoret & Appl Limnol; Sigma Xi; AAAS; Int Asn Landscape Ecol. *Res:* Lake-drainage basin interaction; paleolimnology; ecology of littoral (lake) entomostraca, especially chydoridae (crustacea-cladocera). *Mailing Add:* Dept Landscape Archit Grad Sch Design Harvard Univ 48 Quincy St Cambridge MA 02138

BING, ARTHUR, b Springfield, Mass, Apr 18, 16; m 54; c 1. PLANT PHYSIOLOGY. *Educ:* Univ Conn, BS, 39; Cornell Univ, PhD, 49. *Prof Exp:* Asst bot, Cornell Univ, 46-49; prof floricult, Long Island Hort Res Lab, 49-83, consult, 83-86. *Mem:* Weed Sci Soc Am; Am Soc Hort Sci; Int Soc Hort Sci; Sigma Xi. *Res:* Weed control and factors influencing flowering of ornamentals. *Mailing Add:* Long Island Hort Res Lab 39 Sound Ave Riverhead NY 11901

BING, DAVID H, b East Cleveland, Ohio, Aug 3, 38; m 61; c 3. IMMUNOLOGY, PROTEIN CHEMISTRY. *Educ:* Wesleyan Univ, BA, 60; Case Western Reserve Univ, PhD(immunol), 66. *Prof Exp:* Asst prof microbiol & pub health, Mich State Univ, 68-72, assoc prof microbiol & pub health & human develop, 72-73; sr investr, Ctr Blood Res, Boston, 73-; assoc dir biol sci, -86, SCI DIR, CBR LABS, 86- *Concurrent Pos:* Am Cancer Soc fel immunochem, Univ Calif, Berkeley, 66-68; foreign res worker, Nat Inst Health & Med Res, 70-71; res assoc, Div Immunol, Childrens Hosp Res Ctr, Boston, 72-73; prin res assoc, Dept Biol Chem, Harvard Med Sch, 73- *Mem:* Am Chem Soc; Am Asn Immunologists. *Mailing Add:* CBR Labs 800 Huntington Ave Boston MA 02115

BING, FRANKLIN C, nutrition, for more information see previous edition

BING, GEORGE FRANKLIN, b Barberton, Ohio, Dec 16, 24; m 51; c 5. THEORETICAL PHYSICS. *Educ:* Oberlin Col, BA, 48; Case Inst Technol, MS, 51, PhD(theoret physics), 54. *Prof Exp:* Physicist, Battelle Mem Inst, 51-52; instr, Case Inst Technol, 52-54; mem staff, Lawrence Radiation Lab, Univ Calif, 54-61; asst dir nuclear test detection off, advan res proj agency, US Dept of Defense, DC, 61-63; mem staff, Lawrence Radiation Lab, Univ Calif, 63-67; sci adv to SACEUR, Supreme Hq Allied Powers Europe, 67-70; mem staff, Lawrence Livermore Nat Lab, Univ Calif, 70-88; RETIRED. *Mem:* Am Phys Soc. *Res:* Weapons physics; hydrodynamics; controlled thermonuclear process; problems of national defense. *Mailing Add:* 4128 Colgate Way Livermore CA 94550

BING, KURT, b Cologne, Ger, Apr 30, 14; nat US; m 52; c 2. MATHEMATICAL LOGIC. *Educ:* Hebrew Univ, Jerusalem, MSc, 46; Harvard Univ, PhD(math), 53. *Prof Exp:* Assoc math, Univ Calif, 52-53; from asst prof to prof math, 53-79, EMER PROF MATH SCI, RENSSELAER POLYTECH INST, 79- *Mem:* Am Math Soc; Math Asn Am; Asn Symbolic Logic. *Res:* Mathematical logic; foundations of mathematics. *Mailing Add:* 22 Kuhl Blvd Wynantskill NY 12198

BING, OSCAR H L, b New York, NY, July 13, 35. CARDIAC MUSCLE MECHANICS. *Educ:* Univ Md, MD, 61. *Prof Exp:* ASSOC CHIEF OF STAFF, RES & DEVELOP, BOSTON VET ADMIN MED CTR, 81-; PROF MED, SCH MED, TUFTS UNIV, 81- *Mem:* Am Heart Asn; Int Soc Heart Res; Am Physiol Soc. *Mailing Add:* Dept Med Boston Vet Admin Med Ctr Tufts Univ Sch Med 150 S Huntington Ave Boston MA 02130

BING, RICHARD F, b Sandusky, Ohio, Dec 9, 41; m 66; c 2. VETERINARY MEDICINE. *Educ:* Ohio State Univ, DVM, 66. *Prof Exp:* Animal sci rep, Lilly Res Labs, 66-72, tech specialist, 72-74, mgr nat acct res, 74-79, mgr animal sci res & develop, Europe, Lilly Res Ctr, 79-82, PROD REGIST MGR, LILLY RES LABS, ELI LILLY & CO, 82- *Res:* Antibiotics and growth promotants for livestock and poultry. *Mailing Add:* Dow Elanco 9002 Purdue Rd Quad III-2 Indianapolis IN 46268

BING, RICHARD JOHN, b Nuremburg, Bavaria, Oct 12, 09; nat US; m 38; c 4. INTERNAL MEDICINE. *Educ:* Univ Munich, MD, 34; Univ Bern, MD, 35. *Prof Exp:* Instr physiol, Col Physicians & Surgeons, Columbia Univ, 39-41; instr, NY Univ, 41-43; instr med, Med Sch, Johns Hopkins Univ, 43-44, asst prof surg, 45-47, assoc prof surg & asst prof med, 47-51; prof exp med & clin physiol, Med Col Ala, 51-56; prof med, Washington Univ & dir univ med serv, Vet Admin Hosp, 56-59; prof med & chmn dept, Col Med, Wayne State Univ, 59-69; prof med, Univ Southern Calif, 69-80; DIR EXP CARDIOL & SCI DEVELOP, HUNTINGTON MEM HOSP & HUNTINGTON INST, PASADENA, 69-; EMER PROF MED, UNIV SOUTHERN CALIF, 80- *Concurrent Pos:* Res assoc, Calif Inst Technol, 69-; Harvey Lectr, Royal Soc Med, 54-55. *Honors & Awards:* Res Accomplishment Award, Am Heart Asn, 74. *Mem:* Am Soc Clin Invest; Am Physiol Soc; Harvey Soc; Soc Exp Biol & Med; Asn Am Physicians. *Res:* Physiology and biochemistry of heart muscle; clinical cardiology and internal medicine. *Mailing Add:* Dept Exp Cardiol & Intramural Med Huntington Mem Hosp 100 Congress St Pasadena CA 91105

BINGEL, AUDREY SUSANNA, b Bronx, NY. PHARMACOLOGY, REPRODUCTIVE ENDOCRINOLOGY. *Educ:* Hunter Col, AB, 63; Univ Ill, Chicago, PhD, 68. *Prof Exp:* Assoc prof, 67-81, PROF PHARMACOL, COL PHARM, UNIV ILL, 81- *Mem:* Am Inst Biol Sci; Am Soc Pharmacog; Endocrine Soc; Am Fertil Soc; Soc Study Reproduction; Sigma Xi. *Res:* Antifertility screening; ovulation timing. *Mailing Add:* Col Pharm Univ Ill Rm 310N 883 S Wood St M/C877 Chicago IL 60612

BINGER, WILSON VALENTINE, b Greenwich, NY, Feb 28, 17; m 47, 86; c 3. CIVIL ENGINEERING, ARBITRATION. *Educ:* Harvard Univ, AB, 38, MSCE, 39. *Prof Exp:* Soils engr, US Army Engrs, Wilmington, Del, 39-40; soils & found engr, Gatun 3d Locks Proj, Panama Canal, 40-43; soils engr & resident engr, Parsons Brinckerhoff, Hogan & MacDonald, Caracas, Venezuela, 45-46, chief soils engr, Buenos Aires, Arg, 48-50; chief soils & found sect, Isthmian Canal Studies, Panama Canal, 46-47; chief soils br, Mo River Div, US Army Engrs, Omaha, 47-48; vpres, Porterfield-Binger Construct Co, Youngstown, Ohio, 50-52; regional mgr, Tams Engrs & Architects, Bogota, Colombia, 52-56, assoc partner, NY, 57-61, partner, 62-84, chmn, 75-84; CONSULT ENGR, 85- *Concurrent Pos:* Trustee, Robert Col, Istanbul, Turkey, 70-; mem US comt large dams, Am Arbit Asn. *Honors & Awards:* Steinmetz Award, 85. *Mem:* Nat Acad Eng; fel Am Soc Civil Engrs; fel Inst Civil Engrs; Am Consult Engrs Coun (vpres, 73-75); Int Fedn Consult Engrs (pres 81-83); foreign mem Fel Engrs Brit. *Mailing Add:* 420 Lexington Ave Suite 300 New York NY 10170

BINGGELI, RICHARD LEE, b Sioux Falls, SDak, May 16, 37; m 62; c 3. NEUROPHYSIOLOGY. *Educ:* Univ Calif, Los Angeles, AB, 59, PhD(anat), 64. *Prof Exp:* From instr to asst prof, 64-70, ASSOC PROF ANAT, SCH MED, UNIV SOUTHERN CALIF, 70- *Concurrent Pos:* Consult, Astropower, Inc, Douglas Aircraft Co, Inc, 63-66. *Mem:* Soc Neurosci; Am Asn Anatomists. *Res:* Neurophysiological and behavioral studies of the avian, amphibian, and mammalian visual systems; electrobiology of cancer cells. *Mailing Add:* Dept Anat Sch Med Univ Southern Calif 2025 Zonal Ave Los Angeles CA 90033

BINGHAM, BILLY ELIAS, b Fallston, NC, July 31, 31; m 57; c 2. NUCLEAR ENGINEERING. *Educ:* NC State Univ, BEng, 55; Univ Va, PhD(nuclear eng), 70. *Prof Exp:* Head radiol safety, Army Biol Warfare Lab, Fort Detrick, Frederick, Md, 56-58; unit mgr, Nuclear Power Div, Babcock & Wilcox, 68-79, adv engr, adv reactor proj, 79-86, space power & propulsion, 86-88, ADV ENGR SPACE & DEFENCE SYSTS, BABCOCK & WILCOX, 88- *Mem:* Am Nuclear Soc; Am Astronaut Soc. *Res:* Advanced reactor concepts for space and terrestrial applications; design of the reactor concept and their supporting, safety and propection systems. *Mailing Add:* Babcock & Wilcox 3315 Old Forest Rd PO Box 10935 Lynchburg VA 24506-0935

BINGHAM, CARLETON DILLE, b Washington, DC, Mar 25, 29; m 58; c 5. ANALYTICAL CHEMISTRY, RADIOCHEMISTRY. *Educ:* San Diego State Col, AB, 50; Univ Calif, Los Angeles, PhD(phys chem), 59. *Prof Exp:* Radiol engr div radiation safety, Univ Calif, Los Angeles, 53-54, sr radiol engr, 54-59; sr res chemist atomics int div, NAm Aviation, Inc, 59-61, supvr anal chem, 61-68; proj engr fast breeder reactor chem, NAm Rockwell Corp, 68-71; DIR, NEW BRUNSWICK LAB, US DEPT ENERGY, 71- *Concurrent Pos:* Alt US rep, Int Stand Org Tech Comm 85-, Nuclear Energy, 76-; mem adv comt nuclear ref mat, Int Atomic Energy Agency, 76- *Honors & Awards:* Award of Merit, Am Soc Testing & Mat, 79. *Mem:* AAAS; Am Chem Soc; Am Nuclear Soc; Health Physics Soc; Am Soc Test & Mat; Sigma Xi. *Res:* Analytical chemistry of nuclear materials, nondestructive assay, nuclear materials safeguards measurements, reference standards for nuclear materials. *Mailing Add:* 1512 Windsor Ct Naperville IL 60565

BINGHAM, CARROL R, b Fallston, NC, May 22, 38; m 61; c 3. NUCLEAR PHYSICS. *Educ:* NC State Univ, BSNE, 60, MS, 62; Univ Tenn, PhD(nuclear physics), 65. *Prof Exp:* Res assoc nuclear spectros, Oak Ridge Nat Lab, 65-66; asst prof, 66-71, assoc prof, 71-77, PROF PHYSICS, UNIV TENN, KNOXVILLE, 77- *Concurrent Pos:* Consult, Union Carbide Corp, 66-83 & Martin Marietta, 83- *Mem:* Am Phys Soc. *Res:* Investigation of high-spin nuclear state produced by heavy-ion bombardment of nuclei; nuclear spectroscopy from decay properties of short-lived nuclei; study of nuclear shapes and scattering potentials for heavy ions; measurement of nuclear ground state and isomer moments and isotope shifts utilizing laser spectroscopy. *Mailing Add:* Dept Physics Univ Tenn Knoxville TN 37996-1200

BINGHAM, CHRISTOPHER, b New York, NY, Apr 16, 37; m 67. STATISTICS. *Educ:* Yale Univ, BA, 58, MA, 60, PhD(math), 64. *Prof Exp:* Res assoc math, Princeton Univ, 64-66; asst prof statist, Univ Chicago, 66-72; assoc prof, 72-80, PROF APPL STATIST, UNIV MINN, ST PAUL, 80- *Mem:* Biomet Soc; Soc Indust & Appl Math; fel Am Statist Asn; fel Inst Math Statist; Royal Statist Soc. *Res:* Probability distributions of directions; chronobiometry; time series analysis. *Mailing Add:* Dept Appl Statist Univ Minn 1994 Buford Ave St Paul MN 55108

BINGHAM, EDWIN THEODORE, b Ogden, Utah, Nov 4, 36; m 62; c 2. PLANT BREEDING. *Educ:* Utah State Univ, BS, 59, MS, 61; Cornell Univ, PhD(genetics), 64. *Prof Exp:* NIH fel genetics, Univ Minn, 64-65; from asst prof to assoc prof 65-70, PROF AGRON, UNIV WIS-MADISON, 74- *Mem:* Am Soc Agron; Am Genetic Asn; Can Genetic Asn; Crop Sci Soc Am. *Res:* Genetics and breeding of alfalfa; genetics and cytogenetics of autotetraploids; germ plasma transfer in alfalfa; plant tissue culture. *Mailing Add:* Dept Agron Univ Wis-Madison Madison WI 53706-1575

BINGHAM, EULA, b Lexington, Ky, July 9, 29; m 58; c 3. CHEMICAL CARCINOGENESIS, PULMONARY DEFENSE. *Educ:* Eastern Ky Univ, BS, 51; Univ Cincinnati, MS, 54, PhD(zool), 58. *Hon Degrees:* DSc, Eastern Ky Univ, 79; LLD, Mt St Joseph Col, 81. *Prof Exp:* Analytical chemist, Hilton-Davis Chem Co, 51-52; res assoc, Univ Cincinnati, 53-54, res asst, 55-57, res assoc, 57-61, from asst prof to assoc prof environ health, 61-77, assoc dir grad studies, 72-77; asst secy labor, Occup Safety & Health Admin, US Dept Labor, 77-81; vpres & univ dean grad studies & res, 82-90, PROF ENVIRON HEALTH, COL MED, UNIV CINCINNATI, 77- *Concurrent Pos:* Consult, Subcomt on Carcinogenesis of Threshold Limits Comt, Am Conf Indust Hyg, 72, Panel on Vapor Phase Organic Air Pollutants, Nat Res Coun, 72-75; mem, Safety & Occup Health Study Sect, Nat Inst Occup Safety & Health, 72-76; mem, Nat Toxicol Prog, 78-81, chair, 79-81; mem, Comt on Priority Mechanisms for Res on Agents Potentially Hazardous to Human Health, Assembly Life Sci, Nat Res Coun, 81-84, vchmn, Comt Toxicol, Comn Life Sci, 86-, mem, Comt on Toxicol Subcomt Guidelines for Spacecraft Maximum Allowable Concentrations for Space Station Contaminants, 89-91; mem working group, Int Agency for Res on Cancer, 83, chair, 84; vchmn, Comt on Methods for In Vivo Toxicity Testing of Complex Mixtures from the Environ, Comn Life Sci, Nat Acad Sci, 84-86, mem, Comt on Ground-Water Protection, Water Sci & Technol Bd, 85-86; mem, Comt on the Future of Pub Health, Inst Med-Nat Acad Sci, 86-87. *Honors & Awards:* Rockefeller Found Pub Serv Award, 80; Julia Jones Award, Am Lung Asn, 80; Homer N Calver Award, Am Pub Health Asn, 80, Alice Hamilton Award, 84; Haliburton Distinguished Lectr, Tex Tech, 81; Foard Mem Lectr, Univ NC, 81; William Lloyd Award for Occup Safety, US Steel Workers, 84. *Mem:* Inst Med-Nat Acad Sci; Soc Toxicol; Soc Occup & Environ Health (vpres, 75, pres-elect, 79); Am Asn Cancer Res; Sigma Xi; AAAS; Am Col Toxicol (pres, 81). *Res:* Toxicology; chemical carcinogenesis; pulmonary defense mechanisms; regulatory toxicology; occupational/environmental health. *Mailing Add:* Kettering Bldg ML056 Univ Cincinnati Med Ctr Cincinnati OH 45267

BINGHAM, FELTON WELLS, b Greenville, Miss, Aug 18, 35; m 59; c 4. DISPOSAL OF RADIOACTIVE WASTE. *Educ:* Tulane Univ, BS, 57; Ind Univ, MS, 59, PhD(physics), 62. *Prof Exp:* Assoc physics, Ind Univ, 57-59, res asst, 59-62; res assoc, Univ Ill, 62-64; mem tech staff, 64-83, DISTINGUISHED MEM TECH STAFF & SUPVR, SANDIA NAT LABS, 83- *Res:* Disposal of radioactive waste and uranium-mill tailings; collision and radiation processes important in high-power gas lasers; collisions between heavy ions and atoms at energies above 10 kiloelectron volts; studies of repositories for radioactive waste. *Mailing Add:* 12608 Loyola Ave Albuquerque NM 87112

BINGHAM, GENE AUSTIN, VETERINARY MEDICINE. *Educ:* Auburn Univ, DVM, 58; Am Bd Vet Pub Health, dipl, 66. *Prof Exp:* Vet med office, Path Div, US Army Biol Lab, Ft Detrick Md, 58-59, US Army Trop Res Med Lab, San Juan, PR, 59-60; asst chief, Animal Hosp Sect & head, Primated Quarantine Unit, Vet Resources Br, Res Resources Div, NIH, Bethesda, Md, 60-62; chief, Animal Prod & Conditioning Br, Animal Farm Div, US Army, Ft Detrick, Md, 62-67; dir, Lab Animal Facil, Sch Med, Univ Pittsburgh, 67-78, asst prof, Dept Surg, 67-73, adj assoc prof, 73-78; head, Lab Animal Resources Sect, Biol Div, Oak Ridge Nat Lab, Tenn, 78-89; INST VET & CHIEF VET MED & RESOURCES BR, NIMH, NIH, 89- *Concurrent Pos:* Bd dir, Am Asn Lab Animal Sci, 65, 71-74, Am Col Lab Animal Med, 70-72; mem var comt, Am Asn Lab Animal Sci, 68-75, Subcomt Rodent Standards, Inst Lab Animal Resources, Nat Acad Sci, 67-68; mem, Coun Accreditation, Am Asn Accreditation Lab Animal Care, 71-74, consult, 67- *Mem:* Vet Med Asn; Am Asn Lab Animal Sci (pres, 73); Am Col Lab Animal Med. *Res:* Veterinary medicine. *Mailing Add:* Vet Med & Resources Br NIMH NIH Bldg 10 Rm 4N206 Bethesda MD 20892

BINGHAM, HARRY H, JR, b Chicago, Ill, May 25, 31; m 60; c 3. PHYSICS. *Educ:* Princeton Univ, AB, 52; Calif Inst Technol, PhD(physics), 60. *Prof Exp:* Physicist, Electro-Optical Systs, Inc, Calif, 59-60, Ec Polytech Sch, Paris, 60-62; Ford Found fel physics, Europ Orgn Nuclear Res, Switz, 62-64; from asst prof to assoc prof, 64-72, PROF PHYSICS, UNIV CALIF, BERKELEY, 72- *Concurrent Pos:* Prin investr, NSF, 66-, Cern, 70, Rutherford Lab, Eng, 77-78, Tex Accel Ctr, 83-84, Fermilab, 90; rev, Phys Rev Lett, Nuclear Physics. *Mem:* Fel Am Phys Soc; Am Asn Univ Prof. *Res:* Elementary particle physics; bubble chambers; particle beams; computers; particle decays; coherent neutrino pi- and k- nucleus interactions; polarized photoproduction, k-pi scattering; interactions of high energy pi, k, p, neutrino; gamma beams in H, D heavy liquid; electronics engineering. *Mailing Add:* Dept Physics Univ Calif 422 le Conte Hall Berkeley CA 94720

BINGHAM, RICHARD CHARLES, b Middlebury, Vt, May 23, 44; m 72. TECHNICAL MANAGEMENT, ORGANIC CHEMISTRY. *Educ:* Univ Vt, BS, 66; Princeton Univ, PhD(chem), 70. *Prof Exp:* Fel chem, Univ Tex, Austin, 71-72; res chemist, 73-78, res supvr, 78-81, PROD MGR, E I DU

PONT DE NEMOURS & CO, INC, 81- *Mem:* Am Chem Soc. *Res:* Synthesis of novel dyes and pigments; electronic spectra of organic molecules; interpretation of the structures of molecules; theoretical organic chemistry; polymer chemistry. *Mailing Add:* Dupont Polymer Montgomery Bldg Rm 314 E I du Pont de Nemours & Co PO Box 80800 Wilmington DE 19880-0800

BINGHAM, RICHARD S(TEPHEN), JR, b Brooklyn, NY, July 7, 24; m 49; c 4. CHEMICAL ENGINEERING, TECHNICAL MANAGEMENT. *Educ:* Carnegie Inst Technol, BSChE, 48; Univ Wis-Oshkosh, MBA, 73. *Prof Exp:* Chem engr, explosives res, Hercules Powder Co, 48-50 & Pilot Plant Div, 50-51; supvr in charge, Qual Control Dept, Atlas Powder Co, 51-52, tech asst & statist consult, 52-56; sr engr qual control br, Res & Develop, Carborundum Co, 56-57, supv engr, Process Control Dept, 57-59, mgr qual control dept, Coated Abrasives Div, 59-62; mgr qual control, Consolidated Papers, Inc, 62-66, dir tech serv, 66-71, dir, mgt systs, 71-87; RETIRED. *Concurrent Pos:* Lectr, Univ Del, 51-52, Univ Chattanooga, 53-54 & Rochester Inst Technol, 55-72; adv conf qual control, Rutgers Univ, 57-59; ed, Have You Read Dept, Am Soc Qual Control, 61-; qual control inst, Univ Wis, 63; consult, 87- *Honors & Awards:* Brumbaugh Award, Am Soc Qual Control, 64. *Mem:* Fel Am Soc Qual Control (vpres, 66-69); Biomet Soc; Opers Res Soc Am; Am Inst Chem Engrs; Am Statist Asn. *Res:* Application of statistical methods to research, development and process control problems; operations research; data processing; industrial engineering. *Mailing Add:* 7349 Edinburgh Way Brooksville FL 34613

BINGHAM, ROBERT J, b Blackfoot, Idaho, Feb 17, 32; m 59; c 5. BIOCHEMISTRY, FOOD CHEMISTRY. *Educ:* Utah State Univ, BS, 59; Univ Wis, MS, 62, PhD(biochem, food sci), 64. *Prof Exp:* Health physicist, Phillips Petrol Co, AEC, Idaho, 59; res asst food sci, Univ Wis, 59-63; asst prof food chem, NC State Univ, 63-68; head agr-prod res, Beatrice Foods Co, 68-70; vpres, Nutrico, 70-74; vpres, Promarkco, 74-76; PRES, WIN HY FOODS, INC, 76- *Concurrent Pos:* Dir, Banfield of Tulsa, 73-80 & Prime Western, Inc, 75-80. *Mem:* Fel Am Inst Chem; Am Chem Soc; Inst Food Technol; Am Dairy Sci Asn. *Res:* Nutritional properties and means of protecting or enhancing nutritional value during processing and storage; utilization of by-products of food industry for high quality animal feeds. *Mailing Add:* Win-Hy Foods Inc 8620 Regency Dr Tulsa OK 74131

BINGHAM, ROBERT LODEWIJK, b Amsterdam, Netherlands, Sept 13, 30; US citizen; m 60; c 3. NUCLEAR PHYSICS, PLASMA PHYSICS. *Educ:* Williams Col, BA, 52; Harvard Univ, MA, 53; Columbia Univ, PhD(physics), 59. *Prof Exp:* Res asst meson physics, Nevis Cyclotron Lab, Columbia Univ, 55-59; res assoc plasma physics, Princeton Univ, 59-63; res staff mem, Gen Elec Co, NY, 63-68, tech counsr large lamp dept, Ohio, 68-69; Sloan exec fel grad sch bus, Stanford Univ, 69-70; gen mgr, Tektran, Air Prods & Chem Inc, 70-71; consult, R L Bingham, 72-73; coordr plans div controlled thermonuclear res, US Energy Res & Develop Admin, 74-78; prog anal officer, Off Tech Prog Eval, 78-80, INT TECHNOL TRANSFER POLICY & ANALYSIS, DEFENSE PROGS, US DEPT ENERGY, 80- *Mem:* Am Phys Soc. *Res:* Gaseous discharges; controlled thermonuclear fusion; particle detectors and analyzers; energy conversion. *Mailing Add:* 6500 Pyle Rd Bethesda MD 20817-5452

BINGHAM, SAMUEL WAYNE, b Fallston, NC, Apr 6, 29; m 55; c 3. PLANT PHYSIOLOGY, WEED SCIENCE. *Educ:* NC State Col, BS, 54, MS, 56; La State Univ, PhD(weed sci), 60. *Prof Exp:* Res agronomist, Crops Div, USDA, 56-58; asst specialist, La State Univ, 60-61; assoc prof, 61-72, PROF PLANT PHYSIOL, VA POLYTECH INST & STATE UNIV, 72- *Mem:* Weed Sci Soc Am; Int Turfgrass Soc. *Res:* Weed control in turf, penetration, translocation and fate of herbicides in plants; effects of herbicides on physiological process in plants; fate of herbicides in surface water. *Mailing Add:* Dept Plant Physiol Va Polytech Inst & State Univ Blacksburg VA 24061

BINGLER, EDWARD CHARLES, b Philadelphia, Pa, Nov 4, 35; m 54; c 3. GEOLOGY. *Educ:* Lehigh Univ, BA, 59; NMex Inst Mining & Technol, MS, 61; Univ Tex, PhD, 64. *Prof Exp:* Geologist, NMex Bur Mines & Mineral Resources, 64-67; asst prof geol & geol eng, SDak Sch Mines & Technol, 67-69; from assoc mining geologist to mining geologist, Nev Bur Mines, Univ Nev, Reno, 69-78; dep dir, Mont Bur Mines & Geol, Mont Col Mineral Sci & Technol, Butte, 78-85; BUR ECON GEOL, UNIV TEX, 85- *Mem:* Fel Geol Soc Am; Soc Econ Geol; Asn Prof Geol Scientists. *Res:* Structural geology; volcanic stratigraphy. *Mailing Add:* Econ Geol Univ Tex PO Box Univ Sta Austin TX 78713-7508

BINHAMMER, ROBERT T, b Watertown, Wis, Apr 28, 29; m 52; c 3. ANATOMY. *Educ:* Kalamazoo Col, BA, 51; Univ Tex, MA, 53, PhD(anat), 55. *Prof Exp:* From instr to prof anat, Col Med, Univ Cincinnati, 55-79, from asst dean to assoc dean, 67-79; assoc dean acad affairs, 79-86, PROF ANAT, COL MED, UNIV NEBR, 79- *Mem:* Assoc mem Radiation Res Soc; Am Asn Anatomists. *Res:* Radiation biology; parabiosis in physiological studies; pituitary physiology; cytology. *Mailing Add:* 3608 S 101 St Omaha NE 68124

BINKLEY, ROGER WENDELL, b Newark, NJ, Mar 9, 41; m 62; c 2. ORGANIC CHEMISTRY, PHOTOCHEMISTRY. *Educ:* Drew Univ, BA, 62; Univ Wis, PhD(org chem), 66. *Prof Exp:* From asst prof to assoc prof, 66-74, PROF CHEM, CLEVELAND STATE UNIV, 74- *Concurrent Pos:* Res collab grant, 68-69. *Mem:* Am Chem Soc. *Res:* Photochemistry of carbohydrates. *Mailing Add:* Dept Chem Cleveland State Univ Euclid & 24th St Cleveland OH 44115

BINKLEY, SUE ANN, b Dayton, Ohio, May 19, 44; c 1. BIOLOGICAL RHYTHMS, ENDOCRINOLOGY. *Educ:* Univ Colo, BA, 66; Univ Tex, Austin, PhD(zool), 71. *Prof Exp:* NIH fel physiol, Nat Inst Child Health & Human Develop, 71-72; asst prof biol, 73-76, ASSOC PROF BIOL, TEMPLE UNIV, 76- *Concurrent Pos:* Mem regulatory biol panel, NSF, 80-82. *Mem:* Endocrine Soc; AAAS. *Res:* Biological basis of circadian rhythms in vertebrates and in the function of the pineal gland. *Mailing Add:* Dept Biol Temple Univ Philadelphia PA 19122

BINKOWSKI, FRANCIS STANLEY, b Elizabeth, NJ, 1937; m 61; c 4. METEOROLOGY. *Educ:* Rutgers Univ, AB, 60; NY Univ, MS, 66, PhD(meteorol), 72. *Prof Exp:* Weather officer, Air Weather Serv, US Air Force, 60-64; asst prof meteorol, Rutgers Univ, 67-75; METEOROLOGIST, NAT OCEANIC & ATMOSPHERIC ADMIN, 75- *Mem:* Am Meteorol Soc; Sigma Xi. *Res:* Numerical modeling for air pollution meteorology; atmospheric turbulence and diffusion. *Mailing Add:* 717 Tinkerbell Rd Chapel Hill NC 27514

BINN, LEONARD NORMAN, b Lithuania, Nov 6, 27; nat US; m 80; c 3. MEDICAL BACTERIOLOGY. *Educ:* NY Univ, BA, 49; Univ Mich, MS, 51, PhD(bact), 55. *Prof Exp:* MED BACTERIOLOGIST, VIROL, WALTER REED ARMY INST RES, 55- *Mem:* Am Soc Microbiol; US Animal Health Asn; Am Soc Trop Med & Hyg; Am Soc Virol. *Res:* Hepatitis viruses; stable smallpox vaccine; immunology of adenoviruses; immunity in arthropod-borne virus diseases; laboratory animal virus diseases and canine viruses; Sigma Xi. *Mailing Add:* Dept Virus Dis Walter Reed Army Inst Res Washington DC 20307

BINNICKER, PAMELA CAROLINE, b Denmark, SC, July 17, 38. NEUROSCIENCES, CYTOLOGY. *Educ:* Coker Col, BA, 60; Columbia Univ, PhD(anat), 70. *Prof Exp:* Instr human biol, DeKalb Community Col, 71-72; asst prof biol sci, 72-78, ASSOC PROF PHARMACEUT SCI, SCH PHARM, MERCER UNIV, 78- *Concurrent Pos:* Instr anat & physiol, Sch Anestaesia, Ga Bapt Hosp, 75-86; lectr, Mercer Sch Nursing, 85-88. *Mem:* NY Acad Sci; Am Asn Cols Pharm. *Res:* Anatomy. *Mailing Add:* 1385 Pasadena Ave NE Atlanta GA 30306

BINNIE, WILLIAM HUGH, b Glasgow, Scotland, Dec 24, 39; US citizen; m 63, 83; c 3. ORAL PATHOLOGY, ORAL MEDICINE. *Educ:* Glasgow Univ, BDS, 63; McGill Univ, DDS, 65; Ind Univ, MSD, 67; Royal Col Physicians & Surgeons, Glasgow, FDS, 72; Royal Col Pathologists, London, MRC, 75. *Prof Exp:* Dent surgeon, Int Grenfell Asn, 63-64; from lectr to sr lectr oral path, Guy's Hosp Med Sch, 67-79; prof & chmn oral path, 79-82, PROF & CHMN PATH, BAYLOR COL DENT, 82- *Concurrent Pos:* Hon consult, Guy's Hosp, 75-79; consult, WHO, 75- & Baylor Univ Med Ctr, Vet Admin & Methodist Hosp, Dallas, 81-; chmn, path sect, Am Asn Dent Schs, 82-83; mem, Test Construction Comt, Nat Bd Oral Path & Dent Radiol, 84- *Mem:* Am Acad Oral Path; Int Asn Oral Path; Am Dent Asn; Am Asn Dent Schs; Int Asn Dent Res; Am Acad Oral Med. *Res:* Epidemiology, etiology, pathogenesis, diagnosis and management of oral mucosal disease; oral squamous cell carcinoma and salivary gland neoplasms. *Mailing Add:* Dept Path Baylor Col Dent 3302 Gaston Ave Dallas TX 75246

BINNIG, GERD, b Frankfurt, WGer, July 20, 47. PHYSICS. *Educ:* JW Goethe Univ, dipl, 73, PhD(physics), 78. *Prof Exp:* Res staff mem, Zurich Res Lab, 78-86, group leader, 84-86, IBM FEL, IBM PHYSICS GROUP, MUNICH UNIV, 86- *Concurrent Pos:* Vis prof, Univ Munich, WGer, 87. *Honors & Awards:* Nobel Prize in Physics, 86; Physics Prize, Ger Phys Soc; Otto Klung Prize, 83; King Faisal Int Prize for Sci, 84. *Res:* Superconductivity of semiconductors; scanning tunneling microscopy. *Mailing Add:* IBM Physics Group Munich Univ Schelling Strasse 4 Munich 40 8000 Germany

BINNING, LARRY KEITH, b Fond du Lac, Wis, Aug 29, 42; m 64; c 2. HORTICULTURE. *Educ:* Univ Wis, BS, 65; Mich State Univ, MS, 67, PhD(herbicide physiol, crop sci), 69. *Prof Exp:* From asst prof to assoc prof, 69-78, PROF HORT, UNIV WIS-MADISON, 78- *Mem:* Weed Sci Soc Am; Hort Soc; Potato Asn Am. *Res:* Herbicide use and physiology. *Mailing Add:* Dept Hort 385 Hort Univ Wis Madison 1575 Linden Dr Madison WI 53706

BINNING, ROBERT CHRISTIE, b Baton Rouge, La, Feb 11, 21; m 44; c 4. PHYSICAL CHEMISTRY, CHEMICAL ENGINEERING. *Educ:* La State Univ, BS, 48, MS, 50, PhD(phys chem), 52. *Prof Exp:* Res chemist, Am Oil Co, 52-55, group leader, 55-58; group leader, Monsanto Co, 58-61, proj mgr, 61-64, res mgr, Monsanto Res Corp, 64-65, asst dir res, 65-75, dir environ res & develop, 75-78, DIR CONTRACT RES & DEVELOP, MONSANTO RES CORP, 78- *Mem:* Fel Am Inst Chem Engrs; Int Soc Technol Assessment; fel Am Inst Chemists; World Future Soc. *Res:* Direction of multidisciplinary research and development programs in environmental control and energy. *Mailing Add:* 3473 Tall Timber Trail Dayton OH 45409

BINNINGER, DAVID MICHAEL, b Kenosha, Wis. MOLECULAR BIOLOGY, GENETICS. *Educ:* Univ SFla, BA, 79, MA, 81; Univ NC, PhD(genetics), 87. *Prof Exp:* Postdoctoral res molecular genetics, Univ Calif, Davis, Calif, 87-90; ASST PROF MOLECULAR GENETICS, FLA ATLANTIC UNIV, 90- *Mem:* Sigma Xi; Sci Res Soc Am. *Res:* Molecular mechanisms associated with DNA metabolism of eukaryotic cells; particular emphasis is on genetic recombination and DNA repair. *Mailing Add:* Biol Dept Fla Atlantic Univ Boca Raton FL 33431

BINNS, WALTER ROBERT, b Williamsburg, Kans, Nov 7, 40; m 62; c 2. COSMIC RAY PHYSICS, RADIOLOGY. *Educ:* Univ Ottawa, BS, 62; Colo State Univ, MS, 66, PhD(physics), 69. *Prof Exp:* scientist, Res Labs, McDonnell Douglas Corp, 69-80; sr res assoc, 80-89, RES PROF, WASHINGTON UNIV, 89- *Mem:* Am Phys Soc; Am Asn Physicists Med. *Res:* Abundances of very heavy cosmic rays in the primary cosmic radiation; dosimetry and treatment verification in radiology. *Mailing Add:* Dept Physics Washington Univ St Louis MO 63166

BINSTOCK, MARTIN H(AROLD), metallurgical engineering, for more information see previous edition

BINTINGER, DAVID L, b South Bend, Ind. COMPUTER DATA ANALYSIS, DRIFT CHAMBER CONSTRUCTION. *Educ:* Univ Notre Dame, BS, 64; Univ Chicago, PhD(physics), 75. *Prof Exp:* Res assoc, Fermi Nat Accelerator Lab, 75-; AT BROOKHAVEN NAT LAB. *Mem:* Am Phys Soc. *Res:* Experimental high energy physics: design and supervision of construction of experimental equipment, and analysis of data. *Mailing Add:* Lawrence Berkeley Lab One Cyclotron Rd Berkeley CA 94720

BINTZ, GARY LUTHER, b Manchester, Iowa, Feb 25, 41. COMPARATIVE PHYSIOLOGY, ENVIRONMENTAL PHYSIOLOGY. *Educ:* Cornell Col, BA, 63; Univ NMex, MS, 66, PhD(zool), 68. *Prof Exp:* Asst prof physiol, 68-74, ASSOC PROF BIOL, EASTERN MONT STATE COL, 74- *Res:* Water metabolism, including the effects of negative water balance on whole body and tissue levels; studies on several rodents, especially ground squirrels capable of hibernation. *Mailing Add:* Dept Biol Eastern Mont Col Billings MT 59101

BINZ, CARL MICHAEL, b Elgin, Ill, June 29, 47; m 75. ANALYTICAL CHEMISTRY. *Educ:* Loras Col, BS, 69; Purdue Univ, MS, 71, PhD(chem), 74. *Prof Exp:* Res assoc, Purdue Univ, 74-75; instr anal chem, 75-80, ASST PROF CHEMN, LORAS COL0, 80- *Res:* Determination of trace materials in air and water and their accumulated effects. *Mailing Add:* Dept Chem Loras Col Dubuque IA 52001

BIOLEAU, LUC J R, b Schefferville, Que. ENVIRONMENTAL HEALTH. *Educ:* Laurentian Univ, Ont, BSc, 79, MSc, 82. *Prof Exp:* Res assoc biol & chem, Laurentian Univ & McMaster Univ, 78-82; teacher biol, Laurentian Univ, 80-81; res coordr occup health & safety, Northeastern Ont Occup Health & Safety Resource Ctr, 82-85; CONSULT COORDR COMPUT, COMPUT DATA PROCESSING SOLUTIONS, 83-; CONSULT COORDR OCCUP HEALTH & SAFETY, OCCUP SAFETY HEALTH & ENVIRON CONSULTS, 85- *Concurrent Pos:* Researcher histochem, Carleton Univ, 80; technician sampling, Ont Res Found, 80; researcher x-ray fluorescences, Atmospheric Environ Servs, 82. *Res:* Monitor, literature search, analyses of major designated substances in Ontario; asbestos monitoring and analysis on site after sampling; substances on workplace; occupational health and safety and environmental subjects. *Mailing Add:* Occup Safety Health & Environ Consults 230 Ridgemount Sudbury ON P3B 3W5 Can

BIOLSI, LOUIS, JR, b Port Jefferson, NY, Aug 21, 40; m 62; c 3. THEORETICAL CHEMISTRY. *Educ:* State Univ NY Albany, BS, 61, MS, 63; Rensselaer Polytech Inst, PhD(theoret chem), 67. *Prof Exp:* Proj assoc, Theoret Chem Inst, Univ Wis, 66-67; res assoc theoret chem, Brown Univ, 68; From asst prof, to assoc prof, 68-84, PROF CHEM, UNIV MO-ROLLA, 84- *Mem:* Am Phys Soc; Am Chem Soc. *Res:* Scattering and kinetic theory of polyatomic molecules and the specification of nonequilibrium states. *Mailing Add:* Dept Chem Univ Mo Rolla MO 65401

BIONDI, MANFRED ANTHONY, b Carlstadt, NJ, Mar 5, 24; m 52; c 2. ATOMIC PHYSICS, AERONOMY. *Educ:* Mass Inst Technol, SB, 44, PhD(physics), 49. *Prof Exp:* Res assoc physics, Mass Inst Technol, 46-49; from res physicist to adv physicist, Westinghouse Res Lab, 49-57, mgr physics dept, 57-60; prof physics, 60-86, EMER PROF PHYSICS, UNIV PITTSBURGH, 86- *Concurrent Pos:* Mem adv panel physics, NSF; vis fel, Joint Inst Lab Astrophys, 72-73; exchange prof, Univ Paris, 76-; mem, Army Basic Res Steering Comt, Nat Res Coun, 85-88, chmn, 87-88. *Honors & Awards:* Davisson-Germer Prize, Am Phys Soc, 84. *Mem:* Fel AAAS; fel Am Phys Soc; Am Geophys Union; Nat Res Coun. *Res:* Interactions and reactions involving electrons, ions and excited atoms; plasma physics; electromagnetic properties of metals at liquid helium temperatures; upper atmospheric dynamics; thermosphere and ionosphere. *Mailing Add:* Dept Physics & Astron Univ Pittsburgh Pittsburgh PA 15260

BIONDO, FRANK X, b Brooklyn, NY, Mar 25, 27. MICROBIOLOGY, MYCOLOGY. *Educ:* Temple Univ, BA, 48; Athenaeum Ohio, MS, 51, PhD(exp med), 55. *Prof Exp:* Res dir bovine mastitis, Agr Res Inst, NJ, 53-54; instr microbiol, NY Med Col, 56-60; asst prof microbiol, Long Island Univ, 60-68 & C W Post Col, 68-72, assoc prof, 72-80, PROF HEALTH SCI, C W POST COL, LONG ISLAND UNIV, 80- *Concurrent Pos:* Part time med technologist, Ital Hosp, New York, 44-60; La State Univ fel parasitol & trop med, PR Dominican Repub & Haiti, 58. *Mem:* AAAS; Am Soc Microbiol; Am Soc Trop Med & Hyg. *Res:* Medical mycology; tranplantable tumors; tissue extracts and their effect on Micrococcus pyogenes var aureus infections in mice and transplantable tumors; biochemistry of dematiaceous fungi and Candida species. *Mailing Add:* Dept Health Sci C W Post Campus Col Long Island Univ Greenvale NY 11548

BIRCH, ALBERT FRANCIS, b Washington, DC, Aug 22, 03; m 33; c 3. GEOPHYSICS. *Educ:* Harvard Univ, BS, 24, MA, 29, PhD(physics), 32. *Hon Degrees:* DSc, Univ Chicago, 70 & Harvard Univ, 82. *Prof Exp:* Engr, NY Tel Co, 24-26; assist physics, 28-29, instr & tutor , 31-32, res assoc geophys, 32-37, from asst prof to prof geol, 37-49, Sturgis Hooper Prof Geol, 49-74, EMER PROF GEOL, HARVARD UNIV, 74- *Concurrent Pos:* Staff mem radiation lab, Mass Inst Technol, 42-44; chmn div earth sci, Nat Res Coun, 53-4. *Honors & Awards:* Day Medal, Geol Soc Am, 50; William Bowie Medal, Am Geophys Union, 60; Legion of Merit Award, Am Acad Arts & Sci; Nat Medal Sci, 68; Vetlesen Prize, 69; Penrose Medal, Geol Soc Am, 69; Gold Medal, Royal Astron Soc Gt Brit, 73; Sherman Fairchild Distinguished Scholar, 75; Bridgman Medal, Inst Asn Advan High Pressure Sci & Technol, 83. *Mem:* Nat Acad Sci; fel Am Phys Soc; fel Geol Soc Am (pres, 64); Seismol Soc Am; Am Geophys Union; Am Philos Soc; Am Acad Arts & Sci; Royal Astron Soc. *Res:* Properties of materials at high pressures and high temperature; geothermal studies; elasticity. *Mailing Add:* 987 Memorial Dr Cambridge MA 02138

BIRCH, MARTIN CHRISTOPHER, b Crewe, Eng, July 14, 44; m 70; c 2. ENTOMOLOGY, HEAD INJURIES & REHABILITATION. *Educ:* Oxford Univ, BA, 66, DPhil(entom), 69. *Prof Exp:* Res asst org chem, Dyson Perrins Lab, Oxford Univ, 69-70; res asst entom, Univ Calif, Berkeley, 70-73; assoc prof entom, Univ Calif, Davis, 73-; AT DEPT ZOOL, OXFORD UNIV. *Concurrent Pos:* Royal Entom Soc fel, 65-; Headway oxford chmn, 87- *Mem:* Royal Entom Soc; Entom Soc Am; Sigma Xi; Int Soc Chem Ecol; Soc Exp Biol; Asn Study Animal Behav. *Res:* Insect behavior and physiology, particularly chemical communication; behavioral mechanisms in pheromone communication of Coleoptera and Lepidoptera. *Mailing Add:* Dept Zool Oxford Univ S Parks Rd Oxford OX1 3PS England

BIRCH, RAYMOND E(MBREE), b McConnelsville, Ohio, Dec 19, 05; m 27; c 1. CERAMICS ENGINEERING. *Educ:* Ohio State Univ, BCerE, 27, CerE, 37. *Prof Exp:* Ceramic engr, Carlyle-Labold Co, 27-28; res engr, Eng Exp Sta, Ohio State Univ, 28-30; res engr, Harbison-Walker Refractories Co, 30-45, dir res, 45-70; res, Dresser Industs Co, 68-70; RETIRED. *Concurrent Pos:* Founding officer, Nat Inst Ceramic Engrs, 35; pres & founding group, St Clair Mem Hosp, Pittsburgh, 53-70; mem adv comt, eng depts, Ohio State Univ, Carnegie Mellon Univ & Ill Univs, 70- *Honors & Awards:* Greaves-Walker Award, Nat Inst Ceramic Engrs; St Louis Award, John Jeppson Medal, Am Ceramic Soc, A V Bleininger Award. *Mem:* Fel AAAS; fel Am Ceramic Soc; fel Am Soc Testing & Mat; fel Geol Soc Am. *Res:* Refractories utilization of magnesia. *Mailing Add:* 2650 Westlake Rd Coral Oaks Apt C102 Palm Harbor FL 34684

BIRCH, RUTH ELLEN, b Chicago, Ill, Mar 7, 52. CLINICAL IMMUNOLOGY. *Educ:* Univ Ill Med Ctr, MS, 77, PhD(immunol), 79. *Prof Exp:* Res asst prof pediat, Washington Univ, 82-86; RES SCI, VITEK SYSTEM, 86- *Mem:* Am Asn Immunol; Am Soc Microbiol; Am Soc Clin Path. *Res:* Immunoregulation and clinical assay development. *Mailing Add:* Birch Consultants 12605 Villa Hill Lane Creve Coeur MO 63141

BIRCHAK, JAMES ROBERT, b Latrobe, Pa, Mar 20, 39; m 63; c 2. ENGINEERING PHYSICS. *Educ:* Carnegie-Mellon Univ, BS, 61; Rice Univ, MA, 64, PhD(physics), 66; Wayne State Univ, MBA, 71. *Prof Exp:* Assoc sr res physicist, Gen Motors Res Lab, 66-71; sr res physicist, Southwest Res Inst, 71-76; res specialist, Babcock & Wilcox Lynchburg Res Ctr, 76-79; staff scientist & mgr, Nl Indust, Reliability Inc, 79-85, mem tech staff, 86-91; SR RES PHYSICIST, HALLIBURTON LOGGING SERV, 91- *Mem:* Am Phys Soc; Inst Elec & Electronic Engrs; Sigma Xi; Am Soc Nondestructive Testing. *Res:* Irreversible magnetization of superconductors; magnet sources of internal friction; ultrasonics; electric and magnetic properties of materials; geophysics; solar energy; electronics. *Mailing Add:* 3902 Cypressdale Spring TX 77388

BIRCHEM, REGINA, b Sisseton, SDak; m 81. CELL & DEVELOPMENTAL BIOLOGY, ELECTRON MICROSCOPY. *Educ:* Univ NDak, BSEd, 64; Univ Ga, MEd, 70, PhD(cell biol), 77. *Prof Exp:* Elem teacher, St Peter & Paul Sch, NDak, 58-61; sci teacher, St Francis Acad, 63-69; instr, Dept Bot, Sch Forest Resources, Univ Ga, 71, res assoc, 71-72, res assoc ultrastruct, 77-79; res assoc molecular, cellular & develop biol, Univ Colo, Boulder, 79-80 & Wash Univ, St Louis, 80-81; asst prof biol, Fontbonne Col, St Louis, 81-84, Univ S Sewanee, Tenn, 84; asst res prof, Dept Neurol, St Louis Sch Med, 85-88; assoc prof biol, Pa State Univ, McKeesport, 88-90; ASSOC PROF & FLORENCE MARIE SCOTT CHAIR DEVELOP BIOL, SETON HILL COL, GREENSBURG, PA, 90- *Honors & Awards:* Presidential Award, Electron Micros Soc Am, 77. *Mem:* Am Soc Cell Biol; Fedn Am Socs Exp Biol; Sigma Xi. *Res:* Development and morphology of Volvox carteri; gametogenesis in Volvox; embryogenesis and organogenesis in pine and hardwood tissue cultures; plant cell ultrastructure; neuron and Schwann cell tissue culture; deyelinating neuropathy studies in tissue culture; education on environmental issues. *Mailing Add:* Dept Biol Seton Hill Col Greensburg PA 15601

BIRCHENALL, CHARLES ERNEST, b Coatesville, Pa, Feb 19, 22; m 72; c 3. THERMODYNAMICS & MATERIAL PROPERTIES. *Educ:* Temple Univ, AB, 43; Princeton Univ, MA, 45, PhD(chem), 46. *Prof Exp:* Asst, Manhattan Proj, Princeton Univ, 43-46; mem staff, Metall Res Lab, Carnegie Inst Technol, 46-52, asst prof metall eng, 51-52; assoc prof chem, Princeton Univ, 52-60; dean sch grad studies, 64-67, distinguished prof metall, 60-87, EMER PROF METALL, UNIV DEL, 87- *Concurrent Pos:* Sci fac fel, NSF, Imp Col, Univ London, 58-59; metallurgist, Lawrence Radiation Lab, Univ Calif, Berkeley, 67-68; mem, Panel Magnetohydrodynamics, Off Sci & Technol, 68-69; guest tutor, Tech Univ, Eindhoven, Neth. *Mem:* Fel AAAS; fel Am Soc Metals; Am Inst Mining, Metall & Petrol Engrs; Sigma Xi. *Res:* Diffusion in metals, oxides and sulfides; heat storage in eutectic alloys; corrosion. *Mailing Add:* Sch Eng Univ Del Newark DE 19711

BIRCHFIELD, GENE EDWARD, b Bartlesville, Okla, Apr 14, 28; m 59; c 3. ATMOSPHERIC PHYSICS, GEOPHYSICS. *Educ:* Univ Chicago, AB, 52, MS, 55, PhD(geophys sci), 62. *Prof Exp:* Asst meteorol, Univ Chicago, 56-62; NSF fel geophys sci, Univ Stockholm & Mass Inst Technol, 62-63; assoc prof, 63-74, PROF ENG & GEOL SCI, NORTHWESTERN UNIV, 74- *Concurrent Pos:* Vis scientist, Inst Oceanog Sci, Wormley, Surrey, Eng, 72-73. *Mem:* Am Meteorol Soc; Am Geophys Union. *Res:* Geophysical fluid dynamics; physical limnology. *Mailing Add:* Dept Geol Northwestern Univ Evanston IL 60208

BIRCHFIELD, WRAY, b Ware Shoals, SC, July 2, 20; m 45; c 3. PLANT PATHOLOGY. *Educ:* Univ Fla, BS, 49, MS, 51; La Sa State Univ, PhD(plant path), 54. *Prof Exp:* Lab asst plant path, Univ Fla, 49-50; asst, La State Univ, 51-54; phytonematologist, Citrus Exp Sta, Fla State Plant Bd, 54-59; res plant pathologist, 59-72, nematologist, USDA, La State Univ, 72-84, CONSULT, 84- *Concurrent Pos:* Adj prof, La State Univ, nematol teaching, 68-84. *Mem:* Am Phytopath Soc; Soc Nematol. *Res:* Diseases of plants caused by nematodes; host-parasite relations of nematode diseases; soil microbiology and fumigation studies of soils; fungal and nematode complexes in relation to root diseases of plants; taxonomy of nematodes, fungi and plants. *Mailing Add:* 9302 Kingcrest Parkway Baton Rouge LA 70810

BIRCHLER, JAMES ARTHUR, b Red Bud, Ill, Feb 7, 50. GENETICS. *Educ:* Eastern Ill Univ, BS, 72; Ind Univ, PhD(genetics), 77. *Prof Exp:* investr genetics, Oak Ridge Nat Lab, 77-81; res affil, Roswell Park Mem Inst, 81-82; res affil, dept genetics, Univ Calif, Berkeley, 82-85; asst prof, 85-87, ASSOC PROF, HARVARD UNIV, 87- *Mem:* Genetics Soc Am. *Res:* Molecular genetics of eukaryotic gene expression. *Mailing Add:* 290 Biol Labs Harvard Univ 16 Divinity Ave Cambridge MA 02138

BIRCKBICHLER, PAUL JOSEPH, b Greenville, Pa, Nov 13, 42; m 65; c 2. BIOCHEMISTRY, CELL BIOLOGY. *Educ:* Duquesne Univ, BS, 64, PhD(biochem), 69. *Prof Exp:* Res fel, Brown Univ, 69-71; vis scientist, Bergen Univ, 71-72; res assoc, 72-77, asst scientist, 77-81, ASSOC SCIENTIST, SAMUEL ROBERTS NOBLE FOUND, 81- *Concurrent Pos:* NSF traineeship, Duquesne Univ, 65-68; instr chem, Pa State Univ, 68-69; clin enzymologist, RI Hosp, 71. *Mem:* Am Asn Cancer Res; Tissue Cult Asn; Am Chem Soc; Am Soc Biol Chem; NY Acad Sci. *Res:* Transglutaminase and isopeptide product in normal and transformed human cells; regulation of transglutaminase activity in transformed cells; membrane proteins of normal and transformed cells. *Mailing Add:* PO Box 2180 Noble Found Ardmore OK 73402

BIRD, CHARLES DURHAM, b Norman, Okla, July 7, 32; Can citizen; m 57; c 4. BOTANY. *Educ:* Univ Man, BS, 56; Okla State Univ, MS, 58, PhD(bot), 60. *Prof Exp:* Nat Res Coun Can fel, 60-62; asst prof bot, Univ Alta, Calgary, 62-65, actg cur herbarium, 62-79, admin off dept, 66-69, from assoc prof to prof bot, 67-79; RETIRED. *Concurrent Pos:* Assoc ed bot, Can Field Naturalist, 74- *Honors & Awards:* Loran L Goulden Mem Award, 78. *Res:* Ecology and wise land use policies. *Mailing Add:* Box 165 Mirror AB T0B 3C0 Can

BIRD, CHARLES EDWARD, b Kingston, Ont, Oct 19, 31; m 56; c 2. ENDOCRINOLOGY, METABOLISM. *Educ:* Queen's Univ, Ont, MD & CM, 56; McGill Univ, PhD(exp med), 67. *Prof Exp:* Clin asst med, Toronto Gen Hosp, Can, 61-63; clin asst, Royal Victoria Hosp, Can, 63-65; from asst prof to assoc prof, 65-75, assoc dean, Fac Med, 77-81, PROF MED, QUEEN'S UNIV, ONT, 75- *Mem:* Can Med Asn; Endocrine Soc. *Res:* Biological disposition of progesterone; progesterone metabolism in the human fetus; androgen metabolism in carcinoma of the prostate; androgen and estrogen metabolism in carcinoma of the breast. *Mailing Add:* Dept Med Queen's Univ Kingston ON K7L 3N6 Can

BIRD, CHARLES NORMAN, b Rolla, Mo, Dec 27, 29. ORGANIC CHEMISTRY. *Educ:* Col St Thomas, BS, 51; Univ Md, PhD(org chem), 57. *Prof Exp:* Coordr polyester develop, Minn Mining & Mfg Co, 3M Ctr, St Paul, 56-71, supvr graphic arts prod, 71-72, supvr photog bases & prod responsibility coordr, 72-88; RETIRED. *Mem:* Am Chem Soc. *Res:* Pyrolysis of esters; amides and lactones; organic synthesis; adhesives; rubber chemicals; polyester films; photographic films and chemicals. *Mailing Add:* 4102 River Rd S Afton MN 55001

BIRD, FRANCIS HOWE, nutrition, for more information see previous edition

BIRD, GEORGE RICHMOND, b Bismark, NDak, Jan 25, 25; m 48; c 3. PHOTOGRAPHIC CHEMISTRY. *Educ:* Harvard Univ, AB, 49, AM, 52, PhD(phys chem), 53. *Prof Exp:* Nat Res Coun fel, Radiation Lab, Columbia Univ, 52-53; asst prof chem, Rice Inst, 53-58; res chemist, Polaroid Corp, 58-61, mgr phys chem lab, 61-69; dir sch, 71-74, PROF CHEM, SCH CHEM, RUTGERS UNIV, 69- *Concurrent Pos:* Guggenheim fel, Photog Inst, Fed Tech Sch, Zurich, 74-75; vis lectr, Soc Photog Scientists & Engrs, 84-85. *Mem:* Am Chem Soc; fel Optical Soc Am; fel Soc Photog Sci & Eng. *Res:* Photovoltaic materials and laser dyes; basic photographic science. *Mailing Add:* Dept Chem Rutgers Univ PO Box 939 Piscataway NJ 08855-0939

BIRD, GEORGE W, b Newton, Mass, June 16, 39; m 67. NEMATOLOGY, PLANT PATHOLOGY. *Educ:* Rutgers Univ, BS, 61, MS, 63; Cornell Univ, PhD(plant path), 67. *Prof Exp:* Res scientist, Res Br, Can Dept Agr, 66-68; from asst prof to assoc prof, Dept Plant Path, Univ Ga, 68-73; integrated pest mgt coordr, 78-87, pesticide educ coordr, 85-87, PROF ENTOM, MICH STATE UNIV, 73- *Honors & Awards:* CIBA-GEIGY Award, Soc Nematologists, 83. *Mem:* Fel Am Phytopath Soc; fel Soc Nematol (pres, 77-78); Intersoc Consortium Plant Protection; AAAS. *Res:* Phytonematology; biology nematodes. *Mailing Add:* Dept Entom Mich State Univ East Lansing MI 48824

BIRD, GORDON WINSLOW, b NB, Can, Jan 30, 43; m 68; c 4. GEOCHEMISTRY, CHEMISTRY. *Educ:* Univ NB, BSc, 65; Univ Alta, MSc, 67; Univ Toronto, PhD(geol), 71. *Prof Exp:* Res fel mineral sci, Victoria Univ Wellington, 71-74; sr res fel, 74-77; assoc res officer chem, Atomic Energy Can, 77-81; assoc res officer chem, Oil Sands Res Ctr, Alta Res Coun, 82-86; sr res officer & mgr fundamental res, Oil Sands & Hydrocarbon Recovery, 87-90; ADJ PROF GEOL, UNIV ALTA, EDMONTON, 86- *Concurrent Pos:* Mgr indust contracts & bus develop, 90- *Mem:* Soc Petrol Engrs. *Res:* Geology; environmental chemistry of radionuclides; clay mineral dissolution, mass transport, low temperature silicate reaction kinetics; thermally enhanced oil recovery. *Mailing Add:* Alta Res Coun-Oil Sands Res Ctr PO Box 8330 Sta F Edmonton AB T6H 5X2 Can

BIRD, HAROLD L(ESLIE), JR, b Pluckemin, NJ, Dec 9, 21; m 45; c 2. BIOCHEMISTRY, ORGANIC CHEMISTRY. *Educ:* Rutgers Univ, BS, 43; Purdue Univ, MS, 49; Univ Ariz, PhD(biochem), 74. *Prof Exp:* Chemist, Metall & Chem Div, Naval Res Lab, 43-46; asst, Purdue Univ, 46-49; biochemist & anal chemist, Eli Lilly & Co, 49-64; asst prof, Longwood Col, 64-67; instr & lectr, Dept Chem, Univ Ariz, 67-74; prof, 74-84, EMER PROF CHEM, COL OF THE DESERT, CALIF, 84- *Concurrent Pos:* Lectr, Purdue Univ, 63-64. *Mem:* AAAS; Am Chem Soc. *Res:* Chromatography; biochemistry and natural products. *Mailing Add:* 47-967 Sun Corral Trail Palm Desert CA 92260

BIRD, HARVEY HAROLD, b Montclair, NJ, Aug 25, 34; m 57; c 2. PLASMA PHYSICS, GEOPHYSICS. *Educ:* Johns Hopkins Univ, BA, 56; Calif Inst Technol, MS, 59; Stevens Inst Technol, PhD(physics), 69. *Prof Exp:* Mem tech staff, Space Technol Labs, Thompson-Ramo-Wooldridge, Calif, 59-62; from instr to asst prof, 62-70, ASSOC PROF PHYSICS, FAIRLEIGH DICKINSON UNIV, 70- *Concurrent Pos:* US del, Int Asn Geomagnetism & Aeronomy, Madrid, Spain, 69. *Mem:* Am Phys Soc; Am Geophys Union. *Res:* Physics of the earth's magnetosphere; instability theory in plasmas. *Mailing Add:* Dept Physics Fairleigh Dickinson Univ Madison NJ 07940

BIRD, HERBERT RODERICK, animal nutrition; deceased, see previous edition for last biography

BIRD, JOHN MALCOLM, b Newark, NJ, Dec 27, 31; div; c 2. GEOLOGY. *Educ:* Union Col, NY, BS, 55; Rensselaer Polytech Inst, MS, 59, PhD(geol), 62. *Prof Exp:* From instr to prof geol, State Univ NY Albany, 61-72; chmn dept geol sci, 69-72; PROF GEOL, COL ENG, CORNELL UNIV, 72- *Concurrent Pos:* Res grants, Geol Soc Am, 62-, NY State Geol Surv, 63-, NSF, 64, 68, 72, 73, 75, 76, 77, 78, 79, 80 & 81, US Geol Surv, 79, 80 & 81, Off Naval Res 78, 79 & 80, Nat Acad Sci Day fund, 69, Petrol Res Found, 75 & Nat Geog Soc, 77, 78, 79, 80, 81, 82, 83, 84 & 86; mem, NY State Mus & Sci Serv, 63-67; Nat Acad Sci exchange vis scientist, Polish Acad Sci, 67; res assoc, Lamont-Doherty Geol Observ, Columbia Univ, 70, sr res assoc, 70-72; distinguished vis scientist, Am Geol Inst, 71-; assoc ed, J Geophys Res, 71-73; chmn, Appalachian Working Group, US Geodynamics Comt, 71-73; Distinguished lectr, Am Asn Petrol Geologists, 77-78. *Mem:* Fel Geol Soc Am; Am Geophys Union; fel Can Geol Soc; Sigma Xi. *Res:* Geotectonics; genesis of ore deposits; Appalachian geology; lithosphere plate tectonics; continent evolution; evolution of mountain belts; ophiolite tectonics; petrology of ultramafic rocks. *Mailing Add:* Dept Geol Sci Cornell Univ Ithaca NY 14853

BIRD, JOHN WILLIAM CLYDE, b Erie, Pa, Nov 10, 32; m 54; c 2. PHYSIOLOGY, MOLECULAR PHYSIOLOGY. *Educ:* Univ Colo, BA, 58, MA, 59; Univ Iowa, PhD(physiol), 61. *Prof Exp:* From asst prof to prof physiol & biochem, 61-78, dir, Bur Biol Res, 76-87, chmn, dept biol sci, 82-86, Distinguished PROF PHYSIOL, RUTGERS UNIV, 78- *Concurrent Pos:* Fulbright prof biochem, Cairo Univ, 64-65; NIH career develop award, 66-71; Fulbright res scholar, Belg, 69-70. *Honors & Awards:* Outstanding Res Award, Muscular Dystrophy Asn, 82 & 85. *Mem:* AAAS; NY Acad Sci; Am Physiol Soc; Fedn Clin Res; fel Nat Acad Sci France. *Res:* Proteinases and proteinase inhibitors; muscle lysosomes; muscular dystrophy. *Mailing Add:* Bur Biol Res Rutgers Univ PO Box 1059 Piscataway NJ 08854

BIRD, JOSEPH FRANCIS, b Scranton, Pa, Feb 17, 30; m 55; c 1. THEORETICAL PHYSICS. *Educ:* Univ Scranton, AB, 51; Cornell Univ, PhD(theoret physics), 58. *Prof Exp:* Sr staff mem, 58-62, PRIN STAFF MEM, THEORET PHYSICS, APPL PHYSICS LAB, JOHNS HOPKINS UNIV, 62- *Mem:* Am Phys Soc. *Res:* Astrophysics, cosmogony, star formation theory; combustion, acoustic instability in solid fuel rockets; vision, psychophysical and physiological analysis; neural noise; oceanic hydromagnetism; theory of color and brightness sensations; electromagnetic levitation; wave scattering theory. *Mailing Add:* Appl Physics Lab Johns Hopkins Univ Laurel MD 20723-6099

BIRD, JOSEPH G, b Waycross, Ga, Oct 6, 15; m 39; c 3. CLINICAL PHARMACOLOGY, MEDICINE. *Educ:* Fla Southern Col, BS, 37; Univ Md, MD, 42, PhD(pharmacol), 49. *Prof Exp:* Rotating intern, Univ Hosp, Baltimore, Md, 42-43; asst resident med, 43-44 & 46-47; instr pharmacol, Sch Med, Univ Md, 47-54, asst med, 50-54; dir, Div Clin Res, Sterling-Winthrop Res Inst, 54-67, dir exp med, 67-70; asst dir, Ciba-Geigy Corp, 70-72, assoc dir, 73-77, dir clin pharmacol, 77-80; RETIRED. *Concurrent Pos:* Staff physician, Univ Hosp & Md Gen Hosp, Baltimore, 50-54; med dir, Radio Div, Bendix Corp, 49-54. *Mem:* AAAS; assoc emer Am Chem Soc; emer Am Soc Clin Pharmacol & Exp Therapeut; emer NY Acad Sci. *Res:* Pharmaceutical research; clinical trials of new drugs. *Mailing Add:* 29 Sea Spray Dr Biddeford ME 04005-9204

BIRD, LESLIE V(AUGHN), b Medina, NY, Aug 29, 29; m 54; c 3. ELECTRICAL ENGINEERING. *Educ:* Cornell Univ, BEE, 54, MS, 60; Yale Univ, MEng, 65. *Prof Exp:* Engr, Gen Elec Co, Conn, 55-59; from instr to asst prof elec eng, 59-67, asst dean grad studies, 68-70, ASSOC PROF ELEC ENG, UNIV BRIDGEPORT, 67- *Mem:* Am Soc Eng Educ; Inst Elec & Electronics Engrs; Nat Soc Prof Engrs. *Res:* Communication theory; electronic components. *Mailing Add:* PO Box 42 Monroe CT 06468

BIRD, LUTHER SMITH, b Greenville, SC, Nov 25, 21; m 47; c 2. PLANT PATHOLOGY. *Educ:* Clemson Univ, BS, 48; Tex A&M Univ, MS, 50, PhD(genetics), 55. *Prof Exp:* From instr to prof, 50-86, EMER PROF PLANT PATH, TEX A&M UNIV, 86- *Mem:* Crop Sci Soc Am; Am Phytopath Soc; Am Soc Agron; Am Genetic Asn; Sigma Xi. *Res:* Breeding for and genetics of multi-adversity resistance (diseases, insects and stress) in cotton; symbiosis genetics; seed quality and cold tolerance in cotton; nature of resistance and escape mechanisms. *Mailing Add:* 729 Shady Lane Bryan TX 77802

BIRD, MICHAEL WESLEY, b Grand Rapids, Mich, Sept 24, 42; m 65; c 1. GUIDANCE & CONTROL, NAVIGATION. *Educ:* Mich State Univ, BS, 64, MS, 66, PhD(elec eng), 69. *Prof Exp:* Sr staff engr, 69-80, mgr flight mgt technol, Instrument Div, 80-87, DIR, MIL & COM SYSTS, LEAR SIEGLER INC, 87-; DIR, ELECTRONIC SYSTS & SOFTWARE, SMITHS INDUSTRIES, GRAND RAPIDS. *Mem:* Inst Elec & Electronic Engrs. *Res:* Applied research in performance optimization control theory; artificial intelligence with application to vehicle guidance systems. *Mailing Add:* Smiths Industs Grand Rapids 4141 Eastern Ave SE Grand Rapids MI 49518

BIRD, PETER, b Cambridge, Mass, Sept 29, 51; m 72; c 1. TECTONOPHYSICS, FINITE ELEMENT MODELS. *Educ:* Harvard Univ, BA, 72; Mass Inst Technol, PhD(geophys), 76. *Prof Exp:* From asst prof to assoc prof, 76-85, PROF EARTH & SPACE SCI, UNIV CALIF, LOS ANGELES, 85- *Mem:* Fel Am Geophys Union; fel Geol Soc Am. *Res:* Finite-element and analytic modeling of lithosphere deformation to test theories of orogeny and determine the rheology of lithosphere in situ. *Mailing Add:* Dept Earth & Space Sci Univ Calif Los Angeles CA 90024

BIRD, R(OBERT) BYRON, b Bryan, Tex, Feb 5, 24. CHEMICAL ENGINEERING, THEORETICAL CHEMISTRY. *Educ:* Univ Ill, BS, 47; Univ Wis, PhD(chem), 50. *Hon Degrees:* DEng, Lehigh Univ, 72 & Wash Univ, 73; DEng Sci, Technische Hogeschool Delft, Holland, 77; DSc, Clarkson Col Technol, 80; DEng, Colo Sch Mines, 86. *Prof Exp:* Fulbright fel, Amsterdam, 50-51; proj assoc chem, Univ Wis, 51-52; asst prof, Cornell Univ, 52-53; proj assoc chem eng, Univ Wis-Madison, 53-55, assoc prof, 55-57, chmn dept, 64-68, Charles F Burgess distinguished prof, 68-72, PROF CHEM ENG, UNIV WIS-MADISON, 57-, VILAS RES PROF, 72-, JOHN D MACARTHUR PROF, 82- *Concurrent Pos:* Mem adv panel, Eng Sci Div, NSF, 61-64; Fulbright lectr, Kyoto & Nagoya Univ, 62-63; Petrol Res Fund grant, Am Chem Soc, 63; consult, Union Carbide Corp, 64-77; mem adv bd, Int Chem Eng, 65, 66; US ed, Appl Sci Res, 68-87 & 89-; mem adv bd, Indust & Eng Chem, 70-72 & J Non-Newtonian Fluid Mech, 76- *Honors & Awards:* Curtis McGraw Award, 59 & Westinghouse Award, 60, Corcoran Award, 87, Am Soc Eng Educ; William H Walker Award, 62, Prof Prog Award, 65 & W K Lewis Award, 74, Am Inst Chem Engrs; Bingham Medal, Soc Rheology, 74; Eringen Medal, Soc Eng Sci, 83; Nat Medal of Sci, 87. *Mem:* Nat Acad Sci; Nat Acad Eng; fel Am Inst Chem Engrs; fel Am Phys Soc; Am Acad Arts & Sci; foreign mem, Royal Dutch Acad Arts & Sci. *Res:* Transport phenomena; viscoelastic fluid mechanics; polymer fluid dynamics and kinetic theory; molecular theory of gases and liquids; technical Japanese translation. *Mailing Add:* Dept Chem Eng Univ Wis Madison WI 53706-1691

BIRD, RICHARD PUTNAM, b Durango, Colo, July 19, 38; m 62; c 3. BIOPHYSICS. *Educ:* Univ Colo, BS, 60, MS, 62; Univ Calif, Berkeley, MBioradiol, 68, PhD(biophys), 72; Am Bd Toxicol, cert. *Prof Exp:* Res asst physics, Univ Colo, 61-62; health physicist, Idaho Opers Off, 62-64; phys sci adminr, San Francisco Opers Off, US AEC, 64-66; res assoc physics, Kans State Univ, 71-74; res assoc biophysics, Columbia Univ, 74-88; TECH WRITER, SCI TYPOGRAPHERS, 88- *Mem:* Radiation Res Soc. *Res:* Biological effect of radiation of different quality, particularly accelerator-produced radiations, using eucharyotic cell systems. *Mailing Add:* 110 Jennings Ave Patchogue NY 11772

BIRD, ROBERT EARL, b San Antonio, Tex, Nov 4, 43; m 64. MOLECULAR BIOLOGY. *Educ:* Kans State Univ, BS, 65, PhD(genetics), 69. *Prof Exp:* Investr molecular biol, Biol Div, Oak Ridge Nat Lab, 69-71; res dir, Dept Molecular Biol, Univ Geneva, Switz, 71-74; sr staff fel, Lab Molecular Biol, Nat Inst Arthritis, Metab & Digestive Dis, NIH, 74-77; asst prof, Dept Molecular Biol, Vanderbilt Univ, 77-; res dir, Genex Corp, 88-89; DIR, MOLECULAR BIOL LAB, MOLECULAR ONCOL, INC, 89- *Mem:* Genetics Soc Am; Am Soc Microbiol. *Res:* Genetics, microbiology and biochemistry as related to the regulation of chromosome replication and cell division; control and mechanism of DNA replication in vivo and in vitro. *Mailing Add:* Molecular Oncol Inc 19 Firstfield Rd Gaithersburg MD 20878

BIRD, SAMUEL OSCAR, II, b Charleston, WVa, Feb 26, 34; m 55. INVERTEBRATE PALEONTOLOGY, SYSTEMATICS. *Educ:* Marshall Col, BS, 55; Univ Wis, MS, 58; Univ NC, PhD(geol), 62. *Prof Exp:* Asst prof geol & geog, 62-67, assoc prof, 67-68, chmn dept, 70-79, PROF GEOL, MARY WASHINGTON COL, 68-; GEOLOGIST-ED, VA DIV MINERAL RESOURCES, 79- *Concurrent Pos:* Exten grant for res partic, Col Teachers Prog, 65-67; proj dir, Co SIP, NSF, 70-75. *Mem:* Fel Geol Soc Am; Am Geol Inst. *Res:* Quantitative systematics; molluscan ecology and community structure; quantitative approaches to stratigraphic correlation; Appalachian stratigraphy and structural geology. *Mailing Add:* 4400 Bromley Lane Richmond VA 23221

BIRD, STEPHANIE J, b Los Angeles, Calif, Oct 5, 48; m 76. NEUROSCIENCE, HEALTH POLICY. *Educ:* Univ Calif, Los Angeles, BA, 71; Yale Univ, MS, 72, PhD(physiol), 75. *Prof Exp:* Fel, Johns Hopkins Univ, 75-76; fel, Case Western Reserve Univ, 76-78, sr res assoc, 78-79; staff scientist, Neurosci Res Prog, Mass Inst Technol, 79-82, Mellon fel, 82-83, vis scholar, 83-86, RES ASSOC & LECTR, MASS INST TECHNOL, 86- *Concurrent Pos:* Consult, Sen Edward M Kennedy Comt Policy, 82; Social Issues Comt, Soc Neurosci, 83-86, Pub Info Comt, 84-87; nominations comt, Asn Women Sci, 83-85, counr, 86-88. *Mem:* Soc Neurosci; Asn Women Sci (pres-elect, 88-89, pres, 90-91); AAAS. *Res:* Neuroscience, and it's ethical, legal & social policy implications, especially screening tests and premenstrual syndrome; health and science policy relating to neuroscience and or biotechnology issues. *Mailing Add:* 41 Sleepy Hollow Wrentham MA 02093

BIRD, THOMAS JOSEPH, b Scranton, Pa, Nov 15, 27; m 52; c 6. MEDICAL MICROBIOLOGY. *Educ:* Univ Scranton, BS, 51; Univ Pa, MS, 54, PhD(med microbiol), 56. *Prof Exp:* Instr microbiol, Med Sch, Northwestern Univ, 56-58; instr, Stritch Sch Med, Loyola Univ, Ill, 58-59, asst prof, 59-60; res microbiologist, Hines Vet Admin Hosp, Hines, Ill, 64-72, chief microbiologist, 72-88; RETIRED. *Concurrent Pos:* Consult, Ill Dept Pub Health, 65-; vis prof microbiol, Chicago Med Sch, 70- *Mem:* Fel Am Soc Microbiol; Sigma Xi; NY Acad Sci; Brit Soc Gen Microbiol; fel Am Acad Microbiol. *Res:* Virus relationships to host; bacteriophages; lysogeny; tissue culture; drug effects on host virus system; pseudomonas infections; plasmid epidemiology; pyocins; Legionnaire's disease. *Mailing Add:* 1500 Victoria Ave Berkeley IL 60163

BIRDSALL, CHARLES KENNEDY, b New York, NY, Nov 19, 25; m 49, 81; c 4. ELECTRICAL ENGINEERING. *Educ:* Univ Mich, BSE, 46, MSE, 48; Stanford Univ, PhD(elec eng), 51. *Prof Exp:* Res Physicist, Hughes Res & Develop Labs, 51-55, group leader electron physics, Gen Elec Microwave Lab, 55-59; PROF ELEC ENG, UNIV CALIF, BERKELEY, 59- *Concurrent Pos:* Lectr, Univ Calif, Los Angeles, 53-55, Stanford Univ, 55 & Univ Calif exten, 56; prof, Miller Inst Basic Res in Sci, Univ Calif, Berkeley, 63-64, chmn, Energy & Resources Prog, 72-74; assoc ed, Inst Elec & Electronics Engrs Transactions on Electron Devices, 64; NSF grantee, US-Japan Coop Sci Prog, 66; sr vis fel, Univ Reading, Eng, 76; consult, Lawrence Livermore Nat Lab, 60-; foreign res assoc, Inst Plasma Physics, Nagoya Univ, Japan, 81 & 82, vis prof, 88; vis prof Chevron energy, Calif Inst Technol, 82; vis res scholar, Max Planck Inst Plasma Physics, Ger & Inst Theoret Physics, Univ Innsbruck, Austria, 85; Fulbright grant, Innsbruck, Austria, 91. *Honors & Awards:* Plasma Sci & Applns Award, Inst Elec & Electronics Engrs Nuclear & Plasma Sci Soc, 88. *Mem:* Fel Inst Elec & Electronics Engrs; fel Am Phys Soc; fel AAAS. *Res:* Plasmas; computer simulation of plasmas; plasma assisted processing of materials. *Mailing Add:* Dept Elec Eng & Comput Sci Univ Calif Berkeley CA 94720

BIRDSALL, MARION IVENS, b Toronto, Ont, Nov 7, 40; m 69; c 2. FOOD SCIENCE, SOCIAL SCIENCE. *Educ:* Univ Toronto, BHSc, 62; Pa State Univ, MS, 67, PhD(nutrit), 72. *Prof Exp:* Nutrit specialist, Ont Dept Agr, 62-64; nutrit counselor, Donwood Found, 67; instr, Pa State Univ, 70-71; asst prof nutrit, Albright Col, 72-75; consult curric, Pa State Univ, 76-78; asst prof home econ, 79-85, CHAIR, HUMAN ECOL, ALBRIGHT COL, 83-, ASSOC PROF NUTRITION, 85- *Concurrent Pos:* Vis prof, Univ Guelph, Ont, 68; instr nutrit, Berks, Allentown Campuses, 72-78; visitor, Univ Oxford, 78-79; consult nutrit, Reading Sch Dist & educ training grants, Oley Valley Sch Dist, Pa, 79-81; plan IV rep, Albright Col, Am Dietetic Asn, 79-; liaison staff, Nutrit Info & Resource Ctr, 81-84. *Mem:* Am Dietetic Asn; Soc Nutrit Educ; Am Home Econ Asn. *Res:* Social-behavioral influences on preschool food patterns; attitude measurement and food behavior. *Mailing Add:* 1780 Acorn Dr Reading PA 19608

BIRDSALL, THEODORE G, b Lakewood, Ohio, Oct 30, 27; m 51; c 6. UNDERWATER ACOUSTIC PROPAGATION, SIGNAL PROCESSING. *Educ:* Univ Mich, BSE, 50, MS, 52, PhD(commun sci), 66. *Prof Exp:* Res engr, 50-66, assoc prof, 67-70, PROF COMMUN, PROCESSING & DETECTION THEORY COMMUN & SIGNAL PROCESSING LAB, UNIV MICH, 70-, DIR, 74- *Concurrent Pos:* Prin Investr Off Naval Res, Univ Mich, 60- *Mem:* Inst Elec & Electronics Engrs; fel Acoust Soc Am. *Res:* Stochastic detection and estimation theory applied to underwater acoustic propagation and underwater acoustic tomography; design and conducting of experiments, applications of results. *Mailing Add:* Communications & Signal Processing Lab Univ Mich 4242 EECS Bldg N Campus Ann Arbor MI 48109-2122

BIRDSALL, WILLIAM JOHN, b Waterbury, Conn, Oct 28, 44; m 69; c 2. INORGANIC CHEMISTRY. *Educ:* Univ Maine, BA, 66; Pa State Univ, PhD(chem), 71. *Prof Exp:* from asst prof to assoc prof, 71-83, PROF CHEM, ALBRIGHT COL, 83- *Concurrent Pos:* Fel, Pa State Univ, 71; vis scientist, Oxford Univ, 78-79 & Danforth Assoc; vis prof, Lehigh Univ, 87-88. *Mem:* AAAS; Am Chem Soc; Sigma Xi; NY Acad Sci. *Res:* Isolation of natural products with emphasis on inorganic constituents; synthesis and structural determination of metal-purine complexes and alkaloid-metal adducts; study of bilayer phospholipid vesicles. *Mailing Add:* Dept Chem Albright Col Reading PA 19604

BIRDSELL, DALE CARL, b Spokane, Wash, Feb 9, 40; m 62; c 2. MICROBIOLOGY, PHYSIOLOGY. *Educ:* Wash State Univ, BS, 62, MS, 64; Univ Calif, Riverside, PhD(biol), 67. *Prof Exp:* Res assoc microbiol, Scripps Clin & Res Found, 67-68; asst prof, Sch Med, Univ Louisville, 69-70; asst prof, Sch Dent, Loyola Univ, 70-74; ASSOC PROF BASIC DENT SCI, UNIV FLA, 74- *Concurrent Pos:* NIH fel microbiol, Scripps Clin & Res Found, 68-69. *Mem:* AAAS; Am Soc Microbiol; Int Asn Dent Res. *Res:* Structure and function of the synthesis of macromolecular components of cell surfaces; membrane genesis in microorganisms; role of surfaces in diseases of oral cavity. *Mailing Add:* Dept Surg Univ Calgary Faculty Med 3330 Hosp Dr NW Calgary AB T2N 4N1 Can

BIRDSONG, RAY STUART, b Naples, Fla, June 18, 35; m 61; c 2. BIOLOGY, ICHTHYOLOGY. *Educ:* Fla State Univ, BS, 62, MS, 63; Univ Miami, PhD(marine sci), 69. *Prof Exp:* Asst prof biol, 68-73, ASSOC PROF BIOL & OCEANOG, OLD DOMINION UNIV, 73- *Concurrent Pos:* Asst prog dir, Syst Biol & Ecol Progs, NSF, 70-71. *Mem:* Am Soc Ichthyol & Herpet; Soc Syst Zool. *Res:* Ecology, distribution and classification of fishes, especially marine fishes; functional morphology of fishes in general. *Mailing Add:* 5000 Welleston Ct Virginia Beach VA 23462

BIRDWHISTELL, RALPH KENTON, b Columbus, Ohio, May 11, 24; m 43; c 3. PHYSICAL INORGANIC CHEMISTRY. *Educ:* Ohio State Univ, BSc, 49; Univ Kans, PhD(chem), 53. *Prof Exp:* Fel, Mich State Univ, 53-54, asst prof, 54-58; assoc prof chem, Butler Univ, 63-67; PROF CHEM & CHMN DEPT, UNIV W FLA, 67- *Concurrent Pos:* Vis lectr, Univ Ill, 62-63. *Mem:* Am Chem Soc; Sigma Xi; AAAS; NY Acad Sci. *Res:* Synthesis and thermodynamic properties of inorganic complex compounds; species occurring in both aqueous and nonaqueous media. *Mailing Add:* 35 Blithewood Rd Pensacola FL 32504

BIRECKA, HELENA M, b Poland, May 13, 21; m 48; c 1. PLANT PHYSIOLOGY, BIOCHEMISTRY. *Educ:* Univ Perm, MS, 44; Timiriazev Acad, Moscow, PhD(plant physiol), 48. *Prof Exp:* Asst prof agr chem, Agr Univ, Warsaw, 49-51, assoc prof plant physiol, 53-61, chmn dept, 54-68, prof, 61-68; assoc prof agr chem, Univ Poznan, 51-53; consult, Int Atomic Energy Agency, Vienna, 68-69; res assoc plant physiol, Yale Univ, 69-70; assoc prof plant physiol & biochem, 70-74, PROF BIOSCI, UNION COL, NY, 74- *Concurrent Pos:* head plant metab lab, Polish Acad Sci, 60-68, chmn, Comt Use Isotopes & Nuclear Energy in Agr & Biol Sci, 62-68; prof, Isotope Lab, Inst Plant Cultivation, Warsaw, 61-68; Food & Agr Orgn fel, 64. *Mem:* Polish Bot Soc; Polish Biochem Soc; Am Soc Plant Physiol; Am Soc Agron. *Res:* Mineral nutrition of plants; alkaloid biosynthesis and metabolism; photosynthesis; long distance translocation in plants; polyamine biosynthesis; scientific papers published over 100. *Mailing Add:* Dept Biol Sci Union Col Schenectady NY 12308

BIRECKI, HENRYK, b Warsaw, Poland, Nov 19, 48; m 71; c 2. LIQUID CRYSTAL TECHNOLOGY, COMPUTER DATA STORAGE. *Educ:* Mc Gill Univ, BSc, 70 & MSc, 72; Mass Inst Technol, PhD(physics), 76. *Prof Exp:* Fel & mem tech staff physics liquid crystals, Lawrence Berkeley Labs,

76-79; mem tech staff displays, 79-81, optical data storage, 81-83, proj mgr, 83-85, technol eval & optical computing, 85-87, PROJ MGR, OPTICAL DATA STORAGE, HEWLETT PACKARD LABS, 87- *Mem:* Am Phys Soc; Sigma Xi. *Res:* Optical data storage concentrating on magneto-optic technology for high performance systems. *Mailing Add:* 3001 Ross Rd Palo Alto CA 94303

BIRELY, JOHN H, b Glen Ridge, NJ, Oct 17, 39; m 82; c 1. CHEMICAL PHYSICS. *Educ:* Yale Univ, BS, 61; Univ Calif, Berkeley, MS, 63; Harvard Univ, PhD(phys chem), 67. *Prof Exp:* Asst phys chem, Univ Calif, Berkeley, 61-63 & Harvard Univ, 63-66; NIH fel, Cambridge Univ, 66-67; asst prof chem, Univ Calif, Los Angeles, 67-69; mem tech staff, Aerospace Corp, 69-74; group leader, asst dir & dep div leader, 74-81, assoc dir chem, Earth & Life Sci, 81-86, DEP DIR, DEFENSE, LOS ALAMOS NAT LAB, 86- *Mem:* Am Phys Soc; Am Chem Soc; AAAS. *Res:* Atomic and molecular collision physics; molecular spectroscopy; gas phase chemical kinetics and energy transfer; laser photochemistry. *Mailing Add:* Los Alamos Nat Lab PO Box 1663 MSA 105 Los Alamos NM 87545

BIRGE, ANN CHAMBERLAIN, b San Francisco, Calif, Jan 20, 25; m 48; c 3. RADIATION BIOPHYSICS. *Educ:* Vassar Col, AB, 46; Harvard Univ, AM, 47, PhD(physics), 51. *Prof Exp:* Physicist, Donner Lab, Univ Calif, Berkeley, 51-59, consult, 59-62, lectr, Eng & Sci Exten, 63-65; assoc prof, 65-75, PROF PHYSICS, CALIF STATE UNIV, HAYWARD, 75- *Mem:* AAAS; Fedn Am Scientists; Am Asn Physics Teachers; Sigma Xi. *Mailing Add:* 1 Greenwood Common Berkeley CA 94708

BIRGE, ROBERT RICHARDS, b Washington, DC, Aug 10, 46; m 68; c 2. LASER SPECTROSCOPY, MOLECULAR ELECTRONICS. *Educ:* Yale Univ, BS, 68; Wesleyan Univ, PhD(chem), 72. *Prof Exp:* Researcher chem air pollution res div, Environics Br, Kirtland Air Force Base, NMex, 72-73; NIH fel chem, Harvard Univ, 73-75; from asst prof to assoc prof, Univ Calif, Riverside, 75-84; prof chem & head dept, Carnegie-Mellon Univ, 85-87, dir, Ctr Molecular Electronics, 85-87; PROF CHEM, SYRACUSE UNIV, 88-, DIR, CTR MOLECULAR ELECTRONICS, 88- *Concurrent Pos:* Consult molecular spectros, Air Pollution Res Div, Environics Br, Kirtland Air Force Base, NMex, 73-75; mem & sci consult, Bice Comn Atmospheric Pollution, US Senate, 76-77; chmn, res comt, Univ Calif, 81-82; Weingart fel, Calif Inst Technol, 82-83; treas, Carnegie Inst Natural Hist, 85-86; mem, Molecular & Cellular Biophys Chem Study Sect, NIH, 84-88 & Adv Grant Rev Panel, Biophys Prog, NSF, 90; bd dir, West Penn Hosp Res Found, 87-88; Univ Calif Regents Fac Fel, 76. *Honors & Awards:* Nat Sci Award, Am Cyanamid Co, 64. *Mem:* Am Chem Soc; Sigma Xi; Am Phys Soc; Biophys Soc. *Res:* Theory and application of nonlinear (two-photon) laser spectroscopy to the study of biological molecules and protein bound chromophores; development and application of molecular orbital theory and molecular dynamics in biophysics; laser photocalorimetry and solvent effects. *Mailing Add:* Dept Chem Syracuse Univ Syracuse NY 13244-4100

BIRGE, ROBERT WALSH, b Berkeley, Calif, Jan 30, 24; m 48; c 3. HIGH ENERGY PHYSICS. *Educ:* Univ Calif, AB, 45; Harvard Univ, AM, 47, PhD(physics), 50. *Prof Exp:* Jr physicist, 42-45, lectr, 58-70, physicist, Lawrence Radiation Lab, Univ Calif, Berkeley, 50-87, assoc dir physics, comput sci & math, 73-81, dep assoc dir, 82-87, EMER ASSOC DIR, LAWRENCE BERKELEY LAB, UNIV CALIF, 87- *Concurrent Pos:* Orgn Europ Econ Coop sr vis fel, 60; Guggenheim fel, 61; consult, Arms Control & Disarmament Agency, 63; NATO sr fel sci, 71; vis scientist, Europ Orgn Nuclear Res, 71-72; mem sci policy comt, Stanford Linear Accelerator Ctr, 70-74. *Mem:* Am Phys Soc; Italian Phys Soc. *Res:* Experimental high energy particle physics. *Mailing Add:* One Greenwood Common Berkeley CA 94708

BIRGE, WESLEY JOE, b Pomeroy, Wash, Mar 11, 29; m 69. ZOOLOGY, AQUATIC TOXICOLOGY. *Educ:* Eastern Wash State Col, BA, 51; Ore State Col, MS, 53, PhD(zool), 55. *Prof Exp:* Asst zool, Ore State Col, 52-53; instr, Univ Ill, 55-58, asst prof, 58-62; assoc prof, Univ Minn, Morris, 62-68; PROF BIOL SCI, UNIV KY, LEXINGTON, 68-, PROF, GRAD CTR TOXICOL, 77-, ACTG DIR, SCH BIOL SCI, 80- *Mem:* Soc Environ Toxicol & Chem; Am Soc Zoologists; Soc Develop Biol; Am Soc Cell Biol; Int Soc Develop Biologists. *Res:* Environmental toxicology; reproductive and developmental biology; neuroembryology; electron microscopy, histology. *Mailing Add:* Sch Biol Sci Univ Ky Lexington KY 40506

BIRGENEAU, ROBERT JOSEPH, b Toronto, Ont, Mar 25, 42; m 64; c 4. CONDENSED MATTER PHYSICS. *Educ:* Univ Toronto, BSc, 63; Yale Univ, PhD(physics), 66. *Prof Exp:* Instr, Dept Eng & Appl Sci, Yale Univ, 66-67; fel, Oxford Univ, 67-68; mem tech staff, Bell Lab Inc, 68-74, head res dept, 75; prof physics, 75-82, CECIL & IDA GREEN PROF PHYSICS, MASS INST TECHNOL, 82-, HEAD DEPT, 88- *Concurrent Pos:* Vis scientist, Riso Nat Lab, Roskilde, Denmark, 71 & 79; mem solid state sci adv panel, Nat Res Coun, 74-80; guest sr scientist, Brookhaven Nat Lab, 68-; mem external rev comt, Solid State & Mat Sci Div, Argonne Nat Lab, 78-80; co-chmn, Gordon Conf Quantum Solids & Fluids, 79; chmn policy & adv bd, Cornell High Energy Synchrotron Source, 80-84; mem rev panel neutron scattering, US Dept Energy, 80; assoc ed, Condensed Matter Physics, Phys Rev Lett, 80-; consult, Los Alamos Nat Lab, 82-, Sandia Lab, 85-; assoc dir, Res Lab Electronics, Mass Inst Technol, 83-86; mem, Major Mat Facil Comt, Nat Acad Sci, 84, Prog Comt, High Flux Beam Reactor, Brookhaven, 84, Dept Energy - NSF Synchrotron Adv Comt, 85, NSF Mat Sci Adv Comt, 89-90 & bd trustees, Assoc Univs, Inc, 90- *Honors & Awards:* Morris Loeb lectr, Harvard Univ, 86; Oliver E Buckley Prize for Condensed Matter Physics, Am Phys Soc, 87; Bertram Eugene Warren Award, Am Chem Asn, 88; Mat Sci Outstanding Accomplishment Award, Dept Energy, 88; 48th Richtmyer Mem Lectr, 89. *Mem:* Fel Am Phys Soc; fel AAAS; fel Am Acad Arts & Sci. *Res:* Neutron and x-ray scattering spectroscopy of condensed matter, especially near magnetic, liquid crystal, and melting phase transitions; transport and scattering studies of high-technetium superconductors. *Mailing Add:* Dept Physics 6-113 Mass Inst Technol 77 Massachusetts Ave Cambridge MA 02139-4307

BIRINGER, PAUL P(ETER), b Marosvasarhely, Hungary, Oct 1, 24; Can citizen; m 52; c 2. ELECTRICAL ENGINEERING. *Educ:* Univ Budapest, Dipl, 47; Royal Inst Technol, Sweden, MSc, 51; Univ Toronto, PhD(elec eng), 56. *Prof Exp:* Asst elec eng, Royal Inst Technol, Sweden, 48-52; spec lectr, 52-56, res assoc, 56-57, from asst prof to assoc prof, 57-65, PROF ELEC ENG, UNIV TORONTO, 65- *Concurrent Pos:* Consult, several Can, US & Europ Co, 53; grants, Nat Res Coun Can & Defence Res Bd Can, 60-; mem, Int Elec Comn, 63-; sr res fel, Nat Res Coun Can, 65-; pres, Elec Eng Assocs, Ltd, Res Consults, 68-72; chmn, Can Nat Comt TC27 & Magnetic Soc; fel, Canadian Acad Eng; sr res fel, Japanese Soc Promo Sci; Erskin fel, Univ, Canterbury, New Zealand. *Honors & Awards:* Pleyel Res Award; Inst Elec & Electronics Engrs Award; Centennial Medal, Inst Elec & Electronic Engrs; Achievement Award, Magnetic Soc Inst Elec & Electronic Engrs. *Mem:* Asn Prof Engrs; Int Elec Comm; fel Inst Elec & Electronic Engrs. *Res:* Static frequency multipliers; speed control of motors; induction heating; electric furnaces; static inverters and regulated power supplies. *Mailing Add:* Dept Elec Eng Univ Toronto Toronto ON M5S 1A4 Can

BIRITZ, HELMUT, b Vienna, Austria, Oct 4, 40; m 66. THEORETICAL PHYSICS. *Educ:* Univ Vienna, PhD(theoret physics), 62. *Prof Exp:* Res assoc theoret physics, Univ Vienna, 62-63 & Max Planck Inst Physics & Astrophys, 63-67; asst prof, 68-74, ASSOC PROF THEORET PHYSICS, GA INST TECHNOL, 74- *Mem:* Am Phys Soc. *Res:* Theoretical high energy physics. *Mailing Add:* Sch Physics Ga Inst Technol Atlanta GA 30332

BIRIUK, GEORGE, b Sdolbunov, Russia, Apr 1, 28. MATHEMATICS. *Educ:* Ind Univ, MA, 57; Univ Calif, Los Angeles, PhD(math), 63. *Prof Exp:* From asst prof to assoc prof, 62-70, PROF MATH, CALIF STATE UNIV, NORTHRIDGE, 70- *Concurrent Pos:* Assoc ed, Math Rev, 67-68. *Mem:* Am Math Soc. *Res:* Functional analysis, especially self adjoint and normal operators in Hilbert spaces; applications to differential operators. *Mailing Add:* Dept Math Calif State Univ Northridge CA 91330

BIRK, JAMES PETER, b Cold Spring, Minn, Aug 21, 41; m 74; c 1. CHEMICAL EDUCATION, ARTIFICIAL INTELLIGENCE. *Educ:* St John's Univ, Minn, BA, 63; Iowa State Univ, PhD(phys chem), 67. *Prof Exp:* Res assoc inorg chem, Univ Chicago, 67-68; asst prof, Univ Pa, 68-73; assoc prof, 73-79, PROF CHEM, ARIZ STATE UNIV, 79- *Concurrent Pos:* Asst chmn Chem Dept, Ariz State Univ, 84-86; vis prof, Eastern Mich Univ, 87-88; ed, "Filtrates & Residues", J Chem Educ, 88-89; bd dir, Odyssey Sch Gifted & Talented, 85-87; adv bd, Ctr Acad Precocity; SERAPHIM fel, 87-88. *Honors & Awards:* Nat Calalyst Award, 90. *Mem:* Am Chem Soc; Soc Col Sci Teachers; Nat Sci Teachers Asn; fel Am Inst Chemists; Asn Computers in Sci & Math Teaching; Sigma Xi. *Res:* Kinetics and mechanisms of reactions of transition metal complexes in solution; oxidation-reduction reactions; substitution reactions; linkage isomerism; computers in chemical education; artificial intelligence in chemical education. *Mailing Add:* Dept Chem Ariz State Univ Tempe AZ 85287

BIRKBY, WALTER H, b Gordon, Nebr, Feb 28, 31; m 55; c 2. FORENSIC ANTHROPOLOGY. *Educ:* Univ Kans, BA, 60, MA, 63; Univ Ariz, PhD(anthrop), 73. *Prof Exp:* Physical anthropologist, 68-85, CURATOR PHYSICAL ANTHROP, ARIZ STATE MUSEUM, 85-; DIR, FORENSIC ANTHROPOLOGY GRAD PROG, UNIV ARIZ, 83-, ADJ RES PROF ANTHROPOLOGY, 89- *Concurrent Pos:* Consult, Inst Forensic Sci 71-76; chmn, anthrop sect, Am Acad Forensic Sci, 76-77; consult, Off Med Examr, Pima County, Ariz, 81-, lectr, Univ Ariz, 81-89; consult, USAF Hosp, Davis-Monthan, AFB, Tuscon, Ariz, 84-; adv, forensic anthropology, Chief Dept Army, Armed Serv Graves Regist Off, Wash, DC, 86-; pres, Am Bd Forensic Anthropology. *Honors & Awards:* T Dale Stewart Award, Am Acad Forensic Sci, 90. *Mem:* Fel Am Acad Forensic Sci; Am Asn Phys Anthrop; Int Asn Human Biologist. *Res:* Analysis of unidentified recent human remains; establishment of positive identity for law enforcement agencies; medical examiners, criminal and civil attorneys. *Mailing Add:* Dept Anthropology Univ Ariz Tucson AZ 87521

BIRKE, RONALD LEWIS, b St Louis, Mo, Jan 4, 39; m 62, 75; c 4. ELECTROCHEMISTRY, ANALYTICAL CHEMISTRY. *Educ:* Univ NC, BS, 61; Mass Inst Technol, PhD(anal chem), 65. *Prof Exp:* Res asst chem, Mass Inst Technol, 62-63, 65; vis fel, Free Univ Brussels, 65-66; instr, Harvard Univ, 66-69; from asst prof to assoc prof, Univ South Fla, 69-74; assoc prof, 74-80, PROF CHEM, CITY COL NEW YORK, 80- *Concurrent Pos:* William F Milton Fund grant, Harvard Univ, 66-67; consult, US Army Electronic Components Commmand, NJ, 68 & US Army Harry Diamond Labs, DC, 69; NIH grant, 70-73 & 75-82, NSF grant, 79-82, 83-85, 87-90. *Mem:* AAAS; Am Chem Soc; Electrochem Soc; NY Acad Sci. *Res:* Thermodynamics and kinetics of electrochemical reactions in solution and interfaces; principles of electrochemical measurements; electroanalytical chemical analysis; biological redox processes and raman spectroscopy at electrodes. *Mailing Add:* Dept Chem City Col New York New York NY 10031

BIRKEBAK, RICHARD C(LARENCE), b St Paul, Minn, Mar 14, 34; m 53; c 4. MECHANICAL ENGINEERING. *Educ:* Univ Minn, BS, 55, MSME, 56, PhD(mech eng), 62. *Prof Exp:* Instr mech eng, Univ Minn, 56-62, US Pub Health Serv fel life sci, Mus Natural Hist, 62-64; asst prof mech eng, Ga Inst Technol, 64-66; assoc prof mech eng, Univ Ky, 66-70, prof, 70-84; PROF SCI & TECHNOL, UNIV NATIONS, 85- *Concurrent Pos:* Res prof, Univ Ky, 78-79; prin investr, USAF Aircraft Icing, 82-86. *Mem:* Am Soc Mech Engrs. *Res:* Heat transfer in icing conditions; thermal physical properties of biological materials; heat transfer in biological systems; thermal radiation property measurements and instrumentation; thermal physical property measurements at low temperatures and pressures. *Mailing Add:* Box 922 Winfield BC V0H 2C0 Can

BIRKEDAL HANSEN, HENNING, COLLAGEN BREAKDOWN. *Educ:* Royal Dent Col, Denmark, PhD(biochem), 77. *Prof Exp:* PROF DENT, SCH DENT, UNIV ALA, 83- *Mailing Add:* Dept Oral Biol Univ Ala Sch Dent BHM University Sta Birmingham AL 35294

BIRKELAND, CHARLES JOHN, b Warwick, NDak, Apr 16, 16; m 41; c 3. HORTICULTURE. *Educ:* Mich State Col, BS, 39; Kans State Col, MS, 41; Univ Ill, PhD(hort), 47. *Prof Exp:* Res asst, Kans State Col, 39-41, asst horticulturist, 41-46; from asst to asst prof, 46-49, actg head dept, 49-50, prof hort & head dept, Univ Ill, Urbana-Champaign, 50-77; RETIRED. *Mem:* AAAS; fel Am Soc Hort Sci; fel Am Pomol Soc; Am Genetic Asn; Bot Soc Am; Sigma Xi. *Res:* Anatomy and photosynthesis of fruit-tree leaves; internal structure of apple leaves of varieties, species and hybrids, with special reference to growth and fruitfulness; fruit breeding. *Mailing Add:* 2111 Zuppke Dr Urbana IL 61801

BIRKELAND, PETER WESSEL, b Seattle, Wash, Sept 19, 34; m 59; c 2. GEOLOGY, SOIL SCIENCE. *Educ:* Univ Wash, BS, 58; Stanford Univ, PhD(geol), 62. *Prof Exp:* Asst prof soil morphol, Univ Calif, Berkeley, 62-67; assoc prof, 67-72, PROF GEOL SCI, UNIV COLO, BOULDER, 72- *Honors & Awards:* Kirk Bryan Award, Geol Soc Am. *Mem:* Geol Soc Am; Soil Sci Soc Am; Am Quaternary Asn; Int Union Quaternary Res. *Res:* Geomorphology; Pleistocene geology; soil morphology and genesis; soil stratigraphy and mineralogy; soil development rates in alpine and arctic environments. *Mailing Add:* Dept Geol Sci Univ Colo Boulder CO 80309-0250

BIRKEMEIER, WILLIAM P, b Evanston, Ill, Nov 10, 27; m 51; c 3. ELECTRICAL ENGINEERING. *Educ:* Northwestern Univ, BS, 51; Univ Ark, MS, 55; Purdue Univ, PhD(elec eng), 59. *Prof Exp:* Elec engr, Collins Radio Co, Iowa, 51-54; assoc prof elec eng, Univ Ark, 59-60; assoc prof, 60-65, PROF ELEC ENG, UNIV WIS-MADISON, 65-, CHMN DEPT, 80- *Concurrent Pos:* Consult, Emerson Elec Co, Mo, 60-61, Collins Radio Co, Iowa, 63-64 & Harris Corp, 77-78. *Mem:* Sr mem Inst Elec & Electronics Engrs; Int Union Radio Scientists. *Res:* Radio probing of the atmosphere; radar technology; biomedical engineering; coherent optics. *Mailing Add:* Dept Elec & Comput Eng Univ Wis 1415 Johnson Dr Madison WI 53706

BIRKEN, STEVEN, b Brooklyn, NY, Dec 17, 45; m 72; c 4. ENDOCRINOLOGY, PROTEIN STRUCTURE-FUNCTION. *Educ:* NY Univ, BA, 67; Hofstra Univ, MA, 69; St Johns Univ, PhD(biol), 72. *Prof Exp:* Res assoc protein chem, 72-84, res scientist, 85-88, DIR, PROTEIN CORE FACIL, COLUMBIA UNIV, 86-, SR RES SCIENTIST, 88- *Concurrent Pos:* NSF fel, 70-72; NIH fel, 73-74. *Mem:* Endocrine Soc; Am Chem Soc; AAAS; Am Soc Biol Chemists. *Res:* Protein structural and immunochemical studies of glycoprotein hormones, chiefly human choriogonadotropin. *Mailing Add:* Dept Med Columbia Univ 630 W 168th St New York NY 10032

BIRKENHAUER, ROBERT JOSEPH, b Toledo, Ohio, Feb 7, 16; m 41; c 2. TEXTILE CHEMISTRY. *Educ:* Univ Detroit, BS, 37; Univ Notre Dame, MS, 39, PhD(phys chem), 41. *Prof Exp:* Chemist, Gulf Refining Co, Tex, 37-38; asst, Univ Notre Dame, 38-41; res chemist, process develop viscose rayon, 41-44, prod develop, 44-46, res group leader equip develop, 46, res supvr, 46-47, plant res surv, Textile Fibers Dept, 47-53, process supt rayon, 53-58, dacron, 58-60, mgr tech serv, 60-62, asst to prod mgr dacron, E I du Pont de Nemours & Co, 62-78; RETIRED. *Res:* Process and development of viscose rayon and of dacron and polyester fibers. *Mailing Add:* 2518 Deepwood Dr Foulk Woods Wilmington DE 19810

BIRKENHOLZ, DALE EUGENE, BIOLOGY. *Educ:* Iowa State Univ, BS, 56; Southern Ill Univ, MA, 58; Univ Fla, PhD(biol), 62. *Prof Exp:* Instr biol, Univ Fla, 61-62; assoc prof biol sci, 62-70, PROF ECOL, ILL STATE UNIV, 70- *Mem:* AAAS; Am Soc Mammalogists; Ecol Soc Am; Am Soc Zoologists; Am Ornith Union; Sigma Xi. *Res:* Ecology and behavior of small mammals; ornithology. *Mailing Add:* Dept Biol Sci 1 Ill State Univ Normal IL 61761

BIRKES, DAVID SPENCER, b Portland, Ore, Dec 1, 42. STATISTICS. *Educ:* Stanford Univ, BS, 64; Univ Chicago, MS, 66; Univ Wash, PhD(math), 69; Ore State Univ, MS, 72. *Prof Exp:* Res staff mem math, Inst Defense Anal, 70-71; trainee, 72-74, asst prof, 74-82, ASSOC PROF STATIST, ORE STATE UNIV, 82- *Mem:* Am Statist Asn; Inst Math Statist. *Res:* Linear models. *Mailing Add:* Dept Statist Ore State Univ Corvallis OR 97331

BIRKETT, JAMES DAVIS, b Norwalk, Conn, Sept 30, 36; m 60; c 3. PHYSICAL CHEMISTRY. *Educ:* Bowdoin Col, AB, 58; Yale Univ, MS, 60, PhD(chem), 63. *Prof Exp:* Chemist res & develop, Arthur D Little, Inc, 62-88; PROPRIETOR & CONSULT, W NECK STRATEGIES, 88- *Mem:* AAAS; Am Chem Soc; Water Supply Improv Asn; Int Desalination Asn; Am Water Works Asn; NAm Membrane Soc; NY Acad Sci. *Res:* Membrane permeation systems; desalination systems; water treatment technology; strategic management of technology. *Mailing Add:* PO Box 193 Nobleboro ME 04555-0193

BIRKHAHN, RONALD H, b 39; m 69; c 2. SYNTHETIC ENERGY NUTRIENTS, FAT METABOLISM. *Educ:* Purdue Univ, PhD(phys chem), 67. *Prof Exp:* PROF SURG & BIOCHEM, MED COL OHIO, 76- *Mem:* Am Inst Nutrit; Am Soc Clin Nutrit; Am Col Nutrit; Nutrit Soc Gt Brit; Am Physiol Soc; AAAS; Am Soc Parenteral & Enternal Nutrit. *Res:* Improved parenteral nutritional support of the critically ill; ketosis and its influence on wound healing in surgical patients; nutritional interaction with tumor growth. *Mailing Add:* Med Col Ohio PO Box 10008 Toledo OH 43699

BIRKHEAD, PAUL KENNETH, geology, invertebrate paleontology; deceased, see previous edition for last biography

BIRKHOFF, GARRETT, b Princeton, NJ, Jan 10, 11; m 38; c 3. MATHEMATICS. *Educ:* Harvard Univ, AB, 32. *Hon Degrees:* Dr, Nat Univ Mex, 51, Univ Lille, 59, Case Inst Technol, 64, Tech Univ, Munich, 86. *Prof Exp:* From instr to prof math, Harvard Univ, 36-69, Putnam prof pure & appl math, 69-81; CONSULT, 51- *Concurrent Pos:* Guggenheim fel, 48; consult, Los Alamos Sci Lab, 51-86, Gen Motors Corp, 59-79, Argonne Nat Lab, 63-79, Brookhaven Nat Lab, 64-86 & Rand Corp, 65-75; chmn, Conf Bd Math Sci, 69-70; Fairchild distinguished scholar, 81. *Mem:* Nat Acad Sci; Soc Indust & Appl Math (pres, 67-68); Math Asn Am (vpres, 70-71); Am Math Soc (vpres, 58-59); Am Acad Arts & Sci (vpres, 66-67); Am Philos Soc. *Res:* Modern algebra; fluid mechanics; scientific computing; reactor theory; differential equations, history of math. *Mailing Add:* 45 Fayerweather St Cambridge MA 02138

BIRKHOFF, ROBERT D, b Chicago, Ill, Jan 29, 25; m 45. EXPERIMENTAL SOLID STATE PHYSICS. *Educ:* Mass Inst Technol, BS, 45; Northwestern Univ, PhD(physics), 49. *Prof Exp:* Consult, Health Physics Div, Oak Ridge Nat Lab, 50-55, physicist, 55-67, sect chief, Radiation Physics Sect, 67-74, mem res staff, Health & Safety Res Div, 74-81; prof physics, Univ Tenn, 67-81; RETIRED. *Mem:* Fel Am Phys Soc; Health Physics Soc. *Res:* Beta ray spectroscopy; interaction of radiation with matter; health physics; electronic and optical properties of liquids; electronic structure of submicron objects. *Mailing Add:* PO Box 1255 Linville NC 28646

BIRKIMER, DONALD LEO, b New Lexington, Ohio, Sept 6, 41; m 62; c 3. STRUCTURES, APPLIED MECHANICS. *Educ:* Ohio Univ, BSCE, 63; Univ Cincinnati, MS, 65, PhD(struct, appl mech), 68; Harvard Univ, PMD, 73. *Prof Exp:* Civil engr, Wright Patterson AFB, 63-64; res civil engr, US Army Corps Engrs, 64-68; res struct engr, Battelle Mem Inst, 68-69; actg chief res & develop, US Army Construct Eng Res Lab, 69-71; asst dir, Naval Surface Weapons Ctr, 71-75; tech dir res & develop, Coast Guard Res & Develop Ctr, Groton, 75-81; tech dir, Naval Civil Eng Lab, Port Hueneme, Calif, 81-85; RETIRED. *Concurrent Pos:* Tech consult, Battelle Mem Inst, Nat Standard Co & Technovation Mgt, Inc, 69-71. *Honors & Awards:* Wason Medal, Am Concrete Inst, 73; Coast Guard Meritorius Unit Commendation Award, 76. *Mem:* AAAS; Am Mgt Asn; Inst Mgt Sci; Am Soc Testing & Mat; Nat Soc Prof Engrs. *Res:* Oil identification; vessel traffic services; aid to navigation; domestic and polar ice technology; marine fire safety; search and rescue. *Mailing Add:* 1291 Seybolt Ave Camarillo CA 93010

BIRKLE, A(DOLPH) JOHN, b Chicago, Ill, Aug 26, 30; m 57; c 4. PHYSICAL METALLURGY. *Educ:* Univ Ill, BS, 57; Univ Wis, MS, 58. *Prof Exp:* Sr res metallurgist, Appl Res Lab, US Steel Corp, 58-65; res supvr new prod develop & mech metall, Res Dept, Youngstown Sheet & Tube Co, 65-68; mgr new prod res & develop, CF&I Steel Corp, 68-71; SECT HEAD, ELEC PLANT PROJ DEPT, CONSUMER POWER CO, 71- *Concurrent Pos:* Mem, Bd Nuclear Codes & Standards. *Honors & Awards:* Howe Gold Metal Award, Am Soc Metals, 67. *Mem:* Am Soc Metals; Am Welding Soc; Am Soc Mech Engrs. *Res:* Alloy development; high-strength and high temperature steels; enameling steels; fracture; fractography; electron metallography; sheet steels; welding research on steels, pipe and tube; research and development; product development; mechanical metallurgy; nuclear codes and standards. *Mailing Add:* 10771 Peninsula Dr Stanwood MI 49346

BIRKS, JOHN WILLIAM, b Vinita, Okla, Dec 10, 46; m 72, 90; c 3. ATMOSPHERIC CHEMISTRY, PHOTOCHEMISTRY. *Educ:* Univ Ark, BS, 68; Univ Calif, Berkeley, MS, 70, PhD(chem), 74. *Prof Exp:* Asst prof chem, Univ Ill, Urbana-Champaign, 74-77; assoc prof, 77-84, PROF CHEM, UNIV COLO, BOULDER, 84- *Concurrent Pos:* Fel, Coop Inst Res Environ Sci, 77-85, actg dir, 85-86; Alfred P Sloan Found fel, 79-81; vis scientist, Max Planck Inst, Mainz, Germany, 81-82; Guggenheim fel, 86. *Honors & Awards:* Leo Szilard Award, Am Phys Soc, 85. *Mem:* Am Chem Soc; Sigma Xi. *Res:* Environmental effects of nuclear war, especially nuclear winter; gas phase kinetics of reactive species in the atmosphere; development of sensitive and selective detectors for chemical analysis based chemiluminescence and photochemical reactions; development of methods whole column detection in chromatography. *Mailing Add:* Dept Chem & CIRES Univ Colo Campus Box 216 Boulder CO 80309-0216

BIRKS, NEIL, b Sheffield, Eng, Oct 16, 35; m 58; c 2. METALLURGY, PHYSICAL CHEMISTRY. *Educ:* Sheffield Univ, BMet, 57, PhD(metall), 60. *Prof Exp:* NATO res fel phys chem, Max Planck Inst for Phys Chem, Gottingen, 60-62, res fel, United Steel Co, 62-64; sr lectr metall, Sheffield Univ, 64-68; PROF MAT SCI & ENG, UNIV PITTSBURGH, 78- *Mem:* Am Soc Metals; fel Inst Metallurgists (London); Metals Soc (London); Sheffield Metall & Eng Asn; Am Inst Metall Eng. *Res:* Iron and steel making; deoxidation, desulfurization, high temperature oxidation of metals; hot corrosion, scaling and decarburization; erosion-corrosion of metals. *Mailing Add:* Dept Mat Sci & Eng Univ Pittsburg Pittsburgh PA 15261

BIRKY, CARL WILLIAM, JR, b Urbana, Ill, June 5, 37; m 60, 84; c 1. EVOLUTION. *Educ:* Ind Univ, BA, 59, PhD(embryol), 63. *Prof Exp:* From instr to asst prof zool, Univ Calif, Berkeley, 63-70; assoc prof, 70-76, PROF GENETICS, OHIO STATE UNIV, 76- *Concurrent Pos:* NIH fel, 69, 87, 88; Hargitt fel cell biol, Duke Univ, 77-78; mem, genetics study sect, NIH, 84-87; regional ed, Current Genetics, 85-; coun mem, Am Genetic Asn, 88-91. *Mem:* Am Inst Biol Sci; Am Soc Cell Biol; Soc Study Evolution; Int Soc Develop Biol; Genetics Soc Am; Am Genetic Asn (pres-elect, 91). *Res:* Genetics and evolution, especially of mitochondria and chloroplasts. *Mailing Add:* Dept Molecular Genetics Ohio State Univ Columbus OH 43210

BIRLE, JOHN DAVID, b Columbus, Ohio, Sept 11, 39. MATERIALS SCIENCE. *Educ:* Ohio State Univ, BChE, 62, MS, 63, PhD(mineral), 67. *Prof Exp:* Develop engr, 67-69, prog engr, 69-72, MGR, SPECIALTY MAT DEPT, GEN ELEC CO, 72- *Res:* High pressure research; x-ray crystal structure analysis; acid mine drainage abatement; fossilized bacteria in sulfide minerals; crystal growth; coal and crude oil analysis; computerized automation. *Mailing Add:* 6325 Huntly Rd PO Box 568 Worthington OH 43085

BIRMAN, JOAN SYLVIA, b New York, NY, May 30, 27; m 50; c 3. MATHEMATICS. *Educ:* Barnard Col, Columbia Univ, BA, 48, Columbia Univ, MA, 50; NY Univ, PhD(math), 68. *Prof Exp:* Asst prof math, Stevens Inst Technol, 68-73; PROF MATH, COLUMBIA UNIV, 73- *Concurrent Pos:* Vis asst prof, Princeton Univ, 71. *Mem:* Am Math Soc. *Res:* Topology; group theory; Riemann surfaces. *Mailing Add:* Columbia Univ New York NY 10027

BIRMAN, JOSEPH HAROLD, b West Hartford, Conn, June 2, 24; m 45; c 2. GEOLOGY. *Educ:* Brown Univ, AB, 48; Calif Inst Technol, MSc, 50; Univ Calif, Los Angeles, PhD, 57. *Prof Exp:* Chmn dept, 50-70, PROF GEOL, OCCIDENTAL COL, 50- *Concurrent Pos:* Pres, Geothermal Surveys Inc, 61-; NSF grant glacial geol studies, Turkey, 63. *Mem:* Geol Soc Am; Am Inst Mining, Metall & Petrol Engrs; Sigma Xi. *Res:* Minerals exploration and development; glacial geology; shallow earth temperatures in ground water exploration. *Mailing Add:* 1617 Silverwood Dr Los Angeles CA 90041

BIRMAN, JOSEPH LEON, b New York, NY, May 21, 27; m 50; c 3. THEORETICAL PHYSICS. *Educ:* City Col New York, BS, 47; Columbia Univ, MA, 50, PhD(theoret chem), 52. *Hon Degrees:* Dr es Sci, Univ Rennes, France, 74. *Prof Exp:* Asst, Columbia Univ, 48-52; from physicist to head luminescence sect, Gen Tel & Electronics Res Labs, NY, 52-62; from assoc prof to prof physics, NY Univ, 62-74; HENRY SEMAT PROF PHYSICS, CITY COL NEW YORK, 74- *Concurrent Pos:* Mary Amanda Wood vis lectr, Univ Pa, 60; vis prof, Univ Paris, 69-70. *Mem:* Fel Am Phys Soc. *Res:* Solid state and many-body quantum theory; group theory. *Mailing Add:* Dept Physics-City Col CUNY 138 St Convent Ave New York NY 10031

BIRMINGHAM, BRENDAN CHARLES, b London, Eng, Nov 30, 45; m 69; c 3. ENVIRONMENTAL TOXICOLOGY. *Educ:* Univ Col, Ireland, BSc, 67, HDipEd, 68; McGill Univ, MSc, 71; Univ Windsor, PhD(biol), 80. *Prof Exp:* Fel, Biol Dept, York Univ, 75-79, res assoc, 80-82; pesticide res specialist, 82-83, STANDARDS COORDR, ONT MINISTRY ENVIRON, 83- *Mem:* Can Soc Plant Physiologists. *Res:* Environmental physiology and toxicology measuring behavior and impact of pesticides in the aquatic ecosystem; plant productivity: measured photosynthesis and photorespiration of freshwater algae using sensitive assay of dissolved inorganic carbon; scientific risk analysis of hazardous contaminants. *Mailing Add:* Hazardous Contaminants Coord Br Ministry Environ 135 St Clair Ave W Toronto ON M4V 1P5 Can

BIRMINGHAM, MARION KRANTZ, b Munich, Ger, Nov 2, 17; Can citizen; m 42; c 2. BIOCHEMISTRY, ENDOCRINOLOGY. *Educ:* Bennington Col, BA, 41; McGill Univ, MSc, 46, PhD(exp med), 49. *Prof Exp:* Res biochemist, Geront Unit, 49-54, DIR LAB STEROID BIOCHEM, ALLAN MEM INST PSYCHIAT, 54-; PROF PSYCHIAT, McGILL UNIV, 72- *Concurrent Pos:* Demonstr, McGill Univ, 53-61, lectr, 54-61, hon lectr biochem, 59, from asst prof to assoc prof psychiat, 61-72; med res assoc, Med Res Coun Can. *Honors & Awards:* Ann Eduardo Brown Menendez Lectr. *Mem:* Am Soc Biol Chemists; Endocrine Soc; Can Physiol Soc; Can Biochem Soc. *Res:* Adrenal steroid biogenesis and its control. *Mailing Add:* Allan Mem Inst Psychiat 1033 Pine Ave W Montreal PQ H3A 1A1 Can

BIRMINGHAM, THOMAS JOSEPH, b Milford, Mass, Sept 28, 38; m 65; c 3. PLASMA PHYSICS, SPACE PHYSICS. *Educ:* Boston Col, BS, 60; Princeton Univ, MA, 62, PhD(physics), 65. *Prof Exp:* Res assoc, Nat Acad Sci, 65-66, STAFF SCIENTIST SPACE PLASMA PHYSICS, GODDARD SPACE FLIGHT CTR, 66- *Concurrent Pos:* Ed, J Geophys Res & Space Physics, 86- *Mem:* Fel Am Phys Soc; Am Geophys Union; Sigma Xi. *Res:* Theoretical plasma physics. *Mailing Add:* Code 695 Goddard Space Flight Ctr Greenbelt MD 20771

BIRNBAUM, DAVID, b New York, NY, Mar 30, 40; m 65; c 3. PHYSICS, COMPUTER SCIENCE. *Educ:* Cornell Univ, BEngPhys, 61; Univ Rochester, MA, 63, PhD(physics), 67, MBA, 81. *Prof Exp:* Res assoc physics, Univ Rochester, 67; res physicist, Carnegie-Mellon Univ, 67-69, asst prof physics, 69-71; asst prof physics, Hamline Univ, 71-76; sr res engr, Gen Rwy Signal Co, 76-81; PRIN SCIENTIST, XEROX CORP, 81- *Mem:* Am Phys Soc; Am Asn Physics Teachers; Inst Elec & Electronic Engrs. *Res:* Experimental particle physics; applications of computers to experimental physics; transportation systems research; physics education; astronomy. *Mailing Add:* 800 Phillips Rd Webster NY 14580

BIRNBAUM, EDWARD ROBERT, b Brooklyn, NY, Oct 28, 43; m 68; c 3. INORGANIC CHEMISTRY, BIOINORGANIC CHEMISTRY. *Educ:* Brooklyn Col, BS, 64; Univ Ill, Urbana, MS, 66, PhD(inorg chem), 68. *Prof Exp:* Asst prof, 68-73, assoc prof, 73-78, PROF INORG CHEM, NMEX STATE UNIV, 78- *Concurrent Pos:* Vis prof, Univ Alta, 76-77. *Mem:* AAAS; Am Chem Soc; Am Soc Biol Chem; Biophys Soc. *Res:* Fluorescence and nuclear magnetic resonance studies of inorganic and biological complexes of the lanthanide ions. *Mailing Add:* Dept Chem NMex State Univ Las Cruces NM 88003

BIRNBAUM, ERNEST RODMAN, b Newark, NJ, Oct 4, 33. INORGANIC CHEMISTRY. *Educ:* Univ Calif, Berkeley, BA, 55; Univ Southern Calif, MS, 58; Univ Fla, PhD(chem), 61. *Prof Exp:* Sr res engr, Rocketdyne Div, NAm Aviation, Inc, 61-62; fel chem, Univ Tex, 62-63; from instr to asst prof, 63-71, assoc prof, 71-77, PROF CHEM, ST JOHN'S UNIV, NY, 77- *Mem:* Am Chem Soc; Royal Soc Chem. *Res:* Inorganic reactions in nonaqueous solvents; kinetics and mechanisms of inorganic reactions; Lewis acid-base reactions; hydrido complexes of transition metals. *Mailing Add:* Dept Chem St John's Univ Jamaica NY 11439

BIRNBAUM, GEORGE, b July 16, 19; m; c 2. NONDESTRUCTIVE EVALUATION. *Educ:* Brooklyn Col, 41; George Washington Univ, MS, 49, PhD(physics), 56. *Prof Exp:* Head microwave dielectric measurement group, Nat Bur Standards, Washington, DC, 46-54, head microwave frequency & spectros sect, Boulder Labs, 54-56; sr scientist & head, Quantum Physics Sect, Hughes Res Labs, Calif, 56-62; group leader, Spectros Sci Ctr, NAm Rockwell Corp, 62-75; SR SCIENTIST, NAT INST STANDARDS & TECHNOL, 75- *Concurrent Pos:* Mem USA comn I & chmn, Int Radio Sci Union, 73-76; mem & chmn, Nat Acad Sci adv panel to radio standards lab, Nat Bur Standards, 65-68; vis prof, Univ Paris, 79 & 83-85; distinguished vis lectr, Univ Tex, Austin, 80. *Honors & Awards:* Gold Medal, Dept Com, 89. *Mem:* Fel Am Phys Soc; fel Inst Elec & Electronics Engrs. *Res:* Microwave and far infrared spectroscopy, collision induced absorption and light scattering. *Mailing Add:* Rm B344 Bldg 223 Nat Inst Standards & Technol Gaithersburg MD 20899

BIRNBAUM, GEORGE I, b Cracow, Poland, July 14, 31; Can citizen; m 67; c 2. STRUCTURAL CHEMISTRY. *Educ:* Columbia Univ, BS, 54, MA, 55, PhD(org chem), 61. *Prof Exp:* Res worker x-ray crystallog, Columbia Univ, 59-61, res assoc, 61-64; fel, Nat Res Coun Can, 64-65; NIH spec fel, Glasgow Univ, 65-66; assoc res officer, 66-76, SR RES OFFICER, NAT RES COUN CAN, 76- *Mem:* Am Chem Soc; Am Crystallog Asn; Chem Inst Can. *Res:* X-ray structure analyses of proteins and biologically significant organic molecules. *Mailing Add:* Div Biol Sci Nat Res Coun Montreal Rd Ottawa ON K1A 0R6 Can

BIRNBAUM, H(OWARD) K(ENT), b New York, NY, Oct 18, 32; m 54; c 3. METALLURGY. *Educ:* Columbia Univ, BS, 53, MS, 55; Univ Ill, PhD(metall), 59. *Prof Exp:* From instr to asst prof metall, Univ Chicago, 59-61; assoc prof, 61-63, PROF METALL, UNIV ILL, URBANA-CHAMPAIGN, 63-, DIR, MAT RES LAB. *Mem:* Am Inst Mining, Metall & Petrol Engrs; fel Am Phys Soc. *Res:* Physical metallurgy; physics of solids; dislocation theory; mechanical properties of solids; hydrogen in solids; interstitial diffusion; hydrogen embrittlement. *Mailing Add:* Mat Res Lab Univ Ill 104 S Goodwin Urbana IL 61801

BIRNBAUM, HERMANN, b Gera, Ger, Apr 30, 05; nat; m 44; c 1. ORGANIC CHEMISTRY. *Educ:* Univ Leipzig, PhD(chem), 34. *Prof Exp:* Asst to Dr Berl, Carnegie Inst Technol, 35-36; instr chem, Duquesne Univ, 36-37; chemist res & qual control, Hachmeister, Inc, H J Heinz Co, 37-70; CONSULT, EMULSIFIERS, 70- *Mem:* Am Asn Cereal Chemists; Am Chem Soc; Am Oil Chemists Soc; Inst Food Technologists. *Res:* Oils and fats; food emulsifiers; cereal chemistry and molecular distillation. *Mailing Add:* 5701 Munhall Rd Pittsburgh PA 15217

BIRNBAUM, JOEL S, ARCHITECTURE. *Educ:* Cornell Univ, BS, 60; Yale Univ, MS, 61, PhD(nuclear physics), 65. *Prof Exp:* Dir computer serv, T J Watson Res Lab, IBM Corp, Yorktown Heights, NY, 65-80; vpres & dir, Hewlett-Packard Labs, 80-86, vpres & gen mgr, Info Technol Group, 86-88, VPRES & GEN MGR, INFO ARCHIT GROUP, HEWLETT-PACKARD CO, 88- *Concurrent Pos:* Mem, Res Roundtable, Nat Acad Sci, Nat Acad Eng & Inst Med-Nat Acad Sci; mem eng coun, Cornell Univ, Univ Southern Calif & Beckman Inst Univ Ill. *Mem:* Nat Acad Eng; Inst Elec & Electronics Engrs; Asn Comput Mach. *Res:* Information architecture. *Mailing Add:* Hewlett-Packard Co 19046 Pruneridge Ave Cupertino CA 95014

BIRNBAUM, LEON S, b New York, NY, Apr 10, 16; m 41; c 2. MATERIALS SCIENCE. *Educ:* City Col New York, BS, 36; Am Univ, MA, 69. *Prof Exp:* Chemist indust test lab, Philadelphia Naval Shipyard, US Navy, 38-49, mat engr, Bur Ships, 49-65, head chem, coatings & corrosion eng br, Naval Ship Eng Ctr, 65-73; CONSULT MAT ENGR, 73- *Concurrent Pos:* US Navy proj officer coatings & corrosion control, Info Exchange Progs-Foreign Centered, 65-73; lectr, Cath Univ, 71-73. *Mem:* Hon mem Am Soc Testing & Mat; Am Chem Soc; Fedn Socs Coatings Technol; Nat Asn Corrosion Engrs. *Res:* Protective coatings; chemistry; corrosion and allied technologies. *Mailing Add:* 3214 Arrowhead Circle Fairfax VA 22030

BIRNBAUM, LINDA SILBER, b Passaic, NJ, Dec 21, 46; m 67; c 3. TOXICOLOGY. *Educ:* Univ Rochester, AB, 67; Univ Ill, Urbana, MS, 69, PhD(microbiol), 72. *Prof Exp:* Asst prof microbiol, Univ Ill, Urbana, 72; fel biochem, Univ Mass, 73-74; asst prof biol, Kirkland, 74-75; res assoc biochem, Masonic Med Res Lab, 75-79; sr staff fel, 79-80, RES MICROBIOLOGIST, SYST TOXICOLOGY BR, NAT INST ENVIRON HEALTH SCI, 80- *Concurrent Pos:* adj prof, Sch Pub Health, Univ NC, Chapel Hill, 81- *Mem:* Am Soc Pharmacol Exp Therapeut; AAAS; Soc Toxicol; Geront Soc; Am Aging Asn; Int Soc Study Xenobiotics. *Res:* Biochemistry of drug metabolism and carcinogenesis in liver and extra-hepatic tissues from organism of different ages; mechanisms of toxicity; chemical disposition. *Mailing Add:* 726 Shadylawn Ct Chapel Hill NC 27514

BIRNBAUM, MILTON, b Brooklyn, NY, Nov 27, 20; m 57; c 2. LASERS. *Educ:* Brooklyn Col, AB, 42; Univ Md, MS, 48, PhD(physics), 53. *Prof Exp:* Nuclear physicist, Naval Res Lab, DC, 46-53; head dept physics, Bulova Res & Develop Co, 53-55; assoc prof, Polytech Inst Brooklyn, 55-61; sect head, Aerospace Corp, 61-70, head quantum optics dept, 70-72, sr scientist, 72-86; PROF RES, DEPT ELEC ENG, UNIV SOUTHERN CALIF, 86- *Concurrent Pos:* Consult, Brookhaven Nat Lab, 57-58, Repub Aviation Corp, 60, Loral Electronics Corp, 60-61, Hughes Aircraft Co, 87-89, Northrop Corp, 87-89, Aerospace Corp, 86- *Mem:* Fel Am Phys Soc; fel Optical Soc Am; fel Inst Elec & Electronic Engrs; Int Soc Optical Eng. *Res:* Accelerator design; neutron binding energies; electron paramagnetic resonance; laser physics; optics; semiconductors; solid state lasers. *Mailing Add:* 4904 Elkridge Dr Ranch Palos Verdes CA 90274

BIRNBAUM, SIDNEY, b New York, NY, Sept 22, 28; m 54; c 3. MATHEMATICS. *Educ:* NY Univ BA, 48, MS, 49; Univ Colo, PhD(math), 65. *Prof Exp:* Engr, Martin Co, 53-56, res scientist, 56-65; assoc prof math, Univ SC, 65-70, actg head dept, 67-68; prof math & chmn dept, 70-73, assoc dean, Sch Sci, 73-80, PROF MATH, CALIF STATE POLYTECH UNIV, POMONA, 80- *Concurrent Pos:* Consult, Sandia Corp, NMex, 66- *Mem:* AAAS; Am Math Soc; Math Asn Am; Soc Indust & Appl Math; Sigma Xi. *Res:* Functional analysis; spectral theory; probability; aeroelasticity. *Mailing Add:* Calif State Polytech Univ Pomona CA 91768

BIRNBAUMER, LUTZ, b Vienna, Austria, Feb 6, 39; Arg citizen; m 65. BIOCHEMISTRY, ENDOCRINOLOGY. *Educ:* Univ Buenos Aires, BA, 62, BS, 64, PhD(glycogen metab), 66. *Prof Exp:* Nat Inst Arthritis, Metab & Digestive Dis fel, 68-69; staff fel endocrinol, Nat Inst Arthritis, Metab & Digestive Dis, 69-71; assoc prof physiol, Sch Med, Northwestern Univ, Chicago, 71-75; PROF CELL & REPRODUCTIVE BIOL, BAYLOR COL MED, 75- *Concurrent Pos:* Nat Inst Child Health & Human Develop res grant, Northwestern Univ, 71-, Nat Inst Arthritis, Metab & Digestive Dis & Nat Heart, Lung & Blood Inst grants, 76- *Mem:* Soc Study Reproduction; Endocrine Soc; Am Soc Biol Chemists. *Res:* Cellular, biochemical and molecular bases of hormone and catecholamine action; role of cyclic nucleotides in cellular regulation; plasma membrane biochemistry; reproductive physiology of the ovary. *Mailing Add:* Dept Cell Biol Baylor Col Med Houston TX 77030

BIRNBOIM, HYMAN CHAIM, b Winnipeg, Man, Feb 29, 36; m 59; c 2. CANCER RESEARCH, MOLECULAR BIOLOGY. *Educ:* Univ Man, MD, 60, MSc, 63; FRCP(C), 85. *Prof Exp:* Intern, St Boniface Hosp, Winnipeg, 60-61; asst resident med, 61-62; fel biochem, Nat Cancer Inst Can, 62-63, Med Res Coun, Can, 63-67; assoc res officer, Dept Biol & Health Physics, Chalk River Nuclear Lab, Atomic Energy Can, Ltd, 67-78, sr res officer, Health Sci Div, 78-84; sr lectr, dept biochem, assoc prof, dept med & microbiol-immunol, 84-87, PROF, DEPT MED & MICROBIOL-IMMUNOL UNIV OTTAWA, 87-, PROF, DEPT BIOCHEM, 89-; CAREER SCIENTIST, GEN DIV, OTTAWA REGIONAL CANCER CTR, 84- *Concurrent Pos:* Assoc ed, Can J Biochem, 73-78. *Mem:* Can Biochem Soc; AAAS; Am Asn Cancer Res. *Res:* Activation of oncogenes by DNA strands breaks and oxyradicals. *Mailing Add:* Gen Div 501 Smyth Rd Ottawa Regional Cancer Ctr Ottawa ON K1H 8L6 Can

BIRNEY, DAVID MARTIN, b Washington, DC, May 12, 56; m 80. MOLECULAR STRUCTURE CALCULATIONS. *Educ:* Swarthmore Col, BA, 78; Yale Univ, PhD(chem), 87. *Prof Exp:* Ed, Sadtler Res Labs, 79-81; ASST PROF ORG CHEM, TEX TECH UNIV, 89- *Mem:* Am Chem Soc. *Res:* Ab initio calculations of transition states for organic reactions; massively parallel molecular mechanics calculations; quantitative studies of binding in molecular recognition; multiphoton infrared photochemistry. *Mailing Add:* Dept Chem & Biochem Tex Tech Univ Lubbock TX 79409-1061

BIRNEY, DION SCOTT, JR, b Washington, DC, Apr 23, 26; m 55; c 3. ASTRONOMY. *Educ:* Yale Univ, BS, 50; Georgetown Univ, MA, 56, PhD(astron), 61. *Prof Exp:* Engr, Eng Res Corp, Md, 50-52, Am Instrument Co, 52-54 & Stone Straw Corp, DC, 54-56; asst astron, Georgetown Univ, 56-60; from instr to asst prof astron, Univ Va, 60-68; from assoc prof to prof astron, Wellesley Col, 74-91; RETIRED. *Mem:* Am Astron Soc; Sigma Xi. *Res:* Photometry of Cepheid variables and eclipsing binaries. *Mailing Add:* Whitin Observ Wellesley Col Wellesley MA 02181

BIRNEY, ELMER CLEA, b Satanta, Kans, Mar 26, 40; m 61; c 2. MAMMALOGY, ECOLOGY. *Educ:* Ft Hays State Univ, BS, 63, MS, 65; Univ Kans, PhD(zool), 70. *Prof Exp:* Instr biol, Kearney State Col, 65-66; from asst prof to assoc prof, 70-81, PROF ECOL, UNIV MINN, MINNEAPOLIS, 81-, CUR MAMMAL, JAMES FORD BELL MUS NATURAL HIST, 70- *Concurrent Pos:* From managing ed to ed, J Mammal, 78-84, ed spec publ mammalogy, 84- *Mem:* Ecol Soc Am; Am Soc Mammal (1st vpres, 86-); Soc Study Evolution; Soc Syst Zool; Am Soc Naturalists. *Res:* Evolution, ecology and behavior of mammals. *Mailing Add:* J F Bell Mus Nat Hist Univ Minn Minneapolis MN 55455

BIRNHOLZ, JASON CORDELL, b Newark, NJ, Dec 11, 42; m 71; c 3. ULTRASONIC IMAGING, FETAL PHYSIOLOGY. *Educ:* Union Col, NY, BS, 63; Johns Hopkins, MD, 67. *Prof Exp:* Intern internal med, Bellevue Hosp, 67-68; fel, NIH, 68-70; resident physician radiol, Mass Gen Hosp, 70-73; adv acad fel, biomed eng, Mass Inst Technol, 73-75; asst prof radiol, Stanford Univ, 75-76; assoc prof, Med Sch, Harvard Univ, 76-82; prof radiol, Rush Med Col, 82-88; PVT PRACT, 89- *Concurrent Pos:* Var int vis professorships. *Mem:* Fel Am Col Radiol; assoc fel Am Col Obstetricians & Gynecologists; fel Royal Col Radiol; Sigma Xi; AAAS; AMA; Am Inst Ultrasound Med. *Res:* Applied research in ultrasonic imaging including technical development and clinical applications; human fetal development. *Mailing Add:* 440 Moraine Road Highland Park IL 60035

BIRNIE, RICHARD WILLIAMS, b Boston, Mass, Dec 8, 44; m 73; c 3. MINERALOGY. *Educ:* Dartmouth Col, AB, 68, AM, 71; Harvard Univ, PhD(geol), 75. *Prof Exp:* From asst prof to assoc prof, 75-87, PROF GEOL, DARTMOUTH COL, 87-, DEAN, GRAD STUDIES, 90- *Mem:* Geol Soc Am; Mineral Soc Am; Soc Econ Geologists; Explorers Club; Sigma Xi. *Res:* Infrared radiation thermometry of volcanoes; remote sensing applied to prospecting for ore deposits; sulfide mineralogy and crystal chemistry. *Mailing Add:* Dept Earth Sci Dartmouth Col Hanover NH 03755

BIRNIR, BJÖRN, b Reykjavik, Iceland, Aug 19, 53; m 80; c 1. APPLIED MATHEMATICS, GEOMETRY. *Educ:* Union Col, Schenectady, NY, BS, 76; Courant Inst, NY Univ, MS, 78, PhD(math), 81. *Prof Exp:* Asst prof math, Univ Ariz, Tucson, 81-83; res assoc, Univ Calif, Berkeley, 83-84; res fel math & comput, Univ Iceland, Reykjavik, 84-85; ASSOC PROF MATH, UNIV CALIF, SANTA BARBARA, 85- *Mem:* Am Math Soc; Soc Indust & Appl Math; Icelandic Math Soc; Icelandic Phys Soc; Icelandic Eng Soc. *Res:* Nonlinear partial differential equations, describing a variety of phenomena in physics and engineering; perturbation theory for integrable partial differential equations; planetary motion and string theories in physics; numerical analysis in computer science. *Mailing Add:* Univ Calif Santa Barbara Santa Barbara CA 93106

BIRNSTIEL, CHARLES, b New York, NY, Dec 6, 29. MOVABLE BRIDGE MACHINERY ENGINEERING. *Educ:* NY Univ, BCE, 54, MCE, 57, EngScD, 62. *Prof Exp:* From instr to prof civil eng, NY Univ, 54-73; prof, Poly Inst NY, 73-74; CONSULT ENGR, 74- *Concurrent Pos:* Adj prof, Columbia Univ, NY, 88- *Mem:* Fel Am Soc Civil Engrs; Int Asn Bridge & Struct Engrs; Am Soc Eng Educ; Am Concrete Inst; Struct Stability Res Coun. *Res:* Elastic and inelastic stability of metal structures; cable-suspended structures; mechanisms. *Mailing Add:* 108-18 Queens Blvd Forest Hills NY 11375

BIRO, GEORGE P, b Budapest, Hungary, Jan 14, 38; Can citizen; m 64; c 2. CARDIOVASCULAR PHYSIOLOGY, BLOOD SUBSTITUTES & TRANSFUSION MEDICINE. *Educ:* Queen's Univ, Kingston, Ont, MS, 63; PhD(physiol), 70. *Prof Exp:* Prof, 72-91, ASSOC DEAN, FAC MED, UNIV OTTAWA, 90- *Concurrent Pos:* Vis prof, Univ Calif, San Diego, 81-82. *Mem:*

Can Physiol Soc; Am Physiol Soc; Int Soc Heart Res; Int Soc Oxygen Transport Tissue. *Res:* Blood substitutes including stroma-free hemoglobin solution and perfluorocarbon; oxygen supply to the myocardium; blood rheology; possible ultimate clinical applications. *Mailing Add:* Fac Med Univ Ottawa 451 Smythe Rd Ottawa ON K1H 8M5 Can

BIRON, CHRISTINE A, b Woonsocket, RI, Aug 8, 51. VIRAL IMMUNOLOGY. *Educ:* Univ NC, Chapel Hill, PhD(immunol), 80. *Prof Exp:* asst prof path, Med Sch, Univ Mass, 83-87; asst prof, 87-90, ASSOC PROF MED SCI, DIV BIOL & MED, BROWN UNIV, 90- *Concurrent Pos:* Scholar, Leukemia Soc Am. *Mem:* Am Asn Immunol; AAAS; Sigma Xi. *Res:* Immune response to viral infections; NK cells; T lymphocytes. *Mailing Add:* Div Biol Box G Brown Univ Providence RI 02912

BIRREN, JAMES E(MMETT), b Chicago, Ill; m 42; c 3. GERONTOLOGY, PSYCHOLOGY. *Educ:* Chicago State Univ, BE, 41; Northwestern Univ, MA, 43, PhD(psychol), 47. *Hon Degrees:* PhD, Univ Gothenburg, Sweden, 82, DSc, Northwestern Univ, 86; LLD, St Thomas Univ, Fredericton, Can, 90. *Prof Exp:* Chief sect on aging, NIMH, 51-64; head prog on aging, Nat Inst Child Health & Human Develop, 64-65; DIR GERONT CTR, UNIV SOUTHERN CALIF, 65-, DEAN SCH GERONT, 75- *Concurrent Pos:* Adv, Aging Soc Proj, Carnegie Corp, 83- *Honors & Awards:* Brookdale Award, Brookdale Found, 80. *Mem:* Am Physiol Soc; Am Psychol Asn; Geront Soc (pres, 61). *Res:* Psychophysiology of aging; changes in attention. *Mailing Add:* 647 Toyopa Dr Pacific Palisades CA 90272

BIRRENKOTT, GLEN PETER, JR, b Feb 6, 51; US citizen. POULTRY SCIENCE. *Educ:* Univ Wis-Madison, BS, 73, MS, 75, PhD(poultry sci), 78. *Prof Exp:* ASSOC PROF POULTRY SCI, CLEMSON UNIV, 78- *Mem:* AAAS; Poultry Sci Asn. *Res:* Avian endocrinology; avian reproductive physiology; comparative endocrinology; animal physiology; stress; hyperthermia. *Mailing Add:* Dept Poultry Sci Clemson Univ Clemson SC 29631

BIRSS, FRASER WILLIAM, theoretical chemistry; deceased, see previous edition for last biography

BIRSS, VIOLA INGRID, b Coleman, Alta. ELECTROCHEMISTRY, SURFACE CHEMISTRY. *Educ:* Univ Calgary, BSc Hons 72; Univ Audelord, NS, PhD(chem), 78. *Prof Exp:* Postdoctoral fel, Univ Ottawa, 78-80; res chemist, Continental Group Inc, 80-81; res scientist, Alcon Int Ltd, 82; from asst prof to assoc prof 83-90, PROF, UNIV CALGARY, 91- *Concurrent Pos:* Consult, CEDA Mfg & Sales, Calgary, 84-86, Alcon Int Ltd, Kingston, Ont, 91 & Allied-Signal Inc, South Bend, Ind, 91; vis scientist, Allied-Signal Inc, Morristown, NJ, 89 & Los Alamos Labs, NMex, 90; mem, Phys & Anal Chem Grant Selection Comt, 89-91, chairperson, 91-92. *Honors & Awards:* Lash Miller Award, Electrochem Soc Inc, 86. *Mem:* Electrochem Soc Inc; Can Inst Chem; Nat Asn Corrosion Engrs; Sigma Xi. *Res:* Rates and mechanisms of the adsorption/desorption of organics and the formation/removal of films at electrode surfaces, properties of these films/conductivity, morphology, composition, etc, which range from sub-monolayer to microns in thickness. *Mailing Add:* Dept Chem Univ Calgary Calgary AB T2N 1N4 Can

BIRSTEIN, SEYMOUR J, b Brooklyn, NY, May 1, 27; div; c 1. SURFACE CHEMISTRY, CLOUD PHYSICS. *Educ:* NY Univ, BA, 47; Mont State Col, MS, 49. *Prof Exp:* Res chemist res labs, Air Reduction Co, Inc, 49-50; res chemist, Air Force Cambridge Res Labs, 51-59, proj scientist, 59-68, chief aerosol interaction br, 68-76; PRES, SJB ASSOCS, INC, 77- *Mem:* Am Chem Soc; Am Meteorol Soc; Electron Micros Soc Am; Sigma Xi; fel Am Inst Chemists. *Res:* Application of aerosols in controlling the aerospace environment; basic research and its application to meteorological and weather modification processes. *Mailing Add:* 24 Pippen Rd Marlborough MA 01752

BIRT, ARTHUR ROBERT, b Winnipeg, Man, May 19, 06; m 52; c 3. MEDICINE, DERMATOLOGY. *Educ:* Univ Man, MD, 30; Royal Col Physicians & Surgeons, Can, cert dermat & syphilol, 43, FRCPS(C), 65. *Prof Exp:* From lectr med to assoc prof dermat, Univ Man, 43-76, chmn dept, 64-76, emer prof med, 76-89; RETIRED. *Mem:* Am Acad Dermat; Am Dermat Asn; Can Dermat Asn (secy, 52-57, pres, 62). *Res:* Medical mycology and cause of cutaneous flushing in human mast cell disease; hereditary multiple fibrofolliculomas with trichodiscomas and achrochordons; photodermatitis in North American Indians. *Mailing Add:* 409-3161 Grant Ave Winnipeg MB R3R 3R1 Can

BIRT, DIANE FEICKERT, b Petaluma, Calif, Oct 12, 49. BIOCHEMISTRY, HUMAN NUTRITION. *Educ:* Whittier Col, BA, 72; Purdue Univ, PhD(nutrit), 75. *Prof Exp:* Asst prof human nutrit, Iowa State Univ, 75-76; from asst prof to assoc prof, Eppley Inst Res Cancer, Univ Nebr Med Ctr, 76-88, assoc prof, Dept Pharaceut Sci, Col Pharm, 82-88, assoc prof biochem, Col Med, 85-88, PROF, EPPLEY INST RES CANCER & DEPT PHARMACEUT SCI, COL PHARM & PROF BIOCHEM, COL MED, UNIV NEBR MED CTR, 88- *Concurrent Pos:* Prin investr, numerous res grants, Nat Cancer Inst, 78-94; mem, Chem Path Ad Hoc Study Sect, NIH, 84-87 & Metab Path Study Sect, 87-91, chmn, 89-91. *Mem:* Am Inst Nutrit; Sigma Xi; Am Asn Cancer Res; Am Home Econ Asn. *Res:* Author of numerous technical publications. *Mailing Add:* Eppley Inst Res Cancer Univ Nebr Med Ctr Omaha NE 68105-1065

BIRTCH, ALAN GRANT, b La Porte, Ind, Feb 3, 32; m 54; c 4. SURGERY. *Educ:* Johns Hopkins Univ, BA, 54; Johns Hopkins Med Sch, MD, 58. *Prof Exp:* Resident surg, Peter Bent Brigham Hosp, 58-63, asst, 63-66; instr surg, Harvard Med Sch, 66-67, assoc, 68-70, asst prof, 70-72; PROF SURG & ASST CHMN, SCH MED, SOUTHERN ILL UNIV, 72-, CHMN GEN SURG, 72-, CHIEF SECT TRANSPLANTATION, 72-, COORDR RES AFFAIRS, 79- *Concurrent Pos:* Res fel surg, Harvard Med Sch, 63-64; Arthur-Tracey Cabot fel surg, Peter Bent Brigham Hosp, 64-65, chief resident surg, 64-65; res fel physiol, Harvard Med Sch, 65-66; consult surg, Robert

Breck Brigham Hosp, 68-72; coordr res affairs, Southern Ill Univ, 79- *Mem:* Am Surg Asn; Transplantation Soc; Am Col Surgeons; Cent Surg Asn; Soc Univ Surgeons. *Res:* Clinical and basic transplantation immunology. *Mailing Add:* Southern Ill Univ Sch Med PO Box 3926 Springfield IL 62708

BIRTEL, FRANK T, b New Orleans, La, Apr 4, 32; m 64; c 2. MATHEMATICS, SCIENCE & RELIGION. *Educ:* Loyola Univ, La, BS, 52; Univ Notre Dame, MS, 53, PhD(math), 60. *Prof Exp:* Sr mathematician, US Naval Nuclear Power Sch, 55-57; instr math, Conn Col, 56-57; asst prof, Ohio State Univ, 60-62; from asst prof to assoc prof, Tulane Univ, 62-67, provost & dean, Grad Sch, 78-81, PROF MATH, TULANE UNIV, 67-, UNIV PROF, 82. *Concurrent Pos:* Lectr, res assoc & Off Naval Res fel, Yale Univ, 61-62; vis prof, Cath Univ Nijmegen, 68-69. *Mem:* Math Asn Am; Am Math Soc. *Res:* Uniform algebras; several complex variables, harmonic analysis, topological rings and algebras; higher education; science, ethics and religion. *Mailing Add:* 1229 Cadiz New Orleans LA 70115

BIRX, DEBORAH L, b Baltimore, Md, Apr 4, 56; m; c 2. ALLERGY & IMMUNOLOGY. *Educ:* Houghton Col, BS, 76; Pa State Univ, MS & MD, 80; Am Bd Internal Med, dipl, 83; Am Bd Allergy & Immunol, 85. *Prof Exp:* Intern, Walter Reed Army Med Ctr, 80-81, resident internal med, 81-83, fel allergy & clin immunol, 83-85, asst chief allergy & immunol serv, 85-89, ASST CHIEF, DEPT RETROVIRAL RES, WALTER REED ARMY INST RES, 89-; ASST PROF MED, UNIFORMED SERV UNIV HEALTH SCI, 85-, ASST PROF PEDIAT, 86- *Concurrent Pos:* Teaching fel, Uniformed Serv Univ Health Sci, 80-85; Acad Allergy & Immunol young invest award & travel grant, 85; sr investr cellular immunol, Walter Reed Army Med Ctr, 86-89, consult allergy immunol serv, 89- *Mem:* Am Asn Clin Res; Am Col Physicians; Am Asn Immunologists; Am Acad Allergy & Immunol. *Res:* Cation effect on lymphocyte activation; calcium and protein kinase regulation in polymorphonuclear activation; B cell dysfunction in AIDS patients; Epstein-Barr virus infection and immunoregulation in man; immunology of human imunodeficiency virus; analysis of T and B cell specific responses; author of numerous technical publications. *Mailing Add:* Dept Retroviral Res Walter Reed Army Inst Res 13 Taft Ct Suite 200 Rockville MD 20850

BIS, RICHARD F, b Woonsocket, RI, Mar 9, 35; m 57; c 2. SOLID STATE PHYSICS. *Educ:* Worcester Polytech Inst, BS, 57; Univ Md, MS, 61, PhD(physics), 69. *Prof Exp:* PHYSICIST, US NAVAL ORD LAB, 57- *Concurrent Pos:* Instr, Univ Md, 64- *Res:* Optical and electrical properties of tin tellurium and its alloys with lead; bulk and film samples. *Mailing Add:* 55513 Ridge Rd Airy MD 21771

BISAILLON, ANDRE, b Montreal, Que, Aug 9, 43; m 78; c 2. VETERINARY ANATOMY, MAMMALOGY. *Educ:* Col Bourget, BA, 65; Univ Montreal, DMV, 69, MScV, 73. *Prof Exp:* Instr gross vet anat, Univ Sask, 71-73; from asst prof to assoc prof, 73-83, PROF GROSS VET ANAT, UNIV MONTREAL, 83- *Concurrent Pos:* Secy, Fac Med Vet, Univ Montreal. *Mem:* Am Asn Vet Anatomists; World Asn Vet Anatomists; Can Asn Vet Anatomists. *Res:* Morphology of aquatic mammals, mainly pinnipeds; study of aquatic locomotion of pinnipeds and gross anatomy of the blood vessels of semi-aquatic and aquatic mammals; morphology of the newborn pig. *Mailing Add:* Fac Vet Med Univ Montreal St Hyacinthe PQ J2S 7C6 Can

BISALPUTRA, THANA, b Bangkok, Thailand, Jan 6, 32; m 64. BOTANY. *Educ:* Univ New Eng, Australia, BSc, 59, MSc, 61; Univ Calif, Davis, PhD(bot), 64. *Prof Exp:* Demonstr bot, Univ New Eng, Australia, 59-61; res asst bot, Univ Calif, Davis, 61-64; from asst prof to assoc prof, 64-75, PROF BOT, UNIV BC, 75- *Res:* Light and electron microscopic study of the vascular anatomy of higher plants and the ultrastructure of algae. *Mailing Add:* Dept Bot Univ BC Suite 3529 at 6270 University Blvd Vancouver BC V6T 1Z4

BISBY, MARK A, b Malvern, Eng, Aug 8, 46; Can citizen; m 68; c 3. MEDICAL EDUCATION, SCIENCE POLICY. *Educ:* Oxford Univ, Eng, MA, 69, DPhil, 72. *Prof Exp:* Prof physiol, Univ Calgary, 73-89, asst dean med, 79-84 & 86-89; HEAD PHYSIOL, QUEEN'S UNIV, 89- *Mem:* Can Physiol Soc (secy, 84-87); Soc Neurosci; Int Soc Neurochem; Am Physiol Soc; Can Fedn Biol Socs (pres, 89-90). *Res:* Nerve regenerations; axonal transport; neuroplasticity. *Mailing Add:* Dept Physiol Queen's Univ Kingston ON K7L 3N6 Can

BISCAYE, PIERRE EGINTON, b New York, NY, Nov 24, 35; m 58; c 4. MARINE GEOCHEMISTRY, DEEP-SEA SEDIMENTATION. *Educ:* Wheaton Col, Ill, BS, 57; Yale Univ, PhD(geochem), 64. *Prof Exp:* Geochemist, Jersey Prod Res Co, Okla, 64-65; scientist, Isotopes, Inc, NJ, 65-67; res assoc, 67-71, SR RES SCIENTIST, LAMONT-DOHERTY GEOL OBSERV, COLUMBIA UNIV, 71- *Concurrent Pos:* Nat Acad Sci-Nat Res Coun resident res assoc, Inst Space Studies, NY, 67-69. *Mem:* AAAS; Am Geophys Union; Clay Minerals Soc; Am Geochem Soc; fel Geol Soc Am. *Res:* Isotope-, major- and trace-element geochemistry, and mineralogy to study atmospheric dusts, suspended particulates, deep-sea and continental shelf sediments; sources, sinks and transport mechanisms of marine sedimentation. *Mailing Add:* Lamont-Doherty Geol Observ Palisades NY 10964

BISCHOFF, ERIC RICHARD, b Joliet, Ill, June 16, 38; c 4. DEVELOPMENTAL BIOLOGY, CELL BIOLOGY. *Educ:* Knox Col, Ill, AB, 60; Wash Univ, PhD(zool), 66. *Prof Exp:* Fel anat, Univ Pa, 65-68, instr med sch, 68-69; asst prof, 69-76, ASSOC PROF ANAT, MED SCH, WASH UNIV, 76- *Mem:* Am Soc Cell Biol; Am Soc Zoologists. *Res:* Differentiation of cultured skeletal muscle cells; skeletal muscle regeneration. *Mailing Add:* Dept Anat & Neurobiol Wash Univ Sch Med 660 S Euclid St Louis MO 63110

BISCHOFF, HARRY WILLIAM, b Evansville, Ind, May 15, 22; m 47; c 5. ZOOLOGY, BOTANY. *Educ:* Evansville Col, BA, 49; Vanderbilt Univ, MS, 52; Univ Tex, PhD(phycol), 63. *Prof Exp:* From instr to assoc prof, 50-65, dean men, 53-54, PROF BIOL, TEX LUTHERAN COL, 65- *Concurrent Pos:* NSF fel, 59-60. *Mem:* Bot Soc Am; Phycol Soc Am. *Res:* Phycology; culturing and systematics. *Mailing Add:* 1107 Zunker Seguin TX 78155

BISCHOFF, JAMES LOUDEN, b Los Angeles, Calif, Mar 20, 40; m 65; c 2. GEOCHEMISTRY. *Educ:* Occidental Col, AB, 62; Univ Calif, Berkeley, PhD(geochem), 66. *Prof Exp:* Fel marine geochem, Woods Hole Oceanog Inst, 66-67, asst scientist, 67-69; from asst prof to assoc prof geol sci, Univ Southern Calif, 69-74; consult prof, Dept Geol, Stanford Univ, 74-78; GEOLOGIST, MARINE GEOL BR, US GEOL SURV, 74- *Honors & Awards:* US Dept Interior Award for Meritorious Serv; Hon Fel, Calif Acad Scis. *Mem:* AAAS; Geochem Soc; Geol Soc Am; Int Asn Geochem & Cosmochem; Am Geophys Union. *Res:* Chemistry of water-rock interaction; metal deposits in the oceans; uranium-series dating. *Mailing Add:* Br of Marine Geol US Geol Surv Menlo Park CA 94025

BISCHOFF, KENNETH BRUCE, b Chicago, Ill, Feb 29, 36; m 59; c 2. CHEMICAL & BIOLOGICAL ENGINEERING. *Educ:* Ill Inst Technol, BS, 57, PhD(chem eng), 61. *Prof Exp:* NSF res fel, State Univ Ghent, 60-61; from asst prof to assoc prof chem eng, Univ Tex, Austin, 61-67; from assoc prof to prof, Univ Md, College Park, 67-70; dir sch chem eng, Cornell Univ, 70-75, Walter R Read prof eng, 70-76; chmn Dept Chem Eng, 78-82, UNIDEL PROF BIOMED & CHEM ENG, UNIV DEL, 76- *Concurrent Pos:* Consult, Exxon Res, 64-, Biomed Eng Br, NIH, 67-, Exxon Res & Eng Co, 77-, Gen Foods Corp & W R Grace Co, 80-; mem eng & urban health sci study sect, Environ Protection Agency, 66-71, chmn planning comt, 68-71; meeting chmn, 1st Int Symp Chem Reaction Eng, Am Inst Chem Engrs-Am Chem Soc-Europ Fedn Chem Eng, 70; mem, Artificial Kidney Chronic Uremia Contracts Rev Group, 70-76; assoc ed, Advan Chem Eng; mem, Bd Chem Sci & Technol, Nat Res Coun, 83-86; chmn, Coun Chem Res, 85, Food Pharmaceut Bioeng Div, Am Inst Chem Engrs, 85; numerous lectrs, US & foreign. *Honors & Awards:* Ebert Prize, Acad Pharmaceut Sci, 71; Prof Progress Award, Am Inst Chem Engrs, 76, Food, Pharmaceut & Bioeng Div Award, 82; R H Wilhelm Award, Am Inst Chem Engrs, 87. *Mem:* Nat Acad Eng; fel AAAS; fel Am Inst Chem Engrs; Am Chem Soc; Am Soc Eng Educ; fel Am Inst Chem; NY Acad Sci; Sigma Xi. *Res:* Chemical reaction, biological and medical engineering; author or co-author of 4 books and over 100 publications. *Mailing Add:* Dept Chem Eng Univ Del Newark DE 19716

BISCHOFF, ROBERT FRANCIS, b Beech Grove, Ind, June 28, 43; m 67; c 4. PHYSICAL CHEMISTRY. *Educ:* Ind Univ, AB, 65; Purdue Univ, MS, 67. *Prof Exp:* Prod regist mgr, Dow Chem Co, 67-89, PROD REGIST MGR, DOW ELANCO, 89- *Mem:* Am Chem Soc. *Res:* Chemical and physical properties of agricultural pesticides, optimization of these properties via formulation research; surface and colloid chemistry; registration of pesticide chemicals. *Mailing Add:* 1602 Continental Dr Zionsville IN 46077-9083

BISE, CHRISTOPHER JOHN, b Philadelphia, Pa, Aug 16, 50. UNDERGROUND COAL MINING, MAINTENANCE ENGINEERING. *Educ:* Va Polytech Inst & State Univ, BS, 72; Pa State Univ, MS, 76, PhD(mining eng), 80. *Prof Exp:* Resident engr, Consolidation Coal Co, 72-74; asst prof, 80-85, ASSOC PROF MINING ENG, PA STATE UNIV, 85- *Concurrent Pos:* Instr, Pa State Univ, 76-80. *Mem:* Soc Mining Engrs; Inst Indust Engrs; Nat Acad Forensic Engrs. *Res:* Improvement of productivity and safety in underground coal mining; mining methods; mine power systems; mine subsidence. *Mailing Add:* 123 Mineral Sci Bldg University Park PA 16802

BISEL, HARRY FERREE, b Manor, Pa, June 17, 18; m 54; c 3. CANCER. *Educ:* Univ Pittsburgh, BS, 39, MD, 42. *Prof Exp:* Intern, Med Ctr, Univ Pittsburgh, 42-43; res physician, Knoxville Gen Hosp, 50-51; resident, Mem Sloan-Kettering Cancer Ctr, 51-53; cancer coordr med, Sch Med, Univ Pittsburgh, 53-63; sr consult med oncol, 63-83; from assoc prof to prof med, 70-85, EMER PROF ONCOL, MAYO MED SCH, 83- *Concurrent Pos:* Asst med, Harvard Med Sch, Boston City Hosp, 49-50; lectr, Grad Sch Pub Health, Univ Pittsburgh, 53-56; consult & exec secy, Clin Studies Panel, Cancer Chemother Nat Serv Ctr, 55-56; consult, Nat Cancer Inst, 56-; chmn, Div Med Oncol, Mayo Clin, 64-83. *Mem:* Am Cancer Soc; AMA; Am Fedn Clin Res; Am Asn Cancer Res; Am Soc Clin Oncol (pres, 65); Sigma Xi; Am Asn Cancer Educ; Soc Surg Oncol. *Res:* Cancer chemotherapy. *Mailing Add:* 1223 Skyline Dr Rochester MN 55902-0940

BISGAARD, SØREN, b Umanak, Greenland, June 17, 51; Danish citizen. QUALITY IMPROVEMENT, MANAGEMENT. *Educ:* Copenhagen Col Eng, BS, 75; Tech Univ Denmark, MS, 79; Univ Wis-Madison, MS, 82, PhD(statist), 85. *Prof Exp:* Engr, exp toolmaking, WGer Quartz Smelting Works, 75-76; consult opers res, Int Serv Syst, 79 & statist, Burmeister & Wain, Copenhagen, 79-80; res asst, 80-85, RES ASSOC STATIST, UNIV WIS-MADISON, 86- *Honors & Awards:* Shewell Prize, Am Soc Qual Control, 8, Brumbaugh Award 87. *Mem:* Am Statist Asn; AAAS. *Res:* Industrial and physical science application of statistics; quality control, quality improvement, design of experiments; manufacturing and computerized manufacturing; statistical inference; management; design of environmental regulations; design of laws and standards. *Mailing Add:* Dept Indust Eng 390 Mech Eng Univ Wisc Madison 1513 University Ave Madison WI 53706

BISGARD, GERALD EDWIN, b Denver, Colo, Aug 4, 37; m 61; c 3. VETERINARY PHYSIOLOGY. *Educ:* Colo State Univ, BS, 59, DVM, 62; Purdue Univ, MS, 67; Univ Wis, PhD(physiol), 71. *Prof Exp:* From instr to asst prof vet clins, Purdue Univ, 62-69; asst prof, 71-74, assoc prof, 74-77, prof vet physiol, 77-80, PROF VET SCI, DEPT COMP BIOSCI, UNIV WIS-MADISON, 80-, PROF & CHMN DEPT COMP BIOSCI, 80- *Mem:* Am Physiol Soc; Am Vet Med Asn. *Res:* Control of respiration at high altitude, role of peripheral and central chemoreceptors. *Mailing Add:* Dept Comp Biosci Univ Wis 2015 Linden Dr W Rm 2015 Madison WI 53706

BISH, DAVID LEE, b Arlington, Va, Mar 5, 52; m 81; c 1. X-RAY POWDER DIFFRACTION, CLAY & ZEOLITE MINERALOGY. *Educ:* Furman Univ, Greenville, BS, 74; Pa State Univ, PhD(mineral), 77. *Prof Exp:* Res fel clay mineral, Dept Geol Sci, Harvard Univ, 77-80; STAFF MEM, LOS ALAMOS NAT LAB, 80- *Concurrent Pos:* Counr, Clay Minerals Soc, 86-89; assoc ed, Mineral Soc Am, 88-91. *Mem:* Fel Mineral Soc Am; Clay Minerals Soc; Int Asn Study Clays; Am Geophys Union. *Res:* Quantitative phase analysis of mixtures using x-ray powder diffraction methods; Rietveld refinement of crystal structures using x-ray and neutron powder diffraction data; importance of zeolites and clays in a high-level radioactive waste repository. *Mailing Add:* 394 Richard Ct Los Alamos NM 87544

BISHARA, MICHAEL NAGEEB, b Alexandria, Egypt, Apr 26, 33; US citizen; m 59; c 1. ATOMIC & MOLECULAR PHYSICS, TECHNICAL MANAGEMENT. *Educ:* Cairo Univ, BAE, 57; Univ Va, MAE, 64, DSc, 69. *Prof Exp:* Eng trainee, Egyptian Ministry Indust, 59-60, asst dir tech training, 60-61; res scientist eng, Univ Va Res Labs, 64-69; eng specialist, AiResearch Mfg Co, 69-70; dir eng, S & S Corp, 79-82; CHMN ENG, SOUTHWEST VA COMMUNITY COL, 70- *Concurrent Pos:* Mem, adv comt, Va Ctr Coal & Energy Res, 78-, Va Gov's Adv Comt Mined Land Reclamation, 79-; chmn, VCCS Chancellors Task Force Microcomput, 83; mem, Va Task Force Joint Training Partnership Act, 83; consult, Va Commonwealth Univ, 83- & Nat Independent Coal Operators' Asn, 83. *Honors & Awards:* Minta Martin Award, Am Asn Aeronaut & Astronaut, 64. *Mem:* Inst Elec & Electronics Engrs; Soc Automotive Engrs; Soc Mining Eng; Nat Soc Staff & Prof Develop; Nat Comput Graphics Asn; Sigma Xi. *Res:* Low density rarefied gas flow, including the interaction between inert gas atoms or molecules and solid surfaces; molecular and atomic beam experimentation; vacuum technology and space simulation; tribology in space environments. *Mailing Add:* 175 Valley View Dr Abingdon VA 24210

BISHARA, RAFIK HANNA, b Cairo, Egypt, Mar 21, 41; US citizen; m 68; c 3. ANALYTICAL CHEMISTRY. *Educ:* Cairo Univ, BSc, 62; Butler Univ, MSc, 70; Purdue Univ, PhD(bionucleonics), 72. *Prof Exp:* Phamaceut chemist & head biochem unit, Chem Indust Develop Co, Egypt, 62-67; anal chemist, 67-69, sr anal chemist, 71-75, RES SCIENTIST, ELI LILLY & CO, 76- *Mem:* Am Chem Soc; Am Pharmaceut Asn; Acad Pharmaceut Sci. *Res:* Analytical methods, specifications, reference standards; investigational new drug and new drug applications for new drug substances; corresponding precursors and degradation products; chromatographic techniques and compilation of analytical data. *Mailing Add:* 329 Fourth Ct E Carmel IN 46032

BISHARA, SAMIR EDWARD, b Cairo, Egypt, Oct 31, 35; m 75. ORTHODONTICS. *Educ:* Univ Alexandria, BChD, 57, dipl, 66; Univ Iowa, MS, 70, DDS, 72. *Prof Exp:* From intern to mem med staff dent, Moassat Hosp, Alexandria, Egypt, 57-68; asst prof orthod, 70-73, assoc prof, 73-76, PROF ORTHOD, UNIV IOWA, 76- *Concurrent Pos:* Fel, Guggenheim Dent Clin, 59-60. *Mem:* Int Asn Dent Res; Am Asn Orthodontists; Am Cleft Palate Asn; Int Dent Fedn; AAAS; fel Am Col Dentists; fel Int Col Dentists. *Res:* Changes in facial and dental relations in normal orthodontic and cleft populations. *Mailing Add:* Univ Iowa Col Dentistry Iowa City IA 52240

BISHIR, JOHN WILLIAM, b Joplin, Mo, Sept 23, 33; m 54, 77; c 4. APPLIED MATHEMATICS. *Educ:* Univ Mo, AB, 55; State Univ Iowa, MS, 57; NC State Univ, PhD(statist), 61. *Prof Exp:* From instr to asst prof math, NC State Univ, 57-62; asst prof statist, Fla State Univ, 62-63; from asst prof to assoc prof, 63-68, PROF MATH, NC STATE UNIV, 68- *Mem:* Math Asn Am; Soc Industr Appl Math. *Res:* Mathematical modeling in wildlife biology, ecology, and forest evolution. *Mailing Add:* NC State Univ Box 8205 Raleigh NC 27607

BISHOP, A(VERY) A(LVIN), irrigation engineering, water management; deceased, see previous edition for last biography

BISHOP, ALBERT B, b Philadelphia, Pa, Apr 7, 29; m 51; c 3. INDUSTRIAL ENGINEERING, MANUFACTURING SYSTEMS. *Educ:* Cornell Univ, BEE, 51; Ohio State Univ, MS, 53, PhD(indust eng), 57. *Prof Exp:* Mem tech staff, Bell Tel Labs, 54; res assoc opers res & instr indust eng, 54-57, from asst prof to assoc prof, 57-65, chmn dept indust & systs eng, 74-82, PROF INDUST ENG, OHIO STATE UNIV, 65-, PROJ SUPVR, SYSTS RES GROUP, 66- *Concurrent Pos:* Consult, NAm Aviation, Inc, 58-71, Indust Nucleonics Corp, 59-73, Chem Abstr Serv, 64-70, Stanford Res Inst, 65-67, US Army Combat Develop Command, 70-71, US Army Sci Adv Panel, 75-77, Galion Div, Dresser Indust, 76-, US Army Mats Systs Anal Activ, 78-79, Aerojet Electrosyst, 79-81, Ohio State Int, 79-81, Perfection Coby, 82-83, Container Recovery Corp, 83-84, Accu Ray, 85-86; vis scholar sci technol & soc & mech eng, Mass Inst Technol, 84. *Mem:* Opers Res Soc Am; fel Am Inst Indust Engrs; Inst Elec & Electronics Engrs; fel Am Soc Qual Control; Inst Mgt Sci; Soc Mfg Engrs. *Res:* Formulation and manipulation of analytic models of man-machine systems, including production and industrial systems; control-theory and mathematical programming models; product liability; principles for machine/human compatibility in manufacturing systems. *Mailing Add:* Dept Indust & Systs Eng Ohio State Univ 1971 Neil Ave Columbus OH 43210-1271

BISHOP, ALLEN DAVID, JR, b Meridian, Miss, Sept 5, 38; m 62; c 2. INORGANIC CHEMISTRY, PHYSICAL CHEMISTRY. *Educ:* Mills Col, BS, 60; La State Univ, Baton Rouge, MS, 63; Univ Houston, PhD(inorg chem), 67. *Prof Exp:* Chemist, Res & Develop Lab, Texaco, Inc, Tex, 63-65; fel chem, Univ Houston, 67; assoc prof, 67-74, prof chmn, 74-78, CHMN DEPT COMPUT SCI & DIR ACAD COMPUT, MILLSAPS COL, 78- *Mem:* Am Chem Soc; Geochem Soc; Am Soc Oceanog; Asn Comput Mach. *Res:* Polymerization of silicic acid under conditions similar to geologic conditions, effect of various ions on the polymerization process; complexes of oxomolybdenum V, complexes of molybdenum in the plus-5 oxidation state in an attempt to stabilize this oxidation state; thixotropic properties of clays, effect of various organic chelating agents on the thixotropic properties of clays; large data base management; micro processor-instrument. *Mailing Add:* Dept Comput Sci Millsaps Col Jackson MS 39210

BISHOP, ASA ORIN, JR, b Richmond, Va, June 30, 38; m 64; c 3. ELECTRICAL ENGINEERING, COMPUTER SYSTEM DESIGN. *Educ:* Va Mil Inst, BS, 59; Clemson Univ, MS, 66, PhD(elec eng), 69. *Prof Exp:* Instr elec eng, Va Mil Inst, 62-64; teaching asst instr, Clemson Univ, 64-69, fel & res assoc, 69-70; from asst prof to assoc prof 70-80, PROF ELEC ENG, UNIV TENN, KNOXVILLE, 80-, ASSOC DIR, COMPUT CTR, 75- *Concurrent Pos:* Consult, Duke Med Ctr, 69- & Vet Admin Hosp, 69-; spec asst financial mgt to asst secy of Navy, 78-79. *Mem:* Inst Elec & Electronics Engrs; Am Soc Engr Educ; Indust Elec & Control Inst Soc. *Res:* Feature extraction of information from the electroencephalogram and electrocardiogram using digital filter techniques and the on-line, real-time analysis of bioelectric phenomena; long range planning for data processing installations. *Mailing Add:* 1112 Venice Rd NW Knoxville TN 37923

BISHOP, BEVERLY PETTERSON, b Corning, NY, Oct 19, 22; m 44; c 1. PHYSIOLOGY. *Educ:* Syracuse Univ, BA, 44; Univ Rochester, MA, 46; State Univ NY Buffalo, PhD(physiol), 58. *Prof Exp:* Asst psychol, Univ Rochester, 44-46; teacher, Ohio Pub Sch, 46-48; instr bus admin, State Univ NY Buffalo, 49-50; asst physiol, Glasgow Univ, 56-57; from instr to assoc prof, 58-75, PROF PHYSIOL, STATE UNIV NY BUFFALO, 75- *Concurrent Pos:* NIH Study Sect, Respiration & Appl Physiol, 82-84; mem coun, Am Physiol Soc, 89-91. *Honors & Awards:* Golden Pen Award, Am Phys Ther Asn, 76. *Mem:* Am Physiol Soc; Soc Neurosci; Am Cong Rehab Med; Am Asn Electrodiag Med; hon mem Am Phys Ther Asn. *Res:* Neurophysiology; neural regulation of respiration, mastication and other motor systems; integration of the diverse functions of the abdominal muscles. *Mailing Add:* Dept Physiol Sch Med Biomed Sci & Dent Med State Univ NY 124 Sherman Hall Buffalo NY 14214

BISHOP, CHARLES (WILLIAM), b Elmira, NY, June 30, 20; m 44; c 1. BIOCHEMISTRY. *Educ:* Syracuse Univ, BS, 42, MS, 44; Univ Rochester, PhD(biochem), 46; Am Bd Clin Chem, dipl. *Prof Exp:* Asst chem, Syracuse Univ, 42-44; res chemist, Manhattan Proj, Univ Rochester, 44-46; instr chem, Kent State Univ, 46-47; asst prof, 47-64, ASSOC PROF BIOCHEM, SCH MED, STATE UNIV NY BUFFALO, 64-, ASSOC PROF MED, 65- *Concurrent Pos:* NIH spec res fel, Glasgow Univ, 55-56; supvr med res lab, Buffalo Gen Hosp, 47-51, asst lab dir, Chronic Dis Res Inst, 51-55; founder, Blood Info Serv; head chem lab, Buffalo Gen Hosp, 67-80. *Mem:* AAAS; Am Chem Soc; Am Soc Biol Chemists; Am Asn Clin Chemists; Soc Exp Biol & Med. *Res:* Red cell metabolism; carbohydrate and purine metabolism; blood storage systems; clinical chemistry; medical informatics. *Mailing Add:* Dept Med State Univ NY 508 Getzville Rd Buffalo NY 14226

BISHOP, CHARLES ANTHONY, b Rochester, NY, May 28, 34; m 57; c 6. ORGANIC CHEMISTRY, PHYSICAL CHEMISTRY. *Educ:* Rochester Inst Technol, BS, 57; Iowa State Univ, PhD(org chem), 61. *Prof Exp:* Res chemist, 61-63, sr res chemist, 63-68, lab head, 68-74, sr lab head, 75-84, qual adv to gen mgr, Eastman Kodak Co, 84-86; PRES, BISHOP CONSULT SERV, 87- *Res:* Displacement and elimination reactions; quinonoid compounds; photographic chemistry, dyes; interfacial reactions. *Mailing Add:* 1109 Cypress Trace Dr Melbourne FL 32940

BISHOP, CHARLES FRANKLIN, b Doylestown, Pa, June 29, 18; m 43; c 4. PLANT PATHOLOGY. *Educ:* Goshen Col, BA, 40; Univ WVa, MS, 42, PhD(plant path), 48. *Prof Exp:* Exten specialist plant path & entom, Univ WVa, 44-56; prof agr, 56-63, chmn div natural sci, 68-73, prof biol, 63-86, EMER PROF BIOL, GOSHEN COL, 86-, PLANT PEST CONSULT, 88- *Concurrent Pos:* Vis prof, Univ S Fla, 73 & Univ Fla, 75-76; open fac fel, Lilly Endowment Fund, 75. *Mem:* Sigma Xi; Am Inst Biol Sci. *Res:* Purpletop of potatoes. *Mailing Add:* Dept Biol Goshen College Goshen IN 46526

BISHOP, CHARLES JOHNSON, b Sask, Can, Jan 6, 20; m 51; c 1. GENETICS. *Educ:* Acadia Univ, BSc, 41; Harvard Univ, AM & PhD(cytogenetics), 47. *Hon Degrees:* DSc, Acadia Univ, 82. *Prof Exp:* Assoc prof genetics & agr res officer, Acadia Univ, 47-52, res prof & supt exp sta, Kentville, NS, 52-58; dir res sta, Summerland, BC, 58-59; assoc dir prog, Res Br, Can Dept Agr, 59-64, res coordr hort, 64-85; RETIRED. *Honors & Awards:* Jubilee Medal, 77. *Mem:* Fel Am Soc Hort Sci; Am Pomol Soc (vpres, 57); Genetics Soc Can (pres, 58); Can Soc Hort Sci (pres, 57); fel Royal Soc Can; fel Agr Inst Can; hon life mem, Can Hort Coun. *Res:* X-ray and thermal neutron induced mutations in apples; polyploidy and scab resistance in apples; fruit and vegetable breeding; male sterility in tomatoes; editorial practices in plant science reporting; coordination of horticultural research. *Mailing Add:* 1968 Bel Air Dr Ottawa ON K2C 0W9 Can

BISHOP, CHARLES JOSEPH, b Gary, Ind, June 22, 41; m 63; c 2. ENERGY SYSTEMS ENGINEERING, SOLAR ENERGY ENGINEERING. *Educ:* Purdue Univ, BS, 63; Univ Wash, PhD(nuclear chem), 69. *Prof Exp:* Mgr, Boeing Co, 69-77; br chief, Solar Energy Res Inst, 77-81; dir tech servs, 81-84, exec dir corp technol, 84-85, VPRES, A O SMITH CORP, 85- *Concurrent Pos:* Prog comt, Indust Res Inst, 87-89, chmn advan study group, 88-89, mem bd dirs, 89- *Honors & Awards:* Cert of Recognition, NASA. *Mem:* Indust Res Inst. *Res:* Systems engineering; solar energy; energy systems; systems analysis; electrochemistry; photovoltaics. *Mailing Add:* 335 Douglas Lane Cedarburg WI 53012

BISHOP, CLAUDE TITUS, b Liverpool, NS, May 13, 25; m 52; c 1. IMMUNOCHEMISTRY, CARBOHYDRATE CHEMISTRY. *Educ:* Acadia Univ, BSc, 45, BA, 46; McGill Univ, PhD(chem), 49. *Hon Degrees:* DSc, Univ Western Ont, 86. *Prof Exp:* Asst res officer, Carbohydrate Chem Sect, Div Biosci, Nat Res Coun Can, 49-55, assoc res officer, 55-60, sr res officer, 60-67, prin res officer, 67-70, asst dir biochem lab, 70-72, assoc dir Biol Res Div, 72-78, dir, Biol Sci Div, 78-87, secy gen, 87-89; RETIRED. *Concurrent Pos:* Ed-in-chief, Can J Res, 71-90. *Mem:* Can Biochem Soc; Chem Inst Can; Am Chem Soc; fel Royal Soc Can. *Res:* Carbohydrate chemistry; structures and immunochemistry of microbial polysaccharide antigens. *Mailing Add:* 63 Holborn Ave Nepean ON K2C 3H1 Can

BISHOP, DAVID HUGH LANGLER, b London, Eng, Dec 31, 37; m 71; c 3. MOLECULAR BIOLOGY, VIROLOGY. *Educ:* Univ Liverpool, BSc, 59, PhD(biochem), 62. *Prof Exp:* Fel biochem, Nat Ctr Sci Res, Gif-sur-Yvette, France, 62-63; res assoc, Univ Edinburgh, 63-66 & Univ Ill, 66-69; asst prof molecular biol, Columbia Univ, 69-70; assoc prof, Rutgers Univ, 71-75, prof virol, 75; prof microbiol & sr scientist, Comprehensive Cancer Ctr, Univ Ala Med Ctr, Birmingham, 75-; INST VIROL, NAT ENVIRON RES CAN. *Concurrent Pos:* Mem, Virus Cancer Prog Sci Rev Comt A, 75-77. *Mem:* Am Soc Microbiol. *Res:* Molecular aspects of the replication of RNA viruses, viral genetics and host-virus interactions. *Mailing Add:* Nat Environ Res Can Inst Virol Mansfield Rd Mansfield 0X1 3SR England

BISHOP, DAVID MICHAEL, b London, Eng, Sept 19, 36. THEORETICAL CHEMISTRY. *Educ:* Univ London, BSc, 57, PhD(chem), 60, DSc, 72. *Prof Exp:* Asst prof chem, Carnegie Inst Technol, 62-63; from asst prof to assoc prof, 63-67, PROF CHEM, UNIV OTTAWA, 72- *Concurrent Pos:* Am Chem Soc fel, 60-62. *Res:* Molecular quantum chemistry. *Mailing Add:* Dept Chem Univ Ottawa Ottawa ON K1N 9B4 Can

BISHOP, DAVID WAKEFIELD, physiology; deceased, see previous edition for last biography

BISHOP, EDWIN VANDEWATER, b Jamaica, NY, May 17, 35; m 57; c 1. ASTRONOMY, PHYSICS. *Educ:* Swarthmore Col, BA, 58; Yale Univ, MS, 60, PhD(physics), 66. *Prof Exp:* Res staff astronr, Yale Univ, 66-69; PROF ASTRON & PHYSICS, YOUNGSTOWN STATE UNIV, 69- *Mem:* Am Astron Soc; Am Phys Soc. *Res:* Solar-system physics; thermodynamic properties, particularly temperature-profile histories of the major planets' interiors, especially Jupiter's. *Mailing Add:* Dept Physics & Astron Youngstown State Univ Youngstown OH 44555

BISHOP, ERRETT A, b Newton, Kans, July 14, 28; m 56; c 3. MATHEMATICS. *Educ:* Univ Chicago, BS, 48, MS, 50, PhD(math), 55. *Prof Exp:* From instr to prof math, Univ Calif, Berkeley, 54-65; Miller prof, 64-65, PROF MATH, UNIV CALIF, SAN DIEGO, 65- *Concurrent Pos:* Sloan fel, 59-62; fel, Inst Advan Study, 61-62. *Mem:* Am Math Soc. *Res:* Theory of functions of several complex variables; theory of uniform algebras and functional analysis. *Mailing Add:* 6120 Waverly Ave La Jolla CA 92037

BISHOP, EUGENE H(ARLAN), b Ellisville, Miss, Oct 29, 33; m 55; c 3. MECHANICAL ENGINEERING. *Educ:* Miss State Univ, BS, 55; Univ Tex, PhD(mech eng), 64. *Prof Exp:* Asst engr, Eng Exten Serv, Miss State Univ, 60-61; asst heat transfer, Univ Tex, 61-64; assoc prof mech eng, Miss State Univ, 64-67; assoc prof, Mont State Univ, 67-69, prof aerospace & mech eng & head dept, 69-74; PROF MECH ENG & HEAD DEPT, CLEMSON UNIV, 74- *Mem:* Am Soc Mech Engrs. *Res:* Natural convection heat transfer and fluid mechanics. *Mailing Add:* Dept mech Eng Clemson Univ Main Campus Clemson SC 29634

BISHOP, FINLEY CHARLES, b Minneapolis, Minn, May 1, 49. MINERALOGY, PETROLOGY. *Educ:* Carleton Col, AB, 72; Univ Chicago, SM, 74, PhD(geol), 76. *Prof Exp:* From instr to asst prof, 76-82, ASSOC PROF GEOL, NORTHWESTERN UNIV, EVANSTON, 82- *Mem:* AAAS; Am Geophys Union; Geol Soc Am; Geochem Soc; Mineral Soc Am. *Res:* Ion exchange in minerals at high temperatures and pressures; mineralogy of Archean layered igneous complexes; kimberlites and upper mantle rocks; properties of magnesian calcites. *Mailing Add:* Dept Geol Sci Northwestern Univ Evanston IL 60201

BISHOP, GALE ARDEN, b Jamestown, NDak, Dec 10, 42; m 64; c 2. PALEONTOLOGY. *Educ:* SDak Sch of Mines, BS, 65, MS, 67; Univ Tex, Austin, PhD(geol), 71. *Prof Exp:* PROF GEOL, GA SOUTHERN COL, 71- *Concurrent Pos:* Dir, Ga Southern Col Mus, 80-83. *Mem:* Sigma Xi; Crustacean Soc; Paleont Soc. *Res:* Cretaceous crabs; preservation processes; functional morphology; Tertiary fresh-water clams; ammonite biostratigraphic mapping and larval development; predation-produced trace fossils; paleoecology; decapod trace fossils; fossil and living ghost shrimp; coastal geology. *Mailing Add:* Dept Geol Ga Southern Univ Statesboro GA 30460-8149

BISHOP, GUY WILLIAM, b Medford, Ore, May 16, 26; m 48; c 3. ENTOMOLOGY. *Educ:* Ore State Col, BS, 51, MS, 53; State Col Wash, PhD(entom), 58. *Prof Exp:* Asst entom, Ore State Col, 48-52; asst, State Col Wash, 54-56; from asst entomologist to assoc entomologist, Univ Idaho, 57-74, assoc prof, 68-74, prof entom & entomologist, 74-88; RETIRED. *Mem:* Am Entom Soc. *Res:* Virus vectors of plants; potato insects. *Mailing Add:* 3403 N Valley View Rd Ashland OR 97520

BISHOP, HAROLD (OSWALD), b Harrisburg, Pa, Jan 9, 09; m 31; c 2. ELECTRONICS. *Educ:* Univ Pittsburgh, EE; Univ Md, cert, 54. *Prof Exp:* Owner, Radio Eng Labs, 32-42; instr electronics, US Naval Postgrad Sch, 42-46; electronics scientist, Bur Ord, US Navy, DC, 46-51, dir facil eng, 11th Naval Dist, 51-70; consult electronics & cathodic protection eng, 70-78; RETIRED. *Concurrent Pos:* Pres, Keystone Refrig Inc, Pa, 37-42; pres, WABX, Pa, 47-52 & WXNJ, Inc, NJ, 48-52. *Mem:* Sr mem Inst Elec & Electronics Engrs. *Res:* Stable microwave oscillators; electronic and guided missile counter measures; field operation evaluation of guided missile counter measures; field operation evaluation of guided missiles and electronic devices; antenna transmission lines. *Mailing Add:* 6232 La Jolla Blvd La Jolla CA 92037

BISHOP, JACK BELMONT, b Seymour, Tex, Aug 26, 43; m 71; c 2. GENETICS, TOXICOLOGY. *Educ:* McMurry Col, BA, 67; La State Univ, Baton Rouge, MS, 70, PhD(genetics), 74. *Prof Exp:* Res geneticist, Bee Breeding & Stock Ctr Res Lab, Agr Res Serv, USDA, 72-75; res geneticist, Nat Ctr Toxicol Res, Dept Health Educ & Welfare, Food & Drug Admin, 75-85; RES GENETICIST, NAT INST ENVIRON HEALTH SCI, DEPT HEALTH & HUMAN SERV, NIH, 85- *Concurrent Pos:* Adj prof med sci, Grad Sch, Univ Ark, 80-88; spec appt, Univ NC Med Sch, Chapel Hill, 89- *Honors & Awards:* Commendable Serv Award, Food & Drug Admin, 80. *Mem:* Environ Mutagens Soc; Sigma Xi; Europ Environ Mutagen Soc. *Res:* Chemical mutagenesis including studies of chemically induced heritable translocations and biochemical mutations and aneuploidy in the mouse; studies in dosimetry and spermatogenic cell stage response specificity with chemical mutagens in experimental organisms such as the mouse, drosophila and the honeybee. *Mailing Add:* Exp Carcinogenesis & Mutagenesis Br Nat Inst Environ Health Sci PO Box 12233 Raleigh NC 27612

BISHOP, JACK GARLAND, b Ft Worth, Tex, Sept 12, 19; m 46; c 1. PHYSIOLOGY. *Educ:* NTex State Col, BS, 46, MS, 49; Ind Univ, PhD(physiol), 55. *Prof Exp:* Instr, NTex State Col, 49-51; asst physiol, Ind Univ, 51-53, res assoc, 53-54; from asst prof to prof, Col Dent, Baylor Univ, 54-85, chmn dept, 56-69, dir res, 61-69, actg dir grad studies, 68-69, assoc dean grad sch, 70-85; RETIRED. *Mem:* AAAS; Am Physiol Soc; Int Asn Dent Res. *Res:* Cardiovascular physiology. *Mailing Add:* Dept Physiol Baylor Dent Sch Dallas TX 75246

BISHOP, JACK LYNN, b Abilene, Kans, Nov 15, 29; m 58; c 2. ENTOMOLOGY. *Educ:* Kans State Univ, BS, 56, MS, 57, PhD, 59. *Prof Exp:* From asst prof to assoc prof entom, Va Polytech Inst & State Univ, 58-67; entomologist, Shell Develop Co, Calif, 67-70, toxicologist-environmentalist, Shell Chem Co, 70-77, MEM STAFF, SHELL DEVELOP CO, CALIF, 77- *Mem:* Entom Soc Am; Sigma Xi. *Res:* Forage entomology and insect toxicology. *Mailing Add:* PO Box 267 Bethel Island CA 94511

BISHOP, JAMES MARTIN, b Dodge City, Kans, Mar 13, 36; m 64; c 2. PHYSICS. *Educ:* Kans State Teachers Col, AB, 58; Univ Wis, MS, 60, PhD(physics), 67. *Prof Exp:* Instr physics, Ohio Univ, 66-67, asst prof, 67-72; FAC FEL, DEPT PHYSICS, UNIV NOTRE DAME, 72- *Mem:* AAAS; Am Phys Soc; Asn Comput Mach; Sigma Xi. *Res:* Elementary particle physics; computer applications in physics. *Mailing Add:* Dept Physics Univ Notre Dame Notre Dame IN 46556

BISHOP, JAY LYMAN, b Salt Lake City, Utah, July 7, 32; m 58; c 9. INDUSTRIAL CHEMISTRY, PALEOGRAPHY. *Educ:* Univ Utah, BS, 53, PhD(org chem), 62. *Prof Exp:* Res assoc org chem, Ariz State Univ, 62-63, instr & res assoc chem, 64-67; sr chemist, Ciba Pharmaceut Corp, 67-71; chief chemist & metallurgist, Assoc Smelters Int, 73 & United Refinery Inc, 73-75; chief chemist & metallurgist, US Nat Metals Inc, 75-76; PRES, WESTERN CONSULT, 76-; CHEM & ENVIRON ENGR, US ARMY, 82- *Concurrent Pos:* Pres, Bishop Mfg Co, 71-, consulting chemist, 72-; lectr, Univ Utah, 60-62; traveling sci lectr, Ariz Acad Sci, 62-64; sr pres, 481st Quorum of 70, 70-71. *Res:* Heterocyclics; polymers; amine-formaldehyde alkylation; industrial processes; chemotherapeutic agents; precious metals refining and fabricating; energy research; fuel alcohol production; paleography and translation; munitions deactivation; EPA laws. *Mailing Add:* 1663 S 75 East Bountiful UT 84010-5218

BISHOP, JOHN MICHAEL, b York, Pa, Feb 22, 36; m 59; c 2. VIROLOGY, BIOCHEMISTRY. *Educ:* Gettysburg Col, AB, 57; Harvard Univ, MD, 62. *Hon Degrees:* DSc, Gettysburg Col, 83. *Prof Exp:* Intern internal med, Mass Gen Hosp, Boston, 62-63, resident, 63-64; res assoc virol, NIH, 64-66, sr investr, 66-68; from asst prof to assoc prof, 68-72, PROF MICROBIOL, UNIV CALIF, SAN FRANCISCO, 72-, PROF BIOCHEM & BIOPHYS, 82-; DIR, G W HOOPER RES FOUND, 81- *Concurrent Pos:* Res grants, NIH, 68-, Cancer Res Coord Comt, Univ Calif, 68- & Calif Div, Am Cancer Soc, 69-; vis scientist, Heinrich-Pette Inst, Ger, 67-68; assoc ed, Virol, 70-89; mem bd trustees, Leukemia Soc Am, 73-78; co-chair, Gordon Conf Animal Cells & Viruses, 79; chair, Virol Sect, NIH, 80-82, mem adv comt to dir, 86-88; non-resident fel, Salk Inst, 90- *Honors & Awards:* Nobel Prize in Physiol or Med, 89; Biomed Res Award, Am Asn Med Cols, 81; Albert Lasker Basic Med Res Award, 82; Armand Hammer Cancer Award, 84; Gen Motors Found Cancer Res Award, 84; Gairdner Found Int Award, Can, 84; Medal of Honor, Am Cancer Soc, 85. *Mem:* Nat Acad Sci; Inst Med-Nat Acad Sci; Am Soc Biol Chemists; Am Soc Microbiol; Am Soc Cell Biol; Am Acad Arts & Sci; Fed Am Scientists; Am Asn Univ Prof; fel AAAS. *Res:* Biochemistry of animal viruses; replication of nucleic acids; viral oncogenesis; molecular genetics. *Mailing Add:* Dept Microbiol Med Ctr Univ Calif PO Box 0552 San Francisco CA 94143

BISHOP, JOHN RUSSELL, b Pa, Dec 6, 20; m 45; c 3. AGRICULTURAL CHEMISTRY. *Educ:* Ursinus Col, BS, 42. *Prof Exp:* Dir res & develop, Am Chem Prod Inc, 75-78; assoc dir explor & develop, Herb & Plant Growth Reg, Union Carbide Agr Prod Co Inc, 78-81; RETIRED. *Mem:* Am Chem Soc; Weed Sci Soc Am; Int Weed Sci Soc. *Res:* Organic chemical research pertaining to agricultural chemicals; preparation of chemicals for use as herbicides, plant hormones, plant growth regulators; defoliators; formulation of herbicides and plant growth regulators. *Mailing Add:* Food Sci Va Polytechnic Inst & State Univ Blacksburg VA 24061

BISHOP, JOHN WATSON, b Glenridge, NJ, Mar 21, 38; m 62; c 2. ECOLOGY, AQUATIC BIOLOGY. *Educ:* Rutgers Univ, BA, 60; Cornell Univ, MS, 62, PhD(oceanog), 66. *Prof Exp:* Asst prof biol, Univ Del, 64-66; from asst prof to assoc prof, 66-83, PROF BIOL, UNIV RICHMOND, 83- *Concurrent Pos:* Univ res found grant, Univ Del, 65-66; Off Water Resources grant, 68-70. *Mem:* Am Soc Limnol & Oceanog; Ecol Soc Am. *Res:* Zooplankton ecology; trophic relationships in aquatic ecosystems. *Mailing Add:* 4701 Cosby Rd Powhatan VA 23139

BISHOP, JOHN WILLIAM, b Jefferson Co, Ind, June 12, 16; m 39; c 2. PETROLEUM CHEMISTRY. *Educ:* DePauw Univ, BA, 38; Western Reserve Univ, MA, 39, PhD(org chem), 42. *Prof Exp:* Chemist, Tidewater Oil Co, 42-44, group leader, res dept, 44-56, mgr opers anal, 56-68; mgr econ & eval, Getty Oil Co, 68-79, mgr strategic crude oil planning, 80-84; RETIRED. *Mem:* Am Chem Soc; Am Nuclear Soc; Sigma Xi; AAAS. *Res:* Hydrocarbon isomerization; lubricants; additives for lubricants; electronic computer applications in petroleum industry; planning and economic evaluations; nuclear power technology. *Mailing Add:* 982 Marisa Lane Encinitias CA 92024

BISHOP, JONATHAN S, b Cleveland, Ohio, Apr 20, 25. GLYCOGEN METABOLISM. *Educ:* Yale Univ, MD, 49. *Prof Exp:* ASSOC PROF INTERNAL MED, MED SCH, UNIV MINN, 69-, LECTR PHARMACOL, 79- *Mailing Add:* Med Sch 3-260 Millard Hall Univ Minn 420 Delaware St SE Minneapolis MN 55455

BISHOP, JOSEPH MICHAEL, b Syracuse, NY, Jan 6, 43; m 69; c 3. APPLIED OCEANOGRAPHY, MARINE POLICY. *Educ:* State Univ NY Maritime Col, BS, 66; NY Univ, MS, 70, PhD(oceanog), 73. *Prof Exp:* Asst dean, State Univ NY Maritime Col, 71-72, instr meteorol, 72-74; oceanogr, US Coast Guard, 74-76 & Nat Oceanic Atmospheric Admin, 76-80; policy adv meteorol, Fed Emergency Mgt Agency, 80-84; SR SCI ADV, NAT ADV COMT OCEANS & ATMOSPHERE, 84- *Concurrent Pos:* Instr, George Washington Univ, 77-82 & Catholic Univ, 79- *Mem:* Marine Technol Soc. *Res:* Coastal circulation; wind waves; air-sea interaction; applied oceanography; meteorology; ocean affairs and policy. *Mailing Add:* 16748 Tin Ct Dumfries VA 22026

BISHOP, KEITH C, III, b Portland, Ore, Jan 31, 47; c 2. ORGANOMETALLIC CHEMISTRY, CATALYSIS. *Educ:* Univ Calif, Santa Barbara, BS, 69; Yale Univ, PhD(chem), 73. *Prof Exp:* Fel chem, Stanford Univ, 73-74; res chemist, 74-81, sr environ engr, 81-85, POLICY COORDR CHEM, CHEVRON RES CO. *Concurrent Pos:* Bd dir, Main Municipal Water Dist, 79-83. *Mem:* Am Chem Soc; Sigma Xi; AAAS. *Res:* Environmental science. *Mailing Add:* 576 Standard Ave Richmond CA 94802

BISHOP, MARGARET S, b Lewiston, Mich, June 21, 06; m 37; c 2. GEOLOGY. *Educ:* Univ Mich, AB, 29, MS, 31, PhD(geol), 33. *Prof Exp:* Geologist, Pure Oil Co, 29-30, asst to chief geologist, 33-38, consult geologist, 38-53; from asst prof to prof, 53-71, EMER PROF GEOL, UNIV HOUSTON, 71- *Mem:* Geol Soc Am; Am Asn Petrol Geologists. *Res:* Subsurface geology and earth science. *Mailing Add:* PO Box 2567 Texas City TX 77590

BISHOP, MARILYN FRANCES, b Sacramento, Calif, Jan 19, 50. THEORETICAL SOLID STATE PHYSICS. *Educ:* Univ Calif, Irvine, BA, 71 & 72, MA, 73, PhD(physics), 76. *Prof Exp:* Res asst physics, Univ Calif, Irvine, 72-76; res assoc physics, Purdue Univ, 76-79; ASST PROF PHYSICS, DREXEL UNIV, PHILADELPHIA, 79- *Mem:* Am Phys Soc; Sigma Xi. *Res:* Interacting fermi systems; charge density waves; transport properties of metals; electron-phonon interactions; superconductivity; light scattering; surface electromagnetic waves; optical properties of dielectrics. *Mailing Add:* Dept Physics Va Commonwealth Univ 1020 W Main St Richmond VA 23284-2000

BISHOP, MARVIN, b New York, NY, Feb 17, 45; m 90; c 1. CHEMICAL PHYSICS, COMPUTER SCIENCE. *Educ:* City Col New York, BS, 66; Columbia Univ, PhD(chem), 71; New York Univ, MS Comput Sci, 83. *Prof Exp:* Res assoc chem, Nat Bur Standards, 71-73 & Columbia Univ, 73-74; vis asst prof, Howard Univ, 74-75; res asst prof, State Univ NY Albany, 75-76; asst prof chem, Fordham Univ, Lincoln Ctr, 76-81; vis scientist, Vanderwaals Lab, Neth, 81-82; res scientist, Courant Inst Math, 82-83; ASSOC PROF, MANHATTAN COL, 83- *Concurrent Pos:* Nat Res Coun fel, Nat Bur Standards, 71-73; Am Chem Soc-Petrol Res Fund grant, Fordham Univ, 76-; comput grant & fac res grant, Nat Res Coun Can, 78-; type B grants, PRF-Am Chem Soc, 79-81, 84-86 & 87-89; PRF-ACS type B grants, 79-81, 84-86, 87-89; IBM grants, 85-86, 87-; vis scientist, UMIST, Univ Manchester Inst Sci & Technol, 88-91. *Mem:* Am Chem Soc; Am Phys Soc; Asn Comput Mach. *Res:* Statistical mechanics; computer simulation; polymers; liquid-crystals; fluids. *Mailing Add:* Dept Math Comput Sci Manhattan Col Riverdale NY 10471

BISHOP, MURIEL BOYD, b Billingsley, Ala, Oct 7, 28; m 60; c 1. BIOCHEMISTRY, ORGANIC CHEMISTRY. *Educ:* Huntingdon Col, BA, 52; Emory Univ, MS, 55; Mich State Univ, PhD(chem), 58. *Prof Exp:* NSF grant pharmacol, Yale Univ, 58-59; asst prof, 59-74, ASSOC PROF CHEM, CLEMSON UNIV, 74- *Mem:* Am Chem Soc. *Res:* Nucleic acid chemistry, especially enzymes associated with pyrimidine biosynthesis; structure of nucleic acids. *Mailing Add:* Dept Chem Clemson Univ Clemson SC 29631

BISHOP, NANCY HORSCHEL, b Yakima, Wash, Mar 15, 36. MICROBIOLOGY, IMMUNOLOGY. *Educ:* Wash State Univ, BS, 58, MS, 61; Univ Calif, Los Angeles, PhD(microbiol & immunol), 75. *Prof Exp:* Res asst bact, Wash State Univ, 60-62; microbiologist, Yakima Valley Mem Hosp, 62 & Greater Bakersfield Mem Hosp, 63-64; sr microbiologist, Santa Barbara Cottage Hosp, 64-68; microbiologist, Los Robles Hosp, 69-70; scholar microbiol & immunol, Univ Calif, Los Angeles, 75-77; from asst prof to assoc prof, 77-84, PROF BIOL, CALIF STATE UNIV, NORTHRIDGE, 84- *Mem:* Am Soc Microbiol; Am Venereal Dis Asn; AAAS. *Res:* Mechanisms of immunity in syphilis. *Mailing Add:* Dept Biol Calif State Univ Northridge CA 91330

BISHOP, NORMAN IVAN, b Silverton, Colo, June 29, 28; m 48; c 4. PLANT PHYSIOLOGY. *Educ:* Univ Utah, BS, 51, MS, 52, PhD(physiol), 55. *Prof Exp:* Res asst biol, Univ Utah, 52-55; res assoc biochem, Univ Chicago, 55-57, asst prof, 57-60; assoc prof, Fla State Univ, 60-63; assoc prof, 63-65, PROF PLANT PHYSIOL, ORE STATE UNIV, 65-, DEPT CHMN, 86- *Concurrent Pos:* Lalor Found fel, 56; Guggenheim fel, 71. *Mem:* AAAS; Am Soc Plant Physiol; Biophys Soc; Am Soc Biol Chemists; Am Soc Photobiol. *Res:* Photosynthesis and metabolism of microorganisms; photochemistry; general biochemistry. *Mailing Add:* Dept Bot Ore State Univ Corvallis OR 97331

BISHOP, PAUL EDWARD, b Portland, Ore, Feb 12, 40; m 68; c 1. MICROBIOLOGY, GENETICS. *Educ:* Wash State Univ, BS, 64; Ore State Univ, MS, 70, PhD(food sci), 73. *Prof Exp:* Technician retinal metab, Sch Med, Univ Wash, 66-67; res assoc biochem nitrogen fixation, Ore State Univ,

73-75; res assoc genetics nitrogen fixation, Univ Wis-Madison, 75-77; asst prof, 77-83, ASSOC PROF MICROBIOL, NC STATE UNIV, 83- *Concurrent Pos:* Prin investr, Competitive Grants Prog, Sci & Educ Admin, USDA, 78-80, 81-83, 84-86, 86-88, 88-90, 90-92; Agr Res Serv fel, Agr & Food Nitrogen , Univ Sussex, Brighton, UK, 84-85. *Mem:* Am Soc Microbiol. *Res:* Genetics and molecular biology of alternative nitrogen fixation systems. *Mailing Add:* Dept Microbiol NC State Univ Raleigh NC 27650

BISHOP, PAUL LESLIE, b Hyannis, Mass, Nov 27, 45; m 71; c 2. HAZARDOUS WASTE TREATMENT, WASTEWATER TREATMENT. *Educ:* Northeastern Univ, BSCE, 68; Purdue Univ MSCE, 70, PhD(environ eng), 72. *Prof Exp:* From asst prof to prof environ eng, Univ NH, 72-87, head, 76-82; WILLIAM THOMS PROF & HEAD CIVIL & ENVIRON ENG, UNIV CINCINNATI, 88- *Concurrent Pos:* Environ eng consult, 72-; vis prof, Heriot-Watt Univ, Edinburgh, Scotland, 80 & Tech Univ Denmark, Lyngby, 86-87; bd trustees, Am Acad Environ Engrs, 88-93; liaison, Asn Environ Eng Professors, 88-93; prog comt vchair, Water Pollution Control Fedn; mem, Hazardous Waste Comt, Am Soc Civil Engrs. *Mem:* Int Asn Water Pollution Res & Control; Am Soc Eng Educ. *Res:* Hazardous waste treatment; biological waste treatment; solidification & stabilization of hazardous wastes; leaching mechanisms; use of biofilms for treatment; solid waste composting. *Mailing Add:* Dept Civil & Environ Eng Univ Cincinnati Cincinnati OH 45221-0071

BISHOP, RICHARD LAWRENCE, b Lake Odessa, Mich, Aug 12, 31; m 54; c 5. GEOMETRY. *Educ:* Case Inst Technol, BS, 54; Mass Inst Technol, PhD(math), 59. *Prof Exp:* PROF MATH, UNIV ILL, URBANA-CHAMPAIGN, 59- *Mem:* Am Math Soc. *Res:* Relations between the topological and analytical properties of riemannian manifolds. *Mailing Add:* Univ Ill 1409 W Green St Urbana IL 61801

BISHOP, RICHARD RAY, b Ft Worth, Tex, Mar 23, 38; m 62. ELECTRICAL ENGINEERING, BIOENGINEERING. *Educ:* Southern Methodist Univ, BSEE, 61, MSEE, 65; Univ Tex, Austin, PhD(elec eng), 68. *Prof Exp:* Technician, Bell Helicopter Corp, Tex, 57-61; engr, Tex Instruments, 61-62; instr elec eng, Southern Methodist Univ, 62-63; asst prof elec eng, Pa State Univ, University Park, 68-80; ASST PROF ELEC ENG & CHMN DEPT, UNIV NEW ORLEANS, 80- *Mem:* Inst Elec & Electronics Engrs; Asn Comput Mach. *Res:* Biomedical electronics; biomedical signal processing, particularly electroencephalogram analysis by computer; biomedical simulation. *Mailing Add:* Dept Elec Eng Univ New Orleans New Orleans LA 70122

BISHOP, RICHARD STEARNS, b Dowagiac, Mich, April 14, 45; m 71; c 2. EXPLORATION TECHNOLOGY. *Educ:* Tex Christian Univ, BS, 67; Univ Mo, MA, 69; Stanford Univ, PhD(geol), 77. *Prof Exp:* Prod geologist, Univ Oil Co, Calif, 69-71; sr res specialist, 75-81, GEOL ASSOC, EXXON PROD RES CO, 81- *Concurrent Pos:* Instr, La State Univ, 71; chmn, Nat Convention, Am Asn Petrol Geologists, 88,. *Honors & Awards:* Sproule Award, 80. *Mem:* Am Asn Petrol Geologists (secy, 91); Geol Soc Am; Soc Petrol Engrs; Am Inst Mech Engrs. *Res:* Quantitative model of the accumulation and dispersion of hydrocarbons in the subsurface; mechanics of diapir movement and origins of abnormal pressures. *Mailing Add:* Exxon Prod Res Co Box 2189 Houston TX 77252

BISHOP, ROBERT H, b Vicenza, Italy; US citizen. CONTROL THEORY, SPACECRAFT SYSTEMS. *Educ:* Tex A&M Univ, BS, 79, MS, 80; Rice Univ, PhD(elec), 90. *Prof Exp:* Res staff, Charles Stark Draper Lab, 80-90; ASST PROF AEROSPACE, UNIV TEX, 90- *Concurrent Pos:* Mem, Guidance, Navigation & Control Tech Comt, Am Inst Aeronaut & Astronaut, 91. *Res:* Spacecraft guidance, navigation, and control; nonlinear control of evolutionary spacecraft; manned Mars mission navigation analysis; adaptive control; robust control. *Mailing Add:* Aerospace Eng & Eng Mech Univ Tex Austin TX 78712

BISHOP, ROY LOVITT, b Wolfville, NS, Sept 22, 39;; m 61; c 3. ASTRONOMY. *Educ:* Acadia Univ, BSc, 61; McMaster Univ, MSc, 63; Univ Man, PhD(physics), 69. *Prof Exp:* Head dept, 76-83, PROF PHYSICS, ACADIA UNIV, 63-, HEAD DEPT, 86- *Concurrent Pos:* Ed Observer's Handbk, Royal Astron Soc Can, 81- *Mem:* Royal Astron Soc Can (pres, 84-86); Can Astron Soc; Int Astron Union; Am Asn Physics Teachers. *Res:* History of astronomy with an emphasis on Canada. *Mailing Add:* Dept Physics Acadia Univ Wolfville NS B0P 1X0 Can

BISHOP, SANFORD PARSONS, b Springfield, Vt, Aug 28, 36; m 57; c 3. COMPARATIVE PATHOLOGY, CARDIOVASCULAR DISEASES. *Educ:* NY State Col Vet Med, Cornell Univ, DVM, 60; Ohio State Univ, MSc, 65, PhD(vet path), 68; Am Col Vet Pathologists, dipl; Am Col Vet Internal Med, dipl & cert cardiol. *Prof Exp:* Pvt pract vet med, NY, 60-62; instr cardiol, Sch Vet Med, Univ Pa, 62-63; from asst prof to assoc prof path, Col Med & Dept Vet Path, Ohio State Univ, 68-75; PROF PATH, UNIV ALA, BIRMINGHAM, 75- *Concurrent Pos:* Res career develop award, NIH, 72-75; counr, Int Soc Heart Res. *Mem:* AAAS; Am Asn Pathologists; Am Vet Med Asn; Am Heart Asn; Int Acad Path; Int Soc Heart Res. *Res:* Comparative cardiovascular pathology; hypertrophy; congestive heart failure; atherosclerosis; myocardial infarction. *Mailing Add:* Dept Path Univ Ala at Birmingham University Sta Birmingham AL 35294

BISHOP, STEPHEN GRAY, b York, Pa, Jan 26, 39; m 63; c 2. SOLID STATE PHYSICS. *Educ:* Gettysburg Col, BA, 60; Brown Univ, PhD(physics), 65. *Prof Exp:* Res assoc physics, Brown Univ, 65-66; Nat Acad Sci-Nat Res Coun res assoc, 66-68, res physicist semiconductor physics, 68-73, SUPVRY RES PHYSICIST SEMICONDUCTOR PHYSICS, NAVAL RES LAB, 74- *Concurrent Pos:* Vis scientist semiconductor physics, Max Planck Inst, Stuttgart, Ger, 73-74; exchange scientist, Royal Signals & Radar Estab, Great Malvern, Eng, 78-79. *Honors & Awards:* Pure Sci Award, Res & Eng Soc Am, 77. *Mem:* AAAS; fel Am Phys Soc; Sigma Xi. *Res:* Optical and electronic properties of crystalline and amorphous semiconductors; optical spectroscopy, luminescence and magnetic resonance techniques. *Mailing Add:* Code 6870 Naval Res Lab Washington DC 20375

BISHOP, STEPHEN HURST, b Philadelphia, Pa, June 22, 36; m 62; c 3. BIOCHEMISTRY, COMPARATIVE PHYSIOLOGY. *Educ:* Gettysburg Col, BA, 58; Duke Univ, MA, 60; Rice Univ, PhD(biol), 64. *Prof Exp:* Res assoc biochem, Univ Kans Med Ctr, 64-67; asst prof biochem, Baylor Col Med, 67-77; assoc prof, 77-79, PROF ZOOL, IOWA STATE UNIV, 79- *Concurrent Pos:* Fel biochem, Univ Kans, 64-67; mem, Inst Summer Prog, MBL, NSF, 75-77, Regulatory Biol Grant Rev Panel, 78-81, prog officer, Regulatory Biol, 86-88; coordr molecular cell develop prog, Iowa State Univ, 80-83. *Mem:* Fel AAAS; Soc Complex Carbohydrates; Am Chem Soc; Am Soc Biochem & Molecular Biol; Am Soc Zool; NY Acad Sci. *Res:* Phosphoglycoprotein structure-function; nitrogen metabolism; regulation of aminoacid metabolism; osmoregulation; enzymology. *Mailing Add:* Dept Zool & Genetics Iowa State Univ Ames IA 50011

BISHOP, THOMAS PARKER, b Richland, Ga, July 22, 36; m 62; c 2. SOLID STATE PHYSICS. *Educ:* Carson-Newman Col, BS, 59; Emory Univ, MS, 63; Clemson Univ, PhD, 68. *Prof Exp:* From asst prof to assoc prof, 67-79, PROF PHYSICS, GA SOUTHERN UNIV, 79- *Mem:* Am Asn Physics Teachers; Am Phys Soc. *Res:* Color centers in solids; electron paramagnetic resonance; health physics. *Mailing Add:* Dept Physics Ga Southern Univ Statesboro GA 30460-8031

BISHOP, VERNON S, CARDIOVASCULAR PHYSIOLOGY. *Prof Exp:* PROF PHARMACOL, UNIV TEX, 68- *Mailing Add:* Dept Pharmacol Health Sci Ctr Univ Tex 7703 Floyd Curl Dr San Antonio TX 78284-7764

BISHOP, WALTON B, b LeRoy, Ill, Aug 25, 17; m 40, 59; c 3. RELIABILITY, SEQUENTIAL ANALYSIS. *Educ:* Ill State Univ, BEd, 39; Boston Univ, MA, 50; Northeastern Univ, MSEE, 54; Univ Md, MA, 83, PhD(pub commun), 90. *Prof Exp:* Instr pub schs, Ill, 39-42; radio engr, Scott Field, Ill, 42-43; electronics engr & mathematician, Air Force Cambridge Res Labs, 46-66; res electronics engr, 66-80, CONSULT, US NAVAL RES LAB. *Concurrent Pos:* Lectr eve classes, Northeastern Univ, 60-66; consult, 85- *Mem:* Sigma Xi; AAAS; Inst Elec & Electronics Engrs; Soc ProfJournalist. *Res:* Electronic means of identifying aircraft; reliability; science communication and science education; medical electronics. *Mailing Add:* 6 Balmoral Dr E Oxon Hill MD 20745

BISHOP, WILLIAM P, b Lakewood, Ohio, Jan 18, 40; m 63. SPACE & ENVIRONMENTAL SCIENCE. *Educ:* Col Wooster, BA, 62; Ohio State Univ, PhD(phys chem), 67. *Prof Exp:* Vis res assoc chem, Ohio State Univ, 67-69; mem staff nuclear fuel cycle, Sandia Labs, 69-75; chief waste mgt br, US Nuclear Regulatory Comn, 75-76, asst dir, 76-78; dep dir Environ Observations Div, NASA, 78-81, dep dir, Life Sciences Div, 81-83; dep asst admin satellites & actg asst admin environ satellite, Data & Info Serv, Nat Oceanic & Atmospheric Admin, 83-87; vpres, SAIC, 87-89; VPRES RES, DESERT RES INST, 90- *Concurrent Pos:* Vpres tech, Am Inst Aeronaut & Astronaut, 86-87. *Honors & Awards:* Silver Medal, Nat Res Coun & Dept Com. *Mem:* AAAS; Am Nuclear Soc; Am Geophys Union; Sigma Xi; Am Inst Aeronaut & Astronaut. *Res:* Radiation chemistry, radiation dosimetry; underground testing; nuclear fuel cycle research; nuclear waste management; remote sensing; atmospheric and oceanographic science. *Mailing Add:* 1855 Quarley Pl Henderson NV 89014

BISHOP, YVONNE M M, b Eng, Jan 12, 25; US citizen. BIOSTATISTICS, MATHEMATICAL STATISTICS. *Educ:* Univ London, BA, 47; Harvard Univ, MSc, 61, PhD(statist), 67. *Prof Exp:* Math ed, Ministry Agr & Fisheries, Eng, 49-51; ed asst J Iron & Steel Inst, 51-52; sr ed asst, Brit Coal Utilization Res Asn, 52-53; statistician, Pac Biol Sta, BC, 53-54, Int Pac Salmon Fisheries Comn, Can, 54-56, Inter-Am Trop Tuna Comn, 56-57 & St Paul's Rehab Ctr for Newly Blinded Adults, 58-61; instr statist, Sch Pub Health, Harvard Univ, 61-63, asst prof biostatist, 63-71, assoc prof, 71-80; dep asst adminr, Energy Data Opers, Energy Info Admin, 79-81, DIR OFF STATIST STANDARDS, 81- *Concurrent Pos:* Head, Div Biostatist & Data Processing, Sidney Farber Cancer Ctr, 66-75; mem bd dirs, Am Statist Asn, 81-83. *Mem:* Biomet Soc; fel Am Statist Asn; Int Statist Inst; fel Am Pub Health Asn. *Res:* Statistics applied to energy; development of statistical methodology, particularly quality assessment of survey data. *Mailing Add:* 5432 39th St NW Washington DC 20015

BISIO, ATILLIO L, b New York, NY, Aug 21, 30; m 55; c 3. SCALEUP, POLYMERIZATION PROCESSES. *Educ:* Columbia Col, BA, 52, BS, 53, MS, 54. *Prof Exp:* Engr & mgr, Exxon Res & Eng Co, 57-81, eng adv, 81-86; PRIN, ATRO ASSOCS, 86- *Concurrent Pos:* Adj lectr, Dept Chem Eng, Pa State Univ, 85-; ed, Chem Eng Progress, 86-90. *Mem:* AAAS; Am Chem Soc; Am Inst Chem Engrs; Asn Consult Chemists & Chem Engrs; Asn Res Dirs; Nat Asn Sci Writers. *Res:* Chemical and energy engineering; scaleup emphasis. *Mailing Add:* 1509 Woodacres Dr Mountainside NJ 07092

BISKEBORN, MERLE CHESTER, b Scotia, Nebr, Sept 22, 07; m 35; c 3. ENGINEERING, MATERIALS SCIENCE. *Educ:* SDak Sch Mines & Technol, BS, 30. *Hon Degrees:* DEng, SDak Sch Mines & Technol, 61. *Prof Exp:* Mem tech staff cable develop, Bell Tel Labs, Inc, Am Tel & Tel Co, 30-38, supvr, 38-42, engr radar mfg, Western Elec Co, Md, 42-44, supvr cable develop, Bell Tel Labs, Inc, 44-51, asst toll cable engr, 51-52, toll cable engr, 52-54, sub-dept head, 54-62, head cable dept, 62-72; dir res & develop, Phelps Dodge Commun Co, 72-83; RETIRED. *Concurrent Pos:* Consult, 83- *Mem:* Inst Elec & Electronics Engrs. *Res:* Carrier telephone cable; crosstalk and high frequency primary constants of cable; first automatic radar; microwave resonant cavities; microwave coaxials; first transatlantic telephone cable; anti-submarine cable; wire connectors; cable reclamation compound; waterproof telephone cable. *Mailing Add:* 34 Linden Lane Chatham NJ 07928

BISMANIS, JEKABS EDWARDS, b Svitene, Latvia, Dec 28, 11; m 37; c 3. BACTERIOLOGY. *Educ:* Latvian Univ, MD, 35; Med Coun Can, LMCC, 50;. *Prof Exp:* Asst bact, Latvian Univ, 35-39; in charge, Pasteur Inst, 39-44; in charge bact lab, Dept Health, Riga, Latvia, 39-44; practicing physician, Ger, 44-48; lectr bact, Univ Ottawa, 48-51; med bacteriologist, Lab Pub Health, Regina, Sask, 51-52; from asst prof to prof bact, 52-77, EMER PROF MED MICROBIOL, UNIV BC, 77- *Concurrent Pos:* Family physician, 77-88, 89-90. *Mem:* Emer mem Can Soc Microbiol; fel Royal Col Physicians Can, 73. *Res:* Virology; activation of latent infections by immunosuppression. *Mailing Add:* 4306 W Eighth Ave Vancouver BC V6R 2A1 Can

BISNO, ALAN LESTER, b Memphis, Tenn, Sept 28, 36; m 63; c 2. INTERNAL MEDICINE, INFECTIOUS DISEASES. *Educ:* Princeton Univ, AB, 58; Wash Univ, MD, 62. *Prof Exp:* Intern internal med, Vanderbilt Univ Hosp, 62-63, resident, 63-65; med epidemiologist, Ctr Dis Control, USPHS, 65-68; from asst prof to prof, Univ Tenn, Memphis, 69-86, chief sect infectious diseases, 71-86, assoc dean acad & fac affairs, 81-87 Univ Tenn, Memphis, 81-84; PROF MED & VCHMN VA AFFAIRS, DEPT MED, UNIV MIAMI, 87-; CHIEF MED SERV, MIAMI VA MED CTR, 87- *Concurrent Pos:* NIH training grant infectious dis, Med Sch, Univ Tenn, Memphis, 68-69; teaching & res scholar Am Col Physicians, 69-72; Comt Rheumatic Fever & Endocarditis, Am Health Asn, 75-89. *Mem:* Infectious Dis Soc Am; fel Am Col Physicians; Cent Soc Clin Res; Am Fedn Clin Res; Am Heart Asn. *Res:* Host-parasite relations in infectious diseases and clinical epidemiology as it relates to infectious diseases; special interest and emphasis in streptococcal diseases and their sequelae. *Mailing Add:* Med Serv Miami Vet Admin Med Ctr 1201 NW 16 St Miami FL 33125

BISQUE, RAMON EDWARD, b Stambaugh, Mich, Sept 1, 31; m 54; c 5. GEOCHEMISTRY. *Educ:* St Norbert Col, BS, 53; Iowa State Univ, MS, 56 & 57, PhD(geochem), 59. *Prof Exp:* From asst prof to assoc prof geochem, 59-69, prof chem & head dept, 69-77, adj prof, 77-90, EMER PROF CHEM & GEOCHEM, COLO SCH MINES, 90- *Concurrent Pos:* Consult, Arthur D Little, Inc, Mass, 62-64 & Am Geol Inst, DC, 64-66; co-founder, Earth Sci, Inc, Colo, 62; assoc dir, Earth Sci Curric Proj, Colo Sch Mines, 65, dir, 65- *Mem:* Am Geochem Soc. *Res:* Geochemistry and analytical chemistry as applied to mineral exploration, benefication and exploitation; curriculum development in earth science. *Mailing Add:* 9113 Fern Way Coal Creek Canyon Golden CO 80401

BISSELL, CHARLES LYNN, b Cumberland, Iowa, Mar 22, 39; m 59; c 3. PHYSICAL CHEMISTRY. *Educ:* Tarkio Col, BA, 61; Iowa State Univ, MS, 63; Tex A&M Univ, PhD(phys chem), 65. *Prof Exp:* From asst prof to assoc prof, 65-74, PROF CHEM, NORTHWESTERN STATE UNIV, LA, 74- *Mem:* Am Chem Soc. *Res:* Electrochemistry of fused salts with conductivity and electrochemical cells as the primary concern. *Mailing Add:* Dept Chem & Physics Northwestern State Univ La Natchitoches LA 71497

BISSELL, DWIGHT MONTGOMERY, b San Francisco, Calif, July 11, 40; m 67; c 2. HEPATOLOGY, GASTROENTEROLOGY. *Educ:* Harvard Univ, AB, 62, MD, 67. *Prof Exp:* Assoc prof, 80-86, PROF MED, UNIV CALIF, SAN FRANCISCO, 86- *Concurrent Pos:* Coun, Am Asn Study Liver Dis, 90- *Mem:* Asn Am Physicians; Am Soc Clin Invest; Am Soc Cell Biol. *Res:* Cellular and molecular basis for liver fibrosis and cirrhosis; iron-related hepatic injury; liver disease of alcohol abuse; porphyrias. *Mailing Add:* Bldg 40 Rm 4102 San Francisco Gen Hosp San Francisco CA 94110

BISSELL, EUGENE RICHARD, b San Francisco, Calif, June 24, 28; m 63. ORGANIC CHEMISTRY. *Educ:* Univ Calif, BS, 49; Mass Inst Technol, PhD(org chem), 53. *Prof Exp:* Chemist, Lawrence Livermore Nat Lab, 53-83; chemist, Enzyme Systs Prod, 83-85; CONSULT, PROTOTEK INC & ENZYME SYSTS PROD, 87- *Mem:* Am Chem Soc. *Res:* Organic synthesis; fluorine and polymer chemistry; synthesis of biologically active materials; dye synthesis. *Mailing Add:* 101 Via Lucia Alamo CA 94507

BISSELL, GLENN DANIEL, b Rochester, NY, Sept 21, 35; m 73. VETERINARY MEDICINE. *Educ:* Univ Calif, BS, 57, DVM, 59, MPH, 83. *Prof Exp:* Vet I diag lab, Calif Dept Agr, 59-62; med researcher, US Army Biol Labs, 62-63; vet II & III diag lab, Calif Dept Agr, 63-66; Miller prof, Univ Calif, San Diego, 64-65, prof meth 65-; RETIRED. *Concurrent Pos:* Consult, Hayward State Univ, 71- & Cutter Labs, 74-; asst prof, Sch Vet Med, Univ Calif, Davis, 76. *Mem:* Am Asn Lab Animal Sci; Am Vet Med Asn. *Res:* Veterinary medicine in the study of the care, breeding, utilization and diseases of laboratory animals. *Mailing Add:* 6250 Lambie Rd Suisun City CA 94585

BISSELL, GROSVENOR WILLSE, b Buffalo, NY, Mar 17, 15; wid. INTERNAL MEDICINE. *Educ:* Univ Buffalo, MD, 39. *Prof Exp:* Intern, Buffalo Gen Hosp, 39-40, asst res med & pediat, 40-41, chief res med, 41-42; assoc, Thorndike Mem Lab, 43-44; instr internal med, sch med Univ Buffalo, 44-46, assoc, 46-48, asst prof, 48-51; asst med dir, Armour Labs, 51-52; asst prof internal med, Sch Med, Univ Buffalo, 52-56, assoc prof med, 56-62; chief med serv, Vet Admin Hosp, Buffalo, 52-62; chief endocrine sect & assoc chief staff for res, Sunmount Vet Admin Hosp, Tupper Lake, NY, 62-65; from assoc prof to prof internal med, Sch Med, Wayne State Univ, 65-76; chief med serv, Vet Admin Hosp, Allen Park, 65-76 & Saginaw, 76-81; prof med, Mich State Univ, East Lansing, 76-81; clin prof internal med, Sch Med, Wayne State Univ, Detroit, 76-86; RETIRED. *Concurrent Pos:* Res fel, Thorndike Mem Lab, 42-43; instr clin path, Sch Med, Harvard Univ & vis physician, Boston City Hosp, 43-44; asst internal med, E J Meyer Mem Hosp, 44-51, attend, 52-62; chmn, Vet Admin Coop Study Oral Hypoglycemic Agents in Treatment of Diabetes Mellitus, 51-64; consult internal med, endocrinol & addictology, Vet Admin Med Ctr, Saginaw, 81. *Mem:* AAAS; AMA; Am Thyroid Asn; Endorine Soc. *Res:* Endocrinology; metabolism. *Mailing Add:* 5685 Noel Ct Saginaw MI 48603

BISSELL, HAROLD JOSEPH, b Springville, Utah, Feb 9, 13; m 40; c 3. GEOLOGY. *Educ:* Brigham Young Univ, BS, 34; Univ Iowa, MS, 36, PhD(sedimentation), 48. *Prof Exp:* Teacher pub sch, Utah, 37-38; instr geol, Brigham Young Univ, 38-40; jr geologist, US Army Corps Engrs, Miss, 41-42; engr, Geneva Steel Co, Utah, 42-43; chmn dept geol geog, Br Agr Col, Utah State Agr Col, 43-44; asst geologist, US Geol Surv, Utah, 44-45; from asst prof to assoc prof geol, Brigham Young Univ, 46-52, chmn dept, 54-56, prof, 52-80; RETIRED. *Concurrent Pos:* Assoc geologist, US Geol Surv, Utah,

47-50; vis prof, Univ Wash, 56-66 & Univ NC, Chapel Hill, 66-67. *Mem:* Am Asn Petrol Geologists; fel Geol Soc Am; Paleont Soc; Soc Econ Paleontol & Mineral; Int Asn Sedimentologists; Sigma Xi. *Res:* Lake Bonneville sedimentation; Pennsylvanian Fusulinidae of Utah; eolian deposits of Utah; Pleistocene sedimentation in southern Utah Valley, Utah; lower Triassic southern Nevada; sedimentary petrographic carbonates; Permo-Triassic eastern Great Basin. *Mailing Add:* 452 S 850 E Orem UT 84058

BISSELL, MICHAEL G, b Ridgecrest, Calif, Mar 5, 47. CLINICAL CHEMISTRY. *Educ:* Univ Ariz, BS(chem) & BS(math), 69; Stanford Univ, MD, 75, PhD(neuro & biobehav sci), 77; Univ Calif, Berkeley, MPH, 78; Am Bd Path, dipl clin path, 81. *Prof Exp:* Resident clin path, Martinez Vet Admin Med Ctr, Univ Calif, Daivs, 78-81; sr staff assoc fel, Lab Neurochem, NIMH, 81-84; asst prof path, Pritzker Sch Med, Univ Chicago, 84-88; dir clin path, City Hope Nat Med Ctr, 88-91; CLIN ASST PROF PATH, SCH MED, UNIV CALIF, LOS ANGELES, 89-; VPRES & MED DIR, NICHOLS INST, 91- *Concurrent Pos:* Asst dir clin chem, Med Ctr, Univ Chicago, 84-86, dir gen clin chem, Hosps, 86-88; consult, Bulk Power Facil Siting Comt, State NH Pub Utilities Comn, 85-86, Health Care Financing Admin, 87, Biomed & Environ Consults, 88-, Clin Lab Mgt Asn, 89-, Nat Cancer Inst, 89, Nat Comt Clin Lab Standards & Acad Med Arts & Sci, 90-; ed, Clin Lab Mgt Asn Newslett, 89-; mem, Coun Continuing Educ, Am Soc Clin Pathologists, 90- *Mem:* Am Asn Clin Chem; Am Pub Health Asn; Am Soc Clin Pathologists; Clin Lab Mgt Asn; Col Am Pathologists; Sigma Xi. *Res:* Clinical pathology; clincial chemistry; author of more than 30 technical publications. *Mailing Add:* Nichols Inst Ref Labs 26441 Via de Anza San Juan Capistrano CA 92675

BISSELL, MINA JAHAN, b Tehran, Iran, May 14, 40; m 67; c 2. MOLECULAR BIOLOGY. *Educ:* Radcliffe Col, AB, 63; Harvard Univ, MA, 65, PhD(molecular genetics), 69. *Prof Exp:* Milton fel, Harvard Univ, 69-70; Am Cancer Soc fel molecular biol, Univ Calif, Berkeley, 70-72, sr biochemist, 72-76, SR STAFF MEM, LAWRENCE BERKELEY LAB, UNIV CALIF, BERKELEY, 76-, CO-DIR, LAB CELL BIOL, 80-, DIR, DIV CELL & MOLECULAR BIOL, 88- *Mem:* Am Soc Biol Chemists; Am Soc Cell Biol; AAAS; Asn Women in Sci; Tissue Culture Asn. *Res:* Regulation of gene expression in normal and malignant cells. *Mailing Add:* Lab Cell Biol Lawrence Berkeley Lab Univ Calif Bldg 83 Berkeley CA 94720

BISSETT, J(AMES) R(OBERT), b Junction, Tex, May 9, 10; m 35; c 1. CIVIL ENGINEERING. *Educ:* Univ Tex, BSCE, 41; Univ Ill, MSCE, 50. *Prof Exp:* Field engr, Hwy Dept, Tex, 32-39, bridge design engr, 44-47; construct engr, US Engrs, 41-44; instr, Univ Ill, 49-50; asst prof civil eng, 47-49, from assoc prof to prof, 50-75, assoc dir end, Exp Sta, 53-61, EMER PROF CIVIL ENG, UNIV ARK, 75- *Mem:* Am Soc Civil Engrs; Nat Soc Prof Engrs; Am Soc Eng Educ. *Res:* Highway engineering and soil mechanics. *Mailing Add:* 14300 Chenal Pkwy No 3210 Little Rock AR 72211-5811

BISSETT, MARJORIE LOUISE, b Miami, Ariz, Sept 21, 25. MICROBIOLOGY. *Educ:* Univ Calif, Los Angeles, AB, 49; Univ Mich, PhD(epidemiol sci), 65. *Prof Exp:* Microbiologist, Calif State Dept Pub Health, 49-55, asst microbiologist, 55-61; trainee epidemiol sci, Univ Mich, 61-65; RES MICROBIOLOGIST, CALIF STATE DEPT PUB HEALTH, 65- *Concurrent Pos:* Mem, Diag Prod Adv Comt, Food & Drug Admin, 72-76. *Mem:* Soc Gen Microbiol; Am Soc Microbiol; NY Acad Sci; Conf State & Prov Pub Health Lab Dirs; fel Am Acad Microbiol. *Res:* Clinical and public health microbiology; microbiology and laboratory diagnosis of Neisseria gonorrheae, Yersinia species, Legionellaceae, Enterobacteriaceae, Borrelia burgdorferi and other medically important organisms. *Mailing Add:* 1751 Spruce St Berkeley CA 94709

BISSEY, LUTHER TRAUGER, b New Britain, Pa, Aug 2, 12; m 40. CHEMISTRY, PETROLEUM ENGINEERING. *Educ:* Pa State Univ, BS, 34, MS, 42. *Prof Exp:* Analyst, Mineral Indust Exp Sta, Col Mineral Indust, 36-38, instr, Petrol & Natural Gas Dept, 38-50, from asst prof to assoc prof, 50-73, EMER ASSOC PROF, PETROL & NATURAL GAS DEPT, PA STATE UNIV, 73- *Res:* Water-flooding; household refuse incineration; natural gas pipeline deposits; gas flow in porous media; gas reservoir mechanics; instrumentation and communications. *Mailing Add:* 930 N Atherton St State College PA 16803-3540

BISSHOPP, FREDERIC EDWARD, b Beloit, Wis, Oct 2, 34; m 58; c 1. APPLIED MATHEMATICS. *Educ:* Ill Inst Technol, BS, 54; Univ Chicago, MS, 56, PhD(physics), 59. *Prof Exp:* Asst prof, 61-65, ASSOC PROF MATH, BROWN UNIV, 65- *Res:* Fluid mechanics; electromagnetic theory; partial differential equations; numerical analysis. *Mailing Add:* c/o Dept Appl Math Brown Univ Providence RI 02912

BISSING, DONALD EUGENE, b McCook, Nebr, Sept 10, 34; m 54; c 2. ORGANIC CHEMISTRY. *Educ:* Fort Hays State Col, BA, 59; Univ Kans, PhD(org chem), 62. *Prof Exp:* Mgr process develop, Monsanto Co, 62-74; dir chem eng res & develop chem, BASF Wyandotte Corp, 74-76, gen mgr agr chem div, 76-78; DIR, RES & DEVELOP, AGR CHEM GROUP, FMC CORP, 78- *Mem:* Am Chem Soc; Res & Eng Soc Am. *Res:* Chemistry of agricultural products, polyether polyols, graft polyols; environmental control chemistry and engineering; residue, metabolism of agricultural products. *Mailing Add:* FMC Corp 2000 Market St Philadelphia PA 19103

BISSING, DONALD RAY, b Poplar Bluff, Mo, Feb 18, 43; m 69; c 2. BOTANY, VASCULAR PLANT ANATOMY. *Educ:* Univ Md, BS, 71; Claremont Grad Sch, MA, 73, PhD(bot), 76. *Prof Exp:* ASST PROF BOT, SOUTHERN ILL UNIV, CARBONDALE, 76- *Mem:* AAAS; Int Asn Wood Anatomists; Bot Soc Am. *Res:* Systematic and functional aspects of plant anatomy and morphology. *Mailing Add:* Dept Bot Southern Ill Univ Carbondale IL 62901

BISSINGER, BARNARD HINKLE, b Lancaster, Pa, Jan 27, 18; m 50; c 2. MATHEMATICS, ENGINEERING. *Educ:* Franklin & Marshall Col, BS, 38; Syracuse Univ, MA, 40; Cornell Univ, PhD(math), 43. *Prof Exp:* Instr math, Cornell Univ, 43; asst prof, Mich State Col, 44; opers res analyst, US Army Air Force, 44 & Chennault Flying Tigers, China, 45; gen mgr, Athletic Shoe Factory, 46-52; Lehman prof math & chmn dept, Lebanon Valley Col, 53-68; prof on leave, 68-69, PROF MATH SCI, PA STATE UNIV, CAPITOL CAMPUS, 70- *Concurrent Pos:* Assoc actuary, Pa State Univ, 68; NSF sci fac fel, Princeton Univ, 58-59. *Mem:* Am Math Soc; Math Asn Am; fel Royal Statist Soc. *Res:* Operations research; mathematical analysis; functional analogues of continued fractions; group physical mortality. *Mailing Add:* 281 W Main St Middletown PA 17057

BISSON, EDMOND E(MILE), b East Barre, Vt, July 16, 16; m 47; c 3. TRIBOLOGICAL RESEARCH. *Educ:* Univ Fla, BS, 38, ME, 54. *Prof Exp:* Res & proj engr lubrication & wear res, NACA/NASA, Va, 39-43, head piston ring res sect, 43-45, head lubrication & wear res sect, 45-55, asst chief, Fluid Syst Components Div, 55-68, assoc chief, 68-73; CONSULT ENGR, 73- *Concurrent Pos:* Teacher short courses, Univ Calif, Los Angeles, Univ Tenn Space Inst, Va Polytech Inst & State Univ; ed-in-chief, Soc Tribologists & Lubrication Engrs, 76-90. *Honors & Awards:* Hunt Mem Medal, Am Soc Lubrication Engrs, 54, Nat Award, 67; Jacques de Vaucanson Medal, Groupement pour l'Avancement de la Mecanque Industrielle, 66; Medal Except Sci Achievement, NASA, 68; P M Ku Medal, Am Soc Lubrication Engrs, 80; Hersey Award, Am Soc Mech Engrs, 81. *Mem:* Fel Am Soc Mech Engrs; fel Soc Tribologists & Lubrication Engrs (pres, 63-64); NY Acad Sci. *Res:* Friction, lubrication and wear, tribology; unconventional lubricants; bearings and seals under extreme operating conditions. *Mailing Add:* 805 Windrush Dr Unit A-1 Westlake OH 44145

BISSON, MARY A, b Fairfield, Calif, Oct 24, 48; m 68. ELECTROPHYSIOLOGY, ECOPHYSIOLOGY. *Educ:* Univ Chicago, BA, 70; Duke Univ, PhD(bot), 76. *Prof Exp:* Trainee physiol, Univ NC, Chapel Hill, 76-77; res fel biophys, Univ Sydney, New South Wales, Australia, 78-80; asst prof, 80-86, ASSOC PROF, DEPT BIOL SCI, STATE UNIV NY, BUFFALO, 87- *Concurrent Pos:* Guest lectr, Marine Biol Lab, 82. *Mem:* AAAS; Am Soc Plant Physiologists; Asn Women Sci; Sigma Xi. *Res:* Control of ion transport in plants; control of cytoplasmic pH by active and passive proton movements, and mechanisms of salt tolerance in charophyte algae. *Mailing Add:* Dept Biol Sci Cooke 109 State Univ NY Buffalo NY 14260

BISSON, PETER ANDRE, b Dover, Del, Aug 28, 45; c 4. AQUATIC BIOLOGY. *Educ:* Univ Calif, Santa Barbara, BA, 67; Ore State Univ, MS, 69, PhD(fisheries), 75. *Prof Exp:* Sr scientist, 75-80, PROJ MGR AQUATIC BIOL, WEYERHAEUSER CO, 80- *Mem:* Ecol Soc Am; Am Fisheries Soc; Am Soc Limnol Oceanog; Sigma Xi; NAm Benthological Soc. *Res:* Biology of streams; systematics and zoogeography of freshwater fishes; ecology of fish populations. *Mailing Add:* Technol Ctr Weyerhaeuser Co Tacoma WA 98477

BISSONETTE, JOHN ALFRED, b Colchester, Vt, July 9, 41; m 66; c 2. WILDLIFE ECOLOGY & MANAGEMENT. *Educ:* Univ Vt, BA, 64; Yale Univ, MFS, 70; Univ Mich, PhD(wildlife mgt & resource ecol), 76. *Prof Exp:* Vis asst prof vert ecol, Ariz State Univ, 75-76, asst prof, 76-77; asst leader wildlife, Okla Coop Wildlife Res Unit, 77-81, ASST LEADER WILDLIFE, MAINE COOP WILDLIFE RES UNIT, US FISH & WILDLIFE SERV, 81- *Concurrent Pos:* Adj assoc prof wildlife ecol, Okla State Univ, 77-; coop assoc prof wildlife resource, Univ Maine, 81- *Mem:* Wildlife Soc. *Res:* Mammals, especially ungulate ecology & behavior; collared peccary (Dicotyles tajacu) biology population dynamics. *Mailing Add:* Dept Wildlife Mgmt Utah State Univ Logan UT 84322-5210

BISSONNETTE, GARY KENT, b Claremont, NH, May 8, 47; m 72; c 2. ENVIRONMENTAL MICROBIOLOGY, AQUATIC ECOLOGY. *Educ:* Univ NH, BA, 69; Mont State Univ, MS, 71, PhD(microbiol), 74. *Prof Exp:* From asst prof to assoc prof bact Div Plant Sci, 74-82, PROF AGR MICROBIOL, WVA UNIV, 82- *Mem:* Am Inst Biol Sci; Sigma Xi; Am Soc Microbiol; Soc Indust Microbiol; fel Am Acad Microbiol; Am Water Works Asn. *Res:* Recovery of organisms of public health significance from aquatic environments; effects of environmental stress on microorganisms. *Mailing Add:* Div Plant Sci 401 Brooks Hall WVa Univ PO Box 6057 Morgantown WV 26506-6057

BISSONNETTE, HOWARD LOUIS, b Detroit, Mich, Aug 28, 27; c 3. PLANT PATHOLOGY. *Educ:* Col St Thomas, BS, 52; Univ Minn, MS, 58, PhD, 64. *Prof Exp:* Res asst, Univ Minn, 53-56, plant pathologist, USDA, 57-62; plant pathologist, USDA, 57-62; exten plant pathologist, NDak State Univ, 62-68; prof plant path & plant patholoogist, agr exten, 68-87, EMER PROF, UNIV MINN, MINNEAPOLIS, 87- *Concurrent Pos:* Consult, crop prod, potato storage. *Mem:* Am Phytopath Soc. *Res:* Soil mycroflora; root rot diseases of sugar beets and peas; general pathology; potato diseases; aerial application of fungicides. *Mailing Add:* 3456 Milton Shoreview MN 55126

BISSONNETTE, JOHN MAURICE, b Montreal, Can, Feb 11, 39. OBSTETRICS & GYNECOLOGY. *Educ:* Loyola Col Montreal, BA, 60; McGill Univ, MD CM, 64. *Prof Exp:* Rotating intern, Montreal Gen Hosp, 64-65, asst resident, 65-66; resident obstet & gynec, Johns Hopkins Hosp, Baltimore, 66-69; fel pulmonary physiol, Sch Pub Health, Johns Hopkins Univ, 69-70; asst prof obstet & gynec, Univ Calif, Irvine, 70-73; from asst prof to assoc prof, 73-79, PROF OBSTET & GYNEC, ORE HEALTH SCI UNIV, 79-, PROF CELL BIOL & ANAT, 81- *Honors & Awards:* Duncan Reid Lectr, Harvard Univ, 84. *Mem:* Am Physiol Soc; Soc Gynec Invest; AAAS. *Res:* Respiratory and cardiovascular physiology of the mammalian fetus and placenta, and membrane transport in the placenta especially placental glucose transport; control of fetal breathing movements in utero. *Mailing Add:* Rm 426 Res Bldg Ore Health Sci Univ Portland OR 97201-3098

BISSOT, THOMAS CHARLES, b Grand Rapids, Mich, Apr 20, 30; m 51; c 8. INORGANIC CHEMISTRY, PHYSICAL CHEMISTRY. *Educ:* Aquinas Col, BS, 52; Univ Mich, MS, 53, PhD(chem), 56. *Prof Exp:* Mem staff, E I Du Pont de Nemours & Co, 55-82, sr res fel, 82-90; CONSULT, 90- *Mem:* Am Chem Soc. *Res:* Boron hydrides; high vacuum techniques; catalysis; polymer and surface chemistry; membrane development for chloralkali cells; barrier polymers for packaging. *Mailing Add:* 25 Deer Run Rd Newark DE 19711

BISTRIAN, BRUCE RYAN, b Southhampton, NY, Oct 22, 39; m 64; c 3. NUTRITION, MEDICINE. *Educ:* NY Univ, AB, 61; Cornell Univ, MD, 65; Johns Hopkins Sch Pub Health, MPH, 71; Mass Inst Technol, PhD(nutrit biochem), 76. *Prof Exp:* Intern, Bellevue Hosp, Cornell Div, 65-66; fel metab, Univ Vt Med Sch, 68-69, resident, 69-70; from clin asst prof to assoc prof, 76-90, PROF MED, HARVARD MED SCH, 90- *Concurrent Pos:* Res assoc, Mass Inst Technol, 76-81, lectr nutrit, 81-84; mem, Nutrit Study Sect, NIH, 85-89. *Mem:* Sigma Xi; Am Soc Clin Nutrit; Fedn NAm Soc Exp Biol; Am Col Nutrit; fel Am Col Physicians; Am Soc Parental & Enteral Nutrit (pres, 89-90). *Res:* Optimal compostion of diets for the treatment of obesity; nutritional assessment and nutritional support of the hospitalized patient; dynamic aspects of protein metabolism in semistarvation, injury or infection; novel fats in clinical nutrition; effect of cytokines on intermediary metabolism. *Mailing Add:* Cancer Res Inst 194 Pilgrim Rd Boston MA 02215

BISWAL, NILAMBAR, b Khamar, Orissa, India, Feb 20, 34; m 67; c 3. VIROLOGY. *Educ:* Punjab Col Vet Med, BVSc & AH, 58; Mich State Univ, MS, 63, PhD(microbiol), 65. *Prof Exp:* Asst res virologist, Virus Lab, Univ Calif, Berkeley, 65-67; fel virol, 67-68, asst prof, 68-72, ASSOC PROF VIROL, BAYLOR COL MED, HOUSTON, 72- *Mem:* Am Soc Microbiol; AAAS; Sigma Xi. *Res:* Basic mechanism of biosynthesis of herpes virus DNA; analysis of structure and replication of herpes virus DNA. *Mailing Add:* 9704 Kerrigan Ct Randalls MD 21133

BISWAS, ASIT KUMAR, b Balasore, India, Feb 25, 39. ENVIRONMENTAL MANAGEMENT, AGRICULTURAL DEVELOPMENT. *Educ:* Indian Inst Technol, Kharagpur, BTech, 60, MTech, 61; Univ Strathclyde, PhD(water management), 67. *Hon Degrees:* DTech, Univ Lund, Sweden, 84. *Prof Exp:* Asst civil engr, Ward, Ashcroft & Parkman, UK, 61-62; res fel hydraul eng, Loughborough Univ Technol, Eng, 62-63; lectr water resources eng, Univ Strathclyde, 63-67; prof water resource planning, Queen's Univ, Ont, 67-68; sr res officer, Dept Energy, Mines & Resources, 68-70; dir, Dept Environ, Ottawa, Can, 70-76; DIR, BISWAS & ASSOCS, OTTAWA, 76- *Concurrent Pos:* Vis prof, Univ Ottawa, 68-70; sr consult, UN Environ Prog, 74-; Rockefeller Found Int Relations fel, 74; sr res scientist, Int Inst Appl Syst Anal, 78-79; ed, J Ecol Modeling, Water Resources Develop J & Mazingira J; sr sci adv to exec dir, UN Environ Prog, pres, Int Fund Agr Develop, UN Univ & UN Indust Develop Orgn. *Honors & Awards:* Walter L Huber Res Medal, Am Soc Civil Engrs, 74. *Mem:* Am Soc Civil Engrs; Int Soc Ecol Modelling (pres, 76-91); Int Water Resource Asn (vpres, 79-82, pres, 88-91); Int Asn Hydraulic Res; hon mem Indian Water Resources Soc. *Res:* Resources management; irrigated agriculture; environmental sciences; climate and society; hydrology and water resources; rural and agricultural development. *Mailing Add:* 28 Elvaston Ave Nepean ON K2G 3T4 Can

BISWAS, CHITRA, b Oct 16, 36; c 2. CONNECTIVE TISSUE BIOCHEMISTRY. *Educ:* Univ Ill, Urbana, PhD(nutrit biochem), 63. *Prof Exp:* Asst prof, 81-88, ASSOC PROF ANAT, CELL BIOL & MED HISTOL, SCH MED, TUFTS UNIV, 88- *Concurrent Pos:* Mem biomed sci study sect, NIH, 82-85. *Mem:* Am Soc Cell Biol; Fedn Am Soc Exp Biol; Am Asn Cancer Res. *Res:* Mechanism of tumor invasion. *Mailing Add:* Dept Anat & Cellular Biol Sch Med Tufts Univ 136 Harrison Ave Boston MA 02111

BISWAS, DIPAK R, b India, Feb 3, 49; US citizen. HIGH TEMPERATURE MATERIAL PROCESSING, COATINGS ON OPTICAL FIBER CHEMICAL VAPOR DEPOSITION THIN FILM. *Educ:* Univ Calcutta, India, BTech, 70; Univ Calif, Berkeley, MS, 74, PhD(mat sci & eng), 76. *Prof Exp:* Postdoctoral fel mat sci & ceramics, Lawrence Berkeley Lab, 76-78; res assoc ceramics, Univ Utah, 78-79; mat scientist mat sci, Inst Gas Technol, 79-81; mgr & prin staff scientist optical fiber, ITT Electro Optical Prod Div & Alcatel Cable Systs, 81-88; dir res & develop optical fiber, Spectran Corp, 88-91; DEPT HEAD RES, DEVELOP & ENG, FIBERGUIDE, 91- *Mem:* Am Ceramic Soc; Soc Photo-Optical Engrs; Optical Soc Am. *Res:* Development of specialty coatings on optical glass fibers, fiber optic sensors; high strength hermetically coated optical fibers for adverse environment; optical fibers embedded in smart skin or structures. *Mailing Add:* Fiberguide One Bay St Stirling NJ 07980

BISWAS, NRIPENDRA NATH, b Calcutta, India, Aug 1, 30; m 59; c 2. PHYSICS. *Educ:* Univ Calcutta, BSc, 49, MSc, 51, DrPhil(physics), 55. *Prof Exp:* Res asst cosmic rays, Bose Res Inst, Calcutta, 52-55; res scholar physics, Max Planck Inst Physics, Gottingen, 55-57; res assoc, Physics Inst, Univ Bologna, 57-58; vis scientist high energy physics, Lawrence Radiation Lab, Univ Calif, Berkeley, 58-60; res scientist bubble chamber anal, Max Planck Inst Physics, Munich, 61-65; from asst prof to assoc prof, 65-76, PROF PHYSICS, UNIV NOTRE DAME, IND, 76- *Mem:* Sigma Xi. *Res:* Elementary particle physics, theoretical and experimental; study of cosmic rays. *Mailing Add:* Dept Physics Univ Notre Dame Notre Dame IN 46556

BISWAS, PROSANTO K, b Calcutta, India, Mar 1, 34; US citizen; m 62; c 1. HORTICULTURE, PLANT PHYSIOLOGY. *Educ:* Univ Calcutta, BS, 58; Univ Mo, MS, 59, PhD(hort), 62. *Prof Exp:* Asst prof hort & head dept, 62-69, PROF HORT & CHMN, DEPT PLANT & SOIL SCI, TUSKEGEE UNIV, 69- *Mem:* Am Soc Plant Physiol; Am Soc Hort Sci; Weed Sci Soc Am. *Res:* Physiological responses of plants to different growth regulators; metanolism and mode of action of herbicides by plants. *Mailing Add:* Dept Agric Sci Tuskegee Inst Tuskegee AL 36083

BISWAS, ROBIN MICHAEL, b Oxford, Eng, May 31, 42; US citizen; m 73; c 2. HEPATITIS VIRUSES, MOLECULAR BIOLOGY. *Educ:* Univ G06ttingen, Ger, MD, 71, PhD(med), 75. *Prof Exp:* Intern med & surg, Univ Hosps, Göttingen, Ger, 71-72; res assoc viral hepatitis, Dept Med Microbiol, Univ Göttingen, Ger, 72-78; resident path, Univ Tex, San Antonio, Bexar County Hosp, 80-81 & Univ Md Hosp, Baltimore, 81-84; staff fel med microbiol, Dept Clin Path, Clin Ctr, NIH, 84-87; sr staff fel viral hepatitis, 87-89, LAB CHIEF VIRAL HEPATITIS, LAB HEPATITIS, CTR BIOLOGICS EVAL & RES, FOOD & DRUG ADMIN, 89- *Concurrent Pos:* Instr med microbiol, Dept Med Microbiol, Univ Göttingen, Ger, 72-78. *Mem:* Soc Exp Biol & Med; Am Asn Blood Banks. *Res:* Development of tests for viral hepatitis; relationships between viral hepatitis markers and prevalence of clinical disease and the search for possibly existing unidentified viral agents of hepatitis. *Mailing Add:* Ctr Biologics Eval & Res Food & Drug Admin 8800 Rockville Pike Bethesda MD 20892v

BISWAS, SHIB D, b Agra, India, Feb 1, 40; m 69. CHEMISTRY, BIOCHEMISTRY. *Educ:* Univ Allahabad, BSc, 57, MSc, 60, PhD(chem), 64. *Prof Exp:* Fel chem, Univ Cambridge, 64; fel biochem, 64-65, lectr, 65-67 & Sch Dent Hyg, 65, ASST PROF BIOCHEM & ORAL BIOL, UNIV MANITOBA, 68- *Res:* Lignin complexes in metal ions; etiology of periodontal and caries diseases processes in relation to biochemical standpoint; application of radiochemicals in the study of metabolic pathways in oral bacteria. *Mailing Add:* Dept Dent Univ Manitoba Winnipeg MB R3T 2N2 Can

BITCOVER, EZRA HAROLD, b New York, NY, Jan 16, 20; m 44; c 4. LEATHER CHEMISTRY. *Educ:* Rutgers Univ, BS, 41; Univ Mass, PhD(agron), 51. *Prof Exp:* Jr chemist, Tenn Valley Authority, 41-43; res chemist, Prophylactic Brush Co, 43-46; soils chemist, Citrus Exp Sta, Univ Fla, 48; anal res supvr, Lindsay Chem Co, 50-52; res chemist food processing, Miner Labs, 53-56; res biochemist, Eastern Utilization Res Div, USDA, 56-80; RETIRED. *Mem:* AAAS; Am Leather Chem Asn; fel Sigma Xi; Am Chem Soc; fel Am Inst Chemists (secy, 58-63). *Res:* Nitrogen chemicals; processing of vegetables; tannery pollution; leather defects; collagen by-products; hide processing. *Mailing Add:* 12 Nora Lane Plainview NY 11803

BITENSKY, MARK WOLFE, b Brooklyn, NY, Apr 5, 34; m 57; c 2. PATHOLOGY. *Educ:* Yale Univ, BA, 55, MD, 59. *Prof Exp:* Intern, Univ Div Bellevue Hosp, 59-60, asst res med, 60-61, asst res path, 61; clin assoc, Nat Inst Arthritis & Metab Dis, 61-63, spec res fel, 63-64; instr, dept med, NY Univ Sch Med, 64-65, asst prof med, 65-70; assoc prof path, 70-76, dir student educ, Dept Path, 77-81, PROF PATH, SCH MED, YALE UNIV, 76-; SR FEL, LOS ALAMOS NAT LAB. *Concurrent Pos:* Consult, NIH extramural progs; consult NIH grants & contracts off cancer ctrs grants & contracts; assoc ed, J Cyclic Nucleotide Res; dir, Div Life Sci, Los Alamos Nat Lab, 81-86. *Mem:* AAAS; fel, NY Acad Sci; Endocrine Soc; Harvey Soc; Am Asn Pathologists. *Res:* Cyclic nucleotide chemistry especially in the vertebrate rod; membrane biochemistry and enzyme regulatory mechanisms, especially as related to the study of cell surface regulatory mechanism and cyclic nucleotide effects on normal metbolism and disease states; genetic disorders of metabolism and regulation; diabetes mellitus and protien glucation. *Mailing Add:* Los Alamos Nat Lab PO Box 1663 MS D434 Los Alamos NM 87545

BITGOOD, J(OHN) JAMES, b Agawam, Mass, Feb 26, 40; m 64; c 3. CYTOGENETICS. *Educ:* Univ Mass, BVA, 61; Univ Minn, MS, 75, PhD(animal sci), 80. *Prof Exp:* Res asst, Dept Animal Sci, Univ Minn, 72-80, fel, 80-81; asst prof, 81-87, ASSOC PROF POULTRY MGT & GENETICS, DEPT POULTRY SCI, UNIV WIS, 87- *Mem:* Poultry Sci Asn; AAAS; Sigma Xi. *Res:* Avian cytogenetics; chromosome behavior; single gene inheritance in the chicken; development of linkage maps in the chicken. *Mailing Add:* 3105 Nightingale Lane Middleton WI 53562

BITHA, PANAYOTA, b Greece; US citizen. COMPUTERS. *Educ:* Polytech Sch Athens, dipl chem & eng, 58; Columbia Univ, MS, 60. *Prof Exp:* SR RES CHEMIST, LEDERLE LABS, DIV AM CYANAMID CO, 60- *Mem:* Am Chem Soc; Sigma Xi. *Res:* Medicinals used for infectious diseases such as tetracyclines, B-lactams, matural antibiotics and quinolones; drugs for cancer chemotherapy; author of 22 publications; awarded 36 patents. *Mailing Add:* 287 Treetop Circle Nanuet NY 10954

BITLER, WILLIAM REYNOLDS, b Nyack, NY, Nov 25, 27; m 54; c 2. PHYSICAL METALLURGY. *Educ:* Carnegie Inst Technol, BS, 53, MS, 55, PhD(physics), 57. *Prof Exp:* Asst prof metall eng, Carnegie Inst Technol, 56-62; assoc prof, 62-69, chmn metall sect, 69-73, PROF METALL, PA STATE UNIV, 69- *Concurrent Pos:* Consult, Allegheny-Ludlum Steel Corp, 57-, carrier, 79-80; sr Fulbright res prof, Inst Physics, Univ Oslo, Norway, 85-86. *Mem:* Am Phys Soc; Am Soc Metals; Am Inst Mining, Metall & Petrol Engrs; Am Electroplaters Soc. *Res:* Physical metallurgy; kinetics; statistical thermodynamics; magnetism; superconduction. *Mailing Add:* 2006 N Oak Lane State College PA 16803

BITMAN, JOEL, b Elizabeth, NJ, Nov 29, 26; m 50; c 3. BIOCHEMISTRY. *Educ:* Cornell Univ, AB, 46; Univ Minn, MS, 48, PhD(physiol chem), 50. *Prof Exp:* Asst physiol chem, Univ Minn, 47-50, res fel, 50; chemist, Div Nutrit & Physiol, Dairy Husb Res Br, USDA, 51-55; res assoc, Columbia Univ, 55-56; chemist & leader physiol invests, Dairy Cattle Res Br, 56-74, chief biochem lab, Animal Physiol & Genetics Inst, Agr Res Ctr-East, 74-79, CHEMIST, MILK SECRETION & MASTITIS LAB, USDA, 79- *Mem:* Am Chem Soc; AAAS; Am Dairy Sci Asn; Am Oil Chemists Soc; Am Soc Animal Sci; Fedn Am Soc Exp Biol. *Res:* Chemistry and metabolism of sex hormones; determination of estrogens in blood; vitamin A deficiency and cerebrospinal fluid mechanics; body temperature during the estrous cycle; chemical composition of the mammary gland during mastitis; mechanism of estrogen action in the uterus; DDT effect on carbonic anhydrase and egg-shell thickness; estrogenic effects of DDT; lipid composition of human and bovine milk; thyroid hormones in cattle. *Mailing Add:* Milk Secretion & Mastitis Lab Animal Sci Inst USDA Beltsville MD 20705

BITO, LASZLO Z, b Budapest, Hungary, Sept 7, 34; US citizen. OCULAR PHYSIOLOGY. *Educ:* Bard Col, BA, 59; Columbia Univ, PhD, 63. *Prof Exp:* From instr to asst prof, 65-74, sr res assoc, 74-77, assoc prof & head ocular physiol lab, 77-80, PROF OCULAR PHYSIOL, COLUMBIA UNIV, 80- *Concurrent Pos:* NIH fel, Univ Louisville, 63-64 & Univ Col, London, Eng, 64-65. *Honors & Awards:* Semmelweis Sci Soc Award, 74. *Mem:* Am Soc Pharmacol & Exp Therapeut; Am Physiol Soc; Int Soc Eye Res; Asn Res Vision & Ophthal. *Res:* Prostagland in transport and its physiological role; nature of blood-ocular and blood-brain barriers and the composition and homeostasis of ocular and cerebrospinal fluids; mechanisms and pathophysiology of inflammation; mechanism of accommodation and development of presbyopia. *Mailing Add:* Dept Ophthal Res Columbia Univ Col P&S 630 W 168th St New York NY 10032

BITONDO, DOMENIC, b Welland, Ont, June 7, 25; US citizen; m 49; c 4. AERONAUTICAL ENGINEERING. *Educ:* Univ Toronto, BASc, 47, MASc, 48, PhD(aerophys, aeronaut), 50. *Prof Exp:* Res engr, Nat Res Coun Can, 47; aeronaut engr, US Naval Ord Lab, Md, 48, Defense Res Bd Can, 49 & NAm Aviation, Inc, 50-51; prin investr, Aerophys Develop Corp, 52-56, chief aerodyn, 55-58, proj engr, 58-59; proj engr, Norair Div, Northrop Corp, 59; head test planning & anal sect, Space Tech Labs, 59-60; head dept aeromech, Aerospace Corp, 60-61, dir advan ballistic re-entry systs prog, 61-63, dir vehicle systs, 63; tech dir & head eng, Bendix Systs Div, Bendix Corp, 63-67, dir eng, Apollo Lunar Exp Prog, Aerospace Syst Div, 67-69, dir & gen mgr, Bendix Res Labs, Bendix Ctr, 69-79, exec dir res & develop, Bendix Corp, 79-80; PRES, BITONDO ASSOC, INC, 81- *Concurrent Pos:* Consult, Inst Defense Anal & mem systs panel, defender prog, res planning & mgt, 63-65; lectr technol planning, Univ Mich, 83-87. *Mem:* Am Inst Aeronaut & Astronaut. *Res:* Development of shock tube as aerodynamic research tool; non-stationary fluid flow; re-entry vehicles for ballistic missiles; research management, planning and control. *Mailing Add:* 5 Manchester Court Ann Arbor MI 48104-6562

BITSIANES, GUST, b Virginia, Minn, Sept 24, 19; m 56; c 2. METALLURGICAL ENGINEERING. *Educ:* Univ Minn, BChE, 41, PhD(metall), 51. *Prof Exp:* Asst metall, Univ Minn, 41-43, instr, 43; res metallurgist, Manhattan Proj, Mass Inst Technol, 44-46; from instr to prof metall, 45-69, PROF CIVIL & MINING ENG, UNIV MINN, MINNEAPOLIS, 69- *Concurrent Pos:* Res metallurgist, Nuclear Energy Propulsion Aircraft Proj, Oak Ridge Nat Lab, 48-51. *Honors & Awards:* J E Johnson Jr Award, Am Inst Mining, Metall & Petrol Engrs, 55, Jour Metals Award, 56, Robert H Hunt Medal, 63. *Mem:* Am Inst Mining, Metall & Petrol Engrs. *Res:* Mechanisms of heterogenous reactions pertinent to the field of iron and steel-making, equilibria and kinetic studies of the same; agglomeration and preparation of raw materials for metal-extraction processes; transport mechanisms in high-temperature metallurgical systems; role of basic reaction mechanisms in specific engineering applications. *Mailing Add:* Civil Engr 122 Civ Min Bldg Univ Minn 500 Pillsbury Dr SE Minneapolis MN 55455-0116

BITTAR, EVELYN EDWARD, b Jaffa, Israel, Oct 12, 28; m 61; c 4. PHYSIOLOGY. *Educ:* Colby Col, BA, 51; Yale Univ, MD, 55. *Prof Exp:* Instr med, Med Sch, George Washington Univ, 59-61; chief med officer geriat, St Elizabeth's Hosp, Washington, DC, 61-63; Fulbright lectr nephrology, Univ Damascus, 63-64; vis scientist cell physiol, Bristol Univ, 64-65; Wellcome Trust investr, Cambridge Univ, 67-68; from vis assoc prof to assoc prof, 68-72, PROF PHYSIOL, UNIV WIS-MADISON, 72- *Concurrent Pos:* NIH fel, DC Gen Hosp, 60-61; Wellcome Trust fel, Oxford Univ, 65-67; Fulbright lectr, Univ Algiers, 80; Burroughs-Wellcome vis prof, Oxford Univ, 82. *Honors & Awards:* Osler Medal, Am Asn Hist Med. *Mem:* Brit Biochem Soc; Brit Biophys Soc; Brit Soc Exp Biol; Am Physiol Soc; Am Biophys Soc; NY Acad Sci. *Res:* Cell physiology with special reference to membrane transport and ion transport in single cell preparations; mode of action of hormones and drugs. *Mailing Add:* Dept Physiol Univ Wis Madison WI 53706

BITTER, GARY G, b Hoisington, Kans, Feb 2, 40; m 62; c 3. MATHEMATICS EDUCATION, COMPUTER EDUCATION. *Educ:* Kans State Univ, BS, 62; Emporia State Univ, MA, 65; Univ Denver, PhD(comput & math educ), 70. *Prof Exp:* Teacher math & sci, Derby High Sch, 62-65; teacher math, Forsythe Jr High Sch, 65-66; instr math, Washburn Univ, 66-67; instr comput sci & math, Colo Col, 67-70; from asst prof to assoc prof, 70-77, PROF MATH & COMPUT EDUC, ARIZ STATE UNIV, 77- *Concurrent Pos:* Lectr, Univ Colo, 68-69; consult, Kaman Sci, 67-70; supvr student teachers & comput educ res assoc, Univ Denver, 69-70; ed, Sch Sci & Math J. *Mem:* Int Asn Computers Educ; Nat Coun Teachers Math; Math Asn Am; Sch Sci & Math Asn; Am Educ Res Asn. *Res:* Incorporation of computer into elementary and secondary school curriculum; futuristic mathematics curricula; mathematics laboratory; the application of the hand held calculator in the school; computer assisted mathematics. *Mailing Add:* Sch Sci & Math Jrnl Ariz State Univ FMC Payne 146 Tempe AZ 85287-0111

BITTERMAN, MORTON EDWARD, b New York, NY, Jan 19, 21; m 67; c 3. LEARNING, INTELLIGENCE. *Educ:* New York Univ, BA, 41; Columbia Univ, MA, 42; Cornell Univ, PhD(psychol), 45. *Prof Exp:* Asst prof, psychol, Cornell Univ, 45-50; assoc prof psychol, Univ Tex, 50-55; mem Instit for Adv Study, 55-57; prof psychol, Bryn Mawr Col, 57-70; prof psychol, Univ S Fla, 70-71; PROF PSYCHOL, UNIV HAWAII, 71-; DIR, BEKESY LAB NEUROBIOL, 90- *Concurrent Pos:* Ed, Am Jour Psychol, 55-73; assoc ed, Animal Learning & Behavior, 73-76 & 84-88; prin investr, ONR, Nat Sci Found, Nat Instit Mental Health, 57-; consult ed, Journ Comp Psychol, 88- *Honors & Awards:* Humbolt Prize, Alexander von Humboldt-Stiftung, WGer, 81. *Mem:* Soc Exp Psychologists; Psychonomic Soc; Am Psychol Asn. *Res:* Comparative analysis of learning in vertebrates and invertebrates; evolution of intelligence. *Mailing Add:* Bekesy Lab Neurobiol 1993 East-West Rd Honolulu HI 96822

BITTERS, WILLARD PAUL, b Eau Claire, Wis, June 4, 15; m 40; c 2. HORTICULTURE. *Educ:* St Norbert Col, BA, 37; Univ Wis, MA, 40, PhD(plant physiol), 42. *Prof Exp:* Lab asst, Univ Wis, 39-40; res asst, Univ Ariz, 40-42, asst horticulturist, 42-46; from asst horticulturist to assoc horticulturist, 46-58, horticulturist & prof hort, 58-82, EMER HORTICULTURIST & PROF HORT, CITRUS EXP STA & AGR RES CTR, UNIV CALIF, RIVERSIDE, 82- *Res:* Effect of rootstocks on long term yields; fruit quality and size; effect of rootstocks in relation to disease resistance; clonal varieties of citrus; ornamental citrus; dwarfing rootstocks. *Mailing Add:* 1185 La Subida Ct Riverside CA 92507

BITTINGER, MARVIN LOWELL, b Akron, Ohio, Aug 9, 41; m 65; c 1. MATHEMATICS. *Educ:* Manchester Col, BA, 63; Ohio State Univ, BS, 65; Purdue Univ, PhD(math educ), 68. *Prof Exp:* Asst prof math, Ind Univ-Purdue Univ, Indianapolis, 68-74, ASSOC PROF MATH, PURDUE UNIV, WEST LAFAYETTE, 74- *Concurrent Pos:* Consult, Addison-Wesley Publ Co, 70- *Mem:* Math Asn Am. *Res:* Mathematics education; trigonometry; logic and proof. *Mailing Add:* 3011 Whispering Tr Carmel IN 46032

BITTKER, DAVID ARTHUR, b Rochester, NY, Sept 22, 27; m 53; c 2. CHEMICAL KINETICS. *Educ:* Cornell Univ, BA, 49; Univ Rochester, PhD(phys chem), 53. *Prof Exp:* RES SCIENTIST, LEWIS RES CTR, NASA, 52- *Concurrent Pos:* Lectr math, Cleveland State Univ, 67- *Mem:* Am Inst Aeronaut & Astronaut. *Res:* Chemical mechanism of hydrocarbon fuel thermal degradation; complex kinetic computations of hydrocarbon combustion and pollutant formation for various combustor models. *Mailing Add:* 4095 Charlton Rd South Euclid OH 44121

BITTLE, JAMES LONG, b Norristown, Pa, Mar 7, 27; m 54; c 2. VETERINARY MICROBIOLOGY, VIROLOGY. *Educ:* Univ Calif, BS, 51, DVM, 53; Am Col Vet Microbiol, dipl. *Prof Exp:* Virologist, Virus Res Lab, Pitman-Moore Co, 58-60; supt tissue cult prod, head dept polio & virus vaccine testing & mgr biol testing, Lederle Labs, NY, 60-65; dir virus res, Pitman-Moore Co, Ind, 65-66, head dept infectious dis, Pitman-Moore Div, Dow Chem Co, 67-69, dir res, 69-76, exec vpres, Pitman-Moore, Inc, 76-77, pres, 77-79; ADJ MEM, SCRIPPS RES INST, 80- *Mem:* Am Vet Med Asn; Soc Exp Biol & Med. *Res:* Viral infectious diseases; vaccine development; production; testing. *Mailing Add:* 5353 Calle Vista San Diego CA 92109

BITTMAN, ROBERT, b New York, NY, Mar 19, 42; m 75; c 2. BIOPHYSICAL CHEMISTRY. *Educ:* Queens Col, NY, BS, 62; Univ Calif, Berkeley, PhD(chem), 65. *Prof Exp:* NSF fel chem, Max Planck Inst Phys Chem, 65-66; from asst prof to prof, 66-88, DISTINGUISHED PROF CHEM, QUEENS COL, NY, 88- *Concurrent Pos:* Res grant, Am Chem Soc, 67-71; asst secy, Org Reactions, 68, secy, 69-73, co-secy, 74-; res grants, NIH, 71-; vis res scientist, Nat Ctr Sci Res, France, 80-81. *Mem:* NY Acad Sci; Sigma Xi; Am Soc Biol Chemists; Biophys Soc; Am Chem Soc. *Res:* Fast reactions in solution; molecular interactions between lipids and sterols in membranes; biomembrane structure; mode of action of membrane-active agents; mechanisms of biochemical and organic reactions; chemical synthesis of lipids. *Mailing Add:* Dept Chem Queens Col Flushing NY 11367

BITTNER, BURT JAMES, b Ft Collins, Colo, Feb 8, 21; m 46; c 4. ELECTROMAGNETICS. *Educ:* Colo State Univ, BSEE, 43. *Prof Exp:* Jr engr, Submarine Signal Co, 46-48; div supvr, Sandia Corp, 48-56; vpres & res dir, Gulton Indust Div, 57-58; sr staff scientist, Kaman Sci Corp, 58-74; GROUP LEADER, INTERSTELLAR, LTD, 65-, DIR, 74-; SR STAFF ENGR, MARTIN MARIETTA, DENVER, 77- *Concurrent Pos:* Task engr, Pac Nuclear Tests, 62; secy & pres, Air Acad Bd Educ, 63-75; mem, Group Six, Int Radio Consult Comt, 64-; consult, Army Res Off, Duke Univ, 68-73, Nat Ctr Atmospheric Res, 74 & Shock Tube Lab, Univ NMex; sr consult, Kaman Sci Corp & Kaman Aerospace, 74-77; res physicist, Univ Colo, Colorado Springs, 74-75. *Mem:* Sr mem Inst Elec & Electronics Engrs; Prof Soc Protective Design. *Res:* Antennas; nuclear electromagnetic plasma. *Mailing Add:* 1033 Grove Denver CO 80219

BITTNER, HARLAN FLETCHER, b Eugene, Ore, Mar 26, 51. PHYSICAL CHEMISTRY, SOLID STATE CHEMISTRY. *Educ:* Univ Ore, BA, 73; Univ Calif, San Diego, PhD(chem), 77. *Prof Exp:* Res staff mat chem, Oak Ridge Nat Lab, Union Carbide Corp, 78-80; MEM TECH STAFF, AEROSPACE CORP, 80- *Mem:* Am Chem Soc; Electrochem Soc. *Res:* Reactive metal electrode passivation; lithium battery electrochemistry; oxidation of metals and alloys; hydrogen absorption by and permeation through metals and alloys; alloy and intermetallic compound formation. *Mailing Add:* M2/275 PO Box 92957 Los Angeles CA 90009

BITTNER, JOHN WILLIAM, b Iowa, Mar 6, 26; m 64; c 2. GENERAL PHYSICS. *Educ:* Univ Western Ont, BSc, 48, MS, 50, PhD, 54. *Prof Exp:* Assoc physicist, 54-63, PHYSICIST, BROOKHAVEN NAT LAB, 63- *Mem:* Am Phys Soc; Inst Elec & Electronics Engrs; Sigma Xi. *Res:* Nuclear accelerators; scanning transmission electron microscope. *Mailing Add:* Brookhaven Nat Lab Bldg 725 Upton NY 11973

BITTON, GABRIEL, b Marrakech, Morocco, Sept 8, 40; m 70; c 2. ENVIRONMENTAL MICROBIOLOGY. *Educ:* Univ Toulouse, BS & Agr Eng, 65; Laval Univ, MS, 67; Hebrew Univ, PhD(microbiol), 73. *Prof Exp:* Teacher biol, Acad Quebec, 66-68; res fel appl microbiol, Harvard Univ, 72-74; PROF ENVIRON MICROBIOL, UNIV FLA, 74- *Concurrent Pos:* Prof, Univ Perpignan, France. *Mem:* Am Soc Microbiol; Am Soc Limnol Oceanog; Soc Indust Microbiol. *Res:* Microbiology of surfaces; environmental virology; environmental toxicology. *Mailing Add:* Dept Environ Eng Univ Fla Gainesville FL 32611

BITZER, DONALD L, b East St Louis, Ill, Jan 1, 34; m 55; c 1. ELECTRICAL ENGINEERING. *Educ:* Univ Ill, BS, 55, MS, 56, PhD(elec eng), 60. *Hon Degrees:* PhD, MacMurray Col, 85. *Prof Exp:* from res asst prof to assoc prof radar, 60-67, PROF ELEC ENG & DIR, 67-, EMER PROF ELEC ENG & EMER DIR, COMPUTER-BASED EDUC RES LAB, UNIV

ILL, URBANA-CHAMPAIGN; DISTINGUISHED UNIV RES PROF, NC STATE UNIV. *Concurrent Pos:* Mem high resolution radar comt, US Dept Defense, 58-59; comput consult, US Agency Int Develop, 64. *Honors & Awards:* IR 100 Award, 66; Vladimir K Zworykin Award, Nat Acad Eng, 73; Chester F Carlson Award, Am Soc Eng Educ, 81; Educ Award, Am Fedn Info Processing Socs, 89. *Mem:* Nat Acad Eng; Am Soc Eng Educ; fel Inst Elec & Electronics Engrs; fel AAAS; Sigma Xi; Data Processing Mgt Asn. *Res:* Generation of large synthetic radar antennas; automatic instruction with high speed digital computers; measurement of small signals in a plasma; numerous patents, including Plato and Novanet systems. *Mailing Add:* Comput-Based Educ Res Lab Univ Ill Urbana-Champaign Urbana IL 61801

BITZER, MORRIS JAY, b Huntington, Ind, Mar 3, 36; m 56; c 3. AGRONOMY. *Educ:* Purdue Univ, BS, 58, MS, 65, PhD(plant breeding, genetics), 68. *Prof Exp:* Teacher, high schs, Ind, 59-63; res asst hybrid wheat, Purdue Univ, 63-68; asst prof agron, Univ Ga, 68-71; EXTEN SPECIALIST GRAIN CROPS, UNIV KY, 71- *Mem:* Am Soc Agron; Crop Sci Soc Am. *Res:* Hybrid vigor and gene action in hybrid wheat; pollen dispersal and seed set for hybrid wheat production; effect of nitrogen on yields of dwarf wheats; vernalization of soft winter wheats. *Mailing Add:* Dept Agron ASC North Univ Ky Lexington KY 40546

BITZER, RICHARD ALLEN, b Buenos Aires, Arg, May 11, 39; US citizen; m 64; c 3. ELECTRICAL & MICROWAVE ENGINEERING, AEROSPACE ENGINEERING. *Educ:* Bucknell Univ, BSEE, 61; Rensselaer Polytech Inst, MEE, 63. *Prof Exp:* Planning engr, Western Elec Co, Inc, 65-67; proj engr, Eastern Tube Div, Varian Assocs, 67-71; chmn dept elec eng, DeVry Tech Inst, 71-73; proj engr, Dataram Corp, 73-74; mem engr & sci staff, Princeton Univ, 74-82; STAFF ENGR, ASTRO SPACE DIV, GENERAL ELECTRIC CORP, 82- *Mem:* Sr mem Am Inst Aeronaut & Astronaut. *Res:* Millimeter microwave diagnostic systems in thermonuclear research, particularly radiometer and interferometer techniques; spacecraft electrical power; control and telemetry systems. *Mailing Add:* Astro Space Div G E Corp PO Box 800 Princeton NJ 08543-0800

BIVENS, RICHARD LOWELL, b Denver, Ill, Aug 19, 39; m 63; c 2. PHYSICAL CHEMISTRY. *Educ:* Monmouth Col, Ill, BA, 61; Case Western Reserve Univ, MS, 63, PhD(chem), 69. *Prof Exp:* From asst prof to assoc prof, 65-79, assoc dean, 84-88, PROF CHEM, ALLEGHENY COL, 79- *Concurrent Pos:* Dana vis prof, Univ Rochester, 91. *Mem:* Am Chem Soc; Sigma Xi. *Res:* Thermodynamics and statistical mechanics of phase transitions; structure of ion-solvent clusters. *Mailing Add:* Dept Chem Allegheny Col Meadville PA 16335

BIVER, CARL JOHN, JR, b Owensboro, Ky, July 1, 32; m 58; c 4. APPLIED PHYSICS, MATERIALS SCIENCE. *Educ:* Univ Notre Dame, BS, 54, MS, 56. *Prof Exp:* Asst scientist metall, Argonne Nat Lab, 56-59; from physicist to sr physicist, X-ray Dept, Gen Elec Co, 59-66, sr physicist, Neutron Devices Dept, 66-90; HEAD SCI DEPT & TEACHER MATH & PHYSICS, ST PETERSBURG JR COL, ST PETERSBURG CAMPUS, 91- *Honors & Awards:* Gen Elec Co Inventors Award, 72. *Mem:* Am Vacuum Soc; AAAS; Am Soc Metals. *Res:* Gaseous permeation phenomena; high pressure physics; thermoelectric devices physics; photoconductivity, heat transfer, transport phenomena, metallurgical and chemical kinetics; mathematical modeling. *Mailing Add:* 4 Westwood Lane Belleair Clearwater FL 33516

BIVINS, BRACK ALLEN, b Nashville, Tenn, Nov 28, 43; m 73; c 2. TRAUMA, METABOLISM. *Educ:* Western Ky Univ, BS, 66; Univ Ky, MD, 70. *Prof Exp:* Residency surg, Univ Ky, Med Ctr, Lexington, 70-77, from asst prof to assoc prof, 77-82; HEAD, DIV TRAUMA SURG, HENRY FORD HOSP, DETROIT, 83- *Concurrent Pos:* Clin scholar, Am Cancer Soc, 72 & 73 & Frederick A Collier Soc, 76; clin assoc prof surg, Med Sch, Univ Mich, Ann Arbor, 83-; adj assoc prof pharm, Col Pharm, Univ Ky, Lexington & Wayne State Univ, Detroit, 83-; assoc vpres clin serv, Henry Ford Hosp, 87- *Honors & Awards:* Res Award, Parenteral Drug Asn, 76; Res Award, Am Soc Hosp Pharmacists, 76. *Mem:* Am Col Surgeons; Am Asn Surg Trauma; Soc Surg Alimentary Tract; Am Soc Parenteral & Enteral Nutrit; Soc Critical Care Med. *Res:* Physiology and metabolism of the critically ill and injured. *Mailing Add:* Henry Ford Hosp Detroit MI 48202

BIXBY, W(ILLIAM) HERBERT, b Indianapolis, Ind, Dec 28, 06; m 63. ELECTRICAL ENGINEERING. *Educ:* Univ Mich, BSE, 30, MS, 31, PhD(elec eng), 33; Chrysler Inst Technol, MME, 35. *Prof Exp:* Res engr, Detroit Edison Co, 28-32; spec probs engr, Chrysler Corp, 33-36; from instr to prof elec eng, Wayne State Univ, 36-56; vpres appl res, Power Equip Co Div, North Elec Co, 56-80; RETIRED. *Concurrent Pos:* Develop engr, Power Equip Co, Detroit, 37-80. *Mem:* Fel AAAS; fel Inst Elec & Electronics Engrs. *Res:* Automatic control; dielectric materials; general and industrial electronics; electricity and magnetism; applied mathematics; time-lag in breakdown of liquid insulations as a function of voltage application; breakdown and time-lag of dielectric materials; breakdown of liquid carbon tetrachloride. *Mailing Add:* 2308 SE 21st St Ft Lauderdale FL 33316-3620

BIXBY, WILLIAM ELLIS, physics, for more information see previous edition

BIXLER, DAVID, b Chicago, Ill, Jan 7, 29; m 51; c 6. GENETICS, DENTISTRY. *Educ:* Ind Univ, AB, 50, PhD(zool), 56, DDS, 59; Am Bd Med Genetics, dipl, 82. *Prof Exp:* Asst prof dent sci, 59-66, asst prof med genetics, 66-70, assoc prof dent sci & med genetics, 70-74, PROF MED GENETICS & ORAL-FACIAL GENETICS & CHMN DEPT ORAL-FACIAL GENETICS, SCHS DENT & MED, IND UNIV, INDIANAPOLIS, 74- *Concurrent Pos:* USPHS fel, Sch Dent, Ind Univ, Indianapolis, 56-58; USPHS career develop award, 67-72. *Mem:* AAAS; Am Soc Human Genetics; Sigma Xi; Am Cleft Palate Asn; Soc Craniofacial Genetics. *Res:* Genetics of cleft lip and palate; hereditary anomalies of oral and facial structures. *Mailing Add:* Ind Univ Sch Dent 1121 W Michigan St Indianapolis IN 46202-5186

BIXLER, DEAN A, b Gretna, Nebr, Oct 24, 19; m 41; c 2. TEXTILE CHEMISTRY. *Educ:* Univ Nebr, Lincoln, BS, 40, MS, 46; Purdue Univ, PhD(agr biol), 53. *Prof Exp:* DIR RES, CONE MILLS CORP, 69- *Mailing Add:* 2029 SE 13th St Cape Coral FL 33990

BIXLER, JOHN WILSON, b Eau Claire, Wis, July 27, 37. ANALYTICAL CHEMISTRY. *Educ:* Lakeland Col, BS, 59; Univ Minn, PhD(anal chem), 63. *Prof Exp:* Instr chem, Lake Forest Col, 63-65, asst prof, 65-69; assoc prof, 69-77, PROF CHEM, STATE UNIV NY COL BROCKPORT, 77- *Concurrent Pos:* Vis prof, Purdue Univ, 75-76; vis scientist, Melbourne Univ, 77-78 & Deakin Univ, 84-85. *Mem:* Am Chem Soc. *Res:* Electroanalytical chemistry; chemical instrumentation. *Mailing Add:* Dept Chem State Univ NY Col Brockport NY 14420

BIXLER, OTTO C, b Morenci, Ariz, May 9, 16; m 38; c 2. ELECTRONICS ENGINEERING, MANAGEMENT SCIENCES. *Educ:* Univ Southern Calif, BS, 37; Kennedy Western Univ, PhD, 85. *Prof Exp:* Engr, Valuation Dept, Southern Calif Edison Co, Ltd, 37-42; field engr radio & sonar systs, Radio Div, Western Elec Co with Bur Ships, 42-43, sr engr, 43-45, systs engr, Elec Res Prod Div, 45-49; develop engr, AiRes Mfg Co, 49-51; dir eng & res, Magnecord, Inc, 51-53, vpres & works mgr, 53-54; tech dir, Helipot Corp Div, Beckman Instruments, 54-58; mgr prod design & mfg computer oper, Aeronutronic Div, Ford Motor Co, 58-62; asst gen mgr, Space & Systs Div, Packard Bell Electronics Corp, 62-63; vpres & mgr defense opers, Univac Div, Sperry-Rand Corp, 63-66; vpres & mgr opers, Wanlass Elec Co, 66-67; vpres mfg & distribution, Rexall Drug Co, 67-68; pres, Eng Res Prod Co & Ad Corp, 68-74; gen mgr, Waimea Dispensary & Clin, 74-84; Adminr, Hawaiian Eye Ctr, 85-86; RETIRED. *Concurrent Pos:* Mem bd, Info Systs Design, Inc, 70-74; exec vpres, Info Serv Industs, 70-72; prog mgr, QCM Res, 90-91. *Mem:* AAAS; Inst Elec & Electronic Engrs; a Xi; Soc Motion Picture & TV Engrs; NY Acad Sci; Med Group Mgt Asn. *Res:* Microfilm systems; management information systems; first magnetic tape recorder and stereo tape recorder, computer mass memory system; first multicolor computer display terminals and microfilm automated indentured parts control system; re-entry decoy systems. *Mailing Add:* PO Box 633 Koloa HI 96756

BIZIOS, RENA, b Larissa, Greece, July 7, 43. BLOOD COAGULATION, TRANSPORT PHENOMENA IN PULMONARY PATHOPHYSIOLOGY. *Educ:* Univ Mass, BS, 68; Calif Inst Technol, MS, 71; Mass Inst Technol PhD(biomed eng), 79. *Prof Exp:* Jr scientist, Dept Lab Med, Med Sch, Univ Minn, 70-71, assoc scientist proj coordr, Dept Surg, 72-73; fel spec coagulation lab, Roger Williams Gen Hosp, 79-80; ASST PROF BIOMED ENG, RENSSELAER POLYTECH INST, 81- *Concurrent Pos:* Adj asst prof physiol, Albany Med Col, 82- *Mem:* Biomed Eng Soc; Soc Women Engrs; Am Inst Chem Engrs; AAAS; NY Acad Sci; Am Soc Eng Educ; Sigma Xi. *Res:* Blood coagulation; platelet and leukocyte function in health and disease; transport properties of endothelium; transport phenomena in pulmonary pathophysiology; modeling physiological process and systems. *Mailing Add:* Dept Biomed Eng Rensselaer Polytech Inst Troy NY 12180-3590

BIZZI, EMILIO, b Rome, Italy, Feb 22, 33; US citizen. NEUROPHYSIOLOGY, NEUROBIOLOGY. *Educ:* Univ Rome, MD, 58, Docenza, 68. *Prof Exp:* Res assoc, Dept Psychol, Mass Inst Technol, 66-67, lectr, 67-68, from assoc prof to prof neurophysiol, 69-80, dir, Whitaker Col, 83-89, EUGENE MCDERMOTT PROF BRAIN SCI & HUMAN BEHAV, 80-, CHMN, DEPT BRAIN & COGNITIVE SCI, 86- *Concurrent Pos:* Res assoc, Zool Neurophysiol Lab, Washington Univ, Mo, 63-64; vis assoc, Psychol Lab Clin Sci, NIMH, 64-66; sr investr, Inst Cardiovasc Res, Univ Milan, 68-69; fel, Found Fund Res Psychiat, 68; mem, Study Sect Vision B, NIH, 73-77; Whitaker health sci award, 78-79 & 82-83; NIH Javits neurosci investr award, 88; counr, Soc Neurosci, 88; corp mem, Boston Mus Sci, 89- *Honors & Awards:* Alden Spencer Award, 78; Bartlett Lectr, 88. *Mem:* Nat Acad Sci; Soc Neurosci; AAAS; Am Physiol Soc; Italian Physiol Soc; Int Brain Res Orgn; Am Acad Arts & Sci. *Res:* author and co-author of numerous publications and books relating to neurophysiology and neurobiology. *Mailing Add:* Dept Brain & Cognitive Sci Mass Inst Technol 25-526 Cambridge MA 02139-4307

BIZZOCO, RICHARD LAWREMCE WEISS, b New York, NY, Dec 28, 40; div; c 2. CELL BIOLOGY, BOTANY. *Educ:* Univ Conn, BA, 64; Calif State Univ, Long Beach, MS, 70; Ind Univ, PhD(microbiol), 72. *Prof Exp:* Res fel bot, Univ Calif, Berkeley, 73-74; res fel cell biol, Harvard Univ, 74-76; res assoc med microbiol, Univ Calif, Irvine, 76-77; from asst prof to assoc prof, PROF BOT CALIF STATE UNIV, SAN DIEGO, 85- *Concurrent Pos:* Am Cancer Soc fel, Univ Calif, Berkeley, 73-74; NIH fel cell biol, Harvard Univ, 75-76; NIH grant, Calif State Univ, San Diego, 78-, NIH biomed res, 85-87; NSF, 87. *Mem:* Sigma Xi. *Res:* Mechanism of cell fusion; cell-cell interaction; cell differentiation cell structure; biological electron microscopy; immunoelectron micros. *Mailing Add:* Dept Biol Calif State Univ San Diego CA 92182

BJARNGARD, BENGT E, b N Akarp, Sweden, Oct 2, 34; m 58; c 3. RADIOLOGICAL PHYSICS. *Educ:* Univ Lund, Fil Mag, 58, Fil Lic(radiation physics), 62. *Prof Exp:* Asst radiation physics, Univ Lund, 59-61; res physicist, Atomic Energy Co, Sweden, 61-65; dir radiation physics, Controls for Radiation, Inc, 65-68; from asst to assoc prof, 68-80, PROF RADIOTHER, HARVARD MED SCH, 80-, DIR PHYSICS, 68- *Concurrent Pos:* Lectr, Harvard Sch Pub Health. *Mem:* Am Asn Physicists in Med; Health Physics Soc; Radiation Res Soc; Am Soc Therapeut Radiologists. *Res:* Dosimetry; medical radiological physics. *Mailing Add:* 51 Samoset Rd Winchester MA 01890

BJERKAAS, ALLAN WAYNE, b Alexandria, Minn, Nov 25, 44; m 65; c 4. DYNAMICAL OCEANOGRAPHY, REMOTE SENSING. *Educ:* Univ NDak, BS, 66; Univ Ill, Urbana, MS, 67, PhD(physics), 71. *Prof Exp:* Res assoc physics, Univ Pittsburgh, 71-73; SR PHYSICIST, APPL PHYSICS LAB, JOHNS HOPKINS UNIV, 73- *Concurrent Pos:* Lectr, Johns Hopkins Univ. *Mem:* Sigma Xi. *Res:* Signal processing; image processing; remote sensing. *Mailing Add:* 4922 Snowy Reach Columbia MD 21044

BJERREGAARD, RICHARD S, b Gunnison, Utah, Jan 9, 43; m 62; c 5. AGRICULTURAL FIELD RESEARCH, ECOPHYSIOLOGY. *Educ:* Brigham Young Univ, BS, 67; Utah State Univ, PhD(ecol), 72. *Prof Exp:* Asst prof biol sci, Calif State Polytech Univ, 71-72; exten specialist, Exten Serv, Tex A&M Univ, 72-76; plant sci rep, Lilly Res Labs, 76-84, proj mgr, 84-90; REGULATORY SERVS, DOW ELANCO, 90- *Res:* Agricultural research projects; agricultural extension. *Mailing Add:* 542 Shady La Greenwood IN 46142

BJORHOVDE, REIDAR, b Harstad, Norway, Nov 6, 41; US citizen; m 72; c 2. STRUCTURAL ENGINEERING, STEEL & COMPOSITE STRUCTURES. *Educ:* Norweg Inst Technol, Trondheim, MS, 64, DrIng(civil eng), 68; Lehigh Univ, Bethlehem, Pa, PhD(civil eng), 72. *Prof Exp:* Res fel struct eng, Norweg Inst Technol, 65-66, asst prof, 66-68; res asst struct eng, Fritz Eng Lab, Lehigh Univ, 68-72; regional engr steel construct, Am Inst Steel Construct, Boston, 72-74, res engr, NY, 74-77; prof civil eng, Univ Alta, Edmonton, Can, 77-81; prof civil eng, Univ Ariz, Tucson, 81-87; PROF & CHMN CIVIL ENG, UNIV PITTSBURGH, PA, 87- *Concurrent Pos:* Tech secy, Struct Stability Res Coun, 69-71; chmn & mem, Nat & Internat Tech Comt, 72-, Nat & Internat Tech Conf, 82-; lectr, Northeastern Univ, Boston, Mass, 73; ed, J Constructional Steel Res, 86-; vis prof, Univ Paris VI, France, 87; NATO sr guest scientist fel, 87; mem related fac, Latin Am Studies & W Europ Studies, Univ Pittsburgh, 88-; vis prof & steel indust fel, Univ Witwatersrand, Johannesburg, SAfrica, 89; dir, Bridge & Struct Info Ctr, 89- *Honors & Awards:* Duggan Medal, Eng Inst Can, 80; T R Higgins Lectr, 87. *Mem:* Fel Am Soc Civil Engrs; Am Inst Steel Construct; Int Asn Bridge & Struct Eng; Asn Bridge Construct & Design; Struct Stability Res Coun; Sigma Xi; Am Welding Soc; Mex Soc Struct Eng. *Res:* Strength and stability of all types of steel structures; load and resistance factor design; steel columns; composite structures; steel connections; fracture of welded and other joints; metallic materials. *Mailing Add:* 2025 Murdstone Rd Upper St Clair PA 15241-2242

BJORK, CARL KENNETH, SR, b Burlington, Iowa, Mar 16, 26; m 47; c 7. CHEMISTRY. *Educ:* Augustana Col, AB, 47; DePauw Univ, MA, 50; Univ Ky, PhD(inorg chem), 53. *Prof Exp:* Res chemist, Int Mining & Chem Co, 53-54; res & develop engr, 54-59, patent agt & group leader, 59-70, mgr bioprod sect, Patent Dept, 70-78, mgr int sect, Patent Dept, 78-87, ADMIN MGR, PATENT OPERS, DOW CHEM CO, 87- *Concurrent Pos:* Chmn, Chem & Law Div, Am Chem Soc, 85 & 86, mem, Comt Patents & Related Matters, 78- *Mem:* Am Chem Soc; Sigma Xi. *Res:* Patent law; management; US and international patent applications and prosecution. *Mailing Add:* 2712 Lambros Midland MI 48640

BJORK, PHILIP R, b Wyandotte, Mich, Sept 14, 40; m 64; c 2. VERTEBRATE PALEONTOLOGY. *Educ:* Univ Mich, BS, 62, PhD(geol), 68; SDak Sch Mines & Technol, MS, 64. *Prof Exp:* From asst prof to assoc prof geol, Univ Wis-Stevens Point, 68-75; assoc prof, 75-80, PROF GEOL, SDAK SCH MINES & TECHNOL, 80-, DIR MUS GEOL, 75- *Mem:* Soc Vert Paleont; Am Paleont Soc. *Res:* Blancan carnivora in North America; Oligocene faunas of North America; Blancan faunas of North America; dinosaurs of western North America. *Mailing Add:* Mus Geol SDak Sch Mines & Technol Rapid City SD 57701

BJORKEN, JAMES D, b Chicago, Ill, June 22, 34. PHYSICS. *Educ:* Mass Inst Technol, BS, 56; Stanford Univ, PhD, 59. *Prof Exp:* Res assoc, Stanford Univ, 60, from asst prof to prof, 61-79; sr physicist, Fermi Nat Accelerator Lab, 79-84, assoc dir physics, 84-88; STAFF MEM, THEORET PHYSICS DEPT, STANFORD LINEAR ACCELERATOR CTR, 89- *Honors & Awards:* E O Lawrence Award, US Dept Energy, 78. *Mem:* Nat Acad Sci; fel Am Phys Soc; fel Am Acad Arts & Sci. *Res:* Author and coauthor of several books. *Mailing Add:* Stanford Linear Accelerator Ctr PO Box 4349 Stanford CA 94309

BJORKHOLM, JOHN ERNST, b Milwaukee, Wis, Mar 22, 39; m 64; c 2. EXPERIMENTAL PHYSICS, SOFT X RAY OPTICS. *Educ:* Princeton Univ, BSE, 61; Stanford Univ, MS, 62, PhD(appl physics), 66. *Prof Exp:* mem tech staff, 66-83, DISTINGUISHED MEM TECH STAFF, ELECTRONICS RES LAB, BELL TEL LABS, 83- *Concurrent Pos:* Chmn, Gordon Res Conf Nonlinear Optics & Lasers, 77; mem, Exec Comt of Tech Coun, Optical Soc Am. 86-87; comptroller, CLEO/QELS, 88, 89, 90, 91; bd dirs, Optical Soc Am, 90. *Mem:* Fel Am Phys Soc; sr mem Inst Elec & Electronics Engrs; fel Optical Soc Am (treas, 92). *Res:* Quantum electronics; nonlinear optics; interaction of resonant light with atoms; atomic physics; high resolution two-photon spectroscopy; x-ray projection lithography. *Mailing Add:* Silicon Electronics Res Lab Bell Tel Labs 4B-423 Holmdel NJ 07733

BJORKHOLM, PAUL J, b Milwaukee, Wis, May 27, 42; m 67; c 2. DIGITAL MEDICAL IMAGING. *Educ:* Princeton Univ, BA, 64; Univ Wis, MA, 65, PhD(low energy nuclear physics), 69. *Prof Exp:* Sr scientist x-ray astron & lunar geol, 69-73, sr staff scientist, 73-83, vpres res, 83-86, SR VPRES RES, AM SCI & ENG INC, 86- *Mem:* AAAS; Am Phys Soc; Am Asn Physicists in Med. *Res:* Design and development of x-ray imaging techniques as applied to diagnostic digital radiography; digital mammography; lunar geophysics; remote sensing of elemental composition, natural and induced radioactivity. *Mailing Add:* Am Sci & Eng Inc Ft Washington Cambridge MA 02139

BJORKLAND, JOHN A(LEXANDER), b Chicago, Ill, Oct 16, 23. ENGINEERING PHYSICS. *Educ:* Univ Chicago, BS, 46, MS, 50; NC State Col, PhD(eng physics), 56. *Prof Exp:* Sr engr, Cook Res Labs Div, Cook Elec Co, 56-60; assoc physicist, Armour Res Found, 60-61; assoc engr, Argonne Nat Lab, 61-85; RETIRED. *Res:* Nuclear reactor instrumentation and electronic pulse circuitry. *Mailing Add:* 1005 Harvard Terr Evanston IL 60202

BJORKLUND, GARY CARL, b Passaic, NJ, Nov 17, 46; m 69. QUANTUM ELECTRONICS, OPTICS. *Educ:* Mass Inst Technol, BS, 68; Stanford Univ, MS, 69, PhD(appl physics), 74. *Prof Exp:* Mem tech staff, Bell Tel Labs, 74-79; res staff mem, 79-80, MGR, IBM ALMADEN RES CTR, 80- *Mem:* Fel Am Phys Soc; fel Optical Soc Am; fel Inst Elec & Electronics Engrs. *Res:* Organic nonlinear optical materials and devices, frequency domain optical memories and FM spectroscopy. *Mailing Add:* Dept K95-801 IBM Almaden Res Ctr 650 Harry Rd San Jose CA 95120

BJORKLUND, RICHARD GUY, b Milwaukee, Wis, Feb 10, 28; div; c 5. ECOLOGY, CONSERVATION. *Educ:* Mont State Univ, BS, 51, MS, 53; Univ Mich, PhD, 58. *Prof Exp:* Res asst, Univ Mich, 54-55; from instr to prof biol, 57-68, chmn dept, 64-70, dean lib arts & sci, 73-78, PROF BIOL, BRADLEY UNIV, 68- *Concurrent Pos:* Asst blister rust control, US Forest Serv, 48; aide, Inst Fish Res, 53-54; consult teach & res proj, 78-81 & 84-91; contract res, Avian Ecol, Ill Dept Conserv, 81 & 85. *Mem:* Am Inst Biol Sci; Sigma Xi; Colonial Waterbird Group; Wilson Ornith Soc; Am Ornithologists Union; Am Fisheries Soc. *Res:* Environmental biology; temperature sense; photoperiodism; endocrine functions; aquatic biology; ecology and natural history of herons; ecology and natural history of other birds. *Mailing Add:* Dept Biol Bradley Univ Peoria IL 61625

BJORKMAN, OLLE ERIK, b Jonkoping, Sweden, July 29, 33; m 55; c 2. PHYSIOLOGICAL ECOLOGY, PHOTOBIOLOGY. *Educ:* Univ Stockholm, MS, 57; Univ Uppsala, PhD, 60. *Prof Exp:* Asst scientist, Dept Genetics & Plant Breeding, Royal Agr Col Sweden & Univ Uppsala, 57-61, res fel, Swed Natural Sci Res Coun, 61-63; res fel, 64-65, STAFF BIOLOGIST, DEPT PLANT BIOL, CARNEGIE INST WASH, CALIF, 66- *Concurrent Pos:* Assoc courtesy prof biol, Stanford Univ, 67-, courtesy prof biol, 75- *Honors & Awards:* Linneus Prize, Royal Physiog Soc, Sweden; The Selby Award, Australian Acad Sci; Stephen Hales Award, Am Soc Plant Physiologists. *Mem:* Nat Acad Sci; Am Soc Plant Physiologists; fel AAAS; Am Acad Arts & Sci; Australian Acad Sci; Royal Swed Acad Sci. *Res:* Physiological and biochemical mechanisms of plant response and adaptation to ecologically diverse environments; environmental and biological control of photosynthesis. *Mailing Add:* Dept Plant Biol Carnegie Inst Wash 290 Panama St Stanford CA 94305

BJORKMAN, PAMELA J, b Portland, Ore, June 21, 56; m 88; c 1. PROTEIN CRYSTALLOGRAPHY, PROTEIN EXPRESSION. *Educ:* Univ Ore, BA, 78; Harvard Univ, PhD(biochem & molecular biol), 84. *Prof Exp:* Postdoctoral fel, Harvard Univ, 84-86; postdoctoral fel, Dept Microbiol & Immunol, Stanford Univ, 86-89; ASST PROF BIOL & CHEM, DIV BIOL, CALIF INST TECHNOL, 89-; ASST INVESTR BIOL & CHEM, HOWARD HUGHES MED INST & DIV BIOL, CALIF INST TECHNOL, 89. *Concurrent Pos:* Young investr award, Am Soc Histocompatibility & Immunogenetics, 88; mem, Nat Res Coun Comt, Investigate Funding Young Investigators, 90- *Honors & Awards:* William B Coley Award, Distinguished Res Fundamental Immunol, Cancer Res Inst, 91. *Mem:* Am Crystallog Asn. *Res:* Structure and function of all surface recognition molecules found in the immune and nervous systems by x-ray crystallographic, biochemical and molecular biological techniques. *Mailing Add:* Div Biol 156-29 Calif Inst Technol Pasadena CA 91125

BJORKSTEN, JOHAN AUGUSTUS, b Tammerfors, Finland, May 27, 07; nat US; m 61; c 5. CHEMISTRY. *Educ:* Univ Helsinki, MS, 27, PhD(protein chem), 31. *Prof Exp:* Int Educ Bd fel, Univ Minn, 31-32; res chemist, Felton Chem Co, 33-34, chief chemist, 34-35; res chemist, Pepsodent Co, 35, in-chg develop, 36; chief chemist, Ditto, Inc, 36-41; chem dir, Quaker Chem Prod Corp, Pa, 41-44; CHMN, BD DIRS, BJORKSTEN RES LABS, 44- *Concurrent Pos:* Vpres, ABC Packaging Mach Corp, Fla, 40- & Bee Chem Co, Ill, 44-57; pres, Bjorksten Res Found, 53-; dir, Reef Indust Col, Tex, 56- *Mem:* AAAS; Am Chem Soc; Am Asn Cereal Chemists; Am Geriat Soc; fel Am Inst Chemists (pres, 62-63); Sigma Xi. *Res:* Synthetic resins and plastics; paper and fibrous materials; special coating compositions; proteins; industrial problems; chemical gerontology. *Mailing Add:* 4455 SW 34th St No 4A141 Gainesville FL 32608-6530

BJORNDAHL, JAY MARK, b Lafayette, Ind, Nov 14, 55; m 82; c 2. SIGNAL TRANSDUCTION, T-LYMPHOCYTES. *Educ:* Calif State Univ, BA, 79; Univ Kans, MS, 85, PhD(pharmacol), 87. *Prof Exp:* Assoc res scientist immunol, Okla Med Res Found, 85-88; DIR IMMUNOL, FLOW CYTOMETRY LAB, HARRINGTON CANCER CTR, 88- *Concurrent Pos:* Clin asst prof, Dept Internal Med, Health Sci Ctr, Tex Tech Univ, Amarillo, 88- *Mem:* Sigma Xi; AAAS; Am Soc Microbiol; Am Asn Immunologists. *Res:* Signal transduction in human T lymphocytes; ligand-receptor interactions; surface antigen expression; flow cytometry; protein isolation/purification. *Mailing Add:* Harrington Cancer Ctr 1500 Wallace Blvd Amarillo TX 79106

BJORNDAL, ARNE MAGNE, b Ulstein, Norway, Aug 19, 16; m 52; c 3. DENTISTRY. *Educ:* Volda State Col, BS, 39; Univ Oslo, DDS, 47; Univ Iowa, MS & DDS, 56; Am Bd Endodont, dipl, 64. *Prof Exp:* Instr pedodont, 48-52; from instr to prof operative dent & endodont, Col Dent, Univ Iowa, 55-88; RETIRED. *Concurrent Pos:* Mem, Coun Med TV, 65-; vpres, Clyde Davis Midwestern Endodontic Study Group. *Mem:* Fel Am Col Dentists; AAAS; Am Asn Endodont; Int Asn Dent Res; Norweg Dent Asn. *Res:* Endodontics in relation to hemophiliacs and blood dyscrasias. *Mailing Add:* 2510 Bluffwood Circle Iowa City IA 52245

BJORNERUD, EGIL KRISTOFFER, b Hamar, Norway, May 3, 25; nat US; m 59; c 4. PHYSICS. *Educ:* Univ Wash, BS, 49, MBA, 69; Calif Inst Technol, PhD(physics), 54. *Prof Exp:* Staff mem physics dept, Rand Corp, Calif, 54-55; reactor dept, Inst Atomic Energy, Norway, 55-56; res assoc, Gen Atomic Div, Gen Dynamics Corp, 56-59, asst export mgr, 59-62; assoc dir sci div, Nat Eng Sci Co, Calif, 62-64; chief, Nuclear Technol Aerospace Div, 64-67, dir laser develop, Res Div, 70-75, dir nuclear projs, eng & construct div, Boeing Co, 75-84; MANAGING DIR, BRAATHEN INDUST SERV, OSLO, NORWAY, 84- *Concurrent Pos:* Consult, AEC, India, 59-60; deleg, Int

Atomic Energy Agency, Vienna, 80. *Mem:* Am Phys Soc; Am Nuclear Soc; Am Inst Aeronaut & Astronaut; fel Royal Astron Soc. *Res:* Cloud-chamber studies of elementary particle interactions; experimental studies of flame emission spectra; neutron diffusion and criticality experiments; reactor physics and nuclear technology. *Mailing Add:* Elisenberg Van 32 Oslow 0265 Norway

BJORNSON, AUGUST SVEN, b Reykjavik, Iceland, May 10, 22; nat US; m 44; c 3. ORGANIC CHEMISTRY. *Educ:* Univ Wis, BSc, 44, MSc, 45, Univ Kans, PhD(chem), 48. *Prof Exp:* Res chemist, 48-51, group leader, 51-53, supvr, 53-54, supvr develop dept, 54-56, res mgr, 56-60, res mgr new prod develop, 60-62, mgr develop planning, 62-66, MGR PATENTS & LICENSING, E I DU PONT DE NEMOURS & CO, INC, 66- *Mem:* AAAS; Am Chem Soc. *Res:* Negotiations and administration of technical agreements. *Mailing Add:* 4616 Weldin Rd Wilmington DE 19803

BJOTVEDT, GEORGE, b Brooklyn, NY, Oct 7, 29; m 51; c 4. VETERINARY MEDICINE, ANIMAL PATHOLOGY. *Educ:* Pa Mil Col, BS, 51; Univ Pa, VMD, 58; Am Col Lab Animal Med, dipl, 65. *Prof Exp:* Dir, Div Lab Med, Sch Med, Univ Pa, 59-63; asst prof vet surg, Mt Sinai Sch Med, 64-69; owner, Ariz Animal Welfare League Vet Clin, Phoenix, Ariz, 70-74; UNIV VET, ARIZ STATE UNIV, 74- *Concurrent Pos:* NIH fel, Wistar Inst Anat & Biol, 58-59; coun mem, Gen Med Sci Div, NIH, 61-62. *Mem:* Sigma Xi. *Res:* Applied research in the nature and variety of laboratory animal diseases and pathology. *Mailing Add:* Div Biosci Ariz State Univ Tempe AZ 85287-0608

BJUGSTAD, ARDELL JEROME, range science; deceased, see previous edition for last biography

BLACHERE, JEAN R, b Autun, France, Dec 25, 37; m 59; c 2. CERAMICS. *Educ:* Nat Sch Advan Indust Ceramics, Dipl eng, 59; Alfred Univ, MS, 61, PhD(ceramics), 69. *Prof Exp:* Ceramic technologist, Cent Res Lab, Domtar Ltd, Can, 62-65; asst prof, 68-74, ASSOC PROF CERAMIC ENG, UNIV PITTSBURGH, 75- *Mem:* Am Ceramic Soc; Nat Inst Ceramic Engrs; Ceramic Educ Coun; Am Inst Mining, Metall & Petrol Engrs. *Res:* High temperature materials; surfaces, grain boundaries and hot corrosion; solid state reactions; mechanical properties; refractories; electromagnetic properties. *Mailing Add:* Dept Metall Eng Univ Pittsburgh Main Campus 848 Benedum Pittsburgh PA 15260

BLACHFORD, CAMERON W, b Colonsay, Sask, July 10, 31; m 56, 87; c 4. ELECTRICAL ENGINEERING. *Educ:* Univ Sask, BSc, 53; Univ Ill, MS, 59, PhD(elec eng), 63. *Prof Exp:* Salesman, English Elec Co, Ltd, 56-58; instr elec eng, Univ Ill, 60-63; from asst prof to assoc prof, Queen's Univ, Ont, 63-67; PROF ELEC ENG, UNIV REGINA, 67-, DEAN, GRAD STUDIES & RES, 77- *Concurrent Pos:* Bd mem, Sask Res Coun. *Mem:* Inst Elec & Electronics Engrs; fel Eng Inst Can; Asn Prof Engrs Sask; Can Soc Elec & Computer Eng. *Res:* Automatic controls; electro-mechanical energy conversion. *Mailing Add:* Fac Grad Studies & Res Univ Regina Regina SK S4S 0A2 Can

BLACHMAN, ARTHUR GILBERT, b Cleveland, Ohio, June 29, 26; m 53; c 3. PHYSICS. *Educ:* Western Reserve Univ, BS, 50; Ohio State Univ, MS, 56; NY Univ, PhD(physics), 66. *Prof Exp:* Physicist res dept, Erie Resistor Corp, Pa, 50-51; physicist digital comput lab, Mass Inst Technol, 51-52; teacher high sch, Colo, 52-54; res staff mem, Watson Lab, NY, 56-65, RES STAFF MEM RES CTR, IBM CORP, YORKTOWN HEIGHTS, 65- *Mem:* AAAS; Am Phys Soc; Am Vacuum Soc. *Res:* Semiconductor physics; hyperfine structure of atomic metastable states; structure and properties of sputtered thin films. *Mailing Add:* 215 Cedar Dr E Briarcliff Manor NY 10510

BLACHMAN, NELSON MERLE, b Cleveland, Ohio, Oct 27, 23; m 53; c 2. PHYSICS. *Educ:* Case Sch Appl Sci, Cleveland, BS, 43; Harvard Univ, AM, 47, PhD(appl physics), 47. *Prof Exp:* Spec res assoc, Underwater Sound Lab, Harvard Univ, 43-45, res assoc, Ctr Commun Res, 45-46, Cruft Lab, 46; assoc physicist, Brookhaven Nat Lab, 47-51; physicist, Math Sci Div, Off Naval Res, 51-54, liaison scientist, London, 58-60 & 76-78, Naval Res Lab, Wash, 85-86; PRIN SCIENTIST, GOVT SYSTS CORP, GEN TEL & ELECTRONICS CORP, 54-58, 60-76, 78-85, 86- *Concurrent Pos:* Lectr, Univ Md, 51-52; lectr, Eng Exten, Univ Calif, 61-63; sr Fulbright lectr, Madrid, 64-65; lectr, Eng Exten, Univ Calif, 61-63; lectr Stanford Univ, 67; mem Info Theory Group Bd Govs, Inst Elec & Electronics Engrs, 69-75 (vpres, chmn com, 70). *Mem:* Fel AAAS; Math Soc Am; fel Inst Elec & Electronics Engrs; Soc Indust & Appl Math; fel Inst Elec Engrs London; Am Statist Asn; Union Radio Sci Int. *Res:* Statistical communication theory; information theory. *Mailing Add:* GTE Sylvania Gov't Systs Corp Mail Stop 6209 Box 7188 Mountain View CA 94039-7188

BLACHUT, T(HEODORE) J(OSEPH), b Czestochowa, Poland, Feb 10, 15; nat US; m 48; c 3. PHOTOGRAMMETRY. *Educ:* Inst Technol, Lwow, Poland, Dipl Eng, 38. *Prof Exp:* From asst to assoc scientist photogrammetry, Swiss Fed Inst Technol & lectr cartog, Polish Univ, Camp, Switz, 41-44; geodesist & photogrammetrist, Sci Dept, Wild Co, Switz, 45-51; head photogrammetric res, Nat Res Coun Can, 51-80; RETIRED. *Concurrent Pos:* Vis prof, Laval Univ; NSF vis scientist; Can deleg & officer, Pan Am Inst Geog & Hist, Orgn Am States. *Mem:* Fel Royal Soc Can; Am Soc Photogram; Can Inst Surv; Int Soc Photogram; hon mem Brazilian Cartog Soc. *Res:* Photogrammetric equipment and methods; introduction of aerial triangulation method based on the use of radar auxiliary data; long distance bridging; precise large-scale photogrammetry; non-cartographic photogrammetry; development of stereo-orthophoto mapping system. *Mailing Add:* 61 Ruthwell Dr Gloucester ON K1J 7G7 Can

BLACK, ALEX, b Richmond, Ky, Nov 16, 06; m 29. ANIMAL NUTRITION, BIOCHEMISTRY. *Educ:* Univ Ky, AB, 29; Pa State Univ, MS, 33; Univ Rochester, PhD(nutrit), 38. *Prof Exp:* Asst, 29-34, from instr to prof, 34-69, asst dir agr exp sta, 53-60, assoc dir, 60-69, EMER PROF ANIMAL

NUTRIT, PA STATE UNIV, 69- *Concurrent Pos:* Asst dir, Agr Exp Sta, Pa State Univ, 53-60, assoc dir, 60- 69. *Mem:* Am Soc Animal Sci; Am Inst Nutrit; NY Acad Sci. *Res:* Human and animal nutrition; energy metabolism; protein, fat and mineral metabolism. *Mailing Add:* 5695 Balkan Ct SW Ft Myers FL 33913-2741

BLACK, ARTHUR HERMAN, b Toledo, Ohio, Aug 16, 19; m 45; c 3. ANALYTICAL CHEMISTRY. *Educ:* Univ Toledo, BS, 41, MS, 48; Univ Mich, MS, 55. *Prof Exp:* From instr to assoc prof chem, 46-61, asst dean Col Arts & Sci, 61-64, dean men, 64-68, assoc dean, 68-84, prof chem, 70-84, EMER PROF CHEM, UNIV TOLEDO, 84- *Mem:* AAAS; NY Acad Sci; Am Chem Soc. *Res:* Volumetric and colorimetric methods involving various chelating agents. *Mailing Add:* 3209 Kylemore Rd Toledo OH 43606

BLACK, ARTHUR LEO, b Redlands, Calif, Dec 1, 22; m 45; c 3. BIOCHEMISTRY. *Educ:* Univ Calif, BS, 48, PhD(comp physiol), 51. *Prof Exp:* From instr to assoc prof, 51-62, PROF BIOCHEM, UNIV CALIF, DAVIS, 62- *Honors & Awards:* Borden Award, Am Inst Nutrit, 63. *Mem:* AAAS; Am Soc Biol Chemists; Am Inst Nutrit; Am Physiol Soc. *Res:* Gluconeogenesis and its control; metabolism of amino acids in animals; study of intermediary metabolism in intact animals, including ruminants. *Mailing Add:* Dept Physiol Sch Vet Med Univ Calif Davis CA 95616

BLACK, ASA C, JR, b Clarksville, Tenn, Jan 2, 43; m. ANATOMY. *Educ:* Vanderbilt Univ, BA, 65, PhD(anat), 74. *Prof Exp:* Instr anat, Vanderbilt Univ, 72; res assoc, Univ Iowa, 73-75, asst prof anat, 75-85; ASSOC PROF ANAT, MERCER MED SCH, 85- *Mem:* AAAS; Am Asn Anatomists; Soc Neurosci; Am Soc Neurochem; Res Soc Alcoholism. *Res:* Role of nucleotides in neuronal function; mechanisms of action of ethanol on the nervous system; mechanisms responsible for the fetal alcohol syndrome. *Mailing Add:* Sch Med 841 Mercer Univ 1400 Coleman Ave Macon GA 31211

BLACK, BETTY LYNNE, b St Louis, Mo, Mar 28, 46. EMBRYOLOGY, CELL PHYSIOLOGY. *Educ:* Lindenwood Col, BA, 67; Vanderbilt Univ, MS, 69; Wash Univ, St Louis, PhD(biol), 76. *Prof Exp:* Instr biol, Harris Teachers Col, 69-72; res fel, Sch Med, Wash Univ, 77-79; ASSOC PROF DEVELOP BIOL, NC STATE UNIV, 79- *Honors & Awards:* Res Career Develop Award, Pub Health Serv. *Mem:* AAAS; Sigma Xi; Soc Develop Biol; Am Soc Cell Biol. *Res:* Regulatory role of hormones and ions in differentiation of embryonic intestine; development of intestinal transport systems. *Mailing Add:* Dept Zool NC State Univ Campus Box 7617 Raleigh NC 27695-7617

BLACK, BILLY C, II, b Beatrice, Ala, Feb 1, 37; m 61; c 1. BIOCHEMISTRY, ANALYTICAL CHEMISTRY. *Educ:* Tuskegee Inst, BS, 60; Iowa State Univ, MS, 62, PhD(biochem), 64. *Prof Exp:* Res asst biochem, Iowa State Univ, 60-64; PROF CHEM, ALBANY STATE COL, 64-, CHMN DEPT, 61- *Mem:* Am Chem Soc; Am Oil Chem Soc; Int Food Technologists; Am Inst Chemists. *Res:* Fatty acid metabolism; molecular dynamics; glyceride structure. *Mailing Add:* Pres/Chancellor Albany State Col Albany GA 31705

BLACK, CARL (ELLSWORTH), polymer science, for more information see previous edition

BLACK, CHARLES ALLEN, b Lone Tree, Iowa, Jan 22, 16; m 39; c 3. SOIL FERTILITY. *Educ:* Colo State Univ, BS, 37; Iowa State Univ, MS, 38, PhD(soil fertil), 42. *Prof Exp:* Asst land classifier, Resettlement Admin, 37; jr soil surveyor, Soil Conserv Serv, 38; instr soils, 39-42, res assoc, 43-44, from res asst prof to res assoc prof, 44-49, prof, 49-67, distinguished prof, 67-79, adj prof, 79-85, EMER PROF SOILS, IOWA STATE UNIV, 85- *Concurrent Pos:* Vis prof, Cornell Univ, 55-56; Kearney Found lectr, Univ Calif, 60; NSF fel, 64-65; consult, USDA, 64; chmn bd dirs, Coun Agr Sci & Technol, 72, pres, 73, exec vpres, 74-85, exec chmn bd dirs, 85-88. *Honors & Awards:* Soil Sci Award, Am Soc Agron, 57, Agron Serv Award, 86; Bouyoucos Soil Sci Distinguished Career Award, Soil Sci Soc Am, 81; Distinguished Serv Award, Am Agr Ed Asn, 79; Henry A Wallace Award, Iowa State Univ, 80; Nat Award Agr Excellence, Nat Agr-Mkt Asn, 83; Charles A Black Award, Coun Agr Sci & Technol, 86. *Mem:* Fel AAAS, 76; fel Am Soc Agron (pres, 70-71); fel Soil Sci Soc Am (pres, 62); fel Am Inst Chemists, 69; Int Soc Soil Sci; Am Soc Agr Engrs. *Res:* Soil phosphorus and soil-plant relationships; selection by sequential elimination based on concordant judgments. *Mailing Add:* Dept Agron Iowa State Univ Ames IA 50011

BLACK, CLANTON CANDLER, JR, b Tampa, Fla, Nov 27, 31; m 52; c 3. BIOCHEMISTRY. *Educ:* Univ Fla, BSA, 53, MSA, 57, PhD(agron), 60. *Prof Exp:* NIH fel biochem, Cornell Univ, 60-62; fel, Kettering Res Lab, 62-63, staff scientist, 63-67; asst prof biol, Antioch Col, 63-67; from assoc prof to prof biochem, 67-74, prof bot & head dept, 74-77, prof biochem, 77, RES PROF, BIOCHEM, UNIV GA, 77- *Mem:* Am Soc Biol Chemists; Am Soc Plant Physiologists (pres-elect, 77-78, pres, 78-79). *Res:* Photosynthesis; enzymology; electron transport; pyrophosphate & sucrose metabolism. *Mailing Add:* Dept Biochem GSRC Univ Ga Athens GA 30602

BLACK, CRAIG C, b Peking, China, May 28, 32; US citizen; m 54, 67; c 2. VERTEBRATE PALEONTOLOGY. *Educ:* Amherst Col, AB, 54, MA, 57; Harvard Univ, PhD(biol), 62. *Prof Exp:* Assoc cur vert paleont, Carnegie Mus, 60-63, cur, 63-70; assoc prof ecol & systs, Mus Natural Hist, Univ Kans, 70-72; dir mus & prof geosci, Tex Tech Univ, 72-75; dir, Carnegie Mus Natural Hist, 75-82; DIR, COUNTY MUS NATURAL HIST, LOS ANGELES, 82- *Concurrent Pos:* Mem, Mus Panel, Nat Endowment Arts, 74-77; mem, Nat Sci Bd, 85-90. *Mem:* Soc Vert Paleont (pres, 70-71); fel Geol Soc Am; Soc Study Evolution; fel Linnean Soc London; fel AAAS. *Res:* Tertiary mammals; paleoecology and biogeography; rodent evolution. *Mailing Add:* County Mus Natural Hist 900 Exposition Blvd Los Angeles CA 90007

BLACK, CRAIG PATRICK, b Greeley, Colo, Apr 25, 46; m 69. ANIMAL PHYSIOLOGY, ANIMAL ECOLOGY. *Educ:* Tufts Univ, BS, 68; Dartmouth Col, PhD(biol), 75. *Prof Exp:* Lectr, Dept Biol Sci, 74-75, NIH FEL, DEPT PHYSIOL, SCH MED, DARTMOUTH COL, 75- *Concurrent Pos:* Vis asst prof, Dept Biol Sci, Dartmouth Col, 75-76. *Mem:* Ecol Soc Am; Cooper Ornith Soc; Am Ornithologists Union; Sigma Xi. *Res:* Avian energetics; avian respiration physiology; high altitude physiology. *Mailing Add:* Dept Biol Univ Toledo Toledo OH 43606

BLACK, CURTIS DOERSAM, b Toledo, Ohio, Mar 23, 51; m 74; c 1. NUTRITION, CLINICAL PHARMACY. *Educ:* Univ Toledo, BSPharm, 74; Purdue Univ, MS, 76, PhD(clin pharm & pharmaceuts), 78. *Prof Exp:* Asst prof, 78-84, ASSOC PROF CLIN PHARM, PURDUE UNIV, 84-, ASSOC DEPT HEAD, 86- *Concurrent Pos:* Consult, Prof Educ Bur, Searle Labs, 83-, Clin Nutrit, Travenol Labs, 84-85; site eval team, Am Coun Pharm Ed, 88. *Mem:* Sigma Xi; Am Asn Col Pharm; Am Soc Parenteral & Enteral Nutrit; Am Col Clin Pharm; AAAS. *Res:* Biochemical and physiological effects of total parenteral nutrition as an adjunct to cancer chemotherapy; pharmaceutical considerations for intravenous drug and nutrient delivery; cost effectiveness of clinical pharmacy services. *Mailing Add:* Dept Pharm Pract Univ Toledo Toledo OH 43606

BLACK, DAVID CHARLES, b Waterloo, Iowa, May 14, 43; m 67. THEORETICAL ASTROPHYSICS, METEORITICS. *Educ:* Univ Minn, BS, 65, MS, 67, PhD(physics), 70. *Prof Exp:* Nat Acad Sci fel, 70-72, RES SCIENTIST THEORET ASTROPHYS, NASA AMES RES CTR, 72-, CHIEF SCIENTIST SPACE STA, 85- *Mem:* AAAS; NY Acad Sci. *Res:* Theoretical studies of the formation and evolution of stars and planetary systems; interpretation of rare gas isotopic data from meteorites and lunar samples. *Mailing Add:* 3303 NASA RD 1 Houston TX 77058

BLACK, DEAN, b Salt Lake City, Utah, May 6, 42; m 65; c 9. HEALTH, FAMILY RELATIONS. *Educ:* Brigham Young Univ, BA, 66, MS, 69; Pa State Univ, PhD(human relations), 71. *Prof Exp:* Asst prof sociol, Univ Southern Calif, 71-74; assoc prof, instr Brigham Young Univ, 74-79; pres, Sunrider Naturalfife Corp, 79-84; sr mkt spec, Comput Resources & Info Corp, 84-85; PRES, BIORES FOUND, TAPESTRY PRESS, 85- *Concurrent Pos:* Consult, Action, Washington, DC, 72, 73, US State Dept, USAID, 77. *Mem:* Sigma Xi. *Res:* Scientific, philosophical and political relationships between conventional medicine and natural healing therapies; the influence of social and psychological factors on health; physiological correlates of various coping styles and their manifestation in physical, mental and emotional parameters. *Mailing Add:* Tapestry Press PO Box 653 Springville UT 84663

BLACK, DONALD K, b Ely, Eng, Feb 23, 36. ORGANIC CHEMISTRY. *Educ:* Univ London, BSc, 60, PhD(org chem), 63. *Prof Exp:* Res chemist, Middlesex Hosp Med Sch, London, 64-66; Nat Acad Sci resident res assoc org chem, Naval Stores Res Sta, USDA, 66-68; RES CHEMIST, RES CTR, HERCULES, INC, 68- *Mem:* The Chem Soc. *Res:* Diterpene, resin acid and pesticide chemistry. *Mailing Add:* Hercules Inc Hercules Res Ctr 1313 N Market St Wilmington DE 19894-0001

BLACK, DONALD LEIGHTON, b Portland, Maine, Aug 3, 28; m 50; c 3. ANIMAL PHYSIOLOGY. *Educ:* Univ Maine, BS, 54; Cornell Univ, MS, 57, PhD(animal physiol), 59. *Prof Exp:* Assoc prof animal physiol, 59-68, PROF VET & ANIMAL SCI, UNIV MASS, AMHERST, 68- *Res:* Oviduct physiology; physiology of ovarian function. *Mailing Add:* Paige Lab Univ Mass Amherst MA 01003

BLACK, EMILIE A, medical administration, pediatrics, for more information see previous edition

BLACK, FRANCIS LEE, b Taipei, Formosa, Jan 2, 26; Can citizen; m 52; c 3. VIROLOGY, IMMUNITY. *Educ:* Univ BC, BA, 47, MA, 49; Univ Calif, PhD(biochem), 52. *Prof Exp:* Fel virus chem, Rockefeller Found, 52-54; chemist viruses, Lab Hyg, Dept Nat Health & Welfare, Can, 54-55; res assoc & lectr, 55-61, from asst prof to assoc prof, 61-73, PROF EPIDEMIOL, SCH MED, YALE UNIV, 73- *Concurrent Pos:* Consult, Pan Am Health Orgn, 79. *Mem:* Am Asn Immunol; Soc Exp Biol & Med; Am Epidemiol Soc; Sigma Xi; Am Asn Anthropology. *Res:* Chemistry and functions of viruses; measles virus; infectious disease patterns in isolated populations; South American Indians; population genetics. *Mailing Add:* 109 Mill Plain Rd Branford CT 06405

BLACK, FRANKLIN OWEN, b St Louis, Mo, Aug 8, 37. OTOLARYNGOLOGY. *Educ:* Southeast Mo State Univ, BA, 59; Univ Mo-Columbia, MD, 63. *Prof Exp:* Intern, Mobile Gen Hosp, 63-64; resident gen surg, Bataan Mem Hosp & Lovelace Found, 64-65 & otolaryngol, Med Ctr, Univ Colo, 65-68, instr, 68-69, assoc prof surg, 69-71 & Col Med, Univ Fla, 71-74; assoc prof & dir, div vestibular dis, Eye & Ear Hosp, Pittsburgh, 74-, vchmn otolaryngol, 76-; CHIEF DIV NEUROTOLOGY, GOOD SAMARITAN HOSP. *Concurrent Pos:* NIH spec res fel, Univ Colo, 68-69; attend physician otolaryngol, Denver Vet Admin Hosp & Denver Gen Hosp, 68-69; NIH res career develop award, Univ Fla, 72-74; pres, Pittsburgh Neurosci Soc, 77-78 & Asn for Res Otolaryngol, 78-79; chmn, Comt for Res, Am Acad Otolaryngol, 78-83; ad hoc reviewer, NIH & Am Inst Biol Sci. *Mem:* Asn for Res Otolaryngol; Am Inst Biol Sci; Am Col Surgeons; Am Acad Otolaryngol; Am Bd Otolaryngol; Sigma Xi. *Res:* Objective clinical vestibular examination methods; computerized posturography; control characteristics of human postural sway trajectories; clinical vestibulo-occular responses to rotational acceleration. *Mailing Add:* Chief of Div Neurotology Good Samaritan Hosp Med Ctr 1040 NW 22nd Ave Portland OR 97210

BLACK, GRAHAM, physical chemistry, for more information see previous edition

BLACK, H(AROLD) S(TEPHEN), electrical engineering; deceased, see previous edition for last biography

BLACK, HENRY M(ONTGOMERY), mechanical engineering, for more information see previous edition

BLACK, HOMER SELTON, b Port Arthur, Tex, Sept 9, 35; m 68; c 2. CANCER. *Educ:* Agr & Mech Col Tex, BS, 56; Sam Houston State Teachers Col, MEd, 60; La State Univ, PhD(bot), 64; Univ Houston, MSA, 77. *Prof Exp:* Res chemist, Seed Protein Pioneering Res Lab, Southern Utilization Res & Develop Div, USDA, 64-65; res fel biochem, Univ Tex-M D Anderson Hosp & Tumor Inst, 65-66; asst prof biol, Sam Houston State Univ, 66-68; physiologist, Vet Admin Hosp, Houston, Tex, 68-76; res asst prof dermat, 68-74, adj asst prof biochem, 69-74, assoc res prof, 74-80, RES PROF DERMAT, BAYLOR COL MED, 80- *Concurrent Pos:* Nat Acad Sci-Nat Res Coun res assoc, 64-; dir, Photobiol Lab, Vet Admin Hosp, Houston, Tex, 76- *Honors & Awards:* Res Career Scientist, Vet Affairs, 82- *Mem:* Am Oil Chemists Soc; Am Soc Photobiol; Europ Soc Photobiol; Am Asn Cancer Res; Admin Mgt Soc. *Res:* Photobiology; photo-carcinogenesis; lipid metabolism of skin; chemical carcinogenesis. *Mailing Add:* Vet Affairs Med Ctr 2002 Holcombe Blvd Houston TX 77030

BLACK, HOWARD CHARLES, b Warsaw, Ind, Sept 20, 12; m 38. ORGANIC CHEMISTRY. *Educ:* DePauw Univ, AB, 34; Univ Ill, AM, 36, PhD(biochem), 38. *Prof Exp:* Res chemist, Swift & Co, 37-41, in chg fat & oil res div, Res Labs, 41-50, asst dir res, 50-54, assoc dir, 54-63, dir, 63-70, gen mgr sci res, 70-73; RETIRED. *Mem:* Am Chem Soc; Am Oil Chem Soc (pres, 57). *Res:* Amino acid metabolism; fats and oils; autoxidation; antioxidants; emulsifying agents; industrial oils; fatty acids; drying oils; oil seed extraction; solvent extraction of oils; soaps and detergents; adhesives; proteins; plastics. *Mailing Add:* 12 Concordia Dr Bella Vista AR 72714-2401

BLACK, HUGH ELIAS, b Mindemoya, Ont, Oct 30, 38; m 61; c 3. VETERINARY PATHOLOGY. *Educ:* Ont Vet Col, DVM, 63; Ohio State Univ, MSc, 69, PhD(vet path), 72. *Prof Exp:* Pvt pract vet med, Wellesley Vet Clin, Wellesley, Ont, 63-67; path resident vet path, Ohio State Univ, 67-72; res pathologist, Procter & Gamble Co, 72-78; mem staff toxicol lab, Warner-Lambert, 78-79; dir, Path Lab, 79-89, VPRES DRUG SAFETY & METAB, SCHERING CORP, 89- *Concurrent Pos:* Adj asst prof path, Dept Vet Pathobiol, Ohio State Univ, 74- *Mem:* Am Col Vet Pathologists; Int Acad Path; Am Vet Med Asn; Can Vet Med Asn. *Res:* Parathyroid gland ultrastructure and secretory activity; calcium homeostatic mechanisms; pathophysiology of skeletal disease; toxicologic pathology. *Mailing Add:* Schering Corp Toxicol Lab PO Box 32 Rte 94 Lafayette NJ 07848

BLACK, IRA B, NEUROLOGY. *Prof Exp:* PROF & CHMN NEUROL, ROBERT WOOD JOHNSON MED SCH. *Mailing Add:* Dept Neurosci & Cell Biol Robert Wood Johnson Med Sch Univ Med & Dent NJ 675 Hoes Piscataway NJ 08854

BLACK, JAMES, PHARMACOLOGY. *Prof Exp:* PROF, ANALYTICAL PHARMACOL UNIT, SCH MED, RAYNE INST, KINGS COL, 80- *Mem:* Nat Acad Sci. *Mailing Add:* Anal Pharmacol Unit Sch Med Rayne Inst Kings Col 123 Coldharbour Lane London SE5 9NU England

BLACK, JAMES FRANCIS, hydrology & water resources, atmospheric dynamics; deceased, see previous edition for last biography

BLACK, JAMES H(AY), chemical engineering; deceased, see previous edition for last biography

BLACK, JEFFREY HOWARD, b Prairie City, Ore, Feb 11, 43; m 66; c 2. HERPETOLOGY, ECOLOGY. *Educ:* Ore State Univ, BS, 65; Univ Mont, MS, 70; Univ Okla, PhD(zool), 73. *Prof Exp:* Asst biol, Univ Mont, 65-67; asst zoologist, Univ Okla, 70-72; asst prof, 72-74, ASSOC PROF BIOL, OKLA BAPTIST UNIV, 74- *Concurrent Pos:* Adv biol, Cent Okla Grotto, Nat Speleol Soc, 69-; res assoc herpet, J Willis Stovall Mus, Univ Okla, 70-; mem res coun, Okla City Zoo, 75; res assoc herpet, Okla Biol Surv, 75; dir, Webster Natural Hist Mus, Okla Baptist Univ, 75. *Mem:* Herpetologist's League; Am Soc Ichthyologists & Herpetologists; Soc Study Amphibians & Reptiles; Sigma Xi; Nat Speleol Soc. *Res:* Amphibians of Oklahoma; cavelife of Oklahoma; ecology of temporary pools. *Mailing Add:* Dept Biol E Cent Univ Ada OK 74820

BLACK, JESSIE KATE, b Hogansville, Ga. DEVELOPMENTAL BIOLOGY. *Educ:* Tuskegee Inst, BS, 66; Atlanta Univ, MS, 72, PhD(develop biol), 75; Harvard Sch Pub Health, MPH, 81. *Prof Exp:* High sch biol teacher, 68-71; fel biol, Emory Univ, 76-77; INSTR, SPELMAN COL, 77- *Concurrent Pos:* Res fel, Boston Med Ctr; Consult, IBM & Apple. *Mem:* Soc Develop Biol; Am Zoologist Soc; Soc Study Reproduction; Orgn Minority Women in Sci. *Res:* Role of renin-angiotensin system in the regulation of regional contractile function during acute myocardial ischemia. *Mailing Add:* 2633 Wood Valley Dr East Point GA 30344

BLACK, JOE BERNARD, b Noxapater, Miss, Nov 1, 33; m 55; c 1. INVERTEBRATE ZOOLOGY. *Educ:* Miss Col, BS, 55; Tulane Univ, MS, 57, PhD(zool), 63. *Prof Exp:* Asst prof biol, La Col, 57-60; assoc prof, Miss Col, 62-64; from assoc prof to prof, McNeese State Univ, 64-74; PROF BIOL & CHMN DEPT, LA COL, 74- *Mem:* Am Soc Zool; Int Soc Astacology; Am Sci Affil; Sigma Xi. *Res:* Development, taxonomy, ecology and genetics of crawfishes. *Mailing Add:* Dept Biol La Col Col Sta Pineville LA 71360

BLACK, JOHN ALEXANDER, b Kettle, Scotland, June 15, 40. BIOCHEMISTRY. *Educ:* Glasgow Univ, BSc, 61, PhD(biochem), 64. *Prof Exp:* Res assoc biochem, Univ BC, 64-67; from asst prof to prof biochem & med genetics, Ore Health Sci Univ, 66-85; RETIRED. *Res:* Structural, functional and genetic relationships of proteins. *Mailing Add:* 3181 SW Sam Jackson Park Rd Portland OR 97201

BLACK, JOHN B, b Ann Arbor, Mich, Apr 11, 39; m 65; c 1. ENDOCRINOLOGY, ANATOMY. *Educ:* Mercer Univ, AB, 61; Med Col Ga, PhD(endocrinol), 69. *Prof Exp:* Asst prof, 68-75, actg chmn dept, 75-77, PROF BIOL, AUGUSTA COL, 75- *Concurrent Pos:* Consult, Augusta Lab Hormone Assays. *Mem:* Am Fertil & Steril Soc; Am Soc Andrology; Int Fedn

Fertil Soc; Am Asn Tissue Banks Sperm Coun; Sigma Xi. *Res:* Reproductive endocrine physiology; effect of androgens on hypothalamic-hypophyseal-ovarian axis; separation and storage of sperm fractions. *Mailing Add:* Dept Biol Augusta Col Augusta GA 30904

BLACK, JOHN DAVID, b Winslow, Ark, July 2, 08; m; c 2. VERTEBRATE ZOOLOGY. *Educ:* Univ Kans, BA, 35; Ind Univ, MA, 37; Univ Mich, PhD(zool), 40. *Hon Degrees:* DSc, Kirksville Col Osteop Med, 60. *Prof Exp:* Instr zool, Univ Ark, 37-38; prof sci, Anderson Col, Ind, 40-41 & 45-47; asst prof zool, Ala Polytech Inst, 41-43; biologist, State Conserv Comn, Wis, 43-45; assoc prof zool, Eastern Ill State Col, 47-48; from prof to emer prof zool, Northeast Mo State Teachers Col, 48-73; RETIRED. *Concurrent Pos:* Sr sci fac fel, NSF, 59. *Res:* Taxonomy of Mississippi Valley fishes; fisheries biology; vertebrate distribution and ecology; biological conservation. *Mailing Add:* 1065 Casitis Past Rd Apt 105 Carpenteria CA 93013-1847

BLACK, JOHN HARRY, b Indianapolis, Ind, May 7, 49; m 77. ASTRONOMY. *Educ:* Harvard Univ, BA, 71, AM, 73, PhD(astron), 75. *Prof Exp:* Asst prof astron, Univ Minn, 75-78; res assoc astron & aeronomy, Harvard Col Observ, 78-83; assoc prof astron & assoc astronom, 83-89, PROF & ASTRONR, STEWARD OBSERV, UNIV ARIZ, 89- *Concurrent Pos:* Lectr astron, Harvard Univ, 80-83. *Honors & Awards:* Robert Trumpler Award, Astron Soc Pac, 77. *Mem:* Am Astron Soc; Astron Soc Pac; Royal Astron Soc; Sigma Xi; Int Astron Union. *Res:* Theoretical studies of diffuse matter; ultraviolet space astronomy; atomic and molecular processes in a variety of astronomical contexts; spectroscopy; interacting galaxies. *Mailing Add:* Steward Observ Univ Ariz Tucson AZ 85721

BLACK, JOHN WILSON, communication, phonetics; deceased, see previous edition for last biography

BLACK, KENNETH ELDON, b Adrian, Mo, Feb 2, 22; m 45; c 2. WILDLIFE MANAGEMENT. *Educ:* Wash State Univ, BS, 49. *Prof Exp:* Regional supvr river basin studies, Ore, US Fish & Wildlife Serv, 63-71, dep assoc dir environ, Washington, DC, 71-74, assoc dir, 74, regional dir, 74-81; RETIRED. *Concurrent Pos:* Am Polit Sci Asn cong fel, 66-67. *Mem:* Wildlife Soc; Am Fisheries Soc. *Res:* Waterfowl population dynamics. *Mailing Add:* 5940 Hwy 2297 Panama City FL 32404

BLACK, KIRBY SAMUEL, b Salinas, Calif, Mar 29, 54. SURGICAL RESEARCH. *Educ:* Univ Calif-Los Angeles, BS, 76; Univ Calif-Irvine, PhD(develop biol), 90. *Prof Exp:* Res asst, biomech res sta, Div Orthop Surg, Univ Calif, Los Angeles, 73-77; res assoc, div neonatology, Dept Pediat, 77-78, res assoc, div plastic surg, Dept Surg, 78-84, DIR RECONSTRUCT MICROSURG LABS & ASSOC DEVELOP ENGR, DEPT SURG, UNIV CALIF, IRVINE, 78-, DIR TRANSPLANTATION LABS & BURN RES LABS, DIV PLASTIC SURG & UROL, 81-; PRES & CHIEF EXEC OFFICER, EPIX PHARMACEUT INC, 90- *Concurrent Pos:* Consult, Sandoz Pharmaceut, Beckman Instruments, Inc & Am Edwards, 82-83; clin instr, div plastic surg, Univ Calif, Irvine, 83- *Mem:* Asn Advan Med Instrumentation; AAAS; Am Burn Asn; Inst Elec & Electronics Engrs; Transplantation Soc; Am Fedn Clin Res; Sigma Xi. *Res:* Safe manipulation of the immune system so it will recognize new antigens as self; applications in transplantation immunology & autoimmune disease; Immunosuppression of rejection of transplanted tissues. *Mailing Add:* Dept Surg Med Sci Univ Calif Irvine C \ 92717

BLACK, LINDSAY MACLEOD, b Edinburgh, Scotland, Apr 20, 07; US citizen; m 36; c 3. PLANT VIRUS TUMORS, INSECT VECTORS. *Educ:* Univ BC, Can, BSA, 29; Cornell Univ, PhD(plant path), 36. *Prof Exp:* Teaching asst plant path, Univ BC, 29-30; teaching asst, Cornell Univ, 30-34, res asst, 34-36; res fel biol, Nat Res Coun, 36-37; res asst plant virol, Rockefeller Inst Med Res, 37-41, res assoc, 41-46; res cur, Brooklyn Bot Garden, 46-52; prof, dept bot, 52-73 & dept genetics & develop, 73-75, EMER PROF PLANT VIROL, UNIV ILL, URBANA, 75- *Concurrent Pos:* Prin investr res grants, Am Cancer Soc, NIH & NSF, 47-75; mem growth comt, morphogenesis panel, Nat Res Coun, 50-52; ed, Virol, 55; mem etiology comt, Am Cancer Soc, 60-64; spec NIH fel, Markham's Virus Res Unit, Cambridge, Eng, 61-62; vis prof plant path, Univ Calif, Berkeley, 66-67. *Honors & Awards:* Wright lectr, Purdue Univ, 67; Ruth Allen Award, Am Phytopath Soc, 78. *Mem:* AAAS; Am Phytopath Soc. *Res:* Insect transmission of plant viruses; multiplication of plant viruses in insect vectors; vector cell monolayer cultures; immunofluorescent assay of infections; loss of transmissibility by deletion mutations in viruses; plant virus tumors; virus-double stranded RNA; plant phloem Rickettsias; associated evolution of vector and virus; overwintering of virus in vector; genetic variation in vector's ability to transmit virus; infection of vector organs. *Mailing Add:* 550 Stratford Lane Ridge NY 11961-2038

BLACK, LOWELL LYNN, b Norman, Ark, Oct 9, 38; m 60; c 3. PLANT PATHOLOGY. *Educ:* Univ Ark, BS, 60, MS, 62; Univ Wis, PhD(plant path), 65. *Prof Exp:* Asst prof plant path, WVa Univ, 65-68; from asst prof to assoc prof, 68-76, PROF PLANT PATH, LA STATE UNIV, 76- *Mem:* AAAS; Am Phytopath Soc; Am Soc Plant Physiologists; Am Inst Biol Sci; Sigma Xi. *Res:* Physiology of diseased plants; virus diseases; diseases of vegetables. *Mailing Add:* Dept Plant Path La State Univ Baton Rouge LA 70803

BLACK, MARK MORRIS, m; c 1. NEURONAL CYTOSKELETON, DEVELOPMENTAL NEUROSCIENCE. *Educ:* Case Western Reserve Univ, PhD(anat), 78. *Prof Exp:* Asst prof, 80-86, ASSOC PROF NEUROANAT, TEMPLE UNIV, 86- *Mailing Add:* Dept Anat Temple Univ Med Sch 3420 N Broad St Philadelphia PA 19140

BLACK, MARTIN, MEDICINE, PHARMACOLOGY. *Prof Exp:* PROF HEPATOLOGY & PHARMACOL, MED SCH, TEMPLE UNIV, 73- *Mailing Add:* Dept Med & Pharmacol Med Sch Temple Univ 3420 N Broad St Philadelphia PA 19140

BLACK, OTIS DEITZ, industrial chemistry, for more information see previous edition

BLACK, PAUL H, b Boston, Mass, Mar 11, 30; m 62; c 3. MEDICINE, VIROLOGY. *Educ:* Dartmouth Col, AB, 52; Columbia Univ, MD, 56. *Prof Exp:* Intern med, Mass Gen Hosp, 56-57, asst resident, 57-58, clin & res fel, 58-60, resident, 60-61; sr surgeon, NIH, 61-67, consult, Nat Inst Allergy & Infectious Dis, 64-67; from asst prof to assoc prof med, Harvard Med Sch, 67-79; PROF MICROBIOL & MED, RES PROF SURG & CHMN, DEPT MICROBIOL, SCH MED, BOSTON UNIV, 79- *Concurrent Pos:* Asst physician, Mass Gen Hosp, 67-70, assoc physician, 70-79, hon physician, 79- *Mem:* Am Soc Clin Invest; NY Acad Sci. *Res:* Oncogenic viruses and the mechanisms by which these viruses induce malignancy; cell surface changes in malignant cells; cell surface shedding; psychoneuroimmunology (effect of the mind on the immune system). *Mailing Add:* 21 Dawes Rd Lexington MA 02173

BLACK, PERRY, b Montreal, Que, Oct 2, 30; m 63; c 3. NEUROSURGERY, NEUROPHYSIOLOGY. *Educ:* McGill Univ, BSc, 51, MD, CM, 56; Am Bd Neurol Surg, dipl, 66. *Prof Exp:* Intern & asst resident med & gen surg, Jewish Gen Hosp, Montreal, 56-58; asst resident neurol, Montreal Neurol Inst, 58-59; resident neurosurg, Johns Hopkins Hosp, 59-63, from instr to asst prof, Sch Med, 64-69, asst prof psychiat, 67-70, assoc prof neurosurg, Sch Med, Johns Hopkins Univ, 69-80, assoc prof psychiat, 70-80, neurosurgeon, 64-80; PROF & CHMN DEPT NEUROSURG, HAHNEMANN MED COL PHILADELPHIA, 80- *Concurrent Pos:* NIH fel physiol, Johns Hopkins Univ, 61-62; dir lab neurol sci, Friends Med Sci Res Ctr, Baltimore, Md, 64-; vis neurosurgeon, Baltimore City Hosps, 65-; neurosurg consult, NCharles Gen Hosp, Baltimore, 75-, & Good Samaritan Hosp, 70-; mem neurol A study sect, NIH, 73-77; assoc ed, Neurosurg, J Cong Neurol Surgeons, 76-; neurosurg rep, Johns Hopkins Med Sch Coun, 77-79, chmn tenure policy comt, 77-78, vchmn coun, 78-79, mem adv bd med fac, Sch Med, 78-79; chmn int comt, Cong Neurol Surgeons, 75-79. *Mem:* Int Asn Study Pain; coordr, Epilepsy Found Am, 73- Mem: Cong Neurol Surg; Am Asn Neurol Surg; Soc Neurosci; Am Neurol Asn. *Res:* Neurosurgery, particularly primate model of spinal cord injury; experimental study of recovery of motor function; clinical and experimental epilepsy. *Mailing Add:* Hahnemannu Med Sch Broad & Vine St Philadelphia PA 19102

BLACK, PETER ELLIOTT, b New York, NY, Nov 19, 34; m 56, 87; c 4. WATERSHED MANAGEMENT, ENVIRONMENTAL IMPACT ANALYSIS. *Educ:* Univ Mich, BS, 56, MF, 58; Colo State Univ, PhD(watershed mgt), 61. *Prof Exp:* Res forester, Coweeta Hydrol Lab, Southeastern Forest & Range Exp Sta, US Forest Serv, 56-59; asst, Coop Watershed Mgt Unit, Colo State Univ, 59-60; asst prof forestry & watershed mgt, Humboldt State Col, 61-65; PROF WATER & RELATED LAND RESOURCES, COL ENVIRON SCI & FORESTRY, STATE UNIV NY, 65- *Concurrent Pos:* Watershed mgt consult, Nat Park Serv, Washington, DC, 63-64; prin, Impact Consults, Syracuse, 74-86; vis prof, dept geog, Univ Colo, 85-86. *Honors & Awards:* President's Award Outstanding Serv, Am Water Resources Asn, 88. *Mem:* Am Geophys Union; Soil Conserv Soc Am; Nat Asn Environ Professionals; Am Water Resources Asn; Am Soc Prof Consults. *Res:* Forest influences; wildland hydrology; water law; environmental impact analysis; conservation policy. *Mailing Add:* Col Environ Sci & Forestry State Univ NY Syracuse NY 13210

BLACK, RICHARD H, b Deerfield, Wis, Apr 24, 25. APPLIED MATHEMATICS. *Educ:* Univ Wis, BS, 47, MS, 51, PhD(appl math), 63. *Prof Exp:* Instr math, Univ Wis-Milwaukee, 56-63, asst prof math & dir, Comput Ctr, 63-68; ASSOC PROF MATH, CLEVELAND STATE UNIV, 68- *Mem:* Asn Comput Mach; Am Math Soc. *Res:* Pattern recognition; computer science. *Mailing Add:* 11800 Edgewater Cleveland OH 44107

BLACK, ROBERT CORL, b Philadelphia, Pa, June 16, 41; m 63; c 1. BOTANY, PLANT PHYSIOLOGY. *Educ:* Pa State Univ, BS, 63, MS, 65, PhD(bot), 70. *Prof Exp:* Food technologist, Gen Foods Corp, 65-67; ASSOC PROF BIOL, PA STATE UNIV, DELAWARE COUNTY CAMPUS, 70- *Mem:* AAAS; Am Soc Plant Physiologists; Int Soc Plant Molecular Biol. *Res:* Mechanism and control of indole acetic acid biosynthesis in plants; initiation of plant tumors. *Mailing Add:* Dept Biol Pa State Univ Delaware County Campus Media PA 19063

BLACK, ROBERT EARL LEE, b Cassville, Mo, Nov 20, 28; m 53; c 3. EMBRYOLOGY. *Educ:* William Jewell Col, BA, 51; Univ Wash, PhD(zool), 57. *Prof Exp:* USPHS fel, Calif Inst Technol, 57-59; assoc prof biol, 59-65, PROF BIOL & MARINE SCI, COL WILLIAM & MARY, 65- *Mem:* AAAS; Am Soc Zool. *Res:* Embryology of marine invertebrates. *Mailing Add:* 98 Gilley Dr Williamsburg VA 23185

BLACK, ROBERT L, b Los Angeles, Calif, Aug 25, 30; m 53; c 3. PEDIATRICS. *Educ:* Stanford Univ, BA, 52, MD, 55. *Prof Exp:* Instr pediat, Stanford Univ, 61-62, from asst clin prof to assoc clin prof, 62-75, CLIN PROF PEDIAT, STANFORD UNIV, 75- *Concurrent Pos:* Mem, State Maternal, Child, Adolescent Health Bd, Calif, 83; mem, Inst Med, Nat Acad Sci, 83. *Mem:* Nat Acad Sci; Am Acad Pediat. *Res:* Health planning for children. *Mailing Add:* 920 Cass St Monterey CA 93940

BLACK, SAMUEL HAROLD, b Lebanon, Pa, May 1, 30; m 61; c 2. MEDICAL MICROBIOLOGY & MICROBIAL CYTOLOGY. *Educ:* Lebanon Valley Col, BS, 52; Univ Mich, MS, 58, PhD(microbiol), 61. *Prof Exp:* NSF fel, Delft Technol Univ, 60-61; instr microbiol, Univ Mich, Ann Arbor, 61-62; from asst prof to assoc prof, Baylor Col Med, 62-71; from assoc prof to prof microbiol & pub health, Mich State Univ, 71-75; PROF MICROBIOL & HEAD, DEPT MED MICROBIOL & IMMUNOL, COL MED, TEX A&M UNIV, 75- *Concurrent Pos:* Lectr, Univ Mich-Flint, 61-62 & Univ Houston, 64-66; guest prof, Swiss Fed Inst Technol, 69-70; asst dean curric & undergrad med educ, Col Med, Tex A&M Univ, 85-87, intern dean med, 87-88, assoc dean acad affairs, 88- *Mem:* Soc Gen Microbiol; Am Soc Microbiol; fel Am Acad Microbiol; Am Soc Cell Biol; Soc Invert Path; Electron Micros Soc Am; Sigma Xi. *Res:* Medical microbiology; microbial physiology; cytology of morphogenesis in prokaryotic and eukaryotic cells; biology of Bacillus species, of Hansenula species, and of Treponema pallidum. *Mailing Add:* Col Med Tex A&M Univ College Station TX 77843-1114

BLACK, SAMUEL P W, medicine; deceased, see previous edition for last biography

BLACK, SIMON, b Deerfield, Wis, Aug 9, 17; m 44; c 3. CHEMISTRY. *Educ:* Univ Wis, BS, 38, MS, 41, PhD(biochem), 42. *Prof Exp:* Asst pharmacol, Univ Chicago, 42-45, res assoc, 45-46, from instr to asst prof biochem, Dept Med, 46-51; fel, Mass Gen Hosp, 51-52; chemist, 52-58, chief, Sect Bichem Amino Acids, 58-83, RES CHEMIST, NIH, 83- *Concurrent Pos:* Consult, Off Sci Res & Develop, 42-45; mem exec coun, Space Res Comt, Int Coun Sci Unions, 74-83; course dir, short course for col teachers, NSF, 79-80. *Mem:* AAAS; Am Soc Biol Chemists; Am Chem Soc. *Res:* Biochemistry of amino acids; enzymes of amino acid metabolism and activation; origin of life and the genetic code. *Mailing Add:* NIH Bldg 8 Bethesda MD 20892

BLACK, SYDNEY D, b Newcastle-on-Tyne, Eng, Aug 17, 15; US citizen; m 45; c 1. ENGINEERING TECHNICAL TRAINING, PHYSICS & MATHEMATICS TRAINING. *Educ:* Wash Univ, BS, 38; Ill Inst Technol, MS, 40; Univ Cincinnati, DSc(appl eng sci), 46. *Prof Exp:* Res assoc eng, Riverside Res Inst, 66-76; sr res engr, Grumman Data Systs, 77-78; assoc prof elec eng, US Merchant Marine Acad, 78-80; proj engr, Norden Systs, United Technol, 80-81; supvr, 81-83, prin engr, 83-87, CONSULT COMPUTER MFG, DIGITAL EQUIP CORP, 87- *Concurrent Pos:* Dir, Exp Physics Lab, assoc prof aero eng & chief propeller designer, USAF; consult aerophy, Goodyear Aircraft Co, 52-55; supvr, Small Aircraft Engines, Gen Elec, 55-60; sr specialist, Repub Aviation Corp, 60-62; dept head, Vehicle Eng, Gen Precision Inc, 62-65; pres, S Black & Assoc, 64-; adj assoc prof mech eng, Columbia Univ, Fairleigh Dickinson Univ & Stevens Inst Technol, 67-72; adj prof mech eng, City Col NY. *Honors & Awards:* Distinguished Serv Award, Am Inst Aeronaut & Astronaut, 87. *Mem:* Sigma Xi; Am Inst Aeronaut & Astronaut; Soc Mfg Engrs. *Res:* Systematic method of aircraft propeller design and stress analysis; theory of ferromagnetic fluids; supersonic aeroelasticity; exotic propulsion systems; laser weaponry development. *Mailing Add:* 33 Stonegate Lane Derry NH 03038

BLACK, TRUMAN D, b Houston, Tex, Sept 30, 37; m 57; c 1. SOLID STATE PHYSICS. *Educ:* Univ Houston, BS, 59; Rice Univ, MA, 62, PhD(physics), 64. *Prof Exp:* Mem tech staff, Tex Instruments, 64-65; asst prof physics, 65-69, ASSOC PROF PHYSICS, UNIV TEX, ARLINGTON, 69- *Mem:* Am Phys Soc; Am Asn Physics Teachers; Int Soc Magnetic Resonance; AAAS; Sigma Xi. *Res:* Spin lattice interactions in paramagnetic solids; low temperature magnetic effects; laser optics; holography. *Mailing Add:* Physics Dept Univ Tex Box 19059 Arlington TX 76019

BLACK, VIRGINIA H, b Detroit, Mich, Sept 24, 41; div; c 2. CELL & MOLECULAR BIOLOGY. *Educ:* Kalamazoo Col, AB, 63; Sacramento State Col, MA, 66; Stanford Univ, PhD(anat), 68. *Prof Exp:* Instr anat, 67-68, from instr to asst prof, 68-73, ASSOC PROF CELL BIOL, SCH MED, NY UNIV, 73- *Concurrent Pos:* USPHS res grants, 69-87 & 90-, Am Cancer Soc res grants, 87-90; lectr, Univ Oporto, Port, 83. *Mem:* Endocrine Soc; Am Soc Cell Biologists; Am Asn Anat; NY Acad Sci; Soc Develop Biol; Sigma Xi. *Res:* Cytodifferentiation and functional modulation of steroid-secreting cells; regulation of cytochrome P450s in the adrenal cortex; role of peroxisomes in cholesterol metabolism; influence of hormones on adrenal and reproductive system development. *Mailing Add:* Dept Cell Biol NY Univ Sch Med New York NY 10016

BLACK, WALLACE GORDON, b Arlington, Mass, Feb 24, 22; m 49; c 5. REPRODUCTIVE PHYSIOLOGY. *Educ:* Univ Wis, BS, 48, MS, 49, PhD(genetics), 52. *Prof Exp:* Instr reproductive physiol, Univ Wis, 52-53; asst prof animal husb, Univ Nev, 53-54; from assoc prof to prof animal sci, Univ Mass, Amherst, 54-80; RETIRED. *Res:* Embryo transplantation; uterine defense mechanisms; role of progesterone in maintenance of pregnancy and estrous cycle control; laboratory animal technology. *Mailing Add:* PO Box 126 Pittsburg NH 03592-0126

BLACK, WAYNE EDWARD, b East St Louis, Ill, Sept 15, 35; m 85. PUBLIC HEALTH, ENVIRONMENTAL HEALTH. *Educ:* Ill Col, BS, 57; Univ St Louis, PhD, 64; Nat Registry Clin Chem, cert. *Prof Exp:* Chief radiol health res, 63-64, chief radiol health res & lab serv, 64-67, DIR LABS, ST LOUIS COUNTY HEALTH DEPT, 67- *Concurrent Pos:* Lectr physiol chem, Wash Univ, 66-70, instr, Dept Physiol Chem, Sch Dent, 69-79; mem, Biohazard & Environ Safety Bd, St Louis Univ Med Ctr, 79 & Environ Hazard Response Team, St Louis Co, Mo, 79. *Mem:* Fel Am Inst Chemists; AAAS; Am Chem Soc; Health Physics Soc; Sigma Xi. *Res:* Laboratory work, including environmental health. *Mailing Add:* Dept Commun Health & Med Care 111 S Meramec Clayton MO 63105

BLACK, WILLIAM BRUCE, b Indianapolis, Ind, Feb 25, 23; m 45; c 2. ORGANIC CHEMISTRY, POLYMER CHEMISTRY. *Educ:* Univ Va, BA, 50, MS, 53, PhD(org chem), 54. *Prof Exp:* Res chemist, Chemstrand Res Ctr, Monsanto Textiles Co, 54-61, group leader, 61-62, Chemstrand Co Div, Monsanto Co, 62-65, mgr polymer sci, 65-69, sr group leader, 69-80, sr fel, Monsanto Tech Ctr, 80-85; RETIRED. *Concurrent Pos:* Co-ed, High-Modulus Wholly Aromatic Fibers, 73, Stress Induced Crystallization Part II, 79. *Mem:* Am Chem Soc; Fiber Soc. *Res:* Synthesis, spinning and evaluation of novel polymers and fibers, especially very high modulus, very high strength aromatic fibers; the attainment of unusually high strength and high modulus fibers from conventional polymers; technical management. *Mailing Add:* 2300 N Whaley Ave Pensacola FL 32503

BLACK, WILLIAM Z(ACHARY), b Champaign, Ill, Oct 9, 40; m 62; c 2. MECHANICAL ENGINEERING. *Educ:* Univ Ill, BS, 63, MS, 64; Purdue Univ, PhD(mech eng), 68. *Prof Exp:* Asst prof, 67-70, assoc prof, 70-75, PROF MECH ENG, GA INST TECHNOL, 75- *Honors & Awards:* Recipient of Meriam, Am Soc Eng Educ, 86. *Mem:* Am Soc Mech Engrs; Am Soc Eng Educ; Inst Elec & Electronics Engrs; Sigma Xi; Am Soc Heating Refrig & Air Conditioning Engrs; Nat Soc Prof Engrs. *Res:* Heat transfer; fluid mechanics; thermodynamics. *Mailing Add:* Sch Mech Eng Ga Inst Technol Atlanta GA 30332

BLACKADAR, ALFRED KIMBALL, b Newburyport, Mass, July 6, 20; m 46; c 3. METEOROLOGY. *Educ:* Princeton Univ, AB, 42; NY Univ, PhD(meteorol), 50. *Prof Exp:* Res assoc prof meteorol, NY Univ, 46-56; assoc prof, 56-60, head dept, 67-81, prof, 60-85, EMER PROF METEOROL, PA STATE UNIV, 85- *Concurrent Pos:* Ed, Meteorol Monogr, 61-65; mem, Nat Acad Sci-US Nat Comt Global Atmospheric Res Prog, 68-71; mem, NSF Adv Panel Atmospheric Sci, 68-71; mem comt basic sci adv to US Army, Nat Acad Sci-Nat Res Coun, 70-73, mem exec comt, Earth Sci Div. *Honors & Awards:* Charles F Brooks Award, Am Meteorol Soc, 69; Humboldt Found Sr Scientist Award, 73; Cleveland Abbe Award, Am Meteorol Soc, 86. *Mem:* Fel AAAS; fel Am Meteorol Soc (secy, 65-69, pres, 71-72); fel Am Geophys Union; Deutsche Meteorol Gesellschaft. *Res:* Atmospheric energy transformation; atmospheric turbulence; atmospheric boundary layer modelling. *Mailing Add:* 805 W Foster Ave State College PA 16801

BLACKADAR, BRUCE EVAN, b Nyack, NY, Oct 22, 48. MATHEMATICS. *Educ:* Princeton Univ, AB, 70; Univ Calif, Berkeley, MA, 74, PhD(math), 75. *Prof Exp:* ASST PROF MATH, UNIV NEV, RENO, 75- *Mem:* Am Math Soc. *Res:* Tensor products and crossed products of von Neumann algebras; induced representations of locally compact groups; representations of products and extensions of groups. *Mailing Add:* Dept Math Univ Nev Reno NV 89557

BLACKADAR, ROBERT GORDON, b Ottawa, Ont, Mar 18, 28. GEOLOGY. *Educ:* Univ Toronto, BA, 50, MA, 51, PhD(geol), 54. *Prof Exp:* chief sci ed, Geol Surv Can, 70-79, geologist, 53-89, dir, Geol Info Div, 79-89; RETIRED. *Concurrent Pos:* Sci ed, Geol Surv Can, 65-70. *Mem:* Geol Soc Am; fel Arctic Inst NAm. *Res:* Precambrian geology of Arctic regions; geography and geology of the Arctic. *Mailing Add:* 65 Lock Isle Rd Nepean ON K2H 8G7 Can

BLACKARD, WILLIAM GRIFFITH, b Baltimore, Md, July 14, 33; m 60; c 3. INTERNAL MEDICINE, ENDOCRINOLOGY. *Educ:* Duke Univ, MD, 57; Am Bd Internal Med, dipl, 65. *Prof Exp:* Intern med, New York Hosp-Cornell Med Ctr, 57-58, resident, 58-59; fel endocrinol & metab, Med Ctr, Duke Univ, 59-60, resident med, 60-61, fel endocrinol & metab, 63-64; from instr to assoc prof med, Sch Med, La State Univ, New Orleans, 64-75; PROF & DIR CLIN RES UNIT & DIV ENDOCRINOL & METAB, MED COL VA, 75- *Concurrent Pos:* Markle scholar acad med, 68-73. *Mem:* Am Fedn Clin Res; Endocrine Soc; fel Am Physicians; Am Soc Clin Invest; Am Diabetes Asn; Sigma Xi. *Res:* Carbohydrate and lipid metabolism; insulin and growth hormone homeostasis; diabetes; calcium and phosphorous metabolism. *Mailing Add:* PO Box 155 MCV Sta Richmond VA 23298

BLACKBURN, ARCHIE BARNARD, b Austin, Tex, July 6, 38; m 67; c 2. PSYCHIATRY. *Educ:* Baylor Univ, BA, 60, MS & MD, 65; Am Bd Psychiat & Neurol, dipl psychiat, 74. *Prof Exp:* Resident psychiatrist, Vet Admin Hosp & Univ Tex Southwestern Med Sch Dallas, 69-70 & Vet Admin Hosp & Univ Okla, 70-72; asst prof, Health Sci Ctr, Univ Okla, 72-73; ASST PROF PSYCHIAT, BAYLOR COL MED, 73-; STAFF PSYCHIATRIST, VET ADMIN HOSP, HOUSTON, 73- *Concurrent Pos:* Staff psychiatrist & med dir, Drug Abuse Treatment Unit, Vet Admin Hosp, Oklahoma City, 72-73; med dir, Houston Vet Admin Therapeut Community Ward, 76-84; med staff, Mem Southwest Hosp, W Oaks Hosp & Belle Park Hosp, Houston, Tex, 83- *Mem:* AAAS; Am Psychiat Asn. *Res:* Clinical hypodynamics; biorhythms; methadone maintenance; application of research methodology in clinical disorders of sleep. *Mailing Add:* 3120 Southwest Freeway No 555 Houston TX 77098-4510

BLACKBURN, DALE WARREN, b La Porte, Ind, Sept 12, 26; m 52; c 2. SYNTHETIC ORGANIC & NATURAL PRODUCTS CHEMISTRY. *Educ:* Purdue Univ, BS, 48, MS, 51, PhD(pharmaceut chem), 54. *Prof Exp:* Anal chemist, Whitehall Pharmacal Co, 50; instr pharmaceut chem, Purdue Univ, 51-52, asst prof, 54-56; group leader, 56-66, asst dir, Org Chem Sect, 66-78, ASSOC DIR, RADIOCHEM, SMITH KLINE CORP, 79- *Concurrent Pos:* Founder, Org Reactions Catalysis Soc, 65, Int Isotope Soc, 86. *Mem:* Am Chem Soc; Am Pharmaceut Asn; Health Physics Soc; Int Isotope Soc; Org Reaction Catalysis Soc. *Res:* Organic and radiochemical synthesis; catalysis and high pressure reactions; instrumental analysis; separation and purification; radiological defense. *Mailing Add:* 144 Ramblewood Rd Moorestown NJ 08057

BLACKBURN, DANIEL GLENN, b Pittsburgh, Pa, Aug 27, 52; m 85. REPRODUCTIVE BIOLOGY, FUNCTIONAL MORPHOLOGY. *Educ:* Univ Pittsburgh, BSc, 75; Cornell Univ, MSc, 78, PhD(zool), 85. *Prof Exp:* Teaching asst biol, Cornell Univ, 75-85, lectr histophysiol, Dept Physiol, 85; teaching asst vet anat, NY State Col Vet Med, 81-84; res assoc, Cell Biol Dept, Vanderbilt Univ Sch Med, 86-88; ASST PROF ZOOL, HISTOPHYSIOL & EVOLUTION, DEPT BIOL, TRINITY COL, HARTFORD, 88- *Mem:* Am Soc Zoologists; Herpetologists League; Soc Study Amphibians & Reptiles; Am Soc Ichthyologists & Herpetologists. *Res:* Placentation and fetal nutrition of reptiles and mammals; evolution of reproductive patterns, especially viviparity and lactation; functional morphology of skeletomuscular systems of reptiles amd amphibians; sexual dimorphism of fish and amphibians. *Mailing Add:* Dept Biol Trinity Col Hartford CT 06106

BLACKBURN, EDWARD VICTOR, b Skegness, Eng, Sept 18, 44; m 73; c 3. ORGANIC CHEMISTRY. *Educ:* Univ London, BSc, 66; Univ Nottingham, PhD(org chem), 69. *Prof Exp:* Fel, 69-70, lectr, 70-72, assoc prof chem, 72-78, PROF CHEM & ASSOC DEAN, FAC ST JEAN, UNIV ALTA, 83- *Mem:* The Chem Soc; assoc Royal Inst Chem; Am Chem Soc; Chem Inst Can. *Res:* Photochemistry of olefins with particular reference to stilbene analogues; polar effects in radical reactions; computer assisted and computer based instructional techniques. *Mailing Add:* Dept Chem Fac St Jean 8406 91st St Edmonton AB T6C 4G9 Can

BLACKBURN, ELIZABETH HELEN, b Hobart, Australia, Nov 26, 48; m 75; c 1. MOLECULAR BIOLOGY, BIOCHEMISTRY. *Educ:* Univ Melbourne, BS, 70, MS, 71; Cambridge Univ, PhD(molecular biol), 75. *Prof Exp:* Fel biol, Yale Univ, 75-77; from asst prof to assoc prof molecular biol, Univ Calif Berkeley, 78-90; fel biochem, 77-78, PROF MOLECULAR BIOL, UNIV CALIF, SAN FRANCISCO, 90- *Honors & Awards:* Eli Lilly Award Microbiol, 88; Molecular Biol Award, Nat Acad Sci, 90. *Mem:* Am Soc Cell Biol. *Res:* Repetitive DNA sequences and chromosomal structure in eukaryotes; studies on structure and function synthesis of telomeres chromosome ends; ribonucleoprotein enzyme that synthesizes telomgres has been discovered and characterized. *Mailing Add:* Dept Microbiol Univ Calif San Francisco CA 94143

BLACKBURN, GARY RAY, b Shickshinny, Pa, May 13, 46; m 75. BIOCHEMISTRY, ONCOLOGY. *Educ:* Wilkes Col, BS, 68; Univ Wis, PhD(oncol), 74. *Prof Exp:* Res assoc radiation biol, Colo State Univ, 74-77; res assoc oncol, Inst Cancer Res, Fox Chase Cancer Ctr, 77-81; GENETIC TOXICOLOGIST, MOBIL OIL CORP, 81- *Concurrent Pos:* Nat Cancer Inst fel, Colo State Univ, 75-77. *Mem:* Sigma Xi; Am Asn Cancer Res; Soc Exp Biol & Med. *Res:* Biochemical mechanisms by which chemical carcinogens effect transformation of susceptible cells in target organs, with an emphasis on these agents' alteration of cell architecture and division. *Mailing Add:* Toxicol Div Mobil Oil Corp PO Box 1029 Princeton NJ 08540

BLACKBURN, GEORGE L, b McPherson, Kans, Feb 12, 36. NUTRITIONAL BIOCHEMISTRY, SURGERY. *Educ:* Univ Kans, MD, 65; Mass Inst Technol, PhD(nutrit biochem), 73. *Prof Exp:* MEM, DEPT NUTRIT, CANCER RES INST, BOSTON, 75-; ASSOC PROF SURG, MED SCH, HARVARD UNIV, 76- *Mailing Add:* Dept Hyperaliment New Eng Cancer Res Inst 1194 Pilgrim Rd Boston MA 02215

BLACKBURN, HENRY WEBSTER, JR, b Miami, Fla, Mar 22, 25; m 51; c 3. EPIDEMIOLOGY, PUBLIC HEALTH. *Educ:* Univ Miami, BS, 47; Tulane Univ, MD, 48; Univ Minn, MS, 57. *Hon Degrees:* Dr, Univ Kuopio, Finland, 82. *Prof Exp:* Intern, Chicago Wesley Mem, 48-49; intern, Am Hosp of Paris, 49-50; med officer-in-chg, USPHS, Austria & Ger, 50-53; chief resident med, Ancker Hosp, Minn, 56; from asst prof to assoc prof, Univ Minn, Minneapolis, 56-68, dir physiol hyg, Sch Pub Health & prof med, Sch Med, 72-83, chmn, Div Epidemiol, 83-90, PROF PHYSIOL HYG, SCH PUB HEALTH, UNIV MINN, MINNEAPOLIS, 68-, MAYO CHAIR PUB HEALTH, 90- *Concurrent Pos:* Med dir, Underwriting Mutual Serv Ins Co, Minn, 56-; temporary consult, WHO, 61, 64, 65, 70 & 71, consult, 81-; chmn coun epidemiol, Am Heart Asn, 71-73; med consult, Retail Credit Co, 69-72; vis prof, Univ Geneva, 70-71; dir electrocardiographic ctr & mem steering comt, Coronary Drug Proj, Nat Heart & Lung Inst, nat vchmn multiple risk factor intervention trial, 72-75, mem adv comt, Cardiol Br, 75-; chmn, Coun Epidemiol, Int Soc Cardiol, 86-90 & Med Sect AAAS, 91-92; mem, Study Sect Epidemiol & Dis Control, NIH & adv bd, Nat Heart, Lung & Blood Inst, 89-92. *Honors & Awards:* Thomas Francis Award Epidemiol, 75; Naylor Dana Award Prev Med, 76; Louis Bishop Award Cardiol, 79. *Mem:* Fel Am Col Epidemiol; Am Heart Asn; fel Am Col Cardiol; fel Am Epidemiol Soc; Belgian Royal Acad Med; Soc French Cardiol. *Res:* Epidemiology of cardiovascular disease; electrocardiography; internal medicine; chronic disease epidemiology and prevention; promotion of health in population-wide strategies with community-based programs and experimental design. *Mailing Add:* Div Epidemiol Univ Minn Minneapolis MN 55455

BLACKBURN, JACK BAILEY, b Sterling, Okla, Oct 19, 22; m 49; c 2. CIVIL ENGINEERING. *Educ:* Univ Okla, BS, 47; Purdue Univ, MS, 49, PhD(civil eng), 55. *Prof Exp:* Res asst joint hwy res proj, Purdue Univ, 47-49, res engr, 49-55; assoc prof civil eng, Univ Md, 55-58; transp planning engr, Harland Bartholomew & Assocs, 58-60; prof transp eng & dir Ariz transp & traffic inst, Univ Ariz, 60-63; prof & head civil eng dept, Kans State Univ, 63-72; PROF ENG, ARIZ STATE UNIV, 72- *Concurrent Pos:* Mem, Hwy Res Bd, Nat Acad Sci-Nat Res Coun, 48-70, Comts Durability of Concrete, 53-65, Qual Traffic Serv, 65-70, Theory Traffic Flow, 65-70 & Characteristics Traffic Flow, 65-70; contact rep, Kans State Univ, 63-72; mem bldg res adv bd, Nat Acad Sci-Nat Acad Eng, 68-71. *Mem:* Am Soc Civil Engrs; Am Arbit Asn; Nat Soc Prof Engrs. *Res:* Materials of construction; urban and regional planning; highway and airport planning, economics and design; traffic and transportation engineering; engineering statistics and design of experiments. *Mailing Add:* 4343 N 84th Pl Scottsdale AZ 85251

BLACKBURN, JACOB FLOYD, b Newton, NC, Nov 27, 18; m 45; c 1. COMPUTER SCIENCES. *Educ:* Lenoir-Rhyne Col, Hickory, NC, AB, 40; Duke Univ, MA, 47; Univ NC, Chapel Hill, PhD(math), 53. *Prof Exp:* Meteorologist, US Army & Air Force, 41-45 & 51-56; asst prof math, The Citidel, Charleston, SC, 47-50; assoc prof math, USAF Acad, 55-56; mgr, IBM, Armonk, NY, 56-69, dir, IBM-Europ, Paris, France, 70-77; dir, technol policy, State Dept, Washington, DC, 77-80; exec dir comput sci bd, Nat Acad Sci, 80-82; liaison scientist comput sci, US Off Naval Res, London, 82-84; INDUST ASSESSMENT OFFICER COMPUT SCI & TELECOMMUN, US DEPT COM, LONDON, 84- *Concurrent Pos:* Comput Pioneer, Nat Comput Conf, 75. *Mem:* Asn Comput Mach; Am Math Soc. *Res:* Partial differential equations; numerical weather forecasting; European computers, robots and telecommunications. *Mailing Add:* Foreign Com Serv Embassy US 24/31 Grosvenor Sq London W1A 1AE England

BLACKBURN, JOHN GILL, b Lake Charles, La, June 25, 35; m 56; c 2. PHYSIOLOGY. *Educ:* Tulane Univ, BS, 59, PhD(physiol), 65. *Prof Exp:* From instr to assoc prof, 64-86, PROF PHYSIOL, MED UNIV SC, 86- *Concurrent Pos:* USPHS res grants, 65-68, 74-87 & 88- *Mem:* Sigma Xi; Soc Gen Physiologists; Am Physiol Soc. *Res:* Epithelial transport. *Mailing Add:* Dept Physiol Med Univ SC Charleston SC 29425

BLACKBURN, JOHN LEWIS, b Kansas City, Mo, Oct 2, 13; m 43; c 3. ELECTRICAL ENGINEERING. *Educ:* Univ Ill, BSEE, 35. *Prof Exp:* Engr, Westinghouse Elec Corp, 36-52, supv eng, 52-55, eng sect mgr, 55-69, consult eng, 69-78; lectr & instr, Westinghouse Grad Prog, 48-54; adj prof, Polytech Inst NY, 49-65, power protection relaying, NJ Polytech Inst, 58-71; prog dir & lectr protective relaying, Westinghouse Elec Corp, 59-81; RETIRED. *Concurrent Pos:* Lectr, Westinghouse Grad Prog, NY Univ, Stevens Inst Technol, Brooklyn Polytech Inst, Newark Col Eng, 48-54; adj prof, power syst transmission, distrib & protective relaying, Polytech Inst NY, 49-65; spec lectr, relaying, symmetrical components, power transmission, Inst Elec & Electronics Engrs, Can, Brazil, Taiwan, Venezuela, 52-; adj prof, protective relaying & symmetrical components, NJ Inst Technol, 58-71; dir, Westinghouse Relay Sch, 59-81, Can Westinghouse Relay Sch, 79-87. *Honors & Awards:* Centennial Medal, Inst Elec & Electronics Engrs, 84, Distinguished Serv Award, 78, Outstanding Serv Award, 79; Attwood Assoc Award, CIGRE US Nat Comt, 86. *Mem:* Fel Inst Elec & Electronic Engrs. *Res:* Design, application, technical assistance and instruction, fault analysis, power system performance for both utility and industrial systems. *Mailing Add:* 21816 Eighth Pl W Bothell WA 98021

BLACKBURN, JOHN ROBERT, b Berea, Ky, Jan 30, 45; m 69; c 2. FORENSIC & ENVIRONMENTAL CHEMISTRY. *Educ:* Westminster Col, BS, 67; Vanderbilt Univ, PhD(inorg chem), 72. *Prof Exp:* Res assoc, Vanderbilt Univ, 72; from asst prof to assoc prof, 72-84, PROF CHEM, GEORGETOWN COL, 84-, CHMN DEPT, 72-, CHMN DIV NATURAL SCI, 82- *Concurrent Pos:* Vis asst prof chem, Vanderbilt Univ, 74; vis res assoc, Univ Ky, 79, 82. *Mem:* Am Chem Soc; Water Pollution Control Fedn. *Res:* Chemical education; chemical dynamics; crystallography. *Mailing Add:* Georgetown Col No 205 Georgetown KY 40324

BLACKBURN, MICHAEL N, b Oak Park, Ill, June 30, 44. PHYSICAL BIOCHEMISTRY OF PROTEINS. *Educ:* Univ Calif, Riverside, PhD(biochem), 71. *Prof Exp:* ASSOC PROF BIOCHEM, LA STATE UNIV, 75- *Mem:* Am Soc Biol Chem; Am Chem Soc; Am Heart Asn. *Mailing Add:* Dept Micromolecular Sci Smith Kline & French Labs PO Box 1539 King of Prussia PA 19406-0939

BLACKBURN, PAUL EDWARD, b West Branch, Iowa, May 24, 24; m 48; c 2. HIGH TEMPERATURE CHEMISTRY. *Educ:* Ohio Wesleyan Univ, BA, 48; Ohio State Univ, PhD(chem), 54. *Prof Exp:* Res assoc, Ohio State Univ, 48-51; res chemist, Westinghouse Res Labs, 54-61; proj dir, Arthur D Little, Inc, 61-65; ASSOC CHEMIST, ARGONNE NAT LAB, 65- *Mem:* AAAS; Am Phys Soc; Am Chem Soc; Electrochem Soc. *Res:* High temperature physical chemistry; vapor pressures; thermodynamics; crystal structures; gas-solid reactions. *Mailing Add:* Argonne Nat Lab Bldg 205 Rm R135 Argonne IL 60439

BLACKBURN, THOMAS HENRY, b Newton, NC, June 1, 23; m 44; c 2. MATHEMATICS. *Educ:* Lenoir-Rhyne Col, AB, 45; Western Reserve Univ, MA, 48. *Prof Exp:* From instr to asst prof, 46-55, PROF MATH, LENOIR-RHYNE COL, 58- *Mem:* Math Asn Am. *Res:* Functions of a complex variable. *Mailing Add:* Dept Math Lenoir-Rhyne Col Hickory NC 28603

BLACKBURN, THOMAS ROY, b St Louis, Mo, Nov 30, 36; m 60; c 3. ANALYTICAL CHEMISTRY, PHYSICAL CHEMISTRY. *Educ:* Carleton Col, BA, 58; Harvard Univ, MA, 59, PhD(anal chem), 62. *Prof Exp:* From instr to asst prof chem, Carleton Col, 62-64; asst prof, Wellesley Col, 64-67; from asst prof to prof chem, Hobart & William Smith Cols, 67-78; MCGAW DISTINGUISHED PROF, ST ANDREWS PRESBY COL, 78- *Concurrent Pos:* Vis prof, Swiss Fed Inst Water Pollution Control & Water Conserv, 72-73; res assoc geochem, Harvard Univ, 77; vis prof, Lab Planetary Studies, Cornell Univ, 78; vis prof, Dept Geol, Univ St Andrews, Scotland, 86. *Mem:* AAAS. *Res:* Aquatic geochemistry; geochemistry of aqueous solutions. *Mailing Add:* Dept Math & Sci St Andrews Presby Col Laurinburg NC 28352

BLACKBURN, WILBERT HOWARD, b Loa, Utah, Mar 23, 41; m 65; c 5. WATERSHED MANAGEMENT. *Educ:* Brigham Young Univ, BS, 65; Univ Nev, Reno, MS, 67, PhD, 73. *Prof Exp:* Lab instr bot, Brigham Young Univ, 63-65; res aide range ecol, Intermountain Forest & Range Exp Sta, 64-65; res asst, Univ Nev, Reno, 65-66, range ecologist, 66-74, asst prof range & watershed mgt, 73-75; from assoc prof to prof watershed mgt, Tex A&M Univ, 75-87; SUPVRY HYDROLOGIST, USDA-AGR RES SERV, BOISE, IDAHO, 87. *Concurrent Pos:* mem, Comt on Wild & Free-Roaming Horses & Burros, & Livestock Phase, Capital Reef Nat Park, Nat Res Coun & Nat Acad Sci; mem adv panel, assessment water-related technol sustaining agr semi-arid lands, Off Technol Assessment. *Honors & Awards:* Outstanding Achievement Award, Soc Range Mgt, 87. *Mem:* Soc Range Mgt. *Res:* Pinyon-juniper invasion, condition and trend; vegetation and soil relationships; range hydrology; infiltration studies of selected vegetation; soil units; range and forest practices and water quality; water yield improvement from rangeland water sheds; soil erosion. *Mailing Add:* USDA-Agr Res Serv 800 Park Blvd Plaza IV No 105 Boise ID 83712-7716

BLACKBURN, WILL R, b Durant, Okla, Nov 4, 36. PEDIATRIC-PERINATAL PATHOLOGY, TROPICAL MEDICINE. *Educ:* Univ Okla, BS, 57; Tulane Univ, MD, 61. *Prof Exp:* Intern med, 61-62, resident path, 62-64, instr, Col Physicians & Surgeons, Columbia Univ, 63-64; resident & instr, Univ Colo, 64-66; from asst prof to assoc prof, Col Med, Pa State Univ, 69-73; PROF PATH, COL MED, UNIV S ALA, 73- *Concurrent Pos:* Fel, Univ Colo, 64-66; res pathologist, Div Geog Path, Armed Forces Inst Path, 66-68; bd dirs, Ctr for Birth Defects Info Serv, Dover, Mass. *Mem:* AAAS; Am Asn Path & Bact; Soc Invert Path; Soc Pediat Res; Am Soc Exp Biol; Teratology Soc. *Res:* Human nutrition; protozoology; developmental biology; pediatric pathology; cellular immunology; electron microscopy; parasitology; experimental teratology; human development; birth defects. *Mailing Add:* Dept Path Univ S Ala Mobile AL 36617

BLACKBURN, WILLIAM HOWARD, b Ottawa, Ont, July 30, 41; m 83; c 2. PETROLOGY, GEOCHEMISTRY. *Educ:* St Francis Xavier Univ, BSc, 64; Mass Inst Technol, PhD(geol), 67. *Prof Exp:* From asst prof to prof geol & geochem, Univ Ky, 67-88; PROF & CHMN MINERAL & PETROL, UNIV WINDSOR, 89- *Mem:* Geochem Soc; Geol Asn Can; Mineral Asn Can. *Res:* Equilibration of metamorphic rocks; mobility of chemical elements during metamorphism and kinetics of metamorphic reactions; variation of major and minor element chemistry of granites; tectonics and basalt geochemistry. *Mailing Add:* Dept Geol Univ Windsor Windsor ON N9B 3P4 Can

BLACKERBY, BRUCE ALFRED, b Colon, Repub Panama, Nov 14, 36; US citizen; m 57; c 4. GEOLOGY. *Educ:* Univ Calif, Riverside, AB , 59; Univ Calif, Los Angeles, PhD(geol), 65. *Prof Exp:* From asst prof to assoc prof, 63-71, PROF GEOL, CALIF STATE UNIV, FRESNO, 71-, CHMN DEPT, 69- *Concurrent Pos:* NSF instrl grant, 64-65; geologist, US Geol Surv, 65- *Mem:* Geol Soc Am. *Res:* Tertiary and Mesozoic volcanics; regional geology; geology of Sierra Nevada, especially batholith. *Mailing Add:* Dept Geol Calif State Univ Fresno Fresno CA 93740

BLACKETT, DONALD WATSON, b Boston, Mass, May 2, 26; m 51. MATHEMATICS. *Educ:* Harvard Univ, AB, 47; Princeton Univ, MA, 48, PhD(math), 50. *Prof Exp:* Instr math, Princeton Univ, 50-51, res assoc, 51-53; from asst prof to prof, 53-90, EMER PROF MATH, BOSTON UNIV, 90- *Mem:* Am Math Soc; Math Asn Am. *Res:* Algebra; operations research. *Mailing Add:* 97 Eliot Ave West Newton MA 02165

BLACKETT, SHIRLEY ALLART, b Greenport, NY, Jan 24, 28; m 51. MATHEMATICS. *Educ:* Univ Rochester, AB, 48; Pa State Univ, EdM, 51. *Prof Exp:* Test specialist, Math Test Develop Dept, Educ Testing Serv, 49-52; from instr to assoc prof math, Northeastern Univ, 53-89; RETIRED. *Mem:* Math Asn Am. *Res:* Testing and educational measurement. *Mailing Add:* 97 Eliot Ave West Newton MA 02165

BLACKHAM, ANGUS UDELL, b East Ely, Nev, Apr 16, 26; m 46; c 6. PHYSICAL ORGANIC CHEMISTRY. *Educ:* Brigham Young Univ, AB, 49; Univ Cincinnati, MS, 50, PhD, 52. *Prof Exp:* From asst prof to assoc prof, 52-60, PROF CHEM, BRIGHAM YOUNG UNIV, 60- *Mem:* Am Chem Soc; Sigma Xi. *Res:* Electroorganic chemistry; azoalkanes and aliphatic azines; transition metal complexes; catalytic reforming of hydrocarbons; coal combustion and gasification. *Mailing Add:* Dept Chem Brigham Young Univ ESC 120 Provo UT 84602

BLACKKETTER, DENNIS O, b Westville, Okla, Aug 30, 32; m 55; c 3. MECHANICAL & DESIGN ENGINEERING. *Educ:* Fresno State Col, BS, 55; Univ Idaho, MS, 60; Univ Ariz, PhD, 66. *Prof Exp:* Instr mech eng, Univ Idaho, 58-60; asst prof, Sacramento State Col, 60-63; prof & dept head mech eng, Mont State Univ, 66-84; PROF MECH ENG, CALIF STATE UNIV, CHICO, 84- *Concurrent Pos:* Consult, Power Policy Bd, Am Soc Mech Engrs, 75-, vchmn, Energetics Div, 79. *Res:* Low frequency sound transmission in the human body. *Mailing Add:* Dept Mech Eng Calif State Univ Chico CA 95929-0930

BLACKLER, ANTONIE W C, b Portsmouth, Eng, Oct 19, 31; m 70; c 2. DEVELOPMENTAL BIOLOGY. *Educ:* Univ London, BSc, 53, PhD(embryol), 56. *Prof Exp:* Sr biologist, Loughborough Col Technol, Eng, 56-57; asst lectr zool, Queen's Univ, Belfast, 57-59; Brit Empire Cancer Campaign res asst embryol, Oxford Univ, 59-61; extraordinary prof zool, Univ Geneva, 61-64; assoc prof, 64-73, PROF ZOOL, CORNELL UNIV, 73- *Concurrent Pos:* Div dir, NSF, 81-82. *Mem:* Soc Develop Biol; Int Soc Develop Biol. *Res:* Origin and differentiation of amphibian sex cells; nucleocytoplasmic interactions in amphibian development. *Mailing Add:* 14 Nottingham Dr Ithaca NY 14850-8704

BLACKLOW, NEIL RICHARD, b Cambridge, Mass, Feb 26, 38; m 63; c 2. VIROLOGY. *Educ:* Harvard Univ, BA, 59; Columbia Univ, MD, 63. *Prof Exp:* Resident med, Beth Israel Hosp, Harvard Univ, 63-65; res virologist, NIH, 65-68; sr scientist, 69-71; clin fel med, Mass Gen Hosp, Harvard Univ, 68-69, NIH sr scientist, 69-71; from asst to assoc prof med, Sch Med, Boston Univ, 71-76; prof med & microbiol, 76-90, CHMN DEPT MED, SCH MED, UNIV MASS, 90- *Concurrent Pos:* Assoc ed, Reviews of Infect Dis, 82-; lectr med, Harvard Med Sch, 84- *Mem:* Am Soc Clin Invest; Infectious Dis Soc Am; Am Soc Microbiol; Am Asn Immunol; Am Epidemiol Soc. *Res:* Viral enteritis; defective viruses; clinical virology; clinical infectious disease. *Mailing Add:* Univ Mass Med Ctr 55 Lake Ave N Worcester MA 01655

BLACKMAN, CARL F, b Annapolis, Md, Mar 16, 41; m 67. MOLECULAR BIOLOGY, PHYSICAL CHEMISTRY. *Educ:* Colgate Univ, AB, 63; Pa State Univ, MS, 67, PhD(biophys), 69. *Prof Exp:* Fel, Brookhaven Nat Lab, 68-70; biologist, Bur Radiol Health, 70; BIOLOGIST, ENVIRON PROTECTION AGENCY, 70- *Mem:* Am Soc Photobiol; NY Acad Sci; Am Inst Biol Sci; Biophys Soc; Bioelectromagnetic Soc. *Res:* Interactions of chemical agents and non-ionizing electromagnetic radiation - DC to 300 GHz - with biological systems at the molecular and cellular level. *Mailing Add:* MD-68 Environ Protection Agency Research Triangle Park NC 27711

BLACKMAN, J(AMES) S(AMUEL), engineering mechanics; deceased, see previous edition for last biography

BLACKMAN, JEROME, b New York, NY, Apr 22, 28; m 55. MATHEMATICS. *Educ:* Mass Inst Technol, BS, 48; Cornell Univ, PhD(math), 51. *Prof Exp:* From instr to asst prof, 52-59, ASSOC PROF MATH, SYRACUSE UNIV, 59- *Res:* Mathematical analysis, geometric quantization. *Mailing Add:* Dept Math Syracuse Univ Syracuse NY 13210

BLACKMAN, MARC ROY, b Boston, Mass, Mar 2, 46; m 71; c 3. PITUITARY GLYCOPROTEINS, CELL PHYSIOLOGY & BIOCHEMISTRY. *Educ:* Northeastern Univ, AB, 68; NY Univ, MD, 72. *Prof Exp:* Intern jr & sr asst resident internal med, Bronx Munic Hosp Ctr & Albert Einstein Col Med, 72-75; fel endocrinol, Clin Endocrinol Br, Nat Inst Arthritis, Diabetes & Digestive & Kidney Dis, NIH, Bethesda, Md, 75-77; ASSOC PROF ENDOCRINOL & METAB, DEPT MED, FRANCIS SCOTT KEY MED CTR & JOHNS HOPKINS UNIV SCH MED, 77-; GUEST SCIENTIST ENDOCRINOL & METAB, GERONT RES CTR, NAT INST AGING, NIH, BALTIMORE, MD, 80- *Concurrent Pos:* Consult endocrinol, Johns Hopkins Hosp, 85- *Mem:* Endocrine Soc; Am Fedn Clin Res; Geront Soc Am; AAAS. *Res:* Clinical and laboratory studies of cellular mechanisms for age-related derangements in anterior pituitary hormone secretion. *Mailing Add:* Geront Res Ctr Rm 2B-20 Baltimore MD 21224

BLACKMAN, SAMUEL WILLIAM, b New York, NY, Aug 25, 13; m 39; c 3. PATENT ATTORNEY. *Educ:* Cornell Univ, AB, 34; NY Univ, MS, 37; Polytech Inst Brooklyn, PhD(chem), 60. *Prof Exp:* Dir control, Premo Pharmaceut Labs, Inc, 35-42; microanalyst, Hoffmann-La Roche, 42-44; sr chemist, Burroughs Wellcome & Co, Inc, 44-70; from assoc prof to prof chem, Yeshiva Univ, 68-80; CONSULT INDUST, CHEM PATENTS & REGULATORY AFFAIRS, 69- *Mem:* Am Chem Soc. *Res:* Polynuclear condensed heterocycles; quaternary ammonium compounds. *Mailing Add:* 1349 Lexington Ave New York NY 10128

BLACKMON, BOBBY GLENN, b Rodessa, La, May 17, 40; m 58; c 3. FOREST SOILS. *Educ:* La Tech Univ, BS, 62; Duke Univ, MF, 63; La State Univ, PhD(forest soils), 69. *Prof Exp:* Res asst soil fertility, Univ Ark, Stuttgart, 63-65; soil scientist, Southern Hardwoods Labs, USDA Forest Serv, Stoneville, Miss, 67-72, proj leader forest soils, 72-79; PROF & HEAD, DEPT FOREST RESOURCES, UNIV ARK, MONTICELLO, 79- *Mem:* Soil Sci Soc Am; Am Soc Agron; Soc Am Foresters; Nat Asn Prof Forestry Schs & Cols. *Res:* Forest soil fertility and tree nutrition including nutrient cycling, fertilization and nutrient changes in forest soils. *Mailing Add:* Dept Forestry State Univ NY Col Environ Sci Syracuse NY 13210

BLACKMON, CLINTON RALPH, b Timmonsville, SC, Aug 13, 19; m 59; c 5. PLANT BREEDING. *Educ:* Clemson Col, BS, 41; Univ Mass, MS, 49; Rutgers Univ, PhD(farm crops), 55. *Prof Exp:* Instr bot, Univ Mass, 48; assoc prof agron, Nat Agr Col, 48-55, head dept, 49-55; from asst prof to assoc prof, Univ Maine, 56-62; prof & agr adv, Panama, 62-64; assoc prof plant sci, 64-71, chmn dept hort, 69-75, PROF PLANT SCI, DELAWARE VALLEY COL SCI & AGR, 71-, DEAN, 75- *Concurrent Pos:* Partic, Comn Undergrad Educ Biol Sci Teaching Conf, NSF, Washington, DC, 68 & Biol Inst, Williams Col, 68; UN agr consult to Paraguay, 75; Eastern regional dir, NACTA, 78-80. *Mem:* Am Soc Hort Sci; Am Genetic Asn; Am Soc Agron. *Res:* Plant breeding techniques; fungi spores in the plant and animal environment; varietal improvement; effect of sludge on crop and weed species in forages; pollen viability in economic plants. *Mailing Add:* RD 2 Box 864 Waldoboro ME 04572

BLACKMON, CYRIL WELLS, plant pathology, for more information see previous edition

BLACKMON, JAMES B, b Charlotte, NC, Dec 6, 38; m; c 4. SPACE POWER, DIGITAL IMAGE PROCESSING. *Educ:* Calif Inst Technol, BS, 61; Univ Calif, Los Angeles, MS, 65, PhD(eng & appl sci), 72. *Prof Exp:* Engr & scientist, Douglas Aircraft Co, 61-65, McDonnell Douglas Astronaut Co, 65-73; staff engr, 73-86, mgr, 86-90, SR MGR, ADVAN SPACE SYSTS, MCDONNELL DOUGLAS SPACE SYSTS CO, 90- *Concurrent Pos:* Adj prof & lectr ecol, Calif State Univ, Northridge, 73-76; adj prof, Orange Coast Col, 76-85; chmn, Subcomt E44-08 Solar Thermal, Am Soc Testing & Mat, 84- *Mem:* Am Inst Aeronaut & Astronaut; Am Soc Testing & Mat; fel Inst Advan Eng. *Res:* Advanced space systems; terrestrial and space solar power; space chemical and nuclear power; advanced thermal management systems; digital imaging system applications for optical evaluation of solar concentrators, spacecraft docking sensors, and large structural alignment; artificial intelligence based system design; micro gravity cryogenic fluid management. *Mailing Add:* McDonnell Douglas Space Systs Co 050852750a Ave Huntington Beach CA 92647

BLACKMON, JOHN R, b Canton, Ohio, Jan 28, 30. INTERNAL MEDICINE. *Educ:* Mt Union Col, BS, 52; Western Reserve Univ, MD, 56; Am Bd Internal Med, dipl, 63; Am Bd Cardiovasc Dis, dipl, 70. *Prof Exp:* Intern, Univ Hosps, Cleveland, 56-57, resident, 57-59, fel cardiol, 59-60; from instr to asst prof cardiol, 62-69, assoc prof med, 69-76, PROF MED, SCH MED, UNIV WASH, 76-, DIR HOSP CARDIOL LABS, 64- *Mem:* AMA; Am Fedn Clin Res. *Res:* Total and regional blood flow in normal and cardiac subjects. *Mailing Add:* Univ Hosp Seattle WA 98105

BLACKMON, MAURICE LEE, b Beaumont, Tex, Aug 24, 40; m 70. ELEMENTARY PARTICLE PHYSICS. *Educ:* Lamar State Col, BS, 62; Mass Inst Technol, PhD(physics), 67. *Prof Exp:* Appointee physics, Argonne Nat Lab, 67-69; asst prof, Syracuse Univ, 69-74; mem staff, Nat Ctr Atmospheric Res, 74-75, chmn, 76-80, scientist, Advan Study Prog, 81-87; DIR CLIMATE RES DIV, US DEPT COM, NAT OCEANIC & ATMOSPHERIC ADMIN, 88- *Mem:* Am Meteorol Soc. *Res:* General circulation of the atmosphere. *Mailing Add:* 1535 Kendall Dr Boulder CO 80307

BLACKMORE, DENIS LOUIS, b Jamaica, NY, July 20, 43; c 1. NONLINEAR DYNAMICS & CHAOS, MATHEMATICAL FLUID MECHANICS. *Educ:* Polytech Univ NY, BS, 65, MS, 66 & PhD(math), 71. *Prof Exp:* Res asst fluid mech, Polytech Univ NY, 64-65; instr math, 68-71; actg chmn, NJ Inst Technol, 85, assoc chmn, 86-88, from asst prof to prof, 71-82, assoc dir, Ctr Non Linear Sci & Math Comput, 86-90, PROF MATH, NJ INST TECHNOL, 82- *Concurrent Pos:* Statistician, Elmhurst Hosp, 70-71; prin investr, NSF Grant, 72-73, NJ Inst Technol, SBR, Res Grant,

81-83; consult, Raymar Graphic Design Co, 85. *Mem:* Am Math Soc; Soc Natural Philos; Sigma Xi. *Res:* Qualitative theory of differential equations; differential invariants of manifold; control systems; fluid mechanics; differential geometry; analysis of vortex breakdown phenomena; characterization of chaos in discrete dynamical systems; topological classification of isolated singularities and triangulation of manifolds. *Mailing Add:* Dept Math NJ Inst Technol Newark NJ 07102

BLACKMORE, PETER FREDRICK, HORMONE MECHANIZATION, ENDOCRINOLOGY. *Educ:* Univ New South Wales, Australia, PhD(biochem), 76. *Prof Exp:* Asst prof endocrinol & assoc investr, Howard Hughes Med Inst, Vanderbilt Univ, 78-88; ASSOC PROF PHARMACOL, EASTERN VA MED SCH, 88- *Mailing Add:* Eastern Va Med Sch PO Box 1980 Norfolk VA 23501

BLACKSHAW, GEORGE LANSING, b Bay Shore, NY, May 9, 36; m 55; c 3. NUCLEAR ENGINEERING. *Educ:* Hofstra Col, BA, 58; NC State Univ, MS, 62, PhD(nuclear eng), 66. *Prof Exp:* Physicist, Atomics Int Div, NAm Aviation, Inc, 58-60; from asst prof to assoc prof chem & nuclear eng, 65-73, asst dean eng, 79-80, prof chem & nuclear eng, WVa Univ, 73-, assoc dean eng, 80-; DEAN SCI & ENG, FAIRLEIGH DICKINSON UNIV. *Concurrent Pos:* Res grants, NSF, 67-79, WVa Dept Hwy, 68-70, AEC, 68, Radiation Mach Corp, 68-71 & US Environ Protection Agency, 71-77. *Mem:* Am Nuclear Soc; Am Soc Eng Educ; Am Inst Chem Engrs. *Res:* Neutron activation analysis; neutron scattering theory; process radiation; reverse osmosis; mine drainage purification and treatment; educational methodology. *Mailing Add:* Col Eng Sci Fairleigh Dickenson Univ 1000 River Rd Teaneck NJ 07666

BLACKSHEAR, PERRY L(YNNFIELD), JR, b Atlanta, Ga, July 19, 21; m 48; c 6. MECHANICAL ENGINEERING. *Educ:* Ga Inst Technol, BS, 43, MS, 47; Case Inst Technol, PhD(mech eng), 56. *Prof Exp:* Aeronaut res scientist, Nat Adv Comt Aeronaut, 47-57, head combustion dynamics sect, 57; assoc prof mech eng, 57-59, dir, Bioeng Training Prog, 69-76 & Ctr for Studies Phys Environ, 71-77, PROF MECH ENG, UNIV MINN, MINNEAPOLIS, 59-, DIR BIOENG, GRAD PROG, 80- *Concurrent Pos:* Fulbright lectr, Stuttgart Tech Univ, 62; consult eng sect, Mayo Clin, 65-; mem comt fire res, Nat Acad Sci; consult, Aerospace Corp, Roanoke, Va, 68- & Artificial Kidney Prog, NIH, 70- *Honors & Awards:* Res Award, Am Soc Artificial Internal Organs, 72; Herbert R Lissner Award Contrib Biomed Eng, Am Soc Mech Eng, 84; Clemson Award Appl Res Biomat, 86. *Mem:* Am Inst Aeronaut & Astronaut; Combustion Inst; Sigma Xi; Am Soc Artificial Internal Organs. *Res:* Fluid mechanics of chemically reacting systems as found in power and propulsion equipment; fluid mechanical aspects of artificial internal organs. *Mailing Add:* Dept Mech Eng Rm 455A Univ Minn Minneapolis MN 55455

BLACKSTEAD, HOWARD ALLAN, b Minot, NDak, Feb 24, 40; m 64. MAGNETIC RESONANCE, LOW TEMPERATURE PHYSICS. *Educ:* NDak State Univ, BS, 62; Dartmouth Col, MA, 64; Rice Univ, PhD(physics), 67. *Prof Exp:* Res assoc physics, Univ Ill, Urbana-Champaign, 67-69; asst prof, 69-75, ASSOC PROF PHYSICS, UNIV NOTRE DAME, 75- *Mem:* Am Phys Soc; Sigma Xi. *Res:* Superfluid helium flow properties; ferromagnetic resonance in the heavy rare earth metals; nuclear quadrupole resonance; phonon spectroscopy. *Mailing Add:* Dept Physics Univ Notre Dame Notre Dame IN 46556

BLACKSTOCK, DAVID THEOBALD, b Austin, Tex, Feb 13, 30; m 55; c 4. MECHANICAL ENGINEERING. *Educ:* Univ Tex, BS, 52, MA, 53; Harvard Univ, PhD(appl physics), 60. *Prof Exp:* Physicist, Bioacoust Br, Wright Air Develop Ctr, Ohio, 54-56; sr res asst, Acoust Res Lab, Harvard Univ, 60; sr res staff mem nonlinear acoust, Res Dept, Gen Dynamics/Electronics NY, 60-63; assoc prof elec eng, Univ Rochester, 63-70; lectr mech eng, 71-81, sr lectr, 81-87, FAC RES SCIENTIST, APPL RES LABS, UNIV TEX AUSTIN, 70-, PROF MECH ENG, 87- *Concurrent Pos:* Vis prof, Elec Eng Dept, Univ Rochester, 87, 88. *Honors & Awards:* Silver Medal Phys Acoust, Acoust Soc Am, 85. *Mem:* Fel Acoust Soc Am (vpres, 78-79, pres, 82-83); Am Asn Physics Teachers; Inst Noise Control Eng; Sigma Xi; Int Comn Acoust. *Res:* Nonlinear acoustics; wave motion. *Mailing Add:* Appl Res Labs Univ Tex Austin PO Box 8029 Austin TX 78713-8029

BLACKSTOCK, REBECCA, b Oklahoma City, Okla, Feb 13, 43. IMMUNOLOGY, MICROBIOLOGY. *Educ:* Okla State Univ, BS, 65; Univ Okla, MS, 68, PhD(immunol), 72. *Prof Exp:* Res technician med microbiol, Univ Okla Health Sci Ctr, 67-68; res technician infectious dis, Univ Tex Southwestern Med Sch, 68-69; fel tumor immunol, Sch Med, Univ Fla, 72-76; asst prof, 76-82, ASSOC PROF PEDIAT, UNIV OKLA HEALTH SCI CTR, 82- *Concurrent Pos:* NIH fel, 72-75; Nat Res Serv-Nat Cancer Inst fel, 75-76; adj asst prof, 77-83, adj assoc prof microbiol & immunol, Univ Okla Health Sci Ctr, 83- *Mem:* Sigma Xi; AAAS; Am Soc Microbiol; Am Asn Immunologists; Soc Leukocyte Biol; Soc Exp Biol. *Res:* Cellular immunology; tumor immunology; immune response to C neoformans; cryptococcal immunity, involving the study of macraphase regulation by suppressor T-lymphocytes; suppressor cell to down regulate the phagocytic activity in macrophages. *Mailing Add:* Dept Pediat PO Box 26901 Oklahoma City OK 73190

BLACKWELDER, BLAKE WINFIELD, b Buffalo, NY, June 25, 45; m 67. COASTAL PLAIN GEOLOGY, PALEOENVIRONMENTAL ANALYSIS. *Educ:* Duke Univ, BA, 67; George Washington Univ, MS, 71, PhD(geol), 72. *Prof Exp:* Geologist, US Geol Surv, 72-81; geologist, Tenneco Oil Co, 81-89; SR GEOLOGIST, BRIT GAS, 89- *Concurrent Pos:* Res fel assoc, Nat Acad Sci, 73. *Mem:* Am Asn Petrol Geologists. *Res:* Biostratigraphy; lithostratigraphy; structural evolution of the United States atlantic coastal plain. *Mailing Add:* Brit Gas 1100 Louisiana Suite 2500 Houston TX 77002

BLACKWELDER, RON F, b Pratt, Kans, July 16, 41; m 65; c 2. TURBULENT FLUID MECHANICS. *Educ:* Univ Colo, BS, 64; Johns Hopkins Univ, PhD(fluid dynamics), 70. *Prof Exp:* Teaching asst, Johns Hopkins Univ, 67-68, res asst, 68-70; from asst prof to assoc prof, 70-80, PROF AEROSPACE ENG, UNIV SOUTHERN CALIF, 80- *Concurrent Pos:* Vis scientist, Max Planck Inst Fluid Mech, Ger, 76-77; vis prof, Inst Nat Polytech, Grenoble, France, 83-84. *Mem:* Fel Am Phys Soc; assoc fel Am Inst Aeronaut & Astronaut; Sigma Xi. *Res:* Turbulent fluid dynamics; boundary layers and different methods for controlling turbulent mixing and for reducing the drag of objects in a moving fluid. *Mailing Add:* Dept Aerospace Eng Univ Southern Calif Los Angeles CA 90089-1191

BLACKWELL, ALAN TREVOR, b Cobourg, Ont, July 8, 42. INTERNATIONAL CO-OPERATION IN SCIENCE, PATENT LAW. *Educ:* Univ Western Ont, BSc, 64; Queen's Univ, Kingston, Ont, MSc, 67; Univ Sask, MBA, 72, LLB, 83; London Sch Econ, LLM, 84. *Prof Exp:* Astronr meteors, Dominion Observ, 67-70; astronr meteors, Herzberg Inst Astrophys, 70-85, ADV, INT SCI, BUR INT RELS, NAT RES COUN CAN, 85- *Concurrent Pos:* Lectr, Col Com, Univ Sask, 72-80; consult, 78- *Mem:* Int Astron Union; Can Astron Soc; Meteoritical Soc; Royal Astron Soc Can. *Res:* Meteors and meteorites; international co-operation and exchanges of scientists; patent law. *Mailing Add:* Legal Off Nat Res Coun Ottawa ON K1A 0R6 Can

BLACKWELL, BARRY M, b Birmingham, Eng, July 5, 34; m 58; c 3. PSYCHOPHARMACOLOGY. *Educ:* Cambridge Univ, BA, 57, MB, BChir, 60, MA & MD, 66; Univ London, DPM, 67. *Prof Exp:* Intern med & surg, Guys Hosps, Cambridge, 60-61; resident neurol, Whittington Hosp, London, 61-62; resident psychiat, Maudsley Hosp & Inst Psychiat, 62-68; asst prof psychiat, Univ Cincinnati, 68-70; prof psychiat & assoc prof pharmacol, 70-75; prof psychiat & pharmacol & chmn Dept Psychiat, Sch Med, Wright State Univ, 75-; GROUP DIR CLIN PSYCHIAT RES, MERRELL-NAT CO, 68-; PROF & CHMN, UNIV WIS MED SCH. *Mem:* Royal Col Psychiat; Am Col Neuropsychopharmacol. *Res:* Interactions between foodstuffs and monoamine oxidase inhibitors; various aspects of clinical and psychopharmacology research; clinical research. *Mailing Add:* Univ Wis Med Sch Milwaukee Clin Campus Sinai Samaritan Med Ctr 2000 W Kilbourn Ave Milwaukee WI 53233

BLACKWELL, BENNIE FRANCIS, b Magnolia, Ark, Nov 11, 41; m 60; c 2. MECHANICAL ENGINEERING. *Educ:* Univ Ark, BS, 65; Univ NMex, MS, 67; Stanford Univ, PhD(mech eng), 73. *Prof Exp:* Mem tech staff, Sandia Lab, 65-77; asst prof mech eng, La Tech Univ, 77-79; MEM TECH STAFF, SANDIA NAT LAB, 79- *Concurrent Pos:* Consult, Nat Insulation Contractors Asn, 77-79, Pabco Div, La Pac Corp & Thermal Insulation Mfg Asn, 78-79. *Mem:* Fel Am Soc Mech Engrs; Am Inst Aeronaut & Astronaut. *Res:* Heat transfer; solar energy; energy conservation. *Mailing Add:* Orgn 1553 Bldg 634 Rm 157 Sandia Nat Lab PO Box 5800 Albuquerque NM 87185-5800

BLACKWELL, CHARLES C, JR, b Houston, Tex, Nov 25, 33; m 56; c 3. SYSTEMS CONTROL, DYNAMIC SYSTEMS SCIENCE. *Educ:* Rice Univ, BA, 55, BSME, 56; Southern Methodist Univ, MSME, 60; Univ Ariz, PhD(systems control), 66. *Prof Exp:* Propulsion engr, Gen Dynamics/Ft Worth, 56-60; asst prof mech eng, NMex State Univ, 64-66; from asst prof to assoc prof, 66-71, PROF MECH ENG, UNIV TEX, ARLINGTON, 71- *Concurrent Pos:* Consult, Off Saline Water Res Prog, NMex State Univ, Life Sci Inc, Ling-Temco-Vought, Inc, Gen Dynamics Corp. *Mem:* Am Soc Mech Engrs; assoc NY Acad Sci; Inst Elec & Electronics Engrs. *Res:* Optimal, robust and adaptive control, mathematical modeling, and estimation of dynamic systems. *Mailing Add:* Dept Mech Eng Univ Tex PO Box 19023 Arlington TX 76019

BLACKWELL, CRIST SCOTT, b Winston-Salem, NC, Feb 17, 45; m 90. VIBRATIONAL SPECTROSCOPY, NUCLEAR MAGNETIC RESONANCE SPECTROSCOPY. *Educ:* Univ NC, Chapel Hill, BS, 67; Mass Inst Technol, PhD(phys chem), 71. *Prof Exp:* Res assoc, Univ Southern Calif, 71-72, & Mich State Univ, 72-75; proj scientist, Union Carbide Corp, 75-79, group leader, 79-85, sr res scientist, 85-91, CORP RES FEL, UNION CARBIDE CORP, 91- *Mem:* Am Chem Soc; Soc Appl Spectros; Am Phys Soc; Am Soc Testing & Mat; Coblenz Soc. *Res:* Analytical spectroscopy with particular emphasis on use of infrared, Raman and nuclear magnetic resonance spectroscopy to solve structural chemical problems; solid state nuclear magnetic resonance; molecular sieves; catalysts. *Mailing Add:* Union Carbide Corp Old Sawmill River Rd Tarrytown NY 10591

BLACKWELL, DAVID (HAROLD), b Centralia, Ill, Apr 24, 19; m 44; c 7. MATHEMATICS, STATISTICS. *Educ:* Univ Ill, AB, 38, AM, 39, PhD(math), 41. *Hon Degrees:* DSc, Univ Ill, 66, Mich State Univ, 69, Southern Ill Univ, 71, Carnegie-Mellon Univ, Nat Univ Lesotho, 87, Amherst Col & Harvard Univ, 88, Howard Univ, Yale Univ & Univ Warwick, 90. *Prof Exp:* Rosenwald fel, Inst for Advan Study, 41-42; asst statistician, Off Price Admin, 42; instr math & physics, Southern Agr & Mech Col, 42-43; instr math, Clark Univ, 43-44; prof, Howard Univ, 44-54; prof statist, Univ Calif, Berkeley, 54-73, dir, Study Ctr, UK & Ireland, 73-75, prof, 73-89, EMER PROF STATIST & MATH, UNIV CALIF, BERKELEY, 89- *Concurrent Pos:* Res fel, Brown Univ, 43; mathematician, Rand Corp, 48-50; vis prof, Stanford Univ, 50-51 & Univ Calif, Berkeley, 54-55. *Honors & Awards:* W W Rouse Ball Lectr, Univ Cambridge, 74; Wald Lectr, Inst Math Statist, 77; John Von Neumann Theory Prize, Inst Mgt Sci & Opers Res Soc Am, 79. *Mem:* Nat Acad Sci; AAAS; Am Math Soc (vpres, 68-71); fel Inst Math Statist (pres, 55); Am Acad Arts & Sci; Am Statist Asn (vpres, 78-). *Res:* Markoff chains; sequential analysis. *Mailing Add:* Dept Statist Univ Calif Berkeley CA 94720

BLACKWELL, FLOYD ORIS, b Mayfield, Idaho, Feb 27, 25; m 51; c 4. ENVIRONMENTAL HEALTH, PUBLIC HEALTH ADMINISTRATION. *Educ:* Wash State Univ, BS, 50; Univ Mass, MS, 54; Univ Calif, Berkeley, MPH, 65, DrPH, 69; Am Acad dipl. *Prof Exp:* Sanitarian, Benton-Franklin Dist Health Dept, 50-53; health & sanit adv, US AID, Pakistan, 54-56, sr sanit adv, 56-59; asst prof environ & pub health, Am Univ Beirut,59-64; assoc prof, Rutgers Univ, 67-71; assoc prof, Col Med, Univ Vt, 71-74; prof environ & pub health, E Carolina Univ, 74-; PROF & CHMN DEPT, ENVIRON HEALTH SCI, EASTERN KY UNIV, 81- *Concurrent Pos:* Mem pub health rev comt, Bur Health Manpower Educ, NIH, 71-73; mem bd dirs, Am Acad Sanitarians, 72-; led pub health deleg, Peoples Repub China, 87. *Mem:* Am Pub Health Asn; Nat Environ Health Asn (pres, 76). *Res:* Environmental control of communicable diseases, particularly food and water borne diseases, considering biological, microbiological, physical and social aspects in complex interrelationships; ecology of human well-being; health administration. *Mailing Add:* Dept Environ Health Sci Eastern Kentucky Univ Richmond KY 40475

BLACKWELL, HAROLD RICHARD, b Harrisburg, Pa, Jan 16, 21; m 43; c 2. VISION. *Educ:* Haverford Col, BS, 41; Brown Univ, MA, 42; Univ Mich, PhD(psychol), 47. *Prof Exp:* Res psychologist, Polaroid Corp, 43; res assoc, L C Tiffany Found, NY, 43-45; dir vision res labs, Univ Mich, 46-58; prof biophys, 65-78, DIR INST RES IN VISION, OHIO STATE UNIV, 58-, PROF PHYSIOL OPTICS & RES PROF OPHTHALMOL, 58-, PROF & CHAIRPERSON, DIV SENSORY BIOPHYS, 78- *Concurrent Pos:* Exec secy vision comt, Nat Res Coun, US Armed Forces, 45-55. *Honors & Awards:* Lomb Medal, Optical Soc Am, 50; Gold Medal, Illum Eng Soc, 72. *Mem:* Optical Soc Am; Asn Res Vision & Ophthal; Am Acad Optom; Psychonomic Soc; Illum Eng Soc. *Res:* Psychophysics and psychophysiology of vision; psychophysical methodology and sensory theory; physiological optics. *Mailing Add:* 4485 Gulf Mexico Dr Apt 703 Longboat Key FL 34228-3407

BLACKWELL, JOHN, b Sheffield, Eng, Jan 15, 42; m 65; c 2. POLYMER SCIENCE, BIOPHYSICS. *Educ:* Univ Leeds, BSc, 63, PhD(biophys), 67. *Prof Exp:* Fel chem, State Univ NY Col Forestry, Syracuse Univ, 67-69; vis asst prof, 69-70, asst prof, 69-74, assoc prof, 74-77, PROF MACROMOLECULAR SCI, CASE WESTERN RESERVE UNIV, 77-, CHMN DEPT, 85-, F ALEX NASON PROF, 91- *Concurrent Pos:* Prin investr, NSF Grants, 70, NIH Grant, 70, NIH res career develop award, 73-77; mem exec comt, Biopolymer Subgroup, Biophys Soc, 74-77, chmn, 75-76; mem exec comt, High Polymer Physics Div, Am Phys Soc, 86-90, vchmn, 87-88, chmn, 88-89; bd mem, Edison Polymer Innovation Corp, 85-, secy-treas, 90- *Honors & Awards:* Distinguished Achievement Award, Fiber Soc, 81. *Mem:* Biophys Soc; Soc Complex Carbohydrates; AAAS; Am Chem Soc; fel Am Phys Soc; Fiber Soc. *Res:* Study of the structure and interactions of macromolecules, using x-ray crystallography and other physical-chemical techniques; liquid-crystalline polymers; structure and morphology of cellulose, chitin and other polysaccharides; polysaccharide-protein interactions; structures of polyurethane elastomers. *Mailing Add:* Dept Macromolecular Sci Case Western Reserve Univ Cleveland OH 44120

BLACKWELL, JOHN HENRY, b Melbourne, Australia, July 9, 21; m 46; c 4. APPLIED MATHEMATICS. *Educ:* Univ Melbourne, BSc, 41; Univ Western Ont, MSc, 47, PhD(physics), 52. *Prof Exp:* Demonstr physics, Univ Western Ont, 46-47, instr, 47-49, lectr, 49-52, from asst prof to prof, 52-62, appl math, 62-63, sr prof, 63-67, prof, head & chmn dept appl math, 67-76, exec asst to pres, 76-79, prof, 81-86, EMER PROF APPL MATH, UNIV WESTERN ONT, 86- *Concurrent Pos:* Vis res fel, Australian Nat Univ, 55-56; vis Commonwealth fel, Oxford Univ, 64-65; mem, Can Nat Comt, Int Union Theoret & Appl Mech, 70-74. *Mem:* Fel Inst Math & Appln, UK; fel Can Soc Mech Eng; Can Soc Appl Math; Soc Appl Math. *Res:* Heat flow and diffusion theory; non-linear mathematics. *Mailing Add:* Dept Appl Math Univ Western Ont London ON N6A 5B8 Can

BLACKWELL, LAWRENCE A, b Houston, Tex, Nov 23, 31; m 57; c 2. SAFETY ENGINEERING. *Educ:* Rice Inst, BA, 54; Duke Univ, PhD(physics), 58. *Prof Exp:* Engr microwave, Tex Instruments, 58-62; vpres res & develop, Microwave Physics Inc, 62-66; acad adminr, Rice Univ, 66-76; staff mem, Los Alamos Nat Lab, 76-88; CONSULT, 89- *Mailing Add:* 11430 Game Preserve Rd Gaithersburg MD 20878

BLACKWELL, LEO HERMAN, b Austin, Tex, June 2, 34; m 63; c 3. PHYSIOLOGY. *Educ:* Univ Tex, BA, 55; ETex State Col, MA, 60; Mich State Univ, PhD(physiol), 64. *Prof Exp:* Res technician physics, Univ Tex M D Anderson Hosp & Tumor Inst, Houston, 56-59, res assoc, 59-60; asst biol, ETex State Col, 59; asst physiol, Mich State Univ, 62-63; resident student assoc, Argonne Nat Lab, 63-64; radiation safety officer, Univ Tenn, Memphis, 64-68, actg chmn div radiation biol, 67, from asst prof to assoc prof radiation biol & clin physiol, 64-75, asst to vchancellor acad affairs, 73-77, asst prof physiol & biophys & prof radiation oncol, 75-78; PROF PHYSIOL, SCH DENT, UNIV DETROIT, 78- *Concurrent Pos:* Consult, City of Memphis Hosps, 67-78. *Mem:* AAAS; Radiation Res Soc; Int Soc Exp Hemat; Health Physics Soc; Am Physiol Soc. *Res:* Cell proliferation; biological effects of radiation; radiation dosimetry; tracer physiology. *Mailing Add:* Sch Dent Univ Detroit 2985 Jefferson Ave Detroit MI 48207

BLACKWELL, LYLE MARVIN, b Charleston, WVa, Jan 29, 32; m 53; c 5. ELECTRICAL ENGINEERING. *Educ:* WVa Univ, BSEE, 54; Chrysler Inst, MAE, 56; Ohio State Univ, PhD(elec eng), 66. *Prof Exp:* Studio engr, Styling Div, Chrysler Corp, Mich, 54-58; asst prof elec eng, WVa Inst Tehcnol, 60-61, assoc prof & chmn dept, 61-63; res assoc, Commun & Control Systs Lab, Ohio State Univ, 64-66; DEAN, SCH ENG, WVA INST TECHNOL, 66- *Mem:* Am Soc Eng Educ; Inst Elec & Electronics Engrs. *Res:* Automatic control systems; application of automatic control to automotive design. *Mailing Add:* WVa Inst Technol Montgomery WV 25136

BLACKWELL, MEREDITH, b Abbeville, La, Mar 27, 40; c 1. BIOLOGY. *Educ:* Univ Southwestern La, BS, 61; Univ Ala, MS, 63; Univ Tex, Austin, PhD(bot), 73. *Prof Exp:* Asst prof biol, Hope Col, 75-81; assoc prof, 85-88, PROF BOT, LA STATE UNIV, 88- *Honors & Awards:* Alexopoulos Prize, Mycol Soc Am. *Mem:* Mycol Soc Am (secy, 86-88, vpres, 90-91, pres-elect, 91-92); Sigma Xi; Brit Mycol Soc; Soc Invert Pathol. *Res:* Biology of myxomycetes and entomogenous fungi; studies of fungal spore dispersal primarily by arthropod assemblies; other interest includes biology of desert fungi and myxomycetes; phylogeny of fungi infered from DNA sequences. *Mailing Add:* Dept Bot La State Univ Baton Rouge LA 70803

BLACKWELL, PAUL K, II, b Miami, Fla, July 10, 31; m 49; c 3. COMPUTER SCIENCE, MATHEMATICS. *Educ:* Univ Chicago, BA, 52; Syracuse Univ, MS, 61, PhD(math), 68. *Prof Exp:* Assoc prof math, Milwaukee Sch Eng, 55-57; analyst, Systs Develop Corp, 57-58; analyst defense systs, Gen Elec Co, NY, 58-61, consult, 61-63, analyst defense systs, 63-65; mathematician, Syracuse Univ Res Corp, 65-68; assoc prof, 68-, PROF COMPUT SCI, UNIV MO, COLUMBIA. *Mem:* Am Math Soc; Asn Comput Mach; Inst Elec & Electronics Engrs; Math Asn Am; Soc Indust & Appl Math. *Res:* Data base systems; medical computing; combinatorics. *Mailing Add:* Dept Comput Sci Univ Mo Columbia MO 65211

BLACKWELL, RICHARD QUENTIN, b Wichita, Kans, Sept 5, 18; m 45, 60; c 4. BIOCHEMISTRY. *Educ:* Univ Wichita, AB, 42; Northwestern Univ, PhD(biochem), 49. *Prof Exp:* Lectr chem, Northwestern Univ, 49-50, from instr to assoc prof, 50-61; head, biochem dept, US Naval Med Res Unit 2, Taipei, Taiwan, Repub of China, 58-74; mem, Div Cancer Biol & Diag, Nat Cancer Inst, 74-81; RETIRED. *Mem:* AAAS. *Res:* Clinical biochemistry; human hemoglobin variants; nutrition. *Mailing Add:* 3425 Imperial Valley Dr Little Rock AR 72212

BLACKWELL, ROBERT JERRY, b Clovis, NMex, Dec 31, 25; m 55; c 3. PHYSICS. *Educ:* Tex Christian Univ, BA, 47; Univ NC, PhD(physics), 53. *Prof Exp:* Instr physics, Univ NC, 47-52, asst prof, 53-54; mem, Prod Res Div, Humble Oil & Refining Co, 54-64 & Esso Prod Res Co, 64-74, RES SCIENTIST, EXXON PROD RES CO, 74- *Mem:* Am Phys Soc; Am Asn Physics Teachers; Am Inst Chem Eng; Am Inst Mineral, Metall & Petrol Eng; Sigma Xi. *Res:* Cosmic rays; flow through porous media; petroleum production; enhanced oil recovery processes. *Mailing Add:* 8904 Memorial Houston TX 77024

BLACKWELL, ROBERT LEIGHTON, b Clayton, NMex, Nov 24, 24; m 50; c 1. ANIMAL SCIENCE, ANIMAL GENETICS. *Educ:* NMex State Univ, BS, 49; Ore State Univ, MS, 51; Cornell Univ, PhD(animal genetics), 53. *Prof Exp:* From asst prof to assoc prof animal husb, NMex State Univ, 53-58; dir, Sheep Exp Sta, Agr Res Serv, USDA, 59-66; prof animal & range sci & head dept, Mont State Univ, 66-78, prof animal sci, 78-86; RETIRED. *Concurrent Pos:* Prin investr, Collab Res Support Prog, Small Ruminants, Peru, 79-86; actg assoc prog dir, Small Ruminant Collab Res Support Prog, Univ Calif-Davis. *Mem:* Am Soc Animal Sci. *Res:* Genetics and breeding of domestic animals. *Mailing Add:* 109 Sourdough Ridge Rd Bozeman MT 59715

BLACKWELL, WILLIAM A(LLEN), b Ft Worth, Tex, May 17, 20; m 49; c 2. ELECTRICAL NETWORK THEORY, ELECTRICAL ENGINEERING EDUCATION. *Educ:* Tex Tech Col, BS, 49; Univ Ill, MS, 52; Mich State Univ, PhD(elec eng), 58. *Prof Exp:* From instr to asst prof elec eng, Tex Tech Col, 49-55; test engr, Gen Elec Co, 49-50; instr elec eng, Univ Ill, 55-56 & Mich State Univ, 56-59; proj engr, Gen Dynamics/Ft Worth, 59-61; prof elec eng, Southern Methodist Univ, 61 & Okla State Univ, 61-66; head dept, 66-81, prof elec eng, 66-88, EMER PROF ELEC ENG, VA POLYTECH INST & STATE UNIV, 88- *Concurrent Pos:* Consult, aerosysts sect, Gen Dynamics/Ft Worth, 61; vis prof elec eng, US Military Acad, 81-82; res dir, Southeastern Ctr Elec Eng Educ, 84-85. *Honors & Awards:* Centennial Medal, Inst Elec & Electronics Engrs, 84. *Mem:* Am Soc Eng Educ; fel Inst Elec & Electronics Engrs. *Res:* System theory; author of three textbooks. *Mailing Add:* Bradley Dept Elec Eng Va Polytech Inst & State Univ Blacksburg VA 24061-0111

BLACKWOOD, ANDREW W, b Newton, NJ, Apr 1, 42; m 65; c 3. MATERIALS SCIENCE ENGINEERING. *Educ:* Rensselaer Polytech Inst, BMetE, 63 & PhD(mat eng), 70. *Prof Exp:* Res metallurgist, Am Smelting & Refining Co, 67-71, sect head, Tech Sales Serv, 71-73; supt, Metals Appln, ASA RCO Inc, 74-79; tech dir, Enthone Inc, 80-82; sr mat engr, 82-87, DIR ANAL SERV, STRUCT PROBE INC, 88- *Mem:* Metall Soc; Am Soc Metals; Am Soc Testing & Mat; Am Electroplaters Soc; Nat Asn Corrosion Engrs; Int Soc Hybrid Microelectronics; Soc Die Casting Engrs; Soc Advan Mat & Process Eng; Sigma Xi. *Res:* Materials science, particularly the application of electron microscopy; surface analysis to materials characterization; problem solving for government, industrial academic and legal clients. *Mailing Add:* Struct Probe Inc 63 Unquowa Rd Fairfield CT 06430-5015

BLAD, BLAINE L, b Cedar City, Utah, Apr 2, 39; m 64; c 7. AGRICULTURAL METEOROLOGY, REMOTE SENSING. *Educ:* Brigham Young Univ, BS, 64; Univ Minn, MS, 68, PhD(soil sci), 70. *Prof Exp:* From asst prof to assoc prof, 70-82, PROF AGR METEOROL, UNIV NEBR, 82-, HEAD, DEPT AGR METEOROL, 87- *Concurrent Pos:* Prin investr, NASA grant, 72-75, 81-92, NSF grant, Standard Oil grant, 84-86; co-investr, Off Water Res & Technol grant, 74-75, prin investr, 77-79; Prin investr, USDA-SEA grant, 78-81; consult, NASA, 78-79, 81-84 & Standard Oil Co, Ohio, 84-88. *Mem:* Fel Am Soc Agron; Crop Sci Soc Am; AAAS; Am Meteorol Soc; Sigma Xi. *Res:* Crop temperature; plant water relations; improving crop water use efficiency; evapotranspiration; radiation balance of crops; field photosynthesis; remote sensing of crop and rangeland productivity; global climate change. *Mailing Add:* Dept Agr Meteorol Univ Nebr Lincoln NE 68583-0728

BLADES, ARTHUR TAYLOR, b Milton, Ont, July 20, 26; m 50; c 3. PHYSICAL CHEMISTRY. *Educ:* Univ Western Ont, BSc, 48; Univ Wis, PhD(phys chem), 52. *Prof Exp:* Photochemist, Nat Res Coun Can, 52-54; fel nuclear chem, McMaster Univ, 54-55; res chemist, Res Coun Alta, 55-87, RES ASSOC, UNIV ALTA, 87- *Concurrent Pos:* Mem sci adv comt, Environ Conserv Authority, Alta, 74-80. *Mem:* Chem Inst Can. *Res:* Chemical kinetics; kinetic isotope effects in gas phase reactions; chemistry in flames; ion-molecule chemistry. *Mailing Add:* 7204 119 St Edmonton AB T6G 1V6 Can

BLADES, JOHN DIETERLE, b Cincinnati, Ohio, Nov 1, 24; m 49; c 3. PHYSICS. *Educ:* Western Md Col, BS, 49; Univ Cincinnati, MS, 51, PhD(physics), 54. *Prof Exp:* Sr staff scientist, Burroughs Corp, 54-65; mgr solid state physics lab, Franklin Inst Res Labs, 65-71; mgr solid state physics, Develop Labs, Addressograph-Multigraph Corp, 71-77; staff engr, Burroughs Corp, 77-88; RETIRED. *Concurrent Pos:* Vis scientist, CENG, Grenoble, France, 63-64. *Mem:* Am Phys Soc; Mat Res Soc; Int Soc Hybrid Microelectronics; Soc Photo Scientists & Engrs; Am Ceramic Soc. *Res:* Engineering physics; solid state physics; mechanics; electromagnetism; materials science engineering. *Mailing Add:* 18243 Verano Dr San Diego CA 92128

BLADH, KATHERINE LAING, b Pendleton, Ore, Feb 3, 47; m 71. IGNEOUS PETROLOGY. *Educ:* Ore State Univ, BS, 69; Univ Ariz, MS, 72, PhD(geochem), 76. *Prof Exp:* Jr geologist, Amerada-Hess Oil Corp, 69; teaching asst geol, Univ Ariz, 71-74, lectr, 76-78; asst prof, 78-85, ASSOC PROF GEOL , WITTENBERG UNIV, 85- *Mem:* Geochem Soc; Asn Women Geoscientists; Nat Asn Geol Teachers; Sigma Xi. *Res:* Silicic volcanics: their textures and chemistry; rapakivi texture in a volcanic environment. *Mailing Add:* Dept Geol Wittenberg Univ Box 720 Springfield OH 45501

BLADH, KENNETH W, b Chicago, Ill, Nov 7, 47. DESCRIPTIVE MINERALOGY. *Educ:* Wittenberg Univ, BA, 69; Univ Ariz, MS, 73, PhD(geol), 78. *Prof Exp:* ASST PROF GEOL, WITTENBERG UNIV, 78- *Mem:* Mineral Soc Am; Geol Soc Am; Mineral Asn Can; Sigma Xi. *Res:* Descriptive mineralogy; low temperature geochemistry of weathering. *Mailing Add:* Geol Dept Wittenburg Univ PO Box 720 Springfield OH 45501

BLAEDEL, WALTER JOHN, b New York, NY, May 26, 16; m 42; c 3. ANALYTICAL CHEMISTRY. *Educ:* Univ Calif, Los Angeles, BA, 38, MA, 39; Stanford Univ, PhD(chem), 42. *Prof Exp:* Instr chem, Northwestern Univ, 41-44; res assoc, Nat Defense Res Comt, 42-44; res assoc, Manhattan Proj, Univ Chicago, 44-46; res assoc, Radiation Lab, Univ Calif, 46-47; from asst prof to prof chem, Univ Wis-Madison, 47-81; RETIRED. *Concurrent Pos:* Mem comt postdoctoral fels in chem, Nat Acad Sci, 63-66; mem anal chem adv comt, Nat Bur Standards-Nat Acad Sci, 67-69; mem comt anal chem, Nat Res Coun, 75-78. *Mem:* Am Chem Soc; Sigma Xi. *Res:* Analysis in membrane systems; separations; electrochemistry; ion exchange; equipment and procedures for the electrochemical characterization of surface absorbed substances. *Mailing Add:* 20 Heritage Circle No 5 Madison WI 53711-2747

BLAESE, R MICHAEL, b Minneapolis, Minn, Feb 16, 39; m 62; c 2. IMMUNOLOGY, PEDIATRICS. *Educ:* Gustavus Adolphus Col, BS, 61; Univ Minn, MD, 64. *Prof Exp:* Intern, Parkland Hosp, Dallas, 64-65; resident pediat, Hosp, Univ Minn, 65-66; clin assoc immunol, 66-68, sr investr, 68-73, HEAD CELLULAR IMMUNOL SECT, NAT CANCER INST, NIH, 73-, DEP CHIEF, METAB BR, 85- *Concurrent Pos:* Wellcome vis prof, Royal Col Med, 80. *Honors & Awards:* Mead Johnson Award, Am Acad Pediat, 80. *Mem:* Am Soc Clin Invest; Am Asn Immunologists; Soc Pediat Res; Am Fedn Clin Res; fel AAAS. *Res:* Fundamental mechanisms of host defense by studying patients with immunodeficiency diseases, allergy, malignancy and autoimmunity; cellular immunology; macrophage function; cytotoxicity; antibody formation; neonatal immunology and phagocytosis; immunoregulation; gene therapy. *Mailing Add:* Nat Cancer Inst Bldg 10 Rm 6B05 Bethesda MD 20892

BLAGER, FLORENCE BERMAN, b Wheeling, WVa, July 2, 28; m 73. SPEECH PATHOLOGY. *Educ:* Ohio Univ, BA, 50; Univ Denver, MA, 66, PhD(speech path), 70. *Prof Exp:* ASSOC PROF, DEPT OTOLARYNGOL & PSYCHIAT, UNIV COLO MED CTR, DENVER, 70- *Concurrent Pos:* Chief speech path & audiol, JFK Child Develop Ctr, Univ Colo Med Ctr, Denver & Dept Commun Dis, Univ Colo, Boulder; chief speech path & audiol serv, Nat Jewish Ctr Immunol & Respiratory Med; res assoc, Recording & Res Ctr, Denver Ctr Performing Arts. *Mem:* Am Speech Lang & Hearing Asn; NY Acad Sci. *Res:* Laryngeal dysfunction, chronic cough, professional voice, speech, and language development in Downs infants; hearing impaired Downs infants; language of abused children. *Mailing Add:* Univ Colo Med Ctr 4200 E Ninth Ave Box B210 Denver CO 80262

BLAHA, ELI WILLIAM, b Collinsville, Ill, Oct 30, 27; m 48; c 2. ORGANIC CHEMISTRY. *Educ:* Univ Ill, BS, 51; Univ Iowa, PhD(org chem), 54. *Prof Exp:* Chemist, Standard Oil Co, Ind, 54-60, proj chemist, Am Oil Co Div, 60-64, sr proj chemist, Amoco Chem Corp, 64-70, group leader polystyrene process develop, 70-73, RES SUPVR, AMOCO CHEM CORP, STANDARD OIL CO, INC, 73- *Mem:* Am Chem Soc; Soc Plastics Engrs. *Res:* Discovery, development and production of motor oil additives, including detergents, dispersants, oxidation inhibitors and antiwear agents; research and development of improved polystyrene resins and new condensation polymers. *Mailing Add:* 803 Howard St Wheaton IL 60187

BLAHA, GORDON C, b Chicago, Ill, July 21, 34. ANATOMY, HISTOLOGY. *Educ:* Northern Ill Univ, BS, 57; Univ Ill, MS, 61, PhD(anat), 63. *Prof Exp:* Instr anat, Univ Ill Col Med, 62-63; from instr to asst prof, 63-72, assoc prof, 72-78, PROF ANAT, COL MED, UNIV CINCINNATI, 78- *Mem:* AAAS; Am Asn Anatomists; Geront Soc; Soc Study Reproduction; Sigma Xi. *Res:* Reproductive physiology; senescent decline in reproductive function; senescence in golden hamster. *Mailing Add:* 885 E Orleans St Paxton IL 60957

BLAHD, WILLIAM HENRY, b Cleveland, Ohio, May 11, 21; m 45; c 3. NUCLEAR MEDICINE, INTERNAL MEDICINE. *Educ:* Tulane Univ, MD, 45; Am Bd Internal Med, dipl, 53; Am Bd Nuclear Med, dipl, 72. *Prof Exp:* Ward officer metab res, Vet Admin, 51-52, asst chief radioisotope serv, 52-56; CHIEF NUCLEAR MED SERV, WADSWORTH VET ADMIN HOSP, 56- *Concurrent Pos:* Prof med, Univ Calif, Los Angeles, 66-; mem bd dirs, Am Bd Nuclear Med, 78-, chmn, 82; mem bd govs, Am Bd Internal Med, 81-83. *Mem:* Fel Am Col Physicians; AMA; Health Physics Soc; Soc Exp Biol & Med; Soc Nuclear Med (pres, 77-78). *Res:* Development of nuclear medicine techniques, study of body composition; treatment of thyroid disease; clinical applications of positron emission tomography; author of 3 textbooks on nuclear medicine and 239 scientific papers and published abstracts. *Mailing Add:* Nuclear Med Serv 691/W115 Vet Admin Wadsworth Med Ctr Los Angeles CA 90073

BLAHUT, RICHARD E, b Orange, NJ, June 9, 37. ENGINEERING. *Educ:* Mass Inst Technol, BS, 60; Stevens Inst Technol, MS, 64; Cornell Univ, PhD(elec eng), 72. *Prof Exp:* Staff mem navig systs, Kearfott Div, Gen Precision, 60-64; staff mem, Fed Systs Div, 64-80, FEL, IBM CORP, 80- *Concurrent Pos:* Courtesy prof elec eng, Cornell Univ, 73-; adj prof, Univ Ill, distinguished vis prof elec & computer eng, 84-85, vis res prof, Coord Sci Lab; consult prof, SChina Univ Technol; lectr, Int Ctr Mech Sci, Udine, Italy, 78 & 82, NATO Advan Study Inst, Bonas, France, 83, Il Ciocco, Italy, 86; fel, Japan Soc Propagation Sci, 82; assoc ed, Inst Elec & Electronics Engrs Trans Info Theory, 85-89; vis lectr, Princeton Univ, 89. *Mem:* Nat Acad Eng; fel Inst Elec & Electronics Engrs; AAAS. *Res:* Mathematical aspects of statistical information processing including information theory, communications theory, surveillance theory, error-control codes and digital signal processing. *Mailing Add:* IBM Corp Rte 17C Owego NY 13827

BLAIKLOCK, ROBERT GEORGE, b Newark, NJ, July 6, 40; c 4. CARDIOVASCULAR TOXICOLOGY, PHYSIOLOGY. *Educ:* Univ Vt, BA, 63; Rutgers Univ, MS, 65, PhD(physiol), 69. *Prof Exp:* Res assoc pharmacol, Med Sch, Univ Wis, 69-71; instr, Mt Sinai Sch Med, 71-73; res fel, NY Med Col, 73-74, res asst prof psychiat, 74-75, asst prof pharmacol, 75-80; sr res pharmacol, Lederle Labs, Am Cyanamid Co, 80-84; ASSOC PROF PHARMACOL, ARNOLD & MARIE SCHWARTZ COL PHARM & HEALTH SCI, LONG ISLAND UNIV, 84- *Concurrent Pos:* Adj asst prof pharmacol, NY Med Col, 80- *Mem:* AAAS; NY Acad Sci; Sigma Xi. *Res:* Cardiovalcular pharmacology; myocardial necrosis; superoxides and membrane damage, structure and function; reperfusion injury, photooxidation and bacterial exotoxins; lipid peroxidation, hemolysis and Na-K Atpase. *Mailing Add:* Six Adams Pl Glen Ridge NJ 07028

BLAINE, EDWARD HOMER, b Farmington, Mo, Jan 30, 40; m 63; c 2. CARDIOVASCULAR, RENAL. *Educ:* Univ Mo, AB, 62, MA, 67, PhD(physiol), 70. *Hon Degrees:* DSc, Univ Mo, 89. *Prof Exp:* Fel, Howard Florey Inst, 71-73; asst prof physiol, Univ Pittsburgh Sch Med, 73-77; sr res fel, Merck Inst Therapeut Res, 77-78, sr investr, 78-80, dir renal pharmacol, 80-86; DIR, CARDIOVASC DIS RES, SEARLE RES & DEVELOP, 86-, SR DIR, WASH UNIV PROG, 88- *Concurrent Pos:* Res career develop award, NIH, 75; Adj prof physiol, Temple Univ Sch Md, 77-86; mem study sect, NIH, 78-85; adj prof pharmacol, Sch Med, Wash Univ, 86- *Honors & Awards:* Nelson Lectr, Univ Mo, 86. *Mem:* Am Physiol Soc; Am Soc Pharmacol Exp Therapeut; AMA; Am Soc Hypertension (vpres, 91-). *Res:* Salt and water homeostasis and hypertension; salt and water intake; cardiovascular and renal physiology and pharmacology; hormonal control of blood pressure; appetitive behavior and physiological ecology. *Mailing Add:* G D Searle Res & Develop 800 N Lindbergh St Louis MO 63167

BLAIR, ALAN HUNTLEY, b Vancouver, BC, Oct 20, 33; m 60; c 2. ENZYMOLOGY. *Educ:* Univ BC, BSc, 56, MSc, 57; Univ Calif, Berkeley, PhD(biochem), 64. *Prof Exp:* Damon Runyon Cancer Res Fund fel, Biophys Res Lab, Harvard Med Sch, 64-65; assoc staff med, Peter Bent Brigham Hosp, Boston, Mass, 65; from asst prof to assoc prof, 66-80, PROF BIOCHEM, DALHOUSIE UNIV, 81- *Mem:* Can Biochem Soc. *Res:* Targeting of drugs by linkage to antitumor antibodies; enzymology; mechanism of action of alcohol and aldehyde dehydrogenases. *Mailing Add:* Dept Biochem Dalhousie Univ Halifax NS B3H 4H7 Can

BLAIR, BARBARA ANN, b Gastonia, NC, Oct 21, 26. PROTEIN CHEMISTRY, ENZYMOLOGY. *Educ:* Agnes Scott Col, BA, 48; Univ Tenn, MS, 53, PhD(chem), 56. *Prof Exp:* Res assoc, Med Sch, Univ Buffalo, 56-57 & Med Sch, Univ Va, 57-61; asst prof chem, Wilson Col, 61-62; from asst prof to assoc prof, 62-79, PROF CHEM, SWEET BRIAR COL, 79- *Concurrent Pos:* Vis lectr & actg head, chem dept, Women's Christian Col, Madras, India, 68-69; asst dean, Sweet Briar Col, 69-74, acad dean, 74-77. *Mem:* Fel AAAS; Am Chem Soc; Sigma Xi. *Res:* Amino acid metabolism; protein structure. *Mailing Add:* PO Box P Sweet Briar VA 24595

BLAIR, CAROL DEAN, b Salt Lake City, Utah, Jan 31, 42; m 68; c 3. VIROLOGY, MOLECULAR BIOLOGY. *Educ:* Univ Utah, BA, 64; Univ Calif, Berkeley, PhD(molecular biol), 68. *Hon Degrees:* MA, Dublin Univ, 74. *Prof Exp:* Fel biochem, Dublin Univ, 68-70, from jr lectr to lectr, 70-75; from asst prof to assoc prof microbiol, 75-85, asst dean, Col Vet Med Biomed Sci, 83-88, PROF MICROBIOL, COLO STATE UNIV, 85-, DEPT HEAD MICROBIOL, COLO STATE UNIV, 88- *Concurrent Pos:* Fel, Am Cancer Soc, 68-69, Irish Dept Educ, 69-70 & Int Union Against Cancer, Am Cancer Soc, 72-73; vis prof, Univ Colo Health Sci Ctr, 82. *Mem:* Am Soc Microbiol; Soc Gen Microbiol; AAAS; Am Soc Trop Med & Hyg; Sigma Xi; Am Soc Virol. *Res:* Mechanisms of animal virus replication; mechanisms of virus pathogenesis; latent and oncogenic virus infections; virus genetics. *Mailing Add:* Dept Microbiol Colo State Univ Ft Collins CO 80523

BLAIR, CHARLES EUGENE, b New York, NY, Dec 30, 49. OPERATIONS RESEARCH. *Educ:* Mass Inst Technol, BS, 71; Carnegie-Mellon Univ, MA, 72, PhD(math), 75. *Prof Exp:* Vis asst prof, Dept Math, NC State Univ, 75-76; From asst prof to assoc prof, 76-85, PROF, DEPT BUS ADMIN, UNIV ILL, 88- *Concurrent Pos:* Vis asst prof, Ga Inst Technol, 80. *Mem:* Am Math Soc; Opers Res Soc. *Res:* Integer programming, combinatorics; game theory. *Mailing Add:* 1206 S Sixth St Champaign IL 61820

BLAIR, CHARLES MELVIN, JR, b Vernon, Tex, Oct 24, 10; m 36; c 2. SURFACE CHEMISTRY, SYNTHETIC ORGANIC CHEMISTRY. *Educ:* Rice Univ, AB, 31, AM, 32; Calif Inst Technol, PhD(chem), 35. *Prof Exp:* Chemist, Petrolite Corp, 35-43, dir res, 43-53, pres, 53-64; vchancellor, Wash Univ, 64-67; vpres & tech dir, 67-70, CHMN BD DIRS & SR SCIENTIST, MAGNA CORP, 70- *Mem:* AAAS; Am Chem Soc; Nat Asn Corrosion Eng; Soc Petrol Engrs. *Res:* Colloid and surface chemistry; demulsifiers; corrosion inhibitors; chemical processing of petroleum; chemical agents for enhancement of petroleum production. *Mailing Add:* 5320 Buck Hill Buena Park CA 90670

BLAIR, DAVID W(ILLIAM), b Santa Barbara, Calif, Oct 5, 29; m 54; c 6. THERMODYNAMICS, COMBUSTION. *Educ:* Ore State Univ, BS, 52; Columbia Univ, MS, 54, PhD(combustion), 61. *Prof Exp:* Teaching asst mech eng, Columbia Univ, 52-54, instr, 54-58; res assoc aeronaut eng, Princeton Univ, 58-61; sr res engr, Aerochem Res Labs, 61-62; fel combustion & propulsion, Royal Norweg Inst Indust & Sci Res, 62-63; assoc prof mech eng, Polytech Inst Brooklyn, 63-69; sr res engr, Exxon Res & Eng Co, 69-74, eng assoc, 74-83; sr res engr, M L Energia, 83-85; PRES, PRINCETON SCI ENTERPRISES, 85- *Concurrent Pos:* Lectr, Columbia Univ, 59, adj assoc prof mech eng, 75-; lectr, Stevens Inst Technol, 61-62. *Mem:* Combustion Inst; Am Phys Soc; Am Soc Mech Engrs; Sigma Xi; Am Inst Chem Engrs. *Res:* Solid propellant rocket propulsion; fluid mechanics; energy conversion; plasma physics; combustion emissions and pollutant control; coal combustion and gasification; high temperature test apparatus; tribology test apparatus; automated instrumentation. *Mailing Add:* 1108 Kingston Rd Princeton NJ 08540

BLAIR, DONALD GEORGE RALPH, b Lloydminster, Sask, Nov 5, 32; m 59; c 2. BIOCHEMISTRY, CANCER RESEARCH. *Educ:* Univ Alta, BSc, 55, MSc, 56; Univ Wis, PhD, 61. *Prof Exp:* Lectr cancer res, 61-63, lectr biochem, 63-67, from asst prof to assoc prof cancer res, 63-76, ASSOC PROF BIOCHEM, UNIV SASK, 76- *Concurrent Pos:* Res assoc, Nat Cancer Inst Can, 71-75. *Mem:* AAAS; Am Soc Biochem & Molecular Biol; Sigma Xi; Am Asn Cancer Res; Can Biochem Soc; Can Oncol Soc. *Res:* Cell biology; rat intestinal sucrase; mechanisms and regulation of the biosynthesis of pyrimidine nucleotides; mechanisms of the resistance of mammalian cells to purine and pyrimidine analogues; RNA synthesis in neoplastic cells; RNA polymerase; polyamines and RNA synthesis. *Mailing Add:* Dept Biochem Univ Sask Saskatoon SK S7N 0W0 Can

BLAIR, EMIL, b Satu-mare, Rumania, Oct 20, 23; US citizen. THORACIC SURGERY, CARDIOVASCULAR SURGERY. *Educ:* Med Col Ga, MD, 46. *Prof Exp:* Instr med & NIH fel, Med Ctr, Duke Univ, 52-54; Halsted fel surg, Univ Colo Med Ctr, Denver, 56-57; asst prof surg, Sch Med, Univ Md, Baltimore, 60-66; prof cardiothoracic surg, Col Med, Univ Vt, 66-69; prof, 69-88, EMER PROF SURG, UNIV COLO MED CTR, ENVER, 88- *Concurrent Pos:* Vis prof, Univ Uppsala, Guys Hosp, Univ Oslo, Univ Helsinki, 65-66 & Hodossoh Univ Hosp, Jerusalem, 77-80; consult, Cent Off, Vet Admin, 65-68; Career res award, 61 & fel Fogarty Int Ctr, NIH, 61. *Honors & Awards:* Hektoen Award, AMA, 55. *Mem:* Am Asn Thoracic Surg; Am Col Surg; Sigma Xi; Soc Vascular Surg; Oriental Inst; Am Inst Archeol. *Res:* Hypothermia; hyperbaric oxygen; cardiovascular pulmonary physiology and surgery; shock; biomechanics of trauma; emergency medical services; education, health and hospital planning. *Mailing Add:* 6020 California Circle Suite 211 Rockville MD 20852

BLAIR, ETCYL HOWELL, b Wynona, Okla, Oct 16, 22; m 49; c 3. ORGANIC CHEMISTRY. *Educ:* Southwestern Col, Kans, AB, 47; Kans State Col, MS, 49, PhD(chem), 52. *Hon Degrees:* DSc, Southwestern Col, Kans, 74. *Prof Exp:* Instr org chem, Southwestern Col, Kans, 47-48; res asst, Kans State Col, 48-51; res chemist, 51-56, group leader, 56-65, div leader, 65-67, dir res, Britton Lab, 67-68, mgr res & develop, Agr Dept, 68-71, dir res & develop, Agr-Org Dept, 71-73, dir health & environ res, US Area Res & Develop, 73-78, VPRES & DIR HEALTH & ENVIRON SCI, DOW CHEM CO, 78- *Concurrent Pos:* mem bd dirs, Chem Indust Inst Toxicol, 77- *Mem:* Int Acad Environ Safety; AAAS; Am Chem Soc; NY Acad Sci; Sigma Xi. *Res:* Isolation and characterization of natural products; synthesis of biological active compounds; organic phosphorus compounds. *Mailing Add:* Corp Res & Develop Four Crescent Ct Midland MI 48640

BLAIR, GEORGE RICHARD, b San Bernardino, Calif, Aug 17, 20; m 51; c 2. INORGANIC CHEMISTRY. *Educ:* Univ Redlands, BA, 42. *Prof Exp:* Chief anal chemist, Metal Hydrides, Inc, Mass, 43-46; anal group leader, Mineral Eng Lab, Am Cyanamid Co, 50-52; chief res chemist, McMillan Lab, Inc, Mass, 52-59; head space mat group, Hughes Aircraft Co, Culver City, 59-68, head chem process, Electron Dynamics Div, 68-72, head anal servs, Hughes Aircraft Res Labs, 72-74, qual engr, 74-77, res chemist, 77-84; RETIRED. *Honors & Awards:* NASA Award for Contribution to Surveyor Prog. *Mem:* Am Chem Soc. *Res:* High vacuum studies; clean surfaces; optical fibers; instrumental analysis; fiber optics. *Mailing Add:* 10748 Cranks Rd Culver City CA 90230

BLAIR, GRANT CLARK, b Elderslie, Scotland, Dec 1, 65. IMAGING COMPUTER SYSTEMS, IMAGE DATA RETRIEVAL & INTERCHANGE. *Educ:* Univ Kent, Canterbury, UK, BS, 86. *Prof Exp:* Commun scientist, Govt Commun Hq, 86-87; software engr, Digital Appln Int, 87-88; software consult, Bell Labs, 89; SOFTWARE CONSULT, E I DU PONT DE NEMOURS, INC, 89- *Mem:* Inst Physics. *Res:* Investigation of problems related to producing photographic quality hardcopies of color images displayed on computer monitors. *Mailing Add:* 523 Spruce Ave Maple Shade NJ 08052

BLAIR, JAMES BRYAN, b Waynesburg, Pa, May 26, 44; m 66; c 2. BIOCHEMISTRY. *Educ:* WVa Univ, BS, 66; Univ Va, PhD(biochem), 70. *Prof Exp:* Lab asst chem, WVa Univ, 64-66; fel, Inst Enzyme Res, Univ Wis, 69-72; from asst prof to assoc prof, 72-81, interim chmn, 84-85, PROF BIOCHEM, WVA UNIV, 81- *Concurrent Pos:* NIH fel, 70; NIH res career

develop award, 78- *Mem:* AAAS; Am Soc Biol Chemists; Biophys Soc; Am Chem Soc; Sigma Xi. *Res:* Regulation of carbohydrate and lipid metabolism in mammals; kinetic and regulatory properties of hepatic pyruvate kinase. *Mailing Add:* 1924 Elvin Dr Stillwater OK 74074

BLAIR, JAMES EDWARD, b Cedar Rapids, Iowa, Dec 6, 35; m 61; c 2. PHYSICAL CHEMISTRY. *Educ:* Rockhurst Col, BS, 58; Univ Wis, PhD(chem), 63. *Prof Exp:* Res chemist, Photo Prod Dept, E I du Pont de Nemours & Co, 63-65; from asst prof to assoc prof, Milton Col, 65-80; assoc prof, Edgewood Col, 80-81; scientist II, Raltech, 81-86; CHEMIST 4, WIS STATE LAB HYG, MADISON, 87- *Concurrent Pos:* Res assoc, Enzyme Inst, 65-81; NSF fel. *Mem:* Sigma Xi. *Res:* Sedimentation studies of polymer solutions; proteins and DNA. *Mailing Add:* 2210 Ravenswood Rd Madison WI 53711

BLAIR, JOHN, b Budapest, Hungary, Dec 5, 29; US citizen; m 55; c 2. ELECTRICAL ENGINEERING. *Educ:* Mass Inst Technol, BS, 54, MS, 55, ScD(elec eng), 60. *Prof Exp:* Physicist, Pac Semiconductors Inc, 55-57; prof, dept elec eng, Mass Inst Technol, 57-66; CORP DIR RES, RAYTHEON CO, 66- *Concurrent Pos:* Ford Found fel, 60-61; consult, Raytheon Co, 58-61, Electro Optical Syst, Inc, 59-60 & Esso Res Corp, 60; dir, Energy Conversion, Inc, 61-64; mem, Army Sci Bd, US Dept Army, 78-84 & 86-90; mem, Sea Grant Res Panel, US Dept Com, 79-85. *Mem:* Inst Elec & Electronics Engrs; Am Phys Soc; Indust Res Inst. *Res:* Corporate responsibility for company wide applied research activities in electronics and associated fields. *Mailing Add:* Exec Offs Raytheon Co Lexington MA 02173

BLAIR, JOHN MORRIS, b Russellville, Ark, May 24, 19; m 47; c 2. EXPERIMENTAL NUCLEAR PHYSICS. *Educ:* Okla Agr & Mech Col, BS, 40; Univ Wis, PhM, 42; Univ Minn, PhD(physics), 47. *Prof Exp:* Jr scientist, Los Alamos Sci Lab, NMex, 43-45; res assoc physics, Univ Minn, 47-48; assoc scientist, Argonne Nat Lab, 48-50; from asst prof to assoc prof, 50-62, prof, 62-89, EMER PROF PHYSICS, UNIV MINN, MINNEAPOLIS, 89- *Mem:* Am Phys Soc; Am Asn Physics Teachers; Sigma Xi. *Res:* development of equipment to aid in teaching physics; development and use of Van de Graaf machines; reaction involving heavy ions. *Mailing Add:* Sch Physics Univ Minn Minneapolis MN 55455

BLAIR, JOHN SANBORN, b Madison, Wis, Apr 28, 23; m 51; c 3. PHYSICS. *Educ:* Yale Univ, BS, 43; Univ Ill, MA, 49, PhD(physics), 51. *Prof Exp:* Jr physicist, Manhattan Proj, 44-45; res assoc, Univ Ill, 51-52; from instr to assoc prof, Univ Wash, 52-61; prof physics, 61-; AT RAYTHEON CO. *Concurrent Pos:* Res assoc, Princeton Univ, 57-58; NSF sr fel, Inst Theoret Physics, Copenhagen, 61-62; vis prof, Univ Surrey, 68-69; vis scientist, Ctr Nuclear Studies, Saelay, France, 75. *Mem:* Am Phys Soc. *Res:* Theoretical nuclear physics. *Mailing Add:* Raytheon Co 141 Spring St Lexington MA 02173

BLAIR, LOUIS CURTIS, b Worcester, Mass, Oct 2, 54; m 78; c 2. PLANT TISSUE CULTURE, PHYTOTOXICOLOGY. *Educ:* Univ Mass, BS, 77; Univ Ill, PhD(agron), 83. *Prof Exp:* Res asst herbicides, 77-83, res assoc genetic toxicol, 83-85, RES ASSOC TISSUE CULT & GENETIC ENG, UNIV ILL, URBANA, 85- *Mem:* AAAS. *Res:* Photosynthetic tissue cultures of important crop plant species; discovering the factors limiting photosynthesis in tissue cultured cells; identification and regulation of genes involved in photosynthesis. *Mailing Add:* Dept Agron PABL Univ Ill 1201 W Gregory Dr Urbana IL 61801

BLAIR, MARTHA LONGWELL, b 1948; m 78; c 2. BIOMEDICAL RESEARCH, ANIMAL PHYSIOLOGY. *Educ:* Wellesley Col, BA, 69; Univ Wash, PhD(physiol & psychol), 74. *Prof Exp:* ASSOC PROF PHYSIOL, UNIV ROCHESTER, 77- *Mailing Add:* Dept Physiol Univ Rochester Med Ctr Box 642 Rochester NY 14642

BLAIR, MURRAY REID, JR, b Somerville, Mass, July 13, 28; m 51. PHARMACOLOGY. *Educ:* Tufts Univ, BS, 49, PhD(pharmacol), 53. *Prof Exp:* From instr to asst prof pharmacol, Tufts Univ, 53-58; vis asst prof, State Univ NY, Buffalo, 56-57, asst prof, 58-60; consult pharmacologist, Arthur D Little, Inc, 60-61; assoc prof pharmacol, Med Col Va, 61-66, from asst dean to assoc dean sch med, 61-66; dir res, Astra Pharmaceut Prod, Inc, 66-77, assoc dir sci affairs, 74-77; assoc dean admin, Sch Med, Tufts Univ, 77-80, lectr biochem & pharmacol,77-, actg dean, 80; ASSOC VPRES ACAD AFFAIRS, HSC, TEX TECH UNIV. *Concurrent Pos:* USPHS fel, Tufts Univ, 53-54. *Mem:* Am Soc Pharmacol & Exp Therapeut; NY Acad Sci. *Res:* Pharmacology and physiology of the smooth muscle of the gastrointestinal tract; biological action of polypeptids; action of atropine; medical education. *Mailing Add:* Health Sci Ctr Rm 2B 154 Vpres Off Tex Tech Univ Lubbock TX 79430

BLAIR, PAUL V, b Kimball, Nebr, Nov 11, 29; c 4. BIOCHEMISTRY. *Educ:* Utah State Univ, BS, 55; Purdue Univ, PhD(genetics & chem), 61. *Prof Exp:* Fel biochem, Inst Enzyme Res, Univ Wis, 61-64; asst prof, Univ, 64-66; asst prof, Sch Med, Univ Ind, Indianapolis, 66-69, assoc prof biochem, 69-81; pub rels & sales, Engineered Models Corp, 82-91; PROF BIOCHEM, SCH MED, UNIV IND, INDIANAPOLIS, 91- *Concurrent Pos:* Sabbatical, Cambridge Univ, 72-73; Univ Alaska, 79. *Mem:* Am Soc Biol Chemists. *Res:* Mitochondrial metabolism; correlation of metabolic functions with ultrastructure; metabolic regulation. *Mailing Add:* Dept Biochem & Molecular Biol Ind Univ Sch Med Indianapolis IN 46202-5122

BLAIR, PHYLLIS BEEBE, b Buffalo, NY, May 17, 31; m 55. IMMUNOLOGY. *Educ:* Cornell Univ, BS, 53; Univ Calif, Berkeley, PhD(zool), 58. *Prof Exp:* PROF IMMUNOL, UNIV CALIF, BERKELEY, 72- *Mem:* AAAS; Am Asn Cancer Res; Am Asn Immunologists; Am Soc Zoologists; Clin Immunol Soc. *Res:* Immune regulation. *Mailing Add:* Dept Bact 3573LSB Univ Calif Berkeley CA 94720

BLAIR, ROBERT, b May 29, 33; UK & Can citizen; m 58; c 2. ANIMAL & POULTRY NUTRITION. *Educ:* Univ Glasgow, BSc, 56; Univ Aberdeen, PhD(animal & swine nutrit), 60; DSc, Univ Sask, 83. *Prof Exp:* Sr res fel swine prod, Univ Aberdeen, 60-66; prin sci officer nutrit, Agr Res Coun, 66-75; res assoc, Cornell Univ, 69-70; dir nutrit, Swift Can Co Ltd, 76-78; from assoc prof to prof animal & poultry sci, Univ Sask, 78-84; PROF & HEAD ANIMAL SCI, UNIV BC, 84- *Concurrent Pos:* Exec dir, Prairie Swine Ctr, 78-84; pres, World Soc Animal Prod, 88-93. *Mem:* Can Soc Animal Sci; Am Soc Animal Sci; Nutrit Soc; Poultry Sci Asn; Am Inst Nutrit; Brit Soc Animal Prod. *Res:* Vitamin utilization in poultry and swine; swine nutrition; poultry nutrition; nutritional toxicology. *Mailing Add:* Dept Animal Sci Univ BC Suite 248 2357 Main Mall Vancouver BC V6T 2A2 Can

BLAIR, ROBERT G, b Toledo, Ohio, Dec 6, 36. ECONOMIC GEOLOGY. *Educ:* Dartmouth Univ, BA, 58; Univ Mich, MA, 60. *Prof Exp:* CHIEF GEOLOGIST, COEURD ALENE, 87- *Mem:* Fel Geol Soc Am; Soc Econ Geologists; Geol Soc Can. *Mailing Add:* 7764 S Elm Ct Littleton CO 80122

BLAIR, ROBERT LOUIE, general topology; deceased, see previous edition for last biography

BLAIR, ROBERT WILLIAM, b Kansas City, Mo, Apr 20, 17; m 41; c 2. PETROLEUM GEOLOGY. *Educ:* Univ Colo, BA, 38. *Prof Exp:* Recorder, US Geol Surv, 38-39; engr, Gravity Serv Co, 39; geologist, US Geol Surv, 46-47; geologist, Continental Oil Co, 47-49; div geologist, 49-54; chief surface geologist, Sahara Petrol Co, Egypt, 54-57; asst to regional geologist, Continental Oil Co, 57-61; chief geologist, San Jacinto Oil & Gas Co, 61-62; geologist, Continental Oil Co, 62-67; consult geologist, 67-69; geologist, Geophoto Serv, Inc, 69-70; CONSULT GEOLOGIST, 71- *Mem:* Am Asn Petrol Geologists; fel Geol Soc Am. *Res:* Domestic and foreign petroleum exploration. *Mailing Add:* 1704 N Chestnut St Colorado Springs CO 80907

BLAIR, ROBERT WILLIAM, NEUROPHYSIOLOGY. *Educ:* Univ Tex, PhD(pharmacol), 79. *Prof Exp:* ASST PROF PHYSIOL, OKLA UNIV, 80- *Mailing Add:* Dept Physiol PO Box 26901 Oklahoma City OK 73190

BLAIR, WILLIAM DAVID, b Baltimore, Md, Apr 20, 43; m 65; c 2. PURE MATHEMATICS. *Educ:* Johns Hopkins Univ, AB, 65; Univ Md, PhD(math), 71. *Prof Exp:* From asst prof to assoc prof, 71-86, PROF MATH, NORTHERN ILL UNIV, 86-, DEPT CHAIR, 90- *Concurrent Pos:* Vis fel, Univ Leeds, 78-79 & Univ Calif, San Diego, 85-86. *Mem:* Am Math Soc; Math Asn Am; London Math Soc; Sigma Xi. *Res:* Ring theory, noetherian rings; localizations and quotient rings. *Mailing Add:* Dept Math Sci Northern Ill Univ DeKalb IL 60115

BLAIR, WILLIAM EMANUEL, b Elmira, NY, Apr 10, 34; m 59; c 2. ELECTRICAL ENGINEERING, GEOPHYSICS. *Educ:* Cornell Univ, BSEE, 58, MSEE, 60; Univ NMex, DSc, 64. *Prof Exp:* Res asst elec eng, Cornell Univ, 58-60; assoc, Univ NMex, 60-64; res engr, Stanford Res Inst, 64-68; sr res engr, 68-76; proj mgr, Elec Power Res Inst, 76-84; LEAD CONSULT, ENERGY & CONTROL CONSULTS, 84- *Mem:* Sr mem Inst Elec & Electronics Engrs; Int Electrotech Comn. *Res:* Very-low-frequency propagation and antenna theory; ionospheric propagation theory; satellite data processing; underground and underwater antenna theory; microwave tube theory; electric utility communications, especially power-line carrier, telephone and microwave systems. *Mailing Add:* Energy & Control Consults 1530 Alameda Suite 300 San Jose CA 95126

BLAIS, BURTON W, b Montreal, Que, June 2, 63; m 88. IMMUNODIAGNOSTICS, IMMUNOTECHNOLOGY & BIOTECHNOLOGY. *Educ:* Univ NB, BSc, 84; Carleton Univ, MSc, 86, PhD(biol & biochem), 90. *Prof Exp:* Vis fel molecular biol, Nat Lab Sexually Transmitted Dis, Lab Ctr Dis Control, Ottawa, 90-91; BIOLOGIST & HEAD MOLECULAR MARKERS SECT, PATHOGENS, LAB CTR DIS CONTROL, OTTAWA, DEPT HEALTH & WELFARE CAN, 91. *Concurrent Pos:* Sci consult, Ricoh Kyosan, Inc, 89-; lectr, Dept Biol, Inst Biochem, Carleton Univ, Ottawa, 90- *Mem:* Can Biochem Soc; Can Fedn Biol Sci. *Res:* Development of novel enzyme immunoassay strategies for the diagnosis of human, animal and plant diseases; detection of microbial pathogens in foods and feeds; molecular characterization of enteric pathogens as applied in epidemiology. *Mailing Add:* RR 2 Finch ON K0C 1K0 Can

BLAIS, J A RODRIGUE, b Rimouski, Que, Dec 17, 41. ESTIMATION, INFORMATION. *Educ:* Loyola Col, Montreal, BSc Hons, 63; Univ Toronto, MA, 65; Univ NB, PhD(geodesy), 79. *Prof Exp:* Mathematician, Topographical Surv, Surv & Mapping Br, Dept Energy, Mines & Resources, 68-74 & Geodetic Surv, 75-79; assoc prof, 79-87, PROF SURV ENG, UNIV CALGARY, 87- *Concurrent Pos:* Consult, Dept Energy, Mines & Resources, Ottawa, 79-, Alta Provincial Govt, 80- *Mem:* Can Math Soc; Can Asn Physicists; Can Inst Surv; Can Geophys Union; Int Soc Gen Relativity & Gravitation; Am Math Soc. *Res:* Estimation, numerical methods, gravitation, and geodetic applications; stochastic processes and applications; information theory and applications; earth's gravity field. *Mailing Add:* Dept Surv Eng Univ Calgary Calgary AB T2N 1N4 Can

BLAIS, NORMAND C, b Springfield, Mass, Jan 4, 26; m 51; c 3. CHEMICAL PHYSICS. *Educ:* Union Col, BS, 52; Yale Univ, MS, 53, PhD(physics), 56. *Prof Exp:* MEM RES STAFF, LOS ALAMOS NAT LAB, UNIV CALIF, 56- *Mem:* Am Phys Soc. *Res:* Collision interactions between atoms or molecules; measuring transport properties of gases; reactive scattering in molecular beams; theoretical calculations; laser induced chemical reactions; photoionization of molecular clusters. *Mailing Add:* 56 Coyote Los Alamos NM 87544

BLAIS, ROGER A, b Shawinigan, Que, Feb 4, 26; m 50; c 2. INDUSTRIAL INNOVATION, ECONOMIC GEOLOGY. *Educ:* Laval Univ, BASc, 49, MSc, 50; Univ Toronto, PhD(geol), 54. *Prof Exp:* Geologist, Que Dept Natural Resources, 53-56; chief develop engr, Iron Ore Co, Can, 56-61; assoc

dean res, 70-71, assoc prof, 61-66, dean sci res, 70-80, PROF, ECOL POLYTECH, UNIV MONTREAL, 66-, DIR INNNOVATION CTR, 80. *Concurrent Pos:* Dir, Eldorado Nuclear Ltd, 68-; chmn nat study group solid earth sci in Can, Sci Coun Can, 68-70; mem earth sci comt, Nat Res Coun Can, 67-69, chmn, 69; mem exec comt, Nat Adv Comt Res Geol Sci, 68-; gov, Int Develop Res Ctr, Ottawa, Ont, 74-, chmn exec comt, 76-; Arctic Inst NAm fel, 75; chmn study group mat res, Nat Res Coun Can, 76-77. *Honors & Awards:* Gold Medal, Can Inst Mining & Metall, 67, Barlow Mem Gold Medal, 62; Bancroft Award, Royal Soc Can, 76; Archambault Gold Medal & Alcan Prize, Fr Can Asn Advan Sci, 77. *Mem:* AAAS; Geol Asn Can (pres, 69-70); Can Inst Mining & Metall (past pres); Mineral Asn Can; Soc Econ Geol. *Res:* Applied geostatistics; applications of the theory of regionalized variables to ore reserve estimation, grade control and mineral exploration; origin of iron ores; geology of base metal sulphide deposits and their relations to volcanic and carbonate rocks; industrial innovation. *Mailing Add:* Ecol Polytech Univ Montreal Montreal PQ H3C 3A7 Can

BLAIS, ROGER NATHANIEL, b Duluth, Minn, Oct 3, 44; m 71; c 2. MULTIPHASE FLOW INSTRUMENTATION. *Educ:* Univ Minn, BA, 66; Univ Okla, PhD(physics), 71. *Prof Exp:* Instr physics, Westark Community Col, 71-72; asst prof physics & geophys sci, Old Dominion Univ, 72-77; asst prof, 77-81, chmn physics, 86-88, vice provost, 89, ASSOC PROF ENG PHYSICS, UNIV TULSA, 81-, ASSOC DIR, ARTIFICIAL LIFT PROJS, 83- *Concurrent Pos:* Consult, Brill Eng, ARCO & LaBarge Electronics; bd dirs, Aerospace Industs, Div Instrument Soc Am; actg provost & vpres acad affairs, Univ Tulsa, 90- *Mem:* Am Phys Soc; AAAS; Am Geophys Union; Nat Soc Prof Engrs; Soc Petrol Engrs; Instrument Soc Am; Am Asn Physics Teachers. *Res:* Fluid dynamics; artificial lift technology, multiphase flow in pipes, atmospheric pollution diffusion modeling, remote sensing of atmospheric pollution, instrumentation, gaseous electronics. *Mailing Add:* Off Provost Univ Tulsa 600 S College Ave Tulsa OK 74104-3189

BLAISDELL, ERNEST ATWELL, JR, b Brewer, Maine, Jan 10, 40; m 58; c 2. EXPERIMENTAL DESIGN. *Educ:* Univ Maine, BA, 62, MA, 64; Temple Univ, PhD(statist), 79. *Prof Exp:* Instr math, Clarkson Col Technol, 64-68; PROF & CHMN MATH DEPT, ELIZABETHTOWN COL, 68- *Concurrent Pos:* Adj lectr math, Pa State Univ, 78-79; lectr statist, Temple Univ, 79- *Mem:* Am Statist Asn; Am Asn Univ Professors. *Res:* Partially balanced change-over designs on m-associate class PBIB designs; development of new series of these designs and a study of their efficiencies. *Mailing Add:* 606 Aspen Lane Lebanon PA 17042

BLAISDELL, FRED W(ILLIAM), b Goffstown, NH, July 21, 11; m 35; c 3. HYDRAULIC ENGINEERING. *Educ:* Univ NH, BS, 33, CE, 52; Mass Inst Technol, MS, 34. *Prof Exp:* Town engr, NH, 33; asst engr, H K Barrows, Mass, 34; tech asst, Mass Inst Technol, 35; sta engr, Soil Conserv Serv, USDA, Pa, 36, from jr soil conservationist to asst hydraul engr, Bur Standards, Washington, DC, 36-40, from asst to res hydraul engr, Soil Conserv Serv, 40-53 & Agr Res Serv, 53-78, hydraul engr sci & educ admin agr res, 78-81, res hydraul engr, 81-86, COLLABR, AGR RES SERV, USDA, 86- *Honors & Awards:* Rickey Medal, Am Soc Civil Engrs, 49, J C Stevens Award, 69. *Mem:* Am Soc Civil Engrs; Am Soc Agr Engrs; Int Asn Hydraul Res. *Res:* generalized engineering research on soil conservation structures; design of SAF Stilling basin to dissipate energy below dams, culverts and chutes; type D artificial rainfall apparatus for use in hydrologic research; hydraulics of closed conduit spillways; local scour at pipe outlets and plunge pool energy dissipator design criteria. *Mailing Add:* USDA Agr Res Ctr Water Conserv Struct Lab 1301 N Western St Stillwater OK 74075

BLAISDELL, JAMES PERSHING, b Holbrook, Idaho, June 29, 18; m 41; c 5. RANGE ECOLOGY. *Educ:* Utah State Univ, BS, 39; Univ Idaho, MS, 42; Univ Minn, PhD(bot), 56. *Prof Exp:* Field asst, Grazing Serv, US Dept Interior, 37-39; agr aid, US Forest Serv, USDA, 40-42, range conservationist, Soil Conserv Serv, 46, range conservationist, Forest Serv, 46-66, asst sta dir res, Intermountain Forest & Range Exp Sta, 66-80. *Mem:* Soc Range Mgt; Ecol Soc Am. *Res:* Management of livestock and big game ranges; effects of fire on range vegetation; plant growth in relation to climate. *Mailing Add:* 2664 Shamrock Dr Ogden UT 84403

BLAISDELL, JOHN LEWIS, b Norwood, Mass, Feb 27, 35; m 57; c 5. FOOD SCIENCE, CHEMICAL ENGINEERING. *Educ:* Univ Mass, BSChE, 56, MS, 58; Mich State Univ, PhD(food sci), 63. *Prof Exp:* Asst instr agr eng, Univ Mass, 56-58; asst food sci, Mich State Univ, 58-63, NIH fel, 63-64; from asst prof to assoc prof food sci & nutrit & agr eng, 64-74, PROF FOOD SCI & NUTRIT & AGR ENG, OHIO STATE UNIV, 74- *Concurrent Pos:* Ed, Food Eng Div, Am Soc Agr Engrs. *Mem:* Am Soc Eng Educ; Am Soc Heat, Refrig & Air-Conditioning Engrs; Inst Food Technol; Am Inst Chem Engrs; Am Soc Agr Engrs; Sigma Xi. *Res:* Application of engineering to food preservation problems; natural convection in confined space; transient conduction and convection; dairy processing waste treatment; rheology; packaging; food process controls curriculum development. *Mailing Add:* Ohio State Univ Agr Engr Bldg 590 Woody Hayes Dr Columbus OH 43210

BLAISDELL, RICHARD KEKUNI, b Honolulu, Hawaii, Mar 11, 25; m 62; c 2. INTERNAL MEDICINE, HEMATOLOGY. *Educ:* Univ Redlands, AB, 45; Univ Chicago, MD, 47. *Prof Exp:* Intern med, Johns Hopkins Hosp, 48-49; asst resident, Charity Hosp, New Orleans, 49-50; instr path, Duke Univ, 54-55; resident med, Univ Chicago, 55-57, instr, 57-58; res assoc, Atomic Bomb Casualty Comn, 59-61; asst prof, Univ Chicago, 61-66; chmn dept, 66-70, PROF MED, SCH MED, UNIV HAWAII, 66- *Mem:* AAAS; Am Fedn Clin Res; Am Soc Hemat; Am Col Physicians; Am Asn Hist Med. *Res:* Hemoproliferative disorders; platelet-subendothelial interactions; Polynesian medicine. *Mailing Add:* 2230 Liliha St Honolulu HI 96817

BLAKE, ALEXANDER, b Poland, Sept 24, 20; m 52; c 1. MECHANICAL & STRUCTURAL ENGINEERING. *Educ:* Polish Univ, London, Dipl, 51, Univ London, 55. *Prof Exp:* Design engr indust & tractors, Vickers Armstrongs Ltd, Eng, 51-53; sr stress engr aircraft eng, de Havilland Eng, Co,

Eng, 53-54; head eng sect distrib & generation elec, Elec Auth Eng , 54-56; design anal indust heating, Westinghouse Elec Corp, 56-58; sect head res appl & theoret mech, Assoc Spring Corp, 58-60; sr res engr, Kaiser Aluminum & Chem Corp, 60-61; mgr stress & Wright control, Aerojet Gen Corp, 61-66; tech adv, Boeing Airplain Co, 66-67; SR STRESS ANALYST, NUCLEAR TEST ENG DIV, LAWRENCE RADIATION LAB, 66- *Mem:* Am Soc Mech Engr; Brit Inst Mech Eng. *Res:* Stredd and structural analysis; Machine disign; engineering mechanics. *Mailing Add:* 2358 Westminster W Livermore CA 94550

BLAKE, CARL, b Sarasota, Fla, Dec 17, 25; m 48; c 4. ELECTRONICS ENGINEERING. *Educ:* Mass Inst Technol, BS & MS, 49. *Prof Exp:* Asst prof elec eng, Univ Maine, 49-57; staff mem microwave components, Mass Inst Technol, 57-63, asst group leader, 63-65, assoc group leader, Phased Array Radars, Lincoln Lab, 65, group leader, 65-86; CONSULT, 86- *Mem:* Inst Elec & Electronics Engrs. *Res:* Microwave theory and techniques, especially noise theory, low noise receivers, phased array radar systems, components and costing, and microwave integrated circuits. *Mailing Add:* Carl Blake Assoc One Pacific Lane Westford MA 01886

BLAKE, CARL THOMAS, b Jacksonville, NC, Apr 14, 26; m 49; c 2. AGRONOMY. *Educ:* NC State Col, BS, 52, MS, 57; Pa State Univ, PhD(agron), 63. *Prof Exp:* Asst county agr agt, NC Agr Exten Serv, 52-53; spec acct, Carolina Tel & Tel Co, 53-54; agron exten specialist turf mgt, NC Agr Exten Serv, 54-60,; EXTEN PROF CROP SCI, NC STATE UNIV,61-; AGRON EXTEN SPECIALIST TURF MGT NC AGR EXTEN SERV, 61- *Mem:* Am Soc Agron. *Res:* Ecological and physiological aspects of plant communities, particularly production of cultivated species. *Mailing Add:* 4509 Leaf Ct Raleigh NC 27612

BLAKE, DANIEL BRYAN, b New York, NY, Apr 4, 39; m 66. PALEOBIOLOGY. *Educ:* Univ Ill, Urbana, BS, 60; Mich State Univ, MS, 62; Univ Calif, Berkeley, PhD(paleont), 66. *Prof Exp:* Actg instr paleont, Univ Calif, Berkeley, 66-67; from asst prof to assoc prof, 67-79, PROF GEOL, UNIV ILL, URBANA, 79- *Mem:* Soc Econ Paleontologists & Mineralogists; Paleont Soc; Int Paleont Union; Brit Paleont Asn; Sigma Xi; AAAS. *Res:* Paleozoic Bryozoa; fossil and modern asteroids. *Mailing Add:* Rm 245 NHB Dept Geol Univ Ill 1301 W Green St Urbana IL 61801

BLAKE, DANIEL MELVIN, b Miami, Fla, Oct 23, 43; m 65; c 2. PHOTO CHEMISTRY, ORGANOMETALLIC. *Educ:* Colo State Univ, BS, 65; Wash State Univ, PhD(inorg chem), 69. *Prof Exp:* Vis asst prof chem, Harvey Mudd Col, 69-70; from asst prof to prof chem, Univ Tex, Arlington, 70-81; sr scientist & mgr raw mat & chem alumina, Arco Metals Co, 81-85; mgr mat br, 86-89, PRIN SCIENTIST, SOLAR ENERGY RES INST, 87- *Mem:* Am Chem Soc; Am Ceramic Soc; Metall Soc; AAAS; Am Solar Energy Soc. *Res:* Coordination chemistry; platinum metals chemistry; organotransition metal chemistry; photochemistry; alumina chemicals; materials science; photochemistry; environmental chemistry. *Mailing Add:* Solar Energy Res Inst 1617 Cole Blvd Golden CO 80401

BLAKE, DAVID ANDREW, b Baltimore, Md, Aug 26, 41; m 63; c 3. PHARMACOLOGY, TERATOLOGY. *Educ:* Univ Md, Baltimore City, BS, 63, PhD(pharmacol), 66. *Prof Exp:* Res assoc, Lab Chem Pharmacol, Nat Heart Inst, 66-67; from asst prof pharmacol to assoc prof pharmacol & toxicol, Sch Pharm, Univ Md, Baltimore City, 67-73, chmn dept, 69-73; ASSOC PROF GYNEC-OBSTET & PHARMACOL, SCH MED, JOHNS HOPKINS UNIV, 73-, SR ASSOC DEAN, 80- *Concurrent Pos:* Mem, Human Embryol & Develop Study Sect, NIH, 84- *Mem:* Am Soc Pharmacol & Exp Therapeut. *Res:* Drug disposition; chemoteratogenicity; biochemical toxicology; biotransformation of volatile fluorocarbons; hepatotoxicity of halothane; mechanism of action of thalidomide; bioactivation and mutagenesis anesthetics carcinogenesis; toxicology. *Mailing Add:* Johns Hopkins Univ Sch Med 101 Admin Bldg 720 Rutland Ave Baltimore MD 21205

BLAKE, EMMET REID, b Abbeville, SC, Nov 29, 08; m 47; c 2. ORNITHOLOGY. *Educ:* Presby Col, SC, BA, 28; Univ Pittsburgh, MS, 33. *Hon Degrees:* DSc, Presby Col, SC, 66. *Prof Exp:* Asst zool, Univ Pittsburgh, 31-32; tech asst, Nat Geog Soc exped, Brazil & Venezuela, 30-31; leader, Mandel-Field Mus exped, Orinoco River, 31-32; ornithologist, Field Mus exped, Guatemala, 33-34; leader, Carnegie Mus exped, Brit Honduras, 35; asst ornithologist, 35-37, from asst cur to cur, 37-73, EMER CUR, DIV BIRDS, FIELD MUS NATURAL HIST, 73- *Concurrent Pos:* Mem Field Mus Natural Hist zool exped, Brit Guiana & Brazil, 37-38, Brit Guiana exped, 38-39, Southwestern zool exped, 41 & Mex field studies, 53; leader, Conover Peru Exped, 58. *Mem:* Fel Am Ornith Union; Cooper Ornith Soc; Wilson Ornith Soc. *Res:* Origin, distribution and variation of neotropical birds. *Mailing Add:* Field Mus of Natural Hist Roosevelt Rd & Lake Shore Dr Chicago IL 60605

BLAKE, GEORGE HENRY, JR, b Faunsdale, Ala, June 21, 22; m 46; c 4. ENTOMOLOGY. *Educ:* Auburn Univ, BS, 47, MS, 49; Univ Ill, PhD, 58. *Prof Exp:* Assoc entomologist, Auburn Univ, 50-64, assoc prof, 64-65, prof entom & zool, 65-84, EMER PROF ENTOM & ZOOL, AUBURN UNIV, 84- *Mem:* Entom Soc Am. *Res:* Teaching and graduate research. *Mailing Add:* PO Box 2543 Auburn AL 36831

BLAKE, GEORGE MARSTON, b Los Angeles, Calif, Jan 21, 32; m 52; c 4. FORESTRY. *Educ:* Univ Idaho, BS, 57; Univ Minn, MS, 59, PhD, 64. *Prof Exp:* Asst, Univ Minn, 57-62; from asst prof silvicult to assoc prof forestry, 62-72, PROF FORESTRY, UNIV MONT, 72-, RES COORDR FORESTRY, 73- *Mem:* AAAS; Am Soc Foresters. *Res:* Genetic variation; western larch and ponderosa pine; tissue culture of western larch; selection and breeding of drought resistance in ponderosa pine for mine spoil reclamation. *Mailing Add:* 3019 Martinwood Rd Missoula MT 59802

BLAKE, GEORGE ROWLAND, b Provo, Utah, Mar 14, 18; m 41; c 4. SOIL PHYSICS. *Educ:* Brigham Young Univ, AB, 43; Ohio State Univ, PhD, 49. *Prof Exp:* Asst asst prof soil physics & asst res specialist, Rutgers Univ, 49-55; from assoc prof to prof soil physics, 55-85, dir, Water Resources Ctr, 79-85, EMER PROF SOIL PHYSICS, UNIV MINN, ST PAUL, 85- *Concurrent Pos:* Coop agt, USDA, 49-55; NSF sr fel, Ger, 62-63; consult, Ford Found, Chile, 67; Fulbright guest prof & lectr, Univ Hohenheim, Stuttgart, Ger, 70-71; guest prof & consult, Univ Kesthely, Hungary, 74; consult, Univ Minn-USAID, Inst Agronomique Hassan II, Morocco, 78-; guest prof, Univ Warsaw, Poland, 81, Humboldt Univ, Berlin, Ger, 86. *Mem:* Fel Soil Sci Soc Am; fel Am Soc Agron; Soil Conserv Soc Am; Int Soc Soil Sci; Sigma Xi. *Res:* Soil structure and erosion control; water resources; soil water relations; soil modification. *Mailing Add:* 1579 Burton St St Paul MN 55108

BLAKE, IAN FRASER, b Saskatoon, Sask, June 7, 41; m 66; c 2. ELECTRICAL ENGINEERING. *Educ:* Queen's Univ, BSc, 62, MSc, 64; Princeton Univ, PhD(elec eng), 67. *Prof Exp:* Res assoc telecommun, Jet Propulsion Lab, 67-69; chmn dept, 78-84, PROF ELEC ENG, UNIV WATERLOO, 69- *Concurrent Pos:* IBM Res Lab, Yorktown Heights, NY, 75-76; M/A-COM Linkabit Inc, San Diego, 84-85. *Mem:* Inst Elec & Electronics Engrs. *Res:* Communication theory; coding theory; combinatorial theory. *Mailing Add:* Dept Elec Eng Univ Waterloo Waterloo ON N2L 3G1 Can

BLAKE, J BERNARD, b New York, NY, Dec 14, 35; m 60; c 2. ASTROPHYSICS, SPACE PHYSICS. *Educ:* Univ Ill, BS, 57, MS, 58, PhD(physics), 62. *Prof Exp:* Res assoc physics, Univ Ill, 62; DEPT HEAD SPACE PARTICLES & FIELDS DEPT, SPACE SCI LAB, AEROSPACE CORP, 62- *Mem:* Fel Am Phys Soc; Am Geophys Union; Am Astron Soc; Sigma Xi. *Res:* Space plasma physics; cosmic-ray physics; nuclear astrophyics; nuclear physics. *Mailing Add:* Aerospace Corp M2/259 PO Box 92957 Aerospace Corp PO Box 92957 Los Angeles CA 90009

BLAKE, JAMES ELWOOD, b St George, Utah, Oct 4, 38; m 63; c 4. INFORMATION SCIENCE. *Educ:* Brigham Young Univ, BA, 64. *Prof Exp:* Nat Defense Educ Act fel chem, Utah State Univ, 64-67, asst, 67-68; assoc ed org chem, 68-72, sr assoc ed, 72-76, SR ED CHEM INFO, CHEM ABSTR SERV, 76- *Mem:* Am Chem Soc. *Res:* Applications and uses of chemical information; development of new systems to handle chemical information. *Mailing Add:* Dept 67 Chem Abstr Serv PO Box 3012 Columbus OH 43210

BLAKE, JAMES J, b New York, NY, June 10, 37. ORGANIC CHEMISTRY, BIOCHEMISTRY. *Educ:* Mass Inst Technol, BS, 58; Univ Calif, PhD(chem), 65. *Prof Exp:* Develop engr, rheology, Aerojet Gen Corp, 58-60; asst res biochemist, Hormone Res Lab, Univ Calif, San Francisco, 65-85; AT ONCOGEN, SEATTLE, 86- *Concurrent Pos:* NIH fel, 65-67. *Mem:* AAAS; Am Chem Soc. *Res:* Synthesis of peptides. *Mailing Add:* 201 Galer Apt 243 Seattle WA 98109-3192

BLAKE, JAMES NEAL, b Great Bend, Kans, July 18, 33; m 63. SPEECH & HEARING SCIENCES. *Educ:* Kans State Univ, BS, 55, MS, 60; Univ Southern Miss, PhD(speech & hearing sci), 68. *Prof Exp:* Speech therapist, Charlotte-Mecklenburg Schs, 60-64; asst prof speech & hearing, Western Carolina Univ, 66-70; ASSOC PROF AUDIOL & SPEECH PATH, UNIV LOUISVILLE, 70- *Concurrent Pos:* Consult, Rauch Ctr Retarded Children, 70- *Mem:* Ling Soc Am; Speech Asn Am; Am Speech & Hearing Asn; Am Psychol Asn; Coun Except Children. *Res:* Relationship between normal and abnormal verbal behavior and various cognitive states; effects of variable verbal loadings on cognitive. *Mailing Add:* Dept Spec Educ Univ Louisville Louisville KY 40292

BLAKE, JOHN ARCHIBALD, b Tobago, WI; m 61; c 2. SCIENCE EDUCATION. *Educ:* Howard Univ, BS, 63, MS, 64; George Peabody Col Teachers, Vanderbilt Univ, EdS, 74; Univ Tenn, Knoxville, EdD, 78. *Prof Exp:* DEPT HEAD MATH, OAKWOOD COL, 67-, DIR, SCI LEARNING CTR, 78- *Mem:* Nat Coun Teachers Math; Nat Asn Math; Math Asn Am. *Res:* Problem solving in mathematics and students' thought processes. *Mailing Add:* 2628 Brookline Dr NW Huntsville AL 35810

BLAKE, JOSEPH THOMAS, b Provo, Utah, Jan 24, 19; m 41; c 6. VETERINARY MEDICINE. *Educ:* Brigham Young Univ, BS, 49; Iowa State Univ, MS, 50, PhD, 55, DVM, 56. *Prof Exp:* Instr animal husb, Ariz State Univ, 50-51; instr dairy husb, Iowa State Univ, 51-53; from asst prof to assoc prof vet sci, 56-69, EMER PROF, UTAH STATE UNIV. *Mem:* Am Vet Med Asn; Am Soc Vet Physiol & Pharmacol; Am Soc Animal Sci; Am Asn Vet Med Cols. *Res:* Physiology and nutrition in relation to veterinary medicine. *Mailing Add:* 1314 N Oquirrh Dr Provo UT 84604

BLAKE, JULES, b New York, NY, July 7, 24; m 49; c 3. ORGANIC CHEMISTRY, TECHNOLOGY ACQUISITION. *Educ:* Univ Pa, BS, 49, MS, 51, PhD(org chem), 54. *Prof Exp:* Res chemist, Marshall Lab, E I du Pont de Nemours & Co, Philadelphia, 54-59, res supvr, 60-65, develop supvr, 65-66; res dir, Indust Chem Div, Mallinckrodt Chem Works, 66-69, dir res & develop & gen mgr res & develop div, Chem Group, 69-71; vpres res & develop, Kendall Co, 71-72; vpres res & develop, Colgate-Palmolive Co, 73-87, vpres Corp Sci Affairs, 87-89; Off Sci & Technol Policy, Exec Off of the President, 89- *Concurrent Pos:* Mem, Indust Res Inst, 67-, pres, 80-81; chmn adv comt anal, Nat Bur Standards, 70-73; mem Res & Develop Coun, Am Mgt Asn, 75-79; mem ad hoc adv comt, President's Sci Advisor, 80; bd overseers, Sch Dental Med, Univ Pa, 80-; dir, Cosmetic Toiletries Fragrance Asn, 85-; adv comt, Nat Sci Asn, Univ Pa, 85-, Res & Develop Adv Comt Conf Bd, 88; bd dirs, Immune Response Corp & Marter Corp; chmn, Dirs Indust Res. *Mem:* Am Chem Soc; Asn Res Dirs (pres); Sigma Xi. *Res:* Research management; consumer products. *Mailing Add:* 867 Sunset Ridge Bridgewater NJ 08807

BLAKE, LAMONT VINCENT, b Somerville, Mass, Nov 7, 13; m 38, 57, 80; c 2. MICROWAVE PHYSICS, RADIO ENGINEERING. *Educ:* Mass State Col, BS, 35; Univ Md, MS, 50. *Prof Exp:* Radio interference investr, Ark Power & Light Co, 37-40; physicist, Naval Res Lab, 40-67, consult physicist, Radar Div, 67-69, Head Radar Geophys Br, 69-72; sr scientist, Technol Serv Corp, Silver Spring, Md, 72-84; consult, 84- *Concurrent Pos:* Free lance tech writer, 60- *Honors & Awards:* Sci Res Soc Am Appl Sci Award, 63. *Mem:* AAAS; Am Phys Soc; Sigma Xi; Inst Elec & Electronics Engrs; Am Geophys Union. *Res:* Design of radar systems and components; radar maximum range theory; atmospheric absorption of radio waves; radio noise theory. *Mailing Add:* 800 Copley Lane Silver Spring MD 20904

BLAKE, LOUIS HARVEY, mathematics; deceased, see previous edition for last biography

BLAKE, MARTIN IRVING, b Paterson, NJ, Oct 20, 23; m 48; c 4. PHARMACY. *Educ:* Brooklyn Col Pharm, BS, 47; Rutgers Univ, MS, 50; Ohio State Univ, PhD(pharm), 51. *Prof Exp:* Asst prof pharmaceut chem, Duquesne Univ, 51-55; prof, NDak State Univ, 55-59; resident res assoc, Argonne Nat Lab, 59-60; head dept, 61-80, PROF PHARMACEUT, COL PHARM, UNIV ILL-CHICAGO, 60- *Concurrent Pos:* Consult & fac assoc, Chem Div, Argonne Nat Lab, 60-81; consult pharm serv, Westside Vet Admin Hosp, Chicago, 65-66 & Hines Vet Admin Hosp, Ill, 66-; mem rev comt, Nat Formulary, 65-75 & US Pharmacopeia, 70-; mem, Ill Bd Pharm, 75-82, chmn, 77-79; travel fel, WHO, 82; hon pres, Ill Pharm Asn, 84. *Mem:* Sigma Xi; Am Pharmaceut Asn; Am Asn Cols Pharm; AAAS. *Res:* Nonaqueous titrimetry; drug analysis; isotope effects; biosynthesis; ion exchange. *Mailing Add:* Dept Pharm Univ Ill Med Ctr 833 S Wood St Chicago IL 60612

BLAKE, MILTON CLARK, JR, b San Francisco, Calif, Feb 20, 32; m 59; c 1. GEOLOGY. *Educ:* Univ Calif, Berkeley, AB, 58; Stanford Univ, PhD(geol), 65. *Prof Exp:* Jr geologist, 58-64, GEOLOGIST, BR WESTERN ENVIRON GEOL, US GEOL SURV, 64- *Concurrent Pos:* Vis prof, Auckland Univ, 70, Univ Paris, 78; GK Gilbert fel, 88. *Mem:* Geophys Union; Geol Soc Am; Mineral Soc Am. *Res:* Igneous and metamorphic petrology; low grade metamorphic rocks of Western United States; Blueschist-facies rocks of the world. *Mailing Add:* US Geol Surv 345 Middlefield Rd Menlo Park CA 94025

BLAKE, RICHARD D, b Greenfield, Mass, Sept 24, 32; m 58; c 2. PHYSICAL BIOCHEMISTRY. *Educ:* Tufts Col, BS, 58; Rutgers Univ, MS, 63; Princeton Univ, PhD(biochem), 67. *Prof Exp:* Res asst chem, Harvard Univ, 58-60; res asst chem, Princeton Univ, 61-63, res assoc biochem, 67-68, res staff, 68-73; asst prof, 73-77, assoc prof, 77-82, PROF BIOCHEM, UNIV MAINE, 82- *Concurrent Pos:* Dir, Worthington Biochem Corp, 68-77; prin investr, NIH, 75-; vis prof, Med Sch Hannover, Fed Repub Ger, 80. *Mem:* Am Soc Biol Chemists; Biophys Soc; Sigma Xi. *Res:* Structure and interactions of DNA; helix-coil transitions of DNA; evolution of DNA sequences; effect of water on the stability of DNA. *Mailing Add:* Dept Biochem Univ Maine Orono ME 04473

BLAKE, RICHARD L, b Berkeley Springs, WVa, Mar 8, 37; m 86; c 1. PHYSICS, GENERAL. *Educ:* Rensselaer Polytech Inst, 59; Univ Colo, PhD(astro-geophys), 68. *Prof Exp:* Physicist, Naval Res Lab, 59-63; asst prof astron & astrophys, Enrico Fermi Inst Nuclear Studies, Univ Chicago, 68-74; staff physicist, Los Alamos Nat Lab, 74-84, x-ray sect leader, 85-90; CONSULT, 90- *Mem:* Am Phys Soc; Sigma Xi. *Res:* X-ray spectroscopy of ions and atoms; x-ray physics; solar physics; x-ray astronomy. *Mailing Add:* Los Alamos Nat Lab Mail Stop D-410 Los Alamos NM 87545

BLAKE, ROBERT GEORGE, b Cornell, Ill, May 4, 06; m 29; c 1. MATHEMATICS. *Educ:* Univ Fla, AB, 38, MA, 45, PhD(math), 53. *Prof Exp:* Teacher pub schs, Fla, 26-43; from instr to assoc prof math, Univ Fla, 43-76; RETIRED. *Mem:* Am Math Soc; Math Asn Am; Soc Indust & Appl Math. *Res:* Distributions and generalized transforms; partially ordered algebraic systems. *Mailing Add:* 222 NW 27th Terr Gainesville FL 32607

BLAKE, ROBERT L, b Claremont, NH, Mar 21, 33; m 61; c 2. BIOCHEMISTRY, PHARMACOLOGY. *Educ:* Bates Col, BS, 55; Univ Rochester, MS, 57; Univ Calif, San Francisco, PhD(biochem), 62. *Prof Exp:* Res assoc pharmacol, Univ Rochester, 55-57; asst biochem, Univ Calif, Berkeley & San Francisco, 57-59; USPHS fel, McArdle Mem Lab, Univ Wis-Madison, 62-64; res chemist, Med Ctr, Univ Calif, San Francisco, 64-65, Cardiovasc Res Inst fel, 65-66; from assoc to staff scientist, Jackson Lab, 66-80; ASSOC PROF FAMILY MED, UNIV MO, COLUMBIA, 80- *Concurrent Pos:* NSF travel fel, NATO Advan Study Inst, Bergen, Norway, 68. *Mem:* AAAS; NY Acad Sci; Brit Biochem Soc. *Res:* Biochemical control mechanisms of mammalian genetic expression; mitochondrial biogenesis; hormonal induction of enzymes; inborn errors of metabolism; pharmacogenetics. *Mailing Add:* Health Sci M222 Univ Mo Columbia MO 65211

BLAKE, ROBERT WESLEY, b Long Beach, Calif, Oct 23, 45; m 71; c 2. ANIMAL SCIENCE, ANIMAL GENETICS. *Educ:* Univ Minn, BS, 71; NC State Univ, MTech, 73, PhD(animal sci), 77. *Prof Exp:* Dairy extensionist, US Peace Corps, Peru, 68-71; res asst animal genetics, NC State Univ, 72-76; from asst prof to assoc prof dairy sci, Tex A&M univ, 77-86; ASSOC PROF INT ANIMAL SCI, CORNELL UNIV, 86- *Concurrent Pos:* Livestock consult, Bahamas, Colombia, El Salvador, Brazil, Mex, Costa Rica, Honduras, Trinidad, Saudi Arabia, Venezuela, Nigeria, Kenya & Ethopia, 73,77 & 78-88. *Mem:* Am Dairy Sci Asn; Am Soc Animal Sci; AAAS. *Res:* Dairy cattle breeding, production economics, and management; correlated responses in nutrient utilization, physiological mechanisms, and tropical ruminant production. *Mailing Add:* 149 Morrison Hall-Animal Sci Cornell Univ Main Campus Ithaca NY 14853

BLAKE, ROLAND CHARLES, horticulture; deceased, see previous edition for last biography

BLAKE, ROLLAND LAWS, b Minneapolis, Minn, Jan 16, 24; m 53; c 3. GEOLOGY. *Educ:* Univ Minn, BS, 50, MS, 51, PhD(geol), 58. *Prof Exp:* Geologist, Cleveland-Cliffs Iron Co, Mich, 51-52, Minn, 52-54; petrol & mineral res geologist, 58-71, supvry geologist, 71-88, GEOLOGIST, US BUR MINES, 88- *Mem:* Soc Mining Engrs. *Res:* Mineralogy; iron silicates; carbonates; iron oxides associated with iron ore deposits; manganese minerals; petrology; pre-Cambrian in Canadian shield area. *Mailing Add:* 6701 Southdale Rd Minneapolis MN 55435

BLAKE, THOMAS LEWIS, b Wilmington, NC, Nov 9, 46; m 68; c 2. HORTICULTURE, AGRICULTURE. *Educ:* Clemson Univ, BS, 70. *Prof Exp:* Agr res tech, 70-82, SUPT, HORT CROPS RES STA, NC STATE UNIV, 82- *Res:* Blueberries, grapes, strawberries, potatoes, cabbage, cucumbers, asparagus, & woody ornamentals. *Mailing Add:* Hort Crops Res Sta 3800 Castle Hayne Rd Castle Hayne NC 28429

BLAKE, THOMAS MATHEWS, b Sheffield, Ala, Aug 4, 20. CARDIOLOGY. *Educ:* Univ Ala, BA, 41; Vanderbilt Univ, MD, 44; Am Bd Internal Med, dipl. *Prof Exp:* Intern med, Vanderbilt Univ Hosp, 44-45, intern path, 47-48, asst resident med, 48-49, fel cardiovasc res, Med Sch, 50-52, instr, 52-54, instr clin med, 54-55; from asst prof to assoc prof, 55-70, PROF MED, SCH MED, UNIV MISS, 70- *Concurrent Pos:* Asst resident, Strong Mem Hosp, Rochester, NY, 49-50; chief clin physiol sect, Res Lab, Vet Admin Hosp, 54-55; mem coun arteriosclerosis, Am Heart Asn. *Mem:* AMA; Am Fedn Clin; fel Am Col Cardiol; fel Am Col Chest Physicians; fel Am Col Physicians. *Res:* Electrocardiography; cardiovascular physiology. *Mailing Add:* Heart Sta Univ Miss Med Ctr Jackson MS 39216-4505

BLAKE, THOMAS R, b New York, NY, Apr 5, 38; m 63; c 3. FLUID MECHANICS, RHEOLOGY. *Educ:* Polytech Inst Brooklyn, BME, 60; Rennselaer Polytech Inst, MS, 65; Yale Univ, PhD(eng), 71. *Prof Exp:* Engr, Pratt & Whitney Aircraft Div, United Technol Corp, 65-65, res scientist, Res Lab, 65-68; scientist & mgr, S-Cubed, Inc, 70-80; PROF MECH ENG, UNIV MASS, 80-, HEAD DEPT, 81- *Concurrent Pos:* Vis prof, dept physiol & adj prof, dept chem eng, Univ Ariz, 80-84. *Mem:* Am Inst Chem Engrs; Microcirculatory Soc. *Res:* Fluid mechanics with an emphasis upon the mathematical description of multiphase flow processes with chemical reaction. *Mailing Add:* Dept Mech Eng Univ Mass Amherst MA 01002

BLAKE, TOM, US citizen. QTL ANALYSIS, MOLECULAR EVOLUTION. *Educ:* Univ Calif, BS, 76; SDak State Univ, MS, 79; Wash State Univ, PhD(genetics), 83. *Prof Exp:* Res assoc genetics, Univ Wis, 83-84; asst prof barley breeding, 84-89, ASSOC PROF BARLEY BREEDING, MONT STATE UNIV, 89- *Res:* Breeding of barley; construct genome maps and identify QTL's; genetics and forensics. *Mailing Add:* Plant & Soil Sci Dept Mont State Univ Bozeman MT 59717

BLAKE, WESTON, JR, b Boston, Mass, Feb 26, 30; m 60; c 2. GLACIAL GEOLOGY. *Educ:* Dartmouth Col, AB, 51; McGill Univ, MSc, 53; Ohio State Univ, PhD(geol), 62; Univ Stockholm, Fil Lic(geog), 64, Fil Dr(phys geog), 75. *Prof Exp:* res scientist & head paleoecol & geo-chronol sect, Terrain Sci Div, 69-86; RES SCIENTIST, GEOL SURV CAN, 62- *Concurrent Pos:* Mem Can nat comt, Int Geol Correlation Proj, 74-77; docent, Dept Phys Geog, Univ Stockholm, 75-; assoc ed, Can J Earth Sci, 75-81. *Mem:* Geol Asn Can; Arctic Inst NAm; Glaciol Soc; Geol Soc Finland; Swed Soc Anthrop & Geog; Am Quaternary Asn. *Res:* Glacial history and geomorphological processes in arctic regions; radiocarbon dating; pleistocene marine faunas. *Mailing Add:* 19 Madawaska Dr Ottawa ON K1S 3G5 Can

BLAKE, WILLIAM KING, b Teaneck, NJ, Sept 19, 42; m 67; c 2. ACOUSTICS, FLUID MECHANICS. *Educ:* Univ Notre Dame, BS, 64; Mass Inst Technol, PhD(naval archit), 69. *Prof Exp:* Instr acoust, Dept Naval Archit & Marine Eng, Mass Inst Technol, 66-69; physicist, 69-73, RES PHYSICIST, TAYLOR NAVAL SHIP RES & DEVELOP CTR, 73- *Concurrent Pos:* NSF trainee, Mass Inst Technol, 64-66; consult, Bolt, Beranek and Newman Inc, 66-69; lectr, Univ Minn, Cath Univ Am & von Karman Inst, Brussels. *Mem:* Fel Acoust Soc Am; Sigma Xi. *Res:* Hydroacoustics; flow-induced noise and vibration; turbulence; structural acoustics; cavitation; hydroelasticity. *Mailing Add:* 8607 Hempstead Ave Bethesda MD 20817

BLAKE, WILSON, b San Francisco, Calif, Aug 29, 34; m; c 5. MINING ENGINEERING. *Educ:* Univ Calif, Berkeley, BA, 57, MS, 62; Colo Sch Mines, PhD, 71. *Prof Exp:* Mining geologist, Malachite Corp, 57-58; sr sci data analyst, Lawrence Radiation Lab, Univ Calif, 61-65; res civil engr, Denver Mining Res Ctr, US Bur Mines, 65-71; res mining dir mining res, Gecamines, 72-74; CONSULT MINING ENGR, 74- *Concurrent Pos:* Adj asst prof, Colo Sch Mines, 71-72. *Honors & Awards:* Peele Award, Am Inst Mining Engrs, 70; Interdisciplinary Award, US Nat Comt Rock Mech, 72. *Mem:* Am Inst Mining, Metall & Petrol Engrs; Int Soc Rock Mech. *Res:* Rock mechanics; development and application of the broad-band microseismic method of determining rock stability; development of mathematical models of rock behavior; mechanics of rock bursts; development of microseismic monitoring systems. *Mailing Add:* PO Box 928 Hayden Lake ID 83835

BLAKELY, ELEANOR ALICE, b Quantico, Va, July 29, 47; m 86; c 1. RADIATION PHYSIOLOGY, BIOPHYSICS. *Educ:* Univ San Diego, AB, 69; Univ Ill, MS, 71, PhD(physiol), 75. *Prof Exp:* BIOPHYSICIST, LAWRENCE BERKELEY LAB, UNIV CALIF, 75- *Concurrent Pos:* Co-investr, Nat Cancer Inst, 75-; assoc ed, Radiation Res, 84-88 & Space Solar Power Rev, 81-; mem, Diag Radiol Study Sect, Div Res Grants, NIH, 87-; mem, Diag Radiol Study Sect, Div Res Grants, NIH, 87-91, Nat Coun Radiation Protection & Measurements, sub-comt No 75, 90- *Mem:* Radiation Res Soc; AAAS; Am Soc Therapeut Radiologists; Ny Acad Sci. *Res:* Radiobiology of accelerated charged particle beams for radiotherapy; mechanisms of radiation damage in mammalian and insect cells in vitro. *Mailing Add:* Lawrence Berkeley Lab Univ Calif Bldg 70A-1120 Berkeley CA 94720

BLAKELY, J(OHN) M, b Scotland, Apr 8, 36; m 60; c 2. MATERIALS SCIENCE. *Educ:* Univ Glasgow, BS, 58, PhD(physics), 61. *Prof Exp:* Res fel appl physics, Harvard Univ, 61-63; from asst prof physics & mat sci to assoc prof mat sci, 63-70, PROF MAT SCI, CORNELL UNIV, 70- *Concurrent Pos:* Guggenheim fel, Cavendish Lab, Cambridge Univ, UK, 70-71; vis scientist, Argonne Nat Lab, 76; NSF fel, Lawrence Berkeley Lab, Calif, 77; fel, Sci & Eng Res Coun, York Univ, Eng, 84. *Mem:* Am Inst Mining, Metall & Petrol Engrs; fel Am Phys Soc; fel Brit Inst Physics & Phys Soc. *Res:* Properties and structure of interfaces; surface reactions; crystalline defects. *Mailing Add:* Dept Mat Sci Bard Hall Cornell Univ Ithaca NY 14853

BLAKELY, LAWRENCE MACE, b Los Angeles, Calif, Nov 12, 34; m 60; c 2. PLANT PHYSIOLOGY. *Educ:* Mont State Univ, BA, 56, MA, 58; Cornell Univ, PhD(plant physiol), 63. *Prof Exp:* Asst bot, Mont State Univ, 56-58; asst plant physiol, Cornell Univ, 58-62, res assoc, 62-63, instr, 63; PROF BOT, CALIF STATE POLYTECH UNIV, POMONA, 63- *Mem:* AAAS; Bot Soc Am; Am Soc Plant Physiologists; Plant Growth Regulator Soc Am; Sigma Xi; Am Inst Biol Sci. *Res:* Physiology of plant growth; lateral root initiation; plant tissue culture. *Mailing Add:* 415 Sierra Grande Rte 1 Bishop CA 93514

BLAKELY, ROBERT FRASER, b Newark, NJ, Apr 1, 21; m 43; c 2. GEOPHYSICS. *Educ:* Miami Univ, AB, 46, MA, 48; Ind Univ, PhD, 74. *Prof Exp:* Instr physics, Miami Univ, 47-49; spectrographer, Dept Conserv, State Geol Surv, Ind, 49-53, geophysicist, 53-74; from assoc prof to prof geophys, Ind Univ, 74-86; RETIRED. *Mem:* AAAS. *Res:* Spectrographic determination of major constituents in limestones; physical properties of Indiana sediments; computers in geology; solid earth geophysics and meteorology. *Mailing Add:* 116 Meadowbrook Ave Bloomington IN 47408-4107

BLAKEMORE, JOHN SYDNEY, b London, Eng, May 25, 27; m 53; c 2. SOLID STATE PHYSICS. *Educ:* Univ London, BSc, 48, PhD(physics), 51,. *Hon Degrees:* DSc, Univ London,77. *Prof Exp:* Res physicist, Stand Telecommun Labs, London, 50-54; res fel, Univ BC, 54-56; staff scientist, Honeywell, Inc, 56-63; sr staff scientist, Lockheed Res Labs, 63-64; prof physics, Fla Atlantic Univ, 64-78; prof appl physics, Ore Grad Ctr, 78-89; RES ASSOC, WESTERN WASH UNIV, 89- *Concurrent Pos:* Fulbright-Hays sr fel, Univ New South Wales, 71; dir, Fla Solar Energy Ctr, 76-77. *Mem:* Fel Am Phys Soc; fel Brit Inst Physics. *Res:* Semiconducting materials; photoconductivity, recombination and trapping; charge carrier statistics. *Mailing Add:* Physics & Astron Bond Hall 152-MS 9064 Western Wash Univ Bellingham WA 98225-9064

BLAKEMORE, RICHARD PETER, b Schenectady, NY, July 16, 42; m 67; c 2. MICROBIOLOGY. *Educ:* State Univ NY Albany, BS, 64, MS, 65; Univ Mass, Amherst, PhD(microbiol), 76. *Prof Exp:* Asst prof biol, North Adams State Col, Mass, 66-70; investr, Woods Hole Oceanog Inst, 75-76; res assoc microbiol, Univ Ill, Urbana, 76-77; asst prof microbiol, 77-80, assoc prof, 81-85, PROF, UNIV NH, 86- *Concurrent Pos:* NSF grants, 77-80 & 80-83, 83-85, 85-; Off Naval Res contracts, 79-81, 81-84, 85-; Nat Geog Soc grant, 80; lectr, Am Soc Microbiol found, CSM traveling, Am Soc Microbiol Div. *Honors & Awards:* Lectr, ASM Found, CSM Traveling, ASM Div. *Mem:* Am Soc Microbiol; AAAS; Biophys Soc; Can Soc Microbiol. *Res:* Magnetism in bacteria; denitrification. *Mailing Add:* Four Davis Ave Durham NH 03824

BLAKEMORE, WILLIAM STEPHEN, b Stockdale, Pa, June 22, 20; m 49; c 6. SURGERY. *Educ:* Washington & Jefferson Col, BS, 42; Univ Pa, MD, 45; Am Bd Surg, dipl, 52; Am Bd Thoracic Surg, dipl, 55. *Prof Exp:* Intern, Hosp Univ Pa, 45-46, asst resident, 48-51; asst instr, Sch Med, Univ Pa, 48-51, instr, 51-52, assoc, 52-54, from asst prof to prof, 54-73, asst chief surg clin, Hosp, Univ Pa, 55-62, from asst dir to assoc dir, Harrison Dept Surg Res, Univ, 56-73, from J William White asst prof to J William White assoc prof surg res, 56-62, chmn dept surg, 62-73, surgeon-in-chief, Grad Hosp, 62-73, chmn dept grad surg, 62-73; chmn dept, 73-79, chief surg, hosp, 73-79, prof surg, Med Col Ohio, 73-79; DIR SURG, BAPTIST MED CTR, BIRMINGHAM, 79- *Concurrent Pos:* Fel, Harrison Dept Surg, Univ Pa, 48-51, Damon Runyon cancer fel, 50-51; scholar, Am Cancer Soc, 52-57; Ravdin traveling fel, 53-54; vis fel, Karolinska Hosp, Sweden, 53-54; mem study sect pharmacol, NIH, 47, mem study sect surg, 65-69; attend physician, Philadelphia Gen Hosp, 52-73; assoc surg, Hosp Univ Pa, 52-73; chief surgeon, Emergency Am Med Team to Algeria, 62; mem comt blood & transfusion probs, Nat Res Coun Dean's Comt, Vet Admin Hosp, Wilmington, Del, 62-68; mem cent adv comt, Coun Cardiovasc Surg, Am Heart Asn, 66-73; life trustee, Washington & Jefferson Col. *Mem:* AAAS; Asn Cancer Res; Am Asn Thoracic Surg; Am Col Surg; Int Cardiovasc Soc (secy, 73-77, pres-elec, 77-78, pres, 78-). *Res:* Cardiopulmonary physiology; cancer immunology. *Mailing Add:* Dir Surg Educ Baptist Med Ctr 840 Montclair Rd Suite 310 Birmingham AL 35213

BLAKER, J WARREN, b Wilkes-Barre, Pa, May 25, 34; m 57, 74; c 3. OPTICS. *Educ:* Wilkes Col, BS, 55; Mass Inst Technol, PhD(chem), 58. *Prof Exp:* Asst prof physics, Fairleigh Dickinson Univ, 58-61; mem staff, G C Dewey Co, 61-62 & John Wiley & Sons, Inc, 62-64; prof physics, Vassar Col, 64-76; vpres res & develop, Danker & Wohlk, Inc, 76-79; consult, Optical Sci, 79-82; chief exec officer, Univ Optical, 82-85; PROF & CHMN, ELEC ENG DEPT, FAIRLEIGH DICKINSON UNIV, 87- *Concurrent Pos:* Consult, Optical Sci & Regulatory Affairs, 58- *Honors & Awards:* Eugene S Farley Award. *Mem:* AAAS; Am Phys Soc; Am Chem Soc; Am Asn Physics Teachers; fel Optical Soc Am. *Res:* Optics and optical image processing. *Mailing Add:* 3117 Palisades Ave Bronx NY 10463-4001

BLAKER, ROBERT HOCKMAN, b Meadow Bridge, WVa, Apr 6, 20; m 45; c 2. PHYSICAL CHEMISTRY, INFORMATION SCIENCE. *Educ:* Berea Col, AB, 42; Calif Inst Technol, PhD(chem), 49. *Prof Exp:* Instr chem, Berea Col, 43-44 & Off Naval Res, Calif Inst Technol, 46-49; Imp Chem Industs fel, Univ Liverpool, 49-50; res chemist, Jackson lab, E I du Pont de Nemours & Co, Inc, 50-53, res supvr, Petrol Lab, 53-56, asst dir lab, 56-60, res mgr, Orchem Dept, 60-64, head patent serv, 64-70, mgr cent patent & report serv, Info Systs Dept, 70-85; PRES, TECH INFO SERVS, INC, 85- *Honors & Awards:* Am Dyestuff Reporter Award, 53. *Mem:* Am Chem Soc; Am Soc Info Sci. *Res:* Size and shape of high polymer molecules; theory of dyeing synthetic fibers; fuels and lubricants for internal combustion engines. *Mailing Add:* 1105 N Rodney St Wilmington DE 19806

BLAKESLEE, A EUGENE, b Sayre, Pa, June 20, 28; m 54; c 2. PHYSICAL CHEMISTRY, ELECTRONIC MATERIALS SCIENCE. *Educ:* Pa State Univ, BS, 50; Cornell Univ, PhD(phys chem), 55. *Prof Exp:* Investr photoconductor develop, NJ Zinc Co, 55-57; mem tech staff semiconductor mat, Bell Tel Labs, 57-60; engr, Advan Semiconductor Lab, Gen Elec Co, 60-61; chemist, Components Div Lab, IBM Corp, 62-64 & World Trade Lab, 64-65, chemist, Res Div, 66-79; sr scientist, Solar Energy Res Inst, 79-89; RETIRED. *Concurrent Pos:* Mgt govt res contracts. *Mem:* Am Asn Crystal Growth. *Res:* X-ray crystallography; photoconductive zinc oxide powder; electrophotography; vapor and liquid phase epitaxial growth and diffusion in semiconductors; III-V compound semiconductor materials and devices; superlattices; solar cells; defects in semiconductor crystals; heteroepitaxy on silicon; metallic-organic chemical vapor deposition growth of superconductors. *Mailing Add:* 152 S Holman Way Golden CO 80401

BLAKESLEE, DENNIS L(AUREN), b Wilmington, Del, July 2, 38; m 68; c 1. SCIENCE WRITER. *Educ:* Middlebury Col, BA, 59; Columbia Univ, MS, 60; Univ Wis, MS, 67, PhD(genetics), 70. *Prof Exp:* Trainee, Dept Path, Sch Med, Univ Pa, 70-72; fel, 72-74, res scholar & asst prof, Nat Cancer Inst, Can, Dept Path, Queens Univ, 74-78; res assoc, dept genetics, Univ Wis, 78-80; dir sci info, Salk Inst, 80-83; sci writer, Lawrence Berkeley Lab, 83; Sci Writer, Scripps Clin & Res Found, 84-88; SCI ED, RES INST SCRIPPS CLIN, 89- *Mem:* AAAS; Genetics Soc Am; Am Asn Immunologists. *Res:* Non-technical articles on biomedical research and medical practice; editing and production of research publications. *Mailing Add:* Res Inst of Scripps Clin 10666 N Torrey Pines Rd La Jolla CA 92037

BLAKESLEE, GEORGE M, b Pa, Aug 30, 46; m 67; c 3. FOREST PATHOLOGY. *Educ:* Albright Col, BS; Duke Univ, MF, 70, PhD(forest path), 75. *Prof Exp:* Asst prof forest path, Univ Vt, 75-76; asst prof forest path, 76-81, ASSOC PROF FOREST PATH, UNIV FLA, 81- *Mem:* Am Phytopath Soc; Soc Am Foresters. *Res:* Etiology, epidemiology and control of diseases of intensively managed forest trees. *Mailing Add:* Sch Forest Resources & Conserv Univ Fla Gainesville FL 32611

BLAKEY, LEWIS HORRIGAN, b Burlington, NC, Nov 28, 33. HYDRAULICS, HYDROLOGY. *Educ:* Univ Notre Dame, BSCE, 54; George Washington Univ, MSE, 62; Catholic Univ, PhD(civil eng), 71; Univ Chicago, MBA, 74. *Prof Exp:* Asst prof civil eng, 65-67, dep chief, Spec Eng Div, Engr Studies Ctr, 68-70, chief, Eng Div, NCent Div, 71-76, dep dir, Facil Eng, 76-78, chief off policy, 78-80, CHIEF, PLANNING DIV, CIVIL WORKS, US ARMY CENGR, 80- *Mem:* Am Soc Civil Engrs; Soc Am Mil Engrs. *Res:* Operations research; numerical analysis. *Mailing Add:* Soc Am Mil Engrs 607 Prince St Alexandria VA 22314

BLAKLEY, BARRY RAYMOND, b Saskatoon, Sask, Mar 7, 49; m 82; c 3. IMMUNOTOXICOLOGY, VETERINARY TOXICOLOGY. *Educ:* Univ Sask, BSc, 70, DVM, 75, MSc, 77; Univ Cincinnati, PhD(toxicol), 80. *Prof Exp:* Assoc prof, Vet Med, Univ Sask, 80-86, acad coordr toxicol, 85-91, PROF VET MED, UNIV SASK, 86-, CHMN, TOXICOL GROUP, 89- *Mem:* Can Vet Med Asn; fel Am Acad Vet Comp Toxicol; Soc Toxicol Can; Am Asn Vet Lab Diagnosticians. *Res:* Effects of chemicals and altered nutritional states (deficiency) on immune function; applied veterinary toxicology and nutritional toxicology. *Mailing Add:* Dept Vet Physiol Sci Univ Sask Saskatoon SK S7N 0W0

BLAKLEY, GEORGE ROBERT, b Chicago, Ill, May 6, 32; m 57; c 3. MATHEMATICS, COMPUTER SCIENCE. *Educ:* Univ Md, PhD(math), 60. *Prof Exp:* Off Naval Res res assoc, Cornell Univ, 60-61; Nat Acad Sci-Nat Res Coun fel, Harvard Univ, 61-62; asst prof math, Univ Ill, 62-66; assoc prof, State Univ NY Buffalo, 66-70; head dept, 70-78, PROF MATH, TEX A&M UNIV, 70- *Concurrent Pos:* Mathematician, Nat Inst Standards & Technol, 56-58. *Mem:* AAAS; Am Math Soc; Inst Elec & Electronics Engrs; Asn Comput Mach; Int Asn Cryptologic Res. *Res:* Information theory; combinatorial theory; cryptography; secret sharing technology. *Mailing Add:* Dept Math Tex A&M Univ College Station TX 77843-3368

BLAKLEY, RAYMOND L, b Christchurch, NZ, May 14, 26; m; m 49; c 4. ANTICANCER DRUGS, ACTIVE SITE STRUCTURE. *Educ:* Canterbury Univ Col, BSc, 46, MSc, 47; Univ Otago, NZ, PhD(biochem), 51;. *Hon Degrees:* DSc, Austialian Nat Univ, 65. *Prof Exp:* Actg head dept biochem, Australian Nat Univ, 65-66; prof biochem, Col Med, Univ Iowa, 68-81; CHMN PHARMACOL & MEM, ST JUDE CHILDREN'S RES HOSP, 81- *Concurrent Pos:* Vis prof, Univ Calif, 62-63; prof fel, Australian Nat Univ, 65-68; Mayre guest prof, Univ Queensland, 83. *Mem:* Am Soc Biol Chemists; Am Chem Soc; Am Asn Cancer Res. *Res:* Folic acid metabolism; dihydrofolate reductase structure function relations; mechanism of cytotoxicity of deoxyadenosine analogues; binding of antifolates to dihydrofolate reductase. *Mailing Add:* Dept Pharmacol St Jude Children's Hosp 332 N Lauderdale Memphis TN 38101

BLAKNEY, WILLIAM G G, b Moncton, NB, Apr 4, 26; m 51; c 2. CIVIL ENGINEERING, PHOTOGRAMMETRY. *Educ:* NS Tech Col, BE, 49; Ohio State Univ, MSc, 54. *Prof Exp:* Engr surv mapping, Topog Surv Div, Mines & Tech Surv, Can Govt, 50-58; ASSOC PROF INDUST ENG, AUBURN UNIV, 58-, SR DANFORTH ASSOC, 65- *Mem:* Am Soc Photogram & Remote Sensing; Am Cong Surv & Mapping; Am Soc Eng Educ; Can Inst Surv. *Res:* Surveying; photo interpretation; geology; computer graphics; numerical control. *Mailing Add:* Dept Indust Eng Auburn Univ Auburn AL 36849

BLALOCK, J EDWIN, b Madison, Fla, Sept 29, 49. VIROLOGY, IMMUNOLOGY. *Educ:* Univ Fla, Gainesville, BS, 71, PhD(virol & immunol), 76. *Prof Exp:* Postdoctoral virol & immunol, Univ Tex Med Br, Galveston, 76-77; PROF, DEPT PHYSIOL & BIOPHYS, UNIV ALA, BIRMINGHAM. *Res:* Bidirectional communication between the immune and neuroendocrine systems; immune system derived neuroendocrine hormones. *Mailing Add:* Dept Physiol & Biophys Univ Ala UAB Sta Birmingham AL 35294-0005

BLALOCK, JAMES EDWIN, b Madison, Fla, Sept 29, 49. VIROLOGY, CELL BIOLOGY. *Educ:* Univ Fla, BS, 72, PhD(immunol & med microbiol), 76. *Prof Exp:* Fel, 76-77, asst prof virol, 77-79, assoc prof, 79-84, PROF MICROBIOL, UNIV TEX MED BR, GALVESTON, 84- *Concurrent Pos:* Prin investr, US Army Med Res & Develop Command Contract, 78--81; NIH grants, 81-; Off Naval Res grant, 84- *Mem:* Am Soc Microbiol; Am Asn Immunologists. *Res:* Interferon; vitamin A; cell biology; immunology; physiology. *Mailing Add:* Dept Physiol & Biophysiol Univ Ala Sch Med Birmingham AL 35294

BLALOCK, THERON VAUGHN, b Fayette Co, Tenn, Jan 9, 34; m 52; c 3. ELECTRICAL ENGINEERING. *Educ:* Univ Tenn, BS, 57, MS, 61, PhD(eng sci), 65. *Prof Exp:* Develop engr, Oak Ridge Nat Lab, 58-59, consult, 59-63, res engr, 63-68; instr elec eng, 59-63, PROF ELEC ENG, UNIV TENN, KNOXVILLE, 68- *Mem:* Inst Elec & Electronics Engrs. *Res:* Measurement science; nuclear instrumentation; Johnson noise thermometry; electronics for inertial sensors. *Mailing Add:* Dept Elec Eng Univ Tenn Knoxville TN 37996

BLANC, FRANCIS LOUIS, b Knoxville, Tenn, Aug 22, 16; m 38; c 2. ENTOMOLOGY. *Educ:* Univ Calif, BS, 38. *Prof Exp:* Inspector entom & plant quarantine, State Dept Agr, Calif, 40-43 & 45-47, syst entomologist, 47-58, supvr insect pest detection & surv, 58-71, prog supvr pest prevention, 71-75; CONSULT, 75- *Mem:* Entom Soc Am. *Res:* Systematics of diptera, tephritidae. *Mailing Add:* 5309 Spilman Ave Sacramento CA 95819

BLANC, FREDERIC C, b New York, NY, June 21, 39; m 63; c 2. WASTEWATER TREATMENT, INDUSTRIAL WASTE TREATMENT. *Educ:* City Col, City Univ New York, BCE, 62; NY Univ, MS, 66, PhD(civil eng), 69. *Prof Exp:* Engr, John Thatcher & Son Engrs, 62; sanitary engr, US Pub Health Serv, 62-64; lectr, City Col, City Univ New York, 64-65; fel, NY Univ, 65-69; engr, Bowe Walsh & Assocs, 69; PROF CIVIL ENVIRON ENG, NORTHEASTERN UNIV, 69- *Concurrent Pos:* Consult, 69-; water comnr, Town in Harvard, 75-; bd mem, Town of Harvard, Bd Health, 75- *Mem:* Am Soc Civil Engrs; Water Pollution Control Fedn. *Res:* Environmental engineering: methane fermentation, rotating biological contactor applications, river modeling, combine sewer management studies and synfuel waste treatment. *Mailing Add:* Dept Civil Eng Northeastern Univ 360 Huntington Ave Boston MA 02115

BLANC, JOSEPH, b Cernauti, Romania, Mar 12, 30; US citizen; m 49; c 3. SOLID STATE SCIENCE, PHYSICAL CHEMISTRY. *Educ:* Columbia Col, AB, 54; Columbia Univ, MA, 55, PhD(phys chem), 59. *Prof Exp:* Res asst infrared spectros, Columbia Univ, 56-57 & 58-59; mem tech staff, RCA Labs, 59-87, INDEPENDENT CONSULT, 87-; RES SCIENTIST, PRINCETON X-RAY LASER, INC. *Concurrent Pos:* Res assoc, Optic Electronics Lab, CNRS, Toulouse, France, 74-75. *Mem:* Mat Res Soc; Am Phys Soc. *Res:* Chemical physics of defects in solids and in the spectroscopy of the lanthanide ions; study of defects in semiconductors; scanning capacitance microscopy; composite materials; non linear phenomena; oxidation kinetics; novel microscopies and non-destructive testing. *Mailing Add:* 12 Willow St Princeton NJ 08542

BLANC, ROBERT PETER, b Pittsburgh, Pa, June 8, 46; m 69; c 2. COMPUTER SCIENCE. *Educ:* Univ Pittsburgh, BS, 67, MS, 69. *Prof Exp:* Asst dir, Comput Ctr, Univ Pittsburgh, 69-72; proj mgr comput sci & technol, Nat Bur Standards, 72-74, sci prog analyst, 74-75, staff asst, Comput Utilization Prog, 75-78, actg chief, Prog Develop Off Comput Sci & Technol, 78-79, CHIEF, SYSTS & NETWORK ARCHIT DIV, NAT BUR STANDARDS, 79- *Concurrent Pos:* Mem task force, President's Comt on Right to Privacy, 73-74; course dir, AAAS, 76-77; mem, President's Reorgn Proj, 78-79. *Honors & Awards:* Silver Medal, Dept Com, 78. *Mem:* AAAS; Inst Elec & Electronics Engrs; Asn Comput Mach. *Res:* Computer network protocols; network access; network measurement; computer architecture; distributed processing; local area networking; automated office systems. *Mailing Add:* 20920 Lockhaven Ct Gaithersburg MD 20879

BLANC, WILLIAM ANDRE, b Geneva, Switz, Sept 28, 22; nat US; m 54; c 1. PATHOLOGY. *Educ:* Univ Geneva, BA, 40, MD, 47, MedScD, 52; Am Bd Path, dipl. *Prof Exp:* Resident path, Univ Geneva, 47-50, assoc, 50-53; fel pediat path, Harvard Univ, 53-54; instr, 54-57, assoc prof, 61-66, PROF PATH, COL PHYSICIANS & SURGEONS, COLUMBIA UNIV, 66- *Concurrent Pos:* From asst attend pathologist to assoc attend pathologist, Columbia-Presby Med Ctr, 57-67, attend pathologist, 67-; dir path, Babies Hosp; consult, USPHS. *Honors & Awards:* Laureate, Fac Med, Univ Geneva, 51. *Mem:* AAAS; Am Pediat Soc; fel Am Soc Clin Path; Int Acad Path; AMA. *Res:* Pathology in pediatrics and obstetrics; placenta diseases of prematurity; fetal medicine; morphometrics of cardiovascular adaptation at birth. *Mailing Add:* Eight Robert-de-Traz CH-1206 Geneva Switzerland

BLANCH, HARVEY WARREN, b Sydney, Australia, Jan 17, 47. BIOCHEMICAL ENGINEERING, CHEMICAL ENGINEERING. *Educ:* Univ Sydney, BSc, 68; Univ New SWales, PhD(chem eng), 71. *Prof Exp:* Asst I tech microbiol, Swiss Fed Inst Technol, Zurich, 71-73; res investr, E R Squibb & Co, 73-74; assoc prof chem eng, Univ Del, 75-78; assoc prof, 78-82, PROF CHEM ENG, UNIV CALIF, BERKELEY, 82-, SR FAC SCIENTIST, LAWRENCE BERKELEY LAB, 84- *Concurrent Pos:* Ed, Bioprocess Eng & Biochem Eng J. *Mem:* Am Inst Chem Engrs; Am Chem Soc. *Res:* Fermentation systems; reaction kinetics and mass transfer in biological systems; enzyme engineering. *Mailing Add:* Dept Chem Eng Univ Calif Berkeley CA 94720

BLANCHAER, MARCEL CORNEILLE, b Antwerp, Belg, Apr 4, 21; nat Can; c 2. BIOCHEMISTRY. *Educ:* Queen's Univ, Ont, BA, 45, MD & CM, 46. *Prof Exp:* Assoc prof physiol, 48-64, head dept biochem, 64-75, PROF BIOCHEM, UNIV MAN, 64- *Concurrent Pos:* Dir biochem, St Boniface Hosp, 54-64; Markle scholar, 48-53. *Mem:* Can Biochem Soc; Can Soc Clin Chem. *Res:* Medical education. *Mailing Add:* Dept Biochem Univ Man Fac Med 753 McDermot Ave Winnepeg MB R3E 0W3 Can

BLANCHAR, ROBERT W, b Sherburn, Minn, June 30, 37; m 58; c 2. SOIL CHEMISTRY, BIOCHEMISTRY. *Educ:* Macalester Col, BA, 58; Univ Minn, MS, 61, PhD(soil sci), 64. *Prof Exp:* Res agronomist soil chem, Int Minerals & Chem Corp, 64-68; PROF AGRON, UNIV MO-COLUMBIA, 68- *Mem:* Am Soc Agron; Sigma Xi; Fel Soil Sci Am. *Res:* Polyphosphate chemistry in soils; ion products of iron, aluminum and manganese in soils; arsenic transformations in soils; chemical transformations in coal mine spoils; chemistry of the rhizosphere. *Mailing Add:* Sch Nat Resources Univ Mo 144 Mumford Columbia MO 65211

BLANCHARD, BRUCE, b Ft Stotsenburg, Phillippines, Dec 26, 32; US citizen. ENVIRONMENTAL SCIENCES. *Educ:* Mass Inst Technol, SB, 57, SM, 64. *Prof Exp:* Teaching asst surv, Mass Inst Technol, 57-59; hydraul engr, chief engr off, US Bur Reclamation, Colo, 59-61, Phoenix develop off, Ariz, 61-63, civil engr, 63-64, proj planning engr, Cent Ariz Proj, 65-66; staff specialist, Water Resources Coun, 66-69; staff asst, Off Secy of Interior, 70-71, dir off environ proj rev, Off Secy, 71-89, DEP DIR, US FISH & WILDLIFE SERV, DEPT INTERIOR, 89- *Mem:* Am Soc Civil Engrs; Am Geophys Union; Am Water Resources Asn; AAAS. *Res:* Analysis of resource allocation and environmental problems in the management of natural resources and the protection and enhancement of environmental quality. *Mailing Add:* US Fish & Wildlife Serv Dept Interior Washington DC 20240

BLANCHARD, CONVERSE HERRICK, b Boston, Mass, Sept 25, 23; m 46; c 4. NUCLEAR PHYSICS. *Educ:* Harvard Univ, AB, 45; Univ Wis, PhD(physics), 50. *Prof Exp:* Nuclear physicist, Nat Bur Standards, 50-53; assoc prof physics, Pa State Univ, 53-61; assoc prof, 61-63, PROF PHYSICS, UNIV WIS-MADISON, 63-, ASSOC CHMN DEPT, 75- *Mem:* Fel Am Phys Soc. *Res:* Theoretical nuclear and quantum physics. *Mailing Add:* Dept Physics Univ Wis Sterling Hall Madison WI 53706

BLANCHARD, DUNCAN CROMWELL, b Winterhaven, Fla, Oct 8, 24; m 57; c 4. AIR-SEA INTERACTION, AEROBIOLOGY. *Educ:* Tufts Univ, BNavalSc, 45, BS, 47; Pa State Univ, MS, 51; Mass Inst Technol, PhD(meteorol), 61. *Prof Exp:* Res assoc, Res Lab, Gen Elec Co, 48-49; assoc scientist, Woods Hole Oceanog Inst, 51-68; sr res assoc, Atmospheric Sci Res Ctr, State Univ NY, Albany, 68-89; RETIRED. *Concurrent Pos:* Mem joint panel air & air sea interaction, Nat Acad Sci, 63-65; mem steering comt, Ocean Sci Comt Study Ocean Dumping, Nat Acad Sci-Nat Res Coun, 74-76; consult, NJ Cent Power & Light Co & McDonnell Douglas Astronaut Co; chmn, electorate nominating comt, AAAS, 80-84. *Mem:* Fel AAAS; Sigma Xi; fel Am Meteorol Soc; Am Geophys Union. *Res:* Aerosol production by the sea, bubble production in the sea, and flux of charged particles across the surface; ejection of dissolved organic matter and bacteria from water surfaces into the atmosphere. *Mailing Add:* 32 Highland Dr Albany NY 12203

BLANCHARD, FLETCHER A(UGUSTUS), JR, b Cooperstown, NY, Nov 13, 24; m 47; c 2. ELECTRICAL ENGINEERING. *Educ:* Union Col, BA, 48; Lehigh Univ, MS, 50. *Prof Exp:* Instr elec eng, Lafayette Col, 48-50; from instr to prof elec eng, Univ NH, 68-88; RETIRED. *Mem:* Inst Elec & Electronics Engrs; Am Soc Eng Educ; Marine Technol Soc. *Res:* Instrumentation and control; ocean engineering. *Mailing Add:* PO Box 668 Northville NY 12134

BLANCHARD, FRANK NELSON, b Ann Arbor, Mich, May 24, 31; m 53; c 4. GEOLOGY. *Educ:* Univ Mich, AB, 53, MS, 54, PhD(geol), 60. *Prof Exp:* From asst prof to assoc prof, 56-76, PROF GEOL, UNIV FLA, 76- *Mem:* Mineral Soc Am; Nat Asn Geol Teachers; Mineral Asn Can. *Res:* Mineralogy; crystallography; petrology; petrography; x-ray analysis; x-ray powder diffraction standards for phosphate and arsenate minerals; mineralogy and geochemistry of magnesium-rich phosphorites from Florida. *Mailing Add:* Dept Geol Univ Fla Gainesville FL 32611

BLANCHARD, FRED AYRES, b Cleveland, Ohio, May 27, 23; m 48; c 3. BIOPHYSICS. *Educ:* La State Univ, BS, 44; Univ Cincinnati, MS, 48, PhD(physics), 51. *Prof Exp:* Radiochemist tracers, Dow Chem Co, 51-74, Environ Scientist, 74-86; RETIRED. *Mem:* Soc Environ Toxicol & Chem; Am Chem Soc; Sigma Xi. *Res:* Applications of radioisotopes to biological, chemical and industrial tracer problems; liquid scintillation counting; computer processing of counting data; low level counting; environmental science. *Mailing Add:* 39 Rosemary Ct Midland MI 48640

BLANCHARD, GORDON CARLTON, b Monroe, NH, Dec 5, 32; m 53; c 2. MICROBIOLOGY. *Educ:* Univ Vt, BS, 56, MS, 58; Syracuse Univ, PhD(microbiol), 61. *Prof Exp:* Sr scientist microbiol, Melpar, Inc Div, Westinghouse Air Brake Co, 61-62, br head, 62-67; res microbiologist, 67-78, CHIEF, CLIN IMMUNOL, VET ADMIN MED CTR, 78- *Mem:* AAAS; Am Soc Microbiol; Am Asn Immunologists. *Res:* Characterization of ragweed pollen allergens and study of hypersensitive reactions; automated methods for microbiology; mechanism of nonresponsiveness of ragweed aerosols. *Mailing Add:* Vet Admin Med Ctr 150 S Huntington Ave Boston MA 02130

BLANCHARD, JAMES, b Toronto, Ont, Mar 12, 40; m 62; c 1. PHARMACEUTICS, PHARMACOKINETICS. *Educ:* Univ Toronto, BScPharm, 63, MScPharm, 66; Univ Calif, San Francisco, PhD(biopharmaceut), 71. *Prof Exp:* Teaching asst, Univ Toronto, 63-65, Univ Calif, San Francisco, 66-68; assoc prof, 71-87, PROF PHARMACEUT SCI, COL PHARM, UNIV ARIZ, 87- *Concurrent Pos:* Prin investr res grants, 78-81; vis prof, Dept Clin Pharmacol, Royal Infirmary, Scotland, 80-81,

Nestle Res Ctr, Lausanne, Switz, 87- 88; Fogarty Sr Int Fel Award, Fogarty Int Ctr & Nat Inst Aging, 80-81; chmn-elect, Acad Pharmaceut Res & Sci Sect, Am Pharmaceut Asn, 90. *Mem:* Fel Am Pharmaceut Asn; Am Asn Col Pharm; Sigma Xi; fel Am Col Clin Pharmacol; fel Am Col Nutrit; Acad Pharmaceut Res & Sci. *Res:* Drug absorption; protein binding; drug-complexation interactions; effect of nutrition on drug metabolism; pharmacokinetics of drugs and nutrients in the elderly; formulation of controlled-release ocular drug-delivery systems. *Mailing Add:* Col Pharm Univ Ariz Tucson AZ 85724

BLANCHARD, JONATHAN EWART, b Truro, NS, Mar 22, 21; m 58; c 2. GEOPHYSICS. *Educ:* Dalhousie Univ, BSc, 40; Univ Toronto, MA, 47, PhD(geophys), 52. *Prof Exp:* Dir geophys, 49-66, vpres, 66-68, PRES, NS RES FOUND CORP, 68- *Concurrent Pos:* From asst prof to prof geophys, Dalhousie Univ, 52-66. *Mem:* Soc Explor Geophysicists; Am Geophys Union; Can Asn Physicists; Can Inst Mining & Metall; fel Royal Soc Can. *Mailing Add:* 6470 Cobourg Rd Halifax NS B3H 2A7 Can

BLANCHARD, RICHARD LEE, b Hutchinson, Kans, May 2, 33; m 55, 73; c 2. RADIOCHEMISTRY, ECOLOGY. *Educ:* Ft Hays Kans State Col, BS, 55; Vanderbilt Univ, MS, 58; Washington Univ, PhD(chem), 63. *Prof Exp:* Chemist, Div Radiol Health, USPHS, Oak Ridge Nat Lab, 56-58, health physicist, Utah, 58-59, chemist, Robert A Taft Sanit Eng Ctr, 62-64, chg physics & bioassay group, Radiol Health Res Activities, 64-68, dep chief radiol eng lab, Bur Radiol Health, Environ Control Admin, Ohio, 68-74; dir, Radiochem & Nuclear Eng Facil, US Environ Protection Agency, 74-77; CHIEF, RADIOCHEM SPEC STUDIES BR, EASTERN ENVIRON RADIATION FACIL, MONTGOMERY, ALA, 77- *Concurrent Pos:* Adj prof nuclear eng, Univ Cincinnati, 68-77; mem N 13 subcomt, Am Nat Standards Inst, 74-76; adj prof, Auburn Univ, Montgomery, 78- *Mem:* Health Physics Soc; Sigma Xi. *Res:* Radiation dose to man from naturally occurring radionuclides and from radionuclides in the environment of nuclear power reactors; movement and concentration of radionuclides in aquatic environments. *Mailing Add:* 102 Saccapatoy Dr Montgomery AL 36117

BLANCHARD, ROBERT OSBORN, b Cumberland Center, Maine, July 5, 39; m 65; c 2. BOTANY, PHYTOPATHOLOGY. *Educ:* Southern Maine Voc Tech Inst, AAS, 59; Univ Maine, Portland-Gorham, BS, 64; Univ Ga, MEd, 69, PhD(plant path), 71. *Prof Exp:* Fel plant path, Univ Ga, 71-72; from asst prof to assoc prof, 72-84, PROF BOT & PLANT PATH, UNIV NH, 84- *Concurrent Pos:* Dept chmn bot & plant path, Univ NH, 73-76, chmn, Intercol Biol Sci Orgn, 76-78, assoc dean, Col Life Sci & Agr, 82- *Mem:* Sigma Xi; Mycol Soc Am; Am Phytopath Soc; Smithsonian Assocs. *Res:* Light and electron microscopic observations of fungi; host parasite interactions associated with tree diseases using electrophysiological techniques. *Mailing Add:* Dean's Off Taylor Hall Univ NH Durham NH 03824

BLANCHE, ERNEST EVRED, b Passaic, NJ, Oct 22, 12; m 38; c 2. MATHEMATICS, STATISTICS. *Educ:* Bucknell Univ, AB & AM, 38; Univ Ill, PhD(math), 41. *Prof Exp:* Asst math, Univ Ill, 38-41; instr math & statist, Mich State Univ, 41-42; statist dir, Curtiss-Wright Corp, NY, 42-43, head math & statist lab, 43-44; prin statistician, Foreign Econ Admin, 44-45; instr, US Army Univ, Italy, 45, prin admin analyst, Army Serv Forces, 46-47; chief statist br, Res & Develop Div, Gen Staff, US War Dept, 47-48; chief statistician, Logistics Div, US Dept Army, 48-54, vpres for res, Frederick Res Corp, Bethesda, Md, 54-55; pres & chief res scientist, Blanche & Assocs, Inc, Kensington, 55-70; chmn bd trustees, Capitol Inst Technol, 70-74, actg pres, 71-72; PRES, ERNEST E BLANCHE CONSULTS, CHEVY CHASE, 74- *Concurrent Pos:* Army mem subpanel math & statist, Comt Basic Phys Sci, Res & Develop Bd, 49; consult logistics div, US Dept Army, 54-; adj prof math & statist, Am Univ, 46-66; prof math & statist & trustee, Capitol Inst Technol, 69-70. *Mem:* Am Math Soc; Math Asn Am; Am Statist Asn; Inst Math Statist. *Res:* Probability; choice and chance; work simplification and work measurement; logistics; computer research and applications; system design; programming engineering computations; traffic projections. *Mailing Add:* 14818 Carrolton Rd Manor Club Rockville MD 20853

BLANCHET, WALDO W E, b New Orleans, La, Aug 6, 10; m 43; c 2. SCIENCE EDUCATION. *Educ:* Talladega Col, AB, 31; Univ Mich, MS, 36, PhD(sci educ), 46. *Prof Exp:* Head sci dept, Ft Valley Normal & Indust Sch, 32-35, head dept & dean sch, 36-38; prof sci & admin dean, 39-66, pres, 66-73, EMER PRES, FT VALLEY STATE COL, 73- *Concurrent Pos:* Consult & lectr, NSF sci insts, Atlanta Univ & Albany State Col; consult spec proj, Tenn Valley Authority, 63-64; mem, Nat Adv Coun on Educ Disadvantaged Children, 70-; consult sci educ & higher educ, 73- *Mem:* AAAS; Nat Inst Sci; Nat Asn Res Sci Teaching. *Res:* Curriculum problems of general science education on college level. *Mailing Add:* 508 Camelot Dr College Park GA 30349

BLANCHET-SADRI, FRANCINE, b Trois-Rivieres, Que, July 25, 53; m 79; c 3. GAME THEORY, DISCRETE MATHEMATICS. *Educ:* Univ Que, Trois-Rivieres, BSpSc, 76; Princeton Univ, MS, 79; McGill Univ, PhD(math), 89. *Prof Exp:* Lectr, Univ Technol Isfahan, Iran, 82-84; teaching asst, McGill Univ, 84-88, lectr, 88-89, researcher, 89-90; res fel, Nat Sci & Eng Res Coun Can, 90; ASST PROF, UNIV NC, GREENSBORO, 90- *Concurrent Pos:* Lectr, Univ Que, Trois-Rivieres, 76; prin investr, NSF, 91- *Mem:* Am Math Soc. *Res:* Decidability problem of the dot-depth hierarchy, more specifically on some congruences known to characterize the different levels of the hierarchy. *Mailing Add:* Dept Math Univ NC Greensboro NC 27412

BLANCHETTE, ROBERT ANTHONY, b Lowell, Mass, July 30, 51; m 75; c 2. FOREST PATHOLOGY, PLANT PATHOLOGY. *Educ:* Merrimack Col, BA, 73; Univ NH, MS, 75; Wash State Univ, PhD(plant path), 78. *Prof Exp:* Lab asst forest path, USDA Forest Serv, Northeastern Sta, 74-75; res asst plant path, Wash State Univ, 75-77, res assoc, 77-80; from asst prof to assoc prof, 80-88, PROF, DEPT PLANT PATH, UNIV MINN, 88- *Mem:* Sigma Xi; Am Phytopath Soc; Mycol Soc Am; Soc Am Foresters; Am Soc Microbiol; Int Soc Arboricult; fel Int Soc Wood Sci. *Res:* Discoloration and decay of living trees; forest products deterioration; diseases of urban and forest trees; biological delignification; dutch elm disease; canker diseases; stem rusts of hard pines; pinewood nematode; immunocytochemistry; host-pathogen interactions. *Mailing Add:* Dept Plant Path 495 Borlang Hall Univ Minn 1919 Buford Circle St Paul MN 55108

BLANCK, ANDREW R(ICHARD), b NY, Feb 7, 25; m 50; c 2. MATERIALS SCIENCE, ELECTRICAL ENGINEERING. *Educ:* NY Univ, BA, 48; Polytech Inst Brooklyn, MA, 53; Sussex Inst Technol, ScD. *Prof Exp:* Chem engr, Interchem Corp, 48-49; chemist, Crownoil Chem Corp, 50-51; consult, Centro Res Labs, 51-53; tech serv engr, Celanese Corp Am, 53-56; chief chemist, Applications Lab & head Dielectrics Lab, Polymer Chem Div, W R Grace & Co, 56-58; mat engr, electromagnetic interference engr & coordr & consult, Ord Corps, Picatinny Arsenal, Dover, NJ, 57-66; PRES, RUTHERFORD RES PROD CO, 58-; PRES, PRINCETON RES PROD CO, 64-, CONSULT, 50-; DIR PUB RELS, CLEARING HOUSE FOR CONSULT, NY, 66- *Concurrent Pos:* Dir pub rels, Assoc Consult Chem & Chem Engr, Inc, 66-; trustee, Hotel Salisbury, 66-68; fac mem, NY Univ, 68- *Mem:* Am Soc Testing & Mat; Inst Elec & Electronics Engrs; Am Chem Soc; Soc Plastic Engrs; Chem Industs Asn. *Res:* Plastics; electrical insulating systems and materials; surface coatings; plasticizers; instrumentation design for research and control purposes; adhesives. *Mailing Add:* Rutherford Res Corp Drawer 249 Rutherford NJ 07070

BLANCK, HARVEY F, JR, b North Ridgeville, Ohio, Apr 4, 32; m 60; c 2. PHYSICAL CHEMISTRY, ANALYTICAL CHEMISTRY. *Educ:* Miami Univ, BA, 54; Ohio State Univ, MS, 62, PhD(chem), 67. *Prof Exp:* PROF CHEM, AUSTIN PEAY STATE UNIV, 64- *Mem:* Am Chem Soc. *Res:* Microcomputer interfacing. *Mailing Add:* Dept Chem Austin Peay State Univ Clarksville TN 37044

BLANCO, BETTY M, b New York, NY, Dec 16, 27; m 71; c 2. ASTRONOMY. *Educ:* Cornell Univ, BA, 49, MA, 55. *Prof Exp:* Astronr, US Naval Observ, 60-71; ASTRONR, CERRO TOLOLO INTER-AM OBSERV, 77- *Mem:* Am Astron Soc; Int Astron Union. *Res:* Search for and discover, determine periods and magnitudes of RR hyvae variables. *Mailing Add:* Cerro Tololo Inter-Am Observ Casilla 603 La Serena Chile

BLANCO, VICTOR MANUEL, b Guayama, PR, Mar 10, 18; m 43, 71; c 3. ASTRONOMY. *Educ:* Univ Chicago, BS, 43; Univ Calif, MA, 47, PhD(astron), 49. *Hon Degrees:* DSc, Univ PR, 72. *Prof Exp:* From asst prof to assoc prof physics, Univ PR, 49-51; from instr to prof astron, Case Inst Technol, 51-65; dir astrometry & astrophys div, US Naval Observ, Washington, DC, 65-67; dir, 65-80, ASTRONR, CERRO TOLOLO INTER-AM OBSERV, 81- *Concurrent Pos:* Mem comn galactic struct, photomet and spectros, Int Astron Union; corresp mem, Chilean Acad Sci. *Honors & Awards:* Distinguished Pub Serv Award, US NSF, 81. *Mem:* Am Astron Soc (vpres, 72-75). *Res:* Galactic structure and Magellanic clouds research. *Mailing Add:* Cerro Tololo Inter-Am Observ Casilla 603 La Serena Chile

BLAND, BRIAN HERBERT, b Calgary, Alta, July 10, 43; m 64; c 2. NEUROBIOLOGY, NEUROPSYCHOLOGY. *Educ:* Univ Calgary, BSc, 66, MSc, 68; Univ Western Ont, PhD(neuropsychol), 71. *Prof Exp:* NATO sci fel, Inst Neurophysiol, Oslo, Norway, 71-73; asst prof, Univ Sask, 73-74; asst prof neuropsychol, 74-77, assoc prof, 77-81, PROF PSYCHOL, UNIV CALGARY, 81- *Concurrent Pos:* mem, Psychol Grants Comt, Natural Sci & Eng Res Coun, 82-85; Killam res fel, 85. *Honors & Awards:* Int Sci Exchange Award, Nat Sci & Eng Res Coun Can, 90. *Mem:* Can Physiol Soc; AAAS; NY Acad Sci; Soc Neurosci; Int Brain Res Orgn; World Fedn Neuroscientists. *Res:* Electrophysiological and pharmacological investigation of hippocampal formation function. *Mailing Add:* Behav Neurosci Res Group Dept Psychol Univ Calgary 2500 University Dr Calgary AB T2N 1N4 Can

BLAND, CHARLES E, b Raleigh, NC, Aug 17, 43; m 63; c 2. MYCOLOGY. *Educ:* Univ NC, Chapel Hill, AB, 64, PhD(mycol), 69. *Prof Exp:* From asst prof to assoc prof, 69-75, PROF BIOL, ECAROLINA UNIV, 76-, CHMN DEPT, 81- *Mem:* Mycol Soc Am; Soc Invert Path; World Maricult Soc. *Res:* Fungi; parasitic in marine and freshwater invertebrates. *Mailing Add:* Dept Biol ECarolina Univ Greenville NC 27858

BLAND, CLIFFORD J, b London, Eng, June 22, 36; m 63; c 2. COSMIC RAY PHYSICS, RADIATION PHYSICS. *Educ:* Univ London, BSc, 58, PhD(physics), 61; Imp Col, Univ London, DIC, 61. *Prof Exp:* Res fel physics, Lab Cosmic Physics, La Paz, 62-63; res fel, Inst Physics, Milan, 64-68; assoc prof, 68-77, PROF PHYSICS & EXEC DIR TO DEAN FAC SCI, UNIV CALGARY, 77- *Concurrent Pos:* Lectr, La Paz, 62-63 & Milan, 64-68; consult, Nat Inst Nuclear Physics, Italy, 64-68; Nat Res Coun Can res grants, 68-75. *Mem:* Am Geophys Union; assoc Royal Col Sci. *Res:* Geomagnetic effects; primary electrons; modulation theory; terrestrial radioactivity. *Mailing Add:* Dept Physics Univ Calgary 2500 University Dr NW Calgary AB T2N 1N4 Can

BLAND, HESTER BETH, b Sullivan, Ind, June 22, 06. HEALTH SCIENCES. *Educ:* Ind State Univ, BS, 42; Butler Univ, MS, 49; Ind Univ, HSD, 56. *Hon Degrees:* LLD, Ind State Univ, 73. *Prof Exp:* Supvr & teacher pub schs, Ind, 30-41, teacher, 42-47; supvr plant protection, Allison Div, Gen Motors Corp, 41-42; consult health educ, Ind State Bd Health, 47-72; ADJ PROF, IND STATE UNIV, 72- *Concurrent Pos:* Lectr, Div Allied Sci, Sch Med, Ind Univ, 50-72 & Dept Health Educ, 58-; dir, Am Sch Health Asn & Pharmaceut Mfrs Asn Drug Educ Curric Proj, 69-70; consult ed, J Sch Health, 74-; consult, Med Arts Publ Found, 77-78 & McGovern Allergy Clin, Houston, 79- *Honors & Awards:* William A Howe Award, Am Sch Health Asn, 75. *Mem:* Fel Am Sch Health Asn (pres, 72-73); Am Asn Health, Phys Educ & Recreation. *Res:* Critical areas of health education, especially venereal disease, drug misuse, smoking and alcohol; school and public health. *Mailing Add:* 2511 Parkwood Indianapolis IN 46224

BLAND, JEFFREY S, b Peoria, Ill, Mar 21, 46; m; c 3. BIO-ORGANIC CHEMISTRY. *Educ:* Univ Calif, Irvine, BS, 67; Univ Ore, PhD(chem), 71. *Prof Exp:* Asst prof, 71-77, ASSOC PROF CHEM, UNIV PUGET SOUND, 77- *Concurrent Pos:* Vis prof, Univ Hawaii, 74; sr res fel, Univ Ore, 74 & 75; clin lab dir, Nutrit Biochem, Bellevue-Redmond Med Lab, Evergreen State Col, 77- *Honors & Awards:* Henry Schroeder Trace Mineral Res Award, 80. *Mem:* NY Acad Sci; Nat Acad Clin Biochem; AAAS; Am Chem Soc; Am Asn Clin Chemists. *Res:* Investigation of the effects of biological antioxidants upon the rate of photohemolysis of erythrocytes; examination of mechanisms of photodynamic processes and relation to membrane structure; trace minerals in human health and disease. *Mailing Add:* 5800 Soundview Dr Gig Harbor WA 98335

BLAND, JOHN (HARDESTY), b Globe, Ariz, Nov 7, 17; m 44; c 4. MEDICINE. *Educ:* Earlham Col, AB, 40; Jefferson Med Col, MD, 44; Am Bd Int Med, dipl, 52 & cert, internal med, 84. *Prof Exp:* Resident, Burlington County Hosp & Pa Hosp, Philadelphia, 44-46; chief cardiovasc sect, Sta Hosp, Ft Hood, Tex, 46-48; res fel cardiol, Col Med, Univ Vt, 48-49, from instr to prof med, 49-89, dir, Rheumatism Res Unit, 50-69, EMER PROF MED & RHEUrMATOLOGY, COL MED, UNIV VT, 89- *Concurrent Pos:* Fel rheumatic dis, Mass Gen Hosp & New Eng Ctr Hosp, Boston, 50; hon res fel rheumatol, Univ Manchester, 58-59; vis sci worker, NIH, 72-73. *Mem:* AAAS; fel Am Col Physicians; Am Fedn Clin Res; NY Acad Sci; Am Rheumatism Asn; Sigma Xi; master Am Col Rheumatology. *Res:* Rheumatic disease; connective tissue metabolism; cervical spine anatomy; osteoarthritis; reversibility. *Mailing Add:* Rheumatology & Clin Immunol Unit Univ Vt Col Med Given Med Bldg Rm C302 Burlington VT 05401

BLAND, KIRBY ISAAC, b Dothan, Ala, Feb 6, 42; m 66; c 3. SURGERY, ONCOLOGY. *Educ:* Auburn Univ, BS, 64; Univ Ala, Birmingham, MD, 68. *Prof Exp:* Intern surg, Univ Fla Col Med, 69, resident, 76; fel & res assoc, surg oncol, Univ Tex M D Anderson Hosp, 76-77; from asst prof to assoc prof surg, Univ Louisville Sch Med, 78-83; PROF & ASSOC CHMN SURG, UNIV FLA COL MED, 83-, DIR SURG RESIDENCY PROG, DEPT SURG, 83- *Concurrent Pos:* Prin investr, Univ Fla, 84-; dir, Am Bd Surg, 86-92; exec coun, Soc Surg Oncol, 88-90; comt surg educ, Am Col Surgeons, 88-90, chmn, comt young surgeons, 85-87; comt surg educ, Soc Univ Surgeons, 84-86; dir, Am Bd of Colon & Rectal Surg. *Honors & Awards:* Lester R Dragstedt Physician Award. *Mem:* Am Col Surgeons; Asn Acad Surg (pres, 87-88); Am Surg Asn; Soc Univ Surgeons; Soc Surg Oncol; Am Soc Clin Oncol; Int Soc Surg. *Res:* Tumor biology and metabolism; surgical oncology, especially melanoma, gastro-intestinal, breast neoplasms. *Mailing Add:* Dept Surg Univ Fla Box J-286 JHMHC 1600 Archer Rd Gainesville FL 32610

BLAND, RICHARD DAVID, b New Rochelle, NY, Nov 16, 40; c 3. CARDIOPULMONARY PHYSIOLOGY, NEWBORN MEDICINE. *Educ:* Yale Univ, BA, 62; Boston Univ, MD, 66. *Prof Exp:* Intern & resident pediat, Johns Hopkins Hosp, Baltimore, 66-69; chief newborn nurseries & staff pediatrician, Tripler Army Med Ctr, Honolulu, 69-72; fel pulmonary dis & newborn med, Cardiovasc Res Inst, 74, asst prof, 75-80, ASSOC PROF NEONATAL PEDIAT, UNIV CALIF, SAN FRANCISCO, 80- *Concurrent Pos:* Staff mem, Cardiovasc Res Inst, Univ Calif, San Francisco, 75-; vis prof pediat, Univ Hawaii, Honolulu, 76; NIH grant, NIH, Bethesda, Md, 76-; mem planning comt, Ross Labs Seminars Perinatal Med, 77-; estab investr, Am Heart Asn, 79-84. *Mem:* Soc Pediat Res; Am Thoracic Soc; Am Fedn Clin Res; Am Acad Pediat. *Res:* Lung fluid balance in newborn animals and human infants; microcirculation in the developing lung; intensive care of newborn infants. *Mailing Add:* Dept Pediat Univ Utah Med Ctr 50 N Medical Dr Salt Lake City UT 84132

BLAND, RICHARD P, b Oklahoma City, Okla, Nov 23, 28; m 53. MATHEMATICAL STATISTICS. *Educ:* Univ Okla, BS, 52; Univ NC, PhD(math statist), 61. *Prof Exp:* Anal engr, Chance Vought Aircraft, Inc, 53-56; res asst statist, Univ NC, 60-61; asst prof, Johns Hopkins Univ, 61-63; from asst prof to assoc prof statist, Southern Methodist Univ, 63-89; RETIRED. *Mem:* Am Statist Asn; Am Inst Math Statist; Royal Statist Soc. *Res:* Statistical decision theory, statistical inference, Bayerian statistics. *Mailing Add:* 1436 N Cheyenne Richardson TX 75080

BLAND, ROBERT GARY, b New York, NY, Feb 25, 48; m 74. OPERATIONS RESEARCH, APPLIED MATHEMATICS. *Educ:* Cornell Univ, BS, 69, MS, 72, PhD(oper res), 74. *Prof Exp:* Asst prof math sci, State Univ NY Binghamton, 74-78; asst prof opers res, 78-81, ASSOC PROF OPERS RES, CORNELL UNIV, 81- *Concurrent Pos:* Res assoc oper res, Ctr Oper Res & Economet & lectr appl sci, Cath Univ Louvain, 75-78; vis prof, Europ Inst Advan Studies Mgt, 76-77; Alfred P Sloan Found Res fel, 78-; NSF graduate fel, 69-72. *Mem:* Am Math Soc; Oper Res Soc Am. *Res:* Mathematical programming; combinatorial optimization; matroids; graph theory; networks. *Mailing Add:* Sch Opers Res & Indust Eng Cornell Univ Upson Hall Ithaca NY 14853

BLAND, ROGER GLADWIN, b Los Angeles, Calif, Dec 28, 39; m 64; c 2. NATURAL RESOURCE EDUCATION, SCANNING ELECTRON MICROSCOPY. *Educ:* Univ Calif, Davis, BS, 61; Ore State Univ, MS, 64; Univ Ariz, PhD(entom), 67. *Prof Exp:* From asst prof to assoc prof, 67-77, PROF BIOL, CENT MICH UNIV, 77- *Mem:* Entom Soc Am; Orthopterists Soc. *Res:* Behavior, ecology and sense organ morphology of insects. *Mailing Add:* Dept Biol Cent Mich Univ Mt Pleasant MI 48859

BLAND, WILLIAM M, JR, b Portsmouth, Va, July 23, 22; m 49; c 2. AEROSPACE ENGINEERING & TECHNOLOGY. *Educ:* NC State Col, BS, 47. *Prof Exp:* Res scientist aerodyn, Langley Res Ctr, Nat Adv Comt Aeronaut, 47-58; head, Systs Test Br, NASA, 58-60, asst chief, Eng Div, 60-62, dep mgr, Mercury Proj, 62-63, chief, Apollo Test Div, 63-64, chief, Apollo Test & Checkout, 65-66 & Apollo Reliability, Qual Assurance & Test Div, 67, dep mgr, reliability & qual assurance, Manned Spacecraft Ctr, 68-72, dep dir, safety, reliability & qual assurance, Johnson Space Ctr, 72-79; MGT & TECH CONSULT & PRES, GEEBS, INC, 80- *Concurrent Pos:* Mem tech eval teams, Mercury Little Joe booster, Mercury & Apollo spacecraft & Little Joe II booster & mem mission & flight readiness rev bd, Mercury missions; vis prof space activ, Rice Univ, Univ Houston & La State Univ, 63-64; mem tech staff, President's Comn on Accident Three Mile Island, 79; consult, Nat Acad Sci, 80 & Nuclear Regulatory Comn, 84-85; expert witness, 80-; ed, Nuclear Safety Dept, J Hazard Prev, 81-82. *Mem:* Am Nuclear Soc; Syst Safety Soc; Am Inst Aeronaut & Astronaut. *Res:* Aerodynamics, aircraft, audits, bioengineering, contracts, data analysis, health services, engineering design, expendable launch vehicles, nonmetallic materials, nuclear operations, risk management, reliability; lessons learned, manned space flight, safety, spacecraft, testing, quality assurance, management techniques; nuclear waste, technical editing and investigations and analysis of failures and accidents; author of over 70 technical publications. *Mailing Add:* 18575 Martinique Dr Houston TX 77058

BLANDER, MILTON, b Brooklyn, NY, Nov 1, 27; m 76; c 2. PHYSICAL CHEMISTRY. *Educ:* Brooklyn Col, BS, 50; Yale Univ, PhD(phys chem), 53. *Prof Exp:* Res assoc, Cornell Univ, 53-55; sr chemist, Oak Ridge Nat Lab, 55-62; mem tech staff, NAm Rockwell Sci Ctr, 62-67, group leader, 67-71; GROUP LEADER, ARGONNE NAT LAB, 71- *Honors & Awards:* Mat Chem Award, Dept Energy; Max Bredig Award, Electrochem Soc. *Mem:* Fel AAAS; Am Chem Soc; fel Meteoritical Soc; Electrochem Soc; Am Inst Mining, Metall & Petrol Engrs. *Res:* Statistical mechanics and thermodynamic measurements of molten ionic systems; vapors and alloys; pyrometallurgical research; welding chemistry and the origin of meteorites; neutron studies of high temperature materials. *Mailing Add:* Argonne Nat Lab 9700 S Cass Ave Argonne IL 60439

BLANDFORD, ROBERT ROY, b Columbus, Ga, Jan 1, 37; m 66; c 1. PHYSICAL OCEANOGRAPHY, SEISMOLOGY. *Educ:* Calif Inst Technol, BS, 59, PhD(geophys), 64. *Prof Exp:* Res scientist, NY Univ, 64-66; DIR RES, SEISMIC DATA LAB, GEOTECH, TELEDYNE, INC, 66- *Concurrent Pos:* Consult, Rand Corp, 57-66. *Mem:* Am Geophys Union. *Res:* Rossby waves; thermocline; internal waves; Gulf Stream dynamics; signal detection; arrays; time-series. *Mailing Add:* 1809 Paul Spring Rd 1809 Paul Spring Rd Alexandria VA 22307

BLANDFORD, ROGER DAVID, b Grantham, Eng, Aug 28, 49; m 72; c 2. ASTROPHYSICS. *Educ:* Cambridge Univ, BA, 70, MA & PhD(astrophys), 74. *Prof Exp:* Bye fel, Magdalene Col, Cambridge Univ, 72-73, res fel, St John's Col, 73-76; asst prof, 76-79, PROF THEORET ASTROPHYS, CALIF INST TECHNOL, 79-, RICHARD CHACE TOLMAN PROF THEORET ASTROPHYS, 89- *Concurrent Pos:* Mem, Inst Advan Study, Princeton Univ, 74-75; Parisot fel, Univ Calif, Berkeley, 75. *Mem:* Am Astron Soc; fel Royal Astron Soc. *Res:* Theoretical astrophysics. *Mailing Add:* Theoret Astrophys 130-33 Calif Inst Technol Pasadena CA 91125

BLANE, HOWARD THOMAS, b Deland, Fla, May 10, 26; div; c 2. PSYCHOLOGY, ALCOHOL RESEARCH. *Educ:* Harvard Univ, BA, 50; Clark Univ, MA, 51, PhD(psychol), 57. *Prof Exp:* Instr psychol, Harvard Med Sch, 57-66, asst clin prof, 66-70; from assoc prof to prof educ & psychol, Univ Pittsburgh, 70-86; DIR, RES INST ALCOHOLISM, 86- *Concurrent Pos:* Vis res fel, Univ Hawaii, 68-69; prin investr numerous grants, NIMH, Nat Inst Alcoholic Abuse & Alcoholism, NIH, Japan & Nat Inst Drug Abuse; consult, Nat Inst Alcohol Abuse & Alcoholism, 70-; res prof psychol, State Univ NY, Buffalo, 86-; mem bd dirs, Res Found Ment Hyg. *Mem:* Am Pub Health Asn; fel Am Psychol Asn; AAAS; Res Soc Alcoholism; Soc Psychologists Addictive Behav; Health Educ Found (vpres, 75-); NY Acad Sci. *Res:* Psychological study of development of alcoholism in late adolescence and early adulthood; prevention of alcoholic disorders; substance abuse in the developmentally disabled. *Mailing Add:* 1021 Main St Buffalo NY 14203

BLANEY, DONALD JOHN, b Cincinnati, Ohio, May 18, 26; m 57; c 3. BIOCHEMISTRY. *Educ:* Xavier Univ, Ohio, BS, 49; Univ Iowa, PhD(biochem), 53; Univ Mich, MD, 57. *Prof Exp:* Res asst, Univ Mich, 53-55; intern, Philadelphia Gen Hosp, 57-58; fel dermat, Cincinnati Gen Hosp, 58-61, instr, 61-63; assoc prof, Univ Cincinnati, 65-67; asst prof, 63-77, ASSOC PROF DERMAT, CINCINNATI GEN HOSP, 77-; ASSOC CLIN PROF DERMAT, UNIV CINCINNATI, 67- *Mem:* AAAS; Am Chem Soc; Am Acad Dermat; AMA. *Res:* Immunology. *Mailing Add:* 4966 Glenway Ave Price Hill Cincinnati OH 45238

BLANFORD, GEORGE EMMANUEL, JR, b Lebanon, Ky, Sept 16, 40; div; c 2. PHYSICS, PLANETARY SCIENCES. *Educ:* Cath Univ Am, BA, 64; Univ Louisville, MS, 67; Wash Univ, PhD(physics), 71. *Prof Exp:* Assoc instr physics, Univ Clermont-Ferrand, 71-73; Nat Res Coun res assoc, L B Johnson Space Ctr, 73-75; asst prof, 75-78, assoc prof, 78-86, PROF PHYSICS, UNIV HOUSTON, CLEAR LAKE CITY, 86- *Concurrent Pos:* Prin investr, NASA, 76-83. *Mem:* Am Phys Soc; Am Asn Physics Teachers; AAAS; Am Geophys Union; Am Astron Soc. *Res:* Cosmic ray physics, planetary science, solar and planetary relationships. *Mailing Add:* Dept Phys Sci 2700 Bay Area Blvd Houston TX 77058

BLANK, ALBERT ABRAHAM, b New York, NY, Nov 29, 24; m 50; c 4. MATHEMATICAL PHYSICS, VISION. *Educ:* Brooklyn Col, AB, 44; Brown Univ, MS, 46; NY Univ, PhD(math), 51. *Prof Exp:* Math analyst physiol optics, Columbia Univ, 51-52; assoc info & control systs, Control Systs Lab, Univ Ill, 53-54; asst prof math, Univ Tenn, 54-59; from assoc prof to prof, Courant Inst Math Sci, NY Univ, 59-69; PROF MATH, CARNEGIE-MELLON UNIV, 69- *Concurrent Pos:* Scientist, Magneto-Fluid Dynamics Div, Courant Inst Math Sci, NY Univ, 57-69, sr res scientist, 83-; assoc ed, Am Math Monthly. *Mem:* Fel AAAS; Am Phys Soc; Am Math Soc; Math Asn Am; Sigma Xi. *Res:* Magnetofluid dynamics; theory and computational studies of equilibrium, stability and diffusion in connection with problems in controlling thermonuclear reactions; partial differential equations and applications; binocular space perception. *Mailing Add:* Dept Math Carnegie-Mellon Univ Pittsburgh PA 15213

BLANK, BENJAMIN, b Philadelphia, Pa, May 12, 31; m 58; c 3. ORGANIC CHEMISTRY, MEDICINAL CHEMISTRY. *Educ:* Temple Univ, BA, 53, PhD(chem), 58. *Prof Exp:* Chemist, Muscular Dystrophy Asn, Vet Admin Hosp, Philadelphia, 56; sr med chemist, Smith Kline & French Labs, 58-74, Sr investe, 74-79, Sr proj admin, 79-80, mgr sci & admin affairs, New Compound Eval & Licensing, 81-84, sci dir, Res & Develop Acquisitions, 84-88; dir, CPD Acquisitions, Eastman Pharmaceuts, 88-89; VPRES, TECHNOL ASSESSMENT & ACQUISITION, CAMPBELL SOUP CO, 89- *Concurrent Pos:* Biochemist, Johnson Res Found, Sch Med, Univ Pa, 69-70. *Mem:* AAAS; NY Acad Sci; Am Chem Soc; Sigma Xi. *Res:* Thyromimetics; peptides; nonsteroidal anti-inflammatories; inhibitors of steroidal biosynthesis and gluconeogenesis; cephalosporins; biotechnology; drug delivery. *Mailing Add:* 502 Bruce Rd Trevose PA 19053

BLANK, CARL HERBERT, b Toledo, Ohio, Mar 16, 27; m 50; c 1. PUBLIC HEALTH, MICROBIOLOGY. *Educ:* Univ Toledo, BS, 50; Utah State Univ, MS, 57; Univ NC, MPH, 65, DrPH(lab practice & admin), 67. *Prof Exp:* Microbiologist, Div Labs, Utah State Health Dept, 51-58, actg asst dir, 58-62, asst dir, 62-64; dep dir, Bur Labs, Utah State Div Health, 67-72; chief, Exam & Documentation Br, Ctr Dis Control, 72-81, chief, Lab Training Br, 81-83, chief, Lab Consult Br, 83-85, dep dir & consult, 85-86, SR LAB CONSULT, PUB HEALTH PRACT PROG, OFF, CTR DIS CONTROL, 87- *Concurrent Pos:* Clin instr, Univ Utah, 68-70; consult, Ctr Dis Control, 71. *Mem:* Am Pub Health Asn; Am Soc Microbiol; Conf Pub Health Laboratorians. *Res:* Incidence of Q fever; improved methods in isolation and identification of actinomyces species; toxo plasmosis; fluorescent antibody techniques; laboratory evaluation and improvement programs. *Mailing Add:* Ctr Dis Control G-25 1600 Clifton Rd Atlanta GA 30333

BLANK, CHARLES ANTHONY, b Brooklyn, NY, Apr 15, 22; div; c 3. PHYSICAL INORGANIC CHEMISTRY, ENVIRONMENTAL SCIENCES GENERAL. *Educ:* Brooklyn Col, BA, 46, MA, 49; Syracuse Univ, PhD(phys inorg chem), 54. *Prof Exp:* Lab technician res, Int Flavors & Fragrances, Inc, 41-43; asst chem, Syracuse Univ, 47-48; res chemist, Film Dept, E I du Pont de Nemours & Co, 51-52, Bell Tel Labs, 54-56 & Int Bus Mach Corp, 56-59; supvr tech staff, Phys Sci Dept, Melpar, Inc, Va, 59-60; phys scientist, US Dept Defense, 60-78; CONSULT, 81- *Concurrent Pos:* Teacher chem, Assumption Col, Worcester, Mass, 84-85, physics, Northeastern Univ, Boston, 85-86; physics Worchester State Col, Worchester, Mass, 86-87, chem, Curry Col, Milton Mass, 86-87; adj sci facil, Manatee Community Col, Bradenton, Fl, 89- *Mem:* Am Chem Soc; fel Am Inst Chemists; NY Acad Sci; New Eng Asn Chem Teachers. *Res:* Coordination compounds; organic and inorganic synthesis; solid state chemistry; chemistry of electronic devices and materials; physical instrumentation; geophysics; nuclear weapon effects; chemistry and physics of ionosphere; research administration; environmental sciences. *Mailing Add:* 7085 46th Ave W Apt 173 Bradenton FL 34210

BLANK, GREGORY SCOTT, b Nov 20, 54; m; c 1. CELL BIOLOGY, PROTEIN CHEMISTRY. *Educ:* Univ Southern Calif, PhD(biol), 81. *Prof Exp:* RES ASSOC, BECTON DICKINSON MONOCLONAL CTR, 85-; SCIENTIST PROCESS RECOVERY, GENENTECH, INC. *Mailing Add:* Scientist Process Recovery Genentech 460 Point San Burno Blvd South San Francisco CA 94080

BLANK, HARVEY, b Chicago, Ill, June 21, 18; m 50; c 1. DERMATOLOGY. *Educ:* Univ Chicago, BS, 38, MD, 42; Am Bd Dermat & Syphil, dipl, 48. *Prof Exp:* Intern, Harper Hosp, Mich, 42-43; fel dermat, Univ Pa, 46, assoc, Grad Sch Med, 47-52; assoc med dir, Squibb Inst Med Res, 51-56; PROF DERMAT & MED, SCH MED, UNIV MIAMI, 56-, CHMN DEPT DERMAT, 60- *Concurrent Pos:* Mem res dept, Children's Hosp, Philadelphia, 46-56; Nat Res Coun fel, Univ Pa, 47-48; consult, Surgeon-Gen, US Dept Army, 48-; assoc dermat, Col Physicians & Surgeons, Columbia Univ, 51-55; mem adv comt Qm res & develop, Nat Res Coun, 52-; chief ed, Arch Dermat, 62-64; chmn comn cutaneous dis, Armed Forces Epidemiol Bd, 62-; consult, Vet Admin; mem dermat training grant comt, NIH. *Mem:* Soc Invest Dermat; Am Acad Dermat; Am Dermat Asn; fel Am Col Physicians; fel AMA. *Res:* Virus diseases of the skin; mycology; experimental cytology; investigative and clinical dermatology. *Mailing Add:* 1550 NW Tenth Ave Miami FL 33101

BLANK, IRVIN H, b Mt Carmel, Ill, Mar 20, 02; m; c 2. DERMATOLOGY, PHOTO MEDICINE. *Educ:* Univ Cincinnati, PhD(bacteriol), 28. *Prof Exp:* ASSOC BIOCHEMIST, WELLMAN LAB, MASS GEN HOSP, 55- *Honors & Awards:* Rothman Award, Soc Invest Dermat. *Mem:* Fel AAAS; NY Acad Sci; hon mem Soc Invest Dermat; Am Acad Dermat; Am Physiol Soc; Dermat Found. *Res:* Percutaneous absorption; mechanism of action of emollients. *Mailing Add:* Wellman Lab Mass Gen Hosp Boston MA 02114

BLANK, JOHN EDWARD, b Cleveland, Ohio, Dec 5, 42; m 66; c 2. ARCHAEOLOGY, BIOLOGICAL ANTHROPOLOGY. *Educ:* Case Inst Technol, BS, 64, MA, 67; Univ Mass, PhD(anthrop), 70. *Prof Exp:* Asst prof, 69-73, ASSOC PROF ANTHROP, CLEVELAND STATE UNIV, 73- *Mem:* Soc Am Archaeol; fel Am Anthrop Asn. *Res:* Cultural paleoecological adaptational systems of the Ohio Valley; conservation resource management, and public archaeology. *Mailing Add:* Dept Anthrop Cleveland State Univ Euclid Ave at E 24th Cleveland OH 44115

BLANK, LELAND T, b San Antonio, Tex, Jan 15, 44. ENGINEERING ECONOMICS, RESEARCH CENTER DEVELOPMENT. *Educ:* St Mary's Univ, BS, 67; Okla State Univ, MS, 68, PhD(indust eng), 70. *Prof Exp:* Prof indust eng, Univ Tex, El Paso, 70-76; mgt scientist, indust eng, GTE Data Serv, 76-78; head dept, 84-86, PROF INDUST ENG, TEX A&M UNIV, 78-, DIR, INST MFG SYSTS, 86-, ASST DEAN GRAD STUDIES, 88- *Concurrent Pos:* Coordr, mfg systs, Tex A&M Univ, 85-86; automated mfg equip tech adv comn, Int Trade Admin, US Dept Commerce, 85-; asst dir, Tex Eng Exp Sta, 86- *Mem:* Inst Indust Engrs; Nat Soc Prof Engrs; Sigma Xi; Soc Mfg Engrs; Soc Comput Simulation; Am Soc Eng Educ. *Res:* Cost modeling and quality systems analysis for manufacturing systems; analysis of indirect costs in a computer integrated manufacturing environment. *Mailing Add:* Asst Dean Eng Tex A&M Univ College Station TX 77843

BLANK, MARTIN, b US, Feb 28, 33; m 55; c 3. PHYSIOLOGY, BIOPHYSICS. *Educ:* City Col New York, BS, 54; Columbia Univ, PhD(phys chem), 57; Univ Cambridge, PhD(colloid sci), 59. *Prof Exp:* From instr to asst prof, 59-68, ASSOC PROF PHYSIOL, COL PHYSICIANS & SURGEONS, COLUMBIA UNIV, 68- *Concurrent Pos:* NIH fels, 60-70; res chemist, Unilever Res Lab, Eng, 64 & 67, Netherlands, 69; vis scientist, Weizmann Inst, 67, Univ Calif, 68 & Hebrew Univ,70-,; liaison scientist, Off Naval Res, London, 74-75; vis lectr, Monash Univ, 82; biologist, Off Naval Res, Arlington, 84-85, 86-88; ed, Bioelectrochem & Bioenergetics, J Electrochem Soc & Colloids & Surfaces & J Colloid Interface Sci; distinguished vis prof, Univ Western Australia, 82; distinguished lectr, Physiol, Wayne State Univ, 84. *Honors & Awards:* Yasnda Award, Bioelec Repair & Growth Soc, 90. *Mem:* AAAS; Am Chem Soc; Biophys Soc; Electrochem Soc; Bioelectrochem Soc; Bioelectromagnetics Soc; Bioelec Repair & Growth Soc. *Res:* Physical chemistry of ion transport and excitation in natural membranes; permeability of surface films; electrical effects at interfaces and membranes; transport mechanisms in natural membranes, such as NaKATPase, sperm cell, lung surfactant system; theoretical aspects of membranes and biopolymers; biotechnology, biosensors, electroporation, electromagnetic stimulation of biosynthesis and enzyme activity. *Mailing Add:* Col Physicians & Surgeons Columbia Univ New York NY 10032

BLANK, ROBERT H, b USA, Feb 14, 26; m 49; c 2. BIOLOGY. *Educ:* NY Univ, BA, 51. *Prof Exp:* Biochemist fermentation, 51-66, mgr microbiol res serv, 66-70, MGR BIOTHERAPEUT RES SERV, LEDERLE LABS, AM CYANAMID CO, 70-, RADIATION SAFETY OFFICER, 70- *Mem:* Health Physics Soc. *Res:* Fermentation biochemistry; antibiotics and microbiological transformation of steroids. *Mailing Add:* Lederle Labs Seven South Rd Wading River NY 11792

BLANK, STUART LAWRENCE, b Brooklyn, NY, Mar 4, 42; m 58; c 2. MATERIALS SCIENCE, CERAMIC ENGINEERING. *Educ:* Alfred Univ, BS, 62; Univ Calif, Berkeley, MS, 64, PhD(mat sci), 67. *Prof Exp:* Res asst mat sci, Lawrence Radiation Labs, Univ Calif, 62-67; res scientist, 69-77, supvr, Epitaxial Mat & Processes Group, 77-80, DEPT HEAD, ADVAN TECHNOL DEVELOP DEPT, BELL TEL LABS, 80- *Mem:* Am Ceramic Soc; Nat Inst Ceramic Engrs; Sigma Xi. *Res:* Solid state reactions; diffusion phenomena in refractory materials such as oxides; electro-optic materials research; crystal growth mechanisms; mechanical property effects of cation and anion impurities on crystals. *Mailing Add:* 125 Greenwood Ave Madison NJ 07940

BLANK, ZVI, b Tel-Aviv, Israel; m; c 3. SOLID STATE & POLYMER CHEMISTRY, ENVIRONMENTAL SCIENCES. *Educ:* Israel Inst Technol, BS, 60; NY Univ, MChE, 66, PhD(chem eng), 70. *Prof Exp:* Asst res scientist crystal growth, NY Univ, 65-69; staff engr solid state chem, Mat Res Ctr, Allied Chem Corp, 69-75; mgr chem & mat dept, Corp Res & Develop Lab, Singer Co, 75-80; vpres tech & eng, Radiation Tech Inc, 80-83, Enviro-Sci, Inc, 84-88; PRES, ECRA LABS, INC, 88- *Mem:* Am Chem Soc; Am Asn Crystal Growers; Int Solar Energy Soc; Adhesion Soc; NY Acad Sci. *Res:* Liquid crystals; electrochemical materials; acousto-optic materials; solid state chemistry related to energy storage and conversion; metal hydrides; crystal growth; structure-properties relationships in inorganics and in polymers; hybrid microelectronics; adhesives; garment fabrication technology; environmental analysis; environmental remediation. *Mailing Add:* 103 Littleton Rd Morris Plains NJ 07950

BLANKE, JORDAN MATTHEW, b New York, NY, Sept 16, 54; m 82; c 2. SOFTWARE SYSTEMS. *Educ:* State Univ NY Stony Brook, BS, 76, MS, 77; Emory Univ, JD, 80. *Prof Exp:* Assoc atty, Otterbourg, Steindler, Houston & Rosen, 81-83; law clerk, Justice Joseph Cohen, Bronx, NY, 83; asst prof computer sci & bus law, St John's Univ, Jamaica, NY, 83-85; ASSOC PROF & DEPT CHAIR COMPUTER SCI & COMPUTER INFO SYSTS, MERCER UNIV, ATLANTA, 85- *Concurrent Pos:* Adj prof, Emory Univ, Sch Law, 86-87. *Mem:* Asn Comput Mach. *Res:* Computers and the law; programming languages; database systems. *Mailing Add:* Mercer Univ Atlanta 3001 Mercer University Dr Atlanta GA 30341

BLANKE, ROBERT VERNON, b Leavenworth, Kans, Dec 3, 24; m 50; c 3. TOXICOLOGY. *Educ:* Northwestern Univ, BS, 49; Univ Ill, MS, 53, PhD(pharmacol), 58. *Prof Exp:* Res asst, Univ Ill, 50-53, instr, 53-57; asst toxicologist, Med Examr Off, Md, 57-60; toxicologist, Ill, 60-63; toxicologist, Med Examr Off, Commonwealth of Va, 63-72; prof path, pharmacol & toxicol, Med Col Va, 72-87; RETIRED. *Concurrent Pos:* Sr chemist, Cook County Coroner's Lab, Chicago, 50-53; assoc prof, Med Col Va, 63-74, prof, 87-; consult, Vet Admin Hosp, Richmond, 64-70 & 74-87; mem toxicol study sect, Div Res Grants, NIH, 70-74; dir, Toxicol Lab, Med Col Va Hosp, 72-87. *Honors & Awards:* Harger Award, Toxicol Sect, Am Acad Forensic Sci, 86. *Mem:* AAAS; Am Acad Forensic Sci; Am Chem Soc; NY Acad Sci; Soc Toxicol; Soc Forensic Toxicol; Sigma Xi; Am Asn Clin Chem; Int Asn Forensic Toxicol. *Res:* Analytical toxicology; environmental toxic substances; general toxicology. *Mailing Add:* 4222 Croatan Rd Richmond VA 23235-1116

BLANKENBECLER, RICHARD, b Kingsport, Tenn, Feb 4, 33; wid; c 2. OPTICS. *Educ:* Miami Univ, AB, 54; Stanford Univ, PhD(physics), 58. *Hon Degrees:* DSc, Miami Univ, Ohio, 90. *Prof Exp:* NSF fel physics, Princeton Univ, 58-63, assoc prof, 63-67; prof, Univ Calif, Santa Barbara, 67-69; PROF, STANFORD LINEAR ACCELERATOR CTR, STANFORD UNIV, 69- *Concurrent Pos:* fel, NSF & Sloan Found. *Res:* Particle physics, optics and theoretical high energy physics. *Mailing Add:* Stanford Linear Accelerator Ctr Stanford Univ Stanford CA 94305

BLANKENHORN, DAVID HENRY, b Cleveland, Ohio, Nov 16, 24; m 48; c 4. CARDIOLOGY. *Educ:* Univ Cincinnati, MD, 47. *Prof Exp:* Res asst, Rockefeller Inst, 52-54; chief resident internal med, Cincinnati Gen Hosp, 54-55; instr med, Univ Cincinnati, 55-57; from asst prof to assoc prof med, 57-67, head cardiol sect, 63-80, PROF MED, UNIV SOUTHERN CALIF, 67- *Mem:* Am Fedn Clin Res; Am Soc Exp Biol; Am Soc Clin Invest. *Res:* Lipid metabolism; vascular disease. *Mailing Add:* Sch Med Dept Med Univ Southern Calif 2025 Zonal Ave Los Angeles CA 90033

BLANKENHORN, PAUL RICHARD, b Shenandoah, Pa, Apr 16, 44; m 68; c 2. WOOD SCIENCE & TECHNOLOGY, MATERIALS ENGINEERING. *Educ:* Pa State Univ, BS, 66, MS, 68, PhD(wood sci, mat sci), 72. *Prof Exp:* Eng aide aerospace eng, Johnsville Naval Air Sta, 66-67; res asst nuclear eng, Breazeale Nuclear Reactor, 67-68, res asst wood sci, Sch Forest Resources, 70-72, res assoc mat sci, Pa Transp Inst, 72-75, from asst prof to assoc prof wood technol, 75-83, PROF WOOD TECHNOL, SCH FOREST RESOURCES, PA STATE UNIV, 83- *Honors & Awards:* Wood Award, 2nd, Forest Prod Res Soc, 72. *Mem:* Forest Prod Res Soc; Soc Wood Sci & Technol. *Res:* Physical and mechanical properties of natural occurring polymers and polymer base composites; conversion of wood into energy. *Mailing Add:* Sch Forest Resources Univ Pa 310 Forest Resources Lab University Park PA 16802

BLANKENSHIP, FLOYD ALLEN, b Atlanta, Ga, Nov 21, 30. PHYSICAL CHEMISTRY. *Educ:* Univ Ga, BS, 57; Univ Ill, PhD(chem), 62. *Prof Exp:* Chemist, Celanese Chem Corp, 57; res chemist, Rohm & Haas Co, 62-66; asst prof, 66-68, ASSOC PROF CHEM, TOWSON STATE COL, 68- *Res:* Computer utilization in chemistry; electronic structures of transition metal complexes; properties of non-ionic surfactants. *Mailing Add:* Dept Chem Towson State Univ Towson MD 21204

BLANKENSHIP, JAMES EMERY, b Sherman, Tex, Mar 19, 41; m 60; c 2. NEUROPHYSIOLOGY. *Educ:* Austin Col, BA, 63; Yale Univ, MS, 65, PhD(neurophysiol), 67. *Prof Exp:* Fel physiol, Med Sch, NY Univ, 67-69; NIH fel neurophysiol, NIMH, 69-70; PROF PHYSIOL, ANAT & NEUROSCI & MEM, DIV COMP MARINE NEUROBIOL, MARINE BIOMED INST, UNIV TEX MED BR GALVESTON, 70-, ASSOC DEAN, GRAD SCH BIOMED SCI, 80- *Concurrent Pos:* NIH res grant, Nat Inst Neurol Dis & Stroke, 70, 72, 73, 86 & 89; Off Naval Res contract, 73; NSF res grant, 77, 79, 83 & 87-; dir, MD-PhD prog, Univ Tex Med Br Galveston, 82-, dir, Neurosci Grad Prog, 87- *Mem:* AAAS; Soc Neurosci; Am Physiol Soc; Sigma Xi; Soc Gen Physiol. *Res:* Cellular electrophysiology and neuroendocrinology; comparative neurobiology, synaptic and interneuronal properties; neuropeptides and reproductive peptides. *Mailing Add:* Marine Biomed Inst 200 University Blvd Galveston TX 77550

BLANKENSHIP, JAMES LYNN, b Knoxville, Tenn, Mar 26, 31; m 51; c 4. EXPERIMENTAL SOLID STATE PHYSICS, SEMICONDUCTORS. *Educ:* Univ Tenn, BS, 54, MS, 55, PhD(physics), 73. *Prof Exp:* Circuit develop engr, 55-58, res engr, 58-65, educ assignment, 65-67, physicist, 68-72, SR RES STAFF MEM, OAK RIDGE NAT LAB, 72- *Mem:* Am Phys Soc. *Res:* Nuclear medicine image processing; heavy ion physics instrumentation and research; gaseous and solid state detector development. *Mailing Add:* Oak Ridge Nat Lab Bldg 6000 MS 6368 PO Box 2008 Oakridge TN 37831

BLANKENSHIP, JAMES W, b Lafayette, Ind, Jan 15, 28; m 51; c 3. BIOCHEMISTRY. *Educ:* Southern Missionary Col, BA, 51; Univ Ark, MS, 53; Univ Wyo, PhD(biochem), 69. *Prof Exp:* Instr biochem, Univ Wyo, 67-68, asst prof, 69-70; from asst prof to assoc prof, Sch Allied Health Professions, PROF NUTRIT, SCH HEALTH, LOMA LINDA UNIV, 77- *Mem:* Am Chem Soc; Am Oil Chem Soc. *Res:* Lipid metabolism in insects; minor lipids and their effects on normal and abnormal tissue development. *Mailing Add:* Nutrit Sch Health Loma Linda Univ Loma Linda CA 92354

BLANKENSHIP, JAMES WILLIAM, b San Diego, Calif, June 8, 43; m 68; c 2. BIOCHEMICAL PHARMACOLOGY. *Educ:* Tex A&M, BS, 65, MS, 67; Univ Utah, PhD(pharmacol), 72. *Prof Exp:* Res assoc cell biol, Med Col, Ga, 72-74; asst prof pharmacol, Med Univ SC, 74-77; PROF PHARMACOL, SCH PHARM, UNIV PAC, 77-, CHMN, 90- *Concurrent Pos:* Mem & consult, competency comt, Calif State Bd Pharm, 80-; prin investr, NIH res grant, 80-83 & NSF, 84-87, 87- *Mem:* Am Soc Pharmacol & Exp Therapeut. *Res:* Biochemistry and cellular functions of polyamines; metabolism of polyamines and effects on DNA synthesis and on cell growth. *Mailing Add:* Sch Pharm Univ Pac Stockton CA 95211

BLANKENSHIP, LYTLE HOUSTON, b Campbellton, Tex, Mar 1, 27; m 54; c 4. WILDLIFE RESEARCH, CONSERVATION EDUCATION. *Educ:* Tex A&M Univ, BS, 50; Univ Minn, MS, 52; Mich State Univ, PhD(physiol), 57. *Prof Exp:* Biologist, Game Div, Mich Dept Conserv, 54-56; res biologist, Minn Div Game & Fish, 56-61; wildlife biologist, Wildlife Res Br, Bur Sport Fisheries & Wildlife, US Fish & Wildlife Serv, 61-69; res scientist, Tex A&M Univ-E African Agr & Forestry Res Orgn, 69-72; from assoc prof to prof, Agr Exp Sta, Tex A&M Univ, 72-88; RETIRED. *Concurrent Pos:* Consult, The World Bank, Kenya, 71 & 75, Orgn Am States, Dominican Republic, 75, US Fish & Wildlife Serv, India, 81 & 82, Exxon, 82-; vis lectr, Univ Dar es Salaam, Tanzania, 78; owner & pres, Safari Adventures. *Honors & Awards:* Outstanding Serv Award, Wildlife Soc, 83. *Mem:* Hon mem Wildlife Soc (pres elect, 85-86, pres, 86-88); E Africa Wildlife Soc; Wildlife Protection & Conserv Soc S Africa; Wildlife Soc Southern Africa; S African Wildlife Mgt Asn; Nat Audubon Soc; Nat Wildlife Fedn; Ducks Unlimited; World Wildlife Fund; Nature Conservancy. *Res:* Ecology, behavior and diseases of doves, waterfowl, woodcock, moose and deer; physiology of deer; ecology, reproduction and meat production of large game animals in East Africa; range nutrition of plants and animals. *Mailing Add:* PO Drawer 5220 Uvalde TX 78802-5220

BLANKENSHIP, ROBERT EUGENE, b Auburn, Nebr, Aug 25, 48; m 71; c 2. PHOTOSYNTHESIS. *Educ:* Nebr Wesleyan Univ, BS, 70; Univ Calif, Berkeley, PhD(chem), 75. *Prof Exp:* Postdoctoral fel, Univ Wash, 76-79; asst prof chem, Amherst Col, 79-85; assoc prof, 85-88, PROF CHEM, ARIZ STATE UNIV, 88- *Concurrent Pos:* Dir, Ctr Study Early Events Photosynthesis, 88-; ed-in-chief, Photosynthesis Res, 88-; chmn, Gordon Res Conf Photosynthesis, 91. *Mem:* Am Chem Soc; Biophys Soc; Am Soc Plant Physiologists; AAAS; Union Concerned Scientists. *Res:* Excitation and electron transfer in photosynthetic systems; evolution of biological energy conserving systems; mechanisms of electron transfer reactions in biological systems. *Mailing Add:* Dept Chem & Biochem Ariz State Univ Tempe AZ 85287-1604

BLANKENSHIP, VICTOR D(ALE), b Topeka, Kans, Feb 9, 34; m 56; c 2. FLUID MECHANICS, HEAT TRANSFER. *Educ:* Univ Kans, BS, 56; Univ Notre Dame, MS, 59; Univ Mich, PhD(mech eng), 62. *Prof Exp:* Engr, Bendix Aviation Corp, 56-59; dir Off Res mark 18 & reentry systs concepts, Reentry Systs Div, Aerospace Corp, 62-68; mgr reentry systs eng, 68-83, mgr penetration systs, 83-88, MGR SPEC PROJS, TRW SYSTS GROUP, 88- *Concurrent Pos:* Lectr, Univ Southern Calif, 63-66 & Univ Redlands, 68, 75, 78, 84 & 85; mem, Strategic Arms Limitation Talk Support Group, 69 & 71; trustee, Valley Prep Sch, 70-71. *Mem:* Am Inst Aeronaut & Astronaut; NY Acad Sci; Sigma Xi. *Res:* Effects of oscillating boundary layers; rarefied gas dynamics; partially ionized gases; antenna breakdown; boundary layer; heat transfer; shock-shock interactions; nuclear effects; systems analysis; reentry systems; optics; microelectronics. *Mailing Add:* 1740 Canyon Rd Redlands CA 92373

BLANKESPOOR, HARVEY DALE, b Boyden, Iowa, Oct 15, 39; m 64; c 2. INVERTEBRATE ZOOLOGY, PARASITOLOGY. *Educ:* Westmar Col, BA, 63; Iowa State Univ, MS, 67, PhD(invert zool), 70. *Prof Exp:* Asst prof biol, Univ Northern Iowa, 70-71 & Trinity Christian Col, 71-72; asst prof zool, Univ Mich, Ann Arbor, 72-77; ASSOC PROF BIOL, HOPE COL, 77- *Honors & Awards:* Chester A Herrick Award, Midwest Conf Parasitologists, 70. *Mem:* Am Soc Parasitol; Am Micros Soc; Am Soc Trop Med & Hyg. *Res:* Host-parasite relationships of parasitic flatworms. *Mailing Add:* Dept Biol Hope Col Holland MI 49423

BLANKESPOOR, RONALD LEE, b Hull, Iowa, Feb 22, 46; m 69; c 1. ORGANIC CHEMISTRY. *Educ:* Dordt Col, AB, 68; Iowa State Univ, PhD(org chem), 71. *Prof Exp:* Asst prof chem, Univ Wis-Oshkosh, 71-73; asst prof chem, Wake Forest Univ, 73-77; assoc prof, 77-79, PROF CHEM, CALVIN COL, 79- *Mem:* Am Chem Soc; Sigma Xi. *Res:* Studies of radical ion reactions using electron spin resonance and electrochemical methods. *Mailing Add:* Dept Chem Calvin Col Grand Rapids MI 49546

BLANKFIELD, ALAN, b Passaic, NJ, July 31, 32; m 65. PHYSICS, MATHEMATICS. *Educ:* Univ Ill, BS, 54, MS, 55. *Prof Exp:* Res assoc vacuum technol, Coord Sci Lab, Univ Ill, 55-60; mgr advan prog mil aerospace, Bell Aerospace-Textron, 60-72; prin scientist optics, Cornell Aeronaut Lab, 72-75; dir, Dept Defense Consult, Falcon Res & Develop, 75-80; sr assoc, Ketron, Inc, 81-83; SPEC PROJ ENGR, DEFENSE COMMUN AGENCY, 84- *Concurrent Pos:* Consult, Syst Planning Corp, 72-73, Inst Defense Anal, 74-75 & 80-81, USAF, 76-77, Inst Defense Anal, 78 & Opers Res, Inc, 78-79. *Mem:* Am Defense Preparedness Asn. *Res:* Optical and radio instrumentation. *Mailing Add:* 4704 Warren St NW Washington DC 20016

BLANKLEY, CLIFTON JOHN, b Chicago, Ill, Apr 21, 42; m 76. MEDICINAL CHEMISTRY, COMPUTER DRUG DESIGN. *Educ:* Stanford Univ, BS, 63; Mass Inst Technol, PhD(org chem), 67. *Prof Exp:* Sr res chemist, 67-80, sr res assoc, 80-89, ASSOC RES FEL, PARKE DAVIS RES DIV, WARNER-LAMBERT CO, 89- *Mem:* Am Chem Soc; AAAS; NY Acad Sci. *Res:* Synthesis of potential anti-inflammatory agents, antibacterial agents, antitumor agents and antihypertensive agents and hypolipidemic agents; study of quantitative structure activity relationships; computer-assisted drug design and molecular modeling. *Mailing Add:* Parke Davis Pharm Res Div Warner-Lambert Co 2800 Plymouth Rd Ann Arbor MI 48106-1047

BLANKS, JANET MARIE, b Berkeley, Calif, Sept 25, 44; m 67. NEUROCYTOLOGY. *Educ:* Humboldt State Col, BA, 66; Univ Calif, Los Angeles, PhD(anat), 73. *Prof Exp:* Feldman fel neurocytol, Jules Stein Eye Inst, Univ Calif, Los Angeles, 73-74; vis scientist, Max Planck Inst Brain Res, 74-75; instr, Harvard Med Sch, Boston, 76-78; ASST PROF OPHTHALMOL & ANAT, SCH MED, UNIV SCALIF, 78-; DIR EM LAB, ESTELLE DOHENY EYE FOUND, 78- *Mem:* Asn Res Vision & Ophthal. *Res:* Correlation of biochemical and morphological changes which occur during synaptogenesis in the mammalian central nervous system; light and electron microscopic autoradiography and Golgi light microscopy. *Mailing Add:* Dept Anat Univ SCalif Sch Med 2025 Zonal Ave Los Angeles CA 90033

BLANKS, ROBERT F, b Denver, Colo, June 11, 36; m 57; c 3. CHEMICAL ENGINEERING. *Educ:* Univ Colo, BS(chem eng) & BS(bus admin), 59; Univ Calif, PhD(chem eng), 63. *Prof Exp:* Process res engr, plastics div, Union Carbide Corp, 62-69; prof chem eng, Mich State Univ, 69-78; SR RES ENGR, AMOCO CHEMS CORP, 78- *Mem:* Am Chem Soc; Am Inst Chem Engrs. *Res:* Polymer processing technology and polymer structure and properties as related to processing behavior; molecular rheology and thermodynamics of polymer systems; polymer reaction engineering and process design. *Mailing Add:* 1514 Inverrary Dr Naperville IL 60563

BLANKSCHTEIN, DANIEL, b Buenos Aires, Arg, Dec 1, 51; Israeli citizen; m 78; c 2. COLLOID & INTERFACE SCIENCE, BIOTECHNOLOGY. *Educ:* Tel-Aviv Univ BSc, 76, MSc, 79, PhD (physics), 83. *Prof Exp:* Weizmann postdoctoral fel, Mass Inst Technol, 82-84, Bantrell postdoctoral fel, 84-86, Texaco-Mangelsdorf asst prof, 86-89, TEXACO-MANGELSDORF ASSOC PROF, MASS INST TECHNOL, 89- *Concurrent Pos:* Vis scientist, Israel-Norway Cult Exchange, Oslo, 85; consult, Texaco & Oculon Corp, 89-; NSF presidential young investr, 89-94. *Mem:* Am Phys Soc; Am Chem Soc; Am Inst Chem Engrs; Soc Petrol Engrs; Mat Res Soc. *Res:* Statistical mechanics; thermodynamics and physical chemistry of structured fluids including micellar solutions, polymers and protein mixtures; molecular simulations and liquid state theory applications to predict the phase behavior of complex fluids; bioseparations using two-phase aqueous polymer and micellar systems. *Mailing Add:* Dept Chem Eng Rm 66-448 Mass Inst Technol Cambridge MA 02139

BLANN, H MARSHALL, b Los Angeles, Calif, Aug 22, 35; m 59; c 1. NUCLEAR CHEMISTRY. *Educ:* Univ Calif, Los Angeles, BS, 57; Univ Calif, Berkeley, PhD(chem), 60. *Prof Exp:* From instr to prof chem, Univ Rochester, 60-80; CONSULT, 80- *Concurrent Pos:* Res Corp grant, 60- *Mem:* Am Phys Soc. *Res:* Nuclear reaction mechanisms at moderate excitation energy; nuclear fission. *Mailing Add:* 2471 Sheffield Livermore CA 94550

BLANPAIN, JAN E, b Diest, Belg, Feb 24, 30; m; c 2. MEDICINE. *Prof Exp:* Dir, Leuven Univ Hosp, 60-70, PROF, MED SCH & SCH PUB HEALTH, LEUVEN UNIV, 65- *Concurrent Pos:* Lectr health serv mgt & health policy, Cornell Univ, Ottawa Univ, Montreal Univ, Leeds Univ, Manchester & others; consult health affairs, WHO, UN Develop Prog, Pan Am Health Orgn, World Bank, Rockefeller Found & others, 61-; vis scientist, US Nat Ctr Health Serv Res & Develop, 71; chmn & co-founder, Europ Health Policy Forum, 81- *Mem:* Inst Med-Nat Acad Sci; hon fel Am Col Healthcare Execs. *Res:* Comparative health care organization and health policy; quality assurance in health care. *Mailing Add:* Sch Pub Health Katholieke Univ Leuven Naamsestraat 22 Louvain 3000 Belgium

BLANPIED, GARY STEPHEN, b Corpus Christi, Tex, Nov 25, 49; m 74; c 2. INTERMEDIATE ENERGY NUCLEAR PHYSICS. *Educ:* Auburn Univ, BS, 73; Ga Inst Technol, MS, 74; Univ Tex, PhD(physics), 77. *Prof Exp:* Res asst physics, Univ Tex, 74-77; res assoc, NMex State Univ, 77-79; from asst prof to assoc prof, 79-88, PROF PHYSICS, UNIV SC, 88- *Concurrent Pos:* Fulbright res scholar, 86-87. *Mem:* Am Phys Soc. *Res:* Intermediate energy proton-nucleus scattering; pion-nucleus scattering; distribution of matter in spherical and deformed nuclei; multistep excitations in inelastic scattering; electromagnetic measurements; polarized photon induced reactions. *Mailing Add:* Dept Physics Univ SC Columbia SC 29208

BLANPIED, GEORGE DAVID, b Ridgewood, NJ, June 29, 30; m 52; c 3. POMOLOGY, PHYSIOLOGY. *Educ:* Dartmouth Col, BA, 52; Cornell Univ, MS, 55; Mich State Univ, PhD(pomol), 59. *Prof Exp:* Asst prof, 55-57 & 59-64, assoc prof, 65-76, PROF POMOL, CORNELL UNIV, 76- *Concurrent Pos:* Vis prof, Univ Dublin, 65-66, Agr Can, 74-75 & Fao Iran, 75; sabbatical leave, UK, 81-82. *Mem:* Am Soc Hort Sci. *Res:* Biological and commercial aspects of fruit maturity, storage, handling and marketing. *Mailing Add:* Dept Promology Cornell Univ Ithaca NY 14853

BLANPIED, WILLIAM ANTOINE, b Rochester, NY, May 11, 33; m 59, 73; c 3. SCIENCE POLICY, HISTORY OF SCIENCE. *Educ:* Yale Univ, BS, 55; Princeton Univ, PhD(physics), 59. *Prof Exp:* Instr physics, Princeton Univ, 58-59; NSF fel, Synchrotron Lab, Frascati, Italy, 59-60; from instr to asst prof, Yale Univ, 60-66; assoc prof, Case Western Reserve Univ, 66-69; staff scientist, NSF, New Delhi, India, 69-71; assoc prof, Case Western Reserve Univ, 71-72; sr res fel, Harvard Univ, 72-74; head div pub sector progs, AAAS, 74-76; prog dir, Ethics & Values in Sci & Technol, 76-79, Off Spec Proj, 79-84, SR PROG MGR, DIV INT PROGS, NSF, 84- *Concurrent Pos:* Yale Univ jr fac fel, Cambridge Electron Accelerator, Harvard Univ, 63-64; vis scholar, Grad Sch Int Rels & Pac Studies, Univ Calif, San Diego, 87-89. *Mem:* Am Phys Soc; Am Hist Sci Soc; AAAS. *Res:* Science and public policy; science curriculum development; public understanding of science; history of science in Asia. *Mailing Add:* Div Int Progs Nat Sci Found Washington DC 20550

BLANQUET, RICHARD STEVEN, b Brooklyn, NY, Jan 17, 40; m 67. INVERTEBRATE PHYSIOLOGY, BIOCHEMISTRY. *Educ:* City Col New York, BS, 61; Duke Univ, PhD(zool), 66. *Prof Exp:* Res assoc zool, Lab Quant Biol, Univ Miami, 65-67; instr physiol, 67-69, asst prof, dept biol, 69-73, ASSOC PROF BIOL, GEORGETOWN UNIV, 73- *Mem:* Am Soc Cell Biol; Am Soc Zool; Sigma Xi. *Res:* Physiology and biochemistry of lower invertebrates; problems of coelenterate nematocyst discharge and chemical nature of capsule and enclosed toxin; algal-cnidarian symbioses. *Mailing Add:* Dept Biol Georgetown Univ Washington DC 20057

BLANTON, CHARLES DEWITT, JR, b Kings Mountain, NC, Jan 11, 37; m 59; c 1. ORGANIC CHEMISTRY, MEDICINAL CHEMISTRY. *Educ:* Western Carolina Col, BS, 59; Univ Miss, PhD(org chem), 63. *Prof Exp:* Fel, Dept Chem, Ind Univ, 63-64; assoc res prof pharmaceut & med chem, Sch Pharm, Auburn Univ, 64-68; assoc prof med chem, 68-75, dir res & grad studies, 81-85, PROF MED CHEM, SCH PHARM, UNIV GA, 75- *Mem:* Acad Pharmaceut Sci; Am Chem Soc; Am Pharmaceut Asn; Int Soc Heterocyclic Chem; The Chem Soc. *Res:* Synthesis of quinazolines as potential anticonvulsants; pyrrolopyridines and c-nucleosides as potential antineoplastic agents; potential prophylactic and radical curative antimalarial; indole and benzodiazepine analogues as potential psychopharmacological and antineoplastic agents. *Mailing Add:* Sch Pharm Univ Ga Athens GA 30601

BLANTON, JACKSON ORIN, b Atlanta, Ga, Oct 28, 39; m 62; c 2. PHYSICAL OCEANOGRAPHY, LIMNOLOGY. *Educ:* Univ Fla, BSCE, 62; Ore State Univ, MS, 64, PhD(oceanog), 68. *Prof Exp:* Consult coastal eng, Marine Adv Inc, 64-65; res assoc, Univ Fla, 65-66; asst prof phys oceanog, Marine Lab, Duke Univ, 68-70; res scientist phys limnol, Can Ctr Inland Waters, 70-74; prog mgr phys oceanog, US Energy Res & Develop Admin, 74-76; Assoc prof ocean, 76-, PROF OCEAN, SKIDAWAY INST OCEANOG, STATE UNIV SYST GA. *Concurrent Pos:* Consult phys oceanog, Res Triangle Inst, NC, 68-69, NSF, 77-80 & US Fish & Wildlife Serv, 80. *Mem:* Am Geophys Union; Int Asn Great Lakes Res. *Res:* Energy dissipation in lakes, estuaries and shallow seas; processes of upwelling and water mass exchanges; maintenance and dissipation of thermoclines or frontal zones in oceans and lakes; continental shelf circulation. *Mailing Add:* Skidaway Inst Oceanog PO Box 13687 Savannah GA 31416

BLANTON, JOHN DAVID, b St Louis, MO, Jan 23, 27; m 71; c 4. LOW DIMENSIONAL TOPOLOGY, MORSE THEORY. *Educ:* St Louis Univ, AB, 53, MS, 55; Univ Ill, Urbana, PhD(math), 70. *Prof Exp:* Res asst, Inst Advanced Study, 67-68; instr math, Rockhurst Col, 68-70; PROF MATH, ST JOHN FISHER COL, 70-88. *Mem:* Am Math Soc; Math Asn Am. *Res:* Translation from the Latin of L Euler's "Introduction to Analysis of the Infinite". *Mailing Add:* Dept Math Saint John Fisher Col Rochester NY 14618

BLANTON, PATRICIA LOUISE, b Clarksville, Tex, July 9, 41. GROSS ANATOMY, PERIODONTICS. *Educ:* Hardin-Simmons Univ, BA, 62; Baylor Univ, MS, 64, PhD(anat), 67; Baylor Dent Col, DDS, 74, cert periodont, 75. *Prof Exp:* From asst prof to assoc prof gross anat, Baylor Dent Col, 67-75, chmn, Dept Gross Anat, 83-85, PROF GROSS ANAT & PERIODONT, BAYLOR DENT COL, 75-, ADJ PROF ANAT, 85- *Concurrent Pos:* Mem, Nat Adv Coun Health Prof Educ, NIH, 73-74 & State Anat Bd, 82-85; consult, Am Dent Asn & Coun Dent, NIH Oral Biol-Med-Study Sect, 83-86 & Am Dent Asn, Coun Dent Educ & Comt Dent Accreditation, 81-85. *Honors & Awards:* First Place Award, Student Clinicians, Am Dent Asn, 72. *Mem:* Am Asn Anatomists; Am Acad Periodont; Int Soc Electromyographic Kinesiology; Am Dent Asn; Sigma Xi; fel Am Col Dent; Int Col Dent; Am Acad Periodont. *Res:* Electromyographic evaluation of the human head and neck musculature; masticatory under normal conditions and relative to temporomandibular joint dysfunction and occlusal discrepancies and adjustments. *Mailing Add:* 10614 Creekmore Dallas TX 75218

BLANTON, RONALD EDWARD, b Houston, Tex, Mar 11, 52. MOLECULAR BIOLOGY, GENERAL MEDICAL SCIENCES. *Educ:* Harvard Univ, BA, 74; Case Western Reserve Univ, MD, 79. *Prof Exp:* ASST PROF GEOG MED, UNIV HOSPS, 86- *Mem:* Am Col Physicians; Am Fed Clin Res. *Res:* Molecular biology of schiztosomes with particular emphasis on gene regulation during development and vaccine production. *Mailing Add:* 2884 Warrington Rd Cleveland OH 44120

BLANTON, WILLIAM GEORGE, b Bowie, Tex, Jan 9, 30; m 61; c 1. BIOLOGICAL OCEANOGRAPHY. *Educ:* Tex Wesleyan Col, BS, 58; NTex State Univ, MS, 59; Tex A&M Univ, PhD(biol oceanog), 66. *Prof Exp:* Res scientist, Off Naval Res Proj, Inst Marine Sci, Univ Tex, 59-61; dir corrosion res, Nutro Prod Corp, Tex, 62-65; Dow Chem Co co-investr, Off Saline Waters grant, Tex A&M Univ, 65-66; assoc prof biol, Tex Wesleyan Col, 66-74, prof, 74-80; DR CHIROPRACTICS, WIPF CHIROPRACTIC CTR. *Concurrent Pos:* Dow Chem Co co-investr, Environ Studies Group, Tex Christian Univ, 67-69; co-investr, Ecol Study Cedar Bayou, US Steel Corp, 69; prin investr, Ecol Study Lavaca Bay, Alcoa Aluminum grant, 70-71; group leader, Environ Response Team, Oil & Hazardous Chem Div, Environ Protection Agency, 71-73; staff scientist, Biologische Anstalt Helgoland, 73-74; mem bd consult, USACE, 77-78. *Mem:* AAAS; Am Soc Microbiol; Am Soc Oceanog. *Res:* Microbial ecology; cell physiology; intermediate metabolism of marine bacteria. *Mailing Add:* 4402 Glassock No 612 Harlingen TX 78550

BLANTZ, ROLAND C, NEPHROLOGY & HYPERTENSION. *Prof Exp:* HEAD DEPT & PROF NEPHROLOGY-HYPERTENSION, VET ADMIN MED CTR, 72- *Mailing Add:* Dept Med Div Nephrology-Hypertension Univ Calif San Diego & Vet Admin Med Ctr 3350 La Jolla Village Dr V-111H San Diego CA 92161

BLASBALG, HERMAN, b Poland, June 17, 25; nat US; m 56; c 3. ELECTRICAL ENGINEERING. *Educ:* City Col New York, BEE, 48; Univ Md, MS, 52; Johns Hopkins Univ, DEE, 56. *Prof Exp:* Engr res & develop, Melpar, Inc, 48-51; consult & res scientist commun theory, Radiation Lab, Johns Hopkins Univ, 51-56; res scientist, Res Div, Electronic Commun, Inc, 56-60; sr engr & mgr advan modulation tech group, Commun Systs Dept, Fed Systs Div, Int Bus Mach Corp, 60-66, mgr digital satellite transmission dept, Eng Lab, 66-67, satellite commun tech mgr, Ctr Explor Studies, 67-69; dept head commun technol, Mitre Corp, 69-70, dept head spec purpose commun, 70-76; SR ENGR, IBM, 76- *Concurrent Pos:* Lectr, Drexel Inst Technol, 57-60. *Mem:* Inst Elec & Electronics Engrs; NY Acad Sci. *Res:* Communication theory; detection theory; observation and automatic observing systems; communication systems. *Mailing Add:* Inst Defense Analyses 1801 N Beauregard St Syst Eval Div Alexandria VA 22311

BLASCHEK, HANS P, b Bruchsal, WGer, Mar 16, 52; US citizen; m 75; c 2. PLASMID BIOLOGY. *Educ:* Rutgers Col, BA, 74; Rutgers Univ, MS, 77, PhD(food sci), 80. *Prof Exp:* Res intern, Dept Food Sci, Rutgers Univ, 74-77, fel, 77-78, res asst, 78-80; asst prof food microbiol, 80-86, ASSOC PROF FOOD MICROBIOL, DEPT FOOD SCI, UNIV ILL, URBANA, CHAMPAIGN, 86- *Concurrent Pos:* Consult, Food Indust. *Honors & Awards:* New Investr Res Award, NIH. *Mem:* Am Soc Microbiol; Inst Food Technologists; AAAS; Soc Indust Microbiol. *Res:* Genetic and physiological manipulation of the clostridia. *Mailing Add:* Dept Food Sci 580 Bevier Hall Univ Ill 905 S Goodwin Urbana IL 61801

BLASDELL, ROBERT FERRIS, b Kuala Lumpur, Malaya, May 12, 29; US citizen; m 63; c 3. SYSTEMATIC BOTANY, PLANT MORPHOLOGY. *Educ:* Ohio Wesleyan Univ, BA, 51; Univ Mich, MA, 54, PhD(bot), 59. *Prof Exp:* Asst prof biol, La State Univ, 59-64; asst prof, 64-69, chmn dept, 78-81, ASSOC PROF BIOL, CANISIUS COL, 69- *Mem:* Am Fern Soc; Am Soc Plant Taxonomists; Am Inst Biol Sci. *Res:* Taxonomy and morphology of the vascular plants; plant tissue and cell culture. *Mailing Add:* Dept Biol Canisius Col Buffalo NY 14208

BLASE, EDWIN W(ILLIAM), b St Charles, Mo, Sept 22, 23; m 45; c 3. CHEMICAL ENGINEERING. *Educ:* Univ Kans, BSChE, 44; NY Univ, MBA, 65. *Prof Exp:* Chem engr, Chas Pfizer & Co, 46-52, develop supvr, 52-56, pharmaceut develop supvr, 56-60, asst to vpres prod, 60-61, mgr res co-ord, Pfizer Int, 61-65, mgr pharmaceut prod serv, 65-66, asst prod mgr, Brooklyn Plant, Cordis Corp 66-68, mgr pharmaceut develop, 68-69; mgr, biochem prod, 69-84; CONSULT 84- *Mem:* Am Chem Soc; Am Inst Chem Engrs. *Res:* Technical administration; biochemical reagents; pharmaceuticals and fine chemicals. *Mailing Add:* 908 N Doral Lane Venice FL 34293-7101

BLASER, DWIGHT A, b Bucyrus, Ohio, Oct 5, 43; m 82; c 1. NOISE & VIBRATION CONTROL, FLUID MECHANICS. *Educ:* Ohio State Univ, BSME, 66, MS, 67, PhD(mech eng), 71. *Prof Exp:* Res asst thermodynamics, Exp Sta, Ohio State Univ, 65-66, res assoc, 67-71; assoc sr res engr noise &

vibration, Gen Motors Res Labs, 71-74, sr res engr, 74-80, sect mgr, Signatur Anal Group, 88-89, STAFF RES ENGR NOISE & VIBRATION, GEN MOTORS RES LABS, 80-, SECT MGR METAL CUTTING MECH & PWTRN NOISE & VIBRATION, 90- *Concurrent Pos:* Lectr, Dept Mech Eng, Ohio State Univ, 67-71. *Mem:* Soc Automotive Engrs; Sigma Xi; Soc Mfg Engrs; Am Soc Mech Engrs. *Res:* Mechanics of manufacturing systems specializing in metal removal processes, quality inspection methods through signature analysis, fourier analysis techniques and duct acoustics. *Mailing Add:* Gen Motors Res Labs Dept 15 Gen Motors Tech Ctr Warren MI 48090-9055

BLASER, MARTIN JACK, b New York, NY, Dec 18, 48; m 79; c 3. CAMPYLOBACTER & HELICOBACTER, ENTERIC BACTERIOLOGY. *Educ:* Univ Pa, BA, 69; NY Univ, MD, 73. *Prof Exp:* Intern internal med, Univ Colo Sch Med, 73-74, resident, 74-77, fel infectious dis, 77-79; EIS officer bacterial dis, Ctr Dis Control, 79-81; from asst prof to assoc prof infectious dis, Univ Colo Sch Med, 81-89; SCOVILLE PROF MED & DIR DIV INFECTIOUS DIS, VANDERBILT UNIV SCH MED, 89- *Concurrent Pos:* Chief, Infectious Dis Sect, Vet Admin Med Ctr, Denver, 81-86, clin investr, 86-; clin investr bacteriol & immunol, Rockefeller Univ, 87-88; nat counr, Am Fedn Clin Res, 86-89; assoc ed, Am J Epidemiol, 85-; young investr award, Western Soc Clin Invest, 89-; mem, Bact & Mycol Study sect, NIH, 90- *Mem:* Fel Am Col Physicians; fel Am Acad Microbiol; fel Am Col Epidemiol; Am Soc Microbiol; Campylobacter Soc (secy, 79-83); Am Soc Clin Invest. *Res:* Normal bacterial flora of the gastrointestinal tract; exogenous enteric pathogens in the epidemiology, pathogenesis, and molecular biology of gastrointestinal and extraintestinal disease. *Mailing Add:* Div Infectious Dis Vanderbilt Univ Rm A3310 Med Ctr N Nashville TN 37232-2605

BLASER, ROBERT U, b Akron, Ohio, Sept 1, 16; m 40; c 1. RESEARCH ADMINISTRATION, RESOURCE MANAGEMENT. *Educ:* Univ Akron, BS, 37. *Prof Exp:* Res engr, Babcock & Wilcox Co, 37-47, chief eng, Physics Sect, 47-51, chief, Nuclear Eng Sect, 49-51, asst supt, Mat Dept, Res Ctr, 51-61, staff asst, 61-64, chief plans, Res & Develop Div, 64-72, tech mgr acct, 72-81; RETIRED. *Concurrent Pos:* Consult, 81-; Serv Corps Ret Execs, 82- *Honors & Awards:* Charles B Dudley Medal, 57. *Mem:* Am Phys Soc; Am Soc Mech Engrs; Am Nuclear Soc; Score. *Res:* Steam generating equipment; nuclear power; materials research; environmental testing; nuclear materials for fuel elements and structural parts; forecasting and planning of research, development, and allocation of resources. *Mailing Add:* 115 Vincent Blvd Alliance OH 44601-3945

BLASIE, J KENT, b Flint, Mich, Apr 30, 43; m 65. BIOPHYSICS, PHYSICS. *Educ:* Univ Mich, BS, 64, PhD(biophys), 68. *Prof Exp:* From asst prof to assoc prof, 72-78, PROF BIOPHYS & CHEM, UNIV PA, 78- *Concurrent Pos:* NIH fel, 64-68, career develop award, 71-76; assoc biophysicist, Brookhaven Nat Lab, 73-78, biophysicist, 78- *Mem:* Biophys Soc. *Res:* Structural and dynamical study of isolated and reconstituted biological membrane systems via static and time-resolved x-ray and neutron diffraction utilizing synchrotron and laser plasma sources, spectroscopy, optical and infrared linear dichroism. *Mailing Add:* Dept Chem Univ Pa Philadelphia PA 19104

BLASING, TERENCE JACK, b Waukesha, Wis, Dec 16, 43; m 67; c 2. CLIMATOLOGY, STATISTICAL METEOROLOGY. *Educ:* Univ Wis-Madison, BS, 66, MS, 68, PhD(meteorol), 75. *Prof Exp:* Res assoc dendrochronology, Lab Tree-Ring Res, Univ Ariz, 71-77; res assoc, 77-84, RES STAFF MEM, ENERGY DIV, OAK RIDGE NAT LAB, 84- *Mem:* Am Meteorol Soc; Am Geophys Union; Sigma Xi; Am Statist Assoc; Air & Waste Mgt Asn. *Res:* Statistical meteorology; paleoclimatology; air quality. *Mailing Add:* Bldg 4500N Oak Ridge Nat Lab MS 6200 Oak Ridge TN 37831

BLASINGAME, BENJAMIN P(AUL), b State College, Pa, Aug 1, 18; m 42; c 4. ASTRONAUTICS. *Educ:* Pa State Univ, BS, 40; Mass Inst Technol, ScD, 50. *Prof Exp:* Officer, US Air Force, 41-59, dir, Titan Intercontinental Ballistic Missile Develop, 56-58; prof astronaut & head dept, US Air Force Acad, 58-59; dir eng & res, AC Spark Plug-electronics Div, Gen Motors Corp, 59-62, mgr, Milwaukee Opers, 62-65, gen mgr, AC Electronics Div, 65-70, mgr, Delco Electronics Div, Milwaukee Opers, 70-72, mgr, Delco Electronics Div, Santa Barbara Opers, 72-79; RETIRED. *Concurrent Pos:* Dir, Santa Barbara Cottage Hosp, 77-88; Santa Barbara Bank & Trust, 84- *Mem:* Nat Acad Engrs. *Res:* Instrument design and development; accelerometers; gravimeters; gyroscopes. *Mailing Add:* 517 Carriage Hill Ct Santa Barbara CA 93110

BLASINGHAM, MARY CYNTHIA, b Indianapolis, Ind, Aug 27, 48. PHYSIOLOGY. *Educ:* Duke Univ, BA, 70; Ind Univ, MS, 72, PhD(physiol), 76. *Prof Exp:* Asst, Ind Univ Med Ctr, 71-76, res assoc physiol, 76; res assoc pharmacol, Univ Tenn Ctr Health Sci, 76-79; CONSULT, PHYSIOL, 79- *Mem:* Am Physiol Soc. *Res:* Metabolism of prostaglandin E by the lung during hemorrhagic shock; interrelationship among renal hormone systems, especially prostaglandin, kallikrein-kinin, and renin angiotensin in different states of electrolyte balance. *Mailing Add:* 4150 N Meridian St Indianapolis IN 46208

BLASS, ELLIOTT MARTIN, b New York, NY, Sept 10, 40; m 76; c 4. PSYCHOPHYSIOLOGY, ANIMAL BEHAVIOR. *Educ:* Brooklyn Col, BS, 63; Univ Conn, MS, 64; Univ Va, PhD(psychol), 67. *Prof Exp:* Fel zool, Univ Pa, 67-69; from asst prof to assoc prof psychol, Johns Hopkins Univ, 69-79, prof psychol & psychiat, 79-90; PROF PSYCHOL & NUTRIT, CORNELL UNIV, 90- *Concurrent Pos:* Consult ed, J Comp & Physiol Psychol, 74- & Develop Psychol, 78-; John Simon Guggenheim Mem fel; Fulbright Hayes res prof, 83; Nat Acad Sci Exchange fel, 85. *Mem:* AAAS; Neurosci Soc; Psychonomic Soc; Develop Psychobiol Soc. *Res:* Development of ingestive behavior in infants, neurology and physiology of ingestion; maternal-infant relationships. *Mailing Add:* Dept Psychol Cornell Univ Ithaca NY 14853

BLASS, GERHARD ALOIS, b Chemnitz, Germany, Mar 12, 16; nat US; m 45; c 5. THEORETICAL PHYSICS. *Educ:* Univ Leipzig, Dr rer nat(theoret physics). *Prof Exp:* Sci collabr solid state theory, Siemens & Halske, Berlin, 43-45; lectr math & physics, OhmPolytechnicum, Nuremberg, 46-49; asst prof physics, Col St Thomas, 49-51; from asst prof to prof, 51-81, EMER PROF MATH PHYSICS & ASTROPHYS, UNIV DETROIT, 81 - *Concurrent Pos:* Chmn dept physics, Univ Detroit, 62-71; guest prof & adv, Univ Baroda, India, 67. *Mem:* Fel AAAS; Soc Asian & Comparative Philos. *Res:* Field theory; solid state physics; philosophy of science. *Mailing Add:* 4441 Stewart Rd Metamora MI 48455

BLASS, JOHN P, b Vienna, Austria, Feb 21, 37; US citizen; m 60; c 2. NEUROPSYCHIATRY, BIOCHEMISTRY. *Educ:* Harvard Univ, AB, 58; Univ London, PhD(biochem), 60; Columbia Univ, MD, 65. *Prof Exp:* Intern, Mass Gen Hosp, 65-66, asst resident med, 66-67; res assoc, Molecular Dis Br, Nat Heart Inst, 67-70; from asst prof to assoc prof biochem & psychiat, Sch Med, Univ Calif, Los Angeles, 70-76; BURKE PROF NEUROL & CHIEF DIV CHRONIC & DEGENERATIVE DIS, DEPT NEUROL, CORNELL UNIV MED COL, 78- *Mem:* Am Soc Biol Chemists; Int Brain Res Orgn; Int Soc Neurochem; Am Fedn Clin Res; Am Soc Clin Invest. *Res:* Metabolic disorders affecting the nervous system. *Mailing Add:* Burke Rehab Dementia Res Serv 785 Mamaroneck Ave White Plains NY 10605

BLASS, JOSEPH J(OHN), b Jersey City, NJ, Dec 22, 40; m 62; c 1. ENGINEERING MECHANICS OF SOLIDS. *Educ:* Manhattan Col, BME, 62; Brown Univ, ScM, 64, PhD(mech of solids), 68; Univ Tenn, MBA, 81. *Prof Exp:* Mat res engr, US Air Force Mat Lab, 67-70, vis instr mech of solids, US Air Force Inst Technol, 68-70; asst prof mech eng sci, Wayne State Univ, 70-74; MEM RES STAFF, OAK RIDGE NAT LAB, 74- *Mem:* Am Soc Mech Engrs; Soc Exp Mech. *Res:* Mechanics of solids; elasticity, plasticity; mechanical behavior of metals; fatigue; creep; fracture; high strain rate effects; wave propagation. *Mailing Add:* 236 Gum Hollow Rd Oak Ridge TN 37830

BLASS, WILLIAM ERROL, b Minneapolis, Minn, Aug 5, 37; m 59, 74, 85; c 3. MOLECULAR SPECTROSCOPY, COMETARY ASTROPHYSICS. *Educ:* St Mary's Col, Minn, BA, 59; Mich State Univ, PhD(physics), 63. *Prof Exp:* Asst prof physics, St Mary's Col, Minn, 63-67; asst prof physics & astron, Univ Tenn, 67-71, assoc prof, 71-78, vdir opers, Molecular Spectros Lab, 71-78, PROF PHYSICS & ASTRON, UNIV TENN, KNOXVILLE, 81-, DIR, MOLECULAR RES LAB, 78-, ASSOC DIR, UNIV COMPUT CTR, 90- *Concurrent Pos:* Co-prin investr, AEC res contract, 64-67; prin investr, Univ Tenn-NASA grants, 68-70 & 81; mem sr sci staff, NASA grant, 69-77; NASA/Am Soc Eng Educ summer fac fel, Goddard Space Flight Ctr, 86, 87, 90 & 91; consult, Corp Res, Honeywell, Inc, 81, transp serv, Tenn Valley Authority, 88, Martin Marietta Energy Systs, Inc, 89-; nuclear res assoc, Dept Energy, Metals & Ceramic Div, Oak Ridge Nat Lab, 89. *Mem:* Am Phys Soc; Optical Soc Am; Planetary Soc; Soc Appl Spectros; Coblentz Soc; Int Neural Network Soc; Inst Elec & Electronic Engrs Computer Soc. *Res:* Theoretical and experimental molecular spectroscopy including laser spectroscopy; planetary astrophysics especially physics of comets; neural network computational systems; methods of large system data analysis; digital signal processing particularly deconvolution of experimental data; non-equilibrium thermal physics. *Mailing Add:* Dept Physics & Astron Univ Tenn Knoxville TN 37996-1200

BLASZKOWSKI, THOMAS P, b Mount Pleasant, Pa, Nov 9, 34. PHARMACOLOGY. *Educ:* Univ RI, PhD(pharmacol), 69. *Prof Exp:* PHARMACOLOGIST, NAT HEART, LUNG & BLOOD INST, NIH, 72- *Mailing Add:* 14342 Chesterfield Rd Rockville MD 20853

BLATCHFORD, JOHN KERSLAKE, b Chicago, Ill, Sept 6, 25; m 60; c 1. ORGANIC CHEMISTRY. *Educ:* Univ Colo, BA, 49, MA, 56; Univ Cincinnati, PhD(org chem), 63. *Prof Exp:* Res chemist, Chicago Copper & Chem Co, 50-51; chief chemist, Edcan Labs, 57-61; res chemist, Whirlpool Corp, 63-68, staff chemist, Res & Eng Ctr, 68-71, staff scientist, 71-79; pres, Quanta Labs, Inc, 79-85; MGR CHEM, PHILLIPS ENG CO, 85- *Concurrent Pos:* Adj fac chem, Southwestern Mich Col, 83-84. *Mem:* AAAS; Am Chem Soc; Sigma Xi; Int Soc Hybrid Microelectronics. *Res:* Organometallic compounds; polycyclic-aromatic hydrocarbons; organic synthesis; cyclopropanes; carbene chemistry; snake venoms; detergency; thick film hybrid electronics; polymer synthesis; chromatography; absorption refrigeration; corrosion. *Mailing Add:* 411 Wallace Ave St Joseph MI 49085

BLATSTEIN, IRA M, b Hackensack, NJ, Apr 18, 44; m 68; c 2. UNDERWATER ACOUSTICS. *Educ:* Drexel Univ, BS, 67; Cath Univ Am, MS, 71, PhD(physics), 73. *Prof Exp:* Res physicist underwater explosion effects, 67-81, HEAD, RADIATION DIV, RES & TECHNOL DEPT, WHITE OAK LAB, NAVAL SURFACE WEAPONS CTR, 81- *Mem:* Am Phys Soc; Acoustical Soc Am. *Res:* Long range propagation of underwater explosion energy; normal mode theory; ray theory and modified ray theory; ocean basin reverberation. *Mailing Add:* 600 Avon Square Ct Silver Spring MD 20904

BLATT, CARL ROGER, b Summit, NJ, July 16, 38; m 62; c 4. SOIL SCIENCE, PLANT PHYSIOLOGY. *Educ:* Del Valley Col Agr & Sci, BSc, 60; Rutgers Univ, MSc, 63, PhD(soils), 64. *Prof Exp:* res scientist plant nutrit, Can Dept Agr, 64-80; RES SCIENTIST PLANT NUTRIT, KENTVILLE AGR CTR, 80-, AGRON PROG LEADER, 87- *Mem:* Am Soc Hort Sci; Can Soc Hort Sci; Coun Soil Testing & Plant Anal. *Res:* Nutritional requirements and management studies of berry crops, such as strawberry, raspberry, high and lowbush blueberry, and vegetables, such as tomatoes, cabbage, cauliflower, broccoli, and cereal crops (wheat, etc). *Mailing Add:* Kentville Agr Ctr Kentville NS B4N 1J5 Can

BLATT, ELIZABETH KEMPSKE, b Baltimore, Md, Dec 2, 36; m 58; c 3. MEDICAL PHYSIOLOGY, NEUROENDOCRINOLOGY. *Educ:* Goucher Col, BA, 56; Vassar Col, MA, 58; Union Univ, PhD(life sci & syst), 71. *Prof Exp:* Instr physiol, Mt Holyoke Col, 58-59; res asst zool, Kans State

Univ, 64-65; from instr to asst prof biol, Concord Col, 67-73; assoc prof physiol & chmn dept, 74-78, prof physiol & assoc dean acad affairs & basic sci, WVa Sch Osteop Med, 78-84; dean, Lake Erie Col, 84-88; DEP DIR, SCI MUS, VA, 88- *Concurrent Pos:* Fac fel, US Dept Health, Educ & Welfare, 69-70; grant-in-aid, Sigma Xi, 70; consult health planning, Appalachian Regional Comn, 73-74. *Mem:* AAAS; Sigma Xi; assoc mem Am Osteop Asn. *Res:* Influence of pineal on reproductive rhythmicity; electrophysiological correlates of melatonin activity. *Mailing Add:* Sci Mus Va 2500 W Broad St Richmond VA 23220

BLATT, FRANK JOACHIM, b Vienna, May 1, 24; nat US; m 46; c 2. PHYSICS. *Educ:* Mass Inst Technol, BSEE, 46, MSEE, 48; Univ Wash, PhD(physics), 53. *Prof Exp:* Asst elec eng, Mass Inst Technol, 46-48; assoc, Univ Wash, 48-50, asst physics, 50-53; res assoc, Univ Ill, 53-55, res asst prof, 55-56; from asst prof to prof physics, Mich State Univ, 56-87, chmn dept, 69-73; PROF PHYSICS, UNIV VT, 87- *Concurrent Pos:* Consult, Naval Res Lab, 55-59, Eng Res Inst, Univ Mich, 58-59 & Boeing Aircraft Co, 59; NSF fel, Oxford Univ, 59-60; consult, Argonne Nat Lab, 61-64, E I du Pont de Nemours & Co, Inc, 62-, Argonne Nat Lab, 67-75, Simon Fraser Univ, 72-78, Univ de Louvan, 81, Leed Univ, 82 & Univ N Sumatra, 85 & 86; guest prof, Swiss Fed Inst Technol, 63-64; mem prog rev comt, Solid State Sci Div, Argonne Nat Lab, 66-72; vis prof, ITM, Shah Alam, Malaysia, 86-87. *Mem:* Am Phys Soc; Biophys Soc. *Res:* Physics of solids; transport theory; thermoelectric and thermomagnetic effects. *Mailing Add:* Dept Physics Univ Vt Burlington VT 05405

BLATT, HARVEY, b New York, Sept 11, 31; m 84; c 3. PETROLOGY. *Educ:* Ohio State Univ, BS, 52; Univ Tex, MA, 58; Univ Calif, Los Angeles, PhD(geol), 63. *Prof Exp:* From instr to assoc prof geol, Univ Houston, 62-68; assoc prof, 68-76, PROF GEOL, UNIV OKLA, 76- *Concurrent Pos:* Fulbright sr fel, 74-75. *Mem:* Geol Soc Am; Soc Econ Paleont & Mineral. *Res:* Sedimentary petrology; geochemistry and genesis of sandstones, limestones, dolmites and cherts; provenance studies; petrographic stratigraphy. *Mailing Add:* Sch Geol & Geophys Univ Okla Norman OK 73019

BLATT, JEREMIAH LION, b Ft Worth, Tex, Oct 27, 20; m 59; c 3. MOLECULAR BIOLOGY, SCIENCE EDUCATION. *Educ:* Univ Calif, AB, 43; Univ Calif, Los Angeles, PhD(biochem), 55. *Prof Exp:* Degaussing physicist, USN, 43-45; assoc, Agr Exp Sta, Univ Calif, 45-47; res assoc & instr biochem, Univ Chicago, 55-57; asst prof physiol, Mt Holyoke Col, 57-59; asst prof biophys, Kans State Univ, 59-65; tutor, St Johns Col, 65-67; chmn div natural sci, Concord Col, 67-84, prof natural sci, 67-84; RETIRED. *Mem:* AAAS; Am Chem Soc; Sigma Xi. *Res:* Theory of chromatography; protein structure and biosynthesis; information transfer in biological systems; general education in natural science. *Mailing Add:* 1831 Glencove Lane Richmond VA 23225

BLATT, JOEL HERMAN, b Washington, DC, Mar 30, 38; m 68. APPLIED OPTICS, INSTRUMENTATION. *Educ:* Harvard Univ, BA, 59; Univ Ala, Huntsville, MS, 67; Univ Ala, Tuscaloosa, PhD(physics), 70. *Prof Exp:* Physicist, Missile Support Command Calibration Ctr, Redstone Arsenal, US Army, 62-64, res physicist, Missile Command Propulsion Lab, 64-66; sr scientist, Hayes Int Corp, Ala, 66-67; sr res asst physics, Univ Ala, Huntsville, 68-70; from asst prof to assoc prof, 70-82, PROF PHYSICS, FLA INST TECHNOL, 82- *Concurrent Pos:* Consult eng instrumentation, electromigration, acousto- optics. *Mem:* Int Soc Optical Eng; Nat Speleol Soc; Optical Soc Am; Laser Inst Am. *Res:* Mass spectrometry of organic radicals; optical spectroscopy and fluorometry; radiometry; general and computer interfaced instrumentation in optical and engineering areas; solid state physics; electromigration; acousto-optics; moire profilometry. *Mailing Add:* Dept Physics & Space Sci Fla Inst Technol Melbourne FL 32901

BLATT, JOEL MARTIN, b Brooklyn, NY, Oct 28, 42; m 75; c 1. BIOCHEMISTRY. *Educ:* Brooklyn Col, BS, 63; Univ Wis-Madison, PhD(physiol chem), 68. *Prof Exp:* NIH fel, Purdue Univ, West Lafayette, 68-70, res assoc molecular biol, 70-72; from asst prof to assoc prof biochem, Meharry Med Col, 72-84; staff scientist, 84-88, SR STAFF SCIENTIST, DIAG DIV, MILES INC, ELKHART, IND, 88- *Concurrent Pos:* NSF res grant, Meharry Med Col, 74-75. *Mem:* AAAS; Am Chem Soc; Am Asn Clin Chem; Am Soc Microbiol; Sigma Xi; NY Acad Sci. *Res:* Interlocking control mechanisms for gene expression and enzyme activity; regulation of branched-chain amino acid biosynthesis in bacteria; dry chemistries for medical diagnostic testing; film coating and surface chemistry. *Mailing Add:* 17100 Barryknoll Way Granger IN 46530

BLATT, S LESLIE, b Philadelphia, Pa, June 10, 35; m 59; c 3. RADIATIVE CAPTURE REACTIONS, INSTRUMENTATION. *Educ:* Princeton Univ, AB, 57; Stanford Univ, MS, 59, PhD(physics), 65. *Prof Exp:* Res assoc physics, Van de Graaff Accelerator Lab, Ohio State Univ, 64-66, from asst prof to prof physics, 66-87, chmn dept, 80-86; PROF PHYSICS, DEAN GRAD SCH & RES COORDR, CLARK UNIV, 87- *Concurrent Pos:* Vis researcher, Nuclear Res Ctr, Strasbourg, France, 72-73; chmn, Ind Univ Cyclotron Facil, Nat User's Group, 76-77. *Mem:* Am Asn Univ Prof; Am Phys Soc; Am Asn Physics Teachers; Sigma Xi. *Res:* Studies of charged particle radiative capture reactions and other nuclear reactions; gamma-ray detector systems; design and applications of nuclear instrumentation and computer-based data acquisition systems. *Mailing Add:* Grad Sch Clark Univ 950 Main St Worcester MA 01610

BLATT, SYLVIA, b New York, NY, July 10, 18. CLINICAL CHEMISTRY, SCIENCE ADMINISTRATION. *Educ:* Hunter Col, AB, 38; City Col New York, MS, 40. *Prof Exp:* Asst chemist, City Col New York, 43-46 & Morrisania City Hosp, 46-49; chemist, Bd of Water Supply Lab, New York, 49-53; supvr biochem unit, 53-60, asst chief div supv clin labs & blood banks, 60-62, chief div lab field serv, 62-71, from asst dir lab improv to dir, 71-85, CONSULT CLIN LAB MATTERS, BUR LABS, NEW YORK CITY DEPT HEALTH, 85- *Concurrent Pos:* Consult, Prof Exam Serv, Am Pub Health Asn, 65-; mem adv comts, Queensboro & New York Community Cols. *Mem:* NY Acad Sci; AAAS; Am Asn Clin Chemists; Am Chem Soc; Am Pub Health Asn. *Res:* Clinical chemistry, especially the technical and administrative aspects; quality control and proficiency testing programs in clinical chemistry; laboratory improvement programs. *Mailing Add:* 352 W 46th St New York NY 10036

BLATTEIS, CLARK MARTIN, b Berlin, Ger, June 25, 32; nat US; m 58; c 3. NEUROSCIENCE. *Educ:* Rutgers Univ, BA, 54; Univ Iowa, MSc, 55, PhD(physiol), 57. *Prof Exp:* Asst physiol, Univ Iowa, 54-57; res physiologist, US Army Med Res Lab, Ft Knox, Ky, 58-61 & US Army Res Inst Environ Med, Natick, Mass, 63-66; assoc prof, 55-74, PROF PHYSIOL & BIOPHYS, COL MED, UNIV TENN, MEMPHIS, 74- *Concurrent Pos:* USPHS res fel, Inst Andean Biol, Lima, Peru, 61-62 & Nuffield Inst Med Res, Oxford, Eng, 62-63; Orgn Am States exchange sci award, 68; vis prof physiol, Sch Med, Univ Guanajuato, 68-78; Fulbright-Hays sr scholar, 75; vis prof, Philipps-Univ, Marburg, WGer, 79, 81 & Commonwealth Sci & Indust Res Orgn, Sydney, Australia, 83. *Mem:* AAAS; Soc Neurosci; Am Physiol Soc; Soc Exp Biol & Med; corresp mem Peruvian Soc Physiol Sci; Soc Levkocyte Biol. *Res:* Central nervous system; temperature regulation; host-defense mechanisms. *Mailing Add:* Dept Physiol & Biophys Univ Tenn Col Med Memphis TN 38163

BLATTNER, FREDERICK RUSSELL, b St Louis, Mo, Oct 6, 40; m 71; c 2. BIOPHYSICS, SOFTWARE SYSTEMS. *Educ:* Oberlin Col, BA, 62; Johns Hopkins Univ, PhD(biophys), 68. *Prof Exp:* Res assoc oncol, McArdle Lab, 68-73, asst scientist genetics, 73-74, from asst prof to assoc prof, 74-81, PROF GENETICS, UNIV WIS-MADISON, 81- *Concurrent Pos:* Pres, DNASTAR, Inc; cur, GenBank Adv Comt, NIH, & mem, Protein Identification Adv Comt. *Mem:* Sigma Xi. *Res:* Molecular biology, DNA sequencing, bacteriophage lambda, DNA cloning, recombination, replication, immunology, structure of immunoglobulin genes in mouse and man. *Mailing Add:* 1547 Jefferson Ave Madison WI 53711

BLATTNER, MEERA MCCUAIG, b Chicago, Ill, Aug 14, 30; m 85; c 3. COMPUTER SCIENCES. *Educ:* Univ Chicago, BA, 52; Univ Southern Calif, MA, 66; Univ Calif, Los Angeles, PhD(eng), 73. *Prof Exp:* Asst prof math, Los Angeles Harbor Col, 66-73; instr comput sci, Univ Mass & res fel comput sci, Harvard Univ, 73-74; asst prof math sci, Rice Univ, 74-81; ASSOC PROF APPL SCI, UNIV CALIF, DAVIS & COMPUT SCIENTIST, LAWRENCE LIVERMORE NAT LAB, 80- *Concurrent Pos:* Adj prof biomath, M D Anderson Cancer Res Hosp, Univ Tex Med Ctr, 77- *Mem:* Soc Women Engrs; Asn Comput Mach; Inst Elec & Electronics Engrs. *Res:* Multimedia; user interfaces; software environments; data and format conversion. *Mailing Add:* Dept Appl Sci Univ Calif Davis PO Box 808 Livermore CA 94550

BLATTNER, ROBERT J(AMES), b Milwaukee, Wis, Aug 6, 31; m 56, 83; c 3. LIE GROUPS, HARMONIC ANALYSIS. *Educ:* Harvard Col, AB, 53; Univ Chicago, PhD(math), 57. *Prof Exp:* From instr to assoc prof, 57-71, chmn dept, 81-84, PROF MATH, UNIV CALIF, LOS ANGELES, 71- *Concurrent Pos:* NSF fel, Mass Inst Technol, 61-62; mem, Inst Advan Study, 64-65; prof math & head dept, Univ Mass, 73-75; vis scholar, Mass Inst Technol, 81; visitor, Univ Warwick, 81; hon fel, Univ Wis-Madison, 85; mem, Math Sci Res Inst, Berkeley, 85. *Mem:* Am Math Soc. *Res:* Representations of lie groups; theory of lie algebras and superalgebras; Hopf algebras. *Mailing Add:* Dept Math Univ Calif Los Angeles CA 90024-1555

BLATTNER, WILLIAM ALBERT, b St Louis, Miss, Oct 16, 43; m 74; c 5. EPIDEMIOLOGY, ONCOLOGY. *Educ:* Wash Univ St Louis, BA, 66, MO, 70; diplomate, Am Bd Internal Med, 74, Subspecialty Bds Med Oncol, 77. *Prof Exp:* Intern, Strong Mem Hosp, 70-71, asst resident, 71-72; assoc res, New York Cornell Med Ctr, 72-73; staff assoc, Nat Cancer Inst, 73-75; clin invest, 75-76, sr clin invest, 76-81, chief, Family Studies Sect, 81-87, CHIEF, VIRAL EPIDEMIOL SECT, NAT CANCER INST, 87- *Concurrent Pos:* mem, Epidemiol Subcomt PHS AIDS Exec Task Force, 85-87, co-chmn, AID Epidemiol & Surveillance Subcomt, 85-; mem, div Cancer Etiol, Sr Promotions Review Comt, 86- *Honors & Awards:* Fifth Annual Russel J Blattner Lectr, Texas Childrens Hosp, 84. *Mem:* Am Soc Clin Oncol; Am Fedn Clin Res; Am Assoc Adv Sci; Am Assoc Cancer Res; Int Soc Comparative Cancer Res. *Res:* Pioneering studies undertaken to define the epidemiology of human retroviruses from a national and international perspective; major focus on role of these viruses in human malignancy, neurologic disease and AIDS. *Mailing Add:* 6130 Executive Blvd Exec Plaza N Rm 434 Rockville MD 20852

BLATZ, PAUL E, b Pittsburgh, Pa, Aug 29, 23; m 65; c 5. BIOCHEMISTRY. *Educ:* Southern Methodist Univ, BS, 51; Univ Tex, PhD(chem), 55. *Prof Exp:* Sr res chemist org synthesis, Dow Chem Co, 55-59; res assoc polymer chemist, Socony Mobil Oil Co, 59-63; assoc prof chem, NMex Highlands Univ, 63-64; from assoc prof to prof, Univ Wyo, 64-71; chmn dept, 71-76, PROF CHEM, UNIV MO-KANSAS CITY, 71- *Concurrent Pos:* Vis prof, Fla State Univ, 79, Pan Am Univ, 80. *Mem:* AAAS; Am Chem Soc; Biophys Soc; Am Asn Biol Chemists; Sigma Xi. *Res:* Chemistry of vision, particularly the spectroscopy and chemistry of visual pigments, retinal and related polyenes. *Mailing Add:* Dept Chem Univ Mo 5100 Rockhill Kansas City MO 64110

BLAU, HARVEY ISAAC, b Los Angeles, Calif, Aug 31, 42; m 68; c 2. ALGEBRA. *Educ:* Reed Col, BA, 63; Yale Univ, MA, 65, PhD(math), 69. *Prof Exp:* Asst prof math, Southern Conn State Col, 67-69; from asst prof to assoc prof, 69-86, PROF MATH, NORTHERN ILL UNIV, 86- *Mem:* Am Math Soc; Sigma Xi; Math Asn Am. *Res:* Representations of finite groups. *Mailing Add:* Dept Math Northern Ill Univ DeKalb IL 60115

BLAU, HELEN MARGARET, b London, Eng, May 8, 48; US citizen; m 76; c 2. DEVELOPMENTAL BIOLOGY, PHARMACOGENETICS. *Educ:* Univ York, Eng, BA, 69; Harvard Univ, MA, 70, PhD(biol), 75. *Prof Exp:* Resident tutor, Harvard Univ, 72-74; fel biochem, med, & genetics, Univ

Calif, San Francisco, 75-78; ASST PROF PHARM, STANFORD UNIV SCH MED, 78- Concurrent Pos: Mem bd dir, Northern Calif Burn Coun, 77-; Mellon fel, 79-80; mem, Medfly Health Adv Comt, Calif, 81; Hume fel, 81-82; NIH res career develop award. Mem: Am Soc Cell Biol; Fedn Am Scientists; fel AAAS; Sigma Xi; Hastings Inst Soc, Ethics & Life Sci. Res: Control of gene expression during human muscle development; studies of the differentiative transition; muscle gene expression in somatic cell hybrids; etiology of human muscular dystrophies and pharmacogenetic disorders of muscle. Mailing Add: Stanford Univ Sch Med Dept Pharmacol Stanford CA 94305

BLAU, HENRY HESS, JR, b Speers, Pa, Feb 16, 30; m 58; c 3. ATMOSPHERIC PHYSICS. Educ: Yale Univ, BS, 51; Ohio State Univ, MSc, 52, PhD(physics), 55. Prof Exp: Physicist, Arthur D Little, Inc, 55-66, leader physics sect, Res & Develop Div, 62-66; physicist, Space Sci Lab & group leader geophys, Missile & Space Div, Gen Elec Co, 66-67; physicist, Arthur D Little, Inc, 67-71; vpres & dir, Environ Res & Technol, Inc, 71-80; consult, 80-82; PRES, WAYLAND RES INC, 82- Mem: Fel Optical Soc Am; Am Phys Soc; AAAS; Sigma Xi. Res: Optical physics; molecular spectroscopy; air quality sensors. Mailing Add: Center Hill Rd, RR No 1 Box No 595, Mt Blue Weld ME 04285-9710

BLAU, JULIAN HERMAN, mathematics; deceased, see previous edition for last biography

BLAU, LAWRENCE MARTIN, b New York, NY, Jan 6, 38; m 59; c 2. MEDICAL PHYSICS. Educ: Princeton Univ, BA, 59; Univ Rochester, MA, 63, PhD(physics), 65. Prof Exp: Res assoc physics, Columbia Univ, 65-69; assoc scientist, Hosp Spec Surg & asst prof physics in radiol, Med Col, Cornell Univ, 69-87, dir comput serv, Hosp Spec Surg, 73-87; DIR, MGT INFO SYSTS, HOSP JOINT DIS ORTHOP INST, 87- Concurrent Pos: Consult. Mem: Soc Nuclear Med; Am Phys Soc; Am Asn Physics Teachers; Orthop Res Soc; Sigma Xi. Res: Investigation and application of physics, mathematics and computers in nuclear medicine; application of computer technology in hospitals. Mailing Add: 18 Channing Pl East Chester NY 10709

BLAU, MONTE, b New York, NY, June 17, 26; m 46; c 2. NUCLEAR MEDICINE. Educ: Polytech Inst Brooklyn, BS, 48; Univ Wis, PhD(chem), 52. Prof Exp: Res asst, Geochronomet Lab, Yale Univ, 52-53; mem staff, Radioisotope Lab, Montefiore Hosp, 53-54; mem staff, dept biochem res, Roswell Park Mem Inst, 54-57, mem staff, dept nuclear med, 57-75; prof nuclear med & biophys, State Univ NY, Buffalo, 70-83, chmn, dept nuclear med, 75-83; VIS PROF, RADIOL DEPT, HARVARD MED SCH, 83- Honors & Awards: Hevesy Nuclear Pioneer Award, Soc Nuclear Med, 79 600. Mem: Am Soc Nuclear Med (pres, 72-73); Am Chem Soc; Am Asn Physicists Med. Res: Radiopharmaceuticals; radioisotope instrumentation; nuclear magnetic resonance imaging. Mailing Add: Radiol Dept Harvard Med Sch Boston MA 02115

BLAU, WILLIAM STEPHEN, ecology, evolutionary biology, for more information see previous edition

BLAUER, AARON CLYDE, b Burley, Idaho, Apr 26, 39; m 62; c 7. PLANT TAXONOMY, RANGE ANALYSIS. Educ: Brigham Young Univ, BS, 64, MS, 65. Prof Exp: From instr to assoc prof, 66-86, PROF BOT & MICROBIOL, SNOW COL, 86- Concurrent Pos: Range res technician, Intermountain Forest & Range Exp Sta, USDA, 67-73; botanist, 74-; pres, Snow Col Fac Asn, 70-72; mem bd dirs, Dept Higher Educ, Utah Educ Asn, 72; deleg, Int Bot Sci, SAfrica, 84. Mem: Bot Soc Am; AAAS; NY Acad Sci; Am Hort Soc; Am Inst Biol Sci. Res: Taxonomic and genetic research on Western range browse shrubs, primarily of the Chenopodiaceae, Rosaceae and Asteraceae. Mailing Add: Snow Col 150 E College Ave Ephraim UT 84627

BLAUFOX, MORTON D, b New York, NY, July 19, 34; m 59; c 3. INTERNAL MEDICINE, NUCLEAR MEDICINE. Educ: State Univ NY, MD, 59; Univ Minn, PhD(med), 64. Prof Exp: Asst, Peter Bent Brigham Hosp, 64-66; instr med & dir sect nuclear med, 66-68, from asst prof to assoc prof radiol, 66-76, from asst prof to assoc prof med, 68-78, PROF RADIOL & NUCLEAR MED, ALBERT EINSTEIN COL MED, 76-, PROF MED, 78-, CHMN, DEPT NUCLEAR MED, 82- Concurrent Pos: Am Heart Asn advan res fel, 64-66; scholar, Harvard Col & res fel, Harvard Med Sch, 64-66; attend physician & dir, Div Nuclear Med, Hosp Albert Einstein Col Med, Montefiore Hosp & Med Ctr, Bronx Munic Hosp Ctr & NCent Bronx Hosp; mem, High Blood Pressure Res Coun & Coun Cardiovasc Radiol, Am Heart Asn; prin investr hypertension detection & follow-up prog, New York Ctr, Nat Heart & Lung Inst, 72-80; mem hypertension detection & follow-up prog, Nat Heart, Lung & Blood Inst, 79-, dietary intervention study hypertension, 80- Honors & Awards: Edgar Cigelman Award, 52; Edward Noble Found Award, 63; Albert Lasker Award, NIH, 80. Mem: Fel Am Col Physicians; Am Col Nuclear Physicians; fel Coun High Blood Pressure Res; fel NY Acad Med; Am Heart Asn; Am Fedn Clin Res; AMA; Soc Nuclear Med. Res: Nephrology and nuclear medicine. Mailing Add: Albert Einstein Col Med 1300 Morris Park Ave Bronx NY 10461

BLAUNSTEIN, ROBERT P, b New York, NY, Nov 9, 39; m 62; c 2. ADMINISTRATION, RESEARCH & DEVELOPMENT. Educ: City Col New York, BS, 61; Western Reserve Univ, MS, 64; Univ Tenn, PhD(physics), 68. Prof Exp: Res scientist radiation physics, Oak Ridge Nat Lab, 68; asst prof physics, Univ Tenn, Knoxville, 68-73; prof appt, USERDA, 73-76; prog mgr, 76-80, mgr, Div Technol Assessment, 80-83, MGR, OFF NUCLEAR SAFETY, US DEPT ENERGY, 83- Concurrent Pos: Consult, Oak Ridge Nat Lab, 68-73. Mem: Am Phys Soc; Sigma Xi. Res: Interaction of low energy electrons with polyatomic molecules; electron attachment and scattering processes; interaction of ultra-violet light with polyatomic molecules; excimer formation inter- and intra-molecular energy transfer; environmental assessment. Mailing Add: 3907 Virgilia Chevy Chase MD 20815

BLAUSCHILD, ROBERT ALAN, b New York, NY, Dec 2, 48; m 79. ELECTRICAL ENGINEERING. Educ: Columbia Univ, BS, 71; Univ Calif, Berkeley, MS, 73. Prof Exp: MGR ANALOG RES INTEGRATED CIRCUITS, SIGNETICS CORP, 73-; PRES, LINEAR DESIGN INC. Concurrent Pos: Mem prog comt, Int Solid State Circuits Conf, 79- Mem: Inst Elec & Electronics Engrs. Res: New areas of analog integrated circuits including high speed data acquisition and conversion and linear metal-oxide-semiconductor circuits. Mailing Add: Linear Design Inc 22351 Hartman Dr Los Altos CA 94022

BLAUSTEIN, ANDREW RICHARD, b New York, NY, July 17, 49; m 71. POPULATION BIOLOGY, BEHAVIORAL ECOLOGY. Educ: Southampton Col, BA, 71; Univ Nev, Reno, MS, 73; Univ Calif, Santa Barbara, PhD(biol sci), 78. Prof Exp: Teaching asst & assoc biol, Univ Calif, Santa Barbara, 73-78; ASST PROF ZOOL, ORE STATE UNIV, 78- Mem: AAAS; Ecol Soc Am; Animal Behav Soc; Am Soc Naturalists; Am Soc Mammalogists. Res: Population, community, behavioral and evolutionary ecology. Mailing Add: Dept Zool Ore State Univ Corvallis OR 97331

BLAUSTEIN, BERNARD DANIEL, b Philadelphia, Pa, Apr 26, 29; m 54; c 3. PHYSICAL CHEMISTRY, COAL SCIENCE. Educ: Univ Pa, BSChE, 50; Johns Hopkins Univ, MA, 51, PhD(chem), 57. Prof Exp: Jr instr chem, John Hopkins Univ, 50-54; instr, Sch Pharm, Univ Md, 55-56; asst technologist, Res Ctr, US Steel Corp, 56-58; from instr to adj assoc prof, Univ Pittsburgh, 57-77; supvry phys chemist, 58-62, res chemist, 62-68, supvry res chemist, Bur Mines, US Dept Interior, 68-75; prog coordr, 77-79, div dir, 79-85, dep assoc dir, 85-87, SR SCIENTIST, PITTSBURGH ENERGY TECHNOL CTR, US DEPT ENERGY, 87- Mem: Sigma Xi; AAAS; Am Chem Soc. Res: Chemistry of coal; coal liquefaction; bioprocessing of coal. Mailing Add: DOE-Perc PO Box 10940 Pittsburgh PA 15236-0990

BLAUSTEIN, ERNEST HERMAN, b Boston, Mass, June 29, 21; m 42; c 2. EDUCATIONAL ADMINISTRATION, MICROBIOLOGY. Educ: Boston Col, AB, 41; Mass Inst Technol, MPH, 42; Boston Univ, PhD(microbiol), 52. Prof Exp: From instr to assoc prof, 46-57, chmn, Dept Sci, 64-69, Col Lib Arts, 69-73, assoc dean, Col Lib Arts, 69-86, coordr, Six Yr Lib Arts Med Ed Prog 73-86, PROF BIOL, BOSTON UNIV, 57-, ASSOC VPRES DEVELOP, 86- Concurrent Pos: Spec res fel, NIH, 60-61; bk ed, J Col Sci Teaching; dir, Washington Legis Internship Prog, co-dir, Modular Med Curric; dir, Combined Degree Progs Med, Law & Mgt, Boston Univ; vis dean, Hebrew Univ, Hadassah Med Sch, 84; consult, NJ Dept Educ. Mem: Fel Am Acad Microbiol; fel Royal Soc Health; Am Conf Acad Deans; Asn Am Med Cols; Asn Adv Health Prof. Res: Virology; tissue culture; medical microbiology. Mailing Add: 855 Commonwealth Ave Boston MA 02215

BLAUSTEIN, MORDECAI P, b New York, NY, Oct 19, 35; m 59; c 2. CELL PHYSIOLOGY, NEUROPHYSIOLOGY. Educ: Cornell Univ, BA, 57; Washington Univ, MD, 62. Prof Exp: Intern internal med, Boston City Hosp, 62-63; med res officer, US Navy, 63-66; NIH spec fel physiol, Cambridge Univ, 66-68; assoc prof physiol & biophysics, Med Sch, Washington Univ, 68-75, prof, 75-80; PROF PHYSIOL & CHMN DEPT, MED SCH, UNIV MD, 79-, PROF MED, 83-, SCI DIR, HYPERTENSION CTR, 85- Concurrent Pos: NATO sr fel, Univ Bern, Switz, 71; guest scientist, Lab Marine Biol Assoc, Plymouth, Eng, 73; vis prof, Univ Nat Autonomous Mex, 79, Univ Padua, Italy, 87, Univ La Plata, Argentina, 90. Mem: Am Physiol Soc; Biophys Soc; Soc Gen Physiologists; Soc Neurosci. Res: Regulation of cell calcium in nerve and muscle; physiology of presynaptic nerve terminals; role of salt in the etiology of essential hypertension. Mailing Add: Dept Physiol Univ Md Sch Med 655 Baltimore St Baltimore MD 21201

BLAW, MICHAEL ERVIN, b Gary, Ind, Nov 14, 27; m 50; c 4. PEDIATRIC NEUROLOGY. Educ: Univ Chicago, PhB, 49, MD, 54; Am Bd Psychiat & Neurol, dipl. Prof Exp: Resident pediat, Univ Chicago, 54-56, instr pediat, 59-60; resident neurol, Univ Mich, 56-59; from asst prof to assoc prof pediat neurol, Univ Minn, Minneapolis, 60-68; PROF PEDIAT & NEUROL, UNIV TEX SOUTHWESTERN MED SCH DALLAS, 68-, VCHMN, DEPT NEUROL, 80- Concurrent Pos: Nat Inst Neurol Dis & Blindness spec trainee grant, 58 & pediat neurol training grant, 65; NIH grant, 64; consult, Scottish Rite Hosp, Dallas; mem bd trustees, United Cerebral Palsy Asn Dallas County. Mem: AAAS; Int Child Neurol Asn (vpres); Child Neurol Soc; fel Am Acad Neurol. Res: Clinical studies related to neurological disease in childhood. Mailing Add: Health Sci Ctr Univ Tex Dallas TX 75235

BLAY, GEORGE ALBERT, b Buenos Aires, Arg, Mar 25, 32; US citizen; m 65; c 2. ELECTROANALYTICAL CHEMISTRY. Educ: Eng Technol Inst, Arg, BS, 56; Univ Buenos Aires, MS, 58, PhD(anal chem), 60. Prof Exp: Analyst, City Water Works Lab, Arg, 54-55; res chemist, AEC, 55-60; res chemist, 60-63, sr res chemist, 63-68, RES ASSOC, HOECHST-CELANESE, 68- Concurrent Pos: Asst inorg & radio chem, Univ Buenos Aires, 57-60. Mem: Am Chem Soc; Arg Chem Asn. Res: Analytical chemistry; uranium processing; radiochemistry; electroanalysis; polarography; analytical instrumentation; computer control of bioreactors; catalytic oxidation; nylon chemistry; polyesters, terephthalic acid, wacker acetaldehyde; monstanto acetic acid, acrylic acid, ethyl acrylate chemistry; ethylene glycol/ethylene oxide. Mailing Add: Hoechst-Celanese PO Box 9077 Corpus Christi TX 78469

BLAYDEN, LEE CHANDLER, b Opportunity, Wash, Aug 30, 41; m 67; c 3. METALLURGICAL ENGINEERING, ENVIRONMENTAL ENGINEERING. Educ: Mich State Univ, BS, 64; Univ Wis, MS, 65. Prof Exp: Engr powder metall, 65-68, engr cast metals, 68-71, sr engr metal fluxing, 71-73, sr engr recycling, 73-76, staff engr metal forming, 76-78, supvr, 78-80, MGR, ENVIRON CONTROL LABS, ALUMINUM CO AM, 80- Concurrent Pos: Consult, Off Technol Assessment, 75. Honors & Awards: Arthur Vining Davis Award. Mem: Am Soc Metals; Sigma Xi. Res: Metal forming; municipal refuse processing; scrap recycling and remelting; non-polluting molten metal purification; continuous ingot casting; environmental control systems; environmental monitoring. Mailing Add: 247 Whiteoak Dr New Kensington PA 15068

BLAYDES, DAVID FAIRCHILD, b Columbus, Ohio, Aug 17, 34; m 61; c 2. PLANT PHYSIOLOGY, TISSUE & CELL CULTURE. *Educ:* Ohio State Univ, BS, 56; Univ Wis, MS, 57; Ind Univ, PhD(bot), 62. *Prof Exp:* Res assoc, Mich State Univ, 62-65, NIH fel, 64-65; from asst prof to assoc prof, 65-80, PROF BIOL, WVA UNIV, 80- *Mem:* Tissue Cult Asn; Am Soc Plant Physiologists; NY Acad Sci; Scand Soc Plant Physiologists; Japanese Soc Plant Physiologists. *Res:* Physiology and metabolism of plant effectors; tissue and cell culture; culture media; invitro techniques and manipulations. *Mailing Add:* Dept Biol WVa Univ Box 6057 Morgantown WV 26506-6057

BLAYLOCK, LYNN GAIL, animal nutrition, for more information see previous edition

BLAYLOCK, W KENNETH, b Bristol, Va, Sept 6, 31; m; c 3. DERMATOLOGY. *Educ:* King Col, BS, 53; Med Col Va, MS, 58. *Prof Exp:* Intern, Duke Hosp, Durham, NC, 58-59, resident dermat, 59-61 & internal med, 61-62; from asst prof to assoc prof med, 64-67, asst dean grad med educ, 72-80, PROF DERMAT, MED COL VA, 67-, CHMN DEPT, 77-, ASSOC DEAN GRAD MED EDUC, 80- *Concurrent Pos:* Dermat rep, Fed Drug Admin, 74-; chmn workshop comt eczema, NIH, 74-; mem resident eval comt, Am Asn Prof Dermat, 74- *Mem:* Am Acad Allergy; Am Acad Dermat; Am Dermat Asn; Soc Invest Dermat; Am Asn Prof Dermatologists. *Res:* Allergy and immunology. *Mailing Add:* Dept Dermat Va Commonwealth Univ Sch Med MCV Sta Box 565 Richmond VA 23298

BLAZAR, BEVERLY A, b Providence, RI, Jan 1, 34; m 56; c 3. IMMUNOLOGY. *Educ:* Brown Univ, PhD. *Prof Exp:* ASST PROF BIOL, WELLESLEY COL. *Res:* Epstein-Barr virus. *Mailing Add:* 265 Laurel Ave Providence RI 02906

BLAZER, DAN GERMAN, b Nashville, Tenn, Feb 23, 44; m 66; c 2. GERIATRIC PSYCHIATRY, PSYCHOSOCIAL EPIDEMIOLOGY. *Educ:* Vanderbilt Univ, BA, 65, Univ Tenn, MD, 69; Univ NC-Chapel Hill, MPH, 79, PhD(epidemiol), 80. *Prof Exp:* Intern, Univ Tenn, 69-70; dir Christian Mobile Clin, W Cameroun, Africa, 71-72; resident psychiat, Duke Univ Med Ctr, 73-75; fel consult & liaison psychiat, Montefiore Hosp & Med Ctr, 75-76; from asst prof to assoc prof, 76-84, PROF PSYCHIAT, DUKE UNIV MED CTR, 84-, DIR AFFECTIVE DISORDERS PROG, 85-, J P GIBBONS PSYCHIAT, 90- *Concurrent Pos:* Assoc dir Ctr Study Aging, Duke Univ Med Ctr, 76-80, head Div Soc & Community Psychiat, 80-85, hon teacher, 82-83, interim chmn, Dept Psychiat, 90-; prin investr, Epidemiol Catchment Area Proj, Duke Univ, 80-84, Estab Populations Epidemiol Studies Elderly, 80-84, Clin Res Ctr Study Depression in Elderly, 84-; mem, Coun Int Affairs, Am Psychiat Asn, 82-86, MacArthur Found Proj Successful Aging, 85-; assoc ed, J Am Geriat Soc, 84-86, J Gerontol & Med Sci, 88-; mem, Epidemiol & Dis Control Study Sect, 86-88, chmn, 88- *Mem:* Am Geriat Soc (vpres, 88); Am Psychiat Asn; fel Psychiat Res Soc; Am Col Psychiat; Soc Epidemiol Res; Am Pub Health Asn. *Res:* Psychiatric disorders elderly community populations; depression in late life; cross sectional and longitudinal perspectives; social factors as they relate to the onset and outcome of psychiatric disorders. *Mailing Add:* Dept Psychiat Duke Univ Med Ctr Box 3215 Durham NC 27710

BLAZEVIC, DONNA JEAN, b Wyandotte, Mich, Nov 8, 31. MEDICAL MICROBIOLOGY. *Educ:* Univ Mich, BS, 54, MPH, 63. *Prof Exp:* Res assoc virol, Dept Microbiol, Univ Mich, 63-65; from instr to assoc prof, 65-75, prof clin microbiol, 75-80, prof lab med & path, Dept Lab Med, Univ Minn, Minneapolis, 80-81, lectr, dept microbiol, 70-81, assoc dir microbiol, Univ Minn Hosps, 71-81; PARTNER, ATTORNEYS BASSFORD, HECKT, LOCKHART, TAUESOBLL & BRIGGS, 81- *Concurrent Pos:* Mem subcomt microbiol reagents, Nat Comt Clin Lab Stand, 72-; trustee, Am Soc Med Technol Educ & Res Fund, Inc, 73-78. *Mem:* Am Soc Microbiol; Am Soc Med Technol; Acad Clin Lab Physicians & Scientists. *Res:* Development of rapid tests for diagnosis of infectious disease and identification of microorganisms of medical importance. *Mailing Add:* 3550 Multifoods Tower Minneapolis MN 55402

BLAZEY, LELAND WAY, chemical engineering; deceased, see previous edition for last biography

BLAZEY, RICHARD N, b Rochester, NY, Aug 23, 41; m 64; c 3. ELECTROOPTICS, ELECTROPHOTOGRAPHY. *Educ:* Cornell Univ, BEE, 64; Purdue Univ, MSEE, 66. *Prof Exp:* Proj engr laser syst, Sperry Gyroscope Div, Sperry Rand, 66-69; RES ASSOC LASER SCANNING, RES LABS, EASTMAN KODAK CO, 69- *Mem:* Optical Soc Am; Inst Elec & Electronics Engrs. *Res:* Electrooptic systems, particularly laser systems and laser scanning systems; acoustooptic modulators and deflectors; ultrasonics, especially high frequency systems; color electrophotography; medical imaging-radiology. *Mailing Add:* Eastman Kodak Co B82A-Res Labs Rochester NY 14650-2123

BLAZKOVEC, ANDREW A, b Kewaunee, Wis, Dec 9, 36. IMMUNOBIOLOGY. *Educ:* Univ Wis, BS, 58, MS, 61, PhD(biol), 63. *Prof Exp:* NIH fel immunol, Swiss Nat Res Inst, 63-64; res assoc, Swiss Nat Res Inst & Univ London, 64-66; asst prof, 67-71, ASSOC PROF MED MICROBIOL, UNIV WIS-MADISON, 71- *Mem:* Am Asn Immunologists; AAAS. *Res:* Role and significance of delayed-type cutaneous hypersensitivity in acquired cellular resistance to infectious agents. *Mailing Add:* 1114 Division St Algoma WI 54201

BLAZQUEZ Y SERVIN, CARLOS HUMBERTO, b Mexico City, Mex, Oct 9, 26; nat US; m 65; c 2. PLANT PATHOLOGY. *Educ:* Univ Calif, BS, 55; Univ Fla, MS, 57, PhD(plant path), 59. *Prof Exp:* Jr plant pathologist virus diseases, Bur Plant Path, Calif State Dept Agr, 55; plant pathologist, Firestone Rubber Plantations, Brazil, 59-61; asst plant pathologist, United Fruit Co, Honduras, 62-63; plant pathologist, Dow Chem Co, Mich, 64; plant pathologist citrus res, Univ West Indies, 64-66; asst prof plant path & asst plant pathologist, Agr Res Ctr, Univ Fla, 66-72, assoc prof plant path & assoc plant pathologist, 72-; AGR ADV TO TECH UTIL OFF, 80-, PRES, PROJ LEADER, SENSOR SYSTS, KENNEDY SPACE CTR, 76- *Concurrent Pos:* Hon consult, Univ West Indies, 67; proj leader aerial appl grant, Univ Fla, 69-70; partic, Plant Protection Cong, Paris, 70; prin investr, Inst Food & Agr Sci, Univ Fla-NASA ERTS-B Prog, 73- *Mem:* Mycol Soc Am; Am Phytopath Soc; Asn Appl Biol; Am Chem Soc; Am Asn Hort Sci. *Res:* Pesticide residue analysis; remote sensing; low volume and ultra low volume aerial application methods for disease control; effect of fungicide degradation on biological activity. *Mailing Add:* 63 Westview Lane Cocoa Beach FL 32931

BLAZYK, JACK, b New York, NY, Sept 22, 47; m 72; c 3. BIOLOGICAL MEMBRANES, ANTIMICROBIAL PEPTIDES. *Educ:* Hamilton Col, AB, 69; Brown Univ, PhD(biochem), 75. *Prof Exp:* Vis asst prof chem, Univ SC, 74-75; asst prof chem, 75-81, ASSOC PROF CHEM, OHIO STATE UNIV, 81- , ASSOC DEAN BASIC SCI, 90- *Concurrent Pos:* Fel gen med div, NIH, 75. *Mem:* Biophys Soc; Protein Soc; Am Chem Soc. *Res:* Structure and function of biological membranes; interactions with antimicrobial peptides by spectroscopic techniques. *Mailing Add:* Dept Chem Col Osteop Med Ohio Univ Athens OH 45701

BLEACKLEY, ROBERT CHRISTOPHER, b Manchester, Eng, Feb 29, 52. BIOCHEMISTRY. *Educ:* Univ Birmingham, BSc, 72, PhD(chem), 75. *Prof Exp:* Postdoctoral, Dept Biochem, Univ Alta, 75-77, prof asst, 77-81, sessional instr, 78-80, from asst prof to assoc prof, 81-89, assoc prof, Dept Immunol, 86-89, PROF, DEPT BIOCHEM, UNIV ALTA, 89-, PROF, DEPT IMMUNOL, 89- *Concurrent Pos:* Sci officer, Panel A, Nat Cancer Inst, 86 & 87, chmn, 88-; grants, Nat Cancer Inst, 88-93, Med Res Coun, 90-93 & Can Diabetes Asn, 91-93. *Mem:* Can Biochem Soc; AAAS; NY Acad Sci; Can Soc Immunologists; Am Asn Immunologists. *Res:* T-cell activation; human cytotoxic T-cells. *Mailing Add:* Dept Biochem Univ Alta Med Sci Bldg Edmonton AB T6G 2H7 Can

BLEAKNEY, JOHN SHERMAN, b Corning, NY, Jan 14, 28; m 52; c 2. MALACOLOGY. *Educ:* Acadia Univ, BSc, 49, MSc, 51; McGill Univ, PhD(zool), 56. *Prof Exp:* Cur ichthyol & herpet, Nat Mus Can, 52-58; from assoc prof to prof biol, 58-88, UNIV FEL, ACADIA UNIV, 88- *Concurrent Pos:* Nat Res Coun Can sr res fel, Eng & Denmark, 67-68. *Mem:* Am Malacol Union; Malacol Soc London; Sigma Xi. *Res:* Herpetology and nudibranch molluscs of Eastern Canada; fauna of the Minas Basin, Nova Scotia. *Mailing Add:* Dept Biol Acadia Univ Wolfville NS B0P 1X0 Can

BLEAKNEY, WALKER, b Elderton, Pa, Feb 8, 01; m 31. PHYSICS. *Educ:* Whitman Col, BS, 24; Univ Minn, PhD, 30. *Hon Degrees:* DSc, Whitman Col, 55. *Prof Exp:* from instr to prof, 32-70, chmn dept, 60-67, emer class 1909 prof, 60-69, EMER PROF PHYSICS, PRINCETON UNIV, 69- *Mem:* Nat Acad Sci; fel Am Phys Soc; Am Acad Arts & Sci; Sigma Xi. *Mailing Add:* 900 Calle De Los Amigos N-14 Santa Barbara CA 93105

BLECHER, F(RANKLIN) H(UGH), b Brooklyn, NY, Feb 24, 29. ELECTRICAL ENGINEERING, COMMUNICATIONS. *Educ:* Polytech Inst Brooklyn, BEE, 49, MEE, 50, DEE, 55. *Prof Exp:* Mem tech staff, Polytech Res & Develop Co, 50-52; mem tech staff, Bell Tel Labs, 52-60, dir, Carrier Transmission Lab, 60-68, Electron Device Lab, 68-74, Mobile Commun Lab, 74-82, exec dir, Integrated Circuit Design Div, 82-87, exec dir, Tech Info Systs Div, 87-89, CONSULT TELECOMMUN, AT&T BELL LABS, 89- *Concurrent Pos:* Vchmn, Eng Found, 90-; pres, Polytech Univ Alumi Asn, 90-92. *Honors & Awards:* Thompson Award, Inst Elec & Electronics Engrs, 59. *Mem:* Nat Acad Eng; fel Inst Elec & Electronics Engrs (vpres, 77). *Res:* Design of solid state carrier systems for transmission over cables; design of single & multiple loop feedback amplifiers. *Mailing Add:* 6039 Collins Ave Apt 1526 Miami Beach FL 33140

BLECHER, MARVIN, b New York, NY, Sept 12, 40; m 62; c 3. NUCLEAR PHYSICS, ELEMENTARY PARTICLE PHYSICS. *Educ:* Columbia Univ, BA, 62; Univ Ill, Urbana-Champaign, PhD(physics), 68. *Prof Exp:* From asst prof to assoc prof, 68-81, PROF PHYSICS, VA POLYTECH INST & STATE UNIV, 82- *Mem:* Am Phys Soc. *Mailing Add:* Dept Physics Va Polytech Inst & State Univ Blacksburg VA 24061

BLECHER, MELVIN, b Rahway, NJ, July 19, 22; m 50, 63, 73; c 2. BIOCHEMISTRY, BIOTECHNOLOGY PATENT LAW. *Educ:* Rutgers Univ, BS, 49; Univ Pa, PhD(biochem), 54; Georgetown Univ, JD, 86. *Prof Exp:* Asst instr biochem, Sch Med, Univ Pa, 49-51; res assoc, Col Physicians & Surgeons, Columbia Univ, 54-56; from instr to asst prof, Albert Einstein Col Med, 56-61; assoc prof, 61-68, prof biochem, Schs Med & Dent & Grad Sch, Georgetown Univ, 68-88; ASSOC ATTY, LYON & LYON, 88- *Concurrent Pos:* Vis prof, Med Clin, Univ Rome; vis prof biochem, Univ Paris. *Mem:* Fel AAAS; Am Soc Biochem & Molecular Biol; Am Soc Pharmacol & Exp Therapeut; Endocrine Soc. *Res:* Mechanism of action of hormones; regulatory mechanisms; hormone receptors; membrane biochemistry; autoimmune diseases; immunology; molecular and cell biologies. *Mailing Add:* Lyon & Lyon 1225 Eye St NW Washington DC 20005

BLECHMAN, HARRY, b Brooklyn, NY, Aug 22, 18; m 43; c 2. MICROBIOLOGY. *Educ:* City Col New York, BS, 38; NY Univ, DDS, 51; Hunter Col, MA, 59; Am Bd Endodont, dipl, 68. *Prof Exp:* Clin bacteriologist, St John's Long Island Col Hosp & Rockaway Beach Hosp, 38-41; clin bacteriologist, Tilton Gen Hosp, 41-43; sr bacteriologist, Harlem Hosp, 46-47; chmn dept microbiol, NY Univ, 51-54, from asst prof to prof, 51-68, chmn dept microbiol, prev med & hyg, 54-68, asst dean, Col Dent, 64-68, dean, 68-75, prof endodont & chmn dept, 75-80, Samuel & Hannah Holzman prof & chmn dept endodont, Col Dent, 80-89; RETIRED. *Concurrent Pos:* Exec secy, Guggenheim Inst Dent Res, 54-64; mem & chmn, Nat Bd Dent Examr; pres, Am Bd Endodont, 75; consult, Vet Admin Hosps, USPHS & NY State Narcotic Addiction Comn. *Mem:* Am Dent Asn; fel Am Col Dent; Am Soc Microbiol; Am Asn Endodontists (pres, 78); Int Asn Dent Res. *Res:* Oral microbiology; antibiosis and interactions among oral microorganisms in dental caries and periodontal disease. *Mailing Add:* 122 E 78th St New York NY 10021

BLECHNER, JACK NORMAN, b New York, NY, Jan 7, 33; m 57; c 3. OBSTETRICS & GYNECOLOGY, FETAL PHYSIOLOGY. *Educ:* Columbia Col, BA, 54; Yale Univ, MD, 57; Am Bd Obstet & Gynec, dipl & cert, 80. *Prof Exp:* Intern, Bronx Munic Hosp Ctr, 57-58; Josiah Macy Jr Found fel physiol, Yale Univ, 58-60, 62-63; asst resident obstet & gynec, Columbia Presby Med Ctr, 60-63, chief resident, 63-65; from asst prof to assoc prof, Col Med, Univ Fla, 65-70; PROF OBSTET & GYNEC & HEAD DEPT, SCH MED, UNIV CONN, 70-, ASSOC DEAN, CLIN AFFAIRS, 85- *Concurrent Pos:* Am Cancer Soc fel, 64; chief dept obstet & gynec, New Britain Gen Hosp, 70- & Dempsey Teaching Hosp, 75-; physician consult, dept obstet & gynec, Vet Admin Hosp, 70-, St Francis Hosp, 70-, Mt Sinai Hosp, Hartford, Conn, 70-, Hartford Hosp, 72-, Bristol Hosp, 73-, St Mary's Hosp, 78- & Meriden-Wallingford Hosp, 80-; examr, Am Bd Obstet & Gynec; vchmn, Coun Resident Educ in Obstet & Gynec, 78-79; mem coun, Soc Gynec Invest, 67. *Mem:* Am Col Obstetricians & Gynecologists; Am Gynec & Obstet Soc; Asn Univ Prof Gynec & Obstet; Soc Gynec Invest; Sigma Xi. *Res:* Fetal physiology; membrane transfer; experimental intrauterine surgery; respiratory gas exchange. *Mailing Add:* Dept Obstet & Gynec Univ Conn Sch Med Farmington CT 06030

BLECK, RAINER, b Danzig, Ger, Nov 4, 39; m 66; c 2. METEOROLOGY, PHYSICAL OCEANOGRAPHY. *Educ:* Free Univ Berlin, MS, 64; Pa State Univ, PhD(meteorol), 68. *Prof Exp:* Res asst meteorol, Inst Theoret Meteorol, Free Univ Berlin, 65; Scientist, Nat Ctr Atmospheric Res, 67-75; assoc prof, 75-80, PROF METEOROL, UNIV MIAMI, 80- *Mem:* Am Meteorol Soc. *Res:* Weather analysis and prediction; ocean circulation modeling. *Mailing Add:* Rosenstiel Sch Marine & Atmospheric Sci Univ Miami Miami FL 33149

BLECKER, HARRY HERMAN, b Philadelphia, Pa, Apr 10, 27; m 51; c 4. ORGANIC CHEMISTRY. *Educ:* Bucknell Univ, BS, 51; Rutgers Univ, MS, 54, PhD(chem), 55. *Prof Exp:* Res assoc chem, Univ Mich, 54-56; asst prof, Bucknell Univ, 56-57; from asst prof to prof, 57-89, chmn dept, 65-88, EMER PROF CHEM, UNIV MICH, FLINT, 89- *Mem:* Am Chem Soc. *Res:* Bioanalytical chemistry. *Mailing Add:* Dept Chem Flint MI 48502-2186

BLECKMANN, CHARLES ALLEN, b Evansville, Ind, Nov 24, 44; m 68; c 1. BOTANY. *Educ:* Univ Evansville, BA, 67; Incarnate Word Col, MS, 71; Univ Ariz, PhD(bot), 77. *Prof Exp:* Res assoc bot, Sch Renewable Natural Resources, Univ Ariz, 77-78; RES SCIENTIST BIOL, CONOCO, INC, 78- RES GROUP LEADER, 85- *Mem:* Sigma Xi; AAAS; Am Inst Biol Sci. *Res:* Waste treatment and other environmental problems of the petroleum industry; botanical ultrastructure. *Mailing Add:* Res & Develop Dept Conoco Inc Ponca City OK 74603

BLEDSOE, HORACE WILLIE LEE, b Blue Ridge, Tex, Feb 8, 44; m 71; c 2. PHYSICS, MATHEMATICS. *Educ:* Univ Tex, Austin, BS, 66, PhD(physics), 71. *Prof Exp:* Instr physics, State Univ NY, Stony Brook, 71-73; asst prof, 73-78, ASSOC PROF PHYSICS, UNIV TEX, 78- *Mem:* Am Phys Soc. *Res:* Low energy nuclear reaction theory; the origin of quantum mechanics and the connection between mass and energy. *Mailing Add:* 11301 Santa Cruz Dr Austin TX 78759

BLEDSOE, JAMES O, JR, b New Bern, NC, Mar 22, 38; m 62; c 2. ORGANIC CHEMISTRY. *Educ:* Univ NC, Chapel Hill, BS, 60; Univ Okla, PhD(org chem), 65. *Prof Exp:* Proj leader org synthesis, Int Flavors & Fragrances, Inc, 65-66; res chemist, 66-71, mgr new prod res, Org Chem Group, Glidden Durkee Div, SCM Corp, 71-78; head develop lab, 78-88, SR SCIENTIST, UNION CAMP CORP, JACKSONVILLE, FL, 88- *Mem:* Am Chem Soc. *Res:* Research and development of terpene products especially for uses in the perfumery and flavor fields. *Mailing Add:* Union Camp PO Box 37617 Jacksonville FL 32236

BLEDSOE, LEWIS JACKSON (SAM), b Los Angeles, Calif, Dec 13, 42; m 62; c 1. RESOURCE ECOLOGY, APPLIED MATHEMATICS. *Educ:* Univ Tenn, AB, 64; Colo State Univ, MS, 68, PhD(systs ecol), 76. *Prof Exp:* Math consult, Oak Ridge Nat Lab, 61-66; instr systs ecol, Colo State Univ, 67-71; res assoc prof fisheries, Univ Wash, 72-91; CONSULT SCIENTIST, COMPUTER SCI CORP, 90- *Concurrent Pos:* Prin investr, Norfish Marine Resource Proj, Univ Wash Sea Grant Off & Nat Marine Fisheries Serv, 73-82; landscape ecology, Nat Sci Found, Mt St Helens, 86-90; nitrogen fixation, NSF, High Arctic, 87-89 Salmon Run Trends Proj, US Fish & Wildlife Serv, 78-81. *Mem:* Ecol Soc Am; Am Fisheries Soc; AAAS. *Res:* Systems ecology; application of mathematics, statistics and computer technology to analysis of environmental systems or managed resources. *Mailing Add:* Computer Sci Corp 10000A Aerospace Rd Lanham MD 20706

BLEDSOE, WOODROW WILSON, b Maysville, Okla, Nov 12, 21; m 44; c 3. MATHEMATICS. *Educ:* Univ Utah, BS, 48; Univ Calif, MA, 50, PhD(math), 53. *Prof Exp:* Lectr math, Univ Calif, 51-52; mathematician & mgr systs anal dept, Sandia Corp, 53-60; mathematician & pres, Panoramic Res, Inc, Calif, 60-66; actg chmn dept math, 67-69, chmn, 73-75, PROF MATH, UNIV TEX, AUSTIN, 66- *Concurrent Pos:* Vis prof, Mass Inst Technol, 70-71 & Carnegie Mellon Univ, 78. *Mem:* Am Math Soc; Asn Comput Mach; Am Asn Artificial Intel; Sigma Xi. *Res:* Real variables; topological product measures; systems analysis; artificial intelligence; automatic theorem proving. *Mailing Add:* Dept Computer Sci Univ Tex Austin TX 78712

BLEECKER, EUGENE R, b New York, NY, Jan 23, 43. PULMONARY & CRITICAL CARE MEDICINE. *Educ:* NY Univ, BA, 64; State Univ NY, MD, 68; Am Bd Internal Med, dipl, 74, dipl pulmonary dis, 76. *Prof Exp:* Intern, State Univ NY, Kings County Hosp, 68-69, asst med resident, 69-70; clin assoc, Geront Res Ctr, NIH, 70-72; sr asst med resident, Johns Hopkins Hosp, 72-73; fel pulmonary dis, Dept Med, Univ Calif, San Francisco, 73-74 & Dept Med & Cardiovac Res Inst, 74-75, asst res physiologist, 75-76; from asst prof to assoc prof med, Sch Hyg & Pub Health, Johns Hopkins Univ, 76-82; PROF MED, DIV PULMONARY & CRITICAL CARE MED, SCH MED, UNIV MD, 89-; CHIEF, PULMONARY SECT, BALTIMORE VET ADMIN MED CTR, 89- *Concurrent Pos:* Prin investr, Nat Heart, Lung & Blood Inst, 75-76, 78-83 & 84-92, Nat Inst Aging, 83-88; clin dir pulmonary med, Francis Scott Key Med Ctr, 76-89; mem, Grants & Awards Comt, Am Lung Asn, 82-85, Bronchoprovocation Comt, Am Acad Allergy & Immunol, 82-, Comt Disability Determination, Am Thoracic Soc, 85-86, Steering Comt Sect Respiratory Pathophysiol, Am Col Chest Physicians, 85-87, Res Coun, Am Acad Allergy & Immunol, 89- & Expert Panel Diag & Treat Asthma, NIH, 89-; guest scientist, Geront Res Ctr, NIH, 86-; chmn, Comt Bronchoalveolar Lavage, Am Acad Allergy & Immunol, 88- & Panel Investigative Use Bronchoscopy Asthma, Nat Heart, Lung & Blood Inst, NIH, 90-; consult, Childhood Asthma Mgt Prog, Nat Heart Lung & Blood Inst, NIH, 91- *Mem:* Am Thoracic Soc; Am Fedn Clin Res; Am Col Chest Physicians; fel Am Acad Allergy & Immunol; Am Physiol Soc; Am Heart Asn; Geront Soc Am. *Res:* Pathogenesis of asthma and chronic airflow obstruction; airways hyperreactivity and bronchial inflammation; cardiopulmonary and exercise function during aging. *Mailing Add:* Div Pulmonary & Critical Care Med Sch Med Univ Md Ten S Pine St Baltimore MD 21224

BLEECKER, MARGIT, b Vienna, Austria, Nov 20, 43; US citizen; m 64. NEUROCHEMISTRY, NEUROENDOCRINOLOGY. *Educ:* NY Univ, BA, 64; State Univ NY Downstate Med Ctr, PhD(anat), 70. *Prof Exp:* Instr anat, Sch Med, Johns Hopkins Univ, 70-72; instr anat, Sch Med, Temple Univ, 72-76; FEL NEUROL, SCH MED, JOHNS HOPKINS UNIV, 76- *Mem:* Am Soc Neurochem; Int Soc Psychoneuroendocrinol; Am Asn Anat; Soc Neurosci; Sigma Xi. *Res:* In vitro studies on Taurine in the rat brain. *Mailing Add:* 6000 Hunt Club Lane Baltimore MD 21210

BLEEKE, JOHN RICHARD, b Port Washington, Wis, Sept 16, 54; m 83; c 2. HOMOGENEOUS CATALYSIS. *Educ:* Carthage Col, BA, 76; Cornell Univ, MS, 79, PhD(chem), 81. *Prof Exp:* Assoc, Dept Chem, Univ Calif, Berkeley, 81; asst prof, 81-87, ASSOC PROF, DEPT CHEM, WASHINGTON UNIV, ST LOUIS, 87- *Mem:* Am Chem Soc. *Res:* Synthesis of transition metal organometallic complexes; development of selective homogeneous arene hydrogenation and alkane functionalization catalysts; mechanistic studies of transition metal-catalyzed reactions; chemistry of metallacyclohexadienes and metallabenzenes. *Mailing Add:* Dept Chem Washington Univ St Louis MO 63130

BLEFKO, ROBERT L, b July 5, 32; US citizen. TOPOLOGY. *Educ:* Pa State Univ, PhD(math), 65. *Prof Exp:* Asst prof math, Univ RI, 65-68; ASSOC PROF MATH, WESTERN MICH UNIV, 68- *Mem:* Am Math Soc; Math Asn Am. *Res:* General topology, specializing in continuous function theory. *Mailing Add:* Dept Math Western Mich Univ Kalamazoo MI 49008

BLEI, IRA, b Brooklyn, NY, Jan 19, 31; m 52; c 2. BIOPHYSICAL CHEMISTRY. *Educ:* Brooklyn Col, BS, 52, MA, 54; Rutgers Univ, PhD(biophys chem), 57. *Prof Exp:* Instr chem, Rutgers Univ, 53-57; prin res chemist, Lever Bros Co Res Ctr, 57-61; supvr biophys & biodetection sect, Melpar, Inc, 61-66, mgr environ sci lab, 66-67; from asst prof to assoc prof, 67-74, PROF CHEM, COL STATEN ISLAND, 74- *Mem:* Am Chem Soc; NY Acad Sci. *Res:* Detection of extra-terrestrial life; optical rotatory dispersion; polarization of fluorescence; physical chemistry of non-aqueous phosphatide solutions; biological differentiation of sodium and potassium; mechanism of action of psychoactive drugs. *Mailing Add:* Col Staten Island St George 130 Stuyvesant Pl Staten Island NY 10301

BLEI, RON CHARLES, b Prague, Czech, July 15, 45; US citizen; m 69; c 2. ANALYSIS, FUNCTIONAL ANALYSIS. *Educ:* Univ Calif, Berkeley, BA, 67, PhD(math), 71. *Prof Exp:* From asst prof to assoc prof, 71-81, PROF MATH, UNIV CONN, 81- *Concurrent Pos:* Vis NATO prof, Inst Math, Italy, 73; vis assoc prof, Uppsala Univ, 77, Hebrew Univ, 78; vis fel, Inst Mittag Leffler, Sweden, 77; prin investr grants, NSF, 76-77, 77-79, 80-82 & 83-86; vis prof, Univ BC, Vancouver, 83-84; vis scholar, Cambridge Univ, 87. *Mem:* Am Math Soc. *Res:* Combinatorial dimension as a continuous parameter and its impact in analytic and probabilistic frame works. *Mailing Add:* Univ Conn Storrs CT 06269

BLEIBERG, MARVIN JAY, b Brooklyn, NY, Feb 19, 28; m 60; c 2. BIOCHEMISTRY, BIOMATHEMATICS. *Educ:* Col William & Mary, BS, 49; Med Col Va, PhD(pharmacol), 57; Am Bd Toxicol, dipl, 80, 85, 90. *Prof Exp:* USPHS trainee steroid biochem, Worcester Found Exp Biol, 57-58; from instr to asst prof pharmacol, Jefferson Med Col, 58-61; pharmacologist, Food & Drug Admin, 61-62; sr scientist res div, Melpar, Inc, 62-64; pharmacologist, Div Environ Toxicol, Woodard Res Corp, 64-68; dir div pharmacol, 68-70; sect head pharmacol, Spec Projs Dept, 70-75; PHARMACOLOGIST & TOXICOLOGIST, ADDITIVES EVAL BR, DIV TOXICOL, CTR FOOD SAFETY & APPL NUTRIT, US FOOD & DRUG ADMIN, WASHINGTON, DC, 75- *Concurrent Pos:* Chmn, DC Sect, Soc Exp Biol & Med, 84-85; vpres, Nat Capital Area Sect, Soc Toxicol. *Mem:* AAAS; Biomet Soc; Am Soc Pharmacol; Am Chem Soc; Soc Toxicol; Am Col Toxicol; NY Acad Sci. *Res:* Toxicology; biochemical and autonomic nervous system pharmacology; behavioral pharmacology; HPLC analysis of drugs in serum, blood; pharmacokinetics. *Mailing Add:* 3613 Old Post Rd Fairfax VA 22030

BLEIBTREU, HERMANN KARL, b Darmstadt, Ger, Aug 31, 33; US citizen; m 63; c 2. PHYSICAL ANTHROPOLOGY. *Educ:* Harvard Univ, BA, 56, PhD, 64. *Prof Exp:* Tutor dept anthrop, Harvard Univ, 61; res asst biol, Case Western Reserve Univ, 62; asst prof anthrop, Univ Calif, Los Angeles, 64-66; from asst prof to assoc prof, Univ Ariz, 66-71, prof anthrop & dean col lib arts, 71-75; DIR, MUS NORTHERN ARIZ, 75- *Concurrent Pos:* NIH grant pop characteristics Pima-Papago, Univ Ariz, 67-68, NSF grants, 68 & 69-; mem bd, Prescott Col, 71- *Mem:* Fel AAAS; Am Asn Phys Anthrop; fel Am Anthrop Asn; Am Eugenics Soc. *Res:* Human microevolution; multivariate methods in anthropometrics. *Mailing Add:* 1931 N Calle Delsuerte Tucson AZ 85745

BLEICHER, MICHAEL NATHANIEL, b Cleveland, Ohio, Oct 2, 35; m 57, 80; c 6. DISCRETE GEOMETRY, ELEMENTARY NUMBER THEORY. *Educ:* Calif Inst Technol, BS, 57; Tulane Univ, MS, 59, PhD(math), 61; Univ Warsaw, PhD(math), 61. *Prof Exp:* NSF fel math, Univ Calif, Berkeley, 61-62; from asst prof to assoc prof, 62-69, chmn dept, 73-75, PROF MATH, UNIV WIS-MADISON, 69- *Concurrent Pos:* Spec asst to dir energy res, US Dept Energy, 79-81; prof math, Inst Tech Mara, Shah Alam, Malaysia, 87-90; adj prof polit sci, INTI Inst Higher Learning, Malaysia, 87-88. *Mem:* Am Math Soc; Math Asn Am; Polish Math Soc; Sigma Xi; Southeast Asia Math Soc. *Res:* Geometry of numbers; convexity; number theory; foundations of mathematics; universal algebra. *Mailing Add:* Dept Math Univ Wis Madison WI 53706

BLEICHER, SHELDON JOSEPH, b New York, NY, Apr 9, 31. INTERNAL MEDICINE, METABOLISM. *Educ:* NY Univ, BA, 51; Western Ill Univ, MS, 52; State Univ NY Downstate Med Ctr, MD, 56. *Prof Exp:* Res fel med, Thorndike Mem Lab, Boston City Hosp & Harvard Med Sch, 60-63; dir metab res unit, Jewish Hosp & Med Ctr Brooklyn, 63-69; from instr to assoc prof, 63-75, PROF MED, STATE UNIV NY DOWNSTATE MED CTR, 75-, CHMN DEPT INTERNAL MED, 77- CHIEF DIV ENDOCRINOL & METAB & ATTEND PHYSICIAN, JEWISH HOSP & MED CTR BROOKLYN, 69- *Concurrent Pos:* USPHS fel, 61-63; NIH res career develop award; consult-expert in med use of radioisotopes, Int Atomic Energy Agency. *Mem:* Am Col Physicians; Am Fedn Clin. *Res:* Am Diabetes Asn; AMA; Endocrine Soc. Res: Diabetes; hypoglycemia; carbohydrate metabolism in pregnancy; fat metabolism. *Mailing Add:* 175 Jerchio Turnpike Syosset NY 11791

BLEICK, WILLARD EVAN, b Newark, NJ, Dec 8, 07; m 36; c 3. MATHEMATICAL PHYSICS. *Educ:* Stevens Inst Technol, ME, 29; Johns Hopkins Univ, PhD(appl math), 33. *Prof Exp:* Mem, Sch Math Inst Advan Study, 33-34 & 35-36; actuarial student, Prudential Ins Co, 36-38; engr, Gibbs & Cox, Inc, 38-40; instr math, Cooper Union, 39-40; from instr to asst prof, US Naval Acad, 40-46, assoc prof math & mech, 46-50, prof math & mech, Naval Postgrad Sch, 50-74; RETIRED. *Concurrent Pos:* Contractor, US Navy Optimum Track Ship Routing, 74-82. *Mem:* Am Math Soc; Math Asn Am. *Res:* Quantum mechanics; stress analysis; crystal lattice energies; digital computers; missile dynamics; control optimization. *Mailing Add:* Hacienda Carmel Apt 181 Carmel CA 93921

BLEIDNER, WILLIAM EGIDIUS, analytical biochemistry; deceased, see previous edition for last biography

BLEIER, ALAN, b Rochester, NY, Mar 12, 48. CERAMIC PROCESSING, COLLOID & SURFACE CHEMISTRY. *Educ:* Clarkson Univ, BS, 70, MA, 73, PhD, 76. *Prof Exp:* DEVELOP STAFF MEM, OAK RIDGE NAT LAB, 83- *Concurrent Pos:* Corning asst prof ceramics, Mass Inst Technol, 80-83, abstractor, Ceramics Abstracts, 85-; chmn comt, Proctor & Gamble fel, 87-88. *Mem:* N Y Acad Sci; Sigma Xi; Am Chem Soc; Am Phys Soc; Int Asn Colloid Interface; Am Ceramic Soc. *Res:* colloid and physical chemistry of processing ceramic composites; colloid and physical chemistry of processing composite ceramics. *Mailing Add:* Oak Ridge Nat Lab Metal & Ceramics Div PO Box 2008 Oak Ridge TN 37831-6068

BLEIER, FRANK P(HILIPP), b Vienna, Austria, Jan 26, 13; nat US; m 55; c 1. AERODYNAMICS. *Educ:* Tech Univ, Vienna, Dipl Ing, 38. *Prof Exp:* Res engr, DeBothezat Fan Co, 40-42; dir res, Ilg Elec Ventilating Co, 42-46; CONSULT ENGR, FAN DESIGN & DEVELOP WORK, 46- *Mem:* Am Soc Heating, Refrig & Air Conditioning Engrs. *Res:* Increase of capacity and efficiency on various types of fans; design of axial flow fans and centrifugal fans for tank engine cooling; ventilation of invasion boats and pressure blowers; roof ventilators; fans for agricultural crop drying; multi-stage blower to pressurize flotation bags. *Mailing Add:* 536 W Cornelia Ave Chicago IL 60657-2742

BLEIER, WILLIAM JOSEPH, b Jourdanton, Tex, July 8, 47; m 73. REPRODUCTIVE BIOLOGY. *Educ:* Univ Tex, Austin, BA, 69; Tex Tech Univ, MS, 71, PhD(zool), 75. *Prof Exp:* FROM ASST PROF TO ASSOC PROF ZOOL, NDAK STATE UNIV, 75- *Mem:* Sigma Xi; Am Soc Mammalogists. *Res:* Factors controlling embryonic development in mammals. *Mailing Add:* Dept Zool NDak State Univ Fargo ND 58105

BLEIFUSS, RODNEY L, b Rochester, Minn, May 13, 28; m 50; c 4. GEOLOGY, METALLURGICAL ENGINEERING. *Educ:* Univ Minn, Minneapolis, BE, 51, MS, 53, PhD(geol), 66. *Prof Exp:* Mineral engr, Appl Res Lab, US Steel Corp, 52-58; res assoc, Mineral Resources Res Ctr, Univ Minn, Minneapolis, 61-66, from asst prof to assoc prof metall eng, 66-77; assoc res consult, 77-80, div chief, Appl Res Lab, US Steel Corp, 80-; PROG DIR, METALL MINERALS DIV, NATURAL RESOURCES RES INST. *Mem:* Geol Soc Am; Am Inst Mining, Metall & Petrol Engrs; Soc Mining Engrs. *Res:* Iron ore beneficiation, agglomeration and direct reduction. *Mailing Add:* Natural Resources Res Inst Univ Minn Duluth MN 55811

BLEIL, CARL EDWARD, b Detroit, Mich, Oct 7, 23; m 79; c 4. MATERIAL SCIENCE, RESEARCH ADMINISTRATION. *Educ:* Mich State Univ, BS, 47; Univ Okla, MS, 50, PhD(physics), 53. *Prof Exp:* Asst, Univ Okla, 51-53; dept res scientist, Res Labs, Gen Motors Corp, 53-84; PRES, ENERGY MAT RES, 84- *Concurrent Pos:* Mem solid state sci panel, SSS Comt, Nat Res Coun, 74-84; adj prof, Okla Univ, 87- *Mem:* AAAS; Am Phys Soc; Sigma Xi; Mat Res Soc. *Res:* Compound semiconductors; surface physics; crystal growth; coherent optics; film deposition and modification. *Mailing Add:* 132 Chalmers Dr Rochester Hills MI 48309

BLEIL, DAVID F, b Detroit, Mich, Dec 4, 08; c 4. SOLID STATE PHYSICS, ELECTRIC & MAGENTICS. *Educ:* Univ Mich, BS, 34, MS, 37; Mich State Univ, PhD(physics), 48. *Prof Exp:* Dir res & assoc tech dir, Naval Ordinance Lab, 43-73; RETIRED. *Concurrent Pos:* Instr physics & math, Mich State Univ, Montgomery Jr Col, Univ Md & Bluefield State Col. *Mem:* Am Phys Soc; Am Inst Physics; Sigma Xi. *Res:* Induced polarization-Geophy prospecting; magnetic properties of materials. *Mailing Add:* PO Box 429 Rupert WV 25984

BLEISTEIN, NORMAN, b New York, NY, June 29, 39; m 61; c 2. APPLIED MATHEMATICS, MATHEMATICAL ANALYSIS. *Educ:* Brooklyn Col, BS, 60; NY Univ, MS, 61, PhD(math), 65. *Prof Exp:* Res assoc math, Courant Inst Math Sci, NY Univ, 65-66; asst prof, Mass Inst Technol, 66-69; assoc prof math, Univ Denver, 69-74, prof, 74-; DIR, MATH DEPT, CTR WAVE PHENOMENA, COLO SCH MINES, 83- *Concurrent Pos:* Assoc ed, Soc Indust & Appl Math J, 73-; partner, Denver Appl Anal, Colo. *Mem:* Soc Explor Geophysicists; Soc Indust & Appl Math. *Res:* Asymptotic methods; geometrical theory of diffraction; inverse problems in acoustics; electromagnetics; seismology. *Mailing Add:* 2555 S Ivanhoe Pl Colo Sch Mines Denver CO 80222

BLEIWEIS, ARNOLD SHELDON, b Brooklyn, NY, Aug 8, 37; m 83. MICROBIOLOGY, DENTISTRY. *Educ:* Brooklyn Col, BS, 58; Pa State Univ, MS, 60, PhD(microbiol), 64. *Prof Exp:* Lectr biol, Brooklyn Col, 58; asst microbiol, Pa State Univ, 58-63; fel epidemiol, Med Sch, Washington Univ, 64-65; asst biochem, Albert Einstein Med Ctr, 66-67; asst prof bact, 67-72, assoc prof microbiol, 72-76, prof microbiol & cell sci, microbiologist, inst food & agr sci, 76-85, PROF & CHMN, ORAL BIOL, UNIV FLA, 86-, GRAD RES PROF, ORAL BIOL, 88- *Concurrent Pos:* NIH training grant epidemiol, 64-65, fel, 66-67, res grant sr investr, 69-; consult, Nat Inst Dent Res, 70-; mem, Nat Adv Dent Res Coun, NIH, 78-81; mem, US Army Basic Res Comt, Nat Res Coun, 79-81; mem, Dent Res Progs Adv Comt, NIH, 89- *Mem:* Am Soc Microbiol; Int Asn Dent Res; Am Asn Dent Res. *Res:* Structural and antigenic aspects of streptococcal cell walls and membrane; immunologic studies of cariogenic streptococci. *Mailing Add:* J H Miller Health Ctr Univ Fla Box J424 Gainesville FL 32610

BLEIWEISS, MAX PHILLIP, b Nogales, Ariz, June 23, 44; m 63; c 2. IMAGE PROCESSING, REMOTE SENSING. *Educ:* Calif State Polytechnic Col, BS, 66; Calif State Col, Los Angeles, MS, 69. *Prof Exp:* Physicist astron, US Navy Naval Weapons Ctr, 66-70, Naval Ocean Syst Ctr, 70-77; vpres bus, Deseret Dent Supply, 77-85; physicist, Dugway Proving Ground, 85-89, PHYSICIST REM SENSING, US ARMY, ATMOSPHERIC SCI LAB, 89- *Mem:* Am Astron Soc; Am Geophys Union. *Res:* Remote sensing of smoke obscurent clouds using multispectral imaging; atmospheric characterization using digital image processing. *Mailing Add:* 1215 Villita Loop Las Cruces NM 88005

BLEJER, HECTOR P, toxicology, epidemiology; deceased, see previous edition for last biography

BLEJWAS, THOMAS EDWARD, b Plainfield, NJ, Sept 17, 46; m 69; c 2. MECHANICAL ENGINEERING, STRUCTURAL ENGINEERING. *Educ:* Princeton Univ, BS, 68; Univ Southern Calif, MS, 72; Univ Colo, PhD(civil eng), 78. *Prof Exp:* Flight test engr, US Air Force Flight Test Ctr, Edwards AFB, 68-72; sr engr anal mech, Martin Marietta Aerospace, Denver Div, 72-77; res engr civil eng, Univ Calif, Berkeley, 77-78; ASST PROF MECH ENG, OKLA STATE UNIV, 78- *Concurrent Pos:* Instr, Univ Colo, 76-77. *Mem:* Sigma Xi. *Res:* Structural dynamics; earthquake engineering; vehicle and structure interaction; mechanism dynamics. *Mailing Add:* 2116 Bluecorn Maiden Ct Albuquerque NM 87112

BLEM, CHARLES R, b Dunkirk, Ohio, Apr 18, 43; m 65; c 2. VERTEBRATE ECOLOGY. *Educ:* Ohio Univ, BS, 65; Univ Ill, Urbana, MS, 68, PhD(zool), 69. *Prof Exp:* assoc prof, 69-80, PROF BIOL, VA COMMONWEALTH UNIV, 80- *Mem:* Fel AAAS; Am Soc Ichthyologists & Herpetologists; Am Soc Mammal; fel Am Ornith Union; Cooper Ornith Soc; Wilson Ornith Soc. *Res:* Energetics and ecology of vertebrates, especially birds. *Mailing Add:* Dept Biol Sci 816 Park Ave Va Commonwealth Univ Richmond VA 23284-2012

BLENDEN, DONALD C, b Webster Groves, Mo, Aug 13, 29; m 51; c 2. INFECTIOUS DISEASES, EPIDEMIC INVESTIGATION. *Educ:* Univ Mo, BS, 51, MS, 53, DVM, 56. *Prof Exp:* Med technician, Boone County Hosp, 52-56; clin vet pvt pract, 56-57; prof, 57-90, EMER PROF EPIDEMIOL & PUB HEALTH, UNIV MO-COLUMBIA, 90- *Concurrent Pos:* emer consult, US Air Force, 78; consult, WHO, Pan-Am Health Orgn, Food & Agr Orgn int pub health; dir, World Health Orgn Collab Ctr for Training Ref in Enteric Zoonoses, 87- *Mem:* Fel Am Pub Health Asn; Am Vet Med Asn; Am Soc Microbiol; Sigma Xi; Am Col Vet Prev Med; Am Col Epidemiol. *Res:* Comparative medicine; epidemiology, pathogenesis and microbiology of zoonoses; leptospirosis; listeriosis; rabies; enteric infections; antigen detection; disaster preparedness. *Mailing Add:* Univ Mo Col Vet Med Columbia MO 65211

BLENDON, ROBERT JAY, b Philadelphia, Pa, Dec 19, 42; m 77. HEALTH ECONOMICS, PUBLIC HEALTH PLANNING. *Educ:* Marietta Col, Ohio, BA, 64; Univ Chicago, MBA, 66; Johns Hopkins Univ, MPH, 67, ScD(health serv admin & res), 69. *Hon Degrees:* ScD, Sch Pub Health, John Hopkins Univ, 69; MA, Harvard Unv, 88. *Prof Exp:* Asst assoc dean, Health Care Progs, Sch Med, Johns Hopkins Univ, 69-70; asst dir planning & develop, Off Health Care Progs, Johns Hopkins Med Insts & from instr to asst prof, dept med care & hosp, John Hopkins Sch Hyg & Pub Health, 70-71; spec asst health affairs to dep under secy policy coord & for policy develop to asst secy health & sci affairs, Dept Health, Educ & Welfare, 71-72; vpres, Robert Wood Johnson Found, 72-80, sr vpres & dir prog, 80-87; PROF & CHMN, DEPT HEALTH POLICY & MGT, SCH PUB HEALTH & DEP DIR, DIV HEALTH POLICY RES & EDUC, HARVARD UNIV, 87- *Concurrent Pos:* Vis lectr, Woodrow Wilson Sch Pub & Int Affairs & course coordr, Princeton Univ, 80-87; sr policy analyst, comt health serv indust, Cost Living Coun, Washington, DC, 71; consult, Fiber Comn Priv Philanthropy & Pub Needs, 77, sr consult, NBC, 78, comt corp soc responsibility, Inst Life Insurance, 75-79; partic, White House Conf Balanced Nat Growth & Econ Develop, 78; comt US China rel, Coun Foreign Rel, 77-78, adv comt priv philanthropy & pub needs to secy treas, 77-79, steering comt, Int Conf Health Care Year 2000, Inst Med, 80-82, spec blue ribbon comt, corp restructuring NJ Hosp, Statewide Health Coord Coun, 82-84; bd overseers, Sch Dent Med, Univ Pa

& Leonard Davis Inst, Wharton Sch Bus, Philadelphia, 79-86; adv panel long term planning to vpres health serv, Am Hosp Asn, 80-82; chmn, Inter Am Workshop Legislative Approach Prev Alcohol Rel Prob, Inst Med, Nat Acad Sci & Pan Am Health Orgn, 82-84; mem, Health Adv Coun, John Hopkins Univ Sch Hyg & Pub Health, 84-86; bd mem, Am Bd Med Spec & Am Pharmaceut Inst, 88-90; chmn, Comt Access Health Care, Inst Med, 89-; chair, Health Prog Comt, Am Asn Pub Opinion Res, 89-; sr adv, Subcomt Health, Nat Gov Asn, 89-; sr consult, Pub Opinion & Health Care, US Ways & Means Comt, 90- *Mem:* Inst Med-Nat Acad Sci; Coun Foreign Rels; Am Pub Health Asn; Am Hosp Asn; Asn Health Serv Res; Am Asn Pub Opinion Res. *Res:* Author of over 40 publications in medical & public health care. *Mailing Add:* Dept Health Policy & Mgt Sch Pub Health Harvard Univ 677 Huntington Ave Boston MA 02115

BLENKARN, KENNETH ARDLEY, b Amarillo, Tex, May 17, 29; m 52; c 2. MECHANICAL ENGINEERING. *Educ:* Rice Univ, BA, 51, BS, 52, MS, 54, PhD(mech eng), 60. *Prof Exp:* Instr mech eng, Rice Univ, 52-54; res engr, Amoco Prod Co, 54-57, sr res engr, 60-63, staff res engr, 63-68, res assoc, 68-72, res supvr, 72-76, res dir, 76-86; CONSULT, 86- *Mem:* Nat Acad Eng; Am Inst Mining, Metall & Petrol Engrs; Soc Petrol Engrs. *Res:* Applied mechanics treatment of oilwell drilling and production; effects of arctic ice on structures and vessels; ocean waves and vessel response; design of offshore structures; deepwater systems. *Mailing Add:* 9115 E 37 Ct S Tulsa OK 74145

BLESS, ROBERT CHARLES, b Ithaca, NY, Dec 3, 27; m 70. ASTRONOMY. *Educ:* Univ Fla, BS, 47; Cornell Univ, MS, 55; Univ Mich, PhD(astron), 58. *Prof Exp:* Physicist, Naval Res Lab, 47-48; lectr astron, Univ Mich, 58; res assoc, 58-61, from asst prof to assoc prof, 61-69, chmn dept, 72-76, PROF ASTRON, UNIV WIS-MADISON, 69- *Concurrent Pos:* Nat Res Coun sr fel, Goddard Space Flight Ctr, 67-68; prin investr, High Speed Photometer, Hubble Space Telescope, 77- *Mem:* Int Astron Union; Am Astron Soc; fel AAAS. *Res:* Photoelectric spectrophotometry; space astronomy. *Mailing Add:* Washburn Observ Univ Wis Madison WI 53706

BLESSER, BARRY ALLEN, b Brooklyn, NY, Apr 3, 43. ELECTRICAL ENGINEERING. *Educ:* Mass Inst Technol, SB, 64, SM, 65, PhD(elec eng), 69. *Prof Exp:* From instr to asst prof elec eng, Mass Inst Technol, 66-76, assoc prof, 76-80; MEM STAFF, BLESSER ASSOCS, 80-, PRES. *Concurrent Pos:* Vinton-Hayes fel; NSF trainee; assoc radiol, Harvard Med Sch, 71- *Mem:* AAAS; Audio Eng Soc; Acoust Soc Am; Inst Elec & Electronics Engrs. *Res:* Communications with emphasis on perceptual psychology; compression systems and signal processing in broadcasting; character recognition, speech perception and related issues in psychology. *Mailing Add:* Blesser Assocs RFD2 Box 335 Raymond NH 03077

BLESSER, WILLIAM B, b Warren, Pa, Feb 19, 24; m 52, 79; c 2. ELECTRICAL & MECHANICAL ENGINEERING, ENGINEERING EDUCATION. *Educ:* Rensselaer Polytech Inst, BME, 50; Polytech Inst Brooklyn, MEE, 58. *Hon Degrees:* ScD, Ind Northern Univ, 74. *Prof Exp:* Plant engr, Beaunit Mills, Inc, 50-52; chief mech engr, Anton Electronic Lab, 52-53; engr, Bulova Res & Develop Lab, 53-54; instr mech eng, Polytech Inst NY, 54-57, from asst prof to assoc prof elec eng, 58-66, prof bioeng & mech eng & dir, Bioeng Prog, 66-90; RETIRED. *Concurrent Pos:* Lectr, navig & guid course, Bendix Aircraft, 58-59 & Rockefeller Inst, 59-60; mem fac, State Univ NY, Downstate Med Ctr, 62-; tech adv, US Army Res Off, 62-72; consult, Vet Admin & Cardiac Test Lab, Montefiore Med Ctr, Cardiac Ambulatory Testing, 80- *Mem:* Inst Elec & Electronics Engrs; NY Acad Sci; Asn Advan Med Instrumentation; Robotics Inst Am; Soc Comput Simulation. *Res:* Electromechanical and biomedical engineering; inertial guidance, navigation and control; rehabilitation engineering; bioengineering; robotics; simulation. *Mailing Add:* Polytech Inst New York 333 Jay St Brooklyn NY 11201

BLESSING, GERALD VINCENT, b Cincinnati, Ohio, July 7, 42; m 67; c 2. ULTRASONIC NONDESTRUCTIVE TESTING, ULTRASONIC MATERIAL PROPERTIES. *Educ:* Xavier Univ, Ohio, BS, 64; Col William & Mary, Va, MS, 66; Cath Univ Am, PhD(physics), 73. *Prof Exp:* Physicist, NASA, Langley, Va, 66; instr physics, Randolph-Macon Col, Va, 66-68; res physicist, Naval Surface Weapons Ctr, Md, 73-80; PHYSICIST, NAT INST STANDARDS & TECHNOL, 80- *Mem:* Am Phys Soc; Am Soc Nondestructive Testing; Inst Electronics & Elec Engrs; Sigma Xi; Am Soc Testing & Mat. *Res:* Materials research of the elastic properties of metals, ceramics and composites by ultrasonic methods; nondestructive evaluation of material defect properties by ultrasonics; surface roughness and thickness measured ultrasonically. *Mailing Add:* Nat Inst Standards & Technol Sound Bldg 233 A147 Gaithersburg MD 20899

BLESSING, JOHN A, b Wilkes-Barre, Pa, June 12, 46; m 68; c 1. CANCER CLINICAL TRIALS, GYNECOLOGIC ONCOLOGY. *Educ:* King's Col, BS, 68; State Univ NY, Buffalo, MA, 70, PhD(statist), 74. *Prof Exp:* Asst statistician cancer res, Acute Leukemia Group B, 74; GROUP STATISTICIAN CANCER RES & PRIN INVESTR, GYNEC ONCOL GROUP, 74-; ASST PROF, STATE UNIV NY, BUFFALO, 75-; CANCER RES SCIENTIST V, ROSWELL PARK MEM INST, 79- *Mem:* Soc Gynec Oncologists; Am Soc Clin Oncol; Am Radium Soc; Biomet Soc; Inst Math Statist. *Res:* Clinical trials investigating the treatment of gynecologic cancer. *Mailing Add:* GOG Statist Off Roswell Park Cancer Inst Tower Apts 2nd Floor Elm & Carlton Sts Buffalo NY 14263

BLESSING, ROBERT HARRY, b Wilkes-Barre, Pa, Apr 7, 41; m 62; c 2. STRUCTURAL CHEMISTRY, CRYSTALLOGRAPHY. *Educ:* King's Col, BS, 62; Ohio Univ, PhD(chem), 71. *Prof Exp:* Teacher math & physics, Riverside High Sch, NJ, 62-63; fel chem, Lund Univ, 69-70; fel crystallog, Univ Pittsburgh, 70-72 & State Univ NY Buffalo, 72-74; asst prof chem, Mercyhurst Col, 74-78; RES SCIENTIST, MED FOUND BUFFALO, 78- *Mem:* Am Chem Soc; Am Crystallog Asn. *Res:* Relationship of measurable macroscopic properties of molecules and crystals to measurable microstructure. *Mailing Add:* 291 Nassau Ave Buffalo NY 14217

BLESSINGER, MICHAEL ANTHONY, b New Albany, Ind, Sept 14, 56. OPTOELECTRONICS, SOLID STATE ELECTRONICS. *Educ:* Purdue Univ, BS, 78; Calif Inst Technol, MS, 80. *Prof Exp:* Mem tech staff, Jet Propulsion Lab, 80-84; SR RES SPECIALIST, ROCKWELL INT, 85- *Mem:* Am Phys Soc; Inst Elec & Electronics Engrs; Soc Photo-Optical Engrs. *Res:* Solid state infrared sensing electronics; design of integrated circuits and characterization of infrared sensors. *Mailing Add:* 3480 Greenfield Ct Maineville OH 45039

BLETNER, JAMES KARL, b Mason, WVa, July 29, 12; m 35; c 2. POULTRY HUSBANDRY. *Educ:* WVa Univ, BS, 34, MS, 50; Ohio State Univ, PhD, 58. *Prof Exp:* Asst exten poultryman, Agr Exp Sta, WVa Univ, 35-38, asst prof poultry husb, Univ & asst poultry husbandman, Agr Exp Sta, 46-55; instr poultry husb, Ohio State Univ, 55-58; from assoc prof to prof, 58-78, EMER PROF POULTRY NUTRIT, UNIV TENN, KNOXVILLE, 78- *Concurrent Pos:* Feed serviceman, Quaker Oats Co, 38-44 & 46. *Mem:* AAAS; Am Inst Nutrit; Poultry Sci Asn; World Poultry Sci Asn; Sigma Xi. *Res:* Poultry nutrition and management. *Mailing Add:* 300 Essex Dr Knoxville TN 37922

BLEUER, NED KERMIT, b Oak Park, Ill, Nov 21, 43; m 65; c 2. GLACIAL GEOLOGY, ENVIRONMENTAL GEOLOGY. *Educ:* Univ Wis-Madison, BS, 65, PhD(geol), 70; Univ Ill, MS, 67. *Prof Exp:* GLACIAL GEOLOGIST, IND GEOL SURV, BLOOMINGTON, IND, 68- *Concurrent Pos:* Adj prof, Dept Geol, Indiana Univ. *Mem:* Am Quaternary Asn. *Res:* Glacial stratigraphy and history of Indiana and the Midwest; landsat and shuttle imaging radar studies of Midwestern glacial morphology. *Mailing Add:* Ind Geol Surv 908 S Sheridan Dr Bloomington IN 47401

BLEVINS, CHARLES EDWARD, b Ritzville, Wash, Sept 5, 24; m 48; c 2. ANATOMY. *Educ:* Stanford Univ, BA, 47, MA, 48; Univ Calif, San Francisco, PhD(anat), 61. *Prof Exp:* Instr anat, physiol & zool, Glendale Col, 49-51; instr anat physiol, Contra Costa Col, 51-58, chmn dept life sci, 61-62; res anatomist, Div Otolaryngol, Univ Calif, San Francisco, 59-60; from instr to asst prof anat, Univ Wash, 62-65; asst prof, Baylor Col Med, 65-67; assoc prof, Med Sch, Northwestern Univ, 67-74; PROF & CHMN DEPT ANAT, SCH MED, IND UNIV, INDIANAPOLIS, 74- *Mem:* AAAS; Am Asn Anat; Am Asn Clin Anatomists. *Res:* Gross and microscopic anatomy; histology; histochemistry and ultrastructure of muscles, nerves, connective tissue and joints of the middle ear and larynx; ultrastructure of the acoustic system. *Mailing Add:* Dept Anat Ind Univ Sch Med 1120 South Dr Indianapolis IN 46223

BLEVINS, DALE GLENN, b Ozark, Mo, Aug 29, 43; m 67; c 1. PLANT NUTRITION, NITROGEN METABOLISM. *Educ:* Southwest Mo State Univ, BS, 65; Univ Mo, MS, 67; Univ Ky, PhD(plant physiol), 72. *Prof Exp:* Res assoc, Dept Bot, Ore State Univ, 72-74; asst prof plant nutrit, Univ Md, College Park, 74-77; from asst prof to assoc prof, 78-85, PROF AGRON, UNIV MO-COLUMBIA, 85- *Mem:* Am Soc Plant Physiologists; Am Soc Agron. *Res:* Ureide metabolism in nodulated leguminous plants; mineral element metabolism. *Mailing Add:* Dept Agron Univ Mo Columbia MO 65211

BLEVINS, GILBERT SANDERS, b Smyth Co, Va, Oct 31, 27; m 50; c 2. EXPERIMENTAL PHYSICS, SYSTEMS ENGINEERING. *Educ:* Va Polytech Inst, BS, 53; Duke Univ, PhD(physics), 57. *Prof Exp:* Asst, Duke Univ, 53-57; sr res physicist, Nat Cash Register Co, 57-58; tech consult, Sperry Farragut Co, 58-60; mgr data acquisition systs, Guided Missile Range Div, Pan Am World Airways, 60-62; assoc dir appl physics, Surface Div, Westinghouse Elec Corp, 62-65; consult, US Dept Defense, 65-69; tech dir, US Navy Security Group, 69-83; SR TECH STAFF, ANAL SCI CORP, 83- *Res:* Systems engineering; applied microwave and solid state physics; communications systems; data processing and storage systems. *Mailing Add:* 11986 Hall Shop Rd Clarksville MD 21029

BLEVINS, MAURICE EVERETT, b Franklinville, NJ, Mar 30, 28; m 52; c 1. APPLIED PHYSICS, COMPUTER SCIENCES. *Educ:* Duke Univ, BS, 52, PhD(physics), 58. *Prof Exp:* Asst physics, Duke Univ, 52-56, res assoc, 56-59; prof, Wofford Col, 59-64; fac mem, Spartanburg Tech Educ Ctr, 64-72, head electronics dept, 66-72; sr res engr, Steel Meddle Mfg Co, Greenville, 72-89; sr res engr, Cinn Milicron, 89-90; RETIRED. *Concurrent Pos:* Consult, 68- *Mem:* Am Phys Soc; Am Asn Physics Teachers; Sigma Xi. *Res:* Computer controlled systems design-robots; computer programming; textile technology; electronics; low temperature physics; high energy physics; superconductivity. *Mailing Add:* 98 Canterbury Rd Spartanburg SC 29302

BLEVINS, R(ALPH) W(ALLACE), b Washington, DC, Mar 30, 23; m 46; c 4. CIVIL ENGINEERING. *Educ:* Northwestern Univ, BS, 45; Univ Tex, MS, 49. *Prof Exp:* Detailer, Bridge Div, City of Chicago, 45-46; instr civil eng, Univ Tex, 46-49 & Cornell Univ, 49-51; assoc engr, Appl Physics Lab, Johns Hopkins Univ, 51-52, sr engr, 52-53, proj supvr, 53-59, asst group supvr, 59-78, group supvr, 78-86; RETIRED. *Mem:* Am Soc Civil Engrs; Am Inst Aeronaut & Astronaut. *Res:* Missile and satellite structural design, development and environmental protection; transportation systems design and development; ocean energy systems design and development. *Mailing Add:* 10901 Swansfield Rd Columbia MD 21044

BLEVINS, RAYMOND DEAN, b Butler, Tenn, Apr 3, 39; m 62; c 4. PHYSIOLOGY, BIOCHEMISTRY. *Educ:* ETenn State Univ, BS, 60, MA, 61; Univ Tenn, PhD(physiol, biochem), 71. *Prof Exp:* Teacher & chmn dept sci, Va High Sch, 61-62; assoc prof biol, King Col, 62-69, chmn dept, 62-64; asst prof human physiol & anat, Univ Tenn, Knoxville, 69-71; assoc prof human physiol & anat, 71-81, PROF BIOL SCI & BIOCHEM, EAST TENN STATE UNIV, 81- *Concurrent Pos:* Consult, Biol Div, Oak Ridge Nat Lab, 74-; Environ Mutagen Info Ctr, 78- *Mem:* AAAS; Am Soc Zoologists; Am Inst Biol Sci; Am Chem Soc; Sigma Xi. *Res:* Effects of specific chemicals on the nucleic acids, protein and lipids on human cells grown in culture. *Mailing Add:* 1401 Buffalo St Johnson City TN 37601

BLEVINS, ROBERT L, b Finley, Ky, Oct 24, 31; m 55; c 2. SOIL SCIENCE. *Educ:* Univ Ky, BS, 52, MS, 58; Ohio State Univ, PhD(agron), 67. *Prof Exp:* Soil scientist, Soil Conserv Serv, 58-61; mem staff, Eastern Ky Resource Develop Prog, 61-62; asst prof soil surv, Univ Ky, 62-65; res assoc agron, Ohio State Univ, 65-67; ASST PROF AGRON, UNIV KY, 67- *Mem:* Am Soc Agron; Soil Sci Soc Am. *Res:* Forest soils; soil classification; genesis and land use, especially problems involving soil management. *Mailing Add:* Dept Agron Univ Ky Lexington KY 40506

BLEVIS, BERTRAM CHARLES, b Toronto, Ont, Feb 26, 32; m 59; c 3. SPACE TECHNOLOGY, COMMUNICATIONS. *Educ:* Univ Toronto, BA, 53, MA, 54, PhD(physics), 56. *Prof Exp:* Sci officer radio physics, Defence Res Telecommun Estab, 56-57, group leader ultra high frequency propagation, 57-62, sci officer satellite commun, 62-66; sci liaison officer, Can Defence Res Staff, Washington, DC, 66-69; leader satellite commun sect, Commun Res Ctr, 69-71, mgr space systs prog, 71-75, dir int arrangements, 75-77, dir gen space technol & appln, Commun Res Ctr, 77-87, exec dir res, Dept Commun, 85-86; DIR, GOVT LIAISON, CAN ASTRONAUT LTD, 87- *Concurrent Pos:* Mem, Can Nat Orgn, Int Radio Consult Comt, 61-74, deleg, Int Telecommun Union, 63-77; mem Can comn II, Inst Sci Radio Union, 65-75; vis prof, Indian Space Res Orgn, 83; mem, Space Adv Bd, Brit Satellite Bd Broadcasting, 87-90; chmn, 42nd Cong Int Astronaut Fedn, 91. *Honors & Awards:* Emmy Award, Nat Acad Television Arts & Sci, 87. *Mem:* Sr mem Inst Elec & Electronic Engrs; Can Asn Physicists; fel Can Aeronaut & Space Inst. *Res:* Gaseous discharge electronics; lunar radar; radio auroral backscattering; radio propagation, particularly application to space communications; satellite communications and broadcasting applications and space systems studies. *Mailing Add:* 1997 Neepawa Ave Ottawa ON K2A 3L4 Can

BLEVISS, ZEGMUND O(SCAR), b Calgary, Alta, Aug 19, 22; US citizen; m 51; c 4. SYSTEMS ENGINEERING. *Educ:* Univ Calif, Berkeley, BS, 44; Calif Inst Technol, MS, 47, PhD(aeronaut math), 51. *Prof Exp:* Aeronaut res scientist, Ames Aeronaut Lab, Nat Adv Comt Aeronaut, 44-46; design specialist, Douglas Aircraft Co, Inc, 51-59; mem tech staff, Space Tech Labs, Thompson-Ramo-Wooldridge, Inc, 59-60; head dept magnetohydrodyn, Aerospace Corp, 60-64; asst prog mgr, BRMS data correlation facility, Heliodyne Corp, 64-66; chief sci, Space & Commun Group, Hughes Aircraft Co, 66-89; RETIRED. *Mem:* AAAS; Am Inst Aeronaut & Astronaut; Am Phys Soc. *Res:* Systems engineering and analysis; aeronautics; gas dynamics; re-entry physics; magnetohydrodynamics; plasma physics. *Mailing Add:* 9106 Monte Mar Dr Los Angeles CA 90035

BLEWETT, CHARLES WILLIAM, b Norfolk, Va, Aug 15, 43; m 64; c 5. ORGANIC CHEMISTRY. *Educ:* Thomas More Col, AB, 63; Univ Cincinnati, PhD(org chem), 69. *Prof Exp:* Chemist, 68-73, group leader, 73-84, MGR, EMERY CHEMICAL, 84- *Concurrent Pos:* Instr, Univ Cincinnati, 75- *Mem:* Am Chem Soc; Am Oil Chemists Soc; Sigma Xi; Clay Minerals Soc. *Res:* Fatty acid synthesis and applications; organometallic chemistry; oxo chemistry; dimer acids. *Mailing Add:* Henkel Corp/Emery Group 4900 Este Ave Cincinnati OH 45232

BLEWETT, JOHN P, b Toronto, Can, Apr 12, 10; m 36, 83. NUCLEAR PHYSICS, PARTICLE ACCELERATORS. *Educ:* Univ Toronto, BAC & MAS, 32; Princeton Univ, PhD(physics), 36. *Prof Exp:* Physicist, Gen Elec Res Lab, 37-46; sr physicist, Brookhaven Nat Lab, 47-48; RETIRED. *Mem:* Fel Am Phys Soc; fel Inst Elec & Electronics Engrs; fel NY Acad Sci; fel AAAS. *Res:* Nuclear fusion; synchotron light sources; co-author book of particle accelerators; atomic physics; thermionic emission; microwave generation; particle accelerator design. *Mailing Add:* 310 W 106th St Apt 7D New York NY 10025

BLEWETT, MYRTLE HILDRED, b Toronto, Ont, May 28, 11; nat US; div. PHYSICS. *Educ:* Univ Toronto, BA, 35. *Prof Exp:* Engr appl physics, Gen Elec Co, 42-47; physicist, Brookhaven Nat Lab, 47-64; sr physicist, Argonne Nat Lab, Ill, 64-69; sr physicist, Europ Orgn Nuclear Res, 69-77; RETIRED. *Mem:* Fel Am Phys Soc. *Res:* Particle accelerators; high energy physics. *Mailing Add:* 1666 Pendrell St Apt 1503 Vancouver BC V6G 1S9 Can

BLEYL, ROBERT LINGREN, b Salt Lake City, Utah, Mar 25, 36; m 61; c 4. TRANSPORTATION ENGINEERING, TRAFFIC ENGINEERING. *Educ:* Univ Utah, BS, 61, MS, 62; Pa State Univ, PhD(transp eng), 71. *Prof Exp:* Traffic engr, Utah Hwy Dept, 61-63, dep traffic engr, 64-65; res assoc hwy traffic, Yale Univ, 65-68; from instr to assoc prof traffic eng, Pa State Univ, 68-72; assoc prof transp eng, Univ NMex, 72-81; CONSULT HWY TRAFFIC, ROBERT L BLEYL & ASSOCS, 81- *Concurrent Pos:* Consult, 66-81. *Mem:* Inst Transp Engrs; Nat Acad Forensic Engrs (treas, 91-); Nat Soc Prof Engrs; Soc Automotive Engrs. *Res:* Author of numerous publications on traffic engineering and accident reconstruction. *Mailing Add:* 7816 Northridge Ave NE Albuquerque NM 87109

BLEYMAN, LEA KANNER, b Halle, Ger, Nov 9, 36; US citizen; c 1. GENETICS, CELL BIOLOGY. *Educ:* Brandeis Univ, BA, 58; Columbia Univ, MA, 60; Ind Univ, Bloomington, PhD(genetics of ciliates), 66. *Prof Exp:* Res assoc zool, Univ Ill, Urbana, 64-69; res assoc, Labs for Reproductive Biol, Univ NC, Chapel Hill, 69-72; assoc prof biol, 73-78, dep chmn, 77-79, chmn, Dept Natural Sci, 81-84, PROF BIOL, BARUCH COL, CITY UNIV NY, 79- *Concurrent Pos:* City Univ NY Chancellor's fel, 85-86; vis prof, Dept Anat, Albert Einstein Col Med, 84, 85-89. *Mem:* AAAS; Genetics Soc Am; NY Acad Sci; Am Soc Cell Biol; Soc Protozoologists. *Res:* Genetics of ciliates, especially inheritance and determination of mating type specificity and expression and nuclear-cytoplasmic relations in cellular development. *Mailing Add:* Box 502 17 Lexington Ave New York NY 10010

BLICHER, ADOLPH, b Warsaw, Poland; nat US; m 34; c 1. SOLID STATE SCIENCE. *Educ:* Polytech Inst Poland, DSc(tech sci), 38. *Prof Exp:* Dir eng, Polish Broadcasting Co, 44-46; sr physicist, Radio Receptor Co, NY, 51-55; mgr spec prod develop & applns dept, RCA Corp, 55-56, mgr comput devices develop dept, Solid State Div, 56-62, mgr spec prod develop & applns dept, 63-64, mgr advan devices & applns dept, Solid State Technol Ctr, 64-73, consult, RCA Labs, Princeton, 73-74; CONSULT & SCI WRITER, 74- *Mem:* Inst Elec & Electronics Engrs. *Res:* Solid state device physics and electronics. *Mailing Add:* 345 Judges Lane North Plainfield NJ 07063

BLICK, EDWARD F(ORREST), b Covington, Ky, Mar 9, 32; m 55; c 2. AEROSPACE & MECHANICAL ENGINEERING. *Educ:* Univ Okla, BS, 58, MS, 59, PhD(eng sci), 63. *Prof Exp:* Aerodynamicist, McDonnell Aircraft Corp, 58-59; instr aeronaut eng, 59-63, from asst prof to assoc prof aerospace & mech eng, 63-69, asst dean grad col, 66-68, assoc dean, Col Eng, 68-69, prof aerospace & mech eng, 69-81, PROF PETROL & GEOL ENG, UNIV OKLA, 81- *Concurrent Pos:* Res assoc, Res Inst, Univ Okla, 59-; consult, Gulf Coastal Aircraft Co, 62-63 & F W Dodge Co, 63- & Channel Wing Corp, 69- *Mem:* Am Soc Mech Engrs; Soc Petrol Engrs. *Res:* Drilling; production; fluid flow. *Mailing Add:* 7006 Lago Ranchero Dr Norman OK 73071

BLICKENSDERFER, PETER W, b Cincinnati, Ohio, Nov 28, 32; m 56; c 4. ANALYTICAL CHEMISTRY. *Educ:* Col Wooster, BA, 54; Univ Mo, PhD(phys chem), 63. *Prof Exp:* Instr chem, Univ Mo-Rolla, 61-62; assoc res chemist, Res Labs, Whirlpool Corp, 62-70, sr res chemist, 70-71; from assoc prof to prof chem, 71-87, HEAD CHEM DEPT, KEARNEY STATE COL, 87- *Concurrent Pos:* Vis prof, Univ Nebr, Lincoln, 85-86. *Mem:* Am Chem Soc; Sigma Xi. *Res:* Adsorption of gases on solids; surface chemistry of silicate minerals; high vacuum techniques; analytical methods development for trace inorganic constituents and pesticide residues. *Mailing Add:* Dept Chem Kearney State Col Kearney NE 68849

BLICKENSTAFF, ROBERT THERON, b North Manchester, Ind, July 3, 21; m 44, 78; c 2. SYNTHETIC ORGANIC & NATURAL PRODUCTS CHEMISTRY. *Educ:* Manchester Col, AB, 44; Purdue Univ, MS, 46, PhD(org chem), 48. *Prof Exp:* Res chemist, Procter & Gamble Co, 48-55; res assoc, Med Units, Univ Tenn, Memphis, 55-58; assoc prof, 64-75, ADJ PROF BIOCHEM, SCH MED, IND UNIV, INDIANAPOLIS, 75- *Concurrent Pos:* Res chemist, Vet Admin Hosp, Indianapolis, 58-, exec secy res, 70-73. *Mem:* Am Chem Soc; Sigma Xi. *Res:* Steroids; anti tumor compounds; radioprotective compounds. *Mailing Add:* Vet Admin Hosp 1481 W Tenth St Indianapolis IN 46202

BLICKSTEIN, STUART I, b Brooklyn, NY, Mar 16, 39; m 67; c 2. NP-COMPLETENESS, PEDAGOGICAL PROBLEM IN MATHEMATICS. *Educ:* Brooklyn Col, BS, 61; Fairleigh Dickinson Univ, MBA, 76. *Prof Exp:* Sr appl analyst, Serv Bur Subsid, Int Bus Mach, 64-66; mgr biostatist, Geigy Pharmaceut, 66-68; mgr, Data Anal, Med Studies Inc, 68-70; dir, Prod Discovery, Warner Lambert, 70-80; asst prof quant anal, Fairleigh Dickinson Univ, 80-86; ASST PROF MGT SCI, RIDER COL, 86- *Concurrent Pos:* Chair, Pharmaceut Indust Statisticians, 72-75 & Biostatist subsect, Am Statist Asn, 75; adj prof, Fairleigh Dickinson Univ, 76-80 & 86-; adj fac mem, Grad Sch Mgt, Rutgers Univ, 84-; proj res coordr, Writing Across Curric Prog, 90- *Mem:* Inst Mgt Sci; Am Statist Asn; Pharmaceut Indust Statisticians. *Res:* Heuristic development for machine scheduling problems; application of writing techniques as an aid to teaching mathematics. *Mailing Add:* Stuart I Blickstein & Assoc 18 Brownstone Pl Flanders NJ 07836

BLICKWEDE, D(ONALD) J(OHNSON), b Detroit, Mich, July 20, 20; m 43; c 2. METALLURGY. *Educ:* Wayne State Univ, BS, 43; Mass Inst Technol, ScD (metall), 48. *Prof Exp:* Metallurgist, Curtis-Wright Corp, NJ, 43-45; asst metall, Mass Inst Technol, 45-48; head high temperature alloys sect, Naval Res Lab, 48-50; res engr, Bethlehem Steel Co, 50-53, asst head div, 53-56, head div, 56-60, dir appln res, 60-63, mgr res, 63-64, vpres, 64-84; CONSULT, 84- *Concurrent Pos:* Trustee, Am Soc Metals, 63-65. *Mem:* Nat Acad Eng; Am Soc Metals (pres, 84-85); Am Inst Mining, Metall & Petrol Engrs; Am Iron & Steel Inst; Brit Iron & Steel Inst; Brit Inst Metals. *Res:* Raw materials, reduction, refining, forming, finishing and application of ferrous materials. *Mailing Add:* 891 W Placita Cotonia Green Valley AZ 85614

BLIESE, JOHN C W, b Waterloo, Iowa, Mar 10, 13; m 39; c 2. ZOOLOGY, ORNITHOLOGY. *Educ:* Univ Northern Iowa, BA, 35; Columbia Univ, MA, 36; Iowa State Univ, PhD(zool), 53. *Prof Exp:* High sch instr, Iowa, 36-41; sci supvr, Campus Sch, Univ Northern Iowa, 41-42; civilian instr, US Air Force, 42-44 & Tenn Eastman Corp, 44-45; sci supvr, Campus Sch, Univ Northern Iowa, 45-47; instr biol, Cornell Col, 47-49; instr zool, Iowa State Univ, 49-53; from assoc prof to prof biol, Kearney State Col, 53-78, head dept, 62-66; RETIRED. *Concurrent Pos:* Prof, Kearney State Col, 79-83. *Honors & Awards:* Zool Lab named in honor, John C W Bliese Zool Lab, Kearney State Col, 87. *Mem:* Am Sci Affil; Nat Audubon Soc; Am Ornithologists Union; Int Crane Found; Nature Conservancy. *Res:* Photoperiodism of aquarium plants; roosting of bronzed grackles and associated birds; diurnal ecology of sandhill cranes. *Mailing Add:* 107 E 27 St Kearney NE 68847

BLIGH, JOHN, b London, Eng, July 18, 22; m 52; c 2. PHYSIOLOGY. *Educ:* Univ London, BSc, 50, PhD(physiol), 52, DSc(physiol), 77. *Prof Exp:* From sci officer to sr sci officer physiol, Hannah Res Inst, Ayr, Scotland, 52-56; from prin sci officer to sr prin sci officer, Inst Animal Physiol, Cambridge, Eng, 57-77; DIR, INST ARCTIC BIOL, UNIV ALASKA, FAIRBANKS, 77-, PROF PHYSIOL, 80- *Concurrent Pos:* Permanent deleg, Int Sci Comt Antarctic Res, Int Union Physiol Sci & chmn bd definition terms, 77-; prof animal physiol & dir, Div Life Sci, Univ Alaska, 77- *Mem:* Am Physiol Soc; Brit Physiol Soc; Peruvian Physiol Soc; Int Biometeorol Soc. *Res:* Mammalian thermoregulation; the neurophysiology of homeothermy in particular, and homeostasis generally. *Mailing Add:* Little Garth off High St Harston Cambridge CB2 5QB England

BLIGH, THOMAS PERCIVAL, b Grahamstown, SAfrica, Jan 18, 41; UK citizen; m 66; c 2. MECHANICAL ENGINEERING, PHYSICS. *Educ:* Univ Witwatersrand, BSc, 62, MSc, 64, PhD(physics), 72. *Prof Exp:* Sr res officer mech eng, Chamber Mines, SAfrica Res Labs, 65-71; asst prof civil & mineral eng, Univ Minn, 72-76, assoc prof mech eng, 76-78; ASSOC PROF

MECH ENG, MASS INST TECHNOL, 79- *Concurrent Pos:* Consult various industs, 74- *Res:* Energy conservation; energy production; heat transfer; thermodynamics; combustion; controlled explosives; earth sheltered buildings; underwater photographic equipment; solar energy; computer aided design. *Mailing Add:* Dept Mech Eng Cambridge Univ Trumpington St Cambridge CB2 1PZ England

BLIM, RICHARD DON, PEDIATRICS. *Prof Exp:* DIR MED AFFAIRS, ST LUKES HOSP, 89- *Mem:* Inst Med-Nat Acad Sci. *Mailing Add:* St Lukes Hosp Wornall Rd & 44th St Kent City MO 64111

BLINCOE, CLIFTON (ROBERT), b Odessa, Mo, Nov 21, 26; m 49; c 3. PHYSICAL BIOCHEMISTRY. *Educ:* Univ Mo, BS, 47, MA, 48, PhD, 55. *Prof Exp:* From instr to asst prof dairy husb, Univ Mo, 50-56; asst prof & asst res chemist, Univ Nev, Reno, 56-60, assoc prof, 60-69, prof biochem, 69-85, RES CHEMIST, UNIV NEV, RENO, 60-, EMER PROF BIOCHEM, 85- *Concurrent Pos:* Vis prof, US-Ireland Scholar Exchange Prog, Trinity Col, Dublin, 69-70; Nat Feed Ingredients Asn travel fel, 74; vis prof phys biol, Cornell Univ, 76-77; mem, Dose Assessment Adv Group, app by US Secy Energy & Gov Nev, 85-86. *Mem:* AAAS; Am Chem Soc; Am Soc Animal Sci; Biophys Soc; Brit Soc Chem Indust. *Res:* Physical biochemistry of domestic animals, especially in relation to nutrition, as studied with radiotracers; computer simulation of biological processes. *Mailing Add:* Div Biochem Univ Nev 1041 Univ Terr Reno NV 89503-2722

BLINDER, SEYMOUR MICHAEL, b New York, NY, Mar 11, 32; m 68; c 5. THEORETICAL CHEMISTRY, MATHEMATICAL PHYSICS. *Educ:* Cornell Univ, AB, 53; Harvard Univ, AM, 55, PhD(chem physics), 59. *Prof Exp:* Physicist, Appl Physics Lab, Johns Hopkins Univ, 58-61; asst prof, Carnegie Inst Technol, 61-62; res assoc, Harvard Univ, 62-63; from asst prof to assoc prof, 63-70, PROF CHEM, UNIV MICH, 70- *Concurrent Pos:* Guggenheim fel, Univ Col, Univ London, 65-66; NSF sr fel, Ctr de Mecanique Ondulatoire Appliquee, Paris, 70-71 & Math Inst, Oxford, 70-71; Rackham Res fel, 77; mem, bd ed, J Am Chem Soc, 78-80. *Mem:* Am Phys Soc; AAAS. *Res:* Green's functions and propagators; applications of quantum mechanics to atomic and molecular problems. *Mailing Add:* Dept Chem Univ Mich Ann Arbor MI 48109

BLINK, JAMES ALLEN, b Kenosha, Wis, Apr 21, 48; m 72; c 2. HEAT TRANSFER, LASER PHYSICS. *Educ:* Univ Nev, Reno, BS, 70; Ga Inst Technol, MS, 71; Univ Calif, Davis, PhD, 84. *Prof Exp:* Mil res assoc, Lawrence Livermore Nat Lab, 76-78, consult, 78-79, engr, 79-84, physicist, 85-90, MECH ENGR, LAWRENCE LIVERMORE NAT LAB, 90- *Mem:* Am Nuclear Soc. *Res:* Design of inertial fusion reactors with emphasis on neutronics, activation, safety, tritium processing, and hydrodynamic response of fluids to pulsed fusion energy; design of high average power lasers; design of high level nuclear waste repositories. *Mailing Add:* L-204 Lawrence Livermore Nat Lab PO Box 808 Livermore CA 94550

BLINKS, JOHN ROGERS, b New York, NY, Mar 21, 31; m 53; c 3. PHARMACOLOGY, PHYSIOLOGY. *Educ:* Stanford Univ, AB, 51; Harvard Univ, MD, 55. *Prof Exp:* Med house officer, Peter Bent Brigham Hosp, Boston, 55-56; res assoc, Nat Heart Inst, Bethesda, Md, 56-58; instr pharmacol, Harvard Med Sch, 58-61, assoc, 61-64, asst prof, 64-68; from assoc prof to prof pharmacol, Mayo Grad Sch Med, Univ Minn, 73-90; head, Dept Pharmacol, Mayo Found, 68-88, distinguished investr, 85-90; PROF PHYSIOL/BIOPHYS, UNIV WASH, 90- *Concurrent Pos:* Markle scholar med sci, 61-66; hon res asst, Univ Col, Univ London, 62-63; estab investr, Am Heart Asn, 65-70, chmn, 79, mem, res comt, 75-79 & sci publ comt, 83-87; mem, prog proj comt, Nat Heart & Lung Inst, 68-71 & cardiol adv comt, 75-79; field ed, J Pharmacol & Exp Therapeut, 69-71; vis lectr, dept physiol, Univ Auckland, 74; Edmond A & Marion F Guggenheim Prof, Mayo Found, 81-87; mem, pharmacol test comt, Nat Bd Med Examrs, 84-86. *Honors & Awards:* Otto Krayer Award, Am Soc Pharmacol & Exp Therapeut, 87; Res Achievement Award, Am Heart Asn, 89. *Mem:* Biophys Soc; Cardiac Muscle Soc; Am Physiol Soc; Am Soc Pharmacol & Exp Therapeut; Soc Gen Physiologists; fel AAAS. *Res:* Physiology and pharmacology of heart muscle; the role of calcium in regulation of cell function. *Mailing Add:* Friday Harbor Labs Univ Wash Friday Harbor WA 98250

BLINKS, LAWRENCE ROGERS, plant physiology; deceased, see previous edition for last biography

BLINN, DEAN WARD, b Carroll, Iowa, May 30, 41; m 65; c 2. PHYCOLOGY, AQUATIC ECOLOGY. *Educ:* Simpson Col, BA, 64; Univ Mont, MA, 66; Univ BC, PhD(phycol), 69. *Prof Exp:* Asst prof biol, Univ NDak, 69-71; asst prof, 71-77, assoc prof, 77-81, PROF BIOL, NORTHERN ARIZ UNIV, 81- *Concurrent Pos:* Dir, Ariz Pub Serv, 74-; dir, Nat Park Serv, 75- *Mem:* Sigma Xi. *Res:* Ecology, nutrition and life histories of freshwater species of algae; ecology of invertebrates; predator-prey interactions. *Mailing Add:* Dept Biol Sci Northern Ariz Univ Box 5640 Flagstaff AZ 86011

BLINN, ELLIOTT L, b Pittsburgh, Pa, Sept 25, 40; m 69. INORGANIC CHEMISTRY. *Educ:* Univ Pittsburgh, BS, 62; Ohio State Univ, MS, 64, PhD(inorg chem), 67. *Prof Exp:* Asst chem, Ohio State Univ, 61-62; fel, State Univ NY Buffalo, 67-68; from asst prof to assoc prof, 68-77, PROF CHEM, BOWLING GREEN STATE UNIV, 77- *Mem:* Am Chem Soc; Sigma Xi. *Res:* Coordination chemistry; reaction mechanism of metal complexes. *Mailing Add:* 631 Crestview Dr Bowling Green OH 43402

BLINN, JAMES FREDERICK, b Detroit, Mich, Feb 23, 49. COMPUTER GRAPHICS, SCIENCE EDUCATION. *Educ:* Univ Mich, BS, 70, MSE, 72; Univ Utah, PhD(comput sci), 78. *Prof Exp:* Res asst, Univ Mich, 69-74; fel, 77-79, MEM TECH STAFF, JET PROPULSION LAB, CALIF INST TECHNOL, 77- *Concurrent Pos:* Lectr, Univ Calif, Berkeley, 81. *Res:* Computer animation for educational purposes; modelling techniques and computer systems development. *Mailing Add:* 305 S Hill Ave Pasadena CA 91106

BLINN, LORENA VIRGINIA, b East Chicago, Ind, Sept 8, 39; m 85; c 2. SCIENCE EDUCATION. *Educ:* Univ Ga, BS, 61, MS, 64; Mich State Univ, PhD(sci ed & geol), 71. *Prof Exp:* Teacher biol, MacIntyre Park Sr High Sch, Thomasville, Ga, 61-62; from instr to assoc prof, 64-85, PROF NATURAL SCI, MICH STATE UNIV, 85- *Concurrent Pos:* Vis prof biol, Inst Technol, MARA, Malaysia, 86-87. *Mem:* Nat Sci Teachers Assn; AAAS. *Res:* History of Michigan Archaeological Society; college science teaching. *Mailing Add:* 117 N Kedzie Hall Mich State Univ East Lansing MI 48824-1031

BLINN, WALTER CRAIG, b Belleville, Ill, May 16, 30; m 85; c 4. SCIENCE EDUCATION, NATURAL SCIENCE. *Educ:* Ill State Univ, BS, 51; Okla State Univ, MS, 58; Northwestern Univ, PhD(biol), 61. *Prof Exp:* Instr biol, Northwestern Univ, 60-61; from instr to prof natural sci, 61-89, PROF, CTR INTEGRATED STUDIES GEN SCI, MICH STATE UNIV, 89- *Concurrent Pos:* Consult, Welch Sci Co, 61-64; NSF res grant, 63-65; vis prof physics, Indiana Univ, Cooperative Prog, Malaysia, 86-87. *Mem:* Hist Sci Soc; Sigma Xi. *Res:* History and philosophy of science; metaphysical foundations of scientific thought. *Mailing Add:* Ctr Integrated Studies - Gen Sci Mich State Univ East Lansing MI 48824

BLINT, RICHARD JOSEPH, b Fairmont, Minn, Feb 23, 45; m 66; c 3. CHEMICAL PHYSICS. *Educ:* St Mary's Col, Minn, BA, 67; Calif Inst Technol, PhD(chem), 72. *Prof Exp:* Presidential internship, Brookhaven Nat Lab, 72-73; assoc chem, Nat Res Coun, 73-74; res assoc physics, Univ Ill, 74-76; assoc sr res scientist, 76-80, staff res scientist, 80-85, SR STAFF RES SCIENTIST, DEPT PHYS CHEM, GEN MOTORS RES, GEN MOTORS CORP, 85- *Honors & Awards:* John M Campbell Award. *Mem:* Am Chem Soc; Combustion Inst. *Res:* Molecular quantum mechanics is utilized to investigate problems in combustion physics; molecular simulations are used to study polymer electrolytes. *Mailing Add:* Dept Phys Chem Gen Motors Res Warren MI 48090

BLISCHKE, WALLACE ROBERT, b Oak Park, Ill, Apr 20, 34; wid; c 5. RELIABILITY. *Educ:* Elmhurst Col, BS, 56; Cornell Univ, MS, 58, PhD(statist), 61. *Prof Exp:* Res fel biomath, NC State Col, 61-62; mem tech staff, Appl Math Dept, TRW Space Tech Labs, 62-64; sr statistician & proj dir, C-E-I-R, Inc, 64-69; CONSULT STATIST ANAL, 69-; ASSOC PROF, SCH BUS, UNIV SOUTHERN CALIF, 72- *Mem:* Inst Math Statist; fel Am Statist Asn; Biomet Soc; Inst Mgt Sci; Am Inst Aeronaut & Astronaut. *Res:* Mathematical statistics, especially estimation, mixtures of distributions and discrete distributions; applied statistics, especially experimental design, sample surveys, regression and related data analysis techniques; economic and statistical analysis of warranties; reliability; litigation support. *Mailing Add:* Decision Systs Dept Univ Southern Calif Los Angeles CA 90089-1421

BLISS, ARTHUR DEAN, b Waterbury, Conn, Jan 7, 27; m 57; c 2. POLYMER ALLOYS, COMPOSITES. *Educ:* Yale Univ, BS, 49, PhD(org chem), 55. *Prof Exp:* Res chemist, Olin Mathieson Chem Corp, 55-57, sr res chemist, 57-63, res assoc, 63-69; sect head nylon res & develop, Am Hoechst Corp, 69-74, mgr res lab, 74-79, dir res plastics div, 79-85; tech dir, Texapol Corp, 86-87; DIR, RES & DEVELOP, PLASTICS DIV, WELLMAN INC, 87- *Mem:* Am Chem Soc. *Res:* Synthesis and rearrangements of amino-alcohols; preparation of boron hydrides and organic thiophosphates; preparation and reaction of aliphatic sultams; condensation polymerization; nylon 6; nylon 12; copolyamides; addition polymerization; polystyrene; acrylonitrile butadiene styrene; nylon 66 alloys; thermoplastic polyesters; plastics recycling. *Mailing Add:* PO Box 1696 Litchfield Plantation Pawleys Island SC 29585

BLISS, DAVID FRANCIS, b Durham, NC, Oct 18, 46; m 80; c 2. SOLID STATE PHYSICS. *Educ:* Case Western Reserve Univ, BA, 68; Mass Inst Technol, SM, 81. *Prof Exp:* Technician crystal growth, Mobil Solar Energy Corp, 73-79; res asst, Dept Mat Sci, Mass Inst Technol, 79-81; eng mgr, M/A-COM, Inc, 81-89; PHYSICIST, SOLID STATE SCI DIRECTORATE, ROME LAB, USAF, 89- *Concurrent Pos:* Sr lectr, Northeastern Univ, State of the Art Eng, Inc, 82-; indust fel, Lawrence Berkeley Lab, 87-88; ed, J Electronic Mat, 91. *Mem:* Am Asn Crystal Growth; Mat Soc; Electrochem Soc. *Res:* Research in capital growth and characterization of semi conductors including silicon gallium arsenide, and indium phosphate; synthesis and growth of high melting temperature compounds at high pressure. *Mailing Add:* 14 Coolidge Rd Arlington MA 02174

BLISS, DOROTHY CRANDALL, b Westerly, RI, Feb 20, 16; m 69; c 1. ECOLOGY. *Educ:* Univ RI, BS, 36, MS, 38; Univ Tenn, PhD, 57. *Prof Exp:* Teacher high sch, RI, 36-37; instr sci & math, Greenbrier Jr Col, 41-43, Southern Sem & Jr Col, 43-47; instr bot, Univ Wyo, 47-49; from asst prof to assoc prof, 49-68, PROF BIOL, RANDOLPH-MACON WOMAN'S COL, 68-, EMER PROF, 83- *Concurrent Pos:* Southern Fel Fund scholar, 55-56. *Mem:* Ecol Soc Am; Am Inst Biol Sci; Sigma Xi. *Res:* Ferns of Rhode Island; ecological problem in the spruce-fir region of the Smoky Mountains; floristic survey of wilderness areas in Blue Ridge Mountains. *Mailing Add:* 322 Sumpter St Lynchburg VA 24503-4430

BLISS, ERLAN S, b Chicago, Ill, Jan 18, 41. LASER PHYSICS & ALIGNMENT, DIAGNOSTIC & CONTROL SYSTEMS. *Educ:* Lawrence Univ, Wis, 63; Carnegie-Mellon Univ, MS, 65, PhD(physics), 69. *Prof Exp:* Res physicist laser physics, Air Force Cambridge Res Lab, 68-72; group leader laser physics, 72-75, laser proj engr, Shiva, 76-77 & Nova, 78-84, SR SYST ENGR, LAWRENCE LIVERMORE LAB, UNIV CALIF, 85- *Mem:* Am Phys Soc. *Res:* Electronic properties of metals; laser-induced damage in dielectrics; propagation of intense laser beams in dielectrics, alignment and diagnostic systems for large lasers. *Mailing Add:* 357 Conway Dr Danville CA 94526

BLISS, EUGENE LAWRENCE, b Pittsburgh, Pa, Feb 18, 18; m 47; c 3. MEDICINE, PSYCHIATRY. *Educ:* Yale Univ, BA, 39; NY Med Col, MD, 43. *Prof Exp:* Lectr psychiat, Col Med, Univ Utah, 50-51, asst prof, 51-53; asst clin prof, Sch Med, Yale Univ, 53-56; assoc prof, 56-61, chmn dept, 70-

79, FOUND FUND PROF PSYCHIAT, COL MED, UNIV UTAH, 61- *Concurrent Pos:* Consult, Vet Admin Hosp, Salt Lake City, 50-53 & 56- *Honors & Awards:* Moses Award, 55. *Mem:* Am Psychiat Asn; AMA. *Res:* Neurochemistry; neuroendocrinology; multiple personalities. *Mailing Add:* Univ Utah Med Ctr Salt Lake City UT 84132

BLISS, FREDRICK ALLEN, b Inavale, Nebr, Dec 5, 38; m 63; c 3. PLANT GENETICS, PLANT BREEDING. *Educ:* BSc, Univ Nebr, 60; Univ Wis, PhD(hort, genetics), 65. *Prof Exp:* Fel genetics, Univ Minn, St Paul, 65-66; from asst prof to assoc prof, 66-76, PROF HORT, UNIV WIS-MADISON, 76- *Concurrent Pos:* Lectr, Univ Ife, Nigeria, 66-68; mem vis fac, Inst Agron & Plant Genetics, Univ Gottingen, 68-69. *Honors & Awards:* Assinsel Grand Prize, 86. *Mem:* Am Genetics Asn; fel Am Soc Hort Sci; fel Crop Sci Soc Am. *Res:* Plant breeding and genetical research dealing with Phaseolus vulgaris; quantitative genetics studies; genetic control of protein synthesis; genetic control of nitrogen fixation; disease and insect resistance edible legumes. *Mailing Add:* Dept Pomol Univ Calif Davis CA 95616

BLISS, JAMES C(HARLES), b Ft Worth, Tex, Oct 21, 33; m 62; c 2. BIOENGINEERING, ELECTRICAL ENGINEERING. *Educ:* Northwestern Univ, BS, 56; Stanford Univ, MS, 58; Mass Inst Technol, PhD(elec eng), 61. *Prof Exp:* Res engr character recognition, Stanford Res Inst, 56-58, res engr bioeng, 61-71; PRES, TELESENSORY SYSTS INC, 71- *Concurrent Pos:* Lectr, Stanford Univ, 66-; assoc prof elec eng, Stanford Univ, 68-71. *Honors & Awards:* Isabelle & Leonard H Goldenson Award, 77; John Scott Medal, 80. *Mem:* Inst Elec & Electronics Engrs. *Res:* Studies on tactile and kinesthetic perception, visual information processing in anthropods; development of sensory aids for the blind; electrooptical techniques. *Mailing Add:* Telesensory Systs Inc 455 N Bernardo Mountain View CA 94043

BLISS, LAURA, b Davenport, Iowa, Apr 10, 16. CHEMISTRY. *Educ:* Iowa State Univ, BS, 38, PhD(enzyme chem), 43. *Prof Exp:* Patent chemist, Hercules Powder Co, Del, 43-45; from instr to asst prof, 45-57, assoc prof chem, 57-84, Registr, 74-84, EMER PROF CHEM, RANDOLPH-MACON WOMAN'S COL, 84- *Concurrent Pos:* NSF sci fac fel, Inst Enzyme Res, 61-62; researcher, Dept Biochem, Univ Queensland, 68-69. *Mem:* Am Chem Soc. *Res:* Enzyme chemistry; phosphorylase; hydrolysis of beta-D-glucosides; phosphorylase of waxy maize; acetylesterases. *Mailing Add:* 203 Rowland Dr Lynchburg VA 24503

BLISS, LAWRENCE CARROLL, b Cleveland, Ohio, Nov 29, 29; m 52; c 2. ECOLOGY. *Educ:* Kent State Univ, BS, 51, MS, 53; Duke Univ, PhD(bot), 56. *Prof Exp:* Instr biol, Bowling Green State Univ, 56-57; from instr to prof bot, Univ Ill, Urbana, 57-68; prof bot & dir controlled environ facil, Univ Alta, 68-78; chmn dept, 78-87, PROF BOT, UNIV WASH, 78- *Concurrent Pos:* Fulbright res fel, NZ, 63-64; assoc ed, Oecologia, 69-75; dir Can tundra terrestrial ecosyst study, Int Biol Prog, 70-75; assoc ed, Arctic & Alpine Res, 69-71; mem environ protection bd, Can Arctic Gas Pipeline Ltd, 71-75; mem consultative comt, Nat Mus Can, 75-77; dir revegetation modeling study, Alta Oil Sands, 75-78; dir ecosyst study, Mt St Helen, 81-; mem ecol panel, Nat Sci Found, 81-83. *Mem:* Fel AAAS; Am Inst Biol Sci; fel Arctic Inst NAm; Ecol Soc Am (vpres, 75-77, treas, 76-81, pres elect, 81-82, pres, 82-83); Can Bot Asn. *Res:* Arctic ecosystem and manipulation studies; plant production; plant-soil relationships; microenvironmental studies; eco-physiology, including plant growth chamber research; tundra boreal forest and mountain ecology. *Mailing Add:* Dept Bot Univ Wash Seattle WA 98195

BLITSTEIN, JOHN, b Bucharest, Romania, May 28, 38; US citizen. POLYMER CHEMISTRY. *Educ:* Univ Bologna, Italy, PhD(indust chem), 66. *Prof Exp:* Res Chemist, Toni Co, 67-68; sr res chemist, Eugene Dietzgen Co, 68-70 & Lake Chem, 70-76; VPRES RES, TAPECOAT CO, 76- *Mem:* Am Chem Soc; Am Soc Testing & Mat; Nat Asn Corrosion Engrs; Soc Plastic Engrs. *Res:* Tapes; adhesives coatings; corrosion; polymers; plastic and alloys extrusion. *Mailing Add:* 1527 Lyons Evanston IL 60204

BLITZER, LEON, b New York, NY, Dec 13, 15; m 42; c 2. CELESTIAL MECHANICS. *Educ:* Univ Ariz, BS, 38, MS, 39; Calif Inst Technol, PhD(physics), 43. *Prof Exp:* Instr physics, Calif Inst Technol, 43-45, res assoc, Off Sci Res & Develop Proj, 42-46; from asst prof to prof, 46-83, EMER PROF PHYSICS, UNIV ARIZ, 84- *Concurrent Pos:* Consult, Space Technol Labs, 55-70 & Jet Propulsion Lab, Calif Inst Technol, 71-86. *Mem:* AAAS; Am Phys Soc; Am Asn Physics Teachers. *Res:* Exterior ballistics of rockets; stellar spectroscopy; spectroscopy; satellite orbit perturbations; satellite and planetary perturbation theory. *Mailing Add:* Dept Physics Univ Ariz Tucson AZ 85721

BLITZSTEIN, WILLIAM, b Philadelphia, Pa, Sept 3, 20; m 44; c 2. ASTRONOMY. *Educ:* Univ Pa, PhD, 50. *Prof Exp:* Physicist with Dr J Razek, 41-47; res engr, Franklin Inst Labs, 47-52; electronic scientist, Frankford Arsenal, 52-54; res assoc, Univ Pa, 47-54, asst prof astron & elec eng, 54-60, assoc prof astron, 60-64, prof astron, 64-86, prof astron & astrophysics & chmn dept, 80-85, dir, Observ, 80-86, EMER PROF ASTRON & ASTROPHYS, UNIV PA, 86- *Concurrent Pos:* Consult, Princeton Observ, 48-50, Decker Corp, Radio Corp Am & Spitz Labs, 58-64 & Frankford Arsenal, 72-77. *Mem:* Int Astron Union; Am Astron Soc; Optical Soc Am; Royal Astron Soc. *Res:* Electronic instrumentation; astronomical radiometry and polarimetry; eclipsing binaries; celestial mechanics. *Mailing Add:* 2329 Poplar Rd Havertown PA 19083-1643

BLIVAISS, BEN BURTON, b Chicago, Ill, July 4, 17; m 46; c 3. PHYSIOLOGY. *Educ:* Univ Chicago, BS, 38, MS, 40, PhD(zool), 46. *Prof Exp:* Lab asst biol, Wright Jr Col, 38-41; lab asst zool, Univ Chicago, 43-44, asst zool, Toxicity Lab, 45-46, Dept Zool, 46; instr physiol, Univ Ill, 46-47; guest investr zool, Univ Chicago, 47-48; from instr to assoc prof, 48-66, coordr, Basic Med Sci Educ, 75-77, PROF PHYSIOL, CHICAGO MED SCH, 66-, ASSOC DEAN CURRIC, 77- *Mem:* AAAS; Aerospace Med Asn; Am Soc Zoologists; Endocrine Soc; Soc Exp Biol & Med. *Res:* Cryptorchidism; thyroid-gonad relations in fowl; mating behavior of fowl; thiouracil on pigment; pharmacology of respiratory poisons; testicular tumors; adrenal-gonad relations; steroid hormone metabolism; experimental arthritis; aerospace physiology; nutritional effects on implantation and gestation; experimental carcinogenesis. *Mailing Add:* 3333 Green Bay Rd North Chicago IL 60064

BLIVEN, FLOYD E, JR, b Erie, Pa, May 20, 21; m 50; c 6. ORTHOPEDIC SURGERY. *Educ:* Univ Rochester, AB & MD, 42; Am Bd Orthop Surg, dipl, 56. *Prof Exp:* From instr to asst prof, Sch Med, Univ Rochester, 52-56; chief orthop, Talmadge Mem Hosp, Augusta, 56-79; from assoc prof to prof orthop, Med Col Ga, 56-86; RETIRED. *Concurrent Pos:* Consult, Battery State Hosp Tuberc, 56-65, Vet Admin Hosp, Augusta, 56-, Crippled Children Serv, Ga Dept Pub Health, 60-86, Milledgeville State Hosp, 63-78, US Army Hosp Spec Treatment Ctr, Ft Gordon, 65-80 & Eisenhower Med Ctr, US Army, Augusta, 80- *Mem:* Am Acad Orthop Surg; Orthop Res Soc; Am Col Surg; AMA. *Res:* Infection of bone and joint; circulation and developmental studies of the hip; disability evaluation. *Mailing Add:* 619 Scotts Way Augusta GA 30909

BLIX, SUSANNE, b Crawfordsville, Ind, Oct 29, 49. CHILD PSYCHIATRY. *Educ:* DePauw Univ, BA, 71; Ind Univ, MD, 75. *Prof Exp:* Internship, 75-76, resident, 77-79, fel child psychiat, 79- 81, ASST PROF PSYCHIAT, IND UNIV SCH MED, 81-, ACTG DIR SECT CHILD PSYCHIAT, 88- *Concurrent Pos:* Consult, Ind Pastoral Coun Ctr, 78-79, Cummins Ment Health Ctr, 79-84, Midtown Ment Health Ctr, 84-85; dir, Univ Hosp, Riley Hosp Children, 83-85, asst dir, 83-85, dir psychiat serv, 85- *Mem:* Am Acad Child & Adolescent Psychiat; Am Psychiat Asn; Asn Acad Psychiat; Am Med Asn. *Res:* Psychiatric problems involving children including Munchausen's syndrome by proxy; multiple personality; eating disorders; affective disorders; effects on staff caring for seriously disturbed youth. *Mailing Add:* Riley Child Psychiat Clin 702 Barnhill Dr Indianapolis IN 46223

BLIZNAKOV, EMILE GEORGE, b Kamen, Bulgaria, July 28, 26; m 54. MEDICINE, MICROBIOLOGY. *Educ:* Acad Med, Fac Med, Sofia, MD, 53. *Prof Exp:* Dir, Regional Sta Hygiene & Epidemiol, Pirdop, Bulgaria, 53-55; staff scientist-microbiologist, Res Inst Epidemiol & Microbiol, Sofia, 55-59; sr staff scientist, New Eng Inst, 61-81, vpres, 74-76, pres, 76-81; PRES & SCI DIR, LUPUS RES INST, 81- *Concurrent Pos:* Vis scientist, Res Inst Epidemiol & Microbiol, Moscow, USSR, 58-59; exec dir res & develop, Libra Res, 81-83; mem, Sci Adv Bd, Coalition Immune Systs Disorders. *Mem:* Fel Royal Soc Trop Med & Hyg London; AMA; Am Fedn Clin Res; Am Soc Microbiol; NY Acad Sci; Am Soc Neurochem; Am Col Toxicol; Reticuloendothelial Soc; Bioelectromagnetic Soc; AAAS; Int Am Soc Chemother. *Res:* Public health, preventive medicine and epidemiological control; administration of research; specific and nonspecific resistance to experimental bacterial, parasitic and viral infections and tumors; phagocytosis of particles by the reticuloendothelial system; bacterial viruses; neuropharmacology; mediators and their metabolism in allergy and shock; experimental and clinical screening for drugs with anti-cancer and anti-infectious effect; autoimmunity. *Mailing Add:* 2801 N Course Dr H-205 Pompano Beach FL 33069

BLIZZARD, ROBERT M, b East St Louis, Ill, June 20, 24. PEDIATRICS, PEDIATRIC ENDOCRINOLOGY. *Educ:* Northwestern Univ, BS, 48, MD, 52; Am Bd Pediat, dipl. *Prof Exp:* Rotating intern, Iowa Methodist Hosp & Raymond Blank Mem Hosp Children, Des Moines, Iowa, 52-53; from asst resident physician to resident physician, Raymond Blank Mem Hosp Children, 53-55; fel pediat endocrinol & pediatrician, Harriet Lane Home Children, Johns Hopkins Hosp, 55-57; asst prof pediat & med, Ohio State Univ, 57-59, assoc prof pediat, 59-60, assoc, Div Endocrinol & Metab, Univ Hosp, 57-60; from assoc prof to prof pediat, Johns Hopkins Hosp, 60-73, chief div pediat endocrinol, 60-73, actg chmn dept pediat, 72-73; CHMN DEPT PEDIAT, SCH MED, UNIV VA, 73- *Concurrent Pos:* Consult, NIH, 60-; Dir, Nat Pituitary Agency, 63-67; pres, Human Growth Found. *Honors & Awards:* Ayerst Award, Endocrine Soc. *Mem:* Am Fedn Clin Res; Soc Pediat Res; Endocrine Soc; Am Pediat Soc; Am Fedn Clin Res; L Wilkins Pediat Endocrine Soc; Europ Pediat Endocrine Soc. *Mailing Add:* Dept Pediat Univ Va Sch Med Charlottesville VA 22901

BLOBEL, GUNTER, b Waltersdorf, Silesia, Germany, May 21, 36; m. CYTOLOGY. *Educ:* Univ Tubingen, Germany, MD, 60; Univ Wis, PhD(oncol), 67. *Prof Exp:* From asst prof to assoc prof, 69-76, PROF CELL BIOL, ROCKEFELLER UNIV, 76-; INVESTR, HOWARD HUGHES MED INST, 86- *Concurrent Pos:* Internship in various Ger Hosp, 60-62; postdoctoral fel, Lab Cell Biol, Rockefeller Univ, 67-69. *Honors & Awards:* US Steel Award in Molecular Biol, Nat Acad Sci, 78; Warburg Medal, German Biochem Soc, 83; Richard Lounsbery Award, Nat Acad Sci, 83; V D Mattia Award, Roche Inst Molecular Biol, 86; Wilson Medal Am Soc Cell Biol, 86; Louisa Gross Horwitz Prize, 87; Waterford Bio-Med Sci Award, 89. *Mem:* Nat Acad Sci; Am Acad Arts & Sci; hon mem Japan Biochem Soc; Am Soc Cell Biol (pres, 90); assoc mem Europ Molecular Biol Orgn; hon mem Ger Soc Cell Biol; Am Philos Soc. *Mailing Add:* Cell Biol Lab Rockefeller Univ 66th St & York Ave New York NY 10021-6339

BLOCH, AARON N, b Chicago, Ill, Feb 28, 42; m 67; c 3. TECHNOLOGY TRANSFER, SCIENCE POLICY. *Educ:* Yale Univ, BS, 63; Univ Chicago, PhD(chem physics), 68. *Prof Exp:* Postdoctoral res assoc, Mass Inst Technol, 68-69; from asst prof to prof chem, Johns Hopkins Univ, 69-81; sr res assoc, Exxon Res & Eng Co, 80-84, dir, Phys Sci Lab, 84-88; VPROVOST, COLUMBIA UNIV, 88- *Concurrent Pos:* Consult, Bell Labs, 71, 73 & 77, IBM Thomas J Watson Res Ctr, 74-79, Energy Conversion Devices, Inc, 78-79; vis assoc prof, Dept Chem, James Franck Inst, 76-77; vis prof, Dept Chem, Johns Hopkins Univ, 81-83; mem, Numerical Data Adv Bd, Nat Res Coun, 84-87, Corp Assocs Adv Comt, Am Inst Physics, 84-88; adj prof chem, Columbia Univ, 88-; mem, Comt Pub Policy, Am Inst Physics, 88-, Panel Pub Affairs, Am Phys Soc, 89-, Technol Exec Coun, New York City Partnership, 90- *Mem:* Fel Am Phys Soc; Mat Res Soc. *Res:* Theoretical and experimental condensed matter physics and chemistry; organic electronic materials; microscopic basis of global trends in the properties of solids; complex fluids and interfaces. *Mailing Add:* 205 Low Mem Libr Columbia Univ New York NY 10027

BLOCH, AARON NIXON, b Chicago, Ill, Feb 28, 42; m 67; c 3. CHEMICAL PHYSICS, SOLID STATE CHEMISTRY. *Educ:* Yale Univ, BS, 63; Univ Chicago, PhD(chem physics), 68. *Prof Exp:* Res assoc chem, Mass Inst Technol, 68-69; from asst prof to assoc prof, Johns Hopkins Univ, 69-77, prof chem, 77-80; MEM STAFF, EXXON RES & ENG CO, 80- *Concurrent Pos:* Consult, Bell Labs, 71-73, 77; fac visitor, Thomas J Watson Res Labs, IBM Corp, 74-75; A P Sloan Found fel, 74-76; vis assoc prof, Univ Chicago, 76-77; consult, Energy Conversion Devices, 78- *Mem:* Am Chem Soc; Am Phys Soc. *Res:* Organic conductors; electronic processes in nearly one-dimensional structures; metal-insulator transitions; pseudopotential theory of chemical trends in solids; chemical physics of surfaces. *Mailing Add:* Exxon Res & Eng Co Clinton Township Corp Res Lab Annandale NJ 08801

BLOCH, ALAN, b New York, NY, Nov 28, 15; m; c 2. PHYSICS, MATHEMATICS. *Educ:* Swarthmore Col, BA, 38; Oberlin Col, MA, 39. *Prof Exp:* Teaching fel physics, Iowa State Col, 40-41; head comput develop sect, Arma Corp, 46-49; chief develop engr, Audio Instrument Co, Inc, 51-55; head syst anal dept, GPL Div, Gen Precision Inc, 55-60, staff consult to vpres res & eng, 60-61, regional mkt mgr, Librascope Div, 61-63, staff physicist, 63-65; staff analyst, Gyrodyne Co Am, Inc, 65-70; SCI CONSULT & WRITER, 70- *Mem:* AAAS; Am Phys Soc; Asn Comput Mach; Audio Eng Soc; NY Acad Sci; Sigma Xi. *Res:* Hard and social sciences, particularly system analysis and synthesis; interdisciplinary problem solving; microprocessor application design. *Mailing Add:* 333 E 30th St New York NY 10016

BLOCH, ALEXANDER, b Freiburg, Germany, Oct 6, 23; US citizen. BIOCHEMICAL PHARMACOLOGY, CANCER BIOLOGY. *Educ:* City Col NY, BS, 54; Long Island Univ, MS, 58; Cornell Univ, PhD(microbiol, biochem), 62. *Prof Exp:* Asst biol, Long Island Univ, 57-58; asst microbiol, Sloan-Kettering Inst, 58-62; from scientist to assoc scientist, 62-74, PRIN SCIENTIST, CANCER RES, ROSWELL PARK MEM INST, 74- *Concurrent Pos:* Res prof, Niagara Univ, 68- & Canisius Col, 68-; assoc res prof, State Univ NY Buffalo, 68-75, res prof, 75- *Honors & Awards:* Aaron Bendich Award, Sloan-Kettering Inst. *Mem:* Soc Exp Biol & Med; fel NY Acad Sci; AAAS; Am Chem Soc; Am Asn Cancer Res; Am Soc Biol Chem; Am Soc Pharmacol & Exp Therapeut. *Res:* Experimental cancer chemotherapy; tumor cell proliferation and differentiation. *Mailing Add:* Seven Forsythia Ct Hamburg NY 14075

BLOCH, DANIEL R, b Brownsville, Wis, Dec 13, 40; c 2. CHEMISTRY. *Educ:* Univ Wis, BS, 62; Univ Ill, MS, 67, PhD(chem), 70. *Prof Exp:* Instr inorg chem, Ctr Syst, Univ Wis, 62-65; res chemist, S C Johnson & Son, Inc, 70-74, sr res chemist, 74-79, mgr, Polymer Res Dept, 79-85, asst to vpres res & develop, 85-88, MGR EXTERNAL TECHNOL, S C JOHNSON & SON, INC, 88- *Concurrent Pos:* Vis prof, Univ Wis, Parkside, Univ Wis-Madison & Univ Wis- Milwaukee. *Mem:* Sigma Xi; Adhesion Soc; Fedn Soc Coatings Technol; Am Chem Soc. *Res:* Acrylic latex (emulsion) polymers; monomer sequencing; reaction kinetics; thermosetting acrylic polymers; copolymerizable surfactant-like monomers; latex particle stability parameters; electron beam polymerization kinetics; properties of aqueous amylose gels; modification of natural polysaccharides. *Mailing Add:* S C Johnson & Son 1525 Howe St Racine WI 53403-5011

BLOCH, EDWARD HENRY, b Berlin, Ger, Feb 1, 14; US citizen. ANATOMY. *Educ:* Univ Chicago, BSc, 39, PhD(microphysiol), 49; Univ Tenn, MD, 45. *Prof Exp:* Asst, Univ Tenn, 40-44; intern, Michael Reese Hosp, Chicago, 46; assoc, Univ Chicago, 46-48; from asst prof to prof, 53-84, actg chmn dept, 80-82, EMER PROF ANAT, CASE WESTERN RESERVE UNIV, 84- *Concurrent Pos:* Estab investr, Am Heart Asn, 50-55. *Mem:* Am Asn Trop Med & Hyg; Am Asn Immunol; Microcirculatory Soc (pres, 54 & co-founder); Am Heart Asn; Am Asn Anatomists; Europ Microcirculatory Soc; NY Acad Sci. *Res:* Scanning microspectrophotometry of organs in vivo; microvascular physiology and pathology of the liver, spleen, gastrointestinal tract, muscle and lung; high speed cinephotography; systems analysis of organs; instrumentation for the microscopic study of living organs in situ; pathophysiology of Schistosomiasis; in vivo microscopy of alografts in transparent chambers. *Mailing Add:* Dept Anat Case Western Reserve Univ Cleveland OH 44106

BLOCH, ERIC, b Munich, Ger, Apr 4, 28; nat US; m 61; c 2. NUTRITION, DEVELOPMENTAL BIOCHEMISTRY. *Educ:* City Col New York, BS, 48; Univ Tex, AM, 50, PhD(biochem), 53. *Prof Exp:* Asst, Univ Tex, 50-52; res biochemist, Worcester Found Exp Biol, 52-57; res assoc, Children's Cancer Res Found, 57-58; asst prof, 58-65, assoc prof, 65-76, PROF BIOCHEM, OBSTET & GYNEC, ALBERT EINSTEIN COL MED, 76- *Concurrent Pos:* Asst, Med Sch, Harvard Univ, 57; lectr, steroid training prog, Worcester Found Exp Biol, 57, 58; res career develop award, 60-69; mem, NIH Study Sect, 78-82. *Mem:* Fel AAAS; Am Soc Biol Chemists; Endocrine Soc; Soc Study Reproduction; Soc Gynec Invest; Sigma Xi. *Res:* Control of and drug effects on fetal adrenal and gonadal functional development; testicular control of reproductive tract differentiation; effects of toxins and nutrients in testis function. *Mailing Add:* Dept Biochem Gynec & Obstet Albert Einstein Col Med Bronx NY 10461

BLOCH, ERICH, b Sulzburg, WGer, Jan 9, 25; US citizen; m 48; c 1. COMPUTER SCIENCES. *Educ:* Univ Buffalo, BSEE, 52. *Hon Degrees:* DEng, Colo Sch Mines, 84; DSc, Univ Mass, George Washington Univ & State Univ NY, Buffalo, 85, Oberlin Col, 87; Notre Dame, 88 & Univ SC, 90. *Prof Exp:* Mgr stretch comput syst, IBM, Armonk, NY, 58-62, mgr memory develop, 63-66, lab dir, 66-69, vpres, components div, 69-72, div vpres & gen mgr, 75-80, asst group exec technol, 80-81, corp vpres, 81-84; dir, NSF, 84-90; RETIRED. *Honors & Awards:* US Medal Technol; Founders Medal, Inst Elec & Electronics Engrs. *Mem:* Nat Acad Eng; fel Inst Elec & Electronics Engrs; fel AAAS. *Res:* Computer system design of advanced CPU's and memory subsystem; semiconductor and packaging development; manufacturing management. *Mailing Add:* Former Dir NSF 2801 New Mexico Ave NW Washington DC 20007

BLOCH, HERMAN SAMUEL, petroleum chemistry, catalysis; deceased, see previous edition for last biography

BLOCH, INGRAM, b Louisville, Ky, Aug 27, 20; m 45, 60; c 2. THEORETICAL PHYSICS. *Educ:* Harvard Univ, BA, 40; Univ Chicago, MS, 41, PhD(physics), 46. *Prof Exp:* Asst math biophys & lab asst physics, Univ Chicago, 41-42, asst instr electronics eng, sci & mgt war training, 42-43, jr scientist, Metall Lab, 43-44; asst scientist, Los Alamos Sci Lab, Univ Calif, 44-46; assoc nuclear physics, Univ Wis, 46-47; asst, Yale Univ, 47-48; from asst prof to prof physics, Vanderbilt Univ, 48-86; RETIRED. *Mem:* Fel Am Phys Soc; AAAS; Am Asn Physics Teachers. *Res:* Quantum theory; theory of particles; Author of 290 scientific journals. *Mailing Add:* 926 Cantrell Ave Nashville TN 37215

BLOCH, KONRAD EMIL, b Neisse, Ger, Jan 21, 12; nat US; m; m. BIOLOGICAL CHEMISTRY. *Educ:* Tech Hochschule, Munich, Chem E, 34; Columbia Univ, PhD(biochem), 38. *Prof Exp:* Instr biochem, Columbia Univ, 41-44, assoc, 44-46; from asst prof to prof, Univ Chicago, 46-54; Higgins prof biochem, 54-82, HIGGINS EMER PROF, HARVARD UNIV, 82- *Concurrent Pos:* Guggenheim fel, 53, 68 & 75. *Honors & Awards:* Nobel Prize med physiol, 64; Fritsche Award, Am Chem Soc, 64; Nat Medal Sci, NSF, 88. *Mem:* Nat Acad Sci; fel Am Acad Arts & Sci; Am Chem Soc; Am Soc Biol Chem (pres, 67); Am Philos Soc; foreign mem Royal Soc London; hon mem Bavarian Acad Sci; Jap Biochem Soc. *Res:* Structure, biosynthesis and function of lipids; enzyme inhibitors. *Mailing Add:* Dept Chem Harvard Univ Cambridge MA 02138

BLOCH, KURT JULIUS, b Ger, Oct 17, 29; US citizen; m 53; c 2. IMMUNOLOGY. *Educ:* City Univ New York, BS, 51; New York Univ Sch Med, MD, 55. *Prof Exp:* Asst physician med, Mass Gen Hosp, 65-70, assoc physician, 70-74; from asst prof to assoc prof, 68-76, PROF MED, HARVARD MED SCH, 76-; PHYSICIAN MED, MASS GEN HOSP, 74- *Concurrent Pos:* Dir, Am Bd Allergy & Immunol, 76-81. *Mem:* Am Soc Clin Invest; Am Rheumatism Asn; Am Acad Allergy. *Res:* Biologic properties of antibodies. *Mailing Add:* Dept Med Harvard Univ Sch Med & Chief Clin Immunol & Allergy Units Mass Gen Hosp Fruit St Bulfinch 422 Boston MA 02114

BLOCH, ROBERT JOSEPH, b New York, NY, Oct 18, 46; m 73; c 2. CELLULAR NEUROBIOLOGY, DEVELOPMENTAL NEUROBIOLOGY. *Educ:* Columbia Univ, AB, 67; Harvard Univ, PhD(biochem), 72. *Prof Exp:* Res fel, Bioctr, Univ Basel, Switz, 72-76; res assoc, Neurobiol Lab, Salt Inst, La Jolla, Calif, 76-80; asst prof, 80-85, ASSOC PROF, DEPT PHYSIOL, SCH MED, UNIV MD, BALTIMORE, 85- *Concurrent Pos:* McKnight Scholar's Award, 81-84. *Honors & Awards:* Career Develop Award, NIH, 82. *Mem:* AAAS; Soc Neurosci; Am Soc Cell Biol; NY Acad Sci. *Res:* Formation of synapses; formation of neuromuscular junction, to learn how the postsynaptic region of this synapse is assembled. *Mailing Add:* Dept Physiol Sch Med Univ MD 660 W Redwood St Baltimore MD 21201

BLOCH, SALMAN, b Afula, Israel, May 22, 45; US citizen. SEDIMENTARY PETROLOGY, LOW TEMPERATURE GEOCHEMISTRY. *Educ:* Wroclaw Univ, MS, 68; George Washington Univ, PhD(geochem), 77. *Prof Exp:* Ed, Am Geol Inst, 71-72; geologist, US Geol Surv, 74-78, Okla Geol Surv, 78-82; sr res geologist, 82-86, PRIN RES GEOLOGIST, ARCO OIL & GAS CO, 86- *Mem:* Asn Explor Geochemists; Geochem Soc; Soc Econ Paleontologists & Mineralogists; Soc Environ Geochem & Health. *Res:* provenance and reservoir quality of sandstones; porosity prediction in sandstones. *Mailing Add:* Arco Oil & Gas Co 2300 Plamo Pkwy Plano TX 75075

BLOCH, SYLVAN C, b Vicksburg, Miss, Nov 9, 31; m 58; c 2. PHYSICS. *Educ:* US Coast Guard Acad, BS, 54; Univ Miami, MS, 59; Fla State Univ, PhD(physics), 62. *Prof Exp:* Physicist microwave spectros, Martin Co, Fla, 59; pres, Recon, Inc, 61, res dir, 61-63; asst prof physics, Fla State Univ, 63; from asst prof to assoc prof, 63-70, PROF PHYSICS, UNIV S FLA, 70- *Concurrent Pos:* Consult, Recon, Inc, 63- *Mem:* Am Phys Soc; Inst Elec & Electronics Eng; Am Asn Physics Teachers. *Res:* Theoretical and experimental plasma physics; atmospheric physics. *Mailing Add:* 11718 Lipsey Rd Univ S Fla Tampa FL 33618

BLOCHER, JOHN MILTON, JR, b Baltimore, Md, Jan 6, 19; m 41; c 5. PHYSICAL CHEMISTRY. *Educ:* Baldwin-Wallace Col, BS, 40; Ohio State Univ, PhD(phys chem), 46. *Prof Exp:* Asst chem, Ohio State Univ, 40-43, 45, 46-47; instr physics, Capital Univ, 43-44; jr chemist, Tenn Eastman Corp, 44-45; res engr, 46-50, asst supvr, 50-58, div chief, 58-69, FEL, BATTELLE-COLUMBUS LABS, 69- *Mem:* Am Chem Soc; Electro-chem Soc; Am Nuclear Soc; Sigma Xi. *Res:* Thermodynamics of halide vapor equilibria; mass spectrometry; photovoltaic effect; high purity metals; refractory coatings; chemical vapor deposition. *Mailing Add:* 915 Silvoor Ln Oxford OH 45056

BLOCK, A JAY, b Baltimore, Md, Apr 11, 38; m 60; c 2. PULMONARY MEDICINE, CRITICAL CARE. *Educ:* Johns Hopkins Univ, BA, 58, MD, 62. *Prof Exp:* Intern, Johns Hopkins Hosp, 62-63, resident, 64-67; fel pulmonary, Johns Hopkins Univ, 63-66, instr med, 67-69, asst prof, 69-70; asst prof, 70-73, assoc prof, 73-75, PROF MED, UNIV FLA, 75- *Concurrent Pos:* Chief pulmonary, Vet Admin Med Ctr, Gainesville, 70-77 & Col Med, Univ Fla, 73-; prog specialist, Vet Admin, 79-82. *Mem:* Am Thoracic Soc; Am Fedn Clin Res; Am Col Chest Physicians. *Res:* Breathing and blood oxygenation during sleep in patients with disease and in the general population; identification and modification of dermographic and exogenous factors which promote sleep apnea and desaturation. *Mailing Add:* E R Squibb & Sons PO Box 4000 Princeton NJ 08540

BLOCK, ARTHUR MCBRIDE, environmental chemistry, quantum chemistry, for more information see previous edition

BLOCK, BARTLEY C, b Chicago, Ill, Apr 12, 33; m 63; c 3. BIOACOUSTICS, OLFACTION. *Educ:* Northwestern Univ, BS, 54, MS, 55. *Prof Exp:* Asst prof biol, Lycoming Col, 59-62, Drexel Inst Technol, 63-64, Southern Conn State Col, 65-67; asst prof, 67-74, ASSOC PROF BIOL, UNIV BRIDGEPORT, 74- *Concurrent Pos:* Entomologist, Agr Res Ctr, USDA, Beltsville, 58-61; consult, Williamsport, Pa, 61-62, Maimonides Med Ctr, 67-68, Wethersfield, Conn, 76-77. *Mem:* AAAS; Am Inst Biol Sci; Am Soc Zoologists; Animal Behav Soc. *Res:* Sensory physiology and behavior of insects, birds and mammals. *Mailing Add:* Dept Biol Univ Bridgeport Bridgeport CT 06601

BLOCK, DAVID L, b Davenport, Iowa, July 10, 39; m 61; c 2. SOLAR ENERGY, ENGINEERING MECHANICS. *Educ:* Univ Iowa, BSCE, 62; Va Polytech Inst, MS, 64, PhD(eng mech), 66. *Prof Exp:* Engr, NASA Langley Res Ctr, 62-66; staff engr, Martin Marietta Corp, 66-68; assoc dean eng, Fla Technol Univ, 68-77; DIR, FLA SOLAR ENERGY CTR, 77- *Concurrent Pos:* Fla rep, Southern Solar Energy Ctr, 77-, Solar Energy Task Force, Nat Gov Conf, 77- *Mem:* Int Solar Energy Soc; Am Soc Testing & Mat; Nat Soc Prof Engrs; Am Soc Eng Educ. *Res:* Solar air conditioning systems, energy policy, solar energy standards and certification and marketing of energy systems. *Mailing Add:* Fla Solar Energy Ctr 300 State Rd 401 Cape Canaveral FL 32920

BLOCK, DOUGLAS ALFRED, b Rockford, Ill, July 12, 21; m 60; c 2. GEOLOGY, STRATIGRAPHY. *Educ:* Wheaton Col, Ill, AB, 43; Univ Iowa, MS, 52; Univ NDak, PhD(Pleistocene geol), 65. *Prof Exp:* From instr to assoc prof geol, Wheaton Col, Ill, 45-67; prof geol, 67-71, PROF GEOL & EARTH SCI, ROCK VALLEY COL, 71- *Mem:* Geol Soc Am; fel Am Sci Affil; Nat Asn Geol Teachers. *Res:* Glacial geology and geomorphology; sedimentary petrology; Pleistocene stratigraphy. *Mailing Add:* Dept Phys Sci Rock Valley Col 3301 N Mulford Rockford IL 61101

BLOCK, DUANE LLEWELLYN, b Madison, Wis, Dec 27, 26; m 49; c 2. OCCUPATIONAL MEDICINE. *Educ:* Univ Wis, BS, 49, MD, 51; Am Bd Prev Med, dipl & cert occup med, 61. *Prof Exp:* Plant physician, Cadillac Div, Gen Motors Corp, 52-54, med dir, Gen Motors Tech Ctr, 54-55; physician-in-chg, Rouge Med Serv, 55-70, MED DIR, FORD MOTOR CO, 70- *Concurrent Pos:* Mem vis staff, William Beaumont Hosp, 60-; clin asst prof occup & environ health, Med Sch, Wayne State Univ, 60-; lectr occup med, Sch Pub Health, Univ Mich, 60-; chmn bd, Occup Health Inst, 72-78; mem, Mich Task Force Prev Health Serv & Mich Occup Health Stand Comn, 75-77; mem, Am Bd Prev Med, 75-, chmn, 77-; vchmn air pollution res adv comt, Coord Res Coun, 75-78. *Honors & Awards:* William S Knudsen Award, Am Occup Med Asn, 75; Robert A Kehoe Award, 81. *Mem:* Am Occup Med Asn (pres, 69-70); fel Am Acad Occup Med; Int Asn Occup Health. *Res:* Health effects of automotive emissions; effects of the work environment on worker health. *Mailing Add:* 3699 Brookside Dr Bloomfield MI 48013

BLOCK, EDWARD R, b Baltimore, Md, May 27, 44; c 2. LUNG RESEARCH. *Educ:* Johns Hopkins Univ, MD, 68. *Prof Exp:* PROF MED, COL MED, UNIV FLA, 75-; ASSOC CHIEF STAFF RES, VA MED CTR, GAINESVILLE, FLA,. *Concurrent Pos:* A Blaine Brower travelling scholar, Am Col Physicians; Edward Livingston Trudeau fel, Am Lung Asn. *Mem:* Fel Am Col Physicians; Am Physiobiol Soc; Am Thoracic Soc. *Res:* Mechanisms by which oxygen excess and oxygen deprivation damage mammaliam lungs and cause pulmonary injury, dysfunction and disease. *Mailing Add:* Dept Med Univ Fla Vet Admin Med Ctr 151 Gainesville FL 32608

BLOCK, ERIC, b New York, NY, Jan 25, 42; m 66; c 2. ORGANIC CHEMISTRY. *Educ:* Queens Col, NY, BS, 62; Harvard Univ, AM, 64, PhD(chem), 67. *Prof Exp:* Fel, Harvard Univ, 67; from asst prof chem to prof, Univ Mo-St Louis, 67-81; PROF CHEM & DEPT CHMN, STATE UNIV NY, ALBANY, 81- *Concurrent Pos:* Vis assoc prof chem, Harvard Univ, 74; vis prof, Univ Frankfurt, 78 & Univ Bologna, 84; consult, Elf Aquitaine, France, 83-; John Simon Guggenheim fel, 84-85; assoc ed, Phosphorus & Sulfur, 86-; ed, Advances in Sulfur Chem, 88- *Honors & Awards:* Am Chem Soc Award, Advancement Appln Agr & Food Chem, 87. *Mem:* Am Chem Soc; Royal Soc Chem. *Res:* Organic sulfur chemistry; organic photochemistry; novel methods of organic synthesis; organosilicon chemistry; garlic and onion chemistry. *Mailing Add:* Chem Dept State Univ NY Albany NY 12222

BLOCK, GENE DAVID, b New York, NY, Aug, 17, 48; m 70; c 2. CIRCADIAN RHYTHMS PHYSIOLOGY. *Educ:* Stanford Univ, BS, 70; Univ Ore, MS, 73, PhD(psychol), 75. *Prof Exp:* Fel, Stanford Univ, 75-78; from asst prof to assoc prof, 78-88, PROF BIOL, UNIV VA, 89- *Concurrent Pos:* Dir, Sci & Technol Ctr Biol Timing, NSF. *Mem:* Soc Neurosci; AAAS. *Res:* Cellular and molecular basis of circadian rhythms including the retina of a marine gastropod containing a circadian pacemaker which is accessible for electrophysiological study. *Mailing Add:* 244 Turkey Ridge Rd Charlottesville VA 22901

BLOCK, GEORGE E, b Joliet, Ill, Sept 16, 26; m; c 3. SURGERY. *Educ:* Univ Mich, MD, 51, MS, 58; Am Bd Surg, dipl, 59, recert, 80. *Prof Exp:* Intern, Univ Hosp, Univ Mich, 51-52, resident, 51-53 & 55-58, from instr to asst prof surg, 58-60; from asst prof to assoc prof, 60-67, PROF SURG, UNIV CHICAGO, 67-, PRES MED STAFF, MED CTR, 78- *Concurrent Pos:* Coordr clin oncol, Univ Chiago; attend surgeon, Cook County Hosp; consult, US Naval Hosp, Great Lakes, Ill; Am Cancer Soc fel, Ben May Lab Cancer Res, Univ Chicago, 55. *Honors & Awards:* McClintock Award, Univ Chicago, 65; Pybus Medal, 78; Hamilton Award, 79. *Mem:* Fel Am Col Surgeons; Am Surg Asn; Soc Surg Alimentary Tract (vpres, 77-78); Frederick A Coller Surg Soc (pres, 73-74); Sigma Xi. *Res:* Inflammatory diseases of the bowel; thyroid cancer; malignant melanoma, breast cancer; cancer of the colon. *Mailing Add:* 7401 Block Rd Yorkville IL 60560

BLOCK, HENRY WILLIAM, b Newark, NJ, Sept 9, 41; m 68. MATHEMATICAL RELIABILITY THEORY. *Educ:* Carnegie-Mellon Univ, BS, 63; Ohio State Univ, MS, 64, PhD(math), 68. *Prof Exp:* Asst math, Ohio State Univ, 63-68; from asst prof to assoc prof, 68-82, PROF MATH STATIST, UNIV PITTSBURGH, 82- *Concurrent Pos:* Vis assoc prof, Rensselaer Polytech Inst, 74-75; vis assoc prof, Univ Calif, Berkeley, 75-76; vis assoc prof, Univ Sao Paulo, Brazil, 80; vis prof, Univ Rome, Italy, 81. *Mem:* Fel Inst Math Statist; Oper Res Soc Am; Am Statist Asn; Int Statist Inst. *Res:* Reliability theory; statistical distribution theory; concepts of statistical dependence. *Mailing Add:* Dept Math & Statist Univ Pittsburgh Pittsburgh PA 15260

BLOCK, I(SAAC) EDWARD, b Philadelphia, Pa, Aug 8, 24; m 57; c 3. MATHEMATICS. *Educ:* Haverford Col, BS, 44; Harvard Univ, MS, 47, PhD(math), 52. *Prof Exp:* Res asst, Harvard Univ, 50-51; math consult, Philco Corp, 51-54; coordr comput appln, Burroughs Corp, 54-55, mgr comput ctr, 56-59; mgr, Univac Eng Comput Ctr, Remington Rand Univac, 59-61, mgr appl math dept, 61-64; tech adv, Info Sci Div, Auerbach Corp, 64-65, mgr planning, 65-66, mgr sci info civ, 66-68, vpres, Auerbach Publ, Inc, 71-76; MANAGING DIR, SOC INDUST & APPL MATH, 76- *Concurrent Pos:* Mem, Conf Bd Math Sci, 57-59 & 63-64 & Franklin Inst; ed, J Soc Indust & Appl Math. *Mem:* Am Math Soc; Math Asn Am; Soc Indust & Appl Math (secy, 52-53); Asn Comput Mach. *Res:* Singular integrals; harmonic analysis; applications of mathematics and automatic computation; technical publishing. *Mailing Add:* Soc Indust & Appl Math 3600 University City Sci Ctr Philadelphia PA 19104-2688

BLOCK, IRA, b Brooklyn, NY, Nov 11, 41; div; c 2. TEXTILE SCIENCE, CONSERVATION SCIENCE. *Educ:* Univ Md, BS, 63, PhD(nuclear eng), 71. *Prof Exp:* ASSOC PROF TEXTILES & CONSUMER ECON, UNIV MD, 78- *Concurrent Pos:* Mem, Nat Adv Comt for Flammable Fabrics Act, 80-; Smithsonian Inst fel, 80-81. *Mem:* Sigma Xi; Am Inst Conserv; Int Inst Conserv. *Res:* Performance of textile products; textiles conservation; conservation science; textile fabric flammability. *Mailing Add:* Dept Textiles & Consumer Econ Univ Md College Park MD 20742

BLOCK, IRVING H, b New York, NY, July 31, 16; m 43; c 1. PODIATRIC RADIOLOGY, RADIATION SAFETY. *Educ:* NY Col Podiatric Med, POD, 42, DPM, 70; Am Col Podiatric Radiologists, 57; Am Soc Podiatric Med, 69; Am Podiatric Hosp Asn, 79; Am Bd Podiatric Pub Health, dipl, 87. *Hon Degrees:* APWA, Am Podiatric Writers Asn, 62-63. *Prof Exp:* Pres, Fla Podiatric Med Asn, 62-63; pres radiation safety, Am Col Podiatric Radiologists, 71-73; mem & pres, Fla Bd Podiatry Examiners, 73-84; pres, 69-72, SECY PODIATRIC MED, AM SOC PODIATRIC MED, 82-; SECY RADIOL, AM COL PODIATRIC RADIOLOGISTS, 68-; CHMN PUB HEALTH, PODIATRIC HEALTH SECT, AM PUB HEALTH ASN, 84- *Concurrent Pos:* Chmn, radiation safety, Am Col Podiatric Radiologists, 67-; credentials, NY Col podiatric Med, 68-72, Am Soc Podiatric Med, 78-; comnr, Am podiatric Med Asn, 70-80; podiatric staff, St Francis Hosp, Miami Beach, 71-; mem podiatric health, Am Podiatric Med Asn, 75-; consult, Nat Podiatry Bds, Princeton, 82-84; Fla Bd Podiatry Examr, 73-83; assoc prof, Barry Col Podiatric Med. *Honors & Awards:* President's Award, Fla Podiatric Med Asn. *Mem:* Am Podiatric Med Asn; Am Pub Health Asn; fel Am Asn Hosp Podiatrists; Am Col Podiatric Radiologists; Am Soc Podiatric Med. *Res:* Radiation safety. *Mailing Add:* One Lincoln Rd Bldg Suite 308 Miami Beach FL 33139

BLOCK, JACOB, b Brooklyn, NY, Mar 8, 36; m 63; c 2. SOL-GEL PROCESSES, THERMALLY CONDUCTIVE COMPOSITES. *Educ:* Brooklyn Col, BS, 56; Case Inst Technol, PhD(anal chem), 61. *Prof Exp:* Sr res analyst, Olin Mathieson Chem Corp, 61-65; res chemist, 65-68, sr res chemist, 68-83, RES ASSOC, W R GRACE & CO, 83- *Mem:* Am Chem Soc; Am Ceramic Soc. *Res:* Ceramics; sol-gel processes; rheology; colloidal chemistry; thermally conductive composites; hollow glass spheres. *Mailing Add:* Washington Res Ctr W R Grace & Co Columbia MD 21044

BLOCK, JAMES A, b Evanston, Ill, Feb 23, 44; m 65; c 2. FLUID DYNAMICS, HEAT TRANSFER. *Educ:* Northwestern Univ, BS, 66; Harvard Univ, MS, 67, PhD(mech eng), 70. *Prof Exp:* Engr, 70-73, head, Multiphase Div, 73-80, VPRES, CREARE INC, 76-, PRES, CREARE RES & DEVELOP INC, 80- *Mem:* Am Soc Mech Engrs; Am Inst Chem Engrs; Am Nuclear Soc; Combustion Inst. *Res:* Multiphase flow and heat transfer; energy systems; fluid dynamics; nuclear safety; experimental techniques and instrumentation; product development, trouble shooting and improvement; direct contact condensation; foam technology. *Mailing Add:* Creare Inc Box 71 Hanover NH 03755

BLOCK, JEROME BERNARD, b New York, NY, Jan 7, 31; m 56; c 3. ONCOLOGY, IMMUNOLOGY. *Educ:* Stanford Univ, BA, 52; NY Univ, MD, 56. *Prof Exp:* Intern med, Mass Mem Hosp, Boston, 56-57, NIH fel, 57-59; resident, Univ Wash, 59-60; clin assoc oncol, Nat Cancer Inst, 60-64; staff physician, 64-66, chief, Baltimore Res Ctr, Nat Cancer Inst, 66-76; MEM FAC, DIV MED ONCOL UNIV CALIF, HARBOR GEN HOSP, 76- Instr med, George Washington Univ, 62-63. *Honors & Awards:* Karger Int Prize Biochem of Leukemia, 64. *Mem:* AAAS; Am Asn Cancer Res; Am Fedn Clin Res; Am Soc Hemat. *Res:* Chemotherapy; internal medicine. *Mailing Add:* Harbor Gen Hosp 1000 W Carson St Torrance CA 90509

BLOCK, JOHN HARVEY, b Yakima, Wash, May 11, 38; m 64; c 3. MEDICINAL CHEMISTRY. *Educ:* Wash State Univ, BS & BPharm, 61, MS, 63; Univ Wis, PhD(pharmaceut chem), 66. *Prof Exp:* From asst prof to assoc prof, 66-78, PROF PHARMACEUT CHEM, ORE STATE UNIV, 78- *Concurrent Pos:* NIH spec fel & vis scholar, Stanford Univ, 72-73. *Mem:* Am Pharmaceut Asn; Am Chem Soc; AAAS. *Res:* Drug design; quantitative structure activity relationship. *Mailing Add:* Col Pharm Ore State Univ Corvallis OR 97331-3507

BLOCK, LAWRENCE HOWARD, b Baltimore, Md, Nov 24, 41; m 74; c 2. CLINICAL PHARMACOLOGY. *Educ:* Univ Md, BS, 62, MS, 66, PhD(pharmaceut), 69. *Prof Exp:* Asst prof pharm, Univ Pittsburgh, 68-70; from asst prof to assoc prof, 70-75, PROF PHARMACEUT, SCH PHARM, DUQUESNE UNIV, 75-, CHMN DEPT PHARMACEUT CHEM & PHARMACEUT, 85- *Concurrent Pos:* Consult, Thrift Drug Co, Div JC Penney, Inc, 76-; mem, Tech Adv Bd, Dept Health, Commonwealth Pa, 78- *Mem:* Am Asn Pharmaceut Sci; Am Pharmaceut Asn; Soc Cosmetic Chem; NY Acad Sci; Am Asn Col Pharm; Sigma Xi. *Res:* Mass transport; biopharmaceutics and pharmacokinetics; drug and cosmetic delivery system design; drug interactions. *Mailing Add:* Sch Pharm Duquesne Univ Mellon H Pittsburgh PA 15282

BLOCK, M SABEL, b Louisville, Ky, June 2, 17; m 39; c 2. MECHANICAL ENGINEERING. *Educ:* Univ Louisville, BME, 39. *Prof Exp:* Develop engr, Park Aerial Serv, 39-40; engr, US Navy Bur Ord, 40-44; res engr, Aeronca Aircraft Corp, 46-47; chief engr, P Lorillard Co, 47-64, dir eng, Lorillard Div, Loew's Theatres, Inc, 64-82; RETIRED. *Mem:* Am Soc Mech Engrs; Nat Soc Prof Engrs; NY Acad Sci. *Res:* Engineering research on tobacco products; design and construction of special machinery; construction of manufacturing, processing and storage facilities and installation of equipment. *Mailing Add:* 818 Rollingwood Dr Greensboro NC 27410

BLOCK, MARTIN M, b Newark, NJ, Nov 29, 25; m 49; c 2. PHYSICS. *Educ:* Columbia Univ, BS, 47, MA, 48, PhD(physics), 51. *Prof Exp:* Instr math, Newark Col Eng, 47-48; asst physics, Columbia Univ, 47-51, scientist, 51; assoc, Duke Univ, 51, from asst prof to assoc prof, 52-61; chmn dept, 61-68, PROF PHSYICS, NORTHWESTERN UNIV, 61- *Concurrent Pos:* Guggenheim fel, 58-59. *Mem:* Fel Am Phys Soc; Italian Physics Soc. *Res:* High energy physics; liquid helium bubble chamber; neutrino physics; total cross sections at high energies. *Mailing Add:* Dept Physics Northwestern Univ Evanston IL 60201

BLOCK, MICHAEL JOSEPH, b Chicago, Ill, June 29, 42; m 70; c 2. ORGANIC CHEMISTRY, CARBON CHEMISTRY. *Educ:* Univ Mich, BS, 64; Harvard Univ, MA, 65, PhD(chem), 69. *Prof Exp:* Res chemist, Unocal Corp, 68-76, sr res chemist, 76-80, supvr carbon & nitrogen chemicals res, 81-86, mgr, chem res, 86-90, MGR POLYMER RES, SCI & TECHNOL DIV, UNOCAL CORP, 90- *Mem:* AAAS; Am Chem Soc. *Res:* Physical-organic and inorganic chemistry; petroleum coke and graphite research; agricultural chemicals research. *Mailing Add:* Unocal Sci & Technol Div 376 S Valencia Ave Brea CA 92621

BLOCK, RICHARD B, b Xenia, Ohio, Feb 27, 34; m 58; c 4. TRANSFER OF TECHNOLOGY FROM LAB TO END USER. *Educ:* Kenyon Col, AB, 56, Case Inst Technol, MS, 59, PhD(appl physics), 63. *Prof Exp:* Asst prof elec eng, Case Inst Technol, 63-67; sr physicist & mgr, Physics Int Co, 67-72 & 77-79; sr staff assoc, NSF, 72-77; chief exec officer, Approtech Inc, 79-82; pres, Block Develop Co, 82-84; chief exec officer & pres, Interlab Robotics Inc, 83-87; mgr, Elec Power Res Inst, 87-90; MANAGING PRIN, THRESHOLD GROUP INC, 90- *Concurrent Pos:* Consult, TRW, 58-63, NSF, 71-72, Small Bus Admin, 81-82 & SRI Int, 80-82; dir, Technol Transfer Soc, 77-78 & 87-90; dir small bus, High Tech Inst, 90- *Mem:* Sigma Xi; AAAS; Inst Mgt Sci; Technol Transfer Soc; NY Acad Sci; Inst Elec & Electronics Engrs. *Res:* Applied physics; new technology markets; strategic planning; new venture planning; science and technology policy. *Mailing Add:* 810 Magellan Lane Foster City CA 94404

BLOCK, RICHARD EARL, b Chicago, Ill, Mar 8, 31; m 64; c 2. MATHEMATICS, LIE ALGEBRAS. *Educ:* Univ Chicago, PhD(math), 56. *Prof Exp:* Instr math, Univ Ind, 56-58; Off Naval Res res assoc, Yale Univ, 58-59; Bateman res fel, Calif Inst Technol, 59-60, asst prof, 60-64; assoc prof, Univ Ill, 64-68; PROF MATH, UNIV CALIF, RIVERSIDE, 68- *Concurrent Pos:* Consult, Jet Propulsion Lab, 62-68; prin investr, Nat Sci Found Grants, 66-; res assoc, Yale Univ, 67; vis prof, Rutgers Univ, 80-81; fac res lectr, Univ Calif, Riverside, 86. *Mem:* Am Math Soc; Math Asn Am. *Res:* Algebra; Lie algebras; representations; associative and non-associative ring theory; combinatorial designs; hopf algebras; determined differentiably simple algebras; classification of restricted simple lie algebras. *Mailing Add:* Dept Math Univ Calif Riverside CA 92521

BLOCK, ROBERT CHARLES, b Newark, NJ, Feb 11, 29; m 52; c 2. NUCLEAR PHYSICS, NUCLEAR ENGINEERING. *Educ:* Newark Col Eng, BS, 50; Columbia Univ, MA, 53; Duke Univ, PhD, 56. *Prof Exp:* Physicist, Oak Ridge Nat Lab, 55-66; PROF NUCLEAR ENG, RENSSELAER POLYTECH INST, 66-, DIR, GAERTTNER LINEAR ACCELERATOR LAB, 75- *Concurrent Pos:* Lectr, vis scientists prog, Am Inst Physics & Am Asn Physics Teachers, 58-; exchange scientist, Atomic Energy Authority Res Estab, Harwell, Eng, 62-63; guest lectr, Rensselaer Polytech Inst, 65-66; vis prof, Kyoto Univ, Japan, 73-74; consult, Knolls Atomic Power Lab; mem, Nuclear Cross-sect Adv Comt, US AEC; vpres & treas, Becker, Block & Harris Inc, 80-; mem vis comt, Dept Nuclear Energy, Brookhaven Nat Lab, 82-85. *Mem:* Am Phys Soc; fel Am Nuclear Soc; AAAS. *Res:* Basic and applied neutron physics research; industrial applications of radiation; radiation effects in electronics. *Mailing Add:* Dept Nuclear Eng Rensselaer Polytech Inst Troy NY 12181

BLOCK, ROBERT JAY, b Brooklyn, NY, Feb 18, 35; m 60; c 3. PHYSICAL METALLURGY, FAILURE ANALYSIS. *Educ:* Mass Inst Technol, BSc, 56; Columbia Univ, MSc, 58; Univ Ill, PhD(metall), 63. *Prof Exp:* Asst metall, Mass Inst Technol, 56; asst metall eng, Columbia Univ, 56-58 & Univ Ill, 58-63; from asst prof to assoc prof, 63-72, PROF METALL ENG, UNIV OKLA, 72- *Mem:* Am Soc Metals; Am Inst Mining, Metall & Petrol Engrs. *Res:* Plastic deformation of metals; failure analysis; physical metallurgy; crystal plasticity; dislocation arrangements in metals; products liability. *Mailing Add:* Dept Chem Eng & Metall Eng Univ Okla Main Campus Norman OK 73019

BLOCK, RONALD EDWARD, b Charleston, SC, Apr 29, 41; m 65; c 2. PHYSICAL CHEMISTRY, BIOPHYSICAL CHEMISTRY. *Educ:* Col Charleston, BS, 63; Clemson Univ, MS, 66, PhD(phys chem), 69. *Prof Exp:* Fel chem, Univ Fla, 69-70; res scientist, Papanicolaou Cancer Res Inst, 72-73, assoc scientist biochem, Nuclear Magnetic Resonance Studies, 74-84, sr scientist, 84; nuclear magnetic resonance scientist, 84-90, PHYSICIST & ASSOC DIR MAGNETIC RESONANCE RES, MT SINAI MED CTR, 90- *Concurrent Pos:* Interim vis asst prof chem, Univ Fla, 70; adj asst prof neurol, Sch Med, Univ Miami, 74-76, adj assoc prof chem, 76-, adj asst prof biochem, 79-84, adj assoc prof, 84-; consult nuclear magnetic resonance proj, NIH, 75-88; sabbatical res chemist USEPA, 80-81; mem NIH Diag Radiol Study Sect, 86-88. *Mem:* Am Chem Soc; AAAS. *Res:* Application of nuclear magnetic resonance spectroscopy to studies of the structure of polymers, biological membranes, proteins, and tissue components; magnetic resonance imaging and spectroscopy studies in vivo. *Mailing Add:* 545 NE 112th St Miami FL 33161-7161

BLOCK, SEYMOUR STANTON, b New York, NY, May 16, 18; m 42; c 2. DISINFECTION, APPLIED MICROBIOLOGY. *Educ:* Pa State Univ, BS, 39, MS, 41, PhD(agr biochem), 42. *Prof Exp:* Res chemist fermentation, Joseph E Seagram & Sons, Inc, 42-44; prof chem eng, 44-85, EMER PROF CHEM ENG, UNIV FLA, 85- *Concurrent Pos:* Consult, Charles Pfizer Co, 55-60, Hooker Chem Co, 60-62, USDA, 65-66 & Martin-Marietta Co, 72; mem adv comt recommendations on basic res, US Army, 76-79; ed bk series on hazardous & toxic wastes, Marcel Dekker Publ, 78; consult, Quaker Oats Co, 78 & Recovery Tech Inc; numerous grants & contracts; res assoc, Frankford Arsenal, US Army, Philadelphia, 63, PR Nuclear Ctr, San Juan, 64 & Inst Nuclear Studies, Oak Ridge, Tenn, 68. *Mem:* Soc Indust Microbiol (treas, 64); Am Chem Soc; Sigma Xi. *Res:* Antimicrobial agents; mushroom research; coal investigations; toxic and hazardous wastes; developed culture of wild mushroom (pleurotus ostreatus) now grown commercially worldwide; editor and co-author of foremost book on disinfection and sterilization, in fourth edition, Biography of Benjamin Franklin; treatment and disposal of medical wastes. *Mailing Add:* Dept Chem Eng Univ Fla Gainesville FL 32611

BLOCK, STANLEY, b Baltimore, Md, Feb 14, 26; m 50; c 3. PHYSICAL CHEMISTRY, CRYSTALLOGRAPHY. *Educ:* Johns Hopkins Univ, MA, 52, PhD(chem), 55. *Prof Exp:* Chemist, Crown Cork & Seal Co, Ind, 43-50; physicist solid state, Nat Bur Standards, 54-67, chief, Crystallog Sect, 67-87, res scientist, 87-90; RETIRED. *Concurrent Pos:* Lectr, Am Univ, 58 & Howard Univ, 67; tech expert x-ray crystallog, UNESCO, 59-61; consult, 90- *Honors & Awards:* S B Meyer, Jr Award, Am Ceramic Soc, 59; Gold Medal Award, US Dept Com, 74. *Mem:* Am Chem Soc; Am Crystallog Asn. *Res:* Materials science, particularly the effects of high pressure on the structure and properties of inorganic and organic materials; synthesis of organic compounds, sintering and toughening of ceramics, reaction kinetics and structural transformations. *Mailing Add:* Nat Bur Standards Bldg 223 Rm A219 Gaithersburg MD 20899

BLOCK, STANLEY L, b Cincinnati, Ohio, Oct 10, 23; m 47; c 2. PSYCHIATRY. *Educ:* Univ Cincinnati, BS, 50, MD, 47; Chicago Inst Psychoanal, cert, 64. *Prof Exp:* From instr to asst prof, 53-64, assoc clin prof, 64-71, assoc prof, 71-74, PROF PSYCHIAT, UNIV CINCINNATI, 75- *Concurrent Pos:* Consult, Vet Admin Hosp, 57- & Jewish Family Serv, 62-; lectr, Hebrew Union Col, 62-; vis assoc prof, Med Ctr, Univ Ky, 64-69, vis prof, 69-; dir dept psychiat, Jewish Hosp, Cincinnati, Ohio, 67-77, vpres, 77- *Mem:* Fel Am Psychiat Asn; Am Psychoanal Asn; Am Asn Hist Med; Am Acad Sci; Hist Sci Soc. *Res:* Psychoanalysis; history of psychiatry; group psychotherapy. *Mailing Add:* 2451 Brookwood Lane Amberly OH 45237

BLOCK, STANLEY M(ARLIN), b Minneapolis, Minn, Feb 4, 22; m 47; c 4. INDUSTRIAL ENGINEERING. *Educ:* Univ Minn, BME, 43, MBA, 50, PhD(mech eng), 56. *Prof Exp:* Indust engr, Minn Mining & Mfg Co, 47-49; from instr to asst prof indust eng, Univ Minn, 49-59; assoc prof prod mgt, Univ Chicago, 59-62; prof mgt sci, Ill Inst Technol, 62-80; RETIRED. *Concurrent Pos:* Consult, US Air Force, 54-55; labor arbitrator, 55-; fac adv & lectr, Span Study Group, SAfrica, 57; lectr, Japan Mgt Asn, Tokyo, 64 & Nat Develop & Mgt Found, Africa, 65. *Mem:* Am Inst Indust Engrs; Nat Acad Arbitrators; Soc Prof Dispute Resolution. *Res:* International education; labor-management relations; hand-motion, eye hand coordination and simultaneous hand motions; wage incentives; managerial financial rewards. *Mailing Add:* 6105 Palo Verde Dr Rockford IL 61111

BLOCK, TOBY FRAN, b Brooklyn, NY, Oct 5, 49; m 77; c 3. GENERAL CHEMISTRY, PHYSICAL & INORGANIC CHEMISTRY. *Educ:* Brooklyn Col, BS, 70; Univ Wis-Madison, PhD(phys chem), 76. *Prof Exp:* Asst prof chem, Univ Wis-Stevens Points, 76-80; LAB COORDR, GEN CHEM, SCH CHEM, GA INST TECHNOL, 80- *Mem:* Am Chem Soc; Sigma Xi. *Res:* Chemical education. *Mailing Add:* 3044 Stantondale Dr Atlanta GA 30341

BLOCK, WALTER DAVID, b Dayton, Ohio, Oct 16, 11; m 41; c 2. BIOCHEMISTRY, NUTRITION. *Educ:* Univ Dayton, BS, 33; Univ Mich, MS, 34, PhD(biochem), 38. *Prof Exp:* From instr to asst prof biochem, Univ Mich, Ann Arbor, 38-48; pvt consult, 48-52; PROF NUTRIT, UNIV MICH, ANN ARBOR, 66-, ASSOC PROF BIOCHEM, 52- *Concurrent Pos:* Res consult, Caylor-Nickel Res Found, Bluffton, Ind, 53-80, dir, 81. *Mem:* AAAS; Soc Exp Biol & Med; Am Inst Nutrit; Am Soc Biol Chemists. *Res:* Lipid metabolism; amino acid metabolism; protein research. *Mailing Add:* 1335 Glendaloch Circle Ann Arbor MI 48104

BLOCKER, HENRY DERRICK, b Waterboro, SC, May 18, 32; m 58; c 2. SYSTEMATIC ENTOMOLOGY, CICADELLIDAE. *Educ:* Clemson Univ, BS, 54, MS, 58; NC State Univ, PhD(entom), 65. *Prof Exp:* From asst prof to assoc prof, 65-76, PROF ENTOM, KANS STATE UNIV, 76- *Mem:* Entom Soc Am. *Res:* Taxonomy and biology of leafhoppers. *Mailing Add:* Dept Entom Kans State Univ Waters Hall Manhattan KS 66506

BLOCKSTEIN, DAVID EDWARD, b Pittsburgh, Pa, Jan 1, 56. DEVELOPMENT OF NATIONAL INSTITUTES FOR THE ENVIRONMENT, POLICIES TO CONSERVE BIOLOGICAL DIVERSITY. *Educ:* Univ Wis-Madison, BS, 78; Univ Minn, MS, 82, PhD(ecol), 86. *Prof Exp:* Postdoctoral researcher, Univ Wis-Madison, 86-87; congressional fel, Am Inst Biol Sci, 87-88, prog assoc, 89-90; asst prof zool, Conn Col, 88-89; RESEARCHER BIOL, PRINCETON UNIV, 90- *Concurrent Pos:* Consult, Comt Sci, Space & Technol, US House Rep, 88-89 & Forest Serv, Bur Land Mgt, US Environ Protection Agency, 88-; exec dir, Comt Nat Insts Environ, 90-; lectr, Univ Minn, Minneapolis, 90-; chair, Pub Responsibility Comt, Am Ornith Union, 90- *Honors & Awards:* Ernest P Edwards Award, Wilson Ornith Soc, 90. *Mem:* Am Ornithologists Union; Soc Conserv Biol; Animal Behav Soc; Cooper Ornith Soc; Wilson Ornith Soc; AAAS. *Res:* Assessment of national needs for environmental research and federal spending on environment research; policies to conserve biodiversity; population ecology of forest nesting birds; conservation of endangered birds. *Mailing Add:* Comt Nat Insts Environ 730 11th St NW Washington DC 20001-4521

BLOCKSTEIN, WILLIAM LEONARD, b Irwin, Pa, Nov 11, 25; wid; c 3. HEALTH SCIENCES, ACADEMIC ADMINISTRATION. *Educ:* Univ Pittsburgh, BS, 50, MS, 53, PhD(pharm), 59. *Prof Exp:* Asst, Sch Pharm, Univ Pittsburgh, 52-53, admin asst to dean & instr pharm, 53-58; asst to dean & assoc prof, Sch Pharm, Wayne State Univ, 59-64; from assoc prof to prof pharm, Med Sch, Univ Wis-Madison, 64-88, chmn, Health Sci Unit, Univ Exten, 69-80, clin prof prev med, 72-91, chmn, Health & Human Serv Area, 80-86, Edward Kremers prof pharm, 88-91, EMER PROF, MED SCH, UNIV WIS-MADISON, 91- *Concurrent Pos:* Vis scientist & lectr, Am Asn Cols Pharm, 69-72; chmn, Wis Comn on Prev & Wellness, 78-79. *Honors & Awards:* Remington Honor Medal Award, Am Pharmaceut Asn, 85. *Mem:* Fel Am Col Apothecaries; fel Acad Pharmaceut Res Sci; fel Am Col Pharmaceut Scientists. *Res:* Continuing education in pharmacy; education and educational evaluation in health sciences; health care delivery systems. *Mailing Add:* Univ Wis Sch Pharm Madison WI 53703

BLODGETT, FREDERIC MAURICE, b Bucksport, Maine, July 3, 20; m; c 5. PEDIATRICS. *Educ:* Bowdoin Col, BS, 42; Yale Univ, MD, 45; Am Bd Pediat, dipl, 55. *Prof Exp:* Instr pediat, Harvard Med Sch, 52-54, assoc, 54-57; from asst prof to assoc prof, Sch Med, Yale Univ, 57-66; PROF PEDIAT, MED COL WIS, 66- *Concurrent Pos:* Chmn comt instr clin years, Sch Med, Yale Univ, 63-65. *Honors & Awards:* Children's Bur Award, Interchange Experts Prog, Int Health Res Act, 66. *Mem:* Fel Am Acad Pediat; Am Bd Pediat. *Res:* General care of children; lead poisoning. *Mailing Add:* 1700 W Wells St Milwaukee WI 53233

BLODGETT, ROBERT BELL, b Milwakee, Wis, Feb 2, 16; m 39; c 2. CHEMISTRY. *Educ:* Northwestern Univ, BS, 37; Univ Wis, PhD(phys chem), 40. *Prof Exp:* Res chemist, Rayon Dept, E I du Pont de Nemours & Co, 40-46, res supvr cellophane res, 46-48; chief chemist, Robbins Mills, Inc, 48-54; res mgr, Okonite Co, NJ, 54-74; consult res, Anaconda-Enisson, Inc, 74-80, chief consult eng, Wire & Cable Div, 81-86; RETIRED. *Mem:* AAAS; Am Chem Soc; Inst Elec & Electronics Engrs. *Res:* High-voltage insulation; cable technology; high-tenacity rayon for fabric uses; plasticization and characterization of high polymers; fabric finishing and dyeing. *Mailing Add:* 12000 Southall Court Richmond VA 23233-1681

BLODI, FREDERICK CHRISTOPHER, b Vienna, Austria, Jan 11, 17; nat US; m 46; c 2. MEDICINE. *Educ:* Univ Vienna, MD, 40. *Prof Exp:* Instr ophthal, Univ Vienna, 46-47; res fel, Inst Ophthal, Columbia Univ, 47-52; assoc prof, 52-66, PROF OPHTHAL, UNIV IOWA, 66- *Concurrent Pos:* Head, Dept Ophthal, Univ Iowa, 67-86. *Mem:* Am Ophthal Soc; Asn Res Vision & Ophthal; Am Acad Ophthal & Otolaryngol; Sigma Xi. *Res:* Morphologic pathology of the eye; electromyography of the eye muscles. *Mailing Add:* Dept Ophthal Univ Iowa Iowa City IA 52242

BLOEDEL, JAMES R, b Minneapolis, Minn, Apr 7, 40; c 3. NEUROPHYSIOLOGY. *Educ:* St Olaf Col, BA, 62; Univ Minn, Minneapolis, PhD(physiol), 67, MD, 69. *Prof Exp:* Asst Prof, 68-73, assoc prof, 73-78, prof neurosurg & physiol, Univ Minn, Minneapolis, 78-, dir neurosurg res lab, 72-; AT NEUROBIOL, BARROW NEUROL INST, MED CTR, ST JOSEPH'S HOSP. *Concurrent Pos:* NIH awards, Univ Minn, Minneapolis, 70-76; vis scientist, Lab Neurophysiol, Good Samaritan Hosp, Portland, Ore, 72. *Mem:* Am Physiol Soc; Soc Neurosci; Sigma Xi. *Res:* Cerebellar physiology; pathophysiology of pain. *Mailing Add:* Neurobiol Barrow Neurol Inst Med Ctr St Joseph's Hosp 350 W Thomas Rd Phoenix AZ 85013

BLOEMBERGEN, NICOLAAS, b Dordrecht, Neth, Mar 11, 20; nat US; m 50; c 3. QUANTUM OPTICS. *Educ:* State Univ Utrecht, Phil Can, 41, Phys Drs, 43; State Univ Leiden, Dr Phil(physics), 48. *Hon Degrees:* AM, Harvard Univ, 51; DSc, Laval, Que, 87, Univ Conn, 88. *Prof Exp:* Asst physics, Harvard Univ, 46-47; assoc, State Univ Leiden, 47-48; jr fel, Soc Fels, 49-51, from assoc prof to Gordon McKay Prof appl physics, 51-74, Rumford prof physics, 74-80, GERHARD GADE UNIV PROF, HARVARD UNIV, 80- *Concurrent Pos:* Guggenheim fel, 57, von Humboldt sr fel, 80; vis prof, Univ Paris, 57; Univ Calif, Berkeley, 64; Univ Leiden, 73 & Raman Inst, Bangalore, India, 80; consult, United Technols Res Ctr, 78, DARPA, 72- *Honors & Awards:* Nobel Prize in physics, 81; Buckley Prize, Am Phys Soc, 58; Liebmann Prize, Inst Elec & Electronics Engrs, 59, Medal of Honor, 83; Ballantine Medal, Franklin Inst, 61; Nat Medal Sci, President US, 74; Lorentz Medal, Royal Dutch Acad Sci, 78; Frederic Ives Medal, Oht Soc Am, 79. *Mem:* Nat Acad Sci; fel Am Phys Soc; fel Inst Elec & Electronics Eng; Am Acad Arts & Sci; fel Optical Soc Am. *Res:* Nuclear and electronic magnetic resonance; solid state masers; nonlinear optics. *Mailing Add:* Pierce Hall Harvard Univ Cambridge MA 02138

BLOEMER, WILLIAM LOUIS, b Covington, Ky, Nov 26, 47; m 69; c 1. PHYSICAL CHEMISTRY, CHEMICAL INSTRUMENTATION. *Educ:* Thomas More Col, BA, 67; Univ Ky, PhD(chem), 72. *Prof Exp:* Teaching asst chem, Univ Ky, 68-73; from asst prof to assoc prof, 73-84, PROF CHEM & MED TECHNOL, SANGAMON STATE UNIV, 84- *Concurrent Pos:* Dean, Sch Lib Arts & Sci, Sangamon State Univ, 88- *Mem:* Am Chem Soc; AACC; ADCIS. *Res:* Molecular orbital calculations; educational applications of computers. *Mailing Add:* Sangamon State Univ Springfield IL 62794-9243

BLOGOSLAWSKI, WALTER, b New Britain, Conn, Feb 8, 43; m 77; c 3. MARINE MICROBIOLOGY. *Educ:* Fairfield Univ, BS, 66; Long Island Univ, MS, 68; Fordham Univ, PhD(marine microbiol), 71. *Prof Exp:* Res microbiologist water treatment, Monsanto Biodize, St Louis, Mo, 69-71; RES MICROBIOLOGIST SHELLFISH DIS, NAT MARINE FISHERIES SERV, 71- *Concurrent Pos:* Adj prof biol, Fairfield Univ, 72-; adv mem biol & climate effects stratospheric ozone reduction, Environ Protection Agency, Washington, DC, 77-78; adj assoc prof marine sci, Long Island Univ, 77- *Mem:* Int Ozone Inst (secy, 74-78, chmn, 80-84, pres, 88-89); AAAS; Int Ozone Inst; Sigma Xi; NY Acad Sci. *Res:* Marine microbial research; elimination of shellfish pathogens; disinfection of seawater; detoxification of biotoxins by ozone and other oxidants. *Mailing Add:* Milford Lab 212 Rogers Ave Milford CT 06460-6499

BLOHM, THOMAS ROBERT, b South Bend, Ind, Oct 28, 20; m 48; c 1. BIOCHEMISTRY. *Educ:* Univ Notre Dame, BS, 42; Northwestern Univ, PhD(biochem), 48. *Prof Exp:* From instr to assoc prof biochem, Dent Br, Univ Tex, 48-52; head sect, Pharmacol Dept, Richardson-Merrell Inc, 52-56, head, Dept Biochem, 56-61, prin res chemist, Merrell-Nat Labs Div, 61-90; CONSULT, 90- *Concurrent Pos:* Fel, Coun Arteriosclerosis, Am Heart Asn. *Mem:* Fel AAAS; Am Soc Biol Chem; fel Am Inst Chem; Int Soc Biochem Pharmacol; Am Chem Soc. *Res:* Biochemical pharmacology, lipid metabolism; atherisclerosis; metabolic diseases. *Mailing Add:* 6468 Oldbarn Ct Cincinnati OH 45243

BLOIS, MARSDEN SCOTT, JR, dermatology; deceased, see previous edition for last biography

BLOM, CHRISTIAN JAMES, b Berkeley, Calif, Dec 3, 28; m 53; c 3. GEOLOGY. *Educ:* Calif Inst Technol, BS, 50; Innsbruck Univ, PhD(geol), 53. *Prof Exp:* Geologist, Standard Oil Co Calif, 55-64; geologist, Occidental Petrol Corp, 64-65, explor mgr & vpres, Occidental of Libya, Inc, 66-67, sr vpres & resident mgr, 67-69, sr vpres & coordr Libyan affairs, 69, explor mgr Eastern hemisphere, Occidental Explor & Prod Co, 69-78, vpres explor Eastern hemisphere, 78-83; CONSULT, 84- *Mem:* Geol Soc Am; Am Asn Petrol Geol; Soc Petrol Engrs. *Res:* Fifteen patents. *Mailing Add:* 203 Fairway Dr Bakersfield CA 93309

BLOM, GASTON EUGENE, b Brooklyn, NY, Mar 29, 20; m 46; c 4. CHILD PSYCHIATRY, EDUCATIONAL PSYCHOLOGY. *Educ:* Colgate Univ, BA, 41; Harvard Univ, MD, 44; Am Bd Psychiat & Neurol, dipl & cert psychiat, 55; Boston Psychoanal Inst, dipl psychoanal, 56, dipl child anal, 58; Am Bd Psychiat & Neurol, cert child psychiat, 59. *Prof Exp:* Resident psychiat, USPHS Hosp, Ft Worth, Tex, 46-47; vet resident, Univ Mich Hosp, Ann Arbor, 47-48; fel child psychiat, Mass Gen Hosp, Boston, 48-49 & Judge Baker Guid Ctr, 49-50; asst psychiat, Harvard Med Sch, 51-53, instr, 54-58; assoc prof, Med Sch & chief, Child Psychiat Div, Univ Colo Med Ctr, Denver, 58-63, dir, Day Care Ctr, 63-70, actg chmn, Dept Psychiat, 70-71, prof psychiat, Med Sch, 63-75, prof educ, Denver Ctr, 66-75; PROF PSYCHIAT & ELEMENTARY & SPEC EDUC, MICH STATE UNIV, 75-, ADJ PROF PSYCHOL, 80- *Concurrent Pos:* Dir, Child Psychiat Unit, Mass Gen Hosp, 56-58; consult, Ment Health Study Sect, Div Res Grants, NIH, 60-64, mem, Conf Social Competence & Clin Pract, Nat Inst Ment Health, 65; mem, Conf on Adaptation to Change, Found Fund for Res in Psychiat, 68; consult, Nat Adv Comn on Dyslexia & Related Reading Dis, Dept Health, Educ & Welfare, 68-69 & Denver Pub Schs, 69-75; training & supvry analyst, Denver Psychoanal Inst, 69-75 & supvry child analyst, 70-75; mem, Conf on Child Psychiat in Gen Psychiat Training, 71; mem vis staff, Gentofte Hosp, Copenhagen, Denmark, 72-73; med consult, US Gen Acct Off, 74-75; consult, Ingham Intermediate Sch Dist, 76-; Med coordr, Univ Ctr Int Rehab, Mich State Univ, 79-, fac coordr, Univ Resource Network on Youth Advocacy, 80-; Mich State Spec Educ Adv Coun, 81- *Mem:* Fel Am Psychiat Asn; Am Acad Psychoanal; fel Am Acad Child Psychiat; Coun Except Children; Int Reading Asn. *Res:* Numerous publications on child and youth advocacy; developmental aspects of language acquisition; mental health intervention in collective disasters; stress and coping behaviors in handicapped children and adults. *Mailing Add:* Dept Psychiat Mich State Univ East Lansing MI 48824

BLOMBERG, BONNIE B, b Richmond, Calif, Oct 16, 48. MICROBIOLOGY, IMMUNOLOGY. *Educ:* Univ Calif, San Diego, PhD, 77. *Prof Exp:* ASST PROF MICROBIOL & IMMUNOL, SCH MED, UNIV MIAMI. *Mailing Add:* Univ Miami PO Box 016960 R-138 Miami FL 33101

BLOMGREN, GEORGE EARL, b Chicago, Ill, Apr 15, 31; m 56; c 3. PHYSICAL CHEMISTRY. *Educ:* Northwestern Univ, BS, 52; Univ Wash, PhD(chem), 56. *Prof Exp:* Boese fel, Columbia Univ, 56-57; res chemist, Nat Carbon Co, 57-62, group leader, Consumer Prod Div, 62-73, res group leader, Battery Prod Div, 73-75, sr technol assoc, 76-81, corp technol fel, Union Carbide Corp, 81-87; technol fel, 87-89, SR TECHNOL FEL, EVEREADY BATTERY CORP, 89- *Mem:* AAAS; Am Chem Soc; Electrochem Soc. *Res:* Infrared spectra; theory of liquids; electron spin resonance; quantum chemistry; electrolyte transport; electrode processes. *Mailing Add:* Technol Lab Eveready Battery Co PO Box 45035 Westlake OH 44145

BLOMMERS, ELIZABETH ANN, b Pittsburgh, Pa, Nov 21, 23. ORGANIC POLYMER CHEMISTRY. *Educ:* Bryn Mawr Col, AB, 45, MA, 46, PhD(chem), 50. *Prof Exp:* Instr chem, Wheaton Col, Mass, 50-51; instr, Wells Col, 51-52; chemist org res, Signal Corps, US Army, 52-53; chemist textile res, Quaker Chem Prods, 53-56; res chemist struct adhesives, Borden Chem Co, 56-62; sr chemist, Thiokol Chem Corp, 62-65; sr chemist, Koppers Co, Inc, 65-74; sr res scientist, Arco Polymers, Inc, 74-80 & Arco Chem Co, 80-85; CONSULT, 85- *Mem:* Am Chem Soc. *Res:* Mechanisms of organic reactions; adhesives; analysis of thermoplastic polymers; utilization of polymers with coal and/or oil. *Mailing Add:* 109 Hunt Club Lane Newton Square PA 19073

BLOMQUIST, CHARLES HOWARD, b Rockford, Ill, Oct 27, 33; m 62. BIOCHEMISTRY. *Educ:* Univ Ill, BS, 55; Univ Minn, PhD(biochem), 64. *Prof Exp:* Asst biochem, Univ Wash, 55-56; Olson Mem Found fel, Karolinska Inst, Sweden, 64-65; mem staff H-4 div, Los Alamos Sci Lab, 65-69; mem staff biochem, 69-77, MEM STAFF OBSTET & GYNEC, UNIV MINN, MINNEAPOLIS, 77-; RES DIR, DEPT OBSTET-GYNEC, ST PAUL-RAMSEY HOSP, 69- *Mem:* AAAS; Am Chem Soc; Am Soc Biol Chem. *Res:* Steroid enzymology; kinetics and mechanisms of dehydrogenase enzymes; fluorometric methods as applied to enzyme and protein chemistry. *Mailing Add:* Dept Obstet & Gynec St Paul-Ramsey Hosp St Paul MN 55101

BLOMQUIST, GARY JAMES, b Iron Mountain, Mich, Apr 1, 47; m 70; c 3. BIOCHEMISTRY. *Educ:* Univ Wis, La Crosse, BS, 69; Mont State Univ, PhD(chem), 73. *Prof Exp:* Asst prof chem, Univ Southern Miss, 73-77; from asst prof to assoc prof, 77-83, PROF BIOCHEM, UNIV NEV, RENO, 83- *Mem:* AAAS; Am Soc Biol Chemists; Entom Soc Am. *Res:* Lipid biochemistry; insect biochemistry; insect hydrocarbon and pheromone biosynthesis. *Mailing Add:* Dept Biochem Univ Nev Reno NV 89557-0014

BLOMQUIST, RICHARD FREDERICK, b Mankato, Minn, Nov 25, 12; m 37; c 3. ADHESIVES, WOOD UTILIZATION. *Educ:* Coe Col, AB, 34; Univ Iowa, MS, 36, PhD(org chem), 37. *Hon Degrees:* DSc, Coe Col, 82. *Prof Exp:* Prof chem & chmn dept, Doane Col, 37-42; from asst chemist to assoc chemist, Div Wood Preserv, Forest Prod Lab, 42-45, proj leader glues & gluing res, 47-60 & glued prod res, 60-66, proj leader housing res, Forestry Sci Lab, US Forest Serv, 66-76; CONSULT, 76- *Concurrent Pos:* Marburg lectr, Am Soc Testing & Mat, 63; sr res fel, NZ Forest Serv, 65; adj prof, Sch Forest Resources, Univ Ga, 68-76; chmn ad hoc comt on aerospace struct adhesives, Nat Adv Bd, Nat Acad Sci, 72-74; adv, Mex Forestry Serv, 73 & INDUPERU, Peru, 78; consult, 76- *Honors & Awards:* Adhesives Award, Am Soc Testing & Mat, 62; Gottshalk Award, Prod Ressoc, 72. *Mem:* Am Chem Soc; Am Soc Testing & Mat; Forest Prod Res Soc (pres, 74-75); fel Am Inst Chemists; fel Int Acad Wood Sci. *Res:* Better utilization of wood products in building construction; adhesives for wood, metal and plastics; development of composite lumber products for building construction. *Mailing Add:* 156 S Stratford Dr Athens GA 30605

BLOMQVIST, CARL GUNNAR, b Bararyd, Sweden, Dec 31, 31; US citizen; m 61; c 2. CARDIOLOGY, CARDIOVASCULAR PHYSIOLOGY. *Educ:* Univ Lund, BM, 54, MD, 60; Karolinska Inst, Sweden, PhD(med), 67. *Prof Exp:* Res fel cardiovasc epidemiol, Physiol Hyg Lab, Col Med Sci, Univ Minn, Minneapolis, 60-61; resident dept med, Karolinska Inst, Sweden, 62-65; from instr to assoc prof med & physiol, 66-76, PROF MED & PHYSIOL, UNIV TEX HEALTH SCI CTR DALLAS, 76- *Concurrent Pos:* Sr attend physician & dir EKG & exercise labs, Parkland Mem Hosp, 68-; estab investr, Am Heart Asn, 68-72; consult, Dept Med, Vet Admin Hosp, Dallas, Tex, 72-; mem, ed bd Circulation, 69-73, Med & Sci in Sports, 79-86, Clin Cardiol, 77-90, J Cardiac Rehab, 79-, Appl Physiol, 87-; mem study sect appl physiol & bioeng, NIH, 74-78, prin investr grants NIH, NASA; mem Nat Acad Sci Comt on Space Biol & Med, 86-90. *Mem:* Am Fedn Clin Res; fel Am Col Cardiol; Int Soc Cardiol; AAAS; Aerospace Med Asn; Int Acad Astronaut; Am Heart Asn (Council Epodemiology). *Res:* Cardiovascular regulation in man; space physiology and medicine. *Mailing Add:* Southwest Med Ctr Univ Tex Dallas TX 75235

BLOMSTER, RALPH N, b Lynn, Mass, May 18, 31; m 62; c 2. PHARMACOGNOSY. *Educ:* Mass Col Pharm, BS, 53; Univ Pittsburgh, MS, 57; Univ Conn, PhD(pharmacog), 63. *Prof Exp:* Asst prof pharmacog, Univ Pittsburgh, 63-68; prof pharmacog & chmn dept, 68-80, PROF MED CHEM & PHARMACOG & CHMN DEPT, SCH PHARM, UNIV MD, BALTIMORE, 80- *Concurrent Pos:* Am Found Pharmaceut Educ fel, 59-61; Lederle Labs res fel, 62-63. *Mem:* AAAS; Am Soc Pharmacog; Am Pharmaceut Asn; Sigma Xi. *Res:* Phytochemical investigations; isolation, purification and identification of plant constituents with biological activity, mainly the genera Heimia and Catharanthus; plant tissue culture and biotransformations of natural products. *Mailing Add:* Sch Pharm Univ Md 20 N Pine St Baltimore MD 21201

BLOMSTROM, DALE CLIFTON, b Lincoln, Nebr, Dec 13, 27; m 56; c 3. PROTEIN PURIFICATION, PROTEIN STRUCTURE. *Educ:* Univ Nebr, BS, 48, MS, 50; Univ Ill, PhD(chem), 53. *Prof Exp:* RES ORG & BIOCHEMIST, EXP STA, E I DU PONT DE NEMOURS & CO, 53- *Mem:* Am Chem Soc. *Res:* Heterocyclic compounds; restricted rotation; quinone diimides; polymers; organic reaction mechanisms; protein structures; hormone controls in adipose tissue; mechanism of herbicide action; plant hormone metabolites; isolation and characterization of interferon-induced proteins from human cells. *Mailing Add:* 3333 Rockfield Dr S Wilmington DE 19810

BLONDIN, JOHN MICHAEL, b Worcester, Mass, Oct 29, 60; m 88; c 2. THEORETICAL HIGH ENERGY ASTROPHYSICS. *Educ:* Univ Wis-Madison, BA, 82; Univ Chicago, MS, 84, PhD(astron & astrophys), 87. *Prof Exp:* Nat Acad Sci-Nat Res Coun res assoc, Goddard Space Flight Ctr, NASA, 87-89; RES ASSOC, DEPT ASTRON, UNIV VA, 89- *Mem:* Am Astron Soc. *Res:* Theoretical high-energy astrophysics and numerical hydrodynamics. *Mailing Add:* Dept Astron Univ Va Charlottesville VA 22903

BLOOD, BENJAMIN DONALD, b Wabash, Ind, June 11, 14; m 39; c 5. PRIMATOLOGY, SCIENCE ADMINISTRATION. *Educ:* Colo State Univ, DVM, 39; Harvard Univ, MPH, 49; Am Col Vet Prev Med, dipl. *Prof Exp:* Chief adv vet affairs, Ministry Pub Health, Govt Korea, 46-48; chief vet pub health, Pan Am Sanit Bur, Regional Off WHO, 49-53, consult, Zones V & VI, Buenos Aires, 53-57; founder & dir, Pan Am Zoonoses Ctr, Pan Am Health Orgn, Azul, Arg, 57-64; int health rep, Off Int Health, USPHS, DC, 64-66, assoc dir int orgn affairs, 66-69, int health attache, US Mission to UN & Other Int Orgn in Geneva, 69-74; exec dir, Fed Interagency Primate Steering Comt, NIH, 75-79; consult, WHO, 80-81; CONSULT, US AGENCY INT DEVELOP, PAN AM HEALTH ORGN & OTHER PUB & PRIVATE RES INST, 81- *Concurrent Pos:* Mem WHO-Food & Agr Orgn expert comt on zoonoses, Geneva, Switz, 50 & Stockholm, Sweden, 58; secy directing coun, Pan Am Cong Vet Med, Lima, Peru, 51, Sao Paulo, Brazil, 54 & Kansas City, Mo, 59; mem US del to directing coun, Pan Am Health Orgn, 65, 67 & 68, US rep, Exec Comt, 66-69, vchmn, 69; mem US del, Pan Am Sanit Conf, 66; mem US del to World Health Assemblies, 66-71; alt mem exec bd, WHO, 69-72; secy, Am Bd Vet Pub Health, 51-53. *Honors & Awards:* Twelfth Int Vet Cong Prize, Am Vet Med Asn, 78; Order of Bifuracted Needle, WHO, 76. *Mem:* Fel Am Pub Health Asn; Am Vet Med Asn; Int Primatological Soc; NY Acad Sci; Am Soc Primatologists. *Res:* International aspects of public health; epidemiology of zoonoses and other diseases common to man and animals; comparative medicine. *Mailing Add:* 1210 Potomac School Rd McLean VA 22101-2327

BLOOD, CHARLES ALLEN, JR, b White River Junction, Vt, Sept 6, 23; m 48; c 1. ORGANIC CHEMISTRY. *Educ:* Univ Vt, BS, 50, MS, 52; Lehigh Univ, PhD(chem), 56. *Prof Exp:* Asst inorg & org chem, Univ Vt, 50-52; asst inorg chem, Lehigh Univ, 52-53, res fel org chem, 53-56; res chemist, E I du Pont de Nemours & Co, Inc, 56-62; assoc prof, 62-69, chmn dept, 68-74 & 76-88, PROF CHEM, STATE UNIV NY COL PLATTSBURGH, 69- *Mem:* Am Chem Soc; Sigma Xi. *Res:* Mechanisms and kinetics of organic reactions; cellophane development chemistry. *Mailing Add:* Dept Chem State Univ NY Col Plattsburgh NY 12901

BLOOD, ELIZABETH REID, b Glasgow, Scotland, Apr 1, 51; US citizen. BIOGEOCHEMISTRY, ECOSYSTEM SCIENCE. *Educ:* Va Commonwealth Univ, BS, 74, MS, 75; Univ Ga, PhD(ecol), 81. *Prof Exp:* Res assoc ecol, Va Commonwealth Univ, 76; lab mgr, J R Reed & Assoc, Inc, 77; proj mgr, Off Water Res & Technol, Dept Interior grants, Univ Ga, 79-80, NSF grants, 80-81; asst prog dir ecol, NSF, 81-82; Baruch res assoc & prog mgr N Inlet LTER, 82-86; ASST PROF, DEPT ENVIRON HEALTH SCI, UNIV SC, 86- *Mem:* NAm Benthological Soc; AAAS; Sigma Xi. *Res:* The biogeochemistry of major, minor and trace elements and their role in the structure and functioning of terrestrial wetland and aquatic ecosystems; development of analytical methodology, instrumentation and quality control protocol. *Mailing Add:* Dept Environ Health Sci Univ South Carolina Columbia SC 29208

BLOODGOOD, ROBERT ALAN, b Holyoke, Mass, Sept 24, 48; m 70; c 2. CELL BIOLOGY. *Educ:* Brown Univ, BS, 70; Univ Colo, Boulder, PhD(biol), 74. *Prof Exp:* Fel & res assoc biol, Yale Univ, 74-77; asst prof anat, Albert Einstein Col Med, 77-80; from asst prof to assoc prof, 80-91, PROF ANAT & CELL BIOL, SCH MED, UNIV VA, 91- *Concurrent Pos:* Am Cancer Soc res fel, Yale Univ, 75-76, NIH fel, 76-77. *Mem:* Am Soc Cell Biol; fel AAAS. *Res:* Mechanisms of cell motility; structure and function of microtubules; flagella; dynamics of the cell surface; trans-membrane signaling. *Mailing Add:* Dept Anat & Cell Biol Univ Va Sch Med Box 439 Charlottesville VA 22908

BLOODSTEIN, OLIVER, b New York, NY, Dec 2, 20; m 41; c 2. STUTTERING. *Educ:* City Col NY, BA, 41; State Univ Iowa, MA, 42, PhD(speech path), 48. *Prof Exp:* PROF SPEECH PATH, BROOKLYN COL, NY, 48- *Concurrent Pos:* Prof, doc prog speech & hearing scis, City Univ NY, 66- *Mem:* Fel Am Speech-Lang-Hearing Asn. *Res:* Conditions under which stuttering varies in frequency; factors related to the distribution of stutterings in the speech sequence; features of early stuttering; developmental stages of stuttering. *Mailing Add:* Dept Speech Brooklyn Col Bedford Ave & Ave H Brooklyn NY 11210

BLOODWORTH, JAMES MORGAN BARTOW, JR, b Atlanta, Ga, Feb 21, 25; m 47, 72; c 5. CLINICAL ENDOCRINOLOGY, MEDICINE. *Educ:* Emory Univ, MD, 48, Am Bd Path, cert anat path & forensic path. *Prof Exp:* From intern to asst resident path, Presby Hosp, NY, 48-50; asst resident internal med, Univ Iowa Hosps, 50-51; from instr to prof path, Col Med, Ohio State Univ, 51-62, chief div path anat, 54-61, attend staff, Univ Hosp, 51-62; PROF PATH, SCH MED, UNIV WIS-MADISON, 62- *Concurrent Pos:* Instr, Col Physicians & Surgeons, Columbia Univ, 49-50; attend path, Columbus State Hosp, 54-57; chief labs, Madison Vet Admin Hosp, 62-89, pathologist, 89-; ed, Endoc Path, 68 & 82. *Honors & Awards:* Lilly Award, Am Diabetes Asn, 63. *Mem:* Am Asn Pathologists; Int Acad Path; Soc Exp Biol & Med; Am Soc Cell Biol; Endocrine Soc; Am Diabetes Asn. *Res:* Degenerative vascular disease, especially diabetic microangiopathy; related aspects of diabetes mellitus, renal disease electron microscopy, enzyme histochemistry and radioautography. *Mailing Add:* Dept Path Univ Wis Sch Med Madison WI 53706

BLOODWORTH, M(ORRIS) E(LKINS), b Axtell, Tex, Sept 18, 20; m 60; c 2. SOIL PHYSICS, AGRICULTURAL ENGINEERING & ADMINISTRATION. *Educ:* Agr & Mech Col Tex, BS, 41, MS, 53, PhD(soil physics), 58. *Prof Exp:* Agr engr soil & water, Soil Conserv Serv, USDA, 41-42 & 46-48; agr engr irrig & drainage, Tex Agr Exp Sta, 48-54, assoc soil physicist, 54-56; assoc prof agron, Tex A&M Univ, 56-63, prof soil physics & head, dept soil & crop sci, 63-80, prof agron, dept soil & crop sci & dir, int progs, 80-84; RETIRED. *Mem:* Fel AAAS; Am Soc Agr Engrs; fel Am Soc Agron; Soil Sci Soc Am; Int Soil Sci Soc. *Res:* Soil-plant-water relationships; interrelated physical and chemical factors affecting plant growth and production; engineering aspects concerning irrigation and drainage of agricultural soils; large scale international agricultural developments and operations; agricultural teaching; experiments in agriculture. *Mailing Add:* 801 N Rosemary Dr Bryan TX 77802

BLOOM, ALAN S, b Buffalo, NY, May 28, 47; m 69; c 2. DRUG ABUSE, NEUROTOXICOLOGY. *Educ:* State Univ NY, Buffalo, BA, 69; Univ Louisville, MA, 71, PhD(neuropsychopharmacol), 74. *Prof Exp:* Fel pharmacol, Med Col, Va, 74-76; from asst prof to assoc prof, 76-87, PROF PHARMACOL, MED COL WIS, 88- *Mem:* Am Soc Pharmacol & Therapeut; Soc Neurosci; AAAS. *Res:* Neurochemical measures of effects of drugs of abuse and insecticides, particularly their effects on brain membranes and their associated receptors and enzymes; aging and the brain. *Mailing Add:* Dept Pharmacol Med Col Wis 8701 Watertown Plank Rd Milwaukee WI 53226

BLOOM, ALLEN, b New York, NY, Sept 9, 43; m 67; c 3. ORGANIC CHEMISTRY, PHOTOCHEMISTRY. *Educ:* Brooklyn Col, BS, 65; Iowa State Univ, PhD(org chem), 71; NY Law Sch, JD, 80. *Prof Exp:* Res assoc x-ray crystallog, Cornell Univ, 69-72; mem tech staff org chem & photochem, RCA Labs, 72-77, mem patent coun, patent opers, David Sarnoff Res Ctr, 77-82; patent atty, Pfizer Inc, 82-85; VPRES, GEN COUN & SECY, LIPOSOME CO, 85- *Mem:* Am Chem Soc. *Res:* Organic materials for optically recording and storing information; liquid crystal phenomena; electrical and photochemical properties of organic molecules. *Mailing Add:* Liposome Co Inc One Res Way Princeton Forrestal Ctr Princeton NJ 08540

BLOOM, ARNOLD LAPIN, b Chicago, Ill, Mar 7, 23; m 51; c 1. LASERS, THIN FILMS. *Educ:* Univ Calif, PhD(physics), 51. *Prof Exp:* Asst physics, Univ Calif, 48-51; physicist, Varian Assocs, 51-61; dir theoret res, Spectra-Physics, Inc, 61-68, chief scientist, 68-71; PHYSICIST, COHERENT, INC, 72- *Concurrent Pos:* Lectr, Stanford Univ, 69-70, res assoc physics, 71-74. *Mem:* Fel AAAS; fel Am Phys Soc; fel Optical Soc Am. *Res:* Optical pumping; lasers; physical optics; gas lasers; organic dye lasers; optical thin films design and computation. *Mailing Add:* 1955 Oak Ave Menlo Park CA 94025-5843

BLOOM, ARTHUR DAVID, b Boston, Mass, Oct 4, 34; m 82; c 4. GENETICS, ENVIRONMENTAL SCIENCES. *Educ:* Harvard Univ, AB, 56; NY Univ, MD, 62. *Prof Exp:* Intern pediat, Bellevue Hosp, New York, 60-61; asst resident, Johns Hopkins Hosp, Baltimore, Md, 61; surgeon, USPHS, NIH, 62-64; sr asst resident pediat, Johns Hopkins Hosp, 64-65; cytogeneticist, Atomic Bomb Casualty Comn, Nat Res Coun, Hiroshima, Japan, 65-68; from asst prof to assoc prof, human genetics & res assoc pediat, Med Sch, Univ Mich, Ann Arbor, 68-70; prof pediat, Human Genetics & Develop, Col Physicians & Surgeons, Columbia Univ, 78-88; PRES, ENVIRON HEALTH INST, PITTSFIELD, MA, 86-; ATTEND PHYSICIAN & GENETICIST, BERKSHIRE MED CTR, 86- *Concurrent Pos:* Consult, Bur Radiol Health, USPHS, 68-70; Atomic Bomb Casualty Comn, Nat Res Coun, 68-74; mem, Pan Am Health Orgn, 76-79, coun, Environ Mutagen Soc, 77-80, bd sci counrs, NIH, USPHS, 82-86; prog comt chmn, Am Soc Human Genetics, 80 & bd dirs, 81-83; chmn, adv panel Genetic Screening Workplace, Off Technol Assessment, US Cong, 81-83; med dir, Diagenetic Labs Inc, Teaneck, NJ, 83-88; ed, Bull Berkshire Med Ctr. *Mem:* AAAS; Am Soc Human Genetics; Soc Pediat Res; Radiation Res Soc; Environ Mutagen Soc. *Res:* Biological and clinical effects of environmental chemicals and radiations. *Mailing Add:* Environ Health Inst Berkshire Med Ctr 725 North St Pittsfield MA 01201

BLOOM, ARTHUR LEROY, b Genesee, Wis, Sept 2, 28; m 53; c 3. TECTONICS. *Educ:* Miami Univ, AB, 50; Victoria Univ, NZ, MA, 52; Yale Univ, PhD(geol), 59. *Prof Exp:* Instr geol, Yale Univ, 59-60; from asst prof to assoc prof, 60-76, PROF GEOL SCI, CORNELL UNIV, 76- *Concurrent Pos:* Fulbright sr scholar, James Cook Univ NQueensland, Townsville, Australia, 73; vis scientist, Japan Soc Prom Sci, 80. *Mem:* AAAS; Geol Soc Am; Int Union Quaternary Res; Am Quaternary Asn; Sigma Xi. *Res:* Glacier controlled sea level changes in the quaternary period and tectonic geomorphology of island arcs in western Pacific coral reef zone; Pleistocene geology; tectonic geomorphology of the Andes. *Mailing Add:* Dept Geol Sci Cornell Univ 2122 Snee Hall Ithaca NY 14853-1504

BLOOM, BARRY MALCOLM, b Roxbury, Mass, Aug 12, 28; m 56; c 3. ORGANIC CHEMISTRY. *Educ:* Mass Inst Technol, SB, 48, PhD(org chem), 51. *Prof Exp:* Nat Res Coun fel, Univ Wis, 51-52; res chemist, Pfizer Inc, 52-58, res supvr, 58-59, mgr, 59-63, dir med chem res, 63-68, Med Prod Res & Develop Div, 68-71, vpres res & develop, Pfizer Pharmaceut, 69-71, pres cent res, 71-90, MEM BD DIRS, PFIZER, INC, 73-, SR VPRES RES & DEVELOP, 90- *Concurrent Pos:* US Cong Off Tech, Assessment, 76-77; mem, Conn High Tech Coun, 83, vis comn, Univ Conn, Inst Mat Sci; mem, Cong Comn Fed Drug Approval Process, Comt Drugs Rare Dis, Pharmaceut Mgr Asn, chmn, Res & Develop Sect, 76; Polytech Inst NY fel; mem bd dirs, Oncogene Sci Inc, 86-, Southern New Eng Telecommun Corp, 87-; mem bd mgrs, Lawrence & Mem Hosp, 89- *Mem:* Am Chem Soc; NY Acad Sci. *Res:* Publication on drug research, government regulatory policy; administration of research and development of human therapeutics, agricultural and fine specialty chemical products. *Mailing Add:* Pfizer Inc Eastern Point Rd Groton CT 06340

BLOOM, BARRY R, b Philadelphia, Pa, Apr 13, 37; m 63. IMMUNOLOGY. *Educ:* Amherst Col, AB, 58; Rockefeller Univ, PhD(immunol), 63. *Hon Degrees:* DSc, Amherst Col, 90. *Prof Exp:* NSF fel, Wright-Fleming Inst, London, 63-64; from asst prof to prof, Dept Microbiol & Immunol, 64-73, chmn dept, 78-90, WEINSTOCK PROF MICROBIOL & IMMUNOL, ALBERT EINSTEIN COL MED, 73-, INVESTR, HOWARD HUGHES MED INST, 90- *Concurrent Pos:* Vis asst prof, Dept Microbiol, Meharry Med Col, 64-65; sect ed cellular immunol, J Immunol, 69-75; consult immunol, WHO, 71-, int health, White House, 77-78, Nat Acad Sci, 83; chmn, Steering Comt Immunol Leprosy, WHO, 77-86, Basic Sci Study Sect, Nat Mult Sclerosis Soc, 79, US-Japan Leprosy Panel, 79-83, Comt Immunol Tuberculosis, WHO, 86- & Sci & Tech Adv Comt Res & Training Trop Dis, WHO, 89-; mem, Bd Int Health, Inst Med, Nat Acad Sci, 85-88, Nat Adv coun, Nat Inst Allergy & Infectious Dis, 88 & Bd Sci & Technol Develop, Nat Res Coun, 90-; vis scientist, Lab Molec Biol, Imp Cancer Res Fund, 84-85. *Honors & Awards:* Oscar Besredka Prize for Immunol, WGer, 84; Shipley Lectr, Harvard Univ, 88; Wellcome Distinguished Lectr, Univ Conn, 89. *Mem:* Nat Acad Sci; Inst Med-Nat Acad Sci; Am Asn Immunologists (vpres, 84, pres, 85-86); French Soc Immunol; Fedn Am Socs Exp Biol (vpres, 85-86); fel AAAS; Am Acad Arts & Sci. *Res:* Cell-mediated immunity; virus-lymphocyte interactions; leprosy and tuberculosis; vaccines. *Mailing Add:* Dept Microbiol Albert Einstein Col Med Bronx NY 10461

BLOOM, EDA TERRI, b Los Angeles, Calif, May 11, 45; m 83; c 1. MICROBIOLOGY, CYTOLOGY. *Educ:* Univ Calif, Los Angeles, AB, 66, PhD(med microbiol & immunol), 70. *Prof Exp:* USPHS postdoctoral fel, Sloan Kettering Inst Cancer Res, 70-71; lectr med microbiol & immunol, Univ Calif, Los Angeles, 71-72, asst res immunologist, 72-78 & 80-81, assoc res immunologist, 81-88, adj prof, 88-89; asst prof res med, Univ Southern Calif, 78-80; RES BIOLOGIST & SR INVESTR, LAB CELLULAR IMMUNOL, DIV CYTOKINE BIOL, CTR BIOLOGICS EVAL & RES, FOOD & DRUG ADMIN, 89-; ADJ PROF MICROBIOL & IMMUNOL, SCH MED, GEORGE WASHINGTON UNIV, 90- *Concurrent Pos:* Mem, Immunother Comt Rev Panel B, Nat Cancer Inst, 77-79 & Immunobiol Study Sect, NIH, 91-; res affil, Vet Admin Wadsworth Med Ctr, Geriat Res, Educ & Clin Ctr, 80-83; prin investr, USPHS, Dept Health & Human Serv, 81-89; chief, Lab Tumor Immunol, Geriat Res, Educ & Clin Ctr & supvry res biologist, Vet Admin W Los Angeles Med Ctr, 83-89; adj scientist, Cell Biol & Metab Br, Nat Inst Child Health & Human Develop, NIH, 88. *Mem:* AAAS; Sigma Xi; Transplantation Soc; Am Asn Immunologists; Am Asn Cancer Res; NY Acad Sci; Soc Exp Biol & Med; Geront Soc Am; Am Pub Health Asn; Soc Leukocyte Biol. *Mailing Add:* Div Cytokine Biol Ctr Biologics Eval & Res Food & Drug Admin Bldg 29A Rm 2D-16 8800 Rockville Pike Bethesda MD 20892

BLOOM, ELLIOTT D, b New York, NY, June 11, 40; m 62; c 2. PARTICLE PHYSICS. *Educ:* Pomona Col, BA, 62; Calif Inst Technol, PhD(physics), 67. *Prof Exp:* Res assoc physics, 67-70, asst prof, 70-74, assoc prof, 74-80, PROF PHYSICS, LINEAR ACCELERATOR CTR, STANFORD UNIV, 80- *Concurrent Pos:* Consult, Sci Appln, Inc, 75-82, Space Appln Inc, 85-86, Theta Corp, 85-87. *Honors & Awards:* Alexander Von Humboldt Sr Scientist Award, 82. *Mem:* Fel Am Phys Soc. *Res:* Study of heavy flavor production and decay, high energy electron positron colliding beams. *Mailing Add:* Stanford Linear Accelerator Ctr Stanford Univ Stanford CA 94305

BLOOM, EVERETT E, MECHANICAL ENGINEERING. *Educ:* SD Sch Mines & Technol, BS, 63; Univ Tenn, MS, 64, PhD(metall eng) 70. *Prof Exp:* Res staff mem, Mech Prop & Electron Micros Groups, 64-79, group leader, Radiation Effects & Microstruct Anal Group, 79-82, MGR MAT SCI SECT & FUSION MAT PROG, METALS & CERAMICS DIV, OAK RIDGE NAT LAB, 82- *Mem:* Fel Am Soc Metals Int; Am Nuclear Soc. *Res:* Mechanical and physical metallurgy of materials with emphasis on materials for fission and fusion reactor applications; radiation effects in materials; development of improved materials for reactor applications. *Mailing Add:* Metals & Ceramics Div Oak Ridge Nat Lab PO Box 2008 Bldg 4500-S Oak Ridge TN 37831

BLOOM, FLOYD ELLIOTT, b Minneapolis, Minn, Oct 8, 36; m; c 2. NEUROSCIENCES, CYTOCHEMISTRY. *Educ:* Southern Methodist Univ, AB, 56; Wash Univ, MD, 60. *Hon Degrees:* Dr, Southern Methodist Univ, 83, Hahnemann Univ, 85 & Univ Rochester, 85. *Prof Exp:* Intern & asst resident med, Barnes Hosp, St Louis, Mo, 60-62; NIMH res assoc, St Elizabeth's Hosp, Washington, DC, 62-64; USPHS spec fel anat, Sch Med, Yale Univ, 64-66, asst prof anat & pharmacol, 66-67, assoc prof pharmacol, 68; actg chief, Lab Neuropharmacol, Div Spec Ment Health Res Progs, NIMH, St Elizabeth's Hosp, 68-69 & chief, 69-75, chief, sect cytochem pharmacol, 68-75, actg div dir, 73-75; dir, Arthur V Davis Ctr Behav Neurobiol, Salk Inst, San Diego, Calif, 75-83; dir & mem, Div Preclin Neurosci & Endocrinol, 83-89, CHMN & MEM, DEPT NEUROPHARMACOL, SCRIPPS CLIN & RES FOUND, LA JOLLA, CALIF, 89- *Concurrent Pos:* Asst med, Sch Med, Wash Univ, 60-62; assoc physiol, Sch Med, George Washington Univ, 62-64; res career develop award, Nat Inst Neurol Dis & Blindness, 66-68; consult, Merck, Sharpe & Dohme, 76-82, Nova Pharmaceut, 82-, Comt Sci, Eng & Pub Policy, Nat Acad Sci, 82- & McNeil Pharmaceut, 83-; adj prof, depts neurosci & psychiat, Univ Calif, San Diego, 77, depts pharmacol & psychiat, Col Med, Univ Calif, Irvine, 82-; vis prof, Col France, 79; chmn, Neurobiol Sect, Nat Acad Sci, 79-83 & 85-86, & Comt Policies Allocation Resources Biomed Rest, Inst Med, 88-90; Wellcome vis prof, Royal Soc Med, 84; co-chmn, Gordon Conf Reptiles, 88. *Honors & Awards:* A E Bennett Award, Soc Biol Psychiat, 71; A Cressy Morrison Award, NY Acad Sci, 71; Arthur S Flemming Award, 72; Mathilde Solowey Award, 73; McAplin Award, Nat Asn Ment Health, 80; Harold Cummins Mem lectr, 81; Paul Lamson lectr, Vanderbilt Univ, 82; Marjorie Guthrie lectr, NIH, 84; Albert Dorfman Lectr, Univ Chicago, 87; George Klenk Lectr & Medal, Univ Cologne, 87; John D DeLuca Lectr, Nat Libr Med, 88; Erspamer Lectr, Georgetown Univ, 89; Janssen Award, 89; Pasarow Award, 89; Hermann von Helmholtz Award, 90. *Mem:* Nat Acad Sci; Inst Med-Nat Acad Sci; Soc Neurosci (secy, 73-75, pres, 76-77); Am Soc Cell Biol; Am Soc Pharmacol & Exp Therapeut; Sigma Xi; Am Asn Anatomists; fel Am Col Neuropsychopharmacol; Am Physiol Soc; fel AAAS. *Res:* Chemical control of neuronal activity; neuronal basis of behavior and drug action; author of various technical publications and books. *Mailing Add:* Dept Neuropharmacol Scripps Clin & Res Found La Jolla CA 92037

BLOOM, FRANK, PATHOLOGY. *Educ:* Cornell Univ, DVM, 30; Am Col Vet Pathologists, dipl, 49; Am Col Lab Animal Med, dipl, 61. *Prof Exp:* STAFF MEM, PAPANICOLAOU COMPREHENSIVE CANCER CTR, SCH MED, UNIV MIAMI, 79-, CLIN PROF PATH, 83- *Honors & Awards:* Practr Res Award, Am Vet Med Asn, 64. *Mem:* Vet Med Asn; Fedn Am Socs Exp Biol; Am Asn Pathologists; Am Vet Med Asn. *Res:* Veterinary clinical pathology; pathologic anatomy; experimental pathology. *Mailing Add:* Consult Pathologist Towers Oceanview 800 Parkview Dr Bldg 2 No 121 Hallandale FL 33009

BLOOM, JACK, food chemistry, for more information see previous edition

BLOOM, JAMES R, b Clearfield, Pa, Feb 20, 24; m 47; c 4. PLANT PATHOLOGY. *Educ:* Pa State Univ, BS, 50; Univ Wis, PhD, 53. *Prof Exp:* Asst plant path, Univ Wis, 50-53; from asst prof to prof, 53-87, EMER PROF, PLANT PATH, PA STATE UNIV, 87- *Mem:* Am Phytopath Soc; Soc Nematol. *Res:* Phytonematology and disease of vegetable crops; interactions between plant pathogens. *Mailing Add:* Dept Plant Path Buckhout Lab Pa State Univ University Park PA 16802

BLOOM, JEROME H(ERSHEL), b Boston, Mass, May 20, 24; m 53; c 2. ELECTRICAL ENGINEERING. *Educ:* Northeastern Univ, BS, 45. *Prof Exp:* Test engr, Sylvania Elec Prod, Inc, 45-46; electronic scientist & chief vacuum electronics sect, Air Force Cambridge Res Ctr, 46-63, phys scientist, 63-84; RETIRED. *Concurrent Pos:* Pvt consult vacuum deposition for semiconductor prod & res with thin films. *Mem:* AAAS; Inst Elec & Electronics Engrs; Am Vacuum Soc; Int Union Vacuum Sci Tech & Appln. *Res:* Vacuum tube materials and techniques; semi-conductor thin films; infrared sensors; electron beam lithography; ion implantation. *Mailing Add:* 37 Montclair Rd Waban MA 02168-1122

BLOOM, JOSEPH MORRIS, b Indianapolis, Ind, Apr 26, 37; m 63; c 2. FRACTURE & APPLIED MECHANICS. *Educ:* Purdue Univ, BS, 59; Harvard Univ, SM, 60, PhD(appl math), 64. *Prof Exp:* Asst elasticity, Harvard Univ, 62-64; mathematician, US Army Mat Res Agency, Watertown, 64-66; prin scientist, Aeronutronic Div, Philco-Ford, 66-68; scientist, Mat Sci Lab, Lockheed-Ga Co, 68-70, sr tech specialist, McDonnell-Douglas Corp, 70-72; RES SPECIALIST, RES CTR, BABCOCK & WILCOX CORP, 72- *Mem:* Am Soc Mech Engrs; Am Soc Testing & Mat; Sigma Xi. *Res:* Research and development in fracture mechanics of pressure vessel materials; elastic and plastic probability. *Mailing Add:* 908 22nd St NE Canton OH 44714

BLOOM, KERRY STEVEN, b Washington, DC, Dec 28, 53; m 83; c 1. CELL BIOLOGY, CHROMOSOME MECHANICS. *Educ:* Tulane Univ, BS, 75; Purdue Univ, PhD(molecular biol), 80. *Prof Exp:* Teaching fel, Jane Coffin Childs Mem Fund Med Res, Univ Calif, Santa Barbara, 80-82; asst prof, 82-85, ASSOC PROF BIOL, UNIV NC, CHAPEL HILL, 85- *Concurrent Pos:* Symp lectr ann meeting, Am Soc Cell Biologists, 83; instr physiol, Marine Biol Lab, Woods Hole, Mass, 85-89. *Mem:* Am Soc Cell Biologists; Sigma Xi. *Res:* Chromosome mechanics in yeast; DNA-protein interaction at the centromere; structural organization of eukaryotic chromosomes; microtubule associated proteins. *Mailing Add:* Dept Biol Univ NC Wilson Hall CB No 3280 Chapel Hill NC 27599-3280

BLOOM, L(OUIS) R(ICHARD), b Chicago, Ill, May 9, 14; m 42; c 4. TELECOMMUNICATION SCIENCES. *Educ:* Univ Ill, BS, 38, MS, 41, EE, 50. *Prof Exp:* Asst physics, Univ Ill, 36-39, asst elec eng, 40-43, asst prof, 46-53; res engr, Radio Corp Am, 43-46; head microwave tube sect, Sylvania Res Labs, 53-58, mgr phys electronics lab, 58-60, sci adv to dir res, 61-63; prog mgr mat sci & electronic devices, Gen Tel & Electronics Lab, 64-67, mgr phys electronics & electronic mat, 67-69, dir phys electronics lab & assoc dir res, 69-73; assoc div chief, Electromagnetic Div, Dept Commerce, Boulder, 73-75; GROUP CHIEF NEW TECHNOL, NAT TELECOMMUN & INFO ADMIN, INST TELECOMMUN SCI, BOULDER, 75- *Mem:* Fel AAAS; Am Phys Soc; fel Inst Elec & Electronics Engrs; Sigma Xi; Optical Soc Am. *Res:* Microwaves; laser devices; electro-optical behavior of materials; optical communications; microwaves; electron beam devices; telecommunications; interactive information systems; fiber optical communication systems. *Mailing Add:* 2240 Linden Ave Boulder CO 80302

BLOOM, MIRIAM, b New York, NY, Dec 25, 34; m 60; c 2. GENETIC TOXICOLOGY, ENVIRONMENTAL MUTAGENESIS. *Educ:* Brooklyn Col, BS, 57; Univ Utah, PhD(genetics), 72. *Prof Exp:* Instr bot & hort, Brooklyn Bot Gardens, 63-66; vis scholar chem, Univ Calif, Los Angeles, 72; asst res prof biol, Univ Utah, 73; fel, Univ Southern Fla, 74, res assoc, 75; adj prof, Hillsborough Community Col, 75-77; USPHS res training awardee, Col Med, Georgetown Univ, 77-79; GENETICIST, US CONSUMER PROD SAFETY COMN, 79- *Concurrent Pos:* Mem, Interagency Regulatory Liaison Group, 79-81; liaison, HHS Subcomt Environ Mutagenesis, Comt Coord Environ Related Probs, 80-86; appointee, Interagency Radiation Res Comt, 80-82; ed, Environ Mutagen Soc Newsletter, 85-; counr, Environ Mutagen Soc, 87-, & Fed Asbestos Task Force, 88- *Mem:* AAAS; Genetic Toxicol Asn; Environ Mutagen Soc; Asn Women Sci; Sigma Xi. *Res:* Biochemical genetics of floral flavonoid patterning; pattern evolution; genetic toxicology of environmental chemicals. *Mailing Add:* 4433 Wedgewood St Jackson MI 39211

BLOOM, MYER, b Montreal, Que, Dec 7, 28; m 54. PHYSICS. *Educ:* McGill Univ, BSc, 49, MSc, 50; Univ Ill, PhD(physics), 54. *Prof Exp:* Nat Res Coun Can traveling fel, Kamerlingh Onnes Lab, Univ Leiden, Holland, 54-56; res assoc physics, 56-57, from asst prof to assoc prof, 57-63, PROF PHYSICS, UNIV BC, 63- *Concurrent Pos:* Alfred P Sloan fel, 61-65; Guggenheim fel, 64-65. *Honors & Awards:* Steacie Prize, 67; Gold Medal, Can Asn Physicists, 73; Izaak Walton Killam Mem Scholar, 78; Jacob Biely Prize, 69. *Mem:* Am Phys Soc; Can Asn Physicists; Royal Soc Can. *Res:* Nuclear magnetic resonance; spin relaxation; biological membranes. *Mailing Add:* Dept Physics Univ BC Vancouver BC V6T 1W5 Can

BLOOM, SANFORD GILBERT, b Columbus, Ohio, Apr 11, 37; m 60, 79; c 1. CHEMICAL ENGINEERING. *Educ:* Ohio State Univ, BChE & MSc, 59, PhD(chem eng), 64. *Prof Exp:* Sr res engr, Petrol Prod, Atlantic Richfield Co, 64-66; assoc fel, Ecol & Environ Systs Div, Battelle Mem Inst, 66-77, prin scientist, Columbus Labs, 77-83, staff chem engr, Battelle Proj Mgt Div, 83-87; ENG SPECIALIST, MARTIN MARIETTA, 87- *Mem:* AAAS; Am Inst Chem Engrs; Soc Petrol Engrs. *Res:* Mathematical analysis of engineering problems of heat and mass transfer; atmospheric dispersion; fluid flow in porous media; hydraulic fracturing; ecosystem modeling; mathematical analysis of biological systems. *Mailing Add:* 7104 Delbourne Ave Knoxville TN 37919

BLOOM, SHERMAN, b Brooklyn, NY, Jan 26, 34; m 60; c 2. CARDIOVASCULAR PATHOLOGY, CELL BIOLOGY. *Educ:* NY Univ, AB, 55, MD, 60. *Prof Exp:* Res trainee exp path, NY Univ, 61-65, instr path, 65-66; from asst prof to assoc prof, Univ Utah, 66-73; from assoc prof to prof path, Univ South Fla, 73-77, actg chmn path, 77; PROF PATH, GEORGE WASHINGTON UNIV, 77- *Concurrent Pos:* Vis assoc prof physiol, Med Ctr, Univ Calif, Los Angeles, 72; consult, Nat Heart & Lung Inst, 74 & Drug Toxicol Br, Food & Drug Admin, 78-; bd dir, Scientists Ctr Animal Welfare, 81-; bd dir, Am Col Nutrit, 87- *Mem:* NY Acad Sci; Int Study Group Res Cardiac Metab; Am Asn Path; Am Physiol Soc; Soc Cardiovasc Path (pres, 86-87). *Res:* Determinants of myocardial necrosis; role of dietary metals in cardiovascular resistance to stress; control of myocardial contractility; ultrastructure of myocardial cells and interstitium; role of cell structure in passive mechanical properties; cardiac toxicology. *Mailing Add:* 7604 Hemlock St Bethesda MD 20817

BLOOM, STANLEY, b Plainfield, NJ, Oct 18, 24; m 48; c 4. ELECTRONICS, ELECTRON OPTICS. *Educ:* Rutgers Univ, BS, 48; Yale Univ, MS, 51, PhD(physics), 52. *Prof Exp:* Sr mem staff, RCA Labs, 52-87; RETIRED. *Concurrent Pos:* RCA Corp res fel, Cambridge Univ, 66-67. *Mem:* Inst Elec & Electronics Engrs; Am Phys Soc. *Res:* Microwave electronics; plasma physics; noise; quantum electronics; solid state; electron optics. *Mailing Add:* 1185 Sherlin Dr Bridgewater NJ 08807

BLOOM, STEPHEN ALLEN, b Tucson, Ariz, Mar 9, 47; m 70. QUANTITATIVE MACROBENTHIC ECOLOGY. *Educ:* Univ SFla, BA, 69; Univ Wash, PhD(zool), 74. *Prof Exp:* Asst, Dept Biol, Univ SFla, 74-75, vis asst prof marine ecol, 75-76; ASST PROF MARINE ECOL & INVERT ZOOL, DEPT ZOOL, UNIV FLA, 76-; CHIEF SCIENTIST, ECOL DATA CONSULT INC, 79- *Concurrent Pos:* Vis instr marine ecol, Duke Marine Lab, Beaufort, NC, 76; adj sr res scientist, Mote Marine Labs, Sarosota, Fla, 82- *Mem:* Ecol Soc Am; Sigma Xi; AAAS; Am Zool Soc. *Res:* Investigation of community structure, population dynamics and autecology of marine macroinfauna by descriptive and experimental approaches; development of analytical techniques for community ecology. *Mailing Add:* PO Box 760 Archer FL 32618

BLOOM, STEPHEN EARL, b Brooklyn, NY, Oct 18, 41; m 63; c 2. TOXICOLOGY, CYTOGENETICS. *Educ:* Long Island Univ, BS, 63; Pa State Univ, MS, 65, PhD(genetics), 68. *Prof Exp:* Asst genetics, Pa State Univ, 63-68; from asst prof to assoc prof, 68-81, PROF CYTOGENETICS, CORNELL UNIV, 81- *Concurrent Pos:* Vis prof biol, Univ Tex M D Anderson Hosp & Tumor Inst Houston, 75; vis staff scientist, Worcester Found Exp Biol, 81. *Honors & Awards:* Merck Award Sci Achievement, 89. *Mem:* AAAS; Am Genetic Asn; Genetics Soc Am; Poultry Sci Asn; Am Inst Biol Sci; Environ Mutagen Soc. *Res:* Gene mapping in avian species; gene-dosage in growth and development; genetic toxicology; avian cytology and cytogenetics; chromosome aberrations and neoplasia. *Mailing Add:* Dept Poultry & Avian Sci Cornell Univ Ithaca NY 14853

BLOOM, STEWART DAVE, b Chicago, Ill, Aug 22, 23; m 49; c 3. NUCLEAR PHYSICS. *Educ:* Univ Chicago, PhD(physics), 52. *Prof Exp:* Assoc physicist, Brookhaven Nat Labs, 52-56; Guggenheim res fel, Cambridge Univ, 56-57; staff physicist, 57-69, PROF APPL SCI, LAWRENCE RADIATION LAB, 69- *Concurrent Pos:* Fulbright fel, France, 63. *Mem:* Fel Am Phys Soc. *Res:* Beta ray; nuclear reactions. *Mailing Add:* Lawrence Radiation Lab Livermore CA 94550

BLOOM, VICTOR, b New York, NY, Aug 17, 31; m 73; c 5. PSYCHIATRY. *Educ:* Univ Mich, BS, MD, 57; Am Bd Psychiat & Neurol, dipl & cert psychiat, 63. *Prof Exp:* Intern med, Sinai Hosp Detroit, 57-58; resident psychiat, Lafayette Clin, 58-61, staff psychiatrist, 61-67, chief adult inpatient serv, 68-72; CLIN ASSOC PROF PSYCHIAT, SCH MED, WAYNE STATE UNIV, 72- *Concurrent Pos:* Instr, Sch Med, Wayne State Univ, 61-63, asst prof, 64-72; teacher, Lafayette Clin, 61-; lectr psychiat, Mich State Dept Ment Health, 61-69; consult group psychother, Neuropsychiat Inst, Med Sch, Univ Mich, 68-69 & Kidney Ctr, Mt Carmel Mercy Hosp, 68-82; consult group ther, Harper Hosp Sch Nursing, 72-76. *Mem:* Fel Am Psychiat Asn; fel Am Group Psychother Asn; hon assoc mem Inst Bioenergetic Anal Asn; psychiat assoc, Am Acad Psychoanal. *Res:* Psychoanalysis; psychotherapy; group psychotherapy; suicide prevention; schizophrenia; milieu therapy; bioenergetic analysis; bioenergetics in group therapy. *Mailing Add:* 1007 Three Mile Dr Grosse Pointe Park MI 48230

BLOOMBERG, WILFRED, b Pittsburgh, Pa, Mar 25, 05; m 35, 78. PSYCHIATRY, NEUROLOGY. *Educ:* Harvard Univ, SB, 24, MD, 28. *Prof Exp:* Chief neuro-psychiat serv, Cushing Vet Admin Hosp, Framingham, 46-52; chief neuro-psychiat serv, Vet Admin Hosp, Boston, 52-58; assoc prof psychiat & neurol, Sch Med, Boston Univ, 58; assoc clin prof psychiat & pub health, Sch Med, Yale Univ, 58-69; ment health adminr, Region VI, Commonwealth of Mass, 69-71, dep comnr, Dept Ment Health, 71-75; RETIRED. *Concurrent Pos:* Vis neurologist, Boston City Hosp, 34-49; hon consult, Surgeon Gen, US Dept Army; dir ment health, Southern Regional Educ Bd, Ga, 57-58; comnr ment health, State of Conn, 58-69; lectr, Harvard Med Sch, 70-; consult, Mass Gen Hosp, 70- *Mem:* Am Psychiat Asn; AMA; Am Neurol Asn; Am Acad Neurol. *Mailing Add:* 16 Farrar St Cambridge MA 02138

BLOOMER, JAMES L, b Knoxville, Tenn, Jan 6, 39. ORGANIC CHEMISTRY. *Educ:* Univ Tenn, BS, 59, MS, 61; Univ London, PhD(org chem), 64. *Prof Exp:* NSF fel org chem, Harvard Univ, 64-65; asst prof, 65-70, ASSOC PROF ORG CHEM, TEMPLE UNIV, 70- *Concurrent Pos:* NSF res grant, 66-68; NIH res grant, 69-71. *Mem:* Am Chem Soc; Brit Chem Soc. *Res:* Synthetic organic chemistry. *Mailing Add:* Dept Chem 016-00 Temple Univ Philadelphia PA 19122

BLOOMER, OSCAR T(HEODORE), b St Joseph, Mo, Aug 21, 23; m 48; c 3. CHEMICAL ENGINEERING. *Educ:* Univ Kans, BS, 44; Ill Inst Technol, MS, 51, PhD(chem eng), 54. *Prof Exp:* Jr engr, chem eng design, Eastman Kodak Co, 44-48; sr res engr, gas technol, Inst Gas Technol, 48-53; from sr develop engr to sect supvr, coating & drying of photog films, 53-69, asst dir, 69-71, DIR MFG TECH DIV, EASTMAN KODAK CO, 71- *Mem:* AAAS; Am Chem Soc. *Res:* Process development; coating and drying of photographic films and papers; pressure-volume-temperature relationships; thermodynamic properties; phase equilibrium of gases at low temperatures. *Mailing Add:* 83 Coronado Dr Rochester NY 14617

BLOOMFIELD, DANIEL KERMIT, b Cleveland, Ohio, Dec 14, 26; m 55; c 3. MEDICINE, BIOCHEMISTRY. *Educ:* US Naval Acad, BS, 47; Western Reserve Univ, MS & MD, 54. *Prof Exp:* From intern to asst resident, Beth Israel Hosp, 54-56; resident, Mass Gen Hosp, 56-57; fel biochem, Harvard Univ, 57-59; hon asst registr, Inst Cardiol, 59-60; asst med, Western Reserve Univ, 60-64; dir cardiovasc res, Community Health Found, 64-66; assoc med, Mt Sinai Hosp, 66-69; dean, Sch Clin Med, 74-78, dean, Col Med, 78-84, PROF MED, UNIV ILL, URBANA-CHAMPAIGN, 70- *Mem:* Fel Am Col Physicians; Am Fedn Clin Res; Sigma Xi. *Res:* Biosynthesis of unsaturated fatty acids; cardiac arrhythmias; gas chromatography of steroids; cholesterol metabolism; congenital heart disease; hypertension anticoagulant therapy; mass cardiac screening; medical education. *Mailing Add:* 103 E Michigan Urbana IL 61801

BLOOMFIELD, DAVID PETER, b New York, NY, Mar 27, 38; m 58; c 2. FUEL CELLS, ELECTROCHEMICAL ENGINEERING. *Educ:* Polytech Inst NY, BME, 61; Rensselaer Polytech Inst, MS, 70. *Prof Exp:* Asst proj engr, Power Systs Div, United Technologies Corp, 65-77; sr proj engr, Giner, Inc, 77-78; vpres systs div, Phys Sci Inc, 78-84; PRES, ANAL POWER CORP, 84- *Mem:* Am Soc Mech Engrs; Am Chem Soc; Electrochem Soc. *Res:* Power plant and chemical process systems analysis; electrochemical and heat transfer design and analysis; microcomputer software development. *Mailing Add:* Analytic Power Corp PO Box 1189 Boston MA 02117

BLOOMFIELD, DENNIS ALEXANDER, b Perth, Western Australia, May 4, 33; m 64; c 3. CARDIOLOGY, INTERNAL MEDICINE. *Educ:* Univ Adelaide, MB, BS, 56. *Prof Exp:* Australian Heart Found traveling fel cardiol, Vanderbilt Univ Hosp, 62-65; fel, Inst Cardiol, London, Eng, 65-66; dir, cardiol cath lab, Maimonides Med Ctr, 67-75; DIR MED & CHIEF CARDIOL CATH LAB, ST VINCENTS MED CTR, STATEN ISLAND, 75-; PROF CLIN MED, NY MED COL, 79- *Mem:* Fel Royal Col Physicians; fel Am Col Chest Physicians; fel Am Col Physicians; fel Am Col Cardiol; fel Australasian Col Physicians. *Res:* Application of indicator-dilution techniques to the measurement of regurgitant and regional blood flows; author or coauthor of over 50 publications. *Mailing Add:* 355 Bard Ave Staten Island NY 10310

BLOOMFIELD, HAROLD H, b New York, NY, Oct 8, 44; m 82; c 1. TRANSCENDENTAL MEDITATION, HOLISTIC MEDICINE. *Educ:* Univ Pittsburg, BS, 65; State Univ NY, MD, 69. *Prof Exp:* Intern, Kaiser Hosp, San Francisco, 70; resident & post doc fel, Dept Psychiat, Yale Univ Sch Med, 73; dir, Inst Psychophysiol Med, San Diego, 75-80; DIR PSYCHIAT, N COUNTY HEALTH CTR, DEL MAR, 80- *Honors & Awards:* David Berger Award, 78; Golden Apple Award, 82. *Mem:* Am Psychiat Asn; Asn Humanistic Psychol; Am Holistic Med Asn; Asn Holistic Health. *Res:* Medical & psychiatric benefits of meditation and holistic health; family therapy; group behavior therapy with out-patient chronic schizophrenics. *Mailing Add:* 1011 Camino del Mar Suite 234 Del Mar CA 92014

BLOOMFIELD, JORDAN JAY, b South Bend, Ind, Feb 25, 30; m 60; c 3. ORGANIC CHEMISTRY. *Educ:* Univ Calif, Los Angeles, BS, 52; Mass Inst Technol, PhD(org chem), 58. *Prof Exp:* Asst chem, Mass Inst Technol, 52-53, res, 55-57; instr, Univ Tex, 57-60; res assoc, Univ Ill, 60-61; res assoc, Univ Ariz, 61-62; asst prof, Univ Okla, 62-66 & assoc prof on leave, 66-67; sr res specialist, 67-81, fel, Monsanto Co, 81-85; vis prof, Dept Chem, Univ Mich, 85-86; VPRES RES, ORGANIC CONSULTS INC, 87- *Concurrent Pos:* Vis prof & sr res group leader, 66-67; vis assoc prof, Univ Mo-St Louis, 75, adj prof, 78-87. *Honors & Awards:* St Louis Award, Am Chem Soc, 80. *Mem:* AAAS; Am Chem Soc; Royal Soc Chem; Am Soc Photobiol. *Res:* Reaction mechanisms; transannular reactions; molecular rearrangements; small rings; bicyclic and polycyclic systems of unusual structure; synthetic photochemistry. *Mailing Add:* 88587 Ellmaker Rd Veneta OR 97487

BLOOMFIELD, LOUIS AUB, b Boston, Mass, Oct 11, 56; m 83; c 2. ATOMIC & MOLECULAR PHYSICS. *Educ:* Amherst Col, BA, 79; Stanford Univ, PhD(physics), 83. *Prof Exp:* Fel physics, Stanford Univ, 83; fel physics, AT&T Bell Labs, 83-85; ASST PROF PHYSICS, UNIV VA, 85- *Concurrent Pos:* Sloan fel, 89. *Honors & Awards:* Apker Award, Am Phys Soc; Young Investr Award, 88. *Mem:* Am Phys Soc; Sigma Xi; Optical Soc Am. *Res:* Clusters of atoms and the origins of solid state behavior; atomic and molecular physics; laser physics and nonlinear optics. *Mailing Add:* Dept Physics Univ Va Charlottesville VA 22901

BLOOMFIELD, PHILIP EARL, b Erie, Pa, Aug 28, 34; m 58; c 2. SOLID STATE PHYSICS, THEORETICAL PHYSICS. *Educ:* Univ Chicago, BA, 56, BS, 57, MS, 59, PhD(physics), 65. *Prof Exp:* Technician, Nat Bur Standards, Washington, DC, 56-57; physicist, Lab Appl Sci, Univ Chicago, 58-60; res asst solid state physics, Inst Study Metals, 60-64; asst prof physics, Univ Pa, 64-69; assoc prof, Drexel Univ, 69-70 & City Col New York, 70-73; physicist, Nat Bur Standards, Gaithersburg, Md, 74-77; SR RES SCIENTIST, PENNWALT TECHNOL CTR, PA, 77- *Concurrent Pos:* Instr, Univ Chicago Col, 58-59; res assoc, John Crerar Libr, Ill, 57; lectr, Ill Inst Technol, 62; ed, Accelerated Instr Methods, Ill, 62-63; consult, Pittman Dunn Labs, Frankford Arsenal, Pa, 67-74; adj prof, Drexel Univ, 83- *Honors & Awards:* Army Res & Develop Award, 76. *Mem:* Am Phys Soc; Acoust Soc Am; Inst Elec & Electronics Engrs. *Res:* Thermodynamics; statistical mechanics; electromagnetic phenomena in solids; ferroelectrics; magnetic materials; general relativity; optical properties of metals; superconductivity; electrical properties of polymers. *Mailing Add:* Mat Eng Dept Drexel Univ Philadelphia PA 19104

BLOOMFIELD, SAUL, b Montreal, Que, June 30, 25; m 49; c 3. CLINICAL PHARMACOLOGY. *Educ:* Univ Geneva, MD, 53. *Prof Exp:* PROF MED & PHARMACOL, UNIV CINCINNATI, 77- *Concurrent Pos:* Consult drug eval, AMA, 74-; vis prof & prin investr, Oxford Univ Churchill Hosp, 80; mem orphans prod develop, initial rev group & arthritis adv comt liason, 83- *Mem:* Am Soc Clin Pharmacol & Therapeut; Am Fedn Clin Res; NY Acad Sci. *Res:* Clinical pharmacology and clinical testing of new analgesic agents for efficacy and safety; methodology of clinical trials of investigational new drugs, phases 1,2 &3. *Mailing Add:* Dept Internal Med Univ Cincinnati Med Ctr Div Clin Pharmacol (K-4) ML 578 Cincinnati OH 45267-0578

BLOOMFIELD, VICTOR ALFRED, b Newark, NJ, June 10, 38; div. BIOPHYSICAL CHEMISTRY. *Educ:* Univ Calif, Berkeley, BS, 59; Univ Wis, PhD(phys chem), 62. *Prof Exp:* NSF fel phys chem, Univ Calif, San Diego, 62-64; from asst prof to assoc prof, Univ Ill, 64-70; assoc prof, 70-74, prof biochem & chem, 74-79, PROF & HEAD, BIOCHEM DEPT, UNIV MINN, ST PAUL, 79- *Concurrent Pos:* Alfred P Sloan Found fel, 69-71; NIH career develop award, 71-76; cochmn, Gordon Res Conf Physics & Phys Chem of Biopolymers, 74; mem NIH study sect BBCB, 75-78; mem biophys panel, NSF, 80; Fogarty scholar, NIH, 85-86. *Mem:* Fel AAAS; Am Chem Soc (secy, Biol Chem Div, 81-83); Am Soc Biol Chem; Biophys Soc (pres, 87); Sigma Xi. *Res:* Physical chemistry of biological macromolecules, viruses and other high polymers; biological hydrodynamics; quasi-elastic laser light scattering; scanning tunneling microscopy. *Mailing Add:* Dept Biochem Col Biol Sci Univ Minn 140 Gortner Lab St Paul MN 55108

BLOOMQUIST, EUNICE, b Worcester, Mass, Sept 9, 40; m 70; c 2. PHYSIOLOGY. *Educ:* Simmons Col, BS, 63; Boston Univ, PhD(physiol), 68. *Prof Exp:* From instr to assoc prof physiol, 69-76, actg chair, 80-84, PROF PHYSIOL, SCH MED, TUFTS UNIV, 84- *Mem:* Am Physiol Soc; Soc Gen Physiologists; NY Acad Sci. *Res:* Airway smooth muscle and inflammation; neuropeptides. *Mailing Add:* Dept Physiol Tufts Univ Med Sch 136 Harrison Ave Boston MA 02111

BLOOMSBURG, GEORGE L, b Salmon, Idaho, Mar 28, 31; m 55; c 3. AGRICULTURAL & CIVIL ENGINEERING. *Educ:* Univ Idaho, BS, 57, MS, 59; Colo State Univ, PhD(civil eng), 64. *Prof Exp:* Agr engr, agr res serv, US Dept Agr, 58-59; hydraul engr, Wash State Univ, 60-61; asst res prof hydraul, 61-62, assoc prof irrig, 64-68, prof agr eng & eng sci & chmn dept eng sci, 68-76, eng res coordr, 76-79, PROF AGR ENG & ENG SCI, UNIV IDAHO, 79-; DIR, IDAHO WATER RESOURCES RES INST, 84- *Mem:* Am Soc Agr Engrs; Nat Soc Prof Engrs; Nat Water Well Asn. *Res:* Fluid mechanics; hydrology; irrigation and drainage. *Mailing Add:* Dept Agr Eng Univ Idaho Moscow ID 83843

BLOOR, COLIN MERCER, b Sandusky, Ohio, May 26, 33; m 59; c 3. PATHOLOGY, PHYSIOLOGY. *Educ:* Denison Univ, BS, 55; Yale Univ, MD, 60. *Prof Exp:* From intern to resident path, Yale Univ, 60-62; from asst prof to assoc prof, 68-74, co-chmn dept, 76-78, PROF PATH, UNIV CALIF, SAN DIEGO, 74-, DIR, MOLECULAR PATH GRAD PROG, 89- *Concurrent Pos:* USPHS trainee path, 60-62; Life Ins Med Res Fund fel physiol, Nuffield Inst Med Res, 62-63 & fel path, Yale Univ, 63-64. *Mem:* Am Physiol Soc; Am Asn Pathologists & Bacteriologists; Am Soc Exp Path; Am Col Cardiol; Int Acad Pathologists. *Res:* Cardiovascular physiology and pathology, particularly the coronary circulation and coronary artery disease; effects of exercise on the development of coronary collateral circulation; cardiovascular genetics; exercise physiology. *Mailing Add:* Dept Path Sch Med 0612 Univ Calif at San Diego La Jolla CA 92093-0612

BLOOR, JOHN E, b Stoke on Trent, Eng, Aug 31, 29; m 61; c 2. CHEMISTRY. *Educ:* Oxford Univ, BA, 52; Univ Manchester, PhD(chem), 59. *Prof Exp:* Res chemist, Clayton Aniline Co, Ciba Corp, 52-55; instr org chem, Univ Manchester, 55-57; assoc res chemist, BC Res Coun, Can, 60-63; sr res chemist, Franklin Inst Labs, 64-67; prof chem, Univ Va, 67-69; ASSOC PROF CHEM, UNIV TENN, KNOXVILLE, 69- *Mem:* Am Chem Soc. *Res:* Molecular and far infrared spectroscopy; organic metallic compounds; organic reactivity; quantum chemistry; application of computers to chemical problems; quantum biochemistry and theory of drug action. *Mailing Add:* Dept Chem Univ Tenn Knoxville TN 37916

BLOOR, W(ILLIAM) SPENCER, b Trenton, NJ, Oct 16, 18; c 2. AUTOMATIC CONTROL SYSTEMS. *Educ:* Lafayette Col, Pa, BS, 40. *Hon Degrees:* DrEng, Lafayette, 81. *Prof Exp:* Elec engr, Leeds & Northrop Co, 40-55, var mgt positions, 55-75, mgr steam & nuclear power systs, 75-81; CONSULT ENGR, 81- *Concurrent Pos:* Adj prof, Drexel Univ, 81-87; consult, Univ Pa, 82-, St Hudson Int, Macro Corp, Network Systs Develop Ed, ISA Transactions. *Mem:* Nat Acad Eng; Instrument Soc Am (pres, 74); Inst Elec & Electronics Engrs. *Mailing Add:* 1904 Jody Rd Meadowbrook PA 19046

BLOSS, FRED DONALD, b Chicago, Ill, May 30, 20; m 46; c 3. MINERALOGY. *Educ:* Univ Chicago, BS, 47, MS, 49, PhD(geol), 51. *Prof Exp:* Asst, Univ Chicago, 47-50; instr phys sci, Univ Ill, 50-51; from asst prof to assoc prof geol & mineral, Univ Tenn, 51-57; assoc prof geol, Southern Ill Univ, 57-61; prof, 61-72, ALUMNI DISTINGUISHED PROF MINERAL, VA POLYTECH INST & STATE UNIV, 72- *Concurrent Pos:* Consult, Tenn Valley Authority, 55-; Geol Soc Am Penrose Fund grant-in-aid, 57-58; NSF sr fel, Cavendish Lab, Cambridge & Swiss Fed Inst Technol, 62-63; ed, The Am Mineralogist, 72-76; Caswell Silver distinguished vis prof geol, Univ NMex, 81-82. *Mem:* Fel Mineral Soc Am (vpres, 75-76, pres, 76-77); Mineral Asn Can; Mineral Soc Gt Brit & Ireland. *Res:* Crystallography. *Mailing Add:* Dept Geol Sci Va Polytech Inst & State Univ Blacksburg VA 24061

BLOSS, HOMER EARL, b Cumberland, Md, Dec 3, 33. PLANT PATHOLOGY, PLANT BIOCHEMISTRY. *Educ:* Univ Md, BS, 59; Univ Del, MS, 61; Univ Ariz, PhD(plant path), 65. *Prof Exp:* Res technician plant physiol, USDA, Md, 57-59; plant pathologist, Univ Ariz, 63-65, asst prof, 66-70, assoc prof plant path, 70-; RETIRED. *Mem:* Am Phytopath Soc; Am Mycol Soc; British Mycol Soc; fel Am Inst Chemists. *Res:* Physiology of fungi and fungus diseases of plants; biochemistry of disease resistance in plants and mycorrhizae for plant productivity. *Mailing Add:* 3431 N Grannen Tucson AZ 85745

BLOSSER, HENRY GABRIEL, b Harrisonburg, Va, Mar 16, 28; m 51, 73; c 4. NUCLEAR PHYSICS. *Educ:* Univ Va, BA, 51, MS, 52, PhD(physics), 54. *Prof Exp:* Physicist, Cyclotron Nuclear Res Group, Oak Ridge Nat Lab, 54-56, group leader, Cyclotron Analogue Group, 56-58; from assoc prof to prof, 58-90, dir, Nat Superconducting Cyclotron Lab, 58-89, UNIV DIST PROF PHYSICS, MICH STATE UNIV, 90- *Concurrent Pos:* NSF sr fel, 66-67; Guggenheim fel, John Simon Guggenheim Found, 73-74. *Mem:* Fel Am Phys Soc. *Res:* Design and construction of sector focused cyclotrons of medium and high energy; nuclear reaction studies at medium energies. *Mailing Add:* Cyclotron Lab Mich State Univ East Lansing MI 48824-1321

BLOSSER, JAMES CARLISLE, b Goshen, Ind, July 5, 45; m 74; c 2. RECEPTOR PHARMACOLOGY. *Educ:* Ind Univ, BA, 67; Mich State Univ, PhD(biochem), 72. *Prof Exp:* Lt res & neurobiol, Armed Forces Radiobiol Res Inst, Bethesda, MD, 72-76; res assoc biochem, Duke Univ Med Ctr, 76-77; res assoc neurosci, Baylor Col Med, 77-79, asst prof, 79-82; prin investr pharmacol, Pennwalt Pharmaceut, 82-88; SR PRIN INVESTR BIOCHEM, FISONS PHARMACEUT, 88- *Mem:* Am Soc Pharmacol & Exp Therapeut; Neurosci Soc. *Res:* Identification of sites for therapeutic intervention and discovery of drugs for the treatment of central nervous system diseases. *Mailing Add:* Dept Biol Fisons Pharmaceut 755 Jefferson Rd Rochester NY 14623

BLOSSER, TIMOTHY HOBERT, b Nappanee, Ind, Jan 7, 20; m 45; c 3. DAIRY SCIENCE, ANIMAL SCIENCE. *Educ:* Purdue Univ, BSA, 41; Univ Wis, MS, 47, PhD, 49. *Prof Exp:* Asst, Univ Wis, 41-42, 46-48; asst prof, Wash State Univ, 48-52, assoc prof dairy sci & assoc dairy scientist, 52-58, prof dairy sci & dairy scientist, 58-74, chmn dept dairy sci, 61-64, chmn dept animal sci, 64-74; staff scientist, Nat Prog Staff, 74-78, prog coordr, Prog Planning Staff, 78-81, prog coordr, Prog Planning & Anal Group, Nat Prog Staff, 81-82, res animal scientist, Ruminant Nutrit Lab, Beltsville Agr Res Ctr, Agr Res Serv, USDA, 82-87 & Leader, Joint Coun Reports Staff, Off Grants & Prog Systs, 84-87; RETIRED. *Mem:* Am Dairy Sci Asn (pres, 79, treas, 81-84); fel, Am Soc Animal Sci. *Res:* Parturient paresis and ketosis in dairy cows; nutritive value of dehydrated forages for ruminants; physiological changes associated with parturition in dairy cows; factors affecting forage quality; forage quality prediction with near infrared reflectance spectroscopy. *Mailing Add:* Rm 10 Bldg 005 NRS USDA Barc W Beltsville MD 20705

BLOSSEY, ERICH CARL, b Toledo, Ohio, June 10, 35; m 79; c 3. BIOORGANIC CHEMISTRY. *Educ:* Ohio State Univ, BS, 57; Iowa State Univ, MS, 59; Carnegie-Mellon Univ, PhD(chem), 63. *Prof Exp:* NIH fel chem, Stanford Univ, 62-63; res fel org chem, Syntex, S A, Mex, 63-64; teaching intern, Wabash Col, 64-65; PROF CHEM, ROLLINS COL, 65- *Concurrent Pos:* Vis prof, Univ NMex, 72-73; Arthur Vining Davis Found Fel, 78; A G Bush prof sci, 80-87; vis prof, Okla State Univ, 85-86. *Mem:* Am Chem Soc; Royal Soc Chem; AAAS; NY Acad Sci. *Res:* Application of polymer-supported reagents to problems in organic and bioorganic chemistry; use of 13C NMR spectroscopy in polymeric reagent structures. *Mailing Add:* Dept Chem Rollins Col Winter Park FL 32789

BLOSTEIN, MAIER LIONEL, b Montreal, Que, Oct 21, 32; m 57; c 2. ELECTRONICS ENGINEERING. *Educ:* McGill Univ, BEng, 54, MEng, 59; Univ Ill, PhD(elec eng), 63. *Prof Exp:* Electronic engr, Bell Tel Co Can, 54-55 & Can Marconi Co, 55-58; demonstr elec eng, McGill Univ, 58-59, lectr, 59-60; asst, Univ Ill, 60-61, instr, 61-63; from asst prof to assoc prof, 63-77, PROF ELEC ENG, McGILL UNIV, 77- *Mem:* Inst Elec & Electronics Engrs. *Res:* Solid state devices and circuits; circuit and system theory. *Mailing Add:* Dept Elec Eng McGill Univ Montreal PQ H3A 2T5 Can

BLOSTEIN, RHODA, b Montreal, Que, Nov 4, 36; m 57; c 1. BIOCHEMISTRY. *Educ:* McGill Univ, BS, 56, MS, 57, PhD(biochem), 60. *Prof Exp:* Res assoc biochem, Univ Ill, 60-63; asst prof, McGill Univ, 63-73, assoc prof med, 71-, assoc prof med, 73-; PROF, DEPT MED, MONTREAL GEN HOSP RES INST, McGILL UNIV,. *Mem:* Biophys Soc; Am Biochem Soc; Can Biochem Soc. *Res:* Membrane biochemistry in relation to cellular functions. *Mailing Add:* Montreal Gen Hosp Res Inst Dept Med & Biochem McGill Univ 1650 Cedar Ave Rm 3823 L H Montreal PQ H3G 1A4 Can

BLOT, WILLIAM JAMES, b New York, NY, Aug 30, 43; m 65; c 2. BIOSTATISTICS, EPIDEMIOLOGY. *Educ:* Univ Fla, BS, 64, MS, 66; Fla State Univ, PhD(statist), 70. *Prof Exp:* Assoc sci researcher comput prog, Oak Ridge Nat Lab, 66-67; statistician med res, Atomic Bomb Casualty Comn, 70-72; asst prof, Johns Hopkins Sch Hyg & Pub Health, 72-74; BIOSTATISTICIAN, RES EPIDEMIOL BR, NAT CANCER INST, 74- *Mem:* Am Pub Health Asn; Am Statist Asn. *Res:* Study of the patterns and determinants, particularly environmental, of cancer in the United States and abroad. *Mailing Add:* 6130 Executive Blvd Suite 431 Rockville MD 20892

BLOTCKY, ALAN JAY, b Omaha, Nebr, July 5, 30; m 53; c 2. APPLICATION OF RESEARCH REACTOR TO MEDICAL RESEARCH, MANAGEMENT & OPERATION OF RESEARCH REACTOR. *Educ:* Carnegie Inst Technol, BS, 52; Creighton Univ, MS, 71. *Prof Exp:* Head, Mass Spec Maintenance & instr elec maintenance, Goodyear Atomic Corp, 53-55; pres, consult & mfg rep, Radiation Eng Org, 55-57; REACTOR MGR, VET ADMIN MED CTR, 57-, RADIOL SAFETY OFF, 57- *Concurrent Pos:* Asst clin instr med, Creighton Univ Sch Med, 58-84; instr radiol, Univ Neb Col Med, 62-73; lectr physics, Creighton Univ, 64-65,

Univ Omaha, 66-67; consult, Int Atomic Energy Agency, 85; fel Am Nuclear Soc, 83; co-tech prog chmn, Int Conf Nuclear Methods Environ & Energy Res, 84. *Mem:* Sigma Xi; Am Nuclear Soc; Am Asn Physicists Med; Soc Nuclear Med; Health Physics Soc. *Res:* Development of sensitive neutron activation analysis methodology for trace elements and molecules in biological samples; the application of these methods in correlating trace element quantity with disease states. *Mailing Add:* Omaha Vet Admin Med Ctr Med Res 4101 Woolworth Ave Omaha NE 68114

BLOUET, BRIAN WALTER, b Darlington, Eng, Jan 1, 36; US citizen; m 70; c 3. EDUCATION ADMINISTRATION. *Educ:* Univ Hull, Eng, BA, 60, PhD(geog), 64. *Prof Exp:* Lectr geog, Univ Sheffield, Eng, 64-69; from assoc prof to prof geog, Univ Nebr, 69-83, chmn dept, 76-80, dir, Ctr Great Plains Studies, 79-83; PROF GEOG & HEAD DEPT, TEX A&M UNIV, 83- *Concurrent Pos:* Vis assoc prof geog, Univ Nebr, 66-67. *Mem:* Fel Royal Geog Soc; Asn Am Geographers. *Res:* Development of settlement patterns in the Americas and Europe. *Mailing Add:* Dept Geog Tex A&M Univ College Station TX 77843

BLOUGH, HERBERT ALLEN, b Philadelphia, Pa, Dec 18, 29; m 54; c 4. VIROLOGY. *Educ:* Pa State Col, BSc, 51; Chicago Med Sch, MD, 55. *Prof Exp:* Intern, Cincinnati Gen Hosp, 55; 56; univ fel, Univ Minn, 58-60, USPHS fel, 60-61; USPHS res fel virol, Univ Cambridge, 61-63; asst prof microbiol, 63-69, assoc prof ophthal, 70-75, PROF OPHTHAL, UNIV PA, 75-, HEAD DIV BIOCHEM VIROL & MEMBRANE RES, 70- *Concurrent Pos:* Consult, US Naval Hosp, Philadelphia, Pa; symp mem & lectr, Ciba Found, 64; assoc mem, Comn Influenza, Armed Forces Epidemiol Bd; Nat Multiple Sclerosis Soc sr fel, Oxford Univ, 71-72; vis prof, Univ Helsinki, 72, Katholieke Universiteit, Leuven, 85-86; mem virol study sect, NIH, 75-77, site vis, 81-; USPHS grant, 64-67, 74-; Damon Runyon-Walter Winchell Mem Fund Cancer Res grant, 67-70, 73-75; US Army Med Res & Develop Command grant, 67-76; chmn, Subcomt Virol, Univ Pa Cancer Ctr, 77-79; Mary Jennifer Selznick Fel, Nat Hereditary Disease Found, 78-; charter mem, Med & Sci Adv Bd, Herpes Resource Ctr, 79-82. *Honors & Awards:* Most Distinguished Alumnus Award, Chicago Med Sch, 75. *Mem:* AAAS; NY Acad Sci; Am Asn Immunol; Am Soc Microbiol; Am Soc Biol Chem; Am Soc Virol. *Res:* Structure and assembly of enveloped viruses; attachment of viruses to artificial membranes; lipid biochemistry; development of antivirals. *Mailing Add:* US Army Med Res Inst Infect Dis SGRD-UIV-S Ft Detrick Frederick MD 21701

BLOUIN, ANDRE, b Montreal, Que, Dec 18, 45; m 68; c 1. PHARMACOLOGY. *Educ:* Univ Montreal, DVM, 68, MSc, 70, PhD(pharmacol), 73. *Prof Exp:* Fel anat, Univ Bern, 73-74; asst prof, 75-78, ASSOC PROF PHARMACOL, FAC VET MED, UNIV MONTREAL, 78- *Mem:* Fr-Can Asn Advan Sci; Can Asn Anatomists; Can Asn Vet Anatomists; World Asn Vet Physiologists & Pharmacologists. *Res:* Quantitation by stereological methods of the ultrastructural modifications following the use of different drugs. *Mailing Add:* Fac of Vet Med PO Box 5000 Univ Montreal St Hyacinthe PQ J2S 7C6 Can

BLOUIN, FLORINE ALICE, b New Orleans, La, July 3, 31. NATURAL PRODUCTS CHEMISTRY. *Educ:* La State Univ, BS, 53, MS, 56; Univ Akron, PhD(polymer chem), 75. *Prof Exp:* CHEMIST, SOUTHERN REGIONAL RES LAB, 53-, RES CHEMIST, 80- *Mem:* Am Chem Soc; Sigma Xi; Inst Food Technologists; Am Asn Cereal Chemists. *Res:* Textile, cellulose, polymer, protein, pigment, and food chemical research. *Mailing Add:* 8133 Spruce St New Orleans LA 70118

BLOUIN, GLENN M(ORGAN), b Houston, Tex, Nov 10, 19; m 48; c 1. CHEMICAL ENGINEERING. *Educ:* Rice Univ, BS, 42. *Prof Exp:* Jr chem engr, 42-43, asst chem engr, 46-50, chem engr, 52-62, proj engr, 62-72, CHEM RES SUPVR, TENN VALLEY AUTHORITY, NAT FERTILIZER DEVELOP CTR, 72- *Mem:* Am Chem Soc. *Res:* Development of new nitrogenous fertilizers; chemical nitrogen fertilizer processes and products. *Mailing Add:* 1202 Sorrento Rd Florence AL 35630

BLOUIN, LEONARD THOMAS, b Detroit, Mich, June 15, 30; m 67; c 3. CARDIOVASCULAR PHYSIOLOGY, ANALYTICAL CHEMISTRY. *Educ:* Mich State Univ, BS, 51, MS, 56; Univ Tenn, PhD(physiol), 59. *Prof Exp:* Anal develop chemist, Gen Elec Co, Wash, 51-55; lab asst, Mich State Univ, 55; lab asst, Univ Tenn, 56-59; instr clin physiol, 59; from assoc res physiologist to res physiologist, Parke, Davis & Co, 59-65, sect leader cardiovasc-renal physiol, 65-66, sr res physiologist & head cardiovasc-renal sect, 66-67, assoc lab dir, 67-70, dir cardiovasc-renal sect, 70-74, sr cardiovasc physiologist, Clin Res Dept, 74-81; pres, LFK Assocs, Inc, 81-86; clin pharmaceut res consult, 81-; sr cardiovasc physiologist res, Dept Internal Med, Cardiol Sect, Univ Ariz, 86-; RETIRED. *Concurrent Pos:* Solar energy & passive residential design, 81- *Mem:* AAAS; NY Acad Sci; Asn Advan Med Instrumentation. *Res:* Renal physiology; hypertensive cardiovascular disease; diuretics; radioprotective compounds; electrocardiography; radio and clinical chemistry; congestive heart disease and edema; peripheral vascular disease; angina pectoris; cardiac arrhythmias; radiofrequency energy ablation of cardiac arrhythmias. *Mailing Add:* 4071 W Hardy Rd Arizona Health Sciences Centre Tucson AZ 85741

BLOUNT, CHARLES E, b Hobbs, NMex, Apr 8, 31; m 53; c 2. PHYSICS. *Educ:* Tex Western Col, BS, 54; Tex A&M Univ, MS, 60, PhD(physics), 63. *Prof Exp:* Instr math, Tex Western Col, 56-58; asst prof, 63-68, assoc prof, 68-72, PROF PHYSICS, TEX CHRISTIAN UNIV, 72- *Concurrent Pos:* Consult, El Paso Nat Gas Co, 56-58 & Gen Dynamics Ft Worth, 64- *Mem:* Am Phys Soc; Am Asn Physics Teachers; Am Soc Metals; Am Inst Mining, Metall & Petrol Eng. *Res:* Electronic spectra of molecular solids; imperfections in molecular solids; color centers in alkali halides. *Mailing Add:* Dept Physics Tex Christian Univ Ft Worth TX 76129

BLOUNT, DON HOUSTON, b Cape Girardeau, Mo, Mar 25, 29; m 52; c 3. PHYSIOLOGY. *Educ:* Univ Mo, BA, 50, MA, 56, PhD(physiol), 58. *Prof Exp:* Asst instr med physiol, Col Med, Univ Mo, 56-58; from instr to asst prof physiol & biophys, Col Med, Univ Vt, 58-61; asst prof, Dept Physiol, Med Ctr, WVa Univ, 61-67; head physiol br, Life Sci Labs, Melpar, Inc, 67-69; scientist adminr, Spec Progs Br, Nat Heart & Lung Inst, 69-72, chief cardiac functions br, div heart & vascular dis, 72-; GRAD DEAN & VPROVOST RES, GRAD SCH, UNIV MO, COLUMBIA. *Mem:* Am Physiol Asn. *Res:* Cardiovascular physiology; relationship between the biochemical and mechanical events of the heart. *Mailing Add:* Dept Physiol Univ Mo Sch Med MA415 Med Sci Bldg Columbia MO 65212

BLOUNT, EUGENE IRVING, b Boston, Mass, May 26, 27; m 55; c 3. THEORETICAL SOLID STATE PHYSICS. *Educ:* Harvard Univ, AB, 47; Univ Chicago, MS, 51, PhD, 59. *Prof Exp:* Res physicist, Int Harvester Co, 50-54; res physicist, Westinghouse Elec Corp, 57-60; res physicist, AT&T Bell Labs, 60-89; CONSULT, 89- *Concurrent Pos:* Guggenheim fel, 63-64. *Mem:* Fel Am Phys Soc. *Res:* Behavior of electrons in crystals; symmetry. *Mailing Add:* 13 Woodcliff Dr Madison NJ 07940

BLOUNT, FLOYD EUGENE, inorganic chemistry, physical chemistry; deceased, see previous edition for last biography

BLOUSE, LOUIS E, JR, b San Antonio, Tex, Nov 29, 31; m 79; c 3. BACTERIOLOGY, MICROBIOLOGY. *Educ:* Univ Tex, BA, 53, MA, 55; La State Univ, New Orleans, PhD(microbiol), 71. *Prof Exp:* Asst immunologist, Tex State Dept Health, 54-55; dir lab, El Paso City-County Health Dept, 57-59; chief bact br, USAF Epidemiol Lab, 59-68; res microbiologist & chief exp methods function, 71-72, microbiologist & chief, Dis Surveillance Br, 73-75, microbiologist & chief, Microbiol Br, 76-83, CHIEF, LAB SERV, EPIDEMIOL DIV, USAF SCH AEROSPACE MED, 83- *Concurrent Pos:* Mem, adv comt, United Cath Cols; mem, biomed minority res prog, NIH, 72-; mem adv comt, Ment Health Ment Retardation, Bexar Co. *Honors & Awards:* Fisher Sci Award, 72. *Mem:* AAAS; Am Soc Microbiol; Electron Micros Soc Am; Sigma Xi; fel Am Acad Microbiol; Clin Lab Mgt Asn. *Res:* Isolation and characterization of bacteriophages; epizootiology of staphylococcal infections in canines; comparative studies on staphylococcal typing phages; epidemiology of coagulase-negative staphylococcal infections; combined viral-bacterial infection. *Mailing Add:* 206 Palo Grande San Antonio TX 78232

BLOUT, ELKAN ROGERS, b New York, NY, July 2, 19; m 85; c 3. BIOCHEMISTRY. *Educ:* Princeton Univ, AB, 39; Columbia Univ, PhD(chem), 42. *Hon Degrees:* AM, Harvard Univ, 62; DSc, Loyola Univ, 76. *Prof Exp:* Nat Res Coun fel, Harvard Univ, 42-43, res assoc, Med Sch, 50-52 & 56-60, lectr biophys, 60-62, prof biol chem, 62-64, Edward S Harkness prof, 64-90, chmn dept, 65-69, dean acad affairs, Sch Pub Health, chmn, Dept Environ Sci & Physiol, 86-88, DIR, DIV BIOL SCI, PUB HEALTH, HARVARD UNIV, 87-, EMER EDWARD S HARKNESS PROF BIOL CHEM, 90- *Concurrent Pos:* Mem staff, Polaroid Corp, 43-48, assoc dir res, 48-58, vpres & gen mgr res, 58-62, consult, 62-; consult, Children's Med Ctr, 52- & NSF, 63-64; mem, Sci Adv Comt, Ctr Blood Res, Inc, 67-72 & mem Bd Dirs, 72-; mem, Dept Chem Vis Comt, Carnegie-Mellon Univ, 68-70; mem, Sci Adv Comt, Mass Gen Hosp, Boston, 68-71; mem, Bd Visitors, Fac Health Sci, State Univ NY Buffalo, 68-70; trustee, Boston Biomed Res Inst, 72-; mem, Corp Mus of Sci, Boston, 74-; mem, Adv Coun, Dept Biochem Sci, Princeton Univ, 74- & chmn, adv coun, dept molecular biol, 83-; trustee, Bay Biochem Res, 73-83; mem, Bd Dirs, CHON Corp, 74-83, ESA, Inc, 85- & Nat Health Res Fund, 85-; mem, Conseil de Surveillance, Compagnie Financiere du Scribe, 75-81; mem finance comt, Nat Acad Sci, 76- & mem coun, 80-; mem, Assembly Math & Phys Sci, Nat Res Coun, 79-82; mem, Adv Comt USSR & Eastern Europe, Nat Acad Sci, 79-; mem, Investments Adv Comt, Fedn Am Soc Biol, 81-; mem coun, Int Orgn Chem Sci Develop, 81-; mem bd gov, Weizmann Inst Sci, Rehovot, Israel, 77- & sci adv coun Am Comt, 79-; mem gov bd, Nat Res Coun, 80- & mem comn phys sci, math & resources; vpres & treas, Int Orgn Chem Sci Develop, 85-; gen partner, Gosnold Investment Fund Ltd, 85-; mem bd dirs & investment mgr, Auburn Investment Mgt Corp, 85-; sci adv, Affymax Res Inst, 88- *Honors & Awards:* Nat Medal Sci, 90. *Mem:* Nat Acad Sci (treas, 80-); Inst Med-Nat Acad Sci; fel Am Acad Arts & Sci; fel NY Acad Sci; fel Optical Soc Am; fel AAAS. *Res:* Polypeptides and proteins; spectroscopy and rotatory phenomena; synthetic organic chemistry. *Mailing Add:* Div Biol Sci Sch Pub Health Harvard Univ 677 Huntington Ave Bldg 2 Rm 213 Boston MA 02115

BLOXHAM, DON DEE, CARDIOVASCULAR BIOLOGY. *Educ:* La State Col, PhD(physiol), 73. *Prof Exp:* PROF BIOL, BRIGHAM YOUNG UNIV, 78- *Mailing Add:* Dept Zool Brigham Young Univ Provo UT 84602

BLOXHAM, LAURENCE HASTINGS, b Sandusky, Ohio, Mar 27, 45; m 68. SIGNAL PROCESSING, SIGNALS WARFARE. *Educ:* Bowling Green State Univ, BS, 78, MS, 80. *Prof Exp:* Analyst, Commun Electronics, Inc, 67-69; planner, Control Data Corp, 69-73, Digital Commun Corp, 74-76; teaching asst physics, Bowling Green State Univ, 78-80; programmer, Ill Inst Technol Res Inst, 80-81; task leader, Computer Sci Corp, 81-82; software engr, Watkins Johnson Co, 82-84; ADVAN ENG SPECIALIST, LOCKHEED AIRCRAFT SERV CO, 84- *Concurrent Pos:* Prin investr advan signal processing, Lockheed Aircraft Serv, 84- & focal plane signal processing for A-O receivers, 88- *Honors & Awards:* Robert E Gross Award, 88. *Mem:* Sigma Xi; Inst Elec & Electronics Engrs; Int Soc Optical Eng. *Res:* Advanced signal processing systems for radio frequencies; spread spectrum signal intercept, automatic identification, and tracking; acousto-optic radio frequency spectrum analysis; high performance photo detection techniques; design and develop parallel digital signal processing architectures and algorithms. *Mailing Add:* 5918 Birdie Dr LaVerne CA 91750

BLOZIS, GEORGE G, b Chicago, Ill, Nov 3, 29; m 56; c 3. PATHOLOGY. *Educ:* Ohio State Univ, DDS, 55; Univ Chicago, SM, 61; Am Bd Oral Path, dipl, 66. *Prof Exp:* Intern, Billing's Hosp, Univ Chicago, 55-56, instr oral path, Walter G Zoller Mem Dent Clin, 61-64, asst prof, 64-66; prof & chmn dept, 66-88, EMER PROF ORAL PATH/DIAG, COL DENT, OHIO STATE UNIV, 88- *Concurrent Pos:* Proj dir, Oral Cancer Demonstration grant, 63-66; consult oral path, Vet Admin. *Mem:* Fel AAAS; Am Dent Asn; fel Am Acad Oral Path; Sigma Xi; Int Asn Dent Res. *Res:* Parotid cytology, candida infections in the rat. *Mailing Add:* Ohio State Univ Col Dent 1370 Reymond Columbus OH 43220-3833

BLUDMAN, SIDNEY ARNOLD, b New York, NY, May 13, 27; m 49, 88; c 3. THEORETICAL PHYSICS, ASTROPHYSICS. *Educ:* Cornell Univ, AB, 45; Yale Univ, MS, 48, PhD(physics), 51. *Prof Exp:* Instr physics, Lehigh Univ, 50-52; mem staff, Lawrence Radiation Lab, lectr, Univ Calif, Berkeley, 52-61; PROF PHYSICS, UNIV PA, 61- *Concurrent Pos:* Assoc, Oceanog Inst, Woods Hole, 51; vis mem, Inst Advan Studies, 56-57, Inst Theory Phys, Univ Calif, Santa Barbara; lectr, Univ Calif, 59-60; vis res fel, Imp Col, Univ London, 67-68; vis prof, Tel-Aviv Univ, 71; Lady Davis vis prof, Hebrew Univ, 76, 83; Guggenheim fel, 83-84; Ctr Particle Astrophysics, Univ Calif, Berkeley, 90-91. *Mem:* Am Phys Soc; Sigma Xi; Am Astron Soc; Int Astron Union. *Res:* Elementary particle physics; theoretical astrophysics and cosmology. *Mailing Add:* Dept Physics Univ Pa Philadelphia PA 19104

BLUE, E(MANUEL) M(ORSE), b Spokane, Wash, June 27, 12; m 38, 81; c 6. CHEMICAL ENGINEERING. *Educ:* Univ Calif, BS, 35; Mass Inst Technol, SM, 37. *Prof Exp:* Sr eng assoc, Calif Res Corp, 38-64; mgr invention develop div, Chevron Res Co, 64-77; consult chem engr, 77-79; pres, E M Blue & Assoc, 79-88; RETIRED. *Concurrent Pos:* Lectr chem eng, Univ Calif, Berkeley, 59-89. *Honors & Awards:* Prof Progress Award, Am Inst Chem Engrs, 90. *Mem:* Fel Am Inst Chem Engrs; Am Chem Soc; Licensing Execs Soc. *Res:* Catalytic processes in petroleum refining; hydrocracking; hydrofining; isomerization and reforming. *Mailing Add:* 2642 Saklan Indian Dr 2 Walnut Creek CA 94595

BLUE, JAMES LAWRENCE, b Pekin, Ill, Sept 20, 40; m 66; c 2. MATHEMATICAL MODELING, APPLIED MATHEMATICS. *Educ:* Occidental Col, AB, 61; Calif Inst Technol, PhD(physics), 66. *Prof Exp:* Mem tech staff, Bell Tel Labs, 66-79; MEM STAFF, COMPUT & APPL MATH LAB, NAT INST STANDARDS & TECHNOL, 79- *Honors & Awards:* Gold Medal, US Dept Com, 83. *Mem:* Inst Elec & Electronics Engrs; Soc Indust Appl Math; Asn Comput Mach. *Res:* Mathematical modeling of physical systems, scientific software, magnetic microstructure, computer science algorithms, neural networks. *Mailing Add:* Comput & Appl Math Lab Nat Inst Standards & Technol Gaithersburg MD 20899

BLUE, JOSEPH EDWARD, b Quitman, Miss, Sept 29, 36; m 62; c 2. ACOUSTICS. *Educ:* Miss State Univ, BS, 61; Fla State Univ, MS, 66; Univ Tex, Austin, PhD(acoust), 71. *Prof Exp:* Res physicist acoust, Mine Defense Lab, US Navy, 61-68; res physicist, Appl Res Labs, Univ Tex, 68-71; head measurements br, 71-81, SUPT, UNDERWATER SOUND REFERENCE DETACHMENT, NAVAL RES LAB, 81- *Mem:* Fel Acoust Soc Am. *Res:* Methods and systems for characterizing transducers and other devices for the generation and reception of underwater acoustic signals and noise; sonar transduction and underwater acoustic materials. *Mailing Add:* Naval Res Lab PO Box 8337 Orlando FL 32856

BLUE, MARTS DONALD, b Des Moines, Iowa, Feb 4, 32; m 60; c 4. SOLID STATE PHYSICS. *Educ:* Iowa State Univ, BS, 52, MS, 54, PhD(physics), 56. *Prof Exp:* Asst physics, Iowa State Univ, 53, asst, Inst Atomic Res, 55-56; res scientist, Honeywell Res Ctr, 56-62, head res sect, Res Ctr, Honeywell Inc, 63-69, sci attache, Honeywell Europe, Inc, 69-72, mem opers staff, Honeywell Radiation Ctr, 72-74; PRIN RES SCIENTIST, GA TECH RES INST, GA INST TECHNOL, 74- *Concurrent Pos:* Fulbright fel, Univ Paris, 56-57; consult, US Army Missile Res, Develop & Eng Lab, Huntsville, Ala, 75, Lockheed Missiles & Space Co Inc, Rexham Aerospace & Defense Group Huntsville, Ala & Fairchild Weston Systs NY. *Mem:* Am Phys Soc; Sigma Xi. *Res:* Photographic emulsions; electron microscopy; transport properties of metals; semiconductors; thin films; infrared and submillimeter physics; radiation detection; semiconductors. *Mailing Add:* Ga Tech Res Inst Ga Inst of Technol Atlanta GA 30332

BLUE, MICHAEL HENRY, b Ovid, Colo, Nov 25, 29; div; c 2. COSMIC RAY PHYSICS. *Educ:* Colo State Univ, BS, 50; Univ Wash, PhD(physics), 60. *Prof Exp:* Fel physics, Colo State Univ, 53-54; engr, Boeing Co, Wash, 54-56; res asst physics, Univ Wash, 56-59, res instr, 60; physicist, Lawrence Radiation Lab, Univ Calif, 60-64; ASSOC PROF PHYSICS, UNIV TEX, EL PASO, 64- *Mem:* AAAS; Am Phys Soc; Am Asn Physics Teachers; Italian Phys Soc; Am Mgt Asn; Sigma Xi. *Res:* Proton-proton scattering in the Bev energy region; low and medium energy nuclear physics; Monte Carlo simulation of nucleon-nucleon collisions. *Mailing Add:* Box 36 Univ Tex El Paso TX 79968

BLUE, RICHARD ARTHUR, b Le Mars, Iowa, Nov 12, 36. EXPERIMENTAL NUCLEAR PHYSICS, VACUUM SYSTEM ENGINEERING. *Educ:* Cornell Col, BA, 58; Univ Wis, MS, 60, PhD(physics), 63. *Prof Exp:* Fel physics, Ohio State Univ, 63-66; asst prof physics, Univ Fla, 66-73, assoc prof, 73-79; SR LIAISON PHYSICIST, NAT SUPERCONDUCTING CYCLOTRON LAB, MICH STATE UNIV, EAST LANSING, 79- *Concurrent Pos:* Vis assoc prof, Univ Wis, 77-78. *Mem:* Am Phys Soc. *Res:* Low energy experimental nuclear physics; giant resonances observed via radiative capture reactions. *Mailing Add:* Nat Superconducting Cyclotron Lab Mich State Univ East Lansing MI 48824

BLUE, WILLIAM GUARD, b Poplar Bluff, Mo, Nov 7, 23; m 47; c 3. SOILS. *Educ:* Univ Mo, PhD(soils), 50. *Prof Exp:* Asst, Univ Mo, 47-50; from asst biochemist to assoc biochemist, 50-63, prof soil chem & fertil, 73-79, BIOCHEMIST, UNIV FLA, 63-, EMER PROF SOIL CHEM & FERTIL,

89- *Concurrent Pos:* Pasture agronomist, Univ Fla-Costa Rican Contract, San Jose, 58-60; consult, Guyana, 67; lectr trop soils, Orgn Trop Studies, Costa Rica, 69; consult, Ecuador, 74 & Cameroor, 80; res ext adv, Dschang, Cameroun, 85-87. *Mem:* Fel Soil Sci Soc Am; Am Soc Agron; Am Forage & Grassland Coun; Int Grassland Asn. *Res:* Chemistry, fertility and biological studies of soils, especially chemistry and fertility of tropical soils in British Honduras, Costa Rica, Panama, Venezuela, Guyana and phosphate mine waste reclamation in Florida. *Mailing Add:* 106 Newell Hall Dept Soils Agr Exp Sta Univ Fla Gainesville FL 32611-0313

BLUEFARB, SAMUEL M, b St Louis, Mo, Oct 15, 12; m 45; c 1. DERMATOLOGY. *Educ:* Univ Ill, BS, 35, MD, 37. *Prof Exp:* From instr to assoc prof, 42-57, PROF DERMAT & CHMN DEPT, MED SCH, NORTHWESTERN UNIV, CHICAGO, 57- *Concurrent Pos:* Consult, Res Vet Hosp, 64- *Mem:* AMA; Soc Invest Dermat; Am Acad Dermat; Am Dermat Asn; Am Col Physicians. *Res:* Syphilology; reticuloendothelial system in cutaneous disorders. *Mailing Add:* 1695 Second St Highland Park IL 60035

BLUEMEL, VAN (FONKEN WILFORD), b Freeport, Ill, Feb 8, 34; m 63; c 2. QUANTUM OPTICS. *Educ:* Univ Mich, BS, 56; Univ Ill, MS, 60, PhD(nuclear physics), 67. *Prof Exp:* Asst prof, 66-73, ASSOC PROF PHYSICS, WORCESTER POLYTECH INST, 73- *Mem:* Am Phys Soc; Am Asn Physics Teachers; Sigma Xi. *Res:* Statistics of scattered light; applications of atomic coherent states. *Mailing Add:* Dept Physics Worcester Polytech Inst 100 Institute Rd Worcester MA 01609

BLUEMLE, LEWIS W, JR, b Williamsport, Pa, Mar 9, 21; m 53; c 4. INTERNAL MEDICINE. *Educ:* Johns Hopkins Univ, AB, 43, MD, 46. *Hon Degrees:* DSc, Philadelphia Col Pharm & Sci; LHD, Washington & Jefferson Col. *Prof Exp:* From asst instr to instr med, Univ Pa, 50-52, assoc, 52-55, from asst prof to assoc prof, 55-68, dir clin res ctr, Univ Hosp, 61-68; prof med & pres, State Univ NY Upstate Med Ctr, 68-74; prof med & pres, Univ Ore Health Sci Ctr, 74-77; PROF MED JEFFERSON MED COL & PRES, THOMAS JEFFERSON UNIV, 77- *Concurrent Pos:* Markle scholar, 55-60; vis consult, Vet Hosp, Philadelphia, 55-68; consult, Nat Inst Arthritis & Metab Dis, 67-74; dir, Asn Acad Health Ctr, 80-83. *Mem:* Am Col Physicians; Am Fedn Clin Res; Am Clin & Climat Asn; Royal Col Physicians Edinburgh. *Res:* Renal disease. *Mailing Add:* Thomas Jefferson Univ Philadelphia PA 19107

BLUESTEIN, ALLEN CHANNING, b Lynn, Mass, Aug 25, 26; m 47; c 5. ORGANIC CHEMISTRY. *Educ:* Univ Mass, BS, 49. *Prof Exp:* Jr chemist, Advan Res & Develop Measurements Lab, Gen Elec Co, 52-55; rubber chemist, Anaconda Wire & Cable Co, 55-56, sr develop chemist, 56-61; res dir, Aerovox Corp, 61-63; tech dir, Cooke Color & Chem Co, Inc, Reichhold Chem Inc, Hackettstown, 63-71, vpres, 68-71; pres, Polymer Serv Inc, 71-79; PRES, BERLINGTON ASSOCS, 79- *Concurrent Pos:* Mem, Underwriters Lab Tech Adv Comt. *Mem:* Am Chem Soc; sr mem Inst Elec & Electronics Engrs; fel Am Inst Chemists. *Res:* Rubber compounding technology; cross-linking of thermoplastics; reactions between ozone and polymers; electrical properties of polymers; plastics technology. *Mailing Add:* 232 Burlington Ave Spotswood NJ 08884

BLUESTEIN, BEN ALFRED, b Chicago, Ill, July 5, 18; m 47; c 3. ORGANOMETALLIC CHEMISTRY, POLYMER CHEMISTRY. *Educ:* Univ Chicago, BS, 38, PhD(org chem), 41. *Prof Exp:* Res assoc chem, Univ Chicago, 41-43, asst, Metall Lab, 43-44; asst, Los Alamos Sci Lab, 44-46; asst, Res Lab, 46-50, ASST, SILICONE PROD DEPT, GEN ELEC CO, 50- *Concurrent Pos:* Pittsburgh Plate Glass Co fel, Univ Chicago, 41-43. *Mem:* AAAS; Am Chem Soc; NY Acad Sci; fel Am Inst Chem; Sigma Xi. *Res:* Reactions of 9-formylfluorene; synthesis of organo-metallic resins; plastics; radiochemistry; organo-silicon chemistry. *Mailing Add:* 2037 Hoover Rd Schenectady NY 12309

BLUESTEIN, BERNARD RICHARD, b Philadelphia, Pa, Oct 7, 25; m 47; c 4. ORGANIC CHEMISTRY. *Educ:* Univ Pa, BS, 46; Univ Ill, MS, 47, PhD(org chem), 49; Fairleigh Dickinson Univ, MBA, 67. *Prof Exp:* Assoc chem, Rutgers Univ, 49-51; fel, Purdue Univ, 51-52; asst prof, Coe Col, 52-55; sr res chemist, Sonneborn Chem & Refining Co, 55-59, asst supt, 59-62; supvr chem res, Cent Res Labs, Oakland, NJ, 62-66, from asst dir to dir, Corp Res & Develop, 67-83, VPRES RES & DEVELOP, ALLIED-KELITE DIV, WITCO CORP, 83- *Mem:* Am Oil Chemists Soc; Am Chem Soc; Royal Soc Chem; Am Inst Chem; Asn Res Dirs. *Res:* Process development; synthetic organic chemistry; vapor phase catalysis; detergents; surface active agents; polymers; organophosphorus chemistry; petroleum waxes; metal finishing; electroless plating. *Mailing Add:* Witco Corp Allied-Kelite Div 29111 Milford Rd New Hudson MI 48165

BLUESTEIN, CLAIRE, b Philadelphia, Pa, May 3, 26; m 47; c 4. INDUSTRIAL ORGANIC CHEMISTRY. *Educ:* Univ Pa, AB, 47; Univ Ill, MS, 48, PhD(org chem), 50. *Prof Exp:* Fel, Purdue Univ, 52-53; instr, Ext Serv, Pa State Univ, 58-59; sr res chemist, Sonneborn Chem & Refining Co Div, 59-63, group leader, Cent Res Labs, 63-75, mgr new technol, Witco Chem Corp, 75-76; CONSULT & PRES, CAPTAN ASSOCS, INC, 76-; VPRES, EPOLIN INC, 85- *Mem:* Am Chem Soc; Soc Plastics Engrs; Soc Advan Mat & Process Eng. *Res:* Polyurethane coatings, particularly radiation curable; high solids; aqueous emulsions; polyurethane elastomers; polyurethane chemistry; packaging coatings and adhesives technology; fabric coating technology; polymer analytical and physical characterization; plastics processing aids; petroleum sulfonates and waxes; alkaline earth salt dispersions; bicyclospirane polymerization; coatings and adhesives based on ring opening expansion polymerization. *Mailing Add:* Captan Assoc Inc PO Box 1563 Rutherford NJ 07070

BLUESTEIN, HARRY GILBERT, b New York, NY, Aug 8, 39; m 62; c 2. RHEUMATOLOGY, CLINICAL IMMUNOLOGY. *Educ:* Mass Inst Technol, BS, 61; Albert Einstein Col Med, MD, 65. *Prof Exp:* Resident internal med, Georgetown Univ Hosp, 65-67; res assoc biochem, Lab Gen Comp Biochem, NIMH, NIH, 67-69, res fel immunol, Lab Immunol, 69-70; res fel, dept path, Sch Med, Harvard Univ, 70-71; from asst prof to assoc prof, 71-84, PROF MED, SCH MED, UNIV CALIF, SAN DIEGO, 84- *Concurrent Pos:* Attend physician, Univ Hosp, Univ Calif, San Diego Med Ctr, 71- & San Diego Vet Admin Hosp, 73- *Mem:* Am Soc Clin Invest; Am Asn Immunologists; Am Rheumatism Asn; Am Asn Physicians. *Res:* Elucidating the mechanisms responsible for the disordered immune function that characterizes rheumatoid arthritis and other autoimmune rheumatic diseases; diminished immunity in AIDS and other immunodeficiency diseases. *Mailing Add:* Univ Calif Med Ctr Dept Med 225 Dickinson St Mail Code H-811-G San Diego CA 92103

BLUESTEIN, HOWARD BRUCE, b Chelsea, Mass, Oct 8, 48; m. SYNOPTIC & MESOSCALE METEOROLOGY. *Educ:* Mass Inst Technol, BS, 71, MS(elec eng) & MS(meteorol), 72, PhD(meteorol), 76. *Prof Exp:* Vis asst prof, 76-79, from asst prof to assoc prof, 79-90, PROF METEOROL, UNIV OKLA, 90- *Mem:* Am Meteorol Soc; Sigma Xi. *Res:* Synoptic dynamic meteorology; severe convection; mesoscale meteorology; tropical meteorology. *Mailing Add:* Sch Meteorol Univ Okla Rm 217 200 Felgar St Norman OK 73019

BLUESTEIN, THEODORE, b New York, NY, July 7, 34; m 57; c 4. MISSILE AERODYNAMICS. *Educ:* Pa State Univ, BS, 56; Univ Southern Calif, MS, 58. *Prof Exp:* Sr engr, Missile Syst Design, Raytheon Co, 61-65, missile mgr, Patriot Prog Off, 65-68, prin engr, 68-69, prin engr, missile eng, 71-76; rep, Piedmont Capital, 69-71; prog mgr, Govt Electronics, Motorola, 76-79; mem tech staff, Hughes Aircraft Co, 56-61, head syst engr, Roland Missile Syst, 79-80, Tow Missile sr proj engr, 81-88; RETIRED. *Honors & Awards:* Challenger Flag Award, Am Inst Aeronaut & Astronaut, 85. *Mem:* Am Inst Aeronaut & Astronaut. *Res:* Missile aerodynamics; missile guidance and control; program management; missile and missile system design; systems engineering; design of Patriot missile. *Mailing Add:* 1531 W Sendero Cuatro Tucson AZ 85704

BLUESTONE, HENRY, b Chicago, Ill, May 17, 14; m 38; c 2. ORGANIC CHEMISTRY. *Educ:* Univ Chicago, SB, 38, SM, 39. *Prof Exp:* Instr chem, Chicago City Jr Col, 39-42; res chemist, Velsicol Corp, 42-47; asst res dir, Julius Hyman & Co, 47-51; from res group leader to sr res group leader, Diamond Shamrock Corp, 51-73, staff chemist, 73-78, staff sci liaison, 78-82. *Mem:* Am Chem Soc. *Res:* Organic synthesis; diene reactions; heterocyclic compounds; polycyclic compounds; stereochemistry; bactericides; fungicides; herbicides; insecticides; plant growth regulators; nematocides. *Mailing Add:* 325 Buckingham Way Apt No 301 San Francisco CA 94132

BLUESTONE, JEFFREY A, b Ft Sill, Okla, Apr 8, 53. TRANSPLANTATION BIOLOGY. *Educ:* Cornell Univ, PhD(immunol), 80. *Prof Exp:* HEAD LAB, NAT CANCER INST, NIH, 83- *Mailing Add:* Dept Path Ben May Inst Univ Chicago Box 242 5841 S Maryland Ave Chicago IL 60637

BLUFORD, GUION STEWART, JR, b Philadelphia, Pa, Nov 22, 42; m 64; c 2. AEROSPACE ENGINEERING. *Educ:* Pa State Univ, BS, 64; USAF, Inst Technol, MS, 74, PhD(aerospace eng), 78; Univ Houston, MBA, 87. *Hon Degrees:* Several from US univs. *Prof Exp:* F4C pilot, 12th Tactical Fighter Wing, USAF, 66-67, T-38 pilot instr, 3630th Flying Training Wing, 67-72, dep advan concepts, Air Force Flight Dynamics Lab, 75-77, chief aerodyn & airframe br, 77-78; astronaut cand, 78-79, ASTRONAUT, JOHNSON SPACE CTR, NASA, 79- *Honors & Awards:* Distinguished Nat Scientist Award, Nat Soc Black Engrs, 79, Ebony Black Achievement Award, 83. *Mem:* Am Inst Aeronaut & Astronaut; Nat Tech Asn. *Res:* Computational fluid dynamics; aerodynamics. *Mailing Add:* 16439 Brookvilla Dr Houston TX 77059

BLUHM, AARON LEO, b Boston, Mass, May 26, 24; m 48; c 1. ORGANIC CHEMISTRY. *Educ:* Northeastern Univ, BS, 51; Boston Univ, PhD(chem), 56. *Prof Exp:* RES CHEMIST ORG CHEM, US ARMY NATICK LABS, 56- *Mem:* Am Chem Soc; Sigma Xi. *Res:* Synthesis; reaction mechanisms and kinetics; polymer structural research; infrared spectroscopy of organic and polymeric materials; photochromic materials; electron spin resonance studies of free radicals; photochemistry. *Mailing Add:* Five Wentworth Rd Canton MA 02021

BLUHM, HAROLD FREDERICK, b Williamsport, Pa, Mar 12, 27. ORGANIC CHEMISTRY, EXPLOSIVES. *Educ:* Bucknell Univ, BS, 52; Pa State Univ, MS, 54. *Prof Exp:* Asst chem, Pa State Univ, 52-54; res chemist, Hercules Powder Co, 54-57; res chemist, Atlas Chem Industs, Inc, 57-66, supvr chem develop lab, 66-76, ASSOC MGR, ATLAS RES & DEVELOP LAB, ATLAS POWDER CO, 76- *Mem:* AAAS; Am Mgt Asn; Am Chem Soc. *Res:* Organometallic compounds; plasticizers; synthetic resins; polymers; new explosive compounds and propellants; development of new commercial explosives. *Mailing Add:* 303 Union Ave Williamsport PA 17701

BLUHM, LESLIE, b Winthrop, Mass, May 15, 40; m 64; c 3. FOOD FERMENTATION, ALCOHOLIC BEVERAGES. *Educ:* Univ Mass, BSc, 63; Univ Ill, MS, 67, PhD(food sci), 68. *Prof Exp:* Res scientist, Miles Labs, Inc, 68-72, sect head, 72-78, sr res scientist, 76-78, dir food sci res, 78-79; mgr, distillery res & develop, 79-81, vpres res & develop, Joseph E Seagram & Sons, 81-90; CONSULT, 90- *Mem:* Am Soc Microbiol; Am Soc Food Technologists; Am Soc Brewing Chemists; Soc Indust Microbiol. *Res:* Food fermentations; sanitary microbiology; alcohol production; enzymology, microbiology, biochemistry and analytical aspects associated with food products. *Mailing Add:* 150 Columbus PL-D Stamford CT 06907

BLUHM, TERRY LEE, b Niagara Falls, NY, June 22, 47; m 71; c 1. POLYMER STRUCTURE & MORPHOLOGY, X-RAY DIFFRACTION. *Educ:* State Univ NY Col Environ Sci & Forestry, BS, 70, MS, 73, PhD(polymer chem), 76. *Prof Exp:* Fel, Inst Molecular Chem, WGer, 76-78; RES SCIENTIST, XEROX RES CTR CAN, 78- *Mem:* Am Chem Soc; NY Acad Sci; Mat Res Soc. *Res:* Solid state structure and morphology of natural and synthetic polymers using x-ray scattering techniques and electron microscopy; molecular modelling of polymers and small molecules; structural studies of photoactive pigments. *Mailing Add:* Xerox Res Ctr Can 2660 Speakman Dr Mississauga ON L5K 2L1 Can

BLUM, ALVIN SEYMOUR, b Boston, Mass, Jan 23, 26; m 51; c 1. BIOPHYSICS. *Educ:* Brown Univ, ScB, 48. *Prof Exp:* Radioisotope scientist, Vet Admin Hosp, Calif, 50-52, biochemist, La, 52-53, Fla, 53-54, prin scientist, 55-61; chief scientist, Broward Gen Hosp, Ft Lauderdale, 61-86; RETIRED. *Concurrent Pos:* Instr, Sch Med, Univ Miami, 56-77; consult, Broward Gen Hosp, 58-61, Mt Sinai Hosp, 60-81, Vet Admin Hosp, Miami, Miami Heart Inst, 71-81 & Doctor's Gen Hosp Plantation, 71-77; regist patent agent, 81- *Mem:* Soc Nuclear Med. *Res:* Circulation; kidney function; biochemistry of thyroid; iron metabolism; pulmonary function; radio pharmaceuticals; instrumentation development. *Mailing Add:* 2350 Del Mar Pl Ft Lauderdale FL 33301

BLUM, BARTON MORRILL, b Newark, NJ, May 17, 32; m 58; c 3. SILVICULTURE. *Educ:* Rutgers Univ, BS, 54; Yale Univ, MF, 57; State Univ NY Col, PhD, (forestry). *Prof Exp:* Res forester, Northeastern Forest Exp Sta, US Forest Serv, 57-64, from assoc silviculturist to silviculturist, 64-72, PROJ LEADER, NORTHEASTERN FOREST EXP STA, US FOREST SERV, 72- *Mem:* Soc Am Foresters. *Res:* Silviculture and ecology of northern hardwood forests; individual tree growth; physiology of maple sap flow and attendant problems; silviculture of northern conifers; northern conifer (spruce and fir) silviculture and ecology; spruce bud-worm. *Mailing Add:* US Forest Serv NEFES Libby Hall Univ Maine Orono ME 04469

BLUM, BRUCE I, b New York, NY; m 54; c 2. SOFTWARE ENGINEERING, CLINICAL INFORMATION SYSTEMS. *Educ:* Rutgers Univ, BS, 51; Columbia Univ, MA, 55; Univ Md, MA, 64. *Prof Exp:* Mathematician, Appl Physics Lab, Johns Hopkins Univ, 62-67; vpres, Wolf Res & Develop Corp, 67-73; sr systs eng, TRW Systs, 73-74; PRIN STAFF, APPL PHYSICS LAB, JOHNS HOPKINS UNIV, 74-, ASSOC PROF BIOMED ENG, 75- *Concurrent Pos:* Dir, Clin Info Systs Div, John Hopkins Univ, 76-83. *Mem:* Asn Comput Mach; Inst Elec & Electronics Engrs; Am Asn Artificial Intel. *Res:* Software engineering and the application of knowledge-based techniques; development, implementation and evaluation of clinical imformation systems. *Mailing Add:* 5605 Vantage Point Rd Columbia MD 21044

BLUM, EDWARD H(OWARD), b Washington, DC, Jan 1, 40; m 65; c 2. CHEMICAL ENGINEERING, SYSTEMS ANALYSIS. *Educ:* Carnegie Inst Technol, BS, 61; Princeton Univ, MA, 63, PhD(chem eng), 65. *Prof Exp:* Sloan fel chem eng, Princeton Univ, 64-65, asst prof, 65-67; asst to vpres res, Rand Corp, 67-68, proj leader & mem sr prof staff, New York-Rand Inst, 68-72, vpres, 72-74; sr scientist, Int Inst Appl Systs Anal, 74-76; dir advan technol, US Dept Energy, 76-80; vpres & exec dir, Merrill Lynch Capital Markets, 80-86; pres & chief exec officer, Md Nat Investment Banking Co, 86-89; PRES & CHIEF EXEC OFFICER, BLUM & CO, INC, 89- *Concurrent Pos:* Consult, Union Carbide Corp & Rand Corp, 65-67 & Bur Budget, Exec Off President, 66-67; vis assoc prof, State Univ NY, Stony Brook, 69-72; trustee, Univ Detroit, 70-79; mem adv comt, NBS Fire Res Prog, Nat Acad Sci-Nat Res Coun-Nat Acad Eng, 71-74; mem adv bd, Solar Energy Res Inst, 83-89; mem Fed Private Sector Partnership Bd, 85-89; mem bd dirs, Thames Water Holdings Inc, 90- *Honors & Awards:* Int Prize in Mgt Sci, NATO, 76; Appln Mgt Sci Prize, Int Inst Mgt Sci, 74. *Mem:* AAAS; Am Inst Chem Engrs. *Res:* Applied mathematics; engineering and public policy; systems analysis; energy; combustion and fire protection; management science. *Mailing Add:* 2417 Luckett Ave Vienna VA 22180-6818

BLUM, FRED A, b Austin, Tex, Nov 30, 39; m 61; c 2. SOLID STATE ELECTRONICS, LASERS. *Educ:* Univ Tex, BS, 62; Calif Inst Technol, MS, 64, PhD(physics), 68. *Prof Exp:* Res scientist, Appl Sci Lab, Gen Dynamics/Ft Worth, 63-64; staff mem solid state div, Lincoln Lab, Mass Inst Technol, 68-73; prog mgr, Cent Res Labs, Tex Instruments, 73-75; dir solid state electronics dept, Sci Ctr, Rockwell Int, 75-80, vpres, 80-; PRES & CHMN, GIGABIT LOGIC. *Mem:* Am Phys Soc; Inst Elec & Electronics Engrs; AAAS; Optical Soc Am; Soc Photo-Optical Instrumentation Engrs. *Res:* Semiconductor electronic and electro-optical devices; integrated optics; fiber optics. *Mailing Add:* 14582 Tam OShanter Dr Westlake Village CA 91362

BLUM, HAROLD A(RTHUR), b New York, NY, Apr 19, 21; m 43; c 4. CHEMICAL ENGINEERING. *Educ:* Rensselaer Polytech Inst, BChE, 42; Northwestern Univ, MS, 48, PhD(chem eng), 50. *Prof Exp:* Asst engr, procurement dist, Philadelphia Sig Corps, 42-43; asst chem, Northwestern Univ, 47, res asst chem eng, 48-49; group leader petrol prod res, Atlantic Ref Co, 49-55; lectr heat transfer & chem eng, Southern Methodist Univ, 51-55; assoc prof petrol eng, Tex Tech Col, 55-57; chmn mech eng & aerospace dept, 68-72, PROF MECH ENG, SOUTHERN METHODIST UNIV, 57-; PRES, DEVICES & SERVS CO, 77- *Concurrent Pos:* NSF grant, 61-64; Nat Aeronaut & Space Admin grant, 64-68; res grants, Alcoa Found, 72-74, Fair found, 73-76 & Dallas Power & Light Co, 73-79. *Mem:* Am Chem Soc; Am Soc Eng Educ; Am Soc Heating, Refrig & Air Conditioning Engrs; Int Solar Energy Soc; fel Am Soc Mech Engrs. *Res:* Heat transfer; thermodynamics; engineering education; information; solar energy. *Mailing Add:* 3501 Milton Ave Dallas TX 75205

BLUM, HARRY, b New York, NY, Jan 30, 24; m 49; c 3. VISUAL SHAPE, BIOLOGICAL SHAPE. *Educ:* Cornell Univ, BEE, 50; Syracuse Univ, MEE, 58. *Prof Exp:* Electronic engr & supv scientist, Mil Electronics, USAF, Rome Air Develop Ctr, NY, 50-58; systems anal, Mil Systs, Supreme Hq Allied Power Europ Air Defense Tech Ctr, NATO, Hague, The Neth, 58-60; electronic scientist, Animal & Mach Vision, USAF Cambridge Res Labs, Mass, 60-67; gen phys scientist biol shape, Div Comput Res & Technol, NIH, Bethesda, Md, 67-82; RETIRED. *Concurrent Pos:* Writer & consult, 82- *Mem:* AAAS; Sigma Xi; NY Acad Sci. *Res:* Expanding and applying a new growth geometry to understanding, describing and quantifying biological shape in morphology, animal development and vision (including computer); developing brain models and experiments for understanding nature of perceptual-cognitive organization. *Mailing Add:* 11409 Rokby Ave Kensington MD 20895

BLUM, JACOB JOSEPH, b New York, NY, Oct 3, 26; m 50, 60; c 4. PHYSIOLOGY, BIOPHYSICS. *Educ:* NY Univ, BA, 47; Univ Chicago, MS, 50, PhD(physiol), 52. *Prof Exp:* Merck fel, Calif Inst Technol, 52-53; asst prof biochem, Univ Mich, 56-58; chief biophys sect, Geront Br, Nat Heart Inst, 58-62; assoc prof to prof physiol, 62-80, JAMES B DUKE PROF PHYSIOL, DUKE UNIV, 80- *Concurrent Pos:* Guggenheim fel, Weizmann Inst Sci, Israel, 69-70. *Mem:* Am Physiol Soc; Biophys Soc; Soc Cell Biologists; Soc Gen Physiol; Protozool Soc; Sigma Xi. *Res:* Control and structural organization of intermediary metabolism. *Mailing Add:* Dept Physiol Duke Univ Durham NC 27706

BLUM, JOHN BENNETT, b Boston, Mass, Dec 22, 52; m 75; c 2. ELECTRONIC CERAMICS, POWDER PROCESSING. *Educ:* Brown Univ, BS, 74; Mass Inst Technol, MS, 77, PhD(ceramics), 79. *Prof Exp:* Sr exp ceramist, AC Spark Plug Div, Gen Motors, 79-80; asst prof ceramics, Rutgers Univ, 80-86; SR RES ENGR, NORTON CO, 86- *Concurrent Pos:* Dir, Glenn N Howatt Lab, Electronic Ceramics Res, Rutgers Univ, 80-86. *Honors & Awards:* Ralph R Teetor Award, Soc Automotive Engrs, 83. *Mem:* Am Ceramic Soc; Am Phys Soc; Sigma Xi; AAAS; Nat Inst Ceramic Engrs. *Res:* Electronic ceramics; fundamental electrical properties of the materials used and the processing needed to produce high quality devices from these materials; electronics packaging; ceramics processing. *Mailing Add:* Ceratronics 112 Turnpike Rd Suite 303 Westboro MA 01581

BLUM, JOHN LEO, b Madison, Wis, May 2, 17; m 47; c 3. PHYCOLOGY. *Educ:* Univ Wis, BS, 37; MS, 39; Univ Mich, PhD(bot), 53. *Prof Exp:* Instr biol, Canisius Col, 41-43, asst prof, 46-51, from assoc prof to prof, 53-63; assoc prof, Univ Wis-Milwaukee, 63-66, assoc dean, Col Lett & Sci, 67-69, prof, 66-84, chmn, Dept Bot, 80-84, EMER PROF BOT, UNIV WIS-MILWAUKEE, 85- *Mem:* Int Palm Soc; Sigma Xi; Int Phycol Soc. *Res:* Taxonomy of algae, especially of fresh and brackish water brackish water; aquatic plants; palms. *Mailing Add:* Dept Biol Sci Univ Wis Milwaukee WI 53201

BLUM, JOSEPH, b New York, NY, Mar 7, 19; m 50; c 3. MATHEMATICS, COMPUTER SCIENCE. *Educ:* City Col New York, BS, 40; George Washington Univ, AM, 48, PhD(math), 58. *Prof Exp:* Photogram engr, US Geol Surv, 41-47; res analyst cryptol, US Dept Army, 47-49; staff mem math, Los Alamos Sci Labs, 49-50; mathematician, Nat Bur Standards, 50-51 & Nat Security Agency, 51-75; from assoc prof to prof computer sci, Am Univ, 75-87; GROUP LEADER, INST LEARNING IN RETIREMENT, 89- *Concurrent Pos:* Consult, Nat Security Agency, 75-80. *Mem:* Sigma Xi; Am Math Soc; Soc Indust & Appl Math; Asn Comput Mach; Inst Elec & Electronics Engrs. *Res:* Signal analysis in communications; computer system design and analysis; applications of queueing theory to computer systems. *Mailing Add:* 8101 Connecticut Ave Chevy Chase MD 20815

BLUM, KENNETH, b Brooklyn, NY, Aug 8, 39; m 63; c 2. PHARMACOLOGY. *Educ:* Columbia Univ, BS, 61; NJ Col Med, MS, 65; NY Med Col, PhD(pharmacol), 68. *Prof Exp:* Res assoc biol res, Southwest Found Res & Educ, 68-70, asst found scientist clin sci, 70-71; from asst prof to assoc prof, 71-82, PROF PHARMACOL & CHIEF, DIV DRUG & ALCOHOL ABUSE, UNIV TEX HEALTH SCI CTR, SAN ANTONIO, 82- *Concurrent Pos:* Pharmacol consult, Regional Ctr Drug Educ, 70-72, Lackland Air Force Drug Rehab Ctr, 71-74, Drug Educ Training & Resource Ctr, Region VII, US Off Educ, 72-73 & Southern Calif Res Inst, US Dept Hwy Traffic Safety, 85-; consult, Nat Drug Abuse Ctr, 77; chairperson, Nat Drug Abuse Conf Pharmacol Task Force, 77-79; chmn, Gordon Res Conf Alcohol & Int Neurotoxicol Conf Alcoholism, 79, 82; founder & ed-in-chief, J Substance & Alcohol Actions & Misuse, 79-85; chief, sci affairs coun, Matrix Technol Inc. *Mem:* AAAS; Am Soc Col Professors; Am Soc Pharmacol & Exp Therapeut; Acad Med Educr & Substance Abuse; fel Am Found Pharmaceut Educ; Neurosci Soc. *Res:* Neuropharmacology; psychopharmacology; drug and alcohol abuse; drug identification; monomine correlates of drug-induced behavior; endorphins and alcohol; psychogenetics of drug seeking behavior; inventor of SAAVE and Tropamine nutritional adjuncts to alcohol and drug abuse treatment. *Mailing Add:* Dept Pharmacol Univ Tex Med Sch 7703 Floyd Curl Dr San Antonio TX 78284

BLUM, LEE M, b Philadelphia, Pa, Sept 17, 56; m 81; c 2. TOXICOLOGY, FORENSIC SCIENCE. *Educ:* Pa State Univ, BS, 78; Northeastern Univ, MS, 80; Thomas Jefferson Univ, PhD(pharmacol), 85; Am Bd Forensic Toxicol, dipl. *Prof Exp:* Post doctoral fel, Cuyahoga County Coroner's Off, 85-86; res assoc, Case Western Univ Sch Med, 85-86; TOXICOLOGIST, NAT MED SERVS, INC, 86- *Mem:* Sigma Xi; Am Acad Forensic Sci; AAAS; Am Chem Soc. *Res:* Analysis and interpretaton of exogenous substances in biological matrices. *Mailing Add:* Nat Med Servs Inc 2300 Stratford Ave Willow Grove PA 19090

BLUM, LENORE CAROL, b New York, NY, Dec 18, 42; m 61; c 1. MATHEMATICAL LOGIC, APPLIED MATHEMATICS. *Educ:* Simmons Col, BS, 63; Mass Inst Technol, PhD(math), 68. *Prof Exp:* Air Force Off Sci Res fel, 68-69; lectr math, 69-71, RES ASSOC MATH, UNIV CALIF, BERKELEY, 71-; ASSOC PROF MATH, MILLS COL, 77-, LETTS-VILLARD CHAIR MATH & COMPUTER SCI & LETTS-VILLARD RES PROF, 78- *Concurrent Pos:* Head, Dept Math & Computer Sci, Mills Col, 74-; coun mem, Conf Bd Math Sci, 76-; res scientist, Int Computer Sci Inst, Berkeley, 88- *Mem:* Asn Women in Math (pres, 75-77); Asn Women in Sci;

Am Math Soc (vpres, 90-); fel AAAS; Sigma Xi. *Res:* Applications of mathematical logic to mathematics, in particular, applications to differential algebra; mathematical theory of inductive inference; information theory; complexity theory. *Mailing Add:* 700 Euclid Ave Berkeley CA 94708

BLUM, LEON LEIB, b Telsiai, Lithuania, May 4, 08; nat US; m 36; c 2. MEDICINE, PATHOLOGY. *Educ:* Univ Berlin, MD, 33. *Prof Exp:* Res pathologist, Mt Sinai Hosp, Ill, 34-36; instr path bact & trop dis, Sch Nursing, Union Hosp, 37-57, dir, Sch Med Technol, 54-68, pathologist & dir labs, 37-83; RETIRED. *Concurrent Pos:* Pathologist & dir labs, Assoc Physicians & Surgeons' Clin, 36-83, St Anthony Hosp, 44-54, Clay County Hosp, Brazil, 50-73, Putnam County Hosp, Greencastle, 53-82, Greene County Gen Hosp, Linton, 55-82 & Vermillion County Hosp, Clinton, 56-83; sr partner, Terre Haute Med Lab, 47-83; consult, Charles Pfizer & Co, 58-65; adj prof life sci & clin dir, Ctr Med Technol, Ind State Univ, 67-83; med dir, Med Lab Asst Prog, Ind Voc Tech Col, 73-80; clin prof path, Sch Med, Ind Univ, 80-83. *Mem:* Am Soc Cytol; fel Col Am Path; Am Soc Clin Path; AMA; Pan-Am Med Asn. *Res:* Hematology. *Mailing Add:* 806 Wintergreen Ct Marco Island FL 33937-3464

BLUM, LESSER, b Berlin, Ger, Mar 6, 34; US citizen; m 64; c 3. CHEMICAL PHYSICS, STATISTICAL MECHANICS. *Educ:* Univ Buenos Aires, PhD(chem), 56. *Prof Exp:* Asst prof phys chem, Univ Buenos Aires, 58-66; res assoc, Rice Univ, 67-68; PROF PHYSICS, UNIV PR, 69- *Concurrent Pos:* Fel, Free Univ Brussels, 59-60, Nat Res Coun Can, 64 & J S Guggenheim Found, 78. *Mem:* Am Chem Soc; fel Am Phys Soc. *Res:* Statistical mechanics and thermodynamics of fluids and solutions; ionic solutions and electrodes in equilibrium. *Mailing Add:* Dept Physics Univ PR Rio Piedras PR 00931

BLUM, MANUEL, b Caracas, Venezuela, Apr 26, 38; m 61; c 1. COMPUTER SCIENCE. *Educ:* Mass Inst Technol, BS & MS, 61, PhD(math), 64. *Prof Exp:* Res assoc elec eng, Mass Inst Technol, 64-65, asst prof math, 65-69; assoc prof, 69-72, assoc chairperson comput sci, 77-80, PROF ELEC ENG & COMPUT SCI, UNIV CALIF, BERKELEY, 72- *Concurrent Pos:* Sloan Found fel, 71-72. *Mem:* Am Math Soc; fel Inst Elec & Electronics Engrs; fel AAAS; Asn Comput Mach. *Res:* Theory of algorithms; mathematical theory of inductive inference, automata in 2 and 3 dimensional space; cryptography and number theory. *Mailing Add:* 700 Euclid Ave Berkeley CA 94708

BLUM, MARVIN, b Manhattan, NY, June 18, 28; m 49; c 1. APPLIED MATHEMATICS, RADAR DATA PROCESSING. *Educ:* Brooklyn Col, BS, 48. *Prof Exp:* Physicist, Naval Proving Grounds, Va, 48-49; elec engr, Ord Div, Nat Bur Standards, 49-53; proximity fuze studies, Dept Army, 53-54; design specialist, Gen Dynamics, 54-59; sr opers res scientist, Syst Develop Corp, 59-62; sr res mathematician, Conductron Corp, 62-64; res mathematician, Rand Corp, 64-68; sr res scientist, Syst Develop Corp, 68-71; mem tech staff, Inst Defense Anal, 71-72; mem sr tech staff, Technol Serv Corp, 72-73; mem res & tech staff systs anal & high energy laser applns, Northrop Res & Technol Ctr, 73-77; mem tech staff precision satellite ephemeris generation & estimation, Aerospace Corp, 77-82; STAFF ENGR, RADAR SYSTS GROUP, HUGHES AIRCRAFT CO, 82- *Mem:* Inst Elec & Electronics Engrs. *Res:* Detection; parameter estimation; statistical filter theory; real-time and recursive smoothing techniques; high energy laser applications; system analysis; laser propagation; scenario simulation; vulnerability and system effectiveness evaluation; precision satellite ephemeris and orbit estimation; algorithm development and verification radar data processing. *Mailing Add:* Swerling Manassa Smith 22801 Ventura Blvd Suite 100 Woodland Hills CA 91364

BLUM, MURRAY SHELDON, b Philadelphia, Pa, July 19, 29; m 53; c 4. ENTOMOLOGY, PHYSIOLOGY. *Educ:* Univ Ill, BS, 52, MS, 53, PhD(entom), 55. *Prof Exp:* Entomologist, USDA, 57-58; assoc prof entom, La State Univ, Baton Rouge, 58-67; prof, 67-78, RES PROF ENTOM, UNIV GA, 78- *Concurrent Pos:* consult, NSF, 74-76; Fulbright fel, Univ Paris, 84-85; ed, J Chem Ecol, 1972- *Mem:* Fel AAAS; Entom Soc Am; Acad Applied Sci. *Res:* Chemistry of insect pheromones and defensive secretions; biochemistry of the insect reproductive system; regulation of insect behavior by chemical releasers; chemistry and functions of arthropod natural products; biochemical strategies of insects feeding on toxic plants. *Mailing Add:* Dept Entom Univ Ga Athens GA 30602

BLUM, NORMAN ALLEN, b Boston, Mass, Dec 29, 32; m 57; c 3. SOLID STATE SCIENCE, MICROELECTRONICS. *Educ:* Harvard Univ, AB, 54; Brandeis Univ, MA, 59, PhD(physics), 64. *Prof Exp:* Sr res scientist, Res & Advan Develop Div, Avco Corp, 58-60; res staff mem, Nat Magnet Lab, Mass Inst Technol, 60-66 & NASA Electronics Res Ctr, 66-70; PRIN STAFF PHYSICIST, APPL PHYSICS LAB, JOHNS HOPKINS UNIV, 70- *Mem:* Am Phys Soc; Inst Elec & Electronics Engrs; Int Soc Hybrid Microelectronics; Am Vacuum Soc. *Res:* Solid state physics; magnetism; Mossbauer effect; instrumentation; amorphous films; photovoltaics; optical and electronic properties of solids; microelectronics; biomedical sensors. *Mailing Add:* Appl Physics Lab Johns Hopkins Univ Laurel MD 20723

BLUM, PATRICIA RAE, b Dayton, Ohio, July 21, 48; m 80; c 1. HYDROCARBON PROCESSING. *Educ:* Wright State Univ, BS, 70, MS, 72; Ohio State Univ, PhD(inorg chem), 77. *Prof Exp:* Teaching asst, Wright State Univ, 70-72; contract chemist, Wright Patterson AFB, 72-73; teaching asst, Ohio State Univ, 73-77; PROJ LEADER, STANDARD OIL CO, OHIO, 77- *Mem:* Am Chem Soc. *Res:* Catalytic processes for the selective oxidation of hydrocarbons. *Mailing Add:* 4440 Warrensville Center Rd Cleveland OH 44128

BLUM, RICHARD, b Cernauti, Romania, Sept 26, 13; Can citizen; m 36; c 2. MATHEMATICS. *Educ:* Univ Cernauti, Romania, Lic, 34; Univ Bucharest, PhD(math), 46. *Prof Exp:* Asst math, Univ Bucharest, 45-49; lectr, Univ Cluj, 49-50; asst prof, Acadia Univ, 53-54; lectr, 54-56, assoc prof, 56-63, PROF MATH, UNIV SASK, 63- *Mem:* Am Math Soc; Can Math Cong. *Res:* Differential geometry, particularly imbedding properties of Riemannian spaces in Euclidean spaces. *Mailing Add:* 2510 Munroe St Saskatoon SK S7J 1S9 Can

BLUM, ROBERT ALLAN, b Philadelphia, Pa, May 16, 38; m 59; c 3. PSYCHIATRY, PSYCHOANALYSIS. *Educ:* Mass Inst Technol, BS, 59, MS, 60; Univ Pa, MD, 64. *Prof Exp:* Engr, Gen Atronics Corp, 59-61; intern med, Mem Hosp Long Beach, Calif, 64-65; resident psychiat, Mass Ment Health Ctr, 65-66, chief resident, Clin Res Ctr, 66-67; staff assoc clin psychopharm, Div Spec Ment Health Res Progs, NIMH, 67-69, res psychiatrist, 69-71; res psychiatrist, Nat Inst Alcoholism & Alcohol Abuse, 71-74; ASST PROF PSYCHIAT, GEORGETOWN UNIV, 72-, CLIN ASSOC PROF PSYCHIAT, UNIFORMED SERVS, UNIV HEALTH SCIS, 87- *Concurrent Pos:* Teaching fel, Harvard Med Sch, 65-67; instr, Sch Med, George Washington Univ, 69-74; instr, Sch Med, Johns Hopkins Univ, 69-; consult, US Govt, 74-; teaching analyst, Baltimore-DC Inst Psychoanal, 78- *Mem:* AAAS; fel Am Psychiat Asn; Int Psychoanal Asn; Am Psychoanal Asn; Soc Gen Systs Res; Soc Clin Exp Hypn; Int Soc Hypn; Am Soc Indust Security. *Res:* Information theory; error correcting codes; cognitive and attentional mechanisms in schizophrenia; electrophysiologic correlates of behavior; mathematical models of behavior using automata theory; psychoanalytic models; psychobiographical analysis; behavioral aspects of terrorism. *Mailing Add:* 9819 Hill St Kensington MD 20895

BLUM, SAMUEL EMIL, b New York, NY, Aug 28, 20; c 3. PHYSICAL CHEMISTRY, ATMOSPHERIC SCIENCES. *Educ:* Rutgers Univ, BS, 42, PhD(phys chem), 50. *Prof Exp:* Prod chemist, US Rubber Co, 42-43; asst, Rutgers Univ, 46-50; prin res chemist, Battelle Mem Inst, 52-59; mem res staff, Thomas J Watson Res Lab, IBM Corp, 59-90; CONSULT, 90- *Mem:* AAAS; Sigma Xi; fel Am Inst Chem. *Res:* Semiconductors; material science. *Mailing Add:* 18-1 Granada Crescent White Plains NY 10603

BLUM, SEYMOUR L, b New York, NY, Jan 10, 25; m 48; c 1. CERAMICS, ENVIRONMENTAL SCIENCE. *Educ:* Alfred Univ, BS, 48; Mass Inst Technol, ScD(ceramics), 54. *Prof Exp:* Plant engr, Commercial Decal, Inc, 48-51; res assoc, Mass Inst Technol, 51-54; mgr high temperature mat, Raytheon Co, 54-63; asst dir metals & ceramics res, IIT Res Inst, 63-65, dir ceramics res, 65-69, vpres, 69-71; dir energy & resource planning, Mitre Corp, 71-78; vpres, Northern Energy Corp, Boston, 78-83; vpres, Charles River Asn, Boston, 83-86; PRES, S L B ASSOC, 86- *Concurrent Pos:* chmn, Nat Mat Adv Bd, Nat Acad Sci, 72-76; mem vis comt, Mass Inst Technol, 73-79; mem mat comt, Off Technol Assessment, US Cong, 74-; mem adv comt, Eng & Tech Systs, Nat Acad Sci, 83- *Honors & Awards:* Norton Award, Am Ceramic Soc, 73. *Mem:* AAAS; fel Am Ceramic Soc (vpres, 76-77); Am Soc Metals; Nat Soc Prof Engrs; fel Am Inst Chem. *Res:* Materials research in ceramics, especially electronic ceramics, graphite, refractories and solid state behavior; environmental studies in pollution and waste utilization; environmental research in solar energy and recycling; studies on United States materials policy and availability of materials. *Mailing Add:* 39 Pilgrim Path Sudbury MA 01776

BLUM, STANLEY WALTER, b Sheboygan, Wis, July 3, 33; div; c 2. ORGANIC CHEMISTRY, BIOCHEMISTRY. *Educ:* Univ Wis, BS, 55; Univ Ill, PhD(org chem), 60. *Prof Exp:* Res fel org chem, Univ Minn, 60-61; res assoc chem microbial prod, Rutgers Univ, 61-62; scientist, Warner-Lambert Res Inst, 62-66, sr scientist, 66-77; MEM STAFF, FOOD & DRUG ADMIN, 78- *Mem:* Am Chem Soc. *Res:* Biochemical pharmacology; drug metabolism; biochemistry. *Mailing Add:* 532 Meadow Hall Dr Rockville MD 20851

BLUM, UDO, b Lüdenscheid, Ger, Nov 29, 39; US citizen; m 68. BOTANY, ECOLOGY. *Educ:* Franklin Col, BA, 63; Ind Univ, Bloomington, MA, 65; Univ Okla, PhD(bot), 68. *Prof Exp:* Vis asst prof bot, Univ Okla, 68-69; from asst prof to assoc prof, 69-81, PROF BOT, NC STATE UNIV, 81- *Concurrent Pos:* Fac res grant, 70, NC State Univ & Agr Res Serv, USDA grants 69-91 & Carolina Power & Light Co, 73-79, competitive res grant prog, 81-85 & Lisa-USDA, 88-92; ed, Int J Biometeorol, 81-84. *Mem:* Brit Ecol Soc; Sigma Xi; Ecol Soc Am; Int Soc Biometerol; Phytochem Soc NAm. *Res:* Plant plant interactions, such as competition and allelopathy; effects of natural and man-made stresses on the physiology of plants. *Mailing Add:* Dept Bot NC State Univ Raleigh NC 27695-7612

BLUMBERG, ALAN FRED, b Ancon, Canal Zone, Oct 16, 48; US citizen; m 76; c 1. COASTAL CIRCULATION, NUMERICAL SIMULATION. *Educ:* Fairleigh Dickinson Univ, BS, 70; Johns Hopkins Univ, MA, 73, PhD(phys oceanog), 75. *Prof Exp:* Asst gen oceanog, Johns Hopkins Univ, 73-75; res scientist estuarine & coastal circulation, Princeton Univ, 75-80; sr scientist & sr vpres, Dynalysis of Princeton, 80-85; PRIN SCIENTIST, HYDROQUAL INC, 85- *Concurrent Pos:* Consult, State Md, 78-79, State Calif, 87. *Mem:* Am Meteorol Soc; Am Geophys Union. *Res:* Fundamental research in physical oceanography and application of results to solution of practical engineering and environmental problems. *Mailing Add:* 22 East 14 Barnegat NJ 08006

BLUMBERG, AVROM AARON, b Albany, NY, Mar 3, 28; m 55, 69; c 3. PHYSICAL CHEMISTRY, POLYMER CHEMISTRY. *Educ:* Rensselaer Polytech Inst, BS, 49; Yale Univ, PhD(phys chem), 53. *Prof Exp:* Lab instr, Rensselaer Polytech Inst, 49; Plate Glass fel, Mellon Inst, 53-59, res chemist polymer studies, 59-63; from asst prof to assoc prof, 63-75, head, Div Natural Sci & Math, 66-82, PROF CHEM, DEPAUL UNIV, 75-, CHMN, CHEM DEPT, 86- *Concurrent Pos:* Lectr, Univ Pittsburgh, 57-58. *Mem:* Am Chem Soc; Royal Inst Chem; Arms Control Asn; Sigma Xi. *Res:* Reaction kinetics; solid-liquid reactions; glassy state; differential thermal analysis; x-ray diffraction structure calculations; polymer configurations; transport phenomena in polymers and through films; kinetics of organic growth; arms control; polymer degradation; interfaces. *Mailing Add:* Dept Chem DePaul Univ Chicago IL 60614

BLUMBERG, BARUCH SAMUEL, b New York, NY, July 28, 25; m 54; c 4. EPIDEMIOLOGY, GENETICS. *Educ:* Union Col, NY, BS, 46; Columbia Univ, MD, 51; Oxford Univ, PhD(biochem), 57; FRCP, 84. *Hon Degrees:* DSc, Univ Pittsburgh, Dickinson Col, Hahnemann Univ, 77, Franklin &

Marshal Col, 78, Yeshiva Univ, 80, Ursinus Col, 82, Bard Col, 85, Elizabethtown Col, 88, Ball State Univ, 89, Univ Pa & Columbia Univ, 90; LLD, Union Col, 77, La Salle Univ, 82; LHD, Jewish Theol Sem Am, 77, Thomas Jefferson Univ & Rush Univ, 83; MA, Oxford Univ, 84. *Prof Exp:* Intern & asst resident, First Med Div, Bellevue Hosp, 51-53; teaching fel, Dept Med, Col Physicians & Surgeons & Presby Hosp, NY, 53-55; mem, Dept Biochem, Oxford Univ, 55-57; chief, geog med & genetics sect & attend physician, Clin Ctr, NIH, Bethesda, Md, 57-64; assoc dir clin res & sr mem, Fox Chase Cancer Ctr, Philadelphia, 64-86; clin prof, Dept Epidemiol, Sch Pub Health, Univ Wash, Seattle, 83-88; UNIV PROF MED & ANTHROP, UNIV PA, PHILADELPHIA, 77-; MASTER, BALLIOL COL, OXFORD UNIV, ENG, 89- *Concurrent Pos:* Res collabr, Brookhaven Nat Lab, 60-; assoc prof clin med, Georgetown Univ Sch Med, 62-64; prof human genetics, Univ Pa, Philadelphia, 70-75, prof med, 70-, prof anthropol, 75-, mem grad comt, 68-, mem grad group history & sociol sci, 79-; vis fel, Trinity Col, Oxford, England, 72-73; sr attend physician, Philadelphia Gen Hosp, 75-76; Fox Chase Cancer Ctr rep, US Nat Comt for Int Union Against Cancer, 76-80; fel, Balliol Col, Oxford Univ, 76, George Eastman vis prof, 83-84; mem, sci adv bd, Leonard Wood Mem, 77- & Am Comt for Weizmann Inst, 79-; consult, Franklin Inst, Philadelphia, 80-, Univ Wash Med Res Unit, Taipei, Taiwan, 82-, Off Surg Gen, 89-; Ashland vis prof, Univ Ky, 86-87; vpres pop oncol, Fox Chase Cancer Ctr, Philadelphia, 86-89, sr adv to pres, 89-, Fox Chase distinguished scientist, 89-; mem bd dirs, Zool Soc Philadelphia, 79-89, emer mem, 89-; mem sci adv bd, Stazione Zool, Naples, Italy, 87-, Fogarty Int Ctr, NIH, Bethesda, 89-; mem bd dirs, Inst Sci Info, Philadelphia, 89- *Honors & Awards:* Nobel Prize for Physiol & Med, 76; Passano Award, 74; Gairdner Found Int Award, 75; Karl Landsteiner Mem Award, Am Asn Blood Banks, 75; Richard & Hinda Rosenthal Found Award, Am Col Physicians; Pa Med Soc Distinguished Serv Award, 82; Molly & Sidney N Zubrow Award, Pa Hosp, Philadelphia, 86; Achievement Award, Sammy Davis Jr Nat Liver Inst, 87; John P McGovern Award, Am Med Writers Asn, 88; James Blundell Award, Brit Blood Transfusion Soc, 89; Gold Medal Award, Can Liver Found & Can Asn for the Study of the Liver, 90. *Mem:* Nat Acad Sci; Inst Med-Nat Acad Sci; Am Soc Human Genetics; fel Am Col Physicians; Am Soc Clin Invest; Asn Am Physicians; Am Fedn Clin Res; hon fel, Am Gastroenterol Asn; hon fel, Am Col Gastroenterol; Am Philos Soc. *Res:* Discovery of hepatitis B virus and its vaccine; etiological role of hepatitis B virus and prevention of primary hepatocellular carcinoma. *Mailing Add:* Fox Chase Cancer Ctr 7701 Burholme Ave Philadelphia PA 19111

BLUMBERG, HAROLD, b Fairmont, WVa, June 19, 09; m 34; c 1. PHARMACOLOGY. *Educ:* Johns Hopkins Univ, ScD(biochem), 33. *Prof Exp:* Asst pediat, Sch Med, Johns Hopkins Univ, 33-35, instr, 35-36; assoc biochemist, Off Child Hyg, USPHS, 36-38; res biochemist, Dept Biochem, Sch Hyg & Pub Health, Johns Hopkins Univ, 38-42; assoc toxicologist, Army Indust Hyg, US War Dept, 42-44; sr biologist, Sterling-Winthrop Res Inst, NY, 44-47; dir biol labs, Endo Labs, Inc, 47-60, assoc dir res, 60-74; res prof, 74-81, EMER RES PROF PHARMACOL, NEW YORK MED COL, 81- *Concurrent Pos:* Nat Res Coun fel, Sch Hyg & Pub Health, Johns Hopkins Univ, 34-35; consult. *Honors & Awards:* John Scott Medal Award, 82. *Mem:* AAAS; Soc Toxicol; Am Soc Pharmacol & Exp Therapeut; Soc Exp Biol & Med; NY Acad Sci. *Res:* Vitamin E; lead poisoning; argyremia; manganese rickets; choline and liver cirrhosis; mercurial diuretics; theophylline derivatives; antispasmodics; naloxone and naltrexone narcotic antagonists; molindone tranquilizer; nalbuphine analgesic. *Mailing Add:* 1731 Beacon St Apt 213 Brookline MA 02146

BLUMBERG, JEFFREY BERNARD, b San Francisco, Calif, Dec 31, 45; m 71; c 2. PHARMACOLOGY, TOXICOLOGY. *Educ:* Wash State Univ, BPharm & BSc(psychol), 69; Vanderbilt Univ, PhD(pharmacol), 74. *Prof Exp:* Res psychopharmacol, Vanderbilt Univ, 74; asst prof pharmacol, Northeastern Univ, 75-78, dir toxicol prog, 78-79, assoc prof pharmacol & toxicol, 79-81, head sect pharmacol, 79-81; asst dir, 81-87, actg assoc dir, 85-86, assoc prof nutrit, 81-88, PROF NUTRIT, USDA HUMAN NUTRIT RES CTR, TUFTS UNIV, 88- , ASSOC DIR, 88- *Concurrent Pos:* NIH fel biochem, Univ Calgary, 74-75; co-prin investr, Am Heart Asn grant, 78-81; prin investr, USDA grant, 81-; Surgeon Gen Workshop Health Prom Aging; DHHS year 2000 Health Object Comt. *Honors & Awards:* Cert of Merit, USDA, 82. *Mem:* Soc Neurosci; Sigma Xi; AAAS; Am Col Toxicol; Am Inst Nutrit; Am Col Nutrit; Am Soc Clin Nutrit; Am Aging Asn; Am Soc Parenteral Enteral Nutrit; NY Acad Sci; Union Concerned Scientists. *Res:* Neurotoxicology; nutrition; immunology biochemistry. *Mailing Add:* USDA Human Nutrit Res C Tufts Univ Boston MA 02111

BLUMBERG, LEROY NORMAN, b Atlantic City, NJ, June 22, 29; m 57; c 3. PARTICLE ACCELERATORS. *Educ:* Mass Inst Technol, BSc, 51; Columbia Univ, MA, 55, PhD(physics), 62. *Prof Exp:* Physicist hydrodyn, Los Alamos Sci Lab, Univ Calif, 55-57, nuclear physics, 57-60; physicist, Union Carbide Nuclear Div, Oak Ridge Nat Lab, 62-65; physicist, Cambridge Electron Accelerator, Harvard Univ, 65-66; PHYSICIST, BROOKHAVEN NAT LAB, 66- *Concurrent Pos:* Fulbright fel, 51-52; vis prof, Japan Nat High Energy Phys Lab, 75; vis scholar, Stanford Univ, 82-83; vis scientist, US Dept Energy. *Mem:* AAAS; fel Am Phys Soc. *Res:* Shock wave hydrodynamics; fission fragment angular distributions; polarized proton scattering; pion production; K-zero decay; accelerator physics; electromagnetism; atomic and molecular physics; optics; design of electron storage rings as synchrotron radiation sources; orbit calculations and computer modeling. *Mailing Add:* Nat Synchrotron Light Source Brookhaven Nat Lab Bldg 725 Upton NY 11973

BLUMBERG, MARK STUART, b New York, NY, Nov 16, 24; m 52, 74; c 2. MEASURING OUTCOMES OF CARE. *Educ:* Harvard Sch Dent Med, DMD, 48; Harvard Med Sch, MD, 50. *Prof Exp:* Opers analyst, Opers Res Off, Johns Hopkins Univ, 51-54; commissioned officer, Occup Health Prog, USPHS, 54- 56; from opers analyst to dir health economics, Stanford Res Inst, 56-66; dir health planning, Off Pres, Univ Calif, 66- 70; CORP PLANNING ADV TO DIR, KAISER FOUND HEALTH PLAN, INC, 70- *Concurrent Pos:* Consult, Pan Am Health Orgn, 66; mem, Res Adv Comt, Calif State Dept Ment Hyg, 67-70; lectr, Sch Pub Health, Univ Calif, Berkeley, 70-75; consult, Robert Wood Johnson Found, 83-87. *Mem:* Inst Med-Nat Acad Sci; Health Care Info & Mgt Systs Soc; Am Pub Health Asn; Opers Res Soc Am. *Res:* Quantitative analyses of data on health care service utilization delivery, organization, outcomes and costs; measuring risk-adjusted outcomes of health care. *Mailing Add:* 6150 Pinewood Rd Oakland CA 94611

BLUMBERG, PETER MITCHELL, b Glen Ridge, NJ, Jan 6, 49; m 49. CARCINOGENESIS, TUMOR PROMOTION. *Educ:* Harvard Univ, AB, 70, AM, 70, PhD(biochem & molecular biol), 74. *Prof Exp:* Asst prof pharmacol, Harvard Med Sch, 75-81, assoc prof, 81; CHIEF, MOLECULAR MECHANISMS OF TUMOR PROM SECT, LAB CELLULAR CARCINOGENESIS & TUMOR PROM, DIV CANCER CAUSE & PREV, NAT CANCER INST, 81- *Concurrent Pos:* Assoc ed, Cancer Res, 81-; J Nat Cancer Inst, 85-; Phytotherapy Res, 87. *Honors & Awards:* Rhoads Award, 87. *Mem:* Am Asn Cancer Res; Am Soc Cell Biol; Soc Neurosci; AAAS. *Res:* Analysis of the mode of action of tumor promoters, with particular emphasis on characterization of receptors for the phorbol esters. *Mailing Add:* Lab Cellular Carcinogenesis & Tumor Prom Nat Cancer Inst NIH Bldg 37 Rm 3B25 Bethesda MD 20892

BLUMBERG, RICHARD WINSTON, b Winston-Salem, NC, Nov 10, 14; m 69. PEDIATRICS. *Educ:* Emory Univ, BS, 35, MD, 38. *Prof Exp:* Res assoc, Children's Hosp, Cincinnati, 47-48; assoc, Emory Univ, 48-51, from asst prof to assoc prof, 51-59, actg chmn dept, 51-59, prof pediat & chmn dept, Sch Med, 59-81; DIR, DIV PHYS HEALTH, DEKALB COUNTY HEALTH DEPT, 81- *Mem:* Am Acad Pediat; AMA; Am Pediat Soc. *Res:* Infectious diseases. *Mailing Add:* 2419 Woodward Way NW Atlanta GA 30305

BLUMBERG, WILLIAM EMIL, b College Station, Tex, Dec 23, 30. PHYSICS, BIOPHYSICS. *Educ:* Univ Tex, BS, 52; Univ Calif, Berkeley, PhD(physics), 59. *Prof Exp:* Phys scientist, US Army Biol Labs, Md, 53-55; MEM TECH STAFF BIOPHYS, BELL LABS, INC, 59- *Concurrent Pos:* Assoc prof, Albert Einstein Col Med, 70-; adj assoc prof, Rockefeller Univ, 71-72. *Mem:* AAAS; fel Am Phys Soc; Biophys Soc; Am Soc Biol Chem; Am Soc Photobiol; Sigma Xi. *Res:* Magnetic interactions and magnetic resonance of electrons and nuclei; role of paramagnetic ions in biologically important reactions. *Mailing Add:* AT&T Bell Labs IC427 600 Mountain Ave Murray Hill NJ 07974

BLUME, ARTHUR JOEL, b New York, NY, May 19, 41; m 68; c 2. NEUROCHEMISTRY. *Educ:* Univ Rochester, BA, 63; Syracuse Univ, MS, 66, PhD(molecular biol), 68. *Prof Exp:* USPHS fel neurochem, Lab Biochem Genetics, NIH, 68-71; ASST MEM NEUROCHEM, DEPT PHYSIOL CHEM, ROCHE INST MOLECULAR BIOL, 71- *Mem:* Sigma Xi; Fedn Am Soc Exp Biol. *Res:* Storage and transfer of information in the nervous system; emphasis on properties of neuronal membranes. *Mailing Add:* Am Cyanamid Co Neuro- Molecular Bio Pearl River NY 10965

BLUME, FREDERICK DUANE, b Mishawaka, Ind, Aug 27, 33; m 55; c 8. ENVIRONMENTAL PHYSIOLOGY. *Educ:* Wabash Col, AB, 55; Univ Calif, Berkeley, PhD(altitude metab), 65. *Prof Exp:* Asst res physiologist, White Mountain Res Sta, Univ Calif, 64-69, asst dir, 69-72; PROF BIOL, CALIF STATE COL, BAKERSFIELD, 72- *Mem:* Fedn Am Soc Exp Biol. *Res:* Environmental physiology; carbohydrate metabolism at altitude; growth and development and cardiovascular changes at altitude. *Mailing Add:* Dept Biol Calif State Col 9001 Stockdale Hwy Bakersfield CA 93309

BLUME, HANS-JUERGEN CHRISTIAN, b Braunschweig, WGer, May 10, 26; m 52; c 4. ELECTRONICS ENGINEERING. *Educ:* Tech Univ Braunschweig, WGer, dipl eng, 56, DEng, 66. *Prof Exp:* Patent engr, Siemens, 55-58; SR ENGR, NASA, 59- *Honors & Awards:* Apollo Achievement Award, NASA, 69, Spec Achievement Award, 77, NASA Award, 67. *Mem:* Inst Elec & Electronics Engrs. *Res:* Low-noise preamplification of microwaves with negative resistance amplifiers such as superconductive tunneling junction; development of microwave radiometer systems for the remote sensing of ocean surface parameters from aircrafts and satellites. *Mailing Add:* NASA Langley Res Ctr MS 490 Hampton VA 23665

BLUME, JOHN A(UGUST), b Gonzales, Calif, Apr 8, 09; m 85. DYNAMICS, EARTHQUAKE ENGINEERING. *Educ:* Stanford Univ, AB, 32, Engr, 34, PhD(civil eng), 67. *Prof Exp:* Mem staff, Seismol Div, US Coast & Geod Surv, 34-35; engr, Div Hwy, State of Calif, 35-36; engr, Standard Oil Co Calif, 36-40, design engr, 40-45; pvt practice, 45-57; pres & chmn bd, John A Blume Assoc Engrs, 57-84, consult, URS/Blume Engrs, 84-87; CONSULT, 87- *Concurrent Pos:* UNESCO expert, earthquake risk; mem interoceanic canal study air blast & ground shock technol working groups, seismic & dynamic consult, Div Reactor Licensing, US Atomic Energy Comn; mem & past pres, Earthquake Eng Res Inst; consult, 45-57; mem organizing comt, World Conf Earthquake Eng, 56; lectr, Stanford Univ, 57-84, consult prof, 80-86. *Honors & Awards:* Moisseiff Award, Am Soc Civil Engrs, 53, 61 & 69, Howard Award, 62; Medal of the Seismol Soc of Am, 87. *Mem:* Nat Acad Eng; hon mem Am Soc Civil Engrs; Seismol Soc Am; hon mem Am Concrete Inst; Soc Am Mil Engrs; Sigma Xi; hon mem Int Asn Earthquake Eng; hon mem Earthquake Eng Res Inst. *Res:* Structural-dynamic properties of buildings and other structures and the response of structures to earthquake, underground nuclear explosion, windstorm, blast, sonic boom and ocean swell. *Mailing Add:* 85 El Cerrito Ave Hillsborough CA 94010

BLUME, MARTIN, b New York, NY, Jan 13, 32; m 55; c 2. PHYSICS. *Educ:* Princeton Univ, AB, 54, Harvard Univ, AM, 56, PhD(physics), 59. *Prof Exp:* Fulbright res fel physics, Univ Tokyo, 59-60; res assoc theoret physics, Atomic Energy Res Estab, Eng, 60-62; from assoc physicist to physicist, 62-70, assoc dir, 81-84, SR PHYSICIST, BROOKHAVEN NAT LAB, 70-; PROF PHYSICS, STATE UNIV NY, STONY BROOK, 72-, DEP DIR, 84- *Honors & Awards:* E O Lawrence Award, 81. *Mem:* Fel Am Phys Soc; fel AAAS; fel NY Acad Sci. *Res:* Theoretical solid state and atomic physics; neutron scattering; magnetism; magnetic resonance; synchrotron radiation. *Mailing Add:* Dirs Off Brookhaven Nat Lab Upton NY 11973

BLUME, SHEILA BIERMAN, b Brooklyn, NY, June 21, 34; m 55; c 2. PSYCHIATRY. *Educ:* Harvard Med Sch, MD, 58. *Prof Exp:* Intern pediat, Med Ctr, Children's Hosp, 58-59; Fulbright fel psychiat biochem, Sch Med, Tokyo Univ, 59-60; resident psychiat, Cent Islip Psychiat Ctr, 62-65, chief alcoholism unit, 64-79; dir, NY State Div Alcoholism & Alcohol Abuse, 79-83; med dir, Nat Coun Alcoholism, 83; MED DIR, ALCOHOLISM CHEM DEPENDENCY & COMPULSIVE GAMBLING PROGS & DIR S OAKS INST, S OAKS HOSP, AMITYVILLE, NY, 84-; CLIN PROF PSYCHIAT STATE UNIV NY, STONY BROOK, 84- *Concurrent Pos:* Clin asst prof psychiat, Sch Med, State Univ NY Stony Brook, 71-84; co-dir, Caribbean Inst Alcoholism, 75-87; bd dirs, Nat Comn Confidentiality Health Rec, 76-80; Nat Coun Alcoholism, 79- & Children of Alcoholics Found, 82-; mem panel Alcoholism, Am Med Asn, 83-85; mem, Coun on Addiction Psychiat, Am Phychiat Asn, 90- & Gov Coun Pychiat & Substance Abuse Facil, Am Hosp Asn, 88-90. *Mem:* Fel Am Psychiat Asn; Am Med Soc Alcoholism (pres, 79-80); Am Acad Psychiat in Alcoholism & Addictions; Nat Coun Compulsive Gambling. *Res:* Clinical aspects, treatment, public policy, family aspects and confidentiality of records of alcoholism; drug dependence; compulsive gambling. *Mailing Add:* South Oaks Hosp 400 Sunrise Hwy Amityville NY 11701

BLUMEN, WILLIAM, b Hartford, Conn, Aug 7, 31; m 60; c 2. FLUID DYNAMICS, METEOROLOGY. *Educ:* Fla State Univ, BS, 57, MS, 58; Mass Inst Technol, PhD(mountain-wave drag), 63. *Prof Exp:* Res assoc, Univ Oslo, 63-64; fel meteorol, Nat Ctr Atmospheric Res, 64-66; from asst prof to assoc prof, 66-74, PROF ASTROPHYS, PLANETARY & ATMOS SCI, UNIV COLO, BOULDER, 74- *Concurrent Pos:* Lectr, Univ Colo, 65-66; res grants, NATO, 63-65, NSF, 68-70; NATO sr vis fel, Cambridge Univ, 70-71; Univ Colo fac fel & sci vis, Europ Centre Medium Range Weather Forecasts, Eng, 76-77; ed, J Atmospheric Sci, 78-; partic, The Alpine Exp Observ Prog, 82; sci visitor, Oxford Univ, 85; vis prof, Naval Postgrad Sch, 86-; hon prof, Dept Atmospheric Sci, Nanjing Univ, 88; ed, Meteorol Monographs Am Meteorol Soc, 84- *Mem:* Am Meteorol Soc. *Res:* Theoretical aspects of geophysical fluid dynamics, with application to the dynamics of atmospheric and oceanic circulations; wave motion, atmospheric fronts and hydrodynamic instability. *Mailing Add:* Dept Astrophys Sci Univ Colo Boulder CO 80309

BLUMENFELD, HENRY A, b Amsterdam, Neth, May 31, 25; US citizen; m 56; c 2. PHYSICS, PARTICLE DETECTION. *Educ:* Harvard Univ, AB, 48; Columbia Univ, PhD(physics), 56. *Prof Exp:* Res assoc physics, Mass Inst Technol, 55-56 & Duke Univ, 56-57; mem physics staff, Princeton Univ, 58-67 & Europ Orgn Nuclear Res, Geneva, Switz, 67-70; MEM PHYSICS STAFF, SACLAY NUCLEAR RES CTR, FRANCE, 70- *Mem:* Am Phys Soc. *Res:* High energy nuclear physics; experimental, elementary particles; large and rapid cycling bubble chambers; scintillation and photon dectors; data analysis. *Mailing Add:* DPhPE-STIPE Saclay Nuclear Res Ctr BP 2 Gif-sur-Yvette 91191 France

BLUMENFELD, MARTIN, b Baltimore, Md, Sept 4, 41; m 66; c 2. CELL BIOLOGY, MOLECULAR BIOLOGY. *Educ:* Johns Hopkins Univ, BA, 63; Case Western Reserve Univ, PhD(biol), 68. *Prof Exp:* Fel zool, Univ Mich, 67-68; res assoc, Univ Tex, 68-72; fel genetics, Univ Wis, 72-73, res assoc, 73-74; asst prof zool, Univ Minn, Minneapolis, 74-80. *Concurrent Pos:* USPHS res grant, Inst Gen Med, 74. *Res:* Satellite DNA and molecular organization of chromosomes of drosophila. *Mailing Add:* Genetics 250 Bio Sci Ctr Univ Minn 1445 Gortner Ave St Paul MN 55108

BLUMENFELD, OLGA O, b Lodz, Poland, Apr 6, 23; nat US; m 44; c 1. BIOCHEMISTRY. *Educ:* City Col New York, BS, 46; Univ Colo, MS, 48; NY Univ, PhD(biochem), 57. *Prof Exp:* Fel, Nat Res Coun & Rockefeller Inst, 57-61; from asst prof to assoc prof, 61-72, PROF BIOCHEM, ALBERT EINSTEIN COL MED, 72- *Concurrent Pos:* Sr investr, Arthritis Found, 61-67; NIH career develop award, 67. *Mem:* Am Chem Soc; Fedn Am Soc Exp Biol. *Res:* Protein structure. *Mailing Add:* 11 Highland Ave North Tarrytown NY 10593

BLUMENFIELD, DAVID, b Philadelphia, Pa, Mar 7, 28; m 59; c 2. HORTICULTURE, PLANT PHYSIOLOGY. *Educ:* Del Valley Col, BS, 50; Rutgers Univ, MS, 56, PhD(hort, plant physiol), 59. *Prof Exp:* Spec rep tech sales, Wyeth Labs, Pa, 53-55; res asst hort, Rutgers Univ, 55-59; from asst prof to assoc prof, 59-71, PROF HORT, DEL VALLEY COL, 71- *Mem:* AAAS; Am Soc Hort Sci; fel Nat Asn Col & Teachers Agr. *Res:* Plant nutrition; agricultural climatology. *Mailing Add:* Dept Hort Del Valley Col Doylestown PA 18901

BLUMENSON, LESLIE ELI, b New York, NY, Mar 20, 34; c 2. MATHEMATICAL BIOLOGY, BIOSTATISTICS. *Educ:* NY Univ, AB, 55, MS, 56, PhD(math), 62. *Prof Exp:* Sr res mathematician, Electronics Res Lab, Columbia Univ, 56-63; USPHS trainee, Univ Chicago, 63-65; from sr cancer res scientist to assoc cancer res scientist, 65-71, PRIN CANCER RES SCIENTIST, ROSWELL PARK MEM INST, 72- *Concurrent Pos:* Res asst prof, Dept Biomet, State Univ NY, Buffalo, 86- *Honors & Awards:* George W Snedcor Award, Am Statist Asn, 77. *Mem:* Soc Math Biol. *Res:* Mathematical models in biological sciences; cancer research; biostatistics. *Mailing Add:* Dept Biomath Roswell Park Mem Inst Buffalo NY 14263

BLUMENSTEIN, MICHAEL, b New York, NY, Nov 16, 47. BIOCHEMISTRY. *Educ:* City Col New York, BS, 68; Calif Inst Technol, PhD(chem), 72. *Prof Exp:* Mem tech staff biochem, Bell Labs, 73-74; res assoc, Univ Ariz, 74-75; ASST PROF BIOCHEM, SCH MED, TUFTS UNIV, 75- *Concurrent Pos:* NIH fel, Univ Ariz, 75. *Mem:* AAAS; Am Chem Soc. *Res:* Nuclear magnetic resonance studies of biochemical systems; protein-hormone interactions; mechanism of action of adenine nucleotide coenzymes. *Mailing Add:* Dept Chem Hunter Coll 695 Park Ave New York NY 10021

BLUMENSTOCK, DAVID A, b Newark, NJ, Feb 14, 27; m 52; c 3. MEDICINE. *Educ:* Union Col, NY, BS, 49; Cornell Univ, MD, 53; Am Bd Surg, dipl, 59; Am Bd Thoracic Surg, dipl, 60. *Prof Exp:* Resident surgeon, Mary Imogene Bassett Hosp, 54-60, NIH res career develop award, 62-63, surgeon-in-chief, 63-86; PROF CLIN SURG, COL PHYSICIANS & SURGEONS, COLUMBIA UNIV, 74- *Mem:* Soc Univ Surgeons; Am Asn Thoracic Surg; Soc Thoracic Surgeons; Am Surg Asn; Am Fedn Clin Res; Int Transplantation Soc. *Res:* Transplantation of tissues and organs; preservation of living tissues. *Mailing Add:* Mary Imogene Bassett Hosp Atwell Rd Cooperstown NY 13326

BLUMENTHAL, GEORGE RAY, b Milwaukee, Wis, Oct 20, 45; m 77. ASTROPHYSICS. *Educ:* Univ Wis-Milwaukee, BS, 66; Univ Calif, San Diego, PhD(physics), 71. *Prof Exp:* Sr scientist astrophys, Am Sci & Eng, 71-72; from asst prof to assoc prof, 72-83, PROF ASTRON & ASTROPHYS, UNIV CALIF, SANTA CRUZ, 80- *Mem:* Am Astron Soc. *Res:* Galaxy formation; nature of active galaxies and quasistellar objects; compact x-ray sources; cosmology; cosmic rays; clusters of galaxies. *Mailing Add:* Dept Astron Univ Calif Santa Cruz CA 95064

BLUMENTHAL, HAROLD JAY, b New York, NY, Jan 21, 26; m 50; c 3. MICROBIOLOGY, MICROBIAL BIOCHEMISTRY. *Educ:* Ind Univ, BS, 47; Purdue Univ, MS, 49, PhD(bact), 53. *Prof Exp:* Teaching asst chem, Purdue Univ, 47-49; Am Cancer Soc fel, Inst Cancer Res, 53-54; from instr to assoc prof microbiol, Univ Mich, 54-65; chmn dept, 65-86, PROF MICROBIOL, STRITCH SCH MED, LOYOLA UNIV CHICAGO, 65- *Concurrent Pos:* AEC fel, 52-53; res assoc, Rackham Arthritis Res Unit, 54-56; NIH spec res fel, 63-64; mem adv comt study, NIH Res Training Grant Progs, Nat Acad Sci, 66-75; consult, Argonne Univs Asn-Argonne Nat Lab, 71-74; Am Acad Microbiol vis prof, Cath Univ Chile, 72. *Mem:* AAAS; Am Soc Microbiol; Asn Am Med Cols; Am Chem Soc; Am Asn Univ Profs. *Res:* Pathways of carbohydrate catabolism in microorganisms; metabolism of hexaric acids; staphylococcal virulence. *Mailing Add:* Dept Microbiol & Immunol Loyola Univ Stritch Sch Med Maywood IL 60153

BLUMENTHAL, HERBERT, b New York, NY, May 1, 25; m 50; c 2. BIOCHEMISTRY, TOXICOLOGY. *Educ:* City Col New York, BS, 48; Univ Southern Calif, PhD(biochem), 55. *Prof Exp:* Actg chief, Pharmaco-Dynamics Br, Div Pharmacol, Food & Drug Admin, 55-58; biochemist, Nat Inst Dent Res, 58-61; chief petitions res br, 61-71, dep dir, Div Toxicol, 70-74, actg dir, 74-76, DIR, DIV TOXICOL, FOOD & DRUG ADMIN, 76- *Mem:* Fel AAAS; NY Acad Sci; Soc Toxicol; Environ Mutagen Soc. *Res:* Toxicology of food additives, pesticides, colors, drugs. *Mailing Add:* 12545 Two Farm Dr Silver Spring MD 20904

BLUMENTHAL, HERMAN T, b New York, NY, Apr 8, 13; m 40; c 2. PATHOLOGY. *Educ:* Rutgers Univ, BS, 34; Univ Pa, MS, 36; Washington Univ, PhD(path), 38, MD, 42; Am Bd Path, dipl. *Prof Exp:* Asst path, Sch Med, Washington Univ, 39-42; resident, Jewish Hosp, St Louis, 42-43; lab dir, Jewish Hosp, Louisville, 46; dir div labs, Jewish Hosp, St Louis, 50-57, inst exp path, 57-62; gerontol res assoc, 62-68, RES PROF GERONT, BIOPSYCHOL RES LAB, DEPT PSYCHOL, WASHINGTON UNIV, 68- *Concurrent Pos:* Attend pathologist, Vet Admin Hosp, Jefferson Barracks, Mo, 47-50; dir clin res prog aging, 60-66; co-chmn endocrinol sect, Int Cong Gerontol, London, 54, prog biol & clin med, San Francisco, 60; consult pathologist, Cochran Vet Admin Hosp, 64-; mem coun arteriosclerosis, Am Heart Asn. *Honors & Awards:* Berg Prize, 41. *Mem:* AAAS; Soc Exp Biol & Med; Am Asn Cancer Res; Am Asn Path & Bact; Col Am Path; Sigma Xi. *Res:* Cancer; endocrinology; aging; transplantation; virus diseases; amebiasis; arteriosclerosis. *Mailing Add:* Dept Psychol Washington Univ St Louis MO 63130

BLUMENTHAL, IRWIN S(IMEON), b Derby, Conn, June 6, 25; c 2. ELECTRICAL ENGINEERING. *Educ:* Yale Univ, BE, 45, ME, 47, PhD(elec eng), 51. *Prof Exp:* Test engr, Gen Elec Co, 45-46; lab asst elec eng, Yale Univ, 47-48, instr, 48-51; res engr, Northrop Aircraft, Inc, 51-53; res engr, Rand Corp, 53-73; mem staff, Off Naval Res, NATO, 73-76; RES ENGR, RAND CORP, 76- *Honors & Awards:* Second Prize, Inst Elec & Electronics Engrs, 52. *Mem:* Sigma Xi; Inst Elec & Electronics Engrs; Mil Opers Res Soc; Asn Unmanned Vehicle Safety; Am Defense Preparedness Asn. *Res:* Servomechanism; control system analysis and design; defense systems, concepts, analysis and evaluation; sonar; medical electronics; astronautics; cost-benefit health analysis, positive aspects of pollution; concepts and systems which will minimize the chance of nuclear war occurring. *Mailing Add:* The Rand Corp 1700 Main St Santa Monica CA 90406

BLUMENTHAL, KENNETH MICHAEL, b Chicago, Ill, Aug 24, 45; m 67; c 2. BIOCHEMISTRY. *Educ:* Univ Wis, BSc, 67; Univ Chicago, PhD(biochem), 71. *Prof Exp:* Fel biochem, Univ Calif, Los Angeles, 71-74; asst prof, Univ Fla, 74-76; from asst prof to assoc prof, 76-86, PROF BIOCHEM & MOLECULAR BIOL, COL MED, UNIV CINCINNATI, 86- *Concurrent Pos:* NIH fel, 71; NIH res career develop award, 77; Biochem Study Sect, NIH, 86- *Mem:* Am Chem Soc; Am Soc Biol Chemists; Sigma Xi. *Res:* Protein chemistry; polypeptide neurotoxin structure and function; structure and action of protein cytolysins. *Mailing Add:* Dept Biol Chem Col Med Univ Cincinnati Cincinnati OH 45267

BLUMENTHAL, RALPH HERBERT, b New York, NY, Feb 24, 25; m 48; c 3. PHYSICS. *Educ:* Brooklyn Col, BA, 45, MA, 49; NY Univ, PhD(sci educ), 56. *Prof Exp:* Radio engr, Hamilton Radio Corp, 45; lectr physics, Brooklyn Col, 46-48; physicist & proj leader, US Naval Supply Activ, 48-52; physicist & supvr test group, Picatinny Arsenal, 52; lectr physics, Brooklyn Col & Queens Col, NY, 52-54; tutor, City Col New York, 54-56, instr physics, 58-61; sr engr & group leader, Sperry Gyroscope Corp, NY, 58-62, assoc mem tech & mgt staff, 62-63; staff engr, Grumman Aerospace Corp, 63-70; physics teacher, Sewanhaka Cent High Sch Dist, Elmont, NY, 70-88; ADJ ASSOC PROF NATURAL SCI, HOFSTRA UNIV, 88- *Concurrent Pos:* Adj asst prof physics, Queensborough Community Col, 70-78, adj assoc prof, 78-81; adj asst prof physics, Adelphi Univ, 76-80, adj assoc prof, 80-87. *Mem:* Fel AAAS; Am Phys Soc. *Res:* Laser communication systems; electrooptical light modulation; space systems; microwave electronics; transistor physics; pyrotechnic radiation; heat transfer; programmed physics instruction. *Mailing Add:* 15 Bonnie Dr Westbury NY 11590-2803

BLUMENTHAL, ROBERT ALLAN, b New York, NY, Sept 24, 51. DIFFERENTIAL GEOMETRY, FOLIATION THEORY. *Educ:* Univ Rochester, BA, 73; Washington Univ, PhD(math), 78. *Prof Exp:* Maitre asst math, Univ de Lille, France, 78-79; asst prof, 79-82, ASSOC PROF MATH, ST LOUIS UNIV, 82- *Mem:* Am Math Soc. *Res:* Differential geometry of foliations; effects of various targential and transverse structures on the global structure of a foliated manifold. *Mailing Add:* 1110 Bonsella St Walla Walla WA 99362

BLUMENTHAL, ROBERT MCCALLUM, b Chicago, Ill, Feb 7, 31; m 52; c 2. MATHEMATICS. *Educ:* Oberlin Col, BA, 52; Cornell Univ, PhD, 56. *Prof Exp:* Instr, 56-57, from asst prof to assoc prof, 57-65, PROF MATH, UNIV WASH, 65- *Mem:* Inst Math Statist; Am Math Soc. *Res:* Probability theory; mathematical statistics. *Mailing Add:* Univ Wash Seattle WA 98195

BLUMENTHAL, ROBERT MARTIN, b Lafayette, Ind, Nov 3, 51. MOLECULAR GENETICS, COMPUTER SOFTWARE. *Educ:* Ind Univ, AB, 72; Univ Mich, MSc, 75, PhD(microbiol), 77. *Prof Exp:* Am Cancer Soc fel, Univ BC, 77-79; fel, Cold Spring Harbor Lab, 79-81; ASST PROF MICROBIOL, MED COL OHIO, 81- *Mem:* Am Soc Microbiol. *Res:* Mechanism of protein-DNA interaction; regulation of synthesis and activity of restriction endonucleases; development of computer software for processing nucleic acid sequence information. *Mailing Add:* Dept Microbiol Med Col Ohio CS 10008 Toledo OH 43699

BLUMENTHAL, ROBERT N, b Oak Park, Ill, Aug 31, 34; m 57; c 2. PHYSICAL METALLURGY. *Educ:* Northwestern Univ, BS, 57, MS, 61, PhD(mat sci), 62. *Prof Exp:* Asst prof metall eng, Mich Technol Univ, 62-64; assoc prof mech eng, 64-69, PROF MECH ENG, MARQUETTE UNIV, 69- *Res:* Role of departures from stoichiometry of compounds on their thermodynamic and electrical properties. *Mailing Add:* Dept Mech Eng Marquette Univ Milwaukee WI 53233

BLUMENTHAL, ROBERT PAUL, b Djakarta, Indonesia, July 28, 38; US citizen; m 66; c 3. BIOPHYSICS, BIOCHEMISTRY. *Educ:* Univ Leyden, Holland, BSc, 60, MSc, 62; Weizman Inst, Israel, PhD(phys chem), 67. *Prof Exp:* Fel neurobiol, Inst Pasteur, 68-70; res assoc, Columbia Univ Col Physicians & Surgeons, 70-71; vis assoc biophys, Nat Inst Arthritis & Metab Dis, 71-72; vis scientist, 72-75, sr investr biophys, 78-80, CHIEF, SECT MEMBRANE STRUCT & FUNCTION, LAB THEORET BIOL, NAT CANCER INST, 80- *Mem:* Biophys Soc; AAAS. *Res:* Cell membrane structure and function; membrane transport; receptors; liposomes; mobility of cell surface receptors; reconstitutions; membrane fusion; viral entry. *Mailing Add:* NIH Bldg 10 Rm 4B56 Bethesda MD 20892

BLUMENTHAL, ROSALYN D, b New York, NY, Mar 7, 59; m 85; c 1. TUMOR BIOLOGY, EXPERIMENTAL THERAPEUTICS. *Educ:* City Col NY, BS, 70; City Univ NY, MPhil, 82, PhD(biomed sci), 84. *Prof Exp:* Postdoctoral fel, Pop Coun, Rockefeller Univ, 84-85; res assoc, Dept Radiother, Mt Sinai Med Ctr, NY, 85-87; sr res assoc, 87-90, ASST MEM, CTR MOLECULAR MED & IMMUNOL, 90-, DIR TUMOR BIOL, 91- *Concurrent Pos:* Lectr, Baruch Col, 83-85 & Hunter Col, 84-86; adj asst prof, Dept Biol, St John's Univ, 85-86 & Queensborough Community Col, 85-86. *Mem:* Am Asn Cancer Res; NY Acad Sci; AAAS. *Res:* Antibody-targeted tumor therapy; tumor physiology affecting therapy; animal modeling of human neoplasms; characterization of tumor biological charges post therapy. *Mailing Add:* 77-44 166th St Flushing NY 11366

BLUMENTHAL, SAUL, b Philadelphia, Pa, Oct 5, 35; m 59; c 3. MATHEMATICAL STATISTICS. *Educ:* Cornell Univ, BME, 58, PhD(statist), 62. *Prof Exp:* Res assoc statist, Stanford Univ, 61-62; asst prof indust eng, Univ Minn, 62-64; asst prof statist, Rutgers Univ, 64-66; assoc prof indust eng, NY Univ, 66-71, prof, 71-73; prof statist, Univ Ky, 73-78; prof math, Univ Ill, 78-84; PROF STATIST, OHIO STATE UNIV, 83- *Concurrent Pos:* Consult, Avionics Div, Minneapolis-Honeywell Regulator Co, 63-64, US Naval Appl Sci Lab, 68-70, Touche, Ross & Co, 69- & US Army Res Off, Durham; NSF res grant, 65-71; proj dir res contract, US Naval Appl Sci Lab, 68-69; reviewer, Math Rev; vis prof opers res, Cornell Univ, 72-73; USAF Off Sci Res grant, 75-78 & 84-91, NSF res grant, 79-83. *Mem:* Fel Inst Math Statist; fel Am Statist Asn; Opers Res Soc Am; Int Statist Inst. *Res:* Applied statistics; sequential methods in ranking and selection problems; reliability of complex systems; renewal processes; estimation of ordered parameters; estimation of population size with truncated data; analysis of incomplete data. *Mailing Add:* Dept Statist 1958 Neil Ave Columbus OH 43210

BLUMENTHAL, THOMAS, TRANSCRIPTIONAL REGULATION, GENE STRUCTURE. *Educ:* Johns Hopkins Univ, PhD(genetics), 70. *Prof Exp:* PROF BIOL, IND UNIV, 73- *Mailing Add:* Dept Biol Ind Univ Bloomington IN 47405

BLUMENTHAL, WARREN BARNETT, industrial chemistry, for more information see previous edition

BLUMER, JEFFREY L, b Philadelphia, Pa, 1951; m. PHARMACOLOGY, PEDIATRIC MEDICINE. *Educ:* Northwestern Univ, PhD(pharmacol), 77; Case Western Reserve, MD, 79. *Prof Exp:* Asst prof pharmacol, Med Sch, Case Western Reserve, 80-86; ASSOC PROF PEDIAT, RAINBOW BABIES & CHILDREN'S HOSP, 86-, ASSOC PROF ONCOL, 86- *Concurrent Pos:* Prof pediat, Univ Va, 89, Case Western Res Univ, 91. *Mem:* AAAS; Soc Pediat Res; Am Asn Pediat; Am Soc Microbiol; Soc Critical Care Med. *Res:* Pediatric Clinical Pharmacology; genetics of drug metabolism; pediatric critical care. *Mailing Add:* Dept Pediat Case Western Res Univ Rainbow Babies & Childrens Hosp 2101 Adelbert Rd Cleveland OH 44106

BLUMHAGEN, VERN ALLEN, b Ronan, Mont, July 18, 29; m 54; c 3. ELECTRICAL & SYSTEMS ENGINEERING. *Educ:* Iowa State Univ, BS, 54. *Prof Exp:* Asst elec eng, Iowa State Univ, 54-55; engr electronic instruments, x-ray div, Gen Elec Co, 55-58; sr res engr guid anal, Rockwell Int Corp, 58-60, supvr cruise systs anal, 60-62, chief systs eval, 62-63, guid anal, 63-66, mgr guid & control anal, 66-67, dir advan anal, 67-70, mgr systs requirements & anal, NAm Rockwell Corp, 70-72, mgr oper syst technol, 72-85, mgr, systs anal, Autonetics Div, 85-87; RETIRED. *Mem:* Inst Elec & Electronics Engrs. *Res:* System analysis; operations analysis; computer applications; optimal estimation and control applied to aerospace systems; system concept development, definition, design and evaluation. *Mailing Add:* 47182 Goodpasture Rd Vida OR 97488

BLUMSTEIN, ALEXANDRE, b Grodno, Poland, Jan 13, 30; m 59; c 2. POLYMER CHEMISTRY. *Educ:* Univ Paris, BS, 51; Univ Toulouse, Chem Engr, 52; Univ Strasbourg, PhD(polymer chem), 60. *Prof Exp:* Res trainee polymer chem, Nat Sci Res Ctr, France, 52-54, researcher, 54-57 & 59-60, res asst, 60; sr proj engr, Instruments Div, Budd Co, 61, eng supvr, 61-62; NSF res assoc, Univ Del, 62-64; from asst prof to assoc prof chem, Lowell Technol Inst, 64-72; PROF CHEM, UNIV LOWELL, 72- *Mem:* Am Chem Soc. *Res:* Polymer structure; characterization and application of polymers; photoelasticity; matrix polymerization; polymerization in liquid-crystalline media; liquid crystalline order in polymers; functional polymers. *Mailing Add:* Two Regina Dr Chelmsford MA 01824

BLUMSTEIN, ALFRED, b New York, NY, June 3, 30; m 58; c 3. OPERATIONS RESEARCH. *Educ:* Cornell Univ, BEP, 51, PhD(opers res), 60; Univ Buffalo, MA, 54. *Prof Exp:* Prin opers analyst, Cornell Aeronaut Lab, 51-61; res staff, Sci & Technol Div, Inst Defense Anal, 61-69; prof, 69-79, J ERIK JONSSON PROF & DEAN, URBAN SYSTS & OPERS RES, SCH URBAN & PUB AFFAIRS, CARNEGIE-MELLON UNIV, 79-, PROF, DEPT ENG & PUB POLICY & DIR URBAN SYSTS INST, 79- *Concurrent Pos:* Mem staff, Off Naval Res, 54-55 & 57; vis assoc prof, Cornell Univ, 63-64; dir, Sci & Technol Task Force, President's Comn Law Enforcement & Admin of Justice, 66-67; dir, Off Urban Res, Inst Defense Anal, 68-69; chmn, deterrence & incapacitation panel, Nat Res Coun, 76-78, CRLEAJ, 79-83, sentencing res panel, 80-82, criminal careers panel, 83-86; chmn, Pa Comn Crime & Delinq, 79-90; panel mem, Decision & Mgt Sci Prog, NSF, 82-85; overseas fel, Churchill Col, Univ Cambridge, 83; sci comn, Int Soc Criminol. *Honors & Awards:* Kimball Medal, Opers Res Soc Am, 85; Sutherland Award, Am Soc Criminol, 87. *Mem:* Fel AAAS; Opers Res Soc Am (pres, 77-78); Am Statist Asn; Inst Mgt Sci (pres, 87-88); fel Am Soc Criminol (pres, 91-92). *Res:* Quantitative and policy research in law enforcement and criminal justice; urban transportation; family planning. *Mailing Add:* Sch Urban & Pub Affairs Carnegie-Mellon Univ Pittsburgh PA 15213

BLUMSTEIN, CARL JOSEPH, b Minneapolis, Minn, June 7, 42; m; c 4. ENERGY CONSERVATION & INFORMATION. *Educ:* Reed Col, BA, 66; San Diego State Univ, MS, 68; Univ Calif, San Diego, PhD(chem), 74. *Prof Exp:* Staff scientist, Energy & Environ Div, Lawrence Berkeley Lab, 74-81; RES POLICY ANALYST, UNIVERSITYWIDE ENERGY RES GROUP, UNIV CALIF, BERKELEY, 81- *Concurrent Pos:* Spec consult, Calif Energy Comn, 78-81. *Mem:* AAAS; Am Coun Energy-Efficient Econ (pres). *Res:* Energy use and conservation, including development of computer models of energy demand, analysis of energy conservation policies, assessment of energy conservation technologies. *Mailing Add:* Universitywide Energy Res Group Bldg T-9 Univ Calif Berkeley CA 94720

BLUMSTEIN, RITA BLATTBERG, b Krakow, Poland, Jan 11, 37; French citizen; m 59; c 2. POLYMER CHEMISTRY. *Educ:* Univ Paris, BS, 59; Univ Strasbourg, MS, 60; Univ Del, PhD(phys chem), 65. *Prof Exp:* Asst prof chem, Merrimack Col, 65-69; from asst prof to assoc prof, 75-81, PROF CHEM, UNIV LOWELL, 81- *Concurrent Pos:* Mem polymer educ comt, Am Chem Soc, 76-, legis counr, 77- *Mem:* Am Chem Soc. *Res:* Polymerization within ordered systems; liquid crystalline order in polymers; morphology of blends; synthesis of oligomers. *Mailing Add:* Two Regina Dr Chelmsford MA 01824

BLUNDELL, GEORGE PHELAN, HEMATOLOGY, IMMUNOLOGY. *Educ:* McGill Univ, MD, 48. *Prof Exp:* chief lab serv, Vet Admin Med Ctr, 80-88; RETIRED. *Mailing Add:* PO Box 1253 Shepherdstown WV 25443

BLUNDELL, JAMES KENNETH, b Prescot, Eng, Oct 21, 49; div; c 1. MANUFACTURING SYSTEMS & ROBOTICS. *Educ:* Univ Salford, BS, 72; Loughborough Univ Technol, MS, 74; Univ Nottingham, PhD(mech eng), 77. *Prof Exp:* Res asst, Hawker Siddeley Aviation & Loughborough Univ, 73; asst lectr, Univ Nottingham, 74-76; sr proj engr, Tecquipment Ltd, Nottingham, 76-77; lectr, Univ West Indies, 77-79; from asst prof to assoc prof mech eng, Univ Mo, Columbia, 82-88; mem staff, Truman Eng Labs, 81-88, assoc dir, Design Productivity Ctr, 85-88, DIR, CTR INTELLIGENT DESIGN, DEPT COMPUTER SCI, UNIV MO, KANSAS CITY, 88-, ASSOC PROF, COMPUTER SCI, 88- *Concurrent Pos:* Consult, Western Elec, Mo, 81-; mem, Trinidad & Tobago Comn Prod Eng, 78; Pres, James K. Blundell Asn, 83-; mem comt, Unified Life Cycle Eng, Nat Acad Sci. *Mem:* Am Soc Mech Engrs; Am Soc Metals; Soc Mfg Engrs & Robotics Inst; Soc Logistics Engrs; Brit Inst Prod Engrs. *Res:* Planning and introduction of automation into manufacturing planks including basic robotics research and the simulation of industrial processes using digital computers; lubrication in metalworking processes. *Mailing Add:* Ctr Intelligent Design Dept Computer Sci Univ Mo Kansas City MO 64111

BLURTON, KEITH F, b Grays, Eng, Apr 11, 40; m 66; c 1. ELECTROCHEMICAL SYSTEMS, RESEARCH MANAGEMENT. *Educ:* Southampton Univ, BSc, 61, PhD(chem), 66. *Prof Exp:* Lectr, Portsmouth Polytech, 61-66; sr chemist, Leesona Corp, 66-68; res dir, Energetics Sci Inc, Elmsford, 68-76; dir energy conversion & storage res, Inst Gas Technol, Chicago, 76-80; vpres chem res, PCK Technol Div, Kollmorgen

Corp, Mellville, 80-86; GEN MGR, CTR INNOVATIVE TECHNOL, HERNDON, 86- *Honors & Awards:* Silver Medal Award, Am Electroplaters & Surface Finishers Soc, 86. *Mem:* Am Chem Soc; Electrochem Soc; AAAS; Royal Soc Chem. *Res:* Printed circuits; batteries; electrolytic processes; fuel cells; electroplating; electrochemical sensors; electrolysis; electroless deposition; materials selection and stability. *Mailing Add:* 1151 Clinch Rd Herndon VA 22070

BLUZER, NATHAN, b Tomashow Lubelsky, Poland, Jan 1, 47; US citizen. DEVICE PHYSICS, ELECTRONICS. *Educ:* Univ Md, BSEE, 67, PhD(physics), 74. *Prof Exp:* Assoc engr radar, Westinghouse, Cockeysville, 67-68, engr design, Westinghouse Aerospace & Electronics Systs Div, 68-71; res asst physics, Univ Md, 71-74; sr engr, Solid State Devices, Westinghouse Advan Technol Labs, 74-77, fel physicist, 77-81, adv physicist, IR Focal Plane Detectors, 81-84; HEAD INDUST RELS PROG, WESTINGHOUSE ADVAN TECHNOL DIV, 85- *Mem:* Am Phys Soc; Inst Elec & Electronics Engrs. *Res:* Infrared detector and charge coupled devices; radiation effects in metal-nitrate-oxide silicon. *Mailing Add:* Westinghouse Advan Technol Labs Elect Corp PO Box 1521 MS-3531 Baltimore MD 21203

BLY, CHAUNCEY GOODRICH, b Honan, China, June 14, 20; m 46; c 5. MEDICAL ADMINISTRATION, PATHOLOGY. *Educ:* Univ Rochester, MS & MD, 46, PhD(exp path), 53, Am Bd Path, cert, 53. *Prof Exp:* Pathologist, Army Med Nutrit Lab, 47-49; Atomic Energy Comn fel, Univ Rochester, 49-51; asst prof path & oncol, Med Ctr, Univ Kans, 51-54, assoc prof, 54-57; chief path, Highland Hosp, 57-59; assoc prof path, Duke Univ, 59-60; prof path, Sch Med, Wake Forest Univ, 60-61; pres & adj prof biochem, Thiel Col, 61-74; assoc dir & sr path dir, Armed Forces Inst Path, 74-76; PATHOLOGIST, NAT CANCER INST, 76- *Concurrent Pos:* Consult path & radiosotopes, Vet Admin, 52-61; pres, Kans Soc Pathologists, 56-57; consult, Nat Cancer Inst, 60-61, Public Health Serv, 71-72; dir, Med Mus, Armed Forces Inst Path, 74-76; clin prof path, Uniformed Serv Sch Med, Univ Health Sci, 74- *Mem:* AAAS; Am Asn Path; Int Acad Path; Asn Am Med Col; Am Soc Cytol. *Res:* Protein metabolism; cancer; experimental pathology; nutrition. *Mailing Add:* Four Cedar Dr Greenville PA 16125

BLY, DONALD DAVID, b Bryan, Ohio, Sept 24, 36; m 59; c 3. ANALYTICAL CHEMISTRY, POLYMER CHEMISTRY. *Educ:* Kenyon Col, BA, 58; Purdue Univ, MS, 61, PhD(anal chem), 62. *Prof Exp:* Eli Lilly fel, Purdue Univ, 62-63; res chemist, Carothers Res Lab, 63-67, sect supvr, 67-77, MGR ANAL SCI, CENT RES DEPT, E I DU PONT DE NEMOURS & CO, INC, 77- *Concurrent Pos:* Chmn, Anal Div, 85, Am Chem Soc & Delaware Sect, 87. *Mem:* Am Chem Soc; Am Soc Testing & Mat; AAAS; Soc Appl Spectros. *Res:* Polymer physical chemistry; gel permeation chromatography; spectroscopy. *Mailing Add:* 409 Brentwood Dr Carrcroft Crest Wilmington DE 19803

BLY, ROBERT STEWART, b Lakeland, Fla, Aug 10, 29; m 56. PHYSICAL ORGANIC CHEMISTRY, ORGANOMETALLIC CHEMISTRY. *Educ:* Fla Southern Col, BS, 51; Northwestern Univ, MS, 56; Univ Colo, PhD(org chem), 58. *Prof Exp:* Res chemist, Nylon Res Lab, E I du Pont de Nemours & Co, 57-59; NIH fel, Mass Inst Technol, 59-61; from asst prof to assoc prof chem, 61-70, head dept, 70-73, PROF CHEM, UNIV SC, 70- *Mem:* Am Chem Soc; The Royal Soc Chem. *Res:* Mechanisms of reactions; carbonium ions; pi-complexes; transition-metal alkylidenes. *Mailing Add:* Dept Chem Univ SC Columbia SC 29208

BLY, SARA A, b Arkansas City, Kans, June 22, 46. COMPUTER SCIENCE, COMPUTING HUMAN INTERACTIONS. *Educ:* Univ Kans, BS, 68; Stanford Univ, MS, 69; Univ Calif, Davis, PhD(computer sci), 82. *Prof Exp:* User interface designer, 84-87, MEM RES STAFF, SYST SCI LAB, XEROX PALO ALTO RES CTR, 87- *Mem:* Asn Comput Mach (secy, 88-90). *Res:* Computer-supported cooperative work. *Mailing Add:* Xerox Palo Alto Res Ctr 3333 Coyote Hill Rd Palo Alto CA 94304

BLYE, RICHARD PERRY, b West Chester, Pa, Nov 11, 32; m 56; c 2. REPRODUCTIVE BIOLOGY. *Educ:* Trinity Coll, Conn, BS, 55; Rutgers Univ, PhD, 60. *Prof Exp:* Asst zool & gen biol, Rutgers Univ, 55-59; assoc scientist, Div Endocrinol, Ortho Pharmaceut Corp, 59-62, sr scientist, 62-63, Ortho res fel, 64-70; HEALTH SCIENTIST ADMINR, CDB, CPR, NAT INST CHILD HEALTH & HUMAN DEVELOP, 71- *Concurrent Pos:* Consult, Bur Drugs, Food & Drug Admin, 74-79. *Mem:* AAAS; NY Acad Sci; Soc Study Reprod; Am Chem Soc. *Res:* Endocrinology of reproduction; contraceptive development; contraceptive steroids pharmacology and safety. *Mailing Add:* 7352 Minkhollow Rd Highland MD 20777

BLYHOLDER, GEORGE DONALD, b Elizabeth, NJ, Jan 10, 31; m 55; c 3. PHYSICAL CHEMISTRY. *Educ:* Valparaiso Univ, BA, 52; Purdue Univ, BS, 53; Univ Utah, PhD(chem), 56. *Prof Exp:* Fel, Univ Minn, 56-57; assoc, Johns Hopkins Univ, 57-59; from asst prof to assoc prof chem, 59-67, vchmn dept, 68-72, PROF CHEM, UNIV ARK, FAYETTEVILLE, 67- *Concurrent Pos:* Fel, Oxford Univ, Eng, 65; vis prof, Hokkaido Univ, Japan, 72, Univ Munich, Ger, 75, Cornell Univ, 80, Cambridge Univ, Eng, 88. *Mem:* Am Chem Soc; Sigma Xi; Am Asn Univ Prof. *Res:* Kinetics; catalysis; surface chemistry; molecular spectroscopy; molecular orbital theory. *Mailing Add:* Dept Chem Univ Ark Fayetteville AR 72701

BLYLER, LEE LANDIS, JR, b New Brunswick, NJ, Oct 22, 38; m 62; c 2. RHEOLOGY, POLYMER ENGINEERING. *Educ:* Princeton Univ, BSE, 61, MSE, 62, PhD(mech eng), 66. *Prof Exp:* Mem tech staff, 65-71, GROUP SUPVR, BELL LABS, 71- *Honors & Awards:* Fred O Conley Award in Eng/Technol, Soc Plastics Eng. *Mem:* Am Chem Soc; Soc Rheol; Soc Plastics Eng; fel Am Inst Chem. *Res:* Polymer mechanics; flow and processing behavior of polymer melts; relationships between polymer properties and molecular structure; electrical charge storage in polymer dielectrics; polymeric materials for optical fiber waveguides. *Mailing Add:* 55 Kensington Rd Basking Ridge NJ 07920

BLYSTONE, ROBERT VERNON, b El Paso, Tex, July 4, 43; m 64; c 1. CELL BIOLOGY, ELECTRON MICROSCOPY. *Educ:* Univ Tex, El Paso, BS, 65; Univ Tex, Austin, MA, 68, PhD(zool), 71. *Prof Exp:* From asst prof to assoc prof, 71-84, chmn, Dept Biol, 84-86, DIR ELECTRON MICROS, TRINITY UNIV, 71-, PIPER PROF BIOL, 84- *Concurrent Pos:* Consult var publ, 86-91; prin investr, Off Sci Res, USAF, 89-91. *Mem:* Fel AAAS; Electron Micros Soc Am; Nat Sci Teachers Asn; Am Inst Biol Sci; Am Soc Cell Biol; Am Micros Soc; Sigma Xi. *Res:* Fine structure and cytochemistry of developing lung alveoli; growth of macrophage cells. *Mailing Add:* Dept Biol Trinity Univ San Antonio TX 78212

BLYTAS, GEORGE CONSTANTIN, b Cairo, Egypt, Dec 20, 30; US citizen; m 63; c 2. PHYSICAL CHEMISTRY. *Educ:* Am Univ Cairo, BSc, 56; Univ Wis, PhD(phys chem), 61. *Prof Exp:* Chemist, 61-72, sr staff res chemist, 72-81, RES ASSOC, SHELL DEVELOP CO, 81- *Concurrent Pos:* Lectr electrochem, Univ Calif, Berkeley, 66-67 & 69; chmn, Indust Eng Chem Div, Am Chem Soc, 81-82. *Mem:* Sigma Xi; Am Chem Soc; Am Inst Chem Eng. *Res:* Research and development in novel separation systems and processes; adsorbents, zeolites, ion exchange, solvent extraction, membrane separations, hydrometallurgy, chemistry in nonaqueous solvents, energy systems, environmental chemistry and engineering; supercritical systems, surface science and emulsions. *Mailing Add:* 14323 Apple Tree Houston TX 77079

BLYTH, COLIN ROSS, b Guelph, Ont, Oct 24, 22; m 55; c 6. MATHEMATICAL STATISTICS. *Educ:* Queen's Univ, Ont, BA, 44; Univ Toronto, MA, 46; Univ Calif, PhD(statist), 50. *Prof Exp:* From asst prof to prof math, Univ Ill, Urbana, 50-74; PROF MATH, QUEEN'S UNIV, KINGSTON, 72- *Concurrent Pos:* Assoc ed, J Am Statist Asn, 67-71. *Mem:* fel Inst Math Statist; fel Am Statist Asn; Math Asn Am; Statist Soc Can. *Res:* Statistical inference. *Mailing Add:* Dept of Math & Statist Queen's Univ Kingston ON K7L 3N6 Can

BLYTHE, DAVID K(NOX), b Georgetown, Ky, May 18, 17; m; c 3. CIVIL ENGINEERING. *Educ:* Univ Ky, BS, 40, CE, 48; Cornell Univ, MCE, 50. *Prof Exp:* Jr civil engr, US Forest Serv, 40-41; mat engr, Ohio River Div Labs, Corps Engrs, US Army, 46-47; from instr to prof civil & hwy eng, Univ Ky, 47-87, chmn dept, 57-69, assoc dean continuing educ & exten, Col Eng, 69-87; RETIRED. *Concurrent Pos:* Consult, Ky Dept Hwys, 54-57 & Spindletop Res, Inc, 64-; OEEC & NSF sr vis fel, Univ Durham, 61; Fulbright lectr, Ecuador, 69. *Mem:* Am Soc Civil Engrs; Am Soc Eng Educ; Am Cong Surv & Mapping; Nat Soc Prof Engrs; Sigma Xi. *Res:* Highway planning; engineering interpretation of aerial photographs; control surveys; weighing highway vehicles in motion. *Mailing Add:* 975 Edgewater Dr Lexington KY 40502

BLYTHE, JACK GORDON, b Kansas City, Mo, July 15, 22; m 48; c 4. PHYSICAL GEOGRAPHY, GENERAL EARTH SCIENCES. *Educ:* Wichita State Univ, BA, 47; Northwestern Univ, MS, 50; Univ Okla, PhD(geol), 57. *Prof Exp:* Instr geol, Wichita State Univ, 49-51; assoc prof, Okla City Univ, 53-57; from assoc prof to prof geol, Wichita State Univ, 57-85, chmn dept, 65-70; RETIRED. *Mem:* Geol Soc Am; Am Asn Petrol Geol. *Res:* Stratigraphy; structural geology. *Mailing Add:* 557 N Broadview Wichita KS 67208

BLYTHE, PHILIP ANTHONY, b Dewsbury, Eng, Mar 30, 37; m 63; c 2. APPLIED MATHEMATICS, FLUID MECHANICS. *Educ:* Univ Manchester, BSc, 58, PhD(fluid mech), 61. *Prof Exp:* Res fel aerodyn, Nat Phys Lab, Eng, 61-63, sr sci officer, 63-64; lectr, Imp Col, Univ London, 64-68; assoc prof, 68-70, PROF APPL MATH, LEHIGH UNIV, 70- *Concurrent Pos:* Vis prof, Univ Newcastle, Eng, 74-75. *Res:* Nonlinear wave propagation; nonequilibrium flows; thermal convection; combustion; asymptotic methods. *Mailing Add:* Dept Mech Eng & Mech Lehigh Univ Packard Lab No 19 Bethlehem PA 18015

BLYTHE, WILLIAM BREVARD, b Huntersville, NC, Sept 23, 28; m 56; c 4. INTERNAL MEDICINE. *Educ:* Univ NC, AB, 48; Washington Univ, MD, 53. *Prof Exp:* Life Ins Med Res Fund fel, 58-60; from instr to assoc prof med, 60-70, assoc dir clin res unit, 65-66, dir clin res unit, 66-76, PROF MED, SCH MED UNIV NC, CHAPEL HILL, 70-, HEAD DIV NEPHROLOGY, 72-, MARIEN COVINGTON DISTINGUISHED PROF MED, 88- *Mem:* Am Fedn Clin Res; AMA; Am Physiol Soc; Am Soc Artificial Internal Organs; fel Am Col Physicians; Am Clin & Climat Asn. *Res:* Renal physiology and disease; hemodialysis. *Mailing Add:* Dept Med Univ NC Sch Med Chapel Hill NC 27599

BLYTHE, WILLIAM RICHARD, b Martinez, Calif, Aug 8, 31; m 55; c 3. FAILURE ANALYSIS, AUTOMOBILE HANDLING. *Educ:* Univ Calif, Berkeley, AB, 55, MS, 57; Stanford Univ, PhD (eng mech), 62. *Prof Exp:* from assoc prof civil eng to prof civil eng, San Jose State Univ, 60-65, assoc dean grad studies, 68-69, chmn dept civil eng & appl mech, 69-81; PRES, WILLIAM BLYTHE INC, 82- *Concurrent Pos:* Consult, many eng firms, govt agencies. *Mem:* Am Soc Civil Engrs; Am Acad Mech; Int Soc Terrain-Vehicle Systs. *Res:* Analysis of hyperstatic structures; shells; vehicle dynamics; accident reconstruction. *Mailing Add:* Dept Civil Eng & Appl Mech San Jose State Univ San Jose CA 95114

BO, WALTER JOHN, b Chisholm, Minn, Aug 12, 23; m 48; c 3. ANATOMY, ENDOCRINOLOGY. *Educ:* Marquette Univ, BS, 46, MS, 47; Univ Cincinnati, PhD(anat), 53. *Prof Exp:* Instr zool, Xavier Univ, 47-49; from asst prof to assoc prof anat, Sch Med, Univ NDak, 53-60; assoc prof, 60-63, PROF ANAT, BOWMAN GRAY SCH MED, 63- *Mem:* Am Asn Anat; Asn Cancer Res; Histochem Soc; Soc Exp Biol & Med; Soc Study Reproduction; Sigma Xi. *Res:* Gross histochemistry and abnormal growth of the uterus, cervix and vagina; endocrinology; female reproductive system; intra uterine device effect on uterus mode of action. *Mailing Add:* Dept Anat Bowman Gray Sch Med Winston-Salem NC 27103

BOACKLE, ROBERT J, IMMUNOLOGY. *Educ:* Univ Ala, Birmingham, PhD(biochem), 74. *Prof Exp:* PROF IMMUNOL, MED UNIV SC, 75- *Mailing Add:* Dept Immunol & Microbiol MUSC 171 Ashley Ave Charleston SC 29425

BOADE, RODNEY RUSSETT, b Armstrong, Iowa, Aug 17, 35; m 65; c 2. PHYSICS, STRUCTURAL GEOLOGY. *Educ:* Augustana Col, SDak, BS, 57; Iowa State Univ, PhD(physics), 64. *Prof Exp:* Staff mem physics, Sandia Corp, 64-81; SR RES PHYSICIST, PHILLIPS PETROL CO, 81- *Mem:* Am Phys Soc; Soc Petrol Engrs. *Res:* Energy transfer processes in gases and their influence on the propagation of sound; propagation of shock waves in porous and composite solids; explosive fracturing and in situ processing of oil shale; rock mechanics; petroleum reservoir fundamentals; subsidence. *Mailing Add:* 5410 SE Nowata Rd Bartlesville OK 74006

BOADWAY, JOHN DOUGLAS, b Collingwood, Ont, Apr 5, 22; m 49; c 3. FLUID MECHANICS, ENVIRONMENTAL ENGINEERING. *Educ:* Univ Toronto, BASc, 45; Queens Univ, MSc, 65, PhD(civil eng), 66. *Prof Exp:* Res engr, Nat Res Coun, 45-46; res physicist, Shawinigan Chem Ltd, 46-52; mem res staff eng physics, Consol Paper Corp, 52-63; from asst prof to assoc prof civil eng, Queens Univ, Ont, 66-87; RETIRED. *Mem:* Can Pulp & Paper Asn; Chem Inst Can; Can Soc Chem Engrs. *Res:* Vortex separating devices; computer simulations in hydraulics; paper making, especially hydraulics; sedimentation of effluents; electron microscopy. *Mailing Add:* 35 Holland Cresent Kingston ON K7M 2V7 Can

BOAG, DAVID ARCHIBALD, b Edmonton, Alta, Jan 24, 34; m 63; c 2. ANIMAL ECOLOGY. *Educ:* Univ Alta, BSc, 57, MSc, 58; Wash State Univ, PhD(zool), 64. *Prof Exp:* Lectr zool, Univ Alta, 58-59, asst prof, 59-60; researcher, Zool Sta, Naples, Italy, 61; asst prof, 63-69, assoc prof, 69-76, PROF ZOOL, UNIV ALTA, 76- DIR, R B MILLER BIOL STA, 63- *Mem:* Am Ornith Union; Wildlife Soc; Ecol Soc Am; Can Soc Zool; Can Soc Wildlife & Fishery Biol. *Res:* Population ecology, especially waterfowl and grouse. *Mailing Add:* Dept Zool Univ Alta Edmonton AB T6G 2M7 Can

BOAG, THOMAS JOHNSON, b Liverpool, Eng, Apr 11, 22; Can citizen; m 50; c 4. PSYCHIATRY, PSYCHOANALYSIS. *Educ:* Univ Liverpool, MB, ChB, 44; McGill Univ, dipl, 53; FRCP(C). *Prof Exp:* Asst psychiat, Royal Victoria Hosp, 52-53; lectr, McGill Univ, 53-55; from asst psychiatrist to assoc psychiatrist, Royal Victoria Hosp, 55-61; prof psychiat & chmn dept, Col Med, Univ Vt, 61-67; chief serv, DeGoesbriand Mem & Mary Fletcher Hosps, 61-67; prof psychiat & head dept, Queen's Univ, Ont, 67-75, dean fac med, 75-82, vprin health sci, 83-88; RETIRED. *Concurrent Pos:* Asst prof, McGill Univ, 57-61; asst dir, Allan Mem Inst, 59-61; psychiatrist-in-chief, Kingston Gen Hosp, Ont, 67-75; consult, Hotel Dieu, Kingston Psychiat & St Mary's of the Lake & Kingston Gen Hosp, 67- *Mem:* AMA; Am Psychiat Asn; Can Psychiat Asn. *Res:* Human adaptation in Arctic; interaction recording and study of interview situation; psychiatric treatment milieu; history of psychiatry. *Mailing Add:* Fac Med Queen's Univ Kingston ON K7L 3N6 Can

BOAK, RUTH ALICE, medicine, for more information see previous edition

BOAKE, WILLIAM CHARLES, cardiology, internal medicine, for more information see previous edition

BOAL, DAVID HAROLD, b Toronto, Ont, Jan 12, 48; m 73; c 2. PROPERTIES OF SURFACES & INTERFACES. *Educ:* Univ Toronto, BSc, 70, MSc, 71, PhD(physics), 75. *Prof Exp:* Vis asst prof physics, Univ Alta, 76-78; from asst prof to assoc prof chem, 78-86, PROF PHYSICS, SIMON FRASER UNIV, 86- *Concurrent Pos:* Pres Res Univ Alta grant, 76-77; Nat Res Coun Can grant, 77-; vis assoc prof, Michigan State Univ, 83-84, Univ Ill, 86. *Mem:* Can Asn Physicists; Am Phys Soc. *Res:* Statistical mechanics; properties of surfaces and interfaces; computational physics. *Mailing Add:* Dept Physics Simon Fraser Univ Burnaby BC V5A 1S6 Can

BOAL, JAN LIST, b Canton, Ohio, Oct 20, 30; m 53; c 3. APPLIED MATHEMATICS. *Educ:* Ga Inst Technol, BME & MS, 54; Mass Inst Technol, PhD(math), 59. *Prof Exp:* Instr math, Mass Inst Technol, 59-60; from asst prof to assoc prof, Univ SC, 60-69; chmn dept, 69-77, PROF MATH, GA STATE UNIV, 69- *Concurrent Pos:* Vis lectr cols, Math Asn Am, 64 & 68. *Mem:* Math Asn Am; Am Soc Indust & Appl Math; Sigma Xi. *Res:* Numerical analysis; differential equations; function approximation. *Mailing Add:* Ga State Univ Atlanta GA 30303

BOARD, JAMES ELLERY, b Sacramento, Calif, Sept 7, 48; m 87. GROWTH DYNAMICS, ROW SPACING. *Educ:* Univ Calif, Davis, BS, 70, MS, 71, PhD(plant physiol), 78. *Prof Exp:* Postdoctoral fel, dept agron, Univ Calif, Davis, 79-80; PROF, DEPT AGRON, LA STATE UNIV, 80- *Mem:* Am Soc Agron; Crop Sci Soc Am. *Res:* Photoperiod effects in soybeans as affecting planting date in the southeastern USA; growth dynamics of soybeans in narrow, wide row culture in southeastern USA. *Mailing Add:* 9721 Siegen Lane Baton Rouge LA 70810

BOARD, JOHN ARNOLD, b Altavista, Va, June 30, 31; m 59; c 3. OBSTETRICS & GYNECOLOGY. *Educ:* Randolph-Macon Col, BS, 53; Med Col Va, MD, 55. *Prof Exp:* Intern, Louisville Gen Hosp, Ky, 55-56; from jr asst resident to resident obstet & gynec, Med Col Va Hosps, 56-59; USPHS fel reprod physiol & infertil, Sch Med, Yale Univ, 61-62; from instr to assoc prof, 62-69, PROF OBSTET & GYNEC, MED COL VA, VA COMMONWEALTH UNIV, 69- *Mem:* Am Col Obstet & Gynec; Am Fertil Soc; Soc Study Reproduction; Endocrine Soc; Soc Reprod Endocrinologists. *Res:* Gynecologic endocrinology and infertility. *Mailing Add:* Dept Obstet & Gynec Med Col Va Richmond VA 23298

BOARDMAN, GREGORY DALE, b Montpelier, Vt, Dec 12, 50; m 70, 87; c 5. ENVIRONMENTAL MICROBIOLOGY, ENVIRONMENTAL TOXICOLOGY. *Educ:* Univ NH, BSCE, 72, MSCE, 73; Univ Maine, PhD(sanit eng), 76. *Prof Exp:* ASSOC PROF ENVIRON ENG, VA POLYTECH INST & STATE UNIV, 76- *Mem:* Am Soc Civil Engrs; Water Pollution Control Fedn; Am Water Works Asn; Sigma Xi; Soc Environ Toxicol & Chem. *Res:* Inactivation of microorganisms; treatment of industrial wastes; early assessment of toxicity. *Mailing Add:* 107 Cherokee Dr Blacksburg VA 24060

BOARDMAN, JOHN, b Turlock, Calif, Sept 8, 32; m 63; c 2. THEORETICAL PHYSICS. *Educ:* Univ Chicago, AB, 52; Iowa State Univ, MS, 56; Syracuse Univ, PhD(physics), 62. *Prof Exp:* Lectr physics, Queens Col, NY, 61-62; instr, 62-65, asst prof, 65-80, ASSOC PROF PHYSICS, BROOKLYN COL, 80- *Mem:* Am Phys Soc; Fedn Am Sci. *Res:* General theory of relativity. *Mailing Add:* Dept Eng CUNY Brooklyn Col Brooklyn NY 11210

BOARDMAN, JOHN MICHAEL, b Manchester, Eng, Feb 13, 38; div; c 2. HOMOTOPY THEORY, DIFFERENTIAL TOPOLOGY. *Educ:* Cambridge Univ, BA, 61, PhD(math), 65. *Prof Exp:* Fel, Dept Sci & Indust Res, Eng, 64-66; instr math, Univ Chicago, 66-67; lectr, Univ Warwick, 67-68; visitor, Haverford Col, 69; assoc prof, 69-72, PROF MATH, JOHNS HOPKINS UNIV, 72- *Mem:* Am Math Soc. *Res:* Algebraic topology, particularly homotopy theory, stable homotopy theory and singularities of differentiable maps. *Mailing Add:* Dept Math Johns Hopkins Univ Baltimore MD 21218

BOARDMAN, MARK R, b Albany, NY, Mar 14, 50; div; c 2. MARINE GEOLOGY, CARBONATE SEDIMENTOLOGY. *Educ:* Princeton Univ, AB, 72; Univ NC, Chapel Hill, MS, 76, PhD(marine sci), 78. *Prof Exp:* Prof geochem, Fed Univ Fluminense, Rio de Janeiro, 78-81; asst prof, 81-86, ASSOC PROF GEOL, MIAMI UNIV OHIO, 86- *Mem:* Sigma Xi; Soc Econ Paleontologists & Mineralogists; Int Asn Sedimentologists. *Res:* Geochemical and sedimentologic characteristics of modern carbonate sediments; evaluation of facies and benthic communities; holocene sea level. *Mailing Add:* Dept Geol Miami Univ Oxford OH 45056

BOARDMAN, SHELBY JETT, b Akron, Ohio, Nov 7, 44; m 66; c 2. MINERALOGY, PETROLOGY. *Educ:* Miami Univ, BA, 66; Univ Mich, MS, 69, PhD(geol), 71. *Prof Exp:* Petrol geologist, Mobil Oil Corp, 66; explor geologist, Bear Creek Mining Co, 67; res asst stratig, Mobil Res & Develop Corp, 68; from asst prof to prof, 71-88, chmn dept, 77-83, CHARLES L DENISON PROF GEOL, CARLETON COL, 88- *Concurrent Pos:* Res assoc, Univ Glasgow, Scotland, 78-79; dir geol prog, Assoc Cols of Midwest, 80; courtesy res assoc, Univ Kans, 81-82; pres, Coun Undergrad Res, Geol Coun, 87-89. *Mem:* Geol Soc Am; Sigma Xi; Nat Asn Geol Teachers; AAAS; Mineral Soc Am. *Res:* Geochemistry, petrology and geochronology of Precambrian rocks of Central Colorado; metamorphic petrology, northern Sierra Nevada, California. *Mailing Add:* Dept Geol Carleton Col Northfield MN 55057

BOARDMAN, WILLIAM JARVIS, b Akron, Ohio, Aug 19, 39; m 65; c 3. PHYSICS, ASTRONOMY. *Educ:* Miami Univ, AB, 61, MS, 63; Univ Colo, PhD(astrogeophys), 68. *Prof Exp:* Asst prof physics, 68-74, assoc prof math & physics, 74-82, PROF MATH & PHYSICS, BIRMINGHAM-SOUTHERN COL, 82-, DIR MEYER PLANETARIUM, 73- *Mem:* Am Astron Soc; Am Phys Soc. *Res:* Radiative processes in the solar corona; temperature and density of coronal enhancement over regions of solar activity. *Mailing Add:* Dept Math & Sci Birmingham-Southern Col 900 Arkadelphia Rd Birmingham AL 35254

BOARDMAN, WILLIAM WALTER, JR, b Rugby, NDak, June 2, 16; m 41; c 6. PHYSICAL INORGANIC CHEMISTRY. *Educ:* Coe Col, BA, 38; Univ Iowa, MS, 40, PhD(phys chem), 42. *Prof Exp:* Res chemist, Gaseous Diffusion Isotope Separation Plant, Carbide & Carbon Chem Corp, 45-49; supvr, Control Lab, Twin Cities Arsenal, 50-55; res chemist, Lithium Corp Am, Inc, 55-67; head dept chem, Biola Col, 69-70; res assoc chem, Creation-Sci Res Ctr, 70-71; chief chemist, Gen Monitors, Inc, 71-74; staff chemist, Occidental Res Corp, 74-83; CONSULT, 83- *Mem:* Am Chem Soc; Sigma Xi; Creation Res Soc. *Res:* Synthesis of inorganic compounds; development of catalytic and semiconductor devices for gas detection; analytical methods for oil shale and coal; thermal analytical methods. *Mailing Add:* c/o Tanksely 1927 Lincoln St Oceanside CA 92054

BOAS, MARY LAYNE, b Prosser, Wash, Mar 10, 17; m 41; c 3. PHYSICS. *Educ:* Univ Wash, BS, 38, MS, 40; Mass Inst Technol, PhD(physics), 48. *Prof Exp:* Fel math, Univ Wash, 38-40; instr, Duke Univ, 40-43 & Tufts Col, 43-48; lectr, Wellesley Col, 49-50; consult, Northwestern Nuclear Res Lab, 50-52; vis lectr physics, Northwestern Univ, 52-53; lectr, 55-56, from asst prof to assoc prof, 57-75, prof physics, 75-86, EMER PROF PHYSICS, DEPAUL UNIV, 86- *Mem:* Am Phys Soc; Am Asn Physics Teachers; Math Asn Am. *Res:* Theoretical nuclear physics; photo-disintegration of H-3; mathematical physics; special relativity. *Mailing Add:* 3540 NE 147th St Seattle WA 98155-7822

BOAS, NORMAN FRANCIS, b New York, NY, Aug 4, 22; m 45; c 3. INTERNAL MEDICINE, RHEUMATOLOGY. *Educ:* Harvard Univ, MD, 45. *Prof Exp:* Intern & med resident, Michael Reese Hosp, 45-47; fel path, Mt Sinai Hosp, NY, 47-48, fel med, 48-51; sr asst surgeon, NIH, 51-54; res dir, Norwalk Hosp, 54-78, sr attend physician, 67-78; mem staff, Lawrence & Mem Hosps, New London, 76-85, consult staff, 80-84; RETIRED. *Concurrent Pos:* Asst clin prof med, Yale Univ, 75-78. *Res:* Clinical medicine. *Mailing Add:* Six Brandon Lane Mystic CT 06355

BOAS, RALPH PHILIP, JR, b Walla Walla, Wash, Aug 8, 12; m 41; c 3. MATHEMATICS. *Educ:* Harvard Univ, AB, 33, PhD(math), 37. *Prof Exp:* Instr math, Harvard Univ, 36-37; Nat Res fel, Princeton & Cambridge Univs, 37-39; instr math, Duke Univ, 39-42; asst instr, US Navy Pre-Flight Sch, NC, 42-43; vis lectr, Harvard Univ, 43-45; res assoc, Brown Univ, 45-50; prof, 50-80, EMER PROF MATH, NORTHWESTERN UNIV, 80- *Concurrent Pos:* Exec ed, Math Revs, 45-50; lectr, Mass Inst Technol, 48, Guggenheim fel, 51-52. *Mem:* AAAS; Am Math Soc (vpres, 59-60); Math Asn Am (pres, 73-75); London Math Soc; Soc Indust & Appl Math. *Res:* Laplace integrals; moment problems; Fourier series and integrals; entire functions; power series; approximation of functions. *Mailing Add:* 3540 NE 147th St Seattle WA 98155

BOAST, CHARLES WARREN, b Ames, Iowa, Nov 12, 43; m 66; c 2. SOIL PHYSICS, AGRONOMY. *Educ:* Iowa State Univ, BS, 66, MS, 69, PhD(agron), 70. *Prof Exp:* From asst prof to assoc prof, 70-87, PROF AGRON, UNIV ILL, 87- *Mem:* Soil Sci Soc Am; Am Soc Agron; Am Geophys Union. *Res:* Retention and movement of water in soil; surface and shallow sub-surface hydrology; plant-water relationships; transport of solutes and gases in soil. *Mailing Add:* Dept Agron Univ Ill 1102 S Goodwin Ave Urbana IL 61801

BOAST, W(ARREN) B(ENEFIELD), electrical engineering; deceased, see previous edition for last biography

BOATMAN, EDWIN S, b London, Eng, July 24, 21; m 47. MICROBIOLOGY, ELECTRON MICROSCOPY. *Educ:* Univ London, BSc, 53; Univ Wash, MSc, 61, PhD(marine microbiol), 67. *Prof Exp:* Res asst bact, Royal Postgrad Med Sch, 51-52; res asst, St George's Hosp Med Sch, London, 52-55; head microbiologist, Royal Columbian Hosp, BC, 55-56; res asst microbiol, 56-58, res assoc prev med, 58-61, res instr, 61-67, assoc prof environ health, 67-77, PROF ENVIRON HEALTH, SCH PUB HEALTH, UNIV WASH, 77- , HEAD, DIV HEALTH EFFECTS RES, 79- *Concurrent Pos:* Fel, Inst Med Lab Sci, London, 53-; Josiah Macy Found fac award, Univ Bern, 74; consult, Nat Heart, Lung & Blood Inst, 74- *Mem:* Am Soc Microbiol; Electron Micros Soc Am; NY Acad Sci; fel Royal Soc Health; Am Acad Microbiol; Sigma Xi. *Res:* Relationship of ultra-structure to function in Mycoplasmatales and the anatomy of bacteria and viruses; growth and development of the lungs and the effects of air pollutants on the lungs; analysis of drinking water for asbestos. *Mailing Add:* Dept Pathobiol & Environ Health/SC-34 Univ Wash Sch Pub Health Seattle WA 98195

BOATMAN, JOE FRANCIS, b St Cloud, Minn, Apr 16, 49; m 85; c 1. CLOUD PHYSICS, AEROSOL PHYSICS. *Educ:* St Cloud State Univ, BS, 71; Univ Wyo, MS, 74, PhD(atmospheric sci), 81. *Prof Exp:* Grad asst, Univ Wyo, 72-74 & 77-81; meteorologist, NAm Weather Consults, 74 & State Mont, 75-77; postdoctoral, Nat Res Coun, 81-83; res assoc, Univ Colo, 83-85; SUPV METEOROLOGIST, NAT OCEANIC & ATMOSPHERIC ADMIN, 85- *Concurrent Pos:* Mem, Atmospheric Chem Comt, Am Meteorol Soc, 88-90. *Mem:* Am Meteorol Soc; Sigma Xi. *Res:* Aerosols in the atmosphere as they relate to air quality, acid deposition, cloud processes, and climate change. *Mailing Add:* Nat Oceanic & Atmospheric Admin R-E-ARX1 325 Broadway Boulder CO 80303

BOATMAN, RALPH HENRY, JR, b Carlinville, Ill, Apr 20, 21; m 43; c 2. PUBLIC HEALTH. *Educ:* Southern Ill Univ, BS, 43; Univ NC, MPH, 47, PhD(pub health), 54. *Prof Exp:* Health coordr & instr health educ & dir sch-community health proj, Southern Ill Univ, 47-54, chmn dept health educ, 48-54; dir community health serv, Chicago & Cook County Tuberc Inst, Ill, 54-60; chmn dept pub health educ, 61-70, dir continuing educ & field serv, 68-70, dean allied health prof, 70-86, PROF PUB HEALTH EDUC, SCH PUB HEALTH, UNIV NC, CHAPEL HILL, 60-, DIR OFF CONTINUING EDUC HEALTH SCI, 74- *Concurrent Pos:* Lectr, Prof Schs, Univ Ill, 55-56, asst prof, 56-60; consult, Radiol Health Conf, WHO, 62; mem nat adv coun allied health prof, HEW, 68-; consult, NIH & Am Hosp Asn, 78-88; consult, univs & state bd of educ; gen col adv, Univ NC, Chapel Hill, 80- *Mem:* Fel Am Pub Health Asn (pres, 76-77); Am Soc Allied Health Prof (secy, 73-75, pres-elect, 75-76, pres, 77); Soc Pub Health Educr (pres, 66-67). *Res:* Public, school and allied health education; continuing education. *Mailing Add:* Off Continuing Educ, Div Health Sci Miller Hall #8160 Univ NC Chapel Hill NC 27599-8160

BOATMAN, SANDRA, b Tampa, Fla, Nov 1, 39. ORGANIC CHEMISTRY, BIOCHEMISTRY. *Educ:* Rice Univ, BA, 61; Duke Univ, PhD(chem), 65. *Prof Exp:* Fel, Univ NC, 65-66, NIH fel, 65-67; asst prof, 67-74, assoc prof, 74-79, PROF CHEM, HOLLINS COL, 79- *Mem:* AAAS; Am Chem Soc; The Chem Soc; Sigma Xi. *Res:* Synthetic studies on polyanions of B-ketoaldehydes, pyridones; B-ketoenamines and related compounds; condensations of polyketo compounds; chemical modification of proteins and enzymes, especially at the active sites; stabilizing interactions in viruses, writing in science courses. *Mailing Add:* 1947 Laurel Mountain Dr Salem VA 24153

BOATNER, LYNN ALLEN, b Clarksville, Tex, Aug 3, 38; m 61; c 3. MAGNETIC RESONANCE, MATERIALS SCIENCE. *Educ:* Tex Tech Univ, BS, 60, MS, 61; Vanderbilt Univ, PhD(physics), 66. *Prof Exp:* Asst physics, Tex Technol Col, 60-61; asst, Vanderbilt Univ, 61-66; res scientist, LTV Res Ctr, 66-70; prin scientist, Advan Technol Ctr, Inc, 71-74; sr res scientist, Lab Exp Physics, Fed Polytech Sch, Switz, 75-77; staff scientist, 77-80, SR RES STAFF SCIENTIST & SECT HEAD, SOLID STATE DIV, OAK RIDGE NAT LAB, 80- *Concurrent Pos:* Vis prof, Unam, Mex, 74. *Honors & Awards:* IR-100 Award, 82 & 85; Mat Sci Res Competition, Dept Energy, 84; Jacquet-Lucas Award, Am Soc Metals Int, 88. *Mem:* fel AAAS; fel Am Phys Soc; NY Acad Sci; assoc fel Am Inst Aeronaut & Astronaut; Mat Res Soc; Am Asn Crystal Growth; Am Ceramic Soc; Am Soc Metals Int. *Res:* Electron paramagnetic resonance investigations of transition metal ions in single crystals; studies of rare-earth and actinide ions; single crystal growth; superconductivity; investigations of the Jahn-Teller effect; ferroelectrics; nuclear waste disposal; ion-solid interactions; glass properties. *Mailing Add:* Solid State Div PO Box 2008 Oak Ridge TN 37831

BOAZ, DAVID PAUL, b Wichita, Kans, Oct 19, 44; m 65; c 3. ORGANIC CHEMISTRY, BIOCHEMISTRY. *Educ:* Univ Dubuque, BA, 66; Univ Iowa, MS, 69, PhD(org chem), 72. *Prof Exp:* Fel, Dept Pediat, Univ Iowa, 71-72; res assoc hypertension, Dept Med, 72-73, asst res scientist, 73-76, assoc res scientist biochem, 76; sr res chemist adhesives, 76-80, chemist specialist, 80-84, mgr, Specialty Chem Div, 84-87, EUROP TECH MGR, SPECIALTY CHEM, 3M CO, 87- *Mem:* Am Chem Soc; Sigma Xi. *Res:* Pressure sensitive adhesives; polymer chemistry; aminimids. *Mailing Add:* Ruiter Sdreef 21 2230 Schilde St Paul Belgium

BOAZ, PATRICIA ANNE, b Chester, Pa, May 13, 22; div; c 5. PHYSICAL CHEMISTRY, ANALYTICAL CHEMISTRY. *Educ:* Vassar Col, AB, 44; Univ Iowa, PhD(phys chem), 51. *Prof Exp:* Chemist, Nat Aniline Div, Allied Chem & Dye Corp, 44-45; asst chem, Univ Iowa, 45-48; asst phys chem, Northwestern Univ, 48; instr chem, Smith Col, 49-50; asst prof, 67-70, ASSOC PROF CHEM, IND UNIV-PURDUE UNIV, INDIANAPOLIS, 70- *Mem:* Geol Soc Am; Am Chem Soc; Sigma Xi. *Res:* Water chemistry; environmental chemistry and geology. *Mailing Add:* IUPUI 925 Agnus St Indianapolis IN 46202

BOBALEK, EDWARD G(EORGE), b Chicago, Ill, Oct 13, 15; m 42; c 7. CHEMICAL ENGINEERING. *Educ:* St Mary's Col, Minn, BS, 38; Creighton Univ, MS, 40; Ind Univ, PhD(phys chem), 42. *Prof Exp:* Res chemist, Dow Chem Co, Tex, 42-43, Mich, 43-45; mgr resin res, Arco Co, 45-49; prof chem eng, Case Inst Technol, 49-53; chmn, Dept Chem Eng, 66-76, Gottesman Res Prof, 63-80, EMER PROF, UNIV MAINE, ORONO, 80- *Concurrent Pos:* Sr chem engr adv, US Environ Protection Agency, Res, 76-80; dir, Essof Corp, 49- *Honors & Awards:* Matiello Award, Fed Socs Coatings Technol. *Mem:* Am Chem Soc; fel Am Inst Chem Engrs. *Res:* Theory of non-aqueous solution; theoretical electrochemistry; reaction rates; colloid chemistry; plastics; chemical process theory; paper technology; environmental engineering and technology. *Mailing Add:* 120 Shirley Dr Cary NC 27511

BOBB, MARVIN LESTER, entomology; deceased, see previous edition for last biography

BOBBIN, RICHARD PETER, b Northampton, Mass, Dec 25, 42; m 66; c 2. AUDITORY PHYSIOLOGY. *Educ:* Northwestern Univ, Boston, BS, 64; Tulane Univ, MS, 67, PhD(pharmacol), 69. *Prof Exp:* asst prof pharmacol, Dept Otorhinolaryngol, Kresge Hearing Res Lab, Sch Med & Dept Pharmacol & Exp Therapeut, 71-76, ASSOC PROF PHARMACOL, LA STATE UNIV, 76- *Concurrent Pos:* Guest worker, Lab Neuro-Otolaryngol, Nat Inst Neurol & Commun Disorders & Stroke, NIH, 81. *Mem:* Am Soc Pharmacol & Exp Therapeut; Acoust Soc Am; Asn Res Otolaryngol; AAAS; Sigma Xi. *Res:* Pharmacology and biochemistry of neurotransmitters, drug toxicity and cell death using the peripheral auditory and acousticolateralis organs as model systems; auditory biochemistry and pharmacology. *Mailing Add:* Kresge Hearing Res Lab La State Univ Med Ctr 1100 Florida Ave New Orleans LA 70119

BOBBITT, DONALD ROBERT, b Philadelphia, Pa, Oct 21, 56; m 80; c 3. LASERS, ANALYTICAL CHEMISTRY. *Educ:* Univ Ark, BS, 80; Iowa State Univ, PhD(chem), 85. *Prof Exp:* Chemist, E I Dupont, 79-80; asst prof, 85-89, ASSOC PROF ANALYTICAL CHEM, UNIV ARK, 89- *Concurrent Pos:* Dreyfus Teacher/Scholar fel, 88. *Mem:* Am Chem Soc; Sigma Xi. *Res:* Application of lasers to problems in analytical chemistry; development of laser-based detection schemes for high performance liquid chromatography; investigation of non linear Raman technique; Raman induced Kerr effect. *Mailing Add:* Dept Chem & Biochem Univ Ark Fayetteville AR 72701

BOBBITT, JAMES MCCUE, chemistry, for more information see previous edition

BOBBITT, JESSE LEROY, b Yellow Springs, Ohio, Sept 18, 33; m 56. BIOCHEMISTRY. *Educ:* Berea Col, BA, 55; Univ Louisville, PhD(biochem), 64. *Prof Exp:* Teacher pub schs, Ky, 58-60; biochemist, US Army Med Res Labs, Ft Knox, Ky, 63-66; RES BIOCHEMIST, ELI LILLY & CO, 66- *Mem:* AAAS; Sigma Xi; Am Chem Soc; Am Soc Biochem & Molecular Biol. *Res:* Protein isolation and characterization; mechanism of enzyme and hormone action. *Mailing Add:* 8101 Rosemead Lane Indianapolis IN 46240-2750

BOBBITT, OLIVER BIERNE, b Charleston, WVa, Jan 10, 17; m 43; c 3. CLINICAL PATHOLOGY. *Educ:* Univ Ga, BS, 39; Univ Va, MD, 43. *Prof Exp:* From instr to prof, 47-82, chmn dept & dir clin lab, Hosp, 52-71, EMER PROF CLIN PATH, MED SCH, UNIV VA, 82- *Mem:* AMA; Col Am Path; Am Soc Clin Path. *Res:* Blood groups. *Mailing Add:* Box 6128 Charlottesville VA 22906

BOBEAR, JEAN B, b Schenectady, NY, 1922. SYSTEMATIC BOTANY, ETHNOBOTANY. *Educ:* Col St Rose, BS, 48; Cornell Univ, MS, 54; Trinity Col, Univ Dublin, PhD(taxon bot), 64. *Prof Exp:* Eng asst, Control Dept, Knolls Atomic Power Lab, Gen Elec Co, 48-53; herbarium asst, Cornell Univ, 54-56; from asst prof to prof biol, State Univ NY Col Brockport, 56-83; RETIRED. *Concurrent Pos:* Grant review panelist & consult, NSF. *Mem:* Am Inst Biol Sci; Am Soc Plant Taxon. *Res:* Experimental taxonomic treatment of critical genera; genus Euphrasia; floristics Monroe Co NY. *Mailing Add:* 28 College Brockport NY 14420

BOBECHKO, WALTER PETER, b Toronto, Ont, Aug 22, 32; m 57; c 5. ORTHOPAEDIC SURGERY. *Educ:* Univ Toronto, MD, 57, BSc, 59; Royal Col Physicians & Surg Can, FRCS(C), 63. *Prof Exp:* res scientist, Res Inst, Hosp Sick Children, 64-87; from asst prof to assoc prof orthop, Univ Toronto, 70-88; DIR, HUMANA ADVAN SURG INST, 88- *Concurrent Pos:* Orthop surgeon, Hosp Sick Children, 63-87 & chief orthop surg, 77-87. *Mem:* Can Orthop Soc; Paediat Orthop Soc; Scoliosis Res Soc; Int Soc Orthop Surg & Traumatology. *Res:* Designer of electro-spinal instrumentation--spinal pacemaker to correct spinal scoliosis deformities of children; designer of Double Hook spinal instrumentation for correction of severe spinal deformities without casts or braces. *Mailing Add:* Humana Advan Surg Inst Humana Hosp-Med City Dallas 7777 Forest Lane Dallas TX 75230

BOBECK, ANDREW H, b Tower Hill, Pa, Oct 1, 26; m 49; c 3. APPLIED PHYSICS, COMPUTER SCIENCE. *Educ:* Purdue Univ, BS, 48, MS, 49. *Hon Degrees:* EngD, Purdue Univ, 72. *Prof Exp:* Mem tech staff & supvr, Bell Labs, 49-89; RETIRED. *Honors & Awards:* Stuart Ballentine Medal, Franklin Inst, 73; M N Liebmann Award, Inst Elec & Electronics Engrs, 75;

Valdemar Poulsen Gold Medal, Danish Acad Sci, 76. *Mem:* Nat Acad Eng; Inst Elec & Electronics Engrs. *Res:* Magnetic logic and storage; development of very high density magnetic bubble devices. *Mailing Add:* 41 Ellers Dr Chatham NJ 07928

BOBER, WILLIAM, b New York, NY, Mar 16, 30; m 54; c 3. FLUID MECHANICS, APPLIED MATHEMATICS. *Educ:* City Col New York, BCE, 52; Pratt Inst, MS, 60; Purdue Univ, PhD(eng sci), 64. *Prof Exp:* Asst eng sci, Pratt Inst, 58-60; instr aerospace & eng sci, Purdue Univ, 60-64; assoc eng physicist, Cornell Aeronaut Lab, 64-68; vis asst prof eng mech, Va Polytech Inst, 68-69; assoc prof mech eng, Rochester Inst Technol, 69-81; ASSOC PROF MECH ENG, FLA ATLANTIC UNIV, 82- *Concurrent Pos:* Ford Found fel, 60- 63; lectr, State Univ NY, Buffalo, 65-67. *Mem:* Am Soc Mech Engrs. *Res:* Sub-sonic aerodynamics; heat transfer; numerical methods. *Mailing Add:* Dept Mech Eng Fla Atlantic Univ Boca Raton FL 33431

BOBISUD, LARRY EUGENE, b Midvale, Idaho, Mar 16, 40; m 63. DIFFERENTIAL EQUATIONS & APPLICATIONS. *Educ:* Col Idaho, BS, 61; Univ NMex, MS, 63, PhD(math), 66. *Prof Exp:* Vis mem, Courant Inst Math Sci, NY Univ, 66-67; from asst prof to assoc prof, 67-74, chmn dept, 78-82, PROF MATH, UNIV IDAHO, 74- *Mem:* Am Math Soc. *Res:* Singular perturbation problems for partial differential equations; mathematical biology; nonlinear boundary value problems. *Mailing Add:* Dept Math Univ Idaho Moscow ID 83843

BOBKA, RUDOLPH J, b Akron, Ohio, July 27, 28; m 55; c 1. PHYSICAL CHEMISTRY. *Educ:* Marietta Col, AB, 50; Miami Univ, MS, 51; Western Reserve Univ, PhD(phys chem), 60. *Prof Exp:* Anal chemist, Army Chem Ctr, Md, 51-52; asst, Western Reserve Univ, 54-58; chemist, Res Lab, Union Carbide Corp, 58-68; assoc prof, 68-72, PROF CHEM, STATE UNIV NY COL PLATTSBURGH, 72- *Mem:* Am Chem Soc; Sigma Xi. *Res:* Low temperature calorimetry; surface chemistry; carbon technology. *Mailing Add:* Dragon Rd RR 1 Box 459 Altoona NY 12910

BOBKO, EDWARD, b Cleveland, Ohio, Feb 15, 25; m; c 3. ORGANIC CHEMISTRY. *Educ:* Western Reserve Univ, BS, 49; Northwestern Univ, PhD(org chem), 52. *Prof Exp:* Instr chem, Northwestern Univ, 52-53; res chemist, Olin Industs, Inc, 53-54; asst prof chem, Washington & Jefferson Col, 54-55, instr, 55-56; from asst prof to assoc prof, 56-69, PROF CHEM, TRINITY COL, CONN, 69- *Mem:* AAAS; Am Chem Soc. *Res:* Organometallic compounds; pyrimidines; physical organic chemistry. *Mailing Add:* Dept Chem Trinity Col Hartford CT 06106

BOBLITT, ROBERT LEROY, b Springfield, Ohio, Nov 1, 25; m 44; c 4. MEDICINAL CHEMISTRY. *Educ:* Ohio Northern Univ, BS, 48; Univ Minn, PhD(pharmaceut chem), 53. *Prof Exp:* Asst, 53-55, assoc prof, 56-70, asst dean, 72-76, PROF PHARMACEUT CHEM, COL PHARM, UNIV HOUSTON, 70-, ASSOC DEAN, 76- *Mem:* Am Chem Soc; Am Pharmaceut Asn; Sigma Xi. *Res:* Synthesis of amidines; physical and chemical properties of amidines; phytochemistry; synthesis of beta-adrenergic blocking agents. *Mailing Add:* Pharmaceut Chem Univ Houston Houston TX 77004

BOBO, EDWIN RAY, b Oklahoma City, Okla, May 17, 38. MATHEMATICS. *Educ:* Pa State Univ, BA, 60; Univ Va, MA, 62, PhD(math), 65. *Prof Exp:* Asst prof, 65-70, ASSOC PROF MATH, GEORGETOWN UNIV, 70- *Mem:* Am Math Soc; Math Asn Am. *Res:* Algebra; finite groups; automorphism groups of Jordan algebras; group representations. *Mailing Add:* Dept Math Georgetown Univ Washington DC 20057

BOBO, MELVIN, b Blair, Tex, Feb 13, 24. AERONAUTICAL ENGINEERING. *Educ:* Tex Univ, BS, 49. *Prof Exp:* Test prog, Gen Elec, 49, aircraft eng group, 50-85, chief engr, 85-91; RETIRED. *Mem:* Nat Acad Eng; Aerospace Industs Asn Am. *Mailing Add:* 5629 Oak Vista Dr Cincinnati OH 45227

BOBONICH, HARRY MICHAEL, b Ashland, Pa, Sept 13, 24; m 49; c 2. INORGANIC CHEMISTRY. *Educ:* Susquehanna Univ, BA, 50; Bucknell Univ, MS, 56; Syracuse Univ, PhD(chem), 65. *Prof Exp:* Chemist, Barrett Div, Allied Chem Corp, 52-55; prof chem, Pa State Univ, 55-61 & 64-65; chmn dept, 69-74, actg dean grad studies, 74-75, assoc dean grad studies, 75-79, PROF CHEM, SHIPPENSBURG STATE COL, 65-, DEAN GRAD SCH, 80- *Res:* Synthesis of coordination compounds; use of magnetic measurements; infrared and ultraviolet spectra; conductance and x-rays to determine structure; development of new experiments for freshman chemistry programs. *Mailing Add:* 138 Park Pl W Shippensburg PA 17257

BOBONIS, AUGUSTO, b Humacao, PR, June 20, 07; m 36; c 4. MATHEMATICS. *Educ:* Univ PR, AB, 27, MS, 34; Univ Chicago, PhD(math), 39. *Hon Degrees:* LLD, Inter-Am Univ PR, 75. *Prof Exp:* Instr math & physics, St Augustine Mil Acad, PR, 27-30; critic teacher physics, 30-34, from instr to prof, 34-48, chmn dept, 45-47, dir city col, 47-52, tech adv to secy educ, 52-53, dir sec educ, 53-57, prof math, 57-58, from actg dean to dean col educ, 58-70, prof math, 70-75, EMER PROF MATH & EMER DEAN, UNIV PR, RIO PIEDRAS, 75- *Concurrent Pos:* Vpres acad affairs, Inter-Am Univ PR, 70-75. *Mem:* Am Math Soc. *Res:* Boundary value problems. *Mailing Add:* 62 Marbella Santurce PR 00907

BOBROW, DANIEL G, b New York, NY, Nov 29, 35; div; c 3. INTELLIGENT SYSTEMS. *Educ:* Rensselaer Polytech Inst, BS, 57; Harvard Univ, SM, 58; Mass Inst Technol, PhD(math), 64. *Prof Exp:* Mem staff, Bolt, Beranek & Newman Inc, 62-64; asst prof elec eng, Mass Inst Technol, 64-65; vpres, Info Sci Div, Bolt, Beranek & Newman, Inc, 65-69, dir comput sci div, 69-72; RES FEL, XEROX PALO ALTO RES CTR, 72- *Concurrent Pos:* Fulbright fel, 72-73; lectr, Stanford Univ, 73-; ed-in-chief, Artificial Int, 77-; chair, Cognitive Sci Soc; pres, Am Asn Artificial Intel. *Honors & Awards:* Prog Lang Award, Asn Comput Mach, 74. *Mem:* AAAS; Asn Comput Mach; Asn Comput Ling; Cognitive Sci Soc; Am Asn Artificial

Intel. *Res:* Development of knowledge representation languages and programming systems; development of a computer program that understands natural language; artificial intelligence; development of a designers assistant for very large scale integrated systems. *Mailing Add:* Xerox Palo Alto Res Ctr 3333 Coyote Hill Rd Palo Alto CA 94304

BOBROW, LEONARD S(AUL), b Newark, NJ, Nov 27, 40. ELECTRICAL ENGINEERING. *Educ:* Univ Miami, BSEE, 62; Northwestern Univ, MS, 64, PhD(elec eng), 68. *Prof Exp:* Res asst elec eng, Northwestern Univ, 66-67; electronic engr, Beltone Electronics Corp, 63-65; asst prof, 68-73, ASSOC PROF ELEC ENG, UNIV MASS, AMHERST, 73- *Res:* Active network synthesis; coding theory; graph theory and the application of graph theory to coding. *Mailing Add:* Dept Elec Eng Univ Mass Amherst MA 01002

BOBST, ALBERT M, b Zurich, Switz, Sept 10, 39; nat US; m 70; c 3. BIOPHYSICAL CHEMISTRY, ORGANIC BIOCHEMISTRY. *Educ:* Univ Zurich, PhD(chem), 65. *Prof Exp:* Res fel, Inst Biophys & Biochem, Paris, 66; Swiss Nat Found fel, Univ Calif, Berkeley, 67-68; res assoc biochem, Princeton Univ, 68-69; asst prof, 69-74, assoc prof chem, 74-78, PROF CHEM, UNIV CINCINNATI, 78- *Concurrent Pos:* Vis scientist, Nat Inst Arthritis, Metab & Digestive Dis, 75-76 & 83-84. *Mem:* AAAS; Am Soc Biol Chem; Biophys Soc; Am Chem Soc. *Res:* Application of electron spin resonance and circular dichroism spectroscopy to biological systems; biological redox mechanisms; synthesis of nucleic acid analogs and hybridization probes. *Mailing Add:* Dept Chem 172 Univ Cincinnati Cincinnati OH 45221

BOCARSLY, ANDREW B, b Los Angeles, Calif, Apr 23, 54; m 77; c 2. PHOTOCHEMISTRY, ELECTROCHEMISTRY. *Educ:* Univ Calif, Los Angeles, BS, 76; Mass Inst Technol, PhD(inorg chem), 80. *Prof Exp:* Asst prof, 80-87, ASSOC PROF CHEM, PRINCETON UNIV, 87- *Concurrent Pos:* A P Sloan fel, 86-88. *Honors & Awards:* Solid State Chem Award, Am Chem Soc, 84. *Mem:* Am Chem Soc; Electrochem Soc. *Res:* Inorganic photochemistry and electrochemistry; chemically modified electrode interfaces; solar energy conversion; semiconducting electrodes; electrocatalysis and mechanism of interfacial charge transfer; sensors. *Mailing Add:* Dept Chem Princeton Univ Princeton NJ 08544

BOCCABELLA, ANTHONY VINCENT, b Brooklyn, NY; m 51; c 2. ANATOMY. *Educ:* Seton Hall Univ, BS, 50, JD, 84; Univ Iowa, MS, 56, PhD(anat), 58. *Prof Exp:* Instr anat, Univ Iowa, 58-59; from instr to assoc prof, 59-67, PROF ANAT, COL MED & DENT NJ, 67-, CHMN DEPT, 71- *Mem:* Endocrine Soc; Am Asn Anat; Am Soc Andrology; Am Soc Law & Med; Asn Anat Chmn; Sigma Xi; Soc Study Reproduction. *Res:* Thyroid physiology; male and female reproductive systems; muscular dystrophy; hormonal factors regulating spermatogenesis; histophysiological evaluation of the male reproductive system in various states of hormonal deprivation and repletion. *Mailing Add:* Dept Anat Col Med & Dent NJ Newark NJ 07103

BOCHENEK, WIESLAW JANUSZ, b Radymno, Poland, Nov 11, 38; US citizen; m 69; c 2. GASTROENTEROLOGY. *Educ:* Akad Medyczna, MD, 62, PhD(physiol). *Prof Exp:* Assoc prof med, Akad Medyczna, Wroclaw, 76-81; assoc prof med & gastrointestinal nutrit, Albany Med Col, 82-88; assoc dir, 88-91, DIR RES & DEVELOP, METAB DIV, WYETH-AYERST RES, 91- *Concurrent Pos:* mem comt physiol, gastrointestinal tract & physiol, Polish Acad Sci, 75-81. *Mem:* Am Heart Asn; Am Soc Clin Nutrit; Am Diabetic Asn. *Res:* Development of antidiabetic drugs; research on lipid absorption, intestinal lipoprotein secretion. *Mailing Add:* Wyeth-Ayerst Clin Res & Develop Dept 145 King of Prussia Rd Radnor PA 19087

BOCHNER, BARRY RONALD, FERMENTATION DEVELOPMENT, BIOTECHNOLOGY. *Educ:* Univ Mich, PhD(bioeng), 76. *Prof Exp:* SR SCIENTIST, GENENTECH, INC, 82- *Mailing Add:* R & D Biol Inc 3447 Investment Blvd No 3 Hayward CA 94545

BOCK, ARTHUR E(MIL), thermodynamics, fuels & combustion; deceased, see previous edition for last biography

BOCK, CARL E, b Berkeley, Calif, Mar 29, 42; m 64. ZOOLOGY, ECOLOGY. *Educ:* Univ Calif, Berkeley, AB, 64, PhD(zool), 68. *Prof Exp:* Asst prof biol, 68-77, assoc chmn dept, 74-77, assoc prof, 74-80, PROF ENVIRON, POP & ORGANISMIC BIOL, UNIV COLO, 80- *Mem:* Ecol Soc Am; Soc Study Evolution; Cooper Ornith Soc; Am Ornith Union; Brit Ecol Soc. *Res:* Animal ecology; evolution and ecology of birds; fire ecology. *Mailing Add:* EPO Biol Box 334 Univ Colo Boulder CO 80309

BOCK, CHARLES WALTER, b Philadelphia, Pa, Jan 19, 45; m 68; c 2. QUANTUM & COMPUTATIONAL CHEMISTRY, MOLECULAR MODELING. *Educ:* Drexel Univ, BS, 68, MS, 70, PhD(physics), 72. *Prof Exp:* ASST PROF CHEM & COMPUT SCI, PHILADELPHIA COL TEXTILES & SCI, 72- *Mem:* Am Phys Soc. *Res:* Investigation of the most appropriate definitions of stabilization and destabilization energies; determination of molecular geometries for various organic and inorganic compounds using ab initio quantum mechanical calculations; determination of transition states for various chemical processes; solid state chemistry. *Mailing Add:* Philadelphia Col Textiles & Sci Henry Ave Philadelphia PA 19144

BOCK, ERNST, b Charkow, USSR, Feb 6, 29; nat Can; m 55; c 4. PHYSICAL CHEMISTRY. *Educ:* Univ Man, BSc, 56, MSc, 57, PhD, 60. *Prof Exp:* Lectr, 58-61, from asst prof to assoc prof, 61-77, PROF CHEM, UNIV MAN, 77- *Mem:* Sigma Xi. *Res:* Nuclear magnetic resonance spectroscopy. *Mailing Add:* Dept Chem Univ Man Winnipeg MB R3T 2N2 Can

BOCK, FRED G, b St Louis, Mo, Aug 28, 23; m 44, 65; c 4. BIOCHEMISTRY. *Educ:* Univ Minn, BA, 47, MS, 48; Univ Buffalo, PhD, 58. *Prof Exp:* Assoc scientist, Univ Minn, 52-53; from sr to assoc cancer res scientist, 53-66, prin cancer res scientist, Roswell Park Mem Inst, 66-, dir, Orchard Park Labs, 64-;

assoc res prof, State Univ NY Buffalo 70-; at Papanicolaou Cancer Res Inst, Miami; RETIRED. *Concurrent Pos:* Asst res prof, State Univ NY Buffalo, 63-68; res prof, Niagara Univ, 77- *Mem:* Fel AAAS; Am Chem Soc; Am Asn Cancer Res; Soc Exp Biol & Med. *Res:* Carcinogenesis. *Mailing Add:* 8128 SW 103 Ave Miami FL 33173

BOCK, JANE HASKETT, b Rochester, Ind, Oct 3, 36; m 64; c 1. GENETICS. *Educ:* Duke Univ, BA, 58; Ind Univ, Bloomington, MA, 60; Univ Calif, Berkeley, PhD(bot), 66. *Prof Exp:* Asst prof biol, Calif State Col, Hayward, 67-68; from asst prof to assoc prof, 68-79, PROF BIOL, UNIV COLO, 79- *Concurrent Pos:* Co-dir, Audubon Res Ranch, Elgin, Ariz, 80-91. *Mem:* Bot Soc Am; Brit Ecol Soc; Ecol Soc Am; Soc Study Evolution; Am Inst Biol Sci. *Res:* Evolution and ecology of flowering plants; ecology of fire; conservation of native flora of the western US; herbivory in grassland and alpine; forensic botany. *Mailing Add:* EPO Dept Biol Univ Colo Box 334 Boulder CO 80309

BOCK, MARK GARY, b Munich, Ger, Oct 10, 46; US citizen; m 68; c 2. PEPTIDE CHEMISTRY, HETEROCYCLIC CHEMISTRY. *Educ:* Rutgers Univ, BA, 69; Univ Ga, PhD(chem), 74. *Prof Exp:* Postdoctoral fel chem, Harvard Univ, 74-76; sr res chemist, Merck & Co, Inc, 76-82, res fel med chem, 82-87, sr res fel, 87-91, SR INVESTR, MERCK & CO, INC, 91- *Mem:* Am Chem Soc. *Res:* Design and synthesis of hormone antagonists for therapeutic application. *Mailing Add:* Merck Sharp & Dohme Res Labs W26A-4044 West Point PA 19486

BOCK, PAUL, b Baltimore, Md, Sept 3, 26; m 49; c 3. HYDROLOGY & WATER RESOURCES. *Educ:* Mass Inst Technol, 47; Johns Hopkins Univ, MS, 51, DE, 58. *Prof Exp:* Jr engr, Off Spec Studies, Pa Water & Power Co, 47-48; jr bridge engr, Md State Rd Comn, 50; proj engr, Inst Co-op Res, Johns Hopkins Univ, 51-57; from asst prof to assoc prof civil eng, Univ Del, 58-60; sr res scientist, Travelers Res Ctr, Inc, 61-62, dir, Hydrol Water Resources Div, 62-68, res fel, 67-68; PROF HYDROL & WATER RESOURCES, UNIV CONN, 68- *Concurrent Pos:* Consult, Massey & Assoc, 57; lectr, Univ Conn, 61-68; mem comt int prog atmospheric sci & hydrol, Hydrol Task Force, Nat Acad Sci, 62-63; mem, US Nat Comt Int Hydrol Decade, Nat Acad Sci-Nat Res Coun, 65-70, chmn hydrol panel, 65-71, mem comt space appln adv to NASA, 66-70, mem comt remote sensing progs earth resources surv adv to Dept Interior, 71-; consult, Space Sci & Appln Steering Comt, Off Space Sci & Appln, NASA, 66- *Mem:* Sigma Xi; Am Meteorol Soc; Am Geophys Union; NY Acad Sci; Am Soc Photogram. *Res:* Hydrometeorology; urban drainage; hydrological investigations; physical hydromechanics; remote sensing of environment; atmospheric water vapor; water resources planning. *Mailing Add:* Dept Civil Eng Univ Conn Main Campus U-37 261 Glenbrook Rd Storrs CT 06268

BOCK, ROBERT MANLEY, biophysical chemistry; deceased, see previous edition for last biography

BOCK, S ALLAN, b Baltimore, Md, Apr 28, 46; m; c 1. ALLERGY, PEDIATRICS. *Educ:* Washington Univ, St Louis, AB, 68; Univ Md, MD, 72. *Prof Exp:* ASST PROF PEDIAT & SR STAFF PHYSICIAN ALLERGY, IMMUNOL & PEDIAT, NAT JEWISH HOSP & UNIV COLO MED CTR, 76- *Concurrent Pos:* Pediat allergist, Dept Health & Hosps, Denver, 76-; clin prof pediat, Sch Med, Univ Colo, 90. *Mem:* Am Acad Allergy; Am Acad Pediat. *Res:* Food allergy and immunologic sensitivity to food in children; immunology of the gastrointestinal tract. *Mailing Add:* Dept Pediat 1600 Jackson St Denver CO 80206

BOCK, WALTER JOSEPH, b New York, NY, Nov 20, 33; m 57; c 3. VERTEBRATE MORPHOLOGY, EVOLUTION. *Educ:* Cornell Univ, BS, 55; Harvard Univ, MA, 57, PhD(biol), 59. *Prof Exp:* From asst prof to assoc prof zool, Univ Ill, 61-65; from asst prof to assoc prof, 65-73, PROF EVOLUTIONARY BIOL, COLUMBIA UNIV, 73- *Concurrent Pos:* NSF fel, Univ Frankfurt, 59-61; res assoc ornith, Am Mus Natural Hist, 65- *Honors & Awards:* Coues Award, Am Ornithologists Union, 75. *Mem:* Fel AAAS; fel Am Ornithologists Union; Soc Study Evolution; Am Soc Naturalists (treas, 78-81); corresp mem Ger Ornith Soc. *Res:* Functional and evolutionary morphology; morphology and classification of birds; general theories of major evolutionary changes. *Mailing Add:* Dept Biol Columbia Univ Main Div New York NY 10027

BOCK, WAYNE DEAN, b Truman, Minn, Nov 15, 32; div; c 4. MICROPALEONTOLOGY, ENVIRONMENTAL SCIENCE. *Educ:* Univ Wis, BS, 58, MS, 61; Univ Miami, PhD(oceanog), 67. *Prof Exp:* Instr geol, Sch Marine & Atmospheric Sci, Univ Miami, 64-67, res scientist, 67-69, from asst prof to assoc prof micropaleont, 67-81; pres, Appl Eco-Tech Serv, Inc, 78-83; partner, Bock-Joensuu Ent, 84-87; VPRES, SPECTRO SCAN, 86-; PRES, SPECTROANALYTICA, INC, 87- *Concurrent Pos:* Prin investr, NSF res grants, 67- *Mem:* AAAS; Paleont Soc; Paleont Res Inst. *Res:* Paleoecology of ocean sediment cores; use of Foraminifera as indicators of pollution: environmental assessment & monitoring. *Mailing Add:* 8005 SW 52nd Ave Miami FL 33145

BOCKELMAN, CHARLES KINCAID, b San Francisco, Calif, Nov 29, 22; div; c 1. NUCLEAR PHYSICS. *Educ:* Univ Wis, PhD(physics), 51. *Prof Exp:* Res assoc physics, Mass Inst Technol, 51-55; from asst prof to assoc prof, 55-65, dep provost sci, 69-78, dep provost, 78-87, PROF PHYSICS, YALE UNIV, 65- *Concurrent Pos:* Fel, Univ Inst Theoret Physics, Copenhagen, 58-59; Guggenheim fel, Oxford Univ, 70. *Mem:* Fel Am Phys Soc; Sigma Xi. *Res:* Nuclear structure; electron scattering; nucleon transfer reactions. *Mailing Add:* A W Wright Nuclear Struct Lab Yale Univ New Haven CT 06511

BOCKELMANN, JOHN B(URGGRAF), b Union City, NJ, July 24, 15; m 42; c 3. CHEMICAL ENGINEERING. *Educ:* Columbia Univ, AB, 37, BS, 38, ChE, 39, PhD(chem eng), 41. *Prof Exp:* Sr proj chem engr, Soaps Div, Colgate-Palmolive-Peet Co, NJ, 41-45; asst tech dir & process engr, Tech Dept, F & M Schaefer Brewing Co, 45-49, tech dir, 49-70, vpres, 70-77;

RETIRED. *Mem:* AAAS; Am Chem Soc; Am Soc Brewing Chem (pres, 63-64); Am Inst Chem Engrs; Inst Food Techologists; Sigma Xi. *Res:* Liquid extraction employing solvents near their critical temperatures; process control variables in brewing and their relation to final taste and taste testing. *Mailing Add:* 40 Sussex Rd Tenafly NJ 07670-2512

BOCKEMUEHL, R(OBERT) R(USSELL), b Omaha, Nebr, Oct 17, 27; m 52; c 1. ELECTRICAL ENGINEERING. *Educ:* Univ Mich, BSE, 52. *Prof Exp:* Jr engr, physics dept, 52-54, res engr, 54-56, sr res engr, 56-62, asst head electronics & instrumentation dept, 62-73, HEAD INSTRUMENTATION DEPT, GEN MOTORS RES LABS, 73- *Mem:* Sigma Xi. *Res:* Advanced electronic instruments; solid state materials and device research. *Mailing Add:* 4800 N Harsdale Bloomfield Hills MI 48302

BOCKHOFF, FRANK JAMES, b Tiffin, Ohio, Mar 26, 28; m 51; c 4. PHYSICAL CHEMISTRY, POLYMER CHEMISTRY. *Educ:* Case Inst Technol, BS, 50; Case Western Reserve Univ, MS, 52, PhD(phys chem), 59. *Prof Exp:* Res chem, Western Reserve Univ, 50-51; instr chem & chem eng, Fenn Col, 51-54, from asst prof to assoc prof chem, 54-64, chmn dept, 62-64; chmn dept, 64-83, PROF CHEM, CLEVELAND STATE UNIV, 64- *Concurrent Pos:* Consult, Am Agile Corp, 53-61, Apex Reinforced Plastics Div, White Sewing Mach Co, 61-63 & Signal Chem Mfg Co, 60-62; tech consult, Re-Ox Corp, 62-67; consult, Incar, Inc, Am Smelting and Refining Co, 64-67, Oatey Co, 80-84; res assoc, Cleveland Mus Natural Hist, 74- *Mem:* AAAS; Am Chem Soc; Soc Plastics Eng; Soc Plastics Indust; Am Inst Chem. *Res:* Plastics and resins development; complex-ion; thermodynamics and quantum theory; kinetics and mechanisms of anionic polymerizations; polymer development and applications. *Mailing Add:* Dept Chem Cleveland State Univ Cleveland OH 44115

BOCKHOLT, ANTON JOHN, b Westphalia, Tex, Sept 23, 30; m 52; c 2. PLANT BREEDING, GENETICS. *Educ:* Tex A&M Univ, BS, 52, MS, 58, PhD(plant breeding), 67. *Prof Exp:* Res asst corn breeding, 54-55, instr, 55-66, asst prof corn breeding & leader, corn improv prog, 66-72, ASSOC PROF AGRON, TEX A&M UNIV, 72- *Mem:* Am Soc Agron; Crop Sci Soc Am; Sigma Xi. *Res:* Plant breeding, genetics, pathology, biology-environment, entomology and nutrition and metabolism. *Mailing Add:* Dept Soil & Crop Sci Tex A&M Univ College Station TX 77843

BOCKMAN, DALE EDWARD, b Winona, Mo, Feb 4, 35; m 53; c 3. ANATOMY. *Educ:* Southwest Mo State Col, BS, 56; Los Angeles State Col, MA, 58; Univ Ill, PhD(anat), 63. *Prof Exp:* Instr anat, Univ Ill, 62-63; from instr to asst prof, Univ Tenn Med Units, 63-68; from assoc prof to prof, Med Col Ohio, 68-75; PROF ANAT & CHMN DEPT, MED COL GA, 75- *Res:* Fine structure; pancreas and gastrointestinal tract; development of lymphoid system; immunology. *Mailing Add:* Dept Anat Med Col Ga Augusta GA 30912

BOCKMAN, RICHARD STEVEN, b New York, NY, Sept 14, 41; m 73; c 2. MOLECULAR BIOLOGY. *Educ:* Johns Hopkins Univ, BA, 62; Yale Univ, MD, 67; Rockefeller Univ, PhD(biochem), 71. *Prof Exp:* assoc prof med, Med Ctr, 82-90, ASSOC PROF BIOCHEM, CORNELL UNIV, 71-, PROF MED, MED CTR, 90- *Concurrent Pos:* Assoc attending physician, Mem Hosp, 81-88, Hosp Special Surg; attend physician, NY Hosp, 88- *Mem:* Am Col Physicians; Am Fedn Clin Res; NY Acad Sci; Sigma Xi; Endocrine Soc; Am Soc Clin Invest; Am Soc Bone & Mineral Res. *Res:* Endocrinology; mechanisms of normal and pathological bone resorption; molecular biology of bone formation. *Mailing Add:* Hosp Spec Surg 535 E 70th St New York NY 10021

BOCKO, MARK FREDERICK, b Utica, NY, May 21, 56; m 78; c 1. ELECTRICAL ENGINEERING. *Educ:* Colgate Univ, BA, 78; Univ Rochester, MA, 80, PhD(physics), 84. *Prof Exp:* Res assoc physics, 84-85, ASST PROF ELEC ENG, UNIV ROCHESTER, 85- *Mem:* Am Phys Soc; Inst Elec & Electronics Engrs. *Res:* Superconductivity and tunneling; low noise detection; quantum non-demolition measurement techniques; measurement science; quantum measurement theory. *Mailing Add:* Dept Elec Eng Univ Rochester Hopeman Hall Rochester NY 14627

BOCKOSH, GEORGE R, b Charlerio, Pa, Jan 28, 51; div. RESEARCH ADMINISTRATION. *Educ:* Univ Pittsburgh, BS, 72, MS, 76. *Prof Exp:* Elec eng, Repub Steel Corp, 72-73, Corning Glass Works, 73-74; elec eng, 74-79, RES SUPVR, US BUR MINES, PITTSBURGH RES CTR, 79- *Res:* Program in underground mining in the following areas: equipment design, material handling, human factors, life support (breathing apparatus development), noise control and advanced mining technology. *Mailing Add:* 22 Rosemary Lane Bureau of Mines Cochrans Mill Rd PO Box 18070 Library PA 15129

BOCKRATH, BRADLEY CHARLES, b Hamburg, NY, Dec 11, 42; m 65; c 1. ORGANIC CHEMISTRY. *Educ:* Union Col, BS, 65; State Univ NY Albany, PhD(chem), 71. *Prof Exp:* Chemist, Eastman Kodak Co, 65-67; fel phys chem, Ohio State Univ, 71-74; supvry res chemist coal chem, 74-85, CHIEF, COAL CHEM BR, PITTSBURGH ENERGY TECHNOL CTR, US DEPT ENERGY, 85- *Honors & Awards:* Storch Award, 90. *Mem:* Am Chem Soc. *Res:* Physical organic chemistry; mechanisms of coal liquefaction; coal chemistry; kinetics of fast organic reactions; radical anion chemistry; solvated electrons; radiation chemistry; bioconversion of coal. *Mailing Add:* Pittsburgh Energy Technol Ctr PO Box 10940 Pittsburgh PA 15236

BOCKRATH, RICHARD CHARLES, JR, b St Louis, Mo, May 7, 38; div; c 2. MOLECULAR GENETICS, BIOLOGY. *Educ:* Yale Univ, BS, 61; Pa State Univ, MS, 63, PhD(biophys), 65. *Prof Exp:* Vis res fel microbial genetics, Univ Sussex, 65-67; asst prof microbiol & med biophysics, 68-77, assoc prof microbiol, immunol & med biophysics, 77-81, PROF MICROBIOL & IMMUNOL, SCH MED, IND UNIV, INDIANAPOLIS, 81- *Concurrent Pos:* NIH fel, Univ Sussex, 66-67; vis researcher, Aarhus Univ Inst Molecular Biol, 72; res consult radiobiol, Stanford Univ, 74, vis scholar biol sci, 78-79; fac res partic, Biol & Med Res Div, Argonne Nat Lab, 77; vis worker, MRC Cell Mutation Unit, Univ Sussex, 85-86. *Mem:* Am Soc Photobiol. *Res:* Effects of low energy beta particles on cells; molecular biology and control; cellular accumulation of streptomycin; mutagenesis and DNA repair induced by ultraviolet radiation; radiation mutagenesis; nucleoid structure/function. *Mailing Add:* Dept Microbiol Ind Univ Med Sch Indianapolis IN 46202-5120

BOCKRIS, JOHN O'MARA, b Johannesburg, SAfrica, Jan 5, 23; nat US; div; c 2. PHYSICAL CHEMISTRY. *Educ:* Univ London, BSc, 43, PhD, 45, DSc(electrochem), 52. *Prof Exp:* Lectr phys chem, Imperial Col, Univ London, 45-56; prof chem, Univ Pa, 56-62 & electrochem, 62-71; chmn dept phys chem, Flinders Univ, 71-74, prof phys chem, Sch Phys Sci, 72-78; RETIRED. *Concurrent Pos:* Founder, Inst Solar & Electrochem Energy Conversion. *Honors & Awards:* Medaille d'honneur, Univ Louvain, 53; Faraday Medal, Chem Soc London, 79; Lectr Award, Swedish Acad, 79. *Mem:* Fel Royal Inst Chem; Int Soc Hydrogen Energy; Int Soc Electrochem. *Res:* Hydrogen on and in metals; metal deposition and dissolution; adsorption at solid electrolyte surfaces; constitution of molten salts and silicates; electrode reactions in molten salts; quantum mechanics; electrode processes; hydrogen economy; energy science; photoelectrochemistry; bioelectrochemistry. *Mailing Add:* Dept Chem Texas A&M College Station TX 77843-1248

BOCKSCH, ROBERT DONALD, b Detroit, Mich, May 5, 31; m 58; c 5. ORGANIC CHEMISTRY. *Educ:* Wayne State Univ, BS, 54, Univ Wis, PhD, 60. *Prof Exp:* From asst prof to assoc prof chem, Whitworth Col, 58-68, chmn dept chem, 63-75, chmn dept health sci, 75-78, chmn dept chem, 78-90, PROF CHEM, WHITWORTH COL, 68- *Mem:* Am Chem Soc; Brit Chem Soc. *Res:* Synthetic organic chemistry; natural product chemistry; polynuclear hydrocarbons. *Mailing Add:* Dept Chem Whitworth Col Spokane WA 99251

BOCKSTAHLER, LARRY EARL, b La Junta, Colo, May 13, 34; m 65; c 1. VIROLOGY, BIOCHEMISTRY. *Educ:* Mich State Univ, BS, 56; Univ Wis-Madison, MS, 60, PhD(biochem), 64. *Prof Exp:* NIH fel biochem, Max Planck Inst Biol, 64-67, biochemist & Humboldt fel biochem, 67-69; BIOCHEMIST, FOOD & DRUG ADMIN, USPHS, 69- *Concurrent Pos:* Mem, Foreign Work/Study Prog, USPHS, Munich, Ger, 89-90. *Mem:* Am Soc Virol; Am Soc Photobiol; Biophys Soc; Radiation Res Soc Am; Sigma Xi; AAAS; Am Soc Testing & Mat; Soc Biol Chem. *Res:* Molecular biology of viruses; photobiology; radiation virology. *Mailing Add:* 511 Carr Ave Rockville MD 20850

BOCKSTAHLER, THEODORE EDWIN, b Monticello, Ill, Nov 22, 20; m 43; c 2. INDUSTRIAL ORGANIC CHEMISTRY. *Educ:* Ind Univ, AB, 42; Univ Ill, MS, 48, PhD(chem), 49. *Prof Exp:* Instr, Chem Warfare Sch, Edgewood Arsenal, 45; group leader org chem res, Rohm & Haas Bayport Inc, 49-66, mgr labs, 66-80, tech dir, Rohm & Haas Co, 80-83; RETIRED. *Mem:* Am Chem Soc. *Res:* Process development for plastic chemicals in laboratory and plant. *Mailing Add:* 792 Fairgate Dr Hendersonville NC 28739

BOCQUET, PHILIP E(DMUND), b Beeville, Tex, Oct 6, 18; c 4. CHEMICAL ENGINEERING. *Educ:* Agr & Mech Col Tex, BS, 40; Univ Mich, MSE, 47, PhD (chem eng), 53. *Prof Exp:* Assoc physiol, Univ Mich, 49-52; group supvr, Continental Oil Co, Okla, 53-57; assoc prof chem eng, Univ NMex, 57-61; prof & head dept, 61-69, assoc dean eng, 69-77, prof, 77-84, EMER PROF CHEM ENG, UNIV ARK, FAYETTVILLE, 84- *Concurrent Pos:* Consult, 57-; Fulbright lectr, Univ Madrid, 68-69. *Mem:* AAAS; Am Inst Chem Engrs; Am Chem Soc; Am Soc Eng Educ; Am Inst Mining, Metall & Petrol Engrs; Electrochem Soc. *Res:* Fluid mechanics; electrokinetics; static electricity in fluids; reservoir mechanics; miscible phase behavior; hydraulic fracturing; well logging in petroleum production; cryogenics; natural gas liquefaction. *Mailing Add:* 555 E North Fayetteville AR 72701

BOCTOR, MAGDY, aerodynamics, computer science; deceased, see previous edition for last biography

BOCZAR, BARBARA ANN, b E Orange NJ, May 22, 51. MOLECULAR BIOLOGY. *Educ:* Univ Ore, BS, 74; Univ Calif, Santa Barbara, MA, 81, PhD(biol sci), 85. *Prof Exp:* Instr biochem, Rutgers Univ, 77; FEL BOT, UNIV WASH, 85- *Mem:* AAAS; Phycol Soc Am; Bot Soc Am; Asn Woman Sci; Am Soc Limnol & Oceanog; Sigma Xi. *Res:* Photosynthetic processes in marine algae; application of biochemical and molecular biological techniques to questions concerning adaptive processes in marine algae. *Mailing Add:* Procter & Gamble Co Ivorydale Tech Ctr 5299 Spring Grove Ave Cincinnati OH 45217

BOCZKOWSKI, RONALD JAMES, b Bridgeport, Conn, Nov 14, 43; m 70; c 2. ANALYTICAL CHEMISTRY. *Educ:* Canisius Col, BS, 65; Univ Notre Dame, PhD(chem), 71. *Prof Exp:* Res fel, Univ Cincinnati, 71-72; lab mgr textiles, Armtex, Inc, 72-73; METHODS DEVELOP GROUP LEADER, UNION CAMP CORP, 73- *Mem:* Am Chem Soc; Anal Am Chem Soc; Am Asn Textile Chemists & Colorists. *Res:* Analytical application of electrochemical techniques; preconcentration techniques in trace analysis; analysis of minor components in paper, pulp and process liquors; analysis of chemicals from wood. *Mailing Add:* Union Camp PO Box 3301 Princeton NJ 08543

BODA, JAMES MARVIN, b Mt Victory, Ohio, July 3, 24; m 53; c 1. COMPARATIVE PHYSIOLOGY. *Educ:* Univ Calif, BS, 48, PhD, 53. *Prof Exp:* Actg instr animal husb, univ & animal husbandman, Exp Sta, Univ Calif, 52-54, from asst prof to assoc prof, 59-68, prof animal physiol, 65-80, chmn dept 68-73, EMER PROF, DEPT ANIMAL PHYSIOL, UNIV CALIF, DAVIS, 80- *Concurrent Pos:* Fulbright res scholar, 60-61. *Mem:* Am Physiol Soc; Am Soc Zoologists; Endocrine Soc. *Res:* Comparative endocrinology; physiological adaptations. *Mailing Add:* PO Box 385 North San Juan CA 95960

BODALY, RICHARD ANDREW, b Vancouver, BC, June 24, 49; m 77; c 2. FISHERIES, EVOLUTION. *Educ:* Simon Fraser Univ, BSc, 72; Univ Man, PhD(zool), 77. *Prof Exp:* RES SCIENTIST, FRESHWATER INST, DEPT FISHERIES & OCEANS, CAN, 77- *Concurrent Pos:* Adj prof, dept zool, Univ Man, 86-; vis scientist, environ biol unit, Loughborough Univ, Eng, 87-88. *Mem:* Can Soc Zoologists; Am Fisheries Soc; Can Soc Environ Biologists; Sigma Xi. *Res:* Reservoir fisheries problems related to northern areas; fish zoogeography; speciation and evolution; mercury contamination of fish. *Mailing Add:* Freshwater Inst 501 University Crescent Winnipeg MB R3T 2N6 Can

BODAMER, GEORGE WILLOUGHBY, b Philadelphia, Pa, Apr 12, 16; m 44; c 4. CHEMISTRY. *Educ:* Univ Pa, BS, 38; Cornell Univ, PhD(org chem), 41. *Prof Exp:* Chemist, Rohm and Haas, Philadelphia, 41-59; sr scientist, ESB, Inc, 59-61, mgr phys chem labs, 61-64, assoc dir res, 64-70; chief chemist, Int Hydronics Corp, Princeton, NJ, 70-76; chemist, C&D Batteries, 78-80; CONSULT, ROB SMITH ASSOCS, 81. *Mem:* Am Chem Soc; Electrochem Soc. *Res:* Electrochemistry; ion exchange; industrial waste treatment. *Mailing Add:* 311 Franklin Ave Cheltenham PA 10912

BODAMMER, JOEL EDWARD, b Chicago, Ill, Feb 18, 42. ENVIRONMENTAL TOXICOLOGY. *Educ:* Univ Ill, BS, 64, MS, 66; Univ Wis, PhD(anat), 74. *Prof Exp:* Res assoc, Armed Forces Inst Path, 70-72; RES PHYSIOLOGIST CYTOL, NAT MARINE FISHERIES SERV, NAT OCEANIC & ATMOSPHERIC ADMIN, 73- *Concurrent Pos:* NIH trainee, Anat Dept, Univ Wis, 66-70. *Mem:* Soc Protozoologists. *Res:* Ultrastructural studies on the sensory systems of marine fish larvae; the effects of heavy metal contaminants on sensory systems of fish; fine structural studies on teleost blood and hemopoietic tissues. *Mailing Add:* Box 158 Easton MD 21601

BODANSKY, DAVID, b New York, NY, Mar 10, 24; m 52; c 2. PHYSICS. *Educ:* Harvard Univ, BS, 43, MA, 48, PhD(physics), 50. *Prof Exp:* Instr physics, Columbia Univ, 50-52, assoc, 52-54; chmn dept, 76-84, PROF PHYSICS, UNIV WASH, 54- *Concurrent Pos:* Alfred P Sloan res fel, 59-63; Guggenheim fel, 66-67 & 74-75; vis scholar, Energy & Environ Policy Ctr, Harvard Univ, 81-82. *Mem:* Am Nuclear Soc; fel AAAS; Am Asn Physics Teachers; fel Am Phys Soc; Health Physics Soc. *Res:* Nuclear physics, nuclear astrophysics and energy policy. *Mailing Add:* Dept Physics Univ Wash Seattle WA 98195

BODANSZKY, MIKLOS, b Budapest, Hungary, May 21, 15; US citizen; m 50; c 1. BIO-ORGANIC CHEMISTRY. *Educ:* Budapest Tech Univ, dipl chem eng, 39, DSc(org chem), 49. *Prof Exp:* Lectr med chem, Budapest Tech Univ, 50-56; res assoc biochem, Med Col, Cornell Univ, 57-59; sr res assoc peptide chem, Squibb Inst Med Res, 59-66; prof chem & biochem, 66-78, Charles Frederic Mabery prof res in chem, 78-83, EMER PROF CHEM, CASE WESTERN RESERVE UNIV, 83- *Concurrent Pos:* USPHS grants, Case Western Reserve Univ, 66-; Alexander von Humboldt Found prize, 79. *Honors & Awards:* Alan Pierce Award, Am Peptide Symposia, 77; Morley Medal, Am Chem Soc, 78. *Mem:* Am Chem Soc; Am Soc Biol Chemists; Hungarian Acad Sci. *Res:* Peptide synthesis, structure and activity relationships of biologically active peptides; conformation of peptide hormones and peptide antibiotics. *Mailing Add:* One Markham Rd Princeton NJ 08540

BODDY, DAVID EDWIN, b Erie, Pa, Aug 31, 36; m 65. MECHANICAL ENGINEERING. *Educ:* Purdue Univ, BS, 58, MS, 60, PhD(elec eng), 66. *Prof Exp:* ASSOC PROF ENG, OAKLAND UNIV, 65- *Concurrent Pos:* NSF res initiation grant, 67-68. *Mem:* Asn Comput Mach; Sigma Xi. *Res:* Computer science. *Mailing Add:* 616 Sorbonne Rochester MI 48063

BODDY, DENNIS WARREN, b Portland, Ore, Aug 26, 22. ENTOMOLOGY, ZOOLOGY. *Educ:* Univ Wash, BS, 47, PhD, 55. *Prof Exp:* Instr zool, Columbia Basin Col, 55-59, head sci dept, 59-60, dean instr, 60-62; from asst prof to assoc prof biol, Portland State Univ, 64-87; RETIRED. *Mem:* AAAS. *Res:* Taxonomy of the Coleoptera Tenebrionidae. *Mailing Add:* 1867 SW 14th Portland OR 97201

BODDY, PHILIP J, physical chemistry; deceased, see previous edition for last biography

BODE, ARTHUR PALFREY, b Lima, Ohio, Oct 12, 53; m 77; c 2. CLINICAL COAGULATION, BLOOD BANKING. *Educ:* Univ NC, Chapel Hill, BA, 75, PhD(exp path), 82. *Prof Exp:* Postdoctoral fel blood bank, Univ NC, Chapel Hill, 82-84, res asst prof path, 84-86; asst prof path, 86-91, ASSOC PROF PATH, ECAROLINA UNIV MED SCH, 91- *Concurrent Pos:* Sci dir, Clin Coagulation Lab, Pitt County Mem Hosp, Greenville, 86-, co-dir, Clin Flow Cytometry Lab, 86-89; prin investr, Am Heart Asn, 88-89 & Naval Res & Develop Command, 89-92. *Mem:* Am Asn Blood Banks; Am Heart Asn; AAAS; NY Acad Sci; Int Soc Thrombosis & Hemostasis; Am Asn Pathologists. *Res:* Storage of blood platelets for transfusion; vesiculation of platelets in vitro and in vivo; interference by platelets in heparin monitoring during open-heart surgery; post-operative bleeding in cardiac surgery; blood compatibility of foreign surfaces. *Mailing Add:* Dept Path & Lab Med E Carolina Univ Brody 1S-08 Greenville NC 27858

BODE, DONALD EDWARD, b Merrill, Wis, Aug 26, 22; m 44; c 2. SOLID STATE PHYSICS. *Educ:* Syracuse Univ, AB, 48, MS, 49, PhD(physics), 53. *Prof Exp:* Res assoc, Syracuse Univ, 48-54; assoc physicist, Int Bus Mach Corp, 54-56, proj physicist, 56-57, physicist, 57; head physics sect, Hughes Aircraft Co, 57-74, mgr detector res & develop dept, 72-74, mgr Components Lab, Santa Barbara Res Ctr, 74-79, dir technol, 79-86; CONSULT, SEMICONDUCTOR TECHNOL, 86- *Concurrent Pos:* Chmn, IRIS Detector Specialty Group, 67-72; vchmn Sci & Eng Coun, Santa Barbara, 75-76. *Res:* Photoconductivity; semiconductor properties; xerography; transistors, infrared detectors. *Mailing Add:* 940 Chandler Santa Barbara CA 93110

BODE, JAMES DANIEL, b Pittsburgh, Pa, Sept 6, 23. PHYSICAL CHEMISTRY. *Educ:* Univ Pittsburgh, BS, 46, PhD(chem), 52. *Prof Exp:* Grad asst, Univ Pittsburgh, 46-51; proj engr, Fisher Sci Co, 51-52, head anal chem, 52-55; sr res chemist, Jones & Laughlin Steel Corp, 55-58, res supvr, 58-60; mem tech staff, Bell Tel Labs, 60-63; sr res chemist, FMC Corp, 63-66; prin scientist, Bendix Res Labs, Bendix Corp, 66-81; RETIRED. *Mem:* AAAS; Am Chem Soc. *Res:* Instrumental methods; physics and chemistry of surfaces. *Mailing Add:* 70 Kings Ct Apt 4-C Santurce PR 00911

BODE, VERNON CECIL, b Marion County, Mo, Feb 19, 33; m 56; c 4. MAMMALIAN GENETICS, MAMMALIAN DEVELOPMENT. *Educ:* Univ Mo, BS, 55; Univ Ill, PhD(biochem), 61. *Prof Exp:* USPHS fel biochem, Sch Med, Stanford Univ, 61-64; from asst prof to assoc prof biochem, Med Sch, Univ Md, 64-70; PROF BIOL, KANS STATE UNIV, 70- *Concurrent Pos:* Am Cancer Soc scholar, Oxford Univ, 77-78; Eleanor Roosevelt, Int Cancer fel, Pasteur Inst, Am Cancer Soc, 84-85. *Mem:* Am Soc Biol Chem; Am Soc Microbiol; Genetics Soc Am; Soc Develop Biol. *Res:* Mouse genetics; regulation of embryological development in the mouse; selection of new mouse mutants after ethylnitrosourea mutagenesis; genetic analysis of the mouse t-complex; isolation of mouse mutants useful as models for human disease; preservation of mouse mutants by sperm freezing. *Mailing Add:* Div Biol Kans State Univ Manhattan KS 66506

BODE, WILLIAM MORRIS, b Wooster, Ohio, Mar 18, 43; m 66; c 2. ENTOMOLOGY. *Educ:* Col Wooster, BA, 65; Ohio State Univ, PhD(entom), 70. *Prof Exp:* ASST PROF ENTOM, PA STATE UNIV, 70- *Mem:* Entom Soc Am; Soc Invert Path; Am Inst Biol Sci. *Res:* Biology and management of insect pests of field corn and vegetables; develop sampling methods for determining population density and extent of injury; establish economic injury thresholds for crops; pest control materials and methods of application. *Mailing Add:* Dept Entomol Pa State Univ Main Campus University Park PA 16802

BODEK, ARIE, b Tel-Aviv, Israel, May 11, 47; div; c 3. PARTICLE PHYSICS. *Educ:* Mass Inst Technol, BS, 68, PhD(physics), 72. *Prof Exp:* Res assoc particle physics, Mass Inst Technol, 72-74; Millikan res fel particle physics, Calif Inst Technol, 74-76; from asst prof to assoc prof, 76-85, PROF PHYSICS, UNIV ROCHESTER, 86- *Concurrent Pos:* Sloan fel, 80-81; spokesman exp FNAL E595 & SLAC E140; Japan fel, NSF, 86, Dept Eng, 90. *Mem:* Fel Am Phys Soc. *Res:* Experimental particle physics; neutrino interactions; direct production of muons by hadrons; production of new states (charm, bottom); deep inelastic electron and muon scattering; scattering from nuclear targets; neutrino oscillations; e plus and e minus reactions; proton-antiproton collisions. *Mailing Add:* Dept Physics Univ Rochester Rochester NY 14627

BODELL, WILLIAM J, CANCER CHEMOTHERAPY, DRUG RESISTANCE. *Educ:* Univ Nebr, Lincoln, PhD(biochem), 77. *Prof Exp:* ASST ADJ PROF BIOL, DEPT NEUROL SURG, BRAIN TUMOR RES CTR, UNIV CALIF, 81- *Mailing Add:* Dept Neurol Surg Univ Calif HSE-710 San Francisco CA 94143

BODEN, GUENTHER, b Ludwigshafen, WGer, Jan 8, 35; m 70; c 4. METABOLISM, ENDOCRINOLOGY. *Educ:* Univ Heidelberg, MS, 56; Univ Munich, MD, 59. *Prof Exp:* Asst biochem, Univ Tubingen, 63-65; res fel med, E P Joslin Res Lab, Harvard Sch Med, 65-67; resident, Rochester Gen Hosp, 67-70; from asst prof to assoc prof, 70-77, PROF MED, TEMPLE UNIV HOSP, 77- *Concurrent Pos:* Asst med, Peter Bent Brigham Hosp, Boston, 65-67; gen med A, Study Sect, NIH, 78-82; chief, Div Metab/Endocrinoloss, Sch Med, Temple Univ, 87-; prog dir, Gen Clin Res Ctr, Temple Univ Hosp, 89- *Mem:* Am Soc Clin Invest; Am Col Physicians; Am Endocrine Soc; Am Diabetes Asn; Am Fed Clin Res. *Res:* Metabolism; diabetes. *Mailing Add:* Temple Univ Hosp 3401 N Broad St Philadelphia PA 19140

BODEN, HERBERT, b Staten Island, NY, Apr 5, 32; m 57. ORGANIC CHEMISTRY. *Educ:* Univ Vt, BS, 54, MS, 56; Ohio State Univ, PhD(org chem), 60. *Prof Exp:* Instr gen & org chem, Ohio State Univ, 56-58; res chemist, Org Chem Dept, 60-65, res chemist, Freon Prod Div, 65-70, tech assoc, 70, territory mgr, 70-72, prod & develop mgr, 72-74, nat acct mgr, 74-77, econ anal & distrib mgr, Freon Prod Div, 77-78, mkt mgr, Fire Extinguishants, 78-79, mgr, Sales Support, 79-80, REGIONAL SALES MGR, E I DU PONT DE NEMOURS & CO, INC, 81- *Mem:* Am Chem Soc. *Res:* Biguanide, benzo phenanthrene, fluorine and polymer chemistry; chemistry of tetra substituted amino ethylenes. *Mailing Add:* 904 Barnstable Hockessin DE 19898

BODENHEIMER, PETER HERMAN, b Seattle, Wash, June 29, 37; m 65; c 2. ASTROPHYSICS. *Educ:* Harvard Univ, AB, 59; Univ Calif, Berkeley, PhD(astron), 65. *Prof Exp:* NSF fel, Princeton Univ, 65-66, res assoc astrophys, 66-67; asst astronr & asst prof, 67-70, assoc astronr & assoc prof astron, 70-76, ASTRONR & PROF ASTRON, UNIV CALIF, SANTA CRUZ, 76- *Concurrent Pos:* chmn, Bd Studies Astron & Astrophics, Univ Calif, Santa Cruz, 75-78, 82-83, 88-89; vis scientist, Max Planck Inst Physics & Astrophys, 76, 78, 80, 85, 87-88; prin investr, NSF grant, 77-; mem, Comt Planetary & Lunar Explor, Space Sci Bd, Nat Acad Sci, 85- 87; Alexander von Humboldt Found US sr scientist award; 87. *Mem:* Am Astron Soc; Int Astron Union; Royal Astronomical Soc. *Res:* Theoretical calculations of stellar structure and evolution, including problems involving hydrodynamics and rotation; star formation; evolution of the giant planets. *Mailing Add:* Lick Observ Univ Calif Santa Cruz CA 95064

BODENLOS, ALFRED JOHN, geology; deceased, see previous edition for last biography

BODEY, GERALD PAUL, SR, b Hazelton, Pa, May 22, 34; m 56; c 3. ONCOLOGY, INFECTIOUS DISEASES. *Educ:* Lafayette Col, AB, 56; Johns Hopkins Univ, MD, 60; Am Bd Internal Med, dipl, 72, cert oncol, 73 & cert infectious dis, 74. *Prof Exp:* Intern, Osler Med Serv, Johns Hopkins Hosp, 60-61, resident, 61-62; clin assoc, Nat Cancer Inst, 62-65; resident, Univ Wash, 65-66; asst internist & asst prof, 66-72, assoc internist & asst prof med, 72-75, chief, chemother, 75-83, med dir, Cancer Clin Res Ctr, 77-81, dir, Off Protocol Res, 81-83, CHIEF, SECT INFECTIOUS DIS, UNIV TEX M D ANDERSON HOSP, & TUMOR INST, 71-, INTERNIST & PROF MED, 75-, CHMN, DEPT MED SPECIALTIES, 87- *Concurrent Pos:* From adj asst prof & clin asst prof to adj assoc prof & assoc clin prof, Baylor Col Med, 69-75, adj prof microbiol, immunol & med, 75-; Leukemia Soc scholar, 69-74; from asst prof to assoc prof, Grad Sch Biomed Sci, 69-75, assoc prof, Med Sch, 73-76, prof med & pharmacol, Univ Tex, Houston, 76-; consult prev med div, Space Flight Biotechnol Specialty Team, Manned Spacecraft Ctr, 70-71; consult, Brooke Gen Hosp, Ft Sam Houston, Tex, 71-; consult hemat-oncol, Wilford Hall USAF Med Ctr, Lackland AFB, Tex, 75-76, Comm Develop Comprehensive Cancer Care Ctr, Repub Panama, Orphan Prod Develop Initial Rev Group, FDA, 84-; mem, Am Soviet meetings cancer chemother, Collab Cancer Treat Res Prog, Pan Am Health Orgn, med/sci adv bd, SC Oncol Found, 86-, Janssen Res Coun, 86-; fac mem, Union Int Contre le Cancer, mem, Tex Med Asn Comt Cancer, 84-, vis fac, Hoffman LaRouche adv bd, sci adv bd, Liposome Co, 85-, adv comt, Immunocompromised Host Soc, 87- *Honors & Awards:* Am Chem Soc, Am Inst Chemists, Merck Chem & Robert B Youngman Greek Prizes, 56. *Mem:* Fel Am Col Physicians; AMA; Am Asn Cancer Res; fel Royal Col Med; fel Royal Soc Prom Health. *Res:* Cancer research; hematology; infectious diseases. *Mailing Add:* Univ Tex M D Anderson Hosp Tex Med Ctr Houston TX 77030

BODFISH, RALPH E, b Denver, Colo, Jan 24, 22; m 52; c 2. MEDICINE, NUCLEAR STRUCTURE. *Educ:* Duke Univ, BA, 49; Emory Univ, MD, 54; Am Bd Internal Med, dipl, 61; Am Bd Nuclear Med, dipl. *Prof Exp:* Chief radioisotope serv, Vet Admin Hosp, Long Beach, 56-70, dep dir res, 63-70, dep chief of staff, 68-70; head dept med, Rancho Los Amigos Hosp, 70-74; asst dean, Irvine-Calif Col Med, Univ Calif, 74-84; chief staff, Vet Admin Hosp, Long Beach, 74-84, assoc chief staff qual assurance, 84-90; RETIRED. *Concurrent Pos:* Asst clin prof, Sch Med, Univ Southern Calif, 60-70, prof med & radiol, 70-74; assoc clin prof, Univ Calif, Irvine-Calif Col Med, 67-70, clin prof radiol sci, 74-; consult, Community Hosp, Long Beach, 66-74, Scripps Clin & Inst, 66-74 & Vet Admin Hosp, Long Beach, 70-74. *Mem:* Sr mem Am Fedn Clin Res; Soc Nuclear Med; AMA; fel Am Col Physicians. *Res:* Radiation physics; control of thyroid function; renal disease; relationship of metabolic state to cardiopulmonary and bone marrow function; hospital administration. *Mailing Add:* 2682 Copa de Oro Dr Rossmoor Los Alamitos CA 90720-4910

BODI, LEWIS JOSEPH, b Racine, Wis, Dec 31, 24; m 50; c 4. PHYSICAL CHEMISTRY. *Educ:* DePauw Univ, BS, 50; Univ Wis, PhD(chem), 55. *Prof Exp:* Asst prof chem, Brooklyn Col, 54-60; eng specialist, Gen Tel & Electronics Labs, 60-67; dean sci & math, 67-70, PROF CHEM, CITY UNIV NEW YORK, 67-, DEAN FAC, YORK COL, 70-, DEAN ACAD AFFAIRS, 74-, VPRES, DEP TO THE PRES, 85- *Res:* Luminescence; solid state chemistry and physics; science education. *Mailing Add:* Provost/Dir City Univ New York York Col 94-20 Guy Brewer Blvd Jamaica NY 11451

BODIAN, DAVID, b St Louis, Mo, May 15, 10; m 44; c 5. NEUROBIOLOGY. *Educ:* Univ Chicago, ScB, 31, PhD(anat), 34, MD, 37. *Hon Degrees:* LLD, Johns Hopkins Univ, 87. *Prof Exp:* Asst anat, Univ Chicago, 35-38; Nat Res Coun fel med, Univ Mich, 38; res fel anat, Johns Hopkins Univ, 39-40; asst prof anat, Western Reserve Univ, 40-41; lectr epidemiol, 41, res fac, 42-57, assoc prof Sch Hygiene & Pub Health, 46-53, prof anat & dir dept, 57-75, prof neurobiol, dept otolaryngol, Johns Hopkins Univ Sch Med, 75-85; RETIRED. *Concurrent Pos:* Managing ed, Am J Hygiene, 47-57, assoc ed, Virol, 56-60, Exp Neurol, 71-75 & Anat Rec, 68-72; consult to Surgeon Gen, USPHS, Polio Hall of Fame & Warm Springs Found, 58; mem tech comt poliomyelitis vaccine, USPHS, 57-64; mem bd sci counrs, Div Biol Standards, NIH, 57-59 & Nat Inst Neurol Dis & Stroke, 68-74. *Honors & Awards:* E Mead Johnson Award, 41; USPHS Award, 56; Harvey lectr, 56; Lashley Award, Am Philos Soc, 85. *Mem:* Nat Acad Sci; Am Acad Arts & Sci; Am Philos Soc; Soc Neurosci; Am Asn Anat (pres, 72); AAAS; Am Physiol Soc; Int Brain Res Orgn. *Res:* Histology of nervous system; pathology and pathogenesis of poliomyelitis; development of nervous system; neurohypophysis; fine structure of synapses. *Mailing Add:* 4100 N Charles St Apt 913 Baltimore MD 21218

BODIG, JOZSEF, b Gonc, Hungary, Jan 20, 34; US citizen; m 62; c 3. WOOD SCIENCE, WOOD TECHNOLOGY. *Educ:* Univ BC, BSF, 59; Univ Wash, MF, 61, PhD(wood technol), 63. *Prof Exp:* Proj leader, Fiber Res, Inc, Wash, 60-61; asst mech properties for wood, Univ Wash, 61-63; asst prof & asst forester, Agr Exp Sta, 63-66, from assoc prof to prof wood sci & technol, 66-79, prof wood sci & civil eng, 79-89, EMER PROF WOOD SCI & CIVIL ENG, COLO STATE UNIV, 89-; SR VPRES, ENG DATA MGT INC, FORT COLLINS, COLO, 82- *Concurrent Pos:* Vis prof wood physics, Univ Wash, 69-70; mem tech adv comt, Am Inst Timber Construct, 70-, spec adv comt, Forest Prod Asn, 79-; vis res prof archit, Univ Forestry & Wood Technol, Sopron, Hungary, 79; vis prof, Capricornia Inst Adv Educ, 85; sr vpres, Eng Data Mgt Inc, Fort Collins, Colo, 82-88, exec vpres, 88-90, pres, 90-; tech dir, LRFD Engineered Wood Construct, 88-; dir, NATO Adv Res Workshop, RBD Engineered Wood Struct, 91. *Honors & Awards:* L J Markwardt Wood Eng Award, Forest Prod Res Soc, 82 & Am Soc Testing & Mat, 85. *Mem:* Forest Prod Res Soc; Soc Wood Sci & Tech; Nat Forest Prod Asn; Am Soc Testing & Mat; fel Int Acad Wood Sci; Am Nat Standards Inst. *Res:* Mechanical and rheological properties of wood and wood composites; nondestructive testing of wood; wood engineering; reliability-based design. *Mailing Add:* Eng Data Mgt Inc Oak Ridge Bus Park 4700 McMurray Ave Ft Collins CO 80525

BODILY, DAVID MARTIN, b Logan, Utah, Dec 16, 33; m 58; c 4. PHYSICAL CHEMISTRY, FUEL TECHNOLOGY & PETROLEUM ENGINEERING. *Educ:* Brigham Young Univ, BA, 59, MA, 60; Cornell Univ, PhD(phys chem), 64. *Prof Exp:* Res assoc, Northwestern Univ, 64-65; asst prof chem, Univ Ariz, 65-67; from asst prof to assoc prof fuels eng in mining, metall & fuels eng, Col Mines & Earth Sci, Univ Utah, 69-77, prof mining & fuels eng & chmn dept, 77-83, assoc dean col, 83-89. *Mem:* Am Chem Soc; Asn Inst Mining Engrs; Sigma Xi. *Res:* Radiation chemistry; coal science and cleaning. *Mailing Add:* Dept Fuels Eng Univ Utah Salt Lake City UT 84112

BODIN, JEROME IRWIN, b New York, NY, July 2, 30; m 58; c 2. PHARMACEUTICAL CHEMISTRY. *Educ:* Columbia Univ, BS, 52, MS, 54; Univ Wis, PhD(pharmaceut chem), 58. *Prof Exp:* Res analyst pharmaceut chem, Chas Pfizer & Co, Inc, 58-61; dir pharmaceut anal, Drug Standards Lab, Am Pharmaceut Assoc Found, 62-63; chemist, Div of New Drugs, US Food & Drug Admin, 63-64; dir qual control, Carter Prod Inc, 64-69, dir anal res & develop, 69-78, 82-83, dir develop res, 78-82, mgr tech admin, 83-85, ASSOC DIR PHARMACEUT CHEM, WALLACE LABS DIV, CARTER-WALLACE, INC, 85- *Concurrent Pos:* Mem, Am Found Pharmaceut Educ. *Mem:* Am Chem Soc; Am Pharmaceut Asn; Acad Pharmaceut Res & Sci; Am Asn Pharmaceut Sci; Asn Off Anal Chemists. *Res:* Chemical kinetics and drug stability; drug availability and absorption; pharmaceutical quality control; federal drug regulation; organic chemistry; pharmaceutics. *Mailing Add:* Three Wickman Lane Hightstown NJ 08520

BODINE, PETER VAN NEST, b Syracuse, NY, Mar 14, 58; m 86. STEROID HORMONE RECEPTORS, SIGNAL TRANSDUCTION. *Educ:* Syracuse Univ, BS, 80; Temple Univ, PhD(biochem), 88. *Prof Exp:* Res asst biol, Syracuse Univ, 80-82; postdoctoral fel cancer res, Fels Inst Cancer Res, 88-91; RES ASST PROF CANCER RES, JEFFERSON CANCER INST, 91- *Concurrent Pos:* Distinguished grad res award, NY Acad Sci, 88. *Mem:* AAAS; Sigma Xi; Am Soc Biochem & Molecular Biol; Endocrine Soc; Am Chem Soc; Am Asn Cancer Res. *Res:* Steroid hormone receptor biochemistry; regulation of steroid hormone action; regulation of protein kinase C; isolation and characterization of endogenous steroid hormone receptor and protein kinase C regulators. *Mailing Add:* Jefferson Cancer Inst Thomas Jefferson Univ Phildelphia PA 19107

BODINE, RICHARD SHEARON, b Norman, Okla, Dec 10, 46. ORGANIC SYNTHESIS, PROCESS DEVELOPMENT. *Educ:* BA, 68; MS, 77; PhD, 78. *Prof Exp:* Instr math, Parks Col, Denver, Colo, 69-70; Comput programmer, McGraw-Hill Inc, Englewood, Colo, 70-73; Chemist, Midwest Res Inst, Kansas City, MO,79-84; CHEM DEVELOP TEAM LEADER, MARION LABS INC, KANSAS CITY, MO, 84- *Concurrent Pos:* Prin invesr, NIH & Environ Protection Agency, 81-84. *Mem:* Am Chem Soc; Sigma Xi. *Res:* Synthesis of selected unlabeled and isotopically labeled carcinogens and their metabolites; synthesis of nitro-substituted polycyclic aromatic hydrocarbons; synthesis of drug metabolites and analytical standards for pharmaceutical research and development. *Mailing Add:* 4340 Harrison Kansas City MO 64110

BODINE, ROBERT Y(OUNG), b Peoria, Ill, Sept 7, 33. ENGINEERING MECHANICS, HISTORY & PHILOSOPHY SCIENCE. *Educ:* Bradley Univ, BS, 54; Univ Kans, MS, 58; Univ Wis, PhD(eng mech), 61. *Prof Exp:* Engr, Westinghouse Elec Corp, 54-58; instr eng mech, Univ Wis, 58-61; consult mech, A O Smith, 61-66, mgr eng systs, 66-86; PRES CAD COMP INC, 87- *Concurrent Pos:* Dir, MacNeal-Schwendler Corp, 88-; lectr, Milwaukee Sch Eng, 86- *Mem:* Am Soc Mech Engrs; Sigma Xi. *Res:* Computer methods in engineering analysis; history of technology. *Mailing Add:* 2228 E Woodstock Pl Milwaukee WI 53223

BODKIN, NORLYN L, b Harrisonburg, Va, Jan 9, 37; m 58; c 2. PLANT BIOSYSTEMATICS, MID-APPALACHIAN FLORISTICS. *Educ:* WVa Univ, AB, 60, MS, 62; Univ Md, PhD(bot), 78. *Prof Exp:* Instr biol, Westminster Col, 62-64; from asst prof to assoc prof, 64-85, PROF BIOL, JAMES MADISON UNIV, 85- *Concurrent Pos:* Teaching asst bot, Univ Md, 70-71; curator, James Madison Univ Herbarium, 73-; dir, James Madison Univ Arboretum. *Mem:* Am Asn Plant Taxon; Int Asn Plant Taxon. *Res:* Shale barren endemics; monographic and revisionary research on the genus Melanthium; biosystematics of the varieties of Trillium pusillum; Mid-Appalachian floristic research. *Mailing Add:* Dept Biol James Madison Univ Harrisonburg VA 22807

BODLEY, HERBERT DANIEL, II, b Cleveland, Ohio, Nov 6, 39; m 61; c 2. ANATOMY. *Educ:* Hamline Univ, BS, 61; Univ Minn, MS, 68, PhD(anat), 70. *Prof Exp:* From instr to asst prof anat, Univ Ariz, 70-75; assoc prof, 75-80, actg chmn dept, 78, PROF ANAT, CHICAGO COL OSTEOP MED, 80-, CHMN DEPT, 79- *Mem:* Am Asn Anatomists; Electron Micros Soc Am; Am Soc Zoologists; Am Soc Cell Biol. *Res:* Ultrastructure. *Mailing Add:* Dept Anat Chicago Col Osteop Med 555 31st St Downers Grove IL 60515

BODLEY, JAMES WILLIAM, b Portland, Ore, Oct 5, 37; m 60; c 3. BIOCHEMISTRY. *Educ:* Walla Walla Col, BS, 60; Univ Hawaii, PhD(biochem), 64. *Prof Exp:* Prod control chemist, Dole Corp, 60; fel biochem, Sch Med, Univ, Wash, 64-66, actg asst prof, 66; asst prof, 67-71, assoc prof, 71-77, PROF BIOCHEM, MED SCH, UNIV MINN, MINNEAPOLIS, 77-, ASSOC HEAD, 84- *Concurrent Pos:* Am Cancer Soc fel, 65-66 & sr fel, 73-74. *Mem:* AAAS; Am Soc Biochem & Molecular Biol. *Res:* Mechanism and control of biological reactions involving nucleic acids. *Mailing Add:* Dept Biochem Univ Minn Med Sch Minneapolis MN 55455

BODMAN, GEOFFREY BALDWIN, b Barking, Essex, Eng, Oct 11, 94; US citizen; m 20; c 3. SOIL CLASSIFICATION, ENGINEERING PROPERTIES OF SOILS. *Educ:* Univ Sask, BSA, 19; Univ Minn, Minneapolis, MS, 24, PhD(soils), 27. *Prof Exp:* Instr physics, chem & soils, Schs Agr, Prov Alta, 19-20, prin, 20-22; asst soils, dept soils, Col Agr, Univ Minn, Minneapolis-St Paul, 22-27, asst prof soil chem, 27; from instr to assoc prof soils, Div Soil Technol, 27-39, prof soil physics, dept soils, 39-62, prof soils & chmn, 48-55, EMER PROF SOILS, UNIV CALIF, BERKELEY, 62- *Concurrent Pos:* Soil scientist, US Geol Surv consult to US Army Corps Engrs, Washington DC & SW Pac, 43-45, Int Coop Admin, Univ Taiwan, Taipei, 56; Ford Found prof soil physics, Univ Alexandria, Egypt, 62-63. *Mem:* AAAS; fel Am Soc Agron; Soil Sci Soc Am (pres, 58); fel Crop Sci Soc Am; Am Geophys Union. *Res:* Rate of water entry into, conductivity for water by, and retention of water in soils, with affecting factors; physical and chemical properties of soils. *Mailing Add:* 346 Santa Clara St Stockton CA 95207

BODMAN, SAMUEL WRIGHT, III, b Chicago, Ill, Nov 26, 38; m 61; c 3. CHEMICAL ENGINEERING. *Educ:* Cornell Univ, BChE, 61; Mass Inst Technol, ScD(chem eng), 64. *Prof Exp:* From asst prof to assoc prof, 64-76, VIS LECTR CHEM ENG, MASS INST TECHNOL, 76- *Concurrent Pos:* Tech adv, Am Res & Develop Corp, Mass, 64-70; vpres, FMR Develop Corp, Mass, 70- *Mem:* Am Inst Chem Engrs. *Res:* Heat and mass transfer; process design; applied mathematics. *Mailing Add:* 950 Winter St Waltham MA 02254

BODMER, ARNOLD R, b Frankfurt, Germany, May 23, 29; US citizen; m 56; c 4. PHYSICS, THEORETICAL NUCLEAR PHYSICS. *Educ:* Univ Manchester, BSc, 49, PhD(physics), 53. *Prof Exp:* Sci officer, Royal Armament Res & Design Estab, Eng, 53-56; res fel theoret physics, Univ Manchester, 56-57; res fel, Theoret Nuclear Physics, Europ Orgn Nuclear Res, Switz, 57-58; asst lectr theoret physics, Univ Manchester, 58-59, lectr, 60-65; PROF PHYSICS, UNIV ILL, CHICAGO, 65- *Concurrent Pos:* Resident res assoc, Argonne Nat Lab, 63-65, assoc physicist, 65-72, sr physicist, 72-85; vis prof, Nuclear Physics Lab, Oxford Univ, 70-71 & Justus-Liebig Univ, Giessen, 86. *Mem:* AAAS; Fedn Am Scientists; fel Am Phys Soc; Brit Inst Physics & Phys Soc. *Res:* Theoretical nuclear physics; nuclear structure and many-body problems; hypernuclei and baryon-baryon interactions; heavy-ion collisions. *Mailing Add:* Argonne Nat Lab B203 9700 S Cass Ave Argonne IL 60439

BODMER, WALTER FRED, b Frankfurt-Am-Main, Ger, Jan 10, 36; Brit citizen; m 56; c 3. CANCER. *Educ:* Cambridge Univ, BA, 56, PhD(pop genetics), 59. *Hon Degrees:* DSc, Univ Bath, 88, Univ Oxford, 88. *Prof Exp:* Res fel, Clare Col Cambridge Univ, 58-61; demonstr, Cambridge Univ, 60-61, fel, 61-62; from asst prof to prof, Stanford Univ, Calif, 62-70; prof, Univ Oxford, 70-79; dir res, 79-91, GEN DIR, IMP CANCER RES FUND, 91- *Concurrent Pos:* Coun mem, Royal Soc, 78-80 & Int Union Against Cancer, 82; hon fel, Keble Col, Oxford, 81; mem, Gen Adv Coun, Brit Broadcasting Corp, 81-; dir, Imp Cancer Res Technol Ltd, 82-; chmn, Exec Bd, Orgn Europ Cancer Insts, 90-; pres, Human Genome Orgn, 90- *Honors & Awards:* William Allan Mem Award, Am Soc Human Soc, 80; Conway Evans Prize, Royal Col Physicians, Royal Soc, 82; Rabbi Shai Shacknai Mem Prize Lectr in Immunol & Cancer Res, 83; Rose Payne Distinguished Scientist Lectr, 85. *Mem:* Nat Acad Sci; hon mem Am Acad Arts & Sci; fel Royal Soc; Royal Inst (vpres, 83); Int Union Against Cancer; Royal Statist Soc. *Res:* Genetics and biology of colorectal cancer human leukocyte- antigen system and immune response to cancer; human genome analysis. *Mailing Add:* Imp Cancer Res Fund PO Box 123 44 Lincoln's Inn Fields London WC2A 3PX England

BODNAR, DONALD GEORGE, b Ft William, Ont, Apr 30, 41; m 74; c 3. ANTENNAS, ELECTROMAGNETIC THEORY. *Educ:* Ga Inst Technol, BEE, 63, PhD(elec eng), 69; Mass Inst Technol, MSEE, 64. *Prof Exp:* Eng staff consult, Microwave Electronics Div, Sperry Rand, 69-70; res engr consult, US Army White Sands Missile Range, 71, Res Engr, 71-73; sr res engr, 73-81, chief, advan technol div, 86-89, PRIN RES ENGR, GA INST TECHNOL, 81-, DIR, ADVAN TECHNOL LAB, 89- *Concurrent Pos:* Lectr, Sch Electrical Eng, Ga Inst Technol, 71-78; paper reviewer, Inst Elec & Electronics Engrs Antenna & Propagation Soc, 75-; assoc ed, Inst Elec & Electronics Engrs Antennas & Propagation Soc Newsletter, 79-82; mem, Antenna Standards Comt, Inst Elec & Electronics Engrs, 79-, chmn, 83-, admin comt, Antennas & Propagation Soc, 84-87, 91- *Honors & Awards:* Certificate of Recognition for creative technical innov, NASA, 87. *Mem:* Inst Elec & Electronics Engrs Antennas & Propagation Soc (Admin Comt, 84-88); Int Test & Evaluation Asn; Antenna Measurement Techniques Asn; Asn Old Crows. *Res:* Design and evaluation of microwave antennas with primary emphasis on specially shaped reflector antennas; waveguide slotted arrays; electromechanical scanners; millimeter antennas; near-zone and far-zone radar cross section analyses and measurements. *Mailing Add:* 3568 W Hampton Dr NW Marietta GA 30064

BODNAR, LOUIS EUGENE, b Prince Albert, Sask, Jan 29, 29. CHEMICAL ENGINEERING, PHYSICAL CHEMISTRY. *Educ:* Univ Sask, BA, 50, MA, 51; McMaster Univ, PhD(chem), 57. *Prof Exp:* Lectr chem, Royal Mil Col, Can, 54-58; asst prof chem eng, 60-62, assoc prof & actg chmn dept, 62-64, assoc prof, 64-76, dir first year studies, 70-73, FAC ADMISSIONS OFFICER, UNIV WATERLOO, 75-, MEM FAC CHEM ENG, 76- *Res:* Mechanisms of chemical reactions; thermodynamics. *Mailing Add:* Dept Chem Eng Univ Waterloo Waterloo ON N2L 3G1 Can

BODNAR, ROBERT JOHN, b McKeesport, Pa, Aug 25, 49; m 79; c 2. FLUID INCLUSION ANALYSIS, GENESIS OF MINERAL DEPOSITS. *Educ:* Univ Pittsburgh, BS, 75; Univ Ariz, MS, 78; Pa State Univ, PhD(geochem), 85. *Prof Exp:* Geologist, US Geol Surv, 78-80; res geochemist, Chevron Oil Field Res Co, 84-85; asst prof, 85-87, ASSOC PROF GEOL, VA POLYTECH INST & STATE UNIV, 87- *Concurrent Pos:* NSF presidential young investr award, 86. *Honors & Awards:* Lindgren Award, Soc Econ Geologists, 86. *Mem:* Soc Econ Geologists; Am Mineral Soc; Am Geophys Union; Geochem Soc. *Res:* Experimental and theoretical studies of the volumetric and phase equilibrium properties of geologic fluids; fluid evolution in hydrothermal systems; magma-volatile interactions; vibrational spectroscopic analysis of materials. *Mailing Add:* Dept Geol Sci Va Polytech Inst & State Univ Blacksburg VA 24061

BODNAR, STEPHEN J, b Carteret, NJ, July 6, 25; m 47; c 3. CHEMICAL ENGINEERING, POLYMER CHEMISTRY. *Educ:* Lafayette Col, BA, 51; Univ Ill, Urbana-Champaign, PhD (chem), 54. *Prof Exp:* Chem engr, ESSO Standard Oil Co, 54-60; mgr process eng sect, Res & Develop Dept, Tex-US Chem Co, 60-83; Gmelin Inst, Max Planck Soc, 84-85; CONSULT, 86- *Mem:* Am Chem Soc; AAAS; Sigma Xi. *Res:* Process development of emulsion and solution polymerization; petro chemicals. *Mailing Add:* 408 W Nimitz St Fredericksburg TX 78624

BODNARYK, ROBERT PETER, b Kamsack, Can, Nov 4, 40; m 67. BIOCHEMISTRY, PHYSIOLOGY. *Educ:* McMaster Univ, BA, 63; Univ Waterloo, MSc, 65, PhD(physiol), 67. *Prof Exp:* RES SCIENTIST, RES BR, CAN AGR, 68- *Concurrent Pos:* NIH fel, 67-68; vis scientist, Oxford Univ, 72-73. *Honors & Awards:* C Gordon Hewitt Award, Entom Soc Can, 75. *Mem:* Entom Soc Can. *Res:* Biochemistry of insects; neurochemistry of insects; integrated pest management. *Mailing Add:* Can Agr Res Sta 195 Dafoe Rd Winnipeg MB R3T 2M9 Can

BODNER, GEORGE MICHAEL, b Rochester, NY, Mar 8, 46. CHEMICAL EDUCATION. *Educ:* State Univ NY Buffalo, BS, 69; Ind Univ, PhD(inorg chem), 72. *Prof Exp:* Vis asst prof gen chem, Univ Ill, 72-75; mem fac chem, Stephens Col, 75-77; from asst prof to assoc prof, 77-86, PROF CHEM EDUC, PURDUE UNIV, 87- *Mem:* Am Chem Soc; AAAS; Sigma Xi; The Chem Soc. *Res:* Application of educational technology in chemical education; how students learn chemistry. *Mailing Add:* Dept Chem Purdue Univ Lafayette IN 47907

BODNER, SOL R(UBIN), b New York, NY, Mar 7, 29; m 52; c 4. SOLID & STRUCTURAL MECHANICS. *Educ:* Polytech Inst Brooklyn, BCE, 50, PhD(appl mech), 55; NY Univ, MS, 53. *Prof Exp:* Res assoc appl mech, Polytech Inst Brooklyn, 51-55; sr mathematician, Repub Aviation Corp, 55-56; sr scientist, Avco Res & Advan Develop Div, 56; assoc prof eng, Brown Univ, 56-64; prof mech, 64-68; prof mat eng & head dept, 68-72, HEAD, MAT MECH LAB, MECH ENG, TECHNION-ISRAEL INST TECHNOL, 65- *Concurrent Pos:* Consult, Avco Res & Advan Develop Div, 56-64; Southwest Res Inst, 68, 73, 80-; vis prof, Cambridge Univ, 62, Weizmann Inst, 62-63, Univ Calif, 70, Univ Ill, 75, 79 & ETH Zurich, 83, 90. *Honors & Awards:* Guggenheim fel, 62; Rothschild Prize, 81; fel, Japan, SPS, 88. *Mem:* Soc Rheol; fel Am Soc Mech Engrs; Soc Eng Sci; fel AAAS; Sigma Xi. *Res:* Applied mechanics, dynamic plasticity, constitutive equations, penetration mechanics. *Mailing Add:* Dept Mech Eng Technion-Israel Inst Technol Haifa Israel

BODNER, STEPHEN E, b Rochester, NY, June 20, 39; m; c 2. THEORETICAL PHYSICS. *Educ:* Univ Rochester, BS, 61; Princeton Univ, MA, 62, PhD(physics), 65. *Prof Exp:* Physicist, Lawrence Radiation Lab, Univ Calif, 64-74; BR HEAD, NAVAL RES LAB, 74- *Concurrent Pos:* Instr appl sci, Univ Calif, Davis-Livermore, 66-68; OMB adv, 81-82. *Mem:* AAAS; Am Phys Soc; Optical Soc Am. *Res:* Plasma physics; turbulence; laser fusion controlled thermonuclear reaction research. *Mailing Add:* 8304 Fenway Rd Bethesda MD 20817

BODONYI, RICHARD JAMES, b Cleveland, Ohio, Nov 26, 43; m 66; c 3. AERONAUTICAL ENGINEERING, APPLIED MATHEMATICS. *Educ:* Ohio State Univ, BS, 66, PhD(aero eng), 73; Case Western Reserve Univ, MS, 70. *Prof Exp:* Aerospace engr fluid mech, Lewis Res Ctr, NASA, 67-68; asst prof aeronaut eng, Va Polytech Inst & State Univ, 74-76; from asst prof to prof appl math, Ind Univ-Purdue Univ, Indianapolis, 76-86; PROF AERO ENG, OHIO STATE UNIV, COLUMBUS, 86- *Mem:* Sigma Xi; Am Phys Soc; Am Inst Aeronaut & Astronaut; Am Acad Mech. *Res:* Theory of viscous flows; high Reynolds number flows; rotating flows; hydrodynamic stability. *Mailing Add:* Dept Aero/Astro Eng Ohio State Univ 2036 Neil Ave Mall Columbus OH 43210-1276

BODRE, ROBERT JOSEPH, b Philadelphia, Pa, Mar 6, 21; m 45. ANALYTICAL CHEMISTRY. *Educ:* St Josephs Col, Pa, BS, 43; Univ Louisville, PhD(chem), 61. *Prof Exp:* Chemist, Publicker Industs, Pa, 43-44; res chemist, Whitaker Wool Co, 44-47; res chemist, Girdler Div, Chemetron Corp, Ky, 47-55, group leader, 55-57; sr res chemist, Monsanto Co, 60-67, process specialist, 67-79, sr process specialist, 79-85, RETIRED. *Mem:* Am Chem Soc. *Res:* Development of analytical methods for industrial processes; monomer syntheses and polymerization reactions; catalyst development and applications; production of ethylbenzene and styrene. *Mailing Add:* 2501 Meadow Lane La Marque TX 77568

BODVARSSON, GUNNAR, applied mathematics, geophysics; deceased, see previous edition for last biography

BODWELL, CLARENCE EUGENE, b Woodward, Okla, July 30, 35. NUTRITION. *Educ:* Okla State Univ, BS, 57; King's Col, Cambridge Univ, MSc, 59; Mich State Univ, PhD(food sci), 64. *Prof Exp:* NIH staff fel, Lab Biophys Chem, Nat Inst Arthritis & Metab Dis, 64-65; USPHS res assoc, 65-67; res chemist, Human Nutrit Res Div, 67-72; res chemist nutrit inst, 72-80, chief, Energy & Protein Nutrit Lab, 80-86, RES CHEMISTS, HUMAN NUTRIT RES CTR, USDA, 86- *Mem:* AAAS; Am Chem Soc; Inst Food Technologists; Am Soc Clin Nutrit; Am Inst Nutrit; Am Asn Cereal Chem. *Res:* Evaluation of protein nutritive value in humans; human energy requirements, protein-energy interactions, chemical and biochemical methods for estimating protein nutritional value and amino acid bioavailability for humans. *Mailing Add:* Rm 213 Bldg 308 Human Nutrit USDA Barc-East Beltsville MD 20705

BOE, ARTHUR AMOS, b Westfield Twp, Minn, Apr 27, 33; m 65; c 4. HORTICULTURE. *Educ:* Utah State Univ, BS, 62, PhD(plant sci), 66. *Prof Exp:* Teaching asst hort, Utah State Univ, 62-65, agr specialist proj, Rural Indust Tech Assistance, Brazil, 66-67; from asst prof to prof plant sci & plant physiologist, Univ Idaho, 67-83; CHMN DEPT HORT, NDAK STATE UNIV, 83- *Concurrent Pos:* Assoc plant physiologist, Univ Idaho, 72-77.

Mem: Am Soc Hort Sci; Am Inst Biol Sci. *Res:* Fruit development and maturation as affected by plant hormones, light, temperature and magnetic fields; dormancy of woody plants and seeds; plant tissue culture; biotechnology. *Mailing Add:* Dept Hort NDak State Univ State Univ Sta Fargo ND 58105

BOECK, WILLIAM LOUIS, b Buffalo, NY, June 23, 39; m 62; c 4. ATMOSPHERIC PHYSICS, SOFTWARE SYSTEMS. *Educ:* Canisius Col, BS, 61; Univ Notre Dame, PhD(physics), 65. *Prof Exp:* From asst prof to assoc prof physics, Niagara Univ, 65-73; sr res assoc, State Univ NY, Albany, 73-74; prof, Niagara Univ, 74-77; vis scientist, Nat Ctr Atmospheric Res, 77-78; prof physics, 78-85, PROF COMPUT & INFO SCI & PHYSICS, NIAGARA UNIV, 85- *Concurrent Pos:* Chmn, Krypton 85 Working Group, Int Comn Atmospheric Elec, 74-, mem, Applications Atmospheric Elec, 75-; sci adv, Ecumenical Task Force, 80-88, Indust Waste Siting Bd, 88-90; vis scientist, Marshall Space Flight Ctr, NASA, 90-91. *Mem:* Am Geophys Union; Asn Comput Mach; Am Asn Physics Teachers. *Res:* Relationships between atmospheric radioactivity and cosmic rays and atmospheric electrical fields and currents; environmental radon measurements; stratospheric lightning. *Mailing Add:* Dept Comput & Info Sci Niagara Univ Niagara University NY 14109

BOECKER, BRUCE BERNARD, b Aurora, Ill, July 9, 32; m 60; c 2. RADIOBIOLOGY, HEALTH PHYSICS. *Educ:* Grinnell Col, BA, 54; Univ Rochester, MS, 60, PhD(radiobiol), 62; Am Bd Health Physics, dipl, 84. *Prof Exp:* Radiobiologist, 61-64, asst head dept radiobiol, 64-75, ASST DIR INHALATION TOXICOL RES INST, LOVELACE BIOMED & ENVIRON RES INST, 75- *Concurrent Pos:* Radiation biologist div biol & med, USAEC, 70-71; assoc ed, Radiation Res, 84-88; mem, bd dirs, Health Physics Soc, 85-88; chmn, ORAU/NCI Working Group Bioassy, Detecting Previous Radiation Exposures, 86; mem, NAS/NRC Comt, Biol Effects Ionizing Radiation (BEIR V), 86-89; mem, Nat Coun Radiation Protection & Measurements, 87-, Environ & Health Standards Steering Group, US Dept Energy, 90- *Mem:* Health Physics Soc; Radiation Res Soc; AAAS. *Res:* Deposition, retention and dosimetry of inhaled materials; inhalation toxicology; biological effects of inhaled materials on the respiratory tract and bioassay for internally deposited radionuclides. *Mailing Add:* Inhalation Toxicol Res Inst PO Box 5890 Albuquerque NM 87185-5890

BOECKMAN, ROBERT K, JR, b Pasadena, Calif, Aug 26, 44; m 76. SYNTHETIC METHODOLOGY, TOTAL SYNTHESIS. *Educ:* Carnegie Inst Technol, BS, 66; Brandeis Univ, PhD(org chem), 71. *Prof Exp:* Asst prof chem, Wayne State Univ, 72-75, assoc prof, 76-78, prof, 79-80; PROF CHEM, UNIV ROCHESTER, 80- *Concurrent Pos:* Fel, NIH, 70-72, Career Develop Award, 75-81; fel, A P Sloan Found, 76-80 & Japanese Soc Prom Sci, 90. *Mem:* Am Chem Soc; Royal Soc Chem. *Res:* Methodology for applications of organic synthesis to the construction of stereochemically complex molecules, primarily natural products possessing biological activity of some type. *Mailing Add:* Dept Chem Hutchinson Hall River Sta Univ Rochester Rochester NY 14627

BOEDEKER, RICHARD ROY, b St Louis, Mo, Apr 8, 33; m 61; c 3. PHYSICS. *Educ:* St Louis Univ, BS, 55, MS, 57, PhD(physics), 59. *Prof Exp:* Res assoc physics, St Louis Univ, 59-61; res scientist, McDonnell Aircraft Corp, 61-62; from asst prof to assoc prof, 62-74, PROF PHYSICS, SOUTHERN ILL UNIV, EDWARDSVILLE, 74- *Mem:* Am Phys Soc; Sigma Xi. *Res:* Neutron scattering in solids and liquids; temperature determination in gases by neutron scattering; lasers; nonlinear optics; scattering theory. *Mailing Add:* 639 Harvard Dr Edwardsville IL 62025

BOEDTKER, OLAF A, b Colombo, Ceylon, Feb 10, 24; US citizen; m 52; c 4. SOLID STATE PHYSICS. *Educ:* Swiss Fed Inst Technol, BS, 49; Calif Inst Technol, MS, 58, PhD(eng physics), 61. *Prof Exp:* Res fel phys sci, Calif Inst Technol, 60-62; res scientist metals physics, Lawrence Radiation Lab, Univ Calif, 62-63; assoc prof physics, Ore State Univ, 63-90, dir eng physics, 66-90; RETIRED. *Mem:* Am Phys Soc; Am Soc Metals. *Mailing Add:* Col Sci Ore State Univ Corvallis OR 97331

BOEDTKER DOTY, HELGA, b July 14, 24; m 54; c 3. REGULATION OF COLLAGEN GENE EXPRESSION. *Educ:* Radcliffe Col, PhD(phys chem), 54. *Prof Exp:* SR RES ASSOC BIOCHEM, HARVARD UNIV, 76- *Mem:* Sigma Xi. *Res:* Chondrocyte cell culture; structure and expression of collagen genes. *Mailing Add:* Fairchild Biochem Bldg Harvard Univ Seven Divinity Ave Cambridge MA 02138

BOEHLE, JOHN, JR, b New York, NY, July 5, 32; m 55; c 2. AGRONOMY, WEED SCIENCE. *Educ:* Univ NH, BS, 54; Pa State Univ, MS, 61, PhD(agron), 64. *Prof Exp:* Agronomist, Chevron Chem Co, 62-64; mgr agron serv, Kerr McGee Chem Co, 64-68, dir field res & res farms, Agr Div, Ciba-Giegy, 68-75, dir biol res, 75-81, vpres planning & admin, Plastics & Additives Div, 81-83, vpres pigments, 83-86, vpres additives, 86-88, PRES PIGMENTS DIV, CIBA-GEIGY, 88- *Mem:* Am Soc Agron; Weed Sci Soc Am; Soil Sci Soc Am. *Res:* Weed control; plant protectants; micronutrient nutrition of plants; growth regulators for plants. *Mailing Add:* Ciba-Geigy Pigments Seven Skyline Dr Hawthorne NY 10532

BOEHLER, GABRIEL D(OMINIQUE), b Paris, France, Aug 6, 26; nat US; m 57; c 5. MECHANICAL & AEROSPACE ENGINEERING. *Educ:* Univ Paris, BS, 48; Cornell Univ, MS, 51; Cath Univ Am, PhD(aeronaut eng), 53. *Prof Exp:* Res asst, Cornell Univ, 50-51; chief aerodyn & asst to pres, Thieblot Aircraft Co, 51-56; from asst prof to assoc prof aeronaut eng, 55-62, PROF MECH ENG, CATH UNIV AM, 62- *Concurrent Pos:* Consult, Aerophysics Co, 57-80, chmn bd, 80-; mem adv panel aeronaut, Off Secy Defense, 59-61. *Mem:* Am Inst Aeronaut & Astronaut; Am Helicopter Soc; Am Soc Heating, Refrig & Air-Conditioning Engrs; Soc Automotive Engrs. *Res:* Aerospace technology; aerodynamics; low speed flight; systems engineering; heating, ventilation and air conditioning; energy conservation. *Mailing Add:* 3010 Ordway St NW Washington DC 20008

BOEHLER, ROBERT A, b Chicago, Ill, Nov 18, 24; m 51; c 5. CHEMICAL ENGINEERING, CHEMISTRY. *Educ:* Loyola Univ, Los Angeles, BS, 50; Ill Inst Technol, MS, 66. *Prof Exp:* Area supt, Pharmaceut Div, Armour & Co, 50-53, supvr & process engr, Chem Div, 53-57; process eng, Nalco Chem Co, 57-67, group leader, microbiol dept, 67-69, sr prod engr, 69-81, mgr int opers, 81-87; RETIRED. *Mem:* Am Chem Soc; Am Inst Chem Engrs; Sigma Xi. *Res:* Pharmaceutics; fatty acid derivatives; process engineering and design. *Mailing Add:* Rte 4 Box 159 Jerseyville IL 62052-9434

BOEHLERT, GEORGE WALTER, b Staten Island, NY, May 10, 50; m 74; c 2. ECOLOGY, ZOOLOGY. *Educ:* Univ Calif, Santa Barbara, BA, 72, San Diego, PhD(marine biol), 77. *Prof Exp:* Nat Res Coun-Nat Oceanic & Atmospheric Admin res assoc, Northwest & Alaska Fisheries Ctr, Nat Marine Fisheries Serv, Seattle, 77-78; assoc marine scientist & asst prof, Va Inst Marine Sci, Col William & Mary, 78-79; from asst prof to assoc prof, Dept Fisheries & Wildlife, Sch Oceanog, Ore State Univ, 79-83; chief insular resources, 83-88, DIR, HONOLULU LAB, NAT MARINE FISHERIES SERV, 88- *Concurrent Pos:* Affil grad fac, Dept Oceanog, Univ Hawaii, 83- *Mem:* Am Fisheries Soc; Am Geophys Union; Am Soc Limnol & Oceanog; Oceanog Soc; Ichthyol Soc Japan. *Res:* Ecology and physiology of the early life stages of fishes, age, growth and reproduction in fish; recruitment dynamics; biological oceanography around islands and seamounts. *Mailing Add:* NOAA Nat Marine Fisheries Serv Honolulu Lab 2570 Dole St Honolulu HI 96822-2396

BOEHLJE, MICHAEL DEAN, b Sheffield, Iowa, Aug 6, 43. FARM & AGRICULTURAL BUSINESS FINANCE, AGRICULTURAL POLICY. *Educ:* Iowa State Univ, BS, 65; Purdue Univ, MS, 68, PhD(agr econ), 71. *Prof Exp:* Asst prof agr econ, Okla State Univ, 70-73; from asst prof to prof econ, Iowa State Univ, 73-80, asst dean, Col Agr, 81-85; dept head, 85-90, PROF AGR ECON, UNIV MINN, 91- *Mem:* Am Agr Econ Asn; Southern Agr Econ Asn. *Res:* Strategic planning for firm in the banking, agribusiness and farm sectors to assist those firms in adapting to and thriving in a rapidly changing economic and political climate. *Mailing Add:* 130 COB Univ Minn St Paul MN 55372

BOEHM, BARRY WILLIAM, b Santa Monica, Calif, May 16, 35; m 61. COMPUTER SCIENCE. *Educ:* Harvard Univ, BA, 57; Univ Calif, Los Angeles, MA, 61, PhD(math), 64. *Prof Exp:* Res engr, Gen Dynamics, Astronaut, 57-59; head comput systs anal group, Rand Corp, 59-71, head, Info Sci Dept, 71-73; dir software res & technol, 73-86, CHIEF SCIENTIST, TRW DEFENSE SYSTS GROUP, TRW SYSTS & ENERGY, INC, 86- *Concurrent Pos:* Lectr, Exten, Univ Calif, Los Angeles, 65- & Grad Bus Sch, 69-; mem, NASA adv comt instrumentation & data processing, 68-70; mem, Southern Calif Sci Adv Comt to Selective Serv Syst, 69-, chmn, 70-; chmn conf comt, Am Fedn Info Processing Socs, 69-; guest lectr, USSR Acad Sci, 70; mem Nat Acad Sci panel comput sci develop in Brazil, 70-; sector ed, Comput Rev, 70-; chmn, NASA adv comt guid, control & info syst sci dept, 73-76; vis prof comput sci, Univ Southern Calif, 78-79 & Univ Calif, Los Angeles, 81- *Honors & Awards:* Cert of Merit, Air Force Systs Command, 72; Info Systs Award, Am Inst Aeronaut & Astronaut, 79. *Mem:* Assoc fel Am Inst Aeronaut & Astronaut; Asn Comput Mach; Int Acad Astronaut; Inst Elec & Electronics Engrs. *Res:* Man-computer interaction; information systems; computer applications in education; aerospace; urban services; developing nations; software engineering; information systems; software economics; software reliability. *Mailing Add:* 1400 Wilson Blvd Arlington VA 22209

BOEHM, FELIX H, b Basel, Switz, June 9, 24; nat US; m 56; c 2. NUCLEAR PHYSICS, ATOMIC PHYSICS. *Educ:* Swiss Fed Inst Technol, Dipl physics, 48, Dr nat sci(physics), 51. *Prof Exp:* Res asst, Swiss Fed Inst Technol, 48-51; Boese fel, Columbia Univ, 52-53; res fel physics, Calif Inst Technol, 53-57, asst prof, 58; vis prof, Univ Heidelberg, 58-59; assoc prof, 59-61, PROF PHYSICS, CALIF INST TECHNOL, 61- *Concurrent Pos:* Sloan fel, 61-63; NSF sr fel, Univ Copenhagen, 65-66 & Europ Orgn Nuclear Res, Geneva, Switz, 71-72; mem, comt, Rev of US Medium Energy Sci, US AEC & NSF, 74, nuclear sci adv comt, US Dept Energy/NSF, 81-83; mem, prog adv comt, Los Alamos Meson Facil, 76-79, physics div adv comt, Los Alamos Nat Lab, 82-; trustee, Aspen Ctr for Physics, 76-79; fel, Laue Langevin Inst, Grenoble, France, 79-80, Univ Munich, 80 & Swiss Int Nuclear Res, 81; mem, bd on physics & astron, Nat Res Coun. *Honors & Awards:* Alexander von Humboldt Award, 80. *Mem:* Nat Acad Sci; fel Am Phys Soc; AAAS. *Res:* Nuclear structure; weak interaction; x-rays; neutrino physics; meson physics. *Mailing Add:* 161-33 Calif Inst Technol Pasadena CA 91125

BOEHM, JOHN JOSEPH, b Libertyville, Ill, May 27, 29; m 59; c 4. PEDIATRICS. *Educ:* Univ Notre Dame, BS, 51; Northwestern Univ, MD, 55. *Prof Exp:* Intern, St Luke's Hosp, Chicago, 55-56; resident pediat, Children's Mem Hosp, Chicago, 56-58; res fel newborn & premature pediat, Med Ctr, Univ Colo, 60-62; from instr to asst prof pediat, Med Sch, Univ Ky, 62-65; asst prof & vchmn dept, Univ Calif, Irvine, 65-66; asst prof pediat, Med Sch, Northwestern Univ, Chicago, 66-69, asst dean med sch, 70-72, chief pediat, Northwestern Mem Hosp, 72-90, ASSOC PROF PEDIAT, MED SCH, NORTHWESTERN UNIV, CHICAGO, 69-, ASSOC CHMN, NORTHWESTERN MEM HOSP, 90- *Mem:* Am Acad Pediat; Am Soc Clin Invest. *Res:* Neonatology. *Mailing Add:* Northwestern Mem Hosp Superior St & Fairbanks Ct Chicago IL 60611

BOEHM, PAUL DAVID, b New York, NY, Dec 26, 48; m 71; c 3. ORGANIC CHEMISTRY. *Educ:* Univ Rochester, BS, 70; Univ RI, MS, 73, PhD(chem oceanog), 77. *Prof Exp:* Engr, Western Elec Co, 70-71; prin scientist, Energy Resources Co, Inc, 76-80, chief scientist, 81-83; res leader, Battelle Ocean Sci, 83-89; VPRES, ARTHUR D LITTLE INC, 89- *Concurrent Pos:* Vis prof, Coastal Resources Ctr, Woods Hole Oceanog Inst, 81- *Mem:* AAAS; Am Chem Soc. *Res:* Investigations of organic chemical distributions in the ocean with emphasis on the chemical oceanography and biogeochemistry of marine organic pollutants. *Mailing Add:* 11 Ridgewood Rd Concord MA 01742

BOEHM, ROBERT FOTY, b Portland, Ore, Jan 16, 40; m 61; c 2. HEAT TRANSFER, THERMODYNAMICS. *Educ:* Wash State Univ, BS, 62, MS, 64; Univ Calif, Berkeley, PhD (mech eng), 68. *Prof Exp:* Heat transfer engr, aerospace elec dept, Westinghouse Elec Co, Ohio, 61; design engr, Lawrence Radiation Lab, 62; weights anal engr, Boeing Co, Wash, 63; engr, atomic power equip dept, Gen Elec Co, Calif, 64-66; res engr, Jet Propulsion Lab, Calif, 67; from asst prof to prof mech eng, Univ Utah, 68-90, chmn, Mech & Indust Eng Dept, 81-84; PROF MECH ENG & CHMN, UNIV LAS VEGAS, 90- *Concurrent Pos:* Mem tech staff Saudia Labs, 84-85. *Honors & Awards:* Ralph R Teetor Award, Soc Automotive Engrs, 71. *Mem:* Fel Am Soc Mech Engrs; Am Soc Eng Educ; Sigma Xi; Int Solar Energy Soc. *Res:* Applied thermodynamics, including design of thermal system; biotechnology studies; heat transfer; solar energy; geothermal energy; bioengineering heat transfer. *Mailing Add:* Dept Mech Eng Univ Las Vegas 4505 Maryland Pkwy Las Vegas NV 89154-4027

BOEHM, ROBERT LOUIS, b Buffalo, NY, Nov 17, 25; m 51; c 6. PAPER CHEMISTRY, PULP CHEMISTRY. *Educ:* Colgate Univ, AB, 48; Lawrence Col, MS, 50, PhD(chem), 53. *Prof Exp:* Res chemist, Mead Corp, 53-57, tech asst to mgr, Mass, 57, asst tech dir, Chillicothe Div, 57-60, mgr, Leominster Div, 60-67, bus planner, Gilbert Paper Div, 67-69; staff mgr paper & paperbd opers, Am Can Co, 69-76; SR CONSULT ENGR, H A SIMONS (INT) LTD, VANCOUVER, 76- *Mem:* Am Chem Soc; Tech Asn Pulp & Paper Indust. *Res:* Introduction of chlorine atoms into cellulose; research in pulping and bleaching of wood pulps; paper machines. *Mailing Add:* 1221 Nanton Ave Vancouver BC V6B 2J6 Can

BOEHME, DIETHELM HARTMUT, pathology, neuropathology, for more information see previous edition

BOEHME, HOLLIS, b Liberty, Okla, Nov 28, 33; m 54; c 3. ACOUSTICS. *Educ:* Tex A&M Univ, BA, 60, MS, 61, PhD(physics), 67. *Prof Exp:* Engr, Boeing Airplane Co, 60; res asst, Tex A&M Univ, 60-61; engr, Tex Instruments Inc, 61-62; instr physics, 62-67; engr scientist, Tracor, Inc, 67-68, sr scientist, 68-70; pres, Synergy Inc, 70-73; RES SCIENTIST, APPLIED RES LAB, UNIV TEX, 73- *Concurrent Pos:* Consult, Tex Instruments Inc, 62-63. *Mem:* Am Phys Soc. *Res:* Nuclear magnetic resonance; acoustics. *Mailing Add:* 320 King Arthur Ct Austin TX 78746

BOEHME, WERNER RICHARD, b Englewood, NJ, Jan 15, 20; m 46; c 2. ORGANIC CHEMISTRY. *Educ:* Polytech Inst Brooklyn, BS, 42; Univ Md, PhD(chem), 48; Univ Chicago, MBA, 71. *Prof Exp:* Asst chemist, Gen Dyestuff Corp, 37-41; res chemist, Winthrop Chem Co, 42-44; res chemist, Naval Med Res Inst, 45-46; asst, Univ Md, 46; res chemist, Nat Drug Co, 48-52; sr chemist, Ethicon, Inc, 52-55, head med chem sect, 56-57, mgr dept org chem, 58-61; dir cent res, Shulton, Inc, 61-62; gen mgr & pres, A M Arnold Labs, Inc, 62-67; mgr chem opers & dir res develop, Dawe's Labs, Inc, 67-73; tech dir, Fats & Proteins Res Found, Inc, 73-81; CONSULT, 82- *Mem:* Am Chem Soc; NY Acad Sci; fel Am Inst Chem; Inst Food Technologists; Am Oil Chem Soc. *Res:* Pharmaceuticals and animal nutrition; fatty acid and protein chemistry; animal by-products utilization; pollution control; triglycerides; chemistry in agriculture. *Mailing Add:* 1856 River Ct Jensen Beach FL 34957

BOEHMS, CHARLES NELSON, b Nashville, Tenn, Feb 1, 31; m 52; c 2. PHYSIOLOGICAL ECOLOGY, FRESHWATER BIOLOGY. *Educ:* George Peabody Col, BS, 56, MA, 57; Univ NC, Chapel Hill, PhD(zool), 71. *Prof Exp:* From instr to asst prof biol, Austin Peay State Univ, 57-61; instr zool, Univ NC, Chapel Hill, 62-64; from asst prof to prof biol, 65-87 Austin Peay State Univ, 65-71, dean students, 68-71, vpres student affairs, 71-82; PROF & CHMN BIOL DEPT, GEORGETOWN COL, 87- *Concurrent Pos:* NSF fac fel, 64. *Mem:* Am Zool Soc; Am Soc Biol; Am Inst Biol Sci; NAm Benthological Soc. *Res:* Influence of temperature and photoperiod on seasonal regulation in fresh water invertebrates. *Mailing Add:* Dept Biol Georgetown Col Georgetown KY 40324

BOEHNE, JOHN WILLIAM, b Evansville, Ind, Feb 6, 21; m 46; c 2. NUTRITION. *Educ:* Ind Univ, BS, 42. *Prof Exp:* Chemist, Lederle Labs Div, Am Cyanamid Co, NY, 46-48; asst, Western Reserve Univ, 48-57; biochemist div nutrit, Bur Foods, Food & Drug Admin, 57-59, asst to dir, 59-65, chief spec dietary foods br, 66-70, asst dir, 71-74, asst assoc dir nutrit & consumer sci, 74-78; RETIRED. *Mem:* Am Inst Nutrit. *Res:* Bioanalytical nutrition; mineral metabolism; nutrient value of foods; food additives. *Mailing Add:* 10763 Kinloch Rd Silver Spring MD 20903

BOEHNKE, DAVID NEAL, b Dayton, Ky, July 25, 39. PHYSICAL CHEMISTRY, THEORETICAL CHEMISTRY. *Educ:* Georgetown Col, BS, 61; Univ Ky, MS, 64; Univ Cincinnati, PhD(theoret chem), 68. *Prof Exp:* Instr chem & physics, Ky State Col, 64-65; res assoc, Univ NC, Chapel Hill, 67-69; asst prof, 69-74, ASSOC PROF CHEM, JACKSONVILLE UNIV, 74- *Mem:* Am Chem Soc. *Res:* Theory of critical viscosity; properties of critical fluids; kinetics of bimolecular reactions; synthesis and properties of detergents. *Mailing Add:* 3152 Cesery Blvd Jacksonville FL 32211

BOEKELHEIDE, VIRGIL CARL, b Chelsea, SDak, July 28, 19; m 45; c 3. ORGANIC CHEMISTRY. *Educ:* Univ Minn, AB, 39, PhD(org chem), 43. *Prof Exp:* Instr chem, Univ Ill, 43-46; asst prof org chem, Univ Rochester, 46-51, from assoc prof to prof, 51-60; PROF ORG CHEM, UNIV ORE, 60- *Concurrent Pos:* Guggenheim fel, Swiss Fed Inst Technol, Zurich, 53-54; Swiss-Am Found Lectr, 60; Frontiers of Sci & Welch Symposium lectr, 68; Karl Pfister lectr, Mass Inst Technol, 69; mem adv coun, Nat Inst Environ Health Sci; chmn div chem & chem technol, Nat Res Coun-Nat Acad Sci; mem, Coun Int Exchange Scholars, 75-78. *Honors & Awards:* Fulbright-Hays Distinguished Prof Award, Yugoslavia, 72; Alexander von Humboldt Award, WGer Govt; Coover Award, 81. *Mem:* Nat Acad Sci; Am Chem Soc; Swiss Chem Soc; Ger Chem Soc; The Chem Soc. *Res:* Natural products, heterocycles and aromatic character in polycyclic molecules. *Mailing Add:* Dept Chem Univ Ore Eugene OR 97403

BOEKER, ELIZABETH ANNE, biochemistry; deceased, see previous edition for last biography

BOELL, EDGAR JOHN, b Rudd, Iowa, Oct 30, 06; m 32; c 2. BIOLOGY. *Educ:* Univ Dubuque, AB, 29; Univ Iowa, PhD(zool), 35. *Hon Degrees:* MA, Yale Univ, 46; DSc, Univ Dubuque, 52. *Prof Exp:* Res assoc zool, Univ Iowa, 34-37; from instr to prof biol, 38-46, Ross G Harrison prof exp zool, 47-75, chmn dept, 55-62, actg dean, Yale Col, 2nd term, 68-69, ROSS G HARRISON EMER PROF EXP ZOOL & SR BIOLOGIST, YALE UNIV, 75- *Concurrent Pos:* Rockefeller Found fel, Cambridge Univ, 37-38; Fulbright award, Carlsberg Lab, Univ Copenhagen, 53-54; Guggenheim Found fel, Univ Rome, 63-64; ed, J Exp Zool, 65-85; master, Jonathan Edwards col, 75-77. *Honors & Awards:* Josiah Willard Gibbs Medal, 75. *Mem:* AAAS; Am Physiol Soc; Am Soc Zool; Soc Develop Biol; Am Soc Cell Biol; Am Acad Arts & Sci. *Res:* Chemical embryology; developmental physiology; reproductive biology; physiology of spermatozoa. *Mailing Add:* Dept Biol Yale Univ Osborn Mem Lab PO Box 6666 New Haven CT 06511

BOELLSTORFF, JOHN DAVID, b Johnson, Nebr, May 14, 40; m 64; c 2. STRATIGRAPHY, GEOCHRONOLOGY. *Educ:* Univ Nebr, BS, 63, MS, 68; La State Univ, PhD(geol), 73. *Prof Exp:* Geologist, Western Labs, Inc, 64-65; res geologist stratig, Univ Nebr, 65-68, res geologist & prof, 71-80; SR RES SCIENTIST, AMOCO PROD CO, TULSA, OKLA, 80- *Mem:* Geol Soc Am; Am Quaternary Asn; Am Asn Petrol Geologists. *Res:* Chronology and stratigraphy of Cenozoic deposits of the Great Plains; petrography of glacial deposits; fission track dating; seismic stratigraphy. *Mailing Add:* Amoco Prod Co PO Box 591 Tulsa OK 74102

BOELTER, DON HOWARD, b Sanborn, Minn, Sept 1, 33; m 56; c 4. FORESTRY RESEARCH, SOIL PHYSICS. *Educ:* Iowa State Univ, BS, 55; Univ Minn, MS, 58, PhD(soil physics), 62. *Prof Exp:* Researcher soil physics, NCent Forest Exp Sta, Minn, 59-77, watershed mgt researcher, Forest Serv, 77-80, ASST STA DIR, PAC NORTHWEST FOREST EXP STA, USDA, ORE, 80- *Mem:* Am Soc Agron; Soil Sci Soc Am; Int Soc Soil Sci; Soil Conserv Soc Am; Soc Am Foresters. *Res:* Physical and hydrologic properties of peats and organic soil and relationship to the watershed management of bog and swamp areas in north central United States. *Mailing Add:* 4724 Laura Lane Shoreview MN 55126

BOELTER, EDWIN D, JR, b Greeley, Colo, Mar 3, 28; m 54; c 2. CHEMICAL ENGINEERING. *Educ:* Univ Wash, BS, 49, PhD(chem eng), 52. *Prof Exp:* Tech investr, Electrochem Dept, E I du Pont de Nemours & Co Inc, 52-56, group leader, 57, sr supvr, 57-62, mgr pilot plants, 62-63, lab dir, 63-69, mgr works technol, 69-70, dir, Electronic Prod Div, 70-74, asst dir Cent Res Dept, 74-77, gen mgr, Elastomers Dept, 77-80, vpres, Polymer Prods Dept, 81-85; RETIRED. *Mem:* Am Chem Soc. *Res:* Electrokinetics; electroplating; molten salts; sodium; polymers. *Mailing Add:* Nine Point North Dr Salem SC 29676

BOEN, JAMES ROBERT, b Fergus Falls, Minn, July 21, 32; m 57; c 1. BIOSTATISTICS. *Educ:* Dartmouth Col, AB, 56; Univ Ill, MS, 57, PhD(math), 59. *Prof Exp:* Asst prof math, Southern Ill Univ, 59-60; res assoc, Univ Chicago, 60-61; lectr, Univ Mich, 61-62, fel biostatist, Stanford Univ, 62-64; from asst prof to assoc prof, 64-75, PROF BIOSTATIST, SCH PUB HEALTH, UNIV MINN, MINNEAPOLIS, 75- *Mem:* Am Statist Asn; Biomet Soc; Am Pub Health Asn. *Mailing Add:* Pub Health Box 197 Mayo Univ Minn Med Sch 420 Delaware St SE Minneapolis MN 55455

BOENIGK, JOHN WILLIAM, b Braddock, Pa, July 1, 18; m 47; c 4. PHARMACEUTICAL CHEMISTRY. *Educ:* Duquesne Univ, BS, 40; Western Reserve Univ, MS, 42; Purdue Univ, PhD(pharmaceut chem), 48. *Prof Exp:* Asst prof pharm, Univ Tex, 48-50; assoc prof, Med Col Va, 50-56; assoc prof, Univ Pittsburgh, 56-61; group leader pharm prod develop, Mead Johnson & Co, 61-67, sect leader tablet develop, 67-69, sr res assoc, 69-70, res assoc, 70-78, prin res scientist, 79-83; RETIRED. *Mem:* AAAS; Am Pharmaceut Asn. *Res:* Sustained release; stability; vitamins; antacids. *Mailing Add:* 2511 Bayard Park Dr Evansville IN 47714

BOENING, PAUL HENRIK, b 1953. ANAEROBIC DIGESTION, INDUSTRIAL WASTE TREATMENT. *Educ:* Univ Minn, BS, 75; Univ Ill, MS, 76, PhD(civil eng), 80. *Prof Exp:* Fel, biotechnol dept, Massey Univ, 80-81; VIS ASST PROF, DEPT CIVIL ENG, STATE UNIV NY BUFFALO, 8- *Mem:* Am Water Works Asn; Am Soc Civil Engrs; Water Pollution Control Fedn. *Res:* Anaerobic digestion, especially applied to industrial and municipal waste water treatment; anaerobic fixed film processes and mixing in digesters. *Mailing Add:* 5962 Hobe Lane St Paul MN 55110

BÖER, KARL WOLFGANG, b Berlin, Germany, Mar 23, 26; m; c 2. SOLID STATE PHYSICS, CARRIER TRANSPORT IN SOLIDS. *Educ:* Humboldt Univ, Berlin, dipl physics, 49, Dr rer nat(electronic noise), 52, Dr rer nat habil(photoconductivity), 54. *Prof Exp:* Asst physics, Humboldt Univ, Berlin, 49-56, docent, 56-58, prof & head dept, 58-61; res prof, NY Univ, 61-62; from assoc prof to prof physics, 62-72, dir, Inst Energy Conversion, 72-75, PROF PHYSICS & ENG, UNIV DEL, 72-, CHIEF SCIENTIST, INST ENERGY CONVERSION, 73-, ADV TO PRESIDENT, 77- *Concurrent Pos:* Ed, Fortschritte der Physik, 53-54; consult, Askania Corp Berlin Teltow, 53-55 & US Govt & industs, 62-; dir, lab elec breakdown, German Acad Sci, Berlin, 56-61; ed-in-chief, Physica Status Solidi, 60-61, ed, 72-, ed, Solar Energy Mat, ed-in-chief, Advan Solar Energy; chmn bd, SES, Inc, 73-79, chief scientist, 78-83. *Honors & Awards:* Charles G Abbot Award. *Mem:* Fel Am Phys Soc; Am Solar Energy Soc (secy, 78); sr mem Inst Elec & Electronics Engrs; Int Solar Energy Soc; Elec Chem Soc. *Res:* Solid state electronics; noise; radiation damage; semiconductivity; photoconductivity; photo-electromotive force; dielectric breakdown; field and current instabilities; surface and electrodes; reaction kinetics; electro-optics; semiconducting glasses; photovoltaic; CdS solar cell; solar energy conversion; Solar One house; electron transport in solids; defect chemistry; high Tc superconduct. *Mailing Add:* 239 Buck Toe Hills Kennett Square PA 19348

BOERBOOM, LAWRENCE E, CARDIOVASCULAR DISEASES, PHYSIOLOGY. *Educ:* Univ NDak, Grand Forks, PhD(physiol), 75. *Prof Exp:* ASST PROF PHYSIOL, DEPT CARDIOVASC DIS, MED COL WIS, 77- *Mailing Add:* Dept Cardiothoracic Surg Med Col Wis 8700 W Wisconsin Ave Milwaukee WI 53226

BOERCKER, DAVID BRYAN, b St Louis, Mo, Nov 30, 49; m 71; c 2. KINETIC THEORY, STATISTICAL PHYSICS. *Educ:* Austin Peay State Univ, BA, 70; Univ Fla, PhD(physics), 78. *Prof Exp:* Res assoc, Air Force Weapons Lab, 78-80; PHYSICIST, LAWRENCE LIVERMORE NAT LAB, 80- *Mem:* Am Phys Soc. *Res:* Evaluation of equilibrium time correlation functions and their associated transport coefficients for dense plasmas; broadening of spectral lines in plasmas. *Mailing Add:* Lawrence Livermore Lab-L296 Univ Calif PO Box 808 Livermore CA 94550

BOERÉ, RENÉ THEODOOR, b Wageningen, Neth, May 31, 57; Can citizen; m 85; c 2. MAIN-GROUP SULFER NITRIDES, TRANSITION-METAL SULFUR NITRIDES. *Educ:* Dalhousie Univ, BSc, 79; Univ Western Ont, PhD(chem), 84. *Prof Exp:* Res fel, Guelph-Waterloo Centre Grad Work Chem, 84-86; vis scientist & Inst für Anorganic Anal Chemie, Free Univ Berlin, 86-87; asst prof chem, Univ Victoria, 87-88; ASST PROF CHEM, UNIV LETHBRIDGE, 88- *Honors & Awards:* Silver Medal, Chem Inst Can, 78. *Mem:* Chem Inst Can; Can Soc Chem; Am Chem Soc. *Res:* Chemistry of the P-block elements at transition metal centres; the structures, reactivity and electronic nature of such compounds; use of these compounds in the design of new inorganic materials. *Mailing Add:* Dept Chem Univ Lethbridge Lethbridge AB T1K 3M4 Can

BOERNER, RALPH E J, b Brooklyn, NY, Oct 2, 48; m 82; c 1. FOREST ECOLOGY, NUTRIENT CYCLING. *Educ:* Col Cortland, BS, 70; Adelphi Univ, MS, 72; Rutgers Univ, MPh, 74, PhD(bot), 80. *Prof Exp:* Asst prof biol, Burlington County Col, 74-80; asst prof, 80-86, ASSOC PROF PLANT BIOL, OHIO STATE UNIV, 86-, CHMN DEPT, 90- *Mem:* Ecol Soc Am; Bot Soc Am; Sigma Xi. *Res:* Structure-function relationships in forest ecosystems; successional patterns and land use; forest understory dynamics; plant environmental physiology; mycorrhizae. *Mailing Add:* Dept Plant Biol Ohio State Univ 1735 Neil Ave Columbus OH 43210

BOERNKE, WILLIAM E, b Uniontown, Pa, Feb 7, 44; m 85. ENZYMOLOGY, PHYSIOLOGICAL ECOLOGY. *Educ:* Univ Minn, BA, 66, PhD(zool), 72. *Prof Exp:* PROF BIOL, NEBR WESLEYAN UNIV, 71- *Concurrent Pos:* Vis scientist, Argonne Nat Lab, 81- *Mem:* AAAS; Am Soc Zoologists; Am Inst Biol Sci; Am Soc Biochem & Molecular Biol. *Res:* Metabolic or biochemical adaptations in vertebrate organisms; kinetic and physiochemical studies of enzymes; regulation of enzymes involved in nitrogen metabolism. *Mailing Add:* Dept Biol Nebr Wesleyan Univ 50th & St Paul Lincoln NE 68504

BOERS, JACK E, b Kalamazoo, Mich, Feb 14, 35; c 2. PARTICLE BEAM ACCELERATORS. *Educ:* Univ Mich, BSc, 57, MSc, 58, PhD(physics), 68. *Prof Exp:* Mem tech staff, Sandia Nat Lab, 67-84; Sr engr, Varian/Extrion, 84-90; PRES, THUNDERBIRD SIMULATIONS, 90- *Concurrent Pos:* Consult, Los Alamos Nat Labs, EG&G. *Mem:* Am Phys Soc; Inst Elec & Electronics Engrs; Asn Comput Mach. *Res:* Computer simulation of particle beams. *Mailing Add:* 626 Bradfield Dr Garland TX 75042

BOERSMA, P DEE, b Mt Pleasant, Mich, Nov 1, 46. ZOOLOGY, ECOLOGY. *Educ:* Cent Mich Univ, BS, 69; Ohio State Univ, PhD(zool), 74. *Prof Exp:* Lectr zool, Ohio State Univ, 74; from asst prof to assoc prof, 74-87, PROF ENVIRON SCI, INST ENVIRON STUDIES, UNIV WASH, 88-, ASSOC DIR, 87- *Concurrent Pos:* Adv, US Del, Status of Women Comn & World Pop Conf, UN, 74; mem, Off Endangered Species Sci Adv Comt, Dept Interior; Kellogg Nat Leadership fel, 82-85. *Mem:* AAAS; Ecol Soc Am; Am Ornith Union; Cooper Ornith Soc; Sigma Xi; Wilson Ornith Soc. *Res:* Adaptations of organisms to predictable and unpredictable environments, reproductive strategies, sexual dimorphism, temperature regulation; seabird growth rates in relationship to food supply and population dynamics; seabirds as indicators of marine pollution. *Mailing Add:* Eng Annex FM-12 Environ Studies Univ Wash Seattle WA 98195

BOERTJE, STANLEY, b Pella, Iowa, Aug 1, 30; m 53; c 6. ZOOLOGY, PARASITOLOGY. *Educ:* Calvin Col, AB, 51; Univ Iowa, MS, 57; Iowa State Univ, PhD(zool), 66. *Prof Exp:* Teacher high sch, Iowa, 54-60; assoc prof biol, Dordt Col, 60-64, 65-67 & Midwestern Col, 67-70; PROF BIOL, SOUTHERN UNIV, NEW ORLEANS, 70- *Mem:* Am Soc Parasitol; Am Inst Biol Sci; Am Sci Affil; Soc Neurosci. *Res:* Life cycle and host-parasite relationships of the cestode, Schistotaenia tenuicirrus. *Mailing Add:* 7000 Morrison Rd Apt 102 New Orleans LA 70126

BOES, ARDEL J, b Wall Lake, Iowa, Sept 24, 37; m 63; c 2. MATHEMATICS. *Educ:* St Ambrose Col, BA, 59; Purdue Univ, MS, 61, PhD(math), 66. *Prof Exp:* Instr math, Marycrest Col, 63-64; asst prof, 66-69, ASSOC PROF MATH, COLO SCH MINES, 69-, HEAD DEPT, 78- *Res:* Foundations of probability; measure theory. *Mailing Add:* Dept Math Colo Sch Mines Golden CO 80401

BOESCH, FRANCIS THEODORE, b New York, NY; m; c 1. COMBINATORICS, FINITE MATHEMATICS. *Educ:* Polytech Inst New York, BS, 57, MS, 60, PhD(elec eng), 63. *Prof Exp:* Mem tech staff, Bell Labs, 58-59; instr elec eng, Polytech Inst New York, 59-62, asst prof, 62-63; mem tech staff, Bell Labs, 63-68; prof comput sci, Univ Calif, Berkeley, 68-69; mem tech staff, Bell Labs, 69-79; prof & dept head elec eng & computer sci, 79-88, PROF MATH, STEVENS INST TECHNOL, 80-, DEAN OF FAC, 88- *Concurrent Pos:* Consult, Exec Off of Pres, 68-69; ed in chief, J Networks, 71-; ed, J Graph Theory, 79-; prin investr grants, Nat Sci Found, 81-84. *Mem:* Fel Inst Elec & Electronics Engrs; fel NY Acad Sci; Sigma Xi; Asn Comput Mach. *Res:* Theory and design of reliable large-scale networks, particularly computer networks, using graph theory and combinatorial optimization. *Mailing Add:* Dean Fac Stevens Inst Technol Hoboken NJ 07030-5991

BOESCH, WILLIAM J, b Peoria, Ill, Nov 25, 23. MATERIALS SCIENCE ENGINEERING, HIGH TEMPERATURE ALLOYS. *Prof Exp:* Assoc dir technol, Spec Metals Corp, 85; RETIRED. *Mem:* Sigma Xi. *Mailing Add:* 16 Lin Rd Utica NY 13501

BOESE, GILBERT KARYLE, b Chicago, Ill, June 24, 37; m 59; c 2. ANIMAL BEHAVIOR. *Educ:* Carthage Col, BA, 59; Northern Ill Univ, MS, 65; Johns Hopkins Univ, PhD(pathobiol), 73. *Prof Exp:* Teacher sci, pub sch, 60-62, teacher sci & head dept, 62-65; instr biol, Thornton Community Col, 65-67; asst prof biol, Elmhurst Col, 67-69; res assoc, Brookfield Zoo, 69-71, cur educ & res, 71-72, assoc dir, 72-77, dep dir, 77-80; dir, Milwaukee Co Zool Gardens, 80-89; PRES, ZOOL SOC MILWAUKEE CO, 89- *Concurrent Pos:* Primate adv comt mem, Baltimore Zool Soc, 71-72; adj prof, Northern Ill Univ, 72-73; vis lectr biol, Elmhurst Col, 71-80; adj prof, Univ Wis-Milwaukee, 80. *Mem:* Am Asn Zool Parks & Aquariums; Animal Behav Soc; EAfrican Wildlife Soc; Int Union Dirs Zool Gardens. *Res:* Social behavior and ecology of the Guinea baboon and comparison of social behavior with other species of the genus Papio. *Mailing Add:* 10005 W Bluemound Rd Milwaukee WI 53266

BOESGAARD, ANN MERCHANT, b Rochester, NY, Mar 21, 39; m 66. ASTRONOMY. *Educ:* Mt Holyoke Col, AB, 61; Univ Calif, Berkeley, PhD(astron), 66. *Hon Degrees:* DSc, Mount Holyoke Col, 81. *Prof Exp:* Res fel astrophys, Calif Inst Technol & Mt Wilson & Palomar Observ, 66-67; from asst prof to assoc prof physics & astron, 67-77, PROF PHYSICS & ASTRON, INST ASTRON, UNIV HAWAII, 77- *Concurrent Pos:* Sabbatical leave, researcher, Nat Ctr Sci Res, France, 73-74; res assoc, Hale Observ, 74; counr, Am Astron Soc, 78-81; vis scientist, Harvard-Smithsonian Ctr Astrophysics, 80-81; Morrison visitor, Lick Observ, 85; vis assoc, Caltech, 85-87, vis prof, 87; NATO sr sci fel, 83; Guggenheim fel, 86. *Honors & Awards:* Muhlmann Prize, Astron Soc Pac, 90. *Mem:* AAAS; Am Astron Soc; Int Astron Union; Astron Soc Pac (vpres, 74-76, pres, 76-78). *Res:* Stellar spectroscopy, nucleosynthesis and chemical composition of stars; stellar evolution. *Mailing Add:* Inst Astron Univ Hawaii 2680 Woodlawn Dr Honolulu HI 96822

BOESHAAR, PATRICIA CHIKOTAS, b Butler Twp, Pa, Sept 25, 47; m 70; c 1. ASTRONOMY. *Educ:* Northwestern State Univ, La, BS, 69; Ohio State Univ, PhD(astron), 76. *Prof Exp:* Instr res assoc astron, Univ Wash, 75-77; asst prof physics & astron, Univ Ore, 77-80; res assoc, Univ Ariz, 80-81; prof physics, Rider Col, 81-88; ASSOC PROF PHYSICS, DREW UNIV, 88- *Concurrent Pos:* NSF grants astron, 78, 84, 85 & 86-88; consult, AT&T Bell Labs; vis scientist, Can-France-Hawaii Telescope Corp, 89 & 90. *Mem:* Am Astron Soc; Sigma Xi; Astron Soc Pac. *Res:* Stellar spectroscopy; effects of changes in temperature, surface gravity, and chemical composition on the spectra of the coolest main sequence; subluminous and chemically peculiar evolved stars; galactic structure. *Mailing Add:* PO Box 165 Pottersville NJ 07979

BOETTCHER, ARTHUR LEE, b Glasgow, Mont, Apr 27, 35; c 2. GEOCHEMISTRY, PETROLOGY. *Educ:* Mont Sch Mines, BS, 61; Pa State Univ, MS, 63, PhD(geol), 66. *Prof Exp:* Fel geochem, Univ Chicago, 66-67; from asst prof to assoc prof petrol, 67-72, chmn geochem & mineral, 71-75, prof petrol, Pa State Univ, 72-76; PROF GEOCHEM & GEOPHYS, UNIV CALIF, LOS ANGELES, 76-, CHMN DEPT EARTH & SPACE SCIS, 87- *Concurrent Pos:* Vis scientist, US Geol Surv, 75. *Honors & Awards:* Award, Mineral Soc Am, 72. *Mem:* Fel Mineral Soc Am; fel Geol Soc Am. *Res:* High-pressure, high-temperature investigations of phase equilibria and mineral synthesis and stability; igneous and metamorphic petrology; planetary interiors; andesitic volcanism and plate tectonics; hydrous minerals and ultramafic xenoliths; volatiles in planetary interiors; volcanic history of the Mojave Desert. *Mailing Add:* Inst Geophys & Planetary Physics Univ Calif Los Angeles CA 90024-1567

BOETTCHER, F PETER, b Berlin, Ger, Aug 23, 32; US citizen; m 64; c 4. POLYMER CHEMISTRY, ORGANIC CHEMISTRY. *Educ:* T H Darmstadt, DiplIng, 58, DrIng(org chem), 61. *Prof Exp:* Res assoc org chem, Univ Mich, Ann Arbor, 61-63; res chemist polymer chem, 64-68, sr res chemist textile technol, 68-74, res supvr textile technol, 74-79, sr supvr polymer chem, 79-81, RES MGR POLYMER CHEM, CENT RES DEPT, E I DU PONT DE NEUMORS & CO, INC, 81-, RES MGR POLYMER SCI, POLYMER PROD DEPT. *Mem:* Am Chem Soc. *Res:* Polymer and intermediate synthesis; synthetic fibers; high performance polymers; liquid crystalline polymers. *Mailing Add:* 204 Hullihen Dr Newark DE 19711

BOETTCHER, HAROLD P(AUL), b Eagle, Wis, July 24, 23; m 48; c 3. ELECTRICAL ENGINEERING, MECHANICS. *Educ:* Univ Wis, BS, 47, MS, 50, PhD(elec eng & mech), 54. *Prof Exp:* Instr mechs, Univ Wis, 46-54; supvr long range res lab, A O Smith Corp, 54-61; from assoc prof to prof, 61-88, EMER PROF ELEC ENG, UNIV WIS, 88- *Concurrent Pos:* Mem Nat Acad Sci-Nat Res Coun Conf Insulation. *Mem:* Inst Elec & Electronics Engrs; Am Soc Eng Educ; Sigma Xi. *Res:* Electric motor design and application; application of computers to electrical design; analysis and synthesis of networks; power semiconductor applications. *Mailing Add:* Dept Elec Eng Univ Wis PO Box 784 Milwaukee WI 53201

BOETTGER, SUSAN D, b Two Rivers, Wis, Oct 6, 52. ORGANIC CHEMISTRY. *Educ:* Univ Wis-Madison, BS, 74; Cornell Univ, MS, 77, PhD(org chem), 79. *Prof Exp:* Fel, Cornell Univ, 79 & Univ Rochester, 79-81; res scientist, Indust Div, 81-83, SR RES SCIENTIST, PHARMACEUT RES & DEVELOP DIV, BRISTOL-MEYERS, 84- *Mem:* Am Chem Soc. *Res:* Synthesis of B-lactam antibiotics and antiviral compounds; development of automated systems for optimization of organic reactions. *Mailing Add:* Apt I-3 126 Jamesville Ave Syracuse NY 13210

BOETTINGER, WILLIAM JAMES, b Baltimore, Md, July 22, 46; m 70; c 2. METALLURGY. *Educ:* Johns Hopkins Univ, BES, 68, PhD(metall), 72. *Prof Exp:* Nat Res Coun-Nat Acad Engrs, res assoc alloy solidification, Nat Bur Standards, 72-74; METALLURGIST ALLOY SOLIDIFICATION & PHASE TRANSFORMATIONS, NAT INST STANDARDS & TECHNOL, 74- *Concurrent Pos:* Prof lectr phys metall, George Washington Univ, 77. *Honors & Awards:* Mat Sci Div Award, Am Soc Mat, 89. *Mem:* Am Inst Mining, Metall & Petrol Engrs; Am Soc Mat. *Res:* Alloy solidification; x-ray diffraction. *Mailing Add:* Mat Sci & Eng Lab Nat Inst Standards & Technol Gaithersburg MD 20899

BOETTNER, EDWARD ALVIN, b Mich, Aug 28, 15; m 38; c 2. PHYSICS. *Educ:* Lawrence Inst Technol, BSc, 37; Univ Mich, MSc, 51. *Prof Exp:* Spectroscopist, Wyandotte Chem Corp, 37-46; physicist eng res inst, 46-59, res assoc, Inst Indust Health, 59-65, from asst prof to assoc prof indust health, 59-69, PROF INDUST HEALTH, UNIV MICH, 69- *Mem:* Am Indust Hyg Asn; Optical Soc Am. *Res:* Applied spectroscopy; application of physics instrumentation to the assessment of the environment. *Mailing Add:* 1271 Leisure World Mesa AZ 85206

BOETTNER, FRED EASTERDAY, b Murphysboro, Ill, Dec 10, 18; m 42; c 3. PHARMACEUTICAL CHEMISTRY. *Educ:* Carthage Col, AB, 40; Tulane Univ, MS, 42; Univ Ill, PhD(org chem), 47. *Prof Exp:* Asst chem, Tulane Univ, 40-42; res chemist, Monsanto Chem Co, Ala, 42-43, Higgins Indust, Inc, La, 43-44 & Armour Res Found, 44; asst org chem, Univ Ill, 44-47; sr scientist, Rohm and Haas Co, 47-76; sr scientist, Polysciences, Inc, 76-86; RETIRED. *Mem:* Am Chem Soc; fel Am Inst Chemists; AAAS. *Res:* Lignin utilization; plastics and resins; synthesis of new anion exchange resins; structure of amino amide formaldehyde resins; insecticides; fungicides; herbicides; surface active compounds; antineoplastic agents from plants; pharmaceuticals. *Mailing Add:* 1604 Ternberry Rd High Point NC 27262-7412

BOFFA, LIDIA C, b Monza, Italy, Aug 23, 45; div; c 1. MOLECULAR CARCINOGENESIS, GENES MODULATION. *Educ:* Univ Padova, BS, 65, PhD(org chem), 69. *Prof Exp:* Postdoctoral cell biol & biochem, Inst Path, Univ Padova, 69-71; res asst cell biol, Med Sch, Rockefeller Univ, NY, 71-75, asst prof carcinogenesis & cell biol, 76-82; RES ASSOC PROF MOLECULAR CARCINOGENESIS, NAT CANCER INST, GENOA, ITALY, 82- *Concurrent Pos:* Adj fac, Rockefeller Univ, NY, 82-; vis prof, Med Sch, Univ CapeTown, SAfrica, 87. *Mem:* Europ Asn Cancer Res; Am Asn Cancer Res; Am Asn Cell Biol; NY Acad Sci. *Res:* Determination of macromolecular targets of alkylating agents, with different degree of genotoxicity, in cells; gene sequence and the control of their expression in carcinogenesis. *Mailing Add:* Dept Chem Carcinogenesis Nat Cancer Inst 1st Viale Benedetto XV n10 Genoa 16132 Italy

BOFINGER, DIANE P, b Philadelphia, Pa, Sept 26, 51; m 88. PROTEIN CHEMISTRY, CLINICAL CHEMISTRY. *Educ:* East Stroudsburg State Col, BA, 73; State Univ NY, Buffalo, PhD(biochem), 82. *Prof Exp:* Res health scientist, Vet Admin Med Ctr Buffalo, 84-87; instr chem, Erie County Community Col, 88; res instr, 82-83, clin asst prof clin biochem, 89-90, ASST PROF CLIN BIOCHEM, STATE UNIV, NY, BUFFALO, 90- *Mem:* Protein Soc; Am Asn Clin Chem; Am Chem Soc; AAAS; Asn Women Sci. *Res:* intracellular signal transduction pathways invloving calmodulin and protein phosphorylation; mapping drug and protein binding sites in calmodulin; identify proteins which are phosphorylated in various cells in response to a variety of external cell stimuli. *Mailing Add:* Dept Med Technol 107 AA Clin Ctr State Univ Buffalo NY 462 Grider St Buffalo NY 14215

BOGAN, DENIS JOHN, b Winchester, Mass, Aug 10, 41; m 69; c 2. CHEMICAL KINETICS, MOLECULAR SPECTROSCOPY. *Educ:* Northeastern Univ, AB, 65; Carnegie Mellon Univ, MS, 70, PhD(phys chem), 73. *Prof Exp:* Fel phys chem, Kans State Univ, 72-74, Nat Res Coun fel; res chemist phys chem, Naval Res Lab, 76-85; ASSOC PROF CHEM, CATH UNIV AM, 85- *Concurrent Pos:* Assoc, Nat Res Coun-Nat Acad Sci, 74 & 75; adj assoc prof chem, Cath Univ Am, 81-85. *Mem:* Am Chem Soc; Sigma Xi; Am Soc Photobiol; Inter-Am Photochem Soc; Combustion Inst. *Res:* Physical chemistry of reactive systems, particularly gas phase and gas surface catalytic behavior of molecules containing carbon, hydrogen, oxygen, and halogen. *Mailing Add:* 5110 Althea Dr Annandale VA 22003

BOGAN, RICHARD HERBERT, b Portland, Ore, June 11, 26; m 47, 70; c 5. CIVIL ENGINEERING. *Educ:* Univ Wash, Seattle, BSCE, 49; Mass Inst Technol, SM, 52, ScD(eng), 54. *Prof Exp:* Engr, Pac Water Works Supply Co, Wash, 49-51; asst, Mass Inst Technol, 52-54; from asst prof to assoc prof civil eng, 54-65, PROF CIVIL ENG, UNIV WASH, 65- *Concurrent Pos:* Consult, eng firms, indust & govt, 55-; prog mgr, Div Advan Energy Res & Technol, NSF, 73-75; mem, Interagency Task Force Synthetic Fuel From Coal, Washington, DC, 73-75; mem, proj adv bds various alt energy proj including Batelle (Columbus), MITRE Corp, Elec Power Res Inst, US Dept Energy, 74-; mem, Solar Energy Task Force, Energy Res & Develop Admin, Washington, DC, actg br chief, Div Solar Energy, 75. *Mem:* Am Soc Civil Engrs; Am Soc Eng Educ; Sigma XI; Am Acad Environ Engrs; AAAS. *Res:* Control, measurement and removal of physical, chemical and biological entities contained in air, water and waste waters; computer simulation, analysis and design of treatment processes, water supply and waste management systems. *Mailing Add:* Dept Civil Eng FX-10 Univ Wash Seattle WA 98105

BOGAN, ROBERT L, b Anderson, Ind, July 28, 26; m 48; c 3. DENTISTRY. *Educ:* Butler Univ, BS, 50; Ind Univ, DDS, 54, MSD, 67. *Prof Exp:* Asst prosthetics, Sch Dent, Ind Univ, Indianapolis, 54-55, instr crown & bridge, 55-62, from asst prof to prof partial prosthodontics, 62-77, asst to dean, 64-67, asst dean, 67-73, assoc dean, 67-90; RETIRED. *Mem:* Am Asn Dent Schs; Int Col Dentists. *Res:* Clinical evaluation of factors contributing to stability in removable partial prosthodontics; laboratory study in soldering of high fusing gold alloys used in dental restorations. *Mailing Add:* Rte 3-11 WS-24S Nineveh IN 46164

BOGAR, LOUIS CHARLES, b Bluefield, WVa, Aug 22, 32; m 54; c 3. RADIOCHEMISTRY, HEALTH PHYSICS. *Educ:* Mass Inst Technol, SB, 54. *Prof Exp:* Resident radiochemist, Bettis Atomic Power Lab Off, Chalk River Nuclear Labs, Can, 57-60; scientist, Bettis Atomic Power Lab, Westinghouse Elec Corp, 60-63, supvr mat activation, 63-64, supvr radiation sources, 65-67, mgr radiation eng, 68-69, mgr radiation control, 69-71, mgr radiation & safety, 71-74, mgr steam generators, 78-85; spec assignment, Duquesne Light Co, Shippingport Atomic Power Sta, 74-77,; vpres & mgr, environ safety & health, 85-87, vpres site remediation, 87-88, VPRES GOVT OPERS BUS UNIT, ENVIRON AFFAIRS, FEED MAT PROD CTR, WESTINGHOUSE MAT CO, OHIO, 88- *Mem:* AAAS; Am Nuclear Soc; Am Chem Soc; Am Defense Preparedness Asn; Am Soc Testing & Mat; fel Am Inst Chem. *Res:* Fission product chemistry; fuel element release; radioactivation and transport processes in reactor coolants; low level radioactivity measurements; high temperature aqueous corrosion mechanisms; radiological controls; in-vivo dosimetry; environmental radioactivities. *Mailing Add:* Feed Mat Prod Ctr Westinghouse Mats Co of Ohio 7400 Willey Rd Fernald OH 45239

BOGAR, THOMAS JOHN, b Welland, Ont, Oct 3, 47; US citizen; m 69; c 2. EXPERIMENTAL FLUID DYNAMICS. *Educ:* Kalamazoo Col, BA, 69; Univ Mich, MS, 70, PhD(physics), 75. *Prof Exp:* Asst prof physics, Westminster Col, 75-79; RES SCIENTIST, MCDONNELL DOUGLAS RES LABS, 79- *Mem:* Am Phys Soc; Am Inst Aeronaut & Astronaut. *Res:* Unsteady transonic internal flows with emphasis on instrumentation, data acquisition and processing, computer automation. *Mailing Add:* 13164 Weatherfield Dr Creve Coeur MO 63146

BOGARD, ANDREW DALE, b Savannah, Mo, June 18, 15; m 46; c 2. RADIOCHEMISTRY. *Educ:* William Jewell Col, AB, 37; Univ Ark, MA, 39. *Prof Exp:* Asst chem, Univ Ark, 37-39; chemist coal prod & clay utilization, US Bur Mines, 39-42; chemist, Oak Ridge Nat Lab, 46-48; chemist liquid metals res, Naval Res Lab, 48-56; adv engr, Bettis Atomic Power Lab, Westinghouse Elec Corp, 56-80; RETIRED. *Concurrent Pos:* Consult, 80- *Mem:* Am Chem Soc. *Res:* Coal utilization; liquid metals for nuclear reactors; fallout; fission product analysis; heavy element chemistry; gamma ray spectrometry; reactor coolant radiochemistry. *Mailing Add:* 4765 Baptist Rd Pittsburgh PA 15227

BOGARD, DONALD DALE, b Fayetteville, Ark, Feb 6, 40. GEOCHEMISTRY. *Educ:* Univ Ark, Fayetteville, BS, 62, MS, 64, PhD(isotope geochem), 66. *Prof Exp:* Res fel geol sci, Calif Inst Technol, 66-68; cur, Antarctica Meteorite Collection, 78-84, STAFF SCIENTIST, SOLAR SYST EXPLOR DIV, JOHNSON SPACE CTR, NASA, 68-, DISCIPLINE SCIENTIST PLANETARY MAT PROG, 84- *Concurrent Pos:* NSF fel, Calif Inst Technol, 66-67; assoc ed, J Geophys Res, Am Geophys Union, 75-77. *Mem:* AAAS; Am Geophys Union; Geochem Soc; fel Meteoritical Soc (secy, 81-87). *Res:* Age and origin of planetary objects in the solar system; geochronology; isotope geochemistry nuclear reactions in nature; meteorites. *Mailing Add:* NASA Johnson Space Ctr Code SN Houston TX 77058

BOGARD, TERRY L, b Bicknell, Ind, Aug 8, 36; m 62; c 3. ORGANIC CHEMISTRY, COMPUTER SCIENCE. *Educ:* Ohio State Univ, BSc, 58; Univ Calif, Los Angeles, PhD(org chem), 63. *Prof Exp:* Asst chem, Univ Calif, Los Angeles, 58-62; NIH fel, Swiss Fed Inst Technol, 62-64; fel, Brandeis Univ, 64-65; res chemist, Lederle Labs Div, Am Cyanamid Co, 65-70, mgr, Process Develop & Clin Mat, Cyanamid Int Div, 71-80, mgr, Clin Mat, 81-83, mgr, Proj Anal, Cyanamid Med Res Div, 83-87, mgr, area opers, Int Regulatory Affairs, 88-89, MGR, CL FILE & RES INFO SERV, AM CYANAMID CO, 89- *Mem:* Am Chem Soc; Royal Soc Chem. *Res:* Theoretical, synthetic and natural products organic chemistry; polycyclic bridged compounds, alkaloids, terpenes and reaction mechanisms. *Mailing Add:* 102 W Prospect St Nanuet NY 10954

BOGARDUS, CARL ROBERT, JR, b Hyden, Ky, June 26, 33; m 57; c 2. MEDICINE, RADIOLOGY. *Educ:* Hanover Col, BA, 55; Univ Louisville, MD, 59; Am Bd Radiol, cert therapeut radiol & nuclear med, 64; Am Bd Nuclear Med, dipl, 72. *Prof Exp:* Intern & resident therapeut radiol, Penrose Cancer Hosp, Colorado Springs, Colo, 63; from asst prof to assoc prof, 64-69, PROF RADIOL & VCHMN DEPT, MED CTR, UNIV OKLA, 69-, DIR DIV RADIATION THER, 66- *Concurrent Pos:* Res fel radiation physics, Mallinckrodt Inst Radiol, Sch Med, Wash Univ, 64. *Mem:* Am Soc Therapeut Radiol; Radiation Res Soc; Soc Nuclear Med; AMA; fel Am Col Radiol. *Res:* Cancer; radiation therapy, physics and biology; radioisotopes. *Mailing Add:* Okla Mem Hosp Radiation Ther Treatment Ctr PO Box 26307 Oklahoma City OK 73126

BOGARDUS, EGBERT HAL, b New York, NY, Feb 20, 31; m 53; c 3. SOLID STATE SCIENCE. *Educ:* Univ Tex, BS, 55; Pa State Univ, PhD(mat sci), 64. *Prof Exp:* Res assoc physics, Cornell Univ, 64-66; mem tech staff, Tex Instruments, Inc, Dallas, 66-68; staff physicist, IBM Components Div, IBM T J Watson Res Ctr, 69-73, adv physicist, 73-80, mem corp tech staff, 80-83, mgr advan mfg eng, Gen Prod Div, 83-85, mgr tech staff, Gen Technol Div, 85-86, mgr Process Control Systs, 86-89, MFG TECH STAFF, GEN TECH DIV, IBM T J WATSON RES CTR, 89- *Mem:* Am Phys Soc; Sigma Xi; Inst Elec & Electronics Engrs. *Res:* Optical characterization of III-V materials; ion implantation; magnetic properties of ferrofluids; surface properties of thin insulating films; electrical properties of ion-implanted devices. *Mailing Add:* Box 323 Sharon CT 06069

BOGART, BRUCE IAN, b New York, NY, Sept 11, 39; m 64; c 2. CELL BIOLOGY. *Educ:* Johns Hopkins Univ, BA, 61; NY Univ, PhD(basic med sci), 66. *Prof Exp:* From instr to asst prof, 66-74, ASSOC PROF CELL BIOL, MED CTR, NY UNIV, 74-, COURSE DIR, GROSS ANAT, 79- *Mem:* AAAS; Am Asn Anatomists; Am Soc Cell Biol. *Res:* Cell biology of secretion; salivary and lacrimal glands; cytophysiology, pharmacology and cytopathology of exocrine secretion; cystic fibrosis; signal transduction. *Mailing Add:* Dept Cell Biol NY Univ Med Ctr 550 First Ave New York NY 10016

BOGART, KENNETH PAUL, b Cincinnati, Ohio, Oct 6, 43; m 66; c 2. COMBINATORICS. *Educ:* Marietta Col, BS, 65; Calif Inst Technol, PhD(math), 68. *Prof Exp:* Asst instr, Marietta Col, 65-66; teaching asst, Calif Inst Technol, 66-68; from asst prof to assoc prof, 68-80, PROF MATH & COMPUTER SCI, DARTMOUTH COL, 80- *Mem:* Math Asn Am; Am Math Soc; Soc Indust & Appl Math; NY Acad Sci. *Res:* Abstract commutative ideal theory, noether lattices; lattice theory; ring theory; partically ordered sets; application of algebraic and combinatorial techniques to social and behavioral science problems; algebraic combinatorics; combinatorial algorithms; theoretical computer science. *Mailing Add:* Dept Math & Computer Sci Dartmouth Col Hanover NH 03755

BOGART, MARCEL J(EAN) P(AUL), b Paris, France, Apr 14, 13; nat US; m 47, 62; c 2. CHEMICAL ENGINEERING. *Educ:* Cooper Union, BS, 33; Univ Mich, MSE, 34. *Prof Exp:* Chem engr, Lummus Co, 35-52, sr develop engr, 52-57, dir process develop, 57-62, sr process specialist, 62-67; supv develop engr, Fluor Engrs & Constructors, Inc, 67-83; RETIRED. *Concurrent Pos:* Guest lectr, Columbia Univ, 47-50; adj prof, Polytech Inst Brooklyn, 50-61. *Mem:* Fel Am Inst Chem Engrs; Sigma Xi; AAAS. *Res:* Reduction of pilot plant results to commercial practice; adaption of foreign processes to American industrial practice; aqua-ammonia absorption refrigeration. *Mailing Add:* 10602 Cordoba Ct Whittier CA 90601

BOGASH, R(ICHARD), b Philadelphia, Pa, Dec 26, 22; m 44; c 2. CHEMICAL ENGINEERING. *Educ:* Univ Pa, BS, 43, MS, 47, PhD(chem eng), 49. *Prof Exp:* Asst, Manhattan Proj, Columbia Univ, 43-44; test engr, Insinger Mach Co, 44-46; head, Prod Develop Lab Div, Wyeth Labs, 47-52, dir, Prod Develop Div, 52-59, asst vpres res & develop, 59-68, vpres res & develop, 68-76, PRES RES & DEVELOP, WYETH LABS, 76-; CHMN & VPRES CORP DEVELOP, AM HOME PRODS, 78- *Mem:* Am Chem Soc. *Res:* Physiochemical properties of selected solvents near their critical temperatures; selective precipitation of enzymes; phenothiazines; antibiotics. *Mailing Add:* 101 Cheswold Lane Apt 4-F Haverford PA 19041

BOGATY, HERMAN, b New York, NY, Apr 9, 18; m 38; c 2. RESEARCH ADMINISTRATION, TECHNOLOGY. *Educ:* City Col NY, BS, 40. *Prof Exp:* From sci aide to textile technologist, Bur Standards, 37-47; res assoc, Harris Res Labs, Inc, 47-58; assoc res dir, Toni Div, Gillette Co, 58-62, lab dir & vpres res & develop, 62-68; vpres res & develop consumer prod group, Warner Lambert Co, 68-75; pres, 75-81, CONSULT RES MGT, HERMAN BOGATY CONSULT CORP, 81- *Mem:* Am Chem Soc; Fiber Soc; Soc Cosmetic Chem; AAAS; Inst Food Technol. *Res:* Research and development in toiletries, proprietaries, chewing gum and confections. *Mailing Add:* Laurel Hill HL No R Chapel Hill NC 27514

BOGDAN, VICTOR MICHAEL, b Kiev, Ukrainia, Jan 4, 33; nat US; m 68; c 2. MATHEMATICS. *Educ:* Univ Warsaw, BS, 53, MS, 55; Polish Acad Sci, PhD(math), 60. *Prof Exp:* Asst math, Univ Warsaw, 52-53, asst, 53-55, from instr to asst prof, 55-61; res assoc, Univ Md, 61-62; from asst prof to assoc prof, Georgetown Univ, 62-64; assoc prof, 64-66, PROF MATH, CATH UNIV AM, 66- *Concurrent Pos:* Mem res staff math inst, Polish Acad Sci, 55-61; NSF res grant, 64-65; vis scientist, Univ Montreal, 70-71; sr resident res assoc, Johnson Space Ctr, NASA, Houston, 78. *Mem:* Am Math Soc; Polish Math Soc; Soc Appl & Indust Math. *Res:* Functional analysis; linear methods of summability; integration theory; theory of distributions; differential equations in Banach spaces; almost periodic solutions of ordinary and partial differential equations; theory of random processes; computer mathematical software systems. *Mailing Add:* Dept Math Cath Univ Am Washington DC 20064

BOGDANOFF, JOHN LEE, b East Orange, NJ, May 25, 16; m 45; c 2. ENGINEERING. *Educ:* Syracuse Univ, BME, 38; Harvard Univ, MSE, 39; Columbia Univ, PhD(appl mech), 50. *Prof Exp:* Asst proj engr appl mech, Wright Aero Corp, 39-46; instr, Columbia Univ, 46-50; from assoc prof to prof eng sci, 50-86, assoc head, Sch Aeronaut, Astronaut & Eng Sci, 67-72, EMER PROF AERONAUT & ASTRONAUT ENG, PURDUE UNIV, 86- *Concurrent Pos:* Consult, Chatham Electronic Co, 47-49, Houdaille-Hersey, 53-54 & Allison Div, Gen Motors Corp, 53-63; dir & consult, Midwestern Appl Sci Corp, 57- *Mem:* Nat Acad Engrs; Am Phys Soc; fel AAAS; fel Am Soc Mech Engrs. *Res:* Application of stochastic process in engineering; dynamics; vibration; elasticity; cumulative damages, fatigue and wear. *Mailing Add:* 327 Laurel Dr West Lafayette IN 47906

BOGDANOVE, EMANUEL MENDEL, endocrinology; deceased, see previous edition for last biography

BOGDANSKI, DONALD FRANK, b Port Chester, NY, Oct 31, 28; m 59; c 4. PHARMACOLOGY. *Educ:* Columbia Univ, BS, 50; Georgetown Univ, PhD, 54. *Prof Exp:* Lab asst pharmacol, Georgetown Univ, 52-53, lab asst biochem, 53; pharmacologist, Chem Pharmacol Lab, Nat Heart Inst, 54-58; spec res assoc from Med Sch, to Dept Neurophysiol, Walter Reed Army Inst Res, 58-64; PHARMACOLOGIST, LAB CHEM PHARMACOL, NAT HEART INST, 64- *Mem:* AAAS; Am Soc Pharmacol & Exp Therapeut. *Res:* Binding and release of chemical transmitters in the peripheral nervous system; phylogenetic distribution of amines; transport, storage and release of chemical transmitters in the peripheral and central nervous system. *Mailing Add:* Nat Heart Lung & Blood Inst NIH Lab Biochem & Pharmacol Bethesda MD 20892

BOGDEN, ARTHUR EUGENE, b DuBois, Pa, Oct 6, 21; m 48; c 2. EXPERIMENTAL ONCOLOGY. *Educ:* Univ Pa, BA, 52, MS, 53, PhD(med microbiol), 56. *Prof Exp:* Fel, Wistar Inst, 56-58, chief res serv, 58; chief div cancer immunol, Biochem Res Found, 58-63; dir immunobiol & exp oncol, 63, VPRES, MASON RES INST, 68-; SCI DIR, EG & G BOGDEN LAB, 84- *Concurrent Pos:* Sr Res Assoc, Med Sch, Univ Mass, 80- *Mem:* Am Asn Cancer Res; Am Asn Immunol; NY Acad Sci. *Res:* Immunobiology; immunogenetics; rat and rhesus blood groups; tumor cryopreservation; radiobiology; anti-inflammatory and hormone immunoassay viral-hormonal cocarcinogenesis; cancer chemotherapy and experimental oncology. *Mailing Add:* Dept Biomeasure Inc 11-15 East Ave Hopkinton MA 01748

BOGDEN, JOHN DENNIS, b Jersey City, NJ, Sept 5, 45; m 67; c 2. TRACE ELEMENT NUTRITION, ENVIRONMENTAL TOXICOLOGY. *Educ:* Brown Univ, ScB, 67; Seton Hall Univ, MS, 70, PhD(chem), 71. *Prof Exp:* Fel, NJ Med Sch, 71-72, from clin instr to instr, 73-75, from asst prof to assoc prof, 75-81, PROF PREV MED, UNIV MED & DENT, NJ MED SCH, 87- *Concurrent Pos:* Prin investr, NIH, NJ Med Sch, 84-87; mem, bd dirs, Soc Environ Geochem & Health, 84-86. *Mem:* Am Chem Soc; Am Asn Univ Profs; Am Pub Health Asn; Soc Environ Geochem & Health; Am Inst Nutrit; Am Soc Clin Nutrit. *Res:* Trace element nutrition; trace element toxicology; drug-nutrient interactions; zinc and immune function; antitumorigenic activity of trace elements. *Mailing Add:* Dept Prev Med & Community Health 185 S Orange Ave Newark NJ 07103-2714

BOGDONOFF, MORTON DAVID, b New York, NY, Dec 8, 25; m 51, 75; c 4. MEDICINE. *Educ:* Cornell Univ Med Col, MD, 48. *Prof Exp:* From asst prof to prof, Duke Univ, 56-70, assoc dean, 68-70; prof & chmn dept med, Univ Ill, 70-75; exec assoc dean, 75-78, PROF MED, CORNELL UNIV, 75- *Concurrent Pos:* Nat Counr, Am Fedn Clin Res, 60-65; pres, Am Fedn Clin Res, 64-65; mem, Presidential Adv Heart Panel, 71-72; ed, Clin Res, 58-64; chief ed, Archives Internal Med, 67-77, Drug Therapy, 77-, New Develop in Med. *Mem:* Asn Am Physicians; Am Soc Clin Invest; Am Fedn Clin Res; Endocrine Soc; Harvey Soc. *Res:* Metabolic processes; diabetes; autonomic nervous system; medical care; medical education. *Mailing Add:* Dept Med 1300 York Ave New York NY 10021

BOGDONOFF, PHILIP DAVID, JR, b Lowell, Mass, June 11, 27; m 53; c 3. ANIMAL PHYSIOLOGY, ANIMAL NUTRITION. *Educ:* Univ Md, BS, 51, MS, 53, PhD(physiol), 55. *Prof Exp:* Res analyst animal & human sci, Nat Acad Sci, 55-57; res nutritionist, Com Solvents Corp, 57-59, head animal nutrit res dept, 59-63; res assoc, Bio-Res Inst, Inc & Bio-Res Consult, Inc, 63-67; prof, N Shore Community Col, 70-72; vpres, Wheeler Assocs, Inc, 67-79; PRES, PEER GROUP, INC, 76- *Res:* Administration, research and educational development; consulting for personal and business asset management. *Mailing Add:* 17 Liberty Lane Exeter NH 03833

BOGDONOFF, SEYMOUR (MOSES), b New York, NY, Jan 10, 21; m 44; c 3. MECHANICAL ENGINEERING. *Educ:* Rensselaer Polytech Inst, BAE, 42; Princeton Univ, MSE, 48. *Prof Exp:* Nat Adv Comt Aeronaut, 42-46; from asst prof to prof aeronaut eng, Princeton Univ, 46-61, chmn, Dept Mech & Aerospace Eng, 73-81, Robert Porter Patterson prof aeronaut eng, 61-89, head, Gas Dynamics Lab, 53-89, EMER SR RES SCHOLAR & ROBERT PORTER PATTERSON PROF AERONAUT ENG, PRINCETON UNIV, 89- *Concurrent Pos:* Consult, US Air Force & various industs; mem, Fluid Mech Panel, Adv Group Aerospace Res & Develop, NATO, 67-76; mem, Aeronaut Adv Comt, NASA, 68-; mem, French Nat Academie of Air & Space, 90. *Honors & Awards:* Fluid & Plasmadynamics Award, Am Inst Aeronaut & Astronaut, 83; Dryden Lectureship Res, Am Inst Aeronaut & Astronaut, 90; Except Civilian Serv Award, US Air Force. *Mem:* Nat Acad Eng; fel Am Inst Aeronaut & Astronaut; Supersonic Tunnel Asn; Corresp mem Int Acad Astronaut; Am Soc Mech Engrs. *Res:* High speed gas dynamics; hypersonic and supersonic aerodynamics; energy; development of high speed research facilities, experimental research in supersonics and hypersonic low density flows; shock wave boundary layer interactions in two and three-dimensional configurations. *Mailing Add:* 39 Random Rd Princeton NJ 08540

BOGENSCHUTZ, ROBERT PARKS, b Ponca City, Okla, Mar 25, 33. PHYSIOLOGY, BIOLOGY. *Educ:* Okla State Univ, BS, 54, MS, 56; Cent State Col, Okla, BS, 61; Univ Okla, PhD(zool), 66. *Prof Exp:* Teacher, Am Sch London, 62-63; res assoc, Okla Univ Res Inst, 66; teacher, Am Int Sch, Vienna, 66-67; asst prof, 67-70, assoc prof, 70-77, PROF BIOL & PHYSIOL, CENT STATE UNIV, OKLA, 77- *Mem:* Am Inst Biol Sci; Am Soc Zool. *Res:* Effects of environmental factors on pituitary proteins and histology. *Mailing Add:* Dept Biol Cent State Univ 100 N University Dr Edmond OK 73034

BOGER, DALE L, b Hutchinson, Kans, Aug 22, 53. ORGANIC MEDICINAL CHEMISTRY. *Educ:* Univ Kans, BS, 75; Harvard Univ, PhD(org chem), 80. *Prof Exp:* From asst prof to assoc prof med chem, Univ Kans, 79-85; from assoc prof to prof chem & med chem, Purdue Univ, 85-91; RICHARD & ALICE CRAMER CHAIR CHEM, SCRIPPS RES INST, 91- *Concurrent Pos:* Cottrell res grant, Res Corp, 80-81; Searle scholar, Chicago Comn Trust Co, Searle Family Found, 81-84; petrol res fund, Am Chem Soc, 81-83; career develop award, NIH, 83-88; Alfred P Sloan fel, 85-89. *Honors & Awards:* A C Cope Scholar Award, Am Chem Soc, 88. *Mem:* Am Chem Soc; AAAS. *Mailing Add:* Dept Chem Scripps Res Inst 10666 N Torrey Pines Rd La Jolla CA 92037

BOGER, EDWIN AUGUST, SR, b Providence, RI, Oct 7, 23; m 46, 61; c 2. PHYCOLOGY, MICROBIOLOGY. *Educ:* Valparaiso Univ, BS, 61; Univ Conn, Storrs, MS, 68; Clark Univ, PhD(biol), 76. *Prof Exp:* From instr to assoc prof, 68-80, PROF BIOL, WORCESTER STATE COL, 80- *Concurrent Pos:* Danforth Assoc, 77-83. *Mem:* AAAS; Sigma Xi; Am Soc for Microbiol; Int Phycol Soc; PhycolSoc Am. *Res:* Chemical constituents of marine algae; nematode and other host-parasite relationships in algae; innovative science teaching and testing methods. *Mailing Add:* Dept Biol Worcester State Col Worcester MA 01602-2597

BOGER, PHILLIP DAVID, b Xenia, Ohio, July 12, 43; m 75. GEOLOGY. *Educ:* Ohio State Univ, BSc, 66, PhD(geol), 76. *Prof Exp:* Grad teaching asst geol, Dept Mineral & Geol, Ohio State Univ, 70-71, grad admin assoc acad advisor, Univ Col, 71-73; ASST PROF, DEPT GEOL, STATE UNIV NY, GENESEO, 76- *Concurrent Pos:* Res asst, Dept Mineral & Geol, Ohio State Univ, 73-74; co-prin & prin investr, NSF grants 74-76; fel, Grant-In-Aid, State Univ NY, 78-79 & Kodak, Geneseo, 78-79. *Mem:* Geol Soc Am. *Res:* Interpretation of the variation of strontium isotope ratios of various materials in terms of the geologic history of these materials. *Mailing Add:* Dept Geol Sci State Univ NY at Geneseo Greene Sci Bldg Geneseo NY 14454

BOGER, ROBERT SHELTON, b Richmond, Va, Jan 21, 47; m 72; c 3. INTERNAL MEDICINE, NEPHROLOGY. *Educ:* Amherst Col, BA, 68; Harvard Med Sch, MD, 72. *Prof Exp:* Asst prof med, Med Col Ga, 78-80; CONSULT NEPHROLOGY, AUGUSTA VET ADMIN HOSP & DUBLIN VET ADMIN HOSP, 78- *Mem:* Am Soc Nephrology; Int Soc Nephrology; Am Col Physicians. *Res:* Renal pathophysiology; effects of hyperuricemia on renal function; hypertension; clinical nephrology. *Mailing Add:* 7800 N Kings Hwy Suite 4 Myrtle Beach SC 29577

BOGER, WILLIAM PIERCE, b Johnstown, Pa, July 23, 13; m 43; c 2. MEDICINE. *Educ:* Bucknell Univ, BS, 34; Harvard Univ, MD, 38; Am Bd Internal Med, dipl, 45. *Prof Exp:* Intern, Philadelphia Gen Hosp, 38-40; asst resident med, Mass Gen Hosp, 41-42; chief resident & instr, Med Col Va, 42-43; dir dept internal med & assoc, St Luke's Hosp, WVa, 43-44; assoc med dir, Sharp & Dohme, Inc, 45-51, med dir, 51-56; instr, Grad Sch & Sch Med, Univ Pa, 46-50, assoc med, 50-62; dir res & chief infectious dis, Montgomery Hosp, Norristown, 62-67; asst chief staff, Med Serv, 66-72, chief med serv, Vet Admin Hosp, Coatesville, 72-84; assoc prof psychiat & human behav & clin assoc prof med, 77-86, EMER PROF, JEFFERSON MED SCH, 86- *Concurrent Pos:* Dir dept res therapeut, Norristown State Hosp, 52-62; dir int med affairs, McNeil Labs, 62-64; corp med dir, Miles Labs, Inc, 64-65, vpres res & med affairs, Dome Labs Div, 65-66. *Mem:* Am Col Physicians; fel Am Soc Pharmacol & Exp Therapeut; fel Endocrine Soc; fel Am Soc Clin Invest; fel Royal Soc Med. *Res:* Chemotherapy; antibiotics; internal medicine; metabolic disease. *Mailing Add:* 1675 Glenhardie Rd Wayne PA 19087

BOGERT, BRUCE PLYMPTON, b Waltham, Mass, Sept 26, 23; m 49; c 2. PHYSICS. *Educ:* Mass Inst Technol, BS, 44, MS, 46, PhD(math), 48. *Prof Exp:* Staff mem, Radiation Lab, Mass Inst Technol, 44-45; mem tech staff, Bell Labs, 48-81; RETIRED. *Honors & Awards:* Biennial Award, Acoustical Soc Am, 58. *Mem:* Fel Acoustical Soc Am. *Res:* Digital transmission; seismic wave instrumentation and processing; underwater acoustics; digital computer analysis of time series. *Mailing Add:* Eight Penryn Way Rockport MA 01966

BOGGESS, ALBERT, b Dallas, Tex, Jan 30, 29; m 52; c 3. ACTIVE GALAXIES, SPACE INSTRUMENTATION. *Educ:* Univ Tex, BA, 50; Univ Mich, MA, 52, PhD(astron), 54. *Prof Exp:* Fel, Johns Hopkins Univ, 54-55, consult, 55-56; physicist, Naval Res Lab, 55-58; astrophysicist interstellar medium sect, 59-69, head observational astron br, 69-72, head astron systs br, 72-77, head advan syst develop br, 77-83, ASSOC DIR SCI, GODDARD SPACE FLIGHT CTR, NASA, 83- *Mem:* Am Astron Soc; Int Astron Union. *Res:* Active galaxies; galactic nebulae; interstellar medium; high altitude research; stellar atmospheres; airglow. *Mailing Add:* 319 Stonington Rd Silver Spring MD 20902

BOGGESS, GARY WADE, b Hardin, Ky, Aug 31, 36; m 54; c 2. INORGANIC CHEMISTRY, CHEMICAL EDUCATION. *Educ:* Murray State Univ, BS, 62; Purdue Univ, MS, 66; Univ Tenn, PhD(chem), 73. *Prof Exp:* Teacher math & chem, Murray High Sch, 62-66; from instr to assoc prof chem, 66-78, PROF & DEAN COL SCI, MURRAY STATE UNIV, 78- *Concurrent Pos:* Dir, NSF Integrated Sci Math Educ Proj, 77-; dir, Ctr Excellence & Reservoir Res. *Mem:* Am Chem Soc; AAAS; Sigma Xi. *Res:* High-temperature photoelectron spectroscopy; science and mathematics education for pre and in-service teachers. *Mailing Add:* RR1 Box 196 Hardin KY 42048

BOGGESS, NANCY WEBER, b Philadelphia, Pa, Apr 25, 29; m 52; c 3. ASTRONOMY. *Educ:* Wheaton Col, Mass, BA, 47; Wellesley Col, MA, 49; Univ Mich, PhD, 67. *Prof Exp:* Res asst astrophys, Haverford Col, 49-50; astronr, Hq, NASA, 67-87; CONSULT, 87- *Mem:* Int Astron Union; Am Astron Soc; Sigma Xi. *Res:* Theoretical transition probabilities for elements of astrophysical interest; infrared astronomy; galactic structure of dwarf irregular galaxies; analysis of space data from infrared astronomical satellite; observational-cosmology from COBE satellite. *Mailing Add:* Goddard Space Flight Ctr-NASA Code 685-0 Nancy W Boggess Greenbelt MD 20771

BOGGESS, WILLIAM RANDOLPH, b Oakvale, WVa, Apr 9, 13; m 38; c 4. FORESTRY. *Educ:* Concord Col, AB, 33; Duke Univ, MF, 40. *Prof Exp:* Teacher pub sch, WVa, 33-35; res assoc forester, Ala Agr Exp Sta, Ala Polytech Inst, 39-48; prof & assoc forest res, Dixon Springs Exp Sta, 48-58, prof forestry, 58-73, head dept, 69-73, emer prof forestry, Univ Ill, Urbana, 73-; environ consult, Lab Tree-Ring Res, Univ Ariz, 73-78, res assoc, 78-83; RETIRED. *Concurrent Pos:* Ed, Am Water Resources Asn Bull, 65-69 & 76-81. *Mem:* Soc Am Foresters; Am Soc Agron; Soil Sci Soc Am; Ecol Soc Am; Am Water Resources Asn (pres, 70, gen secy, 77-82). *Res:* Forest ecology; soils; forest influences. *Mailing Add:* 61 E Mediterranean Dr Tucson AZ 85704

BOGGIO, JOSEPH E, b Holyoke, Mass, Mar 20, 36; m 58; c 2. PHYSICAL CHEMISTRY. *Educ:* Worcester Polytech Inst, BS, 58, MS, 60, PhD(phys chem), 63. *Prof Exp:* Res assoc surface chem, Brown Univ, 62-64; instr chem, 64-65, from asst prof to assoc prof, 65-75, PROF CHEM, FAIRFIELD UNIV, 75- *Mem:* Am Phys Soc. *Res:* Theories of formation of very thin oxide films; contact potential studies of oxidizing metal surfaces; adsorption on the dropping mercury electrode; ellipsometry of metals and thin films. *Mailing Add:* Dept Chem Fairfield Univ Fairfield CT 06430

BOGGS, DANE RUFFNER, b Orton, WVa, Apr 21, 31; m; c 4. HEMATOLOGY. *Educ:* Univ Va, BA, 52, MD, 56; Am Bd Internal Med, dipl, 63. *Prof Exp:* Intern, Univ Va Hosp, 57; clin assoc, Nat Cancer Inst, 57-59; asst resident med, Col Med, Univ Utah, 59-60, fel clin hemat, 60-61, from instr to asst prof med, 61-67; assoc prof, Med Sch, Rutgers Univ, 67-69; PROF MED, SCH MED, UNIV PITTSBURGH, 69- *Concurrent Pos:* Leukemia Soc scholar, 62-65; fac res assoc, Am Cancer Soc, 65-70. *Mem:* AAAS; Am Soc Clin Res; Am Soc Hemat; Reticuloendothelial Soc; Soc Exp Biol & Med. *Mailing Add:* 4055 Rodeo News Way Missoula MT 59803

BOGGS, GEORGE JOHNSON, b Beckley, WVa, Mar 2, 49. PSYCHOACOUSTICS, ENGINEERING PSYCHOLOGY. *Educ:* Marshall Univ, BS, 72, MS, 74; Purdue Univ, PhD(psychophysics), 81. *Prof Exp:* Psychologist, WVa Dept Health, Nicholas Ct Ment Health, 74-75, Cabell Ct Bd Educ, 75-77; mem tech staff, 81-84, SR MEM TECH STAFF, GTE LABS, INC, 85- *Concurrent Pos:* Sr Fulbright res fel, Univ Nottingham, Eng, 84-85; Adj fac eng psychol, Northeastern Univ, 84- *Mem:* Acoust Soc Am; Soc Eng Psychologists; Am Psychol Asn; Asn Comput Mach; Sigma Xi; NY Acad Sci. *Res:* Mathematical and computer models of auditory signal processing; psychophysics of complex sounds and speech. *Mailing Add:* Bldg No 40 2850 Golf Rd Rolling Meadows IL 60008

BOGGS, JAMES ERNEST, b Cleveland, Ohio, June 9, 21; m 48; c 3. THEORETICAL CHEMISTRY, STRUCTURAL CHEMISTRY. *Educ:* Oberlin Col, BA, 43; Univ Mich, MS, 44, PhD(phys chem), 53. *Prof Exp:* Res chemist, Linde Air Prod Co, NY, 44-46; asst prof chem, Eastern Mich Univ, Mich, 48-52; instr, Univ Mich, 52-53; PROF CHEM, UNIV TEX, AUSTIN, 53- *Mem:* Am Chem Soc; Am Phys Soc. *Res:* Molecular structure and dynamics; microwave spectroscopy; theoretical chemistry. *Mailing Add:* Dept Chem Univ Tex Austin TX 78712

BOGGS, JAMES H(ARLOW), b Beaumont, Tex, Dec 10, 21; m 43; c 3. MECHANICAL ENGINEERING. *Educ:* Okla State Univ, BS, 43, MS, 48; Purdue Univ, PhD(thermo & heat transfer), 53. *Prof Exp:* Jr engr, Airplane Div, Curtiss-Wright Corp, 43; from instr to prof mech eng, Okla State Univ, 43-57, head sch mech eng, 58-65, dean grad sch, 64-66, vpres acad affairs, 66-77, PROF MECH ENG, OKLA STATE UNIV, 77-, RES COORDR, 69-, VPRES ACAD AFFAIRS & RES, 77- *Concurrent Pos:* Mem, Nat Adv Res Resources Coun, 70-74; ed, Mech Eng News. *Mem:* fel Am Soc Mech Engrs. *Res:* Graduate education. *Mailing Add:* Whitehurst Hall Okla State Univ Stillwater OK 74074

BOGGS, JOSEPH D, b Bellefontaine, Ohio, Dec 31, 21. PATHOLOGY. *Educ:* Ohio Univ, AB, 41; Jefferson Med Col, MD, 45. *Prof Exp:* Instr & assoc path, Harvard Univ, 46-50; asst prof, 50-55, PROF PATH, NORTHWESTERN UNIV, CHICAGO, 55- *Concurrent Pos:* Dir labs, Children's Hosp, Chicago. *Mem:* Fel Am Soc Clin Pathologists; fel Col Am Pathologists. *Res:* Pediatric pathology. *Mailing Add:* Children's Mem Hosp 2300 Childrens Plaza Chicago IL 60614

BOGGS, LAWRENCE ALLEN, b Spokane, Wash, June 14, 10; m 47; c 1. CHEMISTRY OF CARBOHYDRATES, ISOLATION OF COMPONENTS FROM MIXTURES. *Educ:* Univ Hawaii, BS, 38, MS, 39; Univ Minn, PhD(biochem), 51. *Prof Exp:* Asst chemist, Pac Chem & Fertilizer Co, Honolulu, 39-42; chemist, Pearl Harbor Navy Yard, Hawaii, 42-43; res chemist, 51-74; RES CHEMIST, INST PAPER CHEM, APPLETON, WI, 74- *Mem:* Am Chem Soc; AAAS. *Res:* Composition of fructosan from Hawaii Ti root; analyses of sugar components of spent sulfite liquor (paper pulp by-products); reactive solvent extraction; selective precipitation of non-sugars; isolation of sugar components of spent sulfite liquor. *Mailing Add:* 1705 Ravinia Pl Appleton WI 54915

BOGGS, PAUL THOMAS, b Columbus, Ohio, Apr 14, 44; m 66; c 2. NONLINEAR OPTIMIZATION, INTERIOR-POINT METHODS. *Educ:* Univ Akron, BS, 66; Cornell Univ, PhD(computer sci), 70. *Prof Exp:* Asst prof computer sci, Univ Kans, 70-72; asst prof math, Rensselear Polytech Inst, 72-75; mathematician, Math Div, US Army Res Off, 75-78, assoc dir, 78-82; computer scientist, 82-86, CHIEF, APPL & COMPUTATIONAL MATH DIV, NAT INST STANDARDS & TECHNOL, 86- *Concurrent Pos:* Adj prof, Univ NC, 76-82, NC State Univ, 76-82, Duke Univ, 78-82, Univ Colo, 82-85; assoc ed, Trans Math Software, Asn Comput Mach, 81-, Appl Math Letters, 87-, J Optimization, Soc Indust & Appl Math, 91-; chmn, activ group optimization, Soc Indust & Appl Math, 86-89. *Mem:* Soc Indust & Appl Math; Asn Comput Mach; Math Prog Soc. *Res:* Numerical methods for the solution of optimization problems, in particular, methods for constrained and unconstrained nonlinear optimization problems, and interior-point methods for large-scale linear and quadratic programming problems. *Mailing Add:* Nat Inst Standards & Technol A-238 Gaithersburg MD 20899

BOGGS, ROBERT WAYNE, b St Helena, Calif, Sept 17, 41; m 67; c 2. FOOD SCIENCE & TECHNOLOGY. *Educ:* Fresno State Col, BS, 64; Univ Calif, Davis, PhD(nutrit & biochem), 70. *Prof Exp:* Mem staff nutrit res, 70-73, sect head, 73-76, assoc dir food & nutrit res, 76-83, ASSOC DIR CLIN & PRECLIN RES, MIAMI VALLEY LABS, PROCTER & GAMBLE CO, 78-, DIR HEALTH CARE, 83- *Concurrent Pos:* Vis prof, Med Sch, Univ Cincinnati, 72-73. *Mem:* Am Inst Nutrit; AAAS; Am Chem Soc; Sigma Xi. *Res:* Metabolic bone research; diabetes; exploratory biological research; animal science. *Mailing Add:* Miami Valley Labs Procter & Gamble Co PO Box 398707 Cincinnati OH 45239-8707

BOGGS, SALLIE PATTON SLAUGHTER, b Roanoke, Va, Nov 25, 37; m 69; c 4. RADIOLOGY. *Educ:* Emory Univ, BA, 59; Univ Calif, Berkley, MS, 63, PhD(biophys), 67. *Prof Exp:* Fel, Nat Cancer Inst, Nat Inst Health, 67-69; res asst prof med environ, 69-76, asst prof, 76-77, res assoc prof, 77-78, ASSOC PROF, MED & RADIOTHERAPY, UNIV PITTSBURGH, 78- *Concurrent Pos:* Fel health physics, Atomic Energy Comn, 63-67; guest worker, Nat Inst Health, 64-67; fel, Leukemia Soc, 72-64, scholar, 75-80; mem adv comt blood diseases & resources, Nat Heart, Lung & Blood Inst, NIH, 76-80; mem NIH Hematology Study Sect, 86-90. *Mem:* Radiation Res Soc; Int Soc Exp Hemat; Am Soc Hemat. *Res:* Use of the murine transplant model and genetic modification to delineate mechanisms involved in leukemic transformation and regulation of normal hematopoietic stem cells in vivo. *Mailing Add:* RC514 Scaife Hall Dept Radiotherapy Sch Med Univ Pittsburgh Pittsburgh PA 15261

BOGGS, SAM, JR, b King's Creek, Ky, July 26, 28; m 52; c 3. GEOLOGY. *Educ:* Univ Ky, BS, 56; Univ Colo, PhD(geol), 64. *Prof Exp:* Explor geologist, Phillips Petrol Co, 56-61; res geologist, Jersey Prod Res Co, Exxon Res Co, NJ, 64-65; head dept Geol, 74-80, PROF GEOL SEDIMENTATION, UNIV ORE, 65- *Concurrent Pos:* Vis prof, Inst Oceanog, Nat Taiwan Univ, 72-73; res geologist, US Geol Survey, 66-81; vis prof, Res Scientists Foreign Specialists, Ocean Res Inst, Univ tokyo, Japan, 81-82; scientist residence, Argonne Nat Lab, Univ Chicago, 82. *Mem:* Fel Geol Soc Am; Am Asn Petrol Geologists; Soc Econ Paleontologists & Mineralogists. *Res:* Petrology of sedimentary rocks; sedimentology and modern sedimentation processes; marine geology. *Mailing Add:* Dept Geol Sci Univ Ore Eugene OR 97403

BOGGS, STEVEN A, b Mar 15, 46. SULFUR HEXAFLUORIDE, HIGH VOLTAGE ENGINEERING. *Educ:* Reed Col, Portland, Ore, BA, 68; Univ Toronto, MS, 69, PhD(physics), 72, MBA, 88. *Prof Exp:* Sr res physicist, Ont Hydro Res, 75-87; DIR RES, UNDERGROUND SYSTS, INC, 87- *Concurrent Pos:* Prog comt, Conf Elec Insulation & Dielectric Phenomena, Inst Elec & Electronics Engrs, 80-, mem, Substa Comt & contrib ed, Elec Insulation Mag, 90- *Mem:* Optical Soc Am; Inst Elec & Electronics Engrs Power Eng Soc; Inst Elec & Electronics Engrs Elec Insulation Soc; Inst Elec & Electronics Engrs Eng Mgt Soc. *Res:* High voltage systems for the transmission and control of electric power, especially those insulated with sulfur hexafluoride gas; underground transmission, including effects of soil thermal properties and the use of high temperature superconductors. *Mailing Add:* Underground Systs Inc 63 Pears Ave Toronto ON M5R 1S9 Can

BOGGS, WILLIAM EMMERSON, b Zanesville, Ohio, May 9, 24; m 53; c 2. PHYSICAL CHEMISTRY, ANALYTICAL CHEMISTRY. *Educ:* Denison Univ, AB, 48; Carnegie Inst Technol, BS, 49, MS, 62. *Prof Exp:* Anal chemist, State Health Dept, Ohio, 50; microchemist appl res lab, US Steel Corp, 51-58, sr technologist, 59-65, sr res engr, 65-66, assoc res consult, 66-82, res consult, 82-83; RETIRED. *Mem:* Am Chem Soc. *Res:* Oxidation of metals and alloys, tin and iron; instrumental and micro methods of chemical analysis; combustion of variable composition of gaseous fuels (coke oven gas & blast furnace gas). *Mailing Add:* 3761 Windover Rd Murrysville PA 15668

BOGITSH, BURTON JEROME, b Brooklyn, NY, Feb 9, 29; m 51; c 4. PARASITOLOGY, CYTOCHEMISTRY. *Educ:* NY Univ, AB, 49; Baylor Univ, MA, 54; Univ Va, PhD(biol), 57. *Prof Exp:* From asst prof to prof biol, Ga Southern Col, 57-63; asst prog dir, NSF, 63-64; assoc prof, 64-71, chmn dept, 72-75, 78-84, PROF BIOL, VANDERBILT UNIV, 71- *Concurrent Pos:* Instr, Med Field Serv Sch, US Army, 52-54; NIH fel, Hammersmith, London, Eng, 73; Fulbright-Hayes fel, Egypt, 83; coun mem, Am Soc Parasitol, 88-90; vis researcher, Univ Leiden, Neth, 90; prog chmn, Am Soc Zoologists, 90-92. *Mem:* Am Soc Parasitol; Am Micros Soc (vpres, 78, pres, 85); Histochem Soc; Am Soc Zoologists. *Res:* Histochemistry of host-parasite relationships; ultrastructural localization of enzymes in helminths. *Mailing Add:* Dept Biol Vanderbilt Univ Box 1733 Sta B Nashville TN 37235

BOGLE, MARGARET L, CLINICAL NUTRITION. *Educ:* Tex Women's Univ, PhD, 90. *Prof Exp:* DIR CLIN NUTRIT, ARK CHILDREN'S HOSP, 76- *Concurrent Pos:* Asst prof, Dept Pediat, Children's Hosp; chairperson, Am Dietitian Asn. *Mem:* Am Dietitians Asn. *Mailing Add:* Ark Children's Hosp 800 Marshall St Little Rock AR 72202-3519

BOGLE, ROBERT WORTHINGTON, b Chicago, Ill, Dec 25, 18; m 42; c 3. EXPERIMENTAL PHYSICS. *Educ:* Univ Mich, BS, 41, MS, 42, PhD(physics), 48. *Prof Exp:* Asst, Off Sci Res & Develop, Nat Defense Res Comt, US Navy & Manhattan Dist contracts, Univ Mich, 42-43, assoc, 43-44, physicist, 44-48; physicist, Appl Physics Lab, Johns Hopkins Univ, 48-58; sr scientist, Santa Barbara Div, Curtiss-Wright Corp, 58-60; mgr Pac Range eng, Gen Elec Co, 60-64; mem, Appl Sci Staff, Gen Res Corp, 64-69; prog dir, Astro Res Corp, 69-71; mem, Radar Anal Staff, US Naval Res Lab, 71-79; sect chief, US Dept Com Inst Telesci, 79-81; PRES, WORTH RES ASSOCS, 81- *Concurrent Pos:* Parsons fel, Johns Hopkins Univ, 56-57. *Mem:* Am Phys Soc; Am Inst Aeronaut & Astronaut; Am Geophys Union. *Res:* Experimental research in radar geophysics; applied research in physics; electronics; hypersonic fluid mechanics; system development and analysis. *Mailing Add:* 991 Skylark Dr La Jolla CA 92037

BOGLE, TOMMY EARL, b Logansport, La, Sept 4, 40; m 62; c 1. SOLID STATE PHYSICS. *Educ:* Va Polytech Inst, BS, 62; La State Univ, PhD(physics), 68. *Prof Exp:* Design engr, Tex Instruments, Inc, 62-63; from asst prof to assoc prof, 68-77, PROF PHYSICS, MCNEESE STATE UNIV, 77- *Mem:* Am Phys Soc; Am Inst Physics. *Res:* Investigation of Fermi surface of metals at liquid helium temperatures using the magneto-acoustic effect. *Mailing Add:* Dept Physics McNeese State Univ Lake Charles LA 70609

BOGNER, FRED K, b Mansfield, Ohio, July 7, 39; m 59; c 2. STRUCTURAL MECHANICS, FINITE ELEMENT STRESS ANALYSIS. *Educ:* Case Inst Technol, BS, 61, MS, 64, PhD(eng mech), 67. *Prof Exp:* Mem tech staff anal mech, Bell Tel Labs, 67-69; from asst prof to prof mech eng & res engr, 69-84, PROF & CHMN DEPT CIVIL ENGR & ENG MECH, UNIV DAYTON, 84- *Mem:* Am Inst Aeronaut & Astronaut; Am Acad Mech; Am Soc Civil Engrs; Soc Exp Stress Anal; Soc Eng Sci. *Res:* Finite element analysis; nonlinear structural analysis; plates and shells; structural dynamics; elastic-plastic structures; composite structures analysis. *Mailing Add:* Dept Civil Engr & Eng Mech Univ Dayton Dayton OH 45469-0243

BOGNER, PHYLLIS HOLT, b Middleboro, Mass, Mar 26, 30; m 56, 77; c 2. PHYSIOLOGY. *Educ:* Tufts Univ, BS, 52; Brown Univ, MSc, 54; Univ Md, PhD, 57. *Prof Exp:* Asst, Tufts Univ, 50, Brown Univ, 52-54, Columbia Univ, 54-55 & Univ Md, 55-57; instr physiol & pharmacol, Sch Med, Univ Pittsburgh, 57-61, from asst prof to assoc prof pharmacol, 61-70; assoc prof, 70-78, asst dean student affairs, 76-81, assoc, Educ Systs Sect, Ctr For Educ Develop, 70-73, PROF PHARMACOL, SCH BASIC MED SCI, COL MED, UNIV ILL, 78-, ASSOC DEAN CURRIC & STUDENT AFFAIRS, 81- *Mem:* AAAS; Am Physiol Soc; Am Soc Zoologists; NY Acad Sci. *Res:* Development of mechanisms responsible for transport of monosaccharides across the intestinal wall; developmental physiology of the gastrointestinal tract; effectiveness of audiotutorials versus the lecture-mode of teaching at the medical school level. *Mailing Add:* 13623 Gemstone Dr Sun City West AZ 85375

BOGOCH, SAMUEL, b Saskatoon, Sask, Jan 13, 28; m 53. BIOCHEMISTRY, PSYCHIATRY. *Educ:* Univ Toronto, MD, 51; Harvard Univ, PhD(biochem), 56. *Prof Exp:* Intern, Toronto Gen Hosp, Ont, 51-52; resident psychiat, Crease Clin, Vancouver, BC & McLean Hosp, Mass, 52-53; resident psychiat & asst psychiatrist, Boston Psychopath Hosp Community Clin, 56-57, dir neurochem res lab, 56-61; asst res prof psychiat, 61-65, ASSOC RES PROF BIOCHEM & PSYCHIAT, SCH MED, BOSTON UNIV, 65-, MEM FAC GRAD SCH, 64-; DIR, FOUND RES NERV SYST, 61- *Concurrent Pos:* Fel, Harvard Med Sch, 56-57; sr psychiatrist, Mass Ment Health Ctr, 57-60; from asst to instr, Harvard Med Sch, 57-61; gen dir, Dreyfus Med Found, 68-, exec vpres. *Mem:* Am Psychiat Asn; Am Acad Neurol; Int Soc Neurochem; Am Soc Neurochem. *Res:* Biochemistry, physiology and clinical disorders of central nervous system. *Mailing Add:* AOMega A Bermuda Bio Sta St George 00100 Bermuda

BOGORAD, LAWRENCE, b Tashkent, Russia, Aug 29, 21; nat US; m 43; c 2. PLANT PHYSIOLOGY, MOLECULAR BIOLOGY. *Educ:* Univ Chicago, SB, 42, PhD(plant physiol), 49. *Hon Degrees:* MA, Harvard Univ, 67. *Prof Exp:* Instr bot, Univ Chicago, 48-51, from asst prof to assoc prof, 53-61, prof, 61-67; chmn dept, 74-76, prof biol, 67-80, MARIA MOORS CABOT PROF BIOL, HARVARD UNIV, 80- *Concurrent Pos:* Nat Res Coun-Merck fel, Rockefeller Inst, 51-53; Fulbright res scholar & NSF sr fel div plant indust, Commonwealth Sci & Indust Res Orgn, Australia, 60; NSF sr fel biochem, Nobel Inst Med, Sweden, 61; NIH career award, 63-67; chmn sect molecular bot, Int Bot Cong, Seattle, 68; chmn, Gordon Conf Pyrrolic Compounds, 70 & Gordon Conf Plant Molecular Biol, 80; chmn bot sect, Nat Acad Sci, 74-77; dir, Maria Moors Cabot Found, 75-87; ed-in-chief, Proceedings Nat Acad Sci, 91; coun mem, Nat Acad Sci, 89-92, Am Soc Cell Biol, 71-74. *Honors & Awards:* Quantrell Award, Univ Chicago, 60; Stephen Hales Award, Am Soc Plant Physiol, 82. *Mem:* Nat Acad Sci; Am Soc Plant Physiol (pres, 68); fel Am Acad Arts & Sci; Am Soc Cell Biol; Am Soc Biol Chem; Soc Devel Biol (pres, 83-84); AAAS (bd dirs, 82-85, pres, 86-87, chmn bd dirs, 87-88); Am Philos Soc; foreign mem Royal Danish Acad Arts & Sci. *Res:* Plant molecular developmental biology; chloroplasts; eukaryotic organelle biology; molecular biology of chloroplasts and blue-green algae. *Mailing Add:* Biol Labs Harvard Univ 16 Divinity Ave Cambridge MA 02138

BOGOROCH, RITA, b Montreal, Que. TOXICOLOGY. *Educ:* McGill Univ, BS, MS, 51; Univ Ottawa, PhD(anat & histol), 72. *Prof Exp:* Prof & res, Harvard Med Sch, 51-58, Sch Med, Univ Calif, 58-73; res mgr toxicol, Nat Health & Welfare, Ottawa, Can, 73-79; dir health effects, Nat Coun Paper Indust, 79-83; prin toxicologist, Arco Chem Co, 83-87; CONSULT, OCCUP HEALTH & MED RES, 87- *Mem:* Am Soc Cell Biol. *Mailing Add:* 300 E 75th St No 18H New York NY 10021

BOGOSIAN, GREGG, GENETICS. *Educ:* Purdue Univ, PhD(biochem), 83. *Prof Exp:* Postdoctoral fel, Univ NC, 83-85; sr res biologist, 85-87, res specialist, 87-90, RES FEL, MONSANTO CORP, 90- *Mem:* Fed Am Soc Exp Biol; Am Soc Microbiol. *Mailing Add:* Monsanto Corp BB3M 700 Chesterfield Village Pkwy Chesterfield MO 63198

BOGUCKI, RAYMOND FRANCIS, b Wallingford, Conn, Mar 6, 28; m 54; c 5. INORGANIC CHEMISTRY. *Educ:* Col Holy Cross, BS, 53, MS, 54; Clark Univ, PhD(chem), 59. *Prof Exp:* Instr chem, Lafayette Col, 54-56; res fel, Clark Univ, 59; from asst prof to assoc prof, Boston Col, 59-68; chmn dept, 68-78, PROF CHEM, UNIV HARTFORD, 68- *Concurrent Pos:* Vis prof, Middle East Tech Inst, Ankara, 64-65, Purdue Univ, 66-68, Tex A&M Univ, 74-75 & NC State Univ, 82. *Mem:* Am Chem Soc; Sigma Xi. *Res:* Reactions of metal chelate compounds in aqueous solution; polynuclear metal chelates; metal coordination polymers. *Mailing Add:* Dept Chem Univ Hartford West Hartford CT 06117

BOGUE, DONALD CHAPMAN, b St Petersburg, Fla, Jan 9, 32. CHEMICAL ENGINEERING. *Educ:* Ga Inst Technol, BChE, 53; Univ Del, PhD(chem eng), 60. *Prof Exp:* Process engr, Esso Standard Oil Co, La, 53-54; from asst prof to prof chem eng, 60-76, ALUMNI DISTINGUISHED SERV PROF CHEM ENG, UNIV TENN, KNOXVILLE, 76- *Concurrent Pos:* Vis lectr, Kyoto Univ; consult, Tenn Eastman Co, 70- *Mem:* Soc Rheol; Am Inst Chem Engrs; Am Chem Soc; Am Soc Eng Educ. *Res:* Non-Newtonian fluids and general rheological problems. *Mailing Add:* Dept Mat Sci Chem Eng Univ Tenn Knoxville TN 37996

BOGUSLASKI, ROBERT CHARLES, b Grand Rapids, Mich, Feb 22, 41; m 71; c 2. IMMUNOCHEMISTRY, CLINICAL CHEMISTRY. *Educ:* Aquinas Col, BS, 62; Univ Notre Dame, PhD(org chem), 66. *Prof Exp:* Fel, Univ Calif Med Ctr San Francisco, 66-67; res scientist, 67-72, sr res scientist, 72-75, mgr Immunochem Lab, 75-76, dir, 76-81, VPRES RES & DEVELOP, AMES DIV, MILES LABS, INC, 81- *Honors & Awards:* IC 100 Award, Indust Res Coun, 81. *Mem:* AAAS; Am Chem Soc; Sigma Xi. *Res:* Development of novel, convenient analytical methods for use in the clinical laboratory, with special emphasis on homogeneous, now-radioisotopic immunoassay methods. *Mailing Add:* 51783 Winding Waters Lane N Elkhart IN 46514

BOGYO, THOMAS P, b Budapest, Hungary, July 14, 18; US citizen; m 46; c 5. GENETICS, STATISTICS. *Educ:* Jozsef Nador Univ, Budapest, DSc(agr bot) 41; NC State Univ, MES, 63. *Prof Exp:* Plant breeder, Hungarian Plant Breeding Co, 41-45; asst prof genetics, Jozsef Nador Univ, Budapest, 45-46; asst dir, Hungarian Inst Genetics, 46-47; sr prof officer genetics, SAfrican Dept Agr, 56-58, chief prof officer genetics & agron, 56-60; asst statistician, Wash State Univ, 62-65, assoc statistician, 65-70, from asst prof to assoc prof bot & info sci, 64-66, from assoc prof to prof genetics, 66-82, statistician, 70-82, EMER PROF GENETICS & STATISTICIAN, WASH STATE UNIV, 82- *Mem:* Biomet Soc; Genetics Soc Am. *Res:* Genotype-environment interaction; simulation of genetic systems on electronic computers; experimental designs; selection. *Mailing Add:* 1600 NE Upper Dr Pullman WA 99163

BOHACEK, PETER KARL, b Olomouc, Czech, Mar 4, 36; US citizen; m 60; c 4. ENGINEERING. *Educ:* Yale Univ, BE, 58, MEng, 59, PhD(elec eng), 63. *Prof Exp:* Instr elec eng, Mass Inst Technol, 62-63, asst prof, 63-64, Ford fel, 62-64; mem tech staff, Bell Tel Labs, Inc, 64-68, dept head, 68-80, dir, 80-83; dir, 83-86, VPRES AT&T-NJ, 86- *Mem:* Inst Elec & Electronics Engrs; Sigma Xi. *Res:* Synchronization of coherent detectors; distortion of signals passing through plasma; signal processing; detection theory; real time data processing and system design; switching systems engineering; network planning. *Mailing Add:* AT&T Crawfords Corners Rd Rm 4M612 Holmdel NJ 07733

BOHACHEVSKY, IHOR O, b Sokal, Ukraine, Sept 7, 28; US citizen; div; c 9. FLUID MECHANICS, APPLIED MATHEMATICS. *Educ:* NY Univ, BAE, 56, PhD(math), 61. *Prof Exp:* Assoc res scientist, NY Univ, 60-63, adj asst prof aeronaut, 61-63; aerodyn res engr, Cornell Aeronaut Lab, 63-66; prin res scientist, Avco-Everett Res Lab, 66-68; staff mem, Bell Tel Lab, Inc, 68-75; staff mem, Los Alamos Nat Lab, 75-87; STAFF SCIENTIST, ROCKETDYNE, DIV ROCKWELL INT, 87- *Honors & Awards:* W O Bryans Medal, 56; Shewell Prize, 86; T Saaty Prize, 89. *Mem:* Soc Indust & Appl Math; Am Inst Aeronaut & Astronaut; Sigma Xi; NY Acad Sci. *Res:* Magnetohydrodynamics; numerical methods in fluid mechanics; nonlinear wave propagation; geophysical fluid mechanics; mathematical economics; gas dynamics; problems related to laser applications and to laser initiated fusion reactor development; advanced weapons concepts; nonlinear optimization theory. *Mailing Add:* 22204 Victory Blvd, Apt B-314 Woodland Hills CA 91367

BOHACS, KEVIN MICHAEL, b Port Chester, NY, Oct 2, 54. PALEOENVIRONMENTAL ANALYSIS. *Educ:* Univ Conn, BSc, 76; Mass Inst Technol, DSc, 81. *Prof Exp:* RES GEOLOGIST, EXXON PROD RES CO, 81- *Mem:* Soc Econ Paleontologists & Mineralogists; Sigma Xi. *Res:* Process oriented sedimentology; physical and sequence stratigraphy; operation of sediment-transport processes in depositional environments: record they leave, and how the record in the rocks may be used to interpret paleoenvironments. *Mailing Add:* 9550 Elle Lee Lane-2157 Houston TX 77063

BOHANDY, JOSEPH, b Barnesville, Ohio, July 9, 38; m 59; c 2. SOLID STATE PHYSICS, OPTICAL SPECTROSCOPY. *Educ:* Ohio State Univ, BS, 60, MS, 61, PhD(physics), 65. *Prof Exp:* SR PHYSICIST, APPL PHYSICS LAB, JOHNS HOPKINS UNIV, 65- *Mem:* Am Phys Soc. *Res:* Electron paramagnetic resonance used for study of magnetism and magnetic properties of crystalline solids; optical spectroscopy of biologically significant molecules. *Mailing Add:* Appl Physics Lab Johns Hopkins Univ Johns Hopkins Rd Laurel MD 20707

BOHANNON, RICHARD WALLACE, b Salisbury, NC, Dec 29, 53; m 89; c 1. PHYSICAL THERAPY. *Educ:* Univ NC, Chapel Hill, BS, 77, MS, 82; NC State Univ, EdD(adult educ), 88. *Prof Exp:* Chief phys therapist, Univ Ill Hosp, 82; chief phys therapist, SE Regional Rehab Ctr, 83-87; ASSOC PROF PHYS THER, UNIV CONN, 87- *Concurrent Pos:* Ed-in-chief, J Human Muscle Performance, assoc ed, Phys Ther & asst ed, Clin Rehab. *Honors & Awards:* Res Award, Neurol Sect, Am Phys Ther Asn, 90. *Mem:* Am Phys Ther Asn; Soc Res Rehab; Am Acad Cerebral Palsy & Develop Med; Am Congress Rehab Med. *Res:* Human movement studies; muscular function in patients with stroke; measurement of muscular performance. *Mailing Add:* Sch Allied Health U-101 Univ Conn Storrs CT 06269

BOHANNON, ROBERT ARTHUR, b Holton, Kans, Feb 5, 22; m 46; c 3. SOIL SCIENCE. *Educ:* Mich State Univ, BS, 49; Kans State Univ, MS, 51; Univ Ill, PhD, 57. *Prof Exp:* County agent, Nemaha County, Kans, 51-52; exten specialist & agronomist crops & soils, 52-55, 56-61, asst to dean agr, 61-65, actg head dept agron, 65-66, dir int agr progs, 66-68, assoc prof, 51-61, dir exten serv, 68-76, PROF AGRON, KANS STATE UNIV, 61-, EXTEN SPECIALIST, SOIL & WATER CONSERV, 76- *Concurrent Pos:* Fel, Univ Mich, 64-65; chmn exten comt, Great Plains Agr Coun, 71. *Mem:* Am Soc Agron; Soil Sci Soc Am. *Res:* Soil testing, soil fertility, and reduced tillage. *Mailing Add:* 210 York Rd Greenville NC 27858-5601

BOHANNON, ROBERT GARY, b Hollywood, Calif, Feb 17, 49; m 70. GEOLOGY. *Educ:* Glendale Col, AA, 69; Calif State Univ, Northridge, BS, 71; Univ Calif, Santa Barbara, PhD(geol), 76. *Prof Exp:* Teaching asst geol, Univ Calif, Santa Barbara, 71-74; GEOLOGIST, DEPT INTERIOR, US GEOL SURV, 74- *Mem:* Geol Soc Am. *Res:* Structural geology, geochemistry and sedimentology of the Muddy Mountains, southern Nevada; occurences of lithium in the western United States. *Mailing Add:* Geol Survey Mail Stop 999 345 Middlefield Rd Menlo Park CA 94025

BOHART, RICHARD MITCHELL, b Palo Alto, Calif, Sept 28, 13; m 39. ENTOMOLOGY. *Educ:* Univ Calif, BS, 34, MS, 35, PhD(entom), 38. *Prof Exp:* Asst, Univ Calif, 35-37; assoc entom, Univ Calif, Los Angeles, 38-39, instr & jr entomologist, Exp Sta, 39-46; from asst prof to assoc prof, 46-58, from vchmn to chmn dept, 57-67, prof, 58-80, EMER PROF ENTOM, COL AGR, UNIV CALIF, DAVIS, 80- *Mem:* Am Entom Soc. *Res:* Taxonomy of Strepsiptera Vespidae Chrysididae and Sphecidae; biology; taxonomy and control of mosquitoes. *Mailing Add:* Div Entom Univ Calif Davis CA 95616

BOHEN, JOSEPH MICHAEL, b Philadelphia, Pa, Feb 8, 46; m 69; c 2. ORGANIC CHEMISTRY. *Educ:* Temple Univ, AB, 68; Univ Pa, PhD(chem), 73. *Prof Exp:* Proj leader, 73-80, group leader, 80-83, VENTURE MGR, PENNWALT CORP, 83- *Concurrent Pos:* Instr, Univ Pa, 74-75. *Mem:* Am Chem Soc; Soc Plastics Engrs. *Res:* Organometallic compounds; heterocumulenes; cycloaddition reactions of heterocumulenes; photochemistry. *Mailing Add:* 540 Morwyck Dr King of Prussia PA 19406

BOHINSKI, ROBERT CLEMENT, b Wilkes-Barre, Pa, Apr 11, 40; m 61; c 3. BIOCHEMISTRY. *Educ:* Scranton Univ, BS, 61; Pa State Univ, MS, 63, PhD(biochem), 65. *Prof Exp:* Res biochemist, Charles Pfizer & Co, Inc, 65-66; from asst prof to assoc prof chem, 71-74, PROF CHEM, JOHN CARROLL UNIV, 74- *Honors & Awards:* Distinguished Fac Award, John Carroll Univ, 73. *Mem:* AAAS; Sigma Xi. *Res:* Growth and survival characteristics of bacteria under conditions of phosphate, sulfate and nitrogen deficiencies with specific emphasis on sulfur metabolism and turnover of ribosomes. *Mailing Add:* Dept Chem John Carroll Univ University Heights Cleveland OH 44118

BOHL, ROBERT W, b Peoria, Ill, Sept 29, 25; m 47; c 4. METALLURGICAL & NUCLEAR ENGINEERING. *Educ:* Univ Ill, BS, 46, MS, 49, PhD(metall eng), 56. *Prof Exp:* From instr to assoc prof metall eng, 46-59, EMER PROF METALL & NUCLEAR ENG, UNIV ILL, URBANA-CHAMPAIGN, 59- *Concurrent Pos:* Partic Atomic Energy Comn-Am Soc Eng Educ nuclear eng inst, Argonne Nat Lab, 57, 58, 60 & 62, Ames Lab, 58 & Hanford Lab, 62; res assoc, Argonne Nat Lab, 61, 68 & 69 & Battelle Mem Inst, 63, 64, 66 & 67. *Mem:* Fel Am Soc Metals; Am Inst Mining, Metall & Petrol Engrs; Am Soc Eng Educ. *Res:* Thermodynamics of metal alloy systems; physical metallurgy of materials for nuclear engineering; failure analysis. *Mailing Add:* 206 MMB Univ Ill 1304 W Green St Urbana IL 61801

BOHLEN, HAROLD GLENN, b Raleigh, NC, Oct 11, 46; m 70; c 2. CARDIOVASCULAR PHYSIOLOGY. *Educ:* Appalachian State Univ, BA, 68; Bowman Gray Sch Med, PhD(physiol), 73. *Prof Exp:* Am Heart Asn fel physiol, Univ Ariz, 73-75; ASST PROF PHYSIOL, IND UNIV, INDIANAPOLIS, 75- *Mem:* Am Heart Asn; Fedn Am Soc Exp Biol. *Res:* Neural and local vascular control of the intestinal microcirculation. *Mailing Add:* Dept Physiol Ind Univ Sch Med 635 Barnhill Dr Indianapolis IN 46223

BOHLEN, STEVEN RALPH, b Indianapolis, Ind, July 12, 52. METAMORPHIC PETROLOGY, EXPERIMENTAL PETROLOGY. *Educ:* Dartmouth Col, AB, 74; Univ Mich, MS, 77, PhD(geol), 79. *Prof Exp:* Res geochemist, Univ Calif, Los Angeles, 79-82; assoc prof geochem, Dept Earth & Space Sci, State Univ NY, Stony Brook, 82-89; CONSULT PROF, DEPT GEOL, STANFORD UNIV, 89- *Concurrent Pos:* Ed, Am Mineralogist, 90-95. *Mem:* Mineral Soc Am; Am Geophys Union; Sigma Xi. *Res:* Metamorphic, experimental and field petrology; thermodynamic properties of minerals; phase equilibria; mineralogy, crystallography, geochemistry and evolution of the earth's crust. *Mailing Add:* US Geol Surv 345 Middlefield Rd Menlo Park CA 94025

BOHLEN, WALTER FRANKLIN, b Tarrytown, NY, June 21, 38; m 67; c 4. PHYSICAL OCEANOGRAPHY. *Educ:* Univ Notre Dame, BSEE, 60; Mass Inst Technol, PhD(oceanog), 69. *Prof Exp:* Res asst geophys, Woods Hole Oceanog Inst, 62-63; engr electronics, Robert Taggert Inc, 63-65; res asst oceanog, Woods Hole Oceanog Inst, 65; res asst geol, Mass Inst Technol, 65-69; asst prof, 69-77, ASSOC PROF OCEANOG, UNIV CONN, 77- *Concurrent Pos:* Mem res planning adv comt, Long Island Sound Study, New Eng River Basins Comn, 72-75. *Mem:* Am Geophys Union; Marine Technol Soc; Sigma Xi. *Res:* Experimental investigations of sediment transport in coastal waters with particular emphasis on the factors governing variability, the relationship between laboratory and field data and the development of quantitative transport models. *Mailing Add:* Marine Sci Inst Univ Conn Avery Point Groton CT 06340

BOHLIN, JOHN DAVID, b Valparaiso, Ind, Mar 19, 39; m 66; c 2. SOLAR PHYSICS. *Educ:* Wabash Col, AB, 61; Univ Colo, Boulder, PhD(solar physics), 68. *Prof Exp:* Res fel astron, Calif Inst Technol, 67-70; astrophysicist solar physics, Naval Res Lab, 70-76; prog scientist, 77-78, chief, Solar Physics Disc Br, 78-90, CHIEF SCIENTIST, SPACE PHYSICS DIV, NASA, HQ, 90- *Honors & Awards:* Except Serv Award, NASA, 89. *Mem:* Am Astron Soc; Int Astron Union; Am Geophys Union. *Res:* Physics of the outer atmosphere of the sun, especially as inferred from ultraviolet wave lengths observed from space vehicles; structure and dynamics of the solar corona. *Mailing Add:* Code SS NASA Hq Washington DC 20546

BOHLIN, RALPH CHARLES, b Iowa City, Iowa, Nov 26, 43. ASTROPHYSICS. *Educ:* Univ Iowa, BA, 66; Princeton Univ, MA, 69, PhD(astrophysics), 70. *Prof Exp:* Nat Res Coun fel, 69; res assoc astrophysics, Lab Atmospheric & Space Physics, Univ Colo, 70-73; ASTRONR, GODDARD SPACE FLIGHT CTR, GREENBELT, 73- *Concurrent Pos:* Co-investr, Astron Spacelab Payloads Proj, 74-77, Space Telescope Proj Faint Object Spectrog, 74- & Spacelab Ultraviolet Imaging Telescope, 79- *Mem:* Am Astron Soc; Int Astron Union; Sigma Xi. *Res:* Ultraviolet spectroscopy and ultraviolet imaging of astronomical objects, including early-type stars, planetary nebulae, diffuse nebulae and galaxies using data from sounding rockets and orbiting satellites; emphasis on the gas and dust in the interstellar medium. *Mailing Add:* 7662 Sweet Hours Way Columbia MD 21046

BOHLMANN, EDWARD GUSTAV, b Milwaukee, Wis, July 19, 17; m 47; c 3. GEOCHEMISTRY, CORROSION. *Educ:* Univ Wis, BS, 39, MS, 41. *Prof Exp:* Asst sanit chem, Univ Wis, 39-40, asst phys chem, 40-41; chemist, Solvay Process Co, NY, 41-42; res chemist metal lab, Manhattan Proj, Univ Chicago, 43; res chemist, Oak Ridge Nat Lab, 43-48, group leader, 45-48; res engr, Battelle Mem Inst, 48-49; sr res chemist, Oak Ridge Nat Lab, 49-52, chief corrosion sect reactor exp eng div, 52-55, from asst dir to assoc dir, 55-61, assoc dir reactor chem div, 61-73, mem sr res staff chem div, 73-80;

RETIRED. *Mem:* AAAS; Am Chem Soc. *Res:* Physical chemistry; inorganic chemistry; aqueous corrosion; materials and corrosion in nuclear power reactors; molten salt reactors; silica and scaling in geothermal brines. *Mailing Add:* 27 Ona Rd Lake Tansi Village Crossville TN 38555

BOHM, ARNO, b Stettin, Poland, Apr 26, 36; m 65; c 3. HIGH ENERGY PHYSICS, THEORETICAL PHYSICS. *Educ:* Tech Univ, Berlin, DiplPh, 62; Dr rer nat, 66. *Prof Exp:* Sci asst theoret physics, Univ Karlsruhe, 63-64; Int Atomic Energy Agency res fel particle physics, Int Ctr Theoret Physics, Trieste, Italy, 64-66; res assoc, Syracuse Univ, 66-68; assoc prof particle physics, 68-78, PROF PHYSICS, UNIV TEX, AUSTIN, 78- *Concurrent Pos:* Vis prof, Chalmers Inst Technol, Sweden, Univ Brussels, Belg, Max Planck Inst, Munich, Univ Wurzburg, Ger, Los Alamos Nat Lab, NMex, Inst High Energy Physics, Serpukhov, USSR; IAEA fel, Italy, 65; Mortimer & Raymond Sackler fel, Israel, 80; Fulbright Award, 87. *Honors & Awards:* Humboldt Prize, 81. *Mem:* Fel Am Phys Soc. *Res:* Theoretical and elementary particle physics; mathematical physics; quantum physics. *Mailing Add:* Dept Physics Univ Tex Austin TX 78712

BOHM, BRUCE ARTHUR, b Burlington, Wis, Apr 9, 35; Can citizen; m 61; c 2. BOTANY. *Educ:* Alfred Univ, BS, 56; Univ RI, MS, 58, PhD(chem), 60. *Prof Exp:* Fel bot, McGill Univ, 60-61; asst prof chem, Univ Sask, 61-63; asst res prof pharmacog, Univ RI, 63-66; from asst prof to assoc prof, 66-73, PROF BOT, UNIV BC, 73- *Mem:* Am Bot Soc; Am Soc Plant Taxonomists; Phytochem Soc NAm (pres, 69); AAAS. *Res:* Chemical plant taxonomy utilizing secondary metabolites and comparative protein structure. *Mailing Add:* Dept Bot Univ BC 2075 Wesbrook Pl Vancouver BC V6T 1W5 Can

BOHM, GEORG G A, b Brunn, Czech, Oct 7, 35; m 61; c 2. POLYMER PHYSICS. *Educ:* Univ Vienna, PhD(physics), 62. *Prof Exp:* Staff mem, Max Planck Inst Phys Chem, Gottingen, 62-64; vis prof, Int Atomic Energy Agency, Vienna, 64-65; vis scholar, Northwestern Univ, Ill, 65-67; ASST DIR RES, BRIDGESTONE/FIRESTONE INC, 67- *Mem:* Am Phys Soc. *Res:* Polymer physics: interrelationship between polymer structure morphology and physical properties of heterogenous systems; radiation chemistry; processes occurring in polymers on exposure to radiation, polymer processing. *Mailing Add:* Bridgestone/Firestone Inc Cent Res Labs 1200 Firestone Pkwy Akron OH 44317

BOHM, HENRY VICTOR, b Vienna, Austria, July 16, 29; US citizen; m 50; c 2. CONDENSED MATTER PHYSICS. *Educ:* Harvard Univ, AB, 50; Univ Ill, MS, 51; Brown Univ, PhD(physics), 58. *Prof Exp:* Jr physicist, Gen Elec Co, 51 & 53-54; staff mem, Arthur D Little, Inc, 58-59; assoc prof physics, 59-64, actg chmn dept, 62-63, vpres grad studies & res, 68-71, vpres spec projs, 71-72, provost, 72-75, dean fac, 84, interim dean, Col Lib Arts, 84-86, PROF PHYSICS, WAYNE STATE UNIVERSITY, 64- *Concurrent Pos:* Vis prof physics, Cornell Univ, 66-67, Purdue Univ, 77; consult, Aerospace Syst Command, US Air Force, 60-62; mem bd dir, Ctr Res Libr, Chicago, Ill, 70-75, chmn, 73; consult NCent Asn Schs & Cols, 70-80; comnr, Comn Inst High Educ, 74-78; pres, Argonne Univ Asn, 78-83. *Mem:* Am Phys Soc; AAAS. *Res:* Superconductivity. *Mailing Add:* Dept Physics Wayne State Univ Detroit MI 48202

BOHM, HOWARD A, b New York, NY, July 28, 43; m 67; c 1. MICROBIAL PESTICIDES, DRUG FORMULATIONS. *Educ:* NY Univ, BA, 66; Adelphi Univ, PhD(chem), 71. *Prof Exp:* Sr res chemist, Pennwalt Corp, 74-81; res chemist, IKapharm Div, Teva Corp, 81-85; sr res scientist, Lim Technol, 85-90; RES CHEMIST, CHEVRON CORP, 90- *Mem:* Am Chem Soc; Controlled Release Soc; Entom Soc Am. *Res:* Time release systems for food, drugs, and pesticides; microbial pesticides; microencapsulations; photochemistry; cyclodextrins; tableting; drug synthesis; analysis; polymers; sunscreening agents; liposomes. *Mailing Add:* 4339 Solano Rd Fairfield CA 94533

BOHMAN, VERLE RUDOLPH, b Peterson, Utah, Dec 29, 24; m 45; c 5. ANIMAL NUTRITION. *Educ:* Utah State Agr Col, BS, 49, MS, 51; Cornell Univ, PhD(animal nutrit), 52. *Prof Exp:* From asst prof & asst animal nutritionist to assoc prof animal husb & assoc animal nutritionist, 52-63, chmn div animal sci, 60-78, PROF ANIMAL HUSB & NUTRIT, UNIV NEV, 63- *Concurrent Pos:* Ed-in-chief, J Animal Sci, 70-72. *Honors & Awards:* Distinguished Serv Award, Am Soc Animal Sci, 84. *Mem:* Fel AAAS; Am Inst Nutrit; fel Am Soc Animal Sci (pres-elect, 73-74, pres, 74-75); Am Dairy Sci Asn. *Res:* Range livestock nutrition; antibiotics; fat, hormones, protein and mineral metabolism; radioactive fallout. *Mailing Add:* 916 Sbragia Way Sparks NV 89431

BOHME, DIETHARD KURT, b Boston, Mass, June 20, 41; m 66; c 3. PHYSICAL CHEMISTRY, CHEMICAL PHYSICS. *Educ:* McGill Univ, BSc, 62, PhD(phys chem), 65. *Prof Exp:* Nat Res Coun Can Overseas fel & hon res asst physics, Univ Col, Univ London, 65-67; resident res assoc aeronomy, Aeronomy Lab, Environ Sci Serv Admin, 67-69; from asst prof to assoc prof, 70-77, dir, Grad Prog Chem, 79-85, chmn dept chem, 85-90, PROF CHEM, YORK UNIV, 77- *Concurrent Pos:* Res assoc, Nat Acad Sci, Nat Res Coun, 67-69; A P Sloan Found fel, 74-76; sr vis fel, Sci Res Coun Eng, 78-79; mem, Grant Selection Comt, Nat Sci Eng Res Coun Can, 83-86; A von Humboldt res award, 90. *Honors & Awards:* Rutherford Mem Medal, Royal Soc Can, 81; Noranda Lect Award, Chem Inst Can, 83. *Mem:* Fel Chem Inst Can; Combustion Inst; Am Soc Mass Spectrometry. *Res:* Gas-phase kinetics and energetics of ion-molecule reactions; silicon-ion chemistry; ion chemistry in dense interstellar gas clouds; circumstellar envelopes, hydrocarbon flames and planetary ionospheres; gas-phase ion-induced polymer chemistry. *Mailing Add:* Dept Chem York Univ North York ON M3J 1P3 Can

BOHMER, HEINRICH EVERHARD, b Munster, Ger, May 6, 31; m 62; c 1. PHYSICS. *Educ:* Univ Munster, dipl, 59, PhD(diffusion in metals), 61. *Prof Exp:* Asst solid state physics, Inst Metal Res, Univ Munster, 62; res assoc surface physics, 62-63 & plasma physics, 63-65, res asst prof, 65-69, asst prof physics, Coord Sci Lab, Univ Ill, Urbana, 69-76; ASSOC RES PHYSICIST, UNIV CALIF, IRVINE, 76-; MEM TECH STAFF, TRW SYSTS, REDONDO BEACH, CALIF, 76- *Concurrent Pos:* Vis physicist, Dept Plasma Physics, Fontenay Nuclear Res Ctr, Fontenay, France, 72-73 & Dept Physics, Univ Calif, Irvine, 75- *Mem:* Am Phys Soc. *Res:* Plasma physics. *Mailing Add:* Dept Physics Attn Bohmer Univ Calif Irvine CA 92717-4575

BOHMONT, DALE W, b Wheatland, Wyo, June 7, 22; m 69; c 2. PLANT SCIENCE, ACADEMIC ADMINISTRATION. *Educ:* Univ Wyo, BS, 48, MS, 50; Univ Nebr, PhD, 52; Harvard Univ, MPA, 59. *Prof Exp:* Asst agronomist & asst prof agron, Univ Wyo, 48-53; assoc agronomist & head dept, 53-55, agronomist, prof & head dept, 55-60, head, Div Plant Sci, 59-61, assoc dir, Agr Exp Sta, Univ Colo, 61-63; prof, dean & dir, Exp Sta & Coop Exten Serv, Max C Fleischmann Col Agr, Univ Nev, 63-82; PRES BOHMONT CONSULT INC, 82- *Concurrent Pos:* Asst agronomist, USDA, Univ Nebr, 50-51; conserv fel, Harvard Univ, 58-59; sr agr consult, Develop & Resources Corp NY, 68-74; pres, Enide Corp Nev, 74-80; consult, TAMS Agr Develop Group, 75-78, Nev Agr Found, 82-, Brucheum-Wayneborough, Va, 84- *Mem:* AAAS; fel Am Soc Agron; fel Am Acad Sci; fel Plant Sci Soc; Weed Soc. *Res:* Crops, soils, weed control and public administration. *Mailing Add:* Box 3032 Reno NV 89505

BOHN, HINRICH LORENZ, b New York, NY, Aug 25, 34; m 62; c 3. SOIL CHEMISTRY. *Educ:* Univ Calif, Berkeley, BS, 55, MS, 57; Cornell Univ, PhD(soil chem), 63. *Prof Exp:* fel, Univ Calif, Riverside, 63-64; res chemist, Tenn Valley Authority, 64-66; assoc prof agr chem, 66-76, PROF SOIL CHEM, UNIV ARIZ, 74-, PROF SOILS, WATER & ENG & RES SCIENTIST AGR CHEM, 76- *Mem:* Am Chem Soc; Soil Sci Soc Am. *Res:* Soil-pollutant reactions. *Mailing Add:* Dept Soil Sci Univ Ariz Tucson AZ 85721

BOHN, RANDY G, b Toledo, Ohio, Feb 11, 41; m 65; c 2. LOW TEMPERATURE PHYSICS, CRYOGENICS. *Educ:* Univ Toledo, BS, 63; Ohio State Univ, PhD(physics), 69. *Prof Exp:* From asst prof to assoc prof, 69-81, PROF PHYSICS, UNIV TOLEDO, 81- *Mem:* Am Asn Physics Teachers; Am Phys Soc. *Res:* Thermal conductivity in solid hydrogen; vitreous solids; thermal properties of solids at low temperatures. *Mailing Add:* 2801 W Bancroft Toledo OH 43606

BOHN, ROBERT K, b New York, NY, Jan 24, 39; m 64, 83; c 4. MICROWAVE SPECTROSCOPY, AIR POLLUTANT DISPOSAL. *Educ:* Univ Calif, Berkeley, BS, 59; Cornell Univ, PhD(phys chem), 64. *Prof Exp:* NATO fel electron diffraction, Univ Oslo, 64-65; asst prof, 65-71, assoc prof, 71-78, PROF CHEM, UNIV CONN, 78- *Concurrent Pos:* Vis scientist, Univ Tokyo, 71-72 & Columbia Univ, 79-81; vis prof, Univ Mich, 87-88. *Mem:* Am Chem Soc; Am Phys Soc. *Res:* Molecular structure; electron diffraction; microwave spectroscopy. *Mailing Add:* Dept Chem Univ Conn Storrs CT 06269-3060

BOHN, SHERMAN ELWOOD, b New England, NDak, Mar 11, 27; m 52; c 3. MATHEMATICS. *Educ:* Concordia Col, Moorhead, Minn, BA, 49; Univ Nebr, MA, 51, PhD(math, physics), 61. *Prof Exp:* Asst math, Univ Nebr, 49-51; res scientist, Nat Adv Comt Aeronaut, Va, 51-52; asst, Univ Minn, 52-53; from instr to asst prof, Concordia Col, Moorhead, Minn, 53-56; asst, Univ Nebr, 56-57, instr, 57-59; assoc prof, Wartburg Col, 59-61; from asst prof to assoc prof, Bowling Green State Univ, 61-64; assoc prof, 64-66, chmn dept, 65-71, PROF MATH, MIAMI UNIV, 66-, chmn dept, 73-82. *Concurrent Pos:* Chmn math sect, Ohio Acad Sci, 69-70, Math Asn Am, Ohio Sect, 71-72; bd govs, Math Asn Am, 73-76,; joint comt, Dept Chairs, Math Asn Am & Am Math Soc, 84-85. *Mem:* Math Asn Am; Am Math Soc; Soc Indust & Appl Math. *Res:* Mathematical analysis, function spaces and partial differential equations; semi-topological groups. *Mailing Add:* Dept Math Miami Univ Oxford OH 45056

BOHNENBLUST, KENNETH E, b Bala, Kans, Dec 9, 23; m 45; c 3. PLANT BREEDING, PLANT PATHOLOGY. *Educ:* Kans State Univ, BS, 59; Univ Wyo, MS, 61; Univ Minn, St Paul, PhD(plant path), 66. *Prof Exp:* Instr agron, Univ Wyo, 59-61; asst prof biol, Wis State Univ, River Falls, 66-67; Asst prof, 67-78, ASSOC PROF PLANT BREEDING, UNIV WYO, 78- *Mem:* Sigma Xi. *Res:* Potato diseases; Actinomycetes in plant disease; potato breeding; alfalfa breeding and pest resistance; virus diseases of plants; breeding cereals; morphology of cereal seedlings. *Mailing Add:* Plant Sci Div-Agr Col Box 3354 Univ Wyo Univ Sta Laramie WY 82071

BOHNERT, JANICE LEE, BIOCHEMISTRY. *Educ:* Bucknell Univ, BA, 70; Univ Wash, PhD(biochem), 81. *Prof Exp:* Chemist, NIH, Bethesda, 70-84; res asst, Dept Biochem, Univ Wash, 74-81, fel, 81-82, res scientist II, Dept Chem Eng, 84, res assoc, 84-86, res assoc, 86- 87, BIOCHEM CONSULT, CTR BIOENG, UNIV WASH, 89- *Mem:* Am Soc Biochem & Molecular Biol. *Res:* Stabilize proteins loaded in hydrogels for controlled release applications; coordinated studies of bioadhesion with new polymer development; growth of marine bacteria for incorporation into polymer gels being developed for nonfouling surfaces; one US patent; numerous publications. *Mailing Add:* 3538 NE 85th St Seattle WA 98115

BOHNING, DARYL EUGENE, b Belmond, Iowa, Sept 21, 41; c 3. BIOMATHEMATICS, PHYSIOLOGY. *Educ:* Iowa State Univ, BS, 62, PhD(physics), 66. *Prof Exp:* Fel physics, Rutgers Univ, 66-69; NIH trainee biophysics, Mass Gen Hosp & Harvard Univ, 69-70; asst prof physics, Miami Univ, 70-71; res scientist environ med, NY Univ, 71-75; asst prof & res assoc med, State Univ NY Stony Brook, 75-78; staff scientist biomath & med, Brookhaven Nat Lab, 78-; PHILIPS MED SYSTS, INC. *Mem:* Am Phys Soc; Am Asn Physicists in Med; Soc Magnetic Resonance Med. *Res:* Research and development on cardiac applications of magnetic resonance imaging and spectroscopy. *Mailing Add:* 177 Town Hill Rd Cornwall Bridge CT 06754

BOHNING, JAMES JOEL, b Cleveland, Ohio, Apr 11, 34; m; m; c 3. PHOTOCHEMISTRY, HISTORY OF CHEMISTRY. *Educ:* Valparaiso Univ, BS, 56; NY Univ, MS, 59; Northeastern Univ, PhD(phys chem), 65. *Prof Exp:* Instr chem, Wilkes Col, 59-62; res fel, Northeastern Univ, 62-64; from asst prof to prof chem, Wilkes Col, 67-90, chmn dept, 70-86, prof, 87-90, chmn, Dept Earth & Environ Sci, 87-90, EMER PROF CHEM, WILKES COL, 90-; ASST DIR, ORAL HIST, BECKMAN CTR HIST CHEM, 90- *Concurrent Pos:* Vis prof, Univ Ky, 80-81 & Lehigh Univ, 85, 86; chmn, Div Hist of Chem, Am Chem Soc, 86. *Mem:* Am Chem Soc. *Mailing Add:* Beckman Ctr Hist Chem 3401 Walnut St Philadelphia PA 19104-6228

BOHNSACK, KURT K, b Stuttgart, Ger, Mar 23, 20; nat US; m 46; c 3. ECOLOGY, ACAROLOGY. *Educ:* Ohio Univ, BS, 46; Univ Mich, MS, 47, PhD(zool), 54. *Prof Exp:* Instr zool, Univ Mich, 50-51; from instr to asst prof biol, Swarthmore Col, 51-56; from asst prof to prof zool, San Diego State Univ, 56-90, chmn dept, 64-67 & 80; RETIRED. *Concurrent Pos:* Investr, Oak Ridge Nat Lab, 56-57 & Arctic Res Lab, 62-64; mem bd dirs, San Diego Natural Hist Mus, 66-79. *Mem:* Acarol Soc Am; Sigma Xi; AAAS; Arctic Inst NAm; Audubon Soc; Soc Vector Ecol. *Res:* Entomology; acarology. *Mailing Add:* 4309 Mt Helix Highlands Dr La Mesa CA 91941

BOHON, ROBERT LYNN, b Decatur, Ill, July 20, 25; m 47; c 3. ENVIRONMENTAL SCIENCES, ANALYTICAL CHEMISTRY. *Educ:* Univ Ill, BS, 46, PhD(phys chem), 50. *Prof Exp:* Asst phys chem, Univ Ill, 49-50; phys chemist & head infrared lab, Anderson Phys Lab, 50-56; sr chemist, Minn Mining & Mfg, Co, 56-66, supvr, 66-69, mgr, 69-85, dir, Anal & Properties Res, Corp Res, 3M Environ Lab, 85-90; RETIRED. *Mem:* Am Chem Soc; Am Sci Affil. *Res:* General physical chemistry; thermodynamics of solutions; thermal analysis; solid state reaction kinetics; physical properties of polymers; material science; environmental chemistry and engineering; technical management. *Mailing Add:* 5960 Hobe Lane White Bear Lake MN 55110

BOHOR, BRUCE FORBES, b Chicago, Ill, May 4, 32; m 53; c 4. SHOCKED MINERALS, CLAY MINERALOGY. *Educ:* Beloit Col, BS, 53; Univ Ind, MS, 55; Univ Ill, PhD(geol), 59. *Prof Exp:* Res geologist, Prod Res Div, Res & Develop Dept, Continental Oil Co, Okla, 57-63, Explor Res Div, 63-65; geologist, Clay Resources & Clay Mineral Technol Sect, Ill State Geol Surv, 65-74; GEOLOGIST, US GEOL SURV, BR COAL RESOURCES, 74- *Concurrent Pos:* Ill Clay fel; G K Gilbert fel, US Geol Surv, 85-86; NASA grant T-5175P, 88. *Mem:* Soc Econ Paleont & Mineral; Meteoritical Soc; Clay Minerals Soc; AAAS; Geochem Soc; Meteoritical Soc. *Res:* Clay minerals and geochemistry in coal exploration; industrial applications of clays and clay minerals; development of clay resources; tonsteins; shocked minerals, K/T boundary and other impact events. *Mailing Add:* US Geol Surv Br Coal Resources Fed Ctr Box 25046 Mail Stop 901 Denver CO 80225-0046

BOHR, AAGE, b Copenhagen, Denmark, June 19, 22. THEORETICAL PHYSICS. *Educ:* Univ Copenhagen, PhD(physics), 54. *Prof Exp:* Res asst physics, Inst Theoret Physics, 46, PROF PHYSICS, NIELS BOHR INST, UNIV COPENHAGEN, 56- *Honors & Awards:* Nobel Prize in Physics, 75; Dannie Heineman Prize, 60; Rutherford Medal, 72; John Price Wetherill Medal, 74. *Mem:* Finnish Acad Sci; Yugoslavia Acad Sci; Am Philol Soc. *Mailing Add:* Niels Bohr Inst Univ Copenhagen Blegdamsgade 17 Copenhagen 2100 Denmark

BOHR, DAVID FRANCIS, b Zurich, Switz, June 22, 15; US citizen; m 40; c 3. CARDIOVASCULAR PHYSIOLOGY. *Educ:* Univ Mich, MD, 42. *Prof Exp:* Life Inst Med Res fel, Univ Calif, San Francisco, 46-48, res fel path, 47-48; from asst prof to assoc prof physiol, Med Sch, Univ Mich, Ann Arbor, 48-55; vis prof pharmacol, Univ Calif, San Francisco, 55-56; PROF PHYSIOL, MED SCH, UNIV MICH, ANN ARBOR, 56- *Concurrent Pos:* Vis prof physiol, Univ Heidelberg, 61-62; mem coun high blood pressure, Am Heart Asn, chmn, 78-80; distinguished prof, Univ Mich, 73. *Honors & Awards:* Ciba Award, 84. *Mem:* Am Physiol Soc (pres, 78-79). *Res:* Vascular smooth muscle; hypertension. *Mailing Add:* Univ Mich 7710 Med Sci II Ann Arbor MI 48109

BOHREN, BERNARD BENJAMIN, b Omaha, Nebr, Aug 15, 14; m 38; c 3. ANIMAL GENETICS. *Educ:* Univ Ill, BS, 37; State Col Wash, MS, 40; Kans State Col, PhD(genetics), 42. *Prof Exp:* Instr voc agr schs, Ill, 37-38; asst, State Col Wash, 38-39; asst, poultry husb, Kans State Col, 39-43; asst poultry husb & agr chem, Purdue Univ, 50-83, from assoc prof to prof genetics, 46-83; RETIRED. *Honors & Awards:* Res Award, Poultry Sci Asn, 44. *Mem:* Poultry Sci Asn; Am Genetic Asn; Genetics Soc Am. *Res:* Genetics of and selection for quantitative traits in fowl. *Mailing Add:* 2236 Camelback Blvd WL Lafayette IN 47906

BOHREN, CRAIG FREDERICK, b San Francisco, Calif, Feb 24, 40; m 64. APPLIED PHYSICS. *Educ:* San Jose State Univ, BS, 63; Univ Ariz, MS, 65 & 71, PhD(physics), 75. *Prof Exp:* Res asst nuclear eng, Gulf Gen Atomic, 65-67; res assoc hydrol, Sch Renewable Natural Resources, Univ Ariz, 72-75, lectr, Inst Atmospheric Physics, 74-75; res asst, Dept Appl Math & Astron, Univ Col Cardiff, 75-78; vis scholar, Inst Atmospheric Physics, Univ Ariz, 78-80; from asst prof to prof, 80-90, DISTINGUISHED PROF, DEPT METEOROL, PA STATE UNIV, 90- *Concurrent Pos:* Vis staff mem, Los Alamos Nat Lab, 80 & 86-87; vis scholar, Univ Ariz, 81, Dartmouth Col, 84, 85; res physicist, Cold Regions Res & Eng Lab, 84, 85; vis prof, Dartmouth Col, 86-87; ed, Appl Optics. *Honors & Awards:* Louis J Batten Author's Award, Am Meteorol Soc, 89. *Mem:* Int Glaciological Soc; Am Asn Physics Teachers; fel Optical Soc Am. *Res:* Light scattering by small particles; atmospheric physics; meteorological optics; physics of everyday phenomena. *Mailing Add:* Dept Meteorol Penn State Univ University Park PA 16802

BOHRER, JAMES CALVIN, b Evansville, Ind, Oct 27, 23; m 49; c 5. ORGANIC CHEMISTRY. *Educ:* Evansville Col, AB, 44; Cornell Univ, PhD, 51. *Prof Exp:* Sr proj chemist, Colgate-Palmolive Co, Piscataway, 50-83; RETIRED. *Concurrent Pos:* Instr, Univ Col, Rutgers Univ, 63-78. *Mem:* Am Chem Soc. *Res:* Many-membered ring compounds; detergents. *Mailing Add:* 19 Peach Orchard Dr East Brunswick NJ 08816

BOHRER, ROBERT EDWARD, b St Paul, Minn, July 24, 39; m 60; c 1. MATHEMATICS, STATISTICS. *Educ:* SDak Sch Mines & Technol, BS, 61; Univ NC, Chapel Hill, PhD(statist), 66. *Prof Exp:* Statistician, Statist Res Div, Res Triangle Inst, 64-65; vis lectr statist, Univ Col Wales, 65-66; statistician, Statist Res Div, Res Triangle Inst, 66-68; from asst prof to assoc prof math, 68-79, PROF MATH, UNIV ILL, URBANA-CHAMPAIGN, 79- *Concurrent Pos:* Res fel, Univ Kent at Canterbury, 72; vis assoc prof, Univ Calif, Berkeley, 74-75; vis scholar, Grad Theol Union, Berkeley, 74-75; acad vis, Imp Col, Univ London, 77; vis res assoc statistician, Univ Calif, Berkeley, 78-81; vis prof statist, 82-83. *Mem:* Math Asn Am; Inst Math Statist; Royal Statist Soc; fel Am Statist Asn. *Res:* Decision theory; confidence bounds; multiple inference; time series. *Mailing Add:* Dept Statist Univ Ill Urbana IL 61801

BOHRER, THOMAS CARL, b Brooklyn, NY, Nov 8, 39; m 66; c 2. CHEMICAL ENGINEERING. *Educ:* Rensselaer Polytech Inst, BChE, 60. *Prof Exp:* Jr proj engr, Org Chem Div, Am Cyanamid Co, 60-62, proj engr, 62; chem engr, Celanese Res Co, 62-64, res engr, 64-67, sr res engr, 67-69, group leader fibers res, Celanese Fibers Co, 69-71, mgr tire cord & indust yarns develop, Celanese Fibers Mkt Co, 71-73, dir spec prod develop, 73-76, dir planning, 76, vpres tech & planning, 77-78, dir corp planning, 78-79, vpres tech, 79-81, VPRES & GEN MGR ENG RESINS, CELANESE PLASTICS & SPECIALTIES CO, 81- *Mem:* Am Chem Soc; Fibers Soc. *Res:* Exploratory fiber research and development; fibers application research and development in tire cord, industrial fibers, cigarette tow, floor coverings, fiberfill and civil engineering applications of fabric; plastics research and development in engineering resins, films, plastic pipe, structural composites, emulsions, water soluble polymers, and automotive coatings. *Mailing Add:* Hoechst Celanese Corp One Main St Chatham NJ 07928

BOHRER, VORSILA LAURENE, b Chicago, Ill, Jan 22, 31. ETHNOBOTANY. *Educ:* Univ Ariz, BA, 53, PhD(bot), 68; Univ Mich, MS, 54. *Prof Exp:* Lectr biol, Univ Mass, Boston, 69-71, asst prof, 71-73; from res assoc prof to res prof, Paleoindian Inst, Eastern NMex Univ, 73-85; DIR, SOUTHWEST ETHNOBOTANICAL ENTERPRISES, 85- *Concurrent Pos:* Consult, Sch Am Res, Santa Fe, 72-79, Univ Ariz Field Sch at Grasshopper, 69-75, Puerco River Valley Archaeol Proj, 70-77, San Juan Valley Archaeol Proj, 72-80, Cent Ariz Ecotone Proj, 74-78, La Civdad, Phoenix, 82-87, Chambers Group Inc, 84- 85 & Okla Archeol Surv, Norman, 84-; head workshop in archeobotany, Univ Ariz, 84 & Southern Methodist Univ, 88. *Mem:* Soc Am Archaeol; Soc Econ Bot; Soc Range Mgt; Sigma Xi; AAAS. *Res:* Inter-relationship of plants and pre-historic man in the American Southwest; ethnobotany, palynology, plant microfossils, plant macrofossils, paleoecology. *Mailing Add:* 220 New Mexico Dr Portales NM 88130

BOHRMAN, JEFFREY STEPHEN, b Easton, Pa, Jan 19, 44; m 70; c 2. PHARMACOLOGY, PHYSIOLOGY. *Educ:* Dickinson Col, Pa, BS, 67; Univ Pittsburgh, BS, 70; Univ Ill Med Ctr, MS, 72; Univ Pac, Stockton, Calif, PhD(pharmacol, physiol), 77. *Prof Exp:* investr, Oak Ridge Nat Lab, 77-80; res pharmacologist, Nat Inst Occup Safety & Health, Cincinnati, 80-89. *Mem:* AAAS; Found Sci & Handicapped; Asn Handicapped Student Servs Prog; NY Acad Sci. *Res:* Occupational chemical carcinogenesis; mechanism of tumor promotion; occupational photosensitization agents; role of prostaglandins and other inflammatory mediators both in initial and promotional phase of tumorigenesis; mechanism of inflammation and anti-inflammation; drug screening. *Mailing Add:* 3763 Earls View Ct Cincinnati OH 45226

BOICE, JOHN D, JR, b Brooklyn, NY, Dec 20, 45; m 74; c 4. PUBLIC HEALTH & EPIDEMIOLOGY. *Educ:* Univ Tex, El Paso, BS, 67; Rensselaer Polytech Inst, MS, 68; Harvard Univ, ScM, 74, ScD, 77. *Prof Exp:* CHIEF, RADIATION EPIDEMIOL BR, NAT CANCER INST, 84- *Concurrent Pos:* Lectr epidemiol, Harvard Sch Pub Health, 79- *Honors & Awards:* Meritorious Serv Medal, USPHS, 87, Distinguished Serv Medal, 91. *Mem:* Health Physics Soc; Soc Epidemiol Res; Am Pub Health Asn; Radiation Res Soc; Am Epidemiol Soc; Am Asn Cancer Res. *Res:* Cancer-producing effects of exposure to ionizing radiation in human populations. *Mailing Add:* Nat Cancer Inst Radiation Epidemiol Br Exec Plaza N Rm 408 NIH 6130 Executive Blvd Rockville MD 20892

BOIKESS, ROBERT S, b Brooklyn, NY, Feb 2, 37; m 69; c 2. ORGANIC CHEMISTRY. *Educ:* Columbia Univ, BA, 57, MA, 60, PhD(org chem), 61. *Prof Exp:* Res fel org chem, Univ Calif, Los Angeles, 61-62; asst prof chem, State Univ NY Stony Brook, 63-68; from asst prof to assoc prof, 68-84, PROF CHEM, RUTGERS UNIV, 84- *Concurrent Pos:* NSF fel, 62-63. *Mem:* Am Chem Soc. *Res:* Small rings; valency tautomerism; bicyclics; photochemistry; chemical education. *Mailing Add:* Dept Chem Rutgers Univ New Brunswick NJ 08903

BOILEAU, OLIVER C, b Camden, NJ, Mar 31, 27; m; c 4. ELECTRICAL ENGINEERING. *Educ:* Univ Penn, BS, 51, MS, 53; Mass Inst Technol, MS, 64. *Prof Exp:* Vpres, Boeing Aerospace Co, 68-73, pres, 73-80; pres & mem bd dirs, Gen Dynamics Corp, 80-88; PRES & GEN MGR, B2 DIV, NORTHROP CORP, 89- *Concurrent Pos:* Mem, Lawrence Inst Technol Corp & Lincoln Lab Adv Bd, Mass Inst Technol. *Mem:* Nat Acad Eng; Nat Aeronaut Asn; fel Am Inst Aeronaut & Astronaut; Am Defense Preparedness Asn. *Mailing Add:* B2 Div Northrop Corp 8900 E Washington Blvd Pica Rivera CA 90660

BOIME, IRVING, b St Louis, Mo, Feb 5, 41; m 61; c 2. MOLECULAR PHARMACOLOGY. *Educ:* St Louis Col Pharm, BS, 64; Purdue Univ, MS, 66; Washington Univ, PhD(pharmacol), 70. *Prof Exp:* Fel molecular biol, NIH, 70-72; assoc prof, 72-80, PROF OBSTET, GYNEC & PHARMACOL, WASH UNIV, 80- *Concurrent Pos:* Am Cancer Soc fel, 70; NIH res grant, 73; Pop Coun res grant, 73. *Mem:* Fedn Am Soc Exp Biol. *Res:* Biosynthesis of peptide hormones; proinsulin and the protein hormones of human placenta. *Mailing Add:* Dept of Pharmacol Univ Wash Sch Med 660 S Euclid St Louis MO 63110

BOIS, PIERRE, anatomy, for more information see previous edition

BOISSE, NORMAN ROBERT, b Biddeford, Maine, Aug 18, 47. NEUROPHARMACOLOGY. *Educ:* Univ Conn, Storrs, BS, 70; Cornell Univ, PhD(pharmacol), 76. *Prof Exp:* Res asst pharmacol, Cornell Univ Med Col, 76-77; ASST PROF PHARMACOL, NORTHEASTERN UNIV, 77- *Mem:* AAAS; NY Acad Sci; Soc Neurosci; Sigma Xi. *Res:* Barbiturate, alcohol and benzodiazepine physical dependence and tolerance. *Mailing Add:* Pharmacol Dept Col Pharm-Allied Health Northeastern Univ Boston MA 02115

BOISSY, RAYMOND E, b July 8, 54. CELL BIOLOGY, MORPHOLOGY. *Educ:* Univ Mass, PhD(vet & animal sci), 82. *Prof Exp:* Assoc res scientist, Sch Med, Yale Univ, 82-86; ASST PROF CELL BIOL, UNIV CINCINNATI, 86- *Mem:* AAAS; Am Soc Cell Biol. *Res:* Pigment cells. *Mailing Add:* Mermatology U Dept ML592 Univ Cincinnati Col Med Cincinnati OH 45267

BOISVENUE, RUDOLPH JOSEPH, b Windsor, Ont, Aug 30, 26; nat US; m 51; c 1. MICROBIOLOGY, PARASITOLOGY. *Educ:* Univ Western Ont, BSc, 49; Univ Detroit, MSc, 51; Mich State Univ, PhD(microbiol), 55. *Prof Exp:* Wildlife biologist, Kellogg Bird Sanctuary & Forest, 52-55; mgr parasitol, Ralston Purina Co, 55-62; RES SCIENTIST, LILLY RES LABS, 62- *Concurrent Pos:* Microtechnician, Sparrow Hosp, 53-55; asst, Mich State Univ, 54-55; mem fac, Dept Zool, Eve Div, Ind Cent Col, 62; mem, Livestock Insect Workshop Conf. *Mem:* Am Soc Parasitol; Am Asn Vet Parasitol; Wildlife Dis Asn; Am Heartworm Soc; Entom Soc Am; World Asn Advan Vet Parasitol. *Res:* Veterinary helminthology and entomology; biological and chemical control of major external and internal parasite species infecting domestic animals; in vitro cultivation of nematodes; antigenic nematode research. *Mailing Add:* PO Box 708 Greenfield IN 46140

BOISVERT, RONALD FERNAND, b Manchester, NH, July 26, 51; m 72; c 1. MATHEMATICAL SOFTWARE, NUMERICAL ANALYSIS. *Educ:* Keene State Col, BS, 73; Col William & Mary, MS, 75; Purdue Univ, MS, 77, PhD (computer sci), 79. *Prof Exp:* COMPUTER SCIENTIST, COMPUT & APPL MATH LAB, NAT INST STANDARDS & TECHNOL, 79- *Concurrent Pos:* Assoc ed, Asn Comput Mach Trans Math Software, 87-; consult, NSF, 87- *Mem:* Asn Comput Mach; Soc Indust & Appl Math; Inst Elec & Electronic Engrs, Computer Soc. *Res:* Development and maintenance of mathematical software; knowledge-based systems for software selection; scientific computing; algorithms for high-performance computers; numerical analysis; numerical solution of elliptic boundary-value problems. *Mailing Add:* 9631 Shadow Oak Dr Gaithersburg MD 20879

BOIVIN, ALBERIC, b Baie St Paul, Que, Feb 11, 19; m 45; c 4. PHYSICS, OPTICS. *Educ:* Laval Univ, BA, 40, BSc, 44, MSc, 47, PhD(diffraction optics), 60. *Prof Exp:* Asst, Laval Univ, 44, lectr, 45, assoc prof, 50, dir Lab d'Optique & Hyperfrequences, 68-70, prof optics, 55-84, EMER PROF OPTICS, LAVAL UNIV, 84- *Concurrent Pos:* Guggenheim fel, Univ Rochester, 62-63. *Honors & Awards:* Laureate, Prix David, Prov Que, 65; Pariseau Medal, Fr-Can Asn Advan Sci, 67. *Mem:* Fel Optical Soc Am; Can Asn Physicists; Royal Soc Can. *Res:* Scalar diffraction theory applied to image formation, circular gratings and optical resonators; electromagnetic diffraction theory applied to wide-angle antennas and optical systems; holography, with applications to interferometry and image deconvolution; microwave optics. *Mailing Add:* 834 Eymard Quebec PQ G1S 4A1 Can

BOIVIN, LOUIS-PHILIPPE, b Ottawa, Ont, Mar 12, 44; m 70; c 2. OPTICAL RADIOMETRY. *Educ:* Univ Ottawa, BSc, 65, PhD(physics, optics), 70. *Prof Exp:* Res scientist optical commun, Bell-Northern Res, 70-73; RES OFFICER, INST NAT MEASUREMENT STANDARDS, NAT RES COUN CAN, 73- *Res:* Radiometry and photometry. *Mailing Add:* Rm 110 Inst Nat Measurement Standards Montreal Rd Ottawa ON K1A 0R6 Can

BOJALIL, LUIS FELIPE, b Merida, Yucatan, Mar 12, 25; m 58; c 3. MEDICAL MICROBIOLOGY. *Educ:* Univ Campeche, BSc, 43; Nat Polytech Inst, Mex, PhD(microbiol), 49, DSc(biol), 64. *Prof Exp:* From assoc prof to prof gen microbiol, Nat Polytech Inst, Mex, 50-55; prof microbiol & chmn dept, Nat Univ Mex, 55-76, prof, Postgrad Med Sch, 60-62; dean, Div Biol & Health Sci, 74-78, rector, 78-82, head postgrad study social med, 84-85, HEAD ADVAN TRAINING PROG COL TEACHERS, AUTONOMOUS METROP UNIV, XOCHIMILCO, MEX, 89-, DISTINGUISHED PROF. *Concurrent Pos:* Consult, Nat Campaign Against Tuberc, 57; WHO fel, 58; lectr, Univ PR, 62; mem permanent comt, Nat Coun Med Invest, 64; vis prof, Postgrad Sch Sci, Nat Polytech Inst, Mex, 64-; vis prof, Inst Educ, London Univ, 86. *Honors & Awards:* Mex Acad Sci Invest Award, 63; Pasteur Medal, 90. *Mem:* Am Soc Microbiol; Am Acad Microbiol; Mex Acad Sci Invest; Mex Asn Microbiol (vpres, 58-60, pres, 61-62); Brit Soc Gen Microbiol; Mex Nat Acad Med. *Res:* Mycobacteria and Nocardia; epidemiology, taxonomy, physiology and immunology; higher education. *Mailing Add:* Div Biol & Health Sci Univ Autonoma Metrop Xochimilco Mexico 23 DF Mexico

BOJANIC, RANKO, b Breza, Yugoslavia, Nov 12, 24; m 58; c 2. MATHEMATICAL ANALYSIS. *Educ:* Serbian Acad Sci, PhD(math), 53. *Prof Exp:* Docent math anal, Univ Skoplje, 54-56 & Sch Mech Eng, Univ Belgrade, 56-58; vis mem, Tata Inst Fundamental Res, India, 58-59; vis lectr math, Stanford Univ, 59-60; asst prof, Univ Notre Dame, 60-63; assoc prof, 63-66, PROF MATH, OHIO STATE UNIV, 66- *Mem:* Am Math Soc; Math Asn Am. *Res:* Asymptotic analysis; Fourier series and the theory of approximation; numerical analysis. *Mailing Add:* Dept Math Ohio State Univ Columbus OH 43210

BOJAR, SAMUEL, b New York, NY, Jan 23, 15; m 47; c 2. PSYCHIATRY. *Educ:* Brown Univ, AB, 36; Univ Rochester, MA, 37; Johns Hopkins Univ, MD, 41; Am Bd Psychiat & Neurol, cert, 52; Am Psychoanal Assoc, cert, 60. *Prof Exp:* Asst psychol, Univ Rochester, 36-37; asst, Harvard Med Sch, 47; res fel, 48-51, instr, 51-61, clin assoc, 61-67, from asst prof to assoc clin prof psychiat, 67-85. *Concurrent Pos:* Internship, Jewish Hosp Brooklyn, 41-42 asst res path, 43; res, Boston Psychopath Hosp, 47-48; asst, Peter Bent Brigham Hosp, 48-49, from jr assoc to assoc, 50-58, sr assoc, 58-83, physician, 83-89; candidate, Boston Psychoanal Inst, 51-56; psychiatrist, Harvard Med Area Health Serv, 53-79; lectr, Harvard Divinity Sch, 63-81, William James lectr, 71-72; hon physician, 89- *Mem:* Sigma Xi; AMA; fel Am Psychiat Asn; Am Psychoanal Asn. *Res:* Psychoanalysis; psychosomatic and psychophysiological correlates of personality and behavior; motivational studies in higher education. *Mailing Add:* Ten Aston Rd Chestnut Hill MA 02167

BOK, P DEAN, b Douglas Co, SDak, Nov 1, 39; m 64; c 3. ANATOMY, CELL BIOLOGY. *Educ:* Calvin Col, AB, 60; Calif State Univ, Long Beach, MA, 65; Univ Calif, Los Angeles, PhD(anat), 68. *Prof Exp:* From asst prof to assoc prof, 68-76, assoc dir, Jules Stein Eye Inst, 72-78, PROF ANAT, UNIV CALIF, LOS ANGELES, 76- *Concurrent Pos:* Nat Council Combat Blindness fel, Univ Calif, Los Angeles, 68-69, USPHS grant, 68-81; mem bd sci counselors, Nat Eye Inst, 80-, Retinal & Choroidal Dis Panel, 80-81. *Honors & Awards:* Gladys Shea Award, Univ Calif, Los Angeles, 68. *Mem:* Fel AAAS; Asn Res Vision & Ophthal; Am Asn Anatomists; Am Soc Cell Biol. *Res:* Cell biology of retinal photoreceptors and pigment epithelium including interactions between these cell types that involve circadian and diurnal phenomena, vitamin A transport, membrane receptors and inherited photoreceptor degenerations. *Mailing Add:* Jules Stein Eye Inst Univ Calif Ctr Health Sci Los Angeles CA 90024

BOKE, NORMAN HILL, b Mobridge, SDak, Mar 14, 13; m 48; c 1. PLANT ANATOMY, MORPHOLOGY. *Educ:* Univ SDak, AB, 34; Univ Okla, MS, 36; Univ Calif, PhD(bot), 39. *Prof Exp:* Asst bot, Univ SDak, 33-34 & Univ Okla, 34-36; asst bot, Univ Calif, 36-39, instr, 39-40, exp work in genetics & cytol, Bot Gardens, 40-41; inst biol, Univ NMex, 41-42; Johns Hopkins Univ, 42-44; opers analyst, US Army Air Force Pac Ocean Area & Seventh Air Force, 44-45; PROF BOT, UNIV OKLA, 45- *Concurrent Pos:* Guggenheim fel, 53-54; ed-in-chief, Am J Bot, Bot Soc Am, 70-75. *Mem:* Bot Soc Am; Torrey Bot Club; Bot Soc Mex; Mex Soc Cactology; Explorers' Club. *Res:* Developmental anatomy and morphology of vascular plants; development of foliar organs; floral histogenesis; developmental morphology of Cactaceae. *Mailing Add:* 610 Broad Lane Norman OK 73069

BOKELMAN, DELWIN LEE, b Linn, Kans, Sept 13, 34; m 58; c 4. PATHOLOGY. *Educ:* Kans State Univ, BS & DVM, 58; Cornell Univ, PhD(path), 64. *Prof Exp:* Resident vet immunol, Plum Island Animal Dis Lab, USDA, NY, 58-59; instr path, NY State Vet Col, Cornell Univ, 60-61, NIH trainee, 61-64; pathologist, Merck Inst Therapeut Res, 64-67, dir path, 67-77, dir toxicol, 70-77, sr dir safety assessment, 77-81, exec dir safety assessment, 81-87, VPRES SAFETY ASSESSMENT, MERCK, SHARP & DOHME RES LABS, WEST POINT, PA, 87- *Mem:* Fel Am Vet Med Asn. *Res:* Pathological changes associated with drug toxicity; safety assessment of drugs. *Mailing Add:* 81 Woodview Dr Doylestown PA 18901

BOKOCH, GARY M, b Erie, Pa, Apr 15, 54; m 77; c 2. SIGNAL TRANSDUCTION, GTP BINDING PROTEINS. *Educ:* Pa State Univ, BS, 76; Vanderbilt Univ, PhD(pharmacol), 81. *Prof Exp:* Asst mem, 85-90, ASSOC MEM, RES INST SCRIPPS CLIN, 91- *Concurrent Pos:* Postdoctoral fel, Southwestern Med Sch, Univ Tex, 81-85. *Honors & Awards:* Estab Investr, Am Heart Asn, 86, Louis N Katz Award, 88, John J Sampson Res Award, 89. *Mem:* Am Heart Asn; Am Soc Immunologists; Am Soc Biochem & Molecular Biol. *Res:* Research involves signal transduction in phagocytic cells and white blood cells by GTP-binding proteins of both the heterotrinone (receptor-coupled) and low molecular weight (vas- lihc) variety. *Mailing Add:* Dept Immunol Res Inst Scripps Clin 10666 N Torrey Pines Rd La Jolla CA 92037

BOKROS, J(ACK) C(HESTER), b Park Falls, Wis, Feb 24, 31; m 58; c 3. BIOMATERIALS. *Educ:* Univ Wis, BS, 54, MS, 55; Univ Calif, Berkeley, PhD(metall), 63. *Hon Degrees:* Rose-Hulman Inst Technol, DEng, 86. *Prof Exp:* Sr res engr, Atomics Int Div, N Am Aviation Inc, 55-58; res staff mem, John Jay Hopkins Lab, Pure & Appl Sci, Div Gen Dynamics Corp, 58-69; dir, Med Prod Div, Gen Atomic Co, 69-78; PRES, CARBO MEDICS, INC, 79- *Concurrent Pos:* Consult, Gen Atomic Div, Gen Dynamics Corp, 60-62 & Aerojet Nuclear Div, Gen Tire & Rubber Co, 61-62. *Honors & Awards:* IR-100 Award, Indust Res, 71; Charles E Pettinos Award, Am Carbon Soc, 75; Clemson Univ Award, Soc Biomat, 76. *Mem:* Am Soc Testing & Mat; Am Carbon Soc; Soc Biomat; Am Heart Asn; Am Soc Artificial Internal Organs. *Res:* Physical and mechanical metallurgy; structure and properties of pyrolytic carbon; irradiation effects in pyrolytic carbon and graphite; crystallography and kinetics transformation; biomaterials; medical devices. *Mailing Add:* 2204 Manana St Austin TX 78730

BOKSAY, ISTVAN JANOS ENDRE, b Ungvar, Hungary, June 22, 40; Ger citizen; wid; c 1. MEDICINE, CLINICAL PHARMACOLOGY. *Educ:* Univ Med Sch, Budapest, MD, 64; Inst Med Sch, Frankfurt, PhD(pharmacol), 72. *Prof Exp:* Intern, Budapest, Hungary, 63-64; resident, Hosp Neurol & Psychiat, 64-65; pharmacologist, Chem Fabrik v Heyden/Squibb, 65-68; sr pharmacologist, Hoechst AG, 68, exec res asst head pharmaceut div, 70, head sect clin pharmacol & drug develop, 72-75; assoc dir clin res & internal med, Hoechst-Roussel Pharmaceut Inc, 75-78; DIR CLIN RES, AYERST LABS, DIV AM HOME PROD CORP, 78- *Concurrent Pos:* Fel, Coun Stroke, Am Heart Asn. *Mem:* Int Soc Clin Pharmacol & Therapeut; AAAS; Int Soc Neurochem; Am Soc Clin Pharmacol & Therapeut; Int Union Biochem. *Res:* Basic research; research coordination; clinical pharmacology and clinical evaluation of new drugs in Phase I-IV analgesic, antihypertensive, antiinflammatory, beta-blocking, diuretics, hormones, psychotropic and vasoactive agents; anti-anginal drugs, immunosuppressives, and immunostimulants; biostatistics. *Mailing Add:* Seven N Crescent Maplewood NJ 07040

BOKUNIEWICZ, HENRY JOSEPH, b Chicago, Ill, July 29, 49. GEOPHYSICS, MARINE GEOLOGY. *Educ:* Univ Ill, BA, 71; Yale Univ, MPhil, 73, PhD(geophysics), 76. *Prof Exp:* Res assoc geophysics, Yale Univ, 76-77; asst prof, 77-82, ASSOC PROF GEOL OCEANOG, MARINE SCI RES CTR, STATE UNIV NY, STONY BROOK, 82- *Concurrent Pos:* Res affil, US Army Corps Engrs, 71-75, 79-, Waterways Exp Sta, 75-77, Geol Soc Am, 74-75, NY Sea Grant Inst, 77-78 & 80, NY Off Gen Serv, 77-78 & Nat Oceanog & Atmospheric Admin, 80-82. *Mem:* Am Geophys Union; AAAS; Geol Soc Am; NY Acad Sci. *Res:* Sediment transport and coastal oceanography; dredging specialist; coastal groundwater hydrology. *Mailing Add:* Marine Sci Res Ctr State Univ NY Stony Brook NY 11794-5000

BOL, KEES, b Netherlands, June 16, 25; nat US; m 47; c 4. PLASMA PHYSICS. *Educ:* Stanford Univ, PhD(physics), 50. *Prof Exp:* Proj engr electron beams res, Sperry Gyroscope Co, 49-54; asst prof physics, Adelphi Univ, 54-57; NSF fel, Harvard Univ, 57-59; mem res staff, Proj Matterhorn, 59-64, mem res staff, 64-83, mgr exp projs, Plasma Physics Lab, Princeton Univ, 83-88; RETIRED. *Mem:* fel Am Phys Soc. *Res:* Measurement of velocity of light; electron beam problems; problems of tokamak plasmas. *Mailing Add:* 37 Stouts Rd Skillman NJ 08558

BOLAFFI, JANICE LERNER, b Boston, Mass. ENDOCRINOLOGY, NEUROENDOCRINOLOGY. *Educ:* Brandeis Univ, BA, 54; Boston Univ, MA, 73, PhD(biol), 78. *Prof Exp:* Res fel endocrinol, New Eng Med Ctr Hosp, 78-81; vis asst prof physiol, Sch Med, Tufts Univ, 81-83; RES BIOCHEMIST, UNIV CALIF, SAN FRANCISCO, 84- *Mem:* AAAS; Endocrine Soc; Am Soc Zoologists; NY Acad Sci; Sigma Xi; Am Diabetes Asn. *Res:* Regulation of secretion of and synthesis of islet hormones, especially insulin. *Mailing Add:* 2331 Bush St San Francisco CA 94115

BOLAND, JAMES P, b Philadelphia, Pa, Mar 6, 31; m 63; c 6. CARDIOVASCULAR SURGERY, THORACIC SURGERY. *Educ:* St Joseph's Col, Pa, BS, 52; Jefferson Med Col, MD, 56. *Prof Exp:* From assoc prof to prof surg, Med Col Pa, 64-76; PROF SURG, W VA UNIV, CHARLESTON, 76-; CHIEF SURG, CHARLESTON AREA MED CTR, 76- *Mem:* Soc Thoracic Surg; Am Col Surgeons. *Mailing Add:* 3110 MacCorkle Ave Charleston WV 25304

BOLAND, JOSEPH S(AMUEL), III, b Montgomery, Ala, Sept 23, 39; m 62; c 2. ELECTRICAL ENGINEERING. *Educ:* Auburn Univ, BEE, 61, MSEE, 65; Ga Inst Technol, PhD(elec eng), 69. *Prof Exp:* Instr elec eng, Auburn Univ, 61-62, res asst, missile attitude control systs, 64-65; instr elec eng, Ga Inst Technol, 65-68; from asst prof to assoc prof, 68-79, PROF ELEC ENG, AUBURN UNIV, 79-, ASSOC DEAN ENG, 82- *Concurrent Pos:* Proj leader, NASA contract, 68-72; consult & proj leader, US Army Missile Command, 73- *Honors & Awards:* Western Elec Award, 77. *Mem:* Inst Elec & Electronics Engrs; Am Soc Eng Educ. *Res:* Modern control theory and applications; attitude control of space vehicles; adaptive systems; correlation techniques. *Mailing Add:* 202 Ramsay Hall Auburn Univ Auburn AL 36830

BOLAND, W ROBERT, b Winter Haven, Fla, Sept 11, 37; m 60; c 2. NUMERICAL MATHEMATICS, MATHEMATICAL SOFTWARE. *Educ:* Davidson Col, BS, 59; Col William & Mary, MA, 63; Univ Colo, PhD(appl math), 68. *Prof Exp:* Mathematician, Langley Res Ctr, NASA, 59-63; instr math, Randolph-Macon Col, 63-64; NSF trainee, Univ Colo, 65-67; asst prof math, Drexel Univ, 67-70; from asst prof to prof, Clemson Univ, 70-81; MEM STAFF, LOS ALAMOS NAT LAB, 81- *Mem:* Soc Indust & Appl Math. *Res:* Integral equations; operations research; computer science. *Mailing Add:* Los Alamos Nat Lab C-10 MS B296 PO Box 1663 Los Alamos NM 87545

BOLANDE, ROBERT PAUL, b Chicago, Ill, Apr 16, 26; m 54; c 4. PATHOLOGY. *Educ:* Northwestern Univ, BS, 48, MS & MD, 52; Am Bd Path, dipl, 58. *Prof Exp:* Intern, Chicago Wesley Mem Hosp, 52-53; resident pediat path, Children's Mem Hosp Chicago, 53-54; from resident to chief resident path, Inst Path, Case Western Reserve Univ, 54-56, from instr to assoc prof path, Sch Med, 56-72, sr instr pediat, 58-72, chg pediat path, Inst & Hosp, 56-72; prof path, McGill Univ, 72-82, prof pediat, 75-82; PROF PATH, E CAROLINA SCH MED, 82- *Concurrent Pos:* USPHS grants, 58 & 64; consult, Inst Path, Case Western Reserve Univ, 66-; chief pathologist & dir labs, Akron Children's Hosp, 66-72; co-ed, Perspectives in Pediat Path, 70-; dir path, Montreal Children's Hosp, 72-80. *Honors & Awards:* Sidney Farber Mem lectr, Soc Pediat Path, 85. *Mem:* Am Asn Path & Bact; Int Acad Path; Soc Pediat Path. *Res:* Application of electron microscopy, histochemistry and tissue culture to the study of disease in early life, especially neoplasia and natural host resistance in neoplasia. *Mailing Add:* Dept Path E Carolina Sch Med Greenville NC 27834

BOLANDER, RICHARD, b Parsons, Kans, Mar 26, 40; m 59; c 4. ENGINEERING EDUCATION, ACTUARIAL SCIENCE. *Educ:* Mo Sch Mines, BS, 61; Tex Christian Univ, MS, 63; Univ Mo, Rolla, PhD(physics), 69. *Prof Exp:* Assoc engr, Gen Dynamics, Ft Worth, 62; instr physics, Tex Woman's Univ, 63-65; asst prof, La Polytech Inst, 65-66; from instr to asst prof ceramic eng, Univ Mo, Rolla, 67-69, res assoc cloud physics, 69; from asst prof to prof math, 69-80, prof mech eng, 81-86, asst head, Dept Mech Eng & Mgt Inst, 82-86, PROF ENG PHYSICS & MATH, GEN MOTORS INST, 86- *Mem:* Am Soc Eng Educ. *Res:* Engineering education. *Mailing Add:* Dept Sci Math GMI Eng & Mgt Inst Flint MI 48504-4898

BOLAÑOS, BENJAMIN, b Cali, Columbia, Apr 22, 57; m 87. MEDICAL MYCOLOGY, AIR MICROBIOLOGY. *Educ:* Univ Valle, BS, 73; Univ Antioquia, MS, 77; Duke Univ, PhD(microbiol & immunol), 83. *Prof Exp:* Res asst mycol, Int Ctr Med Res, 73-75; instr microbiol, Univ Valle, 77-78; ASSOC PROF MICROBIOL, UNIV PR MED SCI CAMPUS, 83- *Concurrent Pos:* Med mycol consult, Vet Admin Med Ctr, 85-; consult, Occup & Safety Health Admin, 88 & PR Dept Health, 90-; lectr trop med, USN, 90- *Res:* Cryptococcosis in AIDS patients; interaction between the fungus Cryptococcus neoformans and the host which involves multiple virulence factors of the fungus; virulence factors of cryptococcus neoformans and purification of antigens for detection of cell mediated immunity in cryptococcosis. *Mailing Add:* Microbiol & Med Zool Univ PR PO Box 365067 San Juan PR 00936

BOLAR, MARLIN L, b Lincoln, Nebr, Dec 10, 27; m 51; c 5. PLANT PHYSIOLOGY. *Educ:* Colo State Univ, BS, 49; Columbia Bible Col, BD, 52; Univ Nebr, PhD(bot), 63. *Prof Exp:* Instr, Ben Lippen High Sch, NC, 52-57, asst headmaster, 55-57; chmn dept, 68-74, PROF BIOL SCI, CALIF STATE UNIV, SACRAMENTO, 62- *Mem:* AAAS; Am Soc Plant Physiol; Sigma Xi. *Res:* Physiological relationship between plant hosts and obligate parasites, especially rust-producing organisms. *Mailing Add:* Dept Biol Sci Sacramento State Col Sacramento CA 95819

BOLCH, WILLIAM EMMETT, JR, b Lenoir, NC, Oct 27, 35; m 59; c 2. ENVIRONMENTAL ENGINEERING, RADIOLOGICAL HEALTH. *Educ:* Univ Tex, Austin, BS, 59, MS, 64; Univ Calif, Berkeley, PhD(sanit eng), 67. *Prof Exp:* PROF ENVIRON ENG, UNIV FLA, 66- *Concurrent Pos:* Consult engr, Fla Power & Light, Miami, Fla, Environ Sci & Eng, Inc, Gainesville & Ardaman & Assoc, Orlando. *Mem:* Am Nuclear Soc; Health Physics Soc. *Res:* Radiotracer techniques; radiological health; radioactive waste disposal; radiological health and health physics; radon. *Mailing Add:* Dept Environ Eng Univ Fla Gainesville FL 32611

BOLDEN, THEODORE EDWARD, b Middleburg, Va, Apr 19, 20; wid. DENTISTRY, PATHOLOGY. *Educ:* Lincoln Univ, Pa, AB, 41; Meharry Med Col, DDS, 47; Univ Ill, Chicago Circle, MS, 51, PhD(path), 58; Am Bd Oral Med, dipl, 70; Am Bd Oral Path, dipl, 72. *Hon Degrees:* LLD, Lincoln Univ, 81. *Prof Exp:* Instr periodont & oper dent, Meharry Med Col, 48-49; instr oral path, Sch Dent, Univ Ill, 55-57; asst prof oral diag & path, Seton Hall Col, 57-60, assoc prof, 60-62; prof dent & oral med, Sch Dent, Meharry Med Col, 62-69, dir res, 62-73, assoc dean, Sch Dent, 69-73, prof oral path & chmn dept, 69-77; prof path, Col Med & Dent NJ, NJ Dent Sch, 77-90, dean, 77-78; RETIRED. *Concurrent Pos:* Lectr, Meharry Med Col & Univ Ill, 56; attend, Med Ctr, NJ, 58-62; chmn adv health comt, Town of Montclair, NJ, 59-60, comnr urban develop, 61-62; abstractor, NY State Dent J, 60 & J Oral Therapeut & Pharmacol, 64; consult, Vet Admin Hosp, Tuskegee, Ala, 62-, Nashville, Tenn, 68-77 & Brooklyn, NY, 79-; consult, Inst Advan Educ Dent Res, Colgate-Palmolive Co, 67-70; ed, Quart Nat Dent Asn, 75-82, consult, US Dept Health & Human Serv, 80 & 81; trustee, Am Found Dent Health, 77-85, Neighborhood Coun, 85- *Mem:* Nat Dent Asn; Int Asn Dent Res; fel Am Acad Oral Path; NY Acad Sci; fel Am Acad Oral Med; Sigma Xi. *Res:* Histology; periodontal disease; salivary gland pathology; lacrimal gland pathology; pigmentation; effect of aging; experimental carcinogenesis; oral neoplasia. *Mailing Add:* 29 Montague Pl Montclair NJ 07402

BOLDRIDGE, DAVID WILLIAM, b Richmond, Va, Dec 28, 54; m 79; c 1. AEROSOL SCIENCE, AEROSOL DYNAMICS. *Educ:* Davidson Col, BS, 77; Univ Calif, Riverside, MS, 80, PhD(chem), 84. *Prof Exp:* CHEMIST RES & DEVELOP, R J REYNOLDS TOBACCO CO, 84- *Mem:* Am Chem Soc; Am Phys Soc; Am Asn Aerosol Res. *Res:* Aerosol properties and dynamics; fast reaction kinetics and mechanisms; laser spectroscopy. *Mailing Add:* 3154 Shannon Dr Winston-Salem NC 27106

BOLDRIDGE, WILLIAM FRANKLIN, b Culpeper Co, Va, Sept 3, 17; m 42; c 2. PHYSICAL CHEMISTRY. *Educ:* Randolph-Macon Col, BS, 39; WVa Univ, MS, 41, PhD(chem), 53. *Prof Exp:* Asst prof physics, Randolph-Macon Col, 43-44; asst, WVa Univ, 39-41; chemist, Am Viscose Corp, Va, 41-43 & Clinton Labs, Oak Ridge, 44-45; from asst prof to prof chem, Randolph-Macon Col, 49-83; RETIRED. *Mem:* Am Chem Soc. *Res:* Organic reagents as inorganic precipitants; radiochemistry; physics; radioactive tracer technique in the study of adsorption. *Mailing Add:* Rte 1 Box 2220 Ashland VA 23005

BOLDT, CHARLES EUGENE, forestry; deceased, see previous edition for last biography

BOLDT, DAVID H, HEMATOLOGY, IMMUNOLOGY. *Educ:* Tufts Univ, MD, 69. *Prof Exp:* PROF MED & CHIEF DIV HEMAT, HEALTH SCI CTR, UNIV TEX, SAN ANTONIO, 80- *Res:* Oncology. *Mailing Add:* Dept Med Health Sci Ctr Univ Tex 7703 Floyd Curl Dr San Antonio TX 78284

BOLDT, ELIHU (AARON), b New Brunswick, NJ, July 15, 31; m 71; c 3. PHYSICS. *Educ:* Mass Inst Technol, SB, 53, PhD(physics), 58. *Prof Exp:* Guest scientist, Brookhaven Nat Lab, 55-57; res scientist physics, Mass Inst Technol, 58; asst prof, Rutgers Univ, 58-64; Nat Acad Sci sr res assoc cosmic ray physics, 64-66, GROUP LEADER X-RAY ASTRON, GODDARD SPACE FLIGHT CTR, 66- *Concurrent Pos:* Res assoc physics lab, Polytech Sch, Paris, 60-62. *Mem:* Fel Am Phys Soc; Am Astron Soc; Int Astron Union. *Res:* High energy astrophysics; elementary particles. *Mailing Add:* NASA Code 666 Goddard Space Flight Ctr Greenbelt MD 20771

BOLDT, ROGER EARL, b Milwaukee, Wis, Dec 23, 28; m 54; c 3. BIOCHEMISTRY. *Educ:* Univ Wis, BS, 51, MS, 53, PhD(biochem), 58. *Prof Exp:* Asst, Univ Wis, 51-58, proj assoc pediat, 58; USPHS cancer trainee, Univ Minn, 58-60; asst prof biochem, Univ Pittsburgh, 61; res biochemist, Directorate Med Res, Edgewood Arsenal, Md, 61-66; chief, biochem br, 66-81, non-metals anal br, 81-89, CHIEF, RADIOL & INORG CHEM DIV, US ARMY ENVIRON HYG AGENCY, 89- *Concurrent Pos:* Adj prof chem, Univ Md, 64-66. *Mem:* Sigma Xi. *Res:* Analysis of pollutants in environmental and tissue samples. *Mailing Add:* 113 Duncannon Rd Bel Air MD 21014

BOLDUC, REGINALD J, b Lac Drolet, Frontenac, Que, July 28, 39; m 64; c 1. PLANT PHYSIOLOGY, HISTOCHEMISTRY. *Educ:* Laval Univ, Bes A, 60, Bsc A, 64; Purdue Univ, PhD(plant physiol & biochem), 68. *Prof Exp:* Res asst plant physiol & biochem, Purdue Univ, 64-68; ASST PROF, LAVAL UNIV, 68-; MEM STAFF, RES STA, CAN AGR, 72- *Concurrent Pos:* Researcher for Minister of Nat Ctr Sci Res, France, 69; vis prof, Nauchnii Sotroudnik, KAT Fiziologiya Rastenii Inst, Moscow, 79; Nat Res Coun Can exchange scholar to USSR Acad Sci, Moscow, 79; Res assoc, Univ York, UK, 80. *Mem:* Can Soc Plant Physiol; Agr Inst Can; Can Soc Agron; Int Asn Plant Physiol; Sigma Xi. *Res:* Physiology and histochemistry of stresses, cold, drought and salinity in plants. *Mailing Add:* 31 Robitaille Breakeyville PQ G0S 1E0 Can

BOLE, GILES G, b Battle Creek, Mich, July 28, 28; m 85; c 2. INTERNAL MEDICINE, RHEUMATOLOGY. *Educ:* Univ Mich, BS, 49, MD, 53. *Prof Exp:* Resident physician, Univ Hosp, Univ Mich, 54-56, instr med, Med Sch, 58-61, from asst prof to assoc prof, 61-70, chief, Rheumatol Div, 76-86, sr assoc dean, 86-90, PROF INTERNAL MED, RACKHAM ARTHRITIS RES UNIT, UNIV MICH, ANN ARBOR, 70-, INTERIM DEAN, MED SCH, 90- *Concurrent Pos:* Arthritis Found fel, 61-63, sr investr, 63-68; dir, Multipurpose Arthritis Ctr, 77-86; mem, Nat Arthritis Adv Bd, 81-87; Am Bd Internal Med, 62-; Subspeciality Bd Rheumatology, 74- *Mem:* Am Rheumatism Asn (pres, 80). *Res:* Biochemistry and biomechanic studies of osteoarthritis. *Mailing Add:* Univ Mich Med Sch M7324 Med Sci I Ann Arbor MI 48109-0624

BOLEF, DAN ISADORE, b Philadelphia, Pa, June 10, 21; m 44; c 3. SOLID STATE PHYSICS. *Educ:* Columbia Univ, AM, 48, PhD(physics), 52. *Prof Exp:* Instr physics, Stevens Inst Technol, 48-49; from instr to asst prof math & physics, State Univ NY, 49-53; res physicist, Westinghouse Res Labs, 53-63; PROF PHYSICS, WASH UNIV, 63- *Concurrent Pos:* Vis prof, Empire State Col, 74-75. *Mem:* Fel Am Phys Soc. *Res:* Nuclear and electron spin-phonon interactions in solids; elastic properties of solids; acoustic magnetic resonance; acoustic absorption and dispersion in metals, ferromagnets; ultrasonic high frequency techniques. *Mailing Add:* Dept Physics Wash Univ Skinker & Lindell St Louis MO 63130

BOLEMON, JAY S, b Atlanta, Ga, Oct 14, 41; m; c 3. PHYSICS. *Educ:* Univ SC, BS, 64, PhD(physics), 69. *Prof Exp:* Asst prof, 68-73, ASSOC PROF PHYSICS, UNIV CENT FLA, 73- *Mem:* Am Phys Soc; Am Asn Physics Teachers. *Res:* Teaching of physics and astronomy. *Mailing Add:* Dept Physics Univ Cent Fla Box 25000 Orlando FL 32816

BOLEN, DAVID WAYNE, b Whitby, WVa, May 23, 42; m 64; c 3. PHYSICAL BIOCHEMISTRY. *Educ:* Concord Col, BS, 64; Fla State Univ, PhD(biophys chem), 69. *Prof Exp:* NIH res fel biochem, Johns Hopkins Univ, 69-70 & Univ Minn, 70-71; from asst prof to assoc prof, 71-81, PROF BIOCHEM, SOUTHERN ILL UNIV, CARBONDALE, 77- *Concurrent Pos:* Sabbatical leave, Univ Chicago, 78-79 & Johns Hopkins Univ, 85-86. *Mem:* Am Chem Soc; Biophys Soc; Am Soc Biochem & Molecular Biol. *Res:* Thermodynamics; kinetics; mechanisms of interacting systems; protein folding. *Mailing Add:* Dept of Chem & Biochem Southern Ill Univ Carbondale IL 62901

BOLEN, ERIC GEORGE, b Plainfield, NJ, Nov 24, 37; m; c 2. WILDLIFE ECOLOGY. *Educ:* Univ Maine, BS, 59; Utah State Univ, MS, 62, PhD(wildlife), 67. *Prof Exp:* Instr biol, Tex A&I Univ, 65-66; from asst prof to prof, Texas Tech Univ, 66-73, prof range & wildlife mgt & assoc dean, grad sch, 78-88, Paul Whitfield Horn prof, 81-88; DEAN, GRAD SCH, UNIV NC, 88- *Concurrent Pos:* Asst dir, Welder Wildlife Found, 73-78; adj prof, Corpus Christi State Univ, 74-; mem, Fish & Wildlife & Park Natural Sci Adv Comt for Secy Interior, 75-78; ed, Wildlife Soc Bull, 75-78; ed bd, Conserv Biol. *Mem:* Wildlife Soc; Wilson Ornith Soc; Am Ornith Union; Am Soc Mammal; Cooper Ornith Soc; Soc Range Mgt; Forest & Conserv Hist Soc; Nat Leaders Am Conserv. *Res:* Waterfowl ecology and management; wetland ecology; non-game ecology and management; ornithology; plant ecology; environmental history; co-author of one textbook. *Mailing Add:* Dean Grad Sch Univ NC Wilmington NC 28403

BOLEN, LEE NAPIER, JR, b Memphis, Tenn, June 11, 37; m 60; c 1. ACOUSTICS. *Educ:* Univ Miss, BS, 59; Univ Va, MS, 61, PhD(physics), 64. *Prof Exp:* Res assoc physics, Univ Va, 63-64; from asst prof to assoc prof, 64-70, PROF PHYSICS, UNIV MISS, 70- *Mem:* Am Phys Soc; Am Asn Physics Teachers; Sigma Xi; Acoust Soc Am; AAAS; Nat Asn Acad Sci. *Res:* Instrumentation design; physical acoustics. *Mailing Add:* Dept of Physics Univ of Miss University MS 38677

BOLEN, MAX CARLTON, b Waynetown, Ind, Sept 23, 19; m 42; c 2. PHYSICS. *Educ:* Wabash Col, AB, 41; Univ Chicago, cert meteorol, 43; Purdue Univ, MS, 50; Tex A&M Univ, PhD, 61. *Prof Exp:* Teacher high sch, Ind, 41-42; asst physics, Purdue Univ, 46-48; from asst prof to assoc prof & chmn dept, Millikin Univ, 48-59; assoc prof & chmn dept, Trinity Univ, Tex, 59-63; assoc prof, Oklahoma City Univ, 63-65; head dept, Univ Tex, El Paso, 65-69, prof physics, 65-87, coordr sci educ, Col Sci, 69-87, EMER PROF PHYSICS, UNIV TEX, EL PASO, 87- *Concurrent Pos:* Actg asst prof, Tex A&M Univ, 57-58, res fel, 58-59 Consult, Prog Develop Div, Tex Educ Agency, 68-87 Meteorologist, WTVP-TV Sta, Decatur, Ill, 53-55, KOCO, Oklahoma City, Okla, 64-65; sr res physicist, Southwest Res Inst, Tex, 63. *Honors & Awards:* Fel, AAAS, 68- *Mem:* Fel AAAS; Am Asn Physics Teachers; Am Phys Soc; Am Meteorol Soc; Nat Sci Teachers Asn. *Res:* Science education and how students learn science; high polymer physics. *Mailing Add:* 6625 Westwind Dr El Paso TX 79912-3831

BOLENDER, CARROLL H, b Cincinnati, Ohio, Nov 2, 19; m 42; c 2. AERONAUTICAL SCIENCE. *Educ:* Wilmington Col, Ohio, BS, 41; Ohio State Univ, MBA, 49. *Prof Exp:* US Air Force, 41-, chief B-58 weapon systs proj, Detachment 1 Air Res & Develop Command, 58-61, prog dir, Skybolt, Aeronaut Systs Div, Air Force Systs command, 61-62, asst prog dir, 62, Skybolt test dir, Air Proving Ground Ctr, 62-63, dep chief staff opers, 63-64, adv study group, Off VChief Staff, Hq, Washington, DC, 64-65, Apollo mission dir, Off Manned Space Flight, hq, NASA, 65-67, lunar module prog mgr, Manned Spacecraft Ctr, Houston, 67-69, dept dir develop & acquisition, Dep Chief Staff res & develop, Hq, US Air Force Pentagon, 69-72; dep prog mgr, Hampton Opers, Syst Develop Corp, Va, 72-73; prog mgr, 73-79; consult, Eng Inc, Va, 79-81; consult, 81-84; prog mgr, Syst Develop Corp & Unisys, Va, 84-88; RETIRED. *Honors & Awards:* Apollo Achievement Award, Apollo Mgt Team Award & Group Group Achievement Award, NASA, 69. *Mailing Add:* 128 Randolph's Green Williamsburg VA 23185-6537

BOLENDER, CHARLES L, b Iowa City, Iowa, June 2, 32; m 55; c 3. DENTISTRY. *Educ:* Univ Iowa, DDS, 56, MS, 57. *Prof Exp:* From instr to assoc prof, 59-68, chmn dept, 63-89, PROF PROSTHODONTICS, SCH DENT, UNIV WASH, 68- *Mem:* Am Prosthodontic Soc; Am Dent Asn; fel Am Col Dent; fel Acad Denture Prosthetics; Acad Osseointegration; Am Asn Dent Schs. *Res:* Tissue response to artificial dentures; dental implants. *Mailing Add:* Sch Dent Univ Wash Seattle WA 98105

BOLENDER, DAVID LESLIE, b Uniontown, Pa, June 11, 47; m 69; c 3. DEVELOPMENTAL & CELL BIOLOGY. *Educ:* Bethany Col, WVa, BS, 69; WVa Univ, PhD(anat), 75. *Prof Exp:* Asst prof anat, Sch Med, Tex Tech Univ, 74-81; asst prof biol, Bethany Col, 81-85; ASST PROF RES, DEPT ANAT & BIOL, MED COL WIS, 85- *Mem:* Soc Study Reproduction; Soc Develop Biol; Am Asn Anatomists. *Res:* Developmental and cell biology, specifically the role of extracellular matrix in mediation of cell differentiation during development of craniofacial areas with emphasis on neural crest cells. *Mailing Add:* Med Col Wis 8701 Watertom Plank Rd Milwaukee WI 53226

BOLENDER, ROBERT P, b Mt Kisco, NY, Sept 29, 38. CELL BIOLOGY. *Educ:* State Univ NY Albany, BS, 60; Columbia Univ, MA, 65; Harvard Univ, PhD(anat), 70. *Prof Exp:* Fel anat, Univ Bern, 70-71; from asst to head asst, 71-74; lectr, Harvard Univ, 74-75; asst prof, 75-81, ASSOC PROF BIOL STRUCT, UNIV WASH, 81- *Concurrent Pos:* Vis scholar biochem, Univ Louvain, 71-72. *Mem:* Am Soc Cell Biol; Union Swiss Socs Exp Biol; Int Soc Stereology; Am Asn Anatomists; NY Acad Sci. *Res:* Integration of stereological and biochemical information as a way of analyzing relationships of structure to function mathematically; biological membrane kinetics associated with drugs, toxins, and carcinogens; mathematical models for the liver and lung. *Mailing Add:* Dept Biol Struct Univ Wash Sch Med Seattle WA 98195

BOLES, JAMES RICHARD, b Richmond, Ind, Sept 3, 44; m 67; c 2. SEDIMENTARY PETROLOGY. *Educ:* Purdue Univ, BS, 66; Univ Wyo, MS, 68; Univ Otago, NZ, PhD(geol), 72. *Prof Exp:* Fel geol, Univ Wyo, 72-73; res geologist, Geosci Group, Atlantic Richfield Co, 73-75; from asst prof to assoc prof, 75-81, PROF GEOL, UNIV CALIF, SANTA BARBARA, 81- *Mem:* Am Asn Petrol Geologists. *Res:* Diagenesis of clastic sediments including sandstone cementation, clay diagenesis and zeolitization of volcanogenic sediments; sedimentary basin reconstructions. *Mailing Add:* Dept Geol Univ Calif Santa Barbara CA 93106

BOLES, ROBERT JOE, b Brandsville, Mo, Nov 7, 16; m 49; c 2. AQUATIC BIOLOGY. *Educ:* Southwestern Col, Kans, AB, 38; Kans State Univ, MS, 49; Okla State Univ, PhD(zool), 60. *Prof Exp:* Teacher high schs, 39-42 & 45-58; from asst prof to assoc prof, 60-70, prof biol, Emporia State Univ, 70-; RETIRED. *Concurrent Pos:* Ed, Kans Sch Naturalist. *Res:* Farm pond dynamics and fish parasitology; productivity and management of small impoundments; parasites of freshwater fishes of Kansas. *Mailing Add:* 2007 Briarcliff Lane Emporia KS 66801

BOLEY, BRUNO ADRIAN, b Italy, May 13, 24; US citizen; m 49; c 2. THEORETICAL MECHANICS, APPLIED MECHANICS. *Educ:* City Col New York, BCE, 43; Polytech Inst Brooklyn, MAeE, 45, ScD, 46. *Hon Degrees:* ScD, City Col NY, 82. *Prof Exp:* Eng specialist, Goodyear Aircraft Corp, 48-50; assoc prof aeronaut eng, Ohio State Univ, 50-52; prof civil eng, Columbia Univ, 52-68; Joseph P Ripley prof eng & chmn dept theoret & appl mech, Cornell Univ, 68-72; dean, Technol Inst & Walter P Murphy prof, 73-87, WALTER P MURPHY EMER PROF & DEAN ENG, NORTHWESTERN UNIV, EVANSTON, 87-; PROF CIVIL ENG & ENG MECH, COLUMBIA UNIV, 87- *Concurrent Pos:* Vis prof, 64-65; mem cong comt, Int Union Theoret & Appl Mech, 70-; NATO sci fel, 72; ed, Mech Res Commun, 74-; chmn, US Nat Comt Theoret & Appl Mech, 74-76, vchmn, 76-78; gen chmn, Int Cong Struct Mech Reactor Technol, 77, adv general, 79-; chmn, Midwest Prog Minorities Educ, Comt Instnl Coop, 77-82; chmn, Nat Acad Eng Task Force on Eng Educ, 79-81; chmn, Ill Council Res & Dev, 80-84; bd gov, Argonne Nat Lab, 83, Am Soc Mech Exp, 84-86; proj bd, Am Soc Eng Educ, 87. *Honors & Awards:* Townsend Harris Medal, 81; Distinguished Serv Medal, Am Acad Mech & Soc Eng Sci, 87. *Mem:* Nat Acad Eng; hon mem Am Soc Mech Engrs; fel Am Inst Aeronaut & Astronaut; fel Am Acad Mech (pres, 73-74); fel Soc Eng Sci (pres, 75); fel AAAS; Int Union Theoret & Appl Mech; Am Soc Eng Educ; NY Acad Sci. *Res:* Thermal stresses; structural theory; elasticity and inelasticity; wave propagation in solids; heat conduction and change of phase; vibrations and buckling; aerostructures and nuclear reactor structural mechanics; micromechanics of solids. *Mailing Add:* Dept Civil Eng & Eng Mech Columbia Univ New York NY 10027-6699

BOLEY, CHARLES DANIEL, b Urbana, Ill, Dec 15, 43; m 82; c 2. LASER PHYSICS, COMPUTATIONAL PHYSICS. *Educ:* Mass Inst Technol, SB, 66, PhD(physics), 71. *Prof Exp:* Fel Dept Physics, Univ Toronto, 71-73; res fel appl sci, Calif Inst Technol, 73-75; physicist, Argonne Nat Lab, 75-85; mem tech staff, AT&T Bell Labs, 85-86; PHYSICIST, LAWRENCE LIVERMORE NAT LAB, 86- *Mem:* Am Phys Soc; Sigma Xi. *Res:* Plasma physics; laser physics; computational physics; statistical mechanics. *Mailing Add:* Lawrence Livermore Nat Lab L-464 Livermore CA 94550

BOLEY, FORREST IRVING, b Ft Madison, Iowa, Nov 27, 25; m 46, 69; c 4. PLASMA PHYSICS, ASTROPHYSICS. *Educ:* Iowa State Univ, BS, 46, PhD(physics), 51. *Hon Degrees:* MA, Wesleyan Univ, 59; Dartmouth Col, 64. *Prof Exp:* Elec engr, Raytheon Mfg Co, 46-47; from asst prof to prof physics, Wesleyan Univ, 51-61; physicist, Lawrence Radiation Lab, Univ

Calif, 61-64; chmn dept, 75-78, PROF PHYSICS & ASTRON, DARTMOUTH COL, 64- *Concurrent Pos:* Ed, Am J Physics, Am Asn Physics Teachers, 66-72. *Mem:* AAAS; Am Astron Soc; Am Phys Soc; Int Astron Union. *Res:* Cosmic rays; plasma phenomena; astrophysics. *Mailing Add:* Dept Physics & Astron Dartmouth Col Hanover NH 03755

BOLEY, ROBERT B, b Big Spring, Tex, Aug 19, 28; m 60; c 1. BACTERIOLOGY, IMMUNOLOGY. *Educ:* Sam Houston State Univ, BS, 49; Tex A&M Univ, MS, 60; Ohio State Univ, PhD(microbiol), 63. *Prof Exp:* From instr to assoc prof biol, Tarleton State Col, 58-65; ASSOC PROF BIOL, UNIV TEX, ARLINGTON, 65- *Mem:* Am Soc Microbiol; Am Inst Biol Sci; NY Acad Sci; Brit Soc Gen Microbiol. *Res:* Intestinal flora and immune responses in small mammals. *Mailing Add:* 907 Pontiac Ct Arlington TX 76010

BOLEY, SCOTT JASON, b Brooklyn, NY. PEDIATRIC SURGERY. *Educ:* Wesleyan Univ, AB, 46; Jefferson Med Col, MD, 49. *Prof Exp:* Assoc prof, 67-71, PROF SURG & PEDIAT & CHIEF PEDIAT SURG SERV, ALBERT EINSTEIN COL MED, 71-; DIR PEDIAT SURG, MONTEFIORE HOSP & MED CTR, 67- *Mem:* Fel Am Acad Pediat; fel Am Col Surgeons; Am Pediat Surg Asn; Soc Surg Alimentary Tract; Am Med Writers' Asn; Am Physiol Soc; Am Gastroenterol Asn; Soc Cardiovasc & Interventional Radiol. *Res:* Vascular diseases of the intestines; Hirschprung's disease. *Mailing Add:* Albert Einstein Col Med Montefiore Med Ctr 111 E 210th St Bronx NY 10467

BOLGER, EDWARD M, b Jersey City, NJ, Jan 9, 38; m 60; c 2. MATHEMATICS. *Educ:* St Peter's Col, NJ, BS, 59; Pa State Univ, MA, 61, PhD(probability, statist), 64. *Prof Exp:* Asst prof math, Bucknell Univ, 64-67; asst prof math, 67-69, asst prof math & statist, 69-75, PROF MATH & STATIST, MIAMI UNIV, 75- *Mem:* Soc For Indust & Appl Math; Math Asn Am; Inst Math Statist. *Res:* N-person game theory. *Mailing Add:* Dept Math Miami Univ Oxford OH 45056

BOLGER, P MICHAEL, b Corpus Christi, Tex, July 30, 49; m; c 2. TOXICOLOGY OF METALS. *Educ:* Georgetown Univ, PhD(physiol & biophys), 76; Am Bd Topicol, cert gen topicol. *Prof Exp:* TOXICOLOGIST & PHYSIOLOGIST, FOOD & DRUG ADMIN, 80-, SUPVR, 86- *Mem:* Am Physiol Soc; Am Soc Neprology. *Mailing Add:* Div Toxicol HFF-156 Food & Drug Admin 200 C St SW Washington DC 20204

BOLGIANO, LOUIS PAUL, JR, b Baltimore, Md, June 20, 22; m 46; c 3. ELECTRICAL ENGINEERING. *Educ:* Haverford Col, BS, 43; Johns Hopkins Univ, MA, 48, PhD(physics), 53. *Prof Exp:* Jr instr physics, Johns Hopkins Univ, 48-52; asst prof elec eng, US Naval Acad, 52-55; assoc prof, 55-66, PROF ELEC ENG, UNIV DEL, 66- *Concurrent Pos:* Teaching internship, Ford Found, 53-54; prin investr, Res Corp, Univ Del, 56-57; res assoc, Stanford Univ, 60, 61; expert, US Army Ballistic Res Labs, 62; prin investr, Air Force Off Sci Res, 62-73, Univ Del Res Found, 73-75, NSF, 77-79. *Mem:* Int Elec & Electronics Engrs. *Res:* Communication theory and signal processing. *Mailing Add:* Dept Elec Eng Univ Del Newark DE 19711

BOLGIANO, NICHOLAS CHARLES, b Baltimore, Md, Oct 26, 23; m 55; c 8. ORGANIC POLYMER CHEMISTRY. *Educ:* Loyola Col, Md, BS, 44; Catholic Univ, MS, 49; Univ Md, PhD(chem), 54. *Prof Exp:* Res fel, Mellon Inst, 54-56; res assoc, Armstrong World Industs, 54-89; RETIRED. *Mem:* Am Chem Soc. *Res:* Monomer and polymer synthesis; foamed plastics; coatings; binders; radiation curing. *Mailing Add:* 824 Dorsea Rd Lancaster PA 17601

BOLGIANO, RALPH, JR, b Baltimore, Md, Apr 1, 22; m 48; c 4. RADIO PHYSICS, ATMOSPHERIC TURBULENCE. *Educ:* Cornell Univ, BSEE, 44, BEE, 47, MEE, 49, PhD(radio physics), 58. *Prof Exp:* Develop engr, Electronics Lab, Gen Elec Co, 49-50, proj engr, 50-52, sect supvr, Advan Electronics Ctr, 53-54; from assoc prof to prof, 58-90, EMER PROF ELEC ENG, CORNELL UNIV, 90- *Concurrent Pos:* Consult, Gen Elec Co, 54-63, ADCOM, Inc, 62-63 & Rome Air Develop, Ctr, US Air Force, 63-64; mem, US nat body, Int Sci Radio Union, Inter-Union Comt Radiometeorol, Int Sci Radio Union-Int Union Geod & Geophys, 60-70, pres, 67-70; trustee, Univ Corp Atmos Res, 61-64; mem rep, 65-71; Guggenheim fel & Fulbright travel grant, 64-65; mem, Adv Group Aerospace Res & Develop, NATO, 68-78; gen chmn, Int Colloq Spectra of Meteorol Variables, Stockholm, 69; vis scientist, Radio & Space Res Sta, Sci Res Coun, UK & Inst Mech Statist of Turbulence, Univ Aix Marseille, 71-72; sr res assoc, Coop Inst Res Environ Sci, Univ Colo, Boulder, 79-80. *Mem:* Fel AAAS; Inst Elec & Electronics Engrs; Am Meteorol Soc; Am Geophys Union; Sigma Xi. *Res:* Scatter propagation of radio waves; radio scatter measurement of atmospheric structure; study of turbulent mixing in the atmosphere; theory of turbulence in stably stratified atmosphere. *Mailing Add:* Sch Elec Eng 207 Philips Hall Cornell Univ Ithaca NY 14853

BOLHOFER, WILLIAM ALFRED, b Jamaica, NY, May 18, 20; m 42; c 4. ORGANIC CHEMISTRY. *Educ:* Mass Inst Technol, BS, 42, PhD(chem), 50. *Prof Exp:* Chemist, Merck & Co, 42-47; asst, Mass Inst Technol, 47-49; sr res fel, Merck Sharp & Dohme Res Lab, 49-74, sr investr, 74-82; RETIRED. *Mem:* AAAS; Am Chem Soc. *Res:* Medicinal chemistry; synthetic organic chemistry. *Mailing Add:* PO Box 278 Frederick PA 19435-0278

BOLHUIS, REINDER L H, b Winschoten, Neth, July 2, 45; m; c 1. TCR GAMMA DELTA LYMPHOCYTES, IMMUNOTHERAPY. *Educ:* V Univ, Amsterdam, BS, 69; Erasmus Univ, Rotterdam, MD, 77. *Prof Exp:* RES ASSOC, ROTTERDAM RADIO THERAPEUT INST, 74-, STAFF MEM, 76- *Concurrent Pos:* Guest staff mem, Radiobiol Orgn, TNO Health Orgn, Rijswijk, 72, dept head, 76; mem adv bd, Comprehensive Cancer Ctr, Rotterdam, 76-, Nat Working Party, Immunostimulantia, 77-; coordr, Tumor-Immunol, Erasmus Univ, Radiobiol Inst, TNO Health Orgn & Daniel den Hoed Cancer Ctr, 76-; numerous grants, Dutch Cancer Soc, 76-, Dept Econ

Affairs, Leiden & Centocor, 89-91, Concerted Action Prog, 90-92; mem comt, Qual Assurance Clin Immunol, 81-; mem coun, Sci Adv, Daniel den Hoed Cancer Ctr, Nat Health & Med Res, Commonwealth Australia, 88-; extramural rev cand acad, Nat Cancer Inst, NIH, 86-; chmn, Standardization Immunodiag Leukemia & Lymphoma, 86-; founder & chmn, co-operation Immunophenotyping, Haematological Malignancies, Neth, 86-89; ad hoc rev, Europ J Cancer & Clin Oncol, Human Immunol, Hybridoma, Immunol Letters, Urologia Internationalis, J Cancer Res & Clin Oncol. Mem: Soc Qual Control Clin Immunol Assays; Am Asn Immunol; Am Asn Cancer Res. Res: Head and neck tumors; MHC-unrestricted cytotoxicity; NK cell receptor; bispecific monoclonal antibodies against renal and ovarian cancer, small cell lung carcinoma and human immune-deficiency Virus Syndrome. Mailing Add: PO Box 5201 Rotterdam 3008 AE Netherlands

BOLICK, MARGARET RUTH, b Winston-Salem, NC, Sept 8, 50; m 78. BOTANY, SYSTEMATICS. Educ: Duke Univ, BS, 72, MA, 74; Univ Tex, Austin, PhD(bot), 78. Prof Exp: CURATOR & ASST PROF BOT, UNIV NEBR STATE MUS, 78- Mem: Bot Soc Am; Am Soc Plant Taxonomists; Int Asn Plant Taxon; Sigma Xi. Res: Plant systematics, especially Asteraceae; pollen morphology, biology and evolution; mechanics of pollen walls. Mailing Add: W-532 Nebr Hall Univ Nebr State Mus Lincoln NE 68588

BOLIE, VICTOR WAYNE, b Silverton, Ore, July 23, 24; m 45. ELECTROPHYSICS, NEURAL NETWORKS. Educ: Iowa State Univ, BS, 49, MS, 50, PhD, 52; Coe Col, BA, 57; Stanford Univ, MS, 59. Prof Exp: Eng res & admin, Collins Radio Co, 52-57; prof & chmn biomed eng, Iowa State Univ, 57-63; eng res & admin, NAm Rockwell Corp, 63-66; prof elec eng, Univ Ariz, 66-67; chaired prof, Okla State Univ, 67-71; chmn dept elec & comput engr, 71-76, PROF ELEC & COMPUT ENGR, UNIV NMEX, 76- Concurrent Pos: Prin investr, aircraft collision avoidance, Collins Radio Co, 55-57, NIH & Iowa State Univ bioeng facil, 57-63, NIH & Iowa State Univ/ Rockwell artificial hearts, 60-66, atmospheric phenomena, US Dept Defense, Okla State Univ, 67-69, automatic speech recognition, US Air Force, Okla State Univ, Univ NMex, 70-77; nat chmn, joint comt Eng Med & Biol, 64-65; mem, Accreditation Bd Eng & Technol eng col accreditation team, 69-76; consult, NIH, NSF, US Navy, Space Labs, Aerojet, Fisher, Airesearch, Los Alamos. Mem: Fel Inst Elec & Electronics Engrs; Am Physiol Soc; Nat Soc Prof Engrs. Res: Electrophysics particle beams, free election lasers; neutral networks, adaptive control systems. Mailing Add: Dept Elec & Computer Sci Univ NMex Albuquerque NM 87131

BOLIEK, IRENE, b Hickory, NC, Apr 18, 07. ZOOLOGY. Educ: NC Col for Women, AB, 29; Univ NC, AM, 33, PhD(zool), 36. Prof Exp: Teacher high sch, NC, 29-31; asst zool, Univ NC, 32-35; asst prof biol, Ala Col, 35-36; from instr to prof zool, Fla State Univ, 36-65; prof biol & chmn dept, Coker Col, 65-73; EMER CUR, CATAWBA SCI CTR, 74- Mem: Sigma Xi. Res: Fresh water and marine invertebrates. Mailing Add: 223 Ninth Ave NW Hickory NC 28601

BOLIN, HAROLD R, b Lefors, Tex, Jan 16, 30; m 45; c 2. FOOD CHEMISTRY, BIOCHEMISTRY. Educ: San Jose State Col, BS, 59; Utah State Univ, PhD(food sci & technol), 70. Prof Exp: Lab supvr canned fruits & veg, Dole Corp, 59-60; res chemist collabr dried fruits, 60-65, FOODS CHEMIST, WESTERN REGIONAL RES LAB, USDA, 65- Concurrent Pos: Res prof, CSIRO, Sydney, Australia, 83-84. Mem: Am Chem Soc; Inst Food Technol; Sigma Xi. Res: Dried fruits, antimicrobial protection, water activity; crystal development in fruits; solar energy in food preservation; intermediate moisture foods; cellular turgor changes in fruits; nonenzymatic browning; minimum processed fruits & vegetables. Mailing Add: Western Utilization Res & Develop Div USDA 800 Buchanan St Albany CA 94710

BOLING, JAMES A, MAGNESIUM RELATED PROBLEMS IN LACTATING BEEF COWS. Educ: Univ Wis, PhD(animal nutrit & biochem), 67. Prof Exp: PROF ANIMAL SCI, UNIV WIS, 67- Res: Protein and mineral nutrition of ruminant animals. Mailing Add: Assoc Dean Res/ Col Agr Univ Ky S-107 Agr Sci Bldg N Lexington KY 40546

BOLINGER, ROBERT E, b Independence, Kans, May 31, 19. MEDICINE. Educ: Univ Kans, AB, 40, MD, 43. Prof Exp: From instr to assoc prof, 49-60, PROF MED & DIR CLIN RES CTR, MED CTR, UNIV KANS, 60-, LECTR HIST MED, 77- Res: Insulin metabolism; computerized health care. Mailing Add: Univ Kans Med Ctr Kansas City KS 66103

BOLKER, ETHAN D, b Brooklyn, NY, June 11, 38; m 60; c 2. MATHEMATICS. Educ: Harvard Univ, AB, 59, AM, 61, PhD(math), 65. Prof Exp: Instr math, Princeton Univ, 64-65; from asst prof to assoc prof, Bryn Mawr Col, 65-72; PROF MATH, UNIV MASS, BOSTON, 72- Concurrent Pos: Res fel, Univ Calif, Berkeley, 67-68; consult, BGS Systs, 82- Mem: Math Asn Am; Am Math Soc; Asn Women Math; Asn Comput Mach. Res: Combinatorics; geometry; queueing theory. Mailing Add: Dept Math & Comput Sci Univ Mass Boston MA 02125-3393

BOLKER, HENRY IRVING, b Montreal, Que, Feb 19, 26; m 53; c 1. WOOD CHEMISTRY. Educ: Queen's Univ, Ont, BA, 48, MA, 50; Yale Univ, PhD(chem), 52. Prof Exp: Fel biochem, McGill Univ, 51-54; chemist, Dominion Tar & Chem Co, 54 & DuPont of Can, Ltd, 54-60; chemist, Pulp & Paper Res Inst Can, 60-63, asst head, Wood Chem Div, 63-67, assoc div head chem, 67-69, head, Org Chem Sect, 69-72, head, Chem Processes Sect, 72-77, dir, Process Chem Div, 77-80; dir res, 80-87, dir acad affairs, Pulp & Paper Res Inst Can, 87-90; RES ASSOC CHEM, MCGILL UNIV, 63-; EXEC DIR, MECH WOODPULP NETWORK EXCELLENCE, 90- Concurrent Pos: Mem exec comt, Youth Sci Found, 61-64, pres, 65-66, mem admin coun, 69-71; chmn, Sci Affairs, 67-71; mem admin coun, Int Coord Comt, UNESCO, 68-70; chmn educ comt, Can Pulp & Paper Asn, 71-73; mem indust liaison comt, Ecole de Technol Superieure, 73-76; counr, Chem Inst Can, 74-77, bd dir, 77-81, chmn, 79-81. Honors & Awards: Montreal Medal Chem, Chem Inst Can, 84. Mem: Chem Inst Can (vpres, 86-87, pres, 87-88); Can Pulp & Paper Asn; Am Chem Soc; The Chem Soc; fel Int Acad Wood Sci (secy-treas, 90-). Res: Lignin; cellulose; chemical pulping and bleaching. Mailing Add: Mech & Chemimech Wood Pulps Network 570 Blvd St Jean Pointe Claire PQ H9R 3J9 Can

BOLL, H(ARRY) J(OSEPH), b St Bonifacius, Minn, Mar 5, 30; m 56; c 6. ELECTRICAL ENGINEERING, INTEGRATED CIRCUITS. Educ: Univ Minn, BS, 56, MS, 58, PhD(elec eng), 62. Prof Exp: Supvr, Bell Tel Labs, Inc, Murray Hill, 67-; RETIRED. Mem: Inst Elec & Electronics Engrs. Res: Secondary electron emission from magnesium oxide thin films; semiconductor devices. Mailing Add: 2780 70th St SW Naples FL 33999

BOLL, WILLIAM GEORGE, b Manchester, Eng, Dec 5, 21; m 54; c 3. PLANT TISSUE CULTURE, BIOCHEMISTRY. Educ: Univ Manchester, BSc, 48, PhD(plant physiol), 51. Prof Exp: Fel, Plant Res Inst, Univ Tex, 52-54, res scientist, 54-55; from asst prof to prof bot, McGill Univ, 55-90; RETIRED. Mem: Can Soc Plant Physiol. Res: Physiological and biochemical study of growth and morphogenesis of plants; excised root, tissue and cell suspension culture. Mailing Add: 34 Eighth St Roxboro PQ H8Y 1G5 Can

BOLLA, ROBERT IRVING, b Dansville, NY, Aug 18, 43; m 66; c 2. MOLECULAR BIOLOGY, NEMATODE BIOCHEMISTRY. Educ: State Univ NY Buffalo, BS, 65; Univ Mass, Amherst, MSc, 67, PhD(zool), 70. Prof Exp: NIH res fel biol, Univ Notre Dame, 70-73; res fel, Roche Inst Molecular Biol, 73-76; from asst prof to prof biol, Univ Mo, St Louis, 76-88; PROF & CHAIR BIOL, ST LOUIS UNIV, 88- Concurrent Pos: Vis asst prof microbiol, Wash Univ Sch Med, 82-87. Mem: AAAS; Sigma Xi; NY Acad Sci; Soc Nematologists; Am Aging Asn; Am Soc Microbiol. Res: Pinewood nematodes; population biology; toxin production in Bursaphelenchus xylophilus caused rapid pine wilt disease; metabolism Bursaphelenchus xylophilus; plant resistance to nematodes. Mailing Add: Dept Biol St Louis Univ St Louis MO 63103-2010

BOLLAG, JEAN-MARC, b Basel, Switz, Feb 19, 35; m 59; c 4. SOIL BIOCHEMISTRY. Educ: Univ Basel, PhD(plant physiol), 59. Prof Exp: Lectr plant physiol, Swiss Trop Inst, Basel, 58-59; Julius Baer fel, Switz, 63; fel plant genetics, Weizmann Inst Sci, Rehovot, Israel, 63-65; res assoc soil microbiol, Cornell Univ, 65-67; from asst prof to assoc prof, 67-77, PROF SOIL MICROBIOL, PA STATE UNIV, 77- Concurrent Pos: mem, Biochem Sect, CIBA-GEIGY, Basel, Switz, 75-76. Honors & Awards: Badge of Merit, Polish Ministry Agr. Mem: Am Soc Microbiol; Am Soc Agron; fel Soil Sci Soc Am; Int Soc Soil Sci; Int Humic Substances Soc; AAAS; affil mem Am Chem Soc. Res: Bioremediation, microbial control of the environment; degradation of pesticides by microbes; incorporation of xenobiotics into humic material; environmental application of immobilized enzymes. Mailing Add: Lab Soil Biochem Pa State Univ University Park PA 16802

BOLLARD, R(ICHARD) JOHN H, b Hamilton, NZ, July 13, 27; m 55; c 4. AERONAUTICS, ASTRONAUTICS. Educ: Univ NZ, BS, 51. Prof Exp: Lectr aeronaut, Univ NZ, 54-56; from asst prof to prof, Purdue Univ, 56-61, assoc dir aerospace sci lab, 60-61; head dept aeronaut & astronaut, 61-77, PROF AERONAUT & ASTRONAUT, UNIV WASH, 61- Concurrent Pos: Consult, NZ Govt, 54-56, James Aviation, Ltd, 54-, Midwest Appl Sci Corp, Ind, 56- & Math Sci Corp Calif, 60-; dir, Math Sci Northwest Inc. Mem: Assoc fel Am Inst Aeronaut & Astronaut; Am Soc Eng Educ. Res: Systems analysis; solid mechanics; structural analysis and design; vibration theory; photothermoviscoelasticity. Mailing Add: 6810 51st NE Seattle WA 98115

BOLLE, DONALD MARTIN, b Amsterdam, Netherlands, Mar 30, 33; US citizen; m 57; c 4. ELECTROMAGNETISM. Educ: Univ Durham, BSc, 54; Purdue Univ, PhD(elec eng), 61; Brown Univ, MA, 66. Prof Exp: Res engr, EMI Industs, Eng, 54-55; asst elec eng, Purdue Univ, 56, from instr to asst prof, 56-62; sr visitor appl math, Cambridge Univ, 62-63; from asst prof to assoc prof eng, Brown Univ, 63-70, prof, 70-80; chmn & Chandler-Weaver prof, Dept Elec & Comput Eng, 80-81, dean, Col Eng & Phys Sci, Lehigh Univ, 81-88; SR VPRES ACAD AFFAIRS, POLYTECH UNIV NY, 88- Concurrent Pos: Consult, Md Electronics Mfg Co, 56-58, Gen Elec Co, 58, Western Elec Co, 59, Bell Tel Labs, 60, P R Mallory & Co, 61 & Naval Underwater Systs Ctr, 69-80; NSF fel, 62-63. Honors & Awards: Distinguished Serv Award, Oceanic Eng Soc, 80; Centennial Medal, Inst Elec & Electronics Engrs, 84. Mem: Am Soc Eng Educ; fel Inst Elec & Electronics Engrs; fel AAAS. Res: Electromagnetic theory, propagation, radiation, scattering. Mailing Add: Sr VPres Acad Affairs Polytech Univ 333 Jay St Brooklyn NY 11201

BOLLER, FRANCOIS, b Montreux, Switz, July 17, 38; US citizen; m 65. BEHAVIORAL NEUROLOGY, NEUROPSYCHOLOGY. Educ: Univ Pisa, Italy, MD, 63; Case Western Reserve Univ, Cleveland, PhD(exp psychol), 82. Prof Exp: Asst prof neurol, Boston Univ, 71-73; assoc prof neurol, Case Western Reserve Univ, 73-79; ASSOC PROF NEUROL, MED SCH, UNIV PITTSBURGH, 79- Mem: Am Neurol Asn; Int Neuropsychol Soc; Behav Neurol Soc (secy-treas, 82-). Res: Disorders of higher cortical functions (aphesia, amnesia, dementia); Alzheimer disease. Mailing Add: Inserm Unit 324 Centre Paul Broca Two Ter Rue D'Alesia Paris 75014 France

BOLLES, THEODORE FREDERICK, b LaCrosse, Wis, Nov 25, 40; m 62; c 5. INORGANIC CHEMISTRY, PHYSICAL CHEMISTRY. Educ: Univ Wis, BS, 62; Univ Ill, PhD(inorg chem), 66. Prof Exp: Res chemist, 3M Co, 66-71, supvr, Nuclear Med Res Lab, Nuclear Prod Proj, 71-76, mgr, Graphic Arts Lab, 77, tech dir, Transp & Com Graphics Div, 81-, Printing & Publishing Systs Div, TECH DIR SCI RES LAB, 3M CO. Mem: AAAS. Soc. Res: Organometallic chemistry of group IV, carbon, silicon, germanium, tin and lead; thermodynamics; calorimetry; coordination chemistry; nuclear medicine; radiopharmaceuticals; biochemistry; drug metabolism. Mailing Add: 7249 Courtly Rd St Paul MN 55119

BOLLET, ALFRED JAY, b New York, NY, July 15, 26; m 54. MEDICINE. Educ: NY Univ, BS, 45, MD, 48; Am Bd Internal Med, dipl. Prof Exp: Resident physician, Chest Serv, Bellevue Hosp, New York, 50, asst resident physician, 52-53; clin res assoc, Nat Inst Arthritis & Metab Dis, 53-55; asst prof med, Col Med, Wayne State Univ, 55-59; assoc prof prev med & med, Sch Med, Univ Va, 59-65, prof internal & prev med, 65-66; prof med & chmn dept, Med Col Ga, 66-74; prof med & chmn dept, State Univ NY Downstate Med Ctr, 74-80; clin prof, 80-90, VPRES ACAD AFFAIRS, DANBURY HOSP & SCH MED, YALE UNIV, 90- *Concurrent Pos:* Asst, Sch Med, Johns Hopkins Univ, 54-55; consult physician, Baltimore City Hosp, 54-55; Markle scholar med sci, 56-61; mem, Am Bd Intern Med, 71-78. *Mem:* AAAS; Am Soc Clin Invest; Am Fedn Clin Res (pres, 63-64); Am Rheumatism Asn; Harvey Soc. *Res:* Clinical and laboratory studies in the rheumatic diseases. *Mailing Add:* Sch Med Yale Univ Danbury Hosp Danbury CT 06810

BOLLETER, WILLIAM THEODORE, b Lufkin, Tex, June 29, 27; m 52; c 3. ANALYTICAL CHEMISTRY, PROPELLANT CHEMISTRY. *Educ:* Tex Col Arts & Indust, BS, 48, MS, 50; Univ Tex, PhD(anal chem), 55. *Prof Exp:* Instr chem, San Antonio Col, 50-52; res chemist, Monsanto Chem Co, 55-62; group supvr, Hercules Inc, Utah, 62-64, asst supt, 64-66, sr tech specialist, Va, 66-68, asst tech dir, 68-74, tech dir, 74-85, qual assurance & tech dir, 85-89; RETIRED. *Mem:* Am Chem Soc. *Res:* Spectroscopy; propellant chemistry. *Mailing Add:* 220 Price St Blacksburg VA 24060

BOLLING, GUSTAF FREDRIC, b Toronto, Ont, Apr 20, 31; c 2. METALLURGICAL ENGINEERING, ENGINEERING PHYSICS. *Educ:* Univ Toronto, BASc, 54, MASc, 55, PhD(metall eng), 57. *Prof Exp:* Nat Res Coun Can fel, Univ Birmingham, Eng, 57-58; res engr, Westinghouse Res Labs, Pa, 58-62; res scientist, 62-72, res & planning mgr, 72-74, mat planning & strategy mgr, 74-76, exec engr, Vehicle Mat Develop & Planning, 76-81, assoc dir & dir, North Am Res Liaison Off, 81-84, DIR, MFG DEVELOP CTR, FORD MOTOR CO, 84- *Concurrent Pos:* Demonstr appl physics, Univ Toronto, 54-55, lectr physics of metals, 55-57; fel engr, Westinghouse Res Labs, Pa, 61-62; expert consult, Orgn Am State, 68-72 & Space Shuttle Producibility Assessment, NASA, 81. *Mem:* Sigma Xi; Am Inst Mining, Metall & Petrol Engrs. *Res:* Solidification; structure sensitivity; economics of research; development of lightweight materials and new processes; directed advanced development from research through commercialization in the core manufacturing technologies, manufacturing systems and operations engineering, automation and machining, emphasizing applied computer technologies in manufacturing and close partnerships with customers. *Mailing Add:* 5000 Town Ctr Dr Apt 1602 Southfield MI 48075

BOLLINGER, EDWARD H(ARRY), b Milwaukee Wis, Aug 30, 26; m 56; c 4. CHEMICAL ENGINEERING. *Educ:* Rose Polytech Inst, BS, 49; Univ Ill, MS, 51; Ohio State Univ, PhD(chem eng), 58. *Prof Exp:* Res engr, Battelle Mem Inst, 51-52, consult, 53-58; asst, Ohio State Univ, 53; sr res engr, res ctr, B F Goodrich, 58-63, prog mgr, aerospace & defense div, 63-69, tech mgr corp new prod, 69-70, sect leader, aerospace res, 70-73, sr planning analyst, 73-78, mgr planning systs, Tire Div, 78-83, mgr, Com Anal, Res & Develop Ctr, 83-87; RETIRED. *Concurrent Pos:* Chm, Cuyahoga River Basin Water Qual Comt; chm, Summit County Energy Task Force, CO Naval Res Reserve Co. *Mem:* Am Chem Soc; Am Inst Chem Engrs. *Res:* Transport phenomenon; thermochemistry; industrial chemistry; management systems; acoustics. *Mailing Add:* 442 Mowbray Rd Akron OH 44333

BOLLINGER, FREDERICK W(ILLIAM), b Tyndall, SDak, Aug 5, 18; m 45; c 3. ORGANIC CHEMISTRY. *Educ:* Univ SDak, BS, 39; State Col Wash, MS, 41; Calif Inst Technol, MS, 43; Ill Inst Technol, PhD(chem), 51. *Prof Exp:* Asst chemist, State Chem Lab, SDak, 39; asst chem, Ill Inst Technol, 48-50; post-doctoral fel, Col Med, Univ Ill, 50-51; sr chemist, Merck & Co, Inc, 51-87; RETIRED. *Honors & Awards:* Thomas Emery McKinney Prize, 39. *Mem:* Am Chem Soc. *Res:* Stereochemistry; parkinsonism; natural products and azido compounds. *Mailing Add:* 607 Lawrence Ave Westfield NJ 07090

BOLLINGER, GILBERT A, b Edwardsville, Ill, Jan 4, 31; m 55; c 3. GEOPHYSICS, SEISMOLOGY. *Educ:* St Louis Univ, PhD(geophys), 67. *Prof Exp:* Geophysicist, Calif Oil Co, 56-61; res asst seismol, St Louis Univ, 61-67; PROF GEOPHYS, VA POLYTECH INST & STATE UNIV, 67- *Mem:* Seismol Soc Am (pres, 83-84); Am Geophys Union; AAAS. *Res:* Earthquake and engineering seismology. *Mailing Add:* Dept Geol Sci Va Polytech Inst & State Univ Blacksburg VA 24061

BOLLINGER, JOHN G(USTAVE), b Grand Forks, NDak, May 28, 35; m 58; c 3. MECHANICAL, ELECTRICAL & MANUFACTURING ENGINEERING. *Educ:* Univ Wis, BS, 57, PhD(mech eng), 61; Cornell Univ, MS, 58. *Prof Exp:* From instr to prof mech eng, 60-85, PROF, DEPT ECE & IE, COL ENG, UNIV WIS-MADISON, 85-, BASCOM PROF ENG, 73-, DEAN, 81- *Concurrent Pos:* Fulbright res fel, Aachen, Ger, 62-63; US Educ Comt intercountry lectr grant, Univs Birmingham & Manchester, 63, Cranfield Inst, 80; chmn bd, Unico, Inc, 83-89; mem, vis comt, Nat Inst Standards & Technol, 88-; mem technol adv bd, US Postal, 88-; assoc ed, J Mfg Systs, 89- *Honors & Awards:* Gustus L Larson Mem Award, Am Soc Mech Engrs, 76, Centennial Award, 80; Res Medal, Soc Mfg Engrs, 78. *Mem:* Nat Acad Eng; fel Am Soc Mech Engrs; Am Soc Eng Educ; fel Soc Mfg Engrs; Nat Soc Prof Engrs. *Res:* Computer control of machines and processes; environmental noise control; control and design of machines; structural dynamics. *Mailing Add:* Col Eng Univ Wis Madison WI 53706

BOLLINGER, LOWELL MOYER, b Greene Co, Va, Apr 28, 23; m 44; c 3. EXPERIMENTAL NUCLEAR PHYSICS. *Educ:* Oberlin Col, AB, 43; Cornell Univ, PhD, 51. *Prof Exp:* Physicist, Nat Adv Comt Aeronaut, 43-46; from res asst to res assoc physics, Cornell Univ, 46-51; dir physics div, 63-72 & 73-74, PHYSICIST, ARGONNE NAT LAB, 51-, DIR SUPERCONDUCTING LINAC PROJ, 72- *Honors & Awards:* Bonner Prize, Am Phys Soc. *Mem:* Am Phys Soc; AAAS. *Res:* Nuclear physics of neutron-induced reactions; development of the technology and design, and construction of the first superconducting heavy-ion linear accelerator; cosmic rays; combustion research. *Mailing Add:* Argonne Nat Lab Bldg 203 Argonne IL 60439

BOLLINGER, LYNN DAVID, plasma physics, for more information see previous edition

BOLLINGER, RALPH R, b Oct 3, 44; m; c 2. TRANSPLANTATION IMMUNOLOGY. *Educ:* Tulane Univ, MD, MS(biochem) 70; Duke Univ, PhD(immunol), 77. *Prof Exp:* CHIEF SURG TRANSPLANTATION, DUKE UNIV MED CTR, 80- *Mem:* Am Col Surgeons; Transplantation Soc; Soc Univ Surgeons; Am Soc Transplant Surgeons. *Res:* Transplantation immunology. *Mailing Add:* Dept Surg Duke Univ Box 2910 Durham NC 27710

BOLLINGER, RICHARD COLEMAN, b Johnstown, Pa, Feb 1, 32; m 54, 79; c 5. APPLIED MATHEMATICS. *Educ:* Univ Pittsburgh, BS, 54, MS, 57, PhD(math), 61. *Prof Exp:* Eng specialist, Westinghouse Res Labs, 54-59, res mathematician, 59-62; from asst prof to prof math, Pa State Univ, 62-88, prof-in-charge, Behrend Col Grad Ctr, 74-79. *Mem:* AAAS; Math Asn Am; Soc Indust & Appl Math. *Res:* Numerical analysis; mathematical analysis and applied mathematics; combinatorics and finite mathematics. *Mailing Add:* Dept Math Behrend Col Pa State Univ Station Rd Erie PA 16563

BOLLINGER, ROBERT OTTO, b Detroit, Mich, June 15, 39; m 62; c 2. CELL BIOLOGY, PATHOLOGY. *Educ:* Wayne State Univ, BS, 62, MS, 65, PhD(biol), 72. *Prof Exp:* Asst prof zool, Eastern Ill Univ, 69-71; res assoc, Child Res Ctr, Mich, 71-76; asst prof pediat & assoc path, 76-89, DIR CONTINUING MED EDUC, SCH MED, WAYNE STATE UNIV, 89- *Concurrent Pos:* Adj asst prof, Col Pharm & Allied Health Prof, Wayne State Univ, 73-76. *Mem:* Soc Appl Learning Technol; Electron Microscope Soc Am. *Res:* Continuing medical education; applications of computer science in medicine and education. *Mailing Add:* Univ Health Ctr 4H Wayne State Univ 4201 St Antoine Detroit MI 48201

BOLLINGER, WILLIAM HUGH, b Los Angeles, Calif, Aug 18, 47. REFORESTATION, REVEGETATION. *Educ:* Univ Calif, BA, 69; Univ Colo, MA, 72, PhD(ecol), 74. *Prof Exp:* Teaching asst biol, Univ Colo, 69-74; mem staff field ecol, Continuing Educ, Univ Colo, 72-74; sr mem staff ecol, VTN Corp, 74-76; vpres res, Native Plants, Inc, 76-89; VPRES, ENVIRON DIV, TERRATEK, 89- *Concurrent Pos:* Mem staff bot res, Univ Calif Bot Gardens, 66-67; reforestation prog leader, US Agency Int Development, Sri Lanka, 78, reforestation proj staff, Kenya, 80; adv comt tech consult food & renewable resources prog, US Congress Off Technol Assessment, 81-82. *Mem:* Am Mus Natural Hist; Sigma Xi. *Res:* Relation between energy, environment and development: technical developments of solutions to vegetation problems, particularly the restoration of primary vegetation on severely damaged landscapes. *Mailing Add:* Univ Res Park 400 Wakara Way Salt Lake City UT 84108

BOLLS, NATHAN J, JR, b De Witt, Mo, May 16, 31; m 62; c 1. VERTEBRATE PHYSIOLOGY. *Educ:* Kans State Univ, BS, 59, MS, 61, PhD(vert physiol), 63. *Prof Exp:* From instr to asst prof, 63-68, assoc prof, 68-80, PROF BIOL, WITTENBERG UNIV, 80- *Concurrent Pos:* Author & proj dir, NSF Matching Funds grant for undergrad physiol lab equip, 65-67. *Mem:* Am Soc Mammalogists. *Res:* Role of glucocorticoids in energy utilization during bodily dehydration in mammals; ecological physiology of tree squirrels. *Mailing Add:* Dept Biol Wittenberg Univ Springfield OH 45501

BOLLUM, FREDERICK JAMES, b Ellsworth, Wis, June 14, 27; div; c 4. BIOCHEMISTRY. *Educ:* Univ Minn, BA, 49, PhD(physiol chem), 56. *Prof Exp:* Asst physiol chem, Univ Minn, 53-55, instr, 55-56; USPHS fel, Univ Wis, 56-58; biochemist, Biol Div, Oak Ridge Nat Lab, 58-65; prof biochem, Med Sch, Univ Ky, 65-77; chmn dept, Uniformed Servs Univ Health Sci, 77-81, prof biochem, 77-89; PRES, SUPERTECHS INC, 89- *Concurrent Pos:* Consult radiation study sect, NIH, 67-68, biochem study sect, 69-73 & cancer treat rev panel, 75-77; vis prof, Univ Chile, 69, Johns Hopkins Univ, 83; consult, Am Cancer Soc, 71-72; USPHS spec fel, Med Ctr, Univ Calif, 72-73. *Honors & Awards:* K A Forster Award & Lectureship, Mainz Acad Sci, Ger, 74. *Mem:* Am Soc Biol Chem; Am Chem Soc; Sigma Xi. *Res:* Enzymes of nucleic acid synthesis; structure and function of nucleic acid; differentiation of lymphoid cells. *Mailing Add:* Dept Biochem Uniformed Serv Univ Health Sci 4301 Jones Bridge Rd Bethesda MD 20814

BOLLYKY, L(ASZLO) JOSEPH, b Bolyok, Hungary, Dec 9, 32; US citizen; m; c 3. CHEMICAL ENGINEERING. *Educ:* Budapest Tech Univ, dipl, 56; Columbia Univ, MA, 61, PhD(chem), 64. *Prof Exp:* Chem engr, Nitrokemia Co, Hungary, 56; chemist, Am Cyanamid Co, NJ, 57-60, sr res chemist & proj leader, Cent Res Div, 63-70; vpres & tech dir, Pollution Control Industs, Inc, 70-77, pres subsid, PCI Ozone Corp, 73-77; PRES, BOLLYKY ASSOC, INC, 77- *Concurrent Pos:* Found & ed-in-chief, Ozone Sci & Eng, 79-84; dir Int Ozone Asn. *Mem:* Am Chem Soc; Water Pollution Control Fedn; Am Waterworks Asn; Am Inst Chem Engrs; hon mem Int Ozone Asn (vpres, pres-elect). *Res:* Water, waste water treatment and air pollution control; ozone generation and application for pollution control; chemiluminescent reactions; photochemistry; Ozone-UV reactions and reactor design. *Mailing Add:* Bollyky Assoc Inc 83 Oakwood Ave Norwalk CT 06850

BOLMAN, WILLIAM MERTON, b Elyria, Ohio, Oct 8, 29; m 55; c 2. CHILD PSYCHIATRY. *Educ:* Harvard Univ, BA, 51, MD, 55; Am Bd Psychiat & Neurol, dipl, 63 & 69. *Prof Exp:* NIMH career develop award, Boston Univ & Boston City Hosp, 63-64; fel community psychiat, Harvard Lab Community Psychiat, 64-65; from asst prof to assoc prof psychiat, Med Sch, Univ Wis, 65-69; dir, Westside Community Ment Health Ctr, Inc, 69-70; PROF PSYCHIAT, SCH MED, UNIV HAWAII, 70-, PROF PUB HEALTH, 78- *Concurrent Pos:* Consult psychiat, Bur Testing & Guid, Univ Hawaii, 57-59 & Mass Div Legal Med, 60-63; consult child psychiat, Dane County Ment Health Ctr, 64-69, St Coletta's Sch, 67-69 & Joint Comn Ment Health Children, 67-68; consult prev psychiat, Wis State Comprehensive Ment Health Comt, 66-67; mem juv probs res rev comt, NIMH, 71-; med dir, Child Guid Clin, Kauikeolani Children's Hosp, 71-80; Consult, comt psychiat, WHO, Indonesia, 79. *Mem:* Fel Am Psychiat Asn; fel Am Orthopsychiat Asn; fel Am Pub Health Asn; fel Royal Soc Health. *Res:* Community mental health; child psychiatry. *Mailing Add:* Kapiolani Children's Hosp 1319 Punahou St Rm 636 Honolulu HI 96826

BOLMARCICH, JOSEPH JOHN, b Philadelphia, Pa, Dec 22, 42; div; c 3. OPERATIONS RESEARCH, APPLIED MATHEMATICS. *Educ:* Drexel Univ, BS, 65; Univ Pa, PhD(appl math), 72. *Prof Exp:* Res physicist, Frankford Arsenal, 65-72; consult opers res & math, Daniel H Wagner Assocs, 72-76; OPERS RES CONSULT & TREAS, QUANTICS INC, 77- *Honors & Awards:* David Rist Prize, Mil Opers Res Soc, 75. *Mem:* Inst Mgt Sci; Soc Indust & Appl Math; Mil Opers Res Soc; Math Asn Am. *Res:* Bayeslan information processing; combinatorial probability; mathematical logistics. *Mailing Add:* Quantics Five Country View Rd Suite 240 Malvern PA 19355

BOLME, DONALD W(ESTON), b Louisville, Ky, Apr 17, 28; m 52; c 3. CHEMICAL ENGINEERING. *Educ:* Mich Col Mining & Technol, BS, 51; Univ Wash, MS, 55, PhD(chem eng), 57. *Prof Exp:* Chem engr, Dow Chem Co, Mich, 57-59 & Stanford Res Inst, 59-60; asst prof chem eng, Wash State Univ, 60-65; sr develop engr, Pac Northwest Labs, Battelle Mem Inst, 65-70; CONSULT CHEM & PROCESS ENG, 70- *Mem:* Am Chem Soc; Am Inst Chem Engrs. *Res:* Reduction and control of industrial pollution, air and water; evaluations and definition of hazards including explosions, pollutants and radiochemicals; development and analysis of process systems; kinetics; mechanisms and heat transfer; process kinetics and industrial chemistry of nitrogen oxides and nitric acid. *Mailing Add:* Bolme Eng 1001 Fourth Ave Suite 3200 Seattle WA 98154-1075

BOLME, MARK W, b Ft Belvoir, Va, Dec 10, 52. SOFTWARE SYSTEMS, CHEMICAL ENGINEERING. *Educ:* Mich State Univ, BS, 75; Univ Wis, PhD(chem eng), 81. *Prof Exp:* Pilot Plant eng, Celanese Chem Corp, 81-84; CONSULT, 84- *Mem:* Am Inst Chem Eng; Apple Programmer & Developer Asn. *Mailing Add:* PO Box 3746 Bellevue WA 98009

BOLMER, PERCE W, b Springfield, Ohio, Nov 24, 28; m 56; c 1. ELECTROCHEMISTRY, SURFACE CHEMISTRY. *Educ:* Ohio Univ, BS, 50; Purdue Univ, MS, 52, PhD(anal chem), 54. *Prof Exp:* Sr res technologist, Field Res Lab, Mobil Oil Co Inc, 54-66; sr res technologist, Kaiser Aluminum & Chem Corp, 66-73, head finishing sect, 73-87, consult ctr technol, 87-90; RETIRED. *Mem:* Am Chem Soc; Am Electroplaters & Surface Finishers. *Res:* Electrochemistry of corrosion reactions; aluminum and other metal finishing; solid waste disposal; industrial pollution control; anodic oxides; electrodeposition; paint application technology; anodizing. *Mailing Add:* Seven Pasillo Way Hot Springs Village AR 71909

BOLOGNESI, DANI PAUL, b Forgaria, Italy, Mar 19, 41; US citizen; m 64; c 2. VIROLOGY, IMMUNOLOGY. *Educ:* Rensselaer Polytech Inst, BS, 63, MS, 65; Duke Univ, PhD(virol), 67. *Prof Exp:* Res assoc, Dept Surg, Duke Univ Med Ctr, 67-68; NIH postdoctoral fel, Max-Planck Inst Virus Res, WGer, 68-71; from asst prof to prof surg, Duke Univ Med Ctr, 71-84, from asst prof to assoc prof microbiol & immunol, 71-80, dep dir, Duke Comprehensive Cancer Ctr, 81-87, PROF, MICROBIOL & IMMUNOL, DUKE UNIV MED CTR, 81-, JAMES B DUKE PROF, 84-, DIR, DUKE CTR AIDS RES, 87- *Concurrent Pos:* Mem tumor virus detection segment rev bd, Nat Cancer Inst, 73-76; consult, M D Anderson Hosp & Tumor Inst, Tex & Frederick Cancer Res Ctr, Md, 73-74; Nat Cancer Inst grant, Med Ctr, Duke Univ, 74-, USPHS grant, 74-75; Am Cancer Soc grant, 74-76 & Am Soc Fac Res award, 76-81; mem med & sci adv comt, Leukemia Soc Am; mem study sect, Am Cancer Soc, 77-80 & NIH Virol, 77-81; ed, AIDS Res & Human Retroviruses; mem, AIDS Task Force, Nat Cancer Inst, chmn subcomt, AIDS, adv, Vaccine Task Force; mem sci adv bd, Retrogenes, Inc; mem adv comt, AIDS Policy, NIH; mem sci adv comt, Am Found AIDS Res, 85-; Mem selection comt, AIDS Vaccine, Nat Inst Allergy & Infectious Dis. *Mem:* Am Soc Microbiol. *Res:* Tumor virology; tumor immunology; human cancer; representative of over two hundred full length publications. *Mailing Add:* Dept Surg Duke Univ Med Ctr PO Box 2926 Durham NC 27710

BOLON, ALBERT EUGENE, b Montgomery, Ala, July 19, 39; m 65; c 2. NUCLEAR ENGINEERING. *Educ:* Mo Sch Mines, BS, 61, MS, 62; Iowa State Univ, PhD(nuclear eng), 65. *Prof Exp:* Asst prof, 65-71, ASSOC PROF NUCLEAR & METALL ENG, UNIV MO, ROLLA, 71-, REACTOR DIR, 80- *Concurrent Pos:* Dept chair, Univ Mo, Rolla, 87- *Mem:* Am Nuclear Soc; Soc Prof Engrs; Sigma Xi. *Res:* Basic or applied research on materials related to applications of nuclear power; safety, environmental impacts and licensing of fission and fusion power plants. *Mailing Add:* Fulton Hall Univ Mo Rolla MO 65401

BOLON, DONALD A, b Cleveland, Ohio, Sept 7, 34; m 59; c 4. ORGANIC CHEMISTRY. *Educ:* Allegheny Col, BS, 56; Univ Minn, PhD(org chem), 60. *Prof Exp:* Fel, Harvard Univ, 60-62; RES CHEMIST, GEN ELEC CO, 62- *Mem:* Am Chem Soc. *Res:* Carbene chemistry; free radical rearrangements and reactions; autoxidation and catalytic oxidative coupling reactions; solvolitic reactions; photochemical crosslinking; acetylene chemistry; ultraviolet light cured coatings; electrical insulation coatings; epoxy resin cure and use. *Mailing Add:* Gen Elec Co Res & Develop Ctr PO Box 8 Schenectady NY 12301

BOLON, ROGER B, b Cleveland, Ohio, May 19, 39; m 63; c 2. MATERIALS CHARACTERIZATION, SURFACE ANALYSIS TECHNIQUES. *Educ:* Pa State Univ, BS, 61; Col Advan Sci, MS, 63. *Prof Exp:* Assoc mem staff, 66-74, mem res staff & supvr, 74-80, MGR RES & DEVELOP, GEN ELEC CORP, 80- *Honors & Awards:* IR-100 Award, 69; Corning Award, Microbeam Anal Soc, 77. *Mem:* Microbeam Anal Soc (treas, 79-80, pres, 82-83). *Res:* New and or improved instrumentation methods and techniques for scanning electron microscopy; electron microprobe analysis; auger electron spectroscopy; photo electron spectroscopy. *Mailing Add:* Corp Res & Develop Gen Elec Corp One River Rd Schenectady NY 21345

BOLOTIN, HERBERT HOWARD, b New York, NY, Jan 11, 30; m 51; c 2. NUCLEAR PHYSICS. *Educ:* City Col New York, BS, 50; Ind Univ, MS, 52, PhD(exp nuclear physics), 55. *Hon Degrees:* DSc, Univ Melbourne, 80. *Prof Exp:* Physicist, US Naval Radiol Defense Lab, 55-58; res fel nuclear physics, Brookhaven Nat Lab, 58-60; asst physicist, 60-61; asst prof physics, Mich State Univ, 61-62; assoc physicist, Argonne Nat Lab, 62-71; CHAMBER MFRS PROF PHYSICS, UNIV MELBOURNE, 71- *Concurrent Pos:* Vis scientist, Univ Calif, Berkeley, 62 & Los Alamos Sci Lab, 66; chmn, Conf Slow-Neutron-Capture Gamma Rays, Argonne Nat Lab, 66; vis scientist, Brookhaven Nat Lab, 68; mem adv comt, Int Conf Neutron-Capture Gamma Rays Studies, Sweden, 69 & Int Conf Nuclear Structure Study Neutrons, Hungary, 72; vis prof nuclear physics, Osaka Univ Res Ctr, 77; consult nuclear physics, Rutgers Univ, 79; vis prof, Lawrence Berkeley Lab, 79; consult, Chinese Acad Sci, 84; chmn, Int Conf on Nuclear Structure, Univ Melbourne, 87; ed, Proc Int Conf on Nuclear Structure, Melbourne, 87. *Mem:* fel Australian Inst Physics. *Res:* Experimental nuclear physics; beta and gamma ray, neutron and nuclear spectroscopy; lifetimes of excited nuclear states; neutron capture gamma rays; nuclear structure; interdisciplinary research in nuclear medicine investigations; static magnetic moments excited nuclear states. *Mailing Add:* Sch Physics Univ Melbourne Parkville Victoria 3052 Australia

BOLOTIN, MOSHE, b Jerusalem, Palestine, July 3, 22; US citizen; m 60; c 2. BOTANY, FORESTRY. *Educ:* Univ Wash, BS, 51; Univ BC, MF, 58; Smith Col, PhD(bot), 74. *Prof Exp:* Forester growth study, State Sustained Yield Forest, Olympia, Wash, 51-52; state forester, Israel Forest Serv, 52-56, silviculture, 58-65; ASST PROF BIOL, HOLYOKE COMMUNITY COL, 69- *Mem:* Sigma Xi. *Res:* Juvenility in woody plants; flowering in woody plants; tree improvement. *Mailing Add:* Div Biol Holyoke Community Col 303 Homestead Ave Holyoke MA 01040

BOLS, NIELS CHRISTIAN, b Revelstoke, BC, Feb 3, 48. FISH CELL CULTURING, BIOTECHNOLOGY. *Educ:* Simon Fraser Univ, BS, 70; Univ BC, MS, 72; Univ Toronto, PhD(zool), 75. *Prof Exp:* Fel, Karolinska Inst, Sweden, 76-77; from asst prof to assoc prof, 77-90, PROF CELL BIOL, REPRODUCTIVE PHYSIOL & BIOTECHNOL, UNIV WATERLOO, 90- *Concurrent Pos:* Vis scientist, Univ Calif, San Diego, 83-84. *Mem:* Tissue Cult Asn; Can Soc Cell Biol; Can Fedn Biol Soc. *Res:* Development of fish cell lines and culturing technology for the study of the growth and differentiation of fish cells in vitro; contribution of animal cell cultures to biotechnology. *Mailing Add:* Dept Biol Univ Waterloo Waterloo ON N2L 3G1 Can

BOLSAITIS, PEDRO, b Daugavpils, Latvia, July 12, 37; m 67. PHYSICAL METALLURGY, THERMODYNAMICS. *Educ:* Calif Inst Technol, BS, 60, MS, 61; Univ Del, PhD(chem eng), 64. *Prof Exp:* Res asst, Max Planck Inst Phys Chem, Ger, 64-65; res engr, Esso Res & Eng Co, 65-66; from asst prof to assoc prof chem eng, 66-76, PROF CHEM ENG, GRAD SCH, UNIV MD, 76- *Concurrent Pos:* Consult, Econ Assocs, Inc, 67- & Bur Mines, 68- *Mem:* Am Chem Soc; Am Soc Metals; Am Inst Mining, Metall & Petrol Engrs. *Res:* Theory of alloys; thermodynamics of condensed systems; solid electrolytes. *Mailing Add:* Mass Inst Technol Rm 8-115 Cambridge MA 02139-4301

BOLSTAD, LUTHER, b Floodwood, Minn, Feb 11, 18; m 54; c 2. PLASTICS CHEMISTRY. *Educ:* St Olaf Col, BA, 41; NDak State Univ, MS, 43; Purdue Univ, PhD(chem), 49. *Prof Exp:* Chemist, Plastics Res & Develop, Bakelite Corp, 43-45 & Honeywell, Inc, Minn, 51-65; chemist, Stackpole Carbon Co, 65-78; CONSULT, 78- *Mem:* Sigma Xi; Am Chem Soc. *Res:* Aerospace materials; atmospheric contamination within manned spacecraft; carbon composition resistors; electrical properties of carbon dispersions in thermosetting resins and ceramics; epoxy resins. *Mailing Add:* 1222 MD Hwy Kersey PA 15846

BOLSTEIN, ARNOLD RICHARD, b New York, NY, Sept 2, 40; m 63; c 1. SURVEY SAMPLING, STATISTICAL ESTIMATION. *Educ:* Wagner Col, BA, 62; Purdue Univ, MS, 64, PhD(math), 67. *Prof Exp:* Grant, Univ NC, Chapel Hill, 68, asst prof math, 67-73; assoc prof math, 73-84, ASSOC PROF STATIST, GEORGE MASON UNIV, 84- *Concurrent Pos:* Consult, Acad Interscience Methodology, 77-80; prin investr statist anal surv data, Nat Marine Fisheries, 81-82, 90-; design work sampling methodology, Va Dept Social Serv, 84-87; design survey, Alexandria, Va, 90- *Mem:* Am Statist Asn; Opers Res Soc Am. *Res:* Measurement error in surveys; non-response error in surveys; effect of incentives on survey response; pre-election polls. *Mailing Add:* Dept Opers Res & Statist George Mason Univ Fairfax VA 22030

BOLSTERLI, MARK, b New Haven, Conn, Oct 3, 30; m 74. THEORETICAL PHYSICS. *Educ:* Wash Univ, AB, 51, PhD(physics), 55. *Prof Exp:* Res assoc, Argonne Nat Lab, 55; Fulbright student physics, Univ Birmingham, 55-56; res assoc, Wash Univ, 56-57; from asst prof to prof physics, Univ Minn, 57-68; STAFF MEM, LOS ALAMOS SCI LAB, 68- *Concurrent Pos:* Vis prof, Copenhagen Univ, 61-62; consult, Los Alamos Sci Lab, 63-64; Guggenheim fel, Oxford Univ, 64-65; vis staff mem, Swiss Inst Nuclear Res, 81. *Mem:* Fel Am Phys Soc. *Res:* Theoretical nuclear physics, especially quantum field theory for systems of mesons and nucleons. *Mailing Add:* Los Alamos Nat Lab T-2 MS B243 Los Alamos NM 87545

BOLT, BRUCE A, b Largs, NSW, Australia, Feb 15, 30; m 56; c 4. SEISMOLOGY, APPLIED MATHEMATICS. *Educ:* Univ New Eng, Australia, BSc, 52; Univ Sydney, MSc, 54, PhD(math), 56. *Hon Degrees:* DSc, Univ Sydney, 72. *Prof Exp:* Lectr appl math, Univ Sydney, 54-61, sr lectr, 62; PROF SEISMOL & DIR SEISMOG STA, UNIV CALIF, BERKELEY, 63- *Concurrent Pos:* Fulbright res fel, Columbia Univ, 60; vis res scientist, Dept Geod & Geophys, Cambridge Univ, 61; consult seismologist, UK Atomic Energy Authority, 61; mem, Comt Seismol, Nat Acad Sci, 65-72 & Earthquake Eng Res Inst, 67-; ed, Bull Seismol Soc Am, 65-; consult, Vet Admin seismic adv comt, 71-; mem, Gov's Earthquake Coun, Calif, 72-74 & Seismic Safety Comn, Calif, 78- *Honors & Awards:* H O Wood Award for Res in Seismol. *Mem:* Nat Acad Eng; fel Am Geophys Union; Seismol Soc Am (pres, 74-75); assoc Royal Astron Soc; Australian Math Soc. *Res:* Classical mechanics as applied to the propagation of waves and structure of the earth; statistical and computational methods of data reduction. *Mailing Add:* Seismog Sta Dept Geol & Geophys Univ Calif Berkeley CA 94720

BOLT, DOUGLAS JOHN, b Isabel, Kans, Mar 2, 39; m 62. REPRODUCTIVE PHYSIOLOGY, ENDOCRINOLOGY. *Educ:* Kans State Univ, BS, 61, MS, 66, PhD(animal breeding), 68. *Prof Exp:* From res asst to res assoc animal husb, Kans State Univ, 62-67; fel obstet & gynec, Sch Med, Univ Mo, 67-68; RES PHYSIOLOGIST, DAIRY CATTLE RES BR, AGR RES SERV, USDA, 68- *Mem:* Am Asn Animal Sci; Soc Study Reproduction. *Res:* Endocrinology and physiology of reproduction, especially as related to corpus luteum function. *Mailing Add:* Agr Res Ctr USDA Beltsville MD 20705

BOLT, JAY A(RTHUR), b New Era, Mich, Oct 31, 11; m 38; c 2. MECHANICAL ENGINEERING. *Educ:* Mich State Univ, BSME, 34, ME, 47; Chrysler Inst, MME, 37; Purdue Univ, MS, 40. *Prof Exp:* Asst prof mech eng, Univ Ill, 38-40 & Univ Notre Dame, 40-41; supvr aircraft carburetor res, Bendix Aviation Corp, 41-47; from assoc prof to prof mech eng, 54-82, prof-in-charge, Automotive Eng Lab, 70-80, EMER PROF MECH ENG, UNIV MICH, 82- *Concurrent Pos:* Consult, indust, govt & legal prof. *Mem:* Fel Am Soc Mech Engrs; fel Soc Automotive Engrs; Nat Soc Prof Engrs; Inst Mech Engrs, London, Eng. *Res:* Internal combustion engine research; induction systems; fuel metering; injection and combustion for piston and turbine engines; air pollution; mechanical analysis and expert testimony in casualty cases involving internal combustion engines in vehicles and aircraft. *Mailing Add:* 2125 Nature Cove Ann Arbor MI 48104

BOLT, JOHN RYAN, b Grand Rapids, Mich, May 28, 40; m 64. VERTEBRATE ZOOLOGY, PALEOZOOLOGY. *Educ:* Mich State Univ, SB, 62; Univ Chicago, PhD(paleozool), 68. *Prof Exp:* Asst prof anat, Med Ctr, Univ Ill, 68-72; asst cur, 72-77, assoc cur, 77-89, chmn, 81-90, CUR FOSSIL REPTILES & AMPHIBIANS, FIELD MUS NATURAL HIST, 89- *Concurrent Pos:* Lectr, Comt Evolutionary Biol, Univ Chicago, 73-; assoc prof, Geol Sci, Univ Ill, Chicago, 85- *Mem:* Soc Vert Paleont; Soc Study Amphibians & Reptiles; Paleont Soc; Soc Syst Zool. *Res:* Evolution and origin of Paleozoic amphibia; evolution of Paleozoic vertebrate faunas. *Mailing Add:* Field Mus Natural Hist Roosevelt Rd at Lake Shore Dr Chicago IL 60605

BOLT, RICHARD HENRY, b Peking, China, Apr 22, 11; m 33; c 3. PHYSICS, ACOUSTICS. *Educ:* Univ Calif, AB, 33, MA, 37, PhD(physics), 39. *Prof Exp:* Nat Res Coun fel physics, Mass Inst Technol, 39-40; assoc, Univ Ill, 40; dir, Underwater Sound Lab, Mass Inst Technol, 41-43; sci liaison officer, Off Sci Res & Develop, London, 43-44; chief tech aide, Div 6, Nat Defense Res Comt, New York, 44-45; assoc prof physics, Mass Inst Technol, 46-54, dir acoustics lab, 46-57, prof acoustics, 54-63, lectr polit sci, 63-70; chmn bd, 53-76, EMER CHMN BD, BOLT BERANEK & NEWMAN, INC, 76- *Concurrent Pos:* Pres, Int Comn Acoust, 51-57; prin consult, NIH, 57-59; assoc dir, NSF, 60-63; fel, Ctr Advan Study Behav Sci, 63-64; mem gov bd, Am Inst Physics; adj prof archit, Mass Inst Technol, 83-85; guest lectr, Acad Sinica, Peking, China, 81. *Honors & Awards:* Biennial Award, Acoust Soc Am, 42, Gold Medal, 76. *Mem:* Nat Acad Eng; fel Acoust Soc Am (pres, 48); Am Phys Soc; fel Inst Elec & Electronics Engrs; fel Am Acad Arts & Sci. *Res:* Theoretical and experimental studies of sound distribution in enclosures; acoustics of rooms; bioacoustics; effects and control of noise; physics of sound; biophysics; science and public policy. *Mailing Add:* Tabor Hill Rd Lincoln MA 01773

BOLT, ROBERT JAMES, b Grand Rapids, Mich, Feb 23, 20; m; c 4. INTERNAL MEDICINE. *Educ:* Calvin Col, AB, 42; Univ Mich, MD, 45. *Prof Exp:* Instr internal med, Univ Mich, 51-52, instr & res assoc, Inst Indust Health, 52-54, from asst prof to prof internal med, 54-66; chmn dept, 66-79, PROF MED, UNIV CALIF, DAVIS, 66- *Concurrent Pos:* Am Cancer Soc fel, Univ Mich, 51-52; attend physician, Vet Admin Hosp, 53; lectr, Univ Medellin, Colombia, 56; lectr, Univ Louvain, 59-60; vis prof, St Francis Hosp, Honolulu; consult, Health Serv, Univ Mich, Vet Admin Hosp, Wayne County Hosp, 63- & David Grant Hosp, Travis AFB. *Mem:* AAAS; Am Soc Gastrointestinal Endoscopy; Am Gastroenterol Asn; Asn Am Med Cols. *Res:* Gastroenterology; small intestinal epithelial cell enzymes and histochemical aspects of the small bowel mucosa in normal and abnormal states. *Mailing Add:* Dept Med Med Ctr Rm 4012 2315 Stockton Blvd Sacramento CA 95817

BOLT, ROBERT O'CONNOR, b Detroit, Mich, Aug 31, 17; m 43, 89; c 3. ORGANIC CHEMISTRY. *Educ:* James Millikin Univ, BS, 39; Purdue Univ, MS, 42, PhD(chem), 44. *Prof Exp:* From res chemist to sr res assoc, 46-69; mgr Mkt Serv, Chevron Res Co, 69-85; CONSULT, 85- *Concurrent Pos:* Mem Air Force Aircraft Nuclear Propulsion Radiation Damage Adv Group, 53-55, Mat Adv Group, 55-58 & AEC Org Reactor Working Group, 61-63; lectr, Calpoly State Univ, San Luis Obispo, Ca, 75- *Mem:* Am Chem Soc; Soc Tribologists & Lubrication Engrs; Sigma Xi. *Res:* Radiolysis of materials; development of radiation resistant lubricants, fuels, nuclear reactor coolants, and other fluids; development of industrial lubricants; preparation and properties of synthetic oils; organic fluoride chemistry. *Mailing Add:* 55 Culloden Park Rd San Rafael CA 94901

BOLTAX, ALVIN, b Brooklyn, NY, Oct 20, 30; m 56, 85; c 3. REACTOR FUELS TECHNOLOGY, RADIATION EFFECTS. *Educ:* Mass Inst Technol, SB, 51, ScD(phys metall), 55. *Prof Exp:* Group leader, Nuclear Metals, Inc, 55-58, proj mgr, 58-60; mgr fuel develop, Astronuclear Lab, 60-67, mgr fuels technol, Advan Energy Syst Div, Westinghouse, 67-90; CONSULT, 90- *Concurrent Pos:* Lectr, Carnegie Mellon Univ, 70-81. *Mem:* Am Nuclear Soc; Am Inst Metall Eng; Am Soc Metals. *Res:* Fast reactor fuels (oxide and carbide); materials development; irradiation testing; irradiation effects (creep, swelling, embrittlement); computer modeling of fuels and materials behavior; compact space reactor fuels and refractory metals; nerva graphite; coated fuel particles and carbide coatings; metals and ceramics fabrication. *Mailing Add:* 272 Baywood Ave Pittsburgh PA 15228

BOLTE, EDOUARD, b Montreal, Que, July 23, 31; m 65; c 5. REPRODUCTIVE PHYSIOLOGY, MEDICINE. *Educ:* Univ Montreal, BA, 51, MD, 56. *Prof Exp:* Resident med, Univ Hosps, Columbia Univ, 58-60; Med Res Coun Can fel steroids, Columbia Univ, 60-62 & Univ Stockholm, 62-63; from asst prof to prof physiol, 64-75, PROF MED, UNIV MONTREAL, 75- *Concurrent Pos:* Mem endocrine serv, Hotel-Dieu Hosp, 64- *Mem:* Endocrine Soc; Can Soc Clin Invest; Int Soc Res Reproduction. *Res:* Various aspects of androgen and estrogen metabolism in human beings in vivo; placental and fetal steroidogenesis. *Mailing Add:* Univ Montreal Hosp 3840 St Urbank Montreal PQ H2W 1T8 Can

BOLTE, JOHN R, b Waterloo, Iowa, Nov 19, 29; m 48; c 3. OPTICS, SCIENCE EDUCATION. *Educ:* Univ Northern Iowa, BA, 51, MA, 56; Okla State Univ, MS, 57; Univ Iowa, PhD(sci educ, physics), 62. *Prof Exp:* Instr pub schs, Iowa, 51-59; instr physics & sci educ, Univ Iowa, 59-62; from asst prof to assoc prof physics, San Diego State Col, 62-68; asst dean, 68-87, PROF PHYSICS, UNIV CENT FLA, 68-, VPRES ADMIN & FINANCE AFFAIRS, 86- *Concurrent Pos:* Consult, Educ Testing Serv, NJ, 66-68; workshops, educ planning & budgeting. *Mem:* Am Asn Physics Teachers. *Res:* Carbon dioxide laser; use of lasers in communications applications; solar energy application; use of computers in science education. *Mailing Add:* Dept Physics Univ Cent Fla Box 25000 Orlando FL 32816

BOLTER, ERNST A, b Mengen-Wuertt, Ger, Feb 16, 35. GEOCHEMISTRY, GEOLOGY. *Educ:* Univ Gottingen, PhD(mineral), 61. *Prof Exp:* Fel, Yale Univ, 61-63; fel, Rice Univ, 63-65; asst prof geol, 65-67, assoc prof geochem, 67-76, PROF GEOCHEM, UNIV MO-ROLLA, 76- *Mem:* Mineral Soc Am; Geochem Soc; Am Geophys Union. *Res:* Geochemical prospecting; environmental geochemistry; water and soil pollution. *Mailing Add:* Dept Geol Univ Mo Rolla MO 65401

BOLTINGHOUSE, JOSEPH C, b Mt Vernon, Ohio, Nov 16, 19; m 55; c 2. INSTRUMENTS, CONTROL THEORY. *Educ:* Ohio Wesleyan Univ, BA, 40; Mass Inst Technol, SB, 42, Calif State Univ, Fullerton, MS, 82. *Prof Exp:* Proj engr, Sperry Gyroscope Co, Inc, 42-46; control engr, Gen Mach Corp, 46-47; sr res engr, Curtiss-Wright Corp, 48-50; electromech res specialist, NAm Aviation Autonetics Div, Rockwell Int, 50-67, staff scientist inertial instruments, Autonetics Marine Systs Div, 67-88; RETIRED. *Honors & Awards:* Thurlow Award, Inst Navig, 74. *Res:* Inertial platform design and instrument research; gas spin bearing gyros; development of electric bearing gyroscopes. *Mailing Add:* 16057 E Arbela Dr Whittier CA 90603

BOLTON, BENJAMIN A, b Wellsboro, Pa, Sept 5, 28; m 54; c 4. ORGANIC CHEMISTRY. *Educ:* Lafayette Col, BS, 50; Northwestern Univ, MBA, 58. *Prof Exp:* Chemist, Atlas Powder Co, 50-52; chemist, Amoco Chem Corp, 54-61, group leader org chem, 61- 70, sect leader polymers & plastics, Res & Develop Dept, 70-72, res supvr, 72-76, div dir, 76-81, res supvr, 81-83, bus develop mgr, 83-86; RETIRED. *Mem:* Am Chem Soc. *Res:* Organic chemicals; product applications; high temperature polymers; foamed polymers; vinyl thermoplastics; process and product development on polystyrenes; condensation polymers. *Mailing Add:* Five Fern Lane Chesterton IN 46304

BOLTON, CHARLES THOMAS, b Camp Forest, Tenn, Apr 15, 43; m 86. ASTRONOMY, STELLAR SPECTROSCOPY. *Educ:* Univ Ill, BS, 66; Univ Mich, MS, 68, PhD(astron), 70. *Prof Exp:* From asst prof to assoc prof, 73-80, assoc dir, 78-88, PROF ASTRON, DAVID DUNLAP OBSERV, UNIV TORONTO, 80-, ASSOC DIR, 89- *Concurrent Pos:* Asst prof, David Dunlap Observ, Univ Toronto, 70-71, instr Scarborough Col, 71-72, asst prof, Erindale Col, 72-73. *Mem:* Am Astron Soc; Can Astron Soc; Int Astron Union; Astron Soc Pac; fel Royal Soc Can. *Res:* Spectroscopy and photometry of variable and multiple stars; spectroscopic studies of stellar atmospheres; light pollution control; mass loss from stars. *Mailing Add:* David Dunlap Observ Box 360 Richmond Hill ON L4C 4Y6 Can

BOLTON, DAVID C, b Great Lakes, Ill, Sept 6, 52. NEUROVIROLOGY. *Educ:* Univ Calif, Davis, BS, 76, PhD(microbiol), 81. *Prof Exp:* Postdoctoral fel, Univ Calif, San Francisco, 81-84; HEAD, LAB MOLECULAR STRUCT & FUNCTION, NY STATE INST BASIC RES, 84- *Concurrent Pos:* Res grant, NIH, 86-90. *Mem:* AAAS; Am Soc Microbiol; Protein Soc; Am Soc Biochem & Molecular Biol; Sigma Xi. *Mailing Add:* Dept Molecular Biol NY State Inst Basic Res 1050 Forest Hill Rd Staten Island NY 10314

BOLTON, ELLIS TRUESDALE, b Linden, NJ, May 4, 22; m 43; c 2. BIOPHYSICS. *Educ:* Rutgers Univ, BSc, 43, PhD, 50. *Prof Exp:* Res Coun fel, Rutgers Univ, 46, teaching asst genetics & biol, 46-47, instr, 47-50, dir res, Serol Mus, 47-50; mem staff, Dept Terrestrial Magnetism, Carnegie Inst, 50-64, assoc dir, 64-66, dir, 66-74; prof, 74-, DIR CTR MARICULT RES, COL MARINE STUDIES, UNIV DEL, 77-, EMER PROF MARINE STUDIES. *Mem:* World Maricult Soc; AAAS; Biophys Soc. *Res:* Molecular biology; mariculture; mass culturing of algae; biosynthesis in microorganisms. *Mailing Add:* Col Marine Studies Attn: Connie Edwards 700 Pilottown Rd Lewes DE 19958

BOLTON, JAMES R, b Swift Current, Sask, June 24, 37; m 59; c 2. PHYSICAL CHEMISTRY. *Educ:* Univ Sask, BA, 58, MA, 60; Cambridge Univ, PhD, 63. *Prof Exp:* Boese fel & res assoc, Columbia Univ, 63-64; asst prof teaching & res, Univ Minn, 64-66, assoc prof, 66-69, prof assoc chmn, 69-70; dir, Photochemistry Unit, 79-85, PROF TEACHING & RES, UNIV WESTERN ONT, 70-; CONSULT, SOLAR ENERGY RES INST, 78- *Concurrent Pos:* Alfred P Sloan Found fel, 66-68; grants, NSF, 64-70, Petrol Res Fund, 70-73, Nat Res Coun, 70-77; Nat Sci Eng Res Coun Can, 77-; consult, Solar Energy, Res Inst, 78-86; distinguished lectr chem, Simon Fraser Univ, 80. *Honors & Awards:* Noranda Lectr Award, Chem Inst Can, 76. *Mem:* Fel Chem Inst Can; Am Chem Soc; Am Soc Photobiol; Inter Am Photochem Soc; Solar Energy Soc Can. *Res:* Transient free radical intermediates in photochemistry and in model systems of photosynthesis using optical and electron spin resonance techniques; photochemical reactions with the potential for conversion and storage of solar energy; mechanism of photodegradation of aqueous pollutants with UV light. *Mailing Add:* Dept Chem Univ Western Ont London ON N6A 5B7 Can

BOLTON, JOSEPH ALOYSIUS, b Attleboro, Mass, Sept 18, 39; m 63; c 5. HYDRAULICS, FILTRATION. *Educ:* Northeastern Univ, BSME, 63, MS, 65. *Prof Exp:* Res & develop mgr, Bird Mach Co, 65-74; DIR NEW TECHNOL, ALBANY INT, 74- *Concurrent Pos:* Teacher hydraul, Adirondack Community Col, 82-85. *Mem:* Tech Asn Pulp & Paper Indust. *Res:* Equipment used in the pulp and paper industry relating to vacuum pumps, filters, screening systems, control valves and new energy saving process control systems; recipient of 23 patents. *Mailing Add:* Albany Int Quaker Rd Glen Falls NY 12801

BOLTON, LAURA LEE, b Royal Oak, Mich, Oct 13, 44; m 68; c 3. EXPERIMENTAL BIOLOGY. *Educ:* Univ Ill, BA, 66; Stanford Univ, MS, 69; Rutgers Univ, PhD(psychobiol), 73. *Prof Exp:* Sr scientist skin biol, Johnson & Johnson Res, 72-86; mgr wound care, 87-90, DIR WOUND CARE, CONVATEC DIV, BRISTOL-MYERS SQUIBB, 91- *Concurrent Pos:* Adj asst prof, Dept Surgery, Col Med & Dent NJ, Rutgers Med Sch, 80-87, adj assoc prof, 88-; Fulbright scholar, 67-68. *Mem:* Sigma Xi; Surg Infection Soc; Soc Investigative Dermatol; AAAS; NY Acad Sci. *Res:* Exploration of variables affecting soft tissue repair. *Mailing Add:* 15 Franklin Pl Metuchen NJ 08840

BOLTON, RICHARD ANDREW ERNEST, b Montreal, Can, Oct 16, 39; m 74; c 3. NUCLEAR FUSION. *Educ:* McGill Univ, BSc, 60, MSc, 63; Univ Montreal, PhD(plasma physics), 66. *Prof Exp:* Asst prof physics, Univ Montreal, 66-68; res assoc fusion, Culham Lab, UK Atomic Energy Authority, 68-73; sr scientist, Hydro-Que Res Inst, 73-82, mgr, 82-88; DIR GEN, CAN CTR MAGNETIC FUSION, 88- *Concurrent Pos:* Mem, Sci Coun Can, 87- & Nat Coun, Can Nuclear Soc, 84-87. *Mem:* Can Asn Physicists; Can Nuclear Soc. *Res:* Magnetic confinement of hot plasma in toroidal geometry for controlled fusion research and in view of eventual energy production. *Mailing Add:* 507 Argyle Wsmt Montreal PQ H3Y 3B6 Can

BOLTON, W KLINE, IMMUNONEPHROLOGY, GLOMERULONEPHRITIS. *Educ:* Univ Va, MD, 69. *Prof Exp:* PROF MED, DEPT NEPHROLOGY, SCH MED, UNIV VA, 73- *Res:* Cellular immunity. *Mailing Add:* Dept Internal Med Box 133 Univ Va Health Sci Ctr Charlottesville VA 22908

BOLTON, WESSON DUDLEY, b Hardwick, Vt, Dec 30, 22; m 46; c 5. ANIMAL PATHOLOGY. *Educ:* Mich State Col, DVM, 44; Univ Vt, MS, 50. *Prof Exp:* Practicing vet, Vt, 44-47; asst animal pathologist, 47-50, chmn dept, 51-81, PROF ANIMAL PATH, UNIV VT, 51- *Mem:* Am Vet Med Asn; Sigma Xi. *Res:* Causes of infertility in dairy cattle. *Mailing Add:* 9 Pine Tree Terrace South Burlington VT 05401

BOLTRALIK, JOHN JOSEPH, b Mt Carmel, Pa, June 24, 26; m 51; c 2. BIOCHEMISTRY. *Educ:* Univ Pittsburgh, BS, 50, MS, 53. *Prof Exp:* Asst trop testing electronic equip, Univ Pittsburgh, 50-52; jr fel food technol, Mellon Inst Sci, 52-58; asst microbiol, Sch Med, Univ Pittsburgh, 58-60; jr biochemist, Ciba Pharmaceut Co, 60-67; res assoc, St Barnabas Med Ctr, 67-68; scientist, 68-70, sr res scientist biochem, 70-74, HEAD BIOCHEM, ALCON LABS INC, 74- *Mem:* Am Chem Soc; Acad Pharmaceut Sci. *Res:* Methods development for analysis of drugs from biological fluids; studies on absorption, distribution, metabolism and excretion of topically applied ophthalmic drugs; bioavailability of drugs and drug products. *Mailing Add:* 5017 Whistler Ft Worth TX 76133

BOLTZ, ROBERT CHARLES, JR, b Aug 4, 45; US citizen. IMMUNOLOGY, CYTOLOGY & BIOPHYSICS. *Educ:* Pa State Univ, BS, 67 & 68, MS, 72, PhD(biophys), 78. *Prof Exp:* Vis scientist cell separation electrophoresis, Univs Space Res Asn, 77-78; sr res, 78-89, RES FEL IMMUNOLOGIST CELL SEPARATION & ANALYSIS, MERCK SHARP & DOHME RES LABS, 89- *Mem:* Soc Anal Cytol. *Res:* Flow cytometry and separation of mammalian cells. *Mailing Add:* Merck Sharp & Dohme Res Labs Box 2000 R80-107W Rahway NJ 07065

BOLZ, HAROLD A(UGUST), b Cleveland, Ohio, May 27, 11; m 37; c 3. MECHANICAL ENGINEERING. *Educ:* Case Inst Technol, Case Western Reserve Univ, BSME, 33, MSME, 35. *Hon Degrees:* DEng, Purdue Univ, 64; DEng, Tri-State Univ, 73; DHL, Ohio Northern Univ, 80. *Prof Exp:* Asst mech eng, Case Inst Technol, 33-35; develop engr, Weatherhead Co, 35-38; instr mach design, Purdue Univ, 38-40, from asst prof to assoc prof mech eng, 40-46, head dept gen eng, 46-54; assoc dean, Col Eng, 54-58, dean, 58-76, dir, Eng Exp Sta, 58-76, EMER DEAN, COL ENG, OHIO STATE UNIV, 76- *Concurrent Pos:* Ed, Mat Handling Handbk, 58; interim pres, Ohio North Univ, 79. *Mem:* AAAS; Am Soc Mech Engrs; Am Soc Eng Educ (pres, 71-72); Nat Soc Prof Engrs; Sigma Xi. *Res:* Machine design; industrial relations; educational administration. *Mailing Add:* 3097 Herrick Rd Columbus OH 43221

BOLZ, R(AY) E(MIL), b Cleveland, Ohio, Oct 24, 18; m 44; c 4. ACADEMIC ADMINISTRATION, MECHANICAL ENGINEERING. *Educ:* Case Inst Technol, BS, 40; Yale Univ, MS, 42, DEng(mech eng), 49. *Hon Degrees:* DEng, Worcester Polytech Inst, 84. *Prof Exp:* Head jet engine combustion sect, Lewis Flight Propulsion Lab, Nat Adv Comt Aeronaut, 42-46; asst prof, Rensselaer Polytech Inst, 47-50; assoc prof aeronaut eng, Case Western Reserve Univ, 50-52, prof & coordr res, 52-55, head mech eng dept, 56-60, head eng div, 61-67, Leonard S Case Prof, 61-73, dean eng, 67-73; vpres & dean fac, Worcester Polytech Inst, 73-84; RETIRED. *Mem:* AAAS; Am Inst Aeronaut & Astronaut; Am Soc Eng Educ; Am Soc Mech Engrs. *Res:* Applied fluid dynamics and thermodynamics to fields of propulsion, energy conversion and fluid machinery; fluid mechanics; rigid body dynamics and propulsion. *Mailing Add:* 21 Lantern Lane Worcester MA 01609

BOMAR, LUCIEN CLAY, b Birmingham, Ala, Aug 24, 47; m 69; c 3. RADAR SYSTEMS, MILLIMETER WAVES. *Educ:* Ga Inst Technol, BEE, 69. *Prof Exp:* Sr engr elec eng, Sperry Microwave Electronics Co, 69-76; res engr, 76-81, SR RES ENG, ENG EXP STA, GA INST TECHNOL, 81-, ASSOC DIV CHIEF, 80- *Concurrent Pos:* Radar consult, Martin Marietta Corp, 78; radar syst consult, Millimeter Wave Technol Inc, 81. *Mem:* Sr mem Inst Elec & Electronics Engrs; Sigma Xi. *Res:* Radar reflectivity characterization of cultural and man made objects; development of specialized radar equipments. *Mailing Add:* 2712 Chestnut Ridge Way Marietta GA 30062

BOMBA, STEVEN JAMES, b Chicago, Ill, June 28, 37; m 59; c 2. INDUSTRIAL AUTOMATION SYSTEMS, APPLIED PHYSICS & MATHEMATICS. *Educ:* Univ Wis-Madison, BS, 59, MS, 61, PhD(physics), 68. *Prof Exp:* Res physicist, Field Res Lab, Mobil Res & Develop Corp, 68-72; mem tech staff, Central Res Lab, Tex Instruments, Inc, 72-75; systs engr, Collins Radio Group, Rockwell Int, 75-78; dir, Corp Technol Develop, Allen-Bradley Co, 78-85, vpres, 85-87; vpres, Adv Mfg Technol, Rockwell Int, 87-90; VPRES TECHNOL, JOHNSON CONTROLS INC, 90- *Concurrent Pos:* Consult, S J Bomba & Assocs, 71- *Mem:* Am Asn Physics Teachers; Am Phys Soc; Inst Elec & Electronic Engrs; Sigma Xi; Soc Mfg Engrs; Am Soc Mech Engrs. *Res:* Industrial automation with emphasis on communications and control; materials, sensors and processes for materials conversion and assembly mostly regarding discrete parts and batch industries; miscellaneous applied classical physics; applied solid state physics in device and process development; systems science and engineering. *Mailing Add:* 875 E Birch Ave Whitefish Bay WI 53217

BOMBECK, CHARLES THOMAS, III, b Dallas, Tex, June 19, 37; m 62; c 3. SURGERY, ANIMAL PHYSIOLOGY. *Educ:* Rockhurst Col, BS, 58; St Louis Univ Sch Med, MD, 63; Am Bd Surg, cert, 72. *Prof Exp:* Intern surg, Univ Ind, 63-64; resident dept surg, Univ Wash Sch Med, 64-67, asst surg, 64-65; resident dept surg, Univ Ill Col Med, 67-70; from instr to assoc prof surg, Univ Ill Col Med, 68-73, DIR SURG GASTROINTESTINAL RES LABS, UNIV ILL GROUP HOSPS, 67-; CHIEF SURG SERVS, WEST SIDE VETS HOSP, CHICAGO, 87- *Concurrent Pos:* Instr gastrointestinal endoscopy, Cook County Grad Sch Med, 72-; assoc mem, Grad Col Fac, Univ Ill Med Ctr, 75, chief sect gastroenterol, 75-87, sr fel dept surg, Univ Ill Col Med, 67-68; prof surg, West Side Vets Hosp, Chicago, 77; fel surg, Univ Wash Sch Med, Seattle, 65, Am Cancer Soc fel, 65, sr fel dept surg, 66-67; fel, Nat Inst Arthritis & Metab Dis, 65-67, Nat Inst Gen Med Sci, 68-69; vis prof, Univ Basel, Switz, 76, Univ Goteberg, Sweden, 80, Univ Gottingen, Ger, 80, Stanford Univ, 80, Med Col Wis, 81 & 85, Tech Univ Munich, Ger, 83, Univ Lancaster, Eng, 84, Nagasaki Univ & Kyoto Univ, Japan, 86, Univ Fla Sch Med, 87, Ohio State Univ, 88; spec lectr, Japanese Surg Soc, Tokyo, 86 & Ube Univ, Yamaguchi City, 86; Mead Johnson Award, Am Col Surgeons, 69-72. *Mem:* Sigma Xi; Soc Univ Surgeons; Asn Acad Surg; Am Col Surgeons; Soc Surg Alimentary Tract; Europ Soc Surg Res; Europ Soc Gastrointestinal Endoscopy; Am Gastroenterol Asn; Int Soc Surg; Soc Am Gastrointestinal Endoscopic Surgeons (vpres, 87-88); Asn Surg Educ; Am Surg Asn; AAAS; Am Motility Soc; Int Soc Dis Esophagus. *Res:* Gastrointestinal physiology, specifically physiology and pathophysiology of the esophagus, stomach, duodenum, pancreas and liver; complete applications in medicine. *Mailing Add:* 820 S Damen 112 Chicago IL 60612

BOMBERGER, H(OWARD) B(RUBAKER), JR, b Lebanon, Pa, Feb 8, 22; m 43; c 3. METALLURGY. *Educ:* Pa State Univ, BS, 42; Ohio State Univ, MS, 50, PhD(metall), 52. *Prof Exp:* Metallurgist, Wright Aeronaut Corp, 43-45 & Buckeye Steel Castings Co, 46-47; self employed, 47-49; supvr fundamental res, Crucible Steel Co Am, Pa, 52-58; prod develop, 65-67, dir res & develop, 67-73, DIR METALL & RES, RMI CO, 73- *Concurrent Pos:* Mem, US Mat Adv Bd, titanium, 68; grad instr metall, Youngstown State Univ, 68-71. *Honors & Awards:* US Cong Award Metall, 80. *Mem:* Electrochem Soc; Nat Asn Corrosion Engrs; fel Am Soc Metals; Am Inst Mining, Metall & Petrol Engrs; Brit Inst Metals. *Res:* Physical metallurgy; titanium alloy development; alloy properties; kinetics; corrosion and electrochemical phenomena; author of fifty-five technical papers. *Mailing Add:* 385 Bradford Dr Canfield OH 44406

BOMKE, HANS ALEXANDER, b Berlin, Ger, May 26, 10; nat US; m 32; c 1. PHYSICS. *Educ:* Univ Berlin, PhD(physics, math), 31; Tech Univ Berlin, Dr habil, 35. *Prof Exp:* Asst prof, Radiation Lab, Berlin, 31-33; mem sci staff, Ger Bur Stand, 33-37 & Kaiser Wilhelm Co, 38-41; mem, Aerial Navig Inst, 41-45; assoc prof, Med Sch, Univ Munich, 46-52; scientist, US Army, Ft Monmouth, NJ, 52-58, dir, Explor Res Div, Surveillance Dept, 58-61, prin scientist, Exp Res Inst, Signal Res & Develop Labs, 61-77, prin scientist, Electronics Res & Develop Command, 77-81; RETIRED. *Concurrent Pos:* Adj prof, Fordham Univ, 61- *Mem:* Am Phys Soc; Optical Soc Am; Am Geophys Union; Am Inst Aeronaut & Astronaut; Ger Phys Soc. *Res:* Atomic and nuclear physics; geophysics; space. *Mailing Add:* 408 Central Ave Spring Lake NJ 07762

BOMPART, BILLY EARL, b Dallas, Tex, Dec 5, 33; m 57; c 3. DIOPHANTINE EQUATIONS. *Educ:* Univ Tex, Austin, BS, 56, PhD(math educ), 67; Southwestern Baptist Theol Sem, MRE, 60; NTex State Univ, MEd, 64. *Prof Exp:* Teacher high schs, Tex, 58-62; youth dir, Polytech Baptist Church, Ft Worth, 62-63; teacher high schs, Tex, 63-65; from assoc prof to prof math, 67-90, dept chmn, 83-88, VPRES ACAD AFFAIRS, AUGUSTA COL, 90- *Mem:* Nat Coun Teachers Math; Sch Sci & Math Asn; Am Educ Res Asn; Math Asn Am; Am Math Assoc Two-Year Cols. *Res:* Theory of numbers. *Mailing Add:* VPres Acad Affairs Dept Augusta Col Augusta GA 30910

BOMSE, FREDERICK, b New York, NY, Jan 24, 39; m 65; c 3. PHYSICS. *Educ:* Antioch Col, BS, 61; Johns Hopkins Univ, PhD(elem particle physics), 65. *Prof Exp:* Instr physics, Johns Hopkins Univ, 65-66; from instr to asst prof, Vanderbilt Univ, 66-71; MEM PROF STAFF, CTR NAVAL ANALYSES, 71- *Concurrent Pos:* mem staff, Appl Physics Lab, Johns Hopkins Univ, 76

& Inst Defense Anal, 80-81. *Mem:* Am Phys Soc; Sigma Xi. *Res:* Interactions of fundamental particles; acoustics; fluids; optics; directed energy weapons. *Mailing Add:* Ctr Naval Analyses 4401 Ford Ave PO Box 16268 Alexandria VA 22302-0268

BONA, JERRY LLOYD, b Little Rock, Ark, Feb 5, 45; m 66; c 2. FLUID MECHANICS, PARTIAL DIFFERENTIAL EQUATIONS. *Educ:* Wash Univ, St Louis, BS, 66; Harvard Univ, PhD(math), 72. *Prof Exp:* Res fel, Fluid Mech Res Inst, 70-72; L E Dickson instr, 72-73, from asst prof to prof math, Univ Chicago, 73-86; PROF MATH, PA STATE UNIV, 86-, CHMN DEPT, 90- *Concurrent Pos:* Sr res fel, Sci Res Coun, UK, 73-78; vis res assoc, Brookhaven Nat Lab, 76 & 77; vis prof, Inst Budownictwa Wodnego, Polish Acad Sci, 77, Ctr de Pesquisas Fisicas, Brazil, 80, Math Res Ctr, Madison, 80-81, Numerical Anal Lab, Orsay, France, 82-91, INRS-Oceanology, Univ Quebec, 82-91, Appl Math Physics Lab, Denmark, 82, Inst Math & Applications, Univ Minn, 85-91 & US Navy Appl Res Lab, 86-; lectr, Int Cong Math, 78, Am Phys Soc, 79, Am Math Soc, 84. *Honors & Awards:* Britton Lectr, 86. *Mem:* Am Math Soc; Soc Indust & Appl Math; Soc Comput Simulations; AAAS; Math Asn Am; Asn Women Math. *Res:* Nonlinear waves; partial differential equations; functional analysis; mathematical economics; probability theory. *Mailing Add:* Dept Math Pa State Univ University Park PA 16802

BONACCI, JOHN C, b Utica, NY, Aug 11, 35; m 57; c 2. CHEMICAL ENGINEERING, CATALYSIS. *Educ:* Clarkson Col Technol, BChE, 57, MChE, 59; Univ Pa, PhD(chem eng), 66. *Prof Exp:* Engr, Yerkes Res Lab, E I Du Pont de Nemours & Co, Inc, 58-61, res engr, 61, supvr prod, 61-63; res eng appl res div, Mobil Res & Develop Corp, 66-68, sr res engr, 68-75; res sect head, Englehard Corp, 75-77, res mgr, Res & Develop Dept, 77-81, vpres, 82-83, vpres, Ventures Dept, 84, gen mgr, Systs Dept, 81-83, dir, mkt & technol Environ Catalyst Dept, 85-87; sr scientist, Sci Appl Int Corp, 88-89; PRES, FIBONACCI, INC, 88-; MGR, COM DEVELOP, H-R INT, 89- *Concurrent Pos:* Dir, NJ Energy Res Inst, 80-86, chmn, 83-84; rep, World Environ Ctr, 83-86; vis prof, Rutgers Univ, 88-, adj prof, 91-; licensed patent agent, 88- *Mem:* Fel Am Inst Chem Engrs; Am Chem Soc; Sigma Xi; NY Acad Sci. *Res:* Gas-solid mass transfer studies; mass transfer in evaporating, condensing water; pilot plant, films, polymers, catalytic converters, refining processes; petrochemical ventures; commercial development; air pollution control; technology transfer; patent services. *Mailing Add:* 156 Gallinson Dr Murray Hill NJ 07974

BONAKDARPOUR, AKBAR, b Esfahan, Iran. RADIOLOGY. *Educ:* Univ Tehran, MD, 53; Temple Univ, MS, 58. *Prof Exp:* From instr to assoc prof, 64-72, PROF RADIOL, HEALTH SCI CTR, TEMPLE UNIV, 72- *Mem:* Asn Univ Radiologists; fel Am Col Radiol; Radiol Soc NAm; Am Roentgen Ray Soc; Int Skeletal Soc. *Res:* Skeletal radiology. *Mailing Add:* Temple Univ Health Sci Ctr 3401 N Broad St Philadelphia PA 19140

BONAN, EUGENE J, b Pottsville, Pa, Jan 19, 23; m 48; c 2. AERONAUTICAL ENGINEERING, APPLIED MATHEMATICS. *Educ:* Cath Univ Am, BS, 49. *Prof Exp:* Design engr, 49-52, group leader heat transfer, 52-54, group leader armament control systs, 54-56, head weapon syst anal, 56-58, weapon syst proj engr, 58-59, asst dept head weapons syst, 59, dept head, 59-63, chief systs anal sect, 63-64, mgr systs eng, 64-66, dep dir systs technol, 66-70, DIR SYSTS TECHNOL, GRUMMAN AIRCRAFT ENG CORP, BETHPAGE, 70- *Concurrent Pos:* Mem, US Navy Spec Weapons Long Range Planning Comn, 58-60. *Mem:* Am Inst Aeronaut & Astronaut; Am Ord Asn. *Res:* Selection, procurement, technical direction and monitoring of electronic hardware systems included in major weapon and space systems. *Mailing Add:* Quaker Path Box 151 Stony Brook NY 11790

BONAR, DANIEL DONALD, b Murraysville, WVa, July 7, 38; m 66; c 1. MATHEMATICS. *Educ:* WVa Univ, BSChE, 60, MS, 61; Ohio State Univ, PhD(math), 68. *Prof Exp:* From instr to asst prof math, Denison Univ, 65-67; teaching assoc, Ohio State Univ, 67-68; asst prof, Wayne State Univ, 68-69; from asst prof to assoc prof, 69-77, chmn dept, 71-77, PROF MATH, DENISON UNIV, 77- *Concurrent Pos:* Teacher, Ohio State Univ, 85. *Mem:* Math Asn Am; Sigma Xi; Am Math Soc; Nat Coun Teachers Math. *Res:* Complex and real analysis; annular functions; topological idea of cluster sets. *Mailing Add:* Dept Math Sci Denison Univ Granville OH 43023

BONAR, ROBERT ADDISON, b Kalamazoo, Mich, Aug 23, 25; m 51; c 4. CHEMISTRY. *Educ:* Univ Calif, AB, 49, PhD(biochem), 53. *Prof Exp:* Res fel, Med Sch, Univ Calif, Los Angeles, 53-55; res assoc, Dept Surg, Duke Univ, 55-59, asst prof, 59-63, assoc pr of biophys, Dept Surg, Sch Med; res chemist, Vet Admin Hosp, 72-87; RETIRED. *Concurrent Pos:* Nat Cancer Inst fel, 53-55. *Mem:* AAAS; Am Soc Biol Chemists; Sigma Xi. *Res:* Biochemical cytology; carcinogenesis; oncology. *Mailing Add:* Indian Banks Rte 1 Box 387-E Farnham VA 22460

BONAVENTURA, CELIA JEAN, b Silver City, NMex, June 19, 41; m 60; c 2. PROTEIN CHEMISTRY, MARINE BIOMEDICINE. *Educ:* Calif State Univ, San Diego, BA, 64; Univ Tex, Austin, PhD(zool), 68. *Prof Exp:* Fel chem, Calif Inst Technol, 68-70; fel molecular biol, Regina Elena Inst Cancer Res, 71-72; assoc prof, 72-90, PROF CELL BIOL, DUKE UNIV MED CTR & MARINE LAB, 90-, ASSOC DIR, MARINE BIOMED CTR, 78- *Concurrent Pos:* Asst dir, Marine Biomed Affairs, Duke Univ. *Mem:* Biophys Soc; Am Chem Soc; Am Soc Biol Chemists; Sigma Xi. *Res:* Allosteric control in respiratory proteins; protein engineering; pollutant interactions with hemoglobins. *Mailing Add:* Dept Cell Biol Duke Univ Med Ctr & Marine Lab Beaufort NC 28516

BONAVENTURA, JOSEPH, b Oakland, Calif, Feb 15, 42; m 60; c 2. PROTEIN CHEMISTRY. *Educ:* Calif State Univ, San Diego, BA, 64; Univ Tex, Austin, PhD(zool), 68. *Prof Exp:* Fel chem, Calif Inst Technol, 68-70; fel molecular biol, Regina Elena Inst Cancer Res, 71-72; assoc prof, 72-90, PROF CELL BIOL, DUKE UNIV MED SCH & MARINE LAB, 90-, DIR, MARINE BIOMED CTR, 78- *Concurrent Pos:* Estab investr, Am Heart

Asn, 75-80. *Mem:* Biophys Soc; Sigma Xi; Am Chem Soc; Am Soc Biol Chemists; Am Soc Zoologists. *Res:* Protein structure-function relationships; properties of insolubilized proteins; respiratory proteins as models for large multisubunit macromolecular complexes; biochemical basis for adaptation and acclimation. *Mailing Add:* Dept Cell Biol Duke Univ Med Ctr & Marine Lab Beaufort NC 28516

BONAVIDA, BENJAMIN, b Cairo, Egypt, Jan 29, 40. IMMUNOBIOLOGY OF CANCER. *Educ:* Univ Calif, Berkeley, PhD(immunochem), 68. *Prof Exp:* Assoc prof, 78-82, PROF IMMUNOL & BIOCHEM, SCH MED, UNIV CALIF, LOS ANGELES, 82- *Mem:* Am Asn Immunologists; Am Asn Cancer Res; Transplantation Soc. *Mailing Add:* Dept Microbiol & Immunol Univ Calif Sch Med Health Sci Ctr Rm A2-084 Los Angeles CA 90024

BONAVITA, NINO LOUIS, b White Plains, NY, Sept 5, 31; m 65; c 2. PHYSICS, MAXIMUM ENTROPY FORMALISM. *Educ:* Fordham Univ, BS, 54; Cath Univ, MS, 63, PhD(physics), 71. *Prof Exp:* Physicist, Aberdeen Proving Ground, 54-55 & Naval Res Lab, 57-59; PHYSICIST, SPACE DATA DIV, GODDARD SPACE FLIGHT CTR, NASA, 59- *Mem:* Am Phys Soc; Sigma Xi. *Res:* Statistical mechanics; theoretical solid state physics and magnetism; celestial mechanics; physics of condensed matter; stochastic processes and self-organization; information theory; intelligent systems; maximum entropy formalism; artificial neural networks; signal processing; Bayesian probability theory. *Mailing Add:* Goddard Space Flight Ctr NASA Code 636 Greenbelt MD 20771

BOND, ALBERT F, b Clarksburg, WVa, July 27, 30; m 51; c 3. MATHEMATICS. *Educ:* WVa Univ, AB & MS, 56. *Prof Exp:* Develop engr, Goodyear Aircraft Corp, Ohio, 56-57; staff engr, Bendix Systs Div, 57-63; mgr systs eng dept, Brown Eng Co, 63-68; mgr Huntsville opers, Planning Res Corp, 68-70; gen mgr, Huntsville Div, Syst Sci, Inc, 70-72; dept mgr, Comput Sci Corp, Huntsville Opers, 72-77; SR PROJ MGR, UNISYS, 77- *Concurrent Pos:* Guest lectr, Univ Mich, 59. *Mem:* Math Asn Am; Opers Res Soc Am. *Res:* Systems engineering; operations research; management sciences; applied mathematics; software engineering. *Mailing Add:* 2908 Gallalee Rd SE Huntsville AL 35801

BOND, ALBERT HASKELL, JR, b Williamson, WVa, Sept 13, 40; m 68. NUCLEAR PHYSICS. *Educ:* Harvard Univ, AB, 61; Univ Wis, MS, 63, PhD(physics), 68. *Prof Exp:* Found for Aid to Res res assoc nuclear physics, Van Der Graaff Lab, Univ Sao Paulo, 68-74; ASSOC PROF PHYSICS, BROOKLYN COL, 74-, DIR INST NUCLEAR THEORY, 77- *Res:* Fast neutron physics; fabrication techniques of lithium-germanium detectors; particle-gamma angular correlations. *Mailing Add:* Dept Physics CUNY Brooklyn Col Brooklyn NY 11210

BOND, ALECK C, b Columbus, Ga, Jan 11, 22; m 48; c 2. AERONAUTICAL ENGINEERING. *Educ:* Ga Inst Technol, BS, 43, MS, 48. *Prof Exp:* Aeronaut res scientist, Nat Adv Comt Aeronaut, Langley Field, Va, 48-58, aeronaut res scientist & proj engr, Mercury Manned Space Prog, Space Task Group, 58-59, aeronaut res scientist & head performance br, 59-60, asst chief, Flight Syst Div, 60-62; chief, Syst Eval & Develop Div, 62-63, mgr syst test & eval, 63-67, reliability & qual off & flight safety off, 67-68, asst dir chem & mech systs, 68-75, asst dir prog support, 75-77, assoc dir prog support, 77-80, DEP DIR ENG & DEVELOP DIRECTORATE, JOHNSON SPACE CTR, NASA, HOUSTON, 80- *Concurrent Pos:* Mem large launch vehicle planning group NASA & US Dept Defense, 61. *Mem:* Assoc fel Am Inst Aeronaut & Astronaut. *Res:* Ram jet performance; flight evaluation of various aircraft and missile configurations; aerodynamic heating; high temperature materials research; development and testing of manned spacecraft systems. *Mailing Add:* 234 Driftwood Dr Seabrook TX 77586

BOND, ANDREW, b Brownsville, Tenn, Aug 8, 27. BIOCHEMISTRY, AGRONOMY. *Educ:* Tenn Agr & Indust State Col, BS, 48; Tenn State Univ, MS, 52; Univ Minn, PhD(biochem), 65. *Prof Exp:* Instr agron, 54-58, assoc prof biochem, 65-69, assoc prof animal sci, 69-75, PROF BIOCHEM & DEAN ALLIED HEALTH PROFESSIONS, TENN STATE UNIV, 75- *Mem:* Am Chem Soc; Nat Soc Allied Health; Am Soc Allied Health. *Res:* Carbohydrate transformations in germinating cereal seeds. *Mailing Add:* Sch Allied Health Tenn State Univ Nashville TN 37209-1561

BOND, ARMAND PAUL, metallurgy; deceased, see previous edition for last biography

BOND, ARTHUR CHALMER, b Salem, WVa, Feb 14, 17. PHYSICAL CHEMISTRY. *Educ:* Mich State Univ, BS, 39; Univ Mich, MS, 40, PhD(chem), 51. *Prof Exp:* Res chemist, Nat Defense Res Comt, Univ Chicago, 42-43, instr & asst, 43-46; from instr to asst prof phys chem, Univ Rochester, 51-57; from asst prof to assoc prof chem, Rutgers Univ, 57-67, prof chem, 67-82; RETIRED. *Concurrent Pos:* Asst ed, J Phys Chem, 51-57. *Mem:* AAAS; Am Chem Soc; Sigma Xi. *Res:* Molecular structure; inorganic reaction mechanisms. *Mailing Add:* 13 Dansfield Dr Talley Hill Wilmington DE 19803

BOND, CARL ELDON, b Culdesac, Idaho, Sept 11, 20; m 42; c 2. ICHTHYOLOGY. *Educ:* Ore State Univ, MS, 48; Univ Mich, PhD(fisheries), 63. *Prof Exp:* Res asst zool, Univ Calif, 49; from asst prof to assoc prof fish & game mgt, 49-64, asst dean grad sch, 69-74, prof, 64-84, EMER PROF FISHERIES, ORE STATE UNIV, 84- *Concurrent Pos:* NSF sci fac fel, 60 & 71; vis res prof, Tokyo Univ Fisheries, 72. *Mem:* Am Soc Ichthyol & Herpet; fel Am Inst Fishery Res Biol; Am Fisheries Soc; Sigma Xi. *Res:* Distribution, systematics and ecology of northwest fishes. *Mailing Add:* Dept Fisheries & Wildlife Ore State Univ Corvallis OR 97331

BOND, CHARLES EUGENE, b Royston, Ga, Feb 1, 30; m 71; c 7. SOLAR ENERGY, WIND POWER. *Educ:* Ga Inst Technol, BS, 51; Univ Mich, MSE, 56, PhD(aeronaut & astronaut eng), 64. *Prof Exp:* Proj engr, Arnold Eng Develop Ctr, Aro, Inc, 51-54 & Jet Propulsion Lab, 56-57; asst aeronaut

& astronaut eng, Univ Mich, 54-56 & 57-58; sr scientist, Res & Advan Develop Div, Avco Corp, 59-60, lead scientist, 60-61; assoc res engr, Univ Mich, 61-64, res engr, Aerodyn Lab, 64; from asst prof to assoc prof, 64-66, PROF AERONAUT & ASTRONAUT ENG, UNIV ILL, URBANA, 71- *Mem:* Am Soc Heating Refrig & Air-Conditioning Engrs; Am Soc Mech Engrs; Int Solar Energy Soc. *Res:* Aerodynamics; wind tunnel research; electric arc technology; solar energy, technology and society; magnetogasdynamics. *Mailing Add:* 210 Bliss Ave Urbana IL 61801

BOND, CLIFFORD WALTER, b Buffalo, NY, Apr 7, 37; m; c 1. VIROLOGY. *Educ:* State Univ NY Buffalo, BA, 66; Univ Ky, PhD(microbiol), 73. *Prof Exp:* Technician virol, Case Western Reserve Univ, 66-67, res fel, 67-69; res fel, Univ Ky, 69-74; Leukemia Soc Am fel virol, Univ Calif, San Diego, 74-76, USPHS trainee, 76-77, asst res pathologist, 77-78; asst prof, 78-84, ASSOC PROF MICROBIOL, MONT STATE UNIV, 84-, ACTG HEAD, DEPT MICROBIOL, 90- *Mem:* Am Soc Microbiol; AAAS; Am Soc Virol. *Res:* Elucidation of molecular interactions between pathogenic viruses and their host that result in disease. *Mailing Add:* Dept Microbiol Mont State Univ Bozeman MT 59717

BOND, EDWIN JOSHUA, b Delta, Ont, June 9, 27; m 51; c 3. ZOOLOGY. *Educ:* Univ Toronto, BSA, 50; Univ Western Ont, MSc, 55; Univ London, DIC & PhD(insect toxicol), 59. *Prof Exp:* Tech officer forest entom, 50-51, res scientist insect toxicol, Can Agr, 51-87; CONSULT, 87- *Concurrent Pos:* Mem toxicol res unit, Med Res Coun, London, Eng, 67-68; Food & Agr Orgn consult, Cent Food Technol Res Inst, Mysore, India, 71; tech adv, Intergovt Maritime Consult Orgn, London, Eng, 78. *Mem:* Entom Soc Am. *Res:* Toxicology; application of fumigants for control of insects and other organisms; food storage and preservation. *Mailing Add:* Res Inst Can Agr Univ Sub PO London ON N6A 5B7 Can

BOND, ELIZABETH DUX, b Altoona, Pa, June 17, 23. CLINICAL CHEMISTRY. *Educ:* Pa State Univ, BS, 44, MS, 51; Rutgers Univ, PhD(chem), 73; NY Univ, MBA, 80. *Prof Exp:* Instr chem, Orlando Col, 63-67; asst chem, Rutgers Univ, 68-73, asst prof, 73-74; res chemist, Cities Serv Co, 74-76; sr scientist, Hoffman LaRoche, 76-79; SR CHEMIST MED PRODS, DUPONT CO, 80- *Mem:* Sigma Xi; Am Chem Soc; AAAS. *Res:* Chemistry of coal and its products; electron spin resonance spectroscopy; infrared and raman spectroscopy. *Mailing Add:* 13 Dansfield Dr Talley Hill Wilmington DE 19803

BOND, FREDERICK E, b Philadelphia, Pa, Jan 10, 20; m 46; c 1. ELECTRICAL ENGINEERING, COMMUNICATIONS. *Educ:* Drexel Inst, BS, 41; Rutgers Univ, MS, 50; Polytech Inst Brooklyn, PhD(elec eng), 56. *Prof Exp:* Asst chief, radio commun br, US Army Sig Res & Develop Labs, Ft Monmouth, NJ, 46-56, dir, Radio Commun Div, 56-57; co-mgr, commun dept, Ramo-Wooldridge Div, Calif, 57-60; pres, Ryan Commun, Inc, 60-61; chief scientist, Astroelectronics Lab, Westinghouse Elec Corp, 61-64; assoc prog dir, Aerospace Corp, 64-73; dep dir MILSATCOMM systs, Defense Commun Agency, 73-76; sr staff engr, 77-80, prin engr, Aerospace Corp, 81-88; RETIRED. *Mem:* Fel Inst Elec & Electronic Engrs; Sigma Xi; assoc fel Am Inst Aeronaut & Astronaut. *Res:* Communication systems; communication satellite technology; signal processing and modulation methods. *Mailing Add:* 6124 Buckingham Pkwy 301 Culver City CA 90230

BOND, FREDERICK THOMAS, b Brooklyn, NY, Oct 26, 36; m 77; c 2. ORGANIC CHEMISTRY. *Educ:* Mass Inst Technol, BS, 58; Univ Calif, Berkeley, PhD(chem), 62. *Prof Exp:* Asst prof chem, Ore State Univ, 62-67; asst prof, 67-70, ASSOC PROF CHEM, UNIV CALIF, SAN DIEGO, 70-, PROVOST, REVELLE COL, 83- *Concurrent Pos:* NIH spec fel, 72-73. *Mem:* Am Chem Soc; Sigma Xi; AAAS. *Res:* Chemistry of natural products; structure and reactivity relationships in simple and complex molecules; chemistry of divalent carbon; synthetic organic chemistry. *Mailing Add:* Dept Chem Univ Calif at San Diego La Jolla CA 92093

BOND, GARY CARL, b Lawrence, Kans, July 28, 42; m 65; c 1. MEDICAL PHYSIOLOGY. *Educ:* Univ Kans, BS, 65, PhD(physiol), 70. *Prof Exp:* Res physiologist, St Luke's Hosp, 70-73; asst prof physiol, Med Ctr, Univ Ark, Little Rock, 73-76; asst prof physiol, 76-78, ASSOC PROF PHYSIOL, MED COL GA, 78- *Concurrent Pos:* Lectr physiol, Med Ctr, Univ Kans, 71-73; clin asst prof, Sch Med, Univ Mo-Kansas City, 72-73. *Mem:* AAAS; Am Heart Asn; Am Physiol Soc. *Res:* Cardiovascular physiology with emphasis on cardiovascular control mechanisms involved in blood pressure and blood volume regulation. *Mailing Add:* Dept Physiol & Endocrinol Med Col Ga Augusta GA 30912

BOND, GEORGE WALTER, b Lancaster, Pa, Mar 28, 44; m 65; c 4. BIOLOGY, ICHTHYOLOGY. *Educ:* Dartmouth Col, AB, 66; Univ RI, MS, 69, PhD(biol sci), 74. *Prof Exp:* Res asst ichthyol, Div Fishes, Smithsonian Inst, 72-73; from asst prof to assoc prof, 73-84, PROF BIOL, FITCHBURG STATE COL, 84- *Mem:* Am Soc Ichthyologists & Herpetologists. *Res:* Biology of deepsea fishes, especially the family gonostomatidae. *Mailing Add:* Dept Biol Fitchburg State Col Fitchburg MA 01420-2697

BOND, GERARD C, b Altus, Okla, May 20, 40; c 3. TECTONICS. *Educ:* Univ Wis, PhD(geol), 70. *Prof Exp:* Asst prof geol, Williams Col, 70-72 & Univ Calif, Davis, 72-80; SR RES SCIENTIST, LAMONT-DOHERTY GEOL OBSERV, 80- *Mem:* Fel Geol Soc Am; Am Geophys Union; Am Asn Prof Geol. *Res:* Tectonics of continens, especially in basins; paleoclimate; long-term history of sea level. *Mailing Add:* Lamont-Doherty Geological Observat Rte Nine West Palisades NY 10964

BOND, GUY HUGH, b Newark, NJ, Apr 2, 34; m 75; c 1. PHYSIOLOGY, BIOCHEMISTRY. *Educ:* Rutgers Univ, BS, 56, MS, 61, PhD(physiol, biochem), 65. *Prof Exp:* Instr physiol & biochem, Rutgers Univ, 64-65; physiol fel, Vanderbilt Univ, 65-67; asst prof, Med Col Va, 67-75; assoc prof physiol, 75-85, PROF PHYSIOL, KIRKSVILLE COL OSTEOP MED 85- *Mem:* Sigma Xi; Soc Gen Physiologists; Am Physiol Soc. *Res:* Membrane transport; human red cell membrane sodium plus potassium adenosine triphosphatase and calcium plus magnesium adenosine triphosphatase. *Mailing Add:* Dept Physiol Kirksville Col Kirksville MO 63501

BOND, HOWARD EDWARD, b Roseville, Calif, Mar 9, 30; m 59. BIOPHYSICS, CANCER. *Educ:* Univ Calif, BS, 53, DVM, 55; Cornell Univ, PhD(physiol), 59. *Prof Exp:* Asst physiol, State Univ NY Vet Col, Cornell Univ, 55-59; res assoc cellular physiol, Biol Div, Oak Ridge Nat Lab, 59-61; res biologist, US Naval Radiol Defense Lab, 61-65; res biologist, Nat Cancer Inst, Md, 65-66, head molecular separations unit, 66-69; dir res, Electro-Nucleonics Labs, Inc, 69-77, dir technol, Corp Develop Group, 77-79; mgr res & develop, Hynson Westcott & Dunning, Becton Dickinson & Co, 79-82; MGR RES & DEVELOP, DIAG DIV, ABBOTT LABS, 82- *Concurrent Pos:* Clin assoc prof, Sch Med, Georgetown Univ, 71- *Mem:* Am Soc Vet Physiol & Pharmacol; Biophys Soc; Am Soc Microbiol. *Res:* Nutritional and metabolic diseases of ruminants; biochemistry and hormonal regulation of cellular proteins; cancer; ultracentrifugal analysis. *Mailing Add:* Abbott Labs D908AP-20 Abbott Park IL 60064-3500

BOND, HOWARD EMERSON, b Danville, Ill, Oct 11, 42; m 67; c 1. ASTRONOMY. *Educ:* Univ Ill, BS, 64; Univ Mich, MS, 65, PhD(astron), 69. *Prof Exp:* Res assoc, La State Univ, Baton Rouge, 69-70, from asst prof to assoc prof, 70-79, prof astron, 79-84; ASTRONR, SPACE TELESCOPE SCI INST, 84- *Concurrent Pos:* Vis prof, Univ Wash, 75-76; prin investr, NSF & NASA grants; Vis prof, Univ Calif, 82-83. *Honors & Awards:* Shapley lectr, Am Astron Soc, 81- *Mem:* Am Astron Soc; Int Astron Union; Astron Soc Pac. *Res:* Astronomical spectroscopy and photometry; nucleosynthesis; population II stars; cataclysmic variable stars; nuclei of planetary nebulae. *Mailing Add:* Space Telescope Sci Inst 3700 San Martin Dr Baltimore MD 21218

BOND, JAMES ANTHONY, b Santa Monica, Calif, Sept 11, 52; m 85. METABOLISM, PHARMACOKINETICS. *Educ:* Pomona Col, BA, 75; Univ Wash, PhD(pharmacol), 79; Am Bd Toxicol, dipl, 83. *Prof Exp:* Res fel, Chem Indust Inst Toxicol, 80-81; toxicologist, Lovelace Inhalation Res Inst, 81-88, group supvr, 88-89; DEPT HEAD, BIOCHEM TOXICOL, CHEM INDUST INST TOXICOL, 89- *Concurrent Pos:* Clin asst prof, Col Pharm, Univ NMex, 85-89; mem, extrapolation models subcomt, sci adv bd, Environ Protection Agency & consult, sci adv bd, 85-; mem, Toxicol Study Sect, NIH, 90- *Honors & Awards:* Kenneth Morgareidge Award, Int Life Sci Inst, 90. *Mem:* Soc Toxicol; Am Soc Pharmacol & Exp Therapeut; Am Asn Cancer Res. *Res:* Biological fate of xenobiotics in various biological systems; role that metabolism plays in development of xenobiotic-induced toxicity and carcinogenicity. *Mailing Add:* Chem Indust Inst Toxicol PO Box 12137 Research Triangle Park NC 27709

BOND, JAMES ARTHUR, II, b Orangeburg, SC, July 11, 17; m 56; c 1. ECOLOGY. *Educ:* J C Smith Univ, BS, 38; Univ Kans, MA, 42; Univ Chicago, PhD, 61. *Prof Exp:* Teacher, Fla Pub Sch, 38-40; instr biol, Langston Univ, 46-48; John M Prather fel, Univ Chicago, 53-54; asst prof, 57-77, ASSOC PROF ZOOL, UNIV ILL, CHICAGO CIRCLE, 77- ASST DEAN COL ARTS & SCI, 74- *Mem:* AAAS; Am Soc Zoologists; Ecol Soc Am; Biomet Soc; Soc Study Evolution. *Res:* Population ecology; problems of spacing in insect populations; aging phenomena. *Mailing Add:* Univ Ill Box 4348 Chicago IL 60680

BOND, JENNY TAYLOR, b Kinston, NC, Sept 9, 39; m 76; c 2. NUTRITION. *Educ:* Meredith Col, AB, 61; Mich State Univ, PhD(nutrit, physiol), 72. *Prof Exp:* Exten agt youth, NC Agr Exten Serv, 61-64; from instr nutrit to instr pharmacol, 70-72, fel develop nephrology, 72-74, assoc prof 74-79, assoc prof Dept Food Sci & Human Nutrit & Dept Pediat & Human Develop, 79-84, PROF, DEPT FOOD, SCI & HUMAN NUTRIT, MICH STATE UNIV, 84- *Concurrent Pos:* Specialist foods, NC Agr Exten Serv, 64. *Honors & Awards:* George Harrar Res Coun Award. *Mem:* Am Soc Nephrology; Am Fedn Clin Res; Nutrit Soc Brit; Int Soc Nephrology; Am Inst Nutrit; Am Dietetic Asn; Soc Nutrit Educ. *Res:* Nutrition prenatally and postnatally; subsequent effects on development, specifically nutrition; drug interactions; xenobiotic metabolism; clinical nutrition. *Mailing Add:* Dept Food Sci & Human Nutrit Mich State Univ East Lansing MI 48824-1224

BOND, JUDITH, b New York, NY, Sept 27, 40; m 74; c 5. BIOCHEMISTRY, PHYSIOLOGY. *Educ:* Bennington Col, BA, 61; Rutgers Univ, MS, 62, PhD(biochem & physiol), 66. *Prof Exp:* NIH fel, Vanderbilt Univ, 66-68; from instr to prof biochem, Med Col Va, Va Commonwealth Univ, 68-88; PROF & HEAD BIOCHEM & NUTRIT, VA POLYTECH INST & STATE UNIV, 88- *Concurrent Pos:* Vis scientist, Strangeways Res Lab, Cambridge, Eng, 78, Commonwealth Sci & Indust Res Orgn, Adelaide, Australia, 86. *Honors & Awards:* Merit Award, NIH, 89. *Mem:* Am Soc Biochem & Molecular Biol; Am Chem Soc; Am Soc Microbiol; Biochem Soc; Am Physiol Soc. *Res:* Intracellular protein degradation; regulation of enzyme activity and concentration; diabetes; hormonal regulation of metabolism; lysosomes; mechanisms of intracellular proteinases. *Mailing Add:* Dept Biochem & Nutrit Va Polytech Inst & State Univ Blacksburg VA 24061-0308

BOND, LORA, b Bryan, Tex, May 17, 17. BOTANY. *Educ:* Univ Tenn, BA, 38; Wellesley Col, MA, 41; Univ Wis, PhD(bot), 45. *Prof Exp:* Asst bot, Wellesley Col, 38-41 & Univ Wis, 41-43; actg instr, Univ Tenn, 43; instr biol & chem, Drury Col, 43-45 & bot, Wellesley Col, 45-48; from assoc prof to prof biol, 48-82, head dept, 49-74, chmn, Div Sci & Math, 52-55, 64-67 & 79-82, EMER PROF, DRURY COL, 82- *Concurrent Pos:* Sci fac fel, NSF, 61. *Mem:* Fel AAAS; Bot Soc Am; Sigma Xi; Am Inst Biol Sci. *Res:* Morphology of angiosperms; colchicine induction of polyploidy in petunia; edaphic factors of Hempstead Plains, Long Island; plant morphology. *Mailing Add:* 1512 N Washington Ave Springfield MO 65803-2849

BOND, MARTHA W(ILLIS), b St Louis, Mo, Oct 10, 46. PROTEIN CHEMISTRY, MOLECULAR IMMUNOLOGY. *Educ:* Univ Rochester, AB, 69; Univ Pa, PhD(chem), 79. *Prof Exp:* Res asst, Rockefeller Univ, 69-72; fel, Calif Inst Technol, 79-81; STAFF SCIENTIST, DNAX RES INST MOLECULAR & CELLULAR BIOL, INC, 82- *Concurrent Pos:* Vis assoc, Calif Inst Technol, 82. *Mem:* Am Chem Soc; Fedn Am Scientists; Sigma Xi; AAAS; Protein Soc; Am Peptide Soc. *Res:* Primary structure of proteins;

analysis of antibody diversity; structure-function correlations; sequencing methodology; biochemical characterization of cytokines and other proteins involved in immune responses and hematopoesis. *Mailing Add:* DNAX Res Inst 901 California Ave Palo Alto CA 94304-1104

BOND, MARVIN THOMAS, b Canton, Miss, Jan 17, 30; m 53; c 1. PORTABLE WATER SUPPLY, WASTE TREATMENT. *Educ:* La Tech Univ, BS, 57, MS, 61; Tulane Univ, PhD(environ eng), 69. *Prof Exp:* Head, Civil Eng Dept, Hong Kong Baptist Col, 61-66; assoc prof, environ eng technol, Western Ky Univ, 68-73; PROF, WATER TREATMENT, DEPT CIVIL ENG, MISS STATE UNIV, 73-; EXEC DIR, WATER RESOURCES, MISS WATER RESOURCES RES INST, 78- *Concurrent Pos:* Lectr Water Resources Res Doc Ctr, 84-88. *Mem:* Am Water Works Asn; Am Soc Civil Engr; Nat Soc Prof Engr; Am Asn Water Inst; Am Acad Environ Engr. *Res:* Water (portable) treatment, waste water treatment; industrial waste treatment. *Mailing Add:* 208 S Washington Starkville MS 39759

BOND, NORMAN T(UTTLE), b Beverly, Mass, Aug 19, 20; m 52; c 4. ELECTRICAL & INDUSTRIAL ENGINEERING. *Educ:* Univ Ill, BS, 42; Ill Inst Technol, BS, 48; Clarkson Col Technol, MS, 61. *Prof Exp:* Elec engr, Delta Star Elec Co, Ill, 47-48; res engr, 48-68, sr res scientist, Elec Eng Div, 68-78, staff scientist, Elec Products Div, Alcoa Labs, Aluminum Co Am, 78-86; RETIRED. *Mem:* Inst Elec & Electronics Engrs; Sigma Xi. *Res:* Electric connectors and contacts; test techniques for electric connections; data analysis. *Mailing Add:* Box 181 Massena NY 13662

BOND, PETER DANFORD, b Providence, RI, Jan 30, 40; m 68; c 2. EXPERIMENTAL NUCLEAR PHYSICS. *Educ:* Harvard Univ, BA, 62; Western Reserve Univ, MA, 63; Case Western Reserve Univ, PhD(physics), 69. *Prof Exp:* Res assoc physics, Stanford Univ, 69-72; from asst physicist to assoc physicist, 72-76, physicist, 76-86, SR PHYSICIST, BROOKHAVEN NAT LAB, 86- *Concurrent Pos:* Mem, Div Nuclear Physics Prog Comt, 77-79, 90-, Super Heavy-Ion Linear Accelerator Prog Adv Comt, 77-82, chmn, 81-82; vis scientist, Kernfysisch Versneller Inst, Neth, 83-84; assoc chmn, Physics Dept, Brookhaven Nat Lab, 86-87, dep chmn, 87, chmn, 87. *Mem:* Fel Am Phys Soc; fel AAAS. *Res:* Relativistic heavy ion collisions; the reaction mechanism in heavy ion induced transfer reactions. *Mailing Add:* Bldg 510A Brookhaven Nat Lab Upton NY 11973

BOND, RICHARD RANDOLPH, b Lost Creek, WVa, Dec 1, 27; m 49; c 4. ACADEMIC ADMINISTRATION, ZOOLOGY. *Educ:* Salem Col, WVa, BS, 48; Univ WVa, MS, 49; Univ Wis, PhD(zool), 55. *Hon Degrees:* LHD, Salem Col, 78. *Prof Exp:* Asst prof biol, Milton Col, 49-51; assoc prof, Salem Col, WVa, 55-58, dean men, 57-58; fel col admin, Univ Mich, 58-59; dean fac, Elmira Col, 59-63; prof acad admin, Cornell Univ & arts & sci consult, Univ Liberia, 63-64, chief party, Cornell Univ Proj Liberia, 64-66; prof zool, vpres acad affairs & dean fac, Ill State Univ, 66-71; pres, 71-81, EMER PRES & PROF ZOOL, UNIV NORTHERN COLO, 81- *Concurrent Pos:* Mem, Bd Trustees, Salem Col; comnr, Educ Comn States, 74-81, 85; mem bd, Am Asn State Cols & Univs, 79-81; Colo State Legislator, 85-91; interim pres, Front Range Community Col, 91- *Mem:* AAAS; Ecol Soc Am; Sigma Xi; Am Ornith Union. *Res:* Ecology and behavior of birds; autonomic responses; problems of higher education; international education, biology and public policy. *Mailing Add:* 1954 25th Ave Greeley CO 80631

BOND, ROBERT FRANKLIN, b Pullman, Wash, Apr 9, 37; m 60; c 3. PHYSIOLOGY. *Educ:* Ursinus Col, BS, 59; Temple Univ, MS, 61, PhD(physiol), 64. *Prof Exp:* Assoc physiol & cardiovasc trainee, 64-65, from instr to assoc prof physiol, Bowman Gray Sch Med, 65-73; prof physiol, chmn dept & dir, Cardiovasc Res Labs, Kirksville Col Osteop Med, 73-81; prof physiol & chmn dept, Sch Med, Oral Roberts Univ, 81-88; PROF PHYSIOL, SCH MED, UNIV SC, 88- *Concurrent Pos:* Am Heart Asn advan res fel, 68-70; consult, Carolina Med Electronics & Electromagnetic Probe Co. *Mem:* Am Heart Asn; Am Physiol Soc; Microcirc Soc; Shock Soc; Am Inst Biol Sci; AAAS. *Res:* Autoregulation of blood flow; control of cardiovascular reflexes in shock and hypertension; control of regional vascular beds; myocardial metabolism; hyperbaric physiology; endotoxic shock. *Mailing Add:* Dept Physiol Sch Med Univ SC Sch Med Columbia SC 29208

BOND, ROBERT HAROLD, b Denver, Colo, July 27, 36; m 58; c 3. ELECTRICAL ENGINEERING. *Educ:* Colo State Univ, BSEE, 58; Calif Inst Technol, MSEE, 59, PhD(plasma physics), 65. *Prof Exp:* From asst prof to assoc prof elec eng, Va Polytech Inst & State Univ, 65-71; sr engr, AT&T, 71-79, dept chief, Advan Develop Test Facil, 79-86, eng mgr, Mfg Develop Ctr, 86-89, CURRIC MGR CORP EDUC, AT&T, 90- *Mem:* Inst Elec & Electronics Engrs. *Res:* Use of Langmuir probes for determining directed electron velocity distributions in gaseous plasmas. *Mailing Add:* 139 Nancy Lane Ewing NJ 08638-1543

BOND, ROBERT LEVI, b Ozark, Ark, Feb 23, 40. INSTRUMENTATION, PHYSICS. *Educ:* State Col Ark, BS, 61; Univ Ark, MS, 64, PhD(instrumental sci), 68. *Prof Exp:* Sr res physicist, Southwest Res Inst, 68-71; asst prof, 71-74, ASSOC PROF, DEPT ELECTRONICS & INSTRUMENTATION, UNIV ARK, 74- *Mem:* AAAS; Instrument Soc Am; Optical Soc Am; Sigma Xi; Inst Elec & Electronics Engrs. *Res:* Particle optics; modern photon optics; mass spectrometry; vacuum technology; holography; interferometry; optical data processing; biophysics; bioengineering. *Mailing Add:* Univ Ark Grad Inst Technol Etas Bldg Rm 575 2801 S University Little Rock AR 72204

BOND, ROBERT SUMNER, b Cambridge, Mass, Oct 12, 25; m 50; c 2. FOREST ECONOMICS, FOREST POLICY. *Educ:* Univ Mass, BS, 51; Yale Univ, MF, 52; State Univ NY Col Forestry, Syracuse Univ, PhD(forestry econ), 66. *Prof Exp:* Res forester, Fordyce Lumber Co, Ark, 52-54; dist forester, Div Forests & Parks, Mass Dept Natural Resources, 54-56; instr gen forestry, Univ Mass, Amherst, 56-63, from asst prof to prof forestry econ, 63-78; prof forest sci & dir, Sch Forest Resources, Pa State Univ, 78-89; RETIRED. *Mem:* Soc Am Foresters; Forest Hist Soc; Am Forestry Asn. *Res:* Forest oriented outdoor recreation; forest industry organization and labor; forest resource policy research. *Mailing Add:* 499 Hubbardston Rd Princeton MA 01541

BOND, STEPHON THOMAS, b Lost Creek, WVa, Jan 25, 34; m 60; c 4. INORGANIC CHEMISTRY. *Educ:* Salem Col, BS, 60; WVa Univ, MA, 62; Kent State Univ, PhD(chem), 71. *Prof Exp:* Teacher chem, Belpre Pub Schs, 62-65; ASSOC PROF CHEM, SALEM COL, 65-68 & 71- *Mem:* Am Chem Soc; Sigma Xi; Am Indust Hyg Asn. *Res:* Organometallic complexes; computer control of mass spectroscopy. *Mailing Add:* Rte 1 Box 74 Jane Lew WV 26378

BOND, VICTOR POTTER, b Santa Clara, Calif, Nov 30, 19; m 46, 56; c 4. RADIATION BIOPHYSICS, EXPERIMENTAL MEDICINE. *Educ:* Univ Calif, AB, 43, MD, 45, PhD(med physics), 51. *Hon Degrees:* DSc, Long Island Univ, 65. *Prof Exp:* Asst, Med Sch, Univ Calif, 45; res group leader, Naval Radiol Defense Lab, Brookhaven Nat Lab, 48-55, mem staff, 55-58, head, Div Microbiol, 58-62, chmn, Med Dept, 62-67, assoc dir, 67-85, SR SCIENTIST, MED DEPT, BROOKHAVEN NAT LAB, 85- *Concurrent Pos:* Prof med, State Univ NY, Stony Brook, 68-; prof radiol, Columbia Univ, 69-; pres, 5th Int Cong Radiation Res, 71-74; distinguished assoc award, US Dept Energy, 90. *Honors & Awards:* Sr Alexander von Humbolt Award, 84-85; Distinguished Sci Achievement Award, Health Physics Soc, 86. *Mem:* Am Phys Soc; Radiation Res Soc (pres, 73-74); Soc Exp Biol & Med; Soc Nuclear Med; Am Environ Mutagen Soc. *Res:* Biological effects of radiation; effects of radiation on mammalian systems; the use of particulate radiations in radiotherapy. *Mailing Add:* Brookhaven Nat Lab Upton NY 11973

BOND, WALTER D, b Lebanon, Tenn, Mar 21, 32; m 55; c 3. PHYSICAL CHEMISTRY, CHEMICAL ENGINEERING. *Educ:* Mid Tenn State Univ, BS, 53; Vanderbilt Univ, PhD(phys chem), 57. *Prof Exp:* staff mem res & develop, 57-61, GROUP LEADER, OAK RIDGE NAT LAB, 61- *Concurrent Pos:* Guest scientist, Swiss Fed Inst Reactor Res, 71-72. *Honors & Awards:* IR-100 Award, Indust Res Mag, 85. *Mem:* Am Chem Soc; Am Nuclear Soc. *Res:* Chemistry of separations of transuranium elements; chemical kinetics; nuclear fuel cycle chemistry; nuclear fuel reprocessing; colloidal synthesis of ceramic materials. *Mailing Add:* 6704 Stone Mill Rd Knoxville TN 37919

BOND, WILLIAM BRADFORD, b Ithaca, NY, Oct 11, 29; m 52; c 4. PHYSICAL CHEMISTRY, ORGANIC CHEMISTRY. *Educ:* Cornell Univ, BA, 51; Fla State Univ, PhD(chem), 55. *Prof Exp:* Res Chemist, 55-63, res supvr, 63-72, SR RES ASSOC, SPECIALITY PROD DIV, FABRICATED PRODUCTS DEPT, E I DU PONT DE NEMOURS & CO, 72- *Mem:* AAAS; Am Chem Soc; Sigma Xi. *Res:* Free radical chemistry; polyamidation kinetics; polymer chemistry; fibers. *Mailing Add:* 101 Hillside Way Marietta OH 45750

BOND, WILLIAM H, b Toronto, Ont, Sept 20, 16; US citizen; m 49; c 4. INTERNAL MEDICINE, HEMATOLOGY. *Educ:* Univ Chicago, SB, 40, MD, 42. *Prof Exp:* From instr to prof, 52-85, EMER PROF, INTERNAL MED, SCH MED, IND UNIV-PURDUE UNIV, INDIANAPOLIS, 85- *Concurrent Pos:* Consult, Vet Admin Hosp, 55-85. *Mem:* Am Fedn Clin Res; fel Am Col Physicians; Am Soc Hemat; Am Asn Cancer Res; Cent Soc Clin Res. *Res:* Cancer chemotherapy; clinical hematology. *Mailing Add:* 4525 W 59th St Indianapolis IN 46254

BOND, WILLIAM PAYTON, b Franklinton, La, Nov 18, 41; m 69; c 2. PLANT PATHOLOGY. *Educ:* Southeastern La Col, BS, 63; La State Univ, MS, 66, PhD(plant path), 68. *Prof Exp:* Asst prof bot & plant path, 67-74, assoc prof, 74-80, PROF BIOL, SOUTHEASTERN LA UNIV, 80- *Mem:* AAAS; Am Phytopath Soc. *Res:* Plant virology; soil transmission of plant viruses; serology. *Mailing Add:* Dept Biol Sci Southeastern La Univ Hammond LA 70402

BONDAR, RICHARD JAY LAURENT, b New York, NY, Sept 4, 40; m 61; c 2. CLINICAL BIOCHEMISTRY, CLINICAL CHEMISTRY. *Educ:* McGill Univ, BSc, 62; Calif Inst Technol, MS, 65; Univ Calif, Riverside, PhD(biochem), 69. *Prof Exp:* Instr biol, La State Univ, New Orleans, 64-65; res assoc biochem, Mich State Univ, 69-71; res scientist, 71-73, sr res scientist, 73-77, head, Anal Enzymology Dept, 77-80, head develop oper, Worthington Biochem Corp, 80-83; prin develop chemist, Beckman, 83-85; Mgr Process Control, Abbott Labs, 86-89; vpres, R & D Int Medication Systs, Ltd, 89-90; RETIRED. *Concurrent Pos:* Mem subcomt on glucose ref methods, Standards Comt, Am Asn Clin Chem, 72-78; chmn, subgroup reagent composition, subcomt on enzyme assay conditions, standards comt, Nat Comt on Clin Lab Standards, 73-, subcomt on methodological principles, 78-89. *Mem:* Am Chem Soc; Am Asn Clin Chemists; Sigma Xi; fel Am Inst Chemists. *Res:* Enzymology and protein chemistry; mechanisms of enzyme catalysis and mechanism of reactions of biological interest; analytical biochemistry; enzymology; analytical clinical chemistry. *Mailing Add:* 1661 S Tiara Way Anaheim CA 92802

BONDAREFF, WILLIAM, b Washington, DC, Apr 29, 30; m 58; c 2. ANATOMY. *Educ:* George Washington Univ, BS, 51, MS, 52; Univ Chicago, PhD(anat), 54; Georgetown Univ, MD, 62. *Prof Exp:* USPHS res fel, Nat Cancer Inst, 54-55; cytologist, Sect Aging, NIMH, 55-64; assoc prof, Med Sch, Northwestern Univ, Chicago, 64-70, prof anat & chmn dept, 70-; AT GERONT CTR, UNIV SOUTHERN CALIF. *Mem:* Fel AAAS; Electron Micros Soc; Geront Soc; Am Asn Anatomists; Am Acad Neurologists. *Res:* Microscopic and ultrastructure of nervous system. *Mailing Add:* Clare Hall Cambridge CB3 9AL London England

BONDE, ERIK KAUFFMANN, b Odense, Denmark, Oct 19, 22; nat US; m 53; c 3. PLANT PHYSIOLOGY. *Educ:* Univ Colo, BA, 47, MA, 48; Univ Chicago, PhD(bot), 51. *Prof Exp:* Res assoc bot, Univ Calif, Los Angeles, 51-53; NSF fel, Calif Inst Technol, 53-54; asst prof bot, Univ Mo, 54-55; from asst prof to assoc prof, 55-67, PROF BIOL, UNIV COLO, 67- *Mem:* AAAS; Ecol Soc Am; Am Bot Soc; Am Soc Plant Physiol; Scandinavian Soc Plant Physiol; Sigma Xi. *Res:* Plant growth substances and inhibitors; photoperiodism; ecology. *Mailing Add:* Dept EPO Biol Univ Colo Box 334 Boulder CO 80309

BONDE, MORRIS REINER, b Presque Isle, Maine, Aug 7, 45; m 73. PLANT PATHOLOGY. *Educ:* Univ Maine, BS, 67; Cornell Univ, MS, 69, PhD(plant path), 75. *Prof Exp:* RES PLANT PATHOLOGIST, AGR RES SERV, USDA, 74- *Mem:* Am Phytopath Soc; Sigma Xi. *Res:* Epidemiology and control of diseases of major food crops. *Mailing Add:* Fort Detrick Bldg 1301 Frederick MD 21701

BONDELID, ROLLON OSCAR, b Grand Forks, NDak, Jan 8, 23; m 73; c 3. NUCLEAR PHYSICS. *Educ:* Univ NDak, BS, 45; Washington Univ, MS, 48, PhD(physics), 50. *Prof Exp:* Mem staff, Los Alamos Sci Lab, 50-52; physicist, Naval Res Lab, 52-63, head, Cyclotron Br, 63-80; RETIRED. *Mem:* Fel Am Phys Soc; Sigma Xi; Am Nuclear Soc. *Res:* Development of sector-focusing cyclotron research facility; correlation studies of few-nucleon systems; neutron cancer therapy; absolute energy measurements of nuclear reactions. *Mailing Add:* 13218 Rollie Rd E Bishopville MD 21813

BONDER, SETH, b Bronx, NY, July 14, 32; div; c 2. OPERATIONS RESEARCH, RESEARCH MANAGEMENT. *Educ:* Univ Md, BSME, 60; Ohio State Univ, PhD, 65. *Prof Exp:* Res asst indust eng, Ohio State Univ, 61, res assoc, 61-62, proj supvr & prin investr, 62-65; from asst prof to assoc prof indust eng, Univ Mich, Ann Arbor, 65-72, dir systs res lab, 67-71, ADJ PROF INDUST & OPERS ENG, UNIV MICH, ANN ARBOR, 72-; PRES, VECTOR RES, INC, 72- *Concurrent Pos:* Opers res consult, Joint Chiefs Staff, 73-77, Army Sci Bd, 74-78 & Dept HEW, 75-77; comptroller gen, GAO, 80; Jet Propulsion Lab, 81. *Mem:* AAAS (pres, 78-79); Opers Res Soc Am (vpres, 77-78, pres, 78-79); Nat Acad Scis; Inst Mgt Scis. *Res:* Methodology for long range planning; theories of research and development allocation strategies; studies of large-scale systems; organization of municipal governments; research on theories of search; modeling of population processes. *Mailing Add:* 2900 Fuller Ann Arbor MI 48105

BONDERMAN, DEAN P, b Primghar, Iowa, July 6, 36; m 68; c 3. CLINICAL PATHOLOGY, TOXICOLOGY. *Educ:* Westmar Col, BA, 62; Univ Iowa, PhD(inorg chem), 68. *Prof Exp:* Sr chemist & instr pesticide toxicol, Dept Prev Med, Univ Iowa, 66-70; from asst prof to assoc prof clin path, Med Ctr, Ind Univ, Indianapolis, 76-84; PRES & CHIEF EXEC OFFICER, ANALYTICAL CONTROL SYSTS INC, 84- *Mem:* Am Chem Soc; Am Soc Clin Path; Am Asn Clin Chem. *Res:* High temperature thermodynamics of inorganic compounds; enzyme response to environmental stimuli; pesticide detection; enzyme kinetics and thermodynamics; development of quality control materials in clinical chemistry and hematology. *Mailing Add:* Analytical Control Systs Inc 9001 Technology Dr Suite A Fishers IN 46038

BONDESON, STEPHEN RAY, b Zion, Ill, June 9, 52; m 78. MAGNETIC RESONANCE. *Educ:* Univ Wis, Stevens Point, BS, 74; Duke Univ, PhD(phys chem), 78. *Prof Exp:* Res assoc, Princeton Univ, 78-80; ASST PROF CHEM, UNIV WIS, STEVENS POINT, 80- *Mem:* Am Phys Soc; Am Chem Soc; Sigma Xi. *Res:* Experimental and theoretical aspects of low-dimensional charge-transfer solids; solid state magnetic resonance and relaxation phenomena; stochastic processes; random walk theory. *Mailing Add:* Dept Chem Univ Wis Stevens Point WI 54481

BONDI, AMEDEO, b Springfield, Mass, Dec 13, 12; m 39; c 4. MICROBIOLOGY. *Educ:* Univ Conn, BS, 35; Univ Mass, MS, 37; Univ Pa, PhD(microbiol), 42. *Prof Exp:* Instr bact, Sch Med, Temple Univ, 38-40, asst prof, 42-47; chmn, microbiol dept, 47-78, dean, grad sch, 73-84, PROF MICROBIOL, HAHNEMANN MED COL, 47-73 & 84- *Honors & Awards:* Becton-Dickinson Award Clin Microbiol, 80. *Mem:* AAAS; Am Soc Microbiol; NY Acad Sci; Am Soc Trop Med & Hygiene; Sigma Xi. *Res:* Immunology of Brucellosis and whooping cough; studies on the antibiotic agents; studies on staphylococcal infections. *Mailing Add:* 2064 Horace Ave Abington PA 19001

BONDINELL, WILLIAM EDWARD, b Passaic, NJ, March 26, 42; m 70; c 1. MEDICINAL CHEMISTRY. *Educ:* Fairleigh Dickinson Univ, NJ, BS, 63; Univ Calif, Berkeley, PhD(org chem), 68. *Prof Exp:* Fel, dept biochem, Columbia Univ, 69-71; res assoc, Am Health Found, 71-72; assoc sr investr, 73-75, sr investr, 76-87, ASST DIR MED CHEM, SMITH KLINE & FRENCH LABS, 87- *Mem:* AAAS; Am Chem Soc. *Res:* Design and synthesis of receptor agonists and antagonists, uptake and enzyme inhibitors as potential agents for cardiovascular, gastrointestinal and CNS diseases. *Mailing Add:* SmithKline Beecham Pharmaceut 709 Swedeland Rd PO Box 1539 King of Prussia PA 19406-0939

BONDURANT, BYRON L(EE), b Lima, Ohio, Nov 11, 25; m 44; c 3. AGRICULTURAL ENGINEERING ADMINISTRATION. *Educ:* Ohio State Univ, BAE, 49; Univ Conn, MS, 53. *Prof Exp:* Eng trainee surv & inspection, Bur Reclamation, US Dept Interior, 48; lab asst agr eng, Ohio State Univ, 48-49; dist agr engr, Cornell Univ, 49-50; instr & exten engr, agr eng, Univ Conn, 50-53; assoc prof, Univ Del, 53-54; prof & head dept, Univ Maine, 54-64; PROF AGR ENG, OHIO STATE UNIV, 64- *Concurrent Pos:* Adv to dean, Col Agr Eng, Punjab Agr Univ, India, 65-67, 69-71 & 72; team leader, Midwestern Consortium Int Activity, 76-77; proj mgr, UN Develop Prog, Food & Agr Orgn, 76-77; Fulbright prof, Univ Nairobi, 79-80. *Mem:* Fel AAAS; fel Am Soc Agr Engrs; Nat Soc Prof Engrs; Am Soc Civil Engrs; NY Acad Sci; Am Soc Eng Educ. *Res:* Soil and water conservation engineering; hydrology; domestic sewage disposal. *Mailing Add:* Dept Agr Eng Ohio State Univ 590 Woody Hayes Dr Columbus OH 43210

BONDURANT, CHARLES W, JR, polymer chemistry, analytical chemistry; deceased, see previous edition for last biography

BONDURANT, JAMES A(LLISON), b Ness City, Kans, Oct 10, 26; m 51; c 3. AGRICULTURE, ENGINEERING. *Educ:* Kans State Col Agr, BSc, 49; Univ Nebr, MSc, 51. *Prof Exp:* Asst agr eng irrig res, Univ Nebr, 51-55; AGR ENGR, AGR RES, USDA 58- *Concurrent Pos:* Mem, US nat comt, Int Comn Irrig & Drainage. *Mem:* Fel AAAS; fel Am Soc Agr Engrs; Sigma Xi. *Res:* Automatic irrigation gates methods and practices in surface irrigation; sediment and erosion control in surface irrigation. *Mailing Add:* Route 2 Kimberly ID 83341

BONDURANT, STUART, b Winston-Salem, NC, Sept 9, 29; m 54; c 3. INTERNAL MEDICINE. *Educ:* Duke Univ, BS, 52, MD, 53. *Hon Degrees:* DSc, Indiana Univ, 80. *Prof Exp:* Intern, Duke Hosp, Durham, NC, 53-54, resident internal med, 54-56; resident, Peter Bent Brigham Hosp, Boston, 58-59; from asst prof to prof med, Sch Med, Ind Univ, Indianapolis, 59-67, assoc dir, Med Ctr, 61-67; prof med & chmn dept, Albany Med Col, 67-74, pres & dean, 74-79; PROF MED & DEAN, SCH MED, UNIV NC, 79- *Concurrent Pos:* Chmn, Res Study Comt, Coun Circulation, Am Heart Asn, 64-67, res comt, 64-69, chmn, 67-68, vpres-at-large, 69-72; physician-in-chief, Albany Med Ctr Hosp, 67-74; mem adv coun, Nat Heart & Lung Inst, NIH, 72-75; mem adv comt to dir, NIH, 78-81, Panel Phys Med & Rehab Res, 89; mem coun, Inst Med-Nat Acad Sci, 90-, Comt Sci, Eng & Pub Policy, 90- *Honors & Awards:* Citation Distinguished Serv Res, Am Heart Asn, 69, Lyman Duff Mem lectr, Award of Merit, 75; Alfred E Stengel Award, Am Col Physicians, 86. *Mem:* Inst Med-Nat Acad Sci; master Am Col Physicians (vpres, 78, pres-elect, 79-80, pres, 80-81); Am Soc Clin Invest (vpres, 74); Asn Am Physicians (treas, 74-79, pres, 85-86); Am Physiol Soc; fel Royal Col Physicians; fel AAAS; Am Clin & Climat Asn; Am Col Cardiol; AMA. *Res:* Author or co-author of over 45 books & papers. *Mailing Add:* CB 7000 125 MacNider Bldg Univ NC Sch Med Chapel Hill NC 27599-7000

BONDY, DONALD CLARENCE, b Harrow, Ont, June 5, 32; m 56; c 2. GASTROENTEROLOGY. *Educ:* Univ Western Ont, MD, 56; FRCPS(C), 61 & 71. *Prof Exp:* Instr med, 62-63, lectr, 63-65, lectr physiol, 65-67, assoc prof, 67-74, from asst prof med to assoc prof med & physiol, 68-73, PROF MED & PHYSIOL, UNIV WESTERN ONT, 73- *Concurrent Pos:* Consult gastroenterol, Westminster Hosp, 62-67, dir clin invest unit & chief of serv-gastroenterol, 67-72; consult gastroenterol, Victoria & St Joseph's Hosps, 62-67; chief of serv, Dept Gastroenterol, Univ Hosp, 73- *Mem:* Can Soc Clin Invest; Can Med Asn; Can Asn Gastroenterol. *Res:* Clinical gastroenterology. *Mailing Add:* Univ Hosp London ON N6A 5A5 Can

BONDY, MICHAEL F, b Vienna, Austria, Dec 6, 23; nat US; m 50; c 3. ELECTRICAL ENGINEERING. *Educ:* George Washington Univ, BEE, 43; Univ Buffalo, MS, 60. *Prof Exp:* Res assoc, Nat Defense Res Comt, Off Sci Res & Develop, George Washington Univ, 43-45; jr engr, Raytheon Mfg Co, 46-48; intermediate scientist, Westinghouse Elec Corp, 49-50; electromech engr, Glenn L Martin Co, 50-53; res engr, Cornell Aeronaut Lab, Inc, 53-54; head autopilot & stability augmentation group, Bell Aircraft, 54-58; leader systs group, microelectronic prog, Commun Systs Div, Radio Corp Am, 58-65; pres, Dytonis, Inc, 65-90; PRES, MICHAEL BONDY INC, 90- *Concurrent Pos:* Vpres, Butler & Smith, Inc, 65-70; pres, Hap Jones Atlantic, Inc, 70-73; consult, Singer - Kearfott, 83-88. *Mem:* Sigma Xi; sr mem Inst Elec & Electronic Engrs. *Mailing Add:* Michael Bondy Inc PO Box 123 Haworth NJ 07641-0123

BONDY, PHILIP K, b New York, NY, Dec 15, 17; m 49; c 3. FUNCTION OF POLYAMINES IN CELL DIFFERENTIATION, ENDOCRINE FUNCTION OF ADRENAL GLAND. *Educ:* Columbia Univ, BA, 38; Harvard Univ, MD, 42. *Hon Degrees:* MS, Yale Univ, 60. *Prof Exp:* Assoc med, Sch Med, Emory Univ, 48-52; from asst prof to prof med, Sch Med, Yale Univ, 52-74, chmn dept, 65-72; prof med, Inst Cancer Res, London, UK, 76-77; assoc chief of staff res, 77-83, chief staff, Vet Admin Med Ctr, W Haven, 83-89; assoc dean, Vet Admin Affairs, 83-89, prof med, 77-78, EMER PROF, SCH MED, YALE UNIV, 88- *Concurrent Pos:* Ed in chief, J Clin Invest, Am Soc Clin Invest, 56-62 & Yale J Biol Med, 79- *Honors & Awards:* Brainerd Prize, Columbia Univ, 38. *Mem:* Asn Am Physicians; Am Col Physicians; Royal Col Physicians; Am Soc Clin Invest; Endocrine Soc; Royal Soc Med; Sigma Xi. *Res:* Endocrinology, metabolism and control of differentiation; diabetes mellitus and adrenal cortical function; pattern of formation of ectopic hormones by neoplasms; polyamine metabolism, investigating the influence of these substances on growth and differentiation, especially of immune cells and cancers. *Mailing Add:* 9 Chestnut Lane Woodbridge CT 06525

BONDY, STEPHEN CLAUDE, b Hilversum, Holland, Jan 10, 38; US citizen; m; c 3. BIOCHEMISTRY, NEUROBIOLOGY. *Educ:* Cambridge Univ, BA, 59, MA, 62; Univ Birmingham, MSc, 61, PhD(biochem), 62. *Prof Exp:* Res biochemist, Unilever Ltd, Bedford, Eng, 59; res asst, Univ Birmingham, 59-62; res scientist, NY State Psychiat Inst, 63-65; res biol chemist, Sch Med, Univ Calif, Los Angeles, 65-70; asst prof neurol, Med Ctr, Univ Colo, Denver, 70-75, assoc prof neurol & pharmacol, 75-78; head, Neurochem Sect, Lab Behav & Neurol Toxicol, Nat Inst Environ Health Sci, 78-85; PROF TOXICOL, DEPT COMMUNITY & ENVIRON MED, UNIV CALIF, IRVINE, 85- *Concurrent Pos:* NIH res career develop award, 71-75; NIMH res scientist, 75-78. *Mem:* Toxicol Soc; Int Soc Neurochem; Am Soc Neurochem; Int Soc Develop Neurosci. *Res:* Neurobiology of cerebral development; neurotoxicology. *Mailing Add:* Dept Comm & Environ Med Univ Calif Col Med Irvine CA 92717-1825

BONDYBEY, VLADIMIR E, b Prague, Czech, Jan 4, 40; US citizen; m 66; c 3. LASER CHEMISTRY & SPECTROSCOPY, MATRIX ISOLATION. *Educ:* Univ Rostock, BSc, 65; Charles Univ, Prague, MSc, 68; Univ Calif, Berkeley, PhD(chem), 72. *Prof Exp:* Asst prof chem, Charles Univ, Prague, 66-69; postdoctoral fel chem, Ore State Univ, 72-73; tech staff chem, AT&T Bell Labs, 73-86, distinguished mem tech staff, 84; PROF CHEM, OHIO STATE UNIV, 86- & TECH UNIV MUNICH, 88- *Concurrent Pos:* Mem, Steering Comt Int Conf Matrix Isolation, 83-; adv ed, J Chem Physics, 84-86; consult, NSF, 86-87; chmn, Spectros Group, Optical Soc Am, 88-89 & Gordon Conf Matrix Isolation, 91. *Mem:* Am Chem Soc; fel Am Phys Soc; Optical Soc Am; AAAS; NY Acad Sci; Ger Soc Phys Chem. *Res:* Photochemistry, spectroscopy and structure of molecular ions, free radicals and other reactive intermediates; techniques for their production and low temperature studies; clusters and complexes, their formation, dynamics and chemical reactions. *Mailing Add:* Inst Phys & Theoret Chem Tech Univ Munich Lichtenbergstr 4 Garching 8046 Germany

BONE, JACK NORMAN, b Montrose, Colo, Feb 10, 19; m 40; c 2. PHARMACY. *Educ:* Univ Colo, BS, 41, MS, 48; Univ Wash, PhD(pharm), 53. *Prof Exp:* Pharm res asst, Cutter Labs, Berkeley, Calif, 41-42, supvr blood fractionation, 42-45; instr pharm, Univ Colo, 45-48; from asst prof to prof, 48-83, Univ Wyo, dean, Col Pharm, 66-83 & Col Health Sci, 68-83; RETIRED. *Concurrent Pos:* Fulbright lectr, Univ Baghdad, 63-64. *Res:* Synthesis and biotesting of antibacterials and antifungals. *Mailing Add:* 808 Dellwood Dr Ft Collins CO 80524

BONE, JESSE FRANKLIN, b Tocoma, Wash, June 15, 16; m 50; c 4. VETERINARY MEDICINE. *Educ:* State Col Wash, BA, 37, BS, 49, DVM, 50; Ore State Col, MS, 53. *Prof Exp:* Lab technician vet med, State Col Wash, 47-50; instr & res asst, 50-53, from asst prof to prof, 53-80, EMER PROF VET MED, ORE STATE UNIV, 80- *Concurrent Pos:* Fulbright lectr vet path, Univ Assuit, 64-65, lectr vet anat, Univ Nairobi, 80-82; consult, Univ Zimbabwe, 82; prof vet anat, Ross Univ, 84. *Mem:* Am Asn Vet Toxicologists; Am Asn Lab Animal Sci; Royal Soc Health. *Mailing Add:* 3017 Brae Burn Sierra Vista AZ 85635

BONE, LARRY IRVIN, b Perry, Iowa, Aug 31, 35; m 57; c 2. CHEMICAL WASTE TREATMENT, PHYSICAL CHEMISTRY. *Educ:* Coe Col, BA, 57; Ohio State Univ, MSc, 64, PhD, 66. *Prof Exp:* Res chemist, Aerospace Res Labs, Wright-Patterson Air Force Base, US Air Force, 64-67, nuclear scientist, US Dept Defense, 67-68; from asst prof to prof chem, ETex State Univ, 68-78, asst dean sci & technol, 73-78; mem staff, 79-80, group leader, 80-84, ENVIRON PROJ MGR, ENVIRON SERV DEPT, DOW CHEM CO, 84- *Concurrent Pos:* Chief, Chem Div, Nat Ctr Toxicol Res, 78-79; vis prof, Appalachian State Univ, 79-80. *Mem:* Am Chem Soc; Am Phys Soc; Am Soc Mass Spectrometry. *Res:* Mass spectrometric ion-molecule reactions; analytical mass spectrometry; chemical waste treatment and disposal. *Mailing Add:* Dow Chem Co 2030 Willard H Dow Ctr Midland MI 48674

BONE, LEON WILSON, b Memphis, Tenn, Aug 19, 45; m 67; c 2. INVERTEBRATE PHYSIOLOGY. *Educ:* Memphis State Univ, BS, 70, MS, 72; Univ Ark, Fayetteville, PhD(zool), 76. *Prof Exp:* Asst res biologist, Div Toxicol & Physiol, Dept Entom, Univ Calif, Riverside, 77-78; asst prof zool, Div Biol Sci, Ark State Univ, 78-80; MEM FAC, PHYSIOL DEPT, SOUTHERN ILL UNIV, 80- *Concurrent Pos:* Res asst, Div Toxicol & Physiol, Dept Entom, Univ Calif, Riverside, 76-77, co-prin investr, 77- *Mem:* Sigma Xi; Am Soc Parasitologists; Soc Protozoologists; Wildlife Dis Asn. *Res:* Pheromone communication of nematodes. *Mailing Add:* Regional Parasite Res Lab USDA Box 952 Auburn AL 36830

BONE, ROGER C, b Bald Knob, Ark, Feb 8, 41; m 63; c 2. INTERNAL MEDICINE, PULMONARY DISEASE. *Educ:* Hendrix Col, BA, 63; Univ Ark Med Sch, BMS, 65, MD, 67. *Prof Exp:* Asst prof, Univ Kans Med Ctr, 74-77; assoc prof, 77-79, PROF MED, UNIV ARK MED CTR, 79-; AT DEPT MED, RUSH PRESBYTERIAN ST LUKE'S MED CTR. *Concurrent Pos:* Chief pulmonary div, Univ Ark Med Ctr, 77- *Mem:* Am Thoracic Soc; Am Col Chest Physcians; Sigma Xi. *Res:* Basic mechanisms of lung injury in the adult respiratory distress syndrome. *Mailing Add:* Rush Presbyterian St Luke's Med Ctr 1753 W Congress Pkwy Chicago IL 60612

BONEHAM, ROGER FREDERICK, b Boston, Mass, July 26, 35; m 63; c 2. GEOLOGY. *Educ:* Mich Technol Univ, BS, 60; Wayne State Univ, MS, 65; Univ Mich, PhD(geol), 68. *Prof Exp:* From asst prof to assoc prof geol & bot, 66-78, PROF GEOL, IND UNIV, KOKOMO, 78- *Res:* Radon in soils; computer modeling of geologic structures. *Mailing Add:* Dept Geol Ind Univ Kokomo PO Box 9003 Kokomo IN 46904

BONEM, RENA MAE, b Tucumcari, NMex, May 6, 48. INVERTEBRATE PALEONTOLOGY, PALEOECOLOGY. *Educ:* NMex Inst Mining & Technol, BS, 70, MS, 71; Univ Okla, PhD(geol), 75. *Prof Exp:* Lectr geol, Univ Tex, Arlington, 73; asst prof geol, Hope Col, 75-79; asst prof geol, Tex Christian Univ, 79-81; assoc prof, 81-90, PROF GEOL, BAYLOR UNIV, 90- *Concurrent Pos:* Chmn, Coral Reef Ecol Specialty Protection, Nat YMCA Scuba Prog, 78-, mem adv bd, 89- *Mem:* Paleont Soc; Int Paleont Asn; Geol Soc Am; Am Asn Petrol Geologists; Soc Econ Paleontologists & Mineralogists; Sigma Xi. *Res:* Development and succession of modern and ancient reefs systems; human impact on coral reefs. *Mailing Add:* Dept Geol Baylor Univ Waco TX 76798

BONESS, MICHAEL JOHN, b London, Eng, Oct 2, 40; US citizen; m 65; c 3. PHYSICS. *Educ:* Univ London, BSc, 63, PhD(atomic physics), 67. *Prof Exp:* Res assoc atomic physics, Dept Eng & Appl Sci, Yale Univ, 67-69, lectr, 69-73; PRIN RES SCIENTIST ATOMIC PHYSICS, AVCO EVERETT RES LAB, INC, 73- *Mem:* Am Phys Soc. *Res:* Study of fundamental processes involving electron-molecule and electron-excited state interactions; role of fundamental kinetic processes in gas discharges and the application to high energy electric discharge lasers; high power IR; visible and UV laser research and development. *Mailing Add:* Avco Everett Res Lab 2385 Revere Beach Pkwy Everett MA 02149

BONEWITZ, ROBERT ALLEN, b Ottawa, Kans, Dec 18, 43. ELECTROCHEMISTRY, CORROSION. *Educ:* Lawrence Univ, BA, 65; Univ Fla, PhD(chem), 70. *Prof Exp:* Scientist corrosion, Alcoa Labs, 70-73, sr scientist, 73-75, sect head mat & corrosion sci, 75-79, mgr, surface technol, 79-81, mgr, Alloy Technol Div, 81-82, mgr, Prod Eng Div, 82-83, opers dir, Mill Prod Res & Develop, 83-84, DIR, PROD DESIGN & MFG, ALCOA LABS, 84- *Mem:* Sigma Xi; Am Soc Metals. *Mailing Add:* Alcoa Labs Alcoa Center PA 15069

BONEY, WILLIAM ARTHUR, JR, b Iola, Tex, May 13, 16; m 41, 61; c 1. VETERINARY MEDICINE, IMMUNOLOGY. *Educ:* Agr & Mech Col Tex, BS, 40, DVM, 42, MS, 48. *Prof Exp:* Exten poultry vet, Exten Serv, Agr & Mech Col Tex, 42-44, poultry vet, Exp Sta, 44-46, assoc prof poultry dis & poultry pathologist, 46-54; pres & dir, Boney Labs, 54-59; mgr vet technol, Int Div, Schering Corp, 59-62; res vet & virol proj leader, Nat Animal Dis Lab, USDA, 63-80; RETIRED. *Mem:* Am Vet Med Asn; Am Asn Avian Path; US Animal Health Asn. *Res:* Microbiology; virology; bacteriology; avian diseases. *Mailing Add:* 310 O'Neil Drive Ames IA 50010

BONFIGLIO, MICHAEL, b Milwaukee, Wis, Apr 3, 17; m 43; c 3. BIOLOGY, ORTHOPEDIC SURGERY. *Educ:* Columbia Univ, AB, 40; Univ Chicago, MD, 43. *Prof Exp:* Instr orthop surg, Univ Chicago, 49-50; assoc, 50-51, from asst prof to assoc prof, 51-87, EMER PROF ORTHOP SURG, UNIV IOWA, 87- *Mem:* AMA; Am Acad Orthop Surg; Orthop Res Soc; Am Col Surg; Am Orthop Asn; Sigma Xi; Int Skel Soc. *Res:* Pathology of irradiation fractures; aseptic necrosis of bone; bone transplant studies, histological and immunological; bone and skin transplant studies. *Mailing Add:* Univ Hosp Iowa City IA 52242

BONGA, JAN MAX, b Bern, Switz, Dec 5, 29; Can citizen; m 59; c 3. PLANT PHYSIOLOGY. *Educ:* State Univ Groningen, BSc, 54, MSc, 57; McMaster Univ, PhD(biol), 60. *Prof Exp:* Res officer forest path, Can Dept Forestry, 60-72, RES OFFICER FOREST PATH, FORESTRY CAN MARITIMES REGION, 72- *Mem:* Can Soc Plant Physiologists. *Res:* Physiology of tree growth. *Mailing Add:* Forestry Can Maritimes Region PO Box 4000 Fredericton NB E3B 5P7 Can

BONGARTEN, BRUCE C, b Pittsburgh, Pa, Mar 4, 51; m 75. FOREST GENETICS. *Educ:* State Univ NY Col Environ Sci & Forestry, BS, 73; Mich State Univ, PhD(forest genetics), 78. *Prof Exp:* Asst prof forestry, Univ Ga, 78-80. *Mem:* Soc Am Foresters. *Res:* Genetic improvement of southern hardwoods; pine hybridization; advanced generation breeding; population genetics of tree species. *Mailing Add:* Dept Forestry Univ Ga Athens GA 30602

BONGIORNI, DOMENIC FRANK, b Ivoryton, Conn, June 19, 08. PHYSICAL ORGANIC CHEMISTRY. *Educ:* Univ Conn, BS, 33; Johns Hopkins Univ, PhD(chem), 36; Fordham Univ, LLB, 44. *Prof Exp:* Res chemist, Gen Chem Co, Allied C & D, 36-39, spec consult, 40; patent dept, Union Carbide & Carbon Res Labs, Inc, 41-42; res chemist, Bakelite Corp, 42; chem res asst, Off Sci Res & Develop, War Res Div, Columbia Univ, 42-44; assoc patent lawyer, Campbell, Brumbaugh, Free & Graves, 44- 51; patent lawyer, Gen Anil & Film Corp, 51-52; counr, Sci Design Co, Inc, 52-55; patent counr, Johnson & Johnson, 55-60; CONSULT, 60- *Mem:* Am Chem Soc. *Res:* Reaction kinetics; methylene free radical and methyl, ethyl, acetonyl free radical studies; chain reactions; synthetic organic chemistry in drugs, insecticides; plastics; scientific-legal counsel. *Mailing Add:* PO Box 291 Ivoryton CT 06442-0291

BONGIORNO, SALVATORE F, b Mt Vernon, NY, Aug 16, 39; m 68; c 1. VERTEBRATE ECOLOGY. *Educ:* Fordham Univ, BS, 61; Rutgers Univ, MS, 63, PhD(ecol), 67. *Prof Exp:* Asst prof ecol, La State Univ, New Orleans, 67-71; ASST PROF BIOL, FAIRFIELD UNIV, 71- *Mem:* AAAS; Ecol Soc Am; Am Ornith Union; Am Inst Biol Sci; Sigma Xi. *Res:* Nesting ecology of avifauna; Avifauna of Northwestern Spain. *Mailing Add:* Dept Biol Fairfield Univ Fairfield CT 06430

BONGIOVANNI, ALFRED MARIUS, biology, medical science; deceased, see previous edition for last biography

BONHAM, CHARLES D, b Cleburne, Tex, Feb 4, 37; m 55; c 5. PLANT ECOLOGY, BIOSTATISTICS. *Educ:* Abilene Christian Col, BS, 59; Utah State Univ, MS, 65; Colo State Univ, PhD(plant ecol), 66. *Prof Exp:* Grassland ecologist, Colo State Univ, 62-64; plant ecologist, Cornell Aeronaut Labs, 66-67; from asst prof to assoc prof statist ecol, Univ Ariz, 67-70; assoc prof, 70-75, PROF QUANT PLANT ECOL, COLO STATE UNIV, 75- *Concurrent Pos:* Consult, Mining Ind & Inst, Range Veg. *Mem:* Am Soc Range Mgt; Ecol Soc Am; Sigma Xi; Biomet Soc. *Res:* Quantitative plant ecology studies using statistical models and computer graphic techniques; vegetation description and plant community analyses using statistics and mathematical procedures; vegetation mapping by computer graphics. *Mailing Add:* Dept Range Sci Colo State Univ Ft Collins CO 80523

BONHAM, HAROLD F, JR, b Los Angeles, Calif, Sept 1, 28; m 52; c 3. GEOLOGY. *Educ:* Univ Calif, Los Angeles, BA, 54; Univ Nev, MS, 62. *Prof Exp:* Field geologist, Southern Pac Co, 55-57, geologist, 57-61; from asst mining geologist to assoc mining geologist, Nev Bur Mines, Univ Nev, 63-74, MINING GEOLOGIST, NEV BUR MINES & GEOL, UNIV NEV, 74- & PROF GEOL, MACKAY SCH MINES, UNIV NEV, RENO, 74- *Concurrent Pos:* Co-prin investr, Off Water Resources Res grant, 67-68; co-convenor, invitational res conf on zoning in volcanic & subvolcanic mineral deposits, Lake Tahoe, 81 & brecciation & ore deposits, Colorado Springs, 83; counr, Asn Explor Geochemists, 88-90. *Mem:* Fel Geol Soc Am; Soc Econ Geol; Asn Explor Geochemists. *Res:* Geology and geochemistry of metalliferous ore deposits; petrology and petrography of volcanic rocks; environmental geology; geology of epithermal precious metal deposits. *Mailing Add:* Nev Bur Mines Univ Nev Reno NV 89507-0088

BONHAM, KELSHAW, b Seattle, Wash, Mar 19, 09; m 38; c 4. RADIOBIOLOGY. *Educ:* Univ Wash, BS, 31, MS, 35, PhD(zool), 37. *Prof Exp:* Asst prof fisheries, Agr & Mech Col Tex, 38-41; asst biologist, Fish & Wildlife Serv, US Dept Interior, Univ Wash, 41-42; biologist, State Fisheries Dept, Wash, 42-44; asst prof zool, Univ Hawaii, 44-45; res assoc prof lab radiation biol, Univ Wash, 45-68, res prof, Col Fisheries, 68-76; RETIRED. *Res:* Monogenea taxonomy; commercial biology of dogfish shark; low-level gamma-irradiation of salmon. *Mailing Add:* 2230 NE46 Seattle WA 98105-5709

BONHAM, LAWRENCE COOK, b Springfield, Mo, June 9, 20; m 46; c 4. GEOLOGY. *Educ:* Drury Col, BS, 42; Washington Univ, MS, 48, PhD, 50. *Prof Exp:* Photogram engr, US Coast & Geod Surv, 42-44; instr geol, Washington Univ, 46-48; supvry res geologist, Calif Res Corp, 50-64; explor geologist, Standard Oil Co, Calif, 64-66; sr res assoc, Chevron Oil Field Res Co, 66-80, mgr geol div, 80-85; RETIRED. *Concurrent Pos:* Instr, Eve Div, East Los Angeles Col, 65-72; adj prof, Univ Southern Calif, 78- & Univ Calif, Los Angeles, 80- *Mem:* Soc Econ Paleont & Mineral; Am Asn Petrol Geologists. *Res:* Sedimentary petrology; organic geochemistry; computer applications in geology. *Mailing Add:* 14169 Bronte Dr Whittier CA 90602

BONHAM, RUSSELL AUBREY, b San Jose, Calif, Dec 10, 31; m 57; c 3. CHEMICAL PHYSICS. *Educ:* Whittier Col, BA, 54; Iowa State Univ, PhD(phys chem), 58. *Prof Exp:* Res assoc chem, Iowa State Col, 57-58; from instr to assoc prof, 58-65, PROF CHEM, IND UNIV, BLOOMINGTON, 65- *Concurrent Pos:* Res assoc, Nat Acad Sci-Nat Res Coun & US Naval Res Lab, 60; asst prof, Univ Col, Univ Md, 60; NSF grants, Conf Magnetism & Crystallog, Japan, 61 & Winter Inst Quantum Chem & Solid State Physics, Fla, 62; Alfred P Sloan Found fel, 64-66; Guggenheim Mem Found fel, 64-65; Fulbright Found res scholar, Univ Tokyo, 64-65; Royal Norweg Sci & Indust res grant, 65; vis scientist, NSF grant under US-Japan Coop Sci Prog, Univ Tokyo & Hokkaido Univ, Japan, 69-70; mem, Nat Acad Sci Joint Brazilian-US Comt Grad Teaching & Res in Chem, 69-76; assoc ed, J Chem Physics, 74-77; Mem, Comn Spin & Momentum Density, Int Union Crystallog, 75. *Honors & Awards:* Alexander von Humboldt Sr Scientist Award, WGer, 77. *Mem:* Am Chem Soc; Am Crystallog Asn; fel AAAS; fel Am Phys Soc; Sigma Xi. *Res:* Low energy electron spectroscopy using pulsed electron beam time of flight techniques; measurement of accurate absolute total electron impact collision cross sections for atoms and molecules; measurement of secondary electron cross sections for electron impact. *Mailing Add:* 4750 E Heritage Woods Rd Bloomington IN 47401

BONHORST, CARL W, b Van Metre, SDak, Dec 31, 17; m 45; c 1. BIOCHEMISTRY. *Educ:* SDak State Univ, BS, 43; Pa State Univ, MS, 47, PhD(biol chem), 49. *Prof Exp:* Instr chem, Univ Portland, 49-51; res assoc pharmacol, Univ Va, 51-52; assoc biochemist, SDak State Univ, 52-56; assoc prof chem, 56-66, head dept, 68-73, PROF CHEM, UNIV PORTLAND, 66- *Concurrent Pos:* Mem teacher prep-cert study, Nat Asn State Dirs Teacher Educ, 61; vis prof & NSF fac fel, Univ Wis, 62-63. *Mem:* Am Chem Soc; Sigma Xi. *Res:* Physical properties of esters, peptide synthesis, selenium metabolism, active transport and Pasteur effect. *Mailing Add:* 4506 N Amherst St Portland OR 97203

BONI, ROBERT EUGENE, b Canton, Ohio, Feb 18, 28; m 52; c 2. METALLURGICAL ENGINEERING. *Educ:* Univ Cincinnati, MetE, 51; Carnegie Inst Technol, MS & PhD(metall eng), 54. *Prof Exp:* Eng asst, Timken Roller Bearing Co, 47-51, metall engr, 54; chief ferrous metall res sect, gun tube develop, Watertown Arsenal Lab, 55-56; res engr, 56-61, sr res engr, 61-68, mgr appl sci, metall res dept, 68-70, dir metall res, res & technol, 70-76, asst vpres, 76, vpres, 76-78, group vpres, Steel Oper & Oil Gas Subsid, 78-82, exec vpres & chief oper officer, 82-83, pres & chief oper officer, 83-86, CHMN & CHIEF EXEC OFFICER, ARMCO STEEL CORP, 86- *Honors & Awards:* Distinguished Eng Alumnus Award, Univ Cincinnati, 70; Benjamin F Fairless Award, Am Inst Mining, Metall & Petrol Engrs, 87. *Mem:* Am Soc Metals; Am Inst Mining, Metall & Petrol Engrs. *Mailing Add:* 420 Devon St Kearny NJ 07032

BONICA, JOHN JOSEPH, b Filicudi, Italy, Feb 16, 17; US citizen; m 42; c 4. ANESTHESIOLOGY. *Educ:* NY Univ, BS, 38; Marquette Univ, MD, 42; Am Bd Anesthesiol, dipl, 48; Royal Col Surgeons Eng, FFARCS, 73; Med Col Wis, DSc, 77. *Hon Degrees:* DMSc, Univ Siena, 72; DSc, Northwestern Univ, 89. *Prof Exp:* Dir anesthesiol, Tacoma Gen & Pierce County Hosps, Tacoma, Wash, 47-63; anesthesiologist in chief, Med Ctr, 60-77, chmn dept, Sch Med, 60-77, dir anesthesia res ctr, 68-77, PROF ANESTHESIOL, SCH MED, UNIV WASH, 60- *Concurrent Pos:* Consult, US Army Med Corps, 47-77, Vet Admin Hosp, 60-, Ministry of Health, Italy, 54 & 60, Brazil, 59, Arg, 55, Sweden, 69 & Ministry of Educ, Japan, 69; mem, Anesthesiol Training Comt, Nat Inst Gen Med Sci, 65-69, chmn, Gen Med Res Prog-Proj Comt, 70-72; chmn, Sci Adv Comt, World Fedn Socs Anesthesiol, 68-72, secy-gen, 72-80, pres, 80-84; foreign ed, Minerva Anestesiologica, Italy. *Honors & Awards:* Silver Medal, Swedish Med Soc, 69. *Mem:* Am Soc Anesthesiol (pres, 66); Am Acad Anesthesiol; Asn Am Med Cols; Int Anesthesia Res Soc; World Fedn Socs Anesthesiol. *Res:* Pharmacology of local anesthetics; neurophysiology; physiopathology and psychology of pain; obstetric anesthesia and perinatal biology; effects of regional and general anesthesia on human cardiovascular and respiratory functions. *Mailing Add:* Dept Anesthesiol RN-10 Univ Wash Sch Med Seattle WA 98195

BONIFAZI, STEPHEN, b Hartford, Conn, Oct 31, 24; m 59; c 1. CHEMISTRY. *Educ:* Trinity Col, BS, 47. *Prof Exp:* Asst chem, Trinity Col, 47-49; sr chemist, Pratt Whitney Aircraft Group, West Palm Beach, Fla, 49-56, supvr chem, 56-58, proj chemist, 58-63, gen supvr chem, 63-78, fuels & lubricants specialist 78-86; CONSULT, 86- *Mem:* Am Chem Soc; Am Soc Lubrication Engrs; Int Asn Hydrogen Energy; Am Soc Testing & Mat; Coord Res Coun; Am Soc Mech Eng. *Res:* High performance; pocket propellants; aerospace fuels technology; alternatives fuels research; aerospace lubrication; solid lubrication; protective coatings. *Mailing Add:* 516 Kingfish Rd North Palm Beach FL 33408-4316

BONILLA, MANUEL GEORGE, b Sacramento, Calif, July 19, 20; m 49; c 3. GEOLOGY. *Educ:* Univ Calif, AB, 43; Stanford Univ, MS, 60. *Prof Exp:* Geologist, US Bur Reclamation, 46-47; GEOLOGIST, US GEOL SURV, 47- *Honors & Awards:* Claire P Holdredge Award, Asn Eng Geologists, 71. *Mem:* Fel Geol Soc Am; Seismol Soc Am; Asn Eng Geologists; Am Geophys Union; Int Asn Eng Geol. *Res:* Engineering geology; earthquake faults in relation to land use. *Mailing Add:* US Geol Surv MS 977 345 Middlefield Rd Menlo Park CA 94025

BONIN, JOHN H(ENRY), b Buzet, France, May 18, 19; nat US; m 46; c 3. MECHANICAL ENGINEERING. *Educ:* Ill Inst Technol, BS, 47, MS, 48. *Prof Exp:* Supvr heat power equip, Armour Res Found, 48-56; dir eng div, Chicago Midway Labs, 56-59; mgr eng res div, Lockheed Propulsion Co, 59-65, dir res br, 65-77; CONSULT SCIENTIST, LOCKHEED MISSILES & SPACE CO, 77- *Mem:* Am Soc Mech Engrs; Am Inst Aeronaut & Astronaut; Sigma Xi. *Res:* Heat transfer and thermodynamics; high temperature phenomena. *Mailing Add:* 1257 Sargent Dr Sunnyvale CA 94087

BONIN, KEITH DONALD, b Haverhill, Mass, Dec 15, 56; m 78; c 2. PROPERTIES OF ATOMS, NONLINEAR OPTICS. *Educ:* Loyola Univ, BS, 78; Univ Md, College Park, PhD(physics), 84. *Prof Exp:* Res assoc physics, 84-85, instr, 85-86, ASST PROF RES, DEPT PHYSICS, PRINCETON UNIV, 86- *Concurrent Pos:* Consult physicist, Tracor-Jitco, Inc, Rockville, Md, 83-, Columbus Div, Battelle, 88-89. *Mem:* Optical Soc Am; Am Phys Soc; Inst Elec & Electronics Engrs. *Res:* Nonlinear optics, specifically generation of coherent extreme-ultraviolet radiation; developed new techniques for measuring fundamental properties of atoms and molecules, for example, excited-state oscillator strengths, absolute photoionization cross sections and atomic polarizabilities. *Mailing Add:* Dept Physics Princeton Univ Princeton NJ 08544-0708

BONINI, WILLIAM E, b Washington, DC, Aug 23, 26; m 54; c 4. GEOLOGY, GEOPHYSICS. *Educ:* Princeton Univ, BSE, 48, MSE, 49; Univ Wis, PhD(geol), 57. *Prof Exp:* From instr to assoc prof geol eng, 53-66, prof geophysics & geol, 66-70, MAGEE PROF GEOPHYSICS & GEOL ENG, PRINCETON UNIV, 70- *Concurrent Pos:* NSF sr fel, 63-64; vis scientist, Yugoslavia, Nat Acad Sci Eastern Europ Exchange Prog, 74, Spain, 81, Italy, 88; chmn geophys div, Geol Soc Am, 85-86. *Mem:* Am Asn Petrol Geol; Am Geophys Union; Geol Soc Am; Soc Explor Geophys; Nat Asn Geol Teachers (pres, 84-85). *Res:* Gravity and magnetic anomalies and major tectonic features of the northwestern United States and Venezuela. *Mailing Add:* Dept Geol & Geophys Sci Princeton Univ Princeton NJ 08544-1003

BONK, JAMES F, b Menominee, Mich, Feb 6, 31. CHEMISTRY. *Educ:* Carroll Col, Wis, BS, 53; Ohio State Univ, PhD(chem), 58. *Prof Exp:* From asst prof to assoc prof, 59-74, PROF CHEM, DUKE UNIV, 74-, SUPVR FRESHMAN INSTRUCTION, 77-, DIR UNDERGRAD STUDIES, 89- *Mem:* Am Chem Soc. *Res:* Physical-inorganic chemistry. *Mailing Add:* Dept Chem Duke Univ Durham NC 27706

BONKALO, ALEXANDER, psychiatry, neurophysiology, for more information see previous edition

BONKOVSKY, HERBERT LLOYD, b Cleveland, Ohio. MEDICINE, BIOCHEMISTRY. *Educ:* Earlham Col, AB, 63; Western Reserve Univ, MD, 67. *Prof Exp:* Intern, Duke Univ Med Ctr, 67-68; jr asst resident, Cleveland Metro Gen Hosp, 68-69; clin assoc, Nat Cancer Inst, 69-71; res fel & chief resident, Dartmouth-Hitchcock Med Ctr, 71-73; clin & res fel liver dis, Yale Liver Study Unit, 73-74; from asst prof to prof, Dartmouth Med Sch, 74-85, chief hepatol, Metab Lab, Dartmouth-Hitchcock Med Ctr, 77-85; chief div Digestive Dis, Vet Admin Hosp, White River Junction, Vt, 76-85; PROF & DIR CLIN RES FACIL, SCH MED, EMORY UNIV, 85- *Concurrent Pos:* Consult, Hitchcock Clin & Mary Hitchcock Mem Hosp, 74-; res assoc, Vet Admin, 74-76; prin investr, Vet Admin Merit Rev grant, 76-; co-prin investr, Nat Cancer Inst res grant, 76- *Honors & Awards:* Borden Res Award. *Mem:* Am Fedn Clin Res; Am Asn Study Liver Dis; fel Am Col Physicians; Am Soc Clin Invest; Am Gastroent Asn; Am Soc Biol Chemists. *Res:* Liver physiology, biochemistry and disease; hepatic heme metabolism; hepatic drug metabolism and drug-induced liver injury; iron storage diseases, experimental and clinical; porphyria. *Mailing Add:* Emory Univ Sch Med Emory Univ Hosp Clifton Rd NE Atlanta GA 30322

BONN, ETHEL MAY, b Cincinnati, Ohio, Oct 14, 25. ADMINISTRATIVE PSYCHIATRY, QUALITY ASSURANCE. *Educ:* Univ Cincinnati, BA, 47; Univ Chicago, MD, 51; Am Bd Psychiat & Neurol, cert, 58; Am Psychiat Asn, cert, 65; Am Bd Qual Assurance & Utilization Rev, cert, 82. *Prof Exp:* Rotating intern med, pediat, psychiat, surg & obstet-gynec, Strong Mem Hosp, Univ Rochester, NY, 51-53; psychiat residency, Menninger Sch Psychiat, Vet Admin Hosp, Topeka, Kans, 53-55, chief, Women's Neuropsychiat Sect Psychiat, 57-61, chief, North Psychiat Serv, Women's & Geriatric Psychiat, 61-62; asst hosp dir clin serv, Ft Logan Ment Health Ctr, Denver, Colo, 62-67, hosp dir, 67-76; field rep, Joint Comn Accreditation Hosps, Chicago, Ill, 76-78; chief qual assurance, Vet Admin Med Ctr, Brentwood, Calif, 78-81, chief, Psychiat Serv, Vet Admin Med Ctr, Albuquerque, NMex, 81-89; RETIRED. *Concurrent Pos:* Clin instr psychiat, Menninger Sch Psychiat, Topeka, Kans, 56-62; clin prof psychiat, Sch Med, Univ Colo, Denver, Colo, 62-76; consult psychiat, Fitzsimmons Army Hosp, 63-76, Vet Admin Hosps, Sheridan, Wyo, Tuscaloosa, Ala & Ft Lyon, Colo, 63-76; chmn, Psychiat Sect, Am Hosp Asn, 72-73; mem, Comt Hosp & Community Psychiat, Am Psychiat Asn, 77-81, chair, 80-81; clin prof psychiat, Sch Med, Univ Calif Los Angeles, 78-81; assoc prof psychiat, Sch Med, Univ NMex, 81-89. *Honors & Awards:* President's Award, Admin Psychiat, Am Asn Psychiat Adminr, 76. *Mem:* Am Hosp Asn; fel Am Psychiat Asn; fel Am Asn Psychiat Adminr (pres, 75-76); fel Am Col Ment Health Admin; fel Am Col Psychiatrists; AMA; Am Col Utilization Rev Physicians. *Res:* Published papers in psychiatric residency supervision, regressive electroshock treatment, psycho-active medications, administrative psychiatry and quality assurance. *Mailing Add:* 32 Altadena Dr Pueblo CO 81005

BONN, FERDINAND J, b Erstein, France, Oct 7, 43; Can citizen; m 85; c 3. REMOTE SENSING, GEOGRAPHIC INFORMATION SYSTEMS. *Educ:* Lynée Kléber, Strasbourg, France, Lèsl, 62; Univ Louis Pasteur de Strasbourg, BSc, 68; Univ Strasbourg, France, MSc, 69, PhD(geog), 75. *Prof Exp:* Asst lectr geog, Univ Sherbrooke, 69-71, teaching asst geog, 71-73, adj prof remote sensing & geog, 73-78, assoc prof remote sensing & geog, 78-85, PROF REMOTE SENSING, UNIV SHERBROOKE, 82-, DIR CARTEL, 85- *Concurrent Pos:* Vis prof, Univ Calif, Santa Barbara, 79-80; consult var industs & agencies, 82-; chmn, Car Remote Sensing Soc, 85-86; comt mem, Car Earth Sci, Natural Sci & Eng Res Coun Can, 85-88; coun mem, Car Adv Coun Remote Sensing, 87-93; vis scientist, Inst Nat Res Agr, France, 88-89. *Mem:* Am Soc Photogram & Remote Sensing; Car Remote Sensing Soc (pres, 85-86). *Res:* Remote sensing applied to natural resources management and environmental modeling; soils; geomorphology; desertification; author of various publications. *Mailing Add:* Ctr Remote Sensing Res Appln Sherbrooke Univ 2500 Blvd Univ Sherbrooke PQ J1K 2R1 Can

BONN, T(HEODORE) H(ERTZ), b Philadelphia, Pa, May 27, 23; m 46; c 3. APPLIED COMPUTER RESEARCH. *Educ:* Univ Pa, BS, 43, MS, 47. *Prof Exp:* Res asst commun & noise, Moore Sch Elec Eng, Univ Pa, 43-47; develop engr, magnetic tape develop, magnetic recording, servo mechanisms, Eckert-Mauchly Comput Corp, 47-49; chief sonar components, mil elec equip, Naval Air Develop Ctr, 49-50; proj engr magnetic mat, memory devices & amplification devices, Eckert-Mauchly Div, Remington Rand, Inc, 50-57; chief engr res & peripheral develop, Philadelphia Labs, Remington Rand Univac Div, Sperry Rand Corp, 57-64, mgr comput electronics res, 64-; spec asst for res & develop to tech vpres, Electronic Data Processing Div, Honeywell, Inc, 64-68, dir Standards & Appl Res Div, 68-71; dir comput res lab, Sperry Corp Res Ctr, 71-84, mgr C3I systs dept, Sperry Electronic Systs Div, 84-85; INDEPENDENT CONSULT, 85- *Concurrent Pos:* Vpres publ, Inst Elec & Electronics Engrs, 80-82 & bd dirs, 80-84. *Honors & Awards:* Centennial Medal, Inst Elect & Electronics Engrs, 84. *Mem:* Fel Inst Elec & Electronics Engrs. *Res:* Applied research and engineering administration; computer technology; computer architecture; digital magnetic recording; data bases; software; command control communications and intelligence systems. *Mailing Add:* 1601 Pelican Point Dr Apt H213 Sarasota FL 34231

BONN, WILLIAM GORDON, b Toronto, Ont, Sept 28, 46; m 69; c 2. PLANT PATHOLOGY, BACTERIOLOGY. *Educ:* McGill Univ, BSc, 67; Rutgers Univ, MS, 69; Univ Wis, PhD(plant path), 73. *Prof Exp:* Res asst plant path, Rutgers Univ, 67-69 & Univ Wis, 69-73; asst plant path, Purdue Univ, 73; RES SCIENTIST PLANT PATH, CAN DEPT AGR, 73- *Mem:* Can Phytopath Soc; Am Phytopath Soc; Sigma Xi. *Res:* Bacterial diseases of plants; epidemiology of fire blight disease; control of bacterial diseases. *Mailing Add:* Can Dept Agr Harrow Res Sta Harrow ON N0R 1G0 Can

BONNE, ULRICH, b London, Eng, Mar 10, 37; US citizen; m; c 3. PHYSICAL CHEMISTRY, COMBUSTION. *Educ:* Univ Freiburg, BS, 57; Univ Goettingen, MS, 60, PhD(phys chem), 64. *Prof Exp:* Res assoc, Dept Phys Chem, Univ Göttingen, 60-65; prin res scientist combustion, Honeywell Inc, Corp Tech Ctr, 65-67; prin res scientist fire res, Factory Mutual Res Corp, 67-68; STAFF SCIENTIST, HONEYWELL INC, 68-, PRIN RES FEL THERMOCHEM, SENSOR & SYST DEVELOP CTR. *Honors & Awards:* H W Sweatt Award, Honeywell Inc, 71. *Mem:* Optical Soc Am; Combustion Inst; Am Soc Heating, Refrig & Air-Conditioning Engrs. *Res:* Efficiency improvement of energy conversion systems; mathematical modeling and simulation of combustion and heat pump systems; thermochemistry of coal gasification; optical, thermochemical and electrochemical gas sensing methods. *Mailing Add:* Honeywell Inc Corp Tech Ctr 4936 Shady Oaks Rd Meka MN 55420

BONNELL, DAVID WILLIAM, b Wichita, Kans, Sept 29, 43; m 75; c 2. PHYSICAL INORGANIC CHEMISTRY, HIGH TEMPERATURE CHEMISTRY. *Educ:* Rice Univ, BA, 68, PhD(phys chem), 72. *Prof Exp:* Res chemist, Rice Univ, 64-68, mass spectrometrist, 68-69, res assoc chem, 72-73; dir res, MCR-Houston, Inc, 73-75; RES CHEMIST INORG CHEM, NAT INST STANDARDS & TECHNOL, 75- *Concurrent Pos:* Consult, Rice Univ, 74-75; vice chair, IUPAC, HTMC Conf, 89. *Honors & Awards:* IR 100, 80. *Mem:* AAAS; Sigma Xi. *Res:* Electron spectroscopic chem analysis instrumentation development; laser ablation; levitation calorimetry property measurements of high temperature materials; high temperature, high pressure molecular beam mass spectrometry; thermodynamic modeling; ultra high temperature; mass spectrometry. *Mailing Add:* Mat Sci & Eng Lab Nat Inst Standards & Technol Div 450 Metall MS B106 223 Gaithersburg MD 20899

BONNELL, JAMES MONROE, b Chicago, Ill, Apr 5, 22; m 49. ORGANIC CHEMISTRY, CHEMICAL ENGINEERING. *Educ:* Northwestern Univ, BS, 44, BChE, 48, MS, 49; Univ Fla, PhD(chem), 55. *Prof Exp:* Asst dir res, Snow Crop, Fla, 51-55; tech dir, Orange Crystals, Inc, 55-57; DIR RES, TROPICANA PROD, INC, 57- *Concurrent Pos:* Vpres, Prof Eng Serv, Inc, 55- *Mem:* AAAS; Am Chem Soc; Am Inst Chem Eng; fel Am Inst Chemists; Sigma Xi. *Res:* Citrus products and by-products. *Mailing Add:* 7308 18th Ave NW Bradenton FL 34209

BONNER, BILLY EDWARD, b Oak Grove, La, Dec 12, 39. ELEMENTARY PARTICLE PHYSICS. *Educ:* La Polytech Inst, BS, 61; Rice Univ, MA, 63, PhD(physics), 65. *Prof Exp:* Fel physics, Rice Univ, 65-66; sr sci officer, Rutherford High Energy Lab, Sci Res Coun, Eng, 66-70; assoc res physicist, Crocker Nuclear Lab, Univ Calif, Davis, 71-72; mem staff physics, Los Alamos Sci Lab, Univ Calif, 72-85; PROF, RICE UNIV, 85-, CHMN, PHYSICS DEPT, 86-, DIR, BONNER NUCLEAR LABS, 87- *Concurrent Pos:* Vis prof physics, Rice Univ, 81-82; sci assoc at CERN, 83-84. *Mem:* Am Phys Soc; Sigma Xi; AAAS. *Res:* Intermediate and high energy particle physics; studies of hadron polarization, jets and spin dependent structure functions of nucleons. *Mailing Add:* Bonner Nuclear Labs Rice Univ Houston TX 77251-1892

BONNER, BRUCE ALBERT, b Jamestown, NDak, Apr 14, 29; m 53; c 4. PLANT PHYSIOLOGY. *Educ:* Univ Calif, AB, 52, PhD(bot) 57. *Prof Exp:* USPHS fel physiol genetics lab, Nat Cent Sci Res, France, 56-58; res assoc bot, Yale Univ, 58-61; instr, Harvard Univ, 61-63; asst prof, 63-71, ASSOC PROF BOT, UNIV CALIF, DAVIS, 71- *Mem:* Am Soc Plant Physiol; Bot Soc Am. *Res:* Mechanism of photomorphogenesis; growth and development. *Mailing Add:* Dept Bot Univ Calif Davis CA 95616

BONNER, DANIEL PATRICK, b Bayonne, NJ, Oct 9, 45; m 72; c 2. MICROBIOLOGY, CHEMOTHERAPY. *Educ:* Fairleigh Dickinson Univ, BS, 67; Rutgers Univ, MS, 69, PhD(microbiol), 72. *Prof Exp:* Biol sci asst microbiol, Dept Rickettsial Dis, Walter Reed Army Inst Res, Walter Reed Army Med Ctr, 69-71; res assoc microbiol, Waksman Inst Microbiol, Rutgers Univ, 72-77, asst res prof, 77-78; res invest, Squibb Inst Med Res, 78-81, res group leader, 81-84, asst dept dir, 84-87; DIR MICROBIOL, BRISTOL-MYERS SQUIBB PHARM RES INST, 87- *Concurrent Pos:* Div A chmn, Am Soc Microbial, 89-90. *Mem:* AAAS; Am Soc Microbiol; Sigma Xi; Japan Antibiotics Res Asn; Am Acad Microbiol. *Res:* Chemotherapy of infectious diseases; beta-lactam antibiotics; polyene macrolide antibiotics. *Mailing Add:* Dept Micro F2117 Bristol-Myers Squibb Box 4000 Princeton NJ 08540

BONNER, DAVID CALHOUN, b Port Arthur, Tex, Nov 20, 46; m 73; c 2. STATISTICAL MECHANICS SOLUTIONS, POLYMER PHASE SEPARATION. *Educ:* Univ Tex, Austin, BSChE, 67, MSChE, 69; Univ Calif, Berkeley, PhD(chem eng), 72. *Prof Exp:* Res engr, Union Carbide Corp, 68-69; assoc prof chem eng, Tex Tech Univ, 72-76 & Tex A&M Univ, 76-77; tech support mgr, Shell Oil Co, 77-86; dir corp res, 87-88, VPRES RES & DEVELOP, B F GOODRICH CO, 88- *Mem:* Am Inst Chem Eng; Am Chem Soc; Am Soc Compts. *Res:* Polymer phase equilibrium and solution theory. *Mailing Add:* B F Goodrich Co 9921 Brecksville Rd Brecksville OH 44141

BONNER, FRANCIS JOSEPH, b Wilmington, Del, Jan 4, 34; m 76; c 2. PHYSICAL CHEMISTRY, CHEMICAL ENGINEERING. *Educ:* Mass Inst Technol, SB, 55, SM, 56; Univ Del, PhD(chem eng), 64; Univ Uppsala, Fil Lic & Fil Dr(phys chem), 67. *Prof Exp:* Swed Int Develop Agency lectr phys chem, Univ Uppsala, 66; sr res chem engr, Mobil Res & Develop Corp, 68-72; oberassistent, Swiss Fed Inst Technol, 72-81; PROF CHEM ENG, VANDERBILT UNIV, 81- *Concurrent Pos:* Swed Govt fel, Univ Uppsala, 64-65, Swed Natural Sci Res Coun grants, 65-68. *Mem:* Am Chem Soc; Am Inst Chem Engrs; NY Acad Sci; Sigma Xi. *Res:* Tribology, rheology, metallurgy, material science and engineering; physical-surface-interfacial chemistry, including mass transport of synthetic and natural macromolecules and emulsifiers; polymer science and engineering; applied surface chemistry. *Mailing Add:* Chem Eng Dept Univ Lowell One Univ Ave Lowell MA 01854

BONNER, FRANCIS TRUESDALE, b Salt Lake City, Utah, Dec 18, 21; m 46; c 3. CHEMISTRY. *Educ:* Univ Utah, BA, 42; Yale Univ, MS, 44, PhD(phys chem), 45. *Prof Exp:* Res chemist, Manhattan Proj, Columbia Univ, 44-46; res chemist, Clinton Nat Lab, Oak Ridge, 46-47; res chemist, Brookhaven Nat Lab, NY, 47-48; asst prof chem, Brooklyn Col, 48-54; Carnegie Found fel, Harvard Univ, 54-55; phys chemist, Arthur D Little, Inc, 55-58; chmn dept, 58-70, dean, Int Progs, 83-86, PROF CHEM, STATE UNIV NY STONY BROOK, 58- *Concurrent Pos:* NSF sr fel, Saclay Nuclear Res Ctr, France, 64-65; consult ed, Addison-Wesley Publ Co, 56-77; res collabr, Brookhaven Nat Lab, 58-88; Rockefeller Found consult, Univ Del Valle, Colombia, 61, 64 & 68; Ford Found consult, Univ Antioquia, Colombia, 62, 63 & 68; vis scientist, EAWAG/ETH, Zurich, Switz, 73; Nat Acad Sci exchange vis scientist, Romania, 75; mem bd dirs, Res Found, State Univ NY, 76-88; vis prof, Kings Col, Univ London, 87. *Mem:* Fel AAAS; Am Chem Soc. *Res:* Inorganic chemistry of nitrogen; kinetic and isotopic studies of reactions in solution; inorganic chemistry of nitrogen in aqueous solution; mechanisms of reaction; reactive intermediate species; isotopic and kinetic studies. *Mailing Add:* Box 2063 Setauket NY 11733

BONNER, HUGH WARREN, b Chicago, Ill, Oct 27, 44; m 67. EXERCISE PHYSIOLOGY. *Educ:* Univ Minn, BS, 67; Calif State Col, Hayward, MS, 70; Univ Calif, Berkeley, PhD(phys educ), 72. *Prof Exp:* Asst prof, 72-76, ASSOC PROF PHYS EDUC, UNIV TEX, AUSTIN, 77-, DIR EXER PHYSIOL LABS, 73- *Mem:* AAAS; Am Asn Health Phys Educ & Recreation; Am Col Sports Med. *Res:* Determination of the influence of exercise training upon calcium transport of mitochondria and sarcoplasmin reticulum in cardiac and skeletal muscle and morphometric changes of subcellular organelles. *Mailing Add:* 6909 Danwood Dr Austin TX 78759

BONNER, JAMES (FREDRICK), b Ansley, Nebr, Sept 1, 10; m 39; c 2. DEVELOPMENTAL BIOLOGY, BIOCHEMICAL GENETICS. *Educ:* Univ Utah, AB, 31; Calif Inst Technol, PhD(biology, genetics), 34. *Prof Exp:* Nat Res Coun fel, State Univ Utrecht & Swiss Fed Inst Technol, 34-35; asst biol, 35-36, from instr to assoc prof plant physiol, 36-46, prof, 46-81, EMER PROF BIOL, CALIF INST TECHNOL, 81- *Concurrent Pos:* Eastman prof, Oxford Univ, 63-64; mem, Malaysian Rubber Bd, 66-90; chmn & chief exec, Phytogen Inc, Pasadena, Calif, 81-84, chief sci, 84-86. *Mem:* Nat Acad Sci; Bot Soc Am; Am Chem Soc; Am Soc Plant Physiol (pres, 49-50); Am Soc Biol Chem. *Res:* Molecular biology of chromosomes; control of genetic activity. *Mailing Add:* Div Biol Calif Inst Technol Pasadena CA 91125

BONNER, JAMES JOSE, b Los Angeles, Calif, May 1, 50; m 74; c 1. MOLECULAR BIOLOGY, DEVELOPMENTAL BIOLOGY. *Educ:* Univ Calif, Berkeley, BA, 72; Mass Inst Technol, PhD(biol), 77. *Prof Exp:* Teaching asst biol, Mass Inst Technol, 72-76; scholar biochem, Univ Calif, San Francisco, 77-78; asst prof, 79-85, ASSOC PROF BIOL, IND UNIV, BLOOMINGTON, 85-, ASSOC CHMN DEPT, 87- *Concurrent Pos:* Jr fac res award, Am Cancer Soc. *Mem:* Am Soc Cell Biol; AAAS; Genetics Soc Am; Soc Develop Biol; Sigma Xi. *Res:* Developmental regulation of genetic expression; genetic and biochemical analysis of molecular aspects of metabolic stress. *Mailing Add:* Dept Biol Ind Univ Bloomington IN 47405

BONNER, JILL CHRISTINE, b Cottingham, UK, Aug 20, 37. SOLID STATE PHYSICS, THERMAL PHYSICS. *Educ:* Univ London, BSc, 59, PhD(physics), 68, DSc, 84,. *Prof Exp:* Asst lectr physics, Royal Holloway Col, Univ London, 62-64, lectr, 64-67; res assoc, Carnegie-Mellon Univ, 68-70; eng & appl sci res fel, Yale Univ, 70-71; res physicist, Carnegie-Mellon Univ, 72-73; vis asst prof, Univ Utah, 73-74; sr res assoc math, Brookhaven Nat Lab, 74-76; assoc prof, 76-81, PROF PHYSICS, UNIV RI, 81- *Concurrent Pos:* Mem, Div Sci Manpower Improvement, NSF Rev Panel, 76, Adv Comt, Am Inst Physics, 75-78, 80-86 & 90-93, Prog Comt Ann Conf Magnetism & Magnetic Mat, 78, 80 & 89; NATO sci res grant, 77-80 & 84-88; consult, Mitre Corp, 78-79 & Gen Elec Corp Res & Develop Ctr, 79-82; fac fel, Bunting Inst, Radcliffe Col, 78-80; NSF sci res grant, 78-80, 80-83, 85-88; vis prof, Women Sci & Technol, Mich State Univ, 84; vis fel, Sci & Eng Res Coun UK, 85. *Mem:* Fel Am Phys Soc; Brit Inst Physics; fel AAAS; Asn Women Sci; Fedn Am Scientist; sr mem Inst Elec & Electronic Engrs. *Res:* Theoretical solid state physics; cooperative phenomena; magnetism and magneto chemistry; low-dimensional systems; interaction with experimentalists. *Mailing Add:* Dept Physics Univ RI Kingston RI 02881

BONNER, JOHN FRANKLIN, JR, b Smullton, Pa, Sept 11, 17. BIOCHEMISTRY. *Educ:* Rensselaer Polytech Inst, BChE, 38; Univ Rochester, MS, 40, PhD(biophys), 48. *Prof Exp:* Res fel radiol, Univ Rochester, 39-47, from instr toxicol to asst prof radiation biol, 47-55; asst chief med res br, AEC, 55-61; PROF BIOCHEM, SCH MED, IND UNIV, INDIANAPOLIS, 61- *Concurrent Pos:* Fulbright lectr & guest prof, Univ Groningen, 54-55; vis prof molecular & genetic biol, Univ Utah, 67-68; USPHS spec fel, Univ Utah, 67-68. *Mem:* AAAS; Am Chem Soc; Biophys Soc; Sigma Xi. *Res:* Modification of biomolecules by ionizing radiation; radioprotective compounds. *Mailing Add:* Dept Biochem Ind Univ Sch Med Indianapolis IN 46223

BONNER, JOHN TYLER, b New York, NY, May 12, 20; m 42; c 4. DEVELOPMENTAL BIOLOGY. *Educ:* Harvard Univ, BS, 41, MA, 42, PhD(biol), 47. *Hon Degrees:* DSc, Middlebury Col, 70. *Prof Exp:* From asst prof to prof biol, Princeton Univ, 47-90, chmn, Dept Biol, 65-77 & 83-84, actg chmn, Dept Biochem Sci, 83-84, George M Moffett prof, 66-90, EMER PROF BIOL, PRINCETON UNIV, 90- *Concurrent Pos:* Instr embryol, Marine Biol Labs, Woods Hole, 51-52; Rockefeller traveling fel, France, 53; spec lectr, Univ Col, London, 57 & Brooklyn Col, 66; Guggenheim fel, Scotland, 58 & 71-72; NSF sr fel, Cambridge Univ, 63; ed of two sect, Biol Abstr, 57-82, trustee, 58-63; mem bd trustees, Princeton Univ Press, 64-68, 71 & 75-82. *Honors & Awards:* Waksman Medal, 55. *Mem:* Nat Acad Sci; Am Philos Soc; fel AAAS; fel Am Acad Arts & Sci; Soc Develop Biol; Am Soc Naturalists. *Res:* Development in lower organisms; cellular slime molds; evolution. *Mailing Add:* Dept Ecol & Evolutionary Biol Princeton Univ Princeton NJ 08544-1003

BONNER, LYMAN GAYLORD, b Kingston, Ont, Sept 16, 12; m 37, 73; c 3. CHEMISTRY. *Educ:* Univ Utah, AB, 32; Calif Inst Technol, PhD(chem), 35. *Prof Exp:* Nat Res fel chem, Princeton Univ, 35-37; from instr to asst prof physics, Duke Univ, 37-44; res assoc, George Washington Univ, 44-45; tech dir, Hercules Powder Co, 45-55, mgr, Explosives Res Div, 55-58, dir develop, Explosives & chem Propulsion Dept, 58-65; dir found rels, Calif Inst Technol, 65-67, asst to pres, 67-69, dir student rels, 69-80, assoc chem, 66-87, registr, 77-89; RETIRED. *Res:* Infrared and Raman spectroscopy; ballistics. *Mailing Add:* 439 S Catalina No 302 Pasadena CA 91106

BONNER, NORMAN ANDREW, b San Francisco, Calif, Aug 3, 20; m 45; c 1. CHEMISTRY, PSYCHOLOGY. *Educ:* Univ Calif, BS, 42; Princeton Univ, MA, 44, PhD(chem), 45; Calif State Univ, Hayward, MS, 76. *Prof Exp:* Res chemist, Manhattan Proj, Univ Calif, 41-42, Princeton Univ, 42-45, Los Alamos Sci Lab, Univ Calif, 45-46 & Gen Elec Co, NY, 46; instr chem, Wash Univ, 47-48, asst prof, 48-51; asst prof, Cornell Univ, 51-53; chemist, Lawrence Livermore Lab, Univ Calif, 53-87; RETIRED. *Concurrent Pos:* Mem adv coun, Bay Area Air Pollution Control Dist, 70-78. *Mem:* AAAS; Am Chem Soc. *Res:* Trace element analysis of environmental samples; radiochemistry; kinetics of reactions in solution; inorganic chemistry; chemistry of air pollution. *Mailing Add:* 1093 Nielsen Lane Livermore CA 94550

BONNER, OSCAR DAVIS, b Jackson, Miss, May 9, 17; m 40; c 3. PHYSICAL CHEMISTRY. *Educ:* Millsaps Col, BS, 39; Univ Miss, MS, 48; Univ Kans, PhD(chem), 51. *Prof Exp:* With Miss Testing Lab, 40-41 & Filtrol Corp, 46-47; res partic, Oak Ridge Nat Lab, 51; from asst prof to prof, 51-70, from actg head to head dept, 59-70, Robert L Sumwalt chair chem, 70-82, DISTINGUISHED EMER PROF CHEM, UNIV SC, 82- *Concurrent Pos:* Fulbright advan res award, Germany, 57-58; Russell fac res award, 60; Fulbright teaching-res award, Korea, 83. *Mem:* Sigma Xi; Am Chem Soc. *Res:* Ion exchange; thermodynamics of solutions; polyelectrolytes; structure of water; interaction of water and salts with biological systems. *Mailing Add:* Dept Chem Univ of SC Columbia SC 29208

BONNER, PHILIP HALLINDER, MYOGENESIS, NEURO-BIOLOGY. *Educ:* Univ Calif, San Diego, PhD(biol), 71. *Prof Exp:* ASSOC PROF CELL BIOL, UNIV KY, 75- *Mailing Add:* Univ Ky Sch Biol Sci Lexington KY 40506

BONNER, R ALAN, b Leominster, Mass, Dec 1, 39; m 64; c 3. MEDICAL RESEARCH. *Educ:* Mich State Univ, BS, 62, DVM, 64; Boston Univ, PhD(physiol), 72. *Prof Exp:* Vet intern, Rowley Mem Hosp, Springfield, Mass, 64-65; pvt pract, Dewitt Animal Hosp, North Attleboro, Mass, 65-67; fel, Boston Univ, 67-69; from instr to asst prof physiol, Hahnemann Med Col, 69-74; res assoc cardiovasc physiol, Biomed Res Inst, Ctr Res & Advan Study, Univ Maine, Portland, 72-78; DIR, DEPT VET SCI, BRISTOL-MYERS SQUIBB, 82- *Concurrent Pos:* NASA fel, Biospace Technol Training Prog, 68; fel, Pa Heart Asn, 71; consult, Maine Med Ctr, 71. *Mem:* Am Vet Med Asn; AAAS. *Res:* Development of animal model for sudden infant death syndrome studies; pathophysiologic mechanisms involved in sudden infant death syndrome; pharmacologic agents in an attempt to reverse irreversible hemorrhagic shock. *Mailing Add:* Dept Vet Sci Bristol-Myers Squibb PO Box 4000 Princeton NJ 08543-4000

BONNER, ROBERT DUBOIS, b Camden, Ala, Apr 3, 26; m 57; c 3. BOTANY. *Educ:* Howard Univ, BS, 49, MS, 52; Pa State Univ, PhD, 58. *Prof Exp:* Asst Pa State Univ, 55-57; asst prof bot, Tex Southern Univ, 57-63; assoc prof, 63-69, PROF BIOL, HAMPTON INST, 69-, CHMN DEPT, 63-, DIR, SCH PURE & APPL SCI, 74- *Mem:* AAAS; Nat Inst Sci; Bot Soc Am; Mycol Soc Am. *Res:* Taxonomy and ecology of animal parasitic fungi; marine microbiology. *Mailing Add:* Sch Pure & Appl Sci Hampton Univ Hampton VA 23668

BONNER, TOM IVAN, b Boston, Mass, Mar 1, 42; m 66; c 3. MOLECULAR BIOLOGY. *Educ:* Rice Univ, BA, 63; Univ Wis, PhD(physics), 68. *Prof Exp:* Res assoc nuclear physics, Ctr Nuclear Studies, Univ Tex, Austin, 67-70; Carnegie fel, Carnegie Inst, 70-73, res assoc biophys, Dept Terrestrial Magnetism, 73-76; expert, Nat Cancer Inst, 76-82; scientist, Frederick Cancer Res Ctr, 82; res biophysicist, 83-87, spec expert, 87-91, RES BIOLOGIST, NIMH, 91- *Concurrent Pos:* Mem, Comt Receptor Nomenclature & Drug Classification, Int Union Pharmacol, 90- *Honors & Awards:* Boehringer Ingleheim Award 1st Prize, 88. *Mem:* AAAS; Am Phys Soc; Biophys Soc. *Res:* Nucleic acid reassociation techniques; nucleic acid sequence organization and regulation; evolution of nucleic acids; RNA tumor viruses; recombinant DNA; molecular neurobiology; molecular biology of neurotransmitter receptors. *Mailing Add:* Lab Cell Biol Bldg 36 Rm 3A-17 NIMH Bethesda MD 20892

BONNER, WALTER DANIEL, JR, b Salt Lake City, Utah, Oct 22, 19; m 44; c 2. PLANT BIOLOGY. *Educ:* Univ Utah, BS, 40; Calif Inst Technol, PhD(bio-org chem), 46. *Prof Exp:* Asst, US Bur Mines, Utah, 41 & Calif Inst Technol, 41-46; res fel, Harvard Univ, 46-49; Am Cancer Soc & USPHS fel, Molteno Inst, Cambridge Univ, 49-52; from asst prof to assoc prof bot, Cornell Univ, 53-59; prof phys biochem & plant physiol, Johnson Res Found, 59-75, PROF BIOCHEM & BIOPHYS, SCH MED, UNIV PA, 75- *Concurrent Pos:* Guggenheim & overseas fel, Churchill Col, Cambridge Univ, 67-68; prog dir, Molecular Biol Sect, NSF, 74-76. *Mem:* Am Soc Plant Physiol; Am Soc Biol Chem; Am Chem Soc; Biophys Soc; Brit Biochem Soc. *Res:* Bioenergetics; cellular physiology; mechanisms of energy transfer and conservation in plant mitochondria and chloroplasts; metabolic control mechanism in plant and protist cells. *Mailing Add:* Dept Biochem & Biophys Univ Pa Sch Med Philadelphia PA 19104-6059

BONNER, WILLARD HALLAM, JR, b New Haven, Conn, May 22, 28; m 48; c 2. ORGANIC CHEMISTRY. *Educ:* Wesleyan Univ, BA, 47; Univ Buffalo, MA, 48; Purdue Univ, PhD(org chem), 52. *Prof Exp:* Res chemist, Humble Oil & Refining Co, 52-54; RES CHEMIST, E I DU PONT DE NEMOURS & CO, INC, 54- *Mem:* Am Chem Soc. *Res:* Reaction mechanisms; organic synthesis; petrochemicals; polymer chemistry; synthetic fibers. *Mailing Add:* Star Rte 2 Box 583 Bath ME 04530

BONNER, WILLIAM ANDREW, b Chicago, Ill, Dec 21, 19; m 44; c 4. CHEMISTRY. *Educ:* Harvard Univ, BS, 41; Northwestern Univ, PhD(org chem), 44. *Prof Exp:* Res chemist, Corn Prod Ref Co, Ill, 42; asst org chem, Northwestern Univ, 42-44, Nat Defense Res Comt res chemist & instr chem, univ col, 44-46; from instr to assoc prof, 46-59, PROF ORG CHEM, STANFORD UNIV, 59- *Concurrent Pos:* Guggenheim fel, 53; Nat Res Coun res assoc, 70. *Mem:* Am Chem Soc. *Res:* Carbohydrate and synthetic organic chemistry; application of radioactive carbon to elucidation of organic reaction mechanisms; isotope effects in organic reactions; mechanisms of heterogeneous catalytic reactions; natural products. *Mailing Add:* Dept Chem Stanford Univ Stanford CA 94305

BONNER, WILLIAM PAUL, b Bowdon, Ga, July 7, 31; m 56; c 2. SANITARY SCIENCE, CHEMISTRY. *Educ:* Univ Ga, BS, 52; Univ Fla, MSEng, 65, PhD(sanit sci), 67. *Prof Exp:* Jr engr, Union Carbide Nuclear Div, 52-54, supvr anal lab, 56-59, assoc chemist, 56-62; assoc chemist, Oak Ridge Nat Lab, 62-64, res staff mem water pollution, 67-72; PROF, TENN TECHNOL UNIV, 72- *Mem:* Am Chem Soc; Health Physics Soc; Am Water Works Asn; Water Pollution Control Fedn; Sigma Xi. *Res:* Water pollution; radioactive waste disposal; radiological health; land-water interactions. *Mailing Add:* Dept Civil Eng Tenn Technol Univ Cookeville TN 38501

BONNET, JUAN A, JR, b Santurce, PR, Apr 22, 39; m 53; c 6. GENERAL ENVIRONMENTAL SCIENCES. *Educ:* Univ Mich, BS, 60, PhD(nuclear eng), 71; Univ PR, MS, 61. *Prof Exp:* Res asst rum aging processes, Dept Chem Eng, Univ PR, 60-61; organizer & dir, health physics prog, Boiling Nuclear Superheat Reactor, Power Plant, PR Elect Power Authority, 62-64, shift supvr, 64-65, nuclear safety & anal engr, 65-67, head nuclear eng dept & dep proj dir, Aguirre Nuclear Plant, 71-73, head, environ protection, quality assurance & nuclear divs, 73-75, asst exec dir, planning & eng, 75-77; prof, Civil Eng Fac, evening div, Polytech Univ PR, 74-77; dir, Ctr Energy & Environ Res, Univ PR, 77-89; ADMINR, PRODUCTIVITY OFF, PR ELEC POWER AUTHORITY, 90- *Concurrent Pos:* Consult combustion eng, Palisades Nuclear Plant, 70; gen secy, InterAm Confedn Chem Engrs, 77-79; mem, PR Engrs, Architects & Survrs Exam Bd, 78-, pres, 79-; US comn to Unesco Comt on Rational Use of Island Ecosystems, 79-; prin investr, develop of alt energy sci & eng coop in the Caribbean, Phases I & II, NSF, 81-82, Phase III, 83; dir, Syst Energy Conservation Prog, Univ PR, 81-; mem, US Deleg to first US-China Conf on energy resources & environ, Peking, 82, Int Solar Storage Workshop, Jeddah, Saudi Arabia, 82. *Honors & Awards:* Mobil Energy Award, 81. *Mem:* Sigma Xi; Int Solar Soc; AAAS; NY Acad Sci; Nat Soc Prof Engrs; Asn Energy Eng; PR Acad Art & Sci (vpres, 85-); Pan Am Union Eng Soc(vpres, 86-); PR Chemist Asn (pres, 88-89). *Res:* Caribbean energy assessment; renewable energy technologies. *Mailing Add:* PR Elec Power Authority GPO Box 4267 San Juan PR 00936-4267

BONNETT, RICHARD BRIAN, b Jamestown, NY, Feb 7, 39; m 64; c 3. GEOMORPHOLOGY, ENGINEERING & ENVIRONMENTAL GEOLOGY. *Educ:* Allegheny Col, BS, 61; Univ Maine, MS, 63; Ohio State Univ, PhD(geol), 70. *Prof Exp:* Teaching assoc geol, Univ Maine, 61-63 & Ohio State Univ, 63-68; from asst prof to assoc prof, 68-82, PROF GEOL, MARSHALL UNIV, 82-, CHMN DEPT, 78- *Concurrent Pos:* Consult, landslides, groundwater and blast damage. *Mem:* AAAS; Geol Soc Am; Am Quaternary Asn; Sigma Xi. *Res:* Pleistocene-glacial geology; environmental geology; Pleistocene stratigraphy and geomorphology in tri-state area of Ohio, West Virginia and Kentucky; engineering geology. *Mailing Add:* RR 4 Garden Terr Box 393 Proctorville OH 45669

BONNETTE, DELLA T, b Bartlett, Kans, Mar 22, 36. OPERATING SYSTEMS. *Educ:* Univ Southwestern La, MS, 69. *Prof Exp:* Dir, Comput Ctr, 75-83, ASST PROF COMPUTER SCI, UNIV SOUTHWESTERN LA, 71-, DIR, COMPUT & INFO SERV, 83- *Concurrent Pos:* Mem, Spec Interest Group Bd, Asn Comput Mach, 85-, Finance Comt, 88-, alt, Computer Sci Accreditation Bd, 91- *Mem:* Asn Comput Mach. *Mailing Add:* Univ Southwestern La PO Box 41690 Lafayette LA 70504

BONNEVILLE, MARY AGNES, b Pittsfield, Mass, May 30, 31. CELL BIOLOGY. *Educ:* Smith Col, BA, 53; Amherst Col, MA, 55; Rockefeller Inst, PhD(cytol), 61. *Prof Exp:* Trainee anat, Col Physicians & Surgeons, Columbia Univ, 61-63; res assoc dermat, Sch Med, Tufts Univ, 63-64; asst prof biol, Brown Univ, 64-66; instr, Harvard Univ, 66-70; asst prof, 70-73, ASSOC PROF MOLECULAR, CELLULAR & DEVELOP BIOL, UNIV COLO, BOULDER, 73- *Concurrent Pos:* Mem panel for eval of fel applns, Nat Acad Sci & NIH, 73-74. *Mem:* Am Soc Cell Biol; Electron Micros Soc Am; Am Asn Anat. *Res:* Structure and formation of the cell surface. *Mailing Add:* Dept Molecular Cell & Develop Biol Univ Colo Campus Box 347 Boulder CO 80309

BONNEY, ROBERT JOHN, b Rumford, Maine, Feb 6, 42; m 62; c 3. BIOCHEMISTRY. *Educ:* Univ Maine, BS, 64; State Univ NY Buffalo, MS, 67, PhD(biochem), 71. *Prof Exp:* Fel biochem, McArdle Lab Cancer Res, Univ Wis, 71-73; sr res scientist biochem, Dept Health, NY State, 73-76; res fel, Dept Immunol, Merck Inst Therapeut Res, 76-84, sr investr, 85-89; EXEC DIR, BRISTOL-MYERS SQUIBB, 89- *Mem:* Am Soc Cell Biol; Retiuloendothelial Soc. *Res:* The study of the role of the mononuclear phagocytes in inflammatory disease. *Mailing Add:* Dept Preclin Dermat Bristol-Myers Squibb Res Inst 100 Forest Ave Buffalo NY 14213

BONNEY, WILLIAM WALLACE, ONCOLOGY, FEMALE UROLOGY. *Educ:* Univ Calif, Los Angeles, MD, 61. *Prof Exp:* ASSOC PROF UROL, DEPT UROL, COL MED, UNIV IOWA, 69-; STAFF PHYSICIAN, VET ADMIN MED CTR, 69- *Res:* Biology of the R3377 rat prostatic tumor. *Mailing Add:* Dept Urol Col Med Univ Iowa Iowa City IA 52242

BONNICHSEN, BILL, b Twin Falls, Idaho, Sept 25, 37; m 59; c 4. PETROLOGY, ECONOMIC GEOLOGY. *Educ:* Univ Idaho, BS, 60; Univ Minn, Minneapolis, PhD(geol), 68. *Prof Exp:* Instr econ geol, Univ Minn, Minneapolis, 66-67; res assoc geol, Minn Geol Surv, 67-69; asst prof petrol, Cornell Univ, 69-76; SUPVRY RES GEOLOGIST, IDAHO BUR MINES & GEOL, 76- *Mem:* Geol Soc Am; Int Asn Genesis Ore Deposits; Soc Econ Geologists. *Res:* Igneous and metamorphic petrology; economic geology of metals. *Mailing Add:* 927 E Seventh St Moscow ID 83843

BONNICHSEN, MARTHA MILLER, b Salem, Mass, May 2, 41; m 89; c 1. PETROLOGY, VOLCANOLOGY. *Educ:* Wellesley Col, BA, 62; Univ Ore, PhD(geol), 69. *Prof Exp:* Fel geol, Yale Univ, 69-70; asst prof, 70-76, ASSOC PROF GEOL, MT HOLYOKE COL, 76- *Mem:* Geol Soc Am; Am Geophys Union; Sigma Xi. *Res:* Volcanology and eruption mechanisms; chemical nature and evolution of volcanic rocks; metamorphism of ultramafic rocks; structural geology. *Mailing Add:* Dept Geog & Geol Mt Holyoke Col South Hadley MA 01075

BONO, VINCENT HORACE, JR, b Brooklyn, NY, July 23, 33; m 57; c 2. INTERNAL MEDICINE, BIOCHEMISTRY. *Educ:* Columbia Univ, AB, 53, MD, 57. *Prof Exp:* Intern internal med, Duke Univ Hosps, 57-58, univ res fel hemat, 58-59; resident internal med, Strong Mem Hosp, Univ Rochester, 59-60; clin assoc cancer chemother, Med Br, Nat Cancer Inst, Md, 60-63; trainee biochem, Dartmouth Med Sch, 63-64; sr investr cancer chemother, Nat Cancer Inst, 64-73, head molecular biol & methods develop sect, Lab Chem & Biol, 73-76, CHIEF, INVEST DRUG BR, DIV CANCER TREATMENT, NAT CANCER INST, 76- *Res:* Mechanisms of action and clinical evaluation of new drugs for the therapy of human neoplasia. *Mailing Add:* 6192 Oxon Hill Rd Suite 506 Oxen Hill MD 20745

BONORA, ANTHONY CHARLES, b San Francisco, Calif, 43. MECHANICAL ENGINEERING. *Educ:* Univ Calif, Berkeley, BS, 64, MS, 66. *Prof Exp:* Design engr, Lockheed Missiles & Space Co, 66-69; eng mgr, Semimetals, Inc, 69-73, opers mgr, 73-75; mgr eng, 75-76, VPRES CORP TECHNOL, SILTEC CORP, 77-, GEN MGR, INSTRUMENT DIV, 80- *Mem:* Electrochem Soc; Am Soc Eng Educ; Solar Energy Indust Asn. *Res:* Advanced fabrication of silicon materials; development of wafer manufacturing and processing equipment; crystal growth; wafer polishing. *Mailing Add:* 300 Felton Dr Menlo Park CA 94025

BONSACK, JAMES PAUL, b Westminster, Md, Nov 30, 32; m 57; c 5. INORGANIC CHEMISTRY, PHYSICAL CHEMISTRY. *Educ:* Wash Col, BS, 53; Lehigh Univ, MS, 55. *Prof Exp:* Teaching asst chem, Lehigh Univ, 53-55; chemist, J T Baker Chem Co, 55-62; chemist, 62-69, scientist chem, 69-81, SR SCIENTIST, CHEM DIV, SCM CHEM, 81- *Mem:* Am Chem Soc; AAAS; Am Ceramic Soc. *Res:* High temperature chemical processing; mineral chlorination; plasma-arc reaction processes; surface chemistry of silica; titanium chemistry. *Mailing Add:* SCM Chem 3901 Ft Armistead Rd Baltimore MD 21226

BONSACK, WALTER KARL, b Cleveland, Ohio, Apr 14, 32; m 57. ASTROPHYSICS. *Educ:* Case Inst Technol, BS, 54; Calif Inst Technol, PhD(astron), 59. *Prof Exp:* Res fel astron, Calif Inst Technol, 58-60; from asst prof to assoc prof, Ohio State Univ, 60-66; assoc prof, Univ Hawaii, 66-72, prof physics & astron, 72-87, asst dir, Inst Astron, 83-86; RETIRED. *Concurrent Pos:* Vis prof, Univ Vienna, 79-80; pres, Comn 29, Int Astron Union, 79-82. *Mem:* AAAS; Astron Soc Pac. *Res:* Composition and physical conditions in stellar atmospheres; spectrum variations; spectroscopic instrumentation for astronomy. *Mailing Add:* 2696 Glen Fenton Way San Jose CA 95148

BONSI, CONRAD K, b Ghana, Dec 11, 50. AGRICULTURAL RESEARCH. *Educ:* Univ Ghana, BSc 75; Tuskegee Univ, MS, 78; Cornell Univ, PhD(plant path), 82. *Prof Exp:* Res & teaching asst, Univ Ghana, Legon, 75-76; teaching asst, Tuskegee Univ, 76-78; grad res asst, Cornell Univ, 78-82; res assoc, Pa State Univ, 82-83; asst dir, Carver Agr Exp Sta, 86-87, asst prof dept Agr Sci, 83-87, ASSOC DIR RES GEORGE WASHINGTON CARVER AGR EXP STA & ASSOC PROF, DEPT AGR SCI, TUSKEGEE UNIV, ALA, 87- *Concurrent Pos:* Co-prog leader, Farmers Home Admin, USDA, 84-86; proj leader develop tomatoe for dis & environ stress resistance for limited resource farmers, USDA/CSRS, 83-, Tuskegee Univ Summer High Sch Apprenticeship Prog, USDA, 84-, eval Fungacides for control of Southern Blight of Veg Crops, Ciba-Geigy Corp, 84-88, Sweet Potatoe Cultivar & Nutrit Studies, Gerber Food Corp, 85-87, Kudzu-Goat Interactions, Forest Serv, USDA, 90-92; prin investr, Eval Tech Control Mold Walla Walla Sweet Onions, Batelle Pac NW Lab, 84-85, Sweet Potatoe Res Prog, USDA/CSRS, 84-, preservation sweet potatoes Phase Land III, US Dept energy, 85-88; proj co-dir, Sweet Potato Prod Soilless Culture Controlled Environ Life Support Syst, NASA, 85-; co-proj leader, Controlling Storage Root Develop Sweet Potatoe, US Agency Int Develop, 89- *Mem:* Am Phytopath Soc; Am Soc Hort Sci; Sigma Xi; Int Soc Hort Sci; Soc Nematologists. *Res:* Development of high yielding, disease resistant and environmental stress tolerant sweet potatoe cultivars with high nutritive value and production for controlled ecological life support systems; development of tomatoe cultivars adaptable for home gardens and limited resource; over 70 technical writings and publications. *Mailing Add:* George Washington Carver Agr Exp Sta Farm Mechanization Bldg Tuskegee Univ Tuskegee AL 36088

BONSIB, STEPHEN M, SURGICAL & RENAL PATHOLOGY. *Educ:* Ind Univ, MD, 78. *Prof Exp:* ASST PROF PATH, UNIV IOWA HOSP & CLIN, 83- *Res:* Glomerulonephritis. *Mailing Add:* Univ Iowa Hosps & Clins Path Iowa City IA 52242

BONSIGNORE, PATRICK VINCENT, b New York, NY, May 10, 29; m 58; c 2. POLYMER CHEMISTRY. *Educ:* Queens Col, NY, BS, 51; Polytech Inst Brooklyn, PhD(org & polymer chem), 58. *Prof Exp:* Sr scientist plastics, Columbia Southern Div, PPG Industs, 58-59; sr scientist acrylic plastics, Rohm and Haas, 59-70; sr scientist hydrogels, Hydron Labs, 70; staff scientist reprographics, Memorex, 70-71; sci assoc fire retardancy, Aluminum Co Am, 72-82; sr scientist ion-exchange membrane technol, Electrochem Int, 84-88; POLYMER CHEMIST, DEGRADABLE PLASTICS, ARGONNE NAT LAB, 88- *Concurrent Pos:* Fel, Univ Ill, 57-58; adj prof, Sch Med Chem, Univ Pittsburgh, 79-80; mem, Gov Task Force Degradable Plastics, State Ill. *Mem:* Soc Plastics Engrs; Am Chem Soc; Sigma Xi; Fire Retardant Chem Asn. *Res:* Characterization and improvement of fire retardancy and smoke suppression in plastics; development of controlled pore alumina (CPA) as supports for biocatalysts in bioengineering applications; ion exchange membrane technology; poly(lactic acid) degradable plastics. *Mailing Add:* 4426 Sussex Rd Joliet IL 60436-9333

BONTADELLI, JAMES ALBERT, b St John, Wash, Sept 27, 29; m 51; c 2. INDUSTRIAL & SYSTEMS ENGINEERING. *Educ:* Ore State Univ, BS, 51; Stanford Univ, MS, 58; Ohio State Univ, PhD(indust eng), 72. *Prof Exp:* Asst chief, Systs Res Br, US Army Eng Maintenance Ctr, 59-60, chief, 60-62; proj leader, Systs Anal Div, Battelle Mem Inst, 62-66, assoc chief, 66-70, group leader mgt systs group, 70-74; mgr indust eng, Tenn Valley Authority, 74-89; PROF INDUST ENG, UNIV TENN, 90- *Concurrent Pos:* Adj prof indust eng, Univ Tenn, 75-89. *Mem:* Fel Inst Indust Engrs. *Res:* Productivity and quality improvement; engineering economic analysis; decision analysis; application of industrial engineering in construction, power production and service industry operations. *Mailing Add:* 11517 N Monticello Dr Knoxville TN 37922

BONTE, FREDERICK JAMES, b Bethlehem, Pa, Jan 18, 22; m 52, 68; c 6. RADIOLOGY. *Educ:* Western Reserve Univ, BS, 42, MD, 45. *Prof Exp:* Fel radiobiol, Atomic Energy Med Res Proj, Western Reserve Univ, 48-49; resident physician, Univ Hosps, Cleveland, 49-52; instr radiol, Med Sch, Western Reserve Univ, 52-56; prof radiol & chmn dept, 56-73, dean, Southwestern Med Sch, Univ Tex Health Sci Ctr, 73-80, DIR, NUCLEAR MED RES CTR, UNIV TEX, 80- *Concurrent Pos:* Dir dept radiol, Parkland Mem Hosp, 56-73; consult, US Vet Admin Hosp Syst, 56-; mem bd, Nat Coun Radiation Protection & Measurements, 65-71; mem radiol training comt, Nat Inst Gen Med Sci, 65-70; mem med adv comt, Oak Ridge Assoc Univs, 68-70; trustee, Am Bd Radiol, 69-75 & Am Bd Nuclear Med, 71-73; chmn, Am Bd Nuclear Med, 77-80; Effie & Wofford Cain Distinguished Chair Diag Imaging. *Mem:* AAAS; fel Am Col Radiol; Am Roentgen Ray Soc; Radiol Soc NAm; Sigma Xi; fel Am Col Nuclear Physicians; Soc Nuclear Med. *Res:* Clinical and experimental nuclear medicine; radiobiology. *Mailing Add:* Southwestern Med Ctr Univ Tex Dallas TX 75235-9061

BONTEMPO, JOHN A, b Pesco, Italy, Sept 13, 30; US citizen; m 56; c 3. MICROBIOLOGY. *Educ:* Heidelberg Col, AB & BS, 56; Fairleigh Dickinson Univ, MS, 69; Rutgers Univ, PhD(microbiol), 75. *Prof Exp:* Asst scientist, Warner-Lambert Res Inst, NJ, 56-62; microbiologist, 62-70, sr microbiologist, 70-72, group leader, 72-75, TECH FEL, HOFFMANN-LA ROCHE, INC, 75- *Mem:* NY Acad Sci; Am Soc Microbiol; Sigma Xi. *Res:* Antimicrobial agents, analytical microbiology. *Mailing Add:* Hoffmann-La Roche Inc 340 Kingsland St Nutley NJ 07110-1199

BONTING, SJOERD LIEUWE, b Amsterdam, Holland, Oct 6, 24; nat US; m 51, 87; c 4. BIOCHEMISTRY. *Educ:* Univ Amsterdam, BSc, 44, MSc, 50, PhD(biochem), 52. *Hon Degrees:* LTh, St Mark's Inst Theol, London, 75. *Prof Exp:* Instr anal chem, Univ Amsterdam, 47-49; biochemist, Neth Inst Nutrit, 50-52; res assoc physiol, Col Med, Univ Iowa, 52-55; asst prof physiol chem, Sch Med, Univ Minn, 55-56; asst prof biol chem, Col Med, Univ Ill & asst attend biochemist, Presby-St Luke's Hosp, Chicago, 56-60; head sect cell biol, Ophthal Br, Nat Inst Neurol Dis & Blindness, NIH, 60-65; prof biochem & chmn dept, Univ Nijmegen, 65-85; RES SCIENTIST, NASA AMES RES CTR, MOFFETT FIELD, CA, 85- *Concurrent Pos:* Instr biochem, Univ Amsterdam, 51-52; fel, USPHS & Nat Cancer Inst, 52-54; guest scientist, Inst Animal Physiol, Cambridge, Eng, 64-65. *Honors & Awards:* Fight for Sight Award, Nat Coun Combat Blindness & Asn Res Ophthal, 61 & 62; Merit Award, Int Rescue Comt, 63; Arthur S Fleming Award, 64; Karger Mem Found Prize, 64; Pro Mundi Beneficis Award, Brazilian Acad Humanities, 75. *Mem:* AAAS; Am Soc Biol Chem; Am Soc Cell Biol. *Res:* Membrane and transport biochemistry; quantatative histochemistry; visual mechanism; space biology. *Mailing Add:* 1006 E Evelyn Ave Sunnyvale CA 94086

BONURA, THOMAS, b New York, NY, Jan 27, 47; div. RADIATION BIOLOGY, BIOCHEMISTRY. *Educ:* Duquesne Univ, BS, 68; State Univ NY, Buffalo, MA, 71, PhD(biol), 73. *Prof Exp:* Res assoc, Dept Radiol, Radiobiol Res Div, 72-77, res asst DNA repair, 77-79, RES ASSOC, LAB EXP ONCOL, DEPT PATH, SCH MED, STANFORD UNIV, 79- *Mem:* AAAS; Radiation Res Soc; Sigma Xi. *Res:* Enzymology of DNA repair; interaction of ionizing radiation and DNA; cellular inactivation by DNA damaging agents. *Mailing Add:* 606 El Salto Dr Capitola CA 95010

BONVENTRE, PETER FRANK, b New York, NY, Aug 18, 28; m 63. MICROBIOLOGY. *Educ:* City Col New York, BS, 49; Univ Tenn, MS, 55; Univ Mich, PhD, 57. *Prof Exp:* NSF fel, Higher Inst Sanit, Italy, 57-58; from asst prof to assoc prof, 58-68, PROF MICROBIOL, COL MED, UNIV CINCINNATI, 68- *Concurrent Pos:* Vis res assoc, dept pharmacol, Royal Vet Col Sweden, 60; career develop award, USPHS, 63; vis scientist, Clin Res Ctr, London, UK, 75 & Scripps Res Inst, La Jolla, Calif, 87; ed, Infection & Immunity, 79-90 & J Leukocyte Biol, 82-; Path & Immunopath Res, 83-90. *Mem:* AAAS; Am Soc Microbiol; Sigma Xi; Soc Leukocyte Biol. *Res:* Host-parasite relationships in infectious processes; mechanisms of pathogenesis; in vivo action of bacterial toxins; biochemistry and functions of phagocytic cells; biology of leishmanial protozoa; toxic shock syndrome. *Mailing Add:* Dept Molecular Genetics, Biochem & Microbiol Univ Cincinnati Col Med Cincinnati OH 45267-0524

BONVICINO, GUIDO EROS, b Castrovillari, Italy, Feb 9, 21; nat US; m 52; c 1. ORGANIC CHEMISTRY, SYNTHETIC ORGANIC. *Educ:* Fordham Univ, BS, 47, MS, 49, PhD(org chem), 52. *Prof Exp:* Instr org chem & anal, Fordham Univ, 47-52; res chemist org, Lederle Labs, 52-63; assoc prof org chem, 63-68, PROF ORG CHEM, C W POST CTR, LONG ISLAND UNIV, 68- *Honors & Awards:* Am Inst Chemists Medal, 47. *Mem:* Sigma Xi; fel Am Inst Chemists; sr mem Am Chem Soc; NY Acad Sci. *Res:* Thiamine; chemical reactions; antibiotics and synthetic chemotherapeutic agents; chemopsychiatric agents; reaction mechanisms. *Mailing Add:* 73 Hill Dr Oyster Bay NY 11771

BONVILLIAN, JOHN DOUGHTY, b Caldwell, Idaho, Sept 4, 48. CHILD PSYCHOLOGY, PSYCHOLINGUISTICS. *Educ:* Johns Hopkins Univ, BA, 70; Stanford Univ, PhD(psychol), 74. *Prof Exp:* Asst prof psychol, Vassar Col, 74-78; vis assoc prof, 78-79, asst prof, 79-84, ASSOC PROF PSYCHOL, UNIV VA, 84- *Concurrent Pos:* Asst ed, Sign Language Studies, 78-; mem, Comt Linguistics, Univ Va, 79-, Inst Clin Psychol, 81-87; vis scholar, Gallaudet Col, 84-85. *Mem:* Am Psychol Asn; Am Speech Language & Hearing Asn; Linguistic Soc Am; Soc Res Child Develop. *Res:* Longuitudinal investigation of how words and manual signs acquire meaning for normal and language-disabled (deaf, autistic, aphasic) individuals. *Mailing Add:* Dept Psychol Univ Va 102 Gilmer Hall Charlottesville VA 22903-2477

BOODMAN, DAVID MORRIS, b Pittsburgh, Pa, July 4, 23; m 48; c 2. SYSTEMS SCIENCE. *Educ:* Univ Pittsburgh, BS, 44, PhD(phys chem), 50. *Prof Exp:* Sr staff mem, Opers Eval Group, Mass Inst Technol, 50-60; sr staff mem, Opers Res Sect, Arthur D Little Inc, 60-72, vpres, 72-89; RETIRED. *Concurrent Pos:* Opers analyst, Staff, Comdr in Chief, Pac Fleet, US Navy, 51, Oper Develop Force, 53-54 & staff, Comdr First Fleet, 58. *Mem:* Fel AAAS; Math Asn Am; Opers Res Soc Am. *Res:* Systems science-operations; military and industrial operations research. *Mailing Add:* Four Linmoor Terrace Lexington MA 02173

BOODMAN, NORMAN S, b Pittsburgh, Pa, Feb 13, 27; m 63; c 3. FUEL SCIENCE. *Educ:* Univ Pittsburgh, BS, 49, MLitt, 57. *Prof Exp:* Asst chem, Univ Pittsburgh, 49-50; anal chemist, Vitro Mfg Co, Pa, 50-51; chemist, Koppers Co Res Ctr, 51-55; res consult, US Steel Tech Ctr, 56-86; ARISTECH CHEM CORP, 86- *Res:* Recovery and purification of chemicals from high-temperature carbonization coal tar; production of phenolics, alcohols and monomers via hydroperoxidation of alkylaromatic hydrocarbons; coal conversion via hydrogenation and fluidized-bed carbonization. *Mailing Add:* Aristech Chem Corp Monroeville PA 15146

BOOE, J(AMES) M(ARVIN), b Austin, Ind, Nov 12, 06; m 38; c 3. INDUSTRIAL ELECTROCHEMISTRY. *Educ:* Butler Univ, BS, 28. *Prof Exp:* Chemist, Indianapolis Plating Co; chief chemist, P R Mallory & Co, Inc, 29-45, charge electrochem res, 45-51; exec chem engr, 51-53, charge cent chem & metall res labs, 53-55, dir mat res labs, 55-57, cent chem res labs, 57-63, dir chem labs, Mallory Capacitor Co, 63-72, consult, 72-79; consult, Capacitor Co, Emhart Co, 79-86; RETIRED. *Honors & Awards:* Naval Ord Develop Award, 45; Army-Navy E Award, 45; Prof Chemist Award, Am Inst Chem, 69. *Mem:* Am Chem Soc; Electrochem Soc; fel Am Inst Chem. *Res:* Aluminum and tantalum electrolytic capacitors; dry rectifiers; primary batteries; electrostatic capacitors; electroplating; analytical methods; semiconductors and transistors; high temperature materials; 38 US patents in above fields. *Mailing Add:* 548 N Audubon Rd Indianapolis IN 46219

BOOHAR, RICHARD KENNETH, b Philadelphia, Pa, Mar 7, 35; m 59; c 5. DEVELOPMENTAL BIOLOGY. *Educ:* Drew Univ, BA, 57; Univ Wis, MA, 60, PhD(zool), 66. *Prof Exp:* Asst zool, Univ Wis, 64; from instr to asst prof, Butler Univ, 64-67; asst prof, 67-71, ASSOC PROF LIFE SCI, UNIV NEBR, LINCOLN, 71- *Mem:* AAAS; Am Inst Biol Sci; Sigma Xi. *Res:* Bioethics. *Mailing Add:* Sch Biol Sci Univ Nebr Lincoln NE 68588-0118

BOOK, DAVID LINCOLN, b Boston, Mass, Aug 4, 39. PLASMA PHYSICS. *Educ:* Yale Univ, BA, 59; Princeton Univ, MA, 61, PhD(physics), 64. *Prof Exp:* Staff assoc, Gen Atomic Div, Gen Dynamics Corp, 64-67; staff physicist, Lawrence Radiation Lab, 67-74; MEM STAFF, NAVAL RES LAB, 74- *Mem:* Am Phys Soc. *Res:* Theoretical plasma physics; controlled thermonuclear fusion; kinetic theory. *Mailing Add:* US Naval Res Lab Code 4405 Washington DC 20375

BOOK, LINDA SUE, b Youngstown, Ohio, Oct 8, 46; m 70; c 2. PEDIATRIC GASTROENTEROLOGY. *Educ:* Univ Cincinnati, MD, 71. *Prof Exp:* Intern, Univ Cincinnati, 71-72; resident, 72-74, instr, 74-76, ASST PROF PEDIAT & NURSING, UNIV UTAH, 76- *Concurrent Pos:* NIH fel pediat gastroenterol, Univ Utah, 74-76, fel premature infants res, 76- & Ross Lab award res in infant nutrit, 76- *Honors & Awards:* Res Award, Western Soc for Pediat, 74. *Mem:* Fedn Am Soc Exp Biol. *Res:* Infant nutrition; effect of various feeding practices on development of necrotizing enterocolitis; gastrointestinal immunology. *Mailing Add:* Univ Utah Med Ctr 50 N Medical Dr Salt Lake City UT 84132

BOOK, STEPHEN ALAN, b Newark, NJ, Dec 11, 41; m 68; c 5. STATISTICS. *Educ:* Georgetown Univ, AB, 63; Cornell Univ, MA, 66; Univ Ore, PhD(math), 70. *Prof Exp:* from asst prof to assoc prof, 70-79, PROF MATH, CALIF STATE UNIV, DOMINGUEZ HILLS, 79- *Concurrent Pos:* Statist consult. *Mem:* Math Asn Am; Inst Math Statist; Am Math Soc; London Math Soc; Am Statist Asn. *Res:* Limit theorems of probability theory; large deviation probabilities; laws of large numbers; almost sure convergence. *Mailing Add:* Dept Math Calif State Univ Dominguez Hills CA 90747

BOOK, STEVEN ARNOLD, b Albany, Calif, Dec 13, 45; m 68; c 2. RADIATION BIOLOGY, PHYSIOLOGY. *Educ:* Univ Calif, Berkeley, AB, 67; Univ Calif, Davis, MA, 69, PhD(physiol), 73. *Prof Exp:* Res radioecologist, 69-73, asst res physiologist, Radiobiol Lab, 73-79, lectr radiol sci, 74-81, assoc res physiologist, 79-81, ASSOC ADJ PROF, LAB ENERGY-RELATED HEALTH RES, SCH VET MED, UNIV CALIF, DAVIS, 81- *Mem:* AAAS; Health Physics Soc; Radiation Res Soc. *Res:* Metabolism and effects of radioiodine; radiation biology; risk assessment; radioecology. *Mailing Add:* Univ Calif Lab Energy Related Health Res Davis CA 95616-2908

BOOKCHIN, ROBERT M, b New York, NY, Oct 14, 35. HEMATOLOGY. *Educ:* Wash Univ, MD, 59. *Prof Exp:* PROF MED, ALBERT EINSTEIN COL MED, 67- *Concurrent Pos:* Macy fac scholar award. *Mem:* Asn Am Physicians; Am Soc Clin Invest; Am Soc Hemat. *Res:* Erythrocyte physiology and pathophysiology, hemoglobin-opathies red cell membrane transport. *Mailing Add:* Albert Einstein Col Med 1300 Morris Park Ave Bronx NY 10461

BOOKE, HENRY EDWARD, b Brooklyn, NY, Jan 14, 32; m 59; c 1. ICHTHYOLOGY. *Educ:* Cornell Univ, BS, 60; Mich State Univ, MS, 62; Univ Mich, PhD(ichthyol), 68. *Prof Exp:* Conserv biologist, Cortland Exp Hatchery, NY State Conserv Dept, 61-63; US Bur Com Fisheries res fel, 63-65; res asst ichthyol, Mus Zool, Fish Div, Univ Mich, 65-66, teaching asst, Dept Wildlife & Fisheries, 66-67; instr biol, Yale Univ, 68-69; asst prof, Boston Univ, 69-72; assoc prof fishery biol, Univ Wis-Stevens Point, 73-79, asst leader, Wis Coop Fishery Unit, US Dept Interior Fish & Wildlife Serv, 73-80; prof biol & unit leader, Ma Coop Fishery Res Unit, Univ Ma, Amherst, 80-88; DIR, NE ANADROMOUS FISH LAB, US FISH & WILD SERV, 88- *Concurrent Pos:* NSF res traineeship, Sport Fishing Inst Res grant & Theodore Roosevelt Mem fund grant, Am Mus Natural Hist, 66-67; Nat Oceanic & Atmospheric Admin sea grant, 75-80. *Honors & Awards:* Unit Gulf Conser Award, 83. *Mem:* AAAS; Am Fisheries Soc; Am Soc Ichthyologists & Herpetologists; Am Soc Zoologists. *Res:* Fish nutrition and hematology; biochemical systematics, cytotaxonomy population genetics and ecological physiology of fishes. *Mailing Add:* NE Anadromous Fish Res Lab One Migratory Way Turners Falls MA 01376

BOOKER, HENRY GEORGE, radio physics; deceased, see previous edition for last biography

BOOKER, J(OHN) F, b Chicago, Ill, Dec 13, 34; m 57; c 4. MECHANICAL ENGINEERING. *Educ:* Yale Univ, BE, 56; Chrysler Inst Eng, MAE, 58; Cornell Univ, PhD, 61. *Prof Exp:* Engr, Chrysler Corp, 56-58; asst prof mech & aerospace eng, 61-65, assoc dir, 77-80, PROF MECH & AEROSPACE ENG, CORNELL UNIV, 79- *Concurrent Pos:* Vis engr, Glacier Metal Co, 67; Ford Found Eng Residency, IBM Corp, 67-68; consult, Gen Motors Res Labs, 69-; sr vis fel, Univ Leeds, Eng, 74-75 & Twente Univ Technol, Netherlands, 75; mem, Am Soc Mech Engrs Res Comt Lubrication; assoc ed, J Lubrication Technol, 76-79; Fulbright fel, Eindhoven Univ Technol, Netherlands, 82; vis prof, Univ Durham, Eng, 82-83; vis prof, Colo State Univ, 88-89; assoc ed, Tribology Trans, 88-91. *Honors & Awards:* Henry Hess Award, Am Soc Mech Engrs, 65; Starley Premium Prize, Brit Inst Mech Engrs, 67; B P Oil Tribology Award, Brit Inst Mech Engrs, 87. *Mem:* Fel Am Soc Mech Engrs; Soc Tribologists & Lubrication Engrs. *Res:* Hydrodynamic lubrication; dynamics of mechanical systems, especially those containing fluid film bearings; finite element methods. *Mailing Add:* Sch Mech & Aero Eng Upson Hall Cornell Univ Ithaca NY 14853

BOOKER, JOHN RATCLIFFE, b Swanage, Eng, Apr 28, 42; US citizen; m 86; c 2. GEOPHYSICS. *Educ:* Stanford Univ, BS, 63; Univ Calif, San Diego, MS, 65, PhD(earth sci), 68. *Prof Exp:* Fulbright-Hays fel, Meteorol Inst, Univ Stockholm, 68-69; res assoc & lectr geophys, Stanford Univ, 69-71; from asst prof to assoc prof, 71-82, PROF GEOPHYS, UNIV WASH, 82- *Concurrent Pos:* Mem, Earth Sci Proposal Rev Panel, NSF, 81-85; mem, Earth & Ocean Sci Adv Comts, 85-87, Continental Dynamics Proposal Rev Panel, 89- *Mem:* Fel Am Geophys Union. *Res:* Geomagnetic induction; inverse theory; thermal instabilities in viscosity stratified fluids and porous media; ocean acoustic tomography. *Mailing Add:* Geophys Prog Univ Wash Seattle WA 98195

BOOKHOUT, CAZLYN GREEN, b Gilboa, NY, Jan 28, 07; m 36; c 2. MARINE ZOOLOGY. *Educ:* St Stephens Col, AB, 28; Syracuse Univ, AM, 29; Duke Univ, PhD(zool), 34. *Prof Exp:* Instr biol, Women's Col, Univ NC, 29-31; asst zool, Duke Univ, 31-32; assoc prof biol, Elon Col, 34-35; from instr to prof zool, 35-76, EMER PROF ZOOL, DUKE UNIV, 76- *Concurrent Pos:* Dir marine lab, Duke Univ, 50-63 & 64-68, actg dir oceanog prog, 67-70. *Mem:* AAAS; Am Soc Zool; Am Soc Limnol & Oceanog. *Res:* Germ cells of mammals; embryology of invertebrates and polychaetes; growth in relation to molting in Crustacea; larval development of Portuninae; effect of pesticides on development of crabs; effect of drilling fluids on development of crabs. *Mailing Add:* Duke Univ Marine Lab Beaufort NC 28516

BOOKHOUT, THEODORE ARNOLD, b Salem, Ill, June 11, 31; m 52; c 2. WILDLIFE RESEARCH. *Educ:* Southern Ill Univ, BA, 52, MS, 54; Univ Mich, PhD(wildlife mgt), 63. *Prof Exp:* Asst wildlife mgmt, Univ Mich, 56-59; game biologist, Mich Dept Conserv, 59-62; asst wildlife mgmt, Univ Mich, 62, instr, 62-63; res asst prof vert ecol, Univ Md, 63-64; assoc prof, 68-70, PROF ZOOL, OHIO STATE UNIV, 70-; BIOLOGIST & LEADER, OHIO COOP FISH AND WILDLIFE RES UNIT, US FISH & WILDLIFE SERV, 64- *Mem:* Sigma Xi; Wildlife Soc (vpres, 78-79, pres-elect, 79-80, pres, 80-82); Am Soc Mammal. *Res:* Ecology and population dynamics of terrestrial game birds and mammals; waterfowl and wetland ecology. *Mailing Add:* Ohio Coop Fish & Wildlife Res Unit Ohio State Univ 1735 Neil Ave Columbus OH 43210

BOOKMYER, BEVERLY BRANDON, b Glen Head, NY; c 2. ASTRONOMY. *Educ:* Chestnut Hill Col, AB, 46; Univ Pa, MS, 61, PhD(astron), 64. *Prof Exp:* Asst prof astron & math, Villanova Univ, 64-67; res asst prof, Univ Ariz, 67-71; from assoc prof physics & astron to prof, Clemson Univ, 71-87; RETIRED. *Concurrent Pos:* NSF res grant, 66 & int travel grant, 73 & 75. *Mem:* Am Astron Soc; Am Phys Soc; Int Astron Union; fel AAAS; Astron Soc Pac. *Res:* Photoelectric photometry of variable stars; analysis of observations of eclipsing binary systems; applications of computerized synthetic light curve techniques to contact binary systems; mass exchange; stellar evolution. *Mailing Add:* 505A Eunice St N Sequim WA 98382

BOOKSTEIN, ABRAHAM, b New York, NY, Mar 22, 40; m 67. INFORMATION SCIENCE. *Educ:* City Col New York, BS, 61; Univ Calif, Berkeley, MS, 66; Yeshiva Univ, PhD(physics), 69; Univ Chicago, MA, 70. *Prof Exp:* Asst prof libr sci, 71-75, asst prof behav sci, 74-75, assoc prof 75-82, PROF LIBR & BEHAV SCI, UNIV CHICAGO, 82- *Concurrent Pos:* Prin investr, NSF grants & organizing & prog comt conf; vis prof, Royal Inst Technol, Stockholm, 82, Grad Sch Libr & Info Sci, Univ Calif, Los Angeles, 85. *Mem:* Am Soc Info Sci; Asn Comput Mach. *Res:* Mathematical modelling in the information sciences; information storage and retrieval; data compression (computer science); research methods. *Mailing Add:* Ctr Info Studies Univ Chicago Chicago IL 60637

BOOKSTEIN, FRED LEON, b Detroit, Mich, July 27, 47; m; c 2. MORPHOMETRICS, COMPUTER GRAPHICS. *Educ:* Univ Mich, BS, 66, PhD(statist & zool), 77; Harvard Univ, MA, 71. *Prof Exp:* Asst res scientist, Univ Mich, Ann Arbor, 77-80, assoc res scientist, 80-84, res scientist, 84-89, DISTINGUISHED RES SCIENTIST HUMAN GROWTH & DEVELOP, INST GERONT & BIOSTATIST, UNIV MICH, 89- *Concurrent Pos:* Consult, Sch Eng, Columbia Univ, NY Univ Med Ctr & Soc Sci Res Coun & others. *Mem:* Inst Math Statist; Inst Elec & Electronics Engrs. *Res:* New methods for morphometrics, the measurement of biological shape and shape change; applications involve radiology, evolutionary biology, orthodontics, plastic surgery and growth prediction. *Mailing Add:* Ctr Human Growth Univ Mich Ann Arbor MI 48109-0406

BOOKSTEIN, JOSEPH JACOB, b Detroit, Mich, July 25, 29; m 54; c 3. RADIOLOGY. *Educ:* Wayne State Univ, BS, 50, MD, 54. *Prof Exp:* Am Cancer Soc fel, Stanford Univ, 59-60; from instr to prof radiol, Med Ctr, Univ Mich, Ann Arbor, 60-74; PROF RADIOL, HOSP, UNIV CALIF, SAN DIEGO, 74- *Concurrent Pos:* Consult, Vet Admin Hosp, Ann Arbor, Mich, 61-74; mem coun cardiovasc radiol, Am Heart Asn. *Mem:* Asn Univ Radiol; Radiol Soc NAm; NAm Soc Cardiac Radiol. *Res:* Cardiovascular radiology; interventional radiology; renovascular hypertension; magnification angiography; translumind onycoplasty. *Mailing Add:* Dept Univ Hosp 225 W Dickinson St San Diego CA 92103

BOOKWALTER, GEORGE NORMAN, b Springfield, Ill, Nov 25, 24; m 51; c 2. FOOD PROCESSING, CEREAL FOOD FORMULATIONS. *Educ:* Univ Ill, BS, 51. *Prof Exp:* Food technologist, Gen Foods Corp, 51; chemist, Armour & Co, 52-55 & Pillsbury Co, 55-61; appln chemist, A E Staley Co, 61-65; RES FOOD TECHNOLOGIST & LEAD SCIENTIST, NATIONAL CTR AGR UTIL RES, AGR RES SERV, USDA, 66-, TECH ADV FOOD FOR PEACE PROG, 84- *Concurrent Pos:* Mem, NC-120 Food Serv Comt, USDA, 75-; res collabr, Col Agr, Univ Ill, 81- *Honors & Awards:* Babcock-Hart Award, Nutrit Found, Inc & Inst Food Technologists, 86. *Mem:* Sigma Xi; fel Inst Food Technologists; Am Asn Cereal Chemists; Am Oil Chemists' Soc. *Res:* Development of processed food blends from cereals, oilseeds, and dairy products that are utilized in international food relief; cone milk product; whipped topping formulations. *Mailing Add:* 1808 W Teton Dr Peoria IL 61614

BOOLCHAND, PUNIT, b Varanasi, India. EXPERIMENTAL SOLID STATE PHYSICS, EXPERIMENTAL NUCLEAR PHYSICS. *Educ:* Punjab Univ, BS, 64, MS, 65; Case Western Reserve Univ, PhD(nuclear & solid state physics), 69. *Prof Exp:* Teaching asst physics, Case Western Reserve Univ, 65-67, grad fel, 67-69; from instr to asst prof, 69-75, ASSOC PROF PHYSICS, UNIV CINCINNATI, 75- *Concurrent Pos:* Cottrel res grant, Res Corp, 72; res assoc & vis scholar, Stanford Univ, 73-74; vis prof, Katholieke Univ, Leuven, Belg, 74-75; NSF res grant, 75; Am Philos Soc res grant, 75. *Mem:* AAAS; Am Phys Soc. *Res:* Hyperfine interactions studies by Mossbauer effect, perturbed angular correlations, nuclear magnetic resonance and nuclear orientation; structure of amorphous solids; lattice location of implanted ions in semiconductors and metals. *Mailing Add:* Dept ECE Mail Loc 30 Univ Cincinnati Cincinnati OH 45221

BOOLE, JOHN ALLEN, JR, b Exmore, Va, Dec 14, 21; m 49; c 3. PLANT MORPHOLOGY, PLANT PHYSIOLOGY. *Educ:* Univ Va, BA, 49; Va Polytech Inst, MS, 51; Univ NC, PhD(bot), 55. *Prof Exp:* Assoc prof, 55-58, head dept, 55-60, chmn div sci & math, 60-74, PROF BIOL, GA SOUTHERN COL, 58-, DIR ADVISEMENT, 74- *Mem:* AAAS; Bot Soc Am; Am Inst Biol Sci; Sigma Xi. *Res:* Systematic anatomy of the family Celastraceae; wood anatomy. *Mailing Add:* Ga Southern Col PO Box 8042 Statesboro GA 30458

BOOLOOTIAN, RICHARD ANDREW, b Fresno, Calif, Oct 17, 27; m 67; c 3. INVERTEBRATE ZOOLOGY. *Educ:* Fresno State Col, BA, 51, MA, 53; Stanford Univ, PhD(biol), 57. *Prof Exp:* Asst comp physiol, Hopkins Marine Sta, Stanford Univ, 54-57, res assoc, 58; from asst prof to assoc prof zool, Univ Calif, Los Angeles, 57-67; staff consult biol sci curric study, Univ Colo, 67-68; pres, Visual Sci Prod, Los Angeles, 68-69; PRES, SCI SOFTWARE SYSTS, INC, 69- *Concurrent Pos:* NIH spec fel, 65; dir, Inst Visual Med, Los Angeles, 68-79; pres, Instrnl Systs for Health Sci, 70-75. *Honors & Awards:* Lalor Found Award, 59, 61. *Mem:* Fel AAAS; Soc Gen Physiologists; NY Acad Sci; Am Soc Zoologists; Animal Behav Soc. *Res:* Comparative physiology of marine invertebrates. *Mailing Add:* 3576 Woodcliff Rd Sherman Oaks CA 91403

BOOM, ROGER WRIGHT, b Bladen, Nebr, Mar 10, 23; m 51. LOW TEMPERATURE ENGINEERING PHYSICS. *Educ:* Univ Nebr, AB, 44; Univ Minn, MS, 51; Univ Calif, Los Angeles, PhD(physics), 58. *Prof Exp:* Res engr space physics, NAm Aviation, La, 51-52; electronic engr product design, Hoffman Electronic Corp, La, 52-54; fel nuclear physics, Univ Bonn, Ger, 58-60; res engr appl superconductivity, Oak Ridge Nat Lab, 60-63; supvr appl superconductivity, Atomics Int Div, Rockwell Int, 63-68; PROF NUCLEAR & METALL ENG, UNIV WIS-MADISON, 68- *Concurrent Pos:* Superconductive magnetic energy storage dir, Appl Superconductivity Ctr, 72- *Mem:* Am Phys Soc; Inst Elec & Electronic Engrs; Am Soc Mining Engrs; Sigma Xi. *Res:* Applied superconductivity systems for energy storage, fusion reactor designs, magnetic ore separators, particle detection and composite conductors of aluminum-niobium titanium; including electrical, mechanical, thermal and magnetic experimental research topics in above systems. *Mailing Add:* 917 Eng Res Bldg Univ Wis Madison WI 53706

BOOMAN, KEITH ALBERT, b Bellingham, Wash, Jan 15, 28; m 53; c 3. TIME SERIES ANALYSIS, PHYSICAL ORGANIC CHEMISTRY. *Educ:* Univ Wash, BS, 50; Calif Inst Technol, PhD(chem), 56. *Prof Exp:* Res chemist, Res Labs, Rohm and Haas Co, Philadelphia, 56-59; head res lab, Redstone Arsenal Res Div, 59-63, head res lab, 63-70; TECH DIR, SOAP & DETERGENT ASN, 71- *Mem:* Am Chem Soc; Water Pollution Control Fedn; Am Water Resources Asn; Am Statist Asn. *Res:* Time series analysis; detergents. *Mailing Add:* Soap & Detergent Asn 475 Park Ave New York NY 10016

BOOMS, ROBERT EDWARD, b Cleveland, Ohio, Mar 25, 45; m 68; c 1. PHOTOGRAPHIC CHEMISTRY. *Educ:* Ohio State Univ, BS, 67; Univ Calif, Los Angeles, PhD(org chem), 71. *Prof Exp:* CHEMIST, EASTMAN KODAK CO, 71- *Mem:* Am Chem Soc; Soc Photog Scientists & Engrs; Sigma Xi. *Res:* Color photographic systems. *Mailing Add:* 8 Maple Grove Dr Churchville NY 14428

BOOMSLITER, PAUL COLGAN, speech pathology, audiology; deceased, see previous edition for last biography

BOON, JOHN DANIEL, JR, b Ft Worth, Tex, Dec 29, 14; m 39; c 2. GEOLOGY. *Educ:* Southern Methodist Univ, BS, 37. *Prof Exp:* Instr geol, Southern Methodist Univ, 41; computer, Seismograph Party, Nat Geophys Co, 41-42; assoc prof geol & physics, Univ Tex, Arlington, 42-43; prof physics & actg head dept, 43-45, head, Dept Geol, 45-71, prof geol, 71-80; RETIRED. *Mem:* Fel Geol Soc Am; Am Asn Petrol Geologists. *Res:* Field geology of regions in trans-Pecos Texas; structural, general and economic geology; petroleum engineering; seismology. *Mailing Add:* 1708 Venetian Circle Arlington TX 76013

BOONE, BRADLEY GILBERT, b Sept 27, 50; c 2. ELECTROOPTICS, SIGNAL PROCESSING. *Educ:* Wash & Lee Univ, BS, 72; Univ Va, PhD(physics), 77. *Prof Exp:* Physicist solid state, US Army Foreign Sci & Tech Div, 77; SR PHYSICIST & SECT SUPVR, ELECTRO-OPTICS SYSTS GROUP, APPL PHYSICS LAB, JOHNS HOPKINS UNIV, 77- *Concurrent Pos:* Lectr physics, Univ Va, 77; lectr elec eng, Johns Hopkins Univ, 83-, vis prof, Elec & Computer Eng Dept, 90- *Mem:* Soc Photo-optical Instrumentation Engrs. *Res:* Experimental and analytical evaluation of optical processing architectures; radar image and signal processing; radar signal analysis and modeling; microelectronic applications of high temperature superconductivity; pattern recognition. *Mailing Add:* Appl Physics Lab Johns Hopkins Univ Laurel MD 20723-6099

BOONE, CHARLES WALTER, b Berkeley, Calif, Dec 21, 25; m 65; c 2. PATHOLOGY, BIOCHEMISTRY. *Educ:* Univ Calif, San Francisco, MD, 51; Univ Calif, Los Angeles, PhD(biochem), 64; Am Bd Path, dipl, 60. *Prof Exp:* Gen pract med, Los Angeles, 52-56; resident path, 56-60; HEAD CELL BIOL SECT, LAB VIRAL CARCINOGENESIS, NAT CANCER INST, NIH, 65- *Concurrent Pos:* Res fel, Albert Einstein Col Med, 64-65. *Mem:* Am Soc Exp Path; Am Soc Biol Chemists. *Res:* Isolation of pure tumor cells; production of tumor immunity; cellular mechanisms of carcinogenesis. *Mailing Add:* 8705 Hempstead Ave Bethesda MD 20817

BOONE, DONALD H, metallurgical engineering, for more information see previous edition

BOONE, DONALD JOE, b Spokane, Wash, May 12, 43; m 73. CLINICAL CHEMISTRY, CLINICAL PATHOLOGY. *Educ:* Cent State Univ, BS, 65; Iowa State Univ, MS, 67; Wash State Univ, PhD(chem), 70. *Prof Exp:* NSF fel biophys, Albert Einstein Col Med, 71-72; fel clin chem, Univ Iowa, 72-74;

asst prof path, Univ Ky, 74-76; chief, Clin Chem & Toxicol Sect, Training & Lab Improv Prog, 76-86, HEALTH SCI ADMINR, CTR DIS CONTROL, ATLANTA, 87- *Concurrent Pos:* Consult, Lexington Clin, 74-76; clin chemist, Vet Admin Hosp, Lexington, 74-76. *Mem:* Am Chem Soc; Am Asn Clin Chemists. *Res:* Trace element pathology; toxicology and clinical chemistry methodology; kidney and liver disease. *Mailing Add:* Pub Health Prac Prog Off Ctr for Dis Control Atlanta GA 30333

BOONE, DONALD MILFORD, b Phoenix, Ariz, Jan 7, 18; m 47; c 3. PLANT PATHOLOGY. *Educ:* Marion Col, Ind, BA & BS, 40; Univ Wis, PhD(plant path), 50. *Prof Exp:* Instr biol, Marion Col, Ind, 40-41; res asst, 46-49, proj assoc, 49-56, from asst prof to assoc prof, 56-69, PROF PLANT PATH, UNIV WIS, MADISON, 69- *Mem:* Am Phytopath Soc; Bot Soc Am; Mycol Soc Am. *Res:* Inheritance and nature of pathogenicity and disease resistance; genetics of microorganisms; diseases of small fruits. *Mailing Add:* 3621 Spring Tr Madison WI 53715

BOONE, GARY M, b Presque Isle, Maine, July 16, 29; m 52; c 2. PETROLOGY, MINERALOGY. *Educ:* Bowdoin Col, AB, 51; Brown Univ, MA, 54; Yale Univ, PhD(geol), 59. *Prof Exp:* Lectr geol, Univ Western Ont, 57-60, asst prof, 60-63; from asst prof to assoc prof, 63-71, PROF GEOL, SYRACUSE UNIV, 71- *Concurrent Pos:* Nat Res Coun Can res grants, 59-62; Off Naval res contract volcanic rocks, Azores, 65-67. *Mem:* Geol Soc Am; Mineral Soc Am; Mineral Soc Gt Brit & Ireland; Geol Asn Can; Sigma Xi. *Res:* Metamorphism and regional stratigraphy in northern Appalachians. *Mailing Add:* Dept Geol Syracuse Univ Syracuse NY 13244

BOONE, JAMES EDWARD, b Hamilton, Ont, July 31, 27; m 58; c 3. PEDIATRICS, PEDIATRIC RHEUMATOLOGY. *Educ:* Univ Toronto, MD, 51; FRCPS(C), 57. *Prof Exp:* From asst prof to assoc prof pediat, Univ Toronto, 64-73; PROF PEDIAT & CHMN DEPT, UNIV WESTERN ONT, 73- *Mem:* Can Pediat Soc; Royal Col Physicians & Surgeons Can. *Res:* Medical manpower; rheumatic disease of children. *Mailing Add:* 800 Commission Rd E London ON N6C 2V5 Can

BOONE, JAMES LIGHTHOLDER, b San Antonio, Tex, Sept 11, 32; c 4. INORGANIC CHEMISTRY. *Educ:* St Louis Univ, BS, 54; Univ Southern Calif, PhD(chem), 60. *Prof Exp:* Res asst, Univ Southern Calif, 54-59; chemist, 59-75, mgr chem econ, US Borax Res Corp, 75-77, MGR ENG ECON, US BORAX & CHEM CO, 77- *Mem:* Am Chem Soc; Sigma Xi. *Res:* Boron hydrides and their derivatives; organo-boron and organometallic compounds; high temperature boron compounds and refractories; engineering economics. *Mailing Add:* Two Sunridge Irvine CA 92714

BOONE, JAMES ROBERT, b Buffalo, NY, Aug 9, 39; m 61; c 2. MATHEMATICS. *Educ:* Tex A&M Univ, BA, 61, MS, 62; Tex Christian Univ, PhD(math), 68. *Prof Exp:* Opers res analyst, Gen Dynamics/Ft Worth, 63-64, aerodyn engr, 64-65; asst prof math, 68-74, ASSOC PROF MATH, TEX A&M UNIV, 74- *Mem:* Am Math Soc; Math Asn Am. *Res:* General topology. *Mailing Add:* Dept Math Tex A&M Univ College Station TX 77843

BOONE, JAMES RONALD, b Brownsville, Tenn, Nov 10, 46; m 70. ORGANIC CHEMISTRY. *Educ:* David Lipscomb Col, BA, 68; Ga Inst Technol, PhD(chem), 74. *Prof Exp:* Instr chem, Ga Inst Technol, 69-70, fel biochem, 74-75; ASST PROF CHEM, DAVID LIPSCOMB COL, 75- *Mem:* Am Chem Soc. *Res:* Stereoselective reductions of ketones by metal hydrides. *Mailing Add:* Dept Chem David Lipscomb Col Nashville TN 37203

BOONE, LAWRENCE RUDOLPH, III, b Memphis, Tenn, Apr 21, 48; m 75; c 1. VIRAL CARCINOGENESIS, RECOMBINANT DNA. *Educ:* Memphis State Univ, BS, 70; Univ Tenn, PhD(microbiol), 77. *Prof Exp:* fel virus res, Roche Inst Molecular Biol, 77-79; res assoc, Oak Ridge Nat Lab, 79-81; sr staff fel, 81-87, MICROBIOLOGIST, NAT INST ENVIRON HEALTH SCI, 87- *Concurrent Pos:* vis investr, Oak Ridge Nat Lab, 81-83. *Mem:* Am Soc Microbiol. *Res:* Use of recombinant DNA and related techniques to study the mechanism of viral and chemical carcinogenesis; development of in vitro model systems for carcinogen detection and genetic toxicology. *Mailing Add:* 101 Treelawney Lane Rte 2 Apex NC 27502

BOONSTRA, BRAM B, b Arnhem, Neth, June 4, 12; m 45; c 2. ELECTROCHEMISTRY. *Educ:* Univ Amsterdam, PhD(electrochem), 38. *Prof Exp:* Asst prof anal chem, Univ Amsterdam, 36-39; teacher high sch, Neth, 39-40; res chemist, Rubber Found, Neth, 40-43; mgr, Tech Rubber & Plastics Lab, 46-51; develop chemist, Royal Holland Cable Works, 43-46; sect head, Godfrey Cabot, Inc, 51-62, assoc dir res, Rubber & Plastics Lab, 62-77; CONSULT, 77- *Concurrent Pos:* Chmn, Gordon Res Conf Elastomers, 67. *Mem:* Am Chem Soc. *Res:* High polymer physical chemistry and technology; carbon black. *Mailing Add:* 1562 Serenity Lane Sanibel FL 33957

BOOP, WARREN CLARK, JR, b Baltimore, Md, July 27, 33; c 3. NEUROSURGERY. *Educ:* Univ Tenn, BS & MD, 56; Univ Minn, MA, 64. *Prof Exp:* Intern, Baroness Erlanger Hosp, Chattanooga, Tenn, 56-57; resident, St Mary's Hosp, Knoxville, 57-58; resident neurosurg, Univ Minn, Minneapolis, 60-64; staff neurosurgeon, US Naval Hosp, Oakland, Calif, 64-66; head neurosurg serv, US Naval Hosp, Great Lakes, Ill, 66-69; ASSOC PROF NEUROSURG, MED CTR, UNIV ARK, LITTLE ROCK, 70- *Concurrent Pos:* US Navy rep, Nat Res Coun Spinal Cord Injuries, 67; vis prof, Wesley Mem Hosp, Chicago, 68; staff neurosurgeon, Vet Admin Hosp, Little Rock, 70-; consult neurosurg, Ark Children's Hosp & Ark Crippled Children's Div, Dept Pub Welfare Hot Springs Rehab Ctr, 70- *Mem:* Cong Neurosurg; Am Asn Neurol Surg; AMA; Neurosurg Soc Am; Am Col Surg; Sigma Xi. *Res:* Circulatory effects of increased intracranial pressure. *Mailing Add:* Dept Neurosurg Slot 507 Univ Ark Med Ctr Little Rock AR 72205

BOORD, ROBERT LENNIS, b Masontown, Pa, July 29, 26; m 53; c 1. VERTEBRATE MORPHOLOGY, NEUROANATOMY. *Educ:* Washington & Jefferson Col, AB, 50; Univ Md, MS, 58, PhD(zool, comp anat), 60. *Prof Exp:* USPHS fel, 60-61; instr zool, Duke Univ, 61-62; asst prof biol sci, 62-69, assoc prof, 69-77, PROF HEALTH SCI, UNIV DEL, 77- *Honors & Awards:* Jacob K Javits Neurosci Investigator Award, 85. *Mem:* AAAS; Am Asn Anat; Am Soc Zoologists; Sigma Xi. *Res:* Comparative anatomy of vertebrate octavolateralis system; comparative neuroanatomy. *Mailing Add:* Sch Life & Health Sci Univ Del Newark DE 19716

BOORMAN, GARY ALEXIS, b Leonard, Minn, June 30, 42; m 65; c 2. RODENT PATHOLOGY, TOXICOLOGY. *Educ:* Univ Minn, BS, 65, DVM, 67; Univ Mich, MS, 70; Univ Calif, Davis, PhD(path), 78; Am Col Vet Path, dipl, 77. *Prof Exp:* Resident, lab animal med, Univ Mich, 68-70; pathologist, rodent path, Inst Exp Geront, Neth, 70-74; resident path, Univ Calif, 74-75; PATHOLOGIST TOXICOL, NAT INST ENVIRON HEALTH SCI, 78- *Concurrent Pos:* Adj asst prof, Sch Med, Univ NC, 79-; path consult, Interagency Regulatory Liaison Group, Chem Indust Inst Toxicol, 80; mem, comt infectious dis rats & mice, Nat Res Coun, Nat Acad Sci, 82- *Mem:* Am Bd Toxicol; Am Soc Toxicol Pathologists; Am Asn Lab Animal Sci; Am Col Vet Path; Am Col Lab Animal Med. *Res:* Spontaneous and induced neoplasia in rats and mice; potential hazard of environmental chemicals testing to rodents; classification of cancer in rats and mice. *Mailing Add:* Nat Toxicol Prog Nat Inst Environ Health Sci PO Box 12233 Res Triangle Park NC 27709

BOORMAN, PHILIP MICHAEL, b Hitchin, Eng, May 11, 39; m 66; c 4. INORGANIC CHEMISTRY. *Educ:* Univ Nottingham, BSc, 61, PhD(inorg chem), 64. *Prof Exp:* Brit Titan res fel inorg chem, Univ Newcastle, 64-67; from asst prof to assoc prof, 67-80, asst dean fac arts & sci, 70-73, head dept, 82-87, PROF CHEM, UNIV CALGARY, 80- *Concurrent Pos:* Hon vis fel, Dept Chem, Univ Manchester, Eng, 73-74. *Mem:* Fel Chem Inst Can; Royal Soc Chem; Am Chem Soc. *Res:* Interaction of molybdenum and tungsten compounds with sulfur donor ligands; development and chemistry of hydrotreating; upgrading of heavy oil; mechanism of hydrodesulfurization process. *Mailing Add:* Univ Calgary Dept Chem Calgary AB T2N 1N4 Can

BOORN, ANDREW WILLIAM, b London, Eng, July 14, 54; m 80; c 2. ATOMIC SPECTROMETRY, QUADROPOLE MASS SPECTROMETRY. *Educ:* Thames Polytech, BSc, 77; Ga Inst Technol, PhD(anal chem), 81. *Prof Exp:* Sci officer, Lab Govt Chemist, London, Eng, 72-77; res & teaching asst chem, Ga Inst Technol, 77-81; applications engr, Beckman Instruments, Subsidiary of Spectrametrics, 81-83; DIR, ANALYTICAL INSTRUMENTS, SCIEX, 83- *Mem:* Royal Soc Chem; Soc Appl Spectros; Am Chem Soc. *Res:* Development of quantitative analytical methods for fourier transform nuclear magnetic resonance (carbon 13 and phosphorus 31); analysis of foods by x-ray fluorescence; studies of sample introduction techniques for atomic spectroscopy; the last topic and development of atomic spectrometers; computer applications in analytical chemistry; ICP-MS instrument development and applications; LC-MS instrument development. *Mailing Add:* SCIEX 55 Glen Cameron Rd Thornhill ON L3T 1P2 Can

BOORNE, RONALD ALBERT, b Ottawa, Ont, Sept 14, 26; c 1. ENGINEERING PHYSICS. *Educ:* Univ Toronto, BASc, 48. *Prof Exp:* Instr, Univ Toronto, 48-49; theoret aerodynamicist, Aeroelastics Sect, Avro Aircraft Ltd, Can, 49-53; dir sch eng, 56-72, dir eng & comput ctr, 66-72, PROF ENG, MT ALLISON UNIV, 53- *Concurrent Pos:* Sr consult, Seca Co Ltd, 67-72 & P A Lapp Ltd, Ottawa, 72-77. *Mem:* Asn Prof Engrs; Eng Inst Can; Can Asn Univ Teachers. *Res:* Aeroelastic vibrations of aircraft wings and other structures; impact of technology on society. *Mailing Add:* Dept Eng Mt Allison Univ Sackville NB E0A 3C0 Can

BOORSE, HENRY ABRAHAM, b Norristown, Pa, Sept 18, 04; m 31; c 2. LOW TEMPERATURE PHYSICS. *Educ:* Columbia Univ, AM, 33, PhD(physics), 35. *Prof Exp:* Asst, Columbia Univ, 28-31, instr, 31-33; Lydig fel, Cambridge Univ, Eng, 34-35; instr physics, City Col New York, 35-37; from asst prof, 37-70, head dept, 37-43, actg dean fac, 57, dean fac, 59-70, actg pres, 62 & 67, spec lectr & asst to pres, 70-74, EMER PROF PHYSICS & EMER DEAN, BARNARD COL, COLUMBIA UNIV, 70- *Concurrent Pos:* Adams fel, Columbia Univ, 38-40, res physicist, SAM Labs, Manhattan Dist, 41-45, div dir, 45-46; consult, US AEC, 46-58 & Brookhaven Nat Lab, 51-55; chmn, Calorimetry Conf, 56-57; mem, Comn I, Int Inst Refrig & vchmn subcomt basic sci, US Nat Comt, 58-72; chmn, Tenth Int Conf on low Tempertue Physics, USSR Acad Sci, 66. *Mem:* Fel Am Phys Soc. *Res:* Low temperature physics; liquefaction of gases; superconductivity; properties of liquid helium. *Mailing Add:* 338 Summit Ave Leonia NJ 07605

BOORSTYN, ROBERT ROY, b New York, NY, May 28, 37; m 59; c 3. ELECTRICAL ENGINEERING. *Educ:* City Col New York, BEE, 58; Polytech Inst Brooklyn, MS, 61, PhD(elec eng), 66. *Prof Exp:* Engr, Sperry Gyroscope Co, 58-61; from instr to assoc prof, 61-75, PROF ELEC ENG, POLYTECH UNIV, 75- *Concurrent Pos:* Consult. *Mem:* Fel Inst Elec & Electronic Engrs. *Res:* Telecommunication networks. *Mailing Add:* Dept Elec Eng Polytech Univ 333 Jay St Brooklyn NY 11201

BOOS, DENNIS DALE, b Pensacola, Fla, Apr 27, 48; m 74; c 2. MATHEMATICAL STATISTICS. *Educ:* Fla State Univ, BS, 70, MS, 75, PhD(statist), 77. *Prof Exp:* PROF STATIST, NC STATE UNIV, 77- *Mem:* Inst Math Statist; Am Statist Asn. *Res:* Robust estimation; goodness of fit; asymptotic distribution theory. *Mailing Add:* Dept Statist NC State Univ PO Box 8203 Raleigh NC 27695-8203

BOOSALIS, MICHAEL GUS, b Faribault, Minn, Sept 20, 17; m 46; c 1. PLANT PATHOLOGY. *Educ:* Univ Minn, BS, 41, MS, 48, PhD(plant path), 51. *Prof Exp:* From asst prof to prof plant path, Agr Exp Sta, 51-89, chmn dept, 64-89, prof bot, 70-89, EMER PROF PLANT PATH, INST AGR & NATURAL RESOURCES, UNIV NEBR, LINCOLN, 89- *Mem:* Am Phytopath Soc. *Res:* Ecology of fungi, causing root diseases of plants. *Mailing Add:* Dept Plant Path Inst Agr & Natural Resources Univ Nebr 4006 Plant Sci Hall Lincoln NE 68583-0722

BOOSER, E(ARL) R(ICHARD), b Harrisburg, Pa, Jan 7, 22; m 44; c 2. CHEMICAL ENGINEERING. *Educ:* Pa State Col, BS, 42, MS, 44, PhD(chem eng), 49. *Prof Exp:* Instr chem eng, Pa State Col, 43, res asst petrol ref, 44-48; engr, Thomson Lab, 48-55, atomic power equip dept, 55-56, bearing & lubrication engr, Motor Eng Lab, 56-60, sr engr, Electromech Eng, 60-86, CONSULT ENGR, GEN ELEC CO, 87- *Concurrent Pos:* Independent Consult Engr, 87- *Honors & Awards:* Nat Serv Award, Soc Tribologists and Lubrication Engrs, 83. *Mem:* Am Chem Soc; Soc Tribologists & Lubrication Engrs (pres, 56-57); Am Soc Mech Engrs. *Res:* Petroleum refining; analysis of petroleum products; bearings and lubricants; hydraulics; atomic power equipment. *Mailing Add:* 65 St Stephens Lane Scotia NY 12302

BOOSMAN, JAAP WIM, b Amsterdam, Holland, Mar 2, 35; US citizen; m 61. SCIENCE ADMINISTRATION, SEDIMENTOLOGY. *Educ:* Syracuse Univ, BS, 57, MS, 59; George Washington Univ, MPhil, 70, PhD(geol), 73. *Prof Exp:* Arctic proj officer, US Off Naval Res, 62-65; oceanogr, Off Oceanog, US Navy, 65-75; PHYS SCI ADMINR SCI & TECHNOL PLANNING, OFF CHIEF NAVAL OPERS, 75- *Concurrent Pos:* Spec Duty Officer oceanog, Captain, US Naval Reserve, 81-; comndg officer, Naval Oceanog Reserve Activ, Washington, DC, 87- *Mem:* Sigma Xi. *Res:* Clay mineral distribution and source, provenance and transport of sediments on the Argentine continental shelf; relict morphology of the Argentine continental shelf and evidence for sea level changes. *Mailing Add:* 6947 33rd St NW Washington DC 20015

BOOSS, JOHN, b New York, NY, Apr 3, 38; m 72; c 2. NEUROIMMUNOLOGY,NEUROVIROLOGY. *Educ:* Oberlin Col, AB, 61; Albany Med Col, MD, 65. *Prof Exp:* Intern med, Cleveland Clin Hosp, 65-66; resident neurol, NY Univ-Bellevue Med Ctr, 66-69; teaching asst virol, Yale Univ, Sch Med, 69-71; neurologist, Sch Aerospace Med, US Air Force, 71-73; teaching asst virol & immunol, Jefferson Med Col, 73-75; asst prof, 76-81, ASSOC PROF NEUROL & LAB MED, 81-, ASSOC DIR NEUROIMMUNOLOGY CLIN, YALE UNIV SCH MED, NEW HAVEN, CONN, 90-; NEUROVIROLOGIST, VET ADMIN MED CTR, WEST HAVEN, CONN, 76- *Honors & Awards:* Sr Scientist Fel, French Ministry Res & Technol, 90. *Mem:* Am Neurol Asn; Am Acad Microbiol; Am Soc Virol; Am Asn Immunol; Am Asn Neuropathologists. *Res:* Host-viral interaction in the central nervous system; modulation of the immune response by viral infection; organization of the host response in the central nervous system; immunology of brain grafting. *Mailing Add:* Neurol Serv Vet Admin Med Ctr West Haven CT 06516

BOOSTER, DEAN EMERSON, b Salem, Ore, Sept 20, 26; m; c 5. AGRICULTURAL ENGINEERING. *Educ:* Ore State Univ, BS, 54, MS, 56. *Prof Exp:* From instr to prof, 56-87, EMER PROF AGR ENG, ORE STATE UNIV, 87- *Mem:* Am Soc Agr Engrs. *Res:* Mechanical harvesting and handling of horticultural crops; electrostatic separation of field crop seeds; development of range land seeding equipment; nursery and greenhouse mechanization; conservation tillage in vegetable crop production; mechanical harvesting of oil seed crops. *Mailing Add:* 1544 NW 12th St Corvallis OR 97330

BOOTE, KENNETH JAY, b Hull, Iowa, Sept 12, 45; m 69; c 1. CROP PHYSIOLOGY. *Educ:* Iowa State Univ, BS, 67; Purdue Univ, MS, 69, PhD(crop physiol), 74. *Prof Exp:* From asst prof to assoc prof, 74-85, PROF CROP PHYSIOL, UNIV FLA, 85- *Mem:* Am Soc Agron; Crop Sci Soc Am; Am Peanut Res & Educ Asn; Am Soc Plant Physiol; Fel, Am Soc Agron, 90; Fel, Crop Sci Soc Am, 90. *Res:* Relationships of physiological traits to productivity of peanut and soybean genotypes under varying environments; water relations and pest effects on photosynthesis and evapotranspiration of crop canopies; double cropping; soybean crop model development. *Mailing Add:* Dept Agron Univ Fla Gainesville FL 32611

BOOTH, AMANDA JANE, b Banbury, Eng, Mar 22, 62; Can citizen. INTERNAL MEDICINE. *Educ:* Lakehead Univ, BSc, 79; Univ Sask, DVM, 83, MVet Sc, 86; Am Col Vet Internal Med, dipl, 90. *Prof Exp:* Intern large animal med & surg, Univ Pa, 83-84; resident internal med, Western Col Vet Med, 84-86; dir vet serv, Mercury Farm Technol Ctr, 86-89; OWNER & VETERINARIAN, SASCENOS VET SERV LTD, 89- *Mem:* Am Col Vet Internal Med; Can Vet Med Asn. *Res:* Investigation of correction of metabolic acidosis by oral fluid therapy in diarrheic neonatal calves; investigation of efficacy of desensitization therapy for horses with chronic obstructive pulmonary disease. *Mailing Add:* 5317 Sooke Rd RR 1 Sooke BC V0S 1N0 Can

BOOTH, ANDREW DONALD, b East Molesey, Eng, Feb 11, 18; m 50; c 2. PHYSICS, ELECTRICAL ENGINEERING. *Educ:* Univ London, BSc, 40, DSc(physics), 51; Univ Birmingham, PhD(chem), 44. *Prof Exp:* Res physicist, Brit Rubber Producers' Res Asn, 44-46; lectr physics, Birkbeck Col, Univ London, 46-49; prof theoret physics, Univ Pittsburgh, 49-50; dir comput proj, Univ London, 50-54, reader comput methods, Birkbeck Col, 54-57, head, dept comput sci, 57-62; prof elec eng & head dept, Univ Sask, 62-63, dean eng, 63-77; pres & hon prof physics, Lakehead Univ, 72-78; RETIRED. *Concurrent Pos:* Nuffield fel, 46-47; Rockefeller fel & mem, Inst Advan Study, 47-48; consult physicist, Brit Rayon Res Asn, 48-55; sci adv, Int Computs & Tabulators, Ltd, 50-64; European sci adv, Am Mach & Foundry Co, 56-62; sci adv, Inst Aviation Med, 57- & H M Underwater Defense Estab, 59-63; ed, J Royal Micros Soc, 61-65; prof-at-large, Case Western Reserve Univ, 63-; mem, Can proc, Brit Inst Electronics & Radio Engrs, 66-; mem, Nat Res Coun Can, 75-; chmn bd, Autonetics Res Assocs, Victoria, BC, 78-; hon scientist & instr ocean sci, Victoria, BC, 79-; chmn, Vancouver Island Br, Eng Inst Can, 86-, pres, Scholar Found, 87- *Honors & Awards:* Centennial Medal, Can, 67. *Mem:* Fel Inst Physics; hon fel Inst Ling; fel Brit Inst Elec Engrs; Brit Comput Soc; Royal Micros Soc. *Res:* X-ray crystallography; computer design and application; biomedical engineering; machine linguistics; microscopy. *Mailing Add:* Timberlane 5317 Sooke Rd RR1 Sooke BC V0S 1N0 Can

BOOTH, BEATRICE CROSBY, b Minneapolis, Minn, Aug 29, 38; m 60; c 3. PHYTOPLANKTON ECOLOGY, PHYTOPLANKTON TAXONOMY. *Educ:* Radcliffe Col, BA, 60; Harvard Sch Educ, MAT, 62; Univ Wash, MS, 69. *Prof Exp:* Teaching asst oceanog, Univ Wash, 74-75; instr marine plankton, Univ Wash, 75-78; teaching asst oceanog, 74-75, instr marine Plankton, 75-78, res oceanogr, 78-80, sr oceanogr, 80-83, PRIN OCEANOGR, UNIV WASH, 83- *Concurrent Pos:* Oceanogr, Pac Marine Environ Lab, Nat Oceanic & Atmospheric Admin, 78-79. *Mem:* Phycol Soc Am; Am Soc Limnol & Oceanog; AAAS. *Res:* Quantitative investigation, including taxonomy and ultrastructure, of nanoplankton species from the eastern subarctic pacific using electron and fluorescence microscopy on cultures and field collections; differentiation of heterotrophic from autotrophic component. *Mailing Add:* 5521 17th Ave NE Seattle WA 98105

BOOTH, BRUCE L, b Philadelphia, Pa, Mar 22, 38; m 68; c 4. SOLID STATE PHYSICS, LASERS & INTEGRATED OPTICS. *Educ:* Dartmouth Col, BA, 60; Northwestern Univ, PhD(physics), 67. *Prof Exp:* From res physicist to sr res physicist, E I du Pont de Nemours & Co, Inc, 67-73, res supvr, 73-74, sr supvr, 74-79, res mgr, 79-84, prin consult, 84-85, sr assoc, 85-89, RES FEL, E I DU PONT DE NEMOURS & CO, INC, 89. *Mem:* Am Phys Soc; Optical Soc Am; Sigma Xi. *Res:* Low temperature band structure studies using de Haas-Shubnikov effect in semiconductors, mainly gray tin; supervise development of techniques and applications of laser-optical measurements for industry; manage research on medical and clinical instrumentation systems and test methods; integrated optics using polymers. *Mailing Add:* RD No 4 Warpath Rd West Chester PA 19382

BOOTH, CHARLES E, b Cleveland, Ohio, Jan 9, 52; m 85. ANIMAL PHYSIOLOGY. *Educ:* Col Wooster, BA, 74; Col William & Mary, MA, 77; Univ Calgary, PhD(biol), 83. *Prof Exp:* Postdoctoral fel, McMaster Univ, 82-84; asst prof, 84-89, ASSOC PROF BIOL, EASTERN CONN STATE UNIV, 89- *Concurrent Pos:* Vis investr, Huntsman Marine Lab, 83-, Fish Physiol & Toxicol Lab, Univ Wyo, 85- & Rosenstiel Sch Marine & Atmospheric Sci, Univ Miami, 87-; vis fac, Univ Miami, 87- *Mem:* AAAS; Am Soc Zoologists. *Res:* Comparative respiratory, acid-base, and ionoregulatory physiology of aquatic animals; avian hematology. *Mailing Add:* Dept Biol Eastern Conn State Univ Willimantic CT 06226

BOOTH, DAVID LAYTON, b Aurora, Ill, July 20, 39; m 62; c 2. MARKETING, PRODUCTION. *Educ:* Beloit Col, BS, 61; Univ Ore, PhD(org chem), 65. *Prof Exp:* Res chemist, Org Div, Morton Chem Co, 65-69, supvr org res, 69-74, asst dir new prod develop, Morton Int, Inc, 74-79, res dir new prod develop, Morton-Norwich Prod, Inc, 79-83, dir res, 83-89, VPRES, DYES & ORG SPECIALTIES, MORTON INT, INC, 89- *Mem:* Am Chem Soc; Plant Growth Regulator Working Group; Coun Agr Sci & Technol; Soc Appln Mat & Process Eng; Soc Plastics Indust; Synthetic Org Chem Mfrs Asn. *Res:* Alkaloid chemistry; synthetic studies in pesticide research; liquid polysulfide polymers; sulfur chemistry; sulfur polymers; epoxy and polyester polymers. *Mailing Add:* 673 Elsinoor Lane Crystal Lake IL 60014

BOOTH, DONALD CAMPBELL, b Syracuse, NY, May 13, 50; m 86; c 2. ORNAMENTAL PLANT ENTOMOLOGY, INTEGRATED PEST MANAGEMENT. *Educ:* SUNY Col Environ Sci & Forestry, BS, 73, PhD(entomol), 79. *Prof Exp:* Assoc entomologist, United Brands Co, 78-81; res assoc, Univ Ga, 82-83; ENTOMOLOGIST, BARTLETT TREE RES LAB, 83- *Mem:* Entomol Soc Am. *Res:* Product development, pesticide efficacy trials and applications of integrated pest management techniques; pheromone traps, insecticidal soap, spray oils, pyrethoi. *Mailing Add:* Bartlett Tree Res Labs 13768 Hamilton Rd Charlotte NC 28278

BOOTH, FRANK, b July 3, 43. MOLECULAR BIOLOGY, EXERCISE PHYSIOLOGY. *Educ:* Univ Iowa, PhD(exercise physiol), 70. *Prof Exp:* PROF PHYSIOL, UNIV TEX, 75- *Res:* Muscle inactivity; exercise mechanisms. *Mailing Add:* Dept Physiol & Cell Biol Univ Tex Med Sch PO Box 20708 Houston TX 77225

BOOTH, GARY EDWIN, b Campton, Ky, Dec 26, 40; m 62; c 2. ORGANIC CHEMISTRY. *Educ:* Eastern Ky State Col, BS, 62; Ohio State Univ, PhD(org chem), 65. *Prof Exp:* Res chemist, 65-69, SECT HEAD PHYS ORG CHEM, MIAMI VALLEY LABS, PROCTER & GAMBLE CO, 69- *Mem:* Am Chem Soc. *Res:* Conformational analysis; fluorescence of organic molecules; adsorption of dyestuffs onto cellulose; thin-layer chromatography. *Mailing Add:* 27 Weybridge Park Weybridge Park Surrey KT13 8SQ England

BOOTH, GARY MELVON, b Provo, Utah, Oct 9, 40; m 60; c 6. ENTOMOLOGY. *Educ:* Utah State Univ, BS, 63, MS, 66; Univ Calif, Riverside, PhD(entom), 69. *Prof Exp:* NIH fel, Univ Ill, 69-70, asst prof entom, 70-71, asst prof agr entom, 71-72; asst prof zool, 72-73, ASSOC PROF ZOOL, BRIGHAM YOUNG UNIV, 73- *Concurrent Pos:* Consult, Environ Protection Agency, 71; asst entomologist, Ill Nat Hist Surv, 71-72; consult, Thompson-Hayward Chem Co, 72-76; assoc ed environ entom, Entom Soc Am, 75. *Honors & Awards:* First Place Res Award, Sigma Xi, 73. *Mem:* Sigma Xi; Entom Soc Am; Weed Sci Soc Am. *Res:* Metabolism toxicity and environmental behavior of insecticides, herbicides, fungicides and rodenticides in model ecosystems; fish, insects, water and soil; development of pest management research programs. *Mailing Add:* Dept Zool 575 Widstoe Hall Brigham Young Univ Provo UT 84601

BOOTH, IAN JEREMY, b Banbury, Eng, Mar 3, 60; Can citizen. ACOUSTIC OCEANOGRAPHY, LOW TEMPERATURE SOLID STATE PHYSICS. *Educ:* Lakehead Univ, BSc, 77, BSc Hons, 78, MSc, 90; Univ BC, PhD(physics), 85. *Prof Exp:* Postdoctoral fel physics, Simon Fraser Univ, 85-87; DIR PHYSICS, AUTONETICS RES ASSOC, 79- *Concurrent Pos:* Consult, Autonetics Res Assoc, 89-; lectr, Univ Victoria, 90- *Mem:* Can Asn Physicists; Am Phys Soc; NY Acad Sci; Inst Elec Engrs UK. *Res:* Photoluminescence and far infrared transmission studies of various semiconductors at low temperature; investigation of anomolous thermionic emission effects; computer simulation and analysis of underwater sound propagation for research and communication. *Mailing Add:* 5317 Sooke Rd RR No 1 Sooke BC V0S 1N0 Can

BOOTH, JAMES SAMUEL, b LeFlore, Okla, Nov 10, 27; m 62; c 1. BACTERIAL PHYSIOLOGY. *Educ:* Calif State Col, BS, 59; Univ Southern Calif, MS, 62, PhD(staphylocoagulase), 68. *Prof Exp:* Lectr microbiol, Calif State Col, Los Angeles, 63-65; asst prof bact physiol, Univ NMex, 67-74; asst prof biol sci, 74-76, ASSOC PROF BIOL SCI, CALIF POLYTECH STATE UNIV, SAN LUIS OBISPO, 76- *Mem:* Am Soc Microbiol; Sigma Xi. *Res:* Degradation of ergothioneine by alcaligenes faecalis; factors influencing in vitro levels of staphylocoagulase. *Mailing Add:* Rte 2 Box 340 Wilburton OK 74578

BOOTH, JOHN AUSTIN, b DuBois, Pa, Jan 27, 29; m 48; c 1. PLANT PATHOLOGY. *Educ:* Univ Ariz, BS, 51, MS, 58, PhD(plant path), 63. *Prof Exp:* Investr air pollution, Phelps Dodge Corp, Ariz, 51-57; asst prof, 63-69, assoc prof, 69-76, PROF PLANT PATH, NMEX STATE UNIV, 76- *Concurrent Pos:* Consult sulfur oxide injury to plants. *Mem:* Sigma Xi; Air Pollution Control Asn; Am Phytopath Soc. *Res:* Disease physiology; grape diseases; air pollution. *Mailing Add:* Dept Entom & Plant Path NMex State Univ Col Agr University Park NM 88003

BOOTH, LAWRENCE A(SHBY), b Louisville, Ky, Apr 10, 34; m 57; c 3. NUCLEAR ENGINEERING, PHYSICS. *Educ:* Purdue Univ, BS, 55; Univ Louisville, MChE, 56, PhD(chem eng), 60. *Prof Exp:* Asst, Univ Louisville, 56-58; chem engr, Rocket Propulsion Unit, Holloman AFB, 58-60; mem staff, 60-68, asst group leader, 68-70, staff mem, 70-74, GROUP LEADER, LOS ALAMOS SCI LAB, 74- *Concurrent Pos:* Consult, Div Licensing & Regulations, AEC, 67-72. *Mem:* Am Chem Soc; Am Nuclear Soc; fel Am Inst Chemists; Sigma Xi. *Res:* Heat transfer; fluid dynamics; nuclear equipment and materials; gas cooled reactor technology; nuclear propulsion; laser technology; applications of laser fusion. *Mailing Add:* 115 Monte Rey Dr N Los Alamos NM 87544

BOOTH, NEWELL ORMOND, b Pocatello, Idaho, Nov 23, 40; m 65; c 1. UNDERWATER ACOUSTICS, OCEAN ENGINEERING. *Educ:* Univ Calif, Berkeley, BA, 62; Univ Calif, Los Angeles, MS, 66, PhD(physics), 71. *Prof Exp:* Physicist optics, Naval Ord Test Sta, 63-65, RES PHYSICIST UNDERWATER ACOUST, NAVAL OCEAN SYSTS CTR, 71- *Mem:* Acoust Soc Am. *Res:* Adaptive matched-field array processing. *Mailing Add:* 4643 El Cerrito Dr San Diego CA 92115

BOOTH, NICHOLAS HENRY, b Hannibal, Mo, Oct 22, 23; m 44; c 3. TOXICOLOGY, PHARMACOLOGY. *Educ:* Mich State Univ, DVM, 47; Colo State Univ, MS, 51; Univ Colo, PhD, 59. *Prof Exp:* Pvt pract vet med, Hannibal, Mo, 47-48; from asst prof to prof physiol, Colo State Univ, 48-66, head dept, 56-66, dean col vet med & biomed sci, 66-71; dir div vet med res, Bur Vet Med, Food & Drug Admin, 71-74; prof physiol & pharmacol, Univ Ga, 74-85, EMER PROF, 85- *Concurrent Pos:* Chmn vet drug panel, Nat Res Coun-Nat Acad Sci, 66-68; consult, Div Physician Manpower, Bur Health Manpower, Dept Health, Educ & Welfare, 67-68; mem vet med rev comt, Bur Health Physicians Educ & Manpower Training, Dept Health, Educ & Welfare, 69-; mem spec comt vet med educ, Southern Regional Educ Bd, 70. *Mem:* AAAS; Am Physiol Soc; Am Vet Med Asn; Am Heart Asn; Soc Exp Biol & Med; Soc Toxicol. *Res:* Pharmacodynamic studies on d-tubo curarine chloride and succinylcholine chloride in the horse; meperidine studies in the cat; artificial respiration in large animals; electrical defibrillation of the dog heart; electro and phonocardiography studies in the lamb; cardiovascular studies in swine; veterinary pharmacology and toxicology; drug residue studies in food-producing animals; contributing author and co-editor of one publication. *Mailing Add:* 3641 Newcastle Creek Dr Jacksonville FL 32211-2662

BOOTH, NORMAN E, b Toronto, Ont, Mar 7, 30; m 58; c 4. ELEMENTARY PARTICLE PHYSICS. *Educ:* Univ Toronto, BA, 52; Queen's Univ, Ont, MA, 54; Univ Birmingham, PhD(physics), 57. *Prof Exp:* Res fel physics, Univ Birmingham, 57-58; res physicist, Lawrence Radiation Lab, Univ Calif, 59-62; from asst prof to assoc prof physics, Univ Chicago, 62-70; sr res officer, 70-76, LECTR UNIV OXFORD, 76- *Mem:* Fel Am Phys Soc; fel Inst Physics. *Res:* High energy physics; low temperature detectors. *Mailing Add:* Nuclear Physics Lab Univ Oxford Keble Rd Oxford OX1 3RH England

BOOTH, RAY S, b Miami, Fla, Oct 5, 38; m 62; c 3. NUCLEAR ENGINEERING. *Educ:* Univ Fla, BEE, 61, MSNE, 62, PhD(nuclear eng), 65. *Prof Exp:* Asst nuclear eng sci, Univ Fla, 61-65; asst prof physics, US Air Force Inst Technol, 65-68; head, Develop Group, Instrumentation & Controls Div, Oak Ridge Nat Lab, 68-76, head, Develop Group, Controls Sect, 76-81, prog mgr, Clinch River Breeder Reactor Project Off, 81-83, proj mgt, Nuclear Progs, Instrumentation & Controls Div, 83-84, TECH DIR, LIGHT WATER REACTOR PROG, GEN TECHNOL PROG & LIQUID METAL REACTOR PROG, OAK RIDGE NAT LAB, 84- *Mem:* Fel Am Nuclear Soc. *Res:* Neutron wave propagation in nuclear systems; neutron noise analysis in power reactors; nuclear reactors analysis. *Mailing Add:* Oak Ridge Nat Lab Bldg 9102-1 MS 8038 PO Box 2009 Oak Ridge TN 37831-8038

BOOTH, RAYMOND GEORGE, b Utica, NY, Aug 17, 60. NEUROBIOCHEMISTRY, DOPAMINE METABOLISM. *Educ:* Northeastern Univ, BS, 83; Univ Calif, San Francisco, PhD(pharm chem), 89. *Prof Exp:* NIH postdoctoral fel neuro biochem, Harvard Med Sch, 88-90; instr med chem, Northeast Univ, 90; ASST PROF MED CHEM, UNIV NC, CHAPEL HILL, 90- *Concurrent Pos:* Prin investr, Univ NC, Chapel Hill, 90- *Mem:* Soc Neurosci; Am Chem Soc; Am Pharm Asn. *Res:* Research concerning central dopaminergic systems as they relate to Parkinson's disease and schizophernia; regulation of central dopamine synthesis, release, metabolism and receptor-mediated modulators; neurotoxicology of central dopaminergic systems. *Mailing Add:* Sch Pharm CB 7360 Univ NC Chapel Hill NC 27599

BOOTH, RICHARD W, b Cincinnati, Ohio, Mar 17, 24; m 45; c 3. MEDICINE. *Educ:* Univ Cincinnati, MD, 52; Am Bd Internal Med, dipl. *Prof Exp:* Donald L Mahanna res fel, Ohio State Univ Hosp, 56-59; asst prof med & instr aviation & prev med, Ohio State Univ, 59-61; assoc prof, 61-64, PROF MED, SCH MED, CREIGHTON UNIV, 64-, DIR MED SERV, 71- *Concurrent Pos:* Consult, Vet Admin Hosp, Dayton, Ohio, 56-61 & Wright Patterson AFB, 60-61; attend physician, Ohio State Univ, 59-61 & Vet Admin Hosp, Nebr, 63-64; fel coun clin cardiol, Am Heart Asn; med dir & chief staff, St Joseph Hosp, 71-; assoc dean, Sch Med, Creighton Univ, 72-79. *Mem:* Am Fedn Clin Res; Am Col Cardiol; Am Col Angiol; fel Am Col Physicians; Am Heart Asn. *Res:* Cardiovascular diseases. *Mailing Add:* St Joseph Hosp 601 N 30th St Omaha NE 68131

BOOTH, ROBERT EDWIN, b New York, NY, Mar 28, 21; m 42; c 3. ORGANIC CHEMISTRY. *Educ:* NY State Col Forestry, Syracuse Univ, BS, 42, PhD, 50. *Prof Exp:* Asst instr org chem, Syracuse Univ, 49-50; res chemist chem & radiation labs, Chem Corps, US Army, 51-52; res chemist, Solvay Process Div, Allied Chem Corp, 55-68, sr res chemist, Indust Chem Div, 68-84; RETIRED. *Mem:* Am Chem Soc; The Chem Soc. *Res:* Plastics and polymers; vinyl ethers and chloride; polyesters; polyethers; phenolics; inorganic fluorine chemistry; polyurethanes; isocyanates. *Mailing Add:* 4909 South Ave Syracuse NY 13215

BOOTH, SHELDON JAMES, b Marshalltown, Iowa, Feb 21, 45; m 76; c 4. MICROBIOLOGY. *Educ:* Univ Iowa, BA, 68; Univ Nebr, PhD(microbiol), 75. *Prof Exp:* Biol sci asst microbiol, US Army, Dugway, Utah, 68-70; res assoc microbiol, Anaerobe Lab, Va Polytech Inst, 75-77; PROF, DEPT PATHOL & MICROBIOL, UNIV NEBR MED CTR, 77- *Concurrent Pos:* Fel, Va Polytech Inst Anaerobe Lab, 75. *Mem:* Am Soc Microbiol; AAAS; Sigma Xi. *Res:* Ultrastructure, genetics, and biochemistry of bacteriophage and bacteriocins of anaerobic bacteria; bacteriophage and bacteriocin typing of anaerobic bacteria; pathogenicity, ultrastructure, genetics and biochemistry of anaerobic bacteria; water pollution. *Mailing Add:* Dept Path & Microbiol Univ Nebr Med Ctr Omaha NE 68198-6495

BOOTH, TAYLOR LOCKWOOD, computer science, software engineering; deceased, see previous edition for last biography

BOOTH, THOMAS EDWARD, b New York, NY, April 3, 52. MONTE CARLO VARIANCE, REDUCTION TECHNIQUES. *Educ:* Univ Calif Berkeley, BS, 73, MS, 74, PhD(nuclear eng), 78. *Prof Exp:* STAFF MEM, LOS ALAMOS NAT LAB, 74- *Mem:* Am Nuclear Soc. *Res:* Adaptive Monte Carlo techniques for particle transport calculations to achieve an exponential convergence rate rather than one half N by using sophisticated learning techniques. *Mailing Add:* 3472 Questa Dr Los Alamos NM 87544

BOOTHBY, WILLIAM MUNGER, b Detroit, Mich, Apr 1, 18; m 47; c 3. MATHEMATICS, SYSTEMS SCIENCE. *Educ:* Univ Mich, PhD, 49. *Prof Exp:* Swiss-Am Found for Sci Exchange fel, Swiss Fed Inst Technol, 50-51; from instr to asst prof math, Northwestern Univ, 48-59; assoc prof, 59-62, PROF MATH, WASH UNIV, 62- *Concurrent Pos:* NSF sr fel, Inst Advan Study, Princeton, NJ, 61-62; vis mem, Inst Math, Univ Geneva, 65-66; vis prof, Univ Strasbourg, 71-77; Univ Cordoba, Argentina, 81; vis researcher, Univ Rome II, 87. *Mem:* Math Asn Am; Am Math Soc; London Math Soc; Soc Indust Appl Math. *Res:* Differential geometry and topology; geometric control theory. *Mailing Add:* Dept Math Wash Univ St Louis MO 63130

BOOTHE, JAMES HOWARD, b Okanogan, Wash, Nov 2, 16; c 2. ORGANIC CHEMISTRY. *Educ:* State Col Wash, BS, 39; Univ Minn, PhD(pharmaceut chem), 43. *Prof Exp:* res chemist, Lederle Labs, 43-83; RETIRED. *Honors & Awards:* Distinguished Serv Chem Award, Am Chem Soc. *Mem:* Am Chem Soc; NY Acad Sci. *Res:* Chemistry of pteridines; synthesis of folic acid and analogues; syntheses of thiobarbiturates; chemical modification of antibiotics to improve biological properties; chemistry and total synthesis of tetracyclines; chemistry of antibacterial substances. *Mailing Add:* 418 E Longport PO Box 1004 Ocean Gate NJ 08740

BOOTHE, RONALD G, b Havre, Mont, May 25, 47; m; c 4. VISUAL PERCEPTION, VISUAL DEVELOPMENT. *Educ:* Concordia Col, BA, 68; Univ Wash, PhD(psychol), 74. *Prof Exp:* Sr fel, Univ Wash Sch Med, 74-77; res assoc, Univ Wash, 77-79, res asst prof, 79-82, res asst prof psychol, 82-84; ASSOC PROF PSYCHOL & ASST PROF OPHTHAL, EMORY UNIV, 84- *Concurrent Pos:* Adj res prof ophthal, Univ Wash, 80-84; res affiliate, Child Develop & Mental Retardation Ctr, 77-84 & Regional Primate Res Ctr, 78-87; prin investr, NIH res grants, 78-; res assoc prof, Yerks Primate Ctr, Emory Univ, 84- *Mem:* AAAS; Asn Res Vision & Ophthal; Soc Neurosci. *Res:* Behavioral and neuroanatomical research regarding normal and abnormal development of vision in primates. *Mailing Add:* Dept Psychol Emory Univ Atlanta GA 30322

BOOTHROYD, CARL WILLIAM, b Woodsville, NH, Jan 15, 15; m 41, 82; c 2. PLANT PATHOLOGY. *Educ:* Dartmouth Col, AB, 38; State Col Wash, MS, 41; Cornell Univ, PhD, 50. *Prof Exp:* From asst prof to assoc prof, 49-57, assoc head dept, 63-69, prof plant path, Col Agr & Life Sci, 57-80, EMER PROF PLANT PATH, CORNELL UNIV, 80- *Concurrent Pos:* Guggenheim fel, 63. *Res:* Diseases of corn. *Mailing Add:* 350 Plant Sci Cornell Univ Ithaca NY 14850

BOOTHROYD, ERIC ROGER, cytology, for more information see previous edition

BOOTHROYD, GEOFFREY, b Lancashire, Eng, Nov 18, 32; m 54; c 2. MECHANICAL ENGINEERING. *Educ:* Univ London, BSc, 57, PhD(mech eng), 63, DSc, 74. *Prof Exp:* Designer centrifugal pumps, Mather & Platt Ltd, Eng, 56-57; design engr, Eng Elec Co Ltd, 57-58; lectr mech eng, Royal Col Adv Tech, 58-60, sr lectr, 60-64; vis assoc prof, Ga Inst Tech, 64-65; sr lectr, Univ Salford, 65-66, reader, 66-67; prof mech eng, Univ Mass, Amherst, 67-85; PROF INDUST & MFG ENG, UNIV RI, KINGSTON, 85-

Honors & Awards: Nat Medal of Technol, 91. *Mem:* Soc Mfg Engrs; Int Inst Prod Eng Res; Nat Acad Eng. *Res:* Production engineering; manufacturing research including metal machining and mechanized assembly and design for ease of assembly. *Mailing Add:* Dept Indust & Mfg Eng Univ RI Kingston RI 02879

BOOTON, RICHARD C(RITTENDEN), JR, b Dallas, Tex, July 26, 26; m 73. ELECTRICAL ENGINEERING. *Educ:* Agr & Mech Col, Tex, BS & MS, 48; Mass Inst Technol, ScD(elec eng), 52. *Prof Exp:* Asst elec eng, Mass Inst Technol, 48-51, mem staff, Div Indust Coop, 52-54, lectr, 53-54, asst prof elec eng, 54-57; with co, 57-81, mgr electronic info systs opers, TRW Space Technol Labs, 66-78, spec asst to gen mgr, Electronic Systs Div, 78-80, chief scientist, 80-81, CHIEF SCIENTIST, ELECTRONIC SYSTS GROUP, TRW MILCOM, 81- *Honors & Awards:* Thompson Mem Prize, Inst Radio Engrs, 53. *Mem:* Am Inst Aeronaut & Astronaut; assoc fel Inst Elec & Electronics Engrs; Sigma Xi. *Res:* Optimization and analysis of time-varying and nonlinear systems; analogue computation; random processes. *Mailing Add:* 2983 Foothills Ranch Rd Boulder CO 80302

BOOTS, MARVIN ROBERT, b St Louis, Mo, Jan 29, 37. MEDICINAL CHEMISTRY. *Educ:* St Louis Col Pharm, BS, 58; Univ Wis, MS, 60; Univ Kans, PhD(pharmaceut chem), 63. *Prof Exp:* Asst prof pharmaceut chem, Univ Miss, 63-64; res org chemist, Gulf Res & Develop Co, 64-66; asst prof, 66-71, ASSOC PROF MED CHEM, MED COL VA, VA COMMONWEALTH UNIV, 71- *Honors & Awards:* Richardson Award, 63. *Mem:* Am Chem Soc; Am Pharmaceut Asn; Asn Am Univ Professors. *Res:* Design and synthesis of potential therapeutic agents. *Mailing Add:* Dept Med Chem Va Commonwealth Univ Box 581 Richmond VA 23298-0001

BOOTS, SHARON G, b Grand Rapids, Mich, May 5, 39; div; c 1. ORGANIC CHEMISTRY. *Educ:* Univ Wis, BA, 60; Stanford Univ, PhD(org chem), 64. *Prof Exp:* NIH fel org chem, Mass Inst Technol, 64; assoc chemist, Midwest Res Inst, 65-66; res assoc, Am Tobacco Co, 66-67 & Med Col Va, 67-76; res assoc, Dept Chem, Stanford Univ, 78-81; assoc clin toxicol, Nat Inst Environ Health Sci, Univ Wis-Madison, 81-82; ed, J Pharmaceut Sci, Am Pharmaceut Asn, 82-86; MANAGING ED, ANALYTICAL CHEM, AM CHEM SOC 87- *Mem:* Am Chem Soc; AAAS; Am Asn Pharmaceut Scientists; Coun Biol Ed. *Res:* Ecological restoration; study of native plant species, habitats, etc. *Mailing Add:* Am Chem Soc 1155 Sixteenth St NW Washington DC 20036

BOOZER, ALLEN HAYNE, b Orangeburg, SC, July 28, 44; m 68; c 2. PLASMA PHYSICS. *Educ:* Univ Va, BA, 66; Cornell Univ, PhD(physics), 70. *Prof Exp:* Physicist, Princeton Plasma Physics Lab, 74-86; PROF PHYSICS, COL WILLIAM & MARY, 86- *Concurrent Pos:* External sci mem, Max Plank Soc Fed Repub Ger, 89- *Mem:* Fel Am Phys Soc (secy & treas, 88-90); Am Geophys Union. *Res:* Theory of plasmas primarily for application to magnetic fusion; equilibrium of and transport in three dimensional plasmas and methods of non-inductive current drive in plasmas. *Mailing Add:* Dept Physics Col William & Mary Williamsburg VA 23185

BOOZER, CHARLES (EUGENE), b Nacogdoches, Tex, Oct 4, 27; m 53; c 6. RUBBER CHEMISTRY. *Educ:* Austin Col, BS, 49; Rice Inst, MA, 51, PhD(chem), 53. *Prof Exp:* Res assoc oxidation inhibitors, Iowa State Col, 53-54; assoc prof chem, La Polytech Inst, 54-59 & Emory Univ, 59-64; res supvr, Copolymer Rubber & Chem Corp, 64-67, res mgr, 67 -74, appln develop mgr, 74-87; RETIRED. *Concurrent Pos:* Consult, Copolymer Rubber & Chem Corp, 56-64, Houdry Div, Air Prod & Chem Corp. *Mem:* Am Chem Soc. *Res:* Secondary isotope effects; mechanism of chlorosulfite decomposition; mechanism of action of oxidation inhibitors; nitric acid and air oxidations; free radical and Ziegler polymerization; rubber technology. *Mailing Add:* 8635 Gail Dr Baton Rouge LA 70809

BOOZER, REUBEN BRYAN, b Jacksonville, Ala, Aug 19, 25; m 54; c 4. ZOOLOGY, BIOLOGY. *Educ:* Jacksonville State Teachers Col, BS, 49; George Peabody Col Teachers, MA, 52; Auburn Univ, PhD(zool, bot), 69. *Prof Exp:* Teacher high sch, Ala, 50-54; asst prof biol, Jacksonville State Col, 54-63; instr zool, Auburn Univ, 63-68; dean, Col Arts & Sci, Jacksonville State Univ, 71-78, prof biol sci, 68-87, dean, Col Sci & Math, 71-87; RETIRED. *Res:* Histotechnique aging of mammals. *Mailing Add:* 6703 Nisbet Lane Rd Jacksonville AL 36265

BOPE, FRANK WILLIS, b Thornville, Ohio, Oct 30, 18; m 44; c 1. PHARMACEUTICAL CHEMISTRY. *Educ:* Ohio State Univ, BS, 41; Univ Minn, PhD(pharmaceut chem), 48. *Prof Exp:* Asst pharmaceut chem, Univ Minn, 41-42, jr scientist, 46-47; from asst prof to assoc prof pharm, 48-60, prof, 60-84, secy, Col Pharm, 58-84, asst dean, Col Pharm, 70-84, PROF EMER PHARM, OHIO STATE UNIV, 84- *Mem:* Am Pharmaceut Asn; Acad Pharmaceut Sci; Am Asn Cols Pharm; Sigma Xi. *Res:* Chemistry and pharmacology of tannins; antioxidants; preliminary phytochemical and pharmacological investigation of the tannin obtained from Pinus Caribaea Morelet; synthesis of esters of gentisic acid; synthesis of urea analogs. *Mailing Add:* 177 Desantis Dr Columbus OH 43214

BOPP, BERNARD WILLIAM, b New York, NY, Jan 11, 47; m 73; c 2. ASTRONOMY. *Educ:* NY Univ, BA, 68; Univ Tex, Austin, PhD(astron), 74. *Prof Exp:* Res assoc astron, Univ Wyo, 73-74; asst prof, 74-78, ASSOC PROF ASTRON, UNIV TOLEDO, 78- *Concurrent Pos:* Vis prof, Univ Wyo, 80-81. *Mem:* Am Astron Soc; Royal Astron Soc; Int Astron Union; Sigma Xi. *Res:* Spectroscopy of variable and binary stars; study of stellar surface phenomena; stellar chromospheres. *Mailing Add:* Univ Toledo Ritter Astrophy Res Ctr Toledo OH 43606

BOPP, C(HARLES) DAN, b Decatur, Ill, Feb 4, 23; m 71. CHEMICAL ENGINEERING. *Educ:* Purdue Univ, BS, 44. *Prof Exp:* Chemist, Tenn Eastman Co, Clinton Engr Works, 44-48; Develop Engr, 48-90, CONSULT, OAK RIDGE NAT LAB, 90- *Mem:* Am Chem Soc; Am Ceramic Soc; Am Nuclear Soc; Am Inst Chem Engrs; Sigma Xi. *Res:* Materials; corrosion; environmental effects of nuclear plants; water treatment; radiation chemistry. *Mailing Add:* 72 Outer Dr Oak Ridge TN 37830

BOPP, FREDERICK, III, US citizen; c 2. GEOLOGY. *Educ:* Brown Univ, BA, 66; Univ Del, MS, 73, PhD(geol), 80. *Prof Exp:* Pres & prin partner, Earth Quest Assocs, 74-76; res & univ fel, Dept Geol, Univ Del, 70-76; vis asst prof geochem, Univ SFla, 76-77; staff hydrogeologist, Waterways Exp Sta, US Army Corps Engrs, 77-79; geologist, 79-84, mgr geosci dept, 84-87, VPRES & TECH DIR, GEOSCI DEPT, WESTON, 87- *Mem:* Nat Water Well Asn; Geol Soc Am; Estuarine Res Fedn. *Res:* Groundwater resources evaluation; hydrogeologic evaluation of sanitary landfills and other waste disposal methods; detection and abatement of groundwater pollution; digital modelling of groundwater flow and solute transport; statistical analysis of geological and geochemical data; geochemical prospecting. *Mailing Add:* 22 Unchan Ave Downington PA 19335

BOPP, GORDON R(ONALD), b Vancouver, BC, Oct 31, 34; US citizen; m 57; c 4. CHEMICAL ENGINEERING. *Educ:* Univ Colo, BS, 59, MS, 61; Stanford Univ, PhD(chem eng), 64. *Prof Exp:* From asst prof to assoc prof chem eng, Univ Idaho, 63-70; prof chem eng & acad dean, Univ Minn, Morris, 70-78; prof & exec vpres, Eastern NMex Univ, 78-81; prof chem engr & vpres acad affairs, NMex Inst Mining & TEchnol, 81-83; DIR, SOCORRO TECHNOL INNOVATION CTR, 83- *Concurrent Pos:* Res Corp grant, 64-65; consult, Wilbur Ellis Co, 65-70, higher educ adv comt, State of Nev, 69-70, Gattlin Res Co. *Mem:* Am Inst Chem Engrs. *Res:* Electrochemical hydrodynamics; mass transfer in multiphase systems; technology transfer; heat pipe equipment design. *Mailing Add:* Dept Metall & Mat NMex Inst Mining & Technol Socorro NM 87801

BOPP, LAWRENCE HOWARD, b Schenectady, NY, Mar 10, 49; m 71; c 2. PLASMID GENETICS, BIODEGRADATION. *Educ:* State Univ NY, Albany, BS, 71; Rensselaer Polytech Inst, MS, 76, PhD(biol), 81. *Prof Exp:* Assoc staff microbiologist, 77-80, STAFF MICROBIOLOGIST, RES & DEVELOP CTR, GEN ELEC CORP, 80- *Concurrent Pos:* Prof, State Univ NY Med Col, Albany, NY, 85- *Mem:* Am Soc Microbiol; AAAS; Sigma Xi. *Res:* Genetics and biochemistry of bacteria associated with leaching, metal resistance and transformations and the degradation of unusual compounds; genetic manipulation of such organisms. *Mailing Add:* Ten Slater Dr Scotia NY 12302

BOPP, THOMAS THEODORE, b Glendale, Calif, Nov 29, 41; m 62, 73; c 4. PHYSICAL CHEMISTRY. *Educ:* Calif Inst Technol, BS, 63; Harvard Univ, PhD(chem), 68. *Prof Exp:* From asst prof to assoc prof, 67-85, PROF CHEM, UNIV HAWAII, 85- *Mem:* Am Chem Soc. *Res:* Applications of magnetic resonance to liquid structure; molecular motion in liquids; computer simulation of spectra; multiple pulse NMR. *Mailing Add:* Dept Chem 2545 The Mall Univ Hawaii Honolulu HI 96822

BOQUET, DONALD JAMES, b Houma, La, Oct 7, 45. AGRONOMY. *Educ:* Nicholls State Univ, BS, 68; La State Univ, MS, 71, PhD(plant breeding), 74. *Prof Exp:* Asst prof agron, Univ Ark, 74-76; asst prof, 76-79, ASSOC PROF AGRON, LA AGR EXP STA, 79- *Mem:* Am Soc Agron; Crop Sci Soc Am; Sigma Xi. *Res:* Development of varieties and production practices for improvement of crop yields of cotton and soybeans. *Mailing Add:* PO Box 438 St Joseph LA 71366

BORA, SUNDER S, b India, Oct 6, 38; US citizen; m 67; c 2. EDUCATIONAL ADMINISTRATION, TECHNICAL MANAGEMENT. *Educ:* Agra Univ, India, BS, 57, MS, 59; Lucknow Univ, PhD(biochem), 67. *Prof Exp:* Res fel biochem, Lucknow Univ, India, 60-67; res assoc biochem, Columbia Univ, NY, 68-70; lab dir, Clin Lab, Clin Anal Serv, Hamden, Conn, 70-72; asst lab dir, 72-74, TECH DIR CLIN LAB, NORTHERN WESTCHESTER HOSP CTR, MT KISCO, 74- *Concurrent Pos:* Prog dir, Sch Med Technol, Northern Westchester Hosp Ctr, Mt Kisco, 75-; adj prof, Iona Col, New Rochelle, NY, 77-, Mercy Col, Dubbs Ferry, NY, 77-, King's Col, Briarcliff, NY, 78-, adj prof, Pace Univ, Pleasantville, NY, 87-; prin investr, NY Acad Sci, 88- *Mem:* Asn Clin Scientist; Am Asn Clin Chemists; assoc mem Am Soc Clin Pathologists. *Res:* Thyroid-metabolism; immunoglobulins. *Mailing Add:* Northern Westchester Hosp Ctr 400 E Main St Mt Kisco NY 10549

BORAH, KRIPANATH, b Calcutta, India, Mar 1, 31; m 64; c 2. PHARMACEUTICAL CHEMISTRY, ORGANIC CHEMISTRY. *Educ:* Univ Calcutta, BS, 52, MS, 56; Univ Munich, PhD(pharm), 61. *Prof Exp:* Prod chemist, Ciba-Geigy Ltd, Bombay, India, 62-67, dep mgr develop dept, 68-70, mgr, 71-76; res assoc, Dept Chem, Boston Col, 76-77; sr scientist pharm res & develop, William H Rorer, Inc, 77-80; DIR, PHARM RES & DEVELOP, ORGANON INC, 80- *Mem:* Am Chem Soc; Am Pharm Asn; Acad Pharm Sci. *Res:* Synthesis of local anesthetics; 19-norsteroids and total synthesis of Adriamycin; process research and formulation research and development of new drugs. *Mailing Add:* 34 Overlook Trail Morris Plains NJ 07950

BORAKER, DAVID KENNETH, b Los Angeles, Calif, Apr 21, 39; m 61; c 1. IMMUNOLOGY, IMMUNOGENETICS. *Educ:* Univ Calif, Santa Barbara, BA, 62; Univ Calif, Los Angeles, PhD(med microbiol), 67. *Prof Exp:* Immunologist, Dept Serol, Walter Reed Army Inst Res, DC, 67-69; from asst prof to assoc prof med microbiol, Col Med, Univ Vt, 69-85; PRES, CHROMOGEN CO, 85- *Mem:* AAAS. *Res:* Kinetics of induction of high and low dose immunological paralysis; comparative immunology of the double thymus of the South American degu. *Mailing Add:* Chromogen PO 128 Milton VT 05468

BORAX, EUGENE, b New York, NY, June 17, 19; m 50; c 2. GEOLOGY. *Educ:* Univ Calif, Los Angeles, MA, 52. *Prof Exp:* Sr geologist, Union Oil Co, 46-69, sr explor res geologist, Union Carbide Petrol Corp, 69-72; vpres, MOBWINC, Consult Geologists, 72-74; SR EXPLORATIONIST, PENNZOIL CO, 74- *Concurrent Pos:* Explor mgr, Pennzoil do Brazil, Inc. *Mem:* Geol Soc Am; Am Soc Photogram; Am Asn Petrol Geologists; Am Geophys Union. *Res:* Petroleum geology. *Mailing Add:* 5418 El Carro Lane Carpinteria CA 93013

BORCH, RICHARD FREDERIC, b Cleveland, Ohio, May 22, 41; m 62. CANCER PHARMACOLOGY, CANCER DRUG DESIGN. *Educ:* Stanford Univ, BS, 62; Columbia Univ, MA, 63, PhD(chem), 65; Univ Minn, MD, 75. *Prof Exp:* Fel chem, Harvard Univ, 65-66; from asst prof to assoc prof org chem, Univ Minn, Minneapolis, 66-82; DEAN'S PROF PHARMACOL & PROF CHEM, UNIV ROCHESTER, 82- *Concurrent Pos:* Resident internal med, Univ Minn Hosp, 75-76. *Mem:* Am Chem Soc; Am Asn Cancer Res; AAAS; Am Soc Pharmacol & Exp Therapuet. *Res:* Synthesis of biologically active compounds; mechanisms of action of antineoplastic compounds; reduction of antineoplastics toxicity; pharmacology. *Mailing Add:* Dept Pharmacol Univ Rochester Rochester NY 14642

BORCHARD, RONALD EUGENE, b Saginaw, Mich, Feb 16, 39; m 62; c 4. VETERINARY PHARMACOLOGY. *Educ:* Mich State Univ, BS, 66, DVM, 67; Univ Ill, PhD(vet pharm), 75. *Prof Exp:* Assoc vet, Northville Vet Clin, Mich, 67-69; instr vet pharm, Col Vet Med, Univ Ill, 69-73; asst prof, 73-80, ASSOC PROF VET PHARM, COL VET MED, WASH STATE UNIV, 80- *Mem:* Am Acad Vet & Comp Toxicol; Am Acad Vet Pharmacol & Therapeut; Am Vet Med Asn. *Res:* Pharmacokinetics and tissue distribution of chloro-biphenyls in food producing animals; immobilizing drugs for horses; pharmacokinetics of Ketamine in cats. *Mailing Add:* Dept VCAPP Col Vet Med Wash State Univ Pullman WA 99164-6520

BORCHARDT, GLENN (ARNOLD), b Watertown, Wis, July 28, 42; m 65; c 2. SOIL MINERALOGY, SOIL TECTONICS. *Educ:* Univ Wis-Madison, BS, 64, MS, 66; Ore State Univ, PhD(soil mineral), 69. *Prof Exp:* Res asst soils, Int Minerals & Chem Corp, 64; Nat Res Coun res assoc, US Geol Surv, 69-71, geochemist, 72-77, SOIL MINERALOGIST, CALIF DIV MINES & GEOL, 77-; CONSULT, 84- *Concurrent Pos:* Assoc ed, Geol Soc Am, 73-82. *Mem:* AAAS; Soil Sci Soc Am; Clay Minerals Soc; Geol Soc Am; Am Geophys Union; Asn Eng Geol. *Res:* Instrumental neutron activation analysis correlation of volcanic ash; multivariate similarity analysis; quantitative soil clay mineralogy; chemical stabilization of landslides; soil smectites; recency of faulting in soils; scientific worldview. *Mailing Add:* Div Mines & Geol 380 Civic Dr Pleasant Hill CA 94523-1997

BORCHARDT, HANS J, b Berlin, Ger, Jan 26, 30; US citizen; m 53; c 5. FLUORINE CHEMISTRY. *Educ:* Brooklyn Col, BS, 52, MA, 53; Univ Wis, PhD(phys chem), 56. *Prof Exp:* Inorg chemist, Gen Eng Lab, Gen Elec Co, 56-59; sr res chemist, Explosives Dept, E I du Pont de Nemours & Co, 60-64, sr res chemist, Cent Res Dept, 65-69, sr res chemist org chem, 69-75, res assoc, Freon Prod Lab, 75-80; RETIRED. *Mem:* Am Chem Soc; Sigma Xi; Am Soc Heating, Refrig & Air Conditioning Engrs. *Res:* Differential thermal analysis; reactions in the solid state; rare earth solid state chemistry; phase equilibria; oxidation of metals; rare earth luminescence; lasers; crystal growth; fluorocarbon chemistry; chemical and physical properties of fluorocarbons. *Mailing Add:* 4062 DuPont Pkwy Townsend DE 19734

BORCHARDT, JACK A, civil & sanitary engineering; deceased, see previous edition for last biography

BORCHARDT, JOHN KEITH, b Evanston, Ill, June 2, 46. SYNTHETIC ORGANIC CHEMISTRY, PHYSICAL ORGANIC CHEMISTRY. *Educ:* Ill Inst Technol, BS, 68; Univ Rochester, PhD(chem), 73. *Prof Exp:* Lab technician chem, Ill Inst Technol, 68; fel, Univ Notre Dame, 72-74; res chemist, Hercules, Inc, 74-77; sr chemist, 77-80, develop chemist, Halliburton Serv, 80-84; SR RES CHEMIST, SHELL DEVELOPMENT CO, 84- *Concurrent Pos:* Contrib ed, The Hexagon; counr, chmn two local sects, Am Chem Soc; mem fluid mechs & oil recovery process tech comt, Soc Petrol Engrs. *Mem:* Soc Petrol Engrs; Tech Assoc Pulp & Paper Industs; Sigma Xi; Am Chem Soc; NY Acad Sci. *Res:* Mechanism of elimination reactions; mechanism of cycloaddition reactions; synthesis and modification of water soluble polymers; mechanism of cycloaddition reactions; synthesis of small and medium ring compounds; polymer chemistry; surfactant chemistry. *Mailing Add:* 8010 Vista Del Sol Dr Houston TX 77083

BORCHARDT, KENNETH, b Chicago, Ill, Sept 20, 28; m 54; c 3. MICROBIOLOGY. *Educ:* Loyola Univ, Ill, BS, 50; Miami Univ, MS, 51; Tulane Univ, PhD(microbiol), 61. *Prof Exp:* US Army, 53, asst chief vet dept, Med Lab, Ger, 54-56, asstchief bact, Fitzsimons Gen Hosp, 57-58, chief clin microbiol, Letterman Gen Hosp, 61-65, CHIEF CLIN MICROBIOL, US PUB HEALTH SERV HOSP, 65-, SCI DIR, 68- *Mem:* Fel Am Pub Health Asn; fel Am Acad Microbiol; Sigma Xi. *Res:* Clinical research in infectious disease associated with problems in urology, dermatology, ophthalmology and internal medicine. *Mailing Add:* 15 Capilano Dr Novato CA 94949

BORCHARDT, RONALD T, b Wausau, Wis, Feb 18, 44; m 66; c 3. DRUG DELIVERY, DRUG DESIGN. *Educ:* Univ Wis-Madison, BS, 67; Univ Kans, Lawrence, PhD(med chem), 70. *Prof Exp:* Sr res asst, NIH, 69-71; from asst prof to prof biochem, 71-81, COURTESY PROF MED CHEM, UNIV KANS, 79-, SLOAN E SUMMERFIELD PROF BIOCHEM, 81-, SLOAN E SUMMERFIELD PROF CHEM & CHMN DEPT, 83- *Concurrent Pos:* Estab investr, Am Heart Asn, 74-79; consult Merck, Sharp & Dohme Labs, 80-85; Upjohn Co, 84-; Glaxo, Inc, 87-; Victorian prof, Victorian Col Pharm, Melbourne, Australia, 83-; dir, Ctr Biomed Res, 81-88. *Honors & Awards:* Sato Mem Int Award, Pharmaceut Soc Japan & Found Advan Educ Sci, 81; Dolph C Simons Sr Award Biomed Res, Univ Kans, 83. *Mem:* Am Chem Soc; Am Soc for Pharmacol & Exp Therapeut; Am Soc Biol Chemists; Soc Neurosci; Am Pharmaceut Asn; AAAS. *Res:* Biochemistry and chemistry of S-adenosylmethionine-dependent methyltransferases; central nervous system epinephrine and hypertension; mechanism of action of neurotoxins; metabolic and transport properties of isolated gastrointestinal mucosal cells and capillary endothelial cells (blood brain barrier). *Mailing Add:* Dept Pharmaceut Chem Univ Kans 3006 Malott Hall Lawrence KS 66045

BORCHERS, EDWARD ALAN, b Spring Lake, NJ, Feb 26, 25; m 53; c 2. PLANT BREEDING. *Educ:* Cornell Univ, BS, 51; Univ NC, MS, 53; Univ Calif, PhD(plant path), 57. *Prof Exp:* Asst hort, State Col Agr & Eng, Univ NC, 51-53; asst veg crops, Univ Calif, 53-56; plant breeder, Va Truck & Ornamentals Res Sta, 56-72, dir, 73-85, DIR, HAMPTON ROADS AGR EXP STA & PROF HORT, VA TECH, 85- *Mem:* Am Soc Hort Sci. *Res:* Breeding of vegetable crops. *Mailing Add:* 4177 Ewell Rd Virginia Beach VA 23455

BORCHERS, HAROLD ALLISON, b Chicago Heights, Ill, June 6, 35; m 60; c 3. ENTOMOLOGY. *Educ:* Iowa State Univ, BS, 61, MS, 64, PhD(entom, parasitol), 68. *Prof Exp:* Res assoc agr bee lab, Ohio State Univ, 63; instr zool & entom, Iowa State Univ, 65-68; assoc prof, 68-79, PROF BIOL, BEMIDJI STATE UNIV, 79- *Mem:* Entom Soc Am; Sigma Xi; Soc Agr Res. *Res:* Nest-cleaning behavior of honey bees; populations of mites symbiotic on necrophilous beetles; diets of entomophagous fish. *Mailing Add:* Dept Biol Bemidji State Univ Bemidji MN 56601

BORCHERS, RAYMOND (LESTER), b Juniata, Nebr, Apr 13, 16; m 42; c 3. BIOCHEMISTRY. *Educ:* Nebr State Teachers Col, BS, 38; Univ Iowa, MS, 40, PhD(biochem), 42. *Prof Exp:* Asst chem, Nebr State Teachers Col, 36-38; asst biochem, Univ Iowa, 38-42; instr sch med & actg head dept, Creighton Univ, 42-45; asst agr chemist, 45-49, assoc prof, 49-57, PROF BIOCHEM & NUTRIT, COL AGR, UNIV NEBR-LINCOLN, 57- *Mem:* Am Chem Soc; Poultry Sci Asn; Am Inst Nutrit; Am Soc Animal Sci; Fedn Am Socs Exp Biol. *Res:* Amino acid metabolism and nutrition; nutrition value of soybean; unidentified nutrients; rumen metabolism. *Mailing Add:* 6200 Walker Ave Lincoln NE 68507

BORCHERT, HAROLD R, b Fairbault, Minn, Apr 3, 40; m 84; c 4. ADMINISTRATION OF A RADIATION CONTROL PROGRAM, HEALTH PHYSICS IMPACTS FOR PUBLIC HEALTH & SAFETY. *Educ:* NDak State Univ, BS, 64, MS, 74. *Prof Exp:* Grad asst, NDak State Univ, 70-73, asst radiation safety officer, 73-76; chief mat licensing & control sect, Kans Dept Health & Environ, 76-84; radiol assessment specialist, Kans Gas & Elec Co, 84-85; DIR, DIV RADIOL HEALTH, NEBR DEPT HEALTH, 85- *Mem:* Conf Radiation Control Prog Dirs Inc; Health Physics Soc; Am Pharmaceut Asn; Air & Waste Mgt Asn. *Res:* Radon concentrations and dose impacts on humans including the various parameters that affect the concentrations; emergency response activities; dose impacts from radiation sources; radiation control program impacts on public health and safety. *Mailing Add:* Nebr Dept Health 301 Centennial Mall S PO Box 95007 Lincoln NE 68509

BORCHERT, PETER JOCHEN, b Allenstein, Ger, Feb 6, 23; m 58; c 1. ORGANIC CHEMISTRY. *Educ:* Univ Würzburg, BS, 48, MS, 51; Hannover Tech Univ, PhD(org chem), 53. *Prof Exp:* Chemist, Ind Farm Endoquimica, Sao Paulo, Brazil, 54, E F Drew, Sao Paulo, 55 & Fabrica de Pintura Montana, Caracas, Venezuela, 56-57; dir chem res lab, Marschall Div, Miles Labs, Inc, 58-73, dir res & develop, Sumner Div, 73-87; RETIRED. *Mem:* Am Chem Soc. *Res:* Animal poisons; detergents; auxiliary textile products; polymers; paints; carbohydrates; organic synthesis. *Mailing Add:* 28993 Ella Dr Elkhart IN 46514

BORCHERT, ROLF, b Frankfurt-Main, Ger, May 7, 33; c 3. PLANT PHYSIOLOGY. *Educ:* Univ Frankfurt, PhD(bot), 61. *Prof Exp:* Instr bot, Univ Frankfurt, 61-62; prof, Univ Andes, Colombia, 62-68; assoc prof, 68-74, PROF BOT, UNIV KANS, 74- *Mem:* Am Soc Plant Physiol; Scand Soc Plant Physiol; Ecol Soc Am. *Res:* Growth and development of trees; calcium excretion and differentiation of calcium oxalate crystal cells. *Mailing Add:* Dept Physiol & Cell Biol Univ Kans Lawrence KS 66045

BORCHERTS, ROBERT H, b Mt Vernon, NY, Mar 22, 36; m 59; c 2. SYSTEMS SCIENCE, ELECTRONIC CONTROLS. *Educ:* Gen Motors Inst, Eng dipl, 57; Univ Mich, MS, 59, PhD(electron spin resonance), 63. *Prof Exp:* Res asst radiation effects & electron spin resonance, Univ Mich, 61-63; res physicist, Inst Solid State Physics, Univ Tokyo, 63-64; res scientist, Sci Lab, Ford Motor Co, 65-83; STAFF RES ENGR & ACTIVITIES HEAD, GEN MOTORS RES LAB, 83- *Mem:* Am Phys Soc; AAAS; Soc Automotive Engrs; Inst Elec & Electronics Engrs. *Res:* Magnetic resonance; optical spectroscopy; Josephson junctions; magnetic suspension; automotive control systems; suspension ride measurements; engine control, automotive fuel economy and emissions; vehicle systems; electronic controls. *Mailing Add:* Gen Motors Proj Trilby Gen Motors Res Labs Warren MI 48090

BORDAN, JACK, b Montreal, Que, Feb 13, 26; m 49; c 3. EDUCATION ADMINISTRATION. *Educ:* McGill Univ, BEng, 50, MSc, 52. *Hon Degrees:* LLD, Concordia Univ, 82. *Prof Exp:* Res assoc radar weather, McGill Univ, 51-53; lectr & asst prof physics, 52-57, from assoc prof to prof eng, 57-81, dean eng, 63-70, vprin acad, 70-77, vrector acad, 77-80, ADJ PROF ENG, CONCORDIA UNIV, 81- *Mem:* Am Soc Eng Educ; fel Eng Inst Can. *Res:* Educational methods; technical writing. *Mailing Add:* Concordia Univ Montreal PQ H3G 1M8 Can

BORDELEAU, LUCIEN MARIO, b Quebec City, Can, July 6, 40; m 67; c 3. MICROBIOLOGY, RESEARCH MANAGEMENT. *Educ:* Laval Univ, BSc, 65, MSc, 67; Rutgers Univ, PhD(microbiol), 71. *Prof Exp:* Teacher biol, Tilly Regional Sch Bd, Que, 65-67; res officer, 67-71, RES SCIENTIST, CAN AGR, QUE, 71-, PROG HEADER, 76-, GROUP HEAD, 86- *Concurrent Pos:* Consult, CPVQ-Agr Que, 76-77; Nat Acad Sci, Wash, 80-; UNESCO, 81-; assoc ed, Can J Microbiol, 75-78; Mircen J Appl Microbiol Biotech, 85-88. *Mem:* Can Soc Microbiologists; Am Soc Microbiol; Can Col Microbiologists; Agr Inst Can. *Res:* Biological nitrogen fixation in the rhizobium-legumes system; bio-degradation; bio-fertilizers; inoculants technologies. *Mailing Add:* 1491 Gean Charles Cantin Ave Caprouge PQ G1Y 2X7 Can

BORDELON, DERRILL JOSEPH, mathematics, physics, for more information see previous edition

BORDEN, CRAIG W, b Springboro, Ohio, Aug 31, 15; m. INTERNAL MEDICINE. *Educ:* Oberlin Col, AB, 37; Harvard Univ, MD, 41; Am Bd Internal Med, dipl, 50. *Prof Exp:* Staff physician & chief blood bank sect, Vet Admin Hosp, Minneapolis, 47-51; asst prof med, Med Sch, Univ Minn, 50-53; assoc prof, Med Sch, Northwestern Univ, Chicago, 54-60, prof med, 60-77; dir med educ, St Vincent Infirmary, 77-86; prof med, Sch Med, Univ Ark, Little Rock, 77-90; RETIRED. *Concurrent Pos:* Dir adult cardiac clin & cardiac catheterization lab, Variety Club Heart Hosp, Minn, 51-53; asst chief med serv & chief cardiovasc sect, Vet Admin Res Hosp, Chicago, 53-54, chief med serv, 54-74, sr physician, 71-77; mem residency rev comt internal med, AMA Coun Med Educ, 66-71, vchmn, 70-71; secy-treas, Am Bd Internal Med, 67-68, vchmn, 68-70. *Mem:* AAAS; fel Am Col Physicians; fel AMA; Am Fedn Clin Res; Sigma Xi. *Res:* Applied cardio-respiratory physiology. *Mailing Add:* PO Box 56198 Little Rock AR 72215

BORDEN, EDWARD B, b New York, NY, Dec 5, 49; m 76; c 2. INTESTINAL MOTILITY DISORDERS. *Educ:* Polytechnic Inst Brooklyn, BS, 72, MS, 72; Albert Einstein Col Med, MD, 76. *Prof Exp:* ASST PROF SURG, ALBERT EINSTEIN COL MED, 82- *Concurrent Pos:* Asst attend surg, Montefiore Med Ctr, 81- *Mem:* Fel Am Col Surgeons; Inst Elec & Electronics Engrs; Am Motility Soc. *Res:* Clinical and experimental research in intestinal motility; mesenteric circulation; laboratory computer applications especially biological signal processing. *Mailing Add:* 3771 Nesconset Hwy Centereach NY 11720

BORDEN, ERNEST CARLETON, b Norwalk, Conn, July 12, 39; m 67; c 2. ONCOLOGY. *Educ:* Harvard Univ, AB, 61; Duke Univ, MD, 66. *Prof Exp:* Intern, internal med, Duke Univ, 66-67; resident internal med, Univ Pa, 67-68; med officer virol, Ctr Dis Control, 68-70; teaching fel oncol, Johns Hopkins Univ, 70-73; from asst prof to prof oncol, 73-90, AM CANCER SOC PROF CLIN ONCOL, UNIV WIS, 84-; DIR CANCER CTR & PROF MED & MICROBIOL, MED COL WIS, 90- *Concurrent Pos:* Assoc ed, J Interferon Res, 80- & J Biol Response Modifiers, 82-90; mem decision network comt, Biol Response Modifiers Prog, Nat Cancer Inst, 82-86; consult, Triton Biosci, 83-90. *Mem:* Fel Am Col Physicians; Am Asn Cancer Res (pres, 88-89); Am Soc Clin Oncol; Am Asn Immunologists; Int Soc Interferon Res; Soc Biol Ther (pres, 87-88). *Res:* Mechanism of action and clinical application of immunomodulators and interferons. *Mailing Add:* Cancer Ctr Med Col Wis 8701 Watertown Plank Rd Milwaukee WI 53226

BORDEN, GEORGE WAYNE, b Shambaugh, Iowa, June 17, 37; m 59; c 2. ORGANIC CHEMISTRY, CHEMICAL ENGINEERING. *Educ:* Northwest Mo State Univ, BS, 59; Iowa State Univ, PhD(org chem), 63. *Prof Exp:* Proj scientist org chem res, Union Carbide Corp, 63-74; ASST DIR ADMIN PROCESS DEVELOP, PFIZER, INC, 74- *Mem:* Am Chem Soc. *Res:* Administration of an interdisciplinary process development group of chemists and chemical engineers involved in the development of new industrial specialty chemical products and upgrading of current manufacturing processes. *Mailing Add:* Pfizer Inc East Point Rd Groton CT 06340

BORDEN, JAMES B, b Platteville, Wis, Feb 5, 27; m 51; c 3. CHEMICAL ENGINEERING. *Educ:* Univ Wis, BS, 50; Mass Inst Technol, MS, 52. *Prof Exp:* Engr, textile fiber dept, Res Div, E I du Pont de Nemours & Co Inc, 52-55, Dacron Res Lab, 55-56, sr engr, Dacron Tech Div, 56-60, sr res engr, 60-61, tech supvr, Nomex Tech Div, 61-64, sr supvr, 64-69, process supt, Dacron Mfg Div, 69-72, asst to dir Dacron Mfg Div, 72-74, planning mgr, Europe, 74-76, mgr spec studies, 76-78, mgr econ & policy, Mat & Logistics Dept, 78-89, chmn Int Cross Boarder Movement Goods Study 89-90; RETIRED. *Concurrent Pos:* Chmn, CMA Energy Comn, 84-85; vchmn, NAM Fossil Fuels Comt, 88-90; NPRA Tech Subcomt, 89-90; consult, 90- *Mem:* Am Chem Soc. *Res:* Textile fiber and synthetic fiber process and development; economic evaluation of new processes, products and end uses; European petrochemical industry and energy economics legislation and regulation. *Mailing Add:* 19 Wood Rd Wilmington DE 19806

BORDEN, JOHN HARVEY, b Berkeley, Calif, Feb 6, 38; Can citizen; m 62; c 2. CHEMICAL ECOLOGY, PEST MANAGEMENT. *Educ:* Wash State Univ, BSc, 63; Univ Calif, Berkeley, MSc, 65, PhD(entom), 66. *Prof Exp:* Res technician entom, Univ Calif, Berkeley, 63, teaching asst, 63-64; from asst prof to assoc prof, 66-75, PROF BIOL SCI, SIMON FRASER UNIV, 75-, DIR, CHEM ECOL RES GROUP, 81- *Concurrent Pos:* Fel, NSF, 64-66; travel fel, Nat Res Coun, Can, 76-77; assoc mem chem ecol, Plant Biotechnol Inst, Nat Res Coun, Saskatoon, Sask, 85-; mem Premier's Adv Coun Sci & Technol, BC, 87-91; BC Forest Res Adv Coun, 90-; Killam res fel, Can Coun, 90-91. *Honors & Awards:* C G Hewitt Award, Entom Soc Can, 77; J E Bussart Mem Award, Entom Soc Am, 84; Gold Medal Natural & Appl Sci, Sci Coun BC, 85. *Mem:* Fel Entom Soc Can; Entom Soc Am; Can Inst Forestry; Nat Asn Advan Sci; Int Soc Chem Ecol; Sigma Xi; Am Reg Prof Entom. *Res:* Host selection, pheromones, basic biology, sensory perception and the use of semiochemicals in the management of economically important forest and agricultural insects. *Mailing Add:* Dept Biol Sci Ctr Pest Mgt Simon Fraser Univ Burnaby BC V5A 1S6 Can

BORDEN, KENNETH DUANE, b Floyd, NMex, May 4, 40; m 65; c 3. PHYSICAL CHEMISTRY, NUCLEAR CHEMISTRY. *Educ:* Eastern NMex Univ, BS, 62; Univ Ill, MS, 64; Univ Ark, PhD(nuclear chem), 68. *Prof Exp:* from asst prof to assoc prof, 68-79, PROF CHEM & DEPT CHMN, IND CENT UNIV, 79- *Mem:* Am Chem Soc; Sigma Xi. *Res:* Mass distribution from the fast neutron induced fission of heavy elements; use of radioactive tracers in reaction rate studies and reaction mechanisms. *Mailing Add:* Dept Chem Univ Indianapolis Indianapolis IN 46227

BORDEN, ROGER R(ICHMOND), b Fall River, Mass, June 7, 30; m 57; c 2. MECHANICAL ENGINEERING. *Educ:* Mass Inst Technol, BS, 52 & 53; Worcester Polytech Inst, ME, 61; Assumption Col, MA, 76, CAGS, 81. *Prof Exp:* Asst, Sloan Lab Aircraft & Automotive Engines, Mass Inst Technol, 52-53; in charge, mech & mach design dept, Franklin Inst Boston, 53-54; head dept, 54-57; head mech eng technol & mech technol design options, Wentworth Inst, 57-59; asst prof mech eng, 59-67, ASSOC PROF MECH ENG, WORCESTER POLYTECH INST, 67- *Concurrent Pos:* Consult, Children's Med Ctr, Mass, Mass State Dept Pub Health & P K Lindsay Co; fel, Trans Syst Ctr, US Dept Trans, 80-81. *Mem:* Am Mech Engrs; Am Soc Eng Educ; Sigma Xi. *Res:* Thermodynamics; engineering materials; internal combustion engines; electronics; instrumentation. *Mailing Add:* Box 445 Sterling MA 01564

BORDEN, WESTON THATCHER, b New York, NY, Oct 13, 43; m 71. ORGANIC CHEMISTRY. *Educ:* Harvard Univ, BA, 64, MA, 66, PhD(chem), 68. *Prof Exp:* From instr to asst prof chem, Harvard Univ, 68-73; assoc prof, 73-77, PROF CHEM, UNIV WASH, 77- *Mem:* Am Chem Soc; The Chem Soc. *Res:* Theoretical organic chemistry; synthesis of molecules of theoretical interest; stereochemistry of organic reactions; reaction mechanisms. *Mailing Add:* Dept Chem Univ Wash Seattle WA 98195

BORDENCA, CARL, b Birmingham, Ala, Aug 13, 16; m 39; c 2. ORGANIC CHEMISTRY. *Educ:* Howard Col, BS, 36; Ga Inst Technol, MS, 38; Purdue Univ, PhD(org chem), 41. *Prof Exp:* Asst chem, Howard Col, 33-36, Ga Inst Technol, 36-38 & Purdue Univ, 38-41; instr org chem, Ala Polytech Inst, 41-43; res chemist, Visking Corp, 43-45 & Southern Res Inst, 45-56; asst to pres, Newport Industs, 56-57; dir res, Heyden Newport Chem Corp, 57-62; mgr res & develop, Org Chem Div, Glidden Co, 62-68, vpres, Biochem Dept, 68-73, mgr tech liaison, Glidden-Durkee Div, SCM Corp, 73-77; CONSULT, 77- *Concurrent Pos:* Instr, Southern Col Pharm, 37-38. *Mem:* Am Chem Soc. *Res:* Chlorinolysis of hydrocarbons and their partially chlorinated derivatives; utilization of pine products; flavors and flavor components; coal tar by-products; pesticides. *Mailing Add:* PO Box 2064 Daphne FL 36526

BORDER, WAYNE ALLEN, b Bremen, Ind, Feb 6, 43; m. MEDICINE. *Educ:* Purdue Univ, BS, 65; Wash Univ, MD, 68; Am Bd Internal Med, dipl, 78; dipl, nephrol, 80. *Prof Exp:* Med intern, NC Mem Hosp, 68-69, med resident, 69-70; res assoc, Scripps Clin & Res Found, La Jolla, Calif, 73-75; nephrol fel, Harbor Univ Calif Los Angeles Med Ctr, Torrance, 75-76; from asst prof to assoc prof med, Univ Calif Los Angeles Sch Med, 76-82; assoc prof, 83-85, PROF MED, UNIV UTAH SCH MED, 85- *Concurrent Pos:* Dir immunol lab, Div Nephrol & Hypertension, Harbor Univ Calif Los Angeles Med Ctr, 76-82, assoc chief, 82; chief, Div Nephrol & Hypertension, Univ Utah Sch Med, 83-, dir, Dialysis Prog, 83- *Mem:* Am Soc Clin Invest; Am Fedn Clin Res; Am Soc Nephrol; Am Asn Immunologists; fel Am Col Physicians. *Res:* Nephrology. *Mailing Add:* Dept Med Div Nephrol Univ Utah Health Sci Ctr 50 N Medical Dr Salt Lake City UT 84123

BORDERS, ALVIN MARSHALL, b Mays, Ind, Mar 7, 14; m 37; c 3. CHEMISTRY. *Educ:* Ind Univ, AB, 35, PhD(chem), 37. *Prof Exp:* Res chemist, Goodyear Tire & Rubber Co, 37-41, group leader, 41-46; chief polymer res br, Off Rubber Reserve, Reconstruction Finance Corp, DC, 46; res coordr, Goodyear Tire & Rubber Co, 47; assoc, Inst Paper Chem, Lawrence Col, 48-50; group leader res, Minn Mining & Mfg Co, 50-52, assoc dir res, 52-58, res mgr, Reflective Prod Div, 58-62, dir, Appl Res Lab, Cent Res Labs, 62-65, dir, Chem Res Lab, 65-71 & Mat Sci & Process Res, 70-74, dir, Univ related progs, Cent Res Labs, 74-79, consult, 79-80; actg patent adminr, Univ Minn, 81-82; CONSULT, 82- *Mem:* Am Chem Soc. *Res:* Chemistry of rubber and rubber-like materials; styrene-diene resins; paper; fluorochemicals and adhesives; research administration. *Mailing Add:* 536 Eastlake Dr SE Rio Rancho NV 87124-1898

BORDERS, CHARLES LAMONTE, JR, b Lebanon, Ky, June 13, 42; c 3. BIOCHEMISTRY. *Educ:* Bellarmine Col, Ky, BA, 64; Calif Inst Technol, PhD(chem), 68. *Prof Exp:* PROF CHEM, COL WOOSTER, 68- *Mem:* Am Chem Soc; Am Soc Biol Chemists; AAAS. *Res:* Protein chemistry; enzymology. *Mailing Add:* Dept Chem Col Wooster Wooster OH 44691

BORDERS, DONALD B, b Logansport, Ind, Jan 12, 32; m 58; c 2. ORGANIC CHEMISTRY. *Educ:* Ind Univ, BS, 54, MA, 58; Univ Ill, PhD(chem), 63. *Prof Exp:* Res chemist, Rohm and Haas Co, 62-64; res chemist, Lederle Labs, Am Cynamid Co, 64-66, group leader fermentation & isolation dept, 66-75, group leader microbiol & chemother res dept, 75-77, group leader, Microbiol Res Dept, 77-87, DEPT HEAD, MICROBIOL CHEM RES DEPT, MED RES DIV, LEDERLE LABS, AM CYANAMID CO, 87- *Mem:* Am Chem Soc; The Chem Soc. *Res:* Syntheses of biologically active compounds; monomer and polymer syntheses; detection, purification, identification and structure of natural products and other antibiotics. *Mailing Add:* 13 Heatherhill Lane Suffern NY 10901

BORDERS, JAMES ALAN, b Akron, Ohio, Aug 8, 41; m 69. SOLID STATE PHYSICS. *Educ:* Reed Col, BA, 63; Univ Ill, Urbana, MS, 65, PhD(physics), 68. *Prof Exp:* Res asst physics, Univ Ill, Urbana, 63-68; physicist, 68-78, SUPVR SURFACE CHEM & ANALYTICAL DIV, SANDIA LABS, 78- *Mem:* fel Am Phys Soc; Sigma Xi; Electrochem Soc; Am Vacuum Soc. *Res:* Energetic ion backscattering analysis of solid surfaces and thin films; channeling location of ion-implanted impurities; formation of non-equilibrium alloys in metals by ion implantation. *Mailing Add:* Sandia Nat Labs Orig 1823 PO Box 5800 Albuquerque NM 87185

BORDIN, GERALD M, b San Francisco, Calif, Sept 27, 40; m 64; c 2. MEDICINE, PATHOLOGY. *Educ:* Univ Calif, Berkeley, BA, 62; St Louis Univ, MD, 66. *Prof Exp:* Surg pathologist, Med Sch, Univ NMex, 72-73; anat pathologist, Lovelace-Bataan Med Ctr, 73-74; surg pathologist, Med Sch, Univ NMex, 75-76; ANAT PATHOLOGIST, HOSP SCRIPPS CLIN, 76- *Concurrent Pos:* Advan clin fel, Am Cancer Soc, 71-72, jr fac clin fel, 72-73. *Mem:* Am Soc Clin Pathologists; Am Col Pathologists; Int Soc Dermatopath; US-Can Acad Path. *Res:* Cancer epidemiology. *Mailing Add:* Dept Path 10666 N Torrey Pines La Jolla CA 92037

BORDINE, BURTON W, b Monroe, Mich, Nov 28, 34; m 61; c 5. MICROPALEONTOLOGY, PALEOBIOLOGY. *Educ:* Western Mich Univ, BS, 63; Brigham Young Univ, MS, 65; La State Univ, PhD(paleont), 74. *Prof Exp:* Geologist-micropaleontologist, Exxon Corp, 65-68; instr, Eastern Ariz Col, 68-70 & 74-75; asst prof, Middle Tenn State Univ, 75-81; MEM STAFF, PHILLIPS PETROL CO, 81- *Mem:* Am Asn Petrol Geologists. *Res:* Neogene foraminifera of northern South America; relationship of trace fossils to benthic foraminifera; cenozoic foraminifera of California. *Mailing Add:* Dept Geog & Geol Middle Tenn State Univ Murfreesboro TN 37132

BORDNER, CHARLES ALBERT, JR, b Salem, Mass, Nov 19, 37; m 60; c 2. PHYSICS. *Educ:* Colo Col, BS, 59; Harvard Univ, AM, 60, PhD(physics), 64. *Prof Exp:* Res assoc physics, Harvard Univ, 64-66; ASST PROF PHYSICS, COLO COL, 66- *Concurrent Pos:* Vis staff mem, Mass Inst Technol, 65-66. *Mem:* Am Phys Soc; AAAS; Am Asn Physics Teachers; Sigma Xi. *Res:* Experimental high energy physics. *Mailing Add:* Dept Physics Colo Col Colorado Springs CO 80903

BORDNER, JON D B, b Massillon, Ohio, Jan 25, 40; m 69; c 3. ORGANIC CHEMISTRY, BIOCHEMISTRY. *Educ:* Case Inst Technol, BS, 62; Univ Calif, Berkeley, PhD(org chem), 66. *Prof Exp:* Fel x-ray crystallog, Calif Inst Technol, 66-69; from asst prof to prof chem, NC State Univ, 69-85; RES ADV, PFIZER CENT RES, 85- *Concurrent Pos:* Dreyfus teacher-scholar, 72; consult, Pfizer Inc, 74-85. *Mem:* Am Chem Soc; Am Crystallog Asn; Sigma Xi. *Res:* Structure and function of steroids; insect pheromones; molecular neurochemistry; x-ray crystallographics. *Mailing Add:* Three Craig Rd Old Lyme CT 06371

BORDOGNA, JOSEPH, b Scranton, Pa, Mar 22, 33. ELECTRICAL ENGINEERING, EDUCATION. *Educ:* Univ Pa, BSEE, 55, PhD(elec eng), 64; Mass Inst Technol, SM, 60. *Prof Exp:* Engr, RCA Corp, 58-64; asst prof, Univ Pa, 64, Winterstein asst prof, 64-68, from assoc prof to prof, 68-76, master, Stouffer Col House, 72-76, from assoc dean to dean eng & appl sci, 73-90, ALFRED FILTER MOORE PROF ENG, UNIV PA, 76-, DIR, MOORE SCH ELEC ENG, 76- *Concurrent Pos:* Mem, USN, 55-58; consult; chair, ed bd Engr Educ, 87-90; mem bd dirs, Roy F Weston Inc, West Chester, Pa, AOI Systs Inc, Lowell, Mass & Univ City Sci Ctr, Philadelphia, Pa; chair, adv comt eng, NSF, 89-91. *Honors & Awards:* George Westinghouse Award, Am Soc Eng Educ, 74; Centennial Medal Inst Elec Electronics Eng. *Mem:* Fel Inst Elec & Electronics Engrs; Am Soc Eng Educ; fel AAAS; Franklin Inst; Sigma Xi. *Res:* Electro-optics; author of books in electrical engineering and technological literacy and papers on educational innovation. *Mailing Add:* Sch Eng & Appl Sci Univ Pa Philadelphia PA 19104-6391

BORDOLOI, KIRON, b Jorhat, India, Oct 1, 34; m 61; c 2. SOLID STATE PHYSICS. *Educ:* Gauhati Univ, India, BSc, 55; Univ Calcutta, MSc, 57; La State Univ, PhD(physics), 64. *Prof Exp:* Lectr physics, J B Col, India, 57-58; asst, La State Univ, 58-63; asst prof, Univ Southern Miss, 63-69; assoc prof eng physics, 69-78, PROF ELEC ENG, UNIV LOUISVILLE, 78- *Mem:* Am Physiol Soc; sr mem Inst Elec & Electronics Engrs; Soc Photo-Optical Instrumentation Engrs. *Res:* Low temperature physics. *Mailing Add:* Dept Elec Eng Univ Louisville Louisville KY 40292

BORDWELL, FREDERICK GEORGE, b Marmarth, NDak, Jan 17, 16; m 39, 72; c 2. ORGANIC CHEMISTRY. *Educ:* Univ Minn, BS, 37, PhD(org chem), 41. *Prof Exp:* Procter & Gamble fel, 41-42, from instr to prof, 42-54, EMER CLARE HAMILTON HALL PROF CHEM, NORTHWESTERN UNIV, 74- *Concurrent Pos:* Humble lectr, 53 & 57; NSF sr fel, 57; Guggenheim fel, 80; vis prof, Univ Ill, 57, Univ Wales, 75, Univ Western Ont, 76, Univ Calif, Irvine, 79, 84, 85, 86, 87, 88. *Honors & Awards:* Am Chem Soc Award Petrol Chem, 86; H C Brown Lectr, 86; C K Ingold Lectr, 87. *Mem:* Am Chem Soc. *Res:* Mechanisms of organic reactions; chemistry of organic sulfur compounds; acidities of weak acids; organic electrochemistry. *Mailing Add:* Dept Chem Northwestern Univ Evanston IL 60201

BOREI, HANS GEORG, b Stockholm, Sweden, Feb 7, 14; m 38; c 3. INVERTEBRATE MARINE ECOLOGY. *Educ:* Univ Stockholm, Fil Mag, 37, PhD(zool), 40, Fil Dr(biochem), 45. *Prof Exp:* Res asst biophys & exp biol, Univ Stockholm, 37-45, assoc prof, 45-52, head dept develop physiol & genetics, 47-50, head dept bioiphys, 48-52, actg head inst, Wenner-Gren Inst, 48,50; prof zool, 33-84, prof gen physiol, 55-60, CHMN BIOL MAJ PROGS & UNDERGRAD CHMN, 64-, EMER PROF, UNIV PA, 84- *Concurrent Pos:* Researcher, Carlsberg Lab, Copenhagen, Denmark, 46, Kristinebergs Zool Sta, Sweden, 46-52, 61, Millport Marine Sta, Scotland, 48, Molteno Inst, Cambridge, Eng, 48-49, Woods Hole, Mass, 51, 53, Mt Desert Island Biol Lab, Maine, 55-64 & Swans Island Marine Sta, Maine, 65-84; vis prof zool, Calif Inst Technol, 51; pres & dir, Swans Island Marine Sta, 66-84. *Honors & Awards:* Aquist Award, Stockholm, 45, 46. *Mem:* AAAS; Sigma Xi; Am Soc Zoologists; Am Soc Limnol & Oceanog; Int Soc Cell Biol. *Res:* Invertebrate marine ecology and embryology; chemical embryology; changes in cellular metabolism at fertilization; micromethods in respirometry. *Mailing Add:* Long Cove Stanley Point Rd Minturn ME 04659

BOREIKO, CRAIG JOHN, b Milford, Ct, June 5, 51; m 75; c 2. GENETIC TOXICOLOGY, EXPERIMENTAL ONCOLOGY. *Educ:* Clark Univ, BA, 73; Univ Wis-Madison, PhD(exp oncol), 79. *Prof Exp:* Assoc scientist, 79-80, SCIENTIST, CHEM INDUST INST TOXICOL, 80- *Mem:* Environ Mutagen Soc; Genetic Toxicol Asn; Am Asn Cancer Res. *Res:* Chemically induced oncogenic transformation of cultured human cells; cell culture models for initiation and promotion; regulation and characteristics of the transformed state. *Mailing Add:* Environ Health ILZRO PO Box 1206 Research Triangle Park NC 27709

BOREK, CARMIA GANZ, b Tel Aviv, Israel, May 27, 37; US citizen; m 58; c 2. CELL BIOLOGY. *Educ:* Am Univ, BSc, 59; George Washington Univ, MSc, 61; Weizmann Inst Sci, PhD(cell biol, genetics), 67. *Prof Exp:* Res asst biochem, Sch Med, Georgetown Univ, 61-62; res asst, Sect Endocrinol, Nat Inst Arthritis & Metab Dis, 63; res assoc cell physiol, Col Physicians & Surgeons, Columbia Univ, 67-69; instr path, Sch Med, NY Univ, 69-71; from asst prof to assoc prof, 71-81, PROF PATH & RADIOL, COLUMBIA UNIV, 81- *Concurrent Pos:* Nat Cancer Inst fel physiol, Col Physicians & Surgeons, Columbia Univ, 67-69. *Mem:* Am Soc Cell Biol; Am Asn Cancer Res; Sigma Xi; Int Soc Develop Biol; Am Tissue Cult Asn; fel NY Acad Sci. *Res:* cellular & molecular carcinogenesis (chemical radiation), anticarcinogenesis and nutritional chemoprevention. *Mailing Add:* Dept Path & Radiation Oncol Columbia Univ 630 W 168th St New York NY 10032

BOREK, FELIX, b Cracow, Poland, May 5, 26; US citizen; m 57. IMMUNOCHEMISTRY, ORGANIC CHEMISTRY. *Educ:* Hobart Col, BSc, 50; Harvard Univ, MA, 52; Rutgers Univ, PhD(org chem), 56. *Prof Exp:* Res asst chem, Rutgers Univ, 52-55; res fel biochem, Col Physicians & Surgeons, Columbia Univ, 55-57; assoc biochemist, Armed Forces Inst Path, 58-63; vis scientist, Weizmann Inst, 63-65, sr scientist, 65-67; asst prof microbiol & immunol, Albert Einstein Col Med, 67-72; FOUNDING ED, J IMMMUNOL METHODS, 71- *Concurrent Pos:* Lectr, NIH, 60-63, spec res fel, 63-65. *Mem:* AAAS; Am Asn Immunol; NY Acad Sci; Biochem Soc Israel. *Res:* Specificity of delayed hypersensitivity; immunosuppression and immune tolerance in adult hosts; relation between immunogenicity and the molecular size and structure of antigens; localization of labeled antigens in the host; steroid-protein conjugates as antigens; the function of the spleen in immune responses. *Mailing Add:* Dept Microbiol Tel-Aviv Univ Tel-Aviv Israel

BOREL, ARMAND, b Switz, May 21, 23; m 52; c 2. MATHEMATICS. *Educ:* Swiss Fed Inst Technol, MS, 47; Univ Paris, PhD, 52. *Hon Degrees:* Univ Geneva, 72. *Prof Exp:* Supplying prof algebra, Univ Geneva, 50-52; vis lectr, Univ Chicago, 54-55; prof, Swiss Fed Inst Technol, 55-56, 83-86; PROF, INST ADVAN STUDY, 57- *Concurrent Pos:* Vis prof, Mass Inst Technol, 58 & 69, Tata Inst, Bombay, 61, 83 & 90, Univ Paris, 64, Univ Calif, Berkeley, 75 & Yale Univ, 78. *Honors & Awards:* Brouwer Medal, Dutch Math Soc, 78. *Mem:* Nat Acad Sci; Math Soc France; Swiss Math Soc; foreign mem Finnish Acad Sci & Lett; assoc mem French Acad Sci; Am Philos Soc; Am Math Soc; Am Acad Arts & Sci. *Res:* Algebraic topology; lie groups; algebraic and arithmetic groups. *Mailing Add:* Inst Advan Study Princeton NJ 08540

BOREL, YVES, IMMUNOLOGY TOLERANCE, AUTO IMMUNITY. *Educ:* Univ Geneva, MD, 57. *Prof Exp:* ASSOC PROF PEDIAT, RHEUMATOLOGY & IMMUNOL, SCH MED, HARVARD UNIV, 76- *Mailing Add:* Dept Pediat Harvard Med Sch Ctr Recherche Nestle Ave Nestle 55 BP353 Ch-1800 Vevey Switzerland

BORELLA, LUIS ENRIQUE, b Sept 29, 30; Can citizen; m 66; c 2. PHARMACOLOGY. *Educ:* Univ Buenos Aires, BSc, 50, MSc, 54; Univ Conn, MSc, 62, PhD(pharmacol), 64. *Prof Exp:* Pharmacist, 55-58; res asst clin chem, St Raphael's Hosp, New Haven, Conn, 58-60; Anna Bradbury Springer fel pharmacol, Univ Toronto, 64-65; sr res assoc biochem & pharmol, Ayerst Res Labs, 65-75, PRIN SCIENTIST, WYETH-AYERST RES, 75- *Mem:* Assoc AAAS; Can Pharmacol Soc; Sigma Xi. *Res:* Gastrointestinal physiology and pharmacology; drug effects on gastric acid secretion and experimental peptic ulcer; methodology in gastrointestinal research; gastrointestinal drug absorption; pharmacological and biochemical effects of interaction between psychoactive drugs; inflammatory bowel disease; cytoprotection; anti-flammatory agents; osteoarthritis; anorexiants. *Mailing Add:* 42 Slayback Dr Princeton Junction NJ 08550-1915

BOREMAN, JOHN GEORGE, JR, b Bronx, NY, Aug 15, 48; m 70; c 2. BIOLOGY. *Educ:* State Univ NY, BS, 70; Cornell Univ, MS, 72, PhD(fishery sci), 78. *Prof Exp:* Northeast power plant specialist, 74-75, aquatic ecologist, Nat Power Plant Team, US Fish & Wildlife Serv, 75-80; sr assessment scientist, Northeast Fisheries Ctr, Nat Marine Fisheries Serv, 80-85, chief res coord sect, 85-89; DIR, COOP MARINE EDUC & RES PROG, NAT OCEANIC & ATMOSPHERIC ASN, UNIV MASS, 89- *Mem:* Am Fisheries Soc. *Res:* Fish population dynamics; impacts of entrainment and impingement by power plants on fish populations; quantitative approaches to fishery management. *Mailing Add:* 171 Aubinwood Rd Amherst MA 01002

BORENFREUND, ELLEN, b Ger, Mar 15, 22; US citizen. BIOCHEMICAL GENETICS. *Educ:* Hunter Col, BS, 46; NY Univ, MS, 48, PhD(biol), 57. *Prof Exp:* Technician cancer res, Mt Sinai Hosp, 40-47; asst biochem, Col Physicians & Surgeons, Columbia Univ, 48-57; asst prof biochem, Sloan-Kettering Div, Grad Sch Med Sci, Cornell Univ, 61-68; from res assoc to assoc, 57-65, assoc prof biochem, Sloan-Kettering Div, Grad Sch Med Sci, Cornell Univ, 68-81, chairperson Biochem Unit, 72-74, assoc mem, Sloan-Kettering Inst Cancer Res, 65-88, ADJ ASSOC PROF, SLOAN-KETTERING DIV, ROCKEFELLER UNIV, 81- *Concurrent Pos:* NIH res career develop award, 63-66. *Mem:* Am Asn Cancer Res; Am Soc Cell Biol; Am Soc Biol Chem; Tissue Cult Asn; Harvey Soc. *Res:* Nucleic acid biochemistry; biochemical genetics; cell transformation and chemical carcinogenesis in vivo and in vitro; in vitro toxicology. *Mailing Add:* Lab Animal Res Ctr Rockefeller Univ 1230 York Ave New York NY 10021

BORENSTEIN, BENJAMIN, b Jersey City, NJ, Nov 5, 28; m 52; c 2. FOOD CHEMISTRY, ORGANIC CHEMISTRY. *Educ:* Rutgers Univ, BS, 50, MS, 52, PhD(food sci), 54. *Prof Exp:* Res chemist, Fearn Foods, Inc, 54-59; res chemist, Nopco Chem Co, 59-62, head prod appl lab, 61-62; sr chemist, 62-66, mgr food indust tech serv, 66-77, dir res, 77-80, DIR PROD DEVELOP, ROCHE CHEM DIV, HOFFMANN-LA ROCHE, INC, 80- *Concurrent Pos:* Adj prof food sci, Rutgers Univ, 73- *Mem:* AAAS; Am Chem Soc; Am Dairy Sci Asn; Inst Food Technol; fel Am Inst Chem. *Res:* Product development in food, feed and drug fields; meat and dairy chemistry; stabilization of enzymes and vitamins; nutrition; pectin chemistry and biochemistry. *Mailing Add:* 90 Van Buren Ave Teaneck NJ 07666

BORER, JEFFREY STEPHEN, b Deland, Fla, Feb 22, 45; m 78; c 2. CARDIOLOGY. *Educ:* Harvard Col, BA, 65; Cornell Univ Med Col, MD, 69. *Prof Exp:* Intern med, Mass Gen Hosp, 69-70, resident, 70- 71; clin assoc cardiol, Nat Heart, Lung & Blood Inst, 71-74, sr investr, 75-79; assoc prof, 79-82, PROF MED, CORNELL UNIV MED COL, 82-, GLADYS & ROLAND HARRIMAN PROF CARDIOVASC MED, 83- *Concurrent Pos:* Sr Fulbright-Hays scholar & Glorney Raisbeck fel med sci, Guy's Hosp, Univ London, Eng, 74-75; chmn Cardiac & Renal Drugs Adv Comt, US Food & Drug Admin, 81 & 83-87; mem, Life Sci Adv Comt, NASA, 84-88, Aerospace Adv Comt, 88-; prin investr, NIH, TIMI-Cornell Ctr, 84-90; pres, NY Cardiol Soc, 90-91. *Honors & Awards:* Investigators Award Prize, Europ Soc Cardiol, 78; Spec Recognition Award, Asn Thoracic & Cardiovasc Surgeons India, 85. *Mem:* Am Soc Clin Res; Am Col Physicians; Am Col Cardiol; Am Col Chest Physicians; Am Heart Asn; Soc Nuclear Med. *Res:* Preclinical and clinical investigations in the areas of chronic and acute coronary artery disease and valvular heart disease. *Mailing Add:* 975 Park Ave New York NY 10028

BORER, KATARINA TOMLJENOVIC, b Tuzla, Yugoslavia, Sept 17, 40; US citizen; div; c 3. NEUROENDOCRINOLOGY, PHYSIOLOGY. *Educ:* Univ Pa, BA, 62, PhD(zool), 66. *Prof Exp:* Lectr physiol & exp psychol, Alaska Methodist Univ, 68-69; lectr biol & comp vert anat, Anchorage Community Col, 68-69; NIH fel exp marine biol, Rosenstiel Sch Marine & Atmospheric Sci, Univ Miami, 69-70; res fel physiol psychol, 71-73, asst res scientist, Neurosci Lab, 73-77, lectr, Dept Psychol, 73-77, asst prof, 77-81, ASSOC PROF, DEPT PHYS EDUC, UNIV MICH, 81- *Mem:* Endocrine Soc; Am Physiol Soc; Am Col Sports Med; Soc Neurosci. *Res:* Role of taste in control of feeding behavior; physiological mechanisms in control of octopus feeding behavior; neuroendocrine mechanisms in weight regulation and growth in rodents; neuroendocrine effects of exercise; alteration by exercise of neuroendocrine controls over somatic growth and reproductive function; pharmacological and neurosurgical mainipulations of the central nervous system and measurement of pulsatile release of growth hormone, prolactin, and luteinizing hormone. *Mailing Add:* Dept Phys Educ Univ Mich 401 Washtenaw Ann Arbor MI 48109

BORER, PHILIP N, b Sept 23, 45; m 88; c 3. NUCLEAR MAGNETIC RESONANCE SPECTROSCOPY. *Educ:* Univ Calif, Berkeley, PhD(chem), 72. *Prof Exp:* RES ASSOC PROF CHEM, SYRACUSE UNIV, 83- *Res:* Nucleic acid chemistry. *Mailing Add:* Dept Chem Syracuse Univ Syracuse NY 13244-4100

BORESI, ARTHUR PETER, b Toluca, Ill; m 46; c 3. THEORETICAL MECHANICS, APPLIED MECHANICS. *Educ:* Univ Ill, BA, 48, MS, 49, PhD(theoret & appl mech), 53. *Prof Exp:* Res engr dynamics, NAm Aviation, 50; mat engr, Nat Bur Standards, 51; from instr to prof theoret & appl mech, Univ Ill, Urbana, 52-80; PROF CIVIL ENG, UNIV WYO, 80- *Concurrent Pos:* Pres, BLM Appl Mech Consult, 60-; distinguished vis prof, Clarkson Col Technol, 68-69 & Naval Post-Grad Sch, Monterey, 78-79 & 86-87; ed eng mech sect, Nuclear Eng & Design. *Mem:* Am Acad Mech (treas, 74-77); fel Am Soc Mech Engrs; fel Am Soc Civil Engrs; Am Soc Eng Educ; Sigma Xi. *Res:* Stress analysis; dynamics; stability; energy methods in mechanics; theory of plates and shells; approximation methods in mechanics. *Mailing Add:* Dept Civil Engr Univ Wyo Laramie WY 82071

BORG, DONALD CECIL, b New York, NY, July 26, 26; m 49, 79; c 5. MEDICAL RESEARCH. *Educ:* Harvard Univ, BS, 46, MD, 50. *Prof Exp:* Teaching fel med, Harvard Univ, 51-52; anal officer, Armed Forces Spec Weapons Proj, 52-54; teaching fel med, Washington Univ, 54-55; res assoc, 55-57, assoc scientist, 57-62, scientist, 62-79, chmn dept med, 79-84, SR SCIENTIST, BROOKHAVEN NAT LAB, 79- *Concurrent Pos:* Consult, Armed Forces Spec Weapons Proj, 54-57; vis res assoc prof physiol, Mt Sinai Sch Med, 68-; biophysicist, Div Biol & Med, US AEC, 70-72; mem adv bd, Nat Biomed ESR Ctr, 76-; consult, Resources for Future, 77-79, Merck & Co, 77-79 & Eastman Pharmaceuts, 88- *Mem:* AAAS; Soc Nuclear Med; NY Acad Sci; Am Physiol Soc; Radiation Res Soc; Am Soc Biochem & Molecular Biol; Soc Free Radical Res; Soc Mag Res Med. *Res:* Free radical mechanisms in biological function mechanisms of oxidative cytotoxicity; electron paramagnetic resonance; health effects of energy production and use; radioactivation analysis; radiation effects; clinical research. *Mailing Add:* Med Res Ctr Brookhaven Nat Lab Bldg 490 Upton NY 11973-5000

BORG, IRIS Y P, b San Francisco, Calif, Oct 6, 28; m 50; c 1. EARTH SCIENCES, MINERALOGY. *Educ:* Univ Calif, Berkeley, BS, 51, PhD(geol), 54. *Prof Exp:* Asst geol, Princeton Univ, 54-56, vis fel, 56-59; res assoc, Univ Calif, Berkeley, 60-61; GEOLOGIST, LAWRENCE LIVERMORE LAB, UNIV CALIF, 61- *Concurrent Pos:* Consult, Shell Develop Co, Tex, 55-67; Geol Soc Am proj grant, 57; Guggenheim fel, 68-69; vis lectr, Am Univ of Cairo, Egypt, 71; lunar investr, NASA, 72; counr, Mineral Soc Am, 74-77; mem gov bd, Am Geol Inst, 75-77; US deleg, Int Mineral Asn USSR, 78; mem, vis comt earth sci, Mass Inst Technol, 78-84; affil, Energy Resource Group, Univ Calif, Berkeley, 78- *Mem:* Fel Geol Soc Am; Am Geophys Union; fel Mineral Soc Am; Am Asn Petrol Geolgists. *Res:* Natural and artificial deformation of rocks and minerals; petrofabrics; high pressure x-ray technology; shock metamorphism; energy and natural resource issues. *Mailing Add:* Lawrence Livermore Lab Univ Calif Livermore CA 94550

BORG, RICHARD JOHN, b Calif, Oct 18, 25; m 50; c 1. PHYSICAL CHEMISTRY. *Educ:* Univ Calif, BS, 50, MS, 52; Princeton Univ, PhD(chem), 57. *Prof Exp:* Micro-analyst, Univ Calif, 51-52, res chemist, Radiation Lab, Livermore, 52-53; asst chem, Princeton Univ, 57-58, instr, 58-59; chemist, 59-75, div leader, Lawrence Livermore Lab, Univ Calif, 75-77; lectr, Univ Calif, Davis, 59-65, prof, 65-77; res chemist, Lawrence Livermore Lab, 77-90; CONSULT, 90- *Concurrent Pos:* Ed, Mineral Processing & Extractive Metall Rev. *Mem:* Fel Am Phys Soc; Am Chem Soc. *Res:* Diffusion in ionic and metallic solids; Mossbauer measurements in metals; magnetic materials. *Mailing Add:* Lawrence Livermore Labs Livermore CA 94550

BORG, ROBERT MUNSON, b Smithfield, Pa, May 19, 10; m 33; c 2. ECOLOGY. *Educ:* Univ Chicago, BS, 39, MS, 40. *Prof Exp:* Asst to staff, Harvard Univ, 29-31; asst forester, Dept Conserv, Mass, 33; tech forester, US Forest Serv, 33-35, supvry wildlife mgr, Mass, 35-36; asst dist agent, US Fish & Wildlife Serv, 40-45; fumigant engr, Dow Chem Co, Mich, 45; MGR, LILLIAN P BORG MEM FOREST, OSSIPEE, NH. *Mem:* Soc Am Foresters. *Res:* Industrial and soil fumigation; agricultural chemicals; development of farm-forest ecosystems; Lillian P Borg memorial forest. *Mailing Add:* RFD 1 Box 302 Ossipee NH 03864

BORG, SIDNEY FRED, b New York, NY, Oct 3, 16; m 44; c 4. CIVIL ENGINEERING, EARTHQUAKE ENGINEERING. *Educ:* Cooper Union, BSCE, 37; Polytech Inst New York, MCE, 40; Johns Hopkins Univ, DrEng, 56; Stevens Inst Technol, MEng, 58. *Prof Exp:* Engr, Turner Construct Co, 40-41; struct engr, Eastern Aircraft, 41-43; asst prof civil eng, Univ Md, 43-45; assoc prof aeronaut, Naval Postgrad Sch, 45-51; res engr, Grumman Aircraft Co, 51-52; head civil eng, 52-77, prof, 52-87, EMER PROF CIVIL ENG, STEVENS INST TECHNOL, 87- *Concurrent Pos:* Consult to various pvt firms & govt agencies in eng mech & aeronaut, 50-; Fulbright prof, Danish Tech Univ, 65; vis prof, US State Dept, Polish Acad Sci, 68, Technische Hochschule, Stuttgart, 66 & 81, Israel Inst Technol, 72, Air Force Inst Technol, 73. *Honors & Awards:* Sigma Xi. *Mem:* Fel NY Acad Sci; fel Am Soc Civil Engrs; fel AAAS; Am Soc Eng Educ. *Res:* Solids and fluids; matrix and tensor analysis as applied to engineering; similarity methods in engineering; physical, medical sciences and the humanities; earthquake engineering. *Mailing Add:* Dept Civil & Ocean Eng Stevens Inst Technol Castle Point Hoboken NJ 07030

BORG, THOMAS K, b Chicago, Ill, Sept 12, 43. CARDIAC STRUCTURE FUNCTION. *Educ:* Univ Wis, PhD(entom), 69. *Prof Exp:* PROF PATH & DIR ELECTROMICROS, MED SCH, UNIV SC, 76- *Mem:* Am Soc Cell Biol; Soc Int Heart Res; Electromicros Soc. *Mailing Add:* Dept Path Sch Med Univ SC Columbia SC 29208

BORGAONKAR, DIGAMBER SHANKARRAO, b Hyderabad, India, Sept 24, 32; m 63; c 2. GENETICS, CYTOLOGY. *Educ:* Osmania Univ, India, BScAgr, 53; Okla State Univ, PhD(genetics), 63. *Prof Exp:* Res asst, Indian Dept Agr, Hyderabad & Bombay, 55-59, lectr agr bot, Parbhani, 56-57; asst, Univ Minn, 59; res assoc bot, Okla State Univ, 59-63; asst prof biol, Univ NDak, 63-64; from instr to asst prof med & head chromosome lab, Johns Hopkins Univ, 64-71, assoc prof med, Div Med Genetics, Sch Med, 72-78; dir res, Genetics Ctr & prof biol sci, N Tex State Univ, Denton, 78-80; DIR, CYTOGENETICS LAB, MED CTR DEL, NEWARK, 80- *Concurrent Pos:* Sigma Xi grant, 64; Am Philos Soc grant, 67; NSF grant, 66-68; NIMH grant, 69-73; adj prof, Sch Life & Health Sci, Univ Del; res prof pediat, med genetics, Thomas Jefferson Univ, Philadelphia; Dr Margaret I Handy chair in Human Genetics, Med Ctr Del, 81- *Honors & Awards:* JBS Haldane Award, Soc Bionaturalists, India. *Mem:* AAAS; Am Inst Biol Sci; Am Soc Human Genetics; Am Fedn Clin Res; Sigma Xi. *Res:* Human cytogenetics and genetics; mammalian cytogenetics. *Mailing Add:* Cytogen Lab Christiana Hosp 4755 Ogletown Station Rd PO Box 6001 Newark DE 19718

BORGATTI, ALFRED LAWRENCE, b Medford, Mass, Apr 28, 28. MICROBIOLOGY, ENTOMOLOGY. *Educ:* Tufts Univ, BS, 51, EdM, 52; Mich State Univ, PhD(entom), 61. *Prof Exp:* Teacher math & sci, Everett Pub Schs, Mass, 54-58; instr entom, Mich State Univ, 58-61; PROF BIOL, SALEM STATE COL, 61- *Concurrent Pos:* Consult, Briggs Eng Co, 80-84. *Mem:* Sigma Xi; Entom Soc Am; Am Soc Microbiol. *Res:* Microbial diseases of insects. *Mailing Add:* Dept Biol Salem State Col Salem MA 01970

BORGEAT, PIERRE, b Montreal, Que, May 22, 47; m 71; c 3. LEUKOTRIENES, INFLAMMATION. *Educ:* Col des Jesuites de Que, BA, 67; Laval Univ, BSc, 71, DSc, 74. *Prof Exp:* Sr researcher, 79-85, assoc prof, 85-90, PROF PHYSIOL, FAC MED, LAVAL UNIV, 90- *Concurrent Pos:* Ed, Prostaglandins Leukotrienes & Med, 82- *Honors & Awards:* Leo-Pariseau Prize, Can-French Asn Advan Sci, 88; Distinguished Scientist Award, Can Soc Clin Investr, 89. *Mem:* Clin Res Club Que; Can-French Asn Advan Sci; Can Biochem Soc; Can Soc Clin Invest; Int Soc Immunopharmacol. *Res:* Physiological and pharmacological control of leukotrienes synthesis; role of leukotrienes in allergic reactions and inflammation. *Mailing Add:* Inflammation et Immunologie-Rhumatologie Le Cent Hosp de l'Univ Laval 2705 Boul Laurier Ste-Foy PQ G1V 4G2 Can

BORGENS, RICHARD BEN, b Little Rock, Ark, May 7, 46; c 2. DEVELOPMENTAL BIOLOGY, NEUROBIOLOGY. *Educ:* NTex State Univ, BS, 70, MS, 73; Purdue Univ, PhD(biol), 77. *Prof Exp:* Res assoc biol, Purdue Univ, 77-78; res assoc biol, Yale Univ, 78-80; assoc staff scientist, Jackson Lab, Bar Harbor, Maine, 80-81; staff scientist, Inst Med Res, 81-; AT DEPT ANAT, SCH VET MED, PURDUE UNIV. *Concurrent Pos:* Nat Paraplegia Found fel, 78-80. *Mem:* Am Soc Zoologists; Soc Develop Biol. *Res:* General problems of development, particularly the role of endogenous electrical fields in amphibian limb regeneration; role of endogenous and applied current on spinal cord regeneration. *Mailing Add:* Dept Anat Purdue Univ Lynn Hall West Lafayette IN 47907

BORGES, CARLOS REGO, b Sao Miguel, Portugal, Feb 17, 39; US citizen; m 58; c 3. MATHEMATICS. *Educ:* Humboldt State Col, BA, 60; Univ Wash, PhD(math), 64. *Prof Exp:* Asst prof math, Univ Nev, 64-65; from asst prof to assoc prof, 65-72, PROF MATH, UNIV CALIF, DAVIS, 72- *Concurrent Pos:* NSF grants, 65-68; Fulbright-Hays scholar, 71. *Mem:* Am Math Soc. *Res:* General topology. *Mailing Add:* Dept Math Univ Calif Davis CA 95616

BORGES, WAYNE HOWARD, b Cleveland, Ohio, Dec 8, 19; m 43; c 2. HEMATOLOGY. *Educ:* Kenyon Col, AB, 41; Western Reserve Univ, MD, 44. *Prof Exp:* Fel hemat, Harvard Med Sch, 51-52, instr pediat, 52; sr instr, Western Reserve Univ, 52-53, asst prof, 53-58; assoc prof, Univ Pittsburgh, 58-63; from assoc prof to prof, 63-71, THE GIVEN PROF PEDIAT, NORTHWESTERN UNIV, CHICAGO, 71-; MED DIR EDUC, CHILDREN'S MEM HOSP, 72- *Concurrent Pos:* Chief div hemat, Children's Mem Hosp, 63-72. *Mem:* Soc Pediat Res; Am Pediat Soc. *Res:* Pediatrics. *Mailing Add:* 355 E Baltimore St Carlisle PA 17013

BORGESE, THOMAS A, b New York, NY, Jan 21, 29; m 58; c 3. PHYSIOLOGY, BIOCHEMISTRY. *Educ:* NY Univ, BA, 50; Rutgers Univ, PhD(physiol, biochem), 59. *Prof Exp:* NIH fel, Harvard Med Sch, 59-61, res assoc biochem, 61-62; res assoc inst nutrit sci, Columbia Univ, 62-64, asst prof nutrit biochem, 64-67; from asst prof to assoc prof, 67-78, PROF BIOL SCI, HERBERT H LEHMAN COL, 78- *Mem:* AAAS; Am Chem Soc; Biophys Soc; Am Physiol Soc; Sigma Xi. *Res:* Cell physiology and biochemistry; comparative biochemistry and physiology of hemoglobin. *Mailing Add:* Dept Biol Sci Herbert H Lehman Col Bronx NY 10468

BORGESON, DAVID P, b Muskegon, Mich, Dec 6, 35; m 58; c 4. FISH BIOLOGY. *Educ:* Mich State Univ, BS, 58, MS, 59. *Prof Exp:* Fishery res biologist, Calif State Dept Fish & Game, 59-66; trout-salmon specialist, Mich Dept Conserv, 66-70, chief, Inland Fisheries sect, 70-79, ASST CHIEF, FISHERIES DIV, MICH DEPT NATURAL RESOURCES, 79- *Mem:* Am Fisheries Soc; fel Am Inst Fishery Res Biologists. *Res:* Fisheries research, particularly trout and salmon. *Mailing Add:* Mich Dept Natural Resources Fisheries Div 4th Floor Mason Bldg PO Box 30028 Lansing MI 48909

BORGIA, GERALD, b Montebello, Calif, Apr 9, 49; m 80. BEHAVIOR ECOLOGY. *Educ:* Univ Calif, Berkeley, AB, 70; Univ Mich, MS, 73, PhD(zool), 78. *Prof Exp:* Trainee, Allec Lab Animal Behav, Univ Chicago, 79-80; univ res fel, Dept Zool, Univ Melbourne, 80-81; asst prof, 80-86, ASSOC PROF ZOOL, DEPT ZOOL, UNIV MD, 86- *Concurrent Pos:* Vis researcher, Univ Melbourne, 82-90; res assoc, Smithsonian Inst, 85-90. *Mem:* Animal Behav Soc; Soc Study Evolution; AAAS. *Res:* Use of evolutionary theory to study social behavior, particularly mate choice in animals; field studies of mate choice in bowerbirds; models of anisogamy; studies of adaptation for testing evolutionary hypothesis. *Mailing Add:* Dept Zool Univ Md College Park MD 20742

BORGIA, JULIAN FRANK, b Starke, Fla, Nov 22, 43; m 68; c 1. CARDIOPULMONARY PHYSIOLOGY, IMMUNOLOGY. *Educ:* Calif State Polytech Univ, BS, 66; Univ Calif Santa Barbara, PhD(physiol), 69. *Prof Exp:* Prin investr, Naval Radiol Defense Lab, 69, Naval Med Res Inst, 69-70; res asst, Inst Environ Stress, 71-76, res physiologist, 76-83; asst prof, Univ Calif, Santa Barbara, 78-83; dir, Biomed Res, Baxter Healthcare Co, 83-88; VPRES RES & DEVELOP, AM CYANAMID, 88- *Mem:* NY Acad Sci; AAAS. *Res:* Cardiovascular physiology; cellular immunology; blood gas physiology; aging of the cardiovascular system; medical device development. *Mailing Add:* Davis & Geck Div One Casper St Danbury CT 06810

BORGIOTTI, GIORGIO V, b Rome, Italy, Nov 23, 32; m 64; c 2. STRUCTURAL ACOUSTICS. *Educ:* Univ Rome, EngD, 57. *Prof Exp:* Scientist, Selenia, Rome, 58-65; tech staff mem, Res Triangle Inst, 65-66; sect mgr, Missile Systs Div, Raytheon, 67-71; dept mgr, CSELT, Torino, Italy, 71-72; sr res assoc, Nat Acad Sci, 74-75; tech staff mem, Jet Propulsion Lab, 76-77; res staff mem, Inst Defense Anal, 80-83; chief scientist, Sci Applns Int Corp, 83-88; PROF ELEC ENG, GEORGE WASH UNIV, 88- *Concurrent Pos:* Consult, struct acoust & electromagnetics. *Mem:* Fel Inst Elec & Electronics Engrs; Sigma Xi; Union Radio Sci Int. *Res:* Applied electromagnetic theory, antennas, radar systems and adaptive systems; signal processing, linear system theory and system identification; acoustic radiation and scattering; modal analysis of vibration. *Mailing Add:* 5024 Klingle St NW Washington DC 20016

BORGLUM, GERALD BALTZER, b Penn Yan, NY, June 30, 33; m 54; c 4. BIOCHEMISTRY. *Educ:* Dana Col, BS, 53; Univ Nebr, PhD(dipeptidases), 66. *Prof Exp:* SR STAFF SCIENTIST, MILES LABS, 65- *Mem:* Am Chem Soc. *Res:* Development of diagnostic reagents. *Mailing Add:* Miles Inc PO Box 70 Elkhart IN 46515

BORGMAN, LEON EMRY, b Chickasha, Okla, Feb 16, 28; m 49; c 2. STATISTICS, GEOLOGY. *Educ:* Colo Sch Mines, 53; Univ Houston, MS, 59; Univ Calif, Berkeley, PhD, 63. *Prof Exp:* Oceanog engr, Shell Oil Co, 53-59; asst prof math, Univ Calif, Davis, 62-67, assoc prof eng geosci, Univ Calif, Berkeley, 67-70; PROF GEOL & STATIST, UNIV WYO, 70- *Mem:* Am Soc Civil Engrs; Am Inst Mining & Metall Engrs; Am Geophys Union. *Res:* Statistics and mathematics applied to problems in geology, coastal engineering, hydrology, mining and various other earth science disciplines. *Mailing Add:* Dept Geol Univ Wyo Box 3006 Laramie WY 82071

BORGMAN, ROBERT F, b Bridgeport, Conn, 1926; m 51; c 4. NUTRITION, HYPERTENSION. *Educ:* Mich State Univ, DVM, 47, MS(physiol), 49; Kans State Univ, PhD(nutrit), 59. *Prof Exp:* PROF FOOD SCI & NUTRIT, CLEMSON UNIV, 60- *Mem:* Am Inst Nutrit; Can Soc Nutrit Sci; Royal Soc Health Fel, London; Am Vet Med Asn; Sigma Xi. *Res:* Metabolism. *Mailing Add:* Rte 1 Box 715 Candler NC 28715

BORGMAN, ROBERT JOHN, b Ft Madison, Iowa, Feb 4, 42; m 66; c 4. MEDICINAL CHEMISTRY. *Educ:* Univ Iowa, PhD(med chem), 70. *Prof Exp:* From asst prof to assoc prof med chem, WVa Univ, 70-75; group leader med chem, 75-76, DIR, MED & ORG CHEM, ARNAR-STONE LABS, INC, 76- *Mem:* Am Pharmaceut Asn; Am Chem Soc. *Res:* Design and synthesis of dopamine; receptor agonists in periphery and central nervous system. *Mailing Add:* 1735 Victoria Mundelein IL 60060

BORGMAN, ROBERT P, b Chicago, Ill, Apr 5, 30; m 57; c 3. PLANT PATHOLOGY, BOTANY. *Educ:* Cent Col, Iowa, BA, 57; Kans State Univ, MS, 59; Iowa State Univ, PhD(plant path), 62. *Prof Exp:* From asst prof to assoc prof biol, Univ Omaha, 62-69; prof biol & chmn dept, Div Natural Sci, 69-77, PROF BIOL, DIV NATURAL SCI, BUENA VISTA COL, 69-, ASST TO PRES, 77- *Mem:* Am Phytopath Soc. *Res:* Fungal and vegetable crop diseases; virus diseases of cereal crop. *Mailing Add:* Dept Bot Buena Vista Col Fourth & College St Storm Lake IA 50588

BORGNAES, DAN, b New York, NY, Nov 16, 35; m 59; c 2. STRUCTURAL REACTION INJECTION MOLDING, RESIN TRANSFER MOLDING. *Educ:* Fla State Univ, BS, 61; Univ Notre Dame, PhD(org chem), 69. *Prof Exp:* Sr res chemist, PPG Indust, Natrium, WVa, 65-71; teacher chem & tech eng, Gym Blumenthal, City of Bremen, 71-74; sr group leader, Ashland Chem Co, Dublin, Ohio, 74-84, res assoc polymer chem, 85-90; res & develop lab mgr, Dexter-Hysol Div, 84-85; CONSULT INDUST POLYMERS, ACCESS SCI CO, 90- *Mem:* Am Chem Soc; Asn Consult Chemists & Chem Engrs. *Res:* Product, process and formulation development of polymer systems used in the structural-reaction injection molding industry; unsaturated polyesters, polyurethanes and other thermoset polymers including additives, reinforcements, inhibitors and catalysts; inventor of first structural reaction injection molding chemical system commercialized as Arimax. *Mailing Add:* 8027 Tipperary Ct N Dublin OH 43017

BORGSTEDT, HAROLD HEINRICH, b Hamburg, Ger, Apr 21, 29; US citizen; m 57; c 2. PHARMACOLOGY, TOXICOLOGY. *Educ:* Univ Hamburg, MD, 56. *Prof Exp:* Asst, Dept Pharmacol, Univ Hamburg, 49-50; intern, Rochester Gen Hosp, 56-57; fel pharmacol & anat, Univ Rochester, 57-60, instr pharmacol, 60-63, sr instr pharmacol & res sr instr anesthesiol, 63-66, asst prof pharmacol & res asst prof anesthesiol, 66-81; VPRES, HEALTH DESIGNS INC, 81- *Honors & Awards:* Physician's Recognition Award, AMA, 74. *Mem:* AAAS; Am Soc Pharmacol & Exp Therapeut; Soc Toxicol; NY Acad Sci; Europ Soc Toxicol; Sigma Xi; Am Chem Soc. *Res:* Pharmacology and toxicology of anesthetics; toxicology of plastics; mathematical modeling in toxicology. *Mailing Add:* Health Designs Inc 183 E Main St Rochester NY 14604

BORGSTROM, GEORG ARNE, food science; deceased, see previous edition for last biography

BORIE, BERNARD SIMON, b New Orleans, La, June 21, 24; m 57; c 3. METALLURGY, SOLID STATE PHYSICS. *Educ:* Southwestern La Inst, BS, 44; Tulane Univ, MS, 49; Mass Inst Technol, PhD(physics), 56. *Prof Exp:* Res physicist, Metall Div, Oak Ridge Nat Lab, 49-53; Fulbright fel, Univ Paris, 56-57; group leader x-ray diffraction, Oak Ridge Nat Lab, 57-60, head fundamental res sect, 60-69, sr scientist, metals & ceramics div, 69-85; RETIRED. *Concurrent Pos:* Prof, Univ Tenn, 63 -; vis prof, Cornell Univ, 71-72 & Univ Calif, Berkeley, 80. *Mem:* Inst Mining, Metall & Petrol Engrs; Am Soc Metals; Am Crystallog Asn; Sigma Xi; AAAS. *Res:* X-ray diffraction studies of imperfect solids; order-disorder effects in solid solutions; structure of thin films; diffraction effects of thermal motion. *Mailing Add:* 13 Brookside Dr Oak Ridge TN 37830

BORING, JOHN RUTLEDGE, III, b Gainesville, Fla, July 7, 30; m 59, 78; c 2. MICROBIOLOGY, EPIDEMIOLOGY. *Educ:* Univ Fla, BS, 53, MS, 55, PhD(microbiol), 61. *Prof Exp:* Nat Res Coun Can fel, 61-62; from asst chief to chief epidemic aid lab, Ctr Dis Control, 62-66; PROF PREV MED & ASSOC PROF MED, EMORY UNIV, 66- *Concurrent Pos:* Ctr Dis Control grants, Emory Univ, 68-75. *Mem:* Am Soc Microbiol; Am Epidemiol Soc. *Res:* Host-parasite relationship between children with cystic fibrosis and the bacterium pseudomonas aeruginosa. *Mailing Add:* 69 Butler St SE Atlanta GA 30303

BORING, JOHN WAYNE, b Reidsville, NC, Oct 9, 29; m 57; c 3. PHYSICS. *Educ:* Univ Ky, BS, 52, MS, 54, PhD(nuclear physics), 61. *Prof Exp:* Sr scientist, Res Labs Eng Sci, 60-67, assoc prof eng physics, 67-71, asst dir ctr advan studies, 68-72, PROF ENG PHYSICS, UNIV VA, 71- *Concurrent Pos:* Consult, Int Comn Radiation Units & Measurement; vis prof, Inst Physics, Univ Aarhus, Denmark, 74-75. *Mem:* Am Phys Soc. *Res:* Atomic collision phenomena; atom-surface interactions. *Mailing Add:* Dept Nuclear Eng & Eng Physics Univ Va Charlottesville VA 22901

BORISENOK, WALTER A, b Brooklyn, NY, Nov 10, 23; m 55; c 2. PHARMACY, NUTRITION. *Educ:* Rensselaer Polytech Inst, BS, 50. *Prof Exp:* SR VPRES MFG & OPERS FOOD FORTIFICATION, FORTITECH INC. *Concurrent Pos:* Sect head parenteral res & develop. *Mem:* Am Pharmaceut Asn; Parenteral Drug Asn. *Res:* Parenteral pharmaceutical products. *Mailing Add:* 61 Miller Rd Latham NY 12110

BORISH, IRVIN MAX, b Philadelphia, Pa, Jan 21, 13; m 36; c 1. OPTOMETRY. *Educ:* Northern Ill Col Optom, OD, 34, DOS, 35. *Hon Degrees:* LLD, Ind Univ, 68; DSc, Pa Col Optom, 75, State Univ NY, 84; DOS, Southern Calif Col Optom, 83, Southern Col Optom, 84. *Prof Exp:* From instr to prof optom, Northern Ill Col Optom, 36-44, registr, 40-44; pvt pract, 44-72; prof optom, Ind Univ, Bloomington, 73-83; BENEDICT PROF, UNIV HOUSTON, 83- *Concurrent Pos:* Asst chief clins, Northern Ill Eye Clin, 36-41, dir clins, 41-44; lectr & vis prof optom, Ind Univ, 53-72; vpres, Ind Contact Lens Inc, 59-; vis lectr & prof var US univs & cols, 65-; consult, Nat Study Optom Educ, Nat Comn Educ, 72-73. *Honors & Awards:* Apollo Award, Am Optom Asn, 68; Wm Feimbloom Award, Am Acad Optom, 85; Distinguished Serv Award, Am Optom Asn, 89. *Mem:* AAAS; Am Optom Asn; hon mem Am Acad Optom; Int Soc Contact Lens Res. *Res:* Contact lenses; refractive techniques, especially development of new instrumentation including use of vectographic design for binocular refractive methods of visual examination. *Mailing Add:* 211 Wildwood Circle Deerfield Beach FL 33442

BORISON, HERBERT LEON, pharmacology; deceased, see previous edition for last biography

BORISY, GARY GUY, b Chicago, Ill, Aug 18, 42; div; c 3. MOLECULAR BIOLOGY. *Educ:* Univ Chicago, BS, 62, PhD(biophys), 66. *Prof Exp:* NSF fel, 66-67; USPHS fel, 67-68; from asst prof to assoc prof, 68-75, PROF MOLECULAR BIOL & ZOOL, UNIV WIS-MADISON, 75-, CHMN LAB MOLECULAR BIOL, 80- *Concurrent Pos:* NATO fel, 68; NIH multidisciplinary biol training rev panel, 71-73; As Hoc rev comts, postdoctoral fellowships, 74; cell biol study sect, 75-79; NIH Res Career Develop Award, 75-80; Perlman-Bascom Professorship of Life Sci, 80-;

NIGMS couns, 82, 84; NY State Doctoral Eval comt, 84, Am Cancer Soc Cellular & Develop Adv comt, 84, 88- *Honors & Awards:* Romnes Award, 75. *Mem:* Am Soc Cell Biol; fel AAAS. *Res:* Principles of biomolecular organization; assembly of macromolecules into functional structures at the cellular level; assembly and function of cytoplasmic microtubules; formation of the mitotic spindle and the mechanism of shromosome movement; cytokinesis; cell motility; the cytoskeleton. *Mailing Add:* 1525 Linden Dr Madison WI 53706

BORK, ALFRED MORTON, b Jacksonville, Fla, Sept, 18, 26; m 48; c 3. PHYSICS, HISTORY OF PHYSICS. *Educ:* Ga Inst Technol, BS, 47; Brown Univ, MS, 50, PhD(physics), 53. *Prof Exp:* From asst prof to prof physics, Univ Alaska, 52-63; from assoc prof to prof, Reed Col, 63-67; vchmn physics dept, 73-78, PROF PHYSICS, INFO & COMPUT SCI, UNIV CALIF, IRVINE, 68- *Concurrent Pos:* NSF fac fel, Harvard Univ, 62-63; chmn spec interest group for comput use & educ, 71, comt instrnl media, Am Asn Physics Teachers, 73- & conduit physics comt, 74- *Mem:* Asn Comput Mach. *Res:* Philosophy and history of science; application of computers to teaching. *Mailing Add:* Dept Info & Comput Sci Univ Calif Irvine CA 92717

BORK, KENNARD BAKER, b Kalamazoo, Mich, Oct 13, 40; m 63; c 1. GEOLOGY, PALEONTOLOGY. *Educ:* DePauw Univ, BA, 62; Ind Univ, MA, 64, PhD(geol, paleont), 66, 67. *Prof Exp:* From asst prof to prof, 66-90, ALUMNI PROF, DENISON UNIV, 90- *Concurrent Pos:* Vis prof, Chapman Col, Orange, Calif, 82; secy, US Comn Hist Geol, 82-85 & corresp mem, Int Comn Hist Geol, 84; pres, E-Cent Sect, Nat Asn Geol Teachers, 85-86; secy, Hist Earth Sci Soc, 87- *Mem:* AAAS; fel Geol Soc Am. *Res:* History of geology; invertebrate paleontology; biostratigraphy; sedimentary petrology. *Mailing Add:* 80 Beechwood Dr Granville OH 43023

BORKAN, HAROLD, b NJ, June 2, 27; m 48; c 3. SOLID STATE ELECTRONICS. *Educ:* Rutgers Univ, BS, 50, MS, 54. *Prof Exp:* Res engr, RCA Labs, 50-65, eng leader, RCA Corp, Somerville, 65-68, mgr custom monolithics, 68-78, mgr spec projs, Solid State Technol Ctr, Somerville, 78-79, mgr progs & planning, Advan Technol Lab, Camden, 79-81; dir, Microelectronics Div, Electronics Technol & Devices Lab, US Army, Electronics Res & Develop Command, 81-85, dep dir, Labcom, Ft Monmouth, 85-90; CONSULT, HARBORTECH, PRINCETON, NJ, 90- *Mem:* Inst Elec & Electronics Engrs. *Res:* Integrated circuits; silicon processing; solid state image sensors; thin film semiconductor devices; television camera tubes. *Mailing Add:* 150 Longview Dr Princeton NJ 08540

BORKE, MITCHELL LOUIS, b Warsaw, Poland, Mar 23, 19; US citizen; m 49; c 2. PHARMACEUTICAL CHEMISTRY, NUCLEAR CHEMISTRY. *Educ:* Univ Ill, BS, 51, MS, 53, PhD, 57. *Prof Exp:* Asst chem, Univ Ill, 51-55, instr, 55-57; from asst prof to assoc prof, 57-64, PROF CHEM, DUQUESNE UNIV, 64- *Concurrent Pos:* Fulbright vis prof, Taiwan, 69-70; partic scientist exchange prog, Nat Acad Sci & Acad Sci Bulgaria, 74, Poland, 77, Yugoslavia, 79, Bulgaria, 81, Hungary, 82. *Honors & Awards:* Elich Prize, 51. *Mem:* Am Pharmaceut Asn; Acad Pharm Pract; Sigma Xi. *Res:* Pharmaceutical and analytical chemistry; radiochemistry; neutron activation analysis using isotopic sources; gas-liquid chromatography; thin layer chromatography. *Mailing Add:* Duquesne Univ Col Pharm Pittsburgh PA 15219

BORKMAN, RAYMOND FRANCIS, b Los Angeles, Calif, Dec 15, 40; m 76; c 4. PHYSICAL CHEMISTRY. *Educ:* Univ Calif, Los Angeles, BS, 62; Univ Calif, Riverside, PhD(chem), 66. *Prof Exp:* Teaching asst chem, Univ Calif, Riverside, 62-63, res asst, 63-66; fel, Johns Hopkins Univ, 66-68; from asst prof to assoc prof, 68-78, PROF CHEM, GA INST TECHNOL, 78- *Concurrent Pos:* NSF fel, Johns Hopkins Univ, 66-67; Sigma Xi res award, Ga Inst Technol, 72. *Mem:* Sigma Xi; Am Soc Photobiol; Asn Res Vision & Ophthal. *Res:* Experimental-theoretical studies in molecular spectroscopy, photochemistry, photobiology and energy transfer; Ab initio computer calculations of molecular electronic wave functions and properties; biomedical sciences, especially spectroscopic studies of cataracts in human and animal eye lenses. *Mailing Add:* Sch Chem Ga Inst Technol Atlanta GA 30332

BORKON, ELI LEROY, b Chicago, Ill, Aug 11, 08; m 37; c 2. INTERNAL MEDICINE, NUCLEAR MEDICINE. *Educ:* Univ Chicago, BS, 31, PhD(physiol), 36, MD, 37; Am Bd Internal Med, dipl. *Prof Exp:* Asst physiol, Univ Chicago, 32-36; assoc prof anat, Chicago Med Sch, 37; assoc prof physiol & Health, Southern Ill Univ, 39-46; fel internal med, Wash Univ, 46; vis lectr physiol, Southern Ill Univ, 48, adj prof, 54; chmn dept med, Carbondale Clin, 54-71; from clin prof to prof med, 71-76, asst to dean, 71-74, ASST DEAN DEVELOP, SCH MED, SOUTHERN ILL UNIV, CARBONDALE, 74-, EMER PROF MED, 76-; dir nuclear med, Mem Hosp, 76-83. *Concurrent Pos:* Vpres, Med Rev Orgn Southern Ill, 80-; mem Ill State Med Disciplinary Bd, 82-, chmn, 84-85. *Mem:* AMA; fel Am Col Physicians; Am Soc Internal Med; Soc Nuclear Med. *Res:* Motor activity of intestine; endocrines; effects of minerals on thyroidparathyroid; compensatory hypertrophy of remaining organs; hazards of anticoagulants. *Mailing Add:* 14 Pinewood Dr Carbondale IL 62901

BORKOVEC, ALEXEJ B, b Prague, Czech, Oct 17, 25; US citizen; m 51. ORGANIC CHEMISTRY, BIOCHEMISTRY. *Educ:* Prague Tech Univ, ChE, 49; Va Polytech Inst, MS, 54, PhD(org chem), 55. *Prof Exp:* Res org chem, Dow Chem Co, Tex, 55-56, sr res chemist, 56-58; asst prof org chem, Va Polytech Inst, 58-60; res chemist, Agr Res Serv, 61-64, invests leader, 64-72, chief, Insect Chemosterilants Lab, 72-79, CHIEF, INSECT REPRODUCTION LAB, AGR RES SERV, USDA, 79- *Concurrent Pos:* Vis prof, Hollins Col, 60-61. *Honors & Awards:* T L Pawlett Award, Univ Sydney, Australia, 71; Bronze Medal, 3rd Int Cong Pesticide Chem, 74; Super Serv Award, USDA, 79; Gold Medal, 11th Cong Plant Protection, 87. *Mem:* AAAS; Am Chem Soc; Entom Soc Am; Czech Soc Arts & Sci Am; NY Acad Sci. *Res:* Insect neuroscience; reaction mechanism of epoxides and aziridines; synthesis and mode of action of insect chemosterilants; biochemistry of reproduction; pest control. *Mailing Add:* Agr Res Ctr USDA Beltsville MD 20705

BORKOVITZ, HENRY S, b Chicago, Ill, Dec 4, 35; m 83; c 3. MICROCOMPUTER CONTROL & MEASURING SYSTEMS. *Educ:* Ill Inst Technol, BS, 58, MS, 62; Univ Chicago, MBA, 70. *Prof Exp:* Engr, Lindberg Eng Co, 58-62; chief develop eng, Electroseal Corp, 62-64; dir eng, Sola Elec Div, Gen Signal, 64-81; vpres eng, Realist Inc, 81-90; AT ASTRONAUT CORP AM, 90- *Mem:* Inst Elec & Electronics Engrs. *Res:* Development level sensing and control systems utilizing microcomputer circuitry; development of software to facilitate the design and programming process including cross assemblers. *Mailing Add:* Destiny Dr W181 N8291 Menomonee Falls WI 53051

BORKOWSKI, RAYMOND P, b Kingston, Pa, Feb 5, 34; m 64; c 2. PHYSICAL CHEMISTRY. *Educ:* King's Col, BS, 55; Cath Univ, PhD(phys chem), 61. *Prof Exp:* Nat Acad Sci-Nat Res Coun res assoc photochem & radiation chem, Nat Bur Stand, 61-63; staff scientist, Aerospace Res Ctr, Gen Precision Inc, 63-67; from asst prof to assoc prof, 67-78, PROF CHEM, KING'S COL, PA, 78- *Concurrent Pos:* Res fel, NASA/Am Soc Elec Engrs, 79-80. *Mem:* Am Chem Soc; Sigma Xi. *Res:* Chemical kinetics; photochemistry; radiation chemistry; phenomena of energy conversion and transfer. *Mailing Add:* Seven Salem Dr Laflin PA 18702

BORKOWSKY, WILLIAM, b Ger, Mar 6, 47; US citizen; m 69; c 3. IMMUNOLOGY, MICROBIOLOGY. *Educ:* City Col New York, BS, 68; NY Univ, MD, 72. *Prof Exp:* Intern pediat, 72-73, resident, 73-75, fel infectious dis & immunol, 75-78, instr pediat, 77-78, asst prof, 78-85, ASSOC PROF PEDIAT, NY UNIV MED CTR & BELLEVUE HOSP, 85-, CHIEF, INFECTIOUS DIS & IMMUNOL, 86- *Mem:* Infectious Dis Soc Am; Am Asn Immunologists; Pediat Res Soc; Pediat Infectious Dis Soc; Am Fedn Clin Res; Harvey Soc. *Res:* Immune regulation and the role of transfer factor as a regulatory molecule; acquired immunity deficiency syndrome; autoimmune diseases, particularly juvenile diabetes. *Mailing Add:* Dept Pediat NY Univ Med Ctr 550 First Ave New York NY 10016

BORLAUG, NORMAN ERNEST, b Cresco, Iowa, Mar 25, 14; m 40; c 2. MICROBIOLOGY. *Educ:* Univ Minn, BS, 37, MS, 41, PhD(plant path), 42. *Hon Degrees:* DSc, Punjab Agr Univ, India, 69; Royal Norweg Agr Col, 70, Luther Col, 70, Kanpur Univ, India, 70, Uttar Pradesh Agr Univ, India, 71; Mich State Univ, 71, Univ La Plata, Arg, 71, Univ Ariz, 72 & Univ Fla, 73; LHD, Gustavus Adolphus Col, 71; LLD, NMex State Univ, 73. *Prof Exp:* With US Forest Serv, 35-36, 37, 38; instr plant path, Univ Minn, 41; microbiologist, E I du Pont de Nemours & Co, Del, 42-44; res scientist in-chg wheat improv, Coop Mex Agr Prog, Mex Ministry Agr & Rockefeller Found, 44-60, assoc dir, Rockefeller Found assigned to Inter-Am Food Crop Prog, 60-63, DIR WHEAT RES & PROD PROG, INT MAIZE & WHEAT IMPROV CTR, MEX, 64- *Concurrent Pos:* Consult, Food & Agr Orgn UN, NAfrica & Asia, 60; consult & collabr, Nat Inst Invest Agr, Mex Ministry Agr, 60-64; mem, Citizen's Comn Sci, Law & Food Supply & Comn Critical Choices for Am, 73- *Honors & Awards:* Nobel Prize for Peace, 70. *Mem:* Nat Acad Sci; Am Phytopath Soc; Soc Am Foresters; Am Chem Soc; hon mem Am Soc Agron. *Res:* Wheat breeding; agronomy; fungicides; weed killers; plant pathology; forestry. *Mailing Add:* Augustin Ahumada 310-F Lomas de Chapultepec Mexico 10 DF Mexico

BORLE, ANDRE BERNARD, b La Chaux-de-Fonds, Switz, May 27, 30; m 66; c 2. PHYSIOLOGY, ENDOCRINOLOGY. *Educ:* Univ Geneva, MD, 55. *Prof Exp:* Intern med, Mt Auburn Hosp, Cambridge, Mass, 56-57; res fel biochem, Harvard Med Sch, 57-59; resident, Cantonal Hosp, Geneva, Switz, 59-61; instr radiation biol, Atomic Energy Proj, Univ Rochester, 61-63; from asst prof to assoc prof, 63-74, PROF PHYSIOL, SCH MED, UNIV PITTSBURGH, 74- *Concurrent Pos:* Ship surgeon, Johnson Line, Stockdm, 56; asst med, Peter Bent Brigham Hosp, Boston, 57-59; Lederle med fac award, 64-67. *Honors & Awards:* Andre Lichtwitz Prize, 70. *Mem:* AAAS; Am Physiol Soc; Endocrine Soc; Biophys Soc. *Res:* Cellular calcium metabolism; general physiology of calcium; mode of action of parathyroid hormone, calcitonin and vitamin D; membrane transport of calcium; calcium and phosphate metabolism; calcium signaling. *Mailing Add:* Dept Physiol Sch Med Univ Pittsburgh Pittsburgh PA 15261

BORM, ALFRED ERVIN, b Houston, Tex, Sept 2, 37; m 58; c 2. ALGEBRA, TOPOLOGY. *Educ:* Univ Tex, BSc, 58, PhD(math), 65; Univ Wash, MA, 61; Univ Tex, MA, 84. *Prof Exp:* Radiation physicist, Col Med, Baylor Univ, 58; assoc res engr, Boeing Co, Wash, 59-62; teaching asst math, Univ Wash, 60-61 & Univ Calif, Berkeley, 61-62; spec instr, Univ Tex, 64-65; from asst prof to assoc prof math, 65-85, assoc prof comput sci, 86-88, ASSOC PROF MATH, SOUTHWEST TEX STATE UNIV, 88- *Concurrent Pos:* Vis assoc prof comput sci, Univ Tex, Austin, 85-87. *Mem:* Math Asn Am. *Res:* Relationships between algebraic structure of set of continuous functions on a topological space and topological structure of the space; model theory; decidability and complexity of algorithms. *Mailing Add:* Dept Math Southwest Tex State Univ San Marcos TX 78666

BORMAN, ALECK, b Toledo, Ohio, July 13, 19; m 46; c 2. BIOCHEMISTRY. *Educ:* Univ Toledo, BS, 41; Univ Ill, PhD(biochem), 45. *Prof Exp:* Sr chemist & sect head biochem div, E R Squibb & Sons, 45-52, head endocrinol res sect, Squibb Inst Med Res, 52-55, head endocrinol res dept, 55-62, dir physiol sect, 62-67, mgr sci personnel rels, 67, personnel mgr res & develop, 67-71, labor rels mgr, 71-86; RETIRED. *Concurrent Pos:* Consult, E R Squibb & Sons, 86-89. *Mem:* AAAS; Soc Exp Biol & Med; Am Asn Cancer Res; Endocrine Soc. *Res:* Protein and steroid hormones; endocrinology. *Mailing Add:* 42 North Dr East Brunswick NJ 08816

BORMAN, GARY LEE, b Milwaukee, Wis, Mar 15, 32; m 71. COMBUSTION, HEAT TRANSFER. *Educ:* Univ Wis-Madison, BS, 54, MS, 56 & MS, 57, PhD(mech eng), 64. *Prof Exp:* Sr engr, Gen Elec Co, 57-60; from asst prof to assoc prof, 64-71, PROF MECH ENG, UNIV WIS-MADISON, 71-, DIR, ENGINE RES CTR, 86- *Honors & Awards:* Ralph Teetor Award, Soc Automotive Engrs, 64, Arch Colwell Award, 66 & Horning Mem Award, 78; Fulbright Lectr, Univ Sarajevo, Yugoslavia, 70;

Fulbright Distinguished Prof, Maribar, Yugoslavia, 90. *Mem:* Nat Acad Eng; Combustion Inst; fel Soc Automotive Engrs. *Res:* Combustion; combustion engines; emissions; unsteady heat transfer, droplet vaporization and sprays in engines; engine cycle simulation and modeling of thermal systems. *Mailing Add:* 4634 Gregg Rd Madison WI 53705

BORMANN, BARBARA-JEAN ANNE, b Bronxville, NY, Oct 5, 58; m 89; c 1. TRANS MEMBRANE ONCOGENES, RECEPTOR OLIGOMERIZATION. *Educ:* Fairfield Univ, BS, 79; Univ Conn, PhD(biomed sci), 85. *Prof Exp:* Fel cytoskeleton, Sch Med, Yale Univ, 85-88, res assoc signal transduction, 88-89; sr scientist, 89-90, PRIN SCIENTIST IMMUNOL, BOEHRINGER INGELHEIM, 91- *Concurrent Pos:* Res affil, Sch Med, Yale Univ, 89- *Honors & Awards:* Gregory Pincas Mem Award, Fedn Am Soc Exp Biologists, 83. *Mem:* Am Soc Cell Biol. *Res:* Analysis of protein-protein interactions in membranes; transmembrane receptors including oncogenes, growth factor receptors and adhesion molecules. *Mailing Add:* Boehringer Ingelheim Dept Immunol 90 E Ridge PO Box 368 Ridgefield CT 06877

BORMANN, FREDERICK HERBERT, b New York, NY, Mar 24, 22; m 52; c 4. ECOLOGY. *Educ:* Rutgers Univ, BS, 48; Duke Univ, MA, 50, PhD(plant ecol), 52. *Prof Exp:* From instr to asst prof bot, Emory Univ, 52-56; from asst prof to prof, Dartmouth Col, 56-66; prof biol, 69-80, OASTLER PROF FOREST ECOL, SCH FORESTRY, YALE UNIV, 66-, PROF FORESTRY & ENVIRON STUDIES, 80-, DIR, ECOSYSTS RES, 82- *Concurrent Pos:* Fel, Ezra Stiles Col, Yale Univ; ecologist, Boston Univ exped, Alaska, 53; vis scientist, Brookhaven Nat Lab, 63-64, Ctr Energy & Environ Studies, Princeton Univ, 84, Air Pollution Info Exchange, Repub China, 85, E-W Environ & Policy Inst, Honolulu, 87 & Univ New Eng, NSW, 90; vis prof, ctr advan studies, Univ Va, 80-81; mem adv comt, Hubbard Brook, 75-, Native Plants Inc, Salt Lake City, 82-84, World Resources Inst, 86-89, Wilderness Soc, 87-; Green prof, Tex Christian Univ, Ft Worth, 87; mem, Nat Comt Scope Fire Prog, Nat Res Coun, Nat Acad Sci, 77-79, Int Environ Progs Comt, 77-80. *Honors & Awards:* George Mercer Award, 54; Cert of Merit, Bot Soc Am. *Mem:* Nat Acad Sci; fel AAAS; Ecol Soc Am (pres, 70-71); Am Acad Arts & Sci; Am Inst Biol Sci. *Res:* Ecology and physiology of Pinus; structure and function of root grafts; structure, function and development of forest ecosystems; author of 162 technical publications. *Mailing Add:* Sch Forestry & Environ Studies Yale Univ 205 Prospect St New Haven CT 06511

BORN, GEORGE HENRY, b Westhoff, Tex, Nov 10, 39. SATELLITE NAVIGATION, SATELLITE OCEANOGRAPHY. *Educ:* Univ Tex, BS, 62, MS, 65, PhD(aeronaut eng), 68. *Prof Exp:* Engr, Ling-Temco-Vought Corp, 62-63; res engr, Tex Ctr Res, 64-67; aerospace technologist, Johnson Space Ctr, 67-70; tech group supvr, Jet Propulsion Lab, 70-83; sr res engr, Ctr Space Res, 83-85; DIR, COLO CTR ASTRODYN RES, 85- *Concurrent Pos:* Lectr, Dept Mech Eng, Univ Houston, 68-70; ed, J Astronaut Sci, 80-83; guest ed, J Geophys Res, 83 & Marine Geod, 84; sci consult, Jet Propulsion Lab, 83-86, Gen Elec, 84-85, Naval Surface Weapons Ctr & Naval Ocean & Atmosphere Res Lab, 85-86. *Mem:* Fel Am Astronaut Soc; fel Am Inst Aeronaut & Astronaut; Inst Navig; Am Geophys Union; Am Meteorol Soc; AAAS. *Res:* Precision orbit determination; satellite navigation; mission design; satellite oceanography; satellite geodesy. *Mailing Add:* Univ Colo Campus Box 431 Boulder CO 80309-0431

BORN, GORDON STUART, b Hammond, Ind, Apr 26, 33; m 57; c 1. HEALTH PHYSICS, PHARMACY. *Educ:* Purdue Univ, BS, 55, MS, 64, PhD(health physics), 66. *Prof Exp:* From teaching asst to teaching assoc, 62-64, from instr to assoc prof, 64-74, PROF MED CHEM & HEALTH SCI, PURDUE UNIV, 74- *Honors & Awards:* Lederle Pharm Fac Award, Lederle Labs, 72. *Mem:* Health Physics Soc. *Res:* Drug and environmental toxicants and application of tracer techniques to analytical problems. *Mailing Add:* Dept Med Chem Purdue Univ West Lafayette IN 47907

BORN, HAROLD JOSEPH, b Evansville, Ind, Nov 22, 22; m 50; c 2. PHYSICS. *Educ:* Rose Polytech Inst, BS, 49; Iowa State Univ, MS, 58, PhD(physics), 60. *Prof Exp:* Equip engr, Phillips Petrol Co, Okla, 52-55; res asst, Ames Lab, AEC, 55-60; res physicist, Whirlpool Corp Res Labs, Mich, 60-61; assoc prof, 61-66, PROF PHYSICS & HEAD DEPT, ILL STATE UNIV, 66- *Mem:* Am Phys Soc; Am Asn Physics Teachers. *Res:* Low temperature thermoelectric effects; thermoelectric refrigeration; solid state physics. *Mailing Add:* Dept of Physics Ill State Univ Normal IL 61761

BORNE, RONALD FRANCIS, b New Orleans, La, Nov 17, 38; m 59; c 3. MEDICINAL CHEMISTRY, ORGANIC CHEMISTRY. *Educ:* Loyola Univ, La, BS, 60; Tulane Univ, MS, 62; Univ Kans, PhD(med chem), 67. *Prof Exp:* Res asst chem, Ochsner Med Found, 56-62; res chemist, C J Patterson Co, Mo, 62-63 & Mallinckrodt Chem Works, 67-68; asst prof pharmaceut chem, 68-70, assoc prof med chem, 70-72, PROF MED CHEM, UNIV MISS, 72-, CHMN DEPT, 79- *Concurrent Pos:* NIH res grant, 75-78. *Mem:* Am Chem Soc; Am Acad Pharmaceut Sci; Int Soc Heterocyclic Chem; Am Asn Cols Pharm. *Res:* Medicinal chemistry, especially organic syntheses and conformational aspects of drug action. *Mailing Add:* Dept of Medicinal Chem Univ of Miss Sch of Pharm University MS 38677

BORNEMANN, ALFRED, b Montclair, NJ, Apr 6, 06; m 86; c 3. METALLURGICAL ENGINEERING. *Educ:* Stevens Inst Technol, ME, 27; Dresden Tech Univ, DEng, 30. *Prof Exp:* From asst prof to assoc prof eng chem, 30-48, prof metall & head dept, 48-74, EMER PROF METALL, STEVENS INST TECHNOL, 75- *Concurrent Pos:* Dir, Peirce Mem Lab, 40-48; trustee, Stevens Acad, 44-74, pres, 69-73; bd mgr, Nantucket Maria Mitchell Assoc, 60-, pres, 74-80. *Honors & Awards:* Award of Merit, Am Soc Metals, 70. *Mem:* Am Inst Mining, Metall & Petrol Engrs; Am Chem Soc; fel Am Soc Metals; Am Soc Testing & Mat. *Res:* Heat treatment and carburizing of steels; properties of metals; corrosion; failure analysis. *Mailing Add:* 63 Cliff Rd Nantucket MA 02554

BORNEMEIER, DWIGHT D, b Limon, Colo, Oct 29, 34; m 63; c 1. PHYSICS. *Educ:* NCent Col, Ill, BA, 56; Kans State Univ, MS, 60, PhD(physics), 65. *Prof Exp:* Res physicist, Naval Ord Test Sta, 56-59; instr physics, Kans State Univ, 64-65; assoc res physicist, Univ Mich, 65-69, asst prof elec eng, 68-69; vpres, Sensors, Inc, 69-72; res physicist, Environ Res Inst Mich, 73-82; PRES, JL, INC, 82- *Mem:* Am Phys Soc; Optical Soc Am. *Res:* Nuclear spectroscopy; magnetism; coherent optics. *Mailing Add:* Park Davis Biochem Sect 2800 Plymouth Rd Ann Arbor MI 48105

BORNMANN, JOHN ARTHUR, b Pittsburgh, Pa, May 1, 30; m 54; c 2. PHYSICAL CHEMISTRY. *Educ:* Carnegie Inst Technol, BS, 52; Technische Hockschule (physics), Stuttgart, Ger, 56-57, Ind Univ, PhD(phys chem), 58. *Prof Exp:* Res chemist, E I du Pont de Nemours & Co, 58-60; res assoc, Princeton Univ, 60-61; asst prof chem, Northern Ill Univ, 61-65; assoc prof, Lindenwood Col, 65-68, chmn, div natural sci & math, 67-71 & 74-81, chmn dept chem, 65-90, prof chem, 68-91; RETIRED. *Concurrent Pos:* Owner, DAS Co, 74-; vis scientist, McDonnell Douglas Res Labs, 84-; Fulbright grantee, Ger, 56-57. *Mem:* AAAS; Am Chem Soc; fel Am Inst Chemists. *Res:* Interdisciplinary applications of physical chemistry. *Mailing Add:* Three Briarwood Lane St Charles MO 63301

BORNMANN, PATRICIA L(EE), b Long Branch, NJ, Feb 22, 58; m 83. SOLAR PHYSICS. *Educ:* Calif Inst Technol, BS, 79; Univ Colo, Boulder, PhD(astrophys), 85. *Prof Exp:* Teaching asst, solar physics lab, astrophys dept, Univ Colo, 79, res asst, Joint Inst Lab Astrophys, 79-85; astrophysicist, Solar Physics Br, Air Force Geophys Lab, 85-86; res assoc, Univ Colo & NOAA, Space Environ Lab, 88-90, resident res assoc, Nat Res Coun, 86-88, PHYSICIST, SPACE ENVIRON LAB, NAT OCEANIC & ATMOSPHERIC ADMIN, 90- *Concurrent Pos:* Res asst, Big Bear Solar Observ, 77-78, Nat Radio Astron Observ, 79 & Sacramento Peak Solar Observ, 80 & 81; res grant, Richter Scholar, Calif Inst Technol, 78 & 79; travel grant, Solar Physics div, Am Astron Soc, 80 & 84; guest invest Solar Maximum Mission, 86- *Honors & Awards:* Billings Award in Astrogeophys, 85. *Mem:* Am Astron Soc; Int Astron Union; Am Geophys Union. *Res:* Interpretation of the evolution of soft x-ray lines observed during solar flares; understanding the evolution of the high temperature flare plasma; analysis of evolution of solar active regimes and the production of solar flares. *Mailing Add:* 4261 Milliken Ct Boulder CO 80303

BORNMANN, ROBERT CLARE, b Pittsburgh, Pa, June 29, 31; m 63; c 4. MEDICINE, PHYSIOLOGY. *Educ:* Harvard Univ, AB, 52; Univ Pa, MD, 56, MS, 63; Am Bd Preventive Med, dipl. *Prof Exp:* Officer-in-chg, Cape Hallett Int Geophys Year Base, Antarctica, 58-59, med officer, Underwater Swimmers Sch, Key West, Fla, 61-62, Washington, DC, 61-65, Deep Sea Divers Sch, Wash Navy Yard, Washington, DC, 63-65 & Exp Diving Unit, 65-68, dep asst med effects, Deep Submergence Systs Proj, 68-70, exchange officer underwater med, Royal Naval Physiol Lab, Alverstoke, Eng, 70-72, head submarine & diving med, Naval Med Res & Develop Command, Bethesda, 73-78, asst med effects Naval Dep, Nat Oceanic & Atmospheric Admin, Dept Com, Washington, 78-81; staff dir, Defense Med Standardization Bd, Frederick, Md, 81-85; CONSULT, LIMETREE CONSULTS, RESTON, VA, 85- *Concurrent Pos:* staff med off, Oceanogr Navy, Washington, DC, 73-78. *Mem:* AMA; AAAS; Aerospace Med Asn; Am Col Occup Med; Royal Soc Med; Undersea Med Soc; Europ-US Biomed Soc; Marine Tech Soc; Am Polar Soc. *Res:* Submarine, diving and occupational medicine; pathogenesis and treatment of decompression sickness and air embolism; development of decompression schedules for divers; design and development of diving equipment; worker's disability. *Mailing Add:* Limetree Med Consults 11569 Woodhollow Ct Reston VA 22091

BORNONG, BERNARD JOHN, chemistry, propellant manufacturing, for more information see previous edition

BORNS, HAROLD WILLIAM, JR, b Cambridge, Mass, Nov 28, 27; m 53, 81; c 4. QUATERNARY GEOLOGY. *Educ:* Tufts Univ, BS, 51; Boston Univ, MA, 55, PhD(geol), 59. *Prof Exp:* From instr to assoc prof, 55-68, chmn dept, 71-74, PROF GEOL, UNIV MAINE, 68-, DIR, INST QUATERNARY STUDIES, 72- *Concurrent Pos:* Fel, Dept Geol, Yale Univ, 63-64; vis prof, Geol Inst, Univ Bergen, Norway, 75; prog mgr glaciol, Div Polar Prog, NSF, 88-90. *Mem:* Geol Soc Am; Am Polar Soc; Glaciol Soc. *Res:* Glacial geology; quaternary history of Northeast North America, Antarctica and Norway; quaternary climates; environments of paleoindians in America. *Mailing Add:* Dept Geol Sci Univ Maine Orono ME 04473

BORNSIDE, GEORGE HARRY, b Wakefield, RI, Oct 8, 25; m 59; c 1. BACTERIOLOGY. *Educ:* Trinity Col, BS, 48; Univ Conn, MS, 50; Univ Iowa, PhD(bact), 55. *Prof Exp:* Instr, Univ Iowa, 55-56; assoc marine microbiol, Univ Ga, 56; microbiologist, Brooklyn Bot Garden, 56-57; from asst prof to prof, 57-86, EMER PROF SURG & MICROBIOL, SCH MED, LA STATE UNIV, NEW ORLEANS, 86- *Mem:* Am Soc Microbiol; Asn Gnotobiotics (pres, 74-75); Soc Exp Biol & Med. *Res:* Surgical bacteriology; bacterial virulence; antibacterial effects of hyperbaric oxygen; intestinal bacteria. *Mailing Add:* 2200 Leo Sea Simon Dr New Orleans LA 70122

BORNSLAEGER, ELAYNE A, b Milwaukee, Wis, Aug 27, 60. MOLECULAR EMBRYOLOGY. *Educ:* Univ Pa, PhD(biol), 86. *Prof Exp:* Postdoctoral neuromuscular develop, Rockefeller Univ, 87-88; POSTDOCTORAL FEL, DEPT MOLECULAR EMBRYOL, SLOAN KETTERING INST, 88- *Concurrent Pos:* Nat res serv award fel, NIH, 89. *Mem:* Am Soc Cell Biol; Soc Develop Biol; AAAS. *Mailing Add:* Dept Molecular Embryol Sloan-Kettering Inst 1275 York Ave New York NY 10021

BORNSTEIN, ALAN ARNOLD, b Brooklyn, NY, Oct 24, 50; m 72; c 2. PHYSICAL CHEMISTRY, TECHNICAL MANAGEMENT. *Educ:* Polytech Inst Brooklyn, BS, 72; Yale Univ, MS, 73. *Prof Exp:* Res scientist, Int Paper Co, 73-74; teaching fel anal chem, Brooklyn Col, 74-75; res chemist,

Vet Admin Res Found, 75-77; sr scientist, Am Can Co, 77-81; res dir, Olin Corp, 81-89; FUJI HUNT PHOTO CHEM, 89- *Mem:* Am Chem Soc; Soc Appl Spectros; Soc Photographic Scientists & Engrs; Nat Asn Photographic Mfrs. *Res:* Analytical chemistry research management; spectroscopy; chromatography; classical techniques. *Mailing Add:* 16 Beech St Westwood NJ 07675

BORNSTEIN, JOSEPH, b Boston, Mass, June 24, 21; div; c 2. AGRICULTURAL ENGINEERING. *Educ:* Univ Mass, BS, 47; Mich State Univ, MS, 49. *Prof Exp:* Asst, Mich State Col, 48; agr engr, Soil Conserv Serv, USDA, 49-54, civil engr, 54-57, agr engr, Agr Res Serv, 57-85; RETIRED. *Concurrent Pos:* Asst prof, Univ Vt, 61-71, adj assoc prof, 71-75; fac assoc agr eng, Univ Maine, 75-85; consult, 85- *Mem:* Am Soc Agr Engrs; fel Soil & Water Conserv Soc; Nat Soc Prof Engrs; Sigma Xi; Am Soc Civil Engrs. *Res:* Agricultural drainage; soil and water conservation engineering and education; improve farming methods for cool climate of Northeast; developing effective land drainage methods to improve environment for survival and growth of quality feed grains and forages on poorly drained cropland soils, clays, silty clay loam and fragipans. *Mailing Add:* 465 North St Burlington VT 05401-1620

BORNSTEIN, JOSEPH, b Boston, Mass, Feb 19, 25; m 54; c 3. SYNTHETIC ORGANIC CHEMISTRY. *Educ:* Boston Col, BS, 46; Mass Inst Technol, PhD(chem), 49. *Prof Exp:* Chemist, Tracerlab, Inc, 49-50; PROF CHEM, BOSTON COL, 50- *Concurrent Pos:* Consult, Qm, US Army. *Mem:* Am Chem Soc; Soc Chem Indust. *Res:* Synthesis of insecticides and synergists; organic fluorine compounds; heterocyclic compounds; carbon-14. *Mailing Add:* 24 Gary Rd Needham Heights MA 02194

BORNSTEIN, LAWRENCE A, b New York, NY, Sept 15, 23; m 45; c 2. HISTORY AND PHILOSOPHY OF SCIENCE. *Educ:* City Col New York, BS, 44; NY Univ, MS, 51, PhD(physics), 57. *Prof Exp:* Tutor physics, City Col New York, 46-50; from instr to assoc prof, NY Univ, 50-68, assoc dean, 68-69, chmn dept, 69-73, assoc chmn dept, 73-82, actg dean, 84-85, PROF PHYSICS, NY UNIV, 68- *Mem:* Hist Sci Soc; AAAS; Am Phys Soc; Am Asn Physics Teachers; Sigma Xi. *Res:* History of physics. *Mailing Add:* 769 Grange Rd Teaneck NJ 07666

BORNSTEIN, MICHAEL, b Zarki, Poland, May 2, 40; US citizen; m 67; c 2. PHYSICAL PHARMACY, ANALYTICAL CHEMISTRY. *Educ:* Fordham Univ, BS, 62; Univ Iowa, PhD(phys anal pharm), 66. *Prof Exp:* Sr pharmacist, Pitman-More Div, Dow Chem Co, 66-67; sr pharmaceut chemist, 67-, SR FORMULATION CHEMIST, ELI LILLY & CO. *Mem:* Am Pharmaceut Asn; Acad Pharmaceut Sci. *Res:* Pilot plant operations in liquid and ointment development; color measurement; drug adjuvant interactions in solid dosage forms observed with diffuse reflectance spectroscopy; development of parenteral cephalosporids. *Mailing Add:* 90 Barchester Way Westfield NJ 07090

BORNSTEIN, PAUL, b Antwerp, Belg, July 10, 34; US citizen. CONNECTIVE TISSUE BIOLOGY, CELL BIOLOGY. *Educ:* Cornell Univ, BA, 54; NY Univ, MD, 58. *Prof Exp:* Res investr, NIH, 65-67; from asst prof to assoc prof, 67-73, PROF BIOCHEM & MED, UNIV WASH, 73- *Concurrent Pos:* Lederle med fac award, 68-71; NIH res career develop award, 69-74; Macy fac scholar award, 75; vis prof, Univ Calif, San Diego & Weizmann Inst, Israel, 75-76 & Fac Med, Inst Biol Chem, France, 85-86; assoc ed, Arteriosclerosis, 80-85; Guggenheim Found fel, 85. *Honors & Awards:* Merit Award, Nat Inst Arthritis, Musculoskeletal & Skin Dis, 89. *Mem:* Am Soc Biochem & Molecular Biol; Am Soc Cell Biol; Am Chem Soc; Am Rheumatism Asn; Asn Am Physicians; Am Soc Clin Invest. *Res:* Biology of the extracellular matrix; role of the extracellular matrix in regulating cell function; regulation of collagen and thrombospondin gene expression. *Mailing Add:* Dept Biochem SJ-70 Univ Wash Seattle WA 98195

BORNSTEIN, ROBERT D, b New York, NY, July 21, 42; m 64; c 2. METEOROLOGY. *Educ:* City Col New York, BS, 64; NY Univ, MS, 67, PhD(meteorol), 72. *Prof Exp:* Asst res scientist, NY Univ, 64-68; instr meteorol, State Univ NY Maritime Col, 68; asst prof, 69-77, ASSOC PROF METEOROL, SAN JOSE STATE UNIV, 77- *Concurrent Pos:* Consult, Ames Res Ctr, NASA, Calif, 69- *Mem:* Am Meteorol Soc. *Res:* Urban meteorology, especially temperature, wind and humidity distribution in an urban area. *Mailing Add:* Dept Meteorol San Jose State Univ San Jose CA 95192

BOROM, MARCUS P(RESTON), b Waynesboro, Ga, Feb 21, 34; m 58; c 4. CERAMICS ENGINEERING, MATERIALS SCIENCE. *Educ:* Ga Inst Technol, BCerE, 56; Univ Calif, Berkeley, MS, 64, PhD(ceramics eng), 65. *Prof Exp:* Develop engr, Ferro Corp, Ohio, 57-60, res engr, 60-61; staff ceramist, Res & Develop Ctr, Gen Elec Co, 65-74, tech coordr, 74-75, prog mgr, 75-84, SR SCIENTIST, GEN ELEC CORP RES & DEVELOP, 84- *Honors & Awards:* Ross Coffin Purdy Award, Am Ceramic Soc, 77. *Mem:* Fel Am Ceramic Soc. *Res:* Physical chemistry of glass-metal interfaces; mechanisms of adherence between dissimilar materials; diffusion kinetics in multicomponent systems; high temperature phase equilibria; properties of glass and enamels; high temperature molds and cores; hard metals and ceramic cutting tool materials; ceramic high temperature structural composites. *Mailing Add:* Gen Elec Corp Res & Develop PO Box 8 Schenectady NY 12301

BORON, WALTER FRANK, b Elyria, Ohio, Nov 18, 49; m 74; c 2. TRANSPORT PHYSIOLOGY, SIGNAL TRANSDUCTION. *Educ:* St Louis Univ, BA, 71; Wash Univ, St Louis, MD & PhD(physiol), 77. *Prof Exp:* From asst prof to assoc prof physiol, 80-87, PROF CELLULAR & MOLECULAR PHYSIOL, YALE UNIV SCH MED, 87- *Concurrent Pos:* Assoc ed, Physiol Reviews, 85; ed J Physiol, 85; young investr award, Am Soc Nephrol & Am Heart Asn, 85- *Mem:* Am Phys Soc; Am Soc Nephrol; Soc Gen Physiologists (treas, 88-). *Res:* Regulation of intracellular acid concentration optical and electrophysiological techniques in single cells;

identifying new acid-base transport processes; their control by growth factors; examining them at a molecular level using biochemical and genetic approaches. *Mailing Add:* Dept Cellular & Molecular Physiol Sch Med Yale Univ 333 Cedar St New Haven CT 06510

BOROS, DOV LEWIS, b Budapest, Hungary, Mar 4, 32; US citizen; m 47; c 2. MICROBIOLOGY, IMMUNOLOGY. *Educ:* Hadassah Med Sch, Hebrew Univ, Jerusalem, MSc, 58, PhD(immunochem), 67. *Prof Exp:* Asst dir, Vaccine & Serum Inst, Jerusalem, 60-68; USPHS grant, Sch Med, Case Western Reserve Univ, 68-70, asst prof immunol, 70-74; assoc prof, 74-80, PROF IMMUNOL & MICROBIOL, SCH MED, WAYNE STATE UNIV, 80- *Concurrent Pos:* Fogarty Sr Int Fel, 81-82; assoc ed, J Immunol, 83-87. *Mem:* Am Soc Trop Med & Hyg; Am Asn Immunologists. *Res:* Delayed hypersensitivity type granulomatous inflammation; parasitic immunity. *Mailing Add:* Dept Immunol & Microbiol Wayne State Univ Sch Med Detroit MI 48202

BOROSH, ITSHAK, b Fes, Morocco, Oct 22, 38; Israel citizen; m 62; c 2. MATHEMATICS, NUMBER THEORY. *Educ:* Hebrew Univ, Israel, MSc, 61; Weizmann Inst Sci, Israel, PhD(math), 66. *Prof Exp:* Lectr, Bar-Ilan Univ, Israel, 66-70; vis lectr, Univ Ill, Urbana, 70-72; vis asst prof, 72-74, asst prof, 74-76, ASSOC PROF MATH, TEX A&M UNIV, 76- *Mem:* Israel Math Soc; Am Math Soc; Math Asn Am. *Res:* Diophantine approximations; relations between number theory and computing. *Mailing Add:* Dept Math Tex A&M Univ College Station TX 77843

BOROVSKY, DOV, b Tel Aviv, Israel, Dec 4, 43; m 72. CHEMISTRY. *Educ:* Univ Calif, Los Angeles, BA, 67; Univ Miami, PhD(biochem), 72. *Prof Exp:* From res instr to res asst prof biochem, Univ Miami, 73-75; BIOCHEMIST, FLA MED ENTOM LAB, 75- *Mem:* AAAS; Am Chem Soc; Sigma Xi. *Res:* Starch and glycogen metabolism and structure; production of proteolytic enzymes and its regulation in insects; hormonal control of insects metabolism and egg development. *Mailing Add:* 135 36th Ct Vero Beach FL 32968

BOROWIECKI, BARBARA ZAKRZEWSKA, b Warsaw, Poland, Nov 20, 24; US citizen; m 71; c 1. ENVIRONMENTAL SCIENCE, GEOMORPHOLOGY. *Educ:* Ind Univ, BS, 56, MS, 57; Univ Wis, PhD(geog), 62. *Prof Exp:* From instr to assoc prof, 60-69, chmn dept, 75-80 & 83-85, PROF GEOG, UNIV WIS-MILWAUKEE, 69- *Concurrent Pos:* Vis prof geog, Ind Univ, 63-64; fel, Nat Res Coun, NSF; ETS consult. *Mem:* Asn Am Geog (secy); Sigma Xi; Am Geog Soc; Int Geog Union; Polish Inst Arts & Sci in Am. *Res:* Pleistocene geomorphology; regional geomorphology of the United States; spatial terrain problems; valley evolution; loess terrain; numerical terrain analysis. *Mailing Add:* Dept Geog Univ Wis Milwaukee WI 53201

BOROWITZ, GRACE BURCHMAN, b New York, NY, Dec 7, 34; m 59; c 2. ORGANIC CHEMISTRY, SYNTHETIC ORGANIC & NATURAL PRODUCTS CHEMISTRY. *Educ:* City Col New York, BS, 56; Yale Univ, MS, 58, PhD(org chem), 60. *Prof Exp:* Asst, Yale Univ, 56-57; res chemist, Am Cyanamid Co, 60-62; lectr gen chem, Yeshiva Col, 67; US Dept Health, Educ & Welfare teaching fel org chem & biochem, Uppsala Col, 67-68, asst prof, 68-73; from asst prof to assoc prof, 73-80, PROF CHEM, RAMAPO COL NJ, 80- *Concurrent Pos:* Sigma Xi grants, 68-71 & 79; NSF URP grant, 70; Ramapo Col NJ State res fac develop grant, 75-85; NSF NMR grant, 85, NJDHE grants, 85, 86 & 87; Sci Enrich HS Students, Women Minorities, NJDHE-IR grant, 86; vis prof, Columbia Univ, 87- *Mem:* Am Chem Soc; fel AAAS; Sigma Xi; NY Acad Sci; Am Asn Univ Professors. *Res:* Mechanisms in organophosphorus chemistry, including nucleophilicities of tricovalent phosphorus compounds and lithium aluminum hydride reduction of phosphoranes; synthesis and studies of new ion-selective molecules. *Mailing Add:* Sch Theoret & Appl Sci 505 Ramapo Valley Rd Mahwah NJ 07430

BOROWITZ, IRVING JULIUS, b Brooklyn, NY, May 15, 30; m 59; c 2. BIO-ORGANIC CHEMISTRY. *Educ:* City Col New York, BS, 51; Ind Univ, MA, 52; Columbia Univ, PhD(org chem), 56. *Prof Exp:* Fel biochem, Columbia Univ, 56-57 & 58-58, res assoc, 59-60; fel org chem, Yale Univ, 57-58; instr, City Col New York, 60-62; from asst prof to assoc prof, Lehigh Univ, 62-66; assoc prof, Yeshiva Col, 66-74, prof chem, Belfer Grad Sch Sci, 74-77; res assoc, Allied Chem Corp, 77-80; PROF, YESHIVA UNIV, 80- *Concurrent Pos:* Fel org chem, Columbia Univ, 60-62; res grants, Sigma Xi, 61-62, Am Philos Soc, 61-64; NSF, 63-72, NIH, 64-68, 69-72 & 74-78, USAF, 65-69 & Petrol Res Fund, 66-69; grants, Res Corp, 73 & Health Res Coun, NY, 73-75; mem comt econ status, Am Chem Soc, 73-74; hon res prof, Ramapo Col, 78- *Mem:* Am Chem Soc; NY Acad Sci; Sigma Xi; Am Asn Univ Professors. *Res:* Organic synthesis; selective ion-chelation; computer graphics. *Mailing Add:* Dept Chem Yeshiva Univ New York NY 10033

BOROWITZ, JOSEPH LEO, b Columbus, Ohio, Dec 19, 32; div; c 3. PHARMACOLOGY. *Educ:* Ohio State Univ, BSc, 55; Purdue Univ, MS, 57; Northwestern Univ, PhD(pharmacol), 60. *Prof Exp:* Asst pharmacol, Purdue Univ, 55-57; pharmacologist, Sch Aerospace Med, Univ Tex, 60-62; fel pharmacol, Harvard Med Sch, 63-64; asst prof, Bowman Gray Sch Med, 64-69; assoc prof, 69-74, PROF PHARMACOL & TOXICOL, PURDUE UNIV, WEST LAFAYETTE, 74- *Concurrent Pos:* Sabbatical, Cambridge Univ Med Sch, 76 & Sandoz A G, Basel, Switz, 84. *Mem:* Am Soc Pharmacol & Exp Therapeut; Sigma Xi. *Res:* Absorption of drugs from gastrointestinal tract; metabolism and release of catecholamines; release of granule bound substances; adenosine; calcium; kidney toxicology; cyanide neuro-toxicology. *Mailing Add:* Dept Pharmacol & Toxicol Purdue Univ West Lafayette IN 47906

BOROWITZ, MICHAEL J, b New York, NY, Dec 9, 50; m; c 2. IMMUNO-PATHOLOGY, HEMATOLOGY. *Educ:* Duke Univ, MD & PhD(physiol), 77. *Prof Exp:* ASSOC PROF PATH, DUKE UNIV, 81- *Mailing Add:* Dept Pathol Med Ctr Duke Univ Box 3712 Durham NC 27710

BOROWITZ, SIDNEY, b Brooklyn, NY, June 12, 18. THEORETICAL PHYSICS. *Educ:* City Col NY, BS, 37; NY Univ, MS, 41 & PhD(physics), 48. *Prof Exp:* Physicist, Nat Bur Standards, 41-43; engr, Western Elec, 43-45; instr physics, NY Univ, 46-48 & Harvard Univ, 48-50; from asst prof to prof, 50-85, EMER PROF PHYSICS, NY UNIV, 85- *Concurrent Pos:* Dean, Univ Col, NY Univ, 69-72, provost, Univ Hts Campus, 71-72, chancellor & exec vice pres, 72-77; exec dir, NY Acad Sci, 77-82; chief exec officer, Cistron Biotechnol, 83-85; consult, NY Univ, 85. *Honors & Awards:* JF Kennedy Res Fel, Weizmann Inst, Rehovath Israel, 65-66. *Mem:* Am Phys Soc; AAAS; NY Acad Sci. *Res:* Theoretical physics; atomic structure; atomic and molecular scattering; quantum electrodynamics. *Mailing Add:* 70 E Tenth St New York NY 10003

BOROWSKY, BETTY MARIAN, b Brooklyn, NY, Jan 24, 43; m 65; c 2. ANIMAL BEHAVIOR, INVERTEBRATE ZOOLOGY. *Educ:* Queens Col, BS, 64, MA, 69; City Univ New York, PhD(biol), 78. *Prof Exp:* NY Zool Soc Noyes fel biol, Osborn Labs Marine Sci, 78-80; asst prof, Yeshiva Univ, 80-82; MEM STAFF, OSBORN LABS MARINE SCI, 82- *Concurrent Pos:* Res serv coordr, Polytech Univ, 87- *Mem:* AAAS; Animal Behav Soc; Am Soc Zoologists; Sigma Xi; Estuarine Res Fedn. *Res:* The behavioral ecology of marine invertebrates; behavioral physiology and ecology of benthic peracarids of the temperate zone. *Mailing Add:* 304 W 114th St New York NY 10014

BOROWSKY, HARRY HERBERT, b New York, NY, Apr 26, 14; m 40; c 2. CHEMISTRY. *Educ:* Brooklyn Col, BSc, 34. *Prof Exp:* Teacher pub schs, New York, 36-41 & 43-44; chemist, Onyx Oil & Chem Co, 44-57; tech dir, Intex Chem Co, 57-61 & Nuvite Chem Co, 61-69; vpres res & prod, Control Chem Corp, 69-73; CONSULT CHEMIST, 73- *Mem:* Am Chem Soc; Tech Asn Pulp & Paper Indust; assoc Am Dairy Sci Asn; assoc Am Pub Health Asn; Chem Specialties Mfrs Asn. *Res:* Detergents; paper chemistry; sanitary chemicals and germicides; paint removers; metal treating compounds; applied, pesticide and surface chemistry. *Mailing Add:* 205B Covered Bridge Manalapan Englishtown NJ 07726-4409

BOROWSKY, RICHARD LEWIS, b New York, NY, Oct 21, 43. EVOLUTION, GENETICS. *Educ:* City Univ New York, BA, 64; Yale Univ, MPhil, 67, PhD(evolutionary biol), 69. *Prof Exp:* Asst prof, 70-75, ASSOC PROF BIOL, NY UNIV, 75- *Mem:* Genetics Soc Am; Soc Study Evolution; Am Soc Human Genetics. *Res:* Population genetics; adaptive significance of biochemical and physiological variability in populations; forensic genetics; human population genetics and the biostatistical evaluation of paternity; DNA fingerprinting. *Mailing Add:* Dept Biol NY Univ New York NY 10003

BORR, MITCHELL, chemistry; deceased, see previous edition for last biography

BORRA, ERMANNO FRANCO, b Gattinara, Italy, Mar 23, 43; Can citizen; m 77; c 2. COSMOLOGY, INSTRUMENTATION. *Educ:* Univ Torino, Italy, DSC, 67; Univ Western Ont, PhD(astron), 72. *Prof Exp:* Carnegie fel astron, Hale Observ, 72-74; from asst prof to assoc prof, 75-84, PROF PHYSICS, LAVAL UNIV, 84- *Concurrent Pos:* Vis scientist astron, Steward Observ, 81-82. *Mem:* Can Astron Soc; Can Asn Physicists; Am Astron Soc; Int Astron Union. *Res:* Observational cosmology; large scale structure of the universe; quasars; instrumentation for astronomy; very large optics. *Mailing Add:* Physics Dept Laval Univ Quebec PQ G1K 7P4 Can

BORRAS, CARIDAD, b Barcelona, Spain, Feb 18, 42. RADIOLOGICAL PHYSICS. *Educ:* Univ Barcelona, MS, 64; Am Bd Radiol, cert, 71; Univ Barcelona, DSc(physics), 74. *Prof Exp:* Hosp physicist, Hosp de la Santa Cruz y San Pablo, Barcelona, 64-66; asst physicist, Thomas Jefferson Univ, Philadelphia, 67-73; radiological physicist, W Coast Cancer Found, 74-88; REGIONAL ADV RADIOL HEALTH, PAN AM HEALTH ORGN, 88- *Concurrent Pos:* Adj asst prof, Sch Med Sci, Univ of the Pac, 75-76; consult, State Calif Dept Health, 75-; clin asst prof, Dept Radiol, Univ Calif, 82- *Mem:* Am Asn Physicists Med; Health Physics Soc; Soc Nuclear Med; Am Col Radiol; Radiol Soc NAm. *Res:* Radioembryopathological effects of high LET radionuclides; radionuclide dosimetry; imaging and radiation dosimetry of computerized tomography scanners; physics of diagnostic radiology and radiation therapy. *Mailing Add:* 4582 Indian Rock Terr NW Washington DC 20007

BORREGO, JOSE M, b Ojuel, Durango, Mex, Mar 21, 31; US citizen; m 58; c 3. MICROWAVE ENGINEERING, SOLID STATE DEVICE PHYSICS. *Educ:* Instit Tecnologico de Monterrey, BSc, 55; Mass Inst Technol, MS, 57, ScD (elec eng), 61. *Prof Exp:* Prof, elec eng, Mass Inst Technol, 61-62; prof, elec eng, Centro de Investigacion del IPN, Mex City, 62-65; PROF, ELEC ENG, RENSSELAER POLYTECH INST, TROY, NY, 65- *Honors & Awards:* Sprague Award, Mass Inst Tech, 59. *Mem:* Inst Elec & Electronics Engr; Sigma Xi. *Res:* Electrical characterization of semiconducture electronic devices; non-destructive measurements in semiconductor materials; microwave measurement techniques; physics of semiconductor electronic devices. *Mailing Add:* Elec Comput Systs Eng Dept Rensselaer Polytech Inst Troy NY 12180-3590

BORREGO, JOSEPH THOMAS, b Tampa, Fla, Sept 30, 39. MATHEMATICS. *Educ:* Univ Fla, BA, 61, MS, 62, PhD(math), 66. *Prof Exp:* Instr math, Univ Fla, 65-66; asst prof, 66-72, ASSOC PROF MATH, UNIV MASS, AMHERST, 73- *Mem:* Am Math Soc. *Res:* Topological semigroups. *Mailing Add:* Dept Math Univ Mass Amherst MA 01002

BORRELLI, NICHOLAS FRANCIS, b Philadelphia, Pa, Nov 30, 36; m 60, 89; c 3. OPTICAL PHYSICS. *Educ:* Villanova Univ, BS, 58; Univ Rochester, MS, 60, PhD(chem eng), 62. *Prof Exp:* Sr res physicist, 62-74, res assoc physics, 74-76, SR RES ASSOC PHYSICS, CORNING GLASS WORKS, 76- *Concurrent Pos:* Lectr math, Elmira Col, NY, 69- *Honors & Awards:* Sullivan Award, Am Chem Soc, 89. *Mem:* Am Chem Soc; Soc Photo Optical Instrumentation Engrs; Sigma Xi. *Res:* Infrared and ultra-violet properties of glass; glass lasers; magneto-optics and electrooptic properties of glasses and glass-ceramics; optical properties of single crystals; photochromic materials; gradient refractive index materials; micro optics; nonlinear optics. *Mailing Add:* Corning Glass Works Sullivan Res Lab Corning NY 14830

BORRELLI, ROBERT L, b Clarksburg, WVa, Mar 4, 32; m 56; c 4. APPLIED MATHEMATICS. *Educ:* Stanford Univ, BS, 53, MS, 54; Univ Calif, Berkeley, PhD(appl math), 63. *Prof Exp:* Mathematician, Gen Tel & Elec Labs, 59-62; asst prof math, US Naval Postgrad Sch, 62-63; mathematician, Philco Corp, 63-64; from asst prof to assoc prof, 64-73, PROF MATH, HARVEY MUDD COL, 73-, CHMN, MATH DEPT, 81-, DIR MATH CLIN, 81- *Mem:* Am Math Soc; Math Asn Am. *Res:* Partial differential equations; non-elliptic boundary problems. *Mailing Add:* Dept Math Harvey Mudd Col Claremont CA 91711

BORROR, ALAN L, b Ambridge, Pa, June 4, 34; m 59; c 1. ORGANIC CHEMISTRY. *Educ:* Drexel Inst Technol, BS, 57; Princeton Univ, PhD(org chem), 61. *Prof Exp:* Fel, Harvard Univ, 61-62; from asst prof to assoc prof chem, Drexel Inst Technol, 62-67; from scientist to sr scientist, Polaroid Corp, 67-70, res group leader, 70-74, lab mgr, 74-80, dir org chem, 80-88; RETIRED. *Mem:* Am Chem Soc. *Res:* Synthesis and reactions of heterocyclic compounds; mechanisms of organic reactions. *Mailing Add:* 19 Cape Woods Dr Cape Elizabeth ME 04107-1250

BORROR, ARTHUR CHARLES, b Columbus, Ohio, May 27, 35; m 57; c 2. PROTOZOOLOGY. *Educ:* Ohio State Univ, BSc, 56, MSc, 58; Fla State Univ, PhD(protozool), 61. *Prof Exp:* From asst prof to assoc prof, 61-74, PROF ZOOL, UNIV NH, 74- *Concurrent Pos:* Assoc dir, Shoals Marine Lab. *Mem:* Am Micros Soc; AAAS; Soc Protozool. *Res:* Morphology, ecology, systematics and distribution of marine ciliated Protozoa. *Mailing Add:* Dept Zool Univ NH Durham NH 03824

BORROWMAN, S RALPH, b Bedford, Wyo, Oct 29, 18; m 41; c 6. CHEMISTRY. *Educ:* Univ Idaho, BS, 41; Univ Utah, 44, PhD(chem), 50. *Prof Exp:* Chem engr, Deseret Chem Warfare Depot, 44; res chemist, US Bur Mines, 50-58, supvr res chemist, 58-80; RETIRED. *Mem:* Am Chem Soc. *Res:* Recovery of metals from low grade domestic ores by hydrometallurgical processes such as solvent extraction and ion exchange. *Mailing Add:* 158 W 2800 S Bountiful UT 84010

BORSA, JOSEPH, b Wakaw, Sask, Aug 5, 38; m 60; c 3. VIROLOGY, CELL BIOLOGY. *Educ:* Univ Sask, BSc, 61, MSc, 63; Univ Toronto, PhD(biophys), 67. *Prof Exp:* Nat Cancer Inst Can fel, Wistar Inst Anat & Biol, Pa, 67-69; res officer, Med Biophys Br, Whiteshell Nuclear Res Estab, Atomic Energy Can Ltd, 69-85; RES OFFICER, RADIATION RES BR, WHITESHELL LABS, AECL RES, 85- *Mem:* Am Soc Microbiol; Can Soc Cell Biol; Radiation Res Soc. *Res:* Animal viruses; cell biology; antimetabolites; radiobiology; radiation processing; mycotoxins. *Mailing Add:* AECL Res Whiteshell Labs Pinawa MB R0E 1L0 Can

BORSARI, BRUNO, b Cesena, Italy, Jan 11, 61; m 89. ZOOLOGY, ANIMAL SCIENCE. *Educ:* Bologna Univ, Italy, Doctorate(agr sci), 86. *Prof Exp:* Instr agron & animal sci, St Joseph Voc Ctr, Lunsar-Sierra Leona, WAfrica, 87-89; INSTR BIOL, LA STATE UNIV, EUNICE, 91- *Res:* Classification and propagation on a protected area of the native plants and flowers of the prairie of southwestern Louisiana. *Mailing Add:* Rte 3 Box 564 Eunice LA 70535

BORSE, GAROLD JOSEPH, b Detroit, Mich, Dec 20, 40; m 63; c 1. THEORETICAL PHYSICS. *Educ:* Univ Detroit, BS, 62; Univ Va, MS, 64, PhD(physics), 66. *Prof Exp:* Asst prof, 66-71, ASSOC PROF PHYSICS, LEHIGH UNIV, 71- *Mem:* Am Phys Soc; Am Asn Physics Teachers. *Res:* Nuclear structure theory; meson-baryon bound states. *Mailing Add:* Dept Physics Lehigh Univ Bethlehem PA 18015

BORSO, CHARLES S, b Ann Arbor, Mich, July 3, 46; m 72; c 3. SMALL ANGLE SCATTERING, ARTIFICIAL INTELLIGENCE. *Educ:* Princeton Univ, PhD(physics), 79. *Prof Exp:* Res consult, Bell Labs, 77-79; asst biophysicist, Argonne Nat Lab, 79-85. *Honors & Awards:* Indust Res-100 Award, 83 & 85. *Mem:* AAAS. *Res:* Artificial intelligence techniques applied to automated programming algorithms; intelligent systems. *Mailing Add:* 345 Cottonwood Naperville IL 60540

BORSOS, TIBOR, b Budapest, Hungary, Mar 12, 27; US citizen; m 50; c 2. CANCER, IMMUNOLOGY. *Educ:* Cath Univ, BA, 54; Johns Hopkins Univ, ScD(hyg), 58. *Prof Exp:* Res fel microbiol, Johns Hopkins Univ, 58-60, asst prof, 60-62; res chemist, 62-66, head immunochem sect, 66-76, assoc chief biol br, 71-76, head humoral immunity sect, 76-80, assoc chief, 76-85, CHIEF LAB IMMUNOBIOL, NAT CANCER INST, 85-; RES PROF PATH, US UNIFORMED HEALTH SERV, 87- *Honors & Awards:* Alexander von Humboldt Award, 88. *Mem:* Am Asn Immunol. *Res:* Tumor immunology; immunochemistry; action, mechanism and characterization of complement components. *Mailing Add:* Dept Path Uniformed Serv Health Sci 4301 Jones Bridge Rd Bethesda MD 20814

BORST, DARYLL C, b Pontiac, Mich, July 8, 40; m 65; c 2. LIMNOLOGY. *Educ:* Ferris State Univ, BS, 62; Cent Mich Univ, MA, 64; Univ Ill, Urbana, PhD(zool), 68. *Prof Exp:* Chemist, Gt Lakes-Ill River Basins Proj, Fed Water Pollution Control Admin, 64; assoc prof, 68-73, PROF BIOL, QUINNIPIAC COL, 73- *Concurrent Pos:* Consult limnol & lake restoration, 75- *Mem:* Am Soc Limnol & Oceanog; Am Micros Soc; NAm Benthological Soc; Ecol Soc Am; Sigma Xi; NAm Lake Mgt Soc. *Res:* Limnological studies of alpine lakes and streams; limnology of lake restoration. *Mailing Add:* Dept Biol Quinnipiac Col Hamden CT 06518

BORST, DAVID W(ELLINGTON), b Jacksonville, Fla; m 41; c 3. ELECTRICAL ENGINEERING. *Educ:* Brown Univ, ScB, 40. *Prof Exp:* Test engr, Gen Elec Co, 40-41, requisition engr, 41-43, proj engr, 43-45, appln engr, 45-58, with low voltage switch gear dept, 58-61; appln engr, 61-62, chief appln engr, 62-65, mgr power devices, Eng Sect, 65-68, mgr devices prod eng, 68-71, mgr customer eng, 71-76, mgr rating & eval 76-78, mgr device rating & nomenclature, 78-75, MGR ENG ADMIN, INT RECTIFIER CORP, EL SEGUNDO, 85- *Concurrent Pos:* Tech mgr, Intercol Broadcasting Syst, 40-

48, opers mgr & eng consult, 48-53, vpres, 53-58, pres, 58, West Coast vpres, 61-71, vchmn-West, 75-, secy, 79- *Mem:* Fel Inst Elec & Electronics Engrs; assoc Sigma Xi. *Res:* Method for supplying power to igniters in mercury arc rectifiers; method for rapid switching of a loading resistor by means of an ignitron. *Mailing Add:* 2104 Chelsea Rd Palos Verdes Estates CA 90274

BORST, LYLE BENJAMIN, b Chicago, Ill, Nov 24, 12; m 39; c 3. PHYSICS. *Educ:* Univ Ill, AB, 36, AM, 37; Univ Chicago, PhD(chem), 41. *Prof Exp:* Instr chem & gen sci, Univ Chicago, 40-41, res assoc, Metall Lab, 41-43; sr physicist, Clinton Lab, Oak Ridge, 43-46; asst prof chem, Mass Inst Technol, 46-51; prof physics, Univ Utah, 51-54; chmn dept, NY Univ, 54-61; master, Clifford Furnas Col, 68-73, prof physics, State Univ NY, Buffalo, 62-83. *Concurrent Pos:* Chmn dept reactor sci & eng, Brookhaven Nat Lab, 46-51. *Mem:* AAAS; fel Am Phys Soc. *Res:* Neutron and general nuclear physics; nuclear reactor design and development; infrared spectroscopy; liquid helium; mathematics and astronomy in pre-history. *Mailing Add:* 17 Twin Bridge Lane Williamsville NY 14221

BORST, ROGER LEE, b Madison, Wis, Apr 14, 30; div; c 5. MINERALOGY. *Educ:* Univ Wis, BS, 56, MS, 58; Rensselaer Polytech Inst, PhD(minerol), 65. *Prof Exp:* Assoc cur geol & mineral, NY State Mus & Sci Serv, 58-66; res mineralogist, 66-69, sr res mineralogist, 69-81, res assoc, Phillips Petrol Co, 81-85; RETIRED. *Mem:* Fel Geol Soc Am; Mineral Soc Am; Clay Minerals Soc; Int Asn Study Clays; Sigma Xi. *Res:* Sedimentary petrology; clay mineralogy. *Mailing Add:* 823 SE Sixth Bend OR 97702

BORST, WALTER LUDWIG, b Prague, Czech, Sept 12, 38; m 64; c 1. ATOMIC PHYSICS, ATMOSPHERIC PHYSICS. *Educ:* Univ Tubingen, BS, 60, MS, 64; Univ Calif, Berkeley, PhD(physics), 68. *Prof Exp:* Res assoc, Space Res Coord Ctr, Univ Pittsburgh, 68-69, res asst prof physics, 70; asst prof, 71-75, assoc prof physics, Southern Ill Univ, 75- *Mem:* Am Phys Soc; Am Geophys Union; Sigma Xi. *Res:* Atomic and molecular collision processes; reactions between electrons, ions and atmospheric gases; metastable spectroscopy; auroral phenomena; surface physics; solar energy research; solar heating and cooling of buildings; thermal design of buildings. *Mailing Add:* Dept Physics Tex Tech Univ Lubbock TX 79409

BORST-EILERS, ELSE, MEDICINE. *Educ:* Univ Amsterdam, 58. *Prof Exp:* Mem, Dept Hemat, Univ Hosp, Utrecht, 66-69, dir, 69-77, med dir, 77-86; VPRES, HEALTH COUN NETH, 86- *Concurrent Pos:* Pres, Dutch Coun Bloodtransfusion; secretariat, Int Soc Technol Assessment. *Mem:* Inst Med-Nat Acad Sci. *Res:* Cause and prevention of Rh-immunization. *Mailing Add:* Health Coun Neth PO Box 90517 The Hague 2507 LM Netherlands

BORSTING, JACK RAYMOND, b Portland, Ore, Jan 31, 29; m 53; c 2. RESOURCE MANAGEMENT, OPERATIONS RESEARCH. *Educ:* Ore State Univ, BA, 51; Univ Ore, MA, 52, PhD(math statist), 59. *Prof Exp:* Instr math, Western Wash State Col, 53-54; from asst prof to prof opers res & statist, US Naval Postgrad Sch, 59-80, chmn, Dept Opers Anal, 64-71, chmn, Dept Opers Res & Admin Sci, 71-74, provost & acad dean, 74-80; asst secy defense, Pentagon, 80-83; sch bus admin, Univ Miami, 83-88; DEAN SCH ADMIN, UNIV SOUTHERN CALIF, 88- *Concurrent Pos:* Consult, Inst Defense Anal, Ctr Naval Anal; IBM lectr, 66-69; vis prof, Univ Colo, 67, 69 & 71; vis distinguished prof, Ore State Univ, 68; mem, Naval Res Adv Bd Personnel Labs, 71-78; mem adv bd unified sci & math for elem schs, NSF Proj; bd dir, Inst Educ Leadership, 81. *Honors & Awards:* Kimbal Medal, Oper Res Soc Am. *Mem:* Am Statist Asn; Inst Mgt Sci; Opers Res Soc Am (pres, 75-76); Mil Opers Res Soc (pres, 71-72); Sigma Xi. *Res:* Statistical classification techniques; reliability technical management. *Mailing Add:* Dean Sch Bus Admin Univ Southern Calif Los Angeles CA 90089-1421

BORSUK, GERALD M, b Newark, NJ, Dec 15, 44; m 71; c 2. MICROELECTRONICS & PHOTODETECTORS, SOLID STATE ELECTRONICS. *Educ:* Georgetown Univ, BS, 66, MS, 69, PhD(physics), 72. *Prof Exp:* Staff physicist, ITT Electro Physics Lab, Columbia, Md, 73-77; mgr, Westinghouse Adv Tech Lab, Baltimore, Md, 77-83; SUPT, NAVAL RES LAB, WASHINGTON, DC, 83- *Concurrent Pos:* Dep Mem, Adv Group Elec Devices, 83-; mem, Microelectronics Bd, Univ Md, 86-89; adj prof electronic eng & computer sci, George Washington Univ, 88-89. *Honors & Awards:* IR 100 Award, 82. *Mem:* Sigma Xi; Inst Elec & Electronics Engrs. *Res:* Electronic research and development; formulates Naval Res Lab, Navy, and Dept Defense policy in electronics tech base. *Mailing Add:* 3547 Appleton St NW Washington DC 20008

BORTIN, MORTIMER M, BONE MARROW. *Educ:* Marquette Univ, MD, 45. *Prof Exp:* PROF MED & SCI DIR, INT BONE MARROW REGISTRY, MED COL WIS, 51- *Mailing Add:* Transplants Med Col Wis Milwaukee WI 53226

BORTNICK, NEWMAN MAYER, b Minneapolis, Minn, May 14, 21; m 43; c 3. SYNTHETIC ORGANIC CHEMISTRY, ORGANIC POLYMER CHEMISTRY. *Educ:* Univ Minn, BA, 41, PhD(chem), 44. *Prof Exp:* Res chemist, Rohm & Haas Co, Bristol, 44-59, head high pressure res lab, 59-66, res supvr, 66-73, mgr explor process res, 73-81 & plastics res, 81-84, corp res fel, 84-90, CONSULT, ROHM & HAAS CO, BRISTOL, 91- *Mem:* Fel AAAS; Am Chem Soc; Royal Soc Chem; Am Inst Chemists; Soc Plastics Engrs; Sigma Xi. *Res:* Synthetic organic chemistry; polymerization. *Mailing Add:* 509 Oreland Mill Rd Oreland PA 19075-2388

BORTOFF, ALEXANDER, b Cleveland, Ohio, Sept 13, 32; m 60; c 2. PHYSIOLOGY. *Educ:* Western Reserve Univ, BS, 53; WVa Univ, MS, 56; Univ Ill, PhD(physiol), 59. *Prof Exp:* Fel, Univ Ill, 59-60; instr ophthalmic res, Western Reserve Univ, 60-62, instr physiol, 61-62; from asst prof to assoc prof, 62-70, PROF PHYSIOL, STATE UNIV NY UPSTATE MED CTR, 70- *Mem:* AAAS; Am Physiol Soc. *Res:* Electrophysiology of smooth muscle. *Mailing Add:* Dept Physiol State Univ NY Health Sci Ctr 750 E Adams St Syracuse NY 13210

BORTOLOZZI, JEHUD, b Rio Claro, Brazil, Nov 29, 40; m 69; c 3. ANIMAL GENETICS, IMMUNOGENETICS. *Educ:* Univ Sao Paulo, BS, 69, PhD(sci), 71. *Prof Exp:* Teaching asst biol, Nobrega Inst Educ, 64-68; asst prof, 69-83, PROF BIOL, BASIC INST BIOL, MED & AGR BIOL, UNIV SAO PAULO, 83- *Mem:* Soc Brasileira de Genetica; Soc Brasileira para o Progresso da Ciencia; Genetics Soc Am; Am Genetics Asn; Int Soc Animal Blood Group Res. *Res:* Animal population genetics, and immunogenetics of cattle. *Mailing Add:* Dept Genetics Caixa Postal 102 Botucatu Sao Paulo 18610 Brazil

BORTON, ANTHONY, b Bryn Mawr, Pa, June 6, 33; m 57; c 2. ANIMAL SCIENCE. *Educ:* Haverford Col, AB, 55; Mich State Univ, MS, 61, PhD(animal sci), 64. *Prof Exp:* From asst prof to assoc prof, 64-86, PROF ANIMAL SCI, UNIV MASS, AMHERST, 86- *Mem:* Am Soc Animal Sci. *Res:* Reproduction in domestic animals; equine nutrition and physiology. *Mailing Add:* Dept Animal Sci Univ Mass Amherst MA 01003

BORTON, THOMAS ERNEST, b Cleveland, Ohio, Sept, 22, 42; m 70; c 2. AUDIOLOGY, HEARING SCIENCE. *Educ:* Univ Ill, BA, 66, MA, 68, PhD(audiol), 73. *Prof Exp:* Asst prof audiol, Murray State Univ, 73-74; asst prof, Boston Univ, 74-75; assoc prof, Auburn Univ, 75-80; assoc prof audiol, Univ Ala, 80-84; assoc prof, Univ SC, 84-86; ASSOC PROF, UNIV ALA, BIRMINGHAM, 86- *Concurrent Pos:* Assoc ed, J Speech & Hearing Asn, Ala, 76-80; ed consult, J Speech & Hearing Dis 79-86; dir, Speech & Hearing Clin, Children's Hosp, Ala, 81-83. *Mem:* Am Speech, Language & Hearing Asn; Am Auditory Soc; Am Acad Audiol. *Res:* Prevention of hearing loss resulting from exposure to intense noise; treatment of patients suffering from tinnitus aurium; development of electrophysiological diagnostic tools for evaluating the auditory system. *Mailing Add:* Univ Ala Dept Surg Div Otolaryngol 1541 Fifth Ave S Birmingham AL 35233

BORTONE, STEPHEN ANTHONY, b Boston, Mass, Sept 3, 46; m 68; c 2. ICHTHYOLOGY, MARINE BIOLOGY. *Educ:* Albright Col, BS, 68; Fla State Univ, MS, 70; Univ NC, Chapel Hill, PhD(marine sci), 73. *Prof Exp:* From instr to assoc prof, 73-84, PROF BIOL, UNIV WFLA, 84- *Mem:* Am Soc Ichthyologists & Herpetologists; Am Fisheries Soc; Fisheries Soc Brit Isles; Ichthyol Soc Japan; Soc Study Evolution. *Res:* Reproductive biology, systematics and life history of hermaphroditic fishes; fish communities; zoogeography of fishes; fishes-parasitic copepod relationships. *Mailing Add:* Dept Biol Univ WFla Pensacola FL 32504

BORTZ, ALFRED BENJAMIN, b Pittsburgh, Pa, Nov, 20, 44; m 67; c 2. APPLIED SCIENCE & TECHNOLOGY, CHILDRENS SCIENCE TRADE BOOKS. *Educ:* Carnegie Inst Technol, BS, 66, MS, 67; Carnegie Mellon Univ, PhD (physics), 71. *Prof Exp:* From instr to asst prof physics, Bowling Green State Univ, 70-73; res assoc, Yeshiva Univ, 73-74; sr engr, Advan Reactors Div, Westinghouse Elec Co, 74-77; staff scientist, Essex Group, Inc, United Technol, 77-79; scientist, Mellon Inst, 79-83, asst dir, Magnetics Technol Ctr, 83-90, DIR SPEC PROJ ENG EDUC, CARNEGIE MELLON UNIV, 90- *Concurrent Pos:* Consult, Childrens Press, 86-, Harper & Row Jr Books Group, 85-, Franklin Watts Inc, 84-, Nat Res Coun Panel Interfaces & Thin Films, 86, Mat Res Coun, Defense Adv Res Proj Agency, 86-87, AAAS Proj 2061, 85; Groliev, 86-89, Millbrook Pres, 89-; Gen Chair Nat Forum on Children Sci Bk, 92. *Mem:* Am Phys Soc; AAAS; Sigma Xi; Soc Children's Bk Writers. *Res:* Applied magnetics, especially data storage technology; childrens science books and articles; pre college science education, with emphasis on middle school age; public understanding of science and technology. *Mailing Add:* 1312 Foxboro Dr Monroeville PA 15146

BORUCKI, WILLIAM JOSEPH, b Chicago, Ill, Jan 26, 39; m 63; c 3. SEARCH FOR OTHER SOLAR SYSTEMS, PHYSICS OF LIGHTNING. *Educ:* Univ Wis-Madison, BS, 60, MS, 62. *Prof Exp:* Res scientist spectros, 62-72, RES SCIENTIST PLANTETARY SCI, PHYSICS, NASA-AMES RES CTR, 72- *Concurrent Pos:* Guest investr, Pioneer Venus Spacecraft, 88. *Mem:* Am Geophys Union; Am Astron Soc. *Res:* Studies of lightning in planetary atmospheres; development of a system to search for other solar systems. *Mailing Add:* Ames Res Ctr MS 245-3 NASA Moffett Field CA 94035

BORUM, OLIN H, b Spencer, NC, Nov 3, 17; div; c 3. RESEARCH ADMINISTRATION, APPLIED CHEMISTRY. *Educ:* Univ NC, BS, 38, AM, 47, PhD(chem), 49. *Prof Exp:* Asst anal chem, Univ NC, 46-49; res chemist, Philadelphia Lab, E I du Pont de Nemours & Co, 49-50; interim res asst prof, Cancer Res Lab, Univ Fla, 50; res adminr chem div, Off Sci Res Hq, Air Res & Develop Command, 51-52; from instr to asst prof chem, US Mil Acad, 52-55; student, USAF Air Command & Staff Col, 55-56; res adminr res div, Chem Corps Res & Develop Command, DC, 56-60 & res & tech div, Wright-Patterson AFB, Ohio, 60-64; sci adminr, Hq, Army Materiel Command, Res, Develop & Eng Directorate, 64-76; RETIRED. *Concurrent Pos:* Consult, Am Chem Soc Petrol Res Fund, 78. *Mem:* Am Chem Soc; fel Am Inst Chemists. *Res:* Formation of acylamido ketones from amino acids; conversion of acylamido ketones to oxazoles and thiazoles; reactions of amino acids; diene synthesis with maleic anhydride and maleimide; synthesis of dioxaspiroheptane; chemistry of chemical warfare agents; materials science; explosives science. *Mailing Add:* 9002 Volunteer Dr Alexandria VA 22309-2921

BORUM, PEGGY R, b 1946. HUMAN NUTRITION, CARNITINE METABOLISM. *Educ:* Univ Tenn, Knoxville, PhD(biochem), 72. *Prof Exp:* ASSOC PROF HUMAN NUTRIT, DEPT FOOD SCI & HUMAN NUTRIT, UNIV FLA, 83- *Res:* Nutritional assessment. *Mailing Add:* Dept Human Nutrit Univ Fla 409 Food Sci Bldg Gainesville FL 32611

BORWEIN, DAVID, b Kaunas, Lithuania, Mar 24, 24; m 46; c 3. MATHEMATICS. *Educ:* Univ Witwatersrand, BSc, 45, BSc, 48; Univ London, PhD(math), 50, DSc(math), 60. *Prof Exp:* Lectr math, Univ St Andrews, 50-63; vis prof, 63-64, head dept, 67-89, PROF MATH, UNIV

WESTERN ONT, 64- *Concurrent Pos:* Adv prob ed, Am Math, 82-85 & co-ed, Anal Int J Anal & Its Appln. *Mem:* Am Math Soc; Can Math Soc (vpres, 73-75, pres, 85-87); London Math Soc; fel Royal Soc Edinburgh; Math Asn Am. *Res:* Theory of summability of series and integrals. *Mailing Add:* Dept Math Univ Western Ont London ON N6A 5B7 Can

BORYSENKO, MYRIN, b Berezhani, Ukraine, June 3, 42; US citizen; m 70; c 3. IMMUNOBIOLOGY. *Educ:* St Lawrence Univ, BS, 64; State Univ NY Upstate Med Ctr, PhD(anat), 68; Univ Calif, Los Angeles, cert immunol, 70. *Prof Exp:* NIH fel, Univ Calif, Los Angeles, 68-70; from asst prof to assoc prof anat & cellular biol, Sch Med, Tufts Univ, 76-89; OWNER, MIND/BODY HEALTH SCI INC, 89- *Concurrent Pos:* Vis assoc prof behav med, Harvard Med Sch, 78-79. *Mem:* Am Asn Anatomists; Transplantation Soc; Am Asn Immunologists; Am Soc Zoologists; NY Acad Sci; AAAS. *Res:* Phylogenetic and developmental aspects of the immune response; effect of stress on the immune system and disease susceptibility. *Mailing Add:* Mind/Body Health Sci Ctr 22 Lawson Terr Scituate MA 02066

BORYSKO, EMIL, b Scranton, Pa, Sept 24, 18; div; c 3. MICROSCOPY, POLYMER CHARACTERIZATION. *Educ:* Brooklyn Col, BA, 40; George Washington Univ, MA, 50; Johns Hopkins Univ, PhD(biol), 55. *Prof Exp:* Supvr, Optical Glass Plant, Nat Bur Stand, 41-46, fiber technologist, 47-50; instr, Johns Hopkins Univ, 55-56; res assoc, NY Univ, 56-58; PRIN RES SCIENTIST, ETHICON, INC, NJ, 58- *Concurrent Pos:* Fel Damon Runyon Fund for Cancer Res, 51-53. *Honors & Awards:* Phillip B Hoffmann Res Scientist Award, 85; J&J Distinguished Anal Chemist Award, 89. *Mem:* Electron Micros Soc Am; NY Acad Sci; Am Soc Cell Biol. *Res:* Comparative optical and electron microscopic studies of the structure of cells and tissues; dynamic aspects of cellular growth and division; connective tissues and collagen; thermal microscopy of organic polymers; transmission and scanning electron microscopy of biomedical materials; chemical milling of surgical needles. *Mailing Add:* 211 Love Rd Bridgewater NJ 08807

BORZELLECA, JOSEPH FRANCIS, b Norristown, Pa, Oct 3, 30; m 55; c 6. PHARMACOLOGY. *Educ:* St Joseph's Univ, Pa, BS, 52; Thomas Jefferson Univ Med Col, MS, 54, PhD(pharmacol), 56. *Prof Exp:* Res asst, Trudeau Found, Jefferson Med Col, 54-55; instr pharmacol, Woman's Med Col Pa, 56-57, assoc, 57-59; from asst prof to assoc prof, 59-67, PROF PHARMACOL, MED COL VA, VA COMMONWEALTH UNIV, 67- *Concurrent Pos:* Pres, Toxicol & Pharmacol, Inc; mem comns, safe drinking water, toxicol & food safety, Nat Acad Sci-Nat Res Coun; consult, US Army, US Environ Protection Agency, indust. *Honors & Awards:* Distinguished Serv Award, US Army, 86; Excellence in Med & Community Serv Award, Nat Italian-Am Found, 87. *Mem:* AAAS; Am Chem Soc; Am Soc Pharmacol & Exp Therapeut; Soc Exp Biol & Med; Soc Toxicol (pres); Sigma Xi; Am Col Toxicol. *Res:* Toxicology of substances of economic importance; drug absorption, distribution and metabolism. *Mailing Add:* Dept Pharmacol & Toxicol Health Sci Div Med Col Va Commonwealth Univ Richmond VA 23298-0613

BOSANQUET, LOUIS PERCIVAL, b Leesburg, Fla, Sept 25, 31; m 78; c 3. REACTION ENGINEERING, PROCESS DEVELOPMENT. *Educ:* Univ Fla, BS, 54; Univ Tenn, MS, 60, PhD(chem eng), 63. *Prof Exp:* Res engr, Shell Oil Co, 54-58; res engr, 63-70, process group leader, 70-78, supt reaction eng, 78-80, ENG MGR, MONSANTO CO, 80- *Mem:* Am Inst Chem Engrs; Sigma Xi; Am Chem Soc. *Res:* Catalytic cracking of gasoline; organic processes research and development; use digital computers binetic modeling; plant start-ups and trouble shooting; reaction engineering; emergency venting of vessels. *Mailing Add:* 1372 Thornwick Dr Manchester MO 63011

BOSART, LANCE FRANK, b New York, NY, Aug 24, 42; m 69. SYNOPTIC-DYNAMIC METEOROLOGY. *Educ:* Mass Inst Technol, BS, 64, MS, 66, PhD(meteorol), 69. *Prof Exp:* Res asst meteorol, Mass Inst Technol, 65-69; asst prof, 69-76, ASSOC PROF ATMOSPHERIC SCI, STATE UNIV NY, ALBANY, 76- *Concurrent Pos:* Res Found NY fac res grant, 70-71; NSF grant, 70-78; vis assoc prof, Mass Inst Technol, 78-79. *Mem:* Am Meteorol Soc; Royal Meteorol Soc. *Res:* Frontogenesis; mesoscale-synoptic scale interaction through diagnostic and prognostic studies; synoptic meteorology; climatology. *Mailing Add:* Dept Atmospheric Sci State Univ NY 1400 Washington Ave Albany NY 12222

BOSCH, ARTHUR JAMES, b Luverne, Minn, Nov 15, 28; m 52; c 4. BIOCHEMISTRY. *Educ:* Cent Col, Iowa, BA, 51; Univ Wis, MS, 52, PhD(biochem), 58. *Prof Exp:* From instr to assoc prof, 58-69, PROF CHEM, CENT COL, IOWA, 69- *Mem:* AAAS; Am Chem Soc. *Res:* Organic, biological and general chemistry. *Mailing Add:* Cent Col Pella IA 50219

BOSCHAN, ROBERT HERSCHEL, b Los Angeles, Calif, Oct 12, 25; m 60; c 2. ORGANIC CHEMISTRY, POLYMER CHEMISTRY. *Educ:* Univ Calif, Los Angeles, BS, 47, PhD(chem), 50. *Prof Exp:* Eli Lilly res fel, 50-51; res chemist synthetic med res, Merck & Co, Inc, 51-52; org chemist, Chem Div, Res Dept, US Naval Weapons Ctr, 52-59; res & develop specialist, Mat Methods Res & Eng Div, McDonnell Douglas Corp, 59-69; mgr chem instrumentation sect, Appl Sci Div, Analog Technol Corp, 69-71; head chem sect & sr staff chemist, Hughes Aircraft Co, 71-81; RES & DEVELOP ENGR, LOCKHEED AERONAUT SYST CO, CALIF, 81- *Concurrent Pos:* Instr, Phys Sci Exten, Univ Calif, Los Angeles, 56- *Mem:* Am Chem Soc; The Chem Soc; Sigma Xi. *Res:* Organic reaction mechanisms; organic synthesis; kinetics; organic fluorine chemistry; chemistry based instrumentation; gas chromatography; mass spectroscopy. *Mailing Add:* 2012 Midvale Ave Los Angeles CA 90025

BOSCHMANN, ERWIN, b Fernheim, Paraguay, Jan 1, 39; m 62; c 3. BIOINORGANIC CHEMISTRY. *Educ:* Bethel Col, BS, 63; Univ Colo, MS, 65, PhD(metal chelates), 68. *Prof Exp:* Asst prof, 68-72, assoc prof, 72-75, PROF INORG CHEM, IND UNIV, INDIANAPOLIS, 76- *Concurrent Pos:* Vis prof, Agrarian Univ, Peru, 68-69, actg chief of party, Ford Found Proj, 69-70; consult, Ford Found, 74, State of Ind, 83 & Asian Develop Bank, 85; Lilly Fac Open fel, 86. *Honors & Awards:* H F Lieber Award, 83. *Mem:* Am Chem Soc. *Res:* Preparation and study of the chemistry of coordination compounds, especially metal chelates; preparation of new chelating agents; sulfoxide ligands. *Mailing Add:* 425 University Blvd Indianapolis IN 46202

BOSCHUNG, HERBERT THEODORE, b Birmingham, Ala, July 16, 25; m 51; c 4. ICHTHYOLOGY. *Educ:* Univ Ala, BS, 48, MS, 49, PhD(zool, bot), 57. *Prof Exp:* From instr to prof biol, Univ Ala, 50-87; RETIRED. *Concurrent Pos:* Dir, Ala Mus Natural Hist, 66-79. *Mem:* Asn Southeastern Biologists; Sigma Xi; Am Soc Ichthyologists & Herpetologists. *Res:* Fishes of Gulf of Mexico; freshwater fishes of southeastern United States. *Mailing Add:* Dept Biol PO Box 870344 University AL 35487-0344

BOSE, AJAY KUMAR, b Silchar, India, Feb 12, 25; m 50; c 6. ORGANIC CHEMISTRY, MEDICINAL CHEMISTRY. *Educ:* Univ Allahabad, India, BSc, 44, MSc, 46, Mass Inst Technol, ScD, 50. *Hon Degrees:* MEng, Stevens Inst Technol, 63. *Prof Exp:* Res fel chem, Harvard Univ, 50-51; lectr & asst prof, Indian Inst Technol, 51-56; res assoc, Univ Pa, 56-57; res chemist, Upjohn Co, 57-59; assoc prof, 59-61, PROF CHEM, STEVENS INST TECHNOL, 61-, GEORGE MEADE BOND PROF CHEM, 83- *Concurrent Pos:* Consult. *Honors & Awards:* Jubilee Gold Medal, Univ Allahabad, 46; Meghnad Saha Mem Prize, 57; Ottens Res Award, Stevens Inst Technol, 68, Davis Res Award, 81. *Mem:* Am Chem Soc; Indian Chem Soc; fel Indian Nat Acad Sci. *Res:* Penicillin chemistry; natural products; synthetic organic chemistry; biogenesis; nuclear magnetic resonance and mass spectrometry; popular science writing. *Mailing Add:* Dept Chem Stevens Inst Technol Hoboken NJ 07030

BOSE, AMAR G(OPAL), b Philadelphia, Pa, Nov 2, 29; m 60; c 2. ACOUSTICS, ELECTRONICS. *Educ:* Mass Inst Technol, SB & SM, 52, ScD, 56. *Prof Exp:* Assoc prof, 56-66, PROF ELEC ENG, MASS INST TECHNOL, 66- *Concurrent Pos:* Fulbright fel, Nat Phys Lab, New Delhi & Indian Statist Inst, Calcutta, 56-57; chmn bd dir, Bose Corp. *Res:* Acoustical transducer design; audio-signal processing and switching techniques for high efficiency power processing. *Mailing Add:* Bose Corp The Mountain Framingham MA 01701

BOSE, ANIL KUMAR, b Calcutta, India, Apr 1, 29; m 59; c 1. PURE MATHEMATICS. *Educ:* Univ Calcutta, BS, 48, MS, 56; Univ NC, PhD(math), 64. *Prof Exp:* Lectr math, Goenka Col, Calcutta, 58-60; part-time instr, Univ NC, 60-63; vis prof, St Augustine's Col, 63-64; asst prof, Univ Ala, 64-68; ASSOC PROF MATH, CLEMSON UNIV, 68- *Mem:* Am Math Soc. *Res:* Differential equations on the properties of mean-value of the solutions of certain class of elliptic equations; mean-values are considered by taking various non-negative weight functions. *Mailing Add:* Dept Math Clemson Univ Clemson SC 29631

BOSE, ANJAN, b Calcutta, India, June 2, 46; m 76; c 3. POWER SYSTEM ENGINEERING. *Educ:* Indian Inst Technol, Kharagpur, BTech, 67; Univ Calif, Berkeley, MS, 68; Iowa State Univ, Ames, PhD(elec eng), 74. *Prof Exp:* Res asst elec eng, Univ Calif, 67-68; engr syst planning, Consolidated Edison Co, 68-70; res asst elec eng, Iowa State Univ, 70-73, instr technol & social change, 73-74; fel, Sci Ctr, IBM, 74-75; asst prof elec eng, Clarkson Univ, 75-76; consult energy mgt systs div, Control Data Corp, 76-78, mgr, 78-81; assoc prof, 81-85, PROF ELEC ENG, ARIZ STATE UNIV, 85- *Concurrent Pos:* Lectr, Univ Minn, 76-81; vpres, Power Math Assocs, Inc, 81-84; consult, 84- *Mem:* Sr mem Inst Elec & Electronics Engrs. *Res:* Analysis, design and control of large electric power grids; power system control centers; economic and secure operation of power grids; training simulators for power system operators. *Mailing Add:* Dept Elec Eng Ariz State Univ Tempe AZ 85287-5706

BOSE, BIMAL K, b Calcutta, India, Sept 1, 32; US citizen; m 61; c 2. POWER ELECTRONICS, AC MACHINE DRIVES. *Educ:* Calcutta Univ, BE, 56, PhD(elec eng), 66; Univ Wis-Madison, 60. *Prof Exp:* Asst engr, Tata Power Co, Bombay, 56-58; asst prof elec eng, Bengal Eng Col, 60-71; assoc prof, Rensselaer Polytech Inst, 71-76; res engr, Gen Elec Res & Develop Ctr, 76-87; PROF ELEC ENG, UNIV TENN, KNOXVILL, 87- *Concurrent Pos:* Adj assoc prof, Rensselaer Polytech Inst, 76-87; chief scientist, Power Electronics Applications Ctr, 87-; chmn, Indust Power Converter Comt, Inst Elec & Electronics Engrs, 90-; consult, Res Triangle Inst, NC, 90-; mem, Nat Power Electronics Comt, 90-; sr adv, Beijing Power Electronics Ctr, 90- *Mem:* Fel Inst Elec & Electronics Engrs. *Res:* Power electronics circuits; microcomputer control; control of AC drives; simulation; fuzzy and expert system application in power electronics. *Mailing Add:* Dept Elec Eng 419 Ferris Hall Univ Tenn Knoxville TN 37922

BOSE, DEEPAK, b Calcutta, India, Aug 24, 41; Can citizen; m 66; c 2. PHARMACOLOGY, MEDICINE. *Educ:* Vikram Univ, MBBS, 63; Univ Indore, India, MD, 66; Univ Man, PhD(pharmacol), 71. *Prof Exp:* Demonstr, MGM Med Col, Indore, 65-67; lectr, 67-71, from asst prof to assoc prof, 71-82, PROF PHARMACOL, UNIV MAN, 82-, PROF INTERNAL MED, 84-, PROF ANESTHESIA, 88- *Concurrent Pos:* Med Res Coun Can fel, Univ Man, 69-71. *Mem:* Am Physiol Soc; Am Pharmacol Soc; Can Pharmacol Soc; Int Soc Heart Res; Can Thoracic Soc. *Res:* Physiology and pharmacology of cardiac, smooth and skeletal muscle; autonomic pharmacology; malignant hyperthermia; cerebral blood flow regulation. *Mailing Add:* Dept Pharmacol & Therapeut 770 Bannatyne Ave Winnipeg MB R3E 0W3 Can

BOSE, HENRY ROBERT, JR, b Chicago, Ill, Sept 20, 40. MICROBIOLOGY. *Educ:* Elmhurst Col, BS, 62; Univ Ind, Indianapolis, MS, 65, PhD(microbiol), 67. *Prof Exp:* NSF fel, 67-69, asst prof, 69-74, assoc prof, 74-80, PROF MICROBIOL, UNIV TEX, AUSTIN, 80- *Res:* Replication and transformation of cells by ribonucleic acid enveloped animal viruses. *Mailing Add:* Dept Microbiol Univ Tex Austin TX 78712

BOSE, NIRMAL K, b Calcutta, India, Aug 19, 40; m 69; c 2. ELECTRICAL ENGINEERING, APPLIED MATHEMATICS. *Educ:* Indian Inst Technol, Kharagpur, BTech, 61; Cornell Univ, MS, 63; Syracuse Univ, PhD(elec eng), 67. *Prof Exp:* From asst prof to prof math, 67-86, PROF ELEC ENG, UNIV PITTSBURGH, 76-; SINGER PROF & DIR SPATIAL & TEMPORAL SIGNAL PROCESSING CTR, PA STATE UNIV, 86- *Concurrent Pos:* Consult, RCA Corp, Pa, 68-69; vis lectr, Col Steubenville, 68-70; vis assoc prof, Dept Archit & Eng, Am Univ Beirut, 71-, Univ Md, 72 & Univ Calif, Berkeley, 73-74; Ger Acad Exchange Serv grant, 78; res & lect grant, Nat Ctr Sci Res-Inst Info & Automatic Res, France, 75. *Mem:* Fel Inst Elec & Electronics Engrs; AAAS; Sigma Xi; NY Acad Sci; Am Math Soc. *Res:* Active and passive network theory; automatic control; multidimensional systems theory; large-scale circuit layout design; computational complexity theory; signal and image processing. *Mailing Add:* Dept Elec Eng Pa State Univ University Park PA 16802

BOSE, SAMIR K, b Decca, Bangladesh, May 1, 34; m 61; c 3. THEORETICAL PHYSICS. *Educ:* Univ Delhi, BS, 56, MS, 58; Univ Rochester, PhD(physics), 62. *Prof Exp:* Mem, Sch Math, Inst Advan Study, 62-63; fel physics, Tata Inst Fundamental Res, India, 63-64; reader, Univ Delhi, 64-66; consult, Int Ctr Theoret Physics, UNESCO, Italy, 66; from asst prof to assoc prof, 66-90, PROF PHYSICS, UNIV NOTRE DAME, 90- *Res:* Theoretical research in elementary particles and their strong and weak interactions; application of group theory and algebraic techniques to the same. *Mailing Add:* Dept Physics Univ Notre Dame Notre Dame IN 46556

BOSE, SHYAMALENDU M, b Dacca, Bangladesh, July 17, 39; m 63; c 2. SOLID STATE PHYSICS, METAL & SEMICONDUCTOR PHYSICS. *Educ:* Calcutta Univ, BS, 58, MS, 60; Univ Md, College Park, PhD(physics), 68. *Prof Exp:* Lectr physics, Midnapore Col, 60-61; fel, Cath Univ Am, 67-70; from asst prof to assoc prof, 70-83, PROF PHYSICS, DREXEL UNIV, 83- *Concurrent Pos:* Res assoc, Cath Univ Am, 70-71; electronic struct in solids fel, Univ Liege, Belg, 76-77; consult engr, United Engrs & Constructors, Philadelphia, 80-84; vis scientist, Schlumberger-Doll Res, Ridgefield, Conn, 82; vis prof, Univ Waterloo, 86; prin investr NASA grants, 83- *Mem:* Sigma Xi; Am Phys Soc; Franklin Inst; Am Asn Univ Professors. *Res:* Many-body problem; electronic and optical properties of metals; dilute magnetic alloys; chemisorption by metals; electronic properties of disordered solids; surface and interface properties of solids and 2 to 6 compounds; high-temperature superconductor; published 80 papers in professional journals. *Mailing Add:* Dept Phys & Atmospheric Sci Drexel Univ Philadelphia PA 19104-9984

BOSE, SUBIR KUMAR, b Calcutta, India, Jan 1, 39; m 65; c 2. THEORETICAL PHYSICS, STATISTICAL MECHANICS. *Educ:* Bihar Univ, BSc, 58; Patna Univ, MSc, 60; Allahabad Univ, PhD(physics), 67. *Prof Exp:* Res fel physics, Univ Delhi, 66-67; res fel, St Louis Univ, 67-68; from asst prof to assoc prof, 68-78, PROF PHYSICS, SOUTHERN ILL UNIV, CARBONDALE, 78- *Concurrent Pos:* Vis scientist, Saha Inst Nuclear Physics, Calcutta, 77; res fel, Ind Univ, Bloomington, 78-79. *Mem:* Am Phys Soc; Sigma Xi. *Mailing Add:* Dept Physics Univ Cent Fla Orlando FL 32816

BOSE, SUBIR KUMAR, b Gaya, India, Sept 3, 31; US citizen; m; c 2. BIOLOGY, MOLECULAR PARASITOLOGY. *Educ:* Univ Lucknow, BSc, 50, MSc, 52; Washington Univ, PhD(molecular biol), 63. *Prof Exp:* Lectr bot, Univ Lucknow, 52-58; Am Cancer Soc res assoc biochem, Med Sch, Univ Mich, 63-65; from asst prof to assoc prof, 65-75, PROF MICROBIOL, SCH MED, ST LOUIS UNIV, 76-, ACTG CHMN, 89- *Concurrent Pos:* NIH res career develop award, 67-76; vis investr, Sch Med, Stanford Univ, 69; vis prof biochem, Indian Inst Sci, 78; vis prof, Inst Pasteur Hellenique, Athens, 86; guest sci, Ctr Dis Control, Atlanta, 87. *Mem:* AAAS; Am Soc Biol Chemists; Am Soc Microbiol; Brit Soc Gen Microbiol. *Res:* Cell growth control; regulation of gene activity; chlamydial host-parasite interactions. *Mailing Add:* Dept Microbiol St Louis Univ Sch Med St Louis MO 63103-1028

BOSE, TAPAN KUMAR, b Calcutta, India, Oct 12, 38; Can citizen; m 66; c 2. DIELECTRIC PROPERTIES MATERIALS, THERMODYNAMIC PROPERTIES GASES. *Educ:* Univ Calcutta, India, BSc Hons, 58, MSc, 61; Univ Louvain, Belg, PhD(physics), 65. *Prof Exp:* Work leader physics, Kammerlingh onnes Lab Holland, 65-67; from asst prof to assoc prof, 69-74, chmn, Dept Physics, 72-75, PROF PHYSICS, UNIV QUE, TROIS-RIVI07RES, 74-, DIR RES PHYSICS, DIELECTRICS RES GROUP, 82- *Concurrent Pos:* Vis prof phys chem, Univ Nancy, France, 76-77 & physics, Univ Bordeaux, France, 84; mem res comn, Fonds pour la formation des chercheurs et l'aide à la recherche, Que, Can, 86-89. *Mem:* Can Asn Physicists; Am Phys Soc; Inst Elec & Electronics Engrs. *Res:* Condensed matter physics-dielectric and optical properties of liquids and gases; dielectric properties of materials over a wide band-microwave and radiofrequency; phase transition. *Mailing Add:* Dielectrics Res Group Univ Que Trois-Rivireres Case Postale 500 Trois-Rivieres PQ G9A 5H7i

BOSEE, ROLAND ANDREW, b Hagerstown, Md, Feb 12, 10; m 41; c 2. BIOCHEMISTRY. *Educ:* Temple Univ, AB, 32; Drexel Inst, BS, 34; Syracuse Univ, PhD(biochem), 37. *Prof Exp:* Chief chemist, Cheplin Biol Labs, NY, 34-37; res chemist, Endo Prod, Inc, 37-38, dir labs, 38-41, vpres & plant supt, 46-47; asst officer in charge physiol br, Air Test Ctr, USN, Md, 47-50, officer in charge, 50-52, test dir parachute unit, Auxiliary Air Sta, 53-56, dir air crew equip lab, Air Mat Ctr, Philadelphia, Pa, 56-64, head aviation med equip br, Bur Med & Surg & Human Factors Off, Bur Naval Weapons, 64-66, dir crew systs div, Naval Air Systs Command Hq, Washington, DC, 66-68; dir res, Philadelphia Gen Hosp, 68-77; CONSULT MED & HEALTH SCI, 77- *Concurrent Pos:* Mem vision comt & hearing & bioacoust comt, Nat Acad Sci-Nat Res Coun. *Honors & Awards:* Mosely Award, Aerospace Med Asn. *Mem:* AAAS; Am Chem Soc; fel Am Inst Chemists; fel Aerospace Med Asn. *Res:* Aviation physiology; carbon monoxide and its effects on aviation personnel; safety and survival equipment. *Mailing Add:* 3084 Pine Valley Rd Port St Lucie FL 34952

BOSHART, CHARLES RALPH, b Lowville, NY, June 9, 32; m 56; c 4. PHARMACOLOGY. *Educ:* Univ Buffalo, BS, 54, MS, 58; Purdue Univ, PhD(pharmacol), 59. *Prof Exp:* Instr pharmacol, Purdue Univ, 57-58; pharmacologist, Am Cyanamid Co, 59-63, sr res scientist & group leader, 63-65, head dept endocrinol res, 65-69, head dept toxicol eval, Lederle Labs, 69-79; asst dir toxicol, 79-86, DIR, DRUG REGULATORY AFFAIRS PRECLIN, STERLING-WINTHROP RES INST, 86- *Mem:* AAAS; Am Soc Pharmacol & Exp Therapeut; Regulatory Affairs Prof Soc. *Res:* Anticonvulsants; central nervous system barriers; biological use of tritium and other radioisotopes; metabolic effects of drugs. *Mailing Add:* Drug Regulatory Affairs Sterling Res Group Nine Great Valley Pkwy Malvern PA 19355

BOSHART, GREGORY LEW, b Lowville, NY, Mar 21, 33. ORGANIC CHEMISTRY. *Educ:* St Lawrence Univ, BS, 55; Mass Inst Technol, PhD(org chem), 60. *Prof Exp:* Chemist, E I du Pont de Nemours & Co, 57, Esso Res & Eng Co, 60-61, 62-64 & Enjay Chem Co, 64-68; EXEC OFFICER, ACAD AIDE & PROF LECTR, UNIV CHICAGO, 68-, ASST TO DEAN, PHYS SCI DIV, 74- *Mem:* AAAS; Am Chem Soc. *Res:* Polypeptide synthesis; carbodiimide chemistry; resolution of amino acids; resin plasticization. *Mailing Add:* Dept Chem Univ Chicago 5735 S Ellis Chicago IL 60637

BOSHELL, BURIS RAYE, b Marion Co, Ala, Oct 9, 26; m 51; c 2. ENDOCRINOLOGY, INTERNAL MEDICINE. *Educ:* Ala Polytech Inst, BS, 47; Harvard Med Sch, MD, 53. *Prof Exp:* Intern med, Peter Bent Brigham Hosp, 53-54, from jr asst resident to sr asst resident, 54-56, asst, Thorn Lab, 56-57, jr assoc physician & asst dir, Diabetic Teaching Unit, 57-58, chief resident, 58-59; from asst prof to prof, Sch Med, Univ Ala, Birmingham, 59-72, asst dir, Dept Med, 63-76, Ruth Lawson Hanson prof med, 72-90, chief, Div Endocrinol & Metab, Diabetes Res & Educ Hosp, 76-90; STAFF MEM, DIABETES CLIN, MED CTR EAST HOSP, 90- *Concurrent Pos:* Am Col Physicians res fel, Thorn Lab, Peter Bent Brigham Hosp, 56-57; asst, Harvard Med Sch, 58-59; clin investr, Vet Admin Hosp, 59-62, chief med serv, 62-; asst physician-in-chief, Univ Hosp & Hillman Clins, 63- *Mem:* Endocrine Soc; Am Diabetes Asn; Am Soc Clin Pharmacol & Therapeut; fel Am Col Physicians; AMA. *Res:* Diabetes mellitus; oral diagnostic test for diabetes; insulin in blood and etiological factors in insulin resistance. *Mailing Add:* 3017 Old Ivy Rd Birmingham AL 35210

BOSHES, BENJAMIN, neurology, psychiatry; deceased, see previous edition for last biography

BOSHES, LOUIS D, b Chicago, Ill, Oct 15, 08; m 42; c 2. NEUROPSYCHIATRY. *Educ:* Northwestern Univ, Evanston, BS, 31, Northwestern Univ, Chicago, MD, 36; Am Bd Psychiat & Neurol, dipl & cert psychiat, 47, cert neurol, 50 & cert pediat neurol, 69. *Hon Degrees:* DHH, Pan Am Med Asn, 76. *Prof Exp:* Fel psychiat, Ill Neuropsychiat Inst, Chicago, 40-42, 46-47; prof neurol & psychiat, Sch Med, Northwestern Univ, Chicago, 47-63; from clin assoc prof to clin prof, 63-78, dir, Consultation Clin for Epilepsy, 63-72, EMER CLIN PROF NEUROL, UNIV ILL COL MED, 78-; PROF NEUROL, COOK COUNTY GRAD SCH MED, 70 - *Concurrent Pos:* Assoc & attend neurologist, Cook County Hosp, Chicago, 47-63, consult neurol, 70-, attend neurol, 84-; neuropsychiat consult, Ill State Psychiat Inst Chicago, 48-70, Columbus Mem Hosp, 48-75, Louis A Weiss Mem Hosp, 50- & Nicholas J Pritzker Ctr for Children, 64-70; sr consult neurol, Vet Admin Hosp, Downey, Ill, 54-60; assoc examr, Am Bd Psychiat & Neurol, 57- & Am Bd Neurol Surg, 63-; assoc ed, Dis Nerv Syst, 60-76, Int J Neuropsychiat, 66-69, Behav Neuropsychiat, 69- & Int Surg Neurol Sci, 77-; neurol consult, Drexel Home for the Aged, 62-70; attend physician, Res & Educ Hosps, Abraham Lincoln Sch Med, Univ Ill Col Med, 63-78; consult ed, Current Med Dig, 65-74 & New Physician, 67 -; sr attend neurologist & psychiatrist, Michael Reese Hosp & Med Ctr, 65-, chief neurol clins, 69-75; assoc attend neurologist, St Joseph Hosp, , 74 -; sr attend neurologist, psychiatrist & child neurologist, Columbus Mem Hosp, 75 -; mem adv bd & coun, Myasthenia Gravis Found, United Parkinson Found, Epilepsy Found Am & Nat Found-March of Dimes; ambassador, Int Bur Epilepsy & League Against Epilepsy; mem sci exhib adv comt, Mus Sci & Indust, Chicago. *Mem:* Fel Am Acad Neurol; fel Am Psychiat Asn; fel AMA; fel Am Col Physicians; fel Pan-Am Med Asn (pres, 73-); Am Epilepsy Soc; Am Med Electroencephalographic Asn; Am Med Writers Asn; Int Col Surgeons; Child Neurol Soc; Sigma Xi. *Res:* Neurology; child neurology; psychiatry; multiple sclerosis; parkinsonism; epilepsy; myasthenia gravis; geriatric neurology; disorders of aging in the neurologic system; history of neurology; convulsive disorders; fungus infections and inflammations of the central nervous system. *Mailing Add:* Univ Ill Col Med 912 S Wood St Chicago IL 60680

BOSHKOV, STEFAN, b Sofia, Bulgaria, Sept 29, 18; nat US; m 43; c 2. MINING ENGINEERING, GEOLOGY. *Educ:* Am Col Sofia, Dipl, 38; Columbia Univ, BS, 41, EM, 42. *Prof Exp:* Asst mining, Columbia Univ, 42-43, assoc, 46-51, from asst prof to prof mining, 51-80, chmn, 67-85, Henry Krumb prof, 80-88, HENRY KRUMB EMER PROF, HENRY KRUMB SCH MINES, COLUMBIA UNIV, 89- *Concurrent Pos:* Consult engr, 49-; mem int orgn comt, Int Mining Cong, 62-; chmn, Int Conf Strata Control, 64; Fulbright distinguished prof & sr scientist, Yugoslavia, 69; mem, Secy Labor adv comt health & safety standards metal & non-metal mining, US Dept Labor, 78. *Honors & Awards:* B Krupinski Medal, State Mining Coun, Poland, 80; Drinov Medal, Acad Sci, Bulgaria, 87- *Mem:* Am Inst Mining, Metall & Petrol Engrs; Am Arbit Asn; Sigma Xi. *Res:* Barodynamics; stress analysis applied to mine design; mine plant, ventilation, valuation and economics. *Mailing Add:* 119 White Plains Ave White Plains NY 10604

BOSIN, TALMAGE R, b Fond du Lac, Wis, Mar 6, 41. ORGANIC CHEMISTRY. *Educ:* Wheaton Col, Ill, BS, 63; Ind Univ, PhD(org chem), 67. *Prof Exp:* NIH res fel org chem, Univ Calif, Berkeley, 67-69; from asst prof to assoc prof, 69-78, PROF PHARMACOL, MED SCI PROG, IND UNIV, BLOOMINGTON, 78- *Concurrent Pos:* Sr int fel, 77-78. *Mem:* Am Chem Soc; Am Soc Exp Pharmacol Therapeut; Soc Toxicol. *Res:* Biologically active indole compounds, particularly serotonin and its role in pulmonary and central nervous system functions. *Mailing Add:* Dean Med Ind Univ Med Sci Prog Myers Hall Bloomington IN 47405

BOSKEY, ADELE LUDIN, b New York, NY, Aug 30, 43; m 70; c 1. STRUCTURAL CHEMISTRY. *Educ:* Columbia Univ, BA, 64; Boston Univ, PhD(phys chem), 70. *Prof Exp:* Instr chem col lib arts, Boston Univ, 69-70; res fel crystallog, Imp Col Sci & Technol, 70-71; res assoc, 74-79, from asst prof to assoc prof, 77-85, assoc scientist, 79-83, SR SCIENTIST, HOSP SPEC SURG, CORNELL UNIV, 85-, PROF BIOCHEM MED COL, 85-, DIR, LAB ULTRASTRL BIOCHEM, 86- *Mem:* Sigma Xi; Am Crystallog Asn; Am Chem Soc; Orthop Res Soc; Am Soc Bone Mineral Res. *Res:* Mechanism of hard tissue mineralization; structure determination by x-ray crystallography and electron microscopy; role of lipids and proteoglyeans in calcification. *Mailing Add:* Four Winding Way North Caldwell NJ 07006

BOSLEY, DAVID EMERSON, b Lundale, WVa, Dec 16, 27; m 52; c 4. PHYSICAL CHEMISTRY. *Educ:* Univ WVa, BS, 50; Mass Inst Technol, PhD(phys chem), 54. *Prof Exp:* Res chemist, dacron plant tech sect, E I du Pont de Nemours & Co, 54-85; CONSULT, TEXTILE, 85- *Res:* Physico-chemical properties of synthetic fibers. *Mailing Add:* Box 531 Grifton NC 28530

BOSLEY, ELIZABETH CASWELL, b Wichita, Kans, July 4, 12; m 39; c 1. SPEECH PATHOLOGY. *Educ:* Friends Univ, AB, 33; Univ Kans, MA, 35. *Prof Exp:* Instr eng & math, Univ Kans, 36; teacher high sch, Kans, 36-38; instr clin logopedics, Inst Logopedics, Wichita State Univ, 39-40; supvr, Inst Logopedics, 40-42, asst prof, 51-67, preceptor & speech clinician, 42-76; RETIRED. *Concurrent Pos:* From instr to asst prof, Wichita State Univ, 40-67; mem, Order of the Tower, hon soc Friends Univ, Gold Key, Wichita State Univ. *Mem:* Am Speech & Hearing Asn. *Res:* Teacher training and clinical work in correcting all types of speech defects; diagnosis of speech defects; physiology of speech structure; anatomy of speech structures; neurology. *Mailing Add:* Box 1 Cascade CO 80809

BOSLOW, HAROLD MEYER, b New York, NY, Apr 30, 15; m 43. PSYCHIATRY. *Educ:* Univ Va, MD, 39. *Prof Exp:* Staff psychiatrist, US Vet Admin, 48-50; instr, 50-66, ASST PROF PSYCHIAT, SCH MED, JOHNS HOPKINS UNIV, 66-, PSYCHIATRIST, OUTPATIENT DEPT, JOHNS HOPKINS HOSP, 50-; ASST PROF PSYCHIAT, SCH MED, UNIV MD, 66- *Concurrent Pos:* Med officer, Supreme Bench, Baltimore, Md, 50-54; consult psychiatrist, Surgeon Gen, US Dept Army, 52-53; dir, Patuxent Inst, 54-75; WHO fel, 66; consult to White House on anti-social behav & delinq, 64, WHO & UN Social Defense Res Inst, 69-70, NIMH, 70, Govts PR & Que, 70 & Off Law Enforcement Assistance Agency; med dir, Forensic Div, St Elizabeth Hosp, Washington, DC, 75- *Mem:* AAAS; AMA; Am Psychiat Asn; NY Acad Sci. *Res:* Personality structure and deviations; forensic psychiatry. *Mailing Add:* Patuxent Inst Jessup MD 20794

BOSMA, JAMES F, b Grand Rapids, Mich, Apr 29, 16; m 42; c 8. PEDIATRICS. *Educ:* Calvin Col, AB, 37; Univ Mich, MD, 41. *Prof Exp:* Intern, Cleveland City Hosp & Western Reserve Hosp, 41-43, resident, 43-44; asst prof pediat, Med Sch, Univ Minn, 48-49; prof, Col Med, Univ Utah, 49-59; CHIEF ORAL & PHARYNGEAL DEVELOP SECT, NAT INST DENT RES, NIH, 61-; LECTR, DEPT RADIOL, JOHNS HOPKINS UNIV, 75- *Concurrent Pos:* Nat Found Infantile Paralysis fel, Univ Minn, 44-46, Kellogg fel, 46-48; NIH fel, Karolinska Inst, Sweden & Wenner Gren Cardiovasc Res Lab, 59-61; pediatrician, Salt Lake Gen Hosp, 55-59; res prof, dept pediat, Sch Med & dept pediat, Sch Dent, Univ Md. *Mem:* AAAS; Soc Pediat Res; Am Pediat Soc; Soc Res Child Develop; Int Soc Cranio-facial Biol. *Res:* Motor coordination of mouth, pharynx and larynx. *Mailing Add:* 3902 Hadley Square W Baltimore MD 21218

BOSMAJIAN, GEORGE, b Fresno, Calif, Apr 21, 21; m 49; c 3. ORGANIC CHEMISTRY, POLYMERS. *Educ:* Fresno State Col, AB, 43; Univ Ill, MA, 47; Wash State Univ, PhD(org chem), 49. *Prof Exp:* Jr chemist, Shell Develop Co, 43-46; res chemist, Assoc Oil Co, 49-50; group leader, Sinclair Res Labs, 50-55; sr scientist, Southern Res Inst, 55-62; res coordr, David Taylor Naval Ship Res & Develop Ctr, 62-68, div dir, 68-84, tech adv, 84-88; RETIRED. *Mem:* Am Chem Soc; Am Soc Metals; Soc Automotive Engrs. *Res:* Tribology, fuel & lubricants, polymers, composites, advanced materials. *Mailing Add:* 17 Emerson Rd Severna Park MD 21146

BOSMANN, HAROLD BRUCE, b Chicago, Ill, June 17, 42; m 66; c 2. BIOPHYSICS, PHARMACOLOGY. *Educ:* Knox Col, AB, 64; Univ Rochester, PhD(biophys), 66. *Hon Degrees:* Cambridge Univ, MA, 67. *Prof Exp:* Asst prof, Sch Med Dent, Univ Rochester, 68-74, assoc prof pharmacol & toxicol, 74-78, assoc prof oncol, 76-78, prof pharmacol, toxicol & oncol, 78-82; assoc dean, 82-85, prof internal med, pharmacol & cell biophys & dir prog on aging & geriatric educ & res, 82-87, vdean, 85-87, prof psychiat, 85-87, PROF PHARM & MED, SR ASSOC DEAN & DIR RES CTR AGING, SCH MED, UNIV CINCINNATI, 87- *Concurrent Pos:* NSF fel biophys, Strangeways Lab, Cambridge Univ, 66-67; USPHS fel, Salk Inst Biol Studies, La Jolla, Calif, 67-68 & career develop award, Univ Rochester, 68-74; scholar award, Leukemia Soc Am, 74-79. *Mem:* Am Chem Soc; Am Soc Pharmacol & Exp Therapeut; Biophys Soc. *Res:* Membranes; pharmacology; drug resistance; synaptosomes; geriatrics; toxicology; receptor analysis. *Mailing Add:* Off Dean Univ Ill Col Med Chicago IL 60612

BOSNIACK, DAVID S, b New York, NY, July 10, 32; m 57; c 3. ORGANIC CHEMISTRY, PETROLEUM CHEMISTRY. *Educ:* City Col New York, BS, 54; NY Univ, PhD(org chem), 61. *Prof Exp:* Chemist, GAF Corp, 60-62; chemist, Exxon Govt Res Lab, Exxon Res & Eng Co, 62-66, sr chemist, Prod Res Div, 66-70, res assoc, Prod Res Div, 70-77, proj head, New Proj Develop, Baytown Petrol Res Lab, 70-77, sr res assoc & sect head indust lubricants & specialties, 77-81, sr res assoc & sect head engine lubricants, Prod Res Div, 81-90, TECH ADVISOR, EXXON CO INT, 90- *Concurrent Pos:* Instr, Cooper Union, 61-63. *Mem:* Am Chem Soc; AAAS; Soc Automotive Engrs. *Res:* Free radical reactions; petroleum chemistry; lubrication technology including synthetic lubricants, lubrication mechanisms and additive synthesis; acetylene chemistry; solid propellants; polymer synthesis. *Mailing Add:* Exxon Co Int 222 Park Ave Florham Park NJ 07932-1002

BOSNIAK, MORTON A, b New York, NY, Nov 13, 29; m; c 1. RADIOLOGY. *Educ:* Mass Inst Technol, BS, 51; State Univ NY, MD, 55. *Prof Exp:* Intern, Mt Sinai Hosp, NY, 55-56; resident radiol, NY Hosp, 56-57, 59-61; instr, Med Sch, Cornell Univ, 60-61; assoc prof, Sch Med, Boston Univ, 64-67; assoc prof, Albert Einstein Col Med, 67-69, sr attend radiologist, Hosp, 67-69, assoc dir dept, 68-69; PROF RADIOL, SCH MED, NY UNIV, 69- *Concurrent Pos:* Radiologist, Montefiore Hosp, 61-62, assoc attend, 63-; sr staff radiologist, Boston City Hosp, Mass, 64-67; assoc radiologist, Univ Hosp, Boston, 64-67; sr attend radiologist, Bronx Munic Hosp Ctr, NY, 67-; attend radiologist, NY Univ Hosp, 69-; vis attend radiologist, Bellevue Hosp; consult radiol, Vet Admin Hosp. *Mem:* Am Col Radiol; Radiol Soc NAm; Asn Univ Radiol; Am Roentgen Ray Soc. *Mailing Add:* Dept Radiol NY Univ Med Ctr Radiol 550 First Ave New York NY 10016

BOSNJAK, ZELJKO J, CARDIAC ELECTRO-PHYSIOLOGY, GENERAL ANESTHETICS. *Educ:* Univ Wis-Milwaukee, PhD(physiol), 79. *Prof Exp:* ASSOC PROF ANAESTHESIOL & PHYSIOL, COL MED, UNIV WIS-MILWAUKEE, 79- *Res:* Calcium antagonists. *Mailing Add:* Dept Anesthesiol Med Col Wis MFRC A100 8701 Watertown Plank Rd Milwaukee WI 53226

BOSOMWORTH, PETER PALLISER, b Akron, Ohio, May 2, 30; m 56; c 4. MEDICINE, ANESTHESIOLOGY. *Educ:* Kent State Univ, BSc, 51; Univ Cincinnati, MD, 55; Ohio State Univ, MMedSc, 58. *Prof Exp:* Intern, Cincinnati Gen Hosp, Ohio, 57-58; asst resident anesthesiol, Ohio State Univ Hosp, 56-57, chief resident & asst instr, 57-58, instr, 58; chief anesthesia div, US Naval Hosp, Great Lakes, Ill, 58-60; dir anesthesia res, Ohio State Univ, 60-62; chmn dept anesthesiol, 62-70, assoc dean col med, 67-70, VPRES, MED CTR, UNIV KY, 70-, CHANCELLOR, 82- *Concurrent Pos:* Consult, Vet Admin Hosp & USPHS, Lexington, 62-, Ireland Army Hosp, Ft Knox, 64-66; chief staff, Univ Ky Hosp, 66-68; mem bd dir, Blue Cross Blue Shield of Ky; vpres, Health Develop Resources Inst, 70; mem, Asn Acad Health Ctr, 70-, chmn bd, 82. *Mem:* Am Soc Anesthesiol; Am Univ Anesthetists; Am Soc Anesthesiologists; AMA; Int Anesthesia Res Soc. *Res:* Cardiovascular, uterine and renal physiology as influenced by anesthetic agents; health care systems and related policy. *Mailing Add:* Univ Ky Med Plaza Chancellor A301 Lexington KY 40536-0223

BOSRON, WILLIAM F, b Dayton, Ohio, Dec 28, 44; m 79. CLINICAL CHEMISTRY, ENZYMOLOGY. *Educ:* Bowling Green Univ, BS, 67; Univ Cincinnati, PhD(biochem), 73. *Prof Exp:* Res fel biochem, Harvard Med Sch, 72-75; asst prof, 75-79, ASSOC PROF, DEPTS MED & BIOCHEM, MED SCH, IND UNIV, 79- *Concurrent Pos:* Assoc prof, Grad Sch, Ind Univ, 81- *Mem:* Am Chem Soc; Res Soc Alcoholism; Am Asn Clin Chemists. *Res:* Regulation of alcohol metabolism; biochemical properties and kinetics of alcohol and aldehyde dehydrogenases; protein turnover. *Mailing Add:* 413 Med Sci Ind Univ Sch Med Indianapolis IN 46223

BOSS, ALAN PAUL, b Lakewood, Ohio, July 20, 51; m 79; c 2. STAR & PLANET FORMATION, NUMERICAL HYDRODYNAMICS. *Educ:* Univ S Fla, BS, 73; Univ Calif, Santa Barbara, MA, 75, PhD(physics), 79. *Prof Exp:* Nat Acad Sci/Nat Res Coun resident res assoc, Space Sci Div, Ames Res Ctr, NASA, 79-81; staff assoc, 81-83, STAFF MEM, DEPT TERRESTRIAL MAGNETISM, CARNEGIE INST WASHINGTON, 83- *Concurrent Pos:* Prin investr, Theory of Star Formation, NSF Grant, 84-; prin investr, Three Dimensional Evolution Solar Nebula, NASA grant, 88-, NASA Planetary Systs Sci Working Group, 88-; mem Comn Planetary & Lunar Explor, Nat Acad Sci/Nat Res Coun, 90-93. *Mem:* Am Astron Soc; Am Geophys Union; AAAS; Astron Soc Pac; Int Astron Union. *Res:* Formation of stellar and planetary sytems; convection in planetary bodies; numerical gravitational hydrodynamics. *Mailing Add:* Carnegie Inst Washington 5241 Broad Br Rd NW Washington DC 20015

BOSS, CHARLES BEN, b Foley, Ala, Dec 27, 45; m 70. ANALYTICAL CHEMISTRY. *Educ:* Wake Forest Univ, BS, 68; Ind Univ, PhD(anal chem), 77. *Prof Exp:* ASST PROF CHEM, NC STATE UNIV, 77- *Mem:* Soc Appl Spectros; Am Chem Soc; Optical Soc Am. *Res:* Atomic spectrometry in flames and plasmas; trace and ultra-trace chemical analysis; on-line computer instrumentation; automated optimization; novel transducers; flame chemistry. *Mailing Add:* Dept Chem Box 8204 NC State Univ Raleigh NC 27695-8204

BOSS, KENNETH JAY, b Grand Rapids, Mich, Dec 5, 35. MALACOLOGY. *Educ:* Cent Mich Col, BA, 57; Mich State Univ, MSc, 59; Harvard Univ, PhD(biol), 63. *Prof Exp:* Res systematist mollusks, US Nat Mus, Dept Interior, 63-66; from asst cur to cur, 66-74, PROF BIOL, HARVARD UNIV, 70- *Mem:* Soc Syst Zool; Am Malacol Union; Marine Biol Asn UK. *Res:* Systematic and evolutionary studies of mollusks; comparative morphology of lamellibranchs; zoogeography. *Mailing Add:* Organ & Evolution Biol Harvard Univ 26 Oxford St Cambridge MA 02138

BOSS, MANLEY LEON, b Atlanta, Ga, Dec 24, 24; m 56; c 3. BIOMETRICS-BIOSTATISTICS, ENVIRONMENTAL SCIENCES. *Educ:* Univ Miami, BS, 49; Inter-Am Inst Agr Sci, MAgr, 51; Iowa State Col, PhD, 55. *Prof Exp:* From instr to assoc prof bot, Univ Miami, 54-63; head macrobiol sect, 63-91, PROF BOT, FLA ATLANTIC UNIV, 63- *Concurrent Pos:* Environ consult, biostatist consult. *Res:* Environmental sciences. *Mailing Add:* 3515 S Ocean Blvd Highland Beach FL 33487

BOSSARD, DAVID CHARLES, b Sellersville, Pa, Mar 23, 40; m 64; c 7. OPERATIONS RESEARCH. *Educ:* Drexel Univ, BSc, 62; Dartmouth Col, AM, 64 & 66, PhD(math), 67. *Prof Exp:* Assoc opers res, Daniel H Wagner Assocs, 67-73, vpres opers res, 73-81; PRES, D C BOSSARD, INC, 81- *Mem:* Soc Indust & Appl Math; Inst Elec & Electronics Engrs; Asn Comput Mach. *Mailing Add:* D C Bossard Inc Two S Park St PO Box 111 Lebanon NH 03766

BOSSARD, MARY JEANETTE, b Keosauqua, Iowa, Nov 25, 54; m 75; c 2. ENZYMOLOGY, PROTEIN CHEMISTRY. *Educ:* Cent Col, BA, 77; Univ Nebr, PhD(chem), 81. *Prof Exp:* Postdoctoral, Univ Calif, Berkeley, 82-83, NIH postdoctoral fel, 83-85; ASSOC SR INVESTR, SMITHKLINE BEECHAM, 85- *Mem:* Am Chem Soc; Am Soc Biochem & Molecular Biol. *Res:* Mechanistic enzymology applied to rational drug design; therapeutic areas: immunosuppression, cardiovascular and antifungal; enzyme categories: prolyl isomerases, proteases, ATPases, P-450, oxygenases. *Mailing Add:* Dept Med Chem SmithKline Beecham Box 1539 L-410 King of Prussia PA 19406

BOSSEN, DOUGLAS C, b Akron, Ohio, Sept 26, 41; m 66; c 2. ELECTRICAL ENGINEERING. *Educ:* Northwestern Univ, BS, 64, MS, 66, PhD(elec eng), 68. *Prof Exp:* Staff engr, IBM Corp, 68-70, adv engr, 70-76, sr engr, 76-88, SR TECH STAFF MEM, IBM CORP, 89- *Mem:* Sigma Xi; fel Inst Elec & Electronics Engrs; Am Fedn Info Processing Soc. *Res:* Applied research in hardware fault-tolerent computer design, including error correcting codes, error detection and recovery techniques. *Mailing Add:* PO Box 1651 Poughkeepsie NY 12601

BOSSHARDT, DAVID KIRN, b Rochester, Minn, Apr 15, 16; m 43; c 1. ANALYTICAL BIOCHEMISTRY. *Educ:* Univ Minn, BS, 38; Rutgers Univ, MS, 40, PhD(dairy chem), 43. *Prof Exp:* Asst, NJ Exp Sta, 38-43; res assoc, Merck Sharp & Dohme Res Lab Div, Merck & Co, Inc, 43-71; res aide, Natural Resources Res Inst, Univ Wyo, 72-73; chemist, Wyo Dept Agr, 73-81; RETIRED. *Res:* Silage preservation; carotinoid pigments in feed and butter; protein and fat metabolism in small animals; energy and vitamin requirements in small animals; nutritive value of end products of silage fermentation; effect of feed and storage on carotinoid pigments in butter; unidentified growth factors; cholesterol and bile acid metabolism. *Mailing Add:* 1707 Mitchell Laramie WY 82070

BOSSHART, ROBERT PERRY, b Orange, NJ, Feb 15, 42; m 75; c 2. FOOD & ESTATE CROPS OF TROPICS. *Educ:* Univ Ill, Urbana, BS, 64; Va Polytech Inst & State Univ, MS, 67, PhD(agron), 69. *Prof Exp:* Res plant physiologist, Vegetation Control Labs, US Army, Fort Detrick, Md, 69-70; food & agr officer, 29th Civil Affairs Co, US Army, Vietnam, 70-71; chief & agronomist, Vietnam Rice Res Proj, Int Rice Res Inst, 72-75; gen mgr for Brazil, IRI Res Inst, 75-78; assoc agronomist, soil fertility & water mgt, 78-80, Hawaiian Sugar Planters Asn, agronomist, 80-85; dep dir Southeast Asia, Potash & Phosphate Inst & Int Potash Inst, 85-89; LAND RECLAMATION SPECIALIST/SOIL SCIENTIST, REMEDIATION SERV DIV, HORSEHEAD RESOURCE DEVELOP CO INC, 90- *Concurrent Pos:* Affiliate fac, Agron & Soil Sci Dept, Univ Hawaii, Honolulu, 79-85; adj prof, Calif State Univ, Fresno, 81-85. *Mem:* Am Soc Agron; Soil Sci Soc Am; Crop Sci Soc Am; Soil Sci Soc Malaysia; Int Soc Sugar Cane Technologists. *Res:* Soil fertility and plant nutrition; irrigation water management; increasing crop yields. *Mailing Add:* Horsehead Resource Develop Co Inc Fourth & Franklin Ave Palmerton PA 18071

BOSSLER, JOHN DAVID, b Johnstown, Pa, Dec 8, 36; m 63; c 1. GEODETIC SCIENCE, GEOPHYSICS. *Educ:* Univ Pittsburgh, BS, 59; Ohio State Univ, MS, 64, PhD(geod sci), 72. *Prof Exp:* Officer, Nat Oceanic & Atmospheric Admin, Ship COWIE, 59-60; chief astron field parties, Dept Com, Nat Oceanic & Atmospheric Admin, Nat Ocean Surv, Nat Geod Surv, 60-63, chief sci experience math formulas, 64-69; exec officer, Nat Oceanic & Atmospheric Admin, Ship Davidson, 69-71; proj mgr, 73-75, dir, Nat Geod Surv & Charting Geod Serv, 83-86, PROJ MGR & DEP DIR, DEPT COM, NAT OCEANIC & ATMOSPHERIC ADMIN, NAT OCEAN SURV, 75- *Concurrent Pos:* Nat Oceanic & Atmospheric Admin grant; dir, Ctr Mapping, Ohio State Univ, 87-90. *Mem:* Am Soc Photo & Remote Sensors; Am Geophys Union; Am Cong Surv & Mapping; Urban & Regional Info Systs Asn. *Res:* Geodetic science; geophysical research; digital mapping. *Mailing Add:* 8500 Davington Dr Dublin OH 43017

BOSSO, JOSEPH FRANK, b New Kensington, Pa, May 18, 31; m 54; c 3. ORGANIC CHEMISTRY. *Educ:* Univ Pittsburgh, BS, 70. *Prof Exp:* Res asst II, 55-58, res asst I, 58-62, sr technician, 62-66, res chemist, 66-72, res assoc, 72-77, SR RES ASSOC, COATINGS & RESINS DIV, PPG INDUSTS INC, 77- *Honors & Awards:* IR-100 Award, Indust Res Mag, 66 & 74; Coating Award, Am Soc Metals, 73. *Res:* Synthesis and development of polymers, especially in the field of cathodic electrodeposition of organic coatings. *Mailing Add:* 428 Wedgewood Dr Lower Burrell PA 15068

BOST, HOWARD WILLIAM, b Robstown, Tex, Sept 29, 24; m 50; c 3. ORGANIC CHEMISTRY. *Educ:* Southwest Tex State Teachers Col, BS, 48; Univ Tex, MA, 50, PhD(org chem), 55. *Prof Exp:* Res chemist, Phillips Petrol Co, 50-51, res chemist & group leader liquid propellants, 54-57, res chemist & sr group leader, 57-60, group leader photochem reactions, 60-68, group leader oxidation processes, 68-70, group leader org sulfur chem, 70-74, tech recruitment rep, 74-82, patent liaison, 83-85; RETIRED. *Concurrent Pos:* Mem joint Army-Navy-Air Force comt on mono-propellant test methods, 59-60,. *Honors & Awards:* Sigma Xi. *Mem:* Am Chem Soc; Sigma Xi. *Res:* Protective coatings; nitrogen chemicals; liquid rocket propellants; photochemical reactions; sulfur reactions; oxidation processes; organic sulfur chemistry; perfume ingredients; flame retardants; technical recruiting. *Mailing Add:* 1334 Quail Dr Bartlesville OK 74006

BOST, ROBERT ORION, b Elida, NMex, Jan 15, 43; m 66; c 2. TOXICOLOGY. *Educ:* Univ Tex, Austin, BS, 65, MA, 67; Univ Houston, PhD(chem), 70; Am Bd Forensic Toxicol, dipl, 80. *Prof Exp:* Lab asst, Southwestern Univ, 62-64; lab asst, Univ Tex, Austin, 65-67; teaching fel, Univ Houston, 67-69, res asst, 69-70; sci res specialist, La State Univ, New Orleans, 70-72; chemist, Vet Admin Hosp, New Orleans, 72-74; assoc toxicologist, Cuyahoga County Coroner's Off, 75-82; CHIEF TOXICOLOGIST, SOUTHWESTERN INST FORENSIC SCI, DALLAS, 82-; PRES, TOXICOL CONSULTS, INC, DALLAS, 87- *Concurrent Pos:* Asst prof, Dept Path, Southwestern Med Ctr Dallas, Univ Tex, Dallas, 82-90, assoc prof, 90-; guest ed, J Anal Toxicol, 88; adv bd, J Forensic Med, Istanbul, Turkey, 85-; dir bd, Soc Forensic Toxicologists, 86-88, vpres, 89, pres, 90. *Mem:* Am Chem Soc; Soc Forensic Toxicologists; fel Am Acad Forensic Sci; Int Asn Forensic Toxicologists. *Res:* Forensic and clinical toxicology; forensic urine drug testing; gas chromatography; derivatives for chromatography; generation and reaction of carbenes; small ring compounds. *Mailing Add:* 5230 Med Ctr Dr PO Box 35728 Dallas TX 75235-0728

BOSTER, THOMAS ARTHUR, b Columbus, Ohio, Sept 28, 36; m 62; c 3. SOLID STATE PHYSICS. *Educ:* Capital Univ, BS, 58; Ohio Univ, MS, 60, PhD(physics), 66. *Prof Exp:* Res engr, Nortronics Div, Northrop Corp, 60-62; fel physics, Ohio Univ, 62-65; from asst prof to assoc prof, Oklahoma City Univ, 65-69; sr physicist, 68-73, group leader, 73-75, group leader, Div Responsibility for Safety & Qual Control, Lawrence Livermore Lab, Univ Calif, 75-80; BOSTER ASSOC INC, 80- *Concurrent Pos:* Consult, Okla Bur Invest, 67-68. *Mem:* Am Soc Testing & Mat; Soc Automotive Engrs; Am Phys Soc; Am Asn Physics Teachers; Am Soc Safety Engrs; Human Factors Soc. *Res:* Absorption of x-rays; radiation damage; cryogenics; microwaves; x-ray diagnostics; x-ray spectrometers; solid state detectors; safety research; accident reconstruction. *Mailing Add:* 2439 Marbury Rd Livermore CA 94550

BOSTIAN, CAREY HOYT, b China Grove, NC, Mar 1, 07; m 29; c 3. GENETICS. *Educ:* Catawba Col, AB, 28; Univ Pittsburgh, MS, 30, PhD(genetics), 33. *Hon Degrees:* DSc, Catawba Col, 53, Wake Forest Col, 54 & Nat Univ Eng, Peru, 57. *Prof Exp:* Asst zool, Univ Pittsburgh, 28-30; asst prof zool, 30-36, assoc prof zool & assoc poultry genetics, 36-45, prof zool & res prof poultry genetics, 45-53, assoc dean agr, 48-53, chancellor, 53-59, prof, 59-88, EMER PROF POULTRY GENETICS, NC STATE UNIV, 88- *Res:* General and human genetics. *Mailing Add:* 2221 Carol Woods Chapel Hill NC 27514

BOSTIAN, CHARLES WILLIAM, b Chambersburg, Pa, Dec 30, 40; m 64; c 2. ELECTRICAL ENGINEERING. *Educ:* NC State Univ, BS, 63, MS, 64, PhD(elec eng), 67. *Prof Exp:* Res engr, Electronics Res Lab, Corning Glassworks, 67; From asst prof to assoc prof, 69-77, PROF ELEC ENG, VA POLYTECH INST & STATE UNIV, 77- *Concurrent Pos:* Consult, US Dept Defense, 69-71 & Electro-Tec Corp, 72-73. *Mem:* Inst Elec & Electronics Engrs; Sigma Xi. *Res:* Satellite communications; radio wave propagation; applied electromagnetics. *Mailing Add:* Elec Eng Dept VPI Va Polytech Inst & State Univ Blacksburg VA 24061

BOSTIAN, HARRY (EDWARD), b Lewisburg, Pa, Jan 16, 33; m 55. MASS TRANSFER SEPARATIONS, CONVERSION PROCESSES. *Educ:* Bucknell Univ, BSChE, 54; Rensselaer Polytech Inst, MChE, 56; Iowa State Univ, PhD(chem eng), 59. *Prof Exp:* Asst chem eng, Rensselaer Polytech Inst, 54-56 & Iowa State Univ, 56-59; asst prof, Univ NH, 59-61; engr, Exxon Res & Eng Co, 61-65; assoc prof, Univ Miss, 65-70; ENGR & RES PROG MGR, US ENVIRON PROTECTION AGENCY, 70- *Concurrent Pos:* Tech session organizer, Am Inst Chem Engrs, 75- *Mem:* Am Inst Chem Engrs; AAAS; Sigma Xi. *Res:* Energy/environmental research; environmental engineering; separations fundamentals and technology; management of residuals from municipal waste water treatment; sludge processing and conversion; sludge management costs. *Mailing Add:* Environ Res Ctr US Environ Protection Agency 26 W Martin Luther King Dr Cincinnati OH 45268

BOSTICK, EDGAR E, b Newville, Ala, Dec 10, 26; m 55; c 5. PHYSICAL ORGANIC CHEMISTRY. *Educ:* Ala Polytech Inst, BS, 50; Univ Akron, PhD(chem), 59. *Prof Exp:* Jr chem engr, Goodyear Tire & Rubber Co, 50-51, chem engr, 53-56; fel, Univ Akron, 59-60; res chemist, 60-69, mgr res & develop, Insulating Mat Dept, 69-76, WITH RES & DEVELOP, LEXO PROD DEPT, GEN ELEC CO, 76- *Mem:* Am Chem Soc; Am Inst Chem Engrs. *Res:* Ionic polymerization of vinyl, diene and heterocyclic compounds; reaction kinetics and mechanisms; characterization of polymers by chemical and physical techniques. *Mailing Add:* 57 Park Ridge Dr Mt Vernon IN 47620

BOSTICK, WARREN LITHGOW, b Dallas, Tex, July 28, 14; m 39; c 2. PATHOLOGY. *Educ:* Univ Calif, AB, 36, MD, 40. *Prof Exp:* From assoc prof to prof path, Sch Med, Univ Calif, San Francisco, 50-64, dir labs, Univ Hosps, 54-64; dean, 64-74, prof, 64-88, EMER PROF PATH, COL MED, UNIV CALIF, IRVINE, 88- *Concurrent Pos:* Trustee, Calif Physicians Serv, 62- *Mem:* AAAS; Col Am Path; Am Soc Exp Path (pres elect); Am Soc Clin Path. *Res:* Hodgkins disease; ameblasis; fibrocystic disease of pancreas; viral carcinogenesis. *Mailing Add:* Univ Calif Col Med Irvine CA 92717

BOSTICK, WINSTON HARPER, b Freeport, Ill, Mar 5, 16; m 42; c 3. PHYSICS. *Educ:* Univ Chicago, BS, 38, PhD(physics), 41. *Prof Exp:* Instr physics, Univ Chicago, 40; staff mem, Radiation Lab, Mass Inst Technol, 41-45, sect chief, 45-46, staff mem, Res Lab Electronics, 46-49; from asst prof to assoc prof physics, Tufts Col, 48-54; staff mem, Radiation Lab, Univ Calif, 54-56; prof, 56-69, head dept, 56-73, GEORGE MEADE BOND PROF PHYSICS, STEVENS INST TECHNOL, 69- *Concurrent Pos:* NSF sr fel, Nuclear Res Ctr, France & Culham Lab. *Honors & Awards:* Prizes, Gravity Res Found, 58 & 61. *Mem:* Fel Am Phys Soc. *Res:* Cosmic ray research with cloud chambers; design of pulse transformers; development of microwave linear accelerator; diffusion of gaseous ions across a magnetic field and plasma waves in presence of a magnetic field. *Mailing Add:* PO Box 1652 Corrales NM 87048

BOSTOCK, JUDITH LOUISE, b Gardiner, Maine. THEORETICAL SOLID STATE PHYSICS. *Educ:* Trinity Col, AB, 60; Dickinson Col, BS, 63; Georgetown Univ, MS, 69, PhD(physics), 71. *Hon Degrees:* LLD, Gannon Univ, 90. *Prof Exp:* Asst physics, Univ Saarland, 70-71; from instr to assoc prof physics, Mass Inst Technol, 72-82, opers res analyst Off Mgt & Budget, 84-89, spec asst to dir & asst dir indust technol, Off Sci & Technol Policy, 89-90; CHIEF OPERATING OFFICER, SC UNIVS RES & EDUC FOUND, 91- *Concurrent Pos:* Consult physicist crystal physics, Naval Res

Lab, 72-75; consult sr physicist, BDM Corp, Va, 75-79; fac develop workshop consult, 75-80; consult ed, Addison Wesley Publ Co, 76-80, Off Mgt & Budget, 82-84; consult-at-large, SURA, Wash DC, 91-, Los Alamos Nat Lab, 91- *Mem:* Am Phys Soc; Sigma Xi. *Res:* Lattice instabilities and superconductivity in metals at low temperatures; low temperature properties in theoretical solid state physics; non-equilibrium conditions in superconductors. *Mailing Add:* Univ SC Columbia SC 29208

BOSTON, ANDREW CHESTER, b Divernon, Ill, Nov 17, 41; m 65; c 2. ANIMAL BREEDING, BEEF CATTLE PRODUCTION & MANAGEMENT. *Educ:* Univ Ill, Urbana-Champaign, BS, 65; Univ Mo, Columbia, MS, 66; Okla State Univ, Stillwater, PhD(animal breeding), 73. *Prof Exp:* Asst prof animal breeding & genetics, La State Univ, Baton Rouge, 73-76; provincial beef prog specialist, Man Agr, Winnipeg, 76-79; chief, rec performance beef cattle, Agr Can, Ottawa, Ont, 79-82; head, Dept Agr & Natural Resources, Morehead State Univ, Ky, 82-85, prof agr & animal sci, 82-87; livestock consult, 87-89; COUNTY EXTEN DIR & AGR AGENT, PURDUE UNIV, PAOLI, IND, 89- *Concurrent Pos:* Researcher, 76-79. *Mem:* Am Soc Animal Sci; Coun Agr Sci & Technol; Can Soc Animal Sci; Nat Asn County Agr Agents; Am Registry Prof Animal Scientists; Sigma Xi. *Res:* Pasture management levels and grazing intensity on beef cow productivity and profitability; utilization of fescue (endophyte infected and clean) with fertilization and clovers by grazing cow-calf pairs; forage species and production levels on beef cow-calif carrying capacity. *Mailing Add:* 214 W Main St Paoli IN 47454-1032

BOSTON, CHARLES RAY, b Bellaire, Ohio, Aug 4, 28; c 3. ENVIRONMENTAL PROGRAM MANAGEMENT, FOSSIL ENERGY. *Educ:* Ohio Univ, BS, 49; Northwestern Univ, PhD(chem), 53. *Prof Exp:* Chemist, 53-73, GROUP LEADER, OAK RIDGE NAT LAB, 73- *Mem:* Fel AAAS; Sigma Xi. *Res:* Molten salt chemistry; physical inorganic; absorption spectra; unusual oxidation states; management of interdisciplinary teams doing environmental impact assessments. *Mailing Add:* Union Carbide Corp Nuclear Div-Oak Ridge Nat Lab PO Box 2008 Bldg 4500 M5200 Oak Ridge TN 37831-6200

BOSTON, JOHN ROBERT, b Evanston, Ill, Oct 16, 42; c 2. CLINICAL NEUROPHYSIOLOGY, SIGNAL PROCESSING. *Educ:* Stanford Univ, BS, 64, MS, 66; Northwestern Univ, PhD(elec eng), 71. *Prof Exp:* Res assoc health care delivery, Hosp Res & Educ Trust, 71-72; asst prof elec eng, Univ Md, 72-75; asst prof biomed eng prog & elec eng, Carnegie-Mellon Univ, 75-80; asst prof otolaryngol, 75-80, RES ASSOC PROF ANESTHESIOL, SCH MED, UNIV PITTSBURGH, 80- *Mem:* Asn Comput Mach; AAAS; Inst Elec & Electronics Engrs. *Res:* Digital signal processing applied to acquisition and analysis of sensory evoked potentials; auditory evoked potentials; computer applications in patient monitoring; real-time computer applications in clinical medicine. *Mailing Add:* 308 High Oaks Ct Wexford PA 15090

BOSTON, ROBERT WESLEY, b Athens, Ont, July 10, 32; m 58; c 2. PEDIATRICS. *Educ:* Queen's Univ, Ont, MD, 57; Royal Col Physicians & Surgeons Can, dipl pediat, 63. *Prof Exp:* Mead Johnson Educ Fund fel pediat, Harvard Med Sch & Boston Lying-In Hosp, 62-64; McLaughlin Found traveling fel, Univ Col Hosp, London, Eng, 64-65; from asst prof to assoc prof pediat, 65-77, assoc prof obstet & gynec, 77-86, PROF PEDIAT, QUEEN'S UNIV, ONT, 77-, PROF, DEPT MED, 87- *Concurrent Pos:* Attend physician, Kingston Gen Hosp, 65- *Mem:* Can Pediat Soc; Can Soc Clin Invest. *Res:* Neonatology; respiratory distress syndrome; assessment of the fetus in uterus and of immediate neonatal adaption. *Mailing Add:* Dept Med Palliative Med Queen's Univ Kingston ON K7L 3N6 Can

BOSTRACK, JACK M, b Stoughton, Wis, Apr 30, 31; m 53; c 3. BOTANY. *Educ:* Wartburg Col, BA, 53; Univ Wis, MS, 59, PhD(bot hort), 62. *Prof Exp:* Asst bot, Univ Wis, 57-61; asst prof biol, Wayne State Col, 61-63; assoc prof, 63-70, PROF BIOL, Univ Wis-River Falls, 70- *Mem:* Bot Soc Am. *Res:* Morphogenesis of plant parts; determination of form in plants, as influenced by exogenous and endogenous factors. *Mailing Add:* Dept Biol Univ Wis River Falls WI 54022

BOSTROM, CARL OTTO, b Port Jefferson, NY, Aug 18, 32; m 54; c 3. SPACE PHYSICS. *Educ:* Franklin & Marshall Col, BS, 56; Yale Univ, MS, 58, PhD(physics), 62. *Prof Exp:* Sr physicist, 60-64, supvr space physics group, 64-74, chief scientist, 74-78, assoc head, 78, head, Space Dept, 79-80, asst dir space systs, 79, dep dir, 79-80, DIR, APPL PHYSICS LAB, JOHNS HOPKINS UNIV, 80- *Concurrent Pos:* Instr, Evening Col, Johns Hopkins Univ, 75- *Mem:* AAAS; Am Phys Soc; Am Geophys Union. *Res:* Space particles and fields; satellite instrumentation; magnetospheric physics; solar physics; satellite systems design; research administration. *Mailing Add:* 12508 Meadowood Dr Silver Spring MD 20904

BOSTROM, ROBERT CHRISTIAN, b Edinburgh, Scotland, July 22, 20; m 52; c 3. GEOPHYSICS. *Educ:* Oxford Univ, BA, 48, MA, 52, DPhil(geophys), 61. *Prof Exp:* Geophysicist, Explor Consults Inc, 50-53 & Standard Oil Co Calif, 53-64; assoc prof, 64-69, PROF GEOL SCI & GEOPHYS, UNIV WASH, 69- *Concurrent Pos:* Seafloor eng comt, Marine Bd, Nat Res Coun, 74-76. *Mem:* Am Geophys Union; Soc Explor Geophys; Am Asn Petrol Geol; Geol Soc London; Seismol Soc Am. *Res:* Mantle circulation and plate motion mechanism; earth tides; interaction of elementary particles and earth's interior; regional mineral exploration. *Mailing Add:* Dept Geol Sci Univ Wash Seattle WA 98105

BOSWELL, DONALD EUGENE, b Durham, NC, Aug 12, 34; m 56; c 3. ORGANIC CHEMISTRY. *Educ:* Duke Univ, BS, 56; Va Polytech Inst & State Univ, MS, 61, PhD(chem), 63. *Prof Exp:* Res chemist, Cent Res Div Lab, Mobil Res & Develop Co, NJ, 63-65, asst supvr catalysis & math res, 65-67; mgr prod planning, Cincinnati Milacron, 67-68, mgr res & develop, Prod Div, 68-76; dir res & develop, Metall Div, Quaker Chem, 76-78, tech dir, 78-91, vpres, res & develop, 85-91; RETIRED. *Mem:* Nat Asn Corrosion Engrs; AAAS; Am Chem Soc; Am Soc Tool & Mfg Eng; Am Soc Lubrication Eng. *Res:* Chemistry of heterocyclic compounds; organometallic chemistry; metalworking synthetic lubricants; corrosion inhibitors; water purification techniques; organic analytical chemistry. *Mailing Add:* 5749 Oak Bluff Lane Wilmington NC 28409

BOSWELL, FRANK WILLIAM CHARLES, b Hamilton, Ont, July 11, 24; m 51; c 2. PHYSICS. *Educ:* Univ Toronto, BA, 46, MA, 47, PhD(physics), 50. *Prof Exp:* Sci officer, Dept Mines & Technol Surv, Can, 50-57, head metal physics sect, 57-60; assoc prof, 60-63, assoc dean sci for grad affairs, 67-76, PROF PHYSICS, UNIV WATERLOO, 63- *Mem:* Electron Micros Soc Am; Am Asn Physics Teachers; Can Asn Physicists. *Res:* Solid state physics; imperfections in crystals; electron microscopy and diffraction. *Mailing Add:* Dept Physics Univ Waterloo Waterloo ON N2L 3G1 Can

BOSWELL, FRED CARLEN, b Monterey, Tenn, Aug 20, 30; m 54; c 2. AGRONOMY, SOILS. *Educ:* Tenn Polytech Inst, BS, 54; Univ Tenn, MS, 56; Pa State Univ, PhD(agron, soils), 60. *Prof Exp:* Asst res agronomist, Agr Exp Sta, Univ Tenn, 55-56 & Exp Sta, Univ Ga, 56-57; asst, Pa State Univ, 58-60; asst soil chemist, Exp Sta, 60-64, assoc soil chemist, 64-78, PROF AGRON & SOILS, EXP STA, UNIV GA, 71- *Mem:* Am Soc Agron; Int Soil Sci Soc; Coun Agr Sci & Technol; Soil Sci Soc Am. *Res:* Soil chemistry, fertility, microbiology and closely related subjects; nitrogen transformations, management and micronutrients as related to plant growth and physiology. *Mailing Add:* Dept Agron Univ Ga 267 Plant Sci Griffin GA 30223-1797

BOSWELL, GEORGE A, b Hayward, Calif, Jan 26, 32; m 52; c 4. ORGANIC CHEMISTRY. *Educ:* Univ Calif, Berkeley, BS, 56, PhD(org chem), 59. *Prof Exp:* With Shell Develop Co, Calif, 59-61; res org chemist, E I du Pont de Nemours & Co, Inc, 61-67, res supvr, 67-79, res mgr, Cent Res Dept, 79-85, res fel, 86-90, RES MGR, MED PROD DEPT, PHARMACEUT DIV, E I DU PONT DE NEMOURS & CO, INC, 85-, SR RES FEL, DU PONT MERCK PHARMACEUT, 91- *Mem:* Am Chem Soc. *Res:* Steroid chemistry; general organic synthesis; organic fluorine and medicinal chemistry. *Mailing Add:* Du Pont Merck Exp Sta Bldg 353 E I du Pont de Nemours & Co Wilmington DE 19898

BOSWICK, JOHN A, b Galatia, Ill, Aug 30, 26; m 49; c 3. SURGERY. *Educ:* Southern Ill Univ, Carbondale, BA, 51; Loyola Univ Chicago, MS, 52, MD, 56. *Prof Exp:* Assoc prof surg, Med Sch, Northwestern Univ, 61-70; prof surg, Univ Ill Med Ctr, 70-72; PROF SURG & CHMN SECT HAND SURG, MED CTR, UNIV COLO, DENVER, 72- *Concurrent Pos:* Dir burn unit & hand surg, Cook County Hosp, Chicago, 61-72; consult, Brooke Army Med Ctr, 70-, Great Lakes Naval Hosp, 70- & Vet Admin Hosp, Denver, 72- *Mem:* Am Asn Surg of Trauma (secy, 74-79); Am Burn Asn (pres, 74-75); Am Soc Surg of Hand (secy-treas, 72-76); Asn Am Med Cols; Int Soc Burn Injuries (secy-gen, 75-). *Res:* Hand surgery. *Mailing Add:* 2005 Franklin St No 660 Denver CO 80205

BOTELHO, LYNNE H PARKER, DIABETES, ENZYMOLOGY. *Educ:* Ind Univ, PhD(chem), 75. *Prof Exp:* GROUP LEADER, CELLULAR BIOCHEM, SANDOZ RES INST, SANDOZ, INC, 79- *Mailing Add:* Dept Pharm Berlex Labs Inc 110 E Hanover Ave Cedar Knolls NJ 07927

BOTELHO, STELLA YATES, b Japan, Jan 14, 19; US citizen. PHYSIOLOGY. *Educ:* Univ Pa, BA, 40; Med Col Pa, MD, 49. *Prof Exp:* From instr to assoc prof, 49-69, prof, 69-81, EMER PROF PHYSIOL, SCH MED, UNIV PA, 81- *Concurrent Pos:* Vis prof, Cambridge Univ, 57-58; lectr, US Naval Hosp, Philadelphia, 64-70; sect chief, Philadelphia Gen Hosp, 63-74; consult, Children's Hosp Philadelphia, 63-81; mem adv panel regulatory biol prog, NSF, 73-76; mem adv comt, Coun Exchange Int Scholars; mem adv bd, Dry Eye Inst, 84- *Honors & Awards:* Lacrima Award, Dry Eye Inst, 84. *Mem:* Am Asn Zool Parks & Aquariums; Am Physiol Soc; Int Soc Dacryology. *Res:* Neuromuscular physiology; secretory mechanisms, especially of the orbital glands; diseases of neuromuscular system; ocular reflexes. *Mailing Add:* Box 1108-C108 Blue Bell PA 19422

BOTERO, J M, b Medellin, Colombia, Nov 12, 29; m 57; c 2. BIOCHEMISTRY. *Educ:* Univ Antioquia, BS, 48, MD, 55. *Prof Exp:* Head dept biochem, Univ Antioquia, 60-61; med & res dir western hemisphere, 61-67, med dir develop mkt, 64-67, med dir, Dome Labs, 67-69, vpres, Ames Co, 69-79, VPRES MED AFFAIRS, PROF PROD GROUP, MILES LABS, INC, 79- *Concurrent Pos:* Kellogg fel, 55-57; Rockefeller Found fel, 57-59. *Honors & Awards:* Cano Award, 48. *Mem:* NY Acad Sci; AMA; Am Chem Soc. *Res:* Synthesis of proteins and nucleic acids; chemical pathology. *Mailing Add:* Div Miles Inc BB07A Diagnostics Div Elkhart IN 46514

BOTEZ, DAN, b Bucharest, Romania, May 22, 48; m 76; c 2. SEMICONDUCTOR DIODE-LASER PHYSICS, PHASE-LOCKED ARRAYS OF DIODE LASERS. *Educ:* Univ Calif, Berkeley, BS, 71, MS, 72 & PhD(elec eng), 76. *Prof Exp:* Postdoctoral fel, IBM Thomas J Watson Res Ctr, 76-77; mem tech staff, RCA David Sarnoff Res Ctr, 77-82, res leader, 82-84; dir device develop, Lytel Inc, 84-86; chief scientist & lab dir, 86-87, SR STAFF SCIENTIST, TRW RES CTR, 87- *Concurrent Pos:* Mem, Tech Comt Semiconductor Lasers, Inst Elec & Electronic Engrs, 82-87 & 90-, chmn, 88-89. *Honors & Awards:* Outstanding Young Engr Centennial-Yr Award, Inst Elec & Electronics Engrs, 84. *Mem:* Fel Inst Elec & Electronics Engrs; Optical Soc Am. *Res:* Developed high-power single-spatial-mode semiconductor diode lasers and high-power phase-locked arrays of diode lasers; Published work on the properties of dielectric waveguides and was co-developer of the coupled-mode theory for phase-locked arrays. *Mailing Add:* 234 Paseo de las Delicias Redondo Beach CA 90277

BOTHAST, RODNEY JACOB, b Union City, Ind, Sept 18, 45; m 65; c 3. MICROBIOLOGY. *Educ:* Ohio State Univ, BS, 67; Va Polytech Inst & State Univ, MS, 70, PhD, 71. *Prof Exp:* Grad res asst food microbiol, Va Polytech Inst & State Univ, 67-71; microbiologist, 71-78, RES LEADER, FERMENTATION BIOCHEM RES UNIT, NORTHERN REGIONAL

RES CTR, AGR RES SERV, USDA, 78- *Concurrent Pos:* Sr res specialist microbiol, Peoria Sch Med, Univ Ill, 71- *Mem:* Inst Food Technologists; Am Soc Microbiol; Sigma Xi; Am Chem Soc. *Res:* Food microbiology; mycotoxins; biodeterioration; grain preservation; grain drying; solid substrate fermentations; the fermentative conversion of agricultural commodities into new and improved products. *Mailing Add:* 311 Indian Circle East Peoria IL 61611

BOTHNER, RICHARD CHARLES, b New York, NY, May 16, 29; m 58; c 3. ECOLOGY, ANATOMY. *Educ:* Fordham Univ, BS, 51, MS, 57, PhD(biol), 59. *Prof Exp:* Asst biol, Fordham Univ, 55-57; mus asst herpet, Am Mus Natural Hist, NY, 57-58; from asst prof to assoc prof, 58-69, PROF BIOL, ST BONAVENTURE UNIV, 69- *Mem:* Am Soc Ichthyol & Herpet; Ecol Soc Am; Sigma Xi. *Res:* Herpetology; phylogenetic significance of ophidian circulatory systems. *Mailing Add:* 788 Main St Olean NY 14760

BOTHNER, WALLACE ARTHUR, b Fitchburg, Mass, May 17, 41; m 63; c 3. STRUCTURAL GEOLOGY, GEOPHYSICS. *Educ:* Harpur Col, BA, 63; Univ Wyo, PhD(geol), 67. *Prof Exp:* Instr geol, Univ Wyo, 66-67; from asst prof to assoc prof, 67-82, PROF GEOL, UNIV NH, 82- *Concurrent Pos:* Br geophys, US Geol Surv, WAE, 77- *Mem:* Geol Soc Am; Fel Geol Soc Am. *Res:* Metamorphic structural geology, regional geophysics and tectonics of the northern Appalachian mountains. *Mailing Add:* Dept Earth Sci Univ NH Durham NH 03824-3589

BOTHNER-BY, AKSEL ARNOLD, b Minneapolis, Minn, Apr 29, 21; m 49; c 2. BIOPHYSICAL CHEMISTRY. *Educ:* Univ Minn, BChem, 43; NY Univ, MS, 47; Harvard Univ, PhD(chem), 49. *Prof Exp:* Assoc scientist chem, Brookhaven Nat Lab, 49-53, scientist, 53; instr, Harvard Univ, 53-55, lectr, 55-58; staff fel, Mellon Inst Sci, 58-59, from asst dir res to dir res, 59-62, mem adv comt, 62-67, chmn dept chem, 67-70, dean, 71-75, prof chem, 67-76; Univ prof chem, 76-86, ACTG HEAD, CARNEGIE-MELLON UNIV, 87- *Concurrent Pos:* Am Cancer Soc fel, Univ Zurich, 52-53; consult, Retina Found, 57; Fulbright lectr, Univ Munich, 62-63; adj prof, Univ Pittsburgh, 64-; vis prof, Univ Calif, San Diego, 76-77. *Honors & Awards:* IR-100 Award. *Mem:* Am Chem Soc; Am Soc Biochem & Molecular Biol. *Res:* Nuclear magnetic resonance; structure of biopolymers. *Mailing Add:* Dept Chem Carnegie-Mellon Univ Pittsburgh PA 15213

BOTHWELL, ALFRED LESTER MEADOR, b Springfield, Mo, Apr 29, 49; m 74, 85; c 2. IMMUNOLOGY. *Educ:* Wash Univ, AB, 71; Yale Univ, MPh, 74, PhD(biol), 75. *Prof Exp:* Fel virol, Cold Spring Harbor Lab, 75-76; fel, Ctr Cancer Res, Mass Inst Technol, 76-82; asst prof, 82-88, ASSOC PROF BIOL & PATH, MED SCH, YALE UNIV, 88- *Concurrent Pos:* Assoc investr, Howard Hughes Med Inst, 82- *Mem:* Am Soc Microbiol; Sigma Xi; Am Asn Immunologists. *Res:* Structure and expression of genes in the immune response. *Mailing Add:* Sect Immunobiol Yale Med Sch 310 Cedar St New Haven CT 06510

BOTHWELL, FRANK EDGAR, b Saginaw, Mich, Feb 25, 18; m 45; c 8. MATHEMATICS, ENGINEERING. *Educ:* Mass Inst Technol, SB(math), 40, SB(elec eng), 41, PhD(math), 46. *Prof Exp:* Staff mem radar develop, Radiation Lab, Mass Inst Technol, 40-47; div dir guided missiles, Ord Res Lab, Univ Chicago, 47-50; assoc prof elec eng, Northwestern Univ, 50-51; dir planning rocket develop, Naval Weapons Ctr, 51-58; dir appl sci, Appl Sci Lab, Univ Chicago, 58-62; dir & chief sci syst anal, Ctr Naval Anal, 62-70; vpres, Honeywell Inc, 70-83; CONSULT, 83- *Mailing Add:* 422 E Jefferson St Falls Church VA 22046-3534

BOTHWELL, MAX LEWIS, b Portsmouth, Va, Dec 5, 46; m 73. LIMNOLOGY. *Educ:* Univ Calif, Santa Barbara, BA, 68, MA, 72; Univ Wis-Madison, PhD(zool), 75. *Prof Exp:* Res assoc limnol, Ctr Gt Lakes Studies, 71-75; res assoc stream ecol, Weyerhaeuser Co, 75-78; res assoc, Ctr Great Lakes Studies, Univ Wis-Milwaukee, 78-80; WITH DEPT ENVIRON, CAN, 80- *Mem:* Am Soc Limnol & Oceanog; Int Soc Theoret & Appl Limnol. *Res:* Phytoplankton ecology; nutrient limitation and nutrient cycling; photosynthetic pigments and carbon fixation rates. *Mailing Add:* 11 Innovation Blvd Saskatoon SK S7N 2X8 Can

BOTKIN, DANIEL BENJAMIN, b Oklahoma City, Okla, Aug 19, 37; m 60, 79; c 2. ECOLOGY. *Educ:* Univ Rochester, AB, 59; Univ Wis-Madison, MS, 62; Rutgers Univ, PhD(biol), 68. *Prof Exp:* Asst prof ecol, Sch Forestry & Environ Studies, Yale Univ, 68-74, assoc prof, 74-75; assoc prof, Ecosysts Ctr, Marine Biol Lab, Woods Hole, Mass, 75-78; chmn, environ studies prog, 78-85, PROF, DEPT BIOL, UNIV CALIF, SANTA BARBARA, 78-, PROF, ENVIRON STUDIES PROG, 78- *Concurrent Pos:* NSF res grants, 69-83; res grants, World Wildlife Fund, 74-76, Nat Oceanic & Atmospheric Admin, 78-80, NASA, 78-79, 80-81, 81-87, Andrew J Mellon Found, 85-88, 88-91, Pew Found, 89-91, Environ Protection Agency, 89-92, Dept Energy, 90-91; mem, Space Sci Bd, Nat Acad Sci, 80-83, Sci Adv Bd, US Marine Mammal Comn, 80-81, Acad Adv Bd, Central Europ Univ, 90; fel, Woodrow Wilson Int Ctr Scholars, Smithsonian Inst Bldg, Washington, 78-79; nat lectr, Sigma Xi, 81-83; fel, Rockefeller Bellagio Study & Conf Ctr, 86, Woodrow Wilson Int Ctr Scholars, 78-79, East-West Ctr, Honolulu, 85-86, 86-87; consult, World Bank, 88, 90-91, Nat Geog Soc, 88-90. *Mem:* Fel AAAS; Ecol Soc Am; Am Inst Biol Sci; Sigma Xi; Am Soc Naturalists. *Res:* Mineral cycling and energy flow in ecosystems; ecosystem theory and models; interactions among plants and animals; population dynamics of long-lived and endangered species. *Mailing Add:* Dept Biol Sci Univ Calif Santa Barbara CA 93106

BOTKIN, MERWIN P, b Fruita, Colo, Sept 19, 22; m 48; c 4. ANIMAL SCIENCE. *Educ:* Univ Wyo, BS, 48, MS, 49; Okla State Univ, PhD(animal breeding), 52. *Prof Exp:* Instr animal prod, Univ Wyo, 48-49; instr animal husb, Okla State Univ, 51-52; PROF ANIMAL BREEDING, UNIV WYO, 52- *Mem:* Am Soc Animal Sci. *Res:* Sheep breeding and physiology. *Mailing Add:* Univ Wyo Agr Bldg Laramie WY 82071

BOTROS, RAOUF, b Mit-Ghamr, Egypt, Aug 28, 32; m 67; c 1. CHEMISTRY. *Educ:* Alexandria Univ, Egypt, BSc, 55; Duquesne Univ, PhD(org chem), 62. *Prof Exp:* Res supvr dyestuff & intermediates, Egypt Dyes & Intermediates Corp, 62-67; SR RES CHEMIST & GROUP LEADER DYESTUFFS & INTERMEDIATES, AM COLOR & CHEM CORP, SUBSID NAM PHILIPS, 68- *Concurrent Pos:* Mgr fluids develop & prod, Eastman Kodak. *Mem:* Am Chem Soc; Am Asn Textile Chemists & Colorists. *Res:* Synthesis of novel disperse, and acid dyes in the azo and anthraquinone classes for polyesters, acetates, nylon, plastic, paper and polypropylene fibers; special interest in making intermediates and developing existing products; development of imaging fluids for inkjet printers. *Mailing Add:* 5647 Hugh Dr Dayton OH 45459

BOTSARIS, GREGORY D(IONYSIOS), b Patras, Greece, Jan 6, 30; m 65. CHEMICAL ENGINEERING. *Educ:* Nat Univ Athens, Dipl, 52; Mass Inst Technol, MS, 59, ChE, 60, PhD(chem eng), 65. *Prof Exp:* From asst to prof anal chem, Univ Athens, 52-54; chem engr, Dennison Mfg Co, 60-61; from asst prof to assoc prof chem eng, 65-76, PROF CHEM ENG, TUFTS UNIV, 76- *Mem:* Am Inst Chem Engrs; Am Chem Soc; Am Asn Crystal Growth; Am Asn Univ Professors. *Res:* Crystal growth; nucleation; industrial crystallizations; stabilization of solid dispersions in liquids; coal slurries. *Mailing Add:* Dept Chem Eng Tufts Univ Medford MA 02155

BOTSFORD, JAMES L, b Spokane, Wash, June 6, 42; m 64; c 2. MICROBIAL PHYSIOLOGY, BIOCHEMISTRY. *Educ:* Univ Idaho, BS, 64; Ore State Univ, PhD(microbial physiol), 68. *Prof Exp:* NIH fel, Univ Ill, 68-70; from asst prof to assoc prof, 70-82, PROF BIOL, NMEX STATE UNIV, 82- *Concurrent Pos:* Inst for Elderhostel Prog Gen Engr; vis prof, Plant Growth Lab, Univ Calif, Davis, 84. *Mem:* AAAS; Am Soc Microbiol; Sigma Xi. *Res:* Cyclic nucleotides in procaryotes; molecular biology of osmoregulation. *Mailing Add:* Dept Biol NMex State Univ Box 3AF Las Cruces NM 88003

BOTSTEIN, DAVID, b Zurich, Switz, Sept 8, 42; US citizen; m 65. MOLECULAR GENETICS, MICROBIAL PHYSIOLOGY. *Educ:* Harvard Univ, AB, 63; Univ Mich, PhD(human genetics), 67. *Prof Exp:* Instr biol, Mass Inst Technol, 67-69, from asst prof to prof genetics, Dept Biol, 69-87, Earle A Griswold prof, 87-88; vpres sci, Genentech, Inc, 88-90; PROF & CHMN, DEPT GENETICS, SCH MED, STANFORD UNIV, 90- *Concurrent Pos:* Mem, Study Sect, NSF, 72-76 & Am Chem Soc, 77-81; NIH career develop award, 72-77; mem sci adv bd, Collab Res, Inc, 78-87; bd dirs, Genetics Soc Am, 84, Cold Spring Harbor, 85-, Life Sci Res Found, 88- & Genomyx, Inc, 88-; mem, Prog Adv Comt on Human Genome, NIH, 88-, Vis Comt Biol Sci, Rice Univ, 88-, Human Genome Ctr Adv Comt, Lawrence Berkeley Lab, 88- & Sci Resource Bd, Genentech, Inc, 89- *Honors & Awards:* Award in Microbiol & Immunol, Eli Lilly & Co, 78; Genetics Soc Am Medal, 88. *Mem:* Nat Acad Sci; fel AAAS; Am Soc Microbiol; Genetics Soc Am; Am Acad Arts & Sci. *Res:* Control of gene expression in temperate phages, bacteria and deoxyribonucleic acid replication. *Mailing Add:* Dept Genetics Stanford Univ Sch Med Stanford CA 94305-5120

BOTT, JERRY FREDERICK, b Tyler, Tex, Feb 27, 36; m 64; c 2. GAS DYNAMICS, SPECTROSCOPY. *Educ:* Rice Univ, AB, 58, SB, 59; Harvard Univ, AM, 60, PhD(eng), 65. *Prof Exp:* Res fel, Harvard Univ, 65-66; res scientist, 66-80, SR SCIENTIST, AEROSPACE CORP, 80- *Mem:* Am Phys Soc; Am Inst Aeronaut & Astronaut. *Res:* High temperature gas dynamics and kinetics; ultraviolet, visible, infrared radiations; shock tube gas kinetics; plasma and spectroscopy; chemical lasers; excimer lasers. *Mailing Add:* Aerospace Corp M5-747 PO Box 92957 Los Angeles CA 90009-2957

BOTT, KENNETH F, b Albany, NY, Dec 19, 36; c 2. MICROBIOLOGY, MOLECULAR GENETICS. *Educ:* St Lawrence Univ, BS, 58; Syracuse Univ, MS, 60, PhD(microbiol), 63. *Prof Exp:* From instr to asst prof microbiol, Univ Chicago, 62-71; assoc prof, 72-80, PROF MICROBIOL, UNIV NC, CHAPEL HILL, 80-, DIR INTERDEPT CURRIC GENETICS, 84- *Concurrent Pos:* Trainee microbial genetics & virol, Univ Chicago, 63-64; Merck & Co Found grant fac develop, 69; Fulbright grant & vis prof, Univ Paris, Orsay, 71; vis prof, Inst Pasteur, Paris, 77; Fulbright grant & vis researcher, Nat Ctr Sci Res, Gif sur Yvette, France. *Mem:* AAAS; Am Soc Microbiol; Sigma Xi. *Res:* Isolation, characterization, cloning and sequencing individual genes from Bacillus including those for ribosome structure and for origins of chromosome replication; translational regulation of genetic expression in prokaryotes; endospore formation in Bacillus species; genomic organization in mycoplasma species; DNA gyrase genes. *Mailing Add:* Dept Microbiol & Immunol Div Health Affairs Univ NC Chapel Hill NC 27599-7290

BOTT, RAOUL H, b Budapest, Hungary, Sept 24, 23; m 47; c 4. MATHEMATICS. *Educ:* McGill Univ, MEng, 46; Carnegie Inst Technol, DrSc(math), 49. *Hon Degrees:* DSc, Notre Dame Univ, 80 & McGill Univ, 87. *Prof Exp:* Mem, Inst Adv Study, 49-51, 55-57, 70-71; from instr to prof math, Univ Mich, 51-59; prof, 59-74, Higgins prof, 74-77, WILLIAM CASPAR GRAUSTEIN PROF MATH, HARVARD UNIV, 77- *Concurrent Pos:* Hon fel, St Catherine's Col, 85; master, Dunsterhouse, Harvard Univ, 78-84. *Honors & Awards:* Veblen prize, Am Math Soc, 64; Nat Medal Sci, 87; Steele Career Prize, Am Math Soc, 90. *Mem:* Nat Acad Sci; Am Math Soc; hon mem London Math Soc. *Res:* Topology; geometry; global analysis. *Mailing Add:* Dept Math Harvard Univ Cambridge MA 02138

BOTT, THOMAS LEE, b Jersey City, NJ, June 15, 40; m 64; c 3. MICROBIOLOGY, HYDROBIOLOGY. *Educ:* Wheaton Col Ill, BS, 62; Univ Wis-Madison, MS, 65, PhD(bact & zool), 68. *Prof Exp:* NIH fel, Ind Univ, 68-69; asst cur, 69-77, assoc cur, 77-86, CUR AQUATIC MICROBIOL, STROUD WATER RES CTR, ACAD NATURAL SCI, PHILADELPHIA, 86- *Concurrent Pos:* Adj asst prof, Biol Dept, Univ Pa, 72-81, adj assoc prof, 81-; consult, NSF, 77-, Environ Protection Res Inst, 81-84, Ill Inst Technol Res Inst, 82- *Mem:* AAAS; Am Soc Microbiol; Am Soc Limnol & Oceanog; NAm Benthological Soc; Sigma Xi. *Res:* Microbial ecology; aquatic microbiology; Clostridium botulinum type E in the Great Lakes; extreme thermophilic bacteria; bacterial and fungal decomposition activity; primary productivity; oil pollution effects on microbes; nitrilotriacetic acid; biological control of algal growths; bacterial utilization of dissolved organics; bacterial productivity, metabolic condition, and biomass in streams and rivers; gems in the environment; cellulose degradation; bioremediation of toxics. *Mailing Add:* Acad Natural Sci Philadelphia Stroud Water Res Ctr RD 1 Box 512 Avondale PA 19311

BOTTEI, RUDOLPH SANTO, b Old Forge, Pa, July 19, 29; m 60; c 4. ANALYTICAL CHEMISTRY, INORGANIC CHEMISTRY. *Educ:* Wilkes Col, AB, 50; Cornell Univ, MS, 52; Princeton Univ, PhD, 56. *Prof Exp:* Instr to assoc prof, 55-71, ASST HEAD DEPT, UNIV NOTRE DAME, 66-, PROF CHEM, 71- *Mem:* Sigma Xi; Am Chem Soc. *Res:* Electroanalytical chemistry; complex ions; organometallic compounds; boron compounds; organic reduction. *Mailing Add:* Dept Chem Univ Notre Dame Notre Dame IN 46556

BOTTGER, GARY LEE, b Los Angeles, Calif, Oct 27, 38; m 65; c 2. CHEMICAL PHYSICS, PHOTOGRAPHIC SCIENCE. *Educ:* Univ Southern Calif, BS, 60; Univ Wash, PhD, 64. *Prof Exp:* Res asst infrared spectros, Univ Wash, 62-63, sr res chemist, 64-71, LAB HEAD, EASTMAN KODAK CO, 71- *Concurrent Pos:* Tech asst to Eastman Kodak dir res, 86. *Mem:* Soc Imaging Sci & Technol; AAAS; Am Chem Soc; Am Phys Soc; Soc Photog Sci & Technol Japan. *Res:* Infrared and Raman spectroscopy; molecular structure; vibrational spectra of solids; theory of the photographic process; imaging science; silver halide research. *Mailing Add:* Eastman Kodak Co Res Labs Kodak Park Bldg 8213 Rochester NY 14650-2131

BOTTICELLI, CHARLES ROBERT, b Springdale, Conn, Feb 14, 28; m 54; c 6. PHYSIOLOGY. *Educ:* Univ Conn, BA, 53; Williams Col, MA, 55; Harvard Univ, PhD(biol), 58. *Prof Exp:* Res fel, Harvard Univ, 58-60, instr, 60-62, lectr, 62-63; asst prof physiol, Boston Univ, 64-69, assoc prof biol, 69-70; prof biol & dir sci & math, Newton Col, 70-76; RES MGR LIFE SCI, EXPLOR RES LAB, GTE LABS, 76- *Mem:* AAAS; Am Soc Zool; Endocrine Soc; Am Soc Agr Consults; Am Soc Hort Sci. *Res:* Physiology of reproduction; photobiology; steroid chemistry; mechanism and action of hormones; field applications of photobiology in agriculture. *Mailing Add:* 63 Westerly Rd Weston MA 02193

BOTTINI, ALBERT THOMAS, b San Rafael, Calif, June 13, 32; m 53; c 3. ORGANIC CHEMISTRY. *Educ:* Univ Calif, BS, 54; Calif Inst Technol, PhD(chem), 57. *Prof Exp:* From instr to assoc prof, 58-68, PROF CHEM, UNIV CALIF, DAVIS, 68- *Mem:* Am Chem Soc; Sigma Xi. *Res:* Phytotoxins; reaction mechanisms. *Mailing Add:* 271 Baja Ave Davis CA 95616

BOTTINO, NESTOR RODOLFO, b La Plata, Arg, Sept 3, 25; m 53; c 2. BIOCHEMISTRY. *Educ:* La Plata Nat Univ, DrChem, 54. *Prof Exp:* Asst prof biochem, La Plata Nat Univ, 56-59; Welch Found fel biochem & biophys, Tex A&M Univ, 59-61; assoc prof biochem, La Plata Nat Univ, 61-63; res assoc, 64-65, from asst prof to assoc prof, 65-74, prof med biochem, 77-78, PROF BIOCHEM & BIOPHYS, TEX A&M UNIV, 75- *Mem:* AAAS; Am Oil Chem Soc; Am Chem Soc; Soc Exp Biol & Med; Am Soc Biol Chemists. *Res:* Lipid metabolism and chemistry. *Mailing Add:* 1405 Post Oak Dr College Station TX 77840

BOTTINO, PAUL JAMES, b Price, Utah, Aug 3, 41; m 63; c 2. PLANT GENETICS. *Educ:* Utah State Univ, BS, 64, MS, 65; Wash State Univ, PhD(genetics), 69. *Prof Exp:* Res assoc, Brookhaven Nat Lab, 69-73; asst prof, 73-76, ASSOC PROF BOT, UNIV MD, COLLEGE PARK, 76- *Mem:* AAAS; Am Genetics Asn. *Res:* In vitro culture of plant cells; genetics of somatic plant cells; induced mutation in plant cells; application of tissue culture techniques to agricultural problems. *Mailing Add:* Dept Bot Univ Md College Park MD 20742

BOTTJER, DAVID JOHN, b New York, NY, Oct 3, 51; m 73. PALEOECOLOGY, BIOTURBATION & TRACE FOSSILS. *Educ:* Haverford Col, BS, 73; State Univ NY-Binghamton, MA, 76; Ind Univ, PhD(geol), 78. *Prof Exp:* Nat Res Coun fel, US Geol Surv, 78-79; RES ASSOC, LOS ANGELES COUNTY MUS NATURAL HIST, 79-; ASSOC PROF PALEONT, UNIV SOUTHERN CALIF, 85- *Concurrent Pos:* Asst prof paleont, Univ Southern Calif, 79-85; vis scientist, Field Mus Natural Hist, Chicago, 86, ed, Palaios, 89- *Mem:* Soc Econ Paleontologists & Mineralogists; Paleont Soc; fel Geol Soc Am, 86; AAAS; Int Paleont Asn; Paleont Res Inst; Geol Soc London. *Res:* Paleoecological and paleobiological trends in the Phanerozoic history of marine metazoan life; bioturbation in marine sedimentary environments; stratigraphy and paleoenvironments of sedimentary rocks, particulary those of Cretaceous age. *Mailing Add:* Dept Geol Sci Univ Southern Calif Los Angeles CA 90089-0740

BOTTJER, WILLIAM GEORGE, b New York, NY, Feb 20, 31; m 58; c 2. INORGANIC CHEMISTRY. *Educ:* Hofstra Univ, BA, 53; Univ Conn, MS, 55; Univ NH, PhD(inorg chem), 63. *Prof Exp:* Res chemist, Gen Chem Res Lab, Allied Chem Corp, NJ, 55-59; res chemist org chem dept, Jackson Lab, 63-65, develop chemist, 65-68, sr res chemist, Photo Prod Dept, 68-72, sr res chemist, Parlin, NJ, 72-75, res assoc, Photo Prod Plant, 75-82, PROCESS ASSOC MFG, E I DU PONT DE NEMOURS & CO, BREVARD, NC, 82- *Mem:* Am Chem Soc; Soc Photog Scientists & Engrs; AAAS. *Res:* Solid state research; magnetic materials; transition metal and fluorine chemistry; metal oxide chemistry; photographic chemistry. *Mailing Add:* E I du Pont de Nemours & Co PO Box 267 Brevard NC 28712

BOTTKA, NICHOLAS, b Huszt, Hungary, Dec 6, 39; US citizen; m 66; c 2. SOLID STATE PHYSICS, OPTICAL PHYSICS. *Educ:* Univ Calif, Los Angeles, BS, 63, MS, 66; Tech Univ, Berlin, PhD(physics), 70. *Prof Exp:* RES PHYSICIST, MICHELSON LAB, NAVAL WEAPONS CTR, US DEPT NAVY, 63- *Mem:* Am Phys Soc. *Res:* Optical properties of solids. *Mailing Add:* 5642 Mt Burnside Way Burke VA 22015

BOTTO, ROBERT IRVING, b Buffalo, NY, Apr 22, 49; m 75; c 2. ANALYTICAL CHEMISTRY, GEOCHEMISTRY. *Educ:* State Univ NY, Buffalo, BA, 71; Cornell Univ, MS, 73, PhD(geochem), 76. *Prof Exp:* SR STAFF CHEMIST, EXXON RES & ENG CO, 75- *Mem:* Am Chem Soc; AAAS; Sigma Xi; Soc Appl Spectros. *Res:* Elemental analysis; instrumental and wet chemical methods; inductively coupled plasma emission spectroscopy; scanning electron microscopy. *Mailing Add:* Baytown Specialty Prod Exxon Res & Eng Co Baytown TX 77522

BOTTOM, VIRGIL ELDON, b Douglas, Kans, Jan 6, 11; m 32; c 3. MATHEMATICS. *Educ:* Friends Univ, AB, 31; Univ Mich, MS, 38; Purdue Univ, PhD, 49. *Hon Degrees:* DSc, McMurry Col, 83. *Prof Exp:* Teacher high sch, Kans, 33-38; prof physics & head dept, Friends Univ, 38-42; from asst prof to assoc prof,Colo state Univ, 42-47; asst instr, Purdue Univ, 47-49; prof, Colo State Univ, 49-53, actg head dept, 52-53; res physicist, Motorola Res Lab, 53-58; prof, 58-73, EMER PROF PHYSICS, MCMURRY COL, 73- *Concurrent Pos:* Chmn working group frequency control devices, US Dept Defense, 58-64; Fulbright lectr, Univ Sao Paulo, 64-65. *Honors & Awards:* Sawyer Award; Cady Award. *Res:* Semiconductors; piezoelectricity and applications to communication; ferromagnetism. *Mailing Add:* 3441 High Meadows Dr Abilene TX 79605

BOTTOMLEY, FRANK, b Hatfield, Eng, May 14, 41. INORGANIC CHEMISTRY. *Educ:* Univ Hull, BS, 63, MS, 65; Univ Toronto, PhD(chem), 68. *Prof Exp:* Fel lectr chem, Univ Toronto, 67-69; from asst prof to assoc prof, 69-78, PROF CHEM, UNIV NB, 78- *Concurrent Pos:* Alexander von Humboldt fel, 75-76 & 77. *Mem:* The Chem Soc; Can Inst Chem; Ges Deutscher Chemiker. *Res:* Reactions of coordinated ligands; transition metal chemistry. *Mailing Add:* Dept Chem Univ NB Bag Serv No 45222 Fredericton NB E3B 6E2 Can

BOTTOMLEY, LAWRENCE ANDREW, b Detroit, Mich, Sept 6, 50; m 73; c 3. ELECTROCHEMISTRY, SPECTROELECTROCHEMISTRY. *Educ:* Calif State Univ, Fullerton, BS, 76; Univ Houston, PhD(anal chem), 80. *Prof Exp:* asst prof chem, Fla State Univ, 80-87; ASSOC PROF CHEM, GA TECH, 88- *Mem:* Am Chem Soc; Electrochem Soc; Sigma Xi; Soc Electroanal Chemists. *Res:* Electroanalytical chemistry of metalloenzyme model compounds; reactivity of technetium and rhenium coordination compounds; cyclic square wave voltametry. *Mailing Add:* Sch Chem Ga Tech Atlanta GA 30332-0400

BOTTOMLEY, PAUL ARTHUR, b Ivanhoe, Australia, Mar 14, 53; m 86; c 1. NUCLEAR MAGNETIC RESONANCE. *Educ:* Monash Univ, Australia, BSc, 75; Univ Nottingham, PhD(physics), 78. *Prof Exp:* Univ demonstrator, Univ Nottingham, 75-78; res assoc, Johns Hopkins Univ Sch Med, 78-80; consult, Nat Cancer Inst, 78, 85, 86; PHYSICIST, RES & DEVELOP CTR, GEN ELEC CO, 80- *Concurrent Pos:* Consult, Nat Heart, Lung & Blood Inst, 79-80, Nat Cancer Inst, 79, 85 & 86, Nat Inst Diabetes & Digestive & Kidney Dis, 90; ed, Magnetic Resonance Imaging, 82, ed, Magnetic Resonance in Med, 83; ed, Rev of Sci Instruments, 86-88, Radiol, 91; trustee, Soc Magnetic Resonance in Med, 86-89; trustee, Soc Magnetic Resonance Imaging, 82-85; Coolidge fel, 90. *Honors & Awards:* Gold Medal, Soc Magnetic Resonance Med, 89. *Mem:* Soc Magnetic Resonance Imaging; Soc Magnetic Resonance Med. *Res:* Applications of nuclear magnetic resonance to medicine and biology, specifically nuclear magnetic resonance imaging, its instrumentation and the application of nuclear magnetic resonance spectroscopy to the study of normal and disease states in vivo. *Mailing Add:* 64 Pico Rd Clifton Park NY 12065-6710

BOTTOMLEY, RICHARD H, b Arkansas City, Kans, Nov 24, 33; m 58; c 2. INTERNAL MEDICINE, ONCOLOGY. *Educ:* Univ Okla, BS, 54, MD, 58; Am Bd Internal Med, dipl. *Prof Exp:* Intern, Salt Lake County Gen Hosp, 58-59; resident internal med, Med Ctr, 59-61, clin asst, Sch Med, 61-64, instr internal med & sr investr oncol, 64-65, instr res med & asst prof res biol in biochem, 65-68, res med, 67-71, assoc prof med, 71-75, PROF MED, SCH MED, UNIV OKLA, 75-, ASSOC PROF RES MOLECULAR BIOL, 68-, HEAD ONCOL DIV, 72- *Concurrent Pos:* Nat Cancer Inst fels, McArdle Mem Lab, 61-62 & Okla Med Res Found, 61-63, asst head cancer sect, 65-72. *Mem:* Am Fedn Clin Res; AMA; Am Soc Clin Oncol; fel Am Col Physicians; Sigma Xi. *Res:* Biology, biochemistry and chemotherapy of human neoplastic disease. *Mailing Add:* PO Box 96798 Oklahoma City OK 73143

BOTTOMLEY, SYLVIA STAKLE, b Riga, Latvia, Mar 9, 34; US citizen; m 58; c 2. INTERNAL MEDICINE, HEMATOLOGY. *Educ:* Okla State Univ, BS, 54; Univ Okla, MD, 58; Am Bd Internal Med, dipl, 65, cert hemat, 72. *Prof Exp:* Intern med, Sch Med, Univ Utah, 59; resident, Med Ctr, 59-61, clin asst, Sch Med, 61-64, instr, 64-67, from asst prof to assoc prof med, 67-75, PROF MED, SCH MED, UNIV OKLA, 75-; ASST CHIEF HEMAT SECT, VET ADMIN HOSP, 69- *Concurrent Pos:* USPHS res fel hemat, 61-64; res assoc, Vet Admin Hosp, Oklahoma City, 64-65, clin investr, 65-68. *Mem:* Sigma Xi; Am Fedn Clin Res; Am Soc Hemat; fel Am Col Physicians; Ctr Soc Clin Res. *Res:* Clinical hematology and erythropoiesis, especially heme synthesis and iron metabolism. *Mailing Add:* Hemat Sect Vet Admin Hosp 921 NE 13th St Oklahoma City OK 73104

BOTTOMS, ALBERT MAITLAND, b Montclair, NJ, Sept 2, 25; m 57; c 2. OPERATIONS RESEARCH, INDUSTRIAL ECONOMICS. *Educ:* Univ Pa, BSChE, 49; Iowa State Univ, MS, 51; Mass Inst Technol, MS, 62. *Prof Exp:* Asst physics, Iowa State Univ, 49-51; res chemist, Am Cyanamid Co, 51-52; opers analyst, Opers Eval Group, Mass Inst Technol, 52-58; opers analyst, Weapons Systs Eval Group, Inst Defense Anal, US Dept Defense, 58-63, opers analyst, Res Anal Corp, 63-64, opers analyst, Weapons Systs Eval Group, Inst Defense Anal, 64-67; head surface weapons, Off Chief Naval Opers, 67-68; dir opers res task force, Chicago Police Dept, 68-69; dir pub safety div, Urban Systs Res & Eng, Mass, 69-70; pres, Fundamental Systs, Inc & res assoc, Mass Inst Technol-Harvard Joint Ctr Urban Studies, 70-72; head, Warfare Anal Dept, USN Underwater Systs Ctr, 73-77; asst for tech strategies, dir Navy Technol, 77-78; asst for Dept Defense Labs, Off Under

Secy Defense Res & Eng, 78-81; dir opers & mgt, Off Asst Comdr Res & Technol, Naval Air Systs Command, 81-86; Navy chair, Defense Systs Mgt Col, Ft Belvoir, Va, 86-88; dir, Ctr Study Indust Competitiveness & Nat Security, George Washington Univ, 88-90; PRES, BOTTOMS & ASSOCS, INC, MELBOURNE BEACH, FLA, 90- *Concurrent Pos:* Cadet, US Mil Acad, 43-45; consult, Nat Comn Causes & Prev Violence, 68-69, Nat Acad Sci, 69- & Fed Exec Inst, 84; sci adv to comdr, US Seventh Fleet, 75-76; adv sci & technol, Ctr Naval Warfare Studies, US Naval War Col, 84-85. *Mem:* Fel AAAS; Opers Res Soc Am. *Res:* Public services systems analysis; military operations research; criminal justice system; police; undersea warfare; strategy and technology; industrial economics. *Mailing Add:* 401 Andrews Dr Melbourne Beach FL 32951

BOTTOMS, GERALD DOYLE, b Holdenville, Okla, Apr 10, 30; m 52; c 3. PHYSIOLOGY, ENDOCRINOLOGY. *Educ:* ECent State Col, BS, 55; Okla State Univ, MS, 58, PhD(physiol), 66. *Prof Exp:* Sci teacher, Holdenville, Okla, 55-63; asst prof, 66-70, assoc prof, 70-74, PROF PHYSIOL, PURDUE UNIV, 74- *Mem:* AAAS; Am Physiol Soc; Sigma Xi. *Res:* Endotoxin shock in domestic animals; correlation of prostaglandin levels with hemodynamic changes and the effects of treatment with antiprostaglandins. *Mailing Add:* 175 Harvest Dr E Lafayette IN 47905

BOTTONE, EDWARD JOSEPH, b New York, NY, Feb 18, 34; m 62; c 2. MICROBIOLOGY. *Educ:* City Col New York, BS, 65; Wagner Col, MS, 68; St John's Univ, PhD(biol), 73. *Prof Exp:* Bacteriologist, Greenpoint Hosp, 62-64; sr bacteriologist, Elmhurst City Hosp, 64-68; from asst microbiologist to assoc microbiologist, 69-73, asst prof microbiol, Sch Med, 73-77, actg dir microbiol, 74-75, DIR MICROBIOL, MT SINAI HOSP, 75-; PROF MICROBIOL, MT SINAI SCH MED, 81-, PROF CLIN PATH, 85- *Concurrent Pos:* Consult, Elmhurst City Hosp of Mt Sinai Serv, 74-; assoc prof microbiol, Mt Sinai Sch Med, 77-81; pres, NYC Br, Am Soc Microbiol, 88-90. *Mem:* Am Soc Microbiol. *Res:* Clinical microbiology. *Mailing Add:* Clin Microbiol Lab Mt Sinai Sch Med New York NY 10029-6574

BOTTORFF, EDMOND MILTON, b Milroy, Ind, Sept 14, 16; m 48; c 2. ORGANIC CHEMISTRY. *Educ:* Hanover Col, AB, 37; Univ Ill, PhD(org chem), 41. *Prof Exp:* Res chemist, Rohm and Haas Co, Philadelphia, 41-47; res chemist, Eli Lilly & Co, 47-80; RETIRED. *Mem:* Am Chem Soc. *Res:* Action of the Grignard reagent on sterically hindered esters. *Mailing Add:* 1043 Carters Grove Indianapolis IN 46260

BOTTS, TRUMAN ARTHUR, b De Land, Fla, Nov 26, 17; m 44, 64, 74; c 2. MATHEMATICS. *Educ:* Stetson Univ, BS, 38; Univ Va, MA, 40, PhD(math), 42. *Prof Exp:* Instr math, Univ Va, 41-42; asst prof, Univ Del, 46-48; from asst prof to assoc prof, Univ Va, 48-68; exec dir, Conf Bd Math Sci, 68-82; RETIRED. *Concurrent Pos:* Ford Found fel, 53-54; fel comput sci, Nat Bur Standards, 59; vis prof, Univ PR, 62-63; exec dir, comt support res math sci, Nat Acad Sci-Nat Res Coun, Columbia Univ, 66-67. *Mem:* Fel AAAS; Am Math Soc; Math Asn Am. *Res:* Real function theory; calculus of variations; convex sets. *Mailing Add:* 3106 N John Marshall Dr Arlington VA 22207-1340

BOTZLER, RICHARD GEORGE, b Detroit, Mich, Jan 27, 42; m 63; c 5. WILDLIFE MANAGEMENT, MICROBIOLOGY. *Educ:* Wayne State Univ, BS, 63; Univ Mich, MWM, 67, PhD(wildlife), 70. *Prof Exp:* Asst res tissue immunity, Univ Mich Hosp, 64-68; PROF WILDLIFE, HUMBOLDT STATE UNIV, 70- *Concurrent Pos:* Fulbright fel, Univ Hohenheim, Stuttgart, Ger, 81-82; asst ed, J Wildlife Dis, 90-91, ed, 92- *Mem:* Wildlife Dis Asn. *Res:* Enzoology of Listeria species and Yersinia species; avian cholera in wildfowl; physiology, reproduction and parasites of black-tailed deer; tissue immunity. *Mailing Add:* Dept Wildlife Humboldt State Univ Arcata CA 95521

BOUBEL, RICHARD W, b Portland, Ore, Aug 1, 27; m 52; c 4. MECHANICAL ENGINEERING. *Educ:* Ore State Univ, BS, 53, MS, 54; Univ NC, PhD(air & indust hyg), 63. *Prof Exp:* From instr to assoc prof, 54-60, PROF MECH ENG, ORE STATE UNIV, 67- *Concurrent Pos:* Consult, US Bur Mines, Ore, 63-64; environ engr, USAF, 78-79; at Defense Environ Leadership Proj, 84-86. *Honors & Awards:* Ripperton Award, Air Pollution Control Asn, 88. *Mem:* Am Soc Mech Engrs; Air Pollution Control Asn (pres, 78-79); Sigma Xi; Am Acad Environ Engrs. *Res:* Combustion of wood residue; air pollutants generated during combustion; instrumentation for field studies in air pollution control. *Mailing Add:* PO Box 3630 Sunriver OR 97707

BOUBLIK, MILOSLAV, b Apr 13, 27; m 59; c 2. BIOLOGICAL MACROMOLECULES, ELECTRON MICROSCOPY. *Educ:* Czech Acad Sci, MS, 52, PhD(biochem), 65. *Prof Exp:* Mem, Portsmouth Polytech, Eng, 68-71; MEM STAFF, ROCHE INST MOLECULAR BIOL, 71- *Mem:* Am Soc Biol Chemists; Electron Micros Soc Am; AAAS; NY Acad Sci. *Res:* Conformation and interactions of proteins and nucleic acids, structure-function relationship of ribosomes by electron spectroscopic techniques. *Mailing Add:* Biochem Dept Roche Inst Molecular Biol Nutley NJ 07110

BOUBOULIS, CONSTANTINE JOSEPH, b Piraeus, Greece, Feb 23, 28; US citizen; m 60; c 3. ORGANIC CHEMISTRY. *Educ:* Nat Univ Athens, dipl, 53; Columbia Univ, MA, 57; Univ Colo, PhD(org chem), 61. *Prof Exp:* Chemist, John Lafis & Co, Greece, 54-55; from res chemist to sr res chemist, Exxon Res & Eng Co, Inc, 61-73, staff chemist, 73-80, sr staff chemist, Exxon Chem Co. 80-86; OWNER-PRES, CHEM-COMP SYSTS INC, 87- *Mem:* Am Chem Soc; Am Inst Chemists; Sigma Xi. *Res:* Steroids; synthesis and rearrangements of bicyclic systems; condensation polymers; oxidation; amines; metal mining by solvent extraction; technical service in the use of solvents; surface protective coatings; addition polymerization; high solids coatings; computer software for coatings. *Mailing Add:* 661 Golf Terr Union NJ 07083

BOUCHARD, MICHEL ANDRÉ, b Montreal, Que, July 26, 48; m 73; c 1. LATE CENOZOIC STRATIGRAPHY, METEORITE CRATERS. *Educ:* Univ Montreal, BA, 67, BSc, 70, MSc, 74; McGill Univ, PhD(geol), 80. *Prof Exp:* Geologist, Ministry Energy Resources, Que, 74-75; lectr geol, 75-80, ASSOC PROF GEOL, UNIV MONTREAL, 80-, ASST PROF GEOL, 86- *Mem:* Can Quaternary Asn; Int Glaciol Soc; Geol Soc Am; Geol Soc Can. *Res:* Pleistocene history and glacial geomorphology applied to the reconstruction of the North American ice sheets; applications of glacial geology to mineral exploration and environmental studies. *Mailing Add:* Dept Geol Univ Montreal PO Box 6128 Montreal PQ H3C 3J7 Can

BOUCHARD, RAYMOND WILLIAM, b Ft Bragg, Calif, May 28, 44; m 73; c 2. FRESH WATER BIOLOGY. *Educ:* Mass State Col, BSEd, 67; Univ Tenn, PhD(zool), 72. *Prof Exp:* Res assoc, Ctr Wetland Resources, La State Univ, 73-74; asst prof, Univ N Ala, 76-78; ACAD NAT SCI PHILADELPHIA, 86- *Concurrent Pos:* Instr ichthyol, Reelfoot Lake Biol Sta, Tenn, 70; postdoctoral fel, Nat Mus Natural Hist Smithsonian Inst, 71; instr zool, Univ Tenn, 73; fel, Nat Mus Natural Hist, Smithsonian Inst, 74-75. *Honors & Awards:* Explorers Club. *Mem:* Crustacean Soc Explorer Club; Am soc Ichthyologists & Herpetologists; Int Asn Astacol; NAm Benthological Soc; Soc Syst Zool; AAAS. *Res:* Systematics; ecology, zoogeography and evolution of freshwater decapod crustaceans and fishes. *Mailing Add:* Div Environ Res Acad Nat Sci Philadelphia PA 19103

BOUCHARD, RICHARD EMILE, b Colchester, Vt, Mar 31, 26; m 56; c 6. CARDIOLOGY, INTERNAL MEDICINE. *Educ:* Univ Vt, MD, 49, MS, 51. *Prof Exp:* Fel physiol, Nat Heart Inst, 50-51, fel cardiol, 54-55; from instr to assoc prof med, 55-77, PROF MED & FAMILY PRACT, COL MED, UNIV VT, 77- *Concurrent Pos:* Fel coun clin cardiol, Am Heart Asn, 68. *Mem:* AMA; fel Am Col Physicians; fel Am Col Cardiol. *Res:* Clinical research in cardiology. *Mailing Add:* Given Health Care Ctr One S Prospect St Burlington VT 05401

BOUCHER, JOHN H, b Kansas City, Mo, Dec 15, 30. MECHANICS OF BLOOD FLOW. *Educ:* Mo Univ, DVM, 56; Ohio State Univ, PhD(physiol), 77. *Prof Exp:* DIR, RHEOTECH LABS, 82- *Mem:* Am Physiol Soc; Int Soc Biorheology; Am Vet Med Asn. *Mailing Add:* Rheotech Labs 2106 Salisbury Rd Silver Spring MD 20910

BOUCHER, LAURENCE JAMES, b Yonkers, NY, Sept 16, 38; m 64; c 2. INORGANIC CHEMISTRY. *Educ:* Mich State Univ, BS, 61; Univ Ill, MS, 62, PhD(inorg chem), 64. *Prof Exp:* Resident res assoc, Chem Div, Argonne Nat Lab, 64-66; from asst prof to assoc prof chem, Carnegie Mellon Univ, 66-76, fel, Mellon Inst, 67-76; prof & head dept inorg chem, Metrop Univ, Iztapalapa, Mexico City, 76-78; prof chem & head dept, Western Ky Univ, 78-85; PROF CHEM & DEAN COL ARTS & SCI, ARK STATE UNIV, 85- *Honors & Awards:* Fulbright Lectr, Colombia, 81. *Mem:* Am Chem Soc; Sigma Xi. *Res:* Chemistry of coordination compounds; role of metal ions in biological systems; homogeneous and heterogeneous catalysis in the conversion of coal and related products. *Mailing Add:* Towson State Univ Col Nat & Math Sci Towson MD 21204

BOUCHER, LOUIS JACK, b Ashland, Wis, May 24, 22; m 49; c 4. DENTISTRY, ANATOMY. *Educ:* Marquette Univ, DDS, 53, PhD(anat), 61. *Prof Exp:* Resident prosthodontics, Vet Admin Hosp, Wood, Wis, 53-55; instr, Sch Dent, Marquette Univ, 55-60, assoc prof, 60-65, dir grad dept, 63-65, dir postgrad dept, 64-65, assoc anat, Sch Med, 60-65; coordr grad studies & res, Col Dent, Univ Ky, 65-66; from asst dean to assoc dean sch dent, Med Col Ga, 66-71; prof prosthodontics & dean sch dent, Fairleigh Dickinson Univ, 71-76; assoc dean, Col Med & Dent NJ, NJ Dent Sch, 76-77; assoc dean, 78-82, PROF & DIR PROSTHODONTICS, SCH DENT MED, STATE UNIV NY, STONY BROOK, 77- *Concurrent Pos:* Spec fel grant, USPHS, 61-62, res career develop award, 62 -65; consult, Vet Admin Hosp, Wood, Wis, 61-64, Nat Inst Dent Res, 64-66, Div Chronic Dis, USPHS, 65-67, Vet Admin Hosp, Augusta, Ga, 67-71, US Army, Ft Jackson, SC, 67-71, Ft Gordon, Ga, 68-71 & USAF Surgeon Gen 69-71; pres, Fedn Prosthodont Orgns, 67. *Honors & Awards:* Int Res Award, Int Asn Dent Res, 70. *Mem:* Am Acad Plastics Res in Dent (pres, 67); Am Equilibration Soc (pres, 76); Am Prosthodontic Soc; Int Asn Dent Res; Am Dent Asn. *Res:* Functional anatomy of temporomandibular joint; response of the stratum corneum to various dental materials; epithelial reaction to various implants; cineradiographic study of the temporomandibular joint. *Mailing Add:* Sch Dent Med State Univ NY Stony Brook NY 11794

BOUCHER, RAYMOND, b Stanbridge, Que, July 21, 06; m 35. HYDRAULIC ENGINEERING. *Educ:* Polytech Sch, Univ Montreal, BS, 33; Mass Inst Technol, MS, 34. *Prof Exp:* From asst prof to prof hydraul eng, 34-72, head div, 44-72, chmn, Dept Civil Eng, 56-66, EMER PROF HYDRAUL ENG, POLYTECH SCH, UNIV MONTREAL, 73- *Concurrent Pos:* Consult hydraul engr, 49-; lifetime individual mem, US Nat Comt, Int Comn Large Dams; mem, Can Nat Comt, Int Comn Large Dams. *Mem:* Emer mem Am Soc Civil Engrs; Am Soc Mech Engrs; Int Asn Hydraul Res; Eng Inst Can. *Res:* Hydraulic model testing and research; cavitation in hydraulic structures and machinery; flow in open channels; dams and hydraulic structures; ice conditions and problems in rivers and hydraulic structures. *Mailing Add:* Pac Power & Light Co 920 SW Sixth Ave Portland OR 97204

BOUCHER, THOMAS OWEN, b Providence, RI, June 25, 42; m 74. PRODUCTION SYSTEM DESIGN. *Educ:* Univ RI, BS, 64; Northwestern Univ, MBA, 70; Columbia Univ, MS, 73, PhD(indust eng), 78. *Prof Exp:* Sr proj engr, Continental Can Co, 67-69; internal consult, ABEX Corp, 70-72; sr financial analyst, Otis Elevator Co, 74-76; asst prof, Sch Opers Res & Indust Eng, Cornell Univ, 78-81; asst prof, 81-87, ASSOC PROF INDUST ENG, COL ENG, RUTGERS UNIV, 87- *Honors & Awards:* Eugene L Grant Award, Am Soc Eng Educ, 83, 86. *Mem:* Sigma Xi; NY Acad Sci; Inst Indust Engrs. *Res:* Production, organization and control; productivity and technical change; production economics; computer aided manufacturing. *Mailing Add:* Dept Indust Eng Rutgers Univ PO Box 909 Piscataway NJ 08854

BOUCHILLON, CHARLES W, b Louisville, Miss, July 17, 31; m 58; c 4. MECHANICAL ENGINEERING. *Educ:* Miss State Univ, BS, 53; Ga Inst Technol, MS, 59, PhD(mech eng), 63. *Prof Exp:* Instr & res assoc mech eng, Ga Inst Technol, 59-62; assoc prof, 62-77, coordr interdisciplinary progs, Off Res, 70-79, dir, Inst Space & Environ Sci & Eng, 69-79, PROF MECH ENG & ASST DEAN, GRAD SCH RES, MISS STATE UNIV, 77-, PROF MECH ENG, 80- *Mem:* Am Soc Mech Engrs; Am Soc Eng Educ. *Res:* Vortex phenomena; particle trajectories in hydrocyclones; turbulent flow; flow in porous media; pressure losses; heat conduction in composite slabs; thermal stresses; combustion. *Mailing Add:* Dept Mech Eng Miss State Univ Mississippi State MS 39762

BOUCK, G BENJAMIN, b New York, NY, Oct 25, 33; m 61; c 3. BOTANY. *Educ:* Hofstra Univ, BA, 56; Columbia Univ, MA, 58, PhD(bot), 61. *Prof Exp:* Training fel, Harvard Univ, 61-62; from instr to assoc prof biol, Yale Univ, 62-71; PROF BIOL, UNIV ILL, CHICAGO, 71- *Concurrent Pos:* Vis assoc prof, Univ Calif, Berkeley, 68-69. *Honors & Awards:* Darbarker Prize, 79. *Mem:* AAAS; Bot Soc Am; Am Soc Cell Biol; Int Phycol Soc. *Res:* Membrane composition and surface assembly in unicellular flagellates. *Mailing Add:* Dept Biol Sci (MC/066) Univ Ill Chicago Box 4348 Chicago IL 60680

BOUCK, GERALD R, b Owosso, Mich, Aug 22, 34; m 56; c 3. AQUATIC ECOLOGY, FISHERIES PHYSIOLOGY. *Educ:* Cent Mich Univ, BS, 60; Mich State Univ, MS, 63, PhD(fisheries physiol), 66. *Prof Exp:* Res asst fisheries physiol, Mich State Univ, 60-62, res instr, 65-66; chief biol effects br, Pac Northwest Water Lab, US Environ Protection Agency, 66-70, dir, Western Fish Toxicol Sta, 70-73, vis scientist, Ore State Univ, 74-75, res physiologist, Western Fish Dis Lab, 76-83; sr fisheries biologist, 83-86, CHIEF, RES SECT, BONNEVILLE POWER ADMIN, 86- *Honors & Awards:* Silver Medal, US Environ Protection Agency, 70 & Gold Medal, 76. *Mem:* Ecol Soc Am; Am Fisheries Soc; Pac Estuarine Res Soc; World Maricult Soc; Fish Health Soc. *Res:* Stress physiology; toxicology/bioassays; hypergaric physiology; gas bubble disease; plasma enzymes; gas supersaturation; estuarine physiology; Pacific salmon water quality requirements. *Mailing Add:* Bonneville Power Admin PO Box 3621 Portland OR 97208

BOUCK, NOEL, b San Francisco, Calif, Oct 23, 36; m 61; c 3. CARCINOGENESIS, GENETICS. *Educ:* Pomona Col, BA, 58; Columbia Univ, MA, 60; Yale Univ, PhD(microbiol), 69. *Prof Exp:* Asst prof, 79-84, ASSOC PROF MICROBIOL, NORTHWESTERN UNIV, 84- *Mem:* Tissue Cult Soc; Am Soc Microbiol; AAAS; Am Soc Cell Biologists; Genetics Soc Am; Am Asn Cancer Res. *Res:* Suppressor gene function & angrogenesis; genetic control of malignant cell characteristics. *Mailing Add:* Dept Microbiol & Immunol Northwestern Univ 303 E Chicago Ave Chicago IL 60611

BOUCOT, ARTHUR JAMES, b Philadelphia, Pa, May 26, 24; m 48; c 4. PALEONTOLOGY, EVOLUTION. *Educ:* Harvard Univ, AB, 48, AM, 49, PhD(geol), 53. *Prof Exp:* Geologist, US Geol Surv, 49-56; Guggenheim fel, Europe, 56-57; from asst prof to assoc prof geol & geophys, Mass Inst Technol, 57-61; assoc prof, Calif Inst Technol, 61-68; prof geol, Univ Pa, 68-69; PROF GEOL, ORE STATE UNIV, 69-, PROF ZOOL, 79- *Concurrent Pos:* Res assoc, Smithsonian Inst, 68-; US rep, Silurian Subcomt, Comt Stratig, IGC, 73-; chmn, Proj Ecostratig, IGCP-IUGS, 74- *Honors & Awards:* Raymond C Moore Medal, Soc Econ Paleontologists & Mineralogists, 85. *Mem:* Geol Soc Am; Mineral Soc Am; Paleont Soc (pres, 80-81); Brit Palaeont Asn; hon corresp mem Swed Geol Soc; Int Paleont Asn (pres, 84-89). *Res:* Silurian and Devonian stratigraphy and paleontology, especially brachiopods, zoogeography and paleoecology; rates of animal evolution and extinction and their controls. *Mailing Add:* Depts Zool & Geol Ore State Univ Corvallis OR 97331-2914

BOUDAKIAN, MAX MINAS, b New York, NY, Sept 8, 25; m 60; c 3. FLUOROAROMATICS CHEMISTRY, PYRIDINE CHEMISTRY. *Educ:* City Col New York, BS, 49; Univ Mich, MS, 50; Purdue Univ, PhD(chem), 55. *Prof Exp:* Res chemist, Olin Mathieson Chem Corp, 50-52, group leader, 55-69, res assoc, 69-77, SR RES ASSOC, OLIN CORP, 77-, LECTR, TECH SEM PROG, 69-, CONSULT SCIENTIST, 89- *Mem:* Am Chem Soc. *Res:* Aromatic fluorine chemistry; halopyridines. *Mailing Add:* 30 Candlewood Dr Pittsford NY 14534

BOUDART, MICHEL, b Brussels, Belg, June 18, 24; m 48; c 4. PHYSICAL CHEMISTRY. *Educ:* Cath Univ Louvain, Cand Ing Civil, 44, Ing Civil Chim, 47; Princeton Univ, PhD(chem), 50. *Hon Degrees:* Dr, Univs Liege & Ghent, Belgium, Notre Dame Univ & Institut Polytech, Lorraine. *Prof Exp:* Res asst, Princeton Univ, 50-53, res assoc, 53, asst to dir proj SQUID, 53-54, from asst prof to assoc prof chem eng, 54-61; prof, Univ Calif, Berkeley, 61-64; prof chem eng & chem, 64-80, WILLIAM M KECK SR PROF CHEM ENG, STANFORD UNIV, 80- *Concurrent Pos:* Co-founder & dir, Catalytica Assocs, Inc, 74-; vis prof, Univ Louvain, 69, Rio de Janeiro, 73, Tokyo, 75 & Paris, 80; vis fel, Clare Hall, Cambridge, 84; mem sci adv bd, Brookhaven Nat Lab & Ctr Study Nuclear Energy, Belg. *Honors & Awards:* Curtis McGraw Res Award, Am Soc Eng Educ, 62; R H Wilhelm Award, Am Inst Chem Engrs, 74; Kendall Award, Am Chem Soc, 77, Murphree Award, 85; Chem Pioneer Award, Am Inst Chemists, 91; Julian Smith Lectr, Cornell Univ, 91. *Mem:* Nat Acad Sci; Nat Acad Eng; Am Inst Chem Eng; Am Chem Soc; fel AAAS; Belgium Royal Acad; Sigma Xi. *Res:* Homogeneous and heterogeneous chemical kinetics; catalysis. *Mailing Add:* Dept Chem Eng Stanford Univ Stanford CA 94305-5025

BOUDETTE, EUGENE L, b Claremont, NH, Aug 24, 26. GEOLOGY, EDUCATION. *Educ:* Univ NH, BA, 51; Dartmouth Col, MA, 59, PhD(geol), 76. *Prof Exp:* Geologist, Army Corps, 51-53, US Geological Surv, 53-85; PROF EARTH SCI, UNIV NEW HAMPSHIRE, 85- *Concurrent Pos:* State geologist, NH, 85- *Mem:* Fel Am Geol Soc. *Res:* Geology. *Mailing Add:* RR 3 PO Box 1060 Pittsfield NH 03263

BOUDINOT, FRANK DOUGLAS, b New Brunswick, NJ, Mar 31, 56. PHARMACOKINETICS, PHARMACODYNAMICS. *Educ:* Springfield Col, BS, 78; State Univ NY, Buffalo, PhD(pharmaceut), 86. *Prof Exp:* Vet technician, Afton Animal Hosp, 78-79; res technician, Millard Fillmore Hosp-Clin Pharmacokinetics Lab, 79-80; grad assist pharmaceut, State Univ NY, Buffalo, 80-85; asst prof, 86-90, ASSOC PROF PHARMACEUT, UNIV GA, 90- *Concurrent Pos:* Consult, Elan Pharmaceut Res Corp, 88-90; mem, Prof Affairs Comt, Am Asn Col Pharm, 90-91. *Mem:* Am Asn Pharmaceut Scientists; Am Asn Cols Pharm; AAAS; Am Soc Microbiol; NY Acad Sci. *Res:* Characterizing the pharmacokinetics, biopharmaceutics, metabolism and pharmacodynamics of drugs; characterizing the pharmacokinetics of anti-HIV agents and assessing the effects of age on drug disposition, metabolism and pharmacodynamics. *Mailing Add:* Col Pharm Univ Ga Athens GA 30602

BOUDJOUK, PHILIP RAYMOND, b New York, NY, Feb 28, 42; m 64; c 4. SYNTHETIC & MECHANISTIC ORGANOMETALLIC CHEMISTRY. *Educ:* St John's Univ, BS, 64; Univ Wis-Madison, PhD(chem), 71. *Prof Exp:* Lectr chem, Univ Calif, Davis, 71-73. *Concurrent Pos:* High sch chem teacher, Farmingdale, NY, 64-67. *Honors & Awards:* Am Cancer Soc Res Award, 79. *Mem:* Am Chem Soc. *Res:* Synthesis of cyclic, bicyclic and tetracyclic organometallics; reactive intermediates containing silicon, germanium and tin; Lewis acid catalyzed reactions of organometallics; bioactive silanes; sonochemistry applied to synthesis; pi-complexes; highly conducting organic and organometallic compounds polysilanes; catalysis of organometallic reactions. *Mailing Add:* Dept Chem NDak State Univ Fargo ND 58105

BOUDREAU, JAMES CHARLES, b Los Angeles, Calif, May 3, 36. NEUROPHYSIOLOGY. *Educ:* Univ Calif, Berkeley, BA, 57, PhD(psychol), 63. *Prof Exp:* Res psychologist, US Army Med Res Lab, 62-65; res physiologist, Vet Admin Hosp, Pittsburgh, 65-72; assoc prof neurol sci, Sensory Sci Ctr, Univ Tex, Houston, 72-90; RETIRED. *Mem:* Soc Neurosci; Am Physiol Soc; Acoust Soc Am; Europ Chemoreception Res Orgn. *Res:* Neurophysiological studies of sensory processes. *Mailing Add:* Univ Tex Grad Sch Biomed Sci 6420 Lamar Flemming Blvd Houston TX 77025

BOUDREAU, JAY EDMOND, b Santa Monica, Calif, Feb 3, 46; div; c 2. NUCLEAR ENGINEERING. *Educ:* Univ Calif, Los Angeles, BS, 67, MS, 69, PhD(eng), 72. *Prof Exp:* Asst nuclear engr, Argonne Nat Lab, 72-73; group leader & staff mem Liquid Metal Fast Breeder Reactor Safety, 73-79, prog mgr, 79-80, DEP ASSOC DIR NUCLEAR PROGS, LOS ALAMOS NAT LAB, 81- *Mem:* Am Nuclear Soc. *Res:* Fast reactor safety; core disruptive accident analysis. *Mailing Add:* Los Alamos Nat Lab 1000 Los Pueblos St Los Alamos NM 87544

BOUDREAU, ROBERT DONALD, b N Adams, Mass, March 9, 31. AVIATION METEOROLOGY, INDUSTRIAL METEOROLOGY. *Educ:* Tex A&M Univ, BS, 62, MS, 64, PhD(meteorol), 68. *Prof Exp:* Res meteorologist, Atmospheric Sci Lab, Ariz, 65-68, Meteorol Div, Desert Test Ctr, Utah, 68-70 & Meteorol Satellite Lab, Washington, DC, 70-71; sr atmospheric scientist, Earth Resources Lab, NASA, Mo, 71-75; asst sci staff, Naval Oceanog Lab, Mo, 75-76; head meteorol, 76-88, PROF METEOROL, METROP STATE COL, DENVER, 76- *Concurrent Pos:* Consult, Solar Energy Res Inst, 80 & Colo Int Corp, Boulder, 81; pres, Int Meteorol, Aviation & Electronics Inst, 81-87; airline transport pilot, 85-; NSF fel, 62; NDEA fel, 62. *Honors & Awards:* Skylab Award, NASA, 74. *Mem:* Am Meteorol Soc; Sigma Xi; Nat Weather Asn; Int Pilot's Asn; US Pilot's Asn; Aircraft Owners & Pilot's Asn. *Res:* Aviation meteorology; media dissemination of weather information; remote sensing from aircraft and satellites. *Mailing Add:* Earth & Atmospheric Sci Dept Metrop State Col PO Box 173362 Campus Box 22 Denver CO 80217-3362

BOUDREAU, WILLIAM F(RANCIS), b Columbus, Ohio, Sept 29, 14; m 34; c 2. MECHANICAL ENGINEERING. *Educ:* Case Inst Technol, BSc, 36; Mass Inst Technol, MSc, 40. *Prof Exp:* With Am Rolling Mill Co, Ohio, 37-43; res scientist, Substitute Alloy Mat Lab, Columbia Univ, 43-46; engr, Union Carbide Nuclear Co, 46-63; mgr compressor develop, York Div, Borg-Warner Corp, 63-73; supvr mech engr, Catalytic, Inc, 73-75; supv mech engr, Union Carbide Nuclear Div, Union Carbide Corp, 75-79; CONSULT, 80- *Mem:* Am Soc Mech Engrs; Nat Soc Prof Engrs. *Res:* Development of special pumps, compressors and mechanical equipment. *Mailing Add:* 102 Norman Lane Oak Ridge TN 37830

BOUDREAUX, EDWARD A, b New Orleans, La, Oct 30, 33; m 55; c 4. INORGANIC CHEMISTRY, CHEMICAL PHYSICS. *Educ:* Loyola Univ, La, BS, 56; Tulane Univ, MS, 59, PhD(chem), 62. *Prof Exp:* Res assoc chem, Tulane Univ, 56-62; asst prof, 62-65, ASSOC PROF CHEM, UNIV NEW ORLEANS, 65- *Concurrent Pos:* Consult, Kalvar Corp, 61-63; Petrol Res Fund grant, 64-66; Cancer Soc Greater New Orleans grant, 66-67 & 77-78; consult, USDA, New Orleans, 68-70; Fulbright teaching fel, Univ Zagreb, 70-71; Biomed Res Ctr grant, 78-79. *Mem:* Am Chem Soc; The Chem Soc; Sigma Xi. *Res:* Inorganic solid compounds and complexes; ligand field theory; molecular orbital calculations; diffuse reflectance electronic spectroscopy; magnetochemical research. *Mailing Add:* Dept Chem La State Univ New Orleans LA 70122

BOUDREAUX, HENRY BRUCE, b Scott, La, Nov 12, 14; m 41; c 1. ENTOMOLOGY. *Educ:* Southwest La Inst, BS, 36; La State Univ, MS, 39, PhD(entom), 46. *Prof Exp:* Asst biol, Southwest La Inst, 33-36, instr, 36-39, asst prof, 39-44; asst entom, La State Univ, Baton Rouge, 44-46; asst prof biol, Southwest La Inst, 46-47; from asst prof to prof, 47-81, EMER PROF ENTOM, LA STATE UNIV, BATON ROUGE, 81- *Mem:* Fel AAAS; Entom Soc Am; Soc Syst Zool; Acarological Soc Am. *Res:* Biology and taxonomy of aphids and spider mites; physiology of hemophilia. *Mailing Add:* 555 Ursuline Dr Baton Rouge LA 70808-4767

BOUFFARD, MARIE ALICE, virology, for more information see previous edition

BOUGAS, JAMES ANDREW, b Bismarck, NDak, Jan 25, 24; m 53; c 2. THORACIC SURGERY, CARDIOPULMONARY PHYSIOLOGY. *Educ:* Harvard Med Sch, MD, 48. *Prof Exp:* Intern, Columbia Univ Serv, Bellevue Hosp, 48-49, resident surg, Presby Med Ctr & Bellevue Hosp, 49-53; dir cardiopulmonary lab, New Eng Deaconess Hosp, 55-65; ASSOC PROF SURG, SCH MED, BOSTON UNIV, 65- *Concurrent Pos:* Fel thoracic surg, Overholt Thoracic Clin, 53-55, assoc thoracic surgeon, 56-65; consult to numerous hosps, 56-; lectr, Tufts Univ & assoc staff, New Eng Med Ctr, 56-; sr active staff, New Eng Deaconess Hosp, 56-; sr active staff New Eng Baptist Hosp, Boston, 61-; vis staff Boston City Hosp, 65-; chief thoracic surg, Boston Univ Hosp, 67-70; chmn biomat, Gordon Res Conf, 67-; mem cardiac adv group, Regional Med Prog, NH, Mass & RI, 69- *Mem:* AAAS; Am Asn Thoracic Surg; Soc Thoracic Surg; Am Col Surg; Am Fedn Clin Res. *Res:* Physical biology; biomaterials; cardiovascular surgery. *Mailing Add:* 125 Parker Hill Boston MA 02120

BOUGH, WAYNE ARNOLD, b Stockton, Mo, Mar 21, 43; div; c 4. POLLUTION CONTROL, WASTE MANAGEMENT. *Educ:* Univ Mo, Columbia, BS, 65; Univ Minn, PhD(biochem), 69. *Prof Exp:* Res dir, Pollution Control, Am Bact & Chem Res Corp, 69-72; asst prof, Pollution Control, Dept Food Sci, Ga Exp Sta, Marine Exten Serv, Univ Ga, 72-76, assoc marine scientist seafood technol, 76-80, assoc dir, 78-80; mgr tech serv, Wastewater Div, Spec Prod, Inc, Mo, 80-84; MGR, ENVIRON AFFAIRS, MID-AM DAIRYMEN, INC, MO, 84- *Concurrent Pos:* Consult, Environ Assoc Inc, 73-75, SCS Eng, 74 & Mid-Am Dairymen, Inc, 79-80; adv comt, Water & Wastewater Tech Sch, Neosho, Mo, 88-; biomass adv comt, Univ Mo, 88- *Mem:* Water Pollution Control Fedn; Inst Food Technologists; NY Acad Sci. *Res:* Recovery and utilization of by-products from food processing wastes and treatment systems; by-products evaluated for feed value; chitosan manufactured from shrimp wastes for coagulation of suspended solids in wastewaters; pretreatment systems for food processing wastes; environmental regulations. *Mailing Add:* PO Box 1837 Springfield MO 65802-4000

BOUGHN, STEPHEN PAUL, b Wheatland, Wyo, Dec 12, 46; m 68; c 1. EXPERIMENTAL PHYSICS, ASTROPHYSICS. *Educ:* Princeton Univ, BA, 69; Stanford Univ, MS, 70, PhD(physics), 75. *Prof Exp:* Physics res assoc, W W Hansen Labs & Dept Physics, Stanford Univ, 75-78; asst prof, dept physics, Joseph Henry Labs, Princeton Univ, 79-86; ASSOC PROF, DEPT ASTRON, HAVERFORD COL, 86- *Mem:* Am Phys Soc; Am Astron Soc. *Res:* Experimental gravitation; gravitational radiation detectors; the interaction of gravitational radiation with matter; low temperature experimental physics; extragalactic astronomy; cosmic background light; primeval galaxies; missing mass; clusters of galaxies. *Mailing Add:* Dept Astron Haverford Col Haverford PA 19041

BOUGHTON, JOHN HARLAND, b Niagara Falls, NY, May 7, 32; m 64; c 4. INORGANIC CHEMISTRY. *Educ:* Ariz State Univ, BS, 53, MS, 61; Univ Colo, PhD(inorg chem), 65. *Prof Exp:* Tech rep petrol ref, Tretolite Co, 55-58; teaching asst chem, Ariz State Univ, 58-60; teaching asst, Univ Colo, 60-61, res asst, 61-65; chemist, Spunbonded div, Textile Fibers Dept, 65-67, chemist, Pigments Dept, 67-72, sr res chemist, 72-76, res assoc, 76-87, SR RES ASSOC, ELECTRONICS DEPT, E I DU PONT DE NEMOURS & CO, 88- *Mem:* Am Chem Soc; Sigma Xi. *Res:* Pseudohalide complexes; solution chemistry; x-ray surface studies; surface characterization of fine particles; ultra-pure chemicals for electronics; contamination control. *Mailing Add:* E I du Pont de Nemours & Co PO Box 80030 Wilmington DE 19880-0030

BOUGHTON, ROBERT IVAN, JR, b Columbus, Ohio, Apr 2, 42; m 66; c 3. LOW TEMPERATURE PHYSICS. *Educ:* Ohio State Univ, BS & MS, 64, PhD(physics), 68. *Prof Exp:* Fel, Swiss Fed Inst Technol, 68-69; vis asst prof physics, Northeastern Univ, 69-70, asst prof, 70-73, assoc prof, 73-80; PROF PHYSICS & CHMN DEPT, BOWLING GREEN STATE UNIV, 80- *Mem:* Am Phys Soc; Am Asn Physics Teachers. *Res:* Transport properties of metals at low temperatures, both normal and superconducting, including measurements of effects of boundary scattering on transport. *Mailing Add:* Dept Physics & Astron Bowling Green State Univ Bowling Green OH 43403

BOUILLANT, ALAIN MARCEL, b Gisors, France, Aug 13, 28; Can citizen; m 65; c 1. VIROLOGY. *Educ:* Advan Sch Agr Engrs & Technicians, Paris, Agr Ing, 50; Univ Montreal, DVM, 58; Univ Wis-Madison, MS, 63; Laval Univ, PhD(microbiol & immunol), 67. *Prof Exp:* Res asst, Dept Vet Sci, Univ Wis-Madison, 60-63; res asst instr & tutor, Dept Microbiol & Immunol, Laval Univ, 64-67; res scientist virol, 67-68, RES SCIENTIST VIROL & TISSUE CULT, ANIMAL DIS RES INST, 68- *Mem:* Can Vet Med Asn; Am Soc Virol; AAAS; Tissue Cult Asn; Sigma Xi. *Res:* Virus diseases of animals; cytopathology; electronmicroscopy; cell transformation; retrovirus. *Mailing Add:* Animal Dis Res Inst PO Box 11300 Sta H Nepean ON K2H 8P9 Can

BOUIS, PAUL ANDRE, b Nice, France, Sept 21, 45; US citizen; m 67; c 3. ANALYTICAL & LANTHANIDE CHEMISTRY. *Educ:* Va Mil Inst, BS, 67; Univ Tenn, PhD(org chem), 74. *Prof Exp:* Res assoc org chem, Rhone Poulenc, USA, 73-76, process mgr, 76-79, anal & process mgr, 79-86; DIR ANALYTICAL RES, J T BAKER, 86- *Concurrent Pos:* Mem, Am Chem Soc & Semichem Reagents Comts, 86- *Mem:* Am Chem Soc. *Res:* Simplification of analytical chemistry of trace analysis in organic and inorganic matrices; novel purification techniques for high purity chemicals; analysis and purification of lanthanides. *Mailing Add:* 3875 Gloucester Bethlehem PA 18017

BOULANGER, JEAN BAPTISTE, b Edmonton, Alta, Aug 24, 22. PSYCHOANALYSIS. *Educ:* Univ Montreal, BA, 41, MD, 48; McGill Univ, MA, 50; Univ Paris, dipl psychopath, 51, lic psychol, 54; FRCP(C), 55. *Prof Exp:* Instr psychiat, 53, lectr, 54, from asst prof to assoc prof, 54-59, coordr behav sci course, 69-77, PROF PSYCHIAT, FAC MED, UNIV MONTREAL, 70- *Concurrent Pos:* Attend physician, neuropsychiat serv, Notre Dame Hosp, 53-61; training analyst, Can Inst Psychoanal, 57 & med dir treatment serv, 59-71; consult & dir group psychother, Ste-Justine Hosp, 59-78, hon consult child psychiat, 78; consult, Inst Albert Prevost, 60-73, Lakeshore Gen Hosp, 65-78, Verdun Gen Hosp, 65-71 & L H Lafontaine Hosp, 71-77; vis scientist, Ctr Bioethics, Clin Res Inst Montreal, 77-78; pvt pract; assoc ed, Can Psychiat Asn J, 60-; dir, Int Centre Comp Criminol, 75-; vis scientist, Dept Educ & Res Forensic Psychiat, Inst Philippe Pinel de Montreal, 78- & Dept Psychiat Educ, L H Lafontaine Hosp, 81; foreign ed, Am J Forensic Psychiat, 80. *Mem:* Can Psychoanal Soc (pres, 58, 59 & 60); fel Am Psychiat Asn; Int Psychoanal Asn; Can Psychiat Asn (pres, 75-76); NY Acad Sci. *Res:* Medical undergraduate education; postgraduate psychiatric training; psychoanalytic education and training; group psychotherapy; forensic psychiatry; adult and child neuropsychiatry. *Mailing Add:* 2162 Sherbrooke W Montreal PQ H3H 1G7 Can

BOULANT, JACK A, NEUROPHYSIOLOGY, TEMPERATURE REGULATION. *Educ:* Univ Rochester, PhD(physiol), 71. *Prof Exp:* Assoc prof, 77-84, PROF PHYSIOL, OHIO STATE UNIV, 84- *Res:* Hypothalamus. *Mailing Add:* Physiol 314A Hamilton Hall Ohio State Univ 1645 Neil Ave Columbus OH 43210

BOULDIN, DAVID RITCHEY, b Sedalia, Mo, Apr 6, 26; m 60; c 2. SOIL FERTILITY. *Educ:* Kans State Univ, BS, 52; Iowa State Univ, MS, 53, PhD(agron), 56. *Prof Exp:* Soil chemist, Tenn Valley Authority, 56-62; assoc prof, 62-69, PROF SOIL SCI, CORNELL UNIV, 69- *Mem:* AAAS; Am Soc Agron; Am Soc Plant Physiol. *Res:* Relationships between properties of the soil-fertilizer system and plant response. *Mailing Add:* 236 Emerson Hall Agron Cornell Univ Main Campus Ithaca NY 14853

BOULDIN, RICHARD H(INDMAN), b Florence, Ala, Feb 23, 42; m 70; c 2. MATHEMATICS. *Educ:* Univ Ala, BS, 64; Univ Chicago, MS, 66; Univ Va, PhD(math), 68. *Prof Exp:* Instr math, Univ Va, 68-69; from asst prof to assoc prof, 69-73, PROF MATH, UNIV GA, 82- , HEAD MATH, 89- *Mem:* Am Math Soc; Math Asn Am. *Res:* Perturbation theory and approximation theory for bounded operators on a Hilbert space. *Mailing Add:* Dept Math Univ Ga Athens GA 30602

BOULDRY, JOHN M(ILLER), b Boston, Mass, Sept 8, 17; m 44; c 2. ELECTRICAL ENGINEERING. *Educ:* Northeastern Univ, ScB, 41; Brown Univ, ScM, 56. *Prof Exp:* Design engr, Holtzer-Cabot Elec Co, 40-43; elec engr, Telechron, Inc, 46; asst prof elec eng, 46-50, ASSOC PROF ELEC ENG, NAVAL POSTGRAD SCH, 55-, CHMN DEPT EW, 85- *Concurrent Pos:* Consult, Hartnell Col, 57-58 & Monterey Peninsula Col, 60-; mem bd trustees, Monterey Inst Foreign Studies; mem, Monterey City Coun; chmn, Monterey Peninsula Area Planning Comn. *Mem:* Sr mem Inst Elec & Electronics Engrs; Am Soc Eng Educ. *Res:* Automatic feedback control systems; radar; EW systems. *Mailing Add:* US Naval Postgrad Sch Monterey CA 93940

BOULET, J LIONEL, b Quebec, Que, July 29, 19; m 49; c 4. ELECTRICAL ENGINEERING. *Educ:* Laval Univ, BScA, 38, BSc, 44; Univ Ill, MSc, 47. *Hon Degrees:* DSc, Sr George Williams Univ, 68 & Ottawa Univ, 71. *Prof Exp:* Jr engr, RCA Victor Co, Ltd, Que, 44-45; asst elec eng, Laval Univ, 45; asst, Univ Ill, 46, res assoc, 57; from asst prof to prof, Laval Univ, 48-54, head dept, 54-64; tech adv, 64-67, dir inst res, Hydro-Quebec, 67-; PRES, FOUND DEVELOP SCI & TECHNOL. *Concurrent Pos:* In charge res prog, RCA Victor Co, Ltd, Que, 48-50; consult, Can Armament Res & Develop Estab, 50-53 & Hydro-Quebec, 64-; mem sci radio comt, Int Sci Radio Union, 63. *Honors & Awards:* Prince of Wales Award, 38; Can Inst Elec & Electronics Engrs Award, 43; Award, Eng Inst Can, 43; Can Medal, 67; Archambault Medal, 70. *Mem:* Can Elec Asn; Eng Inst Can. *Res:* Applied electromagnetization; antennas and waveguides; corona losses and radio interference on extra-high-voltage transmission lines. *Mailing Add:* 755 Muir No 201 St Laurent PQ H4L 5G9 Can

BOULET, MARCEL, b Montreal, Que, Dec 7, 19; m 47; c 4. FOOD SCIENCE, BIOCHEMISTRY. *Educ:* Univ Montreal, BSc, 43; McGill Univ, MSc, 45, PhD(agr chem), 48. *Prof Exp:* Res officer dairy chem, Nat Res Coun Can, 47-62; PROF FOOD SCI, LAVAL UNIV, 62- *Honors & Awards:* Eva Award, Can Inst Food Sci & Technol, 80. *Mem:* Am Dairy Sci Asn; Inst Food Technologists; Can Inst Food Sci & Technol; Fr-Can Asn Advan Sci. *Res:* Inorganic chemistry of milk; control of enzyme activity in foods of vegetable origin; storage of fresh fruits and vegetables; functional properties of proteins. *Mailing Add:* Dept Agr Food Res & Develop Ctr St Hyacinthe PQ J2S 8E3 Can

BOULGER, FRANCIS W(ILLIAM), b Minneapolis, Minn, June 19, 13; m 40. METALLURGY, MECHANICAL ENGINEERING. *Educ:* Univ Minn, MetE, 34; Ohio State Univ, MS, 37. *Prof Exp:* Draftsman, Minn State Dept Hyg, 34; metallurgist, Repub Steel Corp, Cleveland, 37; res engr, Battelle Mem Inst, 38-44, asst supvr steel res, 44-47, supv metallurgist, 47-60, metall div chief, 60-66, sr tech adv, Columbus Div, 66-84; RETIRED. *Concurrent Pos:* Mem metalworking comt, Orgn Econ Coop & Develop. *Honors & Awards:* Hunt Medal, Am Inst Mining, Metall & Petrol Engrs, 55; Gold Medal, Soc Mfg Engrs, 67. *Mem:* Nat Acad Eng; Am Inst Mining, Metall & Petrol Engrs; fel Am Soc Mech Engrs; fel Soc Mfg Engrs; fel Am Soc Metals; hon mem Int Inst Prod Engrs (pres, 75-76); NAm Mfg Res Inst; Int Cold Forging Res Group. *Res:* Steel metallurgy; low temperature properties of metals; improved free-machining steels; prevention of cracking during cold-drawing of steels; casting, forming, forging, fabricating and manufacturing practices for metals. *Mailing Add:* 1816 Harwitch Rd Columbus OH 43221

BOULLIN, DAVID JOHN, b London, Eng, Apr 21, 31; m 77; c 4. CLINICAL PHARMACOLOGY, PHYSIOLOGY. *Educ:* Univ London, BSc, 56 & 58, MSc, 60; St Andrews Univ, PhD(pharmacol), 63. *Hon Degrees:* MA, Oxford Univ, 75. *Prof Exp:* Demonstr pharmacol, St Bartholomew's Hosp Med Sch, London, 56-59; asst lectr, St Andrews Univ, 59-63; lectr, St Thomas Hosp Med Sch, London, 65-67; assoc prof, Univ Vt, 67-68; vis scientist, Nat Inst Ment Health, 68-73; SR SCIENTIST, MED RES COUN, RADCLIFFE INFIRMARY, OXFORD UNIV, 73- *Concurrent Pos:* USPHS fel, NIH, Md, 63-65; res grants, Ciba Ltd, Eng, 65-67, Ciba Inc, 67-68, Brit Nutrit Found, 71-73 & Arnar-Stone Labs, 76-77; spec lectr pharmacol, George Washington Univ, 71- *Mem:* Am Soc Exp Pharmacol & Therapeut; Brit Pharmacol Soc; Brit Physiol Soc; Int Neuro-Psychopharmacological Col. *Res:* Physiology of intestinal peristalsis; mechanisms of uptake, binding and release of neuro-humours and anti-hypertensive drugs; significance of accumulation drugs by blood platelets; clinical pharmacology of trigger factors in carcinoid syndrome. *Mailing Add:* Dept Clin Pharmacol Radcliffe Infirmary Woodstock Rd Oxford 0X2 6HE England

BOULLION, THOMAS L, b Morse, La, Nov 4, 40; m 59; c 5. MATHEMATICS, STATISTICS. *Educ:* La State Univ, BS, 61; Univ Southwestern La, MS, 63; Univ Tex, PhD(math, statist), 66. *Prof Exp:* Asst prof math, Univ Southwestern La, 66-67; from asst prof to assoc prof, 67-74, chmn interdisciplinary statist, 74-77, prof math & statist, Tex Tech Univ, 74-81; dept math, 81-85, HEAD, STATIST DEPT, UNIV SOUTHWEST LA, 85- *Concurrent Pos:* Consult mathematician & statistician, Tex Ctr Res, 66-72. *Mem:* Am Statist Asn; Inst Math Statist. *Res:* Statistical inference; matrix theory. *Mailing Add:* Univ SW La E University Ave Lafayette LA 70504

BOULOS, BADI MANSOUR, b Alexandria, Egypt, July 3, 30; m 64; c 4. PHARMACOLOGY. *Educ:* Univ Alexandria, MB & BCh, 53, DPh, 58, DTM&H, 60; Univ Iowa, MS, 62; Univ Mo, PhD(med pharmacol), 65. *Prof Exp:* Intern med, Univ Alexandria Hosps, 53-54; med officer, UN Relief & Welfare Agency, Gaza Strip, 54-56; instr pharmacol, Univ Alexandria, 56-60; asst radiation biol, Univ Iowa, 60-63; res assoc pharmacol, Univ Mo, 63-65; asst prof pharmacol, Univ Alexandria, 65-66; asst prof, Univ Mo-Columbia & res assoc cancer res & chemother, Ellis Fischel State Cancer Hosp, 66-70; spec postdoctoral fel clin pharmacol & toxicol, NIH, 70-72; ASSOC PROF ENVIRON & OCCUP HEALTH SCI, UNIV ILL MED CTR, 72- *Concurrent Pos:* Asst scientist, Cancer Res Ctr, 68-; Vet Med Res Coun grants, 69-70; USPHS grant, 69-70; USPHS spec fel clin pharmacol & toxicol, Univ Kans Med Ctr, Kansas City, 70-72; mem & assoc dir, Int Cancer Cong, Ill State Environ Health Resource Ctr, 72-78; Univ Ill grant, 73-74; NIHES contract, 73-76; scholarship indust toxicol, Wayne State Univ, 74; NIH fel, occup med, 77; prof commun med & asst dean, Col Med, King Faisal Univ, Saudi Arabia, 78-80; diplomat, Pac Basin Chem Conf, 84; consult, clin & indust toxicol, WHO, 85-; world wide expert toxicol, UNESCO. *Mem:* Soc Toxicol; AMA; Am Soc Vet Physiol & Pharmacol; NY Acad Sci; Egyptian Soc Pharmacol & Exp Therapeut; Am Soc Pharmacol & Exp Therapeut. *Res:* Effect of antioxidant food additives on behavior and changes in biogenic amines; epidemiological studies to determine effect of barium in drinking water on hypertension; effect of polychlorinated biphenyls in water on high risk population, for example, children, using feline species as a model; placental transfer of drugs and pharmacokinetics in vivo; trace metal contaminants in health food supplements, cadmium and hypertension, aluminum, and Alzheimer's disease; approx 250 research articles in scientific journals. *Mailing Add:* Dept Environ & Occup Health Sci Univ Ill MC 922 Chicago IL 60680

BOULOUCON, PETER, b Ottawa, Ont, Sept 26, 35; m 65; c 2. CHEMICAL ENGINEERING. *Educ:* McGill Univ, BEng, 57; Univ Del, MChE, 62, PhD(chem eng), 64. *Prof Exp:* Process engr, Dow Chem Can, Ltd, 57-59; sr chem engr, Cyanamid Can Ltd, 64-68; process engr, 68-69, GROUP SUPVR PROCESS & SYSTS ANALYSIS, AM CYANAMID CO, 69- *Concurrent Pos:* Vis prof, Chem Eng Dept, Univ Va, 77-78. *Mem:* Am Inst Chem Engrs; Am Chem Soc; Sigma Xi. *Res:* Process analysis and applied mathematics in the area of chemical industry. *Mailing Add:* 370 High Crest Dr High Crest Lake West Milford NJ 07480

BOULPAEP, EMILE L, b Aalst, Belg, Sept 15, 38; m 64. PHYSIOLOGY, NEPHROLOGY. *Educ:* Cath Univ Louvain, BS, 58, MD, 62, MMSc, 63. *Hon Degrees:* Dr, Lourain Univ, 87. *Prof Exp:* Res scientist physiol, Cath Univ Louvain, 62-64; res fel, Med Col, Cornell Univ, 64-66; instr-chief asst, Cath Univ Louvain, 66-68; asst prof, Med Col, Cornell Univ, 68-69; asst prof psysiol, 69-72, assoc prof, 72-79, chmn dept & prof physiol, 79-89, PROF CELLULAR & MOLECULAR PHYSIOL, SCH MED, YALE UNIV, 87- *Concurrent Pos:* Belg Govt travel grant, 63-64; Belg Am Educ Found fel, 64-65; res fel, John Polachek Found Med Res, 65-66; mem bd dirs, Belg Am Educ Found, 71-, pres, 77-; fel, Branford Col, Yale Univ, 73-; mem study sect, NIH Gen Med Br, 76-80; overseas fel, Churchill Col, Cambridge, UK, 78-; mem, Coun Kidney Cardiovasc Dis, Am Heart Asn. *Honors & Awards:* Order of the Crown, Belgium, 79; Order of Leopold II, Belgium, 80; Homer W Smith Award, Am Soc Nephrol, NY Heart Asn. *Mem:* Am Soc Nephrol; Am Physiol Soc; NY Acad Sci. *Res:* Electrolyte and water transport across epithelia, particularly renal tubules; electrophysiology of the kidney. *Mailing Add:* Dept Cellular & Molecular Physiol Yale Univ Sch Med 333 Cedar St New Haven CT 06510

BOULTON, ALAN ARTHUR, b New Mills, Eng, Mar 14, 36; m 59; c 4. NEUROPSYCHOPHARMACOLOGY, PSYCHIATRY. *Educ:* Univ Manchester, BSc, 58, PhD(biochem), 62, DSc, 76. *Prof Exp:* Res asst biochem, Univ Manchester, 60-62; mem, Med Res Coun Unit Res Chem Path Ment Dis, Birmingham Univ, 62-68; chief res biochemist & res assoc, 68-69, assoc prof, 69-75, PROF PSYCHIAT, UNIV SASK, 75-, DIR, NEUROPSYCHIAT RES, MED RES BLDG, 71- *Concurrent Pos:* Res fel physiol, Birmingham Univ, 62-68; consult, Can Corrections, 87-; counr, Am Neurochem Soc, 89- *Honors & Awards:* Clark Prize, 70; Heinz Lehman Award, Can Col Neuropsychopharmacol, 89. *Mem:* Can Col Neuropsychopharmacol (pres, 84-86); Can Biochem Soc; Soc Biol Psychiat; Int Neurochem Soc (treas, 80-84, pres, 84-87); Am Neurochem Soc; Col Int Neuropsychopharmacol; Am Col Neuropharmacol. *Res:* Neurobiology of trace amines. *Mailing Add:* Neuropsychiat Res Unit Med Res Bldg Univ Sask Saskatoon SK S7N 0W0 Can

BOULTON, PETER IRWIN PAUL, b Toronto, Ont, May 11, 34; m 67; c 2. COMPUTER SYSTEMS. *Educ:* Univ Toronto, BASc, 60, MASc, 61, PhD(elec eng), 66. *Prof Exp:* Res engr, Electronics Div, Ferranti-Packard Elec Ltd, 61-63; from asst prof to assoc prof, 67-84, PROF COMPUT SYSTS, UNIV TORONTO, 84- *Concurrent Pos:* Consult, Elec Eng Consociates Ltd, 68- *Mem:* Sr mem Inst Elec & Electronics Engrs; Asn Comput Mach. *Res:* Computer systems software and hardware; local area networks; application of graphics; programming languages and language translators. *Mailing Add:* Dept Elec Eng Univ Toronto Toronto ON M5S 1A1 Can

BOULTON, ROGER BRETT, b Melbourne, Australia, July 14, 49; m 71; c 2. WINE FERMENTATION, WINE STABILITY. *Educ:* Univ Melbourne, BEng, 70, PhD(chem), 76. *Prof Exp:* Lectr process control, Dept Chem Eng, Univ Melbourne, 74-76, process simulation, 76; asst prof, 76-82, PROF WINERY DESIGN, UNIV CALIF, DAVIS, 82- *Concurrent Pos:* Enol consult, Robert Mondavi Winery, 80- *Mem:* Am Inst Chem Engrs; Am Chem Soc; Am Soc Enologists. *Mailing Add:* Dept Chem Eng Univ Calif Davis CA 95616

BOULWARE, DAVID G, b Oakland, Calif, Nov 20, 37; c 2. THEORETICAL PHYSICS. *Educ:* Univ Calif, Berkeley, AB, 58; Harvard Univ, AM, 60, PhD(physics), 62. *Prof Exp:* Jr fel, Harvard Univ, 62-65; from asst prof to assoc prof, 65-73, PROF PHYSICS, UNIV WASH, 73- *Mem:* Fel Am Phys Soc; fel AAAS. *Res:* Quantum field theory; quantum electrodynamics; elementary particles and relativity; general relativity and quantum gravity. *Mailing Add:* Dept Physics FM-15 Univ Wash Seattle WA 98195

BOULWARE, RALPH FREDERICK, b Stamps, Ark, Feb 15, 17; m 41; c 1. GENETICS. *Educ:* Okla State Univ, BS, 49; Univ Nebr, MS, 50, PhD(animal genetics), 53. *Prof Exp:* Asst animal husb, Univ Nebr, 49-53; asst prof, Miss State Col, 53-55; mgr, Farrar Farms, La, 55-57; res assoc, 57-58, from instr to assoc prof 58-77, assoc prof animal husb, 66-77, prof, 77-79, EMER PROF ANIMAL SCI, LA STATE UNIV, BATON ROUGE, 79- *Concurrent Pos:* Vis prof, US AID-Nicaraguan Ministry Agr, 68-69. *Mem:* Fel AAAS; Am Soc Animal Sci; Am Meat Sci Asn; Am Genetics Asn; Inst Food Technol; Sigma Xi. *Res:* Animal breeding. *Mailing Add:* 201 Hope Rd Stamps AR 71860

BOUMA, ARNOLD HEIKO, b Groningen, Netherlands, Sept 5, 32; m 60; c 3. MARINE GEOLOGY, SEDIMENTOLOGY. *Educ:* State Univ Groningen, BS, 56; State Univ Utrecht, MS, 59, PhD(sedimentol), 61. *Prof Exp:* Geol asst, State Univ Groningen, 54-56; sedimentol asst, State Univ Utrecht, 57-59, res fel, 60-62, lectr, 63-66; assoc prof, 66-70, prof geol oceanog, Tex A&M Univ, 70-76; marine geologists, US Geol Surv, 76-80; with explor prod div, Gulf Sci Technol Co, 80-, AT GULF RES & DEVELOP CO. *Concurrent Pos:* Fel Scripps Inst, Univ Calif, 62-63; ed-in-chief, Marine Geol, 63-66; mem pub & printing comt, Int Union Geol Sci, 65-69; co-chmn panel sedimentary processes, Gulf Univ Res Corp, 67-70; mem Gulf of Mex panel, Joint Oceanog Insts Deep-Sea Drilling Proj. *Mem:* AAAS; Am Asn Petrol Geol; Geol Soc Am; Soc Econ Paleont & Mineral; Am Geophys Union; Sigma Xi. *Res:* Sediments, ancient and recent; sedimentary facies models; internal characteristics of sedimentary structures with regard to transport and sedimentation; turbidites; techniques used to study sedimentary structures; graphic presentations; pollution. *Mailing Add:* Dept Geol & Geophysics La State Univ Baton Rouge LA 70803

BOUMA, HESSEL, III, b 1950; m 72; c 4. BIO-MEDICAL ETHICS. *Educ:* Univ Tex, PhD(human genetics), 75. *Prof Exp:* asst prof, 78-82, PROF BIOL, CALVIN COL, 82- *Mem:* Sigma Xi; Am Soc Cell Biol; Electron Micros Soc Am; Hastings Ctr; Am Scientific Affil. *Res:* Health care for the poor; human genetics. *Mailing Add:* Dept Biol Calvin Col Grand Rapids MI 49546

BOUMAN, THOMAS DAVID, b Geneva, Ohio, Nov 23, 40; m 66; c 1. THEORETICAL CHEMISTRY. *Educ:* Wash Univ, AB, 62; Univ Minn, PhD(phys chem), 67. *Prof Exp:* Resident res assoc chem, Argonne Nat Lab, 67-69; asst prof, 69-72, assoc prof, 72-77, PROF CHEM, SOUTHERN ILL UNIV, 77- *Concurrent Pos:* Vis assoc prof, Dept Chem, Univ Va, 74-75; vis prof, Chem Lab IV, H C Oersted Inst, Copenhagen Denmark, 75-76; G C Marshall Mem Fund fel, 75; Fulbright-Hays travel fel, 75; Danish NATO Comt sr sci fel, 75; sr res scholar award, Southern Ill Univ, Edwardsville, 77-78. *Honors & Awards:* Eastman Kodak Award, Univ Minn, 65. *Mem:* AAAS; Am Chem Soc; Am Phys Soc. *Res:* Theory of optical activity; molecular orbital calculations on medium-sized molecules; calculations of magnetic properties. *Mailing Add:* Dept Chem Southern Ill Univ Edwardsville IL 62026-1652

BOUNDS, HAROLD C, b Shreveport, La, Aug 13, 40; m 62; c 2. MICROBIAL ECOLOGY. *Educ:* Centenary Col, BS, 63; La State Univ, Baton Rouge, MS, 64, PhD(microbiol), 69. *Prof Exp:* From instr to asst prof, 69-74, ASSOC PROF BACT, NORTHEAST LA UNIV, 74- *Mem:* Am Soc Microbiol. *Res:* Herbicide influence on soil bacteria; microbial flora of ponds and sewage effluent; fungal contaminants in paper mills; antimicrobial properties of plant oils; sulfur oxidation by iron-sulfur autotrophs. *Mailing Add:* Dept Biol Northeast La Univ Monroe LA 71209

BOUNDY, RAY HAROLD, b Brave, Pa, Jan 10, 03; m 26; c 2. CHEMISTRY. *Educ:* Grove City Col, BS, 24; Case Inst Technol, BS, 26, MS, 30. *Hon Degrees:* ScD, Grove City Col, 47. *Prof Exp:* Phys chemist & mgr plastics div, Dow Chem Co, 30-50, vpres & dir res, 50-68; CONSULT MGT OF RES & DEVELOP, 68- *Concurrent Pos:* Mem insulation comt, Nat Res Coun; mem styrene technol comt, US Rubber Reserve; mem rubber adv comt, Army-Navy Munitions Bd; mem, Tech Intel Comt; dir, High Performance Technol Inc; consult to int exec, Govt Iran Serv Corps, 75-76. *Honors & Awards:* Indust Res Inst Medal, 64. *Mem:* Nat Acad Eng; Am Chem Soc; Electrochem Soc; Am Inst Chem Engrs. *Res:* Production and use of metallic sodium; electrometric analysis; automatic control; production of bromine from seawater; utilization of olefin derivatives; plastic development. *Mailing Add:* 3308 N Bent Oak Midland MI 48640

BOUQUOT, JERRY ELMER, b St Paul, Minn, June 23, 45; m 67; c 2. PRE-CANCER EPIDEMIOLOGY. *Educ:* St Olaf Col, Northfield, Minn, BA, 67; Univ Minn, Minneapolis, DDS, 71, MSD, 74; Am Bd Oral Path, dipl, 78. *Prof Exp:* Asst prof oral path, Sch Dent, Univ Minn, 74-75; from asst prof to assoc prof oral path, 75-84, asst prof path, 75-76, PROF ORAL PATH, SCH DENT & PROF PATH, SCH MED, WVA UNIV, 84- , CHMN DEPT ORAL PATH, 76- *Concurrent Pos:* Clin fel surgical path, Mayo Clin, Rochester, Minn, 74-75, vis sci epidemiol & med statist, 85; vis prof Royal Col Dent, Copenhagen, Denmark, 75; consult oral path, Vet Admin Hosps, Clarksburg, WVa, Martinsburg, WVa; Children's Hosp, Pittsburgh, Pa, 76- *Mem:* Int Asn Oral Path; Am Acad Oral Path; Am Dent Asn; Am Asn Cancer Educrs; Am Asn Dent Res; Am Asn Dent Schs. *Res:* Epidemiology of malignant and benign (especially pre-cancerous) lesions of the upper aerodigestive tract; head and neck pathology; methods of professional education and continuing education; health hazards of smokeless tobacco. *Mailing Add:* Dept Oral Path WVa Univ Health Sci Ctr Morgantown WV 26506

BOURBAKIS, NIKLAOS G, b Chania-Crete, Greece, Mar 23, 50; m 77; c 2. COMPUTER SCIENCE. *Educ:* Univ Athens, Greece, BS, 74; Univ Patras, PhD(comput eng image processing), 82, cert elec eng, 84. *Hon Degrees:* PhD, Univ Patras, Greece, 83. *Prof Exp:* Reseacher asst & chair comput sci, dept comput eng, Patras Univ, Greece, 74-75, teaching asst comput orgn & chair comput sci, 77-81, teaching asst assembly lang, 82, lab instr, 83-84, lectr microcomput, 83-84; ASST PROF MULTI-MICROPROCESSOR ARCHIT, DEPT ELECTRONICS & COMPUT ENG, GEORGE MASON UNIV, FAIRFAX, VA, 84- *Concurrent Pos:* Res collabr, NTUA, Athens, Greece, 81; res assoc Comput Technol Inst, Univ Patroas, 87-88; referee, Trans Syst, Man & Cybernet, Inst Elec & Electronics Engrs, 85; Int J, Pattern Recognition & Arttificial Intelligence, Int J, Microcomput Appln, Int J Automatica. *Mem:* Inst Elec & Electronics Engrs; Asn Comput Mach; Euromicro Soc; Pattern Recognition Soc; Greek Math Soc; Eurographics Soc. *Res:* Multimicrorprocessor architectures applied on real-time Robot Vision systems; expert systems, Text Analysis systems; algorithms and languages for computer vision; computer languages and environments; learning systems and artificial intelligence; Knowledge based VLSI design; performance evaluation of multiprocessor architectures; software engineering. *Mailing Add:* 8637 Greenbelt Rd 102 Greenbelt MD 20770

BOURCHIER, ROBERT JAMES, b London, Ont, Dec 28, 27; m 52; c 3. RENEWABLE RESOURCES POLICY. *Educ:* Univ Toronto, BcF, 51; Univ Alta, MSc, 55; State Univ NY Col Forestry, Syracuse, PhD, 60. *Prof Exp:* Mem res staff, Can Forestry Serv, 51-73, dir prog opers, 73-75, dir gen, Can Forestry Serv, 75-80; dir forest res, Ministry Nat Resources, Ont, 80-82; exec dir, Can Inst Forestry, 83-88; exec secy, Forest Res Adv Coun Can, 88-; RETIRED. *Concurrent Pos:* World Bank-FAO consult on forestry res in People's Repub of China, 84. *Mem:* Sigma Xi; Can Inst Forestry. *Mailing Add:* RR 1 PO Box 255 Cantley Quebec PQ J0X 1L0 Can

BOURDEAU, JAMES EDWARD, b Seattle, Wash, Feb 19, 48; m 73; c 2. INTERNAL MEDICINE, RENAL PHYSIOLOGY. *Educ:* Northwestern Univ, BS, 70, PhD(exp path), 73, MD, 74. *Prof Exp:* Intern/resident, Peter Bent Brigham Hosp & Harvard Med Sch, 74-76; res assoc, Lab Kidney & Electrolyte Metabol, Nat Heart, Lung & Blood Inst, 76-79; from asst prof to assoc prof med, Northwestern Univ Med Sch, 79-84; ASST PROF MED, UNIV CHICAGO, 85- *Concurrent Pos:* Prin investr, NIH & Chicago Heart Asn res grants. *Honors & Awards:* Established Investr, Am Heart Asn. *Mem:* Am Fedn Clin Res; Am Soc Nephrol; Am Physiol Soc; Int Soc Nephrol; Central Soc Clin Res. *Res:* Mechanisms and regulation of calcium transport across individual segments of the mammalian nephron. *Mailing Add:* Dept Med Sec Nephrol Univ Okla Health Sci Ctr Oklahoma City OK 73104

BOURDILLON, ANTONY JOHN, b Abercorn, Zambia, Sept 9, 44; Brit citizen; m 89. TEXTURE PROCESSING, MACROSTRUCTURAL CHARACTERIZATION. *Educ:* Univ Oxford, UK, BA, 71, DPhil(physics), 76. *Prof Exp:* Res assoc physics, Cavendish Lab, Cambridge Univ, UK, 76-78; tech officer mat sci, Dept Mat Sci & Metall, Univ Cambridge, UK, 78-86; dir electron micros, Univ New South Wales, Australia, 86-89; PROF MAT SCI, STATE UNIV NY, STONY BROOK, 89- *Concurrent Pos:* Prin investr, Royal Soc & Sci & Eng Res Coun, UK, 81-86, Metal Mfrs & DITAC, Australia, 87-89, NY State Inst Superconductivity, 91- *Mem:* Nat Inst Ceramic Engrs; Am Phys Soc; Am Ceramic Soc; AAAS. *Res:* Spectroscopy of ionic solids using synchrotron radiation; X-ray absorption and structural studies of insulators, metals and semiconductors; analytical electron microscopy; electron energy-loss spectroscopy; processing and characterization of high temmperature superconducting materials; over 100 papers written and book in preparation. *Mailing Add:* Dept Mat Sci & Eng State Univ NY Stony Brook NY 11794-2275

BOURDO, ERIC A, JR, b Muskegon, Mich, Jan 15, 17; m 42; c 3. FOREST MANAGEMENT. *Educ:* Mich Technol Univ, BS, 43; Univ Mich, MA, 51, PhD(forestry, bot), 55. *Prof Exp:* Compounder natural & synthetic rubber prod, Goodyear Tire & Rubber Co, Ohio, 43-46; consult forester, Pomeroy & McGowan Co, Ark, 46; from instr to assoc prof, Mich Technol Univ, 47-58, dir res, Ford Forestry Ctr, 55-68, prof forestry, 58-81, dean sch forestry & wood prod, Ford Forestry Ctr, 68-81, consult forestry, 82; RETIRED. *Mem:* Soc Am Foresters; Ecol Soc Am; Am Soc Agron; Wilson Ornith Soc; fel Soc Am Foresters. *Res:* Management and utilization of northern forest types. *Mailing Add:* PO Box 27 Twin Lake MI 49457

BOURELL, DAVID LEE, b Dallas, Tex, Feb 26, 53; m 73; c 3. DEFORMATION PROCESSING. *Educ:* Tex A&M Univ, BS, 75; Stanford Univ, MS, 76, PhD(mat sci & eng), 79. *Prof Exp:* Asst engr, Vought Aerospace Corp, 75; res asst, Stanford Univ, 76-79; from asst prof to assoc prof, 79-91, PROF MAT SCI, UNIV TEX, AUSTIN, 91- *Concurrent Pos:* Alexander Von Humboldt fel, Ger, 91. *Honors & Awards:* Bradley Stoughton Award, Am Soc Metals, 86. *Mem:* Am Inst Mech Engrs; Am Soc Metals; Sigma Xi; Am Powder Metall Inst. *Res:* Particulate processing of high temperature materials; solid freeform fabrication. *Mailing Add:* Univ Tex ETC 5160 Austin TX 78712

BOURGAULT, PRISCILLA C, b Winooski, Vt, Jan 1, 28. PHARMACOLOGY. *Educ:* Trinity Col, BS, 50; Loyola Univ, Ill, PhD(pharmacol), 66. *Prof Exp:* ASST PROF PHARMACOL, DENT SCH, LOYOLA UNIV, CHICAGO, 66- *Mem:* Sigma Xi. *Res:* Psychopharmacology; neuropharmacology. *Mailing Add:* Sch Dent Loyola Univ 2160 S First Ave Maywood IL 60153

BOURGAUX, PIERRE, b Woluwe-St Pierre, Belg, Feb 25, 34; m 61; c 2. ANIMAL VIROLOGY, MOLECULAR BIOLOGY. *Educ:* Free Univ Brussels, MD, 59; FRCP(C), 80. *Prof Exp:* From lectr to asst prof microbiol, Med Sch, Free Univ Brussels, 60-68; assoc prof, 68-72, PROF MICROBIOL, MED SCH, UNIV SHERBROOKE, 72- *Concurrent Pos:* Med Res Coun res fel, Exp Virus Res Unit, Glasgow Univ, 63-64; Eleanor Roosevelt fel, Salk Inst Biol Studies, 68-69. *Mem:* Am Soc Microbiol; Am Genetic Asn; Can Biochem Soc; Can Soc Microbiologists. *Res:* oncogenic viruses, particularly expression of viral genes in lytic and transformation systems. *Mailing Add:* Dept Microbiol Fac Med Univ Sherbrooke Sherbrooke PQ J1H 5N4 Can

BOURGET, EDWIN, b Senneterre, Que, July 6, 46; m 69; c 2. MARINE ECOLOGY. *Educ:* Univ Laval, Que, BSc, 69, MSc, 71; Univ Col Wales, PhD(marine biol), 74. *Prof Exp:* Prof-researcher, Nat Inst Sci Res, 74-76; asst prof, 76-84, PROF BIOL DEPT, UNIV LAVAL, 84- *Mem:* Am Soc Limnol & Oceanog; Can Soc Zoologists; Crustacean Soc. *Res:* Structure and dynamics of marine littoral benthic communities; structure and formation of barnacle shells. *Mailing Add:* Int Group for Oceanog Res Univ Laval Quebec City PQ G1K 7P4 Can

BOURGET, SYLVIO J, b Quebec, Que, Jan 9, 30; m 53; c 4. SOIL PHYSICS, SOIL CONSERVATION. *Educ:* Laval Univ, BSc, 50; Univ Wis, MS, 51, PhD(soils), 54. *Prof Exp:* Res officer field husb div, Cent Exp Farm, Ottawa, Ont, 54-59, res officer, Soil Res Inst, 59-62; PROF SOILS, LAVAL UNIV, 62-; dir res sta, Can Dept Agr, 68-; RETIRED. *Mem:* Soil Conserv Soc Am; Int Soc Soil Sci; Int Soc Soil Mech & Found Eng; Can Soc Soil Sci. *Res:* Movement and measurement of soil water; measurement and control of soil erosion; measurement of soil structure and temperature. *Mailing Add:* Agr Can Res Sta 2560 Hochelaga Blvd Quebec PQ G1V 2J3 Can

BOURGIN, DAVID GORDON, b New York, NY, Nov 6, 00; c 2. MATHEMATICS. *Educ:* Harvard Univ, PhD(math, physics), 26. *Prof Exp:* Instr, Lehigh Univ, 25-27; assoc math, Univ Ill, 27-37, from asst prof to prof, 37-44; prof, 66-71, M D ANDERSON PROF MATH, UNIV HOUSTON, 71- *Concurrent Pos:* Mem, Inst Adv Study, 40-41, 48-49; prin lectr, NSF col math conf, Univ Ore, 54; Fulbright lectr, 54-55; Fulbright res grant, Rome, 55-56; ed, Ill Jour Math & Houston Jour Math. *Mem:* Am Math Soc. *Res:* Applied mathematics; linear topological spaces; algebraic topology. *Mailing Add:* 1400 Hormann Dr No 16D Houston TX 77004

BOURGOIGNIE, JACQUES J, NEPHROLOGY, HYPERTENSION. *Educ:* Cath Univ Louvaine, Belgium, MD, 58. *Prof Exp:* PROF MED & DIR DIV NEPHROL, SCH MED, UNIV MIAMI, 76- *Res:* Internal medicine of nephrology. *Mailing Add:* Dept Med & Dir Div Nephrol Univ Miami Sch Med PO Box 016960 Miami FL 33101

BOURGON, MARCEL, physical chemistry, inorganic chemistry; deceased, see previous edition for last biography

BOURGOYNE, ADAM T, JR, b Baton Rouge, La, July 1, 44; m 66; c 6. FUEL TECHNOLOGY, PETROLEUM ENGINEERING. *Educ:* La State Univ, BS, 66, MS, 67; Univ Tex, Austin, PhD(petrol eng), 71. *Prof Exp:* Asst drilling engr, Texaco, Inc, La, 66; res assoc reservoir, Chevron Oil Res Lab, Calif, 67; res engr enhanced res, Continental Oil Res Lab, Okla, 68, sr systs engr comput applications, 69-71; asst prof teaching, 71-74, assoc prof, 74-79, CHMN TEACHING, LA STATE UNIV, 77-, PROF, 79- *Concurrent Pos:* Res consult, Baroid Div N L Indust, 72-74; teacher, Blowout Control Sch, La State Univ, 71-; prin investr, Blowout Prev Res, Dept Interior, US Geol Surv, 77- *Mem:* Am Soc Mech & Petrol Engrs; Am Petrol Inst; Sigma Xi. *Res:* Development of improved blowout prevention procedures for use in deepwater drilling operations. *Mailing Add:* Petrol Eng Dept La State Univ Baton Rouge LA 70803

BOURGUIGNON, LILLY Y W, b Shanghai, China, June 16, 46; m 70; c 1. CELL BIOLOGY. *Educ:* Nat Taiwan Univ, BS, 69; State Univ NY, Stony Brook, MA, 71, PhD(develop biol), 73. *Prof Exp:* Fel cell biol, Yale Univ, 73-75; fel membrane biol, Univ Calif, San Diego, 75-77; asst prof biol, Wayne State Univ, 77-81; assoc prof, 81-83, PROF CELL BIOL, SCH MED, UNIV MIAMI, 83- *Concurrent Pos:* Prin investr, NIH grants, 78-79, 78-81 & 81-90, Am Cancer Soc grants, 78-79 & 81-82 & Am Heart Asn grants, 84-89; estab investr, Am Heart Asn. *Mem:* Am Soc Cell Biol; Sigma Xi. *Res:* Biochemical and immunocytochemical studies on the biosynthesis of membrane proteins and transmembrane interactions between cytoskeletal components and surface receptor proteins in lymphocytes, fibroblasts and platelets; hormone-induced signal transduction and cell activation. *Mailing Add:* Dept Anat Sch Med Univ Miami Miami FL 33101

BOURKE, JOHN BUTTS, b Tampa, Fla, Aug 29, 34; m 57; c 3. CHEMISTRY. *Educ:* Colgate Univ, BA, 57; Ore State Univ, MS, 60, PhD(chem), 63. *Prof Exp:* Asst biochem, Ore State Univ, 62-63; res specialist, Cornell Univ, 63-65, from asst prof chem to assoc prof, 65-76, prof chem & dir, Anal Div, Food Sci Dept, 76-90, EMER PROF CHEM, CORNELL UNIV, 90- *Concurrent Pos:* USDA Coop Res Serv, Washington, DC, 86-88. *Mem:* AAAS; Am Chem Soc; Entom Soc Am; Asn Off Anal Chem. *Res:* Metabolism and metabolic fate of pesticides in plants and animals; laboratory data acquisition systems using computers; agricultural regulatory chemistry; analysis toxic substances and agricultural chemicals; environmental chemistry; food chemistry. *Mailing Add:* Anal Div Food Sci Dept Cornell Univ Geneva NY 14456

BOURKE, ROBERT HATHAWAY, b Portsmouth, Va, June 23, 38; m 64; c 2. POLAR OCEANOGRAPHY, UNDERWATER ACOUSTICS. *Educ:* US Naval Acad, BS, 60; Ore State Univ, MS, 69, PhD(phys oceanog), 72. *Prof Exp:* PROF PHYS OCEANOG, NAVAL POSTGRAD SCH, 71-, ASSOC DEAN FAC & GRAD STUDIES, 89- *Concurrent Pos:* Consult, 74-; prin investr, Arctic Submarine Lab, Off Naval Res, 74-; overseas fel, Churchill Col, Cambridge Univ, Eng, 81-82; assoc ed, J Underwater Acoustics, US Navy, 86- *Mem:* Sigma Xi; Am Geophys Union; Oceanog Soc. *Res:* Physical oceanography of the Arctic marginal sea-ice zones and environmental impact on underwater sound propagation. *Mailing Add:* Dept Oceanog Naval Postgraduate Sch Monterey CA 93943

BOURKOFF, ETAN, b Tel-Aviv, Israel, Oct 14, 49; US citizen; c 3. LASER RESEARCH, MICROWAVE ANTENNAS. *Educ:* Mass Inst Technol, BS & MS, 72; Univ Calif, Berkeley, PhD(elec eng), 79. *Prof Exp:* Sr engr, Ford Aerospace & Commun Corp, 72-76; res asst, Univ Calif, Berkeley, 76-79; mem tech staff, Hewlett-Packard Labs, 79-81; from asst prof to assoc prof, Johns Hopkins Univ, 81-89; PROF & CHAIR, UNIV SC, 90- *Concurrent Pos:* NSF presidential young investr award, 85; vis assoc prof, Columbia Univ, 89. *Mem:* Sr mem Inst Elec & Electronic Engrs; NY Acad Sci; Am Phys Soc; Optical Soc Am; Sigma Xi. *Res:* Generation and application of ultrashort laser pulses using novel mode-locking techniques; studies of femto second pulse compression, semiconductor laser modulation, and soliton propagation. *Mailing Add:* Dept Elec & Computer Eng Univ SC Columbia SC 29208

BOURLAND, CHARLES THOMAS, b Osceola, Mo, July 19, 37; m 63; c 1. FOOD MICROBIOLOGY & SPACE FOOD. *Educ:* Univ Mo, Columbia, BS, 59, MS, 67, PhD(food sci & nutrit), 70. *Prof Exp:* Qual control dairy, Adams Dairy, 63-65; asst food sci, Univ Mo, Columbia, 65-69; res scientist food, Technol Inc, 69-79; mgr, Space Food Develop Group, 79-87, FOOD SCIENTIST, NASA, 87- *Honors & Awards:* NASA-Johnson Space Ctr Group Achievement Award, 85. *Mem:* Inst Food Technol; Dairy Sci Asn; Nutrit Today Soc; Int Asn Milk, Food & Environ Sanit; Am Coun Sci & Health. *Res:* Food microbiology; food safety; aerospace food systems; menu planning, food processing and packaging for the space station; advanced design studies for Lunar/Mars missions. *Mailing Add:* 2341 Broad Lawn Dr Houston TX 77058

BOURLAND, FREDDIE MARSHALL, b Blytheville, Ark, May 4, 48; m 69. AGRONOMY, GENETICS. *Educ:* Univ Ark, BS, 70, MS, 74; Tex A&M Univ, PhD(genetics), 78. *Prof Exp:* Res assoc plant sci, Tex Agr Exp Sta, 73-78; from asst prof to assoc prof agron, Miss State Univ, 78-87; PROF AGRON, UNIV ARK, 88- *Mem:* Am Soc Agron; Crop Sci Soc Am; Am Soc Genetics; Coun Agr Sci Technol. *Res:* Breeding cotton for host plant resistance. *Mailing Add:* 115 Plant Sci Bldg Univ Ark Fayetteville AR 72701

BOURN, WILLIAM M, NEUROPHARMACOLOGY, TOXICOLOGY. *Educ:* Univ Ariz, PhD(pharmacol), 74. *Prof Exp:* DIR, DIV PHARMACOL, TOXICOL & NUCLEAR PHARM, COL PHARM & HEALTH SCI, NORTHEAST LA UNIV, 74- *Mailing Add:* Northeast La Univ Sch Pharmacol Monroe LA 71209

BOURNE, CAROL ELIZABETH MULLIGAN, b Rochester, NY, May 4, 48; m 68. PHYCOLOGY, CELL BIOLOGY. *Educ:* Ohio Wesleyan Univ, BA, 70; Miami Univ, MS, 78; Univ Mich, natural resources, 89- *Prof Exp:* Biol lab technician, Forest Serv, USDA, 70-73; grad res asst, Dept Bot, Miami Univ, 73-75; electron microscopist, Dept Nephrology, Col Med, Univ Cincinnati, 75-76; res asst, Dept Epidemiol, Sch Pub Health, Univ Mich, 78-80, res assoc, Dept Anat & Cell Biol, Med Sch, 81-83, grad res asst, Sch Natural Resources, 83-86, grad teaching asst biol, 87; SCIENTIST, INST FOOD & AGR SCI, UNIV FLA, 90- *Mem:* Am Inst Biol Sci; Am Soc Plant Taxonomists; Phycological Soc Am; Soc Study Evolution; Int Soc Plant Molecular Biol; Int Soc Diatom Res. *Res:* Molcular cloning of mycoplasmalike organism DNA for detection of lethal yellowing of coconut palms; characterizing diatom chloroplast DNA for molecular systematics; chloroplast DNA and organelle evolution studies. *Mailing Add:* Univ Fla-IFAS 3205 College Ave Ft Lauderdale FL 33314

BOURNE, CHARLES PERCY, b San Francisco, Calif, Sept 2, 31; m 53; c 2. INFORMATION SCIENCE. *Educ:* Univ Calif, Berkeley, BS, 57; Stanford Univ, MS, 63. *Prof Exp:* Sr res engr, Stanford Res Inst, Menlo Park, Calif, 57-66; vpres, Info Gen Corp, Palo Alto, Calif, 66-70; PRES, CHARLES BOURNE & ASSOC, MENLO PARK, CALIF, 70-; VPRES, GEN INFO DIV, DIALOG INFO SERV INC, PALO ALTO, CALIF, 77- *Concurrent Pos:* Guest lectr, Univ Calif, Berkeley, 63-66, dir, Inst Libr Res, 71-77, prof-in-residence, Sch Libr & Info Studies, 71-77; UNESCO consult, Indonesia & Tanzania; Nat Acad Sci consult, Ghana, 76; mem, US-Egyptian Task Force on Tech Info Probs, 76 & Network Adv Comt, Libr Cong, 87-; bd dirs, Nat Info Standards Orgn, 87-90. *Honors & Awards:* Ann Award of Merit, Am Documentation Inst, 65; Sarada Ranganathan Lectr, India, 78. *Mem:* Am Soc Info Sci (pres, 70). *Mailing Add:* 1619 Santa Cruz Ave Menlo Park CA 94025

BOURNE, EARL WHITFIELD, b Oklahoma City, Okla, July 6, 38. HISTOLOGY, CYTOLOGY. *Educ:* Westminster Col, Mo, AB, 60; Okla State Univ, MS, 62, PhD(zool), 68. *Prof Exp:* Asst prof biol, Bethany Col, WVa, 62-64; ASSOC PROF BIOL, UNIV N MEX, 68- *Mem:* Am Asn Cell Biol; Am Micros Soc; Sigma Xi. *Res:* Effects of carcinogenic hydrocarbons on cells in vitro; cytological effects of synthetic steroids. *Mailing Add:* Dept Biol Univ NMex Albuquerque NM 87106

BOURNE, HENRY C(LARK), JR, b Tarboro, NC, Dec 31, 21; m 53; c 4. ELECTRICAL ENGINEERING. *Educ:* Mass Inst Technol, SB, 47, SM, 48, ScD(elec eng), 52. *Prof Exp:* Asst prof elec eng, Mass Inst Technol, 52-54; from asst prof to assoc prof, Univ Calif, Berkeley, 54-63; prof elec eng, Rice Univ, 63-79, chmn dept, 63-74; vpres acad affairs, 81-86 & 87-88, actg pres, 86-87, MEM FAC ELEC ENG, GA INST TECHNOL, 81- *Concurrent Pos:* Sci fac fel, NSF, 60-61, dir div eng, NSF, 77-81. *Honors & Awards:* Centennial Medal, Inst Elec & Electronics Engrs. *Mem:* AAAS; Am Soc Eng Educ; Inst Elec & Electronics Engrs; Am Phys Soc; Sigma Xi. *Res:* Ferromagnetism and nonlinear magnetics. *Mailing Add:* Elec Eng 0250 Ga Inst Technol Atlanta GA 30332

BOURNE, HENRY R, b Danville, Va, Mar 1, 40; m; c 3. BIOCHEMISTRY. *Educ:* Johns Hopkins Univ, MD, 65. *Prof Exp:* PROF MED & PHARM, MED SCH, UNIV CALIF, SAN FRANCISCO, 80- *Mailing Add:* Dept Pharmacol Box 0450 Univ Calif Med Sch Cardiovasc Res Inst San Francisco CA 94143

BOURNE, JOHN ROSS, b Bryan, Tex, Aug 31, 44; m 68; c 2. ELECTRICAL ENGINEERING. *Educ:* Vanderbilt Univ, BE, 66; Univ Fla, MSE, 67, PhD(elec eng), 69. *Prof Exp:* from asst to prof elec & biomed eng, 77-87, PROF ELEC ENG, VANDERBILT UNIV, 87- *Concurrent Pos:* Vis prof, Chalmers Univ, Gothenburg, Sweden, 82; ed, CRC Critical Reviews in Bioengineering, 79-; assoc ed, Inst Elec & Electronics Engrs Trans Biomed Eng, 79-82. *Honors & Awards:* Samuel A Talbot Award, Inst Elec & Electronics Engrs Med Biol Soc, 76. *Mem:* Inst Elec & Electronics Eng. *Res:* Intelligent systems; pattern recognition; artificial intelligence. *Mailing Add:* Dept Elec Eng Vanderbilt Univ Nashville TN 37235

BOURNE, MALCOLM CORNELIUS, b Adelaide, Australia, May 18, 26; m 53; c 5. FOOD SCIENCE, RHEOLOGY. *Educ:* Univ Adelaide, BSc, 48; Univ Calif, MS, 61, PhD(agr chem), 62. *Prof Exp:* Chief chemist, Brookers Ltd, Australia, 48-58; from asst prof to assoc prof food sci, 62-73, PROF FOOD SCI & TECHNOL, CORNELL UNIV, 74- *Concurrent Pos:* Ed, J Texture Studies; guest lectr, food rheology Ministry Higher & Sec Educ, USSR, 83; team leader, Nat Acad Sci, China, 84; fel, Inst Food Technologists, 85; consult to food indust; Inst Food Technologists Nat Scientific Lectr, 84-87. *Mem:* AAAS; Am Chem Soc; Inst Food Technol; Soc Rheol; Am Asn Cereal Chem; fel Inst Food Sci & Technol (UK). *Res:* Physical measurement of quality of foods, food processing and food rheology; international food science; post harvest food losses, horticultural products; protein beverages; measurement, specification and control of texture and rheology of foods; quality factors and processing of fruits, legumes and vegetables. *Mailing Add:* NY State Agr Exp Sta Cornell Univ Geneva NY 14456-0462

BOURNE, PETER, b Oxford, Eng, Aug 6, 39; US citizen; m 64. MEDICINE, ANTHROPOLOGY. *Educ:* Emory Univ, MD, 62; Stanford Univ, MA, 69. *Prof Exp:* Res psychiatrist, Walter Reed Army Inst Res, 64-67; resident, Dept Psychiat, Sch Med, Stanford Univ, 67-69; asst prof psychiat & community health, Sch Med, Emory Univ, 69-72; dir, Ga Off Drug Abuse, Off Gov, 70-73; asst dir, Spec Action Off Drug Prev, 72-74; pres, Found Int Resources, 75-77; spec asst health to the President, 77-79; asst secy gen, UN Develop Prog & Coordr, Water Decade, 79; from asst prof to assoc prof psychiat, Emory Univ, 79-82; PROF PSYCHIAT, ST GEORGE'S UNIV MED SCH, GRENADA, 78-; PRES, AM ASN WORLD HEALTH, 84- *Concurrent Pos:* Consult, WHO, 73-74, Drug Abuse Coun, 74-77; lectr, Dept Psychiat, Sch Med, Harvard Univ, 74-80; comnr, Presidential Comn White House fels, 77-81; head, US deleg, UN Develop Prog Gov Coun, 78-; mem jury, Lasker Awards, 78-; lectr, Dept Psychiat, Harvard Med Sch, 74-80; assoc attend psychiatrist, MacLean Hosp, 74-80; bd dirs, Save Children Fedn. *Honors & Awards:* William C Menninger Award, Cent Neuropsychiat Asn, 67. *Mem:* Am Psychiat Asn; Royal Soc Med; Am Anthrop Asn; Am Med Soc Alcoholism; World Fedn Ment Health. *Res:* Psychological and biochemical aspects of stress; health service delivery; drug abuse; AIDS author or co-author of 100 technical articles. *Mailing Add:* 2119 Leroy Pl NW Washington DC 20008

BOURNE, PHILIP ERIC, b London, Eng, Mar 22, 53; m; c 1. X-RAY CRYSTALLOGRAPHY. *Educ:* Flinders Univ SAustralia, BSc, 74, MSc, 75, PhD(crystallog), 80. *Prof Exp:* Fel res, Sheffield Univ, 79-81; res assoc, 81-83, dir, Basic Sci Computer Systs, 83-86, SR ASSOC, HOWARD HUGHES MED INST, COLUMBIA UNIV, 87- *Concurrent Pos:* Mgr, Cancer Ctr Computer Facil, Columbia Univ. *Mem:* Am Crystallog Asn; Royal Australian Chem Inst; Asn Comput Mach. *Res:* Use of computers in molecular biology. *Mailing Add:* Biochem Dept Columbia Univ 630 W 168th St New York NY 10032

BOURNE, SAMUEL G, b Liverpool, Eng, Apr 29, 16; US citizen. MATHEMATICS. *Educ:* Rutgers Univ, BS, 38; Johns Hopkins Univ, AM, 44, PhD(math), 50. *Prof Exp:* Instr math, Johns Hopkins Univ, 43-49, Univ Conn, 50-52 & Temple Univ, 52-54; asst prof, Univ Calif, 54-55 & Lehigh Univ, 56-58; res asst, Calif Inst Technol, 58-59; vis scholar, Univ Calif, Berkeley, 59-63; prof, Univ Fla, 63-64; RES MATHEMATICIAN, UNIV CALIF, BERKELEY, 64- *Concurrent Pos:* Mem, Inst Advan Study, 50; lectr, Math Inst, Hungarian Acad Sci, 75 & 81; lectr, Math Inst, Yugoslav Acad Sci, 85. *Honors & Awards:* Bogart Math Prize, 38; Einstein lectr, Hellenic-Am Found, Athens, Greece, 80. *Mem:* Fel AAAS; Am Math Soc; Math Asn Am; Int Platform Asn; Indian Math Soc; Sigma Xi; Smithsonian Inst. *Res:* Structure of semirings and topological semirings; measure theory on locally compact semigroups; history of science. *Mailing Add:* PO Box 4583 Berkeley CA 94704

BOURNIA, ANTHONY, b Warren, Ohio, Jan 30, 25; m 63; c 2. NUCLEAR PROJECT MANAGEMENT, HEAT TRANSFER. *Educ:* Univ Pittsburgh, BS, 50, PhD(chem eng), 61; Univ Idaho, MS, 54. *Prof Exp:* Res chemist, Sherwin Williams Paint Co, 50-51; engr, Hanford Atomic Prod, Gen Elec Co, 51-55; sr engr, atomic power dept, Westinghouse Elec Corp, 55-60, supvr, Astronuclear Lab, 60-71, adv engr, environ syst dept, 71-72; SR PROJ MGR, US NUCLEAR REGULATORY COMN, 72- *Mem:* Am Nuclear Soc; Am Inst Chem Engrs. *Res:* Technical review, analyses, schedule and evaluation of the nuclear safety aspects of applications for construction and operation of central nuclear power plants. *Mailing Add:* 9101 Shad Lane Potomac MD 20854

BOURNIAS-VARDIABASIS, NICOLE, b Athens, Greece, July 24, 54; m 76; c 2. TERATOLOGY, DEVELOPMENTAL BIOLOGY. *Educ:* Univ Calif, Irvine, BSc, 75; Univ Essex, Eng, PhD(develop genetics), 78. *Prof Exp:* Res fel biol, City Hope Res Inst, 78-80; ASST RES SCIENTIST DEVELOP BIOL, CITY HOPE MED CTR, 80- *Concurrent Pos:* adj asst prof, Occidental Col, 80-81; asst prof, Calif State Univ, 81-83. *Mem:* Tissue Cult

Asn; AAAS; Teratology Soc; Differentiation Soc. *Res:* Study of developmental mechanisms, particularly mechanisms of teratogenesis gene control of early embryogenesis and organogenesis; development of in vitro assays for detecting possible teratogens. *Mailing Add:* Dept Biol Calif State Univ 5500 University Pkwy San Bernardino CA 92407

BOURNIQUE, RAYMOND AUGUST, chemistry; deceased, see previous edition for last biography

BOURNS, THOMAS KENNETH RICHARD, b Vancouver, BC, Feb 9, 24; m 51, 83. PARASITOLOGY. *Educ:* Univ BC, BA, 47, MA, 49; Rutgers Univ, PhD(zool), 55. *Prof Exp:* Tech officer, Med & Vet Entom Lab, Can Dept Agr, 49-52; asst zool, Rutgers Univ, 54-55 & bot, 55-56; lectr zool, Queen's Univ, Ont, 56-57; from lectr to prof zool, Univ Western Ont, 57-89; RETIRED. *Honors & Awards:* Wardle lectr, Can Soc Zool, 83. *Mem:* Am Soc Parasitol; Wildlife Dis Asn; Can Soc Zool (treas, 80-84). *Res:* Schistosome trematode-host relationships; trematode life cycles; tick biology; mutidisciplinary curriculum design. *Mailing Add:* 86 Lonsdale Dr London ON N6G 1T6 Can

BOURQUE, DON PHILIPPE, b St Louis, Mo, Nov 23, 42; c 1. MOLECULAR BIOLOGY. *Educ:* Johns Hopkins Univ, AB, 64; Duke Univ, MA, 67, PhD(bot), 69. *Prof Exp:* Res asst plant biochem, Dept Bot, Duke Univ, 64-69; fel molecular genetics, Dept Biol, Univ Calif, Los Angeles, 69-72, res assoc molecular biol, 72-73; asst prof agr biochem, 73-77, ASSOC PROF BIOCHEM, MOLECULAR & CELLULAR BIOL, UNIV ARIZ, 77- *Mem:* AAAS; Am Soc Cell Biol; Am Soc Plant Physiologists; Am Soc Biol Chemists. *Res:* Molecular biology of chloroplasts; molecular basis of rhizobium-plant symbiosis; xenobiotic detoxification by transgenic plants. *Mailing Add:* Dept Biochem Univ Ariz Tucson AZ 85721

BOURQUE, PAUL N(ORBERT), b Haverhill, Mass, June 6, 27; m; c 2. STRUCTURAL ENGINEERING. *Educ:* St Joseph's Univ, NB, BA, 49, BEd, 56; Univ Montreal, BTh, 53; Mass Inst Technol, MS, 61. *Prof Exp:* Prof math & physics, St Joseph's Univ, NB, 53-58; prof civil eng, 61-63; dean fac sci, Univ Moncton, 63-69; dir sch archit, 69-72, asst to director for acad affairs, Laval Univ, 72-81; CONSULT, 88- *Mem:* Int Asn Shell Struct; Eng Inst Can. *Mailing Add:* Sch Archit Laval Univ Quebec PQ G1K 7P4 Can

BOURQUIN, AL WILLIS J, b Castroville, Tex, Mar 2, 43. MARINE MICROBIOLOGY, MICROBIAL ECOLOGY. *Educ:* Univ Houston, BS, 65, MS, 68, PhD(biol), 71. *Prof Exp:* Res assoc microbiol, Lunar Receiving Lab, Brown & Root-Northrup Corp, 67-68; res microbiologist, 72-80, chief, Microbial Ecol & Biotechnol Br, US Environ Protection Agency, 81-; VPRES RES & DEVELOP, ECOVA CORP, 87- *Concurrent Pos:* Adj prof, Univ WFla, 73-75; co-ed, Develop Indust Microbiol, 74-; adj prof, Ga State Univ, 75-; vis prof microbiol, Univ Tex, 76; vis scientist aquatic toxicol, USSR, 77- *Mem:* Am Soc Microbiol; Soc Indust Microbiol; Sigma Xi. *Res:* Fate and effects of toxic chemicals including pesticides and hydrocarbons in aquatic environments; transport, transformation, and bioavailability of toxic pollutants; develop and verify predictive mathematical exposure concentration models. *Mailing Add:* Ecova Corp 3820 159th Ave NE Redmon WA 98052

BOURS, WILLIAM A(LSOP), III, b New York, NY, July 20, 18; m 41; c 4. CHEMICAL ENGINEERING. *Educ:* Princeton Univ, AB, 39, BSE, 40; Columbia Univ, MS, 41. *Prof Exp:* Engr, E I du Pont de Nemours & Co, Inc, 41-69, asst gen mgr indust & biochem dept, 69-75, vpres & gen mgr, Fabrics & Finishes Dept, 75-78, vpres fabrics & finishes, 78-81; RETIRED. *Mem:* Am Inst Chem Engrs; Am Chem Soc; Soc Chem Indust; Nat Paint & Coatings Asn; Sigma Xi. *Mailing Add:* Park Plaza No 1706 1100 Lovering Ave Wilmington DE 19806

BOUSEMAN, JOHN KEITH, b Clinton, Iowa, Aug 11, 36. ENTOMOLOGY. *Educ:* Univ Ill, BS, 60, MS, 62. *Prof Exp:* Exped entomologist, Am Mus Natural Hist Exped, Uruguay, 63 & Bolivia, 64 & 65; instr, Univ Ill, 65-66; asst entomologist, 72-84, ASSOC ENTOMOLOGIST, ECON ENTOM SECT, ILL NATURAL HIST SURV, 84- *Concurrent Pos:* Grant-in-aid res, Sigma Xi, 61; asst entomologist agr entom, Ill Agr Exp Sta, 72-; vis coleopterist, Fla State Collection Arthropods, 80; consult, Zambia Ministry Agr & Water Resources, 84. *Mem:* Am Registry Prof Entomologists; Entom Soc Am; Am Entom Soc. *Res:* Systematic entomology, especially Coleoptera and Hymenoptera; biology of pollination; history of the biological exploration of South America. *Mailing Add:* Econ Entom Natural Hist Surv 607 E Peabody 291 Nat Res Bldg Champaign IL 61820-6970

BOUSFIELD, ALDRIDGE KNIGHT, b Boston, Mass, Apr 5, 41; m 68. TOPOLOGY. *Educ:* Mass Inst Technol, SB, 63, PhD(math), 66. *Prof Exp:* Lectr math, Brandeis Univ, 66-67, asst prof math, 67-72; assoc prof, 72-76, PROF MATH, UNIV ILL CHICAGO, 76- *Concurrent Pos:* Off Naval Res assoc math, 66-67; NSF res grant, 67- *Mem:* Am Math Soc; Sigma Xi; Math Asn Am. *Res:* Homotopy theory; semisimplicial theory; algebraic topology. *Mailing Add:* Dept Math Statist & Comput Sci Univ Ill Box 4348 Chicago IL 60680

BOUSFIELD, EDWARD LLOYD, b Penticton, BC, June 19, 26; m 53, 84; c 4. INVERTEBRATE ZOOLOGY. *Educ:* Univ Toronto, BA, 48, MA, 49; Harvard Univ, PhD(zool), 54. *Prof Exp:* Invertebrate zoologist, 50-64, chief zoologist, 64-74, sr scientist, 74-86, EMER CUR, NAT MUS NATURAL SCI, 86-; emer cur, res assoc, Royal Ont Mus, 84-; RES ASSOC, ROYAL BC MUS. *Concurrent Pos:* Affil prof, Univ Wash, 80-91 & Univ Toronto, 84; adj prof, Carleton Univ, 71-84; mem, Can Comt Oceanog. *Honors & Awards:* Outstanding Achievement Award, Govt Can, 85. *Mem:* Fel Royal Soc Can; Can Soc Zoologists; Soc Syst Zool; Estuarine Res Fedn. *Res:* Marine zoology; Crustacea; Amphipoda; systematics and ecology; intertidal and estuarine ecology; marine fouling; stream biology; aquatic entomology; marine malacology. *Mailing Add:* Royal BC Mus 675 Belleville St Victoria BC V8V 1X4 Can

BOUSH, GEORGE MALLORY, b Norfolk, Va, June 5, 26; m 45; c 3. ENTOMOLOGY. *Educ:* Va Polytech Inst, BSc, 48; Ohio State Univ, MSc, 51, PhD, 55. *Prof Exp:* Asst entomologist, Va Agr Exp Sta, 49-50 & Rockefeller Found, 52-54; assoc prof entom, Univ Ky, 55-57 & Va Polytech Inst, 57-64; assoc prof, 64-68, chmn dept, 76-80, PROF ENTOM, UNIV WIS-MADISON, 68- *Concurrent Pos:* Smith-Mundt fel, Iraq, 60-61; consult, USDA, 61. *Mem:* Entom Soc Am. *Res:* insect microbial symbiotic associations; insect pathology; insect transmission of plant diseases. *Mailing Add:* 237 Russell Labs Dept Entom Univ Wis 1630 Linden Dr Madison WI 53706

BOUSQUET, WILLIAM F, b Milford, Mass, Sept 23, 33; m 57; c 2. PHARMACOLOGY, BIOCHEMISTRY. *Educ:* Mass Col Pharm, BS, 55; Purdue Univ, MS, 57, PhD(pharmacol), 59. *Prof Exp:* Asst prof bionucleonics, Purdue Univ, West Lafayette, 59-61, from asst prof to prof pharmacol, 61-77; DIR BIOL RES, SEARLE LABS, 77- *Concurrent Pos:* NIH res grant, 60; vis biologist, Am Inst Biol Sci, 61-63; Lederle Pharm fac res award, 63. *Mem:* AAAS; Am Chem Soc; Am Pharmaceut Asn; Am Soc Pharmacol & Exp Therapeut; NY Acad Sci. *Res:* Biochemical pharmacology interaction of drugs the cellular and subcellular structures; drug effects on carbohydrate and lipid metabolism; drug metabolism; cellular mechanisms in liver controlling enzyme synthesis. *Mailing Add:* 712 Roslyn Terr Evanston IL 60201

BOUSTANY, KAMEL, b Aleppo, Syria, Mar 22, 41; US citizen; m 64; c 4. RUBBER CHEMISTRY, FUEL TECHNOLOGY. *Educ:* Neuchatel Univ, Switz, chem eng, 64, PhD(organometallics), 67. *Prof Exp:* Sr res chemist rubber, 67-71, proj mgr, 71-74, sr proj mgr prod develop, 74-78, managing dir, Bombay, India, 78-81, bus develop dir, bus div, Monsanto Co, 81-85; PRES, PERMEA INC, 85- *Mem:* Am Chem Soc; Soc Plastic Engr. *Res:* The effect of short fibers on the reinforcement of elastomers and plastics to improve their physical properties, processibility and performances in tires and mechanical good products; enhanced oil recovery - technology, market size, economics, gas processing; membranes for separation and filtration; adsorbtion of gas. *Mailing Add:* Fisher Controls SA2RUE Louis Armand Asnieres Cedex 92607 France

BOUTILIER, ROBERT FRANCIS, b Lawrence, Mass, Nov 28, 37; m 63; c 3. GEOLOGY. *Educ:* Boston Univ, AB, 59, MA, 60, PhD(geol), 63. *Prof Exp:* Instr geol, Boston Univ, 62-65; from asst prof to assoc prof, 65-73, PROF EARTH SCI, BRIDGEWATER STATE COL, 73- *Mem:* AAAS; Geol Soc Am; Soc Econ Mineralogists & Paleontologists; Nat Asn Geol Teachers. *Res:* Origin of igneous and metamorphic rocks of eastern Massachusetts. *Mailing Add:* Dept Earth Sci Bridgewater State Col Bridgewater MA 02324

BOUTON, THOMAS CHESTER, b Milwaukee, Wis, Dec 5, 39; m 62; c 3. POLYMER SCIENCE. *Educ:* Univ Wis, BS, 52; Univ Akron, MS, 67. *Prof Exp:* Jr engr, Firestone Res, 62-64, supvr pilot plant, 65-67, mgr process develop, 67-76; MGR PROD DEVELOP, BRIDGESTONE-FIRESTONE, 76- *Mem:* Am Chem Soc; Am Inst Chem Engrs. *Res:* Structure-property relationships; process simulation; kinetics solution thermodynamics. *Mailing Add:* 2152 Glengary Rd Akron OH 44333

BOUTROS, OSIRIS WAHBA, b Beni-Suef, U A R, Aug 16, 28; US citizen; m 65; c 4. PLANT PHYSIOLOGY. *Educ:* Cairo Univ, BSc, 52; Fla State Uniiv, MS, 62; Univ Pittsburgh, PhD(plant physiol), 68. *Prof Exp:* Instr, El-Mahallah El Kobra Sec Sch, Egypt, 52-53; instr chem, Ministry of Educ, Addis Abeba, Ethiopia, 53-59; tech asst biol & chem, Fla State Univ, 59-61; instr, Col Steubenville, 62-63; asst plant physiol, Univ Pittsburgh, 64-68, Am Cancer Soc fel, 68-69, ASSOC PROF BIOL, UNIV PITTSBURGH, BRADFORD, 69- *Mem:* Am Soc Plant Physiol; Am Inst Biol Sci; AAAS; Bryol Soc Am. *Res:* Physiological properties, modes of action and chemical nature of plant growth regulators; pollution indicators. *Mailing Add:* Dept Natural Sci Univ Pittsburgh Bradford PA 16701

BOUTROS, SUSAN NOBLIT, b Lock Haven, Pa, May 22, 42; m 65; c 4. ENVIRONMENTAL HEALTH. *Educ:* Dickinson Col, BS, 64; Univ Pittsburgh, PhD(plant cytol), 67. *Prof Exp:* Instr biol, 67-68, NIH traineeship human genetics, Grad Sch Pub Health, Univ Pittsburgh, 68-69; auth gen biol lab book for col level, 69-71, FROM ASST PROF TO PROF BIOL, UNIV PITTSBURGH, BRADFORD, 71- *Concurrent Pos:* pres, Environ Assocs Ltd. *Mem:* Am Soc Parasitologists; Am Water Works Asn; Electron Micros Soc Am. *Res:* Giardia lamblia, cryptosporidium and other water borne diseases; improved identification techniques for waterborne pathogens; differentiation of ground water from surface water. *Mailing Add:* Dept Biol Univ Pittsburgh Bradford Bradford PA 16701

BOUTTON, THOMAS WILLIAM, b Lakewood, Ohio, Sept 9, 51; m 79; c 2. GRASSLAND ECOLOGY, STABLE ISOTOPE METHODOLOGY. *Educ:* St Louis Univ, AB, 73; Univ Houston, MS, 76; Brigham Young Univ, PhD(bot), 79. *Prof Exp:* Fel biol, Augustana Col, 80-81, asst prof, 81-82; fel pediat, Baylor Col Med, 82-83, instr, 83-85, asst prof pediat, 85-87; ASSOC PROF RANGELAND ECOL & PLANT PHYSIOL, TEX A&M UNIV, 87- *Concurrent Pos:* Res assoc bot, Univ Nairobi, Kenya, 80-81. *Mem:* AAAS; Brit Ecol Soc; Ecol Soc Am; Sigma Xi; Soil Sci Soc Am. *Res:* The application of stable isotope techniques to problems in grassland ecology; quantification of energy flow and nutrient cycling in grasslands; plant-soil relationships; prehistoric human and animal diets. *Mailing Add:* Dept Rangeland Ecol & Mgt Texas A&M Univ Mail Stop 2126 College Station TX 77843-2126

BOUTWELL, JOSEPH HASKELL, US citizen; m 43; c 3. CLINICAL BIOCHEMISTRY. *Educ:* Wheaton Col, BS, 39; Northwestern Univ, MS, 41, PhD(biochem), 47; MD, 49; Am Bd Clin Chemists, dipl, 54. *Prof Exp:* Asst prof biochem, Sch Med, Temple Univ, 49-66, prof, 66; dep dir, Bur Labs, 66-80, ASST DIR, CTR ENVIRON HEALTH, CTR DIS CONTROL, USPHS, 80- *Concurrent Pos:* Adj prof biochem, Ga State Univ; assoc prof path, Morehouse Med Sch. *Honors & Awards:* Reinhold Award, Clin Chem, 64. *Mem:* Am Asn Clin Chem; AMA; Am Soc Clin Path. *Res:* Analytical biochemical methods; enzyme systems; intracellular electrolytes; clinical medical education in biochemistry; quality control and proficiency testing in clinical laboratories. *Mailing Add:* 3055 Briarcliff Rd NE B-512 Atlanta GA 30326

BOUTWELL, ROSWELL KNIGHT, b Madison, Wis, Nov 24, 17; m 43; c 3. BIOCHEMISTRY, CARCINOGENESIS. *Educ:* Beloit Col, BS, 39; Univ Wis, MS, 41, PhD(biochem), 44. *Prof Exp:* Asst, McArdle Lab Cancer Res, Univ Wis-Madison, 39-44, instr cancer res, 45-49, from asst prof to assoc prof oncol, 49-67, prof oncol, 67-; chief res, Radiation Energy Res Found, Hiroshima, 84-86; RETIRED. *Concurrent Pos:* Fel, Univ Wis, 44-45; assoc ed, Cancer Res. *Honors & Awards:* Clowes Award, Am Asn Cancer Res, 79. *Mem:* Fel AAAS; Am Asn Cancer Res; Am Soc Biol Chem. *Res:* Interaction of carcinogens with tissue constituents and metabolic consequences; prevention of cancer. *Mailing Add:* 2935 Harvard Dr Madison WI 53705

BOUWKAMP, JOHN C, b Grant, Mich, Apr 20, 42; m 66; c 4. HORTICULTURE, PLANT BREEDING. *Educ:* Mich State Univ, BS, 64, MS, 66, PhD(hort), 69. *Prof Exp:* Asst prof, 69-74, ASSOC PROF HORT, UNIV MD, 74- *Mem:* Am Genetic Asn; Am Soc Hort Sci. *Res:* Genetics and physiology of vegetable crops. *Mailing Add:* Dept Hort Univ Md 1123 Holzapfel Hall College Park MD 20742

BOUWSMA, WARD D, b Lansing, Mich, Jan 11, 35; m 60; c 2. MATHEMATICS. *Educ:* Calvin Col, AB, 56; Univ Mich, MA, 57, PhD(math), 62. *Prof Exp:* Asst prof math, Pa State Univ, 60-67; assoc prof math, Southern Ill Univ, 67-80; CONSULT, 80- *Mem:* Am Math Soc; Math Asn Am; Sigma Xi. *Res:* Complex variables. *Mailing Add:* 2525 Londonderry Ann Arbor MI 48104

BOUYOUCOS, JOHN VINTON, b Lansing, Mich, Nov 9, 26; m 53; c 3. APPLIED PHYSICS. *Educ:* Harvard Univ, AB, 49, AM, 50, PhD(appl physics), 55. *Prof Exp:* Asst, Harvard Univ, 51-55, asst to dir, Acoust Res Lab, 55-59; mgr, Acoust Dept, Gen Dynamics Electronics Div, 59-72, PRES & CHIEF SCIENTIST, HYDROACOUST, INC, 72- *Mem:* Fel Acoust Soc Am; fel Inst Elec & Electronics Engrs; Soc Exploration Geophysicists. *Res:* Hydrodynamic power conversion; physical acoustics; underwater sound; fluid mechanics; electric networks; electronics; servomechanisms. *Mailing Add:* PO Box 23447 Rochester NY 14692

BOVARD, FREEMAN CARROLL, b Eugene, Ore, July 18, 21; m 45; c 3. BIO-ORGANIC CHEMISTRY. *Educ:* Pomona Col, AB, 43; Iowa State Col, PhD(biochem), 52. *Prof Exp:* Jr chemist, Shell Develop Co, 43-45; res biochemist, Stine Lab, E I du Pont de Nemours & Co, 51-55; from asst prof to prof chem, Claremont Men's Col, 55-64, prof, Claremont Men's Col, Pitzer Col & Scripps Col, 64-86, EMER PROF CHEM, CLAREMONT MCKENNA COL, PITZER COL & SCRIPPS COL, 86- *Mem:* AAAS; Am Chem Soc; Sigma Xi. *Res:* Amylolytic enzymes; carbohydrates; drug metabolism; peptide and protein synthesis; proteolytic enzymes. *Mailing Add:* 909 W Bonita Claremont CA 91711-4193

BOVARD, KENLY PAUL, b Pittsburgh, Pa, Mar 23, 28; m 50; c 3. ANIMAL HUSBANDRY. *Educ:* Cornell Univ, BS, 50; Iowa State Univ, MS, 54, PhD, 60. *Prof Exp:* Res assoc animal husb, Iowa State Col, 54-57; ASSOC PROF ANIMAL SCI, VA POLYTECH INST, 57- *Mem:* Fel AAAS; Am Soc Animal Sci; Biomet Soc; Am Genetic Asn. *Res:* Beef cattle breeding; problems dealing with heritabilities of economic characteristics of commercial production. *Mailing Add:* Va Technol State Univ 22 Agnew Hall Blacksburg VA 24061

BOVAY, HARRY ELMO, JR, b Big Rapids, Mich, Sept 4, 14; m 77; c 2. ENGINEERING, MATERIALS SCIENCE. *Educ:* Cornell Univ, Civil Engr, 36. *Prof Exp:* Engr, Humble Oil & Refining Co, 37-46; owner, H E Bovay, Jr, Consult Engrs, 46-52, partner, 52-62; pres, Bovay Engrs, Inc, 62-73, chmn bd & chief exec officer, 73-83; pres, Mid-South Telephone Co, 83-89; PRES, MID-SOUTH TELECOMMUN CO, 87-; CHMN & CHIEF EXEC OFF, MID-SOUTH BROADCASTING CO, 87- *Concurrent Pos:* Am Consult Engrs Coun fel, 73- *Honors & Awards:* ASHRAE/ALCO Award, Am Soc Heating, Refrig & Air-Conditioning Engrs, Inc, 71; Nat Soc Prof Engr Award, 87. *Mem:* Nat Acad Eng; Nat Soc Prof Engrs (pres, 75-76); Am Soc Testing & Mat; Am Civil Engrs; Am Soc Heating, Refrig & Air-Conditioning Engrs. *Res:* Expansion of energy field; power generation utilization; petrochemical plant development; advancement of engineering education and ethics through study and practice. *Mailing Add:* 3355 W Alabama No 1140 Houston TX 77098-1718

BOVBJERG, DANA H, b St Louis, Mo. PSYCHONEURO IMMUNOLOGY. *Educ:* Carleton Col, BA, 73; Univ Iowa, BS, 77; Univ Rochester, MS, 83, PhD(neurosci), 83. *Prof Exp:* Asst prof med, 86-89, ASST PROF NEUROSCI, CORNELL UNIV MED COL, 89-; ASST MEM NEUROL, MEM SLOAN-KETTERING CANCER CTR, 89- *Mem:* Am Asn Immunologists; Am Psychosom Soc; Soc Behav Med. *Res:* Psychoneuroimmunology, the regulation of the immune system by the brain. *Mailing Add:* Psychoneuroimmunol Lab Mem Sloan-Kettering Box 457 1275 York Ave New York NY 10021

BOVBJERG, RICHARD VIGGO, b Chicago, Ill, Sept 11, 19; m 42, 60; c 4. ANIMAL ECOLOGY. *Educ:* Univ Chicago, PhD(zool), 49. *Prof Exp:* Asst zool, Univ Chicago, 46-49; from instr to asst prof, Wash Univ, 49-55; from asst prof to assoc prof, 55-61, PROF ZOOL, UNIV IOWA, 61- *Concurrent Pos:* Ford Found fel, 51-52; dir, Iowa Lakeside Lab, 63- *Mem:* Ecol Soc Am; Soc Study Evolution; Am Soc Limnol & Oceanog; Am Soc Zoologists. *Res:* Ecology of aquatic invertebrates. *Mailing Add:* Dept Biol Univ Iowa Iowa City IA 52242

BOVE, ALFRED ANTHONY, CORONARY ARTERY DISEASE. *Educ:* Temple Univ, Philadelphia, MD, 66, PhD(physiol), 70. *Prof Exp:* prof med & physiol & consult cardiovasc dis, Mayo Clin, 81 -86; CHIEF CARDIOL SECT, TEMPLE UNIV MED SCH, 87- *Mailing Add:* Dept Med Temple Univ Med Sch 3400 Broad St Philadelphia PA 19140

BOVE, JOHN L, b New York, NY, Apr 15, 28; m 57; c 2. AZIRIDINE CHEMISTRY. *Educ:* Bucknell Univ, BA, 49, MS, 54; Case Western Reserve Univ, PhD(chem), 73. *Prof Exp:* Asst prof chem, Cooper Union, 57-67; dep dir, New York Dept Air Resources, Bur Tech Serv, 67-70; PROF CHEM & ENVIRON ENG, COOPER UNION, 70- *Res:* Aziridine chemistry; benzyne chemistry; thermolysis reactions; use of FT-IR-computer to pattern recognition of organics; use of sodium perborate in Baeyer-Villiger reactions. *Mailing Add:* Chem Dept Cooper Union 51 Astor Pl New York NY 10003

BOVE, JOSEPH RICHARD, b Orange, NJ, Apr 23, 26; m 52; c 3. HEMATOLOGY, LABORATORY MEDICINE. *Educ:* Univ Md, BS, 51, MD, 53. *Hon Degrees:* MA, Yale Univ, 70. *Prof Exp:* From instr to asst prof med, 59-66, assoc prof lab med, 66-70, PROF LAB MED, YALE UNIV, 70- *Concurrent Pos:* Consult, Vet Admin, 59- *Honors & Awards:* John Elliott Award, Am Asn Blood Banks. *Mem:* AAAS; Am Soc Hemat; Acad Clin Lab Physicians & Sci; Am Fedn Clin; Am Asn Blood Banks; Am Soc Clin Path. *Res:* Blood transfusion serology, especially alterations of red cell antigens and antibodies. *Mailing Add:* Yale-New Haven Hosp New Haven CT 06504

BOVE, KEVIN E, LIVER DISEASE, NEUROMUSCULAR DISEASE. *Educ:* Univ Buffalo, MD, 61. *Prof Exp:* ASSOC DIR PATH, CHILDREN'S HOSP, CINCINNATI, OHIO, 68- *Mailing Add:* Univ Cincinnati Col Med Children's Hosp Med Ctr Cincinnati OH 45229

BOVEE, KENNETH C, b Chicago, Ill, Sept 1, 36; m 58; c 3. VETERINARY MEDICINE, NEPHROLOGY. *Educ:* Ohio State Univ, BSc, 58, DVM, 61; Univ Pa, MMedSci, 69. *Prof Exp:* Intern & resident, Animal Med Ctr, New York, 61-64; from asst prof to assoc prof, 67-78, dept chmn, 79-84, PROF MED, SCH VET MED, UNIV PA, 78-, CORINNE & HENRY BOWER CHAIR MED, 80- *Concurrent Pos:* NIH fel, Grad Sch Med, Univ Pa, 64-67. *Mem:* Am Col Vet Internal Med; Am Soc Nephrology; Int Soc Nephrology. *Res:* Renal structure, function and disease. *Mailing Add:* Univ Pa 3800 Spruce St Philadelphia PA 19104

BOVELL, CARLTON ROWLAND, b New York, NY, Nov 10, 24; m 67. MICROBIAL PHYSIOLOGY. *Educ:* Brooklyn Col, AB, 48, MA, 52; Univ Calif, PhD(microbiol), 57. *Prof Exp:* Instr biol & bact, Brooklyn Col, 50-52; actg instr bact, 54-56, from instr to assoc prof, 57-69, vice chancellor, 81- 84, PROF MICROBIOL, UNIV CALIF, RIVERSIDE, 69- *Concurrent Pos:* Asst vpres, Acad Planning & Prog Rev, Systemwide Admin, Univ Calif, 78-81, chmn Dept Biol, 87-89, vchmn, Assembly Acad Senate, 89-90, chmn, 90-91, Bd Regents, 89-91. *Mem:* Am Soc Microbiologists; AAAS. *Res:* Comparative aspects of the autotrophic and heterotrophic physiology and metabolism of hydrogen bacteria. *Mailing Add:* Dept Biol Univ Calif Riverside CA 92521

BOVEY, FRANK ALDEN, b Minneapolis, Minn, June 4, 18; m 41; c 3. POLYMER CHEMISTRY. *Educ:* Harvard Univ, BS, 40; Univ Minn, PhD(phys chem), 48. *Prof Exp:* Asst chief chemist, Nat Synthetic Rubber Corp, 43-45; head polymer res dept, Minn Mining & Mfg Co, 48-55, res assoc, 55-62; mem tech staff, Bell Labs, 62-71, head, Polymer Chem Res Dept, 67-91; RETIRED. *Honors & Awards:* Polymer Chem Award, Am Chem Soc, 69, Phillips Award, 83, Award in Polymer Sci, 83, Carothers Award, 91; High Polymer Physics Award, Am Phys Soc, 74; Nichols Medal, 78; Silver Medal, Japanese Polymer Soc, 91. *Mem:* Nat Acad Sci; Am Chem Soc; Am Phys Soc; NY Acad Sci. *Res:* Physical chemistry of polymers; nuclear magnetic resonance of polymers; emulsion polymerization; fluorescence; rates of conformational isomerization; optical rotary dispersion. *Mailing Add:* Polymer Chem Res Dept AT&T Bell Labs Murray Hill NJ 07974

BOVEY, RODNEY WILLIAM, b Craigmont, Idaho, July 17, 34; m 56; c 4. WEED SCIENCE. *Educ:* Univ Idaho, BS, 56, MS, 59; Univ Nebr, PhD(agron), 64. *Prof Exp:* Instr & proj leader weed res, Univ Nebr, 59-64; res agronomist, Agr Res Serv, US Dept Agr, Univ Md, 64-67; proj leader defoliation res, Fed Exp Sta, Mayaguez, PR, 67; proj leader, 68-73, RES LEADER, BRUSH CONTROL RES, AGR RES SERV, US DEPT AGR, TEX A&M UNIV, 73- *Concurrent Pos:* Grants, US Dept Defense, 64-67, US Army, Ft Detrick, 67-68 & numerous chem co, 67-; spec adv to Secy Agr for 2,4,5-T hearing called by Environ Protection Agency, US Dept Agr, 72-74 & 78-80; instr, Tex A&M Univ, 88. *Mem:* Am Soc Agron; Soc Range Mgt; Coun Agr Sci & Technol; Weed Sci Soc Am; Sigma Xi; Plant Growth Regulator Soc Am. *Res:* Field evaluation of herbicides for weed and brush control; absorption and translocation of herbicides; effect of herbicides on ultrastructure, physiology, biochemistry, growth and anatomy of plants; herbicide residues in plants and soils. *Mailing Add:* Dept Range Sci Tex A&M Univ College Station TX 77843

BOVILLE, BYRON WALTER, b Ottawa, Ont, Dec 14, 20; m 45; c 3. ATMOSPHERIC SCIENCE. *Educ:* Univ Toronto, BA, 42; McGill Univ, MSc, 58, PhD(meteorol), 61. *Prof Exp:* Meteorologist, Can Meteorol Serv, 42-58; from asst prof to prof meteorol, McGill Univ, 60-72, chmn dept, 68-70; dir, Atmospheric Processes Res Br, Atmospheric Environ Serv, 72-78, dir Can Climate Ctr, 78-80; World Climate Prog, World Meteorol Org, Geneva, 80-82; chmn, 86-88, ADJ PROF, EARTH & ATMOSPHERIC SCI, YORK UNIV, 82- *Concurrent Pos:* Mem, Can Nat Comt, Int Union Geodesy & Geophysics, 56-69, chmn, 69-; mem Int Ozone Comn, Working Group on Stratospheric Pollution; mem, UN Environ Prog Ozone Coord Comt; sabbatical leave, Meteorol Nat France, 70-71; dir, Climate Conf, China, 80, Latin Am, 82. *Honors & Awards:* Patterson Medal, 82; Massey Medal, 89. *Mem:* Am Meteorol Soc; Can Meteorol Soc; fel Royal Meteorol Soc; Sigma Xi. *Res:* Dynamic meteorology, especially on the general circulation and on the stratosphere and interlayer coupling. *Mailing Add:* RR 1 Hawkestone ON L0L 1T0 Can

BOVING, BENT GIEDE, b Washington, DC, Feb 23, 20; m 44, 74; c 3. ANATOMY, REPRODUCTIVE BIOLOGY. *Educ:* Swarthmore Col, AB, 41; Jefferson Med Col, MD, 48. *Prof Exp:* Asst, Swarthmore Col, 41; asst zool, Univ Iowa, 41-42; intern, Wilmington Gen Hosp, Del, 48-49; mem staff dept embryol, Carnegie Inst, 51-70; prof obstet, gynec & anat, Sch Med,

Wayne State Univ, 70-88; RETIRED. *Concurrent Pos:* Nat Cancer Inst fel anat, Yale Univ, 49-51; vis res scientist, Reproductive Sci Prog, Univ Mich, 90-91. *Honors & Awards:* Schering Award, 46. *Mem:* AAAS; Am Asn Anat; Soc Develop Biol; Int Soc Develop Biol; Soc Study Reproduction; Sigma Xi. *Res:* Rabbit blastocyst transport, spacing, orientation and attachment to the uterus; invasive growth chemistry and mechanics. *Mailing Add:* 41901 W Eight Mile Rd Northville MI 48167-1941

BOVOPOULOS, ANDREAS D, b Argos, Greece, Mar 8, 58; m 88. COMMUNICATIONS NETWORKS, COMPUTER NETWORKS. *Educ:* Nat Tech Univ Athens, dipl elec eng, 82; Columbia Univ, MS, 83, PhD(elec eng), 89. *Prof Exp:* ASST PROF COMPUTER SCI, WASH UNIV, ST LOUIS, 88- *Mem:* Inst Elec & Electronic Engrs. *Res:* High speed networks; traffic control framework for automatic teller machine networks; stochastic systems. *Mailing Add:* 647-C W Canterbury St Louis MO 63132

BOVY, PHILIPPE R, b Liege, Belgium, Sept 23, 51; m 80; c 2. MOLECULAR MODELING. *Educ:* Univ Louvain, Belgium, BS, 72, MS, 74, PhD(chem), 78. *Prof Exp:* Res assoc med chem, Continental Pharna, 80-84; res scientist med chem, Monsanto Health Care, 84-86, Searle Res & Develop, 86-91; SR GROUP LEADER MED CHEM, MONSANTO CORP RES, 91- *Concurrent Pos:* Res assoc, Columbia Univ, NY, 78-80. *Honors & Awards:* Stas Award, Royal Acad Sci, 74; Bruylants Award, Chenici Lovanienses, 75, Grad Res Travel Award, 80. *Mem:* Am Chem Soc. *Res:* Design and synthesis of new compounds with therapeutic potential; antihypertensives; antithrombotics; antidiabetics; molecular modeling to interface organic synthesis and molecular biology. *Mailing Add:* Monsanto/AA2I 700 Chesterfield Village Pkwy St Louis MO 63198

BOWDAN, ELIZABETH SEGAL, b Kingston upon Hull, Eng; m 62; c 1. ANIMAL PHYSIOLOGY. *Educ:* Univ Durham, Eng, BSc, 60; Mt Holyoke Col, MA, 62; Univ Mass, PhD, 76. *Prof Exp:* Fel, 76-79, assoc fel, 79-81, VIS ASST PROF, UNIV MASS, 81- *Mem:* AAAS; Am Zool Soc; Asn Chemoreception Sci. *Res:* Insect neurophysiology specifically neurophysiological and behavioral aspects of chemosensory reception. *Mailing Add:* Dept Zool Univ Mass Amherst MA 01003

BOWDEN, CHARLES MALCOLM, b Richmond, Va, Dec 31, 33; m 60; c 3. QUANTUM OPTICS, STATISTICAL MECHANICS. *Educ:* Univ Richmond, BS, 56; Univ Va, MS, 59; Clemson Univ, PhD(physics), 67. *Prof Exp:* Physicist, US Naval Res Lab, 59-61; instr physics, Univ Richmond, 61-64; PHYSICIST, US ARMY MISSILE COMMAND, 67- *Concurrent Pos:* Mem part-time fac, dept physics, Univ Ala, Huntsville. *Mem:* Am Phys Soc; AAAS; Sigma Xi. *Res:* Quantum statistical mechanics of superradiance; cooperative processes in matter-radiation field interactions; laser physics; correlation phenomena in partially coherent radiation fields; laser induced molecular excitation for subsequent chemical reactivity. *Mailing Add:* 716 Versailles Dr Huntsville AL 35803

BOWDEN, CHARLES RONALD, b Senford, Del, Dec 19, 49; m 76. ONCOLOGY. *Educ:* Syracuse Univ, BS, 71; Univ Southern Calif, MS, 74, PhD(biomed eng), 78. *Prof Exp:* Res scientist endocrinol & metab res, McNeil Pharmaceut, 79-80, sr scientist, 80-85, group leader, 85-86, sect head, 86-87; res fel endocrinol, 87-89, SECT HEAD ONCOL & ENDOCRINOL RES, JASSEN RES FOUND, 90- *Concurrent Pos:* Vis scholar & NIH fel, Northwestern Univ, 78-79. *Mem:* Sigma Xi; Am Physiol Soc; Am Diabetes Asn. *Res:* Pharmacology and biochemical mechanism of novel-experimental cancer therapies; numerous publications. *Mailing Add:* Dept Oncol Res Janssen Res Found Welsh & McKean Rds Spring House PA 19477

BOWDEN, DAVID CLARK, b Tekamah, Nebr, Nov 23, 40; m 60; c 3. MATHEMATICAL STATISTICS, BIOMETRY. *Educ:* Colo State Univ, BS, 62, MS, 65, PhD(statist), 68. *Prof Exp:* From instr to asst prof, 65-84, PROF STATIST, COLO STATE UNIV, 84- *Concurrent Pos:* Consult, Colo Div Wildlife, 65-; dir statist workshops for fish & wildlife biologists, 65- *Res:* Simultaneous confidence bands for linear regression models; discrimination and confidence bands on percentiles; sample survey procedures with emphasis on techniques for fisheries and wildlife. *Mailing Add:* Dept Statist Colo State Univ Ft Collins CO 80523

BOWDEN, DAVID MERLE, b Hedley, BC, Sept 2, 29; m 52; c 2. AGRICULTURAL RESEARCH ADMINISTRATION, ANIMAL NUTRITION. *Educ:* Univ BC, BSA, 52, MSA, 57; Ore State Univ, PhD(animal nutrit), 61. *Prof Exp:* Res scientist, Res Br, Agr Can, 52-67; animal prod officer, Food & Agr Orgn, UN, 67-69; res scientist animal nutrit, 69-80, prog specialist, Western Region Res Bur, 80-83, dir, Res Sta, Swift Current, Sask, 83-85, DIR, RES STA, SUMMERLAND, BC, AGR CAN, 85- *Mem:* Can Soc Animal Sci (pres, 65); Am Soc Animal Sci; Agr Inst Can. *Res:* Nutrition of the beef cow and calf; use of forage crops for ruminants; blood constituents as indicators of nutritional status in ruminants; swine nutrition. *Mailing Add:* Res Sta Agr Can Summerland BC V0H 1Z0 Can

BOWDEN, DOUGLAS MCHOSE, b Durham, NC, Apr 7, 37; m 66; c 4. NEUROPSYCHIATRY. *Educ:* Harvard Col, BA, 59; Stanford Univ Sch Med, MD, 65. *Prof Exp:* Exchange fel neurophsiol, Moscow State Univ, 61-62; asst prof psychiat, 69-73, assoc prof, 73-79, PROF PSYCHIAT & BEHAV SCI, SCH MED, UNIV WASH, 79-; CORE STAFF SCIENTIST, REGIONAL PRIMATE RES CTR, UNIV WASH, 69-, DIR REGIONAL PRIMATE RES CTR, 88- *Concurrent Pos:* Res scientist develop award, Nat Inst Mental Health, 70-75; consult, Nat Libr Med Soviet Publ Translation, 73-; adj assoc prof pharmacol, Med Sch, Univ Wash, 75-79, asst dir, Regional Primate Res Ctr, 77-80, assoc dir, adj prof, 79-88; consult, Nat Inst Aging Animal Models Aging Res, 76; subcomt chmn, Comt Animal Models Aging Res, Primate Subcomt, Nat Acad Sci, 78-79; consult ed, Am J Primatol, 80-; mem, Inst Lab Animal Resources, Nat Res Coun, 85- *Mem:* Soc Neurosci; Am Soc Primatologists; Gerontol Soc. *Res:* Analysis of brain mechanisms of neuropsychiatric disorders such as psychosis, depression, dementia, and addiction to psychoactive drugs by application of behavioral, neurophysiological and pharmacological techniques in nonhuman primates; aging in nonhuman primates. *Mailing Add:* Regional Primate Res Ctr SJ-50 Univ Wash Seattle WA 98195

BOWDEN, DRUMMOND HYDE, b Wales, Mar 10, 24; Can citizen; m 48; c 2. PATHOLOGY. *Educ:* Bristol Univ, MB, ChB, 48, MD, 60. *Prof Exp:* Demonstr path, Bristol Univ, 51-52; asst pathologist, Hosp Sick Children, Toronto, 52-54, res assoc, 52-56; from asst prof to assoc prof, Sch Med, St Louis Univ, 56-64; assoc prof, 64-68, PROF PATH, UNIV MAN, 68-, HEAD DEPT, 81- *Concurrent Pos:* Head, Dept Path, Health Sci Ctr, Winnipeg. *Mem:* Am Asn Path; Int Acad Path; Am Thoracic Soc. *Res:* Pulmonary pathology, human and experimental. *Mailing Add:* Dept Path Rm 236 770 Bannatyne Ave Winnipeg MB R3T 2N2 Can

BOWDEN, GEORGE TIMOTHY, b Cincinnati, Ohio, Dec 15, 45; m 66. ONCOLOGY, BIOCHEMISTRY. *Educ:* Ohio Wesleyan Univ, BA, 67; Univ Wis, PhD(exp oncol), 74. *Prof Exp:* Staff fel chem carcinogenesis, Lab Exp Path, Nat Cancer Inst, 76-78; ASST PROF CANCER ETIOLOGY & THER, DEPT RADIOL, DIV RADIATION ONCOL, UNIV ARIZ, 78- *Concurrent Pos:* Am Cancer Soc fel, Inst Biochem, Ger Cancer Res Ctr, Heidelberg, 74-76; dir grant proj heat & radiation cancer ther, USPHS, 78-81. *Mem:* Am Asn Cancer Res; Am Tissue Cult Soc; Environ Mutagen Soc. *Res:* The role of DNA damage and repair in mutagen, carcinogen, chemotherapeutic agent induced cytotoxicity, mutagenesis and malignant transformation of normal cells. *Mailing Add:* Dept Radiation Oncol Univ Ariz Health Sci Ctr Tucson AZ 85724

BOWDEN, JOE ALLEN, b Dolores, Colo, July 13, 40; m 61; c 2. ANALYTICAL CHEMISTRY. *Educ:* Adams State Col, BA, 63; Univ NDak, MS, 65, PhD(biochem), 68. *Prof Exp:* Res assoc lipoproteins, Univ Fla, 68-69, NIH fel, 69-71; asst prof biochem, La State Univ, Baton Rouge, 71-78; PRES & DIR, CDS LABS, 80- *Concurrent Pos:* Consult, Environ Protection Agency, 73-75; consult energy problems, 73- & hazardous materials, 78-81; mem, Solar Energy Steering Comt, La State Univ, 74-78. *Mem:* AAAS; Sigma Xi; Am Soc Biol Chemists; Am Sci Affil; Am Chem Soc. *Res:* Enzymology of metabolic diseases; environmental analysis problems; lipoprotein structure and function; biochemical aspects of mental retardation; solar energy and alternate energy sources; hazardous materials and acid rain. *Mailing Add:* CDS Labs PO Box 2605 Durango CO 81302-2605

BOWDEN, MURRAE JOHN STANLEY, b Brisbane, Australia, Dec 15, 43; m 67; c 2. POLYMER CHEMISTRY. *Educ:* Univ Queensland, Australia, BSc, 64, 1st class hon, 65, PhD(polymer chem), 69. *Prof Exp:* Mem tech staff polymer chem, 71-78, supvr, radiation sensitive mat appln, Bell Labs, 79-83, div mgr, appl chem res, Bellcore, 84-86; div mgr, chem & mat sci res, 87-89, ASST VPRES, NETWORK SCI & TECHNOL RES, BELLCORE, 90- *Honors & Awards:* IR-100 Award, 77. *Mem:* Am Chem Soc. *Res:* Effects of radiation on polymers and its application to resists for electron lithography; polymer coatings for optical fibers. *Mailing Add:* Bellcore 331 Newman Springs Rd Redbank NJ 07701

BOWDEN, ROBERT LEE, JR, b Paris, Tenn, Apr 10, 33; m 68; c 2. MATHEMATICAL PHYSICS. *Educ:* Murray State Univ, AB, 55; Va Polytech Inst & State Univ, MS, 58, PhD(physics), 63. *Prof Exp:* Instr physics & math, Murray State Univ, 55-56; asst prof physics, Va Polytech Inst & State Univ, 63-68; vis assoc prof, Mid East Tech Univ, Turkey, 68-69; ASSOC PROF PHYSICS, VA POLYTECH INST & STATE UNIV, 69- *Mem:* Am Nuclear Soc; Am Phys Soc. *Res:* Linear and nonlinear operator theory analysis arising in neutral particle transport. *Mailing Add:* Dept Physics Va Polytech Inst & State Univ Blacksburg VA 24061

BOWDEN, WARREN W(ILLIAM), b South Penobscot, Maine, Nov 28, 25; m 53; c 3. CHEMICAL ENGINEERING. *Educ:* Univ Maine, BS, 49; Rose Polytech Inst, MS, 59; Purdue Univ, PhD, 65. *Prof Exp:* Res engr, Res Dept, Com Solvents Corp, 49-54; specialist, US Army, 54-56; from instr to assoc prof chem eng, Rose-Hulman Inst Technol, 56-69, prof chem eng, 69-90; RETIRED. *Mem:* Am Chem Soc; Am Inst Chem Engrs. *Res:* Heat transfer; mass transfer; thermodynamics of solutions. *Mailing Add:* 2183 W River Trace Dr No 6 Memphis TN 38134

BOWDISH, FRANK W(ILLIAM), b Kalispell, Mont, June 5, 17; m 44; c 4. MECHANICAL & CHEMICAL REFINING OF GLASS SAND. *Educ:* Mont Sch Mines, BS, 39, MDrE, 55; Mass Inst Technol, SM, 43; Univ Kans, PhD(chem eng), 56. *Prof Exp:* Trainee, Caterpillar Tractor Co, 39-41; res engr, Mass Inst Technol, 43-46; asst metallurgist, Oliver Iron Mining Co, Minn, 46-47; asst prof mining & metall eng, Univ Kans, 48-54; engr, Iron Res Inst, IRSID, France, 56-58; sr chem engr, Colo Sch Mines Res Found, 59; prof metall eng, NMex Inst Mining & Technol, 59-62; assoc prof & assoc mineral technologist, 62-69, prof chem eng & mineral technologist, 69-81, EMER PROF, MACKAY SCH MINES, UNIV NEV, RENO, 81- *Concurrent Pos:* Consult, Owens-Corning Fiberglas Corp & Stewart Sand & Mats Co, 52-55, Arkhola Sand, Gravel Co, 60-71, Texasgulf, Inc, 75-77, Princeton Mining & Refining, 80-83, Dixie Exp Co & BRPM, Morocco, 81-82; process consult, L-BAR Prods Inc, 86- *Mem:* Am Inst Mining, Metall & Petrol Engrs. *Res:* Kinetics of grinding in a ball mill; refining of feldspathic sands for use in glass making; ion exchange in clays; flotation of chrysocolla; beneficiation of iron ores; extraction of metals and minerals. *Mailing Add:* 1531 N 15th Ave Pasco WA 99301

BOWDLER, ANTHONY JOHN, b London, Eng, Oct 16, 28; US citizen; m 55; c 2. HEMATOLOGY, ONCOLOGY. *Educ:* Univ London, BSc, 49, MB & BS, 52, MD, 63, PhD(hemat), 67; FRCP, 75, FACP, 79, FRCPath, 84. *Prof Exp:* House physician, Univ Col Hosp, London, 52-53, Hammersmith Hosp, London, 53, Brompton Hosp, 56-57; casualty med off, Univ Col Hosp, London, 56; sr house officer, Dorking Gen Hosp, Surrey, Eng, 57-58; med registrar, Univ Col Hosp, 58-61; fel, Univ Col Hosp Med Sch, 61-62; sr fel, Sch Med, Univ Rochester, 62-64; sr lectr & mem, Med Res Coun Group Study of Haemolytic Dis, Univ Col Hosp Med Sch, 64-67; assoc prof, Col Human Med, Mich State Univ, 67-71, prof, 71-80; PROF MED, SECT HEMAT/ONCOL, DEPT MED, SCH MED, MARSHALL UNIV, 80- *Concurrent Pos:* Surgeon-Lieutenant, Royal Naval Vol Reserve, 53-55; hon consult, Univ Col Hosp, 67; vis prof, Dept Med, Sch Med & Dent, Univ

Rochester, 74; co-prin, Gen Hemat Serv, Dept Med, Mich State Univ, 75-79; physician, Vet Admin Med Ctr, Huntington, WVa, 80. *Mem:* Med Res Soc; Am Fedn Clin Res; Am Soc Hemat; Cent Soc Clin Res; Am Soc Clin Oncol; Brit Med Asn; Am Asn Cancer Educ. *Res:* Pre-lytic events of the erythrocyte, to define processes interdictable therapeutically; red cell physiology; myeloproliferative disorders; pathophysiology of the spleen; the physical properties of red cells and their variation with age. *Mailing Add:* Dept Med Sch Med Marshall Univ Huntington WV 25701

BOWDON, EDWARD K(NIGHT), SR, b Kansas City, Mo, July 24, 35; m 56; c 3. COMPUTER SCIENCE, ELECTRICAL ENGINEERING. *Educ:* Kans State Univ, BS, 56; Univ NDak, MS, 60; Univ Iowa, PhD(elec eng), 69. *Prof Exp:* Field engr, Western Elec Co, 56-59; asst, Univ NDak, 59-60; develop engr, Minneapolis Honeywell Co, 60-62; design engr, Collins Radio Co, 62-67, mem tech staff comput systs, 67-69; asst prof comput sci, Univ Ill, Urbana-Champaign, 69-76; ASSOC PROF ELEC ENG, UNIV ARK, FAYETTEVILLE, 76- *Concurrent Pos:* Consult, Automation Technol Inc, 70-; Nat Res Coun travel grant, 71; prin investr, NSF, 71- *Mem:* Inst Elec & Electronics Engrs. *Res:* Computer systems analysis, modeling and analysis of network computers employing queueing theory, scheduling theory and other operations research techniques, with emphasis on improving computer throughput; digital network design aids. *Mailing Add:* N Telecom Inc Data Networks Div 1219 Digital Rd Suite 125 Richardson TX 75081

BOWE, JOSEPH CHARLES, b Chicago, Ill, Sept 17, 21; m 53; c 6. PHYSICS. *Educ:* St Procopius Col, AB, 43; DePaul Univ, MS, 46; Univ Ill, PhD(physics), 51. *Prof Exp:* Instr physics, DePaul Univ, 43-44; jr physicist, Metall Lab, Univ Chicago, 44-46; asst physics, Univ Ill, 46-51; assoc physicist, Argonne Nat Lab, 51-67; TEACHER PHYSICS, ILL BENEDICTINE COL, 67- *Mem:* Am Phys Soc; Am Asn Physics Teachers; Am Sci Teachers Assoc. *Res:* Electron mobility in gases; low energy electron-atom interactions; electronic instrumentation. *Mailing Add:* Dept Physics Ill Benedictine Col Lisle IL 60532

BOWE, ROBERT LOOBY, b Worcester, Mass, Jan 25, 25; m 57; c 4. PSYCHOPHARMACOLOGY, PSYCHOPHYSIOLOGY. *Educ:* Boston Col, BS, 50, MS, 57; Univ Tenn, PhD(clin physiol), 60. *Prof Exp:* Instr, New Prep Sch, Mass, 52-56; asst physiol, Boston Col, 56-57; assoc pharmacol, Med Col SC, 60-61, asst prof pharmacol & therapeut, 61-64; asst prof, 64-71, ASSOC PROF PHARMACOL, MED COL VA, 71- *Concurrent Pos:* Vis prof, Columbia State Hosp, 61-64. *Mem:* AAAS. *Res:* Central nervous system; autonomic nervous system; cardiovascular and renal physiology, pathophysiology, and pharmacology; general pathophysiology and clinical physiology; psychophysiology and pathophysiology; drug abuse. *Mailing Add:* Dept Pharmacol Va Commonwealth Univ Sch Med MCV Sta Box 565 Richmond VA 23298

BOWEN, CHARLES E, b St Louis, Mo, Apr 9, 36; c 1. BIOCHEMISTRY. *Educ:* San Jose State Col, AB, 59, BS, 63, MS, 65; Univ Calif, Davis, PhD(biochem), 69. *Prof Exp:* Asst prof, 69-74, assoc prof, 74-80, PROF CHEM, CALIF STATE POLYTECH COL, KELLOGG-VOORHIS, 80- *Mem:* Sigma Xi; Conchologists Am. *Res:* Protein chemistry; enzyme mechanisms; marine biochemistry of invertebrates. *Mailing Add:* Dept Chem Calif State Polytech Univ Pomona CA 91768

BOWEN, DANIEL EDWARD, b Morris, Minn, July 4, 44; m 67; c 2. ECOLOGY, POPULATION BIOLOGY. *Educ:* Rockhurst Col, BA, 66; Kans State Univ, MS, 71, PhD(biol), 76. *Prof Exp:* from asst prof to assoc prof, 76-88, PROF BIOL, BENEDICTINE COL, 88-, CHMN DEPT, 82- *Mem:* Ecol Soc Am; Am Ornithologists Union; Wildlife Soc; Am Soc Zoologists; Am Inst Biol Sci. *Res:* Regulation of prairie bird populations that live in seasonal environments; winter behavior of snowgeese, bald eagles, and non game birds. *Mailing Add:* Dean/Dir Continuing Educ Benedictine Col Atchison KS 66002

BOWEN, DAVID HYWEL MICHAEL, b Gorseinon, Wales, July 1, 39; m 67; c 2. CHEMISTRY, CHEMICAL ENGINEERING. *Educ:* Univ Birmingham, BSc, 60, PhD(chem eng), 63. *Prof Exp:* Res engr, Eng Res Div, E I du Pont de Nemours & Co, Inc, 63-67; from asst ed to managing ed, 67-72, head journals dept, 73-75, dir, Bks & Journals Div, 75-88, DIR, MEM DIV, AM CHEM SOC, 88- *Mem:* Am Chem Soc; Am Inst Chem Eng; AAAS; Soc Scholarly Publ. *Res:* Technical communication; information science. *Mailing Add:* Am Chem Soc 1155 16th St NW Washington DC 20036

BOWEN, DONALD EDGAR, b Brooklyn, NY, Apr 10, 39; m 79; c 3. SOLID STATE PHYSICS. *Educ:* Tex Christian Univ, BA, 61, MA, 63; Univ Tex, PhD(physics), 66. *Prof Exp:* Res assoc high pressure physics, Gen Dynamics/Ft Worth, 62-63; asst, Univ Tex, El Paso, 63-66, from asst prof to assoc prof physics, 66-81, chmn dept, 71-81, asst vpres acad affairs, 81-85; VPRES ACAD AFFAIRS, SOUTHWEST MO UNIV, 85- *Concurrent Pos:* Res Corp Cottrell grant, 67; R A Welch Found grant, 68-71; NSF grant, 77-81. *Mem:* Inst Elec & Electronics Engrs; Sigma Xi; AAAS; Am Asn Physics Teachers; Am Phys Soc. *Res:* Properties of metal-ammonia solutions and the investigations of the acoustical properties of these solutions. *Mailing Add:* Off Vpres Acad Affairs Southwest Mo State Univ Springfield MO 65804

BOWEN, DONNELL, TRANSPORT, PHARMACOKINETICS. *Educ:* Univ NC, PhD(pharmacol), 75. *Prof Exp:* ASST PROF, DEPT PHARMACOL, HOWARD UNIV, 79- *Mailing Add:* 903 Woodmont Ct Mitchellville MD 20716

BOWEN, DOUGLAS MALCOMSON, organic chemistry, for more information see previous edition

BOWEN, GEORGE HAMILTON, JR, b Tulsa, Okla, June 20, 25; m 48; c 5. VARIABLE STARS, STELLAR EVOLUTION. *Educ:* Calif Inst Technol, BS, 49, PhD(biophys), 53. *Prof Exp:* Assoc biologist, Oak Ridge Nat Lab, 52-54; from asst prof to assoc prof, 54-65, PROF ASTROPHYSICS, IOWA STATE UNIV, 65- *Mem:* AAAS; Am Astron Soc; Astron Soc Pacific; Sigma Xi; Int Aston Union. *Res:* Long-period variable stars; atmospheric structure and dynamics; mass loss mechanisms; evolution; stellar pulsation; solar system formation and evolution. *Mailing Add:* Dept Physics & Astron Iowa State Univ Ames IA 50011

BOWEN, H KENT, b Nov 21, 41; m; c 5. TECHNOLOGY TRANSFER. *Educ:* Univ Utah, BS, 67; Mass Inst Technol, PhD, 71. *Prof Exp:* Staff, Dept Mat Sci & Eng, 70-76, prof 76-81, dir Mat Processing Ctr, FORD PROF ENG, 81-, DIR LEADERS FOR MFR, 88- *Honors & Awards:* R M Fulrath Award; F H Norton Award; Schwartzwalder-PACE Award; Ross Coffin Purdy Award; Robert Browning Sosman Award; Gordon Y Billard Award, Mass Inst Technol, Henry B Kane '24 Award. *Mem:* Nat Acad Eng. *Res:* Materials developments; technology transfer; published numerous articles in various journals. *Mailing Add:* Mass Inst Technol 77 Mass Ave Cambridge MA 02139

BOWEN, HENRY DITTIMUS, b Bear Lake, Mich, Oct 16, 21; m 48; c 2. AGRICULTURAL ENGINEERING. *Educ:* Mich State Univ, BS, 49, MS, 51, PhD(agr eng), 53. *Prof Exp:* Asst instr agr eng, Mich State Univ, 52-53; res assoc prof, 53-60, PROF AGR ENG, NC STATE UNIV, RALEIGH, 60- *Mem:* Am Soc Agr Engrs; Sigma Xi. *Res:* Cotton planting; application of electrostatics and servomechanisms to farm mechanization problems. *Mailing Add:* Dept Agr Eng NC State Col Raleigh NC 27650

BOWEN, J HARTLEY, JR, b Camden, NJ, Dec 26, 14; m 43; c 1. CHEMICAL ENGINEERING. *Educ:* Drexel Inst Technol, BS, 38. *Prof Exp:* Chief chemist, Nat Tube & Spool Co, Philadelphia, 38-39; jr mat engr, Naval Aircraft Factory, Philadelphia, 39-41, asst mat engr, 41-43, assoc chemist in charge chem testing & develop, 43, chemist, 43-45; sr chemist, Naval Air Exp Sta, 45-47, head chem eng div, 47-50, head high polymer div, 50-59, tech dir, Aero Mat Lab, Naval Air Eng Ctr, 59-67 & Naval Air Develop Ctr, 67-72; ENG CONSULT, 72- *Mem:* Am Chem Soc; Am Soc Testing & Mat; Am Inst Chemists. *Res:* Development of special finishes; gloss measurement of camouflage lacquer; methods of packaging of aircraft and parts with emphasis on silica gel for dehumidification; engineering applications of high polymer materials; nuclear radiation effects; research on materials for aerospace and hydrospace. *Mailing Add:* 226 Avondale Ave Haddonfield NJ 08033-2625

BOWEN, J(EWELL) RAY, b Duck Hill, Miss, Jan 9, 34; m 56; c 3. CHEMICAL ENGINEERING. *Educ:* Mass Inst Technol, BS, 56, MS, 57; Univ Calif, Berkeley, PhD(chem eng), 63. *Prof Exp:* From asst prof to prof chem eng, Univ Wis-Madison, 63-81, assoc vchancellor, 70-75, chmn dept, 71-73 & 78-81; DEAN COL ENG & PROF CHEM ENG, UNIV WASH, SEATTLE, 81- *Concurrent Pos:* NATO-NSF fel, Cambridge Univ, 62-63, NATO sr fel, Univ London, 68; Richard Merton prof, DF6, Univ Karlsruhe, 76-77. *Mem:* Combustion Inst; Am Phys Soc; Am Inst Chem Engrs; Am Soc Eng Educ. *Res:* Boundary layer techniques applied to heat and mass transfer; flows involving shock and detonation waves and chemical lasers. *Mailing Add:* Off of the Dean Univ Wash Col Eng FH10 Seattle WA 98195

BOWEN, JAMES MILTON, b Graham, Tex, Feb 1, 35; m 62; c 3. VIROLOGY, IMMUNOLOGY. *Educ:* Midwestern Univ, BS, 55; Ore State Univ, MS, 58, PhD(microbiol), 61. *Prof Exp:* Asst res biologist, Sterling-Winthrop Res Inst, 62-64; assoc prof, 64-73, assoc vpres res, 79-82, PROF VIROL, UNIV TEX M D ANDERSON HOSP & TUMOR INST, 73-, VPRES ACAD AFFAIRS, 82- *Concurrent Pos:* USPHS trainee, Univ Tex, M D Anderson Hosp & Tumor Inst, 61-62. *Mem:* Am Asn Cancer Res; Am Soc Microbiol. *Res:* Tumor immunology; oncogenic viruses; experimental immunotherapy. *Mailing Add:* Univ Tex M D Anderson Cancer Ctr 1515 Holcombe Blvd Houston TX 77030

BOWEN, JOHN METCALF, b Quincy, Mass, Mar 23, 33; m 56; c 2. TOXICOLOGY, CELL CULTURE. *Educ:* Univ Ga, DVM, 57; Cornell Univ, PhD(physiol), 60. *Prof Exp:* From asst prof to assoc prof physiol, Kans State Univ, 60-63; fel pharmacol, Emory Univ, 63; assoc prof, 63-69, PROF PHARMACOL, COL VET MED, UNIV GA, 69-, DIR VET MED EXP STA & ASSOC DEAN RES & GRAD AFFAIRS, 76- *Mem:* Soc Neurosci; Am Soc Pharmacol & Exp Therapeut; Am Vet Med Asn; Am Soc Vet Physiol & Pharmacol. *Res:* Development of in vitro excitable cell culture systems for evaluation of mechanisms of pharmacologic effects; toxicologic effects particulary of chemicals accumulating in the environment. *Mailing Add:* Dept Physiol & Pharmacol Univ Ga Col Vet Med Athens GA 30602

BOWEN, JOSHUA SHELTON, JR, b Parran, Md, Aug 29, 19; m 52; c 1. CHEMICAL ENGINEERING. *Educ:* West Md Col, AB, 39; Johns Hopkins Univ, BE, 41, MChE, 48, DrEng, 52. *Prof Exp:* Jr res engr asbestos-magnesia insulation, Keasbey & Mattison Co, 41-44; asst, Inst Co-op Res, Johns Hopkins Univ, 47-51; asst pyrotech compositions, Eng Exp Sta, Va, 51-55, res assoc, 55-57; chem engr eng develop div, Atlantic Res Corp, 57-68; res engr, Nat Air Pollution Control Admin, USPHS 68-73, chief process res sect, Div Process Control Eng, 68-73; chief combustion res br, Indust Environ Res Lab, US Environ Protection Agency, 73-84; RETIRED. *Mem:* Am Chem Soc; Air Pollution Control Asn; Am Inst Chem Eng. *Res:* Permeability and porosity characteristics of polymeric films; thermites and gasless heat sources; solid propellant development and evaluation; protechnic compositions; chemical agent dissemination; polymeric binders; combustion modification pollution control technology. *Mailing Add:* 207 Glasgow Rd Cary NC 27511

BOWEN, KENNETH ALAN, b Boston, Mass, June 14, 42; m 63; c 1. MATHEMATICAL LOGIC. *Educ:* Univ Ill, Urbana-Champaign, BS, 63, MS, 65, PhD(math), 68. *Prof Exp:* Asst prof math 68-77, PROF COMPUT & INFO SCI, SYRACUSE UNIV, 77- *Mem:* Am Math Soc; Math Asn Am; Asn Symbolic Logic. *Res:* Set theory and foundations of mathematics. *Mailing Add:* Dept Math Univ Sta PO Box 90 Syracuse NY 13210

BOWEN, KIT HANSEL, b Grenada, Miss, July 23, 48; m 76; c 1. PHYSICAL CHEMISTRY. *Educ:* Univ Miss, BS, 70; Harvard Univ, PhD, 77. *Prof Exp:* Fel, Harvard Univ, 78-80; PROF CHEM, JOHNS HOPKINS UNIV. *Concurrent Pos:* NSF fel, 78. *Mem:* Am Chem Soc; Am Phys Soc. *Res:* Negative ion photoelectron spectroscopy to study gas phase negatively-charged cluster ions. *Mailing Add:* Dept Chem Johns Hopkins Univ Baltimore MD 21218

BOWEN, LAWRENCE HOFFMAN, b Lynchburg, Va, Dec 20, 34. PHYSICAL CHEMISTRY. *Educ:* Va Mil Inst, BS, 56; Mass Inst Technol, PhD(phys chem), 61. *Prof Exp:* Asst chem, Mass Inst Technol, 56-58 & nuclear chem, 60-61; from asst prof to assoc prof, 61-70, PROF CHEM, NC STATE UNIV, 70- *Concurrent Pos:* Fac res grant tracer chem, 62-63; US AEC nuclear teaching grant, 67; NSF res grant Mossbauer spectros, 68-75, 82-88. *Honors & Awards:* Res Award, Sigma Xi, 70. *Mem:* Am Chem Soc; Am Phys Soc; Sigma Xi. *Res:* Mossbauer spectroscopy structure and bonding in iron oxides and related compounds; nuclear and radiochemistry applied to physicochemical problems; thermodynamics in solutions and solids. *Mailing Add:* Dept of Chem NC State Univ Raleigh NC 27695-8204

BOWEN, PAUL ROSS, b Catline, Ind, July 13, 02; m 44. BOTANY. *Educ:* DePauw Univ, AB, 25; Yale Univ, MA, 29, PhD(path), 31. *Prof Exp:* Teacher high sch, Wash, 25-27 & Mont, 27-28; asst bot, Yale Univ, 29-31, Sterling fel, 31-32; prof biol & head dept, High Point Col, 32-37; prof bot & zool, Beaver Col, 37-42; prof & head dept, Valley Forge Mil Jr Col, 42-46; prof biol, Del Valley Col, 46-76, cur Herbarium, 71-76; RETIRED. *Res:* Mycology; phytopathology of trees; taxonomy; plant materials for landscape design; gardening. *Mailing Add:* 9896 Bustleton Ave Philadelphia PA 19115

BOWEN, PAUL TYNER, b Macon, Ga; m 80; c 2. PHYSICOCHEMICAL WASTE WATER TREATMENT PROCESSES. *Educ:* Mercer Univ, BS,75; Clemson Univ, MS, 76; PhD(environ syst eng), 82. *Prof Exp:* ASST PROF CIVIL ENG, UNIV OKLA, 82- *Mem:* Am Soc Civil Engrs; Int Asn on Water Pollution Res & Control; Water Pollution Fedn; Am Water Works Asn; Am Chem Soc; Asn Environ Eng Prof. *Res:* Application of physicochemical water and wastewater treatment processes to sludge treatment and handling; removal of algal by-products from water; remediation of industrial wastes. *Mailing Add:* Metcalf & Eddy 1201 Peachtree St NE 400 Colony Rd Suite 1101 Atlanta GA 30361

BOWEN, PETER, internal medicine, human genetics; deceased, see previous edition for last biography

BOWEN, RAFAEL LEE, b Takoma Park, Md, Dec 27, 25; m 58; c 2. DENTISTRY, POLYMER CHEMISTRY. *Educ:* Univ Southern Calif, DDS, 53. *Hon Degrees:* DSc, Georgetown Univ, 87. *Prof Exp:* Pvt pract dent, San Diego, 53-55; res assoc, 56-70, assoc dir, Am Dent Asn Res Ont, 70-82; DIR, PAFFENBARBER RES CTR, NAT BUR STANDARDS, 83- *Honors & Awards:* Hollenback Mem Prize, Acad Oper Dent, 81; Mitch Nakayama Mem Award, Japanese Sect, Pierce Fauchard Acad, 84; Wilmer Souder Award, Dent Mat Group, Int Asn Dent Res. *Mem:* Am Dent Asn; Int Asn Dent Res; fel Am Col Dent; Sigma Xi. *Res:* Dental therapeutic materials and prevention of oral diseases; insoluble direct filling material approximating the properties and appearance of the anterior teeth; physical measurements; adhesion of dentin and enamel to dental composite restorations; glass ceramic inserts for dental composite restorations; protective tooth coatings; tooth desensitization. *Mailing Add:* Am Dent Asn Nat Bur Standards Gaithersburg MD 20899

BOWEN, RAY M, b Ft Worth, Tex, Mar 30, 36; m 58; c 2. MECHANICS, APPLIED MATHEMATICS. *Educ:* Tex A&M Univ, BS, 58, PhD(eng), 61; Calif Inst Technol, MS, 59. *Prof Exp:* Asst prof aeronaut eng, US Air Force Inst Technol, 61-64; res fel mech, Johns Hopkins Univ, 64-65; assoc prof eng mech, La State Univ, 65-67; from asst prof to prof mech, Rice Univ, 67-83, chmn dept, 72-77; dean eng, 83-89, PROF MECH, UNIV KY, 83- *Concurrent Pos:* Dep Asst Dir Eng, NSF, 90- *Mem:* Soc Natural Philos; fel Am Soc Mech Engrs; Am Soc Eng Educ. *Res:* Nonlinear continuum mechanics. *Mailing Add:* Dept Eng Univ Ky Lexington KY 40506-0046

BOWEN, RICHARD ELI, b Oskaloosa, Kans, Sept 27, 32; m 55; c 2. PARASITOLOGY. *Educ:* Univ Kans, BA, 54; Kans State Univ, MS, 60, DVM, 61; Univ Mass, PhD(microbiol), 65. *Prof Exp:* Instr vet sci, Univ Mass, 61-65; sr bacteriologist, Vet Res Div, 65-75, DEPT HEAD, PARASITOL RES, ELI LILLY & CO, 75- *Mem:* Am Vet Med Asn; Am Soc Microbiol; Am Asn Avian Path; Wildlife Dis Asn; Indust Vet Asn; Sigma Xi. *Res:* Avian leukosis; avian respiratory diseases; bacterial diseases of swine and wildlife; disinfectants; general diagnostic bacteriology. *Mailing Add:* Eli Lilly & Co Box 708 Dept GL 612 Greenfield IN 46140

BOWEN, RICHARD LEE, b Bunn, NC, July 2, 29; m 57; c 3. PHYSICAL GEOLOGY. *Educ:* Univ NC, AB, 49; Ind Univ, MA, 51; Univ Melbourne, PhD(geol), 60. *Prof Exp:* Teaching fel, Ind Univ, 50; geologist, US Geol Surv, 51; geologist-geophysicist, Standard Oil Co Calif, 52-54; consult, Am Overseas Petrol Corp, 57 & petrol explor, Gulf Oil Corp, Africa, 59-61; US Agency Int Develop vis prof, Univ Rio Grande do Sul, Brazil, 62-64; assoc prof, 64-68, chmn dept, 68-74, PROF GEOL, UNIV SOUTHERN MISS, 68- *Concurrent Pos:* NATO sci conf grants, 68 & 70; instnl rep Gulf Univs Res Corp & mem sci panel. *Mem:* AAAS; Am Asn Petrol Geol; Am Geol Soc; Geophys Union; Am Inst Urban & Regional Affairs; Sigma Xi. *Res:* Petroleum and glacial geology; paleoclimatology; stratigraphy; sedimentology; Gondwana studies; oceanic geochemistry; continental margin tectonics; material balances in geology. *Mailing Add:* Univ Southern Miss Box 8152 Hattiesburg MS 39406

BOWEN, RUTH JUSTICE, b Parkersburg, WVa, Jan 11, 42; m 72; c 1. INORGANIC CHEMISTRY, PHYSICAL CHEMISTRY. *Educ:* Glenville State Col, BA, 63; WVa Univ, PhD(chem), 68. *Prof Exp:* From asst prof inorg chem to assoc prof chem 68-78, PROF CHEM, CALIF STATE POLYTECH UNIV, POMONA, 78- *Mem:* Am Chem Soc; Nat Asn Sci Teachers. *Res:* Chemical education. *Mailing Add:* Dept CheM Calif State Polytech Univ Pomona CA 91768

BOWEN, SAMUEL PHILIP, b Council Bluffs, Iowa, Oct 12, 39; m 62; c 1. THEORETICAL PHYSICS, CONDENSED MATTER PHYSICS. *Educ:* Iowa State Univ, BS, 62; Cornell Univ, PhD(physics), 67. *Prof Exp:* Res physicist, Univ Calif, Berkeley, 67-69; assoc prof physics & asst dir, Va Ctr Coal & Energy, Va Polytech Inst & State Univ, 75-88, prof physics, 89; PROG LEADER, DIV EDUC PROGS, ARGONNE NAT LAB, 89- *Concurrent Pos:* Res adminr, Energy Exten, Va Coop Exten Serv; asst prof physics, Univ Wis, Madison, Admin off campus grad prog. *Mem:* AAAS; Am Phys Soc. *Res:* Dilute magnetic alloys; field theory; biophysics; solid state physics; energy conversion processes; energy conservation technologies; passive solar heating; economic analysis of energy problems; statistical properties of epilepsy. *Mailing Add:* Div Educ Progs Argonne Nat Lab 9700 S Cass Ave Argonne IL 60439

BOWEN, SARANE THOMPSON, b Des Moines, Iowa, Dec 11, 27; m 54. GENETICS. *Educ:* Iowa State Univ, BS, 48, MS, 51, PhD(genetics, physiol), 52. *Prof Exp:* Instr biol, San Francisco Col Women, 53-54; from instr to asst prof, 56-62, assoc prof, 63-68, PROF BIOL, SAN FRANCISCO STATE UNIV, 69- *Mem:* Genetics Soc Am; Am Soc Human Genetics; fel AAAS. *Res:* physiological ecology; genetics of Artemia. *Mailing Add:* Dept Biol San Francisco State Univ San Francisco CA 94132

BOWEN, SCOTT MICHAEL, b Santa Fe, NMex, Dec 7, 56. LANTHANOID ANALYSIS, BERYLLIUM ANALYSIS. *Educ:* Univ NMex, BS, 78, MS, 81, PhD(chem), 83. *Prof Exp:* Teaching asst chem, Univ NMex, 78-79, res asst, 79-83; postdoctoral scientist chem, Idaho Nat Eng Lab, 83, scientist, 84-85; res asst, 78, STAFF MEM CHEM, LOS ALAMOS NAT LAB, 85- *Mem:* Am Chem Soc; AAAS. *Res:* Isolation and analysis of radioactive elements from soil samples; analysis of scandium, yttrium, the lanthanides, and beryllium. *Mailing Add:* Los Alamos Nat Lab MS J514 Los Alamos NM 87545

BOWEN, STEPHEN HARTMAN, b Ann Arbor, Mich, Jan 13, 49; m 73; c 1. AQUATIC ECOLOGY, ICHTHYOLOGY. *Educ:* DePauw Univ, BA, 71; Ind Univ, MA, 73; Rhodes Univ, PhD(zool), 76. *Prof Exp:* Partic, Lake Valencia Proj, Univ Colo, 76-78; ASST PROF ZOOL, MICH TECHNOL UNIV, 78- *Concurrent Pos:* Consult fish ecol & cult, La Salle Found Natural Sci, Caracas, Venezuela, 77; consult subcontract environ assessment, Mich Technol Univ. *Mem:* Ecol Soc Am; AAAS; Am Fisheries Soc; Am Soc Naturalists. *Res:* Ecology of fishes with emphasis on trophic strategies, especially detritivory. *Mailing Add:* Rte 1 Box 120 Houghton MI 49931

BOWEN, THEODORE, b Evanston, Ill, Mar 19, 28; m 61; c 1. PHYSICS, MEDICAL PHYSICS. *Educ:* Univ Chicago, PhB, 47, SM, 50, PhD, 54. *Prof Exp:* From asst to assoc cosmic ray res, Univ Chicago, 53-55; res fel, Brazilian Centre Physics Res, 54-55; from res asst to res assoc physics, Princeton Univ, 56-62, res physicist, 62; from assoc prof to prof physics, 62-75, PROF PHYSICS & RADIOLOGY, UNIV ARIZ, 76- *Concurrent Pos:* Nat Acad Sci sr res assoc, Goddard Space Flight Ctr, NASA, 68 & 78-79. *Mem:* Fel Am phys Soc; fel AAAS; NY Acad Sci; Sigma Xi. *Res:* Detector development; cosmic ray search for rare or hypothetical new particles; properties of hypernuclei; diagnostic ultrasound. *Mailing Add:* 1118 E Fourth St Tucson AZ 85721

BOWEN, THOMAS EARLE, JR, b Decatur, Ala, Oct 16, 38; c 2. PHYSIOLOGY, BIOPHYSICS. *Educ:* Birmingham Southern Col, BS, 61; Univ Ala, cert med technol, 61, PhD(physiol), 68. *Prof Exp:* Chief technologist adult cardiac path lab, Med Col, Univ Ala, Birmingham, 61-63, chief physiol monitoring technologist, 63-65, from instr to assoc prof physiol, 67-72; asst vchancellor, 72-79, ASSOC PROF PHYSIOL & COORDR EDUC RESOURCES, UNIV TENN CTR HEALTH SCI, 72-, VCHANCELLOR MGT SERV, 79- *Concurrent Pos:* Porter Found vis lectr, Am Physiol Asn, 70-71. *Res:* Cardiovascular physiology; ventricular mechanics; cardiac muscle physiology; muscle mechanics. *Mailing Add:* 301A Univ Med Plaza Univ Kentucky Lexington KY 40536

BOWEN, VAUGHAN TABOR, b Buffalo, NY, Aug 23, 15; m 42; c 3. GEOCHEMISTRY, OCEANOGRAPHY. *Educ:* Yale Univ, BA, 37, PhD(zool), 48. *Prof Exp:* assoc biologist, Brookhaven Nat Lab, 48-52, biologist, 52-54; geochemist, 54-63, SR SCIENTIST, WOODS HOLE OCEANOG INST, 63- *Concurrent Pos:* Instr, Yale Univ, 48-52, lectr, 52-68; mem panel on radioactivity in marine environ, Nat Acad Sci Comt Oceanog, 65-71; mem panel on reference methods in marine radioactivity studies, Int Atomic Energy Agency, 68 & 72; mem, Work Group Low-Level Measurement Tech, Int Comt Radionuclide Metrol, 76-81; mem, Sr Adv Comt Environ Res, Ctr Energy & Environ Res, Univ PR, 77-81; mem adv group, Int Atomic Energy Agency, Vienna, 79; consult high-leval radioactive waste disposal, WHO, Bruges, Belgium, 80. *Mem:* Am Soc Nat. *Res:* Analytical and radiotracer studies of heavy metals in marine organisms and insects; radioisotope tracer studies in geochemistry; biogeochemistry of transuranium elements; oceanography; ultimate disposal of radioactive waste. fallout studies. *Mailing Add:* 652 Knox Rd Wayne PA 19087

BOWEN, WILLIAM DONALD, b Oshawa, Ont, Sept 2, 48; Can citizen; m 70; c 2. PINNIPED ECOLOGY. *Educ:* Univ Guelph, Ont, BSc, 71, MSc, 73; Univ BC, PhD(ecol), 78. *Prof Exp:* Res scientist ecol, Can Dept Fisheries & Oceans, 78-84; CHIEF, CAN DEPT FISHERIES & OCEANS, MARINE FISH DIV, BEDFORD INST OCEANOG, 84- *Mem:* Sigma Xi; Soc Marine Mammal. *Res:* Population dynamics of pinnipeds; ecological energetics of reproduction in seals; resource management of marine mammals and marine fishes including studies of competitive interactions between these groups. *Mailing Add:* Bedford Inst Oceanog PO Box 1006 Dartmouth NS B2Y 4A2 Can

BOWEN, WILLIAM H, b Enniscorthy, Ireland, Dec 11, 33; m 58; c 5. ORAL MICROBIOLOGY, IMMUNOLOGY. *Educ:* Nat Univ Ireland, BDentSurg, 55; Univ Rochester, MSc, 59; Univ London, PhD(dent), 65; Univ Ireland, DSc(dent), 74; FFDRCSI, 65; FDSRCS, 74. *Hon Degrees:* DDSc, Univ

Goteberg. *Prof Exp:* Pvt pract, 55-56; res fel dent, Eastman Dent Ctr, Rochester, NY, 56-59; Quinten Hogg fel, Royal Col Surgeons of Eng, 59-62, Nuffield Found fel, 62-65, sr res fel, 65-69, Sir Wilfred Fish fel, 69-73; ACTG CHIEF CARIES PREV & RES BR, NAT CARIES PROG, NAT INST DENT RES, 73- *Concurrent Pos:* Lectr, Univ London, 70; mem sci coun, Europ Orgn Caries Res, 70-73; C L Roberts Mem lectr, Univ Sheffield, 71; guest lectr, Fac Dent, Univ Oslo, 71. *Honors & Awards:* Colgate-Palmolive Prize, Brit Div, Int Asn Dent Res, 63; Int Dent Fedn Prize, 64; John Tomes Prize, Royal Col Surgeons Eng, 66. *Mem:* Europe Orgn Caries Res; Int Asn Dent Res; Am Asn Lab Animal Sci; Am Soc Zool; Int Dent Fedn. *Res:* Microbiology and immunology of dental caries; prevention of dental disease; oral biology of primates. *Mailing Add:* Dept Dent Res PO Box 611 601 Elmwood Ave Rochester NY 14642

BOWEN, WILLIAM R, b Iowa City, Iowa, Oct 15, 36; m 60; c 2. BOTANY, CELL BIOLOGY. *Educ:* Grinnell Col, BA, 60; Univ Iowa, MS & PhD(bot), 64. *Prof Exp:* Asst prof biol, Wis State Univ, Stevens Point, 64-65, Univ Ill, Chicago Circle, 65-66 & Western Ill Univ, 66-70; assoc prof, Ripon Col, 70-75; prof biol, Univ Ark, Little Rock, 75-90; PROF & HEAD BIOL DEPT, JACKSONVILLE STATE UNIV, AL, 90- *Concurrent Pos:* Res investr, Little Rock VEt Admin Med Ctr, 81-85; adj assoc prof path, Univ Ark Med Sci, 85- *Mem:* AAAS; Am Inst Biol Sci; Nat Asn Biol Teachers; Bot Soc Am; Phycol Soc Am; Electron Microscope Soc Am. *Res:* Developmental plant anatomy and ultrastructure of green algae; botanical teaching. *Mailing Add:* Dept Biol Jacksonville State Univ Jacksonville AL 36265

BOWEN, ZEDDIE PAUL, b Rockmart, Ga, Mar 29, 37; m 78; c 3. GEOLOGY, PALEONTOLOGY. *Educ:* Johns Hopkins Univ, AB, 58; Harvard Univ, MA, 60, PhD(geol), 63. *Prof Exp:* From asst prof to prof geol, Univ Rochester, 62-76, chmn, dept geol sci, 74-76; provost, Beloit Col, 76-81; DEAN, FAC ARTS & SCI, COL WILLIAM & MARY, 81- *Concurrent Pos:* NSF sci fac fel, 67-68. *Mem:* Geol Soc Am; Paleont Soc. *Res:* Invertebrate paleontology; evolution, paleoecology, paleobiogeography and taxonomy of the Brachiopoda; Silurian and Devonian biostratigraphy of eastern North America. *Mailing Add:* Provost/Dir Univ Richmond Richmond VA 23173

BOWER, AMY SUE, b Flemington, NJ, Nov 12, 59. PHYSICAL OCEANOGRAPHY, LAGRANGIAN OBSERVATIONS. *Educ:* Tufts Univ, BS, 81; Univ RI, PhD(oceanog), 88. *Prof Exp:* Res assoc, Univ RI, 88; postdoctoral scholar, 88-89, postdoctoral investr, 89-90, ASST SCIENTIST, WOODS HOLE OCEANOG INST, 90- *Mem:* Am Meteorol Soc; Am Geophys Union; Oceanog Soc. *Res:* Large-and meso-scale ocean circulation with emphasis on the structure and dynamics of western boundary currents and mixing processes at oceanic gyre boundaries. *Mailing Add:* Woods Hole Oceanog Inst Clark 3 Woods Hole MA 02543

BOWER, ANNETTE, b Rochester, NY, May 18, 33. NEUROENDOCRINOLOGY, HUMAN PHYSIOLOGY. *Educ:* Mt St Mary's Col, BS, 60; Creighton Univ, MS, 67; Univ Ariz, PhD(biol), 72. *Prof Exp:* Instr physiol, 67-69, from asst prof to assoc prof, 72-75, PROF BIOL SCI, MT ST MARY'S COL, CALIF, 78-, CHAIRPERSON DEPT, 72- *Concurrent Pos:* Res assoc, Univ Ariz, 72-76; dir human serv prog, Mt St Mary's Col, 76-, dir NSF women in sci grant, 77-, dir NSF undergrad res grant, 78-; mem rep, Calif Allied Health Articulation Comt, 77-; Strengthing Developing Inst Prog grant, HEW; dir, NIH-MBRS at MSMC, 87. *Honors & Awards:* NSF grants. *Mem:* AAAS; Am Zool Soc; Intra-Sci Res Found; Nat Coun Aging; Am Geront Soc; Sigma Xi. *Res:* Investigation of the behavioral and neuropharmacological function of specific brain hormones in relation to osmoregulation and hypertension; the control of melanocyte stimulating hormone release and B-endorphin control. *Mailing Add:* Mt St Mary's Col 12001 Chalon Rd Los Angeles CA 90049

BOWER, CHARLES ARTHUR, b Shawnee, Okla, Feb 17, 16; m 41; c 4. SOIL CHEMISTRY. *Educ:* Okla State Univ, BS, 36; Univ Wis, PhD(soils), 41. *Prof Exp:* Soil scientist, US Soil Conserv Serv, 36-38; asst, Univ Wis, 38-41; chemist, NC State Dept Agr, 41-42; asst prof soils, Iowa State Col, 42-45; prin soil scientist, US Salinity Lab, 45-60, dir, 60-72; soil scientist, Univ Hawaii, 72-81. *Concurrent Pos:* Prof soil sci, Univ Calif, Riverside, 65-72; consult, 81- *Mem:* Soil Sci Soc Am; fel Am Soc Agron. *Res:* Chemistry of salt-affected and tropical soils; mineral nutrition of plants; cation exchange; water quality; agricultural development of arid and tropical lands. *Mailing Add:* 1036 B Willow Tree Dr Las Vegas NV 89128

BOWER, DAVID ROY EUGENE, b Port Angeles, Wash, Sept 10, 34; m 55; c 4. FOREST BIOMETRY. *Educ:* Univ Idaho, BS, 58; Duke Univ, MF, 59. *Prof Exp:* Silviculturist, Southern Forest Exp Sta, US Forest Serv, 59-64, math statistician, 66-69; FOREST BIOMETRICIAN, WEYERHAEUSER CO, 69- *Mem:* Soc Am Foresters. *Res:* Biologic models to describe growth of trees and stands; experimental designs for testing effects of silvicultural treatments, effects of measurement error and biological variability on growth estimation. *Mailing Add:* 1114 Rock Creek Rd Hot Springs AR 71913

BOWER, FRANK H(UGO), b Rutherford, NJ, Oct 12, 21; m 47; c 2. CHARGE-COUPLED DEVICE DEVELOPMENT PROGRAMS. *Educ:* Lehigh Univ, BSEE, 43. *Prof Exp:* Prod engr, Western Elec Co, Pa, 46-51, head dept prod eng semiconductor devices, 51-55, head dept automation develop, 55-56; chief prod engr, Motorola Semiconductor Prod Div, Ariz, 56-57, head dept res & develop, 57-59, prod line mgr, 59-60; mgr res & develop contracts & eng admin, Sylvania Semiconductor Div, Mass, 60-63, mgr integrated circuit prod, 62-63; mgr microelectronics mfg, Semiconductor Div, Raytheon Co, Calif, 63-64; microelectronics consult, 64-72; sr staff consult, Integrated Circuit Eng Corp, 67-72; pres, Integrated Technol Assocs, 70-72; mkt mgr, CCD Operation, 72-80, technol planning & admin mgr, Fairchild Semiconductor Div, 80-84, progs mkt mgr, Fairchild-Weston CCD Imaging Div, Advan Res & Develop Lab, 84-86; RETIRED. *Mem:* Sr mem Inst Elec & Electronics Engrs. *Res:* Development, characterization and mass production of semiconductor devices, integrated circuits and microelectronic systems; solid-state electronics. *Mailing Add:* 4256 Manuela Ave Palo Alto CA 94306

BOWER, GEORGE MYRON, b Arbuckle, Calif, May 26, 25; m 51; c 3. ORGANIC POLYMER CHEMISTRY. *Educ:* Ore State Univ, BS, 50; Univ Ore, MS, 54, PhD(chem), 57. *Prof Exp:* SR RES CHEMIST, WESTINGHOUSE RES & DEVELOP CTR, 56- *Mem:* AAAS; Am Chem Soc. *Res:* Polyimides and other thermally stable polymers; composite technology; phenol-formaldehyde polymers and melamine formaldehyde polymers. *Mailing Add:* 2031 Sonny St Pittsburgh PA 15221

BOWER, J(OHN) R(OY), JR, b Helena, Mont, July 12, 15; m 42; c 3. CHEMICAL ENGINEERING, ORGANIC CHEMISTRY. *Educ:* Mont State Col, BS, 37; Univ Idaho, MS, 39; McGill Univ, PhD(chem), 43. *Prof Exp:* Sr res chemist, Am Cyanamid Co, 42-51, accountability dir, 51-53; accountability mgr, Phillips Petrol Co, 53-55, tech plant asst leader, 55-60, sr tech staff specialist, 60-66; sr tech staff specialist, Idaho Nuclear Corp, 66-71; sr tech staff specialist, Allied Chem corp, 71-81; RETIRED. *Mem:* AAAS; Am Chem Soc; Am Nuclear Soc; Am Inst Chem Engrs. *Res:* Radiochemical processing of spent reactor fuel; wood chemistry; high pressure; high temperature catalytic reactions; melamine; guanidines; cyanamide; cyanides; disposition and management of nuclear fuel processing wastes; nuclear fuel safeguards. *Mailing Add:* 2305 Calkins Idaho Falls ID 83402

BOWER, NATHAN WAYNE, b USA, 51. ARCHAEOMETRY, SPECTROSCOPY. *Educ:* Col Wooster, BA, 73; Ore State Univ, PhD(anal chem), 77. *Prof Exp:* PROF ANALYTICAL CHEM, COLO COL, 77- *Concurrent Pos:* Consult, Los Alamos Nat Lab, 83- *Mem:* Am Chem Soc; Soc Appl Spectroscopy; Sigma Xi. *Res:* Atomic spectroscopy and sampling statistics for analysis; precision of analysis and sampling precision in the archaeological, geochemical and environmental fields. *Mailing Add:* Dept Chem Colo Col Colorado Springs CO 80903

BOWER, OLIVER KENNETH, b Hindsboro, Ill, May 12, 02; m; c 2. MATHEMATICS. *Educ:* Univ Ill, AB, 24, AM, 27, PhD(math), 29. *Prof Exp:* Asst instr math, Univ Ill, 25-28, from instr to asst prof, 29-62; ASSOC PROF MATH, UNIV ARK, FAYETTEVILLE, 62- *Mem:* Am Math Soc; Sigma Xi. *Res:* Mathematics of the archery bow; applications of an abstract existence theorem to both differential and difference equations. *Mailing Add:* PO Box 814 Blind River ON P0R 1B0 Can

BOWER, RAYMOND KENNETH, b Kansas City, Mo, Oct 31, 27; m 59; c 2. MICROBIOLOGY, ANIMAL VIROLOGY. *Educ:* Univ Mo-Kansas City, BA, 49; Kans State Univ, MS, 51, PhD(microbiol), 54. *Prof Exp:* Res fel pediat, Sch Med, Univ Kans, 54-55; USPHS fel microbiol, Sch Med, Univ Okla, 57-59; from instr to asst prof, Sch Med, Univ Ark, 59-63; asst prof, Grad Res Inst, Baylor Univ, 63-65, Col Dent & Grad Div, 65-68; asst prof, 68-74, ASSOC PROF, DEPT BIOL SCI, UNIV ARK, FAYETTEVILLE, 74- *Concurrent Pos:* Consult microbiol, Div Pub Health Labs, Ark State Dept Pub Health, 75 & Vet Admin Hosp, 75-85. *Res:* Rous sarcoma virus-host cell relationships; genetic and immunologic interactions that underlie regression and immunity to Rous sarcoma virus-induced neoplasia in the chicken. *Mailing Add:* 1764 Applebury Pl Fayetteville AR 72701

BOWER, SUSAN MAE, b Sioux Lookout, Ont, Nov 5, 51; c 1. PROTOZOOLOGY, SHELLFISH DISEASES. *Educ:* Univ Guelph, BSc, 74, MSc, 76, PhD(parasitol), 80. *Prof Exp:* Res assoc cellular immunol, Mich State Univ, 80-81; res fel, Natural Sci & Eng Res Coun, 81-83, res scientist biol Rhizocephalan parasites, 83-84, RES SCIENTIST SHELLFISH & MARINE FISH DIS, DEPT FISHERIES & OCEANS, CAN GOVT, 84- *Concurrent Pos:* Res scientist cultured abalone dis, Pac Trident Mariculture Ltd, 84. *Mem:* Can Soc Zoologists; Nat Shellfisheries Asn; Aquacult Asn Can; Soc Invert Path. *Res:* Biology of parasites; etiology of diseases that are of concern to the fishery and mariculture of marine fish and shellfish on the west coast of Canada. *Mailing Add:* Pac Biol Sta Dept Fisheries & Oceans Nanaimo BC V9R 5K6 Can

BOWER, WILLIAM WALTER, b Hammond, Ind, Jan 9, 45. FLUID MECHANICS, AERODYNAMICS. *Educ:* Purdue Univ, BS, 67, MS, 69, PhD(mech eng), 71. *Prof Exp:* Sr engr aircraft propulsion, McDonnell Aircraft Co, 71-74, SCIENTIST AERODYN, MCDONNELL DOUGLAS RES LABS, 74- *Mem:* Am Inst Aeronaut & Astronaut; Am Soc Mech Engrs; Sigma Xi. *Res:* Computational fluid mechanics; the aerodynamics of vertical-takeoff-and-landing aircraft. *Mailing Add:* 847 Woodpoint Dr Apt A Chesterfield MO 63017-1751

BOWERFIND, EDGAR SIHLER, JR, b Cleveland, Ohio, May 7, 24; m 56; c 4. INTERNAL MEDICINE. *Educ:* Western Reserve Univ, MD, 49; Am Bd Internal Med, dipl, 58. *Prof Exp:* From intern to resident, Univ Hosps Cleveland, 50-56; demonstr, 56-58, from instr to sr instr, 58-65, ASST PROF MED, SCH MED, CASE WESTERN RESERVE UNIV, 65- *Concurrent Pos:* Consult, Vet Admin Hosp, Cleveland, 56-; asst secy, Citizens Comn Grad Med Educ, 63-64, secy, 64-66. *Mem:* Am Med Soc Alcoholism & Other Drug Dependencies. *Res:* Clinical medical education; chemical dependency. *Mailing Add:* Univ Hosp Cleveland 2074 Abington Rd Cleveland OH 44106

BOWERING, JEAN, b Yonkers, NY, Mar 16, 39; m 77. NUTRITION. *Educ:* Cornell Univ, BS, 60, MNS, 64; Univ Calif, Berkeley, PhD(nutrit), 69. *Prof Exp:* Res chemist, R T French Co, NY, 60-62; trainee, Inst of Nutrit for Cent Am & Panama, 64; res assoc nutrit, Cornell Univ, 64-66; instr dept pediat, Children's Hosp, DC & med sch, George Washington Univ, 69-70; asst prof human nutrit & foods, Cornell Univ, 70-77; ASSOC PROF HUMAN NUTRIT, SYRACUSE UNIV, 77- *Mem:* Sigma Xi; Am Inst Nutrit; Soc Nutrit Educ. *Res:* Iron requirements of humans; dietary & nutritional status of humans. *Mailing Add:* Dept Nutrit & Food Mgt Syracuse Univ Slocum Hall Syracuse NY 13244

BOWERMAN, EARL HARRY, b El Dorado, Ark, Jan 18, 42; c 2. HORTICULTURE, PLANT SCIENCE. *Educ:* Univ Ark, BSA, 65, MS, 72; Rutgers Univ, PhD(hort), 75. *Prof Exp:* Assoc county exten agent, Ark Agr Exten Serv, 65-69; res asst, Univ Ark, 69-72 & Rutgers Univ, 72-75; asst prof hort, Tex A&M Univ, 75-79; ASST DEAN AGR & HOME ECON & PROF PLANT SCI, CALIF STATE UNIV, FRESNO, CALIF, 79- *Mem:* Am Soc Hort Sci; Int Plant Propagators Soc; AAAS. *Res:* Plant reproductive histology. *Mailing Add:* Dept Plant Sci Calif State Univ Fresno CA 93740

BOWERMAN, ERNEST WILLIAM, chemistry; deceased, see previous edition for last biography

BOWERMAN, ROBERT FRANCIS, neurophysiology, behavioral biology, for more information see previous edition

BOWERS, C(HARLES) EDWARD, b Hanna, Wyo, Sept 3, 19; m 46; c 2. CIVIL ENGINEERING. *Educ:* Univ Wyo, BS, 42; Univ Minn, MS, 49. *Prof Exp:* Hydraul engr, David Taylor Model Basin, US Dept Navy, 42-44; res fel, St Anthony Falls Hydraul Lab, Univ Minn, 48-49; hydraul engr, Bur Reclamation, Colo, 49-51; res assoc, St Anthony Falls Hydraul Lab, Univ Minn, Minneapolis, 51-59, from assoc prof to prof hydraul, Univ Minn, 59-59, prof civil eng, 76-90; RETIRED. *Honors & Awards:* Collingwood Award, Am Soc Civil Engrs, 50. *Mem:* Fel Am Soc Civil Engrs; Am Soc Eng Educ; Int Asn Hydraul Res; Sigma Xi. *Res:* Hydrology; hydraulic engineering; storm runoff studies; spillways and related hydraulic structures. *Mailing Add:* 3385 N Oxford St Shoreview MN 55126

BOWERS, DARL EUGENE, b Fresno, Calif, Sept 23, 21; m 44; c 3. ZOOLOGY, ECOLOGY. *Educ:* Univ Calif, Berkeley, AB, 48, MA, 52, PhD(zool), 54. *Prof Exp:* PROF BIOL, MILLS COL, 54- *Concurrent Pos:* NSF grant, 61-66. *Mem:* AAAS; Ecol Soc Am; Am Inst Biol Soc; Sigma Xi. *Res:* Correlation of the distribution and characteristics of animal forms with environmental gradients and physical factor constellations; ornithological ecology; correlation of the distribution of amphipods with physical factors of beaches; salt marsh ecology. *Mailing Add:* 43 Vista Del Mar Ct Oakland CA 94611

BOWERS, FRANK DANA, b Fayetteville, Ark, Mar 21, 36; m 67. BRYOLOGY, PLANT TAXONOMY. *Educ:* Southwest Mo State Univ, BS, 66; Univ Tenn, Knoxville, MS, 68, PhD(bot), 72. *Prof Exp:* Asst cur herbarium, Dept Bot, Univ Tenn, 68-72, fel, 72-74, res assoc, 74-75; asst prof, 75-80, assoc prof, 80-86, PROF BOT, UNIV WIS-STEVENS POINT, 86- *Mem:* Bryological & Lichenological Soc Am; Am Inst Biol Sci; Soc Econ Bot; Am Soc Plant Taxonomists. *Res:* Moss flora of Mexico; mosses of Central America and Wisconsin; wildlife food plants; Heterotheca section Pityopsis, Compositae; edible and poisonous plants. *Mailing Add:* Dept Biol Univ Wis Stevens Point WI 54481

BOWERS, GEORGE HENRY, III, b Philadelphia, Pa, Oct 16, 23; m 49; c 4. ORGANIC CHEMISTRY. *Educ:* Washington & Lee Univ, BS, 44; Univ Pa, MS, 48, PhD(org chem), 51. *Prof Exp:* Asst instr chem, Univ Pa, 48-51; res chemist, E I du Pont de Nemours & Co, Inc, Chestnut Run, 51-74, res assoc, 74-85; RETIRED. *Mem:* Am Chem Soc; Sigma Xi; Inst Elec & Electronics Eng. *Res:* Synthetic elastomers; high temperature organic reactions; synthetic thermo plastic; electrical insulation and systems. *Mailing Add:* 612 Foulkstone Rd Wilmington DE 19803

BOWERS, JAMES CLARK, b South Bend, Ind, Feb 13, 34; m 56; c 3. COMPUTER AIDED DESIGN, ELECTRONIC ANALYSIS. *Educ:* Vanderbilt Univ, BE, 57; Univ Tenn, MS, 58; Wash Univ, ScD(automatic controls), 64. *Prof Exp:* Assoc res engr, Boeing Airplane Co, 58-59; dir eng tech staff, McDonnell Aircraft Corp, 59-65; PROF ELEC ENG, UNIV S FLA, 65- *Concurrent Pos:* Staff consult, Honeywell, Inc, 65-73; sr consult, US Army, 72-, Am Vulcan Corp, 73-76, US Air Force, 76-, US Navy, 77- *Mem:* Inst Elec & Electronics Engrs. *Res:* Computer aided design and analysis--the applications and program development; solid-state modeling; electronics and control systems analysis. *Mailing Add:* Univ SFla Col Eng Eng 118 Tampa FL 33620

BOWERS, JANE ANN (RAYMOND), b Fredonia, Kans, Aug 19, 40; m 63; c 2. FOOD SCIENCE. *Educ:* Kans State Univ, BS, 62, MS, 63, PhD(food, nutrit), 67. *Prof Exp:* Res assoc foods, Iowa State Univ, 63-64; from asst prof to assoc prof, 67-74, actg head dept foods & nutrit, 75-76, PROF FOODS, KANS STATE UNIV, 74-, HEAD, DEPT FOODS & NUTRIT, 76- *Mem:* Inst Food Technologists; Am Meat Sci Asn; Poultry Sci Asn; Coun Agr Sci & Technol; Am Home Econ Asn. *Res:* Chemical characteristics, nutrient and flavor stability in meat and poultry products and sensory analysis of food; over 70 research publications. *Mailing Add:* Dept Foods & Nutrit Kans State Univ Manhattan KS 66506

BOWERS, JOHN DALTON, b West Unity, Ohio, July 15, 15; m 41; c 2. AGRICULTURE. *Educ:* Ohio State Univ, BSc, 40. *Prof Exp:* Asst, Ohio State Univ, 40-42; lab dir, Borden, Morres & Ross, 42-43, supt milk dept, 46-47, lab dir, 47-51, lab dir qual control & res & develop, Mid-West Div, Borden Co, 51-70, dir prod develop, 70-72, dir qual control & res & develop, Northern Div, Borden, Inc, 72-80; RETIRED. *Concurrent Pos:* Mem bd dirs, Am Cultured Prod Inst, 67-70. *Mem:* Am Dairy Sci Asn. *Mailing Add:* 11327 Bob White Lane St Mary's OH 45885

BOWERS, JOHN E, b St Paul, Minn, June 7, 54; m 77; c 2. APPLIED PHYSICS. *Educ:* Univ Minn, BS, 76; Stanford Univ, MS, 78, PhD(appl physics), 81. *Prof Exp:* Scientist res, Honeywell Corp Mat Sci Ctr, 77-78; res asst, Ginzton Lab, Stanford Univ, 76-81, res assoc, 81- 82; mem tech staff, AT&T Bell Labs, 82-87; PROF ELEC ENG, UNIV CALIF, SANTA BARBARA, 87- *Concurrent Pos:* NSF fel, 76; mem adv bd, Int Bd Optoelectronics, 87-; guest ed, J Quantum Electronics, 89; bd dirs, Bach Camerata, 90-; chmn, Lasers & Electro-Optics Soc Subcomt on Semiconductor Lasers, 90- *Honors & Awards:* Presidential Young Investr Award, NSF, 87. *Mem:* Am Phys Soc; sr mem Inst Elec & Electronics Engrs; Optical Soc Am; Int Soc Optical Eng. *Res:* Ultra-fast optoelectronics, including high speed laser photodetectors and systems; fundamental limits to modulation and the femtosecond transport issues that limit the speed of response; granted 16 patents; author of 133 publications. *Mailing Add:* Dept Elec & Computer Eng Univ Calif Santa Barbara CA 93106

BOWERS, KLAUS DIETER, b Stettin, Ger, Dec 27, 29; nat US; m 64; c 2. ELECTRONICS TECHNOLOGY. *Educ:* Oxford Univ, BA, 50, MA, 53, PhD(physics), 53. *Prof Exp:* Res lectr, Christ Church, Oxford, 52-56; mem tech staff, Bell Labs, 56-59; various tech mgt positions, 59-71; managing dir component develop, Sandia Labs, 71-73, vpres components, 73-75; exec dir, AT&T Bell Labs, 75-79, vpres electronics technol, 79-90; RETIRED. *Mem:* Nat Acad Eng; fel Inst Elec & Electronic Engrs. *Res:* Quantum electronics; semiconductor devices and integrated circuits. *Mailing Add:* AT&T Bell Labs 600 Mountain Ave Murray Hill NJ 07974

BOWERS, LARRY DONALD, b York, Pa, Nov 29, 50; m 73; c 2. CLINICAL CHEMISTRY. *Educ:* Franklin & Marshall Col, BA, 72; Univ Ga, PhD(anal chem), 75; Am Bd Clin Chem, dipl. *Prof Exp:* Assoc fel, Univ Ore Health Sci Ctr, 75-77; from asst prof to assoc prof, 78-88, PROF CLIN CHEM, UNIV MINN, 88- *Concurrent Pos:* Vis prof, Cornell Univ. *Honors & Awards:* L S Palmer Award in Chromatography, 53. *Mem:* Am Chem Soc; Am Asn Clin Chem; Sigma Xi; Acad Clin Lab Physicians & Scientists; Am Soc Mass Spectrometry; AAAS. *Res:* Application of high performance liquid chromatography to bioanalytical chemistry; pharmacology of immunosuppressive drugs; toxicology. *Mailing Add:* Dept Lab Med & Path Mayo Bldg Box 198 Univ Minn Minneapolis MN 55455

BOWERS, MARION DEANE, b White Plains, NY, Mar 19, 52. CHEMICAL ECOLOGY, INSECT ECOLOGY. *Educ:* Smith Col, BA, 74; Univ Mass, Amherst, PhD(zool), 79. *Prof Exp:* Asst, Stanford Univ, 79-81; asst prof biol, Harvard Univ, 81-86, assoc prof, 86-; ASST PROF BIOL & CUR ENTOM MUS, UNIV COLO, BOULDER. *Concurrent Pos:* Vis prof chem, Colo State Univ, 86-87. *Mem:* Ecol Soc Am; Entom Soc Am; Lepidopterists Soc; Soc Study Evolution; Chem Energy Soc; AAAS; Phytochem Soc. *Res:* Interactions of insects and their host plants, particularly the role of plant chemistry; predation on insect populations; larval foraging ecology; evolution of unpalatability. *Mailing Add:* Dept EPO Biol Univ Colo Boulder CO 80309

BOWERS, MARY BLAIR, b Jackson, NC, Apr 2, 30. CELL BIOLOGY. *Educ:* Duke Univ, AB, 52, MA, 55; Harvard Univ, PhD(biol), 61. *Prof Exp:* Instr biol, Harvard Univ, 61-63; RES BIOLOGIST, NAT HEART, LUNG & BLOOD INST, 69- *Concurrent Pos:* NIH fel, Harvard Univ, 61-63; staff fel, Nat Heart Inst, 66-69. *Mem:* Electron Micros Soc Am (treas, 83-87; council, 89-91); Histochem Soc (council 88-92); Am Soc Cell Biol. *Res:* Cell ultrastructure as related to membrane function, phagocytosis and cellular transport of macromolecules. *Mailing Add:* Rm B1-22 Bldg Three NIH Bethesda MD 20892

BOWERS, MAYNARD C, b Battle Creek, Mich, Nov 5, 30; m 52, 70; c 4. BOTANY, BIOSYSTEMATICS. *Educ:* Albion Col, AB, 56; Univ Va, MEd, 60; Univ Colo, PhD(bot), 66. *Prof Exp:* Teacher pub sch, Mich, 56-57 & Fla, 57-59; asst prof biol, Towson State Col, 60-62; from asst prof to assoc prof, 66-75, actg head dept biol sci, 69-70, PROF BIOL, NORTHERN MICH UNIV, 75- *Concurrent Pos:* NSF Acad Year Inst, Univ Va, 59-60; researcher, Univ Turku, Finland, 75; Sigma Xi grant res; Peter White Scholar, N Mich Univ, 86-87; mem bd dir & regional dir, N Cent Region, Sigma Xi, 88-94. *Mem:* Am Bryol Soc; Am Inst Biol Sci; Bot Soc Am; Int Asn Plant Taxon. *Res:* Plant biosystematics; cytotaxonomy of bryophytes, especially the family Mniaceae; cytological effects of herbicides on non-target species; physiological effects of herbicides and their reversal using growth stimulators on non-target species. *Mailing Add:* Dept Biol Sci Northern Mich Univ Marquette MI 49855

BOWERS, MICHAEL THOMAS, b Moscow, Idaho, June 6, 39; m 64; c 2. CHEMICAL KINETICS, CHEMICAL PHYSICS. *Educ:* Gonzaga Univ, BS, 62; Univ Ill, Urbana, PhD(phys chem), 66. *Prof Exp:* From asst prof to assoc prof, 68-76, PROF PHYS CHEM, UNIV CALIF, SANTA BARBARA, 76- *Concurrent Pos:* Ed, Int J Mass Spectros & Ion Processes. *Mem:* Am Chem Soc; Am Phys Soc; Am Soc Mass Spectros. *Res:* Gas phase ion chemistry; ion cyclotron resonance spectroscopy. *Mailing Add:* Dept Chem Univ Calif Santa Barbara CA 93106

BOWERS, PAUL APPLEGATE, b Big Run, Pa, Aug 4, 11; m 45; c 3. GYNECOLOGY. *Educ:* Bucknell Univ, BS, 33; Jefferson Med Col, MD, 37; Am Bd Obstet & Gynec, dipl. *Prof Exp:* Intern, Hosp, Jefferson Med Col, 37-39; resident, Chicago Lying-In-Hosp, Ill, 39-42; prof, 46-81, EMER PROF OBSTET & GYNEC, JEFFERSON MED COL, 81- *Concurrent Pos:* Pvt pract, 46-; consult, Valley Forge Gen Hosp, Phoenixville, Pa, 49- *Mem:* AMA; Pan-Am Med Asn; Pan-Pac Surg Asn; Am Col Obstet & Gynec; Am Col Surgeons. *Res:* Physiologic obstetrics; disorder of the breast. *Mailing Add:* 1015 Chestnut St Suite 1321 Philadelphia PA 19107

BOWERS, PETER GEORGE, b Sussex, Eng, May 14, 37. PHYSICAL CHEMISTRY. *Educ:* Cambridge Univ, BA, 61; Univ BC, PhD(chem), 64. *Prof Exp:* NATO res fel, Univ Sheffield, 64-66; vis asst prof chem, Boston Univ, 67-68; from asst prof to assoc prof, 68-79, chmn dept, 79-85, PROF CHEM, SIMMONS COL, 79- *Concurrent Pos:* Vis assoc prof, Univ BC, 74-75, vis prof, Univ Ore, 82 & 89. *Mem:* Am Chem Soc. *Res:* Reaction kinetics; self-organizing processes; nucleation. *Mailing Add:* Dept Chem Simmons Col 300 Fenway Boston MA 02115

BOWERS, PHILLIP FREDERICK, b Huntington, Ind, Dec 14, 47; div. RADIO ASTRONOMY. *Educ:* Ind Univ, BS, 69; Univ Md, PhD(astron), 77. *Prof Exp:* Teaching & res asst astronr, Univ Md, 70-77, res assoc, 77; res assoc astronr, Nat Radio Astron Observ, 77-80; res astronr, 80-82, RES ASTRONR, SACHS & FREEMAN ASSOC, NAVAL RES LAB, 83- *Mem:* Am Astron Soc; Int Astron Union. *Res:* Galactic structure; interstellar medium; radio aspects of stars and circumstellar envelopes. *Mailing Add:* Naval Res Lab Code 4210B Washington DC 20375-5000

BOWERS, ROBERT CHARLES, b Benton Harbor, Mich, Nov 24, 37; m 63. HORTICULTURE, PLANT SCIENCE. *Educ:* Mich State Univ, BS, 59, MS, 60; Univ Ariz, PhD(tree physiol), 66. *Prof Exp:* Res assoc hort, Univ Ariz, 64-66; assoc scientist, 66-68, scientist, Calif, 68-78, RES HEAD, UPJOHN CO, 78- *Mem:* Am Soc Hort Sci. *Res:* Agricultural chemical research and development. *Mailing Add:* 355 Highland Ct Plainwell MI 49080

BOWERS, ROBERT CLARENCE, b Cumberland, Md, Dec 22, 22; m 46; c 3. PHYSICS. *Educ:* Univ Md, BS, 50. *Prof Exp:* RES PHYSICIST, NAVAL RES LAB, 50- *Concurrent Pos:* Ed, ASLE Trans, Am Soc Lubrication Engrs, 77- *Mem:* Am Soc Lubrication Engrs; Sigma Xi; NY Acad Sci. *Res:* Friction; wear; lubrication; explosives; dielectrics. *Mailing Add:* 3705 Deming Dr Suitland MD 20746

BOWERS, ROGER RAYMOND, b Augusta, Ga, Sept 15, 44; m 67; c 4. DEVELOPMENTAL ANATOMY, VERTEBRATE ZOOLOGY. *Educ:* Creighton Univ, BS, 66; Univ Nebr-Lincoln, MS, 68, PhD(zool), 71. *Prof Exp:* Res asst zool, Univ Nebr-Lincoln, 70-71; asst prof, Southern Ill Univ, Carbondale, 71-72; from asst prof to assoc prof, 72-80, PROF BIOL, CALIF STATE UNIV, LOS ANGELES, 80- *Concurrent Pos:* Prin investr NIH minority biomed support grants, 76-81, 84-, area grant, 86-88; vis prof microanat, Rühr Univ Med Sch, Bochum, WGer, 85; vis prof biol, Calif Polytech, San Luis Obispo, 91. *Mem:* Int Pigment Cell Soc; Pan Am Soc Pigment Cell Res; Sigma Xi; Europ Soc Pigment Cell Res; Poultry Sci Asn. *Res:* Ultrastructure and enzymology of pigmentation in fish, amphibia and fowl, fowl models for vitiligo; previous ultrastructure and enzymology of carrageenan-induced granulomas in lung and subcutaneous tissues in the rat. *Mailing Add:* Dept Biol Calif State Univ Los Angeles CA 90032

BOWERS, ROY ANDERSON, b Racine, Wis, May 11, 13; m 40; c 2. PHARMACY. *Educ:* Univ Wis, BS, 36, PhD(pharmaceut chem), 40. *Prof Exp:* Asst instr pharm, Univ Wis, 37-40; asst prof, Univ Toledo, 40-41; from asst prof to assoc prof, Univ Kans, 41-45; prof pharm & dean col pharm, Univ NMex, 45-51; dean, Col Pharm, 51-78, prof, 51-81, EMER PROF PHARM & EMER DEAN, COL PHARM, RUTGERS UNIV, NEW BRUNSWICK, 81- *Mem:* Am Pharmaceut Asn (vpres, 51-52); Am Inst Hist Pharm (exec secy, 78-); Am Asn Cols Pharm (pres, 63-64); NY Acad Sci. *Res:* Chemistry of fats and waxes; pharmaceutical formulation. *Mailing Add:* Col Pharm Rutgers Univ New Brunswick NJ 08903

BOWERS, WAYNE ALEXANDER, b Bilbao, Spain, Mar 1, 19; m 44; c 4. PHYSICS. *Educ:* Oberlin Col, AB, 38; Cornell Univ, PhD(theoret physics), 43. *Prof Exp:* Asst physics, Cornell Univ, 38-42, instr, 42-44; physicist, Manhattan Proj, Los Alamos, NMex, 44-46; res assoc, Mass Inst Technol, 46-47; from assoc prof to prof, 47-84, EMER PROF PHYSICS, UNIV NC, CHAPEL HILL, 84- *Concurrent Pos:* Vis prof, Mass Inst Technol, 68-69; NSF sci fac fel, 63-64; vis scholar, Cavendish Lab, Univ Cambridge, 77-78 & 84-85. *Honors & Awards:* Pegram Award, Am Phys Soc, 90. *Mem:* Am Phys Soc; Am Asn Physics Teachers. *Res:* Theoretical solid state physics. *Mailing Add:* Physics Dept Univ NC Chapel Hill NC 27599-3255

BOWERS, WILLIAM E, b Sunbury, Pa, Oct 29, 38; m 62; c 2. CELL BIOLOGY. *Educ:* Rockefeller Univ, PhD(life sci), 66. *Prof Exp:* From asst prof to assoc prof, Dept Biochem Cytol, Rockefeller Univ, 70-80; res immunologist, May Imogene Bassett Hosp, 80-90, actg sci dir, Med Res Inst, 85-88; PROF & CHMN DEPT MICROBIOL & IMMUNOL, UNIV SC, SCH MED, COLUMBIA, 88- *Mem:* Am Asn Immunol; Am Soc Biol Chemists; Cell Kinetics Soc; Sigma Xi. *Res:* Differentiation and antigen presentation of immune system dendritic cells. *Mailing Add:* Dept Microbiol & Immunol Univ SC Sch Med VA Campus Columbia SC 29208

BOWERS, WILLIAM SIGMOND, b Chicago, Ill, Dec 24, 35; m 58; c 5. INVERTEBRATE PHYSIOLOGY, BIOCHEMISTRY. *Educ:* Ind Univ, AB, 57; Purdue Univ, MS, 59, PhD(entom), 62. *Prof Exp:* Sr insect physiologist, Agr Res Ctr, US Dept Agr, 62-72; prof insect physiol, NY State Agr Exp Sta, Cornell Univ, 72-; PROF ENTOM, UNIV ARIZ. *Mem:* AAAS; Entom Soc Am; Am Chem Soc. *Res:* Insect metabolism and biochemistry; hormonal regulation of invertebrate metamorphosis, reproduction and diapause; isolation identification and synthesis of insect hormones, anti-hormones and pheromones; natural product chemistry; insect-plant interactions. *Mailing Add:* Univ Ariz Forbes Bldg Rm 410 Tucson AZ 85721

BOWERSOX, TODD WILSON, b Lewistown, Pa, Oct 4, 41; m 61; c 3. FORESTRY, SOLAR ENERGY. *Educ:* Pa State Univ, BS, 66, MS, 68, PhD(forest resources), 75. *Prof Exp:* Res asst forestry, 68-75, asst prof, 75-81, ASSOC PROF SILVICULT FORESTRY, PA STATE UNIV, 81- *Mem:* Forest Prod Res Soc; Soc Am Foresters; Sigma Xi. *Res:* Ecological and managerial aspects of culturing natural stands and plantations of hardwood species in eastern United States. *Mailing Add:* Dept Forestry Pa State Univ Main Campus University Park PA 16802

BOWES, GEORGE ERNEST, b London, Eng, May 22, 42; m 70; c 2. PLANT PHYSIOLOGY, PLANT BIOCHEMISTRY. *Educ:* Univ London, BSc Hons, 63, PhD(plant physiol), 67. *Hon Degrees:* MIBiol, Inst Biol, London, 68, CBiol, Inst Biol, London, 88. *Prof Exp:* Lectr biol, Regents St Polytech, London, 66-68; high sch teacher, Inner London Educ Authority, Eng, 68; vis res assoc agron, Univ Ill, Urbana, 68-71; res fel plant biol, Carnegie Inst, Stanford, Calif, 71-72; from asst prof to assoc prof, 72-84, PROF BOT, UNIV FLA, 84- *Concurrent Pos:* Prin investr grants, Fla Dept Natural Resources, 76-93, Ctr Environ Prog, Inst Food & Agr Sci, Fla, 77-78, Sci & Educ Admin, USDA, 78-93, Gas Res Inst, 81-86 & NSF, 84-88; courtesy app prof agron, Univ Fla, 84- *Mem:* Am Soc Plant Physiologists; Brit Soc Exp Biol; fel Inst Biol/Sigma Xi; Aquatic Plant Mgt Soc. *Res:* Regulation of photosynthetic and photorespiratory carbon metabolism, especially in aquatic plants; aquatic weed control; improving crop productivity; photosynthesis in isolated cells and chloroplasts; ribulose bisphosphate carboxylase-oxygenase kinetics; effects of elevated atmospheric CO_2 on photosynthesis and productivity; CO_2 concentrating mechanisms. *Mailing Add:* Dept Bot Univ Fla Gainesville FL 32611

BOWHILL, SIDNEY ALLAN, b Dover, Eng, Aug 6, 27; US citizen; m; c 2. AERONOMY. *Educ:* Cambridge Univ, BA, 48, MA, 50, PhD(physics), 54. *Prof Exp:* Res engr, Marconi's Wireless Tel Co, Eng, 53-55; assoc prof elec eng, Pa State Univ, 55-62; prof elec eng, Univ Ill, Urbana-Champaign, 62-86; prof & head, 86-90, ASSOC DEAN, DEPT ELEC ENG, UNIV LOWELL, 90- *Concurrent Pos:* Fulbright grant, 55-56; ed, Antennas & Propagation Group Trans, 59-62 & Radio Sci, 67-73; assoc ed, Radio Propagation, 64-67; Mem radio stand panel, Nat Acad Sci, 61-64, comt potential contamination & interference from space exp, 63-, polar res comt, 67-70, chmn panel on upper atmospheric physics, 67-70, mem rocket res comt, 67-72, comt on data interchange and data centers, 67-82, adv panel to World Data Ctr A, 69-75, geophys data panel, 75-76, earth sci panel for NSF postdoctoral fels, 68-69, comt on solar-terrestrial res, 69-79, ad hoc panel on Jicamarca Radio Observ, 69-83, chmn, 76-78; mem, US Nat Comt, Int Sci Radio Union, 64-75, 79-83, vice chmn, 85-87, chmn 88-90, chmn, US Comn 3, 64-67, mem, Int Coun Sci Unions comt on space res working group 4 on exp in upper atmosphere, 66-79, co-chmn panel on interactions between neutral & ionized parts of the atmosphere, 66-79, chmn, Sci Comn C, 79-84, Int Coun Sci Unions spec comt on solar-terrestrial physics, 67-, chmn working group 11 on neutral & ion chem, 68-73, chmn atmospheric physics progs steering comt, 74-80, chmn mid atmosphere prog steering comt, 77-; mem, Int Union Radio Sci solar-terrestrial physics comt, 68-72, chmn working group on Int Ref Ionosphere, 67-68, comt on space res, 69-75, vchmn Int Comn 3, 69-72, chmn, 72-75; pres, Aeronomy Corp, 69- *Honors & Awards:* Scostep Serv Award Int, 90. *Mem:* Nat Acad Eng; fel Inst Elec & Electronics Engrs; Am Soc Eng Educ; fel Am Geophys Union; fel AAAS. *Res:* Physics of ionosphere; radio propagation; rocket and satellite studies of upper atmosphere. *Mailing Add:* Dept Elec Eng Univ Lowell One University Ave Lowell MA 01854

BOWICK, MARK JOHN, b Rotogua, NZ, Nov 29, 57; NZ citizen; m 84; c 1. STRING THEORY, TOPOLOGICAL AND GEOMETRICAL ASPECTS OF THEORETICAL PHYSICS. *Educ:* Univ Canterbury, NZ, BSc, 77; Calif Inst Technol, MS, 79, PhD(physics), 83. *Prof Exp:* Res assoc, Yale Univ, 83-86; staff, Mass Inst Technol, 86-87; ASST PROF, SYRACUSE UNIV, 87- *Mem:* Am Phys Soc; Math Asn Am. *Res:* Theoretical particle physics and mathematical physics. *Mailing Add:* Physics Dept Syracuse Univ Syracuse NY 13244

BOWIE, ANDREW J(ACKSON), b McCool, Miss, July 6, 23; m 48; c 3. AGRICULTURAL ENGINEERING. *Educ:* Miss State Univ, BS, 49. *Prof Exp:* Consult engr, Commercial Soils Serv, Miss, 49-58; RES HYDRAULIC ENGR, SEDIMENTATION LAB, AGR RES SERV, USDA, 58- *Mem:* Am Soc Agr Engrs; Nat Soc Prof Engrs. *Res:* Compilation, tabulation and analysis of watershed runoff and sediment yield records; hydrologic and sediment data assembly; preparation of technical research reports and manuscripts for publication. *Mailing Add:* USDA Sedimentation Lab PO Box 1157 Oxford MS 38655

BOWIE, EDWARD JOHN WALTER, b Church Stretton, Eng, Mar 10, 25; m 48; c 4. HEMATOLOGY. *Educ:* Oxford Univ, MA, 50, BM & BCh, 52, DM, 81; Univ Minn, MS, 61. *Prof Exp:* From instr to assoc prof, 66-74, PROF MED & LAB MED, MAYO MED SCH, 74- *Concurrent Pos:* Consult internal med, Mayo Clin, 61- *Honors & Awards:* Trotter Medal, Univ Col Hosp, London, 51. *Mem:* AAAS; Am Soc Hemat; Am Soc Path; Int Soc Thrombosis & Haemostasis; Am Med Asn; Am Heart Asn; Central Soc Clin Res. *Res:* Hemostasis, bleeding and thrombosing disease. *Mailing Add:* Mayo Clin Rochester MN 55905

BOWIE, JAMES DWIGHT, b Sentinel, Okla, Oct 25, 41; m 68; c 2. RADIOLOGY, ULTRASOUND. *Educ:* Univ Okla, BA, 63, MD, 67. *Prof Exp:* Resident radiol, Univ Chicago, 71-73, resident & trainee, 73-74, instr & trainee, 74-75, from asst prof to assoc prof radiol, 75-79; assoc prof, 79-88, PROF RADIOL, DUKE UNIV, 88- *Mem:* Radiol Soc NAm; Am Inst Ultrasound Med; Soc Radiologists Ultrasound. *Res:* Clinical application of ultrasound in particular ultrasound of the pancreas, placenta and pelvic tumors; computer texture analysis; diagnostic cost-benefit analysis. *Mailing Add:* Dept Radiol Box 3808 Duke Univ Med Ctr Durham NC 27710

BOWIE, LEMUEL JAMES, b Vicksburg, Miss, Apr 23, 44; m 71; c 2. CLINICAL CHEMISTRY. *Educ:* Xavier Univ La, BS, 66; Johns Hopkins Univ, PhD(biochem), 72. *Prof Exp:* fel, Vet Admin Hosp, 72-73, staff biochemist clin chem, 73-74, asst chief clin chem lab serv, 74-79; asst prof chem & path, Univ Calif, San Diego, 74-79; DIR CLIN BIOCHEM, EVANSTON HOSP, ILL, 79-; ASSOC PROF CLIN PATH, MED SCH, NORTHWESTERN UNIV, 83- *Concurrent Pos:* Mem subcomt enzyme assay conditions, Nat Comt Clin Lab Stand, 76-78; mem adv comt minority biomed support prog, Univ Calif, San Diego, 77-78, mem med admis comt, Med Sch, 77-79; co-ed, Chicago Clin Chemist, 79-82; consult, Nat Heart, Lung & Blood Inst, NIH, 79-85; mem clin chem sect, Clin Chem & Hematol Devices Panel, Bur Med Devices, Food & Drug Admin, 82-85; mem bd dirs, Am Asn Clin Chem, 87-89, Am Bd Clin Chem, 84-; vpres, Am Bd Clin Chem, 89- *Mem:* Am Asn Clin Chem; Nat Acad Clin Biochem; Am Acad Clin Lab Physicians & Scientists; Int Soc Clin Enzymol; Am Chem Soc; Am Soc Biol Chem. *Res:* Diagnostic enzymology; applied spectroscopy; sickle cell anemia. *Mailing Add:* Evanston Hosp 2650 Ridge Ave Evanston IL 60201

BOWIE, OSCAR L, b Lisbon Falls, Maine, Nov 22, 21; m 44; c 3. APPLIED MATHEMATICS. *Educ:* Am Int Col, BA, 42. *Hon Degrees:* DSc, Am Int Col, 81. *Prof Exp:* Mathematician, Watertown Arsenal, 44-64, mathematician, 64-78, emer sr scientist, Army Mat & Mech Res Ctr, Watertown, 78-85; RETIRED. *Concurrent Pos:* Secy of Army fel, 66-67. *Mem:* Soc Indust & Appl Math. *Res:* Mathematical theory of elasticity, including plane problems; axially symmetric three dimensional problems; theory of plates and shells; fracture mechanics with the application of conformal mapping techniques to plane crack problems. *Mailing Add:* Star Rte 3 Box 317 Phippsburg ME 04562

BOWIE, WALTER C, b Kansas City, Kans, June 29, 25; m 54; c 3. VETERINARY PHYSIOLOGY. *Educ:* Kans State Col, DVM, 47; Cornell Univ, MS, 55, PhD(physiol), 60. *Prof Exp:* From instr to prof, Tuskegee Inst, 47-90, head, Dept Physiol & Pharmacol, 50-90, dean, Sch Vet Med, 80-90; RETIRED. *Concurrent Pos:* Prin investr, Mark L Morris Found Res Grant, 61-62; co-prin investr, NIH Res Grant, 61-; co-prin investr, US Army Res Grants, 61-62, prin investr, 64-71. *Mem:* Am Vet Med Asn; Am Soc Vet Physiologists & Pharmacologists. *Res:* Physiology, diagnosis and evaluation of cerebrospinal fluid of dogs; mechanisms of infection and immunity in listerosis; factors in virulence and immunogenicity in Listeria monocytogenes; cardiovascular and ruminant physiology; movements of the mitral valve. *Mailing Add:* 707 Patterson St Tuskegee AL 36088

BOWIN, CARL OTTO, b Los Angeles, Calif, Jan 30, 34; m 56; c 4. GEOLOGY, GEOPHYSICS. *Educ:* Calif Inst Technol, BS, 55; Northwestern Univ, MS, 57; Princeton Univ, PhD(geol), 60. *Prof Exp:* Instr geol, Princeton Univ, 60-61; res assoc, 61-63, asst scientist, 63-65, assoc scientist geophys, 65-78, SR SCIENTIST, WOODS HOLE OCEANOG INST, 78- *Concurrent Pos:* Consult, NSF, 77-78; adv, North Caribbean geol conf comt, Dominican Republic, 78-80; mem, Numerical Data adv bd, Nat Res Coun, Nat Acad Sci, 80-82; pres, Zarak Corp, 77- *Mem:* AAAS; Geol Soc Am; Am Geophys Union. *Res:* Development and use of geophysical data acquisition and processing systems utilizing digital computers; tectonics and gravity of earth and other planetary bodies. *Mailing Add:* Woods Hole Oceanog Inst Woods Hole MA 02543

BOWKER, ALBERT HOSMER, b Winchendon, Mass, Sept 8, 19; m 42; c 3. MATHEMATICAL STATISTICS. *Educ:* Mass Inst Technol, BS, 41; Columbia Univ, PhD(statist), 49. *Hon Degrees:* LHD, City Univ, New York, 71; LLD, Brandais, 72, Amtech, 80; ScD, Morehouse, 88. *Prof Exp:* Asst statistician, Mass Inst Technol, 41-43; asst dir statist res group, Columbia Univ, 43-45; from asst prof to prof statist, Stanford Univ, 47-53, head dept, 53-59, dean grad div, 58-63, asst to provost, 56-58; chancellor, City Univ New York, 63-71; chancellor, Univ Calif, Berkeley, 71-80; dean, Sch Pub Affairs, 80-83, exec vpres, Univ MD, 83-86; VPRES RES FOUND, CITY UNIV NY, 86- *Concurrent Pos:* Mem-at-large, Div Math, Nat Res Coun-Nat Acad Sci, 62-65; mem adv comt, Off Statist Standards Statist Policy, 63; mem Nat Adv Coun Exten & Continuing Educ; mem, Nat Drug Abuse Coun, 72 & Sloan Comn on Govt and Higher Educ; mem bd trustees, Univ Haifa, 71- & Bennington Col, 77-; asst secy postsecondary educ, US Dept Educ, 80-81. *Honors & Awards:* Frederick Douglass Award, NY Urban League, 69; Shewhart Award, Am Soc Quality Control, 78; Order De Leopold II, 80. *Mem:* Fel AAAS; Inst Math Statist (pres, 61-62; fel Am Soc Qual Control; Opers Res Soc Am; fel Am Statist Asn (vpres, 62-64, pres elec, 63, pres, 64). *Res:* Industrial statistics; multivariate analysis. *Mailing Add:* Univ Calif 1523 New Hampshire Ave NW Washington DC 20036

BOWKER, DAVID EDWIN, b Richmond, Va, Apr 19, 28; m 55; c 4. REMOTE SENSING, ATMOSPHERIC PROPAGATION. *Educ:* Antioch Col, BS, 53; Mass Inst Technol, PhD(geophysics), 60. *Prof Exp:* Sci staff physics, Mitre Corp, 60-64; SR RES SCIENTIST SPACE SCI, NASA-LANGLEY RES CTR, 64- *Concurrent Pos:* Assoc prof lectr, George Washington Univ, 68- *Mem:* Soc Photo-Optical Instrumentation Engrs; AAAS; Am Soc Photogrammetry & Remote Sensing. *Res:* Remote sensing of earth environment; lunar photography; space applications. *Mailing Add:* NASA-Langley Res Ctr MS 473 Hampton VA 23665-5225

BOWKER, RICHARD GEORGE, b Buffalo, NY, Dec 8, 46; m 68; c 2. PHYSIOLOGICAL ECOLOGY, VERTEBRATE ZOOLOGY. *Educ:* Cornell Univ, BS, 68; Northern Ariz Univ, MS, 75, PhD(zool), 78. *Prof Exp:* Res assoc zool, Nat Mus Kenya, 73-74; PROF ZOOL, ALMA COL, 77- *Concurrent Pos:* Vis prof, Univ Coimbra, Portugal, 83-; vis scientist, Los Alamos Nat Lab; Fulbright fel. *Honors & Awards:* Barlow Award; Charles A Dana Prof Biol. *Mem:* Am Soc Ichthyologists & Herpetologists; Ecol Soc Am; Europ Soc Herpetologists; Sigma Xi. *Res:* Thermoregulation and thermal ecology of lizards; population regulation in tropical anurans; metabolism of vertebrates. *Mailing Add:* Dept Biol Alma Col Alma MI 48801

BOWKLEY, HERBERT LOUIS, b Pittston, Pa, July 9, 21; m 47; c 3. INORGANIC CHEMISTRY. *Educ:* Univ Mich, BSc, 50; Mo Sch Mines, MS, 51; Pa State Univ, PhD(inorg chem), 55. *Prof Exp:* Jr res chemist, Titanium Div, Nat Lead Co, 51-52; sr res chemist, Columbia-Southern Chem Co, 55-56 & High Energy Fuels Div, Olin Mathieson Chem Co, 56-57; assoc prof chem, Sch Mines & Metall, Univ Mo, 57-60; group leader, Explosives & Mining Chem Dept, Am Cyanamid Co, 60-63; sr res chemist, Elkton Div, Thiokol Chem Co, 63-64 & Armour Agr Chem Co, 64-65; from assoc prof to prof chem, Appalachian State Univ, 69-85; RETIRED. *Mem:* Am Chem Soc. *Res:* Industrial inorganic chemicals; commercial explosives and rocket propellants research & development. *Mailing Add:* 104 Edora St Boone NC 28607

BOWLAND, JOHN PATTERSON, b Man, Can, Feb 10, 24; m 46; c 2. ANIMAL NUTRITION. *Educ:* Univ Man, BSA, 45; Wash State Col, MS, 47; Univ Wis, PhD(biochem, animal husb), 49. *Prof Exp:* Agr supvr, Exp Farm, Brandon, Man, 45-46; from asst prof to prof animal sci, 49-75, DEAN AGR & FORESTRY, UNIV ALTA, 75- & AT DEPT ANIMAL SCI. *Concurrent Pos:* Vis mem, Nat Inst Res Dairying, Eng, 59-60; guest prof, Animal Nutrit Inst, Swiss Fed Inst Technol, 68-69; mem comt animal nutrit, Nat Acad Sci-Nat Res Coun. *Honors & Awards:* Borden Award, Nutrit Soc Can, 66. *Mem:* Fel AAAS; Am Soc Animal Sci; Can Soc Animal Sci (secy-treas, 51-54); fel Agr Inst Can; Nutrit Soc Can; Sigma Xi. *Res:* Swine and rat nutrition, particularly energy, protein and amino acids. *Mailing Add:* Animal Sci Dept Univ Alta Edmonton AB T6G 2P5 Can

BOWLDEN, HENRY JAMES, b Hamilton, Ont, Apr 5, 25; nat US; m 48; c 2. USER FRIENDLY SYSTEMS, KNOWLEDGE BASE APPLICATIONS. *Educ:* McMaster Univ, BA, 46; Univ Ill, MS, 47, PhD(physics), 51. *Prof Exp:* Asst physics, Univ Ill, 46-50; from asst prof to assoc prof physics & astron, Wayne State Univ, 50-57; res physicist, Parma Res Ctr, Union Carbide Corp, 57-61, group leader, 61-63; adv scientist, Info Systs & Human Sci Dept, Westinghouse Res & Develop Ctr, 63-90; CONSULT, INTELLIGENT INTEGRATED SYSTS, 90- *Concurrent Pos:* Res assoc, Detroit Inst Cancer Res, 52-57; consult, US Naval Res Labs, 56-60; mem working group WG 2-1, Int Fedn Info Processing, 70-, secy, 70-73; tech activ area leader, Intelligent Integrated Systs, W Res & Develop, 82-90. *Mem:* Asn Comput Mach; Am Asn Artificial Intel; Am Defense Preparedness Asn. *Res:* Artifical intelligence; computer languages and programming systems; numerical methods; semiconductor theory; reaction kinetics. *Mailing Add:* 1156 Bucknell Dr Monroeville PA 15146

BOWLER, DAVID LIVINGSTONE, b New York, NY, June 7, 26; m 55; c 2. ELECTRONICS ENGINEERING. *Educ:* Bucknell Univ, BS, 48; Mass Inst Technol, MS, 51; Princeton Univ, MA, 58, PhD, 64. *Prof Exp:* Instr elec eng, Bucknell Univ, 48-49 & 53-54, asst prof, 54-55; engr, Hazeltine Corp, 51-53; from asst prof to prof elec eng, Swarthmore Col, 57-87, Howard N & Ada J Eavenson Prof Eng, 87-89; RETIRED. *Concurrent Pos:* Consult, JPM Co, Pa, 53-55 & Bartol Res Found, 58-64; vis fel, Queen's Univ, Belfast, 66-67; vis res assoc, Univ Fla, 71 & Univ Pa, 89-90; Fulbright lectr, Univ Col, Galway, Ireland, 84. *Mem:* Inst Elec & Electronics Engrs; Sigma Xi. *Res:* Electronic circuits; digital circuits. *Mailing Add:* 505 Yale Ave Swarthmore PA 19081

BOWLES, JEAN ALYCE, b Boston, Mass, June 15, 29. MICROBIOLOGY, BIOCHEMISTRY. *Educ:* Univ Ill, Urbana, BS, 56, MS, 58; Univ Colo Med Ctr, PhD(microbiol), 65. *Prof Exp:* Res assoc, Dept Biophys, Univ Colo Med Ctr, 57-60, res asst, Dept Microbiol, 60-61; from asst prof to prof, 65-90, EMER PROF BIOL, METROP STATE COL, 90- *Mem:* Am Soc Microbiol; Am Inst Biol Sci; Sigma Xi; Fedn Am Scientists; Am Acad Microbiol. *Res:* Kinetics of utilization and the regulation of the metabolism of organic compounds in mycobacterium tuberculosis; development of methods of obtaining axenic cultures of cyanobacteria more quickly and efficiently; bacterial contamination of mattresses. *Mailing Add:* 1045 Olive St Denver CO 80220

BOWLES, JOHN BEDELL, b Karuizawa, Japan, July 29, 33; US citizen; m 58; c 4. ZOOLOGY. *Educ:* Earlham Col, AB, 56; Univ Wash, MS, 63; Univ Kans, PhD(biol), 71. *Prof Exp:* Teacher high sch, Hawaii, 59-61; asst dir, Waikiki Aquarium, Honolulu, 61-62; asst prof biol, William Penn Col, 63-67; from asst prof to assoc prof, 69-75, PROF BIOL, CENT COL, IOWA, 75- *Mem:* AAAS; Am Soc Mammalogists; Sigma Xi. *Res:* Activity and reproductive patterns of bats, especially North and Central American; small mammal distribution in the midwest. *Mailing Add:* Dept Biol Cent Col Pella IA 50219

BOWLES, JOSEPH EDWARD, b Rocky Mount, Va, July 12, 29; m 50. SOIL MECHANICS, STRUCTURAL ENGINEERING. *Educ:* Univ Ala, Tuscaloosa, BSCE, 58; Ga Inst Technol, MSCE, 61. *Prof Exp:* Instr civil technol, Southern Tech Inst, Ga, 58-61; instr civil eng, Univ Wis, Madison, 61-63; from asst prof to prof civil eng, Bradley Univ, 63-80; OWNER, ENG COMPUT SOFTWARE, 80- *Mem:* Am Soc Civil Eng; Am Soc Eng Educ; Am Soc Testing & Mat; Am Concrete Inst. *Res:* Foundation engineering. *Mailing Add:* Eng Comput Software 1605 W Candletree Dr Peoria IL 61614

BOWLES, KENNETH LUDLAM, b Bronxville, NY, Feb 20, 29; m 54; c 3. PHYSICS. *Educ:* Cornell Univ, BEP, 51, MEE, 53, PhD(radio wave propagation), 55. *Prof Exp:* Physicist, Nat Bur Stand, Colo, 55-65; dir comput ctr, Univ Calif, San Diego, 67-72, prof appl physics & computer sci, 65-84; chmn, Telesoft, 81-89; CONSULT. *Concurrent Pos:* Mem US nat comt, Int Sci Radio Union, 62-65; mem Ada Bd, US Dept Defense, 84-91. *Honors & Awards:* Gold Medal, US Dept Com, 62. *Mem:* Inst Elec & Electronics Engrs; Asn Comput Machinery; Inst Elec & Electronics Engrs Computer Soc. *Res:* Plasma physics; physics and radar studies of the upper atmosphere, including measurements of ionization density, temperature, ionized constituents and nature of irregularities; radio and radar astronomy. *Mailing Add:* 13040 Caminito Mar Villa Del Mar CA 92014

BOWLES, LAWRENCE THOMPSON, b Mineola, NY, Sep 23, 31; m 65; c 3. MEDICAL ADMINISTRATION, THORACIC SURGERY. *Educ:* Duke Univ, AB, 53, MD, 57; NY Univ, MS, 64, PhD(higher educ), 71. *Prof Exp:* Intern surg, NY Univ, 57-58, res surg, 58-62; res surg, Triboro Hosp, NY, 62-64; attend surg, George Washington Hosp, Children's Hosp, Nat Cancer Inst, 65-68; vis prof surg, Univ Ceylon, 68-69; assoc prof surg, 73-76,PROF SURG, GEORGE WASHINGTON UNIV, 76- *Concurrent Pos:* vchmn Continuing Educ Comt, Am Col Surgeon, 79-80; mem, 82-86, chmn Nat Libr, Med Bd Regents, 84-86, mem, 83-, Nat Bd Med Examiners, chmn, 87-, mem VA spec Med Adv Group, 84-88, Coun Deans, Admin Assoc Am Med Cols, 85- *Mem:* Soc Thoracic Surgeons; Am Asn Thoracic Surg; Am Col Surgeons; Am Gerontol Soc; Am Acad Chest Physicians. *Res:* Surgery in the elderly; health promotion. *Mailing Add:* George Washington Univ Med Ctr 2300 Pennsylvania Ave NW Washington DC 20037

BOWLES, WILLIAM ALLEN, b Twin Falls, Idaho, Feb 20, 39. POLYMER RHEOLOGY, PHYSICAL-ORGANIC CHEMISTRY. *Educ:* Ariz State Univ, BS, 61, PhD(chem), 64. *Prof Exp:* Sr res chemist polymers, Monsanto Co, 64-66, group leader, 66-69, group supvr, 69-72, mgr com develop polymers, 72-76; mgr polymer develop, 76-80, mgr, 80-81, DIR, TECH CTR, NORTHERN PETROCHEM CO, 81-; tech dir, Tech Ctr, Inter North, Inc; DIR PROCESS RES QUANTUM CHEM, VSI DIV, MORRIS, IL. *Mem:* Am Chem Soc; Soc Plastics Engrs. *Res:* Structure-rheology relationships in polymers; synthesis of polyolefins; polyolefin catalyst development. *Mailing Add:* Process Res Ctr 8935 N Tabler Rd Morris IL 60450-9988

BOWLES, WILLIAM HOWARD, b McCoy, Colo, Aug 1, 36; m 61; c 3. BIOCHEMISTRY. *Educ:* La Sierra Col, BA, 58; Univ Ariz, MS, 60, PhD(biochem), 64; Baylor Col Dent, DDS, 77. *Prof Exp:* Asst prof chem, Southwestern Union Col, 63-65; from asst prof to assoc prof, 65-73, spec instr,

73-77, ASSOC PROF BIOCHEM, COL DENT, GRAD FAC, BAYLOR UNIV, WACO, TEX, 77- *Concurrent Pos:* Pvt pract dent, 77- *Mem:* Am Asn Dent Sch; Am Dent Asn; Int Asn Dent Res; fel Acad Gen Dent; Am Asn Dent Res; Sigma Xi. *Res:* Fat-soluble vitamins, lipids, hormones, dental amalgam, local anesthetics and dental plaque. *Mailing Add:* 426 Beverly Dr Richardson TX 75080

BOWLEY, DONOVAN ROBIN, b Waltham, Mass, May 20, 45. BOTANY, ENVIRONMENTAL SCIENCE. *Educ:* Eastern Nazarene Col, BA, 67; Boston Univ Grad Sch, MA, 70, PhD(bot), 78. *Prof Exp:* Instr sci, Col Basic Studies, Boston Univ, 75-77; asst surv analyst, Mass Dept Environ Qual Eng, 78, dir surface impoundments assessments, 78-80, sr planner, 80, coordr underground water resource protection prog, 80-82, PROG MGR, WATER SUPPLY CONTAMINATION CORRECTION, MASS DEPT ENVIRON PROTECTION, 82-, GIS PROG COORDR, 89- *Concurrent Pos:* Consult ecol, bot & hort, Helden Assocs, 74-77 & Town Eng Assocs, 76-78. *Mem:* NY Acad Sci; Nat Water Well Asn; Asn Eng Geologists. *Res:* Water supply protection; hydrology; mapping of groundwater contamination sources; lichenology, particularly New England, Arctic/Alpine and coastal regions; salt marsh develop in New England; development of environmental protection GIS databases. *Mailing Add:* St Aelreds House 34 Redwing Rd Wellesley MA 02181-3530

BOWLEY, WALLACE WILLIAM, b Burlington, Vt, May 16, 32; m 57; c 3. MECHANICAL ENGINEERING, ORTHODONTICS. *Educ:* Norwich Univ, BS, 55; Rensselaer Polytech Inst, MME, 57; Univ Conn, PhD(mech eng), 65. *Prof Exp:* Instr mech eng, Rensselaer Polytech Inst, 55-57; anal engr res, Pratt & Whitney Aircraft, 58-59; instr mech eng, Norwich Univ, 59-61; from instr to assoc prof mech eng, 61-74, assoc prof orthod, 69-74, PROF MECH ENG & ORTHOD, UNIV CONN, 74- *Concurrent Pos:* Consult, Pratt & Whitney Aircraft, 64-, Sippican Corp, 73-74, Johnson & Johnson Corp, 73-75, Trochoid Corp, 73-75, Aqua-Tech Corp, 74- & Hydra Corp, 75- *Mem:* Am Soc Mech Engrs; Am Soc Eng Educ; Am Inst Aeronaut & Astronaut. *Res:* Heat transfer and fluid mechanics; thermodynamics; biomechanics; medical applications; energy fields, including solar energy. *Mailing Add:* Dept Mech Eng Univ Conn Main Campus U-139 191 Auditorium Storrs CT 06268

BOWLING, ANN L, b Portland, Ore, June 1, 43; m 80; c 1. VETERINARY MEDICINE. *Educ:* Carleton Col, AB, 65; Univ Calif, Davis, PhD(genetics), 69. *Prof Exp:* Asst prof biol, Occidental Col, 69-73; asst res geneticist, 73-81, adj assoc prof, 81-88, ADJ PROF, UNIV CALIF, DAVIS, 88- *Mem:* AAAS; Sigma Xi; Am Genetics Soc; Int Soc Animal Genetics. *Res:* Immunogenetics; genetics in clinical veterinary medicine; cytogenetics. *Mailing Add:* Serol Lab Univ Calif Davis CA 95616

BOWLING, ARTHUR LEE, JR, b Roanoke, Va, May 14, 47; m 70; c 2. THEORETICAL HIGH ENERGY PHYSICS. *Educ:* Col of William & Mary, BS, 69; Univ Ill, MS, 70, PhD(physics), 74. *Prof Exp:* Asst vis prof physics, Swarthmore Col, 74-75; asst prof physics, Ohio State Univ, Mansfield, 75-77; asst prof, 77-80, ASSOC PROF PHYSICS, AGNES SCOTT COL, 81- *Concurrent Pos:* Vis res assoc, Ga Tech Res Inst, 84-85. *Res:* Weak and electromagnetic interactions. *Mailing Add:* Dept Physics & Astron Agnes Scott Col Decatur GA 30030

BOWLING, CLARENCE C, b Salem, Ark, Nov 12, 26; m; c 3. ENTOMOLOGY, AGRICULTURE. *Educ:* Univ Ark, Fayetteville, BS, 54, MS, 55. *Prof Exp:* Asst entomologist, Tex A&M, 55-56, assoc entomologist, 66-67, asst prof, 67-70, assoc prof entom, 70-83; RETIRED. *Concurrent Pos:* Secy & prog chmn, Rice Tech Working Group, 68-70, chmn, 70-72. *Mem:* Entom Soc Am. *Res:* Biology, ecology, economic importance and methods of control of insect pests of the rice plant. *Mailing Add:* 1750 Wooten Rd Beaumont TX 77707

BOWLING, DAVID IVAN, b Los Angeles, Calif, May 26, 40; m 64; c 2. PHYSICS. *Educ:* Univ Calif, Los Angeles, AB, 62; San Diego State Col, MS, 64; Okla State Univ, PhD(physics), 68. *Prof Exp:* Physicist, Naval Electronics Labs, 64; asst prof, 68-74, ASSOC PROF PHYSICS, CENT MO STATE UNIV, 74- *Mem:* Am Asn Physics Teachers. *Res:* Acoustics; macromolecules; impedance of aperture in plates; effect of shipboard noise on job performance; electrooptic effects in solutions of rigid macromolecules. *Mailing Add:* 25315 Tuckahoe Lane Spring TX 77373

BOWLING, FLOYD E, b Elizabethton, Tenn, Aug 28, 11; m 39; c 2. MATHEMATICS. *Educ:* Lincoln Mem Univ, AB, 34; Univ Iowa, MS, 38; Univ Tenn, EdD(math), 53. *Hon Degrees:* DSc, Tenn Wesleyan Col, 76. *Prof Exp:* Teacher & coach high sch, Tenn, 34-36; teacher math & physics, Lincoln Mem Univ, 37-42, prof math & head dept, 45-59; prof, head dept & dean students, 59-78, dir continuing & adult educ, 78-79, emer prof math, Tenn Wesleyan Col, 78-90; RETIRED. *Concurrent Pos:* Lectr, NSF Insts. *Honors & Awards:* Algernon Sydny Sullivan Award, 90. *Mem:* Emer mem, AAAS; emer mem, Math Asn Am; emer mem, Nat Coun Teachers Math; emer mem, Am Asn Univ Prof. *Res:* Development of specific kinds of graph paper as a teaching aid in math; development of a technique for teaching college math, with more meaning and understanding on the part of the student. *Mailing Add:* Dept Math Tenn Wesleyan Col Athens TN 37303

BOWLING, FRANKLIN LEE, b Guyman, Okla, Nov 2, 09; m 52; c 3. MEDICINE. *Educ:* Univ Colo, PhC & BS, 33, MD, 46; Univ Denver, MS, 47; US Air Force Sch Aerospace Med, cert, 47; Harvard Univ, MPH, 52; Mass Inst Technol, cert, 55; US Missile Test Ctr, Cape Kennedy, cert, 60; Univ Mich, cert, 67; Am Bd Prev Med, dipl & cert aerospace med. *Prof Exp:* Intern, Med Ctr, Univ Colo, 46-47; intern Med Corps, US Air Force, 48-67, dir base med serv, Hq, 15th Air Force, Strategic Air Command, Colorado Springs, 47-49, chief med output serv, March AFB, Calif, 49-50, hosp comdr, Saudi Arabia, 50-51, chief flight surgeon, Bolling AFB, DC, 51, chief prev med, Off Inspector Gen, Hq, 52-55, head dept prev med, US Air Force Sch Aviation Med, 55-58, chief prev med, Pac Air Forces Hq, 58-61, chief mil pub health & occup med, Off Surgeon Gen, 62-66, chief epidemiol br, Armed Forces Inst Path, 66-67; med dir, Colo State Dept Social Serv, 67-71, med serv policy dir, 71-74; RETIRED. *Concurrent Pos:* Rep, Armed Forces Epidemiol Bd, 61-66; consult, Am Bd Prev Med, 62-66; pres, Coun Fed Med Dir Occup Health, 65-66; consult to Surgeon Gen, Dept Air Force, 66-67; mem, Permanent Comn & Int Asn Occup Health, 66-; mem, US Pharm Conv, 75-; chmn comt toxicol, Nat Res Coun. *Honors & Awards:* Wisdom Award of Honor, 70. *Mem:* Fel Am Pub Health Asn; fel Am Col Prev Med; fel Indust Med Asn; fel Royal Soc Health; Int Health Soc US (pres, 67-68, secy-treas, 74-83). *Res:* Anatomy, physiology and pharmacology of cystic innervation, Mus norvegicus; epidemiological factors in motor vehicle accidents for Army, Navy and US Air Force. *Mailing Add:* 1001 E Oxford Lane Englewood CO 80110

BOWLING, LLOYD SPENCER, SR, b Newport, Md, Mar 29, 30; m 55; c 2. AUDIOLOGY, SPEECH PATHOLOGY. *Educ:* Univ Md, BS, 43, MA, 57, EdD(human develop), 64. *Prof Exp:* Clin audiologist, West Side Vet Admin Hosp, Chicago, 57-60; clin audiologist, Vet Admin Hosp, Washington, DC, 60-63, supvr audiol clin, 63-66, assoc chief audiol & speech path clin, 66-67; lectr, Sch Med, Georgetown Univ, 67; PROF SPEECH & CHMN DEPT SPEECH & DRAMA, GEORGE WASHINGTON UNIV, 67- *Concurrent Pos:* Consult, Parmly Hearing Inst, Ill, 60-61, Fairfax County Health Dept, Va & Vet Admin Hosp, Washington, DC, 67-; chmn, DC and Brasilia, Brazil Partners Rehab & Educ, 73- *Mem:* Am Speech & Hearing Asn. *Res:* Methods of aural rehabilitation; clinical auditory tests; intelligibility. *Mailing Add:* Dept Speech George Washington Univ Funger Hall 2201 G St NW Washington DC 20052

BOWLING, ROBERT EDWARD, b Pauls Valley, Okla, Aug 9, 26; m 55; c 5. MICROBIOLOGY. *Educ:* Univ Okla, BS, 48, MS, 50, PhD, 57. *Prof Exp:* Asst bacteriologist, Okla State Health Dept, 50-51; asst, Med Ctr, Univ Okla, 51-52, 55-57; from instr to assoc prof microbiol, 57-77, PROF MICROBIOL & IMMUNOL, MED CTR, UNIV ARK, 77-, ASST DEAN COL MED, 73- *Mem:* Am Soc Microbiol; Brit Soc Gen Microbiol; Soc Cryobiol; Sigma Xi. *Res:* Antibiotic resistance; hypothermia; microbial antagonism; ecology; antibody production; pathogenesis and infection. *Mailing Add:* 4301 W Markham Slot 551 Little Rock AR 72205

BOWLING, SUE ANN, b Bridgeport, Conn, Feb 26, 41. METEOROLOGY. *Educ:* Harvard Univ, AB, 63; Univ Alaska, MS, 67, PhD(geophys), 70. *Prof Exp:* Partic, Advan Study Prog, Nat Ctr Atmospheric Res, 71-72; ASST PROF, GEOPHYS INST, UNIV ALASKA, 70- *Mem:* Am Meteorol Soc. *Res:* Local meteorological variations, climatic change, Alaskan climate, paleoclimatology. *Mailing Add:* Geophys Inst Univ Alaska Fairbanks AK 99775-0800

BOWMAN, ALLEN LEE, b Washington, DC, Jan 19, 31; m 52; c 4. PHYSICAL CHEMISTRY, CRYSTALLOGRAPHY. *Educ:* Col William & Mary, BS, 51; Iowa State Univ, PhD(chem), 58. *Prof Exp:* Jr chemist, Electrodeposition Sect, Nat Bur Stand, 51; res asst phys chem, Iowa State Univ, 55-58; staff mem phys chem, Los Alamos Sci Lab, 58-77; staff mem detonation physics, 77-87, SECT LEADER, LOS ALAMOS NAT LAB, 87- *Concurrent Pos:* physics assessment coord, 86- *Mem:* Am Chem Soc; Am Crystallog Asn; NMex Geol Soc. *Res:* Gas-cooled reactor safety; crystal structures; detonation physics. *Mailing Add:* Ten Encino Los Alamos NM 87544

BOWMAN, BARBARA HYDE, b Mineral Wells, Tex, Aug 5, 30. GENETICS. *Educ:* Baylor Univ, BS, 51; Univ Tex, MA, 55, PhD(genetics), 59. *Prof Exp:* Bacteriologist, Tex State Dept Health, 54-55; technician human genetics lab, Dept Zool, Univ Tex, 55-59, res scientist, Genetics Found, 59-64; mem staff, Rockefeller Univ, 64-65, asst prof, 65-67; PROF HUMAN GENETICS & CHMN DEPT, MED BR, UNIV TEX, 67- *Concurrent Pos:* Prof & Chmn Dept Cellular & Struct Biol, Health Sci Ctr, Univ Tex. *Mem:* Harvey Soc; Am Soc Human Genetics; Am Soc Biol Chemists. *Res:* Biochemical genetics of humans; genetic control of protein structure and basic defects occurring in inherited diseases. *Mailing Add:* Dept Anat Univ Tex Health Sci Ctr 7703 Floyd Curl Dr San Antonio TX 78284

BOWMAN, BARRY J, b Pontiac, Mich, May 15, 46; m 71. MEMBRANE BIOLOGY, ION TRANSPORT. *Educ:* Univ Wis-Milwaukee, BA, 68; Univ Mich, PhD(biol), 75. *Prof Exp:* Fel biochem, Yale Univ Sch Med, 75-79; asst prof, 79-84, ASSOC PROF BIOL, UNIV CALIF, SANTA CRUZ, 84- *Mem:* Am Soc Biol Chem. *Res:* Proton-pumping adenosine triphosphatase of plasma membranes and vacuoles in order to understand the role of these enzymes in active transport. *Mailing Add:* Dept Biol Univ Calif Sinsheimer Labs Santa Cruz CA 95064

BOWMAN, BERNARD ULYSSES, JR, b Atlanta, Ga, Oct 1, 26; m 53; c 4. MEDICAL MICROBIOLOGY. *Educ:* Piedmont Col, BS, 50; Emory Univ, MS, 57; Univ Okla, PhD, 63. *Prof Exp:* Instr bact & biol, Ga State Col, 58-60; guest lectr virol, Okla State Univ, 62; instr microbiol, Med Ctr, Univ Okla, 63-64; asst prof path, 64-66, from asst prof to assoc prof med microbiol, 66-72, PROF MED MICROBIOL, COL MED, OHIO STATE UNIV, 72- *Concurrent Pos:* NIH trainee, Univ Okla, 63-64; Am Thoracic Soc grant, Ohio State Univ, 67-68, Nat Tuberc Asn grant, 67-68, NIH career develop award, 68-72 & grant, 69-72; Merck grant, 79 & Bremer grant, 83-84. *Mem:* Am Soc Microbiol; Am Thoracic Soc. *Res:* Etiology of sarcoidosis; lipids of mycobacteria and mycobacteriophages; mechanisms of K cell action; mechanism of action of isonicotinic acid hydradize (INH). *Mailing Add:* Dept Med Microbiol Ohio State Univ Columbus OH 43210

BOWMAN, BRUCE T, b Oshawa, Ont, Jan 21, 42; m; c 2. SOIL CHEMISTRY, PHYSICAL CHEMISTRY. *Educ:* Ont Agr Col, Univ Toronto, BSA, 64; Univ Guelph, MSc, 66; Univ Minn, St Paul, PhD(soil sci), 69. *Prof Exp:* RES SCIENTIST SOIL CHEM, RES INST, CAN DEPT AGR, 69- *Mem:* Am Chem Soc; Am Soc Agron. *Res:* Use of gas-liquid chromatography, high performance liquid chromatography in studies of environmental behavior of insecticides and herbicides; absorption-desorption isotherms,; pesticide leaching-mobility studies; soil-watering persistance-stability; solubility, octanol-water partitioning coefficients; soil tillage-erosion studies; environmental quality studies. *Mailing Add:* London Res Ctr Agr Can 1400 Western Rd London ON N6G 2V4 Can

BOWMAN, C(LEMENT) W(ILLIS), b Toronto, Ont, Jan 7, 30; m 54; c 2. CHEMICAL ENGINEERING. *Educ:* Univ Toronto, BASc, 52, MASc, 58, PhD(chem eng), 61. *Prof Exp:* Lab demonstr chem eng, Univ Toronto, 52-53; tech asst nylon processing, Can Industs, Ltd & Du Pont of Can, 53-57; res engr, Du Pont of Can, 57-58; res chemist, Imp Oil Enterprises, Ltd, 60-62; res mgr tar sand res, Cities Serv Athabasca & Syncrude Can, Ltd, 63-69; chem res mgr, petrochem res, Imp Oil Enterprises, Ltd, 69-71, petrol res mgr, 71-75; chmn, Alta Oil Sands Technol & Res Authority, 75-84; vpres, Esso Petrol Can, Res Dept, 84-86; PRES, ALTA RES COUN, 87- *Concurrent Pos:* Panelist, World Petrol Cong, 67; prog chmn, Can Chem Eng Conf, 69; bd mem coun, Res Coun Alta, 78-; sci secy, UN Inst Training & Res Conf, 79; mem, Fed Govt Adv Group Hydrogen Opportunities, 85; dir, Inst Chem & Technol, 85-86; bd mem, Can Asn World Petrol Congress, 87-88; coun mem, Nat Res Coun Can, 87-89; chmn, Can Res Mgt Asn, 89. *Honors & Awards:* K A Clark Distinguished Serv Award, 89. *Mem:* Asn Prof Engrs; Can Soc Chem Eng (vpres, 73-74, pres, 74-75); fel Chem Inst Can (pres-elect, 81-82, pres, 82-83); fel Can Acad Eng. *Res:* Petrochemical research, particularly olefins to plastics; oil-water-solid separation studies; theoretical analysis of mass tansfer from drops and bubbles; oil sand and heavy oil recovery and upgrading. *Mailing Add:* Alta Res Coun PO Box 8330 Sta F Edmonton AB T6H 5X2 Can

BOWMAN, CARLOS MORALES, b Mexico, DF, Mex, Mar 4, 35; nat US; m 55; c 7. INFORMATION SCIENCE, COMPUTER SCIENCE. *Educ:* Univ Utah, BA, 54, PhD(chem), 57. *Prof Exp:* Chemist, 57-61, info retrieval analyst, 61-64, group leader, 64-67, asst dir, 67-68, res dir, Comput Res Lab, 68-80, mgr health & environ issues, 80-86, PROJ DIR, TECH INFO, DOW CHEM CO, 86- *Concurrent Pos:* Mem comt chem info, 67-74, chmn, 70-74 & toxicol info proj comt, Nat Acad Sci-Nat Res Coun, 88-90, chmn, 90-92; consult, NSF, 69-73; chmn comt div activities, Am Chem Soc, 75-76 & 82-83; chmn, Gordon Res Conf & Sci Info Prob Res, 76; Am Chem Soc, Comt Environ Improv, 84-87, Soc Comt Chem Abstracts, 86-89, Task Force Numerical Data, 87- *Mem:* AAAS; Am Chem Soc; Sigma Xi (pres, 64); Chem Notation Asn (pres, 71); Am Soc Info Sci. *Res:* Information retrieval; chemical notation; computer applications. *Mailing Add:* 1414 Timber Dr Midland MI 48640

BOWMAN, CHARLES D, b Roanoke, Va, May 23, 35; m 56; c 2. NUCLEAR PHYSICS. *Educ:* Va Polytech Inst, BS, 56; Duke Univ, MA, 58, PhD(nuclear physics), 61. *Prof Exp:* Sr physicist, Lawrence Radiation Lab, Univ Calif, 61-68, res prog mgr, Livermore Electron Linac, 68-72; chief, Nuclear Sci Div, Nat Bur Standards, 72-77, leader, Neutron Measurement & Res Group, 77-82; assoc div leader basic physics, 82-85, STAFF MEM, LOS ALAMOS NAT LAB, 85- *Concurrent Pos:* Mem nuclear cross sect adv comt, US AEC, 69-82. *Mem:* Fel Am Phys Soc; Am Nuclear Soc. *Res:* Neutron physics; development and application of techniques for partial neutron cross section studies on very heavy nuclei; resonance neutron spectrometry on medium weight nuclei; photodisintegration of heavy nuclei; inelastic scattering of neutrons; parity and time reversal invariance violation. *Mailing Add:* 1045 Los Puealos Los Alamos NM 87544

BOWMAN, CRAIG T, b Sept 19, 39. MECHANICAL ENGINEERING. *Educ:* Carnegie Inst Technol, BS, 61; Princeton Univ, MA, 64, PhD(aerospace eng), 66. *Prof Exp:* Sr res scientist, United Technologies Res Ctr, 66-76; assoc prof, 76-81, PROF MECH ENG, STANFORD UNIV, 81- *Mem:* Combustion Inst (secy, 88-); Am Soc Mech Engrs; Am Inst Aeronaut & Astronaut. *Res:* Combustion; air pollution; propulsion systems. *Mailing Add:* Mech Eng Dept High Temperature Gasdynamics Lab MC 3032 Stanford CA 94305

BOWMAN, DAVID F(RANCIS), b Erie, Pa, Mar 26, 20; m 45; c 2. ELECTRONICS ENGINEERING. *Educ:* Ohio State Univ, BEE, 42. *Prof Exp:* Engr, Hazeltine Electronics Corp, 42-46; engr, Airborne Instruments Lab, 46-48, asst supv engr, 48-52; chief engr, Develop Eng Corp, 52-56; engr-in-charge r-f lab, ITE Circuit Breaker Co, 56-62; unit mgr design & develop eng, Missile & Surface Radar Div, 62-83; ANTENNA CONSULT, RCA CORP, 83- *Concurrent Pos:* Consult, Air Force Command & Navig Lab, Wright-Patterson AFB, 50-52; consult, RCA Corp, 83-, NASA, 87- *Mem:* Inst Elec & Electronics Engrs. *Res:* Advanced development of antennas and related devices. *Mailing Add:* 51 Wilshire Dr Williamstown MA 01267

BOWMAN, DONALD EDWIN, b Orrville, Ohio, Nov 12, 08; m 34; c 2. BIOCHEMISTRY. *Educ:* Western Reserve Univ, AB, 33, AM, 35, PhD(biochem), 37. *Prof Exp:* Asst biol, Adelbert Col, Western Reserve Univ, 33-35, asst biochem, 35-37, instr biochem, Sch Med, 37-41; from asst prof to prof biochem, Sch Med, Ind Univ, Indianapolis, 41-79, actg chmn, Dept Biochem & Pharmacol, 56-58, chmn, Dept Biochem, 58-66, EMER PROF BIOCHEM, SCH MED, IND UNIV, INDIANAPOLIS, 79- *Mem:* Fel AAAS; Am Chem Soc; Am Soc Biol Chem; Soc Exp Biol & Med; NY Acad Sci. *Res:* Enzymes; enzyme inhibitors. *Mailing Add:* 6845 N Delaware St Indianapolis IN 46223

BOWMAN, DONALD HOUTS, b Osage City, Kans, May 18, 11; m 35; c 2. AGRONOMY. *Educ:* Kans State Univ, BS, 33, MS, 35; Univ Wis, PhD(plant path, agron), 39. *Prof Exp:* Asst pathologist, Div Cereal Crops & Diseases & asst plant path, Ohio Exp Sta, USDA, 39-46, plant pathologist, Tex Agr Exp Sta, 46-48, agronomist, Delta Br Exp Sta, 48-76; RETIRED. *Mem:* AAAS; Am Soc Agron. *Res:* Weed control, fertilization and general culture of rice; corn production. *Mailing Add:* 316 Sycamore Leland MS 38756

BOWMAN, DOUGLAS CLYDE, b St Louis, Mo, Oct 20, 25; m 55; c 3. PHYSIOLOGY, PHARMACOLOGY. *Educ:* Col Puget Sound, BS, 48, BEd, 49; Univ Wash, MS, 57, PhD(zool), 58. *Prof Exp:* Instr physiol, Univ Wash, 58-59; from instr to asst prof, Dent Sch, Northwestern Univ, 59-64; from asst prof to assoc prof, 64-81, prof sch dent, Loyola Univ, Chicago, 81-87; RETIRED. *Mem:* NY Acad Sci; Int Asn Dent Res. *Res:* Inflammation; action of salicylates; functional innervation of peridontal ligament; tongue thrust; hyoid positioning; toxic effects of methyl methacrylate. *Mailing Add:* 120 El Naranjo Grain Valley AZ 85614

BOWMAN, EDWARD RANDOLPH, b Mercer Co, WVa, Feb 26, 27; m 54; c 1. PHARMACOLOGY, PHYSIOLOGY. *Educ:* Concord Col, BS, 52; WVa Univ, MS, 53; Med Col Va, PhD, 63. *Prof Exp:* Bacteriologist, Va State Dept Health, 54-55; res asst pharmacol, 56-58, RES ASSOC PHARMACOL, MED COL VA, 61- *Concurrent Pos:* Vis investr, Royal Vet Col, Stockholm, Sweden, 61 & Inst Physiol, Santiago, Chile, 63. *Mem:* AAAS; Am Chem Soc; Am Soc Pharmacol & Exp Therapeut; Am Soc Exp Biol & Med; Sigma Xi; Nat Heart Inst. *Res:* Drug metabolism; nicotine metabolism; isolation and identification of urinary metabolites of nicotine; synthesis of pyridino compounds; whole body autoradiography; CNS pharmacology evaluation of unknown drugs for abuse potential. *Mailing Add:* 1900 Windsordale Dr Richmond VA 23229

BOWMAN, EUGENE W, b North Powder, Ore, Mar 28, 10; m 38; c 2. MATHEMATICS. *Educ:* Univ Idaho, BS, 35, MS, 36, EdD(admin), 52. *Prof Exp:* Instr math & chem, Coeur d'Alene Jr Col, 36-38; supt pub schs, Wash, 38-43; prof, 47-75, EMER PROF MATH & EDUC, SOUTHERN ORE COL, 75- *Concurrent Pos:* Educ adv, US Agency Int Develop, Ecuador, 60-62. *Res:* Personnel administration; comparative education. *Mailing Add:* 350 Kearney St Southern Ore Col Ashland OR 97520

BOWMAN, H(ARRY) FREDERICK, b Terre Hill, Pa, Sept 16, 41; m 63; c 1. NUCLEAR ENGINEERING, HEAT TRANSFER. *Educ:* Pa State Univ, BS, 63; Mass Inst Technol, MS & NuclE, 66, PhD(nuclear eng), 68. *Prof Exp:* Mech engr, Humble Oil & Refining Co, 63; res asst, Mass Inst Technol, 67-68; asst prof mech eng, 68-73, assoc prof, 73-79, ASSOC PROF BIOMED ENG, NORTHEASTERN UNIV, 79- *Concurrent Pos:* Admin officer, Harvard Univ-Mass Inst Technol prog health sci & technol, 69-74, exec officer biomat sci, 72-77, sr acad adminr, 74-77, sr acad adminr, 77-; lectr, Dept Mech Eng, Mass Inst Technol, 76- *Mem:* Am Soc Mech Engrs; Am Nuclear Soc. *Res:* Cryogenic boiling heat transfer; effects of ionizing radiation induced bubble nucleation in liquid helium; low temperature thermal property measurements; biological thermal property measurements; bioheat transfer, thermal methods to measure perfusion, engineering aspects of hyperthermia. *Mailing Add:* Dept Bioeng Northeastern Univ 360 Huntington Ave Boston MA 02115

BOWMAN, JAMES DAVID, b White Plains, NY, Aug 24, 39; m 65; c 3. NUCLEAR PHYSICS, HIGH ENERGY PHYSICS. *Educ:* Calif Inst Technol, BS, 61, PhD(physics & math), 67. *Prof Exp:* Res fel nuclear & solid state physics, Calif Inst Technol, 67-68; prof nuclear physics, Univ Bonn, 68-70; res fel nuclear chem, Lawrence Berkeley Lab, Univ Calif, 70-73; STAFF MEM NUCLEAR PHYSICS, MESON PHYSICS FACIL, LOS ALAMOS SCI LAB, 73- *Mem:* Fel Am Physics Soc; AAAS. *Res:* Medium energy physics; study of nuclear structure; nuclear stability; pion nucleus interaction; muon decay and symmetry principles. *Mailing Add:* Medium Energy Physics Div Los Alamos Sci Lab MS H846 Los Alamos NM 87545

BOWMAN, JAMES E, b Washington, DC, Feb 19, 23; m 50; c 1. PATHOLOGY, HUMAN GENETICS. *Educ:* Howard Univ, BS, 43, MD, 46. *Prof Exp:* Chmn dept path, Provident Hosp, Chicago, 50-53 & Nemazee Hosp, Shiraz Med Ctr, Iran, 55-61; vis assoc prof path, Fac Med, Pahlavi Univ, 57-59, vis prof & chmn dept, 59-61; hon res asst, Univ Col, Univ London, 61-62; asst prof med & path, 62-67, assoc prof path, med & biol, 64-71, med dir, Blood Bank, 62-80, dir Labs, 71-81, dir Comprehensive Sickle Cell Ctr, 73-83, PROF PATH, MED & GENETICS, UNIV CHICAGO, 71- *Concurrent Pos:* NIH spec res fel, Univ Col, Univ London, 61-62; USPHS res grant, 64-66 & 73-; chief path br, Med Nutrit Lab, Fitzsimmons Army Hosp, Colo; fel, Ctr Adv Study Behav Sci, Kaiser Found Soc, Stanford, 81-82; sr scholar, Ctr Clin Med Ethics, Univ Chicago, 89- *Mem:* Fel Col Am Path; fel Am Soc Clin Path; Am Genetics Asn; Am Soc Human Genetics; NY Acad Sci; Sigma Xi. *Res:* Human and population genetics, particularly blood anthropology, serum and erythrocytic polymorphisms. *Mailing Add:* 4929 Greenwood Chicago IL 60615

BOWMAN, JAMES FLOYD, II, b Orange, NJ, Feb 22, 32; m 68; c 4. SEDIMENTOLOGY, OCEANOGRAPHY. *Educ:* Rutgers Univ, AB, 56, PhD(geol), 66. *Prof Exp:* Lab technician ceramic eng, Rutgers Univ, 58-63; inspector eng geol, Yards Creek Hydroelec Proj, Jersey Cent Power & Light Co, 63-64; lab technician, Bur Mineral Res, Rutgers Univ, 64; lectr geol, Hunter Col, 65-66; instr, 66-69, ASST PROF GEOL, LEHMAN COL, 69-, PROJ DIR, X-RAY DIFFRACTION & FLUORESCENCE LAB, 70- *Concurrent Pos:* NSF matching funds grant USA, 52-54, res, 54-60; col expert to media, Geol, Lehman Col, 72-; mem fac sedimentology, Univ Inst Oceanog, City Univ New York, 75- *Mem:* Geol Soc Am; Soc Econ Paleont & Mineral; Int Asn Math Geol; NY Acad Sci; Sigma Xi. *Res:* Pleistocene and Recent sedimentation; x-ray diffraction studies of weathering; statistical analysis of sedimentary facies; x-ray diffraction and x-ray fluorescence analysis of micrometeorites, clays; Hudson Estuary-Bight sediments; computer mapping; oil spills in the aqueous environment; wetlands environments. *Mailing Add:* Dept Geol Lehman Col City Univ New York Bronx NY 10468

BOWMAN, JAMES SHEPPARD, b Orrville, Ohio, Oct 29, 28; m 56; c 3. ENTOMOLOGY. *Educ:* Ohio State Univ, BS, 51, MS, 54; Univ Wis, PhD(entom), 58. *Prof Exp:* Res asst, Ohio State Univ, 51, 53-54 & Univ Wis, 55-58; entomologist, Hazleton Labs, 58-61; res entomologist, Am Cyanamid Co, 61-70; tech rep, CIBA-Giegy Corp, 70-71; from asst prof to assoc prof, 71-85, PROF ENTOM, EXTEN ENTOMOLOGISTS, UNIV NH, 85-

Mem: Entom Soc Am; Sigma Xi; Am Asn Univ Prof. *Res:* Industrial development of agricultural pesticides; screening, metabolism and residue studies on insecticides, fungicides and herbicides. *Mailing Add:* Dept Entom Univ NH Nesmith Hall Durham NH 03824

BOWMAN, JAMES TALTON, b High Point, NC, Aug 2, 37; div; c 2. GENETICS. *Educ:* Duke Univ, BS, 61; Univ Calif, Davis, PhD(genetics), 65. *Prof Exp:* Assoc prof, 65-77, PROF ZOOL, UTAH STATE UNIV, 77- *Mem:* AAAS; Genetics Soc Am; Sigma Xi. *Res:* Genetic finestructure and function in higher organisms. *Mailing Add:* Dept Biol Utah State Univ Logan UT 84322-5305

BOWMAN, JOEL MARK, b Boston, Mass, Jan 16, 48; m 72. THEORETICAL CHEMISTRY. *Educ:* Univ Calif, Berkeley, AB, 69; Calif Inst Technol, PhD(chem), 74. *Prof Exp:* Teaching asst chem, Calif Inst Technol, 69-74; asst prof, 74-77, Alfred P Sloan fel, 77-79, ASSOC PROF CHEM, ILL INST TECHNOL, 77- *Mem:* Sigma Xi; Am Phys Soc. *Res:* Theoretical aspects of reaction dynamics of molecular systems involving exact and approximate quantum, semiclassical and quasiclassical techniques; theoretical studies of gas-surface interactions. *Mailing Add:* Dept Chem Emory Univ Atlanta GA 30322

BOWMAN, KENNETH AARON, b Alexandria, Va, Mar 20, 48; m 70; c 2. ELECTROCHEMISTRY, ALUMINUM PROCESS METALLURGY. *Educ:* Tusculum Col, BS, 70; Univ Tenn, PhD(chem), 77. *Prof Exp:* Anal chemist qual control, Org Chem Div, Chemetron Corp, 70-71; teaching asst chem, Univ Tenn, 71-72, res asst, Alcoa Found, 72-76; res scientist, 76-78, sr scientist, 78-80, STAFF SCIENTIST, ALCOA LABS, ALUMINUM CO AM, 80- *Mem:* Electrochem Soc; Int Soc Electrochem; Sigma Xi; Am Inst Mining Metall & Petrol Engrs. *Res:* Molten salt chemistry; the production and purification of aluminum, magnesium and calcium; aluminum melting and recycling. *Mailing Add:* 1410 Rt 56 E Apollo PA 15613-9726

BOWMAN, KIMIKO OSADA, b Tokyo, Japan, Aug 15, 27; US citizen;; c 2. MATHEMATICAL STATISTICS. *Educ:* Radford Col, BS, 59; Va Polytech Inst, MS, 61; PhD(statist), 63; Univ Tokyo, Dr Eng, 87. *Prof Exp:* From statistician to math statistician, Comput Technol Ctr, Nuclear Div, Union Carbide Corp, Tenn, 64-70; mem staff, Oak Ridge Nat Lab, 70-73; sr res staff, Comput Sci Div, Nuclear Div, Union Carbide Corp, 73-84; SR RES STAFF, OAK RIDGE NAT LAB, 84- *Mem:* Fel AAAS; Biomet Soc; Int Asn Statist in Phys Sci; fel Am Statist Asn; fel Inst Math Statist; Sigma Xi; Int Statist Inst; Int Asn Statist Comput; Japan Statist Comput. *Res:* Statistical research using high speed computing and estimating parameters. *Mailing Add:* Oak Ridge Nat Lab PO Box 2009 Oak Ridge TN 37831-8083

BOWMAN, LAWRENCE SIEMAN, b Ogden, Utah, Jan 7, 34; m 59; c 5. ELECTRICAL ENGINEERING. *Educ:* Univ Utah, BS, 57, MS, 61, PhD(elec eng), 64. *Prof Exp:* Engr, Utah Power & Light Co, 57-59 & Sperry Utah Eng Lab, 59; mem tech staff, Bell Tel Labs, NJ, 64-67; assoc prof, 67-73, PROF ELEC ENG, BRIGHAM YOUNG UNIV, 73- *Res:* Backward-wave oscillators; silicon avalanche diode millimeterwave oscillators. *Mailing Add:* Dept Elec Eng Brigham Young Univ Provo UT 84602

BOWMAN, LEO HENRY, b Valeda, Kans, May 12, 34; m 52; c 4. ANALYTICAL CHEMISTRY. *Educ:* Ottawa Univ, Kans, BSc, 56; Mich State Univ, PhD(anal chem), 61. *Prof Exp:* Res chemist anal develop, Chem Div, Pittsburgh Plate Glass Co, 60-62; asst prof chem, Parsons Col, 62-65; chmn dept, Midwestern Col, 65-68; prof chem, Southwest Minn State Col, 68-76; DEAN COL PHYS & LIFE SCI, ARK TECH UNIV, 76- *Mem:* Am Chem Soc. *Res:* Environmental science. *Mailing Add:* Ark Technol Univ Russellville AR 72801

BOWMAN, LEWIS WILMER, b Rothsville, Pa, Dec 8, 28; m 51; c 3. ORGANIC CHEMISTRY. *Educ:* Lebanon Valley Col, BS, 50; Univ Del, PhD(org chem), 54. *Prof Exp:* Instr chem, Lebanon Valley Col, 50-51 & Univ Del, 53-54; res chemist, Esso Res & Eng Co, 54-58, sr chemist, 58, sect head, 58-63, asst dir, 63-69, mgr chem planning & coord, 69-70, pres & managing dir, Esso Res SA, 70-73, gen mgr elastomers, 73, vpres elastomers, Essochem Europe, Inc, 73-77, MGT ELASTOMERS TECHNOL, EXXON CHEM CO, 77- *Mem:* Am Chem Soc. *Res:* Petrochemicals; synthetic rubber. *Mailing Add:* One Forest View Dr Chester NJ 07930

BOWMAN, MALCOLM JAMES, b Auckland, NZ, July 30, 42; m 67; c 4. PHYSICAL OCEANOGRAPHY. *Educ:* Univ Auckland, BSc, 65, MSc, 67; Univ Sask, PhD(elec eng), 70. *Prof Exp:* Res oceanogr Defense Sci Estab, Auckland, NZ, 67; mem staff, Univ Sask, 70-71; from asst prof to assoc prof, 71-88, PROF OCEANOG, MARINE SCI RES CTR, STATE UNIV NY STONY BROOK, 88- *Concurrent Pos:* Vis res fel, Univ Auckland, NZ, 78-79; managing ed, Springer Verlag, 78-; vis prof, Univ BC, 85-86. *Mem:* Am Geophys Union; Am Meterol Soc. *Res:* Oceanography of estuarine and coastal waters; shallow sea fronts; coastal upwelling. *Mailing Add:* Marine Sci Res Ctr State Univ NY Stony Brook NY 11794

BOWMAN, MARK (MCKINLEY), JR, engineering, for more information see previous edition

BOWMAN, NEWELL STEDMAN, b Rocky Ford, Colo, Sept 4, 24; m 46; c 3. ORGANIC CHEMISTRY. *Educ:* US Naval Acad, BS, 46; Univ Md, BS, 51; Princeton Univ, AM, 54, PhD(chem), 55. *Prof Exp:* Fel, Purdue Univ, 54-56; from asst prof to assoc prof, 56-70, PROF CHEM, UNIV TENN, KNOXVILLE, 70- *Concurrent Pos:* Instr, Purdue Univ, 55-56; Fulbright lectr, Univ Karlsruhe, 69-70. *Mem:* Am Chem Soc; Sigma Xi. *Res:* Organic fluorine compounds; sterospecific reactions. *Mailing Add:* 2516 Lake Moor Dr Knoxville TN 37920

BOWMAN, PHIL BRYAN, b Tulsa, Okla, July 2, 39; m 62; c 3. ANALYTICAL CHEMISTRY. *Educ:* Kans State Univ, BS, 61; Purdue Univ, PhD(anal chem), 65. *Prof Exp:* SR RES SCIENTIST, UPJOHN CO, 65- *Mem:* Am Chem Soc. *Res:* Gas liquid and high pressure liquid chromatography; electrochemistry; pharmaceutical analysis. *Mailing Add:* Upjohn Co 4822-259-12 Kalamazoo MI 49001

BOWMAN, RAY DOUGLAS, b Indianapolis, Ind, Mar 3, 42; m 69, 85; c 1. BIOCHEMISTRY, ANALYTICAL CHEMISTRY. *Educ:* Ind Univ, BA, 64; Calif Inst Technol, PhD(phys chem), 71. *Prof Exp:* Res fel, Calif Inst Technol, 71-73; asst prof, 73-80, ASSOC PROF NATURAL SCI, UNIV N FLA, 80- *Mem:* AAAS; Biophys Soc. *Res:* Behavior of polymers in hydrodynamic flow fields; measuring aquatic productivity by the oxygen balance technique; science education. *Mailing Add:* Dept Natural Sci Univ NFla Jacksonville FL 32216

BOWMAN, ROBERT CLARK, JR, b Dayton, Ohio, Oct 10, 45; m 65; c 2. CHEMICAL PHYSICS, SOLID STATE CHEMISTRY. *Educ:* Miami Univ, BS, 67; Mass Inst Technol, MS, 69; Calif Inst Technol, PhD(chem), 83. *Prof Exp:* From res chemist to sr res chemist, Monsanto Res Corp, 69-75, res specialist, 75-83, sci fel, Mound Facil, 83-84; mem technol staff, Chem & Phys Lab, Aerospace Corp, 84-89; SR SPECIALIST TECH STAFF, AEROJET ELECTRONIC SYSTS DIV, 90- *Mem:* Am Phys Soc; Mat Res Soc. *Res:* Nuclear magnetic resonance; electron paramagnetic resonance; radiation damage and defects in solids; physical properties of metal hydrides; raman light scattering of semiconductors; ion implantation effects; solid state diffusion. *Mailing Add:* Aerojet Electronic Systs Div PO Box 296 Azusa CA 91702

BOWMAN, ROBERT MATHEWS, b Belfast, Northern Ireland, July 18, 40; m 67; c 2. SYNTHETIC ORGANIC CHEMISTRY, ANALYTICAL CHEMISTRY. *Educ:* Queen's Univ, Belfast, BS, 62, Hons, 63, PhD, 66. *Prof Exp:* Fel nat prod chem, Univ Minn, 66-68; fel photochem, McMaster Univ, 68-70; MGR ANALYTICAL SERV, PHARMACEUT DIV, CIBA-GEIGY CORP, 70- *Mem:* Am Chem Soc. *Res:* Design and application of broad areas of synthetic organic chemistry leading to novel biologically significant agents, particularly those with analgesic, anti-convulsant and anti-tumor activity. *Mailing Add:* 6 Meadowbrook Ct Summit NJ 07901

BOWMAN, ROBERT SAMUEL, b Valley View, Pa, June 26, 17. PHYSICAL CHEMISTRY, ORGANIC CHEMISTRY. *Educ:* Pa State Col, BS, 40; Univ Pittsburgh, MS, 45, PhD(chem), 50. *Prof Exp:* Asst, Inst Animal Nutrit, Pa State Col, 40-41; res chemist, Jones & Laughlin Steel Corp, 41-45 & Gulf Res & Develop Co, 45-51; sr bone prod fel, 51-63, St Joseph Mineral Corp Labs sr fel, 63-78, dir, Mat Technol Div, 78-79, dir res, 79-81, VPRES, MELLON INST, CARNEGIE-MELLON UNIV, 81- *Mem:* AAAS; Am Chem Soc; Am Inst Chemists; Am Inst Chemists; Sigma Xi. *Res:* Surface chemistry; solid phase reactions; coal tar Chemicals and technology; pyridine bases; petroleum hydrocarbons; polymerization; phenol and calcium phosphate chemistry; solid absorbents; catalysis, glass blowing; metal oxides; technical management; research administration. *Mailing Add:* 5 Bayard Rd Apt 615 Pittsburgh PA 15213

BOWMAN, ROGER HOLMES, biochemistry; deceased, see previous edition for last biography

BOWMAN, THOMAS ELLIOT, b Brooklyn, NY, Oct 21, 18; m 43; c 3. ZOOLOGY. *Educ:* Harvard Univ, SB, 41; Univ Calif, Berkeley, MA, 48; Scripps Inst, PhD, 54. *Prof Exp:* Asst zool, Univ Calif, 45-48, res biologist, Scripps Inst, 48-53; asst prof marine biol, Narragansett Marine Lab, RI, 53-54; assoc cur marine inverts, 54-65, CUR DIV CRUSTACEA, NAT MUS NATURAL HIST, SMITHSONIAN INST, 65- *Mem:* Am Soc Limnol & Oceanog; Soc Syst Zool; Asn Trop Biol; Plankton Soc Japan; Crustacean Soc. *Res:* Taxonomy and zoogeography of hyperiid amphipods, calanoid copepods and cymothoid and asellid isopods. *Mailing Add:* Div Crustacea NHB-163 Smithsonian Inst Washington DC 20560

BOWMAN, THOMAS EUGENE, b Darby, Pa, Aug 3, 38; m 61; c 4. THERMAL SCIENCE. *Educ:* Calif Inst Technol, BS, 60, MS, 61; Northwestern Univ, PhD(mech eng), 64. *Prof Exp:* Res scientist, Denver Div, Martin-Marietta Corp, 63-67; lectr appl math, Univ Reading, Eng, 67-68; res scientist, Denver Div, Martin-Marietta Corp, 68-69; assoc prof space technol, 69-71, assoc prof mech eng, 71-75, dept head, 78-86, grad dean, 82-86, dean Sci & Eng, 86-88, dean Eng, 88-89, actg dean Sci & Lib Arts, 88-89, PROF MECH ENG, FLA INST TECHNOL, 75-, GRAD DEAN, 89-, ASSOC VPRES ACAD AFFIL, 89- *Concurrent Pos:* Vis Lectr, Univ Colo, 67; staff engr, Orlando Div, Martin-Marietta Corp, 72; energy consult, Care, Egypt, 79; chmn, Policy Adv Bd, Fla Solar Energy Ctr, 81-83; chmn, Nat Mech Eng Dept Heads Comt, 83-84; chmn, ME Div, Am Soc Eng Educ, 90-91. *Mem:* Am Soc Mech Engrs; Am Soc Eng Educ. *Res:* Fluid mechanics and thermal science, especially with regard to aerospace propulsion systems, low-gravity fluid behavior, and solar energy; developments in appropriate solar technology, especially solar cookers. *Mailing Add:* AVPAA Fla Inst Technol 150 W Univ Blvd Melbourne FL 32901

BOWMAN, WALKER H(ILL), b Louisville, Ky, June 10, 24; m 55; c 3. CHEMICAL ENGINEERING. *Educ:* Princeton Univ, BS, 49; Mass Inst Technol, SM, 51, ScD, 56. *Prof Exp:* Jr proj engr, Standard Oil Co Ind, 55-58, proj engr, 58-60, sr proj engr, 60-63; res assoc, Amoco Chem Corp, 63-69, div dir, 69-73, sr consult engr, 73-84; CONSULT, CHEM TECHNOL, 84- *Mem:* Am Chem Soc; Am Inst Chem Engrs; Sigma Xi. *Res:* Process design; economics of petro chemicals; technological forecasting. *Mailing Add:* 4N011 Thorn Tree St Charles IL 60174

BOWMAN, WILFRED WILLIAM, b Piqua, Ohio, Dec 17, 41; m 59; c 2. CHEMISTRY. *Educ:* Wilmington Col, BA, 63; Univ Rochester, PhD(chem), 68. *Prof Exp:* Res assoc chem, Cyclotron Inst, Tex A&M Univ, 68-73; RES CHEMIST, SAVANNAH RIVER LAB, E I DU PONT DE NEMOURS & CO, INC, 73- *Mem:* Am Chem Soc; Am Phys Soc. *Res:* Gamma-ray spectroscopy; neutron activation analysis. *Mailing Add:* 1008 Holliday Dr North Augusta SC 29841

BOWMAN, WILLIAM HENRY, b Reading, Pa, Apr 8, 38; m 60; c 2. CLINICAL BIOCHEMISTRY. *Educ:* State Col Pa, Kutztown, BS, 60; Pa State Univ, MS, 63, PhD(biochem), 65. *Prof Exp:* Asst prof chem, Ball State Univ, 65-69; res fel enzym, NIH, 69-71; res asst, 71-72, asst dir, 72-76, DIR, BIOCHEM LAB, METROP LIFE INS CO, 76- *Mem:* Am Chem Soc; AAAS; Am Asn Clin Chemists. *Res:* Assay of therapeutic drugs in blood and urine. *Mailing Add:* Biochem Lab Metrop Life Ins Co One Madison Ave New York NY 10010

BOWMER, RICHARD GLENN, b Spokane, Wash, Dec 4, 31; m 57; c 6. PLANT PHYSIOLOGY. *Educ:* Univ Idaho, BS, 53, MS, 57; Univ NC, PhD(bot), 60. *Prof Exp:* Asst prof biol, Lewis-Clark State Col, 60-61; from asst prof to assoc prof, 65-77, PROF BOT, IDAHO STATE UNIV, 77- *Res:* Translocation of solutes in higher plants. *Mailing Add:* Dept Biol Idaho State Univ Pocatello ID 83201

BOWN, THOMAS MICHAEL, b Chariton, Iowa, Jan 4, 46; div. VERTEBRATE PALEONTOLOGY, FLUVIAL SEDIMENTOLOGY. *Educ:* Iowa State Univ, BSc, 68; Univ Wyo, PhD(geol), 77. *Prof Exp:* Res asst paleont, Peabody Mus, Yale Univ, 69-73; res asst geol, Univ Wyo, 73-74, teaching asst, 74-77; GEOLOGIST, US GEOL SURV, 78- *Concurrent Pos:* Consult, US Park Serv, 74, US Bur Land Mgt, 75-77 & US Bur Reclamation, 76-77; Nat Res Coun fel, US Geol Surv, 77-78. *Mem:* Soc Vert Paleont; Soc Mammalogists; Paleont Soc; Explorers Club; Int Asn Sedimentologists. *Res:* Biostratigraphy and sedimentology of Mesozoic and Cenozoic non-marine rocks, paleosols, nonmarine trace fossils, mammalian evolution. *Mailing Add:* US Geol Surv Fed Ctr MS 919 Boulder CO 80225

BOWNDS, JOHN MARVIN, b Delta, Colo, Apr 21, 41; m 62; c 2. MATHEMATICS. *Educ:* Chico State Col, BA, 64; Univ Calif, Riverside, MA, 67, PhD(math), 68. *Prof Exp:* Mathematician, US Naval Weapons Ctr, Calif, 64-65; asst prof math, 68-74, ASSOC PROF MATH, UNIV ARIZ, 74- *Concurrent Pos:* NSF grant, 79-81. *Mem:* Am Math Soc; Soc Indust & Appl Math. *Res:* Differential equations; numerical solution of integral equations. *Mailing Add:* 1000 Misty Springs Rd Knoxville TN 37932

BOWNDS, M DERIC, b San Antonio, Tex, May 16, 42; m 68. NEUROBIOLOGY. *Educ:* Harvard Univ, BA, 63, PhD(biol), 67. *Prof Exp:* Instr neurobiol, Harvard Med Sch, 67-69; from asst prof to assoc prof, 69-75, PROF MOLECULAR BIOL & ZOOL, UNIV WIS-MADISON, 75- *Concurrent Pos:* NIH fel, 67-69 & res grant, 69- *Mem:* Soc Neurosci; Asn Res Vision & Ophthal; Biophys Soc; Soc Gen Physiologists. *Res:* Protein chemistry of visual pigments; chemistry and physiology of photoreceptor and other nerve membranes. *Mailing Add:* Lab Molecular Biol Univ Wis Madison WI 53706

BOWNE, SAMUEL WINTER, JR, b New York, NY, Nov 19, 25; m 54; c 2. CHEMISTRY, GENETICS. *Educ:* Mass Inst Technol, BS, 48; Columbia Univ, MA, 52; Cornell Univ, MS, 53, PhD(genetics), 57. *Prof Exp:* Jr chemist, Merck & Co, Inc, 48-49; res assoc, Wash State Univ, 53-54; asst prof chem, Eastern Wash Col Educ, 57-59; assoc prof biol, Univ Wichita, 59-61; res assoc, Ohio State Univ, 61-62; from asst prof to assoc prof, Edinboro State Col, 62-68, head dept sci, 64-69, prof chem, 68-89; RETIRED. *Mem:* AAAS; Am Chem Soc. *Res:* Biochemical genetics; intermediate metabolism. *Mailing Add:* 5241 Tarpell Rd Edinboro PA 16412

BOWNESS, COLIN, b London, Eng, Oct 26, 29; m 53; c 2. PHYSICS. *Educ:* Univ London, BSc, 50 & 51, PhD(physics), 56. *Prof Exp:* Microwave engr, Elec & Musical Industs, Eng, 51-56, lab mgr, 56-57; tech dir microwaves & lasers, 57-62, MGR SPEC MICROWAVE DEVICES OPER, RAYTHEON CO, 62- *Mem:* Sr mem Inst Elec & Electronics Engrs; Optical Soc Am; Brit Inst Physics & Phys Soc. *Res:* Lasers; microwave component design, especially ferrite devices. *Mailing Add:* 76 Shady Hill Rd Weston MA 02193

BOWNS, BEVERLY HENRY, b Ontario, Calif. COMMUNITY HEALTH, ADOLESCENT HEALTH. *Educ:* Columbia Univ, BSN, 59; Univ Minn, MS, 60; Johns Hopkins Univ, DrPH, 68. *Prof Exp:* Adminr undergrad prog pub health nursing, Univ Calif, San Francisco, 61-63; asst prof, Univ Md, Baltimore, 68-70; assoc prof, Vanderbilt Univ, 70-72; prof & chairperson dept, Med Units, Univ Tenn, Memphis, 72-77; PROF COMMUNITY HEALTH NURSING & DEAN COL NURSING, RUTGERS UNIV, 77- *Concurrent Pos:* Consult primary care nurse clin prog, Tex Women's Univ, Pa State Univ, Yale Univ, Univ Kans, Meharry Med Col, Ariz State Univ & Ind Univ, 68-76. *Mem:* Am Pub Health Asn. *Res:* Adolescent pregnancy in girls age ten to fifteen and mother and daughter relationships; evaluative research of graduate programs and their graduates in family nursing in primary care. *Mailing Add:* Dept Nursing/Grad Prog Rutgers Univ Newark NJ 07102

BOWSER, CARL, b Compton, Calif, Apr 21, 37; m 60; c 2. GEOCHEMISTRY. *Educ:* Univ Calif, Riverside, BA, 59; Univ Calif, Los Angeles, PhD(geol), 65. *Prof Exp:* Asst geol, Inst Geophys, Univ Calif, Los Angeles, 60-64; from asst prof to assoc prof, 64-74, PROF GEOL, UNIV WIS, MADISON, 74-; GEOLOGIST, US GEOL SURV, 80- *Mem:* Geol Soc Am; Geochem Soc; Mineral Soc Am; Am Geophys Union; AAAS. *Res:* Geochemistry of non-marine salt deposits; mineralogy and chemistry of marine and fresh water ferromanganese nodules; chemical sedimentology of lakes; geochemistry of hydrothermal deposits in Red Sea; chemical evaluation of groundwater-lakewater systems. *Mailing Add:* 1802 Camelot Dr Madison WI 53706

BOWSER, JAMES RALPH, b Windber, Pa, March 30, 49; m 81. ORGANOSILICON CHEMISTRY. *Educ:* Carion State Col, BS, 71; Duke Univ, PhD(chem), 76. *Prof Exp:* Instr chem, Univ Va, 76-78; asst prof, Colgate Univ, 78-80; asst prof, 80-84, ASSOC PROF CHEM, COL FREDONIA, STATE UNIV NY, 84-, CHMN DEPT, 83- *Mem:* Am Chem Soc; AAAS; Sigma Xi; NY Acad Sci. *Res:* Synthetic and spectroscopic aspects of organometallic chemistry with emphasis on organosilicon compounds and chromatographic methods for the analysis of such compounds. *Mailing Add:* Chem Dept Univ Notre Dame Notre Dame IN 46556

BOWSHER, ARTHUR LEROY, b Wapakoneta, Ohio, Apr 29, 17; m 67; c 5. ECHINODERMATA, PALEONTOLOGY. *Educ:* Univ Tulsa, BS, 41. *Prof Exp:* Mem staff, Tidewater Assoc Oil Co, 37-38; lab asst, Univ Tulsa, 38-41; lab asst, Univ Kans, 41-42, instr, 46-48; paleontologist, Smithsonian Inst, 48-52; chief lab, Navy Oil Unit, US Geol Surv, 52-57; staff geologist, Sinclair Oil & Gas Co, Okla, 57-69; sr geologist, Atlantic Richfield Oil Co, Tex, 69-70; sr geologist, Arabian Am Oil Co, Saudi Arabia, 70-78; chief explor strategy, Off Nat Petrol Reserve-Alaska, US Geol Surv, 78-81; explorationist, Yates Petrol Corp, Artesia, NMex, 81-86; CONSULT, 86- *Concurrent Pos:* Asst geologist, Kans State Geol Surv, 41-42, geologist, 42; instr, Am Univ, 49. *Mem:* Am Asn Petrol Geol; Paleont Soc; Geol Soc Am. *Res:* Petroleum geology; non-metallic resources; stratigraphy and paleontology. *Mailing Add:* 2707 N Gaye Dr Roswell NM 88201-4328

BOWSHER, HARRY FRED, b Lima, Ohio, Feb 26, 31; m 55; c 3. NUCLEAR PHYSICS. *Educ:* Ohio State Univ, BS, 55, MS, 56, PhD(nuclear physics), 60. *Prof Exp:* Asst prof nuclear physics, Univ Tenn, 60-66; assoc prof & chmn dept, 66-67, chmn dept, 71-74, PROF PHYSICS, AUGUSTA COL, 67- *Concurrent Pos:* Fulbright lectr, Dept Physics, Cheng Kung Univ, Taiwan, 69-70; consult, Oak Ridge Nat Lab, 60- *Mem:* Am Phys Soc. *Res:* Heavy-ion nuclear physics; reaction mechanisms. *Mailing Add:* Dept Physics/Chem Augusta Col Augusta GA 30910

BOWYER, C STUART, b Toledo, Ohio, Aug 2, 34; m 57; c 3. ASTRONOMY, SPACE SCIENCE. *Educ:* Miami Univ, BS, 56; Cath Univ, PhD(physics), 65. *Hon Degrees:* DSc, Miami Univ, 85. *Prof Exp:* Physicist, Nat Bur Standards, 56-58 & Naval Res Lab, 58-67; asst prof space sci, Cath Univ, 66-67; assoc prof, 67-74, PROF ASTRON, UNIV CALIF, BERKELEY, 74- *Concurrent Pos:* Consult, NASA, 68-; sr vis fel, Sci Res Coun, Eng, 74; Humboldt Found sr scientist award, 82. *Honors & Awards:* Tech Achievement Award, NASA, 72, Group Achievement Award, NASA, 75, Except Sci Achievement Award, NASA, 76. *Mem:* Am Geophys Union; Am Astron Soc; Int Astrophys Union; Astron Soc Pac; corresp mem Int Acad Astronaut. *Res:* Extreme ultraviolet astronomy; search for extraterrestrial intelligence; high energy astrophysics. *Mailing Add:* Dept Astron Univ Calif Berkeley CA 94720

BOWYER, J(AMES) M(ARSTON), JR, b Courtland, Kans, Dec 11, 20; m 48; c 3. FLUID MECHANICS. *Educ:* Kans State Col, BS, 42, MS, 49; Univ Calif, PhD(mech eng), 56. *Prof Exp:* Trainee, Convair Div, Gen Dynamics Corp, 46-47, aerodynamics engr, 49-51; asst, Kans State Col, 47-49; jr res engr low pressures proj, Univ Calif, 51-55; sr thermodynamics engr, Convair Div, Gen Dynamics Corp, 55-57, asst res group engr, 57-59, design specialist, 59-60, staff scientist, 60-62, sr staff scientist, Astronaut & Convair Div, 62-63; from assoc prof to prof mech eng, Kans State Univ, 63-69; staff engr, TRW Systs Group, 69-70; staff mem, Gulf Radiation Technol, 70-71, sr staff engr, 71-76; mem tech staff, Jet Propulsion Lab, 76-84; prof & chmn, dept mech eng, Wichita State Univ, 84-88; CONSULT ENG, 89- *Concurrent Pos:* Adj prof, dept mech eng, Wichita State Univ. *Mem:* Assoc fel Am Inst Aeronaut & Astronaut; Am Soc Mech Engrs; Sigma Xi. *Res:* Nucleation kinetics; reacting gas flows; supersonic and hypersonic nonviscous and viscous flows; rarefied gas flows; heat transfer; applied mathematics; solar energy. *Mailing Add:* 3683 Quailview Court Spring Valley CA 92077

BOWYER, KERN M(ALLORY), b Jersey City, NJ, Nov 12, 28; m 80; c 2. ELECTRICAL ENGINEERING, COMPUTER ENGINEERING. *Educ:* NY Univ, BEE, 57. *Prof Exp:* Serv engr commun equip, Winters Radio Lab, 48-49; test engr radar, Gen Elec Co, 51; sr res assoc, E I du Pont de Nemours & Co, Inc, 57-90; RETIRED. *Concurrent Pos:* Mem, Del State Indust Adv Comt. *Mem:* Audio Eng Soc; Inst Elec & Electronics Engrs; NY Acad Sci. *Res:* Image processing; communication electronics; anti-submarine warfare; industrial instrumentation; magnetic recording systems; computer systems applications. *Mailing Add:* 16 Lehigh Ave Wilmington DE 19805

BOX, EDITH DARROW, b Glendale, Md, Jan 3, 22; m 51; c 2. PARASITOLOGY. *Educ:* Iowa State Col, BS, 43; Johns Hopkins Univ, DSc(parasitol), 48. *Prof Exp:* Student asst parasitol, Sch Hyg, Johns Hopkins Univ, 43-48; res assoc, 48-49, from asst prof to assoc prof, 49-85, EMER ASSOC PROF MICROBIOL, MED BR, UNIV TEX, 85- *Mem:* AAAS; Am Soc Trop Med & Hyg; Am Soc Parasitol; Soc Protozool; Wildlife Dis Asn. *Res:* Chemotherapy and biology of avian and rodent malaria; arthropod transmission of blood and tissue protozoa; life cycles of intracellular protozoa; taxonomy and biology of avian coccidia. *Mailing Add:* Dept Microbiol Univ Tex Med Br Galveston TX 77550

BOX, GEORGE EDWARD PELHAM, b Gravesend, Eng, Oct 18, 19; m 59, 85; c 2. INDUSTRIAL & MANUFACTURING ENGINEERING. *Educ:* Univ London, BS, 48, PhD(statist), 52, DSc, 61. *Hon Degrees:* DSc, Univ Rochester, 75; Carnegie Mellon Univ, 89. *Prof Exp:* From statistician to head statist res sect, Dyestuffs Div, Imp Chem Indust Ltd, 48-56; dir res group, Dept Math, Princeton Univ, 56-59; prof, 59-71, Ronald Aylmer Fisher prof, 71-80, VILAS PROF STATIST, UNIV WIS MADISON, 80-, DIR RES, CTR QUALITY IMPROV, 85- *Concurrent Pos:* Vis prof, NC State Col, 53-54 & Univ Essex, 70-71; Ford Found vis prof, Grad Sch Bus Admin, Harvard Univ, 65-66. *Honors & Awards:* Brit Empire Medal, 46; Am Inst Chem Engrs Prof Progress Award, 63; Guy Medal, Royal Statist Soc, 64; Wilks Medal, Am Statist Asn; Shewhart Medal, Am Soc Qual Control, Deming Medal. *Mem:* Fel AAAS; fel Am Statist Asn (pres, 78); Int Statist Inst; fel Inst Math Statist (pres, 79); fel Am Acad Arts & Sci; fel Royal Soc. *Res:* Design and analysis of experiments; time series and forecasting; statistical inference; quality improvement. *Mailing Add:* 2238 Branson Rd Oregon WI 53575

BOX, HAROLD C, b Clarence, NY, Aug 19, 25; m 51; c 5. BIOPHYSICS. *Educ:* Canisius Col, BS, 48; Univ Buffalo, MA, 51, PhD(physics), 54. *Prof Exp:* Asst physicist, Cornell Aeronaut Lab, 51-54; sr cancer res scientist, 54-61, assoc cancer res scientist, 61-70, prin cancer res scientist, 70-76, DIR DEPT BIOPHYS, ROSWELL PARK MEM INST, 76- *Concurrent Pos:* Asst res prof, Roswell Park, Grad Div, State Univ NY Buffalo, 62-74, adj prof physics, 74-76, res prof & chmn, 76- *Mem:* Am Phys Soc; Am Chem Soc; Biophys Soc; Radiation Res Soc; Am Crystallog Asn. *Res:* Magnetic resonance studies of carcinogens and radiation effects in biological compounds. *Mailing Add:* 170 Reist Williamsville NY 14221

BOX, JAMES ELLIS, JR, b Georgetown, Tex, Sept 12, 31; m 56; c 4. SOIL PHYSICS. *Educ:* Tex A&M Univ, BS, 52, MS, 56; Utah State Univ, PhD(soil & irrig), 60. *Prof Exp:* Soil scientist, Soil Conserv Serv, US Dept Agr, 56-57; asst soil physics, Utah State Univ, 57-60; res soil scientist, Soil & Water Conserv Res Ctr, 60-61, supt, US Big Spring Field Sta, 62-65, dir, 65-84, res leader, 72-81, res adv, Southeastern Region, Nat Asn Conserv Districts, 73-84, location leader, lab dir & soil scientist, 65-84, RES SOIL SCIENTIST, SOUTHERN PIEDMONT CONSERV RES CTR, AGR RES SERV, US DEPT AGR, 84- *Concurrent Pos:* US Dept Agr sponsoring scientist, Soil Physics Res, Israel Inst Technol, 66-; vis scientist, Am Soc Agron, 71-72; adv mem, Ga State Soil & Water Comt, Southeast Area Res Comt, Nat Asn Conserv Dist; mem grad fac, Univ Ga, 75-80. *Mem:* Fel Am Soc Agron; Crop Sci Soc Am; Int Soil Sci Soc; fel Soil Conserv Soc Am; Soil Sci Soc Am. *Res:* Basic aspects of plant-soil-water relationships; moisture conservation under dryland conditions. *Mailing Add:* Southern Piedmont Conserv Res Ctr Agr Res Serv PO Box 555 Watkinsville GA 30677

BOX, LARRY, b Chester, Pa, Sept 24, 39; m 60; c 3. CHEMICAL MODIFICATION OF CELLULOSIC FIBER, CHEMISTRY TROUBLE SHOOTING IN PROCESS & PRODUCT AREAS. *Educ:* Cheyney Univ, BA, 65; St Josephs Univ, MS, 77. *Prof Exp:* Lab technician res, Scott Paper Co, 65- 70, promotion series res chem, 70-90, SR PROJ LEADER RES CHEM, SCOTT WORLDWIDE, 90- *Concurrent Pos:* Lectr & tutor chem, Widener Univ, 71-75; founder & vchmn, AGAPE Day Care & Presch Inc, 81-91. *Mem:* Int Soc African Scientists; Tech Asn Pulp & Paper Indust. *Res:* Organic synthesis research on natural polymers; two patents. *Mailing Add:* 90 Adrien Rd Glen Mills PA 19342

BOX, MICHAEL ALLISTER, b Melbourne, Australia, Dec 23, 47; m 72; c 2. RADIATIVE TRANSFER, REMOTE SENSING. *Educ:* Monash Univ, BSc, 70; Sydney Univ, PhD(physics), 75. *Prof Exp:* Res scientist, Inst Atmospheric Optics & Remote Sensing, Hampton, Va, 77-79; res assoc atmospheric radiation, Inst Atmospheric Physics, Univ Ariz, 79-81; SR LECTR, UNIV NEW S WALES, 81- *Mem:* Am Meteorol Soc; Am Geophys Union; Australian Inst Physics; Australian Meteorol & Oceanog Soc. *Res:* Inversion techniques in atmosphere remote sensing, especially aerosol size distribution; radiative transfer calculations; application of rigorous perturbation techniques in radiative transfer. *Mailing Add:* Sch Physics Univ New S Wales Kensington NSW 2033 Australia

BOX, THADIS WAYNE, b Llano Co, Tex, May 9, 29; m 54; c 4. RANGE MANAGEMENT. *Educ:* Southwest Tex State Col, BS, 56; Agr & Mech Col, Tex, MS, 57, PhD, 59. *Prof Exp:* Asst range mgt, Agr & Mech Col, Tex, 57-59; asst prof, Utah State Univ, 59-62; from assoc prof to prof, Tex Technol Col, 62-68, dir, Int Ctr Arid & Semi-Arid Land Studies, 68-70; PROF RANGE SCI & DEAN COL NATURAL RESOURCES, UTAH STATE UNIV, 70- *Concurrent Pos:* Vis res scientist, Commonwealth Sci & Indust Res Orgn, Australia, 68-69, 78; consult, Food & Agr Orgn, UN, EAfrica, 65, 69 & 70. *Mem:* Soc Range Mgt. *Res:* Range ecology; interrelationships between wild and domestic animals; grazing management and applied ecology in range ecosystems. *Mailing Add:* Col Natural Resources Utah State Univ Logan UT 84322

BOX, VERNON G S, b Montego Bay, Jamaica, June 20, 46; m 76; c 2. PHARMACEUTICAL CHEMISTRY. *Educ:* Univ WI, Kingston, Jamaica, BSc,67, PhD(org chem), 71. *Prof Exp:* From asst lectr to sr lectr chem, Univ WI, Kingston, Jamaica, 69-82; sr scientist pharmaceut & prin scientist chem res, Schering-Plough Corp, Bloomfield, NJ, 82-84; ASSOC PROF ORG CHEM, CITY COL, CITY UNIV NY, 84- *Mem:* Am Chem Soc; Inter-Am Photochem Soc. *Res:* Synthesis of natural products from monosaccharides; reaction mechanisms and stereo-electronic effects in monosaccharide chemistry; new protecting groups for monosaccharides. *Mailing Add:* Dept Chem City Col Convent Ave 138th St New York NY 10031

BOXENBAUM, HAROLD GEORGE, b Philadelphia, Pa, Dec 26, 42; m 82; c 1. AGING RESEARCH, INTERSPECIES SCALING. *Educ:* Temple Univ, BS, 65, MS, 69; Univ Calif, San Francisco, PhD(pharmaceut chem), 72. *Prof Exp:* Asst prof pharm, Col Pharm, Ohio State Univ, 72-74; sr scientist, Hoffmann-La Roche Inc, 74-77; owner & mgr, Pennyroyal Apothecary, 77-80; assoc prof pharm, Sch Pharm, Univ Conn, 80-85; RES ASSOC, MARION MERRELL DOW INC, 85- *Concurrent Pos:* Adj assoc prof, Col Pharm, Univ Cincinnati, 88- *Mem:* Fel AAAS; fel Am Asn Pharmaceut Sci; Am Soc Pharmacol & Exp Therapeut; Int Soc Study Xenobiotics; Sigma Xi. *Res:* Pharmacokinetics, biopharmaceutics, clinical pharmacology, toxicology, aging and drug metabolism. *Mailing Add:* Dept Drug Metab Marion Merrell Dow Inc 2110 E Galbraith Rd Cincinnati OH 45215-6300

BOXER, LAURENCE A, b May 17, 40; m 61; c 1. CELL BIOLOGY OF GRANULOCYDES. *Educ:* Univ Colo, BA, 61; Stanford Univ, MD, 66. *Prof Exp:* Intern & resident pediat, Yale Univ Med Ctr, New Haven, 66-68; resident pediat, Stanford Univ Med Sch, 68-69; maj, USAR Tripler Hosp, Honolulu, 69-72; fel hemat, Harvard Univ, 72-75; from asst prof to prof, Ind Univ, 75-82; PROF & DIR PEDIAT HEMAT & ONCOL, UNIV MICH, 82- *Concurrent Pos:* Mem, Hemat Study Sect, 81-86 & 88-92; assoc ed, J Immunol & Blood, J Leukocyte Biol, J Pediat, J Clin Immunol & J Lab Clin Med. *Honors & Awards:* Mead Johnson Award, 83. *Mem:* Soc Pediat Res (pres, 85-86); Am Soc Hemat (coun, 88-92); Am Soc Clin Invest; Am Soc Cell Biol; Am Asn Immunol; Am Asn Pathologists. *Mailing Add:* Dept Pediat Sch Med Univ Mich Box 0238 Ann Arbor MI 48109

BOXER, ROBERT JACOB, b Brooklyn, NY, Apr 9, 35; m 63; c 2. ORGANIC CHEMISTRY. *Educ:* Brooklyn Col, BS, 56; Rutgers Univ, PhD(org chem), 61. *Prof Exp:* From asst prof to assoc prof chem, Oglethorpe Col, 61-64; assoc prof, 64-72, PROF ORG CHEM, GA SOUTHERN UNIV, 72- *Mem:* Am Chem Soc. *Res:* Organic halogenating agents; history of drugs of abuse. *Mailing Add:* Dept Chem Ga Southern Univ Statesboro GA 30460-8064

BOXER, STEVEN GEORGE, b New York City, NY, Oct 18, 47; m 77; c 2. PHYSICAL CHEMISTRY. *Educ:* Tufts Univ, BS, 69; Univ Chicago, PhD(chem), 76. *Prof Exp:* PROF CHEM, STANFORD UNIV, 76- *Concurrent Pos:* A P Sloan Fel. *Honors & Awards:* Presidential Young Investigator Award. *Mem:* Am Chem Soc; Biophys Soc; Am Soc Photobiol. *Res:* Mechanisms of electron energy transfer in biological systems; new magnetic resonance methods; primary photochemistry of photosynthesis; effects of magnetic and electric fields on chemical reactions. *Mailing Add:* Dept Chem Stanford Univ Stanford CA 94305-5080

BOXILL, GALE CLARK, b Indianapolis, Ind, Jan 28, 19; m 46; c 2. PHARMACOLOGY. *Educ:* Washington & Lee Univ, AB, 47; Univ Tenn, MS, 51, PhD(pharmacol), 54. *Prof Exp:* Jr pharmacologist, Wm S Merrell Co, 47-49; asst prof, Sch Pharm, Univ Ga, 52-53; pharmacologist, Res Lab, Mead Johnson & Co, 53-59; sr res assoc, Warner-Lambert Res Inst, 59-62, dir dept toxicol, 62-70; mgr toxicol, Wyeth Labs, 70-78, assoc dir biol res, 78-85; RETIRED. *Mem:* Am Soc Pharmacol & Exp Therapeut; Sigma Xi; Soc Toxicol (secy, 75-79). *Res:* Cardiovascular effect of epinephrine on pressoreceptors; effects of various drugs on blood pressure to epinephrine; toxicological evaluation of new drugs. *Mailing Add:* 34351 Whispering Oaks Blvd Ridge Manor FL 33525

BOXMAN, RAYMOND LEON, b Philadelphia, Pa, June 9, 46; m 72; c 4. ELECTRICAL DISCHARGES & PLASMAS, METALLURGICAL COATINGS & THIN FILMS. *Educ:* Mass Inst Technol, SB & SM, 69, PhD(elec eng), 73. *Prof Exp:* Sr res engr, Gen Elec Co, 73-75; ASSOC PROF ELEC ENG, TEL-AVIV UNIV, 75- *Concurrent Pos:* Vis assoc prof, Univ SC, 81-82; vis scientist, Brown Boveri Corp, 82; assoc ed, Inst Elec & Electronics Engrs Trans Plasma Sci, 87-93; vis prof, Drexel Univ, 89-90. *Honors & Awards:* Boris & Renee Joffe Award, Int Union Electro-Deposition & Surface Finishing, 84. *Mem:* Fel Inst Elec & Electronics Engrs. *Res:* Electrical discharges; high current vacuum arcs, laboratory measurements and formulation of physical models; applications of arcs to high current switching and to the deposition of metallurgical coatings and thin films. *Mailing Add:* Fac Eng Tel Aviv Univ Ramat Aviv Tel Aviv 69978 Israel

BOYARSKY, ABRAHAM JOSEPH, b Baranovitch, Poland, Nov 16, 46; Can citizen; m 74; c 9. DETERMINISTIC DYNAMICAL SYSTEMS. *Educ:* McGill Univ, BEng, 67, MEng, 68, PhD(control theory), 71. *Prof Exp:* From asst prof to assoc prof, 73-82, PROF MATH, CONCORDIA UNIV, MONTREAL, 82- *Mem:* Am Math Soc. *Res:* The limiting behavior of deterministic dynamical systems, including the subject known as "chaos". *Mailing Add:* Dept Math Concordia Univ 7141 Sherbrooke St W Montreal PQ H4B 1R6 Can

BOYARSKY, LILA HARRIET, b Brooklyn, NY, Apr 23, 21; m 41; c 2. BIOLOGY. *Educ:* Hunter Col, BA, 42; Univ Wis, MS, 43, PhD(genetics), 47. *Prof Exp:* Asst physiol reprod, Univ Wis, 42-43, animal husb, Purdue Univ, 43-45 & physiol, Univ Chicago, 45-47; instr biol, George Williams Col, 49-50; asst prof, 55-57, assoc prof, 58-61, PROF BIOL, TRANSYLVANIA COL, 61- *Mem:* AAAS; Am Zool Soc. *Res:* Genetics; endocrinology; physiology of reproduction; genetics of hamsters. *Mailing Add:* Dept Math & Sci Transylvania Univ Lexington KY 40508

BOYARSKY, LOUIS LESTER, b Jersey City, NJ, Sept 5, 19; m 41; c 2. NEUROPHYSIOLOGY. *Educ:* City Col New York, BS, 41; Purdue Univ, MS, 45; Univ Chicago, PhD(physiol), 48. *Prof Exp:* Psychophysiologist, Inst Juv Res, 49-50; from asst prof to assoc prof, 50-59, PROF PHYSIOL & BIOPHYS, COL MED, UNIV KY, 59- *Concurrent Pos:* Fulbright scholar, Univ Milan, 57-58; vis prof, Univ Hawaii, 68-69. *Mem:* Am Physiol Soc; Soc Neurosci; Biophys Soc; Soc Exp Biol & Med. *Res:* Peripheral and central nervous system; biophysics. *Mailing Add:* Dept Physiol & Biophys Univ Ky Med Sch 800 Rose St Lexington KY 40536

BOYARSKY, SAUL, b Burlington, Vt, July 22, 23; m 45; c 3. UROLOGY, PHYSIOLOGY. *Educ:* Univ Vt, BS, 43, MD, 46; Wash Univ, JD, 81. *Prof Exp:* Instr urol, Sch Med, Duke Univ, 54; instr, NY Univ, 55-56; from asst prof to assoc prof, Albert Einstein Col Med, 56-63; prof urol & asst prof physiol, Sch Med, Duke Urologist, Duke Hosp, 63-70; urologist, Barnes Hosp, 70-73; prof genitourinary surg & chmn dept, prof bioeng & assoc prof pharmacol, Sch Med, Wash Univ, 70-73, prof urol, 70-89; CONSULT, 89- *Concurrent Pos:* Fel surg, Col Med, Univ Vt, 47-48; USPHS fel physiol, Sch Med, NY Univ, 54-55; asst attend, Bellevue Hosp, New York, 54-56 & Univ Hosp, 56-63; asst chief urol, Bronx Vet Admin Hosp, 56-57, attend, 60-63; assoc attend, Bronx Munic Hosp, 56-62, vis urologist, 62-63; adj urologist, New Rochelle Hosp, 60-63; chief urol, Vet Admin Hosp, 63-70, dir rehab, 69-70; consult urol, St Louis City Hosp, 70-; consult, John Cochran Vet Admin Hosp, Jefferson Barracks, Jewish Hosp, Food & Drug Admin bur devices, HEW, Vet Admin Spinal Cord Injury Serv & Mo Crippled Childrens Serv, Univ Mo, 70-; mem panel review gastroenterol-urol devices, HEW, Food & Drug Admin; pvt law practice, 85- *Honors & Awards:* William P Burpeau Award; Urodynamics Soc & Int Continence Soc Award, 83. *Mem:* Am Physiol Soc; Am Asn Genitourinary Surgeons; Am Urol Asn; Am Col Surgeons; Soc Univ Urologists; Sigma Xi; Asn Advan Med Instrumentation; Am Soc Testing & Mat; Am Col Legal Med. *Res:* Renal, ureteral and bladder physiology and pharmacology; rehabilitation and bioengineering in urologic surgery; spinal cord injury care; benign prostate hypertrophy; forensic urology. *Mailing Add:* 45 Portland Pl St Louis MO 63108

BOYCE, DONALD JOE, b Duncan, Okla, Dec 30, 31; m 52; c 6. MATHEMATICS. *Educ:* Cent State Col, BS, 56; Okla State Univ, MS, 57, EdD(math educ), 68. *Prof Exp:* Instr math, Cent State Col, 57-58; systs analyst, Tinker AFB, 58-59; from asst prof to assoc prof, 59-71, chmn math, comput sci & stastist dept, 79-83, PROF MATH, CENT STATE UNIV, 71-, PROF MATH & STATIST, 83- *Concurrent Pos:* NSF res grants for col teachers, 67-69. *Mem:* Math Asn Am; Nat Coun Teachers Math. *Res:* Modern differential geometry, particularly the curvature of differentiable manifolds; number theory. *Mailing Add:* Dept Math Cent State Univ Edmond OK 73034

BOYCE, FREDERICK FITZHERBERT, b Barbados, BWI, Sept 22, 03; nat US; m 41; c 3. SURGERY. *Educ:* Harvard Univ, BS, 27; Yale Univ, MD, 30. *Prof Exp:* From instr to asst prof surg, Sch Med, La State Univ, 32-39; from asst prof to prof, 47-76, EMER PROF SURG, SCH MED, TULANE UNIV, 76- *Concurrent Pos:* Pvt pract, 39- *Honors & Awards:* Gross Quinquennial Prize, 40. *Mem:* Fel AMA; fel Am Col Surgeons; Soc Exp Biol & Med; fel Am Col Chest Physicians; Int Soc Surg. *Res:* Liver and biliary tract; liver-kidney death; autolytic peritonitis; liver function tests and burns; acute appendicitis; regional enteritis. *Mailing Add:* 3220 Jefferson Ave New Orleans LA 70125

BOYCE, HENRY WORTH, JR, b Clinton, NC, Sept 21, 30; m 52; c 5. INTERNAL MEDICINE, GASTROENTEROLOGY. *Educ:* Wake Forest Col, BS, 52, MD, 55; Baylor Univ, MS, 61; Am Bd Internal Med, dipl, 62, cert gastroenterol, 65. *Prof Exp:* Intern, Tripler Gen Hosp, Honolulu, Hawaii, US Army, 55-56, resident internal med, Brooke Gen Hosp, San Antonio, Tex, 57-59, resident gastroenterol, 60-61; chief gastroenterol serv, Madigan Gen Hosp, Tacoma, Wash, 61-65, Brooke Gen Hosp, San Antonio, Tex, 65-66, Walter Reed Gen Hosp, 66-75; PROF MED & GASTROENTEROL, COL MED, UNIV SFLA, 75- *Concurrent Pos:* Consult, Surgeon Gen, US Army, 67-75; dir, Ctr for Swallowing Disorders, Univ South Fla; fel, Am Col of Gastroenterol. *Honors & Awards:* Rudolf Schindler Award, ASGE, 82; Rudolf J. and Anita S. Noer Distinguished Prof Award, Univ S Fla, 85. *Mem:* Fel Am Col Physicians; Am Soc Gastrointestinal Endoscopy; Asn Mil Surg US; Am Asn Study Liver Dis; Am Asn Parenteral & Enteral Nutrit; Am Gastroenterol Asn. *Res:* Clinical internal medicine and gastroenterology. *Mailing Add:* Univ SFla Col Med Box 19 12901 Bruce Downs Blvd Tampa FL 33612

BOYCE, MARK S, b Yankton, SDak, May 24, 50; m 71; c 2. LIFE HISTORY EVOLUTION, MATING SYSTEM EVOLUTION. *Educ:* Iowa State Univ, BS, 72; Univ Alaska, MS, 74; Yale Univ, MPhil, 75, PhD(pop biol), 77. *Prof Exp:* From instr to assoc prof, 76-87, PROF ZOOL, UNIV WYO, 87-; DIR, NAT PARK SERV RES CTR, 89- *Concurrent Pos:* NATO res fel, Univ Oxford, 82-83; vis prof, Dept Math, Univ Wyo, 84-85; Fulbright fel, 91. *Mem:* Ecol Soc Am; Am Soc Naturalists; Soc Study Evolution; Am Soc Mammalogists; Brit Ecol Soc; Wildlife Soc. *Res:* Population ecology and evolution of a variety of vertebrate species; sage grouse-parasite coevolution; conservation biology of spotted owls; ecology of Yellowstone. *Mailing Add:* Dept Zool & Physiol PO Box 3166 Univ Sta Laramie WY 82071

BOYCE, PETER BRADFORD, b New York, NY, Nov 30, 36; m 58; c 2. SEARCH FOR EXTRA TERRESTRIAL INTELLIGENCE, STAR FORMATION. *Educ:* Harvard Univ, AB, 58; Univ Mich, MA, 62, PhD(astron), 63. *Prof Exp:* Res astron, Lowell Observ, 63-75; adj prof, Ohio State Univ, 70-75; prog dir astron div, NSF, 73-79; EXEC OFFICER, AM ASTRON SOC, 79- *Concurrent Pos:* Vis prof, Univ Copenhagen, 72, Univ Md, 84; gov bd, Am Inst Phys, 79-; Cong Sci fel, 77-78; mem, Seti sci working group, NASA, 82-90; chair, Panel Status Profession, Nat Acad Sci Astron & Astrophys Surv Comt, 89-91. *Mem:* Fel AAAS; Am Astron Soc; Int Astron Union; Am Phys Soc; Optical Soc Am; Sigma Xi. *Res:* Photoelectric measurement of stellar, nebular and planetary spectra; astronomical instrumentation; microwave search strategies for seti; analysis of federal funding for astronomy. *Mailing Add:* 5700 Sherrier Pl NW Washington DC 20016

BOYCE, RICHARD JOSEPH, b Schenectady, NY, Sept 15, 39; m 57; c 4. RUBBER CHEMISTRY. *Educ:* Rensselaer Polytech Inst, BChE, 61, PhD(phys chem), 66. *Prof Exp:* Res chemist, 65-80, sr res chemist, 80-85, SR RES SCIENTIST, E I DU PONT DE NEMOURS & CO, INC, 86- *Mem:* Am Chem Soc; Soc Advan Mat & Process Eng. *Res:* Polymer physical chemistry; rheology of elastomers. *Mailing Add:* 11 Harvest Lane Glen Farms Elkton MD 21921

BOYCE, RICHARD P, b Pocatello, Idaho, Jan 27, 28; m 50; c 2. BIOPHYSICS. *Educ:* Univ Utah, BA, 55; Yale Univ, PhD(biophys), 61. *Prof Exp:* From res asst to res assoc radiobiol, Sch Med, Yale Univ, 61-64, from asst prof to assoc prof, 64-71; PROF BIOCHEM, SCH MED, UNIV FLA, 71- *Concurrent Pos:* Guggenheim fel, 68-69. *Mem:* Biophys Soc; Am Soc Biol Chemists. *Res:* Enzymatic repair of irradiation damage in bacteria and bacterial viruses; molecular mechanisms of genetic recombination in bacteria. *Mailing Add:* 8800 NW Fourth Pl Gainesville FL 32607

BOYCE, STEPHEN GADDY, b Anson Co, NC, Feb 5, 24; m 51; c 2. PLANT ECOLOGY. *Educ:* NC State Col, BS, 49, MS, 51, PhD(bot), 53. *Prof Exp:* Instr bot, NC State Col, 52-53; asst prof, Univ Ohio, 53-57; silviculturist, Cent States Exp Sta, US Forest Serv, 57-64, asst dir, 64-66, chief genetics res, 66-67, asst to dep chief res, 67-70, dir, 70-73, chief forest ecologist, 73-84; RETIRED. *Concurrent Pos:* Vis lectr, Sch Forestry, Yale Univ, 79-81; vis prof natural resources, Sch Forestry, Duke Univ, 81- *Mem:* Ecol Soc Am; Soc Am Foresters; AAAS. *Res:* Forest and coastal dunes ecology; tree improvement; wood growth relations; ecology of forests; system dynamics. *Mailing Add:* 27 Moytoy Lane Brevard NC 28712

BOYCE, STEPHEN SCOTT, b Indianapolis, Ind, Feb 23, 42; m 65; c 1. MATHEMATICS, QUANTUM MECHANICS. *Educ:* Earlham Col, BA, 64; Univ Wis, MA, 65, PhD(math), 69. *Prof Exp:* Asst prof, 69-77, ASSOC PROF MATH & CHMN DEPT, BEREA COL, 77- *Mem:* Math Asn Am; Inst Mgt Sci. *Res:* Comparison and development of mathematical models for quantum mechanics; sequency and scheduling theory; combinatorial optimization. *Mailing Add:* Dept Math CPO 2334 Berea Col Berea KY 40404

BOYCE, WILLIAM EDWARD, b Tampa, Fla, Dec 19, 30; m 55; c 3. RANDOM DIFFERENTIAL EQUATIONS, EIGENVALUE PROBLEMS. *Educ:* Rhodes Col, BA, 51; Carnegie Inst Technol, MS, 53, PhD(math), 55. *Prof Exp:* Universal Match Found fel appl math, Brown Univ, 55-56, res assoc, 56-57; from asst prof to assoc prof, 57-63, PROF MATH, RENSSELAER POLYTECH INST, 63- *Concurrent Pos:* Managing ed, SIAM Rev, 70-77; mem coun, Soc Indust & Appl Math, 75-77; vpres publ, 78-81. *Mem:* AAAS; Am Math Soc; Soc Indust & Appl Math; Math Asn Am; Am Asn Univ Prof. *Res:* Approximate analytical and numerical methods for random initial value, boundary value, and eigenvalue problems, and application of such problems. *Mailing Add:* Dept Math Rensselaer Polytech Inst Troy NY 12180-3590

BOYCE, WILLIAM HENRY, b Ansonville, NC, Sept 22, 18; m 48; c 4. UROLOGY. *Educ:* Davidson Col, BS, 40; Vanderbilt Univ, MD, 44. *Hon Degrees:* DSc, Davidson Col, 82. *Prof Exp:* Asst res urologist, New York Hosp & Cornell Univ Med Ctr, 48-50; resident urologist, Univ Va, 50-52; from instr to assoc prof, 52-55, PROF UROL, BOWMAN GRAY SCH MED, 60- *Honors & Awards:* Am Urol Asn Prize, 51, 52 & 54. *Mem:* Soc Exp Biol & Med; Am Urol Asn; Am Col Surgeons; Soc Univ Surgeons; Int Soc Urol; Am Asn Genito-Urinary Surgeons. *Res:* Plastic operations for congenital malformations of bladder and external genitalia; physiology of the urinary bladder; proteins, simple and conjugated, of urine in health and in various diseases; urinary calculous disease. *Mailing Add:* Rte 2 Box 150 Stuart VA 24171

BOYD, ALFRED COLTON, JR, b Buffalo, NY, Dec 12, 29; m 59; c 3. INORGANIC CHEMISTRY. *Educ:* Canisius Col, BS, 51; Purdue Univ, MS, 53, PhD(inorg chem), 57. *Prof Exp:* Asst prof chem, 57-68, asst dean col arts & sci, 69-74, ASSOC PROF CHEM, UNIV MD, COLLEGE PARK, 69- *Mem:* Am Chem Soc. *Res:* Boron and organometallic chemistry. *Mailing Add:* Dept Chem Univ Md College Park MD 20742

BOYD, CARL M, b Leavenworth, Kans, Mar 23, 33; m 55; c 4. BIOLOGICAL OCEANOGRAPHY. *Educ:* Univ Ind, AB, 55, MA, 56; Scripps Inst, Univ Calif, PhD(marine biol), 62. *Prof Exp:* From asst prof to assoc prof, 62-70, dir aquatron lab, 70-75, PROF MARINE BIOL, DEPT OCEANOG, DALHOUSIE UNIV, 70- *Concurrent Pos:* Nat Res Coun Can sr res fel, Sci Exchange, France, 68 & 69; sabbatical yr, Univ Paris, 75-76, off sci res & tech, New Caledonia & Allende Inst Mex, 82-83; assoc ed, Progress in Oceanog, 81-83; res assoc, Whitney Lab, Univ Fla, 87. *Mem:* Am Soc Limnol & Oceanog. *Res:* Membrane transport of marine algae; ecology of marine plankton; designs of instruments. *Mailing Add:* Dept Oceanog Dalhousie Univ Halifax NS B3H 4J1 Can

BOYD, CHARLES CURTIS, b Ottawa, Ill, Feb 18, 43; m 70; c 1. FORESTRY, AGRONOMY. *Educ:* Southern Ill Univ, BS, 66; Univ Hawaii, MS, 68; Wash State Univ, PhD(soils), 73. *Prof Exp:* Proj leader nursery agronomist, 74-77, SECT MGR, FOREST BIOL RES, WEYERHAEUSER CO, 77- *Concurrent Pos:* Res tech wheat res, US Agency Int Develop, 72. *Mem:* Agron Soc; Soil Sci Soc Am. *Res:* Development of technology for growing and handling conifer seedlings, rooted cuttings, tissue culture propagules, reforestation, vegetation control and insect control. *Mailing Add:* PO Box 420 Centralia WA 98531

BOYD, CLAUDE ELSON, b Hatley, Miss, Nov 14, 39; m 63; c 3. LIMNOLOGY, PLANT ECOLOGY. *Educ:* Miss State Univ, BS, 62, MS, 63; Auburn Univ, PhD(limnol), 66. *Prof Exp:* Aquatic biologist, Fed Water Pollution Control Admin, 66; asst prof res, Auburn Univ, 67-68; mem res staff, Savannah River Ecol Lab, 69-71; PROF FISHERIES & ALLIED AQUACULT, AUBURN UNIV, 71- *Concurrent Pos:* Consult to various corp, govt agencies & individuals. *Mem:* World Aquacult Soc; Am Fisheries Soc. *Res:* Water quality management in fish ponds; aquatic plant management. *Mailing Add:* Dept Fisheries & Aquacult Auburn Univ Auburn AL 36830

BOYD, DAVID CHARLES, b Pittsburgh, Pa, Dec 31, 42; m 66; c 2. GLASS TECHNOLOGY. *Educ:* Lehigh Univ, BS, 64; Purdue Univ, MS, 66, PhD(metall eng), 69. *Prof Exp:* SR ASSOC, CONSUMER PROD DEVELOP, CORNING GLASS WORKS, 69- *Mem:* Am Ceramic Soc; Sigma Xi. *Res:* Studies relating the properties of glasses and glass-ceramics to their compositions, particularly in the areas of consumer products, tableware and cookware; porcelain enamels. *Mailing Add:* 3669 Hibbard Rd Horseheads NY 14845

BOYD, DAVID WILLIAM, b Toronto, Ont, Sept 17, 41; m 64; c 3. GEOMETRIC PROPERTIES OF ALGEBRAIC NUMBERS, ALGORITHMS IN NUMBER THEORY. *Educ:* Carleton Univ, BSc, 63; Univ Toronto, MA, 64, PhD(math), 66. *Prof Exp:* Asst prof math, Univ Alta, 66-67; from asst prof to assoc prof, Calif Inst Technol, 67-74, PROF MATH, UNIV BC, 74- *Concurrent Pos:* Vis prof, Univ Paris VI, 81 & 84. *Honors & Awards:* E W R Steacie Prize, Nat Res Coun, Can, 78; Coxeter-James Lectr, Can Math Soc, 79. *Mem:* Math Asn Am; Am Math Soc; Can Math Soc (vpres, 79-81); Royal Soc Can. *Res:* Number theory, particularly the theory of the Pisot and Salem numbers; applications of computers to pure mathematics-number theory, algebra and geometry. *Mailing Add:* Dept Math Univ BC Vancouver BC V6T 1Y4 Can

BOYD, DEAN WELDON, b Shreveport, La, July 15, 41. DECISION ANALYSIS, POLICY ANALYSIS. *Educ:* Mass Inst Technol, BS, 63, MS, 65; Stanford Univ, PhD(eng & econ systs), 70. *Prof Exp:* Res engr, Jet Propulsion Lab, 65-67; decision analyst, SRI Int, 67-70; asst prof info sci, Univ Calif, Santa Cruz, 70-75; mgr consult, 75-77; PRIN & PRES, DECISION FOCUS, INC, 77- *Concurrent Pos:* Consult, var pvt & pub orgn, 70-75; adj fac, Col Bus Admin, Univ Ore, 80- *Mem:* Inst Mgr Sci. *Res:* Application of decision analysis to problems in public and private sectors; development of methodologies for logical selection of portfolios of interrelated activities such as research and develop projects, capital investments and so forth; normative theories for the value, use and selection of models. *Mailing Add:* 36245 Wagner Lane Cottage Grove OR 97424

BOYD, DEREK ASHLEY, b Port Elizabeth, SAfrica, Sept 27, 41; US citizen; m 67. FAR INFRARED SPECTROSCOPY, DIFFRACTIVE OPTICS. *Educ:* Univ Cape Town, SAfrica, BS, 64, Hons, 65, MS, 67; Stevens Inst Technol, PhD(physics), 73. *Prof Exp:* Res assoc, 73-74, from asst prof to assoc prof, 74-83, PROF PHYSICS, UNIV MD, COLLEGE PARK, 83-, CHMN DEPT, 90- *Concurrent Pos:* Res assoc, UK Atomic Energy Authority, 67, 69 & 70; vis scientist, Jet Joint Undertaking, UK, 81 & 88. *Mem:* Fel Am Phys Soc. *Res:* Developing experimental and theoretical methods for using electron-cyclotron emission to measure plasma parameters in controlled thermonuclear fusion devices. *Mailing Add:* Dept Physics Univ Md MD 20742

BOYD, DONALD BRADFORD, b Syracuse, NY, Oct 23, 41; m 65; c 3. COMPUTATIONAL CHEMISTRY. *Educ:* Pa State Univ, BS, 63; Harvard Univ, AM, 65, PhD(chem), 68. *Prof Exp:* Res assoc chem, Cornell Univ, 67-68; sr phys chemist, 68-74, res scientist, 75-89, SR RES SCIENTIST, LILLY RES LABS, ELI LILLY & CO, 90- *Concurrent Pos:* Fac mem Quantum Chem Prog Exchange Workshops, Ind Univ, 80-81, 83-84; adj prof chem, Ind Univ-Purdue Univ, Indianapolis, 82-; cochmn, Symp Molecular Mech, 83; founder & cochmn, Gordon Res Conf Comput Chem, 86 & 88; vis scientist, Royal Danish Sch Pharm, Copenhagen, 87; mem, Adv Bd Quantum Chem Prog Exchange, 87; chmn, Symp Molecular Design & Modeling, 90- *Mem:* Am Chem Soc; Sigma Xi. *Res:* Computational chemistry studies of pharmacological, organic, biochemical and inorganic molecules; relation between molecular structure and biological activity of drug molecules; computer-assisted molecular design & molecular modeling. *Mailing Add:* Lilly Res Labs Eli Lilly & Co Indianapolis IN 46285

BOYD, DONALD EDWARD, b Aruba, Neth, Jan 12, 33; US citizen; m 56; c 4. STRUCTURAL & ENGINEERING MECHANICS. *Educ:* Univ Okla, BS, 60; Univ Colo, MS, 63, PhD(civil eng), 65. *Prof Exp:* Engr, Aero Comdr, Inc, 57-60; design specialist, Martin-Marietta Corp, 60-65; mgr design dept, Brunswick Corp, 65; assoc prof civil eng, 65-69, ASSOC PROF MECH & AEROSPACE ENG, 69-; PRES, BOYD & ASSOC, INC, 78- *Mem:* Am Soc Mech Engrs; assoc fel Am Inst Aeronaut & Astronaut; Am Acad Mech; Am Soc Automotive Engrs. *Res:* Static and dynamic analysis of structures. *Mailing Add:* 77 Kensington Dr Bella Vista AR 72714

BOYD, DONALD WILKIN, b Newark, Ohio, Nov 1, 27; m 57. HISTORICAL GEOLOGY, INVERTEBRATE PALEONTOLOGY. *Educ:* Ohio State Univ, BS, 49; Columbia Univ, PhD(geol), 57. *Prof Exp:* Instr geol, Union Col, NY, 53-56; from asst prof to assoc prof, 56-66, PROF GEOL, UNIV WYO, 66- *Concurrent Pos:* Res assoc, Am Mus Natural Hist, 68- *Mem:* Geol Soc Am; Paleont Soc; Soc Econ Paleont & Mineral; Am Asn Petrol Geol. *Res:* Permian and triassic pelecypods; inorganic and organic sedimentary structures; Permian biostratigraphy and depositional environments. *Mailing Add:* Dept Geol & Geophys Univ Wyo Laramie WY 82071-3006

BOYD, EARL NEAL, b Trinity, Ky, Dec 20, 22; m 48; c 1. DAIRY SCIENCE. *Educ:* Eastern Ky State Col, BS, 48; Univ Ky, MS, 49; Ohio State Univ, PhD(dairy sci), 52. *Prof Exp:* Asst prof, Univ Ky, 52-55; res chemist, Swift & Co, 55-57; prin dairy technologist, US Dept Agr, 57-68; head dept food sci & technol, 68-70, asst dean & dir div basic sci, 70-78, ASSOC DEAN, COL AGR & LIFE SCI & ASSOC DIR, VA AGR EXP STA, VA POLYTECH INST & STATE UNIV, 78- *Concurrent Pos:* Vis res scientist, Dept Dairy Sci, Pa State Univ, 63-64; dir, Va Tech Sea Grant Prog, 78- *Mem:* Am Dairy Sci Asn; Inst Food Technol. *Res:* Chemical and bacteriological research in relation to milk and milk products; effects of processing on milk constituents. *Mailing Add:* 813 McBryde Dr Blacksburg VA 24060

BOYD, ELEANOR H, b Philadelphia, Pa, Oct 7, 35; m 64. NEUROPHARMACOLOGY, NEUROPHYSIOLOGY. *Educ:* Wellesley Col, BA, 56; Univ Rochester, PhD(pharmacol), 68. *Prof Exp:* Assoc physiol, Univ Rochester, 71-72, asst prof pharmacol & toxicol, Sch Med, 72-82, asst prof nursing, Sch Nursing, 73-82. *Concurrent Pos:* Prin investr, NIH grant, 78-81. *Mem:* Am Soc Pharmacol & Exp Therapeut; Soc Neurosci; AAAS. *Res:* The relationship of drug therapy to nursing practice. *Mailing Add:* 5225 Serenity Cove Bokeelia FL 33922

BOYD, FRANCIS R, b Boston, Mass, Jan 30, 26; m; c 2. GEOLOGY. *Educ:* Harvard Col, AB, 49; Stanford Univ, MS, 50; Harvard Univ, MS, 51, PhD, 58. *Prof Exp:* Fel, Harvard Univ, 52-53; staff asst, 53-56, STAFF MEM, GEOPHYS LAB, CARNEGIE INST WASH, 56- *Mem:* Nat Acad Sci; fel Mineral Soc Am; fel Geol Soc Am; Geochem Soc (pres, secy); fel Am Geophys Union; Int Asn Geochem & Cosmochem; Sigma Xi. *Res:* Phase equilibrium research at high pressures and temperatures. *Mailing Add:* Geophys Lab 5251 Broad Branch Rd NW Washington DC 20015-1305

BOYD, FRANK MCCALLA, b Canton, Ga, Mar 16, 29. BACTERIOLOGY. *Educ:* NGa Col, BS, 48; Univ Tenn, MS, 50; Univ Wis, PhD(bact), 59. *Prof Exp:* Bacteriologist, Ga Poultry Lab, 50-51; communicable dis ctr, USPHS, 51-54; res asst, Univ Wis, 54-58; asst poultry microbiologist, Univ Ga, 58-67; ASSOC PROF BIOL, NORTHEAST LA UNIV, 67- *Mem:* AAAS; Am Soc Microbiol; Poultry Sci Asn; Soc Gen Microbiol; Sigma Xi. *Res:* Agricultural and medical bacteriology. *Mailing Add:* Dept Biol Northeast La Univ Monroe LA 71209

BOYD, FREDERICK MERVIN, b Brazil, Ind, July 13, 39. CHEMISTRY. *Educ:* Purdue Univ, BS, 61, MS, 63; Univ Calif, Davis, PhD(plant physiol), 67. *Prof Exp:* Chemist, E I DuPont de Nemours & Co, Inc, 67-70, mem staff, Agr Sales, 70-71; dir biol, Res & Develop, Stull Chem Co, 71-72; chemist, 72-73, gen mgr mfg, 73-79, venture analyst, 79-80, dir res & develop, 80-86, GEN MGR, RES & ENGR, HUNTINGTON LABS, INC, 86- *Mem:* AAAS; Chem Specialists Mfrs Asn; Asn Off Anal Chemists; Am Oil Chemists Soc. *Res:* Developed production facilities for quaternary ammonuim chloride disinfectants; formulation research of products directed to the health care field and other speciality chemicals for sanitary maintenance of institutions; engineering; built hazardous materials containment building. *Mailing Add:* 1015 College Ave Huntington IN 46750

BOYD, FREDERICK TILGHMAN, b St Paul, Minn, Mar 9, 13; m 37; c 5. AGRONOMY. *Educ:* Univ Wis, BS, 34, PhD(soils, agron), 38. *Prof Exp:* Asst agron & soils, Univ Wis, 35-38; asst agronomist, Exp Sta, Fla, 38-41; agronomist, Raoul & Haney, Inc, Fla, 41-42; midwest agriculturist, Am Cyanamid Co, NY, 45-48; chief tech serv dept, Crow's Hybrid Corn Co, 48-50; supt plantation field lab & agronomist, Fla Exp Sta, 53-68; agronomist, Inst Food & Agr Sci, 68-74, EMER PROF AGRON, UNIV FLA, 74- *Concurrent Pos:* Consult, S Coast Sugar Corp, La, 48-52. *Mem:* AAAS; Am Soc Agron; Soil Sci Soc Am. *Res:* Fertilizer response on cultivated crops; effect of various cultural practices on growth of crops; chemicals as herbicides; corn, sorghum and grass breeding; nematodes in agronomic crops. *Mailing Add:* 5551 NW Fourth Pl Gainesville FL 32607

BOYD, GARY DELANE, b Los Angeles, Calif, Sept 14, 32; m 64; c 2. ELECTRICAL ENGINEERING, APPLIED PHYSICS. *Educ:* Calif Inst Technol, BS, 54, MS, 55, PhD(elec eng & physics), 59. *Prof Exp:* TECH STAFF, AT&T BELL LABS, 59- *Concurrent Pos:* Head, Systs Elem Res Dept, AT&T Bell Labs, 73-80, & Electronic Device Res Dept, 80-84. *Mem:* Optical Soc Am; fel Inst Elec & Electronics Engrs. *Res:* Quantum electronics; specifically resonator theory, solid state and semiconductor lasers, non-linear optical phenomena; liquid crystal display; surface acoustic waves; optical neural networks; optical bistable devices and photonic switching. *Mailing Add:* AT&T Bell Lab Rm 4B537 Holmdel NJ 07733

BOYD, GEORGE ADDISON, b Ambia, Tex, Nov 17, 07; m 44; c 4. RESEARCH ADMINISTRATION. *Educ:* Austin Col, AB, 29, AM, 30; Univ Iowa, MS, 35. *Prof Exp:* Res chemist, Phillips Petrol Co, Okla, 30-33; mgr info div, Develop & Patent Dept, Stand Oil Co, Ind, 35-37; asst to vpres in charge develop, Sharples Chem Co, 37-39; dir res, Ozone Processes Co, 40-41; res physicist, Biochem Res Found, Del, 41-44; res assoc, Sch Med & Dent, Univ Rochester, 44-50; prof biophys, Univ Tenn, 50-54; dir, Ariz Res Labs, 54-55; dir res grants & contracts, Ariz State Univ, 55-65, assoc dir ctr meteorite studies & managing ed, Meteoritics, 65-77; RETIRED. *Concurrent Pos:* Sr scientist, Oak Ridge Inst Nuclear Studies, 50-54; partic, NATO Int Inst Econ Forecasting on Sci Basis, Portugal, 67; develop officer, Gen Mercury Corp, 69-71; consult, Systs Mgt Corp, Mass, 70. *Mem:* AAAS; Am Soc Exp Biol; Sigma Xi. *Res:* Autoradiography; biophysics; meteoritics; petroleum refining; ozone generation; toxicity of radioactive elements. *Mailing Add:* 18 E 14th St Tempe AZ 85281

BOYD, GEORGE EDWARD, b Evansville, Ind, Sept 1, 11; m 42; c 1. SURFACE CHEMISTRY, POLYMER CHEMISTRY. *Educ:* Univ Chicago, BS, 33, PhD(phys chem), 37. *Prof Exp:* Rockefeller Found fel, Univ Chicago, 37-38, from instr to assoc prof, 38-48; assoc dir, Chem Div, Oak Ridge Nat Lab, 49-54, asst lab dir, 54-70, sr sci adv, 70-73; prof chem, Univ Ga, 73-82; RETIRED. *Concurrent Pos:* Sect chief metall lab, Univ Chicago & Clinton Lab, Manhattan Proj, 43-46; assoc ed, J Phys Chem, 50-54 & Anal Chem, 53-55; Fulbright scholar & Guggenheim fel, Univ Leiden, 52-53; mem adv comt nuclear data, Nat Res Coun, 55-60; mem, Atoms for Peace Mission, AEC, 57; chmn, Fulbright Comt Chem, Nat Acad Sci, 57-63; vis prof, Purdue Univ, 62; Am Nuclear Soc rep, Nat Res Coun, 66-69; co-dir, NATO Advan Study Inst, 73; Am Nuclear Soc rep, Nat Res Coun, 66-69. *Honors & Awards:* Southern Chem Award, Am Chem Soc, 51, Nuclear Appln Award, 69; Reilly lectr, Univ Notre Dame, 54; Welch Found lectr, 75; Charles H Stone Award, 76. *Mem:* Fel AAAS; Am Chem Soc; Am Phys Soc; fel Am Nuclear Soc; Coblentz Soc. *Res:* Physics and chemistry of surfaces; pure and applied nuclear chemistry; laser Raman and infrared spectra of molten salts and aqueous solutions; physical chemistry of polyelectrolyte solutions and gels; physical chemistry of concentrated electrolyte mixtures; technical management. *Mailing Add:* 65 Surfwatch Dr Kiawah Island John's Island SC 29455

BOYD, JAMES BROWN, b Denver, Colo, June 25, 37; m 60; c 2. MOLECULAR GENETICS. *Educ:* Cornell Univ, BA, 59; Calif Inst Technol, PhD(biochem), 65. *Prof Exp:* Helen Hay Whitney fel, Beerman Div, Max Planck Inst Biol, 65-69; from asst prof to assoc prof, 69-77, PROF GENETICS, UNIV CALIF, DAVIS, 77- *Concurrent Pos:* NATO fel, 73-74 & Guggenheim fel, 74-75; vis scholar, Stanford Univ, 88-89. *Honors & Awards:* Alexander von Humboldt US Scientist Prize. *Mem:* Genetics Soc Am; Sigma Xi. *Res:* Characterization of deoxyribonucleases; DNA repair; genetic and biochemical characterization of mutagen-sensitive mutants in Drosophila; molecular and genetic analysis of DNA repair in Drosophila; enzymology of DNA metabolism; molecular cloning and genetic characteristics of mutagen-sensitive mutants. *Mailing Add:* Dept Genetics Univ Calif Davis CA 95616

BOYD, JAMES EDWARD, physiology, pharmacology, for more information see previous edition

BOYD, JAMES EMORY, b Tignall, Ga, July 18, 06; m 34; c 2. NUCLEAR PHYSICS, GENERAL PHYSICS. *Educ:* Univ Ga, AB, 27; Duke Univ, MA, 28; Yale Univ, PhD(physics), 33. *Prof Exp:* Instr physics, Univ Ga, 28-30; head math-sci dept, WGa Col, 33-35; from asst prof to assoc prof physics, Ga Inst Technol, 35-42, prof, 46-61, actg pres, 71-72, res assoc, Eng Exp Sta, 46-50, head physics div, 50-54, from asst dir to dir, 54-61; pres, 61-71, EMER PRES, W GA COL, 74-; EMER PROF PHYSICS, GA INST TECHNOL, 74- *Concurrent Pos:* Vchancellor acad develop, Bd Regents, Univ Syst Ga, 72-74. *Mem:* Sigma Xi; Am Phys Soc; AAAS. *Res:* X-rays and microwave propagation. *Mailing Add:* 3747 Peachtree Rd Apt 1702 Atlanta GA 30319

BOYD, JEFFREY ALLEN, b Chapel Hill, NC, Mar 27, 58; m 87; c 2. MOLECULAR GENETICS. *Educ:* Duke Univ, BS, 80; NC State Univ, MS, 82, PhD(toxicol), 86. *Prof Exp:* Postdoctoral fel, Lineberger Cancer Res Ctr, Dept Path, Univ NC, Chapel Hill, 86-88; staff fel, 88-90, HEAD, GENE EXPRESSION SECT, LAB MOLECULAR CARCINOGENESIS, NAT INST ENVIRON HEALTH SCI, NIH, 90- *Mem:* Am Asn Cancer Res; Am Soc Cell Biol; Am Soc Microbiol; AAAS. *Res:* Molecular genetics and cell biology of human endometrial pathology, particularly carcinoma and endometriosis. *Mailing Add:* Nat Inst Environ Health Sci NIH PO Box 12233 Research Triangle Park NC 27709

BOYD, JOHN EDWARD, b Albany, Ga, Jan 26, 32; m 54; c 2. AGRICULTURAL BIOCHEMISTRY, DRUG METABOLISM. *Educ:* State Teachers Col Pa, BS, 52; Lehigh Univ, MS, 54; Pa State Univ, PhD(agr chem), 57. *Prof Exp:* Asst monomer chem, Lehigh Univ, 52-54; asst pesticide chem, Pa State Univ, 54-57, instr, 55-56; res chemist, 57-62, group leader, Metab Lab, 62-69, mgr, Metab & Anal Chem Sect, 69-81, DIR AGR RES, INFO SERV, AM CYANAMID CO, 81- *Mem:* Am Chem Soc; NY Acad Sci; AAAS. *Res:* Metabolism and mode of action of pesticides; sub-cellular biology; protein structure and synthesis; structure and function of cell wall and plasma membrane; data and information handling by computer. *Mailing Add:* Am Cyanamid Co PO Box 400 Princeton NJ 08540

BOYD, JOHN PHILIP, b Winchester, Mass, Feb 21, 51; m 87. APPLIED MATHEMATICS, ATMOSPHERIC DYNAMICS. *Educ:* Harvard Univ, AB, 73, SM, 75, PhD(appl physics), 76. *Prof Exp:* Asst, Nat Ctr Atmospheric Res, 76; from asst prof to assoc prof meteorol & numerical anal, 77-88, PROF ATMOSPHERIC & OCEANIC SCI, UNIV MICH, 88- *Concurrent Pos:* Vis res assoc, Ctr Earth & Planetary Physics, Harvard Univ, 80; assoc dir, Lab Sci Comput, 86- *Mem:* Am Meteorol Soc; Soc Indust & Appl Math. *Res:* Nonlinear and unstable planetary and equatorial waves in the atmosphere and ocean; Chebyshev and Fourier numerical methods. *Mailing Add:* Dept Atmospheric & Oceanic Sci Univ Mich 2455 Hayward Ave Ann Arbor MI 48109

BOYD, JOHN WILLIAM, b Ft Worth, Tex, Aug 28, 31; m 57; c 3. PHYSICS. *Educ:* Mich State Univ, BS, 57, PhD(physics), 63. *Prof Exp:* Sr scientist, Systs Res Labs, Ohio, 63-66; asst prof, 66-73, ASSOC PROF PHYSICS, KNOX COL, ILL, 73- *Mem:* AAAS; Am Phys Soc; Am Asn Physics Teachers; Sigma Xi. *Res:* Atomic and molecular structure and spectroscopy; optical masers; optics. *Mailing Add:* Dept Physics Knox Col Galesburg IL 61401

BOYD, JOSEPHINE WATSON, b St Paul, Minn, Feb 19, 27; m 56; c 3. WATER POLLUTION. *Educ:* Univ Minn, BChemE & BBusAdmin, 48, MS, 53, PhD(bact), 55. *Prof Exp:* Chemist & bacteriologist, Theo Hamm Brewing Co, 48-55; asst, Univ Minn, 51-55; res assoc bact, Arctic Inst NAm, 56-57, Ohio State Univ, 58-61 & Arctic Inst NAm, Alaska, 62-63 & Can, 64; fel microbiol, Colo State Univ, 68-74; asst dir water pollution assessment facil, City of Fort Collins, 74-78; tech writer, Hach Co, 78-80, mkt publ mgr, 80-84, sr tech ed, 84-90; RETIRED. *Concurrent Pos:* Vis scientist, Norway, 65 & 70; fel, Colo State Univ, 68-73. *Mem:* AAAS; Sigma Xi. *Res:* Microbiology in the arctic environment; waste water and sewage sludge considered as raw materials rather than refuse. *Mailing Add:* 1313 Stover St Ft Collins CO 80521

BOYD, JUANELL N, b Hollywood, Calif, Apr 10, 42. TOXICOLOGY. *Educ:* Westminster Col, BS, 64; Purdue Univ, MS, 66; Cornell Univ, PhD(toxicol, food chem & nutrit), 81; Am Bd Toxicol, dipl, 87. *Prof Exp:* Postdoctoral res, Cornell Univ, 81-86; SR RES TOXICOLOGIST, CORNING INC, 86- *Concurrent Pos:* New investr award, NIH, 83; mem, Comt Pub Commun, Soc Toxicol. *Mem:* Soc Toxicol; Am Chem Soc; NY Acad Sci. *Mailing Add:* Corning Inc HP-ME-03-070 Corning NY 14831

BOYD, L(ANDIS) L(EE), b Iowa, Dec 1, 23; m 46; c 5. AGRICULTURAL ENGINEERING. *Educ:* Iowa State Univ, BSAE, 47, MS, 48, PhD(agr eng, mech), 59. *Prof Exp:* From asst prof to prof agr eng, Cornell Univ, 48-64; prof & head dept, Univ Minn, St Paul, 64-72; asst dir, Agr Exp Sta, 72-78; dir res & assoc dean, Col Agr & Home Econ, Wash State Univ, 78-85; DIR-AT-LARGE, WESTERN ASN AGR EXP STA DIRS, 85- *Concurrent Pos:* Eng design analyst, Allis-Chalmers Mfg Co, 62-63; consult, Food & Agr Orgn, 64 & 69; fel, Univ Mich, 68. *Honors & Awards:* Metal Builders Mfrs Asn-Am Soc Agr Engrs Award, 68. *Mem:* Fel Am Soc Agr Engrs; Am Soc Eng Educ. *Res:* Farm structures and related equipment; engineering mechanics; theory of models. *Mailing Add:* Western Asn Agr Exp Sta Dirs Agr Exp Sta, Colo State Univ Ft Collins CO 80523

BOYD, LEROY HOUSTON, b Arnett, Okla, May 29, 35; m 58; c 2. ANIMAL SCIENCE. *Educ:* Okla State Univ, BS, 57; Univ Ky, MS, 60, PhD(animal husb), 63. *Prof Exp:* Asst animal husbandman, Univ Ky, 57-60; asst prof, 63-66, assoc prof animal husb, 66-80, PROF ANIMAL SCI, MISS STATE UNIV, 80- *Mem:* Am Soc Animal Sci. *Res:* Physiology, production and management of sheep and horses; ruminant nutrition. *Mailing Add:* PO Box 5228 Mississippi State MS 39762

BOYD, LOUIS JEFFERSON, b Lynn Grove, Ky, Mar 14, 28; m 48; c 4. REPRODUCTIVE PHYSIOLOGY. *Educ:* Univ Ky, BS, 50, MS, 51; Univ Ill, PhD(dairy physiol), 56. *Prof Exp:* Field agent, Dairy Exten, Univ Ky, 51-53; assoc prof dairy, Univ Tenn, 56-62; from assoc prof to prof dairy physiol, Mich State Univ, 63-72; prof, Animal Sci & Chmn Div, 72-79, head dept, Animal & Dairy Sci, 74-79, COORDR SPONS PROG, COL AGR, UNIV GA, 79-, COORDR INT AGR, 87- *Concurrent Pos:* Mem staff, Brit Milk Mkt Bd, 70-71; pres, Coun Agr Sci & Technol, 84-85. *Mem:* Fel AAAS; Am Dairy Sci Asn; Am Soc Animal Sci; Brit Soc Study Fertil; Sigma Xi. *Res:* Factors affecting reproductive efficiency in dairy cattle, such as production, extension, preservation and use of bull sperm and conception and gestation in the dairy cow. *Mailing Add:* 201 Conner Hall Col Agr Univ Ga Athens GA 30602

BOYD, MALCOLM R(OBERT), b Boulder, Colo, Jan 23, 24; m 47; c 2. ELECTRICAL ENGINEERING. *Educ:* State Univ, Colo, BS, 47; Stanford Univ, MS, 51, PhD(elec eng), 53. *Prof Exp:* Res assoc, Radio Corp Am Labs, 47-49; res asst microwave lab, Stanford Univ, 49-53; res assoc, Gen Elec Res Lab, 53-64; vpres, Mictron, Inc, 64-72, dir res, 64-68, tech dir, 68-72; pres, 72-76, tech dir, 76-90, PRES, VTM MICROWAVES INC, 90-; VPRES RES & DEVELOP, GENESIS ELECTRON DEVICES, 90- *Mem:* Inst Elec & Electronics Engrs. *Res:* Microwave electronics; microwave circuitry; gaseous electronics; ceramic sealing; vacuum technology. *Mailing Add:* Genesis Electron Devices 9345 Elm Ct Unit B Denver CO 80221

BOYD, MARY K, b Montreal, Que, Oct 16, 60; m 84. PHOTOCHEMISTRY. *Educ:* Univ Toronto, BSc, 84, PhD(chem), 88. *Prof Exp:* Postdoctoral fel org chem, State Univ NY, Stony Brook, 89-90; ASST PROF ORG CHEM, LOYOLA UNIV, CHICAGO, 90- *Mem:* Am Chem Soc; Inter Am Photochem Soc; AAAS; Am Asn Univ Women; Am Women Sci. *Res:* Organic and bioorganic photochemistry; excited state processes particularly involving the generation and reactions of carbocations; fluorescence and phosphorescence techniques. *Mailing Add:* Chem Dept Loyola Univ 6525 N Sheridan Rd Chicago IL 60626

BOYD, MICHAEL R, THERAPEUTICS RESEARCH. *Prof Exp:* CHIEF, LAB DRUG DISCOVERY RES & DEVELOP, NAT CANCER INST, NIH, 90- *Mailing Add:* Nat Cancer Inst Frederick Cancer Res & Development Ctr Bldg 1052 Rm 121 Frederick MD 21702

BOYD, MILTON JOHN, b Jamestown, NY, June 16, 41; m 64; c 3. MARINE ECOLOGY, INVERTEBRATE ZOOLOGY. *Educ:* Univ Calif, Berkeley, BA, 64; Univ Calif, Davis, PhD(zool), 72. *Prof Exp:* Res assoc ecol, Lerner Marine Lab, 71-72; PROF ZOOL, HUMBOLDT STATE UNIV, 72- *Concurrent Pos:* Prin investr, Redwood Nat Park, US Nat Park Serv, 74-76 & San Francisco Dist, US Army Corps Engrs, 75-86. *Mem:* AAAS; Ecol Soc Am; Am Soc Zoologists. *Res:* Community structure of marine intertidal and benthic assemblages; patterns of primary productivity in spermatophytes of Pacific coast salt marshes. *Mailing Add:* Dept Biol Humboldt State Univ Arcata CA 95521

BOYD, RICHARD HAYS, b Columbus, Ohio, Aug 12, 29; m 51; c 2. POLYMER SCIENCE. *Educ:* Ohio State Univ, BSc, 51; Mass Inst Technol, PhD, 55. *Prof Exp:* Res chemist, E I du Pont de Nemours & Co, 55-62; assoc prof chem, Utah State Univ, 62-65, prof, 65-67; chmn dept mat sci & eng, 76-82, PROF CHEM ENG, MAT SCI & ENG & ADJ PROF CHEM, UNIV UTAH, 67- *Concurrent Pos:* Vis fel, Swed Natural Res Coun; chmn, Div High Power Phys, Am Phys Soc, 85-86. *Honors & Awards:* High Polymer Physics Prize, Am Phys Soc, 88. *Mem:* Am Chem Soc; Sigma Xi; fel Am Phys Soc; Am Inst Chem Engrs. *Res:* Physical chemistry and physics of polymers; rheology, dielectric properties and kinetics; thermodynamics of solutions; thermochemistry. *Mailing Add:* 4772 Ichabod Dr Salt Lake City UT 84117

BOYD, RICHARD NELSON, b Omaha, Nebr, Apr 25, 40; m 62; c 3. NUCLEAR PHYSICS, PARTICLE PHYSICS. *Educ:* Univ Mich, BSE(physics) & BSE(math), 62; Univ Minn, PhD(physics), 67. *Prof Exp:* Res assoc physics, Univ Minn, 67-68, Rutgers Univ, 68-71 & Stanford Univ, 71-72; asst prof, Univ Rochester, 72-78; from assoc prof to prof physics, 78-85, PROF PHYSICS & ASTRON, OHIO STATE UNIV, 85- *Concurrent Pos:* Consult, Lawrence Livermore Nat Lab, 86- *Mem:* Fel Am Phys Soc; AAAS; Am Astron Soc. *Res:* Nuclear structure and reactions; nuclear astrophysics; searches for fundamental particles; stellar evolution; primordial nucleasynthesis; gamma-ray astronomy. *Mailing Add:* Dept Physics Ohio State Univ Columbus OH 43210

BOYD, ROBERT B, GROSS & MICROSCOPIC ANATOMY. *Educ:* Univ Okla, PhD(anat sci), 73. *Prof Exp:* ASSOC PROF ANAT, PA COL PODIATRIC MED, 81- *Mailing Add:* Dept Anat Pa Col Podiatric Med Eighth at Race St Philadelphia PA 19107

BOYD, ROBERT EDWARD, SR, b Clarksville, Tenn, Nov 5, 27; m 60; c 3. MEDICINAL CHEMISTRY. *Educ:* Tenn State Univ, BA, 54; Fisk Univ, MA, 57; Univ RI, PhD(med chem), 74. *Prof Exp:* Res chemist, Nat Cancer Inst, NIH, 61-65, chemist, USPHS, 65-67; assoc prof chem, Livingstone Col, 67-; AT DEPT CHEM, BENNETT COL. *Mem:* AAAS; Am Chem Soc. *Res:* Synthesis of potential antidepressant agents based on derivatives of biogenic amines; isolation and characterization of naturally occurring phospholipids. *Mailing Add:* Dept Chem Bennett Col Greensboro NC 27401

BOYD, ROBERT HENRY, b Norristown, Pa, Aug 27, 32; m 56; c 4. PHYSICAL CHEMISTRY. *Educ:* Lebanon Valley Col, BS, 54; Pa State Univ, PhD(chem), 59. *Prof Exp:* Res chemist, 58-72, RES SUPVR, E I DU PONT DE NEMOURS & CO, 72- *Mem:* Am Chem Soc; Soc Photog Sci & Eng. *Res:* Chemistry of photographic systems. *Mailing Add:* 28 Longspur Dr Limestone Hills Wilmington DE 19808-1971

BOYD, ROBERT WILLIAM, b Buffalo, NY, Mar 8, 48; m 71; c 3. NONLINEAR OPTICS. *Educ:* Mass Inst Technol, BS, 69; Univ Calif, Berkeley, PhD(physics), 77. *Prof Exp:* Res asst physics, Univ Calif, Berkeley, 77; from asst prof to assoc prof, 77-87, PROF OPTICS, INST OPTICS, UNIV ROCHESTER, 87- *Concurrent Pos:* Prin investr, NSF, Off Naval Res, Army Res Off, NY SCAOT grants, 77-88; consult of several maj US Corp, 77-88; vis scientist, Jet Propulsion Lab, Calif Inst Technol, 82-83 & Univ Sussex, Brighton, Eng; tech coun, Optical Soc Am, 83-85; topical ed, J Optical Soc Am 86-89; vis lectr, Univ Toronto, 87-88. *Mem:* Fel Optical Soc Am. *Res:* Nonlinear optical interactions in atomic vapors and nonlinear optical properties of materials; optical phase conjugation; deterministic chaos in nonlinear optics; development of new laser systems. *Mailing Add:* Inst Optics Univ Rochester Rochester NY 14627

BOYD, ROGER LEE, b Lawrence, Kans, Nov 2, 47; m 70; c 1. AVIAN ECOLOGY, ZOOLOGY. *Educ:* Baker Univ, BS, 69; Emporia State Univ, MS, 72; Colo State Univ, PhD(zool), 76. *Prof Exp:* From asst prof to assoc prof, 76-87, PROF BIOL, BAKER UNIV, 87-, DIR, BAKER UNIV NATURAL AREAS, 82- *Concurrent Pos:* Frank M Chapman mem fund res award, Am Mus Natural Hist, 75; Louis Agassiz Fuertes grant-in-aid, Wilson Ornith Soc, 75; chmn, Sci Div, Baker Univ, 87- *Honors & Awards:* Paul A Stewart Mem Award, Wilson Ornith Soc, 81; E Alexander Bergstrom Mem Award, NE Bird Banding Asn, 81. *Mem:* Sigma Xi; Am Ornithologists Union; Western Bird Banding Asn (second vpres, 75-77); Wilson Ornith Soc; Cooper Ornith Soc; Tallgrass Prairie Asn (pres, 85-). *Res:* Habitat ordination of avian populations in Kansas deciduous forests; breeding ecology and migration patterns of Kansas shorebirds; behavioral aspects of niche separation in grassland birds; population ecology of Interior Least Tern and Snowy Plover. *Mailing Add:* Dept Biol Baker Univ Baldwin City KS 66006

BOYD, RUSSELL JAYE, b Kelowna, BC, Sept 11, 45; m 68; c 2. THEORETICAL CHEMISTRY, QUANTUM CHEMISTRY. *Educ:* Univ BC, BSc, 67; McGill Univ, PhD(chem), 71. *Prof Exp:* Nat Res Coun fel chem, Math Inst, Univ Oxford, 71-73; Killam fel, Univ BC, 73-75; from asst prof to assoc prof, 75-85, PROF CHEM, DALHOUSIE UNIV, 85- *Concurrent Pos:* Prin investr, Natural Sci & Eng Res Coun Can grant-in-aid, 76-92; chmn, 8th Can Symp Theoret Chem, 83; vis prof, Lehrstuhl fur Theoretische Chemie, Univ Bonn, 84; ed, Can J Chem; vis prof, Eurkal Herriko Unibertsitatea, Spain, 87; chmn, Div Phys & Theoret Chem, Can Soc Chem, 90-93; Sr Killam fel, 89-90. *Honors & Awards:* Fraser Medal, APICS, 83; Can Nat Comt-Int Union of Pure & Appl Chem Award, 86. *Mem:* Fel Chem Inst Can; Can Asn Physicists; Am Chem Soc; Can Asn Theoret Chemists. *Res:* Electronic structure of atoms and molecules; electron correlation; topological aspects of molecular charge distributions; chemical concepts and the electron density; molecular structures and reaction mechanisms. *Mailing Add:* Dept Chem Dalhousie Univ Halifax NS B3H 3J5 Can

BOYD, VIRGINIA ANN LEWIS, b Shreveport, La, Nov 15, 44; m 64; c 2. VIROLOGY, GENETICS. *Educ:* Northwestern State Univ La, BS, 65, MS, 68; La State Univ, PhD(microbiol), 71. *Prof Exp:* Med technician path, Martin Army Hosp, Ft Benning, Ga, 65-66; instr biol, Jacksonville State Univ, 68-69; fel virol, Baylor Col Med, 71-73; scientist virol, Cancer Res Facil, Nat Cancer Inst, Frederick, 73-88; assoc prof biol & dir, Biomed Sci Masters Degree Prog, 82- 88, PROF & CHAIR BIOL, HOOD COL, 88- *Concurrent Pos:* Virol consult, biol abstr; deleg cell biol, People to People Int, Japan & People's Repub China. *Mem:* Am Soc Microbiol; Am Tissue Cult Asn; AAAS; Sigma Xi; NY Acad Sci; Am Soc Virol. *Res:* Microinjection of human epithelial cells with subgenomic DNA clones to map Epstein Barr Virus antigen and transforming regions; comparative studies of oncogene expression in mouse and human cultured cells; pleotrophic effects of oncogene proteins in cultured mouse and human cells; microinjection of proviral clones to determine infectivity and derive clonal retroviral stocks including HIV, BIV, BLV and EIAV; oncogenesis in human cells using various oncogenes and microinjection. *Mailing Add:* Hodson Sci Ctr Rm 121 Hood Col Frederick MD 21701

BOYD, WILLIAM LEE, b Kingsport, Tenn, Apr 20, 26; m 56; c 3. BACTERIOLOGY. *Educ:* Univ Tenn, BA, 50; Univ Minn, MS, 52, PhD(bact), 54. *Prof Exp:* Asst bact, Univ Tenn, 49-50 & Univ Minn, 50-54; asst prof, Univ Ga, 54-55; prin investr, Arctic Res Lab, Point Barrow, Alaska, 55-57; res assoc, Col Dent, Ohio State Univ, 58-59, from asst prof to assoc prof, 59-64; assoc prof, 64-67, actg head dept, 66-67, PROF BACT, COLO STATE UNIV, 67-, DIR WATER QUAL LAB, 82- *Concurrent Pos:* Prin investr, NSF grant, US Antarctica Res Prog, 61-63 & 67-68; vis scientist, Tromsk Mus, Norway, 65, 70, Arctic Res Lab, Barrow, Alaska, 66, 71 & Mus Natural Hist, Reykjavik, Iceland, 71. *Mem:* AAAS; Am Soc Microbiol; Arctic Inst NAm; Ecol Soc Am; NY Acad Sci. *Res:* Ecology and physiology of Arctic, Antarctic and Alpine microorganisms; microbes as indicators of water quality. *Mailing Add:* Dept Environ Health Colo State Univ Ft Collins CO 80523

BOYE, CHARLES ANDREW, JR, b Appalachia, Va, Nov 15, 28; m 50; c 2. POLYMER PHYSICS. *Educ:* Emory & Henry Col, BS, 50; Univ Tenn, MS, 56. *Prof Exp:* Teacher pub sch, Va, 53-54; asst physics, Univ Tenn, 54-56; physicist, 56-70, RES ASSOC, EASTMAN CHEM DIV, 70- *Concurrent Pos:* Asst prof, E Tenn State Col, 56-57. *Mem:* Am Phys Soc; Am Asn Textile Technologists; Sigma Xi. *Res:* X-ray diffraction of polymers; thermal analysis of polymers; polymer morphology; macromolecular structure; man made fibers. *Mailing Add:* Res Lab B-112 Eastman Chem Div Kingsport TN 37662

BOYE, FREDERICK C, b Buffalo, NY, May 30, 23; m; c 3. ORGANIC CHEMISTRY. *Educ:* Univ Buffalo, BA, 49, MA, 51, PhD(org chem), 52. *Prof Exp:* Res chemist, Cornell Aeronaut Lab, 52-53 & Indust Chem Div, Allied Chem Corp, 53-67; sr res chemist, Cowles Chem Div, Stauffer Chem Co, 67-68, Eastern Res Ctr, 68-71; sr res scientist, Paper Prod Res Div, Am Can Co, 71-72 & Consumer Prod Div, 72-78; sr res scientist, M&T Chem, Inc, 78-81; info scientist, Paper Chem Database & Tech Bibliogr, Inst Paper Chem, 81-89; ENVIRON SCIENTIST, GOVT REGULATIONS, PRINTING WKS, PLANT OPERS, BANTA CORP, 86- *Mem:* NY Acad Sci; Fel Am Inst Chemists; Sigma Xi. *Res:* Isocyanate resins; exploratory intermediates for colors; pharmaceuticals; polymers; curing agents; specialty chemicals; process variable studies; process development; plant demonstration; chelating chemicals; amino acids; detergent builders; organic acid anhydrides. *Mailing Add:* 130 Cherry Ct Appleton WI 54915

BOYE, ROBERT JAMES, b Shattock, Okla, Jan 11, 34. NUCLEAR PHYSICS, THEORETICAL PHYSICS. *Educ:* Univ Okla, BS, 55; Johns Hopkins Univ, PhD(physics), 68. *Prof Exp:* ASSOC PROF PHYSICS, MORGAN STATE COL, 61- *Mem:* Am Phys Soc; Am Asn Physics Teachers. *Res:* Molecular structure; nuclear three-body problem. *Mailing Add:* PO Box 209 Morgan State Col Baltimore MD 21239

BOYER, BARBARA CONTA, b Ithaca, NY, Jan 23, 42; m 68; c 2. DEVELOPMENTAL BIOLOGY. *Educ:* Univ Rochester, AB, 63; Univ Mich, MS, 64, PhD(zool), 69. *Prof Exp:* NIH trainee develop biol, Whitman Lab, Univ Chicago, 69-71; vis assoc prof, 73-87, ASSOC PROF BIOL, UNION COL, NY, 87- *Concurrent Pos:* Prin investr, Marine Biol Labs, Woods Hole, 83-; mem, Corp of Marine Biol Lab. *Mem:* Sigma Xi; Am Soc Zool; Western Soc Naturalists; AAAS; Soc Develop Biol. *Res:* Experimental embryology and gametogenesis of marine invertebrates, particularly turbellarian flatworms. *Mailing Add:* 1310 Garner Ave Schenectady NY 12309

BOYER, CINDA MARIE, b Lancaster, Pa, Aug 13, 54. MEDICINE. *Educ:* Dickinson Col, BS, 75; Pa State Univ, DPhil(microbiol), 80; Am Bd Med Lab Immunol, dipl, 86. *Prof Exp:* Grad res asst, Dept Microbiol, M S Hershey Med Ctr, Pa, 75-80; postdoctoral res fel clin immunol, Dept Hosp Labs, NC Mem Hosp, Chapel Hill, 80-82; postdoctoral res fel, Div Immunol, 82-85, ASST MED RES PROF, DEPT MED, MED CTR, DUKE UNIV, DURHAM, NC, 85- *Concurrent Pos:* Young investr award, Acad Clin Lab Physicians & Scientists, 82. *Mem:* Am Soc Microbiol; Am Asn Immunologists; Am Soc Histocompatibility & Immunogenetics. *Res:* Author of numerous publications. *Mailing Add:* Dept Med Duke Univ Med Ctr Box 3843 Durham NC 27710

BOYER, DELMAR LEE, b Salina, Kans, Oct 31, 26; div; c 6. MATHEMATICS. *Educ:* Kans Wesleyan Univ, AB, 49; Univ Kans, MA, 52, PhD(math), 55. *Prof Exp:* from asst prof to assoc prof, Univ Idaho, 61-65; PROF MATH, UNIV TEX, EL PASO, 65- *Mem:* Am Math Soc; Math Asn Am. *Res:* Abelian groups. *Mailing Add:* Dept Math Univ Tex El Paso TX 79968

BOYER, DON LAMAR, b Valley View, Pa, Apr 5, 38; m 60; c 2. OCEANOGRAPHY. *Educ:* Rensselaer Polytech Inst, BS, 60; Johns Hopkins Univ, PhD(mech), 65. *Prof Exp:* Asst prof civil, mech & aerospace eng, Univ Del, 65-69; assoc prog dir fluid mech, NSF, 69-70; assoc prof civil, mech & aerospace eng, Univ Del, 70-71; sci coordr atmospheric sci, NSF, 71-75; PROF MECH ENG & HEAD DEPT, UNIV WYO, 75- *Mem:* Am Soc Mech Engrs; Am Meteorol Soc; Am Geophys Union; AAAS; Europ Geophys Soc. *Res:* Geophysical fluid mechanics. *Mailing Add:* Dept Mech Eng & Aerospace Eng Ariz State Univ Tempe AZ 85287

BOYER, DON RAYMOND, b Lexington, Okla, Mar 31, 29; div; c 3. VERTEBRATE ZOOLOGY. *Educ:* Univ Okla, BS, 50, MS, 53; Tulane Univ, PhD(zool), 58. *Prof Exp:* Instr zool, Tulane Univ, 58; assoc prof, 58-66, PROF BIOL, WASHBURN UNIV, 66- *Mem:* AAAS; Am Soc Ichthyol & Herpet; Am Soc Zool; Am Inst Biol Sci. *Res:* Vertebrate zoology; physiology of temperature regulation and of respiratory and cardiovascular systemic responses in reptiles. *Mailing Add:* Dept Biol Washburn Univ Topeka KS 66621

BOYER, DONALD WAYNE, b Detroit, Mich, Jan 16, 29; m 56, 75; c 3. PHYSICS, AEROPHYSICS. *Educ:* Univ Melbourne, BSc, 53; Univ Toronto, MASc, 54, PhD(aerophys), 60. *Prof Exp:* Asst, Inst Aerospace Studies, Univ Toronto, 53-59; res aerodynamicist, 59-63, PRIN AERODYNAMICIST, CALSPAN CORP, 63- *Mem:* Sigma Xi; Am Inst Aeronaut & Astronaut. *Res:* Nonisentropic flows; viscous flows; hypersonic aerodynamics; chemical and ionization non-equilibrium phenomena; magneto gas dynamics; plasma-microwave interactions; infrared radiation physics; chemical kinetics; propellants and combustion. *Mailing Add:* 92 Sherbrooke Ave Amherst NY 14221

BOYER, ERNEST WENDELL, b Oconto, Nebr, Mar 18, 37; m 58; c 2. MICROBIAL PHYSIOLOGY, MICROBIAL GENETICS. *Educ:* NCent Col, Ill, BA, 58; Iowa State Univ, PhD(bact), 69. *Prof Exp:* Res technician, Mayo Clin, Minn, 62-64; res microbiologist, Miles Labs, Inc, 68-77; SR STAFF SCIENTIST, SOLVAM ENZYMES, INC, 90- *Mem:* AAAS; Am Soc Microbiol; Soc Indust Microbiol; Sigma Xi; Am Chem Soc. *Res:* Regulation of the synthesis of extracellular enzymes by bacteria and fungi; genetic engineering of industrial microorganisms; extracellular enzymes and taxonomy of alkalophilic bacillus species. *Mailing Add:* 1830 Grant Elkhart IN 46514-3949

BOYER, HAROLD EDWIN, b St Lawrence, Pa, Aug 9, 25; div; c 2. ORAL SURGERY. *Educ:* Lebanon Valley Col, 46-48; Univ Pa, DDS, 52, MS, 58; Am Bd Oral Surg, dipl, 59. *Prof Exp:* From asst dean to dean, Sch Dent, 62-72, PROF ORAL SURG, UNIV LOUISVILLE, 59-, VPRES HEALTH AFFAIRS, 72- *Concurrent Pos:* Chief sect oral surg, Louisville Gen & Children's Hosps, 56-71; consult, Jewish & Vet Admin Hosps, 63-, Ireland Army Hosp, Ft Knox, Ky, 63- & Off Econ Opportunity, Proj Head Start, 65-66; pres, Am Bd Oral Surg. *Mem:* Am Soc Oral Surg; fel Am Col Dent; fel Int Col Dent. *Res:* Early recognition of oral cancer; oral exfoliative cytology; evaluation of emergency drugs and development of a practical office emergency kit. *Mailing Add:* Health Sci Ctr Univ Louisville Louisville KY 40292

BOYER, HERBERT WAYNE, b Pittsburgh, Pa, July 10, 36; m 59; c 2. MICROBIOLOGY, MOLECULAR GENETICS. *Educ:* St Vincent Col, AB, 58; Univ Pittsburgh, MS, 60, PhD(bact), 63. *Hon Degrees:* DSc, St Vincent Col, 81. *Prof Exp:* From asst prof to assoc prof microbiol, 66-76, PROF DIV GENETICS, DEPT BIOCHEM & BIOPHYS, UNIV CALIF, SAN FRANCISCO, 76- *Concurrent Pos:* USPHS fel, Yale Univ, 63-66; dir, Grad Prog Genetics, Univ Calif, San Francisco, 76-81; investr, Howard Hughes Med Inst, 76-83; co-founder and mem bd dirs, Genentech, Inc, 77-; mem, President's Circle, Nat Acad Sci, 89. *Honors & Awards:* Nat Medal Technol, 89; Nat Medal Sci, 90; V D Mattia Award, Roche Inst Molecular Biol, 77; Doisy Lectr, Univ Ill, 79; Albert Lasker Basic Med Res Award, 80; Golden Plate Award, Am Acad Achievement, 81; Achievement Award, Indust Res Inst, 82; Moet Hennessy-Louis Vuitton Prize, Biotechnol Prize, 88; First Nat Biotechnol Award, 89. *Mem:* Nat Acad Sci; fel Am Acad Arts & Sci; Am Soc Biol Chemists; fel AAAS. *Res:* Molecular genetics. *Mailing Add:* Dept Biochem & Biophys Univ Calif San Francisco CA 94143-0554

BOYER, JAMES LORENZEN, b New York, NY, Aug 28, 36; m 63; c 2. HEPATOLOGY. *Educ:* Haverford Col, BA, 58; Johns Hopkins Univ, MD, 62. *Prof Exp:* Asst prof med, Sch Med, Yale Univ, 69-72; assoc prof internal med, Univ Chicago & Pritzerk Sch Med, 72-76, dir, Liver Study Unit, 72-78, prof med, 76-78; PROF MED & DIR, LIVER STUDY UNIT, SCH MED, YALE UNIV, 78-, CHIEF, DIV DIGESTIVE DIS, 82-, DIR LIVER CTR, 84- *Concurrent Pos:* Head, Liver Res Lab, Int Ctr Med Res & Training, Johns Hopkins Univ, 64-66; attend physician, Yale-New Haven Hosp & West Haven Vet Admin Ctr, 69-72 & 78-; dept chmn, Nat Digestive Dis Adv Bd, 81-84. *Mem:* Am Asn Study Liver Dis (pres, 80); Am Soc Clin Invest; Asn Am Physicians; Am Gastroenterol Asn; Int Asn Study Liver (pres, 88-90). *Res:* Mechanisms of bile formation and transport; study of liver structure and function; prognosis and treatment of chronic liver disease. *Mailing Add:* Dept Med Yale Sch Med 333 Cedar St New Haven CT 06510

BOYER, JERE MICHAEL, b Lebanon, Pa, Feb 24, 46; m 71; c 3. MICROBIAL PHYSIOLOGY, MYCOLOGY. *Educ:* Univ Scranton, BS, 68; Millersville State Col, MA, 71; St John's Univ, PhD(microbiol), 75. *Prof Exp:* From instr to prof microbiol, Philadelphia Col Osteop Med, 74-87, vchmn, 84-87; DIR RES, AULTMAN HOSP, 87-; ADJ PROF MICROBIOL, NORTHEASTERN OHIO UNIVS COL MED, 89- *Concurrent Pos:* Fel, St John's Univ, 71-74; NSF fel, 74; consult microbiol, Nat Bd Examrs Osteop Physicians & Surgeons, 75-; res grant, Merck, Sharp & Dohme, 78-87; consult microbiol biochem, NIH & NSF, 79-; res grant, Smith, Kline & French, 80-, Am Osteop Asn, 82-87 & Adv Technol Ctr, 84-88, Roche, 88-, Janssen, 89- *Mem:* Am Soc Microbiol; Soc Indust Microbiol; Med Mycol Soc of the Americas; Am Osteop Asn; Sigma Xi; AAAS; Am Asn Bioanalysts; fel Am Acad Microbiol; Am Bd Bioanal. *Res:* Physiological studies of antifungal agents and their effects on fungi; effects of metabolites on fungi; direct fungal diagnostic laboratory. *Mailing Add:* Dir Res Aultman Hosp 2600 6th St SW Canton OH 44710

BOYER, JOHN FREDERICK, b Evanston, Ill, Oct 14, 41; m 68; c 2. POPULATION BIOLOGY. *Educ:* Amherst Col, BA, 64; Univ Chicago, PhD(biol), 71. *Prof Exp:* Fel genetics, Dept Zool, Univ Iowa, 71-73; asst prof, 73-80, ASSOC PROF BIOL, UNION COL, 80- *Mem:* Soc Study Evolution; Ecol Soc Am; Sigma Xi; Am Soc Nat. *Res:* Tribolium populations; evolution of life histories; genetic variation in social behavior. *Mailing Add:* Dept Biol Sci Union Col Schenectady NY 12308

BOYER, JOHN STRICKLAND, b Cranford, NJ, May 1, 37; m 64; c 2. BIOCHEMISTRY, BIOPHYSICS. *Educ:* Univ Wis, MS, 61; Duke Univ, PhD(bot), 64. *Prof Exp:* Vis asst prof bot, Duke Univ, 64-65; asst physiologist, Conn Agr Exp Sta, 65-66; from asst prof to prof bot & agron, Univ Ill, Urbana, 66-78; plant physiologist, USDA, 78-84; prof, Tex A&M Univ, 84-87; DU PONT PROF MARINE BIOCHEM & BIOPHYS, UNIV DEL, 87- *Concurrent Pos:* Mem vis comt, Carnegie Inst Wash, Stanford Univ & Harvard Univ; Climate Lab fel, NZ, 72. *Honors & Awards:* Shull Award, Am Soc Plant Physiol, 77; von Humboldt Sr Scientist Award, Ger, 83. *Mem:* Nat Acad Sci; fel Am Soc Agron; Am Soc Plant Physiol (pres, 81-82); Sigma Xi; fel Crop Sci Soc Am. *Res:* Plant water relations, photosynthesis, plant growth and reproduction. *Mailing Add:* Col Marine Studies Univ Del 700 Pilottown Rd Lewes DE 19958

BOYER, JOSEPH HENRY, b Otto, Ind, Jan 4, 22. ORGANIC CHEMISTRY. *Educ:* Univ Ill, PhD(chem), 50. *Prof Exp:* Prof chem, Univ Ill, Chicago Circle, 66-85; RES PROF DEPT CHEM, UNIV NEW ORLEANS, 85- *Res:* Organic azides; nitro derivatives, nitrogen heterocycles, energetic materialsand laser dyes. *Mailing Add:* Dept Chem Sci Bldg Rm 2015 Univ New Orleans Lakefront New Orleans LA 70148

BOYER, LEROY T, b Ada, Minn, Mar 10, 37; m 61; c 2. CIVIL ENGINEERING. *Educ:* Univ Minn, BS, 60, PhD(civil eng, math), 66. *Prof Exp:* Asst prof, 66-71, assoc prof, 71-76, PROF CIVIL ENG, UNIV ILL, URBANA, 76- *Mem:* Am Soc Civil Engrs. *Res:* Construction engineering; computer assisted instruction and design; estimating, cost control and construction risk. *Mailing Add:* Dept Civil Eng Univ Ill Urbana IL 61801

BOYER, LESTER LEROY, b Hanover, Pa, Apr 6, 37; m 58; c 3. DAYLIGHTING IN BUILDINGS, EARTH SHELTERED BUILDINGS. *Educ:* Pa State Univ, BArchE, 60, MS, 64; Univ Calif, Berkeley, PhD(archit), 76. *Prof Exp:* Instr archit eng, Pa State Univ, 60-64; res engr, Res & Develop Ctr, Armstrong World Industs, 64-68; sr consult, Bolt Beranek & Newman Inc, 68-70; prof archit, Okla State Univ, 70-84; PROF ARCHIT, TEX A&M UNIV, 84- *Concurrent Pos:* Course dir, Nat Soc Prof Engrs, 64-74; prin investr, earth shelter, US Dept Energy & Control Data Corp, 79-81, daylighting, NSF, 85-88; ed proc, nat & int conferences on earth shelter, 80, 81, 83 & 86; nat coordr, Passive Earth Cooling Prog, Am Solar Energy Soc, 81; vis prof archit, Univ New South Wales & Univ Queensland, Australia, 82; Fulbright scholar to Australia, Coun Int Exchange Scholars, 82; gen chmn, Int Conf Earth Sheltered Bldg, Australia, 83, chair, Energy Res Rev Panel Fenestration, US Dept Energy, 88; vis researcher, Solar Energy Res Inst, Colo, 85; mem bd dirs, Am Underground Space Asn, 89-93; guest prof archit eng, Delft Univ Technol, Holland, 92. *Mem:* Am Solar Energy Soc; Am Underground Space Asn; Acoust Soc Am; Am Soc Heating Refrig & Air-Conditioning; Human Factors Soc; Illum Eng Soc. *Res:* Development and validation of preliminary prediction algorithms to evaluate daylighting, solar effects, and energy performance for selected prototypical daylighting configurations; minimum envelope systems for buildings, such as occur with underground and earth integrated buildings; author of publications on earth shelter technology. *Mailing Add:* 811 Camellia Ct College Station TX 77840

BOYER, MARKLEY HOLMES, b Philadelphia, Pa, Dec 29, 32; m 57; c 4. INTERNAL MEDICINE, TROPICAL MEDICINE. *Educ:* Princeton Univ, AB, 55; Univ Pa, MD, 63; Univ Oxford, DPhil(biol), 72; Harvard Univ, MPH, 72. *Prof Exp:* Asst resident, Univ Hosp, Cleveland, 65-67; physician, Hosp Albert Schweitzer, Haiti, 67; med officer, Radcliffe Infirmary, Eng, 68-70; USPHS fel trop pub health, 72, ASST PROF TROP PUB HEALTH, HARVARD SCH PUB HEALTH, 73- *Concurrent Pos:* Govt trustee, Jackson Lab, Maine, 72-; mem adv comt, Dept Biol, Princeton Univ, 76- *Mem:* AAAS; Royal Soc Med. *Res:* Immunology of parasitic diseases. *Mailing Add:* 209 W Canton St Boston MA 02116

BOYER, MICHAEL GEORGE, b Toronto, Ont, May 15, 25; m 52; c 3. PLANT PATHOLOGY. *Educ:* Ont Agr Col, BSA, 52, MS, 53; Iowa State Col, PhD(plant path), 58. *Prof Exp:* Lectr, Kemptville Agr Sch, Ont, 53-54; pathologist, Can Dept Forestry, 54-55, 58-65; asst prof, 65-68, ASSOC PROF BIOL, YORK UNIV, 68- *Concurrent Pos:* Mem, Can Comt Forest Tree Breeding. *Res:* Resistant poplar hybrids. *Mailing Add:* Dept Biol York Univ 4700 Keele St North York ON M3J 1P3 Can

BOYER, PAUL DELOS, b Provo, Utah, July 31, 18; m 39; c 3. BIOCHEMISTRY. *Educ:* Brigham Young Univ, BS, 39; Univ Wis, MS, 41, PhD(biochem), 43. *Hon Degrees:* Doctorate, Univ Stockholm, 74. *Prof Exp:* Asst biochem res, Univ Wis, 39-43; instr chem, Stanford Univ, 43-45, res assoc, 43-45; from asst prof to prof biochem, Univ Minn, 45-56, Hill prof, 56-63; dir, Molecular Biol Inst, 65-83, biotechnol prog, 85-88, PROF BIOCHEM, UNIV CALIF, LOS ANGELES, 63- *Concurrent Pos:* Guggenheim fel, 55-56; mem, Biochem Study Sect, USPHS, 58-62, chmn, 64-66; ed, Annual Rev Biochem, 65-71, assoc ed, 72-88; ed, Biochem Biophys Res Commun, 69-79; ed, The Enzymes, 70- *Honors & Awards:* Award, Am Chem Soc; McCoy Award Chem Res, 76; Tolman Medal, 84; Rose Award, Am Soc Biochem & Molecular Biol, 89. *Mem:* Nat Acad Sci; AAAS; Am Soc Biol Chem (pres, 69-70); Am Chem Soc; Am Acad Arts & Sci (vpres, 85-88). *Res:* Chemistry and mechanism of action of enzymes; oxidative phosphorylation; photophosphorylation; active transport. *Mailing Add:* Molecular Biol Inst Univ Calif Los Angeles CA 90024

BOYER, RAYMOND FOSTER, b Denver, Colo, Feb 6, 10; m 36, 66; c 4. THERMAL PHYSICS. *Educ:* Case Inst Technol, BS, 33, MS, 35. *Hon Degrees:* DSc, Case Inst Technol, 55. *Prof Exp:* Asst physics, Case Inst Technol, 33-35; mem staff, Dow Chem Co, 35-45, asst dir phys res lab, 45-48, dir, 48-52, dir plastics res, 52-68, asst dir corp res & develop polymer sci, 68-75; AFFIL SCIENTIST, MICH MOLECULAR INST, 75- *Concurrent Pos:* Guest, Soviet Acad Sci, 72, 78, 80, 87 & Polish Acad Sci, 73; res fel, Dow Chem Co, 72-75; vis prof, Case Western Reserve Univ, 74; sr vis fel, Univ Manchester, 75; partner, Boyer & Boyer, 75- *Honors & Awards:* Gold Medal & Int Award in Polymer Sci & Eng, Soc Plastics Eng, 68; Borden Award, Am Chem Soc, 70; Swinburne Gold Medal, Plastics Inst Gt Brit, 72. *Mem:* Nat Acad Eng; AAAS; Am Chem Soc; fel Am Phys Soc; NY Acad Sci. *Res:* Physics and physical chemistry of polystyrene family of high polymers; glass transition and related transitions or relaxations in polymers; liquid state transitions in polymers. *Mailing Add:* Mich Molecular Inst 1910 W St Andrews Midland MI 48640

BOYER, ROBERT ALLEN, b Hummels Wharf, Pa, Aug 27, 16; m 39; c 2. PHYSICS. *Educ:* Susquehanna Univ, AB, 38; Syracuse Univ, MA, 40; Lehigh Univ, PhD(physics), 52. *Hon Degrees:* DSc, Muhlenberg Col, 81. *Prof Exp:* Instr physics, Clarkson Col, 40-41; prof & chmn dept, 41-81, EMER PROF PHYSICS, MUHLENBERG COL, 81- *Honors & Awards:* Lindback Award, 61. *Mem:* Am Asn Physics Teachers. *Res:* Ultrasonics. *Mailing Add:* 2000 Cambridge Ave No 129 Wyomissing PA 19610

BOYER, ROBERT ELSTON, b Turners Sta, Ky, Apr 20, 39; m 59; c 2. GEOTECHNICAL, PAVEMENT. *Educ:* Univ Ky, BS, 61; Purdue Univ, MS, 63, PhD(civil eng), 72. *Prof Exp:* Chief, design div, eng & construct, Tachikawa AFB, Japan, 63-66, oper & maintenance div, Pleiku AFB, South Vietnam, 66-67; staff civil engr, Dir Oper, Hq Tactical Air Command, Langley AFB, Va, 67-69; primary researcher, Purdue Univ, W Lafayette, Ind, 69-72; course dir, Civil Eng Sch, Air Force Inst Technol, Wright-Patterson AFB, Ohio, 72-75, chief, dept eng technol, 75-77; dir civil eng, 12th Missile Warning Group, Thule AFB, Greenland, 77-78; chief, eng res div, Air Force Eng & Serv Ctr, Tyndall AFB, Fla, 79-81; comdr, 341st Civil Squadron, Malmstrom AFB, Mont, 81-83; dir eng & environ sci lab, Air Force Eng & Serv Ctr, Tyndall AFB, Fla, 83-; DIST ENG, ASPHALT INST. *Mem:* Fel Am Soc Civil Engrs; Soc Am Mil Engrs; Asn Asphalt Paving Technologists; Nat Soc Prof Engrs. *Res:* Soil mechanics: foundations, drainage, ground water and seepage; structural dynamics; pavements: design, evaluation and maintenance management systems. *Mailing Add:* Dept Geol Sci Univ Tex Austin TX 78712

BOYER, ROBERT ERNST, b Palmerton, Pa, Aug 3, 29; m 51; c 3. GEOLOGY. *Educ:* Colgate Univ, BA, 51; Ind Univ, MA, 54; Univ Mich, PhD, 59. *Prof Exp:* From instr geol to assoc prof geol, 57-67, chmn dept, 71-80, PROF GEOL SCI & EDUC, 67-, DEAN, COL NATURAL SCI, UNIV TEX, AUSTIN, 80- *Concurrent Pos:* Ed, Tex J Sci, 62-64 & J Geol Educ, 65-68. *Honors & Awards:* Neil A Miner Award, 80; Award Outstanding Contrib Pub Understanding Geol, Am Geol Inst, 88, William B Heroy, Jr Award Distinguished Serv, 90. *Mem:* Fel Geol Soc Am; Am Asn Petrol Geol; Am Asn Geol Teachers; fel AAAS; Sigma Xi. *Res:* Structural and field geology; mapping in Precambrian crystalline rocks of Wet Mountains, Colorado and Llano Region, Texas; heavy mineral studies of crystalline rocks; fracture pattern analysis; earth science teaching in secondary schools; space photography and remote sensing. *Mailing Add:* Dept Geol Sci Univ Tex Austin TX 78712

BOYER, RODNEY FREDERICK, b Omaha, Nebr, Aug 25, 42. BIOCHEMISTRY, ORGANIC CHEMISTRY. *Educ:* Westmar Col, BA, 64; Colo State Univ, MS, 67, PhD(chem), 70. *Prof Exp:* Res biochemist, Univ Mich Med Sch, 70-72; asst prof chem, Grand Valley State Cols, 72-74; from asst prof to assoc prof chem, 74-85, chair, 86-89, PROF CHEM, HOPE COL, 85- *Concurrent Pos:* NIH fel biol chem, Univ Mich, 70-72; sabbatical leave, Univ Colo, Boulder, 91. *Mem:* Am Chem Soc; Am Asn Plant Physiol; Am Soc Biochem & Molecular Biol. *Res:* Singlet oxygen in biochemistry; alpha-oxidation of fatty acids; metal complexes of biochemical ligands; cellulase storage and mobilization of iron in plants and animals. *Mailing Add:* Dept Chem Hope Col Holland MI 49423-3698

BOYER, RODNEY RAYMOND, b Aberdeen, Wash, July 4, 39; m 65; c 2. TITANIUM, PROCESS METALLURGY. *Educ:* Univ Wash, BS, 63, MS, 73. *Prof Exp:* Engr, Aerojet Gen Corp, 63-65; PRIN ENGR, BOEING COM AIRPLANE GROUP, 65- *Concurrent Pos:* Mem, Handbk Comt, Am Soc Mat Int, 82-85 & Aeromat Mgt Comt, 89-; asst chmn, Pac Northwest Mat Conf, 88-89; secy, 7th Int Conf Titanium, Am Inst Mining, Metall & Petrol Engrs, 88-; consult, Conoco, Inc, 90- *Mem:* Fel Am Soc Mat Int; Am Inst Mining Metall & Petrol Engrs. *Res:* Relationship between the metallurgy of titanium and its engineering properties; increasing the utilization of titanium by making it more cost effective through advances in proessing technologies; author and co-author of over 100 technical publications. *Mailing Add:* Boeing Com Airplane Group PO Box 3707 MS 74-44 Seattle WA 98124

BOYER, SAMUEL H, IV, b Duluth, Minn, Aug 16, 24; m 52; c 3. MEDICAL GENETICS, EXPERIMENTAL HEMATOLOGY. *Educ:* Stanford Univ, AB, 50, MD, 54. *Prof Exp:* From asst prof to assoc prof, 59-71, PROF MED & BIOL, SCH MED, JOHNS HOPKINS UNIV, 71- *Concurrent Pos:* USPHS fels, Johns Hopkins Univ, 56-58, Univ Mich, 58 & Univ Col, Univ London, 58-59; ed, Johns Hopkins Med J, 75-82; investr, Howard Hughes Med Inst, 76-86. *Mem:* Am Soc Human Genetics (secy, 62-64); Asn Am Physicians; Am Soc Clin Invest. *Res:* Human biochemical genetics; genetic polymorphisms; comparative biochemistry of protein; regulation of protein synthesis; cells; organization and evaluation of reiterated DNA; regulation of erythropoiesis. *Mailing Add:* Traylor Bldg Johns Hopkins Univ Sch Med 720 Rutland Ave Baltimore MD 21205

BOYER, TIMOTHY HOWARD, b New York, NY, Mar 20, 41; m 77; c 2. THEORETICAL PHYSICS. *Educ:* Yale Univ, BA, 62; Harvard Univ, MA, 63, PhD(physics), 68. *Prof Exp:* Fel physics, Univ Md, 68-70; from asst prof to assoc prof, 70-79, PROF PHYSICS, CITY COL NEW YORK, 80- *Mem:* AAAS; Am Phys Soc. *Res:* Stochastic electrodynamics and quantum theories; classical electron theory; zero-point energy; relations between classical and quantum theories; statistical thermodynamics; van der Waals forces. *Mailing Add:* Dept Physics City Col New York New York NY 10031

BOYER, VINCENT S, b Philadelphia, Pa, 1918. NUCLEAR ENGINEERING. *Educ:* Swarthmore Col, BSc; Univ Pa, MSc. *Hon Degrees:* DSc, Spring Garden Col, 79. *Prof Exp:* Eng plant tests, elec opers, Philadelphia Elec Co, 39-44, supvr, power stations, 46-53, asst supt, Cromby Station, 53-56, supt, 56-60, supt, Peach Bottom Atomic Power Station, 60-63, mgr, nuclear power, Elec Oprs Dept, 63-65, general supt, station operating, 65-67, mgr, elec opers, 67-68, vpres, engr res, 68-80, sr vpres, Nuclear Power, 80-87; CONSULT, 87- *Honors & Awards:* James N Landis Medal, Am Soc Mech Engrs, 81. *Mem:* Nat Acad Eng; fel Am Nuclear Soc (pres, 76-77); fel Am Soc Mech Engrs. *Mailing Add:* 1322 Grenox Rd Wynnewood PA 19096

BOYER, WILLIAM DAVIS, b Dayton, Ohio, Sept 27, 24; m 52, 81; c 4. FOREST ECOLOGY. *Educ:* US Merchant Marine Acad, BS, 50; State Univ NY, BS, 51, MS, 54; Duke Univ, PhD, 70. *Prof Exp:* Res asst water resources & power, Comn Orgn of Exec Br of Govt, 54-55; RES FORESTER, USDA, 55-, RES PROJ LEADER, 76- *Concurrent Pos:* Adj assoc prof, Auburn Univ. *Honors & Awards:* Tech Award, Southeastern Sect, Soc Am Foresters, 81. *Mem:* Ecol Soc Am; Soc Am Foresters; Sigma Xi. *Res:* Silvicultural problems of southern pine; vegetation management and weed plant control. *Mailing Add:* 852 Terrace Acres Auburn AL 36830

BOYERS, ALBERT SAGE, b Plainfield, NJ, July 16, 40. ENERGY CONSERVATION, SOLAR ENERGY. *Educ:* Purdue Univ, BS, 62; Univ Ill, MS, 64. *Prof Exp:* Engr, Cornell Aeronaut Lab, 66-68; EXTEN SPECIALIST MECH ENG, NC STATE UNIV, RALEIGH, 68- *Concurrent Pos:* Instr mech eng, State Univ NY, Buffalo, 66-67. *Mem:* Am Soc Heating Refrig & Air-Conditioning Engrs. *Res:* Passive and active solar energy systems for residential buildings; solar industrial process heat; industrial energy conservation; industrial ventilation systems. *Mailing Add:* Mech & Aerospace Eng Box 7910 Raleigh NC 27695

BOYETTE, JOSEPH GREENE, b Colerain, NC, May 10, 29; m 53; c 3. VERTEBRATE ECOLOGY, ACADEMIC ADMINISTRATION. *Educ:* E Carolina Col, BS, 56, MA, 57; NC State Univ, PhD, 66. *Prof Exp:* Chemist, NC State Dept Agr, 55-56; teacher, high sch, NC, 56-57; instr sci, 57-60, from asst prof to assoc prof, 60-69, asst dean grad sch, 70-73, PROF BIOL, E CAROLINA UNIV, 69-, DEAN GRAD SCH, 73- *Mem:* Fel AAAS; Am Soc Mammalogists. *Res:* Behavior of the pine mouse. *Mailing Add:* 1703 Beaumont Dr Greenville NC 27858

BOYKIN, DAVID WITHERS, JR, b Montgomery, Ala, Jan 6, 39; m 59; c 1. ORGANIC CHEMISTRY. *Educ:* Univ Ala, 61; Univ Va, 61, MS, 63, PhD(org chem), 65. *Prof Exp:* Fel org chem, Univ Va, 66-67; asst prof, 65-66, 67-68, assoc prof, 68-72, PROF ORG CHEM, GA STATE UNIV, 72-, CHMN DEPT, 74- *Mem:* Am Chem Soc; Int Soc Heterocyclic Chem. *Res:* Reaction mechanisms of heterocycles, small rings and alpha, beta-unsaturated ketones; oxygen-17 and carbon-13 nmr; synthetic medicinals. *Mailing Add:* Dept Chem Ga State Univ Univerity Plaza Atlanta GA 30303

BOYKIN, LORRAINE STITH, b Crewe, Va, Feb 1, 31; m 53. NUTRITION. *Educ:* Va State Col, BS, 51, MS, 54; NY Univ, MA, 59; Long Island Univ, MS, 64; Columbia Univ, prof dipl nutrit, 67, EdD(nutrit), 70. *Prof Exp:* Dietitian, New York City Dept Hosps, 51-52, A&T Col NC, 52-53 & NY Univ Hosp, 54-55; nutritionist, New York City Dept Health, 55-64; teacher home econ, New York Bd Educ, 64-65; from instr to asst prof nutrit, Pratt Inst, 65-70; from asst prof to assoc prof, 70-81, PROF NUTRIT, HUNTER COL, 81- *Concurrent Pos:* Adj assoc prof, NY Univ, 74-80; expert appointee nutrit, NIH, Bethesda, 78-79. *Mem:* Fel AAAS; fel NY Acad Sci; fel Am Geriat Soc; fel Am Pub Health Asn; fel Royal Soc Health. *Res:* Geriatrics and nutrition; alcoholism and nutrition; stress and nutrition; cultural nutrition. *Mailing Add:* 627 Powells Lane Westbury NY 11590

BOYKIN, WILLIAM H(ENRY), JR, b Birmingham, Ala, July 22, 38; m 61; c 2. CONTROL SYSTEMS, ENGINEERING MECHANICS. *Educ:* Auburn Univ, BSEP, 61, MSME, 63; Stanford Univ, PhD(systs theory), 67. *Prof Exp:* Res scientist, Hayes Int Corp, 61 & 64; instr mech eng, Auburn Univ, 61-64; res asst, Stanford Univ, 64-67; asst prof eng sci & elec eng, Univ Fla, 67-75; prof eng sci & elec eng, 75-80, assoc dir, Ctr Intel Mach & Robotics, 78-80; PRES, SYST DYNAMICS INC, 80- *Concurrent Pos:* Consult, US Army Missile Command, 69-78; Sci Appln, Inc, 77-78 & Naval Training Device Ctr, 77-; Westinghouse Nuclear Serv Div, 78, Martin Marietta, 78-, Comput Sci Corp, 79-80 & Hughes Helicopters, 80. *Mem:* Soc Automotive Eng; sr mem Inst Elec & Electronics Engrs; Sigma Xi; Am Inst Aeronaut & Astronaut; Am Soc Mech Eng. *Res:* Optimal control of complex systems including spacecraft and robots; design of missile guidance and control systems; development of analytical methods for analysis of digital control systems with multiple sampling rates; development of design and testing methods for precision laser pointing systems; optimal design of robot structures including joints and actuators. *Mailing Add:* Systs Dynamics Inc 1030 NW 39th St Gainesville FL 32605

BOYKINS, ERNEST ALOYSIUS, JR, b Vicksburg, Miss, Oct 5, 31; m 55; c 4. VERTEBRATE ZOOLOGY. *Educ:* Xavier Univ La, BS, 53; Tex Southern Univ, MS, 58, Mich State Univ, PhD(zool), 64. *Prof Exp:* Asst prof biol, Alcorn Agr & Mech Col, 54-61, prof, 64-71, dir div arts & sci, 70-71; pres, Miss Valley State Univ, 71-81; ASSOC PROF SCI, UNIV SOUTHERN MISS, 81- *Mem:* AAAS; Am Ornith Union. *Res:* Effects of dichlorodiphenyltrichloroethane on non-target animals. *Mailing Add:* Univ Southern Miss Box 9279 Hattiesburg MS 39406

BOYKO, EDWARD RAYMOND, physical chemistry; deceased, see previous edition for last biography

BOYLAN, D(AVID) R(AY), b Belleville, Kans, July 22, 22; m 44; c 4. CHEMICAL ENGINEERING, PROCESS DESIGN. *Educ:* Univ Kans, BS, 43; Iowa State Col, PhD(chem eng), 52. *Prof Exp:* Proj engr, Gen Chem Co, 43-47; sr engr, Am Cyanamid Co, 47; from asst prof to prof, Iowa State Univ, 48-59, assoc div eng, Exp Sta, 59-66, dean, Col Eng, 70-88, dir, Eng Res Inst, 66-88, PROF CHEM ENG, IOWA STATE UNIV, 88-; DIR ENG RES INST, IOWA STATE UNIV, 66- *Concurrent Pos:* Process consult, Dep Region 7, Nat Defense Exec Res; mem bd dirs, Stanley Consults. *Honors & Awards:* Merit Award, Am Chem Soc. *Mem:* Am Chem Soc; fel Am Inst Chem Engrs; Am Soc Eng Educ; Nat Soc Prof Engrs; Sigma Xi; fel AAAS. *Res:* Fertilizers; fertilizer technology; filtration. *Mailing Add:* 240 Sweeney Hall Col Eng Iowa State Univ Ames IA 50011

BOYLAN, EDWARD S, b New York, NY, Feb 8, 38. MATHEMATICS, COMPUTER SCIENCE. *Educ:* Columbia Univ, AB, 59; Princeton Univ, MA, 61, PhD(math), 62; Stevens Inst Technol, MS, 86. *Prof Exp:* Res assoc math, grad sch sci, Yeshiva Univ, 62-63; asst prof math, Rutgers Univ, 63-66; assoc prof math, Hunter Col, 66-68; chmn dept, 76-82, ASSOC PROF MATH, RUTGERS UNIV, 68- *Concurrent Pos:* Consult, Hudson Inst, NY, 70-82. *Mem:* Am Math Soc; Asn Comput Mach; Inst Mgt Sci. *Res:* Probability and inventory theory; possible connections between ergodic theory and martingale theory; strategic issues. *Mailing Add:* Dept Math Rutgers Univ Newark NJ 07102

BOYLAN, ELIZABETH S, b Shanghai, China, 46; US citizen; m; c 2. HORMONES & DEVELOPMENT, TRANSPLACENTAL CARCINOGENESIS. *Educ:* Wellesley Col, AB, 68; Cornell Univ, PhD(zool), 72. *Prof Exp:* Teaching asst embryol, Cornell Univ, 68-72; trainee fertil & reprod physiol, Marine Biol Lab, Woods Hole, 70-71; res assoc biochem & oncol, Med Ctr, Univ Rochester, 72-73; from asst prof to assoc prof, Queens Col, City Univ NY, 73-82, actg asst provost, 88-89, asst provost, 89-90, PROF BIOL, QUEENS COL, CITY UNIV NY, 83-, ASSOC PROVOST, 90- *Concurrent Pos:* Nat Cancer Inst res grant, 75-83; mem, Breast Cancer Task Force, Nat Cancer Inst, 80-83; mem adv comt, Am Cancer Soc, 81-85; consult, Dept Environ Safety Task Force, NIH, 85; chmn, Acad Sen, Queens Col, City Univ NY, 85-88; vis investr, Sloan Kettering Inst, Cancer Res, NY, 79-80; Grad Fac Biol, City Univ NY, 77-; res grants, Am Inst Cancer Res, 87-90 & Am Fedn Aging Res, 88-89. *Honors & Awards:* Sigma Xi, 73- *Mem:* Soc Develop Biol; Endocrine Soc, Am Asn Cancer Res; Am Soc Zoologists; AAAS; Int Asn Breast Cancer Res; Sigma Xi. *Res:* Effect of hormones on fetal tissue and its relation to mammary carcinogenesis; dietary fat and metastasis; granted one US patent. *Mailing Add:* Off Provost Queens Col City Univ NY Flushing NY 11367

BOYLAN, JOHN W, physiology, medicine; deceased, see previous edition for last biography

BOYLAN, WILLIAM J, b Bozeman, Mont, Dec 25, 29; m 58; c 4. ANIMAL BREEDING, QUANTITATIVE GENETICS. *Educ:* Mont State Univ, BS, 52; Univ Minn, MS, 59, PhD(genetics), 62. *Prof Exp:* From asst prof to assoc prof quant genetics, Univ Man, 62-66; assoc prof, 66-71, PROF ANIMAL SCI, UNIV MINN, ST PAUL, 71- *Concurrent Pos:* NSF fel, Quantitative Genetics, Brown Univ, 66; Sabbatical leave, Animal Breeding Res Orgn, Edinburgh, Scotland, 72-73; Fulbright scholar, Egypt, 82, Turkey, 88. *Mem:* Am Soc Animal Sci; Am Genetic Asn. *Res:* Quantitative inheritance in animals; design and evaluation of systems of selection and crossbreeding with emphasis on genetic improvement of livestock. *Mailing Add:* Dept Animal Sci Univ Minn St Paul MN 55108

BOYLE, A(RCHIBALD) R(AYMOND), b Sutton Coldfield, Eng, Apr 24, 20; m 45; c 2. ELECTRICAL ENGINEERING. *Educ:* Univ Birmingham, BSc, 40, PhD(phys chem), 43. *Prof Exp:* Exp officer, SRE Dept, Brit Admiralty, 45-46; tech dir, Dobbie McInnes (Electronics) Ltd, Scotland, 46-65; PROF ELEC ENG, UNIV SASK, 66- *Mem:* Can Inst Survrs; Am Cong Surv & Mapping. *Res:* Instrumentation; data handling; automatic cartography. *Mailing Add:* Dept Elec Eng Univ Sask Saskatoon SK S7N 0W0 Can

BOYLE, JAMES MARTIN, b Edwardsville, Ill, June 21, 42; m 70; c 2. PROGRAM TRANSFORMATION, PARALLEL PROCESSING. *Educ:* Northwestern Univ, BS, 64, MS, 65, PhD(appl math), 70. *Prof Exp:* Res assoc, 67-69, asst comput scientist, 69-74, COMPUT SCIENTIST MATH & COMPUT SCI DIV, ARGONNE NAT LAB, 74- *Concurrent Pos:* Consult, TRW Inc; mem, High Performance Comput Adv Bd. *Mem:* Asn Comput Mach; Soc Indust & Appl Math; Sigma Xi; AAAS. *Res:* Source-to-source transformations of computer programs; program correctness; mathematical software; abstract programming; parallel programming; automated reasoning. *Mailing Add:* Math & Comput Sci Div Bldg 221 Argonne Nat Lab Argonne IL 60439-4844

BOYLE, JOHN ALOYSIUS, JR, b New Orleans, La, Nov 5, 50; m 73; c 1. VIROLOGY. *Educ:* Univ New Orleans, BS, 72; Duke Univ, PhD(biochem), 78. *Prof Exp:* Fel molecular biophysics & biochem, Yale Univ, 78; asst prof, Miss State Univ, 78-84, assoc prof bicohem, 84-89, PROF & INTERIM HEAD BIOCHEM & MOLECULAR BIOL, MISS STATE UNIV, 89-, ADJ PROF BASIC & APPL SCI, COL VET MED, 89-; EXEC DIR, MISS ACAD SCI, 90- *Honors & Awards:* Young Res Award, Sigma Xi. *Mem:* Am Chem Soc; AAAS; Sigma Xi; Am Soc Microbiol; Am Soc Biochem & Molecular Biol. *Res:* Pathogen detection using nucleic acids; analytical techniques applied to nucleic acids; development of nucleic acid probes; use of polymerase chain reaction in analysis. *Mailing Add:* Dept Biochem PO Box Drawer BB Miss State Univ Mississippi State MS 39762

BOYLE, JOHN JOSEPH, b Middleport, Pa, Nov 23, 30; m 59. VIROLOGY. *Educ:* Pa State Univ, BS, 59, MS, 60, PhD(bact), 62. *Prof Exp:* Res biologist, Sterling-Winthrop Res Inst, 62; microbiologist, US Army Biolabs, 62-65; virologist, Smith Kline & French Labs, 65, sr virologist, 65-70; prod supt viral & bact vaccines, Merck Sharp & Dohme, 70-77; MEM STAFF, NIGHT VISION & ELECTRONICS OPTICS LAB, ELECTRONICS RES & DEVELOP COMMAND, 77- *Mem:* AAAS; Am Soc Microbiol; Sigma Xi; NY Acad Sci. *Res:* Viral chemotherapy; arboviruses; viral and rickettsial diseases; virus vaccines. *Mailing Add:* 12814 Cross Creek Lane Herndon VA 22071

BOYLE, JOSEPH, III, b Glen Ridge, NJ, Mar 10, 34; m 57; c 3. RESPIRATORY PHYSIOLOGY, CARDIOVASCULAR PHYSIOLOGY. *Educ:* Tufts Univ, BS, 56; Seton Hall Col Med, MD, 60. *Prof Exp:* Instr physiol, Seton Hall Col Med, 61-63; flight surgeon aerospace physiol, Brooks AFB, San Antonio, 63-66; assst prof, 66-68, ASSOC PROF PHYSIOL, NJ MED SCH, UNIV MED & DENT NJ, 68- *Mem:* Am Physiol Soc; Am Thoracic Soc; Sigma Xi. *Res:* Lung mechanics; effects of surfactant on lung function; computer based instruction; physiological modeling. *Mailing Add:* Dept Physiol NJ Med Sch Newark NJ 07103

BOYLE, KATHRYN MOYNE WARD DITTEMORE, b Wheaton, Kans; m 76; c 6. ORGANIC CHEMISTRY. *Educ:* Univ Kans, BA, 49; Univ Iowa, PhD(chem), 74. *Prof Exp:* Instr chem, Univ Wash, 74-75; from asst prof to assoc prof, 75-82, PROF CHEM, FRIENDS UNIV, 82- *Mem:* Am Chem Soc. *Mailing Add:* Dept Natural Sci Friends Univ 2100 University Ave Wichita KS 67213

BOYLE, MICHAEL DERMOT, b Belfast, Northern Ireland, Jan 4, 49; m 73; c 2. IMMUNOCHEMISTRY, COMPLEMENT. *Educ:* Univ Glasgow, Scotland, BSc, 71; Univ London, PhD(biochem), 74. *Prof Exp:* Vis fel, Nat Cancer Inst, Bethesda, Md, 74-76, expert, 76-80, vis scientist, 80; assoc prof immunol, Dept Immunol & Med Microbiol & Pediat, Col Med, Univ Fla, 81-85, prof, 85-88; PROF DEPT MICROBIOL, MED COL OHIO, TOLEDO, 88- *Concurrent Pos:* Assoc ed, Molecular & Cellular Biochem, 83-87; assoc ed, Biotechniques, 86- *Mem:* Am Asn Immunologists; Am Asn Cancer Res; Found Advan Educ Sci; Am Asn Vet Immunologists; Am Asn Microbiologists; Sigma Xi. *Res:* Immunochemistry of bacteria; complement and immuno-diagnostic technology. *Mailing Add:* Med Col Ohio CS 10008 Toledo OH 43699

BOYLE, RICHARD JAMES, b Westport, NY, Apr 19, 27; m 55; c 6. SOLVENT EXTRACTION. *Educ:* Univ Del, BS, 50; Univ Notre Dame, PhD(chem), 53. *Prof Exp:* Res chemist, Calco Chem Div, Am Cyanamid Co, 53-59, group leader, Plant Tech Dept, 59-61, tech rep, Res Dept, 61-62, mgr sales develop, Textile Chem Dept, Org Chem Div, 62-67, prod mgr, Mkt Develop Dept, 67-69, prod mgr, Decision Making Systs Dept, 69, field sales supvr, 69-70, tech dir, Indust Prod Dept, Europe-African Region, Cyanamid Int, 70-73, prod mgr, Indust & Mining Chems, 73-74, prod mgr, organic chems, Cyanamid Europe-Mideast-Africa Div, 74-81, mkt develop mgr phosphine chem, Organic Chem Div, 81-84, new prof develop mgr, Cyanamid Int Chem Div, 84-91; RETIRED. *Mem:* Am Chem Soc. *Res:* Analogs of podophyllotoxin; vat dyes; ultraviolet absorbers; textile chemicals; luminescent chemicals; elastomers; phosphine chemicals and solvent extraction. *Mailing Add:* 266 Altamont Pl Somerville NJ 08876

BOYLE, WALTER GORDON, JR, b Seattle, Wash, June 27, 28; m 58; c 2. COMPUTER APPLICATIONS FOR ANALYTICAL CHEMISTRY. *Educ:* Univ Wash, BS, 50, PhD(anal chem), 56. *Prof Exp:* Group leader, Lawrence Livermore Nat Lab, 63-67, res chemist, 77-80, systs mgr chem comput, 80-84, mgr data base methodology, 84-90, computer applns anal chem, 85-90; RETIRED. *Concurrent Pos:* Guest, Lawrence Livermore Nat Lab, 90- *Mem:* Am Chem Soc; Sigma Xi. *Res:* Determination of trace elements in metals and alloys; atomic absorption and flame emission spectroscopy; computer applications in analytical chemistry; nuclear magnetic resonance imaging; data base methods for analytical chemistry. *Mailing Add:* Lawrence Livermore Nat Lab Chem Dept L-311 Livermore CA 94550

BOYLE, WILLARD STERLING, b NS, Aug 19, 24; m 46; c 4. PHYSICS. *Educ:* McGill Univ, PhD(physics), 50. *Hon Degrees:* LLD, Dalhousie Univ, 84. *Prof Exp:* Lectr physics, McGill Univ, 50-51; asst prof, Royal Mil Col, 51-53; mem tech staff, Bell Tel Labs, 53-62, dir explor studies, Bellcomm, Inc, 62-64, dir semiconductor device lab, Bell Tel Labs, 64-68, exec dir, 68-80; CONSULT, 80- *Honors & Awards:* Stuart Ballantine Medal Award, Franklin Inst, 73; Morris N Liebmann Award, Inst Elec & Electronics Engrs, 74. *Mem:* Nat Acad Eng; fel Inst Elec & Electronics Engrs; fel Am Phys Soc. *Res:* Study of electronic devices and systems relation to communications; solid state physics; semiconductors. *Mailing Add:* Box 179 Wallace NS B0Y 1Y0 Can

BOYLE, WILLIAM C(HARLES), b Minneapolis, Minn, Apr 9, 36; m 59; c 4. CIVIL & ENVIRONMENTAL ENGINEERING. *Educ:* Univ Cincinnati, CE, 59, MS, 60; Calif Inst Technol, PhD, 63. *Prof Exp:* From asst prof to assoc prof, 63-70, chmn, dept civil eng, 84-86, PROF CIVIL ENG, UNIV WIS-MADISON, 70- *Concurrent Pos:* Chmn, Oxygen Trans Standards Comt, 78-, aeration comt, Am Soc Civil Eng, 81-90; chmn, Tech Pract Comm on Aeration, Water pollution Control Fedn, 83-90; prin engr, J M Montgomery, 88-89. *Honors & Awards:* Rudolf Hering Medal, Am Soc Civil Engrs, 75; Radebaugh Award, Water Pollution Control Fedn, 78; Founders Award, US Nat Comm Intl Water Pollution Res, 88; USEPA Commendation for Outstanding Leadership, Aeration Technol, 89; Harrison Prescot Eddy, Water Pollution Control Fedn, 90. *Mem:* Am Soc Civil Engrs; dipl Am Acad Environ Engrs; Am Water Works Asn; Water Pollution Control Fedn; Int Asn Water Pollution Res & Control; Am Asn Environ Engr Profs. *Res:* Water pollution control through biological waste treatment; mechanistic reactions in biosystems in waste and water treatment; oxygen transfer; on site waste treatment and disposal; waste water disinfection; leach testing solid wastes; fate of contaminants in soil. *Mailing Add:* 105 Carillon Dr Madison WI 53705

BOYLE, WILLIAM JOHNSTON, JR, b Albany, Ga, Jan 10, 44; m 66, 86; c 1. ORGANIC CHEMISTRY, POLYMER CHEMISTRY. *Educ:* Duke Univ, BS, 66; Northwestern Univ, PhD(org chem), 71. *Prof Exp:* Assoc org chem, Univ Calif, Santa Cruz, 70-73; res chemist, 73-76, sr res chemist, 76-85, RES ASSOC, ALLIED CORP, 85- *Concurrent Pos:* NSF fel, Univ Calif, Santa Cruz, 70-71. *Mem:* Am Chem Soc; AAAS; Sigma Xi. *Res:* Organic reaction mechanisms; synthetic organic chemistry; polymer synthesis. *Mailing Add:* Allied-Signal Inc PO Box 1021 Morristown NJ 07962

BOYLE, WILLIAM ROBERT, b Paterson, NJ, May 27, 32; m 59; c 3. CHEMICAL & NUCLEAR ENGINEERING. *Educ:* Newark Col Eng, BSChE, 54; WVa Univ, MSNuclear E, 61, PhD(chem eng), 65. *Prof Exp:* Instr chem eng, WVa Univ, 63-65, asst prof nuclear eng, 65-67, assoc prof nuclear & chem eng, 67-75, asst dir, Eng Exp Sta, 69-77, prof chem eng, 75-77; chmn, manpower educ, Res & Training Div, 77-89, VPRES & CHMN, ENERGY/ENVIRON SYSTS DIV, OAK RIDGE ASSOC UNIVS, 89- *Concurrent Pos:* Assoc investr, US AEC, 64-67, prin investr, 67-69. *Mem:* Am Soc Eng Educ; Nat Soc Prof Engrs; Am Inst Chem Engrs; Am Nuclear Soc; Eng Manpower Comn. *Res:* Developing and implementing methods and procedures for improving the participation of universities in research relevant to energy, environment and health. *Mailing Add:* Oak Ridge Assoc Univs PO Box 117 Oak Ridge TN 37831-0117

BOYLEN, CHARLES WILLIAM, b Baltimore, Md, Mar 29, 42; m 70; c 2. LIMNOLOGY, AQUATIC ECOLOGY. *Educ:* Ind Univ, AB, 64; Univ Wis-Madison, MS, 67, PhD(microbiol), 69. *Prof Exp:* Postdoctoral microbial ecol, Life Detection Unit, Ames Res Ctr, NASA, 69-71; postdoctoral microbial ecol, Univ Wis-Madison, 71-72; from asst prof to assoc prof, 72-88, DIR FRESH WATER INST, RENSSELAER POLYTECH INST, 85-, PROF BIOL, 89- *Concurrent Pos:* Site coordr, NSF Int Biol Prog, 74-76; appointee, Nat Comt Microbiol, Am Soc Microbiol, 75-78, pres, NY Br, 79-81; vis prof, Woods Hole Oceanog Inst, Mass, 82-83; co-dir environ eng & environ sci, Rensslaer Polytech Inst, 87-91. *Mem:* Am Soc Microbiol; Am Soc Limnol & Oceanog; AAAS; Ecol Soc Am; NAm Lake Mgt Soc. *Res:* Limnology of acid and nutrient impacted lakes; microbial ecology of stressed ecosystems; ecology of invasive aquatic plant species; bioremediation; environmental science; freshwater ecology. *Mailing Add:* Fresh Water Inst Rensselaer Polytech Inst Troy NY 12181

BOYLEN, JOYCE BEATRICE, b Weston, Ont, Aug 15, 26. BIOCHEMISTRY. *Educ:* Univ Toronto, BA, 47, MA, 49; McGill Univ, PhD(biochem), 61. *Prof Exp:* Jr res officer biol, Atomic Energy Can, Ltd, 49-53; Nuffield res asst genetics, Univ Leicester, 53-56; res asst biochem, McGill Univ, 57-60; sect head tissue culture, Frank W Horner, Ltd, Can, 60-63, sect head virol, 63-64; res biochemist, ICI Americas Inc, 64-68, clin info coordr, Stuart Pharmaceut Div, 68-90; RETIRED. *Concurrent Pos:* Mem, Coun Biol Ed. *Mem:* Drug Info Asn; Am Chem Soc; Am Med Writers Asn; AAAS. *Res:* Clinical studies of new drugs. *Mailing Add:* 801 Maple Ave Bellefonte Wilmington DE 19809

BOYLES, JAMES GLENN, b Harrisburg, Pa, Mar 21, 37; m 60; c 2. PHYSICAL CHEMISTRY. *Educ:* Pa State Univ, BS, 59; Rutgers Univ, PhD(chem), 66. *Prof Exp:* Instr chem, Rutgers Univ, 65-66; from asst prof to assoc prof, 66-80, PROF CHEM, BATES COL, 80-, CHMN DEPT, 77- *Mem:* Am Chem Soc. *Res:* Chemical reaction kinetics and mechanisms. *Mailing Add:* Box 637 Poland Range Rd Bates Col Pownal ME 04069

BOYLES, JAMES MCGREGOR, b Birmingham, Ala, Oct 1, 26; m 54; c 2. MEDICAL PARASITOLOGY. *Educ:* Univ Ala, BS, 51, MS, 52, PhD(biol), 66. *Prof Exp:* Instr biol, Mobile Ctr, Univ Ala, 56-64; from instr to asst prof, 64-71, ASSOC PROF BIOL, UNIV S ALA, 71- *Mem:* Am Soc Parasitol; Am Soc Trop Med & Hyg. *Res:* Amebic meningoencephalitis. *Mailing Add:* Dept Biol Univ SAla 307 University Dr Mobile AL 36688

BOYLES, JANET, LIPID TRANSPORT, CYTOSKELETAL STRUCTURE. *Educ:* Univ Calif, Berkeley, PhD(physiol & anat), 78. *Prof Exp:* Fel, Yale Univ Sch Med, 79-82; ADJ ASST PROF PATH, UNIV CALIF, SAN FRANCISCO, 82- *Mem:* Am Soc Cell Biol; fel Am Heart Asn; Am Soc Hemat; Electron Micros Soc Am; NY Acad Sci; AAAS. *Mailing Add:* Gladstone Found Labs Univ Calif PO Box 40608 San Francisco CA 94140

BOYLES, RICHARD QUINN, aeronautical engineering, for more information see previous edition

BOYNE, PHILIP JOHN, b Houlton, Maine, May 1, 24; m 46; c 2. ORAL SURGERY, ANATOMY. *Educ:* Tufts Univ, DMD, 47; Georgetown Univ, MS, 61. *Hon Degrees:* DSc, NJ Col Med & Dent, 75. *Prof Exp:* Instr anat, Tufts Univ, 47-49; resident oral surg, US Naval Dent Sch, Nat Naval Med Ctr, 54-55, staff guest scientist, US Naval Med Res Inst, 55-57, dir dent res dept, 65-68; prof dent, Sch Dent, Univ Calif, Los Angeles, 68-69, prof oral surg & chmn div, 69-74, asst dean, Sch Dent, 71-74; dir bone res lab, Univ Tex Health Sci Ctr, San Antonio, dean dent sch & prof oral surg, Dent & Med Schs, 74-78; DIR ORAL SURG SECT, DEPT SURG, LOMA LINDA UNIV MED CTR, 78- *Concurrent Pos:* Guest lectr, Sch Dent, Georgetown

Univ, 60-62 & US Naval Dent Sch, Nat Naval Med Ctr, 65-68; mem adv comt, Am Bd Oral & Maxillofacial Surg, 66-72, dir, 76-, vpres, 82, pres, 83. *Honors & Awards:* Career Achievement Award Res-Am Soc Maxillo-Facial Surg, 75. *Mem:* Am Dent Asn; Am Soc Oral Surg; fel Am Col Dent; fel Int Col Dentists; Int Asn Dent Res; Am Inst Oral Biol (pres, 70-). *Res:* Oral surgical research; bone healing; response of osseous tissues to trauma bone growth and development; tissue transplantation in maxillo-facial surgery. *Mailing Add:* Dept Oral Surg Attn: Dr Boyne Loma Linda Univ Sch Dent Loma Linda CA 92350

BOYNE, WILLIAM JOSEPH, b Troy, NY, Sept 28, 26; m 52; c 4. CHEMICAL ENGINEERING. *Educ:* Purdue Univ, BS, 48, PhD(chem eng), 52. *Prof Exp:* Asst chem eng, Purdue Univ, 48-50; engr, Am Cyanamid Co, 52-54, sr engr, 54-61; tech mgr, Monochem Inc, 61-66, eng mgr, 66-68, vpres opers, 68-69; sr engr, Borden Chem Div, Borden Inc, 59-61, dir eng, 69-72, dir chem planning, eng & res & develop, 72-79; MGR MFG PETROCHEM, ASHLAND CHEM CO, 80- *Mem:* Am Inst Chem Engrs; Am Chem Soc. *Res:* Acetlyene; air separation; vinyl chloride; ammonia-urea; methanol; coated fabrics; phenol- and urea-formaldehyde resins; UV coatings; hot melt adhesives; aerosols; phosphoric acid and fertilizers; uranium recovery; sulfuric acid; fluosilicic acid. *Mailing Add:* 6008 Kirkwall Ct West Dublin OH 43017-9002

BOYNTON, JOHN E, b Duluth, Minn, June 3, 38. GENETICS. *Educ:* Univ Ariz, BS, 60; Univ Calif, Davis, PhD(genetics), 66. *Prof Exp:* NIH Postdoctoral fel genetics, Univ Calif, Davis, 66 & Inst Genetics, Copenhagen Univ, 66-68; from asst prof to assoc prof, 68-77, PROF BOT, DUKE UNIV, 77- *Concurrent Pos:* Res career develop award, NIH, 72-77. *Honors & Awards:* Campbell Award, Am Inst Biol Sci, 67. *Mem:* Genetics Soc Am; Int Soc Plant Molecular Biol. *Res:* Genetic control of organelle structure and function. *Mailing Add:* Dept Bot Duke Univ Durham NC 27706

BOYNTON, JOHN E, b New York, NY, Dec 16, 50; m 74; c 2. ESTUARINE ECOLOGY, ECOLOGY OF PROTOZOA. *Educ:* Univ Conn, BS, 73; Cent Conn State Col, MS, 76; Univ Md, PhD(marine & estuarine ecol), 85. *Prof Exp:* Teacher biol, Northeast Sch, Bristol, Conn, 74-80; ASST PROF BIOL, ST JOSEPH'S COL, 85- *Mem:* AAAS; Am Soc Limnol & Oceanog; Soc Protozoologists; Nat Sci Teachers Asn; Nat Marine Educrs Asn; Atlantic Estuarine Res Soc. *Res:* Determination of sediment-water column exchange of nutrients, particularly ammonium, in the Patchogue River Estuary; development of nitrogen budget for Patchogue Lake. *Mailing Add:* 36 Clark St Port Jefferson NY 11776

BOYNTON, ROBERT M, b Evanston, Ill, Oct 28, 24; m 47; c 4. PSYCHOLOGY, OPTICS. *Educ:* Amherst Col, AB, 48; Brown Univ, ScM, 50, PhD(psychol), 52. *Prof Exp:* From asst prof psychol to prof psychol & optics, Univ Rochester, 52-74, chmn dept psychol, 71-74; prof psychol, Univ Calif, San Diego, 74-91, assoc dean, grad studies & res, 87-91; RETIRED. *Concurrent Pos:* Mem, Nat Res Coun-Armed Forces Comt Vision, 57-65, Exec Coun, 62-65; mem, US Nat Comt, Int Comn Illumination, 59-; NIH study fel, Eng, 60-61; dir, Ctr Visual Sci, Univ Rochester, 63-71; mem, Visual Sci Study Sect, NIH, 64-67; dir-at-large, Optical Soc Am, 66-69; vis prof, San Francisco Med Ctr, 69-70; chmn, Visual Sci B Study Sect, NIH, 73-77; dir, Asn Vision & Ophthal, 83-88, vpres, 88. *Honors & Awards:* Tillyer Medal, Optical Soc Am, 72; Godlove Award, Inter-Soc Color Coun, 81. *Mem:* Nat Acad Sci; fel Optical Soc Am; fel Am Psychol Asn; Soc Exp Psychologists; fel AAAS. *Res:* Psychology, physics and physiology of human vision. *Mailing Add:* Dept Psychol C-009 Univ Calif-San Diego La Jolla CA 92093

BOYNTON, W(ILLIAM) W(ENTWORTH), b Alameda, Calif, 04; m 40. MINING ENGINEERING. *Educ:* Univ Nev, BS, 29. *Prof Exp:* Res engr, Boeing Airport, Reno, 28-29; construct supt, You Bet & Red Dog Mining Co, 33-34; hydraul model study, US Eng Dept, 34-36; coord engr, Consol Steel Co, 41-43; design engr, US Rubber Co, 44; mech engr, Calif Inst Technol, 45; mech engr physics div, Naval Ord Test Sta, 46-50; prin engr, Cornell Aeronaut Lab, 51; dir, Boynton Assocs, 51-88; RETIRED. *Concurrent Pos:* Researcher, NAm Aviation, 52-56. *Mem:* Am Soc Mech Engrs. *Res:* Neutron and gamma ray attenuation in various soils and materials; atomic blast simulation and application of scientific and engineering principles in solving industrial and agricultural problems. *Mailing Add:* 1508 S Dobson San Pedro CA 90732

BOYNTON, WALTER RAYMOND, b Lawrence, Mass, May 5, 47; m 73. MARINE ECOLOGY. *Educ:* Springfield Col, BS, 69; Univ NC, Chapel Hill, MS, 74; Univ Fla, PhD(environ eng), 75. *Prof Exp:* Res asst, 69-70, RES ASSOC MARINE ECOL, CHESAPEAKE BIOL LAB, UNIV MD, 75- *Mem:* Estuarine Res Fedn; Sigma Xi. *Res:* Population studies of estuarine fish populations; coastal zone management with special emphasis on power plant siting. *Mailing Add:* Chesapeake Bio Lab Box 38 Solomons MD 20688

BOYNTON, WILLIAM VANDEGRIFT, b Bridgeport, Conn, Oct 29, 44; m 74; c 1. GEOCHEMISTRY, COSMOCHEMISTRY. *Educ:* Wesleyan Univ, BA, 66; Carnegie-Mellon Univ, PhD(phys chem), 71. *Prof Exp:* Res assoc geochem, Ore State Univ, 71-74; asst res geochemist, Univ Calif, 74-77; from asst prof to assoc prof, 77-85, PROF, DEPT PLANETARY SCI, UNIV ARIZ, 86- *Concurrent Pos:* Team leader & prin investr, Mars Observer Mission. *Honors & Awards:* NASA Group Achievement Award. *Mem:* AAAS; Meteoritical Soc; Am Chem Soc; Am Geophys Union. *Res:* Theromdynamics of trace element condensation from the solar nebula; trace element solid solution formation; neutron activation analysis; meteorites; comets; planetary surface composition; gamma-ray spectrometry. *Mailing Add:* Lunar & Planetary Lab Univ Ariz Tucson AZ 85721

BOYSE, EDWARD A, b Worthing, Eng, Aug 11, 23; UK & US citizen. IMMUNOGENETICS. *Educ:* Univ London, MB & BS, 52, MD, 57. *Prof Exp:* Staff, Dept Path, Guy's Hosp, London, 55-56, res fel path, 57-60; public health lab serv, Bact & Virol Sect, London, 56-57; res fel path, Sch Med, NY Univ, 60-61, assoc prof, 61-64; assoc prof, 66-69, PROF BIOL, GRAD SCH MED SCI, CORNELL UNIV, 69-; DISTINGUISHED PROF MICROBIOL & IMMUNOL, UNIV ARIZ, 89- *Concurrent Pos:* Casualty Off & Ear, Nose & Throat Off, Essex County Hosp, Colchester, 52, house surgeon, 53; house phys, Metropolitan Hosp, London, 55, med registr, 54- 55; staff, Dept Path, Guy's Hosp, London, 55-56, Pub Health Lab Serv, 56-57; immunobiol study sect, Nat Inst Health, 65-69; mem clin review bd & exec comt, Monell Chem Senses Ctr, Philadelphia; sci adv bd, Showa Univ Res Inst Biomed, St Petersburg, Fla; sci adv coun, Roswell Park Mem Inst, Buffalo, NY; Am Cancer Soc Res prof, Cell Surface Immunogenetics, 77; mem, Am Acad Arts & Sci, 77; mem, Am Acad Sci, 79; assoc scientist, Sloan-Ketting Inst, Can Res, 67; adj assoc prof path, Sch Med, NY Univ, 64-71, adj prof, 71-; Am Cancer Soc res prof cell surface immunogenetics, 77; outstanding investr award, Nat Cancer Inst, 85. *Honors & Awards:* Harvey Lectr, 75; Cancer Res Inst Award, Tumor Immunol, 75; Isaac Adler Award, Rockefeller & Harvard Univ, 76. *Mem:* Nat Acad Sci; fel Royal Soc; Am Acad Arts & Sci. *Mailing Add:* Col Med Univ Ariz Tucson AZ 85724

BOZAK, RICHARD EDWARD, b Aberdeen, Wash, Oct 13, 34; m 59; c 3. ORGANIC CHEMISTRY. *Educ:* Univ Wash, BS, 56; Univ Calif, Berkeley, PhD(org chem), 59. *Prof Exp:* Fel org synthesis, Moscow State Univ, 59-60; fel ferrocene chem, Univ Ill, 60-61; res chemist, Shell Oil Co, Calif, 61-64 & Shell Develop Co, 64; from asst prof to assoc prof org chem, 64-76, coordr res, 66-69, PROF CHEM, CALIF STATE UNIV, HAYWARD, 76- *Concurrent Pos:* Consult, Lawrence Livermore Nat Lab, 70-; Fulbright res fel, Inorg Inst, Munich Tech Univ, 72-73. *Mem:* Am Chem Soc; Sigma Xi; Int Am Photochem Soc. *Res:* Natural products; organic chemistry of ferrocene; photochemistry. *Mailing Add:* Dept Chem Calif State Col Hayward CA 94542

BOZARTH, GENE ALLEN, b Rumsey, Ky, Nov 10, 41; m 68; c 2. BOTANY, PLANT PHYSIOLOGY. *Educ:* Univ Ky, BS, 63; Auburn Univ, MS, 66, PhD(biochem), 69. *Prof Exp:* Fel fungus physiol, Univ Mo-Columbia, 69-71; sr res analyst, Space Sci Lab, Northrop Serv, Inc, 71-74, prin scientist, bot area, Johnson Spaceflight Ctr, 74-75; plant physiologist, 76-80, mgr, Plant Physiol Dept, 81-83, sr res biologist, Shell Develop Co, 84-89; RES ASSOC, E I DU PONT DE NEMOURS CO, 89- *Mem:* Weed Sci Soc Am; Sigma Xi; Aquatic Plant Mgt Soc; Int Weed Sci Soc. *Res:* Herbicide evaluation; weed control; weed biology and plant growth regulators; rice agronomy; rice weed control; rice diseases. *Mailing Add:* Stine-Haskell PO Box 30 Newark DE 19714

BOZARTH, ROBERT F, b Herrin, Ill, Feb 23, 30; m 51; c 3. PLANT VIROLOGY. *Educ:* Univ Fla, BSc, 52, MSc, 57; Cornell Univ, PhD(plant path), 62. *Prof Exp:* Res assoc, Virol Pioneering Lab, USDA, 61-62; virologist, Boyce Thompson Inst Plant Res, 62-76; PROF LIFE SCI, IND STATE UNIV, 76- *Mem:* Sigma Xi; Am Soc Virol; Am Phytopath Soc; Am Soc Microbiol. *Res:* Physico-chemical properties of fungal and plant viruses. *Mailing Add:* Dept Life Sci Ind State Univ Terre Haute IN 47809

BOZDECH, MAREK JIRI, b Wildflecken, WGer, Oct 12, 46; US citizen; m 67; c 3. HEMATOLOGY, ONCOLOGY. *Educ:* Univ Mich, Ann Arbor, AB, 67; Wayne State Univ, Detroit, MD, 72. *Prof Exp:* Res asst otolaryngol, Kresge Hearing Res Inst, Univ Mich, Ann Arbor, 67-69; intern, Univ Wis-Madison, 72-73, resident internal med, 73-75, from asst prof to assoc prof, 78-85, dir, Clin Hemat Lab, 78-82 & bone marrow Transplantation Prog, 84-85; fel hemat & oncol, Univ Calif, San Francisco, 75-78, clin instr internal med, 77-78; ASSOC PROF MED & DIR ADULT BONE MARROW TRANSPLANTATION SERV, UNIV CALIF, SAN FRANCISCO, 85- *Concurrent Pos:* Res assoc, Cancer Res Inst, Univ Calif, San Francisco, 77-78; Nat Cancer Inst fel, NIH, 77-78; vis fel, Fred Hutchinson Cancer Res Ctr, Seattle, Wash, 80. *Mem:* AAAS; Am Soc Hemat; Am Fedn Clin Res; Am Col Physicians. *Res:* Bone marrow transplantation; abrogation of graft-versus-host disease; manipulation of the graft-versus-leukemia effect; in vitro treatment of allogeneic and autologous marrow; study of the immune and hematopoietic systems after transplantation; use of marrow transplants for treatment of genetic defects. *Mailing Add:* Dept 2E Kiser Permanente 401 Bicentennial Way Santa Rosa CA 95403

BOZEMAN, F MARILYN, b Washington, DC, Dec 3, 27; m 64. MICROBIOLOGY. *Educ:* Univ Md, BS, 48, MS, 50. *Prof Exp:* Bacteriologist, Am Type Cult Collection, 48-49; microbiologist, Dept Rickettsial Dis, Walter Reed Army Inst Res, 50-73; microbiologist, Div Virol, Bur Biologics, Food & Drug Admin, 73-84, br chief, 77-84; RETIRED. *Mem:* Am Soc Microbiol; Tissue Cult Asn; Am Asn Immunologists. *Res:* Rickettsiology and virology; cell and tissue culture. *Mailing Add:* 4008 Queen Mary Dr Olney MD 20832

BOZEMAN, JOHN RUSSELL, b Bleckley Co, Ga, Apr 2, 35; m 58; c 3. PLANT ECOLOGY, RESOURCE MANAGEMENT. *Educ:* Ga Southern Col, BS, 61; Univ NC, Chapel Hill, MA, 65, PhD(bot, ecol), 71. *Prof Exp:* Teaching asst bot, Univ NC, Chapel Hill, 61-64; instr biol, Ga Southern Col, 64-66; cur herbarium, Univ NC, Chapel Hill, 66-68, instr bot, 68; from asst prof to assoc prof biol, Ga Southern Col, 68-75; asst chief, Marsh & Beach Sect, Coastal Resources Div, Brunswick, 75-89, PROG MGR, FRESHWATER WETLANDS, HERITAGE INVENTORY, GA DEPT NATURAL RESOURCES, 89- *Concurrent Pos:* Consult veg analyst, Inst Natural Resources, Univ Ga, 72-75; consult resource assessment & mgt, Dept Natural Resources, Off Planning & Res, State Ga, Atlanta, 74-75. *Mem:* Ecol Soc Am. *Res:* Functional role of vegetaion in aquatic and terrestrial ecosystems; stabilizing and disruptive processes, mineral cycling and storage in ecosystems; vegetational surveys of the Coastal Plain, vegetational mapping and remote sensing; resource assessment, dune stabilization, coastal processes; habitat assessment of rare and endangered species. *Mailing Add:* Ga Dept Natural Resources 2117 US Hwy 278 SE Social Circle GA 30279

BOZEMAN, SAMUEL RICHMOND, b Knoxville, Tenn, Nov 15, 15; m 44. BIOCHEMISTRY. *Educ:* Univ Tenn, BA, 39, MS, 40; Va Polytech Inst, PhD(bact), 45. *Prof Exp:* From bacteriologist to asst dir biol prod div, Mich State Dept Health, 44-48; chief lab, Br Pub Health & Welfare, Gen Hq Supreme Comdr Allied Powers, 48-51, UN Command, 50-51; asst dir in chg qual control, Biol Labs, Pitman-Moore Co, 52, dir, 52-57, vpres, 57-63; asst mgr biol prod, Lederle Labs, 63-67, mgr clin lab aids dept, 67-69, gen mgr acquisitions, 69-70; assoc dean, Col Basic Med Sci, 70-72, PROF MICROBIOL, COL BASIC MED SCI, UNIV TENN CTR HEALTH SCI, MEMPHIS, 70-, EXEC ASST TO THE CHANCELLOR, 72- *Concurrent Pos:* Teaching fel bact, State Col Wash, 40-42. *Mem:* Am Pub Health Asn; NY Acad Sci. *Res:* Iso and hetero-haemagglutinogens; autoantibodies; physiology of genus bacillus, antibiotics; methodology of vaccine, toxoid, antitoxin, antiserum production and control. *Mailing Add:* 2398 Kirby Woods Cove Memphis TN 38119

BOZIAN, RICHARD C, b Springfield, Mass, Aug 12, 19; m 51; c 5. BIOCHEMISTRY, INTERNAL MEDICINE. *Educ:* Rutgers Univ, BS, 39; Albany Med Col, MD, 50; Am Bd Internal Med, dipl, 58; Am Bd Nutrit, dipl, 68. *Prof Exp:* Instr med, Sch Med, NY Univ-Bellevue Med Ctr, 53-55, USPHS trainee metab, 54-55, asst prof med & assoc vis physician & dir clins, 55-58; clin investr, Vet Admin Hosp, Nashville, Tenn, 61-63; assoc prof, 63-70, prof med, 70-85, asst prof biochem & dir div nutrit, 63-85, EMER PROF MED, UNIV CINCINNATI, COL MED, 85- *Concurrent Pos:* USPHS res grant, 58 & 64; mem coun arteriosclerosis, Am Heart Asn, 55-; assoc attend physician, Knickerbocker Hosp & consult, St Barnabas Hosp, N Y, 56-58; consult, US Vet Admin Hosp, Jewish Hosp & Epp Mem Hosp & attend, Holmes Hosp, Cincinnati, 63-; Am Cancer Soc res grant, 63 & 64; lectr, Coun Food & Nutrit, 65-80. *Mem:* AAAS; Am Pharmaceut Asn; Harvey Soc; Am Fedn Clin Res; Sigma Xi; Am Inst Nutrit. *Res:* Nutritional biochemistry. *Mailing Add:* 471 W Galbraith Cincinnati OH 45215

BOZLER, CARL O, b Columbus, Ohio, Aug 24, 41; m 66; c 2. ELECTRICAL ENGINEERING. *Educ:* Ohio State Univ, BEE & MSc, 65, PhD(elec eng), 69. *Prof Exp:* Res assoc elec eng, Ohio State Univ, 64-69; res engr electronic devices, F W Bell, Inc, 68-70; res engr microwave devices, Sperry Rand Corp, 70-74; SCIENTIST MICROELECTRONICS, LINCOLN LAB, MASS INST TECHNOL, 74- *Honors & Awards:* Baker Prize Award, Inst Elec & Electronics Engrs. *Mem:* Inst Elec & Electronics Engrs. *Res:* Material properties; characterization; crystal growth; impurity; implantation of gallium arsenide and other semiconductors; new fabrication techniques; new electronic devices made from semiconductors, metals and insulators; inventor and developer of the permeable base transistor. *Mailing Add:* Mass Inst Technol Lincoln Lab 244 Wood St Lexington MA 02173

BOZLER, EMIL, b Steingebronn, Ger, Apr 5, 01; nat US; m 33; c 3. PHYSIOLOGY. *Educ:* Univ Munich, PhD(zool), 23. *Hon Degrees:* DSc, Ohio State Univ, 75. *Prof Exp:* Asst & privat-docent, Zool Inst, Univ Munich, 24-31; Rockefeller Found fel, 28-29; vis fel, Sch Med, Univ Rochester, 29; fel med physics, Johnson Found, Sch Med, Univ Pa, 32-36; from asst prof to prof, 36-71, EMER PROF PHYSIOL, OHIO STATE UNIV, 71- *Concurrent Pos:* Fulbright award, 58. *Mem:* Am Physiol Soc; hon mem Ger Physiol Soc. *Res:* Physiology of primitive nervous systems; energetics; excitability; action potentials; mechanical properties of muscle. *Mailing Add:* Dept Physiol Ohio State Univ 333 W Tenth Ave Columbus OH 43210-1239

BOZNIAK, EUGENE GEORGE, b Radway, Alta, Oct 15, 42; m 65. AQUATIC ECOLOGY, PHYCOLOGY. *Educ:* Univ Alta, BSc, 63, MSc, 66; Wash Univ, PhD(aquatic ecol), 69. *Prof Exp:* From asst prof to assoc prof, 69-80, PROF & CHMN BOT, WEBER STATE COL, 80- *Mem:* AAAS; Water Pollution Control Fedn; Am Soc Limnol & Oceanog; Ecol Soc Am; Phycol Soc Am. *Res:* Natural and synthetic phytoplankton community cultivation and ecology; effects of synthetic organic compounds upon defined phytoplankton communities; algal ecology of the Ogden River drainage system; waste stabilization ponds in the tropics. *Mailing Add:* Dept Bot Weber State Col Ogden UT 84408

BOZOKI, GEORGE EDWARD, b Kunhegyes, Hungary, June 21, 30; US citizen; m 56; c 1. HIGH ENERGY PHYSICS, COSMIC RAY PHYSICS. *Educ:* Eotvos Lorand Univ & Hungarian Acad Sci, PhD(nuclear physics), 63. *Prof Exp:* Sr res group leader high energy physics & cosmic rays & vhead dept cosmic rays, Cent Res Inst Physics, Budapest, 53-70; sr res assoc high energy physics, Inst Nuclear Physics, Paris, 70-71; sr res assoc, Dept Physics, Univ Pa, 71-74; physicist, Breeder Reactor Div, Burns & Roe, Inc, Oradell, NY, 74-78; RES SCIENTIST, BROOKHAVEN NAT LAB, 78- *Concurrent Pos:* From asst prof to assoc prof nuclear physics, Fac Atomic Physics, Eotvos Lorand Univ, 59-64. *Honors & Awards:* Schmid Prize, Roland Eotvos Phys Soc, 65. *Mem:* Am Phys Soc; NY Acad Sci. *Res:* Neutrino related research; nuclear radiation safety at reactors and accelerators; risk and accident analysis of nuclear plants. *Mailing Add:* Nuclear Energy Dept Brookhaven Nat Lab Upton NY 11973

BOZOKI-GOMBOSI, EVA S, b Budapest, Hungary, US citizen; c 1. HIGH ENERGY PHYSICS. *Educ:* Eotvos Lorand Univ, Budapest, MS, 58; Hungarian Acad Sci, PhD(high energy physics), 62. *Prof Exp:* Sr res assoc high energy physics, Cent Res Inst Physics, Budapest, 58-71; vis res assoc, Joint Inst Nuclear Res, Dubna, USSR, 64-65, Lab Accelerateur Lineaire, Orsay, France, 70-71 & Univ Pa, Philadelphia, 71-72; sr systs analyst comput applns, Indust, 72-77; MEM STAFF ACCELERATOR PHYSICS, BROOKHAVEN NAT LAB, 78- *Mem:* Am Phys Soc; Asn Comput Mach. *Res:* Hadronic interactions of high energy elementary particles; application of programmed control and simulation in accelerator physics and in various engineering fields. *Mailing Add:* Six Lantern Ct Stonybrook NY 11790

BOZZELLI, JOSEPH WILLIAM, b East Orange, NJ, Sept 16, 42; m 68; c 3. PHYSICAL CHEMISTRY, ANALYTICAL CHEMISTRY. *Educ:* Marietta Col, BS, 64; Princeton Univ, MS, 70, PhD(chem), 72. *Prof Exp:* Fel chem physics, Univ Pittsburgh, 73-75; DISTINGUISHED PROF CHEM, NJ INST TECHNOL, 75- *Concurrent Pos:* Consult, US Environ Protection agency, 78-91 & NJ Environ Protection Agency, 77-; res grant, US Environ Protection Agency, 78-85, NJ Dept Environ Protection, 77-81 & 86-90, NSF, NJ Inst Technol & Univ Ind Ctr, 84-91. *Mem:* Am Chem Soc; Sigma Xi. *Res:* Chemical kinetics of small molecule and atom reactions in the vapor phase; detection of trace levels of air pollutants; thermal and catalytic reactions of toxic halocarbons with water vapor and hydrogen; kinetics in combustional chloro hydrocarbons, kiv. *Mailing Add:* Dept Chem & Chem Eng Univ Heights Newark NJ 07102

BOZZUTO, CARL RICHARD, b Waterbury, Conn, June 17, 47; m 71; c 3. PROCESS DESIGN, PILOT PLANT DEVELOPMENT & SCALE UP. *Educ:* Mass Inst Technol, BS, 70, MS, 70; Hartford Grad Ctr, MS, 84. *Prof Exp:* Mgr fuel technol, ABB Combustion Eng, 73-74, mgr energy systs, 74-80, dir advan systs, 80-82, chief technologist, 82-84 dir utility mkt, 84-86, gen mgr fluid bed, 86-88, DIR KDL, ABB COMBUSTION ENG, 88- *Concurrent Pos:* Tech opportunities planning, Am Soc Mech Engrs, 76-86, solar energy standards, 77-82; coal comb appln working group, Dept Energy, 81-83; pres, Oxce Fuel Co, 89- *Mem:* Am Soc Mech Engrs; Am Inst Chem Engrs; Combustion Inst. *Res:* Design development, and testing of energy processes and equipment including pulverized coal, fluidized bed, and coal gasification power plants; author of various publications. *Mailing Add:* One Parky Dr Enfield CT 06082-6107

BRAASCH, NORMAN L, b Pierce, Nebr, July 29, 28; m 53; c 1. ENTOMOLOGY. *Educ:* Univ Nebr, BS, 50, MA, 55, PhD(entom), 65. *Prof Exp:* Instr, Nebr Pub Schs, 55-57; res asst entom, Univ Nebr, 58-63; from asst prof to assoc prof zool, 63-77, PROF BIOL, SOUTHEAST MO STATE COL, 77- *Mem:* AAAS; Am Soc Zoologists; Entom Soc Am. *Res:* Invertebrate zoology. *Mailing Add:* Dept Biol Southeast Mo State Col Cape Girardeau MO 63701

BRAATEN, DAVID A, b New York, NY, Nov 22, 55. BOUNDARY LAYER TURBULENCE, AEROSOL PHYSICS. *Educ:* State Univ NY, Oswego, BS, 77; San Jose State Univ, MS, 81; Univ Calif, Davis, PhD(atmospheric sci), 88. *Prof Exp:* Meteorologist, Henningson, Durham & Richardson, Inc, 77-80; res asst, Univ Calif, Davis, 81-88, postgrad researcher, 88-89; ASST PROF ATMOSPHERIC SCI, DEPT PHYSICS & ASTRON, UNIV KANS, 89- *Mem:* Am Meteorol Soc; Am Asn Aerosol Research. *Res:* Laboratory and atmosphere experimental investigations of turbulent structures which are important in physical processes such as aerosol resuspension, aerosol deposition, momentum flux, mass flux and heat flux. *Mailing Add:* Dept Physics & Astron Univ Kans Lawrence KS 66045

BRAATEN, MELVIN OLE, b Greenbush, Minn, Sept 6, 34; m 56; c 2. STATISTICS, ENGINEERING. *Educ:* NDak State Univ, BS, 56, MS, 61; NC State Univ, PhD(statist), 65. *Prof Exp:* Asst prof forest biomet, Duke Univ, 65-69; assoc prof indust eng & radiol sci, Univ Mo-Columbia, 69-76; ENGR, ROCKWELL INT, 76-; ASSOC PROF BUS, UNIV IOWA, 76- *Mem:* Am Statist Asn; Biomet Soc. *Res:* Computer applications in human engineering; operations research systems modeling in medical service; engineering statistical modeling. *Mailing Add:* 350 Collins Rd NE Cedar Rapids IA 52498

BRAATZ, JAMES ANTHONY, b Baltimore, Md, July 17, 43; m 64; c 3. PROTEIN CHEMISTRY, IMMUNOCHEMISTRY. *Educ:* Johns Hopkins Univ, BS, 68, PhD(biochem), 73. *Prof Exp:* Staff fel, NIH, 73-84, head biochem sect, Biol Response Modifiers Prog, Nat Cancer Inst, 81-84; sr res biochemist, Res Div, 84-86, RES ASSOC, RES DIV, W R GRACE & CO, 86- *Concurrent Pos:* Consult, Abbott Labs, North Chicago, Ill, 81-84 & Hoffman-LaRoche, Nutley, NJ, 83-84. *Honors & Awards:* Incentive Award, Dept Com, 84. *Mem:* Am Chem Soc; Am Soc Biol Chemists; AAAS. *Res:* Tumor biochemistry; isolation, purification and characterization of human tumor-associated antigens; new methodology for immobilization of proteins, chemistry of biocompatible polymers. *Mailing Add:* W R Grace & Co 7379 Rte 32 Columbia MD 21044

BRABANDER, HERBERT JOSEPH, b Brooklyn, NY, Apr 17, 32; m 63; c 3. ORGANIC CHEMISTRY. *Educ:* LI Univ, BS, 53; Stevens Inst Technol, MS, 58. *Prof Exp:* RES CHEMIST, LEDERLE LABS, AM CYANAMID CO, 53- *Mem:* Am Chem Soc. *Res:* Pharmaceutical chemical research for the development of new drugs, particularly agents which act upon the central nervous system. *Mailing Add:* Lederle Labs Bldg 65 Pearl River NY 10965

BRABB, EARL EDWARD, b Detroit, Mich, May 27, 29; m 57; c 2. GEOLOGY. *Educ:* Dartmouth Col, AB, 51; Univ Mich, MS, 52; Stanford Univ, PhD(geol), 60. *Prof Exp:* GEOLOGIST, US GEOL SURV, 59-; PROJ CHIEF, SAN FRANCISCO BAY REGION GEOL, 90- *Concurrent Pos:* Geotech adv, City San Jose, 71-74; expert witness, Nuclear Regulatory Comn, 77-81 & comt mudslides, Nat Acad Sci-Nat Res Coun, 81-82; deleg, Am Comn Stratig Nomenclature, US Geol Surv, 78-82; chmn, US-Japan Landslide Conf, 79 & Landslide Proj Third World Countries, UNESCO, 81; chief, US Landslide Hazard Mapping, 81-90, Regional Landslide Res Group, 82-85; coordr, USGS-Ital CNR hazard courses, Perugia, 88- *Honors & Awards:* Group Award, Achievement & Spec Serv, US Geol Surv, 82; Meritorious Award, US Dept Interior, 83; Distinguished Practice Award, Geol Soc Am, 88. *Mem:* Int Asn Eng Geologists; Am Geophys Union. *Res:* Regional landslide hazard mapping; digital cartography for geologic mapping; seismic zonation for land-use planning; air photo interpretation; environmental geology; San Andreas fault. *Mailing Add:* US Geol Surv 345 Middlefield Rd MS975 Menlo Park CA 94025

BRABSON, BENNET BRISTOL, b Washington, DC, July 29, 38; m 63; c 2. ELEMENTARY PARTICLE PHYSICS. *Educ:* Carleton Col, BA, 60; Mass Inst Technol, PhD(physics), 66. *Prof Exp:* Res assoc physics, 66-67, NSF fel, Mass Inst Technol, 67-68; from asst prof to assoc prof, 68-76, sci assoc, 72-78, PROF PHYSICS, IND UNIV, BLOOMINGTON, 76- *Mem:* Sigma Xi; fel Am Phys Soc. *Res:* Study of strong and electromagnetic interactions, including meson spectroscopy, utilizing spark, streamer, bubble, proportional and drift chambers and counter systems; Electron-positron collisions at high energies; exotics, hybrids and gluebulk. *Mailing Add:* Dept Physics Swain Hall W Ind Univ Bloomington IN 47405

BRABSON, GEORGE DANA, JR, b Washington, DC, Feb 18, 35; m 59; c 3. SPECTROSCOPY, INSTRUMENTAL ANALYSIS. *Educ:* Case Inst Technol, BS, 56; Univ Calif, Berkeley, MS, 62, PhD(chem), 65. *Prof Exp:* Chem engr, Dow Chem Co, Mich, 56; engr, Armed Forces Spec Weapons Proj, Sandia Base, NMex, 56-61; from instr to assoc prof chem, USAF Acad, 62-70, dir chem, F J Seiler Res Lab, 71-74; high energy laser prog, Air Force Weapons Labs, 74-78; dep dir, Mat Lab, Wright-Patterson AFB, Ohio, 78-81; dean res & info, Defense Systs Mgt Col. Ft Belvoir, Va, 81-83; asst prof chem, Ind Univ, 83-84; ASSOC PROF CHEM, UNIV VA, 88- *Concurrent Pos:* Vis prof chem, Univ NMex, 84-88. *Honors & Awards:* Tech Achievement Award, Air Force Syst Command, 72. *Mem:* Sigma Xi; Am Phys Soc; Am Chem Soc. *Res:* Spectroscopy of high temperature molecules isolated in frozen inert gas matrices. *Mailing Add:* 9008 Natalie Ave NE Albuquerque NM 87111

BRABY, LESLIE ALAN, b Kelso, Wash, Jan 12, 41; m 66; c 3. RADIOLOGICAL PHYSICS, RADIATION BIOPHYSICS. *Educ:* Linfield Col, BA, 63; Ore State Univ, PhD(radiol physics), 72. *Prof Exp:* RES SCIENTIST RADIOL PHYSICS & BIOPHYS, PAC NORTHWEST LAB, BATTELLE MEM INST, 63- *Mem:* Sigma Xi; Radiation Res Soc; Am Phys Soc; Health Physics Soc; AAAS. *Res:* Investigation of the physical and early biological processes which determine the response of a biological system to ionizing radiation. *Mailing Add:* Battelle-Pac Northwest Lab 3746 Bldg 300 Area Richland WA 99352

BRACCO, DONATO JOHN, b Neresine, Lussino, Italy, Feb 16, 21; nat US; m 50; c 6. PHYSICAL CHEMISTRY, ANALYTICAL CHEMISTRY. *Educ:* City Col New York, BChE, 41. *Prof Exp:* Res chemist, Peter J Schweitzer, Inc, 41-42; chem engr, Titanium Div, Nat Lead Co, 42-45 & AEC, 47; sect head phosphor & cathode ray tube chem, Physics Labs, Sylvania Elec Prod, Inc, 47-55, mgr, Chem Lab, 55-57 & Planning Res Labs, 58-61, mgr mat anal, Gen Tel & Electronics Labs, 61-69, dir mat res lab, GTE Labs, 69-72, dir, Cent Serv Lab, 72-74, dir mat eng lab, 75-78, dir, Elec Equip Technol, 78-83, LECTR CHEM DEPT, GTE LABS, WALTHAM, 83- *Mem:* Fel, AAAS; Electrochem Soc; Inst Elec & Electronics Engrs; Am Soc Qual Control; Solar Energy Soc; Am Chem Soc; fel Am Inst Chemists. *Res:* Instrumental analysis; process monitoring; materials engineering; solar energy thermal conversion. *Mailing Add:* 348 Hayward Mill Rd Concord MA 01742

BRACE, C LORING, b Hanover, NH, Dec 19, 30; m 57; c 3. PHYSICAL ANTHROPOLOGY. *Educ:* Williams Col, BA, 52; Harvard Univ, MA, 58, PhD(anthrop), 62. *Prof Exp:* Instr anthrop, Univ Wis, Milwaukee, 60-61; from asst prof to assoc prof, Univ Calif, Santa Barbara, 61-67; assoc prof, 67-71, PROF ANTHROP, UNIV MICH, ANN ARBOR, 71-, CUR PHYS ANTHROP, MUS ANTHROP, 67- *Concurrent Pos:* Vis prof anthrop, Univ Auckland, 73. *Mem:* AAAS; Am Asn Phys Anthrop; fel Am Anthrop Asn; Soc Study Human Biol; Int Asn Human Biologists; mem Dent Anthropol Asn (pres, 88-90). *Res:* History of biological anthropology; evolutionary theory and study of human evolution as it is expressed in the human fossil record; analysis of the origin of contemporary human physical diversity; dental anthropology. *Mailing Add:* Mus Anthrop Univ Mich Ann Arbor MI 48109

BRACE, JOHN WELLS, b Evanston, Ill, Jan 19, 26; m 50; c 5. PURE MATHEMATICS. *Educ:* Swarthmore Col, BA, 49; Cornell Univ, AM, 51, PhD(math), 53. *Prof Exp:* From instr to prof, 53-64, PROF EMER MATH, UNIV MD, COLLEGE PARK, 85- *Concurrent Pos:* Vis, dept pure math & math statist, Cambridge Univ, Eng, 66-67, 73-74 & 80-81. *Mem:* AAAS; Am Math Soc; Math Asn Am. *Res:* Applications of functional analysis, ranging through theoretical functional analysis. *Mailing Add:* PO Box 227 Cherryfield ME 04622

BRACE, LARRY HAROLD, b Saginaw, Mich, Feb 19, 29; m 53; c 3. IONOSPHERIC PHYSICS. *Educ:* Univ Mich, BS, 58. *Prof Exp:* Res assoc, Space Physics Res Lab, Univ Mich, 58-60, dir, 60-62; res physicist, Goddard Space Flight Ctr, NASA, 62-90; VIS RES SCIENTIST, SPACE PHYSICS RES LAB, UNIV MICH, 90- *Concurrent Pos:* Mem Ionosphere Subcomt, NASA, 65-67 & MARS Sci Working Group, 89-90. *Honors & Awards:* NASA Medal for Exceptional Sci Achievement, 80. *Mem:* Am Phys Soc; Am Geophys Union. *Res:* Rocket and satellite borne instruments in studies of the processes controlling the electron energy balance of the ionosphere; exploration of the ionosphere of Venus by in-situ measurements. *Mailing Add:* Code 914 Goddard Space Flight Ctr Greenbelt MD 20771

BRACE, NEAL ORIN, b Osceola, Wis, Apr 12, 22; m 45; c 7. ORGANIC CHEMISTRY, FLUORINE CHEMISTRY. *Educ:* Univ Minn, BA, 46; Univ Ill, PhD(org chem), 48. *Prof Exp:* Res chemist, Tenn Eastman Corp, 48-49; res chemist, Org Chem Dept, E I du Pont de Nemours & Co, 49- 63; assoc prof chem, North Park Col, 63-66, head dept, 64-66; from assoc prof to prof chem, 66-88, EMER PROF CHEM, WHEATON COL, ILL, 88-; EMER PROF CHEM, WHEATON COL, ILL, 88- *Concurrent Pos:* Consult, Ciba-Geigy Corp, 68-; sr US scientist, Alexander von Humboldt-Stiftung, Ger, 72; Richard Merton prof & hon Fulbright prof, GH Wuppertal, Ger, 78. *Honors & Awards:* Humboldt Prize, Ger, 72- *Mem:* Am Chem Soc; Am Sci Affiliation. *Res:* Perfluoroalkyl-substituted alkenyl radical cyclization, thermal lactonization of iodoalkanoates, perfluoroalkyl-alkanoic acids and alkanols; mechanisms of substitution and elimination reactions of perfluoroalkyl-substituted iodoalkanes; extinguishment of hydrocarbon fuel fires with aqueous fluorocarbon surfactants; synthesis and reactions of alpha, alpha-dihaloalkyl alkanoates. *Mailing Add:* Dept Chem Wheaton Col Wheaton IL 60187

BRACE, ROBERT ALLEN, b Marlette, Mich, May 4, 48; m 69. CARDIOVASCULAR PHYSIOLOGY, CHEMICAL ENGINEERING. *Educ:* Mich State Univ, BS, 70, MS, 71, PhD(chem eng), 73. *Prof Exp:* Instr, Univ Miss, 73-75, asst prof physiol & biophys, Med Ctr, 75-77; assoc prof physics, Marshall Univ, 77-78; assoc prof physiol, Loma Linda Univ, 78-85; PROF REPRODUCTIVE MED, UNIV CALIF, SAN DIEGO, 85- *Mem:* Microcirculatory Soc; Am Physiol Soc. *Res:* Dynamics and control of fluid volumes in fetus and adult, including blood volume control, lymph flow, transcapillary forces, drinking, and urinary output. *Mailing Add:* Dept Reproductive Med Univ Calif San Diego La Jolla CA 92093

BRACE, WILLIAM FRANCIS, b Littleton, NH, Aug 26, 26; m 55; c 3. STRUCTURAL GEOLOGY. *Educ:* Mass Inst Technol, BS, 49, PhD(geol), 53. *Prof Exp:* Fulbright fel, Mass Inst Technol, Austria, 53-54, from asst prof to prof geol, 55-80, Cecil & Ida Green prof geol & head dept, 80-88; RETIRED. *Honors & Awards:* Walter Bacher Medal, Am Geol Soc, 89. *Mem:* Nat Acad Sci; Geol Soc Am; Am Soc Civil Eng. *Res:* Application of mechanics to problems of structural geology. *Mailing Add:* Dept Earth & Planet Sci Mass Inst Technol Cambridge MA 02139

BRACEWELL, RONALD NEWBOLD, b Sydney, Australia, July 22, 21; m 53; c 2. ASTRONOMY, SPACE SCIENCE. *Educ:* Univ Sydney, BSc, 41, BE, 43, ME, 48; Cambridge Univ, PhD(physics), 50. *Prof Exp:* Sr res officer radiophys div, Commonwealth Sci & Indust Res Orgn, Australia, 42-46, 49-55; Lewis M Terman prof, 74-79, PROF ELEC ENG, STANDARD UNIV, 55- *Concurrent Pos:* Fulbright vis prof, Univ Calif, 54-55. *Mem:* Int Astron Union; fel Inst Elec & Electronics Engrs; fel AAAS; Royal Astron Soc; Am Astron Soc; Astron Soc Australia; Int Sci Radio Union. *Res:* Solar physics; tomography; earth's atmosphere; fast alogorithms; radio wave propagation; antennas; radio telescopes; electromagnetic theory; solar activity; image construction; astronomy; fourier analysis; Hartley transform. *Mailing Add:* 836 Santa Fe Ave Stanford CA 94305

BRACH, EUGENE JENŐ, Tokaj, Hungary, Aug 28, 28; Can citizen; m 53; c 4. AGRICULTURAL ENGINEERING. *Educ:* Tech Univ Budapest, cert Eng, 53; Acad Mil Eng, cert mil eng, 53. *Prof Exp:* Develop engr, Defence Res, 53-56; proj engr, Sangamo Elec Co, 57-62; proj engr, Eng Res Serv, 62-65, RES SCIENTIST, ENG & STATIST RES INST, RES BR, CAN AGR, 65-, TECH/SCI ADV, MEAT & POULTRY PROD DIV, FP&I, 88- *Concurrent Pos:* Adj prof, Dept Plant Sci, Fac Agr, Univ Man, 78-84. *Mem:* Inst Elec & Electronics Engrs. *Res:* Grading and inspection of agricultural commodities by machine vision; electro-optics; agricultural research in breeding and disease control using field spectroscopy; instrument development in aid of agriculture. *Mailing Add:* Meat & Poultry Prod Div FP&I Br 2255 Carline Ave Ottawa ON K1A 0Y9 Can

BRACH, RAYMOND M, b Chicago, Ill, Oct 18, 34; m 56; c 7. MECHANICAL ENGINEERING, ACOUSTICS. *Educ:* Ill Inst Technol, BS, 58, MS, 62; Univ Wis, PhD(eng mech), 65. *Prof Exp:* Assoc res engr, IIT Res Inst, 58-62; sr res engr, Continental Can Co, 62-63; asst prof eng sci, 65-70, ASSOC PROF AEROSPACE & MECH ENG, UNIV NOTRE DAME, 70- *Concurrent Pos:* Eng res consult; lic prof eng, State of Ind. *Mem:* Am Soc Mech Engrs; Acoustical Soc Am; Am Soc Eng Educ; Soc Automotive Engrs; Inst Noise Control Engrs. *Res:* Vibrations, mechanical engineering design; acoustics; vehicle dynamics; mechanics of particle and rigid body impact with applications ranging from collision analysis of automotive vehicles to the analysis of erosion and wear data and the rebound and collection of solid aerosol particles. *Mailing Add:* Dept Aerospace & Mech Eng Univ Notre Dame Notre Dame IN 46556

BRACHER, KATHERINE, b San Francisco, Calif, Oct 26, 38. ASTRONOMY. *Educ:* Mt Holyoke Col, AB, 60; Ind Univ, AM, 62, PhD(astron), 66. *Prof Exp:* From instr to asst prof astron, Univ Southern Calif, 65-67; From asst prof to assoc prof, 67-81, PROF ASTRON, WHITMAN COL, 81- *Concurrent Pos:* Chair, Hist Astron Div, Am Astron Soc, 89-91. *Mem:* Sigma Xi; Am Astron Soc; Astron Soc Pac. *Res:* Archaeoastronomy; history of astronomy. *Mailing Add:* Dept Astron Whitman Col Walla Walla WA 99362

BRACHFELD, NORMAN, b New York, NY, Oct 16, 27. BIOCHEMISTRY, CARDIOVASCULAR PHYSIOLOGY. *Educ:* Columbia Univ, AB, 49; Wash Univ, MD, 53. *Prof Exp:* Intern med, Maimonides Hosp, Brooklyn, 53-54; intern surg, Peter Bent Brigham Hosp, Boston, 54-55, asst resident, 55-56, asst med, 57-59; asst resident med, NY Hosp, 59-60; from instr to asst prof, 60-68, ASSOC PROF MED, MED COL, CORNELL UNIV, 68- *Concurrent Pos:* Teaching & res fel, Peter Bent Brigham Hosp, Boston, 56-57; teaching fel, Harvard Univ Med Sch, 56-57, res fel med, 57-59, Samuel A Levine fel, 58-59; res fel physiol, Sloan Kettering Inst, 57; sr res fel, NY Heart Asn, 61-62; career scientist award, City of NY Health Res Coun, 62-68; career develop award, NIH, 68-73; physician, Outpatient Dept, NY Hosp, 60-62, asst attend, 62-68, assoc attend, 68-; asst dir, Comprehensive Care & Teaching Prog, Med Col, Cornell Univ, 60-65, assoc dir, 65-68; res assoc, Inst Muscle Dis Inc, 60-66, head, Div Myocardial Metab & assoc mem, 66-73; med res collabr, Brookhaven Nat Lab, 61-63; asst vis physician, Sec Med Div, Bellevue Hosp, 63-68. *Mem:* Am Fedn Clin Res; Am Heart Asn; Harvey Soc; fel Am Col Physicians; fel Am Col Cardiol. *Res:* Metabolic and hemodynamic changes in the ischemic heart; cardiovascular research; clinical cardiology. *Mailing Add:* 920 Park Ave New York NY 10028

BRACHMAN, MALCOLM K, b Ft Worth, Tex, Dec 9, 26; m 51; c 3. THEORETICAL PHYSICS. *Educ:* Yale Univ, BA, 45; Harvard Univ, MA, 47, PhD(physics), 49. *Prof Exp:* Asst prof physics, Southern Methodist Univ, 49-50; assoc physicist, Argonne Nat Lab, 50-53; res physicist, Tex Instruments, Inc, 53-54; PRES, NORTHWEST OIL CO, 56- *Concurrent Pos:* Chmn and chief exec officer, Pioneer Am Ins Co, 54-79. *Mem:* Am Phys Soc; Inst Elec & Electronics Eng; Am Math Soc; Am Inst Mining; Soc Explor Geophys. *Mailing Add:* Northwest Oil Co Inc 7515 Greenville Ave No 802 Dallas TX 75231

BRACHMAN, PHILIP SIGMUND, b Milwaukee, Wis, July 28, 27; m 50; c 4. MEDICINE. *Educ:* Univ Wis, BS, 50, MD, 53. *Prof Exp:* Intern, Hosp, Univ Ill, 53-54; asst surgeon, Epidemiol Intel Serv, Commun Dis Ctr, Wistar Inst, Univ Pa, 54-58, med resident, Univ Pa Hosp, 58-60; sr surgeon & chief invest sect, Epidemiol Br, Commun Dis Ctr, USPHS, 60-70, dir, Bur Epidemiol Prog Off, 81-82, dir, Global Epidemic Intel Serv Prog, Ctr Dis Control, USPHS, 82-86; PROF, SCH PUB HEALTH, EMORY UNIV, 86- *Concurrent Pos:* Clin prof, Commun Health, Emory Univ, 82-86. *Mem:* Am Pub Health Asn; Am Epidemiol Soc; Int Epidemiol Asn; Infectious Dis Soc; Soc Epidemiol Res; Am Col Epidemiol; Asn Practr Infection Control. *Res:* Epidemiology, public health, preventive medicine, hospital epidemiology, epidemiology training. *Mailing Add:* Emory Univ Sch Pub Health 1599 Clifton Rd NE Atlanta GA 30329

BRACIALE, THOMAS JOSEPH, JR, b Philadelphia, Pa, Oct 22, 46; m 72; c 3. INFLUENZA VIROLOGY, T-LYMPHOCYTE CLONES. *Educ:* St Joseph's Col, BS, 68; Univ Pa, PhD(microbiol), 74, MD, 75. *Prof Exp:* Intern/resident path, Sch Med, Wash Univ, 75-76; vis fel microbiol, Australian Nat Univ, 76-78; asst pathologist, Barnes Hosp, 78-82; from asst prof to assoc prof, 78-88, PROF PATH, SCH MED, WASH UNIV, 88-; ASSOC PATHOLOGIST, BARNES HOSP, 82- *Concurrent Pos:* Fel, Jan Coffin Child's Mem Found, 76-78; assoc ed, J Immunol, 82-86; mem, Exp Virol Study Sect, NIH, 83-87; sect ed, J Immunol, 87- *Mem:* Am Asn Immunologists; Am Asn Pathologists; Am Soc Microbiol; AAAS. *Res:* Analysis of the function and properties of cloned populations of thymus-derived lymphocytes from man and experimental animals. *Mailing Add:* Dept Path Sch Med Wash Univ 660 Euclid Ave St Louis MO 63110

BRACIALE, VIVIAN LAM, b New York, NY, June 5, 48; m 72; c 3. VIRAL IMMUNOLOGY, LYMPHOKINES & RECEPTORS. *Educ:* Cornell Univ, AB, 69; Univ Pa, PhD(microbiol), 73. *Prof Exp:* Fel immunol, Univ Pa Sch Med, 74-75; fel, immunol, Wash Univ Sch Med, 75-76; vis fel immunol, Australian Nat Univ, 76-78; res instr, 78-83, res asst prof, 83-89, ASST PROF PATH, WASH UNIV SCH MED, 89- *Concurrent Pos:* Mem & consult, Clin Sci Study Sect, NIH, 85-89; assoc ed, J Immunol, 89-91. *Mem:* Am Asn Immunologists; AAAS. *Res:* Study of the role of T lymphocytes in the immune response and events in their induction and differentiation, using cloned cell populations. *Mailing Add:* Dept Path Box 8118 Wash Univ Sch Med 660 S Euclid Ave St Louis MO 63110

BRACK, KARL, b Kuttingen, AG, Switz, Nov 22, 23; US citizen; div; c 2. ORGANIC POLYMER CHEMISTRY, PHOTOCHEMISTRY. *Educ:* Swiss Fed Inst Technol, MS, 49, PhD(org chem), 51. *Prof Exp:* Res chemist pesticides, Hercules Inc, 52-57, res chemist polymers, 57-71; sr chemist inks & coatings, Dennison Mfg Co, 71-75, res assoc inks & coatings, 75-78; group leader adhesives & coatings, Polymer Industs, 78-83; mgr adhesives & coatings, Upaco Adhesives, 83-87; INSTR, UNIV LOWELL, MASS, 87- *Mem:* Am Chem Soc; Soc Mfg Engrs; Asn Finishing Processes; Sigma Xi. *Res:* Synthesis of radiation-curable oligomers, prepolymers and polymers; research and development of radiation-curable inks, coatings and adhesives; urethane type adhesives and coatings; polyester adhesives and coatings; biodegradable polymers. *Mailing Add:* PO Box 6711 Holliston MA 01746-6711

BRACKBILL, JEREMIAH U, b Harrisburg, Pa, Jan 19, 41; m 65; c 2. COMPUTER MODELLING. *Educ:* Lehigh Univ, BS, 63; Univ Wis, Madison, MS, 70, PhD(physics), 71. *Prof Exp:* Staff mem, Los Alamos Nat Lab, 71-78 & 80-85; res prof comput physics, Courant Inst, NY Univ, 78-80; prof appl math & numerical modelling, Brown Univ, 85-86; LAB FEL, LOS ALAMOS NAT LAB, 83- *Concurrent Pos:* Assoc ed, J Comput Physics, 87-; spec topics ed, Comput Physics Commun, 87-; vis prof dept math, Univ NMex, 88-89. *Mem:* Fel Am Phys Soc; AAAS; Am Geophys Union. *Res:* Develop implicit and adaptive grid methods for magnetohydrodynamic and plasma kinetic problems with multiple time and space scales; model unstable flow and transport in laboratory and space plasmas. *Mailing Add:* 1874 Camino Manzana Los Alamos NM 87544

BRACKELSBERG, PAUL O, b Mohall, NDak, Aug 27, 39; m 61; c 3. ANIMAL BREEDING. *Educ:* NDak State Univ, BS, 61; Univ Conn, 63; Okla State Univ, PhD(animal breeding), 66. *Prof Exp:* assoc prof, 66-77, PROF ANIMAL SCI, IOWA STATE UNIV, 77- *Mem:* Am Soc Animal Sci. *Res:* Application of genetic principles to applied animal breeding, with emphasis on growth and carcass traits; livestock production. *Mailing Add:* Dept Animal Sci Iowa State Univ Ames IA 50010

BRACKENBURY, ROBERT WILLIAM, b Long Beach, Calif, May 19, 48; m 69; c 1. DEVELOPMENTAL BIOLOGY. *Educ:* Calif Inst Technol, BS, 70; Brandeis Univ, PhD(biol), 76. *Prof Exp:* Jane Coffin Childs fel, 75-77, asst prof, Rockefeller Univ, 78-87; ASSOC PROF, UNIV CINCINNATI MED CTR, 87- *Mem:* Harvey Soc. *Res:* Structure and function of cell-surface molecules and cell-cell interactions during development; role of alterations in adhesion and motility in tumor cell invasiveness and metastasis; developmental neurobiology. *Mailing Add:* Univ Cincinnati Med Ctr 231 Bethesda Ave ML 521 Cincinnati OH 45267-0521

BRACKENRIDGE, DAVID ROSS, b Buffalo, NY, Jan 31, 38; m 64; c 2. ORGANIC CHEMISTRY. *Educ:* Canisius Col, BS, 60; Ohio State Univ, MS, 65, PhD(org chem), 66. *Prof Exp:* chemist, Ethyl Corp, 66-76, sr res chemist res & develop, 76-85, sr res specialist, 85-90, RES ADV, RES & DEVELOP, ETHYL CORP, 90- *Mem:* Am Chem Soc. *Res:* Synthetic, organic chemistry; reaction mechanisms; bromo-organic compounds; flame retardants; lube oil additives. *Mailing Add:* Ethyl Corp 8000 GSRI Ave Baton Rouge LA 70820

BRACKENRIDGE, JOHN BRUCE, b Youngstown, Ohio, Apr 20, 27; m 54; c 4. CLASSICAL MECHANICS. *Educ:* Muskingum Col, BS, 51; Brown Univ, MS, 54, PhD(physics), 59; London Univ, MSc, 74. *Prof Exp:* Asst prof physics, Muskingum Col, 55-59; from asst prof to assoc prof, 59-66, PROF PHYSICS, LAWRENCE UNIV, 66- *Concurrent Pos:* Res grants, Res Corp, 55-60 and NSF, 60-65, 68; vis scholar, Imp Col, Univ London, 74-75; fel, Nat Endowment Humanities, 80-81, 87-88, 90-93. *Honors & Awards:* Chapman Chair of Physics, Lawrence Univ, 63. *Mem:* Am Acoust Soc; Am Asn Physics Teachers; Hist Sci Soc. *Res:* Subaqueous acoustics; fluid dynamics; application of schlieren interferometry systems to measurements of fluid flow; history philosophy of science; 17th and 18th century physics and astronomy with attention to Newton's Principia. *Mailing Add:* Dept Physics Lawrence Univ Appleton WI 54911

BRACKER, CHARLES E, b Portchester, NY, Feb 3, 38; m 63; c 2. FUNGAL CYTOLOGY. *Educ:* Univ Calif, Davis, BS, 60, PhD(plant path), 64. *Prof Exp:* Res asst plant path, Univ Calif, Davis, 60-64; from asst prof to assoc prof, 64-73, PROF BOT & PLANT PATH DEPT, PURDUE UNIV, WEST LAFAYETTE, 73- *Concurrent Pos:* Ann lectr, Mycological Soc Am, 91. *Honors & Awards:* Ruth Allen Award, Am Phytopathol Soc, 83. *Mem:* Fel AAAS; Mycol Soc Am; Brit Mycological Soc; Electron Micros Soc; Am Soc Cell Biol. *Res:* Fungal ultrastructure and development; developmental cytology; cell wall formation; endomembrane system and organelles; cell growth and reproduction; morphogenesis cell ultrastructure. *Mailing Add:* 308 Park Ln West Lafayette IN 47906

BRACKETT, BENJAMIN GAYLORD, b Athens, Ga, Nov 18, 38; m 59; c 3. ANIMAL PHYSIOLOGY. *Educ:* Univ Ga, BSA, 64, DVM, 62, MS, 64, PhD(biochem), 66; Am Col Theriogenologists, dipl. *Hon Degrees:* MA, Univ Pa, 71. *Prof Exp:* Mark L Morris Animal Found fel & Am Vet Med Asn fel, Univ Ga, 62-64, Nat Inst Child Health & Human Develop fel, 64-66; assoc reproductive biol, Dept Obstet & Gynec, Sch Med, 66-68, from asst prof to prof animal reproduction, Dept Clin Studies, Sch Vet Med, Univ Pa, 68-83, managing dir, Primate Colony, 66-74; PROF & HEAD DEPT PHYSIOL & PHARMACOL, COL VET MED, UNIV GA, ATHENS, 83- *Concurrent Pos:* Consult, Nat Inst Child Health & Human Develop, 68-; mem sci adv bd, Mark L Morris Animal Found, 71-74, mem, Primate Res Ctrs Adv Comt, 74-78; consult, Univ Wis Regional Primate Ctr, US Congress Off Technol Assessment, 86-88, Contraceptive Res & Develop Prog, 86- *Mem:* Am Vet Med Asn; Soc Theriogenology; Int Embryo Transfer Soc (pres, 84-85); Am Fertil Soc; Soc Study Reproduction (secy, 82-86). *Res:* Preimplantation stages of mammalian reproduction, especially fertilization of mammalian ova in vitro. *Mailing Add:* Dept Physiol & Pharmacol Col Vet Med Univ Ga Athens GA 30602

BRACKETT, JOHN WASHBURN, computer science, system analysis, for more information see previous edition

BRACKETT, ROBERT E, b Milwaukee, Wis, Nov 28, 53; m 75; c 2. FOOD MICROBIOLOGY, FOOD SAFETY. *Educ:* Univ Wis, BS, 76, MS, 79, PhD(food sci), 81. *Prof Exp:* Res asst, Univ Wis, 76-81; food safety specialist, NC State Univ, 81-; ASSOC PROF, FOOD SCI DIV, UNIV GA. *Mem:* Int Asn Milk Food & Environ Sanitarians; Inst Food Technologists; Am Soc Microbiol. *Res:* Microbiology of fruits and vegetables; survival and behavior of Listeria monocytogenes in foods; psychotropic pathogens in foods. *Mailing Add:* Food Sci Div Univ Ga Griffin GA 30223

BRACKETT, ROBERT GILES, b Nyack, NY, Oct 8, 30; m 51; c 5. IMMUNOLOGY. *Educ:* Rutgers Univ, BS, 53, MS, 57, PhD(bact), 60. *Prof Exp:* Asst bact, Rutgers Univ, 56-59, instr, 59-60; asst res microbiologist cell cult, 60-62, res microbiologist cell cult & virol, 62-66, dir, Virol Sect, 66-70, DIR CLIN IMMUNOL, PARKE, DAVIS & CO, 70- *Concurrent Pos:* Jr microbiologist, Merck & Co, Inc, NJ, 56; bacteriologist, E R Squibb & Sons, 57-58. *Mem:* Am Soc Microbiol; Tissue Cult Asn; Soc Cryobiol. *Res:* Laboratory and/or clinical research and development of bacterial, plasmodial, and viral vaccines, skin test antigens, and blood products with emphasis on the immunology of these biological products. *Mailing Add:* One The Court of Bayview Northbrook IL 60062

BRACK-HANES, SHEILA DELFELD, b Dallas, Tex, Feb 26, 39; m 73. PALEOBOTANY, PALYNOLOGY. *Educ:* Baylor Univ, BA, 61; Univ Ill, Chicago Circle, MS, 69; Ohio Univ, PhD(bot), 75. *Prof Exp:* Res asst bot, Univ Tex, Austin, 61-64; res asst anat, Univ Mo Med Sch, 64-65; res asst bot, Univ Iowa, 65-66, res asst zool, 66-67; asst prof, 76-, ASSOC PROF BIOL, ECKERD COL. *Mem:* Sigma Xi (secy-treas, 78-79); Bot Soc Am; Am Asn Stratig Palynologists; Torrey Bot Club; Int Orgn Paleobot. *Res:* Structure and morphology of living and fossil lycopods; ferns and gymnospermous plants; transmission and scanning electron microscopy of fossil and extant pollen and spores. *Mailing Add:* Div Natural Sci Eckerd Col St Petersburg FL 33733

BRACKIN, EDDY JOE, b Town Creek, Ala, Feb 27, 45; m 66. ALGEBRA. *Educ:* Florence State Univ, BS, 67; Univ Ala, MA, 68, PhD(algebra), 70. *Prof Exp:* From instr to asst prof, 69-74, ASSOC PROF MATH, UNIV NORTH ALA, 74- *Mem:* Am Math Asn. *Res:* Theory of semirings. *Mailing Add:* Dept Math Box 5220 University St Florence AL 35632

BRACKMANN, RICHARD THEODORE, b Kansas City, Mo, Nov 23, 30; m 54; c 3. MASS SPECTROMETRY, ATOMIC PHYSICS. *Educ:* Univ Kans, BS, 53. *Prof Exp:* Asst, Gen Atomic Div, Gen Dynamics Corp, 56-58, mem res staff, 58-63; assoc res prof phys & elec eng, Univ Pittsburgh, 63-72, pres, 66-72; MGR ENG, EXTRANUCLEAR LABS, INC, 73- *Mem:* Am Phys Soc; Am Soc Mass Spectrometry. *Res:* Atomic and upper atmosphere research using modulated crossed beam, mass spectrometric, cryogenic, counting and high vacuum techniques; development of state of the art quadrupole mass spectrometric equipment and techniques used in process control and GC-LC/MS/DS. *Mailing Add:* 106 Sunnyhill Dr Pittsburgh PA 15237

BRADBEER, CLIVE, b Tynemouth, Eng, Feb 20, 33; m 60; c 2. BIOCHEMISTRY. *Educ:* Univ Durham, BSc, 54, PhD(plant biochem), 57. *Prof Exp:* Jr res biochemist, Univ Calif, Berkeley, 57-59; jr res biochemist, Univ Calif, Davis, 59; proj assoc biochem, Univ Wis, 59-60; lectr microbiol, Queen Mary Col, Univ London, 60-62; vis scientist, Nat Heart Inst, 62-64;

from asst prof to assoc prof, 64-79, PROF BIOCHEM, MED SCH, UNIV VA, 79- *Concurrent Pos:* Vis lectr biochem, Univ Otago, Dunedin, New Zealand, 82-83. *Mem:* Am Soc Microbiol; Am Soc Biochem & Molecular Biol. *Res:* Structure-function relationships in bacterial cell envelopes, with special attention to the transport of vitamin B12 and interactions of E colicins and bacteriophage BF23 with the cell envelope of Escherichia coli; transport of vitamin B12 in mammalian cells. *Mailing Add:* Dept Biochem Univ Va Med Sch Charlottesville VA 22908

BRADBURD, ERVIN M, b Philadelphia, Pa, May 29, 20; m 41; c 2. ENGINEERING. *Educ:* Columbia Univ, BS, 41, MS, 43. *Prof Exp:* Asst elec eng, Columbia Univ, 41-43; head dept transmitter eng develop, Fed Telecommun Labs, Inc, 43-58; mgr advan commun technol, Radio Corp Am, 58-66, advan commun lab, 66-68; vpres & dir eng, Defense Commun Div, Int Tel & Tel Corp, Nutley, NAm, 68-75, tech dir telecommunications & electronics group, 75-; RETIRED. *Mem:* Fel Inst Elec & Electronics Engrs. *Res:* Communications engineering; high power transmitter development; communication systems, techniques and technology. *Mailing Add:* White Birch Dr Pomona NY 10970

BRADBURN, GREGORY RUSSELL, b Wichita, Kans, Feb 23, 55; m 77; c 1. CHEMICAL KINETICS. *Educ:* USAF Acad, BSc, 77; Univ Mo, St Louis, MSc, 88; Washington Univ, 90- *Prof Exp:* Laser physicist, Air Force Weapons Lab, 77-80; IR res physicist, Air Force Geophysics Lab, 80-82; res scientist, McDonnell Douglas Res Lab, 82-90. *Concurrent Pos:* Adj prof, Lindenwood Col, 90. *Mem:* Am Chem Soc. *Res:* Measurement of rates and determination of kinetics of gas phase reactions involved in the deposition of diamond films on various substrates. *Mailing Add:* 9001 Olden St Overland MO 63114

BRADBURY, DONALD, b Fall River, Mass, Oct 17, 17; m 42; c 1. MECHANICAL ENGINEERING. *Educ:* Tufts Univ, BS, 39; Harvard Univ, MS, 40, ScD(mech eng), 50. *Prof Exp:* Engr stress anal, Westinghouse Elec Corp, 40-46; instr mech eng, Harvard Univ, 46-49; assoc prof, Univ RI, 50-53, chmn dept, 52-76, prof mech eng & appl mech, 53-80; RETIRED. *Mem:* AAAS; Am Soc Mech Engrs; Am Soc Eng Educ; Hist Sci Soc; Soc Hist Technol. *Res:* Lubrication; vibration; stress analysis. *Mailing Add:* Ten Birchwood Dr Narrangansett RI 02882

BRADBURY, E MORTON, b May 25, 33. CHEMISTRY. *Educ:* Univ London, BSc Hons, 55, PhD(biophys), 58. *Prof Exp:* Res scientist, Courtauld Res Lab, 58-62; head, Dept Molecular Biol & reader biophys, Portsmouth Polytech, 62-79; PROF & CHAIRPERSON, DEPT BIOL CHEM, SCH MED, UNIV CALIF, DAVIS, 79-; DIR, LIFE SCI DIV, LOS ALAMOS NAT LAB, NMEX, 88- *Concurrent Pos:* Mem, Biol Sci Comt, Sci Res Coun UK, 71-75; Neutron Beam Res Comt, 71-76, Brit Nat Comt Biophys, 71-79, Int Coun Magnetic Resonance in Biol, 72-80, sci coun, Inst Laue Langevin, France, 75-80 & prog comt, Argonne & Los Alamos Nat Labs, 83-; chmn, Brit Biophys Soc, 73, Int Coun Magnetic Resonance in Biol, 76-80 & Neutron-Biol Comt, Inst Laue Langevin, 76-80. *Res:* Author of numerous publications. *Mailing Add:* Dept Biol Chem Univ Calif Sch Med Davis CA 95616

BRADBURY, ELMER J(OSEPH), b Kokomo, Ind, June 26, 17; m 46, 73; c 2. CHEMICAL ENGINEERING. *Educ:* Purdue Univ, BS, 42; Ohio State Univ, MS, 49. *Prof Exp:* Pilot plant supvr, Merck & Co, NJ, 42-45; res engr, Battelle Mem Inst, 45-50, prin chem engr, 50-58, proj leader, 58-62, sr res chemist, 62-69, proj leader res chem, 69- 80, prin res scientist, 80-85; RETIRED. *Mem:* AAAS; Am Chem Soc; Am Inst Chem Engrs; Am Soc Testing Mat; Sigma Xi. *Res:* High pressure polymerizations and reactions; processing and fabrication of conventional and high temeprature resins; structural composites, materials and fabrication techniques. *Mailing Add:* 8400 Steitz Rd Powell OH 43065-9474

BRADBURY, JACK W, b Los Angeles, Calif, Sept 28, 41; m 73; c 2. ANIMAL BEHAVIOR, SOCIOBIOLOGY. *Educ:* Reed Col, BS, 63; Rockefeller Univ, PhD(animal behav), 68. *Prof Exp:* NIH training grant to Rockefeller Univ taken at William Beebe Trop Res Sta, Simla, Trinidad, WI, 68-69; asst prof neurobiol & behav, Cornell Univ, 69-72; asst prof, Rockefeller Univ, 72-75; from asst prof to assoc prof, 76-83, PROF BIOL, UNIV CALIF, SAN DIEGO, 83- *Concurrent Pos:* Richard King Mellon fel, Rockefeller Univ, 73-75; mem, Psychobiol Panel, NSF, 78-80; mem, Comt Direction Lab d'Ecol Trop, Ctr Nat Res Sci, France, 78-80; ed, Behav Ecol & Sociobiol, 89- *Mem:* Ecol Soc Am; Sigma Xi; AAAS; Asn Study Animal Behav; Soc Study Evolution. *Res:* Field studies on resource utilization and social evolution in tropical bats and antelopes; evolution of lek behavior in bats and birds. *Mailing Add:* Dept Biol C-016 Univ Calif San Diego La Jolla CA 92093

BRADBURY, JAMES CLIFFORD, b US, July 7, 18; c 3. GEOLOGY. *Educ:* Univ Ill, AB, 41; Harvard Univ, AM, 49, PhD, 58. *Prof Exp:* From asst geologist to geologist, 49-82, head indust minerals sect, 68-81, actg head geol group, 81-82, prin geol & head geol group, 82-84, EMER PRIN GEOLOGIST, ILL STATE GEOL SURV, 84- *Honors & Awards:* Hardinge Award, Am Inst Mining, Metal & Petrol Engrs, 85. *Mem:* Fel AAAS; Am Inst Mining, Metall & Petrol Engrs; fel Geol Soc Am; fel Soc Econ Geologists; Int Asn Genesis Ore Deposits; distinguished mem, Soc Mining Engrs, 85. *Res:* Geology of mineral deposits of Illinois. *Mailing Add:* 101 W Windsor Rd Urbana IL 61801

BRADBURY, JAMES NORRIS, b Palo Alto, Calif, May 25, 35; m 61; c 2. EXPERIMENTAL PHYSICS. *Educ:* Pomona Col, BA, 56; Stanford Univ, PhD(physics), 65. *Prof Exp:* Res scientist, Lockheed Palo Alto Res Lab, 66-73; group leader, 73-85, DEP DIV LEADER, MEDIUM ENERGY PHYSICS DIV, LOS ALAMOS NAT LAB, 86- *Mem:* Am Geophys Union; Am Phys Soc; Am Asn Physicists in Med. *Res:* Applications of LAMPF to biomedical and environmental research. *Mailing Add:* MP-3 MS 844 Los Alamos Sci Lab Los Alamos NM 87545

BRADBURY, JAMES THOMAS, b Cody, Wyo, Apr 7, 06; m 29; c 2. ENDOCRINOLOGY. *Educ:* Mont State Col, BS, 28; Univ Mich, MS, 30, ScD(zool), 32. *Prof Exp:* Asst, Dept Obstet & Gynec, Univ Hosp, Univ Mich, 32-37, instr, 37-40; endocrinologist, Bur Dairy Indust, USDA, 40-44; from asst prof to prof, 44-74, EMER PROF OBSTET & GYNEC, UNIV HOSP, UNIV IOWA, 74-; ADJ PROF, MONT STATE UNIV, 74- *Concurrent Pos:* Assoc prof, Univ Louisville, 49-52. *Mem:* Am Physiol Soc; Am Asn Anat; Endocrine Soc; Soc Exp Biol & Med. *Res:* Physiology, biochemistry and histology in study of the actions of hormones in processes of sex and reproduction. *Mailing Add:* 1020 E Olive St Bozeman MT 59715

BRADBURY, MARGARET G, b Chicago, Ill, July 15, 27. ICHTHYOLOGY, EVOLUTION. *Educ:* Roosevelt Univ, BS, 55; Stanford Univ, PhD(biol), 63. *Prof Exp:* from asst prof to assoc prof, 63-71, PROF BIOL, SAN FRANCISCO STATE UNIV, 71- *Concurrent Pos:* Res assoc, Calif Acad Sci, 68- *Mem:* Am Soc Ichthyologists & Herpetologists; Sigma Xi. *Res:* Systematic ichthyology; monographic treatment of the lophiiform family ogcocephalidae, their distribution and natural history. *Mailing Add:* Dept Biol San Francisco State Univ San Francisco CA 94132

BRADBURY, MICHAEL WAYNE, b Farmington, Maine, Nov 13, 53. MAMMALIAN DEVELOPMENTAL GENETICS, MOLECULAR BIOLOGY EMBRYOS. *Educ:* Rochester Inst Technol, BS, 76; Yale Univ, MPhil, 79, PhD(biol), 82. *Prof Exp:* Postdoctoral fel, Univ Calif, Davis, 82-84; scientist, Biosyne Corp, 84-87; INSTR, MT SINAI MED SCH, 88- *Concurrent Pos:* Assoc ed, J Exp Zool, 91- *Mem:* Sigma Xi; Soc Develop Biol; Am Soc Zoologists; Am Genetics Asn; AAAS. *Res:* Genetic analysis of embryos; transgenic mice and insertional mutagenesis; sex determination; mammalian developmental genetics; mammalian chimeras. *Mailing Add:* Mt Sinai Med Sch One Gustave L Levy Pl Box 1175 New York NY 10029

BRADBURY, NORRIS EDWIN, b Santa Barbara, Calif, May 30, 09; m 33; c 3. PHYSICS. *Educ:* Pomona Col, AB, 29; Univ Calif, PhD(physics, math), 32. *Hon Degrees:* ScD, Pomona Col, 51; LLD, Univ NMex, 53; DSc, Case Inst Technol, 56. *Prof Exp:* Nat Res fel physics, Mass Inst Technol, 32-34; from asst prof to prof, Stanford Univ, 34-51; prof physics, Univ Calif, 51-70, dir, Los Alamos Sci Lab, 45-70; RETIRED. *Honors & Awards:* Enrico Fermi Award, Atomic Energy Comn, 70. *Mem:* Fel Nat Acad Sci; fel Am Phy Soc. *Res:* Conduction of electricity in gases; properties of ions; atmospheric electricity; nuclear physics. *Mailing Add:* 1451 47th St Los Alamos NM 87544

BRADBURY, PHYLLIS CLARKE, b Oakland, Calif. PROTOZOOLOGY. *Educ:* Univ Calif, Berkeley, AB, 58, MA, 61, PhD(zool), 65. *Prof Exp:* USPHS training fel, Rockefeller Univ, 65-67; from asst prof to assoc prof, 67-77, PROF ZOOL, NC STATE UNIV, 77- *Concurrent Pos:* Prin investr, USPHS grant, 68-71; NATO fel, Sta Biologique de Roscoff, 73-74; mem, Trop Med & Parasitol Study Sect, Pub Health Serv, NIH, 77-81; co-ed, J Protozool, 80-, managing ed, 83-88. *Mem:* Soc Protozoologists (pres, 77-78); Am Soc Zoologists; Am Micros Soc. *Res:* Fine structure of morphogenesis in protozoa; taxonomy of ciliated protozoa; differentiation of sexual stages of malarial parasites; genesis of organelles in protozoa; protozoan parasites of crustacea. *Mailing Add:* Dept Zool NC State Univ Box 7617 Raleigh NC 27695-7617

BRADDOCK, JOSEPH V, b Hoboken, NJ, Dec 10, 29; m 65; c 2. PHYSICS. *Educ:* St Peter's Col, BS, 51; Fordham Univ, MS, 52, PhD(physics), 59. *Prof Exp:* Instr physics, Fordham Univ, 53-58; asst prof, Iona Col, 58-60; FOUNDER, BDM CORP, 60-, SR VPRES, BDM INT INC, 85- *Concurrent Pos:* Mem, Weapons Syst Vulnerability Working Groups, Defense Nuclear Agency & Serv Develop Agencies, 65-75; mem, Sci Adv Bd, Nat Security Agency, 76-83, DNA Sci Adv Group on Effects, 77-85, Army Sci Bd, Dept Army, 77-84 & Defense Sci Bd, 84- *Mem:* Am Phys Soc; Inst Elec & Electronics Engrs. *Res:* Weapon system design and analysis; neutron and solid state physics; atomic and molecular spectra. *Mailing Add:* 1101 St Stephens Rd Alexandria VA 22304

BRADDOCK, WILLIAM A, b Rifle, Colo, Feb 3, 29. FIELD GEOLOGY. *Educ:* Univ Colo, BA, 51; Princeton Univ, PhD(geol), 59. *Prof Exp:* Geologist, US Geol Surv, 52-56; from instr to assoc prof, 56-70, PROF GEOL, UNIV COLO, 70- *Concurrent Pos:* Geologist, US Geol Surv, 56- *Mem:* Geol Soc Am; Am Geophys Union. *Res:* Structural geology; igneous and metamorphic petrology; Precambrian geology. *Mailing Add:* Dept Geol Univ Colo Campus Box Boulder CO 80309-0250

BRADEN, CHARLES HOSEA, b Chicago, Ill, Mar 21, 26; m 52; c 2. NUCLEAR PHYSICS, SYSTEM DYNAMICS. *Educ:* Columbia Univ, BS, 46; Wash Univ, PhD(physics), 51. *Prof Exp:* From asst prof to assoc prof, 51-59, PROF PHYSICS, GA INST TECHNOL, 59- *Concurrent Pos:* Assoc dir physics prog, NSF, 59-60. *Mem:* Am Phys Soc. *Res:* Nuclear spectroscopy and structure; socio-economic modeling. *Mailing Add:* Sch Physics Ga Inst Technol Atlanta GA 30332

BRADEN, CHARLES MCMURRAY, b Santiago, Chile, June 9, 18; US citizen; m 43; c 4. MATHEMATICS. *Educ:* Northwestern Univ, BS, 39; Univ Minn, MS, 50, PhD(math), 57. *Prof Exp:* Instr math, Inst Technol, Univ Minn, 46-56; from asst prof to assoc prof, 56-60, dean fac, 69-72, prof math, Macalester Col, 60-; RETIRED. *Concurrent Pos:* NSF fac fel, 59-60. *Honors & Awards:* Thomas Jefferson Award, 70. *Mem:* Am Math Soc; Math Asn Am. *Res:* Partial difference equations; logic; heat engines. *Mailing Add:* 1666 Coffman St No 218 St Paul MN 55108

BRADER, WALTER HOWE, JR, b Beaumont, Tex, Oct 30, 27; m 44; c 3. ORGANIC CHEMISTRY. *Educ:* Rice Univ, BA, 50; Ga Inst Technol, PhD(org chem), 54. *Prof Exp:* Res chemist, Standard Oil Co, Ind, 54-59; sr res chemist, 59-62, sr proj chemist, 62-64, supvr explor res, 64-67, coordr, 67-70, mgr com develop, 70-71, DIR RES & DEVELOP, JEFFERSON CHEM CO, 71- *Mem:* Am Chem Soc. *Res:* Petrochemicals synthesis; application of catalysis to organic synthesis; new process development; economic evaluations; development of new product sales. *Mailing Add:* 8803 Crest Ridge Circle Austin TX 78750

BRADFIELD, JAMES E, b Temple, Tex, Mar 29, 44. PHYSICS. *Educ:* Univ Tex, Austin, BS, 66; Univ Houston, MS, 70, PhD(physics), 72. *Prof Exp:* Physicist, 73-74, spec projs mgr, 74-75, INSPECTION RES MGR, BAKER HUGHES TUBULAR SERVS, 75- *Mem:* Am Phys Soc; Am Soc Testing & Mat; Am Soc Nondestructive Testing. *Res:* Develop scanning methods for the nondestructive examination of oil country tubular goods, using electromagnetic, ultrasonic, & nuclear radiation techniques. *Mailing Add:* 8106 Log Hollow Dr Houston TX 77040

BRADFIELD, ROBERT B, b Columbia, Mo, Nov 15, 28; m 55; c 3. CLINICAL NUTRITION, BIOCHEMISTRY. *Educ:* Cornell Univ, BA, 51, MNS, 53, PhD(biochem, nutrit), 55; Univ Calif, Berkeley, JD, 76; Am Bd Nutrit, dipl, 78. *Prof Exp:* Serv officer, Interam Pub Health Serv, US State Dept, Peru, 55-62; USPHS sr fel, Cornell Univ, 62-63, & London Sch Trop Med, 63-64; clin prof human nutrit, Univ Calif, Berkeley, 64-80; PRES, NUTRITION SCIENCE, POLICY & LAW, ORINDA, CALIF, 80- *Concurrent Pos:* Appt field dir, Latin Am, Harvard Sch Pub Health, Harvard Univ; grants, William Waterman Fund, 57, 58, 60, 62, 64, 68, NIH, 62-67, 69-72, 73-75, Rockefeller Found, 58, 60, 62, 63-64, 67, 68, 69, 71, US Dept Agr, 69-71 & Brit Nutrit Found, 69, 71, 79; fels, Rockefeller Found, 64, 72, 77 WHO, 65, 67, 74, 79 & Guggenheim, 69; consult var worldwide orgns, 64; consult & lectr, Peace Corps, 64-80; ed, Am J Clin Nutrit, 66-78; mem food & nutrit bd subcomt, Nat Res Coun, 77-79; mem, Latin Am Sci Bd, NSF, 78-80, arbitrator, NASP & AAA, 82- & White House Task Force Nutrit, 85; chief party, Child Survival Proj, Cairo, Egypt, 89-90. *Honors & Awards:* Joseph Goldberger Award, AMA, 78, Gold Award. *Mem:* Fel Am Inst Chem; fel Royal Soc Trop Med & Hyg; fel Royal Soc Health; fel Int Col Pediat; fel AAAS; fel Brazilian Soc Health; fel Royal Soc Photog Soc. *Mailing Add:* Nutrit, Sci, Policy & Law 120 Village Sq Suite 122 Orinda CA 94563

BRADFIELD, W(ALTER) S(AMUEL), fluid dynamics, for more information see previous edition

BRADFORD, DAVID S, b Charlotte, NC, Oct 15, 36; c 3. ORTHOPEDIC SURGERY. *Educ:* Davidson Col, BS, 58; Univ Pa, MD, 62. *Prof Exp:* From resident gen surg to resident orthop surg, Columbia-Presby Med Ctr, 62-66, Annie C Kane jr fel, 68-69; from asst prof to prof orthop surg, Univ Minn, Minneapolis, 70-90; PROF & CHMN DEPT ORTHOP SURG, UNIV CALIF, SAN FRANCISCO, 90- *Mem:* Fel Am Col Surgeons; Orthop Res Soc; Am Acad Orthop Surgeons; Scoliosis Res Soc (pres, 86). *Res:* Connective tissue diseases; diseases of growth and development; idiopathic scoliosis; congenital malformation. *Mailing Add:* Dept Orthop Surg Univ Calif 533 Parnassus Ave San Francisco CA 94143

BRADFORD, G ERIC, b Kingsey, Que, Nov 2, 29; m 54; c 4. GENETICS OF REPRODUCTION, INTERNATIONAL AGRICULTURE. *Educ:* McGill Univ, BSc, 51; Univ Wis, MS, 52, PhD(genetics, animal husb), 56. *Prof Exp:* Res asst animal sci, Univ Wis, 51-55; asst prof animal sci & genetics, Macdonald Col, McGill Univ, 55-57; from asst prof to assoc prof, 57-69, PROF ANIMAL SCI, UNIV CALIF, DAVIS, 69-, CHAIR, ANIMAL SCI DEPT, 90- *Concurrent Pos:* Prin investr, Small Ruminants Collab Res Support Prog, USAID-Univ Calif, 78-; ed, Breed & Genetics Sect, J Animal Sci, 79-80; mem, Coun Agr Sci & Technol. *Honors & Awards:* J R Prentice Mem Award, Am Soc Animal Sci, 85. *Mem:* Am Soc Animal Sci; AAAS; Brit Soc Animal Prod. *Res:* Genetics of growth and reproduction in livestock and laboratory animals, with emphasis on genetics of prenatal survival and litter size; sheep breeding; major genes. *Mailing Add:* Dept Animal Sci Univ Calif Davis CA 95616

BRADFORD, HAROLD R(AWSEL), b Salt Lake City, Utah, Jan 30, 09; m 33; c 2. METALLURGY. *Educ:* Univ Utah, BS, 31, MS, 32. *Prof Exp:* Prin clerk, Home Owner's Loan Corp, 33-36; metallurgist, Am Smelting & Refining Co, 36-46; from asst prof to assoc prof metall, 46-76, EMER ASSOC PROF METALL & METALL ENG, UNIV UTAH, 76- *Mem:* Am Inst Mining, Metall & Petrol Engrs. *Res:* Extractive metallurgy; spectrographic analysis. *Mailing Add:* 1318 Logan Ave Salt Lake City UT 84105

BRADFORD, HENRY BERNARD, JR, b Baton Rouge, La, Mar 15, 42; m 64; c 2. BACTERIOLOGY, MICROBIOLOGY. *Educ:* La State Univ, BS, 64, MS, 67; Univ Ala, Birmingham, PhD(microbiol), 75. *Prof Exp:* Microbiologist, 67-69, chief virol unit, 70-71, asst dir Birmingham lab, Bur Labs, State Pub Health, Ala, 71-76, dir, 76-77; DIR BUR LAB SERV, OFF HEALTH SERV & ENVIRON QUAL, LA DEPT HEALTH & HUMAN RESOURCES, 77- *Concurrent Pos:* Adj asst prof pub health & trop med, Tulane Univ, 77- *Mem:* AAAS; Asn State & Territorial Pub Lab Dirs; Am Soc Microbiol; Am Pub Health Asn. *Res:* Molecular biology of molluscum contagiosum virus; environmental microbiology; vibrios and their public health significance. *Mailing Add:* 1112 Field St Metairie LA 70003

BRADFORD, JAMES C, b Wichita Falls, Tex, Aug 28, 30; m 51; c 3. MATHEMATICS. *Educ:* NTex State Col, BS, 51, MS, 52; Univ Okla, PhD(math), 57. *Prof Exp:* Assoc prof math, Abilene Christian Col, 57-61; res mathematician, Teledyne Systs Co, 61-65; prof math & chmn dept, Abilene Christian Col, 65-79; gov, Math Asn Am, 83-85; CONSULT, 86- *Concurrent Pos:* Consult, Chance-Vought Aircraft, 57-61; vis scientist, Tex Acad Sci, 60-61. *Mem:* Am Math Soc; Math Asn Am; Sigma Xi. *Res:* Functions of a real variable; abstract analysis; analysis stationary time series applied to geophysics. *Mailing Add:* 1901 Lincoln Dr Abilene TX 79601

BRADFORD, JAMES CARROW, b Wilmington, Del, Dec 10, 30; m 53; c 3. TERATOLOGY. *Educ:* Univ Del, BS, 54, MS, 60. *Prof Exp:* Sanitarian, Del Dept Pub Health, 54-55; from assoc res biologist to res biologist, 60-74, RES BIOLOGIST & SECTION HEAD, STERLING-WINTHROP RES INST, RENSSELAER, 74- *Mem:* Environ Mutagen Soc; Sigma Xi (past pres); Teratology Soc; Genetic Toxicol Soc. *Res:* Screen and evaluate drugs for teratogenic-embryotoxic potential; evaluation of mutagenic potential; special toxicology studies. *Mailing Add:* Jordan Rd Box 62 RD 2 Nassau NY 12123

BRADFORD, JOHN NORMAN, b Carthage, Mo, May 28, 31; m 63; c 2. ATOMIC PHYSICS, NUCLEAR PHYSICS. *Educ:* Kans State Univ, BS, 54; Iowa State Univ, PhD(physics), 65. *Prof Exp:* Sr physicist, Gen Dynamics & Astronaut, 62-63; asst prof physics, Univ Mont, 65-67 & Kans State Univ, 67-69; res physicist, 69-76, DEP ELECTRONICS TECHNOL, AIR FORCE CAMBRIDGE RES LAB, BEDFORD, 76- *Mem:* Am Phys Soc; Sigma Xi. *Res:* Nuclear physics, particularly photonuclear reactions; chnanneling and energy loss; x-ray induced electron emission; x-ray imaging; radiation effects in electronics. *Mailing Add:* 39 Clarke St Lexington MA 02173

BRADFORD, JOHN R(OSS), b Amarillo, Tex, Nov 26, 22; m; c 2. CHEMICAL ENGINEERING. *Educ:* Tex Tech Col, BS, 42, MS, 48; Case Western Reserve Univ, PhD, 53. *Prof Exp:* Lectr physics, Tex Tech Col, 46-48; dir, Radioisotopes Lab, Case Western Reserve Univ, 50-54; res consult, US Radium Corp, 54-55; PROF CHEM ENG & DEAN COL, TEX TECH UNIV, 55-, CHMN, DEPT SYSTS, 76- *Concurrent Pos:* Mem, US Nat Comn, World Energy Conf, 78- *Mem:* Am Soc Eng Educ; Am Inst Chem Engrs; Am Chem Soc; Am Soc Testing & Mat. *Res:* Industrial applications of radioisotopes; activation analysis; engineering education. *Mailing Add:* PO Box 4650 Lubbock TX 79409

BRADFORD, LAURA SAMPLE, b St Louis, Mo, Dec 5, 45; c 2. APNEA OF NEWBORN, AIDS IN CHILDREN. *Educ:* Grinnell Col, BA, 67; Wash Univ, PhD(biol), 71; Rush Univ, BSN, 81. *Prof Exp:* Fel genetics, dept biol, Univ Chicago, 71-73; asst prof biol, Roosevelt Univ, 73-79; LEVEL D STAFF NURSE, SPEC CARE NURSERY, RUSH-PRESBY-ST LUKE'S MED CTR, 81-, NURSE, NEONATAL APNEA EVAL PROG, 82- *Mem:* Nurses Asn Am Col Obstetricians & Gynecologists; Nat Asn Neonatal Nurses. *Res:* Effects of discharge teaching to parents of infants with breathing abnormalities (apnea, periodic breathing) on family functioning; Effects of a formal screening program for apnea; effects of infant massage on developmental outcome. *Mailing Add:* 5647 S Blackstone Chicago IL 60637

BRADFORD, LAWRENCE GLENN, b Atlanta, Kans, May 24, 39. MOLECULAR BIOLOGY, MICROBIOLOGY. *Educ:* Univ Kans, BA, 61, PhD(microbiol), 89; Northwestern Univ, MAT, 70. *Prof Exp:* High Sch teacher sci & math, Trego Community High Sch, Wakeeney, Kans, 70-73, Raymore-Peculiar High Sch, Peculiar, Mo, 73-76 & Maur Hill Prep Sch, Atchison, Kans, 76-82; ASST PROF BIOL, BENEDICTINE COL, ATCHISON, KANS, 89- *Mem:* Am Soc Microbiol; Sigma Xi. *Res:* Characterization of human epidermal cell lines which had been immortalized by transfection with SV40 T-antigen and with Harvey murine sarcoma virus ras oncogene. *Mailing Add:* Dept Biol Benedictine Col Atchison KS 66002-1499

BRADFORD, MARION MCKINLEY, b Rome, Ga, Oct 28, 46; m 71; c 2. BIOCHEMISTRY. *Educ:* Shorter Col, Ga, BA, 67; Univ Ga, MS, 74, PhD(biochem), 75. *Prof Exp:* Res assoc, Univ Ga, 75-77, asst res scientist biochem, 77-81; res biochemist, A E Staley Mfg Co, Decatur, Ill, 81-83, sr res chemist, 83-90. *Mem:* Am Chem Soc; Sigma Xi. *Res:* Protein functionality; food protein structure and function; enzyme modification of proteins; hydrolyzed vegetable protein flavors; microbial and human nutrition; food starch chemistry. *Mailing Add:* A E Staley Mfg Co 2200 E Eldorado St Decatur IL 62521

BRADFORD, PHILLIPS VERNER, b Washington, DC, June 15, 40; m 64; c 2. ELECTRICAL ENGINEERING, MICROWAVE PHYSICS. *Educ:* Johns Hopkins Univ, BES, 62; Univ Va, MEE, 64; Columbia Univ, EngScD(elec eng), 68. *Prof Exp:* Res asst, Res Labs Eng Sci, Univ Va, 63-64; res asst, Plasma Lab, Dept Elec Eng, Columbia Univ, 65-67; mem tech staff, Bell Tel Labs, Holmdel, 67-70; dir, Precision Optics, Inc, 70-72; indust specialist, Merrill Lynch Pierce Fenner & Smith, 72-75; staff scientist, Ametek, Inc, 75-77; mgr, Phelps Dodge Industs, 77-82; dir corp res, Rutgers Univ, 82-84; dir, Kans Advan Technol Comn, 84-88; EXEC DIR, COLO ADVAN TECHNOL INST, 89- *Mem:* Inst Elec & Electronics Engrs; NY Acad Sci; assoc mem Am Soc Heating, Refrig & Air-Conditioning Engrs; Electrochem Soc. *Res:* Interaction of microwaves with matter; use of microwaves to diagnose some properties of plasma; propagation of microwaves in communication paths to earth satellites; solar energy; economics; technology transfer. *Mailing Add:* 1020 15th St Apt 27-J Denver CO 80202-2327

BRADFORD, REAGAN HOWARD, b Lawton, Okla, Dec 19, 32; m 53; c 2. BIOCHEMISTRY. *Educ:* Univ Okla, BS, 53, PhD(biochem), 57, MD, 61. *Prof Exp:* Res assoc, 59-62, from asst to actg head cardiovasc sect, 62-70, head cardiovasc sect, Okla Med Res Found, 71-76, sci dir, 74-76, assoc prof med, Sch Med, 72-81, prof biochem, 69-81, exec vpres, Okla Med Res Found, 76-81, DIR, OKLA LIPID RES CLIN, UNIV OKLA, 71- *Concurrent Pos:* Res fel, Okla Med Res Found, 57-59; instr res med, Univ Okla, 64-68, assoc prof res biochem, 64-69, asst prof med, 68-72. *Mem:* Am Physiol Soc; AMA; Am Fedn Clin Res; Am Chem Soc; Sigma Xi. *Res:* Lipid transport and metabolism. *Mailing Add:* Okla Med Res 825 NE 13th Oklahoma City OK 73104

BRADFORD, SAMUEL ARTHUR, b Rolla, Mo, July 19, 28; m 57; c 4. METALLURGY, CORROSION. *Educ:* Univ Mo-Rolla, BS, 51, MS, 56; Iowa State Univ, PhD(phys metall), 61. *Prof Exp:* Asst chief chemist, Pillsbury Mills, Inc, NY, 53-55; asst phys metall, Ames Lab, Atomic Energy Comn, 57-61; engr, Homer Res Labs, Bethlehem Steel Co, 61-67; PROF METALL, UNIV ALTA, 67- *Concurrent Pos:* Vis fel, Ohio State Univ, 77-78. *Mem:* Nat Asn Corrosion Engrs; Am Soc Metals; Am Inst Mining, Metall & Petrol Engrs; Can Inst Mining & Metall; Am Soc Testing & Mat. *Res:* Oxidation and corrosion of metals; stability of metal oxides; thermodynamics of solid state diffusion and reaction; effects of impurities on physical and mechanical properties of metals; crystal structures. *Mailing Add:* Dept Mining Metall & Petrol Eng Univ Alta Edmonton AB T6G 2G6 Can

BRADFORD, WILLIAM DALTON, b Rochester, NY, Nov 2, 31; m 61; c 2. PATHOLOGY, PEDIATRICS. *Educ:* Amherst Col, AB, 54; Western Reserve Univ, MD, 58; Am Bd Pediat, dipl, 63; Am Bd Path, dipl anat path, 67. *Prof Exp:* Intern path, Children's Hosp Med Ctr, Boston, 58-59, asst resident pediat, 59-61; resident path, Boston Hosp for Women, 63-64; New Eng Deaconess Hosp, Boston, 64-65; asst prof path & assoc pediat, 66-70, assoc dean grad med educ, 70-71, 74-78 & 84-87, PROF PATH, DUKE UNIV, 70-, ASST TO CHANCELLOR, HEALTH AFFAIRS, 87- *Concurrent Pos:* Teaching fel, Harvard Univ, 63-64; Mead Johnson res fel pediat, 63-64; fel path, Duke Univ, 65-66, Liaison Comt Med Educ, Survey Team Sec, 77- *Mem:* Am Asn Path; Soc Pediat Res; Int Acad Path; Sigma Xi; Soc Pediat Path (pres, 87). *Res:* Iron metabolism; infectious diseases; medical education. *Mailing Add:* 3724 Hope Valley Rd Durham NC 27707

BRADFORD, WILLIS WARREN, b Lincolnton, Ga, Mar 13, 22; m 41; c 2. AGRONOMY. *Educ:* Univ Ga, BSA, 47, MS, 48; Agr & Mech Col Tex, PhD(plant breeding), 54. *Prof Exp:* Assoc agronomist, Univ Ga, 48-50; geneticist, Inst de Fomento Algodonera, Colombia, SAm, 50; agronomist, Coastal Plain Exp Sta, Ga, 52-57; agronomist, Miss, Delta & Pine Land Co, 57-59, mem staff admin cotton breeding, Western Div, 59-89; CONSULT, 89- *Res:* Cotton breeding. *Mailing Add:* 454 West K St Brawley CA 92227

BRADFUTE, OSCAR E, PLANT VIROLOGY, DETECTION OF PLANT VIRUSES. *Educ:* Univ Calif, Berkeley, PhD(plant path), 63. *Prof Exp:* PROF PLANT PATH, OARDC, OHIO STATE UNIV, 81- *Mailing Add:* Dept Plant Path Ohio Res/Develop Ctr Wooster OH 44691

BRADHAM, GILBERT BOWMAN, b Sumter, SC, Aug 5, 31; m; c 2. SURGERY. *Educ:* Med Univ SC, MD, 56. *Prof Exp:* Assoc, 64-65, from asst prof to assoc prof, 65-72, prof, 72-80, CLIN PROF SURG, MED UNIV SC, 80- *Mem:* AMA; Soc Univ Surg. *Mailing Add:* Dept Surg Med Univ Hosp Charleston SC 29425

BRADHAM, LAURENCE STOBO, b Charleston, SC, Dec 3, 29; m 58; c 4. BIOCHEMISTRY. *Educ:* Univ of the South, BS, 51; Univ Tenn, MD, PhD(biochem), 58. *Prof Exp:* Instr pharmacol, Vanderbilt Univ, 58-59; res assoc, Rockefeller Inst, 59-63; res biochemist, Vet Admin Hosp, Little Rock, 63-71; asst prof, 71-74, ASSOC PROF BIOCHEM, UNIV TENN, MEMPHIS, 74- *Mem:* Am Chem Soc; Fedn Am Soc Exp Biol; Am Soc Biol Chemists. *Res:* Metabolism of sulfur compounds; biochemistry of hormone action; enzymology. *Mailing Add:* Dept Biochem Univ Tenn 800 Madison Ave Memphis TN 38163

BRADISH, JOHN PATRICK, b Lawrence, Mass, Nov 12, 23; m 46; c 7. MECHANICAL & NUCLEAR ENGINEERING. *Educ:* Marquette Univ, BSME, 46; Univ Wis, MS, 53, PhD(mech eng), 65. *Prof Exp:* dir dept, Marquette Univ, 58-61, from instr to assoc prof mech eng, 46-88, asst dean, Col Eng, 71-88; RETIRED. *Concurrent Pos:* Consult, Outboard Marine Div, Evinrude Motors, 57-60, Graham Transmission Co, 58-59 & Argonne Nat Lab, 64- *Mem:* Am Soc Eng Educ; Am Soc Mech Engrs. *Res:* Heat transfer in internal combustion engines; nuclear reactor engineering; engineering education; computer-assisted instruction. *Mailing Add:* 17455 Ezure Lane Brookfield WI 53045

BRADLAW, JUNE A, b Norwich, Conn, July 22, 36; div. CELL BIOLOGY, MICROBIOLOGY. *Educ:* Conn Col, AB, 58; Univ Md, College Park, MS, 64; George Washington Univ, PhD(microbiol), 74. *Prof Exp:* Biologist microbial metab, Agr Res Serv, USDA, 58-65; res microbiologist genetic toxicol, Div Toxicol, 65-86, LEADER, IN VITRO TOXICOL TEAM, FOOD & DRUG ADMIN, 85- *Concurrent Pos:* Adj assoc prof microbiol, George Washington Univ, 81-89, lectr, Biochem Dept; lectr, Anat Dept, Univ Sask, Can; chmn, Toxicity, Carcinogenesis & Mutagenesis Eval Comt, Tissue Cult Asn, 82-, ed, J Tissue Cult Methods, 85-; secy, Tissue Cult Asn, 88- *Honors & Awards:* Award of Merit, Food & Drug Admin, 71, Commendable Serv Award, 87. *Mem:* Am Soc Microbiol; Tissue Cult Asn; Am Soc Cell Biol; Environ Mutagen Soc; Sigma Xi; Grad Women Sci; Am Acad Microbiol. *Res:* Mammalian cell culture applications in toxicology and in vitro toxicity testing; chemical mutagenesis; somatic cell genetics; DNA repair mechanisms; enzyme induction; structure-activity relationships. *Mailing Add:* 5706 Ridgway Ave Rockville MD 20851

BRADLEY, A FREEMAN, b Tuskegee, Ala, 32; c 3. RESPIRATION. *Educ:* Lincoln Univ, BA, 53. *Prof Exp:* DIR RES & DEVELOP LAB, UNIV CALIF, SAN FRANCISCO, 77- *Concurrent Pos:* Blood gas analyzer, Smithsonian Inst. *Mem:* Am Asn Med Instrumentation; Am Physiol Soc; Instrument Soc Am. *Res:* Management of engineering; anesthesia. *Mailing Add:* Res & Develop Lab Univ Calif Rm U-10 San Francisco CA 94143

BRADLEY, ARTHUR, b New York, NY, Apr 6, 26; c 3. SURFACE CHEMISTRY, MELT-BLOWN POLYMER FIBERS. *Educ:* Columbia Univ, AB, 48, MA, 50, PhD(chem), 52. *Prof Exp:* Res assoc phys org chem, George Washington Univ, 52-53; org chemist, US Naval Ord Lab, 53-55; radiochemist, Associated Nucleonics, Inc, 55-58; dir res, Radiation Res Corp, 58-69; vpres, Surface Activation Corp, 69-76; APPL RES COORDR, PALL CORP, 76- *Concurrent Pos:* Vpres dir, Solar Conversion Prod, Inc, 78-80. *Mem:* Am Chem Soc. *Res:* Organic reaction mechanisms and analytical techniques; polymer films and fibers; surface grafting; reactions in gas discharges. *Mailing Add:* 146 Beech St Floral Park NY 11001

BRADLEY, DANIEL JOSEPH, b Portage la Prairie, Man, Aug 3, 49; US citizen; m 81; c 3. PHYSICAL CHEMISTRY, PETROLEUM ENGINEERING. *Educ:* Mich State Univ, BSc, 73, PhD(chem), 77; Mont Tech, BSC, 82; Univ Tulsa, MS, 83. *Prof Exp:* Fel, Metals & Ceramics Div, Oak Ridge Nat Lab, 75-77; fel chem, Univ Calif, Berkeley & Lawrence Berkeley Lab, 78-79; dir int progs, 87-89, MEM STAFF, MONT COL MINERAL SCI & TECHNOL, 79-, HEAD PETROL ENG DIV, 89- *Concurrent Pos:* Consult, Daqing Petrol Admin Bur, 86 & 87, Mont Dept State Lands, 90, various attys, petrol eng, 83- *Mem:* Am Chem Soc; Soc Petrol Engrs. *Res:* Thermodynamics of mixtures; high temperature electrolyte solutions; flow through porous material; petroleum reservoir modeling; pressure transient analysis. *Mailing Add:* Dept Petrol Eng Mont Tech Butte MT 59701

BRADLEY, DORIS P, b Laurel, Miss, Oct 21, 32; m 57. SPEECH PATHOLOGY, AUDIOLOGY. *Educ:* Univ Southern Miss, BS, 53; Univ Fla, MA, 57; Univ Pittsburgh, PhD(speech path), 63. *Prof Exp:* Dir speech clin, Miss Soc Crippled Children, 53-54; speech therapist, Gulfport Schs, 54-56; speech clinician, Los Angeles Soc Crippled Children, 57-58; instr speech path, Med Sch, Univ Ore, 58-61; dir speech clin, Children's Hosp, Pittsburgh, Pa, 63-65; asst prof speech path & coordr grad prog speech & audiol, Med Sch, Univ Ore, 65-68; res assoc, Univ NC, Chapel Hill, 68-70, assoc prof oral biol & dir oral facial & commun dis prog, Dent Sch, 70-74; prof, Dept Logopedics, Wichita State Univ & head clin serv, Inst Logopedics, 74-80; chair person, Dept Speech-Hearing Sci, 81-88, PROF SPEECH & HEARING, UNIV SOUTHERN MISS, HATTIESBURG, 88- *Concurrent Pos:* NIH grant, 65-; consult, United Cerebral Palsy, 54-56 & Ore State Dept Educ, 58-61; res assoc, Cleft Palate Res Ctr, Pa, 63-65; prof commun disorders, Univ Cent Florida, 80-81. *Mem:* Am Cleft Palate Asn; Am Speech & Hearing Asn; Soc Res Child Develop. *Res:* Speech problems associated with cleft palate; development of normal human communication skills in children; modification of aerodynamics of speech production through speech therapy; phonological rules in southern dialect; multiple phonemic articulation therapy; development of models of program evaluation; effectiveness of language intervention systems. *Mailing Add:* Dept Speech-Hearing Sci Univ Southern Miss Box 5092 Southern Sta Hattiesburg MS 39401

BRADLEY, EDWIN LUTHER, JR, b Jacksonville, Fla, July 16, 43; m 63; c 3. STATISTICS. *Educ:* Univ Fla, BS, 64, MStat, 67, PhD(statist), 69. *Prof Exp:* Assoc sci programmer, Lockheed-Ga Co, 65; from asst prof to assoc prof biostatist, 70-88, PROF BIOSTATIST & BIOMATH, UNIV ALA, BIRMINGHAM, 88- *Concurrent Pos:* Consult statistician. *Mem:* Am Statist Asn; Biomet Soc. *Res:* Mathematical statistics; application of statistical techniques. *Mailing Add:* Dept Biostatist & Biomath Univ Ala Birmingham Birmingham AL 35294

BRADLEY, ELIHU F(RANCIS), b Allentown, Pa, Oct 26, 17; m 42; c 4. PHYSICAL METALLURGY, MATERIALS ENGINEERING. *Educ:* Yale Univ, BE, 39. *Prof Exp:* Chemist, Seymour Mfg Co, 39-41; metallurgist, 41-45, design metallurgist, 45-57, eng metallurgist, 57-61, CHIEF MAT ENG, PRATT & WHITNEY AIRCRAFT DIV, UNITED AIRCRAFT CORP, 61- *Concurrent Pos:* Mem adv comt aircraft engine mat, NASA, 66-; mem comt titanium, Mat Adv Bd, 68-69. *Mem:* Fel Am Soc Metals; Am Soc Testing & Mat. *Res:* Aircraft engine materials, particularly gamma prime strengthened nickel, high temperature superalloys, cobalt alloys, titanium alloys, protective coatings and refractory metals; columbium alloys and coatings. *Mailing Add:* 71 Crestwood Rd West Hartford CT 06107

BRADLEY, EUGENE BRADFORD, b Georgetown, Ky, Nov 4, 32; m 58; c 2. MOLECULAR PHYSICS, ELECTRICAL ENGINEERING. *Educ:* Georgetown Col, BS, 52; Univ Ky, MS, 57; Vanderbilt Univ, PhD(infrared spectro), 64. *Prof Exp:* Elec engr, Elec Parts Corp, 56; lectr physics, Vanderbilt Univ, 62-63; instr, 63-64; from asst prof to assoc prof, 64-75, univ res prof, 82-83, actg chmn, 84-85, PROF ELEC ENG, UNIV KY, 75- *Concurrent Pos:* From instr to asst prof elec eng, Univ Ky, 56-60. *Mem:* Am Phys Soc. *Res:* Far-infrared spectroscopy, laser Raman spectroscopy. *Mailing Add:* Dept Elec Eng Univ Ky Lexington KY 40506-0046

BRADLEY, FENNIMORE N, ceramic engineering, materials science, for more information see previous edition

BRADLEY, FRANCIS J, b New York, NY, Jan 15, 26; m 51; c 5. HEALTH PHYSICS. *Educ:* Manhattan Col, BEE, 49; Univ Pittsburgh, MS, 53; Ohio State Univ, PhD(physics, chem), 61. *Prof Exp:* Engr, Westinghouse Elec Corp, 50-53; supt radiation safety, Ohio State Univ, 53-61; tech adv, Int Atomic Energy Agency, 61-62; asst prof health physics, Univ Pittsburgh, 62-65; head health physics sect, Isotopes, Inc, 65-66; PRIN RADIOPHYSICIST RADIOL HEALTH SECT, NY STATE DEPT LABOR, 66- *Mem:* AAAS; Am Phys Soc; Health Physics Soc. *Res:* Environmental factors effecting solid state dosimeters; applications of TLD monitors; human radioisotope body burden assessment by bioassay and whole body counting; radiation protection administration and legislation; laser and microwave exposure studies. *Mailing Add:* Radiol Health Sect NY State Dept Labor One Main St Brooklyn NY 11201

BRADLEY, GEORGE ALEXANDER, b Staunton, Va, Dec 4, 26; m 57; c 2. HORTICULTURE. *Educ:* Univ Del, BS, 51; Cornell Univ, MS, 53, PhD(veg crops), 55. *Prof Exp:* Res asst, Cornell Univ, 51-55; from asst prof to assoc prof, 55-63, dir sec sci training prog plant sci, 65, PROF HORT & FORESTRY, UNIV ARK, FAYETTEVILLE, 63-, HEAD DEPT HORT, 68- *Mem:* Fel AAAS; fel Am Soc Hort Sci. *Res:* Mineral nutrition; soils; physiology of vegetable crops. *Mailing Add:* 1700 Viewpoint Fayetteville AR 72701

BRADLEY, H(OWARD) B(ISHOP), b Chicago, Ill, Mar 24, 20; m 40; c 2. PETROLEUM ENGINEERING. *Educ:* Univ Tex, BS, 44, MS, 50. *Prof Exp:* Asst geologist, Bur Econ Geol, Tex, 46-48; res engr, Tex Petrol Res Comt, 48-49; asst prof petrol & natural gas eng, Univ Ala, 49-51; sr res technologist, field res labs, Magnolia Petrol Co, 53-60, supvr tech training, Socony Mobil Oil Co, Tex, 60-68; adv tech educ & training, Mobil Oil Corp, 68-75, corp mgr, tech training, 75-82; CONSULT, SOC PETROL ENGRS, 82- *Concurrent Pos:* Asst prof, Univ Tex, 48, 51. *Honors & Awards:* Distinguished Serv Award, Soc Petrol Engrs, 86. *Mem:* Am Inst Mining, Metall & Petrol Engrs; Sigma Xi; Soc Petrol Engrs. *Res:* Technical training of professional personnel of Mobil; petroleum reservoir and chemical engineering. *Mailing Add:* 215 Fresh Meadow Dr Roanoke TX 76262

BRADLEY, HARRIS WALTON, b Charlotte, NC, Jan 13, 15; m 44; c 4. CHEMISTRY. *Educ:* Davidson Col, BS, 37; Univ Va, PhD(org chem), 42. *Prof Exp:* Res chemist, Jackson Lab, E I du Pont de Nemours & Co, 42-43, supvr plant opers, Chambers Works, 43-45, res chemist, Jackson Lab, 45-57; tech dir, E R Carpenter Co, 57-60; consult, 60-61; mgr res & develop, Carolina Indust Plastics Div, Essex Wire Corp, 61-64; pres, 64-82, ASST TO PRES, NC FOAM INDUST, 82- *Mem:* AAAS; Am Chem Soc. *Res:* Textile chemicals; water-repellents for textiles; sulfoxidation of hydrocarbons; azo dyes; detergents; neoprene latex; rubber chemicals; blowing agents; polyurethanes. *Mailing Add:* NC Foam Industs, Inc Box 1528 Mt Airy NC 27030

BRADLEY, HUGH EDWARD, b Olean, NY. INDUSTRIAL ENGINEERING. *Educ:* Mass Inst Technol, SB & SM, 57; Johns Hopkins Univ, PhD(indust eng), 63. *Prof Exp:* Asst prof indust eng, Univ Mich, 63-67; group mgr, Upjohn Co, 67-79; dir, Corp Info Systs, Syntex Corp, 79-81; assoc dir, Kaiser Aluminum Corp, 81-84; dir, Info Systs, 84-90, dir, MIS, 90, VPRES & CHIEF INFO OFFICER, SHAKLEE US, INC, DIV YAMANOUCHI PHARMACEUT CO, 90- *Concurrent Pos:* Adj assoc prof, Western Mich Univ, 67-73; ed, Int Abstracts Opers Res, 69-79 & Opers Res, 89-91; adj prof, Grad Mgt Sch, Grand Valley Col, Mich, 75-78. *Honors & Awards:* Kimball Medal, Opers Res Soc Am, 90. *Mem:* Opers Res Soc Am (treas, 80-83 & 89-90, pres 85-86); Inst Mgt Sci (vpres, 77-80); Pharmaceut Mgt Sci Asn. *Mailing Add:* 612 Montezuma Ct Walnut Creek CA 94598

BRADLEY, JAMES HENRY STOBART, b London, Eng, Mar 26, 33; m 66. NUMERICAL ANALYSIS, SYSTEMS ANALYSIS. *Educ:* Oxford Univ, BA, 54; Univ Toronto, MA, 60; Univ Mich, PhD(meteorol), 67. *Prof Exp:* Scientist isotope geol, Geol Serv Can, 57-59; scientist instrumentation, Can Meteorol Serv, 59-63; res asst meteorol, Univ Mich, 63-67; prof, Pa State Univ, 67-68; res assoc, McGill Univ, 68-70; prof physics, Drexel Univ, 70-76; sr mem tech staff, Electronic Assocs, Inc, West Long Branch, 76-82; PRES, ATLANTIC CONSULTS, 82- *Concurrent Pos:* Consult, Govt Can, 68-70, US Govt, 68- *Mem:* Soc Indust & Appl Math; Instrument Soc Am; Soc Comput Simulation; Int Asn Math & Comput in Simulation. *Res:* Computational partial differential equations; real-time numerical methods; training simulators; tests of software. *Mailing Add:* 46 Rabbit Run Wallingford PA 19086

BRADLEY, JAMES T, m 72; c 2. INSECT ENDOCRINOLOGY. *Educ:* Univ Wis, Madison, BS, 70; Univ Wash, Seattle, PhD(develop biol), 76. *Prof Exp:* ASSOC PROF ZOOL, AUBURN UNIV, 81- *Mem:* Am Soc Cell Biol; Am Soc Develop Biol; Am Soc Zool; AAAS. *Res:* Multiple aspects of insect and fish vitellogenesis including vitellogenin biochemistry and endocrine regulation of protein synthesis; polyacrylamide gel electrophoresis; monoclonal antibodies; ultracryomicrotomy. *Mailing Add:* Dept Zool Auburn Univ 101 Cary Hall Auburn AL 36849

BRADLEY, JOHN SAMUEL, b US, Feb 23, 23; m 51; c 3. GEOLOGY. *Educ:* Colo Sch Mines, GeolE, 48; Univ Wash, PhD(geol), 52. *Prof Exp:* Geologist, geol res sect, Humble Oil & Refining Co, 50-54; marine geologist, Inst Marine Sci, Tex, 54-56; geologist, res & develop dept, Atlantic Refining Co, 56-64; GEOLOGIST, GEOL RES DEPT, AMOCO PROD CO, 64- *Mem:* Geol Soc Am; Am Asn Petrol Geologists; Soc Econ Paleont & Mineral; Am Geophys Union. *Res:* Rock mechanics; hydrogeology; photogeology. *Mailing Add:* Amoco Prod Res Ctr Box 3385 Tulsa OK 74102-3385

BRADLEY, JOHN SPURGEON, b Gulfport, Miss, Jan 28, 34; m 59; c 3. MATHEMATICS. *Educ:* Univ Southern Miss, BS, 55; George Peabody Col, MA, 56; Univ Iowa, PhD(math), 64. *Prof Exp:* Instr math, Univ Southern Miss, 56-57 & George Peabody Col, 59-60; from asst prof to assoc prof, 64-72, actg head computer sci dept, 84-86, PROF MATH, UNIV TENN, KNOXVILLE, 72-, HEAD DEPT, 80- *Concurrent Pos:* Vis prof, Univ Dundee, 71-72. *Mem:* Am Math Soc; Math Asn Am; Edinburgh Math Soc. *Res:* Boundary value problems; integral inequalities; oscillation theory. *Mailing Add:* Dept Math Univ Tenn Knoxville TN 37996

BRADLEY, JULIUS ROSCOE, JR, b Minden, La, Mar 25, 40; m 63; c 1. ENTOMOLOGY. *Educ:* La Polytech Inst, BS, 62; La State Univ, MS, 64, PhD(entom), 67. *Prof Exp:* Asst prof, 67-72, assoc prof, 72-78, PROF ENTOM, NC STATE UNIV, 78- *Mem:* Entom Soc Am. *Res:* Ecology and control of the boll weevil; pest management of Heliothis species; pesticide interactions. *Mailing Add:* Dept Entom NC State Univ Gardner HL Raleigh NC 27650

BRADLEY, LAURENCE A, b Cleveland, Ohio, Sept 13, 49. CHRONIC PAIN, BEHAVIORAL MEDICINE. *Educ:* Vanderbilt Univ, BA, 71 & PhD(psychol), 75. *Prof Exp:* Intern psychol, Duke Univ Med Ctr, 75-76; asst prof psychol, Univ Tenn, Chattanooga, 76-77 & Fordham Univ, 77-80; from asst prof to assoc prof med psychol, Bowman Gray Sch Med, 80-89; ASSOC PROF, UNIV ALA-BIRMINGHAM, 89- *Concurrent Pos:* Vis behav scientist, Dept Occup Med, Örebro Med Ctr, Sweden, 86-88; assoc prof, Grad Sch, Wake Forest Univ, 84-, adj assoc prof, Univ NC, Greensboro, 85- *Mem:* Sigma Xi; fel Soc Personality Assessment; Int Asn Study Pain; Am Pain Soc; Arthritis Health Prof Asn; fel Am Psychol Asn. *Res:* Study on the measurement and management of chronic pain due to arthritis and gastrointestinal disorders; medical psychology. *Mailing Add:* Dept Psychol 201 Campbell Hall Univ Ala Birmingham Univ Sta Birmingham AL 35294

BRADLEY, LEE CARRINGTON, III, b Birmingham, Ala, May 31, 26. OPTICAL PHYSICS. *Educ:* Princeton Univ, AB, 46; Oxford Univ, DPhil(physics), 50. *Prof Exp:* Instr physics, Princeton Univ, 50-54; asst prof, 54-64, staff physicist, 64-77, SR STAFF PHYSICIST, LINCOLN LAB, MASS INST TECHNOL, 77- *Mem:* Am Phys Soc; Optical Soc Am. *Res:* Laser beam propagation; reentry physics. *Mailing Add:* 201 Somerset St Belmont MA 02178

BRADLEY, MARSHALL RICE, b Hattiesburg, Miss, Jan 11, 51; m 74; c 2. UNDERWATER ACOUSTICS, WAVE PROPAGATION. *Educ:* Univ Miss, BA, 73, MS, 74; Univ Va, PhD(appl math), 77. *Prof Exp:* Instr appl math, Univ Va, 76-77; SR SCIENTIST, PLANNING SYST INC, 77- *Mem:* Acoust Soc Am; Soc Indust & Appl Math. *Res:* Mathematical modeling of acoustical phenomena; elastic wave propagation; statistical analysis of acoustic data; probability theory. *Mailing Add:* Planning Systs Inc 115 Christian Lane Slidell LA 70458

BRADLEY, MATTHEWS OGDEN, b New York, NY, Dec 9, 42; m 63; c 1. CELL BIOCHEMISTRY, CANCER. *Educ:* Univ Pa, BA, 64; Stanford Univ, PhD(biol sci), 71. *Prof Exp:* Sr staff fel carcinogenesis, Nat Cancer Inst, NIH, Bethesda, Md, 74-78, cancer expert, 78-79; SR SCIENTIST MOLECULAR CARCINOGENESIS, 79-, DIR, MOLECULAR BIOL & GENETIC TOXICOL, MERCK SHARP & DOHME RES LABS, 84- *Concurrent Pos:* Royal Norweg Coun Sci Res fel, 71-72; Leukemia Soc Am fel, Stanford Univ, 72-74. *Mem:* Tissue Cult Asn; Am Soc Cell Biol; Geront Soc; Am Asn Cancer Res; Environ Mutagen Soc. *Res:* Molecular mechanisms of carcinogenesis; relationships between DNA damage/repair and toxicity, mutagenicity, and carcinogenicity; short-term assays for carcinogens; alterations in DNA repair during development and aging; recombinant DNA technology; oncogenes and growth factors; early mammalian development; in vitro toxicology; genetic toxicology. *Mailing Add:* Merck Sharp & Dohme PO Box 4 West Point PA 19486

BRADLEY, MICHAEL DOUGLAS, b Wichita, Kans, May 20, 38; m 61. RESOURCE MANAGEMENT, PHYSICAL GEOGRAPHY. *Educ:* Univ NMex, BA, 67; Univ Mich, MPubAd, 68, PhD(resource planning & conserv), 71. *Prof Exp:* Asst res marine sci affairs, Ctr Marine Affairs, Scripps Inst Oceanog, Univ Calif, San Diego, 71-72; asst prof, 72-77, ASSOC PROF WATER RESOURCES, DEPT HYDROL & WATER RESOURCES, UNIV ARIZ, 77- *Concurrent Pos:* Vis prof natural environ & resources, Grad Sch Archit & Urban Planning, Univ Calif, Los Angeles, 80-81; consult water rights, State NMex, 83- *Mem:* AAAS; Am Water Resources Asn; Asn Am Geographers; Am Geophys Union. *Res:* Water resource management in both large scale (large marine ecosystems, multi-state water systems) and small-scale(localized groundwater aquifers); rational use of scientific and technical information in court and adjudicative proceedings. *Mailing Add:* 3301 N Christmas Ave Tucson AZ 85716

BRADLEY, RALPH ALLAN, b Can, Nov 28, 23; nat US; m 46; c 2. STATISTICS. *Educ:* Queen's Univ, Ont, BA, 44, MA, 46; Univ NC, PhD(math statist), 49. *Prof Exp:* Asst prof math, McGill Univ, 49-50; from assoc prof to prof statist, Va Polytech Inst, 50-59; prof, 59-70, head dept, 59-78, distinguished prof statist, 70-83, EMER PROF STATIST, FLA STATE UNIV, 83-; RES PROF, UNIV GA, 82- *Concurrent Pos:* Ed, Biometrics, 57-62; chmn, Gordon Res Conf, 65; Ford Found Prog specialist, Cairo Univ, 66-67, Ford Found consult, 68-76; mem, Comt Statist, Nat Acad Sci, 82-88. *Honors & Awards:* Horsely Award, Va Acad, 57; Brumbaugh Award, Am Soc Qual Control, 56. *Mem:* Fel AAAS; fel Inst Math Statist; fel Am Statist Asn (vpres, 76-79, pres 81); Biomet Soc (pres, 65); Int Statist Inst. *Res:* Mathematical statistics; statistical methods for sensory difference testing. *Mailing Add:* Dept Statistics Univ Georgia Athens GA 30602

BRADLEY, RAYMOND STUART, b Cheshire, UK, June 13, 48; m 72; c 2. CLIMATOLOGY, PALEOCLIMATOLOGY. *Educ:* Univ Southampton, UK, BSc, 69; Univ Colo, Boulder, MA, 71, PhD(geog), 74. *Prof Exp:* From instr to assoc prof, 73-84, PROF GEOG, UNIV MASS, AMHERST, 84- *Concurrent Pos:* Prin investr grants climate dynamics & polar progs, NSF, 75-88 & US Dept Energy, 81-88; Nat Acad Comt Trends Acid Deposition, 83-85; lectr, Acad Sinica, Beijing, 84; Advan Res Workshop, NATO, 85; vis fel, Clare Hall, Cambridge, 87; US-USSR Bilateral Meeting of Experts on Climatic Change, Leningrad, 86. *Mem:* Am Meteorol Soc; Am Quaternary Asn; Int Glaciol Soc. *Res:* Climate and climatic change; paleoclimatology; environmental change in mountain and arctic regions. *Mailing Add:* Dept Geol & Geog Univ Mass Amherst MA 01003

BRADLEY, RICHARD CRANE, b Chicago, Ill, May 14, 22; m 47; c 4. PHYSICS. *Educ:* Dartmouth Col, AB, 43; Univ Calif, PhD(physics), 53. *Prof Exp:* Tech aid, Calif Inst Technol, 43; asst physics, Univ Calif, 47-53; res assoc, Cornell Univ, 53-56, from instr to assoc prof, 56-62; assoc prof physics, Colo Col, 62-66, chmn dept, 70-73, dean, 73-79, prof, 66-87, EMER PROF PHYSICS, COLO COL, 87- *Concurrent Pos:* Consult, Kaman Nuclear, Inc, 63 & Penrose Cancer Res Inst, 64- *Mem:* AAAS; fel Am Phys Soc; Am Asn Physics Teachers. *Res:* Sputtering; secondary ion emissions; surfaces of solids; mass spectrometry; field emission microscopy. *Mailing Add:* 1035 Broadview Pl Colorado Springs CO 80904

BRADLEY, RICHARD CRANE, JR, b Berkeley, Calif, Mar 27, 50. LIMIT THEOREMS. *Educ:* Mass Inst Technol, BS, 72; Univ Calif, San Diego, PhD(math), 78. *Prof Exp:* Asst prof statist, Columbia Univ, 78-80; asst prof math, 80-85, ASSOC PROF MATH, IND UNIV, 85- *Mem:* Sigma Xi; Inst Math Statist; Am Math Soc. *Res:* Properties of strong mixing conditions for sequences of dependent random variables; limit theorems for random sequences under strong mixing conditions. *Mailing Add:* Dept Math Ind Univ Bloomington IN 47405

BRADLEY, RICHARD E, b Decatur, Ill, Oct 18, 27; m 80; c 5. PARASITOLOGY, VETERINARY MEDICINE. *Educ:* Fla State Univ, BS, 49; Univ Ga, MS, 53, DVM, 54; Univ Ga, PhD, 65. *Prof Exp:* Asst zool, Univ Ga, 49-50, lab asst bact & instr zool, 50-54, instr parasitol, 54-55; instr, Univ Ill, 55-56; pvt practr vet med, Ill, 56-62; instr zool, Danville Jr Col, 62 & vet med, Univ Ga, 63; asst parasitologist, Univ Fla, 65-69, assoc parasitologist, 69-74, prof vet med, 75-90, actg chmn dept, 84-90; RETIRED. *Concurrent Pos:* Int consult control of parasitism in food animals. *Honors & Awards:* Merck Found Award, 80; Beecham Res Award, 83. *Mem:* Am Vet Med Asn; Am Soc Parasitol; Am Soc Trop Med & Hyg; Am Asn Vet Parasitol. *Res:* Canine filariasis; ruminant nematode parasites; anthelmintics; parasitic protozoa; gnotobiology; pathology and pathogenesis of parasites; immunology of parasitism. *Mailing Add:* RR 3 Box 441 Open Range Rd Crossville TN 38555

BRADLEY, RICHARD E, b Omaha, Nebr, Mar 9, 26; m 46; c 3. DENTISTRY, PERIODONTICS. *Educ:* Univ Nebr, BSD, 50, DDS, 52; Univ Iowa, MS, 58. *Prof Exp:* Instr periodont, Univ Iowa, 57-58; asst prof, Creighton Univ, 58-59; asst prof periodont, Col Dent, Univ Nebr-Lincoln, 59-61, assoc prof, 61-65, prof, 65-80, chmn dept, 61-80, dean, Col Dent, 68-80; PRES & DEAN, BAYLOR COL DENT, DALLAS, 80. *Concurrent Pos:* Consult, Vet Admin Hosp, Omaha, 60-80, Dallas, 80-, attend, Vet Admin Hosp, Lincoln, 60-80; cent off consult, Vet Admin, 65-72; mem, Nat Adv Coun Health Professions, Pub Health Serv, 82- *Mem:* Am Soc Periodont (secy, 64); Am Acad Periodont; Am Dent Asn; Int Asn Dent Res; AAAS. *Res:* Histochemistry of oral enzymology; clinical periodontics. *Mailing Add:* 6424 Crooked Creek Dr Lincoln NE 68516

BRADLEY, ROBERT FOSTER, b Columbia, SC, Mar 26, 40; m 69; c 3. CHEMICAL ENGINEERING. *Educ:* Univ SC, BS, 62, MS, 64; Vanderbilt Univ, PhD(chem eng), 66; AMA Exten Inst, cert bus mgt, 80. *Prof Exp:* Res engr, Sabine River Works, 65-66, Savannah River Lab, 66-70, res supvr, Savannah River Lab, 70-78, chief supvr, Savannah River Plant, 78-81, res supvr, 81-85, supt, Savannah River Lab, E I du Pont, 86-89, MGR, FACIL SAFETY EVALUATION, WESTINGHOUSE SAVANNAH RIVER CO, 89- *Mem:* Am Inst Chem Engrs. *Res:* Reaction kinetics, radiation chemistry, liquid metal distillation, transplutonium process development, safety analyses, near-term and long-term nuclear waste management, large-scale development of nuclear waste vitrification processes; cost-benefit analyses; nuclear safety oversight. *Mailing Add:* Savannah River Site Bldg 773-57A Aiken SC 29802

BRADLEY, ROBERT LESTER, JR, b Beverly, Mass, Jan 14, 33; m 60; c 2. DAIRY TECHNOLOGY, FOOD SCIENCE. *Educ:* Univ Mass, BS, 58; Mich State Univ, MS, 60, PhD(food sci), 64. *Prof Exp:* PROF FOOD SCI, UNIV WIS-MADISON, 64- *Mem:* Am Dairy Sci Asn; Int Asn Milk, Food & Environ Sanitarians. *Res:* Fate of and antagonism of environmental contaminants in foods on humans and animals; functions of dairy ingredients in food systems. *Mailing Add:* Dept Food Sci Univ Wis 103 Babcock Hall 1605 Linden Dr Madison WI 53706

BRADLEY, ROBERT MARTIN, b Bury, Eng, Oct 15, 39; m 68; c 2. SENSORY PHYSIOLOGY, DEVELOPMENTAL PHYSIOLOGY. *Educ:* Univ London, BDS, 63; Univ Wash, MSD, 66; Fla State Univ, PhD(biol sci), 70. *Prof Exp:* House surgeon dent, Royal Dent Hosp, London, 63; house officer, St Marys Hosp, London, 64; res assoc fetal physiol, Nuffield Inst Med Res, Eng, 70-72; from asst prof to assoc prof dent, 72-79, PROF DENT, SCH DENT, UNIV MICH, 79-, ASSOC PROF, DEPT PHYSIOL, 80- *Mem:* Am Physiol Soc; Soc Neurosci; Europ Chemoreceptor Res Orgn; Asn Chemoreception Soc. *Res:* Functional and anatomical development of taste receptors and central taste pathways in fetal and neonatal mammals; central control of salivary secretion; central control of swallowing. *Mailing Add:* Dept Oral Biol Sch Dent Univ Mich Ann Arbor MI 48109

BRADLEY, RONALD W, b Mansfield, Ohio, Nov 11, 36; m 58; c 2. ADHESIVE FORMULATION, INK FORMULATION. *Educ:* Ashland Col, BS, 58. *Prof Exp:* Asst chemist, Owens Brockway Glass, 60-62, assoc engr, 62- 64, engr, 64-88, SR ENGR, OWENS BROCKWAY GLASS, 88- *Concurrent Pos:* chmn, Test Procedures, Soc Glass Decorators, 70-72, group chmn, 72-76, dir, 80-82. *Mem:* Am Chem Soc; Soc Glass Decorators (treas, 76-80). *Res:* Labeling and decorating of glass bottles; pressure sensitive labeling; electrostatic spraying; pad printing; screen printing; offset printing. *Mailing Add:* 4526 Rambo Lane Toledo OH 43623

BRADLEY, STANLEY EDWARD, b Columbia, SC, Mar 24, 13; m 36; c 1. MEDICINE. *Educ:* Johns Hopkins Univ, AB, 34; Univ Md, MD, 38. *Hon Degrees:* Dr, Univ Strasbourg, 78. *Prof Exp:* Intern, Univ Hosp, Baltimore, 38-40; instr med, Sch Med, Boston Univ, 42-45, asst prof, 45-47; from asst prof to prof, 47-59, chmn dept, 59-70, Bard prof, 60-78, EMER BARD PROF MED, COL PHYSICIANS & SURGEONS, COLUMBIA UNIV, 78- *Concurrent Pos:* Commonwealth Fund fel, Col Med, NY Univ, 40-42; asst clin vis physician, Bellevue Hosp, New York, 40-42; asst vis physician, Evans Mem Hosp, Mass, 42-47 & Presby Hosp, New York, 47-51, assoc vis physician, 51-59, dir med serv, 59-70, attend physician, 59-78, consult, 78-; ed-in-chief, J Clin Invest, 52-57; mem bd sci consult, Sloan Kettering Inst, 62-72; trustee, Mt Desert Island Lab, 71-81; vis prof clin pharmacol, Univ Bern, 78-90. *Honors & Awards:* Gibbs Prize, NY Acad Med, 47. *Mem:* Asn Am Physicians; Am Soc Clin Invest (pres, 57); Am Physiol Soc; fel Am Acad Arts & Sci; Soc Exp Biol & Med; Am Heart Asn; fel AAAS. *Res:* Normal pathologic physiology of the kidney and liver; hemodynamic adjustments. *Mailing Add:* 200 N Wynnewood Ave A516 Wynnewood PA 19096

BRADLEY, STERLING GAYLEN, b Springfield, Mo, Apr 2, 32; m 51, 74; c 5. MICROBIOLOGY, GENETICS. *Educ:* Southwest Mo State Univ, BA & BS, 50; Northwestern Univ, MS, 52, PhD(biol), 54. *Prof Exp:* Instr biol, Northwestern Univ, 54; from instr to assoc prof microbiol, Univ Minn, 56-63, prof microbiol & genetics, 63-68; prof & chmn microbiol, PROF PHARMACOL VA COMMONWEALTH UNIV, 79-, PROF MICROBIOL & DEAN BASIC HEALTH SCI, 82- *Concurrent Pos:* Eli Lilly fel, Univ Wis-Madison, 54-55, NSF fel, 55-56; consult, Upjohn Co, 60-68, Minneapolis Vet Admin Hosp, 61-68 & E R Squibb & Sons, 69-76; mem coun, Soc Exp Biol & Med, 64-66 & Am Soc Microbiol, 68-74; mem exec bd, Int Comt Syst Bacteria, 66-74; mem & chmn bd sci counr, Nat Inst Allergy & Infectious Dis, 68-72; ed, Am Soc Microbiol, 70-77; coordr proj 3, US/USSR Joint Working Group Microbiol, 79-82; pres, US Fed Cult Collections, 84-86. *Honors & Awards:* Charles Porter Award, Soc Indust Microbiol, 82. *Mem:* Soc Indust Microbiol (pres, 64-65); Am Asn Immunol; fel AAAS; Am Acad Microbiol; Soc Exp Biol & Med; Soc Toxicol; Am Soc Microbiol (treas, 85-); Am Soc Pharmacol & Exp Therapeut. *Res:* Proteases and protease inhibitors; biology of actinomycetes and actinophages; regulation of cellular phenotype; Immunotoxicology. *Mailing Add:* MCV Box 110 Va Commonwealth Univ Richmond VA 23298-0110

BRADLEY, STEVEN ARTHUR, b Chicago, Ill, Sept 21, 49; m; c 1. MATERIALS SCIENCE ENGINEERING. *Educ:* Northwestern Univ, BS, 70, PhD(mat sci), 75; Roosevelt Univ, MBA, 79. *Prof Exp:* Res mat scientist, Universal Oil Prod, Inc, 74-76, group leader, 76-81, res mgr, 81-84; RES SCIENTIST, UOP, INC, 84- *Concurrent Pos:* Instr, Sch Bus, Roosevelt Univ, 79-; co-organizer, Frontiers Electron Micros in Mat Sci meetings & co-ed proceedings. *Mem:* Am Soc Metals; Am Chem Soc; Nat Asn Corrosion Engrs; Am Soc Testing Mat; Sigma Xi; Electron Micros Soc Am; Am Ceramics Soc. *Res:* New product development in materials development areas; materials testing and analysis including materials failure analyses; catalyst development; ceramics and composites; application of electron microscopy for characterizing ceramics, catalysts, and metals; polymer characterization. *Mailing Add:* UOP 50 E Algonquin Rd Des Plaines IL 60017

BRADLEY, TED RAY, b Kansas City, Mo, Feb 1, 40. PLANT TAXONOMY, ECOLOGY. *Educ:* Rollins Col, BS, 62; Univ NC, MA, 66, PhD(plant taxon), 67. *Prof Exp:* Asst prof, 67-74, ASSOC PROF BIOL, GEORGE MASON UNIV, 74- *Mem:* Am Soc Plant Taxon. *Res:* Evolution of the genus Triodanis, especially hybridization between species. *Mailing Add:* Dept Biol George Mason Univ 4400 University Dr Fairfax VA 22030

BRADLEY, THOMAS BERNARD, JR, b DuBois, Pa, Dec 2, 28; m 55; c 4. INTERNAL MEDICINE, HEMATOLOGY. *Educ:* Hamilton Col, AB, 50; Columbia Univ, MD, 54. *Prof Exp:* Intern med, Presby Hosp, New York, 54-55, asst resident, 55-56 & 58-59; sr instr, Seton Hall Col Med & Dent, 62-63, asst prof, 63-64; assoc, Albert Einstein Col Med, 64-65, asst prof, 65-69; chief, Hemat Sect, San Francisco Vet Admin Hosp, 69-88, assoc chief staff 76-88, ambulatory care, 88; assoc prof, 69-75, PROF MED, UNIV CALIF, SAN FRANCISCO, 75-; CHIEF STAFF, AUDIE MURPHY VET ADMIN HOSP, SAN ANTONIO, 88-; ASSOC DEAN & PROF MED, SCH MED, UNIV TEX, SAN ANTONIO, 88- *Concurrent Pos:* Fel hemat, Hopkins Hosp, Baltimore, 59-60; Health Res Coun career scientist, 64-69; actg assoc chief of staff res, San Francisco Vet Admin Hosp, 74-76. *Mem:* Am Fedn Clin Res; Am Soc Clin Invest; Am Soc Hemat (secy, 74-79). *Res:* Human hemoglobin variants, structure, function, genetics, pathophysiology; red cell metabolism. *Mailing Add:* 7400 Merton-Minter Blvd San Antonio TX 78284

BRADLEY, W(ILLIAM) E(ARLE), b Pa, Jan 7, 13; wid; c 2. ENGINEERING PHYSICS. *Educ:* Univ Pa, BS, 35. *Prof Exp:* Res engr, Philco Corp, 37-59, tech dir res div, 46-59; assoc dir res, Inst Defense Anal, 59-63, asst vpres, 63-69; pres, Puredesal, Inc, 70-73; CONSULT, 73- *Concurrent Pos:* Mem subcomts, Nat TV Syst Comt; consult to White House, 57-62 & Dept of Defense, 57-; consult, NASA, 80-81; mem, subcomt technol innovation, bd sci & technol int develop, Nat Acad Sci, 81-83. *Mem:* Fel Inst Elec & Electronics Engrs; AAAS; Franklin Inst. *Res:* Color television; microwave systems; physical optics; electron optics; transistors and related semiconducting devices; communications; physical chemistry; water desalination. *Mailing Add:* Willow Hill Farm Post Box 257 RD 2 New Hope PA 18938

BRADLEY, WILLIAM ARTHUR, civil engineering, applied mechanics; deceased, see previous edition for last biography

BRADLEY, WILLIAM CRANE, b Madison, Wis, Feb 22, 25; m 58; c 3. GEOLOGY. *Educ:* Univ Wis, BS, 51; Stanford Univ, MS, 53, PhD(geol), 56. *Prof Exp:* From instr to prof geol, Univ Colo, Boulder, 55-89, chmn dept, 68-72; RETIRED. *Concurrent Pos:* Res scientist, Univ Tex, 65-66; vis lectr, Univ Adelaide, S Australia, 73-74, Univ Wales, Aberystwyth, 82-83. *Mem:* Geol Soc Am; Colo Sci Soc. *Res:* Geomorphology of coasts, rivers, and weathering. *Mailing Add:* 2885 16th St Boulder CO 80304

BRADLEY, WILLIAM ROBINSON, b Minneapolis, Minn, Jan 31, 08; m 31; c 2. INDUSTRIAL HYGIENE, TOXICOLOGY. *Educ:* Cornell Col, BS, 31; Univ Iowa, MS, 40; Am Bd Indust Hyg, dipl. *Prof Exp:* Indust hygienist, Dept Health, Detroit, 37-43 & Fidelity & Casualty Co, NY, 43-44; asst dir environ health, Am Cyanamid Co, NY, 44-60; CONSULT, W R BRADLEY & ASSOC, 60- *Concurrent Pos:* Chmn, Air Pollution Control Comn, NJ, 55-64, expert witness; trustee, Am Indust Hyg Found. *Honors & Awards:* Bordon Award, Am Indust Hyg Asn, 81. *Mem:* AAAS; Am Acad Indust Hyg; Air Pollution Control Asn; hon & emer mem Am Indust Hyg Asn; Am Chem Soc. *Res:* Prevention of occupational diseases; industrial toxicology; radioactivity; air and stream pollution; workmen's compensation; industrial noise and safety. *Mailing Add:* 208 Beth Dr Hendersonville NC 28739

BRADLEY, WILLIAM W(HITNEY), b Brooklyn, Iowa, Nov 9, 22; m 49; c 5. METALLURGY. *Educ:* Univ Chicago, BS, 44; Univ Colo, BS, 48, MS, 49; Mass Inst Technol, ScD, 64. *Prof Exp:* Asst res engr, Univ Colo, 48-49; mem tech staff, Bell Tel Labs, 49-61; Ford fel, Mass Inst Technol, 61-64; assoc prof civil eng, Clarkson Col Technol, 64-86; RETIRED. *Mem:* Am Soc Eng Educ; Electrochem Soc; Sigma Xi. *Res:* Environmental deterioration of materials; corrosion of metals; materials science. *Mailing Add:* 115 Market St Potsdam NY 13676

BRADLOW, HERBERT LEON, b Philadelphia, Pa, Mar 21, 24; m 47; c 3. ORGANIC CHEMISTRY, BIOLOGICAL CHEMISTRY. *Educ:* Univ Pa, BS, 45; Univ Kans, MS, 47, PhD(chem), 49. *Prof Exp:* Fel, Univ Calif, 49-50; assoc, Sloan Ctr, 63-77; prof biochem, Albert Einstein Col Med, 71-77; PROF, ROCKEFELLER UNIV HOSP, 78- & MED CTR, 63-; PROF BIOCHEM, ALBERT EINSTEIN COL MED, 71- *Concurrent Pos:* Assoc prof biochem, Albert Einstein Col Med, 68-71; adj assoc prof, Rockefeller Univ, 71-74, adj prof, 74- *Mem:* Am Chem Soc; Am Soc Biol Chemists; Endocrine Soc; Brit Soc Endocrinol; Royal Soc Chem. *Res:* Fluoraromatic compounds; cholesterol metabolism; steroid hormone metabolism; endocrinology; cancer research. *Mailing Add:* 86-25 Palo Alto St Holliswood NY 11423

BRADNER, HUGH, b Tonopah, Nev, Nov 5, 15; m 44; c 1. EXPERIMENTAL PHYSICS, GEOPHYSICS. *Educ:* Miami Univ, AB, 37; Calif Inst Technol, PhD(physics), 41. *Hon Degrees:* DSc, Miami Univ, 61. *Prof Exp:* Mem res dept, Champion Paper & Fiber Co, 36-37; asst, Calif Inst Technol, 39-41; mem res staff design & test magnetic mines, US Naval Ord Lab, Wash, 41-43; mem res staff, Manhattan Dist, Los Alamos, NMex, 43-46 & Radiation lab, Univ Calif, Berkeley, 46-61; prof, 64-80, EMER PROF ENG PHYSICS & GEOPHYS, UNIV CALIF, SAN DIEGO, 80-, RES PHYSICIST, INST GEOPHYS & PLANETARY PHYSICS, 61- *Mem:* Fel Am Phys Soc; Seismol Soc Am; Am Geophys Union. *Mailing Add:* 0225 Inst Geophys & Planetary Physics Univ Calif San Diego La Jolla CA 92093

BRADNER, MEAD, b Tonopah, Nev, Sept 23, 14; m 43; c 6. ENGINEERING, ENGINEERING MECHANICS. *Educ:* Mass Inst Technol, SB, 38. *Prof Exp:* Res physicist, Champion Paper & Fibre Co, Ohio, 35-36; student, Foxboro Co, 37 & R R Donnelley & Sons, Ill, 38-40; contract employee, Naval Ord Lab, 40-43; consult engr, Westinghouse Elec Co, Pa, 43-45; res engr, Foxboro Corp, 45-50, res dir, 50-62, int eng coordr, 62-68, tech dir res, develop & eng, 66-71, mkt planner, 71-73, tech consult, 73-77, tech dir ideas, 77-84; CONSULT, 84- *Mem:* AAAS; fel Am Soc Mech Engrs; fel Instrument Soc Am; Am Nat Standards Inst. *Res:* Industrial instruments; underwater ordnance. *Mailing Add:* 30 Water St Foxboro MA 02035

BRADNER, WILLIAM TURNBULL, b NJ, Aug 16, 24; m 51; c 3. BACTERIOLOGY, ONCOLOGY. *Educ:* Lehigh Univ, BA, 48, MS, 49, PhD(bact), 52. *Prof Exp:* Assoc bact, Brown Univ, 53-55; asst mem exp chemother, Sloan Kettering Inst, 55-58; sr res scientist cancer chemother screening, Bristol Labs, Bristol-Myers, 58-65, asst dir pharmacol res, 65-73, asst dir microbiol res, 73-77, dir antitumor biol dept, 77-85, dir admin preclin anticancer res, 85-86, PRES RES ADV, BRISTOL-MYERS, 86- *Concurrent Pos:* Assoc, Med Col, Cornell Univ, 56-58, asst prof bact, 58; consult, cancer res. *Mem:* AAAS; Am Soc Microbiol; Am Asn Cancer Res; NY Acad Sci; Am Soc Clin Oncol. *Res:* Amoebicidal agents; bacterial associates of Endamoeba histolytica; bacteriology of radiation infection; in vitro screening of antitumor agents, host defense against cancer; experimental cancer chemotherapy toxicity of anticancer drugs; clinical development of cancer drugs. *Mailing Add:* 4903 Briarwood Circle Manlius NY 13104

BRADSHAW, AUBREY SWIFT, b West Sunbury, Pa, July 12, 10; m 34; c 1. FRESH WATER ECOLOGY. *Educ:* Univ Ky, AB, 34, MA, 44. *Prof Exp:* Asst zool, Univ Ky, 34-35; instr biol, Transylvania Col, 35-43, asst prof biol & geog, 44-51, prof biol, 51-53; from assoc prof to prof, Ohio Wesleyan Univ, 53-75, emer prof, 75; res prof ecol, Univ Tenn, Knoxville, 78-83; RETIRED. *Concurrent Pos:* Asst, Cornell Univ, 49-51; consult, Advan Fossil Energy Prog, Environ Sci Div, Oak Ridge Nat Lab, 76-85. *Mem:* Am Soc Syst Zool; Am Micros Soc; Am Soc Limnol & Oceanog; Int Asn Theoret & Appl Limnol; Int Asn Great Lakes Res; Sigma Xi. *Res:* Life histories, ecology and industrial bioassay of Cladocera. *Mailing Add:* 109 E Price Rd Oak Ridge TN 37830-5138

BRADSHAW, GORDON VAN RENSSELAER, b Kansas City, Mo, Aug 8, 31; m 53; c 2. MAMMALOGY. *Educ:* Cent Mo State Col, BS, 53; Kans State Univ, MS, 56; Univ Ariz, PhD(zool), 61. *Prof Exp:* Asst prof biol, The Citadel, 60-61; prof, 61-80, chairperson dept, 76-80, EMER PROF BIOL, PHOENIX COL, 80- *Mem:* AAAS; Am Soc Mammal; Sigma Xi. *Res:* Natural history of southwestern bats. *Mailing Add:* HC 63 Box 5655 Mayer AZ 86333

BRADSHAW, JERALD SHERWIN, b Cedar City, Utah, Nov 28, 32; m 54; c 2. ORGANIC CHEMISTRY. *Educ:* Univ Utah, BS, 55; Univ Calif, Los Angeles, PhD(chem), 63. *Prof Exp:* NFS fel, Calif Inst Tech, 62-63; chemist, Chevron Res Co, 63-66; assoc prof, 66-73, asst chmn, 80-86, PROF CHEM, BRIGHAM YOUNG UNIV, 73- *Concurrent Pos:* Nat Acad Sci exchange prof, Univ Ljubljana, Yugoslavia, 72-73; chem res, Univ Sheffield, Eng, 78; Sigma Xi ann lectr, Brigham Young Univ, 88- *Honors & Awards:* Maeser Res Award, Brigham Young Univ, 77. *Mem:* Am Chem Soc; Int Soc Heterocyclic Chem. *Res:* Synthesis and complexation properties of macrocyclic compounds; synthesis of polysiloxane stationaty phases for capillary gas and supercritical fluid chromatography. *Mailing Add:* 1616 Oaklane Provo UT 84604

BRADSHAW, JOHN ALDEN, b Ann Arbor, Mich, Dec 9, 19. ELECTROPHYSICS. *Educ:* Harvard Univ, BA, 41, MS, 48, PhD(electrophysics), 54. *Prof Exp:* Res assoc, res lab, Gen Elec Co, 54-62; assoc prof elec eng, Rensselaer Polytech Inst, 62-78; CONSULT, 78- *Mem:* AAAS; Am Asn Physics Teachers; sr mem Inst Elec & Electronics Engrs; Am Solar Energy Soc. *Res:* Electron beam dynamics; wave propagation in overmoded guides and plasma; microwave scattering from waveguide obstacles. *Mailing Add:* 223 Green St Schenectady NY 12305

BRADSHAW, JOHN STRATLII, b High Wycombe, Eng, Oct 15, 27; nat US; m 48; c 4. BIOLOGICAL OCEANOGRAPHY. *Educ:* San Diego State Col, BS, 50; Univ Calif, MS & PhD(oceanog), 58. *Prof Exp:* Jr res biologist, Scripps Inst Oceanog, Univ Calif, San Diego, 58-67; ASSOC PROF BIOL, UNIV SAN DIEGO, 67- *Mem:* Am Soc Limnol & Oceanog; Ecol Soc Am; Sigma Xi. *Res:* Ecology of recent foraminifera and coastal lagoons; physiological ecology. *Mailing Add:* 308 Camino Calafia San Marcos CA 92069

BRADSHAW, LAWRENCE JACK, b Palo Alto, Calif, Dec 3, 24; m 49; c 4. MEDICAL MICROBIOLOGY. *Educ:* Stanford Univ, BA, 50, PhD(med microbiol), 56. *Prof Exp:* Instr microbiol, San Bernardino Valley Col, 54-63, assoc prof, 63-65; assoc prof, 65-68, PROF MICROBIOL, CALIF STATE UNIV, FULLERTON, 68- *Mem:* AAAS; NY Acad Sci; Sigma Xi. *Res:* Immunology; immunochemistry. *Mailing Add:* Dept Biol Calif State Univ Fullerton CA 92631

BRADSHAW, MARTIN DANIEL, b Pittsburg, Kans, June 24, 36; m 83; c 3. ELECTRICAL ENGINEERING. *Educ:* Univ Wichita, BS, 58, MS, 61; Carnegie Inst Technol, PhD(elec eng), 64. *Prof Exp:* Jr engr, Boeing Airplane Co, Kans, 58-59, res engr, 61; instr elec eng, Univ Wichita, 59-61; asst, Carnegie Inst Technol, 61-63; from asst prof to assoc prof elec eng, 63-76, asst dean, Col Eng, 74-76, PROF ELEC & COMPUT ENG, UNIV NMEX, 76- *Concurrent Pos:* Engr, State Elec Comn, Victoria, Australia, 78. *Honors & Awards:* Western Elec Fund Award, Gulf-Southwest Sect, Am Soc Eng Educ, 69; George Westinghouse Award, Am Soc Eng Educ, 73. *Mem:* Sr mem Inst Elec & Electronics Engrs. *Res:* Electromagnetic fields; antennas; electric power; large scale systems; effective teaching techniques; large scale systems with particular emphasis on electromagnetic fields and electric power systems; application of diakoptics techniques to solution of electromagnetics problems. *Mailing Add:* Dept Elec & Comput Eng PO Box 786 Placitas NM 87043

BRADSHAW, RALPH ALDEN, b Boston, Mass, Feb 14, 41; m 61; c 2. BIOCHEMISTRY. *Educ:* Colby Col, BA, 62; Duke Univ, PhD(biochem), 66. *Prof Exp:* Res assoc biochem, Ind Univ, 66-67; sr fel, dept biochem, Univ Wash, Seattle, 67-68, actg res asst prof, 68-69; from asst prof to prof, dept biol chem, Wash Univ, St Louis, 69-82; PROF & CHMN, DEPT BIOL CHEM, UNIV CALIF, IRVINE, 82- *Concurrent Pos:* USPHS fel chem, Ind Univ, 66-67; USPHS fel, Univ Wash, 67-68; USPHS career res develop award, 70-74; Physiol Chem Study Sect, NIH, 75-79, chmn, 78-79, Biomed Sci Study Sect, 80-85; sci adv bd, Hereditary Dis Found, 83-87; Coun Thrombosis, Am Heart Asn, 84-86; external adv bd, Cleveland Clin Prog Hypertension, 85-88; Nat Comt Biochem, Nat Acad Sci-Nat Res Coun, 87-; vis scientist, Univ Zurich, 69; vis lectr, Duke Univ Marine Sta, Beaufort, NC, 72, Univ Sydney, New S Wales, Australia, 77, Univ Ky, 87 & Univ Camerino, Italy, 87; Josiah Macy fac scholar, Howard Florey Inst, Univ Melbourne, 77-78; consult, Durrum Chem Co, Dionex Corp, Sunnyvale, Calif, 70-78, Cortex Pharmaceut, Irvine, Calif, 87-; sci adv, Int Genetic Eng, Santa Monica, Calif, 82-89; sci adv bd, Nucleic Acid Res Inst-ICN, Costa Mesa, Calif, 86-87, bd dirs, 86-91. *Honors & Awards:* Gold Medal, Ital Nat Res Coun, 87. *Mem:* fel AAAS; Am Chem Soc; NY Acad Sci; Am Soc Biol Chem; Am Soc Neurochem; Sigma Xi; Endocrine Soc; Am Soc Cell Biol; Soc Neurosci; Int Soc Neurochem; Protein Soc (pres, 86-87). *Res:* Structure-function relationships in proteins and enzymes; chemistry of growth factors and related hormones. *Mailing Add:* Dept Biochem Col Med Univ Calif Irvine CA 92717

BRADSHAW, WILLARD HENRY, b Orem, Utah, Feb 11, 25. MICROBIOLOGY. *Educ:* Brigham Young Univ, BS, 52, MS, 53; Univ Calif, Berkeley, PhD(comp biochem), 57. *Prof Exp:* Asst biochem, Univ Calif, 56-57; res assoc, Univ Ill, 57-59; res asst, Brigham Young Univ, 59-60; sr biochemist, Wallace Labs, 60-61; asst prof, 61-65, assoc prof bact, 65-80, ASSOC PROF MICROBIOL, BRIGHAM YOUNG UNIV, 80- *Concurrent Pos:* Hon lectr, Stanford Univ, 67-68. *Mem:* Am Soc Microbiol; Am Chem Soc; Brit Soc Gen Microbiol. *Res:* Microbial metabolism and physiology; microbial genetics; enzyme chemistry. *Mailing Add:* Dept Microbiol Brigham Young Univ Provo UT 84602

BRADSHAW, WILLIAM EMMONS, b Orange, NJ, May 16, 42; m 71; c 1. POPULATION BIOLOGY, EVOLUTION. *Educ:* Princeton Univ, AB, 64; Univ Mich, MS, 65, PhD(zool), 69. *Prof Exp:* NIH res fel, Harvard Univ, 69-71; from asst prof to assoc prof, 71-84, res assoc, Tall Timbers Res Sta, 77-78, PROF BIOL, UNIV ORE, 84- *Concurrent Pos:* Guggenheim fel, 86; Fulbright Grantee, 86; acad vis, Imp Col, Silwood Park, 86. *Mem:* Am Soc Naturalists; Ecol Soc Am; Entom Soc Am; Am Mosquito Control Asn; Soc Study Evolution. *Res:* Population, behavioral and community ecology; photoperiodism and seasonal development; biogeography insect diapause; mosquito ecology and physiology; life history evolution. *Mailing Add:* Dept Biol Univ Ore Eugene OR 97403

BRADSHAW, WILLIAM NEWMAN, b Louisville, Ky, Nov 2, 28; m 56; c 2. ENVIRONMENTAL ENGINEERING, SCIENCE ADMINISTRATION. *Educ:* Austin Col, BA, 51; Univ Tex, MA, 56, PhD, 62. *Prof Exp:* Asst prof biol, McMurry Col, 57-61; lectr zool, Univ Tex, 61; from asst prof to prof, WVa Univ, 62-76, dir environ biol prog, 67-71; environ scientist & proj mgr, Stearns-Roger Eng Corp, 76-84, prin environ scientist, Stearns Catalytic Corp, 84-86, prin environ scientist, Stearns-Roger Div, United Engrs & Constructors Inc, 86-89; ENVIRON PROJ SUPVR, WESTERN OPERS, UNITED ENGRS & CONSTRUCTORS INC, 89- *Concurrent Pos:* Vis assoc prof, M D Anderson Hosp & Tumor Inst, Univ Tex, 70-71; NIH spec res fel, 70-71; dir, Acad Assocs, Inc, 73-76. *Mem:* Nat Asn Environ Professionals; Am Soc Mammal; Sigma Xi; Ecol Soc Am. *Res:* Population biology and ecology; cytogenetics, speciation and behavior; terrestial, wildlife ecology, impact analysis and pollution abatement. *Mailing Add:* Western Opers United Engrs & Constructors Inc PO Box 5888 Denver CO 80217-5888

BRADSHAW, WILLIAM S, b Salt Lake City, Utah, Oct 29, 37; m 61; c 5. BIOCHEMISTRY, DEVELOPMENTAL BIOLOGY. *Educ:* Harvard Univ, AB, 63; Univ Ill, PhD(biochem), 68. *Prof Exp:* Res fel biol div, US AEC, Oak Ridge Nat Lab, 68-70; assoc prof, 70-85, PROF ZOOL, BRIGHAM YOUNG UNIV, 85- *Mem:* Soc Develop Biol. *Res:* Biochemistry of development; biochemical effects of environmental teratogens; synthesis of pancreatic enzymes during development; early immediate genes in RSV infected cells. *Mailing Add:* Dept Zool Brigham Young Univ Provo UT 84602

BRADSHER, CHARLES KILGO, b Petersburg, Va, July 13, 12; m 38; c 3. ORGANIC CHEMISTRY. *Educ:* Duke Univ, AB, 33; Harvard Univ, AM, 35, PhD(org chem), 37. *Prof Exp:* Rohm & Haas fel, Univ Ill, 37-38, du Pont fel, 38; from instr to prof, 39-65, chmn dept, 65-70, James B Duke prof chem, 65-79, JAMES B DUKE EMER PROF CHEM, DUKE UNIV, 79- *Concurrent Pos:* Nat Res fel, 41-42; Fulbright lectr, State Univ Leiden, 51-52; NSF sr fel, Swiss Fed Inst Technol, 59-60. *Honors & Awards:* Herty Medal Award, Am Chem Soc, (Ga Sect), 87. *Mem:* Am Chem Soc; Chem Soc London. *Res:* Aromatic cyclodehydration; quinolizinium derivatives; cationic polar cycloaddition. *Mailing Add:* Dept Chem Duke Univ Durham NC 27706

BRADSTREET, RAYMOND BRADFORD, b Salem, Mass, Aug 21, 01; m 30. ANALYTICAL CHEMISTRY, ORGANIC CHEMISTRY. *Educ:* Pratt Inst, dipl, 29; Polytech Inst Brooklyn, BS, 32, MS, 42. *Prof Exp:* Anal chemist, A C Lawrence Leather Co, 24-28; anal res chemist, Gen Labs, US Rubber Co, 29-36; anal res chemist, Standard Oil Develop Co, 36-42; group head anal res, 46; chief chemist, Creole Petrol Corp, Venezuela, 46-49; pres, Bradstreet Labs, Inc, 49-56; res dir, Carbon Solvents Lab, 56-59; supvr chem dept, US Testing Co, 59-61, supvr petrol & leather div, 61-68; CONSULT, 68- *Mem:* Fel AAAS; emer mem Am Chem Soc; fel emer Am Inst Chemists; NY Acad Sci; Royal Soc Chem. *Res:* Kjedahl method for organic nitrogen; organic analysis methods; standard solutions. *Mailing Add:* PO Box 1 Cranford NJ 07016

BRADT, HALE VAN DORN, b Colfax, Wash, Dec 7, 30; m 58; c 2. X-RAY ASTRONOMY. *Educ:* Princeton Univ, AB, 52; Mass Inst Technol, PhD(physics), 61. *Prof Exp:* From instr to assoc prof, 61-72, PROF PHYSICS, MASS INST TECHNOL, 72- *Concurrent Pos:* Assoc ed, Astrophys J Lett, 73-76. *Mem:* Am Phys Soc; Am Astron Soc. *Res:* X-ray astronomy. *Mailing Add:* Mass Inst Technol Rm 37-587 Cambridge MA 02139

BRADT, PATRICIA THORNTON, b Philadelphia, Pa, Oct 4, 30; m 52; c 3. AQUATIC ECOLOGY. *Educ:* Cornell Univ, BA, 52; Lehigh Univ, MS, 70, PhD(biol), 74. *Prof Exp:* Technician, Dept Gynec & Obstet, Med Sch, Yale Univ, 53-55; teaching asst bot, biol & ecol, Lehigh Univ, 73-74, asst prof, 74-77, adj asst prof, 77-81; prin investr, Liv Lakes, Inc, 87-90; res scientist, 84-88, res assoc, 90-91, PRIN RES SCIENTIST, DEPT NAT RESOURCES, CORNELL UNIV, 88-; PRIN INVESTR, PA POWER & LIGHT CO, 81- *Concurrent Pos:* Lectr intro biol & ecol, Lafayette Col, 75-79; prin investr, Off Water Res & Technol, US Dept Interior, 76-77; comt mem, Air & Water Tech Adv Comt, Pa Dept Environ Resources, 79-; consult, Environ Resources Mgt, Inc, 80-82; adj fac assoc, Ctr Social Res, Lehigh Univ, 80-, adj assoc prof, 81-84; res scientist, AAAS, 85. *Mem:* AAAS; Sigma Xi; Am Soc Limnol & Oceanog; Asn Women Sci; NAm Benthological Soc; Am Inst Biol Sci. *Res:* Study of freshwater ecosystems; benthic macroinvertebrate fauna; water chemistry and relation to all components of ecosystem; impact of pesticides on freshwater ecosystems; effect of acid precipitation on freshwater organisms and water chemistry; biological responses to nutrient loadings. *Mailing Add:* ESC Chandler 17 Lehigh Univ Bethlehem PA 18015

BRADT, RICHARD CARL, b St Louis, Mo, Nov 17, 38; m 60; c 2. PHYSICAL CERAMICS, MATERIALS SCIENCE. *Educ:* Mass Inst Technol, BSc, 60; Rensselaer Polytech Inst, MS, 65, PhD(mat eng), 67. *Prof Exp:* Res metallurgist, Fansteel Metall Corp, 60-62, ceramist, V-R/Wesson Div, 62-63; res fel mat eng, Rensselaer Polytech Inst, 63-67; from asst prof to prof ceramic sci, Pa State Univ, 67-78, head, Mat Sci & Eng, 78-83; head, Dept Mat Sci & Eng, Univ Wash, 83-86, Kyocera prof ceramics, 86-89; DEAN, MACKAY SCH MINES, UNIV NEV, 89- *Honors & Awards:* R M Fulrath Award, 80; T J Planje Award, 88. *Mem:* Fel Am Ceramic Soc; Am Soc Metals; Am Inst Mining, Metall & Petrol Engrs; Brit Ceramic Soc; Sigma Xi; Am Asn Univ Professors; Am Soc Eng Educ; Am Soc Testing & Mat; Mat Res Soc; Mineral Soc Am; Nat Inst Ceramic Engrs; Int Acad Ceramics. *Res:* Diffusion, sintering, electrical and mechanical properties and the transmission electron microscopy of ceramics, composite materials and cutting tool materials. *Mailing Add:* Mackay Sch Mines Univ Nev Reno NV 89557-0047

BRADWAY, KEITH E, b Indianapolis, Ind, Dec 31, 26; m 50; c 4. PULP & PAPER TECHNOLOGY, ORGANIC CHEMISTRY. *Educ:* Purdue Univ, BSChE, 48; Inst Paper Chem, MS, 50, PhD(org chem), 53. *Prof Exp:* Res chemist, Camp Mfg Co, 53-54, asst tech dir, 54-56, supt process eng, Union Bag Camp Paper Corp, 56-60, tech control, 60-62, res scientist, 62-77, SR SCIENTIST, UNION CAMP CORP, NJ, 77- *Honors & Awards:* Hugh D Camp Award, 68. *Mem:* Am Tech Asn Pulp & Paper Indust. *Res:* Technology of manufacture of white papers, unbleached paper and paperboard, with emphasis on printability and opacity of white papers and converting characteristics of unbleached products. *Mailing Add:* 30 Pin Oak Dr Lawrenceville NJ 08648

BRADWHAW, SAMUEL LOCKWOOD, JR, b Columbus, Ohio, Jan 30, 37; m 61; c 2. PSYCHIATRY. *Educ:* Univ Southern Calif, BM, 58, MD, 65; Univ Ill, MM, 54. *Prof Exp:* Intern med, Univ Calif, Los Angeles, 65-66; resident psychiat, Menninger Sch Psychiat, 69-72; staff psychiatrist, 72-75, asst chief, 75-77, CHIEF PSYCHIAT, VET ADMIN HOSP, TOPEKA, KANS, 77- *Concurrent Pos:* Dir interdisciplinary students, Menninger Sch Psychiat, 72-74, dir spec student prog, dir med student prog & fac mem, 73- *Mem:* Am Psychiat Asn; AAAS. *Res:* Effects of the use of names on attitudes of hospital employees; the effect of a hospital milieu. *Mailing Add:* 2200 Gage Blvd Topeka KS 66622

BRADY, ALLAN JORDAN, b Fairview, Utah, May 23, 27; m 52; c 2. BIOPHYSICS. *Educ:* Univ Utah, BA, 51, MS, 52; Univ Wash, PhD(biophys), 56. *Prof Exp:* Asst physiol & biophys, Sch Med, Univ Wash, 55-56, Am Heart Asn fel, 56-57; Am Heart Asn fel, Physiol Lab, Univ Cambridge, 57-58; advan res fel, Los Angeles County Heart Asn Cardiovasc Res Lab, Med Ctr, 58-60, estab investr, 60-65, assoc prof in residence, Sch Med, 62-66, PROF PHYSIOL, SCH MED, UNIV CALIF, LOS ANGELES, 66-, CAREER DEVELOP AWARDS, LOS ANGELES COUNTY HEART ASN CARDIOVASC RES LAB, MED CTR, 65- *Concurrent Pos:* Mem basic sci exec comt, Am Heart Asn, 62- *Mem:* Biophys Soc; Am Physiol Soc. *Res:* Electrical and ionic basis of single fiber activity; excitation-contraction coupling; muscle mechanics. *Mailing Add:* Dept Physiol Univ Calif Los Angeles Med Ctr Bri A 3-381 Los Angeles CA 90024

BRADY, ALLEN H, b Sewal, Iowa, Mar 17, 34; m 59; c 3. MATHEMATICS, COMPUTER SCIENCES. *Educ:* Univ Colo, BS, 56; Univ Wyo, MS, 59; Ore State Univ, PhD(math), 65. *Prof Exp:* Jr engr, Mat Res Lab, Collins Radio Co, Calif, 55; physicist, Cent Radio Propagation Lab, Nat Bur Standards, Colo, 56-64; asst prof comput sci, Univ Notre Dame, 65-66; dir, Comput Data Processing Ctr, Ball State Univ, 66-68; assoc res prof comput sci, Desert Res Inst, 68-69, asst dir acad uses, Comput Ctr, Univ Syst, 69-79, PROF MATH, UNIV NEV, RENO, 79- *Mem:* Am Math Soc; Math Asn Am; Asn Comput Mach; Inst Elec & Electronics Engrs; Sigma Xi. *Res:* Artificial intelligence; turing computability; computer organization. *Mailing Add:* Dept Comput Sci Univ Nev Reno NV 89557

BRADY, ALLEN ROY, b Houston, Tex, Feb 7, 33; m 58; c 4. SYSTEMATIC ZOOLOGY. *Educ:* Univ Houston, BS, 55, MS, 59; Harvard Univ, PhD(biol), 64. *Prof Exp:* Res fel arachnology, Mus Comp Zool, Harvard Univ, 63-64; Kettering intern biol, Hope Col, 64-65; asst prof, Albion Col, 65-66; from asst prof to assoc prof, 66-72, PROF ZOOL, HOPE COL, 72-, CHMN DEPT, 80- *Concurrent Pos:* Vis prof zool, Univ Fla, Gainesville, 72-73. *Mem:* Am Arachnol Soc; Am Soc Zoologists; Soc Syst Zool. *Res:* Systematics and zoogeography of spiders. *Mailing Add:* Dept Biol Hope Col Holland MI 49423

BRADY, BARRY HUGH GARNET, b Warwick, Australia, Oct 9, 42; m 71; c 1. MINING ENGINEERING. *Educ:* Univ Queensland, BSc, 64, MSc, 67; Univ London, MSc, 75, PhD(eng), 79. *Prof Exp:* Res scientist, Res Dept, CSR Corp, 64-65; tutor thermodynamics, Univ Queensland, 65-67; res engr, Mt Isa Mines Ltd, 67-71, sr engr, 71-75; lectr eng & rock mech, Univ London, 76-79; assoc prof eng & rock mech, 80-83, ADJ PROF, UNIV MINN, 83-; MGR, DEPT APPL MECH, DOWELL-SCHLUMBERGER, 91- *Concurrent Pos:* Prin consult, Itasca Consult Group, Inc, 81-, 87-91. *Mem:* Soc Mining Engrs; Inst Soc Rock Mech; Inst Mining & Metall; Brit Geotech Soc. *Res:* Development and application of computational schemes for design of surface and understand excavations in rock; investigation of the strength and deformation properties of geologic media. *Mailing Add:* Dowell-Schlumberger PO Box 2710 Tulsa OK 74101

BRADY, BRIAN T, b Cleveland, Ohio, Sept 7, 38; m 65; c 4. MATERIALS SCIENCE, APPLIED MATHEMATICS. *Educ:* Univ Dayton, BSc, 61; Mass Inst Technol, MSc, 64; Colo Sch Mines, PhD(math, metall, mining), 69. *Prof Exp:* Geophysicist, Cities Serv Oil Co, Okla, 64-66; PHYSICIST, US BUR MINES, 67- *Mem:* Am Geophys Union; Am Inst Mining, Metall & Petrol Engrs. *Res:* Analysis of brittle rock failure and applications of results to earthquake seismology and mine failure prediction. *Mailing Add:* US Bur Mines Fed Ctr Denver CO 80225

BRADY, DENNIS PATRICK, b Mitchell, SDak, Feb 18, 49; m 72; c 2. STATISTICAL METHODS TO SOLVE BUSINESS PROBLEMS. *Educ:* St John's Univ, Minn, BA, 71; SDak Sch Mines & Technol, MS, 73; Mont State Univ, PhD(statist), 76. *Prof Exp:* Sr statistician, 76-80, prof scientist, 80-91, DEVELOP SCIENTIST, UNION CARBIDE CORP, 91- *Concurrent Pos:* Adj prof, Col Grad Studies, WVa Univ, 77-80. *Mem:* Am Soc Qual Control; Am Statist Asn. *Res:* Application of statistical methods to solve problems in industrial research, marketing, sales, strategic planning, production, and litigation. *Mailing Add:* Union Carbide Corp PO Box 8361 South Charleston WV 25303

BRADY, DONALD R, b Alamosa, Colo, June 28, 39. CONTINUING EDUCATION. *Educ:* Cornell Univ, PhD(biochem), 70. *Prof Exp:* Dept Chem, Memphis State Univ, 73-80, Ctr Nuclear Studies, 80-85; DEAN ADMIS & SPEC PROGS, AM TECH INST, 85- *Concurrent Pos:* Res grant, Memphis State Univ, NIH, 80-84. *Mem:* Am Soc Biochemists; NY Acad Sci; AAAS. *Mailing Add:* Am Tech Inst PO Box 8 Memphis TN 38014

BRADY, DONNIE GAYLE, b Campbell, Mo, Sept 9, 40; m 62; c 1. ORGANIC CHEMISTRY, POLYMER CHEMISTRY. *Educ:* Southern Ill Univ, BS, 62; Purdue Univ, PhD(org chem), 66. *Prof Exp:* Res chemist, Phillips Petrol Co, 66-71, sr res chemist, 71-77, resin develop engr, Phillips Chem Co, 77-78, sect supvr, Plastics Compounding, 78-80, mgr polymer applications, 80-83, mgr Thermoplastic Composites, 83-86, MGR POLYMERS & MAT DIV, PHILLIPS PETROLEUM CO, 86- *Mem:* Am Chem Soc; Soc Plastics Engrs; Soc Mfg Eng. *Res:* Organo sulfur chemistry; organo halogen chemistry; organo metallic chemistry; flame retardant additives for polymers; engineering thermoplastics; polyolefins; advanced composites. *Mailing Add:* Phillips Petrol Co 258 RF PRC Bartlesville OK 74004

BRADY, DOUGLAS MACPHERSON, b Mt Kisco, NY, Aug 27, 43; m 67; c 2. DIGITAL TRANSMISSION, SIGNAL PROCESSING. *Educ:* Princeton Univ, BSE, 65, MSE, 65. *Prof Exp:* Mem tech staff, Bell Labs, 68-74, supvr, 74-80; CHIEF ENGR, SPECIAL SERV PROD DIV, GEN ELEC LENKURT INC, 80-; VPRES RES & DEVELOP, BASE TWO SYSTS. *Mem:* sr mem Inst Elec & Electronics Engrs. *Res:* Transmission systems. *Mailing Add:* Base Two Systs 5353 Manhattan Circle No 201 Boulder CO 80303

BRADY, EUGENE F, b Philadelphia, Pa, Sept 19, 27; m 54; c 7. MECHANICAL ENGINEERING, MATHEMATICS. *Educ:* Drexel Inst, BSME, 55; Univ Pittsburgh, MS, 58, PhD(mech eng), 63. *Prof Exp:* Mech engr, Bettis Plant, Westinghouse Elec Corp, 55-61; mech engr missile & space div, Gen Elec Co, 61-62; proj leader rolling bearing res, SKF Industs, Inc, Pa, 62-64; assoc prof eng, PMC Cols, 64-67; group mgr mech technol, Univac Div, Sperry Rand Corp, 69-77; SR ENG SPECIALIST, INGALLS SHIPBUILDING CORP, 77- *Mem:* Am Soc Mech Engrs. *Res:* Theory of plasticity; physics of chemical deposition; thermal pollution of estuaries and streams. *Mailing Add:* PO Box 506 Moorestown NJ 08057

BRADY, FRANK OWEN, b Chicago, Ill, Feb 8, 44; m 66; c 3. BIOCHEMISTRY, BIOINORGANIC CHEMISTRY. *Educ:* Carnegie-Mellon Univ, BS, 65; Duke Univ, PhD(biochem), 70. *Prof Exp:* Fel, Inst Cancer Res, Columbia Univ, 69-73; from asst prof to assoc prof, 73-82, PROF BIOCHEM, SCH MED, UNIV SDAK, 82- *Concurrent Pos:* NIH fel, Nat Cancer Inst, 70-72 & trainee, 72-73; NIH career develop award, Nat Inst

Environ Health Sci, 76-81; Vis scientist, Med Res Coun, Toxicol Unit, Carshalton, UK, 79-80; Kelloggs Found Nat Leadership fel, 88-91. *Mem:* Am Soc Biochem & Molecular Biol; NY Acad Sci; Soc Exp Biol & Med; Am Heart Asn. *Res:* Metalloenzymes and metalloproteins; toxic metal metabolism; zinc and copper homeostasis; hormonal control of metal metabolism. *Mailing Add:* Dept Biochem & Molecular Biol Sch Med Univ SDak Vermillion SD 57069

BRADY, FRANKLIN PAUL, b Winnipeg, Man, Aug 14, 31; m 64. NUCLEAR PHYSICS. *Educ:* Univ Man, BS, 53; Princeton Univ, AM, 54, PhD(physics), 60. *Prof Exp:* Res assoc nuclear physics, Princeton Univ, 60-61; physicist, Schlumberger Ltd, 61-62; from asst prof to assoc prof, 62-74, PROF PHYSICS, UNIV CALIF, DAVIS, 74-, ASSOC DIR, CROCKER NUCLEAR LAB, 75- *Concurrent Pos:* Ford Found vis prof prin investr, NSF grant, 72-, Nat Cancer Inst grant, 75-79, Dept Energy grant, 81-; sr Fulbright, Karlsruhe, Fed Repub Ger,82-83. *Mem:* Fel Am Phys Soc. *Res:* Geophysics; experimental nuclear physics; nuclear reactions; cyclotron design. *Mailing Add:* Dept Physics Univ of Calif Davis CA 95616

BRADY, GEORGE W, b Quebec, Que, Jan 22, 21; m 49; c 2. PHYSICAL CHEMISTRY. *Educ:* Laval Univ, BSc, 42; McGill Univ, PhD, 49. *Prof Exp:* Asst, Univ Chicago, 49-51; res fel, Harvard Univ, 51-52; mem tech staff, Bell Labs, 52-78; mem staff, Div Labs-Res, NY State Dept Health, 78-90; RETIRED. *Concurrent Pos:* Adj prof, State Univ NY Albany, 76-; adj prof physics, Rensselaer Polytech Inst. *Mem:* Fel Am Phys Soc; fel NY Acad Sci; Biophys Soc; Am Crystallog Asn. *Res:* Chemical kinetics; surface chemistry; structure of liquids; x-ray diffraction; critical state phenomena; structure of membranes and DNA. *Mailing Add:* Div Labs-Res New Scotland Ave Albany NY 12201

BRADY, JAMES EDWARD, b New York, NY, Jan 26, 38; m 60; c 1. INORGANIC CHEMISTRY. *Educ:* Hofstra Col, BA, 59; Pa State Univ, PhD(inorg chem), 64. *Prof Exp:* Staff scientist chem, Res Ctr, Aerospace Group, Gen Precision, Inc, 64-65; instr, 65-70, ASSOC PROF CHEM, ST JOHN'S UNIV, NY, 70- *Mem:* Am Chem Soc; NY Acad Sci. *Res:* Reaction mechanisms of coordination compounds by means of high pressure techniques; synthesis of inorganic luminescent materials at high temperatures. *Mailing Add:* Dept Chem St John's Univ Grand Cent & Utopia Pkwy Jamaica NY 11439

BRADY, JAMES JOSEPH, b Oregon City, Ore, Nov 24, 04; m 32; c 2. SURFACE PHYSICS. *Educ:* Reed Col, AB, 27; Ind Univ, AM, 28; Univ Calif, PhD(physics), 31. *Prof Exp:* Asst physics, Ind Univ, 27-28; fel, St Louis Univ, 31-32, from instr to asst prof, 32-37; from asst prof to prof, 37-73, actg chmn dept, 65-66 & 69-71, EMER PROF PHYSICS, ORE STATE UNIV, 73- *Concurrent Pos:* Assoc group leader, Radiation Lab, Mass Inst Technol, 42-46; tech consult, Radiation Lab, Univ Calif, 48-; consult physicist, Boeing Airplane Co, 46-54. *Mem:* Fel Am Phys Soc; AAAS; Am Asn Physics Teachers. *Res:* Photoelectric effect of thin films; cyclotron and nuclear physics; microwave optics. *Mailing Add:* 2015 SW Whiteside Corvallis OR 97330

BRADY, JOHN F, b Taunton, Mass, July 12, 28; m 52; c 4. MECHANICAL ENGINEERING. *Educ:* Mass Inst Technol, BS, 48, MS, 50. *Prof Exp:* Asst thermodynamics, Mass Inst Technol, 48-50; mech engr, US Naval Torpedo Sta, 50-51; proj officer, Wright Air Develop Ctr, 51-52; br head proj mgt, Underwater Ord Sta, US Naval Underwater Systs Ctr, 52-56, asst div syst eng, 56-60, div head propulsion, 60-61, head, Dept Appl Sci, 61-64, head, Dept Res, 64-68, assoc dir,, Underwater Weapons Sta, 68-72, assoc dir Newport lab, 70-71, assoc dir weapons res, 71-72, head, Weapons Dept, 72-76; CONSULT TECH STAFF, SUBMARINE SIGNAL DIV, RAYTHEON CORP, 76- *Honors & Awards:* Excellence Sci & Technol Award, Naval Underwater Ord Sta, 63. *Mem:* Am Inst Aeronaut & Astronaut; Am Soc Mech Engrs. *Res:* Underwater propulsion; partial admission turbines; underwater vehicle systems. *Mailing Add:* 206 Immokolee Dr Portsmouth RI 02871

BRADY, JOHN PAUL, b Boston, Mass, June 23, 28; m 63; c 4. PSYCHIATRY. *Educ:* Boston Univ, AB, 51; MD, 55. *Hon Degrees:* MA, Univ Penn, 70. *Prof Exp:* Resident psychiat, Inst Living, Hartford, Conn, 56-59; res psychiatrist, Sch Med, Ind Univ, 59-63; from assoc prof to prof, 63-74, chmn dept, 74-82, KENNETH E APPEL PROF PSYCHIAT, SCH MED, UNIV PA, 74- *Concurrent Pos:* Co-founder & assoc ed, Behav Ther, 69-; consult panel anal of impact of basic res on ment health, Comt Brain Sci, Nat Res Coun, 72-74. *Honors & Awards:* Strecker Award, 72. *Mem:* Asn Advan Behav Ther (pres, 70-71); Soc Biol Psychiat (pres, 79-80); Psychiat Res Soc; Pavlovian Soc NAm. *Res:* Applications of principles of learning to disorders of behavior; biobehavioral approaches to medical disorders. *Mailing Add:* Dept Psychiat Hosp Univ Pa Philadelphia PA 19104

BRADY, JOSEPH VINCENT, b New York, NY, Mar 28, 22. BEHAVIORAL BIOLOGY. *Educ:* Fordham Univ, BS, 43; Univ Chicago, PhD, 51. *Prof Exp:* Chief clin psychologist, Neuropsychiat Ctr, Europ Command, Ger, 46-48; student officer psychol, Univ Chicago, 48-51; chief dept exp psychol, Walter Reed Army Inst Res, 51-64, dep dir, Div Neuropsychiat, 64-70; prof behav biol & dir div, 67-83, PROF NEUROSCI, SCH MED, JOHNS HOPKINS UNIV, 83- *Concurrent Pos:* Prof, Univ Md, 55-69. *Mem:* Fel AAAS; fel Am Col Neuropsychopharmacology. *Res:* Experimental analysis of behavior; behavioral physiology; behavioral pharmacology; comparative psychosomatics. *Mailing Add:* Dept Psych Johns Hopkins Univ Sch Med 720 Rutland Ave Baltimore MD 21205

BRADY, KEITH BRYAN CRAIG, b Dulphin Co, Pa, Apr 23, 54. ACID MINE DRAINAGE, HYDROGEOLOGY. *Educ:* Alaska Methodist Univ, BA, 76; Univ Maine, MS, 82. *Prof Exp:* HYDROGEOLOGIST, DEPT ENVIRON RESOURCES, BUR MINING & RECLAMATION, 80- *Mem:* Geol Soc Am; Asn Groundwater Scientists & Engrs; AAAS; Am Quaternary Asn; Am Sci Affil. *Res:* Prediction of surface coal mine drainage quality; means of preventing acid mine drainage occurrence by considering variables imposed by geologic, geochemical, hydrologic and man induced controls. *Mailing Add:* DER Bureau Mining & Reclamation Box 2357 Harrisburg PA 17105

BRADY, LAWRENCE LEE, b Topeka, Kans, Nov 6, 36; m 60; c 2. GEOLOGY, ECONOMIC GEOLOGY. *Educ:* Kans State Univ, BS, 58; Univ Kans, MS, 67, PhD(geol), 71. *Prof Exp:* Eng geologist, US Army Corps Engr, Kansas City Dist, 58-63; asst prof geol, Okla State Univ, 71; chief Mineral Resources Sect, 75-83, RES ASSOC GEOL, KANS GEOL SURV, 71-, CHIEF GEOL INVESTS, 87- *Mem:* Asn Eng Geologists; Am Asn Petrol Geologists; Soc Econ Paleontologists & Mineralogists. *Res:* Coal geology and resources; quality and distribution of Kansas mineral resources; engineering geology related to mineral development; clastic sedimentology. *Mailing Add:* 1930 Constant Ave Campus West Univ Kans Lawrence KS 66047

BRADY, LEONARD EVERETT, b Brooklyn, NY, Feb 24, 28; m 51; c 1. ORGANIC CHEMISTRY. *Educ:* Wagner Col, BS, 51; NC State Col, MS, 54; Mich State Univ, PhD(org chem), 58. *Prof Exp:* Asst, Mich State Univ, 53-57; res chemist, Abbott Labs, 57-61; res assoc chem, Univ Ill, 61-63, instr, 63-64; assoc prof chem, Ill State Univ, 64-66; assoc prof, 66-70, chmn dept, 75-81, PROF CHEM, UNIV TOLEDO, 71- *Concurrent Pos:* Vis scholar, Cambridge Univ, 73-74; vis prof, Mich State Univ, 85. *Mem:* Am Chem Soc; Am Soc Pharmacog. *Res:* Mechanisms; structure proof; stereochemistry; natural products. *Mailing Add:* Dept Chem Univ of Toledo Toledo OH 43606

BRADY, LUTHER WELDON, JR, b NC, 1925. MEDICINE, RADIOLOGY. *Educ:* George Wash, Univ AB, 46, MD, 48. *Hon Degrees:* DFA, Colgate Univ, 88; DSc, Lehigh Univ, 90. *Prof Exp:* Asst prof med & radiol, 59-66, prof radiol, 66-70, PROF RADIATION ONCOL & NUCLEAR MED & CHMN DEPT, HAHNEMANN UNIV HOSP, 70-, DIR RADIATION ONCOL, 59-, AM CANCER SOC CLIN ONCOL PROF, 75- *Concurrent Pos:* Consult, Vet Admin Hosp, 60-, Crozer-Chester Med Ctr, Pa Hosp, Philadelphia, Lankenau Hosp, Philadelphia, W Jersey Hosp, Camden, NJ, Gaarden State Hosp, Cherry Hill, NJ & St Luke's Hosp, Bethlehem, Pa; asst prof, Harvard Med Sch, 62-63. *Honors & Awards:* Gold Medal, Am Radium Soc, 81; Gold Medal, Am Col Radiol, 83; Maurice Lenz Mem Lectr, Columbia Presby Med Ctr, 83; Medal, Nat Med Soc Ecuador, 80; Gilbert H Fletcher Gold Medal, 84; Swedish Acad Med Medal, 86; Gold Medal, Am Soc Therapeut Radiol & Oncol, 87; John McAfee lect, State Univ NY, Syracuse, 84; Mark Blum lect, Monmouth Med Ctr, 77; Elsa Pardee lect, Midland Med Ctr, 79. *Mem:* Fel Am Col Radiol; Radiol Soc NAm; Am Radium Soc (past pres); Am Soc Therapeut Radiol (past pres); Am Soc Clin Oncol; Am Asn Cancer Res; hon fel, Med Radiol Soc Italy; hon fel, Royal Col Radiologists; hon fel, Ger Roentgenol Soc. *Res:* Effects of radiation on pulmonary function on bone marrow using ferrochromokinetics in patients with established diagnosis of malignancy and on kidney function, monoclonal antibodies in cancer treatment. *Mailing Add:* Radiation Oncol & Nuclear Med Hahnemann Univ Broad & Vine Sts Philadelphia PA 19102

BRADY, LYNN R, b Shelton, Nebr, Nov 15, 33; m 57. PHARMACOGNOSY. *Educ:* Univ Nebr, BS, 55, MS, 57; Univ Wash, PhD(pharmacog), 59. *Prof Exp:* From asst prof to assoc prof, 59-66, chmn, dept pharmaceut sci, 72-80, PROF PHARMACOG, UNIV WASH, 66- *Mem:* Am Pharmaceut Asn; Am Soc Pharmacog; Am Asn Cols Pharm; fel Acad Pharmaceut Sci. *Res:* Constituents of higher fungi; poisonous plants; chemotaxonomy. *Mailing Add:* Col Pharm SC-68 Univ Wash Seattle WA 98195

BRADY, NYLE C, b Manassa, Colo, Oct 25, 20; m 36; c 4. AGRONOMY, SOIL & SOIL SCIENCE. *Educ:* Brigham Young Univ, BS, 41; NC State Col, PhD(soil sci), 47. *Hon Degrees:* PhD, Brigham Young Univ, 79 & Ohio State Univ, 91. *Prof Exp:* Jr agronomist, NC State Col, 42-44, res instr, 46, asst prof agron, 47; from asst prof to prof soil sci, NY State Col Agr, Cornell Univ, 48-73, head dept agron, 55-63, dir agr exp sta, 65-73, assoc dean, 70-73; dir gen, Int Rice Res Inst, Philippines, 73-81, sen asst, Admin Sci & Tech, 81-89; SR CONSULT, UN DEVELOP PROG, 89- *Concurrent Pos:* Dir sci & educ, USDA, 63-65; chmn agr bd, Nat Res Coun, 67-70; ed, Proc Soil Sci Soc Am; ed, Agron J, 65-90. *Honors & Awards:* Int Agron Award, Am Soc Agron, 87; Int Soil Sci Award, Soil Sci Am, 90. *Mem:* AAAS; Soil Sci Soc Am; Am Soc Agron; Soil Conserv Soc Am; Am Inst Biol Sci. *Res:* Physiology of the peanut plant; fundamental effects of lime on soil; influence of fertilizer on the yield of corn, rice and coffee; influence of soil temperature on nutrient uptake. *Mailing Add:* UN Develop Prog 1889 F St Washington DC 20006

BRADY, RICHARD CHARLES, cellular micro tubal associated protein, regulation of calcium function, for more information see previous edition

BRADY, ROBERT FREDERICK, JR, b Washington, DC, July 20, 42; m 65; c 3. POLYMER CHEMISTRY, ORGANIC CHEMISTRY. *Educ:* Univ Va, BSChem, 64, PhD(org chem), 67. *Prof Exp:* Chemist, Nat Bur Stand, 67-72 & US Customs Serv, 72-75; chief, Paints Br, Fed Supply Serv, 75-82; HEAD, COATINGS SECT, NAVAL RES LAB, 82- *Concurrent Pos:* NASA fel, 64-67; contrib ed, J Protective Coatings & Linings, 86-90 & J Coatings Technol, 85-; vis scientist, Mat Res Lab, Melbourne, Australia, 90-91. *Mem:* Sigma Xi; Am Chem Soc; Fed Socs Coatings Technol. *Res:* Synthetic, structural, and physical chemistry of polymers and natural products; trace organic analysis by instrumental methods; coatings technology, testing and specifications. *Mailing Add:* 706 Hope Lane Gaithersburg MD 20878-1883

BRADY, ROBERT JAMES, b Detroit, Mich, Apr 13, 27; m 55; c 4. MICROBIOLOGY. *Educ:* Univ Detroit, BS, 51, MS, 54; Univ Md, PhD(microbiol), 57. *Prof Exp:* Asst biol, Univ Detroit, 53-55; asst microbiol, Univ Md, 54-57; from asst prof to assoc prof, 57-69, chmn dept, 73-78, PROF MICROBIOL, MIAMI UNIV, 69- *Mem:* Am Soc Microbiol; Brit Soc Gen Microbiol. *Res:* Microbial physiology and genetics. *Mailing Add:* Dept Microbiol Miami Univ Oxford OH 45056

BRADY, ROSCOE OWEN, b Philadelphia, Pa, Oct 11, 23; m ; c 2. BIOCHEMISTRY, GENETIC DISORDERS. *Educ:* Harvard Univ, MD, 47. *Prof Exp:* Intern, Univ Hosp, Univ Pa, 47-48; officer-in-chg, Chem Labs, Naval Med Ctr, 52-54; chief lipid chem sect, 54-70, asst chief lab neurochem, 70-72, CHIEF DEVELOP & METAB NEUROL BR, NAT INST NEUROL

& COMMUN DIS & STROKE, 72- *Concurrent Pos:* Nat Res Coun fel med sci, Univ Hosp, Univ Pa, 48-50, fel, 50-51, USPHS fel, 51-52, fel, Endocrine Sect, Med Clin, 52-54; prof lectr, Sch Med, George Washington Univ, 63-73; adj prof biochem, Georgetown Univ, 67- *Honors & Awards:* Gairdner Int Award, 73; Cotsias Award, Am Acad Neurology, 80; Passano Found Award, 82; Lasker Found Award, 82; SACMS Award, Child Neurology Soc, 90; Kovalenko Medal, Nat Acad Sci, 91. *Mem:* Inst Med-Nat Acad Sci; Am Acad Neurol; Am Soc Biol Chem & Molecular Biol; Am Soc Human Genetics. *Res:* Complex lipid metabolism; genetic diseases; viral carcinogenesis; neurochemistry; biosynthesis of fatty acids and complex lipids of the nervous system. *Mailing Add:* Bldg 10 Rm 3D04 NIH Bethesda MD 20892

BRADY, RUTH MARY, b Bridgeport, Conn, Feb 18, 24. PHYSICAL CHEMISTRY. *Educ:* Albertus Magnus Col, AB, 44; Fordham Univ, MS, 55, PhD(phys chem), 65. *Prof Exp:* Asst prof chem, Col St Mary, Ohio, 63-65, Albertus Magnus Col, 65-68; sr scientist, York Res Corp, Conn, 68-69; prof chem, Alcorn State Univ, 69-84, Chairperson Chem & Physics, 73-84. *Mem:* Am Chem Soc; AAAS; Sigma Xi. *Res:* Phase relations in isomeric systems; kinetics and mechanism of reactions of thiolacetic acid; effects of metal ions on growth of seedlings. *Mailing Add:* Alcorn State Univ Box 327 Lorman MS 39096

BRADY, SCOTT T, b San Antonio, Tex, Nov 11, 50. CELLULAR NEUROBIOLOGY, CELL MOTILITY. *Educ:* Mass Inst Technol, ScB (biol) & ScB (physics), 73, Univ Southern Calif, PhD(cellular & molecular biol), 78. *Prof Exp:* Sr res assoc cell biol, Case Western Reserve Univ, 80-85; ASSOC PROF CELL BIOL & ANAT, UNIV TEX SOUTHWESTERN MED CTR, DALLAS, 85- *Honors & Awards:* Jordi Folch-Pi Mem Award, Am Soc Neurochem, 88. *Mem:* Soc Neurosci; Am Soc Cell Biol; Am Soc Neurochem; AAAS; Am Phys Soc; Int Soc Neurochem. *Res:* Indentification and characterization of the molecular mechanisms of axonal transport, resulting in discovery of kinesin-based form of intracellular transport; role of neuronal cytoskeleton in growth and regeneration. *Mailing Add:* Dept Cell Biol & Neurosci Univ Tex Southwestern Med Ctr 5323 Harry Hines Blvd Dallas TX 75235

BRADY, STEPHEN FRANCIS, b New York, NY, Oct 17, 41; m 66; c 1. MEDICINAL CHEMISTRY. *Educ:* Columbia Univ, BA, 63; Stanford Univ, PhD(chem), 67. *Prof Exp:* Res assoc, Dept Chem, Northwestern Univ, 67-70; sr res chemist, 70-79, res fel, 79-87, SR RES FEL, MERCK SHARP & DOHME RES LABS DIV, MERCK & CO, INC, 87- *Mem:* Am Chem Soc; Sigma Xi. *Res:* Synthesis of peptides; development of new methodology and protecting groups for peptide synthesis; chemistry and biology of hormonal peptides and proteins; synthesis and incorporation of novel amino acids in peptides; studies of enzyme action and inhibition. *Mailing Add:* 8803 Crefeld St Philadelphia PA 19118

BRADY, STEPHEN W, b Indianapolis, Ind, Jan 12, 41; m 65. MATHEMATICS. *Educ:* Ind Univ, BA, 63, MA, 65, PhD(math), 68. *Prof Exp:* ASSOC PROF MATH, WICHITA STATE UNIV, 67- *Concurrent Pos:* Consult, Math & Biol Corp, 67-68. *Mem:* Soc Indust & Appl Math; Am Math Soc; Math Asn Am; NY Acad Sci. *Res:* Functional and numerical analysis; differential equations; mathematics applied to physiology. *Mailing Add:* 9823 W Second St Wichita KS 67212

BRADY, THOMAS E, b Elizabeth, NJ, Apr 2, 41; m 70; c 3. SYNTHETIC ORGANIC CHEMISTRY, AGRICULTURAL CHEMISTRY. *Educ:* St Vincent Col, BS, 63; Fordham Univ, PhD(org chem), 68. *Prof Exp:* Res chemist, Org Chem Div, 67-74, sr res chemist, Chem Res Div, 74-83, SR RES CHEMIST, AGR RES DIV, AM CYANAMID CO, NJ, 83- *Mem:* Am Chem Soc; NY Acad Sci. *Res:* Synthesis of dyes and detergent additives; flame retardants; antioxidants, light stabilizers, photo initiators, herbicides, PGRs. *Mailing Add:* Pres Midwest Plastic Res Assoc Inc 333 14th St Toledo OH 43624

BRADY, ULLMAN EUGENE, JR, insect toxicology, physiology; deceased, see previous edition for last biography

BRADY, WILLIAM GORDON, b Zanesville, Ohio, May 31, 23; m 48; c 2. APPLIED MATHEMATICS. *Educ:* Univ Cincinnati, Aero Eng, 50; Brown Univ, MS, 53. *Prof Exp:* Asst res engr appl mech, Cornell Aeronaut Lab, Inc, NY, 52-55, assoc res engr, 55-58, res engr, 58-60, prin engr, 60-70; from asst prof to prof math, Erie Community Col, 70-88; RETIRED. *Honors & Awards:* Sigma Xi. *Res:* Aeroelasticity; low speed aerodynamics. *Mailing Add:* 44 Pryor Ave Tonawanda NY 14150

BRADY, WILLIAM THOMAS, b Ventura, Calif, Sept 25, 33; m 54; c 2. PHYSICAL ORGANIC CHEMISTRY. *Educ:* NTex State Univ, BA, 55, MS, 56; Univ Tex, PhD(chem), 60. *Prof Exp:* Res chemist, Tex Eastman Co, Eastman Kodak Co, 56-57; res scientist, Clayton Found Biochem Inst, 57-60; res chemist, Tex Eastman Co, Eastman Kodak Co, 60-62; from asst prof to assoc prof, 62-68, PROF CHEM, NTEX STATE UNIV, 68- *Mem:* Am Chem Soc. *Res:* Ketene chemistry. *Mailing Add:* Dept Chem NTex State Univ Denton TX 76201

BRADY, WRAY GRAYSON, b Benton Harbor, Mich, July 20, 18; m 43; c 2. MATHEMATICS. *Educ:* Washington & Jefferson Col, BS, 40, MA, 42; Univ Pittsburgh, PhD(math), 53. *Prof Exp:* Instr math, Washington & Jefferson Col, 40-42, prof math, 51-66; asst, Stanford Univ, 46-47; instr, Univ Wyo, 47-50; Bernard prof math & chmn dept, Univ Bridgeport, 66-69; dean, Grad Sch, 69-72, prof, 72-87, EMER PROF MATH, SLIPPERY ROCK STATE COL, 87- *Concurrent Pos:* Consult, Atomic Energy Bd, Westinghouse Elec Bettis Plant. *Mem:* AAAS; Am Math Soc; Math Asn Am; Fibonacci Asn. *Res:* Infinite series; number theory; mathematical statistics. *Mailing Add:* Dept Math Univ Pa 1290 Bestwick Rd Mercer PA 16137

BRADY, YOLANDA J, b Greenwood, Miss, Aug 7, 56. FISH DISEASE DIAGNOSIS, ELECTRON MICROSCOPY. *Educ:* Univ Miss, BS, 78; Univ Southern Miss, MS, 82; Auburn Univ, PhD(fish health), 85. *Prof Exp:* Res assoc diag, 84-85, ASST PROF FISH HEALTH, AUBURN UNIV, 86- *Mem:* Am Fisheries Soc; World Aquacult Soc. *Res:* Bacterial and viral disease mechanisms in warm-water fish, crustaceans, and mollusks. *Mailing Add:* Dept Fisheries & Allied Aquacult Int Ctr Aquacult Auburn Univ Auburn AL 36849-5419

BRAEKEVELT, CHARLIE ROGER, b Sarnia, Ont, Aug 27, 42; m 69; c 2. ELECTRON MICROSCOPY, MORPHOLOGY. *Educ:* Univ Western Ont, BA, 64, MSc, 66, PhD(anat), 69. *Prof Exp:* Demonstr zool, Univ Western Ont, 64-66, demonstr anat, 66-69; from lectr to assoc prof, 60-87, PROF ANAT, UNIV MAN, 87- *Concurrent Pos:* Ont Grad fel, 65-68; Nellie L Farthing mem fel, 68-69; Med Res Coun Can fel, 70-71; vis prof ophthal, Cath Univ Louvain, 76; vis res fel, Univ Western Australia, 84-85; vis scientist, Med Res Coun Can, 84-85; WHO fel, 84-85. *Mem:* Can Soc Zoologists; Am Soc Zoologists; Can Asn Anatomists; Am Asn Anatomists; Asn Res Vision & Ophthal; Int Soc Eye Res. *Res:* Fine structure of normal and abnormal eye; effects of toxic substances on the vertebrate retina; cell renewal within the vertebrate retina. *Mailing Add:* Dept Anat Univ Man 730 William Ave Winnipeg MB R3E 0W3 Can

BRAEMER, ALLEN C, b Woodhaven, NY, May 8, 30; m 81; c 3. LICENSING, BUSINESS DEVELOPMENT. *Educ:* State Univ NY Vet Col, Cornell Univ, DVM, 55. *Prof Exp:* Vet, Pine Tree Vet Hosp, Augusta, Maine, 55-56, Crawford Animal Hosp, Garden City, NY, 56-58 & Mindell Animal Hosp, Albany, 58-61; assoc res biologist, Sterling-Winthrop Res Inst, Rensselaer, NY, 61-62, group leader, 62-64, res biologist, 64-66, res vet & group leader, Exp Farm, 66-67; dir animal clin sect, Norwich Pharmacal Co, 67-69, chief vet bact sect, 69-73; asst dir, Inst Agrisci, 76-85, head dept vet microbiol, 77-80, dir vet biol res, 80-85, HEAD DEPT ANIMAL SCI, SYNTEX RES, 73-, DIR TECHNOL ACQUISITION, 85- *Mem:* AAAS; Am Vet Med Asn; Am Indust Vet Asn. *Res:* Chemotherapy and immunology of microbial diseases; veterinary vaccines; licensing of technology; business development. *Mailing Add:* 1649 Kamsack Dr Sunnyvale CA 94087

BRAENDLE, DONALD HAROLD, b Flushing, NY, Nov 8, 27; m 57. MICROBIOLOGY. *Educ:* Rice Inst, BA, 52; Rutgers Univ, PhD(microbiol), 57. *Prof Exp:* Waksman-Merck fel, Rutgers Univ, 57-58; SR SCIENTIST, ABBOTT LABS, 58- *Mem:* AAAS; Am Soc Microbiol; Brit Soc Gen Microbiol; Sigma Xi. *Res:* Microbial genetics, especially that of filamentous microorganisms, their physiology, and their production of antibiotics and other microbial products. *Mailing Add:* 239 E Washington Lake Bluff IL 60044

BRAGDON, ROBERT WRIGHT, b Marblehead, Mass, Jan 19, 22; m 43; c 3. CHEMISTRY. *Educ:* Bowdoin Col, BS, 43. *Prof Exp:* Anal chemist, Manhattan Proj, Mallinckrodt Chem Works, 43-46; res chemist, United Shoe Mach Corp, 46-47; sr res chemist, Metal Hydrides, Inc, 47-52, assoc dir res, 52-60, dir res, 60-65; sr scientist, Simplex Wire & Cable Co, 65-69; mgr new prod develop, Hampshire Chem, 70-79, dir res, 79-81, vpres new prod develop, 81-84, vpres res & develop, Org Chem Div, W R Grace & Co, 84-89; WRITER, 89- *Concurrent Pos:* Consult, Ventron Corp, 69-70. *Mem:* Am Chem Soc; Com Develop Asn. *Res:* Borohydrides; aluminohydrides; alkali and alkaline earth hydrides; sodium metal cable; chelates; amino acids; hydrogen cyanide chemistry; nitroparaffins; hydrogen cyanide based products. *Mailing Add:* Rte 2 Box 69B South Harpswell ME 04079-9605

BRAGG, DAVID GORDON, b Portland, Ore, May 1, 33; m 55; c 4. MEDICINE, RADIOLOGY. *Educ:* Stanford Univ, AB, 55; Univ Ore, MD, 59; Am Bd Radiol, dipl, 66. *Prof Exp:* Intern, Philadelphia Gen Hosp, 59-60; resident radiol, Col Physicians & Surgeons, Columbia Univ, 62-64, chief resident, 64-65, from instr to asst prof, Med Col, Cornell Univ, 65-70; PROF RADIOL & CHMN DEPT, COL MED, UNIV UTAH, 70- *Concurrent Pos:* Chmn dept diag radiol, Sloan-Kettering Cancer Ctr, New York, 67-70, consult, 70-; res grant, Microwave Methods Lung Water Measurement. *Mem:* AMA; Radiol Soc NAm; Am Roentgen Ray Soc; Soc Surg Oncol; Am Gastroenterol Asn. *Res:* Oncological and gastrointestinal radiology; diagnostic oncology, microwave radiology; quantitation of distribution of Lung Water by nuclear magnetic resonance. *Mailing Add:* Dept Radiol Univ Utah Med Ctr Salt Lake City UT 84132

BRAGG, DENVER DAYTON, b Duffy, WVa, Apr 13, 15; m 41; c 2. POULTRY SCIENCE. *Educ:* WVa Univ, BS, 40; Va Polytech Inst & State Univ, MS, 53. *Prof Exp:* Poultry serviceman, Ralston Purina Co, 40; asst rural rehab supvr, Farm & Home Admin, 40-42; county 4-H Club agent, Agr Exten Serv, WVa, 46, from asst county agent to assoc county agent, 46-49; assoc prof poultry sci & exten specialist, 49-73, EMER ASSOC PROF POULTRY SCI, VA POLYTECH INST & STATE UNIV, 73- *Mem:* Poultry Sci Asn. *Res:* Egg production and processing. *Mailing Add:* 400 Country Club Dr SW Blacksburg VA 24060

BRAGG, GORDON MCALPINE, b Flin Flon, Man, May 13, 39; m 63; c 3. FLUID MECHANICS. *Educ:* Univ Toronto, BASc, 62; Cambridge Univ, PhD(aeronaut eng), 65. *Prof Exp:* Asst prof, 65-70, ASSOC PROF MECH ENG, UNIV WATERLOO, 70- *Concurrent Pos:* Res asst, Univ Toronto, 61-62; mem cent comt, Can Cong Appl Mech. *Mem:* Am Soc Mech Engrs; Can Soc Mech Engrs. *Res:* Subsonic flow; turbulence; air and gas handling equipment; turbomachinery; air pollution control; wind power. *Mailing Add:* Dept Mech Eng Univ Waterloo Waterloo ON N2L 3G1 Can

BRAGG, JOHN KENDAL, b Washington, DC, Nov 12, 19; m 43; c 4. CHEMICAL PHYSICS. *Educ:* Harvard Univ, BS, 41, PhD(chem physics), 48. *Prof Exp:* Asst prof chem, Cornell Univ, 48-50; res assoc, Gen Elec Co, 50-53; pres, Opers Res, Inc, 54-56; proj analyst, Gen Elec Co, 57-63, mgr res consult serv, 64-66, mgr res & develop consult serv, 66-70; asst vpres technol, Singer Co, 70-72, dir res, 72-75, chief tech officer, 75-77; RETIRED. *Concurrent Pos:* Consult, Weapons Syst Eval Group, 53-54 & Opers Res Off, 54; trustee, Textile Res Inst, 72-78. *Mem:* AAAS; Opers Res Soc. *Res:* Applied physics. *Mailing Add:* 22 C Coachman Sq Clifton Park NY 12065

BRAGG, LESLIE B(ARTLETT), b Milford, Mass, Aug 24, 02; m 30; c 3. CHEMICAL ENGINEERING. *Educ:* Mass Inst Technol, BS, 25, MS, 29, ScD(chem eng), 33. *Prof Exp:* Asst chem eng, Mass Inst Technol, 25-27, fuel & gas, 27-28, res assoc, 28-32; lectr, Worcester Polytech Inst, 32-33; asst, Foster Wheeler Corp, 33-34, design engr, 34-37, supvr, Oil Lab & Develop Div, 37-41, supvr Stedman Packing Develop, 41-43, mgr, Stedman Dept, 43-47; head chem eng sect, Knolls Atomic Power Lab, Gen Elec Co, 48-51; pres & treas, Packed Column Corp, 52-70, chmn bd, 71-72. *Mem:* AAAS; Am Chem Soc; Am Inst Chem Engrs. *Res:* Elimination of carbon monoxide from engine exhaust gases; solvent refining; dewaxing and decolorization of lube oil development; development of Goodloe packing; Panapak packing improvements. *Mailing Add:* 260 Ashland Rd Summit NJ 07901

BRAGG, LINCOLN ELLSWORTH, b Buffalo, NY, Jan 25, 36; m 61. MATHEMATICAL PHYSICS. *Educ:* Carnegie Inst Technol, BS, 59, MS, 60, PhD(continuum mech), 64. *Prof Exp:* Asst, US Steel Res Lab, Pa, 60; asst, Mellon Inst, 62 & 64; res asst prof continuum mech, Inst Fluid Dynamics & Appl Math, Univ Md, College Park, 63-66; Dunham Jackson instr math, Univ Minn, Minneapolis, 66-67; asst prof, Univ Ky, 67-71 & Fla Inst Technol, 71-75; MATHEMATICIAN & COMPUT SPECIALIST, CENT NOMIS OFF, 76- *Mem:* Inst Elec & Electronics Engrs; Am Math Soc. *Res:* Electromagnetic behavior of deformable materials; theory of constitutive relations; concepts of electrodynamics and their historical development; gravitational theory; application of exterior calculus in engineering sciences; computer performance prediction and evaluation. *Mailing Add:* 704 Four Mile Rd Alexandria VA 22305

BRAGG, LOUIS HAIRSTON, b Chicora, Miss, Sept 12, 28; m 54; c 3. BOTANY & SEED STRUCTURE, ELECTRON MICROSCOPY. *Educ:* NTex State Col, BS, 53, MS, 57; Univ Tex, PhD(bot), 64. *Prof Exp:* Tutor biol, NTex State Col, 53-54; teacher, Pub Schs, Tex, 54-60; from instr to assoc prof, 60-84, PROF BIOL, UNIV TEX, ARLINGTON, 84- *Concurrent Pos:* Res scientist, Univ Tex, 64. *Mem:* Bot Soc Am; Sigma Xi; Electron Micros Soc Am. *Res:* Experimental plant ecology; seed structure; biochemical systematics; plant cytology; electron microscopy. *Mailing Add:* Dept Biol Univ Tex Arlington TX 76019

BRAGG, LOUIS RICHARD, b Weston, WVa, Aug 5, 31; m; c 2. MATHEMATICS. *Educ:* WVa Univ, AB, 52, MS, 53; Univ Wis, PhD(math), 55. *Prof Exp:* From instr to asst prof math, Duke Univ, 55-59; assoc prof, WVa Univ, 59-61; from asst prof to assoc prof, Case Inst Technol, 61-67; PROF MATH, OAKLAND UNIV, 67- *Concurrent Pos:* Chmn, Oakland Univ, 75-78. *Mem:* Am Math Soc; Math Asn Am; Soc Indust & Appl Math. *Res:* Partial differential equations, transmutations, and special functions; numerical integration. *Mailing Add:* Dept Math Scis Oakland Univ Rochester MI 48309

BRAGG, PHILIP DELL, b Gillingham, Eng, July 2, 32; m 58; c 3. BIOCHEMISTRY, ORGANIC CHEMISTRY. *Educ:* Bristol Univ, BSc, 54, PhD(org chem), 58. *Hon Degrees:* DSc, 86. *Prof Exp:* Fel org chem, Queen's Univ, Ont, 57-59; res assoc biochem, sch med, La State Univ, 59-61; res assoc, 61-64, from asst prof to prof, 64-74, HEAD BIOCHEM DEPT, UNIV BC, 86. *Concurrent Pos:* Med Res Coun Can scholar, Univ BC, 64-69. *Mem:* Royal Soc Chem; Am Soc Microbiol; Can Biochem Soc; Am Soc Biochem & Molecular Biol. *Res:* Biochemistry of microorganisms, especially carbohydrate metabolism; electron transport; oxidative phosphorylation; cellular membranes; control mechanisms; membrane transport. *Mailing Add:* Dept Biochem Univ BC Vancouver BC V6T 1W5 Can

BRAGG, ROBERT H(ENRY), b Jacksonville, Fla, Aug 11, 19; m 47; c 2. CARBON SCIENCE, MATERIALS CHARACTERIZATION. *Educ:* Ill Inst Technol, BS, 49, MS, 51, PhD(physics), 60. *Prof Exp:* Asst physicist, Res Lab, Portland Cement Asn, 51-54; assoc physicist, 54-56; assoc physicist, Ill Inst Technol Res Inst, 56-57, res physicist, 57-59, sr physicist, 59-61; res scientist, Palo Alto Res Lab, Lockheed Missiles & Space Co, 61-63; sr staff scientist, 63-69; chmn dept, 78-81, prof, 69-87, EMER PROF MAT SCI, UNIV CALIF, BERKELEY, 87- *Concurrent Pos:* Consult, Univ Ill & Atomic Energy Comn, Arg, 66, Jamaica Bauxite Inst, 77; prin investr, Mat & Molecular Div, Lawrence Berkeley Lab, 69-; mem, comt Sci & Technol Appl Mat Processing Space, Nat Acad Sci, 77-78, Mat Res Adv Comt, Div Mat Res, NSF, 82-88, Eval Panel, Res Assoc Prog, Nat Res Coun, 84-; vis prof, Howard Univ, 77, Musashi Inst Technol, Japan, 89; prog dir, Div Mat Sci, Dept Energy, 81-82; chair, Eval Panel, Ctr Mat Sci, Nat Bur Standards, 81-84; vis scientist, Centre Nat de Recherche Scientifique, Univ Bordeaux, France, 83. *Mem:* Am Ceramic Soc; Am Phys Soc; Am Crystallog Asn; Am Inst Mining & Metall Eng; Am Carbon Soc; Am Asn Univ Prof. *Res:* Carbon and composite materials; x-ray diffraction, small angle scattering; electron microscopy and diffraction; characterization of structure of materials; correlation of structure with electrical, magnetic, mechanical and thermophysical properties. *Mailing Add:* Dept Mat Sci & Mineral Eng Univ Calif Berkeley CA 94720

BRAGG, SUSAN LYNN, b Wheeling, WVa, Nov 16, 53; m 87. OPTICAL PHYSICS & CHEMISTRY, LASER SPECTROSCOPY. *Educ:* Wash Univ, Mo, BA, 75, MA, 77, PhD(earth & planetary sci), 81. *Prof Exp:* Sr scientist, McDonnell Douglas Res Labs, McDonnell Douglas Corp, 82-90; SR PROJ SCIENTIST, METAPHASE CORP, 90- *Concurrent Pos:* Vis fel, Somerville Col, Oxford Univ, UK, 83-84. *Mem:* Optical Soc Am. *Res:* Experimental high resolution spectroscopy of atoms and small molecules; investigation of atmospheric propagation in visible and near infrared spectral regions; development of laser spectroscopic methods. *Mailing Add:* Metaphase Corp 1266 Andes Blvd St Louis MO 63132

BRAGG, THOMAS BRAXTON, b San Francisco, Calif, June 4, 38; m 63; c 2. PLANT ECOLOGY, ECOLOGY. *Educ:* Calif State Polytech Univ, BS, 63; Kans State Univ, MS, 71, PhD(biol), 74. *Prof Exp:* Asst prof, 74-78, ASSOC PROF BIOL, UNIV NEBR, OMAHA, 78- *Mem:* Ecol Soc Am; Soc Range Mgt; Am Inst Biol Sci; AAAS. *Res:* Fire ecology of native prairie and forest ecosystems with emphasis on prairie vegetation; ecosystem management; grassland restoration. *Mailing Add:* Dept Biol Univ Nebr Omaha NE 68182-0040

BRAGIN, JOSEPH, b Brooklyn, NY, Jan 23, 39; m 68. PHYSICAL CHEMISTRY. *Educ:* Brooklyn Col, BS, 59; Univ Wis, PhD(chem), 67. *Prof Exp:* Fel chem, Univ SC, 67-70; asst prof, 70-74, assoc prof, 74-, PROF CHEM, CALIF STATE UNIV, LOS ANGELES, ASSOC DEAN, SCH LETT & SCI. *Res:* Vibrational spectroscopy as a probe of chemical and biological activity. *Mailing Add:* Sch Natural & Soc Sci King Hall D1052 Calif State Univ 5151 State Univ Dr Los Angeles CA 90032-8200

BRAGINSKI, ALEKSANDER IGNACE, b Warsaw, Poland, Apr 23, 29; US citizen; m; c 1. MAGNETISM, SOLID STATE ELECTRONICS. *Educ:* Wroclaw Tech Univ, MSc, 51; Polish Acad Sci, PhD, 60; Warsaw Tech Univ, DSc, 65. *Prof Exp:* Jr engr, Nat Inst Telecommun, Wroclaw, 51-52; sr engr, Radio Works, Warsaw, 52-53; sect mgr powder cores, 53-54, lab mgr ferrites, 54-55; dept mgr, HF Ceramics Factory, 55-56; tech dir, Polfer Magnetic Mat Factory, 56; head, Polfer Res Lab, 56-66; res engr, Lignes Telegraphiques & Telephoniques, France, 66-67; sr scientist, Westinghouse Elec Co, 67-69, fel scientist, 69-71, adv scientist, 71-72, mgr solid state magnetics, 72-75, mgr superconducting mat, 75-87, consult scientist, Res & Develop Ctr, 87-89; VIS SCIENTIST, RES CTR JUELICH (ISI-KFA), 89- *Concurrent Pos:* Consult, Inst Comput, Polish Acad Sci, 56-60; lectr, Warsaw Tech Univ, 58 & 64-66; from res asst to res leader, Lab Magnetism & Solid Physics, Nat Ctr Sci Res, 60-61. *Honors & Awards:* Heavy Mach Indust Award, Poland, 59; Polish Nat Prize, 64. *Mem:* Sr mem Inst Elec & Electronic Engrs; Am Phys Soc. *Res:* Microwave magnetics; induced anisotropy and aftereffects; ferrites; cylindrical domain memory technology; superconducting materials; Josephson junction technology, HTS; high-temperature superconductor electronics. *Mailing Add:* 1S1 Forschungszentrum Juelich Postfach 1913 Juelich D-5170 Germany

BRAGOLE, ROBERT ANTHONY, b Somerville, Mass, Oct 17, 36; m 62; c 5. ORGANIC CHEMISTRY. *Educ:* Boston Univ, AB, 58; Northeastern Univ, MS, 60; Yale Univ, PhD(org chem), 65. *Prof Exp:* Teaching fel, Boston Univ, 57-58; teaching fel & res asst, Northeastern Univ, 58-60; teaching fel, 60-62, res asst, Yale Univ, 62-64; res chemist, Carwin Res Labs, 65-66; sr res chemist, USM Chem Co, 66-69, mgr appl res, 70-73, lab mgr, Bostik Div, USM Corp, 73-78; tech dir, Bostik Div, Emhart Corp, 78-80, lab mgr, Bostik Div, 80-85; dir res & develop, 85-87, VPRES RES & DEVELOP, UPACO ADHESIVES INC, 87- *Concurrent Pos:* Instr, Southern Conn State Col, 65-66. *Mem:* Sr mem Am Chem Soc; Adhesives & Sealant Coun; Soc Mfg Engrs; Soc Plastics Engrs. *Res:* Adhesion theory and technology; photochemistry; polyolefins; polymer synthesis; liquid adhesives; hot melt adhesives; solvent and water-free adhesives; sheet form sound deadeners; coatings. *Mailing Add:* 5 Innis Dr Danvers MA 01923

BRAGONIER, JOHN ROBERT, b Cedar Falls, Iowa, July 4, 37; m 59; c 3. OBSTETRICS & GYNECOLOGY. *Educ:* Iowa State Univ, BS, 60; Univ Nebr, MS, 63, MD, 64, PhD(med sci, biochem), 67. *Prof Exp:* Intern, Univ Nebr Hosp, 64-65; resident obstet & gynec, 65; resident obstet & gynec & instr, Jefferson Med Col Hosp, 66-68; res assoc, Inst Child Health & Human Develop, NIH, 68-70; head physician, Dept Obstet & Gynec, Harbor Gen Med Ctr, Torrance, 70-85; asst prof obstet & gynec, 70-76, adj assoc prof, 76-82, ADJ PROF, UNIV CALIF, LOS ANGELES, 82-; CHMN, DEPT OBSTET & GYNEC, CIGNA HEALTH PLANS CALIF, LOS ANGELES, 85- *Concurrent Pos:* Adj prof, Univ Calif, Riverside, 82- *Mem:* Fel Am Col Obstetricians & Gynecologyists; Am Fertil Soc; Am Pub Health Asn; AAAS. *Res:* Medical biochemistry; biochemical teratology; social issues in obstetrics and gynecology (family planning, sex education, problem pregnancy); training of paramedical personnel; prevention of preterm birth; health care delivery systems. *Mailing Add:* 1640 Harper Ave Redondo Beach CA 90278

BRAGONIER, WENDELL HUGHELL, b Geneseo Twp, Iowa, Aug 5, 10; m 34; c 3. BOTANY. *Educ:* Iowa State Teachers Col, BA, 33; Iowa State Univ, MS, 41, PhD(plant path), 47. *Prof Exp:* High sch teacher, Iowa, 33-34, Tenn, 34-35 & Iowa, 35-39; instr bot, Iowa State Univ, 40-42, res assoc & asst to dir, Indust Sci Res Inst, 42-47, assoc prof bot, 47-49, prof & head dept bot & plant path, 50-63, assoc dir, Camp Dodge Br, 46-47; prof bot & plant path & dean grad sch, 63-75, asst dir, Exp Sta, 79-82, EMER PROF BOT & PLANT PATH & EMER DEAN GRAD SCH, COLO STATE UNIV, 75-; consult, Colo State Dept Agr, 79-82. *Concurrent Pos:* Chmn sci adv coun, Am Seed Res Found, 59-; consult, Colo State Dept Agr, 79-82. *Mem:* Fel AAAS; Bot Soc Am; Am Inst Biol Sci; Sigma Xi. *Res:* Plant pathology and morphology; teaching of botany; umbrella disease of Rhus glabra L caused by Botryosphaeria ribis G & D; ethanol production from grain and agricultural wastes. *Mailing Add:* 1601 Collindale Dr Ft Collins CO 80525

BRAHAM, HOWARD WALLACE, b Los Angeles, Calif, Dec 4, 42; c 4. VERTEBRATE ZOOLOGY, MARINE ECOLOGY. *Educ:* Calif State Univ, Los Angeles, BS, 66; Calif Polytech State Univ, San Luis Obispo, MS, 72; Ohio State Univ, Columbus, PhD(ecol), 75. *Prof Exp:* Cryptographer, US Dept Justice, Fed Bur Invest, Los Angeles, 63-64; police officer, West Covina, Calif, 64-68; teacher math & phys educ, El Rancho Sch Dist, Calif, 68-69; teaching & res assoc, Calif Tech State Univ, San Luis Obispo, 69-72; teaching & res assoc, Ohio State Univ, 72-75; supvr wildlife biol, 75-84, DIR NAT MARINE MAMMAL LAB, NAT MARINE FISHERIES SERV, NAT OCEANIC & ATMOSPHERIC ADMIN, 84- *Concurrent Pos:* Steering Comt, US-USSR Comn for Environ Protection Marine Mammals, 76-; manuscript ed, Fisheries Bulletin, 77-80; mem, Comt Sci Adv, Int Whaling Comn, 79- *Honors & Awards:* Admin Award, Nat Oceanic & Atomospheric Admin, 83. *Mem:* Soc Marine Mammal. *Res:* Population biology of marine mammals, especially endangered whales, specializing in the ecology of the Bowhead Whale, Gray Whale and Northern Sea Lion and arctic species. *Mailing Add:* Nat Marine Mammal Div Lab NOAA 7600 Sand Point Way NE Bldg 4 Seattle WA 98115

BRAHAM, ROSCOE RILEY, JR, b Yates City, Ill, Jan 3, 21; m 43; c 4. CLOUD PHYSICS, THUNDERSTORMS. *Educ:* Ohio Univ, BS, 42; Univ Chicago, SM, 48, PhD(meteorol), 51. *Prof Exp:* Sr analyst & officer in chg thunderstorm proj, US Weather Bur, 46-49; res asst meteorol, Univ Chicago,

49-50; res meteorologist cloud physics & weather modification, NMex Inst Mining & Technol, 50-51; sr meteorologist, Univ Chicago, 51-56, from assoc prof to prof meteorol, 56-91; SCHOLAR IN RESIDENCE, NC STATE UNIV, 91- *Concurrent Pos:* Assoc ed, J Am Meteorol Soc, 53-69; dir, Inst Atmospheric Physics & assoc prof, Univ Ariz, 54-56; mem bd trustees, Univ Corp Atmospheric Res, 65-67, 73-77 & 79-86; prin investr, Proj Metromex, NSF, 70-78; mem adv comt, Proj Stormfury, Nat Oceanic & Atmospheric Admin, Nat Acad Sci, 72-80; mem weather modification adv comt, US Dept Com, 77-78; consult, Dept Com & Dept Interior. *Honors & Awards:* Losey Award, Inst Aeronaut Sci, 50; Silver Medal, US Dept Com, 50; Rossby Gold Medal, Am Meteorol Soc, 81, C F Brooks Award, 87; Schaefer Award, Weather Modification Asn, 87. *Mem:* Fel AAAS; fel Am Meteorol Soc (pres, 88); Am Geophys Union (pres, Meteorol Sect, 76-77); Sigma Xi; fel Royal Meteorol Soc. *Res:* Cloud physics and weather modification; urban meteorology; thunderstorms and severe weather; understanding the physics of clouds and precipitation and determining the extent to which they can be, and are being, influenced by activities of man. *Mailing Add:* 57 Longcommon Rd Riverside IL 60546

BRAHANA, THOMAS ROY, b Champaign, Ill, June 26, 26; m 51; c 4. MATHEMATICS. *Educ:* Univ Ill, AB, 47; Univ Mich, MA, 50, PhD(math), 55. *Prof Exp:* Instr math, Dartmouth Col, 53-54; from asst prof to assoc prof, 54-68, PROF MATH, UNIV GA, 68- *Concurrent Pos:* Mem, Inst Advan Study, 57-58; Fulbright lectr, Univ Zagreb, Yugoslavia, 71-72. *Mem:* Am Math Soc. *Res:* Algebraic topology and geometry. *Mailing Add:* Dept Math Univ Ga Athens GA 30602

BRAHEN, LEONARD SAMUEL, b Philadelphia, Pa, Nov 28, 21; m 47; c 3. PSYCHIATRY. *Educ:* Temple Univ, BS, 49, MS, 51; Univ Md, PhD(pharmacol), 54; Univ Louisville, MD, 58. *Prof Exp:* Asst pharmacol, Temple Univ, 49-51, res asst, 51; instr pharmacol, Sch Med, Univ Louisville, 54-56; chief investr, Ky Heart Asn grant, 56-58; asst surgeon, USPHS, NY, 58-59; assoc dir clin res, Chas Pfizer & Co, 59-60, dir clin res, 60-62, med dir, Leeming Pacquin Div, 62-64; med dir, Endo Labs, Inc, 64-68, corp med dir, 68-71; DIR MED RES & EDUC, NASSAU COUNTY DEPT DRUG ABUSE & ADDICTION 71- *Concurrent Pos:* NIMH fel, 57; Collins Found scholar, 57; mem drug adv comt, State Sen Dunne; chmn drug abuse subcomt, Nassau County Med Soc, NY; assoc clin prof, New York Med Col, 59-62; clin assoc prof, State Univ NY Stony Brook, 71-; mem staff & consult, Nassau County Med Ctr, 71- *Mem:* AAAS; Am Fedn Clin Res; Am Geriat Soc; fel Am Col Clin Pharmacol & Chemother; Am Therapeut Soc. *Res:* Mechanism of action of drugs on autonomic nervous system; developed quantitative method for determination of irritation; studies in vasodilation; drug action and metabolism in humans. *Mailing Add:* 735 Central Ave Woodmere NY 11598

BRAHMA, CHANDRA SEKHAR, b Calcutta, India, Oct 5, 41; US citizen; m 72; c 2. GEOTECHNICAL ENGINEERING, STRUCTURAL ENGINEERING. *Educ:* Calcutta Univ, India, BE, 62; Mich State Univ, MSc, 65; Ohio State Univ, PhD(civil eng), 69. *Prof Exp:* Asst engr, Govt West Bengal, India, 62-63; proj engr, Frank H Lehr Assoc, East Orange, NJ, 69-70; sr engr, John G Reutter Assoc, Camden, NJ, 70-72; asst prof civil eng, Worcester Polytech Inst, Mass, 72-73; chief geotech engr, Daniel, Mann, Johnson & Mendenhall, Baltimore, Md, 73-79; sr engr, Sverdrup & Parcel, St Louis, Mo, 79-80; PROF CIVIL ENG, CALIF STATE UNIV, FRESNO, 80- *Concurrent Pos:* Res asst, Mich State Univ, 63-65; teaching & res assoc, Ohio State Univ, 65-69; assoc, SAFE Internation Inc, Philadelphia, Pa, 72-74; consult, Expert Resources, Inc, Peoria Heights, Ill, 81- & Twinning Labs, Inc, 82-; spec consult, Sverdrup & Parcel, St Louis, Mo, 80-; secy, Am Soc Civil Engrs, 81-83, vpres, 83-84, pres, 84-85, dir, 85-86; vis prof, eng dept, Univ Republica, Montevidio, Uruguay, 84; Fulbright Cert, Bd Foreign Scholars, 84; Fulbright grant, Coun Int Exchange Scholars, 84; consult, Marderosian & Swanson Law Firm, Fresno, Calif, 85-, & Hurlbutt, Clevenger, Long & Vortman, Visalia, Calif, 88-; Hugh B William fel, Asn Drilled Shaft Contractors, 86. *Mem:* Fel Am Soc Civil Engrs; Am Soc Testing & Mat; Sigma Xi; Int Soc Soil Mech & Found Eng; Int Soc Rock Mech; Am Soc Eng Educ; Nat Soc Prof Engrs. *Res:* Soft-ground and hard rock tunneling; deep excavations; soil-structure interaction; deep foundations; stochastic behavior of soils; physical and creep behavior of soils; specialized engineering constructions; landslides; computer-aided design in foundation engineering; underground engineering; forensic engineering; professional issues in engineering. *Mailing Add:* Dept Civil Eng Calif State Univ Fresno CA 93740-0094

BRAHMI, ZACHARIE, TUMOR IMMUNOLOGY, IMMUNO GENETICS. *Educ:* Ind Univ, PhD(immunol), 70. *Prof Exp:* PROF MED, DEPT IMMUNOL, SCH MED, IND UNIV, 78- *Mailing Add:* Dept Med Riley S-09 Ind Univ Sch Med 702 Barnhill Dr Indianapolis IN 46223

BRAID, THOMAS HAMILTON, b Heriot, Scotland, Dec 21, 25; m 51; c 2. EXPERIMENTAL PHYSICS, INSTRUMENTS. *Educ:* Univ Edinburgh, BSc, 47, PhD(physics), 51. *Prof Exp:* Nat Res Coun Can res fel physics, Atomic Energy Can, Ltd, 50-52; res asst, Princeton Univ, 52-55, instr, 55-56, res assoc, 56; PHYSICIST, ARGONNE NAT LAB, 56- *Concurrent Pos:* Vis physicist, Atomic Energy Res Establishment, Eng, 66-67; assoc ed, Appl Physics Lett, 74-79; ed, Rev Sci Instruments, 79- *Mem:* Fel Am Phys Soc. *Res:* Nuclear reactions and structure; radiation detection and instrumentation; reactor safety; scientific instruments. *Mailing Add:* 7619 S Sussex Creek Dr Apt 301 Darien IL 60559

BRAIDS, OLIN CAPRON, b Providence, RI, Apr 29, 38; m 62; c 1. SOIL CHEMISTRY, WATER CHEMISTRY. *Educ:* Univ NH, BA, 60, MS, 63; Ohio State Univ, PhD(agron, soil chem), 66. *Prof Exp:* Res assoc, Univ Ill, Urbana-Champaign, 66-67, asst prof soil org chem, 67-72; hydrologist, US Geol Surv, 72-75; sr scientist, Geraghty & Miller, Inc, 75-80, assoc, 80-89; prin scientist, 89-90, VPRES, BLASLAND, BOUCK & LEE, 91- *Concurrent Pos:* Abstractor, Chem Abstr Serv, 67-77; NSF guest lectr, Univ Calif, Riverside, 70; guest lectr, Aramco Oil, Dhahran, Saudi Arabia, 86; consult surface clean-up & mobilization processes, US Environ Protection Agency,

90. *Mem:* Am Chem Soc; Am Soc Agron; Soil Sci Soc Am; AAAS; Asn Ground-Water Sci & Eng; Nat Water Well Asn. *Res:* Effect of land disposal of municipal sludges on soils and plants; influence of waste disposal practice on ground-water quality; chemistry of contaminants from waste in aquifers; chemistry of petroleum-related contaminants in aquifers. *Mailing Add:* Blasland Bouck & Lee 3550 W Busch Blvd Suite 100 Tampa FL 33618

BRAIDWOOD, CLINTON ALEXANDER, b Snover, Mich, Nov 14, 14; m 42; c 3. ORGANIC CHEMISTRY. *Educ:* Mich State Univ, BS, 40. *Prof Exp:* Mem staff motor prod develop, US Rubber Co, 40-42; asst dir res, Reichold Chem, Inc, 42-49; vpres mfg & res develop, Schenectady Chems, Inc, 49-71, exec vpres, 71-73, pres, 73-80; BRAIDWOOD CHEM CONSULT, 81- *Mem:* AAAS; Am Inst Chemists; Am Chem Soc. *Res:* Alkylation of phenol and synthesis of phenolic resins from alkyl phenols; polymerization chemistry; terpene resins. *Mailing Add:* 6541 East Shore Rd Traverse City MI 49684

BRAILOVSKY, CARLOS ALBERTO, b Buenos Aires, Arg, Oct 16, 39; Can citizen; m 65; c 3. CELL BIOLOGY, VIROLOGY. *Educ:* Buenos Aires Univ, MD, 61. *Prof Exp:* Buenos Aires Univ fel, Inst Res Sci Cancer, Villejuif, France, 63-65; res assoc virol, 65-68; invited scientist path, Univ Montreal, 69; assoc prof, Univ Sherbrooke, 70-80, prof cell biol, fac med, 80-; UNIV LAVALLE. *Honors & Awards:* Squibb Award, Arg Med Asn, 57. *Mem:* French Soc Microbiol; Am Soc Cell Biol; Am Soc Microbiol; Am Tissue Cult Asn; Can Soc Cell Biol. *Res:* Oncogenic transformation modifications and immune control of tumor growth. *Mailing Add:* Cessul Med Sch Univ Lavalle Quebec ON P4S 1K7 Can

BRAILSFORD, ALAN DAVID, b Mansfield, Eng, Apr 3, 30; m 52; c 3. PHYSICS. *Educ:* Birmingham Univ, Eng, BSc, 53, PhD(math physics), 56. *Prof Exp:* Res physicist, Bell Telephone Labs, 56-57; res physicist, Sci Labs, Ford Motor Co, 57-71; prof fel, Theoret Physics Div Harwell, 71-73; SR STAFF SCIENTIST, SCI LABS, FORD MOTOR CO, 73- *Concurrent Pos:* Lectr, Physics Dept, Wayne State Univ, 64-66; consult, Metals & Ceramics Div, Oak Ridge Nat Lab, 75- *Mem:* Fel Am Phys Soc; fel Inst Physics; Mat Res Soc; Am Soc Metals Int. *Res:* Irradiation effects in metals; electronic and ionic transport properties of materials; point defect and dislocation properties in materials. *Mailing Add:* 1212 Beechmont Dearborn MI 48124

BRAIN, CARLOS W, b Lima, Peru; US citizen. STATISTICAL METHODOLOGY, DESIGN OF EXPERIMENTS. *Educ:* Univ Agraria, Peru, BS, 60, Ing Agr, 64; WVa Univ, MS, 69, PhD(appl statist), 75. *Prof Exp:* ASSOC PROF & CHMN STATIST, FLA INT UNIV, 73- *Concurrent Pos:* Bk rev ed, J Qual Technol, 85-87. *Mem:* Am Statist Asn; Biometric Soc. *Res:* Statistical methodology; development of procedures for testing distributional assumptions. *Mailing Add:* Dept Statist Fla Int Univ Univ Park Miami FL 33199

BRAIN, DEVIN KING, b Mt Vernon, Ohio, Jan 7, 26; m 50, 83; c 3. ORGANIC CHEMISTRY. *Educ:* Univ Ariz, BS, 48, MS, 49; Ohio State Univ, PhD(org chem), 54. *Prof Exp:* Res chemist, 54-88, TECH MGT, PROCTER & GAMBLE CO, 88- *Mem:* Am Chem Soc; Am Soc Mass Spectrometry; AAAS. *Res:* Soap and detergent products; application of, separation and instrumental techniques, especially gas chromatography, gas chromatography and mass spectrometry, mass spectrometry, nuclear magnetic resonance, irridescent radiation and x-ray techniques, to product development; industrial organic chemistry, analysis of waste water; analytical chemistry. *Mailing Add:* Three Walsh Lane Cincinnati OH 45208

BRAIN, JOSEPH DAVID, b Paterson, NJ, Jan 20, 40; m 61; c 3. PHYSIOLOGY, ENVIRONMENTAL HEALTH. *Educ:* Harvard Univ, SM, 62, SMHyg, 63, SDHyg, 66. *Prof Exp:* Res assoc physiol, 66-68, from asst prof to assoc prof, 68-78, actg assoc dean acad affairs, 76-77, DIR, HARVARD PULMONARY SPECIALIZED CTR RES, HARVARD SCH PUB HEALTH, 77-, PROF PHYSIOL, 78-, DIR, RESPIRATORY BIOL PROG, 81- *Concurrent Pos:* NIH res career develop award, 69-74; mem comt, Cardiovascular & Pulmonary Study Sect, NIH, 75-79; mem prog proj res rev comt, Nat Heart, Lung & Blood Inst, 80-83. *Mem:* Am Physiol Soc; Am Thoracic Soc; Reticuloendothelial Soc; Sigma Xi; AAAS. *Res:* Function and structure of macrophages; pulmonary responses to inhaled gases and particles; health effects of air pollution; deposition and clearance of inhaled particles; occupational lung disease; respiratory mechanics. *Mailing Add:* 1427 Great Plain Ave Needham MA 02192

BRAINARD, ALAN J, b Cleveland, Ohio, June 30, 36; m 64; c 2. CHEMICAL ENGINEERING. *Educ:* Fenn Col, BS, 59; Univ Mich, MS, 61, PhD(chem eng), 64. *Prof Exp:* Engr, Esso Res Labs, 64-67; asst prof, 67-71, ASSOC PROF CHEM ENG, UNIV PITTSBURGH, 71- *Concurrent Pos:* NSF res grant, 68. *Mem:* Assoc Am Inst Chem Engrs; Am Chem Soc; Am Soc Eng Educ. *Res:* Phase equilibria of multicomponent systems; behavior of entropy at low temperatures; entropy maximum principle and the conditions of equilibrium; non-equilibrium thermodynamics; anisotropic heat conduction; carbon oxidation catalysis. *Mailing Add:* 2032 Holiday Park Dr Pittsburgh PA 15239

BRAINARD, JAMES R, ISOTOPE. *Educ:* Hope Col, Holland, BA, 71; Ind Univ, PhD(chem), 79. *Prof Exp:* MEM TECH STAFF, ENVIRON CHEM SECT, LOS ALAMOS NAT LAB. *Res:* Bioremediation of contaminated sites and process streams; magnetic resonance imaging contrast agents; applications of nuclear magnetic resonance and stable isotopes to the study of metabolism and structure; numerous publications; one patent. *Mailing Add:* Dept Isotope & Nuclear Chemicals-4 Los Alamos Nat Lab MSC345 Los Alamos NM 87545

BRAINERD, BARRON, b New York, NY, Apr 13, 28; Can citizen. MATHEMATICS, QUANTITATIVE STYLISTICS. *Educ:* Mass Inst Technol, SB, 49; Univ Mich, MS, 51, PhD(math), 54. *Prof Exp:* Instr math, Univ BC, 54-57; asst prof math, Univ Western Ont, 57-59; from asst prof to assoc prof, 59-67, PROF MATH & LING, UNIV TORONTO, 67- *Concurrent Pos:* Vis fel, Australian Nat Univ, 62-63. *Mem:* Modern Lang Asn; Asn Comput & Humanities; Can Ling Asn. *Res:* Theory of partially ordered rings; rings of continuous functions; operators on function spaces; mathematical linguistics; statistics of literary style; stochastic models in linguistics. *Mailing Add:* Dept Math Univ Toronto Toronto ON M5S 1A1 Can

BRAINERD, JEROME J(AMES), b San Angelo, Tex, Sept 15, 32; m 56; c 6. FLUID MECHANICS. *Educ:* Univ Notre Dame, BS, 54, MS, 56; Cornell Univ, PhD(aerospace eng), 63. *Prof Exp:* Sr aerodyn engr, Gen Dynamics/ Convair, 60-62 & space sci lab, Gen Dynamics/Astronaut, 62-63, staff scientist, 63-65; head fluid & thermal sci sect & chmn fluid & thermal eng fac, 68-70, chmn dept mech eng, 73-78, ASSOC PROF AEROSPACE ENG, UNIV ALA, HUNTSVILLE, 65- *Mem:* Am Inst Aeronaut & Astronaut; Sigma Xi; Am Geophys Union; Am Phys Soc. *Res:* Various phases of fluid dynamics including magnetogasdynamics; aerodynamics; hypersonic and high enthalpy flows; perturbation theory; chemically reacting flows; hypervelocity impact; turbomachinery. *Mailing Add:* Dept Mech Aerospace Eng Univ Ala Huntsville AL 35899

BRAINERD, JOHN WHITING, biology, for more information see previous edition

BRAINERD, WALTER SCOTT, b Des Moines, Iowa, May 27, 36; m 58; c 3. MATHEMATICS, COMPUTER SCIENCE. *Educ:* Univ Colo, BA, 58; Univ Md, MA, 61; Purdue Univ, PhD(comput sci), 67. *Prof Exp:* Instr math, Naval Postgrad Sch, 63-65, asst prof comput sci, 67-69; asst prof math statist, Columbia Univ, 69-72; prof math, Harvey Mudd Col, 72-73; sr systs specialist, Burroughs Corp, 73-78; mem staff, Los Alamos Nat Lab, 78-81; prof computer sci, Univ NMex, 81-90; SR SOFTWARE ENGR, SUN MICRO SYSTS, 90- *Concurrent Pos:* Vis asst prof, Univ Calif, San Diego, 68. *Mem:* Asn Comput Mach; Am Nat Standards Inst. *Res:* Programming languages, theory of computation. *Mailing Add:* 235 Mt Hamilton Los Altos CA 94022

BRAITHWAITE, CHARLES HENRY, JR, b Chicago, Ill, Dec 16, 20; m 49; c 2. ORGANIC CHEMISTRY, CHEMICAL & FORENSIC ENGINEERING. *Educ:* Univ Calif, Los Angeles, AB, 41; Univ Mich, BSE, 43; Carnegie Inst Technol, MS, 48, DSc, 49. *Prof Exp:* Insulation engr, Westinghouse Elec Corp, 43-46; sr chemist, Shell Oil Co, 49-51; dir lab, W Vaco ChlorAlkal Div, FMC Corp, 51-57; dir research & develop, Productol Co, 57-58; pres, Cal Colonial Chemsolve, 58-87; BRAITHWAITE CONSULT, 87- *Concurrent Pos:* Consult, Res Dept, US Borax Corp, 58-65, W R Grace Res Corp, 59-70, Piezo Ceramic Div, Gulton Indust, 64-75 & Stratos Div, Fairchild Indust, 76-; dir, Jacksons Prods Corp, 70-88. *Mem:* Am Chem Soc; Am Inst Chem Engrs; Soc Plastics Engrs. *Res:* Coatings; plastics; materials engineering. *Mailing Add:* Braithwaite Consults 11232 Tigrina Whittier CA 90603

BRAITHWAITE, JOHN GEDEN NORTH, b Bishop Auckland, Eng, Oct 5, 20; US citizen; m 47; c 3. OPTICS. *Educ:* Cambridge Univ, BA, 47, MA, 51. *Prof Exp:* Prin sci officer, Royal Radar Estab, Eng, 48-57; sr scientist, Baird Atomic Inc, Mass, 58-61; sr proj leader, Block Assocs, 62; group leader, Res Inst, Ill Inst Technol, 62-63 & Bendix Systs Div, Mich, 64-65; res physicist, Univ Mich, Ann Arbor, 65-72; RES PHYSICIST, ENVIRON RES INST MICH, 73- *Concurrent Pos:* Consult, Manned Spacecraft Ctr, NASA & Denver Div, Martin Marietta Corp. *Mem:* AAAS; Optical Soc Am. *Res:* Infrared physics and technology; optical instrumentation; optical filters; spectrometry. *Mailing Add:* 1332 White St Ann Arbor MI 48104

BRAKEFIELD, JAMES CHARLES, b Janesville, Wis, Nov 28, 44; m 78. COMPUTER THEORY, INTELLIGENT SYSTEMS. *Educ:* Univ Wis-Madison, BS, 71, MS(elec eng) & MS(comput sci), 72. *Prof Exp:* Systs analyst, Tex Instruments, 73-74; RES ENGR, KRUG INT, 75- *Mem:* Inst Elec & Electronics Engrs; Asn Comput Mach; Int Neurol Network Soc. *Res:* Select, integrate, program and support computer equipment for scientific research; image processing and implementation of a facility for image filtering; modeling and implementation of bio-physical models; color video display generator (real time). *Mailing Add:* 1214 Vista Del Rio San Antonio TX 78216-1700

BRAKENSIEK, DONALD LLOYD, b Alexis, Ill, Nov 5, 28; m 56; c 3. AGRICULTURAL, WATERSHED HYDROLOGY. *Educ:* Univ Ill, BS(agr) & BS(agr eng), 51, MS, 52; Iowa State Univ, PhD(agr eng), 55. *Prof Exp:* Agr engr, Agr Res Serv, USDA, Colo, 56, hydraul engr, Ohio, 57-59 & Md, 60-61, res hydraul engr, Hydrograph Lab & chief engr, Plant Indust Sta, 68-72, staff scientist, Watershed Hydrol, Nat Prog Staff, 72-74, location & res leader, Northwest Watershed Res Ctr, 74-85; RETIRED. *Mem:* Am Soc Agr Engrs; Am Geophys Union. *Res:* Analytical hydrology; soil water process modeling and watershed hydrology models; author of 105 publications. *Mailing Add:* 881 E Pennsylvania Dr Boise ID 83706-4435

BRAKKE, MYRON KENDALL, b Preston, Minn, Oct 23, 21; m 47; c 4. PLANT VIROLOGY. *Educ:* Univ Minn, BS, 43, PhD(biochem), 47. *Prof Exp:* Asst biochem, Univ Minn, 43-44, instr, 44-47; res assoc, Brooklyn Bot Garden, 47-52 & Dept Bot, Univ Iil, 52-55; chemist, Agr Res Serv, USDA & prof plant path, Univ Nebr, Lincoln, 55-86; RETIRED. *Honors & Awards:* Superior Serv Award, USDA, 68 & 86; Ruth Allen Award, Am Phytopath Soc, 68; Outstanding Res & Creative Activity Award, Univ Nebr, 82; Sci Hall of Fame, 87; Award of Distinction, Am Phytopath Soc, 88. *Mem:* Nat Acad Sci; fel AAAS; Am Chem Soc; fel Am Phytopath Soc; Am Soc Microbiol. *Res:* Purification and characterization of plant viruses and their nucleic acids; vectors of plant viruses and control of plant virus diseases. *Mailing Add:* 406 Plant Sci Hall Univ of Nebr Lincoln NE 68583-0722

BRALLEY, JAMES ALEXANDER, b Bath Co, Va, Aug 18, 16; m 46; c 3. ORGANIC CHEMISTRY. *Educ:* Univ Va, BS, 36, PhD(org chem), 41. *Prof Exp:* Res chemist, BF Goodrich Co, 41-46; develop chemist, Rohm and Haas Co, 46-56; dir res, AE Staley Mfg Co, 56-61, vpres res & develop, 61-70; vpres res & develop, Puritan Chem Co, 71-76; mgr qual control, 76-79, MGR QUAL ASSURANCE, C H DEXTER DIV, DEXTER CORP, 79- *Mem:* Sigma Xi; AAAS; Am Chem Soc. *Res:* Morphine alkaloids; vinyl and acrylic resins; plasticizers; inhibitors; starch chemistry, processing and industrial uses; vegetable oils and proteins; epoxy resins; fermentation chemistry; amylose and amylopectin; amino acids; organic syntheses; detergents; disinfectants; insecticides; herbicides; waxes; synthetic polymer formulations. *Mailing Add:* 3040 Sawtooth Circle Alpharetta GA 30202-5400

BRALOW, S PHILIP, b Philadelphia, Pa, Aug 28, 21; m 82; c 2. GASTROENTEROLOGY, INTERNAL MEDICINE. *Educ:* Pa State Col, BS, 42; Temple Univ, MD, 45; Univ Ill, MS, 49. *Prof Exp:* From intern to resident gastrointestinal res, Michael Reese Hosp, 46-47; chief, Gastrointestinal Clin & Lab, Temple Univ Hosp, 54-63, from asst prof to assoc prof med, Temple Univ, 59-69; prof, Jefferson Med Col, 69-75; clin prof med, Sch Med, Univ SFla, 75-82; CLIN PROF MED, UNIV PA, 82-, MED DIR, CANCER PREV CTR, GRAD HOSP, 82- *Concurrent Pos:* NIH grant, Temple Univ Hosp; Am Cancer Soc & Nat Cancer Inst grants, Thomas Jefferson Univ Hosp; attend physician gastroenterol, Vet Admin Hosp, 54-63; consult, US Naval Hosp, 69; mem, Coun Cancer, Am Gastroenterol Asn, 72; mem, Nat Sci Adv Bd, Nat Found Ileitis & Colitis, Inc & med & sci comt, Am Cancer Soc, 73; working cadre mem, Nat Large Bowel Cancer Study Group, NIH, 78- *Honors & Awards:* 25 Year Award, Colostomy Ileostomy Rehab Asn, 74. *Mem:* Fel Am Col Gastroenterol; Am Cancer Soc; Am Gastroenterol Asn; Am Asn Cancer Res. *Res:* Experimental carcinogenesis in animal models; radiation injury to small intestines; cancer prevention. *Mailing Add:* One Independence Pl Philadelphia PA 19106

BRAM, JOSEPH, mathematics; deceased, see previous edition for last biography

BRAM, RALPH A, b Washington, DC, Dec 10, 32; m 56; c 4. ENTOMOLOGY. *Educ:* Univ Md, BS, 56, MS, 61, PhD(entom), 64. *Prof Exp:* Instr entom, Univ Md, 63-64; asst prof, Purdue Univ, 64-65; entomologist, Smithsonian Inst, 65-68; res entomologist, Animal Parasite Dis Lab, Entom Res Div, Agr Res Serv, 68-71, res entomologist, Insects Affecting Man & Animals Res Lab, 71-72, vet entomologist, Animal & Plant Health Inspection Serv, 72-81, NAT PROG LEADER, AGR RES SERV, USDA, 81- *Concurrent Pos:* Consult, Food & Agr Orgn, Rome, Italy, 74-76, 82-85, Int Atomic Energy Agency, Vienna, Australia, 82-84; chmn, Sect D, Entom Soc Am, 84-85. *Mem:* AAAS; Entom Soc Am. *Res:* Veterinary and medical entomology; mosquito systematics and zoogeography; general entomology and pathogen-vector relationships; tick and tick-borne disease control. *Mailing Add:* Agr Res Serv USDA Agr Res Ctr West Bldg 005 Beltsville MD 20705

BRAMAN, ROBERT STEVEN, b Lansing, Mich, Aug 31, 30; m 54; c 2. ANALYTICAL CHEMISTRY, INSTRUMENTATION. *Educ:* Mich State Univ, BS, 52; Northwestern Univ, PhD(chem), 56. *Prof Exp:* Group leader anal chem res & develop, Callery Chem Co, 56-59; res chemist, Res Inst, Ill Inst Technol, 59-64, sr chemist, 64-67; from asst prof to assoc prof, 67-73, PROF CHEM, UNIV S FLA, 73- *Mem:* Am Chem Soc; Sigma Xi. *Res:* Development of analytical instrumentation; environmental chemistry. *Mailing Add:* Dept Chem Univ of S Fla Tampa FL 33620

BRAMANTE, PIETRO OTTAVIO, b Rome, Italy, May 21, 20; US citizen; m 57. PATHOLOGY, PHYSIOLOGY. *Educ:* Univ Rome, MD, 44, MS, 50; Drexel Univ, MS, 64. *Prof Exp:* Instr internal med, Sch Med, Univ Rome, 45-46, asst prof, 46-51; res fel, Sch Med, Univ Stockholm & King Gustaf's Res Inst, 51-52; from res assoc to assoc prof physiol, Sch Med, St Louis Univ, 52-65; assoc prof physiol, Univ Ill Col Med, 65-70, prof, 70-75; pathologist, Wm Beaumont Army Med Ctr, El Paso, 78-79; pathologist, Sun Bay Community Hosp, Fla, 81-82; RETIRED. *Concurrent Pos:* Tech med consult, Court of Justice, Rome, 48-51; Fulbright fel, 52; Heart Inst fel, Drexel Univ, 62-64. *Mem:* Am Physiol Soc; Am Soc Clin Path; Col Am Pathologists; Soc Exp Biol & Med. *Res:* Pituitary in uremia; liver function; body electrolyte; hypotension; thyroid function; innervation of nasal mucosa; methodology of energy metabolism; allometry; experimental tumorigenesis; calorigenic drugs; exercise and metabolism; experimental myocardial necrosis; experimental dyslipoproteinemia. *Mailing Add:* 4111 103rd Ave N Clearwater FL 34622

BRAMBL, ROBERT MORGAN, b Ft Smith, Ark, April 26, 42; m 65; c 1. PLANT BIOLOGY. *Educ:* Hendrix Col, BA, 65; Univ Nebr, PhD(bot-biochem), 69. *Prof Exp:* NIH fel & res assoc, Dept Biol Sci & Path, Stanford Univ, 69-71; from asst prof to assoc prof, 71-80, PROF, DEPT PLANT PATH & PLANT BIOL, UNIV MINN, 80- *Concurrent Pos:* Proj leader, Univ Minn Agr Exp Sta, 71- & 81-83; prin investr, Nat Inst Gen Med Sci res grant, 74-; vis scientist, Inst Physiol Chem & Phys Biochem, Univ Munich, 77-80; co-prin investr, McKnight Found res grant, 83-87, USDA res grant, 85- *Mem:* Am Soc Biochem & Molecular Biol; Am Chem Soc; Am Soc Microbiol; AAAS. *Res:* Purification and characterization of mitochondrial respiratory enzymes of fungi; coordination of nuclear and mitochondrial genes in mitochondrial biogenesis; regulation of mitochondrial gene expression; structural characterization and expression of nuclear and mitochondrial genes for mitochondrial proteins in fungi and plants; mechanisms of assembly of multiple-subunit respiratory enzymes; membrane translocation of enzyme sub-units; post-transcriptional regulation of gene expression; regulation of heat shock gene expression; cellular localization and functions of stress proteins; molecular responses of plant pollen and seed embryos to environmental and biotic stresses; role of plant seed lectins in plant seed defense against fungi; biochemistry of fungi spore dormancy and germination; numerous publications. *Mailing Add:* Dept Plant Biol Univ Minn 220 Biol Sci Ctr St Paul MN 55108

BRAMBLE, JAMES H, b Annapolis, Md, Dec 1, 30; m 55, 77; c 4. APPLIED MATHEMATICS, NUMERICAL ANALYSIS. *Educ:* Brown Univ, AB, 53; Univ Md, MA, 55, PhD(math), 58. *Hon Degrees:* DSc, Chalmers Univ Technol, Gothenburg, Sweden, 85. *Prof Exp:* Mathematician, Gen Elec Co, 57-59 & US Naval Ord Lab, 59-60; res asst prof math, Univ Md, 60-63, from res assoc prof to res prof, 63-68; dir, Ctr Appl Math, 74-80, assoc chmn math, 79-84, PROF MATH, CORNELL UNIV, 68- *Concurrent Pos:* Vis lectr, Soc Indust & Appl Math, 62-65; consult, Nat Bur Standards, 60-66; NSF sr fel, 66-67; vis prof, Univ Rome, 66-67, Chalmers Univ Technol, Gothenburg, Sweden, 70, 72-73, 76; vis staf mem, Los Alamos Sci Lab, Univ Calif, 74-76; consult, Brookhaven Nat Lab, 76-, Ecole Poly, Paris, 78, Lausanne, Switzerland, 79, Univ Paris, 81, Australian Nat Univ, 90. *Mem:* Am Math Soc; Soc Indust & Appl Math. *Res:* Partial differential equations; numerical analysis. *Mailing Add:* 24 Tyler Rd Ithaca NY 14850

BRAMBLE, WILLIAM CLARK, b Baltimore, Md, Nov 7, 07; m 39; c 2. FORESTRY. *Educ:* Pa State Univ, BS, 29; Yale Univ, MF, 30, PhD(bot), 32. *Prof Exp:* Instr bot, Carleton Col, 32-37; Nat Res fel, Univ Zurich, 35-36; from asst prof to prof forestry, Pa State Univ, 37-58, head dept forest mgt, 54-58, actg dir sch forestry, 55-58; prof forestry & head dept forestry & conserv, 58-72, EMER PROF FORESTRY, PURDUE UNIV, 72- *Concurrent Pos:* Collabr, USDA, 34-72; mem div biol & agr, Nat Res Coun, 70-73; ed, Hoosier Conserv, 75-78. *Mem:* Soc Am Foresters; Ecol Soc Am; Swiss Bot Soc; Sigma Xi. *Res:* Forest ecology and management; silviculture. *Mailing Add:* Dept Forestry & Conserv Purdue Univ West Lafayette IN 47907

BRAMBLETT, CLAUD ALLEN, b Crystal City, Tex, Oct 8, 39; m 61; c 2. PHYSICAL ANTHROPOLOGY, PRIMATOLOGY. *Educ:* Univ Tex, Austin, BA, 62, MA, 65; Univ Calif, Berkeley, PhD(anthrop), 67. *Prof Exp:* Mgr, Darajani Primate Res Ctr, Southwest Found Res, 63-64; asst prof, 67-73, ASSOC PROF ANTHROP, UNIV TEX, AUSTIN, 73- *Concurrent Pos:* Consult, Southwest Found Res & Educ, 71- & dept path, Univ Tex Med Sch San Antonio, 72-77; mem proj comt, Arashiyama W Japanese Macaque Ranch, 72-; co-investr, Exp Atherosclerosis in Baboons, NIH, 73-78. *Mem:* Fel Am Anthrop Asn; Am Asn Phys Anthrop; Int Primatol Soc; Am Soc Mammalogists; Soc Study Human Biol. *Res:* Behavior as a risk factor in coronary heart disease; acquisition of social signals in developing vervet monkeys; behavioral regulators. *Mailing Add:* Dept Anthrop Univ of Tex Austin TX 78712

BRAMBLETT, RICHARD LEE, b Dallas, Tex, Aug 15, 35; m 57; c 3. APPLIED PHYSICS, NUCLEAR PHYSICS. *Educ:* Rice Univ, BA, 56, MA, 57, PhD(physics), 60. *Prof Exp:* Physicist, Lawrence Radiation Lab, Univ Calif, 60-67; br mgr, Gulf Energy & Environ Systs, Calif, 68-73; SECT MGR, PHOTON LOGGING, SCHUMBERGER WELL SERV, 80- *Mem:* Am Phys Soc; Am Nuclear Soc; Inst Nuclear Mat Mgt. *Res:* Experimental photonuclear and neutron physics using electron accelerator technology; spectroscopy and cross-section measurements for photons and neutrons; application of accelerators, reactors and radioisotopes to non-destructive isotopic assays. *Mailing Add:* 1108 Tall Pines Houston TX 77088

BRAME, EDWARD GRANT, JR, b Shiloh, NJ, Mar 20, 27; m 57. ANALYTICAL CHEMISTRY. *Educ:* Dickinson Col, BS, 48; Columbia Univ, MS, 50; Univ Wis, PhD(anal chem), 57. *Prof Exp:* Asst chem, Columbia Univ, 48-50; res anal chemist, Corn Prod Refining Co, 50-53; asst, Univ Wis, 53-56; res chemist, Plastics Dept, E I Du Pont De Nemours & Co, Inc, 57-64, res chemist, Elastomer Chem Dept, 64-79, res assoc, Polymer Prod Dept, 80-85; CONSULT, CECON GROUP, INC, 85- *Concurrent Pos:* Ed, Appl Spectros Rev; researcher, Pub Exhib Comn, Winterthur Mus, Del; mem postdoctoral res associateships eval panel, Nat Res Coun, 74-76; mem adv bd, Anal Chem, 74-76; chmn, Fedn Anal Chem & Spectros Socs, 76; ed-in-chief, Practical Spectros & ed, Appln Polymer Spectros, 78; Nat Acad Sci exchangee to USSR, 81, 83, 85 & 90; exhib dir, Fedn Anal Chem & Spectros Soc, 86. *Mem:* Sigma Xi; Am Chem Soc; Soc Appl Spectros (pres, 78); NY Acad Sci. *Res:* Infrared spectroscopy; nuclear magnetic resonance spectroscopy; electron paramagnetic resonance spectroscopy; instrumental methods of analysis; nuclear quadrupole resonance spectroscopy; surface spectroscopy. *Mailing Add:* Cecon Group, Inc 242 N James St Wilmington DE 19804

BRAMFITT, BRUCE LIVINGSTON, b Troy, NY, Feb 4, 38; m 63; c 2. FERROUS PHYSICAL METALLURGY. *Educ:* Univ Mo, Rolla, BS, 60, MS, 62, PhD(metal eng), 66. *Prof Exp:* Instr, Univ Mo, Rolla, 60-66; res engr, 66-79, supvr metall eng, 79-84, res fel, 84-87, SR SCIENTIST, BETHLEHEM STEEL CORP, 84- *Honors & Awards:* Joseph Vilella Award, Am Soc Testing & Mat, 74; C D Moore Award, Am Inst Mining, Petrol & Metall Eng, 77 & 79, Michael Jenembaum Award, Iron & Steel Soc, 87; Bradley Stoughton Award, Am Soc Metals Int, 91. *Mem:* Fel Am Soc Metals; Am Inst Mining, Petrol & Metall Eng; Int Metall Soc. *Res:* Physical metallurgy with emphasis on microstructure and its relationship to the properties of steel; design of new alloys based on various strengthening mechanisms. *Mailing Add:* 1647 Pleasant Dr Bethlehem PA 18015

BRAMHALL, JOHN SHEPHERD, b Manchester, Eng, March 30, 50. MOLECULAR BIOLOGY. *Educ:* Manchester Polytech, HND, 71; Ewell Col, MIBiol, 73; Aston Univ, PhD(biochem), 76; Univ Calif, San Diego, MD, 91. *Prof Exp:* Fel, Univ Calif, Los Angeles, 76-79, Max Planck Inst Biol, 79-80, Stanford Univ, 80-81; asst prof, Univ Calif, Los Angeles, 81-87; RESIDENT ANESTHESIOL, VA MASON CLIN, SEATTLE, 91- *Mem:* Biophys Soc. *Res:* Structure and function of biological membranes. *Mailing Add:* 2805 Maple Ave Manhattan Beach CA 90266-2329

BRAMLAGE, WILLIAM JOSEPH, b Dayton, Ohio, Mar 27, 37; m 67; c 3. HORTICULTURE. *Educ:* Ohio State Univ, BS, 59; Univ Md, MS, 61, PhD(hort), 63. *Prof Exp:* Horticulturist, agr mkt serv, USDA, 63-64; asst prof hort, 64-69, assoc prof plant physiol, 69-76, PROF PLANT SCI, UNIV MASS, AMHERST, 76- *Concurrent Pos:* Vis Prof, Univ Neb, Lincoln, 76-77; vis scientist, East Malling Res Sta, Kent, England, 84, DSIR, Fruit & Trees,

Auckland, NZ, 91. *Mem:* Am Soc Plant Physiol; fel Am Soc Hort Sci. *Res:* Post-harvest physiology; physiological disorders of fruit; stress physiology. *Mailing Add:* Dept Plant & Soil Sci Bowditch Hall Univ Mass Amherst MA 01003

BRAMLET, ROLAND C, b Wallowa, Ore, June 11, 21; m 61; c 2. RADIOLOGICAL PHYSICS. *Educ:* Ore State Univ, BS, 48; NY Univ, MS, 61; St John's Univ, NY, PhD(biol), 66. *Prof Exp:* Fel, Sloan-Kettering Inst Cancer Res, 58-59; jr physicist, Queens Gen Hosp, Jamaica, NY, 59-61, radioisotope physicist, 61-69; PHYSICIST, HIGHLAND HOSP, ROCHESTER, NY, 69- *Concurrent Pos:* Clin asst prof radiation oncol, Sch Med & Dent, Univ Rochester, NY, 85. *Mem:* Am Asn Physicists Med. *Res:* Clinical radioisotopic instrumentation; radiation biology; radiation oncology. *Mailing Add:* Radiation Therapy Highland Hosp Rochester NY 14620

BRAMLETT, WILLIAM, b Stephenville, Tex, Sept 09, 42; m 63; c 2. RESEARCH ADMINISTRATION. *Educ:* Texas Tech, BS, 64. *Prof Exp:* VPRES RES & DEVELOP, WICAT SYSTEMS, 83- *Mem:* Inst Elec & Electronics Engrs. *Mailing Add:* 1155 East 100 North Orem UT 84057

BRAMLETTE, WILLIAM (ALLEN), geology; deceased, see previous edition for last biography

BRAMMER, ANTHONY JOHN, b Watford, Eng, May 30, 42; Brit & Can citizen; m 67. ACOUSTICS PHYSICS. *Educ:* Univ Exeter, Eng, BSc, 63, PhD(physics), 68. *Prof Exp:* Engr vacuum tube develop, Standard Tel & Cables, Ltd, 67; res fel, 67-68, asst res officer, 69-76, ASSOC RES OFFICER ACOUSTICS, NAT RES COUN, CAN, 76- *Concurrent Pos:* Lectr physics & math, Exeter Tech Col, 66-67; sessional lectr, Carleton Univ, 69-71; adj assoc prof, Univ Windsor, 72- *Mem:* Acoust Soc Am; Can Acoust Asn. *Res:* Hand-arm vibration; vibration isolation systems; vibration standards; non-linear acoustics applicable to internal combustion engines; industrial noise control. *Mailing Add:* 4792 Massey Lane Ottawa ON K1A 0S1 Can

BRAMMER, FOREST E(VERT), b Mabscott, WVa, July 21, 13; m 42; c 4. ELECTRICAL ENGINEERING. *Educ:* NC State Col, BS, 33; Concord Col, AB, 33; Case Inst Technol, PhD(elec eng), 51. *Prof Exp:* Teacher high sch, NC, 33-36; dist engr geophys, Schlumberger Well Surv Corp, 37-42; supvr underwater res group, Appl Physics Lab, Johns Hopkins Univ, 46-48; prof lectr elec eng, Case Inst Technol, 48-51, assoc prof, 51-60; head dept, 60-70, prof, 60-68, EMER PROF ELEC ENG, WAYNE STATE UNIV, 68- *Concurrent Pos:* Mem, Nat Electronics Conf; consult, Goodyear Aircraft Corp, 54-58 & Repub Steel Corp, 59- *Mem:* Inst Elec & Electronics Engrs. *Res:* Antennas; solid state devices; analog computers. *Mailing Add:* Wayne State Univ Detroit MI 48202

BRAMMER, LEE, b Barnsley, Eng, Apr 23, 63. CRYSTALLOGRAPHY. *Educ:* Univ Bristol, UK, BSc hons, 83, PhD(inorg chem), 87. *Prof Exp:* N Atlantic Treaty Orgn fel chem, Univ New Orleans, 87-88; fel chem, Brookhaven Nat Lab, 88-90; ASST PROF INORG CHEM, UNIV MO, 90- *Mem:* Am Crystallog Asn; Am Chem Soc. *Res:* Single crystal x-ray and neutron diffraction; structure, bonding and electronic properties of organometallic compounds, particularly transition metal cluster compounds and compounds with transition metal-hydrogen interactions. *Mailing Add:* Dept Chem Univ Mo St Louis MO 63121

BRAMS, STEWART L, rubber chemistry; deceased, see previous edition for last biography

BRAMWELL, FITZGERALD BURTON, b Brooklyn, NY, May 16, 45; m 73; c 3. PHYSICAL CHEMISTRY. *Educ:* Columbia Univ, BA, 66; Univ Mich, MS, 67, PhD(chem), 70. *Prof Exp:* Res chemist, Esso Res & Eng Co, 70-71; assoc prof, 71-80, PROF CHEM, BROOKLYN COL, 80- *Concurrent Pos:* Mem tech staff, Bell Tel Labs, 74-75, consult, 77-78; Nat Res Coun, Grad Fel Panel, 81-84; BD ADVS, CHEMTECH, 85-; sci consult, Bd Educ, New York, 85; dir-at-large, NY sect, Am Chem Soc, 86. *Mem:* Am Chem Soc; NY Acad Sci. *Res:* Electron spin resonance of triplets; free radicals in solution; organic metals; charge transfer complexes; Jan-Teller distortions; aryl tin radicals. *Mailing Add:* Dept Chem Brooklyn Col Brooklyn NY 11210

BRANAHL, ERWIN FRED, b St Louis, Mo, Mar 8, 22; m 44; c 2. AERONAUTICAL ENGINEERING. *Educ:* Wash Univ, BS, 43, MS, 51. *Prof Exp:* Stress analyst, Curtiss-Wright Corp, 43-44; res engr, McDonnell Co, 46-47, design engr, 47-51, dynamics engr, 51-52; chief dynamics engr, McDonnell Douglas Astronaut Co, St Louis, 52-54, chief aeromech engr, 54-59, mgr aeromech, 59-60, asst chief engr, Missile Eng Div, 60-61, mgr space & missile eng progs, 61-68, vpres eng, 68-74, vpres & gen mgr, 74-86, exec vpres, 86-87; RETIRED. *Concurrent Pos:* Vpres & dir, McDonnell Douglas Tech Serv Co, 74-; dir, Vitek Syst, Inc, 77-; corp vpres, McDonnell Douglas Corp, 78-, vpres, McDonnell Douglas Ltd, 80- *Mem:* Assoc fel Am Inst Aeronaut & Astronaut. *Res:* Missile aerodynamics; automatic control systems; dy- namics; propulsion; structures; thermodynamics. *Mailing Add:* 14 Lake Pembroke Dr St Louis MO 63135

BRANCA, ANDREW A, b New Rochelle, NY, Aug 20, 50. CELL BIOLOGY. *Educ:* N Adams State Col, BA, 75; Univ NH, MS, 78, PhD(biochem), 80. *Prof Exp:* Asst prof biochem, Albany Med Col, 82-89; SR SCIENTIST, DEPT CELL BIOL, PROCYTE CORP, 89- *Concurrent Pos:* Fel, Nat Cancer Inst, NIH, 81-82; new investr res award, Nat Inst Allergy & Infectious Dis Inst, 85-88. *Mem:* AAAS; Int Soc Interferon Res; Am Soc Biochem & Molecular Biol; Sigma Xi; Tissue Cult Asn. *Mailing Add:* Dept Cell Biol ProCyte Corp 1240 115th Ave NE Suite 210 Kirkland WA 98034

BRANCATO, DAVID JOSEPH, toxicology, physiology, for more information see previous edition

BRANCATO, EMANUEL L, b Brooklyn, NY, Nov 3, 14; m 58; c 2. ELECTRICAL ENGINEERING. *Educ:* Columbia Univ, BA, 36, BS, 37, MS, 38. *Prof Exp:* Jr elec engr, Develop & Design, Boston Naval Shipyard, 38-39; elec engr, Sci Sect, Develop & Res, NY Naval Shipyard, 39-46; sect head, Shipboard Power Systs, US Naval Res Lab, 46-54, head, Insulation Sect, Solid State Div, 54-67, head, Solid State Appl Br, 67-70, consult to dir res, 70-76, head consult staff, 76-78; RETIRED. *Concurrent Pos:* Consult, Elec Power Res Inst & Naval Res Lab, 79. *Honors & Awards:* Centennial Award Medal, Inst Elec & Electronics Engrs, 84. *Mem:* Fel Inst Elec & Electronics Engrs; Sigma Xi. *Res:* Investigation of the effects of ionizing radiation on dielectric and semiconductor materials and devices. *Mailing Add:* 7370 Hallmark Rd Clarksville MD 21029

BRANCATO, LEO J(OHN), b New York, NY, Oct 27, 22; m 48; c 3. MECHANICAL ENGINEERING. *Educ:* Cooper Union Sch Eng, BME, 50; Columbia Univ, MS, 52. *Prof Exp:* Develop engr, Edward Ermold Co, 49-52; mgr eng, Heli-Coil Corp, 52-53, chief engr design & develop, 53-56, vpres, 56-58, exec vpres & mem bd, 58-70; vpres & mem bd, 70-74, Pres, Mite Corp, 75-85; PRES, HELI COIL CORP, 70- *Mem:* Fel Am Soc Mech Engrs; Am Soc Metals; NY Acad Sci. *Res:* Automatic machinery; metal fabrication; screw thread fastener technology. *Mailing Add:* 137 Chambers Rd Danbury CT 06811

BRANCH, CLARENCE JOSEPH, b Jackson, NC, Sept 5, 40; m 72. ANIMAL BIOLOGY, RADIATION BIOLOGY. *Educ:* St Augustine's Col, BS, 65; Univ Tenn, Knoxville, MS, 69; Oak Ridge Assoc Univ, DrIP, 71. *Prof Exp:* Asst researcher, Oak Ridge Inst Nuclear Studies, 68-70; RESEARCHER, ENVIRON BIOL LAB, RESEARCH TRIANGLE PARK, 74- *Concurrent Pos:* Researcher, Atomic Energy Comn, 68-70. *Mem:* Am Inst Sci; Nat Sci Teacher Soc. *Res:* Cancer-radiation; heavy solvent metal; toxicology; tumorigenic and life span effects of low level HTO exposure on pregnant rats. *Mailing Add:* 942 Shelly Rd Raleigh NC 27609

BRANCH, DAVID REED, b Coaldale, Pa, Mar 12, 42; m 69; c 2. ASTROPHYSICS. *Educ:* Rensselaer Polytech Inst, BS, 64; Univ Md, PhD(astron), 69. *Prof Exp:* Res assoc solar physics, Goddard Space Flight Ctr, NASA, 69; res fel astrophys, Calif Inst Technol, 69-70; sr res fel, Royal Greenwich Observ, 70-73; from asst prof to assoc prof, 73-81, chmn dept, 85-90, PROF PHYSICS & ASTRON, UNIV OKLA, 81- *Concurrent Pos:* George Lynn Cross res prof, 87- *Honors & Awards:* Sr US Scientist Award, Alexander Humboldt Found, 88. *Mem:* Int Astron Union; Am Astron Soc. *Res:* Stellar spectroscopy and photometry; stellar and solar chemical composition; supernovae; extragalactic distance scale. *Mailing Add:* Dept Physics & Astron Univ Okla Norman OK 73019-0225

BRANCH, GARLAND MARION, JR, b Plant City, Fla, Apr 16, 22; m 45; c 4. MEDICAL TECHNOLOGY. *Educ:* Stetson Univ, BS, 43; Cornell Univ, PhD(physics), 51. *Prof Exp:* Res asst nuclear reactor physics, Manhattan Proj, 43-46; asst physics, Cornell Univ, 46-48, res asst cosmic ray physics, 48-51; res assoc electron physics, Res Lab, Gen Elec Med Systs Div, Milwaukee, 51-58, liaison scientist, 58-60, consult microwave physicist, Superpower Microwave Tube Lab, 60-63, mgr electron beam & circuit res, Tube Dept, 63-75, sr develop engr, 75-79; prof elec engr & comput sci, Union Col, 79-85; RETIRED. *Concurrent Pos:* Adj prof, Union Col, NY, 67-72. *Mem:* Inst Elec & Electronic Engrs; Am Phys Soc; Am Asn Physics Teachers; NY Acad Acad Sci; Soc Photo-optical Engrs; Sigma Xi. *Res:* Microwave electronics; electron optics; cosmic rays; satellite communication systems; high-energy particle accelerators. *Mailing Add:* 1401 Via Del Mar Schenectady NY 12309-4319

BRANCH, JOHN CURTIS, b Buffalo, Okla, Oct 1, 34; m 60; c 3. PARASITOLOGY. *Educ:* Northwestern State Col, Okla, BS, 59; Univ Okla, MS, 62, PhD(prev med, pub health), 65; Oklahoma City Univ, JD, 80. *Prof Exp:* From asst prof to assoc prof, 64-72, chmn dept, 66-80, PROF BIOL, OKLAHOMA CITY UNIV, 72- *Concurrent Pos:* Atty. *Mem:* AAAS; Am Soc Parasitol; Am Inst Biol Sci; Sigma Xi. *Res:* Parasite metabolism; transfer of learning via brain extracts. *Mailing Add:* 2705 Abbey Rd Oklahoma City OK 73120

BRANCH, LYN CLARKE, b Greenwood, Miss, Sept 25, 53. CONSERVATION BIOLOGY. *Educ:* Miss State Univ, BS, 75; Miami Univ, MS, 77; Univ Calif, Berkeley, PhD(wildland resource sci), 89. *Prof Exp:* ASST PROF CONSERV BIOL, UNIV FLA, 90- *Mem:* Ecol Soc Am; Am Soc Mammalogists; Soc Conserv Biol; Asn Trop Biol; Wildlife Soc. *Res:* Conservation biology; evolution of vertebrate social systems; role of animals in structuring plant communities; tropical ecology. *Mailing Add:* 118 Newins-Ziegler Univ Fla Gainesville FL 32611-0304

BRANCH, ROBERT ANTHONY, b Nairobi, Kenya, Oct 8, 42; m 65; c 3. CLINICAL PHARMACOLOGY. *Educ:* Bristol Univ, MB & ChB, 65; RCOG, 67; MRCP, 68; Bristol Univ, MD, 76. *Prof Exp:* House surgeon, United Bristol Hosp, 65-66, physician, 66, sr house surgeon, 66-67; registrar nephrology, Welsh Nat Sch Med, 68-69; registrar gastroenterol, Univ Bristol, 69-71; lectr clin pharmacol, Univ Bristol, 73-75; res fel, 71-73, from asst prof to assoc prof, 75-80, PROF MED & PHARMACOL, VANDERBILT UNIV, 80- *Mem:* Am Asn Study Liver Dis; Am Fedn Clin Res; Brit Soc Pharmacol; Am Soc Clin Invest; fel Royal Col Physicians. *Res:* Drug disposition in normal and disease states; hepatic and renal disease; regulation of renin release. *Mailing Add:* Vanderbilt Clin 1301 22nd Ave S Nashville TN 37232

BRANCH, WILLIAM DEAN, b Duncan, Okla, Sept 14, 50; m 72; c 2. PEANUT BREEDING, PEANUT GENETICS. *Educ:* Okla State Univ, BS, 72, MS, 74, PhD(crop sci), 76. *Prof Exp:* Res asst, Okla State Univ, 72-77; res assoc, Auburn Univ, 77-78; asst geneticist, 78-84, assoc prof, 85-90, PROF, UNIV GA, 90- *Mem:* Am Soc Agron; Crop Sci Soc Agron; Am Peanut Res & Educ Soc; Am Genetic Asn; Coun Agr Sci & Technol; Nat Peanut Coun. *Res:* Georgia peanut breeding and genetic research program. *Mailing Add:* Univ Ga Dept Agron Coastal Plain Exp Sta Tifton GA 31793-0748

BRANCH, WILLIAM H, electrical engineering, for more information see previous edition

BRANCHINI, BRUCE ROBERT, b New Haven, Conn, Oct 22, 50. ORGANIC CHEMISTRY, BIO-ORGANIC CHEMISTRY. *Educ:* Lehigh Univ, BS, 72; Johns Hopkins Univ, MA, 73, PhD(chem), 75. *Prof Exp:* Fel Biochem, Harvard Univ, 75-76; asst prof, 76-80, ASSOC PROF CHEM, UNIV WIS-PARKSIDE, 81- *Concurrent Pos:* Vis prof chem, Johns Hopkins Univ, 81-82. *Mem:* Am Chem Soc; AAAS. *Res:* Bioluminescence; chemiluminescence; cystic fibrosis; enzyme inhibitors; new substrates. *Mailing Add:* Chem Dept Conn Col Box 5401 New London CT 06320

BRAND, DONALD A, ENGINEERING. *Prof Exp:* SR VPRES & GEN MGR, PAC GAS & ELEC CO. *Mem:* Nat Acad Eng. *Mailing Add:* Pac Gas & Elec Co 77 Beale St F1512 San Francisco CA 94106

BRAND, EUGENE DEW, b CZ, Mar 10, 24; m 53, 72; c 6. PSYCHIATRY, PSYCHOPHARMACOLOGY. *Educ:* Univ Va, BA, 44; Harvard Univ, MD, 50; Am Bd Psychiat & Neurol, cert, 79. *Prof Exp:* Instr physiol, Sch Med, Univ Va, 48, intern med, Univ Hosp, 50-51; resident, Univ Utah, 51-52, asst pharm, 52-53; asst prof, Univ Va Hosp, 53-61, assoc prof pharmacol, Sch Med, 61-74, resident psychiat, Univ Hosp, 70-74; med dir & staff psychiatrist, Mid Peninsula North Neck Ment Health Serv, 74-77; chmn psychiat, Rappahanock Gen Hosp, 77-86; RETIRED. *Mem:* Am Psychiat Asn; AMA. *Res:* Physiology; traumatic shock; psychiatry; ethology. *Mailing Add:* PO Box 26 Wilcomico Church VA 22579

BRAND, JERRY JAY, b Waterloo, Ind, Sept 20, 41; m 67; c 2. PHOTOSYNTHESIS, PHOTOBIOLOGY. *Educ:* Manchester Col, BS, 63; Purdue Univ, PhD(biol sci), 71. *Prof Exp:* Sec sch teacher chem & physics, Nigeria, 63-65; res assoc bot, Ind Univ, 71-74; asst prof, 74-79, ASSOC PROF BOT, UNIV TEX, AUSTIN, 79- *Concurrent Pos:* Vis sci, Carnegie Inst, Stanford, Calif, 82. *Mem:* Sigma Xi; Am Soc Plant Physiologists; Am Chem Soc. *Res:* Mechanism of the light reactions of photosynthesis, particularly in blue-green algae; regulation of photosystem II activity; role of inorganic ions and carotenoids in photosynthetic membranes. *Mailing Add:* Dept Bot Univ Tex Austin TX 78712

BRAND, JOHN S, b Buffalo, NY, May 2, 38; m 60; c 4. BIOPHYSICS. *Educ:* LeMoyne Col, BS, 59; Univ Rochester, MS, 64, PhD, 66. *Prof Exp:* Fel, Univ Rochester, 66-68, asst prof, 68-75, assoc prof radiation biol & biophys, 75-, assoc prof orthop & radiol, 79-, dir musculoskeletal res, 79-; UNIT DIR, EASTMAN KODAK. *Concurrent Pos:* NIH career develop award, 75-80; vis scientist Nat Inst Med Res, Mill Hill, London, Eng, 77-78; mem B Study Sect, NIH Gen Med, 79- *Mem:* AAAS; Sigma Xi; Orthop Res Soc. *Res:* Hormonal regulation of bone cell metabolism. *Mailing Add:* Analytical Technol Div 2nd Floor Bldg 49 Eastman Kodak Kodak Park Rochester NY 14650

BRAND, KARL GERHARD, b Lübeck, Ger, June 10, 22, US citizen; m 49; c 2. MICROBIOLOGY. *Educ:* Univ Hamburg, MD, 49; Trop Inst Hamburg, dipl, 54; Free Univ Berlin, DPH, 56. *Prof Exp:* Resident instr internal med, Med Acad Lübeck, 49-52; asst prof trop med, Trop Inst Hamburg, 52-54; assoc prof pub health & microbiol, Free Univ Berlin, 55-57; from asst prof to prof microbiol, Sch Med, Univ Minn, Minneapolis, 57-85; PHYSICIAN, WGER, 85- *Concurrent Pos:* Vis prof, Univ de Los Andes Mérida Venezuela, 60 & Kawasaki Med Sch, 81. *Mem:* Am Asn Immunol; Soc Exp Biol & Med; Am Soc Microbiol. *Res:* Serological problems in influenza, mumps, psittacosis, viral pneumonia; foreign body carcinogenesis; antigenic structure of cells, tissues, tumors; cancer-aging relationships; asbestos cancer; schistosomiasis cancer. *Mailing Add:* 561 Bay Rd Durham NH 03824

BRAND, LEONARD, b New York, NY, Dec 21, 23; m 51; c 3. MEDICINE, ANESTHESIOLOGY. *Educ:* Yale Univ, BS, 46; Columbia Univ, MD, 49; Am Bd Anesthesiol, dipl, 59. *Prof Exp:* Intern med, Long Island Col Hosp, 49-50; resident gen med, Leo N Levi Mem Hosp, Ark, 50-51; resident anesthesiol, Presby Hosp, 53-55; from instr to assoc prof, 55-72, PROF CLIN ANESTHESIOL, COLUMBIA UNIV, 72- *Concurrent Pos:* Asst anesthesiologist, Presby Hosp, New York, 55-57, asst attend anesthesiologist, 57-66, attend anesthesiologist, 66-, dir, Pain Treat Serv. *Mem:* Am Soc Anesthesiol. *Res:* Pharmacology and physiology of intravenous agents; fate and distribution of barbiturates in man; passage of substances into central nervous system; respiratory physiology; anesthesia for trauma and orthopedic cases; treatment of acute and chronic pain. *Mailing Add:* Presby Hosp 622 W 168th St New York NY 10032

BRAND, LEONARD ROY, b Harvey, NDak, May 17, 41; m 74; c 2. VERTEBRATE ZOOLOGY, PALEONTOLOGY. *Educ:* La Sierra Col, BA, 64; Loma Linda Univ, MA, 66; Cornell Univ, PhD(vertebrate zool), 70. *Prof Exp:* From asst prof to assoc prof, 69-78, chmn dept, 71-86, PROF BIOL & PALEONTOLOGY, LOMA LINDA UNIV, 78-, CHMN DEPT, 88- *Honors & Awards:* A Brazier Howell Award, Am Soc Mammalogists, 67. *Mem:* Sigma Xi; Am Soc Mammalogists; Animal Behav Soc; Geol Soc Am; Soc Vert Paleont. *Res:* Vertebrate typhonomy and fossil trackways; behavior and ecology of mammals. *Mailing Add:* Dept Natural Sci Loma Linda Univ Loma Linda CA 92350

BRAND, LUDWIG, b Vienna, Austria, Jan 3, 32; nat US; m 58. BIOCHEMISTRY. *Educ:* Harvard Univ, BA, 55; Ind Univ, PhD, 60. *Prof Exp:* Asst biochem, Ind Univ, 55-59; Nat Found fel, 59-62; from asst prof to assoc prof, 62-74, PROF BIOL, JOHNS HOPKINS UNIV, 74- *Concurrent Pos:* Fel, Brandeis Univ, 59-61; fel, Weizmann Inst, 61-62. *Mem:* Am Soc Biol Chem; Biophys Soc; Am Soc Photobiol; Am Chem Soc. *Res:* Mechanism of enzyme action; fluorescence studies with proteins; fluorescence lifetimes; fluorescence probe studies of biological membranes. *Mailing Add:* Dept Biol Mudd Hall Johns Hopkins Univ Baltimore MD 21218

BRAND, PAUL HYMAN, b Yonkers, NY, June 19, 40; m 71; c 2. PHYSIOLOGY. *Educ:* City Col New York, BS, 62; Univ Rochester, MSc & PhD(physiol), 72. *Prof Exp:* Fel dept physiol, Univ Rochester, 72-73; fel, Col Med, Univ Ariz, 73-75; asst prof, 75-80, ASSOC PROF PHYSIOL, MED COL OHIO, 80- *Mem:* Am Physiol Soc; Sigma Xi. *Res:* Mechanism of transport of metabolites in the kidney; epithelial transport. *Mailing Add:* CS 10008 Dept Physiol Med Col Ohio Toledo OH 43699

BRAND, RAYMOND HOWARD, b Highland Park, Mich, Sept 22, 28; m 51; c 3. ANIMAL ECOLOGY. *Educ:* Wheaton Col, Ill, BA, 50; Univ Mich, MS, 51, PhD(zool), 55. *Prof Exp:* Asst prof biol, Westmont Col, 55-57, actg chmn div sci, 57-59; assoc prof biol, 59-70, chmn dept, 73-80, PROF BIOL, WHEATON COL, ILL, 73-; RES ASSOC, MORTON ARBORETOM, LISLE, ILL, 88- *Concurrent Pos:* NSF basic res found grant, 62-64; grant, Inst Radiolbiol, Argonne Nat Lab, 65; consult ecol & environ mgt. *Mem:* Ecol Soc Am; Am Sci Affiliation; Sigma Xi; Am Inst Biol Sci; AAAS. *Res:* Population dynamics; animal behavior; temperature adaptations; diversity of species of insects in the order Collembola (springtails) in mid-western native and restored prairies; effects of fire on surface arthropods for different fire management programs. *Mailing Add:* Dept Biol Wheaton Col Wheaton IL 60187

BRAND, RONALD S(COTT), b Springfield, Mass, Jan 31, 19; m 41; c 3. MECHANICAL ENGINEERING. *Educ:* Worcester Polytech Inst, BS, 40; Univ Mich, MS, 49; Brown Univ, PhD(appl math), 60. *Prof Exp:* Instr mech, Rensselaer Polytech Inst, 41-44; from instr to assoc prof mech eng, Univ Conn, 46-57, prof, 57-79, dept head, 68-74; RETIRED. *Concurrent Pos:* NSF res fel, Norweg Inst Technol, 62-64; vis prof, Inst Sound & Vibration Res, Univ Southampton, Eng, 71-72. *Mem:* AAAS; Am Soc Mech Engrs; Sigma Xi. *Res:* Fluid mechanics. *Mailing Add:* 20 Church Rd Box 236 Eastford CT 06242

BRAND, SAMSON, b New York, NY, Apr 30, 43; m 69; c 2. METEOROLOGY, ATMOSPHERIC SCIENCE. *Educ:* City Col New York, BS, 64; Fla State Univ, MS, 66. *Prof Exp:* Navy weather res facil, 66-71, res meteorologist, Naval Environ Prediction Res Facil, 71-89; RES METEOROLOGIST, NAVAL OCANOG & ATMOSPHERIC RES LAB, 89- *Mem:* Am Meteorol Soc; Nat Weather Asn. *Res:* Applied meteorological research with primary emphasis in tropical cyclone and tropical meteorological forecasting and decision making aids; published over 90 articles, mostly technical in nature. *Mailing Add:* Naval Oceanog & Atmospheric Res Lab Atmospheric Dir Monterey CA 93943-5006

BRAND, WILLIAM WAYNE, b Garrett, Ind, Mar 9, 44; m 81; c 5. ORGANIC CHEMISTRY, CHEMOMETRICS. *Educ:* Manchester Col, BA, 66; Purdue Univ, PhD(org chem), 70. *Prof Exp:* Res chemist synthesis-agr, Am Cyanamid Co, 70-76; sr res chemist, Diamon Shamrock Corp, 76-78, res assoc agr, 78-83; res assoc, corp res, 83-87, SUPVR, TESTING SERV, ICI POLYURETHANES, ICI AM INC, 87- *Mem:* Am Chem Soc; Ctr Process Anal Chem. *Res:* Synthetic organic chemistry, particularly in sulfur chemistry and small ring heterocycles; pesticide chemistry. *Mailing Add:* ICI Polyurethanes Mantua Grove Rd West Deptford NJ 08066

BRANDALEONE, HAROLD, b New York, NY, Apr 27, 07; m 47; c 3. MEDICINE, CARDIOLOGY. *Educ:* NY Univ, BS, 28, MD, 31, ScD(med), 38; Am Bd Internal Med, dipl. *Prof Exp:* Intern, Third Med Div, Bellevue Hosp, 31-33; asst prof, 49-57, ASSOC PROF CLIN MED, NY UNIV, 57- *Concurrent Pos:* Assoc chief metab clin, NY Univ Hosp, 33-, asst attend physician, 51-57, assoc attend physician, 57-71, attend physician, 71-; assoc vis physician, Bellevue Hosp, 57-66, vis physician, 66-; med dir, Continental Tel Co. *Mem:* Fel Am Col Physicians; fel Am Col Cardiol; Am Diabetes Asn; AMA; Endocrine Soc. *Res:* Metabolic studies and accident prevention. *Mailing Add:* 815 Park Ave New York NY 10021

BRANDAU, BETTY LEE, b Easton, Pa; m 62; c 1. INORGANIC CHEMISTRY, NUCLEAR CHEMISTRY. *Educ:* Ursinus Col, BS, 53; Carnegie Inst Technol, MS, 55, PhD(inorg chem), 60. *Prof Exp:* Chemist, Sun Oil Co, Pa, 54, Dow Chem Co, Mich, 56, Y-12 Plant, Union Carbide Nuclear Corp, Tenn, 60-64 & Oak Ridge Assoc Univs, 64-65 & 69; res assoc, Rosenstiel Inst Marine & Atmospheric Sci, Univ Miami, 65-68; asst dir, Appl Isotope Studies, Univ Ga, 69-71, assoc dir, 78-81; CONSULT, 81- *Concurrent Pos:* Intergovernmental Personnel Act, Nat Oceanic & Atmospheric Admin, Rockville, Md, 81-82. *Mem:* AAAS; Am Chem Soc; fel Am Inst Chemists; Sigma Xi. *Res:* Marine sediment and radiocarbon dating; geochemistry; natural radioactivity environmental problems; chemical archaeology; marine pollution. *Mailing Add:* 220 Pine Crest Athens GA 30605

BRANDAUER, CARL M(ARTIN), b Peking, China, Oct 23, 24; nat US; m 48, 69; c 4. COMMUNICATIONS SYSTEMS. *Educ:* Columbia Univ, BA, 51, MA, 52, PhD(psychol), 58. *Prof Exp:* Instr psychol, Columbia Univ, 54-55; mem tech staff, Bell Tel Labs, Inc, 55-58; res psychologist, Haskins Labs, 58-60; mem tech staff, Bell Tel Labs, Inc, NJ, 60-66; scientist, Nat Ctr Atmospheric Res, 66-70; MEM TECH STAFF, BELL TEL LABS, 70-; RES ASSOC, UNIV COLO, 77- *Mem:* Acoust Soc Am. *Res:* Speech communications; systems analysis; auditory localization. *Mailing Add:* 1760 Sunset Blvd Boulder CO 80302

BRANDEAU, MARGARET LOUISE, b New York, NY, Aug 28, 55; m 88; c 1. MANUFACTURING SYSTEMS RESEARCH, MANAGEMENT SCIENCE APPLICATIONS. *Educ:* Mass Inst Technol, BS, 77, MS, 78; Stanford Univ, PhD(eng- econ systs), 85. *Prof Exp:* ASST PROF INDUST ENG, STANFORD UNIV, 85- *Concurrent Pos:* NSF presidential young investr award, 88. *Mem:* Sigma Xi; Opers Res Soc Am; Inst Mgt Sci. *Res:* Manufacturing systems design models; policy analysis of Acquired Immune Deficiency syndrome screening and intervention; facility location models. *Mailing Add:* Indust Eng Dept Stanford Univ Stanford CA 94305-4024

BRANDEBERRY, JAMES E, b Toledo, Ohio, Oct 10, 39; m 62; c 2. ELECTRICAL ENGINEERING, SYSTEMS THEORY. *Educ:* Univ Toledo, BSEE, 61, MSEE, 63; Marquette Univ, PhD(elec eng), 69. *Prof Exp:* Sr proj engr, AC Electronics Div, Gen Motors Corp, 63-66; asst prof eng, Wright State Univ, 69-76, assoc prof comput sci & eng & chmn, Dept Comput Sci, 76-80; PRES, SIMULATION TECHNOL, 80- *Concurrent Pos:* Consult, Nat Cash Register Co, 69-70. *Mem:* Inst Elec & Electronics Engrs; Soc Indust & Appl Math; Am Soc Eng Educ. *Res:* Stability and optimal control of stochastic systems; modeling of man-machine systems. *Mailing Add:* 5118 Jackson Rd Enon OH 45323

BRANDELL, BRUCE REEVES, b Detroit, Mich, Oct 4, 26; m 52; c 5. ANATOMY, ZOOLOGY. *Educ:* Univ Mich, BS, 49, MS, 50, PhD(zool), 63. *Prof Exp:* Teaching fel zool, Univ Mich, 54-57; instr biol, Univ Akron, 57-61; asst prof anat, Sch Med, Univ NDak, 62-65; from asst prof to assoc prof, 65-75, PROF ANAT, UNIV SASK, 75- *Concurrent Pos:* USPHS grant, 65-66; Dept Nat Health & Welfare, Can grants, 68-70, Can Med Res Council grants, 72-81 & Sask Med Res Bd grant, 81-82. *Mem:* Am Soc Mammal; Am Asn Anatomists; Can Asn Anat; Int Soc Electromyography & Kinesiology; Can Med & Biol Eng Soc; Am Asn Clin Anatomists. *Res:* Human gross anatomy; electromyographic and anatomical study of the finger moving muscles; comparative anatomy of nerve distribution patterns in mammalian forearm and hand muscles; investigations of locomotion by means of electromyography and cinematography and functional electrical simulation. *Mailing Add:* Dept Anat Univ Sask Saskatoon SK S7N 0W0 Can

BRANDENBERGER, ARTHUR J, b St Gallen, Switz, July 2, 16; m 45; c 2. PHOTOGRAMMETRY, GEODESY & CARTOGRAPHY. *Educ:* Swiss Fed Inst Technol, MS, 40, PhD(photogram), 47. *Prof Exp:* Photogram eng, Switz, 40-44; res assoc & lectr photogram, Swiss Fed Inst Technol, 45-49; expert, Ministry Pub Works, Greece, 49-50; prof photogram, Eng Sch Istanbul, Turkey, 50-54; chief engr, City of Istanbul, 53; vis prof & res assoc photogram, Ohio State Univ, 55-56, assoc prof & chief res assoc, 57-59, prof & res supvr, 59-65; res prof, 75-76, DIR, DEPT PHOTOGRAM, LAVAL UNIV, 65-, PRES ASSEMBLEE OF PROF, GEODESY SECT, FAC FORESTRY & GEOMATICS, 83- *Concurrent Pos:* Mem archaeol exped to Mt Ararat, Turkey, 60; mem adv comt hwy design aids, Hwy Res Coun Ohio, 61-63; leader, Byrd Sta Traverse, Antarctica Exped, 62-63; consult, UN, US Air Force & other orgn, 63-64, world cartog, UN, 75-; hon vis prof, Nat Sci Coun, Repub of China, 69-70; vpres, comt thematic cartog, Panam Inst Geogr & Hist, OSA, 77-81; expert consult world air photo index, Food & Agr Orgn, 80-; invited scientist, Chinese Soc Geod, Photogram & Cartog, 81. *Mem:* Am Soc Photogram & Remote Sensing; hon mem Am Inst Navig; Am Geog Soc; Am Polar Soc; hon mem Indian Soc Photogram; French Soc Photogram & Remote Sensing; hon mem Brazilian Soc Cartog; Can Inst Surv; Am Cong Surv & Mapping. *Res:* Geodesy; cartography; navigation; photogrammetry space science; polar research; archaeology; glaciology; world cartography; space triangulation from satellite imagery. *Mailing Add:* 2590 Plaza No 611 Sillery Quebec PQ G1T 1X2 Can

BRANDENBERGER, JOHN RUSSELL, b Danville, Ill, May 13, 39; m 64; c 2. ATOMIC PHYSICS, LASER SPECTROSCOPY. *Educ:* Carleton Col, AB, 61; Brown Univ, ScM, 64, PhD(physics), 68. *Prof Exp:* Instr physics, Col Wooster, 64-66; from asst prof to assoc prof, 68-83, chmn dept, 73-75, PROF PHYSICS & CHMN DEPT, LAWRENCE UNIV, 83- *Concurrent Pos:* Res fel, Harvard Univ, 75-76; consult, Los Alamos Sci Lab, Univ Calif, 75-78; vis prof, Univ Reading, 81; vis sr scientist, Oxford Univ, 82; counr, Coun Undergrad Res; Fulbright scholar, 90-91. *Mem:* Am Phys Soc. *Res:* Measurement of Sommerfeld fine structure constant by level-crossing in atomic hydrogen; fast atomic beam spectroscopy; time-resolved fluorescence spectroscopy; high resolution laser spectroscopy. *Mailing Add:* Dept Physics Lawrence Univ Appleton WI 54911

BRANDENBERGER, ROBERT H, b Bern, Switz, 1954; US citizen. COSMOLOGY, PARTICLE PHYSICS. *Educ:* Swiss Fed Inst Technol, Zurich, dipl physics, 78; Harvard Univ, MA, 79, PhD(physics), 83. *Prof Exp:* Postdoctoral fel physics, Inst Theoret Physics, Univ Calif, Santa Barbara, 83-85 & Dept Appl Math & Theoret Physics, Univ Cambridge, 85-87; asst prof, 86-91, ASSOC PROF PHYSICS, BROWN UNIV, 91- *Concurrent Pos:* US Dept Energy outstanding jr investr, 88; A P Sloan fel, 88-92. *Res:* Using particle physics and field theory to develop models for the origin of galaxies and other structures in the universe; cosmic string models and inflationary universe scenarios. *Mailing Add:* Physics Dept Brown Univ Providence RI 02912

BRANDENBERGER, STANLEY GEORGE, b Houston, Tex, Jan 18, 30; m 67; c 2. ORGANIC CHEMSTRY. *Educ:* Rice Univ, BA, 52; Univ Tex, PhD(org chem), 56. *Prof Exp:* Res chemist, Houston Res Lab, Shell Oil Co, 56-63, sr res chemist, 63-64, group leader, 64-68, sect head, Royal Dutch/Shell Lab, Holland, 68-69, staff res chemist, Houston Res Lab, Deerpark, 69-72, STAFF RES CHEMIST, SHELL DEVELOP CO, 72- *Mem:* Am Chem Soc; Southwest Catalysis Soc. *Res:* Physical organic chemistry; heterogeneous catalysis. *Mailing Add:* 5726 Kuldell Houston TX 77096

BRANDENBURG, JAMES H, b Green Bay, Wis, July 17, 30; m 54; c 4. OTOLARYNGOLOGY. *Educ:* Univ Wis, BA, 52, MS, 56; Am Bd Otolaryngol, dipl, 63. *Prof Exp:* Intern, William Beaumont Gen Hosp, El Paso, Tex, 57, ear, nose & throat preceptorship, 58; ear, nose & throat resident, Brooke Gen Hosp, San Antonio, 61; from asst prof to assoc prof, 64-72, PROF & CHMN OTOLARYNGOL, MED SCH, UNIV WIS-MADISON, 72- *Concurrent Pos:* Attend consult otolaryngol, Vet Admin Hosp, Madison, 64-; mem fac, Home Study Course, Am Acad Otolaryngol & Ophthal, 67-; prin investr prototype comprehensive network demonstration proj head & neck cancer, NIH Contract, 74- *Honors & Awards:* Cert of Achievement, Armed Forces Inst Path, Walter Reed Hosp, 64. *Mem:* Fel Am Acad Ophthal & Otolaryngol; fel Am Laryngol, Rhinol & Otol Soc; Am Coun Otolaryngol; Am Soc Head & Neck Surg; AMA. *Res:* Combination therapy for epidermoid carcinoma of the head and neck; carcinoma of the larynx; study of the guinea pig cochlea by electron microscopy; study of the changes seen in the cochlea of the guinea pig following noise exposure; traumatic injuries to the larynx. *Mailing Add:* Oto HNS Surg Clin Univ Wis 600 Highland Ave Madison WI 53792

BRANDENBURG, ROBERT O, b Minneapolis, Minn, Aug 5, 18; m 44; c 5. MEDICINE. *Educ:* Univ NDak, BS, 40; Univ Pa, MD, 43. *Prof Exp:* Intern, Presby Hosp, Pa, 44; intern, Nutrit Clin, Hillman Hosp, Ala, 47-48; fel, Mayo Clin, 48-51, consult med, 51-84, from asst prof to assoc prof med, 58-69, coordr continuing med educ, Outreach Prog, 76-84, PROF MED, MAYO MED SCH, 69-; CLIN PROF MED, UNIV ARIZ, 85- *Concurrent Pos:*

Chmn div cardiovasc dis, Mayo Clin, 69-74. *Mem:* AMA; Am Heart Asn; fel Am Col Physicians; fel Am Col Cardiol (vpres, 78); Am Asn Univ Prof. *Res:* Congenital and rheumatic cardiovascular disease. *Mailing Add:* 701 Mission Twin Butts Green Valley AZ 85614

BRANDEWIE, RICHARD A(NTHONY), b Sidney, Ohio, Feb 4, 36; m 59; c 2. ELECTRICAL ENGINEERING. *Educ:* Univ Detroit, BEE, 59; Carnegie Inst Technol, MS, 60, PhD(elec eng), 63. *Prof Exp:* Res asst elec eng, Carnegie Inst Technol, 59-60; res specialist, Autonetics Div, NAm Aviation, Inc, 63-69, supvr laser dept, NAm Rockwell Corp, 69-80, DIR ADV TECH, ROCKETDYNE DIV, ROCKWELL INT, 80- *Mem:* Am Phys Soc; Inst Elec & Electronics Engrs. *Res:* Solid state lasers; laser radar; coherent laser receivers; spectroscopy; plasma physics. *Mailing Add:* 6633 Canoga Ave Canoga Park CA 92803

BRANDFONBRENER, MARTIN, b New York, NY, July 25, 27; m 56; c 4. INTERNAL MEDICINE, CARDIOLOGY. *Educ:* Albany Med Col, MD, 49. *Prof Exp:* Rotating intern, Med Ctr, Univ Ind, 49-50; resident internal med, Boston City Hosp, 50-51; resident, Dept Physiol, Western Reserve Univ, 51-52; res assoc cardiol, Sect Geront, Nat Heart Inst, 53-54; asst resident med, Univ Hosp, Univ Cleveland, 54-55; asst resident cardiol, Presby Hosp, New York, 55-56; fel cardiorespiratory med, Col Physicians & Surgeons, Columbia Univ, 56-58; chief cardiol sect, Vet Admin Res Hosp, Chicago, 58-63; assoc prof med, Univ NMex, 63-67; PROF MED, MED SCH, NORTHWESTERN UNIV, CHICAGO, 67- *Concurrent Pos:* USPHS fel, Sch Med, Western Reserve Univ, 51-52. *Mem:* Am Fedn Clin Res (secy, 62); fel Am Col Physicians; Am Physiol Soc. *Res:* Myocardial metabolism and coronary flow. *Mailing Add:* 707 N Fairbanks Ct Chicago IL 60611

BRANDHORST, HENRY WILLIAM, JR, b Omaha, Nebr, Sept 18, 36; m 56. PERFORMANCE MEASUREMENTS. *Educ:* Univ Okla, BS, 57; Purdue Univ, PhD(chem), 61. *Prof Exp:* Researcher, 61-67, head, Solar Cell Res Sect, 67-79, chief, Space Photovoltaic Br, 79-84, CHIEF, POWER TECHNOL DIV, LEWIS RES CTR, NASA, 84- *Honors & Awards:* William R Cherry Award, Inst Elec & Electronic Engrs, 84; Exceptional Eng Achievement Award, NASA, 84. *Mem:* Sigma Xi. *Res:* Space power technology: photovoltaics, electrochemical storage, power management and distribution, spacecraft environmental interactions; solar nuclear dynamic systems: concentrators, thermal storage, free piston stirling engines, brayton and rankine cycles, advanced radiators. *Mailing Add:* 272 Westbridge Dr Berea OH 44017

BRANDINGER, JAY J, b Bronx, NY, Jan 2, 27; m 51; c 3. ELECTRICAL ENGINEERING, MATHEMATICAL STATISTICS. *Educ:* Cooper Union, BEE, 51; Rutgers Univ, MSEE, 62, PhD(elec eng), 68. *Prof Exp:* Mem tech staff commun res, Res Lab, 51-60, group leader TV systs, Consumer Electronics Res Lab, 60-74, vpres TV eng, Consumer Electronics Div, 74-79, staff vpres & systs engr, RCA Electronic Prods & Labs, 84-85, DIV VPRES, SELECTAVISION VIDEODISC OPER, RCA CORP, 79-, GEN MGR, 84- *Honors & Awards:* David Sarnoff Achievement Award, RCA-David Sarnoff Res Ctr, 52, 70 & 73. *Mem:* Sr mem Inst Elec & Electronics Engrs; NY Acad Sci; Soc Info Display; Inst Math Statist; Sigma Xi. *Res:* Consumer TV systems; statistical communications theory and application in coding, modulation, networks and detection; optical communications; electro-optic techniques; optical information processing; display techniques and systems; video discs. *Mailing Add:* 2 Queens Lane Pennington NJ 08534

BRANDLER, PHILIP, b New York, NY, Feb 2, 43; m 85. FOOD STABILIZATION, FOOD SERVICE EQUIPMENT ENGINEERING. *Educ:* Columbia Col, BA, 64; Brown Univ, MS, 69; Northeastern Univ, MS, 71, MBA, 73. *Prof Exp:* Physicist, NASA Electronics Res Ctr, 67-70; syst analyst, US Army Natick Res & Develop Ctr, 71-73, opers res analyst, 73-75, prin investr, 75-77, supvry opers res analyst, 77-81, dir, Systs Anal, 82-85, asst tech dir prog integration, 85-88, DIR FOOD ENG, US ARMY NATICK RES & DEVELOP CTR, 88- *Honors & Awards:* Rohland Isker Award, Res & Develop Assocs for Food & Packaging Indust, 82. *Mem:* Inst Food Technologists; Mil Opers Res Soc; Opers Res Soc Am. *Res:* Research, development and engineering in military rations, subsistence and related food items ranging from basic food chemistry to product development and testing; food service equipment and systems ranging from basic combustion technology to development and testing of complete mobile kitchens and bakeries. *Mailing Add:* US Army Natick Res & Develop Ctr Code STRNC-W Natick MA 01760

BRANDLI, HENRY WILLIAM, b Boston, Mass, Nov 5, 37; m 59; c 4. METEOROLOGY. *Educ:* Tufts Univ, BSME, 59; Mass Inst Technol, MS(meteorol) & MS(aeronaut & astronaut), 65. *Prof Exp:* Satellite meteorologist, Environ Res Technol, 76-78; CHIEF SCIENTIST, FLA WEATHER SERV, 78- *Concurrent Pos:* Consult, Lockheed Harris. *Mem:* Fel Am Meteorol Soc; Nat Weather Asn; Cousteau Soc. *Res:* Satellite meteorology; remote sensing. *Mailing Add:* 3165 Sharon Dr Melbourne FL 32904

BRANDMAIER, HAROLD EDMUND, b New York, NY, Nov 9, 26; m 51; c 4. MECHANICAL ENGINEERING. *Educ:* Columbia Univ, BS, 47, MS, 48, ME, 62. *Prof Exp:* Asst mech eng, Columbia Univ, 47-48; proj engr, Worthington Corp, 48-54; chief proj engr, Curtiss-Wright Corp, 56-71; proj engr, RCA Astroelectronics, 71-73; MGR ENG SYSTS, BURNS & ROE, INC, 73- *Mem:* Am Inst Aeronaut & Astronaut; Am Soc Mech Engrs. *Res:* Fluid mechanics; thermodynamics; energy conversion; composite materials. *Mailing Add:* 63 Florence Rd Harrington Park NJ 07640

BRANDMAN, HAROLD A, b Newark, NJ, Jan 29, 41; m 64; c 2. ORGANIC CHEMISTRY. *Educ:* Univ Pa, BA, 62; Seton Hall Univ, MS, 66, PhD(chem), 68. *Prof Exp:* Chemist, 68-74, group leader, 75-78, prod chem, 78-80, dir fragrance mfg, 80-82, dir flavor, fragrance mfg, 82-84, DIR CHEM FLAVOR FRAGRANCE MFG, GIVAUDAN CORP, 85- *Mem:* Am Chem Soc; Sigma Xi. *Res:* Synthesis and development of antimicrobial agents used as industrial biocides. *Mailing Add:* Seven Adams Place Glen Ridge NJ 07028

BRANDNER, JOHN DAVID, b Bethlehem, Pa, Mar 23, 10; m 43; c 3. INDUSTRIAL CHEMISTRY. *Educ:* Lehigh Univ, BS, 32, MS, 34; Columbia Univ, PhD(chem), 43. *Prof Exp:* Chemist, Exp Lab, Atlas Powder Co, 34-37; asst, Columbia Univ, 38-40; asst dir control lab, Ravenna Ord Works, 41-42; chemist, Atlas Chem Industs, 42-44, dir res lab, 44-47, asst dir cent res lab, 48-51, mgr, 51-55, mgr phys chem & anal sect, Res Dept, 55-61, assoc dir, Chem Res Dept, 61-70, dir, Chem Res Dept, 70-75; RETIRED. *Concurrent Pos:* Mem comt specifications, Food Chem Codex Proj, Nat Res Coun & Food & Agr Orgn-WHO Expert Comt Food Additives. *Mem:* AAAS; Am Chem Soc; fel Am Inst Chem. *Res:* Calorimetry; kinetics; alkylene oxide reactions; food additives. *Mailing Add:* 2313 E Mall Wilmington DE 19810

BRANDOM, WILLIAM FRANKLIN, b Oklahoma City, Okla, Jan 2, 26; m 52; c 4. DEVELOPMENTAL BIOLOGY, CYTOGENETICS. *Educ:* Stanford Univ, AB, 51, MA, 54, PhD(biol), 59. *Prof Exp:* Asst biol, Stanford Univ, 53-58; Nat Cancer Inst fel, Princeton Univ, 58-59; from instr to asst prof zool, Newcomb Col, Tulane Univ, 59-65, asst prof path, Sch Med, 65-67; asst prof, Delta Primate Res Ctr, 67-68; assoc prof biol sci, 68-74, PROF BIOL SCI, UNIV DENVER, 74- *Concurrent Pos:* Nat Inst Child Health & Human Develop fel, Univ St Andrews, 66. *Mem:* AAAS; Am Soc Zool; Soc Develop Biol; Tissue Cult Asn; Sigma Xi. *Res:* Developmental biology of amphibians; developmental genetics; polyploidy; hybridization; pigment pattern formation in heteroploid hybrids; karyotyptes and idiograms of amphibian chromosomes; lampbrush chromosomes; somatic cell genetics; human radiation cytogenetics; chromosome biology. *Mailing Add:* Dept Biol Sci Univ Denver Denver CO 80210

BRANDON, CLEMENT EDWIN, b Yorkshire, Ohio, Oct 3, 15; m 39; c 2. PULP AND PAPER SCIENCE & TECHNOLOGY. *Educ:* Defiance Col, AB, 36; State Univ NY, Syracuse, MS, 42. *Prof Exp:* Teacher math, Bradford High Sch, Ohio, 36-37; chemist pulp & paper technol, Aetna Paper Co, 37-40; chief chemist, Howard Paper Co, 41-51, asst tech dir, 51-57, tech dir, 57-58; chmn, Dept Pulp & Paper Technol, Miami Univ, 58-81, dir, Coop Prog, Paper Sci & Eng Dept, 81-83; CONSULT, 83- *Concurrent Pos:* Mem steering comt testing div, Tech Asn Pulp & Paper Indust, 54-64, chmn testing div, 58-61; exec dir, Miami Univ Pulp & Paper Found, 60-88; mem exec comt D-6 on paper, Am Soc for Testing & Mat, 60-, chmn, 70-72; chmn US adv comt, Int Orgn Stand Tech Comt 6 on pulp & paper, 63-69, US rep, 65-; mem testing adv bd, Tech Asn Pulp & Paper Indust, 65-80; mem comt P-3 on paper, Am Nat Standards Inst, 69-84. *Honors & Awards:* Testing Div Award of Merit, Tech Asn Pulp & Paper Indust, 68; Merit Award, Am Soc Testing & Mat, 90. *Mem:* Fel Tech Asn Pulp & Paper Indust; Am Soc Testing & Mat; Paper Indust Mgt Asn; Sigma Xi. *Res:* Development of paper testing methods; hygroexpansivity of paper. *Mailing Add:* 121 Oakhill Dr Oxford OH 45056

BRANDON, DAVID LAWRENCE, b New York, NY, Sept 4, 46. IMMUNOLOGY, BIOCHEMISTRY. *Educ:* Harvard Univ, AB, 67, PhD(biochem), 74. *Prof Exp:* Res fel immunol, Nat Inst Med Res, London, UK, 74-76 & Univ Calif, Berkeley, 76-77; RES BIOCHEMIST, WESTERN REGIONAL RES CTR, USDA, 78- *Concurrent Pos:* Nat Multiple Sclerosis Soc fel, 74-76; lectr, Univ Calif, Berkeley, 79. *Mem:* Am Chem Soc; Am Soc Plant Physiologists. *Res:* Immunochemical analysis of foods; regulation of the immune response; cell membrane biochemistry; immunotoxicology; immunological studies of phytohormones and their receptors. *Mailing Add:* USDA West Region Res Ctr 800 Buchanan St Berkeley CA 94710

BRANDON, FRANK BAYARD, b Indiana, Pa, Jan 2, 27; m 48; c 3. BIOPHYSICS, VIROLOGY. *Educ:* Carnegie Inst Technol, BS, 51; Univ Pittsburgh, MS, 54, PhD(biophys), 56. *Prof Exp:* Assoc res microbiologist, Parke, Davis & Co, 56-58, res microbiologist, 58-63, sr res microbiologist, 63-67, sr scientist, 67-86; RETIRED. *Mem:* AAAS; Am Soc Microbiol; Biophys Soc. *Res:* Biophysical and immunologic properties of influenza virus; clinical immunology; biometrics. *Mailing Add:* Watson Blvd Big Pine Key FL 33043

BRANDON, JAMES KENNETH, b Sarnia, Ont, May 21, 40; m 62; c 3. CRYSTALLOGRAPHY, PHYSICS. *Educ:* McMaster Univ, BSc, 62, PhD(physics), 67. *Hon Degrees:* MA, Cambridge Univ, 68. *Prof Exp:* Res asst physics, Cavendish Lab, Cambridge Univ, 68-70; asst prof, 70-76, assoc chmn dept, 78-83, ASSOC PROF PHYSICS, UNIV WATERLOO, 77- *Concurrent Pos:* Res fel, Clare Hall Cambridge Univ & NATO fel, Cavendish Lab, 68-70. *Mem:* Can Asn Physicists; Am Crystallog Asn. *Res:* Inorganic crystal structure analyses using x-ray diffraction; metal and alloy crystal structures. *Mailing Add:* Dept Physics Univ Waterloo Waterloo ON N2L 3G1 Can

BRANDON, RONALD ARTHUR, b Mt Pleasant, Mich, Dec 3, 33; m 54; c 3. VERTEBRATE ZOOLOGY. *Educ:* Ohio Univ, BS, 56, MS, 58; Univ Ill, PhD(zool), 62. *Prof Exp:* Asst prof biol, Univ Ala, 62-63; from asst prof to assoc prof, 63-74, chmn, 80-87, PROF ZOOL, SOUTHERN ILL UNIV, CARBONDALE, 74- *Mem:* AAAS; Soc Study Amphibians & Reptiles; Am Soc Ichthyologists & Herpetologists; Soc Study Evolution; Am Soc Zoologists; Soc Syst Zool. *Res:* Amphibian systematics; biology of salamanders. *Mailing Add:* Dept Zool Southern Ill Univ Carbondale IL 62901

BRANDON, WALTER WILEY, JR, b Gainesville, Ga, Dec 1, 29; m 57; c 3. ROCKET PROPULSION SYSTEMS. *Educ:* Emory Univ, BA, 52, MS, 53. *Prof Exp:* Scientist, Redstone Res Labs, Rohm and Haas Co, 53-65; res specialist, space div, Boeing Co, 65-67; scientist, Redstone Res Labs, Rohm and Haas Co, 67-71; physicist, US Army Missile Command, 72-87; AEROSPACE TECHNOLOGIST PROPULSION SYSTS, MARSHAL SPACE FLIGHT CTR, NASA, 87- *Mem:* Assoc fel Am Inst Aeronaut & Astronaut. *Res:* Nuclear radiation spectroscopy; liquid and solid rocket propulsion; combustion instability of rockets; detonation of rocket propellants and explosives; instrumentation; microwaves; atmospheric optics. *Mailing Add:* 1902 Colice Rd SE Huntsville AL 35801

BRANDOW, BAIRD H, b San Diego, Calif, June 9, 35; m 66; c 2. THEORETICAL PHYSICS. *Educ:* Calif Inst Technol, BS, 57; Cornell Univ, PhD(theoret physics), 64. *Prof Exp:* NSF fel, Niels Bohr Inst, Copenhagen Univ, 63-65; docent theoret nuclear physics, Nordic Inst Theoret Atomic Physics, Denmark, 65-66; instr physics, Cornell Univ, 66-69; fel, Battelle-Seattle Res Ctr, 69-71 & Battelle Mem Inst, Ohio, 71-74; MEM STAFF, LOS ALAMOS NAT LAB, 74- *Mem:* Am Phys Soc. *Res:* Nuclear many-body problem; linked-cluster perturbation theory; theory of liquid and solid helium; theory of magnetic insulators and associated metal-insulator transitions; theory of effective pi-electron hamiltonians; theory of valence fluctuations and heavy-electron behavior in rare earth and actinide compounds; theory of high-temperature superconductors. *Mailing Add:* Los Alamos Nat Lab Group T-11 PO Box 1663 Los Alamos NM 87545

BRANDRETH, DALE ALDEN, b Phoenixville, Pa, Dec 17, 31; m 66; c 2. CHEMICAL ENGINEERING, PHYSICAL CHEMISTRY. *Educ:* Univ Pa, BSc, 53, MSc, 58; Univ Toronto, PhD(chem eng), 65. *Prof Exp:* Tech serv engr plastics processing, Welding Engrs Inc, 55-56; res asst high energy fuels, Inst Coop Res, Univ Pa, 56-57; res engr solid propellant technol, Atlantic Res Corp, 58-61; sr res engr high temperature mat, Westinghouse Astronuclear Lab, 65-66; sr res engr surface chem thermodynamics, E I Du Pont De Nemours & Co, 66-83; ADJ PROF CHEM ENG, WIDENER UNIV, 83- *Mem:* Am Inst Chem Engrs; Am Chem Soc; AAAS. *Res:* Thermodynamics; energy conservation; surface chemistry; chlorofluorocarbons and alternatives. *Mailing Add:* 721 Whitebriar Rd Hockessin DE 19707

BRANDRISS, MARJORIE C, b New York, NY, Apr 10, 49; m 76. GENE EXPRESSION, CELLULAR PHYSIOLOGY. *Educ:* Cornell Univ, AB, 71; Mass Inst Technol, PhD(microbiol), 75. *Prof Exp:* Instr molecular biol, Biol Dept, Mass Inst Technol, 75-76, fel, 76-81; asst prof, Microbiol Dept, 81-87, ASSOC PROF, DEPT MICROBIOL & MOLECULAR GENETICS, NJ MED SCH, UNIV MED & DENT NJ, 87- *Concurrent Pos:* NIH Study Sect Panelist, 88-92. *Mem:* Am Soc Microbiol; Genetics Soc Am; AAAS. *Res:* Regulation of gene expression in the pathways of proline metabolism in the yeast, saccharomyces carevisiae; mitochondrial protein transport. *Mailing Add:* Dept Microbiol & Molecular Genetics NJ Med Sch Univ Med & Dent NJ 185 S Orange Ave Newark NJ 07103-2757

BRANDRISS, MICHAEL W, b Brooklyn, NY, Oct 3, 31; m 55; c 4. IMMUNOLOGY, INFECTIOUS DISEASES. *Educ:* Kenyon Col, AB, 53; NY Univ, MD, 57. *Prof Exp:* Intern med, Johns Hopkins Hosp, 57-58, asst resident, 58-59; asst resident, Baltimore City Hosps, 59-60; clin assoc, Nat Inst Allergy & Infectious Dis, 60-62, clin investr, 62-63, sr investr, 63-65; asst prof, 65-69, ASSOC PROF MED, SCH MED & DENT, UNIV ROCHESTER, 69- *Concurrent Pos:* Nat Inst Allergy & Infectious Dis grants, 66-; consult med, Vet Admin Hosp, Batavia, NY, 65-; attend physician, Rochester Gen Hosp, NY; sr assoc physician, Strong Mem Hosp, Rochester. *Mem:* Am Fedn Clin Res; Am Asn Immunol; Infectious Dis Soc Am. *Res:* Delayed hypersensitivity in animals; antimetabolites in autoimmune states; immunologic studies on penicillin and related antigens; human immune response to haptens and to protein and polysaccharide antigens; human immune response to viral infections. *Mailing Add:* Sch Med & Dent Rochester Gen Hosp Univ Rochester 1425 Portland Ave Rochester NY 14621

BRANDS, ALLEN J, b Kansas City, Mo, Sept 19, 14; m 34; c 1. PHARMACOLOGY. *Educ:* Univ Southern Calif, BS, 41. *Hon Degrees:* DSc, Philadelphia Col Pharm & Sci, 74. *Prof Exp:* Community pharmacist, 41-50; chief, Pharm Br, Indian Health Serv, USPHS, 53-81, chief, Pharm Off, 67-81, asst surg gen, 80-81; RETIRED. *Concurrent Pos:* Mem, Vis Comt, Col Pharm, Wayne State Univ; mem, Pub Health Serv Comt, Nat Asn Retarded Children, 69-72; deleg, US Pharmacopiel Conv, 70, 75 & 80; vis prof, Sch Nursing, Cath Univ Am, 78-79. *Honors & Awards:* Andrew Craigie Award, Asn Mil Surgeons US, 73; Harvey A K Whitney Award, Am Soc Hosp Pharmacists, 78. *Mem:* Fel Am Col Clin Pharmacists; Fedn Int Pharmacists; Am Pharmaceut Asn; Am Soc Hosp Pharmacists. *Res:* Clinical pharmacy; rational drug therapy. *Mailing Add:* 3024 Tilden St NW Washington DC 20008

BRANDS, ALVIRA BERNICE, b Hader, Minn, July 9, 22; m 71; c 2. PSYCHIATRIC NURSING. *Educ:* Univ Minn, BS, 58, MNurse Admin, 60. *Hon Degrees:* DSc, Cath Univ Am, 75. *Prof Exp:* From head nurse to nursing instr, Anoka State Hosp, Minn, 54-63; chief nursing serv, Div Med Serv, Minn Dept Pub Welfare, St Paul, 63-69; psychiat nurse consult, 70-74, prog analyst, 74-79, CHIEF, FACIL & QUAL SECT, MENT HEALTH CARE & SERV FINANCING BR, NIMH, ROCKVILLE, MD, 79- *Concurrent Pos:* Psychiat nurse consult, Jamestown State Hosp, NDak, 67, Woodward State Hosp & Sch, Iowa, 68 & Sch Pharm, Univ Southern Calif, 71; mem, USPHS task force nurse recruit; bd mem, DC League Nursing. *Mem:* Am Col Clin Pharmacol; Am Nurses Asn; Am Pub Health Asn; Am Asn Univ Women; Nat League Nursing; Asn Mil Surg US. *Res:* Drug use for long term nursing home. *Mailing Add:* 3024 Tilden St NW Washington DC 20008

BRANDSTETTER, ALBIN, b Schildorn, Austria, Mar 20, 32; US citizen; m 56; c 2. HYDROLOGY, NUCLEAR WASTE ISOLATION. *Educ:* Col Agr Vienna, Dipl Ing, 58; Univ Calif, Davis, MS, 66, PhD(hydrol), 71. *Prof Exp:* Bridge design engr, Ore State Hwy Dept, 58-59; invests engr, Ore State Water Resources Bd, 59-62; hydrol res engr, Univ Calif, Davis, 65-70; sr res scientist & proj mgr, Pac Northwest Labs, Battelle Mem Inst, 70-80, proj mgr, 80-90; SR HYDROLOGIST, SCI APPLICATIONS INT CORP, 90- *Mem:* Am Geophys Union; Nat Water Well Asn; Am Nuclear Soc; Sigma Xi. *Res:* Hydrologic research, including the mathematical modeling of precipitation-runoff relationships, ground and surface water flow and quality; environmental planning and assessments; mathematical modeling and safety assessments of nuclear waste repositories. *Mailing Add:* 350 Humboldt Dr N Henderson NV 89014-1300

BRANDT, BRUCE LOSURE, b Seattle, Wash, Nov 3, 41; m 63; c 2. PHYSICS, INSTRUMENTATION. *Educ:* Pomona Col, BA, 63; Univ Rochester, PhD, 69. *Prof Exp:* Res assoc neurobiol, Salk Inst Biol Studies, 73-76; staff scientist, Neurosci Res Prog, 76-77, ASSOC HEAD, INSTRUMENTATION & OPERS GROUP, F BITTER NAT MAGNET LAB, 77- *Concurrent Pos:* NIH fel genetics, Sch Med, Stanford Univ; Nat Multiple Sclerosis Soc fel. *Mem:* Am Phys Soc. *Res:* Properties of thin film superconductors; molecular genetics of bacteria and cultured cells; neuromuscular junction formation; electromagnetic measurements in high magnetic fields. *Mailing Add:* F Bitter Nat Magnet Lab 150 Albany St MIT NW14-1313 Cambridge MA 02139

BRANDT, CARL DAVID, b Bridgeport, Conn, Jan 19, 28; m 64; c 2. VIROLOGY. *Educ:* Univ Conn, BS, 49; Univ Mass, MS, 51; Harvard Univ, PhD(bact), 58. *Prof Exp:* Instr vet sci, Univ Mass, 49-52 & 54; res virologist, Charles Pfizer & Co, 58-62; assoc, Dept Epidemiol, Pub Health Res Inst, New York, 62-66; instr, Med Sch, Georgetown Univ, 66-69; asst prof, 69-74, ASSOC PROF PEDIAT, MED SCH, GEORGE WASHINGTON UNIV, 74- *Concurrent Pos:* Res assoc, Virol Sect, Children's Hosp Nat Med Ctr, 66-79, sr res assoc, 79-86, sr scientist, 86- *Mem:* Fel Infectious Dis Soc; Am Soc Microbiol; fel Am Acad Microbiol; Soc Epidemiol Res; fel Am Col Epidemiol; AAAS; Sigma Xi. *Res:* Virus epidemiology; enteric and respiratory tract pathogens; rapid virus diagnosis; electron microscopy; polymerase chain reactions for virus diagnosis. *Mailing Add:* Children's Nat Med Ctr 111 Michigan Ave NW Washington DC 20010

BRANDT, CHARLES LAWRENCE, b Prescott, Ariz, Dec 18, 25; m 54. PHYSIOLOGY. *Educ:* Stanford Univ, AB, 49, PhD(biol sci), 55. *Prof Exp:* Asst gen physiol, Stanford Univ, 50-54; instr zool, Univ Tex, 55-57; assoc prof, 57-66, chmn dept, 67-69, PROF BIOL, SAN DIEGO STATE UNIV, 66- *Concurrent Pos:* Res assoc, Hopkins Marine Sta, 56; NIH fel, Ctr Nat de la Res Sci, Marseille France, 66-67; res assoc, Naval Ocean Systs Ctr, San Diego, CA. *Mem:* AAAS; Soc Gen Physiol; Am Soc Photobiol; Sigma Xi. *Res:* Neurosciences; photophysiology; cellular and general physiology. *Mailing Add:* Biol Dept San Diego State Univ San Diego CA 92182

BRANDT, E J, US citizen. MOLECULAR BIOLOGY, GENETICS. *Educ:* Rosary Hill Col, Buffalo, NY, BS, 62; State Univ NY, Buffalo, MS, 70, PhD(molecular biol & genetics), 75. *Prof Exp:* Res technician, Hematol Lab, Buffalo Gen Hosp, NY, 62-64; res asst, Dept Microbiol, State Univ NY, Buffalo, 64-67; res investr, Div Immunol, Millard Fillmore Hosp, 67-68; asst cancer res scientist, Roswell Park Mem Inst, 68-70; asst res instr & fel human genetics, Dept Pediat, State Univ NY, Buffalo & mem staff, Buffalo Children's Hosp, 70-72; sr res chemist, Corp Res & Develop, 76-77, res specialist, 77-78, group leader, Biomed Res Dept, Monsanto, Co, 79-81, ENVIRON SPECIALIST, MONSANTO AGR PROD CO, 81- *Concurrent Pos:* NIH fel, 75-76; adj asst prof biol, Univ Mo-St Louis, 76- *Honors & Awards:* Certificate of Recognition, Reticuloendothelial Soc, 78. *Mem:* AAAS; Am Asn Clin Chem; Am Asn Immunologists; Am Chem Soc; Am Public Health Asn. *Res:* Immunology. *Mailing Add:* 635 Westborough Pl St Louis MO 63119

BRANDT, EDWARD NEWMAN, JR, b Oklahoma City, Okla, July 3, 33; m 53; c 3. ACADEMIC ADMINISTRATION, MEDICINE. *Educ:* Univ Okla, BS, 54; Okla State Univ, MS, 55, MD, 60, PhD(biostatist), 63. *Hon Degrees:* LHD, Med Univ SC, 84, Rush Univ, 85; DSc, NY Inst Technol, 84. *Prof Exp:* From instr to prof prev med & pub health, Med Ctr, Univ Okla, 61-70, dir biostatist unit & med res comput ctr, 62-70, assoc prof, Dept Internal Med, 63-70, assoc dean sch med & assoc dir med ctr, 68-70; dean grad sch, 70-74, dean med sch, 74-76, exec dean, 76-77, prof prev med & family med, Univ Tex Med Br, Galveston, 70-81, Univ Syst vpres Health Affairs, 77-81; asst secy health, Dept Health & Human Serv, Washington, DC, 81-84; chancellor, Univ Md, Baltimore, 85-88, pres, 88-89; PROF & EXEC DEAN, COL MED, UNIV OKLA, 89- *Concurrent Pos:* Consult, Civil Aeromed Res Inst, Okla, 62-67, Med Res Div, Upjohn Co, 63-67, comn evol foreign Med Schs, Fed State Med Bds, 80-81 & Coun Med Educ & Hosps, Tex Med Asn, 74-81; mem Bd Regents, Am Soc Clin Pharmacol & Therapeut, 68-73; fac adv, Nat Coun Student Sci Forums, 70-77; mem primate res ctrs adv comt, Div Res Resources, NIH, 75-79; mem panel comput technol & physician competence, Off Technol Assessment, Cong US, 78-79; mem Nat Arthritis Bd, Dept Health & Human Serv, Nat Digestive Dis Bd, 81-84; mem Fed Coord Coun Sci, Eng & Technol, US Govt, 81-84; mem working group food safety, US Cabinet Coun Human Resources, 81-84; chmn, Interagency Radiation Res Comt, US Govt, 81-82; mem bd dirs, Nat AIDS Network, 86-88; mem, Gov Coun, Inst Med-Nat Acad Sci, 86-, vchmn, 87-; prin, Ctr Excellence Govt, 86-; mem, Adv Coun Hazardous Substances Res & Training, Nat Inst Environ Health Sci, 87-; ed, AIDS & Pub Policy J, 88-; adj prof, Dept Family Med, Col Med, Univ Okla Health Sci Ctr, 89-, Dept Health Admin, Col Pub Health, 89-, Dept Biostatist & Epidemiol, 89-; pres, Inst Sci Soc, 90- *Honors & Awards:* Cert of Merit, AMA, 68; Am Col Int Physicians Award, 82; R Arnold Griswold Lectr, Univ Louisville, 83; Harvey Wiley Award, Food & Drug Admin, 84; Convocation Medalist, Am Col Cardiol, 85. *Mem:* Inst Med-Nat Acad Sci; AAAS; Am Fedn Clin Res; AMA; Am Acad Family Physicians; Assoc Am Med Cols; Sigma Xi; hon fel Am Col Cardiol. *Res:* Educational administration; published numerous articles in various journals. *Mailing Add:* Col Med Univ Okla PO Box 26901 Oklahoma City OK 73190

BRANDT, G(EORGE) DONALD, b New York, NY, Aug 5, 25; m 69. CIVIL ENGINEERING. *Educ:* City Col New York, BCE, 48; Columbia Univ, MS, 52. *Prof Exp:* Lectr civil eng, City Col New York, 48-52; struct engr, Andrews & Clark, 52-56; from asst prof to assoc prof civil eng, 56-69, chmn dept, 77-90, PROF CIVIL ENG, CITY COL NEW YORK, 70- *Concurrent Pos:* Vis asst prof, Univ Mich, 61. *Honors & Awards:* Spec Citation Award, Am Inst Steel Construct, 83. *Mem:* Fel Am Soc Civil Engrs; Am Concrete Inst; Am Soc Eng Educ; Am Inst Steel Construct. *Res:* Structural engineering; plastic design; computer methods. *Mailing Add:* Dept Civil Eng City Col New York New York NY 10031

BRANDT, GERALD BENNETT, b Pittsburgh, Pa, Apr 20, 38; m 61; c 4. OPTICAL PHYSICS. *Educ:* Harvard Univ, AB, 60; Carnegie Inst Technol, MS, 63, PhD(physics), 66. *Prof Exp:* Engr, Westinghouse Res Labs, 60-66, sr engr, 66-70, fel scientist, 70-74, mgr electrooptics, 74-82, ADV SCIENTIST, APPL PHYSICS DEPT, WESTINGHOUSE SCI & TECHNOL CTR, 82- *Concurrent Pos:* Lectr, Carnegie Mellon Univ, 69 & 70; assoc ed, J Optical Soc Am, 72-77. *Mem:* AAAS; Optical Soc Am; Am Phys Soc; Soc Photo-Optical Instrumentation Engrs. *Res:* Integrated optics; optical and digital signal processing; optical design and instrumentation; coherent optics and holography; low temperature physics; electronic properties of metals; ultrasonics; computer applications. *Mailing Add:* 1600A Trimont Tower One Trimont Lane Pittsburgh PA 15211

BRANDT, GERALD H, b Lompoc, Calif, Feb 25, 33; m 58; c 4. SOIL PHYSICS. *Educ:* Ore State Univ, BS, 55; Mich State Univ, MS, 60, PhD(soil sci), 63. *Prof Exp:* Asst soil sci, Mich State Univ, 60-63; chemist, Dow Chem Co, 63-72, group leader, 73-84, recruiting mgr, 84-86, RES ASSOC, DOW CHEM CO, 86- *Concurrent Pos:* Mem, Comt Soil Physiochem Phenomena, Transp Res Bd, Nat Acad Sci, 71-75, Soil Chem Stabilization Comt, 72-75. *Mem:* Am Soc Agron; Clay Minerals Soc; Soil Sci Soc Am; Int Soc Soil Sci; Sigma Xi; Am Ceramics Soc; Am Chem Soc. *Res:* Dissolved oxygen in soils; clay-chemical phenomena; modifying soil physical properties with chemicals; erosion and sediment control; soil stabilization; frost heave in soils; environmental effects of de-icing salts; chemicals research management; general applications research; ceramics foundation forming and injection molding. *Mailing Add:* 6101 Siebert St Midland MI 48640

BRANDT, H(ARRY), b Amsterdam, Neth, Nov 14, 25; nat US; m 53; c 3. MECHANICAL ENGINEERING. *Educ:* Univ Calif, BS, 49, MS, 50, PhD(mech eng), 54. *Prof Exp:* From asst to assoc, Univ Calif, 50-54, res engr, Inst Eng Res, 51-52; supvr res engr, Chevron Res Corp, 54-64; assoc prof, 64-69, chmn dept, 69-74, PROF MECH ENG, UNIV CALIF, DAVIS, 69-, CHMN DEPT, 86- *Concurrent Pos:* Consult, Lawrence Livermore Nat Lab, 69- & State Calif, 70-; dir, Adaptive Technol Inc, Sacramento, 81-82. *Mem:* Am Inst Aeronaut & Astronaut; Am Soc Mech Engrs. *Res:* Fluid mechanics; heat transfer; refrigeration; solar energy; robot welding machines. *Mailing Add:* 3309 Middle Golf Dr Box 2533 El Macero CA 95618

BRANDT, HOWARD ALLEN, b Philadelphia, Pa, June 18, 46; m 74; c 2. BIOCHEMISTRY, CLINICAL CHEMISTRY. *Educ:* Philadelphia Col Pharm & Sci, BS, 67; Univ Miami, PhD(biochem), 75. *Prof Exp:* Fel biochem, Sch Med, Tufts Univ, 75; FEL BIOCHEM, ORE REGIONAL PRIMATE RES CTR, 76- *Concurrent Pos:* Asst prof biochem, Pac Univ, 78- *Res:* Combining interests in modulating metabolism via protein phosphorylation with the phenomenon of cellular maturation; the initiation of sperm motility in the epididymis appears to involve a protein phosphorylation mechanism. *Mailing Add:* Attn Personnel Pac Univ 2043 College Way Forest Grove OR 97116-1797

BRANDT, IRA KIVE, b New York, NY, Nov 9, 23; m 47; c 4. PEDIATRICS. *Educ:* NY Univ, AB, 42; Columbia Univ, MD, 45; Am Bd Pediat, cert,pediat, 52; Am Bd Med Genetics, cert clin & clin-biochem genetics, 82. *Prof Exp:* Res fel pediat & biochem, Yale Univ, 55-57, from asst prof to assoc prof, pediat, 57-68; chmn dept, Children's Hosp, San Francisco, 68-70; clin prof, Univ Calif, San Francisco, 70; PROF, DEPTS PEDIAT & MED GENETICS, IND UNIV, INDIANAPOLIS, 70- *Mem:* Soc Pediat Res; Am Acad Pediat; Soc Inherited Metabolic Disorders; Am Pediat Soc; Am Soc Human Genetics. *Res:* Metabolic and genetic disorders of children. *Mailing Add:* Riley Childrens Hosp Rm A-36 702 Barnhill Dr Indianapolis IN 46223

BRANDT, J LEONARD, b New York, NY, Aug 3, 19; m 50; c 2. INTERNAL MEDICINE, GERIATRICS. *Educ:* Univ Mich, AB, 40; Long Island Col Med, MD, 43; Am Bd Internal Med, dipl. *Prof Exp:* Intern, Mt Sinai Hosp, Cleveland, 44; res assoc, Sch Trop Med, Columbia Univ, San Juan, PR, 45; asst res pathologist, Montefiore Hosp, New York, 48; from instr to assoc prof med, Col Med, State Univ NY, 49-59; lectr, Sch Med, McGill Univ, 59-70, assoc prof med, 70-79; sr physician, Jewish Gen Hosp, Montreal, 59-78; CHIEF MED, HEBREW HOME & HOSP, HARTFORD, 79- *Concurrent Pos:* Physician-in-chief, Jewish Gen Hosp, Montreal, 59-79; actg chief med, Mt Sinai Hosp, Hartford, 79-80; assoc prof med, Sch Med, Univ Conn, 79-87, prof med, 88. *Mem:* AAAS; Am Fedn Clin Res; Soc Exp Biol & Med; Am Physiol Soc; Harvey Soc; Am Geriat Soc; Geront Soc Am. *Res:* Hepatic and renal physiology; geriatrics. *Mailing Add:* Hebrew Home & Hosp 615 Tower Ave Hartford CT 06112

BRANDT, JOHN T, b Milwaukee, Wis, Sept 15, 48. HEMATOPATHOLOGY. *Educ:* Univ Mo, Columbia, MD, 76. *Prof Exp:* ASST PROF PATH, OHIO STATE UNIV, 81- *Mailing Add:* 400 Riley Ave Worthington OH 43085

BRANDT, JON ALAN, b Sidney, Ohio, Sept 10, 47; m 78; c 2. PRICE ANALYSIS, COMMODITY MODELING. *Educ:* Ohio State Univ, BS & MS, 70; Univ Calif, Davis, PhD(agr econ), 77. *Prof Exp:* Res asst price anal, Dept Agr Econ, Ohio State Univ, 70; res asst commodity anal, Dept Agr Econ, Univ Calif, Davis, 73-77; from asst prof to assoc prof price anal, Dept Agr Econ, Purdue Univ, 77-86; PROF, UNIV MO. *Concurrent Pos:* Vis assoc prof, Dept Agr Econ, Univ Mo, 84-85. *Mem:* Am Agr Econ Asn; Int Inst Forecasters. *Res:* Agricultural price analysis and forecasting research activities related to livestock commodities; livestock modeling and policy analysis. *Mailing Add:* Dept Agr Econ, Mumford Hall Univ Mo Columbia MO 65211

BRANDT, KARL GARET, b Galveston, Tex, Oct 15, 38; m 65; c 2. INSTRUCTIONAL ADMINISTRATION. *Educ:* Rice Univ, BA, 60; Mass Inst Technol, PhD(org chem), 64. *Prof Exp:* NSF fel, Johnson Res Found, Univ Pa, 64-65; NIH fel, Cornell Univ, 65-66; from asst prof to assoc prof, 66-75, asst dean grad sch, 81-84, actg dean agr, 86-87, PROF BIOCHEM, 75-, ASSOC DEAN AGR, PURDUE UNIV, 84- *Mem:* Am Chem Soc; Am Soc Biol Chemists; AAAS. *Res:* Mechanism of enzyme action; structure-function relationships; kinetics of enzyme-catalyzed reactions. *Mailing Add:* Agr Admin Bldg Purdue Univ West Lafayette IN 47907

BRANDT, LUTHER WARREN, b Gradan, Kans, Oct 1, 20; m 42; c 1. PHYSICAL CHEMISTRY. *Educ:* Ft Hays Kans State Col, AB, 41; Kans State Univ, PhD(phys chem), 50. *Prof Exp:* Chemist fission prod anal, Clinton Labs, E I du Pont de Nemours & Co, 44 & Hanford Eng Works, E I du Pont de Nemours & Co & Gen Elec Co, 44-47; res chemist high polymers, E I du Pont de Nemours & Co, 50-52; phys chemist, US Bur Mines, 52-59, chief res div, 59-61, chief helium res ctr, Tex, 61-62, res dir, 62-70, polymer chemist, Spokane Mining Res Ctr, Wash, 70-72, res chemist, Tuscaloosa Mining Res Lab, 72-81; RETIRED. *Mem:* AAAS; Am Chem Soc; fel Am Inst Chemists; Sigma Xi. *Res:* Surface chemistry relating to mineral tailings; electrophoresis of serum proteins of certain animals; phase equilibria and thermodynamics of gases. *Mailing Add:* 22 Riverdale Tuscaloosa AL 35406

BRANDT, MANUEL, b Columbus, Ohio, Jan 7, 15; m 47; c 2. ANALYTICAL CHEMISTRY. *Educ:* Ohio State Univ, BA, 36, MA, 37. *Prof Exp:* Asst physiol chem, Ohio State Univ, 36-38; chemist, US Army Ord, 40-43 & Tenn Eastman Corp, 43-45; chemist, Ethyl Corp, 45-50, group leader, 50-58, asst supvr, 58-66, suprv Res & Develop Dept, 66-81; RETIRED. *Mem:* Am Chem Soc. *Res:* Microchemical analyses; trace and ultratrace metal analyses. *Mailing Add:* 641 Beaker Pl Columbus OH 43213

BRANDT, PHILIP WILLIAMS, b Cleveland, Ohio, Sept 23, 30; m 54; c 2. ANATOMY, PHYSIOLOGY. *Educ:* Swarthmore Col, BA, 52; Univ Pa, MS, 57; Columbia Univ, PhD(anat), 60. *Prof Exp:* Asst instr anat, Univ Pa, 53-57; from asst to asst prof, 57-70, assoc prof, 70-87, PROF ANAT, COL PHYSICIANS & SURGEONS, COLUMBIA UNIV, 88- *Concurrent Pos:* Guggenheim fel, 68-69; NIH career develop award, 68-73; computer prog. *Mem:* Am Soc Cell Biol; Am Asn Anat; Soc Gen Physiol. *Res:* Biophysics; muscle. *Mailing Add:* Col Physicians & Surgeons Columbia Univ 630 W 168th St New York NY 10032

BRANDT, RICHARD BERNARD, b Brooklyn, NY, July 3, 34; m 56; c 3. BIOCHEMISTRY. *Educ:* Queens Col, NY, BS, 56; Brooklyn Col, MA, 61; NY Univ, PhD(biochem), 68. *Prof Exp:* Res asst, Sloan-Kettering Inst Cancer Res, NY, 56-58; substitute instr, Brooklyn Col, 59-61, chem & biol warfare grant proj leader, 61-62; res assoc, Inst Med Res & Studies, NY, 62-69; guest res assoc, Brookhaven Nat Labs, NY, 68-70; asst prof, 70-74, ASSOC PROF BIOCHEM, MED COL VA, VA COMMONWEALTH UNIV, 74- *Concurrent Pos:* Anal chemist, F D Snell, Inc, NY, 58. *Mem:* AAAS; Am Chem Soc; Soc Exp Biol & Med; Am Inst Nutrit; NY Acad Sci; Sigma Xi. *Res:* Vitamin A in trauma; D-lactate metabolism; enzyme systems in cell growth, especially glyoxalase I and II. *Mailing Add:* 3309 Gloucester Rd Richmond VA 23227

BRANDT, RICHARD CHARLES, b Philadelphia, Pa, Dec 18, 40; m 63; c 2. SOLID STATE PHYSICS. *Educ:* Calif Inst Technol, BS, 62; Univ Ill, MS, 64, PhD(physics), 67. *Prof Exp:* Staff mem solid state physics, Lincoln Lab, Mass Inst Technol, 67-71; RES ASST PROF PHYSICS, UNIV UTAH, 71- *Mem:* Am Phys Soc; Asn Comput Mach; Soc Appl Learning Tech. *Res:* Infrared optical properties of solids; development of computer-aided instruction. *Mailing Add:* 3160 Merrill Eng Univ Utah Salt Lake City UT 84112

BRANDT, RICHARD GUSTAVE, b Albany, NY, Nov 2, 36; div; c 3. LOW TEMPERATURE PHYSICS. *Educ:* Yale Univ, BS, 58, MS, 59, PhD(physics), 65. *Prof Exp:* Res staff physicist, Yale Univ, 64-65; mem tech staff, Defense Res Corp, 65-68; PHYSICIST, OFF NAVAL RES, 68- *Mem:* Am Phys Soc; AAAS; Am Vacuum Soc. *Res:* Surface physics; superconductivity; electrooptics; solid state physics; ionospheric physics. *Mailing Add:* Off Naval Res 800 N Quincy St Arlington VA 22217-5000

BRANDT, ROBERT WILLIAM, forestry, for more information see previous edition

BRANDT, STEPHEN BERNARD, b Milwaukee, Wis, May 12, 50; m 74; c 2. BIOLOGICAL OCEANOGRAPHY, FISHERIES ECOLOGY. *Educ:* Univ Wis-Madison, BA, 72, MS, 75, PhD(oceanog & limnol), 78. *Prof Exp:* Res technician, Lab Limnol, Univ Wis-Madison, 70-72, res specialist, 72, res asst, 73-78; res scientist, 78-82, SR RES SCIENTIST, MARINE LAB, DIV FISHERIES RES, COMMONWEALTH SCI & INDUST RES ORGN, 82- *Mem:* AAAS; Am Fisheries Soc; Am Soc Limnol & Oceanog; Int Asn Great Lakes Res; Australian Marine Sci Asn; Sigma Xi. *Res:* Assessment of the effects of oceanic thermal fronts on biological processes; effects of warm-core eddies on microplankton off East Australia; ecological study of oceanic squid. *Mailing Add:* Chesapeake Biol Lab Univ Md PO Box 38 Solomons MD 20688

BRANDT, WALTER EDMUND, b Columbia, Pa, Apr 11, 35; div; c 2. VIROLOGY. *Educ:* Univ Md, BS, 60, MS, 64, PhD(microbiol), 67. *Prof Exp:* Microbiologist, Walter Reed Army Inst Res, 60-67, supvr, 67-70, asst chief virol, 70-85; PROJ MGR, USA MED MAT DEVELOP ACT, 85 - *Concurrent Pos:* Consult, NIH, 71-75; instr, eve div, Univ Md, 66-74, lectr microbiol, 70-78, asst proj dir extramural contracts, 78-84. *Mem:* Am Asn Immunologists; Am Soc Microbiol; Am Soc Trop Med & Hyg; Sigma Xi. *Res:* Dengue fever virus infection of cultured vertebrate and invertebrate cells; isolation of particulate and soluble antigens for study of dengue in humans; attenuated live virus vaccines; antigenic determinants; antibody-mediated infection. *Mailing Add:* USA Med Mat Develop Act Ft Detrick Frederick MD 21701-5009

BRANDT, WERNER WILFRIED, b Friedensdorf, Ger, Feb 13, 27; US citizen; m 52; c 4. PHYSICAL CHEMISTRY. *Educ:* Univ Heidelberg, BS, 49; Polytech Inst Brooklyn, PhD(phys chem), 56. *Prof Exp:* Anal chemist, Merck & Co, Inc, NJ, 50 & 52; res chemist, E I du Pont de Nemours & Co, Del, 55-59; asst prof chem, Ill Inst Technol, 59-65; chmn, Dept Chem, 76-78, ASSOC PROF CHEM, UNIV WIS-MILWAUKEE, 65-, MEM, LAB SURFACE STUDIES, 77- *Mem:* Am Chem Soc; Am Phys Soc; fel Am Inst Chemists. *Res:* Rate processes, especially diffusion; plasma chemistry; solid state electrochemistry. *Mailing Add:* Dept Chem Univ Wis-Milwaukee Milwaukee WI 53201

BRANDT, WILLIAM HENRY, b Great Falls, Mont, May 25, 27; div; c 2. BOTANY, PHYTOPATHOLOGY. *Educ:* Univ Mont, BA, 50; Ohio State Univ, MSc, 51, PhD, 54. *Prof Exp:* Asst oak wilt, Ohio Agr Exp Sta, 52-54; res biologist, B F Goodrich Res Ctr, 54-56; from instr to assoc prof, 56-90, EMER ASSOC PROF BOT, ORE STATE UNIV, 90- *Concurrent Pos:* NIH spec res fel, 69-70; guest investr, Rockefeller Univ, 69-70. *Mem:* AAAS; Am Phytopath Soc; Mycol Soc Am. *Res:* Physiology of fungi; morphogenesis in fungi; vascular wilt diseases. *Mailing Add:* Dept Bot Ore State Univ Corvallis OR 97331

BRANDTS, JOHN FREDERICK, b Celina, Ohio, June 15, 34; m 52; c 6. PHYSICAL CHEMISTRY, BIOCHEMISTRY. *Educ:* Miami Univ, BA, 56; Univ Minn, PhD(phys chem), 61. *Prof Exp:* NIH fel, 61-62; from asst prof to assoc prof, 62-71, PROF CHEM, UNIV MASS, AMHERST, 71- *Concurrent Pos:* NIH res grant, 63-66. *Mem:* Am Chem Soc. *Res:* Physical biochemistry; conformation and stability of globular proteins. *Mailing Add:* Univ Chem Univ Mass Amherst MA 01003

BRANDVOLD, DONALD KEITH, b Maddock, NDak, Aug 12, 36; m 61; c 1. BIOCHEMISTRY, PHYSICAL CHEMISTRY. *Educ:* NDak State Univ, BS, 62, PhD(chem), 65. *Prof Exp:* Technician, NDak State Univ, 59-62; asst prof, 65-74, ASSOC PROF CHEM, NMEX INST MINING & TECHNOL, 74- *Mem:* AAAS; Am Chem Soc. *Res:* Submolecular and electronic biology; general molecular biology; vitamins and hormones mode of action; virus research; pesticide residues; ore leaching by bacteria; water chemistry and pollution. *Mailing Add:* Dept Chem NMex Inst Mining & Technol Socorro NM 87801

BRANDWEIN, BERNARD JAY, b Chicago, Ill, Apr 19, 27; m 51; c 3. BIOCHEMISTRY. *Educ:* Purdue Univ, BS, 48, MS, 51, PhD, 55. *Prof Exp:* Assoc prof, 55-69, PROF CHEM, SDAK STATE UNIV, 69- *Mem:* AAAS; Am Chem Soc; Inst Food Technologists; Sigma Xi. *Res:* Isolation, identification and proof of structure of naturally occurring carbohydrates, glycosides and vitamins; reversion products of carbohydrates. *Mailing Add:* 414 State Ave Brookings SD 57006

BRANEN, ALFRED LARRY, b Caldwell, Idaho, Jan 5, 45; m 72; c 3. FOOD SCIENCE. *Educ:* Univ Idaho, BS, 67; Purdue Univ, PhD(food sci), 70. *Prof Exp:* Asst prof food sci, Univ Wis, 70-73; from asst prof to assoc prof & chmn food sci, Wash State Univ, 74-81; prof food sci & head dept, Univ Nebr, 81-83; assoc dean & dir resident instr, 83-86, DEAN, COL AGR, UNIV IDAHO, 86- *Concurrent Pos:* Regional communicator, Inst Food Technologists, 79-81; mem bd dir, Am Inst Coop, Consortium Int Develop, Idaho Res Found. *Mem:* Inst Food Technologists; Nat Asn Cols & Teachers Agr. *Res:* Toxicology of food additives; preservation of foods with naturally occurring biochemicals; autoxidation and antioxidant function. *Mailing Add:* Col Agr Univ Idaho Moscow ID 83843

BRANGAN, PAMELA J, b Davenport, Iowa, Nov 20, 44. BIOLOGY, ANATOMY. *Educ:* Smith Col, BA, 67; Yale Univ, MFS, 70, PhD(biol), 77. *Prof Exp:* Res assoc biol, Harvard Univ, 74-76; cur mammals, Chicago Zool Soc, 76-79; asst prof ecol & systematics, Cornell Univ, 79-; AT DEPT DENT HYG, OLD DOMINION UNIV. *Concurrent Pos:* Lectr, Univ Chicago, 76-78. *Mem:* Am Soc Zool; Am Soc Mammal; Australian Soc Mammal; Soc Vert Paleont. *Res:* Functional morphology, mammalogy, behavior and evolution; mammalian herbivores and plant communities in the semiarid region of Australia. *Mailing Add:* Dept Community Health Professions Old Dominion Univ 5215 Hampton Blvd Norfolk VA 23508

BRANGES, LOUIS DE, b Paris, France, Aug 21, 32; US & French citizen; m 80. MATHEMATICS. *Educ:* Mass Inst Technol, BS, 53; Cornell Univ, PhD(math), 57. *Prof Exp:* Asst prof math, Lafayette Col, 58-59; mathematician, Inst Advan Study, 59-60; lectr, Bryn Mawr Col, 60-61; mem, Courant Inst Math Sci, 61-62; assoc prof, 62-63, PROF MATH, PURDUE UNIV, 63- *Concurrent Pos:* Sloan Found Fel, 63-65; Guggenheim Found Fel, 67-68. *Honors & Awards:* Alexander von Humboldt Prize, 85; Alexander Ostrowski Prize, 89. *Mem:* Am Math Soc. *Res:* Functional analysis, operator theory. *Mailing Add:* 331 Hollowood Dr West Lafayette IN 47906

BRANHAM, JOSEPH MORHART, b Washington, DC, Jan 31, 32; m 56; c 2. DEVELOPMENTAL BIOLOGY, ECOLOGY. *Educ:* Fla State Univ, BS, 56, MS, 58, PhD(exp biol), 63. *Prof Exp:* Asst prof biol, Oglethorpe Univ, 62-64, assoc prof, 63-65; NIH fel, Inst Animal Genetics, Univ Edinburgh, 65-67; asst prof zool, Univ Hawaii, 69-72; fel biol, Univ Utah, 72-73; sci teacher, Leesburg High Sch, Fla, 74-90; Chmn, Lake County Resource Adv Comt for County Comn, 74-80, LAKE COUNTY WATER AUTHORITY ADV BD, 81- *Concurrent Pos:* Instr invertebrate zool, Marine Resources Ctr, Univ Ga, 79-84. *Mem:* AAAS; Am Soc Zoologists; Am Inst Biol Sci; NY Acad Sci; Sigma Xi. *Res:* Fertilization, sea urchin and rabbit gamete physiology; insect physiology and ecology; coral reef biology. *Mailing Add:* 801 Wilson Ave Leesburg FL 32748

BRANLEY, FRANKLYN M, b New Rochelle, NY, June 5, 15; m 38; c 2. ASTRONOMY. *Educ:* NY Univ, BS, 42; Columbia Univ, MA, 46, EdD(sci educ), 57. *Prof Exp:* Teacher sci, NY Pub Schs, 36-54 & State Teachers Col NY, 54-56; assoc astronr, 56-63, asst chmn, 64-68, astronr 68-72, ASTRONR, AM MUS-HAYDEN PLANETARIUM, 63-; SCI WRITER, 72- *Concurrent Pos:* Mem, Comn Teacher Educ & Cert, AAAS, 61-62 & Adv Comn, Nat Sci Exhibit, Worlds Fair, Wash, 62; proposals referee, NSF, 62; sci writer. *Honors & Awards:* Edison Award. *Mem:* Fel AAAS; Am Astron Soc; Nat Sci Teachers Asn; fel Royal Astron Soc. *Res:* Science education, especially astronomy and full use of planetaria. *Mailing Add:* Harbor Drive Rte 3 Box 436 Sag Harbor NY 11963

BRANN, DARRELL WAYNE, b Dallas, Tex, May 4, 58. NEUROENDOCRINOLOGY, REPRODUCTION. *Educ:* Henderson State Univ, BA, 84; Med Col Ga, PhD(endocrinol), 90. *Prof Exp:* Postdoctoral fel neuroendocrinol, 90-91, ASST RES SCIENTIST, MED COL GA, 91-

Concurrent Pos: Grant, Med Col Ga Res Inst, 91-92. *Honors & Awards:* Nat Res Serv Award, NIH, 89; Distinguished Leadership Award, Am Biog Inst Inc, 91. *Mem:* Endocrine Soc; Int Soc Neuroendocrin; Soc Study Reproduction; NY Acad Sci; AAAS; Sigma Xi; Soc Exp Biol & Med; Am Physiol Soc. *Res:* Neuroendocrine mechanisms involved in the secretion of gonadotropin hormones in the female; numerous publications. *Mailing Add:* Dept Physiol & Endocrinol Med Col Ga Augusta GA 30912-3000

BRANNEN, ERIC, b Manchester, Eng, Sept 25, 21; m 46; c 3. NUCLEAR PHYSICS, QUANTUM ELECTRONICS. *Educ:* Univ Toronto, BA, 44, MA, 46; McGill Univ, PhD(nuclear physics), 48. *Prof Exp:* Demonstr, Univ Toronto, 45-46, McGill Univ, 46-48 & Univ Toronto, 48-49; instr, 49-50, lectr, 50-51, from asst prof to prof, 51-55, SR PROF PHYSICS, UNIV WESTERN ONT, 55- *Mem:* Am Phys Soc; Can Asn Physicists. *Res:* Racetrack microtron electron accelerator; submillimeter radiation; photon correlation; far infrared maser. *Mailing Add:* Dept Physics Univ Western Ont London ON N6A 3K7 Can

BRANNEN, WILLIAM THOMAS, JR, b Chicago, Ill, Sept 30, 36; div; c 5. ORGANIC CHEMISTRY. *Educ:* DePaul Univ, BS, 58; Northwestern Univ, PhD(org chem), 62. *Prof Exp:* Res asst org chem, Northwestern Univ, 59-62; proj chemist, Am Oil Co div, Standard Oil Co, Ind, 62-65; sr proj chemist, Amoco Chem Corp div, 65-67; tech dir, 67-71, vpres res & mfg, 71-77, EXEC V PRES, ELCO CORP, 77- *Concurrent Pos:* Vpres, Detrex Chem Indust, 84- *Honors & Awards:* Outstanding Sr Chem Award, Am Inst Chemists. *Mem:* Am Chem Soc; Nat Lubricating Grease Inst; Sigma Xi. *Res:* Mechanistic studies of organic and enzyme reactions; synthetic organic and inorganic chemistry. *Mailing Add:* 2050 Radcliffe Westlake OH 44145

BRANNIGAN, DAVID, b Cambridge, Mass, Apr 12, 41; m 74; c 1. ZOOLOGY. *Educ:* Suffolk Univ, BS, 68; Univ NH, MS, 72, PhD(zool), 74. *Prof Exp:* Instr biol, Adelphi Univ, 71-74; ASSOC PROF BIOL, STATE UNIV NY, BROCKPORT, 74- *Mem:* Animal Behav Soc; AAAS. *Res:* Effect of behavior on the atherosclerotic process. *Mailing Add:* Dept Biol Sci State Univ NY Col Brockport NY 14420

BRANNON, DONALD RAY, b Fort Peck, Mont, May 30, 39; m 67. ORGANIC CHEMISTRY, TOXICOLOGY. *Educ:* Okla State Univ, BS, 61, PhD(org chem), 65. *Prof Exp:* Res asst, Okla State Univ, 61-65; sr scientist, Lilly Res Labs, 67-71, res scientist, 71-73, head microbiol & fermentation prod red div, 73-77, head toxicity testing, 77-84, dir toxicol studies, 84-86, dir process res & develop div, 86-89, DIR ENVIRON AFFAIRS, LILLY RES LABS, 86- LILLY RES LABS, 86- *Mem:* Am Chem Soc. *Res:* Toxicity testing; hazard evaluation; ecological hazard assessments; process chemistry. *Mailing Add:* Environ Affairs Div Lilly Res Eli Lilly & Co Lilly Corp Ctr Indianapolis IN 46285

BRANNON, H RAYMOND, JR, b Midland, Ala, Jan 23, 26. PETROLEUM ENGINEERING, CIVIL ENGINEERING. *Educ:* Auburn Univ, BS, 50, MS, 51. *Prof Exp:* Engr, 52-73, res scientist, 73-84, sr res scientist, Exxon Prod Res Co, 84-86; OWNER, BRANNON CO, 86- *Mem:* Nat Acad Eng; Am Phys Soc; Soc Explor Geophysicists; Sigma Xi. *Mailing Add:* 5807 Queens Loch Houston TX 77096

BRANNON, PATSY M, NUTRITION. *Educ:* Cornell Univ, PhD(nutrit & biochem), 79. *Prof Exp:* ASST PROF NUTRIT & FOOD SCI, DEPT NUTRIT & FOOD SCI, UNIV ARIZ, 82- *Res:* Nutritional influence on pancreatic functions; regulation of lipid metabolism; adaptation of pancreatic enzymes. *Mailing Add:* 9669 E Moonbeam Dr Tucson AZ 85748

BRANNON, PAUL J, b Flint, Mich, Apr 12, 35; m 65; c 2. SPECTROSCOPY. *Educ:* Univ Mich, BS, 57, MS, 59, PhD(physics, infrared spectros), 65. *Prof Exp:* Mem staff upper atmospheric physics, Sandia Labs, 65-67; fel infrared spectros, Univ Tenn, 67-68; mem staff upper atmospheric physics, 68-70, mem staff, Laser Fusion Physics, 70-79, MEM STAFF, LASER PHYSICS, SANDIA LABS, 79- *Res:* Laser physics and spectroscopy applied to upper atmospheric research, line shape and shifts in gases; optoelectronics; waveguides; xenon infrared laser. *Mailing Add:* Orgn 1128 Bldg MO71 Rm 16 Sandia Labs Albuquerque NM 87185

BRANS, CARL HENRY, b Dallas, Tex, Dec 13, 35; m 57; c 4. MATHEMATICAL PHYSICS. *Educ:* Loyola Univ, La, BS, 57; Princeton Univ, PhD(physics), 61. *Prof Exp:* From instr to assoc prof, 60-70, PROF PHYSICS, LOYOLA UNIV, LA, 70- *Concurrent Pos:* NSF grant, 62-66; vis prof physics, Princeton Univ, 73-74; res grant, Res Corp, 75. *Mem:* Am Phys Soc. *Res:* Varying gravitational constant in general relativity; mathematical methods in general relativity; differential geometry and topology; group theory; application of computers to general relativity. *Mailing Add:* Dept Physics Loyola Univ New Orleans LA 70118

BRANSCOMB, ELBERT WARREN, b Yakima, Wash, Mar 31, 35; m 81; c 1. MOLECULAR BIOLOGY. *Educ:* Reed Col, BA, 57; Syracuse Univ, PhD(physics), 64. *Prof Exp:* Physicist theoret physics, 64-69, PHYSICIST MOLECULAR BIOL, LAWRENCE LIVERMORE LAB, 69- *Res:* Genetic toxicology: assays for in-vivo somatic point mutation injury in humans and in experimental animals; accuracy in the intracellular transfer of genetic information in DNA replication and in gene expression. *Mailing Add:* Lawrence Livermore Lab Biomed Div L-452 PO Box 5507 Livermore CA 94550

BRANSCOMB, LEWIS MCADORY, b Asheville, NC, Aug 17, 26; m 51; c 2. INFORMATION SCIENCE, INFORMATION TECHNOLOGY. *Educ:* Duke Univ, AB, 45; Harvard Univ, MS, 47, PhD(physics), 49. *Hon Degrees:* DSc, Western Mich Univ, 69, Univ Rochester, 71, Univ Colo, 73, Duke Univ, 74, Lycoming Col, 79, Polytech Inst NY, 80, Clarkson Col Technol, 81, Pace Univ, 82, Univ Ala, 83, Pratt Inst, 83, Rutgers Univ, 84, Lehigh Univ, 84, Notre Dame Univ, 86, State Univ NY, Binghamton, 89. *Prof Exp:* Instr physics, Harvard Univ, 50, jr fel, Soc Fels, 49-51; physicist, Nat Bur

Standards, 51-54, chief atomic physics sect, 54-59, chief div, 59-62, chmn joint inst lab astrophys, 62-65, chief lab, Astrophys Div, 62-69, dir, 69-72; chief scientist, IBM Corp, 72, vpres & chief scientist, 72-86; ALBERT PRATT PUB SERV PROF, JOHN F KENNEDY SCH GOVT, HARVARD UNIV, 86- *Concurrent Pos:* Rockefeller pub serv fel, Univ Col, Univ London, 57-58; prof-adjoint physics, Univ Colo, 62-69; mem, President's Sci Adv Comt, 65-68 & President's Comn for Medal of Sci, 70-74; mem, Int Comt Weights & Measures, 69-72; mem bd dirs, Am Nat Standards Inst, 69-72, General Foods Corp, 75-85, Mobil Corp, 78-, Lord Corp, 87-, Mitre Corp, 87-, C S Draper Labs, 88-; ed, Rev Mod Physics, 69-72; mem bd overseers, Harvard Univ, 69-; mem adv comt sci & foreign affairs, US Dept State, 72-74; mem adv comt energy res & develop, Fed Energy Agency, 73-75; trustee, Carnegie Inst, Washington, 74-90, Polytech Inst New York, 74-78, Vanderbilt Univ, 80-, Rand Corp, Woods Hole Oceanog Inst, 85-; mem physics vis comt, Mass Inst Technol, 74-79; mem, Nat Sci Bd, 79-, chmn, 80-; chmn, Panel Appl Sci & Technol, Off Assessment, US Cong, 77-79; chmn, Comt Scholarly Commun with People's Repub China, Nat Acad Sci, 77-80; mem, US-USSR Joint Comn Sci & Technol, 77-80; chmn, Carnegie Form Task Force on Teaching as a Prof, 86, mem, Carnegie Commn Sci Technol & Govt, 89- *Honors & Awards:* Arthur Fleming Award, 58; Samuel Wesley Stratton Award; Proctor Prize, Sci Res Soc Am; Arthur M Bueche Award, Nat Acad Eng, 87. *Mem:* Nat Acad Sci; Inst Med-Nat Acad Sci; Nat Acad Eng; Nat Acad Pub Admin; fel Am Acad Arts & Sci; Sigma Xi (pres, 85-86); fel Am Philos Soc. *Res:* Gaseous electronics; spectra of diatomic molecules; physics of the upper atmosphere; physics of negative ions; science and technology policy; information systems technology; science & technology policy; education reform. *Mailing Add:* J F Kennedy Sch Govt Harvard Univ 79 JFK St Cambridge MA 02138

BRANSCOME, JAMES R, chemical engineering, operations research, for more information see previous edition

BRANSOME, EDWIN D, JR, b New York, NY, Oct 27, 33; m 59; c 2. ENDOCRINOLOGY, BIOCHEMISTRY. *Educ:* Yale Univ, AB, 54; Columbia Univ, MD, 58. *Prof Exp:* House officer, Peter Brent Brigham Hosp, 58-59, asst med, 59-61, asst resident, 61-62; res assoc, Columbia Univ, 62-63; assoc endocrinol, Scripps Clin & Res Found, 64-66; from asst prof to assoc prof exp med, Mass Inst Technol, 66-70; instr med, Harvard Med Sch, 66-70; clin dir 9C MS Lab, 73-88, PROF MED & ENDOCRINOL, MED COL GA, 70- *Concurrent Pos:* NIH res fel, Harvard Med Sch, 59-61; vis fel biochem, Columbia Univ, 62-63; Am Cancer Soc fel, 62-64, fac res assoc award, 65-70; Am specialist, US Dept State, 61; asst physician, Vanderbilt Clin, 62-64; chmn, comt rev, Adv Panel Endocrinol, US Pharmacopeia, 75-90, subcomt biochem microbiol 80-90, mem bd trustees, 90- & deleg, US Pharmacopoeial Conv, Med Asn Ga; mem bd dir, Am Diabetes Asn, 85-88; chmn Gov't Relations Comt, 86-88; chmn coun, Health Care Delivery & Pub Health, 88-90; consult, Stereochemical Genetics Inc, 87- *Mem:* Am Diabetes Asn; Endocrine Soc; Am Physiol Soc; Am Col Clin Pharmacol; Am Chem Soc. *Res:* Internal medicine; mechanism of hormone and drug action on cellular metabolism and growth; drug design. *Mailing Add:* Dept Med Med Col Ga Augusta GA 30912

BRANSON, BRANLEY ALLAN, b San Angelo, Tex, Feb 11, 29; m 64; c 1. ICHTHYOLOGY, MALACOLOGY. *Educ:* Okla State Univ, BS, 56, MS, 57, PhD(zool), 60. *Prof Exp:* Spec instr zool, Okla State Univ, 56-57; asst prof, Kans State Col Pittsburg, 60-66; mem fac, 66-67, assoc prof, 67-70, FOUND PROF BIOL, EASTERN KY UNIV, 70- *Concurrent Pos:* Res grants, NSF, 60-61, Sigma Xi, 62-64 & NIH, 64-66; ed, Ky Acad Sci Trans, 81- *Mem:* AAAS; Am Soc Ichthyologists & Herpetologists; Soc Study Evolution; Soc Syst Zool; Am Micros Soc. *Res:* Comparative morphology of the sensory systems of teleost fishes; ecological adaptations of fishes as regards their sensory system; zoogeography and taxonomy of fishes and mollusks. *Mailing Add:* Dept Biol Eastern Ky Univ Richmond KY 40475

BRANSON, BRUCE WILLIAM, b Athol, Mass, Oct 20, 27; m 51; c 2. SURGERY. *Educ:* Loma Linda Univ, MD, 50; Am Bd Surg, dipl, 63. *Prof Exp:* Asst prof, 62-65, assoc prof, 65-77, PROF SURG, LOMA LINDA UNIV, 77-, CHMN DEPT, 78- *Mailing Add:* Dept Surg Loma Linda Univ Sch Med Loma Linda CA 92350

BRANSON, DAN E(ARLE), b Dallas, Tex, Nov 13, 28; m 55; c 1. STRUCTURAL & CIVIL ENGINEERING. *Educ:* Auburn Univ, BCE, 54, MCE, 56; Univ Fla, PhD(struct eng), 60. *Prof Exp:* Instr civil eng, Auburn Univ, 55-56, asst prof, 57, assoc prof, 62-63; instr eng mech, Univ Fla, 59-60; assoc prof civil eng, Univ Ala, 60-62; PROF CIVIL & ENVIRON ENG, UNIV IOWA, 63- *Concurrent Pos:* Consult, Hayes Int Corp, Ala, 61-65 & concrete & steel struct, 64-; Alexander von Humboldt US sr scientist award, WGer, 80. *Honors & Awards:* Martin P Korn Award, Prestressed Concrete Inst, 71. *Mem:* Fel Am Soc Civil Engrs; fel Am Concrete Inst; Prestressed Concrete Inst; Int Asn Bridge & Struct Eng; Am Soc Eng Educ. *Res:* Creep and shrinkage of different weight concretes; time-dependent deformation and load-deflection response of uncracked and cracked reinforced beams and slabs; prestressed structures; author of numerous papers and books. *Mailing Add:* Dept Civil Eng Univ Iowa Iowa City IA 52242

BRANSON, DEAN RUSSELL, b Livingston, Mont, July 11, 41; m 63; c 3. ENVIRONMENTAL SCIENCE, BIOLOGICAL CHEMISTRY. *Educ:* Univ Redlands, BS, 63; Mont State Univ, PhD(biol chem), 67. *Prof Exp:* Res specialsit insecticide biochem, 67-70, res specialist toxicol, 70-71, res specialist aquatic toxicol, 71-74, RES SPECIALIST ENVIRON SCI, DOW CHEM CO, 74- *Honors & Awards:* IR-100 Award, Indust Res Mag, 76. *Mem:* Inst Elec & Electronics Engrs; Am Soc Testing & Mat; Am Chem Soc; Mfg Chem Asn. *Res:* Bioconcentration of chemicals in fish; fate of chemicals in the aquatic environment; aquatic toxicology and metabolism of chemicals in rats, microbes and fish. *Mailing Add:* 5002 Highridge Midland MI 48640

BRANSON, DOROTHY SWINGLE, b Modesto, Calif, June 17, 21; div; c 4. CLINICAL MICROBIOLOGY. *Educ:* Kans State Univ, BS, 42, MS, 44, PhD(bact), 64. *Prof Exp:* Field secy, Girl Scouts, S Oakland County, Mich, 44; instr zool, Univ Okla, 46-47; med lab technologist, Riverside County Gen Hosp, Calif, 59-60 & St Bernardine's Hosp, San Bernardino, Calif, 61; microbiologist & teaching supr, Columbia Hosp, Milwaukee, Wis, 64-68; microbiologist, Methodist Hosp & Med Ctr, St Joseph, Mo, 68-72; asst prof microbiol, Kirksville Col Osteopath Med, 73-74; microbiologist, Grant Hosp, Columbus, Ohio, 74-82; tech writer Remel Regional Med Labs, Lenexa, Kansas, 82-85; RETIRED. *Concurrent Pos:* lectr clin microbiol, 65-84. *Mem:* Sigma Xi; Am Soc Med Technol. *Res:* Use by soil microorganisms of aromatic compounds; effects of host nutrition on parasitism. *Mailing Add:* 10505 E 42nd St No D Kansas City MO 64133

BRANSON, FARREL ALLEN, b Coats, Kans, May 3, 19; m 47; c 2. PLANT ECOLOGY. *Educ:* Ft Hays State Univ, BS, 42, MS, 47; Univ Nebr, PhD(bot), 52. *Prof Exp:* Instr bot, Ft Hays State Univ, 47-48; asst prof range mgt, Mont State Univ, 51-57; BOTANIST, US GEOL SURV, 57- *Honors & Awards:* Hon Award Superior Serv, US Dept Interior. *Mem:* Fel AAAS; Ecol Soc Am; Am Inst Biol Sci; Soc Range Mgt; Sigma Xi. *Res:* Relationships of vegetation to hydrologic processes; vegetation indicators of quantities and qualities of soil water; effects of land treatment practices on vegetation and hydrology. *Mailing Add:* 906 24th St Golden CO 80401

BRANSON, HERMAN RUSSELL, b Pocahontas, Va, Aug 14, 14; m 39; c 2. PHYSICS. *Educ:* Va State Col, BS, 36; Univ Cincinnati, PhD(physics), 39. *Prof Exp:* Instr math & physics, Dillard Univ, 39-41; prof & head physics dept, Howard Univ, 41-68; pres, Central State Univ, 68-70; pres, Lincoln Univ, Pa, 70-85; RETIRED. *Concurrent Pos:* Chmn, adv comt estab comn minorities in sci & eng, Nat Res Coun, 72-; mem, adv comt major new children's prog in math, Educ Develop Ctr, 73-; mem, adv coun status health sci in Black col, Inst Serv Educ, 74-; mem corp, Mass Inst Technol, 79-84; mem coun, Nat Univ Lesotho, SAfrica, 85-; dir, pre-col sci & math res prog, Howard Univ, 86- *Mem:* Nat Inst Med-Nat Acad Sci; AAAS; Soc Math Biol. *Res:* Mathematical biology; education. *Mailing Add:* 10906 Oakwood St Silver Spring MD 20901

BRANSON, TERRY FRED, b Wichita, Kans, Feb 24, 35. ENTOMOLOGY. *Educ:* Colo State Univ, BS, 57, MS, 64; SDak State Univ, PhD(entom), 70. *Prof Exp:* Entomologist, Agr Res Serv USDA, 64-66, res entomologist, Northern Grain Insects Res Lab, 66-, proj leader, 70-; RETIRED. *Concurrent Pos:* Tech adv, Morelos Inst of Super Agr Studies, Mex, 73. *Mem:* Entom Soc Am. *Res:* Host plant resistance to grain insects; mechanisms of host plant resistance; behavior of insect host selection, feeding, and reproduction. *Mailing Add:* 912 126th St Ct Northwest Gig Harbor WA 98335

BRANSTETTER, DANIEL G, b Cedar Rapids, Iowa, Sept 26, 50. CANCER RESEARCH. *Educ:* Med Col Ohio, Toledo, PhD(path), 85; Am Col Vet Pathologists, dipl. *Prof Exp:* Instr, Med Col Ohio, 79-84; VET PATHOLOGIST & TOXICOLOGIST, UPJOHN CO, 84- *Mem:* Vet Med Asn. *Mailing Add:* Upjohn Co Kalamazoo MI 49001

BRANT, ALBERT WADE, b Isabel, Kans, Mar 28, 19; m 40; c 2. FOOD TECHNOLOGY, POULTRY HUSBANDRY. *Educ:* Kans State Col, BS, 40; Mich State Col, MS, 42; Iowa State Col, PhD, 49. *Prof Exp:* Instr, State Col Wash, 42-44 & 46-47; poultry husbandman, Agr Res Serv, USDA, 49-57, chief, Poultry Res Br, Animal Husb Res Div, 57-59; agriculturist, 59-85, EMER AGRICULTURIST FOOD TECHNOL, COOP EXTEN, DEPT FOOD SCI & TECHNOL, UNIV CALIF, DAVIS, 86- *Mem:* Coun Agr Sci Technol; Poultry Sci Asn; Inst Food Technologists; World Poultry Sci Asn. *Res:* Methods of determining interior quality of eggs; physical properties of meat, eggs and poultry related to quality; microbiology of meat, shell eggs and poultry; technology of animal products; processing of poultry, egg and red-meat products. *Mailing Add:* PO Box 2502 El Maciro CA 95618

BRANT, DANIEL (HOSMER), b St Paul, Minn, Apr 10, 21; m 45, 65, 69; c 3. VERTEBRATE ECOLOGY. *Educ:* Univ Minn, BS, 43; Univ Wash, MS, 49; Univ Calif, PhD(zool), 53. *Prof Exp:* Asst, Univ Calif, 49-52; from asst prof to prof biol sci, 52-74, PROF ZOOL, HUMBOLT STATE UNIV, 74- *Mem:* AAAS; Ecol Soc Am; Am Soc Mammal. *Res:* Behavior of wild small rodents in the field and artificial runway systems. *Mailing Add:* Dept Zool Humbolt State Col Arcata CA 95521

BRANT, LARRY JAMES, b Sept 1, 46; c 2. STATISTICS, COMPUTER PROGRAMMING. *Educ:* Frostburg State Col, BS. 68; Pa State Univ, MA, 72; Johns Hopkins Univ, PhD(biostatist), 78. *Prof Exp:* Chief biostatist br, Alaska Invest Div, Ctrs Dis Control, 78-80; CHIEF STATIST SCI SECT, GERONT RES CTR, NAT INST AGING, NIH, 80- *Concurrent Pos:* Adj clin prof, Sch Pub Health, Univ Hawaii, Manoa, 84-; adj prof Math & Statist, Penn State Univ, 87-; adj assoc prof, Johns Hopkins Sch Hygiene & Pub Health, 89- *Mem:* Am Statist Asn; Biomet Soc; Am Math Soc; Math Asn Am; NY Acad Sci; Geront Soc Am; Int Asn Human Biol. *Res:* Statistical methodology and combinatorial objects applicable to statistical designs; mathematical and statistical modeling; statistical computing; longitudinal data analysis; applications to gerontology, polar populations and biological sciences. *Mailing Add:* Geront Res Ctr Nat Inst Aging 4949 Eastern Ave Baltimore MD 21224

BRANT, RUSSELL ALAN, b Brooklyn, NY, Mar 18, 19; m 45; c 4. GEOLOGY. *Educ:* Univ Mich, BS, 48, MS, 49. *Prof Exp:* Asst chemist rubber chem, Monarch Rubber Co, 42-44; geologist, Fuels Br, US Geol Surv, 49-52; from asst head to head coal sect, Ohio Geol Surv, Columbus, 52-57, areal geol sect, 57-60, asst state geologist, 60-68, sr res geologist, 68; geologist, Ohio River Valley Water Sanit Comn, 68-76; sr geologist & head coal sect, 76-81, res geologist, 81-86, EMER GEOLOGIST, KY GEOL SURV, UNIV KY, 86-; CONSULT GEOLOGIST, 86- *Concurrent Pos:* Mine Drainage Specialist, Water Resources Ctr, Ohio State Univ, 68. *Mem:* Geol Soc Am; Am Inst Prof Geologists. *Res:* Coal and coal resources; areal; geochemistry; water geology; acid mine water; mining; underground injection of waste water; aerial surveillance, air photo interpretation; Lake Erie geology and shore erosion. *Mailing Add:* 309 Pasadena Dr Lexington KY 40503

BRANTHAVER, JAN FRANKLIN, b Davenport, Iowa, Mar 12, 36. PORPHYRIN CHEMISTRY. *Educ:* Millikin Univ, Decatur, Ill, BA, 58; NDak State Univ, Fargo, PhD(chem), 76. *Prof Exp:* Res assoc, Dept Chem, NDak State Univ, 64-76, instr chem, 67; res chemist, Laramie Energy Technol Ctr, US Dept Energy, 76-83; SR RES CHEMIST, WESTERN RES INST, LARAMIE, WYO, 83- *Concurrent Pos:* Adj prof, Dept Chem, Univ Wyo, 81- *Mem:* Am Chem Soc; Sigma Xi. *Res:* Organic geochemistry of petroleum, tar sand, oil shale and coal with emphasis on metal chelates in these materials. *Mailing Add:* Western Res Inst Box 3395 Univ Station Laramie WY 82071

BRANTINGHAM, CHARLES ROSS, b Long Beach, Calif, Feb 14, 17; m 42; c 4. PREVENTIVE PODIATRIC MEDICINE, PUBLIC HEALTH. *Educ:* Calif Col Podiatric Med, DPM, 39; Am Bd Podiatric Pub Health, cert, 88. *Prof Exp:* Dir, Podiatric Med Group, Long Beach, Calif, 46-72; dir pvt pract, Almitos Podiatry Group, Los Alamitos Med Ctr, 72-90; CONSULT PRACT, PODIATRIC ERGONOMICS, 80- *Concurrent Pos:* Lt comdr podiatrist, US Naval Reserve, 41-68; mem bd trustees, Am Podiatric Med Asn, 57-58, Nat Coun, 53-55; podiatrist staff, Pac Hosp, Long Beach, 62-85, hon staff, 85 Woodruff Community Hosp, 72-81, chmn podiatry staff, 74, Lakewood Doctors Hosp, 70-85, Los Alamitos, Med Ctr, 72-90, hon staff, 90-, chief podiatry sect, Dept Orthop, 82-90; clin instr med, Univ Southern Calif, Sch Med, Los Angeles, 65-80, podiatry consult prof staff, 70-, clin asst prof med, 81-; occupational podiatric med consult, Specified Prod Co, El Monte, Calif, 68-90; clin assoc prof, Calif Col Podiatric Med, Southern Campus, Los Angeles, 71-; adj prof, Calif State Univ, Long Beach, 72-88; consult, Armstrong World Industs, Lancaster, Pa, 83-; pres, Int Acad Standing & Walking Fitness, 83-; co-chair, podiatric med, Nat Acad Pract, 83- *Honors & Awards:* Mennen Award, Am Podiatry Asn, 44, Stickel Award, 63, Silver Award, 73; Stephen Toth Award, Am Pub Health Asn, 82. *Mem:* Am Asn Hosp Podiatrists (pres, 58-60); Am Public Health Asn; Asn Military Surgeons US; Am Podiatric Med Asn; Am Soc Podiatric Med; Am Acad Podiatric Sports Med. *Res:* Effect of urban substrata on public and occupational health; stress reducing, varied terrain floor surface; decreasing fatigue, pathology and accidents at the workplace and walkways; foot gear research. *Mailing Add:* Charles R Brantingham DPM Inc 1541 Los Padros Rd Nipono CA 93444-9625

BRANTLEY, LEE REED, b Herrin, Ill, Sept 23, 06; m 84. PHYSICAL CHEMISTRY. *Educ:* Univ Calif, AB, 27; Calif Inst Technol, MS, 29, PhD(chem), 30. *Prof Exp:* Reader physics, Univ Calif, Los Angeles, 26-29; asst chem, Calif Inst Technol, 28-30; instr chem & physics, Occidental Col, 30-36, asst prof chem, 36-40, from assoc prof to prof, 40-67, head dept, 40-62; prof educ, Curric Res & Develop Group, 67-72, consult, 72-73, EMER PROF EDUC, UNIV HAWAII, 72-; PRES, A & R RES ASSOCS, 77- *Concurrent Pos:* Res fel, Calif Inst Technol, 36-43, asst & Nat Defense Res Coun consult, 43-44; teacher schs, Calif, 41-42; Petrol Res Fund award, 58-59; consult, Crescent Eng Co, 44-46, Pac Rocket Soc, 44-51, Nat Bur Standards, 51-54 & Corn Indust Res Found, 57-59; dir contracts, US Off Naval Res, 49-58 & Off Qm Gen, 52-58; vis prof, Lehigh Univ, 58-59; vpres, Alpha Chi Sigma Educ Found, 58-63, trustee, 63-; vis prof, Univ Hawaii, 62-63 & 65-66, res prof, 85-; consult chem engr, 73- *Honors & Awards:* John R Kuebler Award, Alpha Chi Sigma, 73. *Mem:* Fel AAAS; Am Chem Soc; Electrochem Soc. *Res:* Preparation and properties of xenon trioxide; relation of air borne fungal spores and pollen to asthma in Hawaii; x-ray fluoreseence spectroscopy characterization of adhesion,; chemistry of architectural materials. *Mailing Add:* 2908 Robert Pl Honolulu HI 96816-1720

BRANTLEY, SUSAN LOUISE, b winter Park, Fla, Aug 11, 58. AQUEOUS GEOCHEMISTRY, HYDROTHERMAL GEOCHEMISTRY. *Educ:* Princeton Univ, AB, 80, MA, 83, PhD(geol & geophys sci), 87. *Prof Exp:* ASST PROF GEOCHEM, PA STATE UNIV, 86- *Concurrent Pos:* NSF presidential young investr, 87; consult, Sawyer Res Prod, Inc, 88- *Mem:* Geochem Soc; Am Geophys Union; AmChem Soc; AAAS; Asn Women Geosci. *Res:* Thermodynamics and Kinetics of rock-water interaction over the temperature range 25 to 600 degrees centigrade; dissolution and precipitation reactions of minerals, natural brines, mineral surface chemistry, and alteration of porosity and permeability. *Mailing Add:* Dept Geosci 209 Deike Bldg Pa State Univ University Park PA 16802

BRANTLEY, WILLIAM ARTHUR, b Roanoke Rapids, NC, Jan 19, 41; m 67. MATERIALS SCIENCE, DENTAL BIOMATERIALS. *Educ:* NC State Col, BS, 63; Carnegie Inst Technol, MS, 65, Carnegie-Mellon Univ, PhD(metall & mat sci), 68. *Prof Exp:* Res metall engr & lectr, Carnegie-Mellon Univ, 67-68; tech staff mem & group leader, Ceramic Res Lab, US Army Mat & Mech Res Ctr, 68-70; mem tech staff, Bell Labs, 70-74; asst prof, Dept Dent Mat, Marquette Univ, 74-79, assoc prof & actg chmn, 79-80, prof & chmn, 80-; SECT RESTORATIVE & PROSTHETIC DENT, COL DENT, OHIO STATE UNIV. *Mem:* Am Ceramic Soc; Am Soc Metals; Am Inst Mining, Metall & Petrol Engrs; Int Asn Dent Res; Am Asn Dent Res. *Res:* Dental biomaterials; biomechanics; mechanical properties of materials; defect characterization in crystalline materials. *Mailing Add:* Sect Restorative & Prosthetic Dent Col Dent Ohio State Univ 305 W 12th Ave Columbus OH 43210-1241

BRANTLEY, WILLIAM CAIN, JR, b Raleigh, NC, Feb 16, 49; m 70; c 2. COMPUTER ORGANIZATION, ARCHITECTURE. *Educ:* NC State Univ, BS, 71; Univ NMex, MS, 73; Carnegie-Mellon Univ, PhD(elec eng), 78. *Prof Exp:* Staff mem, Los Alamos Sci Lab, 71-74; RES STAFF MEM, IBM RES, 78- *Mem:* Inst Elec & Electronics Engrs; Asn Comput Mach. *Res:* High performance computer organization and architectures; vector, pipeline and multiprocessors; language oriental processors and the design automation of applicators for multiprocessors (graph-partitioning). *Mailing Add:* 8502 Appalachian Dr Austin TX 78759-7928

BRANTLEY, WILLIAM HENRY, b Forsyth, Ga, Aug 23, 38; m 61; c 2. NUCLEAR PHYSICS. *Educ:* Mercer Univ, AB, 60; Vanderbilt Univ, MA, 63, PhD(nuclear physics), 66. *Prof Exp:* Sci cooperator low energy nuclear physics, Tech Univ Delft, 65-66; ASSOC PROF PHYSICS, FURMAN UNIV, 66-, CHMN DEPT, 67- *Mem:* Am Phys Soc. *Res:* Research on the energy concept; educational research; environmental research; low energy nuclear physics, including beta and gamma ray spectroscopy. *Mailing Add:* Dept Physics Furman Univ Greenville SC 29613

BRANTON, DANIEL, b Antwerp, Belgium, Jan 13, 32; US citizen; m 57; c 2. CELL BIOLOGY, BOTANY. *Educ:* Cornell Univ, AB, 54; Univ Calif, Davis, MS, 57; Univ Calif, Berkeley, PhD(plant physiol), 61. *Prof Exp:* NSF fel, Swiss Fed Inst Technol, 61-63; from asst prof to prof bot, Univ Calif, Berkeley, 63-72; PROF BIOL, HARVARD UNIV, 72-, HIGGINS CHAIR PROF BIOL, 72- *Concurrent Pos:* Miller Res Prof, 67-68; Guggenheim fel, 70-71. *Mem:* Nat Acad Sci; Am Acad Arts & Sci; Am Soc Cell Biol (pres, 84-85); Biophys Soc; AAAS. *Res:* Cell and membrane structure; cell biology. *Mailing Add:* Biol Labs Harvard Univ Cambridge MA 02138

BRANTON, PHILIP EDWARD, b Toronto, Ont, June 8, 43; m 70; c 1. VIROLOGY, MOLECULAR BIOLOGY. *Educ:* Univ Toronto, BSc, 66, MSc, 68, PhD(med biophys), 72. *Prof Exp:* Fel biol, Mass Inst Technol, 72-74; asst prof cell biol, Univ Sherbrooke, 74-75; from asst prof to prof path, McMaster Univ, 75-90; PROF & CHMN BIOCHEM, MCGILL UNIV, 90- *Concurrent Pos:* Nat Cancer Inst Can Hunter res fel, 72-74; res scholar, Nat Cancer Inst Can, 74-79, mem fel panel, 76-79, res assoc, mem molecular biol panel, 80-84 & chmn, molecular biol panel, 84-85, mem adv comt res, 88- *Mem:* Am Soc Microbiol; Can Soc Cell Biol. *Res:* Studies on the structure and function of early region 1 proteins of human adenoviruses and on the isolation and characterization of phosphotyrosyl protein phosphatases. *Mailing Add:* McGill Univ 3655 Drummond Rm 802 Montreal ON H3G 1Y6 Can

BRAQUET, PIERRE G, b Dakar, Senegal, June 13, 47; m; c 3. CHEMISTRY. *Educ:* Univ Bordeaux II Med Sch, Pharmacien, 70; Univ Bordeaux I, DSc, 71; Univ Paris V Med Sch, PhD(pharm), 74. *Prof Exp:* Post grad res assoc chem, Col de France, 72; res assoc, Uniformed Army Serv Res Ctr, 72-74; dir, Dept Opers Res, Direction Centrale, Santne01 des Armne01es, 74-78; res dir, Fournier Lab, France, 78-79; dir develop, Merck, Sharp & Dohme, France, 79-82; GEN MGR & DIR RES, IPSEN BEAUFOUR GROUP, 82-; PROF, LA STATE UNIV, NEW ORLEANS, 88- *Concurrent Pos:* Assoc prof, Dept Chem, Univ Bordeaux Med Sch, Bordeaux, France, 69-72; vis prof, Georgetown Univ Med Sch, Wash, 84-86, Univ Sherbrook Med Sch, Can, 84-, Univ Milan Med Sch, Italy, 86-; numerous invited lectrs, US & foreign, 87- *Honors & Awards:* Award by Royal Soc Med London for Discovery of PAF Antagonists, Europ Biol Res Asn. *Mem:* Am Asn Immunologists; Int Soc Heart Res; fel NY Acad Sci; Am Heart Asn; Am Soc Exp Therapeut; Am Thoracic Soc; Int Soc Immunopharmacol; Int Soc Xenobiotics. *Res:* PAF antagonists; antihypertensive agents; antiallergy drugs; antiasthmatic drugs; antibacterials; antiinflammatory agents; antibrain edema agents; antiulcer agents; antivirals/anti AIDS; aspartyl protease antagonists; diuretics; hypdipemic agents; ion transport inhibitors; immunoregulators; leukotriene antagonists; prostaglandins agonists and antagonists; somatostatine analogues. *Mailing Add:* Inst Henri Beaufour 17 Ave Descartes Le Plessis Robinson 92350 France

BRAS, RAFAEL LUIS, b San Juan, PR, Oct 28, 50; m 74; c 2. STOCHASTIC HYDROLOGY. *Educ:* Mass Inst Technol, BS, 72, MS, 74, ScD, 75. *Prof Exp:* Asst prof hydrol & hydraul, Univ PR, 75-76; asst prof, 76-79, assoc prof, 79-84, PROF HYDROL, MASS INST TECHNOL, 84-, HEAD, WATER RESOURCES & ENVIRON ENG DIV, 83-, DIR, RALPH M PARSONS LAB, 83- *Concurrent Pos:* Ad honorem prof, Univ PR, 76-78; Chmn, Gilbert Winslow career develop award, Mass Inst Technol, 79-82 & William E Leonhard, 89-; John Simon Guggenheim Found fel, 82-83; consult govt & private indust, 74-; prin investr basic res sponsored by govt & indust, 76-; assoc ed, Water Resources Res, Am Geophys Union, 81-88; vis assoc prof, Univ Simon Bolivar, Venezuela, 82-83; vis scholar, Int Inst Appl Syst Anal, 83; mem ed bd, J Hydrol, 84-; chmn, Bd J Ed, Am Geophys Union, 84-88 & Budget & Finance, 90-; dir, Mass Inst Technol, MITES Prog, 87; vis prof, Univ Iowa, 89-90; assoc dir, Ctr Global Change Sci, Mass Inst Technol, 90-; chmn bd, Atmospheric Sci & Climate, NRC, 89-; adv bd, eng directorate, NSF, 89- *Honors & Awards:* Horton Award, Am Geophys Union, 81, James B Macelwane Award, 82. *Mem:* Am Soc Civil Engrs; Am Geophys Union; AAAS; Sigma Xi; Am Metrol Soc; Int Asn Hydraulic Res. *Res:* Use of probability theory in geophysics; streamflow and rainfall forecasting; soil moisture estimation and modelling; the estimation of random fields from point data; data collection network design. *Mailing Add:* Dept Civil Eng Rm 48-311 Mass Inst Technol Cambridge MA 02139

BRASCH, FREDERICK MARTIN, JR, b Mexico City, Mex, Sept 12, 43; US citizen; m 66; c 3. ELECTRICAL ENGINEERING, COMPUTER SCIENCE. *Educ:* Rice Univ, BA, 65, BS, 66, PhD(elec eng), 70. *Prof Exp:* Asst prof elec eng, Northwestern Univ, 70-75, asst prof comput sci, 71-75, assoc prof elec eng & comput sci, 75-81; SUPVR, AT&T BELL LABS, 81- *Mem:* Inst Elec & Electronics Engrs; Asn Comput Mach. *Res:* Distributed processing; real time systems; control systems. *Mailing Add:* AT&T Bell Labs 11900 N Pecos 31K70 Denver CO 80234

BRASCH, KLAUS RAINER, b Berlin, WGermany, Dec 19, 40; Can citizen; m 66; c 1. CELL BIOLOGY, ELECTRON MICROSCOPY. *Educ:* Concordia Univ, BS, 65; Carleton Univ, MS, 68, PhD(biol), 71. *Prof Exp:* Asst prof biol, Carleton Univ, Ottawa, 72-73; from asst prof to assoc prof, Queen's Univ, Kingston, 73-83; assoc prof, 83-87, PROF BIOL, UNIV TULSA, 87- *Concurrent Pos:* Mem bd dirs, Can Soc Cell Biol, 76-77; vis scientist, dept med genetics, City Hope Med Ctr, Duarte, Calif, 79-80. *Mem:* AAAS; Am Soc Cell Biol; Can Soc Cell Biol; Planetary Soc. *Res:* Functional organization of eukaryotic chromosomes and cell nuclei; role of nuclear proteins and steroid hormones in gene expression and cellular development; molecular analysis of human chromosomes and genetic disorders. *Mailing Add:* Fac Biol Sci Univ Tulsa 600 S College St Tulsa OK 74104

BRASCHO, DONN JOSEPH, b Syracuse, NY, Jan 9, 33; m 56; c 6. RADIOLOGY. *Educ:* Hobart Col, BS, 54; State Univ NY, MD, 58. *Prof Exp:* Asst chief radiol, Madigan Gen Hosp, Tacoma, Wash, 62-64; chief radiother, 62-64 & 65-66; radiologist, 121st Evacuation Hosp, Korea, 64-65; from asst prof to assoc prof, 66-74, PROF RADIATION THER, UNIV ALA, BIRMINGHAM, 74- *Mem:* AMA; Radiol Soc NAm; fel Am Col Radiol; NY Acad Sci; Am Soc Therapeut Radiol. *Res:* Clinical radiation oncology. *Mailing Add:* Dept Radiation Oncol Univ Ala 619 19th St Birmingham AL 35233

BRASE, DAVID ARTHUR, b Orange, Calif, May 9, 45. PHARMACOLOGY. *Educ:* Chapman Col, BS, 67; Univ Va, PhD(pharmacol), 72. *Prof Exp:* Fel pharmacol, Univ Calif, San Francisco, 72-76; asst prof pharmacol, Eastern Va Med Sch, 76-84; RES ASSOC DEPT PHARMACOL & TOXICOL, MED COL, VA, 84- *Concurrent Pos:* Res starter grantee, Pharmaceut Mfrs Asn Found, 77-78; adj asst prof, Old Dominion Univ, 77-84. *Mem:* Soc Neurosci; Am Soc Pharmacol & Exp Therapeut. *Res:* Pharmacology of narcotic drugs and endogenous opioids; interactions of central nervous system neurotransmitters. *Mailing Add:* Dept Pharmacol & Toxicol Med Col Va Va Commonwealth Univ Box 613 MCV Sta Richmond VA 23298-0001

BRASEL, JO ANNE, b Salem, Ill, Feb 15, 34. PEDIATRICS, ENDOCRINOLOGY. *Educ:* Univ Colo, Boulder, BA, 56; Univ Colo, Denver, MD, 59. *Prof Exp:* From intern to resident pediat, NY Hosp-Cornell Univ Med Ctr, 59-62; fel pediat endocrinol, Sch Med, Johns Hopkins Univ, 62-65, asst prof, 65-68; from asst prof to assoc prof, Med Col, Cornell Univ, 69-72; assoc prof pediat, Dir Div Growth & Develop & asst dir Inst Human Nutrit, Col Physicians & Surgeons, Columbia Univ, 72-79; PROF PEDIAT, SCH MED, UNIV CALIF-LOS ANGELES, 79- & PROF MED, 80- *Concurrent Pos:* Irma T Hirschl trust career scientist award, 73-77; Nat Inst Child Health & Human Develop res career develop award, 74-79; spec consult, Endocrine & Metab Adv Comt, Food & Drug Admin, 71-75; mem nutrit study sect, NIH, 74-78. *Mem:* Soc Pediat Res (secy-treas, 73-77, pres-elect, 77-78, pres, 78-79); Lawson Wilkins Pediat Endocrine Soc; Endocrine Soc; Am Soc Clin Nutrit; Am Pediat Soc. *Res:* Research interests in growth and development, pediatric endocrinology, infant nutrition and adipose tissue development. *Mailing Add:* Dept Pediat Med Ctr Univ Calif Los Angeles 1000 W Carson St Torrance CA 90509

BRASELTON, WEBB EMMETT, JR, b Americus, Ga, July 17, 41; m 64; c 2. ENDOCRINOLOGY. *Educ:* Univ Wis, BS, 63, MS, 66, PhD(endocrinol), 69; Am Bd Toxicol, dipl, 81. *Prof Exp:* Assoc biol chem, Harvard Med Sch, 71-73; asst res prof med, Med Col Ga, 73-77, asst prof endocrinol, 73-77, assoc res prof med, 77-78, dir, Gas Chromatography-Mass Spectrometry Facil, 73-78; assoc prof, 78-84, PROF PHARM & TOXICOL, MICH STATE UNIV, 84-, DIR, ANALYTICAL TOXICOL SECT, 78- *Concurrent Pos:* Res fel biol chem, Harvard Med Sch, 69-71; fel biochem, Mass Gen Hosp, Boston, 69-73. *Mem:* Am Soc Mass Spectrometry; Endocrine Soc; Am Chem Soc; Asn Off Anal Chem; Am Asn Vet Lab Diagnosticians; Am Acad Vet Comp Toxicol. *Res:* Metabolism and pharmacokinetics of drugs and xenobiotics; extraction, purification and physiochemical characterization of the anterior pituitary gonadotropic hormones; enzymes and mechanisms involved in steroid biosynthesis and metabolism; analytical methods development for organics and trace metals. *Mailing Add:* Dept Pharmcol & Toxicol Mich State Univ E Lansing MI 48824

BRASH, JOHN LAW, b Glasgow, Scotland, Mar 8, 37; m 64; c 3. BIOMEDICAL ENGINEERING. *Educ:* Glasgow Univ, BSc, 58, PhD(chem), 61. *Prof Exp:* Nat Res Coun Can fel, Div Pure Chem, Ottawa, Ont, 61-63; res chemist, Yerkes Lab, E I du Pont de Nemours & Co, 63-64; polymer chemist, Stanford Res Inst, 64-69, sr polymer chemist, 69-72; assoc prof chem eng, 72-76, PROF CHEM ENG, McMASTER UNIV, 76- *Concurrent Pos:* Vis scientist, Ctr Res Macromolecules, Strasbourg, France, 78-79; vis prof, Univ Paris-Nord, France, 86-87. *Mem:* AAAS; Am Chem Soc; Am Soc Artificial Internal Organs; Can Biomat Soc; Soc for Biomaterials. *Res:* Biomaterials; blood-surface interactions; adsorption of macromolecules, especially proteins; kinetics of polymerization reactions, segmented polyurethanes, bioactive polymers. *Mailing Add:* Dept Chem Eng McMaster Univ Hamilton ON L8S 4L8 Can

BRASHIER, CLYDE KENNETH, b Marion, La, May 21, 33; m 70; c 3. PLANT TAXONOMY, PLANT MORPHOLOGY. *Educ:* La Polytech Inst, BSc, 55; Univ Nebr, MSc, 57, PhD(bot), 61. *Prof Exp:* Prof bot & bact, Wis State Univ, Superior, 61-67; chmn dept math & sci, 67-74, acad dean, 74-76, prof biol, 76-82 & 85-86, chmn dept, 82-85, CHMN DEPT BIOL, DAKOTA STATE COL, 86- *Concurrent Pos:* Grant, Wis State Univ, 63-68; NSF fel, 63 & 68; dir, NSF Summer Insts, 64-67; grants, Dakota State Col, 68-78, 80-82 & 87-89. *Mem:* Bot Soc Am; Int Soc Plant Morphol. *Res:* Lake pollution research; environmental and career education. *Mailing Add:* Dept Math Sci Dakota State Univ Madison SD 57042

BRASITUS, THOMAS, MEDICINE. *Prof Exp:* MEM STAFF, JOSEPH B KIRSNER CTR STUDY DIGESTIVE DIS, UNIV CHICAGO. *Mailing Add:* Joseph B Kirsner Ctr For THe Study Digestive Diseases Univ Of Chicago 5841 S Maryland Ave Chicago IL 60637

BRASLAU, NORMAN, b Galveston, Tex, July 21, 31; m 55; c 2. PHYSICS. *Educ:* Agr & Mech Col, Tex, BS, 51; Ohio State Univ, MS, 52; Univ Calif, Berkeley, PhD(physics), 60. *Prof Exp:* Physicist, US Air Force Cambridge Res Ctr, Mass, 52-56; staff mem, Lawrence Radiation Lab, Univ Calif, 60; res staff mem, 61-90, ASSOC ED, IBM J RES & DEVELOP, IBM WATSON RES CTR, 90- *Concurrent Pos:* Vis fel, Ecole Normale Superieure, Paris, 60-61, vis prof, 75-76. *Mem:* Am Phys Soc; Sigma Xi. *Res:* Semiconductor device physics; non-linear optics; current instabilities in semiconductors; atmospheric physics. *Mailing Add:* Merritt Ct Katonah NY 10536

BRASSARD, ANDRE, b Quebec City, Que, May 31, 33; m 59; c 2. COMPARATIVE PATHOLOGY, IMMUNOPATHOLOGY. *Educ:* Laval Univ, BA, 55; Univ Montreal, DVM, 60; Univ Pa, MSc, 62, PhD(comp path), 64. *Prof Exp:* Asst comp pathologist, Univ Pa, 60-62, assoc comp pathologist, 62-64; asst prof, 65-70, ASSOC PROF COMP PATH, LAVAL UNIV, 70- *Mem:* NY Acad Sci; Int Union Conserv of Nature. *Res:* Experimental autoimmune glomerulonephritis. *Mailing Add:* 2891 La Promenade Ste-Foy PQ G1W 2J5 Can

BRASTED, ROBERT CROCKER, inorganic chemistry, chemical education; deceased, see previous edition for last biography

BRASTINS, AUSEKLIS, b Riga, Latvia, July 27, 25; US citizen; m 57; c 2. ELECTRICAL ENGINEERING, ELECTRONICS. *Educ:* Carnegie Inst Technol, BS, 54, MS, 55, PhD(elec eng), 60. *Prof Exp:* Res engr, Res Labs, Westinghouse Elec Corp, 60-61, sr engr, 61-64; res engr, Gulf Res & Develop Co, 64-68, sr res engr, 68-83; RETIRED. *Res:* High frequency loss mechanism in ferrites; semiconductor circuitry; seismic instrumentation; computer systems. *Mailing Add:* 108 Rutledge Dr Pittsburgh PA 15215-1920

BRASUNAS, ANTON DES, b Elizabeth, NJ, Mar 11, 19. METALLURGY & PHYSICAL METALLURGICAL ENGINEERING. *Educ:* Antioch Col, BS, 43; Ohio State Univ, MS, 46; Mass Inst Technol, DSc, 50. *Prof Exp:* Prof metall eng, 64-84, assoc dean, 75-80, EMER PROF METALL ENG, UNIV MO, 84- *Mem:* Fel Am Soc Metals; fel US Metric Asn. *Mailing Add:* Dept Metall Eng Univ MO 8001 Natural Bridge Rd St Louis MO 63121

BRASUNAS, JOHN CHARLES, b Elizabeth, NJ, Feb 6, 52; m 86. OUTER PLANET EXPLORATION, SPECTROMETER DEVELOPMENT. *Educ:* Princeton Univ, AB, 74; Harvard Univ, AM, 76, PhD(physics), 81. *Prof Exp:* Mem tech staff, Lincoln Lab, Mass Inst Technol, 81-84; ASTROPHYSICIST, NASA GODDARD SPACE FLIGHT CTR, 84- *Concurrent Pos:* Panel moderator, Workshop Mat Sci High Temperature Superconductors: Magnetic Interactions, 88; steering comt panel moderator, Advances in Mat Sci & Appln High Temperature Superconductors, 90; conf cochair, Superconductivity Appln Infrared & Microwave Devices, Int Soc Optical Eng, 90 & 91; prin investr, high-temperature-superconductor bolometers planetary missions, NASA, 90-92. *Mem:* Am Phys Soc; Am Astron Soc; Optical Soc Am; Am Geophys Union; AAAS. *Res:* Spectroscopic, remote sensing of the Earth's upper atmosphere and of the outer planets; infrared sensor development; development of spectrometer systems; high-temperature superconductors developed for use as sensors. *Mailing Add:* NASA Goddard Code 693 2 Greenbelt MD 20771

BRASWELL, EMORY HAROLD, b Brooklyn, NY, Jan 22, 32; m 52, 71; c 3. BIOPHYSICS, BIOPHYSICAL CHEMISTRY. *Educ:* Polytech Inst Brooklyn, BS, 52, MS, 55; NY Univ, PhD(phys chem), 61. *Prof Exp:* Res chemist, cent lab, Gen Foods Corp, 54-56; lectr chem, Hunter Col, 56-57; NIH fel, Univ Birmingham, 60-62; asst prof chem, Univ Conn, 62-64, bact, 64-68, biol, 68-73, assoc prof biol, 73-87, PROF MOLECULAR & CELL BIOL, UNIV CONN, 87-, HEAD, NAT ANAL ULTRACENTRIFUGATION FAC, 86- *Mem:* Am Chem Soc; Biophys Soc; Sigma Xi. *Res:* Physical biochemistry; physical chemistry of biomacromolecules; studies on dextran, gelatin, DNA and heparin, using techniques such as light scattering, analytical ultracentrifugation, viscosity, conductivity and dye binding; lactic dehydrogenase kinetics and subunit interaction; protein association. *Mailing Add:* U-125 Univ Conn Storrs CT 06268

BRASWELL, ROBERT NEIL, b Boaz, Ala, July 23, 32; m 54; c 2. SYSTEMS ENGINEERING. *Educ:* Univ Ala, BS, 57, MS, 59; Okla State Univ, PhD(eng), 64. *Prof Exp:* Mgr opers anal, Brown Eng Co, Inc, Ala, 59-61, sr proj engr, 61-62, mgr systs eng dept, Cape Kennedy, 62-64; prof indust & systs eng, Univ Fla, 64-73, chmn dept, 65-73, resident dir genesys, 64-66, dir systs res ctr, 68-72; scientist, 72-73, tech dir, Armament Div, Eglin AFB, Fla, 73-85; dir math & comput sci, USAF, Washington, DC, 85-86; DIR COMPUT & PROF, FLA STATE UNIV, 86- *Concurrent Pos:* Asst prof, Redstone Grad Sch, Univ Ala, 59-62; lectr, Ariz State Univ, 61-62; pres, Sole Educ Found Inc, 69- *Honors & Awards:* more than fifty awards for outstanding res and develop. *Mem:* Am Inst Indust Engrs; Am Soc Eng Educ; Nat Soc Prof Engrs; Opers Res Soc Am; Sigma Xi. *Res:* Supercomputing; robotics with applications in large scale design. *Mailing Add:* 804 Tarpon Dr Ft Walton Beach FL 32548

BRATCHER, THOMAS LESTER, b Fortworth, Tex, July 9, 42; c 2. BAYESIAN INFERENCE, SURVIVAL ANALYSIS. *Educ:* Univ Tex, Arlington, BS, 63; Southern Methodist Univ, MS, 65, PhD(statist), 69. *Prof Exp:* Tech staff statist, Sandia Nat Labs, 65-67; assoc prof, Univ Southwestern La, 69-79; ASSOC PROF STATIST & MATH, BAYLOR UNIV, 79- *Concurrent Pos:* Statistician, Nat Ctr Toxicol Res, 75-78. *Mem:* Am Statist Asn; Sigma Xi. *Res:* Statistical inference including estimation, multiple comparisons, selection and interval estimation; biostatistics: models to analyze animal experiment data for tumors and birth defects. *Mailing Add:* Dept Math Baylor Univ Waco TX 76798

BRATENAHL, ALEXANDER, b Washington, DC, June 24, 18; c 3. PLASMA PHYSICS, SPACE SCIENCE. *Educ:* Washington & Lee Univ, BS, 41; Univ Calif, Berkeley, PhD(physics), 52. *Prof Exp:* Chemist, E I du Pont de Nemours & Co Inc, Belle, WVa, 41-42; jr engr, Applied Physics Lab, Johns Hopkins Univ, 42-45; physicist, Lawrence Radiation Lab, Univ Calif, Berkeley, 52-60; sr staff physicist, Avco-Everett Res Lab, Everett, Mass, 60-61; lectr physics, Univ Calif, Santa Barbara, 61-62; physicist space sci, Jet Propulsion Lab, Calif Inst Technol, 62-76; RES PHYSICIST, INST GEOPHYSICS & SPACE PHYSICS, UNIV CALIF, RIVERSIDE, 75- *Mem:* Am Geophys Union; AAAS; Int Solar Energy Soc; Sigma Xi. *Res:* Magneto plasma dynamics; earth's magnetosphere and solar atmosphere; magnetospheric substorms and solar flares and laboratory simulation experiments; controlled thermonuclear fusion experiments. *Mailing Add:* 1111 Armada Dr Pasadena CA 91103

BRATER, D(ONALD) CRAIG, NEPHROLOGY. *Educ:* Duke Univ, MD, 71. *Prof Exp:* Assoc prof pharmacol & internal med, Southwestern Med Sch, Univ Tex, 77-86; PROF INTERNAL MED, SCH MED, IND UNIV, 86- *Mailing Add:* Dept Med Ind Univ Sch Med Wishard Mem Hosp WOP Bldg Rm 316 1001 W 10th St Indianapolis IN 46202

BRATERMAN, PAUL S, b London, UK, Aug 28, 38; c 3. INORGANIC CHEMISTRY. *Educ:* Oxford Univ, BA, 60, MA & DPhil, 63. *Hon Degrees:* DSc, Oxford Univ, 85. *Prof Exp:* Lectr chem, G Langan Univ, 65-77, sr lectr, 77-88, reader, 88; PROF CHEM, UNIV NTEX, 88- *Mem:* Am Chem Soc; fel Royal Soc Chemistry; Int Soc Study Origin Life. *Res:* Spectroscopy and electrochemistry of inorganic and other species in solution; morphology and development, form and pattern in inorganic precipitates. *Mailing Add:* Dept Chem Univ NTex Denton TX 76203

BRATHOVDE, JAMES ROBERT, b Glasgow, Mont, June 8, 26; m 49; c 4. ENVIRONMENTAL MANAGEMENT, ENVIRONMENTAL SCIENCES. *Educ:* Eastern Wash Col, BA, 50; Univ Wash, MS, 55, PhD(phys chem), 56. *Prof Exp:* Teacher, pub sch syst, Wash, 50-51; supvr, freshman labs, Univ Wash, 53-54, instr phys chem lab, 54; assoc prof chem, Whitworth Col, Wash, 56-57, chmn dept, 56-60; mem staff, phys sci res dept, Sandia Labs, NMex, 60-62 & aerospace nuclear safety prog, 62-63; assoc prog dir, undergrad educ sci, NSF, 63-64; prof chem & dir comput ctr, State Univ NY Binghamton, 64-67; prof chem & chmn dept, 67-70, dean col sci & humanistic studies, 70-72, prof, 72-84, EMER PROF ENVIRON SCI & CHEM, NORTHERN ARIZ UNIV, 84- *Mem:* Am Crystallog Asn; AAAS; Sigma Xi. *Res:* X-ray and neutron diffraction; molecular and crystal structures; ambient air and water quality; waste water operations; small scale energy converters; geothermal exploration. *Mailing Add:* 519 N James Flagstaff AZ 86001

BRATINA, WOYMIR JOHN, b Sturie, Yugoslavia, Feb 21, 16; Can citizen; m 55. SOLID STATE PHYSICS. *Educ:* Univ Zagreb, dipl mech eng, 40; Univ Ljubljana, dipl metall eng, 43; Univ Toronto, MASc, 52, PhD(metal physics), 54. *Prof Exp:* Res fel metal physics, Ont Res Found, 54-58, res scientist, 58-66, sr res scientist metal physics, Ont Res Found, 66-80; RETIRED. *Concurrent Pos:* Adj prof mat sci, Univ Toronto, 78- *Mem:* Am Phys Soc; Am Soc Metals. *Res:* Various aspects of cyclic stressing studied by fracture mechanics and materials science methodologies; materials primarily: iron-base alloys and titanium alloys; temperature range: 77 to 400 K; correlation of statistical multiple fatigue crack initiation and growth with microstructural characteristics; parameters: temperature, environment, etc; applications to surgical implants behavior and fracture. *Mailing Add:* Six Ridout St Toronto ON M6R 1Z2 Can

BRATSCHUN, WILLIAM R(UDOLPH), b Chicago, Ill, June 1, 31; m 54; c 3. CERAMIC ENGINEERING. *Educ:* Univ Ill, BS, 53, MS, 57, PhD(ceramic eng), 59. *Prof Exp:* Ceramic engr, Gen Elec Co, NY, 53-54; sr res scientist, Honeywell Res Ctr, 59-64; dir res, Channel Industs, Inc, 64-67; res scientist, IIT Res Inst, 67-72; mgr anal chem & mfg serv, Zenith Radio Corp, 72-87; MGR, MAT LAB, MOTOROLA, 87- *Mem:* Fel Am Ceramic Soc; Nat Inst Ceramic Engrs; Int Soc Hybrid Microelectronics; ASM; Inst Elec & Electronic Engrs. *Res:* Glasses; dielectrics; thick film materials; semiconductors; piezoelectrics; electronic applications of materials; failure analysis, reliability of electrical materials. *Mailing Add:* 127 N Ashland Ave La Grange IL 60525

BRATT, ALBERTUS DIRK, b Bozeman, Mont, Apr 2, 33; m 55; c 4. ENTOMOLOGY. *Educ:* Calvin Col, BS, 55; Mich State Univ, MS, 57; Cornell Univ, PhD(entom), 64. *Prof Exp:* Instr biol, Calvin Col, 58-61; asst entom, Cornell Univ, 61-64; asst prof, 64-70, chmn dept, 73-79, PROF BIOL, CALVIN COL, 70- *Mem:* Am Inst Biol Sci. *Res:* Morphology and taxonomy of Acalyptrate Diptera; fresh-water biology; morphology of immature insects. *Mailing Add:* Dept Biol Calvin Col Grand Rapids MI 49546

BRATT, BENGT ERIK, b Willstad, Sweden, May 18, 22; US citizen; m 66; c 1. OPERATIONS RESEARCH, STATISTICS. *Educ:* Tech Col, Gothenburg, Sweden, ME, 45; West Coast Univ, MSSE, 67; Western Colo Univ, PhD(sci), 77. *Prof Exp:* Systs engr systs res, Gen Dynamics, San Diego, 56-61; res engr space probs, Inst Nat Defense, Stockholm, Sweden, 61-62; res engr rocket propulsion, Northrop Space Labs, Los Angeles, 62-64; res scientist math & statist, Lockheed Res Ctr, Los Angeles, 64-67; systs engr forecasting & statist, Metrop Water Dist, 67-85; CONSULT ENGR, 85- *Concurrent Pos:* Lectr, Calif State Univ, Los Angeles. *Mem:* Svenska Teknolog foreningen, Sweden; AAAS. *Res:* Systems analysis; statistical methods and computer application. *Mailing Add:* 1942 Lemoyne St Los Angeles CA 90026

BRATT, PETER RAYMOND, b Syracuse, NY, Nov 11, 29; m 54; c 5. SOLID STATE PHYSICS. *Educ:* Syracuse Univ, BS, 54, PhD(physics), 65. *Prof Exp:* Sr res scientist, Spencer Labs, Raytheon Co, 60-63; MEM TECH STAFF, SANTA BARBARA RES CTR, 63- *Mem:* Am Phys Soc; Sigma Xi. *Res:* Photoconductivity in solids; semiconductor physics; infrared detection; solid state lasers. *Mailing Add:* 6505 Camino Venturoso Goleta CA 93117

BRATTAIN, MICHAEL GENE, b Ponca City, Okla, Oct 31, 47. BIOCHEMISTRY. *Educ:* Rutgers Univ, BS, 70, PhD(biochem), 74. *Prof Exp:* Res assoc path, 74-76, from instr to asst prof path, Univ Ala, Birmingham, 76-78, asst prof biochem, 78-; ASSOC PROF, DEPT PHARMACOL, BAYLOR COL MED. *Concurrent Pos:* Assoc scientist, Comprehensive Cancer Ctr, Univ Ala, Birmingham, 76- *Mem:* Am Asn Cancer Res. *Res:* Biochemical pathology of human cancer. *Mailing Add:* Detp Pharmacol Baylor Col Med Houston TX 77030

BRATTON, GERALD ROY, b San Antonio, Tex, Sept 25, 42; m 65; c 3. VETERINARY ANATOMY. *Educ:* Tex A&M Univ, BS, 65, DVM, 66, MS, 70, PhD(vet anat), 77. *Prof Exp:* From instr to asst prof vet anat, Tex A&M Univ, 66-75; Assoc prof vet anat in chg Anat progs, dept animal sci & vet med, Univ Tenn, Knoxville, 75-; AT DEPT VET ANAT, TEX A&M UNIV. *Mem:*

Am Asn Vet Anatomists; Am Vet Med Asn. *Res:* Developing an immunological method for tracing anatomical connections of the basal ganglia, and the use of horseradish peroxidase tracer methods to localize the anatomical positions of the medullary respiratory centers. *Mailing Add:* Dept Vet Anat & Pub Health Tex Vet Med Ctr Col Vet Med Tex A&M Univ College Station TX 77843-4458

BRATTON, SUSAN POWER, b Wilmington, Del, Oct 11, 48. PLANT ECOLOGY. *Educ:* Barnard Col, Columbia Univ, AB, 70; Cornell Univ, PhD(bot), 75, Univ Georgia, Athens, cert Environ Ethics, 85; Fuller Seminary, MA, 87. *Prof Exp:* Res biologist & res coordr ecol, Field Res Lab, Great Smoky Mountains Nat Park, US Dept Interior, 74-81; RES BIOLOGIST, US NAT PARK SERV COOP, UNIV GA, 81- *Concurrent Pos:* Instr, AuSable Inst. *Mem:* Ecol Soc Am; Brit Ecol Soc; Torrey Bot Club; Southern Appalachian Bot Club; Soc Conserv Biol. *Res:* Impact of overgrazing and exotic species on native plant communities; ecology of the European wild boar; structure and diversity patterns of herbaceous plant communities; rare and endangered plants; human trampling effects; deer browse; park management, ecotheology; landscape history. *Mailing Add:* Inst Ecol Univ Ga Athens GA 30602

BRATTSTEN, LENA B, b Gothenburg, Sweden. BIOCHEMICAL TOXICOLOGY. *Educ:* Univ Lund, Sweden, Fil kand, 67; Univ Ill, Urbana, PhD(insecticide biochem & toxicol), 71. *Prof Exp:* Fel biochem, Cornell Univ, 71-72, res assoc, 72-77; asst prof biochem & ecol, Univ Tenn, Knoxville, 77-83; sr res biologist, DuPont, 83-87; PROF ENTOM, RUTGERS, 87- *Mem:* Sigma Xi; AAAS; Entom Soc Am; Am Chem Soc; Int Soc Chem Ecol. *Res:* Comparative biochemical toxicology of xenobiotics; insect-plant interactions, chemical ecology; insecticide resistance mechanisms. *Mailing Add:* RD1 Box 58D Jackson NJ 08527

BRATTSTROM, BAYARD HOLMES, b Chicago, Ill, July 3, 29; m 52, 83; c 2. ZOOLOGY, ECOLOGY. *Educ:* San Diego State Col, BS, 51; Univ Calif, Los Angeles, MA, 53, PhD(zool), 59. *Prof Exp:* Asst cur herpet, Natural Hist Mus, San Diego, 49-51; dir educ, 50-51; asst zool, Univ Calif, Los Angeles, 52-55; fel paleoecol, Calif Inst Technol, 55-56; instr biol, Adelphi Univ, 56-60; from asst prof to assoc prof, 60-66, PROF ZOOL, CALIF STATE UNIV, FULLERTON, 66- *Concurrent Pos:* Res assoc, Los Angeles Co Mus, 61-; former pres & bd mem, Fullerton Youth Mus & Natural Sci Ctr; sr fel Monash Univ, Australia, 66-67; vis prof zool, Univ Sydney, 78 & Univ Queensland, 84; consult var govt agencies & pvt orgn. *Mem:* Fel AAAS; Am Soc Ichthyologists & Herpetologists; Ecol Soc Am; Am Soc Mammalogists; Am Ornith Union; fel Herpetologists League. *Res:* Ecology, behavior, paleontology, zoogeography and physiology of reptiles and amphibians; paleoecology and paleoclimates; social behavior of vertebrates; thermoregulation; ecology of tropics; population and repopulation problems. *Mailing Add:* Dept Biol Calif State Univ Fullerton CA 92634-4080

BRATZ, ROBERT DAVIS, b Sherman, Tex, Dec 14, 20; m 47; c 3. BIOLOGY, ECOLOGY. *Educ:* Sam Houston State Teachers Col, BS, 41; Ore State Univ, MS, 50, PhD(zool, bot), 52. *Prof Exp:* From asst prof to assoc prof, 53-63, PROF BIOL, COL IDAHO, 63- *Mem:* AAAS. *Res:* General biology and ecology; biogeography of Mexico and western United States. *Mailing Add:* 1400 Willow Caldwell ID 83605

BRAU, CHARLES ALLEN, b Brooklyn, NY, Nov 4, 38; m 66; c 3. PHYSICAL CHEMISTRY, APPLIED PHYSICS. *Educ:* Cornell Univ, BEP, 61; Harvard Univ, BA, 62, PhD(appl physics), 65. *Prof Exp:* Sr staff phys chem, Avco Everett Res Lab, 65-70; NATO fel molecular physics, Kammerlingh-Onnes Lab, Univ Leiden, Neth, 70-71; sr staff laser physics, Avco Everett Res Lab, 71-76; ASSOC GROUP LEADER LASER DEVELOP, LOS ALAMOS NAT LAB, 76-; NUCLEAR PHYSICS, NUCLEAR PHYSICS LAB. *Res:* Laser chemistry; excimer lasers; free electron lasers. *Mailing Add:* 9308 Crookett Rd Brentwood TN 37027

BRAU, JAMES EDWARD, b Tacoma, Wash, Nov 10, 46; m 69; c 2. EXPERIMENTAL HIGH ENERGY PHYSICS. *Educ:* US Air Force Acad, BS, 69; Mass Inst Technol, MS, 70, PhD(physics), 78. *Prof Exp:* Physicist, Cent Inertial Guid Test Facil, 70-71 & Air Force Weapons Lab, 71-74; res assoc, Stanford Linear Accelerator Ctr, 78-82; dept physics & astron, Univ Tenn, 82-88; PHYSICS DEPT, UNIV ORE, 88- *Mem:* Am Phys Soc; Sigma Xi. *Res:* Experimental investigations of the fundamental particles and their interactions, specifically the production and decay of neutral intermediate vector boson in electron- positron annihilation; emphasis on calorimetric techniques of measurement. *Mailing Add:* Physics Dept Univ Ore Eugene OR 97403

BRAUCHI, JOHN TONY, b Sayre, Okla, Dec 23, 27; m 48; c 3. PSYCHIATRY. *Educ:* Univ Okla, MD, 55; Am Bd Psychiat & Neurol, dipl, 62. *Prof Exp:* Intern, Hillcrest Med Ctr, 55-56; resident, Univ Hosps, Okla, 56-58; resident, Med Ctr, Univ Kans, 58-59, from instr to prof psychiat, 58-75, dir inpatient serv, 60-61, dir residency training, 64-69, dir prof serv psychiat, 64-75, assoc chmn dept psychiat, 68-75; PROF PSYCHIAT, FAMILY MED & CHMN DEPT PSYCHIAT, SCH MED, LA STATE UNIV, 75- *Concurrent Pos:* Dir, Atchison County Guid Clin, Kans, 59-60, consult, 60-61; psychiat consult, Vet Admin Consol Ctr, Wadsworth, 59-61; actg chief psychiat serv, Vet Admin Hosp, Kansas City, Mo, 61-63; psychiat consult, Vet Admin Hosp, Shreveport, La, 75-; chief psychiat, Confed Mem Med Ctr, Shreveport, 75- *Mem:* Fel Am Asn Social Psychiat; Am Asn Chmn Depts Psychiat; fel Am Psychiat Asn; fel Am Col Psychiat; AMA. *Res:* Scintillation measurement of gross locomotor behavior; electroconvulsive shock and retroactive inhibitions; sleep deprivation; concept identification in thinking disorders; psychopharmacology; government; biofeedback. *Mailing Add:* PO Box 33932 Shreveport LA 71130

BRAUD, HARRY J, b Hope Villa, La, Sept 9, 35; m 69; c 2. AGRICULTURAL ENGINEERING. *Educ:* La State Univ, BS, 57, MS, 59; Okla State Univ, PhD(eng), 62. *Prof Exp:* From asst prof to assoc prof, 61-72, PROF AGR ENG, LA STATE UNIV, 72- *Mem:* Am Soc Agr Engrs. *Res:* Environmental control for orchards and fields; hydraulic discharge through slits in an elastic pipe wall; flow of water in porous medium with respect to sub-irrigation. *Mailing Add:* Dept Agr Eng La State Univ EB Doran Bldg Rm 149 Baton Rouge LA 70803

BRAUDE, ABRAHAM ISAAC, medicine; deceased, see previous edition for last biography

BRAUDE, GEORGE LEON, b Samara, Russia, Mar 2, 18; US citizen; m 49. CHEMISTRY, CHEMICAL ENGINEERING. *Educ:* Wilna Univ, Lithuania, dipl, 42; Univ Halle, PhD(org chem), 45, Dr Econ Sc, 46. *Prof Exp:* Res asst chem, Rohm & Haas fel, Univ Halle, 42-45 & Univ Frankfurt, 46-47; res chemist, Prosynthese Co, Paris, 48-52 & Imperial Paper Corp, Collins & Aikman Co, NY, 52-55; dir com res, Alcolac Chem Corp, Baltimore, 56-57; res supvr, W R Grace & Co, Md, 57-66, sect mgr, 66-68, dir spec proj dept, 68-70, mgr org res, 70-71; chief, Chem Indust Pract Br, US Food & Drug Admin, Washington, DC, 72-80, asst dir, Contaminant Coord, 81-83; RETIRED. *Concurrent Pos:* consult, 83- *Honors & Awards:* Award of Merit, Am Chem Soc, 68. *Mem:* Am Chem Soc. *Res:* Environmental contaminants of foods; waste treatment; organic intermediates; foams and aerosols; pharmaceuticals; detergents; pigments; polymers; specialty chemicals. *Mailing Add:* 2410 Parkway Cheverly MD 20785

BRAUDE, MONIQUE COLSENET, b Lisieux, France, Nov 13, 25; US citizen; m 49. PHARMACY, TOXICOLOGY. *Educ:* Univ Paris, dipl, 48; Ohio State Univ, MS, 54; Univ Md, PhD(pharmacol), 63. *Prof Exp:* Res pharmacist, Blaque Labs, Paris, 49-53; res assoc, Sch Med, Univ Md, Baltimore City, 62-65; pharmacologist, US Food & Drug Admin, 66-69; chief biomed sect, Ctr Studies Narcotics & Drug Abuse, NIMH, 70-74; chief preclin pharmacol, Biomed Res Br, Div Res, Nat Inst Drug Abuse, 74-87; CONSULT, 87- *Honors & Awards:* Scientist Emer Award, Soc Exp Biol & Med, 90. *Mem:* AAAS; Am Soc Pharmacol & Exp Therapeut; Soc Toxicol; Soc Exp Biol & Med. *Res:* Neuropharmacology, psychopharmacology, perinatology and toxicology of drugs of abuse, marihuana, other central nervous system agents. *Mailing Add:* 2410 Parkway Cheverly MD 20785

BRAUER, FRED, b Konigsberg, Ger, Feb 3, 32; nat Can; m 58; c 3. MATHEMATICS. *Educ:* Univ Toronto, BA, 52; Mass Inst Technol, SM, 53, PhD(math), 56. *Prof Exp:* From asst to instr, Mass Inst Technol, 53-56; instr, Univ Chicago, 56-58; from lectr to asst prof math, Univ BC, 58-60; from asst prof to assoc prof, 60-66, PROF MATH, UNIV WIS, MADISON, 66- *Mem:* Am Math Soc; Math Asn Am; Can Math Soc; Soc Indust & Appl Math; Soc Theoret Biol; Can Soc Theoret Biol. *Res:* Ordinary differential equations; biomathematics. *Mailing Add:* Dept Math Univ Wis Madison WI 53706

BRAUER, GEORGE ULRICH, b Ger, Mar 18, 27; m 68. MATHEMATICS. *Educ:* Univ Toronto, BA, 49; Univ Mich, MA, 50, PhD(math), 54. *Prof Exp:* Asst, Univ Mich, 50-53; from instr to asst prof, 53-66, ASSOC PROF MATH, UNIV MINN, MINNEAPOLIS, 66- *Mem:* Am Math Soc; Math Asn Am. *Res:* Summation of infinite series; Dirichlet series; functional analysis; real and complex variables. *Mailing Add:* Sch Math Univ Minn Minneapolis MN 55455

BRAUER, GERHARD MAX, b Berlin, Ger, Feb 5, 19; nat US; m 68; c 2. CHEMISTRY, DENTAL RESEARCH. *Educ:* Univ Minn, BS, 41; Univ NC, MS, 48, PhD(chem), 50. *Prof Exp:* Asst chem, Univ NC, 46-47; RES CHEMIST DENT & MED MAT, POLYMER & STANDARDS DIV, NAT BUR STANDARDS, 50- *Concurrent Pos:* Sr res fel & adj prof, Free Univ Berlin, 74-75; mem, biomat adv comt, Nat Inst Dent Res & Comn MD-156 Dent Mat & Devices, Am Nat Standards Comt, subcomt chmn; subcomt chmn, Comn F-4 Med & Surg Mat & Devices, Am Soc Testing & Mat, sect chmn,. *Honors & Awards:* Silver Medal, US Dept Com, 64 & Gold Medal, 75; Souder Award, Int Asn Dent Res, 75; US Sr Scientist Award, Humboldt Found, 74. *Mem:* Am Chem Soc; Int Asn Dent Res; AAAS; Sigma Xi; hon fel Am Col Dentists; Adhesion Soc; Soc Biomats. *Res:* Chemistry of dental materials; reaction mechanisms; polymerization; analysis; reactivity of tooth surfaces; medical implants; adhesion; dental cements. *Mailing Add:* Dent & Med Mat Nat Bur Standards Gaithersburg MD 20899

BRAUER, JOHN ROBERT, b Kenosha, Wis, Apr 18, 43; m 82. FINITE ELEMENT ANALYSIS, COMPUTER-AIDED ENGINEERING. *Educ:* Marquette Univ, BEE, 65; Univ Wis-Madison, MSEE, 66, PhD(elec eng), 69. *Prof Exp:* Res asst elec eng, Univ Wis-Madison, 66-68, teaching asst elec eng, 68-69; res scientist elec mach, A O Smith Corp, 69-75, sr consult engr finite elements, A O Smith Data Systs Inc, 75-88; SR CONSULT ENGR FINITE ELEMENTS, MACNEAL-SCHWENDLER CORP, 88- *Concurrent Pos:* Lectr, Short Course on Finite Elements Analysis, Marquette Univ, 88- *Mem:* Inst Elec & Electronics Engrs; Sigma Xi. *Res:* Area of expertise is finite element analysis of electromagnetic fields; authored computer codes, one book and over 60 technical papers involving calculation of electromagnetic fields of frequencies from O to microwaves. *Mailing Add:* 929 N Astor St Apt 506 Milwaukee WI 53202

BRAUER, JOSEPH B(ERTRAM), b Lawrenceburg, Ind, Feb 12, 30; m 54; c 5. METALLURGY, CHEMISTRY. *Educ:* Purdue Univ, BSMetE, 51; Syracuse Univ, MChemE, 60. *Prof Exp:* Proj officer mat eng, USAF, 51-53, chief Prod Develop Unit, 53-55, Electronic Mat Sect, 55-58, Appl Physics Sect, 58-62, chief, Solid State Appln Sect, 62-79, chief, Microelectronics Br, 79-85, CHIEF, MICROELECTRONICS RELIABILITY DIV, ROME AIR DEVELOP CTR, USAF, 85- *Mem:* Am Soc Metals; fel Am Inst Chemists. *Res:* Metallurgy, chemistry and physics of solid state materials, especially work on magnetics, insulators, thermoelectrics, conductors, photoconductors and semiconductors, design and application of electronic devices and equipment employing these materials. *Mailing Add:* Chief Microelectronics Br USAF Rome Air Develop Ctr (RBR) Rome NY 13441-1500

BRAUER, RALPH WERNER, b Berlin, Ger, June 18, 21; nat US. PHARMACOLOGY. *Educ:* Columbia Univ, AB, 40; Univ Rochester, MSc, 41, PhD(biochem), 43. *Prof Exp:* Res chemist, Wyandotte Chem Co, 43 & Distillation Prod Co, NY, 43-44; instr pharmacol, Harvard Med Sch, 44-47; asst prof, Sch Med, La State Univ, 47-51; head, Pharmacol Br, US Naval Radiol Defense Lab, Calif, 51-66; prof physiol & pharmacol, Sch Med, Duke Univ, 66-71; PROF MARINE PHYSIOL & DIR, INST MARINE BIO-MED RES, UNIV NC, WILMINGTON, 71- *Concurrent Pos:* Mary Scott Newbold lectr, Philadelphia Col Physicians, 62; dir, Wrightsville Marine Bio-Med Lab, 66-71. *Mem:* Marine Technol Soc; Am Soc Pharmacol & Exp Therapeut; Soc Exp Biol & Med; NY Acad Sci; Radiation Res Soc. *Res:* Liver physiology; bile secretion; enzyme and fat chemistry; neutral phosphate esters as cholinesterase inhibitors; bromsulphthalein excretion; liver and plasma proteins; autoxidation of betaeleostearic acid; delayed radiation effects; nutrition and aging; environmental physiology. *Mailing Add:* Inst Res Interrelation Sci & Cult 1601 Doctors Circle Wilmington NC 28401

BRAUER, ROGER L, b Julesburg, Colo. SAFETY ENGINEERING, FACILITIES MANAGEMENT. *Educ:* Valparaiso Univ, BS & BA, 65; Univ Ill, Urbana-Champaign, PhD(mech eng), 72. *Prof Exp:* Engr, Boeing Co, 67-68; sr res engr, Roy C Ingersall Res Ctr, Borg-Warner Corp, 68-69; prin investr, Construct Eng Res Lab, US Army C Engr, 72-86, team leader, 87-90; TECH DIR, BD CERT SAFETY PROFESSIONALS, 90-91. *Concurrent Pos:* Adj asst prof safety eng, Mech & Indust Eng Dept, Univ Ill, Urbana-Champaign, 78-91; adminr, Eng Div, Am Soc Safety Engrs, 83-84, chmn, Acad Accreditation Coun, 86-90, vpres prof develop, 90-92; dir, Bd Cert Safety Professionals, 84-90. *Honors & Awards:* Charles V Culbertson Award, Am Soc Safety Engrs, 85. *Mem:* Am Soc Safety Engrs; Bd Cert Safety Professionals (secy-treas, 86-87, vpres, 88, pres, 89). *Res:* Facilities planning and management; ergonomics. *Mailing Add:* 602 W Austin St Tolono IL 61880

BRAUGHLER, JOHN MARK, b Pittsburgh, Pa, Dec 29, 50; m 72; c 3. NEUROPHARMACOLOGY & BIOCHEMICAL PHARMACOLOGY, FREE RADICAL BIOLOGY. *Educ:* Point Park Col, BS, 72; Univ Pittsburgh, PhD(pharmacol), 77. *Prof Exp:* Res assoc, Sch Med, Univ Pittsburgh, 77; res fel clin pharmacol, Univ Va, 77-79; from asst prof to assoc prof pharmacol, Col Med, Northeastern Ohio Univ, 79-83; asst prof biol & chem, Kent State Univ, 79-82; res scientist, 83-84, sr res scientist, CNS Res, 85-89, assoc dir, Acquisitions Rev, 89-90, DIR, ACQUISITIONS REV, UPJOHN CO, 91- *Mem:* NY Acad Sci; Am Soc Pharmacol & Exp Therapeut; Soc Neurosci. *Res:* Pharmacological treatment of central nervous system injury and ischemia; biochemistry of central nervous system trauma and ischemia; pharmacology and therapeutics of antioxidants and radical scavengers; biochemical mechanisms of lipid peroxidation; membrane damage; drug development. *Mailing Add:* Acquisitions Rev Unit Upjohn Co Kalamazoo MI 49001

BRAULT, JAMES WILLIAM, b New London, Wis, Feb 10, 32; m 52; c 3. SOLAR PHYSICS, ATOMIC SPECTROSCOPY. *Educ:* Univ Wis, BS, 53; Princeton Univ, PhD(physics), 62. *Prof Exp:* Res asst physics, Cornell Univ, 54-55; res asst, Princeton Univ, 55-58, instr, 61-64; physicist, Kitt Peak Nat Observ, 64-84; PHYSICIST, NAT SOLAR OBSERV, 84- *Honors & Awards:* Humboldt Prize, WGer, 86. *Mem:* Am Phys Soc; Int Astron Union; fel Optical Soc Am. *Res:* Solar and laboratory spectroscopy, both atomic and molecular, using a Fourier transform spectrometer. *Mailing Add:* Nat Solar Observ PO Box 26732 Tucson AZ 85726-6732

BRAULT, MARGARET A, b Manistique, Mich. CHEMISTRY. *Educ:* Northern Mich Univ, BA, 68, MA, 73; Ore State Univ, PhD(chem), 80. *Prof Exp:* asst prof chem, Albion Col, 81-; AT NORTHERN MICH UNIV. *Mem:* Am Chem Soc; Sigma Xi. *Mailing Add:* 122 E Hewitt Marquette MI 49855

BRAULT, ROBERT GEORGE, b Hoquiam, Wash, Dec 4, 18; m 48; c 2. POLYMER CHEMISTRY. *Educ:* Whitworth Col, Wash, BS, 42; Mich State Col, PhD(org chem), 48. *Prof Exp:* Asst, Armour & Co, 49-51; asst org chem, Sinclair Res Labs, Inc, 51-54; mkt res engr, Callery Chem Co, 54-59; res chemist, Hughes Aircraft Co, 59-85; RETIRED. *Mem:* Am Chem Soc; Sigma Xi. *Res:* Organic synthesis; fatty acid derivatives; organo-metallics; borane chemistry; organic photochromic compounds; photopolymerization; radiation resists. *Mailing Add:* 924 Princeton St Santa Monica CA 90403

BRAUMAN, JOHN I, b Pittsburgh, Pa, Sept 7, 37; m 64; c 1. CHEMISTRY. *Educ:* Mass Inst Technol, BS, 59; Univ Calif, Berkeley, PhD(chem), 63. *Prof Exp:* NSF fel chem, Univ Calif, Los Angeles, 62-63; from asst prof to prof, 63-80, chmn dept, 79-83, J G JACKSON-C J WOOD PROF CHEM, STANFORD UNIV, 80- *Concurrent Pos:* Mem, Chem Adv Panel, NSF, 74-78; Guggenheim fel, 78-79; mem, Bd Chem Sci & Technol, Nat Res Coun, 78-81 & 83-, Comt to Surv Opportunities in Chem Sci, 82-; Christensen Fel, Oxford Univ, 83-84; dep ed phys sci, Science, 85- *Honors & Awards:* Pure Chem Award, Am Chem Soc, 73 & Harrison Howe Award, 76 & James Flack Norris Award phys org chem, 86; Arthur C Cope Scholar Award, 86. *Mem:* Nat Acad Sci; Am Acad Arts & Sci; Am Chem Soc; The Chem Soc; fel AAAS. *Res:* Physical organic chemistry; gas phase ionic reactions; electron photodetachment spectroscopy; reaction mechanisms. *Mailing Add:* Dept Chem Stanford Univ Stanford CA 94305-5080

BRAUMAN, SHARON K(RUSE), b Elizabeth, NJ, Apr 14, 39; m 64; c 1. PHYSICAL ORGANIC CHEMISTRY. *Educ:* Mt Holyoke Col, BA, 61; Univ Calif, Berkeley, PhD(org chem), 65. *Prof Exp:* Sr chemist, SRI Int, 65-83; AT LOCKHEED RES LAB, 83- *Concurrent Pos:* Lectr chem, Stanford Univ, 75-76. *Mem:* Am Chem Soc; AAAS. *Res:* Polymer degradation, combustion and fire retardance; biocompatible polymers; membrane transport; chemistry of difluoramino compounds. *Mailing Add:* Lockheed Res Lab 93-50 3251 Hanover St Bldg 204 Palo Alto CA 94304-1191

BRAUN, ALVIN JOSEPH, b Chicago, Ill, July 10, 15; m 39; c 3. PLANT PATHOLOGY. *Educ:* Univ Chicago, BS, 37; Univ Wis, PhM, 38; Ore State Col, PhD(plant path), 47. *Prof Exp:* Asst, Univ Wis, 37-38 & Ore State Col, 38-42; anal chemist, Sherwin-Williams Co, Ill, 42-43; asst pathologist, Guayule Res Proj, Bur Plant Indust, Soils & Agr Eng, USDA, Calif, 43-44; pathologist, 44-45; from asst prof to prof, 45-76, EMER PROF PLANT PATH, NY STATE AGR EXP STA, GENEVA, 76- *Concurrent Pos:* Agr off, Food & Agr Orgn, UN, Res & Training Ctr for Rice Prod, Bangkok, Thailand, 66-67; vis prof, Res Inst Pomol, Poland, 75. *Mem:* Am Phytopath Soc; Sigma Xi. *Res:* Fungus and virus diseases; nematode and physiological problems of small fruits and grapes; fungicides; fumigants; spray machinery. *Mailing Add:* Dept Plant Path NY State Agr Exp Sta Geneva NY 14456-1315

BRAUN, CHARLES LOUIS, b Webster, SDak, June 4, 37; m 58; c 2. PHYSICAL CHEMISTRY. *Educ:* SDak Sch Mines & Technol, BS, 59; Univ Minn, PhD(phys chem), 63. *Prof Exp:* Chief process control training sect, Eng Reactors Group, US Army, Ft Belvoir, Va, 64-65; res instr, 65-66, chair, dept chem, 82-85, from asst prof to assoc prof, 66-72, PROF CHEM, DARTMOUTH COL, 77- *Concurrent Pos:* Dartmouth Col fac fel, Phys Inst, Univ Stuttgart, 69-70; vis prof, dept chem, Cornell Univ, 79-80; consult, Eastman Kodak Co, 79- *Mem:* Am Chem Soc; Am Phys Soc. *Res:* Photoionization and photoconductivity in molecular liquids and solids; electronically excited states of organic molecules. *Mailing Add:* Dept Chem Dartmouth Col Hanover NH 03755

BRAUN, CLAIT E, b Kansas City, Mo, Oct 4, 39; m 60; c 3. WILDLIFE BIOLOGY. *Educ:* Kans State Univ, BS, 62; Univ Mont, MS, 65; Colo State Univ, PhD(wildlife biol), 69. *Prof Exp:* Soil scientist, Soil Conserv Serv, 61-69; asst wildlife researcher, Colo Div Game, Fish & Parks, 69-73; wildlife researcher, 73-80, WILDLIFE RES LEADER, COL DIV WILDLIFE, 81- *Concurrent Pos:* Assoc, Mont Forest Conserv Sta, 63-65, Am Mus Natural Hist, 69-70 & Inst Arctic & Alpine Res, 70-79; mem grad fac, Colo State Univ, 69-; ed, J Wildlife Mgt, 81-83. *Honors & Awards:* USDA Soil Conserv Serv Merit Award, 65; Prof Award, CMPS, Wildlife Soc, 89. *Mem:* AAAS; Wildlife Soc; Am Soc Mammal; Cooper Ornith Soc; Wilson Ornith Soc (pres, 83-84); Am Ornith Union. *Res:* Population ecology of grouse, native columbids and alpine ecology. *Mailing Add:* Wildlife Res Ctr 317 W Prospect Rd Ft Collins CO 80526

BRAUN, DONALD E, b Dinuba, Calif, Dec 15, 30; m 56; c 2. ANALYTICAL CHEMISTRY. *Educ:* Fresno State Col, AB, 52, MA, 54; Univ of the Pac, PhD(chem), 65. *Prof Exp:* Lab technician, Biochem & Virus Lab, Univ Calif, Berkeley, 54-55; med technician, Bethel Deaconess Hosp, Newton, Kans, 56-57; PROF CHEM, PAC COL, CALIF, 57- *Concurrent Pos:* NSF lab asst, Univ of the Pac, 63, partic, NSF res participation chem teachers, Ore State Univ, 68-69; consult, Braun, Skaffs & Kevorkian, 66- *Mem:* Am Chem Soc. *Res:* Development and refinement of analytical procedures in water and biochemical samples. *Mailing Add:* Dept Chem Pac Col 1717 S Chestnut Fresno CA 93702

BRAUN, DONALD PETER, b New York, NY, Mar 7, 50; m 74; c 3. CANCER IMMUNOLOGY, MACROPHAGE IMMUNOBIOLOGY IN CANCER. *Educ:* Univ Ill, Urbana, BS, 72; Univ Ill Med Ctr, MS, 74 & PhD(microbiol), 76. *Prof Exp:* Instr immunol, Univ Ill Med Ctr, 76-78; asst prof oncol, 78-82, assoc prof, 83-87, ASSOC PROF ONCOL & IMMUNOL, RUSH MED COL, 87- *Concurrent Pos:* Mem, Small Bus Innovations Rev, Nat Cancer Inst, 83-85, Exp Therapeut Study Sect, 84-, Ariz Dis Control Comn, 86-88. *Mem:* NY Acad Sci; Am Asn Cancer Res; AAAS. *Res:* Elucidating how malignant disease influence immune function in patients with solid tumors; patients with AIDS. *Mailing Add:* Sect Med Oncol Rush Med Col Suite 830 Prof Bldg Chicago IL 60612

BRAUN, ELDON JOHN, b Glen Ullin, NDak, Jan 14, 37. PHYSIOLOGY, COMPARATIVE PHYSIOLOGY. *Educ:* Concordia Col, Moorhead, Minn, BA, 60; Univ Ariz, PhD(zool, biochem), 69. *Prof Exp:* Asst prof, 72-77, ASSOC PROF PHYSIOL, COL MED, UNIV ARIZ, 77- *Concurrent Pos:* NIH fel, Univ Ariz Col Med, 69-72. *Mem:* Am Ornith Union; Cooper Ornith Soc; Am Soc Nephrology; Int Nephrology Soc; Am Physiol Soc. *Res:* Comparative renal physiology; regulation of individual nephron function; measurement of single nephron glomerular filtration rates; function of countercurrent multiplier system changes in intrarenal blood flow patterns. *Mailing Add:* Dept Physiol Univ Ariz Col Med Tucson AZ 85724

BRAUN, JUERGEN HANS, b Hof, Ger, Nov 6, 27; US citizen; div; c 3. INORGANIC CHEMISTRY, PHYSICAL CHEMISTRY. *Educ:* Tech Univ, Berlin, Dipl Ing, 51; Univ Tex, PhD(inorg chem), 56. *Prof Exp:* From res chemist to sr res chemist, 55-71, RES ASSOC, CHEM & PIGMENTS DEPT, E I DU PONT DE NEMOURS & CO, INC, 71- *Honors & Awards:* Roon Found Award, Fedn Socs Coatings Technol, 89. *Mem:* Am Chem Soc. *Res:* Hydrothermal crystallization; crystal growth of semiconductors; chemistry of liquid ammonia; small particle technology; technology of white and colored pigments; pigment optics. *Mailing Add:* DuPont Chemicals Jackson Lab Wilmington DE 19898

BRAUN, LEWIS TIMOTHY, b Sacramento, Calif, Oct 27, 23; m 50; c 3. ECONOMIC GEOLOGY. *Educ:* Univ Calif, Berkeley, BS, 48. *Prof Exp:* Jr geologist, Atlantic Refining Co, Wyo, 48-49; asst mining geologist, Calif Div Mines, 49-51; geologist, Geophoto Serv, Inc, Colo, 51-55, proj mgr, Cagayan Basin, Geophoto Explor, Ltd, Philippines, 55-57; admin geologist, Geophoto Serv, Inc, 57-59, chief geologist, Geophoto Serv, Ltd, Alta, 59-64, proj mgr, Geophoto Serv, Inc, 64-71, dep mgr, Geophoto, Australia, 71-72, proj mgr, Geophoto Serv, Dallas, 72-77; explor geologist, Tex Pac Oil Co, Inc, Denver, 77-80; geol mgr, Hrubetz Oil Co, Denver, 80-85; CONSULT, 85- *Mem:* Am Asn Petrol Geologists; Am Inst Prof Geologists; fel Geol Soc Am; Can Soc Petrol Geologists. *Res:* Regional stratigraphic and structural geology and its relation to petroleum, mineral deposits and geothermal energy. *Mailing Add:* 7950 E Bethany Pl Denver CO 80234

BRAUN, LOREN L, b Waseca, Minn, June 12, 29; m 64; c 3. ORGANIC CHEMISTRY. *Educ:* Mankato State Col, BS, 51; Univ Nebr, MS, 53, PhD(chem), 56. *Prof Exp:* Chemist, Mead Johnson & Co, 56-57; from asst prof to assoc prof, 57-66, PROF CHEM, IDAHO STATE UNIV, 66- *Mem:* Am Chem Soc; Sigma Xi. *Res:* Aliphatic diazo compounds; N-Bromosuccinimide reactions; nitrogen bridgehead compounds. *Mailing Add:* Dept Chem Idaho State Univ Pocatello ID 83209

BRAUN, LUDWIG, b Brooklyn, NY, May 14, 26; m 47; c 4. TECHNOLOGY IN EDUCATION. *Educ:* Polytech Inst Brooklyn, BS, 50, MS, 55, DEE, 59. *Prof Exp:* Elec engr, Allied Control Co, 50-51; elec engr, Anton Electronics Labs, 51-52, head, Electronics Dept, 52-55; from instr to prof elec eng, Polytech Inst Brooklyn, 55-72; dir, Huntington Three Comput Proj, State Univ NY, Stony Brook, 81-82, prof eng, 72-82, dir, Lab Persona Comput Educ, 79-82; prof comput sci, Acad Comput Lab, NY Inst Technol, 82-87; res prof, NY Univ, 87-89; RETIRED. *Concurrent Pos:* Sr lectr, Dept Med, State Univ NY Downstate Med Ctr; mem prof staff, Dept Med, Kings County Hosp; consult, Grumman Aircraft Eng Co, 59-61, Gen Elec Co, 62-65, McGraw-Hill, 70-72, IBM, 78-79, Albert Einstein Med Ctr, Nat Inst Educ & Div Sci Educ Develop & Res, NSF. *Mem:* Inst Elec & Electronics Engrs; Int Soc Technol in Educ; Sigma Xi. *Res:* Feedback control theory; system analysis and design; medical engineering; technology in education; simulation. *Mailing Add:* 11 Parsons Dr Dix Hills NY 11746

BRAUN, MARTIN, b New York, NY, July 26, 41. APPLIED MATHEMATICS. *Educ:* Yeshiva Univ, BA, 63; NY Univ, MS, 65, PhD(math), 68. *Prof Exp:* Instr math, NY Univ, 67-68 & Courant Inst Math Sci, 68-70; asst res prof appl math, Brown Univ, 68-70, asst prof, 70-75; asst prof, 75-80, assoc prof, 80-85, PROF MATH, QUEENS COL CITY UNIV NEW YORK, 86- *Mem:* Soc Indust & Appl Math; Am Math Soc. *Res:* Qualitative theory of ordinary differential equations. *Mailing Add:* Dept Math Queens Col Flushing NY 11367

BRAUN, PETER ERIC, b BC, May 24, 39; m 79; c 2. BIOLOGICAL MEMBRANES, NEUROBIOLOGY. *Educ:* Univ BC, Vancouver, BScP, 61, MSc, 64; Univ Calif, Berkeley, PhD(biochem), 67. *Prof Exp:* Asst prof biochem, Univ Pa, Philadelphia, 69-73; assoc prof, 73-83, PROF BIOCHEM, MCGILL UNIV, MONTREAL, 84- *Mem:* Am Soc Biol Chem; Am Soc Neurochem; Soc Neurosci. *Res:* Biological assembly of neural membranes; structure and function of myelin; demyelinating and dysmyelinating disorders. *Mailing Add:* Dept Biochem McGill Univ 3655 Drummond St Montreal PQ H3G 1Y6 Can

BRAUN, PHYLLIS C, b Bridgeport, Conn, Jan 19, 53; m 75; c 2. MYCOLOGY. *Educ:* Fairfield Univ, BS, 75; Georgetown Univ, PhD(microbiol), 78. *Prof Exp:* Postdoctoral fel, Health Ctr, Univ Conn, 78-80; from asst prof to assoc prof, 80-89, PROF MOLECULAR BIOL, DEPT BIOL, FAIRFIELD UNIV, 89- *Concurrent Pos:* Consult, Int Schoeffel Industs, 84-86 & Miles Pharmaceut, 84-90. *Mem:* Am Soc Microbiol; NY Acad Sci; Med Mycol Asn Am; AAAS; Am Asn Women Sci. *Res:* Determining the mode of action of antifungal compounds and the action of these agents on the regulation of chitin synthelase, a membrane bound enzyme found in most fungii. *Mailing Add:* Dept Biol Fairfield Univ Fairfield CT 06430

BRAUN, ROBERT DENTON, b Santa Ana, Calif, June 28, 43; div. ANALYTICAL & CORROSION CHEMISTRY, ELECTROCHEMISTRY. *Educ:* Univ Colo, BA, 65; Univ Conn, MS, 70, PhD(anal chem), 72. *Prof Exp:* High sch teacher, Conn, 65-66; instr & fel anal chem, Univ Mich, 72-73; vis asst prof anal chem, Univ Ill, 73-74; asst prof chem, Vassar Col, 74-77; from asst prof to assoc prof 77-86, PROF CHEM UNIV SOUTHWESTERN LA, 86- *Mem:* Electrochem Soc; Am Chem Soc; Sigma Xi; NY Acad Sci; Nat Assoc Corrosion Engrs. *Res:* Electrochemistry of biochemical compounds; electrochemistry and electroanalysis in nonaqueous solvents; corrosion. *Mailing Add:* Dept Chem Univ Southwestern La Box 4-4370 Lafayette LA 70504-4370

BRAUN, ROBERT LEORE, b New England, NDak, Dec 25, 36; m 62; c 2. PHYSICAL CHEMISTRY, COMPUTER MODELING. *Educ:* Univ Wash, BS, 59, PhD(phys chem), 66. *Prof Exp:* Chemist, Hanford Labs, Gen Elec Co & Pac Northwest Labs, Battelle Mem Inst, 59-66; CHEMIST, LAWRENCE LIVERMORE NAT LAB, 66- *Honors & Awards:* Robert Peele Mem Award, Am Inst Mining, Metall & Petrol Engrs, 75. *Mem:* Am Inst Chem Engrs. *Res:* Computer modeling of physical and chemical processes related to recovery of natural resources. *Mailing Add:* Lawrence Livermore Lab PO Box 808 L-207 Livermore CA 94550

BRAUN, WALTER G(USTAV), b Springfield, Mass, June 23, 17; m 48; c 2. CHEMICAL ENGINEERING. *Educ:* Cooper Union, BCh, 42; Pa State Univ, MS, 48, PhD(chem eng), 55. *Prof Exp:* Jr chemist, Tidewater Assoc Oil Co, 36-42; asst petrol refining, Pa State Univ, 43-47, from instr to assoc res prof, 47-76, asst dean Col Eng, 70-74, prof, 74-80, EMER PROF CHEM ENG & ASSOC DEAN INSTR, PA STATE UNIV, 80- *Mem:* Fel Am Inst Chemists; Am Chem Soc; Am Inst Chem Engrs; Am Soc Eng Educ. *Res:* Raman and absorption spectroscopy; mass transfer; petroleum refining; thermodynamics. *Mailing Add:* 500 E Marylyn Ave No 74E State College PA 16801-6269

BRAUN, WERNER HEINZ, b Stuttgart, Ger, Mar 29, 45; US citizen; m 66; c 4. PHARMACOKINETICS, TOXICOLOGY. *Educ:* St Edward's Univ, BS, 67. *Prof Exp:* Anal chemist, Dow Life Sci, 67-79, res chemist spectros, 69-72; sr res chemist metab, 72-77, group leader, 77-82, health & Environmental Affairs mgr, 82-88, BIOTRANSFORMATION, DIV TOXICOL, DOW CHEM USA, 77- *Mem:* Soc Toxicol. *Res:* Application of pharmacokinetics and metabolism to hazard assessment. *Mailing Add:* Performance Prod Dept Dow Chem USA 1691 N Swede Rd Larkin Rd Midland MI 48674

BRAUN, WILLI KARL, b Reutlingen, WGer, Sept 22, 31; m 58; c 3. MICROPALEONTOLOGY, STRATIGRAPHY. *Educ:* Univ Tuebingen, WGer, Dr rer nat, 58. *Prof Exp:* From paleontologist to sr paleontologist, Shell Can Ltd, Alta, 58-64; from asst prof to assoc prof, 64-74, PROF PALEONT & STRATIG, UNIV SASK, 74- *Concurrent Pos:* Consult to oil co, 64- *Mem:* Fel Geol Soc Am; Paleont Soc; Brit Palaeont Asn; fel Geol Asn Can. *Res:* Paleozoic and Jurassic-Lower Cretaceous microfaunas and biostratigraphy of Western and Arctic Canada; tertiary to recent microfaunas and biostratigraphy of Mackenzie Delta-Beaufort Sea region, Arctic Canada. *Mailing Add:* Dept Geol Univ Sask Saskatoon SK S7N 0W0 Can

BRAUNDMEIER, ARTHUR JOHN, JR, b Granite City, Ill, Feb 28, 43; m 64; c 1. PHYSICS. *Educ:* Eastern Ill Univ, BSEd, 65; Univ Tenn, Knoxville, MS, 69; Oak Ridge Assoc Univ, PhD(physics), 70. *Prof Exp:* Assoc prof, 70-80, PROF PHYSICS, SOUTHERN ILL UNIV, EDWARDSVILLE, 80- *Concurrent Pos:* Consult, Oak Ridge Nat Lab, 71- *Mem:* Am Phys Soc; Am Asn Physics Teachers; Optical Soc Am; Health Physics Soc; Am Vacuum Soc. *Res:* Vacuum and thin film technology; optical constants of solids in the vacuum ultraviolet; solid state plasma oscillations; spectroscopy as a tool for measuring air pollution; detectors for ultraviolet radiation. *Mailing Add:* 14 Sunrise Ct Highland IL 62249

BRAUNE, MAXIMILLIAN O, b Bethlehem, Pa, Nov 21, 32; m 61; c 1. VIROLOGY, IMMUNOCHEMISTRY. *Educ:* Moravian Col, BS, 55; Univ Maine, MS, 58; Pa State Univ, PhD(bact, biochem), 63. *Prof Exp:* Instr rest vet sci, Pa State Univ, University Park, 58-63, asst prof vet sci, 63-; RETIRED. *Mem:* AAAS; Am Soc Microbiol; Sigma Xi. *Res:* Isolation and serological relationships of pleuropneumonia-like organisms; differential diagnosis of viral respiratory diseases by fluorescent antibody techniques. *Mailing Add:* Rte 1 Box 629 Emigrant MT 59027

BRAUNER, KENNETH MARTIN, b San Francisco, Calif, Dec 29, 27; m 54; c 3. ANALYTICAL CHEMISTRY. *Educ:* Univ Calif, BS, 49; Univ Chicago, PhD, 59. *Prof Exp:* Chemist, A Schilling & Co, 49-50 & Tidewater Assoc Oil Co, 52-53; res chemist, E I du Pont de Nemours & Co, 57-63; CHIEF CHEM DIV, DUGWAY PROVING GROUND, 63- *Mem:* AAAS; Am Chem Soc; Soc Appl Spectros. *Res:* Electrochemistry; trace analyses; chemical separations; physical methods of analysis. *Mailing Add:* 1675 E Granada Dr Sandy UT 84093-3743

BRAUNER, PHYLLIS AMBLER, b Natick, Mass, Oct 2, 16; m 43; c 2. ANALYTICAL CHEMISTRY, INORGANIC CHEMISTRY. *Educ:* Wheaton Col, BA, 38; Wellesley Col, MA, 40; Boston Univ, PhD, 59. *Prof Exp:* Head sci dept, Winnwood Sch, NY, 39; anal chemist, Bellevue Hosp, 41; asst, Purdue Univ, 41-43; anal chemist, Gen Elec Co, Mass, 44-45; asst, Northeastern Univ, 45-46; instr, Swarthmore Col, 46-48; asst chem, Simmons Col, 49-51, from instr to assoc prof, 51-66, prof, 67-83, emer prof chem, 83; RETIRED. *Concurrent Pos:* NSF fel, Switz, 60-61; res assoc, Royal Inst Technol, Sweden, 64-65; vis res assoc, Swiss Fed Inst Water Res & Water Pollution Control, 70-71; vis lectr chem, Framingham-Mass State Col; teacher, Univ Md, overseas prog, Japan & Gaum, 84. *Honors & Awards:* Hill Award, Am Chem Soc. *Mem:* AAAS; Am Chem Soc. *Res:* Physical methods of analysis; compleximetry and homogeneous precipitation; solution equilibria; ionic speciation in natural waters; preparation of complex inorganic compounds. *Mailing Add:* 15 Benton St Wellesley MA 02181

BRAUNFELD, PETER GEORGE, b Vienna, Austria, Dec 12, 30; US citizen; m 59; c 2. MATHEMATICS. *Educ:* Univ Chicago, AB, 49, BS, 51; Univ Ill, Urbana, MA, 52, PhD(math), 59. *Prof Exp:* Res asst prof math & coord sci lab, 59-63, from asst prof to assoc prof math & educ, 63-68, PROF MATH & EDUC, UNIV ILL, URBANA, CHAMPAIGN, 68- *Concurrent Pos:* Dir Math & Computer Educ Prog, 88- *Mem:* Math Asn Am; Nat Coun Teachers Math. *Res:* Mathematics education; mathematics curriculum development. *Mailing Add:* Dept Math Univ Ill Urbana, Champaign 1409 W Green St Urbana IL 61801

BRAUNLICH, PETER FRITZ, b Ger, Feb 25, 37; m 63; c 2. EXPERIMENTAL SOLID STATE PHYSICS, ATOMIC PHYSICS. *Educ:* Univ Marburg, pre-dipl, 58; Univ Giessen, dipl, 61, Dr rer nat(physics), 63. *Prof Exp:* Sci asst physics, Univ Giessen, 63-65; res assoc, mat res lab, Pa State Univ, 65-66; sr physicist, 66-70, sr proj physicist, Bendix Res Labs, 70-75, sr prin scientist, 75-76; from assoc prof to prof physics, 76-90, ADJ PROF PHYSICS, WASH STATE UNIV, 90- *Concurrent Pos:* Adj prof eng sci, Wayne State Univ, 74-76; pres, Int Sensor Technol, 82- *Mem:* Am Phys Soc. *Res:* Luminescence; thermally stimulated processes; photoconductivity; electron kinetics; exolectron emission; laser spectroscopy; interaction of laser light with matter; surface physics; dosimetry. *Mailing Add:* Int Sensor Technol Inc NE 1425 Terrace View Dr Pullman WA 99163

BRAUN-MUNZINGER, PETER, b Heidelburg, Germany, Aug 26, 46. NUCLEAR STRUCTURE. *Educ:* Univ Heidelburg, PhD(physics), 72. *Prof Exp:* From asst prof to assoc prof, 78-82, PROF PHYSICS, STATE UNIV NY, STONY BROOK, 82- *Mem:* Am Phys Soc. *Res:* Nuclear Physics; physics with relativistic heavy ions. *Mailing Add:* Dept Physics State Univ NY Stony Brook NY 11794

BRAUNSCHWEIGER, CHRISTIAN CARL, b Wellsville, NY, Oct 18, 26; m 53; c 5. MATHEMATICS. *Educ:* Alfred Univ, BA, 50; Univ Wis, MS, 51, PhD(math), 55. *Prof Exp:* Teaching asst, Univ Wis, 52-55; instr math, Purdue Univ, 55-57; from asst prof to assoc prof, 57-67, chmn dept, 70-73, PROF MATH, MARQUETTE UNIV, 67- *Concurrent Pos:* NSF fac fel, Univ Heidelberg, 64-65. *Mem:* Am Math Soc; Math Asn Am. *Res:* Functional analysis and topology; geometric models of abstract linear topological spaces. *Mailing Add:* Dept Math Marquette Univ Milwaukee WI 53233

BRAUNSCHWEIGER, PAUL G, b Troy, Ohio, Oct 18, 47; m 69, 88; c 3. CELL & TUMOR BIOLOGY. *Educ:* State Univ NY, Buffalo, BA, 69, PhD(radiation biol), 74. *Prof Exp:* Fel, Allegheny-Singer Res Corp, Allegheny Gen Hosp, 74-75, res assoc, 75-76, asst biologist, 76-77, assoc biologist cell kinetics, 77-81, sr staff scientist, Cancer Res Lab, 81-82, chief, Exp Therapeut AMC Cancer Ctr, 82-90; ASSOC PROF & CHIEF, DIV LAB RES, DEPT RADIOLOGICAL ONCOL, UNIV MIAMI, 90- *Concurrent Pos:* Instr physiol, LaRouche Col, 78-79. *Mem:* AAAS; Radiation Res Soc; Am Asn Cancer Res; Cell Kinetics Soc; Sigma Xi; Soc Magnetic Resonance Imaging. *Mailing Add:* 1720 C Wakeena Dr Miami FL 33133

BRAUNSTEIN, DAVID MICHAEL, b New York, NY, Dec 9, 42; m 65; c 2. ORGANIC CHEMISTRY, POLYMER CHEMISTRY. *Educ:* Polytech Inst Brooklyn, BS, 64; Purdue Univ, PhD(org chem), 68. *Prof Exp:* Assoc chemist, Gen Foods Corp, 64; res chemist, Celanese Corp Am, 68-74, group leader, 74-76, res assoc, 76-77; proj mgr, Shell Develop Co, 77-78; dir res & develop, Certainteed Corp, 78-79; MGR RES & ENG, TECHNOL DEVELOP & PLANNING, ARCO CHEM CO, 79- *Concurrent Pos:* Adj prof, Middlesex County Col, 71, Richmond Col, City Univ New York, 71. *Mem:* Am Chem Soc; Soc Plastics Engrs. *Res:* Polymer product and process development; thermoplastic resins, structural foams; stabilization, fire retardants; regulatory agency approvals; technical liaison with foreign affiliates; organic derivative synthesis; urethane chemistry. *Mailing Add:* 540 Green Hill Lane Berwyn PA 19312

BRAUNSTEIN, GLENN DAVID, b Greenville, Tex, Feb 29, 44; m 65; c 2. ENDOCRINOLOGY. *Educ:* Univ Calif, San Francisco, BS, 65, MD, 68. *Prof Exp:* Med intern, Peter Bent Brigham Hosp, Boston, 68, med resident, 69; clin assoc, Reproduction Res Br, NIH, 70-72; chief resident endocrinol, Harbor Gen Hosp, 72-73; asst prof med, Univ Calif, Los Angeles, 73-77; assoc prof, 77-81, PROF MED, UNIV CALIF, LOS ANGELES, 81- *Concurrent Pos:* Dir, Div Endocrinol, Cedars Sinai Med Ctr, 73-; dir dept med, Cedars Sinai Med Ctr, 86- *Mem:* Am Col Physicians; Endocrine Soc; Am Fedn Clin Res; Am Soc Clin Invest; Am Fert Soc; Am Asn Physicians. *Res:* Ectopic hormone production; placental endocrinology; fertility regulation. *Mailing Add:* Dept Med Rm B118 8700 Beverly Blvd Los Angeles CA 90048

BRAUNSTEIN, HELEN MENTCHER, b NY, Feb 5, 25; m 45; c 3. SCIENCE WRITING, INFORMATION SCIENCE. *Educ:* Univ Maine, BA, 64, MS, 65, PhD(phys chem), 71. *Prof Exp:* Res assoc solution chem, Reactor Chem Div, Oak Ridge Nat Lab, 72-73; DOCUMENT COORDR ENVIRON & HEALTH SCI, ENVIRON RESOURCE CTR, OAK RIDGE NAT LAB, 74- *Mem:* Am Chem Soc; Am Soc Info Sci; AAAS; Soc Tech Info; Sigma Xi. *Res:* Environmental and health literature research. *Mailing Add:* 101 Parsons Rd Oak Ridge TN 37830

BRAUNSTEIN, HERBERT, b New York, NY, Jan 10, 26; m 54; c 4. PATHOLOGY, MICROBIOLOGY & INFECTIOUS DISEASES. *Educ:* City Univ New York, BS, 44; Hahnemann Univ Med Col, MD, 50. *Prof Exp:* Intern, Montefiore Hosp, New York, 50-51; asst resident path, Univ Mich, 51-52; resident, Univ Cincinnati, 52-54, from instr to assoc prof, 54-64, dir, Inter-Dept Path Res Lab, 56-64; prof path, Chicago Med Sch & dir dept path, Michael Reese Hosp & Med Ctr, 64-65; from assoc prof to prof, Univ Ky, 65-70; clin prof path, Sch Med, Univ Calif, Los Angeles, 75-83, prof-in residence, biomed sci, Univ Calif, Riverside, 78-83; dir labs, San Bernardino County Gen Hosp, 70-88; CLIN PROF PATH, MED SCH, LOMA LINDA UNIV, 70- *Concurrent Pos:* Fel gastroenterol, Univ Cincinnati, 52-54; USPHS res career develop award, 58-64; asst & actg chief path, Vet Admin Hosp, 54-56, attend pathologist, 56-64; admin supvr, Cent Labs, Cincinnati Gen Hosp, 56-64. *Mem:* Am Asn Pathologists; Histochem Soc; US-Can Acad Path; Am Soc Clin Path; Sigma Xi. *Res:* Pathology of diseases of liver and cardiovascular systems; histochemistry of neoplasms and cardiovascular system; clinical microbiology and infectious disease. *Mailing Add:* San Bernardino County Med Ctr San Bernardino CA 92404

BRAUNSTEIN, JERRY, b New York, NY, Dec 24, 22; m 45, 77; c 4. PHYSICAL CHEMISTRY. *Educ:* City Col New York, BS, 42; Wesleyan Univ, MA, 47; Northwestern Univ, PhD(chem), 51. *Prof Exp:* Chemist, Manhattan Proj, Columbia Univ, 42-45; fel chem, Univ Wash, 50-52; physicist, Gen Elec Co, 52-54; from asst prof to prof chem, Univ Maine, 54-66; group leader, Chem Div, Oak Ridge Nat Lab, 66-90; RETIRED. *Concurrent Pos:* Res assoc, Oak Ridge Nat Lab, 60-61; lectr, Univ Maine, 66- & Univ Tenn-Oak Ridge Grad Sch Biomed Sci, 71- *Mem:* AAAS; Am Chem Soc; fel Am Inst Chem; Electrochem Soc. *Res:* Molten salts; thermodynamics; electrochemistry; concentrated electrolytes; solution chemistry; theoretical chemistry. *Mailing Add:* Rte 3 Box 272A Clinton TN 37716

BRAUNSTEIN, JOSEPH DAVID, b Long Island City, NY, Sept 19, 43; m 79; c 2. IMMUNOLOGY, CELL BIOLOGY. *Educ:* Cornell Univ, BA, 64; Univ Minn, PhD(chem), 72. *Prof Exp:* Asst Hebrew Univ, Jerusalem, 71-72; res assoc cell immunol, Sloan Kettering Inst, NY, 73-77; res assoc cell biol, Albert Einstein Med Sch, 77-81; ASST PROF PATH, MONTEFIORE HOSP, 81- *Mem:* Am Asn Immunol; Am Asn Clin Chem; NY Acad Sci. *Res:* Investigation of the relationship between the biochemistry, cell biology and endocrinology of cancer to the immune system. *Mailing Add:* Dept Chem Mt Sinai Hosp & Med Sch One Gustave L Levy Pl New York NY 10029

BRAUNSTEIN, JULES, b Buffalo, NY, Nov 4, 13; m 34. STRATIGRAPHY. *Educ:* George Washington Univ, BS, 33; Columbia Univ, MA, 36. *Prof Exp:* Paleontologist, Shell Oil Co, 37-41, stratigrapher, 41-52, spec probs, 52-55, sr stratigrapher, 55-61, area stratigrapher, 61-66, staff geologist, 66-78; ED, GEOL SURVEY ALA, 78- *Concurrent Pos:* Adj prof, Univ New Orleans, 79- *Honors & Awards:* Distinguished Serv Award, Am Asn Petrol Geologists, 75. *Mem:* Fel AAAS; hon mem Am Asn Petrol Geologists; hon mem Soc Econ Paleontologists & Mineralogists; fel Geol Soc Am; hon mem New Orleans Geol Soc. *Res:* Micropaleontology and stratigraphy of the Gulf Coast; lithology of carbonate rocks; petroleum geology of the southeastern Gulf Coast; regional stratigraphy, structure and petroleum geology of the US Gulf Coast. *Mailing Add:* 55 Gull St New Orleans LA 70124

BRAUNSTEIN, RUBIN, b New York, NY, May 6, 22; m 48. PHYSICS. *Educ:* NY Univ, BS, 48; Syracuse Univ, MS, 51, PhD(physics), 54. *Prof Exp:* Asst physics, Syracuse Univ, 48-51, res asst molecular beams, 51-52; res assoc, Columbia Univ, 52-53; mem res staff, Solid State Physics, Radio Corp Am Labs, 53-64; PROF PHYSICS, UNIV CALIF, LOS ANGELES, 64- *Concurrent Pos:* Consult, RCA Labs, 64-; Sci Res Coun fel, Oxford Univ, 74-75. *Mem:* AAAS; fel Am Phys Soc; NY Acad Sci. *Res:* Molecular beams; radiofrequency and microwave spectroscopy; solid state; quantum electronics. *Mailing Add:* Dept Physics Univ Calif Los Angeles CA 90024

BRAUNWALD, EUGENE, b Vienna, Austria, Aug 15, 29; nat US; m 52; c 3. INTERNAL MEDICINE, CARDIOLOGY. *Educ:* NY Univ, AB, 49, MD, 52. *Hon Degrees:* Harvard Univ, MA; Univ Libson, MD. *Prof Exp:* Intern & resident, Mt Sinai Hosp, New York, 52-54; clin assoc physiol, Nat Heart Inst, 55-56, resident med, 56-57; resident, Johns Hopkins Univ, 57-58, chief cardiol sect, Nat Heart Inst, 58-66, clin dir, 66-68; prof med & chmn dept, Sch Med, Univ Calif, San Diego, 68-72; HERSEY PROF MED, HARVARD MED SCH, 72-, CHMN DEPT, 80-; PHYSICIAN IN CHIEF, BRIGHAM & BETH ISRAEL HOSP, 72- *Concurrent Pos:* Fel med, Columbia Univ, 54-55; mem, US/USSR Res Proj Heart Dis, Nat Heart & Lung Inst, 72. *Mem:* Nat Acad Sci; Am Col Physicians; Am Physiol Soc; Am Fedn Clin Res (pres, 69-70); fel Am Acad Arts & Sci; Am Soc Clin Invest (pres, 74-75). *Res:* Cardiovascular hemodynamics and diagnostic techniques; clinical cardiology. *Mailing Add:* Harvard Med Sch Boston MA 02115

BRAUSE, ALLAN R, b New York, NY, July 27, 42; m 70; c 2. FOOD CHEMISTRY. *Educ:* Polytech Inst Brooklyn, BS, 63; Univ Wis, MS, 65; Univ Cincinnati, PhD(org chem), 67. *Prof Exp:* Res chemist, Phys Res Lab, Dow Chem Co, 67-68; chemist, Coatings & Chems Lab, US Army, 68-70; chemist, Res Div, US Indust Chem Co, 70-72; sr food chemist, Kroger Co, 72-84; chief chemist, Unique Anal Serv, 84-85; CHIEF ANALYTICAL CHEMIST, GEN PHYSICS CORP, 85- *Mem:* Am Chem Soc; Asn Official Anal Chemists; Inst Food Technologists; Sigma Xi. *Res:* Verification of authenticity of natural products, especially fruit juices and vanilla, by instrumental analysis; food and flavor chemistry; nutritional analysis; vitamins and fats by gas chromatography and liquid chromatography. *Mailing Add:* PO Box 531 Fulton MD 20759-0531

BRAUTH, STEVEN EARLE, b Trenton, NJ, Apr 12, 47. NEUROPSYCHOLOGY. *Educ:* Rensselaer Polytech Inst, BS, 67; NY Univ, PhD(psychol), 73. *Prof Exp:* Res fel biol, Calif Inst Technol, 72-74; res assoc anat, State Univ NY Stony Brook, 74-75; asst prof, 75-80, ASSOC PROF PSYCHOL, UNIV MD, COLLEGE PARK, 80- *Res:* Comparative neuroanatomy and neurophysiology of the vertebrate brain; study of the evolution of the brain-behavior relationships; comparative neurobiology of the basal ganglia. *Mailing Add:* Dept Psychol Univ Md College Park MD 20742

BRAUTIGAN, DAVID L, b Detroit, Mich, 1950. PROTEIN PHOSPHORYLATION, HORMONE ACTION. *Educ:* Kalamazoo Col, Mich, BA, 72; Northwestern Univ, BS, 73, PhD(biochem), 77. *Prof Exp:* Fel biochem, Univ Wash Med Sch, 77-80, res asst prof, 80-81; ASST PROF MED SCI, BROWN UNIV, 81- *Concurrent Pos:* Consult, Kemin Indust, Des Moines & Merck Sharp & Dohme. *Mem:* AAAS; Am Chem Soc; Am Soc Biol Chemists. *Res:* Mechanism of action of insulin and growth factors; protein phosphorylation of serine and tyrosine residue; protein phosphatases; protein structure and function. *Mailing Add:* Brown Univ Biomed Box G Providence RI 02912

BRAVENEC, EDWARD V, b Temple, Tex, Aug 25, 30; m 56; c 1. MATERIAL FAILURE ANALYSIS. *Educ:* Univ Tex, El Paso, BS, 56; Univ Houston, MS, 61; Pacific Western Univ, PhD(metall), 79. *Prof Exp:* Engr, Eastern State Petrol Co, 56-57; res, mat, Hughes Tool Co, 57-58; mgr, metall, Armco, Inc; VPRES ENGR, DIR & CONSULT, ANDERSON & ASSOC INC, 84- *Concurrent Pos:* Consult, Exxon, NASA, Prudential Life Ins, Union Carbide, et al, 84- *Honors & Awards:* Meritorious Serv, Am Soc Mech Engrs. *Mem:* Fel Am Soc Metals Int; Am Soc Mech Engrs; Am Soc Quality Control. *Res:* Material fracture mechanisms, material high-low temperature properties, corrosion properties, quality control and quality assurance management. *Mailing Add:* 206 Sleepy Hollow Ct Seabrook TX 77586

BRAVERMAN, DAVID J(OHN), b Los Angeles, Calif, Apr 21, 34; m 57; c 2. ELECTRICAL ENGINEERING, COMMUNICATIONS. *Educ:* Univ Calif, Los Angeles, BS, 56, MS, 58; Stanford Univ, PhD(elec eng), 61. *Prof Exp:* From asst prof to assoc prof elec eng, Calif Inst Technol, 61-65; sr scientist, Hughes Aircraft Co, 65-67, dept mgr, 67-68, assoc lab mgr, 68-69; res staff mem sci & technol, Inst Defense Anal, 69-76; lab mgr space & commun group, 76-80, ASST DIV MGR, HUGHES AIRCRAFT CO, 80- *Mem:* Inst Elec & Electronics Engrs; Am Inst Aeronaut & Astronaut. *Res:* Pattern recognition; communications theory; data processing. *Mailing Add:* Space & Commun Group Hughes Aircraft Co PO Box 92919 Los Angeles CA 90009

BRAVERMAN, IRWIN MERTON, b Boston, Mass, Apr 17, 29; m 55; c 3. MEDICINE, DERMATOLOGY. *Educ:* Harvard Univ, BA, 51; Yale Univ, MD, 55. *Prof Exp:* Intern, Med Serv, Yale Hosp, 55-56, asst resident med, 58-59; from asst prof to assoc prof, 62-73, PROF DERMAT, SCH MED, YALE UNIV, 73- *Concurrent Pos:* Helen Hay Whitney Found res fel dermat, Sch Med, Yale Univ, 59-62; vis scientist, Dept Biol Structure, Univ Wash, 69-70. *Mem:* Soc Invest Dermat; Am Fedn Clin Res; Am Acad Dermatologists; Am Dermat Asn. *Res:* Autoimmune disorders; lupus erythematosus; scleroderma and dermatomyositis, both from clinical and research aspects; microcirculation in skin, both from clinical and from electron microscopic aspects. *Mailing Add:* Sect Dermat Yale Univ New Haven CT 06520

BRAVO, JUSTO BALADJAY, b Philippines, Dec 5, 17; nat US; m 42; c 3. INORGANIC CHEMISTRY. *Educ:* Adamson Univ, Manila, BSChE, 40; Univ Kans, PhD(chem), 53. *Prof Exp:* Instr chem, Univ St Tomas, Philippines, 46-49; asst instr, Univ Kans, 49-50, res assoc, 50-53; res chemist, Wyandotte Chem Corp, Mich, 53 & Oldbury Electrochem Co, 54-56; res proj engr, Am Potash & Chem Corp, 56-57; sr res chemist, Foote Mineral Co, Pa, 57-59, group leader, 59-60; res supvr, Sun Oil Co, 60-63; sr res scientist, Glidden Co, Md, 63-64; prof, West Chester State Col, 64-83, chmn dept, 67-68, EMER PROF CHEM, WEST CHESTER UNIV, 83- *Concurrent Pos:* Sr consult, Wastex Industs Co, Inc, 71-77; grad coordr, West Chester State Col, 76-83. *Mem:* Sigma Xi. *Res:* Uninegative rhenium; inorganic pigments and phosphorus compounds; lithium metal and inorganic lithium compounds; electrochemicals; gas-solid reactions; fused system electrolysis; powder metallurgy; electrode processes; fuel cell development; synthesis, structure and properties of inorganic complex compounds. *Mailing Add:* 3000 22nd Ave Rio Rancho NM 87124-1658

BRAWER, JAMES ROBIN, b Patterson, NJ, Dec 15, 44; m 68; c 5. NEUROCYTOLOGY, NEUROENDOCRINOLOGY. *Educ:* Tufts Univ, BS, 66; Harvard Univ, PhD(neurocytol), 71. *Prof Exp:* NIH fel, Harvard Med Sch, 71-72; from asst prof to assoc prof anat, Med Sch, Tufts Univ, 72-75; from asst prof to assoc prof obstet, gynec & anat, 75-81, assoc scientist, 76-78, MED SCIENTIST, MCGILL UNIV-ROYAL VICTORIA HOSP, 78-, PROF OBSTET & GYNEC, MED SCH, 81-, PROF ANAT, 83- *Concurrent Pos:* Med Res Coun Can scholar, 76; assoc mem, Dept Med, Div Exp Med, McGill Univ, 77-, dir, Ctr Study Reproduction, 87-90. *Mem:* Soc Study Reproduction; Am Asn Anatomists; Sigma Xi. *Res:* Cytophysiology of neuroendocrine transducers in the medial basal hypothalamus; regulation of brain, pituitary, ovarian axis, polycystic ovarian disease. *Mailing Add:* Dept Obstet & Gynec Med Sch McGill Univ 3655 Drummond St Montreal PQ H3G 1Y6 Can

BRAWER, STEVEN ARNOLD, b Paterson, NJ, Oct 7, 41; m 67; c 2. PHYSICS. *Educ:* Rutgers Univ, BA, 64; Univ Colo, PhD(physics), 72. *Prof Exp:* Res assoc, Mat Res Lab, Pa State Univ, 72-77; PHYSICIST, LAWRENCE LIVERMORE LAB, 77- *Mem:* Am Phys Soc; Am Ceramic Soc. *Res:* Optical properties of dielectric glasses. *Mailing Add:* Six Dickson Dr Westfield NJ 07090

BRAWERMAN, GEORGE, b Biala Podlaska, Poland, June 12, 27; US citizen; m 53; c 2. BIOCHEMISTRY. *Educ:* Univ Brussels, BS, 48; Columbia Univ, PhD(biochem), 53. *Prof Exp:* Res assoc biochem, Col Physicians & Surgeons, Columbia Univ, 56-60, asst prof, 60-61; from asst prof to assoc prof, Yale Univ Sch Med, 61-70; PROF BIOCHEM, SCH MED, TUFTS UNIV, 70- *Concurrent Pos:* USPHS career develop award, 63-70. *Mem:* AAAS; Am Soc Biol Chemists. *Res:* Biochemistry of gene expression; structure and function of messenger RNA; protein biosynthesis. *Mailing Add:* Dept Biochem Tufts Univ Sch Med Boston MA 02111

BRAWLEY, JOEL VINCENT, JR, b Mooresville, NC, Feb 2, 38; m 59; c 3. ALGEBRA. *Educ:* NC State Univ, BS, 60, MS, 62, PhD(math), 64. *Prof Exp:* Instr math, NC State Univ, 64-65; from asst prof to assoc prof, 65-72, actg head dept, 77-78, PROF MATH SCI, CLEMSON UNIV, 72- *Concurrent Pos:* Vis lectr statist, Southern Regional Educ Bd, 67-69; vis lectr, Math Asn Am, 68-; lectr, Nat High Sch & Jr Col Math Club, 70-; vis assoc prof, NC State Univ, 71-72; sect lectr, Math Asn Am, 75-76; vis prof, Univ Tenn, Knoxville, 79-80. *Mem:* Math Asn Am; Am Math Soc. *Res:* Linear and abstract algebra, combinatorics. *Mailing Add:* Dept Math 201 Sikes Hall Clemson Univ Clemson SC 29631

BRAWLEY, SUSAN HOWARD, b Charlotte, NC, Oct 6, 51; m 90. PHYCOLOGY, COMMUNITY ECOLOGY. *Educ:* Wellesley Col, BA, 73; Univ Calif, Berkeley, PhD(bot), 78. *Prof Exp:* Res assoc, Smithsonian Inst, 79-83 & physiol dept, Univ Conn Health Ctr, 81-83; asst prof, 83-90, ASSOC PROF BIOL, VANDERBILT UNIV, 90- *Concurrent Pos:* NSF traineeship, Univ Calif, 73-74, regents fel, 74-75; Luce scholar, Univ Tokyo, 76-77; Smithsonian fel, 78-79; sci scholar, Bunting Inst, Radcliffe Col, 81-83; prin investr ecol, Nat Geog Soc grant, Qingdao, China, 86, NSF Grant, 84-87 & 88-91; assoc ed, J Phycol, 89-92. *Honors & Awards:* Career Advan Award, NSF, 88-89. *Mem:* Am Soc Cell Biol; Ecol Soc Am; Phycol Soc Am (treas, 89-91); Int Phycol Soc; Soc Develop Biol. *Res:* Early embryogenesis; cell polarization, polyspermy blocks; marine ecol; effects of small herbivores upon algal community structure. *Mailing Add:* Dept Biol Vanderbilt Univ Nashville TN 37235

BRAWN, MARY KAREN, b Ft Campbell, Ky, Oct 22, 55; m 82; c 1. BIOCHEMISTRY. *Educ:* NC State Univ, BS, 77; Duke Univ Med Ctr, PhD(biochem), 84. *Prof Exp:* Res fel, Baylor Col Med, 84-88; res fel, Univ Miami, 89-91; RES FEL, UPJOHN CO, 91- *Mem:* AAAS; Sigma Xi. *Mailing Add:* 6749 N 32nd St Richland MI 49083

BRAWNER, THOMAS A, b Cleveland, Ohio, Dec 8, 45; m 67; c 2. MICROBIOLOGY, VIROLOGY. *Educ:* Albion Col, BA, 67; Univ Tex, Austin, PhD(microbiol), 72. *Prof Exp:* Fel virol, Univ Tex, San Antonio, 71-73; asst prof microbiol, Univ Mo, Columbia, 73-81, assoc prof, 81; mem staff, Abbott Labs, 81-89; CHAIRPERSON, DEPT BIOL, CARTHAGE COL, 89-, DIV NAT SCI, 90- *Mem:* Sigma Xi; Am Soc Microbiol; AAAS. *Res:* Process of viral RNA replication; RNA viral biochemistry and genetics; replication of Hepatitis B virus; Hepatitis B viral markers; Delta virus replication and markers. *Mailing Add:* Dept Biol Carthage Col Kenosha WI 53140

BRAY, A PHILIP, b San Francisco, Calif, Sept 23, 33; m 56; c 6. MECHANICAL ENGINEERING. *Educ:* Univ Calif, Berkeley, BS, 55. *Prof Exp:* Mem staff, Gen Elec Co, 55-79, vpres & gen mgr, Nuclear Power Systs Div, 79-84; sr vpres, 84-85, exec vpres, 84-86, PRES & CHIEF EXEC OFFICER, MGT ANALYSIS CO, 86- *Concurrent Pos:* Chmn, Goose Lake Lumber Co, 87-; chmn, Teal Tech Inc, 87-; mem bd, Sequoia Nat Bank, Energeo, Inc, Thomas Res Corp, Thermal Energy Storage, Inc. *Honors & Awards:* Lawrence Mem Award, US Dept Energy. *Mem:* Nat Acad Eng. *Mailing Add:* Mgt Analysis Co PO Box 85404 San Diego CA 92138

BRAY, BONNIE ANDERSON, b Lincolnton, Ga, Oct 27, 29; m 65. GLYCOPROTEINS, GLYCOSAMINOGLYCANS. *Educ:* Univ Ga, BS, 50; Columbia Univ, PhD(biochem), 63. *Prof Exp:* Asst chemist, Union Carbide Corp, 50-51, jr chemist, 51-52; jr biologist, Oak Ridge Nat Lab, 52-59; res assoc biochem & med, Col Physicians & Surgeons, Columbia Univ, 63-66; NIH fel, Lab Biophys Chem, Nat Inst Arthritis & Metab Dis, 66-67; res assoc, New York Blood Ctr, 67-69; res assoc, 69-83, ASSOC RES SCIENTIST, DEPT MED, COL PHYSICIANS & SURGEONS, COLUMBIA UNIV, 83- *Mem:* Soc Complex Carbohydrates; Am Chem Soc; Am Soc Biochem & Molecular Biol; Biochem Soc; Soc Exp Biol & Med. *Res:* Glycosaminoglycan biochemistry; linkage of glysosaminoglycans to protein; presence of glycosaminoglycans in platelets; glycoproteins: fibrinogen and fibronectin, especially tissue fibronectin; basement membrane biochemistry and immunology; connective tissue of lung and placenta. *Mailing Add:* St Lukes-Roosevelt Hosp Ctr AJA Rm 101 428 W 59 St New York NY 10019

BRAY, BRUCE G(LENN), b Deckerville, Mich, Mar 17, 30; m 54; c 3. CHEMICAL ENGINEERING. *Educ:* Univ Mich, BS, 53, MS, 54, PhD(chem eng), 57. *Prof Exp:* Res assoc, Eng Res Inst, 53-57; res engr, Continental Oil Co, Okla, 57-59, sr res engr, 59-61, res group leader, 61-65, res assoc, 65-66; dir eng, CER Geonuclear Corp, 67-73; DIR SPEC PROJS, CONTINENTAL OIL CO, 73- *Concurrent Pos:* Mem indust adv comt gen eng, Univ Nev, Las Vegas, 71-73 & indust adv comt, Sch Technol, Okla State Univ, 75-76. *Mem:* AAAS; Am Inst Chem Engrs; Sigma Xi; Am Inst Mining, Metall & Petrol Engrs. *Res:* Applied thermodynamics; effects of gamma radiation on several polysulfone reactions; hydrocarbon processing; reservoir engineering; low temperature processing; peaceful uses of nuclear explosives; in situ leaching; coal preparation, treating; hydraulic transport; mining research; coal characterization. *Mailing Add:* 2500 Donner Ponca City OK 74604

BRAY, DALE FRANK, b Paw Paw, Mich, Mar 2, 22; m 47; c 2. ENTOMOLOGY. *Educ:* Mich State Col, BS, 47, MS, 49; Rutgers Univ, PhD(entom), 54. *Prof Exp:* Grad asst entom, Mich State Col, 47-49; from instr to asst prof, Univ Del, 49-55; assoc entomologist, Bartlett Tree Res Labs, 55-58; prof entom & chmn dept, Univ Del, 58-80, emer prof, 80-83; pres & entomologist, Bray Entom Serv, 83-87; RETIRED. *Mem:* Entom Soc Am; hon mem Nat Pest Control Asn; Am Entom Soc. *Res:* Economic entomology; shade tree and ornamental plant insects; microlepidoptera-urban entomology. *Mailing Add:* PO Box 9 18 S Pkwy Dimock PA 18816

BRAY, DONALD JAMES, b Anamosa, Iowa, Nov 8, 23; m 48; c 5. POULTRY NUTRITION. *Educ:* Iowa State Col, BS, 50; Kans State Col, MS, 52, PhD, 54. *Prof Exp:* From asst prof to assoc prof animal sci, Univ Ill, Urbana, Champaign, 54-68, prof, 68-80; mem staff, Coop State Res Serv, USDA, 80-83; RETIRED. *Mem:* AAAS; Poultry Sci Asn; World Poultry Sci Asn; Am Inst Nutrit; Sigma Xi. *Res:* Nutritional and environmental factors as they influence laying hens. *Mailing Add:* 10713 Tenbrook Dr Silver Spring MD 20901

BRAY, GEORGE A, b Evanston, Ill, July 25, 31; c 4. MEDICINE. *Educ:* Brown Univ, AB, 53; Harvard Univ, MD, 57. *Prof Exp:* From asst prof to assoc prof med, Sch Med, Tufts Univ, 64-70; from assoc prof to prof med, Sch Med, Univ Calif, Los Angeles, 70-81; dir clin study ctr, Harbor Gen Hosp, Torrance, Calif, 70-78; prof med & chief, Div Diabetes & Clin Nutrit, Univ Southern Calif, Los Angeles, 81-89; DIR, PENNINGTON BIOMED RES CTR, LA STATE UNIV, BATON ROUGE, 89- *Concurrent Pos:* NSF fel, Nat Inst Med Res, 61-62; NIH fel, New Eng Med Ctr Hosps, 62-64; NIH res grants, Harbor Gen Hosp, 67 & 70; assoc chief, Div Endocrinol & Metab, Harbor Gen Hosp, 70-; chmn, Fogarty Ctr Conf Obesity, 73; consult, Food & Drug Admin, 71 & Dept Health & Welfare, Can, 74; invited witness, Select Comt Nutrit & Human Needs, US Sen, 74. *Honors & Awards:* Osborne & Mendel Award, AIN, 88; McCollum Award, Am Soc Clin Nutrit, 89. *Mem:* Asn Am Physicians; Am Soc Clin Invest; Sigma Xi; Am Physiol Soc; Endocrine Soc; Fel AAAS. *Res:* Etiology, treatment and management of experimental animal and human obesity. *Mailing Add:* Pennington Biomed Res Ctr 6400 Perkins Rd Baton Rouge LA 70808

BRAY, JAMES WILLIAM, b Atlanta, Ga, Feb 6, 48. THEORETICAL SOLID STATE PHYSICS. *Educ:* Ga Inst Technol, BS, 70; Univ Ill, MS, 71, PhD(physics), 74. *Prof Exp:* physicist, Gen Elec Res & Develop Ctr, 74-79; mgr, Phys Sci Br, 79-82, electronic mat sci br, 82-87, SUPERCONDUCTING SYSTS PROG, GE CORP RES & DEVELOP, 87- *Mem:* Am Phys Soc. *Res:* Theoretical study of magnetic and transport properties of quasi-one-dimensional system; superconductivity. *Mailing Add:* Gen Elec Res & Develop Ctr PO Box 8 Schenectady NY 12301

BRAY, JOAN LYNNE, b Yankton, SD, Oct 10, 35. ELECTRON MICROSCOPY. *Educ:* Rosary Col, BA, 64; Purdue Univ, MS, 69, PhD(biol), 75. *Prof Exp:* ASST PROF BIOL, UNIV NORTH FLA, 76- *Mem:* AAAS; Nat Asn Biol Teachers. *Res:* Ultrastructure of algal development. *Mailing Add:* 10561 Hampton Rd Univ NFla Jacksonville FL 32217

BRAY, JOHN ROGER, b Belleville, Ill, June 20, 29; m 61; c 3. ECOLOGY. *Educ:* Univ Ill, BA, 50; Univ Wis, PhD(bot), 55. *Prof Exp:* Asst, Univ Wis, 50-54; vis lectr, Univ Minn, 55-57; asst prof, Univ Toronto, 57-62; prin sci officer, Dept Sci & Indust Res, 63-66; CONSULT ECOLOGIST, 66- *Concurrent Pos:* Vis prof, Univ Minn, 69. *Mem:* AAAS; Ecol Soc Am; Brit Ecol Soc; NZ Ecol Soc. *Res:* Quantitative techniques for sampling and classifying vegetation; productivity and efficiency of terrestrial vegetation; ecologic theory; historical climatology; influence of solar and volcanic activity on climate; forest regeneration dynamics. *Mailing Add:* PO Box 494 Nelson New Zealand

BRAY, JOHN THOMAS, b Terre Haute, Ind, Apr 13, 44; m 67; c 2. ANALYTICAL CHEMISTRY, BIOCHEMISTRY. *Educ:* Rose-Hulman Inst Technol, BS, 67; Johns Hopkins Univ, MA, 70, PhD(chem oceanog, geochem), 73. *Prof Exp:* Res asst dept earth & planetary sci, Chesapeake Bay Inst, Johns Hopkins Univ, 67-77, assoc res scientist dept geog & environ eng, 73-74; chem prog coordr, Ecol Anal, Inc, 74-78; asst prof dept surg & dir, Trace Elements Labs, 78-84, ASSOC PROF SHARED RES RESOURCES LABS & GROUP LEADER, ELEMENTAL ANALYSIS GROUP, SCH MED, E CAROLINA UNIV, 84- *Mem:* Am Chem Soc; Am Soc Limnol & Oceanog; AAAS; Sigma Xi; Soc Appl Spectros; Soc Environ Geochem & Health. *Res:* Analytical chemistry of trace elements in biomedical, environmental and geological matrices; biogeochemical cycling of trace elements in the environment and the role of trace elements in nutrition and other physiological functions. *Mailing Add:* 115 Lakeview Dr Greenville NC 27834

BRAY, NORMAN FRANCIS, b Louisville, Ky, May 27, 38; m 66; c 2. PHYSICAL CHEMISTRY. *Educ:* Univ Louisville, BS, 60, PhD(phys chem), 63. *Prof Exp:* Res scientist, NASA, Ohio, 63-64; from instr to asst prof chem, Hunter Col, 64-69; ASST PROF CHEM, LEHMAN COL, 69- *Mem:* Am Chem Soc. *Res:* Structure determination by nuclear magnetic resonance; solution thermodynamics. *Mailing Add:* Herbert H Lehman Col Dept Chem Bedford Park Blvd W Bronx NY 10468

BRAY, PHILIP JAMES, b Kansas City, Mo, Aug 25, 25; m 51; c 3. SOLID STATE PHYSICS. *Educ:* Brown Univ, ScB, 48; Harvard Univ, MA, 49, PhD(physics), 53. *Prof Exp:* Asst prof physics, Rensselaer Polytech Inst, 52-55; from assoc prof to prof, Brown Univ, 55-85, chmn dept, 63-68, Hazard prof physics, 85-90, EMER HAZARD PROF PHYSICS, BROWN UNIV, 90- *Concurrent Pos:* NSF sr fel, 61-62; John Simon Guggenheim fel, 68-69; vis prof, Dept Glass Technol, Univ Sheffield, 61-62 & 68-69; vis prof, Dept Chem, Univ Exeter, 75-76. *Honors & Awards:* George W Morey Award, Am Ceramic Soc, 70. *Mem:* Fel AAAS; fel Am Phys Soc; Am Acad Arts & Sci; fel Soc Glass Technol; fel Am Ceramic Soc. *Res:* Nuclear magnetic resonance and nuclear quadrupole resonance studies of the structure of glasses and crystalline materials. *Mailing Add:* Dept Physics Brown Univ Providence RI 02912

BRAY, RALPH, b Russia, Sept 11, 21; nat US; m 48; c 3. SOLID STATE PHYSICS. *Educ:* Brooklyn Col, BA, 42; Purdue Univ, PhD(physics), 49. *Prof Exp:* From instr to assoc prof, 45-65, prof, 65-89, EMER PROF PHYSICS, PURDUE UNIV, 89- *Concurrent Pos:* Nat Res Coun fel, Tech Univ Delft, 51-52; consult, Univ Reading, 52-53, Battelle Mem Inst & Nat Cash Register Co; vis scientist, Gen Atomic Div, Gen Dynamics Corp, 60-61; Guggenheim fel, 69-70; vis scientist, Clarendon Lab, Oxford Univ, 69-70 & Becton Ctr, Yale Univ, 70 & Max Planck Inst, 85-86; fac adv, Tex Instruments, Gen Tel Res & Electronics Lab; vis prof, Univ Osaka, 90; von Humboldt Sr Scientist Award, 85-86; Guggenheim Mem Fel Award, 69. *Honors & Awards:* vis Scientist Award, Japan Soc Prom Sci, 78. *Mem:* Fel Am Phys Soc; Sigma Xi. *Res:* Semiconductors; nonequilibrium phenomena; acoustoelectric effects; instabilities; brillouin scattering. *Mailing Add:* 322 Hollowood Dr West Lafayette IN 47906

BRAY, RICHARD NEWTON, b San Diego, Calif, Apr 14, 45; m 76; c 2. FISH ECOLOGY, MARINE ECOLOGY. *Educ:* San Diego State Univ, BS, 67; Univ Calif, Santa Barbara, MA, 74, PhD(biol), 78. *Prof Exp:* Res asst biol, Univ Calif, Santa Barbara, 74-78; lectr, 78-79, asst prof, 79-88, PROF BIOL, CALIF STATE UNIV, LONG BEACH, 88- *Honors & Awards:* Stoye Award, Am Soc Ichthyologists & Herpetologists, 78. *Mem:* AAAS; Am Soc Ichthyologists & Herpetologists; Western Soc Naturalists; Ecol Soc Am; Am Soc Limnol Oceangr. *Res:* Role of fishes in the food web of subtidal reef communities. *Mailing Add:* Dept Biol Calif State Univ Long Beach CA 90840-3702

BRAY, ROBERT S, b Feb 3, 27. METALLURGY, ALLOYS. *Educ:* Yale Univ, BE, 53, ME, 54. *Prof Exp:* EMER PROF CHEM MAT ENG, CALIF STATE POLYTECH UNIV, POMONA, 80- *Mem:* Fel Am Soc Metals. *Mailing Add:* 411 Pinata Pl Fullerton CA 92635

BRAYER, KENNETH, b New York, NY, July 3, 41. COMMUNICATIONS ENGINEERING, COMPUTER NETWORKING. *Educ:* City Col NY, BEE, 64; Columbia Univ, MS, 65. *Prof Exp:* Engr, Eng Res Lab, Columbia Univ, 64; tech staff mem, Mitre Corp, 65-79, dept staff, 79-82, group leader, 82-89, PROJ MGR, MITRE CORP, 89. *Concurrent Pos:* Res referee, Eng Div, NSF, 69-; instr, George Washington Univ, 76-77; ed, Inst Elec & Electronics Engrs Press, 74-75, Trans Commun, 87-90. *Mem:* Fel Inst Elec & Electronics Engrs. *Res:* Digital data communications; high frequency radio; error detection and correction coding; computer communication and networking; adaptive computer systems. *Mailing Add:* Mitre Corp Burlington Rd, Box 208 Bedford MA 01730

BRAYMER, HUGH DOUGLAS, b Oklahoma City, Okla, Mar 28, 33; m 56; c 2. MOLECULAR BIOLOGY. *Educ:* Univ Okla, BS, 55, MS, 57, PhD(biochem), 60. *Prof Exp:* Biochemist, gene action res, radiobiol br, US Air Force Sch Aerospace Med, 60-63; USPHS fel biochem genetics, Stanford Univ, 63-66; from asst prof to assoc prof, 66-74, PROF MICROBIOL, LA STATE UNIV, BATON ROUGE, 74-, VPRES ACAD AFFAIRS. *Concurrent Pos:* Mem fac, Trinity Univ, 60-63. *Mem:* Am Chem Soc; Genetics Soc Am; Am Soc Microbiol; Am Soc Biol Chemists; Brit Biochem Soc. *Res:* Biochemical genetics; gene action; nitrogen fixation; metabolic control mechanisms. *Mailing Add:* Dept Microbiol La State Univ Baton Rouge LA 70803

BRAZEAU, GAYLE A, b Toledo, Ohio, May 27, 57; m 79. PHARMACOKINETICS, PHYSICAL PHARMACY. *Educ:* Univ Toledo, BS, 80, MS, 83; State Univ NY, Buffalo, PhD(pharmaceut), 89. *Prof Exp:* Grad teaching asst pharmaceut, Univ Toledo, 80-83; grad res asst pharmaceut, State Univ NY, Buffalo, 83-89; ASST PROF PHARMACEUT, COL PHARM, UNIV HOUSTON, 89- *Concurrent Pos:* Sigma Xi res award, State Univ NY, Buffalo, 87. *Mem:* Am Asn Pharmaceut Scientists; Am Asn Cols Pharm; assoc mem Soc Exp Biol & Med. *Res:* Mechanisms muscle damage upon intramuscular injection; drug disposition-neuromuscular diseases; lipoproteins of drug delivery systems. *Mailing Add:* Dept Pharmaceut Univ Houston Col Pharm Houston TX 77030

BRAZEE, ROSS D, b Adrian, Mich, Oct 9, 30; m 53; c 4. AEROSOL PHYSICS, TRANSPORT PROCESSES. *Educ:* Mich State Univ, BS, 52, MS, 53, PhD(agr eng), 57. *Prof Exp:* Grad res asst, Mich State Univ, 52-53 & 55-57; agr engr, dept agr eng res, Agr Res Serv, 57-62, leader, Pioneering Lab, Fine Particle Physics, 62-72, res leader, Agr Eng Res Unit, 72-85, RES LEADER, APPLN TECH RES UNIT, AGR RES SERV, USDA, 85- *Concurrent Pos:* Adj prof, Ohio Agr Res & Develop Ctr, 60- & Ohio State Univ, 63- *Mem:* AAAS; NY Acad Sci; Am Geophys Union; Am Soc Agr Engrs; Sigma Xi. *Res:* Agricultural micrometeorology; crop protection; atmospheric turbulence and diffusion in plant populations; image processing; electrostatics; fluid mechanics. *Mailing Add:* Ohio Agr Res & Develop Ctr Agr Res Serv USDA Wooster OH 44691

BRAZEL, ANTHONY JAMES, CLIMATOLOGY, MICROCLIMATOLOGY. *Educ:* Rutgers Univ, BA, 63, MA, 65; Univ Mich, PhD(geog), 72. *Prof Exp:* Res scientist, US Army Corps Eng, 69-70; prof geog, Windsor Univ, Can, 71-74; PROF GEOG, ARIZ STATE UNIV, 74- *Concurrent Pos:* Prin investr, Jet Propulsion, Pasadena, 86-; prin investr, Salt River Proj, Phoenix, Ariz, 87-; state climatologist, Arizona, 79- *Mem:* Asn Am Geographers; Royal Meteorol Soc; Am Meteorol Soc; Nat Weather Asn; Int Mountain Soc. *Res:* Climatology of Arizona; study desertification and acolian processes, energy and moisture budgets of urban areas and arid environments. *Mailing Add:* Dept Geog Univ Del Newark DE 19716

BRAZELTON, WILLIAM T(HOMAS), b Danville, Ill, Jan 22, 21; m 44; c 2. CHEMICAL ENGINEERING. *Educ:* Northwestern Univ, BS, 43, MS, 47, PhD(chem eng), 52. *Prof Exp:* Instr mech & chem eng, 43-47, from asst prof to assoc prof, 47-63, PROF CHEM ENG & ASSOC DEAN TECHNOL INST, NORTHWESTERN UNIV, 63- *Mem:* Am Chem Soc; Am Soc Eng Educ; Am Inst Chem Engrs. *Res:* Heat and mass transfer operations; fluidization. *Mailing Add:* Ten E Willow Rd Prospect Heights IL 60070

BRAZIER, MARY A B, b Eng; US citizen; c 1. NEUROPHYSIOLOGY. *Educ:* Univ London, BSc, 26, PhD(biochem), 29, DSc(neurophysiol), 60. *Hon Degrees:* MD, Univ Utrecht, 76. *Prof Exp:* Neurophysiologist, Mass Gen Hosp, Boston, 40-60; PROF ANAT & PHYSIOL, UNIV CALIF, LOS ANGELES, 61- *Concurrent Pos:* Res fel, Maudsley Hosp, London, 30-40; res assoc, Med Sch, Harvard Univ, 41-60 & Mass Inst Technol, 53-65; ed-in-chief, Electroencephalography & Clin Neurophysiol; secy-gen, Int Brain Res Orgn. *Mem:* Am Physiol Soc; Int Fedn Electroencephalog & Clin Neurophysiol (hon pres); hon mem Am Electroencephalog Soc; Am Neurol Asn; Am Acad Neurologists. *Res:* Brain research. *Mailing Add:* Brain Res Inst Univ Calif Sch Med Los Angeles CA 90024

BRAZINSKY, IRVING, b New York, NY, Oct 27, 36; m 59; c 2. CHEMICAL ENGINEERING. *Educ:* Cooper Union, BSChemE, 58; Lehigh Univ, MS, 60; Mass Inst Technol, ScD(chem eng), 67. *Prof Exp:* Res engr rocket vehicle res, NASA, 59-61; res engr non-Newtonian fluid dynamics, Polaroid Corp, 66-69; sr res engr, Celanese Res Co, 69-76; sr res & develop engr, Halcon Int, 76-81; process mgr, Foster Wheeler Energy Corp, 81-85; assoc prof chem eng, 85-89, CHMN, DEPT CHEM ENG, COOPER UNION SCH ENG, 89- *Concurrent Pos:* Adj prof, Newark Col Eng, 71-81; fel, Petrol Res Fund, Proctor & Gamble & Arthur D Little. *Honors & Awards:* Schweinburg Award. *Mem:* Am Inst Chem Engrs; Sigma Xi; NY Acad Sci. *Res:* Heat transfer and mass transfer; flow of non-Newtonian fluids; fiber spinning; chemical reaction kinetics; thermodynamics; crystal growth from the melt. *Mailing Add:* Six Rustic Lane Matawan NJ 07747

BRAZIS, A(DOLPH) RICHARD, b Bridgeport, Conn, Nov 14, 26; m 52; c 2. MICROBIOLOGY. *Educ:* Norwich Univ, BS, 49; Univ Mo, MS, 51, PhD(dairy husb, bact), 54. *Prof Exp:* Asst instr dairy husb & bact, Univ Mo, 52-54; sr scientist, Robert A Taft Sanit Eng Ctr, 54-69, chmn appl lab methods comt, 64-78; lab cert officer, Div Microbiol, Food & Drug Admin, 69-70, chief lab develop prog, 70-78, actg chief, Lab Qual Assurance Br, Div Microbiol, 77-78; corp microbiologist, Fairmont Foods Co, 78-80; mgr, Sanit Microbiol, W A Golomski & Assoc, 81-84; RETIRED. *Concurrent Pos:* Chmn screening test subcomt, Nat Mastitis Coun, 67-; pres, Int Asn of milk food & environ sanitarians, Inc, 83-84. *Honors & Awards:* USPHS Commendation Medal, 63; Award of Merit, Food & Drug Admin, 76. *Mem:* Sigma Xi; Int Asn Milk, Food & Environ Sanit (pres, 83-84); Inst Food Technol; Am Soc Microbiologists; Am Dairy Sci Asn. *Res:* Water supply and pollution germicidal agents investigations; sanitation of radionuclide removal processes; laboratory evaluation and services; milk and food microbiology; administration and collegiate teaching; industry processing and laboratory quality assurance. *Mailing Add:* Lab Qua Systs Co 1006 Martin Dr W Bellevue NE 68005

BRAZY, PETER C, b 1946; c 2. EPITHELIAL TRANSPORT, CELLULAR METABOLISM. *Educ:* Washington Univ, St Louis, MD, 72. *Prof Exp:* Asst prof & chief nephrol, Duke Univ Med Ctr, 74-88; ASSOC PROF & HEAD NEPHROLOGY SECT, UNIV WIS, 88- *Mailing Add:* Univ Wis Clin Sci Ctr 600 Highland Ave H4/510 Madison WI 53792

BREAKEFIELD, XANDRA OWENS, b Boston, Mass, Oct 6, 42; m 74; c 2. BIOCHEMICAL GENETICS, NEUROSCIENCE. *Educ:* Wilson Col, AB, 65; Georgetown Univ, PhD(microbial genetics), 71. *Prof Exp:* Fel neurobiol, Nat Heart & Lung Inst, 71-73, staff fel, 73-74; asst prof human genetics, Sch Med, Yale Univ, 74-; assoc prof neurol, Harvard Med Sch; dir, Molecular Neurogenetics, E K Shriver Ctr; DEPT NEUROGENETICS, MASS GEN HOSP. *Concurrent Pos:* Prin investr res grants, Nat Inst Neurol Dis & Stroke, 75-78, March of Dimes, 76-79, Dystonia Med Res Found, 77-80 & Nat Inst Gen Med Sci, 78-81; assoc geneticist, neurol sci, Mass Gen Hosp; McKnight Neurosci Develop Award, 82-84; Jants Neurosci Investr Award, 85-92. *Honors & Awards:* Mathilde Solowey Award Neurosci, FAES, NIH, 86. *Mem:* Am Soc Neurochem; Int Soc Neurochem; Am Soc Human Genetics; Soc for Neurosci. *Res:* Genetic control of neurotransmitter metabolism; catecholamines; monoamine oxidase; cell culture; gene mapping; inheritance of monoamine oxidase in human population; inherited neurologic diseases. *Mailing Add:* Dept Neurogenetics Mass Gen Hosp 6th Fl Bldg 149 E 13th St Charlestown MA 02129

BREAKEY, DONALD RAY, b Snohomish, Wash, June 1, 27; m 48; c 2. BIOLOGY. *Educ:* Wilamette Univ, BS, 50; Mich State Univ, MS, 52; Univ Calif, Berkeley, PhD(zool), 61. *Prof Exp:* From instr to assoc prof, 54-67, chmn dept, 68-83, PROF BIOL, WILLAMETTE UNIV, 67- *Concurrent Pos:* Past pres, Malheur Field Sta Consortium, Pac NW Bird & Mammal Soc & Great Basin Soc. *Mem:* Sigma Xi; Am Soc Mammal; Australian Mammal Soc. *Res:* Vertebrate mammalian populations and distribution; ecological habitats and their evaluation in relation to animal forms present; athropod and life histories. *Mailing Add:* Dept Biol Wilamette Univ Salem OR 97301-3922

BREAKSTONE, ALAN M, b Racine, Wis, Oct 13, 50; m 86. PARTICLES AND FIELDS. *Educ:* Calif Inst Technol, BS, 72; Univ Calif, Santa Cruz, MS, 74, PhD(physics), 80. *Prof Exp:* Postdoctoral fel, Ames Lab, Iowa State Univ, 80-84; ASST PHYSICIST, UNIV HAWAII, MANOA, 84- *Mem:* Am Phys Soc. *Res:* Experimental elementary particle physics. *Mailing Add:* PO Box 4349 Stanford CA 94305

BREAKWELL, JOHN, space mechanics, optical control; deceased, see previous edition for last biography

BREAM, CHARLES ANTHONY, b Midland, Pa, Mar 15, 15; m 41; c 4. MEDICINE. *Educ:* Grove City Col, BS, 36; Temple Univ, MD, 40. *Prof Exp:* Intern, Mercy Hosp, Pa, 40-41; vis physician, St Vincent's Hosp, Pa, 46-48; instr radiol, Col Physicians & Surgeons, Columbia Univ, 51-52; from asst prof to assoc prof, 52-58, PROF RADIOL, SCH MED, UNIV NC, CHAPEL HILL, 58- *Concurrent Pos:* Mellon res fel, Mercy Hosp, Pa, 41-42; instr med, Sch Med, Univ Pittsburgh, 40-41; asst vis physician, Hamot Hosp, Pa, 46-48; attend radiologist, NC Mem Hosp, Univ NC, 52-; consult, Watts Hosp, 53- & Womack Army Hosp, 58-; vis attend radiologist, Vet Admin Hosp, Durham, 55- *Mem:* Roentgen Ray Soc; Radiol Soc NAm; Asn Am Med Cols; Asn Univ Radiologists; fel Am Col Radiol. *Res:* Improvement of techniques, methods and scope of investigation of the retioperitoneal space; extend the usefulness of excretory urography in various nephropathies. *Mailing Add:* Dept Radiol Sch Med Univ NC Chapel Hill NC 27514

BREARLEY, HARRINGTON C(OOPER), JR, b Greenville, SC, Jan 17, 26; m 57; c 3. ELECTRICAL ENGINEERING, COMPUTER SCIENCE. *Educ:* Ga Inst Technol, BEE, 46; Univ Ill, MS, 50, PhD(elec eng), 54. *Prof Exp:* Mem tech staff, Bell Tel Labs, 47-49; engr, Gen Elec Co, 53-59; res asst prof elec eng, Univ Ill, 59-65; assoc prof elec eng & comput sci, 65-77, PROF ELEC ENG & COMPUT SCI, IOWA STATE UNIV, 77- *Mem:* Inst Elec & Electronics Engrs; Asn Comput Mach; Sigma Xi. *Res:* Computer science; switching theory. *Mailing Add:* 1537 Linden Dr Ames IA 50010-5533

BREAULT, GEORGE OMER, b Providence, RI, Nov 30, 42; m 67; c 2. DRUG METABOLISM. *Educ:* Univ Notre Dame, AB, 64; Univ RI, MS, 67; Univ Mich, PhD(pharm chem), 70. *Prof Exp:* Sr res biochemist drug metab, Merck Inst Therapeut Res, 70-72, res fel, 72-76, sr res fel, 76-80. *Mem:* AAAS; Am Chem Soc. *Res:* Development of analytical methods for drugs in biological systems. *Mailing Add:* Analytical Develop Corp 4405 N Chestnut St Colorado Springs CO 80907

BREAZEALE, ALMUT FRERICHS, b Berlin, Germany, May 4, 38; US citizen; m 64; c 2. POLYMER CHEMISTRY, ORGANIC CHEMISTRY. *Educ:* Univ Ill, BS, 60; Univ Wash, PhD(org chem), 65. *Prof Exp:* RES ASSOC, E I DUPONT DE NEMOURS & CO, 65- *Mem:* Am Chem Soc; Sigma Xi. *Res:* Structure-property relationships of thermoset and thermoplastic polymers, including elastomers and fluorinated polymers. *Mailing Add:* PPD E323/208 Wilmington DE 19898

BREAZEALE, MACK ALFRED, b Leona Mines, Va, Aug 15, 30; m 52; c 3. PHYSICAL ACOUSTICS, SOLID STATE PHYSICS. *Educ:* Berea Col, BA, 53; Univ Mo, Rolla, MS, 54; Mich State Univ, PhD(physics), 57. *Prof Exp:* Asst physics, Univ Mo, Rolla, 53-54; asst, Mich State Univ, 54-57, asst prof, 57-58 & 59-62; Fulbright res grant, Phys Inst, Stuttgart Tech Inst, 58-59; from assoc prof to prof physics, Univ Tenn, Knoxville, 67-88; DISTINGUISHED RES PROF PHYSICS, UNIV MISS, 88- *Concurrent Pos:* Consult, Oak Ridge Nat Lab, 62-, Naval Res Lab, 72-74 & Leeds & Northrup, 79-81; mem orgn comt, Int Symp Nonlinear Acoust, Moscow, 75, Blacksburg, Va, 76, Paris, 78, Leeds, 79, Kobe, 84, Novosibirsk, 87 & Austin, Tex, 90; guest prof, Tech Univ Denmark, 77; assoc ed, J Acoust Soc Am, 77-; tech prog chmn, Acoust Soc Am, 85. *Mem:* Am Asn Univ Professors; fel Acoust Soc Am; Am Phys Soc; fel Inst Elec & Electronic Engrs; fel Inst Acoust Gt Brit. *Res:* Ultrasonics; nonlinear solid state and liquid state phenomena. *Mailing Add:* Nat Ctr Phys Acoust Univ Miss Coliseum Dr University MS 38677

BREAZEALE, ROBERT DAVID, b Ames, Iowa, Aug 30, 35. ORGANIC CHEMISTRY. *Educ:* SDak State Col, BS, 57; Univ Wash, PhD(org chem), 64. *Prof Exp:* Res chemist, Sinclair Res Labs, Inc, 57-59; res chemist, 64-71, RES SUPVR, AUTOMOTIVE PROD DEPT, RES DIV, E I DU PONT DE NEMOURS & CO, INC, 71- *Mem:* Am Chem Soc. *Res:* Polymer and petroleum chemistry; non-benzenoid aromatic chemistry; finishes and adhesives technology. *Mailing Add:* E I du Pont de Nemours Automotive Prod Marshall Lab 3500 Grays Ferry Ave Philadelphia PA 19146

BREAZEALE, WILLIAM HORACE, JR, b Greensboro, NC, Aug 30, 38; m 63, 85; c 3. PHYSICAL CHEMISTRY, CHEMICAL SAFETY. *Educ:* Univ SC, BS, 61, PhD(phys chem), 66. *Prof Exp:* Asst prof chem, Winthrop Col, 66-70; assoc prof, 70-71, assoc dean, 73-75, PROF CHEM & PHYSICS, FRANCIS MARION COL, 71-, CHMN DEPT, 70- *Mem:* Sigma Xi; Am Chem Soc. *Res:* Chemical education; thermodynamics of solutions; chemical safety. *Mailing Add:* Dept Chem & Physics Francis Marion Col Florence SC 29501

BREAZILE, JAMES E, b Rockport, Mo, Dec 31, 34; m 57; c 3. NEUROANATOMY, NEUROPHYSIOLOGY. *Educ:* Univ Mo, BS, 58, DVM, 58; Univ Minn, PhD(anat, physiol), 63. *Prof Exp:* Pvt pract vet med, 58-60; NIH fel, 61-63; asst prof physiol, Okla State Univ, 63-67; assoc prof vet anat, Univ Mo, Columbia, 67-68, chmn dept, 69-70 & 72-77, dir grad studies vet anat, 72-77, prof vet anat, 68-80; MEM FAC, COL VET MED, OKLA STATE UNIV, 80- *Concurrent Pos:* NIH res grant, 64-; consult, 77- *Mem:* AAAS; Am Vet Med Asn; Am Asn Vet Anat; Am Soc Vet Physiol & Pharmacol; Am Asn Anatomists; Sigma Xi. *Res:* Studies of normal structure of the central nervous system of vertebrates correlated with physiological phenomena, especially pain and autonomic control mechanisms; gross anatomy. *Mailing Add:* Physiol Sci Okla State Univ Stillwater OK 74078

BREBRICK, ROBERT FRANK, JR, b Danvers, Mont, Oct 18, 25; m 57; c 4. SOLID STATE CHEMISTRY. *Educ:* Mont State Col, BS, 47; Cath Univ Am, PhD(chem), 52. *Prof Exp:* Asst, Cath Univ Am, 47-50; phys chemist, Naval Ord Lab, 52-61, group leader, mat group, solid state div, 57-61; phys chemist, solid state div, Lincoln Lab, Mass Inst Technol, 61-70; assoc prof, 70-73, PROF, DEPT MECH ENG MAT SCI PROG, MARQUETTE UNIV, 73- *Honors & Awards:* Meritorious Civilian Serv Award, 56. *Mem:* Am Chem Soc; Am Phys Soc; Electrochem Soc. *Res:* Physical chemistry of semiconductors; theoretical and experimental studies of the homogeneity range and thermodynamic properties of semiconducting compounds; calculation of phase diagrams. *Mailing Add:* Dept Mech Eng Marquette Univ Milwaukee WI 53233

BRECHER, ARTHUR SEYMOUR, b New York, NY, Mar 30, 28; m 66; c 2. BIOCHEMISTRY. *Educ:* City Col New York, BS, 48; Univ Calif, Los Angeles, PhD, 56. *Prof Exp:* Jr res biochemist, NY State Psychiat Inst, 48-52; asst chem, Univ Calif, Los Angeles, 53-56; asst prof biochem, Purdue Univ, 56-58; biochemist pharmacol, Food & Drug Admin, Dept Health, Educ & Welfare, 58-60; assoc res scientist, NY State Dept Ment Hyg, 60-63; asst prof biochem, Sch Med, George Washington Univ, 63-69; assoc prof, 69-75, PROF BIOCHEM, BOWLING GREEN STATE UNIV, 75- *Mem:* AAAS; Am Chem Soc; Soc Exp Biol & Med; Am Soc Biol Chemists; Am Soc Neurochem; Int Soc Neurochem. *Res:* Enzyme chemistry; intermediary metabolism; drug metabolism; hypertension. *Mailing Add:* Dept Chem Bowling Green State Univ Bowling Green OH 43403

BRECHER, AVIVA, b Bucarest, Romania, July 4, 45; m 65; c 2. SPACE PHYSICS, PALEOMAGNETISM. *Educ:* Israel Inst Technol & Mass Inst Technol, BSc & MSc, 68; Univ Calif, PhD(appl physics), 72. *Prof Exp:* Res assoc earth & planetary sci, Mass Inst Technol, 72-77; asst prof physics, Wellesley Col, 77-79; sr staff consult, nuclear & geotech systs safety & environ safety, Arthur D Little, Inc, 79-85; dir acad-corp rels, Boston Univ, 85-87; STRATEGIC PLANNING ANALYST, US DEPT TRANSP, 87- *Concurrent Pos:* Amelia Earhard Zonta fel, 71-72; vis scientist, Mass Inst Technol, 77-79; exchange scientist, Japan Soc for Prom Sci, 79; cong scientist, Am Phys Soc fel & US Senate, 83-84. *Mem:* Am Phys Soc; Am Astron Soc; Am Geophys Union; AAAS; Int Astron Union. *Res:* Laser physics and solid state physics; nature of lunar magnetism; planetary magnetic fields; terrestrial paleomagnetism and continental drift; physical properties of meteorites and models of planetary evolution; risk analysis for nuclear waste disposal in geologic repositories; physico-chemical and mathematical modeling of the long-term performance of nuclear waste packages; probabilistic risk analysis for the mining industry (subsidence, flooding, earthquakes, back failure); technology assesment and strategic planning; university and industry cooperative research. *Mailing Add:* Strategic Planning Analysis Off US Dept Transp-Nat Transp Syst Ctr Cambridge MA 02142

BRECHER, CHARLES, b Brooklyn, NY, Nov 5, 32; m 69; c 1. PHYSICAL CHEMISTRY. *Educ:* Columbia Univ, BA, 54, MA, 55, PhD(phys chem), 59. *Prof Exp:* Res fel phys chem, Columbia Univ, 59-62; RES CHEMIST EXPLOR RES DEPT, GEN TEL & ELECTRONICS LABS, 62- *Mem:* Am Phys Soc; Optical Soc Am; Sigma Xi; Am Chem Soc. *Res:* Optical phenomena in condensed phases, especially spectra of solids and liquids as related to problems of molecular and crystal structure, molecular interactions; energy transfer, laser phenomena and light-generating chemical reactions. *Mailing Add:* Gen Tel & Electronics Lab Nine Skyview Rd Lexington MA 02173

BRECHER, GEORGE, b Olmutz, Czech, Nov 5, 13; US citizen; m 59; c 5. LABORATORY MEDICINE, HEMATOLOGY. *Educ:* Univ Prague, MD, 38; Univ London, DTM&H, 40. *Prof Exp:* Sr asst surgeon, 47, surgeon, 49-51, sr surgeon, 51-53, med dir, 54-56, chief hemat serv, Dept Clin Path, USPHS, 53-66; chmn dept, 66-78, prof lab med, 66-78, EMER PROF, UNIV CALIF BERKELEY, 78- *Concurrent Pos:* Fel, Mayo Clin, 43; NIH fel, 67; assoc fac scientist, Lawrence Berkeley Labs, Donner Lab, Berkeley, 78- *Honors & Awards:* French Legion Hon, 80. *Mem:* Acad Clin Lab Physicians & Scientists (pres, 71); Am Soc Hemat (pres, 73); Asn Am Physicians; Int Soc Hemat. *Res:* Morphologic hematology; kinetics of bone marrow and blood cells; laboratory automation and quality control. *Mailing Add:* LBL-74 Univ Calif Berkeley CA 94720

BRECHER, KENNETH, b New York, NY, Dec 7, 43; m 65; c 2. THEORETICAL ASTROPHYSICS, SCIENCE EDUCATION. *Educ:* Mass Inst Technol, BS, 64, PhD(physics), 69. *Prof Exp:* Res physicist, Univ Calif, San Diego, 69-72; from asst prof to assoc prof physics, Mass Inst Technol, 72-79; assoc prof, 79-81, PROF ASTRON & PHYSICS, BOSTON UNIV, 81-, DIR SCI & MATH EDUC CTR, 90- *Concurrent Pos:* John Simon Guggenheim Mem fel, 79-80; assoc, Harvard Col Observ, 79-; sr res assoc, Nat Res Coun, NASA & Nat Acad Sci, 83-84; W K Kellog fel, 85-88. *Mem:* Fel Am Phys Soc; Am Astron Soc; Int Astron Union; Sigma Xi. *Res:* Theoretical high energy astrophysics; x-ray and gamma-ray astronomy; observational tests and consequences of gravitational theories; cosmology; archaeoastronomy; historical astronomy and archaeoastrophysics. *Mailing Add:* Dept Astron Boston Univ 725 Commonwealth Ave Boston MA 02215

BRECHER, PETER I, b New York, NY, May 19, 40; m 64; c 2. BIOCHEMISTRY. *Educ:* Ohio Univ, BS, 60; Boston Univ, PhD(biochem), 68. *Prof Exp:* Res assoc, Ben May Labs Cancer Res, Univ Chicago, 69-71; from asst prof to assoc prof biochem, 71-81, PROF BIOCHEM, SCH MED, BOSTON UNIV, 81- *Concurrent Pos:* Res fel biochem, Sch Med, Boston Univ, 68-69; NIH trainee physiol, 68-69. *Res:* Lipid metabolism in the arterial wall; cardiovascular diseases. *Mailing Add:* Boston Univ Sch Med 80 E Concord St Boston MA 02218

BRECHNER, BEVERLY LORRAINE, b New York, NY, May 27, 36. MATHEMATICS. *Educ:* Univ Miami, BS, 57, MS, 59; La State Univ, Baton Rouge, PhD(math), 64. *Prof Exp:* Instr math, La State Univ, New Orleans, 62-64, asst prof, 64-68; from asst prof to assoc prof, 68-83, PROF MATH, UNIV FLA, 83- *Concurrent Pos:* Vis assoc prof math, Univ Mich, 77 & Univ Tex, 80. *Honors & Awards:* George Polya Award, Math Asn Am, 89. *Mem:* AAAS; Am Math Soc; Math Asn Am. *Res:* Topology; point set topology; spaces of homeomorphisms; locally setwise homogeneous continua; chainable continua; dimension theory; prime end theory; fixed point problems. *Mailing Add:* Dept Math Univ Fla Gainesville FL 32611-2082

BRECHT, PATRICK ERNEST, b Beaconsfield, Eng, Jan 8, 46; US citizen; m 67; c 1. PLANT PHYSIOLOGY, VEGETABLE CROPS. *Educ:* Whittier Col, BA, 68; Calif State Univ, Los Angeles, MS, 69; Univ Calif, Davis, PhD(plant physiol), 73. *Prof Exp:* Asst prof postharvest physiol, Cornell Univ, 73-76; dir prod res & develop, Sea-Land Serv Inc, 76-78; dir qual control perishables, United Brands Co, 78; DIR SPEC COMMODITIES, AM PRESIDENT LINES. *Mem:* Am Soc Hort Sci; Am Soc Plant Physiologists; Inst Food Technologists; Sigma Xi. *Res:* Postharvest physiology and quality of vegetables and fruits. *Mailing Add:* Am President Lines 1901 Harrison 14th Floor Oakland CA 94612

BRECKE, BARRY JOHN, b Milwaukee, Wis, Jan 16, 47; m 69; c 2. WEED SCIENCE, AGRONOMY. *Educ:* Wis State Univ-River Falls, BS, 69; Cornell Univ, MS, 74, PhD(weed sci), 76. *Prof Exp:* ASSOC PROF WEED SCI, AGR RES CTR, UNIV FLA, 75- *Mem:* Weed Sci Soc Am; Plant Growth Regulator Soc Am; Am Soc Agron. *Res:* Studies of weed management systems; the effects of various cultural practices on weeds; biology of particularly troublesome weeds and and weed population dynamics. *Mailing Add:* Univ Fla Agr Res Educ Ctr Jay FL 32565

BRECKENRIDGE, BRUCE (MCLAIN), b Brooklyn, Iowa, Nov 7, 26; m 49; c 3. PHARMACOLOGY. *Educ:* Iowa State Col, BS, 47; Univ Rochester, MS, 49, PhD(physiol), 52, MD, 56. *Prof Exp:* Instr physiol, Univ Rochester, 53-56; intern med, Barnes Hosp, St Louis, 56-57; from instr to assoc prof pharmacol, Wash Univ, 57-67; PROF PHARMACOL & CHMN DEPT, UNIV MED & DENT, NJ, ROBERT WOOD JOHNSON MED SCH, 67- *Concurrent Pos:* Fel pharmacol, Wash Univ, 57-59; Markle scholar med sci, 59; mem, Pharmacol A Study Sect, NIH, 68-72; adj prof med, Robert Wood Johnson Med Sch, 79- *Mem:* Am Soc Biol Chemists; Am Soc Pharmacol & Exp Therapeut. *Res:* Metabolism of the nervous system; cyclic nucleotides and drug action. *Mailing Add:* Clin Res Ctr Univ Med & Dent, NJ-Robert Wood Johnson Med Sch One Robert Wood Johnson Pl CN 19 New Brunswick NJ 08903

BRECKENRIDGE, CARL, b Asphodel Twp, Ont, July 29, 42; m 65; c 3. ATHEROSCLEROSIS, CLINICAL BIOCHEMISTRY. *Educ:* Queen's Univ, Ont, BSc, 65; Univ Toronto, MSc, 66, PhD(biochem), 70. *Prof Exp:* Med Res Coun Can res fel, Ctr Neurochem, Strasbourg, France, 70-72 & Montreal Neurol Inst, 72; from asst to assoc prof clin biochem, Univ Toronto, 72-80, dir, Core Lab, 72-80; assoc prof, 80-84, PROF BIOCHEM, DALHOUSIE UNIV, 85-, HEAD BIOCHEM, 88- *Concurrent Pos:* Fel, Am Heart Asn. *Mem:* Can Fedn Biol Socs; Am Oil Chemists Soc; Can Soc Clin Chem; Can Soc Clin Invest; Can Atherosclerosis Soc; Am Heart Asn. *Res:* Studies of the structure of lipid and protein constituents of human lipoproteins and their role in hyperlipoproteinemia and the development of atherosclerosis. *Mailing Add:* Dept Biochem Dalhousie Univ Halifax NS B3H 4H7 Can

BRECKENRIDGE, JOHN ROBERT, b Kingston, Pa, Oct 10, 20; m 79; c 1. MEDICAL BIOPHYSICS. *Educ:* Univ Pa, BS, 42. *Prof Exp:* Student engr, Electrometall Co, Union Carbide Corp, Ohio, 42-45; physicist, Qm Res & Eng Ctr, Mass, 46-61; supvry physicist, 61-64, res physicist, 64-74, chief, Biophysics Br, 74-84, CONSULT, MIL ERGONOMICS DIV, US ARMY RES INST ENVIRON MED, 85- *Mem:* Sigma Xi; Am Soc Heat, Refrig & Air-Conditioning Engrs. *Res:* Heat and moisture transfer from clothed man to his environment in terms of clothing parameters and environmental factors. *Mailing Add:* 63 Stonebridge Rd Wayland MA 01778

BRECKENRIDGE, ROBERT A(RTHUR), b Los Angeles, Calif, Mar 29, 24; m 55, 83; c 2. OCEAN ENGINEERING, STRUCTURAL ENGINEERING. *Educ:* Univ Southern Calif, BE, 50, MS, 53. *Prof Exp:* Mat engr, US Naval Civil Eng Lab, 50-51, civil engr, 52-54, res engr, 54-69, ocean eng prog mgr, 69-74, dir found eng div, 74-79; RETIRED. *Mem:* Sigma Xi; fel Am Soc Civil Engrs; Am Concrete Inst; Nat Soc Prof Engrs. *Res:* Structures and structural components; blast-resistant shelters; blast-closure devices; deep-ocean structures and structural plastics; design guidelines, equipment, tools and techniques for ocean installations. *Mailing Add:* 111 S Norton Ave Los Angeles CA 90004-3916

BRECKENRIDGE, WILLIAM H, b Kansas City, Mo, Oct 14, 41; m 84; c 1. CHEMICAL DYNAMICS, ENERGY TRANSFER. *Educ:* Univ Kans, BS, 63; Stanford Univ, PhD(chem), 68. *Prof Exp:* mem tech staff, Bell Labs, NJ, 70-71; from asst prof to assoc prof, 71-80, PROF CHEM, UNIV UTAH, 80- *Concurrent Pos:* vis scholar, Stanford Univ, 77-78, Max Planck Inst, WGer, 84-85; vis prof, Univ Paris-South, 85. *Mem:* Am Chem Soc; Inter-Am Photochem Soc. *Res:* State-to-state dynamics of chemical reactions and energy transfer processes in the gas phase, using laser techniques; spectroscopy and dynamics of van der Waals complexes of metal atoms in supersonic jets. *Mailing Add:* Dept Chem Univ Utah Salt Lake City UT 84112

BRECKINRIDGE, JAMES BERNARD, b Cleveland, Ohio, May 27, 39; m 65; c 2. OPTICAL PHYSICS. *Educ:* Case Inst Technol, BSc, 61; Univ Ariz, MS, 70, PhD(optics), 76. *Prof Exp:* Asst astron, Lick Observ, Univ Calif, 61-64 & Kitt Peak Nat Observ, NSF, 64-66; engr physicist, Rauland Corp, Zenith Radio Corp, 66-67; assoc-in-res, Kitt Peak Nat Observ, NSF, 67-76; co-investr, NASA Spacelab 3 & mem tech staff, 76-80, MGR, OPTICAL SCI

SECT, JET PROPULSION LAB, 80-FAC, CALIF INST TECHNOL, 81- *Mem:* Fel Optical Soc Am; Am Astron Soc; Royal Astron Soc; Int Astron Union; Inst Elec & Electronics Engrs; Sigma Xi; Soc Photo-Optical Instrumentation Eng. *Res:* Optical interferometric techniques for research in astrophysics; image evaluation; optical instruments. *Mailing Add:* 4565 Viro Rd La Canada CA 91011

BREDAHL, EDWARD ARLAN, b Pelican Rapids, Minn, Dec 20, 37; div; c 2. PHYSIOLOGY, ECOLOGY. *Educ:* Concordia Col, Moorhead, Minn, BA, 59; Univ NDak, MST, 64, PhD(physiol ecol), 69. *Prof Exp:* Teacher high sch, Minn, 59-62; instr cell physiol, Univ NDak, 69; asst prof biol & ecol, Stout State Univ, 69-80; mem fac, Moorhead State Col, 80-; AT DEPT BIOL-ZOOL, SEATTLE CENT COMMUNITY COL. *Res:* Physiological investigation of water balance in hibernating and nonhibernating Citellus franklinii and Citellus richardsonii to develop efficiency indices for water balance relative to the environment. *Mailing Add:* Dept Math & Sci Seattle Cent Community Col 1701 Broadway Seattle WA 98122

BREDECK, HENRY E, b St Louis, Mo, Nov 5, 27; m 53; c 6. PHYSIOLOGY. *Educ:* St Louis Univ, BS, 50; Univ Mo, MS, 53, PhD(agr chem), 56. *Prof Exp:* Asst prof chem, Colo State Univ, 56-58, from asst prof to prof physiol, 58-66; res prog mgr, NASA, 66-67; prog dir, Gen Res Support Br, Div Res Resources, NIH, 67-71; assoc dir, Off Res Develop, 71-78, ASST VPRES RES, MICH STATE UNIV, 78- *Mem:* Nat Coun Univ Res Adminr; Soc Univ Patent Adminr; Licensing Execs Soc. *Res:* Endocrinology; physiology of reproduction; cardiovascular physiology. *Mailing Add:* Off Res Develop Mich State Univ East Lansing MI 48824-1046

BREDEHOEFT, JOHN DALLAS, b St Louis, Mo, Feb 28, 33; m 58; c 3. HYDROLOGY, GEOLOGY. *Educ:* Princeton Univ, BSE, 55; Univ Ill, MS, 57, PhD(geol), 62. *Prof Exp:* Geologist, Humble Oil & Refining Co Div, Standard Oil Co, NJ, 57-59; geologist, Desert Res Inst, Nev, 62; res geologist, 62-74, dep asst chief res hydrologist, Water Resources Div, US Geol Surv, 74- *Concurrent Pos:* Vis assoc prof, Univ Ill, 67-68; res assoc, Resources for Future, 68-70. *Honors & Awards:* R E Horton Award, Am Geophys Union & O E Meinzer Award, Geol Surv Am, 75. *Mem:* Am Asn Petrol Geologists; Geol Soc Am; Am Geophys Union. *Res:* Ground water hydrology; physical properties of ground water systems; physics of ground water motion; transport of chemical constituents in ground-water systems. *Mailing Add:* Water Resources Div 345 Middlefield Rd Menlo Park CA 94025

BREDENBERG, CARL ERIC, b San Mateo, Calif, Mar 18, 40; m 66; c 2. THORACIC SURGERY, VASCULAR SURGERY. *Educ:* Johns Hopkins Univ, MD, 64. *Prof Exp:* Intern surg, Johns Hopkins Hosp, 64-65, asst resident, 67-71, resident surgeon, 71-72; asst prof, 72-76, assoc prof, 76-80, PROF SURG, STATE UNIV NY, UPSTATE MED CTR, 80- *Concurrent Pos:* Chief Thoracic Surg Sect, Syracuse Vet Admin Hosp, 73-; surgeon in charge, Vascular Surg Serv, State Univ NY, 75- *Mem:* Soc Univ Surgeons; Asn Academic Surg; Int Cardiovascular Soc; Am Asn Thoracic Surg; Am Col Surgeons. *Res:* Pulmonary and cardiovascular physiology on shock, pulmonary alveolar mechanics, pulmonary circulation and pulmonary edema. *Mailing Add:* Dept Surg Maine Med Ctr 22 Bramhall St Portland ME 04102

BREDER, CHARLES VINCENT, b Atlantic City, NJ, Feb 15, 40; m 62; c 2. ORGANIC POLYMER CHEMISTRY, ANALYTICAL CHEMISTRY. *Educ:* Carson-Newman Col, BS, 62; Vanderbilt Univ, MS, 64, PhD(org chem), 68. *Prof Exp:* Res chemist, E I du Pont de Nemours & Co, 67-71; res chemist, Food & Drug Admin, Dept Health, Educ & Welfare, 71-84; STAFF SCIENTIST, LAW OFF KELLER & HECKMAN, 84- *Honors & Awards:* Award of Merit, FDA. *Mem:* Am Chem Soc; Asn Off Anal Chemists; Am Soc Testing & Mat. *Res:* Determination of the identity and quantity of chemicals that migrate from food packaging into foods. *Mailing Add:* 2209 Senseney Lane Falls Church VA 22043

BREDEWEG, ROBERT ALLEN, b Forest Grove, Mich, Aug 15, 41; m 63; c 2. ANALYTICAL CHEMISTRY. *Educ:* Hope Col, BA, 63; Southern Ill Univ, MA, 65. *Prof Exp:* ASSOC SCIENTIST, DOW CHEM CO, 65- *Honors & Awards:* Vernon A Stenger Award, Dow Chem Anal Scientists, 75; IR 100 Award; Vaaler Award, Chromatography Div, Am Chem Soc. *Mem:* Am Chem Soc. *Res:* Automation of analytical procedures; process optimization. *Mailing Add:* 200 S Eight Mile Rd Midland MI 48640

BREDON, GLEN E, b Fresno, Calif, Aug 24, 32; m 63; c 2. MATHEMATICS. *Educ:* Stanford Univ, BS, 54; Harvard Univ, AM, 55, PhD(math), 58. *Prof Exp:* Mem, Inst Advan Study, 58-60; from asst prof to prof math, Univ Calif, Berkeley, 60-68; PROF MATH, RUTGERS UNIV, NEW BRUNSWICK, 68- *Concurrent Pos:* Sloan fel, 65-67; mem, Inst Advan Study, 66-67. *Mem:* AAAS; Sigma Xi. *Res:* Topological transformation groups; algebraic and differential topology. *Mailing Add:* 521 State Rd Princeton NJ 08540

BREE, ALAN V, b Sydney, Australia, Nov 16, 32; m 61; c 2. PHYSICAL CHEMISTRY. *Educ:* Univ Sydney, BSc, 54, MSc, 55, PhD(phys chem), 58. *Prof Exp:* Nat Res Coun Can fel, 58-59; fel, Chem Dept, Univ Col, Univ London, 60-61; from asst prof to assoc prof, 61-71, PROF CHEM, UNIV BC, 71- *Res:* Electronic and vibrational properties of aromatic molecules in the gas; solution and solid phases. *Mailing Add:* Dept Chem Univ BC 2075 Westbrook Pl Vancouver BC V6T 1W5 Can

BREECE, HARRY T, III, b Huntington, WVa, May 3, 39; m 64; c 4. ELECTRICAL ENGINEERING, COMMUNICATIONS. *Educ:* Univ Cincinnati, BS, 62; Rutgers Univ, MS, 64; Purdue Univ, PhD(elec eng), 69. *Prof Exp:* Engr semiconductor circuit design, Semiconductor Div, RCA, 62-64; mem tech staff digital commun systs, Bell Labs, 69-78; dist mgr strategic planning, AT&T, 78-84; VPRES ENG, SUPERIOR AUTOTRONICS, INC. *Mem:* Inst Elec & Electronics Engrs. *Res:* Digital communications; digital switching; data communications. *Mailing Add:* Superior Autotronics Inc Rm 104 533 Third Ave Huntington WV 25701

BREED, CAROL SAMETH, b New York, NY, Apr 21, 33; m 54, 65; c 5. GEOLOGY, DESERT GEOMORPHOLOGY. *Educ:* Smith Col, BA, 54; Brown Univ, MS, 57. *Prof Exp:* Asst geol, Mus Northern Ariz, 70-72; GEOLOGIST, US GEOL SURV, 72- *Concurrent Pos:* Co-investr desert studies, Nat Air & Space Mus, Smithsonian Inst, 75-; res assoc, Mus Northern Ariz, 76- *Mem:* Sigma Xi; fel Geol Soc Am. *Res:* Desert geologic processes and landforms, especially in southwestern American deserts and Eastern Sahara (NAfrica) with application to problems of geologic history, climate change and soil and water resources in arid regions. *Mailing Add:* 1826 N Beaver St Flagstaff AZ 86001

BREED, ERNEST SPENCER, b Lyndonville, NY, May 27, 13; m 44; c 4. SURGERY, PHYSIOLOGY. *Educ:* Univ Mich, MD, 38; Am Bd Surg, dipl, 49. *Prof Exp:* Intern med & surg, NY Univ Div, Bellevue Hosp, 39-40, asst resident surg & path, 40-41, resident, 41-42; instr surg, 47-50, asst prof clin surg, 50-51 & surg, 51-60, ASSOC PROF SURG, SCH MED, NY UNIV, 60- *Concurrent Pos:* Res fel, Off Sci Res & Develop shock proj, Col Physicians & Surgeons, Columbia Univ, 42-44; Commonwealth Fund res fel physiol, 47-49; attend, Bellevue Hosp, 42-, Univ Hosp, 52- & Vet Admin Hosp, Manhattan, NY, 56-; consult, Misericordia Hosp, Bronx, 65- *Mem:* Soc Univ Surgeons. *Res:* Cardiac and renal vein catheterization in patients in shock; renal function in relation to surgical conditions in patients, including open heart surgery. *Mailing Add:* Dept Surg Sch Med New York Univ 550 First Ave New York NY 10016

BREED, HELEN ILLICK, b New Cumberland, Pa, Mar 12, 25; m 57; c 3. ZOOLOGY. *Educ:* Syracuse Univ, BS, 47, MS, 49; Cornell Univ, PhD(vert zool & biol), 53. *Prof Exp:* Asst, Dept Zool, Syracuse Univ, 47-49; teacher high sch, Lyons, NY, 49-50; instr zool & physiol, Dept Biol, Akron Univ, 54; Ford Found fel & instr physiol, Vassar Col, 54-55; asst prof biol, Russell Sage Col, 55-57; asst dir syst biol, NSF, DC, 57; res assoc prof conserv, Cornell Univ, 57-61; res assoc biol, Rensselaer Polytech Inst, 64-68; CONSULT, ELTICK RES CORP, 68- *Concurrent Pos:* Am-Scand fel, 59-60; mem, Rensselaer Environ Mgt Coun, 71-78. *Mem:* AAAS; Am Soc Zoologists; Soc Syst Zool; Am Soc Ichthyol & Herpet; Am Inst Biol Sci. *Res:* Ichthyology; ecology; conservation; x-ray diffraction of vertebrate hard tissue. *Mailing Add:* RD 3 Box 245 B Troy NY 12180

BREED, HENRY ELTINGE, b New York, NY, Dec 5, 15; m 57. PHYSICS. *Educ:* Colgate Univ, BA, 38; Rensselaer Polytech Inst, MS, 48, PhD, 55. *Prof Exp:* From instr to prof physics, Rensselaer Polytech Inst, 45-81; RETIRED. *Concurrent Pos:* Res asst, Gen Elec Res Lab, 46-48; consult, Ord Corps, US Army, 56-62; Fulbright fel, Inst Teorisk Kjemi, Trondheim, Norway, 59-60; sabbatical, Univ Oslo, 76. *Honors & Awards:* Fulbright lectr optics, Nat Univ Eng, Lima, Peru, 73. *Mem:* Am Phys Soc; Am Asn Physics Teachers; fel Optical Soc Am; Sigma Xi. *Res:* Electron diffraction; optics; acoustics. *Mailing Add:* RD 3 Box 245B Troy NY 12180

BREED, LAURENCE WOODS, b Decatur, Ill, Dec 16, 24; m 48; c 2. ORGANIC POLYMER CHEMISTRY. *Educ:* Park Col, BA, 48; Univ Kans, MA, 50. *Prof Exp:* Asst, Univ Kans, 48-50; res chemist, Battenfeld Grease & Oil Corp, 50-53; assoc chemist, 53-59, sr chemist, 59-62, prin chemist, 62-67, sr adv chem, 67-77, MGR PLAN, MIDWEST RES INST, 77- *Mem:* Am Chem Soc. *Res:* Organosilicon chemistry; organic and semiorganic polymers; organic synthesis; thermal analysis of materials. *Mailing Add:* 6400 W 66th Terr Overland Park KS 66202-4117

BREED, MICHAEL DALLAM, b Kansas City, Mo, Sept 2, 51; m 75; c 1. INSECT BEHAVIOR. *Educ:* Grinnell Col, BA, 73; Univ Kans, MA, 75, PhD(entom), 77. *Prof Exp:* From asst prof to assoc prof, 77-87, PROF BIOL, UNIV COLO, 87- *Concurrent Pos:* Dept chair, Univ Colo, 86-90. *Mem:* Animal Behav Soc; Int Union Study Social Insects (secy & treas, 78-82); AAAS; Int Bee Res Asn. *Res:* Mechanisms of dominance in social insects; kin recognition; the behavioral ecology of aggression; territoriality and spacing behavior; comparative social behavior of bees. *Mailing Add:* Dept Environ Pop & Org Biol Univ Colo Boulder CO 80309-0334

BREED, WILLIAM JOSEPH, b Massillon, Ohio, Aug 3, 28; div; c 1. WRITER, TOUR LEADER. *Educ:* Denison Univ, BA, 52; Univ Ariz, BS, 55, MS, 60. *Prof Exp:* Stratigraphic aide, Shell Oil Co, Tex, 55; geol aide, Ground Water Br, US Geol Surv, Colo, 55-56; cur geol, Mus Northern Ariz, 60-78, head, Geol Dept, 78-81; mem, Dose Assessment Adv Coun, Dept Energy, 80-87; MEM, ARIZ ATOMIC ENERGY COMN, 79- *Concurrent Pos:* Fulbright grant, NZ, 57; naturalist & exped leader, Nature Exped Int, 75-, Betchart Exped, 84- *Honors & Awards:* Gladys Cole Award, Geol Soc Am, 82; NSF Antarctic Serv Medal, 77. *Mem:* AAAS. *Res:* Eolian processes; Proterozoic rocks of Grand Canyon; general geology of northern Arizona; natural history. *Mailing Add:* Box 1424 Flagstaff AZ 86002-1424

BREEDEN, JOHN ELBERT, b Charlotte, Tenn, Nov 11, 31; m 53; c 3. PLANT ECOLOGY. *Educ:* Austin Peay State Univ, BS, 53, MA, 54; Vanderbilt Univ, PhD(plant ecol), 68. *Prof Exp:* PROF BIOL, DAVID LIPSCOMB UNIV, 68- *Mem:* Sigma Xi. *Mailing Add:* David Lipscomb Univ Nashville TN 37203

BREEDING, J ERNEST, JR, b Peoria, Ill, Mar 17, 38; m 70, 90; c 1. OCEAN ACOUSTICS, WATER WAVES. *Educ:* Drake Univ, BA, 60; Columbia Univ, PhD(geophys), 72. *Prof Exp:* Res asst physics, Univ Tenn, Knoxville, 60-61; res physicist, Naval Coastal Systs Ctr, 62-78; vis assoc prof oceanog, Fla State Univ, Tallahassee, 77-79; mem fac, Fla Inst Technol, 79-87; oceanogr, Naval Oceanog Off, 88, RES PHYSICIST, NAVAL RES LAB-STENNIS SPACE CTR, 88- *Concurrent Pos:* Res asst phys oceanog, Lamont-Doherty Geol Observ, Columbia Univ, 65-70. *Mem:* AAAS; Am Asn Physics Teachers; Am Geophys Union; Soc Explor Geophysicists; Sigma Xi; Am Meteorol Soc. *Res:* Modification and prediction of water waves in coastal waters; wave refraction; wave induced beach processes; ambient noise in underwater sound. *Mailing Add:* 115 Blackbeard Dr Slidell LA 70461

BREEDLOVE, CHARLES B, b Chico, Tex, Dec 5, 16; m 42; c 2. SCIENCE EDUCATION. *Educ:* McMurry Col, BS, 37; Southern Methodist Univ, MS, 38; Wayne State Univ, EdD, 63. *Prof Exp:* Teacher pub high schs, Tex, 38-42 & Mich, 43-58; head dept chem, Henry Ford High Sch, 58-64; prof, 64-80, EMER PROF PHYSICS, EASTERN MICH UNIV, 80- *Concurrent Pos:* Mem, State Sci Curriculum Comt, Mich, 64-65; dir, Five Year NSF Proj, 70-75. *Mem:* Nat Asn Sci Teachers; Nat Asn Res Sci Teaching; Am Asn Physics Teachers; Am Chem Soc. *Res:* Science teacher education and science curriculum projects. *Mailing Add:* 555 Pleasant St Birmingham MI 48009

BREEDLOVE, JAMES ROBBY, JR, physics, computer science, for more information see previous edition

BREEN, DALE H, b Mar 10, 25. METALLURGY. *Educ:* Bradley Univ, BS, 49; Univ Mich, MS, 53; Univ Chicago, MBA, 68. *Prof Exp:* Metallurgist, Houdaille-Hershey Corp, 49-52; instr, Univ Mich, 52-53; res metallurgist, Int Harvester Co, 53-56, metal consult, Truck Group, 56-61, mat engr, 62-64, chief res metallurgist, Corp, 64-71, mgr mat res & eng, 71-80, mgr, Metall Tech Ctr, 80-82, mgr strategic planning, 82; DIR, GEAR RES INST, 82- *Mem:* Fel Am Soc Metals; Soc Automotive Engrs; Am Soc Mech Engrs; Soc Tribologists & Lubrication Engrs. *Res:* Metallurgy; gears; materials and tribology; numerous publications. *Mailing Add:* Gear Res Inst PO Box 353 Naperville IL 60566

BREEN, GAIL ANNE MARIE, b Pembroke, Ont, Can. NEUROSCIENCES, MOLECULAR BIOLOGY. *Educ:* Univ Toronto, BSc, 70; Univ Calif, Los Angeles, PhD(neurosci), 74. *Prof Exp:* From res asst pharm to res asst immunol, Univ Toronto, 68-70; res asst neurosci, Univ Calif, Los Angeles, 70-74; Med Res Coun Can fel & res assoc molecular biol, Roswell Park Mem Inst, 74-; AT BIOL PROG DIV, UNIV TEX, DALLAS. *Concurrent Pos:* Achievement Rewards for Col Scientists Found fel, 70-72. *Mem:* Soc Neurosci; Sigma Xi. *Res:* Genetics and biochemistry of various lysosomal enzymes, especially beta-galactosidase; regulation of gene expression during development and differentiation, especially of the nervous system. *Mailing Add:* Biol Programs Div Univ Texas-Dallas, PO Box 688 M-S FO 3 1 Richardson TX 75083-0688

BREEN, JAMES LANGHORNE, b Chicago, Ill, Sept 5, 26; m 51; c 5. OBSTETRICS & GYNECOLOGY. *Educ:* Northwestern Univ, BS, 48, MD, 52; Am Bd Obstet & Gynec, dipl. *Prof Exp:* Rotating intern, Walter Reed Army Hosp, 52-53, resident obstet & gynec, 54-57, asst chief, 57-58; from asst chief to chief, Second Gen Hosp, Landstuhl, Ger, 58-60; assoc prof, Seton Hall Col Med & Dent, 61-69; DIR DEPT OBSTET & GYNEC, ST BARNABAS MED CTR, 69- *Concurrent Pos:* Res fel obstet, gynec & breast path, Armed Forces Inst Path, 60-61; attend obstetrician & gynecologist, Margaret Hague Maternity Hosp, Jersey City, 61-63; gynecologist & obstetrician-in-chief, Newark City Hosp, 63-; attend gynecologist, Med Ctr, Jersey City, 63-; consult, St Elizabeth Hosp & Monmouth Med Ctr, 63- & St Barnabas Med Ctr, 64-; clin prof obstet & gynec, Jefferson Med Col, 76- & NJ Med Sch, 86. *Mem:* Fel Am Col Obstet & Gynec; fel Am Col Surgeons; Asn Mil Surgeons US; NY Acad Sci; Am Soc Cytol; Soc Gynec Surgeons (pres, 88-89). *Res:* Oncology surgery; research cytology; biochemical assays of female genital malignancy; ovarian tumors. *Mailing Add:* St Barnabas Med Ctr Old Short Hills Rd Livingston NJ 07039

BREEN, JOHN E(DWARD), b Buffalo, NY, May 1, 32; m 53; c 7. CIVIL ENGINEERING. *Educ:* Marquette Univ, BCE, 53; Univ Mo, MS, 57; Univ Tex, PhD, 62. *Prof Exp:* Res engr, Univ Mo, 56-57, asst prof civil eng, 57-59; res engr, 59-62, from asst prof to assoc prof civil eng, 62-68, prof, 69-77, J J McKetta prof eng, 77-81, Carol Cocknell Curran Chair Eng, 81-84, NASSER I AL-RASHID CHAIR CIVIL ENG, UNIV TEX, AUSTIN, 84- *Honors & Awards:* Wason Medal, Am Concrete Inst, 72 & 83; Raymond C Reese Res Medal, 72 & 78; T Y Lin Award, Am Soc Civil Engrs, 85 & 89; Arthur A Anderson Award, Am Concrete Inst, 87; J Boase Award, Reinforced Concrete Res Coun, 87; Fedn Internationale de Precontrainte Medal, 90. *Mem:* Nat Acad Eng; fel Am Concrete Inst; fel Am Soc Civil Engrs. *Res:* Structural design and behavior of reinforced and prestressed concrete; codes and standards for building and bridge structures; construction safety in concrete construction. *Mailing Add:* Ferguson Struct Eng Lab 10100 Burnet Rd Austin TX 78758-4497

BREEN, JOSEPH JOHN, b Waterbury, Conn, July 22, 42. ENVIRONMENTAL HEALTH. *Educ:* Fairfield Univ, BS, 64; Duke Univ, PhD(chem), 73. *Prof Exp:* NIH fel, Biophysics & Microbiol Dept, Univ Pittsburgh, 72-74; chemist, Lab Officiel des Analyses, Rech Chimiques, Casablanca, Morocco, 74-75 & Inst Agronomique, Veternaire-Hassan II, Rabat, Morocco, 75-76; supvr chemist, 77-85, BR CHIEF, OFF TOXIC SUBSTANCES, ENVIRON PROTECTION AGENCY, WASHINGTON, DC, 85- *Concurrent Pos:* Consult, The Br Sch, Morocco, 74-76; adj prof, Southeastern Univ, Washington, DC, 84-85. *Honors & Awards:* Silver Medal, Environ Protection Agency, 84; Bronze Medal, Environ Protection Agency, 85, 86 & 87. *Mem:* Am Chem Soc; Sigma Xi; NY Acad Sci; Nat Sci Teachers Asn; Am Pub Health Asn; Asn Off Anal Chem. *Res:* Assess human body burden and environmental levels of agricultural and industrial chemicals; asbestos analytical and quality assurance program; chemometrics applied statistics in chemistry. *Mailing Add:* Environ Protection Agency-OTS-TS798 401 M St SW Washington DC 20460

BREEN, MARILYN, b Anderson, SC, Nov 8, 44; m 75; c 1. GEOMETRY. *Educ:* Agnes Scott Col, BA, 66; Clemson Univ, MS, 68, PhD(math), 70. *Prof Exp:* Vis instr, 70-71; from asst prof to assoc prof, 72-82, PROF MATH, UNIV OKLA, 82- *Mem:* Math Asn Am; Am Math Soc. *Res:* Convexity and combinatorial geometry; convex polytopes; m-convex sets; starshaped sets. *Mailing Add:* Dept Math Univ Okla Norman OK 73019

BREEN, MOIRA, b Madras, India, Dec 18, 23; US citizen. BIOCHEMISTRY. *Educ:* Univ Madras, BS, 44; Vassar Col, MS, 51; Northwestern Univ, PhD(biochem), 60. *Prof Exp:* Asst biochem & physiol, Univ Madras, 44-49; asst physiol, Vassar Col, 49-52; instr, Sch Nursing, Johns Hopkins Univ, 52-53; asst nutrit, Harvard Univ, 53-54; instr biochem, Northwestern Univ, 60-62; res assoc med, Univ Va, 62-63; res assoc physiol, Univ Chicago, 64-65; sr res biochemist, Res-in-Aging Lab, Vet Admin Hosp, North Chicago, Ill, 68-; ASSOC PROF, DEPT PATH, UNIV HEALTH SCI, CHICAGO MED SCH NORTH CHICAGO, ILL, 79- *Mem:* AAAS; Am Chem Soc; Am Soc Biol Chemists; Complex Carbohydrate Soc. *Res:* Complex carbohydrates in central nervous system and other tissues. *Mailing Add:* 139 Woodland Rd Libertyville IL 60048-1701

BREENE, WILLIAM MICHAEL, b Adams, Wis, May 10, 30; m 63. FOOD SCIENCE, BIOCHEMISTRY. *Educ:* Univ Wis, BS, 57, MS, 61, PhD(dairy & food industs), 64. *Prof Exp:* Res asst dairy & food industs, Univ Wis, 58-64, proj asst, 64; res assoc dairy industs, Univ Minn, 64-66, asst prof food sci & industs, 66-67; asst prof, Mich State Univ, 67; from asst prof to assoc prof, 68-80, PROF FOOD SCI & NUTRIT, UNIV MINN, ST PAUL, 80- *Mem:* Inst Food Technologists; Am Soc Hort Sci; Am Asn Cereal Chem. *Res:* Spray drying fruits and vegetables using skim milk as a carrier; carotenoid oxidation, fruit and vegetable texture, soybean foods, adzuki beans, bean sprouts. *Mailing Add:* Dept Food Sci & Nutrit Univ Minn 1334 Eckles Ave St Paul MN 55108

BREESE, GEORGE RICHARD, b Richmond, Ind, Dec 27, 36; m 60; c 2. PHARMACOLOGY, NEUROBIOLOGY. *Educ:* Butler Univ, BS, 59, MS, 61; Univ Tenn, PhD(pharmacol), 65. *Prof Exp:* Instr pharmacol, Med Sch, Univ Ark, 65-66; res assoc pharmacol & toxicol, Nat Inst Gen Med Sci, 66-68; from asst prof to assoc prof, 68-76, PROF PSYCHIAT & PHARMACOL, SCH MED, UNIV NC, CHAPEL HILL, 76- *Concurrent Pos:* NIH grant; Nat Inst Mental Health grant, Study Sect; USPHS grant. *Mem:* AAAS; Am Soc Pharmacol & Exp Therapeut; Soc Neurosci; Am Col Neuropsychopharmacol; Res Soc Alcohol. *Res:* Developmental neuropsychopharmacology; neuropharmacology; depression; alcohol; neurocytotoxic compounds; peptides. *Mailing Add:* 226 Biol Sci Res Ctr Univ NC Sch Med Chapel Hill NC 27514

BREESE, SYDNEY SALISBURY, JR, b New York, NY, Apr 11, 22; m 53; c 3. VETERINARY VIROLOGY. *Educ:* Antioch Col, BS, 44; Univ Md, MS, 53. *Prof Exp:* Biophysicist virol, Walter Reed Army Inst Res, 48-53; physicist, Children's Cancer Res Found, 54-55; physicist, Plum Island Animal Dis Ctr, USDA, 55-78; assoc dir, Cent Electron Micros Facil, Med Sch, Univ Va, 79-84, res asst prof, anat, 80-84; ELECTRON MICROS, MARINE SCI CTR, SUFFOLK COUN COMMUNITY COL, 86- *Concurrent Pos:* Prog chmn & ed, 4th Int Cong Electron Micros, 62; chmn, Gordon Res Conf Immuno-Electron Micros, 72. *Mem:* AAAS; Electron Micros Soc Am (pres, 64); Am Soc Cell Biol; Soc Gen Microbiol Gt Brit. *Res:* Physical properties and purification of viruses; electron microscopy and ultracentrifugation techniques for virology problems; development and identification of viruses; immunochemical techniques in electron microscopy. *Mailing Add:* 1700 Cedar Beach Rd Southold NY 11971

BREG, WILLIAM ROY, b Arlington, Tex, Sept 6, 23; m 48; c 4. PEDIATRICS, CYTOGENETICS. *Educ:* Yale Univ, MD, 47. *Prof Exp:* From clin instr to assoc clin prof, 55-77, PROF HUMAN GENETICS & PEDIAT, YALE UNIV, 77- *Concurrent Pos:* Consult, physician & dir, Res Lab, Southbury Training Sch, 54-77; consult, Nat Inst Child Health & Human Develop, 65-69. *Mem:* Am Acad Pediat; Am Asn Ment Deficiency; Soc Pediat Res; Am Soc Human Genetics; Am Pediat Soc. *Res:* Human chromosomal abnormalities. *Mailing Add:* Dept Human Genetics 333 Cedar St PO Box 3333 New Haven CT 06510

BREGAR, WILLIAM S, b Chicago, Ill, Jan 11, 41; m 63; c 3. PROBLEM SOLVING. *Educ:* Miami Univ, BA, 63; Univ Wis, Madison, MS, 69, PhD(comput sci), 74. *Prof Exp:* Res assoc comput aided instr, Pa State Univ, 74-75; asst prof, 75-81, ASSOC PROF COMPUTER SCI, ORE STATE UNIV, 81- *Mem:* Asn Comput Mach; Sigma Xi; Asn Develop Comput Based Instr Systs. *Res:* Artificial intelligence emphasizing cognitive aspects of problem solving, knowledge representation and their application to computer based instruction. *Mailing Add:* Tektronix Inc PO U Box 500 MS 50-662 Beaverton OR 97077

BREGLIA, RUDOLPH JOHN, b New York, NY, Nov 6, 47; m 74; c 2. TOXICOLOGY. *Educ:* St John's Col Pharm, BS, 70; St John's Univ, MS, 72; New York Med Col, PhD(pharmacol), 79. *Prof Exp:* Toxicologist, Texaco Inc, 79-81, sr toxicologist, 81-85, proj toxicologist, 85- *Honors & Awards:* Chevron USA Award, 84. *Mem:* Assoc Soc Toxicol. *Res:* Designing, monitoring and evaluating toxicity testing for a wide variety of experimental materials and commercial products. *Mailing Add:* British Petrol 200 Public Square-7K Cleveland OH 44114-2375

BREGMAN, ALLYN A(ARON), b Brooklyn, NY, Apr 29, 41; m 65; c 2. CELL BIOLOGY, CYTOGENETICS. *Educ:* Brooklyn Col, BS, 62; Univ Rochester, MS, 64, PhD(biol), 68. *Prof Exp:* From asst prof to assoc prof, 67-85, PROF BIOL, STATE UNIV NY, COL NEW PALTZ, 85- *Concurrent Pos:* NIMH grant; vis scholar, Duke Univ, 75. *Mem:* AAAS; Am Soc Cell Biol. *Res:* Cell biology; auth of 1 book. *Mailing Add:* Dept Biol State Univ of NY Col New Paltz NY 12561

BREGMAN, DAVID, b New York, NY, Apr 24, 40; m 77; c 4. CARDIOVASCULAR SURGERY. *Educ:* NY Univ, BA, 61; State Univ NY, MD, 65. *Prof Exp:* from asst prof to assoc prof, 75-83, ASSOC CLIN PROF SURG, COL PHYSICIANS & SURGEONS, COLUMBIA PRESBY MED CTR, COLUMBIA UNIV, 83-; CHMN, DEPT SURG, ST JOSEPH'S HOSP & MED CTR, NJ, 82- *Concurrent Pos:* Asst attending surgeon, Presby Hosp, New York, 75-80, assoc attending surgeon, 80-; overall chmn, subcomt Formulation Nat Standards, Am Soc Artificial Internal Organs, 77-; mem,

Comt Cardiovasc Surg, Am Col Chest Physicians, 77-; prof surg, Seton Hall Univ, Sch Grad Med Educ, South Orange, NJ, 89- *Mem:* Fel Am Col Cardiol; fel Am Col Chest Physicians; fel Am Col Surgeons; Am Soc Artificial Internal Organs; Soc Thoracic Surgeons; Am Asn Thoracic Surg. *Res:* Artificial organs with emphasis on mechanical circulatory support techniques; pioneered in clinical applications of intra-aortic balloon pumping and pulsatile assist device during open heart surgery. *Mailing Add:* Dept Surg St Joseph's Hosp & Med Ctr 703 Main St Paterson NJ 07503

BREGMAN, JACOB ISRAEL, b Hartford, Conn, Sept 17, 23; m 48; c 3. ENVIRONMENTAL MANAGEMENT. *Educ:* Providence Col, BS, 43; Polytech Inst Brooklyn, MS, 48, PhD(chem), 51. *Prof Exp:* Chemist, Fels & Co, 47-48; sr chemist, Nat Aluminate Corp, 50-52, head dept phys chem, 52-59; supvr phys chem, IIT Res Inst, 59-63, asst dir chem res, 63-65, dir chem sci res & mgr water res ctr, 65-67; dep asst secy, US Dept Interior, 67-69; pres & chmn bd, Wapora, Inc, 69-82; vpres, Dynamic Corp, 83-84; PRES, BREGMAN & CO INC, 84- *Concurrent Pos:* Chmn, Northeast Ill Metrop Area Air Pollution Control Bd, 62-63, Ill Air Pollution Control Bd, 63-67, Adv Comt Saline Water Conversion to Sci & Tech Comt, NATO Parliamentarians Coun, 63 & Water Resources Res Coun, 64-66; ed, Int Series Chem & Allied Sci, Spartan Book Co, 64-67; co-ed, Series on Water Pollution, Acad Press, 69-80. *Mem:* Am Chem Soc; Sigma Xi; Am Inst Chemists. *Res:* Ion exchange; corrosion; water treatment; saline water conversion; air and water pollution; environment. *Mailing Add:* 5630 Old Chester Rd Bethesda MD 20814

BREHM, BERTRAM GEORGE, JR, b Cleveland, Ohio, Nov 26, 26; m 50; c 6. BOTANY, SYSTEMATICS. *Educ:* Western Reserve Univ, BS, 50, MS, 52; Univ Tex, PhD(bot), 62. *Prof Exp:* Instr biol, Lee Col, Tex, 55-61; from asst prof to assoc prof 62-80, PROF BIOL, REED COL, 80- *Concurrent Pos:* NSF grants, 63 & 67. *Mem:* AAAS; Bot Soc Am; Int Asn Plant Taxon; Am Asn Plant Taxon. *Res:* Systematics and evolution of higher plants as determined by comparative chemistry; plant-insect interactions. *Mailing Add:* 727 SW Chestnut Portland OR 97202

BREHM, JOHN JOSEPH, b Memphis, Tenn, Dec 6, 34; m 59; c 4. PHYSICS. *Educ:* Univ Md, BS, 56, PhD(physics), 63; Cornell Univ, MS, 59. *Prof Exp:* Physicist, US Naval Ord Lab, 58-62; NSF fel particle physics, Princeton Univ, 62-63; from asst prof to assoc prof high energy physics, Northwestern Univ, 63-67; assoc prof physics, 67-71, PROF PHYSICS, UNIV MASS, AMHERST, 71- *Mem:* Am Phys Soc. *Res:* Theoretical physics, elementary particle physics; nonlinear dynamics. *Mailing Add:* Dept Physics Univ Mass Amherst MA 01003

BREHM, LAWRENCE PAUL, b Berwyn, Ill, Dec 24, 48; m 87; c 1. HOLOGRAPHY & DIFFERENTIAL HOLOGRAPHY, WAVEGUIDE OPTICS. *Educ:* Manhattan Col, BS, 70; Rensselaer Polytech Inst, MS, 78; Univ Del, PhD(physics), 84. *Prof Exp:* Teacher physics, Barlow Sch, 72-75; teaching asst & res trainee physics, Rensselaer Polytech Inst, 75-78; teacher physics, Simon's Rock Bard Col, 78-79; instr & res fel physics, Univ Del, 79-84; STAFF PHYSICIST, IBM CORP, 85- *Mem:* Sigma Xi; Am Phys Soc; Biophys Soc; Am Asn Physics Teachers. *Res:* Optoelectronics; metrology and analysis of optical digital data communications systems; electronic packaging. *Mailing Add:* Dept T31 Bldg 257-3 IBM Corp 1701 North St Endicott NY 13760

BREHM, RICHARD LEE, nuclear engineering, for more information see previous edition

BREHM, SYLVIA PATIENCE, microbiology, for more information see previous edition

BREHM, WARREN JOHN, b New York, NY, Nov 2, 25; m 49, 78; c 2. CHEMISTRY, PATENTS. *Educ:* NY Univ, AB, 44; Harvard Univ, AM, 46, PhD(org chem), 48. *Prof Exp:* Chemist, Carbide & Carbon Chem Corp, Tenn, 45-46; Jewett fel, Columbia Univ, 48-49; instr, NY Univ, 49-52; chemist, EI Du Pont de Nemours & Co Inc, 52-56, from supvr to sr supvr, 56-68, res labsupt, Plastics Dept, Fluorocarbons Div, Exp Sta, 68, sr supvr, Res & Develop Div, 68-82, patents consult, 82-90; RETIRED. *Mem:* Am Chem Soc; NY Acad Sci; Royal Soc Chem. *Res:* Structure of strychnine; antimalarials; reaction mechanisms; polymers and plastics. *Mailing Add:* 111 W Pembrey Dr Pembrey Wilmington DE 19803

BREHM, WILLIAM FREDERICK, b Franklin, Pa, May 26, 40. RADIOACTIVE MATERIAL TRANSPORT, CORROSION. *Educ:* Mass Inst Technol, BS, 62; Cornell Univ, MS, 64, PhD(mats sci), 67. *Prof Exp:* Res engr, Sci Lab, Ford Motor Co, 62; teaching asst, Cornell Univ, 62-63, res asst, 63-67; sr scientist, Battelle Northwest Labs, 67-70; staff scientist & group leader, Battelle Northwest Labs, 87-88; sr engr & fel engr, 70-76, mgr, Corrosion Technol, 76-87, FEL ENGR, WESTINGHOUSE HANFORD CO, 89- *Mem:* Sigma Xi; Minerals, Metals & Mat Soc; Am Soc Metals Int; Am Nuclear Soc; Nat Asn Corrosion Engrs. *Res:* Materials compatibility and radioactive mass transport in fluid systems especially sodium and lithium; methods for cleaning sodium and lithium from components and requalifying them for continued service; engineering problems associated with fluid systems and cover gas; corrosion of container materials in nuclear waste repository environments; engineering: plant chemist and system cognizant engineer, fast flux test facility-Department of Energy test reactor; chemistry and engineering aspects of high-level nuclear waste storage. *Mailing Add:* 727 Snyder Rd Richland WA 99352

BREHME, ROBERT W, b Washington, DC, Mar 6, 30; m 54; c 4. THEORETICAL PHYSICS. *Educ:* Roanoke Col, BS, 51; Univ NC, MS, 54, PhD(physics), 59. *Prof Exp:* From asst prof to assoc prof, 59-68, PROF PHYSICS, WAKE FOREST UNIV, 68- *Concurrent Pos:* Res assoc, Univ NC, 64-65. *Mem:* Am Asn Physics Teachers; AAAS; Sigma Xi; Am Phys Soc. *Res:* Field theory; astronomy; quantum electrodynamics; relativity. *Mailing Add:* Box 7803 Reynolds Sta Winston-Salem NC 27109

BREHMER, MORRIS LEROY, b Strasburg, Ill, Apr 10, 25; m 54. AQUATIC BIOLOGY. *Educ:* Eastern Ill Univ, BS, 50; Mich State Univ, MS, 56, PhD, 58. *Prof Exp:* Tech asst, Ill Nat Hist Surv, 51; fisheries biologist, Ill Dept Conserv, 51-55; tech asst, Inst Paper Chem, 55; student conserv aide, Mich Inst Fisheries Res, 55-58; assoc biologist, Va Inst Marine Sci, 58-63, sr marine scientist, 63-67, asst dir, 67-71, head div appl marine sci & ocean eng, 69-71; dir environ opers, 71-76, exec mgr environ serv, 76-83, CORP SCIENTIST, VA POWER, 83- *Mem:* Air Pollution Control Asn; Water Pollution Control Fedn. *Res:* Aquatic ecology; toxicology; electric fields. *Mailing Add:* 2103 Fenway Dr Mechanicsville VA 23111

BREHOB, W(AYNE) M, b Indianapolis, Ind, June 19, 36; m 56; c 6. MECHANICAL ENGINEERING. *Educ:* Purdue Univ, BS, 58, MS, 60, PhD(mech eng), 63. *Prof Exp:* Instr mech eng, Purdue Univ, 59-63; res engr, 63-70, prin res engr emission, 70-75, exec engr components & emission res, 75-78, exec engr advanced eng, prod planning Res, Ford Motor Co, 78-81; chmn, Mech Eng Dept, Lawrence Inst Technol, 81-89; PROF MECH ENG, LAWRENCE TECH UNIV, 89- *Mem:* Soc Automotive Engrs; Combustion Inst; Am Soc Mech Engrs. *Res:* Emission control through engine design and after treatment systems; emission regulations and tests; drivetrain and vehicle modifications for fuel economy improvement; alternate engines, engine fuels and fuel sources. *Mailing Add:* 538 N Franklin Dearborn MI 48128-1518

BREIDENBACH, ROWLAND WILLIAM, b Dayton, Ohio, Feb 1, 35; m 57; c 2. PLANT BIOCHEMISTRY, CELL CULTURE. *Educ:* Univ Fla, BS, 59; Univ Calif, Davis, MS, 63, PhD(plant physiol), 66. *Prof Exp:* NIH fel biol sci, Purdue Univ, 66-67; asst prof, 70-71, asst agronomist, Exp Sta, 67-76, assoc agronomist, 76-82, LECTR AGRON & RANGE SCI, UNIV CALIF, DAVIS, 76-, AGRONOMIST, 82-, DIR, PLANT GROWTH LAB, 87- *Concurrent Pos:* NSF fel, 66-67; Du Pont Young fac award, Univ Calif, Davis, 70-71. *Honors & Awards:* NY Bot Garden Award, 68. *Mem:* AAAS; Am Soc Plant Physiol. *Res:* Cell and molecular biology; biomembranes, compartmentation and metabolic regulation; mechanisms of environmental stress effects on plants. *Mailing Add:* Dept Agron Univ Calif Davis CA 95616

BREIG, EDWARD LOUIS, b St Marys, Mo, Oct 6, 32. QUANTUM THEORY, ATOMIC COLLISIONS. *Educ:* Southeast Mo State Col, BS, 54; Univ Okla, MS, 63, PhD(physics), 66. *Prof Exp:* Res engr, Autonetics-NAm Rockwell, 61-62; theoret physicist, Aerospace Corp, Los Angeles, 66-72; RES SCIENTIST, UNIV TEX, DALLAS, 72- *Mem:* Am Phys Soc; Am Asn Physics Teachers; Am Geophys Union. *Res:* Theoretical atomic and molecular physics, with applications to physics and chemistry of planetary atmospheres; atomic processes in lasers; mathematical analysis of ionospheric measurements from satellites. *Mailing Add:* Univ Tex Dallas PO Box 688 Richardson TX 75083-0688

BREIG, MARVIN L, b St Mary's, Mo, Oct 9, 34; m 61; c 1. SOLID STATE PHYSICS. *Educ:* Southeast Mo State Col, BS, 56; Univ Okla, MS, 59, PhD(physics), 63. *Prof Exp:* Asst prof, 63-64, assoc prof, 64-76, PROF PHYSICS, EASTERN ILL UNIV, 76- *Res:* Dislocation study in ionic crystals; pyroelectric effect in triglycene sulfate; electron spin resonance and nuclear magnetic resonance. *Mailing Add:* Dept Physics Eastern Ill Univ Charleston IL 61920

BREIL, DAVID A, b Brockton, Mass, Mar 27, 38; m 63; c 2. BRYOLOGY, PLANT MORPHOLOGY. *Educ:* Univ Mass, Amherst, BS, 60, MA, 63; Fla State Univ, PhD(bot), 68. *Prof Exp:* Instr bot, Pa State Univ, 63-65; Tall Timbers Res Sta grant, Fla, 66-68; PROF BIOL, LONGWOOD COL, 68- *Mem:* Am Bryol & Lichenological Soc; Bot Soc Am. *Res:* Liverwort flora of Florida; bryophytes of Virginia; ecology of bryophytes. *Mailing Add:* Dept Natural Sci Longwood Col Farmville VA 23901

BREIL, SANDRA J, b Springfield, Mass, Apr 30, 37. CELL PHYSIOLOGY, PLANT HISTOCHEMISTRY. *Educ:* Univ Vt, BA, 58; Univ Mass, PhD(bot), 63. *Prof Exp:* Instr biol, Wilson Col, 63-65; fel, Inst Molecular Biophys, Fla State Univ, 66-68; ASSOC PROF BIOL, LONGWOOD COL, 69- *Mailing Add:* Nat Sci Dept Longwood Col Farmville VA 23901

BREILAND, JOHN GUSTAVSON, b Hjelmeland, Norway, Nov 21, 05; nat US; m 42; c 2. METEOROLOGY. *Educ:* Luther Col, Iowa, AB, 33; Univ Iowa, MS, 34; Univ Calif, Los Angeles, PhD(meteorol), 56. *Prof Exp:* Instr math, Luther Col, Iowa, 35-38 & Univ Pittsburgh, 38-39; observer meteorol, US Weather Bur, Iowa, 39-41; from instr to assoc prof physics & meteorol, 42-66, actg chmn dept physics, 57-58, 61-62, prof physics, 66-70, EMER PROF PHYSICS, UNIV NMEX, 70- *Concurrent Pos:* Asst dir, Air Force Meteorol Training Prog, 51-52, dir, 52-54. *Res:* Synoptic meteorology; thunderstorms; atmospheric ozone. *Mailing Add:* 2921 Charleston NE Albuquerque NM 87110

BREILLATT, JULIAN PAUL, JR, b Pensacola, Fla, Mar 2, 38; m 62; c 4. BIOMATERIALS. *Educ:* Univ Calif, Berkeley, AB, 59; Univ Utah, PhD(biochem), 67. *Prof Exp:* Res assoc, Oak Ridge Nat Lab, 67-69, res scientist, 69-74, actg dir, Molecular Anat Prog, 74-76, sr res scientist biol div, 76-77; res supvr, 77-78, sr res chemist, E I Du Pont De Nemours & Co, 78-85; Baxter res scientist, 86-90, RES DIR, BAXTER HEALTH CARE CORP, 90- *Honors & Awards:* IR-100 Award, Indust Res, 77. *Mem:* NY Acad Sci; AAAS; Am Soc Artificial Internal Organs; Biomaterials Soc; Int Soc Artificial Organs. *Res:* Ex vivo blood cell separation; immunodepletion; zonal centrifugation; artificial organs; bioresponse inducing materials. *Mailing Add:* Baxter Healthcare Corp Baxter Technology Park Round Lake IL 60073-0490

BREINAN, EDWARD MARK, physical metallurgy, for more information see previous edition

BREINER, SHELDON, b Milwaukee, Wis, Oct 23, 36; m 62; c 2. GEOPHYSICS. *Educ:* Stanford Univ, BS, 59, MS, 62, PhD(geophys), 67. *Prof Exp:* Res geophysicist, Varian Assocs, 61-68; geophysicist & pres, Geometrics, 69-83; pres & founder, Syntelligence, 83-87; CHMN & CHIEF

EXEC OFFICER, PARA MAGNETIC LOGGIUS INC, 83-; PRES & FOUNDER, QUORUM SOFTWARE, 88- *Concurrent Pos:* Res assoc, NSF grant, 65-67; res assoc, dept geophys, Stanford Univ, 67-, consult prof, dept appl earth sci, 74-78. *Mem:* AAAS; Soc Explor Geophys; Am Geophys Union; Europ Asn Explor Geophysicists; Sigma Xi. *Res:* Remote sensing, magnetic and gamma ray surveys for mineral and petroleum exploration; airborne and marine magnetic gradient studies; earthquake prediction; micropulsations; magnetic surveys for archaeological exploration. *Mailing Add:* 45 Buckeye Portola Valley CA 94025

BREIPOHL, ARTHUR M, b Higginsville, Mo, Nov 14, 31; m 55; c 2. ELECTRICAL ENGINEERING. *Educ:* Univ Mo, BS, 54; Univ NMex, MS, 60, ScD(elec eng), 64. *Prof Exp:* Trainee engr, Westinghouse Elec Corp, 54-55; staff mem, Sandia Corp, 57-62, 63-64; res assoc eng, Univ NMex, 62-63; from asst prof to prof, Okla State Univ, 64-70; chmn dept elec eng, Univ Kans, 70-79; prof, Univ Kans, 80-84; AT DEPT ELEC ENG, UNIV OKLA, 84- *Concurrent Pos:* Consult, Sandia Corp, 62-63 & 64-65; vis prof eng econ systs, Standford Univ, 79; consult, Kans Corp Comn, 79-84. *Mem:* Inst Elec & Electronics Engrs. *Res:* Statistical system studies; electric power pricing; reliability; estimation from data and engineering knowledge. *Mailing Add:* Dept Elec Eng Univ Olka 202 W Boyd Rm 219 Norman OK 73019

BREISCH, ERIC ALAN, b Woodbury, NJ, Nov 5, 50; m 73; c 2. MORPHOMETRY. *Educ:* Temple Univ, BS, 72, PhD(anat), 77. *Prof Exp:* Fel cardiovasc anat/physiol, Dept Med, Temple Univ, 77-79; ASST PROF ANAT, DEPT SURG, UNIV CALIF, SAN DIEGO, 79- *Mem:* Am Asn Anatomists; Am Physiol Soc; AAAS. *Res:* Influence of pressure overload stimuli on the myocradium and the resulting cardiac hypertrophy is then studied from both an ultrastructural and physiologic point of view; morphometric and myocardial blood flow techniques. *Mailing Add:* US Int Univ Sch Osteo Med & Surg 10455 Pomerado Rd Daley Hall San Diego CA 92131

BREITBEIL, FRED W, III, b Cincinnati, Ohio, Sept 25, 31; m 56; c 4. ORGANIC CHEMISTRY. *Educ:* Xavier Univ, Ohio, BS, 53, MS, 57; Univ Cincinnati, PhD(org chem), 60. *Prof Exp:* Res chemist, Proctor & Gamble Co, 60-62; assoc, Iowa State Univ, 62-63; asst prof, 63-75, chmn dept, 69-77, PROF ORG CHEM, DE PAUL UNIV, 75- *Mem:* Am Chem Soc. *Res:* Halogen additions to heterocyclics; chemically bound liquid phases for gas-liquid phase chromatography; reactions of peroxyacetylnitrate with plant cuticle. *Mailing Add:* Dept Chem De Paul Univ Chicago IL 60614

BREITENBACH, E A, b Los Angeles, Calif, Dec 17, 36; m 55; c 2. PETROLEUM ENGINEERING. *Educ:* Stanford Univ, BS, 58, MS, 59, PhD(petrol eng), 64. *Prof Exp:* Petrol engr, Socony Mobil Oil Co, Inc, 59-61; instr formation eval, Stanford Univ, 62-64; ADV RES SCIENTIST, DENVER RES CTR, MARATHON OIL CO, 64-; PRES, SCI SOFTWARE CORP, 70- *Mem:* Am Inst Mining, Metall & Petrol Engrs. *Res:* Reservoir engineering, including primary, secondary and tertiary recovery and formation evaluation. *Mailing Add:* 4491 Marigold Ln Littleton CO 80123

BREITENBACH, ROBERT PETER, b Madison, Wis, Oct 10, 23; m 48; c 3. ZOOLOGY, ENDOCRINOLOGY. *Educ:* Univ Wis, BS, 49, MS, 50, PhD(zool, endocrinol), 58. *Prof Exp:* Instr zool, Univ Wis, Milwaukee, 56-59; from asst prof to assoc prof, Univ Mo, Columbia, 59-67, chmn dept, 64-68, assoc dir, Div Biol Sci, 75-79, prof, 67-87, dir undergrad studies, Biol Div, 79-87, EMER PROF ZOOL, UNIV MO, COLUMBIA, 87- *Mem:* AAAS; Am Physiol Soc; Soc Exp Biol & Med. *Res:* Avian physiology; development of immunological maturity, reproductive physiology, calcium metabolism, endocrines and development and avian behavior. *Mailing Add:* Div Biol Sci 200 Lefevre Hall Univ Mo Columbia MO 65211

BREITENBERGER, ERNST, b Graz, Austria, June 11, 24; c 4. HISTORY & PHILOSOPHY OF SCIENCE, SCIENCE EDUCATION. *Educ:* Univ Vienna, Drphil(math), 50; Cambridge Univ, PhD(physics), 56. *Prof Exp:* Res assoc, Radiuminstitut, Vienna, 50-51; Brit Coun res scholar, Univ Cambridge, 51-54; lectr physics, Univ Malaya, 54-58; from assoc prof to prof, Univ SC, 58-63; PROF PHYSICS, OHIO UNIV, 63- *Concurrent Pos:* Guest prof, Univ Bonn, 69-70, Univ New SWales, Sydney, 88. *Mem:* Am Phys Soc; Brit Math Asn; Math Asn Am; NY Acad Sci; AAAS; Hist Sci Soc; Gauss-Gesellschaft. *Res:* Theoretical physics; probability and statistics; mathematical methods; science history; teaching mathematics to physics students. *Mailing Add:* Dept Physics Ohio Univ Athens OH 45701-2979

BREITHAUPT, LEA JOSEPH, JR, b Natchez, Miss, Sept 16, 29; m 59, 79; c 3. PAPER CHEMISTRY, CELLULOSE CHEMISTRY. *Educ:* St Louis Univ, BS, 51; La State Univ, MS, 56. *Prof Exp:* From res chemist to tech dir, Non-Wovens Div, 69-71, sr res assoc, 72-76, MGR ANALYTICAL SCI, ERLING RIIS RES LAB, INT PAPER CO, 77- *Mem:* Tech Asn Pulp & Paper Indust; Am Chem Soc. *Res:* Pulp processing reactions, especially dissolving pulps, cellulose derivatives and reactions; analytical chemistry. *Mailing Add:* Int Paper Co PO Box 2787 Mobile AL 36652

BREITMAN, LEO, polymer chemistry, physical chemistry, for more information see previous edition

BREITMAN, THEODORE RONALD, b Brooklyn, NY, Feb, 25, 31; m 58; c 2. BIOCHEMISTRY. *Educ:* City Col New York, BS, 53; Ohio State Univ, MS, 56, PhD(biochem), 58. *Prof Exp:* Nat Cancer Inst fel, USPHS, Brandeis Univ, 58-60; res biochemist, Lederle Labs, 60-62; RES BIOCHEMIST, NIH, 63- *Concurrent Pos:* Assoc ed, In Vitro Cellular & Develop Biol. *Mem:* AAAS; Tissue Cult Asn; Am Soc Biol Chemists; Am Asn Cancer Res. *Res:* Hormones and growth factors with particular emphasis on control of cell growth and differentiation. *Mailing Add:* Lab Biochem Nat Cancer Inst Bldg 37 Rm 5D02 Bethesda MD 20892

BREITMAYER, THEODORE, b Grants Pass, Ore, June 21, 22; m 47; c 3. CHEMICAL ENGINEERING. *Educ:* Ore State Univ, BS, 44; Univ Wash, MS, 48. *Prof Exp:* Res engr, Crown Zellerbach Corp, 48-57; process design engr, Monsanto Chem Co, 51-55; sr process engr, C F Braun & Co, 55-64; chief process engr, Bechtel Group Inc, 64-88, consult, 88-91; RETIRED. *Mem:* Fel Am Inst Chem Engrs; Am Nuclear Soc. *Mailing Add:* 148 Draeger Dr Moraga CA 94556

BREITSCHWERDT, EDWARD B, b Baltimore, Md, Oct 25, 48; m 80; c 2. INTERNAL MEDICINE, NEPHROLOGY. *Educ:* Univ Md, BS, 70; Univ Ga, DVM, 74; Am Col Vet Internal Med, cert, 79. *Prof Exp:* Intern, Sch Vet Med, Univ Miss, 74-75, resident, 75-77; from asst prof to assoc prof internal med, Sch Vet Med, La State Univ, 77-82; assoc prof, 82-91, PROF INTERNAL MED & NEPHROLOGY, SCH VET MED, NC STATE UNIV, 91- *Concurrent Pos:* Serv chief, Vet Teaching Hosp, La State Univ, 79-82; assoc ed, J Vet Internal Med, 86-91. *Honors & Awards:* Small Animal Res Award, Ralston Purina, 85. *Mem:* Am Vet Med Asn; Am Animal Hosp Asn; Am Col Vet Internal Med (pres, 85-88); Soc Vet Urol; Am Soc Rickettsial Dis; Am Asn Vet Immunologists. *Res:* Aspects of recognition and characterization of various naturally occurring diseases of companion animals; immunologic response to infectious agents (Rickettsia rickettsii, Borrelia burgdorferi), immunodeficiency diseases and intestinal immunity; clinical immunology. *Mailing Add:* Sch Vet Med NC State Univ 4700 Hillsborough St Raleigh NC 27606

BREITWIESER, CHARLES J(OHN), b Colorado Springs, Colo, Sept 23, 10; m 43; c 2. ELECTRONICS ENGINEERING. *Educ:* Univ NDak, BSEE, 30; Calif Inst Technol, MS, 33. *Hon Degrees:* DSc, Univ NDak, 47. *Prof Exp:* Engr, United Sound Prod Corp, 33-34; owner, C J Breitwieser & Co, 35-37; chief engr & vpres, Caldo Corp, 37-39; secy-treas, Conducto-Therm Corp, 39-42; chief electronics & res labs, Electronic & Guid Sect, Consol-Vultee Aircraft Mfg Co, 42-51; dir eng & exec asst to vpres in chg eng, P R Mallory, 51-54; vpres eng & gen mgr res & develop, Lear Inc, 54-57; pres, Metrolog Corp Div, 57-61; vpres, Air Logistics Corp, 57-61; exec vpres & gen mgr, Cubic Corp, 61-76; DIR, T-SYSTS INC, SAN DIEGO, 76- *Concurrent Pos:* Consult engr, Dept Police & Fire, Los Angeles, 32-35; consult mfg commercial res, 34-37; consult engr & chief engr res, DeForest Labs, 37-41; bd mem, Swan Electronics Corp & US Elevator Corp, chmn bd, 72-76; mgt consult to staff, Calif Pac Univ, San Diego, 76-; consult to trustees, Mesa Col, San Diego, 77- *Mem:* Inst Elec & Electronics Engrs; Aerospace Indust Asn Am; Am Ord Asn; Am Inst Aeronaut & Astronaut. *Res:* Management and direction of applied research in automation; solid state devices; guidance systems for missiles and space vehicles; elevators; computers; business machinery field. *Mailing Add:* 2738 Caminito Prado La Jolla CA 92037

BREKKE, CLARK JOSEPH, b Salem, Ore, May 23, 44; m 67; c 3. FOOD SCIENCE, FUNCTIONAL PROPERTIES. *Educ:* Rutgers Univ, BS, 66; Cornell Univ, MS, 68; Univ Wis, PhD(food sci), 72. *Prof Exp:* Res asst food sci, Cornell Univ, 66-68; res asst meat sci, 68-72; asst prof food sci, Univ Maine, 72-76; from asst prof & asst prof, 76-83, actg chmn, 81-84, PROF, DEPT FOOD SCI & HUMAN NUTRIT, WASH STATE UNIV, 83- *Concurrent Pos:* Vis scientist, Dept Food Sci, Univ BC, Vancouver, BC, 86. *Mem:* Inst Food Technologists; Am Meat Sci Asn; Sigma Xi; Poultry Sci Asn; Coun Agr Sci & Technol; Nat Asn Cols & Teachers Agr. *Res:* Meat and poultry products technology; biochemical, microbial and physical changes influencing product quality; muscle protein biochemistry; protein modification and functionality. *Mailing Add:* Dept Food Sci & Human Nutrit Wash State Univ Pullman WA 99164-6376

BREMEL, ROBERT DUANE, b Eau Claire, Wis, Dec 19, 45; m 66; c 2. ANIMAL PHYSIOLOGY, BIOCHEMISTRY. *Educ:* Univ Wis-Madison, BS, 67; St Louis Univ, PhD(biochem), 72. *Prof Exp:* Fel anat-biophys, Med Ctr, Duke Univ, 72-74; ASST PROF DAIRY SCI, UNIV WIS-MADISON, 74 - *Mem:* Am Dairy Sci Asn; Am Chem Soc; Sigma Xi. *Res:* Mechanistic studies of hormone action and the measurement and control of stress in food producing animals; genetic control of milk secretion; production of transgenic animals which produce alternate proteins in their milk. *Mailing Add:* Dept Dairy Sci Univ Wis Col Agr & Life Sci Madison WI 53706

BREMER, HANS, b Hamburg, Ger, Aug 26, 27; m 57; c 4. MOLECULAR BIOLOGY. *Educ:* Univ Heidelberg, Vordiplom, 49; Univ Gottingen, PhD(biol), 57. *Prof Exp:* Res fel zool, Univ Gottingen, 57-59; asst genetics, Univ Cologne, 59-62; res microbiologist, Univ Calif, Berkeley, 62-65; asst genetics, Univ Cologne, 65-66; from asst prof to assoc prof biol, 66-73, PROF BIOL, SOUTHWEST CTR ADVAN STUDIES, UNIV TEX, DALLAS, 73- *Concurrent Pos:* Ger Res Soc fel, 58-59; Fulbright travel grant, 62. *Res:* Developmental physiology; synthesis of ribonucleic acid; DNA replication. *Mailing Add:* Div Biol Univ Tex PO Box 688 Richardson TX 75083-0688

BREMER, KEITH GEORGE, organic chemistry; deceased, see previous edition for last biography

BREMERMANN, HANS J, b Bremen, Ger, Sept 14, 26; nat US; m 54. EPIDEMIOLOGY OF THE HUMAN IMMUNE DEFICIENCY VIRUS, MATH MODELS OF HIV-IMMUNE SYSTEM INTERACTIONS. *Educ:* Univ Munster, MA & PhD(math), 51. *Prof Exp:* Instr math, Univ Munster, 52; res assoc, Stanford Univ, 52-53, vis asst prof, 53-54; asst prof, Univ Munster, 54-55; staff mem, inst Advan Study, 55-57 &58-59; asst prof, Univ Wash, 57-58; assoc prof math, 59-64, assoc prof math & biophys, 64-66, PROF MATH & BIOPHYSICS, UNIV CALIF, BERKELEY, 66- *Concurrent Pos:* Res fel, Harvard Univ, 53; indust consult; mem, exec comt grad group biophysics & med physics, Univ Calif, Berkeley, 64-68, grad group bioeng, Berkeley, 72-76 & San Francisco, 91- *Mem:* Am Math Soc; Austrian Math Soc; Am Asn Artificial Intel; Ger Math Asn; Biophys Soc; fel AAAS. *Res:* Several complex variables; Schwartz Distributions; physics; dispersion relations; renormalization; information theory; limitations of genetic control; evolution processes; self-organizing systems; biological algorithm; complexity theory; mathematical ethology; pattern recognition; model verification;

optimization; control; medical applications of nonlinear control; mathematical models of epidemics and host-pathogen dynamics; conformation of proteins; artificial intelligence; nerve nets; evolution of sex; HIV epidemiology; mathematical models of the immune system; interactions of HIV with the immune system. *Mailing Add:* Dept Molecular & Cell Biol Univ Calif 102 Donner Berkeley CA 94720

BREMMER, BART J, b Waddinxveen, Netherlands, Sept 4, 30; US citizen; m 53; c 3. ORGANIC CHEMISTRY, POLYMER CHEMISTRY. *Educ:* State Univ Leiden, BS, 50; Mich State Univ, MS, 62. *Prof Exp:* Anal chemist, Gouda Apollo, Netherlands, 50-51; res chemist, Grand Rapids Varnish Co, 54-55; chemist, Kelvinator Div, Am Motors Corp, 55-57; org res chemist, 57-64, proj leader org chem, 64-66, sect head, 66-69, develop mgr, 69-72, res mgr org chem, 72-78, lab dir, 78-84, ELECTRONIC INDUST DEV MGR, DOW CHEM CO, 84- *Mem:* Am Chem Soc; Com Develop Asn. *Res:* Thermoset resins, especially epoxy and phenolic resins with emphasis on fire-retardant resins; episulfide chemistry; chemistry of halogenated aromatic compounds, especially those containing amine groups; pyridine chemistry; materials used in electronics industry. *Mailing Add:* 5609 Woodview Pass Midland MI 48640-3147

BREMNER, JOHN MCCOLL, b Dumbarton, Scotland, Jan 18, 22; nat US; m 50; c 2. AGRICULTURE, ENVIRONMENTAL SCIENCES. *Educ:* Glasgow Univ, BSc, 44; Univ London, PhD(chem), 48. *Hon Degrees:* DSc, Univ Glasgow, 87, Univ London, 59. *Prof Exp:* From sci officer to prin sci officer, Chem Dept, Rothamsted Exp Sta, Eng, 45-59; assoc prof soil biochem, 59-61, PROF AGRON & BIOCHEM, 61-, CHARLES F CURTISS DISTINGUISHED PROF AGR, IOWA STATE UNIV, 75- *Concurrent Pos:* Rockefeller Found fel, 57-58; tech expert, Int Atomic Energy Agency, Yugoslavia, 64-65; Guggenheim Found fel, 68-69. *Honors & Awards:* Soil Sci Achievement Award, Am Soc Agron, 67 & Agron Res Award, 85; Alexander von Humboldt Award, Humboldt Found, 82; Gov's Sci Achievement Medal, Gov Iowa, 83; Harvey Wiley Award, Asn Off Anal Chemists, 84; Spencer Award, Am Chem Soc, 87; Environ Qual Res Award, Soil Sci Soc Am, 89. *Mem:* Nat Acad Sci; fel Soil Sci Soc Am; fel Am Soc Agron; Am Chem Soc; Brit Soc Soil Sci; fel AAAS. *Res:* Soil biochemistry and microbiology. *Mailing Add:* Dept Agron Iowa State Univ Ames IA 50010

BREMNER, RAYMOND WILSON, b Bellingham, Wash, Dec 4, 04; m 35; c 3. SCIENCE EDUCATION. *Educ:* Univ Wash, BS, 28, MS, 32, PhD(chem), 37. *Prof Exp:* Prin & teacher, schs, Wash & Ore, 24-30,; from instr to asst prof, Tex A&M Univ, 37-44; lab proj leader, Dow Chem Co, 44-47; from assoc prof to prof, 47-51, actg chmn, Dept Chem, 58-68, EMER PROF CHEM, CALIF STATE UNIV, FRESNO, 73- *Concurrent Pos:* US Customs Inspector, 25, asst chemist, Oceanog Labs Univ Wash, 32-35; Agent, USDA, 39 & 40; anal res chemist, Stanford Res Inst, 48; fulbright prof & sr lectr, Univ Peshawar, 59-60; researcher, Pakistani Air Force, 60; NSF Award, 62, 64 & 66; mem, policy bd, Moss Landing Marine Labs, Calif State Univs & Cols, 66-68. *Honors & Awards:* Cottrell Awards, Res Corp, 50 & 51; Smith-Mundt Award, US Dept State, 59; NSF Awards, 62, 64 & 66. *Mem:* Am Chem Soc. *Res:* Specific ion electrodes; spectrophotometry; water analysis; pollution; smog; dental caries; radioactive tracers; chemistry of the sea; instrumental and chemical analysis. *Mailing Add:* 1554 W San Ramon Fresno CA 93711

BREMNER, WILLIAM JOHN, b Bellingham, Wash, July 5, 43; m 65; c 4. INTERNAL MEDICINE, REPRODUCTIVE ENDOCRINOLOGY. *Educ:* Harvard Univ, BA, 64; Univ of Wash, MD, 69; Monash Univ, Melbourne, Australia, PhD(physiol), 78. *Prof Exp:* Sr res fel, Univ Wash, 72-74; sr res officer, Nat Health & Med Res Coun, Australia, 74-77; chief, endocrinol, Vet Admin Med Ctr, Seattle, 79-82; asst prof, 77-82, ASSOC PROF MED, OBSTET & GYNEC, UNIV WASH, 82- *Concurrent Pos:* Hon physician endocrinol & diabetes, Prince Henry's Hosp, Melbourne, Australia, 74-77; steering comt, WHO Prog Res Human Reproduction, 84- *Honors & Awards:* Wyeth Award, Pac Coast Fertil Soc, 74. *Mem:* Endocrine Soc; Am Soc Andrology; Am Soc Clin Invest. *Res:* Human and animal reproductive endocrinology. *Mailing Add:* Endocrinol Sect Vet Admin Med Ctr 1660 S Columbian Way Seattle WA 98108

BRENCHLEY, GAYLE ANNE, b Washington, DC, Sept 21, 51. INVERTEBRATE ZOOLOGY, MARINE ECOLOGY. *Educ:* Univ Md, College Park, BS, 73, MS, 75; Johns Hopkins Univ, PhD(ecol), 78. *Prof Exp:* ASST PROF ECOL, UNIV CALIF, IRVINE, 78- *Concurrent Pos:* Scholar, Woods Hole Oceanog Inst; Steps Toward Independence fel, Marine Biol Lab, 80-81. *Mem:* Am Soc Zool; Ecol Soc Am; Am Soc Naturalists; Sigma Xi; AAAS. *Res:* Ecology of marine soft-bottom communities; competition and predation among invertebrates; population ecology of mud snails (Ilyanassa) and mud shrimp (Callianassa, Upogebia). *Mailing Add:* 2222 NE 92nd St Apt 416 Seattle WA 98115

BRENCHLEY, JEAN ELNORA, b Towanda, Pa, Mar 6, 44; m 90. MOLECULAR BIOLOGY. *Educ:* Mansfield State Col, BS, 65; Univ Calif, San Diego, MS, 67, Univ Calif, Davis, PhD(microbiol), 70. *Prof Exp:* Res assoc biol, Mass Inst Technol, 70-71; from asst prof to assoc prof microbiol, Pa State Univ, 71-77; assoc prof biol, Purdue Univ, West Lafayette, 77-81; res dir, Genex Corp, Rockville, MD, 81-84; head, dept molecular & cell biol, 84-87, dir, Bioltechnol Inst, 84-90, PROF MICROBIOL & BIOTECHNOL, PA STATE UNIV, 90- *Concurrent Pos:* Found lectr, Am Soc Microbiol, 75-76; mem, Adv Panel, NSF; mem bd trustees, Biosis; trustee, Found Microbiol; elected nominating comts, AAAS, 87 & 89. *Honors & Awards:* Becton-Dickenson lectr, 80; Waksman Award, 83. *Mem:* Am Soc Microbiol (pres, 86); Am Soc Biol Chemists; Soc Indust Microbiol; Am Chem Soc; Genetics Soc Am. *Res:* Discovery of novel microbial products; microbial genetics; biotechnology. *Mailing Add:* Dept Molecular & Cell Biol Pa State Univ 209 S Frear Lab University Park PA 16802

BRENDEL, KLAUS, b Berlin, Ger, July 14, 33; m 57; c 3. PHARMACOLOGY. *Educ:* Free Univ Berlin, cand chem, 55, dipl chem, 59, Dr rer nat(chem), 62. *Prof Exp:* From asst prof to assoc prof pharmacol, Duke Univ, 67-70; assoc prof pharmacol & toxicol, 70-76, PROF PHARMACOL, COL MED, UNIV ARIZ, 76- *Concurrent Pos:* Fel Ger Chem Indust, Free Univ Berlin, 62-63; NATO fel, Univ of Pac, 63-64 & Duke Univ, 65-67; Am Heart Asn estab investr, Duke Univ & Univ Ariz, 65-70, fel, Coun Arteriosclerosis. *Mem:* Ger Chem Soc; Am Chem Soc; Am Soc Pharmacol & Exp Therapeut; Sigma Xi. *Res:* Bioorganic chemistry; drug metabolism; drugs and intermediary metabolism; toxicology; analytical organic and clinical chemistry. *Mailing Add:* Dept Pharmacol Univ Ariz Col Med Tucson AZ 85724

BRENDER, RONALD FRANKLIN, b Wyandotte, Mich, Sept 22, 43; m 75. INFORMATION SCIENCE. *Educ:* Univ Mich, BSE, 65, MS, 68, PhD(comput sci), 69. *Prof Exp:* CONSULT ENGR & SUPVR COMPUT LANG DEVELOP, SOFTWARE ENG DEPT, DIGITAL EQUIP CORP, 70- *Concurrent Pos:* Mem, Am Nat Stand Comt, X3J3, Fortran Lang, 73-78; vis scientist, Dept Comput Sci, Carnegie-Mellon Univ, 78-79; Ada Bd Fed Adv Comt, 83-88 mem, Int Standards Orgn ISO/IEC JTC1/SC22/WG9, Ada language standards, 83- *Mem:* Asn Comput Mach; Inst Elec & Electronics Engrs. *Res:* Computer language/compiler design and development; implementation languages for system software; software development methodology. *Mailing Add:* Digital Equip Corp ZK02-3/N30 Nashua NH 03062-2698

BRENDLEY, WILLIAM H, JR, b Philadelphia, Pa, Mar 26, 38; m 67. PHYSICAL CHEMISTRY, POLYMER CHEMISTRY. *Educ:* St Joseph's Col, Pa, BS, 60, MS, 62; Univ Pa, PhD(phys chem), 65. *Prof Exp:* Chemist, Rohm and Haas Co, 60-62; NSF fel, Univ Pa, 65-66; res chemist, 66-67, group leader thermoplastic polymer coatings, 67-72, lab head indust coatings, 72-74, DEPT MGR COATINGS TECH SERV, ROHM AND HAAS CO, 74- *Mem:* Am Chem Soc; Fedn Paint Socs; Am Inst Chemists. *Res:* Radiotracer and vacuum line techniques; volume expansion of alkali metals in liquid ammonia; physical chemistry of polymeric coatings and industrial application; polymeric coatings. *Mailing Add:* 2450 Exton Rd Hatboro PA 19040

BRENDLINGER, DARWIN, b Reading, Pa, Nov 29, 34; m 54, 90; c 7. PROSTHODONTIC DENTISTRY, EDUCATION. *Educ:* Temple Univ Sch Dent, DDS, 58; Tufts Univ, MS, 68. *Prof Exp:* Chief prosthodontist, USAF Regional Hosp, 68-72; USAF Med Ctr, Wiesbaden, Ger, 72-76, USAF Regional Hosp, Sheppard AFB, Tex, 76-79; dir dent serv, USAF Med Ctr, Wright Patterson AFB, 79-80, staff officer prof, USAF Dent Serv, 80-82, dep corps chief, 82-86; assoc dir, Dentsply Int Inc, 86-87; prosthodontics consult, 87-89, DIR RES & DEVELOP, 89- *Concurrent Pos:* Air Force rep, NIH adv; consult, Surgeon Gen, USAF, 72-86, Dentsply Int, 86- *Mem:* Am Dent Asn; Am Col Dentists; Am Col Prosthodontists; Am Acad Crown & Bridge Prosthodontics; Am Equilibration Soc; Am Prosthodontics Soc. *Res:* Rate of wear of dissimilar dental materials; clinical research into practical application of castable ceramic material for inlays, onlay, crowns and bridges; utilization of light cured resins in denture construction and relines. *Mailing Add:* 264 Tri-Hill Rd York PA 17403

BRENEMAN, EDWIN JAY, b Orrville, Ohio, Sept 29, 27; m 49; c 3. VISUAL PSYCHOPHYSICS, COLOR SCIENCE. *Educ:* Ohio Wesleyan Univ, BA, 50; Ohio State Univ, MSc, 52. *Prof Exp:* Res physicist visual perception & photog reproduction, Eastman Kodak Co Res Labs, 52-68; res assoc, Macbeth Div, Kollmorgen Corp, 68-73; res assoc visual perception & photog reproduction, Eastman Kodak Co Res Labs, 73-86; RETIRED, 87- *Concurrent Pos:* Consultant, 87- *Mem:* Optical Soc Am. *Res:* Visual perception and photographic color reproduction. *Mailing Add:* Five Tuxford Rd Pittsford NY 14534-1517

BRENEMAN, WILLIAM C, b Braddock, Pa, July 9, 41; m 68; c 1. SILICONE CHEMISTRY, CHEMICAL PLANT DESIGN. *Educ:* Carnegie-Mellon Univ, BS, 63; Ohio Univ, MS, 71, MBA, 75. *Prof Exp:* Res & develop engr, Union Carbide Corp, 66-75, proj scientist, 75-79, develop engr, 79-82, lead process engr, 82-84, technol mgr, chief design engr, 84-85, MGR PROCESS RES & DEVELOP, UNION CARBIDE CORP, 85- *Honors & Awards:* Cert of Recognition, NASA, 81, 82; Chem Pioneer, Am Inst Chemists, 85. *Mem:* Am Inst Chemists; Am Inst Chem Engrs; AAAS. *Res:* Development of a practical large scale process for production of ultra high purity silane used in manufacture of polycrystalline silicon; development of processes for manufacture of organo modified silicones, antifoams and emulsions; new product introduction. *Mailing Add:* 809 Main St Sistersville WV 26175-1221

BRENGELMANN, GEORGE LESLIE, BODY TEMPERATURE REGULATION, SKIN-BLOOD FLOW. *Educ:* Univ Wash, PhD(physiol), 67. *Prof Exp:* PROF PHYSIOL, UNIV WASH, 66- *Mailing Add:* Dept Physiol SJ-40 Univ Wash Seattle WA 98195

BRENHOLDT, IRVING R, ENGINEERING. *Educ:* Univ Minn, BSEE, 41. *Prof Exp:* Sect head, Mass Spectrometer & Electronics, Kellex Corp, 41-45; dir, Refiner Automation, Standard Oil Co, Chicago, 46-52; chief engr, Farrand Optical Co, 52-64; dir res & develop, 70-73; vpres, Pentron Electronics Corp, 64-65; mgr & chief engr, Electro-Optics Div, Gen Precision Corp, 65-70; dir, Instrument & Systs Develop, St Regis Corp, Champion Int Corp, 73-85; PRES, BRENHOLDT INDUSTS INC, 85- *Concurrent Pos:* Mem, Advan Sensors Crit Rev Panel, US Dept Energy. *Honors & Awards:* Vannevar Bush Award, US Off Sci Res & Develop, 43; Henry L Stimson Award, US War Dept, 45. *Mem:* NY Acad Sci. *Res:* Technical problem solving and invention; development, design and engineering utilizing technologies of the applied physical sciences including mechanics, electronics, optics, physical optics, electro-optics, opto-electronics and electrochemistry; holder of 23 patents. *Mailing Add:* Brenholdt Industs Inc PO Box 596 Stratford CT 06497

BRENKERT, KARL, JR, b Detroit, Mich, June 29, 21; m 45; c 5. FLUID MECHANICS. *Educ:* Univ Mich, BS, 44; Stanford Univ, MS, 52, PhD(eng mech), 55. *Prof Exp:* Prod control engr, Packard Motor Car Co, 44; sr engr, Radio Corp Am, Victor Div, Mich, 45-47; gen mgr, TEK Industs, Mich, 47-48 & Bracey Corp, Mich, 48-49; asst, Univ Utah, 49-50; from asst to instr, Stanford Univ, 50-54; asst prof fluid mech, Univ Ala, 54-56; assoc prof appl mech, Mich State Univ, 56-60; asst dean eng, Auburn Univ, 60-63; prog dir, Off Instnl Prog, NSF, 63-64; dean, Sch Eng, 64-79, PROF MECH ENG, UNIV MISS, 64- *Concurrent Pos:* Sr struct engr, Convair, Tex, 57. *Mem:* AAAS; Sigma Xi; Am Soc Mech Engrs. *Res:* Turbulence; theoretical fluid mechanics. *Mailing Add:* Dept Mech Eng Univ Miss University MS 38677

BRENNAN, DANIEL JOSEPH, b South Bend, Ind, Sept 11, 29; m 55; c 7. STRATIGRAPHY, SEDIMENTOLOGY. *Educ:* Univ Notre Dame, BS, 51; SDak Sch Mines & Technol, MS, 53; Univ Ariz, PhD(geol), 57. *Prof Exp:* Geologist, Shell Oil Co, 57-60; from geologist to sr geologist, Sunray DX Oil Co, 60-64; from asst prof to assoc prof geol, Wichita State Univ, 64-68; assoc prof geol, State Univ NY Col Cortland, 68-71, prof, 71-80, chmn dept, 70-80; GEOLOGIST, BORDER EXPLOR CO, 80- *Concurrent Pos:* Consult geologist, 68-; prin investr, NY State Sea Grant Prog, 71-74. *Mem:* Fel Geol Soc Am; Am Asn Petrol Geologists; Soc Econ Paleont & Mineral; AAAS. *Res:* Petroleum and mineral exploration; stratigraphy; sedimentation and sediment-binding mechanism; higher education in marine science. *Mailing Add:* Ariz Oil & Gas Conserv Comn 5150 N 16th St B-141 Phoenix AZ 85016-3803

BRENNAN, DAVID MICHAEL, b Springfield, Mass, Jan 12, 29; m 58; c 2. ENDOCRINE PHYSIOLOGY. *Educ:* Tufts Univ, BS, 50; Purdue Univ, MS, 55, PhD(endocrinol), 57. *Prof Exp:* Asst prof physiol, Univ Notre Dame, 57-58; sr res endocrinologist, Squibb Inst, 58-60; sr endocrinologist, Lilly Res Labs, 60-66, head physiol res, 66-67, asst to vpres res & develop, 67-68, dir biochem & physiol res, 68-72, dir biol res, 72-81, dir pharmaceut proj mgt, 81-89; RETIRED. *Mem:* AAAS; Endocrine Soc; NY Acad Sci. *Res:* Reproductive physiology; endocrine and central nervous system interrelationships; mechanisms of steroid antagonistic action; biochemistry of lipids. *Mailing Add:* 7702 S 775 East Zionsville IN 46077

BRENNAN, JAMES A, b Freeland, Pa, Nov 4, 20; m 43; c 1. ORGANIC CHEMISTRY, PETROLEUM CHEMISTRY. *Educ:* Temple Univ, BA, 50, MA, 52, PhD(org chem), 59. *Prof Exp:* Jr technologist, Mobil Oil Corp, 52-53, technologist, 53-56, sr res chemist, 56-68, res assoc, 68-75, SR RES ASSOC, APPL RES & DEVELOP LABS, MOBIL RES & DEVELOP CORP, 75- *Mem:* AAAS; Am Chem Soc; Am Inst Chemists. *Res:* Synthetic lubricants; combustion; nitrogen heterocyclics; high temperature oxidation; organic synthesis; catalysis. *Mailing Add:* Towers Windsor Toledo 10-T Cherry Hill NJ 08002

BRENNAN, JAMES GERARD, b Hazleton, Pa, Aug 30, 27; m 51; c 5. THEORETICAL PHYSICS. *Educ:* Univ Scranton, BS, 48; Univ Wis, MS, 50, PhD(physics), 52. *Prof Exp:* Asst, Univ Wis, 48-52; from instr to assoc prof, 52-61, chmn dept, 61-74, assoc dean grad prog, Arts & Sci, 74-77, PROF PHYSICS, CATH UNIV AM, 61- *Concurrent Pos:* Consult, Naval Ord Lab, 55-; sci liaison, Off Naval Res, London, Eng, 65-66. *Res:* Theoretical nuclear physics; theoretical interior ballistics; atomic physics; ion implantation studies; electronic stopping power calculations. *Mailing Add:* Dept Physics Cath Univ Am Wash DC 20064

BRENNAN, JAMES ROBERT, b Crawfordsville, Ind, Nov 14, 30; m 53; c 7. PLANT ANATOMY, CYTOLOGY. *Educ:* Va Polytech Inst, BS, 52, MS, 55; Univ Md, PhD(bot), 58. *Prof Exp:* From instr to asst prof biol, Norwich Univ, 58-61; chmn, dept biol, 83-87, actg vpres, acad affairs, 87-88, from instr to assoc prof, 61-69, PROF BOT, BRIDGEWATER STATE COL, 69- *Concurrent Pos:* Dir, NSF Biol Sci Curric Study, Bridgewater State Col, 63-67, NSF, Coop Col-Sch Proj, 66; seasonal ranger-naturalist, Nat Park Serv, 72-80, vis prof, Open Univ, Milton Keynes, Eng, 78, & Australian Nat Univ, Canberra, 86. *Mem:* Int Soc Plant Morphologists; Bot Soc Am; Sigma Xi; AAAS. *Res:* Plant anatomical studies in relation to taxonomy and phylogeny; ultrastructural effects of chemical agents on meristematic plant cells; development of cell shape in differentiating plant cells. *Mailing Add:* Dept Biol Bridgewater State Col Bridgewater MA 02325

BRENNAN, JAMES THOMAS, b St Louis, Mo, Jan 12, 16; m 46; c 2. RADIOTHERAPY, RADIOBIOLOGY. *Educ:* Univ Ill, BA, 39; Univ Minn, MD, 43. *Prof Exp:* Investr radiobiol, Los Alamos Sci Lab, Univ Calif, 48-52; dir dept biophys, Walter Reed Army Inst Res, 52-54, resident radiol, Walter Reed Gen Hosp, 54-57, chief radiation ther, 60-61; consult to surgeon, US Army Europe, 57-60, dir, Armed Forces Radiobiol Res Inst, Md, 61-66; vis lectr radiobiol, 66-67, Wilson prof res radiol, 67-78, EMER WILSON PROF RES RADIOL, UNIV PA, 78- *Concurrent Pos:* USPHS planning grant, 66-; mem main comt, Nat Coun Radiation Protection, 64-72. *Mem:* AMA; Health Physics Soc; Radiol Soc NAm; Am Soc Therapeut Radiol; Radiation Res Soc. *Res:* Neutron radiobiology; experimental radiotherapy. *Mailing Add:* Dept Radiol Hosp Pa 3400 Spruce St Philadelphia PA 19104

BRENNAN, JOHN JOSEPH, b Boston, Mass, Sept 26, 38; m 61; c 3. PHYSICS. *Educ:* Boston Col, BS, 60; Worcester Polytech Inst, MS, 62; Ga Inst Technol, PhD(physics), 68. *Prof Exp:* Teaching asst physics, Ga Inst Technol, 62-66, instr, 66-68; asst prof, 68-71, ASSOC PROF PHYSICS, FLA TECHNOL UNIV, 71- *Mem:* Am Asn Physics Teachers. *Res:* Theoretical nuclear structure; Nilsson model; beta decay; nuclear magnetic resonance; molecular rotations in solids. *Mailing Add:* 549 New Brunswick Rd Somerset NJ 08873

BRENNAN, LAWRENCE EDWARD, b Oak Park, Ill, Jan 29, 27; m 47; c 4. ELECTRONICS. *Educ:* Univ Ill, BA, 48, MA, 49, PhD(elec eng), 51. *Prof Exp:* Res assoc electronic dynamics, Univ Ill, 49-51; specialist systs anal, NAm Aviation, Inc, 51-52; engr, Chicago Midway Labs, 52-55, Lockheed Missile Systs Div, 55, Aeronaut Systs Inc, 56, Off Naval Res, 57 & Rand Corp, 58-67; sr scientist, Technol Serv Corp, 67-80; vpres, Adaptive Sensors, Inc, 80-90; RETIRED. *Mem:* Fel Inst Elec & Electronics Engrs. *Res:* Radar systems; electronic counter-countermeasures. *Mailing Add:* Adaptive Sensors Inc 216 Pico Blvd Suite 8 Santa Monica CA 90405

BRENNAN, MICHAEL EDWARD, b Covington, Ky, July 8, 41; m 65; c 2. ORGANIC CHEMISTRY. *Educ:* Univ Dayton, BS, 62; Univ Fla, MS, 65, PhD(org chem), 67. *Prof Exp:* Sr res chemist, Monsanto Co, 67-69; res assoc chem, Univ Ga, 69-70; sr res chemist, 70-76, proj chemist, 76-85, sr proj chemist, Texaco Chem Co, 85-87, PRIN RES CHEMIST, ARCO CHEM CO. 87- *Mem:* Am Chem Soc. *Res:* Organic synthesis; homogeneous catalysis; reaction mechanisms; bridged polycyclic compounds; small-ring chemistry; photochemistry; nuclear magnetic resonance spectra; polyurethanes; flammability. *Mailing Add:* Arco Chem Co 3801 Westchester Pike Newtown Square PA 19073

BRENNAN, MICHAEL JAMES, b Mt Clemens, Mich, June 11, 21; m 45; c 8. MEDICINE. *Educ:* Univ Detroit, BS, 41; Stritch Sch Med, MD, 47. *Prof Exp:* Assoc physician, Hemat Dept, Henry Ford Hosp, Detroit, 50-52; chief, Med & Lab Sect, US Army Hosp, Ft Monmouth, NJ, 53-54; physician-in-chg, Oncol Div, Henry Ford Hosp, 54-65, chief div, 65-66; prof med, Col Med, Wayne State Univ, 66-76; PRES & MED DIR, MICH CANCER FOUND, 66- *Concurrent Pos:* Dir, Meyer L Prentis Comprehensive Cancer Ctr, Detroit, 78-87. *Honors & Awards:* Horace H Rackam Award. *Mem:* AMA; Am Col Physicians; Am Asn Cancer Res; Am Soc Hematologists; Am Fedn Clin Res. *Res:* Clinical hematology and oncology; host tumor relationships; experimental and clinical chemotherapy. *Mailing Add:* Mich Cancer Found 110 E Warren Ave Detroit MI 48201

BRENNAN, MURRAY F, b Auckland, NZ, Apr 2, 40; US citizen; m 73; c 4. SURGICAL ENDOCRINOLOGY, SURGICAL ONCOLOGY. *Educ:* Univ NZ, BSc, 61; Univ Otago, MD, 64. *Prof Exp:* Sr investr, vis scientist & head surg metab, Nat Cancer Inst, NIH, Washington, DC, 75-81; PROF SURG, CORNELL UNIV, NEW YORK, 81-; CHIEF GASTRIC & MIXED TUMOR SERV, MEM SLOAN-KETTERING CANCER CTR, 81-, MEM, SLOAN-KETTERING INST CANCER RES, 82-, ALFRED P SLOAN PROF SURG, 84-, CHMN DEPT, 85- *Concurrent Pos:* Consult surg, Harvard Med Sch & Peter Bent Brigham Hosp, Boston, 75-79; clin instr surg, Georgetown Univ, 76-81; consult biochem res, Grad Sch Arts & Sci, George Washington Univ, 78-80; attend surgeon, NY Hosp, 81- & gastric & mixed tumor serv, & nutrit & endocrinol serv, Mem Sloan-Kettering Cancer Ctr, 81-; vis physician, Rockefeller Univ, 81- *Mem:* Am Soc Clin Invest; Soc Univ Surgeons; Am Surg Asn; Endocrine Soc; Int Soc Endocrine Surgeons; Soc Surg Oncol; hon fel Brazilian Col Surgeons. *Res:* Intermediary metabolism of the cancer patient and metabolic response to treatment and nutritional support; development of integrated multimodality therapy of human malignancy. *Mailing Add:* Mem Sloan-Kettering Cancer Ctr 1275 York Ave New York NY 10021

BRENNAN, PATRICIA CONLON, b Chicago, Ill, Nov 20, 32; m 58; c 3. MICROBIOLOGY, IMMUNOLOGY. *Educ:* Albion Col, AB, 54; Univ Wis, MS, 57; Loyola Univ, Ill, PhD(microbiol), 68. *Prof Exp:* Technician bact, G D Searle & Co, 54-55; bacteriologist, Animal Dis Diag Lab, Wis Dept Agr, 57-58; microbiologist, Univ Chicago Hosps & Clins, 60-63; sci asst microbiol, Argonne Nat Lab, 63-68, asst microbiologist, 68-72, microbiologist, Div Biol & Med Res, 72-81; sr immunologist, Packard Instruments, 81-85; PROF MICROBIOL & RES, NAT COL CHIROPRACTIC, 85-, DEAN RES, 89- *Mem:* AAAS; Am Soc Microbiol; Am Asn Immunologists. *Res:* Mechanisms of pathogenicity; mechanisms of inflamation; free oxygen radicals; chemiluminescence. *Mailing Add:* 225 Brewster Lombard IL 60148

BRENNAN, PATRICK JOSEPH, b Boyle, Ireland, Mar 16, 38; m 68; c 3. MICROBIOLOGY, BIOCHEMISTRY. *Educ:* Nat Univ Ireland, BSc, 61, MSc, 63; Trinity Col, Univ Dublin, PhD(biochem), 65; fel Royal Acad Med, Ireland, 75. *Hon Degrees:* MA, Trinity Col, Univ Dublin, 67. *Prof Exp:* Fel biochem, Univ Calif, Berkeley, 65-67; jr lectr, Trinity Col, Dublin, 67-71; lectr, Univ Col, Dublin, 71-76; sr res scientist, Nat Jewish Hosp, 76-81, mem staff, 81-; ASSOC PROF MICROBIOL, COLO STATE UNIV. *Concurrent Pos:* USPHS Int fel, Univ Calif, Berkeley, 65-67; Nat Sci Coun grants, Ireland, 68-74; traveling fel, Royal Irish Acad, 69; vis asst prof virol, Baylor Col Med, 72-73; NIH grants & contracts, 78-, USDA grant, 81-; asst prof microbiol, Univ Colo Med Ctr, Denver, 78-81, adj assoc prof, 81-; mem study sect, NIH, 81- *Mem:* Biochem Soc; Am Soc Microbiol; Int Leprosy Asn. *Res:* Immunochemistry of mycoracteria; antigen structure in mycobacteria; pathogenesis of mycoglycoconjugates bacteria; leprosy. *Mailing Add:* Dept Microbiol Colo State Univ Ft Collins CO 80523

BRENNAN, PAUL JOSEPH, b Auburn, NY, June 29, 20; m 51; c 10. CIVIL ENGINEERING. *Educ:* Univ Detroit, BArchE, 43; Yale Univ, MEng, 44, DEng(civil eng), 51. *Prof Exp:* Plant eng dept, Mich Steel Casting Co, 42-43; asst civil eng, Yale Univ, 46-48, asst prof, 48-53; Prof civil eng, Univ Del, 53-58, Chmn dept, 53-55, Chmn dept civil eng & eng mech, 55-88; chmn dept, 58-77, prof civil eng, 58-77, DAVID S RUTTY PROF ENG, SYRACUSE UNIV, 77-, ASSOC DEAN RES & GRAD AFFAIRS, COL ENG, 88- *Mem:* Am Soc Civil Engrs. *Res:* Structures; structural dynamics; steel and reinforced concrete; wind energy. *Mailing Add:* Link Hall L C Smith Col Eng Syracuse Univ Syracuse NY 13244

BRENNAN, THOMAS MICHAEL, b Jacksonville, Ill, June 7, 41@; m 63; c 2. PHOTOBIOLOGY. *Educ:* Univ Ill, BS, 65; Rutgers Univ, MS, 75, PhD(bot), 77. *Prof Exp:* Engr, Southwestern Bell Tel Co, 65-68; computer analyst, Bell Tel Labs, 68-71; res assoc, Princeton Univ, 77-78; asst prof, 78-84, ASSOC PROF BIOL, DICKINSON COL, 84- *Concurrent Pos:* Res assoc, Dept Biol, Univ Ill-Chicago, 79; vis scientist, Dept Biol Sci, Univ, Md Baltimore Co, 85-86; dept chmn biol, Dickinson Col, 86- *Mem:* Am Soc Photobiol; Am Soc Plant Physiologists. *Res:* Light-dependent hydrogen peroxide formation in photosynthetic systems; photosensitizing agents in plant tissues; writing on photosynthesis, photorespiration, and related topics for general readership. *Mailing Add:* Dept Biol Dickinson Col Carlisle PA 17013

BRENNECKE, HENRY MARTIN, b Saratoga, Tex, 24; m 50; c 1. CHEMICAL ENGINEERING. *Educ:* Tex A&M Univ, BS, 47; Univ Tex, PhD(chem eng), 54. *Prof Exp:* Chem engr, Delhi-Taylor, 47-49; sr res engr, Explosives Dept, 54-76, TECH ASSOC, PETROCHEM DEPT, E I DU PONT DE NEMOURS & CO, INC, 76- *Mem:* AAAS; Am Chem Soc; Am Inst Chem Engrs; Am Inst Mining Engrs. *Res:* Research and development of chemical processes. *Mailing Add:* 5261 River Oaks Corpus Christi TX 78413

BRENNECKE, LLEWELLYN F(RANCIS), b St Louis, Mo, Mar 20, 23; m 50; c 2. CHEMICAL ENGINEERING. *Educ:* Wash Univ, BS, 49, MS, 50, DSc(chem eng), 55. *Prof Exp:* Res engr res labs, 54-63, asst chief, Mo, 63-67, asst chief chem eng, Tex, 67-68, proj engr, Pa, 68-70; tech serv mgr, Alcoa Minerals Jamaica, Inc, 70-76; opers chem engr, Aluminum Co Am, 76-82; CONSULT CHEM ENG, 82- *Concurrent Pos:* Vchmn, Am Inst Chem Engrs, 58-59, chmn, 59-60. *Mem:* Am Inst Chem Engrs (secy, 56-57); Am Chem Soc; Sci Res Soc Am; Nat Soc Prof Engrs. *Res:* Refining of bauxite and related chemicals; application of statistics; consultant in crystallization. *Mailing Add:* 605 Limwood Dr Victoria TX 77901

BRENNEMAN, JAMES ALDEN, b Elida, Ohio, Aug 26, 43; m 67; c 2. MYCOLOGY. *Educ:* Goshen Col, BA, 65; WVa Univ, MS, 67; La State Univ, Baton Rouge, PhD(plant path), 70. *Prof Exp:* From asst prof to assoc prof, 70-88, PROF BIOL, UNIV EVANSVILLE, 88- *Mem:* AAAS; Am Inst Biol Sci; Nat Asn Biol Teachers; Sigma Xi; Mycol Soc Am; Nat Audubon Soc. *Mailing Add:* Dept Biol Univ Evansville Evansville IN 47722

BRENNEMANN, ANDREW E(RNEST), JR, b Cincinnati, Ohio, Dec 6, 25. ELECTRICAL ENGINEERING. *Educ:* Univ Cincinnati, MSEE, 50; Syracuse Univ, MS, 61. *Prof Exp:* Res staff mem cathode ray tube storage, IBM Res Lab, 50-54, res staff mem ferroelectric applns, 54-57, res staff mem applns of superconductivity to comput circuitry, 57-64, res staff mem semiconductor surfaces & integrated circuits, 64-68, res staff mem comput commun, 69-72, res staff mem automation res, 72-85, advan robotic technol, 86-91; RETIRED. *Mem:* Inst Elec & Electronics Engrs. *Mailing Add:* Four Morningside Court Ossining NY 10562

BRENNER, ABNER, b Kansas City, Mo, Aug 5, 08; m 36; c 4. ELECTROCHEMISTRY. *Educ:* Univ Mo, AB, 29; Univ Wis, MS, 30; Univ Md, PhD(chem), 39. *Prof Exp:* Asst to Dr Kahlenburg, Univ Wis, 29-30; jr chemist, Nat Bur Standards, 30-35, assoc chemist, 35-50, chief electrodeposition, 50-71; PRES, N-Q ELECTROCHEM RES CORP, 71- *Honors & Awards:* Gold Medal, US Dept Com, 63; Proctor Award, Am Electroplaters Soc, 49, Sci Achievement Award, 62; Hothersall Award, Brit Inst Metal Finishing, 61; Electrochem Eng & Technol Award, Electrochem Soc, 74. *Mem:* Am Electroplaters Soc; Am Chem Soc; Electrochem Soc; Brit Inst Metal Finishing. *Res:* Electrodeposition of metals and alloys; measurement of physical properties and of thickness and metal coatings; electrodeposition of metals from organic and fused salt baths; vapor deposition of tungsten; electroless plating; electrochemical calorimetry and galvanostalametry. *Mailing Add:* 7204 Pomander Lane Chevy Chase MD 20815

BRENNER, ALAN, b Washington, DC, Apr 30, 46; m 68; c 2. INORGANIC CHEMISTRY, PHYSICAL CHEMISTRY. *Educ:* Johns Hopkins Univ, BA, 68; Northwestern Univ, PhD(chem), 75. *Prof Exp:* asst prof, 75-80, ASSOC PROF CHEM, WAYNE STATE UNIV, 80- *Concurrent Pos:* Consult, Dow Chemical Co, 78- *Mem:* Catalysis Soc; Am Chem Soc; AAAS; Sigma Xi. *Res:* Synthesis; characterization and applications of novel heterogeneous catalysts; surface chemistry; organometallic chemistry. *Mailing Add:* 679 Westchester Rd Grosse Pointe Park MI 48230

BRENNER, ALFRED EPHRAIM, b Brooklyn, NY, Sept 11, 31; div; c 3. PHYSICS. *Educ:* Mass Inst Technol, SB, 53, PhD(physics), 58. *Prof Exp:* Ford Found fel physics, Europ Orgn Nuclear Res, 58-59; instr & res fel, Harvard Univ, 59-62, asst prof, 62-66, sr res assoc, 66-70; physicist & dir comput, Fermi Nat Accelerator Lab, 70-85; pres, Consortium Sci Comput, 85-86; DIR APPLICATIONS RES, SUPERCOMPUT RES CTR, 86- *Concurrent Pos:* Consult, Dept Energy, 72-86 & Summagraphics Corp, 73-84; mem, Com Dept, Computer Systs Tech Adv Comt, 74-86; mem subcomt supercomputers, US Activ Bd, Inst Elec & Electronics Engrs, 83-; chmn, Nat Res Coun, Panel Comput & Appl Math, Nat Inst Standards & Technol, 85-; ed, J Supercomput, 86- & Int J High Speed Comput, 89- *Mem:* AAAS; fel Am Phys Soc; Asn Comput Mach; Inst Elec & Electronics Engrs; NY Acad Sci; Soc Indust & Appl Math. *Res:* Computational science; high performance computer architectures and data processing; experimental high energy physics. *Mailing Add:* Supercomput Res Ctr 17100 Science Dr Bowie MD 20715-4300

BRENNER, BARRY MORTON, b Brooklyn, NY, Oct 4, 37; m 60; c 2. INTERNAL MEDICINE. *Educ:* Long Island Univ, BS, 58; Univ Pittsburgh, MD, 62. *Prof Exp:* From asst prof med to prof med & physiol, Univ Calif, San Francisco, 72-76; SAMUEL A LEVINE PROF MED, HARVARD MED SCH & DIR, RENAL DIV & LAB KIDNEY & ELECTROLYTE PHYSIOL, BRIGHAM & WOMENS HOSP, BOSTON, 76- *Concurrent Pos:* Sr staff mem, Cardiovasc Res Inst, Univ Calif, San Francisco, 74-76; physician, Brigham & Womens Hosp, Boston, 76- *Honors & Awards:* Homer Smith Award, Am Soc Nephrol. *Mem:* Am Soc Clin Invest; Am Physiol Soc; Am Soc Nephrol; Asn Am Physicians. *Res:* Renal regulation of ion, water and macromolecule transport; regulation of renal glomerular filtration; pathophysiology of renal disorders. *Mailing Add:* Renal Div Brigham & Womens Hosp 75 Francis St Boston MA 02115

BRENNER, DAEG SCOTT, b Reading, Pa, Aug 9, 39; m 89; c 3. NUCLEAR CHEMISTRY, NUCLEAR PHYSICS. *Educ:* Rensselaer Polytech, BS, 60; Mass Inst Technol, PhD(chem), 64. *Prof Exp:* Ford Fund fel nuclear chem, Niels Bohr Inst, Copenhagen, Denmark, 64, NATO fel, 65; res assoc, Brookhaven Nat Lab, 65-67; from asst prof to assoc prof, 67-79, chmn dept, 78-81, assoc provost, 83-87, chair fac, 88-91, PROF CHEM, CLARK UNIV, 79- *Concurrent Pos:* Consult, Lawrence Livermore Nat Lab; res collabr, Brookhaven Nat Lab & Los Alamos Nat Lab. *Mem:* Am Chem Soc; Am Phys Soc; Sigma Xi. *Res:* Atomic mass measurements; nuclear spectroscopy; nuclear fission; nuclear structure. *Mailing Add:* Chem Dept Clark Univ Worcester MA 01610-1477

BRENNER, DOUGLAS, b Washington, DC, Dec 31, 38; m 68; c 3. POLYMER SCIENCE. *Educ:* Johns Hopkins Univ, BA, 61; Harvard Univ, MA, 63, PhD(physics), 68. *Prof Exp:* Res fel atomic physics, Harvard Univ, 68-69, instr physics, 69; res physicist, 69-73, sr res physicist, corp res lab, Exxon Res & Eng Co, 73-86, MEM TECH STAFF, AT&T BELL LABS, 86- *Mem:* Am Phys Soc; Inst Elec & Electronics Engrs. *Res:* Physical aspects of elastomer and plastic multiphase systems and blends; polymers containing appended ionic groups; filled and reinforced polymers; photoelectron spectroscopy of polymer surfaces; radio frequency spectroscopy; adhesion studies; system engineering of computer and telecommunication systems. *Mailing Add:* 61 E Sherbrooke Pkwy Livingston NJ 07039

BRENNER, DOUGLAS MILTON, chemical dynamics, aerosol molecular physics, for more information see previous edition

BRENNER, EGON, b Vienna, Austria, July 1, 25; US citizen; m 50; c 2. ELECTRICAL ENGINEERING. *Educ:* City Col NY, BEE, 44; Polytech Inst Brooklyn, MEE, 49, DEE, 55. *Prof Exp:* Tutor elec eng, 46-50, from lectr to prof, 50-67, dean grad studies, 61-71, dean, Sch Eng, 71-76, actg vchancellor acad affairs, 76-77, dep chancellor, 77-81, PROF ELEC ENG, CITY COL NY; EXEC VPRES, YESHIVA UNIV, NY, 81- *Mem:* Fel Inst Elec & Electronics Engrs; Am Soc Eng Educ; fel AAAS. *Res:* Higher education; system theory. *Mailing Add:* Yeshiva Univ New York NY 10033

BRENNER, FIVEL CECIL, b Norfolk, Va, Sept 20, 18; m 47; c 2. PERFORMANCE TESTING. *Educ:* Va Polytech Inst, BS, 40; Polytech Inst Brooklyn, MS, 42, PhD(chem), 50. *Prof Exp:* Asst, Polytech Inst Brooklyn, 40-42; teacher, Army Serv Col, Ger, 45-46; asst prof phys chem, Vanderbilt Univ, 49-51; sr res scientist, Johnson & Johnson, 51-60; sr group leader, Chemstrand Res Ctr, 60-66; chief textiles & apparel technol ctr, Inst Appl Technol, Nat Bur Standards, 66-70; chief tire systs div, Safety Res Lab, 70-78, chief consumer info, Nat Hwy Traffic Safety Admin, 78-83; sr statistician statist policy br, Environ Protection Agency, 87-90; RETIRED. *Honors & Awards:* Apparel Res Found Honor Award, 67. *Mem:* Am Chem Soc; Fiber Soc; Am Soc Testing & Mat; Brit Textile Inst; Standards Eng Soc. *Res:* Cellulose chemistry; fiber physics and mechanics; textile mechanics; statistical design; tire test method development; highway-vehicle interaction, tire performance testing, air quality. *Mailing Add:* 1917 Rookwood Rd Silver Spring MD 20910

BRENNER, FREDERIC J, b Warren, Ohio, Dec 25, 36; m; c 2. ECOLOGY, BEHAVIORAL BIOLOGY. *Educ:* Thiel Col, BS, 58; Pa State Univ, MS, 60, PhD(zool), 64. *Prof Exp:* Asst biol, Thiel Col, 57-58; asst zool, Pa State Univ, 58-64; first Kettering teaching intern, Denison Univ, 64-65; asst prof biol, Thiel Col, 65-69; from asst prof to assoc prof, 69-85, PROF BIOL, GROVE CITY COL, 85- *Concurrent Pos:* Grants, Am Philos Soc & Sigma Xi, 65-66, 70, NSF, 66-67, Off Water Resources, 71-72 & 75-77, Nat Wildlife Fedn, 73, Nat Geog Soc, 73-75, Nat Rifle Asn, 79-81, Nat Wild Turkey Fedn & Pa Wild Resources Fund; vis prof, Pa State Univ, 83; consult for coal indust; ed, Newsletter, Pa Acad Sci, 72-76 & 86- *Mem:* AAAS; Am Soc Ichthyol & Herpet; Am Soc Mammal; Am Ornith Union; Am Soc Zool; Ecol Soc Am. *Res:* Hibernation and migration within animal populations; reclamation of surface mine and lands as related to wildlife and fishery management; productivity and fish populations of strip mine lakes; limnology of strip mine lakes and mine drainage streams; wildlife fisheries management of wetland systems. *Mailing Add:* Dept Biol Grove City Col Grove City PA 16127

BRENNER, GEORGE MARVIN, b Ottawa, Kans, Sept 19, 43; m 66; c 2. PHARMACOLOGY. *Educ:* Univ Kans, BS, 66, PhD(pharmacol), 71; Baylor Univ, MS, 68. *Prof Exp:* Asst prof pharmacol, SDak State Univ, 71-76; PROF, COL OSTEOP MED, OKLA STATE UNIV, 76-, CHMN, PHARMACOL DEPT, 89- *Concurrent Pos:* Vis lectr, Oral Roberts Univ; adj prof, N Eastern Okla State Univ. *Mem:* Am Soc Pharm Exp Ther; Sigma Xi; Soc Exp Biol Med. *Res:* Sex hormone effects on hypoxic resistance of myocardium; trace element metabolism; serum lipids. *Mailing Add:* 12520 S Elwood Jenkins OK 74037

BRENNER, GERALD STANLEY, b Brooklyn, NY, July 8, 34; m 59; c 3. PHYSICAL ORGANIC CHEMISTRY, PHARMACEUTICAL. *Educ:* City Col New York, BS, 56; Univ Wis, PhD, 61. *Prof Exp:* Res fel, 72-76, sr res fel, 76-80, DIR, MERCK & CO, INC, 80- *Mem:* Am Chem Soc; Am Asn Pharm Scientist. *Res:* Structure determination; development of organic processes; synthetic organic chemistry; medicinal chemistry; physical organic chemistry; preformulation research; biopharmaceutics; analytical chemistry. *Mailing Add:* Merck Sharp & Dohme Res Labs Div Merck & Co Inc West Point PA 19486

BRENNER, GILBERT J, b New York, NY, Jan 18, 33; m 56; c 2. PALYNOLOGY. *Educ:* City Col New York, BS, 55; NY Univ, MS, 58; Pa State Univ, PhD(geol), 62. *Prof Exp:* Sci asst micropaleont, Am Mus Natural Hist, 56-58; teaching asst geol, Pa State Univ, 58-62; res paleontologist, Stand Oil Co Calif, 62-64; from asst prof to assoc prof geol, 64-70, res grant, 64-65, PROF GEOL, STATE UNIV NY COL NEW PALTZ, 70- *Concurrent Pos:* NSF res grant, 65-68; Am Philos Soc Penrose Fund grant, 70; vis res scientist, Geol Surv Israel, 70-71; NSF res grant, 80. *Mem:* AAAS; Am Paleont Soc; Nat Asn Geol Teachers; Am Asn Stratig Palynologists; Bot Soc Am; Sigma Xi. *Res:* Paleogeography and ecology of recent Foraminifera; Cretaceous angiosperm pollen and spores; origin of angiosperms; plant evolution, paleogene palynomorphs from the Atlantic Coastal Plain; cretaceous floral provinces and paleoclimatology. *Mailing Add:* Dept Geol Sci State Univ NY Col New Paltz NY 12562

BRENNER, HENRY CLIFTON, b Rochester, NY, Nov 21, 46; m 82; c 1. SPECTROSCOPY. *Educ:* Mass Inst Technol, SB, 68; Univ Chicago, MS, 69, PhD(chem), 72. *Prof Exp:* Res assoc chem, Univ Calif, Berkeley, 72-75; asst prof, 75-80, ASSOC PROF CHEM, NY UNIV, 80- *Concurrent Pos:* NSF energy related fel, Univ Calif, Berkeley, 75; Alfred P Sloan fel, 77-79; Fulbright scholar, Soviet Union, 90. *Mem:* Am Chem Soc; AAAS; Am Phys Soc; Biophys Soc; Am Soc Photobiol; Sigma Xi. *Res:* Electron spin resonance and optical spectroscopic studies of molecular crystals and biological systems; excited state energy transfer; coherence and laser spectroscopy; optically detected magnetic resonance; luminescence under high pressure. *Mailing Add:* Dept Chem NY Univ New York NY 10003

BRENNER, HOWARD, b New York, NY, Mar 16, 29; m 51, 81; c 3. CHEMICAL ENGINEERING. *Educ:* Pratt Inst, BChE, 50; NY Univ, MChE, 54, DEngSc, 57. *Prof Exp:* From instr to prof chem eng, NY Univ, 55-66; prof, Carnegie-Mellon Univ, 66-77; prof chem eng & chmn dept, Univ Rochester, 77-81; WILLARD HENRY DOW PROF CHEM ENG, MASS INST TECHNOL, 81- *Concurrent Pos:* Fairchild distinguished scholar, Calif Inst Technol, 75-76; Guggenheim fel, 88-; consult ed & chmn adv bd, Butterworths Series Chem Eng. *Honors & Awards:* Honor Scroll Award, Am Chem Soc, 61; Bingham Medal, Soc Rheology, 80; Alpha Chi Sigma Award, Am Inst Chem Engrs, 75, Walker Award, 85; Kendall Award, Am Chem Soc, 88. *Mem:* Nat Acad Eng; Am Chem Soc; fel Am Inst Chem Engrs; fel Am Acad Mechs; fel AAAS. *Res:* Thermodynamics; transport processes; fluid dynamics; heat transfer; applied mathematics; porous media; Brownian motion; interfacial phenomena; physicochemical hydrodynamics; statistical mechanics. *Mailing Add:* Dept Chem Eng Mass Inst Technol Cambridge MA 02139

BRENNER, JOHN FRANCIS, b Charleston, SC, Sept 13, 41; m 64; c 2. BIOMEDICAL IMAGE ANALYSIS. *Educ:* Mass Inst Technol, SB, 62, SM, 66; Tufts Univ, PhD(physics), 72. *Prof Exp:* Instr therapeut radiol, 72-75, ASST PROF THERAPEUT RADIOL, SCH MED, TUFTS UNIV, 75- *Concurrent Pos:* Physicist spec & sci staff radiation oncol, New Eng Med Ctr Hosp, 72-; consult, Abbott Labs, 74-; mem task force automated differential cell counters, Ctr Dis Control, 76-77; assoc dir image processing res, Dept Therapeut Radiol, New Eng Med Ctr Hosp, 76-77, dir, Image Anal Lab, 77-; assoc ed, Comput in Biol & Med, Cytometry. *Mem:* AAAS; Int Acad Cytol; Soc Anal Cytol; Sigma Xi. *Res:* Computer assisted biomedical image analysis. *Mailing Add:* Tufts New Eng Med Ctr 750 Washington St NEMC 854 Boston MA 02111

BRENNER, LORRY JACK, b Atlanta, Ga, May 16, 23; m 54; c 2. IMMUNOLOGY. *Educ:* Emory Univ, BA, 47; Univ Mich, MS, 49; Western Reserve Univ, PhD(immunol), 55. *Prof Exp:* Asst immunol, Dept Path, Western Reserve Univ, 54-57; instr microbiol, 57-65; immunologist, Highland View Hosp, 57-62; from asst prof to assoc prof, 65-74, PROF BIOL, CLEVELAND STATE UNIV, 74- *Concurrent Pos:* Asst, St Luke's Hosp, Cleveland, 62; consult immunologist, Highland View Cuyahoga County Hosp, 62-78; adj, Div Prof Affairs, Cleveland Clin Found, 76- *Mem:* Am Soc Zoologists; Soc Invert Path; Am Asn Immunologists; Soc Protozoologists; Electron Microscopy Soc Am. *Res:* Ciliates and antibodies; syphilis serology; immunohematology; leukemia. *Mailing Add:* Dept Biol Cleveland State Univ Cleveland OH 44115

BRENNER, MARK, b Boston, Mass, June 19, 42; m 64; c 2. ANALYTICAL CHEMISTRY. *Educ:* Univ Mass, BS, 64, MS, 65; Mich State Univ, PhD(hort), 70. *Prof Exp:* From asst prof to assoc prof, 69-80, PROF HORT SCI, UNIV MINN, ST PAUL, 80-, ASSOC DEAN, GRAD SCH, 89- *Concurrent Pos:* Mem, Plant Growth Regulator Working Group. *Mem:* Fel Am Soc Hort Sci; Am Soc Plant Physiol; Sigma Xi; Int Plant Growth Substances Asn; Crops Soc Am. *Res:* Physiology of growth and development of plants; hormone metabolism in plant tissue; development and use of new methods for plant hormone extraction and identification. *Mailing Add:* Dept Hort Sci Univ Minn St Paul MN 55108

BRENNER, MORTIMER W, b New York, NY, July 28, 12; m 39; c 2. BIOCHEMICAL ENGINEERING, BIOCHEMISTRY. *Educ:* NY Univ, BS, 32, ScM, 38. *Prof Exp:* Asst tech dir & chief engr biochem eng & biochem, Schwarz Labs, Inc, 46-58; vpres, Jacob Ruppert Brewery, 58-60; exec vpres, Schwarz Serv Int Ltd, 60-65, pres, 65-75; CONSULT BIOCHEM ENG, 76- *Mem:* Am Soc Brewing Chemists (pres, 59-60); fel Am Inst Chemists; Inst Food Technologists; Am Inst Chem Engrs; Master Brewers Asn Am; Am Chem Soc. *Res:* Brewing, foods, analytical procedures, process engineering and fermentation industries. *Mailing Add:* 24 Kent Rd Scarsdale NY 10583

BRENNER, RICHARD JOSEPH, b Kankakee, Ill, Aug 22, 53; m 79; c 1. MEDICAL ENTOMOLOGY, VECTOR BIOLOGIST. *Educ:* Univ Ill, Urbana-Champaign, BS, 75, MS, 76; Cornell Univ, Ithaca, PhD(med entom), 80. *Prof Exp:* Grad tech asst med entom, Entom Dept, Univ Ill, Urbana, 75-76 & Cornell Univ, 76-80; postdoctoral res entomologist, Univ Calif, Riverside, 80-82; res entomologist, Screwworm Res Lab, 82-84, RES ENTOMOLOGIST, MED & VET ENTOM RES LAB, AGR RES SERV, USDA, 84- *Honors & Awards:* Tech Transfer Award, Fed Lab Consortium, 89. *Mem:* Entom Soc Am; Am Soc Trop Med & Hyg; Soc Vector Ecologists; Am Mosquito Control Asn; Royal Soc Trop Med & Hyg. *Res:* Behavior and ecology of insects adversely affecting humans, domestic animals, homes, and structures; allergies in humans caused by arthropods; microclimate in structures and the impact on household pests. *Mailing Add:* USDA Agr Res Serv MAVERL 1600 SW 23rd Dr PO Box 14565 Gainesville FL 32604

BRENNER, ROBERT MURRAY, b Lynn, Mass, Feb 6, 29; m 53; c 2. CYTOLOGY. *Educ:* Boston Univ, AB, 50, AM, 51, PhD(biol), 55. *Prof Exp:* From asst scientist to sr asst scientist, USPHS, 55-57; from instr to asst prof biol, Brown Univ, 57-64; assoc scientist & assoc prof, 64-68, SR SCIENTIST & PROF, ORE REGIONAL PRIMATE RES CTR, UNIV ORE HEALTH SCI CTR, 68-, HEAD, DIV REPROD BIOL & BEHAV, 83- *Concurrent Pos:* Res grants, USPHS, 64 -; mem, Reproductive Endocrinol Study Sect, NIH,

85-89. *Honors & Awards:* Wyeth Award, 72. *Mem:* AAAS; Endocrine Soc; Am Soc Cell Biol; Soc Study Reproduction; Am Fertil Soc. *Res:* Electron microscopy; endocrinology; reproductive biology; steroid receptors; immunocytochemistry. *Mailing Add:* Ore Regional Primate Res Ctr 505 NW 185th St Beaverton OR 97006

BRENNER, SIDNEY S(IEGFRIED), b Nuremberg, Ger, Feb 6, 27; US citizen; m 50; c 3. METALLURGY. *Educ:* Mass Inst Tehcnol, BS, 51; Rensselaer Polytech Inst, PhD(metall), 57. *Prof Exp:* Metallurgist, Gen Elec Res Labs, 51-60; sr scientist, Res Lab, US Steel Corp, 60-83; RES PROF, UNIV PITTSBURGH, 84- *Mem:* Am Inst Mining, Metall & Petrol Engrs; Am Phys Soc; Am Vacuum Soc. *Res:* Crystal growth; mechanical properties of crystals; kinetics of reactions; surface chemistry and physics; field-ion microscopy. *Mailing Add:* Mat Engr/Benedum Univ Pittsburgh 4200 Fifth Ave Pittsburg PA 15260

BRENNER, STEPHEN LOUIS, b Brooklyn, NY, Apr 10, 48; m; c 3. BIOPHYSICS. *Educ:* State Univ NY Binghamton, BA, 69; Ind Univ, PhD(chem physics), 74. *Prof Exp:* Asst prof chem, Univ Ky, 72-74; staff fel biophys, NIH, 74-80, res chem, 80-84; group leader, 84-85, res supvr, 86-87, RES MGR, STRUCTURAL BIOL, DUPONT CO, 87- *Concurrent Pos:* Adj Prof chem, Univ Del, 88- *Honors & Awards:* Victor K LaMer Award, Colloid & Surface Chem Div, Am Chem Soc, 74. *Mem:* Biophys Soc; Protein Soc. *Res:* Biophysics of genetic recombination; DNA-protein interactions; biological assembly. *Mailing Add:* DuPont Merck Pharmaceut Co E328/B43 Wilmington DE 19880-0328

BRENNESSEL, BARBARA A, b Brooklyn, NY, Aug 15, 48; m 77; c 3. CELL MEMBRANE IN HORMONE RESPONSES. *Educ:* Fordham Univ, BS, 70; Cornell Univ, PhD(biochem), 75. *Prof Exp:* Instr biochem, Med Sch, Cornell Univ, 73-74; instr & res fel microbiol & immunol, Med Sch, Tulane Univ, 75-77; res fel oral biol & pathophysiol, Harvard Sch Dent Med, 77-80; asst prof, 80-86, ASSOC PROF BIOL, WHEATON COL, 87- *Concurrent Pos:* Vis lectr, Harvard Sch Dent Med, 83- *Mem:* Am Soc Cell Biol; AAAS; Am Soc Zoologists; NY Acad Sci. *Res:* Regulation of prolactin production by thyrotropin-releasing hormone and epidermal growth factor; regulation of hormone production in pituitary tumor cells; effects of artificial sweeteners on cultured cells; relationship of cell morphology to gene expression. *Mailing Add:* Dept Biol Wheaton Col Norton MA 02766

BRENNIMAN, GARY RUSSELL, b Mt Clemens, Mich, Aug 4, 42; m 65; c 2. ENVIRONMENTAL HEALTH. *Educ:* Cent Mich Univ, BS, 64, MS, 66; Univ Mich, Ann Arbor, MPH, 69, PhD(environ health sci), 74. *Prof Exp:* Chief basic sci br, Brooke Army Med Ctr, US Army Med Field Serv Sch, 66-68; asst prof, 74-78, actg dir environ & occup health sci, 86-88, ASSOC PROF, UNIV ILL, CHICAGO, 78- *Concurrent Pos:* Consult, Argonne Nat Lab, 76. *Mem:* Sigma Xi; Water Pollution Control Fedn. *Res:* Health effects of human exposure to metals and chlorinated hydrocarbons in drinking water; characterize emissions from incineration of infections wastes; in-soil diffusion coefficient determination for hydrocarbons. *Mailing Add:* 428 Ethel Lombard IL 60148

BRENOWITZ, A HARRY, b Brooklyn, NY, July 8, 18; m 48; c 2. BIOLOGY. *Educ:* Brooklyn Col, BA, 39; Columbia Univ, MA, 47, EdD, 58. *Prof Exp:* From asst prof to prof, 46-83, dir marine sci, 68-75, EMER PROF BIOL, ADELPHI UNIV, 83- *Concurrent Pos:* NSF fels, 59, 60 & 61; trustee, Affiliated Cols & Univs, NY Ocean Sci Lab. *Mem:* AAAS; Am Soc Zool; Am Soc Limnol & Oceanog; Am Inst Biol Sci. *Res:* Marine invertebrates; environmental studies of Great South Bay, Long Island & New York; biology of Chrysaora Quinquecirrha. *Mailing Add:* Biol Dept Adelphi Univ Garden City NY 11530

BRENT, BENNY EARL, b Alton, Kans, July 3, 37; m 62. ANIMAL NUTRITION. *Educ:* Kans State Univ, BS, 59, MS, 60; Mich State Univ, PhD(animal husb, nutrit), 66. *Prof Exp:* Chemist, Mich State Univ, 64-66; asst prof animal husb, 66-69, assoc prof animal sci & indust, 69-70, PROF ANIMAL SCI & INDUST, KANS STATE UNIV, 76- *Mem:* AAAS; Am Soc Animal Sci; Sigma Xi. *Res:* Mathematical modeling of growth; B-complex vitamins in the rumen; control of feed intake in ruminants; enzymology of the rumen. *Mailing Add:* Dept Animal Sci & Indust Kans State Univ Col Agr Manhattan KS 66502

BRENT, CHARLES RAY, b Hattiesburg, Miss, June 12, 31; m 53; c 3. PHYSICAL CHEMISTRY, ORGANIC CHEMISTRY. *Educ:* Univ Southern Miss, BA, 53; Tulane Univ, MS, 60, PhD(phys chem), 63. *Prof Exp:* Asst chemist, Eagle Chem Co, Ala, 53; assoc prof phys chem, 60-70, asst dean, Col Arts & Sci, 66-70, PROF CHEM, UNIV SOUTHERN MISS, 70-, DIR, INST ENVIRON SCI, 74-, PROF INDUST TECH, 75-, RADIATION SAFETY OFFICER, 81- *Mem:* Am Chem Soc; Sigma Xi; Audubon Soc; Int Solar Energy Soc. *Res:* Thermochemical studies of nitrogen heterocyclic compounds; kinetic solvent effects; environmental measurements; solar energy conversion. *Mailing Add:* SS Box 5156 Hattiesburg MS 39406

BRENT, J ALLEN, b Carmi, Ill, Nov 21, 21; m 46; c 4. SCIENCE ADMINISTRATION. *Educ:* Eastern Ill State Col, BS, 43; Univ Fla, PhD(chem), 49. *Prof Exp:* Naval stores asst, Univ Fla, 46-49; head dept chem, Jacksonville Jr Col, 49-51; chief chem, Sect Res & Develop, Minute Maid Corp, 51-62, from assoc dir res to dir res, Minute Maid Co Div, Coca-Cola Co, 62-66, dir res & develop carbonated beverages, 66-68, vpres, Coca-Cola USA Div, 68-76, VPRES, COCA-COLA CO, 76- *Concurrent Pos:* With Manhattan Proj, 43-46. *Mem:* Am Chem Soc. *Res:* Vapor-liquid equilibrium; vacuum distillation; terpenes; citrus products and by-products. *Mailing Add:* 8900 Huntcliff Trace Atlanta GA 30338

BRENT, MORGAN MCKENZIE, b Evanston, Ill, Jan 31, 23; m 55; c 3. MICROBIOLOGY. *Educ:* Northwestern Univ, BS, 48, MS, 49, PhD(biol sci), 53. *Prof Exp:* Jr res zoologist, Univ Calif, 53-54; instr microbiol, Jefferson Med Col, 54-57; assoc prof biol, Bowling Green State Univ, 57-66, chmn dept, 64-67, dir appl microbiol, Col Health & Community Serv, 75-77, prof biol, 66-55, EMER PROF BIOL SCI, 85- *Mem:* Am Micros Soc; Soc Protozool. *Res:* Effects of environmental factors on morphogenesis and transformation of amoebo-flagellates; nutrition of protozoa. *Mailing Add:* Dept Biol Sci Bowling Green State Univ Bowling Green OH 43403

BRENT, ROBERT LEONARD, b Rochester, NY, Oct 6, 27; m 49; c 4. EMBRYOLOGY. *Educ:* Univ Rochester, AB, 48, MD, 53, PhD, 55. *Hon Degrees:* DSc, Univ Rochester, 88. *Prof Exp:* Res assoc, AEC, Univ Rochester, 47-55, asst physics, 48-49; chief radiation biol, Div Nuclear Med, Walter Reed Army Inst Res, 55-57; assoc prof pediat, 57-60, PROF PEDIAT & RADIOL, JEFFERSON MED COL, 60-, PROF ANAT, 71-, CHMN DEPT PEDIAT, 66-, DIR, STEIN RES CTR, 65-; DIR, ELEANOR ROOSEVELT CANCER RES INST, 62- *Concurrent Pos:* Fel, Fitzwilliam Col, Cambridge Univ, 71; intern, Mass Gen Hosp, 54-55; dir, Stein Res Ctr, Jefferson Med Col, 69-; mem, Embryol Study Sect, NIH, 70-74, chmn, Subcomt Prev Embryol, Fetal & Perinatal Dis, Fogarty Ctr; managing ed, 77-80, ed in chief, Teratology, 76-; fertil & maternal health drug comn, FDA, 81-84; Louis & Bess Stein prof pediat, Jefferson Med Col, 85; dir, Develop Biol Lab, A I du Pont Inst, 89. *Honors & Awards:* Ritchie Prize, 53. *Mem:* Teratology Soc (pres, 68-69); AAAS; Am Asn Anatomists; Am Soc Exp Path; Am Pediat Soc; Sigma Xi; Japan Teratol Soc; Soc Pediat Res; Am Acad Pediat; Radiation Res Soc. *Res:* Experimental embryology; radiation biology; clinical pediatric; immunology. *Mailing Add:* 949 Irvin Rd Huntingdon Valley PA 19006

BRENT, THOMAS PETER, b Leipzig, Ger, Nov 7, 37; Brit citizen; m 66, 76. BIOCHEMISTRY. *Educ:* Cambridge Univ, Univ, BA, 62; Univ London, PhD(biochem), 66. *Prof Exp:* Res fel cell biol, Chester Beatty Res Inst, London, 66-68; asst prof cancer res & biochem, McGill Univ, 68-72; from asst mem to assoc mem, 72-85, MEM, PHARMACOL, ST JUDE CHILDREN'S RES HOSP, 85- *Concurrent Pos:* Assoc prof, Dept Biochem, Univ Tenn Ctr Health Sci, 80-89, assoc prof, Dept Pharmacol, 90-; mem, Radiation Study Sect, NIH, 81-85. *Mem:* Brit Biochem Soc; Biophys Soc; Am Soc Photobiol; Radiation Res Soc; Am Asn Cancer Res; AAAS. *Res:* Repair of DNA in response to radiation and alkylation; regulation of DNA replication and cell proliferation. *Mailing Add:* St Jude Childrens Res Hosp PO Box 318 Memphis TN 38101

BRENT, WILLIAM B, b Ky, June 28, 24. STRUCTURAL GEOLOGY, STRATIGRAPHY. *Educ:* Univ Va, BA, 49, JD, 66; Cornell Univ, MA, 52, PhD(geol), 55. *Prof Exp:* Asst prof geol, Okla State Univ, 55-61; assoc prof, La Tech Univ, 66-68; vis assoc prof, Univ Va, 69-70; GEOLOGIST, TENN DIV GEOL, 70- *Mem:* Geol Soc Am; Sigma Xi. *Res:* Field studies of mountain systems of the United States; chiefly structural and stratigraphic field studies in the complexly folded and faulted Southern Appalachians. *Mailing Add:* 3393 Lake Brook Blvd Knoxville TN 37909

BRENTJENS, JAN R, b Netherlands, Sept 22, 36; m 64; c 3. NEPHROLOGY, IMMUNOPATHOLOGY. *Educ:* Univ Leiden, MD, 64; Univ Amsterdam, PhD(exp diffuse intravascular coagulation), 67; State Univ NY, MD, 79. *Prof Exp:* Resident med, Univ Hosp Groningen, 67-72; asst prof med, Univ Groningen, 72-76; assoc prof, 76-81, PROF PATH, STATE UNIV NY, BUFFALO, 81-, RES PROF MED, 82-, ASSOC PROF MICROBIOL, 76- *Concurrent Pos:* Res asst prof microbiol, State Univ NY, Buffalo, 72-74; fel, Henry & Bertha Buswell Fund, 72-74, Am Nat Kidney Found, 73-74. *Res:* Immunopathology in laboratory animals and in man; nephrology. *Mailing Add:* Dept Path 204 Farber Hall Main St Campus State Univ NY Buffalo NY 14214

BRENTON, JUNE GRIMM, b Wheeling, WVa, Oct 16, 18; m 66; c 1. PHYSICS, MATHEMATICS. *Educ:* Univ WVa, AB, 38; Univ Fla, MS, 48; PhD(physics), 53. *Prof Exp:* Res physicist, Ordnance Eng Corp, 53-55; chief scientist physics, Geo-Sci Inc, 55-66; prog mgr systs anal, Dikewood Indust Inc, 66-81, vpres res, Dikewood Corp, 81-85; RETIRED. *Mem:* Sigma Xi; Am Inst Aeronaut & Astronaut; Am Geophys Union; Asn Old Crows; Air Defense Preparedness Asn. *Res:* Application of technology to military operational requirements. *Mailing Add:* 3301 San Rafael SE Albuquerque NM 87106

BRESCIA, VINCENT THOMAS, b New York, NY, June 2, 30; m 65; c 4. MICROBIAL GENETICS, PROTEIN PURIFICATION. *Educ:* Cent Col, Iowa, BA, 55; Fla State Univ, MS, 65, PhD(genetics), 73. *Prof Exp:* Instr biol, Lees Jr Col, Ky, 60-63; from asst prof to assoc prof biol, The Lindenwood Col, 69-81; DIR QUAL ASSURANCE, LEE SCIENTIFIC, 82- *Mem:* Am Chem Soc; Sigma Xi. *Res:* Purification of clinical enzymes. *Mailing Add:* 912 Wilmington Dr St Charles MO 63301-4761

BRESHEARS, WILBERT DALE, b Eugene, Ore, Apr 25, 39; m 60; c 4. PHYSICAL CHEMISTRY. *Educ:* Portland State Col, BS, 61; Ore State Univ, PhD(phys chem), 65. *Prof Exp:* NSF fel, Sch Chem Sci, Univ EAnglia, 65-66; staff mem, 66-74, GROUP LEADER PHYS CHEM, LOS ALAMOS NAT LAB, 74- *Mem:* AAAS; Am Chem Soc. *Res:* Chemical kinetics; molecular energy transfer; laser isotope separation. *Mailing Add:* Los Alamos Nat Lab PO Box 1663 MS J565 Los Alamos NM 87545

BRESLAU, BARRY RICHARD, b Brooklyn, NY, Feb 10, 42; m 66. MEMBRANE TECHNOLOGY, ULTRAFILTRATION. *Educ:* Polytech Inst Brooklyn, BChE, 63, PhD(chem eng), 69; Univ Mich, MChE, 64. *Prof Exp:* Engr, Pioneer Res Div, E I du Pont de Nemours & Co, Inc, 64-65; res engr, 68-71, sr engr, 71-73; DIR PROCESS RES, ROMICON, INC, 73- *Mem:* Am Chem Soc; Am Inst Chem Engrs; Am Soc Testing & Mat; Am Dairy Sci Asn; Water Pollution Control Fedn. *Res:* Membrane technology; hollow fiber ultrafiltration membranes; mass transfer in membrane systems. *Mailing Add:* Romicon Inc 100 Cummings Park Woburn MA 01801

BRESLER, AARON D, b New York, NY, June 20, 24; m 50; c 4. ELECTRICAL ENGINEERING. *Educ:* City Col New York, BEE, 44; Polytech Inst Brooklyn, MS, 51, PhD(elec eng), 59. *Prof Exp:* Instr elec eng, City Col New York, 48-51; res assoc circuit theory, Microwave Res Inst, Polytech Inst Brooklyn, 51-53; instr elec eng, City Col New York, 53-55; sr res assoc propagation & electromagnetic theories, Polytech Inst Brooklyn, 55-59; sr engr, Jasik Labs, 59-63, chief engr, 63-65, vpres, 65-68; independent consult, 68-72; DEP DIR ANTENNA SYST DIV, CUTLER-HAMMER INC, 72- *Mem:* Sr Inst Elec & Electronics Engrs; Sigma Xi. *Res:* Antennas; electromagnetic theory; propagation of electromagnetic wave in anisotropic media. *Mailing Add:* 12 Helene Ave Merrick NY 11566

BRESLER, B(ORIS), b Harbin, China, Oct 18, 18; nat US; m 46; c 1. STRUCTURAL ENGINEERING. *Educ:* Univ Calif, BS, 41; Calif Inst Technol, MS, 46. *Prof Exp:* Struct designer, Kaiser Shipyards, Calif, 41-43; stress analyst, Convair Co, 43-45; lectr & mem fac, Univ Calif, Berkeley, 46-58, asst dean, Col Eng, 56-59, chmn, Div Struct Eng & Mech, 63-64, dir, Struct Mat Lab, 63-65, prof, 58-78, emer prof civil eng, 78-88. *Concurrent Pos:* Consult engr, 46-; NSF fel, 61; Guggenheim fel, 62; prin, Wiss, Janney, Elstner & Assocs, Inc, 77-; Wiff, Janey, Elstner Assoc, 77-88; consult, 88- *Honors & Awards:* Wason Medal for Res, Am Concrete Inst, 59; State-of-the-Art Civil Eng Award, Am Soc Civil Engrs, 68; J W Kelly Award, Am Concrete Inst, 78. *Mem:* Nat Acad Eng; fel Am Concrete Inst; Int Asn Bridge & Struct Engrs; Reinforced Concrete Res Coun; fel Am Soc Civil Engrs. *Res:* Behavior and distress in structures exposed to complex loading conditions and various special environments; analyses of various means of fire protection of concrete and steel structures; systematic procedures for evaluating seismic hazards of existing buildings. *Mailing Add:* 6363 Christie Ave No 2904 Emeryville CA 94608

BRESLER, JACK BARRY, b New York, NY, May 10, 23; m 52; c 4. HEALTH PLANNING. *Educ:* Univ Denver, BA, 48; Univ Okla, MA, 52; Univ Ill, PhD(biol), 57. *Prof Exp:* Instr biol, Colgate Univ, 54-55; asst prof, Bard Col, 55-57 & Brown Univ, 57-62; assoc prof, Boston Univ, 62-66; assoc prof & dir res, Tufts Univ, 66-76; researcher health sci, Libr Cong, 76-77; consult health, 78-79; sr researcher, 80-86, CONSULT, VET ADMIN CENT OFF, 86- *Concurrent Pos:* Consult Collabr Study Human Reprod, NIH, 57-62; seminar assoc, Columbia Univ, 57-62; mem, New Eng Comt, Dept Health Educ & Welfare, 70-73; mgt consult, Nat Sci Found, 75-76; appln rev, Nat Sci Found, 74-76. *Mem:* fel AAAS; Behav Genetics Soc; Soc Study Social Biol. *Res:* Genetics; environment; scientific productivity. *Mailing Add:* 6901 Stonewood Ct Rockville MD 20852

BRESLIN, ALFRED J(OHN), environmental sciences, for more information see previous edition

BRESLIN, JOHN P, b US, June 9, 19; m 44; c 8. HYDRODYNAMICS. *Educ:* Webb Inst Naval Archit, BS, 44; Univ Md, MA, 51; Stevens Inst Tech, DSc(appl mech), 56. *Prof Exp:* Naval architect & hydraul engr, David Taylor Model Basin, US Dept Navy, DC, 44-51; hydrodynamicist, Gibbs & Cox, Inc, NY, 51-54; sr res eng hydrodynamics, Stevens Inst Technol, 54-56, tech dir, 56-59, dir Hydrodynamics, Davidson Lab, 59-; prof oceanog eng, Inst, 67-; RETIRED. *Concurrent Pos:* Consult, Westinghouse Elec Corp, 58- & Sperry Systs Mgt Div, 69-; J Hydronaut, Am Inst Aeronaut & Astronaut. *Mem:* Soc Naval Archit & Marine Eng; Am Inst Aeronaut & Astronaut. *Res:* Blade-frequency pressures and forces generated by ship propellers; theory of wave resistance of ships. *Mailing Add:* Ciudad de las Comunicacione, 11-3 Stevens Inst of Technol San Miguel de Salinas Alcanta 03193 Spain

BRESLOW, DAVID SAMUEL, b New York, NY, Aug 13, 16; m 46; c 3. POLYMER CHEMISTRY, ORGANIC CHEMISTRY. *Educ:* City Col New York, BS, 37; Duke Univ, PhD(org chem), 40. *Prof Exp:* Lab asst chem, Duke Univ, 37-40; asst org chem, Calif Inst Technol, 40-41; res fel, Radiation Lab, Univ Calif, 42; res assoc, Duke Univ, 42-44, instr, 44-45; from res chemist to sr res chemist, Hercules Inc, 46-63, res assoc, 63-71, sr res assoc, 71-82; PRES, DAVID S BRESLOW ASSOC, 83- *Concurrent Pos:* Vis prof, Univ Munich, 64-65 & Univ Notre Dame, 71; adj prof chem, Univ Del, 72-87, Univ Fla, 88-89. *Honors & Awards:* Am Chem Soc Award, Appl Polymer Sci, 88. *Mem:* Am Chem Soc. *Res:* Polymer synthesis; biologically active polymers; synthetic organic chemistry; organometallic chemistry. *Mailing Add:* Nine Madelyn Ave Wilmington DE 19803

BRESLOW, ESTHER M G, b New York, NY, Dec 23, 31; m 55; c 2. BIOPHYSICAL CHEMISTRY. *Educ:* Cornell Univ, BS, 53; NY Univ, MS, 55, PhD(biochem), 59. *Prof Exp:* Res assoc, 60-64, from asst prof to assoc prof, 64-78, PROF BIOCHEM, MED COL, CORNELL UNIV, 78- *Concurrent Pos:* NIH fel, Med Col, Cornell Univ, 59-61 & NIH res grants, 61-74; mem, Biophys & Biophys Chem Study Sect A, NIH, 73-77; mem, Adv Comt Physiol Cellular Molecular Biol, Biol Inst Subcomt, NSF, 81-84. *Mem:* Fel AAAS; Am Chem Soc; Am Soc Biol Chemists; Harvey Soc; Sigma Xi. *Res:* Relationship between protein structure and biological and chemical reactivity. *Mailing Add:* 275 Broad Ave Englewood NJ 07631

BRESLOW, JAN LESLIE, b New York, NY, Feb 28, 43; m 65; c 2. METABOLISM, PEDIATRICS. *Educ:* Columbia Col, AB, 63; Columbia Univ, MA, 64; Harvard Med Sch, MD, 68. *Prof Exp:* Intern pediat, Boston Children's Hosp, 68-69, jr asst res, 69-70, chief, Metab Lab, 73-84, asst med, 73-75, chief, metab div, 75-84, assoc med, 75-81, sr assoc med, 81-84; staff assoc, molecular dis br, Nat Heart, Lung & Blood Inst, 70-73; instr pediat, Harvard Med Sch, 73-74, from asst prof to assoc prof, 74-84; PROF, ROCKEFELLER UNIV & SR PHYSICIAN, ROCKEFELLER UNIV HOSP, 84-; CLIN AFFIL, PEDIAT NEW YORK HOSP, 84- *Concurrent Pos:* Eugene Higgins fel, Columbia Univ, 63-64; consult med, Brigham & Women's Hosp, 76-84; investr, Am Heart Asn, 81-86, mem, path res study comt, 85- *Honors & Awards:* E Mead Johnson Award, Am Acad Pediat, 84. *Mem:* Am Heart Asn; Soc Pediat Res; Am Soc Clin Invest; Int Atherosclerosis Soc. *Res:* Laboratory and clinical investigation directed at understanding human genetic susceptibility to atherosclerosis; genes that regulate lipoprotein metabolism and mutations in these genes in patients with lipoprotein abnormalities and-or premature heart disease. *Mailing Add:* Lab Biochem Genetics & Metab Rockefeller Univ 1230 York Ave New York NY 10021-6399

BRESLOW, LESTER, b Bismarck, NDak, Mar 17, 15; m 67; c 3. PUBLIC HEALTH, PREVENTIVE MEDICINE. *Educ:* Univ Minn, BA, 35, MD, 38, MPH, 41,. *Hon Degrees:* DSc, Univ Minn, 88. *Prof Exp:* Dist health officer, State Dept Health, Minn, 41-43; chief, Bur Chronic Dis, State Dept Pub Health, Calif, 46-60, chief, Div Prev Med Serv, 60-65, dir, 65-68; prof health serv admin, Sch Pub Health, 68-70, prof prev & social med & chmn dept, Sch Med, 70-72, dean, Sch Pub Health, 72-80, PROF PUB HEALTH, UNIV CALIF, LOS ANGELES, 80- *Honors & Awards:* Lasker Award, 60; Sedgwick Medal, Am Pub Health Asn, 77; Dana Award Health, Dana Found, 88. *Mem:* Inst of Med of Nat Acad Sci; Am Pub Health Asn (pres, 69); Pub Health Cancer Asn Am (pres, 53); Int Epidemiol Asn (pres, 67-68); Asn Schs Pub Health (pres, 73-75). *Res:* Chronic disease epidemiology; health services; health promotion. *Mailing Add:* Sch Pub Health Univ Calif Los Angeles CA 90024

BRESLOW, NORMAN EDWARD, b Minneapolis, Minn, Feb 21, 41; m 63; c 2. MEDICAL STATISTICS. *Educ:* Reed Col, BA, 62; Stanford Univ, PhD(statist), 67. *Prof Exp:* Vis res worker med statist, London Sch Hyg & Trop Med, 67-68; from asst prof to assoc prof, 68-76, PROF BIOSTATIST, UNIV WASH, 76-, CHMN DEPT, 83- *Concurrent Pos:* Statistician, Children's Cancer Study Group, 68-72; mem, Nat Wilm's Tumor Study Comt, 70-; statistician, Int Agency Res Cancer, WHO, 72-74 & 78-79; assoc mem, Fred Hutchinson Cancer Res Ctr, 74-; consult, Nat Inst Environ Health Sci, 75, Nat Ctr Health Servs Res, 75-77, NSF, 75-77; assoc ed, J Am Statist Asn, 77-80 & Int Statist Rev, 81-86; bd sci counselors, Nat Toxicol Prog, 83-87; sr US sci award, Alexander von Humboldt Found, 82-83; mem, Bd Health Prom & Dis Prev, Inst Med, 87-; mem, Comt Appl & Theoret Statist, Nat Acad Sci-Nat Res Coun, 88- *Honors & Awards:* Spiegelman Gold Medal Award, Am Pub Health Asn, 78. *Mem:* Inst Med-Nat Acad Sci; Biomet Soc; fel Am Statist Asn; Soc Epidemiol Res; Int Statist Asn; fel AAAS. *Res:* Statistics of clinical trials, especially Wilm's tumor; cancer epidemiology, particularly in environmental carcinogenesis; statistical methodology for survival time studies and case-control studies. *Mailing Add:* Dept Biostatist SC-32 Univ Wash Seattle WA 98195

BRESLOW, RONALD, b Elizabeth, NJ, Mar 14, 31; m 55; c 2. ORGANIC CHEMISTRY. *Educ:* Harvard Univ, AB, 52, AM, 54, PhD(chem), 56. *Prof Exp:* Nat Res Coun fel, 55-56, from instr to prof chem, 56-67, chmn dept, 76-79, S L MITCHELL PROF CHEM, COLUMBIA UNIV, 67- *Concurrent Pos:* Chmn div chem, Nat Acad Sci, 74-77; bd trustees, Rockefeller Univ, 81-; chmn, bd sci adv, A P Sloan Found, 81-85, chem sect, AAAS, 88-89; mem, sci adv comt, Gen Motors, 82-89, coun, Am Philos Soc, 87- *Honors & Awards:* Nat Medal of Sci, 91; Fresenius Award, 66; Award Pure Chem, Am Chem Soc, 66, Harrison Howe Award, 75; Mark van Doren Medal, Columbia Univ, 69; Baekeland Medal, 69; Remsen Medal, 77; Roussel Prize in Steroid Chem, 78; J R Norris Award, 80; T W Richard Medal, 84; Centenary Medal, Brit Chem Soc, 72; A C Cope Award, 87; Nichols Medal, 89; Chem Sci Award, US Nat Acad Sci, 89; Paracelsus Medal, Swiss Chem Soc, 90; Nat Medal Sci, 91. *Mem:* Nat Acad Sci; Am Acad Arts & Sci; Am Chem Soc; Am Philos Soc; AAAS; Brit Chem Soc. *Res:* Aromaticity and small ring compounds; biochemical model systems; reaction mechanisms. *Mailing Add:* Box 566 Havemeyer Hall Columbia Univ New York NY 10027

BRESNICK, EDWARD, b Jersey City, NJ, Sept 7, 30; m 57. BIOCHEMISTRY. *Educ:* St Peter's Col, BS, 52; Fordham Univ, MS, 54, PhD, 57. *Prof Exp:* Res assoc, Med Br, Univ Tex, 57-58; res biochemist, Burroughs, Wellcome & Co, 58-61; from asst prof biochem to prof pharmacol, Baylor Col Med, 70-72; prof cell & molecular biol & chmn dept, Med Col Ga, 72-78; prof biochem, Col Med, Univ Vt, 78-, CHMN DEPT, 80-; PROF & DIR, EPPLEY INST RES CANCER & ALLIED DIS, MED CTR, UNIV NEBR. *Concurrent Pos:* Consult, Nat Inst Gen Med Sci. *Mem:* AAAS; Am Chem Soc; Am Asn Cancer Res; Am Soc Cell Biol; Am Soc Biol Chemists. *Res:* Cancer research; enzymology; regulatory mechanisms; pyrimidine and nucleic acid metabolism. *Mailing Add:* Eppley Inst Res Cancer & Allied Dis Med Ctr Univ Nebr 42nd & Denver Ave Omaha NE 68105

BRESSLER, BERNARD, b Milan, Mich, May 22, 17; m 48; c 2. MEDICINE. *Educ:* Washington Univ, AB, 38, MD, 42. *Prof Exp:* Assoc prof, 55-59, PROF PSYCHOSOM MED, MED CTR, DUKE UNIV, 59- *Concurrent Pos:* Instr, Washington Psychoanal Inst, 58-60, training analyst, 60-; instr, Univ NC-Duke Univ Psychoanal Inst, 60-; consult, Ft Bragg, NC, 58. *Mem:* Am Psychoanal Asn; Am Psychiat Asn; AMA; Am Psychosomatic Soc; Am Col Psychiat; Sigma Xi. *Res:* Psychosomatic medicine; psychoanalysis. *Mailing Add:* 3600 Floyd Ave Richmond VA 23221

BRESSLER, BERNARD HARVEY, b Winnipeg, Man, Dec 25, 44; m 68; c 2. PHYSIOLOGY, ANATOMY. *Educ:* Sir George Williams Univ, BSc, 66; Univ Man, MSc, 68, PhD(physiol), 72. *Prof Exp:* Asst prof anat, Univ Sask, 73-74 & physiol, 74-76; from asst prof to assoc prof, 76-88, PROF ANAT, UNIV BC, 88- *Concurrent Pos:* Assoc Dean, Res Grad Studies Fac Med, Univ BC, 87-90, assoc vpres res, Health Sci, 90- *Mem:* Biophys Soc; Can Physiol Soc; Can Asn Anatomists (secy, 80-83, vpres, 83-85, pres, 85-87); Can Fedn Biol Soc (vpres, 85-86, pres, 86-87). *Res:* Biophysics of skeletal muscle contraction. *Mailing Add:* Dept Anat Univ BC Vancouver BC V6T 1W5 Can

BRESSLER, DAVID WILSON, b San Francisco, Calif, Sept 7, 23; m 49; c 2. MATHEMATICS. *Educ:* Univ Calif, Berkeley, AB, 49, MA, 51, PhD(math), 57. *Prof Exp:* Instr math, Univ BC, 57-58; instr, Tufts Univ, 58-59; from instr to asst prof, Univ BC, 59-68; assoc prof, 68-74, PROF MATH, CALIF STATE UNIV, SACRAMENTO, 74- *Mem:* Am Math Soc. *Res:* Set theory and measure theory. *Mailing Add:* 401 E Eleventh St Davis CA 95616-2011

BRESSLER, RUBIN, DIABETES, ENDOCRINOLOGY. *Educ:* Duke Univ, MD, 57. *Prof Exp:* PROF INTERNAL MED, HEAD DEPT & CHIEF, CLIN PHARMACOL, UNIV ARIZ, 69- *Mailing Add:* Dept Internal Med Health Sci Ctr Univ Ariz Tucson AZ 85724

BRESSLER, STEVEN L, b St Louis, Mo, June 7, 51; m 81; c 3. NEUROPHYSIOLOGY, PSYCHOBIOLOGY. *Educ:* Johns Hopkins Univ, BA, 72; Univ Calif, Berkeley, PhD(neurophysiol), 82. *Prof Exp:* Systs neurophysiologist, EEG Systs Lab, 82-90; ASSOC PROF NEUROSCI & COMPLEX SYSTS, FLA ATLANTIC UNIV, 90- *Mem:* Soc Neurosci; Soc Psychophysiol Res. *Res:* Recording spatial patterns of the brain's electrical activity from people and animals performing highly controlled tasks; developing mathematical models to describe the operation of the neural circuits generating the brain's electrical activity. *Mailing Add:* Ctr Complex Systs, Fla Atlantic Univ Bldg MT9 500 NW 20th St Boca Raton FL 33431

BRESSON, CLARENCE RICHARD, b Wooster, Ohio, July 30, 25; m 60; c 3. APPLIED CHEMISTRY. *Educ:* Col Wooster, BA, 50; Wash State Univ, MS, 55, PhD(chem), 58. *Prof Exp:* Baking technologist cereal chem, USDA, 51-54; chemist, Phillips Petrol Co, 59-60, 61-65, sr res chemist, 66-90; RETIRED. *Mem:* Am Chem Soc. *Res:* Upgrading refinery by-product streams; oxo-alcohols, primary and secondary plasticizers, fluid coke, pyrolosis gasoline; applications for new products, sulfolanyl ethers; technical service; hydraulic barriers; asphaltic concrete flexure fatigue; rubberized-asphalt roofing-membranes; photopolymers; drilling mud additives; mining chemicals. *Mailing Add:* 118 Harwood Dr Bartlesville OK 74003

BRESSOUD, DAVID MARIUS, b Bethlehem, Pa, Mar 27, 50; m 85. COMBINATORIAL NUMBER THEORY, GENERATING FUNCTIONS. *Educ:* Swarthmore Col, BA, 71; Temple Univ, MA, 75, PhD(math), 77. *Prof Exp:* From asst prof to assoc prof, 77-86, PROF MATH, PA STATE UNIV, 86- *Concurrent Pos:* Mem, Inst Adv Study, 79-80; vis asst prof, Univ Wis-Madison, 80-81; Sloan Found Fel, 82-84; prof assoc, Univ Strasbourg, France, 85-86; Fulbright fel, 85-86. *Mem:* Am Math Soc; Math Asn Am. *Res:* Partition identities; combinatorial enumeration; special functions. *Mailing Add:* Dept Math Pa State Univ University Park PA 16802

BREST, ALBERT N, b Berwick, Pa, Dec 17, 28. MEDICINE. *Educ:* Temple Univ, MD, 53; Am Bd Internal Med, dipl, 61. *Prof Exp:* Intern, Philadelphia Gen Hosp, Pa, 53-54; resident internal med, Temple Univ Hosp, 54 & Albert Einstein Med Ctr, Pa, 56-58; instr med & dir hypertension unit, Hahnemann Med Col, 59-60, assoc, 60, from asst prof to assoc prof, 61-69, head sect hypertension & renology, 59-63, vasc dis & renology, 63-69, cardiol, 68-69; HEAD DIV CARDIOL, JEFFERSON MED COL & HOSP, PA, 69- *Concurrent Pos:* Am Heart Asn res fel, Hahnemann Hosp, 58-59. *Mem:* AMA; Am Heart Asn; NY Acad Sci; Am Fedn Clin Res; fel Am Col Cardiol. *Res:* Cardiovascular diseases. *Mailing Add:* Dept Med Jefferson Med Col 1025 Walnut St Philadelphia PA 19107

BRESTON, JOSEPH N(ORBERT), b Passaic, NJ, June 6, 12; m 41; c 1. CHEMICAL ENGINEERING, PETROLEUM ENGINEERING. *Educ:* Pa State Col, BS, 35, MS, 42, PhD(fuel tech), 44. *Prof Exp:* Asst chem, Pa State Col, 35-36, asst petrol ref res, 37-42; assoc, Pa Grade Crude Oil Asn, 45-48, lab dir, 48-53; vpres & dir res & develop, Bradford Chem Co, 53-56; owner-dir, Teck Labs, 56-75; PRES, BRADFORD CHEM CO, 65-; DIR, J N BRESTON ASSOCS, 75- *Honors & Awards:* Pioneers in Chem Award, Am Chem Soc. *Mem:* AAAS; Am Asn Corrosion Engrs; Am Inst Mining, Metall & Petrol Engrs; Am Chem Soc. *Res:* Petroleum refining; chemical engineering; coal technology and combustion; fuel utilization; petroleum production and secondary recovery; chemicals for oil production; water analyses and treatment; pollution control engineering; environmental engineering. *Mailing Add:* 225 Homan Ave State College PA 16801

BRETHERTON, FRANCIS P, b Oxford, Eng, July 6, 35; m 59; c 3. METEOROLOGY. *Educ:* Univ Cambridge, BA, 58, MA & PhD(fluid dynamics), 62. *Prof Exp:* Instr math, Mass Inst Technol, 61-62; SAR dynamical meteorol, Univ Cambridge, 62-63, ADR, 63-64, univ lectr appl math, 66-69; prof meteorol & oceanog, Johns Hopkins Univ, 69-74; pres, Univ Corp Atmospheric Res, 73-; dir, Nat Ctr Atmospheric Res, 74-88; DIR, SPACE SCI & ENG CTR, 88- *Concurrent Pos:* Vis prof, Univ Miami, 68-69; mem, US Comt for Global Atmospheric Res Prog, 71-76; mem, Comt Atmospheric Sci, Nat Acad Sci, 71-77. *Honors & Awards:* Res Award, Area IV, World Meteorol Orgn, 71; Buchan Prize, Royal Meteorol Soc; Meisinger Award, Am Meteorol Soc, 72. *Mem:* Am Meteorol Soc; Royal Meteorol Soc. *Res:* Mesoscale and large scale dynamics of the atmosphere and ocean; applied mathematics and the general theory of wave propagation. *Mailing Add:* 1225 W Dayton St Madison WI 53706

BRETHOUR, JOHN RAYMOND, b Junction City, Kans, Sept 30, 34; m 64; c 1. RUMINANT NUTRITION, ANIMAL PRODUCTION. *Educ:* Kans State Univ, BS, 55; Okla State Univ, MS, 56. *Prof Exp:* Technician radioisotopes, Univ Tenn, 56-57; asst prof animal sci, Ft Hays Exp Sta, Kans State Univ, 57-66; beef cattle prod, J R Brethour, 66-68; assoc prof, 68-75, PROF ANIMAL SCI, FT HAYS EXP STA, KANS STATE UNIV, 75- *Mem:* Am Soc Animal Sci; Sigma Xi; Am Registry Cert Animal Scientists; Coun Agr Sci & Technol; Am Inst Ultrasound Med. *Res:* Animal nutrition; ultrasound. *Mailing Add:* Ft Hays Exp Sta Kans State Univ Hays KS 67601

BRETON, J RAYMOND, b Hartford, Conn, Nov 14, 31; m 61; c 6. ACOUSTICS, WAVE PROPAGATION. *Educ:* Boston Col, MA, 57; Harvard Univ, AM, 59; Univ RI, MS, 75, PhD(elec eng), 77. *Prof Exp:* Foreign serv officer, Dept State, 61-63; lectr, San Diego State Col, 63-65; prin engr, Naval Underwater Systs Ctr, 67-82; dir, Gen Instrument Corp, 82-89; PRES, BR ENTERPRISES, 89- *Mem:* Inst Elec & Electronics Engrs. *Res:* Sonar system design applied to bathymetry; general propagation and hydrodynamics. *Mailing Add:* BR Enterprises PO Box 71 Norwood MA 02062

BRETSCHER, ANTHONY P, b 1950; m 83; c 1. MICROFILAMENT, BRUSH BORDER. *Educ:* Leeds Univ, Eng, PhD(genetics), 74. *Prof Exp:* ASSOC PROF BIOCHEM, SECT BIOCHEM, MOLECULAR & CELL BIOL, CORNELL UNIV, 81- *Mem:* Am Soc Cell Biol. *Res:* Biochemical and cell biology analysis of the cytoskeleton. *Mailing Add:* Biotechnol Bldg Cornell Univ Ithaca NY 14853

BRETSCHER, MANUEL MARTIN, b River Forest, Ill, May 12, 28; m 55; c 5. NUCLEAR PHYSICS, REACTOR PHYSICS. *Educ:* Wash Univ, AB, 50, PhD(physics), 54. *Prof Exp:* Asst prof physics, Ala Polytech Inst, 54-56; from assoc prof to prof & co-chmn dept, Valparaiso Univ, 56-67; PHYSICIST, APPL PHYSICS DIV, ARGONNE NAT LAB, 67- *Concurrent Pos:* Consult, Oak Ridge Nat Lab, 58-64. *Mem:* Am Phys Soc; Am Nuclear Soc; Am Asn Physics Teachers; Sigma Xi. *Res:* Particle accelerators; stripping reactions; coulomb excitation; scattering; reactor physics; plutonium capture-to-fission ratios; age measurements; mass spectrometry; inhomogeneous magnetic field spectrometers; breeding ratio measurements for liquid metal fast breeder reactors; tritium measurements. *Mailing Add:* 7313 Seminole Ct Darien IL 60559

BRETSCHNEIDER, CHARLES LEROY, b Red Owl, SDak, Nov 9, 20; m 48; c 2. PHYSICAL OCEANOGRAPHY, OCEAN ENGINEERING. *Educ:* Hillsdale Col, BS, 47; Univ Calif, Berkeley, MS, 50; Tex A&M Univ, PhD(phys oceanog), 59. *Prof Exp:* Engr, Waves & Wave Force Proj, 50-51; engr waves & wave forces on pilings, Res Found, Tex A&M Univ, 51-56; hydraul engr & chief, Oceanog Br, Beach Erosion Bd, Corps Engr, US Army, 56-61; dir eastern opers, Nat Eng Sci Co, Washington, DC, 61-64, vpres, Vt, 64-65, geomarine technol, 66-67; PROF & CHMN DEPT OCEAN ENG, UNIV HAWAII, 67- *Concurrent Pos:* Lectr, Mass Inst Technol, 61-63. *Honors & Awards:* Outstanding Contrib Coastal Eng Res Prize, Am Soc Civil Eng, 59; Dept Army, Off Chief Engr, Outstanding Performance & Cash Award. *Mem:* Am Soc Oceanog; Am Soc Eng Educ; Soc Naval Archit & Marine Eng; Am Soc Civil Eng; Am Geophys Union. *Res:* Variability, spectra, forecasting, generation and decay of waves; wave forces on piles; hurricane surge and waves; rubble mound breakwater stability. *Mailing Add:* Dept Ocean Eng Univ Hi at Manoa Honolulu HI 96822

BRETSKY, PETER WILLIAM, b Easton, Pa, Oct 26, 38; m 65; c 1. INVERTEBRATE PALEONTOLOGY. *Educ:* Lafayette Col, AB, 61; Southern Methodist Univ, MS, 63; Yale Univ, PhD, 68. *Prof Exp:* Instr geol, Northwestern Univ, 67-69, asst prof geol & biol sci, 69-70; dir environ paleont prog, 70-76, assoc prof, 70-80, PROF EARTH & SPACE SCI, STATE UNIV NY STONY BROOK, 80- *Mem:* Soc Econ Paleont & Mineral; Paleont Soc; Marine Biol Asn UK; Sigma Xi. *Res:* Paleoecology; biostratigraphy. *Mailing Add:* Dept Earth & Space Sci State Univ NY Stony Brook NY 11790

BRETT, BETTY LOU HILTON, b Hudson, NY, Mar 25, 52; m 74. EVOLUTIONARY GENETICS, ICH. *Educ:* State Univ NY, Buffalo, BA, 74; Univ Mich, MA, 77, PhD(natural resources), 81. *Prof Exp:* Res assoc evolutionary genetics, Univ Rochester, 81-87; VIS ASST PROF, ROCHESTER INST TECHNOL, 87- *Concurrent Pos:* Adj asst prof, State Univ NY, Brockport, 86, 89-90 & Nazareth Col, Rochester, 89-91. *Mem:* Soc Study Evolution; Am Soc Ichthyologists & Herpetologists; Genetics Soc Am; AAAS; Soc Syst Zool; Sigma Xi. *Res:* Population genetics of Genesee River fishes; analyzing transmission of supernummary chromosomes and sex ratios in mealy bugs (Coccidae). *Mailing Add:* 106 Norman Rd Rochester NY 14627

BRETT, CARLTON ELLIOT, b Exeter, NH, June 19, 51; m 74; c 2. TAPHONOMY, PALEOECOLOGY. *Educ:* State Univ NY, Buffalo, BA, 73, MA, 75; Univ Mich, PhD(paleont), 78. *Prof Exp:* From instr to assoc prof, 78-90, PROF GEOL & PALEONT, UNIV ROCHESTER, 90- *Concurrent Pos:* Prin investr, Nat Oceanic & Atmospheric Admin Hydrolab, St Croix, VI, 82; prin investr, Am Chem Soc grant, Univ Rochester, 83-85, 89-91 & 91-, NSF grant, 83-86 & 88-; trustee, Paleont Res Inst, 89- *Honors & Awards:* Chas Schuchert Award, Paleont Soc, 90. *Mem:* Paleont Soc; Paleont Res Inst; Geol Soc Am; Int Paleont Asn; Soc Econ Paleontologists & Mineralogists. *Res:* Paleozoic invertebrate paleontology, taphonomy and paleoecology and sequence-event stratigraphy; systematics and paleoecology of pelmatozoan echinoderms; paleoecology and evolution of hard substrate communities, including epibionts; paleoecology of oxygen limited facies; sequence-event stratigraphy of the Silurian-Devonian of Appalachian Basin. *Mailing Add:* Dept Geol Sci Univ Rochester 227 Hutchison Rochester NY 14627

BRETT, JOHN ROLAND, fisheries; deceased, see previous edition for last biography

BRETT, ROBIN, b Adelaide, S Australia, Jan 30, 35; m 86; c 2. EARTH SCIENCES. *Educ:* Univ Adelaide, BSc, 56; Harvard Univ, AM, 60, PhD(geol), 63. *Prof Exp:* Geologist, US Geol Surv, Washington, DC, 64-69; chief, Geochem Br, Johnson Space Ctr, NASA, 69-74; geologist, US Geol Surv, Reston, Va, 74-78; dir, Earth Sci Div, NSF, 78-82; GEOLOGIST, US GEOL SURV, RESTON, VA, 82- *Concurrent Pos:* Hon fel, Australian Nat Univ, 67-68; assoc ed, J Geophys Res, 72-74, 83-85 & Bd Earth Sci, Nat Acad Sci-Nat Res Coun, 85-87; pres, Planetary Geol Div, Geol Soc Am, 84; pres, Geol Soc Wash, 86; vpres, 28th Int Geol Cong, 86-; chmn, US Nat Comt, Int Union Geol & Geophys, 87-; secy gen, Int Union Geol Sci, 89- *Honors & Awards:* Lindgren Award, Soc Econ Geologists, 64; Exceptional Sci Achievement Medal, NASA, 73. *Mem:* Geochem Soc; Am Geophys Union (pres, Volcanology, Geochemistry & Petrol Sect, 86-88); Soc Econ Geol; fel Meteoritical Soc (pres, 73-74); fel Geol Soc Am. *Res:* Mineralogy, petrology, geochemistry of lunar samples, meteorites, planetary geochemistry; meteorite impact structures, mineral deposits. *Mailing Add:* US Geol Surv 917 Reston VA 22092

BRETT, WILLIAM JOHN, b Chicago, Ill, Mar 23, 23; m 48; c 1. PHYSIOLOGY. *Educ:* Northern Ill Univ, BS, 49; Miami Univ, MS, 50; Northwestern Univ, PhD(biol), 53. *Prof Exp:* Asst, Northwestern Univ, 50-53; asst prof biol, Millsaps Col, 53-56; assoc prof biol, 56-59, chmn dept life sci, 64-67, PROF BIOL, IND STATE UNIV, TERRE HAUTE, 59-, CHMN DEPT LIFE SCI, 77- *Mem:* Am Soc Zool; Am Inst Biol Sci; Animal Behav Soc; Nat Asn Biol Teachers. *Res:* Biological rhythms. *Mailing Add:* Dept Life Sci Ind Univ Terre Haute IN 47809

BRETTELL, HERBERT R, b Rock River, Wyo, Feb 1, 21; m 47; c 4. INTERNAL MEDICINE. *Educ:* Univ Wyo, BS, 42; Univ Rochester, MD, 50; Am Bd Internal Med, cert, 57; Am Bd Family Pract, cert, 74, Am Bd Internal Med, cert, 81. *Prof Exp:* Intern, Vanderbilt Univ, 50-51; resident, 51-55, asst, 54-55, from instr to assoc prof, 56-66, head, Div Family Pract, 69-75, chmn, Dept Family Med, 75-77, PROF MED, UNIV COLO MED CTR, DENVER, 75- *Mem:* Am Col Physicians; Am Col Clin Pharmacol & Chemother; Am Heart Asn; Soc Geol Int Med. *Res:* Cardiovascular disease; thrombosis. *Mailing Add:* Dept Med Univ Colo Med Ctr 4200 E Ninth Ave Denver CO 80262

BRETTHAUER, ROGER K, b Morris, Ill, Jan 4, 35; m 59; c 3. CELL BIOLOGY. *Educ:* Univ Ill, BS, 56, MS, 58; Mich State Univ, PhD(biochem), 62. *Prof Exp:* NIH training grant bact, Univ Wis, 62-64; from asst prof to assoc prof, 64-77, PROF CHEM, UNIV NOTRE DAME, 77- *Mem:* AAAS; Am Chem Soc; Am Soc Biol Chemists; Soc Complex Carbohydrates. *Res:* Complex carbohydrates, biosynthesis, structure and function. *Mailing Add:* Dept Chem & Biochem Univ Notre Dame Notre Dame IN 46556

BRETTON, R(ANDOLPH) H(ENRY), b Wolfeboro, NH, Apr 1, 18; m 44; c 3. CHEMICAL ENGINEERING THERMODYNAMICS. *Educ:* Worcester Polytech Inst, BS, 41, MS, 43; Yale Univ, DEng, 49. *Hon Degrees:* DSc, Worcester Polytech Inst, 57. *Prof Exp:* Chem serv engr, Goodyear Tire & Rubber Co, 43-44; from asst prof to assoc prof chem eng, Yale Univ, 48-67; mgr, Tech Info Serv, 67-73; mgr proc develop, Nat Distillers & Chem Corp, 73-82; RETIRED. *Concurrent Pos:* Vis assoc engr, Brookhaven Nat Lab, 49-50, consult, 50-64; consult, Nat Distillers & Chem Corp, 59-67. *Mem:* Am Chem Soc; Am Inst Chem Engrs. *Res:* Kinetics in flow systems; high pressure thermodynamic equilibria. *Mailing Add:* 1/2 Hepburn Rd Hamden CT 06517

BRETZ, HAROLD WALTER, microbiology, for more information see previous edition

BRETZ, MICHAEL, b Harvey, Ill, June 2, 38; m 64; c 2. ADSORPTION PHYSICS. *Educ:* Univ Calif, Los Angeles, 61; Univ Wash, PhD(physics), 71. *Prof Exp:* Asst scientist space physics, Lockheed Res Labs, 61-65; res assoc physics, Univ Wash, 71-73; asst prof, 73-78, ASSOC PROF PHYSICS, UNIV MICH, ANN ARBOR, 78- *Mem:* Am Phys Soc; AAAS. *Res:* Properties of over-layer quantum films adsorbed on homogeneous substrates; new materials for adsorption and catalysis. *Mailing Add:* Dept Physics Univ Mich Ann Arbor MI 48109

BRETZ, PHILIP ERIC, b Carlisle, Pa, Jan 28, 53; m 79; c 2. PHYSICAL METALLURGICAL ENGINEERING, FAILURE ANALYSIS & FRACTURE MECHANICS. *Educ:* Lehigh Univ, BS, 75, MS, 77, PhD(metall), 80. *Prof Exp:* Welding engr, Chicago Bridge & Iron Co, 74-75; teaching asst & res asst metall, Lehigh Univ, 75-80; staff engr, Aluminum Co Am, 80-84, tech specialist, 84-86, proj mgr, 86-88, div mgr, 88-90, OPERATING SUPT, ALUMINUM CO AM, 90- *Mem:* Am Soc Metals; Am Inst Mining Metall & Petrol Engrs; Sigma Xi. *Res:* Microstructure of materials and the influence of structure on mechanical behavior; fatigue and fracture resistance of materials; aerospace aluminum alloy and aluminum-lithium alloy development; failure analysis and product liability; aluminum alloy development. *Mailing Add:* Aluminum Co Am PO Box 9158 Alcoa TN 37701

BREUER, CHARLES B, b Brooklyn, NY, Aug 7, 31; m 62; c 1. RESEARCH ADMINISTRATION. *Educ:* NY Univ, BA, 53; Rutgers Univ, MS, 59, PhD(biochem), 64. *Prof Exp:* Res scientist, Squibb Inst Med Res, Olin Mathieson Chem Corp, 59-61; res asst biochem, Rutgers Univ, 61-63; res scientist, Lederle Labs, Am Cyanamid Co, 63-66, group leader, 66-70, sr res scientist, 67-70, head biotherapeut dept, 70-71, dir diag res & develop, 71-76, dir sci & regulatory affairs, 76-77; dir sci & regulatory affairs, 77-79, DIR TECH OPERS, FISHER SCI CO, 80- *Mem:* AAAS; Am Chem Soc; Am Soc Biol Chem:. *Res:* Diagnostics; immunochemistry; biology of serum proteins; in new diagnostic agents and principles. *Mailing Add:* Five Turner Rd Pearl River NY 10965-2299

BREUER, DELMAR W, b St James, Mo, June 21, 25; m 49; c 2. AERONAUTICAL ENGINEERING. *Educ:* Iowa State Univ, BS, 47; Mo Sch Mines, MS, 50; Ohio State Univ, PhD(aeronaut & astronaut eng), 61. *Prof Exp:* Instr mech, Mo Sch Mines, 47-50; stress analyst, NAm Aviation, 50-51; from asst prof to assoc prof, USAF Inst Technol, 51-61, from actg head to head dept mech, 55-76, prof mech, 61-80, head dept aeronaut & astronaut, 77-80, EMER PROF AEROSPACE ENG, USAF INST TECHNOL, 80-; CONSULT, 80- *Mem:* Am Inst Aeronaut & Astronaut; Am Soc Eng Educ; Sigma Xi. *Res:* Effect of finite deflections on vibrations of plates; thermal stresses; stability of pressurized conical shells; dynamics of variable mass systems; non-nuclear weapon effects. *Mailing Add:* 5539 Oakshire Circle Dayton OH 45440

BREUER, GEORGE MICHAEL, b Trenton, NJ, Feb 1, 44. PHYSICAL CHEMISTRY, INDUSTRIAL HYGIENE. *Educ:* Univ Mo-Rolla, BS, 66; Univ Calif, Irvine, PhD(chem), 72. *Prof Exp:* Asst res chemist air pollution, Statewide Air Pollution Res Ctr, Univ Calif, Riverside, 74-77; res chemist indust hyg, Nat Inst Occup Safety & Health, 77-; AT UNIV HYG LAB, UNIV IOWA, OAKDALE CAMPUS. *Mem:* Am Chem Soc; Soc Appl Spectros; Am Indust Hyg Asn; Am Conf Govt Indust Hygienists; fel Am Inst Chemists. *Res:* Photochemistry; spectroscopy; analytical chemistry in air pollution. *Mailing Add:* Univ Hyg Lab Univ Iowa Oakdale Campus Iowa City IA 52242

BREUER, MAX EVERETT, b Sperry, Iowa, Sept 29, 38; m 62; c 4. CHEMICAL ENGINEERING. *Educ:* Univ Iowa, BS, 63, MS, 64, PhD(chem eng), 67. *Prof Exp:* Chem engr, Union Carbide Corp, 66-68; CHEM ENGR, UPJOHN CO, 68- *Mem:* Am Chem Soc; Am Inst Chem Engrs. *Res:* Gas mixture separation via selective permeation; crystallization; chemical process development. *Mailing Add:* Upjohn Co 1300-87-2 Upjohn Co 1300-173-1 MI 49001

BREUER, MELVIN A, b Los Angeles, Calif, Feb 1, 38; m 63; c 2. ELECTRICAL ENGINEERING. *Educ:* Univ Calif, Los Angeles, BS, 59, MS, 61; Univ Calif, Berkeley, PhD(elec eng), 65. *Prof Exp:* From asst prof to assoc prof elec eng, 65-79, PROF ELEC ENG & COMPUT SCI, UNIV SOUTHERN CALIF, 79- *Concurrent Pos:* Fulbright-Hays fel, 72; ed-in-chief, J Design Automation & Fault Tolerant Comput, 77-82. *Mem:* Sigma Xi; fel Inst Elec & Electronics Engrs; Int Fedn Info Processing. *Res:* Digital computer design; switching theory; design automation; fault tolerant computing; development of new algorithms for computer aided design of digital circuits, such as the min-cut and forced vector algorithms; development of new approaches to testing, such as a knowledge based system for design for test and fundamental work on automatic test vector generation. *Mailing Add:* Dept Elec Eng Univ Southern Calif Los Angeles CA 90089-0781

BREUHAUS, W(ALDEMAR) O(TTO), b Lowell, Ohio, Apr 26, 18; m 50; c 3. AERONAUTICAL ENGINEERING. *Educ:* Carnegie Inst Technol, BS, 40; Univ Buffalo, MS, 61. *Prof Exp:* Aerodynamics engr, Chance-Vought Div, United Aircraft Corp, 40-42, head, Stability & Control Group, 42-46; head, Eng Sect, Flight Res Dept, Cornell Aeronaut Lab, Inc, 46-48, asst head, 48-56, head, 56-67, dir, Flight Dynamics Div, 67-69, dir, Aerosci Div, 69-72; mem prof staff, Systs Eval Div, Inst for Defense Anal, 73-83; RETIRED. *Concurrent Pos:* Mem, Res & Technol Adv Subcomt Aircraft Flight Dynamics, NASA, 67-70; chmn, Sci Adv Bd C-5A, Performance Subcomt, USAF, 70. *Honors & Awards:* Laura Taber Barbour Air Safety Award, Flight Safety Found, 67; De Florez Training Award for Flight Simulation, Am Inst Aeronaut & Astronaut, 86. *Mem:* Assoc fel Am Inst Aeronaut & Astronaut. *Res:* Aircraft aerodynamics, stability and control, and flying qualities requirements; variable stability aircraft use and development. *Mailing Add:* 661 Mill Rd East Aurora NY 14052

BREUNING, SIEGFRIED M, b Hamburg, Ger, July 25, 24; US citizen; m 51; c 3. TRANSPORTATION ENGINEERING, FUTURES ANALYSIS. *Educ:* Stuttgart Tech, MS, 50; Mass Inst Technol, ScD, 57. *Prof Exp:* Asst transp eng, Stuttgart Tech, 51; eng asst, Montreal Transp Comn, 51-52; res engr, Mass Inst Tech, 52-57; assoc prof civil eng, Univ Alta, 57-59; assoc prof civil eng & res, Mich State Univ, 59-63; chief, Transp Systs Div, Nat Bur Stand, 63-65; vis prof civil eng & dir hwy transp prog, Mass Inst Technol, 65-69; exec vpres, Social Technol Systs Inc, Mass, 69-71; PROF CIVIL ENG & INTERDISCIPLINARY STUDY, SOUTHEASTERN MASS UNIV, 71- *Concurrent Pos:* Res consult, Hwy Intersect Design, 50; comt mem, Hwy Res Bd, Nat Acad Sci, 54-78; consult transp econ, Burma, 58; adv, Traffic Res Corp, 62-63; exec secy panel transp res & develop, US Dept Com, 63-67; consult innovation, transp & social technol, 71-; consult, Cabot Corp, 72-87; vpres & dir, New Eng Energy Policy Coun, 74-78; consult future transportation, 81. *Mem:* Ger Road Res Soc; Sigma Xi. *Res:* Definitions, measurements and research of basic factors in transportation; socio-technical systems design and research; interdisciplinary studies. *Mailing Add:* 129 School St Wayland MA 01778

BREW, DAVID ALAN, b Clifton Springs, NY, Nov 22, 30; m 58; c 4. GEOLOGY. *Educ:* Dartmouth Col, AB, 52; Stanford Univ, PhD(geol), 64. *Prof Exp:* GEOLOGIST, US GEOL SURV, 52- *Concurrent Pos:* Asst, Stanford Univ, 58-59; Fulbright scholar, Univ Vienna, 59-60. *Honors & Awards:* Meritorious Serv Award, US Dept Interior, 72; Career Serv Award for Spec Achievement, Civil Serv League, 73. *Mem:* Geol Soc Am; Am Geophys Union; Am Asn Petrol Geologists; Geol Soc Austria; Sigma Xi. *Res:* Relations of sedimentation and tectonics; structure and structural geometry; tectonic and plutonic history of northeastern Pacific rim; relations of mineral deposits to plutonic rocks. *Mailing Add:* US Geol Surv 345 Middlefield Rd Menlo Park CA 94025

BREW, WILLIAM BARNARD, b Sibley, Ill, Apr 23, 13; m 41; c 5. CHEMISTRY. *Educ:* Wash Univ, BS, 35, MS, 36. *Prof Exp:* Res chemist, 36-38, chief anal chemist, 38-41, mgr org res lab, 41-55, mgr org & inorg res lab, 55-60, mgr spec prod res lab, 60-64, dir cent res labs, 64-73, DIV VPRES & DIR CENT RES LABS, RALSTON PURINA CO, 73- *Mem:* Am Chem Soc; Inst Food Technologists; Am Inst Chemists. *Res:* Vitamin chemistry; amino acid chemistry; chromatography; spectrophotometry. *Mailing Add:* 3227 Hawthorne Blvd St Louis MO 63104

BREWBAKER, JAMES LYNN, b St Paul, Minn, Oct 11, 26; m 54, 70; c 4. PLANT GENETICS. *Educ:* Univ Colo, BA, 48; Cornell Univ, PhD(plant breeding), 52. *Prof Exp:* Asst genetics, Calif Inst Technol, 48, Univ Minn, 50 & Cornell Univ, 49-52; Nat Res Found fel, Univ Lund, Sweden, 52-53; asst prof plant breeding, Univ Philippines & Cornell Univ, 53-55; assoc geneticist, Brookhaven Nat Lab, 56-61; assoc prof, 61-64, PROF HORT, UNIV HAWAII, 64- *Concurrent Pos:* Geneticist, Rockefeller Found, Thailand, 67-68 & Int Atomic Energy Agency, Philippines, 70; consult trop agr, Nigeria, Philippines, Korea, Colombia, Taiwan, Indonesia, India, Dominican Repub & Pakistan. *Honors & Awards:* Crop Sci Res Award, Am Soc Agron, 84; Int Inventors Award, Sweden, 86; Superior Serv, USDA, 90. *Mem:* Bot Soc Am; fel Am Soc Agron; Nitrogen Fixing Tree Asn (pres, 81-90); Nat Sweet Corn Breeders Asn (pres, 87); Soc Am Foresters. *Res:* Maize breeding and genetics; breeding tropical nitrogen-fixing trees. *Mailing Add:* Dept Hort Univ Hawaii 3190 Maile Way Honolulu HI 96822

BREWEN, J GRANT, b Easton, Pa, Oct 2, 39; m 69; c 2. CELL BIOLOGY, MOLECULAR BIOLOGY. *Educ:* Johns Hopkins Univ, BA, 61, PhD(biol), 63. *Prof Exp:* Res assoc cytogenetics, Oak Ridge Nat Lab, 63-66; res assoc, Div Plant Indust, Commonwealth Sci & Indust Res Org, 66-67; sr res staff, Oak Ridge Nat Lab, 67-79; dir genetics, Allied Corp, 79-88; PRES & CHIEF EXEC OFFICER, BIOTECHNOL RES & DEVELOP CORP, 88- *Concurrent Pos:* Adj prof, Sch Biomed Sci, Univ Tenn, 70-78; vis prof, Univ Calif, Berkeley, 77; adv sci comt, Ill Govt, 88- *Honors & Awards:* Tech Achievement Award, Allied Signal, 86. *Res:* Molecular genetics; mechanisms of mutagenesis; genetic toxicology. *Mailing Add:* Biotechnol Res & Develop Corp 1815 N University St Peoria IL 61604

BREWER, ALLEN A, dentistry; deceased, see previous edition for last biography

BREWER, ARTHUR DAVID, b Evesham, Eng, Nov 14, 41; Can citizen; m 68; c 2. AGRICULTURAL CHEMISTRY. *Educ:* Cambridge Univ, BA, 63, MA, 67; Univ Strathclyde, Scotland, PhD(chem), 67. *Prof Exp:* Fel chem, Univ Ariz, 67-68; res chemist, 68-75, res scientist chem, 75-81, SR GROUP LEADER, UNIROYAL RES LAB, 75-81. *Mem:* Fel Royal Soc Chem; Can Inst Chemists; Brit Ornithologists Union; Sigma Xi. *Res:* Basic synthesis research in heterocyclic chemistry with view to discovery of new agricultural and pharmaceutical chemicals. *Mailing Add:* RR 1 Puslinch ON N0B 2J0 Can

BREWER, BEVERLY SPARKS, urban & extension entomology, for more information see previous edition

BREWER, CARL ROBERT, neurobiology; deceased, see previous edition for last biography

BREWER, CURTIS FRED, b Greenville, Tex, July 17, 44; m 89. NUCLEAR MAGNETIC RESONANCE SPECTROSCOPY. *Educ:* Univ Calif, Santa Barbara, BA, 66, PhD(chem), 71. *Prof Exp:* Trainee fel, NSF, 66-69; fel, 71-73, res assoc, 73-74, from asst prof to assoc prof, 74-84, PROF PHARMACOL, ALBERT EINSTEIN COL MED, 84- *Concurrent Pos:* Res Career Develop Award, NIH, 75-80; mem, Biomed Sci Study Sect, NIH, 84-88, chmn, 88. *Honors & Awards:* Meller Res Award, Albert Einstein Col Med, 74. *Mem:* NY Acad Sci; Am Chem Soc; Am Soc Biol Chemists; AAAS. *Res:* Biochemical and biophysical studies of the interactions of carbohydrate binding proteins (lectins) with cell surface carbohydrate molecules; ligand-macromolecule interactions and enzyme catalysis. *Mailing Add:* Dept Molecular Pharmacol 1300 Morris Park Ave Bronx NY 10461

BREWER, DANA ALICE, b Easton, Pa, Nov 6, 50. ATMOSPHERIC CHEMISTRY, PHYSICAL CHEMISTRY. *Educ:* Pa State Univ, BS, 72; Va Polytech Inst & State Univ, PhD(chem), 77. *Prof Exp:* Teaching asst chem, Va Polytech Inst & State Univ, 72-77; chemist, NASA Langley Res Ctr, 75, res assoc, Joint Inst Advan Flight Sci, George Washington Univ, 77-80, asst prof lectr atmospheric chem, 79-81; sr scientist, Systs & Appl Sci Corp, 81-84; res prof chem, Col William & Mary, NASA Langley Res Ctr, 85; res assoc, Vigyan Res Assocs, 86-87; mem tech staff, AT&T Bell Labs, 87; ENVIRON MGR, SPACE STA FREEDOM PROG OFF, NASA, 88- *Concurrent Pos:* Eng mgt, Space Sta Freedom Prog Off, NASA. *Mem:* Am Chem Soc. *Res:* Development of trace contaminant models for space station; atmospheric chemical reaction mechanisms; atmospheric hydrocarbon chemical models; environmental control and life support studies for space station. *Mailing Add:* 11704 Old Englis Dr Apt L Reston VA 22090-3583

BREWER, DOUGLAS G, b Toronto, Ont, Dec 22, 35; m 61; c 1. INORGANIC CHEMISTRY, ANALYTICAL CHEMISTRY. *Educ:* Univ Toronto, BA, 58, PhD(chem), 61. *Prof Exp:* From asst prof to assoc prof anal & inorg chem, 61-72, PROF CHEM & DEAN SCI, UNIV NB, 72- *Concurrent Pos:* Operating grants, Nat Res Coun Can, 61-71 & NB Res & Productivity Coun, 63-64. *Mem:* Chem Inst Can. *Res:* Stability constants of metal complexes as determined by differential potentiometry and related methods; infrared, optical rotary dispersion, ultraviolet-vision, nuclear magnetic resonance and x-ray studies of metal complexes. *Mailing Add:* Dept Chem Univ NB Col Hill-Box 4400 Fredericton NB E3B 5A3 Can

BREWER, FRANKLIN DOUGLAS, b Electric Mills, Miss, Sept 25, 38; m 66, 88; c 3. ENTOMOLOGY. *Educ:* Miss Col, BS, 60; Univ Southern Miss, MS, 63; Miss State Univ, PhD(entom), 70. *Prof Exp:* Instr biol, Miss Woman's Univ, 64-66; from instr microbiol to res assoc biochem, Miss State Univ, 66-71; res entomologist, Screwworm Res Lab, Agr Res Serv, USDA, 71-; AT TUXTLA GUTIERREZ, CHIAPAS, MEXICO. *Mem:* Sigma Xi; Entom Soc Am. *Res:* Insect nutrition, quality control and mass rearing techniques for producing a consistent yet economical number of competitive insects for biological control purposes. *Mailing Add:* Gast Rearing Lab PO Box 5367 Mississippi State TX 39762

BREWER, GARY DAVID, fish biology, for more information see previous edition

BREWER, GEORGE E(UGENE) F(RANCIS), b Vienna, Austria, Jan 23, 09; US citizen; wid; c 3. COATINGS TECHNOLOGY, PAINT SOLVENT EMISSION. *Educ:* State Col Vienna, AB, 28; Univ Vienna, MA, 30, PhD(chem), 33. *Prof Exp:* Asst lectr, Univ Vienna, 34-35; Tech mgr textile dyeing mill, S Wolf & Co, Erlach, Austria, 36-38; lectr textile chem, Inst l'Indust Textile, Brusselles, 39; prof chem, Rosary Col, 40-44; prof & head dept chem, Marygrove Col, 44-67; staff scientist paint technol, Ford Motor Co, 68-72; OWNER, GEORGE E F BREWER COATING CONSULT, 72- *Honors & Awards:* Midgley Medal & Doolittle Award, Am Chem Soc, 69; Mattiello Mem Lectr, Fedn Soc Coatings Technol, 73; Chem Pioneer Award, Am Inst Chemists, 78; Reginald Gower Mem Lectr, Inst Metal Finishing, Eng, 84. *Mem:* Am Chem Soc; fel Am Inst Chemists; Soc Paint Technol; Asn Anal Chemists (pres, 59); Chem Coaters Asn (pres, 76). *Res:* Low polluting, high quality and low cost industrial coatings; over 100 publications in technical and educational journals; 23 patents in electro deposition and other fields of coating. *Mailing Add:* Coating Consult 6135 Wing Lake Rd Birmingham MI 48010-6007

BREWER, GEORGE R, b New Albany, Ind, Sept 10, 22; m 46, 84; c 2. PHYSICS, ELECTRICAL ENGINEERING. *Educ:* Univ Louisville, BS, 43; Univ Mich, MS, 49 & 50, PhD(elec eng), 51. *Prof Exp:* Staff mem, Radiation Lab, Mass Inst Technol, 43-44; res engr, Naval Res Labs, 44-47; res engr res inst, Univ Mich, 47-51; mgr, Electron Device Physics Dept, Hughes Res Lab, 51-81. *Concurrent Pos:* Consult, electron & ion beam syst & microelectronics, 81-86. *Mem:* Fel Inst Elec & Electronics Engrs. *Res:* Electron and ion dynamics; electron gun and beam focusing; ion propulsion; microwave electron tubes; microelectronic processes and devices; ion implantation; electron beam lithography. *Mailing Add:* 217 Peter Pan Rd Carmel CA 93923-9746

BREWER, GLENN A, JR, b New Haven, Conn, Nov 3, 27; m 57; c 4. CHEMISTRY. *Educ:* Hobart Col, BS, 49; Univ Wis, MS, 52, PhD(biochem), 54. *Prof Exp:* Sr res biochemist, Com Solvents Corp, 54-56; sr res microbiologist, E R Squibb & Sons, Inc, 56-60, suprv anal res, 60-67, asst dir, 67-84, dir anal res, 84-90, EXEC DIR, WORLDWIDE ANALYTICAL SYSTS, BRISTOL-MYERS SQUIBB INST PHARMACEUT RES, 90- *Mem:* Am Chem Soc; Am Asn Pharmaceut Scientists; fel AAAS. *Res:* Pharmaceutical analysis. *Mailing Add:* Bristol-Myers Squibb Inst Pharmaceut Res New Brunswick NJ 08903

BREWER, GREGORY J, MEMBRANE BIOCHEMISTRY & BIOPHYSICS, CYTOLOGY. *Educ:* Calif Inst Technol, BS, 68; Univ Calif, San Diego, PhD(biol), 72. *Prof Exp:* ASSOC PROF MED MICROBIOL, SCH MED, SOUTHERN ILL UNIV, 80- *Concurrent Pos:* Dir, Res Imaging Facil. *Res:* Neuron cell culture; reconstitution of gap junctions in model membranes; Alzheimer disease; synaptogenesis; ganglioside function. *Mailing Add:* Sch Med Southern Ill Univ PO Box 19230 Springfield IL 62794

BREWER, H BRYAN, JR, b Casper, Wyo, Aug 17, 38; m 58; c 2. METABOLIC DISEASE, ATHEROSCLEROSIS. *Educ:* Johns Hopkins Univ, BA, 60; Stanford Univ, MD, 65. *Prof Exp:* HEAD SECT PEPTIDE CHEM, NAT HEART, LUNG & BLOOD INST, 70-, CHIEF MOLECULAR DIS BR, 76- *Concurrent Pos:* Mem coun atherosclerosis, Am Heart Asn, 79-81. *Mem:* Fedn Am Soc Biol Chemists; Am Soc Clin Res; Am Soc Clin Invest. *Res:* Molecular structure, function and metabolism of human plasma lipoproteins; the role of plasma lipoproteins in lipid transport in normal individuals and patients with disorders of lipid metabolism and atherosclerosis. *Mailing Add:* Molecular Dis Br Bldg 10 Rm 7N117 Nat Heart Lung & Blood Inst NIH 9000 Rockville Pike Bethesda MD 20892

BREWER, HOWARD EUGENE, b Indianola, Iowa, Apr 30, 10. PLANT PHYSIOLOGY. *Educ:* Simpson Col, AB, 31, BS, 32; Iowa State Col, PhD(plant ecol), 42. *Prof Exp:* Asst bot, Iowa State Col, 37-41; instr plant physiol & ecol, State Col Wash, 41-42; from asst prof to assoc prof bot, Ala Polytech Inst, 46-49, assoc botanist, Exp Sta, 47-49; from asst prof to prof bot, 49-74, coordr gen studies sci, 68-74, EMER PROF BOT, WASH STATE UNIV, 74- *Concurrent Pos:* Coop agent, Soil Conserv Serv, USDA, 38-42. *Mem:* Ecol Soc Am; Am Soc Plant Physiol; Scand Soc Plant Physiol. *Res:* Seed physiology; nitrogen nutrition of cereals; water stress relations of plants. *Mailing Add:* c/o W R Rayburn SE 410 Nebr St Pullman WA 99163

BREWER, JAMES W, b West Palm Beach, Fla, May 29, 42; m 63; c 5. ALGEBRA. *Educ:* Fla State Univ, AB, 64, PhD(math), 68. *Prof Exp:* Asst prof math, Va Polytech Inst, 68-70; from asst prof to prof math, Univ Kans, 70-87; PROF MATH, FLA ATLANTIC UNIV, 85- *Concurrent Pos:* Vis prof, Univ Stellenbosch, SAfrica, 77-78; vis prof, Univ Va, 78. *Mem:* Am Math Soc; SAfrica Math Soc. *Res:* Commutative algebra with special emphasis on polynomial and power series rings; algebraic control theory. *Mailing Add:* Dept Math Fla Atlantic Univ Boca Raton FL 33431

BREWER, JEROME, b Kansas City, Mo, Mar 31, 19; m 49; c 3. CHEMICAL ENGINEERING, THERMODYNAMICS. *Educ:* Iowa State Univ, BS, 40; Univ Kans, MS, 57, PhD(phase equilibria), 60. *Prof Exp:* Asst ord engr, Southwest Proving Ground, 41-42, sect head, 42-43; group leader spectrog, Pratt-Whitney Aircraft Corp, Mo, 43-44; jr engr instrumentation, Metall Lab, Univ Chicago, 44-45; asst physicist indust res, Midwest Res Inst, 45-49, assoc physicist, 49-54, sr chem engr, 54-56; sr scientist thermodyn correlations, Air Prod & Chem, Inc, 59-61; sr physicist, Midwest Res Inst, 61-65, prin physicist, 65-67, prin chem engr, 67-73; process engr, C W Nofsinger Co, 73-80, sr process engr, 80-82; CONSULT, 82- *Mem:* Am Inst Chem Engrs; Sigma Xi. *Res:* Design of continuous fractional crystallization processes; correlation of thermodynamic properties of materials; phase equilibra; process design; chemical kinetics; measurement of second viral coefficients. *Mailing Add:* 9948 Woodson Dr Overland Park KS 66207

BREWER, JESSE WAYNE, b Rives, Mo, Oct 10, 40; m 64; c 2. ENTOMOLOGY, PLANT PATHOLOGY. *Educ:* Cent Mich Univ, BS, 63, MA, 65; Purdue Univ, PhD(entom), 68. *Prof Exp:* Instr biol, Muskegon Community Col, 64-65; from asst prof to assoc prof entom, Colo State Univ, 68-79, prof & coordr entom progs, 79-87, emer prof zool & entom, 87; PROF & HEAD DEPT ENTOM, AUBURN UNIV, 87- *Concurrent Pos:* Vis scientist, Nat Acad Sci, Czech, 78 & 79. *Mem:* AAAS; Entom Soc Am; Can Entom Soc. *Res:* Forest entomology; pollination biology. *Mailing Add:* Auburn Univ 301 Funchess Hall Auburn AL 36849

BREWER, JOHN GILBERT, b Robinson, Ga, May 11, 37. ANALYTICAL CHEMISTRY. *Educ:* Univ Ga, BS, 58, MS, 63, PhD(chem), 66. *Prof Exp:* Anal develop chemist, Chemstrand Co, Monsanto Chem Co, Fla, 58-59; asst chem, Univ Ga, 60-66; asst prof, The Citadel, 66-68; assoc prof, 68-74, PROF CHEM, ARMSTRONG STATE COL, 74- *Mem:* Am Chem Soc; Sigma Xi. *Res:* Analysis of multicomponent mixtures through fitting of their excess functions via computer usage. *Mailing Add:* Dept Chem Armstrong State Col 11935 Abercorn St Savannah GA 31419

BREWER, JOHN MICHAEL, b Garden City, Kans, May 13, 38; m 65; c 2. BIOCHEMISTRY. *Educ:* Johns Hopkins Univ, BA, 60, PhD(biochem), 63. *Prof Exp:* Res assoc chem, Univ Ill, Urbana-Champaign, 63-66; from asst prof to assoc prof, 66-83, PROF BIOCHEM, UNIV GA, 83- *Mem:* Am Chem Soc; Am Soc Biol Chemists; Biophys Soc; Sigma Xi. *Res:* Protein chemistry and structure and the mechanism of enzyme action. *Mailing Add:* Dept Biochem Univ Ga Athens GA 30602

BREWER, KENNETH ALVIN, b Wichita, Kans, Jan 24, 38; m 58; c 3. TRANSPORTATION ENGINEERING. *Educ:* Kans State Univ, BSCE, 60, MS, 61, Tex A&M Univ, PhD(civil eng), 67. *Prof Exp:* Proj engr, Fuller Assocs Consults, 63-65; res asst transp eng, Tex Transp Inst, 66-67; from instr to asst prof, Tex A&M Univ, 67-69; assoc prof, 69-75, PROF CIVIL ENG, IOWA STATE UNIV, 75- *Concurrent Pos:* Res assoc, Tex Transp Inst, 67-68, asst res engr, 68-69. *Mem:* Inst Transp Engrs; Am Soc Civil Engrs; Nat Soc Prof Engrs; Am Soc Eng Educ; Sigma Xi. *Res:* Traffic flow theory; transportation planning; traffic safety; travel behavior; highway maintenance; highway design. *Mailing Add:* 4107 Ross Rd Ames IA 50010-3834

BREWER, LEO, b St Louis, Mo, June 13, 19; m 45; c 3. PHYSICAL CHEMISTRY. *Educ:* Calif Inst Technol, BS, 40; Univ Calif, PhD(chem), 43. *Prof Exp:* Res assoc, Manhattan Dist Proj, Radiation Lab, 43-46, from asst prof to prof, 46-89, EMER PROF, DEPT CHEM, UNIV CALIF, BERKELEY, 89- *Concurrent Pos:* Assoc, Lawrence Berkeley Lab, 47-60, prin investr, 60-, head, Inorg Mat Res Div, 61-75, assoc dir, 67-75; Guggenheim Mem fel, 50; secy comn high temperature, Int Union Pure & Appl Chem, 57-61 & 76-, assoc mem comn thermodyn & thermochem, 74-; fac lectr, Univ Calif, 66 & corn prod lectr, Pa State Univ, 70; assoc ed, J Chem Physics; mem rev comt for reactor chem div, Oak Ridge Nat Lab; mem, Calorimetry Conf; assoc ed, J Electrochem Soc, 76-84; mem comt data needs, Nat Res Coun, 76-78, mem, 75-79, chmn comt high temperature sci & technol, 79-81; ed, Atomic Energy Rev Spec Issue No 7, Atomic Energy Agency, Vienna, 77-80; mem, Int Coun Alloys Phase Diagrams, Am Soc Metals & Nat Bur Standards, 78-, Basic Energy Sci Lab Progs Panel, Dept Energy, 78-82, Coun Mat Sci, 83-85, Res Assistance Task Force, 88 & Steering Comt Workshop Chem Processes & Prod Severe Reactor Accidents, Nuclear Regulatory Comn, 88; lectr, Oak Ridge Nat Lab, 79, Frontiers Chem Res, Tex A&M Univ, 81, res scholar lectr, Drew Univ, 83. *Honors & Awards:* Baekeland Award, 53; E O Lawrence Award, 61; Robert S Williams Lectr, Mass Inst Technol; Henry Werner Lectr, Univ Kans, 63; O M Smith Lectr, Okla State Univ; G N Lewis Lectr, Univ Calif, 64; Palladium Medal, Electrochem Soc, 71, Henry B Linford Award, 88; W D Harkins Lectr, Univ Chicago, 74; William Hume-Rothery Award, Metall Soc, 83, Extractive Metall Sci Award, 91; Louis Jacob Bircher Lectr, Vanderbilt Univ, 86; Eyring Lectr, Ariz State Univ, 89. *Mem:* Nat Acad Sci; fel Am Phys Soc; Am Chem Soc; Royal Soc Chem; Metall Soc; Am Optical Soc; Electrochem Soc; Am Acad Arts & Sci; fel AAAS; fel Am Soc Metals Int. *Res:* High temperature chemistry and thermodynamics; theory of bonding in metallic solutions. *Mailing Add:* Dept Chem Univ Calif Berkeley CA 94720

BREWER, LEROY EARL, JR, b Hagerstown, Md, June 1, 36; m 56; c 1. THERMAL PHYSICS, PHENOMENOLOGY. *Educ:* Univ Fla, BS, 60; Univ Tenn, MS, 65; Univ Brussels, PhD(appl sci), 75. *Prof Exp:* Res engr, Arnold Eng Develop Ctr, Aro, Inc, 60-66; physicist, Space Sci Lab, Gen Elec Co, Pa, 66-69; prin investr, Calspan Inc, 69-77, supvr, Physics Sect, 77-80, mgr, Advan Test Diag Br, Arnold Eng Develop Ctr, 81-85; SR SYST ENGR & PHENOMENOLOGIST, GEN ELEC CO, VALLEY FORGE, PA, 85- *Concurrent Pos:* Adj fac mem math, Motlow Community Col, 76-80. *Mem:* Assoc fel Am Inst Aeronaut & Astronaut; Am Phys Soc. *Res:* Fundamental and applied research in fields of emissions, combustion, infrared physics, atmospheric physics, phenomenology of plumes, backgrounds and space electric propulsion. *Mailing Add:* 317D Howells Rd Malvern PA 19355

BREWER, MAX C(LIFTON), b Blackfalds, Alta, May 7, 24; US citizen; m 54; c 5. GEOLOGICAL ENGINEERING, GEOPHYSICS. *Educ:* Wash Univ, BS, 50. *Hon Degrees:* DSc, Univ Alaska, 65. *Prof Exp:* Geophysicist permafrost, US Geol Surv, 50-71, dir res, Naval Arctic Res Lab, 56-71; comnr, Alaska Dept Environ Conserv, 71-74; Arctic consult, 75-76; CHIEF OPERS, NAT PETROL RESERVE IN ALASKA, 77- *Honors & Awards:* USN Distinguished Pub Serv Award, 62; Edward C Sweeney Medal, Explorer's Club, 70. *Mem:* Fel Arctic Inst NAm; Soc Am Mil Engrs; Am Geophys Union; Brit Glaciol Soc. *Res:* Permafrost; ocean and fresh water ice and their effect on engineering structures through measurements of thermal gradients and physical properties. *Mailing Add:* 3819 Locarno Dr Anchorage AK 99508

BREWER, NATHAN RONALD, b Albany, NY, June 28, 04; m 36; c 3. VETERINARY PHYSIOLOGY. *Educ:* Mich State Univ, BS, 30, DVM, 37; Univ Chicago, PhD(physiol), 36. *Hon Degrees:* DSc, Chicago Col Osteop Med, 77. *Prof Exp:* Vet practr, 37-45; assoc prof & dir, Univ Chicago, 45-69, dir animal quarters, 69-, emer prof physiol, 69-; RETIRED. *Concurrent Pos:* Consult, Chicago Med Sch, Chicago Col Osteop Med, Mercy Hosp & Ill Inst Technol, Chicago, 69-; mem bd, Inst Lab Animal Resources, Nat Res Coun, 53-60; ed, Proc Asn Lab Animal Sci, 50-62, emer ed, Lab Animal Care, 63- *Honors & Awards:* Griffin Award, Asn Lab Animal Sci. *Mem:* Hon mem Am Soc Vet Cardiol; Nat Soc Med Res; Am Asn Lab Animal Sci (pres, 50-55); Am Physiol Soc; Am Vet Med Asn; Am Col Lab Animal Med (pres, 57-59); Wildlife Dis Asn; Am Soc Primatologists. *Res:* Laboratory animal medicine. *Mailing Add:* 5526 Blackstone Ave Chicago IL 60637-1834

BREWER, RICHARD (DEAN), b Murphysboro, Ill, June 17, 33; div. ECOLOGY, ORNITHOLOGY. *Educ:* Southern Ill Univ, BA, 55; Univ Ill, MS, 57, PhD(zool), 59. *Prof Exp:* Res asst zool, Univ Ill, 55-56; from instr to assoc prof, 59-71, PROF BIOL, WESTERN MICH UNIV, 71- *Concurrent Pos:* Asst dir, C C Adams Ctr Ecol Studies, 60-63, dir, 63-68; ed, Jack-Pine Warbler, 76-79. *Honors & Awards:* Edwards Prize, Wilson Ornith Soc. *Mem:* Fel AAAS; Ecol Soc Am; Am Ornith Union; Wilson Ornith Soc. *Res:* Competition, speciation and habitat selection in birds; organization and evolution in plant, animal communities; forest succession. *Mailing Add:* Dept Biol Western Mich Univ Kalamazoo MI 49001

BREWER, RICHARD GEORGE, b Los Angeles, Calif, Dec 8, 28; m 54; c 3. QUANTUM ELECTRONICS, LASER SPECTROSCOPY. *Educ:* Calif Inst Technol, BS, 51; Univ Calif, Berkeley, PhD, 58. *Prof Exp:* Researcher, Aerojet-Gen Corp, 51-53; instr, Harvard Univ, 58-60; asst prof, Univ Calif, Los Angeles, 60-63; res physicist, 63-78, IBM FEL, RES LAB, IBM CORP, 73- *Concurrent Pos:* Consult, Space Technol Labs, Inc, 62-63; vis prof physics, Mass Inst Technol, Cambridge, 68-69; mem comt atomic & molecular physics, Nat Acad Sci-Nat Res Coun, 74-77; Japan Soc Prom Sci vis prof physics & appl physics, Univ Tokyo, 75; vis prof, Univ Calif Santa Cruz, 76; consult prof appl physics, Stanford Univ, 77-; adj prof, Nat Inst Optics, Florence, Italy, 77-; assoc ed, Optics Letters, 77-80; Nat Res Coun Rev Panal, Nat Bureau Standards, 81-84; rev comt, San Francisco Laser Ctr, NSF, 80-; assoc ed, J Optical Soc Am, 80-83; mem bd physics & astron, Nat Acad Sci, Nat Res Coun, 83-86. *Honors & Awards:* A A Michelson Gold Medal, Franklin Inst, 79. *Mem:* Nat Acad Sci; fel Am Phys Soc; fel Optical Soc Am. *Res:* Laser spectroscopy; nonlinear optic effects and nonlinear spectroscopy of atoms, including coherent optical transients and collision phenomena; ultrahigh resolution solid state laser spectroscopy, chaos, ion trapping. *Mailing Add:* IBM Almaden Res Ctr 650 Harry Rd San Jose CA 95193

BREWER, ROBERT FRANKLIN, b Woodbury, NJ, May 25, 27; m 47; c 2. HORTICULTURE. *Educ:* Rutgers Univ, BSc, 50, PhD(hort), 53. *Prof Exp:* Soil chemist, Dept Soils & Plant Nutrit, Citrus Exp Sta, 53-70, HORTICULTURIST, PLANT SCI DEPT, UNIV CALIF, RIVERSIDE, 70- *Mem:* Am Soc Hort Sci; Int Soc Citricult; Air Pollution Control Asn. *Res:* Inorganic plant nutrition; effects of fluorine and ozone on plant growth; cold tolerance; frost protection; environment modification. *Mailing Add:* San Joaquin Valley Res Ctr 9240 S Riverbend Ave Parlier CA 93648

BREWER, ROBERT HYDE, b Richmond, Va, Dec 24, 31; m 60; c 1. ANIMAL ECOLOGY, INVERTEBRATE ZOOLOGY. *Educ:* Hanover Col, AB, 55; Univ Chicago, PhD(zool), 63. *Prof Exp:* Asst prof biol, Ill Col, 63-65; res fel & lectr entom, Waite Agr Res Inst, Univ Adelaide, 65-68; from asst prof to assoc prof, 68-81, PROF BIOL, TRINITY COL, CONN, 82- *Mem:* Fel AAAS; Ecol Soc Am; Am Soc Zool; Am Inst Biol Sci; Sigma Xi; Am Soc Naturalists. *Res:* Physiological and population ecology of aquatic invertebrates. *Mailing Add:* Dept Biol Trinity Col Hartford CT 06106

BREWER, ROBERT NELSON, b Philcampbell, Ala, Feb 24, 34; m 53; c 1. POULTRY PATHOLOGY. *Educ:* Auburn Univ, BS, 55, MS, 60; Univ Ga, PhD(poultry sci), 68. *Prof Exp:* Asst county agr agent, Exten Serv, Auburn Univ, 55-58; poultry specialist, Pillsbury Co, 60-65; asst prof poultry parasitol & path, 68-74, ASSOC PROF POULTRY SCI, AUBURN UNIV, 74- *Mem:* Poultry Sci Asn; Am Inst Biol Sci. *Res:* Poultry disease research, including immunity studies on roundworms, coccidiosis and epidemiological and etiological studies on Marek's disease; environment-disease interrelationships. *Mailing Add:* Dept Poultry Sci Auburn Univ Auburn AL 36830

BREWER, STEPHEN WILEY, JR, b Spartanburg, SC, Jan 18, 41; m 82. ANALYTICAL CHEMISTRY. *Educ:* Univ Fla, BSCh, 62; Univ Wis-Madison, PhD(anal chem), 69. *Prof Exp:* Asst prof, 69-73, assoc prof, 73-79, PROF CHEM, EASTERN MICH UNIV, 79- *Concurrent Pos:* Vis scholar, Univ Mich. *Mem:* Soc Appl Spectros; Spectros Soc Can; Sigma Xi; Am Chem Soc. *Res:* Emission spectroscopy; electron probe microanalysis. *Mailing Add:* Dept Chem Eastern Mich Univ Ypsilanti MI 48197

BREWER, WILLIAM AUGUSTUS, b Oakland, Calif, May 27, 30. REMOTE SENSING, HYDROGEOLOGY OF ARID SITES. *Educ:* Univ Calif, Berkeley, BA, 54, MA, 55, PhD(eng), 65. *Prof Exp:* Sr geologist, Anaconda Co, 55-60; assoc eng, Univ Calif, Berkeley, 60-63; chief engr, Cent Intel Agency, 63-68 & Fed Systs Div, Int Bus Mach, 69-70; consult overseas mining, 70-73; exec dir, Off Gov, Wash, 73-75; prof environ eng, Univ Wash, 75-79; CONSULT NUCLEAR WASTE MGT, 79- *Concurrent Pos:* Consult, numerous US & foreign govt agencies, 73-; sr hydrogeologist, State Wash, 83-89. *Res:* Geotechnical analysis of nuclear-toxic waste disposal sites; regional-national energy policy; environmental legislation; mapping; geophysics. *Mailing Add:* 5131 Fir Tree Rd SE Olympia WA 98501

BREWINGTON, PERCY, JR, b Benton, Pa, Jan 31, 30; m 51; c 3. CIVIL ENGINEERING. *Educ:* Drexel Univ, BS, 54. *Prof Exp:* Staff engr, Mobile Dist Off, US Army CEngr, 58-66, asst chief spec projs br, 66-67, chief proj mgt br, Huntsville Div, 67-73, chief eng div, 73; asst mgr construct eng, AEC/ERDA, 73-76, DEPT MGR, OAK RIDGE OPERS OFF, DEPT ENERGY, 76- *Mem:* Am Soc Civil Engrs; Nat Soc Prof Engrs; Res Officers Asn. *Mailing Add:* 1074 W Outer Dr Oak Ridge TN 37830

BREWSTER, JAMES HENRY, b Ft Collins, Colo, Aug 21, 22; m 54; c 3. ORGANIC CHEMISTRY, STEREO CHEMISTRY. *Educ:* Cornell Univ, BA, 42; Univ Ill, PhD(chem), 48. *Prof Exp:* Chemist, Atlantic Refining Co, 42-43; asst, Univ Ill, 46-47; Fels Fund fel, Univ Chicago, 48-49; from instr to assoc prof, 49-60, PROF CHEM, PURDUE UNIV, 60- *Mem:* AAAS; Am Chem Soc; Royal Chem Soc; Am Asn Univ Professors; Sigma Xi. *Res:* Stereo chemistry of displacement reactions and reductions; synthesis and optical activity of highly symmetrical compounds; stereochemical notation. *Mailing Add:* Dept Chem Purdue Univ West Lafayette IN 47907

BREWSTER, MARCUS QUINN, b Salt Lake City, Utah, Aug 20, 55; m 79; c 3. HEAT TRANSFER, COMBUSTION. *Educ:* Univ Utah, BS, 78, ME, 79; Univ Calif, Berkeley, PhD(mech eng), 81. *Prof Exp:* Assoc mech eng, Kyoto Univ, 81-82; asst prof, Univ Utah, 82-86; asst prof, 86-88, ASSOC PROF MECH ENG, UNIV ILL, URBANA, 88- *Concurrent Pos:* Presidential young investr award, NSF, 84; young investr award, Off Naval Res, 88. *Mem:* Am Soc Mech Eng; Am Inst Aeronaut & Astronaut. *Res:* Radiative heat transfer; solid propellant combustion; laser material processing; author of one book. *Mailing Add:* 49 Glenn Dr White Heath IL 61884

BREWSTER, MARJORIE ANN, b Conway, Ark, Mar 7, 40. CLINICAL BIOCHEMISTRY, ENVIRONMENTAL EXPOSURE. *Educ:* Univ Ark, Little Rock, BS, 64, MS, 66, PhD(biochem), 71; Am Bd Clin Chem, dipl, 74. *Prof Exp:* Clin biochemist, Clin Lab, Univ Ark, 66-67 & 71-77, asst prof path, Med Ctr, 69-79, asst prof biochem, 73-79; dir chem lab, 76-87, DIR METAB LAB, ARK CHILDREN'S HOSP, 76- , PROF PEDIAT, & PATH, 88-; DIR, ARK REPRODUCTIVE HEALTH MONITORING SYST, 82- *Concurrent Pos:* Past pres, Nat Registry Clin Chemists, mem, govt adv comn hazardous waste; mem, Chem Safety Div, Am Chem Soc. *Mem:* Am Asn Clin Chemists; Nat Acad Clin Biochemists (pres, 81-82); Asn Clin Sci; Am Chem Soc; Soc Toxicol. *Res:* Birth defects; environmental, genetic, and nutritional impacts upon health; reproductive monitoring; human exposure detection. *Mailing Add:* Metab Lab Ark Childrens Hosp Little Rock AR 72202

BREY, R(OBERT) N(EWTON), b Philadelphia, Pa, Feb 2, 20; m 48; c 3. CHEMICAL & NUCLEAR ENGINEERING. *Educ:* Univ Pa, BS, 43; Pa State Univ, cert, 43. *Prof Exp:* Head, Nuclear Syst Group, Leeds & Northrup Co, 46-60; mgr opers admin, Ballistic Missile Early Warning Syst Proj, Radio Corp Am, 60-63; spec appln adv space proj dept, Gen Elec Co, 63-69, mgr proj eng, Power Syst Oper Prog & Syst Anal, Info Syst Group, 69-71; PROJ MGR, UNITED ENGRS & CONSTRUCTORS INC, 71- *Mem:* Am Nuclear Soc; Am Inst Chem Engrs; Instrument Soc Am; Franklin Inst. *Res:* Design engineering and construction of nuclear and fossil fueled central station electric generating plants for utilities. *Mailing Add:* 107 W Moreland Ave Philadelphia PA 19118

BREY, WALLACE SIEGFRIED, JR, b Schwenksville, Pa, June 6, 22; m 55; c 2. PHYSICAL CHEMISTRY, MAGNETIC RESONANCE. *Educ:* Ursinus Col, BS, 42; Univ Pa, MS, 46, PhD(chem), 48. *Prof Exp:* Asst chemist, Warner Co, 42-44; res chemist, Publicker Indust, 44; asst prof chem, De Pauw Univ, 48-49; from asst prof to assoc prof, St Joseph's Col, Pa, 49-52; from asst prof to assoc prof, 52-64, PROF CHEM, UNIV FLA, 64- *Concurrent Pos:* Ed, J Magnetic Resonance, 69-; prog dir, NSF, 76-77. *Honors & Awards:* Fla Award, Am Chem Soc. *Mem:* AAAS; Am Chem Soc; Am Phys Soc; Royal Soc Chem; Soc Magnetic Resonance Med; Soc Magnetic Resonance Imaging; Int Soc Magnetic Resonance. *Res:* Magnetic resonance; heterogeneous catalysis and adsorption; molecular interactions in liquids; spectroscopy of biological molecules. *Mailing Add:* 800 NW 37th Dr Gainesville FL 32605

BREYER, ARTHUR CHARLES, b Brooklyn, NY, June 13, 25; m 48; c 3. PHYSICAL CHEMISTRY, INORGANIC CHEMISTRY. *Educ:* NY Univ, BA, 48; Columbia Univ, MA, 50; Rutgers Univ, PhD(chem), 58. *Prof Exp:* Instr chem, Columbia Univ, 48-49; from instr to asst prof, Upsala Col, 49-58; assoc prof, Ohio Wesleyan Univ, 59-63; prof, Harvey Mudd Col, 63-64; PROF CHEM & CHMN DEPT CHEM & PHYSICS, BEAVER COL, 64- *Concurrent Pos:* Chem study proj lectr, Gatlinburg Ion-Exchange Res Conf, 59; dir, NSF Undergrad Res Partic Prog, 60-64, dir, Inst High Sch Chem Teachers, 62-; consult, NSF, 62-, McGraw-Hill Book Co, Inc; Charles E Merrill Books, Inc & Wadsworth Publ Co. *Mem:* AAAS; Am Chem Soc; Am Sci Affil; Franklin Inst. *Res:* Chromatography and electrophoresis of plant pigments, dyes and surfactants; ion exchange catalysis; metal ion complexation. *Mailing Add:* Dept Chem Beaver Col Glenside PA 19038

BREYER, NORMAN NATHAN, b Detroit, Mich, June 21, 21; m 52; c 3. METALLURGICAL ENGINEERING. *Educ:* Mich Tech, BS, 43; Univ Mich, MS, 48; Ill Inst Technol, PhD(metall), 63. *Prof Exp:* Aeronaut res scientist, Nat Adv Comt Aeronaut, 48; chief armor sect, Detroit Tank Arsenal, 48-52; dir res cast steels & irons, Nat Roll & Foundry, 52-54; metallurgist-in-charge armor, Continental Foundry & Mach Div, Blaw-Knox Co, 55-57; mgr tech projs, La Salle Steel Co, 57-64; assoc prof metall, 64-69, prof & chmn dept, 69-84, PROF METALL & MAT ENGR, ILL INST TECHNOL, 84- *Mem:* Am Soc Metals; Am Inst Mining, Metall & Petrol Engrs. *Res:* High strength, strain aging and cold and warm deformation of steels; leaded steels; failure analysis. *Mailing Add:* Dept Metall Ill Inst Technol Ten W 33rd St Chicago IL 60616

BREYERE, EDWARD JOSEPH, b Washington, DC, Apr 25, 27; c 1. IMMUNOGENETICS. *Educ:* Univ Md, BS, 51, MS, 54, PhD(zool), 57. *Prof Exp:* Asst, Univ Md, 54-55; assoc prof, 61-67, PROF BIOL, AM UNIV, 67- *Concurrent Pos:* Res fel, Nat Cancer Inst, 57-61; dir, Immunogenetics Lab, Sibley Mem Hosp, 61-79. *Mem:* AAAS; Am Asn Cancer Res; Transplantation Soc. *Res:* Transplantation immunity; maternal-fetal relationships; immunology of oncogenic viruses. *Mailing Add:* 3355 University Blvd W Apt 101 Kensington MD 20895-1843

BREZENOFF, HENRY EVANS, b New York, NY, July 9, 40; m 64; c 2. NEUROPHARMACOLOGY. *Educ:* Columbia Univ, BS, 62; NJ Col Med & Dent, PhD(pharmacol), 68. *Prof Exp:* From instr to assoc prof, 69-80, asst dean, 78-80, PROF PHARMACOL, COL MED & DENT NJ, 80-, ASSOC DEAN, GRAD SCH BIOMED SCI, 81- *Concurrent Pos:* Fel pharmacol, Sch Med, Univ Calif, Los Angeles, 68-69. *Mem:* Am Soc Pharmacol & Exp Therapeut; Int Soc Hypertension; AAAS. *Res:* Central control of blood pressure; hypertension; brain acetylcholine. *Mailing Add:* Dept Pharmacol Univ Med & Dent NJ 185 S Orange Ave Newark NJ 07103

BREZINSKI, DARLENE RITA, b Chicago, Ill, Oct 9, 41. PAINT & COATINGS, LEAD POISONING. *Educ:* Mundelein Col, BS, 64; Iowa State Univ, MS, 67, PhD(chem), 69; Univ Chicago, cert advan mgt, 76. *Prof Exp:* Chmn & asst prof, Chem Dept, Mundelein Col, 69-73; res chem anal serv, De Soto Inc, 73-75, tech mgr, 75-80, dir anal serv, 80-90; SR CONSULT CHEM COATINGS, CONSOL RES INC, 90- *Concurrent Pos:* Ed, J Coatings Technol, 75- & Fed Ser Coatings Technol, 83-; consult lead poisoning, 75- *Honors & Awards:* George Baugh Heckel Award, Fed Socs Coatings Technol, 83. *Mem:* Am Soc Testing & Mat; Fed Socs Coatings Technol; Nat Paint & Coatings Asn; Inst Elec & Electronics Engrs. *Res:* Analytical laboratory efficiency and technical management. *Mailing Add:* 200 E Evergreen Mt Prospect IL 60056

BREZNAK, JOHN ALLEN, b Passaic, NJ, Aug 28, 44; m 68; c 2. MICROBIAL SYMBIOSES, MICROBIAL ECOLOGY. *Educ:* Rutgers Univ, BS, 66; Univ Mass, PhD(microbiol), 71. *Prof Exp:* NIH fel microbiol, dept bacteriol, Univ Wis, Madison, 71-73; from asst prof to assoc prof, 73-84, PROF MICROBIOL, MICH STATE UNIV, 84- *Concurrent Pos:* Prin investr, NSF Funded Res Proj, 75-; co-dir microbiol course, Marine Biol Lab, Woods Hole, Mass, 90- *Honors & Awards:* Sr US Scientist Award, Alexander von Humboldt Found, 85. *Mem:* Am Soc Microbiol; Soc Gen Microbiol; Entom Soc Am; AAAS; Soc Exp Biol; Sigma Xi. *Res:* Biochemical bases for symbiosis between termites and their intestinal microbiota; anaerobic degradation of plant polysaccharides by microorganisms. *Mailing Add:* Dept Microbiol Mich State Univ East Lansing MI 48824-1101

BREZNER, JEROME, b New York, NY, July 18, 31; m 54; c 2. ENTOMOLOGY, BIOCHEMISTRY. *Educ:* Univ Rochester, AB, 52; Univ Mo, AM, 56, PhD(entom), 59. *Prof Exp:* Asst prof biol, Elmira Col, 59-61; res assoc, 61-62, from asst prof to assoc prof, 62-68, PROF ENTOM, STATE UNIV NY COL ENVIRON SCI & FORESTRY, 68- *Concurrent Pos:* Am Physiol Soc fel biol, Dartmouth Med Sch, 60; NIH res grant, 62-65; US Forest Serv grant, 66-69. *Mem:* AAAS; NY Acad Sci; Entom Soc Am; Soc Environ Toxicol & Chem. *Res:* Insect enzyme proteins, the species specific nature of these molecules; insect toxicology. *Mailing Add:* Dept Biol State Univ NY Col Environ Sci & Forestry Syracuse NY 13210-2788

BREZONIK, PATRICK LEE, b Sheboygan, Wis, July 17, 41; m 65; c 2. WATER CHEMISTRY, LIMNOLOGY. *Educ:* Marquette Univ, BS, 63; Univ Wis, MS, 65, PhD(water chem), 68. *Prof Exp:* Asst prof water chem & environ eng, Univ Fla, 66-70, assoc prof water chem, 70-76, prof environ sci, 76-81; PROF ENVIRON ENG, UNIV MINN, 81-, DIR WATER RESOURCES RES CTR, 85- *Concurrent Pos:* Res grants, Fed Water Pollution Control Admin, 68-71; Off Water Resources Res, 69-71, 73-75 & 78-80, Environ Protection Agency, 74-75 & 77- & NSF, 79-81; NSF sci fac fel, Swiss Fed Inst Technol, Zurich, 71-72; chmn, Panel on Nitrates in the Environ, Nat Res Coun, 75-78; vis prof, Fed Inst for Water Treatment, Zurich, 80. *Mem:* AAAS; Am Chem Soc; Am Soc Limnol & Oceanog; Asn Environ Eng Prof. *Res:* Eutrophication of lakes; nitrogen dynamics in natural waters; nutrient chemistry; acid rain; trace metals in natural waters; organic matter in water. *Mailing Add:* Dept Environ Eng & Sci 308 Black Hall Univ Fla Gainesville FL 32601

BRIAN, P(IERRE) L(EONCE) THIBAUT, b New Orleans, La, July 8, 30; m 52; c 3. CHEMICAL ENGINEERING. *Educ:* La State Univ, BS, 51; Mass Inst Technol, ScD, 56. *Prof Exp:* From asst prof to prof chem eng, Mass Inst Technol, 55-72; VPRES ENG, AIR PROD & CHEM, INC, 72-, MEM BD DIRS, 73- *Concurrent Pos:* Overseas fel, Churchill Col, Cambridge Univ, 69-70. *Honors & Awards:* Prof Progress Award, Am Inst Chem Engrs, 73, Robert L Jacks Mem Award, 89. *Mem:* Nat Acad Eng; Am Chem Soc; Am Inst Chem Engrs. *Res:* Interphase mass transfer; chemical kinetics; heat transfer; numerical mathematics. *Mailing Add:* Air Prod & Chem Inc 7201 Hamilton Blvd Allentown PA 18195

BRIAND, FREDERIC JEAN-PAUL, b Paris, France, Nov 1, 49. ECOLOGY. *Educ:* Univ Paris, BSc, 70; Univ Calif, Irvine, MS, 72, PhD(ecol), 74. *Prof Exp:* Asst prof, 74-80, ASSOC PROF ECOL, DEPT BIOL, UNIV OTTAWA, 80- *Mem:* Ecol Soc Am; Am Soc Limnol Oceanog; Soc Int Limnol; Int Asn Ecol. *Res:* Structure and stability of ecological food webs; biological control of toxic algae. *Mailing Add:* Int Union Conserv Nature & Natural Resources Hq Ave de Mont Blanc Gland CH 1196 Switzerland

BRIANT, CLYDE LEONARD, b Texarkana, Ark, May 31, 48; m 77; c 3. MATERIALS SCIENCE. *Educ:* Hendrix Col, BA, 71; Columbia Univ, BS, 71, DEngSc, 74. *Prof Exp:* Fel metall, Univ Pa, 74-76; STAFF METALLURGIST, GEN ELEC CORP, 76- *Concurrent Pos:* Overseas fel, Churchill Col, Cambridge, Eng, 87-88. *Honors & Awards:* Alfred Noble Prize, Five Eng Soc, 81; Robert Lansing Hardy Gold Medal, Am Inst Mining Metall & Petrol Engrs, 78; R W Raymond Award, Am Inst Mining Metall & Petrol Engrs, 80. *Mem:* Am Inst Mining Metall & Petrol Engrs; Am Soc Metals. *Res:* Grain boundary chemistry; hydrogen embrittlement of metals; nucleation theory; chemical boarding in alloys; corrosion and stress corrosion checking; thermomechanical processing of metals. *Mailing Add:* Four Marvin Dr Charlton NY 12019

BRICE, DAVID KENNETH, b Sulphur Springs, Tex, Apr 15, 33; div. SOLID STATE PHYSICS. *Educ:* ETex State Col, BS, 54; Univ Kans, MS, 56, PhD(physics), 63. *Prof Exp:* DISTINGUISHED MEM TECH STAFF PHYSICS, SANDIA CORP, 63- *Concurrent Pos:* Assoc prof physics, NMex Inst Mining & Technol, 67-73. *Mem:* Am Phys Soc; AAAS. *Res:* Ion implantation; radiation damage; particle-solid interactions and atomic transport in solids. *Mailing Add:* Sandia Labs Orgn 1111 PO Box 5800 Albuquerque NM 87185-5800

BRICE, DONAT B(ENNES), b Alton, Ill, Nov 22, 20; m 48; c 2. CHEMICAL ENGINEERING. *Educ:* Univ Ark, BS, 42; Univ Mo, MS, 47; Ohio State Univ, PhD(chem eng), 50. *Prof Exp:* Instr chem, Mo Sch Mines, 46-47; instr chem eng, Kans State Col, 47-48 & Ohio State Univ, 48-49; chem engr internal ballistics, Naval Ord Test Sta, 50-52; proj leader chem process develop, Food Mach & Chem Corp, 52-56; sr chem engr, Flour Corp, Calif, 56-60; STAFF CHEM ENGR, CALIF DEPT WATER RESOURCES, 60- *Concurrent Pos:* Sr officer, Int Atomic Energy Agency, Vienna, 66-67. *Mem:* Am Inst Chem Engrs; Am Chem Soc. *Res:* Saline water purification; heat transfer; thermodynamics; metallurgy; waste water reclamation and reuse. *Mailing Add:* 3831 Garden Hwy Sacramento CA 95834

BRICE, JAMES COBLE, b Union City, Tenn, Dec 21, 20. GEOMORPHOLOGY. *Educ:* Univ Ala, BS, 42; Univ Calif, PhD(geol), 50. *Prof Exp:* From asst prof to prof geol, Wash Univ, 50-76; HYDROLOGIST, US GEOL SURV, 76- *Mem:* Geol Soc Am; Am Soc Civil Engrs. *Res:* Fluvial geomorphology. *Mailing Add:* US Geol Surv WRD MS-470 1277 Woodland Ave Menlo Park CA 92105

BRICE, LUTHER KENNEDY, b Spartanburg, SC, Jan 29, 28. PHYSICAL CHEMISTRY. *Educ:* Harvard Univ, BA, 49; Dartmouth Col, MA, 51; Duke Univ, PhD(chem), 55. *Prof Exp:* From asst prof to assoc prof, 54-65, PROF CHEM, VA POLYTECH INST & STATE UNIV, 65- *Mem:* Am Chem Soc. *Res:* Chemical education. *Mailing Add:* 1401 17th St NW No 606 Washington DC 20036

BRICK, IRVING B, b Oakland, Calif, Apr 24, 14. INTERNAL MEDICINE, GASTROENTEROLOGY. *Educ:* George Washington Univ, AB, 37, MD, 41; Am Bd Internal Med, dipl, 49; Am Bd Gastroenterol, dipl, 51. *Prof Exp:* From instr to assoc prof, 47-61, actg chmn dept, 68-69 & 70-72, PROF MED, SCH MED, GEORGETOWN UNIV, 61- *Concurrent Pos:* Sr med consult, Rehab & Vet Affairs Comn, Am Legion, 48-84; consult, Clin Ctr, NIH, 57-; actg dir, Georgetown Univ Med Div, Washington, DC Gen Hosp, 64-65. *Mem:* Am Col Physicians; Am Gastroenterol Asn; Am Asn Study Liver Dis; Am Soc Gastrointestinal Endoscopy; Am Fedn Clin Res. *Res:* Radiation effect on the gastrointestinal tract; liver disease; gastrointestinal endoscopy. *Mailing Add:* One Washington Sq Village 10R New York NY 10012

BRICK, ROBERT WAYNE, b Dallas, Tex, May 24, 39; m 68; c 2. ANIMAL PHYSIOLOGY. *Educ:* Tex Technol Col, BS, 62; Univ Hawaii, MS, 70, PhD(zool), 75. *Prof Exp:* Asst prof wildlife & fisheries sci, Tex A&M Univ, 75-80; WITH AQUABIOTICS INC, 80- *Mem:* Am Soc Zoologists; Am Fisheries Soc; Sigma Xi; World Maricult Soc; Am Inst Biol Sci. *Res:* Physiology and culture of crustaceans, especially freshwater shrimps of the genus Macrobrachium and saltwater shrimps of the genus Penaeus. *Mailing Add:* c/o H I M B PO Box 1346 Kaneohe HI 96744

BRICKBAUER, ELWOOD ARTHUR, b Elkhart Lake, Wis. AGRONOMY. *Educ:* Univ Wis, BS, 43, MS, 61. *Prof Exp:* Teacher, Pub Schs, 43-44; mgr, Peck Seed Farms, 44-46; county agent, Univ Wis-Madison, 46-56, exten agronomist, 57-78, from asst prof to assoc prof, 57-70, prof agron, 70- *Mem:* Am Soc Agron. *Res:* Corn, soybeans, small grains and seed certification. *Mailing Add:* 1579 Venture Out Mesa AZ 85205

BRICKER, CLARK EUGENE, b Shrewsbury, Pa, June 17, 18; m 42; c 3. CHEMISTRY. *Educ:* Gettysburg Col, BA, 39; Haverford Col, MS, 40; Princeton Univ, PhD(anal chem), 44. *Hon Degrees:* DSc, Pikeville Col, 70. *Prof Exp:* Res chemist, Heyden Chem Corp, NJ, 43-46; asst prof chem, Johns Hopkins Univ, 46-48; from asst prof to prof, Princeton Univ, 48-61; dean, Col Wooster, 61-63; PROF CHEM, UNIV KANS, 63- *Mem:* AAAS; Am Chem Soc. *Res:* Physicochemistry methods of analysis; electrochemistry; spectrophotometry; photochemistry. *Mailing Add:* 1812 W 21st Terr Lawrence KS 66046

BRICKER, JEROME GOUGH, b Buffalo, NY, Jan 20, 28; m; c 2. MEDICAL PHYSIOLOGY, FEDERAL HEALTH CARE MANAGEMENT. *Educ:* Univ Akron, BSc, 49; Ohio State Univ, MSc, 53, PhD(physiol), 63. *Prof Exp:* Res & develop adminr air force weapons, Air Res & Develop Command, USAF, 54-60, staff scientist, Air Force Systs Command, 63-66, chief life sci div, Off Aerospace Res, 66-69; SR res scientist, Geomet, Inc, Rockville, 69-72; health care consult, 73-74; spec asst for Legis, Off Asst Secy Defense, Health Affairs, Washington, DC, 75-80; dir spec projs, 81-83; sr policy specialist, 84; CONSULT HEALTH CARE MGT & ENVIRON HAZARDS, 85- *Concurrent Pos:* Mem, Sci Panel, Agent Orange Working Group, 81-89. *Mem:* Sigma Xi. *Res:* Endocrinology, especially intermediate metabolism; respiratory and bacterial physiology; toxic effects of herbicides. *Mailing Add:* 6000 Cable Ave Camp Springs MD 20746-3824

BRICKER, NEAL S, b Denver, Colo, Apr 18, 27; m 51; c 3. INTERNAL MEDICINE. *Educ:* Univ Colo, BA, 46, MD, 49; Am Bd Internal Med, dipl, 56. *Prof Exp:* Intern & resident, Post-Grad Med Sch, NY Univ-Bellevue Med Ctr, 49-52; clin asst, Sch Med, Univ Colo, 52-54; sr resident, Peter Bent Brigham Hosp, 54-55; instr, Harvard Med Sch, 55-56; from asst prof to prof med, Sch Med, Wash Univ, 56-72; dir renal div, 56-72; prof med & chmn dept, Albert Einstein Col Med, 72-76; prof med & vchmn dept, Univ Miami Sch Med, 76-78; prof med & dir prog kidney dis, Univ Calif, Los Angeles, 78-86; DISTINGUISHED PROF MED, LOMA LINDA UNIV, 86- *Concurrent Pos:* Fel, Howard Hughes Med Inst, 55-56; jr assoc & assoc dir, Cardiorenal Lab, Peter Bent Brigham Hosp, 55-56; estab investr, Am Heart Asn, 59-64; assoc ed, J Lab & Clin Med, 61-67; investr, Mt Desert Island Biol Labs, 62-66; mem sci adv bd, Nat Kidney Dis Found, 62-69, chmn res & fel grants comt, 64-65, mem exec comt, 68-71; USPHS career res award, 64-72; mem gen med study sect, NIH, 64-68, chmn, 66-68, chmn renal dis & urol teaching grants comt, 69-71; mem drug efficacy comt, Nat Acad Sci, 66-68, mem comt space biol & med, 71-81, mem ad hoc panel renal & metab effects space flight, 72-74, chmn comt, 72-74, mem space sci bd, 77-81; mem nephrol test comt, Am Bd Internal Med, 71-76, chmn, 73-74, bd gov, 70-78. *Honors & Awards:* Gold-Headed Cane Award, Univ Colo, 49; George Norlin Silver Award, 82; Medal, Am Soc Nephrology, 83 & Am Soc Clin Invest, 84. *Mem:* Inst Med-Nat Acad Sci; Am Soc Clin Invest (pres, 72-73); fel Am Col Physicians; Asn Am Physicians; Int Soc Nephrol (vpres, 66-69, treas, 69-81); Am Soc Nephrology (pres, 66-67). *Res:* Nephrology and transport across isolated membranes. *Mailing Add:* Dept Nephrology Rm 1511 Loma Linda Univ 11234 Anderson St Loma Linda CA 92354

BRICKER, OWEN P, III, b Lancaster, Pa, Mar 5, 36; m 58; c 4. GEOCHEMISTRY, HYDROLOGY & WATER RESOURCES. *Educ:* Franklin & Marshall Col, BS, 58; Lehigh Univ, MS, 60; Harvard Univ, PhD(geol), 64. *Prof Exp:* Fel environ sci & eng, Harvard Univ, 64-65; from asst prof to assoc prof geol, Johns Hopkins Univ, 65-75; geologist, Md Geol Surv, 75-79 & Chesapeake Bay Prog, Environ Protection Agency, 79-81; GEOLOGIST, US GEOL SURV, 81- *Mem:* Geochem Soc; Am Chem Soc; Mineral Soc Am; Soc Econ Geol; Geol Soc Am. *Res:* Mineral equilibria in the earth-surface environment; chemistry of weathering and supergene processes; chemistry of natural waters. *Mailing Add:* Water Resources Div US Geol Surv Reston VA 22092

BRICKEY, PARIS MANAFORD, b Denton, Tex, Dec 13, 31; m 59. ENTOMOLOGY, BIOLOGY. *Educ:* Univ Md, 59, MS, 60. *Prof Exp:* Analyst biol, 60-63, res entomologist, 63-77, CHIEF MICROANALYSIS BR, US FOOD & DRUG ADMIN, 77- *Honors & Awards:* Contrib Award, US Food & Drug Admin, 64. *Mem:* Fel Asn Off Anal Chemists; Entom Soc Am; Inst Food Technologists; Am Asn Cereal Chemists; Asn Food & Drug Off. *Res:* Detection and identification of insect, rodent, other animal filth, foreign material and mold in foods and food ingredients; determine regulatory significance of filth in foods. *Mailing Add:* US Food & Drug Admin HFF127 200 C St SW Washington DC 20204

BRICKS, BERNARD GERARD, b Pittsburgh, Pa, Nov 28, 41; m 63; c 2. LASERS. *Educ:* Memphis State Univ, BS, 63; Johns Hopkins Univ, MS, 69, PhD(physics), 71. *Prof Exp:* Fel, New Eng Inst, 71-72; LASER PHYSICIST, SPACE SCI LAB, GEN ELEC CO, 72- *Mem:* AAAS; Am Phys Soc; Am Asn Physics Teachers. *Res:* Research and developmental program in metal vapor lasers and manufacturing of custom laser systems for applications. *Mailing Add:* 32 Elmwood Dr Kennett Square PA 19348

BRIDENBAUGH, PETER R, b Franklin, Pa, July 28, 40; m 65; c 2. METALLURGY, MECHANICAL ENGINEERING. *Educ:* Lehigh Univ, BS, 62, MS, 66; Mass Inst Technol, PhD(metall & mat sci), 68. *Prof Exp:* Group Leader Eng Dept, Aluminum Co Am, Alcoa Ctr, Pa & Evansville, IND, 68-75, mgr, 75-78, opers dir, 78-80,dir, Res & Develop, 83-84, VPRES, ALCOA LABS, ALCOA CENTER, PA, 84- *Concurrent Pos:* Qual Assurance Mgr, Alcoa, Alcoa, Tenn, 78-80; fel Am Soc Metals Int, 87. *Mem:* Am Soc Metals; Am Inst Mining, Metall Engrs; Am Soc Eng Educ; Indust Res Inst; Sigma Xi. *Res:* Productivity of basic deformation processing; manufacturing processes; lubricants for manufacturing processes. *Mailing Add:* Aluminum Co Am Alcoa Labs Alcoa Center PA 15069

BRIDGE, ALAN G, b Eng, Jan 15, 36; m 58; c 4. CHEMICAL ENGINEERING. *Educ:* Univ Birmingham, BSc, 56; Univ Toronto, MASc, 58; Princeton Univ, MA, 61, PhD(chem eng), 65. *Prof Exp:* Process chem engr, Celanese Corp Am, 58-59; res engr, Photo Prods Dept, E I du Pont de Nemours & Co, Inc, 60; res engr, 69-81, SR ENG ASSOC, CHEVRON RES CO, CALIF, 81- *Mem:* Am Inst Chem Engrs. *Res:* Licensing; two phase flow; catalysis; reactor design and development. *Mailing Add:* 7976 Terrace Dr El Cerrito CA 94530

BRIDGE, HERBERT SAGE, b Berkeley, Calif, May 23, 19; m 41; c 3. SPACE PHYSICS. *Educ:* Univ Md, BS, 41; Mass Inst Technol, PhD(physics), 50. *Prof Exp:* Mem res staff, Los Alamos Sci Lab, 43-46; res assoc cosmic ray res, 46-50, mem res staff, 50-55, res physicist, 55-65, assoc dir ctr space res, 65-78, prof, 65-84, dir ctr space res, 78-84, EMER PROF & SR LECTR PHYSICS, MASS INST TECHNOL, 85- *Concurrent Pos:* Vis scientist, Europ Orgn Nuclear Res, Switz, 57-58. *Honors & Awards:* Medal Except Sci Achievement, NASA, 74, 81 & 86. *Mem:* Fel Am Geophys Union; fel Am Acad Arts & Sci; fel AAAS. *Res:* Cosmic rays; solar wind; high energy astrophysics. *Mailing Add:* Ctr Space Res Mass Inst Technol Rm 37-667 Cambridge MA 02139

BRIDGE, JOHN F(LOYD), b Sioux City, Iowa, Dec 27, 33. FLUID MECHANICS. *Educ:* Iowa State Univ, BSME, 55; Ohio State Univ, MS, 59, PhD(fluid mech), 63. *Prof Exp:* Instr, 57-63, ASST PROF MECH ENG, OHIO STATE UNIV, 63- *Concurrent Pos:* Res engr, Boeing Co, 59, 61, res specialist, 63; NSF-Am Gas Asn res grant, 60-61; consult, Battelle Mem Inst, Ohio, 68-70. *Mem:* Am Soc Mech Engrs; Am Inst Aeronaut & Astronaut. *Res:* Biomedical mechanical design and analysis; turbulent and laminar heat transfer; non-uniform subsonic two dimension flow; gas flow and diffusion associated with decompression of humans from hyperbaric environment. *Mailing Add:* Dept Mech Eng Ohio State Univ 206 W 18th Ave Columbus OH 43210

BRIDGE, THOMAS E, b Apr 3, 25; m 47; c 4. GEOLOGY. *Educ:* Kans State Univ, BS, 50, MS, 53; Univ Tex, PhD(geochem), 66. *Prof Exp:* Teacher pub schs, 50-53, prin, 53-56; instr geol, Colo State Univ, 56-59; asst prof, Tex Technol Col, 63-66; PROF GEOL, EMPORIA STATE UNIV, 66- *Mem:* Am Geophys Union; Meteoritical Soc Am. *Res:* Contact metamorphic rocks and ground water geology. *Mailing Add:* Dept Phys Sci Emporia State Univ 1200 Commercial St Emporia KS 66801

BRIDGEFORD, DOUGLAS JOSEPH, b Grand Forks, NDak, Feb 3, 26; m 55; c 4. POLYMER CHEMISTRY, MATERIALS SCIENCE ENGINEERING. *Educ:* Univ Minn, BChem, 47. *Prof Exp:* Chemist, teepak inc, 48-55, res chemist polymers, 55-63, sr res chemist, 58-63, asst res mgr, 63-72, res assoc cellulose, 72-82, adv scientist polymers, 82-91; CONSULT, 91- *Mem:* Am Chem Soc; fel Am Inst Chemists; Tech Asn Pulp & Paper Indust; NY Acad Sci. *Res:* Film technology from amylose, cellulose, polyvinyl alcohol and cellulose derivatives; viscose modifications; lamination; printing; coatings films; organic synthesis; waste treatment; polymer composites; 55 US patents. *Mailing Add:* 1209 W John St Champaign IL 61821

BRIDGEO, WILLIAM ALPHONSUS, b St John, NB, Dec 15, 27; m 55; c 4. ORGANIC CHEMISTRY. *Educ:* St Francis Xavier Univ, BSc, 48; Univ Ottawa, PhD(chem), 52. *Prof Exp:* Instr gen chem, St Mary's Col, Ind, 52-53; assoc prof anal chem, 63-70, dean sci, 67-76, PROF ANALYTICAL CHEM, ST MARY'S UNIV, NS, 70- *Concurrent Pos:* Dir, Tech Serv Div, NS Res Found, Halifax, 53-65, dir, Chem Div, 65-69; pres, Bridco Values Ltd, 71-; dir Can Nat Comt, Int Asn Water Pollution Res & Control, 71-85, pres, 81-82. *Mem:* Can Inst Chem; Am Inst Chemists; Can Asn Water Pollution Res & Control; NS Inst Sci; Int Asn Water Pollution Res & Control. *Res:* Nitrogen, strontium, and sulfur compounds; ceramic glazes. *Mailing Add:* PO Box 3161 Halifax NS B3J 3H5 Can

BRIDGER, WAGNER H, b New York, NY, Jan 9, 28. RESEARCH ADMINISTRATION, PHARMACOLOGY. *Educ:* NY Univ, BA, 46, MD, 50. *Prof Exp:* From assoc prof to prof psychiat, Albert Einstein Col Med, Bronx Munic Hosp Ctr, 65-82, dir res, 65-82; PROF PSYCHIAT & PHARMACOL, MED COL PA, 82-, CHMN DEPT, 82- *Concurrent Pos:* Vis scientist, Instituto Superiore de Sanita, Rome, Italy, 69-70; prof neurosci, Albert Einstein Col Med, 74-82, actg chmn, 76-81, attend psychiatrist, Hosp Albert Einstein Col Med, 76-82; mental health career investr, USPHS, 58-63, res career develop, 63-68; consult, Hastings Ctr, Inst Soc, Ethics & Life Sci, 77- *Mem:* Soc Psychiat Res (secy, 66-69, pres, 70-71, vpres, 86); Soc Biol Psychiat (pres, 88); fel Am Col Neuropsychopharmacol; fel Am Col Psychiatrists; Am Psychopath Asn. *Res:* Neonatal psychophysiology cross-modal transfer in premature infants; galvanic skin response conditioning effect of drugs on animal behavior; role of various neurotransmitters on animal behavior and aggression in children. *Mailing Add:* Dept Psychiat Med Col Pa 3200 Henry Ave Philadelphia PA 19129

BRIDGER, WILLIAM AITKEN, b Winnipeg, Man, May 31, 41; m 83; c 4. BIOCHEMISTRY. *Educ:* Univ Man, BSc, 62, MSc, 63, PhD(biochem), 66. *Prof Exp:* Med Res Coun Can res fel biochem, Univ Calif, Los Angeles, 66-67; from asst prof to assoc prof, 67-77, PROF BIOCHEM, UNIV ALTA, 77-, CHMN, DEPT BIOCHEM, 87- *Concurrent Pos:* Vis prof, Rockefeller Univ, NY, 84-85. *Honors & Awards:* Ayerst Award, Can Biochem Soc, 80. *Mem:* Am Chem Soc; Can Biochem Soc (secy, 74-); Am Soc Biol Chemists; fel Royal Soc Can. *Res:* Study of structure function and assembly of enzymes. *Mailing Add:* Dept Biochem Med Sci Bldg Univ Alta Edmonton AB T6G 2H7 Can

BRIDGERS, WILLIAM FRANK, b Asheville, NC, July 26, 32; m 74; c 4. ACADEMIC ADMINISTRATION. *Educ:* Univ of the South, BA, 54; Wash Univ, MD, 59. *Prof Exp:* From instr to asst prof prev med & med, Sch Med, Wash Univ, 63-66; from asst prof to assoc prof med, Sch Med, Univ Miami, 66-68; assoc prof pediat, biochem & med, 68-70, prof psychiat & dir neurosci prog, Sch Med, 70-73, dir sponsored progs, 73-74, spec asst to vpres health affairs, 75-76, chmn & prof, Dept Pub Health, 76-81, dean & prof, Sch Pub Health, 81-89, dir, Lister Hill Ctr, Health Policy, 87-90, PROF PUB HEALTH, UNIV ALA, 90- *Concurrent Pos:* United Health Found fel, Dept Prev Med, Sch Med, Wash Univ, 63-65; assoc ed, Nutrit Rev, 63-67; mem staff, Nat Acad Sci/Nat Res Coun, 74-75. *Mem:* Am Soc Biol Chemists; Am Inst Nutrit; Am Soc Clin Nutrit; Am Soc Neurochem. *Res:* Neurochemistry and regulatory molecular biology of folic acid metabolism in developing, mammalian brain; health policy. *Mailing Add:* Sch Pub Health Univ Ala Birmingham AL 35294

BRIDGES, C DAVID, b Northiam, E Sussex, UK, Jan 16, 33; US citizen. RECOMBINANT DNA, BIOCHEMISTRY OF RETINOIDS. *Educ:* Univ London, UK, BSc, 53, PhD(physiol & biochem), 56, DSc, 83. *Prof Exp:* Sci staff, Med Res Coun, London, 64-67; from asst prof to prof ophthal, NY Univ Med Ctr, 67-77; prof ophthal, Baylor Col Med, 77-87, prof biochem, 80-87; PROF BIOL SCI PHYSIOL, PURDUE UNIV, 87- *Concurrent Pos:* Dir, Ctr Res Vision, Baylor Col Med, 78-87; distributing ed, Vision Res, Pergamon Press, 79-89; trustee, Asn Res Vision & Opthal, 82-86, vpres, 86; Pfizer traveling fel, Clin Res Inst Montreal, 83; Marc vis scientist, Fedn Am Socs Exp Biol, 83-84; mem, Behav & Neurosci Study Sect, NIH, 83-85 & VISB Study Sect, 75-79; adj prof physiol & biophys, Sch Med, Ind Univ, 90- *Honors & Awards:* Marjorie W Margolin Prize, Retina Res Found, 82, Sam & Bertha Brochstein Award, 86. *Mem:* Physiol Soc; Asn Res Vision & Opthal; Am Soc Biol Chemist; Soc Neurosci; Marine Biol Asn UK. *Res:* Molecular basis of the visual process; biochemistry and transport of vitamin A in light and dark adaptation; biochemical basis of ocular diseases. *Mailing Add:* Dept Biol Sci Purdue Univ West Lafayette IN 47907

BRIDGES, CHARLES H(ENRY), chemical engineering, for more information see previous edition

BRIDGES, CHARLES HUBERT, b Shreveport, La, Feb 23, 21; m 45; c 2. VETERINARY PATHOLOGY. *Educ:* Agr & Mech Col, Tex, DVM, 45, MS, 54, PhD(vet path). *Prof Exp:* Armed Forces Inst Path, cert, 55; Am Col Vet Pathologists, dipl, 56. *Prof Exp:* Res assoc, Agr Exp Sta, La State Univ, 49-51; assoc prof, Tex A&M Univ, 55-59, prof, Agr Exp Sta, 59-60, head dept, 60-78, prof, 60-87, EMER PROF VET PATH, TEX A&M UNIV, 87- *Concurrent Pos:* Adj prof path, Baylor Col Med, Houston, Tex, 78-87. *Res:* Experimental pathology as applied to natural diseases of animals; definition of toxicological factors in good (plant toxins, trace minerals) and their role in producing disease with special evaluation of nutritional, biochemical and pathological interactions. *Mailing Add:* 1502 Glade College Station TX 77840

BRIDGES, DONALD NORRIS, b Shelby, NC, Aug 13, 36; m 59; c 4. NUCLEAR SAFETY. *Educ:* NC State Univ, Raleigh, BCE, 58, MS, 60; Ga Inst Technol, MS, 68, PhD(nuclear eng), 70. *Prof Exp:* Civil engr, DuPont Co, 61-62; naval officer, Civil Eng Corps, USN, 62-66; nuclear engr, AEC, Savannah, 70-74; nuclear engr, Div Operating Reactors, Nuclear Regulatory Comn, 74-76; chief nuclear safety br, 76-81, chief reactors br, 81-85, DIR SAFETY DIV, DEPT ENERGY, SAVANNAH RIVER OPERS OFF, 85- *Mem:* Am Nuclear Soc; Soc Am Mil Engrs; Naval Res Asn; Res Off Asn. *Res:* Spatial and feedback effects resulting from nuclear reactor transients; nuclear methods for measuring soil moisture and density. *Mailing Add:* 1002 Longleaf Ct N Augusta SC 29841

BRIDGES, E(RNEST), b Winnipeg, Man, Aug 4, 31; m 56; c 3. ELECTRICAL ENGINEERING. *Educ:* Univ Man, BSc, 54, MSc, 58. *Prof Exp:* From lectr to assoc prof, 56-70, PROF ELEC ENG, UNIV MAN, 70- *Mem:* Inst Elec & Electronics Engrs; Can Asn Univ Teachers. *Res:* Microwaves; microwave measurement methods; microwave filters and antennae; semiconductor permittivity measurements. *Mailing Add:* Dept Elec Eng Univ Man Winnipeg MB R3T 2N2 Can

BRIDGES, FRANK G, MICROWAVE ABSORPTION SPECTROSCOPY OF ELECTRIC DIPOLE SYSTEMS, EXTENDED X-RAY ABSORPTION FINE STRUCTURE STUDIES OF PURE & SUBSTITUTED CRYSTALS. *Educ:* Univ BC, Can, BS, 62, MS, 64; Univ Calif, San Diego, PhD(physics), 86. *Prof Exp:* Asst res physicist & lectr, Univ Calif, San Diego, 68-70; from asst prof to assoc prof physics, 70-81, chair physics, 82-84, PROF PHYSICS, UNIV CALIF, SANTA CRUZ, 81- *Concurrent Pos:* Prin investr, 71-; consult, Xerox Corp, 85- *Mem:* Am Phys Soc; Mat Res Soc. *Res:* Tunneling behavior of off-center ions in crystals; local structure of defect atoms in high temperature superconductors; investigated low velocity collisions of ice particles, such as occur in Saturn's rings. *Mailing Add:* Dept Physics Univ Calif Santa Cruz CA 95064

BRIDGES, JOHN ROBERT, b Plainview, Ark, Feb 14, 44; m 68; c 2. ENTOMOLOGY, BIOCHEMISTRY. *Educ:* Univ Ark, BS, 66, MS, 68; Univ Wis, PhD(entom), 75. *Prof Exp:* Med entomologist, USN, 69-72; res entomologist, US Forest Serv, 75-89. *Mem:* Entom Soc Am; Am Soc Microbiol; Mycol Soc Am. *Res:* Bark beetle and associated microorganisms; insect physiology and biochemistry. *Mailing Add:* 201 14th & Independence SW PO Box 96090 Washington DC 20090-6090

BRIDGES, KENT WENTWORTH, b Milwaukee, Wis, Aug 25, 41; m 65. ECOLOGY. *Educ:* Univ Hawaii, BA, 64, MS, 67; Univ Calif, Irvine, PhD(biol), 70. *Prof Exp:* Asst prof wildlife resources, Utah State Univ, 70-73; ASST PROF BOT, UNIV HAWAII, 73- *Concurrent Pos:* Chief modeler, Desert Biome, US Int Biol Prog, 70-73. *Mem:* Ecol Soc Am; Japanese Soc Pop Ecol; Asn Comput Mach. *Res:* Computer modeling of biological problems; systems ecology. *Mailing Add:* Dept Bot Univ Hawaii Honolulu HI 96822

BRIDGES, ROBERT STAFFORD, b Northampton, Mass, Aug 5, 47; m 75; c 2. NEUROENDOCRINOLOGY, BEHAVIORAL ENDOCRINOLOGY. *Educ:* Earlham Col, BA, 69; Univ Conn, MS, 72, PhD(endocrinol), 74. *Prof Exp:* Fel biochem/behav, Rutgers Univ, 75-77; res assoc anat, Sch Med, Univ Calif, Los Angeles, 77-78; asst prof anat, Harvard Med Sch, 78-84, assoc prof anat & cell biol, 84-90; PROF COMP MED, SCH VET MED, TUFTS UNIV, 90- *Concurrent Pos:* Vis lectr, Woods Hole Oceanog Inst, 80-83; res scientist develop award, NIMH, 84-89. *Mem:* Soc Study Reproduction; Endocrine Soc; Am Asn Anatomists; Soc Neurosci. *Res:* Relationships among the endocrine, neural and neuroendocrine systems in the regulation of reproductive processes in mammals; control of maternal behavior and maintenance of pregnancy by the endocrine and neuroendocrine systems; neurobiology of aging. *Mailing Add:* Dept Comp Med Sch Vet Med Tufts Univ 200 Westboro Rd North Grafton MA 01536-1895

BRIDGES, THOMAS JAMES, b Gillingham, Eng, Dec 2, 23; UK citizen; m 50, 81; c 2. PHYSICS, ELECTRICAL ENGINEERING. *Educ:* London Univ, BS, 44. *Prof Exp:* Scientist physics & elec eng, Royal Naval Sci Serv, UK, 44-56 & 59-63; mem tech staff, 56-59, MEM TECH STAFF PHYSICS & ELEC ENG, AT&T BELL LABS, 63- *Concurrent Pos:* Vis prof elec eng dept, Mass Inst Technol, 67-68. *Mem:* Inst Elec & Electronics Engrs; Optical Soc Am. *Res:* Microwave and optical electronics. *Mailing Add:* Bell Tel Labs Holmdel NJ 07733

BRIDGES, WILLIAM B, b Inglewood, Calif, Nov 29, 34; m 57, 86; c 3. MILLIMETER WAVE & OPTICAL TECHNOLOGY. *Educ:* Univ Calif, Berkeley, BS, 56, MS, 57, PhD(elec eng), 62. *Prof Exp:* Assoc elec eng, Univ Calif, Berkeley, 57-59; mem tech staff, Res Labs, Hughes Aircraft Co, 60-65, sr mem tech staff, 65-68, mgr laser dept, 69-70, sr scientist, 68-77; prof elec eng & appl physics, 77-83, exec officer elec eng, 78-81, CARL F BRAUN PROF ENG, CALIF INST TECHNOL, 83- *Concurrent Pos:* Res engr, Varian Assocs, 56 & 57; lectr, Univ Southern Calif, 62-64; Sherman Fairchild distinguished scholar, Calif Inst Technol, 74-75; assoc ed, J Quantum Electronics, Inst Elec & Electronics Engrs, 77-82 & J Optical Soc Am, 78-82; dir, Optical Soc Am, 82-84; adv bd, Air Force Sci, 85-88. *Honors & Awards:* Arthur L Schawlow Award, Laser Inst Am, 86; Quantum Electronics Award, Inst Elec & Electronics Engrs, 88. *Mem:* Nat Acad Sci; Nat Acad Eng; fel Optical Soc Am (vpres, 86, pres, 88); fel Inst Elec & Electronics Engrs. *Res:* Laser devices; optical communication and instrumentation; millimeter wave devices and systems. *Mailing Add:* Calif Inst Technol 128-95 Pasadena CA 91125

BRIDGES, WILLIAM G, b Conn, Nov 4, 42. ALGEBRA, COMBINATORICS & FINITE MATHEMATICS. *Educ:* Univ Conn, BA, 64; Syracuse Univ, MS 67; Calif Inst Technol, PhD(math), 69. *Prof Exp:* Res assoc math, City Univ New York, 69-70; PROF MATH, UNIV WYO, 70. *Mem:* Math Asn Am; Am Math Soc. *Res:* Applications of linear algebra to combinatorics matrix theory. *Mailing Add:* Dept Math Univ Wyo Box 3036 Laramie WY 82071

BRIDGEWATER, ALBERT LOUIS, JR, b Houston, Tex, Nov 22, 41. ELEMENTARY PARTICLE PHYSICS. *Educ:* Univ Calif, Berkeley, BA, 63; Columbia Univ, PhD(physics), 72. *Prof Exp:* Fel physics, Lawrence Berkeley Lab, 70-73; staff asst, Physics Sect, 73-76, spec asst for sci planning, 76-81, actg asst dir, 83-85, dep asst dir, 81-86, actg asst dir, 83-85, SR STAFF ASSOC, ASTRON, ATMOSPHERIC, EARTH & OCEAN SCI, NSF, 86- *Concurrent Pos:* Adj asst prof, Howard Univ, 75-76. *Mem:* AAAS; Am Phys Soc; Am Geophys Union. *Mailing Add:* 3713 S George Mason Dr T5W Falls Church VA 22041

BRIDGFORTH, ROBERT MOORE, JR, b Lexington, Miss, Oct 21, 18; m 43; c 2. PHYSICAL CHEMISTRY. *Educ:* Iowa State Univ, BS, 40; Mass Inst Technol, SM, 48. *Prof Exp:* Instr chem, Mass Inst Technol, 40-43, staff mem, Div Indust Coop, 43-48; assoc prof physics & chem, Emory & Henry Col, 49-51; res specialist, Boeing Co, 51-58, chief propulsion syst sect, Syst Mgt Off, 58-59, chief propulsion res unit, 59-60; chmn bd, Rocket Res Corp, 60-69; RETIRED. *Concurrent Pos:* Chmn bd, Explosives Corp Am, 66-69. *Mem:* Am Chem Soc; Am Astronaut Soc; Brit Interplanetary Soc; Am Inst Aeronaut & Astronaut; Am Rocket Soc; Am Inst Physics; Tissue Cult Asn;

Sigma Xi; Soc Leukocyte Biol; Am Inst Chemists; Combustion Inst; NY Acad Sci; Am Defense Preparedness Asn; AAAS; Am Asn Physics Teachers. *Res:* Thermodynamics of rocket propellants; thermodynamics of living systems. *Mailing Add:* 4325 87th Ave SE Mercer Island WA 98040

BRIDGMAN, CHARLES JAMES, b Toledo, Ohio, May 6, 30; m 54; c 6. RESEARCH ADMINISTRATION. *Educ:* US Naval Acad, BS, 52; NC State Univ, MS, 58, PhD(nuclear eng), 63. *Prof Exp:* Asst nuclear eng, USAF Inst Technol, 59-60, from instr to assoc prof, 60-68, chmn, Nuclear Eng Comn, 68-89, chmn, Sch Eng Doctoral Coun, 76-89, PROF NUCLEAR ENG, USAF INST TECHNOL, 68-, DEAN RES, SCH ENG, 89- *Concurrent Pos:* Consult, Air Force Weapons Lab & Nuclear Agency & Defense Nuclear Energy. *Mem:* Fel Am Nuclear Soc; Am Soc Eng Educ; Health Physics Soc; Am Geophys Union; Sigma Xi; Am Asn Physics Teachers. *Res:* Nuclear weapon effects modeling especially fallout and residual radiation; military applications of nuclear power including space nuclear power. *Mailing Add:* 7362 Natoma Pl Dayton OH 45424

BRIDGMAN, GEORGE HENRY, b Minneapolis, Minn, 1940. MATHEMATICS. *Educ:* Hamline Univ, BA, 61; Univ Minn, Minneapolis, MA, 64, PhD(math), 69. *Prof Exp:* Asst prof math, Wartburg Col, 69-76; lectr math, Univ Minn, Duluth, 76-79; asst prof math, Hamline Univ, 79-80; asst prof math, Univ Wis-River Falls, 80-81; programmer/analyst, Cray Res, Inc, St Paul, 84-85; APPLN PROG, DIV EPIDEMIOL, UNIV MINN, MINNEAPOLIS, 86- *Concurrent Pos:* Instr comput prog, Macalester Col, St Paul, 83-84. *Mem:* Math Asn Am. *Res:* Mathematical analysis. *Mailing Add:* 4306 Grimes Ave S Minneapolis MN 55424-1051

BRIDGMAN, HOWARD ALLEN, b Boston, Mass, Nov 9, 44; m 74; c 2. CLIMATOLOGY, AIR POLLUTION. *Educ:* Beloit Col, BA, 66; Univ Hawaii, MA, 68; Univ Wis-Milwaukee, PhD(geog), 77. *Prof Exp:* Consult meteorol, Lawrence Nield & Partners, Architects, Sydney, Australia, 77-80; chmn, Bd Environ Studies, 80-85; sr postdoc consult, GMCC, NOAA, Boulder, Colo, 86-87; SR LECTR, DEPT GEOG, UNIV NEWCASTLE, 77- *Concurrent Pos:* Air qual consult, Hunter Develop Bd, 80-; environ consult, Cleanaway, 90- *Mem:* Sigma Xi; Am Geophys Union; Am Meteorol Soc; Royal Meteorol Soc; Australia-NZ Clean Air Soc. *Res:* Solar radiation and particulate air pollution; air pollution dispersion; air pollution and climatic change; arctic haze; acid rain; author of one book. *Mailing Add:* Dept Geog Univ Newcastle Newcastle NSW 2308 Australia

BRIDGMAN, JOHN FRANCIS, b Kuling, China, Sept 6, 25; US citizen; m 52; c 4. PARASITOLOGY, ANIMAL PHYSIOLOGY. *Educ:* Davidson Col, BS, 49; La State Univ, MS, 52; Tulane Univ, PhD(biol), 68. *Prof Exp:* Instr biol, Delta State Teachers Col, 52 & Univ Tenn, Martin Br, 52-54; lectr, Shikoku Christian Col, 56-67, from asst prof to prof, 57-72; assoc prof, 72-81, PROF BIOL, COL OZARKS, 81- *Honors & Awards:* George Henry Penn Award Biol, 69. *Mem:* AAAS; Am Soc Parasitologists; Japanese Soc Parasitologists; Am Soc Zoologists. *Res:* Invertebrate ecology; life cycles, bionomics and ecology of parasites together with their host-parasite relations. *Mailing Add:* Dept Math & Sci Univ Ozarks 415 Col Ave Clarksville AR 72830

BRIDGMAN, WILBUR BENJAMIN, b New Wilmington, Pa, Jan 28, 13; m 37; c 2. PHYSICAL CHEMISTRY. *Educ:* Wis State Teachers Col, BEd, 33; Univ Wis, PhD(chem), 37. *Prof Exp:* Instr chem, Univ Wis, 37-42, Rockefeller fel, 42-43; from asst prof to prof, 43-78, EMER PROF CHEM, WORCESTER POLYTECH INST, 78- *Mem:* Am Chem Soc. *Res:* Dipole moments; ultracentrifugal determination of particle size and shape; solution thermodynamics; surface chemistry. *Mailing Add:* Dept Chem Worcester Polytech Inst Worcester MA 01609

BRIDLE, ALAN HENRY, b Harrow, Eng, Sept 2, 42; m 68. ASTRONOMY. *Educ:* Cambridge Univ, BA, 63, MA, 67, PhD(radio astron), 67. *Prof Exp:* UK Sci Res Coun fel, 67; from asst prof to prof physics, Queen's Univ, Ont, 72-83; STAFF MEM, NAT RADIO ASTRON OBSERV, 83- *Concurrent Pos:* Vis asst scientist, US Nat Radio Astron Observ, 68, vis scientist, 80-82; adj prof, Univ NMex, 80-81. *Mem:* Am Astron Soc; Can Astron Soc; Int Astron Union. *Res:* Extragalactic radio astronomy at high angular resolution; radio galaxies; galaxy clusters. *Mailing Add:* Nat Radio Astron Observ Edgemont Rd Charlottesville VA 22903-2475

BRIEADDY, LAWRENCE EDWARD, b Syracuse, NY, May 13, 44; m 66; c 2. PHARMACEUTICAL CHEMISTRY, ORGANIC CHEMISTRY. *Educ:* LeMoyne Col, NY, BS, 66; NC State Univ, MS, 69. *Prof Exp:* Res scientist, 71-80, SECT HEAD PHARMACEUT, BURROUGHS WELLCOME CO, 80- *Mem:* Am Chem Soc. *Res:* Prostaglandin synthetase inhibitors; central nervous system; monoamine oxidase inhibitors; anti-inflammatory analgesic fields. *Mailing Add:* 5924 Wintergreen Dr Raleigh NC 27609

BRIEGER, GERT HENRY, b Hamburg, Ger, Jan 5, 32; US citizen; m 55; c 3. HISTORY OF MEDICINE. *Educ:* Univ Calif, Berkeley, AB, 53; Univ Calif, Los Angeles, MD, 57; Harvard Univ, MPH, 62; Johns Hopkins Univ, PhD, 68. *Prof Exp:* Intern med, Univ Calif, Los Angeles, 57-58; med officer, US Army Hosp, Dugway Proving Ground, 58-61; asst prof hist med, John Hopkins Univ Sch Med, 66-70; assoc prof community health sci & hist, Duke Univ, 70-75; prof & chmn, dept hist health sci, Univ Calif, San Francisco, 75-84; WILLIAM H WELCH PROF & DIR, INST HIST MED, JOHN HOPKINS UNIV SCH MED, 84- *Concurrent Pos:* Mem coun, Am Asn Hist Med, 68-81; mem adv bd, Einthoven Found, Neth; mem adv comt, Hist Med, Wellcome Trust, London, Eng; mem vis comt, Hist Sci Dept, Bd Overseers Harvard Univ. *Mem:* Inst Med-Nat Acad Sci; Am Asn Hist Med (vpres, 78-80, pres, 80-82); Int Acad Hist Med. *Res:* History of American medicine and public health; history of medicine in the late 19th and 20th centuries; author of 25 publications in history of medicine. *Mailing Add:* John Hopkins Univ 1900 E Monument Ave Baltimore MD 21201

BRIEGER, GOTTFRIED, b Berlin, Ger, Oct 27, 35; m 59; c 2. ORGANIC CHEMISTRY. *Educ:* Harvard Col, BA, 57; Univ Wis, PhD(chem), 61. *Prof Exp:* Asst prof chem, Univ Calif, Berkeley, 61-63; from asst prof to assoc prof, 63-72, PROF CHEM, OAKLAND UNIV, 72- *Concurrent Pos:* Sr Fulbright fel, Univ Heidelberg, 72-73. *Mem:* AAAS; Am Chem Soc; Royal Chem Soc; NY Acad Sci. *Res:* Synthesis and structural determination of natural products, especially terpenes; chemistry and biochemistry of insect hormones; catalytic hydrogen transfer reactions; clinical chemistry; applied chemistry; microbial degradation of xenobiotics. *Mailing Add:* Dept Chem Oakland Univ Rochester MI 48063

BRIEGLEB, PHILIP ANTHES, b St Clair, Mo, July 23, 06; m 35; c 2. FORESTRY. *Educ:* Syracuse Univ, BS, 29, MF, 30. *Prof Exp:* Jr forester, Pac Northwest Forest Exp Sta, US Forest Serv, 29-34, asst forester, 35, from assoc forester to forester, 36-43, sr forester, Washington, DC, 44, Northeastern Forest Exp Sta, 45-46, chief div forest mgt res, Pac Northwest Forest Exp Sta, 46-51, dir, Cent States Forest Exp Sta, 51-53, Southern Forest Exp Sta, 54-63 & Pac Northwest Forest Exp Sta, 63-71; CONSULT FORESTER, 71- *Concurrent Pos:* Lectr, Univ Calif, 42; mem, US Forestry Mission, Chile, 43-44; guest, Brit Commonwealth Forestry Conf, Australia & NZ, 57; mem adv coun, Cascade Head Scenic Res Area, 75-78. *Mem:* Fel AAAS; fel Soc Am Foresters (pres, 64-65). *Res:* Forest survey; mensuration; management and research administration; ecology; watershed management. *Mailing Add:* 4217 SW Agate Lane Portland OR 97201

BRIEHL, ROBIN WALT, b Vienna, Austria, June 21, 28; US citizen. BIOCHEMISTRY, BIOPHYSICS. *Educ:* Swarthmore Col, BA, 50; Harvard Med Sch, MD, 54. *Prof Exp:* Intern, Montefiore Hosp, New York, 54-55, asst resident med, 55-56; asst resident med & cardiol, Presby Hosp, 56-57, asst physician, 57-60; asst prof physiol, 62-66, assoc med, 65-69, assoc prof, 66-71, PROF PHYSIOL, ALBERT EINSTEIN COL MED, 71-, PROF BIOCHEM, 73-, ASST PROF MED, 69- *Concurrent Pos:* NIH vis fel, Cardiopulmonary Lab, Columbia Univ, 57-60; NIH res fel biol, Harvard Univ, 60-62; career scientist, Health Res Coun, NY, 62-72; asst physician, Bronx Munic Hosp Ctr, 65- *Mem:* Am Soc Biol Chemists; AAAS; Am Fedn Clin Res; Am Soc Clin Invest; Am Physiol Soc; Sigma Xi. *Res:* Allostery; relations between the structure and function of hemoglobin; molecular bases of sickle cell disease; physical chemistry; internal medicine. *Mailing Add:* Dept Physiol Albert Einstein Col Med Bronx NY 10461

BRIEN, FRANCIS STAPLES, b Windsor, Ont, Apr 9, 08; m 37; c 1. INTERNAL MEDICINE. *Educ:* Univ Toronto, BA, 30, MB, 33; FRCP(London) & FACP, 46; FRCP(C), 58. *Hon Degrees:* LLD, Univ West Ont, 79. *Prof Exp:* Instr therapeut, Fac Med, Univ Toronto, 35-36; head dept, 45-70, prof, 45-73, hon prof med & consult to teaching hosps, Univ Western Ont, 73-78; RETIRED. *Concurrent Pos:* Med consult, Northern Life Assurance Co Can; hon mem, Col Family Med Can. *Mem:* Fel Can Geriat Soc; hon fel NY Acad Sci; Can Arthritis & Rheumat Soc; Can Cardiovasc Soc. *Res:* Nutrition; absorption and malabsorption; aging; hypertensive states. *Mailing Add:* 144 Iroquois Ave London ON N6C 2K8 Can

BRIEN, JAMES FREDERICK, b Windsor, Ont, Mar 3, 45; m 67; c 2. PHARMACOLOGY, TOXICOLOGY. *Educ:* Univ Windsor, BSc, 66, PhD(chem), 71. *Prof Exp:* Fel pharmacol, Univ Toronto, 71-73; res scientist, Addiction Res Found Ont, 73-78; from asst prof to assoc prof, 78-86, PROF PHARMACOL & TOXICOL, QUEEN'S UNIV, 86- *Concurrent Pos:* Asst prof, Dept Psychiat, Queen's Univ, 79-; Can J Physiol & Pharmacol, 83-; vis prof, Res Inst, St Joseph's Hosp, Univ W Ont, 85-86; grants, Med Res Coun & Heart & Stroke Found Ont. *Honors & Awards:* Basmajian Award Excellence Med Res, 80. *Mem:* Chem Inst Can; Soc Toxicol Can; Pharmacol Soc Can. *Res:* Effects of prenatal ethanol exposure: acute and chronic studies; mechanism of action of organic nitrate vasodilator drugs; experimental and clinical studies of antiarrhythmic drugs. *Mailing Add:* Dept Pharmacol & Toxicol Queen's Univ Kingston ON K7L 3N6 Can

BRIENT, CHARLES E, b El Paso, Tex, Oct 10, 34; m 59; c 4. NUCLEAR PHYSICS. *Educ:* Univ Tex, BS, 57, MA, 60, PhD(physics), 63. *Prof Exp:* Res assoc nuclear physics, Univ Tex, 63-64; asst prof, 64-68, ASSOC PROF PHYSICS, OHIO UNIV, 68- *Concurrent Pos:* NSF res grant, 65-66; Ohio Univ Fund res grant, 66-67; AEC grant, 68. *Mem:* Am Phys Soc. *Res:* Nuclear reaction mechanisms and structure. *Mailing Add:* Dept Physics Accelerator Lab Ohio Univ Athens OH 45701

BRIENT, SAMUEL JOHN, JR, b Phoenix, Ariz, Mar 24, 30; m 56; c 2. THIN METAL FILMS. *Educ:* Univ Tex, BS, 52, PhD(physics), 59. *Prof Exp:* Res scientist, Defense Res Lab, Univ Tex, Austin, 56, Spectros Lab, 59-60, instr physics, 60-62; assoc prof, 62-74, PROF PHYSICS, UNIV TEX, EL PASO, 74- *Concurrent Pos:* Consult, Air Force Off Sci Res, Holloman AFB, NMex, 63, Braddock, Dunn & McDonald Inc, Tex, 63-64, Los Alamos Sci Lab, 67-70, Harry Diamond Labs, Army Res Off, 69 & NASA, Cleveland, 74-80. *Mem:* Fel AAAS; Am Vacuum Soc; Am Phys Soc; Am Asn Physics Teachers; Sigma Xi. *Res:* Conductive mechanisms in thin films experiments; Green's function equation of motion technique. *Mailing Add:* Dept Physics Univ Tex El Paso TX 79968

BRIENZA, MICHAEL JOSEPH, b Mt Vernon, NY, Aug 20, 39; m 62; c 3. ELECTRO-OPTICS, LASERS. *Educ:* Univ Notre Dame, BS, 60, PhD(physics), 64. *Prof Exp:* Res scientist, Res Labs, 64-67, sr res scientist, 67-68, chief appl laser technol, 68-71, chief electrooptics, Norden Div, United Aircraft Corp, 71-77; dir, Advan Technol, sewing prod group, Singer Corp, 77-81; vpres eng, Diagnostic Retrieval Systs, 81-83; dir, hardware eng, GSSD, Harris Corp, 83-86; DIR HARDWARE ENG, NORDEN SYST UNITED TECHNOL, 86- *Concurrent Pos:* Consult, optical & electronic prod. *Mem:* Am Phys Soc; Optical Soc Am; Sigma Xi. *Res:* Photoelectric and optical properties of metal films; hot electron ranges and devices; surface physics and thin film studies; ultra short laser pulse technology; acousto-optic signal processing; laser applications; electronics; mechanisms; product development; manufacturing engineering. *Mailing Add:* 140 Norton Rd Easton CT 06612

BRIERE, NORMAND, b St-Eustache, Que, Oct 24, 37; m 63; c 1. MICROSCOPY, TISSUE CULTURE. *Educ:* Univ Montreal, BA, 59, BSc, 62, MSc, 64, PhD(cancer), 67. *Prof Exp:* Asst prof dept anat, fac med, 67-73, ASSOC PROF HISTOL, DEPT ANAT & CELL BIOL, UNIV SHERBROOKE, QUE, 73- *Concurrent Pos:* Prin investr, Med Res Coun, 84-86. *Mem:* Am Asn Anatomists; Am Soc Cell Biol; Can Asn Anatomists; Micros Soc Can; Can Soc Cell Biol. *Res:* Culture of fetal kidney in a serum-free medium in order to study the role of various growth factors and hormones on nephrogenesis. *Mailing Add:* Dept Anat & Cell Biol Univ Sherbrooke Fac Med Sherbrooke PQ J1H 5N4 Can

BRIERLEY, CORALE LOUISE, b Shelby, Mont, Mar 24, 45; m 65. INORGANIC WASTE WATER TREATMENT, MICROBIOLOGY. *Educ:* NMex Inst Mining & Technol, BS, 68, MS, 71; Univ Tex, Dallas, PhD, 81. *Prof Exp:* Microbiologist planetary quarantine, Martin Marietta Corp, Denver, 68-69; chem microbiologist bact leaching, NMex Bur Mines & Mineral Resources, 71-82; pres, 82-87, VPRES, ADVAN MINERAL TECHNOL, INC, 87- *Concurrent Pos:* Vis prof, Dept Mining, Metall & Fuels Eng, Univ Utah, 74 & 78; vis researcher, Appl Geochem Res Group, Imp Col, 76 & Salt Lake City Metall Res Ctr, US Bur Mines, 78; vis lectr, Concordia Univ, Montreal, 79; vis scientist, Univ Tex, Dallas, 79-80. *Mem:* Can Soc Microbiologists; Am Inst Mining, Metall & Petrol Engrs; Soc Indust Microbiol; Am Electroplaters Soc. *Res:* Biogenic extractive metallurgy; biological treatment methods for inorganic wastes; thermophilic chemautotrophic micro-organisms. *Mailing Add:* 2872 E Elk Horn Lane Sandy UT 84093-6595

BRIERLEY, GERALD PHILIP, b Ogallala, Nebr, Aug 14, 31; m 54; c 2. BIOCHEMISTRY. *Educ:* Univ Md, PhD(biochem), 61. *Prof Exp:* Fel, Inst Enzyme Res, Univ Wis, 60-62, asst prof biochem, 62-64; asst prof, 64-69, PROF PHYSIOL CHEM, OHIO STATE UNIV, 69- *Concurrent Pos:* Estab investr, Am Heart Asn, 64- *Mem:* Am Soc Biol Chemists; Am Chem Soc. *Res:* Active transport and oxidative phosphorylation in heart mitochondria. *Mailing Add:* Dept Physiol Chem Ohio State Univ Col Med 5166 Graves Hall Columbus OH 43210

BRIERLEY, JAMES ALAN, b Denver, Colo, Dec 22, 38; m 65. GEOMICROBIOLOGY, BIOHYDROMETALLURGY. *Educ:* Colo State Univ, BS, 61; Mont State Univ, MS, 63, PhD(microbiol), 66. *Prof Exp:* Fel microbiol, Mont State Univ, 61-66; asst prof biol, NMex Inst Mining & Technol, 66-68; res scientist, Martin Marietta Corp, Colo, 68-69; asst prof, 69-78, prof & dept chmn biol, NMex Inst Mining & Technol, 78-83; dir, Advan Mineral Technol, 83-88; NEWMOUNT METALL SERV, 88- *Concurrent Pos:* Vis scientist, Univ Warwick, Eng, 76. *Mem:* Am Soc Microbiol; Brit Soc Gen Microbiol; fel AAAS. *Res:* Microbial leaching of metals; microbial processes for metals recovery. *Mailing Add:* Newmount Metall Serv 417 Wakara Way Suite 210 Salt Lake City UT 84108

BRIESKE, THOMAS JOHN, b Milwaukee, Wis, Sept 2, 39; m 63; c 3. MATHEMATICS EDUCATION. *Educ:* St Mary's Univ, Tex, BA, 61, MS, 63; Univ SC, PhD(math educ), 69. *Prof Exp:* Teacher, High Schs, Tex, 61-63; teaching asst math, Univ Tex, Austin, 63-64; sci programmer, A O Smith Corp, Wis, 64-65; instr math, Wis State Univ, Oshkosh, 65-66; teaching asst, Univ SC, 66-69; asst prof, 69-74, ASSOC PROF MATH, GA STATE UNIV, 74- *Mem:* Math Asn Am. *Res:* Learning of mathematics, in particular, the calculus. *Mailing Add:* Dept Math Ga State Univ Univ Plaza Atlanta GA 30303

BRIGGAMAN, ROBERT ALAN, b Hartford, Conn, Aug 14, 34; m 60; c 1. MEDICINE, DERMATOLOGY. *Educ:* Trinity Col, Conn, BS, 56; NY Univ, MD, 60. *Prof Exp:* Intern internal med, Univ Va Hosp, 60-61, asst resident, 61-62; resident, 64-66, from instr to assoc prof, 67-74, PROF DERMAT, SCH MED, UNIV NC, CHAPEL HILL, 74- *Concurrent Pos:* Res fel, Univ NC, 66-67; NIH spec fel, 67-70. *Mem:* AAAS; Am Acad Dermat; Soc Invest Dermat; Tissue Cult Asn; AMA; Sigma Xi. *Res:* Epidermal-dermal interactions; tissue culture; delayed hypersensitivity; electron microscopy. *Mailing Add:* Dermat Dept NC Mem Hosp-137 Chapel Hill NC 27514

BRIGGLE, LELAND WILSON, b Bismarck, NDak, Oct 6, 20; m 44; c 2. PLANT GENETICS. *Educ:* Jamestown Col, BS, 42; NDak State Univ, BS, 49, MS, 51; Iowa State Univ, PhD(plant breeding), 54. *Prof Exp:* Asst agron, NDak State Univ, 49-51; asst, Iowa State Univ, 52-53; geneticist, Agr Res Ctr, USDA, Beltsville, Md, 54-55, res agronomist, 55-74, scientist, Nat Prog Staff, 75-82, res scientist, Cereal Crops Germplasm Prog, 83-89; RETIRED. *Mem:* AAAS; Am Soc Agron; Crop Sci Soc Am; Sigma Xi. *Res:* Wheat and oat breeding and genetics. *Mailing Add:* 701 Copley Lane Silver Spring MD 20904

BRIGGS, ARTHUR HAROLD, b East Orange, NJ, Nov 3, 30; m 53; c 2. PHARMACOLOGY, MEDICINE. *Educ:* Johns Hopkins Univ, BA, 52, MD, 56. *Prof Exp:* Intern internal med, Univ Hosp, Vanderbilt Univ, 56-57, asst resident, 57-58, res assoc pharmacol, Sch Med, 58-59; from asst prof to prof pharmacol, Sch Med, Univ Miss, 59-68, from instr to asst prof med, 60-68; prof pharmacol & chmn dept, Med Sch, Univ Tex, San antonio, 68- *Concurrent Pos:* US Govt fel, 58-60. *Mem:* Am Soc Pharmacol & Exp Therapeut; Soc Exp Biol & Med. *Res:* Smooth muscle and cardiac muscle pharmacology; alcohol. *Mailing Add:* Health Sci Ctr Univ Tex San Antonio TX 78284

BRIGGS, DALE EDWARD, b Alton, Ill, Nov 27, 30; m 53; c 2. CHEMICAL ENGINEERING. *Educ:* Univ Louisville, BChE, 53; Univ Mich, MSE, 58, PhD(chem eng), 68. *Prof Exp:* From instr to assoc prof, 61-79, PROF CHEM ENG, UNIV MICH, ANN ARBOR, 79- *Concurrent Pos:* Coal consult, Consumers Power Co, Jackson, Mich, 74- *Mem:* Am Inst Chem Engrs; Am Chem Soc. *Res:* Colloid chemistry non-aqueous systems; coal liquefaction; coal gasification air pollution control. *Mailing Add:* Dept Chem Eng Univ Mich Ann Arbor MI 48109-2136

BRIGGS, DARINKA ZIGIC, b Belgrade, Yugoslavia, Sept 2, 32; Can citizen; m 68. STRATIGRAPHY, INFORMATION SCIENCE. *Educ:* Univ Belgrade, Dipl Geol, 57. *Prof Exp:* Field geologist, Serbian Acad Sci, Belgrade, 56; res asst paleont, Royal Ont Mus, Toronto, 58; res geologist, Imp Oil Co, Ltd, Toronto, 59; asst geol, Univ Toronto, 58-60; teacher high sch, Tilbury, Can, 62; res assoc geol, Subsurface Lab, Univ Mich, Ann Arbor, 65-68; CONSULT GEOL, SYSTS ANAL RES INFO CO, 68- *Concurrent Pos:* Adj scientist geol, Lab Subsurface Geol, Univ Mich, Ann Arbor, 64- *Mem:* Am Asn Petrol Geologists; Geol Soc Am; Am Asn Univ Professors; Am Soc Info Sci; Int Asn Math Geol. *Res:* Theoretical and mathematical basis for stratigraphic analysis, and the application to historical reconstructions of paleogeographies, ancient environments, and their utilization in the search for fossil energy resources. *Mailing Add:* 3451 Burbank Dr Ann Arbor MI 48105

BRIGGS, DAVID GRIFFITH, b Chicago, Ill, Apr 4, 32; m 56; c 3. MECHANICAL ENGINEERING, THERMODYNAMICS. *Educ:* Dartmouth Col, BA, 54, MS, 55; Univ Minn, PhD(mech eng), 65. *Prof Exp:* From asst prof to assoc prof, 64-88, PROF MECH ENG, RUTGERS UNIV, 88- *Concurrent Pos:* NASA-Am Soc Eng Educ fac fel, Jet Propulsion Lab, Calif Inst Technol, 70. *Mem:* Am Soc Mech Engrs. *Res:* Free convection heat transfer in stratified fluids; computational fluid mechanics. *Mailing Add:* 14 Renfro Rd Somerset NJ 08873

BRIGGS, DAVID R(EUBEN), chemistry; deceased, see previous edition for last biography

BRIGGS, DONALD K, b Northampton, Eng, Apr 10, 24; US citizen; m 58; c 2. MEDICINE, HEMATOLOGY. *Educ:* Cambridge Univ, MA, 44, MB & BCh, 46. *Prof Exp:* ASST PROF MED, NY UNIV, 57- *Mem:* Am Soc Hemat. *Res:* Genetics; hematology. *Mailing Add:* 170 E 79th St New York NY 10021

BRIGGS, EDWARD M, US citizen. OCEAN ENGINEERING. *Educ:* Univ Tex, BS, 63. *Prof Exp:* Res asst, 59-63, asst res engr, 63-65, res engr, 65-67, sr res engr, Dept Struct Res, 68-72, group leader, 72-73, mgr ocean eng, 73-76, DIR DEPT OCEAN ENG & STRUCT DESIGN, SOUTHWEST RES INST, 76- *Mem:* Marine Technol Soc; Am Soc Testing & Mat; Am Soc Mech Engrs. *Res:* Offshore experimental stress analyses of offshore platforms, submersible barges and deep ocean oil production equipment. *Mailing Add:* 6220 Culebra Rd San Antonio TX 78228-0510

BRIGGS, FAYE ALAYE, b Abonnema, Nigeria, Sept 23, 47. ELECTRICAL ENGINEERING, COMPUTER ENGINEERING. *Educ:* Ahmadu Bello Univ, Zaria, Nigeria, BEng, 71; Stanford Univ, MS, 74; Univ Ill, PhD(elec eng), 77. *Prof Exp:* Asst lectr, Col Sci & Technol, Port-Harcourt, Nigeria, 71-72; res asst, Coord Sci Lab, Univ Ill, 74-76; asst prof, Sch Elec Eng, Purdue Univ, 76-; ASST PROF, ELEC ENG DEPT, RICE UNIV. *Mem:* Inst Elec & Electronics Engrs; Asn Comput Mach. *Res:* Computer architecture; multiprocessing systems; memory organizations; microprocessors and applications; operating systems. *Mailing Add:* Sun Microsysts 2550 Garcia Ave Mountain View CA 94043

BRIGGS, FRED NORMAN, b Oakland, Calif, Sept 12, 24; m 49; c 1. PHYSIOLOGY. *Educ:* Univ Calif, AB, 47, MA, 48, PhD(physiol), 53. *Prof Exp:* Radiologist, US Naval Radiol Defense Lab, 48-49; instr pharm, Dent Sch, Harvard Univ, 52-55, assoc pharmacol, Med Sch, 56-58; from asst prof to assoc prof, Med Sch, Tufts Univ, 58-61; prof physiol, Sch Med, Univ Pittsburgh, 61-71; PROF PHYSIOL & CHMN DEPT, MED COL VA, VA COMMONWEALTH UNIV, 71-; PROF VA COMMONWEALTH, 86- *Concurrent Pos:* Res fel, Harvard Univ, 55-56; sect ed, Am J Physiol; NIH-physiol, 67-70; Am Physiol Soc Ed Comt, 71-74; prof & chmn, Va Commonwealth, 71-85; NIH Cardio Pulmonary, 75-78; chmn, 78-79; NIH Manpower Nat Heart Lung & Blood Inst, 87- *Mem:* Am Soc Pharmacol & Exp Therapeut; Am Physiol Soc; Biophys Soc. *Res:* Thyroid; adrenal; muscle; physiology. *Mailing Add:* Dept Physiol Med Col Va Box 551 MCV Sta Richmond VA 23298

BRIGGS, GARRETT, b Dallas, Tex, Dec 31, 34; m 57; c 4. SEDIMENTOLOGY. *Educ:* Southern Methodist Univ, BS, 58, MS, 59; Univ Wis, PhD(geol), 63. *Prof Exp:* Wis Alumni Res Found asst, Univ Wis, 59-60; geologist, Chevron Oil Co, La, 62-65; asst prof geol, Tulane Univ, 65-68; assoc prof, Univ Tenn, 68-74, interim head dept, 72-74, prof geol & head dept, 74-81, assoc dean, Liberal Arts Col, 77-81; dean, Sch Phys & Math Sci, NC State Univ, 81-88; PRES, PEACE COL, 88- *Mem:* Am Geol Soc. *Res:* General geology and sedimentation studies of late Paleozoic rocks in the Ouachita Mountains of Oklahoma; carboniferous sediments of Cumberland Plateau, Tennessee. *Mailing Add:* Sch Phys & Math Sci NC State Univ Raleigh NC 27650

BRIGGS, GEORGE MCSPADDEN, nutrition, animal science; deceased, see previous edition for last biography

BRIGGS, GEORGE ROLAND, b Ithaca, NY, May 21, 24; m 48; c 2. PHYSICS. *Educ:* Cornell Univ, AB, 47; Univ Ill, MS, 50, PhD(physics), 53. *Prof Exp:* Mem staff, Lincoln Lab, Mass Inst Technol, 52-53; mem tech staff, 54-69, MEM TECH STAFF, DIGITAL SYSTS RES LAB, DAVID SARNOFF RES CTR, RCA CORP, 69- *Honors & Awards:* Achievement Awards, RCA Corp, 54 & 60. *Mem:* Am Phys Soc; Inst Elec & Electronics Engrs; Asn Comput Mach. *Res:* Magnetic and ferroelectric computer memory and logic devices; electro-luminescent-magnetic computer and television displays; nuclear-radiation resistant computer memory and logic circuitry. *Mailing Add:* 37 Broadripple Dr Princeton NJ 08540

BRIGGS, HILTON MARSHALL, b Cairo, Iowa, Jan 9, 13; m 35; c 2. ANIMAL HUSBANDRY. *Educ:* Iowa State Univ, BS, 33; NDak State Univ, MS, 35; Cornell Univ, PhD(nutrit), 38. *Hon Degrees:* DSc, NDak State Univ, 63; Dr Higher Educ & Admin, Univ SDak, 74. *Prof Exp:* Asst, NDak State Univ, 34-35 & Cornell Univ, 35-36; from asst prof to prof animal husb, Okla Agr & Mech Col, 36-50, assoc dean agr & assoc dir exp sta, 49-50; dean col

agr & dir exp sta, Univ Wyo, 50-58; pres, 58-75, EMER PRES & DISTINGUISHED PROF AGR, SDAK STATE UNIV, 75- *Mem:* Fel AAAS; fel Am Soc Animal Sci (secy, 47-50, vpres, 51, pres, 52); Sigma Xi. *Mailing Add:* SDak State Univ Rm 241 Libr Box 2115 Univ Sta Brookings SD 57007-1497

BRIGGS, JEFFREY L, US citizen. ENVIRONMENTAL IMPACT ASSESSMENT, WETLANDS DELINEATION & PERMITTING. *Educ:* Univ Denver, BS, 65; Ore State Univ, MA, 68, PhD(ecol), 70. *Prof Exp:* Asst prof biol & zool, Univ Wis-Whitewater, 70-74; cur III, Milwaukee Pub Mus, 74-75; lectr, Univ Wis-Milwaukee, 75-78 & Univ Wis-Waukesha, 78-79; assoc prof biol, Col Med, King Faisal Univ, Damman, Saudi Arabia, 79-84; PROJ MGR ENVIRON STUDIES, SCS ENGRS, 85- *Mem:* Am Soc Ichthyologists & Herpetologists; Soc Wetland Scientists; Ecol Soc Am. *Res:* Environmental impact assessment and wetlands delineation; field ecology studies; environmental risk assessment; flora and fauna assessment and mapping; habitat restoration; hazardous waste impact assessment and remediation. *Mailing Add:* SCS Engrs 11260 Roger Bacon Dr Reston VA 22090

BRIGGS, JOHN CARMON, b Portland, Ore, Apr 9, 20; m 48; c 9. ICHTHYOLOGY, HISTORICAL BIOGEOGRAPHY. *Educ:* Ore State Univ, BS, 43; Stanford Univ, MA, 47, PhD(biol), 52. *Prof Exp:* Aquatic biologist, US Fish & Wildlife Serv, Ore, 45-46; asst gen biol, Stanford Univ, 47-48; instr, Ore Inst Marine Biol, 48; asst gen biol, Hopkins Marine Sta, 49; biologist, Calif State Div Fish & Game, 50-51; res assoc, Stanford Natural Hist Mus, 52-54; from instr to asst prof biol, Univ Fla, 54-57, res assoc anat, 57-58; asst prof, Univ BC, 58-59, asst prof zool, 59-61; asst prof, Hopkins Marine Sta, 61; res scientist, Inst Marine Sci, Tex, 61-64; chmn dept zool, 64-71, prof biol, 64-75, dir grad studies, 71-75, actg chmn, dept marine sci, 76-79, prof marine sci, 79-85, EMER PROF, UNIV SFLA, 85-; RES ASSOC, NAT HIST MUS, UNIV GA, 90- *Mem:* Am Soc Ichthyol & Herpet (vpres, 59, 64, treas, 81-85); fel Am Inst Fishery Res Biologists; Am Soc Naturalists; Am Inst Biol Sci; Am Fisheries Soc; Soc Syst Zool; Soc Study Evolution. *Res:* Ichthyology; historical biogeography; evolutionary significance of zoogeographic patterns. *Mailing Add:* 1260 Julian Dr Watkinsville GA 30677

BRIGGS, JOHN DORIAN, b Santa Monica, Calif, Dec 20, 26; m 51; c 3. INSECT PATHOLOGY. *Educ:* Univ Calif, BS, 51, PhD(entom), 56. *Prof Exp:* Assoc entomologist, Ill Nat Hist Surv, 55-59; head entom dept, Bioferm Corp, 59-62; actg dean col biol sci, 68-69, PROF ENTOM, OHIO STATE UNIV, 62- *Concurrent Pos:* Fulbright sr res fel, Mexico, 80-81. *Mem:* AAAS; Entom Soc Am; Soc Invert Path (vpres, pres, 70-74). *Res:* Insect diseases; entomogenous bacteria, fungi, protozoa and viruses; invertebrate immunology; biological control of invertebrates of public health and agricultural importance; biology of aging. *Mailing Add:* Dept Entom Ohio State Univ Columbus OH 43210-1220

BRIGGS, JOSEPHINE P, b Toronto, Ont, Dec 14, 44; US citizen; m 79; c 2. RENAL PHYSIOLOGY. *Educ:* Harvard Univ, AB, 66, MD, 70. *Prof Exp:* Int resident internal med, Mt Sinai Med Sch, 70-76; chief resident res staff physiol, Yale Univ, 76-80; vis scientist, Univ Munich, 80-85; asst prof, 85-87, ASSOC PROF NEPHROL, UNIV MICH, 87- *Concurrent Pos:* Alexander von Humboldt fel, 80-82; estab investr, Am Heart Asn, 83-88; counr, Am Soc Clin Invest, 89- *Mem:* Women Nephrology (pres-elect, 90-91); Am Soc Clin Invest; Am Soc Nephrol; Am Fedn Clin Res. *Res:* Renal physiology; aiming to elucidate cellular mechanisms operating within the juxtaglomerular apparatus to regulate renin secretion and renal blood flow. *Mailing Add:* Nephrol Res Lab George M O'Brien Renal Ctr 1560 MSRB II 1150 W Medical Center Dr Ann Arbor MI 48109-0676

BRIGGS, PHILLIP D, b Ada, Okla, Dec 10, 37. MATHEMATICS. *Educ:* Univ Okla, BS, 59, MA, 63. *Prof Exp:* PROF MATH, E CENT OKLA STATE UNIV, 65- *Mem:* Math Asn Am. *Mailing Add:* Dept Math E Cent Okla State Univ Ada OK 74820

BRIGGS, REGINALD PETER, b Port Chester, NY, Mar 12, 29; m 56; c 3. GEOLOGY OF PETROLEUM, ENVIRONMENTAL GEOLOGY & HYDROLOGY. *Educ:* Wesleyan Univ, BA, 51. *Prof Exp:* Geologist, 53-76, chief PR Coop Geol Mapping Proj, 64-70, proj dir greater Pittsburgh regional studies, US Geol Surv, 70-76; PRES & CHIEF GEOLOGIST, GEOMEGA, INC, 76- *Concurrent Pos:* Managing ed, Geol Penn proj, 83- *Honors & Awards:* Spec Achievement Award, US Dept Interior, 76. *Mem:* Fel Geol Soc Am; Am Geophys Union; Am Inst Prof Geologists; Asn Eng Geologists. *Res:* Geology of volcanic rocks; Appalachian plateau geomorphology; slope stability; mineral fuels; metallic and nonmetallic mineral resources; reclamation; limestone hydrology; forensic applications. *Mailing Add:* Geomega Inc PO Box 12933 Pittsburgh PA 15241-0933

BRIGGS, RICHARD JULIAN, b Shanghai, China, May 26, 37; US citizen; m 60; c 3. PLASMA PHYSICS. *Educ:* Mass Inst Technol, BS & MS, 61, PhD(elec eng), 64. *Prof Exp:* Instr elec eng, Mass Inst Technol, 61-64; plasma physicist, Lawrence Radiation Lab, Univ Calif, 64-66; from asst prof to assoc prof elec eng, Mass Inst Technol, 66-72; physicist & charged particle beam prog leader, Lawrence Livermore Lab, Univ Calif, 72-88, assoc dir, Beam Res & Magnetic Fusion, 88-; SSC LAB. *Concurrent Pos:* Consult, Lawrence Livermore Lab, 66-72, Microwave Assocs, 67-70 & Avco-Everett Res Lab, 67-72. *Mem:* Fel Am Phys Soc. *Res:* Controlled fusion; relativistic electron beams. *Mailing Add:* SSC Lab Stoneridge Off Park 2550 Beckleymeade Ave MS 1070 Dallas TX 75237-3946

BRIGGS, ROBERT CHESTER, b Iron Mountain, Mich, June 6, 44; m 72; c 2. BIOLOGY, BIOCHEMISTRY. *Educ:* Northern Mich Univ, BS, 66, MA, 72; Univ Vt, PhD(zool), 77. *Prof Exp:* Res assoc, 76-78, res instr, 78-80, RES ASST PROF BIOCHEM, SCH MED, VANDERBILT UNIV, 80-, ASST PROF PATH, 83- *Mem:* Am Soc Cell Biol; AAAS; Am Asn Cancer Res; Sigma Xi; Am Asn Pathologists; Am Soc Biochem & Molecular Biol. *Res:* Nuclear protein function in normal eucaryotic cell differentiation and during malignant transformation. *Mailing Add:* 2706 Sunset Pl Nashville TN 37212

BRIGGS, ROBERT EUGENE, b Madison, Wis, Apr 4, 27; m 48; c 4. AGRONOMY. *Educ:* Univ Wis, BS, 50, PhD(agron), 58; Mich State Univ, MS, 52. *Prof Exp:* Asst farm crops, Mich State Univ, 50-52; asst agron, Univ Wis, 53-56; asst plant breeder, Univ Ariz, 56-57, from asst prof to assoc prof agron, 57-70, prof agron & plant genetics, 70-, agronomist, Agr Exp Sta & Exten Agronomist, Coop Exten Serv, 74-; RETIRED. *Concurrent Pos:* Adv field crops, US AID-Brazil Prog, 64-66. *Mem:* Am Soc Agron. *Res:* Crop production; seed technology; cultural practices; fiber quality; cotton; physiology; management. *Mailing Add:* 3326 E Lester St Tucson AZ 85716

BRIGGS, ROBERT WILBUR, b Delavan, Ill, Mar 27, 34; m 62; c 2. GENETICS, AGRICULTURE. *Educ:* Univ Ill, BS, 56, MS, 58; Univ Minn, PhD(genetics), 63. *Prof Exp:* Res assoc genetics, Brookhaven Nat Lab, 63-65, from asst geneticist to assoc geneticist, 65-69; geneticist, 69-78, PRIN RES SCIENTIST, RES DEPT, FUNK SEEDS INT, 79- *Concurrent Pos:* Fel biol, Brookhaven Nat Lab, 63-64. *Mem:* Genetics Soc Am; Crop Sci Soc Am. *Res:* Chemical mutagens in plants; application of biotechnology to plant breeding and genetics. *Mailing Add:* Ciba-Geigy Seed Div Res Dept PO Box 2911 Bloomington IL 61702-2911

BRIGGS, THOMAS, b New York, NY, May 24, 33; div; c 3. BIOCHEMISTRY. *Educ:* Yale Univ, BS, 54; Univ Pa, PhD(biochem), 60. *Prof Exp:* Res asst med, Yale Univ, 62-63, res assoc, 63-67; asst prof biochem, 67-70, ASSOC PROF BIOCHEM & MOLECULAR BIOL, COLS MED & DENT, UNIV OKLA, 70-, ADJ ASSOC PROF PHARM, 85- *Concurrent Pos:* USPHS fel, Guy's Hosp Med Sch, London, 60-62; vis scientist, NIH, 81. *Mem:* Sigma Xi; AAAS; Am Chem Soc; NY Acad Sci; Brit Biochem Soc; Soc Exp Biol & Med; assoc mem Am Soc Biochem & Molecular Biol. *Res:* Bile salt chemistry and metabolism; cholesterol metabolism; comparative biochemistry; biochemical evolution; metabolism of lipoproteins. *Mailing Add:* Dept Biochem & Molecular Biol Univ Okla Health Sci Ctr Oklahoma City OK 73190

BRIGGS, THOMAS N, m; c 3. ANALYTICAL CHEMISTRY. *Educ:* Inter Am Univ, PR, BS, 69; Case Western Reserve Univ, PhD(phys & anal chem), 76. *Prof Exp:* Anal chemist, plastics div, Gen Elec Co, 76-80, res anal chemist, lamp div, 80-83; mgr anal serv, Fasson Div, 83-87, DIR RES GRAPHICS SYST DIV, AVERY INT, 87- *Mem:* Am Chem Soc; AAAS. *Mailing Add:* Avery Graphics Syst Div 701 E 83rd Ave Suite B Merrillville IN 46410-6239

BRIGGS, WILLIAM EGBERT, b Sioux City, Iowa, Mar 26, 25; m 47; c 4. ALGEBRA, BIOMATHEMATICS. *Educ:* Morningside Col, BA, 48; Univ Colo, MA, 49, PhD(math), 53. *Hon Degrees:* DSc, Morningside Col, 68. *Prof Exp:* Instr math, Univ Colo, 48-53, res asst, 53-54; teacher pub sch, Colo, 54-55; from instr to assoc prof math, Univ Colo, Boulder, 55-64, actg dean, Col Arts & Sci, 63-64, dean, Col Arts & Sci, 64-80, prof math, 64-88; RETIRED. *Concurrent Pos:* Teacher high sch, Iowa, 47-48; dir acad yr inst high sch teachers, NSF, 56-60, actg chmn dept math, 59-60; NSF fel & res assoc, Univ Col, Univ London, 61-62; mem bd dirs, Educ Projs, Inc, 66-; bd dir, Coun Cols Arts & Sci, 69-73 & 74-77; mem regional selection comt, Woodrow Wilson Fel Found, 70-71; mem comn arts & sci, Nat Asn Land Grant Cols & Univs, 70-73 & 76-79. *Mem:* Coun Cols Arts & Sci (pres-elect, 74-75, pres, 75-76); Sigma Xi; Am Math Soc; Math Asn Am; London Math Soc. *Res:* Analytic number theory; prime number theory; dirichlet L-functions and Riemann zeta function; coef of power series; sequences generated by sieve processes; polynomials for protein-ligand binding. *Mailing Add:* Campus Box 426 Univ Colo Boulder CO 80309

BRIGGS, WILLIAM SCOTT, b Shelton, Wash, Aug 15, 41; m 63. ORGANIC CHEMISTRY, ANALYTICAL CHEMISTRY. *Educ:* Univ Wash, BS, 63; Stanford Univ, PhD(org chem), 68. *Prof Exp:* Res chemist, Bellingham Div, Ga-Pac Corp, 67-84; CHEM CONSULT, W S BRIGGS & ASSOCS, 84- *Concurrent Pos:* Lectr chem, Western Wash Univ, 84-88. *Mem:* Am Chem Soc; Am Tech Asn Pulp & Paper Indust. *Res:* Wood chemistry and utilization of pulping wastes; application of chromatographic methods and instrumental techniques to the solution of organic structural problems; mass spectrometry; production of commercial chemicals from biomass by chemical and enzymic methods. *Mailing Add:* 15004 NE Sorrel Dr Vancouver WA 98682

BRIGGS, WINSLOW RUSSELL, b St Paul, Minn, Apr 29, 28; m 55; c 3. PLANT PHYSIOLOGY. *Educ:* Harvard Univ, AB, 51, AM, 52, PhD(biol), 56. *Prof Exp:* From instr to prof biol, Stanford Univ, 55-67; prof, Harvard Univ, 67-73; DIR DEPT PLANT BIOL, CARNEGIE INST WASH, 73- *Honors & Awards:* Alexander von Humboldt US Sr Scientist Award. *Mem:* Nat Acad Sci; AAAS; Bot Soc Am; Am Soc Plant Physiol (pres, 75-76); Am Acad Arts & Sci; Akademie der Naturforscher Leopoldina. *Res:* Plant growth, development; physiology, biochemistry and molecular biology of photomorphogenesis. *Mailing Add:* Dept Plant Biol 290 Panama St Stanford CA 94305

BRIGHAM, KENNETH LARRY, b Nashville, Tenn, Oct 29, 39; m 59; c 1. PULMONARY PHYSIOLOGY. *Educ:* David Lipscomb Col, BA, 62; Vanderbilt Univ, MD, 66. *Prof Exp:* From intern to asst resident internal med, Johns Hopkins Hosp, Baltimore, 66-68; med epidemiologist, Nat Commun Dis Ctr, USPHS Ecol Invest Prog, 68-70; instr internal med, Sch Med, Vanderbilt Univ, 70-71; NIH res fel pulmonary med, Cardiovasc Res Inst, Univ Calif, San Francisco, 71-73; from asst prof to assoc prof, 73-78, assoc biomed eng, 77-78, PROF MED, VANDERBILT UNIV, 78-, PROF BIOMED ENG, 86- *Concurrent Pos:* Dir, Pulmonary Res, Sch Med, Vanderbilt Univ, 73-76, Specialized Ctr Res, 76-, dir pulmonary med, Med Ctr, 78-; mem pulmonary study sect, Nat Heart & Lung Inst, 75-79; mem lung res comt, Vet Admin, 76; estab investr, Am Heart Asn, 75-88; vpres, Nashville Soc Internal Med, 85; pres-elect, Am Thoracic Soc, 88-89. *Honors & Awards:* Simon Rodbard mem-lectr, Am Col Chest Physicians, 87. *Mem:* NY Acad Sci; Am Soc Clin Invest; AAAS; Am Fedn Clin Res; Am Thoracic Soc; Am Physiol Soc; Am Asn Prof; Am Soc Cell Biol; Micros Soc. *Res:* Humoral mechanisms in the pathogenesis of pulmonary edema resulting primarily from changes in the lung circulation. *Mailing Add:* Pulmonary Circulation Ctr Med Ctr N Vanderbilt Univ B-1308 Nashville TN 37232

BRIGHAM, NELSON ALLEN, b Holyoke, Mass, Nov 6, 15. MATHEMATICS. *Educ:* Rutgers Univ, BS, 37, MS, 38; Univ Pa, PhD(math), 48. *Prof Exp:* Asst, Inst Math, Univ Pa, 38-41; sr res engr, Repub Aviation Corp, NY, 47; asst prof math, Swarthmore Ctr, Pa State Col, 47-48; asst prof, Univ Md, 48-51, mathematician, Appl Physics Lab, 51-56; sr scientist, Avco Mfg Corp, 56-59; prin mathematician, Baird-Atomic, Inc, 59-60; staff mem, Lincoln Lab, Mass Inst Technol, 60-62 & Mitre Corp, 62-64; assoc prof math, Southeastern Mass Tech Inst, 65; prof math & astron & head dept, Butler Univ, 65-70; prof math & chmn dept, Bradley Univ, 70-71; pvt tutoring serv, 71-91; RETIRED. *Mem:* Am Math Soc; Math Asn Am. *Res:* Analytic additive theory of numbers; applied mathematics; digital and analog simulation and programming. *Mailing Add:* 753 W Wonderview Dr Dunlap IL 61525

BRIGHAM, RAYMOND DALE, b Stamford, Tex, Apr 1, 26; m 49; c 4. PLANT BREEDING. *Educ:* Tex Tech Col, BS, 50; Iowa State Col, MS, 52, PhD(crop breeding, plant path), 57. *Prof Exp:* Res agronomist, Agr Res Serv, USDA, 57-67; ASSOC PROF AGRON, TEX AGR EXP STA, TEX A&M UNIV, LUBBOCK, 67- *Concurrent Pos:* Prog chmn, Int Sunflower Conf, Memphis, 70; vis prof, India, 81; consult, Ministry of Agr, Pakistan, 82, Agr Corp, Burma, 88. *Mem:* AAAS; Am Phytopath Soc; Crop Sci Soc Am; Am Soc Agron; NY Acad Sci. *Res:* Breeding for disease resistance in oilseed crops; inheritance of quantitative and qualitative characteristics. *Mailing Add:* Tex Agr Exp Sta Rte 3 Box 219 Lubbock TX 79401-9757

BRIGHAM, WARREN ULRICH, b Oak Park, Ill, Feb 27, 42. AQUATIC BIOLOGY, AQUATIC ENTOMOLOGY. *Educ:* Univ Ill, BS, 64, PhD(zool), 72; Tenn Tech Univ, MS, 66. *Prof Exp:* Tech asst, 66-72, asst aquatic biologist, 72-76, ASSOC AQUATIC BIOLOGIST, ILL NATURAL HIST SURV, 76-; PARTNER, MIDWEST AQUATIC ENTERPRISES, 73- *Mem:* Coleopterists Soc; Ecol Soc Am; Am Soc Limnol & Oceanog; Soc Int Limnol; NAm Benthological Soc. *Res:* Streams and their fauna; taxonomy and ecology of aquatic coleoptera; water quality requirements of aquatic insects; threatened and endangered fauna. *Mailing Add:* Aquatic Biol Sect Ill Natural Hist Surv 99 Nat Res Bldg 607 E Peabody Champaign IL 61820

BRIGHAM, WILLIAM EVERETT, b Murphysboro, Ill, Apr 1, 29; m 54; c 5. CHEMICAL ENGINEERING. *Educ:* Iowa State Univ, BS, 50; Univ Okla, MS, 56, PhD(chem eng), 62. *Prof Exp:* Chem engr, S C Johnson & Sons, 50-51, 53-54; instr chem eng, Univ Okla, 56-58; res engr, Res & Develop Dept, Continental Oil Co, 58-60, sr res engr, 60-62, res group leader, 62-65, res sect leader, 65-67, res assoc, Prod Res Div, 67-71; assoc chmn dept petrol eng, 79-87, PROF, STANFORD UNIV, 71- *Concurrent Pos:* Mem, Ed Rev Comt, J Petrol Technol & Soc Petrol Engrs J, 65-69; session co-chmn, World Petrol Cong, 76; mem, Geothermal Resources Coun, 77-; Fulbright-Hays grant, Heriot-Watt Univ, Scotland, 78-79; Los Alamos Lab Adv Comt, 82-; mem, Engr Manpower Comt, Soc Petrol Engrs, 85- *Honors & Awards:* Ferguson Award, Soc Petrol Engrs, 77; John Franklin Carll Award, Soc Petrol Engrs, 88; Homer H Lowry Award Fossil Energy, 90. *Mem:* Soc Petrol Engrs-Am Inst Mining, Metall & Petrol Engrs. *Res:* Geothermal engineering; enhanced recovery of heavy oil; reservoir mixing theories; new cell test techniques; other transient flow, oil and gas recovery phenomena. *Mailing Add:* Dept Petrol Eng Stanford Univ Lloyd Noble Bldg Stanford CA 94305

BRIGHT, ARTHUR AARON, b Hanover, NH, Dec 31, 46; m 69; c 2. EXPERIMENTAL SOLID STATE PHYSICS. *Educ:* Dartmouth Col, AB, 69, MA, 70; Univ Pa, PhD(physics), 73. *Prof Exp:* Res investr physics, Univ Pa, 73-75; res scientist physics, Union Carbide Corp, Carbon Prod, 75-78; RES SCIENTIST PHYSICS, IBM WATSON RES CTR, 78- *Concurrent Pos:* Chmn Thin Films Div, Am Vacuum Soc, 91. *Mem:* Am Physics Soc; Am Vacuum Soc; Mat Res Soc. *Res:* Solid state physics; plasma processing of silicon technology materials. *Mailing Add:* IBM Thomas J Watson Res Ctr Yorktown Heights NY 10598

BRIGHT, DONALD EDWARD, b Columbus, Ohio, Feb 10, 34; m 60; c 4. SYSTEMATIC ENTOMOLOGY. *Educ:* Colo State Univ, Ft Collins, BSc, 57; Brigham Young Univ, MSc, 61; Univ Calif, Berkeley, PhD(entom), 65. *Prof Exp:* Asst res entomologist, dept entom, Univ Calif, 65-66; head, Coleoptera Sect, 72-77, RES SCIENTIST, BIOSYST RES CTR, CAN DEPT AGR, 66-, PROJ LEADER, INSECT PESTS PROJ, 90- *Concurrent Pos:* Mem, syst res comt, Entom Soc Am, 79-87; mem, sci pol comt, Entom Soc Can, 79-84; mem, Biol Coun Can, 79-84; gen chmn ann meeting, Entom Soc Can, 85, mem, Ethics Comt, 89- *Mem:* Entom Soc Am; Sigma Xi; Coleopterists Soc (secy, 84-); Entom Soc Can (treas, 86-90). *Res:* Systematics and biology of Scolytidae coleoptera; general biology of forest coleoptera; forest entomology. *Mailing Add:* Biosyst Res Ctr Cent Exp Farm Ottawa ON K1A 0C6 Can

BRIGHT, GEORGE WALTER, b Port Arthur, Tex, Apr 24, 45. MATHEMATICS EDUCATION. *Educ:* Rice Univ, BA, 68, MA, 69; Univ Tex, Austin, PhD(math educ), 71. *Prof Exp:* From asst prof to assoc prof math, Northern Ill Univ, 71-83; assoc prof, Univ Calgary, 84-85; from assoc prof to prof cur & instr, Univ Houston, 85-90; PROF EDUC, UNIV NC-GREENSBORO, 90- *Concurrent Pos:* Asst prof math, Emory Univ, 74-76; Vis assoc prof math, Univ Wis-Madison, 79-80. *Mem:* Nat Coun Teachers Math; Math Asn Am; Am Educ Res Asn. *Res:* Supplementary instructional techniques, games and micropcomputers in teaching mathematics; psychology of mathematics learning; student performance in algebra. *Mailing Add:* Sch Educ Univ NC-Greensboro Greensboro NC 27412-5001

BRIGHT, GORDON STANLEY, b Smethport, Pa, Apr 29, 15; m 39; c 3. CHEMISTRY. *Educ:* Tusculum Col, AB, 38; Univ Cincinnati, MS, 39. *Prof Exp:* Head dept sci, Pfeiffer Col, 39-40; res chemist, Texaco Inc, 40-54, supvr, Grease Res Dept, 54-62, res assoc, 62-80; RETIRED. *Mem:* Fel AAAS; fel Nat Lubricating Grease Inst; Am Chem Soc; Am Soc Lubrication Eng. *Res:* Grease and other lubricants. *Mailing Add:* PO Box 940 Flat Rock NC 28731

BRIGHT, HAROLD FREDERICK, b Smethport, Pa, Aug 6, 13; m 38; c 2. STATISTICS. *Educ:* Lake Forest Col, BA, 37; Univ Rochester, MS, 44; Univ Tex, PhD(educ psychol), 52. *Prof Exp:* Asst prof math, Denison Univ, 43-44 & Univ Rochester, 44-45; registr & dir guid, San Angelo Col, 45-49; assoc dir res, Am Asn Jr Cols, Univ Tex, 49-52; chief tech serv, Human Resources Res Off, George Washington Univ, 52-54, dept dir, 54-56; specialist opers res & synthesis, Gen Elec Co, 57-58; chmn dept statist, 58-64, assoc dean faculties, 64-66, PROF STATIST, GEORGE WASHINGTON UNIV, 58-, VPRES ACAD AFFAIRS, 66-, PROVOST, 69- *Mem:* AAAS; Am Psychol Asn; Inst Math Statist; Am Statist Asn; Math Asn Am; Sigma Xi. *Res:* Teaching of statistics; computing problems; operations research. *Mailing Add:* 4132 Northwest Blvd Davenport IA 52806

BRIGHT, PETER BOWMAN, b Gallipolis, Ohio, Dec 27, 37; div; c 3. TECHNICAL MANAGEMENT. *Educ:* Antioch Col, BS, 60; Univ Chicago, PhD(math biol), 66; Univ Calif, Los Angeles, MBA, 89. *Prof Exp:* Fel math biol, Univ Chicago, 66; res instr biophys, Ctr Theoret Biol, State Univ, NY Buffalo, 66-69; asst prof bioinfo sci, Dept Surg, Southwestern Med Sch, Univ Tex, 69-70, asst prof biophys & surg, 70-73; mem fac, Dept Biomath, Sch Med, Univ Calif, Los Angeles, 73-75; asst prof math, Calif State Univ, Northridge, 77-80; consult, Aerospace Corp, El Segundo, 75-80; proj dir, Critical Care Inst, Univ Southern Calif, 80-81; mem tech staff, Aerospace Corp, Los Angeles, 81-84, proj engr, 84-88, mgr, 88-90; SR SYSTS ENGR, GEN ELEC CORP, LOS ANGELES, CALIF, 91- *Concurrent Pos:* Adj prof, Dept Statist, Southern Methodist Univ, 70-73. *Mem:* Inst Elec & Electronics Engrs; Soc Math Biol. *Res:* Risk management. *Mailing Add:* 5412 W 124th St Hawthorne CA 90250-3450

BRIGHT, ROBERT C, b Salt Lake City, Utah, Dec 27, 28. GEOLOGY. *Educ:* Univ Utah, BS, 52, MS, 60; Univ Minn, PhD(geol), 63. *Prof Exp:* Instr geol, Univ Minn, 59; NSF fel, Royal Univ Uppsala, 63-64; dir, Wasatch-Uinta Field Geol Camp, 67-72, ASSOC PROF GEOL, UNIV MINN, MINNEAPOLIS, 64-, ASSOC PROF ECOL, 71-, CUR PALEONT, MUS NATURAL HIST, 64-; ASSOC DIR, BELL MUS, 78- *Mem:* AAAS; Am Asn Petrol Geologists; fel Geol Soc Am; Paleont Soc; Am Soc Limnol & Oceanog; Sigma Xi. *Res:* Pleistocene stratigraphy, paleontology; paleoecology and paleolimnology. *Mailing Add:* Mus Natural Hist Univ Minn Minneapolis MN 55455

BRIGHT, THOMAS J, b Millen, Ga, Sept 2, 37; m 60; c 2. BIOLOGICAL OCEANOGRAPHY. *Educ:* Univ Wyo, BS, 64; Tex A&M Univ, PhD(biol oceanog), 68. *Prof Exp:* Asst prof biol & invert zool, Jacksonville Univ, 68-69; from asst prof to assoc prof oceanog, 69-81, PROF OCEANOG, TEX A&M UNIV, 81- *Mem:* Am Soc Limnol & Oceanog. *Res:* Ecology of coral reefs and hard-banks. *Mailing Add:* Dept of Oceanog Tex A&M Univ College Station TX 77843

BRIGHT, WILLARD MEAD, b New York, NY, Mar 26, 14; m 44, 81; c 1. PHYSICAL CHEMISTRY. *Educ:* Univ Toledo, BS, 36; Harvard Univ, MA, 41, PhD(chem), 42. *Prof Exp:* Res chemist, Kendall Co, 41-44; res chemist, Bauer & Black, 44-46, group leader, 46-48; dir, Clark Lab, Kendall Co, 48-52; asst res dir, Lever Bros Co, 52-54, dir res & develop, 54-60, vpres, 60-64, dir, 62-64; vpres & dir, R J Reynolds Tobacco Co, 64-68; sr vpres & dir, Warner-Lambert Pharmaceut Co, 68-70; pres, chief exec officer & dir, Kendall Co, 70-73; pres & dir, Curtiss-Wright Corp, 74-75; pres & dir, Boehringer Mannheim Corp, 75-82; CHMN & DIR, ZMI CORP, 82- *Concurrent Pos:* Mem bd vis, dept chem, Harvard, 59-61; vis com, dept chem engr, Univ Rochester, 62-64; trustee, Corn Indust Res Found, 65-68; mem, Adv Comt on Patents, US Dept Com, 66-69; chmn, Res Comt, Nat Asn Mfrs, 66-68; dir, Am Found Pharmaceut Educ, 69-71; mem bd vis, Dept Chem, Boston Univ, 77- *Mem:* Nat Asn Mfrs; AAAS; NY Acad Sci; Am Chem Soc; Indust Res Inst (pres, 67-68). *Mailing Add:* 134 Prestwick Circle Vero Beach FL 32967

BRIGHTMAN, I JAY, b New York, NY, Oct 28, 09; m 35; c 3. MEDICINE, PUBLIC HEALTH. *Educ:* NY Univ, BS, 30, MD, 34, MedScD(internal med), 40; Am Bd Internal Med, dipl, 41; Columbia Univ, MPH, 43; Am Bd Prev Med, dipl, 55. *Prof Exp:* Dir res proj, Div Syphilis Control, NY State Dept Health, 41-42, dir res proj, Veneral Dis Control, Buffalo, 43-45, asst dir div, 45-46, assoc physician, Div Med Serv, 46-48, asst dir div, 48-50, asst comnr, 50-51, asst comnr welfare med serv, 52-56, exec dir, Health Resources Bd, 56-60, asst comnr chronic dis control, 60-66, dep comnr med serv & res, 66-69; dir, New York Metrop Regional Med Prog, 69-72; CONSULT HEALTH PLANNING, 73- *Concurrent Pos:* Assoc prof, Albany Med Col, 51-; mem, Comt Welfare Serv, AMA, 54-60, consult, 60; mem, Nat Adv Community Health Comt, USPHS, 62-73. *Mem:* Fel Am Col Physicians; Am Col Prev Med. *Res:* Chronic illness, especially natural history of disease, need for specialized resources and personnel, and medical economics. *Mailing Add:* 30 Cole Ave Apt 6E Spring Valley NY 10977

BRIGHTMAN, MILTON WILFRED, b Toronto, Ont, July 13, 23; nat US; m 53; c 2. NEUROANATOMY. *Educ:* Univ Toronto, AB, 45, AM, 48; Yale Univ, PhD(anat), 54. *Prof Exp:* NEUROANATOMIST, LAB NEUROANAT SCI, NAT INST NEUROL DIS & BLINDNESS, 54- *Mem:* Am Asn Anat. *Res:* Perivascular spaces in brain; neurosecretion; blood supply of spinal cord; neurocytology. *Mailing Add:* Bldg 36 Rm 3B-26 NIH Bethesda MD 20205

BRIGHTMAN, VERNON, b Brisbane, Australia, Dec 17, 30; m 62; c 3. DENTISTRY, ORAL MEDICINE. *Educ:* Univ Queensland, BDSc, Hons, 52, MDSc, 56; Univ Chicago, PhD(microbiol), 60; Univ Pa, DMD, 68. *Prof Exp:* Lectr dent, Univ Queensland, 52-56; asst, Zoller Dent Clin, Univ Chicago, 56-60; instr microbiol, Univ Pa, 60-62; sr lectr, Univ Queensland, 62-64; from assoc to assoc prof, 64-70, PROF ORAL MED, SCH DENT MED, UNIV PA, 70-, ACTG CHMN DEPT ORAL MED, 78-, PROF OTORHINOLARYNGOLOGY, 73-, MEM STAFF, CHILDREN'S HOSP, 73- *Concurrent Pos:* Chief dent res, Philadelphia Gen Hosp, 64-72; vis prof dept anat, Sch Vet Med, Cornell Univ, 75-76. *Honors & Awards:* Univ Medal, Univ Queensland, 52. *Mem:* Fel AAAS; fel Am Acad Oral Path; fel Am Acad Oral Med; Int Asn Dent Res; Am Dent Asn. *Res:* Oral medicine and pathology. *Mailing Add:* Gen Clin Res Ctr 4001 Spruce St Philadelphia PA 19104

BRIGHTON, CARL T, b Pana, Ill, Aug 20, 31; m 54; c 4. ORTHOPEDIC SUGERY. *Educ:* Valparaiso Univ, BA, 53; Univ Pa, MD, 57; Univ Ill, Chicago Circle, PhD(anat), 69; Am Bd Orthop Surg, dipl, 65. *Prof Exp:* Staff orthopedist, Philadelphia Naval Hosp, USN, 62-63, Great Lakes Naval Hosp, Ill, 63-66 & USS Sanctuary, South China Sea, 66-67; from asst prof to prof, 68-77, PAUL B MAGNUSON PROF BONE & JOINT SURG, UNIV PA, 77-, CHMN DEPT ORTHOP SURG, 77-, DIR ORTHOP RES, 68- *Concurrent Pos:* NIH res career develop award, 71-76; mem, NAMSD Adv Coun, NIH, 88-92. *Honors & Awards:* Shands Lect Award, Orthop Res Soc, 85; Merit Award, Nat Inst Arthritis & Musculoskeletal & Skin Dis, 87. *Mem:* Orthop Res Soc; fel Am Acad Orhtop Surg; fel Am Col Surgeons; Bioelec Repair & Growth Soc; Am Orthop Asn. *Res:* Epiphyseal plate growth and development; fracture healing; arthritis and articular cartilage preservation and transplantation; bioelectricity; osteoporosis. *Mailing Add:* Dept Orthop Univ Pa Philadelphia PA 19104

BRIGHTON, JOHN AUSTIN, b Gosport, Ind, July 9, 34; m 53; c 2. MECHANICAL & BIOMEDICAL ENGINEERING. *Educ:* Purdue Univ, BS, 59, MS, 60, PhD(mech eng), 63. *Prof Exp:* Lab technician, Schwitzer Corp, 52-57; instr mech eng, Purdue Univ, 60-62; asst prof, Carnegie Inst Technol, 62-66; assoc prof, Pa State Univ, 66-77; CHMN DEPT MECH ENG, MICH STATE UNIV, 77- *Concurrent Pos:* NSF grant, 64-65; consult, Nat Heart & Lung Inst, NIH, 78; tech ed, J Biomech Eng, Am Soc Mech Engrs. *Mem:* Am Soc Mech Engrs; Am Soc Eng Educ; Sigma Xi. *Res:* Fluid mechanics; study of the structure of turbulence in the flow of incompressible fluids; the structure of hypersonic turbulent wakes and boundary layers; biofluid mechanics; cardiovascular dynamics; blood flow. *Mailing Add:* 101 Hammond Bldg Pa State Univ University Park PA 16802

BRIGHTWELL, DENNIS RICHARD, b Des Moines, Iowa, Aug 7, 46; m 83. MEDICINE, PSYCHIATRY. *Educ:* Iowa State Univ, BS, 68; Univ Iowa, MD, 71. *Prof Exp:* Asst prof psychiat, Univ Ky, 74-78; assoc prof psychiat, Univ Mo-Columbia, 78-85; chief psychiat serv, Harry S Truman Mem Vet Hosp, 78-85; PROF PSYCHIAT & BEHAV MED, UNIV SFLA, 85-; CHIEF PSYCHIAT SERV, JAMES A HALEY VET HOSP, 85- *Concurrent Pos:* Staff physician psychiat, Vet Admin Hosp, Lexington, 74-78, chief, Behav Psychiat Sect, 75-78. *Mem:* Am Psychiat Asn; Asn Mil Surgeons US; Soc US Army Flight Surgeons. *Res:* Behavioral treatment of obesity; behavioral factors controlling eating in humans. *Mailing Add:* Psychiat Serv 116A 13000 N 30th St Tampa FL 33612

BRIGNOLI GABLE, CAROL, b New York, NY, Dec 28, 45; div; c 5. RISK ANALYSIS. *Educ:* City Col New York, BS, 68; Univ Md, PhD(chem), 73; Univ Md, MA, 87. *Prof Exp:* Biochemist nutrit, NY Res Inst, 67-68; NSF fel chem, Univ Md, 69-72; lectr chem, Montgomery Col, 72-75; res assoc nutrit, Agr Res Serv, USDA, 74-76; regulatory scientist, Gen Recognized As Safe Rev Br, 77-78 & Food Animal Additive Staff, 78-82, REGULATORY SCIENTIST, CLIN NUTRIT BR, US FOOD & DRUG ADMIN, 82- *Mem:* Am Chem Soc; AAAS; NY Acad Sci. *Res:* Analyzed data from National Heart & Nutrition Examination Survey II on the prevalence of iron deficiency anemia; risk assessment; epidemiology. *Mailing Add:* Systemetrics McGraw Hill Inc 4401 Connecticut Ave NW Suite 400 Washington DC 20008-2324

BRILES, CONNALLY ORAN, b Idabel, Okla, Dec 10, 19; m 45; c 2. IMMUNOGENETICS. *Educ:* Agr & Mech Col Tex, BS, 49, MS, 51; Univ Wis, PhD, 55. *Prof Exp:* Asst poultry husb, Agr Exp Sta, La State Univ, 55-57; asst to head poultry breeding sect, USDA, Md, 57-58; geneticist & dir blood grouping lab chickens, Arbor Acres Farm, Inc, 58-62; assoc prof animal sci & genetics, Macdonald Col, McGill Univ, 63-69; prof biol, 69-71, PROF ANIMAL SCI, TUSKEGEE INST, 72- *Concurrent Pos:* Breeding & incubation expert, Food & Agr Orgn, UN, Karachi, Pakistan, 71-72. *Mem:* Genetics Soc Am; Poultry Sci Asn; World's Poultry Sci Asn; Int Soc Animal Blood Group Res. *Res:* Blood groups in domestic animals and humans; association of blood group genotypes with economic and morphological characteristics in chickens; general poultry breeding; basic and practical interaction of genetic traits and environment in poultry; genetic selection and utilization of nutrients in poultry. *Mailing Add:* 1001 S Dean Rd Auburn AL 36830

BRILES, DAVID ELWOOD, b Hempstead, NY, May 25, 45; m 83; c 2. PATHOGENESIS, INFECTIOUS DISEASE. *Educ:* Univ Tex, Austin, BA, 67; Rockefeller Univ, PhD(immunogenetics), 73. *Prof Exp:* Res instr immunol, dept microbiol, Wash Sch Med, St Louis, Mo, 75-78; from asst prof to assoc prof, 78-85, PROF MICROBIOL, UNIV ALA, BIRMINGHAM, 85-, ASSOC PROF PEDIAT, 84-, SR SCIENTIST, CANCER CTR, 83- *Mem:* Am Asn Immunologists; Am Soc Microbiol. *Res:* Genetic control of bacterial pathogenesis. *Mailing Add:* Dept Microbiol Univ Ala Birmingham AL 35294

BRILES, GEORGE HERBERT, b Neodesha, Kans, Sept 15, 37. ORGANIC CHEMISTRY. *Educ:* Univ Kans, BS, 60; Whittier Col, MS, 61; Univ Mass, PhD(org chem), 66. *Prof Exp:* Chemist, US Borax Res Corp, 61; assoc prof, 65-80, PROF CHEM, SOUTHAMPTON COL, LONG ISLAND UNIV, 80- *Mem:* Am Chem Soc. *Res:* Properties and reactions of pentavalent organoantimony compounds. *Mailing Add:* RR 2 Box 219 Neodesha KS 66757-9407

BRILES, WORTHIE ELWOOD, b Italy, Tex, Jan 31, 18; m 41; c 3. IMMUNOGENETICS, POULTRY GENETICS. *Educ:* Univ Tex, BA, 41; Univ Wis, PhD(genetics, poultry breeding), 48. *Prof Exp:* Res asst genetics, Univ Wis, 41-42, 45-47, instr genetics & zool, 47; res asst genetics, Tex A&M Univ, 47-48, from asst prof to assoc prof poultry husb, 48-57; immunogeneticist, DeKalb AgRes, Inc, 57-70; prof, 70-87, ADJ PROF & EMER PROF IMMUNOGENETICS, NORTHERN ILL UNIV, 87- *Honors & Awards:* Res Prize, Poultry Sci Asn, 51. *Mem:* AAAS; Genetics Soc Am; Am Genetic Asn; fel Poultry Sci Asn; Am Asn Immunol. *Res:* Blood groups of chickens; physiological effects of blood group genes; alloantigens and disease resistance. *Mailing Add:* Dept Biol Sci Northern Ill Univ DeKalb IL 60115

BRILEY, BRUCE EDWIN, b Chicago, Ill, June 24, 36; m 59; c ·4. ELECTRICAL ENGINEERING. *Educ:* Univ Ill, BS, 58, MS, 59, PhD(elec eng), 63. *Prof Exp:* Asst digital comput lab, Univ Ill, 59-63; sr engr, Automatic Elec Res Labs, Gen Tel & Electronics Corp, Northlake, 63-64, mgr device appln res, 64-65, mgr electronic switching systs dept, 65-66; MEM STAFF, BELL TEL LABS, NAPERVILLE, 66- *Concurrent Pos:* Lectr grad sch, Ill Inst Technol, 64- *Mem:* Inst Elec & Electronics Engrs. *Res:* Memory and logic devices, systems and philosophies associated with digital computer design. *Mailing Add:* 5650 S Madison La Grange IL 60525

BRILL, A BERTRAND, b New York, NY, Dec 19, 28; m 50; c 3. NUCLEAR MEDICINE. *Educ:* Grinnell Col, AB, 49; Univ Utah, MD, 56; Univ Calif, Berkeley, PhD(biophys), 61. *Prof Exp:* Med dir, Div Radiol Health, USPHS, 57-64; assoc prof med & radiol, Sch Med, Vanderbilt Univ, 64-71, assoc prof biomed eng & biophys, 69-71, assoc prof physics, 70-71, prof med, radiol, biomed eng, biophys & physics, 71-79; PROF RADIOL, STATE UNIV NY, STONY BROOK, 79- *Mem:* AAAS; Soc Nuclear Med; Radiation Res Soc; Am Thyroid Asn. *Res:* Radiation leukemogenesis; effects of radiation on thyroid function; diagnostic radioisotope studies. *Mailing Add:* Two Rollingwood Dr Worcester MA 01609

BRILL, ARTHUR SYLVAN, b Philadelphia, Pa, June 11, 27; m 57; c 2. BIOPHYSICS, SPECTROSCOPY & SPECTROMETRY. *Educ:* Univ Calif, Berkeley, AB, 49; Univ Pa, PhD(biophys), 56. *Prof Exp:* Res assoc, Electron Micros Lab, Cornell Univ, 50-61; asst prof biophys, Yale Univ, 61-64, assoc prof molecular biol & biophys, 64-68; mem ctr advan studies, 68-71, prof mat, 68-73, PROF PHYSICS, UNIV VA, 73- *Concurrent Pos:* Fel med physics, Univ Pa, 56-58; Donner fel, Med Sci Div, Nat Res Coun, Clarendon Lab, Oxford Univ, 58-59; assoc ed, Biophys J, 78-81; chairperson, Div Biol Physics, Am Phys Soc, 90-91. *Mem:* Fel Am Phys Soc; Biophys Soc; Electron Micros Soc Am; Am Chem Soc. *Res:* Molecular biophysics; mechanisms of enzyme action; transition metal ions in biology; protein structure and dynamics; fundamental limitations of measuring instruments; physical chemistry. *Mailing Add:* Dept Physics Univ Va Charlottesville VA 22901

BRILL, DIETER RUDOLF, b Heidelberg, Ger, Aug 9, 33; m 71; c 1. THEORETICAL PHYSICS. *Educ:* Princeton Univ, AB, 54, MA, 56, PhD(physics), 59. *Prof Exp:* Instr physics, Princeton Univ, 59-60; Flick exchange res fel, Univ Hamburg, 60-61; from instr to assoc prof, Yale Univ, 61-70; PROF PHYSICS, UNIV MD, COLLEGE PARK, 70- *Concurrent Pos:* Vis prof, Univ Wurzburg ,68; vis prof, Max Planck Inst Physics & Astrophys, Munich, 72, 75 & 84; vis lectr, Col de France, Paris, 77. *Honors & Awards:* Humboldt Found Senior US Scientist Award. *Mem:* Sigma Xi; fel Am Phys Soc. *Res:* General relativity and gravitation physics; foundation of quantum mechanics; light, color, perception pedagogy for non-scientists. *Mailing Add:* Dept Physics & Astron Univ Md College Park MD 20742

BRILL, JOSEPH WARREN, b New York, NY, Oct 3, 50; m 85; c 2. SOLID STATE PHYSICS, ACOUSTICS. *Educ:* Columbia Univ, AB, 72; Stanford Univ, MS, 76, PhD(physics), 78. *Prof Exp:* Res assoc physics, Univ Southern Calif, 78-79; from asst prof to assoc prof, 80-89, PROF PHYSICS, UNIV KY, 89- *Mem:* AAAS; Am Phys Soc. *Res:* Properties of low-one-dimensional crystals; transport, thermal, optical and elastic measurements of crystals. *Mailing Add:* Dept Physics & Astron Univ Ky Lexington KY 40506-0055

BRILL, KENNETH GRAY, JR, b St Paul, Minn, Nov 16, 10; m 39; c 2. PALEONTOLOGY. *Educ:* Univ Minn, AB, 35; Univ Mich, MS, 38, PhD(geol), 39. *Prof Exp:* Asst geol, Yale Univ, 36-37; from instr to assoc prof geol & geog, Univ Chattanooga, 39-44; geologist, US Geol Surv, 44-45; from asst prof to prof geol, 46-79, chmn, Dept Geol & Geol Eng, 67-69, EMER PROF GEOL, ST LOUIS UNIV, 79- *Concurrent Pos:* Spec coal consult, Econ Coop Admin, SKorea, 49; hon res assoc, Univ Tasmania, 78. *Honors & Awards:* Fulbright lectr, Univ Tasmania, 53. *Mem:* Fel AAAS; fel Paleont Soc; fel Geol Soc Am; Am Asn Petrol Geol. *Res:* Permo-carboniferous stratigraphy and paleontology; Paleozoic stratigraphy of Rocky Mountain region. *Mailing Add:* Dept Earth & Atmospheric Sci St Louis Univ St Louis MO 63103

BRILL, ROBERT H, b Irvington, NJ, May 7, 29; m 57; c 1. PHYSICAL CHEMISTRY. *Educ:* Upsala Col, BS, 51; Rutgers Univ, PhD(phys chem), 55. *Prof Exp:* Asst prof chem, Upsala Col, 54-60; dir, 72-75, RES SCIENTIST, CORNING MUS GLASS, 60- *Mem:* Am Chem Soc; Am Ceramic Soc. *Res:* Archaeological chemistry; scientific examination of archaeological materials, particularly ancient glass; history of science and technology; glass chemistry; photochemistry. *Mailing Add:* Corning Mus Glass Corning Glass Ctr Corning NY 14830

BRILL, THOMAS BARTON, b Chattanooga, Tenn, Feb 3, 44; m 66; c 2. INORGANIC CHEMISTRY, PHYSICAL CHEMISTRY. *Educ:* Univ Mont, BS, 66; Univ Minn, Minneapolis, PhD(chem), 70. *Prof Exp:* From asst prof to assoc prof, 70-79, actg chmn, 85-86, PROF CHEM, UNIV DEL, 79- *Concurrent Pos:* Prin investr, Air Force Off Sci Res & Army Res Off, 74; vis prof, Univ Ore, 78; consult, Morton-Thiokol Inc, Lawrence Livermore Nat Lab, CDS Instruments, Conoco Oil, 83-; chmn, Gordon res conf energetic mats, 90. *Honors & Awards:* Nat Lectr, Sigma Xi, 84-87. *Mem:* Am Chem Soc; Sigma Xi; Combustion Inst. *Res:* Infrared and Raman spectroscopy; organometallic chemistry; solid state effects; thermal decomposition studies; explosions. *Mailing Add:* Dept Chem Univ Del Newark DE 19716

BRILL, WILFRED G, b Albion, Ind, Aug 11, 30; m 54; c 2. PHYSICS. *Educ:* Manchester Col, BA, 52; Purdue Univ, MS, 55, PhD(physics), 64. *Prof Exp:* Res assoc physics, 64-67, asst prof, 67-74, ASSOC PROF PHYSICS, PURDUE UNIV, NCENT CAMPUS, 74. *Mem:* Am Asn Physics Teachers. *Res:* Atomic spectroscopy. *Mailing Add:* 2905 Maple St Michigan City IN 46360

BRILL, WILLIAM FRANKLIN, b Utica, NY, 1923; m 51; c 3. OXIDATION, CATALYSIS. *Educ:* Univ Conn, BS, 45, MS, 48, PhD(chem), 50. *Prof Exp:* Res assoc org chem, Univ Calif, Los Angeles, 50-52; chemist, Polymers Sect, Olin Industs, 52-55; group leader oxidations, Petro-Tex, 55-60, res assoc, 60-66; dir chem res, Princeton Chem Res, 66-69; sr scientist, Halcon Res, 69-86; Dana fel, Drew Univ, 86-90; RETIRED. *Concurrent Pos:* Consult. *Mem:* Am Chem Soc; NY Acad Sci. *Res:* Reaction mechanisms; oxidations; catalysis; asymmetric oxidations. *Mailing Add:* 914 Rte 518 Skillman NJ 08558

BRILL, WINSTON J, b London, Eng, June 16, 39; US citizen; m 65; c 1. MICROBIOLOGY, BIOCHEMICAL GENETICS. *Educ:* Rutgers Univ, BS, 61; Univ Ill, Urbana, PhD(microbiol), 65. *Prof Exp:* NIH fel biol, Mass Inst Technol, 65-67; prof bact, Univ Wis-Madison, 67-81; dir res, Agracetus, 81-89; PRES, WINSTON J BRILL & ASSOCS, 89. *Concurrent Pos:* Res grants, USPHS, 68-, NIH, 69- & NSF, 69-; panel mem, NSF Metab Biol Prog, 74-77; USDA competitive grants prog, 78; mem, NIH Recombinant DNA Adv Panel, 79-83 & USDA policy adv comt; adj prof, Univ Wis-Madison. *Honors & Awards:* Eli Lilly Award Microbiol & Immunol, 79; Alexander von Humboldt Award, 79. *Mem:* Nat Acad Sci; fel AAAS; Am Soc Microbiol; Am Soc Plant Physiologists. *Res:* Electron transport; ferredoxin; regulation of catabolism; microbial physiology; nitrogen fixation; catabolite repression; soil microbiology; plant molecular biology. *Mailing Add:* 4134 Cherokee Dr Madison WI 53711

BRILL, YVONNE CLAEYS, b Man, Can, Dec 30, 24; US citizen; m 51; c 3. LIQUID ROCKET PROPULSION, SOLID ROCKET PROPULSION. *Educ:* Univ Man, Winnipeg, BSc, 45; Univ Southern Calif, Los Angeles, MS, 51. *Prof Exp:* Mathematician, Aircraft Design Douglas Aircraft Co, 45-46; res analyst propulsion & propellants, RAND Corp, 46-49; group leader igniters & fuels, Marquardt Corp, 49-52; staff engr combustion, United Technol, 52-55; proj engr preliminary design, Wright Aeronaut, 55-58; consult propulsion & propellants, FMC Corp, 58-66; mgr propulsion, 66-81, staff engr preliminary design, RCA Astro Electronics, 83-86; mgr solid rocket motor, NASA Hq, 81-83; staff mem, Space Eng Int Maritime Satellite Orgn, 86-91; CONSULT, 91- *Concurrent Pos:* Mem joint Army, Navy, NASA & Air Force Hydrazine Working Group, 70-77; mem liquid rocket tech comt, Am Inst Aeronaut & Astronaut, 73-76; dir student affairs, Nat Exec Comn, Soc Women Engrs, 79-80 & 83-84, treas, 80-81; mem, USAF Sci Adv Bd, 82-83; educt comt, Int Astronaut Fed, 83-85; mem, Career Enhancement Comt, Am Inst Aeronaut & Astronaut, 84-86; mem, Space Syst Technol Workshop, USAF & Am Inst Aeronaut & Astronaut, 85-86; mem, Comt Advan Solid Rocket Motor Qual & Test Prog, Nat Res Coun, 91. *Honors & Awards:* Marvin C Demlar Award, Am Inst Aeronaut & Astronaut, 83; Achievement Award, Soc Women Engrs, 86. *Mem:* Nat Acad Eng; fel Am Inst Aeronaut & Astronaut; fel Soc Women Engrs; UK Womens Eng Soc; Int Astronaut Fedn; Brit Interplanatary Soc; Sigma Xi. *Res:* Ramjet, turbojet, turbofan and spacecraft rocket propulsion. *Mailing Add:* 914 Rte 518 Skillman NJ 08558

BRILLER, STANLEY A, b New York, NY, Dec 4, 22. BIOMEDICAL ENGINEERING, CARDIOLOGY. *Educ:* Yale Univ, BS, 45; NY Univ, MD, 47. *Hon Degrees:* MA, Univ Pa, 71. *Prof Exp:* PROF ENG & MED, CARNEGIE MELLON UNIV, 74-; PROF MED, MED COL PA, 88- *Concurrent Pos:* Dir, Heart Sta, Allegheny Gen Hosp, 73- *Mem:* Biomed Eng Soc; Biophys Soc; Sigma Xi; Harvey Soc. *Mailing Add:* Allegheny Gen Hosp 320 E North Ave Pittsburgh PA 15212

BRILLHART, DONALD D, b McClure, Ohio, Jan 25, 18; m 48; c 2. CHEMICAL ENGINEERING. *Educ:* Cleveland State Univ, BChE, 48. *Prof Exp:* Chemist, Aluminum Co Am, 38-43 & Stevens Grease & Oil Co, 48-53; lubrication engr, Prof-Chem, Inc, 53-69, chemengr, 69-85; LIPO-TECH, INC CLEVELAND, OH. *Mem:* AAAS; Asn Consult Chemists & Chem Engrs; Am Soc Lubrication Engrs; Am Chem Soc. *Res:* Lubrication; corrosion; industrial processes; lipids; synthetic organic and natural products chemistry. *Mailing Add:* 14713 Rockside Rd Cleveland OH 44137

BRILLIANT, HOWARD MICHAEL, b Baltimore, Md, Aug 15, 45; m 78; c 2. AEROSPACE PROPULSION, AERODYNAMICS. *Educ:* Univ Pittsburgh, BSME, 66; Univ Mich, MSE, 67, PhD(aerospace eng), 71. *Prof Exp:* Proj engr, USAF, Aero Propulsion Lab, Wright-Patterson AFB, 70-75; from instr to asst prof aeronaut, USAF Acad, 75-78, assoc prof, 78-80; chief, Chem Laser Technol Group & Fluid Mech Br, Air Force Weapons Lab, 80-84, from Europ prog mgr to actg dir acquisition suppnt, Dep Propulsion, Aeronaut Systs Div, 84-86; chief spec prof off, Aero Propulsion Lab, Air Force Wright Aeronaut Lab, 86-88, dep dir, Adv Propulsion Div, Aero Propulsion Lab, 88-89, mid prog mgr, 89-90, DIR SYSTS ENG, JOINT TACTICAL INFO DISTRIB SYST, JOINT PROG OFF, ELECTRONICS SYSTS DIV, AIR FORCE WRIGHT RES & DEVELOP LAB, 90- *Concurrent Pos:* Consult, NASA Dryden Flight Res Ctr, 76-78. *Mem:* Am Inst Aeronaut & Astronaut; Sigma Xi. *Res:* Airbreathing propulsion including scramjets; transonic, supersonic and hypersonic aerodynamics; boundary layer flows; laser aerodynamics, kinetics, systems. *Mailing Add:* 32 Biscayne Pkwy Nashua NH 03060-1169

BRILLIANT, LAWRENCE BRENT, b Detroit, Mich, May 5, 44; m 69; c 1. EPIDEMIOLOGY, PUBLIC HEALTH. *Educ:* Wayne State Univ, MD, 69; Univ Mich, MPH, 77; Ctr Dis Control, cert, 77. *Prof Exp:* Med officer epidemiol, WHO, 73-76 & 80-81; asst prof pub health, Univ Mich, 77-81, asst prof epidemiol, 81-89; DIR, SEVA FOUND, SAN RAFAEL, CALIF, 78-; PRES & OWNER, BRILLIANT COLOR CARDS, 89- *Concurrent Pos:* Consult, WHO, New Delhi, 73-; mem, WHO Secretariat, Int Comns Eradicating Smallpox, 77-78; proj dir, Ford Found Proj Environ Health Policy, Mich, 77-78; surv dir, Nepal Blindness Surv, WHO, 80-81; lectr, Dept Epidemiol, Univ Calif, Berkeley, 91- *Honors & Awards:* Health Minister's Award, Govt India, 75; Plaque, India Soc Malaria & Other Commun Dis, 75. *Mem:* Int Epidemiol Asn; Am Pub Health Asn; NY Acad Sci; Am Asn World Health; Fedn Am Scientists. *Res:* Epidemiology of blindness and blindness diseases; health planning and health policy; international health. *Mailing Add:* 524 San Anselmo Ave San Anselmo CA 94960

BRILLIANT, MARTIN BARRY, b Brooklyn, NY, Dec 4, 31; m 55; c 3. TELECOMMUNICATIONS NETWORK PLANNING. *Educ:* Washington & Jefferson Col, BA, 55; Mass Inst Technol, SB & SM, 55, ScD(elec eng), 58. *Prof Exp:* Mem tech staff, Nat Co, Inc, 58-60 & Hazeltine Res Corp, 60-61; asst prof elec eng, Univ Kans, 61-64; mem tech staff, Booz-Allen Appl Res, Inc, 64-66; Mem Tech Staff, Bell Tel Labs, Inc, 66-89. *Concurrent Pos:* Consult, McDonnell Aircraft Corp, 63-64. *Mem:* AAAS. *Res:* Digital network synchronization; telecommunication network objectives. *Mailing Add:* 39 McCampbell Rd Holmdel NJ 07733-2232

BRILLINGER, DAVID ROSS, b Toronto, Ont, Oct 27, 37; m 60; c 2. STATISTICS. *Educ:* Univ Toronto, BA, 59; Princeton Univ, MA, 60, PhD(math), 61. *Prof Exp:* Soc Sci Res Coun fel, London Sch Econ, 61-62; mem tech staff statist, Bell Tel Labs, NJ, 62-64; lectr, London Sch Econ, 64-66, reader, 66-69; chmn dept, 79-80, PROF STATIST, UNIV CALIF, BERKELEY, 69- *Concurrent Pos:* Lectr math, Princeton Univ, 62-64. *Honors & Awards:* Wald Lectr, Inst Math Statist, 82. *Mem:* Fel Am Statist Asn; fel Inst Math Statist; Int Statist Inst; fel Royal Soc Can; fel AAAS. *Res:* Time series; applied probability; data analysis; statistical methods in neurophysiology and seismology. *Mailing Add:* Dept Statist Univ Calif Berkeley CA 94720

BRILLSON, LEONARD JACK, b New York, NY, Dec 15, 45; m 68; c 2. SURFACE PHYSICS, METAL-SEMICONDUCTOR INTERFACES. *Educ:* Princeton Univ, AB, 67; Univ Pa, MS, 69, PhD(physics), 72. *Prof Exp:* Assoc scientist, Xerox Corp Webster Res Ctr, 72-74, scientist, 74-79, sr scientist physics, 79-85, prin scientist, 85-89, mgr interfaces & processing group, 85-89, MGR, MAT RES LAB, XEROX CORP WEBSTER RES CTR, 89- *Concurrent Pos:* Mem adv bd, Ctr Joining Materials, Carnegie-Mellon Univ; adj prof, dept mat sci, Univ Pa, 84-85; dir, Am Vacuum Soc, 85-86, exec chmn, Electronic Mats & processing Div, 85; vis scientist, Tel Aviv Univ, 84; mem, Organizing/Prog Comt, Physics & Chem Semiconductor Interfaces Conf, 89-; mem exec bd, Ctr Photoinduced Charge Transfer, NSF, 89-; secy & treas, Div Condensed Matter Physics, Am Phys Soc, 89-; chair, Interface, Characterization Microelectronic Mat, Int Soc Optical Eng Conf, 89; mem adv bd, Int Conf Formation of Semiconductor Interfaces, 90- *Mem:* Fel Am Phys Soc; Am Vacuum Soc; Mat Res Soc; Electrochem Soc. *Res:* Surface and interface electronic structure of semiconductors and insulators, their dependence on surface chemical composition and bonding, and their influence on charge transport across interfaces with metals. *Mailing Add:* Xerox Webster Res Ctr 800 Phillips Rd W-114 Webster NY 14580

BRIM, CHARLES A, b Spalding, Nebr, Apr 14, 24; m 48; c 2. PLANT BREEDING. *Educ:* Univ Nebr, BS, 48, MS, 50, PhD(agron), 53. *Prof Exp:* Res agronomist, USDA, 53-80, prof, 69-80, EMER PROF CROP SCI & GENETICS, NC STATE UNIV, 80-; VPRES RES, FUNK SEEDS INT. *Mailing Add:* NC State Univ 112 Williams Hall Raleigh NC 27607

BRIM, ORVILLE G, JR, b Apr 7, 23; m 44; c 4. LIFESPAN DEVELOPMENTS. *Educ:* Yale Univ, BA, 47, MA, 49, PhD(sociol), 51. *Prof Exp:* Sociologist, Russell Sage Found, NY, 55-64, asst secy, 60-65, pres, 64-72; pres, Found Child Develop, NY, 74-85; DIR, MAC ARTHUR FOUND RES NETWORK SUCCESSFUL MID-LIFE DEVELOP, 89- *Concurrent Pos:* Chmn, Spec Comt Soc Sci, NSF, 68-69; adv comt, Child Develop, Nat Acad Sci, 71-76; vchair, bd trustees, Am Inst Res, 71-87, chair, 88-90; chmn, Comt Work & Personality Middle Years, Soc Sci Res Coun, 72-79; bd trustees, William T Grant Found, Inc, 75-84. *Honors & Awards:* Kurt Lewin Mem Award, Soc Psychol Study Social Issues, 79; Distinguished Sci Contrib Child Develop Res Award, Soc Res Child Develop, 85. *Mem:* Inst Med-Nat Acad Sci; Am Acad Arts & Sci. *Res:* Mid-life development. *Mailing Add:* MacArthur Found Res Network Successful Mid-Life Develop Vero Beach FL 32960

BRIMACOMBE, ROBERT KENNETH, b Niagara Falls, Ont, Can, Mar 11, 57; m 81; c 1. EXCIMER LASERS, CARBON DIOXIDE LASERS. *Educ:* Royal Mil Col, Can, BEng, 80; McMaster Univ, PhD(physics), 85. *Prof Exp:* Asst prof physics, Royal Mil Col Can, 85-88; SR SCIENTIST, LUMONICS INC, 88- *Mem:* Optical Soc Am; Inst Elec & Electronics Engrs; Quantum Electronics & Appln Soc. *Res:* Excitation and gain mechanisms in carbon dioxide lasers; use of excimer lasers to investigate ultraviolet transmission in optical fibers; development of excimer laser products. *Mailing Add:* Lunonics Inc 105 Schneider Rd Kamata ON K2K 1Y3 Can

BRIMHALL, GEORGE H, b Santa Monica, Calif, Aug 24, 47; m 71; c 2. ORE DEPOSITS, CRYSTAL GEOCHEMISTRY. *Educ:* Univ Calif, Berkeley, BA, 69, PhD(geol), 72. *Prof Exp:* Mine geologist, Anaconda Co, 72-74, proj geologist, 74-76; asst prof, Johns Hopkins Univ, 76-78; from asst prof to assoc prof geol, 78-82, chmn dept, 86-90, PROF GEOL, UNIV CALIF, BERKELEY, 82- *Concurrent Pos:* Mem, Panel Mineral Resources, Continental Sci Drilling Comt, Nat Res Coun, 81-84 & Harvard vis comt, 82- *Honors & Awards:* Lindgren Award, Soc Econ Geologists, 80; McKinstry Lectr, Harvard Univ, 85. *Mem:* Soc Econ Geologists; fel Mineral Soc Am; fel Geol Soc Am; Am Geophys Union; Geochem Soc; Am Inst Mining & Metall. *Res:* Distribution and tranport processes of transition metals in the Earth's crust; fractionation processes involving aqueous and magmatic fluids; hydrochemical transport; rock weathering and soil forming processes. *Mailing Add:* Dept Geol & Geophys Univ Calif Berkeley CA 94720

BRIMHALL, J(OHN) L, b Palo Alto, Calif, Mar 17, 37; m 66. MATERIALS SCIENCE. *Educ:* San Jose State Col, BS, 59; Stanford Univ, MS, 61, PhD(mat sci), 64. *Prof Exp:* Sr engr phys metall, Gen Elec Co, 64-65; SR RES SCIENTIST, PAC NORTHWEST LAB, BATTELLE MEM INST, 65- *Mem:* Am Inst Mining, Metall & Petrol Engrs; Am Soc Metals; Sigma Xi; Mat Res Soc. *Res:* Studies of interaction of radiation with materials and the relationship between crystal lattice defects and structural properties of materials. *Mailing Add:* 2502 Riverside Dr West Richland WA 99352

BRIMHALL, JAMES ELMORE, b Burlington, Iowa, June 5, 36; m 58; c 2. PHYSICS. *Educ:* Hamline Univ, BS, 59; Univ Pittsburgh, MS, 65; Union Grad Sch, PhD, 79. *Prof Exp:* Teacher, WVa State Col, 66-68, chmn dept, 68-74, from assoc prof to prof physics, 81-91, VPRES ADMIN AFFAIRS, WVA STATE COL, 91- *Mem:* Am Phys Soc; Am Asn Physics Teachers; Nat Sci Teachers Asn; Soc Photog Technologists. *Res:* Positronium annihilation in gases; circular polarization analysis of x-rays via Compton scattering; plane polarization analysis of gamma rays via Compton scattering; momentum spectroscopy of charged particles. *Mailing Add:* WVa State Col Campus Box 200 Institute WV 25112

BRIMIJOIN, WILLIAM STEPHEN, b Passaic, NJ, July 1, 42; m 64; c 3. PHARMACOLOGY, NEUROBIOLOGY. *Educ:* Harvard Col, BA, 64, Harvard Univ, PhD(pharmacol), 69. *Prof Exp:* Instr pharmacol, Mayo Grad Sch Med, 71-72; from asst prof to assoc prof, 72-80, assoc dir, res training & degree progs, Mayo Grad Sch Med, 83-87, prof pharmacol, 80-89, WINSTON & IRIS CLEMENT PROF, MAYO MED SCH, 89-, CONSULT, MAYO FOUND, 72- *Concurrent Pos:* NIH pharmacol res assoc training prog, 69-71; NIMH grant, 71-74; NIH grant & career develop award, 75-80, 80-89, 82-; assoc consult, Mayo Found, 71-72; vis scientist, Karolinska Inst, 78-79, Univ Würzburg, 87-88; mem Soc Issues Comt, Soc Neurosci, 87-90; mem Behav & Neurosci Study Sect, NIH, 89- *Honors & Awards:* Jacob Javits Award, Nat Inst Neurol & Commun Dis & Stroke, 87; Sr US Scientist Award, Humboldt Found, 87. *Mem:* Am Soc Pharmacol & Exp Therapeut; Am Soc Neurochem; Soc Neurosci; Int Soc Neurochem. *Res:* Chemical biology of nerve cells; mechanisms and functions of axoplasmic transport; regulation and properties of cholinesterases. *Mailing Add:* Dept Pharmacol Mayo Med Sch Rochester MN 55905

BRIMM, EUGENE OSKAR, b Sheboygan, Wis, July 7, 15; m 40; c 3. EXTRACTIVE METALLURGY, INORGANIC CHEMISTRY. *Educ:* Univ Wis, BS, 38; Univ Ill, MS, 38, PhD, 40. *Prof Exp:* Div head, Tonawanda Lab, Linde Div, Union Carbide Corp, 40-56, asst mgr, 56-63, mgr technol rels group, Union Carbide Europe S A, Switz, 63-68, mgr prod technol, Mat Systs Div, Union Carbide Corp, 68-70; mgr mkt develop, Stellite Div, Cabot Corp, Ind, 70-71, mgr int dept, 71-72; mgr res & develop, Filtrol Corp, 72-78; mgr tech develop, Holmes & Narver, Inc, 78-81; CONSULT, 81- *Mem:* Am Inst Mining & Metall Engrs; Can Inst Mining & Metall; Am Chem Soc. *Res:* Solvent extraction of metals; cracking and hydrotreating catalysts; adsorbents; metal carbonyls; organometallic compounds; molecular sieves. *Mailing Add:* 18121 Allegheny Dr Santa Ana CA 92705

BRIN, MYRON, b New York, NY, July 1, 23; m 44; c 3. BIOCHEMISTRY. *Educ:* NY Univ, BS, 45; Cornell Univ, BS, 47, MS, 48; Harvard Univ, PhD(med sci, biochem), 51. *Prof Exp:* Sr res scientist metab radiation, New Eng Deaconess Hosp, Boston, 51-53; instr biochem, Sch Med, Harvard Univ, 53-56; chief biologist, Food & Drug Res Labs, Inc, 56-58; assoc prof biochem & med, State Univ NY Upstate Med Ctr, 58-68; prof nutrit, Univ Calif, Davis, 68-69; asst dir biochem nutrit, Hoffmann-La Roche Inc, 69-70, dir clin nutrit, 70-88; RETIRED. *Concurrent Pos:* Instr biochem, Sch Med, Tufts Univ, 52-53; res fel & instr, Thorndike Mem Lab, Med Sch, Harvard Univ, 53-56, mem bd tutors, Harvard Univ, 54-56; NIH sr res fel, Hadassah Med Sch, Hebrew Univ Jerusalem, 67-68; adj prof, Columbia Univ, 69- *Honors & Awards:* A Cressy Morrison Award, NY Acad Sci, 62. *Mem:* AAAS; Am Soc Biol Chemists; Am Inst Nutrit; Am Soc Clin Nutrit; Am Chem Soc; Sigma Xi. *Res:* Biochemistry and nutrition; vitamin function in metabolism and health maintenance, particularly vitamin B1, B6 and E adequacy, by functional tests; concept of morgual vitamin deficiency. *Mailing Add:* 129 Coccio Dr West Orange NJ 07052

BRINCH-HANSEN, PER, b Copenhagen, Denmark, Nov 13, 38; m 65; c 2. CONCURRENT PROGRAMMING, OPERATING SYSTEMS. *Educ:* Tech Univ Denmark, MS, 63 & Dr Techn, 78. *Prof Exp:* Software designer, Regnecentralen, Denmark, 63-70; res assoc, Carnegie-Mellon Univ, 70-72; assoc prof comput sci, Calif Inst Technol, 72-76, Henry Salvatori prof, Univ Southern Calif, 76-84, prof, Univ Copenhagen, 84-87; DISTINGUISHED PROF COMPUT SCI, SYRACUSE UNIV, 87- *Concurrent Pos:* Consult, Burroughs, Control Data, Hewlett-Packard, IBM, Jet Propulsion Lab, Los Alamos Sci Lab, Mostek, Tex Instruments, TRW & other var nat & pvt cos. *Mem:* Fel Inst Elec & Electronics Engrs; Int Fed Info Processing; NY Acad Sci. *Res:* Current programs; programming languages; operating systems; computer architecture. *Mailing Add:* 5070 Pine Valley Dr Fayetteville NY 13066

BRINCK-JOHNSEN, TRULS, b Stavanger, Norway, Oct 10, 26; US citizen; m 52; c 3. ENDOCRINOLOGY, BIOCHEMISTRY. *Educ:* Univ Utah, MS, 56, PhD(anat, endocrinol), 59. *Prof Exp:* Res asst anal chem, Univ Oslo, 48-49; res asst, Inst Aviation Med, Norweg Air Force, 49-50; chief chemist, French & Norweg Trop & Antarctic Whaling Expeds, 50-52 & 53-54; res assoc med biochem, Univ Oslo, 52-53, investr steroid chem, Ullevaal Hosp, 59-63, chief biochemist, Endocrine Lab, Rikshosp, 63-64; res assoc steroid chem & asst prof path, 64-69, assoc prof med, 70-77, assoc prof path, 69-83, PROF CLIN PATH, DARTMOUTH MED SCH, 83- *Concurrent Pos:* Mem assoc staff, Mary Hitchcock Mem Hosp, Hanover, NH, 67-; dir, Steroid-Radioimmunoassay Lab, Dartmouth-Hitchcock Med Ctr, 72-; consult, Vet Admin Hosp, 75-; adj assoc prof, Sch Health Studies, Univ NH, 79-83, adj prof, 83. *Mem:* Endocrine Soc; fel Nat Acad Clin Biochem; Royal Soc Med. *Res:* Biochemistry and physiology of steroid hormones. *Mailing Add:* Dept Path Dartmouth Med Sch Hanover NH 03756

BRINCKMAN, FREDERICK EDWARD, JR, b Oakland, Calif, June 24, 28; m 54; c 2. INORGANIC CHEMISTRY, MATERIALS BIOPROCESSING. *Educ:* Univ Redlands, BS, 54; Harvard Univ, AM, 58, PhD(chem), 60. *Prof Exp:* Res chemist & head propellant br, US Naval Ord Lab, Calif, 60-61; sci staff asst head, Anal Chem Div, Gen Res Div, US Naval Propellant Plant, Md, 61-64; actg chief, Inorg Chem Sect, 74-75, RES CHEMIST, CHEM STABILITY & CORROSION DIV, NAT BUR STANDARDS, 64-, GROUP LEADER, CHEM & BIODEGRADATION PROCESSES, 78-,

RES CHEMIST & BIOPROCESS COORDR, POLYMER DIV, 88- *Concurrent Pos:* Richard Merton vis prof geosci, Univ Maine, 78; adj prof Inorg Chem, Univ Md, 85. *Honors & Awards:* Silver Medal, Dept Com, 74; Gold Medal, Dept Com, 80. *Mem:* AAAS; Am Chem Soc; fel Royal Soc Chem; NY Acad Sci; fel Am Inst Chemists; Sigma Xi. *Res:* Synthetic inorganic and organometallic chemistry; metal-nitrogen, metal-fluorine chemistry; coordination chemistry of main group elements; applications of magnetic AA and epifluorescence and mass spectrometry to environmental inorganic chemistry; dynamics and quantative structure-activity relationship of bioactive organometallic systems in aqueous phases; organometallics as intermediates in biotransformations of metals; ultratrace specification of environmental organometals in vivo and in exocellular strategic or waste materials bioprocessing. *Mailing Add:* 5609 Granby Rd Granby Woods RR1 Derwood MD 20855

BRINE, CHARLES JAMES, b Newark, NJ, July 4, 50. BIOPOLYMERS, CELLULOSE & CELLULOSICS. *Educ:* Worcester Polytech Inst, BS, 72; Univ Del, MS, 74, PhD(marine studies), 79. *Prof Exp:* Fel marine chem, Col Marine Studies, Univ Del, 75-79, res assoc, 79-80; res chemist, FMC Corp, 80-82, mgr, explor res, Food & Pharmaceut & Prod Div, 82-88, res assoc, New Technol Food & Pharmaceut Prod Div, 88-91, SR RES ASSOC, NEW TECHNOL, FMC CORP, 91- *Concurrent Pos:* Chmn, Pub Rel, Agr & Food Chem Div, Am Chem Soc-AGFD, 85-89, vchmn, 89-, prog chmn, chair-elect, 90-, chmn, 91-; co-organizer, Joint US-Japan, NSF, coop seminar on chitin, chitosan, & related enzymes, 84-; session chmn Second, Third & Fourth, Int Conf Chitin & Chitosan, 82, 85 & 88; mem, Sci Adv Bd, Fourth Int Conf, Chitin & Chitosan & Sci Adv Bd, 88-89; co-founder & vpres, Am Chito Sci Soc; chmn, Fifth Int Conf, Chitin & Chitosan, 91- *Honors & Awards:* Nat Award grad res in appl marine sci, US Dept Com, 77. *Mem:* Am Chem Soc; AAAS; Sigma Xi; NY Acad Sci; Controlled Release Soc. *Res:* Chitin, chitosan and related polymeric, oligomeric and monomeric derivatives; basic and pharmaceutical application of cellulose and cellulosic derivative polymers in the areas of sustained/controlled drug delivery, drug dosage form development and pharmaceutical latex drug carrier systems; cellulose/hydrocolloid fat substitutes and food additive systems. *Mailing Add:* 28 Tee-Ar Pl Princeton NJ 08540

BRINEN, JACOB SOLOMON, b Brooklyn, NY, Nov 16, 34; m 58; c 2. MOLECULAR SPECTROSCOPY, SURFACE CHEMICAL ANALYSIS. *Educ:* Brooklyn Col, BS, 56; Pa State Univ, PhD(chem & electronic spectros), 61. *Prof Exp:* From res chemist to sr res chemist, 61-74, prin res chemist, 74-82, assoc res fel, 82-86, RES FEL, AM CYANAMID CO, 86- *Mem:* Am Chem Soc; Am Phys Soc; Am Vacuum Soc; AAAS. *Res:* Electronic absorption spectroscopy, fluorescence and phosphorescence; electron spin and photoelectron spectroscopy, auger, SIMS and ISS. *Mailing Add:* Am Cyanamid Co 1937 W Main St Stamford CT 06904

BRINER, WILLIAM WATSON, b Winchester, Ind, Dec 12, 28; m 54; c 4. MICROBIOLOGY. *Educ:* Ohio State Univ, BA, 53, MS, 57, PhD(microbiol), 59. *Prof Exp:* MICROBIOLOGIST, PROCTER & GAMBLE CO, 59- *Concurrent Pos:* Lectr, Eve Col, Univ Cincinnati, 63-70; assoc ed, J Dent Res, 73-75. *Mem:* AAAS; Am Soc Microbiol; Int Asn Dent Res; Am Asn Dent Res; Am Acad Microbiol. *Res:* Oral microbiology; dental diseases in animals; microbial ecology. *Mailing Add:* Procter & Gamble Co 11511 Reed Hartman Hwy Cincinnati OH 45241

BRINEY, ROBERT EDWARD, b Benton Harbor, Mich, Dec 2, 33. COMPUTER SCIENCE. *Educ:* Northwestern Univ, AB, 55; Mass Inst Technol, PhD(math), 61. *Prof Exp:* Instr math, Mass Inst Technol, 61-62; asst prof, Purdue Univ, 62-68; assoc prof, 68-70, prof math, 70-77, chmn dept math, 71-77, PROF COMPUT SCI, SALEM STATE COL, 77-, CHMN DEPT COMPUT SCI, 77- *Mem:* Am Math Soc; Asn Comput Mach; Soc Indust & Appl Math; AAAS; Math Asn Am. *Res:* Algebraic geometry; commutative algebra. *Mailing Add:* Dept Comput Sci Salem State Col Salem MA 01970

BRINGER, ROBERT PAUL, b Peoria, Ill, June 5, 30; c 3. PLASTICS ENGINEERING. *Educ:* Purdue Univ, BS, 52, PhD(chem eng), 56. *Prof Exp:* Chem engr, Dow Chem Co, Mich, 56-57, supvr, 57-59; sr chem engr, 3M Co, 59-62, supvr, 62-66, mgr, Chem Div, 66-72, tech dir, Film Div, 72-84, exec dir, St Paul, 84-85, STAFF VPRES, ENVIRON ENG & POLLUTION CONTROL, 3M CO, 86- *Concurrent Pos:* Mem, environ comt, Nat Asn Mfgrs & Bus Roundtable; mem, corp conserv coun, Nat Wildlife Fedn; chmn, solid & hazardous waste task force, Nat Asn Mfg; bd dirs, Air Pollution Control Asn. *Mem:* Air & Waste Mgt Asn; Water Pollution Control Fedn. *Res:* Plastics, rubbers and resins, particularly fluorocarbon and high temperature; films, particularly oriented. *Mailing Add:* 103 Eastbank Ct N Hudson WI 54016-1079

BRINGHURST, ROYCE S, b Murray, Utah, Dec 27, 18; m 45; c 6. PLANT BREEDING. *Educ:* Utah State Univ, BS, 47; Univ Wis, MS, 48, PhD(agron, genetics), 50. *Prof Exp:* Chmn dept, 50-76, PROF POMOL, UNIV CALIF, DAVIS, 50- *Mem:* Genetics Soc Am; fel Am Soc Hort Sci; Am Phytopath Soc; Bot Soc Am. *Res:* Breeding and genetics of strawberries; evolution, genetics, cytogenetics and breeding of Fragaria polyploids. *Mailing Add:* Dept Pomol Univ Calif Davis CA 95616

BRINIGAR, WILLIAM SEYMOUR, JR, b Wichita, Kans, May 18, 30; m 56; c 3. BIOCHEMISTRY. *Educ:* Univ Kans, BS, 52, PhD(chem), 57. *Prof Exp:* Res assoc, Univ SC, 57-59 & Yale Univ, 59-66; from asst prof to assoc prof, 66-79, PROF CHEM, TEMPLE UNIV, 79- *Mem:* Am Chem Soc. *Res:* Mechanisms of enzyme catalysis; hemoprotein chemistry; site-specific mutagenesis of hemoglobins. *Mailing Add:* Dept Chem Temple Univ Philadelphia PA 19122

BRINK, DAVID LIDDELL, b St Paul, Minn, July 7, 17; m 43; c 3. WOOD CHEMISTRY. *Educ:* Univ Minn, BS, 39, PhD(agr biochem, forestry), 54. *Prof Exp:* Asst, Univ Minn, 39-42; res chemist, Salvo Chem Corp, 43-45; asst, Univ Minn, 47-49; group leader lignin, Mead Corp, 49-53; chemist, Develop Ctr, Weyerhaeuser Timber Col, 54-55, chief chem sect, 55-57, chief org chem sect, Cent Res Dept, 57-58; assoc forest prod chemist & lectr forestry, Univ Calif, Berkeley, 58-62, forest prof chemist & lectr wood chem & prof agr chem, 62-80, grad adv, 72-80. *Mem:* Am Chem Soc; Forest Prod Res Soc; Tech Asn Pulp & Paper Indust. *Res:* Lignin, cellulose, pulping, bark, extractives. *Mailing Add:* 1068 Woodside Rd Berkeley CA 94708

BRINK, FRANK, JR, b Easton, Pa, Nov 4, 10; m 39; c 2. BIOPHYSICS OF NERVE CELLS, ION TRANSPORT. *Educ:* Pa State Univ, BS, 34; Calif Inst Technol, MS, 35; Univ Pa, PhD(biophys), 39. *Hon Degrees:* DSc, Rockefeller Univ, 83. *Prof Exp:* Lalor fel, 39-40; instr physiol, Med Col, Cornell Univ, 40-41; Johnson Found fel & lectr biophys, Univ Pa, 41-47, asst prof, 47-48; assoc prof, Johns Hopkins Univ, 48-53; prof, 53-74, dean grad studies, 57-72, Detlev W Bronk prof, 74-81, EMER PROF, ROCKEFELLER UNIV, 81- *Concurrent Pos:* Dir, Comn for Biol & Med, NSF, 53-59; ed, Biophys J, 60-64; chmn, President's Comt Nat Medal Sci, 63-65. *Mem:* Nat Acad Sci; Am Acad Arts & Sci; Am Physiol Soc; Soc Gen Physiol; Biophys Soc; Soc Neurosci. *Res:* Physical chemistry of nerve cells, in particular, the transport of sodium and potassium ions. *Mailing Add:* Rockefeller Univ New York NY 10021

BRINK, GILBERT OSCAR, b Los Angeles, Calif, May 26, 29; m 57. EXPERIMENTAL ATOMIC PHYSICS. *Educ:* Col of the Pacific, BA, 53; Univ Calif, PhD(chem), 57. *Prof Exp:* From staff res physicist to physicist, Lawrence Radiation Lab, Univ Calif, 57-63; asst res prof, Univ Pittsburgh, 62-63; prin physicist, Cornell Aeronaut Lab, 63-68; chmn, Dept Physics & Astron, 72-74, assoc prof, 68-80, PROF PHYSICS, STATE UNIV NY, BUFFALO, 80- *Concurrent Pos:* Vis McDonnell distinguished prof, Wash Univ, St Louis, 81. *Mem:* Am Phys Soc; Sigma Xi; NY Acad Sci. *Res:* Atomic beam magnetic resonance; hyperfine structure anomaly measurements; nuclear spins and magnetic moments; atomic scattering by crossed beams; molecular structure; laser excitation of atomic states. *Mailing Add:* Dept Physics & Astron State Univ NY Buffalo NY 14214

BRINK, JOHN JEROME, b Secunderabad, India, Mar 18, 34; US citizen; m 60; c 2. NEUROSCIENCES. *Educ:* Univ Orange Free State, SAfrica, BSc, 55; Univ Witwatersrand, BSc(Hons), 56; Univ Vt, PhD(biochem), 62. *Prof Exp:* Biochemist, Stanford Res Inst, 62-64; NIH asst res biochemist, Ment Health Res Inst, Univ Mich, 64-66; PROF BIOCHEM, CLARK UNIV, 66- *Concurrent Pos:* Vis lectr, Harvard Med Sch, 72; res assoc, Harvard Sch Pub Health, 77-79. *Mem:* AAAS; Am Soc Neurochem; NY Acad Sci; Am Soc Biol Chemists; Int Soc Neurochem. *Res:* Neurochemistry of brain proteins and nucleic acids; biochemical pharmacology of nucleic acid precursor antimetabolites. *Mailing Add:* Dept Biol Clark Univ Worcester MA 01110

BRINK, KENNETH HAROLD, b Buffalo, NY, Oct 4, 49. SHELF DYNAMICS, COASTAL UPWELLING. *Educ:* Cornell Univ, BS, 71; Yale Univ, MS, 73, PhD(geol & geophys), 77. *Prof Exp:* Ed, Sch Oceanog, Ore State Univ, 77-80; ASSOC SCIENTIST, WOODS HOLE OCEANOG INST, 80- *Mem:* Am Geophys Union; Am Meteorol Soc; AAAS. *Res:* Physical oceanography of coastal upwelling and subtidal frequency fluctuations over the continental shelf and slope. *Mailing Add:* 53 Broken Bow Lane East Falmouth MA 02536

BRINK, KENNETH MAURICE, b Cleveland, Ohio, May 11, 32; m 56; c 2. HORTICULTURE. *Educ:* Purdue Univ, BS, 54, MS, 58, PhD, 65. *Prof Exp:* From instr to assoc prof hort, Purdue Univ, 58-68; PROF HORT & HEAD DEPT, COLO STATE UNIV, 68- *Mem:* AAAS; Am Soc Hort Sci. *Res:* Marketing horticultural crops; post-harvest physiology and economic factors. *Mailing Add:* Dept Hort Colo State Univ Ft Collins CO 80523

BRINK, LINDA HOLK, b Fairhope, Ala, Jan 13, 44; m 67. BIOCHEMISTRY, IMMUNOLOGY. *Educ:* Univ Ala, BS, 67; Univ London, MSc, 73, PhD(parasitol & immunol), 77; Brunel Univ, London, MSc, 77. *Prof Exp:* Microbiologist cholera, WHO, 70-71; microbiologist tuberculosis, USAID, 71-72; res fel, Rockefeller Found, Univ London, 74-77; RES ASSOC, WHO, ROCKEFELLER FOUND & HARVARD MED SCH, 78- *Concurrent Pos:* Prin investr, WHO, 79-81, Clark Found, 81-82 & Rockefeller Found, 78- *Mem:* Am Soc Trop Med & Hyg; Royal Soc Trop Med & Hyg; Brit Soc Immunol; AAAS; Am Asn Hist Med. *Res:* Effector mechanisms of schistosome immunity and isolation of antigens which are now being tested in pre-vaccine trials. *Mailing Add:* 9518 Kentstone Dr Bethesda MD 20817

BRINK, MARION FRANCIS, b Golden Eagle, Ill, Nov 20, 32. NUTRITION. *Educ:* Univ Ill, BS, 55, MS, 58; Univ Mo, PhD(nutrit), 61. *Prof Exp:* Res biologist, Biomed Div, US Naval Radiol Defense Lab, 61-62; from assoc dir to dir nutrit res, 62-71, pres, 71-85, EXEC VPRES OPERS, NAT DAIRY COUN, UNITED DAIRY INDUST ASN, 85- *Mem:* AAAS; Am Oil Chem Soc; Am Inst Nutrit; Am Dairy Sci Asn; Sigma Xi. *Res:* Mineral requirements; toxicology; nutrition, biochemistry and physiology of humans and animals; effects of ionizing radiation upon nutrient requirements. *Mailing Add:* Nat Dairy Coun 6300 N River Rd Rosemont IL 60018

BRINK, NORMAN GEORGE, b Littleton, Colo, Aug 31, 20; m 47. CHEMISTRY. *Educ:* Princeton Univ, AB, 42, PhD(chem), 44. *Prof Exp:* Res chemist, 44-50, asst dir org & biochem res, 50-56, dir bio-org chem, 56-66, dir, Univ Rels, Merck & Co, Inc, 66-76; dir indust rels, dept chem, Univ Calif, San Diego, 77-80; RETIRED. *Mem:* AAAS; Am Soc Biol Chem. *Res:* Isolation of biologically active natural products; intermediary metabolism in disease; enzyme inhibition; medicinal chemistry; stimulation of academic-industrial cooperation. *Mailing Add:* 7220 York Ave S No 418 Edina MN 55435

BRINKER, RUSSELL CHARLES, b Easton, Pa, Dec 7, 08; m 33, 66. CIVIL ENGINEERING. *Educ:* Lafayette Col, BS, 29; Univ Minn, MS, 33, CE, 39. *Prof Exp:* Struct detailer, Am Bridge Co, Pa, 29-30; instr civil eng, Univ Minn, 30-35, asst prof struct eng, 40-41, assoc prof civil eng, 45-47; instr & asst prof eng, Univ Hawaii, 35-36 & 37-40; exchange teacher civil eng, Worcester Polytech Inst, 36-37; assoc prof, Univ Southern Calif, 47-50; prof, Va Polytech Inst, 50-57, head dept, 51-57; ed dir eng bks, Ronald Press Co, 57-58; prof civil eng, Tex Western Col, 58-61; prof, 61-69, head dept, 62-66, ADJ PROF CIVIL ENG, NMEX STATE UNIV, 69- *Concurrent Pos:* News corresp, Eng News Rec, 38-40 & 45-47; consult ed, Civil Eng Ser, Int Textbook Co, 59-74, Dun-Donnelly Publ Corp, 74-76 & Harper & Row Publ, 76-; instr, Peace Corps, 61, consult, 61-66; prof, Univ Hawaii, 66-67 & 75; Calif-Nev rep, Joint Comt Surv & Mapping. *Honors & Awards:* Barge Math Prize. *Mem:* Am Soc Civil Engrs; Am Soc Eng Educ; Am Cong Surv & Mapping; Sigma Xi. *Res:* Surveying; civil engineering; numerous publications. *Mailing Add:* 13373 Plaza Del Rio Blvd No 7752 Peoria AZ 85381-4873

BRINKHOUS, KENNETH MERLE, b Clayton Co, Iowa, May 29, 08; m 36; c 2. PATHOLOGY. *Educ:* Univ Iowa, AB, 29, MD, 32. *Hon Degrees:* DSc, Univ Chicago, 67. *Prof Exp:* Asst path, Sch Med, Univ Iowa, 32-33, instr, 33-35, assoc, 35-37, from asst prof to assoc prof, 37-46; prof, 46-61, distinguished alumni prof, 61-80, EMER PROF PATH, SCH MED, UNIV NC, CHAPEL HILL, 80- *Concurrent Pos:* Mem, subcomt blood coagulation, Nat Res Coun, 51-54, chmn, 54-62, mem, comt blood, 54-, mem, Thrombosis Task Force, 65-70; chmn, Med Adv Comt, Nat Hemophilia Found, 54-73; mem, sch adv comt, World Fedn Hemophilia, 64-78; mem, Int Comt Haemostasis & Thrombosis, 54-, chmn, 64-66, secy gen, 66-78; consult, Armed Forces Inst Path, 56-72; mem, Univs Asn Res & Educ in Path, 64-; pres, Fedn Am Socs Exp Biol, 66-67; mem, Arteriosclerosis Task Force & Coun, Nat Heart & Lung Inst, 70-71; chief ed, Arch Path and Lab Med, 74-83; ed, Thrombosis Res, 83-86, Year Book Path, 80-91. *Honors & Awards:* Ward-Burdick Awards, Am Soc Clin Path, 41 & 63; O Max Gardner Award, 61; James F Mitchell Found Int Award, 69; Gold Headed Cane Award, Am Asn Path, 81; Maude Abbott lectr, Int Acad Path, 85; Pobert P Grant Medal, Int Soc Thrombosis Haemostasis, 85; Distinguished Serv Award, AMA, 86. *Mem:* Nat Acad Sci; Am Soc Exp Path (secy-treas, 60-63, pres, 65-66); Am Asn Path & Bact (secy-treas, 68-71, pres, 73); Int Soc Thrombosis (pres, 72); AAAS. *Res:* Blood clotting; hemorrhagic and thrombotic diseases; vitamin K; atherosclerosis; blood platelets. *Mailing Add:* Dept Path Univ NC Sch Med Chapel Hill NC 27599-7525

BRINKHUIS, BOUDEWIJN H, marine biology, physiological ecology; deceased, see previous edition for last biography

BRINKHURST, RALPH O, b Croydon, Surrey, UK, Mar 13, 33; Can citizen; m 84; c 2. OLIGOCHAETE BIOLOGY, BENTHOS. *Educ:* Univ London, BSc, 55, PhD(zool), 58, DSc, 72. *Prof Exp:* Lectr zool, Univ Liverpool, 57-65; prof, Univ Toronto, 66-72; dir, St Andrews Biol Lab, 72-75; head, Ocean Ecol Lab, Inst Ocean Sci, 76-87, scientist, 87-92; DIR, AQUATIC RESOURCE CTR, 91- *Concurrent Pos:* Assoc, Great Lakes Inst, Univ Toronto, 67-70, Systs & Ecol, Marine Biol Lab, Woods Hole, 67-70 & Royal Ont Mus, 68-; dir, Pollution Probe, 69-73, Huntsman Marine Lab, 72-75; adj prof, Univ Victoria, 76-; vis prof, Royal Roads Mil Col, 80-81. *Mem:* NAm Benthological Soc (pres, 80-81); Freshwater Biol Asn UK. *Res:* Ecology and systematics of freshwater and marine benthos especially Oligochaeta. *Mailing Add:* Aquatic Resource Ctr 4256 Warren Rd Franklin TN 37064

BRINKLEY, JOHN MICHAEL, b Franklin, NH, Jan 23, 44; m 71; c 4. APPLICATION OF FLUORESCENCE TO ANALYSIS OF BIOLOGICAL MATERIALS, DEVELOPMENT OF UNIFORM FLUORESCENT LATEX MICROSPHERES. *Educ:* Culver-Stockton Col, Canton, Mo, BA, 66; Case Western Reserve Univ, Cleveland, Ohio, PhD(org chem), 70. *Prof Exp:* Sr res chemist, Shell Develop Co, 70-73; sr res chemist, Syva Co, 73-80, clin studies supvr, 80-84; clin studies dir, Sclavo West Coast, Inc, 84-88; PROD DEVELOP DIR, MOLECULAR PROBES, INC, 88- *Concurrent Pos:* Mem bd dirs, Western Biomed Inst, 91- *Res:* Dye-protein conjugation chemistry; fluorescent dyes; development and applications of dyed latex microspheres; development of fluorescence-based analytical techniques, primarily to detect gene transfections. *Mailing Add:* Molecular Probes Inc 4849 Pitchford Ave Eugene OR 97402

BRINKLEY, LINDA LEE, b Glendale, Calif, Aug 19, 43; m 73; c 2. DEVELOPMENTAL BIOLOGY. *Educ:* Calif State Univ, Northridge, BA, 65, MS, 67; Univ Calif, Irvine, PhD(biol), 71. *Prof Exp:* Fel cell biol, Dept Zool, 72, res assoc develop biol, Dept Oral Biol, Sch Dent, 72-74, res investr develop biol, Dept Oral Biol, Sch Dent, 74-76, asst res scientist, Sch Dent, 76-77, asst res scientist & instr, Dept Anat, 77-79, asst prof, 79-84, ASSOC PROF, MED SCH, UNIV MICH, ANN ARBOR, 84- *Concurrent Pos:* Res Career Develop Award, NIH, 81-86. *Mem:* Asn Women Sci; Soc Develop Biol; Tissue Cult Asn; Am Soc Cell Biologists; Int Asn Dent Res. *Res:* Normal and abnormal development of secondary palate; role of extracellular matrix in craniofacial morphogenesis; image analysis as a tool for developmental biology studies. *Mailing Add:* Dept Anat Univ Mich Sch Med Ann Arbor MI 48109

BRINKMAN, CHARLES R, b Salt Lake City, Utah, Oct 21, 37. HIGH TEMPERATURE MATERIALS, METALS & CERAMICS. *Educ:* Univ Utah, BS, 60, MS, 61, PhD(metall eng), 66. *Prof Exp:* GROUP LEADER, MECH PROPERTIES GROUP, OAK RIDGE NAT LAB, 74- *Mem:* Am Soc Mech Engrs; fel Am Soc Metals; Am Soc Testing & Mat; Am Inst Mining Metall & Petrol Engrs; Am Ceramics Soc. *Mailing Add:* Metals & Ceramics Div Oak Ridge Nat Lab PO Box 2008 Oak Ridge TN 37831-6154

BRINKMAN, WILLIAM F, b Washington, Mo, July 20, 38; m 60; c 2. ELECTRONICS ENGINEERING, MATERIALS SCIENCE ENGINEERING. *Educ:* Univ Mo-Columbia, BS, 60, MS, 62, PhD(physics), 65. *Hon Degrees:* DHL, Univ Mo, Columbia, 87. *Prof Exp:* NSF res fel physics, Oxford Univ, 65-66; head, Infrared Physics & Electronics Res Dept, Bell Labs, 72-74, dir, Chem Physics Res Lab, 74-81 & Phys Res Lab, 81-84; vpres res, Sandia Nat Labs, 84-87; EXEC DIR RES, PHYSICS DIV, AT&T BELL LABS, 87- *Mem:* Nat Acad Sci; Am Phys Soc; fel AAAS. *Res:* Solid state theory; itinerant magnetism and spin fluctuations; electron tunneling theory; theory of metal-insulator transitions. *Mailing Add:* AT&T Bell Labs 600 Mountain Ave Murray Hill NY 07974

BRINKMEYER, RAYMOND SAMUEL, b Lima, Ohio, Mar 22, 48; m 71; c 2. ORGANIC CHEMISTRY. *Educ:* Univ Cincinnati, BS, 70; Colo State Univ, PhD(chem), 75. *Prof Exp:* Res scientist, Lilly Res Labs, 78-89; RES ADMINR, DOW ELANCO, 89- *Concurrent Pos:* Fel, Stanford Univ, 75-77 & Nat Cancer Inst, 76-77. *Mem:* Am Chem Soc. *Res:* Natural products; synthetic methods; plant growth regulators. *Mailing Add:* Dow Elanco Greenfield IN 46140

BRINKS, JAMES S, b Plymouth, Mich, Jan 2, 34; m 55; c 9. ANIMAL GENETICS. *Educ:* Mich State Univ, BS, 56, MS, 57; Iowa State Univ, PhD(statist & genetics), 60. *Prof Exp:* Res geneticist, Animal Husb Res Div, Agr Res Serv, USDA, 60-67; PROF ANIMAL SCI, COLO STATE UNIV, 67- *Mem:* Am Soc Animal Sci. *Res:* Estimation of genetic, environmental and phenotypic parameters; estimation of response to selection and inbreeding; comparisons of mating systems; performance testing. *Mailing Add:* Dept Animal Sci Colo State Univ Ft Collins CO 80523

BRINLEY, FLOYD JOHN, JR, b Battle Creek, Mich, May 19, 30; m 55; c 3. BIOPHYSICS. *Educ:* Oberlin Col, AB, 51; Univ Mich, MD, 55; Johns Hopkins Univ, PhD, 61. *Prof Exp:* Intern, Los Angeles County Gen Hosp, 55-56; sr asst surgeon neurophysiol lab, NIH, 57-59; from asst prof to assoc prof physiol, Sch Med, Johns Hopkins Univ, 61-76; prof physiol, Sch Med, Univ MD, 76-79; dir, Neurol Dis Prog, 79-82, DIR, CONVULSIVE, DEVELOP & NEUROMUSCULAR DIS PROG, NAT INST NEUROL & COMMUN DIS & STROKE, NIH, 82- *Mem:* Biophys Soc; Am Physiol Soc; Soc Gen Physiol; Soc Neurosci; Am Neurol Asn; Am Acad Neurol; Am Soc Biol Chemists. *Res:* Membrane phenomena; ionic transport. *Mailing Add:* PO Box 41021 Bethesda MD 20814

BRINN, JACK ELLIOTT, JR, b Norfolk, Va, June 7, 42; m 65; c 3. ENDOCRINOLOGY, ELECTRON MICROSCOPY. *Educ:* ECarolina Col, BA, 64, MA, 66; Univ Wyo, PhD(zool), 71. *Prof Exp:* Instr biol, ECarolina Univ, 66-67; res assoc, Milton S Hershey Med Ctr, 70-72; ASST PROF ANAT, ECAROLINA UNIV, 72- *Mem:* Am Asn Anatomists; Am Soc Zoologists; Am Diabetes Asn; Fedn Am Soc Exp Biol. *Res:* Ultrastructure of polypeptide hormone synthesis and secretion; comparative endocrine cytology and ultracytochemistry. *Mailing Add:* Dept Anat ECarolina Univ Sch Med Greenville NC 27834

BRINSON, MARK MCCLELLAN, b Shelby, Ohio, Oct 6, 43; m 71; c 1. WETLAND ECOLOGY, ESTUARINE ECOLOGY. *Educ:* Heidelberg Col, BS, 65; Univ Mich, Ann Arbor, MS, 67; Univ Fla, PhD(bot), 73. *Prof Exp:* Fisheries biologist, Peace Corps, Costa Rica, 67-69; from asst prof to assoc prof, 73-81, dir biol grad prog, 81-86, PROF BIOL, ECAROLINA UNIV, 81- *Concurrent Pos:* Vis asst prof bot, Univ NC, Chapel Hill, 76; wetland ecologist, US Fish & Wildlife Serv, 79-80. *Mem:* Sigma Xi; Ecol Soc Am; Am Soc Limnol & Oceanog; Soc Wetland Scientists (vpres, 89-90, pres, 90-91); Int Ecol Soc; AAAS. *Res:* Wetland biogeochemistry and hydrology; aquatic macrophyte ecology. *Mailing Add:* Dept Biol ECarolina Univ Greenville NC 27858

BRINSTER, RALPH L, b Montclair, NJ, Mar 10, 32. DEVELOPMENTAL BIOLOGY. *Educ:* Rutgers Univ, BS, 53; Univ Pa, VMD, 60, PhD(physiol), 64. *Prof Exp:* Asst instr physiol, Sch Med, Univ Pa, 60-61, teaching fel, 61-64, instr, 64-65, from asst prof to assoc prof, Sch Vet Med, 65-70, PROF PHYSIOL, SCH VET MED, UNIV PA, 70- *Concurrent Pos:* Lake Placid Conf, 87. *Honors & Awards:* Award in Biol & Med Sci, NY Acad Sci, 83; Harvey Soc Lectr, 84. *Mem:* Nat Acad Sci; Inst Med; Soc Study Reproduction; Brit Soc Study Fertil; Brit Biochem Soc; Am Vet Med Asn; Am Acad Arts & Sci; Am Soc Cell Biol. *Res:* In vitro culture and physiology of the cleavage stages of mammalian embryos; differentiation in the mammalian embryo; gene regulation; transgenic animals. *Mailing Add:* Sch Vet Med Univ Pa 3800 Spruce St Philadelphia PA 19104

BRINTON, CHARLES CHESTER, JR, b Pittsburgh, Pa, Aug 15, 26. MICROBIOLOGY, BIOPHYSICS. *Educ:* Carnegie Inst Technol, BS, 49; Univ Pittsburgh, MS, 52, PhD(biophys), 55. *Prof Exp:* Asst biophys, Univ Pittsburgh, 49-51; res fel, Geneva, Switz, 55-56; res assoc, Univ Pittsburgh, 56-59, from asst res prof to assoc prof, 59-69, PROF MICROBIOL & MEM GRAD FAC, MED SCH, UNIV PITTSBURGH, 69- *Mem:* AAAS; Biophys Soc; Am Soc Microbiol. *Res:* Bacterial and viral genetics, biophysics, physiology and serology; electrophoresis; biochemistry and physical chemistry of proteins and nucleic acids. *Mailing Add:* Dept Biol Sci A-234 Langley Hall Univ Pittsburgh Pittsburgh PA 15260

BRINTON, EDWARD, b Richmond, Ind, Jan 12, 24; m 48; c 4. BIOLOGICAL OCEANOGRAPHY. *Educ:* Haverford Col, BA, 49; Bryn Mawr Col, MA, 50; Scripps Inst Oceanog, Univ Calif, PhD(oceanog), 58. *Prof Exp:* From jr res biologist to res biologist, 53-73, RES BIOL, MARINE LIFE RES GROUP, SCRIPPS INST OCEANOG, UNIV CALIF, 73- *Concurrent Pos:* Partic, AID-Univ Calif Proj, Southeast Asia, 60-61; UNESCO instr, Thailand, 62 & Pakistan, 64; UNESCO cur, Indian Ocean Biol Ctr, Indian Ocean Exped, Ernakulam, S India, 65-67. *Mem:* Am Soc Limnol & Oceanog; Challenger Soc. *Res:* Systematics, population biology and zoogeography of Euphausiacea, including antarctic Krill; ecology of zooplankton. *Mailing Add:* Scripps Inst Oceanog Univ Calif La Jolla CA 92093

BRION, CHRISTOPHER EDWARD, b Eng, May 5, 37; m 61; c 3. PHYSICAL CHEMISTRY. *Educ:* Bristol Univ, BSc, 58, PhD(chem), 61. *Prof Exp:* Res fel chem, Univ BC, 61-63, teaching fel, 63-64, from asst prof to assoc prof, 64-76, PROF CHEM, UNIV BC, 76- *Concurrent Pos:* Sr res fel, Nat Res Coun Can, 69-70; John Simon Guggenheim mem fel, 78-79; Killan res fel, Canada Coun, 84; Sr Killam res fel, Univ BC, 90. *Honors & Awards:* Noranda Award, Chem Inst Can, 77; Herzberg Award, Spectros Soc Can, 83. *Mem:* Fel Chem Inst Can; Can Asn Physicists; Am Chem Soc; fel Royal Soc Can; Spectros Soc Can. *Res:* Excited states of ions; energy loss electron spectroscopy; electron-electron and electron-ion coincidence techniques; photoionization; orbital imaging by electron momentum spectroscopy; synchrotron radiation studies; experimental quantum chemistry. *Mailing Add:* Dept Chem Univ BC Vancouver BC V6T 1Y6 Can

BRISBIN, DOREEN A, b Edmonton, Alta, Dec 19, 26; m 49; c 3. BIOINORGANIC CHEMISTRY. *Educ:* Univ Alta, BSc, 46; Univ Toronto, PhD(chem), 60. *Prof Exp:* Res assoc chem, 60-61, lectr, 61-63, asst prof, 63-66, assoc dean sci, 75-86, ASSOC PROF CHEM, UNIV WATERLOO, 66-, ADV TO VPRES ACAD, ACAD HUMAN RESOURCES, 86- *Res:* Coordination chemistry; stability of metal complexes; kinetics of metalloporphyrin formation; intermolecular complexes with porphyrins. *Mailing Add:* Dept Chem Univ Waterloo Waterloo ON N2L 3G1 Can

BRISBIN, I LEHR, JR, b Drexel Hill, Pa, Apr 2, 40; m; c 1. ZOOLOGY, ECOLOGY. *Educ:* Wesleyan Univ, AB, 62; Univ Ga, MS, 65, PhD(zool), 67. *Prof Exp:* Adj assoc prof vert ecol, 67-68, AEC SR FEL, SAVANNAH RIVER ECOL LAB, 68-; ADJ ASSOC PROF POULTRY SCI & ZOOL, UNIV GA, 69- *Concurrent Pos:* Pop ecologist, US Energy Res & Develop Admin, 73-75. *Mem:* Am Soc Mammal; Am Ornithologists Union; Am Soc Ichthyologists & Herpetologists; Ecol Soc Am; Wildlife Soc. *Res:* Vertebrate ecology and bioenergetics; mammalogy, ornithology and herpetology; ecology of domestication; canine behavior and olfaction; energy-related environmental impacts, radioecology. *Mailing Add:* Savannah River Ecol Lab PO Drawer E Aiken SC 29801

BRISCOE, ANNE M, b New York, NY, Dec 1, 18; m 55. BIOCHEMISTRY, PHYSIOLOGY. *Educ:* Vassar Col, AM, 45; Yale Univ, PhD(physiol chem), 49. *Prof Exp:* From res assoc to asst prof biochem, Med Col, Cornell Univ, 50-56; res assoc, Sch Med, Univ Pa, 56; assoc biochem, 56-72, asst prof med, Col Physicians & Surgeons, Columbia Univ, 72-87; RETIRED. *Concurrent Pos:* NIH fel, Univ Pa, 49-50; lectr, Sch Gen Studies, Hunter Col, 51-64, adj asst prof, Sch Health Sci, 73-; vis asst prof, Temple Univ, 56; consult, Vet Admin Hosp, Castle Point, NY, 57-67; career scientist, New York Health Res Coun, 60-66; lectr, Sch Gen Studies, Columbia Univ, 67-68, Sch Nursing, Harlem Hosp Ctr, 68-77 & Antioch Col Physicians Asst Prog at Harlem Hosp Ctr, 71-73; spec lectr, Columbia Univ, 87- *Mem:* Fel Am Inst Chemists; Am Chem Soc; Am Soc Clin Nutrit; Asn Women Sci (pres, 78-); fel NY Acad Sci. *Res:* Metabolism of calcium and magnesium in human subjects; metabolic acidosis; chronic renal failure. *Mailing Add:* 15 Quarry Hill Rd East Hampton CT 06424

BRISCOE, CHARLES VICTOR, b Abingdon, Va, May 5, 30; m 53; c 3. PHYSICS. *Educ:* King Col, BA, 52; Rice Inst, MA, 57, PhD(physics), 58. *Prof Exp:* Instr math, King Col, 52-53, instr physics, 53-54; res assoc, 58-60, from asst prof to assoc prof, 58-70, PROF PHYSICS, UNIV NC, CHAPEL HILL, 70- *Concurrent Pos:* Vis prof, Tech Univ Karlsruhe, 66 & 67. *Mem:* Am Phys Soc; Am Asn Physics Teachers; Sigma Xi. *Res:* Superconducting properties of metallic films; position annihilation in inert gas liquids. *Mailing Add:* 4814 Oak Hill Rd Chapel Hill NC 27514

BRISCOE, JOHN, b Brakpan, SAfrica, July 30, 48; Ireland citizen; m 79; c 3. SANITARY & ENVIRONMENTAL ENGINEERING. *Educ:* Univ Cape Town, BSc, 69; Harvard Univ, MS, 71, PhD(environ eng), 76. *Prof Exp:* Scientist, Cholera Res Lab, Bangladesh, 76-78; res assoc, Harvard Univ, 79; engr, Govt of Mozambique, 79-81; assoc prof environ eng, Univ NC, 81-85; sr economist & unit chief, World Bank, Washington, DC, 86-91. *Res:* Relationships between engineered water systems and health: viruses and organic chemicals in developed countries; bacteria and helminths in underdeveloped countries; systems analysis and quantitative methods; economics of water systems. *Mailing Add:* World Bank Washington DC 20433

BRISCOE, JOHN WILLIAM, b Westfield, Ill, Feb 24, 17; m 42; c 1. CIVIL ENGINEERING. *Educ:* Univ Ill, BS, 40 & 47, MS, 53. *Prof Exp:* From instr to assoc prof, 48-57, asst head dept civil eng, 57-59, assoc head dept, 59-65, assoc provost, 65-67, assoc chancellor admin, 67-68, vchancellor admin affairs, 68-77, PROF CIVIL ENG, UNIV ILL, URBANA-CHAMPAIGN, 57- *Concurrent Pos:* Consult, Off Civil Defense, 61-65; mem, Proj Harbor Study Comt, 63, Joint Adv Comt Continuing Educ, 64 & USAF Acad, 64. *Mem:* Am Soc Civil Engrs; Am Soc Eng Educ; Soc Hist Technol. *Res:* Behavior of reinforced concrete structures; protective construction. *Mailing Add:* 107 Whitehall Urbana IL 61801-6664

BRISCOE, MELBOURNE GEORGE, b Akron, Ohio, July 17, 41; m 82. INTERNAL WAVES, AIR-SEA INTERACTIONS. *Educ:* Northwestern Univ, BS, 63, PhD(mech engr), 67. *Prof Exp:* Fel, von Karman Inst Fluid Dynamics, Rhode-St Genese, Belg, 67-68; fel air sea interactions, Off Naval Res, NATO Saclant Undersea Res Ctr, La Spezia, Italy, 68-69; scientist Phys Oceanog, Woods Hole Oceanog Inst, 69-89, sci officer, Off Naval Res, 87-89, DIR APPL OCEANOG & ACOUST DIV, OFF NAVAL RES, 89- *Concurrent Pos:* Ed, Oceanic Internal Waves, 74-75; assoc ed, J Geophys Res, 74-77, J Phys Oceanog, 77-79 & 87-, chief ed, J Phys Oceanog, 85-87; consult, Naval Res Adv Comt, 77, Naval Oceanog Res Develop Activ, 78, Sci Appl Inc, 79-85, Joseph Wetzell Assoc, 81, EG&G, 81-, Buoy Technol, 84, Endeco, 85. *Mem:* Am Geophys Union; Am Meteorol Soc; Oceanog Soc; Inst Elec & Electronics Engrs; Acoust Soc Am. *Res:* Field experiments in physical oceanography to understand the pathways of energy between atmosphere and ocean; the role of the ocean in determining the climate of the earth; acoustical methods to investigate oceanographic phenomena. *Mailing Add:* Off Naval Res Arlington VA 22217-5000

BRISCOE, WILLIAM ALEXANDER, physiology, medicine, for more information see previous edition

BRISCOE, WILLIAM TRAVIS, b Tulsa, Okla; m 70; c 2. BIOCHEMISTRY. *Educ:* Rice Univ, BA, 70; Univ Tex, Houston, PhD(biochem), 74. *Prof Exp:* Robert A Welch Found fel biochem, M D Anderson Hosp & Tumor Inst, Univ Tex, Houston, 74; fel microbiol, Scripps Clin & Res Found, Calif, 75-77; ASST PROF BIOCHEM, SCH MED, ORAL ROBERTS UNIV, 77- *Mem:* Am Asn Cancer Res. *Res:* Nucleic acid and protein synthesis, especially the investigation of mutagenic sites in DNA as carcinogenic markers. *Mailing Add:* 2628 E 73rd St Tulsa OK 74136

BRISKE, DAVID D, b Grafton, NDak, Aug 27, 51; m 71, 88; c 2. PHYSIOLOGICAL PLANT ECOLOGY, POPULATION ECOLOGY. *Educ:* NDak State Univ, BA, 73; Colo State Univ, PhD(range sci), 78. *Prof Exp:* Res asst, Colo State Univ, 74-77; asst prof, 78-83, ASSOC PROF RANGELAND ECOL & MGT, TEX A&M UNIV, 83- *Concurrent Pos:* Consult, USAID, Morocco. *Mem:* Soc Range Mgt; Ecol Soc Am; Am Inst Biol Sci. *Res:* Physiology ecology and population biology of perennial grasses including developmental morphology, defoliation physiology, clonal biology, resource allocation; competitive interactions and population structure. *Mailing Add:* Dept Rangeland Ecol & Mgt Tex A&M Univ College Station TX 77843-2126

BRISLEY, CHESTER L(AVOYEN), b Albion, Pa, Apr 3, 14; m 32; c 1. INDUSTRIAL ENGINEERING. *Educ:* Youngstown State Univ, BS, 45; Wayne State Univ, MS, 54, PhD, 57. *Prof Exp:* With Packard Elec Div, Gen Motors Corp, 35-42; supvr job improv & time standards, NAm Aviation, 42-45; indust eng mgr, Wolverine Tube Div, Calumet & Hecla, Inc, 46-58; staff asst to dir prod, Chance Vought Aircraft, Inc, 58-59; sr consult, A T Kearney & Co, Inc, 60; chief indust engr, Allis-Chalmers Mfg Co, 61-62; mgr mgt serv, Touche-Ross, 63-64; assoc chmn dept & prof eng & appl sci, Univ Wis-Milwaukee Ext, 65-83; prof eng, Calif Polytech, San Luis Obispo, 83-85; prof eng, Marquette Univ, 85-86, Lincoln Mem Univ, Harrogate, Tenn, 87-90; DIR INDUST ENG, MARQUETTE UNIV, 90- *Concurrent Pos:* Mem, Mkt & Pub Rels Tech Assistance Team to WBerlin, Ger, 54, Indust Eng Tech Assistance Team to Japan, 57 & Asian Productivity Orgn Assistance Team, 75. *Honors & Awards:* Engr of the Year Award, Am Inst Indust Engrs, Region XI, 78. *Mem:* Fel Am Inst Indust Engrs; Am Soc Eng Educ; Am Soc Mech Eng; Nat Soc Prof Engrs. *Res:* Development of technique of work measurement called work sampling. *Mailing Add:* Dept Mech & Indust Eng Marquette Univ Milwaukee WI 53233

BRISSON, GERMAIN J, b St Jacques, Que, Apr 12, 20; m 48; c 5. NUTRITION. *Educ:* Joliette Sem, BA, 42; Laval Univ, BSc, 46; McGill Univ, MSc, 48; Ohio State Univ, PhD(nutrit), 50. *Prof Exp:* Res officer nutrit, Can Dept Agr, 50-60, dir admin, Lennoxville Exp Sta, 60-62; dir, Nutrit Res Ctr, 68-78, PROF NUTRIT, LAVAL UNIV, 62- *Concurrent Pos:* Sci ed, Can J Animal Sci, 63-66. *Mem:* Am Soc Animal Sci; Am Dairy Sci Asn; Nutrit Soc Can; Agr Inst Can; Can Soc Animal Sci. *Res:* Nutritional requirements of young ruminants; protein and fat utilization by young animals. *Mailing Add:* Nutrit Res Ctr Laval Univ Quebec PQ G1K 7P4 Can

BRISTOL, BENTON KEITH, agriculture; deceased, see previous edition for last biography

BRISTOL, DOUGLAS WALTER, b Rochester, NY, Oct 31, 40; m 64; c 1. ORGANIC CHEMISTRY. *Educ:* St John Fisher Col, BS, 62; Syracuse Univ, PhD(chem), 69. *Prof Exp:* Res assoc chem, Columbia Univ, 69; instr, Univ Utah, 69-71; asst prof pesticides chem, NDak State Univ, 71-77; mem staff, Div Chem & Physics, Food & Drug Admin, 77-80. *Mem:* AAAS; Am Chem Soc. *Res:* Free radical chemistry; reactions and properties of primary and secondary alkoxy-radicals; reactions of pyridinium salts with nucleophiles and dihydropyridine compounds with molecular oxygen; pesticides chemistry. *Mailing Add:* 104 Fieldstone Ct Chapel Hill NC 27514

BRISTOL, JOHN RICHARD, b Mattoon, Ill, Apr 11, 38; m 75; c 2. PARASITOLOGY, PHYSIOLOGY. *Educ:* Cornell Col, BA, 61; Kent State Univ, MA, 66, PhD(physiol), 70. *Prof Exp:* Instr biol, Kent State Univ, 69-; from asst prof to assoc prof biol, Univ Tex, El Paso, 70-85, chmn dept, 77-83, asst dean, 84-85, prof biol & asst vpres acad affairs, 85-88, VPRES ACAD AFFAIRS, UNIV TEX, EL PASO, 88- *Concurrent Pos:* Prog dir, Minority Biomed Support grant, NIH, 74-83; consult, Div Res Resources, NIH, 76-78. *Mem:* Am Soc Parasitologists; Am Micros Soc. *Res:* Animal parasite host interactions; physiology and biochemistry of animal parasites. *Mailing Add:* Dept Biol Sci Univ Tex El Paso TX 79968

BRISTOL, MELVIN LEE, b Hartford, Conn, Dec 3, 36; m 59; c 3. ETHNOBOTANY, HORTICULTURE. *Educ:* Harvard Univ, AB, 60, AM, 62, PhD(bot), 65. *Prof Exp:* USPHS fel, 65-66; asst prof bot & asst botanist, Lyon Arboretum, Univ Hawaii, 66-69; OWNER, BLOOMINGFIELDS FARM, 69- *Mem:* Soc Econ Botanists. *Res:* Ethnobotany and floristics of Colombia and Samoa; taxonomy and ethnobotany of Datura; plant domestication; horticulture; Hemerocallis. *Mailing Add:* Nine Rte 55 W Sherman CT 06784-9722

BRISTOW, JOHN DAVID, b Pittsburgh, Pa, Dec 7, 28; m 50; c 3. MEDICINE. *Educ:* Willmette Univ, BA, 49; Univ Ore, MD, 53; Am Bd Cardiovasc Dis, dipl, 70. *Prof Exp:* Instr med, 60-61, dir cardiol lab, 62-68, from asst prof to assoc prof med, 62-70, chmn dept, 71-75, chief med serv, Univ Hosp, 69-75, prof med, Med Sch, Univ Ore, 70-77, clin res assoc physiol, 65-77; prof med, Univ Calif, San Francisco, 77-81; chief cardiol, Vet Admin Hosp, San Francisco, 77-81; PROF MED & DIR CARDIOVASC RES & TRAINING, ORE HEALTH SCI UNIV, 81- *Concurrent Pos:* Nat Heart Inst spec fel, Cardiovasc Res Inst, Univ Calif, 61-62 & Radcliffe Infirmary, Eng, 67-68; Markle scholar, 64-69; Am Heart Asn Coun Clin Cardiol fel; consult, US Army Hosp, Ft Lewis, Wash, 66-68; mem, Heart & Lung Prog Proj Comt, NIH, 70-74, nat adv coun, Nat Heart, Lung & Blood Inst, 78-82; mem bd gov, Am Bd Internal Med, 78- *Mem:* Fel Am Col Physicians; Asn Am Physicians; fel Am Col Cardiol; Am Soc Clin Invest; Am Fedn Clin Res. *Res:* Clinical cardiology; cardiovascular physiology; medical education. *Mailing Add:* Ore Health Sci Univ Portland OR 97201

BRISTOW, LONNIE ROBERT, b New York, NY, Apr 6, 30; m 61; c 4. INTERNAL MEDICINE, OCCUPATIONAL MEDICINE. *Educ:* Col City NY, BS, 53; NY Univ, MD, 57. *Prof Exp:* Intern, San Francisco City & Co Hosp, 57-58; resident internal med, US Vet Admin Hosp, San Francisco, 58-60 & Bronx, NY, 60-61; resident occup med, Univ Calif, San Francisco, 79-81. *Concurrent Pos:* Pvt pract med, 61-; mem bd trustees, AMA, 85- *Mem:* Inst Med-Nat Acad Sci; Am Soc Internal Med; AMA. *Res:* Medical ethics; socialized medicine as practiced in Great Britain and Canada; health care financing in America; professional liability insurance problems; sickle cell anemia; coronary care unit utilization. *Mailing Add:* 2023 Vale Rd San Pablo CA 94806

BRISTOW, PETER RICHARD, b Detroit, Mich, June 18, 46; m 70; c 1. PLANT PATHOLOGY. *Educ:* Mich State Univ, BS, 68, MS, 72, PhD(plant path), 74. *Prof Exp:* Fel plant path, Univ Mo-Columbia, 74-76; asst plant pathologist, 76-80, ASSOC PLANT PATHOLOGIST, PUYALLUP RES & EXTEN CTR, WASH STATE UNIV, 80- *Mem:* Am Soc Plant Path; Am Phytopath Soc; Sigma Xi. *Res:* Diseases of small fruit crops, especially red raspberry, strawberry, blueberry and cranberry; biology and ecology of soil micro-organisms; root disease pathogens. *Mailing Add:* 7612 Pioneer Way E Wash State Univ Puyallup WA 98371

BRISTOW, WILLIAM WARREN, b Philadelphia, Pa, Jan 14, 40; m 62; c 3. PHYSICAL CHEMISTRY. *Educ:* Temple Univ, BS, 61; Univ Del, PhD(phys chem), 65. *Prof Exp:* Res chemist, Textile Fibers Pioneering Res, E I du Pont de Nemours & Co, 65-67 & Polymer Sect, Atlas Chem Indust, 67-70; res supvr polymer group, Corp Res Lab, ICI US Inc, 70-79, mgr polymer res, 79-87; mgr, Indust Adhesives, Am Cyanamid Co, 87-90. *Concurrent Pos:* AEC fel, 62-65. *Mem:* Am Chem Soc; Sigma Xi; Soc Advan Mat Process Eng. *Res:* Radiation chemistry, effects of radiation on matter; characterization of fibers and foams; solution properties of water soluble polymers and uses; high temperature polymers synthesis and uses; thermosetting resin; science of composites, molding compounds. *Mailing Add:* 3328 Altamont Dr Wilmington DE 19810

BRITT, DANNY GILBERT, b Glasgow, Ky, Sept 19, 46; m 66; c 1. DAIRY NUTRITION. *Educ:* Western Ky Univ, BS, 68; Mich State Univ, MS, 72, PhD(dairy nutrit), 73. *Prof Exp:* Asst prof animal sci, Tex A&I Univ, 73-74; asst prof, 74-80, ASSOC PROF AGR, EASTERN KY UNIV, 80- *Mem:* AAAS; Am Dairy Sci Asn; Am Soc Animal Sci. *Res:* Use of high moisture corn in dairy rations; nutrient requirements of high producing cows. *Mailing Add:* Agr Dept Eastern Ky Univ A B Carter Bldg Richmond KY 40475-3011

BRITT, DOUGLAS LEE, b West Lafayette, Ind, Mar 10, 46; m 70; c 2. LIMNOLOGY, BIOLOGY. *Educ:* Purdue Univ, BS, 68; Univ Toledo, MS, 73. *Prof Exp:* Asst biol & ecol, Univ Toledo, 68-70; assoc instr zool & limnol, Ind Univ, Bloomington, 70-72; sect head fisheries & aquatic ecol, Jack McCormick & Assoc, Wapora Inc, 73-75; proj leader energy & environ prog planning, MITRE Corp, 75-78; dir environ sci & prog planning, Gen Res Corp, 78-84; PRES, INT SCI & TECHNOL INC, STERLING, 84- *Concurrent Pos:* Mem, Fed Interagency Comt Health & Environ Effects Energy Technol, 78-81; mem, Int Expert Working Group on Environ Monitoring & Specimen Banking, 78. *Mem:* AAAS; Nat Audubon Soc; Soc Environ Toxicol & Chem. *Res:* Environmental program planning and environmental assessments of fossil fuel and advanced energy systems; development of monitoring programs for the identification of hazardous materials in aquatic and terrestrial ecosystems. *Mailing Add:* 10802 Midsummer Dr Reston VA 22091

BRITT, EDWARD JOSEPH, b Flagstaff, Ariz, Aug 17, 41; m 65; c 2. PHYSICS, BIOENGINEERING. *Educ:* Univ Ariz, BS, 64, PhD(nuclear eng), 71. *Prof Exp:* Physicist, Am Atomics Corp, 70-71; physicist, 71-80, TECH DIR, RASOR ASSOC INC, 80- *Mem:* Inst Elec & Electronics Engrs; Am Nuclear Soc; Sigma Xi. *Res:* Thermionic energy conversion; bioengineering; laser energy conversion; plasma physics; heart pacemakers; photosynthesis; solar electric power generation; cogeneration; space power; laser development. *Mailing Add:* 20850 Pepper Tree Lane Cupertino CA 95014

BRITT, EUGENE MAURICE, microbiology; deceased, see previous edition for last biography

BRITT, HAROLD CURRAN, b Buffalo, NY, Sept 14, 34; m 56. NUCLEAR PHYSICS. *Educ:* Hobart Col, BS, 56; Dartmouth Col, MA, 58; Yale Univ, PhD(physics), 61. *Prof Exp:* Mem staff, Physics Div, Los Alamos Sci Lab, 61-72; vis prof physics, Univ Rochester, 72-73; group leader, 74-81, fel, Physics Div, Los Alamos Nat Lab, 81-86; mem staff, Physics Dept, 86-88, E DIV LEADER, LAWRENCE LIVERMORE NAT LAB, 88- *Concurrent Pos:* Mem staff, Nuclear Chem Div, Lawrence Radiation Lab, Univ Calif, 64-65; mem subpanel on instrumentation & tech, Nat Acad Sci-Nat Res Coun Nuclear Physics Surv, 70; mem, Lawrence Berkeley Lab, Super Heavy Ion Linear Accelerator Users Exec Comt, 72-73 & 77-79, chmn, 78-79; consult & vis, Niels Bohr Inst, Univ Copenhagen, 76-80; mem, Oak Ridge Nat Lab, Exec Comt Heavy Ion Accelerator Users Group, 78-79; Alexander von Humboldt sr US scientist award, 79; vis prof, Univ Munich, 79-80; mem prog adv comn, Nat Superconducting Cyclotron Lab, Mich State Univ, 81-83; mem Exec Comt Div Nuclear Physics, Am Phys Soc, 86-88. *Mem:* AAAS; fel Am Phys Soc; Am Chem Soc. *Res:* Heavy ion reactions at medium and relativistic energies; development of a multiple logarithmic detector system for the study of exclusive heavy and light particle emission in heavy ion reactions at energies of the order of 100 MeV/nucleon. *Mailing Add:* L289 Lawrence Livermore Nat Lab Livermore CA 94550

BRITT, JAMES ROBERT, physics, computer science, for more information see previous edition

BRITT, N WILSON, b Lucas, Ky, Jan 3, 13; m 41. AQUATIC ENTOMOLOGY, LIMNOLOGY. *Educ:* Western Ky State Col, BS, 39; Ohio State Univ, MS, 47, PhD(hydrobiol, entom), 50. *Prof Exp:* Pub sch teacher, Ky, 33-40; high sch prin, coach & teacher, 40-42; asst ecol, USAAF Weather Sch, Ill, 42-43, USAAF Europ Theatre, 44-46; asst meteorol, Stone Inst Hydrobiol, 47-50, from asst prof to assoc prof, Ohio State Univ, 50-68, prof entom, 68-83; RETIRED. *Res:* Ecology; limnology, aquatic entomology, general entomology. *Mailing Add:* 1914 70th St W Bradenton FL 34209

BRITT, PATRICIA MARIE, b Los Angeles, Calif, May 2, 31. COMPUTER SCIENCE, BIOMATHEMATICS. *Educ:* Univ Chicago, BA, 49; Univ Calif, Los Angeles, BA, 51, PhD(philos), 59. *Prof Exp:* Mathematician, Bendix Comput Corp, Los Angeles, 58-63 & Control Data Corp, 63; sr systs engr, IBM Corp, 63-66; assoc dir health sci, Univ Calif, Los Angeles, 66-78, dir, hosp comput facil, 79-90; CONSULT, 91- *Concurrent Pos:* Consult, Health Serv & Ment Health Admin, 71-73 & NIH, 72-; lectr, Dept Biomath, Univ Calif, Los Angeles, 72-78. *Res:* Application of computers to research in biomedicine. *Mailing Add:* 3240 Flagler Rd Nordland WA 98358

BRITTAIN, DAVID B, b Chicago, Ill, Dec 24, 25; m; c 3. PHYSIOLOGY, GENERAL MEDICAL SCIENCES. *Educ:* DePaum Univ, MA, 51. *Prof Exp:* ASSOC PROF BIOL, RIPON COL, 62- *Mailing Add:* Dept Biol Ripon Col 300 Seward St Box 248 Ripon WI 54971

BRITTAIN, JOHN O(LIVER), b Pittsburgh, Pa, Feb 15, 20; m 45; c 4. METALLURGY. *Educ:* Pa State Univ, BS, 43, PhD(metall), 51. *Prof Exp:* Asst metall, Res Labs, Aluminum Co Am, 43-44; instr, Pa State Univ, 47-48; prin metallurgist, Battelle Mem Inst, 50-51; res assoc, Columbia Univ, 51-55; chmn dept, 68-73, prof, 55-90, dir Mat Res Ctr, 76-79, EMER PROF MAT SCI & ENG, NORTHWESTERN UNIV, 90- *Concurrent Pos:* Air Force Off Sci Res grant, 69-71; NSF grant, 75-; mem, Eng Comt Accreditation Bd Eng Technol; Off Naval Res grant, 78-; NASA grant, 83-; guest scholar, Kyoto Univ, 86. *Honors & Awards:* Ralph A Teetor Educ Award, Soc Automotive Engrs, 77. *Mem:* Am Inst Mining, Metall & Petrol Engrs; fel Am Soc Metals; Am Soc Eng Educ; Sigma Xi. *Res:* Materials science; structure sensitive properties of solids. *Mailing Add:* Technol Inst Northwestern Univ Evanston IL 60201

BRITTAIN, THOMAS M, mechanical engineering, engineering mechanics; deceased, see previous edition for last biography

BRITTAIN, WILLIAM JOSEPH, b Rapid City, SDak, July 12, 55; m 79; c 3. POLYMER SYNTHESIS, PHYSICAL-ORGANIC CHEMISTRY. *Educ:* Univ Northern Colo, BS, 77; Calif Inst Technol, PhD(org chem), 82. *Prof Exp:* Postdoctoral assoc, Dept Chem, Duke Univ, 82-84; res chemist, Cent Res, E I du Pont de Nemours & Co, Inc, 84-88; sr res chemist, Polymer Prod, 88-90; ASST PROF POLYMER SCI, UNIV AKRON, 90- *Mem:* Am Chem Soc. *Res:* Kinetics and mechanism of polymerization; group transfer polymerization; synthesis of biomaterial model surfaces; synthesis of polyamides. *Mailing Add:* Dept Polymer Sci Univ Akron Akron OH 44325-3909

BRITTAN, MARTIN RALPH, b San Jose, Calif, Jan 28, 22; m 47; c 2. VERTEBRATE ZOOLOGY, ECOLOGY. *Educ:* San Jose State Col, AB, 46; Stanford Univ, PhD(biol sci), 51. *Prof Exp:* Teaching asst biol sci, Stanford Univ, 47-49; asst prof biol, SDak Sch Mines & Technol, 49-50; from instr to asst prof zool, San Diego State Col, 50-53; from asst prof to assoc, 53-62, PROF BIOL SCI, CALIF STATE UNIV, SACRAMENTO, 62- *Mem:* Am Fisheries Soc; Am Soc Ichthyologists & Herpetologists; Ecol Soc Am. *Res:* Ichthyology; hydrobiology; taxonomy; evolution. *Mailing Add:* Dept Biol Sci Calif State Univ 6000 J St Sacramento CA 95819

BRITTELLI, DAVID ROSS, b Milwaukee, Wis, Oct 10, 44; m 73; c 1. ORGANIC CHEMISTRY, MEDICINAL CHEMISTRY. *Educ:* Univ Wis, BS, 66; Univ Ill, PhD(org chem), 69. *Prof Exp:* res chemist, Cent Res & Develop Dept, E I du Pont De Nemours & Co Inc, 69-90; SR GROUP LEADER, DUPONT MERCK PHARMACEUT CO, 91- *Mem:* Am Chem Soc; AAAS. *Res:* Chemistry of natural products; beta-lactams; organophosphorus chemistry; heterocyclic chemistry; introduction of fluorine into natural products; new synthetic procedures; antibiotics. *Mailing Add:* 90 Hopewell Rd Nottingham PA 19362

BRITTEN, BRYAN TERRENCE, b Creston, Iowa, Dec 25, 33; m 58; c 3. ECOLOGY, ZOOLOGY. *Educ:* Merrimack Col, AB, 57; Villanova Univ, MS, 62; Univ Wyo, PhD(zool), 66. *Prof Exp:* Assoc prof, 66-80, PROF BIOL, NIAGARA UNIV, 80- *Concurrent Pos:* Environ consult. *Mem:* Wildlife Soc; Ecol Soc Am; Biomet Soc. *Res:* Biogeometry of bird eggs. *Mailing Add:* Dept Biol Niagara Univ Niagara Falls NY 14109

BRITTEN, EDWARD JAMES, b Rouleau, Sask, Mar 5, 15, Australian citizen; m 40; c 3. PLANT CYTOGENETICS. *Educ:* Univ Sask, BSc, 40, MSc, 41; Univ Wis, PhD(genetics), 44. *Prof Exp:* Asst supt forage crops, Dominion Exp Farm, 44-47; asst prof bot, Univ Hawaii, 47-54, asst prof agron, 54-55, from assoc prof agr & assoc agronomist to prof agron, 55-64; head, Dept Agr, 64-73, dean, 65-66, 71, 79-80, prof, 64-80, EMER PROF AGR, UNIV QUEENSLAND, 81- *Concurrent Pos:* Sr res fel, Univ Melbourne, 61-62; Res consult, Thailand Inst Sci Tech Res, 79. *Mem:* Fel AAAS; Am Genetic Asn; Am Soc Agron; fel Australian Inst Agr Sci; Genetic Soc Australia; Trop Grassland Soc Australia. *Res:* Cytogenetics; agricultural education; conservation; tropical agriculture; legumes for semi-arid tropics; essential oil crops. *Mailing Add:* Dept Agr Univ Queensland St Lucia Brisbane 4067 Australia

BRITTEN, MICHEL, b Trois-Rivières, Que, Mar 14, 60; m 82; c 2. DAIRY SCIENCE, PROTEIN CHEMISTRY. *Educ:* Laval Univ, Bsc, 83, PhD(dairy chem), 88. *Prof Exp:* RES SCIENTIST DAIRY PROTEINS, AGR CAN-FOOD RES & DEVELOP CTR, 88- *Concurrent Pos:* Vis scientist, Food Res

Ctr, Reading, Eng, 86-87; auxiliary prof, Laval Univ, 90- *Mem:* Am Dairy Sci Asn. *Res:* Determination of protein structural and functional properties; improvement of functional properties through biochemical or biophysical processes; storage stability of protein-based food systems. *Mailing Add:* Food Res & Develop Ctr-Agr Can 3600 Casavant Blvd W Saint-Hyacinthe PQ J2S 8E3 Can

BRITTEN, ROY JOHN, b Washington, DC, Oct 1, 1919; m; c 2. MOLECULAR BIOLOGY. *Educ:* Univ Va, BA, 40; Princeton Univ, PhD, 51. *Prof Exp:* STAFF MEM BIOPHYS, CARNEGIE INST WASHINGTON DEPT TERRESTRIAL MAGNETISM, 51-; DISTINGUISHED CARNEGIE SR RES ASSOC BIOL, CALIF INST TECHNOL, 81- *Concurrent Pos:* Vis assoc biol, Calif Inst Technol, 71-73; sr res assoc, 73- *Mem:* Nat Acad Sci; AAAS. *Res:* Quadrupole focussing of energetic particle beams; repeated DNA sequences in genomes of higher organisms. *Mailing Add:* Kerckhoff Marine Lab 101 Dahlia St Corona del Mar CA 92625

BRITTIN, DOROTHY HELEN CLARK, b Gadsden Co, Fla, June 27, 38; m 60; c 3. FOOD SCIENCE, NUTRITION. *Educ:* Fla State Univ, BS, 60; Tex Tech Univ, MS, 65, PhD, 74. *Prof Exp:* Elem sch teacher, Mobile, Ala Pub Schs, 60; home economist, NY State & Nassau County Exten Serv, 61-63; from instr to assoc prof, 65-89, PROF FOOD & NUTRIT, TEX TECH UNIV, 89- *Concurrent Pos:* Consult, var textbk publ, 69-; mem Am Dietetic Asn Scholar & Award Comt & Int Fedn Home Econ Res Comt; vis prof, Univ Tenn, 82. *Mem:* Soc Nutrit Educ; Am Soc Animal Sci; Inst Food Technologists; Am Meat Sci Asn; Am Inst Nutrit; Am Home Econ Asn; Int Fedn Home Econ; Am Soc Enol & Viticult. *Res:* Food composition and acceptability, sensory analysis of food; meat science; iron in food; cultural aspects of food and nutrition; food and nutrition education. *Mailing Add:* Food & Nutrit Col Home Econ Tex Tech Univ Lubbock TX 79409

BRITTIN, WESLEY E, b Philadelphia, Pa, Apr 21, 17; m 41; c 4. THEORETICAL PHYSICS. *Educ:* Univ Colo, BS, 42, MS, 45; Princeton Univ, MA, 47; Univ Alaska, PhD, 57. *Prof Exp:* Asst physics, 42-44, instr, 44-45, from asst prof to assoc prof, 47-59, chmn dept, 57-60, chmn dept physics & astrophys, 60-74, PROF PHYSICS & ASTROPHYS, UNIV COLO, BOULDER, 59-, DIR, INST THEORET PHYSICS, 68- *Concurrent Pos:* Asst, Princeton Univ, 49-50. *Mem:* Am Phys Soc; Am Asn Physics Teachers. *Res:* Statistical mechanics. *Mailing Add:* 2425 Vassar Dr Boulder CO 80303

BRITTON, CARLTON M, b Socorro, NMex, Aug 13, 44; m 68; c 4. RANGE SCIENCE, FIRE ECOLOGY. *Educ:* Tex Tech Univ, BS, 68, MS, 70; Tex A&M Univ, PhD(range sci), 75. *Prof Exp:* asst prof range res, Ore State Univ, 75-80; ASSOC PROF RANGE MGT, TEX TECH UNIV, 80- *Mem:* Soc Range Mgt; Ecol Soc Am; Sigma Xi. *Res:* Rangeland productivity; fire ecology; ecosystem analysis. *Mailing Add:* Dept Range Mgt Tex Tech Univ Lubbock TX 79409

BRITTON, CHARLES COOPER, b Whitley City, Ky, Apr 27, 22; m 48; c 2. ELECTRICAL ENGINEERING. *Educ:* Tex Technol Col, BS, 47; Iowa State Univ, MS, 50. *Prof Exp:* Test engr, Gen Elec Co, 47-48; instr elec eng, Iowa State Univ, 48-50; ASSOC PROF ELEC ENG & ASST DEAN ENG, COLO STATE UNIV, 50- *Mem:* Inst Elec & Electronics Engrs; Instrument Soc Am; Am Soc Eng Educ. *Res:* Instrumentation engineering for wind-tunnel and hydraulic research. *Mailing Add:* 804 Birkey Pl Ft Collins CO 80526

BRITTON, DONALD MACPHAIL, b Toronto, Ont, Mar 6, 23; m 50; c 3. CYTOGENETICS. *Educ:* Univ Toronto, BA, 46; Univ Va, PhD(biol), 50. *Prof Exp:* Res scientist, Defense Res Bd Can, 51-52; Nat Res Coun Can fel biol, Univ Alta, 52-53; lectr genetics, 53; asst prof hort, Univ Md, 54-58; from asst prof to prof bot, Univ Guelph, 58-84, prof molecular biol & genetics, 84-88; RETIRED. *Concurrent Pos:* Sigma Xi res grant, 59-61; adj prof, Univ Guelph, 88- *Honors & Awards:* Gilchrist Prize, 44; Fleming Prize, 48. *Mem:* Bot Soc Am; Am Genetic Asn; Am Fern Soc; Genetics Soc Can; Brit Pteridological Soc; Int Asn Pteridologists. *Res:* Genetics; cytotaxonomy; plant breeding; biosystematics of ferns; floristics; pteridophytes of Canada. *Mailing Add:* 81 James St W Univ Guelph Guelph ON N1G 1E7 Can

BRITTON, DOYLE, b Los Angeles, Calif, Mar 6, 30; m 62; c 3. STRUCTURAL CHEMISTRY. *Educ:* Univ Calif, Los Angeles, BS, 51; Calif Inst Technol, PhD(chem), 55. *Prof Exp:* From asst prof to assoc prof, 55-64, PROF INORG CHEM, UNIV MINN, MINNEAPOLIS, 65- *Concurrent Pos:* NSF sr fel, 63-64. *Mem:* Am Crystallog Asn. *Res:* Molecular structure; x-ray crystallography. *Mailing Add:* Dept Chem Univ Minn Minneapolis MN 55455

BRITTON, MARVIN GALE, b Corning, NY, Jan 8, 22; m 48; c 3. MINERALOGY, CERAMICS. *Educ:* Alfred Univ, BS, 43; Ohio State Univ, MS, 49, PhD(mineral), 52. *Prof Exp:* Ceramics engr, Corning Glass Works, 43-46; res engr, Battelle Mem Inst, 48-49; res assoc mineral, Ohio State Univ, 48-52; res assoc ceramics, Corning, Inc, 52-56, develop mgr, 56-59, tech mgr govt serv, 59-61, staff res mgr, 61-63, mgr tech liaison, 63-77, mgr tech & eng educ, 78-87; RETIRED. *Concurrent Pos:* Mem comt aerospace & astronaut adv panel, Mat Adv Bd, Nat Acad Sci, 59-62, mem comt ceramic mat, 62-63, mem comt res-eng interaction, 65-66; instr, Fac Continuing Educ, Corning Community Col, 62-77. *Mem:* Fel Am Ceramic Soc; Sigma Xi. *Res:* Mineralogy of blast furnace slags and refractories; development of refractory ceramics. *Mailing Add:* 121 Hornby Dr Painted Post NY 14870

BRITTON, MAXWELL EDWIN, b Hymera, Ind, Jan 26, 12; m 37, 57. BOTANY. *Educ:* Ind State Univ, AB, 34; Ohio State Univ, MS, 37; Northwestern Univ, PhD(bot), 41. *Hon Degrees:* DSc, Ind State Univ, 67. *Prof Exp:* Jr high sch teacher, Ind, 34-35; asst bot, Ohio State Univ, 35-37; from asst to assoc prof, Northwestern Univ, 37-55; sci officer arctic res, Geog Br, Off Naval Res, 55-66, dir arctic res prog, 66-70, actg dir, Earth Sci Div, 67-69; dir arctic develop & environ prog, Arctic Inst NAm, 70-73; consult environ sci, 73-74; biol scientist, US Geol Surv, 74-80; consult environ sci, 80-82; RETIRED. *Concurrent Pos:* Mem exec comt geog & climatol, Nat Res Coun, 53-54; mem comt environ studies, Proj Chariot, AEC, 59-62; mem panel biol & med, Comt Polar Res, Nat Acad Sci, 71-74; mem, Environ Protection Bd, Winnipeg, Man, 72-74. *Mem:* AAAS; Ecol Soc Am; Am Soc Limnol & Oceanog; Sigma Xi. *Res:* Taxonomy of freshwater algae; microclimatology; ecological relations of bog vegetation; soil temperatures; arctic ecology. *Mailing Add:* 2330 N Vermont St Arlington VA 22207

BRITTON, OTHA LEON, b Portales, NMex, Aug 6, 45; m 68; c 2. NUMERICAL ANALYSIS. *Educ:* Eastern NMex Univ, BS, 67; Drexel Univ, MS, 69, PhD(math), 72. *Prof Exp:* Asst prof math, Univ Miss, 72-77; assoc prof math, 77-80, ASST PROF COMPUT SCI, UNIV TENN, MARTIN, 80- *Mem:* Am Math Soc; Math Asn Am; Sigma Xi. *Res:* Efficient methods of finding approximate solutions to overdetermined systems of equations. *Mailing Add:* Dept Math Univ Tenn Martin TN 38238

BRITTON, STEVEN LOYAL, b Oct 10, 48; m; c 2. CARDIOVASCULAR PHYSIOLOGY, HYPERTENSION. *Educ:* Tex Tech Univ, PhD(physiol), 78. *Prof Exp:* PROF PHYSIOL & BIOPHYS, MED COL OHIO, 81- *Mem:* Am Physiol Soc. *Res:* Regulation of blood flow. *Mailing Add:* Dept Physiol Med Col Ohio Toledo OH 43699

BRITTON, WALTER MARTIN, b Lasker, NC, Aug 10, 39; m 61; c 2. POULTRY NUTRITION. *Educ:* NC State Univ, BS, 61, MS, 63, PhD(nutrit), 67. *Prof Exp:* NIH trainee nutrit, Univ Calif, Berkeley, 67-69; asst prof, 69-76, ASSOC PROF POULTRY SCI, UNIV GA, 76- *Concurrent Pos:* Vis res scientist poultry res, Spelderholt Inst, Beckbergen, Neth, 85. *Mem:* Poultry Sci Asn; World Poultry Sci Asn; Am Inst Nutrit; Sigma Xi. *Res:* Magnesium nutrition; sodium and chloride nutrition. *Mailing Add:* Dept Poultry Sci Univ Ga Athens GA 30602

BRITTON, WILLIAM GIERING, b Wilkes-Barre, Pa, Sept 25, 21; m 43; c 4. PHYSICAL CHEMISTRY, INORGANIC CHEMISTRY. *Educ:* Millikin Univ, BS, 43; Univ Ill, MS, 47; Univ Colo, PhD(phys chem), 56. *Prof Exp:* Prof chem, Bemidji State Univ, 47-82; RETIRED. *Mem:* Am Chem Soc. *Res:* Inclusion of nuclear science in undergraduate instruction. *Mailing Add:* 1624 Bixby Ave Bemidji MN 56601

BRITZ, GALEN C, b St Paul, Minn, Apr 1, 39; m 66. QUALITY ASSURANCE STATISTICS, CHEMICAL ENGINEERING. *Educ:* Univ Minn, BS, 61; Iowa State Univ, MS, 63, PhD(chem eng), 66. *Prof Exp:* Sr chem engr, Cent Res Pilot Plant, 66-71, sr chem engr, New Bus Ventures, 72-74, opers anal specialist, Corp Opers Anal, 75-81, qual assurance mgr, 81-82, process develop mgr, 82-84, qual control mgr, 85-87, TECH MGR, 3M CO, 88- *Mem:* Am Inst Chem Engrs; Am Soc Qual Control. *Res:* Radiation heat transfer; photopolymerization reactor engineering; photographic science and engineering; statistical experimental design; crystallization; statistic quality control. *Mailing Add:* 1255 Rolling Oaks Lane Hutchinson MN 55350

BRITZ, STEVEN J, b Chicago, Ill, July 19, 49. PLANT PHYSIOLOGY, PHOTOBIOLOGY. *Educ:* Johns Hopkins Univ, BS, 71; Harvard Univ, MA, 72, PhD(biol), 77. *Prof Exp:* Res staff biologist, Yale Univ, 77-80; lead scientist, Plant Photobiol Lab, 80-90, RES LEADER, CLIMATE STRESS LAB, BELTSVILLE AGR RES CTR, USDA, 90- *Mem:* Am Soc Plant Physiologists; Am Soc Photobiol. *Res:* Photosynthesis and photosynthate partitioning; photoperiodism and circadian rhythms; photomorphogenesis. *Mailing Add:* Climate Stress Lab USDA Beltsville Agr Res Ctr Beltsville MD 20705

BRITZMAN, DARWIN GENE, b Watertown, SDak, May 13, 31; m 53; c 3. ANIMAL NUTRITION, POULTRY NUTRITION. *Educ:* SDak State Univ, BS, 53, PhD(animal sci), 64; Univ Minn, MS, 62. *Prof Exp:* Asst turkey mgr, Sunshine State Hatchery, Watertown, SDak, 53-54, hatchery mgr, Brookings, 54-55; territory mgr feed sales, McCabe Co, Minneapolis, 55-58; asst nutritionist, Farmer's Union Grain Terminal Asn, St Paul, 58-62; res scientist poultry, Swift & Co, Chicago, 63-65; DIR NUTRIT, GTA FEEDS, HARVEST STATES CO, 65- *Concurrent Pos:* Consult, Mex, China & Japan. *Mem:* Am Soc Animal Sci; Am Dairy Sci Asn; Poultry Sci Asn. *Res:* Applied poultry, livestock and pet nutrition and management. *Mailing Add:* 1900 Edgewood Rd Sioux Falls SD 57103

BRIXEY, JOHN CLARK, mathematics; deceased, see previous edition for last biography

BRIXNER, LOTHAR HEINRICH, b Karlsruhe, Ger, Dec 30, 28; US citizen; m 55; c 1. INORGANIC CHEMISTRY. *Educ:* Univ Karlsruhe, BS, 51, MS, 53, PhD(inorg chem), 55. *Prof Exp:* Asst prof inorg chem, Univ Karlsruhe, 55-56; from res chemist to sr res chemist, E I du Pont de Nemours & Co, 56-66, res assoc, Pigments Dept, 66-69, res fel inorg chem, Cent Res & Develop Dept, 69-90; RETIRED. *Concurrent Pos:* Fel, Mass Inst Technol, 55; vis prof mat sci, Brown Univ, 68. *Mem:* Am Chem Soc; Am Asn Crystal Growth. *Res:* Exploratory as well as crystal growth aspects of ferroelectric, ferroelastic and electrooptic materials; finding and developing novel phosphors, particularly for the application of improved medical x-ray intensifying screens; inventor of optical waveguide on KTP potassium titanyl phosphate. *Mailing Add:* 501 E Lafayette Dr West Chester PA 19382

BRIZIO-MOLTENI, LOREDANA, b Savona, Italy, June 17, 27; US citizen; m 63; c 2. RECONSTRUCTIVE SURGERY. *Educ:* Univ Bologna, Italy, MD, 51, Bd Gen Surg, 56. *Prof Exp:* Asst prof surg, Univ Bologna, Italy, 52-59; instr, State Univ NY, Buffalo, 71-73; asst prof surg, Univ Mo, Kansas City, 74-76; assoc prof clin surg, mem fac, dept surg, Loyola Univ, Chicago, 76-85; ADJ ASSOC PROF, NORTHWESTERN UNIV, CHICAGO, ILL, 88- *Mem:* Fel Am Col Surgeons; AMA; Am Burn Asn; Asn Acad Surg. *Res:* Electron microscope studies of skin and connective tissue in burns; corrective surgery in congenital anomalies; pathogenesis of hypertension in burned patients; role of hormone receptors in malignant tumors; response of the endocrine system to thermal trauma; trace metals in lung pathology. *Mailing Add:* 2664 Sheridan Rd Evanston IL 60201

BRIZZEE, KENNETH RAYMOND, neuroanatomy, neuropathy; deceased, see previous edition for last biography

BROACH, ROBERT WILLIAM, b Fayetteville, Ark, May 19, 49; m 72; c 3. MEASUREMENT SCIENCE. *Educ:* Univ Ark, Fayetteville, BS, 71; Univ Wis-Madison, PhD(chem), 77. *Prof Exp:* Res asst, Univ Wis-Madison, 72-77; res asst, Argonne Nat Lab, 77-78, sr res chemist, 78-81, res scientist, 81-83, group leader, 83-87, mgr, 87-89; SR RES ASSOC, UOP, 89- *Concurrent Pos:* Resident assoc, Argonne Nat Lab, 78-79. *Mem:* Am Chem Soc; Mat Res Soc; Am Crystallog Asn; Catalysis Soc. *Res:* Theoretical and experimental characterization of aerospace, automotive, and engineered materials including catalysts, adsorbents, polymers and ceramics. *Mailing Add:* UOP 50 E Algonquin Rd Des Plaines IL 60017

BROACH, WILSON J, b Atkins, Ark, Aug 14, 15; m 42; c 4. PHYSICAL CHEMISTRY. *Educ:* Henderson State Teachers Col, BA, 37; Univ Ark, MS, 48; PhD(phys chem), 53. *Prof Exp:* Instr chem, Little Rock Jr Col, 46-48 & Univ Ark, 48-52; assoc prof, Southern State Col, 52-54; assoc prof, Northwestern State Col, La, 54-55, actg head div phys sci, 55-57; assoc prof, 57-59, PROF CHEM & DEAN COL SCI, UNIV ARK, LITTLE ROCK, 59- *Mem:* Am Chem Soc; AAAS. *Res:* Reaction rates in aqueous and nonaqueous solutions. *Mailing Add:* 5511 Stonewall Rd Little Rock AR 72207

BROAD, ALFRED CARTER, b Yonkers, NY, Apr 29, 22; m 50; c 1. INVERTEBRATE ZOOLOGY. *Educ:* Univ NC, AB, 43, MA, 51; Duke Univ, PhD, 56. *Prof Exp:* Chief shrimp invests, Inst Fisheries Res, Univ NC, 48-51; res investr, Marine Lab, Duke Univ, 53-57; from asst prof to assoc prof zool, Ohio State Univ, 57-64; chmn dept biol, 64-71, actg dir, Sundquist Marine Lab, 82-85, PROF BIOL, WESTERN WASH UNIV, 64- *Mem:* Am Soc Zoologists; Crustacean Soc. *Res:* Sand dollar biology; arctic littoral ecology. *Mailing Add:* Dept Biol Western Wash Univ Bellingham WA 98225

BROADBENT, FRANCIS EVERETT, b Snowflake, Ariz, Mar 29, 22; m 44; c 6. SOIL MICROBIOLOGY. *Educ:* Brigham Young Univ, BS, 42; Iowa State Col, MS, 46, PhD(soil bact), 48. *Prof Exp:* Instr soils, Iowa State Col, 47-48; jr chemist, Citrus Exp Sta, Univ Calif, 48-50; assoc prof soil microbiol, Cornell Univ, 50-55; assoc prof, Univ Calif, Davis, 55-61, prof soil microbiol, 61-; RETIRED. *Concurrent Pos:* Fulbright sr res scholar, NZ, 62-63. *Mem:* Fel Am Soc Agron; Soil Sci Soc Am. *Res:* Soil organic matter chemistry; metal organic complexes; nitrogen transformations; use of stable tracer isotopes in biological systems. *Mailing Add:* 821 A St Davis CA 95616

BROADBENT, HYRUM SMITH, b Snowflake, Ariz, July 21, 20; m 42; c 8. CHEMISTRY. *Educ:* Brigham Young Univ, BS, 42; Iowa State Univ, PhD(org chem), 46. *Prof Exp:* Lab asst chem, Brigham Young Univ, 40-42; partic, Nat Defense Res Comt & Off Sci Res & Develop Projs, Iowa State Univ, 43-44; Milton Fund fel physico-org chem, Harvard Univ, 46-47; from asst prof to assoc prof, 47-52, prof, 52-85, EMER PROF CHEM, BRIGHAM YOUNG UNIV, 86- *Concurrent Pos:* Group leader med chem, Schering Corp, 58-59; vis scientist, C F Kettering Labs, 62-63; Sigma Xi ann lectr, Brigham Young Univ, 68; res chemist, Eastman Kodak Res Labs, 70-71; vis prof chem, Kuwait Univ & Konstanz Univ, WGer, 80. *Honors & Awards:* Karl G Maeser Awards, Brigham Young Univ, 68 & 70. *Mem:* Am Chem Soc; Royal Soc Chem; Int Soc Heterocyclic Chem. *Res:* Heterocycles; organometallic compounds; physical and general synthetic organic chemistry; contact catalysis; medicinal chemistry. *Mailing Add:* Dept Chem Brigham Young Univ Provo UT 84602

BROADBENT, NOEL DANIEL, b Oakland, Calif, Feb 22, 46; m 73; c 1. RESEARCH ADMINISTRATION. *Educ:* San Diego State Univ, BA, 68; Uppsala Univ, Sweden, FK, 71, MA & PhD(archeol), 79. *Prof Exp:* Lectr archeol, Uppsala Univ, Sweden, 77-81; sr lectr, Umeå Univ, Sweden, 81-83, dir, Ctr Arctic Res, 83-89; DOCENT ARCHEOL, UPPSALA UNIV, SWEDEN, 83-; PROG DIR, DIV POLAR PROGS, NSF, 90- *Concurrent Pos:* Lectr, Int Grad Sch, Univ Stockholm, 75-78; exchange scholar, Royal Swed Acad & Chinese Acad Social Sci, Beijing, 85, Univ Zimbabwe, 88; consult, Swed Ed Cambridge Encycl Archeol & Times Atlas Archeol, 86 & 88; mem, Swed UNESCO man & the Biosphere Comt, 88-90; author, Swed Nat Encycl, 88-93. *Honors & Awards:* Svea Order, Stockholm, Sweden, 74. *Mem:* Am Anthrop Asn; Swed Archeol Asn; Int Union Circumpolar Health; Int Arctic Social Sci Asn; fel Am Scand Found. *Res:* Multidisciplinary analysis of cultural adaptation in northern maritime societies; coastal societies during the prehistoric period in the north Bothian region of Sweden and Finland. *Mailing Add:* Div Polar Progs NSF Washington DC 20550

BROADBOOKS, HAROLD EUGENE, b Wilbur, Wash, Aug 29, 15; m 50; c 6. ECOLOGY. *Educ:* Univ Wash, BA, 37; Univ Mich, MA, 40, PhD(zool), 50. *Prof Exp:* Asst prof biol, Stephen F Austin State Col, 49-50; asst prof zool, Univ Ariz, 50-52; res assoc surg, Sch Med, Univ Wash & Radioisotope Unit, Vet Admin Hosp, 53-54; biologist, State Dept Fisheries, Wash & Ore, 54-56; asst prof zool, Shurtleff Col, 56-57; from asst prof to assoc prof, 57-70, PROF ZOOL, SOUTHERN ILL UNIV, EDWARDSVILLE, 71- *Mem:* Am Soc Mammal; Soc Study Evolution; Ecol Soc Am; Cooper Ornith Soc; Am Behavior Soc; Sigma Xi. *Res:* Mammalogy; vertebrate ecology; behavior; distribution; evolution. *Mailing Add:* 2051 El Monte Dr Thousand Oaks CA 91362

BROADDUS, CHARLES D, b Irvine, Ky, Oct 17, 30; m 57; c 3. ORGANIC CHEMISTRY. *Educ:* Centre Col Ky, AB, 52; Auburn Univ, MS, 54; Univ Fla, PhD(org chem), 60. *Prof Exp:* Res chemist, 60-68, sect head res, 68-72, assoc dir food, paper & coffee technol div, 72-73, dir, 73-80, mgr res, 80-84, VPRES RES, PROCTER & GAMBLE CO, 84- *Res:* Metalation chemistry, base catalyzed exchange reactions. *Mailing Add:* Miami Valley Lab PO Box 398707 Cincinnati OH 45239-8707

BROADFOOT, ALBERT LYLE, b Milestone, Sask, Jan 8, 30; US citizen; m 64; c 2. PHYSICS PLANETARY ATMOSPHERES, SPACE SCIENCES. *Educ:* Univ Sask, BE, 56, MSc, 60, PhD(physics), 63. *Prof Exp:* Engr, Defence Res Bd Can, 56-58; from asst physicist to physicist, Kitt Peak Nat Observ, 63-79; sr res scientist & assoc dir, Earth & Space Sci Inst, Univ Southern Calif, 79-83; SR RES SCIENTIST, LUNAR & PLANETARY LAB, UNIV ARIZ, 83- *Concurrent Pos:* Prin investr, Ultraviolet Spectrometric Exp, Mariner 10 & Voyager Missions. *Honors & Awards:* Exceptional Sci Achievement Medal, NASA, 81, 86. *Mem:* Am Astron Soc; Am Geophys Union; Int Union Geod & Geophys; Int Asn Geomagnetism & Aeronomy. *Res:* Molecular spectroscopy; upper atmospheric physics; planetary atmosphere; airglow, aurora and associated phenomena. *Mailing Add:* Univ Ariz LPL 901 Gould-Simpson Bldg Tucson AZ 85721

BROADHEAD, RONALD FRIGON, b Racine, Wis, July 22, 55. PETROLEUM GEOLOGY. *Educ:* NMex Inst Mining & Technol, BS, 77; Univ Cincinnati, MS, 79. *Prof Exp:* Petrol geologist, Cities Serv Oil Co, 79-81; petrol geologist, 81-88, PETROL GEOLOGIST & HEAD DATA SECT, NMEX BUR MINES & MINERAL RESOURCES, 88- *Concurrent Pos:* Adj asst prof geol, NMex Inst Mining & Technol, 83- *Mem:* Am Asn Petrol Geologists; Soc Econ Paleontologists & Mineralogists; Sigma Xi. *Res:* Investigate and evaluate petroleum resources of New Mexico. *Mailing Add:* NMex Bur Mines & Mineral Resources Socorro NM 87801

BROADHURST, MARTIN GILBERT, b Washington, DC, Apr 28, 32; m 55; c 2. SOLID STATE PHYSICS. *Educ:* Western Md Col, BA, 55; Pa State Univ, MS, 57, PhD(physics), 59. *Prof Exp:* Res assoc physics, Pa State Univ, 59-60; physicist, Nat Inst Standards & Technol, 60-67, chief polymer dielectrics sect, Polymer Div, 76-70, chief Elect Mech & Thermal Properties Sections, 70-82, dep chief, 82-83, asst chief, Polymer Div, 83-88; RETIRED. *Concurrent Pos:* Chmn Conf Elec Insulation, Nat Acad Sci, 76-77; consult, 88- *Honors & Awards:* Silver Medal, US Dept Com, 73, Gold Medal, 79; Samuel Stratton Award, 83; Soc Plastics Engrs Res Award, 84. *Mem:* Am Chem Soc; AAAS; fel Am Phys Soc. *Res:* Experimental and theoretical techniques for electrical, mechanical and thermal properties of organic and polymeric solids; relations between microscopic structure and bulk physical properties; thermodynamics, statistical mechanics, piezoelectric and pyroelectric properties, conduction, aging, electrical breakdown. *Mailing Add:* 116 Ridge Rd Box 163 Washington Grove MD 20880

BROADUS, JAMES MATTHEW, b Mobile, Ala, Feb 24, 47; m 81; c 3. ECONOMICS OF MARINE MINERALS, MARINE RESOURCE ECONOMICS. *Educ:* Oberlin Col, BA, 69; Yale Univ, MA, 72, MPh, 74, PhD, 76. *Prof Exp:* Economist, US Dept Justice, 75-79; vis asst prof, Univ Ky, 78-79 & 81; res fel, 81-82, policy assoc, 82-84, SOCIAL SCIENTIST, MARINE POLICY CTR, WOODS HOLE OCEANOG INST, 84-, DIR, 86- *Concurrent Pos:* Consult, Off Technol Assessment, US Cong, 84; vis assoc prof sci & soc, Wesleyan Univ, 86; mem, Joint Group Experts Sci Aspects Marine Pollution, 86-88, Bd Ocean Sci & Policy, Nat Res Coun; assoc ed, J Coastal Res, 83-; trustee, Bigelow Lab Ocean Sci, 85- *Mem:* Asn Environ & Resource Economists; Int Asn Energy Economists; Marine Technol Soc; Am Asn Adv Sci; Coun Ocean Law. *Res:* Economics of marine science and technology; market organization and public policy, natural resources, energy and environmental economics; structure and behavior of Ocean Industries; marine minerals and seabed mining; role of marine resources and marine policy in developing nations. *Mailing Add:* Marine Policy Ctr Woods Hole Oceanog Inst Crowell House Woods Hole MA 02543

BROADWAY, ROXANNE MEYER, b Maywood, Calif, Mar 31, 51; m 84. INSECT PHYSIOLOGY, INSECT-PLANT INTERACTIONS. *Educ:* Univ Caif, Los Angeles, BA, 73; Univ Calif, Davis, PhD(entom), 85. *Prof Exp:* Res fel entom, Univ Calif, Davis, 85-86; res fel biochem & molecular biol, Wash State Univ, 86-87; ASST PROF ENTOM, NY STATE AGR EXP STA, CORNELL UNIV, 87- *Mem:* Entom Soc Am. *Res:* The biochemistry of insect-host plant interactions; determination of the mechanism of action of plant natural products on insect physiology and the molecular genetics of host plant resistance. *Mailing Add:* Entom Dept NY State Agr Exp Sta Cornell Univ Geneva NY 14456

BROADWELL, JAMES E(UGENE), b Atlanta, Ga, Jan 15, 21; m 43; c 3. AERONAUTICAL ENGINEERING. *Educ:* Ga Inst Technol, BS, 42; Calif Inst Technol, MS, 44; Univ Mich, PhD(aeronaut eng), 52. *Prof Exp:* Asst prof aeronaut eng, USAF Inst Technol, 46-48; instr & asst prof, Univ Mich, 48-55, assoc prof, 55-58; staff engr, Space Technol Labs, Thompson-Ramo-Wooldrige, Inc, 58-75, sr staff engr, 75-84; from res assoc to sr res assoc, 72-85, SR SCIENTIST, CALIF INST TECHNOL, 85- *Concurrent Pos:* Design specialist, Douglas Aircraft Co, 56-57. *Mem:* Nat Acad Eng; Am Phys Soc; fel Am Inst Aeronaut & Astronaut. *Res:* Turbulent mixing and combustion; rarefied gas dynamics; cellular automata; high energy chemical lasers. *Mailing Add:* Calif Inst Technol 1201 E California Blvd 301-46 Pasadena CA 91125

BROADWELL, RICHARD DOW, b Oak Park, Ill, Nov 4, 45. NEUROBIOLOGY & NEUROPATHOLOGY, CYTOLOGY. *Educ:* Knox Col, BA, 67; Univ Wis, MS, 71, DPhil(neurosci), 74. *Prof Exp:* Staff Fel neurocytol, NIH, 74-77, sr staff fel, 77-79, staff biologist, 79-80; asst prof neuropath, 80-84, assoc prof neuropath & neurol surg, 85-89, DIR NEUROL SURG LABS, MED SCH, UNIV MD, BALTIMORE, 84-, PROF NEUROL SURG & NEUROPATH, 89- *Concurrent Pos:* Japan Soc Prom Sci fel, 79-80, NIH grantee, 82-; prog dir molecular & cellular neurobiol, NSF, 87-89. *Mem:* Soc Neurosci; Am Asn Anatomists; Am Soc Cell Biol; Histochem Soc. *Res:* Neurocytology; neuropathology; cell biology; blood-brain barrier; axoplasmic transport; degeneration/regeneration; brain transplants. *Mailing Add:* Div Neurol Surg MSTF Bldg Rm-634 Baltimore MD 21201

BROBECK, JOHN RAYMOND, b Steamboat Springs, Colo, Apr 12, 14; m 40; c 4. PHYSIOLOGY. *Educ:* Wheaton Col, Ill, BS, 32; Northwestern Univ, MS, 37, PhD(neurol), 39; Yale Univ, MD, 43. *Hon Degrees:* LLD, Wheaton Col, Ill, 60; MA, Univ Pa, 71. *Prof Exp:* From instr to assoc prof physiol, Sch Med, Yale Univ, 43-52; prof & chmn dept, 52-70, Herbert C Rorer Prof Med Sci, 70-84, EMER PROF, SCH MED, UNIV PA, 84- *Mem:* Nat Acad Sci; fel Am Acad Arts & Sci; Am Physiol Soc (pres, 71-72); Am Soc Clin Invest; Am Inst Nutrit. *Res:* Physiological controls and regulations; control of energy balance; physiology of hypothalamus. *Mailing Add:* 224 Vassar Ave Swarthmore PA 19081

BROBERG, JOEL WILBUR, b Willmar, Minn, Aug 2, 10; m 37; c 2. INORGANIC CHEMISTRY. *Educ:* Macalester Col, BA, 32; Univ Minn, MA, 40, PhD(chem educ), 62. *Prof Exp:* From asst prof to prof chem, NDak State Univ, 41-81, dir, Inst Teaching Educ, 69-77; RETIRED. *Mem:* Am Chem Soc. *Res:* Molar method of teaching chemistry; synthesis of coordination compounds. *Mailing Add:* 1245 11th St N Fargo ND 58102

BROBST, DONALD ALBERT, b Allentown, Pa, May 8, 25; m 50; c 1. ECONOMIC GEOLOGY. *Educ:* Muhlenberg Col, AB, 47; Univ Minn, PhD(geol), 53. *Prof Exp:* Geologist, US Geol Surv, 48-80, dep chief, Off Mineral Resources, 73-76, chief, Br Eastern Mineral Resources, 78-80; CONSULT GEOLOGIST, 81- *Honors & Awards:* Herbert C Hoover Award, Am Inst Mining Metall & Petrol Engrs, 88. *Mem:* Fel Geol Soc Am; fel Mineral Soc Am; Soc Econ Geologists; Am Inst Mining, Metall & Petrol Eng; Int Asn Genesis Ore Deposits; Soc Mining, Metall & Explor. *Res:* Economic geology of mineral deposits, especially industrial minerals; assessment of mineral resources; petrology. *Mailing Add:* 2268 Wheelwright Ct Reston VA 22091

BROBST, DUANE FRANKLIN, b Medicine Lake, Mont, Oct 8, 23; m 58; c 4. VETERINARY MEDICINE. *Educ:* Univ Southern Calif, AB, 49; Wash State Univ, DVM, 54; Univ Pittsburgh, MPH, 57; Univ Wis, PhD(vet sci), 62. *Prof Exp:* Pvt vet practice, Mont, 54-56; pub health veterinarian, Allegheny County Health Dept, Pa, 57-58; proj asst vet med, Univ Wis, 58-62; asst prof vet path, Sch Vet Sci & Med, Purdue Univ, 62-70; PROF VET MED, WASH STATE UNIV, 70- *Concurrent Pos:* Whitley County Cancer Asn, Ind & Delta Theta Tau res grant, 62-63; consult, Am Med Asn, 63- *Mem:* Am Vet Med Asn; Am Col Vet Path. *Res:* Pathology of animal disease. *Mailing Add:* SW 630 Dawnview Pullman WA 99163

BROBST, KENNETH MARTIN, analytical chemistry; deceased, see previous edition for last biography

BROCARD, DOMINIQUE NICOLAS, b Versailles, France. FLUID MECHANICS, CIVIL ENGINEERING. *Educ:* Mass Inst Technol, MS, 73, PhD(civil eng), 76. *Prof Exp:* Res engr environ fluid mech, Alden Res Lab, Worcester Polytech Inst, 76-78, asst prof civil eng, 76-81, lead engr, 78-82, assoc prof civil engrs, 81-86, asst dir, 82-86; ASSOC METCALF & EDDY, 86- *Concurrent Pos:* Adj prof, Alden Res Lab, Worcester Polytech Inst, 86- *Mem:* Am Soc Civil Engrs; Sigma Xi; Int Asn Hydraulic Res. *Res:* Environmental fluid mechanics; free surface and density flows; turbulent dispersion; waste heat disposal; heat and momentum transfer. *Mailing Add:* 23 Orchard St Wellesley MA 02181

BROCCOLI, ANTHONY JOSEPH, b Newark, NJ, Sept 14, 56; m 84. CLIMATE MODELING, PALEOCLIMATE. *Educ:* Rutgers Univ, BS, 77, MS, 79. *Prof Exp:* Instr meteorol, Rutgers Univ, 79-82; RES METEOROLOGIST, GEOPHYS FLUID DYNAMICS LAB, NAT OCEANIC & ATMOSPHERIC ADMIN, 82- *Mem:* Am Meteorol Soc; Am Geophys Union; AAAS. *Res:* Use of numerical models for study of climate; climate change; paleoclimate. *Mailing Add:* Geophys Fluid Dynamics Lab NOAA Princeton Univ PO Box 308 Princeton NJ 08542

BROCHMANN-HANSSEN, EINAR, b Hvitsten, Norway, June 18, 17; nat US; m 43; c 2. PHARMACEUTICAL CHEMISTRY. *Educ:* Univ Oslo, Cand Pharm, 41; Purdue Univ, PhD(pharmaceut chem), 49. *Prof Exp:* Pharmacist, Flekkefjord Apotek, Norway, 42-44; res assoc, Univ Oslo, 44-46; from asst prof to assoc prof, 49-59, PROF PHARMACEUT CHEM, SCH PHARM, UNIV CALIF, SAN FRANCISCO, 59- *Concurrent Pos:* Instr, Univ Oslo, 45-46; partic scientist, UN Opium Res Prog, 58-; mem rev & exec comts, Nat Formulary, US Pharmacopeia, 60-70; vis prof, Robert Robinson Labs, Liverpool, 65-66; mem, US Pharmacopeial Conv, 70-75; vis prof, Nat Taiwan Univ, 74-75. *Honors & Awards:* Ebert Prize, 62; Powers Award, 63; Silver Medal, Univ Helsinki, 70; Res Achievement Award in Natural Prod, Acad Pharmaceut Sci, 78. *Mem:* Am Soc Pharmacog; Am Chem Soc; Am Pharmaceut Asn; NY Acad Sci; Int Pharmaceut Fedn; fel Acad Pharmaceut Sci. *Res:* Isolation, structure and physiological activity of naturally occurring substances; analytical chemistry; chromatography; alkaloid chemistry and biosynthesis. *Mailing Add:* Sch Pharm Univ Calif San Francisco CA 94143

BROCK, CAROLYN PRATT, b Chicago, Ill, July 25, 46; m 72. STRUCTURAL CHEMISTRY. *Educ:* Wellesley Col, BA, 68; Northwestern Univ, PhD(chem), 72. *Prof Exp:* From asst prof to assoc prof, 72-87, PROF, DEPT CHEM, UNIV KY, 87- *Mem:* Am Chem Soc; Am Crystallog Asn. *Res:* Molecular structure as determined by x-ray diffraction; intra and intermolecular forces; thermal motion and molecular packing in crystals. *Mailing Add:* Dept Chem Univ Ky Lexington KY 40506-0055

BROCK, DONALD R, GYNECOLOGICAL PATHOLOGY. *Educ:* Northwestern Univ, MD, 50. *Prof Exp:* DIR LABS, ST MARY'S HOSP, 61- *Mailing Add:* St Mary's Hosp 36475 Five Mile Rd Livonia MI 48154

BROCK, ERNEST GEORGE, b Detroit, Mich, Apr 7, 26; m 50; c 3. ELECTROPHYSICS. *Educ:* Univ Notre Dame, BS, 46, PhD(physics), 51. *Prof Exp:* Res assoc, Gen Elec Res Lab, 51-56; group leader, Linfield Res Inst, Ore, 56-58; sr physicist, Res Dept, Gen Dynamics/Electronics, 58-59, prin scientist, 59-61; mgr quantum physics lab, 61-66; head quantum electronics

dept, Lab Opers, Aerospace Corp, Calif, 66-68, sr staff scientist, 68-69, sr staff engr, 69-71; sr engr, NAm Rockwell, 72; sr engr specialist, Garrett Corp, 73-74; div off tech staff mem, Laser Div, 74-90, LAB ASSOC, LOS ALAMOS NAT LAB, UNIV CALIF, 90- *Mem:* AAAS; Am Phys Soc; sr mem Inst Elec & Electronics Engrs; NY Acad Sci; Sigma Xi. *Res:* Laser applications to fusion and isotope separation; applications of lasers and electronics technology to advanced strategic systems planning. *Mailing Add:* 1880 Camino Redondo Los Alamos NM 87544

BROCK, FRED VINCENT, b Chillicothe, Ohio, Nov 25, 32; m 60; c 2. METEOROLOGY. *Educ:* Ohio State Univ, BS, 54; Univ Mich, MSE, 60; Univ Okla, PhD(meteorol), 73. *Prof Exp:* Assoc res engr meteorol, Univ Mich, 60-69; spec instr, Univ Okla, 69-73; vis scientist, 73-75, STAFF SCIENTIST METEOROL, NAT CTR ATMOSPHERIC RES, 75- *Concurrent Pos:* Consult, Univ Corp Atmospheric Res & White Sands Missile Range, US Army, 70-73; Nat Ctr Atmospheric Res affil prof, Univ Okla, 75- *Mem:* Am Meteorol Soc. *Res:* Meteorological measurement systems including sensors, data loggers, system analysis and data processing. *Mailing Add:* 200 Felgar Rd No 219 Norman OK 73019

BROCK, GEOFFREY E(DGAR), b Eng, Feb 4, 30; nat US; m 55; c 3. METALLURGY. *Educ:* Univ Wales, BSc, 50; Pa State Univ, MS, 52, PhD(metall), 54. *Prof Exp:* Res metallurgist, NJ Zinc Co, Pa, 54; metallurgist, US Army Frankford Arsenal, 54-56; mem res staff, IBM Corp, 56-64, mgr, Mat Tech Mfg Res Dept, Components Div, 64-68, mfg res lab mgr, 68-70, asst to Components Div pres, 70-71, mgr, Plans & Controls Develop Labs, Components Div, 71-74, mgr silicon process, Technol Data Systs Div, 74-89; RETIRED. *Mem:* Am Soc Metals; Am Inst Mining, Metall & Petrol Engrs; Mat Res Soc. *Res:* Physical metallurgy of semiconductor materials; deformation and x-ray studies of crystal defects; phase equilibria; materials research on electronic computer components. *Mailing Add:* 126 Forest Dr Mt Kisco NY 10549

BROCK, GEORGE WILLIAM, b Grant Co, Ind, Aug 27, 20; m 44; c 3. NUCLEAR ENGINEERING. *Educ:* Nebr State Teachers Col, AB, 39; NY Univ, cert, 43; Air Force Inst Technol, MS, 57; Purdue Univ, MS, 64. *Prof Exp:* Radio engr, Farnsworth TV & Radio Corp, Ind, 41-42; meteorologist, US Army Air Force, 42-46; physicist, Aeronaut Ice Res Lab, 46-48, chief res, 48-50; USAF, 50-75, physicist, Wright Air Develop Ctr, Ohio, 50-59, mem staff, USAF Acad, 59-65, prof physics & actg head dept, 65-66, dir sci, Hq Air Force Systs Command, Andrews AFB, Md, 69-73, dir, F J Seiler Res Lab, USAF Acad, 73-75; sr engr, Ensco Inc, Colorado Springs, 76-81; PRES, ROCKY MOUNTAIN ENG ASSOC, INC, COLORADO SPRINGS, 81- *Concurrent Pos:* Mem subcomt icing, Nat Adv Comt Aeronaut; US deleg, NATO DRG Panel Physics & Electronics, 70-75. *Mem:* AAAS; Am Phys Soc. *Res:* Solid state physics and electronic materials. *Mailing Add:* 6118 Applewood Ridge Circle Colorado Springs CO 80918

BROCK, JAMES HARVEY, b Neubruecke, Ger, July 16, 57; US citizen; m 78. ENGINEERING. *Educ:* Univ Houston, Clear Lake, BS, 84. *Prof Exp:* Technician telemetry, 81-84; engr computer networks, 84-88, ENGR ROBOTICS, NASA-JOHNSON SPACE CTR, 88- *Mem:* Am Astron Soc. *Res:* Design and develop telerobotic components for telerobotic systems; design and develop interface simulators for space craft training. *Mailing Add:* 4222 Royal Manor Pasadena TX 77505

BROCK, JAMES RUSH, b McAllen, Tex, Dec 30, 37; m 64. CHEMICAL ENGINEERING, PHYSICS. *Educ:* Rice Univ, BA, 53, BS, 54; Univ Wis, MS & PhD(chem eng), 60. *Prof Exp:* Res engr, Humble Oil & Refining Co, 54-55; from asst prof to prof chem eng, 60-80, K A KOBE PROF, UNIV TEX, AUSTIN, 80- *Concurrent Pos:* NSF postdoctorate fel, 62-63; res grant, Nat Ctr Atmospheric Res, 64; mem res grants adv comt & prog planning subcomt, Nat Air Pollution Control Admin, 69-; vis prof, Univ Paris, 73-74; Dir, US Environ Protection Agency, Buffalo, NY Field Study & Houston, Tex Field Study, 80, mem Environ Protection Agency chemistry & physics rev comt, 80-; vis prof, Tokyo Inst Technol, 88. *Honors & Awards:* Sci Rev Award, US Environ Protection Agency, 76. *Mem:* AAAS; Am Chem Soc; Am Inst Chem Engrs; Sigma Xi. *Res:* Aerosol physics and chemistry; rarefied gas dynamics; statistical mechanics; plasma physics; author of over 150 research publications. *Mailing Add:* Box 5164 Austin TX 78712-5164

BROCK, JOHN E(DISON), b Chicago, Ill, Mar 7, 18; m 39; c 4. MECHANICAL ENGINEERING. *Educ:* Purdue Univ, BS, 38, MSE, 41; Univ Minn, PhD(math), 50. *Prof Exp:* Asst instr appl mech, Purdue Univ, 38-39; res engr, Westinghouse Elec Co, 39-40; tech engr, Union Elec Co, Mo, 40-42; from instr to assoc prof appl mech, Wash Univ, 42-51; dir res, Midwest Piping Co, Inc, 51-54; prof mech eng, US Naval Post-Grad Sch, 54-62; dir eng sci, Serv Bur Corp, NY, 62-63; prof mech eng, US Naval Postgrad Sch, 63-80; RETIRED. *Concurrent Pos:* Consult; mem bd contrib & consult ed, Heat, Piping & Air Conditioning Mag. *Mem:* Am Soc Mech Engrs; Am Soc Testing & Mat; Am Soc Naval Engrs. *Res:* Applied mechanics; structures theory; piping component and systems theory, design and analysis; computer applications. *Mailing Add:* 4120 Crest Rd Pebble Beach CA 93953

BROCK, KATHERINE MIDDLETON, b Keokuk, Iowa, June 3, 38; m 71; c 2. MICROBIOLOGY, SCIENCE COMMUNICATIONS. *Educ:* Vassar Col, AB, 60; Univ Calif, Berkeley, MA, 63; Univ Mass, Amherst, PhD(microbiol), 67. *Prof Exp:* Res asst biochem, Tech Univ Norway, 63-64; asst prof microbiol, San Francisco State Col, 67-70; res assoc, Ind Univ, Bloomington, 70-71; res assoc bact, Univ Wis-Madison, 71-81; VPRES, SCI TECH PUBL, INC, 81- *Mem:* AAAS; Sigma Xi; Am Soc Microbiol. *Res:* Microbial physiology; study of extreme environments, especially high temperature, high and low hydrogen-ion concentration and saline environments. *Mailing Add:* 1227 Dartmouth Rd Madison WI 53705

BROCK, KENNETH JACK, b Pampa, Tex, Aug 22, 37; m 62; c 1. MINERALOGY. *Educ:* San Jose State Col, BS, 62; Stanford Univ, PhD(geol), 70. *Prof Exp:* Asst prof, 70-74, ASSOC PROF GEOL, IND UNIV NORTHWEST, 74-, CHMN, GEOSCI, 77- *Mem:* Mineral Soc Am; Mineral Soc Gt Brit & Ireland. *Res:* Mineralogy and genesis of skarns; mineralogy of zoned lithium micas. *Mailing Add:* Dept Geol Ind Univ Northwest 3400 Broadway Gary IN 46408

BROCK, LOUIS MILTON, b Davenport, Iowa, Apr 16, 43; m 72. ENGINEERING PHYSICS. *Educ:* Northwestern Univ, BS, 66, MS, 67, PhD(theoret & appl mech), 72. *Prof Exp:* Draftsman, Black & Veatch Consult Engrs, 62; tech asst, Gen Dynamics/Convair, 63-64; tech asst, Sargent-Welch Sci Co, 64; res technician, Am Can Co, 65; PROF ENG MECH, UNIV KY, 71- *Concurrent Pos:* Consult, McDonnell-Douglas Astronaut Co, 72 & Compusport, Inc, 84-87; prin investr, NSF, 72-74, 79-82, 84-87 & 90-92; vis prof, Dept Eng Sci, Univ Oxford, 80-81; chmn, comm elasticity, Am Soc Civil Engrs, 83-85, corresp exec comt, eng mech div; res fel, Naval Res Lab, 83, 85, 87, 90. *Mem:* Sigma Xi; Am Soc Mech Engrs; Am Soc Civil Engrs; Am Asn Univ Professors. *Res:* Wave propagation in solids, with emphasis on problems arising in fracture, composite materials and seismology; dislocation mechanics. *Mailing Add:* Eng Mech 00461 Univ Ky Lexington KY 40506

BROCK, MARY ANNE, b Aurora, Ill, June 29, 32. CELL BIOLOGY, GERONTOLOGY. *Educ:* Grinnell Col, BA, 54; Harvard Univ, MA, 56, PhD(biol), 59. *Prof Exp:* Res assoc med, Harvard Med Sch, 59-60; BIOLOGIST, GERONT RES CTR, NAT INST AGING, NIH, 60- *Concurrent Pos:* Consult; bd gov, Soc Cryobiol, 73-76; vis scientist, Hopkins Marine Sta, Stanford Univ, 77. *Mem:* Am Soc Cell Biol; NY Acad Sci; Int Soc Chronobiol; Soc Cryobiol (secy, 71-72); fel Geront Soc Am; Sigma Xi; Soc Res Biol Rhythms. *Res:* Cell biology of aging mammalian cell types; effect of temperature on cellular aging; biological clocks and aging. *Mailing Add:* Nat Inst Aging NIH 4940 Eastern Ave Baltimore MD 21224

BROCK, PATRICK WILLET GROTE, b Kelowna, BC, Dec 17, 32; m; c 1. PETROLOGY, STRUCTURAL GEOLOGY. *Educ:* Univ BC, BASc, 56; Univ Leeds, PhD(geol), 63. *Prof Exp:* Party leader explor geol, Western Rift Explor Co, 57-59; adv Uganda geol surv, Can External Aid Off, Can Int Develop Agency, 63-70; asst prof, 70-75, ASSOC PROF GEOL, QUEENS COL, NY, 76- *Concurrent Pos:* Hon lectr, Makerere Univ, Uganda, 68-70; Fac Res Award Prog grant, Res Inst, City Univ New York, 74-75; prof staff, Cong-Bd Higher Educ grant, 78-79, 79-80 & 85-87. *Mem:* Geol Asn Can; Sigma Xi; AAAS; Geol Soc Am; Asn Geoscientists Int Develop. *Res:* Structural, metamorphic and age relationships in the rocks of the Manhattan Prong, Southeast New York; regional geology; field geology. *Mailing Add:* 61-18 157th St Flushing NY 11367-1241

BROCK, RICHARD EUGENE, b Astoria, Que, July 1, 43; m 70; c 1. MARINE ECOLOGY, FISHERIES. *Educ:* Univ Hawaii, BS, 68, MS, 72; Univ Wash, Seattle, PhD(fisheries), 79. *Prof Exp:* Dir field studies, Coop Fishery Unit, Univ Hawaii, 72-73; teaching asst, Univ Wash, 73-74; HEAD BENTHIC STUDIES, HAWAII INST MARINE BIOL, 75- *Concurrent Pos:* Consult, Oceanic Inst, 72-; Environ Consult, Inc, 75-; Environ Assessment Co, Pilipili Prod, 76- & Hawaii Planning Design & Res, 77- *Mem:* Am Soc Ichthyologists & Herpetologists; Soc Syst Zool; Am Soc Naturalists; Sigma Xi. *Res:* Ecological relationships of coral reef fishes; tropical marine benthic community structure and function; management of coral reef systems; energy flow through coral reef ecosystems. *Mailing Add:* Hawaii Inst Marine Biol PO Box 1346 Kaneohe HI 96744

BROCK, RICHARD R, b Sewickley, Pa, Mar 7, 38; m 59; c 3. CIVIL ENGINEERING, HYDRAULIC ENGINEERING. *Educ:* Univ Calif, Berkeley, BS, 61, MS, 62; Calif Inst Technol, PhD(civil eng), 68. *Prof Exp:* Civil Engr, Los Angeles County Flood Control Dist, 62-63; asst prof civil eng, Univ Calif, Irvine, 67-73; assoc prof, 73-77, PROF CIVIL ENG, CALIF STATE UNIV, FULLERTON, 77- *Honors & Awards:* J C Stevens Award, Am Soc Civil Engrs, 66. *Mem:* Am Soc Civil Engrs. *Res:* Flow in open channels and through porous media hydrology. *Mailing Add:* Dept Civil Eng Calif State Univ Fullerton CA 92634

BROCK, ROBERT H, JR, b New Rochelle, NY, Feb 15, 33; m 53; c 3. PHOTOGRAMMETRY, REMOTE SENSING. *Educ:* State Univ NY, BS, 58, MS, 59; Cornell Univ, PhD(photogram, geod eng), 71. *Prof Exp:* From instr to asst prof photogram & geod, Syracuse Univ, 59-66; sr systs analyst, CBS Labs, 66-67; assoc prof, 67-77, PROF PHOTOGRAM & REMOTE SENSING, STATE UNIV NY COL ENVIRON SCI & FORESTRY, 77- *Concurrent Pos:* Photogram consult, Syracuse Univ, 67-; proj dir electronic technol lab, US Army Res Inst, 71-72; consult, Syracuse Univ Res Corp, 71- *Mem:* Am Soc Photogram; Int Soc Photogram. *Res:* Photogrammetric engineering; magnitude of image identification errors; film shrinkage and atmospheric refraction studies; application of analytic photogrammetry to experimental structural mechanics; effects of edge information properties on coordinate measurement; image deformation studies. *Mailing Add:* SUNY-CESF 312 Bray Hall Syracuse NY 13210

BROCK, THOMAS DALE, b Cleveland, Ohio, Sept 10, 26; m 52, 71; c 2. MICROBIOLOGY. *Educ:* Ohio State Univ, BSc, 49, MSc, 50, PhD(microbiol), 52. *Prof Exp:* Res microbiologist, Upjohn Co, 52-57; asst prof biol, Western Reserve Univ, 57-59, fel, Sch Med, 59-60; from asst prof to prof bact, Ind Univ, Bloomington, 60-71; E B Fred prof, 71-90, EMER PROF NATURAL BIOL, UNIV WIS-MADISON, 90-; PRES, SCI TECH PUBLISHERS, 90- *Concurrent Pos:* Vis biologist, Am Inst Biol Sci, 61-62; USPHS career develop award, 62-68; lectr, Found Microbiol, 71-72 & 78-79. *Honors & Awards:* Fisher Award, Am Soc Microbiol, 84, Carski Award, 88. *Mem:* Fel AAAS; Am Soc Microbiol. *Res:* Microbial ecology; aquatic microbiology; history of microbiology; computer-assisted publication. *Mailing Add:* Sci Tech Publishers 701 Ridge St Madison WI 53705

BROCK, TOMMY A, CELL CULTURE, ION FLUX. *Educ:* Univ Iowa, PhD(physiol), 78. *Prof Exp:* asst prof path, Brigham & Women's Hosp, Harvard Med Sci, 82-; DEPT MED HYPERTENSION PROG, UNIV ALA. *Mailing Add:* Dept Med Hypertension Prog Univ Ala 1046 Zeigler Bldg Birmingham AL 35294

BROCKE, RAINER H, b Calcutta, India, Nov 24, 33; US citizen; m 57; c 3. ENDANGERED WILDLIFE MANAGEMENT. *Educ:* Mich State Univ, BS, 55, MS, 57, PhD(mammalian bioenergetics), 70. *Prof Exp:* Instr, Mich State Univ, 63-69; sr res assoc, 73-83, ASSOC PROF WILDLIFE ECOL, STATE UNIV NY COL ENVIRON SCI & FORESTRY, 83- *Concurrent Pos:* Consult, Develop Sci Prog, NY, 85-88; var res grants, 71- *Mem:* NY Acad Sci; Am Soc Mammal; Wildlife Soc; Sigma Xi. *Res:* Wildlife ecology and management; physiological ecology of mammals. *Mailing Add:* 249 Illick Hall State Univ NY Col Environ Sci & Forestry Syracuse NY 13210

BROCKEMEYER, EUGENE WILLIAM, b Buckner, Mo, June 22, 29; m 51; c 2. PHARMACY. *Educ:* Univ Kans, BS, 51, MS, 52; Ohio State Univ, PhD(pharm), 54. *Prof Exp:* Res assoc, 57-67, asst dir prod develop, 67, mgr qual control & prod develop, 67-71, dir res develop & qual control, 71-82, VPRES, RES & QUAL ASSURANCE, DORSEY LABS, 83- *Mem:* Am Pharmaceut Asn; Am Soc Qual Control. *Res:* Rate of absorption of drugs from pharmaceutical vehicles and sustained release forms of medication. *Mailing Add:* 1230 Twin Ridge Rd Lincoln NE 68510-5059

BROCKENBROUGH, EDWIN C, b Baltimore, Md, July 24, 30; m 68; c 5. SURGERY. *Educ:* Col William & Mary, BS, 52; Johns Hopkins Univ, MD, 56. *Prof Exp:* Asst surg, Johns Hopkins Univ, 57-58; clin assoc cardiac surg, NIH, 59-61; asst, Sch Med, Univ Wash, 61-64; asst prof, 64-70, clin assoc prof surg, 70-84, CLIN PROF, SCH MED, UNIV WASH, 84-; SR RES SCIENTIST, INST APPL PHYSIOL & MED, SEATTLE, 75-; PVT PRACT, 75- *Concurrent Pos:* Assoc surgeon-in-chief, Harborview Med Ctr, Seattle, 70-75; chief staff, Northwest Hosp. *Mem:* Sigma Xi; Am Col Surgeons; Am Heart Asn. *Res:* Cardiovascular physiology; thoracic surgery; cerebrovascular diseases. *Mailing Add:* 1560 N 115th St Seattle WA 98133

BROCKERHOFF, HANS, b Duisburg, Ger, July 8, 28; m 58; c 3. BIOCHEMISTRY, NEUROCHEMISTRY. *Educ:* Univ Cologne, Dr rer nat, 58. *Prof Exp:* Researcher, Univ Wash, 58-60 & Univ Calif, Berkeley, 60-61; assoc scientist, Fisheries Res Bd, Can, 61-63, sr scientist, 63-73; CHIEF SCIENTIST NEUROCHEM, NY STATE INST FOR BASIC RES IN DEVELOP DISABILITIES, 73- *Mem:* Am Soc Biol Chemists; Am Soc Neurochem. *Res:* Lipid chemistry; structure of phospholipids and triglycerides; protein kinase; lipolytic enzymes; myelin lipids; membrane structure. *Mailing Add:* NY State Inst Res Develop Disabilities 1050 Forest Hill Rd Staten Island NY 10314

BROCKETT, PATRICK LEE, b Monterey Park, Calif, Mar 29, 48; m; c 1. MATHEMATICAL STATISTICS, PROBABILITY. *Educ:* Calif State Univ, Long Beach, BA, 70; Univ Calif, Irvine, MA, 75, PhD(math), 75. *Prof Exp:* Asst prof Math, Tulane Univ, 75-77; asst prof math, Univ Tex, 77-80; vis prof, Dept Statist, Univ Calif, Riverside, 80-81; from asst prof to assoc prof finance, 81-85, PROF FINANCE & ACTUARIAL SCI, UNIV TEX, AUSTIN, 86-, DIR, ACTUARIAL SCI, 86- *Mem:* Inst Math Statist; Am Statist Asn; Royal Statist Soc; Am Risk & Ins Asn. *Res:* Stochastic methodology and theory and application to actuarial science, finance, statistical signal processing and signal detection; math models in finance. *Mailing Add:* Dept Finance Univ Tex Austin TX 78712

BROCKETT, ROGER WARE, b Wadsworth, Ohio, Oct 22, 38; m 60; c 3. APPLIED MATHEMATICS. *Educ:* Case Inst Technol, BS, 60, MS, 62, PhD(eng), 64. *Prof Exp:* From asst prof to assoc prof, Mass Inst Technol, 63-69; PROF APPL MATH, DIV ENG & APPL PHYSICS, AIKEN COMPUT LAB, HARVARD, 69- *Concurrent Pos:* Ford fel, 63-65; vis positions Imperial Col, London, Nagoya Univ, Univ Rome & Australian Nat Univ. *Mem:* Nat Acad Eng; Am Math Soc; fel Inst Elec & Electronics Engrs. *Res:* Automatic control theory including optimal control and stability theory; nonlinear phenomena; mathematical system theory. *Mailing Add:* 29 Oakland St Lexington MA 02173

BROCKHOUSE, BERTRAM NEVILLE, b Lethbridge, Alta, July 15, 18; m 48; c 6. SOLID STATE PHYSICS. *Educ:* Univ BC, BA, 47; Univ Toronto, MA, 48, PhD, 50. *Hon Degrees:* DSc, Univ Waterloo, 69 & McMaster Univ, 84. *Prof Exp:* Lectr physics, Univ Toronto, 49-50; res officer, Atomic Energy Can, Ltd, 50-60, head neutron physics br, 60-62; prof, 62-84, chmn dept, 67-70, EMER PROF PHYSICS, MCMASTER UNIV, 84- *Concurrent Pos:* Guggenheim fel, 62. *Honors & Awards:* Oliver E Buckley Prize, Am Phys Soc, 62; Duddell Medal & Prize, Brit Inst Physics & Phys Soc, 63; Medal Achievement Physics, Can Asn Physicists, 67; Tory Medal, Royal Soc Can, 73. *Mem:* Fel Am Phys Soc; Can Asn Physicists; fel Royal Soc Can; fel Royal Soc London; foreign mem Royal Swed Acad Sci; foreign mem Am Acad Arts & Sci. *Res:* Neutron physics as applied to physics of solid and liquids; philosophy of physics. *Mailing Add:* Dept Physics SSC453 McMaster Univ Hamilton ON L8S 4M1 Can

BROCKINGTON, JAMES WALLACE, b Norfolk, Va, Apr 12, 43; m 63; c 2. ORGANIC CHEMISTRY. *Educ:* Univ Richmond, BS, 65; Univ Miami, MS, 67; Va Commonwealth Univ, PhD(chem), 70. *Prof Exp:* Res chemist, Res & Develop Dept, Texaco, Inc, 70-76 & Atlas Powder Co, 76-78; sr res chemist, Corp Res Dept, 78-79, sect mgr, 79-81, DIR PLANNING & MULTIDISCIPLINARY PROG, CORP SCI CTR, AIR PROD & CHEM INC, 81- *Mem:* Am Chem Soc. *Mailing Add:* 4424 Parkland Dr Allentown PA 18104

BROCKMAN, DAVID DEAN, b Greer, SC, Aug 4, 22; m 50; c 3. PSYCHOANALYSIS, PSYCHIATRY. *Educ:* Furman Univ, BS, 43; Med Col SC, MD, 46. *Prof Exp:* Staff mem psychiat, North Shore Health Resort, Winnetka, Ill, 51-52; asst, Clin & Med Sch, Univ Chicago, 52-53, from instr to asst prof, 53-55, actg chief, Univ, 54-55; prof lectr, 55-56; from clin asst prof to clin assoc prof, 56-85, CLIN PROF PSYCHIAT, COLL MED, UNIV ILL, 85- Mem: Fel Am Psychiat Asn; Am Psychoanal Asn; fel Am Col Psychoanalysts; fel Am Col Psychiatrists; fel Am Soc Adolescent Psychiat. Res: Late adolescence; parent loss; young adulthood. Mailing Add: Inst Psychoanal 180 N Michigan Ave Chicago IL 60601

BROCKMAN, ELLIS R, b St Louis, Mo, Feb 10, 34; m 59; c 3. MICROBIOLOGY. Educ: DePauw Univ, AB, 55; Univ Mo, AM, 60, PhD(bact), 64. Prof Exp: Instr bact, Univ Mo, 63; asst prof biol, Winthrop Col, 63-69; PROF BIOL, CENT MICH UNIV, 69- Concurrent Pos: Consult, Warner Lambert Res Inst, 69-75 & Bristol Labs, 77-78. Mem: AAAS; Am Soc Microbiol; Am Inst Biol Scientists; Nat Asn Biol Teachers; Soc Indust Microbiol. Res: Bacterial taxonomy; biology of the myxobacteria; bacteriology of aquatic ecosystems; screening anti-microbial compounds. Mailing Add: Biol Dept Cent Mich Univ Mt Pleasant MI 48859

BROCKMAN, FRANK ELLIOT, agronomy, for more information see previous edition

BROCKMAN, HAROLD W, b Sidney, Ohio, Mar 31, 22; m 54; c 1. MATHEMATICS. Educ: Capital Univ, BSEd, 48; Ohio State Univ, MA, 50, PhD(math, math educ) 62. Prof Exp: From instr to assoc prof math, 49-64, PROF MATH, CAPITAL UNIV, 64- Concurrent Pos: Ed, Ohio J Sch Math. Mem: Am Math Soc; Math Asn Am; Nat Coun Teachers Math. Res: Algebra; analysis. Mailing Add: Dept Math Capital Univ Columbus OH 43209

BROCKMAN, HERMAN E, b Danforth, Ill, Dec 5, 34; m 56; c 6. GENETICS, ENVIRONMENTAL MUTAGENESIS & ANTIMUTAGENE SIS. Educ: Blackburn Col, BS, 56; Northwestern Univ, MA, 57; Fla State Univ, PhD(genetics), 60. Prof Exp: Res assoc, Biol Div, Oak Ridge Nat Lab, 60-61, Nat Acad Sci-Nat Res Found, Coun fel, 61-62, USPHS fel, 62-63; from assoc prof to prof, 63-82, DISTINGUISHED PROF, ILL STATE UNIV, 82- Concurrent Pos: USPHS spec res fel, 69-70; counr, Environ Mutagen Soc, 84-86; distinguished vis scientist, US Environ Protection Agency, 87-90. Honors & Awards: Environ Mutagenesis Recognition Award, Environ Mutagen Soc, 82. Mem: AAAS; Genetics Soc Am; Sigma Xi; Environ Mutagen Soc. Res: Comparative and environmental mutagenicity and antimutagenicity. Mailing Add: Dept Biol Sci Ill State Univ Normal IL 61761

BROCKMAN, HOWARD LYLE, JR, b Olney, Ill, Sept 5, 44. PHYSICAL BIOCHEMISTRY, ENZYMOLOGY. Educ: DePauw Univ, BA, 66; Mich State Univ, PhD(biochem), 71. Prof Exp: Fel biochem, Univ Chicago, 71-73; from asst prof to assoc prof, 73-81, PROF BIOCHEM, HORMEL INST, UNIV MINN, 81-, BIOPHYS SECT LEADER, 76- Concurrent Pos: Nat Inst Arthritis & Metab Dis fel, 71-73; estab investr, Am Heart Asn, 78-82, coun, atherosclerosis. Mem: AAAS; Am Soc Biol Chemists; Biophys Soc. Res: Interfacial reactions, particularly the influence of interfacial structure on reactions catalyzed by water soluble and membrane bound enzymes. Mailing Add: Hormel Inst Univ Minn Austin MN 55912

BROCKMAN, JOHN A, JR, b Kellogg, Idaho, Apr 4, 20; m 52; c 4. IMMUNOLOGY. Educ: Calif Inst Technol, BS, 42, PhD(org chem), 48. Prof Exp: Asst chem, Calif Inst Technol, 42-46, res fel, 48-49; res chemist, Lederle Labs, Am Cyanamid Co, 49-85; RETIRED. Mem: Am Chem Soc. Res: Analytical chemistry; isolation and structure determination of natural products; alkaloids of dichroa febrifuga; vitamins and growth factors; vitamin B12; leucovorin; thiotic acid; neurochemistry; immuno-chemistry; oncology. Mailing Add: 130 Rose Ave Woodcliff Lake NJ 07675

BROCKMAN, PHILIP, b Boston, Mass. ATMOSPHERIC SCIENCES, ENVIRONMENTAL SCIENCES. Educ: Univ Mass, BS, 59; Col William & Mary, MA, 63. Prof Exp: Researcher, 59-88, ASST BR HEAD, LASER TECHNOL & APPLICATIONS BR, NASA LANGLEY RES CTR, 88- Mem: Optical Soc Am; assoc fel Am Inst Aeronaut & Astronaut. Res: Laser systems for remote sensing of atmosphere including winds, species, and pollutants; specializing in 2 micrometer systems with experience at 10 micrometer and near infrared. Mailing Add: Langley Res Ctr NASA MS 468 Hampton VA 23665

BROCKMAN, ROBERT W, b Chester, SC, Dec 8, 24; m 48; c 2. BIOCHEMISTRY. Educ: Vanderbilt Univ, BS, 47, MS, 49, PhD(chem), 51. Prof Exp: Asst chem, Vanderbilt Univ, 50-51; sr chemist, 51-57, head drug resistance sect, 58-67, HEAD BIOL CHEM DIV, SOUTHERN RES INST, 67-, PRIN SCIENTIST, 64- Concurrent Pos: NIH spec fel, 64-66; mem adv comt, Am Cancer Soc, 66-69 & 75-78; ed, Antibiotics & Chemother, 73-82; exp ther, study sect, NIH, 79-83; assoc ed Cancer Res, 81-87. Mem: Am Chem Soc; Am Soc Biochem & Molecular Biol; Am Asn Cancer Res; Sigma Xi; Soc Exp Biol & Med; NY Acad Sci. Res: Mechanisms of drug inhibition and resistance; cancer research. Mailing Add: 1631 Keith Valley Rd Charlottesville VA 22901

BROCKMAN, WILLIAM WARNER, molecular biology, virology; deceased, see previous edition for last biography

BROCKMANN, HELEN JANE, b Louisville, Ky, Feb 25, 47. ZOOLOGY, ANIMAL BEHAVIOR. Educ: Tufts Univ, BS, 67; Univ Wis-Madison, MS, 72, PhD(zool), 76. Prof Exp: from asst prof to assoc prof, 76-89, PROF ZOOL, UNIV FLA, 89- Concurrent Pos: NSF res fel zool, Oxford Univ, 77-78; Harry Frank Guggenheim Found grant, 77-79; prin investr, NSF grant, 79-82, 89-92; NSF vis prof women, Princeton Univ, 85-86. Mem: Sigma Xi; Animal Behav Soc; Am Soc Zoologists; Soc Study Evolution; Int Union Study Social Insects; Entomol Soc Am. Res: Ethology; social behavior; behavioral ecology and evolution; nesting behavior of digger wasps, sphecidae; origins of sociality in insects; behavior of horseshoe crabs sex ratios; life histories. Mailing Add: Dept Zool Univ Fla Gainesville FL 32611

BROCKMEIER, NORMAN FREDERICK, b Montclair, NJ, Sept 5, 37; m 60; c 3. POLYMER PROCESS DESIGN & ECONOMICS, TECHNICAL LICENSING. Educ: Cornell Univ, BChE, 60; Mass Inst Technol, PhD(chem eng), 66. Prof Exp: Res chem engr, Minn Mining & Mfg Co, 60-63; asst prof chem eng, Univ Tex, Austin, 66-71; RES ASSOC, AMOCO CHEM CO, 71- Mem: Am Chem Soc; Am Inst Chem Engrs; Am Sci Affiliation. Res: Chemical reaction kinetics; molecular transport in polymers; polymer physics and chemistry; polymer process design and economics; polyolefins via transition-metal catalysis; polymer melt and solution viscosity; vapor-polymer thermodynamic equilibrium. Mailing Add: Amoco Chem Co PO Box 3011 Naperville IL 60566

BROCKMEIER, RICHARD TABER, b Grand Rapids, Mich, Apr 13, 37; m 64; c 2. NUCLEAR PHYSICS. Educ: Hope Col, AB, 59; Calif Inst Technol, MS, 61, PhD(physics), 65. Prof Exp: Res fel physics, Calif Inst Technol, 65-66; PROF PHYSICS, HOPE COL, 66-, PROF COMPUT SCI, 80- Mem: Am Phys Soc; Am Asn Physics Teachers; Sigma Xi; AAAS. Res: Nuclear structure physics; x-ray isotope shifts. Mailing Add: Dept Physics Hope Col Holland MI 49423

BROCKWAY, ALAN PRIEST, b Hanover, NH, Aug 21, 36; m; c 3. COMPARATIVE PHYSIOLOGY. Educ: St John's Col, Md, AB, 58; Western Reserve Univ, PhD(biol), 64. Prof Exp: Asst prof zool & entom, Ohio State Univ, 63-67; assoc prof, 67-79, PROF BIOL, UNIV COLO, DENVER, 79-, CHMN, 89- Concurrent Pos: Vis assoc prof oral biol, Sch Dent, Univ Colo, 73-77. Mem: Nat Asn Adv Health Professions. Res: Insect respiration; control of water loss in insects; uses of microcomputers in biology. Mailing Add: Dept Biol Univ Colo Denver CO 80202

BROCKWELL, PETER JOHN, b Melbourne, Australia, Oct 12, 37; m 65; c 3. PROBABILITY, MATHEMATICAL STATISTICS. Educ: Univ Melbourne, BEE, 58, BA, 60, MA, 62; Australian Nat Univ, PhD(math statist), 67. Prof Exp: Asst mathematician appl math, Argonne Nat Lab, 67-70; assoc prof statist, Mich State Univ, 71-73; prof math statist & chmn dept, La Trobe Univ, 73-76; prof statist & chmn dept, 76-77, PROF STATIST, COLO STATE UNIV, 78- Concurrent Pos: Consult, Argonne Nat Lab, 70-73; assoc ed, Math Scientist, 77-84, Stochastic Models, 85-, Theory Probability & Applications, 88-; NSF res grant, 78-87, 90-; ed, Advances Appl Probability, 82-88; vis prof, Kuwait Univ, 82-84; prof statist & chmn dept, Univ Melbourne, 88-89; bk rev ed, J Statist Planning Info, 89- Mem: Australian Statist Soc; Am Statist Asn; fel Inst Math Statist; Cell Kinetics Soc; Sigma Xi; Int Statist Inst. Res: Stochastic processes and their applications in physics, biology and hydrology; time series analysis; inference for stochastic processes. Mailing Add: Dept Statist Colo State Univ Ft Collins CO 80523

BROCOUM, STEPHAN JOHN, b New York, NY, Feb 16, 41; m 69. STRUCTURAL GEOLOGY, TECTONICS. Educ: Brooklyn Col, BS, 63; Columbia Univ, PhD(geol), 71. Prof Exp: Res scientist geol, Lamont-Doherty Geol Observ, Columbia Univ, 71-73; asst prof, Tex Christian Univ, 73-75; proj geologist, E D'Appolonia Consult Engrs, Inc, 75-76; res geologist, Gulf Sci & Technol Co, Gulf Oil Corp, 76-88; AT DEPT OF ENERGY, 88- Honors & Awards: Antarctic Serv Medal, NSF, 73. Mem: AAAS; Geol Soc Am; Am Geophys Union; Sigma Xi. Res: Use of remote sensing for hydrocarbon exploration; improving geological criteria for siting nuclear power plants; relationships between mid-continent tectonics and earthquakes; tectonic and strain history of the Sudbury Basin; tectonic and metamorphic history of the Adirondack lowlands. Mailing Add: US Nuclear Regulatory Comn 1000 Independence Ave Washington DC 20555

BRODASKY, THOMAS FRANCIS, b New London, Conn, Oct 6, 30; m 60; c 2. PHYSICAL CHEMISTRY, ORGANIC CHEMISTRY. Educ: Northeastern Univ, BS, 53, MS, 55; Rensselaer Polytech Inst, PhD(chem), 61. Prof Exp: SR RES SCIENTIST, CHEM & BIOL SCREENING, UPJOHN CO, 87- Mem: Am Chem Soc; Sigma Xi. Res: Gas, high performance liquid, paper and thin-layer chromatographic separations; infrared, ultraviolet and visible absorption spectroscopy; data retrieval; antibiotic metabolism; kinetics of biological reactions; gas liquid chromatography-mass spectroscopy; isolation natural products. Mailing Add: Upjohn Co Res Div 301 Henrietta St Kalamazoo MI 49001

BRODD, RALPH JAMES, b Moline, Ill, Sept 8, 28; m 50; c 3. PHYSICAL CHEMISTRY. Educ: Augustana Col, BA, 50; Univ Tex, MA, 53, PhD(chem), 55. Prof Exp: Electrochemist, Nat Bur Standards, 55-61; sr scientist, Res Ctr, Ling-Temco-Vought, Inc, 61-63; res chemist, Parma Res Ctr, 63-64, group leader, 64-65, tech res mgr, Rechargeable Batteries, Parma Res Ctr, Union Carbide Corp, 65-77, sr scientist, 77-78; staff consult, ESB Technol Ctr, 78-79; dir technol, Inco Electroenergy, 79-80; res fel, Inco Res & Develop Ctr, Inc, 80-82; consult, Broddays Inc, 82-84; proj mgr, Amoco Res Ctr, Naperville, 84-86; MGR, LITHIUM POWER SOURCES, GOULD INC, 86- Concurrent Pos: Instr, USDA Grad Sch, 56-61, Am Univ, 58-59 & Georgetown Univ, 60-61; adj prof, Mich Technol Univ, 85- Mem: AAAS; Am Chem Soc; hon mem Electrom Soc (pres); Chem Soc; Sigma Xi; NY Acad Sci; Int Soc Electrochem. Res: Metal-gas; metal-liquid interfaces; kinetics of electrochemical reactions; lithium batteries. Mailing Add: Gould Inc 35129 Curtis Blvd Eastlake OH 44095-4001

BRODE, WILLIAM EDWARD, b McMinn Co, Tenn, Dec 17, 29; m 60; c 4. MITIGATION OF IMPACTS & LOSSES TO WETLANDS. Educ: Univ Southern Miss, BS, 52, MA, 54, PhD(biol), 69. Prof Exp: Teacher biol, Copiah-Lincoln Jr Col, 55-57, Knox County Schs, Tenn, 59-60; Davidson County Schs, 60-62 & High Sch, Miss, 62-65; prof biol, Wesleyan Col, Ga, 69-74; ENVIRON SPECIALIST V, TENN DEPT TRANSP, 74- Concurrent Pos: Prin investr, WE Brode. Res: Amphibian and reptile taxonomy and ecology; hematological and serological studies on amphibians; impact of proposed highways on wildlife and wetlands. Mailing Add: Environ Planning Off Suite 900, James K Polk Bldg Nashville TN 37243-0334

BRODER, IRVIN, b Toronto, Ont, June 27, 30; m 54; c 3. IMMUNOLOGY. *Educ:* Univ Toronto, MD, 55; FRCP(C), 60. *Prof Exp:* Intern, Toronto Gen Hosp, 55-56, asst resident med, 58-59, resident physician, 59-60; sr intern, Sunnybrook Hosp, Toronto, 56-57; clin instr allergy, Med Ctr, Univ Mich, 60-62; clin teacher med, 63-66, assoc, 66-68, asst prof, 68-71, asst prof pharmacol, 65-75, asst prof path, 65-80, assoc prof med, 71-76, DIR, GAGE RES INST, UNIV TORONTO, 71-, PROF MED, 77-, PROF PATH, 80-, PROF OCCUP & ENVIRON HEALTH, 82- *Concurrent Pos:* Res fel endocrinol, Dept Path, Univ Toronto, 57-58; res fel immunol, Dept Pharmacol, Univ Col, Univ London, 62-63; res scholar, Med Res Coun Can, 63-66; career investr, Med Res Coun Can, 66-; mem, Inst Med Sci, 68-, & Inst Immunol, Univ Toronto, 71-82. *Mem:* Can Med Asn; Can Soc Allergy & Clin Immunol; Can Thoracic Soc; Can Soc Clin Invest; Can Soc Immunol. *Res:* Occupational lung disease research; study of human obstructive airways disease. *Mailing Add:* The Gage Res Inst 223 College St Toronto ON M5T 1R4 Can

BRODER, MARTIN IVAN, b Brooklyn, NY, June 15, 36; m 69; c 4. MEDICINE, CARDIOLOGY. *Educ:* Wayne State Univ, AB, 57; Western Reserve Univ, MD, 61. *Prof Exp:* Chief resident, Cleveland Metrop Gen Hosp, 64-65; fel clin cardiol, Univ London Inst Cardiol, 65-66; health scientist adminr, Nat Heart Inst, 67-69; fel, Vet Admin Hosp, Washington, DC, 69-70; dir cardiac care unit, Cleveland Metrop Gen Hosp, 71-75; from asst clin prof to assoc clin prof med, Sch Med, Case Western Reserve Univ, 76-81; chmn dept med, Fairview Gen Hosp, 76-81; assoc prof, 82-86, PROF MED, SCH MED, TUFTS UNIV, 86-; CHMN, DEPT MED, BAYSTATE MED CTR, 82- *Concurrent Pos:* Consult, Vet Admin, 71-; adj prof clin med, Ohio Col Podiat Med, 73-81; dep chmn, Dept Med, Sch Med, Tufts Univ. *Mem:* Fel Am Col Cardiol; fel Am Heart Asn. *Res:* Coronary artery disease; artificial heart techniques; hypertension treatment; cardiac arrhythmia. *Mailing Add:* 759 Chestnut St Springfield MA 01199

BRODER, SAMUEL, b Feb 24, 45; m; c 2. CANCER RESEARCH. *Educ:* Univ Mich, BS, 66, MD, 70; Am Bd Internal Med, cert, 73, cert oncol, 77. *Prof Exp:* Clin assoc, Metab Br, Nat Cancer Inst, NIH, 72-75, investr, Med Br, 75-81, sr investr, Metab Br, 76-81, assoc dir, Clin Oncol Prog, 81-88, DIR, NAT CANCER INST, NIH, 89- *Honors & Awards:* Upjohn Achievement Award; Arthur S Flemming Award; Augustus B Wadsworth Lectr; DREW Award, Ciba-Geigy; Samuel Rudin Award, Columbia Presby Hosp; Dr Frederick Stohlman Lectr Award; Lifetime Sci Award, Inst Advan Studies in Immunol & Aging; Harvey W Wiley Medal, Food & Drug Admin. *Mem:* Am Soc Clin Invest; Asn Am Physicians; Am Fedn Clin Res; Am Asn Cancer Res; Am Soc Clin Oncol; Am Asn Immunologists; Clin Immunol Soc; fel Am Col Physicians. *Res:* Relationships between cancer and immunodeficiency states; anti-retroviral chemotherapy. *Mailing Add:* NIH Nat Cancer Inst Off Dir Bldg 31 Rm 11A48 Bethesda MD 20892

BRODERICK, GLEN ALLEN, b Chicago, Ill, Oct 15, 45; m 68; c 3. RUMINANT NUTRITION, AMINO ACID METABOLISM. *Educ:* Univ Wis, Madison, BS, 67, MS, 70, PhD(biochem), 72. *Prof Exp:* Asst prof animal nutrit, Tex A&M Univ, 72-78, assoc prof, 78-80; RES SCIENTIST RUMINANT NUTRIT, US DAIRY FORAGE RES CTR, AGR RES SERV, US DEPT AGR, 81- *Mem:* Am Dairy Sci Asn; Am Soc Animal Sci; Am Inst Nutrit; Nutrit Soc. *Res:* Ruminant protein nutrition; protein degradation and protein synthesis of rumen microorganisms; amino acid nutrition and metabolism in the lactating dairy cow. *Mailing Add:* US Dairy Forage Res Ctr USDA-ARS Univ Wis 1925 Linden Dr W Madison WI 53706

BRODERICK, JOHN, b Danville, Pa, Oct 14, 40; m 79; c 4. RADIO ASTRONOMY. *Educ:* Pa State Univ, BS, 62; Brandeis Univ, MA, 64, PhD(physics), 70. *Prof Exp:* Res assoc, Nat Radio Astron Observ, 69-71 & Nat Astron & Ionospheric Ctr, 71-74; from asst prof to assoc prof, 74-85, PROF PHYSICS, VA POLYTECH INST & STATE UNIV, 85- *Mem:* Int Sci Radio Union; Int Astron Union; Am Astron Soc. *Res:* Extragalactic radio astronomy; very long baseline interferometry. *Mailing Add:* Dept Physics Va Polytech Inst & State Univ Blacksburg VA 24061-0435

BRODERIUS, STEVEN JAMES, b Hutchinson, Minn, Oct 23, 43; m 69; c 3. FISHERIES BIOLOGY, AQUATIC TOXICOLOGY. *Educ:* Univ Minn, BS, 65; Ore State Univ, MS, 69, PhD(fisheries), 73. *Prof Exp:* Res assoc aquatic toxicol, Univ Minn, 73-78; RES AQUATIC BIOLOGIST TOXICOL, US ENVIRON PROTECTION AGENCY, 78- *Mem:* Am Fisheries Soc; Sigma Xi; Soc Environ Toxicol & Chem. *Res:* Toxicity of organic and inorganic environmental contaminates to aquatic organisms; use of applied chemistry to better understand the reaction of toxicants in the aquatic environment. *Mailing Add:* Environ Res Lab 6201 Congdon Blvd Duluth MN 55804

BRODERSEN, ARTHUR JAMES, b Fresno, Calif, Aug 31, 39; m 65; c 2. ELECTRICAL ENGINEERING. *Educ:* Univ Calif, Berkeley, BS, 61, MS, 63, PhD(elec eng), 66. *Prof Exp:* From asst prof to prof elec eng, Univ Fla, 66-74; chmn dept elec & bioeng, 74-81, assoc dean, 79-86, PROF ELEC ENG, VANDERBILT UNIV, 74-,. *Mem:* Inst Elec & Electronics Engrs; Am Soc Eng Educr; Am Asn Artificial Intel. *Res:* Electronic and integrated circuits; modeling of semiconductor devices; microelectronics; applied artifical intelligence; intelligent tutorial systems. *Mailing Add:* Dept Elec Eng PO Box 1628 Sta B Nashville TN 37235

BRODERSEN, ROBERT W, DESIGN ANALOG INTERFACE CIRCUITS. *Educ:* Calif State Polytech Univ, BS(elec eng), & BS(math), 66; Mass Inst Technol, MSc & MS, 68 & PhD(eng), 72. *Prof Exp:* Staff, Ctr Res Lab, Tex Instruments, 72-76; PROF, DEPT ELEC ENG & COMPUT SCI, UNIV CALIF, BERKELEY, CALIF, 76- *Concurrent Pos:* Consult, various industr. *Mem:* Nat Acad Eng; fel Inst Elec & Electronics Engrs. *Res:* Speech recognition and synthesis, telecommunications, image processing, control systems and robotics. *Mailing Add:* Dept Elec Eng & Comput Sci Berkeley CA 94720

BRODERSON, STEVAN HARDY, b Lead, SDak, Sept 23, 38; m 64; c 2. ANATOMY, HISTOCHEMISTRY. *Educ:* Ohio State Univ, BS, 60; State Univ NY, PhD(anat), 67. *Prof Exp:* Instr, 67-70, ASST PROF ANAT, UNIV WASH, 70- *Concurrent Pos:* Investr, Am Epilepsy Found, 77-78. *Mem:* Am Asn Anatomists; Histochem Soc. *Res:* Histochemistry and neurochemistry of neurotransmitters and enzymes associated with neurotransmission; histochemistry of lipids associated with biomembrane structure and function. *Mailing Add:* Dept Biol Struct SM-20 Univ Wash Sch Med Seattle WA 98195

BRODEUR, ARMAND EDWARD, b Penacook, NH, Jan 8, 22; m 47; c 6. PEDIATRICS, RADIOLOGY. *Educ:* St Anselm's Col, AB, 45; St Louis Univ, MD, 47, MRd, 52; Am Bd Radiol, dipl, 52. *Hon Degrees:* LLD, St Anselm's Col, 71. *Prof Exp:* Intern, St Mary's Group, 47-49; asst radiologist, City Hosp, St Louis, Mo, 52; instr radiol, 52-57, sr instr, 57-60, asst prof clin radiol, 60-62, assoc prof radiol, 62-70, assoc dean sch med, 62-65, assoc prof pediat, 66-78, chmn dept, 75-78, PROF RADIOL, SCH MED, ST LOUIS UNIV, 70-, VCHMN DEPT, 78- *Concurrent Pos:* Chief radiol sect, USPHS, 52-54, consult, Div Radiol Health, 63-; spec consult, Div Spec Health Serv, Firmin Desloge Hosp, St Louis, 54-, assoc radiologist, Hosp, 57; chief radiologist, Cardinal Glennon Mem Children's Hosp, St Louis, 56; consult x-ray, Cath Health Asn, 56-88; chmn radiol, Shriners Hosp Crippled Children, St Louis, 88-; assoc vpres develop, Cardinal Glennon Childrens Hosp, 88-; health broadcaster, KMOX(CBS) radio, 80. *Mem:* Fel Am Col Radiol; Am Med Asn; Radiol Soc NAm; Soc Nuclear Med; Soc Pediat Radiol; fel Am Acad Pediat. *Res:* Pediatric radiology; reduction of radiation exposure in pediatric diagnostic radiology. *Mailing Add:* 400 Bambury Way St Louis MO 63131

BRODHAG, ALEX EDGAR, JR, b Charleston, WVa, Aug 23, 24; m 63; c 3. ORGANIC CHEMISTRY. *Educ:* Oberlin Col, AB, 48; Duke Univ, PhD(org chem), 54. *Prof Exp:* Asst, Duke Univ, 49-51; res assoc, Off Naval Res, 52; org res chemist, Union Carbide Chem Co, 53-60; asst ed, Chem Abstr Serv, 60-66, assoc ed, 66-70, sr ed, Org Abstr Ed Dept, 70-72, & Chem Technol Dept, 73-85; RETIRED. *Mem:* Am Chem Soc. *Res:* Intramolecular rearrangements; acetylene derivatives; high polymers; dyes. *Mailing Add:* 1760 Ardwick Rd Columbus OH 43220

BRODIE, ANN ELIZABETH, b Chicago, Ill, Jan 21, 43; m 66. CELL BIOLOGY. *Educ:* Purdue Univ, West Lafayette, BS, 65; Univ Calif, Berkeley, PhD(genetics), 70. *Prof Exp:* Fel cell biol, Univ Wis-Madison, 70-72, res assoc neruophysics, 72-75; RES ASSOC BIOCHEM, ORE STATE UNIV, 75- *Concurrent Pos:* N L Tarter res fel, 77. *Mem:* AAAS; Asn Women Sci. *Res:* Cellular synthesis and metabolism of thiol compounds and control mechanisms regulating these processes; controls thiol compounds exert on other systems; oxidative stress; protien thiols. *Mailing Add:* Dept of Biochem & Biophysics Weniger Hall 535 Ore State Univ Corvallis OR 97331-6503

BRODIE, BERNARD BERYL, pharmacology; deceased, see previous edition for last biography

BRODIE, BRUCE ORR, b Allegan, Mich, Apr 19, 24; m 47, 63; c 6. VETERINARY MEDICINE. *Educ:* Mich State Univ, DVM, 51; Univ Ill, MS, 58. *Prof Exp:* Pvt pract, 51-54; from instr to assoc prof, 54-69, PROF VET MED, UNIV ILL, URBANA-CHAMPAIGN, 69- *Mem:* Am Vet Med Asn; Am Asn Vet Clinicians; Am Asn Bovine Practr; Soc Theriogenology. *Res:* Infectious diseases of cattle, trichomoniasis and reproductive herd health programs. *Mailing Add:* 1406 Mayfair Champaign IL 61821

BRODIE, DAVID ALAN, b Albany, NY, June 2, 29; m 53; c 2. PHARMACOLOGY. *Educ:* Philadelphia Col Pharm & Sci, BSc, 51; Ohio State Univ, MSc, 53; Univ Utah, PhD(pharmacol), 56. *Prof Exp:* Instr pharmacol, Sch Med, Johns Hopkins Univ, 56-57; res assoc neuropharmacol, Merck Inst Therapeut Res, 57-62, sr investr gastroenterol, 62-64, sr res fel, 64-66, dir gastroenterol, 66-70; mgr pharmacol, William H Rorer, Inc, Pa, 70-71; assoc dir, Smith Klein & French Labs, 71-72; mgr dept, Abbott Labs, 72-73; dir pharmacol & med chem, 73-84; dir pharmacol, Ayerst Labs Res, Inc, 84-87; dir, CNS & CV Pharmacol, Wyeth Ayerst Res, 87-90; PRES, DISCOVERY RES CONSULTS, 90- *Mem:* Am Soc Pharmacol & Exp Therapeut; Am Physiol Soc; Am Gastroenterol Soc. *Res:* Drug effects on gastric secretion and experimental peptic ulcer. *Mailing Add:* 585 Edison Dr E Windsor NJ 08520

BRODIE, DONALD CRUM, b Carroll, Iowa, Mar 29, 08; m 34. PHARMACY. *Educ:* Univ Southern Calif, BS, 34, MS, 38; Purdue Univ, PhD(pharmaceut chem), 44. *Hon Degrees:* PdD, 81. *Prof Exp:* Lab asst, Col Dent, Univ Southern Calif, 36-38; asst, Col Pharm, Purdue Univ, 38-41, instr pharmaceut chem, 41-44; assoc pharmacol, Sch Med & Dent, Univ Rochester, 44-45; assoc prof pharmaceut chem, Col Pharm, Univ Kans, 45-47; assoc prof pharm, 47-53, lectr, Med Sch, 48-58, dir pharmaceut serv, 58-68, prof pharm & pharmaceut chem, Div Ambulatory & Community Med, Sch Med, 67-73, assoc dean prof affairs, Sch Pharm, 69-73, EMER PROF CHEM & PHARMACEUT CHEM, SCH PHARM, UNIV CALIF, SAN FRANCISCO, 73- *Concurrent Pos:* Res consult, Comn Outpatient Dispensing by Hosps & Related Facilities, 65; spec assignment, dir drug-related studies, Nat Ctr Health Serv Res & Develop, Health Serv & Ment Health Admin, HEW, 70-73; adj prof med & pharm, Univ Southern Calif, 74- *Mem:* AAAS; Am Chem Soc; Am Pharmaceut Asn; Am Soc Hosp Phamacists; Soc Hosp Pharmacists Australia. *Res:* Toxicity of drug agents; antihemorrhagic activity of the naphthoquinones; salicylate analgesics; physiology and pharmacology of vascular smooth muscle; delivery of health care; professional and clinical judgment; education programs for the health care team; health professions education; impact of new technology on pharmaceutical education. *Mailing Add:* 21300 S Heather Ridge Circle Apt 357 Green Valley AZ 85614-5108

BRODIE, DON E, b Bracebridge, Ont, Sept 8, 29; m 55; c 3. SOLID STATE PHYSICS. *Educ:* McMaster Univ, BSc, 55, MSc, 56, PhD(solid state physics), 61; Ont Col Educ, cert, 57. *Prof Exp:* Teacher physics, Humberside Collegiate Inst, 57-58; lectr, Univ Waterloo, 58-59, from asst prof to assoc prof, 61-68, dean sci, 82-90, PROF PHYSICS, UNIV WATERLOO, 68- *Concurrent Pos:* Ed, Physics Can, 69-73; mem, NSERC Strategic Grants Panel; dir, Physics Photovoltaic Group, Univ Waterloo, 79- *Mem:* Can Asn Physicists; NY Acad Sci; Solar Energy Soc Can Inc. *Res:* Physics of thin films, amorphous and crystalline; electronic and optical properties; photovoltaic solar cell development. *Mailing Add:* Univ Waterloo Waterloo ON N2L 3G1 Can

BRODIE, EDMUND DARRELL, JR, b Portland, Ore, June 29, 41; m 62; c 2. HERPETOLOGY, BEHAVIORAL ECOLOGY. *Educ:* Ore Col Educ, BS, 63; Ore State Univ, MS, 67, PhD(zool), 69. *Prof Exp:* Asst prof zool, Clemson Univ, 69-74; from assoc prof to prof biol, Adelphi Univ, 74-84; PROF & CHMN DEPT BIOL, UNIV TEX, ARLINGTON, 84- *Mem:* AAAS; Am Soc Ichthyologists & Herpetologists; Herpetologist's League; Ecol Soc Am; Soc Study Evolution; Sigma Xi. *Res:* Evolution of antipredator mechanisms including behavior, toxins, anatomy and coloration; mimicry in amphibians and reptiles. *Mailing Add:* Dept Biol Univ Tex Arlington TX 76019

BRODIE, HARLOW KEITH HAMMOND, b Stamford, Conn, Aug 24, 39; m 67; c 4. PSYCHIATRY. *Educ:* Princeton Univ, AB, 61; Columbia Univ, MD, 65. *Hon Degrees:* LLD, Univ Richmond, 87. *Prof Exp:* Intern med, Ochsner Found Hosp, New Orleans, 65-66; asst resident psychiat, Columbia-Presby Med Ctr, 66-68; clin assoc psychiat, Lab Clin Sci, NIMH, 68-70; asst prof, Sch Med, Stanford Univ, 70-74, prog dir, Gen Clin Res Ctr, 73-74; James B Duke prof psychiat & law & chmn dept psychiat, Med Ctr & chief psychiat serv, Duke Hosp, 74-82, chancellor, 82-85, PRES, DUKE UNIV, 85- *Concurrent Pos:* Consult, Palo Alto Vet Admin Hosp, Calif, 70-72; assoc ed, Am J Psychiat, 73-80; mem, Nat Adv Coun on Alcohol & Alcohol Abuse, 74-78; examr, Am Bd Psychiat & Neurol, 74-; consult psychiat, Educ Br, NIMH, 73-75, Durham Vet Admin Hosp, NC, 74-82; mem, President's Biomed Panel Interdisciplinary Cluster Pharmacol, Substance Abuse & Environ Toxicol, 75-76; adj prof, Dept Psychol, Duke Univ. *Honors & Awards:* Dean Echols Award, Ochsner Found Hosp, 65; A E Bennett Clin Res Award, Soc Biol Psychiat, 70. *Mem:* Inst Med-Nat Acad Sci; fel Am Psychiat Asn (pres, 82-83); Sigma Xi. *Res:* Psychobiology of mood disorders in man specifically as these are related to changes in endocrine function and neurotransmitter turnover in brain. *Mailing Add:* 63 Beverly Dr Durham NC 27707

BRODIE, HARRY JOSEPH, b New York, NY, Apr 25, 28; m 64; c 2. BIOCHEMISTRY, CHEMISTRY. *Educ:* Fordham Univ, BS, 50, MS, 52; NY Univ, PhD(org chem), 58. *Prof Exp:* Instr chem, Hunter Col, 53-56 & City Col New York, 57-58; chemist, E I du Pont de Nemours & Co, 58-60; fel, Clark & Worcester Found Exp Biol, 60-62; scientist, Worcester Found Exp Biol, 62-67, sr scientist, 67-78; exec secy, physiol chem study sect, 78-85, EXEC SECY, ENDOCRINOL STUDY SECT, NIH, 85- *Mem:* AAAS; Am Chem Soc; Endocrine Soc. *Res:* Mechanism of enzyme catalyzed reactions, especially estrogen biosynthesis and oxidation and reduction of steroids; inhibition of steroid hormone biosynthesis for contraception and cancer therapy; steroid synthesis, including labeling. *Mailing Add:* Div Res Grant NIH WW-218B Bethesda MD 20892

BRODIE, IVOR, b London, Eng, Apr 29, 28; m 54; c 2. PHYSICAL ELECTRONICS. *Educ:* Univ London, BSc, 50, Hons, 51, MSc, 56, PhD(physics), 59. *Hon Degrees:* DSc, London Univ, 76, Eurotechnical Res Univ. *Prof Exp:* Scientist, Res Labs, Gen Elec Co Eng, 51-56 & Nat Coal Bd, Mining Res Estab, 56-59; fel engr, Electronic Tube Div, Westinghouse Elec Corp, 59-62; sr engr, Varian Assocs, 62-65, sr scientist, 65-67, dir res, Vacuum Div, 67-69; pres, Photophysics, Inc, 69-71, chmn bd & chief exec officer, 71-73; DIR, PHYS ELECTRONICS LAB, SRI INT, 73- *Mem:* Fel Am Phys Soc; Soc Info Display; sr mem Inst Elec & Electronics Engrs. *Res:* Physical electronics; electron beam lithography; field emission; microfabrication technologies; electro-photography; electro-radiography. *Mailing Add:* SRI Int 333 Ravenswood Ave Menlo Park CA 94025

BRODIE, JEAN PAMELA, b London, Eng, Dec 25, 53. ASTRONOMY. *Educ:* Univ London, BSc, 75; Univ Cambridge, PhD(astron), 81. *Prof Exp:* Marketer & trader, Shell Int Petrol, Co, 75-77; lectr astron, Bd Extra-Mural Studies, Univ Cambridge, 78-80; FEL, DEPT ASTRON & SPACE SCI LAB, UNIV CALIF, BERKELEY, 80- *Mem:* Fel Royal Astron Soc; fel Royal Col Sci; Am Astron Soc. *Res:* Observation astronomy with particular emphasis on extragulactic research including globulue cluster, active galaxies and jets, BL LAC objects. *Mailing Add:* Dept Astron Univ Calif Santa Cruz CA 95064

BRODIE, JONATHAN D, b New York, NY, June 11, 38; c 3. BRAIN IMAGING, PSYCHIATRY. *Educ:* Univ Wis, PhD(biochem), 62; NY Univ, MD, 75. *Prof Exp:* From asst prof to assoc prof biochem, State Univ, Buffalo, NY, 65-73; from asst prof to assoc prof, 78-87, PROF PSYCHIAT, NY UNIV MED CTR, 87- *Mem:* Am Psychiat Asn; Am Soc Biol Chemists; Am Chem Soc; Am Col Psychiatrists; Soc Nuclear Med. *Res:* Positron emission tomography, application to mental disorders and biochemical pharmacology. *Mailing Add:* Dept Psychiat NY Univ Med Ctr 550 First Ave New York NY 10016

BRODIE, LAIRD CHARLES, b Portland, Ore, Aug 30, 22; m 48; c 3. PHYSICS. *Educ:* Reed Col, BA, 44; Univ Chicago, MS, 49; Northwestern Univ, PhD(physics), 54. *Prof Exp:* Res engr, Lab Div, Radio Corp Am, 53-54; Ford Found intern, Reed Col, 54-55; from instr to prof, 55-86, EMER PROF PHYSICS, PORTLAND STATE UNIV, 86- *Mem:* Am Phys Soc; Am Asn Phys Teachers. *Res:* Nucleation of the vapor phase in liquid helium. *Mailing Add:* Dept Physics Portland State Univ Portland OR 97207

BRODIE, MARK S, ELECTROPHYSIOLOGY, NEUROPHARMACOLOGY. *Educ:* Univ Ill, PhD(pharmacol), 84. *Prof Exp:* FEL, UNIV COLO, 84- *Mailing Add:* Abbott Labs D-47W Apt 10 Abbott Park IL 60064

BRODISH, ALVIN, b Brooklyn, NY, June 11, 25; m 57; c 2. PHYSIOLOGY. *Educ:* Drake Univ, BA, 47; Univ Iowa, MS, 50; Yale Univ, PhD(physiol), 55. *Prof Exp:* Lab instr biol, Drake Univ, 47-48; asst physiol, Univ Iowa, 49-50; asst psychiat, Yale Univ, 51-53, asst physiol, 53-54, NSF fel, 55-57, from instr to assoc prof, 57-68; prof physiol, Col Med, Univ Cincinnati, 68-75; prof physiol & chmn, Dept Physiol & Pharmacol, Bowman Gray Sch Med, Winston-Salem, NC, 75-80; mem staff, dept physiol, Univ Cincinnati, 80-; PROF & CHMN DEPT PHYSIOL & PHARMACOL, BOWMAN GRAY SCH MED. *Concurrent Pos:* Investr, Howard Hughes Med Inst, 57-60. *Mem:* AAAS; Endocrine Soc; Am Physiol Soc. *Res:* Neuroendocrine systems; regulation of anterior pituitary secretions; nerve-muscle regeneration. *Mailing Add:* Wake Forest Univ 300 S Hawthorne Rd Winston-Salem NC 27103

BRODKEY, JERALD STEVEN, b Omaha, Nebr, Jan 20, 34; m 62; c 2. NEUROSURGERY, BIOMEDICAL ENGINEERING. *Educ:* Harvard Univ, AB, 55; Univ Nebr, MS, 59, MD, 60; Am Bd Neurol Surg, dipl, 71. *Prof Exp:* Intern surg, Barnes Hosp, St Louis, Mo, 60-61, asst resident, 61-62; asst resident neurosurg, Mass Gen Hosp, Boston, 62-66, clin asst, 66-67; dir sci comput sect & asst chmn dept bioeng, Presby-St Luke's Hosp, 67-69; asst prof bioeng, Univ Ill, Chicago, 65-69, clin instr neurol & neurosurg, Col Med, 67-69; chief neurosurg, Cleveland Vet Admin Hosp, 69-83; prof, 77-83, CLIN PROF NEUROSURG, CASE WESTERN RESERVE UNIV, 83-; MED DIR, PAIN CTR, ST LUKE'S HOSP, 83-, CHIEF NEUROSURG, 84- *Concurrent Pos:* Clin & res fel neurosurg, Mass Gen Hosp, Boston, 62-63; USPHS spec fel, 65-66; asst attend bioeng, Presby-St Luke's Hosp, 65-67, asst attend neurosurg, 67-69. *Mem:* AAAS; Inst Elec & Electronics Engrs; Soc Cybernet; Asn Comput Mach; NY Acad Sci. *Res:* Embedding of neurophysiological experiments in a neurological control system background; clinical neurosurgery, particularly pituitary and pain surgery. *Mailing Add:* 24755 Chagrin Blvd Suite 205 Beachwood OH 44122

BRODKEY, R(OBERT) S(TANLEY), b Los Angeles, Calif, Sept 14, 28; m 75; c 1. CHEMICAL ENGINEERING. *Educ:* Univ Calif, BS & MS, 50; Univ Wis, PhD(chem eng), 52. *Prof Exp:* Chem engr, E I du Pont de Nemours & Co, 50; res chem engr, Esso Res & Eng Co, 52-56 & Esso Standard Oil Co, 56-57; RETIRED. *Concurrent Pos:* Consult, fluid dynamics; sr fel sci, NATO, 72; US sr scientist award, Alexander von Humboldt Found, 75, 83; Expository lectr, GAMM, Gottingen, 75; sr res award, Col Eng, Ohio State Univ, 83 & 86, Am Soc Eng Educ, 85; distinguished sr res award, Ohio State Univ, 83; numerous lectrs at various American, European & Asian cols & univs; 3m Lecturship Award, Am Soc Eng Educ. *Mem:* Am Chem Soc; fel Am Inst Chem Engrs; fel Am Phys Soc; Soc Rheology; Soc Eng Sci; fel AAAS; fel Am Inst Chemists; Sigma Xi. *Res:* Fundamental structure of turbulence; turbulent motion, mixing, and kinetics; rheology; two-phase flow; bio-fluid flow; author and co-author of 100 scientific publications. *Mailing Add:* Dept Chem Eng Ohio State Univ 140 W 19th Ave Columbus OH 43210-1180

BRODKORB, PIERCE, b Chicago, Ill, Sept 29, 08; m 31; c 1. ZOOLOGY. *Educ:* Univ Ill, AB, 33; Univ Mich, PhD(zool), 36. *Prof Exp:* Asst ornith, Field Mus, 30 & Cleveland Mus, Ohio, 31-32; asst mus zool, Univ Mich, 33-36, asst cur birds, 36-46; from asst prof to assoc prof, 46-55, PROF ZOOL, UNIV FLA, 55- *Concurrent Pos:* Mem exped, Idaho, 31-32, Black Hills, 35, Mex, 37, 39, 41, 53 & Bermuda, 60; consult, Fla Geol Surv, 57- & Govt of Bermuda, 60; Int Comt Avian Anat Nomenclature, 62-; mem Int Ornith Cong, 62- & Int Cong Zool, 63- *Honors & Awards:* Brewster Medal, Am Ornith Union, 78. *Mem:* Cooper Ornith Soc; fel Am Ornithologists Union; Wilson Ornith Soc; Paleont Soc; Soc Study Evolution; Sigma Xi. *Res:* Ornithology; zoogeography; avian paleontology and evolution. *Mailing Add:* Dept Zool Univ Fla 611 Brw Bldg Gainesville FL 32611

BRODMANN, JOHN MILTON, b Savannah, Ga, Aug 20, 33. ORGANIC CHEMISTRY. *Educ:* Lynchburg Col, BS, 55; Emory Univ, MS, 59, PhD(natural prod synthesis), 67. *Prof Exp:* From instr to prof chem, 57-89, chmn, Div Natural Sci, 65-76, ADMIN ASST, CULVER-STOCKTON COL, 89- *Concurrent Pos:* Chmn, Quincy-Keokuk Sect, Am Chem Soc, 87- *Mem:* Sigma Xi; Am Chem Soc. *Res:* Grignard addition to alpha, B-unsaturated ketones. *Mailing Add:* 501 S Monticello Rd Canton MO 63435-1416

BRODOFF, BERNARD NOAH, b New York, NY, Sept 8, 23. DIABETES, OBESITY. *Educ:* NY Univ, BA, 42; MD, 46. *Prof Exp:* CLIN PROF MED & ADJ PROF PHYSIOL, NY MED COL. *Concurrent Pos:* Vis scientist, Brookhaven Nat Lab, 67-69. *Mem:* Fel Am Col Physicians; fel Am Col Cardiol; Am Physiol Soc. *Res:* Hypothalmic mechanisms in obesity and diabetes; lipid metabolism. *Mailing Add:* Dept Physiol NY Med Col Basic Sci Bldg Valhalla NY 10595

BRODOWAY, NICOLAS, b Melfort, Sask, Dec 9, 22; US citizen; m 49; c 6. ORGANIC CHEMISTRY. *Educ:* Univ BC, BSc, 49; Univ Minn, PhD(org chem), 53. *Prof Exp:* Res chemist, Elastomer Chem Div, Polymer Prod Dept, E I du Pont de Nemours & Co, Inc, 53-80, res assoc, 80-84; RETIRED. *Mem:* Am Chem Soc; Sigma Xi. *Res:* Polymer synthesis, evaluation and process development. *Mailing Add:* 215 Dakota Ave Wilmington DE 19803

BRODRICK, HAROLD JAMES, JR, meteorology, for more information see previous edition

BRODSKY, ALLEN, b Baltimore, Md, Nov 5, 28; m 51, 84; c 3. RADIOLOGICAL HEALTH PHYSICS, BIOSTATISTICS. *Educ:* Johns Hopkins Univ, BE, 49, MA, 60; Univ Pittsburgh, ScD(biostatist, radiation health), 66; Am Bd Health Physics, dipl, 60; Am Bd Indust Hyg, dipl, 66; Am Bd Radiol, dipl, 75. *Prof Exp:* Head, Health Physics Unit, Naval Res Lab, Washington, DC, 50-52, physicist, Opers Ivy & Castle, Eniwetok & Bikini, 52-54; pres, Health Physics Servs, Inc, Baltimore, Md, 55-57; radiol defense officer, Fed Civil Defense Admin, Olney, Md, 56-57; health physicist, Health Protection Br, Div Biol & Med, US AEC, Washington, DC, 57-59, radiation physicist, Div Licensing & Regulation, Washington, DC, 59-61; res assoc health physics & epidemiol, Dept Occup Health, Grad Sch Pub Health, Univ

Pittsburgh, 61-66, assoc prof, Dept Occup Health & Dept Radiation Health, 66-71; radiation physicist & safety officer, Dept Radiol, Mercy Hosp, Pittsburgh, Pa, 71-75; sr health physicist, Off Nuclear Regulatory Res, US Nuclear Regulatory Comn, Washington, DC, 75-86; CONSULT, ALLEN B CONSULT INC, 86- *Concurrent Pos:* Chmn standards comt, Health Physics Soc, 60-61 & 67-70, bd dirs, 67-70; consult, Westinghouse, 62-64, Numec, 66-69, Mercy Hosp, 67-70, Case Western Reserve Univ, 68-69, SCA, Inc, 87-89, Brookhaven Nat Lab, 87-88, Dept Energy, 90, Centocor, 90, Fitzpatrick Nuclear Plant, 90-91; tech dir, Dept Radiation Med, Presby Univ Hosp, Pittsburgh, 65-70; mem, AEC Radiation Sci Fel Bd, Oak Ridge Assoc Univs, 68-70; expert witness, J Terasi, 68-71, Dept Justice, 81-83; adj res prof, Sch Pharm, Duquesne Univ, 71-75; adj prof radiation med, Georgetown Univ, 86- *Honors & Awards:* Distinguished Serv Award, Western Pa Chap, Health Physics Soc, 74; Wright Langham lectr award, Radiation Epidemiol, Univ Ky Med Ctr, 79; Failla lectr award, NY Chap Health Physics Soc & Radiol & Med Physics Soc, 87. *Mem:* Am Indust Hyg Asn; Health Physics Soc; Am Asn Physicists Med; Am Pub Health Asn; Am Statist Asn; Am Nuclear Soc; NY Acad Sci. *Res:* Radiation dose measurement and interpretation; radiation hazard evaluation and standards for radiation protection; mathematical models of carcinogenesis and risk estimation from environmental agents; epidemiologic studies of environmental exposures and effects. *Mailing Add:* 2765 Ocean Pines Berlin MD 21811-9127

BRODSKY, CARROLL M, b Lowell, Mass, Dec 23, 22; c 3. PSYCHIATRY, ANTHROPOLOGY. *Educ:* Cath Univ Am, AB, 49, MA, 50, PhD(anthrop), 54; Univ Calif, MD, 56. *Prof Exp:* Lectr anthrop, Cath Univ Am, 52; res assoc, Human Resources Res Off, 53; from instr to assoc prof, 60-70, PROF PSYCHIAT, SCH MED, UNIV CALIF, SAN FRANCISCO, 70- *Concurrent Pos:* NIMH grant, Langley-Porter Neuropsychiat Inst, 57-60. *Mem:* Acad Psychosom Med; Am Antrop Asn; Am Col Psychiat. *Res:* Studies of human disability with emphasis on social and psychiatric factors delaying recovery from illness; problems of work. *Mailing Add:* Dept Psychiat Univ Calif Sch Med San Francisco CA 94143

BRODSKY, FRANCES M(ARTHA), IMMUNOLOGY, ENDOCYTOSIS. *Educ:* Oxford Univ, DPhil, 79. *Prof Exp:* PROG MGR CELL BIOL, BECTON DICKINSON IMMUNOCYTOMETRY SYSTS, 82- *Mailing Add:* Univ Calif Box 0446 Rm S-926 513 Parnassus Ave San Francisco CA 94143

BRODSKY, MARC HERBERT, b Philadelphia, Pa, Aug 9, 38; m 66; c 2. SEMICONDUCTOR TECHNOLOGY. *Educ:* Univ Pa, AB, 60, MS, 61, PhD(physics), 65. *Prof Exp:* Res assoc physics, Univ Pa, 65; res physicist, US Naval Ord Lab, 65-66; physicist, Night Vision Lab, US Army, 66-68; res mem staff, T J Watson Res Ctr, Int Bus Mach Corp, 68-87; prog dir, IBM Advan Gaas Technol Lab, 87-89; DIR, TECH PLANNING, IBM CORP WATSON RES CTR, 89- *Concurrent Pos:* Adj assoc prof, Columbia Univ, 72-74; exchange prof, Univ Paris VI, 74-75. *Mem:* AAAS; fel Am Phys Soc; fel Inst Elec & Electronics Engrs. *Res:* Semiconductors, amorphous semiconductors; lattice vibrations; infrared spectroscopy; Semiconductor devices and technology. *Mailing Add:* IBM Corp Watson Res Ctr PO Box 218 Yorktown Heights NY 10598

BRODSKY, MERWYN BERKLEY, b Chicago, Ill, Mar 4, 30; m 50; c 2. SOLID STATE PHYSICS. *Educ:* Roosevelt Univ, BS, 49; Ill Inst Technol, MS, 51, PhD(phys chem), 55. *Prof Exp:* Asst, Ill Inst Technol, 50-54; assoc chemist, Brookhaven Nat Lab, 54-58; assoc chemist, 58-66, group leader, 66-74, SR SCIENTIST, ARGONNE NAT LAB, 74- *Concurrent Pos:* Sr vis res fel, Imperial Col, Univ London, 71-73. *Mem:* Am Phys Soc; Am Inst Mining, Petrol & Metall Eng. *Res:* Electron transport and magnetism of actinides; fused salt electrorefining; physical chemistry of liquid metal solutions; high temperature thermodynamics; liquid metal reactor systems. *Mailing Add:* 355 Oakwood St Park Forest IL 60466

BRODSKY, PHILIP HYMAN, b Philadelphia, Pa, July 7, 42; m 64; c 3. ADHESIVES, PLASTICS. *Educ:* Cornell Univ, BChE, 65, PhD(chem eng), 69. *Prof Exp:* Sr res engr, Monsanto Co, 69-73, res specialist, 73-76, group leader, 76-79, sr group leader, 79-80, mgr, 80-84, dir res & develop, 84-87, DIR CORP RES, MONSANTO CO, 87- *Mem:* Am Chem Soc; Indust Res Inst. *Res:* Compatibility in polymer blends; manufacturing processes and chemistry of condensation and addition polymers; reprographics and adhesives chemistry; technology organization development; technology management training; plastics research and development. *Mailing Add:* Monsanto Co 800 Lindbergh Blvd St Louis MO 63167

BRODSKY, STANLEY JEROME, b St Paul, Minn, Jan 9, 40; m 62, 86; c 2. HIGH ENERGY PHYSICS, THEORETICAL PHYSICS. *Educ:* Univ Minn, BPhys, 61, PhD(physics), 64. *Prof Exp:* Res assoc theoret physics, Columbia Univ, 64-66; res assoc, 66-68, assoc prof, 75-76, MEM RES STAFF, STANFORD LINEAR ACCELERATOR CTR, STANFORD UNIV, 68-, PROF, 76- *Concurrent Pos:* Vis assoc prof, Dept Physics, Cornell Univ, 69; consult, Particle Data Group, Lawrence Berkeley Lab, 72-78, mem high energy physics rev comt, 78-81; mem comt fundamental constants, Nat Res Coun-Nat Acad Sci, 74-77 & prog comt, Wilson Lab, Cornell Univ, 75-79; mem prog comt, Argonne Nat Lab, 77-79; mem exec bd, Sci Forum, Weizmann Inst, Rehovot, Israel, 77-, vis prof, 78; mem prog adv comt, Fermi Nat Accelerator Lab, 78-; vis prof, Inst Theoret Physics, Univ Calif, Santa Barbara, 81, 85, & 88; mem, Dept Energy & Nat Sci Found Nuclear Sci Adv Subcomt Electromagnetic Interactions, 81-; mem sci adv panel, Southeastern Univ Res Asn, Inc, 81-; vis prof natural sci, Inst Advan Study, Princeton, 82, Max Planck Inst Nuclear Physics, Heidelberg, 87 & 88; mem sci & educ adv comt, Lawrence Berkeley Lab, Univ Calif, 86-; external sci dir, Max Planck Inst Nuclear Physics, 89-; lectr, Colloquium Series, Univ Minn, 89, Inst Nuclear Theory, Univ Wash, 90. *Honors & Awards:* Alexander von Humboldt sr distinguished US scientist. *Mem:* Fel Am Phys Soc. *Res:* Quantum electrodynamics; muonic x-rays; weak interactions; Zeeman structure; Lamb shift; lepton magnetic moments; elementary particles; electromagnetic interactions; colliding beam physics; large transverse momentum reactions; nuclear processes; photon-photon collisions; quark-model; quantum chromodynamics. *Mailing Add:* Stanford Linear Accelerator Ctr Stanford Univ PO Box 4349 Stanford CA 94305

BRODSKY, WILLIAM AARON, b Philadelphia, Pa, Jan 8, 18; m 50; c 3. PHYSIOLOGY. *Educ:* Temple Univ, BS, 38, MD, 41. *Prof Exp:* Intern, Philadelphia Gen Hosp, 41-43; instr pediat, Univ Pa, 43-44; instr, Univ Cincinnati, 48-49; from asst prof to prof pediat, Univ Louisville, 51-68, prof exp med, 60-68; PROF BIOPHYS, MT SINAI SCH MED, 68-, PROF NEPHROLOGY, 83- *Concurrent Pos:* Res fels pediat, Univ Cincinnati, 46-48 & 49-51; USPHS res fel, 62-; estab investr, Am Heart Asn, 55-60; Career Res Award, NIH, 61-68; prof biophys grad div, City Univ NY, 68- *Honors & Awards:* Basic Sci Res Award, Temple Univ, Sch Med, 86- *Mem:* Soc Pediat Res; Am Physiol Soc; Soc Exp Biol & Med; Am Soc Clin Invest; Biophys Soc (secy & exec coun, 67-72); Sigma Xi. *Res:* Osmotic properties of tissue; ion transport mechanisms; acid-base equilibria; renal transport mechanisms. *Mailing Add:* 259 Barnard Rd Larchmont NY 10538

BRODWICK, MALCOLM STEPHEN, b New York, NY, July 26, 44; m 69; c 2. PHYSIOLOGY. *Educ:* San Fernando Valley State Col, Ba, 67; Univ Calif, Los Angeles, PhD(anat), 72. *Prof Exp:* Fel physiol & pharmacol, Duke Univ, 72-74; asst prof, 74-79, ASSOC PROF PHYSIOL, UNIV TEX MED BR, GALVESTON, 79- *Mem:* Soc Gen Physiologists; Biophys Soc. *Res:* Membranes controlling the flux of materials from one side to the other; active transport; excitability; neurotransmitter induced conductance; exocytosis. *Mailing Add:* Dept Physiol & Biophysics Sch Med Univ Tex 301 Univ Blvd Galveston TX 77550

BRODWIN, MORRIS E(LLIS), b New York, NY, July 14, 24; m 49; c 2. ELECTRICAL ENGINEERING. *Educ:* Univ Nebr, BS, 47; Johns Hopkins Univ, MS, 52, DrE, 57. *Prof Exp:* Res scientist, Microwave & Syst Anal, Johns Hopkins Univ, 48-58; from assoc prof to prof microwave eng, 58-69, PROF ELEC ENG, NORTHWESTERN UNIV, EVANSTON, 69- *Concurrent Pos:* Consult, Martin Co, 57-58 & Admiral Corp, 59. *Mem:* AAAS; Inst Elec & Electronics Engrs. *Res:* Microwave physics; propagation in anisotropic media; measurement of physical properties at microwave frequencies. *Mailing Add:* Dept Elec Eng Technol Inst Northwestern Univ Evanston IL 60201

BRODY, AARON LEO, b Boston, Mass, Aug 23, 30; m 53; c 3. FOOD TECHNOLOGY. *Educ:* Mass Inst Technol, BS, 51, PhD(food technol), 57; Northeastern Univ, MBA, 70. *Prof Exp:* Food technologist, Birdseye Fisheries Labs, Gen Foods Corp, 51-52 & Raytheon Mfg Co, 54-55; res food technologist, Whirlpool Corp, Mich, 57-61; mgr pkg & prod develop, M&M's Candies Div, Mars, Inc, 61-67; sr staff mem, Arthur D Little Inc, 67-73; new ventures mgr, Mead Packaging, Atlanta, 73-81; mgr mkt develop, Container Corp Am, 81-85; CONSULT, 85- *Concurrent Pos:* Mem, US Navy Food Serv Adv Comt; sci lect, Inst Food Technologists, 72-75; consult, World Bank, 72-73; mem, sci adv coun, Refrig Res Found, 86- *Honors & Awards:* Willis H Carrier Award, Am Soc Heating, Refrig & Air Conditioning Eng, 60; Riester-Davis Award, Inst Food Technologists, 88. *Mem:* AAAS; fel Inst Food Technologists; Am Soc Heating, Refrig & Air Conditioning Eng; fel Pkg Inst US (vpres, 73-); NY Acad Sci; Inst Packaging Professionals. *Res:* Objective measurements of physical properties of foods; microwave heating of foods; refrigeration and freezing of food products; controlled atmosphere storage; packaging of foods. *Mailing Add:* 733 Clovelly Lane Devon PA 19333

BRODY, ALFRED WALTER, b New York, NY, Feb 20, 20; m 43; c 4. PHYSIOLOGY, PHARMACOLOGY. *Educ:* Columbia Univ, BA, 40, MA, 41; Long Island Col Med, MD, 43; Univ Pa, DSc, 53. *Prof Exp:* Fel, Grad Sch Med, Univ Pa, 51-52, instr physiol & pharmacol, 51-54; from asst prof to prof physiol & pharmacol, Sch Med, Creighton Univ, 54-77, dir pulmonary lab, 54-77, prof med, 60-77, chief, Chest Div, 62-77. *Concurrent Pos:* Consult, pulmonary dis, 77- *Mem:* AAAS; Am Physiol Soc; Am Thoracic Soc; NY Acad Sci; AMA. *Res:* Mechanics of respiration; respiratory physiology; lesser circulation; bronchogenic cancer medicine; internal medicine. *Mailing Add:* Dept Med Creighton Univ Sch Med 2500 California Ave Omaha NE 68178

BRODY, ARNOLD R, b Boston, Mass, March 24, 43; m 67; c 2. EXPERIMENTAL PULMONARY PATHOLOGY. *Educ:* Univ Ill, MS, 67; Colo State Univ, BS, 65, PhD(cell biol), 69. *Prof Exp:* Fel, dept entom, Ohio State Univ, 69-72; res assoc, Univ Vt, 72-74, asst prof, 74-78; sr staff fel, 78-84, res biolist, lab pulmonary pathobiol, 84- 87, HEAD PULMONARY PATH, NAT INST ENVIRON HEALTH SCI, 78-, PROF PATH, 87- *Concurrent Pos:* Vis scientist, Pneumoconiosis Res Ctr, Med Res Coun, Wales, 74 & Dept Path, Hosp Bichat, France, 77; adj prof, dept path, Col Med, Duke Univ, 78-; John P Wyatt Traveling Fel Environ Path, 84; vis prof, dept path, Nat Inst Cardiol, Mexico City, 85. *Mem:* AAAS; Electron Micros Soc Am; Sigma Xi; Nat Wildlife Fedn. *Res:* Defining the mechanisms by which inhaled inorganic particulates cause lung disease; deposition patterns of asbestos and silica and the subcellular interactions of various pulmonary cells with these inhaled particulates; biology and biochemistry of macrophage-derived growth factors. *Mailing Add:* Nat Inst Environ Health Scis Div Intramural Res Research Triangle Park NC 27709

BRODY, BERNARD B, b New York, NY, June 24, 22; m 54; c 2. MEDICINE, CHEMISTRY. *Educ:* Univ Wis-Madison, BS, 43; Univ Rochester, MD, 51. *Prof Exp:* Res assoc chem, Univ Chicago, 43-45; res assoc, Monsanto Chem Corp, Ohio, 45-47; intern med, Strong Mem Hosp, Rochester, NY, 51-52, asst resident, 52-53; chief resident, Genesee Hosp, 55-56; dir med affairs, Genesee Hosp, 74-87; from instr to asst prof med, 56-81, assoc prof path, 77-81, PROF MED & PATH, UNIV ROCHESTER, 81- *Concurrent Pos:* Pvt practr, 56-67; dir clin labs, Genesee Hosp, 67-81. *Mem:* AAAS; AMA; Am Soc Internal Med; Am Asn Clin Chem; Am Col Physicians. *Mailing Add:* 12 Huntington Brook Rochester NY 14625

BRODY, BURTON ALAN, b New York, NY, June 8, 42; m 80. PHYSICS. *Educ:* Columbia Col, BA, 63; Univ Mich, PhD(exp physics), 70. *Prof Exp:* PROF PHYSICS & PHYSICIST, BARD COL, 70-; ADJ RES PROF PHYSICS & PHYSICIST, COLUMBIA UNIV, 81- *Concurrent Pos:* Systs analyst & comput programmer, OLI Systs Inc, 78-80. *Mem:* Am Phys Soc;

Am Asn Physics Teachers; Fedn Am Scientists. *Res:* Superfluid helium; database management systems; acoustics; electronics; photon echoes (laser physics). *Mailing Add:* Dept Physics Bard Col Annandale-on-Hudson NY 12504

BRODY, EDWARD NORMAN, b Chicago, Ill, Mar 2, 39; m 64; c 4. RNA SYNTHESIS, RNA SPLICING. *Educ:* Roosevelt Univ, Chicago, BS, 59; Univ Chicago, MD, 64, PhD(biochem), 65. *Prof Exp:* Fel, Dept Biophysics, Univ Chicago, 65-68 & Dept Molecular Biol, Univ Geneva, 68-70; res, Inst Biol Physics & Chem, Paris, 70-88; DIR RES, CENTRE DE GENETIQUE MOLECULAIRE, GIF-SUR-YVETTE, 88- *Concurrent Pos:* Vis res assoc, Dept Biol, Calif Inst Technol, 84. *Mem:* Europ Molecular Biol Orgn. *Res:* Regulation of transcription in E coli infected by bacteriophage T4; mechanism of RNA splicing in eukaryotes. *Mailing Add:* Centre de Génétique Moldculaire CNRS Gif-sur-Yvette 91190 France

BRODY, EUGENE B, b Columbia, Mo, June 17, 21; m 44; c 3. PSYCHIATRY, PSYCHOANALYSIS. *Educ:* Univ Mo, AB, 41, MA, 41; Harvard Med Sch, MD, 44; Am Bd Psychiat & Neurol, dipl, 50; Am Psychoanal Asn, cert, 60. *Prof Exp:* From intern to asst resident psychiat, Yale Univ, 44-46; chief resident, 48-49, from instr to assoc prof, 49-57, chmn dept, 59-76, dir, Inst Psychiat & Human Behav, 59-76, prof human behav, 76-87, EMER PROF PSYCHIAT & HUMAN BEHAV, UNIV MD, BALTIMORE, 87-; SR ASSOC, JOHNS HOPKINS SCH PUB HEALTH, 86- *Concurrent Pos:* Consult, WHO & Pan Am Health Orgn, 65-, NIMH, 68- & UNESCO, 87-; dir, Interam Ment Health Studies Prog, Am Psychiat Asn, 66- 68; ed-in-chief, J Nerv & Ment Dis, 67-; vis prof, Univ Brazil, 68, Univ West Indies, 72-75 & Univ Otago, New Zealand, 81; fel, Ctr Adv Study Behav Sci, Stanford, 75-76, Sackler Inst Adv Study, Univ Tel Aviv, 86; psychiat epidemiol, US NIMH, 75-79, AIDS, 88-; bd, Hogg Found, Univ Tex, 87-90; mem adv bd, Peruvian NIMH, 84-; secy gen, World Fedn Ment Health, 83-; exec bd, Int Soc Sci Coun, 87-90. *Honors & Awards:* Global Humanitarian, Hasserman Found Int Accord, 90. *Mem:* Fel Am Psychiat Asn; Asn Behav Sch Med Educ (pres, 80-81); World Fedn Mental Health (pres, 81-83, secy gen, 83-); fel Am Anthrop Asn; Am Psychoanal Asn. *Res:* Social psychiatry; culture and fertility; psychoanalytic anthropology; ethics and health. *Mailing Add:* Sheppard & Enoch Pratt Hosp PO Box 6815 Baltimore MD 21285-6815

BRODY, GARRY SIDNEY, b Edmonton, Alta, Sept 21, 32; US citizen; m 57; c 3. PLASTIC SURGERY. *Educ:* Univ Alta, MD, 56; McGill Univ, MS, 59. *Prof Exp:* CHIEF PLASTIC SURG, SCH MED, UNIV SOUTHERN CALIF-RANCHO LOS AMIGOS HOSP, 69-, ASST CLIN PROF SURG, 77- *Concurrent Pos:* Abstract ed, J Plastic & Reconstruct Surg, 74-; chmn, Plastic Surg Res Coun, 76. *Mem:* Am Soc Plastic & Reconstruct Surg; Asn Surg of the Hand; Med Eng Soc. *Res:* Rheology of connective tissue and scar; relationship of breast cancer and prosthetic insects; multiple clinical plastic surgery problems. *Mailing Add:* Downey Comm Hosp 11411 Brookshire Ave Suite 504 Downey CA 90241

BRODY, HAROLD, b Cleveland, Ohio, May 15, 23; m 51; c 2. NEUROGERONTOLOGY. *Educ:* Western Reserve Univ, BS, 47; Univ Minn, PhD(anat), 53; Univ Buffalo, MD, 61. *Prof Exp:* Teaching asst anat, Univ Minn, 47-49, instr, 49-50; asst prof, Univ NDak, 50-54; from asst prof to assoc prof, 54-63, PROF ANAT, SCH MED, STATE UNIV NY BUFFALO, 63-, CHMN DEPT, 71- *Concurrent Pos:* Fulbright sr res scholar, 63; ed-in-chief, J Geront, 75-80; ed, Neurobiol Aging, 82- *Honors & Awards:* Kleemeier Award Res Geront, 79. *Mem:* AAAS; Am Asn Anatomists; Am Geront Soc (pres, 74-75); Int Asn Gerontol; Am Geriatric Soc. *Res:* Neuroanatomy; neuropathology; age changes in the human nervous system. *Mailing Add:* Dept Anat Sci Sch Med State Univ NY Buffalo NY 14214

BRODY, HAROLD D, b Boston, Mass, Apr 13, 39; m 66; c 2. MATERIALS PROCESSING. *Educ:* Mass Inst Technol, SB, 60, SM, 61, ScD, 65. *Prof Exp:* Sr res metallurgist, Monsanto Res Corp, 65-66; from asst prof to assoc prof, 66-74, chmn dept, 70-79, PROF MAT SCI & ENG, UNIV PITTSBURGH, 74-, DIR, CASTING INDUST SCI & ENG INST, 82- *Concurrent Pos:* Res engr, Mass Inst Technol, 65-66, vis prof mat sci & eng, 87- *Honors & Awards:* Foundry Educ Found Prof, 77. *Mem:* Am Foundrymen's Soc; Am Soc Metals; Am Inst Mining, Metall & Petrol Engrs; Mat Res Soc. *Res:* Casting technology; process metallurgy with emphasis on melting, casting and solidification; process modelling and CAD/CAM; composite materials; engineering education. *Mailing Add:* Univ Pittsburgh 848 Benedum Hall Pittsburgh PA 15261

BRODY, HOWARD, b Newark, NJ, July 11, 32; m 54; c 3. PHYSICS. *Educ:* Mass Inst Technol, SB, 54; Calif Inst Technol, MS, 56, PhD(physics), 59. *Prof Exp:* NSF fel, 59; from instr to assoc prof, 59-77, PROF PHYSICS, UNIV PA, 77- *Mem:* Am Phys Soc; Am Asn Physics Teachers. *Res:* Study of elementary particles and their interactions at high energy. *Mailing Add:* Dept Physics Univ Pa Philadelphia PA 19104

BRODY, JACOB A, b Brooklyn, NY, May 5, 31; m 69; c 2. EPIDEMIOLOGY, MEDICAL SCIENCE. *Educ:* Williams Col, BA, 52; State Univ NY Downstate Med Ctr, MD, 56. *Prof Exp:* Intern, Roosevelt Hosp, NY, 56-57; mem staff surveillance of arthropod-borne dis, Surveillance Sect, Epidemiol Br, Commun Dis Ctr, USPHS, 57-59, chief poliomyelitis surveillance unit, 58-59, med officer, Virus Sect, Mid Am Res Unit, Nat Inst Allergy & Infectious Dis, CZ, Panama, 59-61, mem staff, Lab Trop Med, Md, 61-62, chief epidemiol sect, Arctic Health Res Ctr, Alaska, 62-65, chief epidemiol br, Collab & Field Res, Nat Inst Neurol Dis & Stroke, 65-74; res coordr unified prog, Atomic Bomb Casualty Comt, Hiroshima, Japan, 74-75; sr res epidemiol, Nat Inst Neurol & Commun Disorders & Stroke, NIH, 75-76; chief, Epidemiol & Spec Studies Br, Nat Inst Alcohol Abuse & Alcoholism, Alcohol, Drug Abuse & Ment Health Admin, 76-77; assoc dir, Nat Inst Aging, 77-85; PROF EPIDEMIOL & DEAN, SCH PUB HEALTH, UNIV ILL, CHICAGO, 85- *Concurrent Pos:* Exchange scientist to Inst Poliomyelitis & Virus Encephalitis, Moscow, 62; Nat Mult Sclerosis Soc Med Adv Bd, 66-; vpres, Muscular Dystrophy Asn Am, 68-, mem corp, 69-; mem comn geog neurol, World Fedn Neurol, 68-; assoc epidemiol, Sch Hyg & Pub Health, Johns Hopkins Univ, 70-; Amyotrophic Lateral Sclerosis Soc Am Med Advisor, 73-; sci adv comn, Am Found Aging Res, 82-; pres, Gerontol Health Sect, Am Pub Health Asn, 85. *Mem:* Am Epidemiol Soc (pres, 80-81); Am Asn Immunol; Int Epidemiol Asn; fel Am Pub Health Asn; Soc Epidemiol Res. *Res:* Epidemiology of health and disease in aging populations with particular interest in environmental and social influences and the role of genetics, immunity and metabolism. *Mailing Add:* Univ Ill Sch Pub Health 1145 Franklin Ave River Forest IL 60305

BRODY, JEROME IRA, b New York, NY, Jan 24, 28; m 59; c 3. INTERNAL MEDICINE, HEMATOLOGY. *Educ:* NY Univ, AB, 47, AM, 48; Jefferson Med Col, MD, 52; Am Bd Internal Med, dipl, hematol, dipl. *Prof Exp:* Intern, Philadelphia Gen Hosp, 52-53, resident path, 53-54, resident cardiol, 54-55; resident internal med, Grad Hosp, Univ Pa, 57-58; chief hemat sect, Vet Admin Hosp, Coral Gables, Fla, 60-62; from asst prof to assoc prof internal med, Sch Med, Univ Pa, 62-75, dir hemat, Grad Hosp, 62-75; PROF MED, MED COL PA, 75- *Concurrent Pos:* USPHS res fel hemat, Sch Med, Yale Univ, 58-60; asst instr path, Sch Med, Univ Pa, 53-54, asst instr internal med, 54-55, asst physician, Grad Hosp, 62-; asst attend physician, Grace-New Haven Community Hosp, Conn, 58-60 & Jackson Mem Hosp, Miami, Fla, 60-62; asst prof, Sch Med, Univ Miami, 60-62; consult, Variety Children's Hosp, Miami, 61-62; prin investr res grant, USPHS, 62-79, prog dir training grant, 64-74; consult, Walson Army Hosp & Naval Hosp, 63-75; chief med serv, Med Col, Pa Div, Vet Admin Hosp, Philadelphia, 75-79; adj prof med, Med Col, Thomas Jefferson, Univ, 89- *Mem:* AAAS; Am Soc Clin Invest; Am Fedn Clin Res; Am Soc Hemat; Am Asn Pathologists; Am Soc Clin Path. *Res:* Relationship of hemeostasis to cardiovascular disease. *Mailing Add:* Med Col Pa Philadelphia PA 19129

BRODY, JEROME SAUL, b Chicago, Ill, Dec 6, 34; m 55; c 3. PULMONARY DISEASES. *Educ:* Univ Ill, BS, 55, MD, 59. *Prof Exp:* NIH fel, Univ Pa, 65-67, from asst prof to assoc prof med & physiol, 67-73; assoc prof med, 73-78, CHIEF PULMONARY SECT, UNIV HOSP & BOSTON CITY HOSP, 73-; PROF MED, SCH MED, BOSTON UNIV, 78- *Mem:* Fel Am Col Physicians; Am Thoracic Soc; Am Fedn Clin Res; Am Physiol Soc; Am Soc Clin Invest. *Res:* Development, compensatory growth and repair of the mammalian lung viewed from morphologic, physiologic and biochemical points of view. *Mailing Add:* Sch Med Boston Univ 80 E Concord St Boston MA 02118

BRODY, MARCIA, b New York, NY, Dec 3, 29; div; c 3. PHOTOBIOLOGY. *Educ:* Hunter Col, AB, 51; Rutgers Univ, MS, 53; Univ Ill, PhD(biophys), 58. *Prof Exp:* Res asst photosynthesis proj, Univ Ill, 53-58, res assoc, 58-59; res assoc chem, Brandeis Univ, 59-61; from asst prof to assoc prof biol sci, 61-70, chmn dept, 63-69, PROF BIOL SCI, HUNTER COL, 70- *Concurrent Pos:* City Univ New York grant fac award, 61-62, 71-72, 72-73 & 80-81; NSF res grant, 62-65; NSF fel sci, Biochem & Biophysics Panel, 78-80. *Mem:* Biophys Soc; Sigma Xi; NY Acad Sci. *Res:* States of chlorophyll in vivo; light reactions in photosynthesis; low temperature fluorescence spectroscopy; physico-chemical properties of chromoproteins; role of accessory pigments in photosynthesis; organelle structure, enzymology and development; efficiencies of transfer of light energy; phototropism; lectin biochemistry. *Mailing Add:* 111 W 85th St New York NY 10024

BRODY, MICHAEL J, pharmacology; deceased, see previous edition for last biography

BRODY, RICHARD SIMON, b Brooklyn, NY, Oct 20, 50; c 2. ENZYME MECHANISMS. *Educ:* Cornell Univ, BA, 72; Harvard Univ, PhD(chem), 78. *Prof Exp:* Fel, Ohio State Univ, 79-81; asst prof chem, State Univ NY at Buffalo, 81-83; CONSULT, 83- *Mem:* Am Chem Soc; Am Soc Biochem & Molecular Biol. *Res:* Mechanisms of enzymes that synthesize and degrade nucleic acids; protein-nucleic acid interactions; phosphodiesterase inhibitors; model systems. *Mailing Add:* Battelle 505 King Ave Columbus OH 43201

BRODY, SEYMOUR STEVEN, b New York, NY, Nov 29, 27; m 49; c 3. BIOPHYSICS. *Educ:* City Col New York, BS, 51; NY Univ, MS, 53; Univ Ill, PhD(biophys), 56. *Prof Exp:* Res asst photosynthesis, Univ Ill, 53-56, res assoc, 56-59; sect chief photobiol, US Dept Air Force, 59-60; mgr biophys, Watson Lab, Int Bus Mach Corp, 60-65; assoc prof, 66-68, PROF BIOL, NY UNIV, 68- *Concurrent Pos:* Adj assoc prof, Dept Biol, Washington Square Col, NY Univ, 64-65; guest prof, Dept Biol, Jawaharlal Nehru Univ, New Delhi, India, & Dept Biophys, Univ Leiden, Holland, 82; vis scientist, Royal Inst, London, Eng, 79, Carlsberg Res Lab, Copenhagen, 80 & 85, Serv Biophys, Ctr d'Etudes Nucleaires, Saclay, France, 83; guest prof, Inst Basic Biol, Okazaki, Japan, 87; Ecole Normale Superior, Paris, France, 88. *Mem:* AAAS; Biophys Soc; Photo Biol Soc. *Res:* Photobiology; fluorescence and absorption spectroscopy of molecules; heterogeneous photocatalysis; picosecond and nonosecond phenomena in living systems and in molecules; use of monomolecular bilayers and liposomes as films, models of membranes to investigate processes in vision and photosynthesis. *Mailing Add:* Dept Biol NY Univ Wash Square New York NY 10003-6606

BRODY, STUART, b Newark, NJ, June 25, 37; m 65; c 2. BIOCHEMICAL GENETICS. *Educ:* Mass Inst Technol, SB, 59; Stanford Univ, PhD(biol), 64. *Prof Exp:* Guest investr biol, Rockefeller Univ, 64-66, res assoc, 66, asst prof, 66-67; from asst prof to assoc prof, 67-82, PROF BIOL, UNIV CALIF, SAN DIEGO, 82- *Concurrent Pos:* Am Cancer Soc fel, 64-66; prin investr NSF grant, 66-75. *Mem:* AAAS; Am Soc Biol Chemists; Am Soc Microbiol; Genetics Soc Am. *Res:* Biochemical genetics; biochemistry and morphology of microorganisms; circadian rhythms; spore formation and germination; lipid metabolism. *Mailing Add:* Dept Biol Univ Calif San Diego La Jolla CA 92093

BRODY, STUART MARTIN, b Brooklyn, NY, June 25, 36; m 58; c 2. FRACTIONAL DISTILLATION, CHEOMATOGRAPHIC & MASS SPECTROMETRIC TECHNOLOGY. *Educ:* Queens Col NY, BSc, 58. *Prof Exp:* Jr chemist, Pharmaceuticals Co-Ciba, 58-66, lab suprv, 67-68; sr scientist

I, Pharmaceut Div, Ciba-Geigy Corp, 69-80, sr scientist II, 81-83, sr res scientist, 84-85, ASST MGR, PHARMACEUT DIV, CIBA-GEIGY CORP, 85- *Mem:* Fel Am Inst Chemists; NY Acad Sci; Asn Off Anal Chemists; Am Soc Mass Spectrometry. *Res:* Develop methodologies for analytical evaluation of chemical reaction mixtures to elucidate chemical structures; analytical systems for on-line assay of complex reaction mixtures using state-of-the-art instrumentation and computer technology; analytical chemistry research relevant to the drug discovery and development process. *Mailing Add:* Eight Timberlane Dr Colonia NJ 07067

BRODY, THEODORE MEYER, b Newark, NJ, May 10, 20; m 47; c 4. PHARMACOLOGY. *Educ:* Rutgers Univ, BS, 43; Univ Ill, MS, 49, PhD(pharmacol), 52. *Prof Exp:* From instr to prof pharmacol, Med Sch, Univ Mich, 52-66; chmn, Dept Pharmacol & Toxicol, Mich State Univ, 66-86, PROF PHARMACOL, MICH STATE UNIV, 66- *Concurrent Pos:* Mem, NIH Fel Rev Panel Pharmacol & Endocrinol, 64-68 & Nat Acad Sci-Nat Res Coun Pesticide Safety Adv Comt, 64-66; mem, Pharmacol & Exp Therapeut Study Sect, NIH, 69-73; mem, Bd Dirs, Fedn Socs Exp Biol, 73-76; US rep, Int Union Pharmacol, 73; NSF distinguished scholar lectr, Univ Hawaii, 74; consult, Random House Dict Eng Lang, 64-; Basic Sci Coun, Am Heart Asn, 86-; mem, Pharmacol Adv Bd, Pharmaceut Mfrs Asn Found; mem, Drug Abuse Rev Comt, Nat Inst Drug Abuse, 75-79. *Mem:* John Jacob Abel Award in Pharmacol, Am Soc Pharmacol & Exp Therapeuts, 55; Soc Neurosci; Soc Toxicol; Int Soc Biochem Pharmacol; Int Soc Heart Res; Am Heart Asn; Sigma Xi; Int Soc Heart Res. *Res:* Mode of action of drugs and poisons; cardiac glycosides; ion transport. *Mailing Add:* Dept Pharmacol & Toxicol Life Sci Bldg Mich State Univ East Lansing MI 48824

BRODZINSKI, RONALD LEE, b South Bend, Ind, Feb 14, 41; m 67; c 3. NUCLEAR INSTRUMENTATION & APPLICATIONS. *Educ:* Purdue Univ, BS, 63, PhD(nuclear chem), 68. *Prof Exp:* Sr res scientist, Pac Northwest Lab, Battelle Mem Inst, 68-80, staff scientist, 80-86, mgr, nuclear chem, 86-89, SR STAFF SCIENTIST, PAC NORTHWEST LAB, BATTELLE MEM INST, 89- *Concurrent Pos:* Prin investr, NASA, 72- *Mem:* Am Chem Soc. *Res:* High-energy charged particle reactions; cosmic-radiation activation of astronauts and spacecraft; nuclear waste management; environmental research; lunar and space sciences; controlled thermonuclear reactor materials research; nuclear instrument and technique development and application; fundamental particle physics. *Mailing Add:* Pac Northwest Labs PO Box 999 Richland WA 99352

BROECKER, WALLACE, b Chicago, Ill, Nov 29, 31; m 52; c 6. GEOCHEMISTRY. *Educ:* Columbia Col, AB, 53; Columbia Univ, MA, 56, PhD(geol), 58. *Prof Exp:* From instr to prof, 56-77, dir, Geochem Lab, 66-80, NEWBERRY PROF GEOL, COLUMBIA UNIV, 77- *Concurrent Pos:* Sloan fel, 64. *Honors & Awards:* Ewing Medal, Am Geophys Union; Day Medal, Geol Soc Am, 84; Huntsman Award, Bedford Inst Can, 85; Vetlessen Award, 87; Goldschmidt Award, 88. *Mem:* Nat Acad Sci; Am Geophys Union; Geochem Soc; Am Acad Sci. *Res:* Pleistocene geochronology; carbon 14 and thorium 230 dating; chemical oceanography, including oceanic mixing based on radioisotope distribution; author of 10 technical publications and 4 books. *Mailing Add:* Dept Geol Sci Lamont-Doherty Geol Observ Columbia Univ Palisades NY 10964

BROEG, CHARLES BURTON, b Princeton, Ind, Mar 15, 16; m 46. CHEMISTRY, AGRICULTURE & FOOD CHEMISTRY. *Educ:* DePauw Univ, AB, 38; Okla Agr & Mech Col, MS, 40. *Prof Exp:* Asst, Okla Agr & Mech Col, 38-40; chemist, USDA, 41-43 & 46-58; head prod planning & develop, 58, dir tech serv, 58-61, vpres & tech dir, SuCrest Corp, 61-77, VPRES RES & TECH SERV, REVERE SUGAR CORP, 78-, PRES, APPL SUGAR LABS, INC, 71- *Concurrent Pos:* Dir, World Sugar Res Orgn, Adv, Sugar Comn Codex Alimentarius. *Mem:* Am Chem Soc; Am Soc Sugar Beet Technologists; Am Soc Sugar Cane Technologists; Am Inst Chemists; Inst Food Technologists. *Res:* Cane wax; cane juice clarification; polynitro aromatic-benzene addition compounds; color standards for sugar products; direct compression vehicles from sugar; desugaring molasses; agglomeration processes and products; new product development. *Mailing Add:* 101 Victoria Court Houma LA 70360

BROEMELING, LYLE DAVID, b Juneau, Alaska, Mar 17, 39; m 82; c 1. BAYESIAN STATISTICS. *Educ:* Tex A&M Univ, BA, 60, MS, 63, PhD(statist), 66. *Prof Exp:* Asst prof statist, Tex A&M Univ, 66-67; asst prof, Univ Tex, Houston, 67-68; from asst prof to prof, Okla State Univ, 68-85; sci officer, Off Naval Res, 85-87; PROF SURG, MED BR, SCHRINERS BURN INST, UNIV TEX, 87- *Concurrent Pos:* Hon res fel, Univ Col London, 73-74; vis prof, Nat Autonomous Univ Mex, 81 & Tex A&M Univ, 83-84; prin investr, Okla State Univ, 83-85. *Honors & Awards:* H O Hartley Award, Dept Statist, Tex A&M Univ, 85. *Mem:* Am Statist Asn. *Res:* Bayesian statistical inferences for various parametric models, mixed models, dynamic stochastic systems and time series. *Mailing Add:* Schriners Burn Inst Univ Tex 610 Texas Ave Galveston TX 77550

BROENE, HERMAN HENRY, b Grand Rapids, Mich, Dec 28, 19; m 44; c 3. CHEMISTRY. *Educ:* Calvin Col, AB, 42; Purdue Univ, PhD(phys chem), 47. *Prof Exp:* Asst phys chem, Purdue Univ, 42-44; res chemist, Eastman Kodak Co, 46-56; from assoc prof to prof, 56-84, EMER PROF CHEM, CALVIN COL, 84- *Mem:* Am Chem Soc. *Res:* Radio-chemistry; chemistry of water; thermodynamics of aqueous hydrofluoric acid solutions; mechanical properties of high polymers; chemical education. *Mailing Add:* 2601 Golfridge Dr SE Calvin Col Grand Rapids MI 49546-5616

BROER, MATTHIJS MENO, b Madison, Wis, Aug 23, 56. OPTICAL FIBER MATERIALS, SOLID STATE LASER MATERIALS. *Educ:* Tech Univ, Delft, The Netherlands, BSc, 77, Univ Wis, MSc, 79, PhD(physics), 82. *Prof Exp:* Postdoctoral mem tech staff, 83-85, MEM TECH STAFF, RES, AT&T BELL LABS, MURRAY HILL, NJ, 85- *Mem:* Am Phys Soc; Optical Soc Am; Mat Res Soc. *Res:* Nonlinear optics in condensed matter; optical properties of amorphous systems. *Mailing Add:* AT&T Bell Labs Murray Hill NJ 07974

BROERMAN, F S, b Oskaloosa, Iowa, Jan 5, 38; m 61; c 2. RESEARCH ADMINISTRATION, FORESTRY. *Educ:* Iowa State Univ, BS, 60, MS, 65. *Prof Exp:* Res forester, 65-68, sr res forester, 68-72, proj leader, 72-81, MGR WOODLANDS RES DEPT, UNION CAMP CORP, 81- *Concurrent Pos:* Res asst, Iowa Agr Exp Sta, 63-65. *Mem:* Soc Am Foresters; Soil Sci Soc Am. *Res:* Research administration and forestry. *Mailing Add:* Woodlands Res Dept Union Camp Corp PO Box 1391 Savannah GA 31402

BROERS, ALEC N, b Calcutta, India, Sept 17, 38; m 64; c 2. ELECTRICAL ENGINEERING, PHYSICS. *Educ:* Univ Melbourne, BSc, 59; Cambridge Univ, BA, 61, PhD(elec eng), 65. *Prof Exp:* Mem prof staff electron optics, Thomas J Watson Res Ctr, 65-67, mgr electron beam technol, 67-74, mgr photon & electron optics, IBM res, 74-81, MGR LITHOGRAPHY SYST & TECHNOL TOOLS, E FISHKILL FACIL, IBM CORP, 81- *Concurrent Pos:* IBM fel, 77. *Honors & Awards:* Award for Best Tech Paper, Nat Electronics Conf, 69. *Mem:* Electron Micros Soc Am. *Res:* Development of precision electron optical equipment and its application to the production of microcircuitry and microrecording of digital information, especially new cathodes and cathode materials, ion etching and high resolution scanning electron microscopy; general aspects of high resolution lithography for microcircuits. *Mailing Add:* c/o IBM Evelyn Marino T J Watson Res Ctr PO Box 218 Yorktown Heights NY 10598

BROERSMA, DELMAR B, b Lynden, Wash, July 2, 34; m 60; c 3. ENTOMOLOGY. *Educ:* Calvin Col, AB, 56; Syracuse Univ, MS, 63; Clemson Univ, PhD(entom), 65. *Prof Exp:* Teacher, Grand Rapids Christian High Sch, 56-57 & Western Mich Christian High Sch, 58-62; asst entom & zool, Clemson Univ, 63-65; res entomologist, Ill Natural Hist Surv, 65-69; asst prof, 69-71, ASSOC PROF ENTOM, PURDUE UNIV, WEST LAFAYETTE, 71- *Mem:* AAAS; Entom Soc Am. *Res:* Insect pathology; economic entomology. *Mailing Add:* 1836 Locust Lane West Lafayette IN 47906

BROERSMA, SYBRAND, b Harlingen, Netherlands, Sept 20, 19; nat US. MOLECULAR PHYSICS. *Educ:* Leiden Univ, Candidaats, 39, Doctoraal, 41; Delft Inst Technol, DSc(physics), 47. *Prof Exp:* Asst, Leiden Univ & Delft Inst Technol, 39-46; int exchange fel, Northwestern Univ, 47; instr, Univ Toronto, 48; prof exp physics, Univ Indonesia, 49-51; asst prof physics, Northwestern Univ, 52-59; PROF PHYSICS, UNIV OKLA, 59- *Mem:* Am Phys Soc; Netherlands Phys Soc; Europ Phys Soc. *Res:* Magnetism; hydrodynamics; molecular spectroscopy. *Mailing Add:* c/o Paul Hendrickson 512 S Stevenson Olathe KS 66061

BROFAZI, FREDERICK R, b Jersey City, NJ, Apr 21, 33; m 56; c 3. ANALYTICAL CHEMISTRY, PHARMACEUTICAL CHEMISTRY. *Educ:* Rutgers Univ, BS, 54, MS, 57, PhD(chem), 59. *Prof Exp:* Res anal chemist, Nat Cash Register Co, 59-60; res supvr, 60-64, group leader, 64-66, from asst dir to dir qual control, 66-71, EXEC DIR QUAL CONTROL, PHARMACEUT DIV, CIBA-GEIGY CORP, 71- *Mem:* Am Chem Soc; Am Pharmaceut Asn; Am Soc Qual Control; Sigma Xi. *Res:* Development of analytical methods; pharmaceutical analysis; quality control. *Mailing Add:* 72 Alpine Terr Hillsdale NJ 07642

BROGAN, GEORGE EDWARD, b La Jolla, Calif, Feb 18, 44; m 68; c 2. GEOLOGY, ENGINEERING GEOLOGY. *Educ:* San Diego State Univ, BA, 66, MS, 69. *Prof Exp:* Sr proj geologist, Woodward-Clyde Consults, 71-75, dep dir eng geol, 76-, vpres, 78-; AT APPL GEOSCI, INC, TUSTIN. *Concurrent Pos:* Prin investr grants, US Geol Surv & Earthquake Eng Res Inst, 77-81; adv, US Geol Surv Earthquake Hazard Reduction Prog, 77-; mem prof affairs comt, Calif State Bd Regist for Geologists & Geophysicists, 78-; mem, Calif State Mining Bd, 81- *Mem:* Am Geophys Union; Asn Eng Geologists; Earthquake Eng Res Inst; Int Asn Eng Geol; Seismol Soc Am; Sigma Xi. *Res:* Evaluation of youthful seismic activity of faults in diverse geographic and climatic environments, as deduced from the geologic record. *Mailing Add:* Geomatrix Consults 3505 Cadillac Ave No P-201 Costa Mesa CA 92626

BROGAN, WILLIAM L, b Vail, Iowa, Mar 11, 35; m 55; c 6. ELECTRICAL & MECHANICAL ENGINEERING. *Educ:* Univ Iowa, BSME, 58; Univ Calif, Los Angeles, MS, 61, PhD(control theory), 65. *Prof Exp:* Thermodyn engr, Gen Dynamics/Convair, Calif, 58-59; mem tech staff struct dynamics, Hughes Aircraft Co, 59-61, mem tech staff control systs, 61-65; mem tech staff guid & control systs, Aerospace Corp, 65-68; assoc prof systs theory, 68-72, PROF SYSTS THEORY, UNIV NEBR, LINCOLN, 72- *Concurrent Pos:* Lectr, Univ Calif, Los Angeles, 66-68; consult, Brunswick Corp, Nebr, 68-; dir systs sci prog, NSF, Washington, DC, 78-79. *Mem:* Inst Elec & Electronics Engrs. *Res:* Control and systems theory; applied mathematics; dynamics. *Mailing Add:* Dept Elec Eng 243 N WSec Univ Nebr Lincoln NE 68508

BROGDON, BYRON GILLIAM, b Ft Smith, Ark, Jan 22, 29; m 51, 78; c 3. DIAGNOSTIC RADIOLOGY. *Educ:* Univ Ark, BS & BSM, 51, MD, 52. *Prof Exp:* Asst prof radiol, Col Med, Univ Fla, 60-63; assoc prof radiol & radiol sci, Sch Hyg & Pub Health, Sch Med, Johns Hopkins Univ, 63-67; asst dean hosp affairs, Sch Med, Univ NMex, 70-72, prof radiol, 67-78, chmn dept, 67-77; asst dean continuing med educ, 81-89, prof radiol, 78-89, DISTINGUISHED PROF RADIOL, COL MED, UNIV SOUTH ALA, 89-, CHMN DEPT, 85- *Concurrent Pos:* Radiologist-in-chg, Div Diag Roentgenol, Johns Hopkins Hosp, 63-67; med dir, Bernalillo County Med Ctr, 69-72; pres, Southern Radiol Conf, 67-68, secy, 84-; secy-treas, Soc Chm Acad, Radiol Dept, 69-70; mem, House Deleg, Am Med Asn, 88- *Honors & Awards:* Physician Speaker Award, AMA, 79; Gold Medal, Asn Univ Radiologists, 85, Am Col Radiol, 87. *Mem:* Am Roentgen Ray Soc (vpres, 79-80); Radiol Soc NAm; Soc Pediat Radiol; Asn Univ Radiologists (pres, 73-74); Am Col Radiol (pres, 78-79); Int Skeletal Soc; Am Acad Forensic Sci. *Res:* Diagnostic roentgenology; clinical applications; determination of normal and abnormal variations; visual perception, forensic radiology. *Mailing Add:* Dept Radiol Univ S Ala Med Ctr Mobile AL 36617

BROGE, ROBERT WALTER, b Cleveland, Ohio, Oct 27, 20; m 44; c 5. INTERNATIONAL BUSINESS DEVELOPMENT. *Educ:* Harvard Univ, SB, 42; Cornell Univ, PhD(phys chem), 48. *Prof Exp:* Res assoc rocket res, Nat Defense Res Comt, 42-45; res chemist, Procter & Gamble Co, 48-51, sect head basic res, 51-57 & prod res, 57-61, assoc dir, 61-67, dir prod develop, 67-71, dir res, 71-78, int tech dir, 78-83; RETIRED. *Mem:* AAAS; Am Chem Soc; Int Asn Dent Res. *Res:* X-ray crystallography; solutions of surface active agents; dental research; chemistry of keratin; health care product development; pharmaceutical clinical research. *Mailing Add:* 221 Compton Ridge Dr Wyoming OH 45215

BROGLE, RICHARD CHARLES, b Boston, Mass, July 12, 27; m 59; c 3. HEALTH CARE PRODUCTS. *Educ:* Mass Inst Technol, SB, 50, SM, 53, PhD(food sci), 60. *Prof Exp:* Sr chemist, Am Chicle Co, 60-62; group leader contract res, Am Chicle Co, Div Warner-Lambert Co, 63-66; dir clin res, Warner-Lambert Co, 66-69, dir proprietary toiletries res, 69-72, dir clin regulatory affairs, 72-75, vpres res serv, 75-77; vpres lab opers, Block Drug Co, 78-83, vpres-dir res & develop, 83-87; PRES, RICHARD C BROGLE ASSOCS, 87- *Concurrent Pos:* Chmn subcomt, Proprietary Asn, 80-86. *Mem:* AAAS; Inst Food Technologists; Acad Pharmaceut Sci; Soc Cosmetic Chemists; Am Soc Clin Pharmacol & Therapeut; Int Asn Dent Res; Am Asn Pharmaceut Scientists. *Res:* Antacid in vitro-in vivo relationships; objective clinical testing methods; control of dental plaque; research management. *Mailing Add:* Eight Kenneth Rd Upper Montclair NJ 07043

BROIDA, THEODORE RAY, b Louisville, Ky, Dec 6, 28; m 52; c 2. RESEARCH ADMINISTRATION. *Educ:* Univ Calif, BS, 50, MS, 52. *Hon Degrees:* LLD, Transylvania Univ, 86. *Prof Exp:* Physicist, Nucleonics Div, US Naval Radiol Defense Lab, 50-53, Broadview Res Corp, 53-55 & Defense Atomic Support Agency, 55-57; sr opers analyst, Stanford Res Inst, 57-60; dir planning, Broadview Res Corp, 60-62; mgr techno-econ res div, Spindletop Res, Inc, 62-68, pres, 68-73; PRES, QRC RES CORP, 73- *Concurrent Pos:* Consult, US Off Educ, 68-; mem comt remote sensing agr, Nat Acad Sci-Nat Res Coun; consult innovation develop, Off Sci & Technol, US Dept Com, 80-81. *Honors & Awards:* Transylvania Medal, 83. *Mem:* Asn Comput Mach; Nat Soc Prof Engrs. *Res:* High intensity thermal radiation sources and instrumentation; measurement of air temperature; logistics; inventory management; electronic data processing; reconnaissance systems; applied industrial and regional economics; resource development; communications systems; research management. *Mailing Add:* 290 S Ashland Ave Lexington KY 40502-1728

BROIDO, ABRAHAM, b Cherkassi, Russia, Sept 12, 24; nat US; wid; c 2. PHYSICAL CHEMISTRY. *Educ:* Univ Chicago, SB, 43; Univ Calif, PhD(chem), 50. *Prof Exp:* Chemist, Metall Lab, Univ Chicago & Argonne Nat Lab, 43-46; res chemist, Clinton Labs & Oak Ridge Nat Lab, 46-48; chemist, Radiation Lab, Univ Calif, 48-50 & US Naval Radiol Defense Lab, 50-56; chemist, Pac Southwest Forest & Range Exp Sta, US Forest Serv, Berkeley, 56-76; RETIRED. *Concurrent Pos:* Guest scientist, Hebrew Univ, Jerusalem, 64-65; res assoc, Statewide Air Pollution Res Ctr, Univ Calif, 66-, consult, Inst Eng Res, 56-; head thermal radiation br, US Naval Radiol Defense Lab, 51-53; mem, Calif Gov Radiol Defense Adv Comt, 62-66; Nat Acad Sci Adv Comt Civil Defense, 66-70. *Res:* Radiochemistry; ion exchange and solvent extraction; microchemistry; rare earth and transuranic elements; effects of thermal and nuclear radiation; forest fire and air pollution research. *Mailing Add:* 1936 Carquinez Ave Richmond CA 94805

BROIDO, JEFFREY HALE, b New York, NY, Oct 22, 34; m 61; c 2. NUCLEAR PHYSICS, MECHANICAL ENGINEERING. *Educ:* Columbia Univ, AB, 55, BS, 56; Stanford Univ, MS, 60, Engr, 62. *Prof Exp:* Res engr, Beloit Corp, 61-62; PHYSICIST, GEN ATOMIC, INC, SAN DIEGO, 62- *Concurrent Pos:* Consult, Swiss Fed Inst Reactor Res, 68-69. *Mem:* Am Nuclear Soc. *Res:* Economics, dynamics and control of fast breeder nuclear reactors. *Mailing Add:* 8811 Robinhood Lane La Jolla CA 92037

BROIN, THAYNE LEO, b Kenyon, Minn, Sept 18, 22; m 49; c 3. GEOLOGY. *Educ:* St Cloud State Col, BS, 43; Univ Colo, MA, 52, PhD(geol), 57. *Prof Exp:* Asst physics, St Cloud State Col, 42, asst biol, 43; from instr to asst prof geol, Colo State Univ, 50-57; res geologist, Cities Serv Res & Develop Co, 57-65, supvr geol div, 61-65; res coordr, Explor Div, Cities Serv Oil Co, 65-72, chief comput geologist, Explor Technol Dept, 72-78, comput technol mgr, Energy Res Group, Cities Serv Co, 78-83; GEOL CONSULT, 83- *Mem:* Soc Econ Paleontologists & Mineralogists; Geol Soc Am; Am Asn Petrol Geologists; Sigma Xi. *Res:* Sedimentation; stratigraphy; subsurface facies mapping; sedimentary petrology; petroleum geology; computer mapping. *Mailing Add:* 5280 Champagne Dr Colorado Springs CO 80919

BROITMAN, SELWYN ARTHUR, b Boston, Mass, Aug 30, 31; m 53; c 2. MICROBIOLOGY, MEDICAL EDUCATION. *Educ:* Univ Mass, BS, 52, MS, 53; Mich State Univ, PhD(microbiol), 56. *Prof Exp:* Res instr path, 63-64, from asst prof to assoc prof microbiol, 65-75, PROF MICROBIOL & NUTRIT SCI, SCH MED, BOSTON UNIV, 69-, ASST DEAN ADMIS, 81-, PROF PATH, 82- *Concurrent Pos:* Res assoc gastrointestinal res lab, Mallory Inst Path, Boston City Hosp, 56-74, res assoc nutrit path univ, 75-; assoc med, Thorndike Mem Lab, 69-; assoc med, Harvard Med Sch, 69-74; founding mem, Digestive Dis Found, 71 & Nutrit Today Soc, 74; lectr, Div Allied Health Professions, Northeastern Univ, 71-73; lectr, Sargent Col Allied Health Professions, Boston Univ, 72-76, consult, Nat Large Bowel Cancer Proj, 77-; mem comt diet, nutrit & cancer, Nat Acad Sci, 81-83; mem bd dirs, Mentor Corp, 84-87. *Mem:* Am Gastroenterol Soc; Am Soc Microbiol; Brit Soc Appl Bact; Soc Exp Path; NY Acad Sci; Sigma Xi; Am Inst Nutrit; Soc Exp Biol & Med; Am Fedn Clin Res; AAAS; fel Am Col Gastroenterol, 88. *Res:* Experimental gastroenterology; malabsorptive diseases; colon cancer; intestinal flora; bacterial and viral enteric infections. *Mailing Add:* Dept Microbiol Boston Univ Sch Med Boston MA 02118

BROKAW, BRYAN EDWARD, b Pittsfield, Ill, Feb 27, 49; m 70; c 3. ANIMAL BREEDING. *Educ:* Abilene Christian Univ, BS, 71; Ore State Univ, PhD(animal breeding & genetics), 75. *Prof Exp:* asst prof animal sci, 75-80, CHMN & ASSOC PROF, DEPT AGR, ABILENE CHRISTIAN UNIV, 80- *Mem:* Am Soc Animal Sci; Sigma Xi; Nat Asn Cols Teachers Agr. *Res:* Relationships of physiological characteristics with performance traits in beef cattle and their value as selection criteria. *Mailing Add:* Abilene Christian Univ Sta PO Box 7986 Abilene TX 79699

BROKAW, CHARLES JACOB, b Camden, NJ, Sept 12, 34; m 55; c 2. CELL BIOLOGY. *Educ:* Calif Inst Technol, BS, 55; Univ Cambridge, PhD(zool), 58. *Prof Exp:* Res assoc biol div, Oak Ridge Nat Lab, 58-59; asst prof zool, Univ Minn, 59-61; from asst prof to assoc prof biol, 61-68, exec officer, 76-80, & 85-89, assoc chmn, 80-85, PROF BIOL, CALIF INST TECHNOL, 68- *Concurrent Pos:* Guggenheim fel, Univ Cambridge, 70-71; chmn, NIH Cellular & Molecular Dis Rev Comt, 85-87. *Mem:* AAAS; Am Soc Cell Biol; Biophys Soc. *Res:* Motility and behavior of ciliated and flagellated cells; chemotaxis: experimental work with marine invertebrate spermatozoa and Chlamydomonas; computer simulation of theoretical models. *Mailing Add:* Div Biol 156-29 Calif Inst Technol Pasadena CA 91125

BROKAW, GEORGE YOUNG, b St Clairsville, Ohio, Sept 11, 21; m 44; c 4. APPLIED CHEMISTRY. *Educ:* Ohio Wesleyan Univ, BA, 41; Ohio State Univ, MS, 44, PhD(chem), 47. *Prof Exp:* Asst chem, Ohio State Univ, 41-43, asst, Univ Res Found, 43-44; chem engr, Naval Res Lab, Washington, DC, 44-45; res assoc, Distillation Prod Industs Div, Eastman Kodak Co, 47-63, head develop labs, 63-69; sr res assoc, Tenn Eastman Co, 69-76, coordr prod safety, 76-84; RETIRED. *Mem:* AAAS; fel Am Chem Soc. *Res:* Fats and oils; food emulsifiers; fat-soluble vitamins. *Mailing Add:* 4421 Chickasaw Rd Kingsport TN 37664

BROKAW, RICHARD SPOHN, b Orange, NJ, Mar 26, 23; m 47; c 4. PHYSICAL CHEMISTRY. *Educ:* Swarthmore Col, AB, 43; Princeton Univ, AM, 49, PhD(chem), 51. *Prof Exp:* Asst phys chem, Calco Chem Div, Am Cyanamid Co, 43-44 & Princeton Univ, 46-52; aeronaut res scientist phys chem & combustion, Lewis Res Ctr, NASA, 52-55, from assoc head to head, Combustion Fundamentals Sect, 55-57, chief, Phys Chem Br, 57-71, chief, Physics & Chem Div, 71-72, asst chief, Phys Sci Div, 72-73; lectr chem, Baldwin-Wallace Col, 74-82; RETIRED. *Concurrent Pos:* Consult, Am Gas Asn Labs, 75-84. *Honors & Awards:* Medal for Except Sci Achievement, NASA, 72. *Mem:* AAAS; Am Chem Soc; Combustion Inst; Am Inst Aeronaut & Astronaut; Sigma Xi. *Res:* Combustion, especially burning velocity and spontaneous ignition; transport properties of gases, especially heat transport in chemically reacting gases; gas phase chemical kinetics. *Mailing Add:* Seven Beacon Hill Sharon VT 05065-9611

BROKER, THOMAS RICHARD, b Hackensack, NJ, Oct 22, 44; m 74. MOLECULAR GENETICS, TUMOR VIROLOGY. *Educ:* Wesleyan Univ, BA, 66; Stanford Univ, PhD(biochem), 72. *Hon Degrees:* MA, Wesleyan Univ, 81. *Prof Exp:* Res fel chem, Calif Inst Technol, 72-75; sr staff investr, Colo Spring Harbor Lab, 75-79, sr scientist, electron micros, 79-84; CONSULT, 84- *Concurrent Pos:* Res fel, Helen Hay Whitney Found, 72-75. *Mem:* Sigma Xi; Am Soc Microbiol. *Res:* RNA transcription and DNA replication of human papillomaviruses; mechanisms of carcinogenesis; genetic structure, expression and regulation of human papillomavirus; diagnostic and therapeutic approaches to HPV infections of genital tract epithelium. *Mailing Add:* Biochem Dept Sch Med Univ Rochester 601 Elmwood Ave PO Box 607 Rochester NY 14642

BROLIN, ROBERT EDWARD, b Holland, Mich, Apr 12, 48; m 70; c 2. GASTROINTESTINAL SURGERY. *Educ:* DePauw Univ, BA, 70; Univ Mich, Ann Arbor, 74. *Prof Exp:* Asst prof surg, Rutgers Med Sch, Univ Med & Dent NJ, 80-84; assoc prof surg, 84-89, CHIEF GASTROINTESTINAL SURG, UNIV MED & DENT NJ, ROBERT WOOD JOHNSON MED SCH, 88-, PROF SURG, 89- *Mem:* Asn Acad Surg; Am Col Surgeons; Am Soc Bariatric Surg; Am Motility Soc; Soc Univ Surgeons; NAm Asn Study Obesity; Soc Surg Alimentary Tract; Am Soc Parenteral & Enternal Nutrit. *Res:* Gastrointestinal motility in relationship to ischemic bowel disease and postoperative ileus; surgery for morbid obesity. *Mailing Add:* Dept Surg CN19 UMDNJ-Robert Wood Johnson Med Sch New Brunswick NJ 08903-0019

BROLLEY, JOHN EDWARD, JR, b Chicago, Ill, Jan 15, 19; m 46. ELEMENTARY PARTICLE PHYSICS. *Educ:* Univ Chicago, SB, 40; Ind Univ, MS, 48, PhD(physics), 49. *Prof Exp:* Mem staff, Res Elem Particle Physics & Astrophysics, Los Alamos Nat Lab, 49-83; RETIRED. *Mem:* Fel AAAS; fel Am Phys Soc; Archeol Inst Am; Am Astron Soc; Am Geophysical Union; Sigma Xi. *Res:* Nuclear and particle physics; laser physics; artificial intelligence; information theory, timeseries. *Mailing Add:* 1225 Los Pueblos Los Alamos NM 87544

BROLMANN, JOHN BERNARDUS, b Holland, Nov 20, 20; m 52; c 4. PLANT BREEDING. *Educ:* State Agr Univ Wageningen, MS, 52; Univ Fla, PhD(agron), 68. *Prof Exp:* Plant breeder, Exp Sta Mas d'Auge, France, 52-60; ASST PROF AGRON & ASST LEGUME BREEDER, AGR EXP STA, UNIV FLA, 69- *Mem:* Am Soc Agron; Crop Sci Soc Am. *Res:* Selection and breeding of tropical and temperature climate legumes for pasture use. *Mailing Add:* Agr Res Ctr Box 248 Ft Pierce FL 34954-0248

BROM, JOSEPH MARCH, JR, b Petersburg, Va, Oct 8, 42; m 67; c 2. PHYSICAL CHEMISTRY, MOLECULAR SPECTROSCOPY. *Educ:* Col St Thomas, BS, 64; Iowa State Univ, PhD(phys chem), 70. *Prof Exp:* Res assoc chem, Univ Fla, 70-73; asst res physicist, Univ Calif, Santa Barbara, 73-75; asst prof, 75-80, ASSOC PROF CHEM, BENEDICTINE COL, 80- *Mem:* Am Chem Soc; Am Phys Soc. *Res:* Magnetic resonance and optical spectra of trapped radicals; laser chemistry; high temperature chemistry. *Mailing Add:* Col St Thomas Box 5032 2115 Summit St Paul MN 55105-1048

BROMAGE, PHILIP R, anesthesiology, for more information see previous edition

BROMBERG, ELEAZER, b Toronto, Ont, Oct 7, 13; nat US; m 48; c 3. SCIENTIFIC COMPUTING. *Educ:* City Col New York, BS, 33; Columbia Univ, MA, 35; NY Univ, PhD(math), 50. *Prof Exp:* Asst demonstr physics, Univ Toronto, 36-37; instr, Worcester Polytech Inst, 41-42; res mathematician, Appl Math Panel, Nat Defense Res Comt, Columbia Univ, 43-44, NY Univ, 44-45; proj engr, Reeves Instrument Corp, 45-50 & 52; head mech br, Off Naval Res, 50-53; admin dir comput ctr & assoc prof math, 53-57, admin mgr, Courant Inst Math Sci, 57-59, asst dir, 59-66, vchancellor acad affairs, 70-72, dep chancellor, 72-75, prof, 57-79, EMER PROF APPL MATH & COMPUT, NY UNIV, 79-; SCI ADV & MATHEMATICIAN, CTR COMPUT & APPL MATH, NAT INST STANDARDS & TECHNOL, WASHINGTON, DC, 79- *Concurrent Pos:* Res Scientist, Los Alamos Sci Lab, 56-57; mem bd trustees, Soc Indust & Appl Math, 74-77. *Mem:* Fel AAAS; Soc Indust & Appl Math; Asn Comput Mach (secy, 53-56); Am Math Soc; Math Asn Am. *Res:* Nonlinear vibrations; elasticity; computing techniques; wave motion; simulation techniques; computer operations analysis. *Mailing Add:* Admin A-438 Nat Inst Standards & Technol Gaithersburg MD 20899

BROMBERG, J PHILIP, b Brooklyn, NY, Mar 25, 36; m 60; c 2. PHYSICAL CHEMISTRY. *Educ:* Mass Inst Technol, BS, 56; Calif Inst Technol, MS, 59; Univ Chicago, PhD(chem), 64; Duquesne Univ, JD, 77. *Prof Exp:* Engr, Semiconductor Div, Westinghouse Elec Corp, Pa, 59-60; from asst prof to assoc prof chem, Carnegie-Mellon Inst, 64-76; Westinghouse Environ Systs Dept, 76-77; CONSULT, 77- *Concurrent Pos:* Lectr, Univ Pittsburgh, 65-68; practicing atty, 77- *Mem:* Am Chem Soc; Am Bar Asn; Am Phys Soc. *Res:* Nuclear magnetic resonance; paramagnetic susceptibility; electron impact spectra; environmental problems; atmospheric pollution; legal. *Mailing Add:* 906 Grant Bldg Pittsburgh PA 15219

BROMBERG, MILTON JAY, b Brooklyn, NY, Sept 15, 23; div; c 1. INDUSTRIAL ORGANIC CHEMISTRY. *Educ:* City Col NY, BS, 44; Columbia Univ, MS, 48. *Prof Exp:* Asst res chemist, Columbia Univ, 43-44, asst, 46-48; chief chemist, H A Brassert Co, 48-51; chief chemist, Explosive Div, Olin Mathieson Chem Corp, 51-54, chief prod engr, 54-55, tech mgr, 55-57, proj mgr, High Energy Fuels Div, 57-58, mgr qual control, 58-60, staff consult org div, 60-62, mgr qual assurance, Chem Div, Doe Run Plant, 62-72, mgr qual assurance, Chem Group, Lake Charles Complex, Olin Corp, 72-78; RETIRED. *Mem:* Fel Am Inst Chemists. *Res:* Chemical management; quality control programs; fluoro-aromatics; petro chemicals; toluene diisocyanate; polypropylene glycols; explosives; high energy fuels; design of laboratories. *Mailing Add:* PO Box 5274 Lake Charles LA 70606

BROMBERG, R(OBERT), b Phoenix, Ariz, Aug 6, 21; m 43; c 3. ENGINEERING. *Educ:* Univ Calif, BS, 43, MS, 45; Los Angeles, PhD, 51. *Prof Exp:* Jr engr, Univ Calif, 43-49, chmn's rep eng res & assoc engr, Los Angeles, 49-52, chmn's rep eng, res & asst prof, 52-53, asst dir, Inst Industs Co-op & assoc prof eng, 53-54; mem tech staff, Ramo-Wooldridge, Inc, 54-58, assoc dir, Astrosci Lab, TRW Space Technol Labs, 58-59, dir, Propulsion Lab, 59-60, dir, Mech Div, 61-65, vpres, Mech Div, 62, vpres & gen mgr, Appl Technol Div, 71-72, vpres res & eng, TRW Electronics & Defense, 72-82; RETIRED. *Concurrent Pos:* Consult, 82- *Mem:* Nat Acad Eng; Fel AAAS; Fel Am Inst Aeronaut & Astronaut; Am Soc Eng Educ; Sigma Xi. *Res:* Propulsion research and development; heat transfer and fluid mechanics; thermal and luminous radiation; instrumentation; hypersonic boundary layer phenomena; combustion; engineering and research management. *Mailing Add:* 1001 Westholme Ave Los Angeles CA 90024

BROMBERGER, SAMUEL H, b New York, NY, Dec 21, 41; m 66; c 1. EXPLORATION GEOLOGY. *Educ:* City Col New York, BS, 62; Univ Iowa, MS, 65, PhD(geol), 68. *Prof Exp:* From instr to asst prof geol, Cornell Col, 67-69; geologist, Los Angeles Div, Texaco, Inc, 69-73, regional geologist, Texaco Prod Serv Ltd, London, 73-76; med div staff, 76-80, MGR, GULF COAST EXPLOR, OIL & GAS DIV, GULF COAST DIST, J M HUBER CORP, 80- *Mem:* AAAS; Mineral Soc Am; Soc Econ Paleontologists & Mineralogists; Can Soc Petrol Geologists; Am Asn Petrol Geologists. *Res:* Stratigraphy, hydrocarbon potential analysis of sedimentary basins; regional geology. *Mailing Add:* 2321 Westcreek Lane Houston TX 77027

BROMBERGER-BARNEA, BARUCH (BERTHOLD), b Adelnau, Ger, Aug 31, 18; nat US; m 55; c 3. ENVIRONMENTAL MEDICINE, PHYSIOLOGY. *Educ:* Tech Col Berlin, EE, 38; Univ Colo, PhD(physiol), 57. *Prof Exp:* Telecommun engr, Govt Israel, Tel Aviv, 46-51; asst, Med Sch, Univ Colo, 55-57, instr physiol res, 57-60, asst prof, 60-61; from asst prof to assoc prof, 62-74, PROF ENVIRON PHYSIOL, SCH HYG & PUB HEALTH, JOHNS HOPKINS UNIV, 75- *Concurrent Pos:* Electrophysiologist, Nat Jewish Hosp, 57-62. *Mem:* Inst Elec & Electronics Engrs; Am Physiol Soc; Am Heart Asn. *Res:* Electrophysiology; cardiac excitability; ventricular fibrillation; transmembrane potentials; circulation; coronary blood flow; environmental effects on coronary circulation; environmental physiology and health; ozone. *Mailing Add:* Dept Environ Physiol Sch Hyg & Pub Health Johns Hopkins Univ 615 N Wolfe St Baltimore MD 21205

BROMER, WILLIAM WALLIS, b Racine, Wis, Oct 10, 27; m 52; c 3. BIOCHEMISTRY. *Educ:* DePauw Univ, AB, 49; Ind Univ, PhD(chem), 54. *Prof Exp:* Sr biochemist, 53-60, res assoc, 60-70, RES ADV, ELI LILLY & CO, 70- *Mem:* AAAS; Am Diabetes Asn; Am Chem Soc; Am Soc Biol Chemists. *Res:* Metabolism of essential fatty acids and cholesteryl esters; antibiotics isolation; isolation and structural analysis of proteins; glucagon; intrinsic factor and vitamin B12; insulin and proinsulin chemistry; glucagon structure and function; recombinant insulin. *Mailing Add:* 8919 Spicewood Ct PO Box 618 Indianapolis IN 46260

BROMERY, RANDOLPH WILSON, b Cumberland, Md, Jan 18, 26; m 47; c 5. GEOLOGY, GEOPHYSICS. *Educ:* Howard Univ, BS, 56; Am Univ, MS, 62; Johns Hopkins Univ, PhD, 68. *Hon Degrees:* EdD, Western New Eng Col; DSc, Frostburg State Col; LLD, Hokkaido Univ; DPs, Westfield State Col, N Adams State Col. *Prof Exp:* Geophysicist, US Geol Surv, 48-67; assoc prof geophys, 67-69, head dept geol & geog, 69-70, vchancellor, 70-72, sr vpres & chancellor, 72-80, PROF GEOPHYS, UNIV MASS, AMHERST, 69-, EXEC VPRES, 80-, COMMONWEALTH PROF GEOPHYSICS, 80-; PRES, GEOSCI ENG CORP, 81-; CHANCELLOR, BD REGENTS. *Concurrent Pos:* Prof lectr, Howard Univ, 61-65; consult, AID, US Dept State, 66-72; consult, Kennecott Copper & Exxon; mem bd trustees, Johns Hopkins Univ, Mt Holyoke Col; incorporator, Woods Hole Oceanog Inst; mem bd dirs, Exxon Co, NYNEX Corp, Chem Banking Corp, John Hancock Mutual Life Ins Co. *Mem:* Am Geophys Union; Soc Explor Geophys; fel Geol Soc Am; fel AAAS. *Res:* Gravity, magnetic, radioactivity and seismic surveying methods applied to geologic mapping programs. *Mailing Add:* Bd Regents Higher Educ One Ashburton Pl Rm 1401 Boston MA 02108

BROMET, EVELYN J, b New Brit, Conn, Nov 10, 44; c 1. PSYCHIATRY. *Educ:* Smith Col, BA, 66; Yale Univ, PhD(Pub Health & Epidemiol), 71. *Prof Exp:* Trainee, US Pub Health Sev, Yale Univ, 66-71; res assoc, dept psychiat, Stanford Univ, 72-76; from asst prof to assoc prof, dept psychiat, Univ Pittsburgh, 76-86; PROF, DEPT PSYCHIAT, STATE UNIV NY, STONY BROOK, 86- *Concurrent Pos:* Prin investr, psychiat epidemiol training prog, Nat Inst Mental Health, 77-86, three mile island accidents, 79-84, exposure to lead, psychol & behav sequalae, 81-83, psychiat epidemiol study of blue collar women, 84-86, stress & mental health mgrs & prof, 86-89; reviewer, epidemiol & serv res review comt, 83-87, epidemiol of newly diagnosed psychosis, 89-92; mem, consortium res social support, 83-; Vet Admin Sci Adv Comt, Nat Vietnam Vet Readjust Study, 87. *Honors & Awards:* Rema Lapouse Mental Health Res Award, Am Pub Health Asn. *Mem:* Am Pub Health Asn; Soc Epidemiol Res; Am Psychopathol Asn. *Res:* Psychiatric epidemiology which includes studies of affective disorder, alcohol abuse and psychotic disorders. *Mailing Add:* Dept Psychiat State Univ NY Stony Brook Putnam Hall S Campus Stony Brook NY 11794-8790

BROMFIELD, CALVIN STANTON, b Freeport, NY, Feb 4, 23; m 43; c 4. GEOLOGY. *Educ:* Univ Ariz, BS, 48, MS, 50; Univ Ill, PhD, 62. *Prof Exp:* Geologist, Homestake Mining Co, 49-51; GEOLOGIST, US GEOL SURV, 51- *Concurrent Pos:* Asst, Univ Ill, 55-57. *Mem:* Geol Soc Am; Soc Econ Geologists; Sigma Xi. *Res:* Base metal ore deposits and regional geology, especially of Arizona, Colorado and Utah. *Mailing Add:* 6337 W Geddis Dr Littleton CO 80123

BROMFIELD, KENNETH RAYMOND, b Wilkes-Barre, Pa, June 22, 22; m 46; c 5. PLANT PATHOLOGY. *Educ:* Pa State Univ, BS, 49; Univ Minn, PhD(plant path), 57. *Prof Exp:* Plant pathologist epiphytology, US Army Biol Labs, 49-51 & 52-55; res asst, Dept Plant Path, Univ Minn, 51-52; res asst, Dept Plant Path, Univ Minn, 55-56; plant pathologist forest dis, US Forest Serv, Northeastern Forest Exp Sta, NH, 57-59 & US Army Biol Defense Res Ctr, 59-71; RES PLANT PATHOLOGIST & RES LEADER, PLANT DIS RES LAB, USDA, 71- *Concurrent Pos:* Lectr, Frederick Community Col, 61- *Mem:* Am Phytopath Soc. *Res:* Etiology and epiphytology of plant rusts; evaluation of plant pathogens for damage potential; biological control of weeds by plant pathogens; ecology of plant pathogens; weather plant disease relationships; epiphytology; ecology. *Mailing Add:* 5825 Box Elder Ct Frederick MD 21701

BROMLEY, DAVID ALLAN, b Westmeath, Ont, May 4, 26; US citizen; m 49; c 2. NUCLEAR PHYSICS. *Educ:* Queen's Univ, Ont, BSc, 48, MSc, 50, DSc, 81; Univ Rochester, PhD(physics), 52; Yale Univ, MA, 61;. *Hon Degrees:* Numerous from US and foreign univs, 78-91. *Prof Exp:* Demonstr physics, Queen's Univ, Ont, 47; res officer, Nat Res Coun Can, 48; from instr to asst prof physics, Univ Rochester, 52-55; sr res officer, Atomic Energy Can Ltd, 55-60, sect head accelerators, 58-60; assoc prof physics & assoc dir, Heavy Ion Lab, Yale Univ, 60-61, prof physics, 61-72, chmn dept, 70-77, dir, A Wright Nuclear Strut Lab, 61-89, HENRY FORD II PROF, YALE UNIV, 72-; ASST TO PRES SCI & TECHNOL & DIR, OFF SCI & TECHNOL POLICY, WHITE HOUSE, 89- *Concurrent Pos:* Mem Panel Nuclear Struct Physics, NSF, 61; chmn nuclear sci comt, Nat Res Coun, 66-76, mem exec comt, Div Phys Sci & mem-at-lg, 68-78; dir, United Nuclear Corp, 67-, Labcore Inc, 69-71, Extrion Corp, 70-75, United Illum Co, 73-89, Union Trust Co, 76-89, Northeast Bancorp Inc, 76-89, Oak Ridge Assoc Univs, 78-82, Univ Bridgeport, 81-85, Gen Ionex, 80-88 & Barnes Eng Co, 84-86; chmn physics surv comt, Nat Acad Sci, 69-73, mem Naval Studies Bd, 74-78; consult, Oak Ridge Nat Lab, Brookhaven Nat Lab, Acad Press, Bell Tel Labs, NSF, McGraw-Hill, GT&E & Int Bus Mach Corp, 69-89; mem, US Nat Comt, Int Union Pure & Appl Physics, 70-, chmn, 74-, vpres, 75-81, first vpres, 81-84 & pres, 84-87; mem, Joint US-USSR Coord Comt Fundamental Res, 72 & High Energy Physics Adv Panel, Energy Res Develop Admin, 73-78; Guggenheim fel, 77-78; chmn, US Nuclear Scientists vis Peoples Repub China, 78; mem, US Nat Comt, Int Coun Sci Unions, 81-89 & White House Sci Coun, 81-89; mem, Nuclear Sci Adv Panel, NSF & Dept Energy, 81-89; mem adv bd, Inst Nuclear Power Opers, 81-84 & Elec Power Res Inst, 84-89; US co-chmn, India-US Presidential Initiative Sci & Technol, 82-, US chmn, Brazil-US Presidential Initiative Sci & Technol, 87-; mem bd dirs, Southeastern Univ Res Asn; co-chmn, US-Japan joint comn, US-Europ joint consult group, US-Venezuelan comn Sci & Technol, US-Chili comn Sci & Technol, US-China joint comn Sci & Technol. *Honors & Awards:* Humboldt Prize, 79, 82, 85; Benjamin Franklin fel, Royal Soc Arts (London); Presidental Medal, NY Acad Sci, 89; US Nat Medal Sci, 89. *Mem:* Nat Acad Sci; Can Asn Physicists; fel Am Acad Arts & Sci; fel AAAS (pres, 81); fel Am Phys Soc; fel Brazilian Acad Sci; fel Royal Soc SAfrica; fel Am Phys Soc. *Res:* Nuclear structure and reaction mechanisms; heavy ion physics; accelerators. *Mailing Add:* Wright Nuclear Struct Lab Yale Univ New Haven CT 06511

BROMLEY, LE ROY ALTON, b San Jose, Calif, Sept 30, 19; m 41; c 4. CHEMICAL ENGINEERING, PHYSICAL CHEMISTRY. *Educ:* Univ Calif, BS, 41, PhD(chem), 48; Ill Inst Technol, MS, 43. *Prof Exp:* Chemist, Nat Motor Bearing Co, Calif, 41; asst, Ill Inst Technol, 41-43; asst, Manhattan Dist, 43-46, from instr to prof chem eng, 46-76, EMER PROF CHEM ENG, UNIV CALIF, BERKELEY, 76- *Concurrent Pos:* Vis prof, Univ Philippines,

54-56. *Honors & Awards:* Alan P Colburn, Am Inst Chem Engrs, 53. *Mem:* Am Inst Chem Engrs; Am Chem Soc. *Res:* Heat transfer in boiling; condensation; thermodynamic properties of halides, sulfides, carbides and nitrides; high temperature heats reaction; sea water conversion; research on air pollution control by sea water; properties of sea water. *Mailing Add:* PO Box 580 Willits CA 95490

BROMLEY, STEPHEN C, b Los Angeles, Calif, Aug 31, 38; m 67; c 6. DEVELOPMENTAL BIOLOGY. *Educ:* Brigham Young Univ, BS, 60; Princeton Univ, MA, 62, PhD(biol), 65. *Prof Exp:* Res asst ornith & mammal, Los Angeles County Mus, 57, 59-60; instr biol, Princeton Univ, 64-65; asst prof zool, Univ Vt, 65-69; res assoc, 69-70, assoc prof zool & assoc prof biol sci, 70-75, PROF ZOOL & PROF BIOL SCI, MICH STATE UNIV, 75-, DIR PROG, 70- *Mem:* AAAS; Am Soc Zoologists. *Res:* Neural and endocrine factors in regeneration of amphibian limbs; response of amphibian limbs to culture conditions. *Mailing Add:* Dept Biol Scis Mich State Univ East Lansing MI 48823

BROMUND, RICHARD HAYDEN, b Oberlin, Ohio, Apr 28, 40; m 62; c 2. ANALYTICAL CHEMISTRY. *Educ:* Oberlin Col, AB, 62; Pa State Univ, PhD(chem), 68. *Prof Exp:* From asst prof to assoc prof chem, 67-80, PROF CHEM, COL WOOSTER, 80- *Mem:* AAAS; Am Chem Soc. *Res:* Chemical instrumentation; gas chromatography; trace organic analytical chemistry. *Mailing Add:* Dept Chem Col Wooster Wooster OH 44691

BROMUND, WERNER HERMANN, b Duluth, Minn, May 8, 09; m 35; c 2. ANALYTICAL CHEMISTRY, ORGANIC CHEMISTRY. *Educ:* Univ Chicago, BS, 32; Oberlin Col, AM, 35; NY Univ, PhD(chem), 42. *Prof Exp:* Asst chem, NY Univ, 34-37; from instr to prof, 37-75, EMER PROF CHEM, OBERLIN COL, 75- *Mem:* Am Chem Soc; Sigma Xi. *Res:* Amino sugar derivatives; adsorption characteristics of porous solids; macro and microchemical apparatus design; macro and microchemical methods of analysis; methods of analysis of pharmaceuticals; teaching methods in analytical chemistry; Kofler microfusion methods of analysis. *Mailing Add:* 79 Florentine Way Olmsted Township OH 44138

BRON, WALTER ERNEST, b Berlin, Ger, Jan 17, 30; m 52; c 2. SOLID STATE PHYSICS. *Educ:* NY Univ, BME, 52; Columbia Univ, MS, 53, PhD(metal physics), 58. *Prof Exp:* Res assoc, Eng Res & Develop Lab, 54-56; lectr, George Washington Univ, 55-56; res physicist, Watson Lab, IBM Corp, 57-58, staff mem physics group, Res Ctr, 58-66; vis prof, Phys Inst, Univ Stuttgart, 66-67; assoc prof, 67-69, prof physics, Ind Univ, Bloomington, 69-86; PROF PHYSICS, 86-, CHMN, UNIV CALIF, IRVINE, 89- *Concurrent Pos:* Res assoc & lectr, Columbia Univ, 52-58, 64; Guggenheim fel, 66-67; vis prof, Max Planck Inst for Solid State Res, 73-74 & 81-82. *Honors & Awards:* Alexander von Humboldt Found award, 73. *Mem:* Europ Phys Soc; Am Phys Soc; Sigma Xi. *Res:* Defects in solid crystals; optical properties of solids; lattice dynamics; phonon transport. *Mailing Add:* Dept Physics Univ Calif Irvine CA 92717

BRONAUGH, EDWIN LEE, b Salina, Kans, July 22, 32; m 55; c 2. ELECTROMAGNETIC COMPATIBILITY, ANTENNA & INSTRUMENT DESIGN. *Educ:* East Tex State Univ, BA, 55. *Prof Exp:* Res scientist, electromagnetic compatibility, Southwest Res Inst, 68-76, res dir, 76-82; dir res & develop, Electro-Mechanics Div, Penril Co, 82-89; PRIN EMC SCIENTIST, ELECTRO-MECHANICS CO, 89- *Concurrent Pos:* Chmn, SC-1, sect Comt, C 63, Am Nat Standard Inst, 77-85, vchmn, 85-; mem, Electromagnetic Compatability Standards comt, Inst Elec & Electronics Elec, 80-; mem, Electromagnetic Interference Standards Tech, Soc Automotive Engrs, 75-; mem, Electromagnetic Radiation Tech Comt, Soc Automotive Engrs, 87-; dir tech serv, Electromagnetic Compatibility Soc, 81-87, vpres, 88-89, pres, 90- *Honors & Awards:* Richard R Stoddart Award, Electromagnetic Compatibility Soc, Inst Elec & Electronics Engrs. *Mem:* Fel Inst Elec & Electronics Engrs; Soc Automotive Engrs. *Mailing Add:* 11007 Crossland Dr Austin TX 78726-1320

BRONAUGH, ROBERT LEWIS, b Spokane, Wash, Dec, 15, 42; m 70; c 1. TOXICOLOGY, PHARMACOLOGY. *Educ:* Univ NMex, BS, 68; Univ Colo, PhD(pharmacol), 72. *Prof Exp:* Teaching fel, Med Sch, Univ Colo, 72-74 & NY Univ, 74-76; sr pharmacologist, Interx Res Corp, 76-78; RES PHARMACOLOGIST, US FOOD & DRUG ADMIN, 78- *Concurrent Pos:* Adj prof, Univ Cincinnati, 84- *Honors & Awards:* Commendable Serv Award, Food & Drug Admin, 84. *Mem:* Soc Toxicol; Soc Investigative Dermat; Soc Cosmetic Chemists. *Res:* Determination of the absorption of chemicals through skin; mechanisms; methodology effect of vehicles; effect of physico-chemical parameters. *Mailing Add:* 1528 Winding Waye Lane Silver Spring MD 20902

BRONCO, CHARLES JOHN, b Gary, Ind, Jan 25, 28. ATOMIC PHYSICS. *Educ:* Okla State Univ, BS, 49; Univ Tex, MA, 51; Univ Okla, PhD(physics), 59. *Prof Exp:* Sr systs design engr, Chance Vought Aircraft, Tex, 56-59, electronic systs engr, 59-60; prin physicist, Phys Sci Div, Melpar, Inc, Va, 60-61, sr scientist gaseous laser res, 61-62; asst prof math & physics, Grad Inst Technol, Univ Ark, 62-68; assoc prof physics, Ark Tech Univ, 78-82; independent res, 82-89; prof physics, Univ Ozark, 89-91; INDEPENDENT RES, 91- *Concurrent Pos:* Consult, Melpar, Inc, 62. *Mem:* Am Phys Soc. *Res:* Resonance excitation cross-sections; high vacuum physics; electronic systems research; molecular dissociation by electron and photon impact; laser development; focal isolation methods; laboratory techniques. *Mailing Add:* Rte 1 Box 260 Dover AR 72837

BRONDYKE, KENNETH J, b Saulte Ste Marie, Mich, June 16, 22. METALLURGY & PHYSICAL METALLURGICAL ENGINEERING. *Educ:* Univ Mich, BS, 46. *Prof Exp:* Dir, Alcoa Lab, 83; RETIRED. *Mem:* Am Soc Metals; Am Inst Mining Metall & Petrol Engrs; Sci Res Soc Am. *Mailing Add:* 727 14th St Oakmont PA 15139

BRONDZ, BORIS DAVIDOVICH, b Yaroslavle, USSR, June 21, 34; m 56, 80; c 1. EXACT ELUCIDATION OF THE MHC MOLECULE. *Educ:* First Moscow Med Inst, Physician, 57; Gamaleya Inst Epidemiol Microbiol, Cand Sci, 65, Dr Med Sci, 73; Cancer Res Ctr, Prof, 89. *Hon Degrees:* Vet Labour USSR, Moscow Munic Soviet of People's Deputation, 87. *Prof Exp:* Sr laboratorian immunol & oncol, Gamaleya Inst, Acad Med Sci, 61-64, jr sci collabr, 64-68, sr sci collabr, 68-77; sr sci collabr immunol, Lab Immunochem, Cancer Res Ctr, 77-87, chief sci collabr, 87-89, LEADER LAB LYMPHOCYTES, LAB REGULATORY MECH IMMUN, CARCINOGENESIS INST, CANCER RES CTR, 89- *Concurrent Pos:* Lectr, Moscow Univ, Central Inst, 82-; comn AIDS, Acad Med Sci, 88-; comn oncol, Cancer Res Ctr, 88- *Mem:* Immunol Soc; NY Acad Sci; hon mem Am Asn Immunol. *Res:* Specific absorption-elution of different T-cell subsets and their receptors on macrophage monolayers of inbred-mutant strains of mice differing in particular histocompatibility genes; the killer T-cells differ from each other in clonal structure of their receptors; difference in clonal repertoire; fine specificity and epitopes between T-killers and antibodies and between receptors of T-killers, their memory cells and suppressor T-cells; selective markers of suppressor T-cells; mechanism of immunodeficiency. *Mailing Add:* Sci Investment Carcinogenetics Cancer Res Ctr Kashirskoye Shaussae 24 Moscow 115478 USSR

BRONIKOWSKI, THOMAS ANDREW, applied mathematics; deceased, see previous edition for last biography

BRONK, BURT V, b Pittsburgh, Pa, 1935; m 56; c 3. BIOPHYSICS. *Educ:* Pa State Univ, BS, 55; Princeton Univ, PhD(physics), 65. *Prof Exp:* Res assoc physics, State Univ NY, Stony Brook, 64-66; asst prof, Queens Col, NY, 66-69; sr res assoc biol, Brookhaven Nat Lab, 69-71; assoc prof, 71-78, prof physics & microbiol, 78-87, ADJ PROF BIOPHYS, CLEMSON UNIV, 87-; RES PHYSICIST, CHEM RES DEVELOP & ENG CTR, ABERDEEN PROVING GROUND, 87- *Concurrent Pos:* Res collabr, Brookhaven Nat Lab, 71-72; res assoc biol, Univ Tex, Dallas, 81-82; ed, Comments on Molecular & Cellular Biophys. *Mem:* Biophys Soc; Phys Soc; Am Soc Microbiol. *Res:* Synergy of DNA damage with thermal effects; statistical physics as applied to biological processes; synergistic effect of hyperthermia on radiation and drug damage to animal cells; optical physics of micron- sized particles; lightscatering and fluorescence of baterial cells. *Mailing Add:* Dept Physics Clemson Univ Clemson SC 29631

BRONK, JOHN RAMSEY, b Philadelphia, Pa, Dec 20, 29; m 55; c 2. INTESTINAL TRANSPORT, CELL PHYSIOLOGY. *Educ:* Princeton Univ, AB, 52; Oxford Univ, PhD(biochem), 55. *Prof Exp:* From asst prof to prof zool, Columbia Univ, 58-66; PROF BIOCHEM, UNIV YORK, ENG, 66- *Concurrent Pos:* Guggenheim fel, 64-65. *Mem:* Am Soc Biol Chemists; Brit Biochem Soc; Brit Physiol Soc. *Res:* Transport of amino acids, sugars, purines, pyrimidines and vitamins in the small intestine; mode of action of the thyroid hormones; oxidative phosphorylation and mitochondrial structure; urea synthesis and gluconeogenesis in liver. *Mailing Add:* Dept Biol Univ York York YO1 5DD England

BRONNER, FELIX, b Vienna, Austria, Nov 7, 21; nat US; m 47; c 2. CALCIUM METABOLISM, ION REGULATION. *Educ:* Univ Calif, BS, 41; Mass Inst Technol, PhD(nutrit biochem), 52. *Prof Exp:* Asst, Mass Inst Technol, 51-52, assoc, 52-54; vis investr, Rockefeller Inst, 54-56, asst, 56; assoc, Hosp Spec Surg, Med Ctr, Cornell Univ, 57-58, Bicknell assoc mineral metab, 58-61, assoc scientist & asst prof biochem, Cornell Med Col, 61-63; assoc prof, Sch Med, Univ Louisville, 63-69; prof oral biol, 69-86, nutrit sci, 76-89, biostructur & function, 86-89, EMER PRPF, UNIV CONN, FARMINGTON, 89- *Concurrent Pos:* Helen Hay Whitney fel, Rockefeller Inst, 54-55 & Arthritis & Rheumatism fel, 55-56; vis scientist, Pasteur Inst, 62, 76 & Weizmann Inst Sci 65, 76, Univ Capetown, 85, 88; consult, NIH, Acad Press & USDA; organizer chmn, Gordon Res Conf Bones & Teeth 54, Sanger Symp, 24th Int Cong, 65; vis scientist, INSERM, Paris, 72, Lyon, 88; vis prof, Tel-Aviv Univ, 76, Univ Capetown, 85, 88, Caron vis prof, 88; co-chmn, Int Workshop Calcium & Phospate Transport Across Biomembrane, Vienna, Austria, 81, 84, 87. *Honors & Awards:* Andre Lichtwitz Prize, INSERM, Paris, 74. *Mem:* Fel AAAS; Am Inst Nutrit; Am Soc Clin Nutrit; Am Physiol Soc; Soc Exp Biol & Med; Harvey Soc; emer mem Orthop Res Soc; Am Fedn Clin Res; Am Soc Bone & Mineral Res. *Res:* Calcium and mineral metabolism; metabolic bone disease; biologic regulation; ion movement and transport. *Mailing Add:* Dept Bio Structure & Function Sch Dent Med Univ Conn Farmington CT 06030

BRONNES, ROBERT L(EWIS), b Dobbs Ferry, NY, Mar 17, 24. METALLURGY. *Educ:* Univ Ala, BS, 51. *Prof Exp:* From asst engr to engr, Nam Phillips Co, Inc, 51-57, sr engr, 57-66, staff engr metall, 66-85, sr mem res staff, 85-90; RETIRED. *Mem:* Am Soc Metals; Metall Soc; fel Am Inst Chem; Int Metallog Soc. *Res:* Metallurgical and metallographic phases, especially semi-conductors, metal-ceramic seals, ferrites, brazing, electroplating, thin films; surface preparation of semiconductor materials. *Mailing Add:* 21 N Eckar St Irving NY 10533

BRONS, CORNELIUS HENDRICK, b Staten Island, NY, Jan 24, 55; m; c 3. SPECTROSCOPY, GEOLOGY & BIOLOGY RELATED PROJECTS. *Educ:* Fairleigh Dickinson Univ, Madison, NJ, BA, 87. *Prof Exp:* Res technician phys chem, Merck, Sharpe & Dome Res Labs, 76-77; sr res technician mineral sci, ASAR Co, Res & Develop Labs, 77-79; sr res technician phys chem, electro-chem, enhanced oil recovery, RES ASSOC POLYMER & COMPLEX FLUIDS & BIO-REMEDIATION OF CONTAMINATED SOILS, EXXON RES & ENG LABS, 79- *Mem:* Sigma Xi; Am Chem Soc; Geol Sci US. *Res:* Water soluble polymers; encapsulation of products; surfactant research for many applications (household, automotive and soil cleaning and washing); bio-remediation research dealing with microbial populations, nourishment, treating, cleaning and formulating nutrient packages for them; treating waste water and other contaminents; granted five patents; author of various publications. *Mailing Add:* 41 Fairview St Washington NJ 07882

BRONS, KENNETH ALLYN, b Chicago, Ill, July 21, 29; m 55; c 2. COMPUTER SCIENCE. *Educ:* NCent Col, Ill, BA, 51; Univ Ill, Urbana, MS, 52, PhD(math), 56. *Prof Exp:* Appl sci rep, IBM Corp, 56-59; sr methods specialist, RCA Corp, 59-69, mgr mgt sci appln, 69-72; consult, Comput Dynamics Corp, 72-75; SR SYST ANALYST, CRYOVAC DIV, W R GRACE & CO, 75- *Mem:* Asn Comput Mach; Am Math Soc; Math Asn Am; Soc Indust & Appl Math; Sigma Xi. *Mailing Add:* 302 Hermitage Rd Greenville SC 29615

BRONSKY, ALBERT J, b Waterbury, Conn, June 14, 28; m 55; c 2. BIOCHEMISTRY. *Educ:* Boston Col, BS, 54; Purdue Univ, MS, 57, PhD(biochem), 61. *Prof Exp:* Res biochemist, Res Dept, 59-68, head fermentation sect, 68-75, mgr res & develop, 75-78, TECH DIR, JOSEPH E SEAGRAM & SONS, INC, 78- *Mem:* Am Chem Soc; Am Soc Brewing Chemists; AAAS; Sigma Xi; NY Acad Sci. *Res:* Enzymology; mechanism of action of esterases and phosphorylases; metabolism; formation of trace flavor components during fermentation. *Mailing Add:* 330 Dundee Rd Stamford CT 06903

BRONSON, FRANKLIN HERBERT, b Pawnee City, Nebr; m 53; c 2. REPRODUCTIVE PHYSIOLOGY. *Educ:* Kans State Univ, BS, 57, MS, 58; Pa State Univ, PhD(zool), 61. *Prof Exp:* Staff scientist, Jackson Lab, 61-68; assoc prof, 68-72, PROF ZOOL, UNIV TEX, AUSTIN, 72- *Mem:* AAAS; Animal Behav Soc; Soc Study Reprod; Am Soc Zoologists; Ecol Soc Am. *Res:* Environmental control of reporduction. *Mailing Add:* Dept Zool Univ Tex Austin TX 78712

BRONSON, JEFF DONALDSON, b Dallas, Tex, Aug 12, 38. NUCLEAR PHYSICS. *Educ:* Rice Univ, BA, 59, MA, 61, PhD(physics), 64. *Prof Exp:* Res assoc physics, Rice Univ, 64; res fel, Univ Basel, 64-65 & Univ Wis, 65-67; asst prof, 67-74, MEM STAFF, CYCLOTRON INST, TEX A&M UNIV, 74- *Mem:* Am Phys Soc; Sigma Xi. *Res:* Nuclear forces using 3-body nuclear reactions; determination of atomic hyperfine fields and nuclear magnetic moments using perturbed angular correlations. *Mailing Add:* Cyclotron Inst Tex A&M Univ College Station TX 77843

BRONSON, ROY DEBOLT, b Reno, Nev, May 30, 20; m 47; c 3. SOIL SCIENCE. *Educ:* Mich State Univ, BS, 48, MS, 49; Purdue Univ, PhD, 59. *Prof Exp:* Soil technician, Agr Res Dept, Green Giant Co, Minn, 49-52; instr agron & supvr soil testing prog, Purdue Univ, 52-59, assoc prof, 59-60; head dept chem & soils, United Fruit Co, Honduras, 60-61; chief soils & fertile res br, Tenn Valley Authority, Wilson Dam, Ala, 61-63; prof agron, Purdue Univ, 63-68, soil scientist, Purdue-Brazil Proj, Agr Univ, Minas Gerais, 63-65, chief of party, 65-68, PROF AGRON & INT AGR, PURDUE UNIV, WEST LAFAYETTE, 68-, DIR UNDERGRAD PROGS INT AGR, 75-; REGIONAL DEVELOP AGRONOMIST, USAID, IVORY COAST, WEST AFRICA, 79- *Mem:* Fel AAAS; Am Soc Agron; Soil Sci Soc Am; Int Soil Sci Soc; Soc Int Develop. *Res:* Fertility and chemistry of highly weathered soils, particularly of tropics, and relation to agricultural productivity potential; international agricultural education, development assistance planning and evaluation. *Mailing Add:* NIAMEY Agency Int Dev Washington DC 20523

BRONSTEIN, IRENA Y, CHEMISTRY. *Educ:* Bryn Mawr Col, BS, 68; Johns Hopkins Univ, PhD(photochem), 73. *Prof Exp:* Res assoc, dept biol chem, Sch Med, Univ Md, 74-75; scientist photochem & polymer chem res, Polaroid Corp, 76-78, res group leader, 78-80, sr res group leader, photochem res, 80-82, sr res group leader, Land Lab, 81-82, tech mgr, photochem res, 82-84, co-dir biotech, 83-84; SR STAFF SCIENTIST, ALLIED HEALTH & SCIENTIFIC PROD DIAGNOSTICS RES & DEVELOP CO, ALLIED CORP, 84- *Mem:* Am Soc Photobiol; Am Chem Soc; AAAS; Am Asn Clin Chem; Sigma Xi; Soc Photographic Scientists & Engrs. *Mailing Add:* 11 Ivanhoe St Newton MA 02158

BRONZAN, JOHN BRAYTON, b Los Angeles, Calif, Apr 5, 37; m 63; c 2. HIGH ENERGY PHYSICS. *Educ:* Stanford Univ, BS, 59; Princeton Univ, PhD(physics), 63. *Prof Exp:* From instr to assoc prof physics, Mass Inst Technol, 63-71; assoc prof physics, Rutgers Univ, prof, 75-; AT SERIN PHYS LAB. *Mem:* Am Phys Soc. *Res:* Theoretical high energy physics. *Mailing Add:* Dept Physics/Astron Rutgers-The St Univ New Brunswick NJ 08903

BRONZINI, MICHAEL STEPHEN, b Johnstown, Pa, Aug 21, 44; m 67; c 3. TRANSPORTATION ENGINEERING, ECONOMICS. *Educ:* Stanford Univ, BS, 67; Pa State Univ, MS, 69, PhD(civil eng), 73. *Prof Exp:* Res asst civil eng, Pa State Univ, 67-72; asst prof civil eng, Ga Inst Technol, 73-75; from assoc prof to prof civil eng, Univ Tenn, 78-86, from assoc dir to dir, Transp Ctr, 78-86; sr assoc, 75-78, SR CONSULT, CACI INC, 78-; PROF & HEAD, DEPT CIVIL ENG, PA STATE UNIV, 86- *Concurrent Pos:* Consult civil eng, govt, res & develop firms & univs, 72- *Mem:* Am Soc Civil Engrs; Inst Transp Engrs; AAAS; Sigma Xi. *Res:* Development of national freight transportation network data and models; estimation of transport cost and energy use; inland waterway system and lock capacity studies. *Mailing Add:* Dept Civil Eng 212 Sackett Bldg Pa State Univ University Park PA 16802

BRONZINO, JOSEPH DANIEL, b Brooklyn, NY, Sept 29, 37; m 61; c 3. BIOMEDICAL ENGINEERING, SYSTEMS ENGINEERING. *Educ:* Worcester Polytech Inst, BSEE, 59, PhD(elec eng), 68; US Naval Postgrad Sch, MSEE, 61. *Prof Exp:* From instr to asst prof elec eng, Univ NH, 64-67; NSF fac fel, Worcester Found Exp Biol, Shrewsbury, Mass, 67-68; asst prof, Trinity Col, 68-75, chmn, dept eng, 82-90, PROF ENG & VERNON ROOSE PROF APPL SCI, TRINITY COL, 77- *Concurrent Pos:* Mem cooperating staff, Worcester Found Exp Biol, 68-; res assoc, Inst Living, Conn, 68-; dir biomed eng prog, Hartford Grad Ctr, 69-; clin assoc, Dept Surgery, Univ Conn Health Ctr, Farmington, 71-; ed, Newsletter, Am Soc Eng Educ, 77-79; chmn, Biomed Eng Div, Am Soc Eng Educ, 80-81. *Mem:* AAAS; Am Soc Eng Educ; fel Inst Elec & Electronic Engrs (pres, 85); Sigma Xi; Asn Advan Med Instrumentation. *Res:* Application of engineering technology to new or different environments such as medicine, biology and the ocean; neurophysiological research; biomedical engineering. *Mailing Add:* Six Wyngate Simsbury CT 06070

BROODO, ARCHIE, b Wichita Falls, Tex, Feb 3, 25; m 51; c 4. PHYSICAL CHEMISTRY, CHEMICAL ENGINEERING. *Educ:* Agr & Mech Col Tex, BS, 48; Univ Tex, MS, 50, PhD(chem), 54. *Prof Exp:* CHIEF EXEC OFFICER, AID CONSULT ENGRS, INC, 70- *Concurrent Pos:* Ed, Southwest Retort, Am Chem Soc, 57-58. *Mem:* Sigma Xi; Am Chem Soc; Inst Elec & Electronics Engrs; Am Soc Metals. *Res:* Technical investigation of insurance losses; chemical and other product liability; arson determinations; explosion cause analyses; materials studies and analytical chemistry; material flammability studies; corrosion studies; tire defect analyses. *Mailing Add:* 5439 Del Roy Dallas TX 75229

BROOK, ADRIAN GIBBS, b Toronto, Ont, May 21, 24; m 54; c 3. ORGANOSILICON CHEMISTRY. *Educ:* Univ Toronto, BA, 47, PhD(chem), 50. *Prof Exp:* Lectr chem, Univ Sask, 50-51; Nuffield res fel, Imp Col Sci & Technol, 51-52; res assoc, Iowa State Col, 52-53; lectr, Univ Toronto, 53-56, from asst prof to prof chem, 56-89, from assoc chmn to chmn dept, 68-74, EMER PROF CHEM, UNIV TORONTO, 89- *Concurrent Pos:* Vis prof, Sch Molecular Sci, Univs Sussex, 74-75, Cambridge, 81 & Ind, 88. *Honors & Awards:* F S Kipping Award, Am Chem Soc, 73; Chem Inst Can Medal, 85. *Mem:* Am Chem Soc; Chem Inst Can; fel Royal Soc Can. *Res:* Organosilicon, organogermanium chemistry, silenes; organometallic compounds; stereochemistry; reaction mechanisms. *Mailing Add:* Dept Chem Univ Toronto Toronto ON M5S 1A1 Can

BROOK, ITZHAK, b Afula, Israel, June 16, 41; US citizen; m 67; c 5. INFECTIOUS DISEASES, PEDIATRICS. *Educ:* Hebrew Univ, MD, 68; Tel Aviv Univ, MS, 73. *Prof Exp:* Instr microbiol, Hebrew Univ Sch Med, 73-74; instr pediat, Univ Calif, Irvine, 76-77; asst prof, George Wash Univ, 77-80; assoc prof pediat & surg, 80-83, PROF PEDIAT & SURG, UNIFORMED SERV UNIV HEALTH SCI, 83- *Concurrent Pos:* Sr investr, Armed Forces Radiobiol Res Inst, 84-, Naval Med Res Inst, 80-84; attend physician, Naval Hosp, Bethesda, 80-; consult infectious dis, Walter Reed Army Med Ctr, 80-; mem & chmn antinfective adv comt, Food & Drug Admin, 84-88. *Mem:* Fel Am Soc Infectious Dis; fel Soc Pediat Res; fel Soc Pediat Infectious Dis; Am Soc Microbiol; Soc Surg Infectious Dis. *Res:* Infections and pathogenicity of anaerobic bacteria; infections in the immunocompromised host; pathogenesis and therapy; upper respiratory infections. *Mailing Add:* Armed Forces Radiobiol Res Inst Bethesda MD 20814

BROOK, MARX, b New York, NY, July 12, 20; m 47; c 3. PHYSICS. *Educ:* Univ NMex, BS, 44; Univ Calif, Los Angeles, MA & PhD(physics), 53. *Prof Exp:* Asst, Univ NMex, 43-46; res physicist, Univ Calif, Los Angeles, 47-53; res physicist, 54-58, from assoc prof to prof, 58-86, chmn dept, 68-78, 60-86, dir, Res & Develop Div, 78-86, EMER PROF PHYSICS, NMEX INST MINING & TECHNOL, 86- *Concurrent Pos:* John Wesley Powell Lectr, 65; vis scientist, Cavendish Lab, 68-69; vis lectr, Japan Soc Promotion Sci, 75. *Honors & Awards:* Outstanding NMex Scientist of Year, 78. *Mem:* Fel AAAS; fel Am Phys Soc; Am Asn Physics Teachers; fel Am Meteorol Soc; Royal Meteorol Soc; Sigma Xi; fel Am Geophys Union. *Res:* Air quality; cloud physics and lightning; weather radar. *Mailing Add:* Res & Develop Div Rm 129 NMex Tech Socorro NM 87801

BROOK, ROBERT H, New York, NY, July 3, 43; m; c 4. MEDICINE. *Educ:* Univ Ariz, BS, 64; Johns Hopkins Univ, MD, 68, ScD, 72; Am Bd Hyg & Pub Health, dipl, 72. *Prof Exp:* VCHMN & PROF MED, DEPT MED & PROF PUB HEALTH, SCH PUB HEALTH & MED, UNIV CALIF, LOS ANGELES, CALIF, 74-; DIR, DEPT HEALTH PROG, THE RAND CORP, SANTA MONICA, CALIF, 74- *Concurrent Pos:* Prog dir, Robert Wood Johnson Clin Scholar Training Prog, Univ Calif, Los Angeles, 74-; dir, Health Sci Prog, The Rand Corp, Santa Monica, Calif, 74- *Honors & Awards:* Clarke Mem Lectr, Univ Hosp, Nottingham, UK, 85; Richard & Hinda Rosenthal Found Award, Am Col Physicians, 89; Hollister Univ Award lectr, Northwestern Univ, 89; Baxter Health Serv Res Prize, 88; Sonneburn Distinguished lectr, Univ Pa, 88. *Mem:* Inst Med-Nat Acad Sci; fel Am Col Physicians; Am Soc Clin Invest; Am Pub Health Asn; Int Epidemiol Asn; Am Asn Physicians. *Res:* Quality assessment and assurance; development and use of health status measurements in health policy; efficiency and effectiveness of physician behavior and performance, especially as related to academic centers and internal medicine. *Mailing Add:* 1700 Main St PO Box 2138 Santa Monica CA 90406-2138

BROOK, TED STEPHENS, b San Antonio, Tex, Apr 19, 26; m 48; c 2. ENTOMOLOGY. *Educ:* Trinity Univ, Tex, BS, 48; Kans State Univ, MS, 50; Miss State Univ, PhD(entom), 66. *Prof Exp:* Asst entomologist, Tex Agr Exp Sta, 50-54; from asst entomologist to assoc entomologist, Agr Exp Sta, 56-70, from asst prof to assoc prof entom, Miss State Univ, 64-70; plant protection adv, Univ Mo-Columbia-US AID India Prog, Dept Agr, Bihar, 70-72; assoc prof entom, Miss Agr Exp Sta, Miss State Univ, 72-75; pest mgt specialist, Miss Coop Exten Serv, 75-84; RETIRED. *Res:* Household and ornamental plant pests; termite biology and control; insect biology and ecology; general entomology. *Mailing Add:* 239 Pecan St Uvalde TX 78801

BROOKBANK, JOHN WARREN, b Seattle, Wash, Mar 3, 27; m 50; c 3. DEVELOPMENTAL BIOLOGY. *Educ:* Univ Wash, BA, 49, MS, 53; Calif Inst Technol, PhD, 55. *Prof Exp:* Asst biol, Calif Inst Technol, 52-55; from asst prof to prof zool, 55-72, prof zool & microbiol, 72-76, PROF MICROBIOL & CELL SCI, UNIV FLA, 76-, ASSOC GERONT, 76- *Mem:* Geront Soc; Am Soc Zoologists; Soc Develop Biol. *Res:* Regulation of gene expression in development and aging. *Mailing Add:* 406 Central Admin Bldg Div Aquaculture Dept Agr Olympia WA 98504

BROOKE, JOHN PERCIVAL, b Chicago, Ill, Oct 20, 33; m 70. APPLIED GEOPHYSICS. *Educ:* Mich Technol Univ, BS, 54; Univ Utah, MS, 59, PhD(geol eng), 64. *Prof Exp:* Geologist, Am Smelting & Refining Co, 60-61; asst prof geophys & econ geol, 61-70, assoc prof geol, geophys & eng geol, 70-78, PROF GEOL, SAN JOSE STATE UNIV, 78- *Mem:* Soc Explor Geophys. *Res:* Instrumental and field analysis of geochemical alteration halos surrounding ore bodies; mineral exploration; operations research and mineral economics; geophysical and geochemical prospecting; engineering and environmental geology. *Mailing Add:* Dept Geol San Jose State Univ San Jose CA 95192

BROOKE, M(AXEY), b Muskogee, Okla, June 2, 13. CHEMICAL ENGINEERING. *Educ:* Univ Okla, BS, 35. *Prof Exp:* Chemist, Midland Gasoline Co, 36-40; chief chemist, Jefferson Lake Sulphur Co, 40-47; supvry chemist, Phillips Petrol Co, 47-76, chief chemist, 76-78; CONSULT, 78- *Mem:* Am Chem Soc; Nat Asn Corrosion Engrs; hon mem Cooling Tower Inst. *Res:* Water conditioning; cooling towers; industrial waste disposal; corrosion control. *Mailing Add:* 912 Ocean Ave Sweeny TX 77480

BROOKE, MARION MURPHY, b Atlanta, Ga, Dec 6, 13; m 40; c 3. PARASITOLOGY. *Educ:* Emory Univ, AB, 35, AM, 36; Johns Hopkins Univ, ScD(protozool), 42; Am Bd Microbiol, dipl. *Prof Exp:* Instr biol, Emory Jr Col, 36-38; asst protozool, Sch Hyg & Pub Health, Johns Hopkins Univ, 38-40, instr, 40-42, assoc parasitol, 42-44; assoc prof prev med, Col Med, Univ Tenn, 44-45; chief parasitol sect, Lab Br, Ctr Dis Control, USPHS, 45-51, chief parasitol & mycol sect, 51-57, chief microbiol sect, 57-62, chief lab consult & develop sect, 62-69, dep chief licensure & develop br, 69-72, assoc dir Health Lab Manpower Develop, Lab Training & Consult Div, 72-84; AT DEPT MICROBIOL, EMORY UNIV. *Concurrent Pos:* Dir interstate malaria surv, Bd State Health Comn Upper Mississippi River Basin, 42; assoc prof, Med Sch, Emory Univ, 46-; vis investr, Sch Trop Med, Univ PR, 46; mem joint dysentery unit, Armed Forces Epidemiol Bd Korea, 51, assoc mem comn enteric infections, 54-74; mem expert adv panel parasitic dis, WHO, 64- & Conf State & Prov Pub Health Lab Dirs; assoc prof, Sch Pub Health, Univ NC, 63-75, prof, 76-; consult, 85- *Mem:* Am Soc Parasitol; Soc Trop Med & Hyg; Am Pub Health Asn; Am Acad Microbiol. *Res:* Proficiency examinations; health laboratory manpower; malaria; amoebiasis; intestinal parasites; toxoplasmosis; programmed instruction. *Mailing Add:* 1343 Emory Rd NE Atlanta GA 30306

BROOKE, RALPH IAN, b Leeds, UK; Can citizen; m 63; c 2. DENTISTRY. *Educ:* Leeds Univ, BChD, 57; Royal Col Surgeons Eng, MRCS & LRCP, 63; FDSRCS, 67. *Prof Exp:* Prof oral med & chmn dept, 72-82, DEAN FAC DENT & VPROVOST HEALTH SCIS, UNIV WESTERN ONT, 82-; CHIEF DENT, UNIV HOSP, LONDON, ONT, 72- *Mem:* Fel Royal Col Dent; fel Int Col Dent. *Res:* Chronic facial pain. *Mailing Add:* Fac Dent Univ Western Ont London ON N6A 5C1 Can

BROOKER, DONALD BROWN, b Troy Grove, Ill, Dec 5, 16; m 36; c 2. CROP DRYING, MATERIALS HANDLING. *Educ:* Univ Mo, Columbia, BS(agr eng), 47, BS(mech eng), 54, MS, 49. *Prof Exp:* Instr, Purdue Univ, 49-51; from asst prof to prof, 51-82, EMER PROF AGR ENG, UNIV MO, COLUMBIA, 82- *Concurrent Pos:* Dir, Am Soc Agr Engrs, 76-78; Mem, Governor's Adv Bd, Weights & Measures, 81-82. *Honors & Awards:* Massey-Ferguson Medal, Am Soc Agr Engrs, 79; Halliburton Educ Award, 80. *Mem:* Am Soc Agr Engrs; Am Soc Eng Educ; Nat Soc Prof Engrs; Sigma Xi. *Res:* Author of two textbooks. *Mailing Add:* Dept Agr Eng Univ Mo Agr Eng Bldg Columbia MO 65211

BROOKER, GARY, b San Diego, Calif, Mar 24, 42; m 64; c 2. PHARMACOLOGY, BIOCHEMISTRY. *Educ:* Univ Southern Calif, BS, 66, PhD(cardiac pharmacol), 68. *Prof Exp:* Asst prof med, Sch Med, Univ Southern Calif, 68-72, asst prof biochem, 69-72; from assoc prof to prof pharmacol, Med Sch, Univ Va, 72-82; prof & chmn, Dept Biochem, 82-86, PROF, DEPT BIOCHEM & MOLECULAR BIOL, GEORGETOWN UNIV MED CTR, 86- *Concurrent Pos:* Los Angeles County Heart Asn res grants, 68-70; Am Heart Asn res grant, 70-73; USPHS res grants, 71- *Mem:* Am Soc Biol Chemists; AAAS; Am Fedn Clin Res; Am Soc Pharmacol & Exp Therapeut. *Res:* Mechanisms of hormone and drug action as it relates to cardiac metabolism and cardiac contraction; development of methods for rapid analysis of substances of biological interest; microscopic images. *Mailing Add:* Med Ctr Dept Biochem & Molecular Biol Georgetown Univ 3900 Reservoir Rd Washington DC 20007

BROOKER, HAMPTON RALPH, b Columbia, SC, Sept 11, 34; m 62; c 2. NUCLEAR MAGNETIC RESONANCE, MAGNETIC RESONANCE IMAGING. *Educ:* Univ Fla, BS, 56, MS, 58, PhD(physics), 62. *Prof Exp:* Asst prof physics, Abilene Christian Col, 61-63; fel, Univ Fla, 63-64; asst prof, 64-74, ASSOC PROF PHYSICS, UNIV SFLA, 74- *Concurrent Pos:* Univ res fel, Univ Nottingham, Eng, 76-77; consult, dept radiol, Univ Fla, 78- *Mem:* Am Phys Soc; Am Asn Physics Teachers; Am Asn Physicists Med; Soc Magnetic Resonance Med. *Res:* Application of nuclear magnetic resonance to medicine. *Mailing Add:* Dept Physics Univ SFla Tampa FL 33620

BROOKER, ROBERT MUNRO, b Troy Grove, Ill, Jan 3, 18; m 41; c 2. ORGANIC CHEMISTRY. *Educ:* Univ Mo, BS, 47, PhD(chem), 50. *Prof Exp:* Asst, Univ Mo, 46-50; head dept chem, 50-66, Ind Cent Univ, chmn div sci & math, 66-88, Bohn prof chem, 81-88; RETIRED. *Concurrent Pos:* Res Chemist, Pitman-Moore Co, 50-65; consult, Dow Chem Co, 65-; pres, Short Courses, Inc, 73- *Mem:* Am Chem Soc. *Res:* Acenaphthene; veratrum alkaloids; natural products. *Mailing Add:* 1431 Windermire Indianapolis IN 46227

BROOKES, NEVILLE, b London, Eng, Mar 12, 39; m 71; c 2. PHARMACOLOGY, NEUROBIOLOGY. *Educ:* Pharmaceut Soc Gt Brit, MPS, 62; Univ Leeds, PhD(pharmacol), 67. *Prof Exp:* Asst lectr pharmaceut, Sch Pharm, Bristol, 62-63; res fel neuropharmacol, Univ Birmingham, 67-68; instr pharmacol, Hebrew Univ, Jerusalem, 68-70, lectr, 70-74; asst prof, 74-80, ASSOC PROF PHARMACOL, MED SCH, UNIV MD, 80- *Concurrent Pos:* Guest worker, Behav Biol Br, Nat Inst Child Health & Human Develop, 74-76. *Mem:* Am Soc Pharmacol & Exp Therapeut; Soc Neurosci; Pharmaceut Soc Gt Brit. *Res:* Neuropharmacology and neurotoxicology; actions of transmitters drugs and toxic agents on excitable cells. *Mailing Add:* Dept Pharmacol & Exp Therapeut Univ Md Med Sch Baltimore MD 21201

BROOKES, VICTOR JACK, entomology; deceased, see previous edition for last biography

BROOKHART, JOHN MILLS, b Cleveland, Ohio, Dec 1, 13; m 39; c 3. PHYSIOLOGY. *Educ:* Univ Mich, BS, 35, MS, 36, PhD(physiol), 39. *Prof Exp:* Asst physiol, Univ Mich, 35-39; instr, Sch Med, Loyola Univ, Ill, 40-42, assoc, 42-45, asst prof, 45-46; asst prof neurol, Med Sch, Northwestern Univ, 47-49; assoc prof physiol, 49-52, head dept, 52-79, PROF PHYSIOL, MED SCH, ORE HEALTH SCI UNIV, 52-, ACTG VPRES ACAD AFFAIRS, 79- *Concurrent Pos:* Fel, Inst Neurol, Med Sch, Northwestern Univ, 39-40; spec consult physiol study sect, USPHS, 51-55, neurol study sect, 57-60, gen clin res ctr comt, 61-63 & physiol training comt, 63-67; Fulbright res scholar, Univ Pisa, 56-57; mem physiol test comt, Nat Bd Med Exam, 59-62 & 76-79; mem bd sci counr, Nat Inst Neurol Dis & Blindness, 61-65; mem, Int Brain Res Orgn, 62-; ed-in-chief, J Neurophysiol, 64-74; mem cent coun, Int Brain Res Orgn, 64-68 & 74-77; mem adv coun health res facil, NIH, 67-71; US deleg, Gen Assembly, Int Union Physiol Sci, 65, 68, 71 & 74, nat comt chmn, 69-75, treas & mem exec comt, 74-80. *Honors & Awards:* R G Daggs Award, Am Physiol Soc, 74. *Mem:* AAAS; Am Physiol Soc (pres-elect, 64-65, pres, 65-66); Am Acad Arts & Sci; Soc for Neuroscience; Sigma Xi. *Res:* Regional neurophysiology; systems analysis of postural control. *Mailing Add:* 3126 NE 39th Portland OR 97212

BROOKHART, MAURICE S, b Cumberland, Md, Nov 28, 42; m 65; c 2. ORGANIC CHEMISTRY. *Educ:* Johns Hopkins Univ, BA, 64; Univ Calif, Los Angeles, PhD(org chem), 68. *Prof Exp:* NATO fel, Univ Southampton, 68-69; assoc prof, 69-76, PROF ORG CHEM, UNIV NC, CHAPEL HILL, 76- *Concurrent Pos:* Vis prof, Oxford Univ, 82-83. *Mem:* Am Chem Soc. *Res:* Mechanistic and synthetic organometallic chemistry; applications of transition metal complexes in organic synthesis and catalysis. *Mailing Add:* Dept Chem Univ NC Chapel Hill NC 27514

BROOKINS, DOUGLAS GRIDLEY, b Healdsburg, Calif, Sept 27, 36; m 61; c 2. GEOCHEMISTRY. *Educ:* Univ Calif, Berkeley, AB, 58; Mass Inst Technol, PhD(isotope geol), 63. *Prof Exp:* Asst emission spectrog, Mass Inst Technol, 58-59, asst geochronology, 59-63; asst prof geochem, Kans State Univ, 63-66, assoc prof geol, 66-71; chmn dept, 76-79, PROF GEOL, UNIV NMEX, 71- *Concurrent Pos:* Consult, Mass Inst Technol, 63; US Geol Surv, State Surv Maine, Conn & Kans; vis staff mem, Los Alamos Sci Lab, 74-, Lawrence Berkely Lab, 80-87, Oak Ridge Nat Lab, 84-87. *Mem:* Am Geophys Union; fel Geol Soc Am; Geochem Soc; Meteoritical Soc; Am Chem Soc; fel Mineral Soc Am; Am Nuclear Soc; Int Assoc Geochem Cosmodiun; Clay Min Soc; Am Inst Chem; NY Acad Sci; Soc Econ Geol; Sigma Xi. *Res:* Geochronological investigations on problems of petrogenesis and correlation in regionally metamorphosed areas by the rubidium-strontium methods phase equilibrium; kimberlites; radioactive waste disposal; carbonatites; uranium geochemistry; economic geology; chemical waste; environmental geochemistry; author of 5 books and 550 articles for science literature. *Mailing Add:* Dept Geol Univ NMex Albuquerque NM 87131

BROOKMAN, DAVID JOSEPH, b Ft Collins, Colo, Oct 31, 43; m 65. ANALYTICAL CHEMISTRY. *Educ:* Colo State Univ, BS, 65; Univ Calif, Riverside, PhD(anal chem), 68. *Prof Exp:* Asst prof chem, Univ Wis-Madison, 68-70; from res chemist to sr res chemist, Stauffer Chem Co, 70-73, group supvr, 74-78, mgr anal sect, 79-85, mgr, Anal/Metab Dept, 86-87; MGR, ENVIRON SCI DEPT, ICI AMERICAS INC, 88- *Mem:* AAAS; Am Chem Soc. *Res:* Environmental chemistry; analytical separations. *Mailing Add:* ICI Americas Inc 1200 S 47th St PO Box 4023 Richmond CA 94804-0023

BROOKNER, ELI, b New York, NY, Apr 2, 31; m 55; c 2. ELECTRICAL ENGINEERING. *Educ:* City Col New York, BEE, 53; Columbia Univ, MS, 55,. *Hon Degrees:* DSc, COLUMBIA UNIV, 62. *Prof Exp:* Res engr, Electronics Res Lab, Columbia Univ, 53-57, sr res engr, 60-62; proj engr, Fed Sci Corp, 57-60, sr engr, 62-64, prin engr, 64-67, mgr, Info Systs Sect, 67-69, staff engr, Wayland Lab Mgr, 69-70, CONSULT SCIENTIST, RADAR SYSTS LAB, ADVAN SYSTS LAB & DATA ACQUISTION SYSTS, RAYTHEON CO, 70- *Concurrent Pos:* Int lectr radar technol, 78-; mem, Panel Implications Future Space Systs US Navy, Nat Acad Sci, 79-80, Air Force Ad Hoc Comt on Air Mobile MX, Army Space Initiative Panel, 85; mem, US Nat Comn, Int Union Radio Sci; mem, Defense Adv Res Proj Agency assessment of ultra wide band technol, 90, Advan Radar Technol Panel, Naval Studies Bd, Nat Res Coun, 90-91; Distinguished lectr, Inst Elec & Electronics Engrs Antennas & Propagation Soc, 83-85, Aerospace & Electronics Systs, 87-91. *Honors & Awards:* Centennial Medal, Inst Elec & Electronics Engrs, 84, Region One Award, 86, Meritorious Achievement Award, 90. *Mem:* Fel Inst Elec & Electronics Engrs; assoc fel Am Inst Aeronaut & Astronaut. *Res:* Surveillance, defense, space-borne, pulse doppler, synthetic aperture, marine and millimeter radar design; noise studies; signal processing; detection theory; waveform analysis; decoy discrimination; ionospheric propagation; target signatures; characterization of laser and millimeter communication channels; modulation. *Mailing Add:* Raytheon Co Equip Develop Boston Post Rd Wayland MA 01778

BROOKS, ALBERT LAW, biological oceanography, for more information see previous edition

BROOKS, ALFRED AUSTIN, JR, b Swampscott, Mass, Aug 3, 21; m 43; c 2. PHYSICAL CHEMISTRY. *Educ:* Hobart Col, BA, 46; Ohio State Univ, PhD(chem), 50. *Prof Exp:* Chemist, Lake Ontario Ord Works, 43 & Manhattan Proj, 43-45; instr, Univ Chicago, 50-52; res chemist, Standard Oil Co, 52-54 & Upjohn Co, 54-56; phys chemist, Union Carbide Nuclear, 56-62 & Cent Data Processing Div, 62-70; head comput applns dept, Oak Ridge Nat Lab, 70-73; mgr, Comput Appln Dept, 73-81, MGR INFO SYST, COMPUT SCI DIV, NUCLEAR DIV, UNION CARBIDE CORP, 81- *Mem:* Am Chem Soc; Am Phys Soc; Asn Comput Mach; Am Soc Info Sci; Sigma Xi. *Res:* Application of high speed computers to information processing. *Mailing Add:* 100 Wiltshire Dr Oak Ridge TN 37830

BROOKS, ANTONE L, b St George, Utah, July 7, 38; m 63; c 6. GENETIC TOXICOLOGY. *Educ:* Univ Utah, BS, 61, MS, 63; Cornell Univ, PhD(phys biol), 67. *Prof Exp:* MEM SR STAFF, LOVELACE FOUND, 67- *Concurrent Pos:* Tech rep, Dept Energy, 76-78. *Mem:* Radiation Res Soc; Environ Mutagen Soc; Nat Coun Radiation Protection & Measurements. *Res:* Distribution and uptake of radioactive fallout, effects of radiation on the chromosomes of somatic and reproductive tissue; effects of internal emitters on chromosomes; genotoxic effects of pollutants from coal utilization and diesel exhaust; carcinogenic hazards of alpha emitters. *Mailing Add:* 6802 W 13th Ave Kennewick WA 99337-1305

BROOKS, ARTHUR S, b St Johnsbury, Vt, Apr 25, 43; m 66; c 2. PLANKTON ECOLOGY. *Educ:* Wash & Jefferson Col, BA, 65; Univ Vt, MS, 67; Johns Hopkins Univ, PhD(environ biol), 72. *Prof Exp:* Postdoc fel, Ctr Great Lakes Studies, 72-73, from asst prof to assoc prof zool & limnol, Dept Zool, 78-88, PROF LIMNOL, DEPT BIOL SCI & CTR GREAT LAKES STUDIES, UNIV WIS MILWAUKEE, 88- *Concurrent Pos:* Vis scientist, Freshwater Biol Asn, Windermere Lab, Eng, 83-84; fel, Fairbanks Mus Natural Sci, 88. *Mem:* Sigma Xi; AAAS; Am Soc Limnol & Oceanog; Int Asn Great Lakes Res; Int Soc Limnol; Freshwater Biol Asn. *Res:* Ecology of plankton of large lakes; physical and chemical environment in which plankton live; biological interactions and influence of human activities; long-term climactic variability on aquatic ecosystems. *Mailing Add:* Ctr Great Lakes Studies Univ Wis Milwaukee 600 E Greenfield Ave Milwaukee WI 53204

BROOKS, AUSTIN EDWARD, b Ft Wayne, Ind, Aug 10, 38; m 63; c 2. PHYCOLOGY. *Educ:* Wabash Col, AB, 61; Ind Univ, PhD(microbiol), 65. *Prof Exp:* Res assoc biol, Brown Univ, 65-66; from asst prof to assoc prof, 66-80, chair, dept biol, 78-83, PROF BIOL, WABASH COL, 81-, CHAIR, SCI DIV, 82- *Concurrent Pos:* Alexander von Humboldt Found fel, 74-75; consult, fS/Thailand NSF Curric Proj, 84; Hopkins-Dana fel, dept biomed eng, Johns Hopkins Med Sch, 87-88. *Mem:* AAAS; Phycol Soc Am; Int Phycol Soc; Sigma Xi; Found Sci & Handicapped; Nat Sci Teachers Asn. *Res:* Morphology and physiology of green flagellated algae; physiological effects of environmental chemicals on green algae; teaching materials development for the visually impaired; calcium metabolism in vascular smooth muscle. *Mailing Add:* Dept Biol Wabash Col Crawfordsville IN 47933

BROOKS, BARBARA ALICE, b New York, NY, June 18, 34. NEUROSCIENCE. *Educ:* Univ Fla, BA, 57, MA, 59; Fla State Univ, PhD(exp psychol), 64. *Prof Exp:* Vis scientist, Max Planck Inst Brain Res, Frankfurt, 65-66; NIH fel, Univ Freiburg, 66-69; vis scientist, Max Planck Inst, Bad Nauheim, 70-71 & Inst Ophthal, Univ London, 71; NIH sr fel & res asst prof, Univ Wash, 72-73; asst prof, 74-78, ASSOC PROF NEUROSCI, DEPT PHYSIOL, UNIV TEX HEALTH SCI CTR, SAN ANTONIO, 78- *Concurrent Pos:* Prin investr, Eye Inst, NIH res grant, 74-79. *Mem:* Am Physiol Soc; Asn Res in Vision & Ophthal; Neurosci Soc; Sigma Xi. *Res:* Neurophysiology and psychophysics of vision during normal eye movement; sensory-motor integration. *Mailing Add:* Dept Physiol Univ Tex Health Sci Ctr 7703 Floyd Curl Dr San Antonio TX 78284

BROOKS, BRADFORD O, b May 5, 51; m; c 2. IMMUNOTOXICOLOGY, CLINICAL IMMUNOLOGY. *Educ:* Harding Univ, BS, 73; Mont State Univ, MS, 76, PhD(immunol & microbiol), 79. *Prof Exp:* SR IBM SCIENTIST, IBM CORP, 82- *Concurrent Pos:* Pres & chief exec officer, Immunocompetence Inc, 80- *Mem:* Am Asn Immunologists; Int Soc Immunopharmacol; Am Soc Microbiol; Royal Soc Trop Med & Hygiene. *Res:* Influence of chemicals on the immune system; novel bioassays for detection of human chemical exposure; indoor air pollution; parasitology. *Mailing Add:* IBM Corp PO Box 1900 Boulder CO 80302

BROOKS, CHANDLER MCCUSKEY, physiology; deceased, see previous edition for last biography

BROOKS, CHARLIE R(AY), b Pressmen's Home, Tenn, Sept 20, 31; m 57; c 3. METALLURGICAL ENGINEERING. *Educ:* Univ Tenn, BS, 54, MS, 59, PhD(metall eng), 62. *Prof Exp:* Asst metall, 54-56, from instr to assoc prof, 55-69, PROF METALL ENG, UNIV TENN, 69- *Concurrent Pos:* Prof & adv engr, Univ Panama, 63-64. *Mem:* Fel Am Soc Metals; Int Metallog Soc. *Res:* Measurement of specific heats of metals from 0 to 1000 degrees centigrade; adiabatic calorimetry; order-disorder reactions; electron microscopy. *Mailing Add:* Dept Chem & Metall Eng Univ Tenn Knoxville TN 37916

BROOKS, CLYDE S, b Pittsburgh, Pa, May 1, 17; m 46; c 2. PHYSICAL CHEMISTRY. *Educ:* Duke Univ, BS, 40. *Prof Exp:* Koppers Co fel tar synthetics, Mellon Inst, 41-44, assoc fel coal prod phys chem, 46-49; res chemist prod res, Shell Develop Co, 49-61; sr res scientist, Res Labs, United Technologies Res Ctr, 61-81; CONSULT, APPL PHYS CHEM, 81- *Concurrent Pos:* Pres, Recycle Metals, 83- *Mem:* Sigma Xi (secy, 73-75); Am Chem Soc; fel Am Inst Chemists; Metall Soc; NAm Catalysis Soc. *Res:* Surface chemistry; heterogeneous catalysis; colloids; application of physical chemistry to elimination of environmental pollution; applications of catalysis in fossil fuel and chemical conversion processes; fossil fuel desulfurization; recovery of waste metals. *Mailing Add:* 41 Baldwin Lane Glastonbury CT 06033

BROOKS, DANIEL RUSK, b St Louis, Mo, Apr 12, 51; m 87; c 1. SYSTEMATIC BIOLOGY, PARASITOLOGY. *Educ:* Univ Nebr-Lincoln, BS, 73, MS, 75; Univ Miss, PhD(biol), 78. *Prof Exp:* NIH fel, Univ Notre Dame, 78-79; mem staff, Path Dept, Zool Park, Smithsonian Inst, 79-80; from asst prof to assoc prof, Zool Dept, Univ BC, 80-88; assoc prof, 88-91, PROF, ZOOL DEPT, UNIV TORONTO, 91- *Honors & Awards:* H B Ward Medal, Am Soc Parasitologists, 85. *Mem:* Soc Syst Zool; Am Soc Parasitologists; Sigma Xi; Can Soc Zool. *Res:* Phylogenetics of parasitic helminths; historical biogeography; evolutionary biology. *Mailing Add:* Dept Zool Univ Toronto Toronto ON M5S 1A1 Can

BROOKS, DAVID ARTHUR, b St John, NB, May 6, 43; US citizen; m 69; c 2. PHYSICAL OCEANOGRAPHY, ATMOSPHERIC SCIENCE. *Educ:* Univ Maine, Orono, BS, 65; Univ Miami, MS, 71, PhD(phys oceanog), 75. *Prof Exp:* Syst engr elec eng, Gen Elec Co, 65-69; res asst oceanog, Univ Miami, 69-75; res assoc phys oceanog, NC State Univ, 75-78; asst prof phys oceanog, 78-83, ASSOC PROF OCEANOG, TEX A&M UNIV, 83- *Concurrent Pos:* Adj asst prof, NC State Univ, 78-82. *Mem:* Am Geophys Union; AAAS; Am Meteorol Soc. *Res:* Gulf stream dynamics; generation, propagation and dissipation of meanders; continental shelf and slope dynamics; Gulf of Maine circulation. *Mailing Add:* Dept Oceanog Tex A&M Univ College Station TX 77843

BROOKS, DAVID C, b Farmington, NMex, Feb 14, 36. MATHEMATICS. *Educ:* Seattle Pacific Univ, BS, 58; Univ Wash, MS, 60, PhD(math), 78. *Prof Exp:* PROF MATH, SEATTLE PACIFIC UNIV, 67- *Mem:* Am Math Soc. *Mailing Add:* Dept Math Seattle Pacific Univ Seattle WA 98119

BROOKS, DAVID PATRICK, b London, Eng, May 31, 52; US citizen; m 78; c 2. RENAL PHARMACOLOGY, RENAL PHYSIOLOGY. *Educ:* Univ London, BSc, 74; Southampton Univ, MSc, 76, PhD(physiol), 79. *Prof Exp:* Teaching fel physiol, Southampton Univ, 78-79; res fel, Univ Hawaii, 79-81; asst prof, Univ Tenn, Memphis, 81-86; assoc sr investr pharmacol, 86-88, SR INVESTR PHARMACOL, SMITHKLINE & FRENCH LABS, 88-, ASST DIR PHARMACOL, SMITHKLINE BEECHAM PHARMACEUT, 88- *Concurrent Pos:* Adj asst prof, Sch Med, Univ Pa, 87- *Mem:* Am Physiol Soc; Am Soc Pharmacol & Exp Therapeut; Endocrine Soc; Soc Neurosci; Am Soc Nephrology; Am Soc Hypertension. *Res:* Mechanisms involved in hypertension and chronic renal failure; discovery of novel therapeutics for hypertension and chronic renal failure. *Mailing Add:* Smithkline Beecham Pharmaceut L521 King of Prussia PA 19406-0939

BROOKS, DEE W, b Provo, Utah, Mar 2, 52; m 72; c 3. MEDICINAL CHEMISTRY, ENZYME CHEMISTRY. *Educ:* Univ Lethbridge, BSc, 74; Univ Alta, PhD(chem), 78. *Prof Exp:* Fel chem, Mass Inst Technol, 78-79; asst prof, Purdue Univ, 79-84; group leader med chem, 84-87, PROJ LEADER MED CHEM, ABBOTT LABS, 87- *Mem:* Am Chem Soc; Sigma Xi. *Res:* Pharmaceutical discovery in the area of immunosciences, inflammation, allergy, asthma and arthritis; development of new methods in medicinal chemistry research and organic synthesis; applications of microbial and enzyme chemistry. *Mailing Add:* Dept 47K Abbott Labs Abbott Park IL 60064

BROOKS, DERL, b Headrick, Okla, July 25, 30; m 57; c 1. ENTOMOLOGY. *Educ:* Tex Tech Col, BS, 57, MS, 59; Iowa State Univ, PhD(entom), 62. *Prof Exp:* Assoc prof biol, Ark State Col, 62-65; assoc prof, 65-74, PROF BIOL, WTEX STATE UNIV, 74- *Res:* Acarology, chiefly nasal mites of birds. *Mailing Add:* Dept Biol WTex State Univ Canyon TX 79016

BROOKS, DOROTHY LYNN, b Trinidad, Tex, June 4, 35; m 80; c 4. EDUCATION ADMINSTRATION, ALGEBRA. *Educ:* Baylor Univ, BA, 55; Tex Christian Univ, MA, 61; NTex State Univ, PhD(higher educ admin), 81. *Prof Exp:* Teacher math, Univ High Sch, Waco, Tex, 55-58; teaching asst math, Tex Christian Univ, 59-61; instr math, Tex Wesleyan Col, 62-63; from instr to asst prof, 62-82, asst dean sci, 75-82, coordr instnl planning, 82-87, ASST PROF MATH, UNIV TEX, ARLINGTON, 87- *Mem:* Math Asn Am; Asn Women Math; Am Phys Soc; Sigma Xi. *Mailing Add:* 720 Briarwood Blvd Arlington TX 76013-1501

BROOKS, DOUGLAS LEE, b New Haven, Conn, Aug 5, 16; m 41; c 4. SCIENCE POLICY. *Educ:* Yale Univ, BS, 38; Mass Inst Technol, SM, 43, ScD, 48. *Prof Exp:* Instr math, Hill Sch-Hotchkiss Sch, 38-42; pilot & weather officer, US Army Air Corps, 42-46; opers analyst & dir res, Naval Opers Anal, US Navy, 48-62; pres contract res, Travelers Res Ctr, 63-70; spec asst to dir sci policy, NSF, 70-72, exec dir, Nat Adv Coun Oceans & Atmosphere, 72-79; AUTHOR, 79- *Mem:* Fel NY Acad Sci; Am Meteorol Soc; AAAS; Marine Technol Soc; fel NY Acad Sci. *Res:* Atmospheric physics & environmental management; naval operations & defense policies; national, environmental & strategic policy. *Mailing Add:* 40 Loeffler Rd Apt 303P Bloomfield CT 06002

BROOKS, EDWARD MORGAN, b New Haven, Conn, Mar 19, 16; m 41; c 8. METEOROLOGY. *Educ:* Harvard Univ, AB, 37; Mass Inst Technol, SM, 39, ScD(meteorol), 45. *Prof Exp:* Asst observer & asst, Blue Hill Observ, Harvard Univ, 37-39; asst & map analyst, Radiosonde Sta, Mass Inst Technol, 39; map analyst & terminal forecaster, Pan Am Airways, Hawaii, 39-40, tutor meteorol, Calif, 40-41, instr, Fla, 41-42; instr, Mass Inst Technol, 42-46; from asst prof to prof geophys, Inst Technol, St Louis, 46-61; sr meteorologist, Allied Res Assocs, Mass, 61-63; staff scientist, Geophysics Corp Am, 63-65; lectr geophys, Boston Col, 65-67, prof geol & geophys, 68-81 & 85-; guest prof meteorol, King Abdul-Aziz Univ, Jeddah, Saudi Arabia, 81-84; PROF GEOL & GEOPHYS, BOSTON COL, 85- *Concurrent Pos:* Scholar, Milton Acad, 32-33; lectr, Boston Ctr Adult Educ, Mass, 43; ed, Mt Washington Observ News Bull, 65-71; meteorologist, Wallace Howell Assoc, 67 & Edgerton, Germeshausen & Grier, 67; prof physics, State Univ NY Plattsburgh, 67-68; guest prof meteorol, McGill Univ, 67-68; consult, Ocean-Atmosphere Res Inst, 68-69, Fla Power & Light Co, 69-70, Cent Weather Bur, Taiwan, 72-73 & Int Group Training Facil, 79-80; instr & asst prof meteorol & oceanog, div grad & continuing studies, Framingham State Col, Mass, 71-73; vis prof meteorol & chmn dept, Inst Astron & Geophys, Univ Sao Paulo, Brazil, 74-77; extended forecaster, Chicago Bd Trade, 77-80, William E Simpson, Inc, 79, Farmers Export Co, 79-81 & Ralston Purina Co, 80-81. *Honors & Awards:* Soc of the Sigma Xi, 39. *Mem:* AAAS; Am Meteorol Soc; Am Geophys Union; Am Astron Soc. *Res:* Physical oceanography; tropical cyclones; polar front in tropics; tropical cloudiness; solar radiation; microbarography; tornadoes; Coriolis force; pilot balloon accuracy; isentropic analysis; satellite meteorology. *Mailing Add:* Dept Geol & Geophys Boston Col Chestnut Hill MA 02167-3809

BROOKS, ELWOOD RALPH, b Charlevoix, Mich, Aug 10, 34; m 88; c 5. VOLCANOLOGY. *Educ:* Mich Col Mining & Technol, BS, 56; Univ Calif, Berkeley, MS, 58; Univ Wis, PhD(geol), 64. *Prof Exp:* Mine geologist, White Pine Copper Co, Mich, 59-60; lectr geol, Univ Calif, Riverside, 64-65; chmn dept, 76-79 & 85-88, PROF GEOL, CALIF STATE UNIV, HAYWARD, 65- *Concurrent Pos:* Res grants, 64, 66-67, 70-72, 74, 77-82, 86, 88 & 90; mem, Penrose Conf Comt, Geol Soc Am, 81-83; chair, Short Courses Comt, Geol Soc Am, 87-89. *Mem:* Geol Soc Am; Nat Asn Geol Teachers; Am Geophys Union. *Res:* Igneous and metamorphic petrology, particularly petrology of ancient volcanic rocks; optical mineralogy and petrography; field geology. *Mailing Add:* Dept Geol Sci Calif State Univ Hayward CA 94542-3088

BROOKS, FRANK PICKERING, physiology; deceased, see previous edition for last biography

BROOKS, FREDERICK P(HILLIPS), JR, b Durham, NC, Apr 19, 31; m 56; c 3. COMPUTER SCIENCE. *Educ:* Duke Univ, AB, 53; Harvard Univ, SM, 55, PhD(comput sci), 56. *Prof Exp:* Assoc engr, Prod Develop Lab, Int Bus Mach Corp, 56, staff engr, 56-58, adv syst planner, 58-59, mem res staff, 59-60, systs planning mgr, 60-61, corp processor mgr & Syst/360 mgr, 61-64, Operating Syst/360 mgr, 64-65; prof, 64-75, chmn dept, 64-84, KENAN PROF COMPUT SCI, DEPT COMPUT SCI, UNIV NC, CHAPEL HILL, 75- *Concurrent Pos:* Vis instr, Vassar Col, 58; adj asst prof, Columbia Univ, 60-61; mem-at-large, Nat Coun, Asn Comput Mach, 66-70, bd dirs, Spec Interest Group Computer Archit, 73-75; vis prof, Twente Tech Univ, Enschede, Neth, 70; Guggenheim fel, 75; mem, Computer Sci & Technol Bd, Nat Res Coun, 77-80; mem, Artificial Intel Task Force, Defense Sci Bd, 83-84, Computers Simulation & Training Task Force, 86-87, chmn, Mil Software Task Force, 85-87; mem, Nat Sci Bd, 87- *Honors & Awards:* W W McDowell Award, Inst Elec & Electronics Engrs, 70, Comput Pioneer Award, 82; Comput Sci Award, Data Processing Mgt Asn, 70; Nat Medal Technol, 85; Distinguished Serv Award, Asn Comput Mach, 87; Harry Goode Mem Award, Am Fedn Info Processing Socs, 89. *Mem:* Nat Acad Eng; fel Inst Elec & Electronics Engrs; Asn Comput Mach; fel Am Acad Arts & Sci. *Res:* Interactive computer graphics, man-machine interfaces; applications to molecular structure studies, dynamic architectural visualization ("walk-through"); architecture of digital computers, especially design of machine language; software engineering, especially programming-in-the-large and project management; two US patents. *Mailing Add:* Dept Computer Sci Univ NC Chapel Hill NC 27599-3175

BROOKS, GARNETT RYLAND, JR, b Richmond, Va, Nov 25, 36; div. ECOLOGY. *Educ:* Univ Richmond, BS, 57, MS, 59; Univ Fla, PhD(zool), 63. *Prof Exp:* From instr to asst prof, 62-68, assoc prof, 68-73, PROF BIOL, COL WILLIAM & MARY, 73- *Mem:* AAAS; Ecol Soc Am; Am Soc Ichthyologists & Herpetologists. *Res:* Physiological ecology of certain species of reptiles. *Mailing Add:* Dept Biol Col William & Mary Williamsburg VA 23185

BROOKS, GEORGE H(ENRY), industrial engineering; deceased, see previous edition for last biography

BROOKS, GEORGE WILSON, b Warren, Vt, Feb 11, 20; m 43; c 4. PSYCHIATRY. *Educ:* Univ NH, BS, 41; Univ Vt, MD, 44; Am Bd Psychiat & Neurol, dipl psychiat, 55. *Prof Exp:* Intern, Mary Fletcher Hosp, Burlington, 44-45; resident psychiat, Vt State Hosp, Waterbury, 47-49, asst physician, 49-51; exchange resident, NH State Hosp, Concord, 52; sr psychiatrist, Vt State Hosp, Waterbury, 53-56, dir res & staff educ, 57-61, asst supt, 61-68, actg supt, 68; from instr to assoc prof, 53-70, PROF CLIN PSYCHIAT, COL MED, UNIV VT, 70-; SUPT, VT STATE HOSP, 68- *Concurrent Pos:* Smith, Kline & French Found fel, Mass Ment Health Ctr, Boston, 56-57. *Mem:* Fel Am Psychiat Asn; AMA; Asn Med Supt Ment Hosps; fel Am Col Physicians. *Res:* Biological and behavioral research; schizophrenia. *Mailing Add:* Vt State Hosp Waterbury VT 05676

BROOKS, HAROLD KELLY, b Winfield, Kans, Nov 27, 24; m 62; c 4. GEOLOGY. *Educ:* Kans State Univ, BS, 47; Harvard Univ, AM, 50, PhD, 62. *Prof Exp:* Instr geol, Brown Univ, 50-51; instr, Oberlin Col, 51-52; asst prof, Univ Tenn, 52-54 & Univ Cincinnati, 54-56; from asst prof to prof, 56-80, EMER PROF GEOL, UNIV FLA, 80- *Concurrent Pos:* Consult, eng geol, water resources & hazardous waste. *Honors & Awards:* Antarctic Serv Medal, USA. *Mem:* Am Asn Petrol Geologist; Fel Geol Soc Am. *Res:* Marine geology; sedimentology; paleontology; geology and physiology of the state of Florida with published maps of the complete state. *Mailing Add:* PO Box 729 Patillas PR 00723

BROOKS, HARVEY, b Cleveland, Ohio, Aug 5, 15; m 45; c 4. SCIENCE, TECHNOLOGY & PUBLIC POLICY. *Educ:* Yale Univ, AB, 37; Harvard Univ, PhD, 40. *Hon Degrees:* DSc, Yale Univ, 62, Union Col, 62, Harvard Univ, 63, Kenyon Col, 63, Brown Univ, 64. *Prof Exp:* Mem staff, Underwater Sound Lab, Harvard Univ, 42-45; asst dir, Ord Res Lab, Pa State Univ, 45; res assoc, Res Lab & assoc lab head, Knolls Atomic Power Lab, Gen Elec Co, 46-50; dea, Div Eng & Appl Physics, Harvard Univ, 57-75, prof appl physics, 50-89 & technol & pub policy, 75-89, BENJAMIN PIERCE EMER PROF TECHNOL & PUB POLICY, KENNEDY SCH GOVT & GORDON MCKAY EMER PROF APPL PHYSICS, HARVARD UNIV, 89- *Concurrent Pos:* Guggenheim fel, 56-57; ed-in-chief, J Physics & Chem Solids, 56-80; mem adv comts reactor safeguards, progs & policies, AEC, 58; chmn solid state adv panel, Off Naval Res; chmn comt undersea warfare, Nat Res Coun, 57-63; mem, President's Sci Adv Comt, 59-64 & Nat Sci Bd, 62-74; chmn comt sci & pub policy, Nat Acad Sci, 66-72; mem, Adv Comt Sci & Technol Develop, UN, 87. *Mem:* Nat Acad Sci; Nat Acad Eng; sr mem Inst Med-Nat Acad Sci; Am Philos Soc; fel Am Phys Soc; Am Acad Arts & Sci (pres, 70-75). *Res:* Solid state physics; underwater sound; nuclear reactors; science policy; author of numerous technical publications and one book. *Mailing Add:* JFK Sch Govt Harvard Univ 79 JFK St Cambridge MA 02138

BROOKS, JACK CARLTON, b Hornell, NY, May 17, 41; m 65; c 1. PHYSIOLOGY, NEUROSCIENCE. *Educ:* Alfred Univ, BA, 63; Univ Tex, Austin, PhD(zool, physiol), 69. *Prof Exp:* Fel enzym, Univ Wis-Madison, 69-72, fel neurosci, 72-74; from asst prof to assoc prof, 74-91, PROF PHYSIOL, SCH DENT, MARQUETTE UNIV, 91- *Concurrent Pos:* NIH trainee, Univ Wis-Madison, 69-72, NIH fel, 72-73; NSF grants, 83-87; NIH grant, 89-92. *Mem:* Soc Neurosci; Am Soc Neurochem. *Res:* Stimulation; secretion coupling in neurons and neurosecretory cells. *Mailing Add:* Dept Basic Sci Sch Dent Marquette Univ 604 N 16th St Milwaukee WI 53233

BROOKS, JAMES ELWOOD, b Salem, Ind, May 31, 25; m 49; c 3. GEOLOGY, APPLIED, ECONOMIC & ENGINEERING. *Educ:* DePauw Univ, AB, 48; Northwestern Univ, MS, 50; Univ Wash, PhD(geol), 54. *Prof Exp:* From instr to assoc prof geol, 52-62, chmn dept, 61-70, assoc provost & dean humanities & sci, 69-72, provost & vpres ac affairs, 72-80, pres ad interim, 80-81, PROF GEOL, SOUTHERN METHODIST UNIV, 62-, PRES, INST STUDY EARTH & MAN, 81- *Concurrent Pos:* Consult geologist, Gulf Oil Corp, 52-54, DeGolyer & MacNaughton Inc, 54-59 & Egyptian Geol Surv, 67-76. *Mem:* Fel AAAS; Am Asn Petrol Geologists; fel Geol Soc Am. *Res:* Devonian and mid-Paleozoic stratigraphy of Eastern Great Basin and Pennsylvanian sedimentary petrology of north central and west Texas; origin of Qattara Depression, Egypt; regional geology North Yemen, stratigraphy and geomorphology; history of geologic concepts. *Mailing Add:* Inst Study Earth & Man Southern Methodist Univ Dallas TX 75275-0274

BROOKS, JAMES KEITH, b Cleveland, Ohio, Sept 26, 38; m 61; c 2. MATHEMATICS. *Educ:* John Carroll Univ, BS, 59, MS, 61; Ohio State Univ, PhD(math), 64. *Prof Exp:* Asst prof math, Ohio State Univ, 64-66; vis lectr, Univ Southampton, 66-67; from asst prof to assoc prof, 67-74, PROF MATH, UNIV FLA, 74- *Mem:* Am Math Soc; Math Asn Am. *Res:* Functional analysis, measure and integration theory; probability theory. *Mailing Add:* Dept Math Univ Fla Gainesville FL 32611

BROOKS, JAMES LEE, b Toledo, Ohio, Sept 11, 37; m 62; c 3. BIOCHEMISTRY. *Educ:* San Diego State Col, BS, 59, MS, 61; Univ Calif, Davis, PhD(biochem), 65. *Prof Exp:* Res assoc biochem, Univ Calif, Davis, 65, Scripps Inst, Univ Calif, 65-66 & Okla State Univ, 66-67; asst prof, 67-74, asst agr biochemist, 69-74, ASSOC PROF AGR BIOCHEM & ASSOC AGR BIOCHEMIST, WVA UNIV, 74- *Mem:* AAAS. *Res:* Enzymes in fruit ripening; control of carbohydrate metabolism; fatty acid biosynthesis; sulfur metabolism in algae and bacteria; bacterial lipids; lignification in plants. *Mailing Add:* Plant & Soil Sci WVa Univ Morgantown WV 26506-6103

BROOKS, JAMES MARK, b Kansas City, Mo, Dec 29, 47; m 75; c 4. ENVIRONMENTAL CHEMISTRY, GEOCHEMISTRY. *Educ:* Abilene Christian Univ, BS, 69; Tex A&M Univ, MS, 70, PhD(oceanog), 75. *Prof Exp:* Res assoc, 73-76, RES SCIENTIST OCEANOG, TEX A&M UNIV, 76- *Mem:* Am Soc Limnol & Oceanog; Am Geophys Union; AAAS; Am Chem Soc; Sigma Xi. *Res:* Dissolved gas geochemistry; geochemical prospecting; stable isotope geochemistry. *Mailing Add:* Col Geosci Tex A&M Univ 833 Graham Rd College Station TX 77845

BROOKS, JAMES O, b Evanston, Ill, July 7, 30; m 58; c 2. ALGEBRA. *Educ:* Oberlin Col, AB, 52; Univ Mich, MA, 53, PhD(math), 64. *Prof Exp:* Asst prof math, Haverford Col, 59-64; asst prof, 64-66, actg chmn dept, 68-69, chmn dept, 69-77, ASSOC PROF MATH, VILLANOVA UNIV, 66- *Mem:* Am Math Soc; Math Asn Am. *Res:* Algebraic number theory. *Mailing Add:* 441 S Valley Forge Rd Wayne PA 19087

BROOKS, JAMES REED, b Steubenville, Ohio, July 16, 55; m 85; c 1. PLANT PROTEIN FRACTIONATION & CHARACTERIZATION, PROTEIN ALLERGENICITY. *Educ:* Ohio State Univ, BS, 77; Clemson Univ, MS, 81, PhD(plant physiol), 84. *Prof Exp:* Asst prof food sci, Univ Ark, 84-87; GROUP LEADER PROD RES & DEVELOP, ROSS LABS, DIV ABBOTT LABS, 87- *Mem:* Inst Food Technologists; Sigma Xi. *Res:* Plant storage protein fractionation and characterization; plant protein allergen identification; starch hydrolysis and utilization. *Mailing Add:* Dept 104230-S4 Ross Labs 625 Cleveland Ave Columbus OH 43216

BROOKS, JERRY R, b Barboursville, WVa, Sept 9, 30. REPRODUCTIVE ENDOCRINOLOGY. *Educ:* WVa Univ, BS, 53, MS, 58; Univ Mo, PhD(animal husb), 61. *Prof Exp:* Asst dairy husb, Univ Mo, 61-63; Ford Found res fel reproductive physiol, Worcester Found Exp Biol, Mass, 63-65; sr endocrinologist, Merck & Co, Inc, 66-72, res fel, 72-81, sr res fel, 81-85, SR INVESTR, MERCK & CO, INC, 85- *Concurrent Pos:* Coadjutant staff, Rutgers Univ, 77. *Mem:* AAAS; NY Acad Sci; Sigma Xi. *Res:* Causes of and therapy for benign prostatic hyperplasia; anti-acne agents. *Mailing Add:* 111 Knollwood Dr Watchuno NJ 07060-6245

BROOKS, JOHN A(LBERT), b Denver, Colo, Nov 30, 16; m 47; c 1. CHEMICAL ENGINEERING. *Educ:* Ill Inst Technol, BS, 46. *Prof Exp:* Chem engr, res dept, Standard Oil Co, Ind, 46-51, asst proj chem engr, 51-53, group leader, 53-60, process coordr, Am Oil Co Div, 60-65, supt tech serv, Am Oil Co, 65-68, mgr planning, scheduling & tech serv, 68-72; mgr eng & tech, Amoco Oil Co, Texas City, 72-74; dir process & synthetic fuels, Amoco Res Ctr, Amoco Oil Co, 76-78, dir contracts & admin, 78-81; RETIRED. *Concurrent Pos:* Stockbroker, 82-87. *Mem:* Am Chem Soc; Am Inst Chem Engrs. *Res:* Petroleum. *Mailing Add:* 227 Sea Anchor Dr Osprey FL 34229

BROOKS, JOHN BILL, b Hyatt, Tenn, Aug 9, 29; m 51; c 2. BIOCHEMISTRY, MICROBIOLOGY. *Educ:* Western Carolina Univ, BS, 62; Va Polytech Inst, PhD(biochem, microbiol), 69. *Prof Exp:* Chemist, Tenn Copper Co, 59; biol aide, Ctr Dis Control, 61-62; chemist, NIH, 62-63; biochemist, 63-69, RES CHEMIST, CTR DIS CONTROL, 69- *Honors & Awards:* Kimble Metholology Res Award, 78. *Mem:* Am Soc Microbiol; Sigma Xi. *Res:* Develop methods for rapid identification of microorganisms by analysis of their metabolites or cellular constituents both in vivo and in vitro by advanced chemical procedures such as gas chromatography and spectrophotometry; author or coauthor of sixty-six publications in this area. *Mailing Add:* Bldg 5 Rm 308 Ctrs Dis Control 1600 Clifton Rd Atlanta GA 30333

BROOKS, JOHN J, b Philadelphia, Pa, Feb 2, 48; m 69; c 3. HEMATOPATHOLOGY & IMMUNOPATHOLOGY, SOFT TISSUE PATHOLOGY. *Educ:* St Josephs Univ, BS, 70; Thomas Jefferson Med Sch, MD, 74. *Prof Exp:* Resident path, Hosp Univ Pa, 74-78; asst prof, Hahnemann Med Col & Hosp, 78-79; from asst prof to assoc prof, 79-88, PROF PATH, UNIV PA MED SCH, 88- *Concurrent Pos:* Staff pathologist, Hosp Univ Pa, 79-88; vis prof, Royal Marsden Hosp, London, 87-88. *Mem:* Am Asn Pathologists; AAAS; Int Acad Path; NY Acad Sci; Soc Hemotopath. *Res:* Using immunohistochemistry, tumor markers are studied as they apply to the diagnosis of various types of cancer and to our understanding of the growth and differentiation of cancer cells, particularly sarcomas. *Mailing Add:* Div Surg Path Hosp Univ Pa 34 & Spruce St Philadelphia PA 19104

BROOKS, JOHN LANGDON, b New Haven, Conn, Feb 10, 20; m 53; c 1. ECOLOGY, SYSTEMATICS. *Educ:* Yale Univ, BS, 41, MS, 42, PhD(zool), 46. *Prof Exp:* Lab asst, Yale Univ, 42-45, asst instr, 45-46, from instr to asst prof zool, 46-56, assoc prof biol, 56-69; prog dir gen ecol, NSF, 69-75, sect head ecol & pop biol, 75, dep div dir environ biol, 75-81, dir div biotic systs & resources, 81-89; CONSULT, CAPITOL SYSTS GROUP, INC, 90- *Concurrent Pos:* Vis prof, Univ Rangoon, 48-49; ed, Syst Zool, 52-59. *Mem:* Am Soc Zoologists; Soc Syst Zool; Ecol Soc Am; Am Soc Limnol & Oceanog; Soc Study Evolution. *Res:* Ecology and evolution of fresh water organisms; history of evolutionary concepts especially contributions of A R Wallace. *Mailing Add:* 3014 32nd St NW Washington DC 20008

BROOKS, JOHN ROBINSON, b Cambridge, Mass, Nov 15, 18; m 44; c 4. SURGERY. *Educ:* Harvard Univ, AB, 40, MD, 43. *Prof Exp:* prof surg, 70-81, CHIEF SURG, HARVARD UNIV HEALTH SERV, 62-, FRANK SAWYER PROF SURG, HARVARD MED SCH, 81- *Mem:* Am Col Surg; Soc Univ Surg; Am Surg Asn; Int Soc Surg; Soc Surg Alimentary Tract. *Res:* Gastrointestinal surgery; transplantation; endocrine organs. *Mailing Add:* 75 Mt Auburn St Cambridge MA 02138

BROOKS, JOHN S J, b Philadelphia, Pa, Feb 2, 48; m 69; c 3. SURGICAL PATHOLOGY, IMMUNOHISTOCHEMISTRY. *Educ:* St Joseph's Univ, Philadelphia, BS, 70; Thomas Jefferson Med Sch, MD, 74. *Prof Exp:* Asst prof path, Hahnemann Med Col & Hosp, 78-79; from asst prof to assoc prof, 79-88, PROF PATH, UNIV PA MED SCH, 88- *Concurrent Pos:* Vis prof, Royal Marsden Hosp & Inst Cancer Res, London, 87; assoc dir surg path, Hosp Univ Pa, Philadelphia, 88-; chmn sarcoma path, Eastern Coop Oncol Group, 88-; mem, Educ Comt, US-Can Div, Int Acad Path, 89-; prog dir, Arthur Purdy Stout Soc Surg Pathologists, 91-; counr, Am Soc Clin Path, 91- *Mem:* Int Acad Path; AAAS; NY Acad Sci; Soc Hematopath; Am Soc Clin Pathologists. *Res:* Growth and differentiation of tumors of the soft tissues and hematopoietic system; monoclonal antibodies and immunohistochemistry to identify cell types in human tumors. *Mailing Add:* Div Surg Path Hosp Univ Pa Phildelphia PA 19104

BROOKS, KENNETH CONRAD, b Chicago Heights, Ill, Apr 5, 47. GAS CHROMATOGRAPHY, LIQUID CHROMATOGRAPHY. *Educ:* Ill Inst Technol, BS, 69; Univ Colo, Boulder, MS, 73, PhD(anal chem), 81. *Prof Exp:* LECTR, CHEM DEPT, UNIV ILL, URBANA-CHAMPAIGN, 86- *Concurrent Pos:* Instr chem, Univ Colo, Colorado Springs, 74-79; asst prof, Univ Colo, Denver, 81-85. *Mem:* Am Chem Soc. *Res:* Analytical methods development for petrochemicals; chromatography; gel chromatography, high performance liquid chromatography, gas chromatography and mass spectrometry; high temperature super conductors; metal B-diketonates; chemical vapor deposition. *Mailing Add:* Sch Chem Sci 107 Chem Annex, 601 S Matthews Urban IL 61801

BROOKS, MARVIN ALAN, b Trenton, NJ, Jan 28, 45; m 68; c 2. PHARMACEUTICAL ANALYSIS. *Educ:* Lafayette Col, BS, 66; Univ Md, PhD(anal chem), 71. *Prof Exp:* Res group chief drug anal, Hoffmann-La Roche Inc, 71-85; sr res fel, 85-89, DIR PHARMACEUT ANALYSIS, MERCK, SHARP & DOHME RES LABS, 89- *Mem:* Am Chem Soc; Sigma Xi; Am Asn Pharmaceut Sci. *Res:* Analysis of drugs in dosage forms and biological fluids using chromatographic, spectrophotometric, spectrofluorometric and electrochemical methods; use of voltammetry to solve pharmaceutical problems. *Mailing Add:* Pharm Res & Develop Merck Sharp & Dohme Res Labs West Point PA 19486

BROOKS, MERLE EUGENE, b Baldwin, Kans, Feb 8, 16; m 41; c 2. BIOLOGY, BOTANY. *Educ:* Kans State Teachers Col, BS, 46, MS, 47; Univ Colo, PhD(biol), 56. *Prof Exp:* From instr to assoc prof, Kans State Teachers Col, 47-59; assoc prof biol, 59-63, prof, 63-80, EMER PROF BIOL, UNIV NEBR, OMAHA, 80- *Mem:* AAAS; Am Micros Soc; Nat Sci Teachers Asn; Nat Asn Biol Teachers; Bot Soc Am. *Res:* Limnology; Cladocera; plant morphology; science education; microbiology. *Mailing Add:* 8436 Loveland Dr Omaha NE 68124

BROOKS, NORMAN H(ERRICK), b Worcester, Mass, July 2, 28; m 48; c 3. HYDRAULICS & HYDROLOGY. *Educ:* Harvard Univ, AB, 49, MS, 50; Calif Inst Technol, PhD(civil eng, physics), 54. *Prof Exp:* From instr to prof civil eng, 53-76, acad officer environ eng sci, 69-74, DIR, ENVIRON QUAL LAB, 74-, JAMES IRVINE PROF ENVIRON & CIVIL ENG, CALIF INST TECHNOL, 76-, EXEC OFFICER, ENVIRON ENG SCI, 85- *Concurrent Pos:* Hydraul consult govt agencies & elec utilities; vis assoc prof, Grad Sch Eng, Southeast Asia Treaty Orgn, 59-60; vis prof, Mass Inst Technol, 62-63; mem, Univs Coun Water Resources; mem assembly sci & technol adv coun, Calif State legis, 70-73; mem environ studies bd, Nat Acad Sci, 73-76; guest prof, Swiss Fed Inst Technol, 84-85; mem comn eng & tech systs, Nat Acad Sci & Nat Acad Eng, 86- *Honors & Awards:* Rudolph Hering Medal, Am Soc Civil Engrs, 57, 62, Collingwood Prize, J C Stevens Award & Huber Res Prize, 59; Helgard Hydraul Prize, Am Soc Civil Engrs, 70. *Mem:* Nat Acad Sci; Nat Acad Eng; fel AAAS; Am Soc Civil Engrs; Am Geophys Union; Water Pollution Control Fedn. *Res:* Hydraulic engineering; turbulence and diffusion; sediment and pollutant transport in streams and groundwater; dispersion of wastes in ocean; density-stratified flow; environmental policy studies. *Mailing Add:* W M Keck Lab Hydraul Calif Inst Technol Pasadena CA 91125

BROOKS, PHILIP RUSSELL, b Chicago, Ill, Dec 31, 38; div; c 4. DYNAMICS OF CHEMICAL REACTIONS. *Educ:* Calif Inst Technol, BS, 60; Univ Calif, Berkeley, PhD(chem), 64. *Prof Exp:* From asst prof to assoc prof, 64-75, PROF CHEM, RICE UNIV, 75- *Concurrent Pos:* Alfred P Sloan fel, 70-72; Guggenheim Found fel, 74-75; sr US scientist award, Alexander von Humboldt Found, 85. *Mem:* Fel Am Phys Soc; Am Chem Soc. *Res:* Experimental studies on molecules in the process of reacting (transition state spectroscopy); molecular beam reactive scattering of oriented molecules. *Mailing Add:* Dept Chem Rice Univ, P O Box 1892 Houston TX 77251

BROOKS, ROBERT ALAN, b Gloversville, NY, Feb 23, 24; m 51; c 1. ORGANIC CHEMISTRY. *Educ:* Harvard Univ, BS, 44; Yale Univ, MS, 45, PhD(org chem), 49. *Prof Exp:* Instr chem, Yale Univ, 46-48; res chemist, Jackson Lab, E I Du Pont de Nemours & Co Inc, Deepwater, NJ, 48-54, res supvr, 54-57, head div dyes, 57-59, head fluorine chem, 59-61, asst lab dir, 61-62, lab mgr, 62-69, tech mgr, Dyes & Chem Div, Org Chem Dept, 69-70, dir, Dyes & Chem Res, 70-75, tech dir, Freon Prod Div, 75-80,lab dir, Petrochem Dept, Exp Sta, Wilmington, Del, 80-85; RETIRED. *Mem:* AAAS; Am Chem Soc; NY Acad Sci. *Res:* Synthetic and organic dyes; pigments and elastomers; chemistry of fluorine. *Mailing Add:* Seven Stars RR 3 Box 237 Woodstown NJ 08098

BROOKS, ROBERT E, b Los Angeles, Calif, Aug 17, 21; m 50; c 2. EXPERIMENTAL PATHOLOGY. *Educ:* Univ Calif, Los Angeles, BS, 48; Univ Ore, MS, 64, PhD(path), 67. *Prof Exp:* Res asst biochem, Atomic Energy Proj, Univ Calif, Los Angeles, 48-50; res asst path, Sch Med, Univ Calif, San Francisco, 50-60; from instr to asst prof, 60-70, ASSOC PROF PATH, MED SCH, UNIV ORE, 70- *Mem:* Am Asn Cancer Res; Electron Micros Soc Am. *Res:* Ultrastructural analysis of spontaneous and experimental animal tumors; various normal and pathologic animal and human tissues. *Mailing Add:* 11015 SW Collina Ave Portland OR 97219

BROOKS, ROBERT FRANKLIN, b Columbus, Ohio, Mar 17, 28; m 59; c 1. ENTOMOLOGY. *Educ:* Ohio State Univ, BS, 54, MS, 55; Univ Wis, PhD(entom), 60. *Prof Exp:* Asst entom, Ohio State Univ, 54-55; asst, Univ Wis, 56-59, instr, 59-60; asst entomologist, Citrus Exp Sta, Univ Fla, 60-67, assoc prof entom & assoc entomologist, Inst Food & Agr Sci, 67-70, prof, 70-87, emer prof entom & entomologist, Inst Food & Agr Sci, Univ Fla, 70-87; CONSULT, 88- *Concurrent Pos:* Consult citrus, Standard Fruit Co Div, Castle & Cook, 66-, Sun Oil Refining Co, 87- *Mem:* Entom Soc Am; Int Orgn Biol Control; Sigma Xi. *Res:* Integrated control and management of citrus pests; improving and developing more efficient application equipment for use on citrus. *Mailing Add:* Brooks Citrus Consult PO Box 307 Lake Hamilton FL 33851

BROOKS, ROBERT M, b Freeport, Tex, Jan 5, 38; m 60. MATHEMATICS. *Educ:* La State Univ, BS, 60, PhD(math), 63. *Prof Exp:* From instr to asst prof math, Univ Minn, 63-67; assoc prof, 67-72, PROF MATH, UNIV UTAH, 72- *Mem:* Am Math Soc. *Res:* Topological algebras; complex analysis. *Mailing Add:* Dept Math Univ Utah Salt Lake City UT 84112

BROOKS, ROBERT R, b Jersey City, NJ, Nov 8, 44; m 68; c 4, CARDIOVASCULAR DRUGS, ANTIINFLAMMATORY DRUGS. *Educ:* Princeton Univ, AB, 66; Johns Hopkins Univ, PhD(biochem), 71. *Prof Exp:* Postdoctoral fel, NIH, 72; res assoc, Univ Fla, Gainesville, 71-73; res scientist IV, res admin, Norwich Eaton Pharmaceuticals Inc, 73-78, scientist V, 78-83, sci assoc, 83, group leader, 83-84, SECT HEAD RES ADMIN, NORWICH EATON PHARMACEUTICALS INC, 84- *Concurrent Pos:* Instr, State Univ NY, Morrisville, 75, Binghamton, 77-80; consult, I L Richer Co, Inc, New Berlin, NY, 81-83; mem basic sci coun, Am Heart Asn. *Mem:* Am Soc Pharmacol & Exp Therapeut; Soc Exp Biol Med; Am Chem Soc; Am Soc Microbiol. *Res:* General and cardiovascular pharmacology, in vitro models, in vivo models in small and large animals arrhythmias, hypertension, gastric secretion, inflammation, development of drug testing systems. *Mailing Add:* Norwich Eaton Pharmaceut Inc PO Box 191 Norwich NY 13815

BROOKS, RONALD JAMES, b Toronto, Ont, Apr 16, 41; m 84; c 2. ECOLOGY, EVOLUTION. *Educ:* Univ Toronto, BSc, 63, MSc, 66; Univ Ill, PhD(zool), 70. *Prof Exp:* From asst prof to assoc prof 69-88, PROF ZOOL, UNIV GUELPH, 88- *Mem:* Am Soc Mammalogists; Animal Behav Soc; Can Soc Zoologists; Sigma Xi; Am Soc Naturalists; Am Soc Ichthyol & Herpetol. *Res:* Acoustic communication on Zonotrichia albicollis and Dicrostonyx groenlandicus; ontogeny of behavior and analysis of behavioral role in population changes in microtine rodents; life history, evolution and population biology of turtles; Chrysemys sp, Chelydra sp, Clemmys sp; and frogs (Rana sp). *Mailing Add:* Dept Zool Univ Guelph Guelph ON N1G 2W1 Can

BROOKS, SAM RAYMOND, b Austin, Tex, Apr 1, 40. MATHEMATICS. *Educ:* Univ Tex, BA, 62, MA, 64, PhD(math), 69. *Prof Exp:* ASST PROF MATH, MEMPHIS STATE UNIV, 66- *Mem:* Math Asn Am; Am Math Soc. *Res:* Group theory; semigroup theory and generalizations. *Mailing Add:* Dept Math Memphis State Univ Memphis TN 38152

BROOKS, SAMUEL CARROLL, b Winchester, Va, May 12, 28; m 61; c 3. BIOCHEMISTRY, ENDOCRINOLOGY. *Educ:* Carnegie Inst Technol, BS, 51; Univ Wis-Madison, MS, 55, PhD(biochem), 57. *Prof Exp:* Res scientist, John L Smith Mem Lab Cancer Res, Charles Pfizer & Co, Inc, 57-59; from instr to assoc prof, 59-74, PROF BIOCHEM, SCH MED, WAYNE STATE UNIV, 74- *Concurrent Pos:* Res assoc, Detroit Inst Cancer Res, 59-70; Fulbright res scholar, Univ Louvain, 64-65; dir, Dept Chem, Mich Cancer

Found, 70-74. *Mem:* Am Chem Soc; Soc Study Reprod; Endocrine Soc; Am Asn Cancer Res; Am Soc Biol Chem. *Res:* Steroid hormone activity and metabolism. *Mailing Add:* Dept Biochem Wayne State Univ Sch Med 540 Canfield Detroit MI 48201

BROOKS, SHARON LYNN, b Detroit, Mich, Oct 19, 44; m 65. DENTISTRY. *Educ:* Univ Mich, ABEd, 65, DDS, 73, MS, 76, MS, 84. *Prof Exp:* Pvt pract dent, 73-76; clin instr oral diag & radiol, 73-76, asst res scientist herpes virus, Dent Res Inst, 76-77, asst prof oral diag & radiol, 76-80, ASSOC PROF ORAL DIAG & RADIOL, SCH DENT, UNIV MICH, 80- *Mem:* Am Dent Asn; Am Acad Dent Radiol; Am Asn Dent Schools; Orgn Teachers Oral Diag (secy-treas, 85-); Int Asn Dent Res; Health Physics Soc. *Res:* Herpes simplex virus; infective endocarditis; selection criteria for dental radiographs. *Mailing Add:* Dept Oral Diag & Radiol Univ Mich Sch Dent Ann Arbor MI 48109-1078

BROOKS, SHEILAGH THOMPSON, b Tampico, Mex, Dec 10, 23; US citizen; m 51; c 2. PHYSICAL ANTHROPOLOGY. *Educ:* Univ Calif, Berkeley, BA, 44, MA, 47, PhD(phys anthrop), 51. *Prof Exp:* Assoc trop biogeog, Chihuahua & Durango, Mex, 55-58; res assoc physiol, Univ Calif, Berkeley, 58-62; lectr anthrop, Univ Southern Calif, 59-61; lectr, Pasadena City Col, 60-62; asst prof, Univ Colo, Denver & Boulder, 63-65; res asst, Mus, Southern Ill Univ, 65; asst prof, Univ Colo, Boulder, 65-66; assoc prof, 66-69, chmn dept, 73-77, PROF ANTHROP, UNIV NEV, LAS VEGAS, 69- *Concurrent Pos:* NSF fel arch & phys anthrop, Sarawak, Malaysia, 66; consult, Clark County Sheriff's Dept, 67-; mem coord coun, Nev Archaeol Surv, 68-77; grants, Dept Health, Educ & Welfare, mus progs, Univ Nev, Las Vegas, 69-71, fac res comt, Photog Lab, 70 & Nat Endowment Humanities, Preserv Hist Sites, 72-73; mem bd, Gov Comn Nev Lost City Mus, 71-76; consult, Nev Archaeol Surv, 73-77 & Archaeol Res Ctr, 77-; mem bd trustees, Nev State Mus, 77-81; mem adv bd, Div Hist Preserv & Archaeol, Nev. *Mem:* Am Eugenics Soc; Inst Asn Human Biol; Soc Vert Paleont; Am Acad Forensic Sci; Am Asn Phys Anthrop; Sigma Xi. *Res:* Archaeologically recovered skeletal populations with emphasis on demography, paleopathology and morphological differences; analysis and identification of historical burials, forensic physical anthropology. *Mailing Add:* Dept Anthrop Univ Nev 4505 Maryland Pkwy Las Vegas NV 89154

BROOKS, STUART MERRILL, b Cincinnati, Ohio, Apr 28, 36; c 3. PULMONARY DISEASE, OCCUPATIONAL MEDICINE. *Educ:* Univ Cincinnati, BS, 58, MD, 62; Am Bd Internal Med & Am Bd Pulmonary Dis, dipl, 69. *Prof Exp:* Resident internal med, Boston City Hosp, 63-64, 66-67, fel pulmonary dis, 67-69; res fel med, Sch Med, Tufts Univ, 68-69; from asst prof med to asst prof environ health, Col Med, Univ Cincinnati, 69-73, prof environ health & med & chief, Div Clin Studies, 73-86; PROF & CHMN, DEPT ENVIRON & OCCUP HEALTH, UNIV SFLA, 86- *Concurrent Pos:* Mem ad hoc comt case control studies on host factors as determinants of chronic obstructive pulmonary dis susceptibility, Nat Heart & Lung Inst, 70; attend physician, Cincinnati Gen Hosp & Vet Admin Hosp, Cincinnati, 73- *Mem:* Am Fedn Clin Res; Am Thoracic Soc; fel Am Col Physicians; fel Am Col Chest Physicians. *Res:* Pulmonary physiology; occupational lung diseases; pathogenetic mechanisms of bronchial asthma, non-respiratory functions of lung; corticosteroid metabolism. *Mailing Add:* Col Pub Health MH 104 Univ SFla 13301 Bruce B Downs Blvd Tampa FL 33612-3899

BROOKS, THOMAS FURMAN, b Charlotte, NC, Oct 4, 43; m 70; c 2. AEROACOUSTICS. *Educ:* NC State Univ, BS, 68, PhD(acoust & mech eng), 74. *Prof Exp:* Tool engr, Turbine Div, Westinghouse Elec Corp, 70; aerospace technologist acoust, 74-81, AEROSPACE RESEARCHER, LANGLEY RES CTR, NASA, 81- *Mem:* Acoust Soc Am; Am Inst Aeronaut & Astronaut. *Res:* Basic and applied aeroacoustics; rotor, propeller and airframe noise sources; linear acoustics, aerodynamics and turbulence. *Mailing Add:* Mail Stop 461 NASA Langley Res Ctr Hampton VA 23665

BROOKS, THOMAS JOSEPH, JR, b Starkville, Miss, May 23, 16; m 41; c 4. PREVENTIVE MEDICINE. *Educ:* Univ Fla, BS, 37; Univ Tenn, MS, 39; Univ NC, PhD(prev med), 42; Bowman Gray Sch Med, MD, 45; Am Bd Microbiol, dipl. *Prof Exp:* Instr bact & parasitol, Bowman Gray Sch Med, 42-45; intern, Bowman Gray Sch Med & NC Baptist Hosp, 45-46; assoc prof pharmacol, Sch Med, Univ Miss, 47-48; med dir, Fla State Univ Hosp, 48-52; Rockefeller Found travel grant, 52; asst dean in chg student affairs, 56-73, prof & chmn dept, 52-81, EMER PROF PREV MED, SCH MED, UNIV MISS, 81- *Concurrent Pos:* Officer-in-chg res unit, US Naval Training Ctr, Md, 53-55; consult, Vet Admin Hosp, Jackson, Miss, 56-81, UN, India, 66 & epidemiol, Miss State Dept Health, 81-; La State Univ Caribbean travel fel, 56 & 60; vis prof, Sch Med, Univ Costa Rica, 62-63, Keio Univ, Japan, 68 & Kyoto Univ, 68-69; NIH trainee, Univ Wis, 65; Alan Gregg fel, Japan & Southeast Asia, 68-69. *Mem:* Am Pub Health Asn; fel Royal Soc Health Eng; Asn Teachers Prev Med. *Res:* Treatment of filariasis; toxicology of antimony; treatment of canine filariasis with anthiomaline; tuberculosis in university students; epidemiology of streptococcal diseases; mitosis of Entamoeba histolytica; epidemiology of Echinococcus granulosus; biochemistry of schistosomes. *Mailing Add:* Dept Prev Med Univ Miss Med Ctr Jackson MS 39216

BROOKS, VERNON BERNARD, b Berlin, Ger, May 10, 23; nat Can; m 50; c 3. NEUROPHYSIOLOGY. *Educ:* Univ Toronto, BA, 46, PhD(physiol), 52; Univ Chicago, ScM, 48. *Prof Exp:* Lectr physiol, McGill Univ, 50-52, asst prof, 52-56; from asst prof to assoc prof neurophysiol, Rockefeller Inst, 56-64; prof physiol, New York Med Col, 64-71, chmn dept, 64-69; chmn dept, 71-76, prof, 71-89, EMER PROF PHYSIOL, UNIV WESTERN ONT, 89- *Concurrent Pos:* Vis fel, Australian Nat Univ, 54-55; fel, Neural Sci Inst, Rockefeller Univ, NY, 85. *Mem:* Am Physiol Soc; Can Physiol Soc. *Res:* Interaction of neurones; sensorimotor integration; motor control and learning of animals and human subjects. *Mailing Add:* Dept Physiol Univ Western Ont London ON N6A 5C1 Can

BROOKS, WALTER LYDA, b Tazewell, Tenn, Mar 6, 23; m 49; c 1. PHYSICS, ASTRONOMY. *Educ:* Lincoln Mem Univ, BA, 43; NY Univ, MS, 50, PhD(physics), 53. *Prof Exp:* Instr physics, Lincoln Mem Univ, 46-47; asst, NY Univ, 47-53; consult scientist develop div, United Nuclear Corp, 53-71, Gulf United Nuclear Fuels Co, 71-74; NUCLEAR ENGR, US NUCLEAR REGULATORY COMM, 74- *Mem:* Am Nuclear Soc; Am Chem Soc. *Res:* Reactor physics; electronics; computers; nuclear instrumentation. *Mailing Add:* 13218 Lambert Rd Whittier CA 90601

BROOKS, WAYNE MAURICE, b Lynchburg, Va, Mar 11, 39; m 85; c 2. INSECT PATHOLOGY, PROTOZOOLOGY. *Educ:* NC State Univ, BS, 61; Univ Calif, PhD(entom), 66. *Prof Exp:* Res asst entom, Univ Calif, 61-63, jr specialist, 63-66; from asst prof to assoc prof, 66-78, PROF ENTOM, NC STATE UNIV, 78- *Concurrent Pos:* Pres, Soc Invert Path, 82-84. *Mem:* Entom Soc Am; Soc Invert Path; Sigma Xi; Soc Protozoologists; Int Orgn Biol Control. *Res:* Insect pathology with emphasis on entomophilic protozoa. *Mailing Add:* Dept Entom Box 7613 NC State Univ Raleigh NC 27695-7613

BROOKS, WENDELL V F, b Rockford Ill, Mar 7, 25; m 50; c 2. PHYSICAL CHEMISTRY, MOLECULAR SPECTROSCOPY. *Educ:* Swarthmore Col, BA, 48; Univ Minn, PhD, 54. *Prof Exp:* Instr chem, Univ Ariz, 52-54; NSF fel, Yale Univ, 54-55; from asst prof to prof chem, Ohio Univ, 55-67; PROF CHEM, UNIV NB, 67- *Concurrent Pos:* Hon res assoc chem, Harvard, 73-74; vis prof physics, Duke Univ, 80-81. *Mem:* Am Chem Soc. *Res:* Infrared and microwave spectroscopy. *Mailing Add:* 218 Colonial Heights Fredericton NB E3B 5M1 Can

BROOKS, WILLIAM HAMILTON, b Wilkinsburg, Pa, Nov 9, 32; m 69; c 2. RESOURCE MANAGEMENT, TECHNICAL MANAGEMENT. *Educ:* Pa State Univ, BS, 55; NDak State Univ, MS, 64. *Prof Exp:* Asst prof plant sci, Gonzaga Univ, 65-69; develop consult, US Dept Educ, 69-70; prog dir, Natural Resources Mgt, Pima Col, 70-74; proj leader, Ministry Agr/ Water Kingdom Saudi Arabia, 74-76; dir regional develop, Off Arid Land Studies, Univ Ariz, 76-82; dir exploratory technol, Agribusiness Res Corp, 82-83; state watershed planner, 84-90, AREA COORDR, RESOURCE CONSERV & DEVELOP, SOIL CONSERV SERV, USDA, 90- *Concurrent Pos:* Prin investr, Off Arid Lands Studies, Univ Ariz, 79-82; consult, Stanford Res Inst, 80-81, Elec Power Res Inst, 80-81, Riyadh Univ, 81-82, Int Union Conserv Nat Resources, Switz, 82-83, Sultanate of Oman, 83-84; regional expert, United Nations Econ Comn Western Asia, 81-82; consult prod, Desert Realm publ, Nat Geog Soc, 82; mem, bd dirs, Int Tree Crop Inst; fel Grad Study, NSF. *Mem:* Asn Arid Land Studies (pres, 82-83); Sigma Xi; fel Linnaean Soc London. *Res:* Implementation and coordination of regional and international programs in new crop assessment for arid and semiarid lands, environmental assessments and impact analysis, revegetation, afforestation, quantitative plant surveys, resources management, and salinity problems. *Mailing Add:* PO Box 4736 San Luis Obispo CA 93403-4736

BROOKSHEAR, JAMES GLENN, b Denton, Tex, Nov 27, 44; m 68; c 1. MATHEMATICS. *Educ:* NTex State Univ, BS, 67; NMex State Univ, MS, 68, PhD(math), 75. *Prof Exp:* ASST PROF MATH, MARQUETTE UNIV, 75- *Mem:* Am Math Soc; Math Asn Am; Asn Comput Mach. *Res:* Rings of continuous functions and related areas. *Mailing Add:* Dept Math & Statist Marquette Univ Milwaukee WI 53233

BROOKS SPRINGS, SUZANNE BETH, b Ocala, Fla, Aug 6, 35. BIOCHEMISTRY, IMMUNOLOGY. *Educ:* Wheaton Col, Ill, BS, 57; Univ Ill, Urbana, PhD(biochem), 62. *Prof Exp:* Res assoc biochem, Sch Med, Vanderbilt Univ, 62-63; res assoc, Sch Med, Boston Univ, 63-64, instr, 64; res assoc, Am Cancer Ctr, Denver, 70-75; RES ASSOC BIOCHEM, WEBB-WARING INST, MED SCH, UNIV COLO, 75- *Concurrent Pos:* Fel, Nat Cancer Inst, 62-64 & Damon Runyon Found, 64-65. *Res:* Biochemical processes involved in mechanisms of cellular immunology, especially with respect to macrophage vs tumor interactions. *Mailing Add:* Webb-Waring Inst Box C321 Denver CO 80207

BROOM, ARTHUR DAVIS, b Panama, CZ, July 26, 37; m 60; c 3. BIO-ORGANIC CHEMISTRY, MEDICINAL CHEMISTRY. *Educ:* Univ Tex, Austin, BS, 59; Ariz State Univ, PhD(nucleoside methylation), 65. *Prof Exp:* Res assoc, Johns Hopkins Univ, 65-66; res assoc, 66-67, asst res prof, 67-69, from asst prof to assoc prof, 69-75, PROF MED CHEM, COL PHARM, UNIV UTAH, 75-, CHMN DEPT, 78- *Concurrent Pos:* Mem BNP study sect, NIH, 80-84; consult, Egyptian govt, 83; chair, chem sect, Am Col Pharmacol, 85-86; Fulbright sr fel, WGermany, 83; mem, develop therapeut contracts rev comt, NCI, 88-92, chemother hemat grants rev comt, Am Cancer Soc 86-90. *Mem:* AAAS; Am Chem Soc; Am Asn Cancer Res; Am Asn Pharmaaceut Scientists; Am Asn Col Pharm. *Res:* Design and synthesis of multisubstrate analog enzyme inhibitors; conformational analysis by nuclear magnetic resonance; antiviral polynucleotides. *Mailing Add:* Dept Med Chem Col Pharm Univ Utah Salt Lake City UT 84112

BROOMAN, ERIC WILLIAM, b London, Eng, Sept 15, 40; m 62; c 2. ELECTROCHEMISTRY. *Educ:* Univ Surrey, Dipl Tech, 63; Cambridge Univ, PhD(electrochem), 66. *Prof Exp:* Res electrochemist, Columbus Div, Battelle Mem Inst, 66-75, prin res scientist, 75-78, assoc mgr, corrosion & electrochem technol sect, 78-80, assoc mgr, chem process, develop & polymer sci & tech sect, 80-85, assoc mgr, corrosion & electrochem tech dept, 85-89, RES LEADER & BUS AREA MGR, ADVAN MAT DEPT, COLUMBUS DIV, BATTELLE MEM INST, 89- *Concurrent Pos:* Res fel & consult, Univ Salford, 72-73; mem bd dirs, Electrochem Soc, Inc, 78-80 & 88, vchmn for new proj, 88, corp secy & trustee, 88-; mem res bd & public bd, Am Electroplaters & Surface Finishers Soc, 90. *Honors & Awards:* Silver Medal Award, Am Electroplaters & Surface Finishers Soc, 88. *Mem:* Electrochem Soc; Am Electroplaters & Surface Finishers Soc; Sigma Xi. *Res:* Electrochemical energy conversion and storage; primary and secondary batteries; fuel cells and water electrolyzers; hydrogen production and storage; corrosion science and technology; electrodeposition of metals, compounds and alloys. *Mailing Add:* Battelle Mem Inst 505 King Ave Columbus OH 43201-2693

BROOME, CARMEN ROSE, b Miami, Fla, June 19, 39; m 78. SYSTEMATIC BOTANY. *Educ:* Univ Miami, BS, 65; Univ SFla, MA, 68; Duke Univ, PhD(bot), 74. *Prof Exp:* asst prof bot, Univ MD, College Park, 73-79; EXAMR, PLANT VARIETY PROTECT OFF, US DEPT AGR, BELTSVILLE, MD, 79- *Mem:* Sigma Xi; Am Soc Plant Taxonomists; Int Asn Plant Taxon. *Res:* Systematics of angiosperms, particularly Gentianaceae; economic plants; palynology, reproductive biology of angiosperms. *Mailing Add:* USDA-NAL-ISD 10301 Balitmore Blvd 5th Floor Beltsville MD 20705

BROOME, CLAIRE V, b Aug 24, 49; m 88; c 1. INFECTIOUS DISEASES, ACUTE DISEASE EPIDEMIOLOGY. *Educ:* Harvard Univ, BA, 70; Harvard Med Sch, MD, 75; Univ Calif, San Francisco, dipl, Am Bd Internal Med, 81. *Prof Exp:* Dep chief spec pathogens, Bact Dis Div, 79-80, chief spec pathogens, 81-90, ASST DIR SCI, CTR DIS CONTROL, 91- *Concurrent Pos:* Clin asst, Sch Med, Emory Univ, 77-; consult, WHO; infectious dis fel, Mass Gen Hosp, 80-81; mem, Microbiol & Infectious Dis Res Comt, Nat Inst Allergy & Infectious Dis, 87-89, Adv Comt Vaccines & Biologics, Food & Drug Admin, 90-, WHO Prog Vaccine Develop, Steering Comt Encapsulated Bacteria, 89-, Consult Group Vaccine Develop, Nat Vaccine Prog. *Honors & Awards:* Charles C Shepard Award, 86. *Mem:* Fel, Infectious Dis Soc Am; Am Epidemiol Soc; Am Col Physicians; Am Soc Microbiol; Am Col Epidemiol. *Res:* Develop new method for estimating vaccine efficacy of pneumococcal vaccine; responsible for CDC epidemiology investigations of toxic-shock syndrome; haemophilus influenzae, listeriosis and other causes of bacterial meningitis. *Mailing Add:* D39 Off Dir Ctr Dis Control Atlanta GA 30333

BROOME, PAUL W(ALLACE), b Oakdale, Pa, Jan 17, 32; m 57; c 4. ELECTRICAL ENGINEERING, SIGNAL PROCESSING. *Educ:* Carnegie Inst Technol, BS, 54, MS, 55, PhD(elec eng), 60. *Prof Exp:* Res engr, Gulf Res & Develop Co, 55-57; chief analog comput anal, Gen Dynamics-Astronaut, 58-61, design specialist, 61-62; sr staff scientist, Guided Missiles Range Div, Pan Am World Airways, 62-64; sr eng specialist, Earth Sci Div, Teledyne Inc, 64-65, mgr appl res lab, 65-68; CHIEF EXEC OFFICER, ENSCO, 68- *Mem:* Sigma Xi; Inst Elec & Electronics Engrs. *Res:* Signal and noise theory; time series analysis; pattern recognition and spatial filtering; equipment reliability analysis; applications of signal theory. *Mailing Add:* Box 3092 Winter Park CO 80482

BROOMFIELD, CLARENCE A, b Mt Morris, Mich, Sept 18, 30; m 56; c 3. BIOCHEMISTRY. *Educ:* Univ Mich, BS, 53; Mich State Univ, PhD(chem), 58. *Prof Exp:* Res assoc phys chem, Cornell Univ, 58-59; res biochemist, Chem Res & Develop Lab, Edgewood Arsenal, 62-68, chief protein chem, Med Res Dir, Biomed Lab, 68-80; RES CHEMIST, ARMY MED RES INST CHEM DEFENSE, 80- *Concurrent Pos:* NIH fel, Cornell Univ, 60-62. *Mem:* Am Chem Soc; Sigma Xi; NY Acad Sci; Am Soc Biochem & Molecular Biol. *Res:* Secondary and tertiary protein structure; relationship of structure to biological activity; toxic proteins; cholinesterases; biochemistry of nerve transmission; recombinant enzymes; organophosphate hydrolyzing enzymes. *Mailing Add:* 1917 Youngston Rd Jarrettsville MD 21084

BROOTEN, DOROTHY, b Hazleton, Pa, Jan 16, 42; m 63; c 2. NURSING. *Educ:* Univ Pa, BSN, 66, MSN, 70, PhD(educ admin). *Prof Exp:* Assoc prof nursing, Thomas Jefferson Univ, 72-77; from asst prof to assoc prof, 77-88, PROF NURSING & CHAIR, HEALTH CARE OF WOMEN & CHILDBEARING, UNIV PA, 88-, DIR, CTR FOR LOW BIRTHWEIGHT, SCH NURSING, 90- *Concurrent Pos:* Mem gov coun, Am Acad Nursing, 88-91; consult, Sch Med, Univ Utrecht, Neth, 89 & Ministry of Health, Malawi, Africa, 91. *Honors & Awards:* Distinguished Contrib to Nursing Sci, Am Nurse's Asn, 88. *Mem:* Inst Med-Nat Acad Sci; Am Acad Nursing. *Res:* Low birthweight prevention; postdischarge care of low birthweight infants; health care delivery. *Mailing Add:* Box 33 Gradyville Rd Glen Mills PA 19342

BROPHY, GERALD PATRICK, b Kansas City, Mo, Sept 11, 26; m 51; c 3. GEOLOGY, MINERALOGY. *Educ:* Columbia Univ, AB, 51, MA, 53, PhD(geol), 54. *Hon Degrees:* MA, Amherst Col, 68. *Prof Exp:* Res asst geol, Columbia Univ, 51-54; from instr to assoc prof, 54-68, PROF GEOL, AMHERST COL, 68- *Concurrent Pos:* NIH res grants, 62-; Fulbright fel, Univ Baghdad, 65-66; cur, Pratt Mus Geol, 67-; NSF grants, 69-; consult, Bear Creek Mining Co, Cerro Corp, R T Vanderbilt & Co & Conyers Construct Co; prog mgr, Geothermal Explor USA, Div Geothermal Energy, Dept Energy, 78-80; assoc, Dunn Geoscience Inc, 81- *Mem:* Fel Geol Soc Am; Mineral Soc Am; Geochem Soc; Soc Econ Geologists; Yellowstone-Big Horn Res Asn (pres, 75-77). *Res:* Crystals chemistry of phosphates and sulphates; effects of pressure on crystallization of granitic magmas; rock alteration in geothermal systems. *Mailing Add:* Dept Geol Amherst Col Amherst MA 01002

BROPHY, JAMES JOHN, b Chicago, Ill, June 6, 26; m 49; c 3. SOLID STATE ELECTRONICS. *Educ:* Ill Inst Technol, BS, 47, MS, 49, PhD(physics), 51. *Prof Exp:* Res physicist, 51-53, supvr solid state physics, 53-56, asst mgr, Physics Div, 56-61, dir tech develop, 61-63, vpres, Res Inst, 63-67, acad vpres, Ill Inst Technol, 67-76; sr vpres, Inst Gas Technol, 76-80, PROF PHYS & ELEC ENG & VPRES RES, UNIV UTAH, 80- *Mem:* Fel Am Phys Soc; AAAS; Sigma Xi. *Res:* Solid state physics; semiconductors; fluctuation phenomena; secondary emission; magnetism; plasma physics. *Mailing Add:* Univ Utah 210 Park Bldg Salt Lake City UT 84112

BROPHY, JERE H(ALL), b Schenectady, NY, Mar 11, 34; m 56; c 3. METALLURGY. *Educ:* Univ Mich, BSChE & BSMetE, 56, MSMetE, 57, PhD(metall eng), 58. *Prof Exp:* Asst prof metall, Mass Inst Technol, 58-63; sect head, Platinum & Nickel Base Alloys Res Lab, Int Nickel Co, 63-67, non-ferrous metals res mgr, P D Merica Res Lab, 67-73, mgr, 73-77, dir res & develop, Inco Res & Develop Ctr, 78-80, DIR ADVAN TECHNOL INITIATION, INCO LTD, 80- *Mem:* Fel Am Soc Metals; Am Inst Mining, Metall & Petrol Engrs; Brit Inst Metals. *Res:* Refractory metals; powder metallurgy; precious metals; superalloys; superplasticity; microduplex alloys; materials. *Mailing Add:* 29 Brookside Dr No WE Harriman NY 10926

BROPHY, JOHN ALLEN, b Rockford, Ill, Mar 30, 24; m 54; c 2. GEOLOGY. *Educ:* Univ Ill, AB, 48, MS, 49, PhD(geol), 58. *Prof Exp:* Asst instr geol, Univ Ill, 48-49; geologist, Magnolia Petrol Co, 49-51; asst geologist, Ill Geol Surv, 53-59; from asst prof to prof, 59-82, chmn, Div Natural Sci, 67-70, chmn, dept geol, 70-82, EMER PROF GEOL, NDAK STATE UNIV, 82- *Concurrent Pos:* Leverhulme fel, Univ Birmingham, 67-68. *Res:* Pleistocene geology. *Mailing Add:* 702 South Dr Fargo ND 58103

BROPHY, MARY O'REILLY, b New York, NY, Aug 3, 48; m 69; c 3. CANCER RESEARCH, ENVIRONMENTAL TOXICOLOGY. *Educ:* Univ Mich, BS, 70, MS, 72, PhD(human anat), 79. *Prof Exp:* Technician & instr histol, Univ Idaho, 74-75; instr phys ed, Univ Ga, 77-79; fel & instr histol, State Univ NY, Upstate Med Ctr, 79-81, res asst prof, 81-84; res assoc, Syracuse Res Corp, 84-86; SR INDUST HYGIENIST, DIV SAFETY & HEALTH, NY STATE DEPT LABOR, 86- *Concurrent Pos:* Adj asst prof, State Univ NY, Upstate Med Ctr, 84-86, asst prof, Sch Pub Health, 90- *Mem:* Tissue Culture Asn. *Res:* Evaluation of the health effects of environmental pollutants to determine acceptable levels in drinking and ambient water; development of methodology to evaluate the health effects of exposure to several chemical toxicants simultaneously. *Mailing Add:* NY State Dept Labor 677 S Salina St Syracuse NY 13202

BROQUIST, HARRY PEARSON, b Chicago, Ill, Jan 23, 19; m 42; c 2. BIOCHEMISTRY, NUTRITION. *Educ:* Beloit Col, BS, 40; Univ Wis, MS, 41, PhD(biochem), 49. *Prof Exp:* Group leader microbiol, Lederle Lab, Pearl River, NY, 41-46 & 49-58; from assoc prof to prof biol chem, Univ Ill, 58-69; PROF BIOCHEM, VANDERBILT UNIV, 69-, DIR DIV NUTRIT, 72- *Concurrent Pos:* Vis lectr biochem, Mich State Univ, 63; biochem consult interdept comt nutrit nat develop, US Nutrit Surv to Nigeria, 65; NSF sr fel, Karolinska Inst, Sweden, 65-66; mem nutrit study sect, NIH, 66-70; mem food & nutrit bd, Nat Res Coun, 74-; assoc ed, Nutrit Reviews, 79- & Ann Reviews Nutrit, 81- *Honors & Awards:* Borden Award, Am Inst Nutrit, 68. *Mem:* Am Soc Biol Chemists; Am Inst Nutrit; Fedn Am Soc Exp Biol. *Res:* Amino acid metabolism in yeasts, molds and mammalian systems; nutritional biochemistry. *Mailing Add:* Dept Biochem Vanderbilt Univ Nashville TN 37232

BROSBE, EDWIN ALLAN, b Burlington, NJ, Jan 8, 18; m 43; c 2. MEDICAL MICROBIOLOGY. *Educ:* Philadelphia Col Pharm, BS, 40; Univ Colo, MS, 47, PhD(bact), 51. *Prof Exp:* Bacteriologist, Vet Admin Hosp, Ft Logan, Colo, 47-50 & Denver, Colo, 50-53, bacteriologist & actg chief lab serv, NY, 53-55, chief tuberc res lab, Long Beach, Calif, 55-66, microbiologist, Little Rock, Ark, 66-68, clin microbiologist, Long Beach, 68-73; LECTR, CALIF STATE UNIV, LONG BEACH, 73- *Mem:* AAAS; Am Soc Microbiol; Soc Exp Biol & Med. *Res:* Bacteriology and chemotherapy of tuberculosis; biology and chemotherapy of coccidioidomycosis; host-parasite relationship. *Mailing Add:* 300 Granada Ave Long Beach CA 90814

BROSCHAT, KAY O, PROTEIN STRUCTURE & REGULATION. *Educ:* Ohio State Univ, Columbus, PhD(biochem), 82. *Prof Exp:* RES ASST PROF ANAT & CELL BIOL, UNIV MIAMI, 85- *Res:* Myosin mediated motility. *Mailing Add:* Univ Miami R124 PO Box 016960 Miami FL 33101

BROSE, DAVID STEPHEN, b Detroit, Mich, Feb 20, 39; m 65; c 2. PALEO-ANTHROPOLOGY, HUMAN ECOLOGY. *Educ:* Univ Mich, BA, 60, MA, 66, PhD(anthrop), 68. *Prof Exp:* From asst prof to assoc prof anthrop, Case Western Reserve Univ, 68-77, chmn anthrop, 73-35; CHIEF CUR ARCHAEOL, CLEVELAND MUS NATURAL HISTORY, 77- *Concurrent Pos:* Ed, Mid Continental J of Archaeol, 76-91; appointment, Interagency Archaeol Serv, US Dept Interior, 77-78; adj prof, Dept Anthrop, Case Western Reserve Univ, 78-80 & Cleveland State Univ, 80-91; investr, GAO Interagency Task Force, New Melones Archaeol Proj, 80-81; coordr of res, Cleveland Mus Natural History, 81-; chair, Nat Hist Landmarks Comt, 85- *Honors & Awards:* Jared Kirtland Medal, 85. *Mem:* Soc Prof Archaeologists (pres 82-83); Soc Am Archaeol. *Res:* Prehistoric Archaeology of eastern North America; historic sites archaeology of the great lakes and Central Gulf Coast; research focuses upon cultural adaptations to changing social and environmental conditions over time. *Mailing Add:* Dept Archaeol Museum Natural History Cleveland OH 44106

BROSEGHINI, ALBERT L, b Chicago, Ill, Sept 17, 32; m 58; c 2. ZOOLOGY. *Educ:* Northern Ill State Teachers Col, BS, 54; Iowa State Univ, MS, 56, PhD(zool), 59. *Prof Exp:* Asst prof biol, Fresno State Col, 59-62; assoc prof physiol, Col Osteop Med & Surg, 62-64; scientist adminr, Div Res Grants, NIH, 64-65, asst endocrinol prog dir, Nat Inst Arthritis & Metab Dis, 65-66, hemat prog dir, 66-69; dir res admin, Children's Hosp, Boston, 69-87, mgr patents & lic, 76-87; vpres res admin, Mass Eye & Ear Hosp, Boston, 87-90; ASSOC DIR, PATENTS & TECHNOL MKT, CORNELL RES FOUND, 90- *Mem:* AAAS. *Res:* Comparative physiology; histochemistry of insect tissues; hematology; erythropoiesis. *Mailing Add:* 13-4 Deer Path Maynard MA 01754

BROSEMER, RONALD WEBSTER, b Oakland, Calif, Feb 17, 34; m 62; c 2. BIOCHEMISTRY. *Educ:* Univ Calif, BS, 55; Univ Ill, PhD(biochem), 60. *Prof Exp:* NSF res fel, Univ Marburg, 60-62; asst prof biochem, Univ Ill, 62-63; from asst prof to assoc prof, 63-72, assoc dean grad sch, 75-77, PROF BIOCHEM, WASH STATE UNIV, 72-, ASSOC DEAN, DIV SCI, 84- *Concurrent Pos:* Fulbright scholar, Univ Konstanz, 70-71. *Mem:* NY Acad Sci; Am Soc Biol Chemists. *Res:* Molecular basis of neurologic diseases. *Mailing Add:* Biochem/Biophys Prog Wash State Univ Pullman WA 99164-4660

BROSENS, PIERRE JOSEPH, b Uccle, Belgium, Oct 12, 33; US citizen; m 55; c 3. MECHANICAL ENGINEERING. *Educ:* Mass Inst Technol, BS & MS, 56, ScD(mech eng), 59. *Prof Exp:* Asst prof mech eng, Mass Inst Technol, 59-60; proj mgr thermonics, Thermo Electron Eng Corp, 60-67; mech eng consult, 67-69; VPRES, GEN SCANNING, INC, 69-, VCHMN BD. *Mem:* Am Soc Mech Engrs. *Res:* Mechanical vibrations; stress analysis; mechanical design. design; management. *Mailing Add:* Gen Scanning Inc 500 Arsenal St Watertown MA 02172

BROSHAR, WAYNE CECIL, b Boone Co, Ind, May 3, 33; wid; c 4. THERMAL PHYSICS. *Educ:* Wabash Col, AB, 55; Univ Mich, MS, 56; Brown Univ, PhD(physics), 69. *Prof Exp:* Res engr, Convair San Diego, Gen Dynamics-Convair, 56-59; instr physics, Wabash Col, 59-61; from instr to prof physics, Ripon Col, 66-88; VIS PROF, GUSTAVUS ADOLPHUS COL, 88- *Mem:* Am Asn Physics Teachers. *Mailing Add:* 1330 St Paul Ave Apt 3 St Paul MN 55116

BROSILOW, COLEMAN B, b Philadelphia, Pa, Nov 14, 34; m 62; c 2. CHEMICAL & SYSTEMS ENGINEERING. *Educ:* Drexel Inst Technol, BS, 57; Polytech Inst Brooklyn, MChE, 59, PhD(chem eng), 62. *Prof Exp:* Teaching fel chem eng, Polytech Inst Technol, Brooklyn, 57-59, asst rocket propulsion syst eng, 59-60, sr res fel, 60-62; control engr, Am Cyanamid Co, 62-63; from asst prof to assoc prof eng, 63-73, chmn dept chem eng, 80-84 PROF ENG, CASE WESTERN RESERVE UNIV, 73- *Honors & Awards:* Comput in Chem Eng Award, Am Inst Chem Engrs, 89. *Mem:* Am Inst Chem Engrs. *Res:* Control of industrial processes; model predictive control; digital simulation of dynamic systems; inferential control. *Mailing Add:* Dept Eng Case Western Reserve Univ Cleveland OH 44106

BROSIN, HENRY WALTER, b Blackwood, Va, July 6, 04; m 49; c 1. MEDICINE. *Educ:* Univ Wis, BA, 27, MD, 33. *Prof Exp:* Intern, Cincinnati Gen Hosp, 33-34; Commonwealth fel psychiat, Sch Med, Univ Colo, 34-37; Rockefeller fel, Inst Psychoanal, Univ Chicago, 37, from instr to prof psychiat, Sch Med, 37-51; prof, 51-69, EMER PROF PSYCHIAT, SCH MED, UNIV PITTSBURGH, 69-; PROF PSYCHIAT, COL MED, UNIV ARIZ, 69- *Concurrent Pos:* Fel, Ctr Advan Study Behav Sci, Calif, 56 & 66; consult, Off Surgeon Gen; mem comt army med educ, Nat Res Coun; mem nat clearing house, NIMH. *Mem:* Fel Am Acad Arts & Sci; AAAS; fel AMA; Am Psychoanal Asn; Am Psychosom Soc. *Res:* Psychoanalysis; organic cerebral disease; psychosomatic medicine; communication theory. *Mailing Add:* Col Med Univ Ariz Tucson AZ 85724

BROSIOUS, PAUL ROMAIN, physics; deceased, see previous edition for last biography

BROSNAN, JOHN THOMAS, b Kenmare, Ireland, Feb 13, 43; m 70; c 3. METABOLISM. *Educ:* Nat Univ Ireland, BSc, 64, MSc, 66; Oxford Univ, DPhil(biochem), 69. *Prof Exp:* Fel med res, Univ Toronto, 69-71; from asst prof to assoc prof, 72-81, PROF BIOCHEM, MEM UNIV NFLD, 81- *Concurrent Pos:* Mem grants comt metab, Med Res Coun Can, 75-78; mem grants comt, Kidney Found Can, 79-86, Can Heart Found, 82-85. *Honors & Awards:* Borden Award, Nutrit Res, 86. *Mem:* Biochem Soc; Can Biochem Soc; Am Soc Biol Chemists; Can Soc Nutrit Sci; Sigma Xi. *Res:* Metabolic regulation, especially the regulation of amino acid metabolism in kidney and liver; carbohydrate and fatty acid metabolism in heart. *Mailing Add:* Dept Biochem Mem Univ Nfld St Johns NF A1B 3X9 Can

BROSNAN, MARGARET EILEEN, b Tulsa, Okla, Oct 29, 42; m 70; c 3. ENDOCRINOLOGY, PHYSIOLOGY. *Educ:* Univ Toronto, BA, 66, MSc, 67, PhD(physiol), 72. *Prof Exp:* Fel, 72-74, from asst to assoc prof, 74-85, PROF BIOCHEM, MEM UNIV NFLD, 85- *Concurrent Pos:* Assoc ed, Can Jour Physiol & Pharmacol, 78-87. *Mem:* Nutrit Soc Can; Can Fedn Biol Socs (pres, 84-85); Biochem Soc UK; Can Biochem Soc; Am Physiol Soc. *Res:* Control of enzymes of polyamine metabolism in liver and mammary glands; role of polyamines in metabolic regulation and in hormone action; nutritional regulation of metabolism; metabolism of amino acids in diabetes mellitus. *Mailing Add:* Dept Biochem Mem Univ Nfld St John's NF A1B 3X9 Can

BROSNIHAN, K BRIDGET, b Omaha, Nebr, June 3, 41; m 75; c 2. ENDOCRINOLOGY, NEUROBIOLOGY. *Educ:* Col St Mary, BS, 65; Creighton Univ, MS, 70; Case Western Reserve Univ, PhD(physiol), 74. *Prof Exp:* Fel res, 74-76, proj scientist, 76-77, assoc staff, 77-82, STAFF RES, CLEVELAND CLINIC, 82- *Concurrent Pos:* Adj assoc prof physiol, Case Western Reserve Univ Med Sch, 81-82 & Cleveland State Univ. *Mem:* Am Heart Asn; Am Physiol Soc; Endocrine Soc; Soc Neurosci; Sigma Xi; Coun High Blood Pressure Fel. *Res:* Neurohormonal interrelationships of two systems, catecholamines and renin-angiotensin; central and peripheral nervous system. *Mailing Add:* Cleveland Clinic Found 9500 Euclid Ave Cleveland OH 44106

BROSS, IRWIN DUDLEY JACKSON, b Halloway, Ohio, Nov 13, 21; m 49; c 3. STATISTICS, EPIDEMIOLOGY. *Educ:* Univ Calif, Los Angeles, BA, 42; NC State Univ, MA, 48; Univ NC, PhD(exp statist), 49. *Prof Exp:* Asst prof pub health & prev med, Med Col, Cornell Univ, 52-59; dir biostatist, Roswell Park Mem Inst, 59-83; res prof, State Univ NY, BUFFALO, 61-83; PRES, BIOMED METATECHNOL INC, 83- *Concurrent Pos:* Res assoc biostatist, Johns Hopkins Univ, 49-52, assoc epidemiol, 71-; head res design & anal, Sloan-Kettering Inst, 52-59; actg chief epidemiol, Roswell Park Mem Inst, 66-74. *Mem:* AAAS; Am Col Epidemiol; Am Statist Asn; Biometrics Soc. *Res:* Cancer epidemiology aimed at protecting the public against hazardous side effects of technology; effective utilization of biomedical data for better treatment of human diseases. *Mailing Add:* 109 Maynard Dr Buffalo NY 14226

BROSSEAU, GEORGE EMILE, JR, b Berkeley, Calif, July 24, 30; m 78; c 3. GENETICS. *Educ:* Univ Calif, Berkeley, BA, 52, PhD(genetics), 56. *Prof Exp:* Am Cancer Soc fel, Biol Div, Oak Ridge Nat Lab, 56-58, res assoc genetics, 58-59; from asst prof to prof zool, Univ Iowa, 59-70; prog mgr, Nat Sci Found, 70-78; expert consult, Admin Aging, 78-79; prob analyst, 79-81, proj mgr, 81-84, PROG DIR, NSF, 84- *Mem:* Am Soc Human Genetics; AAAS; Robotics Int; Sigma Xi; World Future Soc. *Res:* Administration; applied science; interdisciplinary research; research utilization; robotics; manufacturing engineering. *Mailing Add:* NSF 1800 G St NW Washington DC 20550

BROSTOFF, STEVEN WARREN, b Boston, Mass, Sept 10, 42; m 83; c 2. NEUROCHEMISTRY, NEUROIMMUNOLOGY. *Educ:* Mass Inst Technol, BS, 64, PhD(biochem), 68. *Prof Exp:* Fel biochem, Am Cancer Soc, Eleanor Roosevelt Inst Cancer Res, Univ Colo Med Ctr, 68-70; res assoc neurochem, Salk Inst Biol Studies, 70; sr res biochemist, Dept Exp Biol, Merck Inst Therapeut Res, 71-72; asst prof, dept path, Albert Einstein Col Med, 72-73; assoc prof, 73-77, PROF NEUROCHEM, DEPTS NEUROL & IMMUNOLOGY, MED UNIV SC, 77-, ASSOC DEAN GRAD STUDIES, 77- *Concurrent Pos:* Prin investr, NIH, 77-, mem neurological sci study sect, 79-83; vis scientist, MRC cellular immunol unit, Univ Oxford, 83. *Mem:* AAAS; Am Chem Soc; Am Soc Neurochem; Int Soc Neurochem; Am Soc Biol Chemists; Am Asn Immunologists. *Res:* Immoregulation of autoimmune diseases; chemical and immunological properties of nervous system proteins; autoimmune demyelinating diseases. *Mailing Add:* Immune Response Corp 6455 Nancy Ridge Dr San Diego CA 92121

BROSTOW, WITOLD KONRAD, b Warsaw, Poland, Mar 21, 34; Can citizen; m 73; c 2. MATERIALS SCIENCE ENGINEERING. *Educ:* Univ Warsaw, MS, 55, DrSc, 60; Polish Acad Sci, DSc, 65. *Prof Exp:* Asst prof, Inst Phys Chem, Polish Acad Sci, 65-68; head, Div Phys Chem, Inst Synthetics, Warsaw, 68-70; Nat Acad Sci vis scholar, Stanford Univ, 69-70; Nat Res Coun Can vis scholar, Univ Ottawa, 70-71; full researcher chem, Univ Montreal, 71-76; prof chem, Ctr Advan Studies, Mexico City, 76-79; ASSOC PROF MAT ENG, DREXEL UNIV, PHILADELPHIA, 80- *Concurrent Pos:* Ger Chem Soc vis lectr, Martin Luther Univ, Halle-Wittenberg, 65; French Minister Foreign Affairs vis scholar, Univ Montpellier, 66. *Mem:* Mexico Nat Acad Sci; fel Royal Soc Chem; Int Soc Bioelec; Am Phys Soc; Am Chem Soc. *Res:* Structure and properties of polymer solutions and melts; degradation and failure of polymers; computer simulation of materials and processes; information theory; statistical mechanics and thermodynamics of liquid mixtures; biopolymers. *Mailing Add:* CMC UNT PO Box 5308 Denton TX 76203

BROSTROM, CHARLES OTTO, b Downsville, Wis, Nov 2, 42; m 65; c 1. PHARMACOLOGY, BIOCHEMISTRY. *Educ:* Wis State Univ, River Falls, BS, 64; Univ Ill, PhD(biochem), 69. *Prof Exp:* Asst prof biochem, 71-72, from asst prof to assoc prof pharmacol, 72-84, PROF PHARMACOL, ROBERT WOOD JOHNSON MED SCH, UNIV MED & DENT, NJ, 84-, ACTG CHMN PHARMACOL, 89- *Concurrent Pos:* USPHS fel enzym, Univ Calif, Davis, 68-70; Health Sci Advan Award, 70-71. *Mem:* Am Soc Biochem & Molecular Biol. *Res:* Cyclic nucleotide metabolism in brain; regulation of protein synthesis by calcium. *Mailing Add:* Robert Wood Johnson Med Sch Dept Pharmacol Univ Med & Dent NJ Piscataway NJ 08854

BROSTROM, MARGARET ANN, b Chicago, Ill, Aug 13, 41; m 65; c 1. BIOCHEMISTRY, PHARMACOLOGY. *Educ:* Clarke Col, BA, 63; Univ Ill, PhD(biochem), 68. *Prof Exp:* Sr teaching asst, 71-72, from instr to assoc prof, 72-87, PROF PHARMACOL, ROBERT WOOD JOHNSON MED SCH, UNIV MED & DENT, 87- *Concurrent Pos:* NIH fel, Univ Calif, 69-71; prin investr various grants, 80- *Mem:* Am Soc Pharmacol & Exp Therapeut. *Res:* Calcium and cyclic nucleotides in neural and pituitary function; second messengers and translational control. *Mailing Add:* Dept Pharmacol Univ Med & Dent NJ Robert Wood Johnson Med Sch 675 Hoes Lane Piscataway NJ 08854

BROT, FREDERICK ELLIOT, b Kalamazoo, Mich, May 28, 41; m 65; c 2. BIOCHEMICAL GENETICS. *Educ:* Univ Mich, BS, 62; Stanford Univ, PhD(org chem), 66. *Prof Exp:* NIH fel, Northwestern Univ, 66-68; sr res chemist, Monsanto Co, 68-71; RES INSTR & NIH TRAINEE, SCH MED, WASH UNIV, 71- *Mem:* AAAS; Am Chem Soc. *Res:* Protein purification and characterization; affinity chromatography; enzyme modification; application of enzymes for correction of genetic enzyme defects; synthesis and in vivo fate of site-directed drugs. *Mailing Add:* 8145 Cornell Ct University City MO 63130

BROT, NATHAN, b New York, NY, July 27, 31; m 58; c 3. BIOCHEMISTRY. *Educ:* City Col New York, BS, 53; Univ Calif, PhD(biochem), 63. *Prof Exp:* Res chemist, Med Col, Cornell Univ, 53-58; chemist, Univ Calif, 62-63; USPHS res fel, NIH, 63-65, chemist, 65-67; MEM STAFF, ROCHE INST MOLECULAR BIOL, HOFFMAN-LA ROCHE, INC, 67- *Mem:* AAAS; Am Chem Soc; Am Soc Microbiol; Am Soc Biol Chemists; Sigma Xi. *Res:* Gene expression; regulation of protein synthesis; regulation of leaf sluck genes; systemic lupus erythematosus. *Mailing Add:* Four Greentree Rd West Orange NJ 07052

BROTAK, EDWARD ALLEN, b Trenton, NJ, July 17, 48. SYNOPTIC METEOROLOGY, FOREST METEOROLOGY. *Educ:* Rutgers Univ, BS, 70, MS, 72; Yale Univ, PhD(biometeor), 77. *Prof Exp:* Asst prof meteorol, Kean Col NJ, 77-80; asst prof, Lyndon State Col, 80-81; assoc prof & dir atmospheric sci, Univ NC, Asheville, 81-87; CONSULT, 88- *Concurrent Pos:* Vis prof, Wesleyan Univ, 76, 78-80; AV ed, Weatherwise & meteorel educ ed, Nat Weather Dig. *Mem:* Am Meteorol Soc; Nat Weather Asn; Asn Am Weather Observers; Sigma Xi. *Res:* Synoptic meteorology and climatology; weather forecasting; forest fire weather. *Mailing Add:* Dept Atmospheric Sci Univ NC Asheville Univ Heights Asheville NC 28804

BROTEN, NORMAN W, b Meacham, Sask, Dec 21, 21. ASTRONOMY. *Educ:* Univ Western Ont, BSc, 50. *Prof Exp:* Astronr solar radio, 50-59, RADIO ASTRONR, NAT RES COUN CAN, 59-, HEAD ASTRON SECT, 74- *Concurrent Pos:* Mem comm 40, Int Astron Union, 70-; chmn comn J, Int Union Radio Sci, 75- *Honors & Awards:* Rumford Medal, Am Acad Arts & Sci, 71. *Mem:* Can Astron Soc; Am Astron Soc; Inst Elec & Electronics Engrs; AAAS; Royal Astron Soc Can; 226-5297. *Res:* Radio astronomy in very long baseline interferometry, interstellar molecules, quasars, Seyfert galaxies and extra galactic objects; ionized hydrogen regions. *Mailing Add:* 48 Pineglen Crescent Nepean ON K2E 6X9 Can

BROTHERS, ALFRED DOUGLAS, b Waukegan, Ill, Aug 6, 39; m 72. SOLID STATE PHYSICS. *Educ:* Northern Ill Univ, BS, 62; Iowa State Univ, MS, 65, PhD(physics & educ), 68. *Prof Exp:* Jr physicist, Ames Lab, US AEC, 65-66; asst prof physics, St Benedict's Col, 68-71; actg chmn, 71-72, CHMN DEPT PHYSICS, BENEDICTINE COL, 72- *Mem:* Am Asn Physics Teachers; Am Inst Physics; Sigma Xi. *Res:* Optical qualities of thin films and crystalline solids. *Mailing Add:* Dept Physics Box N66 Benedictine Col Atchison KS 66002

BROTHERS, EDWARD BRUCE, b Brooklyn, NY, Apr 25, 48; m 69; c 2. ICHTHYOLOGY, ECOLOGY. *Educ:* Brooklyn Col, BS, 68; Univ Calif, San Diego, PhD(oceanog), 75. *Prof Exp:* Asst prof zool, Univ Calif, Berkeley, 74; res assoc, Inter-Am Trop Tuna Comn, 75; asst prof ecol & systematics, Cornell Univ, 75-83; ECOL CONSULT, 83- *Concurrent Pos:* Ecol consult, 83-; vis prof, Univ Mont, 86. *Mem:* Am Soc Ichthyologists & Herpetologists; AAAS; Am Soc Zoologists; Am Fisheries Soc; Sigma Xi. *Res:* Ecology and behavior of freshwater and marine fishes; growth dynamics of individuals and populations; fisheries biology. *Mailing Add:* Three Sunset W Ithaca NY 14850

BROTHERS, JOHN EDWIN, b Salt Lake City, Utah, July 6, 37. GEOMETRIC MEASURE THEORY, DIFFERENTIAL GEOMETRY. *Educ:* Univ Utah, BA, 59, MS, 60; Brown Univ, PhD(math), 64. *Prof Exp:* Asst chmn dept, 71-73, PROF MATH, IND UNIV, BLOOMINGTON, 66- *Concurrent Pos:* Co-recipient, NSF grant, 67-81; mem, Inst Advan Study, 73-74; ed, Ind Univ Math J, 81. *Mem:* Am Math Soc. *Res:* Geometric measure theory and applications to the calculus of variations with emphasis on the least area problem and the study of minimal submanifolds. *Mailing Add:* Dept Math Ind Univ Bloomington IN 47401

BROTHERSON, DONALD E, b Chicago, Ill, Sept 4, 32; m 59; c 2. ENERGY & CONSTRUCTION. *Educ:* Univ Ill, BS, 53, MS, 55. *Prof Exp:* DIR, BLDG RES COUNCIL, UNIV ILL, 59- *Concurrent Pos:* Consult, 60- *Res:* Energy conservation. *Mailing Add:* Small Homes Council Bldg Research Council Univ Illinois One East St Mary's Rd Champaign IL 61820

BROTHERSON, JACK DEVON, b Castle Dale, Utah, Sept 18, 38; m 64; c 4. PLANT ECOLOGY. *Educ:* Brigham Young Univ, BS, 64, MS, 67; Iowa State Univ, PhD(plant ecol), 69. *Prof Exp:* PROF ECOL, BRIGHAM YOUNG UNIV, 69- *Concurrent Pos:* Consult, Indian Inst, Brigham Young Univ, 69- & Bur Reclamation, Cent Utah Proj, 72-78; mem, Am Inst Biol Comt Natural Areas State of Utah, 72-74; reviewer, Brown & Co, Publishers, 74. *Mem:* AAAS; Ecol Soc Am; Brit Ecol Soc; Soc Range Mgt; Sigma Xi. *Res:* Ecological adaptation; niche metrics; evolutionary and environmental gradient accomodation of organisms, populations, and biotic communities in the Great Basin of North America. *Mailing Add:* Dept Bot & Range Sci Brigham Young Univ Provo UT 84602

BROTHERTON, ROBERT JOHN, b Ypsilanti, Mich, Aug 4, 28; m 50; c 4. ORGANIC CHEMISTRY, RESEARCH ADMINISTRATION. *Educ:* Univ Ill, BS, 49; Wash State Univ, PhD, 54. *Prof Exp:* Du Pont fel & instr org chem, Univ Minn, 54-55; res chemist, Refinery Res Group, Union Oil Co, 55-57; res chemist, 57-64, VPRES & DIR RES, US BORAX RES CORP, 64- *Mem:* Am Chem Soc. *Res:* Boron chemistry. *Mailing Add:* 424 Panorama Dr Laguna Beach CA 92651

BROTZEN, FRANZ R(ICHARD), b Berlin, Ger, July 4, 15; nat US; m 50; c 2. PHYSICAL METALLURGY. *Educ:* Case Inst Technol, BS, 50, MS, 53, PhD(phys metall), 54. *Prof Exp:* Res metallurgist, US Bur Mines, Md, 50-51; sr res assoc metall, Case Inst Technol, 51-54; assoc prof mech eng, Rice Univ, 54-59, prof mat sci, 59-86, dean eng, 62-66, EMER PROF MAT SCI, RICE UNIV, 86- *Concurrent Pos:* Consult, Naval Res Lab, 55-59; Guggenheim fel, US Sr Scientist Award, Stuttgart Tech Sch, 61-62; vis lectr & consult, Univ Brazil, 63, 65 & 70; vis prof, Swiss Fed Inst Technol, 66-67, Max-Planck Inst, Stuttgart, 73-74 & Univ Lausanne, 81. *Mem:* Fel Am Soc Metals; Am Inst Mining, Metall & Petrol Engrs; Am Inst Physics; Am Soc Eng Educ. *Res:* Lattice imperfection theory; electron emission, electronic materials phenomena; alloy theory; physical properties. *Mailing Add:* Dept Mech Eng & Mat Sci Rice Univ PO Box 1892 Houston TX 77001

BROUGHTON, M(ERVYN) B(LYTHE), b Corbetton, Ont, July 16, 29; m 56; c 4. ELECTRICAL ENGINEERING. *Educ:* Univ Western Ont, BSc, 54; Univ Toronto, MASc, 58; Queens Univ, Can, PhD(elec eng), 71. *Prof Exp:* Defense sci serv officer, Defence Res Bd, Can, 58-59; from asst prof to assoc prof, 60-79, actg head, Elec & Comput Eng Dept, 88-89, PROF ELEC ENG, ROYAL MIL COL CAN, 79- *Concurrent Pos:* Defence Res Bd res grant, 61-63, 66; secy, Can Conf Elec Eng Educ, 65-66; chmn, Can Region Student Activ Comt, Inst Elec & Electronic Engrs, 68-69; exchange prof, Royal Mil Col Sci, Shrivenham, Eng, 76-77; vis res prof, US Naval Acad, Annapolis, Md, 84-85, Australian Defence Force Acad, Canberra, Australia, 91- *Honors & Awards:* Medal Res & Develop, Asn Prof Engrs Ont, 90. *Mem:* Sr mem Inst Elec & Electronic Engrs; Can Soc Elec Engrs; Soc Computer Simulation; Asn Prof Engrs Ont. *Res:* Analysis, design and simulation of circuits and control systems using digital computers; power electronics and instrumentation. *Mailing Add:* Dept Elec & Comput Eng Royal Mil Col Can Kingston ON K7K 5L0 Can

BROUGHTON, ROBERT STEPHEN, b Corbetton, Ont, June 29, 34; m 57; c 4. AGRICULTURAL & HYDRAULIC ENGINEERING. *Educ:* Univ Toronto, BSA, 56, BASc, 57; Mass Inst Technol, SM, 59; McGill Univ, PhD, 72. *Hon Degrees:* LLD, Dalhousie Univ, 89. *Prof Exp:* Res asst fluid mech, Mass Inst Technol, 57-59; hydraul engr, Conserv Br, Ont Civil Serv, 59-61; chmn, Dept Agr Eng, 61-71, coordr, Agr Sci Div, 72-75, from asst prof to assoc prof, 61-74, PROF AGR ENG, MACDONALD COL, MCGILL UNIV, 74-; DIR, CTR DRAINAGE STUDIES, 87- *Concurrent Pos:* External examr agr eng, Trinidad, 70-77, Nigeria, 75-76 & England, 79-81; consult irrig & drainage, Food & Agr Orgn UN Asian Develop Bank & Can Govt Aid Orgns, Trinidad, Haiti, El Salvador, Egypt, India, Pakistan, USA

& Can. *Honors & Awards:* Maple Leaf Award, Can Soc Agr Engrs, James Beamish Award. *Mem:* Am Soc Agr Engrs; fel Can Soc Agr Engrs; Eng Inst Can; Can Water Resources Asn. *Res:* Study of the water balance of flat lands in Canada; mechanics of cavitation; vorticity problems of flood control pumping stations; water resources surveys; irrigation water requirements; drainage and irrigation engineering research. *Mailing Add:* Dept Agr Eng Macdonald Col McGill Univ Quebec PQ H9X 1C0 Can

BROUGHTON, ROGER JAMES, b Montreal, Que, Sept 25, 36; m 59; c 3. NEUROPHYSIOLOGY, PSYCHOBIOLOGY. *Educ:* Queen's Univ, Ont, MD, CM, 60; Univ Aix-Marseille, dipl electroencephalog-neurophysiol, 64; McGill Univ, PhD(neurophysiol), 67. *Prof Exp:* Intern med, Univ Sask Hosp, 60-61; resident, Centre St Paul, Marseille, 62-64; asst prof clin neurophysiol, McGill Univ, 64-68; assoc prof med & pharmacol, 68-74, prof pharmacol, 74-84, PROF MED & PSYCHOL, UNIV OTTAWA, 74-, ASSOC DIR, INST MED ENG, 90-; PHYSICIAN, OTTAWA GEN HOSP, 68- *Concurrent Pos:* Consult, WHO, 64-68; assoc, Med Res Coun Can, 68-; ed, Sleep Rev, 70-80. *Honors & Awards:* Hon Corresp Mem Brazilian Acad Neurologist, Latin Am Sleep Soc, Czech Soc, Clin Neurophysiology, Cuban Neurosci Soc. *Mem:* Am EEG Soc; Am Sleep Disorders Asn; Sleep Res Soc; Int League Against Epilepsy; Can Soc EEG; Sigma Xi. *Res:* Mechanisms of precipitation of epileptic seizures; pharmacology of experimental and clinical epilepsy; clinical disorders of sleep and arousal; phylogeny of sleep; cerebral evoked potentials and their correlates in man; daytime sleepiness narcolepsy. *Mailing Add:* Dept Med Ottawa Gen Hosp Ottawa ON K1H 8L6 Can

BROUGHTON, WILLIAM ALBERT, b Ronan, Mont, Oct 12, 14; m 44; c 4. ECONOMIC GEOLOGY. *Educ:* Univ Wis, AB, 39. *Prof Exp:* Geologist, Wis Geol Surv, 36-41; instr petrol, State Col Wash, 41; geologist, State Div Geol, Wash, 41-45; geologist, Chicago, Milwaukee, St Paul & Pac Rwy, 45-46; consult geologist, 46; prof & head dept, 48-78, EMER PROF GEOL, UNIV WIS-PLATTEVILLE, 79- *Concurrent Pos:* US Geol Surv, 51-54; geologist-in-chg, Mineral Develop Atlas, Wis Geol Surv, 61-, prof & specialist, 79- *Mem:* Fel AAAS; Soc Econ Geologists; Am Inst Mining, Metall & Petrol Eng. *Res:* Sedimentation of Wisconsin lakes; tungsten and magnetic iron deposits of Washington; zinc-lead deposits; meteoritics. *Mailing Add:* 295 Bradford Platteville WI 53818

BROUILLARD, ROBERT ERNEST, b Northbridge, Mass, Aug 16, 15; m 41; c 4. CHEMISTRY, PHYSICS. *Educ:* Bates Col, BS, 38; Clark Univ, MA, 39, PhD(chem), 41. *Prof Exp:* Chemist dyes & pigments, Am Cyanamid Co, 41-49; prod mgr pigments, GAF Corp, 49-56, sales mgr, 56-60; vpres sales carbohydrates, Penick & Ford Ltd, 60-65, vpres tech dir, 65-78, pres, Bedford Labs, 75-82; RETIRED. *Mem:* Am Chem Soc; fel Am Inst Chemists; Tech Asn Pulp & Paper; Am Asn Food Technologists; Am Asn Cereal Chemists. *Res:* Dyes; pigments; carbohydrate chemistry; enzymology. *Mailing Add:* 1723 San Pablo Dr San Marcos CA 92069-4717

BROUILLET, LUC, b L'Assomption, Que, June 18, 54. SYSTEMATICS, PHYTOGEOGRAPHY. *Educ:* Univ Montreal, BSc, 75, MSc, 77; Univ Waterloo, PhD(biol), 81. *Prof Exp:* Prof bot, MacDonald Campus & cur Herbarium, McGill Univ, 80-82; PROF BOT & CUR, MARIE-VICTORIN HERBARIUM, UNIV MONTREAL, 82- *Honors & Awards:* L Cinq-Mars Award, Can Bot Asn, 79. *Mem:* Can Bot Asn (past-pres, 87-88); Can Biol Coun; Bot Soc Am; Soc Study Evolution; Am Soc Plant Taxonomists; Am Inst Biol Sci. *Res:* Systematics, evolution and phytogeography of angiosperms, Compositae and Begoniaceae; floristics and biogeography of Quebec and Newfoundland; cytogeography and flora of North America. *Mailing Add:* Institut de Recherche en Biologie Végétale Univ Montreal 4101 Sherbrooke St E Montreal PQ H1X 2B2 Can

BROUILLETTE, ROBERT T, b Washington, DC, June 27, 47; m 74; c 3. NEONATOLOGY, PEDIATRIC PULMONOLOGY. *Educ:* Providence Col, AB, 69; Wash Univ, St Louis, MD, 74. *Prof Exp:* From asst prof to prof pediat, Med Sch, Northwestern Univ, 79-89; PROF PEDIAT, MCGILL UNIV, 89- *Concurrent Pos:* Attending physician, Children's Mem Hosp, 79-89, Montreal Children's Hosp, 89-, div head, Newborn Med, Dir Sleep Lab. *Mem:* Soc Pediat Res; Am Physiol Soc; Am Acad Pediat; Am Thoracic Soc; Am Perinatal Asn. *Res:* Development of respiratory control and airway patency original models and human diseases; central hyperventilation syndrome and phrenic nerve pacing; clinical neonatology. *Mailing Add:* Montreal Children's Hosp 2300 Tupper St Montreal PQ H3H 1P3 Can

BROUILLETTE, WALTER, b Port Barre, La, Aug 13, 25; m 46. SOLID STATE PHYSICS. *Educ:* La State Univ, BSc, 48, MSc, 50; Univ Mo, PhD(physics), 55. *Prof Exp:* Physicist, Gen Elec Co, Syracuse, 55-80; MEM FAC, CHEM DEPT, UNIV ALA, 80- *Mem:* Am Asn Physics Teachers; Am Phys Soc. *Res:* Physical electronics; electron tube circuitry; information theory; information storage and processing; electron beam lithography. *Mailing Add:* 6675 Woodchuck Hill Rd Jamesville NY 13078

BROUN, THOROWGOOD TAYLOR, JR, b Nashville, Tenn, Apr 4, 23; m 42; c 3. INDUSTRIAL CHEMISTRY. *Educ:* ETex State Univ, BS, 43; Purdue Univ, PhD(chem), 51. *Prof Exp:* Res analyst, Field Res Dept, Magnolia Petrol Co, 43-44; res chemist, Chem Div, PPG Industs, Inc, 51-56, res supvr, 56-67; sr res chemist, 67-75, RES COORDR, HOUSTON CHEMICAL CO DIV, PPG INDUSTS, INC, 75- *Mem:* Am Chem Soc; Sigma Xi. *Res:* Applied product and process research. *Mailing Add:* 5575 Clinton Dr Beaumont TX 77706

BROUNS, RICHARD JOHN, b Osakis, Minn, Oct 2, 17; m 52; c 6. ANALYTICAL CHEMISTRY, NUCLEAR CHEMISTRY. *Educ:* St John's Univ, Minn, BS, 42; Iowa State Col, MS, 44, PhD(anal chem), 48. *Prof Exp:* Plant supvr, Cardox Corp, Okla, 44-45; res chemist, Gen Elec Co, Wash, 48-52, supvr anal chem res, 52-61, opers res analyst, 61-64; res assoc, Pac Northwest Labs, Battelle Mem Inst, 65-86; RETIRED. *Mem:* Am Chem Soc; Inst Nuclear Mat Mgt. *Res:* Analytical chemistry; electrochemistry; radiochemistry; instrumentation; development of nuclear safeguards systems through materials accounting measurements and statistics. *Mailing Add:* 1424 Sanford Richland WA 99352

BROUS, DON W, b Glenside, Pa, Aug 10, 13; m 37; c 4. ARTIFICIAL KIDNEY SYSTEMS. *Educ:* Haverford Col, BS, 36. *Prof Exp:* Engr, Westinghouse Elec Mfg Co, 36-41; exec vpres, Anchor Mfg Co, 41-51; pres, Northeastern Eng Co, 51-61; PRES, ENG ASSOCS, 61-; VPRES, YANKEE INT MED RES LTD, 80- *Mem:* Instrument Soc Am; Am Inst Elec Engrs; Nat Soc Prof Engrs. *Res:* Artificial kidney systems; hardware for kidney therapy. *Mailing Add:* Yankee Int Med Res Ltd Ten Chestnut Dr Bedford NH 03110-5566

BROUS, JACK, b New York, NY, Nov 14, 26; m 51; c 2. INORGANIC CHEMISTRY, PHYSICAL CHEMISTRY. *Educ:* City Col NY, BS, 48; Univ Chicago, SM, 49; Polytech Inst Brooklyn, PhD(inorg chem), 53. *Prof Exp:* Develop engr, Gen Eng Labs, Gen Elec Co, 52-55 & Power Tube Dept, 55-57; develop engr, Radio Corp Am, 57-58, engr leader, Chem & Phys Lab, 58-68; dir process develop, Pyrofilm Corp, NJ, 69-70; mgr chem res & develop, 70-73, SR STAFF SCIENTIST, ALPHA METALS CORP, JERSEY CITY, 73- *Mem:* Am Chem Soc; Sigma Xi; Inst Interconnecting & Packaging Electronic Circuits. *Res:* Inorganic dielectrics; ferroelectrics; x-ray diffraction; thermionic electron emission studies; high vacuum technology; getter materials; thermal measurements; electronic materials; cleaning materials and processes; invented the Ionograph for measurement of residual ionic contamination; surface insulation studies; surface electrical leakage phenomena of electronic insulating surfaces and the effects which different processes and processing materials have on it. *Mailing Add:* 35 Brandon Ave Livingston NJ 07039

BROUSSEAU, NICOLE, b Quebec, Can, Oct 5, 48. OPTICS. *Educ:* Laval Univ, BS, 70, MS, 72, PhD(optics), 75. *Prof Exp:* SCI RESEARCHER OPTICS, DEPT NAT DEFENCE, 75- *Concurrent Pos:* Res asst optics, Laval Univ, 75. *Mem:* Can Asn Physicists. *Res:* Optical processing of synthetic aperture radar data; acousto-optics optical data processing. *Mailing Add:* DREO Shirley's Bay Ottawa ON K1A 0Z4 Can

BROUTMAN, LAWRENCE JAY, b Chicago, Ill, Feb 9, 38; m 61; c 2. MATERIALS & POLYMER SCIENCE. *Educ:* Mass Inst Technol, SB, 59, SM, 61, ScD(mat eng), 63. *Prof Exp:* Res engr, IIT Res Inst, 63-65, sr res engr, 65-66; assoc prof mech & mat, 66-70, PROF MAT ENG, ILL INST TECHNOL, 70- *Concurrent Pos:* Consult, US Air Force Mat Lab, 66-67, Aeronca Inc, 66-69, US Army Mat & Mech Res Ctr, 69- & Atomic Energy Comn, 68- *Mem:* Soc Plastics Engrs; Soc Rheol; Soc Exp Stress Anal; Am Soc Testing & Mat; Sigma Xi. *Res:* Properties and structure of organic polymers; fracture phenomena in nonmetallic materials; structure and properties of composite materials. *Mailing Add:* 950 N Michigan 42C Chicago IL 60611-4503

BROWALL, KENNETH WALTER, b Boston, Mass, May 24, 47; m 72. PHYSICAL CHEMISTRY. *Educ:* Tufts Univ, BS, 69; Rensselaer Polytech Inst, MS, 72, PhD(chem), 76. *Prof Exp:* PHYS CHEMIST, DIV CORP RES & DEVELOP, GEN ELEC CO, 69- *Mem:* Am Chem Soc; Electrochem Soc; Sigma Xi. *Res:* Structure and properties of solid electrolytes; power production-conversion systems; one-dimensional conductors; batteries and fuel cells; thin film deposition. *Mailing Add:* 1265 Sagemont Ct Schenectady NY 12309

BROWDER, FELIX EARL, b Moscow, Russia, July 31, 27; nat US; m 49; c 2. MATHEMATICAL ANALYSIS. *Educ:* Mass Inst Technol, SB, 46; Princeton Univ, MA, 47, PhD(math), 48. *Hon Degrees:* MA, Yale Univ, 62; Dr, Univ Paris, 90. *Prof Exp:* Moore instr math, Mass Inst Technol, 48-51; instr, Boston Univ, 51-53; asst prof, Brandeis Univ, 55-56; from asst prof to prof, Yale Univ, 56-63; prof, 63-72, chmn dept, 72-77 & 80-85, Louis Block Prof, 72-82, Max Mason distinguished serv prof math, Univ Chicago, 82-86; VPRES RES & UNIV PROF MATH, RUTGERS UNIV, 86- *Concurrent Pos:* Vis mem, Inst Advan Study, 53-54 & 63-64; Guggenheim fels, 53-54 & 66-67; NSF sr fel, 57-58; Sloan Found fel, 59-63; ed, Bull Am Math Soc, 59-68 & 77-83, assoc ed, Ann Math, 64- 69; vis prof, Inst Advan Study, 53-54 & 63-64, Inst Pure & Appl Math, Rio de Janeiro, 60, Mass Inst Technol, 61-62 & 77-78, Princeton Univ, 68 & Univ Paris, 72, 75, 78, 81, 83 & 87; vis assoc, Cal Inst Technol, 67, 78, Fairchild distinguished vis scholar, 75-76; sr res fel, Univ of Sussex, 70-76; mem, var sci comts, NSF, Nat Res Coun, Nat Acad Sci, 71- *Mem:* Nat Acad Sci; Am Math Soc; fel AAAS; fel Am Acad Arts & Sci; Sigma Xi. *Res:* Partial differential equations and nonlinear functional analysis. *Mailing Add:* Old Queens Bldg Rutgers Univ New Brunswick NJ 08903

BROWDER, JAMES STEVE, b Goodwater, Ala, Aug 5, 39; m 62; c 3. SOLID STATE PHYSICS, OPTICS. *Educ:* Rollins Col, BS, 61; Univ Fla, MS, 63, PhD(physics), 67. *Prof Exp:* Asst physics, Univ Fla, 64-67, res asst, 67-68, fel, 68; asst prof, Northwestern State Univ, 68-71; from asst prof to assoc prof, 71-77, PROF PHYSICS, JACKSONVILLE UNIV, 77- *Concurrent Pos:* NSF acad year exten grant, 69-71. *Mem:* Am Asn Physics Teachers; Optical Soc Am; Am Phys Soc. *Res:* Use of a three-terminal capacitance dilatometer to study the thermal expansion of solids (single crystals, polycrystals and glasses) in the temperature range from room temperature down to 16 kelvin. *Mailing Add:* Dept Physics Jacksonville Univ Jacksonville FL 32211

BROWDER, LEON WILFRED, b Pueblo, Colo, Apr 19, 40; m 63; c 2. DEVELOPMENTAL BIOLOGY. *Educ:* Univ Colo, BA, 62; La State Univ, MS, 64; Univ Minn, PhD(zool), 67. *Prof Exp:* Res assoc, Univ Colo, 67-69; asst prof biol, 69-72, assoc prof, 72-78, PROF BIOL, UNIV CALGARY, 78- *Concurrent Pos:* NIH fels, 67-69; Nat Res Coun Can grant, 70- *Mem:* Can Soc Cell Biol; Soc Develop Biol; Genetics Soc Can; Am Soc Cell Biol; AAAS; Fedn Am Soc Exp Biol. *Res:* Control of amphibian pigment cell differentiation; control of genic expression during oogenesis and early development in amphibians; nuclear differentiation during development. *Mailing Add:* Dept Biol Univ Calgary 2500 University Dr NW Calgary AB T2N 1N4 Can

BROWDER, LEWIS EUGENE, b McQueen, Okla, Jan 29, 32; m 54; c 3. PLANT PATHOLOGY. *Educ:* Okla State Univ, BS, 54, MS, 56; Kans State Univ, PhD(plant path), 65. *Prof Exp:* Plant pathologist, Okla State Univ, 56-58; plant pathologist, USDA, 58-66, ASST PROF PLANT PATH, KANS STATE UNIV, 58-, RES PLANT PATHOLOGIST, USDA, 66- *Mem:* Am Phytopath Soc. *Res:* Physiologic specialization of cereal rusts. *Mailing Add:* 2012 Ivy Dr Manhattan KS 66502

BROWDER, WILLIAM, b New York, NY, Jan 6, 34. TOPOLOGY. *Educ:* Mass Inst Technol, BS, 54; Princeton Univ, PhD(math), 58. *Prof Exp:* Instr math, Univ Rochester, 57-58; from instr to assoc prof, Cornell Univ, 58-63; mem, Inst Advan Study, 63-64 & 83; chmn, 71-73, PROF, DEPT MATH, PRINCETON UNIV, 64- *Concurrent Pos:* NSF post doc fel, Univ Chicago, Oxford, 59-60; assoc prof, Fac Sci Orsay, Univ Paris, 67-68; vis fel, Harvard Univ, 74; John Simon Guggenheim fel, 74-75; vis lectr, Tata Inst Fundamental Res, Bombay, 75; SRC sr fel, Oxford, 78-79; vis prof, Aarhus Univ, 84, Univ Chicago & North Western Univ, 85; chmn, Off Math Sci & mem, Assembly Math & Phys Sci, Nat Sci Found & Nat Res Coun, 78-84; chmn panel math, Off Sci & Technol Policy, White HOuse, 83; speaker, Int Cong Math, Moscow, 66 & Nice, 70; mem coun, Am Math Soc, 67-69, 72-74 & colloquium lectr, 77; mem, Max Planck Inst Math, 88-89; ed, Annals Math. *Mem:* Nat Acad Sci; Am Acad Arts & Sci; Am Math Soc (vpres, 77-78, pres, 89-90). *Res:* Algebraic topology with special interests in the topology of manifolds and compact transformation groups. *Mailing Add:* Dept Math Princeton Univ Fine Hall Washington Rd Princeton NJ 08544-1000

BROWDY, CRAIG LAWRENCE, b Wash, DC, June 9, 58; m 88; c 1. AQUACULTURE, SHRIMP MARICULTURE. *Educ:* Univ Md, BA, 80, BSc, 81; Tel Aviv Univ, PhD(marine zool), 89. *Prof Exp:* Tech & res asst aquacult & invert zool, Israel Oceanog & Limnol Res, Nat Ctr Maricult, 81-89; RES MARINE SCIENTIST, SC WILDLIFE & MARINE RESOURCES, WADDELL MARICULT CTR, 89- *Concurrent Pos:* Teaching asst, Zool Dept, Tel Aviv Univ, 84-88; consult shrimp maricult, 89- *Mem:* World Aquacult Soc; Crustacean Soc; Am Soc Zoologists. *Res:* Aspects of shrimp mariculture; systems design, experimental to commercial scale for maturation and spawning; larval culture, nurseries and growout; physiology and endocrine control of reproduction in crustaceans; shrimp disease pathology. *Mailing Add:* Waddell Maricult Ctr PO Box 809 Bluffton SC 29910

BROWE, JOHN HAROLD, b Burlington, Vt, Nov 17, 15; m 39; c 4. PUBLIC HEALTH & EPIDEMIOLOGY. *Educ:* Univ Vt, AB, 37, MD, 40; Columbia Univ, MPH, 50. *Prof Exp:* Res assoc med & clin dir nutrit study, Col Med, Univ Vt, 46-49; pub health physician in training, NY State Dept Health, 49-50, actg dir, Bur Nutrit, 50-51, dir, Bur Nutrit, 51-77; RETIRED. *Concurrent Pos:* Consult, USPHS, 47, 56-73 & Vt State Dept Health, 48-49; assoc, Albany Med Col, 52-55, instr, 58-66; mem, Am Bd Nutrit; mem panel nutrit, Life Sci Comt, Space Sci Bd, Nat Acad Sci, 62-70, mem coun epidemiol, Am Heart Asn, 66-77; mem consult group for dietitians, Albany Regional Med Prog, 69-73. *Mem:* Fel Am Pub Health Asn; Am Soc Clin Nutrit; Am Inst Nutrit; Asn State & Territorial Pub Health Nutrit Dirs (pres, 57-58); Latin Am Soc Nutrit. *Res:* Human nutrition. *Mailing Add:* Four Locust Ave Troy NY 12180-5124

BROWELL, EDWARD VERN, b Indiana, Pa, Feb 6, 47; m 69; c 1. LASER REMOTE SENSING, ATMOSPHERIC SCIENCES. *Educ:* Univ Fla, BSAE, 68, MS, 71, PhD(appl optics), 74. *Prof Exp:* Res asst appl optics, Aerospace Eng Dept, Univ Fla, 69-74; SR RES SCIENTIST & HEAD, LIDAR APPLICATIONS GROUP, ATMOSPHERIC SCI DIV, LANGLEY RES CTR, NASA, 74- *Mem:* Optical Soc Am; Am Meteorol Soc; Am Geophys Union; AAAS. *Res:* Remote sensing of atmospheric gases and parameters using pulsed tunable laser systems; studies of atmospheric dynamics and chemistry in troposphere and stratosphere; atmospheric environmental sciences. *Mailing Add:* MS 401-A NASA Langley Res Ctr Hampton VA 23665-5225

BROWER, ALLEN S(PENCER), b Poughkeepsie, NY, Sept 28, 26; m 49; c 3. ELECTRICAL ENGINEERING. *Educ:* Columbia Univ, AB, 47, BS, 48, MSEE, 50. *Prof Exp:* Asst elec eng, Columbia Univ, 48-49, instr, 49-53; mem advan eng prog, 53-56, admin staff, 56-59, mgr tech educ, NY Region, 59-61, metal rolling appln engr, 61-66, systs engr, 66-69, MGR SYSTS ADVAN DEVELOP, DRIVE SYSTS DEPT, GEN ELEC CO, 70- *Mem:* Inst Elec & Electronics Engrs; assoc mem Asn Iron & Steel Engrs. *Res:* Industrial process control; computer technology. *Mailing Add:* Elec Util Syst Bldg 2-520 Gen Elec Co 1 River Rd Schenectady NY 12345

BROWER, FRANK M, environmental management; deceased, see previous edition for last biography

BROWER, JAMES CLINTON, b New Rochelle, NY, June 27, 34; m; c 2. INVERTEBRATE PALEONTOLOGY. *Educ:* Am Univ, BS, 59, MS, 61; Univ Wis, PhD(geol), 64. *Prof Exp:* From asst prof to assoc prof, 63-74, PROF GEOL, SYRACUSE UNIV, 74- *Mem:* Soc Syst Zool; Soc Econ Paleontologists & Mineralogists; Int Asn Math Geol; Paleont Soc; Int Paleont Asn. *Res:* Paleobiology; Paleozoic crinoids; geostatistics. *Mailing Add:* Dept Geol Syracuse Univ Syracuse NY 13244-1070

BROWER, JOHN HAROLD, b Augusta, Maine, June 8, 40. BIOLOGICAL CONTROL, RADIATION ECOLOGY. *Educ:* Univ Maine, BS, 62; Univ Mass, MS, 64, PhD(entom), 65. *Prof Exp:* Res assoc insect radiation ecol, Brookhaven Nat Lab, 61-65; RES ENTOMOLOGIST, STORED PROD INSECTS RES & DEVELOP LAB, AGR RES SERV, USDA, 65- *Concurrent Pos:* Consult, Int Atomic Energy Agency, 74-90. *Mem:* Entom Soc Am; Ecol Soc Am; Radiation Res Soc; Entom Soc Can. *Res:* Effects of increased levels of ionizing radiations on insect populations in nature; radiation effects on insect behavior, population dynamics, physiology, genetics and developmental success; possibilities of radiation control of insect pests; biological control of stored-product pests with insect parasites and predators. *Mailing Add:* Rte 1 Box 441A Rincon GA 31326

BROWER, KAY ROBERT, b Altus, Okla, June 7, 28; m 48; c 2. ORGANIC CHEMISTRY. *Educ:* Mass Inst Technol, BS, 48; Univ Maine, MS, 51; Lehigh Univ, PhD(chem), 53. *Prof Exp:* Instr chem, Lehigh Univ, 53-54; res assoc, Univ Ill, 54-56; assoc prof, 56-70, chmn dept, 70-77, PROF CHEM, NMEX INST MINING & TECHNOL, 70- *Mem:* Fel AAAS; Am Chem Soc. *Res:* Chlorination products of organic compounds of sulfur; coumarins and chromones; kinetics of aromatic nucleophilic substitution reactions; volumes of activation of organic reactions; odor and chemical structure; chemistry of explosive decomposition. *Mailing Add:* Dept Chem NMex Inst Mining & Technol Socorro NM 87801

BROWER, KEITH LAMAR, b South Haven, Mich, Oct 24, 36; m 60; c 3. PHYSICS. *Educ:* Hope Col, AB, 58; Univ Minn, MS, 61; Univ Ill, PhD(physics), 66. *Prof Exp:* Res assoc ion & plasma physics, Electronics Group, Gen Mills, 61-62; MEM TECH STAFF MAGNETIC RESONANCE, SANDIA LABS, 66- *Mem:* Am Phys Soc. *Res:* Radiation damage. *Mailing Add:* Sandia Labs Div 1111 PO Box 5800 Albuquerque NM 87185

BROWER, LINCOLN PIERSON, b Summit, NJ, Sept 10, 31; c 2. ECOLOGY. *Educ:* Princeton Univ, AB, 53; Yale Univ, PhD(zool), 57, Amherst Col, MA, 62. *Prof Exp:* Fulbright scholar, Oxford Univ, 57-58; from instr to prof, Amherst Col, 58-75, Stone prof biol, 75-80; DISTINGUISHED PROF ZOOL, UNIV FLA, 80- *Concurrent Pos:* NIH spec fel, Oxford Univ, 63-64; vpres, Am Soc Naturalists, 84. *Mem:* Fel AAAS; Am Soc Naturalists; Am Soc Zoologists; Am Inst Biol Sci; Soc Study Evolution (pres, 79); Int Soc Chem Ecol (pres, 84). *Res:* Experimental study of ecology; evolution; animal behavior; ecological chemistry; automimicry; migration. *Mailing Add:* Dept Zool Univ Fla 123 Bartram Hall Gainesville FL 32611

BROWER, THOMAS DUDLEY, b Birch Tree, Mo, Mar 15, 24; m 48; c 2. ORTHOPEDIC SURGERY. *Educ:* Cent Col, BS, 45; Wash Univ, MD, 47. *Prof Exp:* Intern surg, Univ Chicago Clins, 47-48; fel orthoped surg, Albany Med Col, 48-49; resident orthoped surg, Univ Chicago Clins, 49-54, instr surg, 54-55; from asst prof to assoc prof orthoped, Univ Pittsburgh, 55-64; PROF ORTHOPED, MED CTR, UNIV KY, 64- *Mem:* Am Acad Orthoped Surg; Am Orthoped Soc. *Res:* Articular cartilage. *Mailing Add:* Div Orthoped Surg Univ Ky Med Ctr Lexington KY 40506

BROWER, WILLIAM B, JR, b Jersey City, NJ, Oct 29, 22; m 46; c 3. FLUID MECHANICS, AERONAUTICAL ENGINEERING. *Educ:* Rensselaer Polytech Inst, BAeroE, 50, MAeroE, 51, PhD(aeronaut eng), 61. *Prof Exp:* Res assoc aerodyn, 52-56, asst prof fluid mech, 56-62, ASSOC PROF FLUID MECH, RENSSELAER POLYTECH INST, 62-, ASSOC PROF AERONAUT & MECH ENG, 68- *Concurrent Pos:* Res Engr, NAm Aviation, Inc, 51-52; fel, Inst Blaise Pascal, Nat Sci Res Ctr, France, 53-54; consult, Curtiss-Wright Corp, 57, Avco Corp, 57-60, NY State Natural Gas Corp, 63-64 & Gen Elec Co, 70. *Mem:* Am Soc Mech Engrs. *Res:* Electrical analogy computation; compressible flow, especially unsteady fluid mechanics; high-speed tube transportation. *Mailing Add:* 47 Second St Troy NY 12180

BROWN, A HAYDEN, JR, b Cookeville, Tenn, Oct 13, 46; m 77. BEEF CATTLE GENETICS & BREEDING, BEEF CATTLE ADAPTATION. *Educ:* Tenn Technol Univ, BS, 68; Univ Tenn, Knoxville, MS, 74, PhD(genetics), 76. *Prof Exp:* From asst prof to assoc prof, 77-88, PROF ANIMAL SCI, UNIV ARK, FAYETTEVILLE, 88- *Concurrent Pos:* Prof, Univ Ark, Fayetteville, 88- *Mem:* Sigma Xi; Am Soc Animal Sci; NY Acad Sci; AAAS. *Res:* Adaptation of beef cattle to climatic and production resources, and genetic improvement through artificial selection. *Mailing Add:* C-125 Animal Sci Bldg Univ Ark Fayetteville AR 72701

BROWN, ACTON RICHARD, b Ash Grove, Kans, Apr 3, 20; m 43; c 2. PLANT BREEDING. *Educ:* Kans State Col, BS, 42; Univ Wis, MS, 47, PhD(agron), 50. *Prof Exp:* From asst prof to assoc prof agron, Univ Ga, 50-87; RETIRED. *Mem:* Am Soc Agron; Sigma Xi. *Res:* Plant production; inheritance of leaf rust resistance in barley; barley loose smut; lysine content of barley; semidwarf wheats; releasing of four barley varieties and one rye variety. *Mailing Add:* Div Agron Univ Ga Athens GA 30602

BROWN, ALAN R, IMMUNO-REGULATION, IDIOTYPE NETWORKS. *Educ:* Univ Fla, Gainesville, PhD(immunol & microbiol), 77. *Prof Exp:* ASST MEM IMMUNOL, ST JUDE'S HOSP, 80- *Res:* Idiotype based assays. *Mailing Add:* St Jude Children's Res Hosp PO Box 318 Memphis TN 38101

BROWN, ALBERT LOREN, b Rochester, NY, Aug 27, 23; m 49; c 3. VETERINARY MICROBIOLOGY. *Educ:* Cornell Univ, BS, 48, MS, 49, PhD(bact), 51. *Prof Exp:* Asst bact, Cornell Univ, 48-51; res assoc virol, Sharp & Dohme Div, Merck & Co, Inc, 51-55, tech asst to dir biol prod, 55-57; res biologist, 57-61, sr res scientist, 61-84, dir molecular microbiol, Res & Develop, 84-87, DISTINGUISHED SCIENTIST, NORDEN LABS, 87- *Concurrent Pos:* NIH spec fel, Sloan-Kettering Inst Cancer Res, 64; res fel, Dept Clin Vet Med, Cambridge Univ, Eng, 79. *Mem:* Am Soc Microbiol; US Animal Health Asn; Conf Res Workers Animal Diseases; Int Asn Biol Standards. *Res:* Veterinary biological products; viruses; canine distemper; rabies; leptospira; tissue culture; immunology. *Mailing Add:* 859 Moraine Dr Lincoln NE 68510

BROWN, ALEX CYRIL, b Petrolia, Ont, July 24, 38; m 64, 74; c 1. ECONOMIC GEOLOGY, ORE GENESIS. *Educ:* Univ Western Ont, BSc, 62; Univ Mich, MS, 65, PhD(geol & mineral), 68. *Prof Exp:* Tech officer, Geol Surv Can, 62-63; dir, Mineral Explor Res Inst, Montreal, 84-85; from asst prof to assoc prof, 70-80, PROF ECON GEOL, ECOLE POLYTECH, UNIV MONTREAL, 80- *Concurrent Pos:* Nat Res Coun Can & NATO fel, Univ Liege, 68-70, Sigma Xi res grant-in-aid, 69; grants, Nat Sci & Eng Res Coun Can, Ecole Polytech, Univ Montreal, 70-, FCAC-FCAR res grants, 82-, Natural Sci & Eng Res Coun Can Strategic grant, 89-; Geol Surv Can res contracts, 74-76; sabbatical leave, Geol Surv Can, 80-81; invited prof, Inst Nat De Recherche Scientifique Georessources, 80-84; adj prof, Derry-Rust Res Unit, Ottawa-Carleton Ctr Geosci Studies, 84-; sabbatical leave, Univ Col, Cardiff, Wales, 87-88. *Honors & Awards:* Gold Medal, Geol, Univ Western Ont, 62; Waldemar Lindgren Citation, Soc Econ Geologists, 71. *Mem:* Geol Asn Can; Mineral Asn Can; Mineral Explor Res Inst; Can Inst Mining & Metall; Soc Geol Appl Mineral Deposits; Soc Econ Geologists; Geol Soc Belg. *Res:* Mineralization of sediment-hosted base metal deposits; Archean gold deposits. *Mailing Add:* Ecole Polytech, Univ Montreal CP 6079 Succursale A Montreal PQ H3C 3A7 Can

BROWN, ALFRED BRUCE, JR, b Oak Park, Ill, May 11, 20; m 56; c 3. ELECTRON PHYSICS. *Educ:* Lehigh Univ, BS, 42; Calif Inst Technol, MS, 47, PhD(physics), 50. *Prof Exp:* Res assoc, Gen Elec Res Lab, 50-56; mem tech staff, 56-62, dept head, 62-75, CONSULT, BELL LABS, 75- *Mem:* AAAS; Am Phys Soc; Inst Elec & Electronics Engrs. *Res:* Data transmission. *Mailing Add:* 109 Edgemont Rd Upper Montclair NJ 07043

BROWN, ALFRED EDWARD, b Elizabeth, NJ, Nov 22, 16; m 44; c 2. ORGANIC CHEMISTRY. *Educ:* Rutgers Univ, BSc, 38; Ohio State Univ, MSc, 40, PhD(chem), 42. *Prof Exp:* Org chemist, USDA, 42-45; res assoc, Off Sci Res & Develop, 45; res assoc & asst dir, Harris Res Labs, 45-53, from asst dir res to dir res, 53-61, from vpres to pres, 53-66; pres, Celanese Res Co, 66-75, corp dir sci affairs, Celanese Corp, 76-81; RETIRED. *Concurrent Pos:* Mem, adv bd mil personnel supplies & mem, comt fire safety aspects polymeric mat, Nat Res Coun. *Mem:* Nat Acad Eng; Asn Res Dirs (pres, 72-73); Am Asn Textile Chemists & Colorists; Am Chem Soc; fel Am Inst Chemists. *Res:* Organic polymers; resins; fibers; coatings; films; carbohydrates and proteins; consumer products; structure-property relationships. *Mailing Add:* 800 25th St NW Washington DC 20037

BROWN, ALFRED ELLIS, b Ingelwood, Calif, Mar 4, 41; m 69; c 1. CELL ULTRASTRUCTURE. *Educ:* Calif State Col, Long Beach, BS, 66; Univ Calif, Los Angeles, PhD(microbiol), 71. *Prof Exp:* Fel, Sch Med, Ind Univ, 71-72; NIH fel & res biologist, Univ Calif, San Diego, 72-76, res chemist, 76-77; asst researcher & lectr, Univ Calif, Los Angeles, 78-80; ASST PROF, DEPT BOT & MICROBIOL, AUBURN UNIV, 80- *Concurrent Pos:* Consult, 3M Corp, 79-80. *Mem:* Am Soc Microbiol; AAAS; Sigma Xi. *Res:* Structure, function and biogenesis of membranes in both eukaryotic and prokaryotic systems; photosynthetic membrane systems. *Mailing Add:* Dept Bot & Microbiol Auburn Univ Auburn AL 36849

BROWN, ALISON KAY, b Edinburgh, Scotland, Jan 1, 57; US citizen. NAVIGATION SYSTEMS, ESTIMATION & CONTROL. *Educ:* Cambridge Univ, Eng, BA & MA, 79; Mass Inst Technol, SM, 81; Univ Calif, Los Angeles, PhD(mech & aerospace eng), 84. *Prof Exp:* Sr engr, Litton Guidance & Control, 81-84, sr mem tech staff, Litton Aero Prod, 84-86; PRES, NAYSYS CORP, 86- *Concurrent Pos:* Chmn, Comt Global Positioning Syst Integrity, 87-89, Ion Satellite Div Int Conf, 88-89; ed, Global Positioning Syst World, 90-, Global Ionospheric Studies World, 90- *Honors & Awards:* Sir George Nelson Prize for Appl Mech, Cambridge Univ, 89. *Mem:* Inst Navig; Inst Elec & Electronics Engrs. *Res:* Research and development related to systems design and applications of the global positioning system. *Mailing Add:* 18725 Monument Hill Rd Monument CO 80132

BROWN, ANTHONY WILLIAM ALDRIDGE, b England, Nov 18, 11; m 38; c 3. INSECT PHYSIOLOGY, TOXICOLOGY. *Educ:* Univ Toronto, BSc, 33, MA, 35, PhD(biochem), 36. *Prof Exp:* Asst biol, Univ Toronto, 33-34; Royal Soc Can res fel, Univ London, 36-37; asst entomologist Can forest insect surv, Dept Agr, Can, 37-42; assoc prof zool, Univ Western Ont, 47-49, prof & head dept, 49-68; vector biol & control, WHO, Switz, 69-73; John A Hannah Distinguished Prof insect toxicol & dir, Pesticide Res Ctr, 73-76, EMER PROF INSECT TOXICOL, MICH STATE UNIV, 76- *Concurrent Pos:* Ed, Can Field-Naturalist, 41-42; biologist, WHO, Geneva, 56-58. *Honors & Awards:* Gold Medal Achievement, Entom Soc Can, 64. *Mem:* Entom Soc Am (pres, 67); fel Royal Soc Can; Entom Soc Can (pres, 62); Can Physiol Soc; Am Mosquito Control Asn (pres, 60). *Res:* Insect biochemistry, especially nitrogenous metabolism; biology and epidemiology of forest insects; large-scale insect control with aircraft and aerosol generators; toxicity and repellency of organic compounds to insects; resistance of insects to insecticides; ecology of pesticides; medical entomology. *Mailing Add:* 1261 Genolier Canton Vaud Switzerland

BROWN, ARLEN, b Goshen, Ind, Sept 5, 26; m 48, 64; c 3. MATHEMATICS. *Educ:* Univ Chicago, PhB, 48, BS, 49, MS, 50, PhD(math), 52. *Prof Exp:* From instr to assoc prof math, Rice Inst, 52-63; assoc prof, Univ Mich, 63-67; PROF MATH, IND UNIV, BLOOMINGTON, 67- *Mem:* Am Math Soc. *Res:* Operators on Hilbert spaces; theory of operators. *Mailing Add:* Dept Math Ind Univ Bloomington IN 47401

BROWN, ARLIN JAMES, b Suffern, NY, Apr 10, 33. PHYSICS, NUTRITION. *Educ:* Franklin & Marshall Col, BS, 55. *Prof Exp:* Instr radiol warfare, Chem Biol Radiol Warfare Sch, US Army, 59-60, physicist liquid crystals, Night Vision Lab, 61-70, physicist infrared detectors, 70-74, physicist image processing, 74-78, physicist lasers, Night Vision & Electro-Optics Lab, 78-88; RETIRED. *Concurrent Pos:* Dir, Arlin J Brown Info Ctr Inc, 63- *Res:* Cancer treatment; preventive health; lasers. *Mailing Add:* PO Box 251 Ft Belvoir VA 22060

BROWN, ARNOLD, b Philadelphia, Pa, July 20, 39; m 64; c 3. INFECTIOUS DISEASES. *Educ:* Univ Calif, Los Angeles, AB, 60; Univ Southern Calif, Los Angeles, MD, 64. *Prof Exp:* Intern, Peter Bent Brigham Hosp, Boston, 64-66; mem staff, Internal Med, US Naval Dispensary, Norfolk, 66-68; sr res & postdoctoral fel, Stanford Univ Hosp, 68-72; postdoctoral fel molecular virol, Univ Geneva, Switz, 72-73; staff physician, Vet Admin Med Ctr, Pittsburgh, 73-75, chief, Infectious Dis Sect & Microbiol Lab, 75-81; ASSOC CHIEF STAFF RES, DORN VET ADMIN HOSP, COLUMBIA, 81- *Concurrent Pos:* Asst prof med, Univ Pittsburgh Sch Med, 73-81, asst prof microbiol,

Grad Sch Pub Health, 75-81, assoc prof infectious dis & microbiol, 84-; assoc prof med & microbiol/immunol, Univ SC Sch Med, Columbia, 81-84, prof, 84-; chief, Prev Med, US Army Med Dept Activ, Ft Jackson, 91- *Mem:* Fel Infectious Dis Soc Am; Am Soc Microbiol; Am Fedn Clin Res; Soc Exp Biol & Med; Asn Mil Surgeons US; AAAS. *Res:* Bacterial taxonomy; control of environmental adaptation; chemical and genetic methods to classify members of the family Legionellaceae; diagnostic probes. *Mailing Add:* Res Serv Dorn Vet Hosp Columbia SC 29201

BROWN, ARNOLD LANEHART, JR, b Wooster, Ohio, Jan 26, 26; m 49; c 5. PATHOLOGY. *Educ:* Med Col Va, MD, 49. *Prof Exp:* Resident path, Presby-St Lukes Hosp, 50-51; instr, Med Sch, Univ Ill, 53-59; from instr to assoc prof path, Mayo Grad Sch Med, Univ Minn, 59-71, chmn, dept path & anat, 68-78, prof path, Mayo Med Sch, 71-78; DEAN & PROF PATH & LAB MED, UNIV WIS, 78- *Concurrent Pos:* Resident, Presby-St Lukes Hosp, 53-56; NIH fel, 56-59; consult, Mayo Clin, 59-; mem, Nat Cancer Adv Coun, 70-72 & Nat Cancer Adv Bd, 72-74; consult, Nat Cancer Adv Bd Carcinogenesis & Nat Organ Site Progs, 74-79, chmn comt determination carcinogenicity of cyclamates, 75-76; chmn coun of deans, Asn Am Med Cols, 84-85. *Mem:* Am Asn Path & Bact; Am Soc Exp Path; Am Soc Clin Path; Am Gastroenterol Asn; Int Acad Path; Sigma Xi. *Res:* Fine structural and biochemical aspects of pathology. *Mailing Add:* 1300 University Ave Madison WI 53706

BROWN, ARTHUR, b New York, NY, Feb 12, 22; m 47; c 4. MICROBIOLOGY. *Educ:* Brooklyn Col, BA, 43; Univ Chicago, PhD(bact) 50. *Prof Exp:* Res assoc bact, Univ Chicago, 51; instr microbiol & immunol, Col Med, State Univ NY, 51-55; microbiologist & br chief, Virus & Rickettsia Div, Ft Detrick, Md, 55-68; PROF MICROBIOL & HEAD DEPT, UNIV TENN, KNOXVILLE, 68-, DISTINGUISHED SERV PROF, 80- *Concurrent Pos:* Sr fel, Inst Molecular Biol, Geneva, 63-64; consult, Virus Cancer Prog, Nat Cancer Inst, 69-77 & Oak Ridge Nat Lab, 69-; mem training grants comt, Nat Inst Allergy & Infectious Dis, 69-73, mem exp virol study sect, NIH, 80-85; macebearer, Univ Tenn, 75-76, chancellor's scholar, 78-79, distinguished serv prof, 80- *Mem:* Soc Exp Biol & Med; Am Asn Immunol; Am Acad Microbiol; Am Soc Microbiol; Am Asn Cancer Res; Am Soc Virol. *Res:* Virology; molecular virology; viral pathogenesis and immunity; mechanism of cross protection among togaviruses. *Mailing Add:* Dept Microbiol Univ Tenn Knoxville TN 37916

BROWN, ARTHUR A(USTIN), b Cleveland, Ohio, Feb 12, 15; m 45; c 2. ENGINEERING PHYSICS. *Educ:* Oberlin Col, AB, 36; Brown Univ, MS, 37, PhD(physics), 41. *Prof Exp:* Installation lab engr, Pratt & Whitney Aircraft Div, United Aircraft Corp, 40-53; vpres & gen mgr, Bowser Tech Refrig Div, Bowser, Inc, 53; vpres eng, Frederick Res Corp, 53-54; proj engr, United Aircraft Corp, 55-68, chief qual engr, Pratt & Whitney Aircraft Div, 68-77; RETIRED. *Concurrent Pos:* Mem spec subcomt induction syst de-icing, Nat Adv Comt Aeronaut, 42-46, chmn, 46, mem subcomt de-icing probs, 47-53, chmn, 51-52. *Mem:* AAAS; Am Nuclear Soc; Am Soc Heating, Refrig & Air-Conditioning Engrs. *Res:* Quality control; aircraft engine performance; icing of aircraft; engine and component test equipment; environmental test equipment; nuclear test facility design. *Mailing Add:* 341 Tall Timbers Rd Glastonbury CT 06033

BROWN, ARTHUR CHARLES, b South Bend, Ind, Feb 7, 29; m 53; c 5. PHYSIOLOGY, NEUROSCIENCES. *Educ:* Univ Chicago, AB, 48, MS, 54, Univ Wash, PhD(physiol, biophys), 60. *Prof Exp:* Asst physics, Univ Chicago, 51-52; asst physics, Univ Wash, 54-55, asst physiol, 55-56; instr, State Univ NY Upstate Med Ctr, 58-60; from instr to assoc prof, Univ Wash, 60-69, prof physiol & biophys, 70-77; PROF PHYSIOL & PHARMACOL, ASSOC DEAN, SCH DENT, PROF PHYSIOL, SCH MED, ORE HEALTH SCI UNIV, 77- *Concurrent Pos:* Vis lectr physiol & actg head dept, Fac Med, Univ Malaya, 69-70; chmn, Physiol Sect, Am Asn Dent Schs, 70-71, 81-82 & 86-87; mem, Comput Biomath Sci Study Sect, NIH, 73-77; hon prof, Facul Med, Univ Hong Kong. *Mem:* Int Asn Study Pain; Am Physiol Soc; Int Asn Dent Res; Soc Neurosci; Am Asn Dent Schs. *Res:* Temperature regulation; dental and oral physiology and biophysics; neurophysiology of pain. *Mailing Add:* Off Dean Sch Dent Ore Health Sci Univ Portland OR 97201-3097

BROWN, ARTHUR LLOYD, b Edmonton, Alta, Feb 16, 15; nat US; m 44; c 4. SOIL SCIENCE. *Educ:* Univ Alta, BSc, 39, MSc, 41; Univ Minn, PhD(soil sci), 46. *Prof Exp:* Agr asst, Can Dept Agr, 39-42; asst soil sci, Univ Minn, 43-45; analyst, Univ Alta, 46; soil surveyor, Res Coun Alta, 46-47; instr soils & jr soil technologist, Univ Calif, Davis, 47-5l, assoc specialist soils, 51-57, lectr soils, Exp Sta, 51-57, specialist, 57-8l; RETIRED. *Mem:* AAAS; Am Soc Agron; Soil Sci Soc Am. *Res:* Soil chemistry, particularly heavy metals, especially zinc and cadmium, in the soil in relation to amounts aquired by food products; methods of diagnosing nutrient deficiencies and excesses. *Mailing Add:* 38 Parkside Dr Davis CA 95616

BROWN, ARTHUR MORTON, b Winnipeg, Man, Mar 3, 32; div; c 4. PHYSIOLOGY, BIOPHYSICS. *Educ:* Univ Man, MD, 56; Univ London, PhD(physiol), 64. *Prof Exp:* Asst resident path, Columbia-Presby Med Ctr, 56-57; asst resident med, Col Med, Univ Utah, 57-59, trainee cardiol, 59-61; res fel, Cardiovasc Res Inst, Univ Calif, San Francisco Med Ctr, 64-66, asst prof med & physiol, 66-69, from assoc prof to prof physiol & internal med, 69-73; prof & chmn dept physiol & biophys, Univ Tex Med Br, Galveston, 73-85; PROF & CHMN DEPT PHYSIOL & MOLECULAR BIOPHYS, BAYLOR COL MED, HOUSTON, 85- *Concurrent Pos:* Chmn physiol study sect, NIH, 74-76; mem Cardiovascular B Res Comt, Am Heart Asn, 76-79. *Mem:* AAAS; Am Physiol Soc; Soc Gen Physiologists; Biophys Soc. *Res:* Membrane biophysics; phototransduction; neurocirculatory control; cardiology. *Mailing Add:* Dept Physiol & Molecular Biophys Baylor Col Med 1 Baylor Plaza Tex Med Ctr Houston TX 77030

BROWN, AUDREY KATHLEEN, b New York, NY, Feb 2, 23; m 54; c 1. PEDIATRICS, HEMATOLOGY. *Educ:* Columbia Univ, BA, 44, MA, 45, MD, 50; Am Bd Pediat, dipl, 56 & cert hematol-oncol, 74. *Prof Exp:* Intern med, Columbia Med Div, Bellevue Hosp, NY, 50-51, intern & asst resident, Children's Med Serv, 51-53; Holt fel pediat, Babies Hosp, Columbia Presby Med Ctr, 53-54; civilian pediatrician, Walter Reed Army Med Ctr, 54-55; from instr to asst prof pediat, Col Med, Wayne State Univ, 55-59; from asst prof to assoc prof, Sch Med, Univ Va, 59-67; prof, Med Col Ga, 67-73, actg chmn dept, 69-73; PROF PEDIAT, STATE UNIV NY DOWNSTATE MED CTR, 74-, VCHMN DEPT, 76- *Concurrent Pos:* Asst & asst pediatrician, Children's Hosp, Mich, 55-57, assoc pediatrician, 58, assoc hematologist, 58-59; sr res assoc, Child Res Ctr Mich, 55-59; assoc dir, Hemat Training Grant; pediatrician, Univ Va Hosp, 59-, dir, Pediat Hemat Clin, 60-; med coordr & dir res, Children's Rehab Ctr, Va, 59-, physician in charge ped, hematol-oncol, 74-; mem perinatal biol & infant mortality training grant rev comt, NIH. *Mem:* Fel Am Acad Pediat; Soc Pediat Res; NY Acad Sci; Am Fedn Clin Res; Am Pediat Soc (secy-treas, 83-); Sigma Xi; Harvey Soc; Soc Woman Geogr. *Res:* Hyperbilirubinemia; neonatal development of the glucuronide conjugating system; Bilirubin bending; infantile pyknocytosis; fetal hemoglobin development; sickle cell anemia; nutrition and development. *Mailing Add:* Downstate Med Ctr 450 Clarkson Ave Box 49 Brooklyn NY 11203

BROWN, AUSTIN ROBERT, JR, b Kansas City, Mo, Dec 13, 25; m 51; c 3. MATHEMATICS. *Educ:* Grinnell Col, BA, 49; Yale Univ, MA, 50, PhD(math) 52. *Prof Exp:* Mathematician, Ballistic Res Labs, 49-53, chief comput methods sect, 53-54; assoc prof, Drury Col, 54-56; chief ballistic comput br, Air Force Armament Ctr, 56-58 & Air Proving Ground Ctr, Eglin AFB, Fla, 58; opers analyst, Hq Air Defense Command, US Air Force, 58-65; from assoc prof to prof math, 65-88, dir comput ctr, 65-81, EMER PROF MATH, COLO SCH MINES, 88- *Concurrent Pos:* Lectr, Univ Del, 52-54; adj prof, Colo Col, 62-63. *Honors & Awards:* Fulbright Lectr, Cyprus, 88. *Mem:* Math Asn Am; Asn Comput Math; Sigma Xi. *Res:* Computer science; numerical analysis; simulation; operations research. *Mailing Add:* 407 Peery Pkwy Golden CO 80403-1539

BROWN, BARBARA ILLINGWORTH, b Hartford, Conn, May 12, 24; m 51; c 3. PHYSIOLOGICAL CHEMISTRY. *Educ:* Smith Col, BA, 46; Yale Univ, PhD, 50. *Prof Exp:* Asst & res asst prof physiol chem, 50-64, res assoc prof biol chem, 64-74, PROF BIOL CHEM, SCH MED, WASH UNIV, 74- *Concurrent Pos:* Estab investr, Am Heart Asn, 66-71. *Mem:* Am Soc Biol Chemists; Brit Biochem Soc; Am Soc Human Genetics. *Res:* Carbohydrate metabolism; phosphorylase; glycogen storage disease. *Mailing Add:* Dept Biochem Wash Univ Sch Med 660 S Euclid Ave St Louis MO 63110

BROWN, BARKER HASTINGS, b Olathe, Colo, June 24, 10; m 37; c 4. BIOLOGICAL CHEMISTRY, AGRICULTURAL CHEMISTRY. *Educ:* Alma Col, AB, 32; Univ Mich, MS, 33, PhD(biol chem), 37. *Prof Exp:* Asst prof chem, Univ Philippines, 37-42; vis asst prof biol chem, Univ Mich, 45-46; sales mgr chem & equip, Philippine Am Drug & Soutraco, 46-48; prof chem, Univ Philippines, 48-52; sales & mkt dir, Macondray & Co, 52-76; dir & tech adv, Vulcan Indust & Mining Corp, 76-86; CHMN & TECH ADV, VULCHEM CORP, 87- *Concurrent Pos:* Fel, Univ Mich, 32-33 & 36-37. *Mem:* Am Chem Soc; AAAS; Philippine AAS; Chem Soc Philippines; Nat Res Coun Philippines; NY Acad Sci. *Res:* Intermediary metabolism sulfur containing amino acids; nutrition in field of utilization of natural occurring inorganic elements, primarily calcium and iron. *Mailing Add:* 126 Matahimik U P Village Diliman 1104 Quezon City Philippines

BROWN, BARRY LEE, b Kingman, Ariz, June 29, 44; m 77; c 2. COMPUTER SCIENCES, REPRODUCTIVE PHYSIOLOGY. *Educ:* Univ Ga, BSA, 66, MS, 68; Purdue Univ, PhD(physiol), 72. *Prof Exp:* Res asst, Dept Animal Sci, Univ Ga, 66-68, Purdue Univ, 68-72; res fel, Dept Pop Dynamics, Johns Hopkins Univ, 72-74; RES CHEMIST ENDOCRINOL, RES UNIT, FED BUR INVEST LAB, 74- *Concurrent Pos:* Asst prof lectr, Dept Biol Sci, George Washington Univ, 75-80, assoc prof lectr, 80- *Mem:* Soc Study Reproduction; Endocrine Soc. *Res:* Biochemistry and physiology of the testis; development and application of techniques, relating principally to endocrinology and forensic science; application of computers to laboratory environment. *Mailing Add:* Res Unit Fed Bur Invest Lab Quantico VA 22135

BROWN, BARRY STEPHEN, b Wilkes-Barre, Pa, Oct 18, 49; m 73; c 2. IN VITRO CENTRAL NERVOUS SYSTEM, CARDIAC ELECTROPHYSIOLOGY. *Educ:* Pa State Univ, BS, 71; Yale Univ, MS, 73; Univ SFla, PhD(pharmacol), 79. *Prof Exp:* NIH trainee, Yale Univ, 71-73; asst, Univ SFla, 74-79; fel, Mich State Univ, 79-80; Mich Heart Asn res fel, Mich State Univ, 80-81; res investr, 81-83, sr res investr, Am Critical Care, Div of Am Hosp Supply Corp, 83-86; group leader, DuPont Critical Care, 86-89, res assoc, Dupont Pharmaceut, 89-90, RES ASSOC, DUPONT MERCK PHAMACEUT CO, 91- *Mem:* AAAS; Am Soc Pharmacol & Exp Therapeut. *Res:* Effects of cardiovascular and central nervous system drugs on membrane potentials and ionic currents in isolated cardiac and brain tissue. *Mailing Add:* CNS Dis Res Dupont Merck Pharmaceut Co PO Box 80400 Exp Sta Wilmington DE 19880-0400

BROWN, BARRY W, b Buffalo, NY, Dec 20, 39; m 63; c 3. MATHEMATICAL BIOLOGY. *Educ:* Univ Chicago, BS, 59; Univ Calif, Berkeley, MA, 61, PhD(math), 63. *Prof Exp:* Res assoc statist & asst prof biol sci, Comput Ctr, Univ Chicago, 63-65; assoc prof, 65-80, PROF, DEPT BIOMATH, UNIV TEX M D ANDERSON HOSP & TUMOR INST, 80- *Concurrent Pos:* Chief sect comput sci, M D Anderson Hosp, 72-76. *Mem:* Asn Comput Mach; Am Statist Asn. *Res:* Mathematical problems arising in biological sciences. *Mailing Add:* Dept Biomath 237 MD Anderson Cancer Ctr Univ Tex 1515 Holcombe Houston TX 77030

BROWN, BENJAMIN LATHROP, b Alton, Ill, Sept 25, 47; m 83; c 1. APPLIED PHYSICS. *Educ:* Principia Col, BS, 72; New York Univ, MS, 84; Brandeis Univ, PhD(physics), 87. *Prof Exp:* Mem tech staff, Physics, AT&T Bell Labs, 84-87; res assoc, Harvard Univ, 87-88; IBM res fel, Harvard Univ, 88-89; ASST PROF, MT HOLYOKE COL, 90- *Concurrent Pos:* Consult, AT&T Bell Lab, 87-88. *Mem:* Am Phys Soc. *Res:* Positron negative work function measurement; search for gravitational waves with a Weber Bar; laboratory simulations of galactic positron annihilation; positron plasma studies. *Mailing Add:* 91 Ferry Hill Rd Granby MA 01033

BROWN, BERT ELWOOD, b The Dalles, Ore, Sept 27, 26. PHYSICS, ATMOSPHERIC PHYSICS. *Educ:* Wash State Univ, BS, 49; Calif Inst Technol, MS, 53; Ore State Univ, PhD(physics), 63. *Prof Exp:* Naval architect, Puget Sound Naval Shipyard, 53-54, physicist, 54-56; from instr to prof, 60-83, chmn, Dept Physics, 70-80, EMER PROF PHYSICS, UNIV PUGET SOUND, 83- *Concurrent Pos:* Consult, 83-84. *Mem:* Am Phys Soc; Am Asn Physics Teachers. *Res:* Atomic physics; atmospheric physics; climatology. *Mailing Add:* Dept Physics 6842 N 11th St WA 98406

BROWN, BEVAN W, JR, b Starr, SC, Oct 19, 28; m 86; c 3. WATER RESOURCES. *Educ:* Clemson Univ, BS, 49, MS, 50; Stanford Univ, MS, 68. *Prof Exp:* Sect supvr, TVA-Flood Protection Br, 50-64, asst br chief, 64-72, asst, div water mgt, 72-74, asst dir, div water mgt, 74-79, dir, Div Nat Resources, 79, deputy mgr, office Nat Resources, 79-82, dir, Div Air & Water Res, 82-88; CONSULT, ENG PLANNING, 88- *Concurrent Pos:* Consult, Agr Int Develop Paraguay, 83, Peoples Republic China, 84, UN, Mozambique govt, 84. *Mem:* Am Soc Civil Engrs; Int Water Resources Assoc. *Mailing Add:* 4638 Cobblestone Circle Knoxville TN 37938-3207

BROWN, BEVERLY ANN, b San Antonio, Tex, July 21, 51; m 72. IMMUNOCHEMISTRY, PROTEIN CHEMISTRY. *Educ:* Incarnate Word Col, BS, 73; Univ Tex, MA, 74; Univ Minn, PhD(biochem), 79. *Prof Exp:* Res assoc, Harvard Med Sch, 79-81; Baxter Health, 89-90; RES ASSOC, NEW ENG NUCLEAR-DUPONT, 81-88, 90- *Mem:* Am Chem Soc; AAAS; Am Soc Med Technol; Am Soc Neurochem; Am Asn Immunologists. *Res:* Investigating the feasibility of using monoclonal antibodies against cancer antigens in the diagnosis and treatment of cancer. *Mailing Add:* Four Dawes Ave Winchester MA 01890

BROWN, BILLINGS, b Seattle, Wash, June 23, 20; m 46; c 7. THERMODYNAMICS. *Educ:* Univ Wash, BS & MS, 51, PhD(phys chem), 53. *Prof Exp:* Assoc prof chem eng, Brigham Young Univ, 53-59; tech specialist thermodynamics, Hercules Inc, 59-63; supvr solid rockets, Boeing Co, 63-64; prof chem eng, Univ Wash, 64-65; sr tech specialist solid rockets, Hercules Inc, 65-66; mem tech staff, Inst Defense Analyses, 66-68; sr tech specialist solid rockets, Hercules, Inc, 68-83; RETIRED. *Concurrent Pos:* Mem joint Army Navy Air Force Thermochem Panel, 58-71 & Interagency Working Group on Safety, 72- *Mem:* Am Inst Chem Engrs. *Res:* Explosion mechanisms of solid rocket fuels. *Mailing Add:* 3501 S 3650 E East Holiday UT 84109

BROWN, BRADFORD E, b Worcester, Mass, Apr 1, 38; m 59, 74; c 4. FISHERIES BIOLOGY, ECOLOGY. *Educ:* Cornell Univ, BS, 60; Auburn Univ, MS, 62; Okla State Univ, PhD(statist zool), 69. *Prof Exp:* Asst fish cult, Auburn Univ, 61-62; fisheries biologist, Biol Lab, US Bur Com Fisheries, 62-65; asst leader, Okla Coop Fisheries Unit, US Bur Sport Fisheries & Wildlife, leader fishery mgt biol invest, Biol Lab, 70-75, chief resource assessment div, Northeast Fisheries Ctr, 76-84, dep dir, 84-89, DIR SOUTHEAST FISHERIES CTR, NAT MARINE FISHERIES SERV, US DEPT COM, 89- *Concurrent Pos:* Asst prof zool, Okla State Univ, 65-70; mem standing comt res & statist, Int Comn Northwest Atlantic Fisheries, 71-; adj prof biol, Bridgewater State Col, 75-; mem, Int Coun Explor of the Sea, Demersal Northern Fish Com, 75-77; mem adv comt fishery management, 78-; adj prof natural sci, Univ Md, Eastern Shore, 78-; mem ocean policy comt, Nat Acad Sci Fisheries Task Force; adj prof, Grad Sch Oceanog, Univ RI & fisheries, Rutgers Univ; consult multispecies fisheries, Food & Arg Orgn, 78; US sci corresp, Int Comn Conserv Atlantic Tunas, 86-, coordr, Enhanced Billfish Res Prog, 87-; adj prof, Rosenstal Sch Marine & Atmospheric Scis, Univ Miami. *Honors & Awards:* Fleming Award, 76. *Mem:* Biomet Soc; Am Fisheries Soc; AAAS; Sigma Xi; fel Am Inst Fisheries Res Biologists; Ecol Soc. *Res:* Population dynamics and ecology of fresh water environments; applications of statistical and computer technology to ecology; population dynamics of exploited marine species; ecosystem approach to marine fisheries management; statistics. *Mailing Add:* Southeast Fisheries Ctr Nat Marine Fisheries Serv NOAA Miami FL 33149

BROWN, BRADFORD S, applied statistics, quality engineering, for more information see previous edition

BROWN, BRUCE ANTONE, forest ecology, silviculture, for more information see previous edition

BROWN, BRUCE CLAIRE, b Burns, Ore, May 10, 44; div; c 2. ACCELERATOR PHYSICS, HIGH ENERGY PHYSICS. *Educ:* Univ Rochester, BS, 66; Univ Calif, San Diego, MS, 69, PhD(physics), 73. *Prof Exp:* Res asst physics, Univ Calif, San Diego, 67-72; res assoc, Univ Mich, 72-73; res assoc, 73-76, mem accelerator div staff, 76-82, MEM MAGNET TEST FACIL, FERMI NAT ACCELERATOR LAB, 82- *Mem:* Am Phys Soc; Inst Elec & Electronics Engrs. *Res:* Development and operation of accelerator facility; structure of protron in hadron and lepton production. *Mailing Add:* Fermi Nat Accelerator Lab MS 316 PO Box 500 Batavia IL 60510

BROWN, BRUCE ELLIOT, b Chicago, Ill, May 15, 30; m 80. GEOLOGY, CRYSTALLOGRAPHY. *Educ:* Wheaton Col, Ill, BS, 52; Univ Wis, MS, 54, PhD(soil sci), 57, MS, 60, PhD(geol), 62. *Prof Exp:* Crystallogr, Allis-Chalmers Mfg Co, 61-65; ASSOC PROF GEOL SCI, UNIV WIS-MILWAUKEE, 65- *Mem:* Mineral Soc Am; AAAS; Am Soc Limnol & Oceanog; Nat Asn Geol Teachers. *Res:* Inorganic structure, especially clay minerals, layer silicates, feldspars and binary chalcogonides; geochemistry of carbonate lakes. *Mailing Add:* Dept Geol Univ Wis Milwaukee WI 53201

BROWN, BRUCE STILWELL, b Chicago, Ill, Mar 15, 45; m 67; c 1. SOLID STATE PHYSICS. *Educ:* Miami Univ, BA, 66; Univ Ill, MS, 68, PhD(physics), 72; Univ Chicago, MBA, 84. *Prof Exp:* Post-doctoral, 72-74, asst physicist, 74-76, physicist, Mat Sci Div, 76-79, PHYSICIST, IPNS DIV, ARGONNE NAT LAB, 79-, DIR, 86- *Mem:* Am Phys Soc; Mat Res Soc; AAAS. *Res:* Neutron scattering. *Mailing Add:* IPNS-360 Argonne Nat Lab 9700 S Cass Argonne IL 60439

BROWN, BRUCE WILLARD, b New York, NY, Nov 25, 27; m 50; c 2. ANALYTICAL CHEMISTRY, INORGANIC CHEMISTRY. *Educ:* Polytech Inst Brooklyn, BS, 49, MS, 52; Univ Wash, PhD(chem), 61. *Prof Exp:* Instr chem, Everett Jr Col, 58-63; from asst prof to assoc prof, Portland State Univ, 63-77, asst dean fac, 68-70, actg dean undergrad studies, 70, asst dean acad affairs, 71, PROF CHEM, PORTLAND STATE UNIV, 77-, CHMN DEPT CHEM, 88- *Mem:* Am Chem Soc; Am Crystallog Asn; Sigma Xi. *Res:* X-ray crystallography of transition metal coordination compounds; computer programing; solid state reactions; lattice defect compounds. *Mailing Add:* Dept Chem Portland State Univ Portland OR 97207-0751

BROWN, BUCK F(ERGUSON), b Monroe, La, Oct 22, 32; m 59; c 3. ELECTRICAL ENGINEERING. *Educ:* Mass Inst Technol, SB, 55; Okla State Univ, MS, 59, PhD(elec eng), 64. *Prof Exp:* Aerophysics engr, Gen Dynamics, Ft Worth, 55-56; instr elec eng, Okla State Univ, 56-64; assoc prof elec eng, La Tech Univ, 64-76, prof, 76-80; dir res & develop, Lincoln Electronics, Inc, 69-80; prog dir, Nat Sci Found, 89-90; head, Elec Eng & Computer Sci & dir, Echo Res Labs, 80-89, CHMN, ELEC & COMPUTER ENG & DIR, RES & GRAD STUDIES, ROSE-HULMAN INST TECHNOL, 90- *Concurrent Pos:* Dir res & develop, Lincoln Electronics, Inc, 69-80; prod develop consult, ITS Indust, 85-86; electronics & computer consult, US Navy, 81- *Mem:* Inst Elec & Electronics Engrs; Am Soc Eng Educ; Sigma Xi; Nat Soc Prof Engrs. *Res:* Linear control system design; linear sample data systems; non-linear control system theory; high-speed electronic interconnections. *Mailing Add:* Dept Elec & Computer Eng Rose-Hulman Inst Technol Terre Haute IN 47803

BROWN, BURTON PRIMROSE, b Denver, Colo, Dec 5, 17; m 44; c 2. ELECTRICAL ENGINEERING, RADIO ENGINEERING. *Educ:* Univ Colo, BSEE, 39; Univ Vt, MS, 41. *Prof Exp:* Mem staff, Electronic Systs Div, Gen Elec Co, 47-62, consult engr, 63-78; RETIRED. *Concurrent Pos:* Asst dir defense res & eng, Dept of Defense, 62-63; asst prof elec eng, Univ Vt, 66-67; mem, US Air Force Sci Adv Bd, Army Sci Adv Panel. *Honors & Awards:* Charles A Coffin Award, Gen Elec Co, 51, Charles P Steinmetz Award, 75. *Mem:* Nat Acad Eng; fel Inst Elec & Electronic Engrs. *Mailing Add:* 50 Brown St Baldwinsville NY 13027

BROWN, BYRON WILLIAM, JR, b Chicago, Ill, Apr 21, 30; m 49; c 6. BIOSTATISTICS. *Educ:* Univ Minn, BA, 52, MA, 55, PhD, 59. *Prof Exp:* Instr biostatist, Univ Minn, 53-54; instr biostatist, Univ Minn, 55-56; asst prof, La State Univ, 56-57; from lectr to prof & head div, Sch Pub Health, Univ Minn, Minneapolis, 57-68; PROF BIOSTATIST & HEAD DIV, STANFORD UNIV MED CTR, 68-, CHMN, DEPT HEALTH RES & POLICY, 88- *Concurrent Pos:* Consult to govt agencies & indust. *Mem:* Inst Med-Nat Acad Sci; Biomet Soc; fel Am Statist Asn; fel Am Heart Asn; Inst Math Statist; Int Statist Inst; fel AAAS. *Res:* Teaching, research and consulting in applied statistics for the biomedical sciences. *Mailing Add:* Div Biostatist Stanford Univ Med Ctr Stanford CA 94305

BROWN, CARL DEE, b New Cambria, Kans, Oct 2, 19; m 44; c 2. ENTOMOLOGY. *Educ:* Okla Baptist Univ, BS, 47; La State Univ, MS, 47; Iowa State Col, PhD(entom), 51. *Prof Exp:* Asst entomologist, Dept Entom & Bot, Agr Exp Sta, Univ Ky, 50-52; from asst prof to assoc prof, 52-58, PROF BIOL, MEMPHIS STATE UNIV, 58-, CHMN DEPT, 62- *Mem:* Am Entom Soc. *Mailing Add:* Dept Biol Memphis State Univ Memphis TN 38152

BROWN, CHARLES ALLAN, b Detroit, Mich, Mar 22, 44. ORGANIC CHEMISTRY, ORGANOMETALLIC CHEMISTRY. *Educ:* Purdue Univ, BSc, 64; Univ Calif, Berkeley, PhD(chem), 67. *Prof Exp:* Chemist mass spectrometry, Stanford Univ, 67-69; chemist pharmaceut chem, Syntex Corp, 69-70; asst prof chem, Cornell Univ, 70-75; RES SCIENTIST CHEM, IBM RES LAB, 75- *Concurrent Pos:* Air Force Off Sci Res fel, Stanford Univ, 67-68, NSF fel, 68-69; consult, Exxon Res & Eng Co, 74-75. *Mem:* Am Chem Soc; AAAS. *Res:* New methods of synthesis of organic and organometallic materials; structural effects on reactivity and selectivity. *Mailing Add:* IBM Dept K33/801 650 Harry Rd San Jose CA 95120-6099

BROWN, CHARLES ERIC, b Spangler, Pa, Nov 23, 46; m 71; c 1. BONE MINERALIZATION & POLYMER COMPOSITE CHARACTERIZATION. *Educ:* State Univ NY, Buffalo, BA, 68; Northwestern Univ, PhD(biochem), 73. *Prof Exp:* Instr biochem, Northwestern Univ, 73-75; res fel physiol chem, Roche Inst Molecular Biol, 75-77; asst prof, 77-83, ASSOC PROF BIOCHEM, MED COL WIS, 84- *Concurrent Pos:* Consult, Nicolet Instrument Corp, 84- *Mem:* Am Chem Soc; Soc for Neurosci; Am Soc Mass Spectrometry; Sigma Xi; Am Soc Pharmacol & Exp Therapeut; Int Soc Magnetic Resonance. *Res:* Noninvasive quantitative determination of bone mineral content during osteoporosis by magnetic resonance techniques; cross polarization, magic angle sample spinning nuclear magnetic resonance and laser desorption, Fourier transform mass spectroscopy of conducting and flame retardant polymers. *Mailing Add:* 8701 Watertown Plank Rd Milwaukee WI 53226

BROWN, CHARLES ERIC, b Spangler, Pa, Nov 23, 46; m 71; c 1. FOURIER TRANSFORM-ION CYCLOTRON RESONANCE MASS SPECTROMETRY, CROSS POLARIZATION. *Educ:* State Univ NY, Buffalo, BA, 68; Northwestern Univ, PhD(biochem), 73. *Prof Exp:* Instr & fel biochem, Northwestern Univ, 73-75; res fel neurosci, Roche Inst Molecular Biol, 75-77; from asst prof to assoc prof biochem, Med Col Wis, 77-88; ANALYTICAL BUS DEVELOP COORDR, CHEM ANALYSIS

SCI, SUDBURY RES CTR, BP RES, 88- *Concurrent Pos:* Consult, Nicolet Instrument Corp, 84-88, Metriflow, Inc, 84-88. *Mem:* Am Chem Soc; Int Soc Magnetic Resonance; Am Soc Pharmacol & Exp Therapeut; Am Soc Mass Spectros. *Res:* Identifying and providing state-of-the-art analytical technology to support of chemical product lines; magic angle sample spinning nuclear magnetic resonance spectrometry. *Mailing Add:* Analytical Res Div BP Res Chertsey Rd Sudbury-on-Thames Middlesex TW16 7LN England

BROWN, CHARLES JULIAN, JR, b Utica, Miss, Dec 31, 32; m 63; c 4. PHYSICAL CHEMISTRY. *Educ:* Miss Col, BS, 54; Univ Tex, PhD(chem), 59. *Prof Exp:* Fel, Univ Tex, 59-60; res chemist, Film Dept, Exp Sta, 60-65 & org chem dept, 65-73, sr res chemist, Plastics Prod & Resins Dept, 73-85, RES ASSOC, POLYMER PROD DEPT, E I DU PONT DE NEMOURS & CO, INC, 85- *Mem:* Am Chem Soc; AAAS; Sigma Xi. *Res:* Polymers; reverse osmosis. *Mailing Add:* Ten Thornberry Dr Hockessin DE 19707

BROWN, CHARLES L(EONARD), b Ranger, Tex, June 30, 20; m 46; c 2. MECHANICAL ENGINEERING. *Educ:* Rice Inst, BS, 40; Purdue Univ, MS, 43; PhD(mech eng), 49. *Prof Exp:* PROF MECH ENG, PURDUE UNIV, 59- *Mem:* Am Soc Mech Engrs; Am Soc Eng Educ. *Res:* Thermodynamics; engineering analysis. *Mailing Add:* Mech Eng & Bldg Purdue Univ Lafayette IN 47907

BROWN, CHARLES MOSELEY, b Kansas City, Mo, July 15, 43. ATOMIC AND MOLECULAR SPECTROSCOPY, SOLAR & LABORATORY SPECTRON OF HIGHLY IONIZED SPECIES. *Educ:* Southern Ill Univ, BA, 65; Univ Md, PhD(physics), 71. *Prof Exp:* Nat Res Coun res assoc atomic & molecular spectros, 71-73; physicist, Naval Res Lab, 73-76; mem sr sci staff, Ball Bros Res Corp, 76-77; PHYSICIST, NAVAL RES LAB, 78- *Concurrent Pos:* Mem comt line spectra elements, Nat Res Coun. *Mem:* Fel Optical Soc Am. *Res:* Vacuum ultraviolet spectroscopy of atomic and molecular species; spectroscopy with synchrotron radiation; spectroscopy of laser heated plasmas; solar spectroscopy. *Mailing Add:* Code 4173B Naval Res Lab Washington DC 20375

BROWN, CHARLES MYERS, b Oswego, SC, Oct 16, 27; m 57; c 2. PLANT BREEDING. *Educ:* Clemson Col, BS, 50; Univ Wis, MS, 52, PhD(agron), 54. *Prof Exp:* Asst agron, Univ Wis, 50-54; first asst, 54-55, from asst prof to assoc prof, 55-68, PROF AGRON, UNIV ILL, URBANA-CHAMPAIGN, 68-, ASSOC HEAD DEPT, 71- *Mem:* Am Soc Agron; Sigma Xi. *Res:* Oat and wheat breeding; genetics; cytogenetics. *Mailing Add:* Dept Agron Univ Ill 1102 S Goodwin Ave Urbana IL 61801

BROWN, CHARLES NELSON, b Victoria, BC, Can, June 3, 41; m 67; c 2. EXPERIMENTAL HIGH ENERGY PHYSICS. *Educ:* Univ BC, BSc, 63; Univ Rochester, AM, 66, PhD(physics), 68. *Prof Exp:* Res fel physics, Harvard Univ, 68-71, asst prof, 71-74; STAFF PHYSICIST, FERMI NAT ACCELERATOR LAB, 74- *Mem:* Am Phys Soc; Fel, Am Physics Soc. *Res:* Experimental study of lepton scattering and lepton production and heavy quark in very high energy collisions. *Mailing Add:* Fermi Nat Accelerator Lab PO Box 500 PO Box 500 Batavia IL 60510

BROWN, CHARLES QUENTIN, b Roanoke Rapids, NC, Sept 12, 28; m 50; c 2. GEOLOGY. *Educ:* Univ NC, BS, 51, MS, 53; Va Polytech Inst, PhD(geol), 59. *Prof Exp:* From instr to assoc prof geol, Clemson Univ, 54-67; dir instnl develop, 69-79, actg dean, Sch Technol, 81-83, chmn dept geol, 67-79 & 79-88, PROF GEOL, EAST CAROLINA UNIV, 67- *Concurrent Pos:* Mem, NC Marine Sci Coun, 71-81 & NC Coastal Resources Ctr Admin Bd, 76-79. *Mem:* Geol Soc Am; Soc Econ Paleont & Mineral; Nat Asn Geol Teachers. *Res:* Sedimentation; clay mineralogy of recent deposits. *Mailing Add:* Dept Geol East Carolina Univ Greenville NC 27834

BROWN, CHARLES S(AVILLE), b Orange Co, Va, Oct 1, 20; m 46; c 3. CHEMICAL ENGINEERING. *Educ:* Va Polytech Inst, BS, 41; Univ Wis, PhD(chem eng), 46. *Prof Exp:* Control engr, Nat Aniline & Chem Co, NY, 41; dept mgr, Fermentation Unit, Abbot Labs, 44-58,dir & supt prod, 58-60, vpres prod, 60-61, US & Can opers, 61-64, pharmaceut opers, 64-66, exec vpres sci & mfg opers, 66-68, sci opers, 68-70, exec vpres admin, 70-81; RETIRED. *Mem:* Fel AAAS; Am Chem Soc; Am Inst Chem Engrs; Sigma Xi. *Res:* Aerobic fermentations; fermentation engineering; pharmaceutical manufacturing. *Mailing Add:* 421 Tiffany Dr Waukegan IL 60085

BROWN, CHARLES THOMAS, b Paterson, NJ, Oct 27, 28; m 53; c 2. PHYSICAL CHEMISTRY. *Educ:* Ohio Wesleyan Univ, BA, 53; Rensselaer Polytech Inst, MS, 55, PhD, 57. *Prof Exp:* Prin chemist, Battelle Mem Inst, 57-60; mgr chem sect, Nuclear Div, Combustion Eng Corp, 60-61; res chemist, United Aircraft Corp Res Labs, United Aircraft Corp, 61-75, sr res chemist, United Technol Res Ctr, 75-80. *Mem:* Am Chem Soc; Electrochem Soc. *Res:* Electrochemistry of non-aqueous systems, including material compatibility. *Mailing Add:* 663 Briarcliffe Rd Lake Hayward Colchester CT 06415

BROWN, CHESTER HARVEY, JR, b Everett, Mass; m 47; c 2. AERONAUTICAL, STRUCTURAL & AEROELASTIC ANALYSIS, SYSTEMS ANALYSIS. *Educ:* Mass Inst Technol, BSc, 37. *Prof Exp:* Res engr, Supersonic Lab, Mass Inst Technol, 47-50, res engr, Aeroelastic & Struct Res Lab, 50-53; sr struct engr, Chas T Main, Inc, 54-57; sr struct engr, Fairchild Engine & Airplane Corp, 57-60; prin struct engr, Atomic Power Develop Assocs, 60-69; sr struct engr, Advan Reactor Div, Westinghouse Elec Corp, 69-78; CONSULT NUCLEAR INDUST, 78- *Concurrent Pos:* Consult proposed mil landing, 84- *Mem:* Am Nuclear Soc; Sigma Xi; Am Soc Qual Control. *Res:* Structural dynamics and impact effects; blast interaction with aircraft and ground structures; solid-fluid interaction; shock wave effects in structures; nuclear plant design and safety analysis. *Mailing Add:* 630 College St No 2 Pittsburgh PA 15232-1955

BROWN, CHRISTOPHER W, b Maysville, Ky, Apr 13, 38; c 3. SPECTROSCOPY. *Educ:* Xavier Univ, BS, 60, MS, 62; Univ Minn, PhD(phys chem), 67. *Prof Exp:* Res fel chem, Univ Md, 66-68; asst prof, 68-72, assoc prof, 72-76, PROF CHEM, UNIV RI, 76- *Honors & Awards:* Award, Chicago Sect, Soc Appl Spectros, 66. *Mem:* Am Chem Soc; Am Phys Soc; Soc Appl Spectros; Am Soc Testing & Mat. *Res:* Infrared, near infrared, visible, ultraviolet, and Raman spectroscopy; fourier transform infrared; spectroscopic applications, environmental and biomedical problems; analysis of air and water pollutants; analysis of petroleum and natural gas; chemical applications of lasers; computerized spectroscopy; fiberoptics. *Mailing Add:* Dept Chem Univ RI Kingston RI 02881

BROWN, CLINTON CARL, psychophysiology, for more information see previous edition

BROWN, COLIN BERTRAM, b London, Eng, Jan 15, 29; m 54; c 2. CIVIL ENGINEERING. *Educ:* Univ London, BSc, 53; Univ Minn, PhD(mech, mats), 62. *Prof Exp:* Eng asst, D H Lee, Eng, 53-55; bridge designer, Lindsey County Coun, 55-57; design engr, Prov of BC, 57-59; res fel eng mech, Univ Minn, 59-61; asst prof civil eng, Univ Calif, Berkeley, 62-66; assoc prof civil eng & appl mech, Columbia Univ, 66-69; PROF CIVIL ENG, UNIV WASH, 69- *Mem:* Am Soc Civil Engrs; Sigma Xi. *Res:* Structural mechanics; theory of interface friction; incremental problems in elasticity; statistical methods in structural engineering. *Mailing Add:* Dept Civil Eng Univ Wash Seattle WA 98195

BROWN, CONNELL JEAN, b Everton, Ark, Mar 6, 24; m 46; c 2. ANIMAL BREEDING. *Educ:* Univ Ark, BSA, 48; Okla State Univ, MS, 50, PhD(animal sci), 56. *Prof Exp:* Asst animal husb, Okla State Univ, 48-50; asst prof, Univ Ark, Fayetteville, 50-53; asst, Okla State Univ, 53-54; from asst prof to prof animal sci, 54-86, sect leader, 76-79, UNIV PROF, UNIV ARK, FAYETTEVILLE, 86- *Concurrent Pos:* Consult, various activities. *Mem:* Fel AAAS; NY Acad Sci; fel Am Soc Animal Sci; Genetics Soc Am; Sigma Xi (pres). *Res:* Performance words of beef cattle for economic records; beef cattle genetics and breeding. *Mailing Add:* Div Animal Sci Univ Ark Fayetteville AR 72701

BROWN, CYNTHIA ANN, b Kansas City, Mo, Oct 9, 45; m 63; c 2. ANALYSIS OF ALGORITHMS. *Educ:* Mich State Univ, BS, 65; Univ Mich, MA, 66, PhD(math), 77. *Prof Exp:* Lectr comput sci, Ind Univ, 76-77, from asst prof to assoc prof, 77-84; ASSOC PROF COMPUT SCI, NORTHEASTERN UNIV, 84-, DEAN, COL COMPUTER SCI, 89- *Concurrent Pos:* Mem tech staff, GTE Labs, 82-84. *Mem:* Asn Comput Mach; Am Asn Artificial Intel; Inst Elec & Electronic Engrs Comput Soc. *Res:* Analysis and development of algorithms, specializing in algorithms for solving search problems. *Mailing Add:* Col Computer Sci Northeastern Univ Boston MA 02115

BROWN, DALE GORDON, b Corvallis, Ore, May 11, 44; m 63; c 2. ORGANIC CHEMISTRY, AGRICULTURAL CHEMISTRY. *Educ:* Abilene Christian Univ, BS, 68; North Tex State Univ, PhD(org chem), 74. *Prof Exp:* Res chemist agr chem, Tenneco Chem Inc, 68-70; res chemist, 74-86, sr res chemist, Agr Chem, 86-89, MGR TECHNOL LICENSING, AM CYANAMID CO, 89- *Concurrent Pos:* Dir Res & Develop, Whitehill Oral Technols, Inc. *Honors & Awards:* Sci Achievement Award, Am Cyanamid, 87. *Mem:* Am Chem Soc; Asn Univ Technol Mgrs. *Res:* Synthesis of pesticide chemicals and pyrethroid analogs. *Mailing Add:* 42 Stoney Brook Rd Hopewell NJ 08525

BROWN, DALE H(ENRY), b Oak Grove, Mo, Apr 30, 22; m 44; c 2. MECHANICAL ENGINEERING. *Educ:* Rensselaer Polytech Inst, BS, 42, MS, 50. *Prof Exp:* Instr mech eng & mech, Rensselaer Polytech Inst, 43-44, asst prof thermodyn & heat eng, 46-50; thermodyn engr, Gen Elec Co, 50-56, thermal systs consult engr, 56-67, systs engr, Res & Develop Ctr, 67-69, thermal & environ systs consult engr, 69-76, energy syst engr, 76-81, energy systs mgt engr, Res & Develop Ctr, 81-85; RETIRED. *Mem:* Am Soc Mech Engrs; Soc Automotive Engrs; NY Acad Sci; Nat Soc Prof Engrs; Sigma Xi. *Res:* Energy production and environmental effects in the biosphere; system dynamics; transient thermodynamics; simulation of power and process apparatus; advanced energy conversion from coal; advanced cogeneration systems to produce power and heat; coal fired locomotive system; advanced diesel engines. *Mailing Add:* Five Berkley Sq Scotia NY 12302

BROWN, DALE MARIUS, b Detroit, Mich; m 58; c 2. SOLID STATE PHYSICS, ELECTRICAL ENGINEERING. *Educ:* Univ Mich, BS, 53, MS, 55; Purdue Univ, PhD(physics), 60. *Prof Exp:* RES PHYSICIST, GEN ELEC RES & DEVELOP CTR, 61- *Concurrent Pos:* Ed, J Electrochem Soc, 77-; Gen Elec Coolidge fel, 83. *Honors & Awards:* Bronze, Silver & Gold Patent Awards, Gen Elec Res Ctr, 66-70; Nat Electronics Award, Electrochem Soc, 90. *Mem:* Am Phys Soc; fel Inst Elec & Electronic Engrs; Electrochem Soc. *Res:* Solid state physics; semiconductors; electronic solid state devices; integrated circuits; materials and processes; solid state imagers; charge transfer devices. *Mailing Add:* Gen Elec Corp PO Box Eight Schenectady NY 12301

BROWN, DANIEL JOSEPH, b Elwood, Ind, June 3, 41; m 66; c 3. PHARMACOLOGY, TOXICOLOGY. *Educ:* Marian Col, BS, 63; Ind Univ, PhD(toxicol), 68. *Prof Exp:* asst prof pharmacol & toxicol, Med Ctr, Ind Univ, Indianapolis, 69- *Concurrent Pos:* Fel pharmacol & toxicol, Univ Wash, 68-69. *Mem:* Am Acad Forensic Sci; Soc Toxicol; Int Asn Forensic Toxicol. *Res:* Pharmacology and toxicology of alcohols and volatile organic substances. *Mailing Add:* 3708 NW Territory Dr Springfield IL 62707

BROWN, DANIEL MASON, b Roanoke, Va, May 26, 39; m 61; c 2. CIVIL ENGINEERING. *Educ:* Duke Univ, BSCE, 61; Univ Ill, Urbana, MS, 63, PhD(civil eng), 65. *Prof Exp:* Asst civil eng, Univ Ill, Urbana, 61-65, asst prof, 65-69; assoc prof civil eng, Vanderbilt Univ, 69- *Concurrent Pos:* Resident res fel, Air Force Off Sci Res, 77-78. *Mem:* Am Soc Civil Engrs. *Res:* Structural engineering; structural optimization; operations research applied to civil engineering; probabilistic aspects of structural engineering. *Mailing Add:* 1113 Bishops Park Denton TX 76205-8065

BROWN, DARRELL QUENTIN, b Manhattan, Kans, May 23, 32; m 56; c 2. RADIATION BIOPHYSICS. *Educ:* Univ Kans, AB, 54, MS, 59; Univ Chicago, PhD(biophys), 64. *Prof Exp:* Mathematician-comput programmer, Lewis Lab, NASA, Ohio, 57; lab instr radiation biophys, Univ Kans, 58; USPHS Trainee biophys, Univ Chicago, 59-64, res assoc, 64; asst prof radiation biol, Univ Tenn, Knoxville, 64-70; RADIOBIOLOGIST, AM ONCOL HOSP, 70-; ASST PROF RADIATION BIOL, UNIV PA, 71-75, 79- *Concurrent Pos:* Consult, Oak Ridge Nat Lab, 68-69. *Mem:* AAAS; Radiation Res Soc; Sigma Xi. *Res:* Chemical and radiation carcinogenesis in vitro; effects of microbeam irradiation of parts of cells; quantitative microscopy; radiation dosimetry; radiobiological basis of radiotherapy; radiation protectors and radiation sensitizers. *Mailing Add:* Am Oncol Hosp Cent & Shelmire Ave Fox Chase Philadelphia PA 19111

BROWN, DAVID CHESTER, b Bay Shore, NY, July 16, 42. PHYSICS. *Educ:* Adelphi Univ, BA, 66; Syracuse Univ, PhD(physics), 73. *Prof Exp:* Engr laser eng, Gen Elec Co, 72-74; res dir thin film/lasers, Inficon Leybold-Heraeus Inc, 74-75; SR SCIENTIST LASER PHYSICS, LAB LASER ENERGETICS, UNIV ROCHESTER, 75- *Concurrent Pos:* Trainee, Nat Aeronaut & Space Admin, 66-69; consult, Gen Elec Co, 76- & Eastman Kodak Co, 78. *Honors & Awards:* VIP Award, Gen Elec Co, 73. *Mem:* Optical Soc Am. *Res:* Physics of glass lasers; glass physics; rare-earth spectroscopy; excimer lasers; efficient high repetition rate glass lasers; nonlinear optics. *Mailing Add:* Laser Technol Assoc Inc 148 Vestal Pkwy E Vestal NY 13850

BROWN, DAVID E, b Ottawa, Ont, Mar 1, 47. ENVIRONMENTAL MANAGEMENT. *Educ:* Univ Ill, Urbana, PhD(animal sci), 81. *Prof Exp:* asst prof & asst animal sci, Univ Nev, 80-; WVA OSTEOP SCH MED. *Mailing Add:* WVa Osteop Sch Med 400 N Lee St Louisburg WV 24901

BROWN, DAVID EDWARD, b Indianapolis, Ind, July 9, 09; m 39; c 2. OTOLARYNGOLOGY. *Educ:* Stanford Univ, AB, 32, MD, 36. *Prof Exp:* From asst to assoc, 44-52, from asst prof to prof, 52-78, chmn dept, 62-71, EMER PROF OTOLARYNGOL, MED CTR, IND UNIV, INDIANAPOLIS, 78- *Mem:* AMA; Am Acad Ophthal & Otolaryngol. *Mailing Add:* 405 Ave Granada No 313 San Clemente CA 92672

BROWN, DAVID FRANCIS, b Seattle, Wash, July 22, 35; m 63; c 3. AGRICULTURAL CHEMISTRY, PLANT BIOCHEMISTRY. *Educ:* NMex Inst Mining & Technol, BS, 58; Wash State Univ, MS, 61; Utah State Univ, PhD(plant nutrit & biochem), 70. *Prof Exp:* Instr chem, Sheridan Col, 61-62 & Yakima Valley Col, 62-65; res assoc, Southern Regional Res Ctr, Sci & Educ Admin-Agr Res, USDA, 69-72; res assoc plant biochem, Food Protein Res & Develop Ctr, Tex A&M Univ, 72-74; RES CHEMIST, SOUTHERN REGIONAL RES CTR, SCI & EDUC ADMIN, USDA, 74- *Mem:* Am Chem Soc; Inst Food Technologists; Sigma Xi. *Res:* Byssinosis; flavor and aroma of foods and agricultural products. *Mailing Add:* 1047 Dana Ave Apt 2 Cincinnati OH 45229

BROWN, DAVID FREDERICK, b Bargoed, Wales, Sept 14, 28; US citizen; m 50; c 2. MEDICINE. *Educ:* Univ Wales, BSc, 47, MB, BCh, 50; Am Bd Internal Med, dipl. *Prof Exp:* House physician, Med Unit, Cardiff Royal Infirmary, 50-51 & Sully Hosp, Wales, 51; med registr, Morriston Hosp, 53-55; resident internal med, Cornell Univ Infirmary & Thompkins County Mem Hops, Ithaca, NY, 55-56; sr pub health officer, NY State Dept Health, 56-57; res assoc, Cardiovasc Health Ctr & asst med, 57-59, from instr to assoc prof med, 59-73, assoc dir, Cardiovasc Health Vtr, 59-69, PROF MED, ALBANY COL MED, 73- *Concurrent Pos:* Nat Heart Inst spec fel physiol chem, Univ Lund, 6364; asst dispensary physician, Clins, Albany Med Ctr, 57-62, assoc dispensary physician, 62-, clin asst, Hosp, 6062, asst attend physician, 62-, asst attend cardiologist, 62-66, attend cardiologist, 66-, dir adult cardiac clin; attend physician, Vet Admin Hosp, 66-; fel, Coun Arteriosclerosis & Coun Epidemiol, Am Heart Asn. *Mem:* Am Fedn Clin Res; Am Soc Clin Nutrit; fel Am Col Physicians. *Res:* Cardiovascular disease; lipid metabolism as related to atherosclerosis. *Mailing Add:* Albany Med Col Albany NY 12208

BROWN, DAVID G, b London, Ont, June 4, 43; m; c 2. DENTAL SCIENCE, TOXICOLOGY. *Educ:* Univ Colo, PhD(biochem), 70. *Prof Exp:* Chmn, dept oral biochem, 81-85, asst dean, 83-89, ASSOC DEAN, COL DENT, UNIV NEBR, 89- *Mem:* Am Asn Dent Res; Am Asn Dent Schs; Sigma Xi; Am Soc Pharmacol & Exp Therapeut. *Mailing Add:* Col Dent Med Ctr Univ Nebr Lincoln NE 68583-0740

BROWN, DAVID HAZZARD, b Philadelphia, Pa, Apr 22, 25; m 55; c 2. CELL BIOLOGY. *Educ:* Rutgers Univ, BA, 51, PhD(plant path), 65; Univ Pa, MS, 54. *Prof Exp:* Res assoc, Rutgers Univ, 65; NIH fel, Biol Div, Oak Ridge Nat Lab, Union Carbide Nuclear Corp, 65-67, staff molecular anat prog, 67-70; scientist, Oak Ridge Assoc Univs, 70-; AT DEPT BIOCHEM, WASHINGTON UNIV SCH MED. *Mem:* AAAS; Am Inst Biol Sci; Am Soc Cell Biol; Sigma Xi. *Res:* Use of radiopharmaceuticals in cancer detection; comparative studies of the cellular and physiological uptake of scintiscanning agents in normal and neoplastic tissue. *Mailing Add:* Dept Biochem Washington Univ Sch Med St Louis MO 63110

BROWN, DAVID HENRY, biological chemistry, for more information see previous edition

BROWN, DAVID LYLE, b Victoria, BC, Mar 2, 43; m 62; c 1. CELL BIOLOGY. *Educ:* Univ BC, BSc, 66; Univ Calif, Davis, MSc, 68, PhD(biol), 69. *Prof Exp:* Fel biol, Yale Univ, 69-70; from asst prof to assoc prof, 75-82, PROF BIOL, UNIV OTTAWA, 82- *Concurrent Pos:* Grants panel, Nat Res Coun Can, 76-79; assoc ed, Can J Biochem & Cell Biol, 85- *Mem:* Can Soc Cell Biol (secy, 71-75, pres, 77-78); Am Soc Cell Biol. *Res:* Synthesis of microtubule protein and the regulation of microtubule assembly; methylmercury effects on cytoskeleton; nuclear matrix in lymphocytes. *Mailing Add:* Dept Biol Univ Ottawa Ottawa ON K1N 6N5 Can

BROWN, DAVID MITCHELL, b Chicago, Ill, Nov 11, 35; c 3. DIABETES MELLITUS, KIDNEY DEVELOPMENT & DISEASE. *Educ:* Univ Ill, BS, 56, MD, 60. *Prof Exp:* Rotating internship, Univ Ill Res & Educ Hosps, 60-61; attend staff pediat endocrinol, Wilford Hall USAF Hosp, 65-67; resident pediat, Univ Minn, 61-62, fel endocrinol & metab, 62-65, from asst prof to assoc prof pediat, Univ Minn Med Sch & Path, 67-73, actg head, 71, assoc head, 71-84, dir clin labs, 71-84, PROF PEDIAT, LAB MED & PATH, UNIV MINN, 73-, DEAN, MED SCH, 84- *Concurrent Pos:* Career develop award, USPHS, 68-73. *Mem:* Am Physiol Soc; Am Soc Path; Endocrine Soc; Am Soc Nephrol; Am Soc Cell Biol; Am Diabetes Asn. *Res:* Biochemical and physiologic bases for kidney complications of diabetes mellitus. *Mailing Add:* Univ Minn Med Sch 420 Delaware St SE Box 293 UMHC Minneapolis MN 55455

BROWN, DAVID P(AUL), b Chicago, Ill, Aug 29, 34. ELECTRICAL ENGINEERING. *Educ:* Univ Ill, BS, 56, MS, 57; Mich State Univ, PhD(elec eng), 61. *Prof Exp:* From instr to asst prof elec eng, Mich State Univ, 60-63; from vis assoc prof to assoc prof, 63-68, PROF ELEC ENG, UNIV WIS, MADISON, 68-; AT DEPT ELEC SCI & SYSTEM ENG, SOUTHERN ILL UNIV. *Concurrent Pos:* Vis prof, Colo State Univ, 67 & 68; consult, Surabaya Inst Tech, Indonesia, 77. *Mem:* Inst Elec & Electronics Engrs; Sigma Xi. *Res:* Circuit theory; network analysis and synthesis; system theory. *Mailing Add:* Dept Elec Sci & Syst Eng Southern Ill Univ Carbondale IL 62901

BROWN, DAVID RANDOLPH, b Los Angeles, Calif, Oct 31, 23; m 44; c 4. COMPUTER SCIENCE, ELECTRICAL ENGINEERING. *Educ:* Univ Wash, BS, 44; Mass Inst Technol, SM, 47. *Prof Exp:* Asst res engr, Univ Wash, 44-46; res assoc, Mass Inst Technol, 46-48, group leader, 51-58; lectr, Univ Calif, 48-51; assoc tech dir res, Mitre Corp, 58-63; ASST DIR COMPUT SCI & TECHNOL DIV, SRI INT, 63- *Honors & Awards:* Naval Ord Develop Award, 46. *Mem:* Fel Inst Elec & Electronics Engrs; Asn Comput Mach. *Res:* Computer sciences; research administration; systems design and systems science; electronics engineering. *Mailing Add:* SRI Int Menlo Park CA 94025-3493

BROWN, DAVID ROBERT, b Johnson City, NY, Oct 1, 54; m 88. GASTRO-INTESTINAL PHARMACOLOGY, MOLECULAR PHARMACOLOGY. *Educ:* Emory Univ, PhD(pharmacol), 81. *Prof Exp:* Fel pharmacol & physiol, Univ Chicago, 81-84; asst prof vet pharmacol, 84-89, asst prof neurosci, 86-89, ASSOC PROF VET BIOL, NEUROSCI & TOXICOL, UNIV MINN, 89- *Concurrent Pos:* Consult, 3M Pharmaceut, 90-; vis assoc prof pharmacol, Univ Naples, Italy, 90. *Mem:* Soc Neurosci; Am Soc Pharmacol & Exp Therapeut; AAAS. *Res:* Neural regulation and pharmacology of epithelial ion transport; molecular pharmacology of neurotransmitter receptors. *Mailing Add:* Dept Vet Biol Univ Minn 1988 Fitch Ave St Paul MN 55108

BROWN, DAVID SMITH, b Bluffton, Ind, Apr 26, 44; m 63; c 1. SOIL CHEMISTRY, ENVIRONMENTAL SCIENCE. *Educ:* Purdue Univ, BSA, 66; Univ Ill, MS, 68, PhD(soil phys chem), 72. *Prof Exp:* res soil scientist, 71-90, ACTG BR CHIEF, ENVIRON RES LAB, US ENVIRON PROTECTION AGENCY, 90- *Honors & Awards:* Bronze Medal, US Environ Protection Agency, 74, 86; Sci Achievement Award, Environ Protection Agency, 90, Ord Managerial Excellence Award, 90. *Mem:* Am Soc Agron; Soil Sci Soc Am; Am Chem Soc. *Res:* Sorption of organic pollutants on natural soils and sediments; speciation and transport of pollutant metals in groundwaters. *Mailing Add:* Environ Res Lab College Station Rd Athens GA 30613

BROWN, DAVID T, b Portland, Ore, May 20, 36; m 60; c 3. MATHEMATICS. *Educ:* Ottawa Univ, BA, 58; Syracuse Univ, MA, 61, PhD(math), 65; WVa Univ, MS(comput sci), 87. *Prof Exp:* Asst prof math, Hiram Col, 65-68 & Univ Pittsburgh, 68-74; from asst prof to assoc prof math, 74-82, prof math & comput sci, Bethany col, WVa, 82-88; ASSOC PROF COMPUT SCI & MATH, ITHACA COL, 88- *Mem:* Math Asn Am; Asn Comput Mach. *Res:* Theory. *Mailing Add:* Dept Math & Comput Sci Ithaca Col Ithaca NY 14850

BROWN, DAWN LARUE, b Frederick, Md, Nov 13, 48; m 80. NEUROSCIENCES. *Educ:* Boston Univ, BA, 70; Drake Univ, MA, 72; Univ NC, Greensboro, PhD(physiol psychol), 76. *Prof Exp:* Instr psychol, NC A&T State Univ, 74-75; fel psychobiol, 75-78, LAB TECHNOLOGIST II, PSYCHOBIOL, FLA STATE UNIV, 78- *Concurrent Pos:* Asn Res Vision & Ophthal travel fel, 75. *Mem:* Soc Neurosci; Asn Res Vision & Ophthal; AAAS. *Res:* Neurophysiology of vision; development of mammalian visual system; correlation of behavioral and electrophysiological properties of visual system; taste preferences and psychophysics. *Mailing Add:* 14219 Otter Run Rd Tallahassee FL 32312

BROWN, DEAN RAYMOND, b Detroit, Mich. MATHEMATICS. *Educ:* Rose Polytech Inst, BS, 60; Rensselaer Polytech Inst, MS, 64; Ohio State Univ, MS, 66, PhD(math), 70. *Prof Exp:* From asst prof to assoc prof, 70-84, chmn dept math & comput sci, 78-83, PROF, YOUNGSTOWN STATE UNIV, 84- *Mem:* Nat Coun Teachers Math; Math Asn Am. *Res:* Functional analysis. *Mailing Add:* Dept Math & Comput Sci Youngstown State Univ Youngstown OH 44555

BROWN, DELOS D, b Mansfield, La, Aug 2, 33; m 59; c 3. FOOD SCIENCE. *Educ:* La State Univ, BS, 56, MS, 63; Univ of Ga, PhD(food sci), 72. *Prof Exp:* Res assoc meat technol, La State Univ, 61-63; sr scientist & asst head sausage develop sect, Food Res Div, Armour & Co, Ill, 63-68; food scientist, Univ Ga, 68-73; DIR & FOUNDER, BAPTIST AGR PROG TRAINING IN SCI & TECHNOL, 73- *Concurrent Pos:* USPHS training grant; Armour Food Res Div grant. *Mem:* Am Meat Sci Asn. *Res:* Development of an agricultural experiment station; research with field crops, livestock and poultry, vegetable crops and food science. *Mailing Add:* BAPTIST Box 120 Petauke Zambia

BROWN, DENNISON ROBERT, b New Orleans, La, May 17, 34; m 56, 87; c 6. MATHEMATICS. *Educ:* Duke Univ, BS, 55; La State Univ, MS, 60, PhD(math), 63. *Prof Exp:* Assoc math, La State Univ, 58-61; from asst prof to assoc prof, Univ Tenn, 63-67; assoc prof, 67-70, PROF MATH, UNIV HOUSTON, 70- *Concurrent Pos:* Vis lectr, Math Asn Am, 65-74; prin investr, NSF grants, 65-71; mem grad coun, Univ Houston, 69-71 & dir grad studies math, 69-72; ed, Semigroup Forum, 70-; sr investr, NASA grants, 72-77; consult, Undergrad Prog Math, 73-81, vis consult, 81-; vis prof, La State Univ, 87-88. *Mem:* Am Math Soc; Sigma Xi; Math Asn Am. *Res:* Topological and algebraic semigroups; semigroups of matrices. *Mailing Add:* Dept Math Univ Houston University Park Houston TX 77204-3476

BROWN, DONALD A, b Roswell, NMex, Dec 23, 16; m 41; c 3. SOIL CHEMISTRY. *Educ:* NMex Col Agr & Mech, BS, 39; Univ Mo, Columbia, MS, 47, PhD(soil chem), 50. *Prof Exp:* Asst prof agron, NMex Col Agr & Mech, 45-46; asst soil chem, Univ Mo, Columbia, 47-50; asst prof, 50-59, PROF AGRON, UNIV ARK, FAYETTEVILLE, 59- *Concurrent Pos:* Consult chem waste disposal, 79-80. *Honors & Awards:* Outstanding Res Award, Univ Ark, 80. *Mem:* Agron Soc; Soil Sci Soc Am; Int Soil Sci. *Res:* Characterization of plant root systems by video camera techniques under field conditions. *Mailing Add:* 7699 Altus Dr Univ of Ark Fayetteville AR 72701

BROWN, DONALD D, b Cincinnati, Ohio, Dec 30, 31; m 57; c 3. DEVELOPMENTAL BIOLOGY. *Educ:* Univ Chicago, MS & MD, 56. *Hon Degrees:* DSci, Univ Chicago, 76, Univ Maryland, 83. *Prof Exp:* Intern, Charity Hosp, New Orleans, La, 56-57; res assoc biochem, NIMH, 57-59; spec fel, Pasteur Inst, Paris, 59-60; spec fel, 60-61, MEM STAFF BIOCHEM, CARNEGIE INST DEPT EMBRYOL, 61-, DIR DEPT EMBRYOL, 76- *Concurrent Pos:* Prof, Johns Hopkins Univ, 69- *Honors & Awards:* US Steel Found Award Molecular Biol, Nat Acad Sci, 73; V D Mattia Lectr, Roche Inst Molecular Biol, 75; Boris Pregel Award, NY Acad Sci, 77; Ross G Harrison Prize, Int Soc Develop Biologists, 81; Feodor Lynen Medal, 87. *Mem:* Nat Acad Sci; Am Soc Biol Chem; Soc Develop Biol; Am Acad Arts & Sci; Am Soc Cell Biol (pres, 92); Am Philos Soc. *Res:* Control of genes during development; isolation of genes. *Mailing Add:* Dept Embryol Carnegie Inst Washington 115 W University Pkwy Baltimore MD 21210

BROWN, DONALD FREDERICK MACKENZIE, b Chicago, Ill, Dec 9, 19; m 45. BIOLOGICAL SCIENCES. *Educ:* Univ Mich, BA, 49, MS, 51, PhD(bot), 58. *Prof Exp:* Asst prof natural sci, Eastern Mich Univ, 56-60, from assoc prof to prof biol, 60-87; RETIRED. *Mem:* AAAS; Wilderness Soc; Am Inst Biol Sci; Nat Audubon Soc; Nat Parks Asn. *Res:* Taxonomy; morphology and ecology of fern genus Woodsia; zoology; botany; applied botany and zoology; pteridology. *Mailing Add:* 15 Northwick Ct Ann Arbor MI 48105

BROWN, DONALD JEROULD, b Salt Lake City, Utah, Nov 4, 26; m 55; c 3. NUCLEAR WASTE ISOLATION. *Educ:* Univ Utah, BS, 55. *Prof Exp:* Engr nuclear waste, Gen Elec Co, 55-64 & Battelle Mem Inst, 64-65; environ engr, Isochem, Inc, 65-67 & Atlantic Richfield Co, 67-77; prin scientist, Rockwell Int, 77-78, proj mgr, 78-79, dept mgr, 79-81, CHIEF SCI ADV NUCLEAR WASTE ISOLATION, ROCKWELL INT, 81- *Mem:* Fel Geol Soc Am; Am Asn Petrol Geologists. *Res:* Environmental safety; evaluating sites for the isolation of radioactive wastes; evaluating radionuclide migration through saturated and unsaturated sediments; directing geological, geophysical and hydrological research and assessing environmental impacts. *Mailing Add:* Westinghouse-Hanford PO Box 1970 Richland WA 99352

BROWN, DONALD JOHN, b Utica, NY, June 1, 33; m 56; c 3. INORGANIC CHEMISTRY. *Educ:* Utica Col, AB, 55; Syracuse Univ, PhD(chem), 63. *Prof Exp:* From instr to assoc prof, 60-87, PROF CHEM, WESTERN MICH UNIV, 87- *Concurrent Pos:* NIH grant, 64; res fel with Prof J Lewis, Univ London, 68; res fel with Prof B Lever, York Univ, 81; dir, Sci Citizens Ctr, Western Mich Univ, 84-; Kellogg Found grants, 85, 88, 89, CS Mott Found, 86. *Mem:* AAAS; Am Chem Soc. *Res:* Magnetic and spectral investigations of transition metal complexes; synthesis and investigation of new chelating agents; ground water chemisty. *Mailing Add:* Dept Chem Western Mich Univ Kalamazoo MI 49008

BROWN, DONALD M, b Canton, Ohio,Feb 11, 20; m; c 4. PETROLEUM GEOLOGY. *Educ:* Mt Union Col, BS, 42; Ohio State Univ, MS, 48. *Prof Exp:* Vpres, Rickelson Oil & Gas, 73-78; explor mgr, White Ellis Drilling, 78-81; PETROL GEOL CONSULT, 81- *Mem:* Fel Geol Soc Am. *Mailing Add:* 7333 E 22 St No 6 Wichita KS 67226-1127

BROWN, DOUGLAS EDWARD, b Mt Morris, Mich, Mar 25, 26; m 50; c 2. OPTICAL PHYSICS. *Educ:* Univ Mich, BS, 49, MS, 51, PhD(physics), 57. *Prof Exp:* Physicist, Gen Elec Co, 51-52; res assoc, Willow Run Res Ctr, 52-53; res fel, Univ Mich, 53-56; physicist, Dow Chem Co, 56-58, Bendix Systs Div, 58-60 & Inst Sci & Technol, Univ Mich, 60-71; dir eng, Syscon Int, Ind, 71-74; PHYSICIST, DEPT DEFENSE, 74- *Res:* Infrared; semiconductors; optics; thermal; optical data processing. *Mailing Add:* 5348 Lightningview Columbia MD 21045

BROWN, DOUGLAS FLETCHER, b Rouleau, Sask, Aug 13, 34. GENETICS. *Educ:* Univ Sask, BSA, 57, MSc, 58; Univ Wis, PhD(genetics), 63. *Prof Exp:* Res asst genetics, Univ Wis, 61-63; asst res officer biol, Atomic Energy Can, Ltd, 63-66; PROF BIOL, BISHOP'S UNIV, 66- *Mem:* Genetics Soc Can; Sigma Xi. *Res:* Mutant isolation in plant cell cultures; isolation of plant genes. *Mailing Add:* Dept Biol Sci Bishop's Univ Lennoxville PQ J1M 1Z7 Can

BROWN, DOUGLAS RICHARD, b Sacramento, Calif, Mar 13, 42; m 66; c 2. PHYSICS. *Educ:* Univ Calif, Berkeley, BS, 66, MA, 68; Univ Wash, PhD(physics), 74. *Prof Exp:* Res asst, Nuclear Physics Lab, Univ Wash, 66-74; res asst nuclear physics, Cyclotron Inst, Tex A&M Univ, 74-77; staff scientist, 77-79, MGR, ADVAN NUCLEONICS DIV, SCI APPLN, INC, 79- *Mem:* Am Phys Soc. *Res:* Study of relatively simple modes of higher lying nuclear excitation; characterization of coal using nuclear and electromagnetic techniques; developed family of on-line coal analysis instruments for process control; applied physics and computer science. *Mailing Add:* Sci Applications Inc 2950 Patrick Henry Dr Santa Clara CA 95054

BROWN, DUANE, b Fredonia, Ariz, Apr 1, 18; m 42; c 2. PHYSICAL CHEMISTRY. *Educ:* Brigham Young Univ, BS, 41; Princeton Univ & Mass Inst Technol, cert, 44; Cornell Univ, PhD(phys chem), 51. *Prof Exp:* Asst chem, Univ Nebr, 41-42 & Cornell Univ, 46-51; assoc prof, 51-58, PROF CHEM, ARIZ STATE UNIV, 58- *Concurrent Pos:* Res chemist, Nat Carbon Res Labs, Ohio, 57 & 58; consult, World Wide Refineries, Phoenix, Ariz & Prof Metal Serv, Inc, 78- *Mem:* AAAS; Am Chem Soc. *Res:* Chemical kinetics; thermodynamics; radiation chemistry; hydrometallurgy. *Mailing Add:* 104 E Huntington Dr Tempe AZ 85282

BROWN, EARL IVAN, II, b Carrolton, Ga, Feb 15, 17; m 41; c 3. CIVIL ENGINEERING. *Educ:* Va Mil Inst, BS, 40; NC State Col, MS, 49; Univ Tex, PhD (struct eng), 53. *Prof Exp:* Instrumentman, Tenn Valley Authority, 40; asst engr, Big Four RR, Ind, 41-42; res engr, NC, 46; instr civil eng, NC State Col, 47-49; asst prof, GA Inst Technol, 50-53; head prof civil eng, Ala Polytech Inst, 54-58, asst dean, 58-60; Jones prof civil eng & Chem Dept, Duke Univ, 60-83; RETIRED. *Concurrent Pos:* Vis prof, Univ Liverpool, 66-67 & Univ Whitewatersrand, SAm, 74. *Mem:* Am Soc Civil Engrs; Am Soc Eng Educ; Am Concrete Inst; Int Asn Bridge & Struct Engrs. *Res:* Concrete design. *Mailing Add:* V237 Carolina Chapel Hill NC 27514

BROWN, EDGAR HENRY, JR, b Oak Park, Ill, Dec 27, 26; m 54; c 2. TOPOLOGY. *Educ:* Univ Wis, BS, 49; Wash State Univ, MA, 51; Mass Inst Technol, PhD, 54. *Prof Exp:* Instr math, Washington Univ, 54-55; instr, Univ Chicago, 55-57; res assoc, Brown Univ, 57-58; from asst prof to assoc prof, 58-63, PROF MATH, BRANDEIS UNIV, 63- *Concurrent Pos:* NSF sr fel, 61-62; Guggenheim fel, 65-66; Sci Res Coun sr fel, Univ Col, London, 73-74; Sci & Eng Res Coun sr fel, Cambridge Univ, 82-83; sr res fel, Jesus Col, Oxford, 86-87. *Mem:* Am Acad Arts & Sci; Am Math Soc. *Res:* Algebraic and differential topology. *Mailing Add:* Dept Math Brandeis Univ Waltham MA 02154

BROWN, EDMOND, b Brooklyn, NY, July 22, 24; m 46; c 3. THEORETICAL SOLID STATE PHYSICS. *Educ:* Univ Ill, BS, 48; Cornell Univ, PhD(physics), 54. *Prof Exp:* From asst prof to assoc prof, 55-65, PROF PHYSICS, RENSSELAER POLYTECH INST, 65- *Concurrent Pos:* Vis prof, State Univ NY Albany, 67-68; consult, Phys Sci Div, Watervliet Arsenal, 73- *Mem:* Fel Am Phys Soc; Am Asn Physics Teachers. *Res:* Electron band theory of solids; theory of two dimensional electron gases. *Mailing Add:* Dept Physics Rensselaer Polytech Inst Troy NY 12181

BROWN, EDMUND HOSMER, b Boston, Mass, June 19, 20; m 51, 71; c 3. ACOUSTIC REMOTE SENSING OCEANS & ATMOSPHERE. *Educ:* Univ Colo, BS, 55, MS, 62. *Prof Exp:* Instr math, Univ Colo, 49-55; physicist, Boulder Labs, Nat Bur Standards, 51-57, educ dir, 57-65; physicist, Wave Propagation Lab, Nat Oceanic & Atmospheric Admin, 65-81; sr res assoc, Coop Inst Res Environ Sci, Univ Colo, 81-86; RETIRED. *Concurrent Pos:* Prof adjoint mech eng, Univ Colo, 57-58, vis lectr, math, 58-61; researcher, Inst Henri Poincaré, 67-68; assoc ed, J Acoust Soc Am, 79-; external examnr, Carleton Univ, Ottawa, 80 & Univ Melbourne, Australia, 85. *Mem:* Fel Acoust Soc Am; Int Soc Acoust Remote Sensing (secy, 83-88). *Res:* Remote sensing of depth or altitude profiles of temperature in oceans and atmosphere; applications of complex geometry in elementary particles. *Mailing Add:* 905 15 St Boulder CO 80302

BROWN, EDWARD HERRIOT, JR, b Dover, NJ, Jan 2, 26; m 52; c 4. FISHERIES SCIENCE, ECOLOGY. *Educ:* Mich State Univ, BS, 50; Ohio State Univ, MS, 61. *Prof Exp:* Fish mgt supvr I & II, Ohio Dept Natural Resources, Div Wildlife, 52-62; asst leader & proj leader fishery res, Little Miami River Proj, Xenia, Ohio, 52-58; proj leader design fishery surv, Olentangy Wildlife Exp Sta, Delaware, Ohio, 58-62; proj leader fishery res, US Dept Interior, Bur of Com Fisheries, Ann Arbor, 62-70; PROJ LEADER FISHERY RES, NAT FISHERIES RES CTR-GREAT LAKES, US FISH & WILDLIFE SERV, 70- *Mem:* Am Inst Fishery Res Biologists; Am Fisheries Soc. *Res:* Population dynamics and ecology of Great Lakes fishes, especially the coregonines, salmonines, and clupieds. *Mailing Add:* Nat Fisheries Res Ctr-Great Lakes 1451 Green Rd Ann Arbor MI 48105-2899

BROWN, EDWARD JAMES, b Litchfield, Minn, Dec 26, 48; m 82; c 2. BIOHYDIOMETALLURGY, ENVIRONMENTAL BIOTECHNOLOGY. *Educ:* Univ Minn, BS, 70, Univ Wis, MS, 73, PhD(bact), 75. *Prof Exp:* From asst prof to assoc prof, 77-88, dir, Water Res Ctr, 86-89, PROF MICROBIOL, UNIV ALASKA, FAIRBANKS, 88- *Concurrent Pos:* vis prof, Univ Minn, 84-85. *Mem:* Am Soc Microbiol; Am Soc Limnology & Oceanog; Am Chem Soc; AAAS. *Res:* Biohydrometallurgy; environmental biotechnology; phytoplankton ecology; bioremediation. *Mailing Add:* Water Res Ctr Univ Alaska Fairbanks AK 99775

BROWN, EDWARD MARTIN, b Philadelphia, Pa, Sept 24, 33; m 59, 74; c 3. MATHEMATICS. *Educ:* Univ Pa, AB, 58, MA, 59; Mass Inst Technol, PhD(math), 63. *Hon Degrees:* AM, Dartmouth Col, 73. *Prof Exp:* Teaching asst math, Univ Pa, 57-59; teaching asst, Mass Inst Technol, 59-62, asst instr, 62-63, instr, 63-64; res instr, 64-65, from asst prof to assoc prof, 65-72, vchmn dept, 74-78, PROF MATH, DARTMOUTH COL, 72- *Concurrent Pos:* Staff mem, Math Res Inst, Univ Warwick, 70-71. *Mem:* Am Math Soc. *Res:* General knotting structure of complexes in manifolds; topology of 3-dimensional manifolds. *Mailing Add:* Dept Math Dartmouth Col Hanover NH 03755

BROWN, EDWIN WILSON, JR, b Youngstown, Ohio, Mar 6, 26; m 52; c 3. COMMUNITY HEALTH, TRAVEL MEDICINE. *Educ:* Harvard Univ, MD, 53, MPH, 57. *Prof Exp:* Intern, E J Meyer Mem Hosp, 54-55; resident pub health, Arlington County Health Dept, Va, 55-56; asst prof prev med, Sch Med, Tufts Univ, 58-61; dep chief staff, Boston Dispensary, 61; vis prof social & prev med, Osmania Med Col, India, 61-63; asst dir, Div Int Med Educ, Asn Am Med Cols, 63-65; dir, Proj Vietnam, Washington, DC, 65-66; dir, Div Int Health, Med Ctr, Ind Univ, 66-69; dir int affairs, 69-77, assoc prof med, Sch Med, Ind Univ-Purdue Univ Indianapolis, 66-86, assoc dean & dir, int serv,

81-86, PRES GLOBAL HEALTH SERV, LTD, 87- *Concurrent Pos:* Field dir, Harvard Epidemiol Proj, Greenland, 56-57; res assoc, Sch Pub Health, Harvard Univ, 59-60; consult, Boston City Health Dept, 58-60 & WHO, 73-; sr adv med educ, King Faisal Univ, Saudi Arabia, 77-78; chmn bd dirs, Med Assistance Progs, Inc, Ill; chmn coun int educ in health prof, Midwestern Univs Consortium Int Activ, Inc; mem med adv comt, Iran Found; mem bd dirs, Paul Carlson Found & CARE-MEDICO, Int Students, Inc; mem bd govs, Int Policy Forum; Pres, Global Health Services, Ltd, 87- *Mem:* Nat Coun Int Health. *Res:* Problems of medical education in developing countries. *Mailing Add:* 8153 Oakland Rd Indianapolis IN 46240

BROWN, ELEANOR MOORE, b East Liverpool, Ohio, Mar 19, 36; m 60; c 1. PHYSICAL BIOCHEMISTRY. *Educ:* Ohio Wesleyan Univ, BA, 58; Drexel Univ, MS, 67, PhD(chem), 71. *Prof Exp:* Chemist, Harshaw Chem Co, 58-61 & Calbiochem, 61-62; res asst biochem, Mich State Univ, 64; res asst phys chem, Drexel Univ, 67-71; RES CHEMIST, EASTERN REGIONAL RES CTR, USDA, 71- *Concurrent Pos:* Assoc, Nat Res Coun, 71-73; fel, Dairy Res Inc, 73-75. *Mem:* Am Chem Soc; AAAS; Am Soc Biochem & Molecular Biol; Protein Soc; Sigma Xi; Am Women Sci. *Res:* Metal-protein and protein-protein interactions; lipid-protein interactions; protein secondary structure; lipoprotein complexes; protein modification. *Mailing Add:* Eastern Regional Res Ctr USDA 600 E Mermaid Lane Philadelphia PA 19118

BROWN, ELISE ANN BRANDENBURGER, b Jacksonville, Fla, Dec 5, 28; m 52; c 3. REGULATORY TOXICOLOGY. *Educ:* George Washington Univ, BS, 49, MS, 50, PhD(pharmacol), 56; Am Bd Toxicol, dipl. *Prof Exp:* Asst prof pharmacol, George Washington Univ, 56-57; pharmacologist, Nat Heart, Lung & Blood Inst, 62-79; STAFF OFFICER TOXICOLOGIST, FOOD SAFETY & INSPECTION SERV, USDA, 79- *Concurrent Pos:* Fel, McCollum-Pratt Inst, Johns Hopkins Univ & Sinai Hosp, Baltimore, Md, 57-59; lectr, Sch Med, Georgetown Univ, 64- *Mem:* Am Chem Soc; Am Soc Pharmacol & Exp Therapeut; Soc Gen Physiol; Int Soc Study Xenobiotics; Int Soc Biochem Pharmacol; Soc Exp Biol Med; Soc Toxicol; Am Col Toxicol. *Res:* Toxicology of food additives; pharmacology of lung tissue; effect of drugs on lipid metabolism; cancer chemotherapy; cystic fibrosis of the pancreas; para-amino-benzoic acid metabolism; Toxicology of Pesticides; environmental Toxicology. *Mailing Add:* 6811 Nesbitt Place McLean VA 22101

BROWN, ELLEN, b San Francisco, Calif, Apr 30, 12. ANIMAL PHYSIOLOGY. *Educ:* Univ Calif, BA, 34, MD, 39. *Prof Exp:* Intern, San Francisco Hosp, 38-39; from asst resident to resident med, Univ Calif Hosp, 39-43; clin instr, 43-44, from instr to prof, 46-79, EMER PROF MED, MED SCH, UNIV CALIF, SAN FRANCISCO, 79- *Concurrent Pos:* Asst physician, Cowell Mem Hosp, 40-42; Commonwealth Fund fel, Harvard Med Sch, 44-46; Guggenheim fel, Oxford Univ, 58. *Mem:* AAAS; Am Physiol Soc; NY Acad Sci; Am Heart Asn. *Res:* Capillary pressure and permeability; blood volume and vascular capacity; cardiac failure; cardiac complications of pregnancy; peripheral circulation in relation to pain syndromes and vascular diseases. *Mailing Add:* Dept Med Univ Calif Sch Med Box 0120 San Francisco CA 94143

BROWN, ELLEN RUTH, b New York, NY, June 15, 47. COMPUTATIONAL PHYSICS, APPLIED THEORETICAL PHYSICS. *Educ:* Mary Washington Col, BS, 69; Pa State Univ, MS, 71; Univ Va, PhD(physics), 81. *Prof Exp:* Physicist, Naval Weapons Lab, 69; teacher math, Va Pub Schs, 70-71; instr physics, Lord Fairfax Community Col, 71-74; engr, EG&G WASC, 79-86, head dept, 82-86, scientist, 86-90, SR SCIENTIST, EG&G, WASC, 90- *Mem:* Am Phys Soc; Sigma Xi. *Res:* Computational physics; band bending in semiconductors; photoionization of open-shell atoms; atmospheric optical-ray tracing; meteorological effects on reentry bodies; electromagnetic pulse calculations; energy technology transfer; physics of music. *Mailing Add:* PO Box 1397 Fredericksburg VA 22402-1397

BROWN, ELLIS VINCENT, b Montreal, Que, Aug 6, 08; m; c 2. ORGANIC CHEMISTRY. *Educ:* Univ Ill, BS, 30; Iowa State Col, PhD(org chem), 36. *Prof Exp:* Asst, Iowa State Col, 30-36, instr chem, 36-37; res chemist, Chas Pfizer & Co, Inc, 37-43, group leader, 43-47; assoc prof org chem, Fordham Univ, 47-53; prof chem, Seton Hall Univ, 53-59, chmn dept, 56-59; PROF CHEM, UNIV KY, LEXINGTON, 59- *Mem:* Fel AAAS; Am Chem Soc; Sigma Xi. *Res:* Chemistry and synthesis of vitamins, medicinals and antibiotics; reactions and synthesis in heterocyclics and carbohydrates; organic reaction mechanisms. *Mailing Add:* Dept Chem Univ Ky Lexington KY 40506

BROWN, ELMER BURRELL, b New York, NY, Apr 1, 26; m 54; c 4. INTERNAL MEDICINE, HEMATOLOGY. *Educ:* Oberlin Col, AB, 46; Wash Univ, MD, 50; Am Bd Internal Med, 57. *Prof Exp:* From intern to asst resident med, Presby Hosp, NY, 50-52; from instr to assoc prof, 55-71, chief hemat clin, 59-73, dir, Div Hemat, 64-73, PROF MED, SCH MED, WASH UNIV, 71-, ASSOC DEAN CONTINUING MED EDUC, 73-, ASSOC DEAN POSTGRAD TRAINING, 83- *Concurrent Pos:* USPHS trainee hemat, Sch Med, Wash Univ, 54-55, Nat Res Coun fel, 55-57; USPHS spec res fel, Enzyme Sect, Lab Cellular Physiol, Nat Heart Inst, 57-59; consult, Wash Univ Clins, 55-57 & St Louis City Hosp, 55-57; Lederle Med Fac Award, 60-63; USPHS res career develop award, 63-70; vis prof, Royal Postgrad Med Sch, London, 69-70. *Mem:* Am Soc Hemat; Am Fedn Clin Res; Soc Exp Biol & Med; Am Soc Clin Invest; Am Soc Exp Path; Am Soc Clin Nutrit; Int Soc Hemat. *Res:* Problems of iron metabolism at the clinical, physiological and biochemical levels. *Mailing Add:* Sch Med Wash Univ Box 8063 660 S Euclid St Louis MO 63110

BROWN, ERIC, b Ann Arbor, Mich, 50; m 83; c 1. COMPLEMENT, PHAGOCYTOSIS. *Educ:* Harvard Univ, MD, 75. *Prof Exp:* Sr investr immunol, NIH, Bethesda, Md, 77-85; ASSOC PROF MED, SCH MED, WASH UNIV, 85- *Mem:* AAAS; NY Acad Sci; Am Soc Clin Immunol; Am Asn Immunol; Am Found Clin Res. *Res:* Macrophage. *Mailing Add:* Sch Med Wash Univ PO Box 8051 St Louis MO 63110

BROWN, ERIC REEDER, b Cortland, NY, Mar 16, 25; m 61; c 4. CHEMISTRY, VIROLOGY. *Educ:* Syracuse Univ, BA, 48, MS, 51; Univ Kans, PhD(virol, immunol biochem), 57. *Hon Degrees:* DSc, Quincy Col, 66. *Prof Exp:* Instr, Col Med, Univ Ill, 57-59; asst prof hemat & immunol, Univ Ala, 59; asst prof, Univ Minn, 60-61; sr res assoc hemat, Hektoen Inst Med Res, 61-65; assoc prof microbiol, Med Sch, Northwestern Univ, 65-68; PROF MICROBIOL & CHMN DEPT, CHICAGO MED SCH, 67- *Concurrent Pos:* Assoc, Northwestern Univ, 63-; res fel, Am Cancer Soc, 60-63; scholar, Leukemia Soc, Inc, 65-70; mem med adv comt, Leukemia Res Found Inc, 70-; med adv, Ill State Dir Selective Serv Syst. *Mem:* NY Acad Sci; Am Asn Cancer Res; fel Am Inst Chemists; fel Am Acad Microbiol; fel Am Pub Health Asn; Sigma Xi. *Res:* Anthrax diagnosis; bacteriophage and transduction genetics; immunochemical tumor antigens; leukemia virus and antibodies; microbiology; epidemiology; pollution. *Mailing Add:* Dept Educ Psychol NY Univ New York NY 10003

BROWN, ERIC RICHARD, b Ithaca, NY, Feb 12, 42; m 66; c 2. ELECTROCHEMISTRY. *Educ:* Mich Technol Univ, BS, 63; Northwestern Univ, PhD(anal chem), 67. *Prof Exp:* RES CHEMIST, RES LABS, EASTMAN KODAK CO, 67- *Mem:* Am Chem Soc. *Res:* Electrochemistry, especially instrumentation and automatic data acquisition; investigation of mechanisms of organic oxidation and reduction reactions. *Mailing Add:* Eastman Kodak Co Res Labs Kodak Park Bldg 82 Rochester NY 14650-2102

BROWN, ERNEST BENTON, JR, physiology, for more information see previous edition

BROWN, ESTHER MARIE, b Birmingham, Mich, May 12, 23; m 63. ANATOMY. *Educ:* Mich State Univ, BS, 46, MS, 51, PhD(path), 55. *Prof Exp:* Bacteriologist, Game Div, State Dept Conserv, Mich, 46-50; bacteriologist, Ft Detrick Proj, Mich State Univ, 50-54, asst vet path, 54-55, from asst prof to prof anat, 55-71, dir sch med technol, 66-71; prof vet anat, Sch Vet Med, Univ Mo-Columbia, 71-; RETIRED. *Concurrent Pos:* Mem vet med rev comt, NIH, 72-74. *Mem:* Am Soc Med Technol; Am Asn Anatomists; Tissue Cult Asn; Am Soc Cytol; NY Acad Sci. *Res:* Microscopic anatomy; cytology and tissue culture, particularly tissue responses of animal diseases; normal histology of the white rat. *Mailing Add:* 10184 Elk Trail Williamsburg MI 49690

BROWN, FARRELL BLENN, b Mt Ulla, NC, Nov 29, 34; m 58; c 2. STRUCTURAL CHEMISTRY. *Educ:* Lenoir-Rhyne Col, BS, 57; Univ Tenn, MS, 59, PhD(phys chem), 62. *Prof Exp:* Robert A Welch Found fel, Tex A&M Univ, 62-63; from asst prof to assoc prof, Clemson Univ, 63-73, asst dean grad studies, 74-77, actg dean, 89-91, PROF CHEM, CLEMSON UNIV, 73-, ASSOC DEAN GRAD STUDIES, 77- *Mem:* Photochem Soc NAm; Am Chem Soc; Sigma Xi. *Res:* Molecular spectroscopy, structure and dynamics; far infrared spectral studies; molecular configurations; quantum mechanics of bonded systems. *Mailing Add:* Grad Sch E 106 Martin Hall Clemson Univ Clemson SC 29634-5120

BROWN, FIELDING, b Berlin, NH, Jan 2, 24; m 44; c 4. PHYSICS. *Educ:* Williams Col, Mass, BA, 47, AM, 49; Princeton Univ, PhD, 53. *Prof Exp:* Res physicist, Sprague Elec Co, Mass, 52-59; from asst prof to assoc prof, 59-67, PROF PHYSICS, WILLIAMS COL, MASS, 67- *Concurrent Pos:* Vis scientist, Lincoln Lab, Mass Inst Technol, 61-62 & Francis Bitter Nat Magnet Lab, 68-; vis prof, Univ Tokyo, 65-66; consult, Mass Inst Technol, 73-74. *Mem:* Am Phys Soc; Sigma Xi. *Res:* Cosmic rays; ferrites; ferroelectricity; nonlinear optics; far infrared; lasers. *Mailing Add:* Dept Physics Williams Col Williamstown MA 01267

BROWN, FORBES TAYLOR, b Newton, Mass, Oct 12, 34; m 63; c 3. MECHANICAL ENGINEERING, FLUID POWER. *Educ:* Mass Inst Technol, SB & SM, 58, ME, 59, ScD, 62. *Prof Exp:* Asst mech eng, Mass Inst Technol, 56-58, from instr to assoc prof, 58-70; head, syst dynamics div, 84-87, PROF MECH ENG, LEHIGH UNIV, 70- *Concurrent Pos:* Mem exec comt, Fluids Eng Div, Am Soc Mech Engrs, 70-75; mem bd gov, Nat Conf Fluid Power, 81- *Mem:* Am Soc Mech Engrs; Fluid Power Soc. *Res:* General techniques and applications of lumped and distributed modeling and simulation of complex engineering systems, plus their design and control; fluid line dynamics and vibration; conception and development of new concepts in fluid power systems. *Mailing Add:* Dept Mech Eng Packard Lab 19 Lehigh Univ Bethlehem PA 18015

BROWN, FOUNTAINE CHRISTINE, b Huffman, Ark, Oct 13, 23. BIOCHEMISTRY. *Educ:* Southeast Mo State Col, BA & BS, 47; Univ Mo, MA, 51; Univ Iowa, PhD(biochem), 55. *Prof Exp:* PROF BIOCHEM, COL MED, UNIV TENN, MEMPHIS, 58-, RES BIOCHEMIST, TENN PSYCHIAT HOSP, 58- *Concurrent Pos:* Fel, M D Anderson Hosp & Tumor Inst, Univ Tex, 55-58. *Mem:* AAAS; Am Soc Biol Chemists; Am Chem Soc; Am Soc Neurochem; Int Soc Neurochem; Sigma Xi. *Res:* Neurochemistry; alcohol and other drug changes in the central nervous system; mechanism of enzyme action. *Mailing Add:* 3084 Rising Sun Rd Memphis TN 38133

BROWN, FRANK MARKHAM, b San Diego, Calif, May 12, 30; m 57; c 3. ELECTRICAL ENGINEERING. *Educ:* US Naval Acad, BS, 53; US Air Force Inst Technol, MS, 58; Ohio State Univ, PhD(elec eng), 67. *Prof Exp:* US Air Force, 54-72, instr electronics, Air Training Command, US Air Force Inst Technol, 54-56, proj scientist, Electronic Technol Lab, Air Res & Develop Command, 58-61, from instr to assoc prof elec eng, 61-72; ASSOC PROF, DEPT ELEC ENG, UNIV KY, 72- *Mem:* Inst Elec & Electronics Engrs. *Res:* Switching theory; boolean equations. *Mailing Add:* Eng Dept Air Force Inst Technol Wright Patterson Air Force Base OH 45433

BROWN, FRED(ERICK) R(AYMOND), b Peoria, Ill, Feb 15, 12; m 36; c 3. CIVIL ENGINEERING. *Educ:* Univ Ill, BS, 34. *Prof Exp:* From hydraul engr to chief, hydrodyn br, hydraul div, US Corps Engrs Waterways Exp Sta, 34-64, chief, nuclear weapons effects div, 64, asst dir sta, 64-69, tech dir sta, 69-85; RETIRED. *Honors & Awards:* Meritorius Civilian Serv, US Dept Army. *Mem:* Hon mem Am Soc Civil Engrs; Nat Soc Prof Engrs; Int Asn Hydraul Res. *Res:* Hydraulic design of dams and appurtenant structures; explosives effects. *Mailing Add:* 105 Stonewall Rd Vicksburg MS 39180

BROWN, FREDERICK CALVIN, b Seattle, Wash, July 6, 24; m 52; c 3. SOLID STATE PHYSICS. *Educ:* Harvard Univ, SB, 45, MA, 47, PhD(physics), 50. *Prof Exp:* Physicist, US Naval Res Lab, 50; assoc physicist, Appl Physics Lab, Univ Wash, 51; from instr to asst prof physics, Reed Col, 51-54; from asst prof to prof, Univ Ill, Urbana-Champaign, 55-73; prin scientist, Xerox Palo Alto Res Lab, 73-74; prof, 74-87, EMER PROF PHYSICS, UNIV ILL, URBANA, 87-; PROF PHYSICS, UNIV WASH, 87- *Concurrent Pos:* NSF sr fel, Clarendon Lab, Oxford Univ, 64-65; assoc, Ctr Advan Study, Univ Ill, 69-70; consult prof appl physics, Stanford Univ, 73-74; Alexander Humboldt sr fel award, 78. *Mem:* Fel Am Phys Soc; Sigma Xi. *Res:* Photoconducting and optical properties of ionic crystals, especially silver and alkali halides; color center phenomena and cyclotron resonance; extreme ultraviolet spectroscopy and synchrotron radiation. *Mailing Add:* 2414 E Discovery Langley WA 98260

BROWN, GARRY LESLIE, b Adelaide, S Australia, May 14, 42; m 67; c 3. FLUID MECHANICS, AERONAUTICS. *Educ:* Univ Adelaide, BE Hons, 64; Oxford Univ, PhD(eng sci), 67. *Prof Exp:* Sr res fel aeronaut, Calif Inst Technol, 67-71; from reader to sr lectr mech eng, Univ Adelaide, 71-78; prof aeronaut, Calif Inst Technol, 78-88; dir aeronaut res labs, Melbourne, Australia, 88-90; CHAIR MECH & AEROSPACE ENG DEPT, PRINCETON UNIV, 90- *Mem:* Am Inst Aeronaut & Astronaut; Inst Engrs Australia. *Res:* Turbulence; structure in turbulent shear plows; combustion in turbulent plows noise and unsteady gas dynamics. *Mailing Add:* Mech & Aerospace Eng Dept Princeton Univ Princeton NJ 08540

BROWN, GARY S, b Jackson, Miss, Apr 13, 40; m 70; c 2. ROUGH SURFACE SCATTERING, REMOTE SENSING. *Educ:* Univ Ill, BS, 63, MS, 64, PhD(elec eng), 67. *Prof Exp:* Consult, Andrew Corp, 65-67; mem tech staff, TRW Systs Group, 67 & 69-70; capt, Elec Eng, US Army, 67-69; sr engr, Res Triangle Inst, 70-73; chief scientist, Appl Sci Assocs, Inc, 73-85; PROF ELEC ENG, VA POLYTECH INST & STATE UNIV, 85- *Concurrent Pos:* Consult, Decision Sci Applications & Johns Hopkins Appl Physics Lab, 89-90, US Naval Res Lab & Dynamics Technol Inc, 89-; mem, comn B & F, Int Union Radio Sci. *Honors & Awards:* R W P King Award, Inst Elec & Electronic Engrs, Soc Antennas & Propagation, 77; Group Achievement Award, NASA, 78 & 79. *Mem:* Inst Elec & Electronic Engrs, Antennas & Propagation Soc (pres, 88); Inst Elec & Electronic Engrs, Geosci & Remote Sensing Soc; Inst Elec & Electronic Engrs, Oceanic Eng Soc; Int Union Radio Sci; Sigma Xi. *Res:* Theoretical modeling of electromagnetic scattering by randomly rough surfaces and the remote sensing applications of the results with particular emphasis on the ocean surface and sea ice. *Mailing Add:* Bradley Dept Elec Eng Va Polytech Inst & State Univ Blacksburg VA 24061-0111

BROWN, GENE MONTE, b Mo, Jan 21, 26; m 54; c 3. BIOCHEMISTRY. *Educ:* Colo State Univ, BS, 49; Univ Wis, MS, 50, PhD(biochem), 53. *Prof Exp:* Assoc, Univ Tex, 52-54; from instr to prof, 54-67, exec officer dept biol, 67-72, assoc head dept biol, 72-77, head dept biol, 77-85, PROF BIOCHEM, MASS INST TECHNOL, 67-, DEAN SCI, 85- *Mem:* Am Soc Biol Chem; Am Soc Microbiol; Am Chem Soc; Am Acad Arts & Sci; Fedn Am Soc Exp Biol. *Res:* Biosynthesis and function of vitamins and coenzymes; enzymology. *Mailing Add:* Dept Biol Sch Sci Rm 6-123 Mass Inst Technol Cambridge MA 02139

BROWN, GEORGE D, JR, b Whiting, Ind, Sept 16, 31; m 59; c 2. GEOLOGY, PALEONTOLOGY. *Educ:* St Joseph's Col, Ind, BS, 53; Univ Ill, Urbana, MS, 55; Ind Univ, PhD(geol), 63. *Prof Exp:* Jr geologist, Pan-Am Petrol Corp, 55-57; sedimentary engr, Youngstown Sheet & Tube Co, 57-58; instr geol, Colgate Univ, 58-59; lectr, Ind Univ, 62; from asst prof to prof geol, Boston Col, 62-86, chmn, Dept Geol & Geophys, 70-80, CONSULT, 88- *Concurrent Pos:* Sigma Xi grant, 66-67; NSF Grant, 68-70. *Mem:* Paleont Soc; Int Bryozool Asn; Sigma Xi. *Res:* Paleozoic Trepostomatous Bryozoa; litho stratigraphy and biostratigraphy of Ordovician strata in the Central States area. *Mailing Add:* 970 Pleasant St Framingham Center MA 01704

BROWN, GEORGE EARL, b Weaubleau, Mo, Sept 4, 06; m 35; c 1. ORGANIC CHEMISTRY. *Educ:* Cent Col, Mo, AB, 31; Iowa State Col, PhD(org chem), 41. *Prof Exp:* Instr chem & math, Cent Col, Mo, 31-32; instr chem & physics, Culver-Stockton Col, 35-36, from asst prof to prof chem, 36-56, head dept phys sci, 40-46, chmn div natural sci, 46-50, head dept chem, 46-56; prof, Southeast Mo State Col, 56-62, chmn dept chem, 58-62; coordr phys sci gen studies, 62-65, prof, 62-74, EMER PROF CHEM, SOUTHERN ILL UNIV, CARBONDALE, 74- *Mem:* Am Chem Soc. *Res:* Orientation studies with dibenzofuran organometallic studies; reactions of organolithium compounds with some organic phosphorus and nitrogen compounds. *Mailing Add:* 911 Glenview Dr Carbondale IL 62901

BROWN, GEORGE LINCOLN, b Brookings, SDak, Dec 27, 21; m 43; c 4. CHEMISTRY. *Educ:* SDak State Univ, BS, 41; Brown Univ, PhD(chem), 47. *Prof Exp:* Chemist, Standard Oil Co, La, 43-45; res chemist, Rohm and Haas Co, 47-51, head lab, 51-57, res supvr, 57-63; vpres res, Polyvinyl Chem, Inc, Peabody, 63-65; tech dir, Chem Coatings Div, Mobil Chem Co, 65-68, mgr, Coatings Res & Develop, 69-80, mgr admin, 80-85; RETIRED. *Honors & Awards:* Potter Prize, Brown Univ, 47. *Mem:* Am Chem Soc. *Res:* Electrochemistry; properties of electrolytic solutions; surface chemistry; polymers; polymer dispersion coatings. *Mailing Add:* 23 Essex Rd Scotch Plains NJ 07076

BROWN, GEORGE MARSHALL, b Rochelle, Ga, Feb 14, 21; m 62; c 1. NEUTRON CRYSTALLOGRAPHY, X-RAY CRYSTALLOGRAPHY. *Educ:* Emory Univ, BA, 42, MS, 43; Princeton Univ, MA, 46, PhD(phys chem), 49. *Prof Exp:* From asst prof to assoc prof chem, Univ Md, 47-59; chemist, Oak Ridge Nat Lab, 59-90; CONSULT, 90- *Concurrent Pos:* Res fel x-ray crystallog, Calif Inst Technol, 57-58; vis scholar protein crystallog, Univ Wash, 74-75. *Mem:* Am Chem Soc; Am Crystallog Asn. *Res:* X-ray and neutron crystal structure analysis; hydrogen atom location and hydrogen bonding; sugars; triarylamines; inorganic complex compounds; heteropoly acids; protein structure refinement; low-temperature apparatus. *Mailing Add:* PO Box 2008 Oak Ridge Nat Lab Oak Ridge TN 37831-6197

BROWN, GEORGE MARTIN, b Evanston, Ill, May 28, 19; m 43; c 5. CHEMICAL ENGINEERING. *Educ:* Univ Mich, BSE, 40; Mass Inst Technol, SM, 41; Northwestern Univ, PhD(chem eng), 49. *Prof Exp:* Asst chem engr, Sun Oil Co, Pa, 39; chem engr, Standard Oil Co, Ind, 41-43; instr chem eng, Northwestern Univ, 43-44; engr, Metall Lab, Univ Chicago, 44-45; engr, Off Sci Res & Develop & USN proj, 45-58, from instr to asst prof chem eng, 45-50, ASSOC PROF CHEM ENG, NORTHWESTERN UNIV, 50- *Concurrent Pos:* Consult Chem Engr Atomic Energy Comn & Manhattan Dist, 44-45. *Mem:* Am Chem Soc; Am Inst Chem Engrs. *Res:* Thermodynamic properties of mixtures; vapor-liquid phase equilibria; distillation; process simulation; application of computers to chemical engineering. *Mailing Add:* 2315 Lincolnwood Dr Northwestern Univ Evanston IL 60201-2048

BROWN, GEORGE STEPHEN, b Santa Monica, Calif, June 28, 45; m; c 1. PHYSICS. *Educ:* Calif Inst Technol, BS, 67; Cornell Univ, PhD(physics), 73. *Prof Exp:* Mem tech staff physics, Bell Labs, 73-77; sr res assoc physics, 77-82, PROF RES, STANFORD UNIV, 82- *Mem:* Fel Am Phys Soc. *Res:* Electron scattering; x-ray diffraction; x-ray scattering; x-ray absorption spectroscopy. *Mailing Add:* SSRL Bin 69 PO Box 4349 Stanford CA 94309

BROWN, GEORGE WALLACE, b Warrensburg, Mo, Jan 31, 39; m 64; c 2. FOREST HYDROLOGY. *Educ:* Colo State Univ, BS, 60, MS, 62; Ore State Univ, PhD(forest hydrol), 67. *Prof Exp:* Head Dept Forest Eng, 73-86, PROF FOREST HYDROL, ORE STATE UNIV, 66-, ASSOC DEAN RES, 86- *Concurrent Pos:* Consult hydrologist, Weyerhaeuser Co, 73; prin res engr, Forest Serv, USDA, 81. *Mem:* Soc Am Foresters; Am Geophys Union; Sigma Xi; AAAS. *Res:* Water quality prediction on small forested streams; temperature, sediment, dissolved oxygen. *Mailing Add:* Col Forestry Ore State Univ Corvallis OR 97331-5704

BROWN, GEORGE WILLARD, JR, b Alameda, Calif, Oct 24, 24; m 48, 61; c 6. BIOCHEMICAL ECOLOGY, ENVIRONMENTAL CHEMISTRY. *Educ:* Univ Calif, Berkeley, BS, 50, MA, 51, PhD(comp biochem), 55. *Prof Exp:* Fel, NIH, 56; res fel & res assoc physiol chem, Univ Wis, 57, from instr to asst prof, 58-61; from asst prof to assoc prof biochem, Med Br, Univ Tex, 61-67; from assoc prof to prof, 67-90, EMER PROF FISHERIES, COL OCEAN & FISHERY SCI, UNIV WASH, 90- *Concurrent Pos:* Consult, NIH, 58, Acad Press, 62-, NASA, 66-67, Highline Col, 67, US Coast Guard, 70-73, Envirodyne, Inc, 77-78, US Fish & Wildlife Serv, 80 & Res & Eval Assocs, Inc, 84-86; vis biologist, Am Inst Biol Sci, 62-73; assoc prog mem, Nat Res Coun, Nat Acad Sci, 80-81. *Honors & Awards:* Belg Fourragere. *Mem:* Am Chem Soc; Sigma Xi; Am Soc Biochem & Molecular Biol; Soc Protection Old Fishes (pres, 67-). *Res:* Intermediary nitrogen metabolism; biochemistry of primitive fishes; enzymology; desert biology; biochemical ecology; water pollution. *Mailing Add:* Sch Fisheries WH-10 Univ Wash Seattle WA 98195

BROWN, GEORGE WILLIAM, b Boston, Mass, June 3, 17; m 41; c 5. MATHEMATICS. *Educ:* Harvard Univ, AB, 37, AM, 38; Princeton Univ, PhD(math), 40. *Prof Exp:* Res statistician, R H Macy & Co, 40-42; res assoc, Princeton Univ, 42-44; res engr, RCA Labs, Princeton, NJ, 44-46; from assoc prof to prof math statist, Iowa State Col, 46-48; mathematician, Rand Corp, 48-52; sr staff engr, Int Telemeter Corp, 52-57; prof bus admin & eng, Univ Calif, Los Angeles, 57-67, dir, Western Data Processing Ctr, 57-64, chmn dept bus admin, 64-65; dean grad sch admin, 67-72, PROF, 67-, EMER PROF ADMIN, UNIV CALIF, IRVINE. *Concurrent Pos:* Dir & consult, Dataproducts Inc, 62-88; dir, Comput Automation Inc, 69-81; dir, Calcomp, 77-80. *Mem:* Fel AAAS; fel Am Statist Asn; fel Inst Math Statist. *Res:* Dynamic decision processes; mathematical and applied statistics; management science; operations research. *Mailing Add:* 2401 Bamboo St Newport Beach CA 92660

BROWN, GERALD E, b Brookings, SDak, July 22, 26; m 53; c 3. PHYSICS. *Educ:* Univ Wis, BA, 46; Yale Univ, MS, 48, PhD, 50; Univ Birmingham, DSc, 57. *Hon Degrees:* DSc, Univ Helsinki, 82 & Univ Birmingham, Eng, 90. *Prof Exp:* Prof physics, Univ Birmingham, 59-60; prof, Nordic Inst Theoret Atomic Physics, 60-85; leading prof, 74-88, DISTINGUISHED PROF PHYSICS, STATE UNIV NY, STONY BROOK, 88- *Concurrent Pos:* Lectr math physics, 55-58, reader, Univ Birmingham, 58-59; prof physics, Princeton Univ, 64-68; prof, State Univ NY, Stony Brook, 68-74; dir nuclear astrophys sect, NSF Inst Theoret Physics, Univ Calif, Santa Barbara, 60; chmn, Gordon Res Conf Nuclear Struct Physics, 80; mem, NSF adv panel physics, 74-76, nuclear sci adv comt, 77-, mem vis comt, Theoret Div, Los Alamos Sci Lab, 76-79; Alexander von Humboldt Found sr distinguished scientist award, 87. *Honors & Awards:* Boris Pregel Award, NY Acad Sci, 76; Tom W Bonner Prize Nuclear Physics, 82. *Mem:* Nat Acad Sci; fel Am Phys Soc; fel Am Acad Arts & Sci; hon mem Finnish Soc Sci & Lett. *Res:* Author of over 240 technical publications. *Mailing Add:* Inst Theoret Physics State Univ NY Stony Brook NY 11794

BROWN, GERALD LEONARD, b New York, NY, May 17, 36; m 67; c 1. APPLIED MATHEMATICS. *Educ:* Univ Miami, BS, 58; Mass Inst Technol, MS, 60; Univ Wis, PhD(math), 65. *Prof Exp:* Tech staff mem, Math Res Dept, Sandia Corp, 65-70; RES STAFF MEM, SYSTS EVAL DIV, INST DEFENSE ANALYSIS, 71- *Mem:* Am Math Soc; Opers Res Soc Am. *Res:* Partial differential equations; systems analysis; operational testing and evaluation; command and control analysis. *Mailing Add:* SED-Inst Defense Analysis 1801 N Beauregard St Alexandria VA 22311

BROWN, GERALD RICHARD, b Poplar Bluff, Mo, Oct 22, 37; m 59; c 2. POMOLOGY, VITICULTURE. *Educ:* Univ Ark, BS, 59, MS, 63, PhD(bot), 74. *Prof Exp:* Res asst hort, Southwest Br Exp Sta, Univ Ark, Hope, 60-61; plant quarantine inspector, Agr Res Serv, USDA, 61-69; res asst genetics, Hort Dept, Univ Ark, 69-74; res pomologist, State Fruit Exp Sta, Southwest Mo State Univ, 74-78; EXTEN PROF POMOL, RES & EDUC CTR, UNIV KY, 78- *Concurrent Pos:* Adv estab demonstration fruit farm, Ecuador, 84; vis prof, Scottish Crop Res Inst, Invergorwie, Dundee, Scotland, Great Britain. *Honors & Awards:* Krezdorn Award, Am Soc Hort Sci, 74. *Mem:* Am

Soc Hort Sci; Sigma Xi. *Res:* Fruit production including the basic relationship of plant growth; especially the use of growth regulators, cultivar evaluation, nutrition, pruning, and irrigation; water relations. *Mailing Add:* Res & Educ Ctr Univ of Ky PO Box 469 Princeton KY 42445

BROWN, GILBERT J, b Bronx, NY, Apr 29, 48; m 69; c 2. TECHNOLOGY & VALUES. *Educ:* Cornell Univ, BS, 69; Mass Inst Technol, MS, 72, PhD(nuclear eng), 74. *Prof Exp:* PROF NUCLEAR ENG, UNIV MASS, LOWELL, 73- *Concurrent Pos:* Vis assoc, Brookhaven Nat Lab, 77; resident assoc, Argonne Nat Lab, 78; consult, Stone & Webster Eng Corp, 81-82, Int Atomic Energy Agency, 84-85 & Yankee Atomic Elec Co, 89-90; vis assoc prof nuclear eng, Mass Inst Technol, 81-82; reactor engr, NH Yankee-Seabrook Reactor, 86; dir, Am Nuclear Soc, 86-89. *Mem:* Am Nuclear Soc; Am Soc Eng Educ; Sigma Xi. *Res:* Nuclear reactor systems; reactor safety; thermal hydraulics; energy economics; cogeneration and alternative energy systems; technology and values; understanding technological risks. *Mailing Add:* Univ Mass One University Ave Lowell MA 01854

BROWN, GILBERT MORRIS, b Valdasta, Ga, Nov 5, 47; m 84; c 3. ELECTROCHEMISTRY, PHOTOCHEMISTRY. *Educ:* Erskine Col, BA, 70; Univ NC, Chapel Hill, PhD(chem), 74. *Prof Exp:* Res assoc, Chem Dept, Stanford Univ, 74-76 & Brookhaven Nat Lab, 76-78; CHEMIST, OAK RIDGE NAT LAB, 78- *Mem:* Am Chem Soc; Int Am Photochem Soc; Electrochem Soc; AAAS. *Res:* Photochemistry of transition metal complexes; kinetics and mechanisms of the reactions of transition metal complexes; materials chemistry; mechanisms of electrode reactions; preparation of refractory and ceramic materials from metallo-organic precursors. *Mailing Add:* Chem Div Oak Ridge Nat Lab PO Box 2008 Oak Ridge TN 37831-6119

BROWN, GLEN FRANCIS, b Graysville, Ind, Dec 14, 11; m 75; c 1. GEOLOGY. *Educ:* NMex Sch Mines, BS, 35; Northwestern Univ, MS, 40, PhD(geol), 49. *Prof Exp:* Geologist, Philippine Bur Mines, 36-38; jr geologist, US Geol Surv, 38-41, from asst geologist to sr geologist, 41-44; sr geologist, Foreign Econ Admin, 44-46; geologist, US Geol Surv, 46-48, actg chief, Mission to Thailand, 49, chief field party, Saudi Arabia, 50-54, chief, Saudi Arabian Proj, Wash, DC, 55-57; geol adv, Kingdom Saudi Arabia, 57-58; mem, World Bank Mission to Saudi Arabia, 60 & Kuwait, 62; chief field party, 63-69, sr staff geologist for mideastern affairs, 69-82, CONSULT, US GEOL SURV, 82- *Concurrent Pos:* US Geol Surv deleg, 22nd & 23rd World Geol Congs; mem, Joint Comn to Saudi Arabia, 74-76. *Honors & Awards:* Distinguished Serv Gold Medal, US Dept Interior, 64. *Mem:* Soc Econ Geologists; fel Geol Soc Am; Am Geophys Union; Am Asn Petrol Geologists; Soc Photogram; fel AAAS; Sigma Xi; fel Explorers Club. *Res:* Geology of metals and coal in the Philippine islands, Thailand and South China; geology of ground water, Mississippi, California, and Saudi Arabia; Precambrian geology in Saudi Arabia; tectonism of Arabian peninsula. *Mailing Add:* Apt 21C 2031 Royal Fern Court Reston VA 22091

BROWN, GLENN A, GEOLOGY. *Educ:* Univ Calif Los Angeles, BA. *Prof Exp:* Investr, Dept Water Resource, 52-62; PRIN HYDROGEOLOGIST & SR VPRES, LAW ENVIRONMENTAL INC, 66- *Concurrent Pos:* Prin geotech consult. *Mailing Add:* 332 N San Franando Blvd Burbank CA 91504

BROWN, GLENN HALSTEAD, b Logan, Ohio, Sept 10, 15; m 43; c 4. CHEMISTRY. *Educ:* Ohio Univ, BS, 39; Ohio State Univ, MS, 41; Iowa State Col, PhD, 51. *Hon Degrees:* DSc, Bowling Green State Univ, 72. *Prof Exp:* Instr chem, Univ Miss, 41-42, asst prof, 43-50; asst prof, Univ Vt, 50-52; from asst prof to assoc prof, Univ Cincinnati, 52-60; prof, Kent State Univ, 60-68, chmn dept, 60-65, dean res, 63-69, dir, Liquid Crystal Inst, 65-83; regents prof chem, 68-85, EMER REGENTS PROF CHEM & EMER DIR, KENT STATE UNIV, 85- *Concurrent Pos:* Instr, Iowa State Col, 45-48; Sigma Xi nat lectr, 70-71; ed, Molecular Crystals & Liquid Crystals; mem, 8th Int Liquid Crystal Conf, Tokyo, 80; fel, Glenn H Brown Liquid Crystal Inst, Kent State Univ, 85- *Honors & Awards:* Morley Award, 77; Bikerman Award, 81; Citation for Res, 9th Int Liquid Crystal Conf, Bangalore, India, 82. *Mem:* Fel AAAS; fel Am Inst Chemists; Am Chem Soc; Am Crystallog Asn; NY Acad Sci; fel Liquid Crystal Inst. *Res:* Structures of solutions; x-ray diffraction; coordination compounds; structure and properties of liquid crystals; role of liquid crystals in life processes; photochromism; structure of concentrated salt solutions. *Mailing Add:* 470 Harvey Ave Kent OH 44240

BROWN, GLENN LAMAR, b Miami, Ariz, Oct 14, 23; m 58; c 4. ELECTRONIC SYSTEMS, LASER TECHNOLOGY DEVELOPMENT. *Educ:* Univ Calif, Los Angeles, BA, 49, MA, 51, PhD(physics), 55. *Prof Exp:* Res assoc geophysics, Univ Calif, Los Angeles, 55-57; staff mem, Ramo-Woolridge Corp, 57-59; div mgr, Space Gen Corp, 59-65; dir res, Am Nucleonics Corp, 65-71; PRES, RESOURCES ENG & PLANNING, 71- *Mem:* Am Phys Soc; Am Geophys Union; Inst Elec & Electronics Engrs. *Res:* Electromagnetics; electronic warfare; laser effects and countermeasures; electromagnetic pulse effects and countermeasures; high power microwave effects; thin film applications to electron counter-countermeasures; multiple quantum well devices. *Mailing Add:* Resource Engineering & Planning Co 5937 Via Robles Lane El Paso TX 79912

BROWN, GLENN R(OBBINS), b Greensburg, Pa, June 1, 30; m 52; c 2. CHEMICAL ENGINEERING. *Educ:* Pa State Univ, BS, 52; Case Inst Technol, MS, 54, PhD, 58. *Prof Exp:* Engr, Standard Oil Co, 53-55, sr engr, 55-56, tech specialist, 56-59, group leader, 59-61, asst mgr mgt methods, 61-63, mgr mgt sci & systs, 63-64; admin, Sohio Chem Co, 64-69; mgr res & develop, Standard Oil Co, 69-75, vpres res & eng, 75-78, vpres technol & planning, 78-79, sr vpres, coal, minerals & planning, 79-80, sr vpres minerals, res & planning, 80-83, dir, 81-86, sr vpres technol, 83-86; dir, strategic planning, 86-87, dean cols, 87-90, VPROVOST, CORP RES & TECHNOL TRANSFER, CASE WESTERN RESERVE UNIV, 90- *Concurrent Pos:* Dir, Nordson Co, 86-, Soc Nat Bank, 86-, Ferro Corp, 88- *Mem:* Am Chem Soc; Am Inst Chem Engrs. *Res:* Petroleum and petrochemical processing; management science; economic evaluation; planning. *Mailing Add:* 150 Greentree Rd Chagrin Falls OH 44022-2424

BROWN, GORDON EDGAR, JR, b San Diego, Calif, Sept 24, 43; m 65; c 2. MINERALOGY, CRYSTALLOGRAPHY. *Educ:* Millsaps Col, BS, 65; Va Polytech Inst & State Univ, MS, 68, PhD(mineral), 70. *Prof Exp:* Res assoc mineral, State Univ NY Stony Brook, 70-71; asst prof, Princeton Univ, 71-73; from asst prof to assoc prof mineral, 73-85, PROF MINERAL & CRYSTALLOG, STANFORD UNIV, 85-, CHMN DEPT GEOL, 86- *Concurrent Pos:* Assoc ed, Am Mineralogist, Mineral Soc Am, 75-79; co-dir, Ctr Mat Res, Stanford Univ, 86. *Mem:* Mineral Soc Am; Am Crystallog Asn; Mineral Soc Can. *Res:* Crystal chemistry of the rock-forming minerals; silicate glasses and melts; high temperature of minerals; paragenesis of pegmatite minerals; mineral surfaces; synchrotron radiation research. *Mailing Add:* Dept Geol Stanford Univ Stanford CA 94305

BROWN, GORDON ELLIOTT, b Ellensburg, Wash, Sept 13, 36. MATHEMATICS. *Educ:* Calif Inst Technol, BS, 58; Cornell Univ, PhD(math), 63. *Prof Exp:* From instr to asst prof math, Univ Ill, Urbana, 62-66; asst prof, 66-69, ASSOC PROF MATH, UNIV COLO, BOULDER, 69- *Mem:* Am Math Soc. *Res:* Non-associative algebras; Lie algebras. *Mailing Add:* Univ Math Campus Box 426 Univ Colo Boulder CO 80309

BROWN, GORDON MANLEY, b Morpeth, Ont, Mar 17, 33; m 64; c 2. ORGANIC CHEMISTRY. *Educ:* Univ Western Ont, BSc, 54, MSc, 56; Laval Univ, DSc(chem), 59; Univ Montpellier, DSc(chem), 61. *Prof Exp:* Nat Res Coun overseas fel, Univ Montpellier, 59-61; lectr, 61-62, from asst prof to assoc prof, 62-75, secy fac sci, 70-76, PROF CHEM, UNIV SHERBROOKE, 75- *Concurrent Pos:* Sabbatical leave, Nat Ctr Sci Res Labs, France, 77-78. *Mem:* Chem Inst Can. *Res:* Configurational and conformational studies of decaline and cyclohexane derivatives; hetero diels-alder reactions. *Mailing Add:* Dept Chem Fac Sci Univ Sherbrooke 2500 Univ Blvd Sherbrooke PQ J1K 2R1 Can

BROWN, GORDON S, ENGINEERING. *Prof Exp:* RETIRED. *Mem:* Nat Acad Eng. *Mailing Add:* 6301 N Calle de Adeltta Tucson AZ 85718

BROWN, GRAYDON L, b Newkirk, Okla, Aug 7, 24; m 54; c 4. SYSTEMS ENGINEERING, ELECTRICAL ENGINEERING. *Educ:* Okla State Univ, BS, 48, MS, 52. *Prof Exp:* Asst proj engr, res found, Okla State Univ, 48-51; res engr, 52-60, res group leader, res & develop dept, 60-83, dir, tech & admin serv sect, Conoco, Inc, 83-86; CONSULT ENGR, 86- *Mem:* Inst Elec & Electronics Engrs; Soc Explor Geophys; Audio Eng Soc; Nat Soc Prof Engrs. *Res:* Exploration; geophysical instrumentation; hydraulic servomechanisms; magnetic recording instrumentation. *Mailing Add:* 1528 Autumn Rd Ponca City OK 74604

BROWN, GREGORY GAYNOR, b Englewood, NJ, Aug 17, 48; m 77; c 1. MITOCHONDRIA, PLANT ORGANELLES. *Educ:* Univ Notre Dame, BS, 70; Mt Sinai Med Sch, PhD(biochem), 77. *Prof Exp:* Res assoc biochem, State Univ NY, Stony Brook, 77-79, fel, 79-81; ASST PROF BIOL, MCGILL UNIV, 81- *Concurrent Pos:* Prin investr, Nat Sci Eng Res Coun, 81- *Mem:* Can Soc Cell Biol; Int Soc Plant Molecular Biol; Am Soc Plant Physiologists. *Res:* The structure, function, expression and evolution of genetic elements in mitochondria, particularly those in higher plants; cytoplasmic male sterility in plants; genetic engineering. *Mailing Add:* Dept Biol McGill 1205 Ave Dr Penfield Montreal PQ H3A 1B1 Can

BROWN, GREGORY NEIL, b Detroit, Mich, Feb 10, 38; m 61, 74; c 5. TREE PHYSIOLOGY, CRYOBIOLOGY. *Educ:* Iowa State Univ, BS, 59; Yale Univ, MF, 60; Duke Univ, DF, 63. *Prof Exp:* Plant physiologist, Oak Ridge Nat Lab, 63-66; from asst prof to prof forest physiol, Univ Mo-Columbia, 66-77; prof forest physiol, Iowa State Univ, 77-78; prof forest resources & head dept, Univ Minn, 78-83; dean forest resources & assoc dir Maine Agr Exp Sta, 83-87, VPRES RES & PUB SERV, UNIV MAINE, ORONO, 87- *Concurrent Pos:* Mem, Nat Tree Physiol Comt, 68-72; NSF grant, 69-71; working group leader, Int Union Forest Res Orgns, 71-83; mem adv bd, Forest Sci, 77-79, ed, 79-82; chair tree physiol working group, Soc Am Foresters, 83-84. *Mem:* Soc Am Foresters; Japanese Soc Plant Physiol; Am Soc Plant Physiologists. *Res:* Membrane structure and function and protein characterization during the induction of cold hardiness in woody plant species. *Mailing Add:* Res & Pub Serv Univ Maine 201 Alumni Hall Orono ME 04469

BROWN, GUENDOLINE, b New Albany, Miss, July 13, 36; m 58; c 3. NUTRITION. *Educ:* Miss State Col Women, BS, 58; Univ Wyo, MS, 69; Utah State Univ, PhD(nutrit & food sci), 78. *Prof Exp:* Asst prof home econ, Univ Wyo, 69-72; asst prof food & nutrit, Weber State Col, 72-74; asst prof food sci & nutrit, Colo State Univ, 78-81; CHAIR & ASSOC PROF FOOD & NUTRIT, NDAK STATE UNIV, 81- *Mem:* Soc Nutrit Educ; Sigma Xi; Inst Fodd Technol. *Res:* Nutrition education; coronary heart disease; nutrient density. *Mailing Add:* 351 Home Econ Bldg NDak State Univ Fargo ND 58105

BROWN, HARLEY PROCTER, b Uniontown, Ala, Jan 13, 21; m 42, 89; c 1. INVERTEBRATE ZOOLOGY, ENTOMOLOGY. *Educ:* Miami Univ, AB & AM, 42; Ohio State Univ, PhD(zool), 45. *Prof Exp:* Asst zool, Ohio State Univ, 42-45; instr, Univ Idaho, 45-47 & Queen's Col, NY, 47-48; from asst prof to prof, 48-84, EMER PROF ZOOL, UNIV OKLA, 84- *Concurrent Pos:* Partic, NSF Inst Marine Biol & Trop Ecol, 64; cur invert, Stovall Mus of Sci & Hist, Okla Mus Nat Hist. *Mem:* Am Soc Zool; AAAS; Am Micros Soc (pres, 75); Am Inst Biol Sci; Entom Soc Am; North Am Benthol Soc; Sigma Xi. *Res:* Aquatic; ecology, life history and systematics of dryopoid beetles, riffle beetles, and water pennies. *Mailing Add:* Dept Zool Univ Okla Norman OK 73019

BROWN, HAROLD, b New York, NY, Sept 19, 27; m 53; c 2. PHYSICS. *Educ:* Columbia Univ, AB, 45, MA, 46, PhD(physics), 49. *Hon Degrees:* 11 honorary degrees. *Prof Exp:* Lectr & mem sci staff, Columbia Univ, 47-50; physicist, Lawrence Radiation Lab, Univ Calif, Berkeley, 50-52, mem staff, Livermore, 52-53, group leader, 53-55, div leader, 55-58, assoc dir, 58-59, from dep dir to dir, 59-61; dir defense res & eng, Off Secy Defense, 61-65, Secy Air Force, 65-69; pres, Calif Inst Technol, 69-76; Secy Defense, DC, 77-81; BUS CONSULT, 81- *Concurrent Pos:* Lectr physics, Stevens Inst Technol, 49-50; mem, Polaris Steering Comt, 56-58; adv, US Deleg Conf Experts Detection Nuclear Weapons Tests, Geneva, 58; sr sci adv, US Deleg Conf Discontinuance Nuclear Weapons Tests, 58-59; mem sci adv comt ballistic missiles, Secy Defense, 58-61; mem, President's Sci Adv Comt, 61; deleg, Strategic Arms Limitation Talks, Helsinki, Vienna & Geneva, 69-75; mem gen adv comt, Arms Control & Disarmament Agency, 69-76; mem exec comt, Trilateral Comn, 73-76; chmn, Technol Assessment Adv Coun, 74-75; Distinguished vis prof, Sch Advan Int Studies, Johns Hopkins Univ, 81-84, chmn, Foreign Policy Inst, 84- *Honors & Awards:* Numerous awards including the Presidential Medal of Freedom, 81. *Mem:* Nat Acad Eng; Nat Acad Sci; Am Acad Arts & Sci; Am Phys Soc. *Res:* Nuclear and neutron physics; nuclear explosives and reactor design; weapons systems; management of research and development; technology and arms control. *Mailing Add:* Johns Hopkins Univ 1619 Massachusetts Ave Washington DC 20036-2297

BROWN, HAROLD, b Lawrence, Mass, Sept 17, 17; m 50; c 3. MEDICINE. *Educ:* Harvard Univ, AB, 39, MD, 43; Am Bd Internal Med, dipl, 50, recert, 73. *Prof Exp:* Asst med, Harvard Med Sch, 46-47; from instr to assoc prof, Sch Med, Univ Utah, 47-61; head metab sect & dir clin res ctr, Baylor Col Med, 62-75, prof med, 62-84, actg chmn dept, 69-70, vchmn dept med, 70-75, dep chmn, 75-84; CLIN PROF MED, UNIV NMEX, COL MED, 85- *Concurrent Pos:* From asst chief to chief med serv, Vet Admin Hosp, Salt Lake City, Utah, 48-61; vis prof, Univ Med Sci, Bangkok, 57-59, vis prof med, W China Univ Med Sci, 84-85, hon prof, 85-; officer, intern & resident, Boston City Hosp, 43-47; external exam med, Univ Hong Kong, 58; consult & dir metab res lab, Vet Admin Hosp, Houston, 62-84. *Mem:* Endocrine Soc; Am Fedn Clin Res; Am Col Physicians; Am Soc Clin Invest. *Res:* Metabolism of corticosteroids; serotonin metabolism; liver disease. *Mailing Add:* Rte 9 Box 90-14 Santa Fe NM 87505

BROWN, HAROLD DAVID, b Mishawaka, Ind, July 12, 40; m 58; c 2. APPLIED MATHEMATICS, COMPUTER SCIENCES. *Educ:* Univ Notre Dame, MSc, 63; Ohio State Univ, PhD(math), 66. *Prof Exp:* Asst to chmn dept math & asst dir, NSF Training Progs Math, Univ Notre Dame, 60-63; instr, Ohio State Univ, 63-66, asst prof, 67; vis mem, Courant Inst Math Sci, 67-68; from asst prof to assoc prof, Ohio State Univ, 68-73; assoc prof comput sci, Stanford Univ, 73-77; SR SCIENTIST, NASA-AMES INST ADVAN COMPUT, 77- *Concurrent Pos:* Dir, NSF Sec Sci Training Prog, Ohio State Univ, 64-71; vis prof, RWTH, Aachen, 71, 73 & 75. *Mem:* Am Math Soc; Math Asn Am; Asn Comput Mach. *Res:* Computational algebra and graph theory; biochemical structure determination; computer algorithms. *Mailing Add:* NASA-Ames Res Ctr Moffett Field CA 94035-2112

BROWN, HAROLD HUBLEY, b Portsmouth, NH, Mar 14, 26; m 57; c 3. CLINICAL CHEMISTRY. *Educ:* Marietta Col, AB, 50; Wayne State Univ, PhD(biochem), 54. *Prof Exp:* Instr biochem, Albany Med Col, 54-56; biochemist, Pawtucket Mem Hosp, RI, 56-59 & Harrisburg Polyclin Hosp, 59-64; asst prof biochem & dir Core Lab Gen Clin Res Ctr, Albany Med Col, 64-65; biochemist, Harrisburg Polyclin Hosp, Pa, 65-70; clin chemist, Berkshire Med Ctr, Pittsfield, Mass, 70-73; ASSOC PROF BIOCHEM, ALBANY MED COL, 73-; DIR, DEPT CLIN CHEM, ALBANY MED CTR HOSP, 73- *Mem:* AAAS; Am Chem Soc; Am Asn Clin Chemists. *Res:* Methodology in clinical chemistry. *Mailing Add:* 28 Sherwood Dr Lenox MA 01240-2346

BROWN, HAROLD MACK, JR, b Salt Lake City, Utah, Mar 6, 36; m 59; c 3. VISUAL PHYSIOLOGY, BIOPHYSICS. *Educ:* Univ Utah, BS, 58, PhD(psychol, molecular biol), 64. *Prof Exp:* Res assoc, Scripps Inst Oceanog, 67-69; asst res prof physiol, Sch Med, Univ Calif, Los Angeles, 69-70; asst prof physiol, 70-72, assoc prof, 72-80, ASST PROF MED, SCH MED, UNIV UTAH, 71-, PROF PHYSIOL, 80- *Concurrent Pos:* Vet Admin fel neurol, Col Med, Univ Utah, 64-66; NIH spec fel, Scripps Inst Oceanog, 67-69. *Honors & Awards:* Flanagan Award, Am Inst Res, 64. *Mem:* AAAS; Am Physiol Soc; Soc Gen Physiol; Biophys Soc; Am Soc Photobiol; Fedn Am Soc Exp Biol. *Res:* Visual biophysics and membrane biophysics. *Mailing Add:* Dept Physiol Res Park Rm 128 Univ Utah Med Ctr 410 Chipeta Way Salt Lake City UT 84108

BROWN, HAROLD PROBERT, organic polymer chemistry; deceased, see previous edition for last biography

BROWN, HAROLD VICTOR, b Los Angeles, Calif, Nov 5, 18; m 46; c 2. ENVIRONMENTAL HEALTH, INDUSTRIAL HYGIENE. *Educ:* Univ Calif, Los Angeles, BA, 40, MPH, 65, DrPH, 70; Am Intersoc Acad Cert Sanit & Am Am Bd Indust Hyg, dipl. *Prof Exp:* Chemist, Eng Corps, US War Dept, 41 & 46; res chemist, Pioneer-Flintkote Co, 47; indust hygienist, State Dept Health, Calif, 47-52; indust hygienist, 52-65, Univ Calif, Los Angeles, environ health & safety officer, Campus Community Safety, 65-82, lectr, Sch Pub Health, 77-82; MEM HEARING BD, SOUTHCOAST AIR QUAL MGT DIST, 85- *Concurrent Pos:* Mem, Calif Occup Safety & Health Standards Bd, 74-76. *Mem:* AAAS; Am Indust Hyg Asn; Health Physics Soc; fel Am Pub Health Asn; Conf Govt Indust Hygienists. *Res:* Cholinesterase inhibitors; laboratory safety. *Mailing Add:* 6008 Chariton Ave Los Angeles CA 90056

BROWN, HARRY ALLEN, b New York, NY, Apr 26, 25. MAGNETISM. *Educ:* Univ Wis, PhD(physics), 54. *Prof Exp:* Instr physics, Oberlin Col, 54-55; asst prof, Miami Univ, 55-59; assoc prof, St John's Univ, 59-61; sr physicist, Lab Appl Sci, Univ Chicago, 61-63; Fulbright lectr, Univ Sao Paulo, 63-64; assoc prof, San Francisco State Col, 64-65; assoc prof, 65-70, PROF PHYSICS, UNIV MO-ROLLA, 72- *Concurrent Pos:* Consult, Nat Carbon Co, Ohio, 55-58 & Am Mach & Foundry Co, Conn, 60-61; Fulbright lectr, Univ Sao Paulo, 71-72. *Mem:* Am Phys Soc; Am Asn Physics Teachers; AAAS. *Res:* Solid state physics; magnetic properties of solids; statistical mechanics; phase transitions. *Mailing Add:* Dept Physics Univ Mo Rolla MO 65401

BROWN, HARRY BENJAMIN, b Chicago, Ill, Apr 11, 13; m 39; c 2. ELECTRONICS ENGINEERING. *Educ:* Auburn Univ, BS, 38; Kans Univ, MS, 52. *Prof Exp:* Engr mining, US Steel, Wylam, Ala, 39-40; engr elec, Dept Interior, US Govt, 40-42; instr radio & TV, Midwest Radio Inst, 45-48; engr crystal, Com Equip Co, 48-50; consult, Lawrence Eng Co, 52-53, Aeronaut Radio Co, 53-63; CHIEF ENGR, APPL ELECTRO MECH INC, 63-, VPRES ENGR, 65- *Res:* High level sound generation and transmission. *Mailing Add:* Appl Electro Mech Inc Rte 28 & Rock Hall Rd Point of Rocks MD 21777

BROWN, HARRY DARROW, b Newark, NJ, July 21, 25; m 45; c 3. BIOCHEMISTRY. *Educ:* Long Island Univ, BS, 50; Columbia Univ, AM, 52, PhD(bot), 57. *Prof Exp:* Asst bot, Columbia Univ, 50-53; lectr biol, Hunter Col, 54-56; asst prof plant physiol, Loyola Univ, 57-61; assoc prof, Southern Ill Univ, 61-63; biochemist, Seed Protein Lab, USDA, 63-64; asst prof biochem, Univ Tex Med Br, 64-68; assoc prof, Sch Med, Univ Mo-Columbia, 68-74; PROF BIOCHEM & DEAN RES, COOK COL, RUTGERS UNIV, NEW BRUNSWICK, 74-; ASSOC DIR, NJ AGR EXP STA, 74- *Concurrent Pos:* Collabr, USDA, 63-64; chmn biochem sect & assoc dir, Cancer Res Ctr, 68-74. *Mem:* Am Asn Cancer Res; fel Am Inst Chemists; Biophys Soc; Biochem Soc. *Res:* Particulate enzymes; biochemical calorimetry; interfacial biochemical reactions. *Mailing Add:* Dept Biochem Rutgers Univ New Brunswick NJ 08903

BROWN, HARRY LESTER, b Brimfield, Mass, June 29, 24; m 46; c 3. MATERIALS SCIENCE & ENGINEERING. *Educ:* Colo Sch Mines, MetE, 50, MSc, 51. *Prof Exp:* Engr, Phillips Petrol Co, Kans, 51-54; staff mem, Los Alamos Sci Lab, 54-67; staff metallurgist, water reactor safety prog off, Atomic Energy Div, Phillips Petrol Co, Idaho, 67; sr scientist, Water Reactor Safety Prog Off, Idaho Nuclear Corp, 67-73; proj engr, Mat Technol Br, Aerojet Nuclear, 73-75, staff specialist, 75-77; staff specialist documentation fuels & mat div, 77-80, sr eng tech reviewer/editor, Mat Sci div, EG&G Idaho, Inc, 80-86; CONSULT, 86- *Res:* Elastic properties of metals, graphites, refractory metal carbides and composites. *Mailing Add:* 1867 Michael St Idaho Falls ID 83402-1735

BROWN, HELEN BENNETT, b Greenwich, Conn, Oct 6, 02; m 28; c 2. NUTRITION, CARDIOVASCULAR DISEASES. *Educ:* Mt Holyoke Col, BA, 24; Yale Univ, PhD(biochem), 30. *Hon Degrees:* ScD, Mt Holyoke Col, 74. *Prof Exp:* Res chem technician, Dept Pediat, New Haven Hosp, 24-28; res asst biochem, Babies & Children's Hosp, Western Reserve Univ, 28-31, instr bact, Sch Med, 42-44; res assoc biochem, Ben Venue Labs, Inc, 44-47; asst, Res Div, Cleveland Clin, 48-61, mem staff, 62-68, EMER RES CONSULT BIOCHEM & NUTRIT, CLEVELAND CLIN FOUND, 68- *Concurrent Pos:* Instr, Cleveland City Health Dept, 42-44; mem coun arteriosclerosis & coun epidemiol, Am Heart Asn; mem, Nat Diet-Heart Study Comt, 60-68; mem adv comt, Epidemiol & Biomet Sect, Nat Heart & Lung Inst, 69-71 & consult nutrit, Multiple Risk Factor Intervention Trial, 72-82; sci dir coop screening prog, Cleveland Chap, Am Heart & Diabetes Asn Greater Cleveland, 70-72; mem adv comt, Human Nutrit, USDA, 80. *Mem:* Am Heart Asn; Am Inst Nutrit; hon mem Am Dietetic Asn; Sigma Xi; NY Acad Sci. *Res:* Health problems in the general population and in the community; diet and heart disease; nutrition of the elderly. *Mailing Add:* Cleveland Clin Found 9500 Euclid Ave Cleveland OH 44106

BROWN, HENRY, b Erie, Pa, Feb 20, 20; m 45; c 5. SURGERY. *Educ:* Univ Mich, AB, 41; Univ Pa, MD, 44. *Prof Exp:* Clin assoc, 63-69, asst clin prof surg, 69-77, ASSOC CLIN PROF SURG, HARVARD MED SCH, 77- *Concurrent Pos:* Consult surgeon & mem clin res ctr, Mass Inst Technol; dir hand surg serv, New Eng Deaconess Hosp; mem surg staff, Cambridge Hosp & Mt Auburn Hosp; hand surg consult, Ganta Hosp & Leprosarium Ganta, Liberia, WAfrica, 69, 71, 73 & 75; asst dir, Harvard Surg Unit, Boston City Hosp, 72-73. *Mem:* Soc Surg Alimentary Tract; Am Soc Surg of Hand; Am Physiol Soc; Am Chem Soc; Am Col Surg; Sigma Xi. *Res:* Surgical physiology of the liver; surgery of the hand. *Mailing Add:* 21 Southwick Rd Waban MA 02168

BROWN, HENRY CLAY, b Gainesville, Fla, Oct 8, 48; m 74; c 2. ORGANIC CHEMISTRY. *Educ:* Emory Univ, BSc, 70; Univ Ga, PhD(chem), 77. *Prof Exp:* SR RES CHEMIST, MONSANTO CO, 76- *Mem:* Am Chem Soc; Sigma Xi. *Res:* Homogeneous and heterogeneous catalysis; new synthetic methods; process design and development; organic electrochemical synthesis. *Mailing Add:* Monsanto Co - T3D 800 N Lindbergh Blvd St Louis MO 63166

BROWN, HENRY CLAY, III, b Carbur, Fla, Mar 7, 19; m 44; c 3. ORGANIC CHEMISTRY. *Educ:* Duke Univ, BS, 41, MA, 42; Univ Fla, PhD(chem), 50. *Prof Exp:* Res chemist, Texas Co, 42-47; prof chem & chem eng, Univ Fla, 52-81; RETIRED. *Mem:* Am Chem Soc; Am Inst Chem Engrs; Am Inst Chemists; Soc Plastics Engrs. *Res:* Fluorine chemistry and chemistry of fluorocarbon derivatives. *Mailing Add:* 2910 NW 13th Ct Gainesville FL 32605-5001

BROWN, HENRY SEAWELL, b Marion, NC, Mar 4, 30; m 51; c 5. ECONOMIC GEOLOGY. *Educ:* Berea Col, BA, 52; Univ Ill, MS, 54, PhD(geol), 58. *Prof Exp:* From asst prof to assoc prof geol, Berea Col, 55-58, head dept geol & geog, 55-58; from asst prof to prof geol, NC State Univ, 58-89, head dept, 86-89; CONSULT, 89- *Concurrent Pos:* Consult, Mineral Res & Develop Corp, 60-, Off Technol Assessment, US Cong, 77-78 & Interstate Mining Compact Comn, 79-81; pres, Geol Resources, Inc, 71- *Mem:* Geol Soc Am; Am Inst Mining, Metall & Petrol Engrs. *Res:* Origin and nature of mineral deposits. *Mailing Add:* 2114 Buckingham Rd Raleigh NC 27607-2114

BROWN, HERBERT, b South Irvine, Ky, Nov 10, 30; m 55; c 1. ANIMAL NUTRITION, ANIMAL HUSBANDRY. *Educ:* Univ Ky, BS, 52, MS, 56; Iowa State Univ, PhD(animal nutrit), 59. *Prof Exp:* Assoc prof animal sci, Ill State Univ, 59-61; area swine specialist, Univ Ky, 61-63; RES & DEVELOP SPECIALIST, LILLY RES LABS, ELI LILLY & CO, 63- *Mem:* Am Soc Animal Sci; Am Registry Prof Animal Scientists. *Res:* Development of products that will stimulate more efficient and economic production of livestock. *Mailing Add:* Lilly Res Labs G21 PO Box 708 Greenfield IN 46140

BROWN, HERBERT ALLEN, b Los Angeles, Calif, Jan 10, 40; m 65; c 2. VERTEBRATE ZOOLOGY. *Educ:* Univ Calif, Los Angeles, BA, 62; Univ Calif, Riverside, PhD(vert zool), 66. *Prof Exp:* Lectr biol, Univ Calif, Riverside, 66-67; asst prof, 67-76, ASSOC PROF BIOL, WESTERN WASH STATE COL, 76- *Mem:* Soc Study Evolution; Am Soc Ichthyologists & Herpetologists; Am Soc Zoologists; Soc Study Amphibians & Reptiles. *Res:* Physiological ecology of amphibians and reptiles; amphibian speciation; reproductive adaptations of amphibians and reptiles. *Mailing Add:* Dept Biol Western Wash State Col 516 High St Bellingham WA 98225

BROWN, HERBERT CHARLES, b London, Eng, May 22, 12; nat US; m 37; c 1. INORGANIC CHEMISTRY, ORGANIC CHEMISTRY. *Educ:* Univ Chicago, BS, 36, PhD(inorg chem), 38. *Hon Degrees:* DSc, Univ Chicago, 68, Wayne State Univ, Lebanon Valley Col, Long Island Univ, Hebrew Univ & Pontificia Univ, 80, Purdue Univ & Univ Wales, 81, Butler Univ & Univ SParis, 82, Ball State Univ, 85. *Prof Exp:* Asst, Univ Chicago, 36-38, Eli Lilly fel, 38-39, instr chem, 39-43, res investr, 41-43; from asst prof to assoc prof, Wayne State Univ, 43-47; prof, 47-59, Wetherill res prof chem, 59-78, EMER WETHERILL RES PROF CHEM, PURDUE UNIV, WEST LAFAYETTE, 78- *Concurrent Pos:* Vis prof & hon lectr, numerous US & foreign univs; mem, Group on Spec Fuels, Defense Dept, 55-60; consult, Exxon Corp & var com orgn; prin investr, NIH, US Army Res Off, Off Naval Res, Ethyl Corp & NSF. *Honors & Awards:* Nobel Prize, 79; Harrison Howe Lectr, 53; Nichols Medal, 59; Am Chem Soc Award, 60; Linus Pauling Medal, 68; Nat Medal Sci, 69; Baker Lectr, 69; Roger Adams Medal, 71; Charles Frederick Chandler Medal, 73; Madison Marshall Award, 75; Ingold Mem Lectr & Medal, 78; Elliot Cresson Medal, 78; Priestly Medal, 81; Gold Medal, Am Inst Chemists, 87; Perkin Medal, 82. *Mem:* Nat Acad Sci; Am Acad Arts & Sci; foreign fel Indian Nat Sci Acad; hon fel Royal Soc Chem; Chem Soc Japan; Pharmacol Soc Japan; Am Chem Soc. *Res:* Hydrides of boron; reactions of atoms and free radicals; nature of Friedel-Craft catalysts; reaction mechanisms; chemistry of addition compounds; steric strains; selective reductions; hydroboration; chemistry of organoboranes; asymmetric synthesis via chiral organoboranes. *Mailing Add:* HC Brown & RB Wetherill Labs Chem Purdue Univ 1393 Brown Bldg West Lafayette IN 47923

BROWN, HERBERT ENSIGN, b Ogden, Utah; m 44; c 2. CHRONOBIOLOGY. *Educ:* Univ Utah, BS, 49, MS, 51, PhD(anat), 55. *Prof Exp:* Instr, Westminster Col, 54-55; res instr, Univ Utah, 55-56; from asst prof to assoc prof, 56-84, EMER PROF ANAT, SCH MED, UNIV MO-COLUMBIA, 85- *Concurrent Pos:* Lt Colonel, US Air Force. *Mem:* Int Soc Chronobiol; Reticuloendothelial Soc; Am Asn Anatomists; Sigma Xi. *Res:* Cell biology; hematology; application of chronobiological principles to endocrinology, biochemistry and immunology for an understanding of the morphology and function of cells of the blood and blood forming organs and connective tissues. *Mailing Add:* Dept Anat Univ Mo Med Ctr Columbia MO 65212

BROWN, HERBERT EUGENE, b Jackson, Ga, Dec 25, 15; div; c 5. MEDICINE. *Educ:* Emory Univ, AB, 36, AM, 37, MD, 46; Univ Calif, PhD(zool), 40. *Prof Exp:* Teaching asst zool, Univ Calif, 37-40; instr, Univ Ga, 40-43; intern internal med, Grady Mem Hosp, Atlanta, 46-47; med officer, US Army, Brooklyn, 47 & Cristobal, CZ, 48-49; asst intermediate & sr resident internal med, Lawson Vet Admin Hosp, Ga, 49-52; internist, Atomic Bomb Casualty Comn, Hiroshima, 52-53; internist & partner, Emory Univ Clin, 54-73, from instr to assoc prof med, Sch Med, Emory Univ, 53-73, phsician, Univ Student Health, 73-76, assoc prof med, Dept Oral Med, Sch Dent, 74-78; RETIRED. *Concurrent Pos:* Physician, Fed Reserve Bank, Atlanta, 73-81; consult int med, Social Security Admin, 81- *Res:* Cytology of termite flagellates; human genetics; amebiasis; antibiotics. *Mailing Add:* 622 Park Lane Decatur GA 30033-5451

BROWN, HOMER E, b Humboldt, Minn, Apr 14, 09; m 32; c 1. ELECTRICAL ENGINEERING. *Educ:* Univ Minn, BSEE, 30. *Prof Exp:* Jr engr, Commonwealth Edison Co, 30-37, engr, 37-50, sr engr, 50-55, staff engr, 55-73; Eletrobras prof power eng, Brazil, 73-76; vis prof, NC State Univ, 76-81; TECH WITNESS, 81- *Concurrent Pos:* Consult, Manhattan Proj, Univ Chicago, 42-45, McDonnell Automation Ctr, St Louis, 62-70, Serv Bur Corp, 64, Comput Sci Corp, 68-69, IBM Corp, 69, Hidroserv, Brazil, 76 & Elec Power Res Inst, 77-78; vis prof, Iowa State Univ, 69-70, Rensselaer Polytech Inst, 70 & 71, Purdue Univ, 70-71 & Univ Ill, Chicago Circle, 71; distinguished lectr, Univ Tex, Arlington, 70. *Mem:* Fel Inst Elec & Electronics Engrs; Int Conf Large Elec Systs; fel Inst Elec & Electronics Engrs Brit; Sigma Xi. *Res:* Network analysis as applied to solution of power system problems using computers. *Mailing Add:* 705 Willow Valley Lakes Dr Willow Street PA 17584-9400

BROWN, HORACE DEAN, b Gainesville, Ga, Feb 13, 19; m 40; c 3. MEDICINAL CHEMISTRY, INFORMATION SCIENCE. *Educ:* Berry Col, BS, 39; Emory Univ, MS, 40; Iowa State Univ, PhD(chem), 47. *Prof Exp:* Chemist, Red Rock Cola Co, Atlanta, 40-41; instr chem, Iowa State Col, 41-43, res chemist, Manhattan Proj, 44-46, asst, Col, 46-47; sr chem & res assoc, Merck & Co, Inc, 47-67, dir sci info, 67-78, sr dir sci info, Merck Sharp & Dohme Res Labs, 78-84; RETIRED. *Concurrent Pos:* Asst, Quaker Oats Co, 46-47. *Mem:* Am Chem Soc; NY Acad Sci; Am Inst Chem; Asn Comput Mach. *Res:* Structure-activity relationships; insecticides; fungicides; antiparasitic agents; scientific information. *Mailing Add:* 7130 Windover Dr Durham NC 27712-9213

BROWN, HOWARD (EARL), b Anaheim, Calif, Nov 1, 16; m 41; c 1. MECHANICAL ENGINEERING. *Educ:* Univ Calif, BS, 39; Univ Tex, MS, 49, PhD, 56. *Prof Exp:* From instr to asst prof mech eng, Univ Tex, Austin, 40-52, assoc prof, 52-80. *Concurrent Pos:* Res engr, Defense Res Lab, 45-50; asst prog dir eng sci, NSF, 60-61. *Mem:* Am Soc Mech Engrs; Am Soc Eng Educ. *Res:* Quick freezing; automatic fuel meters; droplet dispersion in air streams; convective, boiling heat transfer. *Mailing Add:* 3504 Balcones Dr Austin TX 78731

BROWN, HOWARD HOWLAND, JR, b New York, NY, Sept 2, 34. PHYSICS. *Educ:* Mass Inst Technol, BS, 56, PhD(physics), 61. *Prof Exp:* Res asst physics, Res Lab Electronics, Mass Inst Technol, 56-61; res assoc, 61-62, from asst prof to assoc prof, 63-74, PROF PHYSICS, NY UNIV, 75- *Mem:* Am Phys Soc; Sigma Xi; AAAS. *Res:* Plasma physics; physics of electronic and atomic collisions. *Mailing Add:* Dept Physics NY Univ Washington Square New York NY 10003

BROWN, HOWARD S, b Lakewood, Ohio, July 30, 21; m 49; c 4. BIOLOGY, GENETICS. *Educ:* Univ Calif, Los Angeles, BA, 43, MA, 49; Claremont Grad Sch, PhD, 60. *Prof Exp:* Assoc prof, 48-64, prof biol, Calif State Polytech Univ, 64-82. *Concurrent Pos:* Vis prof, Chung Chi Col, Hong Kong, 61-62. *Mem:* AAAS; Bot Soc Am; Sigma Xi. *Res:* Cytogenetics; botanica genetics Vis prof, Chung Chi Col, Hong Kong, 61-62. *Mailing Add:* 29853 Disney Lane Vista CA 92084

BROWN, HUGH KEITH, b Atlanta, Ga, Aug 6, 45; m 66; c 2. RADIOLOGY, NEUROCYTOLOGY. *Educ:* Augusta Col, BS, 67; Med Col Ga, MS, 72; Tulane Univ, PhD(anat), 74. *Prof Exp:* Technician electron micros, Med Col Ga, 67-71; asst prof, 75-80, ASSOC PROF ANAT & RADIOL, COL MED, UNIV S FLA, 80- *Concurrent Pos:* scientist, magnetic resonance imaging, Dept Radiol, Col Med, Univ SFla; pres, Fla Soc Electron Micros, 91. *Mem:* Am Asn Anat; Electron Microscope Soc Am; Am Asn Clin Anat; Soc Magnetic Resonance Imaging. *Res:* Plasticity and regeneration of nervous tissue; quantitative synaptology; tissue characterization and mapping by NMR feature analysis. *Mailing Add:* Dept Anat Box 6 Univ SFla Col Med 12901 N 30th St Tampa FL 33612-4799

BROWN, HUGH NEEDHAM, b Champaign, Ill, Sept 12, 28; m 59; c 2. NUCLEAR PHYSICS. *Educ:* Univ Ill, BS, 50, MS, 52, PhD(physics), 54. *Prof Exp:* PHYSICIST, ACCELERATOR DEPT, BROOKHAVEN NAT LAB, 56- *Mem:* Am Phys Soc. *Res:* Accelerator development and construction; high energy physics experimental facilities. *Mailing Add:* Accelerator Dept Brookhaven Nat Lab Upton NY 11973

BROWN, IAN DAVID, b Edgware, Eng, Apr 11, 32; Can citizen; m 61; c 3. SOLID STATE CHEMISTRY, SCIENTIFIC NUMERICAL DATABASES. *Educ:* Univ London, BSc, 55 & 56, PhD(crystallog), 59. *Prof Exp:* Fel crystallog, 59-63, from asst prof to assoc prof, 63-74, PROF PHYSICS, MCMASTER UNIV, 74- *Mem:* Can Asn Physicists; Brit Inst Physics; Royal Soc Chem; fel Chem Inst Can; Am Crystallog Asn. *Res:* Structure relations in inorganic crystals; information retrieval. *Mailing Add:* Dept Physics McMaster Univ Hamilton ON L8S 4M1 Can

BROWN, IAN MCLAREN, b St Andrews, Scotland, Apr 15, 35; m 62; c 2. MAGNETIC RESONANCE. *Educ:* St Andrews Univ, BSc, 57, PhD(physics), 62. *Prof Exp:* Res assoc chem, Wash Univ, 61-63 & physics, Argonne Nat Lab, 63-65; assoc scientist, 65-73, SCIENTIST, McDONNELL DOUGLAS CORP, 73- *Mem:* Am Phys Soc; Am Chem Soc. *Res:* Magnetic resonance of free radicals in polymers. *Mailing Add:* 2416 Powders Mill Dr Chesterfield MO 63017

BROWN, IAN ROSS, b Douglas, Scotland, UK, Oct 23, 43; Can citizen; m 69; c 3. MOLECULAR NEUROBIOLOGY. *Educ:* Carleton Univ, BSc, 66; Univ Tex, PhD(biol), 69. *Prof Exp:* Med Res Coun fel, Univ Calgary, 69-71; from asst prof to assoc prof biol, 71-81, PROF MOLECULAR & DEVELOP BIOL, UNIV TORONTO, SCARBOROUGH CAMPUS, 81- *Concurrent Pos:* Sr vis investr, dept molecular biol, Res Inst Scripps Clin, 84-85; prof res zool, Univ Toronto, 88-89. *Mem:* Can Soc Cell Biol; AAAS; Am Soc Cell Biol; Am Soc Neurochem; Neurosci Soc; Int Neurochem Soc. *Res:* Analysis of gene expression in the development of the mammalian brain; expression of heat shock (stress) genes in neural tissue; molecular cloning of CDNA for synaptic glycoproteins. *Mailing Add:* Div Life Sci Univ Toronto Scarborough Campus West Hill Toronto ON M1C 1A4 Can

BROWN, IRA CHARLES, hydrology; deceased, see previous edition for last biography

BROWN, IRMEL NELSON, b McAfee, Ky, Feb 18, 12; wid; c 2. ELECTRICAL ENGINEERING. *Educ:* Univ Ky, BS, 33. *Prof Exp:* Prin, high sch, Ky, 36-42; instr elec, Morehead State Teachers Col, 42-44; engr, infrared devices, Radio Corp Am, 44-47, eng leader, 48-50, mgr, 51-55, sysmgr systs studies, 55-57, mgr comput & admin, 58-60, adminr tech coord, 61-63; mgr systs tehcnol, 61-64, leader systs eng RCA Corp, 65-70, mgr data planning & control, 70-74, sr mem eng staff, 75-76; RETIRED. *Mem:* Sr mem Inst Elec & Electronic Engrs. *Res:* Systems engineering techniques applied to major defense systems; development of communication; infrared and radar subsystems for complex military systems; data management. *Mailing Add:* 1221 Glen Terr Glassboro NJ 08028

BROWN, J(OHN) DAVID, b Shawnee, Okla, Feb 20, 57; m 85; c 2. THERMODYNAMICS OF SELF GRAVITATING SYSTEMS, HAMILTONIAN STRUCTURE OF GENERAL RELATIVITY. *Educ:* Okla State Univ, BS, 79; Univ Tex, PhD(physics), 85. *Prof Exp:* Postdoctoral asst, Inst Theoret Physics, Univ Wien, 85-87; postdoctoral assoc, Ctr Relativity & Theory Group, Univ Tex, Austin, 87-88; POSTDOCTORAL ASSOC, INST FIELD PHYSICS, UNIV NC, 88- *Mem:* Am Phys Soc. *Res:* Gravitation theory: gravitational thermodynamics, canonical quantum gravity, quantum cosmology; mathematical physics: BRST quantization, path integral constructions, Hamiltonian structure of gravity, strings and membranes. *Mailing Add:* Physics Dept CB No 3255 Univ NC Chapel Hill NC 27599-3255

BROWN, J(ACK) H(AROLD) U(PTON), b Nixon, Tex, Nov 16, 18; m 43. PATIENT RECORD SYSTEMS, CONSORTIA ORGANIZATION. *Educ:* Southwest Tex State Univ, BS, 36; Rutgers Univ, PhD(biochem), 48. *Prof Exp:* Instr biol, Univ Tex, 39-41; instr & researcher, USAF, 41-43; Gerard Swope Fund fel, Rutgers Univ, 44-50; asst prof physiol, Univ NC Sch Med,

50-52; from assoc prof to prof to actg chmn, Dept Med & Dent Physiol, Emory Univ Sch Med, 52-60; br chief, Nat Inst Gen Med Sci, NIH, 62-65, asst dir, Div Res Resources, 65-68, assoc dir, 68-72 & 72-74, actg dir, 74; prof physiol, Univ Tex Med Ctr, 74-78; assoc provost sci & res, 78-80, prof, 80-89, EMER PROF BIOL, UNIV HOUSTON, 89- *Concurrent Pos:* Lectr, Southwest Tex State Univ, 43-44; Univ Pittsburgh, 48-50, Georgetown Med Sch & George Washington Univ, 60-62, San Antonio Art Inst, 74-78 & Univ Houston, 78-80; dir biol res, Mellon Inst, 48-50; assoc dep adminr, Health Serv & Ment Health Admin & Health Maintenance Orgn, 74-78; adj prof, Trinity Univ, 74-78, Univ Tex & Baylor Col Med, 80-89; mem bd, Biomed Eng Soc, Joint Comt Eng Biol & Med, var comts, Am Physiol Soc, Joint Comt Interagency Technol Transfer, NASA & Dept Health & Human Serv & White House Comt Emergency Med Care. *Mem:* Nat Acad Eng; fel Inst Elec & Electronics Engrs; fel AAAS; sr mem Am Chem Soc; Inst Elec & Electronics Engrs Biomed Eng Soc; Soc Exp Biol & Med; Am Physiol Soc. *Res:* Patient record systems; management in health care; computerization of medical information. *Mailing Add:* Rm 124 Biol Bldg Univ Houston Houston TX 77204

BROWN, J MARTIN, b Doncaster, Eng, Oct 15, 41; m 67; c 2. RADIOBIOLOGY, ONCOLOGY. *Educ:* Univ Birmingham, BSc, 63; Univ London, MSc, 65; Oxford Univ, DPhil(radiation biol), 68. *Prof Exp:* NIH fel radiation biol, Med Ctr, 68-70, res assoc, 70-71, from asst prof to assoc prof, 71-84, PROF & DIR, DIV RADIATION BIOL, 84-, DIR, CANCER BIOL RES LAB, STANFORD UNIV MED CTR, 85- *Concurrent Pos:* Am Cancer Soc Dernham sr fel, 71-74; mem adv comt biol effects of ionizing radiations, Nat Acad Sci, 71- *Honors & Awards:* 9th Res Award, Radiation Res Soc, 80. *Mem:* Brit Inst Radiol; Radiation Res Soc; AAAS; Am Asn Cancer Res; Brit Asn Cancer Res; Am Soc Therapeut Radiol & Oncol. *Res:* Mammalian cellular radiobiology; tumor radiobiology; experimental chemotherapy; bioreductive cytotoxic agents; radiation carcinogenesis. *Mailing Add:* Dept Radiation Oncol Stanford Univ Med Ctr Stanford CA 94305

BROWN, JACK HAROLD UPTON, b Nixon, Tex, Nov 16, 18; m 43. PHYSIOLOGY. *Educ:* Southwest Tex State Col, BS, 39; Rutgers Univ, PhD(biochem), 48. *Prof Exp:* Instr physiol, Univ Tex, 39-41; radio engr, US Air Force, Wis, 41-43; instr physics, Southwest Tex State Col, 43-44; instr phys chem, Rutgers Univ, 44-45, res assoc biochem & nutrit, 45-48; dir, Biol Labs, Mellon Inst, 48-50; asst prof physiol, Sch Med, Univ NC, 50-52; assoc prof, Emory Univ, 52-57, actg chmn dept, 57-59, prof, 58-60; exec secy training grant, Nat Inst Gen Med Sci, NIH, 60-62, chief spec res resources br, 62-63, chief gen clin ctr br, 63, asst chief div res facil & resources, 63-71, dir sci progs, Nat Inst Gen Med Sci, 70-71, from assoc dir to actg dir, 64-70; spec asst to adminr, Health Sci Ment Health Admin, 70-71, assoc dep adminr, 71-73; spec asst to admin, Health Resources Admin, 73-74; coordr, Southwest Res Consortium, 74-78; prof environ sci & physiol, Univ Tex Health Sci Ctr, San Antonio, 74-78 & prof physiol, Univ Tex Med Ctr, 74-78; assoc provost res & advan educ, 78-80, prof biol, 81-88, EMER PROF, UNIV HOUSTON, 88- *Concurrent Pos:* Lectr, Univ Pittsburgh, 48-50, Med Schs, George Washington Univ, 61- & Georgetown Univ, 64-; consult, Vet Admin, Oak Ridge Inst Nuclear Studies, Lockheed Aircraft Co & Nat Libr Med, NASA; adj prof pub admin, Sch Pub Health, Univ Tex, 81-; adj prof health care admin & pub admin, Tex Woman's Univ, 81-; adj prof community health, Baylor Col Med, Houston, 81- *Honors & Awards:* Fulbright Award; Gerald Swore Award. *Mem:* Nat Acad Eng; Am Chem Soc; Soc Exp Biol & Med; Am Physiol Soc; Biomed Eng Soc (pres, 69); fel Inst Elec & Electronics Engrs; fel AAAS; Sigma Xi (pres, 77). *Res:* Research and development in new technology for patient record systems and health care using laser card technology. *Mailing Add:* Dept Biol Univ Houston Houston TX 77004

BROWN, JACK STANLEY, b New Orleans, La, Mar 12, 29; m 48; c 2. FRESHWATER BIOLOGY. *Educ:* Tulane Univ, BS, 48; Univ Ala, MS, 49, PhD, 56. *Prof Exp:* Asst biol, Univ Ala, 48-49 & fel, 52-54; asst prof, Jacksonville State Col, 49-52; instr zool, Auburn Univ, 54-55; med entomologist, US Air Force, 55-59; prof biol & chmn dept, Emory & Henry Col, 59-63; prof biol, Parsons Col, 63-65, dean, 65-68; dir inst freshwater biol, 69-75, head dept, 74-79, PROF BIOL, UNIV N ALA, 68- *Mem:* Am Soc Ichthyol & Herpet; Asn Southeastern Biologists; Ala Acad Sci. *Res:* Herpetology; medical entomology; natural history; freshwater benthos; water quality indicators; macrobenthos. *Mailing Add:* Dept Biol Univ N Ala Florence AL 35632-0001

BROWN, JAMES DOUGLAS, b Hamilton, Ont, July 17, 34; m 57; c 4. PHYSICAL CHEMISTRY, MATERIALS SCIENCE. *Educ:* McMaster Univ, BSc, 57; Univ Md, PhD(phys chem), 66. *Prof Exp:* Asst solubility of plutonium compounds, Atomic Energy Can Ltd, Ont, 56; chemist, Res Dept, Imp Oil Ltd, Ont, 57-60; res chemist, Res Ctr, US Bur Mines, Md, 60-66; from prof to assoc prof, 67-76, chmn mat sci group, 69-75, PROF MAT SCI, UNIV WESTERN ONT, 76-, CHMN, DEPT MAT ENG, 89- *Concurrent Pos:* Vis scientist, Metaalinstituut TNO Apeldoorn, Neth, 75-76; vis prof, Inst Appl Physics, Univ Vienna, Austria, 82-83. *Honors & Awards:* Corning Award, Microbeam Anal Soc, 74, 80, Presidential Award, Significant Tech Achievement, 88. *Mem:* Am Chem Soc; Microbeam Anal Soc (secy, 71, pres, 78); Asn Prof Engrs. *Res:* Structure and properties of vacuum deposited thin films; quantitative intensity; concentration relationships in electron probe microanalysis; x-ray spectroscopy; electrostatics; secondary ion mass spectroscopy. *Mailing Add:* 95 Cumberland Crescent London ON N5X 1B7 Can

BROWN, JAMES EDWARD, b Columbus, Ind, Jan 9, 45; m 71; c 2. BLOOD COAGULATION BIOCHEMISTRY, HAEMOSTASIS. *Educ:* Iowa State Univ, BS, 67, PhD(chem), 71. *Prof Exp:* Res fel, 73-75, ASST RES BIOCHEMIST BIOCHEM BLOOD COAGULATION, DEPT PATH, SCH MED, UNIV CALIF, SAN DIEGO, 75- *Mem:* Int Soc Thrombosis & Haemostasis; Am Chem Soc; AAAS. *Res:* Biochemistry of blood coagulation, particularly the role of factor VIII:C in haemostasis. *Mailing Add:* 1251 van Buren St NW Washington DC 20012

BROWN, JAMES HAROLD, b Richwood, WVa, Nov 9, 31; m 53; c 1. FOREST SOILS & GENETICS. *Educ:* WVa Univ, BSF, 53; Yale Univ, MF, 54; Mich State Univ, PhD(forestry), 67. *Prof Exp:* Res asst forestry, Northeastern Forest Exp Sta, US Forest Serv, 54-55; asst silviculturist, WVa Univ, 57-61, from asst prof to assoc prof forestry, 61-70; assoc prof, 70-74, prof, dept forestry, 74-89, ASSOC DIR, OHIO AGR RES & DEVELOP CTR, 89- *Mem:* Fel Soc Am Foresters; Soil Sci Soc Am. *Res:* Problems of artifical establishment and growth of tree species, including soil, ecological and genetic considerations. *Mailing Add:* Dirs Off Ohio Agr Res & Develop Ctr Wooster OH 44691

BROWN, JAMES HEMPHILL, b Ithaca, NY, Sept 25, 42; m 65; c 2. BIOGEOGRAPHY, EVOLUTION. *Educ:* Cornell Univ, AB, 63; Univ Mich, PhD(zool), 67. *Prof Exp:* Fel zool, Univ Calif, Los Angeles, 67-68, asst prof, 68-71; from asst prof to assoc prof biol, Univ Utah, 71-75; assoc prof, 75-78, prof ecol & evolutionary biol, Univ Ariz, 78-87; PROF BIOL, UNIV NMEX, 87- *Mem:* Ecol Soc Am; Am Soc Naturalists; Soc Study Evolution; Am Soc Mammalogists; Soc Conserv Biol; AAAS. *Res:* Ecology and biogeography; organization of ecological communities and distribution and diversity of species; interspecific interactions of desert ecosystems and diversification of continental biotas; plant-pollinator relationships, interspecific competition and physiological adaptations. *Mailing Add:* Dept Biol Univ NMex Albuquerque NM 87131

BROWN, JAMES HENRY, JR, b Oneco, Conn, July 27, 34; m 56; c 4. FOREST ECOLOGY, SOILS. *Educ:* Univ Conn, BS, 56; Univ RI, MS, 58; Duke Univ, DF(forest ecol), 65. *Prof Exp:* Res asst, 58-60, lectr, 60-62, from asst prof to assoc prof, 64-80, PROF FORESTRY, UNIV RI, 80- *Concurrent Pos:* Actg chmn dept, Univ RI, 67-68; res lectr, Utah State Univ, 69-70, instr, Sch Field Studies, 84. *Mem:* Soc Am Foresters; Sigma Xi. *Res:* Site-growth relationships; impact of fire on forests of southern New England; forest regeneration and forest water budget; recreation impact on forest sites; forest biomass; impact of gypsy moth defoliation. *Mailing Add:* Dept Natural Resources Sci Univ RI Kingston RI 02881

BROWN, JAMES K, b Kalamazoo, Mich, Oct 14, 45. ASTHMA. *Educ:* Johns Hopkins Univ, MD, 72. *Prof Exp:* ASST CHIEF RESPIRATORY CARE, VET ADMIN MED CTR, SAN FRANCISCO, 81- *Mem:* Am Physiol Soc; Am Thoracic Soc. *Mailing Add:* Respiratory Care Sect 111D Vet Admin Med Ctr 4150 Clement St San Francisco CA 94121

BROWN, JAMES KERR, b Osborne, Kans, May 14, 38; m 60; c 2. FORESTRY. *Educ:* Univ Minn, BS, 60; Yale Univ, MF, 61; Univ Mich, PhD(forestry), 68. *Prof Exp:* Res forester fire mgt, NCent Forest Exp Sta, 61-65, fire-fuel appraisal, Northern Forest Fire Lab, 65-79, PROJ LEADER FIRE EFFECTS, INTERMOUNTAIN FOREST & RANGE EXP STA, US FOREST SERV, 79- *Mem:* AAAS; Soc Am Foresters. *Res:* Quantification of the physical properties of forest and range fuels; fuel and fire behavior modeling; effects of fire on fuels and vegetation; biomass estimation. *Mailing Add:* 1504 Woods Gulch Rd Missoula MT 59802

BROWN, JAMES MELTON, b Clayton, Ala, Dec 29, 25; m 47; c 4. AGRONOMY. *Educ:* Auburn Univ, BS, 49, MS, 51; NC State Univ, PhD(soils, plant nutrit), 60. *Prof Exp:* Asst prof soils, Auburn Univ, 51-53, asst agronomist, 53-55; res instr soils, NC State Univ, 55-58, exten specialist, Statewide Soils Educ Prog, 58-60; agronomist, Nat Plant Food Inst, DC, 60-61; agronomist, Nat Cotton Coun Am, 61-68, mgr agron res, 68-69 & agron & pesticide residue res, 69-71, mgr prod technol, 71-91; RETIRED. *Concurrent Pos:* Alt rep, Agr Res Inst, 74-, mem, Pesticide Study Panel, 75-, mem bd dirs, 76-, adv mem, Regional Res Proj S-102, 76-, mem, Technol Transfer Subcomt, 77-; mem, Nat Cotton Indust-Exten Resources Comt, 77-, Agr Metrol Study Panel, 78-, vres, 79-81, pres, 81-; conf coordr & chmn prog & steering comts, Beltwide Cotton Conf, 71-, ed ann proceedings, Beltwide Cotton Prod-Mechanization Conf & Beltwide Cotton Prod Res Conf, 71-; mem, Boll Weevil Eradication Trial Tech Adv Comt, 78-; consult, State-USDA Cotton Res Task Force Rev Comt, 77- *Mem:* Am Soc Agron; Soil Sci Soc Am; Crop Sci Soc; Coun Agr Sci & Technol; Plant Growth Regulation Soc. *Res:* Soil fertility and chemistry; plant nutrition and physiology; weed control; air pollution; pesticide toxic substances; plant diseases; insect control; crop production systems; agricultural weather; legislation-regulations. *Mailing Add:* 4821 White Oak Dr Memphis TN 38116

BROWN, JAMES MICHAEL, b Gallatin, Tenn, Sept 26, 55. INDUSTRIAL WATER TREATMENT, ENHANCED OIL RECOVERY. *Educ:* Western Ky Univ, BS, 77; Vanderbilt Univ, MS & PhD(phys chem), 80. *Prof Exp:* Sr chemist, Halliburton Serv, 80-85, develop chemist, 85-86; PROJ LEADER, BETZ LABS, INC, 86- *Concurrent Pos:* Instr, N Harris County Community Col, 85- *Mem:* Am Chem Soc. *Res:* Development of mineral scale inhibitors for industrial cooling water, reverse osmosis systems and geothermal wells; development of polymeric dispersants for industrial water treatment; development of biodegradable and environmentally friendly additives for water treatment. *Mailing Add:* 23 Tangle Brush Dr The Woodlands TX 77380

BROWN, JAMES RICHARD, b Charleston, Ill, Oct 6, 31; m 60; c 4. SOIL FERTILITY, WASTE MANAGEMENT. *Educ:* Univ Ill, BS, 53, MS, 57; Iowa State Univ, PhD(soil fertil), 63. *Prof Exp:* From asst prof to assoc prof, 63-78, prof agron, 78-90, PROF SOIL SCI, UNIV MO-COLUMBIA, 90- *Concurrent Pos:* Coun Soil Testing & Plant Anal; Am Coun Forage & Grassland Coun. *Mem:* Am Soc Agron; Soil Sci Soc Am; Sigma Xi; Am Forage & Grassland Coun; Coun Soil Testing & Plant Anal. *Res:* Soil fertility and testing; waste management; plant analysis; nutrition of forage crops. *Mailing Add:* Sch Natural Resources Univ of Mo Columbia MO 65211

BROWN, JAMES ROY, b Wilkinsburg, Pa, June 14, 23; m 47; c 3. NUCLEAR PHYSICS. *Educ:* Allegheny Col, BA, 44; Calif Inst Technol, MS, 48, PhD(physics), 51. *Prof Exp:* Asst, Calif Inst Technol & Naval Ord Test Sta, 44-45; instr physics, Allegheny Col, 42-47; sr scientist nuclear physics, Reactors, Westinghouse Bettis Atomic Power Div, 50-57, adv scientist exp reactor physics, 57-59; PRIN PHYSICIST, NEUTRON PHYSICS EXP, LINEAR ACCELERATOR & PHYSICIST CHG CRITICAL EXP, GEN ATOMIC CO, 59- *Mem:* AAAS; Am Nuclear Soc; Am Phys Soc; Sigma Xi. *Res:* Solid propellant rockets; elementary physics; precision spectroscopy of gamma rays using a curved crystal focusing spectrometer; present research on neutron and reactor physics. *Mailing Add:* 12611 Sommerfield Lane Poway CA 92064

BROWN, JAMES RUSSELL, mathematical analysis; deceased, see previous edition for last biography

BROWN, JAMES T, JR, b Jackson Heights, NY, Feb 23, 39; m 59; c 3. THEORETICAL PHYSICS, EXPERIMENTAL LIGHT SCATTERING. *Educ:* Univ Colo, BA, 61, PhD(physics), 68. *Prof Exp:* Physicist, Boulder Labs, Nat Bur Stand, Colo, 58-65; PROF PHYSICS, COLO SCH MINES, 67- *Mem:* Am Asn Aerosol Res. *Res:* Aerosol science and particulate control; condensation and nucleation properties of respirable dust; scattering of light by small particles and surfaces. *Mailing Add:* Dept Physics Colo Sch Mines Golden CO 80401

BROWN, JAMES WALKER, JR, b Hampshire, Tenn, July 10, 30; m 53; c 2. BIOCHEMISTRY. *Educ:* Mid Tenn State Col, BS, 57; NC State Col, MS, 59, PhD(dairy chem), 62. *Prof Exp:* From asst prof to assoc prof, 62-69, PROF CHEM, MID TENN STATE UNIV, 69- *Mem:* Sigma Xi. *Mailing Add:* 1911 Ragland Ave Murfreesboro TN 37130

BROWN, JAMES WARD, b Philadelphia, Pa, Jan 15, 34; m 57; c 2. APPLIED MATHEMATICS. *Educ:* Harvard Univ, AB, 57; Univ Mich, AM, 58, PhD(math), 64. *Prof Exp:* Asst prof math, Univ Mich,Dearborn, 64-66 & Oberlin Col, 66-68; assoc prof, 68-71, PROF MATH, UNIV MICH, DEARBORN, 71- *Concurrent Pos:* NSF res grant, 69; ed consult, Math Rev, 70-84. *Mem:* Sigma Xi; Am Math Soc. *Res:* Special functions of mathematical physics; author of two books. *Mailing Add:* 1710 Morton Ave Ann Arbor MI 48104

BROWN, JARVIS HOWARD, b Goshen, Ind, Aug 7, 37; m 59; c 3. CROP PHYSIOLOGY. *Educ:* Purdue Univ, BSA, 59; Iowa State Univ, MS, 62, PhD(crop physiol), 65. *Prof Exp:* Res assoc agron, Purdue Univ Univ, 65-67; asst prof, 67-71, ASSOC PROF AGRON, MONT STATE UNIV, 71- *Mem:* Am Soc Agron; Crop Sci Soc Am. *Res:* Effects of leaf orientation, leaf width and chlorophyll concentration on the water use, growth, yield and photosynthesis of barley isotypes; cold hardiness of winter wheat. *Mailing Add:* Dept Plant & Soil Sci Mont State Univ Bozeman MT 59717

BROWN, JAY CLARK, b Jersey City, NJ, June 23, 42; m 65; c 3. VIRUS STRUCTURE, ELECRON MICROSCOPY. *Educ:* Johns Hopkins Univ, BS, 64; Harvard Univ, PhD(biochem), 69. *Prof Exp:* Fel molecular biol, MRC Lab Molecular Biol, Cambridge, Eng, 69-71; from asst prof to assoc prof, 71-87, PROF MICROBIOL, SCH MED, UNIV VA, 87- *Concurrent Pos:* Fel sci, NATO, 69-70; instr physiol, Marine Biol Lab, Woods Hole, Mass, 77-80; personnel comt, Am Cancer Soc, 80-86. *Mem:* Am Soc Biol Chemists; Am Soc Microbiol; AAAS. *Res:* Study of controlled sample erosion by ion beams as a method to enhance the usefulness of the electron microscope for analysing biological ultrstructure; applications to virus structure. *Mailing Add:* Dept Microbiol Box 441 Med Ctr Univ Va Charlottesville VA 22908

BROWN, JEANETTE SNYDER, b Rochester, NY, Mar 6, 25; m 50; c 3. PHOTOBIOLOGY. *Educ:* Cornell Univ, BS, 45, MS, 48; Stanford Univ, PhD(biol), 52. *Prof Exp:* Asst chem, NY State Agr Exp Sta, Geneva, 46-47; bacteriologist, Am Inst Radiation, Calif, 48-49; staff biologist, Carnegie Inst Wash Dept Plant Biol, 58-87; RETIRED. *Mem:* Am Soc Plant Physiol; Phycol Soc Am; Am Soc Photobiol; AAAS; Am Inst Biol Sci. *Res:* Spectroscopic investigations of the native state of chlorophyll and other photosynthetic pigments in plant membranes and isolated protein complexes. *Mailing Add:* PO Box 334 Murphys CA 95247-0334

BROWN, JEROME ENGEL, b Buckhannon, WVa, Apr 29, 24; m 62; c 4. INDUSTRIAL CHEMISTRY. *Educ:* WVa Wesleyan Col, BS, 44; Univ Ill, PhD(chem), 49. *Prof Exp:* Asst res supvr, Ethyl Corp, 49-60, commercial develop rep, 60-62, mgr mkt develop, 62-70, sr assoc, 70-87; RETIRED. *Mem:* Am Chem Soc. *Res:* Market development and research; gasoline additives; high-vacuum techniques. *Mailing Add:* PO Box 2332 Baton Rouge LA 70821-2332

BROWN, JERRAM L, b Glen Ridge, NJ, July 19, 30; m 53; c 3. NEUROSCIENCES. *Educ:* Cornell Univ, AB, 52, MS, 54; Univ Calif, Berkeley, PhD(zool), 60. *Prof Exp:* NIH fel, 60-62; from asst prof to prof biol & brain res, Univ Rochester, 62-78; PROF BIOL SCI, STATE UNIV NY ALBANY, 78- *Honors & Awards:* Brewster Medal, Am Ornith Union, 87. *Mem:* Soc Study Evolution; fel Am Ornith Union; fel Animal Behav Soc; fel AAAS; Ecol Soc Am; Soc Behav Ecol. *Res:* Neuro-ethology; evolution of behavior; ecology of aid-giving in animals; behavioral ecology. *Mailing Add:* Dept Biol Sci State Univ NY Albany NY 12222

BROWN, JERRY, b Boundbrook, NJ, Feb 22, 36; m 65; c 2. SOIL SCIENCE. *Educ:* Rutgers Univ, BS, 58, PhD(soils), 62. *Prof Exp:* Soil scientist & chief Earth Sci Br, US Army Cold Regions Res & Eng Lab, 61-85, dir tundra biome, Int Biol Prog, 70-76; HEAD, ARCTIC RES POLICY STAFF, NSF, 85- *Concurrent Pos:* Mem Polar Res Bd, Nat Res Coun, Nat Acad Sci, chmn Comt Permafrost, chmn US Comt, Int Permafrost Asn, 85-90. *Mem:* AAAS; Soil Sci Soc Am; Arctic Inst NAm; Ecol Soc Am; Geol Soc Am; Am Geophys Union; Am Quaternary Asn; Sigma Xi. *Res:* Arctic soil science; arctic ecology; permafrost. *Mailing Add:* Div Polar Progs NSF Washington DC 20550

BROWN, JERRY L, b Malvern, Ark, Oct 3, 35; m 61; c 1. BIOCHEMISTRY. *Educ:* NTex State Univ, BS, 59, MS, 60; Univ Tex, PhD(biochem), 64. *Prof Exp:* Fel biol chem, Univ Calif, Los Angeles, 63-67; asst prof biochem, 67-73, ASSOC PROF BIOCHEM, UNIV COLO, 73- *Mem:* Am Soc Biol Chemists; Fedn Am Soc Exp Biol. *Res:* Reactions resulting in structural amino acid modifications of proteins; reactions leading to blocked amino terminal amino acids are being investigated. *Mailing Add:* Dept Biochem Univ Colo Med Sch 4200 E Ninth Ave B-121 Denver CO 80262

BROWN, JERRY WILLIAM, b Wichita, Kans, July 4, 25; m 50; c 3. ANATOMY, NEUROSCIENCES. *Educ:* Wichita State Univ, AB, 46; Univ Kans, MA, 49, PhD(anat), 51. *Prof Exp:* From asst to instr anat, Univ Kans, 47-51; instr, Sch Med, Univ Pittsburgh, 51-56; from asst prof to assoc prof, Sch Med, Univ Mo, 56-64; assoc prof, 64-70, PROF ANAT, MED CTR, UNIV ALA, BIRMINGHAM, 70- *Concurrent Pos:* Guest worker, Neth Cent Inst Brain Res, Univ Amsterdam, 65-66. *Mem:* AAAS; Am Acad Neurol; Soc Neurosci; Am Asn Anat. *Res:* Embryonic development of the brainstem and telencephalon; development of sensory nuclei of V in human embryos; development of telencephalon of bat embryos; adult bat telencephalon. *Mailing Add:* Dept Cell Biol & Anat Univ Ala at Birmingham, Univ Sta Birmingham AL 35294

BROWN, JESSE J, JR, b Harrisburg, Pa, Oct 1, 39; m 61; c 2. CERAMIC ENGINEERING. *Educ:* Pa State Univ, BS, 61, MS, 62, PhD(ceramic technol), 64. *Prof Exp:* Res & develop engr, Sylvania Elec Prod Inc Div, Gen Tel & Electronics Corp, 64-67; from asst prof to assoc prof ceramic eng, 67-77, PROF MAT ENG & DIR, CTR ADVAN CERAMIC MAT, VA POLYTECH INST & STATE UNIV, 77- *Mem:* Fel Am Ceramic Soc; Am Chem Soc. *Res:* Inorganic chemistry, especially as related to phase equilibria and crystal chemistry; properties including, luminescence, thermal expansion and related fields. *Mailing Add:* Dept Mat Eng Col Eng Va Polytech Inst & State Univ Blacksburg VA 24060

BROWN, JOAN HELLER, b Cleveland, Ohio, Feb 14, 46; m 70; c 1. NEUROPHARMACOLOGY, RECEPTOR MECHANISMS. *Educ:* Cornell Univ, BA, 67; Albert Einstein Col Med, PhD(pharmacol), 73. *Prof Exp:* Fel pharmacol, Univ Colo Med Ctr, 73-75; asst res pharmacologist & lectr med, 75-79, asst prof residence, 79-80, ASST PROF, DIV PHARMACOL, DEPT MED, UNIV CALIF, SAN DIEGO, 80- *Mem:* Am Soc Pharmacol & Exp Therapeut; Soc Neurosci; Am Women Sci; Fedn Am Soc Exp Biol. *Res:* Biochemical mechanisms of neurotransmitter action: receptor pharmacology, cyclic nucleotide and phosphatidylinositol metabolism, regulation of transmitter synthesis and release. *Mailing Add:* Div Pharmacol Dept Med M-013 Univ Calif San Diego La Jolla CA 92093

BROWN, JOE NED, JR, b Bryan, Tex, Jan 24, 47. CRYSTALLOGRAPHY, PHYSICAL CHEMISTRY. *Educ:* Tex A&M Univ, BS, 69; Univ New Orleans, PhD(phys chem), 72. *Prof Exp:* Res fel biochem, Calif Inst Technol, 72-73; asst prof chem, Ill Inst Technol, 73-79; sr res scientist, Celanese Res Co, 79-; AT CIBA-GEIGY, 89- *Concurrent Pos:* Damon Runyon fel, Calif Inst Technol, 72-73; NIH fel, 73; consult, G D Searle & Co, Skokie, Ill, 74-79 & Latticeworks, Inc, 79- *Mem:* Am Crystallog Asn; Am Chem Soc; Am Inst Physics; AAAS. *Res:* Structure-activity correlations among biologically active molecules; peptide structure; hydration effects on structure. *Mailing Add:* Oracal Corp 101 Woods Ave S Iselin NJ 08830

BROWN, JOEL EDWARD, b Middletown, NY, May 23, 37. NEUROPHYSIOLOGY. *Educ:* Mass Inst Technol, BS & MS, 60, PhD(physiol), 64. *Prof Exp:* From asst prof to assoc prof physiol, Mass Inst Technol, 64-71; prof anat, Vanderbilt Univ, 71-76; prof biochem, 74-76; prof physiol, State Univ NY Stony Brook, 76-83; Bernard Becker prof ophthal, Wash Univ, 83-91; TACHNA PROF OPHTHAL, ALBERT EINSTEIN COL MED, 91- *Concurrent Pos:* NIH fel, Sch Med, Univ Chile, 64; corp mem, Woods Hole Marine Biol Lab, trustee, 73-77; McCormick Scholar, Res Prev Blindness, Inc, 84. *Mem:* Asn Res Vision & Ophthal; Am Soc Biochem & Molecular Biol; Biophys Soc; Soc Gen Physiol; fel AAAS. *Res:* Physiology, biophysics and biochemistry of retina and photoreceptors. *Mailing Add:* Dept Ophthal & Visual Sci Albert Einstein Col Med Montefiore Med Ctr 111 E 210th St Bronx NY 10467

BROWN, JOHN ANGUS, b Lake Wales, Fla, Dec 13, 25; m 54; c 3. ORGANIC CHEMISTRY, ANALYTICAL CHEMISTRY. *Educ:* Emory Univ, BA, 45; Ga Inst Technol, MS, 51, PhD(org chem), 54. *Prof Exp:* Engr, Law & Co, Ga, 45-48; teaching asst chem, Ga Inst Technol, 48-54; res chemist, Esso Res & Eng Co, NJ, 54-63, sr chemist, 63-68, staff adv, 68-73; PRES, JOHN BROWN ASSOCS, INC, 73- *Concurrent Pos:* Mem var ad hoc study & planning panels, Dept of Defense; pres, Passaic Valley Chamber Com, 87-90. *Honors & Awards:* Bronze Medallion, Am Defense Preparedness Asn, 82- *Mem:* Am Chem Soc; Sigma Xi; Am Defense Preparedness Asn; Soc Advan Mat & Process Eng; Optical Soc Am; Soc Photo-Optical Instrumentation Engrs. *Res:* Research and development of lubricating oil additives, safety and sensitivity of propellants and explosives; advanced physical-chemical waste reclamation processes; analytical aspects of chemical warfare defense; analytical instrument development; electronics; optics; advanced optical filters for laser eye protection. *Mailing Add:* 15 York Pl Berkeley Heights NJ 07922

BROWN, JOHN BOYER, b Lexington, Ky, Nov 12, 24; m 49; c 3. PHYSICAL CHEMISTRY. *Educ:* Univ Ky, BS, 48; Northwestern Univ, PhD, 56. *Prof Exp:* From instr to prof chem, 52-90, chmn dept, 61-64, 72-73 & 75-76, EMER PROF CHEM, DENISON UNIV, 90-, CONSULT, COMPUTER CTR, 90- *Concurrent Pos:* Vis scientist, Swedish Inst Surface Chem, 73-74; Vis prof, Duke Univ, 81-82. *Mem:* Fel Am Inst Chemists; NY Acad Sci. *Res:* Cationic surfactants; physical properties of micelles; solubilization by detergents. *Mailing Add:* Dept Chem Denison Univ Granville OH 43023

BROWN, JOHN CLIFFORD, b Cullman, Ala, Feb 23, 43; div; c 2. BIOCHEMISTRY, IMMUNOLOGY. *Educ:* Auburn Univ, BS, 65, MS, 67; NC State Univ, PhD(biochem), 73. *Prof Exp:* Fel immunol, Univ Calif, Berkeley, 73-76; asst prof, 76-83, ASSOC PROF MICROBIOL, UNIV KANS, 83-, COURTESY ASSOC PROF BIOCHEM, 90- *Concurrent Pos:* Consult, Biospecific Tech Inc, 86-88, Abbott Labs, 88- & Immtech Int Inc, 89-; mem genetics prog, Univ Kans, 89- *Mem:* Am Asn Immunologists; Fedn Am Soc Exp Biol. *Res:* Rheumatic disease autoantibodies; idiotype and anti-idiotype relationships in rheumatic disease. *Mailing Add:* Dept Microbiol 8041 Haworth Hall Univ Kans Lawrence KS 66045

BROWN, JOHN FRANCIS, JR, b Providence, RI, Oct 11, 26; m 55; c 3. PHYSICAL ORGANIC CHEMISTRY. *Educ:* Brown Univ, BS, 47; Mass Inst Technol, PhD(chem), 50. *Prof Exp:* Res assoc, 50-56, mgr reaction studies unit, 56-61, org chemist, 61-65, mgr life sci br, 65-82, MGR HEALTH RES, RES & DEVELOP CTR, GEN ELEC CO, 82- *Concurrent Pos:* Fel clin path, State Univ NY Upstate Med Ctr, 68-69; mem, Study Critical Environ Problems, 70; mem, Nat Heart, Lung & Blood Inst Working Group on Characterization Stand for Blood-Mat Interactions, 78-79. *Mem:* AAAS; Am Chem Soc; Soc Environ Toxicol Chemists. *Res:* Biomedical surface chemistry; health effects of industrial chemicals; environmental fate of xenobiotics. *Mailing Add:* Gen Elec Corp-Res & Develp PO Box 8 Bldg K1 Rm 3 B29 Schenectady NY 12301-0008

BROWN, JOHN HENRY, b Kalamazoo, Mich, Nov 4, 24; m 46; c 3. WOOD SCIENCE, WOOD TECHNOLOGY. *Educ:* Mich State Univ, BS, 49, MS, 50; Syracuse Univ, PhD(wood prod eng), 62. *Prof Exp:* Trainee, Ga Pac Plywood Co, 50-51; buyer & salesman, Olympia Wood Preserving Co, 51-54; wood technologist, Res & Develop, Potlatch Forests, Inc, 54-56; instr & res asst, Syracuse Univ, 56-62, asst prof, 62-63; technologist, Eng & Tech Serv Dept, Potlatch Corp, 63-88; RETIRED. *Mem:* Soc Wood Sci & Technol; Forest Prod Res Soc; Am Soc Testing & Mat. *Res:* Wood adhesion; electrical properties of wood; wood-moisture relations; energy; wood combined with other materials, plastics, metals, paper and particleboard. *Mailing Add:* 1962 Coulter Lane Clarkston WA 99403

BROWN, JOHN LAWRENCE, JR, b Ellenville, NY, Mar 6, 25. APPLIED MATHEMATICS, COMMUNICATION THEORY. *Educ:* Ohio Univ, BS, 48; Brown Univ, PhD(appl math), 53. *Prof Exp:* From asst prof to prof, 52-88, EMER PROF ELEC ENG, PA STATE UNIV, UNIVERSITY PARK, 88- *Concurrent Pos:* Gen Lew Allen, Jr Res Chair, Air Force Inst Technol, 84-85; Stocker prof electrical & computer eng, Ohio Univ, 88- *Honors & Awards:* Fel, Inst Elec & Electronics Eng. *Mem:* Am Math Soc; Inst Elec & Electronics Engrs; Math Asn Am; Soc Indust & Appl Math; Acoustical Soc Am. *Res:* Statistical communication theory; applied mathematics; underwater acoustics. *Mailing Add:* 1431 Curtin St State College PA 16803

BROWN, JOHN LOTT, b Philadelphia, Pa, Dec 3, 24; m 48; c 4. PSYCHOPHYSIOLOGY, HUMAN FACTORS ENGINEERING. *Educ:* Worcester Polytech Inst, BSEE, 45; Temple Univ, MA, 49; Columbia Univ, PhD(psychol), 52. *Hon Degrees:* DSc, Worcester Polytech Inst, 84. *Prof Exp:* Mem staff, Div Govt Res, Columbia Univ, 51-53, tech dir, US Air Force Res Proj, 53-54; head psychol div, Aviation Med Acceleration Lab, Naval Air Develop Ctr, 54-59; asst prof sensory physiol, Sch Med, Univ Pa, 57-62, assoc prof, 62-65, dir grad training prog physiol, 62-65; dean grad sch, Kans State Univ, 65-66; vpres acad affairs, 66-69; prof psychol & optics, Univ Rochester, 69-78, dir, Ctr Visual Sci, 71-78; pres, 78-88, EMER PROF MED & ENG, UNIV SFLA, 88- *Concurrent Pos:* USPHS sr res fel physiol, Sch Med, Univ Pa, 59-63; mem visual sci study sect, NIH, 67-71; chmn, Comt Vision, Nat Acad Sci-Nat Res Coun, 65-70; staff adv, 68-72; mem bd trustees, Worcester Polytech Inst, 71-83, Asn Res Vision & Ophthal, 74-79, Illum Eng Res Inst, 74-79; assoc ed, J Optical Soc Am, 72-77; mem, Vision Res Prog Comt, Nat Eye Inst, 74-78, chmn, 75-78. *Mem:* fel Am Psychol Asn; fel Optical Soc Am; fel AAAS; Asn Res Vision & Ophthal (pres, 78). *Res:* Psychophysiology and electrophysiology of vision; applications of psychology and physiology to man-machine systems; sensory physiology. *Mailing Add:* Univ SFla Tampa FL 33620

BROWN, JOHN M, CANCER IMMUNOLOGY, MONOCLONAL ANTIBODIES. *Educ:* Wayne State Univ, PhD(biochem), 77. *Prof Exp:* asst prof microbiol surg, Col Med, Univ Ill, 81-; CENTO COR, INC. *Res:* Immunotherapy. *Mailing Add:* Cento Cor Inc 244 Great Valley Pkwy Malvern PA 19355

BROWN, JOHN WESLEY, b St Louis, Mo, Oct 13, 33; m 63; c 4. MATHEMATICS. *Educ:* Univ Mo, BA, 58; Univ Calif, Los Angeles, MA, 64, PhD(math), 66. *Prof Exp:* Mem tech staff, Douglas Aircraft Co, Inc, 58-59, Space Technol Labs, 59-61, Packard Bell Comput Corp, 61-63 & Hughes Aircraft Co, 63-65; specialist, Bell Tel Labs, 65-66; asst prof, 66-71, ASSOC PROF MATH, UNIV ILL, URBANA-CHAMPAIGN, 66- *Mem:* Am Math Soc; Math Asn Am. *Res:* Combinatorial mathematics. *Mailing Add:* 1409 W Green St Urbana IL 61801

BROWN, JOHN WESLEY, b Chicago, Ill, Dec 2, 25; m 50; c 3. BIOLOGICAL CHEMISTRY. *Educ:* Elmhurst Col, BS, 50; Univ Ill, MS, 53, PhD, 56. *Prof Exp:* Tech rep nutrit, E I Du Pont de Nemours & Co, Inc, 56-57; from asst prof to assoc prof biochem, 57-68, chmn dept chem, 68-71, PROF CHEM, UNIV LOUISVILLE, 68-, ASSOC DEAN GRAD SCH & DIR SPONSORED PROG, 71- *Mem:* AAAS; Sigma Xi; Am Soc Biol Chem. *Res:* Animal protein metabolism; amino acid nutrition in relation to neurotransmitter synthesis and behavior. *Mailing Add:* 1960 Meadowcreek Ave Louisville KY 40218

BROWN, JOSEPH M, b Fayetteville, WVa, Jan 9, 28; m 52; c 4. MECHANICAL ENGINEERING. *Educ:* WVa Univ, BS, 46, BSME, 47; Purdue Univ, MSME, 50, PhD(eng design), 52. *Prof Exp:* Struct engr, NAm Aviation, 51-53 & Douglas Aircraft Co, 53-56; owner, Struct Specialties Inc, 56-57; sect head struct, Hughes Aircraft Co, 57-61; assoc dept head vehicle design, Aerospace Corp, 61-67; proj engr, McDonnell Douglas Corp, 67-70; PROF MECH ENG, MISS STATE UNIV, 70- *Res:* Engineering design methodology, all areas of structural analysis and design. *Mailing Add:* Drawer ME-Mech Eng Miss State Univ Mississippi State MS 39762

BROWN, JOSEPH ROSS, b Parkville, Mo, Sept 24, 20; m 48; c 4. MATHEMATICS. *Educ:* Park Col, AB, 41; Univ Kans, MA, 49, PhD(math), 53. *Prof Exp:* Instr math, Park Col, 46-49; prof math, Bradley Univ, 52-87; RETIRED. *Mem:* Math Asn Am. *Res:* Non-Euclidean geometry. *Mailing Add:* Dept Math Bradley Univ Peoria IL 61606

BROWN, JOSHUA ROBERT CALLOWAY, b Switchback, WVa, Jan 6, 15; m 44; c 1. CELL BIOLOGY. *Educ:* Duke Univ, AB, 48, MA, 49, PhD(zool), 53. *Prof Exp:* Instr zool, Duke Univ, 51-52, Army Med Serv trainee biochem, Med Sch, 52-53; from asst prof to prof, 53-82, EMER PROF ZOOL, UNIV MD, 82- *Concurrent Pos:* Asst dir sci teaching improv prog, AAAS, 57-58; NSF fel, Walter Reed Army Inst Res, 63-64. *Mem:* AAAS; Am Soc Cell Biol. *Res:* Cell physiology; isolated cell components; evolution and differentiation of cells in culture; effects of environmental factors on cells in culture. *Mailing Add:* 6513 40th Ave University Park MD 20782

BROWN, JUDITH ADELE, b Providence, RI, Dec 30, 44; m 73. ANIMAL BEHAVIOR, BIOLOGICAL RHYTHMS. *Educ:* Whittier Col, BA, 66; Northwestern Univ, MS, 69, PhD(biol), 73. *Prof Exp:* Asst prof, 69-75, ASSOC PROF BIOL, CALIF STATE COL, STANISLAUS, 75- *Concurrent Pos:* Res assoc, Inst Cult Resources, Calif State Col, Stanislaus, 73-; field res, Calif Fish & Wildlife Agency, 75- *Mem:* AAAS; Animal Behav Soc; Int Audio-Tutorial Cong. *Res:* Field research in animal behavior and laboratory research in patterns of biorhythmicity. *Mailing Add:* Dept Biol Sci Calif State Univ 801 W Monte Vista Stanislaus Turlock CA 95380

BROWN, JULIUS, b New York, NY, Aug 21, 15; m 44; c 3. ELECTRICAL ENGINEERING. *Educ:* City Col New York, BS, 36; St Louis Univ, MS, 52; Wash Univ, DSc(elec eng), 61. *Prof Exp:* Design & develop engr, Hazeltine Elec Corp, 46-49 & Perkin-Elmer Corp, 50; from asst prof to assoc prof elec eng, St Louis Univ, 52-65; chmn dept, 68-74, PROF ELEC ENG, SOUTHERN ILL UNIV, EDWARDSVILLE, 68- *Mem:* Inst Elec & Electronics Engrs. *Res:* Physical electronics; pulse and timing techniques. *Mailing Add:* Dept of Eng & Technol Southern Ill Univ Edwardsville IL 62026

BROWN, KAREN KAY (KILKER), b Manhattan, Kans, July 25, 44; m 66. VACCINE DEVELOPMENT, FERMENTATION. *Educ:* Washburn Univ, BS, 66; Okla State Univ, PhD(molecular biol & microbiol), 72. *Prof Exp:* Res scientist, Cutter Haver Lockhart Labs, 72-75, qual assurance microbiologist, 72; sr res scientist, Bayvet Div, Miles Labs, 75-82, prin scientist, 82-83, mgr biol res, 83-85; DIR BIOL RES, ANIMAL HEALTH DIV, MOBAY CORP, 85- *Concurrent Pos:* Biohazards control officer, Bayvet Div, Miles Labs & Animal Health Div, Mobay Corp, 74-; pres, Pair O'Docs Investment Enterprises, 84- *Honors & Awards:* Miles Sci & Technol Award, 85; Mobay Sci & Technol Awards, 86. *Mem:* AAAS; NY Acad Sci; Am Soc Microbiol; Am Asn Lab Animal Scientists; Sigma Xi. *Res:* Development of veterinary vaccines and drug products; use of biotechnology techniques for development of purified subunit vaccines and diagnostics; fermentation; tissue culture; protein purification; delivery systems. *Mailing Add:* 5501 NW Fox Hill Rd Parkville MO 64155

BROWN, KARL LESLIE, b Coalville, Utah, Sept 30, 25; m 48, 77; c 5. PARTICLE PHYSICS. *Educ:* Stanford Univ, BS, 47, MS, 49, PhD(physics), 53. *Prof Exp:* Res assoc, Stanford Univ, 53-58; consult, Linear Accelerator Lab, Ecole Normale Superieure, France, 58-59; sr res assoc, 59-74, PROF, STANFORD LINEAR ACCELERATOR CTR, 74- *Concurrent Pos:* Consult, Varian Assocs, 66-; vis scientist, European Orgn Nuclear Res, 72-74, 84-85. *Mem:* Fel Am Phys Soc. *Res:* Experimental physics; charged particle accelerators and charged particle optics. *Mailing Add:* Stanford Linear Accelerator Ctr Stanford CA 94305

BROWN, KEITH CHARLES, b Beverley, Eng, Dec 2, 42. PHYSICAL ORGANIC CHEMISTRY. *Educ:* Univ Liverpool, BSc, 64, PhD(chem), 67. *Prof Exp:* Fel, Univ Rochester, 67-69; res officer, Elec Coun Res Ctr, Capenhurst, Eng, 69-73; scientist, Environ Impact Ctr Inc, Mass, 73-74; from res scientist to sr res scientist, 74-77, sect head, 77-78, mgr org chem, 78-83, ASSOC DIR RES, CLAIROL INC, DIV BRISTOL-MYERS CO, 84- *Mem:* Am Chem Soc; Royal Soc Chem; Asn Res Dirs. *Res:* Kinetics and mechanisms of reactions of organic compounds, particularly compounds important in cosmetic chemistry; dye synthesis. *Mailing Add:* Clairol Inc 2 Blachley Rd Stamford CT 06902

BROWN, KEITH H, b Salt Lake City, Utah, Sept 3, 39; m 63; c 3. PLASMA PHYSICS, HIGH PRESSURE PHYSICS. *Educ:* Brigham Young Univ, BA, 64, PhD(physics), 69; Univ Ill, Urbana-Champaign, MS, 65. *Prof Exp:* Asst prof, 68-72, assoc prof, 72-77, PROF PHYSICS, CALIF STATE POLYTECH UNIV, POMONA, 77-, CHMN DEPT, 76- *Concurrent Pos:* Fel, Brigham Young Univ, 74-75. *Mem:* Am Phys Soc; Am Asn Physics Teachers. *Res:* Magnetic confinement of plasmas; x-ray diffraction of liquid metals at high pressures. *Mailing Add:* 204 Jefferson Ave r Pomona CA 91767

BROWN, KENNETH HENRY, b Utica, NY, June 6, 39; m 61; c 4. PHYSICAL CHEMISTRY, ANALYTICAL CHEMISTRY. *Educ:* Hope Col, BA, 60; Rutgers Univ, PhD(phys chem), 66. *Prof Exp:* Asst prof chem, Transylvania Univ, 65-72, consult, T M Regan Inc, 73; asst prof, 73-81, ASSOC PROF CHEM, NORTHWESTERN OKLA STATE UNIV, 81- *Concurrent Pos:* Consult, Ranger Eng, 75-76 & Trex Inc, 84- *Mem:* Am Chem Soc; Sigma Xi; Soc for Col Sci Teachers. *Res:* Chemical education; chemical separations; chemical aspects of energy resource recovery and use; asphalt roofing materials. *Mailing Add:* Dept Chem Northwestern Okla State Univ Alva OK 73717

BROWN, KENNETH HOWARD, b Chicago, Ill, Nov 21, 42; m 69; c 1. ORGANIC POLYMER CHEMISTRY. *Educ:* Univ Ill, BS, 65; Univ Minn, MS, 69; Loyola Univ Chicago, PhD(org chem), 71. *Prof Exp:* Sr res scientist org & polymer chem, Continental Can Co, 70-76; MEM STAFF, RUSTOLEUM CORP, 76- *Concurrent Pos:* Asst prof chem, eve div, Elmhurst Col, 75- *Mem:* Sigma Xi; Am Chem Soc. *Res:* Photopolymerization process of coatings; synthesis of photosensitizers and monomers for polymeric systems; scale-up of organic reactions. *Mailing Add:* Rustoleum Corp 8105 Fergufson Dr Pleasant Prairie WI 53158-0769

BROWN, KENNETH MICHAEL, b Riverside, Calif, Dec 20, 48; m 74; c 1. EVOLUTIONARY ECOLOGY, AQUATIC ECOLOGY. *Educ:* Ariz State Univ, BS, 70; Wash State Univ, MS, 72; Univ Iowa, PhD(zool), 76. *Prof Exp:* Instr zool, Univ Iowa, 76-77; from asst prof to assoc prof biol sci, Purdue Univ, 77-85; ASSOC PROF ZOOL, LA STATE UNIV, 85- *Concurrent Pos:* Prin investr, NSF grants, snail pop ecol, 79-82 & snail Community ecol, 85-88; field sta dir, Crooked Lake Biol Sta, 79-85; prin investr, Off Water Res & Technol grant, zooplankton ecol, 80-83; vis prof, Bodega Marine Lab, 85. *Mem:* Ecol Soc Am; Am Soc Naturalists; Brit Ecol Soc; Soc Study Evolution. *Res:* Population and community ecology of fresh water and marine gastropods; life history evolution; importance of competition; parasitism; predation in structuring snail assemblages. *Mailing Add:* Dept Zool & Physiol La State Univ Baton Rouge LA 70803-1725

BROWN, KENNETH STEPHEN, b Orillia, Ont, May 4, 49; m 71; c 2. BIOSTATISTICS, MEDICAL STATISTICS. *Educ:* Univ Waterloo, BMath, 71, PhD(statist), 74. *Prof Exp:* Lectr, Univ Waterloo, 73-74, from asst prof to assoc prof, 74-88, PROF STATIST, UNIV WATERLOO, 88-, CHAIR, 90- *Concurrent Pos:* Consult, prog smoking & health, Non-Med Use Drugs Directorate, Health & Welfare Can, 74-76; mem working group health hazard appraisal, Non-Med Use Drugs Directorate & Health Prom & Prev Directorate, Health & Welfare Can, 75-78; mem, health care systs grant rev comt, Ont, 87- *Mem:* Am Statist Asn; Can Statist Soc; Royal Statist Soc; Biometrics Soc. *Res:* Mathematical models of disease processes; smoking and disease; biostatistics; smoking prevention. *Mailing Add:* Dept Statist Univ Waterloo Waterloo ON N2L 3G1 Can

BROWN, KENNETH STEPHEN, b Chicago, Ill, Dec 15, 29; m 55; c 3. TERATOLOGY, HUMAN GENETICS. *Educ:* Univ Chicago, BA, 49, MD, 60. *Prof Exp:* Asst zool, Univ Chicago, 52-56; intern med, Blodgett Hosp, 60-61; clin assoc, 61-63, investr, Human Genetics Br, 64-67, sect chief, Develop Genetics Sect, 67-74, INVESTR GENETICS, LAB DEVELOP BIOL & ANOMALIES, NIH, 75- *Concurrent Pos:* Instr, Found Advan Educ Sci, 66-81. *Mem:* Am Soc Human Genetics; AAAS; Teratology Soc; Soc Craniofacial Genetics (pres elect, 81); Sigma Xi. *Res:* Etiology and pathogenesis of birth defects; primary interest in craniofacial malformations, using mouse genetic defects and teratogenic agents as probes for study of genetic control of biochemical processes. *Mailing Add:* 11527 Deborah Dr Potomac MD 20854

BROWN, KENNETH TAYLOR, b Purcellville, Va, Apr 7, 22; m 48; c 2. PHYSIOLOGY. *Educ:* Swarthmore Col, BA, 47; Univ Chicago, MS, 49, PhD(psychol), 51. *Prof Exp:* Res psychologist, Aero-Med Lab, Wright Air Develop Ctr, 50-54; res assoc visual physiol, Brown Univ, 54-55; from instr to asst prof physiol optics, Wilmer Inst, Med Sch, Johns Hopkins Univ, 55-58; from asst prof to assoc prof, 58-66, PROF PHYSIOL, MED SCH, UNIV CALIF, SAN FRANCISCO, 66- *Concurrent Pos:* NSF fel, Brown Univ, 54-55; USPHS spec res fel, Wilmer Inst, Med Sch, Johns Hopkins Univ, 57-58; Nat Eye Inst res grant, Univ Calif, San Francisco, 59-; Commonwealth Fund fel, John Curtin Sch Med Res, Australian Nat Univ, 64-65; fac res lectr, Univ Calif, San Francisco, 69. *Mem:* Fel AAAS; Am Physiol Soc; Soc Neurosci; Asn Res Vision & Ophthal; Int Brain Res Orgn. *Res:* Neurophysiology of the vertebrate retina, with special reference to the photoreceptors; microelectrode techniques. *Mailing Add:* Dept Physiol Univ Calif Sch Med 1459 Willard St San Francisco CA 94117

BROWN, KERMIT EARL, b Haskell, Tex, Nov 2, 23; m 45; c 3. PETROLEUM ENGINEERING. *Educ:* Tex A&M Univ, BS(mech eng) & BS(petrol eng), 48; Univ Tex, MS, 59, PhD, 62. *Prof Exp:* Petrol engr, Stanolind Oil & Gas Co, 48-51; gas lift engr, Garrett Oil Tools, Inc, 51-55; from asst prof to assoc prof petrol eng, Univ Tex, Austin, 55-66, grad adv, 65-66; prof petrol eng & head dept, Univ Tulsa, 66-68, vpres res, 68-88, Halliburton prof petrol eng & chmn, Div Resources Eng, 76-88, Floyd M Stevenson prof, 81-88; CONSULT, 88- *Concurrent Pos:* Consult, Otis Eng Corp, 58-; Convair award, 58-59. *Mem:* Nat Acad Sci; Am Soc Eng Educ; Soc Petrol Engrs. *Res:* Storing reactor fuel waste in salt domes; gas lift applied to oil; vertical two-phase fluid flow. *Mailing Add:* Intl Training & Consult 5429 E 65th Pl Tulsa OK 74104

BROWN, L(AURENCE) C(LAYDON), b Glasgow, Scotland, Mar 23, 37. PHYSICAL METALLURGY. *Educ:* Univ Strathclyde, BSc, 58; Glasgow Univ, PhD(physics), 61. *Prof Exp:* Fel metall, 61-63; from asst prof to assoc prof, 63-72, PROF METALL, UNIV BC, 72- *Mem:* Am Soc Metals. *Res:* Diffusion and phase transformations in binary and ternary metal systems. *Mailing Add:* Dept Metals & Mats Eng Univ BC Vancouver BC V6T 1W5 Can

BROWN, L CARLTON, b Mineral Springs, Ark, Mar 26, 15; m 38; c 1. PHYSICS. *Educ:* Henderson State Teachers Col, AB, 37; Fla State Univ, MA, 52; Ohio State Univ, PhD(physics), 55. *Prof Exp:* Instr math, Pulaski County Schs, Ark, 38-39; eng aide, Ark State Hwy Dept, 39-41 & US Army Corps Engrs, 41-43 & 46-47, US Army Air Force, 43-46; training officer, US Vet Admin, 47-48; asst engr, Ark State Hwy Dept, 48-51; asst physics, Fla State Univ, 51-52; asst, 52-55, res assoc magnetic resonance, 55-57, from asst prof to prof physics, 57-84, CONSULT PHYSICS, OHIO STATE UNIV, 84- *Concurrent Pos:* Co-supvr & chief investr, Ohio State Univ res grant, Off Naval Res & Air Force Off Sci Res, 55-84. *Mem:* Am Phys Soc; Am Asn Physics Teachers. *Res:* Nuclear magnetic resonance; electron paramagnetic resonance; x-ray and gamma ray scattering; electromagnetic theory. *Mailing Add:* Dept Physics Ohio State Univ 2170 Lane Rd Columbus OH 43220

BROWN, LARRY NELSON, mammalogy, ecology, for more information see previous edition

BROWN, LARRY PATRICK, b South Bend, Ind, Mar 17, 42; m 71; c 4. PETROLEUM ENGINEERING. *Educ:* Purdue Univ, BS, 64; Univ Tulsa, MS, 68, PhD(eng), 71. *Prof Exp:* Engr automotive, Chrysler Corp, 64-65; engr aerospace, McDonnell-Douglas Corp, 65-67; res engr petrol, Cities Serv Co, 69-72, sr tech adv, 72-74, sr res engr, 74-77, field technol mgr petrol, 77-84; sr engr adv, Oxy USA Inc, 84-; BROWN & ASSOC. *Mem:* Soc Petrol Engrs. *Res:* Petroleum production; pressure transient testing and analysis; well completions; reservoir engineering. *Mailing Add:* Brown & Assoc 4304 E 98th St Tulsa OK 74137

BROWN, LARRY ROBERT, b Wellsville, NY, Oct 20, 46. PROTEIN STRUCTURE, MEMBRANES. *Educ:* Calif Inst Technol, BS, 68; Australian Nat Univ, PhD(chem), 74. *Prof Exp:* Asst lectr chem, Univ SPac, 69-70; fel, Eidgenossische Tech Hochschule, 75-81, asst, 74-75; assoc res scientist biophysics, Mich Molecular Inst, 81-85; SR FEL & HEAD, UNIV NUCLEAR MAGNETIC RESONANCE CTR, RES SCH CHEM, AUSTRALIAN NAT UNIV, 85- *Mem:* Am Chem Soc; Biophys Soc. *Res:* Nuclear magnetic resonance to investigate the structure and function of synthetic and biological macromolecules. *Mailing Add:* Res Sch Chem Australian Nat Univ Conberra ACT 2601 Australia

BROWN, LAUREN EVANS, b Waukesha, Wis, Sept 4, 39; m 68; c 3. EVOLUTION, BEHAVIOR. *Educ:* Carroll Col, Wis, BS, 61; Southern Ill Univ, Carbondale, MS, 63; Univ Tex, Austin, PhD(zool), 67. *Prof Exp:* from asst prof to assoc prof, 67-77, PROF VERT ZOOL, ILL STATE UNIV, 77- *Concurrent Pos:* Managing ed, Herpetologica, 79-81; consult endangered species; mem, Houston Toad Recovery Team, US Fish & Wildlife Serv. *Mem:* Herpetologists' League; Am Soc Ichthyologists & Herpetologists; Soc Study Amphibians & Reptiles. *Res:* Evolution; ecology; behavior; amphibian speciation; bioacoustics; herpetology; extinction; ecology of flood plains and sand prairies; zoogeography; natural hybridization and reproductive isolation; systematics. *Mailing Add:* Dept Biol Sci Ill State Univ Normal IL 61761

BROWN, LAURIE LIZBETH, b Nyack, NY, July 26, 46; m; c 2. GEOPHYSICS, PALEOMAGNETISM. *Educ:* Middlebury Col, AB, 68; Univ Wyo, MS, 72; Ore State Univ, PhD(geophys), 74. *Prof Exp:* From asst prof to assoc prof, 74-88, PROF GEOL, UNIV MASS, 88- *Mem:* Am Geophys Union; Geol Soc Am; Soc Explor Geophysicists; Asn Women Geoscientists. *Res:* Magnetic stratigraphy; secular variation; plate tectonic reconstructions; magnetic susceptibility anisotropy. *Mailing Add:* Dept Geol & Geog Univ Mass Amherst MA 01003

BROWN, LAURIE MARK, b Brooklyn, NY, Apr 10, 23; m 69; c 4. THEORETICAL PHYSICS. *Educ:* Cornell Univ, AB, 43, PhD(theoret physics), 51. *Prof Exp:* Res assoc, sam labs, Columbia Univ, 43-44; asst physics, Cornell Univ, 46-48, asst, Nuclear Studies Lab, 48-50; from instr to assoc prof, 50-61, PROF PHYSICS, NORTHWESTERN UNIV, EVANSTON, 61- *Concurrent Pos:* NSF fel, Inst Advan Study, 52-53; Fulbright res scholar, Italy, 58-60; consult, Argonne Nat Lab, 60-70; Int Atomic Energy Agency prof, Univ Vienna, 66; vis prof, Univ Rome, 67, Univ Sao Paulo, 72-73, Univ Wash, 78, Keio Univ, 84; mem, NSF Panel Advan Sci Sem, 68, 69; chair, Div Hist Physics, Am Phys Soc, 83. *Mem:* Fel AAAS; fel Am Phys Soc; Hist Sci Soc. *Res:* Theoretical high energy physics; quantum electrodynamics; history of physics. *Mailing Add:* Dept Physics & Astron Northwestern Univ Evanston IL 60208

BROWN, LAWRENCE D, b Dec 16, 40. APPLIED MATHEMATICS. *Educ:* Calif Inst Technol, BS, 61; Cornell Univ, PhD, 64. *Prof Exp:* Asst prof, Dept Statist, Univ Calif, Berkeley, 65-66; from asst prof to assoc prof, Dept Math, Cornell Univ, 66-73; prof, Dept Math, Rutgers Univ, 73-78; PROF, DEPT MATH, CORNELL UNIV, 78- *Concurrent Pos:* Postdoctoral fel, Nat Acad Sci-Nat Res Coun, 64-65, Sloan Found, 69-71; assoc ed, Ann Statist, 70-80 & 83-, Ann Probability, 85-88; dir, Cornell Statist Ctr, 82-83, 86-88 & 89-; mem, Bd Math Sci, Nat Res Coun, 89-; Lady Davis prof, Hebrew Univ, Jerusalem. *Honors & Awards:* Wald Lectr, Inst Math Statist, 85. *Mem:* Nat Acad Sci; fel Inst Math Statist; Am Statist Asn. *Res:* Mathematics and statistics. *Mailing Add:* Dept Math 129 White Hall Cornell Univ Ithaca NY 14853-7901

BROWN, LAWRENCE E(LDON), b Republic Co, Kans, June 19, 14; m 39, 56; c 1. CHEMISTRY. *Educ:* Southwest Tex State Col, AB, 36; Univ Tex, 41; Tulane Univ, 48. *Prof Exp:* Asst chemist in charge microanal lab, US Dept Agr, 42-50, chemist, South Mkt & Nutrit ResS Div, 50-73; RETIRED. *Mem:* Am Chem Soc; Am Oil Chem Soc; fel Am Inst Chem. *Res:* Microchemical analysis; utilization research on oilseeds, meals and oils. *Mailing Add:* 805 Andrews Ave Metairie LA 70005

BROWN, LAWRENCE G, b St Louis, Mo, Feb 6, 43. MATHEMATICS. *Educ:* Harvard Univ, BA, 65, PhD(math), 68. *Prof Exp:* PROF MATH, PURDUE UNIV, 74- *Mem:* Am Math Soc; Am Asn Univ Prof. *Mailing Add:* Dept Math Purdue Univ West Lafayette IN 47907

BROWN, LAWRENCE MILTON, petroleum chemistry; deceased, see previous edition for last biography

BROWN, LAWRENCE S, JR, b Brooklyn, NY, Dec 4, 49; m 78; c 1. INTERNAL MEDICINE, DRUG ADDICTION. *Educ:* Brooklyn Col, BA, 76; Columbia Univ, MPH, 79; NY Univ, MD, 79. *Prof Exp:* Intern med, Harlem Hosp Ctr, 79-80, resident, 80-82, chief resident, 82-83; Fel med & endocrinol, dept med, Columbia Univ, 83-86; ASST ATTEND PHYSICIAN MED & ENDOCRINOL, HARLEM HOSP, 86-; SR VPRES RES, ADDICTION RES & TREAT CORP, 86- *Concurrent Pos:* Adj prof health admin, Columbia Univ, Sch Pub Health, 88-, instr med, dept med, 86-, vis clin fel, 79-82, clin fel, 82-83; mem, Panel IV Drug Use & AIDS, Nat Acad Sci, 88- *Mem:* Am Pub Health Asn; Am Soc Internal Med; Am Col Physicians;

NY Acad Sci; AAAS; Nat Med Asn. *Res:* Clinical investigations of the pathogenesis of diabetes and steroid metabolism; clinical investigations of the endocrinologic and immunological associations with patterns of drug abuse and addiction. *Mailing Add:* Addiction Res & Treat Corp 22 Chapel St Brooklyn NY 11201

BROWN, LEE F(RANCIS), b Elmhurst, Ill, Feb 23, 29; m 67; c 3. CHEMICAL ENGINEERING, PHYSICAL CHEMISTRY. *Educ:* Univ Notre Dame, BS, 51; Univ Del, MChE, 55, PhD(chem eng), 63. *Prof Exp:* Chem engr, Houdry Process Corp, Pa, 51-54; res assoc, Univ Colo, 63-64, actg asst prof, 64, from asst prof to prof chem, 64-81, chmn dept, 80-81; STAFF MEM, LOS ALAMOS NAT LAB, UNIV CALIF, 81- *Concurrent Pos:* Univ Colo fac fel, Tech Univ Delft, Neth, 67-68; vis prof, Yale Univ, 77-78. *Mem:* Am Chem Soc; Am Inst Chem Engrs; Am Soc Eng Educ. *Res:* Transport in and structure of porous materials; hetergeneous kinetics and catalysis; adsorption. *Mailing Add:* MS K557 N-6 Los Alamos Nat Lab Los Alamos NM 87545

BROWN, LEE ROY, JR, b Warren, Ohio, Aug 20, 26; m 46; c 3. MEDICAL MICROBIOLOGY. *Educ:* Univ Ala, BS, 50; George Washington Univ, MS, 56, PhD(microbiol), 61. *Prof Exp:* Bacteriologist, Naval Med Field Res Lab, Camp Le Jeune, NC, 50-53; bacteriologist, Naval Dent Sch, Bethesda, Md, 53-61; asst prof microbiol, Sch Dent, Univ Mo-Kansas City, 61-62, assoc prof & chmn dept, 62-68; PROF MICROBIOL & MEM DENT SCI INST, DENT BR, UNIV TEX HEALTH SCI CTR, HOUSTON, 68- *Concurrent Pos:* Consult, Res Div, Vick Chem Co, NY, 63-73, Procter & Gamble Co, Ohio, 74, 82 & 83, Spectrix Corp, Houston, 75, Vet Admin, Div Dent Res, 78, 80 & 83, Kimberly-Clark Corp, 84-85 & Lyndon B Johnson Space Ctr, 85-; mem, Oral Biol & Med Study Sect, NIH-Nat Inst Dent Res, 79, ad hoc Res Rev Comt, 82, Comt Res & Manpower, 84. *Honors & Awards:* Civil Serv Superior Accomplishment Awards, US Dept Navy, 59 & 61. *Mem:* Am Soc Microbiol; Int Asn Dent Res; Am Asn Dent Schs; Am Asn Lab Animal Sci; Sigma Xi. *Res:* Microbiological, immunological and chemical aspects of the etiology, prevention and treatment of diseases with manifestations in the oral cavity. *Mailing Add:* Inst Dent Sci 6515 John-Freeman Houston TX 77030

BROWN, LELAND RALPH, b Springdale, Ark, Mar 7, 22; m 49; c 3. ENTOMOLOGY. *Educ:* Univ Calif, BS, 46; Cornell Univ, PhD, 49. *Prof Exp:* prof 69-85, EMER PROF ENTOM, UNIV CALIF, RIVERSIDE, 85- *Mem:* AAAS; Entom Soc Am. *Res:* Insecticide applicators; insects affecting ornamental shrubs, shade trees and forests; insect photography; general entomology. *Mailing Add:* 5250 Tower Rd Riverside CA 92506-1036

BROWN, LEO DALE, b Waco, Tex, Apr 11, 48; m 72; c 3. X-RAY CRYSTALLOGRAPHY, OPTICAL MICROSCOPY. *Educ:* Baylor Univ, BS, 70; Univ Calif, Berkeley, PhD(chem), 74. *Prof Exp:* Fel inorg chem, Northwestern Univ, 75-77; RES ASSOC, EXXON RES & ENG CO, 77- *Mem:* Am Chem Soc; Am Crystallog Asn; Sigma Xi. *Res:* Synthetic fuels from coal and oil shale; effects of inorganic constituents on fuel processing; x-ray structural analysis of inorganic materials; optical and acoustic microscopy of oil shale, coal, coke and related materials. *Mailing Add:* Exxon Res & Develop Labs PO Box 2226 Baton Rouge LA 70821-2226

BROWN, LEONARD D, animal nutrition, physiology; deceased, see previous edition for last biography

BROWN, LEONARD FRANKLIN, JR, b Seminole, Okla, June 1, 28; m 56, 82; c 1. GEOLOGY, BASIN ANALYSIS. *Educ:* Baylor Univ, BS, 51; Univ Wis, MS, 53, PhD(geol), 55. *Prof Exp:* Explor geologist, Stand Oil Co Tex, 55-57; from asst prof to assoc prof geol, Baylor Univ, 60-66; dir, Tex Mining & Minerals Resources Res Inst, 81-84; res scientist, 57-60, Bur Econ Geol, Univ Tex, 57-60, res scientist & lectr geol, 66-71, assoc dir res, 71-84, sr res scientist, 84- 89, prof geol, 71-89, EMER PROF GEOL, UNIV TEX, 90- *Concurrent Pos:* Ed, Baylor Geol Studies Bull, 61-66; consult, Int Petrol, 73-; continuing educ lectr, 73- *Honors & Awards:* Distinguished Lectr, Am Assoc Petrol Geologists, 72-73, Petrol Explor Soc, Australia, 85; Cheney Award, Am Asn Petrol Geologist, 90. *Mem:* Am Asn Petrol Geol; Geol Soc Am. *Res:* Stratigraphic studies of the Pennsylvanian system; stratigraphy, environments and paleontology of rocks in Texas; terrigenous depositional systems; environmental geology in Texa; facies analysis; seismic-stratigraphy; basin analysis; sequence stratigraphy. *Mailing Add:* Rte 5 Box 301 Georgetown TX 78626-9805

BROWN, LESTER R, b Bridgeton, NJ, Mar 28, 34; div; c 2. AGRICULTURAL SCIENCE, ECONOMICS. *Educ:* Rutgers Univ, BS, 55; Univ Md, MS, 59; Harvard Univ, MA, 62. *Hon Degrees:* LHD, Dickinson Col, 75, Franklin Col, 77; LLD, Univ Md, 76, Williams Col, 77, Rutgers Univ, 78, Glassboro State Col, 79, Tufts Univ, 85, Clark Univ, 88, Col Wooster, 89, Rippon Col, 90, Otterbein Col, 90. *Prof Exp:* Int agr economist, Econ Res Serv, USDA, 59-64, policy adv secy of agr, 64-66; adminr, Int Agr Develop Serv, 66-69, sr fel, Overseas Develop Coun, 69-74; PRES, WORLDWATCH INST, 74- *Concurrent Pos:* Mem, Overseas Develop Coun, Global Studies Ctr & Global Tomorrow Coalition; MacArthur Found fel, 86. *Honors & Awards:* Arthur S Fleming Award, US Jaycees, 65; A H Boerma Award, UN Food & Agr Orgn, 81; Lorax Award, 85; Spec Conserv Award, Nat Wildlife Fedn, 82; Gold Medal, Worldwide Fund for Nature, 89. *Mem:* Coun Foreign Rels; World Future Soc; Overseas Develop Coun; Global Studies Ctr; Better World Soc; Renew Am; Global Tomorrow Coalition. *Res:* Food; population; environment; resources. *Mailing Add:* Worldwatch Inst 1776 Massachusetts Ave NW Washington DC 20036-1904

BROWN, LEWIS MARVIN, b Brooklyn, NY, June 9, 50; m 72; c 2. PHYCOLOGY, PHYSIOLOGICAL ECOLOGY. *Educ:* State Univ NY, Stony Brook, BS, 72; Univ SFla, MA, 74; Univ Toronto, PhD(bot), 78. *Prof Exp:* Vis fel marine bot, Atlantic Res Lab, Nat Res Coun Can, 78-80; asst prof, Univ Western Ont, 80-87, res assoc, Cancer Res Lab, 87-88; MGR, ALGAL TECHNOL SOLAR ENERGY RES INST, 88- *Concurrent Pos:* Mem bd dirs, Can Bot Asn, 87-88. *Mem:* Phycol Soc Am; Int Phycol Soc; Am Soc Plant Physiologists; Am Soc Limnol & Oceanog. *Res:* Plant molecular biology; physiological ecology of algae; biotechnological applications of microalgae. *Mailing Add:* Solar Energy Res Inst 1617 Cole Blvd Golden CO 80401-3393

BROWN, LEWIS RAYMOND, b Houston, Tex, Aug 11, 30; m 51; c 4. BACTERIOLOGY, MICROBIOLOGY. *Educ:* La State Univ, BS, 51, MS, 53, PhD(bact), 58. *Prof Exp:* Asst br chief, US Civil Serv, Pine Bluff Arsenal, 53-55; instr bact, La State Univ, 55-58; operator, pvt consult & testing lab, 58-61; assoc dean res, Col Arts & Sci, 71-89, PROF MICROBIOL, MISS STATE UNIV, 61- *Honors & Awards:* Charles Porter Award, Soc Indust Microbiol, 90. *Mem:* AAAS; Am Soc Microbiol; Am Acad Microbiologists; Int Water Resources Asn; Sigma Xi; NY Acad Sci; Soc Indust Microbiol; Water Pollution Control Fedn; Am Water Resources Asn. *Res:* Biodeterioration of metals; microbial ecology; geomicrobiological prospecting; sewage and industrial waste disposal; petroleum degradation and water pollution control; isolation, characterization and metabolism of hydrocarbon oxidizing bacteria and other recalcitrant molecules; author of 100 publications in microbiology. *Mailing Add:* PO Drawer CU Miss State Univ Mississipi State MS 39762

BROWN, LINDA ROSE, b Muncie, Ind, Mar 7, 45. MOLECULAR SPECTROSCOPY. *Educ:* Ohio State Univ, BA, 68; Fla State Univ, PhD(physics), 76. *Prof Exp:* Res assoc physics, Nat Res Coun, 76-78; SR SCIENTIST PHYSICS, JET PROPULSION LAB, 78- *Mem:* Soc Appl Spectros; Optical Soc Am; Sigma Xi. *Res:* High resolution infrared spectroscopy of light molecules of atmospheric interest. *Mailing Add:* Jet Propulsion Lab T-1166 4800 Oak Grove Dr Pasadena CA 91109

BROWN, LINDSAY DIETRICH, b Lynchburg, Va, Jan 14, 29; m 52; c 2. AGRONOMY, PLANT PHYSIOLOGY. *Educ:* Lynchburg Col, BA, 51; Va Polytech Inst, MS, 58; Mich State Univ, PhD(hort), 62. *Prof Exp:* Jr instr biol, Johns Hopkins Univ, 55-56; instr hort, Va Polytech Inst, 57-58; instr, Mich State Univ, 61-63; asst prof, Univ Ky, 63-67; supvr agron serv, Southwest Potash Corp, 67-69; assoc prof hort, Va Polytech Inst & State Univ, 69-70; dir mkt serv, Amax Chem Corp, 70-72; dir spec prod mkt, 72-81, DIR NAM MKT, INT MINERAL & CHEM CO, 81- *Concurrent Pos:* Consult, USAID, Southeast Asia, 66. *Mem:* Am Soc Hort Sci; Am Soc Agron. *Res:* Plant breeding; mineral nutrition of higher plants; physiology of greenhouse-grown plants. *Mailing Add:* Int Minerals & Chem Corp 421 E Howley St Mundelein IL 60060

BROWN, LINFIELD CUTTER, b Peterborough, NH, Sept 24, 42; m 64; c 1. SANITARY ENGINEERING. *Educ:* Tufts Univ, BCE, 64, MS, 66; Univ Wis-Madison, PhD(sanit eng), 70. *Prof Exp:* Asst prof, 70-76, assoc prof civil eng, 76-81, PROF CIVIL ENG, TUFTS UNIV, 81- *Mem:* Am Water Works Asn; Water Pollution Control Fedn; Sigma Xi. *Res:* Mass transfer of oxygen into streams; statistical design of experiments; statistical modelling of systems in water pollution control. *Mailing Add:* 106 Damon Rd Medford MA 02155

BROWN, LLOYD LEONARD, b Topeka, Kans, Jan 20, 27; m 54; c 2. PHYSICAL CHEMISTRY. *Educ:* Wichita State Univ, BS, 50, MS, 52. *Prof Exp:* Chemist, Wichita Found Indust Res, 52; chemist, Oak Ridge Sch Reactor Technol, 52-53, chemist, Oak Ridge Nat Lab, 53-91; RETIRED. *Mem:* Am Chem Soc. *Res:* Chemistry of stable isotopes; chemical methods for enrichment of rare isotopes; kinetic effects; nuclear magnetic resonance spectroscopy; electron spin resonance spectroscopy. *Mailing Add:* Chem Div Oak Ridge Nat Lab PO Box 2008 Oak Ridge TN 37831-6201

BROWN, LLOYD ROBERT, b Fayette, Mo, Dec 10, 24; m 49; c 4. ELECTRICAL ENGINEERING. *Educ:* Mo Valley Col, AB, 47; Univ Mo-Columbia, BSEE, 49; Wash Univ, MSEE, 51, DSc(elec eng), 60. *Prof Exp:* Asst elec eng, Wash Univ, 49-51; design engr, Missile Div, McDonnell Aircraft Corp, Mo, 52-54; res engr, White-Rodgers Elec Co, 54-56; instr elec eng, Wash Univ, 56-58; engr, Electro-Mech Res, Inc, Fla, 58-61; from assoc prof to prof, 61-87, EMER PROF ELEC ENG, WASH UNIV, 87- *Concurrent Pos:* Consult, White-Rodgers Elec Co, Mo, 56-58, Universal Match Corp, 62-66, McConnell Aircraft Corp, 64-68 & US Naval Ord Sta, Ill, 66-69; Pvt fire investr, expert witness & electrical specialist, 74- *Mem:* Sr mem Inst Elec & Electronics Engrs; Nat Fire Protection; Int Asn Arson Investrs; Sigma Xi. *Res:* Automatic control and electronic instrumentation; fire and accident investigation. *Mailing Add:* Sch Eng Wash Univ St Louis MO 63130

BROWN, LORETTA ANN PORT, b Kingston, NY, July 30, 45; m 70; c 1. INTERNAL MEDICINE, BIOCHEMICAL GENETICS. *Educ:* State Univ NY Col New Paltz, BS, 67; Univ Mich, MS, 68; Eastern Va Med Sch, MD, 81. *Prof Exp:* NSF res partic, Albion Col, 66; USPHS trainee, Univ Mich, 67-69; lab technologist, Univ Mich, 69-70; res instr biol-med genetics, M D Anderson Hosp & Tumor Inst, 70; res instr obstet & gynec, Baylor Col Med, 71-76; adj prof biol, Christopher Newport Col, 77-78; resident, Internal Med, Eastern Va Grad Sch Med, 81-84; staff physician, Health Am, Hampton, VA, 84-87; CHIEF ADMITTING & SCREENING, VA HOSP, HAMPTON, VA, 88- *Mem:* Am Col Physicians. *Res:* Suppressor mutations of the methionine-1 locus in Neurospora crassa; arginine and polyamine metabolism and control in Neurospora crassa; protein hormones in human reproductive endocrinology; equine color genetics. *Mailing Add:* 21093 Reynolds Dr Carrollton VA 23314

BROWN, LOUIS, b San Angelo, Tex, Jan 7, 29; m 52. NUCLEAR PHYSICS. *Educ:* St Mary's Univ, Tex, BS, 50; Univ Tex, PhD(physics), 58. *Prof Exp:* Res asst physics, Inst Physics, Univ Basel, 58-61; Carnegie fel, 61-64, MEM PHYSICS STAFF, DEPT TERRESTRIAL MAGNETISM, CARNEGIE INST WASHINGTON, 64- *Honors & Awards:* Amerbach Prize, Univ Basel, 64. *Mem:* Fel Am Phys Soc; Am Geophys Union; AAAS. *Res:* Experimental study of the beryllium-10 content of soils, sediments and rocks; experimental study of the isotopic composition of elements in rocks and meteorites; application of resonance ionization mass spectrometry to geochronology. *Mailing Add:* Dept Terrestrial Magnetism Carnegie Inst Washington Washington DC 20015

BROWN, LOWELL SEVERT, b Visalia, Calif, Feb 15, 34; m 56; c 1. PHYSICS. *Educ:* Univ Calif, Berkeley, AB, 56; Harvard Univ, PhD(physics), 61. *Prof Exp:* NSF fel physics, Univ Rome, 61-62 & Imp Col, Univ London, 62-63; res assoc, Yale Univ, 63-64, from asst prof to assoc prof, 64-68; assoc prof, 68-71, PROF PHYSICS, UNIV WASH, 71- *Concurrent Pos:* NSF sr fel, Imp Col, Univ London, 71-72; John Simon Guggenheim Found fel, 79-80; mem, Inst Advan Study, Princeton, NJ, 79-80; consult, Los Alamos Sci Lab; mem, Panel Pub Affairs, Am Phys Soc, 79, exec comt, Div Particles & Fields, 82-83 & 88-91, publ comt, 83-86; chmn, Theoret Physics Panel, US Dept Energy Tech Assessment Comt Univ Progs, 82-83; mem bd trustees, Aspen Ctr Physics, 82-88 & 90-; ed, Phys Rev, 87-; mem sci adv bd, Theoret Study Inst Elem Particle Physics, 84-89; vis prof, Columbia Univ, 90; vis scientist, Los Alamos Nat Lab, 91. *Mem:* Fel Am Phys Soc; fel AAAS. *Res:* Theoretical physics; quantum field theory; elementary particle physics. *Mailing Add:* Dept Physics FM-15 Univ Wash Seattle WA 98195

BROWN, LUTHER PARK, b Bloomington, Ill, Feb 18, 51; div. MATE CHOICE, EVOLUTION. *Educ:* Elmhurst Col, BA, 73; Ohio State Univ, MS, 76, PhD(zool), 78. *Prof Exp:* Asst prof, 78-83, ASSOC PROF BIOL, GEORGE MASON UNIV, 83- *Concurrent Pos:* Asst prof, Mountain Lake Biol Sta, Univ Va, 79. *Mem:* Animal Behav Soc; Ecol Soc Am. *Res:* Behavioral ecology, specifically in the evolution of animal social and mating systems through both natural and sexual selection. *Mailing Add:* Dept Biol George Mason Univ 4400 University Dr Fairfax VA 22030

BROWN, LYNN RANNEY, b Waterloo, Iowa, Sept 26, 28; m 53; c 4. ANIMAL NUTRITION. *Educ:* Iowa State Univ, BS, 50, PhD(dairy nutrit), 59. *Prof Exp:* Teacher, Pub Schs, Iowa, 50-51 & 53-55; asst dairy nutrit, Iowa State Univ, 56-58, assoc, 58-59; from asst prof to assoc prof dairy exten, 60-74, PROF DAIRY EXTEN, UNIV CONN, 74- *Mem:* Am Dairy Sci Asn; Am Soc Animal Sci; Sigma Xi. *Res:* Physiological factors associated with bloat; nutrition of dairy calves and cows. *Mailing Add:* Dept Animal Sci Univ Conn U-40 3636 Horsebarn Rd Ext Storrs CT 06268

BROWN, MARIE JENKINS, b Eldorado, Ill, Sept 26, 09; m 89. INVERTEBRATE ZOOLOGY. *Educ:* Phillips Univ, BA, 29; Cath Univ Am, MA, 51; Univ Okla, PhD(zool), 61. *Prof Exp:* Asst prof biol, Benedictine Heights Col, 54-57, registr, 55-57; instr zool, Univ Okla, 60-62; from assoc prof to prof, 62-75, EMER PROF BIOL, MADISON COL, VA, 75- *Concurrent Pos:* Sigma Xi grant, 64-66; NIH grant, 66-69; writer of sci books for young people, 69- *Mem:* Am Soc Zool; Sigma Xi. *Res:* Planarian physiology, reproductive activity and behavior; life history of sexual versus asexual strains; ageing. *Mailing Add:* 504 Dakota St Norman OK 73069-7013

BROWN, MARK, radiology; deceased, see previous edition for last biography

BROWN, MARK WENDELL, b Riverside, NJ, July 11, 53. FRUIT ENTOMOLOGY, BIOLOGICAL CONTROL. *Educ:* Univ Maine, BSF, 75; Univ Idaho, MS, 77; Pa State Univ, MS & PhD(entom), 81. *Prof Exp:* Res asst, Univ Idaho, 75-77; teaching asst entom, Pa State Univ, 77-81; res assoc, Miss State Univ, 81-88; MEM STAFF, APPALACHIAN FRUIT RES STA, AGR RES SERV, USDA, 88- *Concurrent Pos:* Temp instr entom, Univ Maine, 80. *Mem:* Entom Soc Am; Entom Soc Can; Int Orgn Biol Control; Sigma Xi. *Res:* Analyzing community ecology of insects in deciduous fruit orchards to develop new pest management practices; study biological control of insects in deciduous fruit orchards to use for insect control. *Mailing Add:* Appalachian Fruit Res Sta 45 Wiltshire Rd Kearneysville WV 25430

BROWN, MARVIN ROSS, b Douglas, Ariz, July 25, 47; m 63; c 1. MEDICAL PHYSIOLOGY, INTERNAL MEDICINE. *Educ:* Univ Ariz, BS, 69, MD, 73. *Prof Exp:* Fel, 74-75, res assoc endocrinol, 75-76, ASST RES PROF, PEPTIDE BIOL LAB, SALK INST BIOL STUDIES, 76-; ASST ADJ PROF MED, SCH MED, UNIV CALIF, SAN DIEGO, 77- *Mem:* Soc Neurosci; Am Physiol Soc; Endocrine Soc; AAAS; Am Diabetes Asn. *Res:* Study of the physiology and pharmacology of brain hormones. *Mailing Add:* Salk Inst Bay Hosp Med Ctr 484 Avenida Primavera Del Mar CA 92014

BROWN, MAURICE VERTNER, b Durand, Mich, Jan 31, 08; m 34, 69; c 3. ACOUSTICS, GEOPHYSICS. *Educ:* Univ Mich, BS, 30, MS, 31; NY Univ, PhD(physics), 37. *Prof Exp:* Lab asst physics, Univ Mich, 30-31; from instr to prof, 31-74, EMER PROF PHYSICS, CITY COL NEW YORK, 74- *Concurrent Pos:* Instr, Pa State Col, 41; res physicist, Hudson Labs, Columbia Univ, 52-69; expert in acoust scattering, Naval Res Lab, Washington, DC, 69-73 & Naval Undersea Ctr, San Diego, 74-76. *Mem:* Acoust Soc Am; Am Phys Soc. *Res:* Artificial radioactivity; biophysics of electric fish. *Mailing Add:* 17373 Francisco Dr San Diego CA 92128

BROWN, MAX LOUIS, b Elgin, Ill, Apr 12, 36; m 63; c 4. MECHANICAL & SAFETY ENGINEERING. *Educ:* Univ Cincinnati, ME, 59; Ohio State Univ, MSc, 60, PhD (mech eng), 68. *Prof Exp:* Engr, Magna Am Corp, 53-60; instr math, 60-61, from instr to assoc prof mech eng, 61-75, actg head mech & indust eng dept, 81, PROF MECH ENG, UNIV CINCINNATI, 75-, DIR GRAD STUDIES & ASST HEAD MECH & INDUST ENG DEPT, 80- *Concurrent Pos:* Dir, Mach Tool & Mfg Tech Info Ctr, 68-70 & State Tech Serv Prog Referral Network Off, Cincinnati, 68-69; design, mfg & safety consult, Cincinnati Inc, Lunkenheimer Co, Honeywell Controls Div, Jacobson Co, Aetna Casualty & Surety Co, Liberty Mutual Inc Co, Aluminum Co Am, Ford Motor Co, Mobil Oil Co, Ryder Truck Rental, Taft Broadcasting Co, Ohio Valley Elec Co & numerous others. *Honors & Awards:* Ralph R Teetor Award, Soc Auto Engrs, 75. *Mem:* Am Soc Mech Engrs; Am Soc Eng Educ; Soc Mfg Engrs; Soc Auto Engrs. *Res:* Optimum design of mechanical parts; strength of materials; teaching; safety, engineering economics; machine design; dynamics of machines. *Mailing Add:* Dept Mech & Indust Eng Main Campus Univ Cincinnati Cincinnati OH 45221-0072

BROWN, MELVIN HENRY, b Victor, Iowa, Dec 6, 19; m 47; c 4. CHEMICAL ENGINEERING. *Educ:* Iowa State Univ, BS, 42, MS, 44, PhD(chem eng), 49. *Prof Exp:* Engr, Alcoa Labs, Aluminum Co Am, 49-60, sci assoc, 60-72, sr sci assoc chem eng, 72-80, tech consult, 81-86; RETIRED. *Mem:* Am Inst Chem Engrs; Am Chem Soc; Res Soc Am; NY Acad Sci. *Res:* Corrosion; desalination; waste treatment; inhibitors; heat treatment of aluminum alloys; alloy development; applications of aluminum alloys in the chemical industry; energy systems; aluminum smelting. *Mailing Add:* RR 2 Box 299 Morning Sun IA 52640

BROWN, MERTON F, b Burlington, Vt, Sept 22, 35; m 57; c 3. MYCOLOGY, FOREST PATHOLOGY. *Educ:* Univ Maine, BS, 61, MS, 63; Univ Iowa, PhD(bot), 66. *Prof Exp:* Asst prof biol, Wis State Univ, Superior, 66-68; asst prof, 68-72, assoc prof, 72-80, PROF FORESTRY & PLANT PATH, UNIV MO-COLUMBIA, 80- *Mem:* Mycol Soc Am; Am Phytopath Soc. *Res:* Ultrastructure and composition of fungal cell walls, plant pathogenic fungi and mycorrihizae. *Mailing Add:* Plant Path Univ Mo 108 Waters Columbia MO 65211

BROWN, MEYER, b Chicago, Ill, Oct 16, 10; m 34; c 3. NEUROLOGY, PSYCHIATRY. *Educ:* Univ Chicago, BS, 31, MD, 35; Northwestern Univ, MS, 37, PhD(nervous & ment dis), 39; Am Bd Psychiat & Neurol, dipl, 41. *Prof Exp:* Staff physician, Behav Clin, Chicago, 35-41; attend neuropsychiatrist, Student Health Serv, 39-61, assoc, 40-48, from asst prof to assoc prof neurol & psychiat, 48-73, prof clin neurol & clin psychiat, 73-82, EMER PROF CLIN NEUROL & CLIN PSYCHIAT, MED SCH, NORTHWESTERN UNIV, CHICAGO, 82- *Concurrent Pos:* Attend physician, Evanston Hosp, 47-62, head div psychiat, 62-69, head div neurol, 62-75. *Mem:* Fel AMA; fel Am Psychiat Asn; fel Am Acad Neurol. *Res:* Epilepsy; sensation of vibration; constitutional differences between deteriorated and non-deteriorated epileptics; spinal cord injuries; stroke. *Mailing Add:* 1500 Sheridan Rd Apt 7J Wilmette IL 60091

BROWN, MICHAEL, b Philadelphia, Pa, Jan 8, 39; m 63; c 2. INDUSTRIAL PHARMACY. *Educ:* Temple Univ, BS, 60; Purdue Univ, MS, 62, PhD(phys pharm), 64. *Prof Exp:* Develop chemist pharm, Lederle Lab, Div Am Cyanamid, 64-67; group leader, Carter Prod, Div Carter-Wallace, Inc, 67-74; mgr proprietaries & toiletries res & develop, Div Schering-Plough, Plough Inc, 74-76; dir prod develop & qual assurance, Consumer Prod Div, Becton Dickinson, 76-78; vpres tech affairs, L Perrigo Co, 78-86; PRES, BROWN ASSOCS, 86- *Mem:* Acad Pharmaceut Sci; Am Pharmaceut Asn; Sigma Xi. *Res:* Product research and development of proprietary, toiletries products; Product research and development of pharmaceuticals, proprietary & toiletry products; quality control and quality assurance; physical pharmacy; technical management, business planning. *Mailing Add:* 1235 St James Ct Holland MI 49424

BROWN, MICHAEL, b Hayes, Middx, UK, Mar 19, 47; c 3. METAMORPHIC PETROLOGY, IGNEOUS PETROLOGY. *Educ:* Univ Keele, UK, BA, 69, PhD(geol), 75. *Prof Exp:* Sr lectr, prin lectr & head, Dept Geol & Phys Sci, Oxford Polytech, UK, 72-84; head & prof geol, Sch Geol Sci, Kingston Polytech, UK, 84-90; PROF & CHMN, DEPT GEOL, UNIV MD, 90- *Concurrent Pos:* Subj ed, J Geol Soc London, UK, 81-86; founder & ed, J Metamorphic Geol, Blackwell, Oxford, 82-; asst dean acad affairs, Fac Sci, Kingston Polytechnic, UK, 86-89, vis prof, 90- *Mem:* Geol Soc Am; Am Geophys Union; Mineral Soc Am; Geol Asn Can; Geol Soc London; Mineral Soc Gt Brit. *Res:* Petrology of high-grade metamorphic rocks, their P-T-t histories and their tectonic interpretation; petrogenesis of migmatites and granites and crustal anatexis; origin of calc-alkaline plutonic rocks; strike-slip faults and transpression. *Mailing Add:* Dept Geol Univ Md College Park MD 20742-4211

BROWN, MICHAEL S, b New York, NY, Apr 13, 41. GENETIC ENGINEERING, MOLECULAR GENETICS. *Educ:* Univ Pa, BA, 62, MD, 66. *Hon Degrees:* DSC, Rensselaer Polytech Inst, 82, Univ Chicago, 82, Univ Pa, 86. *Prof Exp:* Clin assoc, NIH, 68-71; asst prof, Southwestern Med Sch, 71-74, PROF MED & GENETICS & DIR, CTR GENETIC DIS, UNIV TEX SOUTHWEST MED CTR, 77-, REGENTAL PROF, 85- *Concurrent Pos:* Estab investr, Am Heart Asn, 74-79; mem, Molecular Cytol Study Sect, NIH, 74-79 & ed adv bd, J Lipid Res, 79-83, J Cell Biol, 81-84 & Arteriosclerosis, 81. *Honors & Awards:* Nobel Prize in Physiol or Med, 85; Pfizer Award, Am Chem Soc, 74; Lounsbery Award, US Nat Acad Sci, 79; Res Achievement Award, Am Heart Asn, 84; US Nat Medal of Sci, 88. *Mem:* Nat Acad Sci; Am Soc Clin Invest; Asn Am Physicians; Fel Am Col Physicians; Am Acad Arts & Sci; Inst Med; Am Soc Biol Chemists. *Res:* Published numerous articles in various journals. *Mailing Add:* Dept Molecular Genetics Univ Tex Southwest Med Ctr Dallas TX 75235

BROWN, MORTON, b New York, NY, May 7, 31; m 57; c 3. POLYMER CHEMISTRY. *Educ:* Cornell Univ, AB, 52; Duke Univ, AM, 54; Mass Inst Technol, PhD(org chem), 57. *Prof Exp:* Fel, Mass Inst Technol, 57; res chemist, Cent Res Dept, 57-62 & Elastomer Chem Dept, 62-79, TECH CONSULT, POLYMER PROD DEPT, E I DU PONT DE NEMOURS & CO, INC, 80- *Mem:* Am Chem Soc; Inst Elec & Electronics Engrs. *Res:* Fluorinated organo-sulfur compounds; cyano-carbons; polyurethanes; interfacial polymerizations; thermoplastic elastomers; effect of polymer melt rheology on processing characteristics; high voltage rubber insulations. *Mailing Add:* 2409 Sweetbriar Rd Wilmington DE 19810-3413

BROWN, MORTON, b New York, NY, Aug 12, 31. MATHEMATICS. *Educ:* Univ Wis-Madison, BS, 53, MS, 55, PhD(math), 58. *Prof Exp:* Instr math, Univ Wis-Madison, 57 & Ohio State Univ, 57-58; Off Naval Res fel, 58-59, from asst prof to assoc prof, 59-64, PROF MATH, UNIV MICH, ANN ARBOR, 64- *Concurrent Pos:* NSF fel, Inst Advan Study, Princeton Univ, 60-62; Sloan Found fel, Cambridge Univ & Univ Mich, 63-65. *Honors & Awards:* Veblen Prize, Am Math Soc, 64. *Mem:* Am Math Soc; Math Asn Am. *Res:* Topology. *Mailing Add:* Dept Math Univ Mich Ann Arbor MI 48109

BROWN, MURRAY ALLISON, b Rochester, NY, Oct 29, 27; m 48; c 3. DAIRY SCIENCE. *Educ:* Mich State Univ, BS, 50; Tex A&M Univ, MS, 53, PhD(animal breeding), 56. *Prof Exp:* Instr dairy prod, Tex A&M Univ, 53 & 55-57, from asst prof to assoc prof dairy sci, 57-65; PROF AGR, SAM HOUSTON STATE UNIV, 65- *Concurrent Pos:* Animal sci consult. *Mem:* AAAS; Am Dairy Sci Asn; Nat Asn Col & Teachers Agr (pres, 69-70, secy-treas, 72-). *Res:* Bovine reproductive physiology; artificial insemination; dairy cattle breeding. *Mailing Add:* Sam Houston State Univ Dept Agr PO Box 2088 Huntsville TX 77341

BROWN, MYRTLE LAURESTINE, b Columbia, SC, June 1, 26. NUTRITION. *Educ:* Bennett Col, BS, 45; Pa State Univ, MS, 48, PhD(nutrit), 57. *Prof Exp:* Nutrit specialist, Human Nutrit Res Div, Agr Res Serv, USDA, 48-63; assoc prof home econ, Univ Hawaii, 63-65, nutrit, 65-66, pub health, 66-69; assoc prof nutrit, Rutgers Univ, 69-72, prof, 72-74; staff officer, Nat Acad Sci, 74-76, exec secy, Food & Nutrit Bd, 77-83; vis prof, Va Polytech Inst & State Univ, 83-84, assoc dean grad sch, 84-86, prof, 86-88, EMER PROF BIOCHEM & NUTRIT, VA POLYTECH INST & STATE UNIV, 88- *Mem:* AAAS; Am Inst Nutrit; Inst Food Technologists; Sigma Xi. *Res:* Nutrition and reproduction; growth and development. *Mailing Add:* 201 Fincastle Dr Blacksburg VA 24060

BROWN, NANCY J, b Erie, Pa, Apr 19, 43; m 74; c 1. CHEMICAL PHYSICS, CHEMICAL DYNAMICS. *Educ:* Va Polytech Inst, BS, 64; Univ Md, MS, 69, PhD(phys chem), 71. *Prof Exp:* Fel, Appl Phys Lab, Johns Hopkins Univ, 71-73; vis scholar combustion, 73, lectr chem, 73-75, res engr combustion, 75-77, staff scientist, 76-84, DEP PROG LEADER, ENVIRON RES GROUP, LAWRENCE BERKELEY LAB, UNIV CALIF, BERKELEY, 79-, SR SCIENTIST, 84- *Concurrent Pos:* Adv comt, Calif State Energy Resources Conserv & Develop Comn, 75-77; affil fac, energy & resources group, Univ Calif, Berkeley, group leader combustion res, 86, actg prog leader environ res, 90; Gov appointee, State Calif Acid Deposition Adv Comt; prof invitee, Univ Pierre & Marie Curie, Paris VI. *Mem:* Am Chem Soc; Am Phys Soc; Combustion Inst; Asn Women Sci. *Res:* Theoretical and experimental chemical kinetics; combustion chemistry; chemical dynamics; theoretical chemistry; energy transfer; air pollution; spectroscopy; theoretical studies of reactivity and energy transfer; functional sensitivity analysis of the interrelationship between dynamics and potential structures; modeling of combustion processes. *Mailing Add:* Bldg 29C Lawrence Berkeley Lab Univ Calif Bldg 29 C Berkeley CA 94707

BROWN, NEAL B, b Moscow, Idaho, Dec 13, 38; m 64, 74, 80; c 5. LOW LIGHT LEVEL SPECTROSCOPY. *Educ:* Wash State Univ, BS, 61; Univ Alaska, MS, 66. *Prof Exp:* Researcher, NASA Ames Res Ctr, Moffett Field, Calif, 61-62 & Am Geophys Soc, 62-63; asst geophysicist, 68-81, ASST PROF PHYSICS, UNIV ALASKA, 81- *Concurrent Pos:* Researcher, Geophys Inst, Univ Alaska, 68-71, dir, Poker Flat Res Range, 71-88. *Mem:* Sigma Xi; AAAS; Am Geophys Union; Am Inst Aeronaut & Astronaut. *Res:* Near-infrared spectral studies of aurora & air glow; low light level color television studies of aurora; low light level imaging of transient phenomena. *Mailing Add:* 1569 La Rue Lane Fairbanks AK 99709

BROWN, NEAL CURTIS, b Vassalboro, Maine, Mar 15, 39; m 64; c 2. BIOCHEMISTRY, PHARMACOLOGY. *Educ:* Cornell Univ, DVM, 62; Yale Univ, PhD(biochem pharmacol), 66. *Prof Exp:* USPHS fel, Dept Chem, Karolinska Inst, Sweden, 66-68, Swed Cancer Soc fel, 68-69; from asst prof to assoc prof cell biol & pharmacol, Sch Med, Univ Md, 69-73; assoc prof biochem, 73-74, PROF PHARMACOL & CHMN DEPT, MED SCH, UNIV MASS, 74- *Mem:* Fedn Am Socs Exp Biol. *Res:* Metabolism of DNA and DNA precursors and its control in vitro and in vivo in prokaryotic and eukaryotic cells. *Mailing Add:* Dept Pharmacol Univ Mass Med Sch Worcester MA 01655

BROWN, NEIL HARRY, b Whitehall, NY, July 12, 40; m 63; c 4. PHARMACEUTICS, PACKAGING. *Educ:* Albany Col Pharm, BS, 65. *Prof Exp:* Lab adminr med technol, Children's Hosp, Albany, 65-66; sr res pharmacist, 67-82, PACKAGING RES & DEVELOP MGR, STERLING RES GROUP, STERLING DRUG INC, 82- *Mem:* Am Soc Clin Pathologists; Packaging Inst Am; Am Soc Testing & Mat; Parenteral Drug Asn. *Res:* Biopharmaceutics; preformulation; dissolution; physical and chemical stability of drug dosage forms; packaging. *Mailing Add:* Sterling Res Group Columbia Turnpike Rensselaer NY 12144

BROWN, NIGEL ANDREW, b Sunderland, Eng, Jan 27, 53. TERATOLOGY, BIOCHEMICAL TOXICOLOGY. *Educ:* Univ Leeds, BSc, 73; Univ Surrey, PhD(biochem), 77. *Prof Exp:* Demonstr biochem, Univ Surrey, Eng, 73-75, 76-77; res fel pharmacol, Pharmakologisches Inst Univ Mainz, Ger, 75-76; vis fel toxicol, Nat Inst Environ Health Sci, NC, 77-78; asst prof pharmacol, George Washington Univ, 78-; AT EXP EMBRYOL & TERATOLOGY UNIT, MED RES COUN LABS. *Mem:* Biochem Soc; AAAS; Fedn Am Socs Exp Biol. *Res:* Mechanisms of teratogenesis; toxicity in reproduction; xenobiotic metabolism; ligand-albumin interactions. *Mailing Add:* Med Res Coun Exp Embryol & Teratology Unit St George's Hosp Med Sch Cranmer Terr London SW17 ORE England

BROWN, NORMAN, b Lynn, Mass, Feb 7, 21; m 43; c 6. METALLURGY. *Educ:* Mass Inst Technol, BS, 43; Stanford Univ, MS, 50; Univ Calif, Berkeley, PhD(metall), 52. *Prof Exp:* Jr metallurgist, Naval Torpedo Sta, RI, 43-44; metallurgist, Ballistics Res Lab, Aberdeen Proving Ground, 46-48; from asst prof to assoc prof metall, 52-60, PROF MAT SCI, UNIV PA, 60- *Concurrent Pos:* Guggenheim fel, Cavendish Lab, 58-59; vis prof, Bristol Univ, 66-67; NIH spec fel, 66-67; consult, Australian Gas Ltd, Czech Gas Co, Philadelphia Gas, Tsaka Gas; vis prof physics dept, Univ Leeds, 81-82. *Mem:* Am Soc Metals; Am Phys Soc; Fel Am Phys Soc. *Res:* Mechanical behavior of polymers. *Mailing Add:* Dept Metall & Mat Sci Univ Pa Philadelphia PA 19104

BROWN, NORMAN LOUIS, b Atlantic City, NJ, Dec 22, 23; m 57; c 3. METHANE GENERATION FROM HUMAN ANIMAL & AGRICULTURAL WASTES, PHYSICAL CHEMISTRY. *Educ:* Mass Inst Technol, SB, 47; Brown Univ, PhD(phys chem), 52. *Prof Exp:* Phys chemist, Electronics Lab, Gen Elec Co, 51-56; physicist, Nat Bur Standards, 57-63 & Nat Ctr Fish Protein Concentrate, Bur Com Fisheries, US Dept Interior, 63-67; prog leader, Fish Protein Concentrate Eng Res & Develop Prog, 67-70; prof assoc, Bd Sci & Technol for Int Develop, Off Foreign Secy, Nat Acad Sci, 70-75; country prog specialist, Off Asst Adminr Int Affairs, ERDA, 75-77, chief tech adv, Develop Country Prog Off, US Dept Energy, 77-78; spec adv energy to asst adminr & Chief, Div Sci, Tech & Environ Problems, Bur for Asia, USAID, 78-81; CONSULT, 81- *Concurrent Pos:* Consult, AID, World Bank & UN projects in Asia, Africa, Caribbean & the Mid East. *Mem:* Fel AAAS; Int Solar Energy Soc; Soc Int Develop; Am Solar Energy Soc. *Res:* Manufacture and use of fish and other protein concentrates; applications of science and technology to problems of development; small-scale renewable-resource energy systems for developing countries. *Mailing Add:* 1746 Q St NW Washington DC 20009

BROWN, OLEN RAY, b Hastings, Okla, Aug 18, 35; m 58; c 3. OXYGEN RADICALS, EXPERT WITNESS. *Educ:* Univ Okla, BS, 58, MS, 60, PhD(microbiol), 64; Am Bd Toxicol, dipl, 89-94. *Prof Exp:* Spec instr microbiol, Univ Okla, 63-64; from instr to assoc prof, 64-77, asst dir, John M Dalton Res Ctr, 74-78, RES INVESTR, UNIV MO-COLUMBIA, 68-, PROF MICROBIOL, 77-, PROF VET BIOMED SCI, 87- *Concurrent Pos:* Peer & site reviewer, NIH & Nat Inst Environ Health Sci; guest lectr, Ross Univ, St Kitts, 84 & 88; expert witness/legal consult, Tech Adv Serv Attorneys; consult, Drug Abuse Policy Off, Washington, DC, 82. *Mem:* Fel Am Inst Chem; Soc Toxicol; Oxygen Soc; Am Chem Soc; Int Soc Study of Xenobiotics; Soc Free Radical Res. *Res:* Mechanisms of toxicity of oxygen and redox-active chemicals at the cellular level in bacteria and tissues; effects of hyperoxia and redox-active chemicals such as paraquat, adriamycin and nitrofurantoin on growth, respiration, permeability, oxidative phosphorylation and on enzymes and coenzymes of lipid metabolism; inhibition of amino acid and nicotinamide adenine dinucleotide biosynthesis, induction of genetic stringency and prevention of these toxicity sites. *Mailing Add:* John M Dalton Res Ctr Univ Mo Columbia MO 65211

BROWN, OLIVER LEONARD INMAN, b Webster City, Iowa, Nov 15, 11; m 43; c 4. PHYSICAL CHEMISTRY. *Educ:* Univ Iowa, AB, 32, MS, 33; Univ Calif, PhD(phys chem), 36. *Prof Exp:* Instr gen & phys chem, Univ Mich, 36-41; from instr to asst prof chem, US Naval Acad, 41-46; asst prof, Syracuse Univ, 46-52; prof & chmn dept, 52-74, Lucretia L Allyn prof, 52-77, EMER PROF CHEM, CONN COL, 77- *Mem:* Am Chem Soc. *Res:* Low temperature heat capacities; heats of solution; free energies from equilibrium measurements; data of state of gases; vapor pressure relationships. *Mailing Add:* Box 162 Quaker Hill CT 06375-0162

BROWN, OLIVER MONROE, b Ithaca, NY, June 11, 44; m 70. NEUROCHEMISTRY, NEUROTRANSMITTERS. *Educ:* Long Beach City Col, AA, 64; Calif State Col, Long Beach, BS, 66; Kans State Univ, PhD(biochem), 70. *Prof Exp:* Fel pharmacol, Mt Sinai Sch Med, 70-71; from instr to asst prof, 71-84, ASSOC PROF PHARMACOL, STATE UNIV NY, UPSTATE MED CTR, 84-; INVESTR, MT DESERT ISLAND BIOL LAB, MAINE, 87- *Concurrent Pos:* Consult, 75-; Markey fel, 87. *Mem:* AAAS; NY Acad Sci; Am Soc Pharmacol & Exp Therapeut; Am Heart Asn; Am Pub Health Asn. *Res:* Chemical approaches to autonomic control of the mammalian heart; drug alterations of neurotransmitters in the brain; mechanisms of neurotransmitter release; toxicology, pharmacology of alcohol. *Mailing Add:* Pharmacol Dept State Univ NY Health Sci Ctr 766 Irving Ave Syracuse NY 13210

BROWN, OTTO G, b San Antonio, Tex, Sept 11, 25; m 50. MECHANICAL ENGINEERING, MATHEMATICS. *Educ:* Univ Okla, BS, 48; Univ Tex, MS, 57, PhD(mech eng), 62. *Prof Exp:* Instr mech eng, Univ Tex, 54-60; PROF MECH ENG & HEAD DEPT, LAMAR UNIV, 62- *Concurrent Pos:* Res ctr grant, Lamar Univ, 62- *Mem:* Am Soc Eng Educ; Am Soc Mech Engrs; Sigma Xi. *Res:* Fluid mechanics; turbulent flow. *Mailing Add:* 14914 Moss Arch San Antonio TX 78232-4655

BROWN, PATRICIA LYNN, b Lafayette, La, Oct 1, 28. INFORMATION SCIENCE. *Educ:* Univ Southwestern La, BS, 47; Univ Tex, MA, 49. *Prof Exp:* Instr anal chem, Smith Col, 49-50; chemist, R&M Labs, 50; res assoc indust toxicol, Albany Med Col, 50-51; lit searcher info serv & res & eng dept, Ethyl Corp, 51-53, reference librn, 53-54, lit specialist, 54, ed asst, 54-55; sr tech writer & ed pub eng, Atomic Power Div, Westinghouse Elec Corp, 55-57, staff engr, 57; supvr info serv, Semiconductor-Components Div, Tex Instruments, Inc, 57-61, mgr info serv, Corp Res & Eng, 61-64, tech info consult, 64-66; sr researcher, Battelle Mem Inst, 66-76; assoc dir, Travenol Labs, Inc, 76-82, tech info counr, 83-86; MGR TECH INFO SERV, STEPAN CO, 87- *Mem:* Am Chem Soc; Spec Libr Asn; Soc Women Engrs (pres, 61-63). *Res:* Analysis and design of information systems; techniques of document preparation, acquisition, indexing, condensation, storage, retrieval, dissemination and use; design and development of methods for useful organization and presentation of information. *Mailing Add:* 8837 Major Ave Morton Grove IL 60053

BROWN, PATRICIA STOCKING, b Cadillac, Mich, Apr 25, 42; m 65; c 2. ZOOLOGY, ENDOCRINOLOGY. *Educ:* Univ Mich, BS, 63, MS, 66, PhD(zool), 68. *Prof Exp:* Res assoc develop biol, State Univ NY Albany, 68-69; from asst prof to assoc prof biol, 69-80, dept head, 84-86 & 88-91, PROF BIOL, SIENA COL, NY, 80- *Concurrent Pos:* Mellon-Wellsley fel, 78-79; NSF fel, Univ Calif, Berkeley, 80-81; vis prof, Univ Bath, Eng, 87-88. *Mem:* Am Soc Zool; Am Soc Soc Icthyologists & Herpetologists; Endocrin Soc; Asn Women Sci; AAAS; History Sci Soc. *Res:* Hormonal control of growth in amphibians and reptiles; salt and water balance in urodeles; function of prolactin in lower vertebrates; history of women scientists. *Mailing Add:* Dept Biol Siena Col Loudonville NY 12211

BROWN, PATRICK MICHAEL, b North Tonawanda, NY, 1938; m 59; c 4. PHYSICAL INORGANIC CHEMISTRY. *Educ:* Univ Miss, BS, 60; Syracuse Univ, PhD, 65. *Prof Exp:* Fel, State Univ NY, Buffalo, 65-66; chemist, W R Grace & Co, 66-73, sr res chemist, 73; pres, Biolytic's Inc, 73-74; dir res, 76-88, MGR CHEM RES, CYPRUS FOOTE MINERAL CO, 74-, SR TECH CONSULT TO PRES, 89- *Mem:* Sigma Xi; Am Chem Soc; Am Ceramic Soc; Am Inst Mining Metall & Petrol Engrs. *Res:* Catalysis; materials science; lithium chemistry. *Mailing Add:* Cyprus Foote Mineral Co 301 Lindewood Dr Malvern PA 19355-1740

BROWN, PAUL B, b Panama City, Panama, Nov 29, 42; US citizen; m 68. NEUROPHYSIOLOGY. *Educ:* Mass Inst Technol, BS, 64; Univ Chicago, PhD(physiol), 68. *Prof Exp:* NIH fel neurophysiol, Cornell Univ, 68-72; res scientist, Boston State Hosp, 72-74; from asst prof to assoc prof, 74-82, PROF PHYSIOL, MED CTR, WVA UNIV, 82- *Concurrent Pos:* Chief ed, J Electrophysiol Tech, 78-88; Fogarty sr int fel, Univ Edinburgh, 80-81. *Mem:* Soc Neurosci; AAAS; Int Soc Neurol Networks. *Res:* Mechanisms of somesthesis; spinal cord neurophysiology; instrumentation for neurophysiology; neuroanatomy; plasticity. *Mailing Add:* Dept Physiol WVa Univ Med Ctr Morgantown WV 26506

BROWN, PAUL EDMUND, b Schuyler Co, Ill, Oct 17, 16; m 52; c 2. CHEMICAL REACTIONS IN HIGH-TEMPERATURE WATER. *Educ:* Western Ill Univ, BEd, 43; Purdue Univ, PhD(phys chem), 49. *Prof Exp:* Engr chem, Atomic Power Div, Westinghouse Corp, 49-51, sr scientist, 51-57, fel engr, 57-59, supv engr, Naval Reactor Facil, 59-61, fel engr, Atomic Power Div, 61-62, supv engr, 62-70, adv sci chem, 70-88; supvr engr chem & thermal hydraulics, Westinghouse Reactor Facil, 59-61; RETIRED. *Concurrent Pos:* Instr, Carnegie-Mellon Univ, 61-63. *Mem:* Fel AAAS; Am Chem Soc; Sigma Xi. *Res:* Nuclear plant decontamination; nuclear plant (primary and secondary) water chemical control; steam generator chemical cleaning. *Mailing Add:* 1531 Redfern Dr Pittsburgh PA 15241-2936

BROWN, PAUL JOSEPH, b St Louis, Mo, April 18, 24; m 51; c 2. BATTERY TECHNOLOGY, ELECTRIC VEHICLE TECHNOLOGY. *Educ:* Wash Univ, BSME, 49, MEA, 64. *Prof Exp:* Prod mgr, Universal Match Corp, St Louis, 56-64; mkt mgr, Am Mach & Foy, York, Pa, 64-66; dir, Off Vehicle Syst Res, Nat Bur Standards, 66-71; dir, Safety Res Lab, US Dept Transp, 71-76; DIR ELEC & HYBRID VEHICLE DIV, US DEPT ENERGY, 76- *Concurrent Pos:* Adv mech engr, Emerson Elec, St Louis, Mo, 48-56 auto safety expert, White House Sci Adv Comn, 71-72; mem adv comt, Stapp Car Crash Conf, 84- *Mem:* Soc Automotive Engrs. *Res:* Management of the federal governments research & development program of electric & hybrid vehicle technologies including batteries, fuel cells, propulsion systems & prototype vehicles. *Mailing Add:* 10611 Glenwild Rd Silver Spring MD 20901

BROWN, PAUL LAWSON, b Ash Grove, Kans, June 1, 18; m 46; c 2. SOIL MANAGEMENT, SOIL FERTILITY. *Educ:* Kans State Univ, BS, 41, MS, 49; Iowa State Univ, PhD(soil fertil), 56. *Prof Exp:* Soil surveyor, Kans Agr Exp Sta, 46-48, soil scientist, Ft Hays Exp Sta, 48-56, res soil scientist, Agr Res Serv, Mont State Univ, 56-73, res soil scientist, Agr Res Serv, Northern Plains Soil & Water Res Ctr, Ft Benton, Mont, 73-80, soil scientist, Agr Res Serv, Mont State Univ, USDA, 81-83; SOIL SCIENTIST, MONT STATE UNIV, 83- *Mem:* Fel Soil Sci Soc Am; fel Am Soc Agron; Can Soc Soil Sci; Int Soc Soil Sci; Sigma Xi; Soil Conserv Soc Am. *Res:* Soil management for dryland crop production; soil, crop and land management for saline seep control. *Mailing Add:* 1624 S Third Ave Bozeman MT 59715

BROWN, PAUL LOPEZ, b Anderson, SC, Feb 18, 19; m 44; c 4. ZOOLOGY. *Educ:* Knoxville Col, BS, 41; Univ Ill, MS, 48, PhD(zool), 55. *Prof Exp:* From instr to assoc prof biol, Southern Univ, 48-58; prof, Fla Agr & Mech Univ, 58-59; prof biol & head dept, Norfolk State Univ, 59-74; acad dean, Clark Col, 74-78; prof biol, Atlanta Univ, 78-82, dean, Sch Arts & Sci, 82-87; RETIRED. *Concurrent Pos:* Mem staff, NSF, 63-64 & 74, consult, 64-; consult, Off Educ, US Dept Health, Educ & Welfare, 64- *Mem:* Am Soc Zoologists; Am Micros Soc. *Res:* Biology of crayfishes including taxonomy, life history, distribution, limnology of waters inhabited; ecology. *Mailing Add:* 855 Flamingo Dr SW Atlanta GA 30311

BROWN, PAUL WAYNE, b Cleveland, Ohio, Oct 31, 35; m 60; c 3. CHEMISTRY, ENGINEERING. *Educ:* Ohio State Univ, BChE, 60. *Prof Exp:* Res engr, M&R Dietetic Labs, Ohio, 60-62, group leader res & develop, 62-65; mgr process eng, Abbott Labs, 65-66, dir qual assurance, 66-69, mgr prod systs & environ control, 69-71, plant eng mgr, Ross Labs Div, 71-76, mgr eng opers, Hosp Prod Div, 76-77, dir eng, Hosp Prod Div, 77-91; CONSULT, 91- *Mem:* Am Inst Chem Engrs. *Res:* Spray drying, especially air-spray jet spatial configuration on drying rate and physical properties; development of automatic batch blend processing techniques for processing infant formulas and IV solutions. *Mailing Add:* 6211 Camino Arturo Tuscon AZ 85718

BROWN, PAUL WHEELER, b Teaneck, NJ, Mar 12, 36; m 57, 82; c 4. NEUROVIROLOGY, NEUROEPIDEMIOLOGY. *Educ:* Harvard Univ, AB, 57; Johns Hopkins Univ, MD, 61; Am Bd Internal Med, cert, 69. *Prof Exp:* Intern & resident, Johns Hopkins Hosp, 61-63; staff assoc virol, USPHS, 63-71, MED DIR USPHS, NAT INST NEUROL & COMMUN DIS & STROKE, NIH, 72- *Concurrent Pos:* Resident, Univ Calif, San Francisco, 65-66 & Johns Hopkins Hosp, 66-67; fel, Nat Multiple Sclerosis Soc, 71-72 & Comt Control Huntington's Dis, 73-74; vis scientist virol, NIH & Med Res, Paris, France, 71-72 & 77-78. *Mem:* Am Col Physicians; Am Epidemiol Soc; Infectious Dis Soc Am; French Neurol Soc. *Res:* Study of slow and latent central nervous system viruses. *Mailing Add:* Bldg 16 Rm 5B20 NIH Bethesda MD 20892

BROWN, PAUL WOODROW, b New York, NY, Dec 28, 19; c 3. HAND SURGERY, ORTHOPEDIC SURGERY. *Educ:* Univ Mich, BS, 42, MD, 50. *Prof Exp:* Resident orthop surg, Letterman Army Hosp, San Francisco, 51-55; fel hand surg, Walter Reed Army Hosp, Washington, DC, 62-63; div chmn orthop, Fitzsimons Gen Hosp, Denver, 66-69; prof surg, Med Ctr, Univ Colo, Denver, 69-72; prof, Sch Med, Univ Miami, 72-74; chmn dept surg, St Vincent's Hosp, 74-83; CLIN PROF ORTHOP PLASTIC & RECONSTRUCT SURG, SCH MED, YALE UNIV, 78- *Concurrent Pos:* Colonel, US Army, retired. *Mem:* Am Col Surg; Am Acad Orthop Surg; Am Soc Surg of the Hand; Int Soc Orthop Surg & Traumatol. *Res:* Reconstructive surgery of the hand; peripheral nerve lesions; mechanics of wound healing; fate of exposed bone. *Mailing Add:* 3101 Main St Bridgeport CT 06606

BROWN, PETER, b Lincoln, Eng, Sept 12, 38. ORGANIC CHEMISTRY. *Educ:* Univ Southampton, BSc, 61, PhD, 64. *Prof Exp:* Res assoc mass spectrometry, Stanford Univ, 64-67; from asst prof to assoc prof, 67-77, PROF ORG CHEM, ARIZ STATE UNIV, 77- *Mem:* Am Chem Soc; Royal Soc Chem. *Res:* Cycloaddition reactions; photochemistry; mass spectrometry. *Mailing Add:* 55 Benslow Rise Hitchin SG4 908 Herts England

BROWN, PETER, b Durham, Eng, Sept 14, 38; m 66; c 2. TEXTILE PHYSICS. *Educ:* Univ Leeds, BSc, 62, PhD(knitting), 65. *Prof Exp:* Sr lectr textile sci, Leicester Polytech, UK, 65-68; asst prof knitting, NC State Univ, Raleigh, 68-75; consult, Inst Textile Technol, 75-76; consult, 76-77; ASSOC PROF TEXTILE PHYSICS, UNIV MINN, ST PAUL, 77- *Concurrent Pos:* Mem US Adv Comt, Textile Inst, 72- *Mem:* Textile Inst; Am Asn Textile Chemists & Colorists. *Res:* Fabric performance; garment performance safety aspects of textile products including garment flammability; psychophysical properties of textile materials. *Mailing Add:* Two Aitkin Close Norwich NR7 8BB England

BROWN, PHILIP EDWARD, b St Louis, Mo, July 12, 52; m 74; c 3. METALLIC MINERAL DEPOSITS, METAMORPHIC PETROLOGY. *Educ:* Carleton Col, BA, 74; Univ Mich, MS, 76, PhD(geol), 80. *Prof Exp:* Teaching asst geol, Univ Mich, 74-79, instr econ geol, 79-80; NSF fel, US Geol Surv, Reston, 80-81; asst prof, 81-87, ASSOC PROF ECON GEOL, UNIV WIS-MADISON, 87- *Mem:* Am Geophys Union; Soc Econ Geologists; Mineral Soc Am; Mineral Soc Can; Sigma Xi; Geol Asn Can. *Res:* Stable isotopic and geochemical studies of metamorphic and hydrothermal fluids and the origin of contact metamorphic and metasomatic ore deposits; evolution of Archean terranes. *Mailing Add:* Dept Geol & Geophys Univ Wis Madison WI 53706

BROWN, PHYLLIS R, b Providence, RI, Mar 16, 24; m 45; c 4. CHEMISTRY, BIOCHEMISTRY. *Educ:* George Washington Univ, BS, 44; Brown Univ, PhD(chem), 68. *Prof Exp:* Res asst chem, Harris Res Lab, 44-45; fel & res assoc, Dept Pharmacol, Brown Univ, 68-71, from instr to asst prof, 71-73; from asst prof to assoc prof, 73-80, PROF CHEM, UNIV RI, 80- *Concurrent Pos:* Vis prof, Hebrew Univ. *Honors & Awards:* Tswett Medal Chromatography, 88; Dal Nogere Award Chromatography, 89; Fulbright, Israel, 87. *Mem:* Am Chem Soc; Am Inst Chemists; Am Asn Clin Chemists; Am Soc Pharmacol & Exp Therapeut; Am Soc Biol Chemists; Sigma Xi; NY Acad Sci. *Res:* Application of high pressure liquid chromatography to biomedical research with special emphasis on the analysis of nucleotides and other nucleic acid components. *Mailing Add:* Dept Chem Pastore Lab Univ RI Kingston RI 02881

BROWN, R(ONALD) A(NDERSON) S(TEVEN), b Winnipeg, Man, June 23, 34; m 62; c 3. CHEMICAL ENGINEERING. *Educ:* Univ Alta, BSc, 56, PhD(chem eng), 63; Calif Inst Technol, MS, 57. *Prof Exp:* Asst res officer, Eng Div, Res Coun Alta, 63-70, assoc res officer, 70-81; consult, Chinook Fuel Innovations Ltd, 81-84; res mgr, Pembine Resources Ltd, 85; INDEPENDENT CONSULT, 86- *Mem:* Am Inst Chem Engrs; Eng Inst Can; Chem Inst Can; Sigma Xi. *Res:* Immiscible fluids; solids pipelining; combustion of solid fuels; high temperature processing; underground coal gasification; fuel technology; coal conversion. *Mailing Add:* 8923 142nd St Edmonton AB T5R 0M6 Can

BROWN, RANDALL EMORY, b Eugene, Ore, May 28, 17; m 50; c 2. WASTE DISPOSAL. *Educ:* Stanford Univ, AB, 38; Yale Univ, MS, 41. *Prof Exp:* Geologist, Glacier Peak Mine, M A Hanna Co, 41, State Dept Geol & Mineral Industs, Ore, 41-42, US Geol Surv, 42-45 & Corps Engrs, US Army, 45-47; sr geologist, Hanford Atomic Prod Opers, Gen Elec Co, 47-65; sr res scientist, Water & Land Resources Dept, Pac Northwest Lab, Battelle Mem Inst, 65-71; asst prof geol, Cent Wash Univ, 71-72; instr, Columbia Basin Col, 72-73; sr res scientist, Pac Northwest Labs, Battelle Mem Inst, 73-82; instr, Joint Ctr Grad Study, Richland, 73-78; SUPVR, FRANKLIN COUNTY CONSERV DIST, 73- *Concurrent Pos:* Adj assoc prof, Cent Wash Univ, 67-69, 79-82; geol consult, 71- *Mem:* Fel AAAS; fel Geol Soc Am; Sigma Xi; Asn Ground Water Scientists & Engrs. *Res:* Ground water geology, geology and hydrology of disposal of radioactive and other wastes; geology of eastern Washington; paleoecology. *Mailing Add:* 504 Rd 49 N Pasco WA 99301

BROWN, RAY KENT, b Columbus, Ohio, Apr 7, 24; m 47; c 3. PROTEIN CHEMISTRY. *Educ:* Ohio State Univ, AB, 44, MD, 47, MS, 48; Harvard Univ, PhD(biochem), 51. *Prof Exp:* Intern, Boston City Hosp, 47-48; sr asst surgeon protein chem, NIH, 51-53; asst dir, Div Labs & Res, State Dept Health, NY, 53-59, assoc dir, 59-63; chmn dept, 63-87, PROF BIOCHEM, SCH MED, WAYNE STATE UNIV, 63- *Concurrent Pos:* From assoc prof to prof, Albany Med Sch, 56-63. *Mem:* Am Chem Soc; Am Asn Immunol; Am Soc Biol Chem; Brit Biochem Soc. *Res:* Plasma proteins; immunologic studies of ribonuclease and thyroglobulin; photolabeling of proteins; role of ds RNA. *Mailing Add:* Dept Biochem Wayne State Univ Detroit MI 48201

BROWN, RAYMOND RUSSELL, b Calgary, Can, Dec 23, 26; US citizen. IMMUNOLOGY & CANCER, AMINO ACID METABOLISM IN INFLAMMATORY DISEASES & EOSINOPHILIA-MYALGIA SYNDROME. *Educ:* Univ Alta, BS, 48, MS, 50; Univ Wis-Madison, PhD(oncol), 53. *Prof Exp:* Instr chem, Univ Alta, Edmonton, 48-50; res asst oncol, Univ Wis-Madison, 50-53, postdoctoral fel clin oncol, 53-55, asst prof clin oncol, 55-60, assoc prof surg, 60-65, PROF CLIN ONCOL, UNIV WIS-MADISON, 65- *Concurrent Pos:* Res collabr, Brookhaven Nat Lab, 61-89;

affil prof, Dept Nutrit Sci, Univ Wis, 68-91; vis prof, McEachern Cancer Lab, Edmonton, 69-70. *Mem:* Am Asn Cancer Res; Am Inst Nutrit; Am Inst Clin Nutrit; Am Soc Biol Chemists. *Res:* Tryptophan metabolism; immunological diseases and tryptophan; interferon and interleukin metabolic effects; eosinophilia/myalgia syndrome and tryptophan metabolism; nutrition and cancer relationships; autoimmune diseases and AIDS. *Mailing Add:* 2817 Van Hise Ave Madison WI 53705

BROWN, RHODERICK EDMISTON, JR, b Hendersonville, NC, Apr 7, 53; m 81; c 1. BIOLOGICAL-MODEL MEMBRANES, LIPID TRANSPORT. *Educ:* Univ NC, AB(chem) & AB(zool), 75; Wake Forest Univ, PhD(biochem), 81. *Prof Exp:* Res assoc, Univ Va, 81-86; ASST PROF, HORMEL INST, UNIV MINN, 86- *Concurrent Pos:* Lectr, Biochem Dept, Mayo Clin, 89- *Mem:* Biophys Soc; Am Soc Cell Biol; AAAS; Soc Exp Biol & Med; Electron Micros Soc Am. *Res:* Structure and function of glycolipids and sphingolipids in biological and model membranes; spontaneous and protein-mediated intermembrane lipid transfer with emphasis on glycolipids and gangliosides; intracellular lipid transport; protein interactions with glycolipids and gangliosides; effects of lyso-glycolipids on membrane properties; structure and function of skin lipids. *Mailing Add:* Hormel Inst Univ Minn 801 16th Ave NE Austin MN 55912-3698

BROWN, RICHARD DEAN, b Mansfield, Ohio, Feb 6, 41; m 63; c 2. BIOACOUSTICS, RAPTOR BIOLOGY. *Educ:* Columbia Union Col, BA, 64; Ohio State Univ, MS, 70, PhD(zool), 75. *Prof Exp:* Chmn dept sci, Garden State Acad, 66-70; asst prof, 75-83, cur, Zool Mus, 76-83, ADJ PROF BIOL, UNIV NC, CHARLOTTE, 83- *Concurrent Pos:* Environ consult, 75-; founder & exec dir, Carolina Rubber Ctr, Inc, 80-86; exec dir, CRC Found, Inc, 86-87. *Mem:* Hawk Migration Asn NAm; Wilson Ornith Soc; Raptor Res Found. *Res:* Animal communication and associated behavior in social systems with emphasis on recording, playing back and analyzing bird songs; migratory bird and hawk studies involving banding; raptor behavior and ecology; wildlife inventories. *Mailing Add:* Dept Biol Univ NC Charlotte NC 28223

BROWN, RICHARD DON, b Alexandria, La, Mar 3, 40; m 63; c 1. PHARMACOLOGY, TOXICOLOGY. *Educ:* La Col, BS, 64; La State Univ, MS, 66, PhD(pharmacol), 68. *Prof Exp:* Instr pharmacol, Sch Med, La State Univ, New Orleans, 68-69; from asst prof to assoc prof 69-76, coordr grad studies, 78-81, PROF PHARMACOL, SCH MED, LA STATE UNIV, SHREVEPORT, 76- *Concurrent Pos:* USPHS gen res support grant, 68-69; Southern Med Asn res grant, 70-71; La State Univ Found res grant, 73-74; Nat Inst Gen Med Sci res grant, 75-78. *Honors & Awards:* Frank R Blood Award, Soc Toxicol, 76. *Mem:* Am Soc Pharmacol & Exp Therapeut; Acoust Soc Am; Soc Neurosci; Soc Toxicol; Am Soc Clin Pharmacol Therapeut. *Res:* Ototoxic action of the loop diuretics and aminoglycosides, pharmacokinetics and computer software development; auditory physiology and pharmacology. *Mailing Add:* Dept Pharmacol La State Univ Sch Med Shreveport LA 71130

BROWN, RICHARD EMERY, b Chatham, NJ, Apr 7, 29; m 51; c 4. ORGANIC CHEMISTRY. *Educ:* Moravian Col, BS, 51; Univ Md, MS, 52, PhD(org chem), 56. *Prof Exp:* Res chemist org chem, Allied Chem Corp, 55-60; assoc dept head, Warner Lambert Res Inst, 60-77; assoc dept head med chem, Revlon Health Care Group, 77-82; SR CHEM PATENT AGENT, BECTON DICKINSON & CO, 83- *Mem:* Am Chem Soc. *Res:* Synthesis of new organic compounds for potential medicinal interest. *Mailing Add:* One Becton Dr Franklin Lakes NJ 07417

BROWN, RICHARD GEORGE BOLNEY, b Wolverhampton, UK, Sept 15, 35; Can citizen. ORNITHOLOGY, MARINE BIOLOGY. *Educ:* Oxford Univ, BA, 57, DPhil(ethology), 62. *Prof Exp:* Researcher ethology, dept zool, Oxford Univ, 62-65; res assoc animal psychol, Dept Psychol, Dalhousie Univ, NS, 65-67; RES SCIENTIST ORNITH MARINE BIOL, CAN WILDLIFE SERV, 67 - *Mem:* Am Ornithologists Union; Brit Ornithologists Union; Brit Trust Ornith; Arctic Inst NAm. *Res:* The pelagic ecology of seabirds; role of seabirds in marine ecosystems. *Mailing Add:* Bedford Inst Oceanog Can Wildlife Serv PO Box 1006 Dartmouth NS B2Y 4A2 Can

BROWN, RICHARD HARLAND, b Gloversville, NY, Nov 9, 21. MATHEMATICS. *Educ:* Columbia Univ, AB, 42, PhD(math), 51. *Prof Exp:* Mathematician, US Dept Navy, 42-46; lectr math, Columbia Univ, 46-49, instr, 50-54; asst prof, Boston Univ, 54-55, mem opers eval group, 55-60; from assoc prof to prof, 60-86, EMER PROF MATH, WASH COL, 86- *Concurrent Pos:* Consult, Opers Res Off, 54-55, Opers Eval Group, 60-63 & Sacandaga Software, 87-88. *Mem:* Am Math Soc; Math Asn Am; Inst Math Statist; Asn Comput Mach; Inst Elec & Electronics Engrs Comput Soc; NY Acad Sci. *Res:* Finite differences; probability. *Mailing Add:* Quatre Rue du Porche Langlade 30980 France

BROWN, RICHARD JULIAN CHALLIS, b Sydney, Australia, Apr 10, 36; m 62; c 3. PHYSICAL CHEMISTRY. *Educ:* Univ Sydney, BSc, 57, MSc, 59; Univ Ill, PhD(phys chem), 62. *Prof Exp:* Asst prof chem, Queen's Univ, Ont, 62-66; mem staff, Australian AEC, 66-69; assoc prof, 68-83, PROF CHEM, QUEEN'S UNIV, ONT, 83- *Mem:* Can Soc Chem. *Res:* Nuclear quadrupole resonance; nuclear spin relaxation and molecular motion in liquids and solids; radio frequency; spectroscopy at high pressure; thermodynamics of solids. *Mailing Add:* Dept Chem Queen's Univ Kingston ON K7L 3N6 Can

BROWN, RICHARD K(EMP), b Miamisburg, Ohio, May 18, 17; m 49; c 1. ELECTRICAL ENGINEERING. *Educ:* Univ Mich, BS, 40, MS, 41, PhD(elec eng), 52. *Prof Exp:* Spec asst, Underwater Sound Lab, Harvard Univ, 41-43; engr electronic prod develop, Physicists Res Co, 46; from instr to prof elec eng, Res Inst, Univ Mich, Ann Arbor, 47-77, prof elec & comput eng, 77-; RETIRED. *Honors & Awards:* Bur Ord Res & Develop Award, 45. *Mem:* Audio Eng Soc; Am Acoust Soc; Inst Elec & Electronics Engrs. *Res:* Underwater sound velocity meters; acoustics; transistors; television; microwaves; radar. *Mailing Add:* 1301 Beale Ave Rm 1128 Univ Mich Ann Arbor MI 48109-2122

BROWN, RICHARD KETTEL, b Long Branch, NJ, Feb 3, 28; m 55. MATHEMATICS. *Educ:* Muhlenberg Col, BS, 48; Rutgers Univ, MS, 50, PhD(math), 52. *Prof Exp:* Asst prof math, Rutgers Univ, 55-58; res mathematician, US Army Electronics Labs, 58-65; chmn dept, 65-75, PROF MATH, KENT STATE UNIV, 65-, DIR OPER INST COMPUT MATH, 86- *Mem:* Am Math Soc; Math Asn Am; Sigma Xi. *Res:* Function of a complex variable; univalent and multivalent functions. *Mailing Add:* 5790 Horning Rd Kent OH 44240

BROWN, RICHARD L, b London, Eng, 64. STRUCTURAL TECTONICS. *Educ:* Edinburgh Univ, PhD(geol), 64. *Prof Exp:* PROF GEOL, CARLETON UNIV, 64- *Mem:* Geol Soc Am. *Mailing Add:* Earth Sci Carleton Univ Colonel Bay Dr Ottawa ON K1S 5B6 Can

BROWN, RICHARD LELAND, b Bryn Mawr, Pa, Dec 2, 12; m 39; c 4. PHYSICS. *Educ:* Amherst Col, AB, 34; Wesleyan Univ, MA, 36; Mass Inst Technol, PhD(physics), 42. *Prof Exp:* Asst physics, Wesleyan Univ, 34-37; teaching fel physics, Mass Inst Technol, 37-41, res assoc, Div Indust Coop, 42-47; res assoc, Div Indust Coop, Mass Inst Technol, 42-47; prof, 47-78, EMER PROF PHYSICS, ALLEGHENY COL, 78- *Mem:* Fel Am Acoust Soc; Inst Elec & Electronics Engrs; Am Phys Soc; Am Asn Physics Teachers. *Res:* Acoustics; sound-absorbing structures. *Mailing Add:* 211 Meadow St Meadville PA 16335

BROWN, RICHARD MCPIKE, b San Diego, Calif, Feb 17, 26; m 60; c 3. PLANT TAXONOMY, PLANT ECOLOGY. *Educ:* Pomona Col, BA, 50; Harvard Univ, MA, 52. *Prof Exp:* Park naturalist, 57-66, RES BIOLOGIST, NAT PARK SERV, 66- *Mem:* Am Soc Plant Taxon; Am Bryol & Lichenological Soc; Sigma Xi. *Res:* Role of man in altering natural vegetation and fauna; restoration of natural biotic conditions and processes in national parks. *Mailing Add:* 490 Estado Way Novato CA 94947

BROWN, RICHARD MALCOLM, JR, b Pampa, Tex, Jan 2, 39; m 61; c 1. BOTANY. *Educ:* Univ Tex, BS, 61, PhD(bot), 64. *Prof Exp:* Fel, Univ Hawaii & Univ Tex, 64-65; asst prof bot, Univ Tex, Austin, 65-68; assoc prof bot, Univ NC, Chapel Hill, 68-73, prof, 73-, dir, Electron Micros Lab, 70-; AT DEPT BOT, UNIV TEX. *Concurrent Pos:* NSF fel, Univ Freiburg, 68-69 & res grant, 70-72. *Mem:* AAAS; Am Soc Cell Biol. *Res:* Airborne algae; algal ecology; ultrastructure of algal viruses; algal ultrastructure; immunochemistry of algae; cytology; Golgi apparatus and cell wall formation; sexual reproduction amon algae; cellulose biogenesis. *Mailing Add:* Dept Bot Univ Tex BIO-308 Austin TX 78712

BROWN, RICHARD MARTIN, b Chicago, Ill, Oct 22, 37; m 60; c 3. CERAMIC ENGINEERING. *Educ:* Univ Ill, BS, 61, MS, 62, PhD(semiconductor glasses), 66. *Prof Exp:* Res asst ceramic eng, Univ Ill, 61-66; MEM TECH STAFF, TEX INSTRUMENTS INC, 66- *Mem:* AAAS; Nat Inst Ceramic Engrs. *Res:* Ferroelectric materials; electro-optics; crystal growth. *Mailing Add:* Tex Asn 13131 Floyd Rd Dallas TX 75243

BROWN, RICHARD MAURICE, b Cambridge, Mass, May 17, 24; m 46, 62; c 4. COMPUTER SCIENCE. *Educ:* Harvard Univ, AB, 44, MA, 47, PhD(physics), 50. *Prof Exp:* Asst, Oceanog Inst, Woods Hole, 44-46; asst prof physics, State Col Wash, 49-54; res prof physics & elec eng, Coord Sci Lab, 55-67, EMER PROF PHYSICS & COMP ENG, UNIV ILL, URBANA-CHAMPAIGN, 67- *Concurrent Pos:* Res asst prof, Control Syst Lab, Univ Ill, 52-54. *Mem:* Fel Am Phys Soc; Inst Elec & Electronics Engr; Asn Comput Mach. *Res:* Electronic computer design and use; scientific data acquisition systems design. *Mailing Add:* Dept Physics Univ Ill 1110 W Green St Urbana IL 61801

BROWN, ROBERT, b Montreal, Que, Can, Dec 13, 41. SPINAL CORD INJURY, CHRONIC LUNG DISEASE. *Educ:* McGill Univ, MDCM, 69. *Prof Exp:* CHIEF PULMONARY MED, VET ADMIN MED CTR, WEST ROXBURY, MASS, 76- *Mem:* Am Thoracic Soc. *Mailing Add:* Vet Admin Med Ctr 1400 VFW Pkwy West Roxbury MA 02132

BROWN, ROBERT A, b San Antonio, Tex, July 22, 51; m 72; c 2. CHEMICAL ENGINEERING. *Educ:* Univ Tex, BS, 73, MS, 75; Univ Minn, PhD(chem eng), 79. *Prof Exp:* Vis researcher, Dept Computer Sci & Appl Math, Lawrence Berkeley Lab, 77; instr, Dept Chem Eng & Mat Sci, Univ Minn, 78; from asst prof to assoc prof, 79-84, PROF CHEM ENG, MASS INST TECHNOL, 84-, ARTHUR DEHUN LITTLE PROF CHEM ENG, 86-, DEPT HEAD, 89- *Concurrent Pos:* Mem, Internal Opers Comt, Mat Processing Ctr, Mass Inst Technol, 84-88, vis comt, Dept Chem Eng, Univ Tex, 89-, Space Sci Bd, Nat Res Coun, 90-; exec officer, Dept Chem Eng, Mass Inst Technol, 86-88; Berkeley lectr chem eng, Univ Calif, Berkeley, 86; co-dir, Mass Inst Technol Supercomputer Fac, 89- *Honors & Awards:* Allan P Colburn Lectr, Univ Del, 86; Allan P Colburn Award, Am Inst Chem Engrs, 86; Robert Vaughan Mem Lectr, Calif Inst Technol, 87; Stanley Corrsin Mem Lectr, Johns Hopkins Univ, 89; Ernest Thiele Lectr, Notre Dame Univ, 89. *Mem:* Nat Acad Eng; Soc Indust & Appl Math; Am Asn Crystal Growth; Am Phys Soc; Soc Rheology; Coun Chem Res; Am Inst Chem Eng. *Res:* Mathematical modelling of the fluid mechanics, heat and mass transfer and interfacial phenomena associated with melt and vapor phase crystal growth; polymer processing and coating deposition; fluid mechanics of viscoelastic fluids; numerical simulation of the flow of complex fluids; prediction and characterization of microscopic changes in interface morphology during directional solidification; efficient numerical solutions of transport problems, especially by flute element and special methods; numerical implementation of bifurcation analysis; solving free- and moving-boundary problems; laser-doppler velocimetry applied to slow flows of viscoelastic fluids and to solidification systems; experimental measurement of microscale morphologies in melt/solids interfaces during solidification; transport processes and control of growth of semiconductors from the melt and vapor, large-scale process simulations; formation of crystalline defects in crystals growth from the melt and vapor; constitutive modelling of defeat dynamics. *Mailing Add:* Dept Chem Eng Mass Inst Technol Rm 66-342 Cambridge MA 02139

BROWN, ROBERT ALAN, b Los Angeles, Calif, June 11, 34; m 57; c 3. ATMOSPHERIC SCIENCE, GEOPHYSICS. *Educ:* Univ Calif, Berkeley, BS, 57, MS, 63; Univ Wash, PhD(geophys, atmospheric sci), 69. *Prof Exp:* Res engr fluid dynamics, Boeing Aircraft Co, Seattle, 64-66; res assoc boundary layer dynamics, Dept Atmospheric Sci, Univ Wash, 69-70; advan study prog fel atmospheric fluid dynamics, Nat Ctr Atmospheric Res, 70-71; prin scientist boundary layer dynamics, Arctic Ice Dynamics Joint Exp, 71-77, PRIN INVESTR, RES SCIENTIST & PROF, DEPT ATMOSPHERIC SCI, UNIV WASH, 77- *Concurrent Pos:* Adj prof, Naval Postgrad Sch, 80; mem, Satellite Sensor Teams, NASA, Numerous Arctic Experiment Teams. *Mem:* Fel Am Meteorol Soc; Am Geophys Union; Sigma Xi. *Res:* Planetary boundary layer modeling in connection with air-sea interaction; instabilities and secondary flows in the atmosphere and ocean; author of texts, analytic planetary boundary layer modelling; fluid mechanics of the atmosphere. *Mailing Add:* Dept Atmospheric Sci-AK-40 Univ Wash Seattle WA 98105

BROWN, ROBERT ALAN, b Hanover, NH, Feb 8, 43; m; c 2. SCIENCE POLICY. *Educ:* Princeton Univ, AB, 65; Harvard Univ, MA, 68, PhD(physics), 71. *Prof Exp:* Res fel, Harvard Univ, 71-75, res assoc earth & planetary physics, 75-79; res fel, Lunar & Planetary Lab, Univ Ariz, 79-82; ASTRONR, SPACE TELESCOPE SCI INST, 82- *Concurrent Pos:* Proj scientist, Hubble Space Telescope, 83-85. *Mem:* Am Astron Soc; AAAS. *Res:* Observational studies of planetary atmospheres and environments; search for other planetary systems; planetary physics; science and exploration public policy. *Mailing Add:* Space Telescope Sci Inst Homewood Campus Baltimore MD 21218

BROWN, ROBERT ALAN, b Columbus, Ohio, Sept 25, 27; m 52; c 4. OPERATIONS RESEARCH, INDUSTRIAL ENGINEERING. *Educ:* US Naval Acad, BSc, 49; Ohio State Univ, MSc, 56, PhD(indust eng), 62. *Prof Exp:* From instr to asst prof indust eng, Ohio State Univ, 56-67; From assoc prof to prof indust eng, 67-90, chmn, Dept Indust & Systs Eng, 76-81 & 84-89, EMER PROF INDUST ENG, UNIV ALA, HUNTSVILLE, 90- *Concurrent Pos:* Consult, NAm Aviation, Inc, 62-67, US Army Missile Command, 69-, Matrix Corp Div, URS Systs Corp, 71-, Huntsville Hosp, 72 & Northrop Serv Corp 74; expert witness testimony. *Mem:* Inst Elec & Electronics Engrs; Sigma Xi. *Res:* Industrial and systems engineering; engineering management; engineering manpower planning; micro computer applications; applied statistics; quality and reliability engineering. *Mailing Add:* Dept Indust & Systs Eng Univ Ala Box 1247 Huntsville AL 35807

BROWN, ROBERT BRUCE, b Portland, Ore, July 19, 38; m 62; c 3. MATHEMATICS. *Educ:* Harvard Univ, AB, 59; Univ Chicago, MS, 60, PhD(math), 64. *Prof Exp:* From instr to asst prof math, Univ Calif, Berkeley, 64-69; assoc prof, Univ Toronto, 69-70; assoc prof, 70-86, assoc dean, 87-90, PROF MATH, OHIO STATE UNIV, 86- *Concurrent Pos:* Asst actuary, Midland Mutual Life Ins, 77-78. *Mem:* AAAS; Am Math Soc; Assoc Soc Actuaries; Am Acad Actuaries. *Res:* Algebra, combinatorics and their applications; actuarial science. *Mailing Add:* Dept Math Ohio State Univ Columbus OH 43210

BROWN, ROBERT CALVIN, b Iredell Co, NC, Jan 7, 33; m 56; c 3. PATHOLOGY, MEDICINE. *Educ:* Erskine Col, AB, 55; Univ NC, MD, 59. *Prof Exp:* From intern to resident path, Sch Med, Univ NC & NC Mem Hosp, 5964, instr, 62-64; head cellular path group, Biol Div, Oak Ridge Nat Lab, 66-69; dir dept path, Rutherford Hosp, 71-; AT DEPT PATH, W C PATH ASN MED LABS. *Concurrent Pos:* Assoc prof path, Sch Med, Univ NC, Chapel Hill, 69-74. *Honors & Awards:* Seard-Sanford Award, Am Soc Clin Path, 59. *Mem:* Am Soc Clin Path; Int Acad Path; Tissue Cult Asn; AMA; Soc Exp Biol & Med. *Res:* Megakaryocyte structure; platelet aggregation; radiation and viral leukemogenesis; ultrastructural pathology; mammalian cytogenetics; nuclear sex chromatin; malacoplakia; viruscell interactions. *Mailing Add:* Dept Path W C Path Asn Med Labs P A PO Box 5686 Statesville NC 28677

BROWN, ROBERT DALE, b Red Bluff, Calif, July 31, 45; m 68, 79; c 3. ANIMAL NUTRITION, PHYSIOLOGY. *Educ:* Colo State Univ, BS, 68; Pa State Univ, PhD(animal nutrit), 75. *Prof Exp:* Asst assoc prof animal nutrit, Tex A&I Univ, 75-81; res sci, Caesar Kleberg Wildlife Res Inst, 81-87; DEPT HEAD, WILDLIFE & FISHERIES, MISS STATE UNIV, 87- *Concurrent Pos:* Dir, Forage Eval Lab, Tex A&I Univ, 75-; prin investr grants, Houston Livestock Show, 75-, Perry Found, 76-, Welder Wildlife Found, 77-, NIH, 77- & Ceaser Kleberg Wildlife Found, 78-; prog admin, Am Asn Fish & Wildlife, 87, SRAC-8 (Fish, Wildlife & Aqua) 88- *Honors & Awards:* Outstanding Teacher Award, Tex Asn Col Teachers, 77. *Mem:* Am Fisheries Soc; Wildlife Soc; Am Inst Nutrit. *Res:* Nutrition; endocrinology; environmental physiology of wild and domestic ruminant animals and bobwhite quail; biomedical aspects of deer antler growth; osteoporosis. *Mailing Add:* Wildlife & Fisheries Dept PO Drawer LW Mississippi State MS 39762

BROWN, ROBERT DALE, b Red Bluff, Calif, July 31, 45; m 80; c 3. FISH & WILDLIFE SCIENCES. *Educ:* Colo State Univ, BS, 68; Pa State Univ, PhD(animal nutrit), 75. *Prof Exp:* From asst prof to assoc prof animal nutrit, Tex A&I Univ, 75-81; assoc res scientist wildlife nutrit, Caesar Kleberg Wildlife Res Inst, Kingsville, Tex, 81-83, res scientist, 83-87; DEPT HEAD, DEPT WILDLIFE & FISHERIES, MISS STATE UNIV, 87- *Concurrent Pos:* Chmn, S Regional Adv Comt, Wildlife, Fisheries & Aquacult, 80-; prin investr, Biomed Appln Deer Antler Growth, NIH, 75-87. *Mem:* Wildlife Soc; Am Fisheries Soc; Am Inst Nutrit. *Res:* Comparative nutrition and physiology of domestic and wild animals to include deer, cattle, javalian, exotic animals, and quail; deer antler growth physiology and human biomedical applications, to include osteoporosis. *Mailing Add:* PO Drawer LW Mississippi State MS 39762

BROWN, ROBERT DILLON, b Paris, Ark, Nov 26, 33; m 68; c 2. MATHEMATICS. *Educ:* Univ Calif, Berkeley, AB, 55, PhD(math), 63. *Prof Exp:* Res mathematician, Univ Calif, Berkeley, 63; from asst prof to assoc prof, 63-88, ASSOC CHAIR MATH, UNIV KANS, 79-, PROF MATH, 88- *Mem:* Am Math Soc; Math Asn Am. *Res:* Eigen value approximations; partial differential equations. *Mailing Add:* Dept Math Univ Kans Lawrence KS 66045

BROWN, ROBERT DON, b Falls City, Nebr, May 30, 42; m 70; c 1. INDUSTRIAL HYGIENE. *Educ:* Northwest Mo State Univ, BS, 64; Univ Mich, MS, 69. *Prof Exp:* Health physicist, 69-72, indust hygienist, 72-74, sr indust hygienist, Kelsey-Seybold Clin, 74-77; MEM STAFF, NASA LANGLEY RES CTR, 77- *Mem:* Health Physics Soc; Am Indust Hyg Asn; Am Acad Indust Hyg. *Res:* Environmental factors that may pose significant health hazards, particularly noise, chemical agents and electromagnetic radiation. *Mailing Add:* 429 Nasa Langley Res Ctr Hampton VA 23665

BROWN, ROBERT EUGENE, b Appleton City, Mo, July 7, 25. PHYSICAL CHEMISTRY. *Educ:* Univ Mo, AB, 49, PhD(phys chem), 57. *Prof Exp:* Res physicist, US Naval Ord Lab, 60-90; CONSULT, 90- *Mem:* Inst Elec & Electronics Engrs. *Res:* Measurements of thermal accommodation coefficients on clean metal surfaces; rare earth magnetism; development of magnetic materials; applications of magnetic materials and superconductivity to devices. *Mailing Add:* 14A E Burnam Rd Columbia MO 65203

BROWN, ROBERT FREEMAN, b Cambridge, Mass, Dec 13, 35; m 57; c 2. FIXED POINT THEORY. *Educ:* Harvard Univ, AB, 57; Univ Wis, PhD(math), 63. *Prof Exp:* From asst prof to assoc prof, 63-73, PROF MATH, UNIV CALIF, LOS ANGELES, 73- *Mem:* Am Math Soc; Math Asn Am. *Mailing Add:* Dept Math Univ Calif Los Angeles CA 90024

BROWN, ROBERT GEORGE, b Montreal, Que, Apr 7, 37; m 63; c 3. MICROBIAL BIOCHEMISTRY. *Educ:* McGill Univ, BSc, 59, MSc, 61; Rutgers Univ, PhD(microbiol), 65. *Prof Exp:* Nat Res Coun Can overseas fel carbohydrate chem, Univ Stockholm, 65-67; from asst prof to assoc prof, 67-75, PROF BIOL, DALHOUSIE UNIV, 75- *Concurrent Pos:* Killam sr fel, 73; vis, Univ Stockholm, 74; Unilever res, Eng, 80-81, Medicarb, Sweden, 87-88. *Mem:* Am Soc Microbiol; Can Soc Microbiol. *Res:* Structure and specificity of lectins from molluscs which bind to bacteria; catabolite repression in yeasts and fungi. *Mailing Add:* Dept Biol Dalhousie Univ Halifax NS B3H 4J1 Can

BROWN, ROBERT GETMAN, b Gloversville, NY, Oct 25, 17; m 37; c 3. INDUSTRIAL CHEMISTRY. *Educ:* Cornell Univ, ChB, 39. *Prof Exp:* Chem microscopist, B F Goodrich Co, 39-44, chem purchasing agent, 44-46; chem salesman, EMERY Industs, 46-47; vpres, Perma Glaze Chem Corp, 47-53; asst to tech dir, Mohasco Industs, 53-57; vpres & tech dir, Perma Glaze Chem Corp, 57-69; pres, Knight Oil Corp & Wells Chem Co Inc, 69-84, treas & tech dir, 84-88; RETIRED. *Mem:* Am Chem Soc. *Res:* Chemical specialties development; emulsion cleaners; polishes; sanitizers; industrial degreasing compounds; chrome tanning compounds; aerosol technology. *Mailing Add:* RD 1 Lakeside Dr Mayfield NY 12117-9776

BROWN, ROBERT GLENN, b Norristown, Pa, Apr 16, 38; m 65; c 2. NUTRITION, PHYSIOLOGY. *Educ:* Drexel Inst, BSc, 60; Univ Tenn, Knoxville, PhD(nutrit & physiol), 64. *Prof Exp:* From asst prof to assoc prof nutrit, Drexel Inst, 64-69, actg chmn dept, 68-69; assoc prof animal sci, Univ Guelph, 69-74, prof, 74-78; PROF DEPT NUTRIT, UNIV MASS, AMHERST, 78- *Honors & Awards:* Dr Francis S W Luken's Award, Del Valley Diabetes Asn, 69. *Mem:* AAAS; Am Chem Soc; Am Soc Animal Sci; Nutrit Soc Can; Am Inst Nutrit; hon mem Can Vet Med Asn. *Res:* Genetic disorders of trace mineral metabolism; collagen and connective tissue metabolism; nutrition & aging; nutrition of companion animals. *Mailing Add:* Dept Nutrit Univ Mass Amherst MA 01002

BROWN, ROBERT GOODELL, b Evanston, Ill, Apr 14, 23; m 67, 88; c 3. OPERATIONS RESEARCH. *Educ:* Yale Univ, BE, 44, MA, 48. *Prof Exp:* Mem staff, Opers Eval Group, US Navy, 48-50; head dept opers res, Willow Run Res Ctr, Univ Mich, 50-53; mem staff, Arthur D Little, Inc, 53-67; vpres & exec dir opers serv, Curtiss-Wright Corp, 67-68; indust consult, Distrib Industs Mkt, Int Bus Mach Corp, 68-71; consult, 71-75, PRES, MAT MGT SYSTS, INC, 75- *Concurrent Pos:* Vis lectr, Amos Tuck Sch Bus Admin, 64; vis prof opers res, math & prod mgt, Yale, Northeastern, Dartmouth, Boston & Lehigh Univs. *Mem:* Int Fedn Opers Res Soc (treas, 62-65); Opers Res Soc Am; Am Soc Mech Engr; Am Math Soc; fel Am Prod Inventory Control Soc. *Res:* Statistical forecasting in discrete, nonstationary, time-series, inventory control, production planning and control systems; development, implementation, evaluation and education; author of 200 works on statistical forecasting and scientific inventory management. *Mailing Add:* PO Box 239 Thetford Center VT 05075-0239

BROWN, ROBERT HARRISON, b Edinburg, Tex, Feb 22, 38; m 60; c 2. VERTEBRATE ZOOLOGY. *Educ:* Univ Ariz, BS, 60, MS, 62, PhD(zool), 65. *Prof Exp:* Asst prof zool, Univ Idaho, 65-67; ASSOC PROF ZOOL, CENT WASH UNIV, 67-, CHMN DEPT, 77- *Mem:* Am Soc Zool. *Res:* Comparative osteology and myology of lizards and rodents. *Mailing Add:* Dept Biol Sci Cent Wash Univ Ellensburg WA 98926

BROWN, ROBERT JAMES SIDFORD, b Lawndale, Calif, Sept 7, 24; m 50, 61; c 5. PHYSICS. *Educ:* Calif Inst Technol, BS, 48; Univ Minn, MS, 51, PhD(physics), 53. *Prof Exp:* Electronic technician & asst, Los Alamos Sci Lab, 44-46; asst, Univ Minn, 48-53; SR RES ASSOC PHYSICS, CHEVRON OIL FIELD RES CO, STANDARD OIL CO CALIF, 53- *Mem:* Am Phys Soc; Soc Explor Geophys; Am Geophys Union. *Res:* Application of nuclear magnetic relaxation phenomena to study of liquids and their interactions with solid surfaces in porous media; application of electromagnetic and seismic techniques to oil exploration; study of elastic properties of porous materials. *Mailing Add:* 5285 Burlingame St Buena Park CA 90621

BROWN, ROBERT LEE, b Ranger, Tex, Nov 23, 32; m 56; c 2. BACTERIAL PHYSIOLOGY, BIOCHEMISTRY. *Educ:* Univ Houston, BS, 62, MS, 64; Univ Tex, PhD(microbiol), 67. *Prof Exp:* Asst biochemist & asst prof, Univ Tex M D Anderon Hosp & Tumor Inst, Houston, 68-69; asst prof microbiol, Dent Br, Univ Mo-Kansas City, 69-71; SR RES INVESTR, ANAL RES & DEVELOP DEPT, E R SQUIBB INST, 71- *Concurrent Pos:* NIH fel, Univ Tex Med Br Houston, 66-67; NIH fel biochem, Univ Tex M D Anderson Hosp & Tumor Inst, Houston, 6768; staff consult, Parkway Hosp, Houston,

69-; consult, BioControl, Inc, Houston, 67-69 & Bioassay, Inc, 68-69. *Honors & Awards:* O B Williams Res Award, Tex Soc Microbiol, 66. *Mem:* AAAS; Am Soc Microbiol; NY Acad Sci. *Res:* Clinical and industrial microbiology; clinical biochemistry and pathology; biochemistry of cancer. *Mailing Add:* 3363 Riviera Dr San Diego CA 92109

BROWN, ROBERT LEE, b Franklin, Pa, Feb 26, 08; m 40; c 3. SURGERY. *Educ:* Univ Mich, AB, 29; Harvard Univ, MD, 33. *Prof Exp:* Instr surg, Sch Med, Univ Rochester, 39-40, assoc, 40-41; from assoc dir to dir, Robert Winship Clin, 45-66, instr, 45-46; assoc, Sch Med, Emory Univ, 46-54, from asst prof to prof, 54-76, dir, Emory Univ Clin, 66-76, emer prof surg, 76-; RETIRED. *Mem:* Am Col Surgeons; AMA; Am Cancer Soc; Am Radium Soc (secy, 58-61, pres, 62); James Ewing Soc (pres, 52). *Res:* Cancer of the cervix, endometrium and melanoma; tumors of the neck; neoplastic diseases. *Mailing Add:* 3747 Peachtree St NE Apt 513 Atlanta GA 30319

BROWN, ROBERT LEE, b Detroit, Mich, Aug 4, 36; m 59; c 4. INDUSTRIAL & MANUFACTURING ENGINEERING. *Educ:* Mich Col Mining & Technol, BS, 58; Mass Inst Technol, SM, 61; Mass Inst Technol, ScD, 64; Boston Univ, MBA, 73. *Prof Exp:* Mgr process res, Vasco Metals, 64-67; eng mgr, Tex Instruments, 67-69; DIR ADVAN PROCESS, RES & DEVELOP, GILLETTE CO, 69- *Mem:* Am Soc Metals; Am Inst Mining & Metall Eng; Am Chem Soc; Am Soc Mfg Eng; Sigma Xi. *Res:* Manufacturing development as related to the design of advanced instrumentation, equipment and manufacturing processes for high volume production. *Mailing Add:* 5E3 Gillette Co Gillette Park Boston MA 02110

BROWN, ROBERT MELBOURNE, b Richmond, Que, Sept 14, 24; wid; c 3. ENVIRONMENTAL ISOTOPE CHEMISTRY. *Educ:* Bishop's Univ, Can, BSc, 44; McGill Univ, BSc, 47, PhD(phys chem), 51. *Prof Exp:* Jr res officer, Atomic Energy Can Ltd, 51-52; sci officer, Defence Res Bd, 52-62; assoc res officer, Environ Res Br, Atomic Energy Can Ltd, 62-70; sci officer, Sect Isotope Hydrol, Int Atomic Energy Agency, Vienna, 70-73; ASSOC RES OFFICER, ENVIRON RES BR, ATOMIC ENERGY CAN LTD, 73-; ADJ PROF, DEPT EARTH SCI, UNIV WATERLOO, 77- *Concurrent Pos:* Mem sub-comt low level radioactivity in mat, Int Comn Radiol Units & Measurements, 60-62. *Mem:* Sr mem Chem Inst Can. *Res:* Low level radiochemistry, including early development of low background B counters; measurement of tritium and deuterium in natural waters and their use as hydrological and meteorological tracers; carbon-14 dating by accelerator mass spectrometry. *Mailing Add:* Atomic Energy Can Ltd Chalk River ON K0J 1J0 Can

BROWN, ROBERT ORDWAY, b Tyler, Minn, July 31, 17; m 43; c 5. CHEMICAL ENGINEERING. *Educ:* Univ Minn, BChE, 39. *Prof Exp:* Chemist, Pillsbury Mills, Inc, 39-43, chem engr, 43-47, sect leader in charge, Cereal Pilot Plant, 47-51, process engr, 51-56; FOOD TECHNOLOGIST & ENGR, R BROWN CO, 56- *Mem:* AAAS; Am Asn Cereal Chem; Am Inst Chem Engrs; NY Acad Sci; Consult Eng Coun; Am Chem Soc. *Res:* Food technology; food, process and chemical engineering. *Mailing Add:* 4500 Morning Side Rd Minneapolis MN 55416

BROWN, ROBERT RAYMOND, b Akron, Ohio, June 30, 22; m 50. ORGANIC CHEMISTRY, POLYMER CHEMISTRY. *Educ:* Univ Akron, BS, 47, MS, 50; Ohio State Univ, PhD(org chem), 55. *Prof Exp:* Chemist, Firestone Tire & Rubber Co, Ohio, 47-50; sr res chemist, Int Latex Corp, 55-57, mgr polymerization res, 57-59, mgr polymer res, 59-65, dir res, Standard Brands Chem Industs, Inc, 65-72, dir res & develop, 72-75; dir res & develop, Reichold Chem, Inc, 76-84, dir technol, Emulsion Polymers Div, 84-88; RETIRED. *Mem:* Am Chem Soc. *Res:* Polymer research, especially elastomers for both latex and dry polymer applications. *Mailing Add:* 45 Shady Lane Dover DE 19901

BROWN, ROBERT REGINALD, b Alameda, Calif, Apr 4, 23; m 48; c 3. PHYSICS. *Educ:* Univ Calif, Berkeley, AB, 44, PhD(physics), 52. *Prof Exp:* Instr physics, Princeton Univ, 52-53; asst prof, Univ NMex, 53-56; asst prog dir, NSF, 56-57; lectr, Univ Calif, Berkeley, 57-58, from asst prof to assoc prof, 58-65, prof, 65-, emer prof physics,; RETIRED. *Concurrent Pos:* Guggenheim fel & Fulbright scholar, 63-64. *Mem:* Am Geophys Union. *Res:* Cosmic ray time variations; auroral and ionospheric physics; geomagnetism. *Mailing Add:* 504 Channel View Dr Anacortes WA 98221

BROWN, ROBERT SCHENCK, chemical engineering, for more information see previous edition

BROWN, ROBERT STANLEY, b High River, Alta, Can, Sept 16, 46; m 68; c 2. PHYSICAL ORGANIC CHEMISTRY. *Educ:* Univ Alta, BSc, 68; Univ Calif, San Diego, MSc, 70, PhD(chem), 72. *Prof Exp:* Nat Res Coun Can fel, Columbia Univ, 72-74; from asst prof to assoc prof, 74-84, PROF CHEM, UNIV ALTA, 84- *Honors & Awards:* Syntex Award, Can Soc Chem, 91. *Mem:* Can Soc Chem; Am Chem Soc. *Res:* Amide hydrolysis mechanistic studies; model enzyme studies; electrophilic halogenation. *Mailing Add:* Dept Chem Univ Alta Edmonton AB T6G 2G2 Can

BROWN, ROBERT STEPHEN, b New York, NY, May 21, 38; m 60; c 2. INTERNAL MEDICINE, NEPHROLOGY. *Educ:* Harvard Univ, AB, 59; Columbia Univ, MD, 63. *Prof Exp:* Intern & asst resident internal med, Bellevue Hosp, NY, 63-65; clin assoc med, Nat Cancer Inst, 65-67; sr asst resident internal med, Yale-New Haven Hosp, 67-68; fel renal med, Med Sch, Yale Univ, 68-69, asst prof, 69-72; asst prof internal med, 72-80, ASSOC PROF MED, HARVARD MED SCH, 80- *Concurrent Pos:* Med dir, Kidney Transplantation & Dialysis Serv, Yale-New Haven Hosp, 69-72 & Boston City Hosp, 72-73; clin chief renal unit, Beth Israel Hosp, Boston, 73- *Mem:* Am Fedn Clin Res; Am Soc Nephrology; fel Am Col Physicians; Int Soc Nephrology; Am Soc Artificial Internal Organs; Am Soc Transplant Physicians; Nat Kidney Found. *Res:* Potassium metabolismin renal disease; acute kidney failure; metabolic aspects of nephrolithiasis; hereditary renal cancer. *Mailing Add:* Dept Med Beth Israel Hosp Boston MA 02215

BROWN, ROBERT THEODORE, b Bay City, Mich, Feb 16, 31; m 56; c 4. PHYSICS. *Educ:* Univ Calif, Riverside, BA, 58; Univ Mich, MS, 59, PhD(physics), 65. *Prof Exp:* Instr physics, Univ Mich, 65-66; asst prof space sci, Rice Univ, 66-68; sci specialist, EG & G, Inc, 68-72; staff mem, Los Alamos Sci Lab, 72-81; sr scientist, Damaskos, Inc, 81-82; sr staff scientist, Grumman Corp Res Ctr, 82-89; GROUP LEADER, LOCKHEED ADVAN DEVELOP CO, 89- *Concurrent Pos:* Res Corp Cottrell grant, 67-68. *Mem:* Inst Elec & Electronics Engrs; Am Phys Soc; Am Astron Soc. *Res:* Computational physics; atomic structure and spectra; x-ray scattering; Monte Carlo charged particle and photon transport; electromagnetic scattering and diffraction. *Mailing Add:* 25838 Vaquero Ct Valencia CA 91355

BROWN, ROBERT THORSON, b Rochester, Minn, Sept 29, 23; m 53; c 4. PLANT ECOLOGY. *Educ:* Univ Wis, BS, 47 & 48, MS, 49, PhD(bot), 51. *Prof Exp:* Assoc prof, 51-62, prof biol sci, Mich Technol Univ, 62-88; RETIRED. *Concurrent Pos:* Radiation biologist, Argonne Nat Lab, 66 & USAID, India, 68; Fulbright res fel, Univ Helsinki, 71-72; lectr, Soviet Acad Sci, 72, Sigma Xi lectr, 75; Fulbright res fel, Univ West Indies, St Augustine, Trinidad, 86-87. *Honors & Awards:* Medal of Finnish Forestry Asn, 79; Fulbright lectr, Univ Adana, Turkey, 80-81. *Mem:* Fel AAAS; Am Inst Biol Sci; Ecol Soc Am; Sigma Xi; Nat Environ Educ Asn. *Res:* Air pollution; plant growth control; seed germination; northern forest ecology; radiation biology; exogenous growth substances; mycorrhizae on conifers; allelopathy; tropical ecology. *Mailing Add:* 500 Garnet St Houghton MI 49931

BROWN, ROBERT WADE, b Dallas, Tex, June 2, 33; m 54; c 4. CLAY STABILIZATION, FOUNDATION REPAIR. *Educ:* NTex State Univ, BA & MS, 55. *Prof Exp:* Process analyst, Temco Aircraft, 55-56; mgr, Indust Res Div, Western Co, 56-63; instr math, Odessa Jr Col, 58-59; instr, API Sch Petrol Technol, 56-63; pres & chmn bd, BPR Construct & Eng, Inc, 64-70; PRES & OWNER, ROBERT WADE BROWN & ASSOCS, 70- *Concurrent Pos:* Pres & chmn bd, Webb Properties, Inc, 68-, Brown Found Repair & Consult, Inc, 73-, Brown Oil Prod, 78- & Soil Sta, Inc, 85-; instr, Massey Realty Col, 73-84; secy & treas, ECI Serv, Inc, 83- *Mem:* Am Soc Civil Engrs; Soc Petrol Engrs. *Res:* Clay stabilization via base exchange chemistry; structural distress relative to differential soil movement; author of over 50 technical publications and four books. *Mailing Add:* Brown Consol Inc 2614 B Industrial Lane Garland TX 75041

BROWN, ROBERT WAYNE, b Atwood, Kans, June 27, 23; m 45; c 2. INTERNAL MEDICINE. *Educ:* Univ Colo, BA, 49; Univ Kans, MD, 55. *Prof Exp:* From asst prof to prof med, Univ Kans, 60-90; dir, Salina Health Educ Found, 77-90; RETIRED. *Concurrent Pos:* Coordr, Kans Regional Med Prog, 69-77; mem staff, Asbury & St Johns Hosps, Salina. *Mem:* Am Fedn Clin Res; Am Col Physicians; AMA. *Res:* Endocrinology and metabolism; diabetes. *Mailing Add:* 910 Marymount Rd Salina KS 67401

BROWN, ROBERT WILLIAM, b St Paul, Minn, Oct 3, 41. PHYSICS. *Educ:* Univ Minn, BS, 63; Mass Inst Technol, PhD(physics), 68. *Prof Exp:* Res assoc physics, Brookhaven Nat Lab, 68-70; from asst prof to assoc prof, 70-81, PROF PHYSICS, CASE WESTERN RESERVE UNIV, 81-, INST PROF, 91- *Mem:* Am Phys Soc. *Res:* High energy physics; electromagnetism. *Mailing Add:* Dept Physics Case Western Reserve Univ Cleveland OH 44106

BROWN, ROBERT ZANES, b Jackson, Mich, July 31, 26; m 48, 71; c 4. ECOLOGY, ANIMAL BEHAVIOR. *Educ:* Swarthmore Col, BA, 48; Johns Hopkins Univ, DSc(vert ecol), 52. *Prof Exp:* Res asst animal behavior, Am Mus Natural Hist, 48-53; sr asst scientist, Commun Dis Ctr, USPHS, 51-54; from assoc prof to prof zool, Colo Col, 54-63; NSF fac fel, Inst Marine Sci, Univ Tex, 63-64; PROF BIOL, DOWLING COL, 64-, EXEC ASST TO PRES, 78- *Concurrent Pos:* Spec staff mem, Rockefeller Found, 68-70. *Mem:* Fel AAAS. *Res:* Role of animal behavior in population dynamics; bioenergetics; ecology of suburban areas. *Mailing Add:* Dept Math & Sci Dowling Col Idle Hour Blvd Oakdale NY 11769

BROWN, RODGER ALAN, b Ellicottville, NY, Mar 24, 37; m 59; c 3. METEOROLOGY, CONVECTIVE STORMS. *Educ:* Antioch Col, BS, 60; Univ Chicago, MS, 62; Univ Okla, PhD, 89. *Prof Exp:* Res meteorologist, Univ Chicago, 60-65; assoc meteorologist, Cornell Aeronaut Lab, Inc, NY, 65-68, res meteorologist, 68-70; res meteorologist, Wave Propagation Lab, Environ Sci Serv Admin, Colo, 70; RES METEOROLOGIST, NAT SEVERE STORMS LAB, 70- *Honors & Awards:* Antarctic Serv Medal, 65. *Mem:* Am Meteorol Soc; Am Geophys Union; Royal Meteorol Soc; Sigma Xi; Nat Weather Asn. *Res:* Thunderstorm kinematics and dynamics; mesometeorology; Doppler weather radar. *Mailing Add:* Nat Severe Storms Lab 1313 Halley Circle Norman OK 73069

BROWN, RODNEY DUVALL, III, b Carbondale, Pa, Aug 28, 31; m 50; c 7. SOLID STATE PHYSICS, BIOPHYSICS. *Educ:* Univ Scranton, BS, 54; Columbia Univ, MA, 61; NY Univ, PhD(physics), 69. *Prof Exp:* Assoc engr xerography, IBM Develop Lab, 54-57, staff mem ferroelec, IBM Watson Lab, 57, staff mem hot electrons in germanium, 57-63, staff mem quantum effects in bismuth, 63-65, staff mem, Automated cytol, 65-67, mem interdiv comt med electronics, 66-70, staff mem quantum effects in bismuth, 67-68, mgr mach develop, 68-70, RES STAFF MEM, T J WATSON RES CTR, IBM CORP, 70- *Mem:* Am Phys Soc; NY Acad Sci. *Res:* NMRD investigation of metal and sugar binding properties of concanavalin A; nuclear magnetic relaxation in protein solutions. *Mailing Add:* T J Watson Res Ctr IBM Corp PO Box 218 Yorktown Heights NY 10598

BROWN, ROGER E, b Marcellus, Mich, Feb 20, 20; m 63; c 3. ANATOMY, SURGERY. *Educ:* Mich State Univ, DVM, 50, MS, 60; Purdue Univ, PhD(anat), 64. *Prof Exp:* Instr surg, Mich State Univ, 50-53, instr anat, 60-62; pvt pract, 53-60; instr anat, Purdue Univ, 62-63; asst prof, Mich State Univ, 64-66, assoc prof anat & asst dir space utilization, 66-69; assoc vet med & surg & dir educ resources, 69-70, prof vet med & surg & chmn dept, 70-80, prof & interim chmn dept vet anat & physiol, 80-82, EMER PROF, UNIV MO, COLUMBIA, 82- *Res:* Microcirculation of the myocardium. *Mailing Add:* 10184 Elk Lake Trail Williamsburg MI 49690

BROWN, RONALD ALAN, b Cleveland, Ohio June 19, 36; m 63; c 3. THEORETICAL PHYSICS, SOLID STATE PHYSICS. *Educ:* Drexel Inst, BS, 59; Purdue Univ, MS, 61, PhD(physics), 64. *Prof Exp:* Asst prof metall & Ford Found fel eng, Mass Inst Technol, 64-67; asst prof physics, Kent State Univ, 67-71; ASSOC PROF PHYSICS, STATE UNIV NY COL OSWEGO, 71- *Mem:* Am Phys Soc; Am Asn Physics Teachers; Sigma Xi; Nat Sci Teachers Asn; Soc Com Seed Technologists. *Res:* Recombination phenomena in semiconductors; electron-donor recombination in n-type germanium and silicon; history and philosophy of physics; physics and art; science teaching in the elementary and secondary grades; development of physics teaching materials. *Mailing Add:* Dept Physics State Univ NY Oswego NY 13126

BROWN, RONALD FRANKLIN, b San Diego, Calif, May 22, 40; m 68. SOLID STATE PHYSICS. *Educ:* Univ Calif, Riverside, BA, 62, MA, 64, PhD(physics), 68. *Prof Exp:* Teaching fel physics, Harvey Mudd Col, 68-69, asst prof, 69-74; ASST PROF PHYSICS, CALIF POLYTECH STATE UNIV, 74- *Concurrent Pos:* Referee, Am J Physics, 75. *Mem:* Am Phys Soc; Am Asn Physics Teachers. *Res:* Low temperature solid state physics; properties of magnetic insulators; electron transport in semiconductors. *Mailing Add:* Dept of Physics Calif Polytech State Univ San Luis Obispo CA 93407

BROWN, RONALD HAROLD, b Dudley, Ga, July 25, 35; m 57; c 4. AGRONOMY, PLANT PHYSIOLOGY. *Educ:* Univ Ga, BS, 57, MS, 59; Va Polytech Inst, PhD(agron), 62. *Prof Exp:* Asst prof agron, Va Polytech Inst, 61-67; assoc prof, Tex A&M Univ, 67-68; assoc prof, 68-71, PROF AGRON, UNIV GA, 72- *Mem:* Am Soc Agron; Crop Sci Soc Am; Am Soc Plant Physiologists. *Res:* Physiology and microclimatology as related to growth of forage crops. *Mailing Add:* Dept Agron Plant Sci Bldg Univ Ga Athens GA 30602

BROWN, ROSS DUNCAN, JR, b Fairmont, WVa, July 21, 35; m 58, 82; c 4. BIOCHEMISTRY. *Educ:* WVa Univ, BS, 57; Univ Wis-Madison, MS, 65, PhD(biochem), 68. *Prof Exp:* Staff chemist, Bjorksten Res Lab, Wis, 60-62; asst prof, Va Polytech Inst & State Univ, 67-80, assoc prof biochem, 80; ASSOC PROF, DEPT FOOD SCI, UNIV FLA, 80- *Mem:* AAAS; Am Chem Soc; Inst Food Technol; Sigma Xi; Am Soc Microbiol; Soc Complex Carbohydrates. *Res:* Microbial biochemistry; cellulose degradation; mode of action of cellulases; glycoprotein structures and biosynthesis. *Mailing Add:* Dept Food Sci & Human Nutrit Univ Fla Gainesville FL 32611

BROWN, RUSSELL VEDDER, b Tulsa, Okla, Mar 20, 25; m 53; c 2. GENETICS, IMMUNOLOGY. *Educ:* Univ Tulsa, BA, 48, MA, 50; Iowa State Univ, PhD(genetics), 62. *Prof Exp:* Geneticist, B-Bar-K Ranch, 53-59; res asst, Iowa State Univ, 59-62; asst prof biol, NTex State Univ, 62-68; assoc prof vet path, community health & med pract & biol sci, Sinclair Comp Med Res Farm, Univ Mo-Columbia, 68-74; PROF BIOL & CHMN DEPT, VA COMMONWEALTH UNIV, 74- *Mem:* AAAS; Genetics Soc Am; Am Genetic Asn; Soc Exp Biol Med; Environ Mutagen Soc; Sigma Xi. *Res:* Immunogenetics, population and biochemical genetics; heritability in mice; blood groups in birds, animals and fish; molecular pathology. *Mailing Add:* Dept Biol Va Commonwealth Univ 816 Park Ave Richmond VA 23298

BROWN, SAMUEL HEFFNER, b Winchester, Va, Dec 27, 33; m 70; c 2. PHYSICS, PHYSICAL CHEMISTRY. *Educ:* George Washington Univ, BS, 58; Cath Univ Am, MS, 70, PhD(chem), 75. *Prof Exp:* RES PHYSICIST, DAVID W TAYLOR NAVAL SHIP RES & DEVELOP CTR, 66- *Concurrent Pos:* Res assoc, Cath Univ Am, 75- *Mem:* Soc Naval Architects & Marine Engrs; Am Chem Soc. *Res:* Theoretical chemical physics; classical electromagnetic theory; theoretical acoustics; fluids. *Mailing Add:* David W Taylor Naval Ship Res & Develop Ctr 2032 Fairfax Rd Annapolis MD 21402

BROWN, SANDRA, b London, Eng, Oct 3, 44; US citizen. TROPICAL FORESTS, WETLANDS. *Educ:* Univ Nottingham, Eng, BSc, 66; Univ S Fla, MS, 73; Univ Fla, PhD(environ eng sci, ecol), 78. *Prof Exp:* Teacher sci, Christopher Wren Sch, Eng, 67-68, Pinellas County Sch Bd, 70-74; res asst environ eng sci, Univ Fla, 74-78, res assoc, Univ Fla & Univ Ill, 78-80; asst prof, 80-86, ASSOC PROF FORESTRY, UNIV ILL, 86- *Concurrent Pos:* Mem staff, President's Coun Environ Qual, 79; co-prin investr, Trop Forests & Global Carbon Cycle, US Dept Energy, 81-90; mem comt, Role Alt Farming Methods, Bd Agr, Nat Res Coun, 85-88; mem, US Man & Biosphere Directorate Trop Ecosysts. *Mem:* Ecol Soc Am; Soc Am Foresters; Int Soc Trop Foresters; Asn Trop Biologists; Soc Wetland Scientists; Sigma Xi. *Res:* Understanding how environmental factors regulate the organic matter dynamics of tropical and wetland forests; role of tropical forests in potential climate change; tropical forest restoration. *Mailing Add:* Dept Forestry 110 Munford Hall Univ Ill 1301 W Gregory Urbana IL 61801

BROWN, SEWARD RALPH, b Glace Bay, NS, Mar 25, 20; m 52; c 4. GEOCHEMISTRY, ECOLOGY. *Educ:* Queen's Univ, Ont, BA, 50 & 51, MA, 52; Yale Univ, PhD(biogeochem), 62. *Prof Exp:* Lectr zool, Yale Univ, 56-58; from asst prof to assoc prof biol, 59-68, dir biol sta, 59-74, PROF BIOL, QUEEN'S UNIV, ONT, 68- *Mem:* Am Soc Limnol & Oceanog; Geochem Soc; Can Soc Zoologists. *Res:* Biogeochemistry; primary productivity of lakes, particularly the relationship between photosynthetic rates and absolute quantities of chlorophylls, carotenoids and their diagenetic derivatives; paleolimnology, using plant pigments from lake sediments as biochemical fossils. *Mailing Add:* Dept Biol Queen's Univ Kingston ON K7L 3N6 Can

BROWN, SHELDON (JACK), b Los Angeles, Calif, Oct 13, 15; wid; c 2. PHYSICS. *Educ:* Univ Calif, Los Angeles, AB, 39, PhD(physics), 51. *Prof Exp:* Assoc physics, Univ Calif, Los Angeles, 42-49 & 50-51; res fel, Univ Strasbourg, 52-54; asst prof physics, DePaul Univ, 54-56; from asst prof to assoc prof, 56-65, PROF PHYSICS, CALIF STATE UNIV, FRESNO, 65- *Concurrent Pos:* Guggenheim fel, 62. *Mem:* Am Phys Soc; Am Asn Physics Teachers. *Res:* Electron inertia effects; gyromagnetism; gravitation; elementary particle physics; superconductivity. *Mailing Add:* Dept Physics M/S 37 Calif State Univ Fresno CA 93740

BROWN, SHERMAN DANIEL, b Salt Lake City, Utah, Sept 10, 29; m 53; c 5. CHEMICAL ENGINEERING, MATERIALS SCIENCE. *Educ:* Univ Utah, BS, 50, PhD(chem eng), 57. *Prof Exp:* Sr res engr high temperature chem, Jet Propulsion Lab, Calif Inst Technol, 57-59; head appl physics sect, Utah Div, Thiokol Chem Corp, 59-60; asst prof ceramic sci, Univ Utah, 60-64; prin scientist ceramic res, Rocketdyne Div, NAm Aviation, Inc, 64-68; assoc prof, 68-71, prof ceramic eng, 71-87, PROF MAT SCI & ENG, UNIV ILL, URBANA-CHAMPAIGN, 87- *Mem:* Fel Am Ceramic Soc; Nat Inst Ceramic Engrs; Ceramic Educ Coun; Am Soc Metals; Electrochem Soc. *Res:* Glass structure; ceramic-to-metal adhesion; high-temperature reaction rates in ceramic systems; brittle fracture; wear-resistant ceramic coatings. *Mailing Add:* Dept Mat Sci & Eng Univ Ill Urbana IL 61801

BROWN, STANLEY ALFRED, b Boston, Mass, Oct 9, 43; m 70. BIOMATERIALS. *Educ:* Mass Inst Technol, BS, 65; Dartmouth Col, DEng, 71. *Prof Exp:* Fel physiol, 70-72, res assoc, 72-74, asst prof surg, 74-80, assoc prof surg, Dartmouth Med Sch, 80, asst prof eng, Dartmouth Col, 78-80, assoc prof, 80; assoc prof bioeng & orthop & dir orthop res, Univ Calif, Davis, 80-, assoc prof eng, 81-; ASSOC PROF BIOMED ENG, DEPT BIOMED ENG, CASE WESTERN RESERVE UNIV. *Concurrent Pos:* Assoc staff, Mary Hitchcock Mem Hosp, 72-; adj asst prof eng, Dartmouth Col, 73-; res fel, Airlift Serv Indust Fund & Inst Straumann, 78. *Mem:* Asn Advan Med Instrumentation; Am Soc Artificial Internal Organs; Inst Elec & Electronics Engrs; Biomat Soc; Orthop Res Soc. *Res:* Biocompatibility of surgical implant materials; implant site infections; fracture healing, materials for fracture fixation and methods of assessment of healing; metal allergy and implants; ultrasonics; orthopaedic materials. *Mailing Add:* Dept Biomed Eng Case Western Reserve Univ Cleveland OH 44106

BROWN, STANLEY GORDON, b Washington, DC, May 17, 39; m 70; c 2. SCIENTIFIC EDITING. *Educ:* Harvard Univ, AB, 60; Univ Pa, MS, 63, PhD(physics), 66. *Prof Exp:* Instr & res assoc physics, Cornell Univ, 66-68; res assoc, Mass Inst Technol, 68-70; adj asst prof, Univ Calif, Los Angeles, 70-72; res assoc, State Univ NY, Stony Brook, 72-74; asst ed, 74-77, assoc ed, 77-81, ED, PHYS REV D, AM PHYS SOC, 81-, ED, PHYS REV LETT, 85- *Concurrent Pos:* Ed, Phys Rev Abstr, 81-85; chmn, Am Inst Physics Publ Bd, 82-84. *Mem:* Am Phys Soc. *Mailing Add:* Am Phys Soc One Research Rd Box 1000 Ridge NY 11961

BROWN, STANLEY MONTY, b New York, NY, Apr 2, 43; m 72; c 2. PETROLEUM CHEMISTRY, SURFACE CHEMISTRY. *Educ:* Queens Col, NY, BA, 64; Northwestern Univ, PhD(org chem), 69. *Prof Exp:* Res chemist, Davison Chem Div, W R Grace & Co, Md, 69-73, res supvr, 73-74, res assoc, 74-75; group leader, Minerals & Chem Div, 75-77, res mgr, Engelhard Minerals & Chem Corp, 78-80, RES DIR, ENGELHARD CORP, 80- *Mem:* Am Chem Soc; Catalysis Soc; Sigma Xi; Soc Advan Educ. *Res:* Catalytic chemistry; organic reaction mechanisms; preparation, characterization and evaluation of heterogeneous catalysts and catalyst supports for industrial processes including fluid cracking, hydrotreating, hydrogenation, isomerization and polymerization; autocatalysts, nox abatement. *Mailing Add:* 1424 Sylvan Lane Scotch Plains NJ 07076

BROWN, STEPHEN CLAWSON, b Caracas, Venezuela, June 16, 41; m 65; c 2. COMPARATIVE ENDOCRINOLOGY. *Educ:* George Washington Univ, BS, 63; Univ Mich, MS, 64, PhD(zool), 66. *Prof Exp:* From asst prof to assoc prof, 67-82, PROF BIOL, STATE UNIV NY, ALBANY, 82- *Mem:* AAAS; Am Physiol Soc; Am Soc Zoologists. *Res:* Hormonal control of hydromineral balance in lower vertebrates. *Mailing Add:* Dept Biol Sci State Univ NY 1400 Washington Ave Albany NY 12222

BROWN, STEPHEN L(AWRENCE), b San Francisco, Calif, Feb 16, 37; m 61; c 3. ENVIRONMENTAL HEALTH, RESEARCH MANAGEMENT. *Educ:* Stanford Univ, BS, 58, MS, 61; Purdue Univ, PhD(physics), 63. *Prof Exp:* Physicist, 63-68, sr opers analyst systs planning, 68-74, prog mgr, 74-76, asst dept dir, 76-77, DEPT DIR, SRI INT, 77-; assoc exec dir, Nat Rec Coun Comn Life Sci, 83-86; PROJ MGR, ENVIRON CORP, 86- *Concurrent Pos:* Mem, comt priority mechanisms, Nat Res Coun,Comt asbestiform fibers; mem, Panels of Sci Adv Bd, Environ Protection Agency. *Mem:* AAAS; Soc Risk Anal; Int Soc Exposure Assessment. *Res:* environmental systems analysis; chemical risk assessment; radiation risk assessment. *Mailing Add:* 9100 Mine Run Dr Great Falls VA 22066

BROWN, STEVEN MICHAEL, b New York, NY, Dec 3, 44; m 69; c 2. ARCHITECTURAL ACOUSTICS, NOISE CONTROL ENGINEERING. *Educ:* Johns Hopkins Univ, BA, 66, MA, 69, PhD(physics, math), 72. *Prof Exp:* Physicist plasma physics, US Naval Res Lab, 66-70; res assoc & instr physics, Johns Hopkins Univ, 66-72; resident & sci secy atomic physics, Ctr Theoret Studies, Univ Miami, 72-73; dir qual control eng, Allegheny Metal Stamping Co, Inc, 73; res physicist acoust, 73-75, res scientist, 75-79, sr res scientist, Armstrong World Industs, Inc, 79-86; sr res eng, 86-89, PRIN ENG, STEELCASE, INC, 89- *Concurrent Pos:* Fel, NSF & NASA; invited guest lectr, PA State Univ, Grand Valley State Univ, Mich, Millersville Univ, PA. *Honors & Awards:* Fel, Acoust Soc Am. *Mem:* Acoust Soc Am; Inst Noise Control Eng; Am Soc Testing & Mat; Am Phys Soc; Am Asn Physics Teachers; Inst Elec & Electronics Engrs. *Res:* Architectural and physical acoustics; noise control engineering; application of applied mathematics; materials science of acoustical materials; software engineering. *Mailing Add:* 599 Rookway SE Grand Rapids MI 49546-9607

BROWN, STEWART ANGLIN, b Peterborough, Ont, Apr 6, 25; m 86. PLANT BIOCHEMISTRY. *Educ:* Univ Toronto, BSA, 47; Mich State Univ, MS, 49, PhD(biochem), 51. *Prof Exp:* From asst res officer to assoc res officer, Prairie Regional Lab, Nat Res Coun Can, 51-62, sr res officer, 62-64; assoc prof chem, Trent Univ, 64-68, dean grad studies, 73- 77, prof chem, 68-87, EMER PROF, TRENT UNIV, 87- *Concurrent Pos:* Vis res worker, Univ Chem Lab, Cambridge Univ, 55-56; vis prof, Col Pharm, Univ Minn, Minneapolis, 70-71. *Mem:* Phytochem Soc NAm. *Res:* Biochemistry of higher plants and plant-parasite relations; chemistry of lignification, biosynthesis of coumarins and related compounds; methodology of biosynthetic investigation in plants; analysis, localization and chemical ecology of plant coumarins. *Mailing Add:* Dept Chem Trent Univ Peterborough ON K9J 7B8 Can

BROWN, STEWART CLIFF, b Philadelphia, Pa, Mar 15, 28; m 50; c 1. PLASTICS CHEMISTRY, POLYMER CHEMISTRY. *Educ:* Philadelphia Col Pharm & Sci, BSc, 50; Univ Del, MSc, 54, PhD(org chem), 57. *Prof Exp:* Res chemist, Tex-US Chem Co, 56-59 & E I Du Pont de Nemours & Co, 59-61; res chemist & tech serv supvr polymer film develop, Avisun Corp, 61-65; mkt mgr & res supvr polymer films, Hercules Inc, 65-80, dir technol film, 81-91, mgr, Mat Sci Dept, 88-91; RETIRED. *Res:* Film forming plastics; effects of orientation on film properties, chemical structure and orientation versus film properties. *Mailing Add:* 1406 Pennsylvania Ave Wilmington DE 19806

BROWN, STUART HOUSTON, b Bryn Mawr, Pa, Mar 25, 41; m 66; c 2. ORGANIC CHEMISTRY. *Educ:* Williams Col, BA, 63; Stanford Univ, PhD(org chem), 68. *Prof Exp:* NIH fel, Univ Liverpool, 68-69; res chemist, Lubricant Res Dept, Chevron Res Co, 69-75, sr res chemist, 75-77, sr res assoc, Greases & Indust Oils Div, 77-80, sr staff specialist, prod mkt dept, 80-84, tech asst to vpres, prod res dept, 84-85, sr res assoc, Lubricants Div, 85-87, unit leader, fuel chem, Fuels Div, 87-90. *Mem:* Am Chem Soc. *Res:* Alkaloid structure and biosynthesis; x-ray crystallography; synthesis of lubricating oil additives; petroleum chemistry; industrial oil product development; business planning; petroleum products marketing; marketing research; performance and quality of petroleum fuels. *Mailing Add:* 85 Main Dr San Rafael CA 94901

BROWN, STUART IRWIN, b Chicago, Ill, Mar 1, 33. OPHTHALMOLOGY. *Educ:* Univ Ill, BMS, 55, MD, 57. *Prof Exp:* Intern, Jackson Mem Hosp, Miami, 57-58; resident ophthal, Harvard Med Sch, 58-59; resident, Eye, Ear, Nose & Throat Hosp, Med Sch, Tulane Univ, 59-61, Heed fel, 61-62; fel ophthal, Cornea Serv, Mass Eye & Ear Infirmary, 62-66; clin asst prof ophthal & asst attend surgeon, New York Hosp-Cornell Univ Med Ctr, 66-70; clin assoc prof & assoc attend surgeon, 70-74, dir, Cornea Serv & Cornea Res Lab, Med Ctr, 66-74; PROF OPHTHAL & CHMN DEPT, EYE & EAR HOSP PITTSBURGH & SCH MED, UNIV PITTSBURGH, 74- *Mem:* Am Acad Ophthal; Soc Contemporary Ophthal; Pan Am Asn Ophthal; Asn Res Vision & Ophthal. *Res:* Cornea. *Mailing Add:* Univ Hosp San Diego UC San Diego-Ophthal La Jolla CA 92093

BROWN, TALMAGE THURMAN, JR, b Raleigh, NC, Apr 18, 39; m 72; c 2. PATHOLOGY, VIROLOGY. *Educ:* NC State Univ, BS, 61; Okla State Univ, DVM, 65; Cornell Univ, PhD(vet path), 73. *Prof Exp:* Res assoc path & virol, Cornell Univ, 72-73; fel comp path, Johns Hopkins Univ, 73-74, asst prof, 74-75; asst prof path, Okla State Univ, 75-78; vet med officer virol, Nat Animal Dis Ctr, Sci & Educ Admin-Agr Res, USDA, 78-81; PROF PATH, NC STATE UNIV, 81- *Concurrent Pos:* Co-investr, Pub Health Serv grant, Nat Inst Child Health & Human Develop, 71-74; fel, Nat Inst Neurol Dis & Stroke. *Honors & Awards:* J Scholar Award, Charles L David DVM Found Adv Vet Path, 75. *Mem:* Am Col Vet Pathologists; Int Acad Path; Am Vet Med Asn; Conf Res Workers Animal Dis. *Res:* Viral diseases affecting respiratory system of domestic animals; congenital and viral-induced leukocyte disorders. *Mailing Add:* Dept Microbiol Path & Parasitol Sch Vet Med NC State Univ 4700 Hillsborough St Raleigh NC 27606

BROWN, THEODORE CECIL, b Paintsville, Ky, May 15, 07; m 27; c 2. MECHANICAL ENGINEERING. *Educ:* Univ Ky, BS, 31, ME, 36; NC State Col, MS, 40. *Prof Exp:* Off engr, US CEngr, Charleston, WVa, 31-32; engr & instr, Drafting Schs, Ky, 32-37; from instr to assoc prof eng, NC State Col, 37-64; consult engr, Booth, Jones & Assocs, 64-70; chief mech engr, Eng Div, NC Dept Ins, 70-78; RETIRED. *Concurrent Pos:* Consult engr, 32- & Cast Iron Soil Pipe Inst, DC, 60- *Mem:* Nat Soc Prof Engrs. *Res:* Design and drafting technique in foreign countries; new methods of radiant heating; drafting room practices; some trends in engineering education; three US patents. *Mailing Add:* 601 Beaver Dam Rd Raleigh NC 27607

BROWN, THEODORE D, b Brooklyn, NY, Apr 25, 44; m 66; c 2. COMPUTER SCIENCE, OPERATIONS RESEARCH. *Educ:* City Col New York, BS, 66; NY Univ, BMOR, 68, PhD(oper res), 70. *Prof Exp:* ASSOC PROF COMPUT SCI, QUEENS COL, NY, 70- *Concurrent Pos:* Bd High Educ-Prof Staff Cong grant, Res Found, City Univ New York, 77. *Mem:* Oper Res Soc Am; Asn Comput Mach; AAAS; Am Statist Asn. *Res:* Computer systems; modeling computer systems; stochastic modeling; simulation methodology. *Mailing Add:* Dept Comput Sci Queens Col NY Flushing NY 11367

BROWN, THEODORE GATES, JR, b Boston, Mass, Sept 29, 20; m 45; c 5. PHARMACOLOGY. *Educ:* Univ Tenn, AB, 48, MS, 50; Med Col SC, PhD(pharmacol), 56. *Prof Exp:* Res asst pharmacol, Sterling-Winthrop Res Inst, 57-58; assoc mem, 58-61, group leader, 61-62, res biologist, 61-63, sr res biologist & sect head, 63-68; dir biomed res, Warren-Teed Pharmaceut, Inc, 68-74; mgr preclin res, Rohm & Haas Res Ctr, 74-77; asst dir preclin res & mgr tech licensing, 77-78, mgr product develop & licensing Adria Labs, 78-83; dir new prods, Ben Venue Lab Inc, 83-87; PRES, THEODORE G BROWN & ASSOCS INC, 83- *Concurrent Pos:* Lectr, Albany Med Col, 58-68; consult, Pharmaceut Res Develop Prod 83- *Mem:* AAAS; Am Soc Pharmacol; Am Chem Soc; Am Heart Asn; Regulatory Affairs Prof Soc. *Res:* Pharmacology and physiology of cardiovascular disease; general pharmacology; neoplastic diseases. *Mailing Add:* 23 Seminary St Middlebury VT 05753

BROWN, THEODORE LAWRENCE, b Green Bay, Wis, Oct 15, 28; m 51; c 5. INORGANIC CHEMISTRY. *Educ:* Ill Inst Technol, BS, 50; Mich State Univ, PhD(chem), 56. *Prof Exp:* From instr to assoc prof, 56-65, PROF INORG CHEM, UNIV ILL, URBANA-CHAMPAIGN, 65-; DIR, BECKMAN INST ADV SCI & TECHNOL, UNIV ILL, URBANA, 87- *Concurrent Pos:* Sloan fel, 62-66; NSF fel, 64-65; assoc ed, Inorg Chem, 68-78; vis scientist, Int Meteorol Inst, Stockholm, 72; Boomer lectr, Univ Alta, Edmonton, 75; Firth vis prof, Sheffield Univ, 77; Guggenheim fel, 79-80; vchancellor res & dean, Grad Col, Univ Ill, Urbana, 80-86. *Honors & Awards:* Inorg Chem Award, Am Chem Soc, 72. *Mem:* Am Chem Soc; Sigma Xi; fel, AAAS. *Res:* Kinetics and mechanisms of organometallic reactions. *Mailing Add:* Beckman Inst Univ Ill 405 N Mathews St Urbana IL 61801

BROWN, THOMAS ALLEN, MUCOSAL IMMUNITY, IGA SUBCLASSES. *Educ:* Univ Fla, PhD(microbiol), 78. *Prof Exp:* ASST PROF ORAL BIOL, UNIV FLA, 84- *Mailing Add:* Dept Oral Biol Col Dent Univ Fla Box J-424 JHMHC Gainesville FL 32610

BROWN, THOMAS EDWARD, b Dayton, Ohio, May 19, 25; m 82; c 3. PLANT PHYSIOLOGY. *Educ:* Antioch Col, BA, 50; Ohio State Univ, MS, 51, PhD(plant physiol), 54. *Prof Exp:* Staff scientist, C F Kettering Res Lab, 55-69; assoc prof biol, 69-70, chmn, Sci Div, 73-79, PROF BIOL, ATLANTIC COMMUNITY COL, 70- *Mem:* AAAS; Am Soc Plant Physiol; Phycol Soc Am; Nat Asn Biol Teachers; Sigma Xi; Am Inst Biol Sci. *Res:* Photosynthesis and algal physiology; algal pollution control. *Mailing Add:* Dept Math & Sci Atlantic Community Col Mays Landing NJ 08330-9888

BROWN, TONY RAY, b Owen Co, Ind, Apr 11, 39; m 59; c 2. HEAT TRANSFER, ENGINEERING. *Educ:* Purdue Univ, BSME, 62, MSME, 64, PhD, 68. *Prof Exp:* Proj engr, Midwest Appl Sci Corp, 64-66, asst mgr & chief engr, Prod Develop Div, 66-67; mem res staff, Sch Mech Eng, Purdue Univ, 67-68; contracting off tech rep, proj engr, & aero-mech engr, US Army Aviation Systs Command, 68-69; vpres, Midwest Appl Sci Corp, 69-72, gen mgr, 71-72; vpres res & develop, 72-77, V PRES RES & BUS DEVELOP, PHARMASEAL DIV, AM HOSP SUPPLY CORP, 77- *Mem:* Am Soc Mech Engrs; Am Mgt Asn; Soc Advan Mgt (int pres, 81-); Sigma Xi; Asn Advan Med Instrumentation. *Res:* Mathematics; physics; cryogenics; mass transfer; thermal radiation heat transfer; environmental engineering; lasers; holography; engineering management; general management; general engineering; hospital and home health care. *Mailing Add:* 6605 Canyon Hills Rd Anaheim Hills CA 92807

BROWN, TRUMAN ROSCOE, b Oct, 4, 1918; c 2. NUCLEAR MAGNETIC RESONANCE. *Educ:* Mass Inst Technol, Cambridge, BS, 64, PhD(physics), 70. *Prof Exp:* Instr physics, Mass Inst Technol, Cambridge, 70-71; mem tech staff, Bell Labs, Murray Hill, NJ, 71-83; CHMN, DEPT NUCLEAR MAGNETIC RESONANCE & MED SPECTROS, FOX CHASE CANCER CTR, PHILADELPHIA, PA, 83- *Concurrent Pos:* NAm ed, Nuclear Magnetic Resonance in Biomed, 87-; chmn, Sci Prog Comt, Soc Magnetic Resonance Med, 88-89; adj prof, Sch Med, Univ Pa, 89- *Mem:* Fel Am Phys Soc; Biophys Soc; Am Col Radiol; Soc Magnetic Resonance Med; Soc Magnetic Resonance Imaging. *Res:* Application of nuclear magnetic resonance to the study in vivo physiology, particularly in medicine. *Mailing Add:* Fox Chase Cancer Ctr 7701 Burholme Ave Philadelphia PA 19111

BROWN, VERNE R, b Dover, NH, Jan 23, 35; m 61; c 3. ELECTRICAL ENGINEERING, SOLID STATE PHYSICS. *Educ:* Univ NH, BSc, 60; Univ Mich, MSc, 61, MSc, 64, DrPhil(elec eng), 66. *Prof Exp:* Res engr, Univ Mich, 64-66; asst prof elec eng, Univ Calif, 66-68; vpres, Transidyne Gen Corp, Mich, 68-70; chmn bd, 70-76, PRES, ENMET CORP, 70-, CHIEF RES EXEC, 76- *Mem:* Instrument Soc Am; Inst Elec & Electronics Engrs; Air Pollution Control Asn; Asn Advan Med Instrumentation; Am Indust Hyg Asn. *Res:* Electrical, optical and infrared properties of solid state materials as related to sensor development, specifically use of materials technology and thin film deposition processes for sensor development; medical research; air pollution monitoring; industrial health and safety instrumentation; industrial safety/confined space entry field. *Mailing Add:* W 28 Southwick Ann Arbor MI 48210

BROWN, VIRGINIA RUTH, b Wollaston, Mass. NUCLEAR THEORY. *Educ:* Northeastern Univ, BS, 57; McGill Univ, PhD(physics), 64. *Prof Exp:* Teaching asst, Ohio State Univ, 57-59, fel, 59-61; RES PHYSICIST, LAWRENCE LIVERMORE LAB, 66- *Concurrent Pos:* Fel, Nat Res Coun Can, 61-63; appointee, Yale Univ, 63-64. *Mem:* Am Phys Soc. *Res:* Nuclear reactions; nuclear structure; nucleon-nucleon interaction in the presence of other fields. *Mailing Add:* Lawrence Livermore Lab Box 808 Theor Dt L-297 Livermore CA 94550

BROWN, W(AYNE) S(AMUEL), mechanical engineering, for more information see previous edition

BROWN, W VIRGIL, b Royston, Ga, Sept 25, 38. METABOLIC DISEASES, LIPOPROTEINS. *Educ:* Yale Univ, MD, 64. *Prof Exp:* PROF MED & CHIEF, DIV METAB DIS, MT SINAI MED CTR, NY, 78- *Concurrent Pos:* Mem, Coun Arteriosclerosis, Am Heart Asn. *Mem:* Am Soc Clin Invest; Am Fedn Clin Res; Am Heart Asn. *Res:* Lipid metabolism. *Mailing Add:* Medlantic Res Found 108 Irving St NW Washington DC 20010

BROWN, WALTER CREIGHTON, b Butte, Mont, Aug 18, 13. BIOLOGY. *Educ:* Col Puget Sound, AB, 35, MA, 38; Stanford Univ, PhD, 50. *Prof Exp:* Head dept sci high, Wash, 38-42; actg instr biol, Stanford Univ, 49-50; instr, Northwestern Univ, 51-53; instr, 53-54, chmn div sci, 55-61, dean div sci & eng, 61-67, dean, col, 67-74, prof, 74-78, EMER PROF BIOL, MENLO COL, 78-; RES ASSOC, CALIF ACAD SCI, 78- *Concurrent Pos:* Fulbright lectr zool, Silliman Univ, Philippines, 54-55; res assoc biol sci, Stanford Univ, 55-74; vis prof, Stanford Univ & Harvard Univ, 62-72. *Mem:* Sigma Xi; Assoc Am Soc Ichthyol & Herpet; assoc Am Soc Zool; assoc Soc Syst Zool. *Res:* Ecology, systematics and zoogeography of amphibians and reptiles in the islands of the Pacific. *Mailing Add:* Dept Herpet Calif Acad Sci San Francisco CA 94118

BROWN, WALTER ERIC, b Butte, Mont, Mar 17, 18; m 47; c 3. CHEMICAL PHYSICS. *Educ:* Univ Wash, BS, 40, MS, 42; Harvard Univ, PhD(chem physics), 49. *Prof Exp:* Physicist electron micros, B F Goodrich Co, 42-45; chemist phosphates, Tenn Valley Authority, 48-62; CHEMIST, AM DENT ASN RES UNIT, NAT BUR STANDARDS, WASHINGTON, DC, 62- *Concurrent Pos:* Rockefeller Found spec grant, Neth, 58-59. *Mem:* AAAS; Am Chem Soc; Am Dent Asn; Int Asn Dent Res. *Res:* Crystallography and physical chemistry of calcium phosphates in tooth and bone. *Mailing Add:* Am Dent Asn Health Found Nat Inst Standards & Technol Bldg 224 Rm A153 Gaithersburg MD 20899

BROWN, WALTER JOHN, b North Providence, RI, Aug 25, 28; m 52; c 3. INTERNAL MEDICINE. *Educ:* Brown Univ, AB, 50; Univ RI, MS, 53; Med Col Ga, MD, 60. *Prof Exp:* Trainee, Clin Cardiovasc Res Prog, Med Col Ga, 62-63, asst prof med & assoc dir cardiovasc res training prog, 64-67, assoc prof, 67-71, prof med, 71-; assoc dir & chief med, Univ Health Serv, Univ Ga, Athens, 68-; prof therapeut med, Sch Pharm & prof med & surg, Inst Comp Med, Sch Vet Med, 69-; AT DEPT MED, MED COL GA. *Concurrent Pos:* Grants, NIH coop study, 66-72, Ga Heart Asn, 66-68 & Cerebrovasc dis, Am Heart Asn, 68; attend physician, Vet Admin Hosp, Augusta, Ga, 6469; consult, Penitentiary Syst, State of Ga, 64-69; dir, Athens High Blood Pressure Ctr; consult, Northeast Ga Health Dist. *Mem:* Am Heart Asn; Sigma Xi. *Res:* Epidemiology and hemodynamics of hypertension and peripheral vascular disease; bioavailability of drugs in humans. *Mailing Add:* PO Box 312 Winterville GA 30683

BROWN, WALTER LYONS, b Charlottesville, Va, Oct 11, 24; m 46; c 4. MATERIALS SCIENCE. *Educ:* Duke Univ, BS, 45; Harvard Univ, AM, 48, PhD(physics), 51. *Prof Exp:* MEM TECH STAFF, BELL TEL LABS, 50- *Honors & Awards:* Von Hipple Award, Mat Res Soc, 84. *Mem:* Nat Acad Sci; fel Am Phys Soc; Nat Acad Eng. *Res:* Semiconductors; space physics; interaction of energetic particles with solids. *Mailing Add:* 138 Cambridge Dr Berkeley Heights NJ 07922

BROWN, WALTER REDVERS JOHN, b Toronto, Ont, Aug 22, 25; m 48; c 5. ENGINEERING PHYSICS. *Educ:* Univ Toronto, BA Sc, 47; Univ Rochester, MS, 49. *Prof Exp:* Sr physicist, Kodak Res Lab, 49-55; res assoc, Phys Res Lab, Boston Univ, 55-57; proj mgr, Itek Corp, 57-62; vpres res & develop, United Carr Inc, 62-69; exec vpres, Ealing Corp, 69-71; PRES, DAEDALON CORP, 71- *Honors & Awards:* Adolph Lomb Medal, Optical Soc Am, 56. *Mem:* Fel Optical Soc Am. *Res:* Comprehensive measurements of the ability of the human eye to perceive small color differences. *Mailing Add:* 35 Congress St PO Box 2028 Salem MA 01970

BROWN, WANDA LOIS, b England, Ark, Oct 30, 22. BIOCHEMISTRY. *Educ:* Ouachita Baptist Col, BS, 43; Sch Med, Univ Ark, cert, 46; Trinity Univ, MS, 60. *Prof Exp:* Chemist anal, Aluminum Ore Co, 43-45; med technologist, Ark Baptist Hosp, 46-50; med technologist hemat, Brooke Army Med Ctr, 51-53, med technologist chem, 53-55; biochemist metab, 55-62, RES BIOCHEMIST PROTEIN METAB, US ARMY INST SURG RES, 62- *Mem:* Am Chem Soc; AAAS; Asn Women Sci. *Res:* Protein metabolism; burn trauma. *Mailing Add:* 212 Brightwood San Antonio TX 78209

BROWN, WARREN SHELBURNE, JR, b Loma Linda, Calif, Sept 8, 44; m 66; c 2. PSYCHOPHYSIOLOGY. *Educ:* Point Loma Col, BA, 66; Univ Southern Calif, MA, 69, PhD(exp psychol), 71. *Prof Exp:* Trainee neurosci, Brain Res Inst, Univ Calif, Los Angeles, 71-73; asst res psychologist, 73-75, asst prof, 75-80, ASSOC PROF PSYCHOL, SCH MED, UNIV CALIF, LOS ANGELES, 80-; PROF PSYCHOL, FULLER GRAD SCH PSYCHOL, PASADENA, 81- *Concurrent Pos:* NIMH spec res fel, 75 & res scientist develop award, 75-80; mem, NSF US-indust countries exchange scientists, 86. *Mem:* AAAS; Am Psychol Asn; Soc Psychophysiol Res; NY Acad Sci; Int Neuropsychol Soc. *Res:* Brain mechanisms of higher mental functions in humans; electroencephalography as a means of studying cortical events; disorders of higher cognitive function in humans. *Mailing Add:* Grad Sch Psychol Fuller Theol Sem Pasadena CA 91182

BROWN, WELDON GRANT, chemistry; deceased, see previous edition for last biography

BROWN, WENDELL STIMPSON, b Pompton Plains, NJ, Apr 4, 43; m 79; c 3. PHYSICAL OCEANOGRAPHY. *Educ:* Brown Univ, BS, 65, MS, 67; Mass Inst Technol, PhD(oceanog), 71. *Prof Exp:* Asst res oceanog, Inst Geophys & Planetary Physics, Univ Calif, San Diego, 70-74; from asst prof to assoc prof, 74-87, PROF OCEANOG, UNIV NH, 87- *Concurrent Pos:* Vis scholar, Scripps Inst Oceanog, 81 & Harvard Univ, 88. *Mem:* Am Geophys Union; AAAS. *Res:* Observational/modelling studies of ocean circulation on the continental shelf, slope and marginal seas in the United States; climate change related studies in the western tropical Atlantic Ocean. *Mailing Add:* Dept Earth Sci Univ NH Durham NH 03824

BROWN, WILBUR K, b Oakland, Calif, July 6, 32; m 54; c 2. GENERAL PHYSICS. *Educ:* Univ Calif, Berkeley, AB, 54, MS, 57, PhD(eng sci), 62. *Prof Exp:* Asst, Lawrence Radiation Lab, Univ Calif, 58-62; mem staff, Res Estab, Riso, Denmark, 62-64 & Los Alamos Sci Lab, 64-70; assoc prof nuclear & mech eng, Univ Wyo, 70-72; mem staff, Los Alamos Nat Lab, 72-87; INSTR MATH SCI, LASSON COL, 87- *Mem:* Am Phys Soc; Am Nuclear Soc; Sigma Xi; Am Astron Soc. *Res:* Cosmogony. *Mailing Add:* 5179 Eastshore Dr Lake Almanor CA 96137

BROWN, WILLARD ANDREW, b Seattle, Wash, Nov 13, 21; m 48; c 2. SCIENCE EDUCATION. *Educ:* Univ Wash, BS, 46; Wash State Col, MAT, 58; Univ Fla, EdD(phys sci), 63. *Prof Exp:* Data analyst, Boeing Co, 48-49; teacher high sch, 51-56; asst prof educ & supvr student teachers, Western Wash State Col, 56-60; asst prof physics, Cent Wash State Col, 60-61; asst prof phys sci & astron, San Francisco State Col, 63-66; coordr sci educ, 66-69, prof, 66-84, EMER PROF PHYSICS & ASTRON, WESTERN WASH UNIV, 84- *Concurrent Pos:* Microcomput consult. *Res:* Data processing; data analysis for implementation of educational objectives. *Mailing Add:* 445 Highland Dr Bellingham WA 98225

BROWN, WILLIAM ANDERSON, b Corpus Christi, Tex, Dec 21, 29; m 52; c 2. SURFACE CHEMISTRY, RUBBER CHEMISTRY. *Educ:* Univ Tex, BS, 51, MA, 53, PhD(chem), 60. *Prof Exp:* Res scientist, Defense Res Labs, Univ Tex, 53-55; res engr, Alcoa Res Labs, Aluminum Co Am, 59-64; res chemist, Houston Res Labs, 64-68; res chemist, Ashland Chem Co, 68-71, sr res chemist, 71-79, mgr carbon black res, 79-87; CONSULT, 87- *Mem:* Am Chem Soc. *Res:* Coordination compounds; catalysis; organic coatings; corrosion; fine particles; elastomers; surface chemistry. *Mailing Add:* 1473 Chelmsford St NW North Canton OH 44720-6030

BROWN, WILLIAM ARNOLD, b New York, NY, Feb 13, 33; m 57; c 2. ATOMIC PHYSICS. *Educ:* Cornell Univ, AB, 55; Univ Mich, MS, 60, PhD(physics), 65. *Prof Exp:* Res assoc, Inst Sci & Technol, Univ Mich, 57-61; RES SCIENTIST, PHYS SCI DEPT, LOCKHEED PALO ALTO RES LAB, LOCKHEED MISSILES & SPACE CO, 64- *Mem:* Am Astron Soc; Optical Soc Am. *Res:* X-ray astronomy and solar physics; statistical analysis and design of experiments; shock tube spectroscopy; vacuum ultraviolet spectroscopy; atomic and molecular collisions. *Mailing Add:* 3525 Greer Rd Palo Alto CA 94303

BROWN, WILLIAM AUGUSTIN, b Mt Vernon, NY, Aug 1, 32; m 59; c 1. STRUCTURAL MECHANICS. *Educ:* Manhattan Col, BCE, 54; Va Polytech Inst, MS, 56; NY Univ, PhD(struct mech), 68. *Prof Exp:* Assoc engr, Sperry Gyroscope Co, NY, 55-56; dir off civil defense prof adv ctr, 69-74, ASSOC PROF CIVIL ENG, MANHATTAN COL, 56-, CHMN DEPT, 77- *Concurrent Pos:* NSF sci fac fel, 63; consult, Off Civil Defense, Washington, DC, 65-83. *Mem:* Sigma Xi; Am Soc Civil Eng. *Res:* Structural analysis. *Mailing Add:* RR 2 Box 41A South Salem NY 10590

BROWN, WILLIAM BERNARD, b Marinette, Wis, Apr 3, 36; m 58; c 3. POLYMER CHEMISTRY. *Educ:* Univ Wis, BS, 58; Univ Akron, PhD(polymer sci), 66. *Prof Exp:* Res chemist, Firestone Tire & Rubber Co, Ohio, 60-61 & Standard Oil Co (Ohio), 65-68; SR RES CHEMIST, GEN MOTORS RES LABS, 68- *Mem:* Am Chem Soc; Electrochem Soc; Am Inst Physics. *Res:* Chemical and physical aspects of providing protective coating to surfaces, including the failures of coatings due to both substrate deterioration and deterioration of the coatings. *Mailing Add:* 1899 Banbury Birmingham MI 48008

BROWN, WILLIAM DUANE, biochemistry, for more information see previous edition

BROWN, WILLIAM E, b Oxford, Pa, May 26, 33; m 56; c 3. MICROBIOLOGY, INDUSTRIAL CHEMISTRY. *Educ:* Univ Del, BS, 56; Univ Ga, MS, 60, PhD(food sci), 64. *Prof Exp:* Bacteriologist, Amerlab Inc, Ga, 56-58; res asst virol, Univ Ga, 58-60, res asst food sci, 60-64; microbiologist, Cornell Univ, 64-66; microbiologist, 67-71, vpres & tech dir, 71-87, SR SCIENTIST, BIO—LAB INC, 87- *Mem:* Poultry Sci Asn; Inst Food Technol; Chem Specialties Mfg Asn; Natl Asn Corrosion Engrs; Soc Indust Microbiol; Sigma Xi. *Res:* Environmental sanitation; enzymology; water treatment; water chemistry. *Mailing Add:* 2046 Meadowbrook Circle Conyers GA 30207

BROWN, WILLIAM E(RIC), b Philadelphia, Pa, Mar 4, 23; m 45, 78; c 2. CERAMICS. *Educ:* Rutgers Univ, BS, 48, MS, 51, PhD(ceramics), 75. *Prof Exp:* Instr ceramics, Rutgers Univ, 50-52; res engr, Kaiser Aliminum & Chem Corp, 52-67, sr res engr, 67-82; RETIRED. *Mem:* Keramos Hon Soc. *Res:* Thermomechanical and thermochemical properties of inorganic materials. *Mailing Add:* 1115 Kolln St Pleasanton CA 94566-5628

BROWN, WILLIAM E, b Cleveland, Ohio, Mar 27, 45; c 3. BIOCHEMISTRY, MICROBIOLOGY. *Educ:* Univ Minn, PhD(biochem), 71. *Prof Exp:* PROF BIOL SCI, CARNEGIE-MELLON UNIV, 73- *Concurrent Pos:* Consult, US Dept Health, Educ & Welfare & Environ Protection Agency; adv, Am Type Cult Collection; partic adv bd, Biotechnol & Bioeng; mem organizing comt, Intersci Conf on Antimicrobial Agents & Chemother, 78-; mem microbiol exchange comt with Soviet Union, NSF. *Honors & Awards:* Can Soc Tech Agr Gold Medal. *Mem:* AAAS; Brit Soc Gen Microbiol; NY Acad Sci; Am Chem Soc; Am Soc Microbiol. *Res:* Production of organic compounds biosynthetically; production of antibiotics, vitamins, dextran and therapeutic steroids; discovery of new anti-infectives and antibiotics; technology transfer; protein biochemistry; structure function; enzymology. *Mailing Add:* Carnegie-Mellon Univ 130 Carnegie Pl Pittsburgh PA 15208

BROWN, WILLIAM ERNEST, dentistry, for more information see previous edition

BROWN, WILLIAM G, b Toronto, Ont, June 13, 38; m 69; c 3. COMBINATORICS & FINITE MATHEMATICS. *Educ:* Univ Toronto, BA, 60, PhD(math), 63; Columbia Univ, MA, 61. *Prof Exp:* Asst prof math, Univ BC, 63-66; assoc prof, 66-76, PROF MATH, McGILL UNIV, 76- *Mem:* Am Math Soc; Math Asn Am; Can Math Soc (treas, 77-80); London Math Soc. *Res:* Combinatorial analysis; graph theory. *Mailing Add:* Dept Math McGill Univ 805 Sherbrooke St W Montreal PQ H3A 2K6 Can

BROWN, WILLIAM HEDRICK, b Yakima, Wash, Mar 18, 33; m 57; c 2. ANIMAL NUTRITION. *Educ:* State Col Wash, BS, 55; Univ Md, MS, 57, PhD(animal nutrit), 59. *Prof Exp:* From asst prof to assoc prof dairy sci, 59-69, PROF ANIMAL SCI, UNIV ARIZ, 69- *Concurrent Pos:* Res grants, Dept HEW, 60-63, 64-67 & NSF, 62-64. *Mem:* Am Dairy Sci Asn; Sigma Xi. *Res:* Roughage utilization by ruminants; lipid and mineral metabolism; biochemistry of the bovine rumen; pesticide residues in milk; physiology of lactation. *Mailing Add:* Dept Animal Sci Univ Ariz Tucson AZ 85721

BROWN, WILLIAM HENRY, b Ogdensburg, NY, Oct 26, 32; m 60, 90; c 4. ORGANIC CHEMISTRY, BIOCHEMISTRY. *Educ:* St Lawrence Univ, BS, 54; Harvard Univ, MA, 55; Columbia Univ, PhD(chem), 58. *Prof Exp:* From instr to asst prof org chem, Wesleyan Univ, 58-64; from asst prof to assoc prof, 64-72, PROF ORG CHEM, BELOIT COL, 73- *Concurrent Pos:* Res fel org chem, Calif Inst Technol, 63-64; res assoc, Univ Ariz, 67-68 & 70-71. *Mem:* Am Chem Soc. *Mailing Add:* Dept Chem Beloit Col Beloit WI 53511

BROWN, WILLIAM HENRY, b Warrensburg, Mo, May 22, 42; m 63. AGRICULTURAL & BIOLOGICAL ENGINEERING. *Educ:* Univ Mo, BS, 64, MS, 66, PhD(agr eng), 69. *Prof Exp:* From asst prof to assoc prof agr & biol eng, Miss State Univ, 68-77; assoc prof, La State Univ, 77-80, PROF AGR ENG & HEAD DEPT, 80- *Mem:* Am Soc Agr Engrs; Am Inst Biol Sci; Am Soc Eng Educ; Inst Elec & Electronics Engrs; Nat Soc Prof Engrs; Sigma Xi. *Res:* Plant and animal responses to environmental factors; engineering control of environmental factors. *Mailing Add:* La Agr Exp Sta PO Box 25055 Baton Rouge LA 70894-5055

BROWN, WILLIAM J, MEMBRANE RECEPTORS. *Educ:* Univ Tex, Dallas, PhD(med sci), 81. *Prof Exp:* Assoc res scientist cell biol, Sch Med, Yale Univ, 81-86; ASST PROF CELL BIOL, CORNELL UNIV, 86- *Mailing Add:* Sect Biochem-Molecular & Cell Biol Cornell Univ Biotechnol Bldg Ithaca NY 14853

BROWN, WILLIAM JANN, neuropathology; deceased, see previous edition for last biography

BROWN, WILLIAM JOHN, b Ashland, Pa, June 7, 40; m 66; c 1. MICROBIOLOGY. *Educ:* Univ Scranton, BS, 63; Duquesne Univ, MS, 65; WVa Univ, PhD(microbiol), 69. *Prof Exp:* From instr to asst prof microbiol, 69-74, ASSOC PROF IMMUNOL & MICROBIOL, SCH MED, WAYNE STATE UNIV, 74-; DIR MICROBIOL, HUTZEL HOSP, DETROIT, 74- *Concurrent Pos:* Mich Kidney Found grant; consult microbiol, Detroit Gen Hosp. *Mem:* Am Soc Microbiol. *Res:* Basic research on lymphocytic choriomeningitis virus, particularly the virus-cell relationship; urinary tract infections from both the viral and bacteriological aspects. *Mailing Add:* 2439 Dorfield Rochester MI 48307

BROWN, WILLIAM LACY, genetics, plant breeding; deceased, see previous edition for last biography

BROWN, WILLIAM LEWIS, b Clover, SC, Nov 23, 28; m 51; c 3. FOOD MICROBIOLOGY. *Educ:* Clemson Col, BS, 49; NC State Col, MS, 51; Univ Ill, PhD(food microbiol), 56. *Prof Exp:* Asst, NC State Col, 49-51, instr meats, 51-54; asst, Univ Ill, 56; res food technologist, John Morrell & Co, 56-57, asst dir res, 57-64, vpres res, 64-67; PRES, AM BACT & CHEM RES CORP, 67- *Mem:* Fel Am Soc Microbiol; Sigma Xi; fel Inst Food Technologists. *Res:* Bacteriology; nutrition; biochemistry. *Mailing Add:* Box 1557 Gainesville FL 32605

BROWN, WILLIAM LOUIS, JR, b Philadelphia, Pa, June 1, 22; m 46; c 3. BIOLOGY. *Educ:* Pa State Col, BS, 47; Harvard Univ, PhD, 50. *Prof Exp:* Asst, Pa State Col, 47; Parker fel from Harvard Univ, Australia, 50-51, Fulbright fel, 51-52; asst cur insects, Mus Comp Zool, Harvard Univ, 52-54, assoc cur, 54-60; from asst prof to assoc prof, 60-68, PROF ENTOM, CORNELL UNIV, 68- *Concurrent Pos:* Assoc, Mus Comp Zool, Harvard Univ, 60-; Guggenheim fel, 73-74. *Honors & Awards:* Donisthorpe Prize, 63. *Mem:* Am Entom Soc; Soc Syst Zool (pres, 84). *Res:* Systematics, ecology and behavior of ants; general evolutionary theory; zoogeography. *Mailing Add:* Dept Entom Cornell Univ Ithaca NY 14853

BROWN, WILLIAM M(ILTON), b Wheeling, WVa, Feb 14, 32; m 63; c 3. SYSTEMS ANALYSIS, APPLIED MATHEMATICS. *Educ:* WVa Univ, BS, 52; Johns Hopkins Univ, MS, 55, DrEng(elec eng), 57. *Prof Exp:* Asst instr physics, WVa Univ, 51-52; assoc elec engr, Westinghouse Elec Corp, 52-54; res engr, Radiation Lab, Johns Hopkins Univ, 54-57, lectr math & elec eng, 55-57; mem tech staff, Inst Defense Anal, 57-58; from asst prof to prof elec eng, Univ Mich, 58-73; dir, Willow Run Labs, 70-73; PRES, ENVIRON RES INST MICH, 72- *Concurrent Pos:* Assoc prof lectr, George Washington Univ, 58; indust consult, 58-; consult, Inst Defense Anal, 63-; spec adv aeronaut systs div, US Air Force, 63-; vis prof, Univ London, 68; vpres, Chain Lakes Res Corp, 69; ed-in-chief, Trans Aerospace & Electronic Systs, Inst Elec & Electronics Engrs. *Mem:* Fel Inst Elec & Electronics Engrs. *Res:* Circuit and information theory; systems analysis; radar and communications; military electronics problems. *Mailing Add:* Environ Res Inst Mich PO Box 8618 Ann Arbor MI 48107-8618

BROWN, WILLIAM PAUL, genetics, bioethics, for more information see previous edition

BROWN, WILLIAM RANDALL, b Staunton, Va, Oct 31, 13; m 42; c 3. GEOLOGY. *Educ:* Univ Va, BS, 38, MA, 39; Cornell Univ, PhD(mineral), 42. *Prof Exp:* From asst to assoc geologist, Va Geol Surv, 42-45; from asst prof to prof, 45-84, EMER PROF GEOL, UNIV KY, 84- *Concurrent Pos:* Consult, Calif Co, Shell Oil Co & US Geol Surv, 64-76; Field Inst Japan, Am Geol Inst, 67. *Mem:* fel Geol Soc Am; Am Asn Petrol Geol. *Res:* Igneous, metamorphic and structural geology of Piedmont Province, Virginia; mica and feldspar deposits of Virginia; Piedmont zinc and lead deposits of Virginia; park basins of Colorado; Wind River Basin, Wyoming; Wyoming overthrust belt; Pennsylvanian of eastern Kentucky. *Mailing Add:* 253 Shady Lane Lexington KY 40503

BROWN, WILLIAM ROY, b Kenora, Ont, June 26, 45; m 84; c 3. IMMUNODERMATOLOGY, CELL PROLIFERATION. *Educ:* Univ Waterloo,BIS, 75; Univ Windsor, MSC, 78; Univ Guelph, PhD(cell biol), 83. *Prof Exp:* Res assoc pathol, Univ Guelph, 82-83; res assoc, 83-87, ASST PROF DERMAT, UNIV TORONTO, 87- *Mem:* Soc Invest Dermat; Am Soc Photobiol; Can Soc Immunol; Can Soc Cell Molecular Biol; Sigma Xi; NY

Acad Sci. *Res:* Cellular adhesion molecules in contact sensitivity reactions; immune suppression by UVB radiation; circadian rhythms in epidermal cell proliferation and aging of the immune system. *Mailing Add:* Div Dermat Univ Toronto 100 College St Toronto ON M5G 1L5 Can

BROWN, WILLIAM SAMUEL, JR, b Pottstown, Pa, Apr 25, 40; m 62; c 2. SPEECH SCIENCE. *Educ:* Edinboro State Col, BS, 62; State Univ NY, Buffalo, MA, 67, PhD(speech sci), 69. *Prof Exp:* Speech & hearing therapist, Crawford County Operated Classes Handicapped, 62-65; res asst speech sci, State Univ NY, Buffalo, 65-68; res assoc & fel, Commun Sci Lab, 68-70, asst prof, 70-75, area head phonetic sci, Inst Advan Study Commun Processes, 74-81, assoc prof speech, 75-81, PROF SPEECH & DIR, INST ADVAN STUDY COMMUN PROCESSES, UNIV FLA, 81- *Mem:* Sigma Xi; fel Int Soc Phonetic Sci; Am Asn Phonetic Sci (secy-treas, 73-79); fel Am Speech & Hearing Asn; fel Acoust Soc Am. *Res:* Experimental phonetics; physiological and aerodynamic study of speech articulatory behavior, laryngeal and respiratory function utilizing both normal and pathological speakers. *Mailing Add:* Inst Advan Study Commun Processes Univ Fla Dauer Hall Gainesville FL 32605

BROWN, WILLIAM STANLEY, b Cleveland, Ohio, Feb 13, 35; m 56; c 2. COMPUTER ALGEBRA & LANGUAGE. *Educ:* Yale Univ, BS, 56; Princeton Univ, PhD(physics), 61. *Prof Exp:* Instr physics, Princeton Univ, 60-61; mem tech staff, Math Physics Dept, 61-66, head comput math res dept, 66-81, head, Info Systs Dept, Bell Labs, 81-86, div mgr venture technologies, AT&T, 86-88, INT PLANNING DIR, BELL LABS, 88- *Concurrent Pos:* Mem, Working Group Numerical Software, Int Fed Info Processing, 76-82 & bd sci counr, Lister Hill Ctr, Nat Libr Med, 81-83; mem, electronic publ comt, Am Inst Physics, 86-89, chmn subcomt numerical databases, 87-89; mem info technol adv panel, Europ Inst Technol, 89-90, sci adv bd, 90- *Res:* Computer algebra and languages; information transfer; software portability; theory of floating-point computation. *Mailing Add:* AT&T Bell Labs Crawfords Corner Rd Rm 2C612 Holmdel NJ 07733

BROWN, WINTON, b Ft Sheridan, Ill, June 12, 12; m 87; c 5. CHEMICAL ENGINEERING. *Educ:* Mass Inst Technol, BS, 34, MS, 35; Univ Pittsburgh, PhD(chem eng), 42. *Prof Exp:* Asst & indust fel, Mellon Inst, 35-41; chem engr, Distillation Prods Industs, 46-69, tech assoc, 69-76; RETIRED. *Concurrent Pos:* Mem, Liberty Amendment Comt, 77-; mem bd dirs, Rochester Comt Sci Info, 78-87. *Mem:* Am Chem Soc; Am Inst Chem Engrs. *Res:* Gas absorption; sulfur technology; processing of vitamins A and E. *Mailing Add:* Lake of Woods HCR 72 Box 551B Locust Grove VA 22508

BROWNAWELL, WOODROW DALE, b Grundy Co, Mo, Apr 21, 42; m 65; c 1. NUMBER THEORY. *Educ:* Univ Kans, BA, 64; Cornell Univ, PhD(math), 70. *Prof Exp:* Asst prof math, Pa State Univ, 70-74; vis assoc prof, Univ Colo, 74-75; assoc prof, 75-80, PROF MATH, PA STATE UNIV, 80- *Concurrent Pos:* Alexander von Humboldt res scholar, Univ Cologne, 77-78; master res, Ecole Polytech, 84; vis prof, Univ Paris, Univ St Etienne, 87. *Mem:* Am Math Soc; Math Asn Am; Deutsch Math Ver. *Res:* Independence properties of numbers and functions arising in classical analysis, including especially the exponential function and the Weierstrass elliptic functions. *Mailing Add:* Dept Math Pa State Univ University Park PA 16802

BROWNE, ALAN LAMPE, b Hamilton, Ohio, Feb 3, 44; m 70; c 3. ENGINEERING MECHANICS. *Educ:* Harvard Univ, AB, 66; Northwestern Univ, PhD(mech eng), 71. *Prof Exp:* Assoc sr res engr lubrication tire traction, Gen Motors Res Labs, 71-76, sr res engr lubrication tire mech, 76-80, staff res eng tire mech, occupant safety, 80-87, SR STAFF RES ENGR, OCCUPANT SAFETY, GEN MOTORS RES LABS, 87- *Concurrent Pos:* Mem adv panel, Nat Coop Highway Res Prog, 74-80; assoc ed, Tire Sci & Technol, 75- *Mem:* Am Soc Mech Engrs; Am Soc Testing & Mat; Soc Automotive Engrs; Sigma Xi. *Res:* Elastohydrodynamic lubrication; pneumatic tire hydroplaning; tire wet traction; tire power loss; vehicle dynamics; occupant safety; impact mechanics. *Mailing Add:* Dept Eng Mech Gen Motors Res Warren MI 48090

BROWNE, CAROLE LYNN, b July 13, 50; m 76; c 2. CELL BIOLOGY. *Educ:* Syracuse Univ, PhD(biol). *Prof Exp:* ASST PROF CELL & MOLECULAR BIOL & ELECTRON MICROS, WAKE FOREST UNIV, 80- *Mem:* Am Soc Cell Biol; Electron Micros Soc Am. *Res:* Cyclic AMP dependent protein kinase; cellular regulation of mitotic process. *Mailing Add:* Wake Forest Univ Reynolda Sta Box 7325 Winston-Salem NC 27109

BROWNE, CHARLES IDOL, b Atlanta, Ga, Feb 8, 22; m 42; c 1. RADIOCHEMISTRY. *Educ:* Drew Univ, AB, 41; Calif Inst Technol, MS, 48; Univ Calif, PhD(chem), 52. *Prof Exp:* Alt leader, Radiochem Group, 55-65, assoc div leader, Weapons Test Div, 65-70, alt div leader, 70-72, div leader, 72-74, from asst dir to assoc dir, Los Alamos Sci Lab, 74-87; RETIRED. *Mem:* Fel Am Phys Soc; fel Am Inst Chem; Sigma Xi. *Res:* Radiochemical diagnostics. *Mailing Add:* 428 Estante Way Los Alamos NM 87544

BROWNE, COLIN LANFEAR, b Buffalo, NY, Apr 11, 28; m 49; c 2. FIBER CHEMISTRY. *Educ:* Lafayette Col, BA, 49; Univ Va, MS, 51, PhD(chem), 53. *Prof Exp:* Textile res chemist, Rohm and Haas Co, 53-57; mgr indust prod develop, 57-66, mgr dyeing & finishing lab, 66-70, mgr dyeing & finishing tech serv & develop, 70-78, RES & DEVELOP MGR, 78-, TECH ASSOC, HOECHST CELANESE CORP, CHARLOTTE, NC, 78- *Mem:* Am Chem Soc; Am Asn Textile Chemists & Colorists; Fiber Soc. *Res:* Development of specialty fibers and films, smoking products and filters. *Mailing Add:* Hoechst Celanese Corp PO Box 32414 Charlotte NC 28232-6085

BROWNE, CORNELIUS PAYNE, b Madison, Wis, Oct 30, 23; m 57; c 2. NUCLEAR PHYSICS. *Educ:* Univ Wis, AB, 46, PhD(physics), 51. *Prof Exp:* Res assoc, Mass Inst Technol, 51-56; from asst prof to assoc prof, 56-64, PROF PHYSICS, UNIV NOTRE DAME, 64- *Concurrent Pos:* Vis prof, Univ Wis, 59-61, Univ Tex, Austin, 72-73; prog officer, Nuclear Physics,

NSF, 80-81. *Mem:* Am Asn Physics Teachers; fel Am Phys Soc; Sigma Xi. *Res:* Measurement of nuclear reaction energies and nuclear energy levels; nuclear reaction mechanisms. *Mailing Add:* 1606 E Washington Ave South Bend IN 46617

BROWNE, DOUGLAS TOWNSEND, b Hamilton, Ohio, Jan 23, 42; m 74; c 2. BIO-ORGANIC CHEMISTRY, BIOCHEMISTRY. *Educ:* Mass Inst Technol, SB, 64; Univ Ill, PhD(chem), 68. *Prof Exp:* NIH fel biochem, Harvard Univ, 68-69; asst prof chem, Univ Calif, Berkeley, 69-74; asst prof, 74-76, ASSOC PROF CHEM, WORCESTER POLYTECH INST, 76- *Concurrent Pos:* Affil scientist, Worcester Found Exp Biol, 75-78; vis prof, Univ Hawaii, 84-85. *Mem:* Am Chem Soc; AAAS; NY Acad Sci. *Res:* Enzyme conformation and mechanism; nuclear magnetic resonance studies of proteins; site-specific enzyme labeling; protein chemical modification; pyrimidine and enzyme photochemistry. *Mailing Add:* Dept Chem Worcester Polytech Inst Worcester MA 01609

BROWNE, EDWARD TANKARD, JR, b Chapel Hill, NC, Dec 17, 26; m 50; c 3. BOTANY. *Educ:* Univ NC, BA, 48, MA, 50, PhD(bot), 57. *Prof Exp:* Head bot slide dept, Carolina Biol Supply Co, Elon Col, 52-55; asst prof bot, Auburn Univ, 56-59 & Univ Ga, 59-60; from asst prof to assoc prof, Univ Ky, 60-67; assoc prof, 67-68, prof, 68-88, EMER PROF BIOL, MEMPHIS STATE UNIV, 88- *Mem:* Am Soc Plant Taxon. *Res:* Contributions of plant embryology to systematic botany; flora of Kentucky; Liliaceae and Aletris systematics. *Mailing Add:* 486 St Nick Dr Memphis TN 38117

BROWNE, JAMES CLAYTON, b Conway, Ark, Jan 16, 35; m 59; c 3. MOLECULAR PHYSICS, COMPUTER SCIENCE. *Educ:* Hendrix Col, BA, 56; Univ Tex, PhD(chem), 60. *Prof Exp:* Asst prof physics, Univ Tex, 60-64; NSF fel, 64-65; prof comput sci, Queen's Univ, Belfast, 65-68; PROF PHYSICS & COMPUT SCI, UNIV TEX, AUSTIN, 68-, RES SCIENTIST, 73- *Concurrent Pos:* Consult, NSF. *Mem:* Am Phys Soc; Asn Comput Mach; Soc Indust & Appl Math. *Res:* Atomic and molecular processes; operating systems; symbolic mathematics. *Mailing Add:* Dept Comp Sci Univ Tex Austin TX 78712

BROWNE, JOHN MWALIMU, b Miami, Fla, Aug 13, 39; m 58; c 3. CELL BIOLOGY. *Educ:* Bethune-Cookman Col, BS, 62; Univ Miami, MS, 68, PhD(cell & develop biol), 70. *Prof Exp:* Instr biol, Bethune-Cookman Col, 62-65; NIH res asst develop biol, Lab Quant Biol, Univ Miami, 65-68; teaching asst, Dept Biol, 68-69; res assoc, Lab Quant Biol, 70; assoc prof biol, Bethune-Cookman Col, 70-71; assoc prof, 71-77, dir res, 83-84, chair, dept biol, 85-89, PROF BIOL, ATLANTA UNIV, 77-, PROF CELL BIOL & DIR, BIOMED RES, 89- *Concurrent Pos:* Res assoc, Rosensteil Sch Marine & Atmospheric Sci, 70-71; adj prof cell biol, Rockefeller Univ, 78- *Mem:* NY Acad Sci; Soc Develop Biol; Am Soc Cell Biol; AAAS; Am Soc Biotechnol; Nato fel, 84. *Res:* Cellular regulation at the translational level in avian, piscean and mammalian systems; effects of carcinogens on the differential synthesis of cell membrane enzymes during mammary gland tumorigenesis. *Mailing Add:* Dept Biol Atlanta Univ Atlanta GA 30314-4391

BROWNE, MICHAEL EDWIN, b Los Angeles, Calif, June 12, 30; m 51; c 7. PHYSICS. *Educ:* Univ Calif, Berkeley, BS, 52, PhD(physics), 55. *Prof Exp:* Staff scientist, Lockheed Res Labs, 55-67; vis res physicist, Physics Inst, Zurich, 63-64; dept chmn, 67-75, PROF PHYSICS, UNIV IDAHO, 67- *Concurrent Pos:* Vis prof physics, Boston Univ, 77. Concurrent. *Mem:* Am Phys Soc; Am Asn Physics Teachers. *Res:* Solid state physics and biophysics. *Mailing Add:* Dept Physics Univ Idaho Moscow ID 83843

BROWNE, ROBERT GLENN, b Brockport, NY, Oct 6, 51; m 84; c 2. MANUFACTURING TECHNOLOGY RESEARCH, APPLIED MACHINE VISION RESEARCH. *Educ:* Univ Rochester, BS, 72, BA, 72; Princeton Univ, MA, 76, PhD(physics), 79. *Prof Exp:* Mem tech staff, RCA Labs, 78-87; MEM TECH STAFF, DAVID SARNOFF RES CTR, INC, 87- *Concurrent Pos:* Adj prof physics, Middlesex County Col, 87-89. *Mem:* Am Phys Soc; sr mem Mach Vision Asn. *Res:* Manufacturing technology research including development of laser-based inspection systems; applications of machine vision systems to industrial uses; laser-materials processing research; modeling and simulation of physical dynamic systems. *Mailing Add:* 93 Remmey St Fords NJ 08863

BROWNE, RONALD K, b Chicago, Ill, Oct 21, 34. SCIENTIFIC COMMUNICATIONS, REGULATORY SUBMISSIONS. *Educ:* Loyola Univ, PhD(pharm), 72. *Prof Exp:* Mgr sci affairs, Sci Commun Inc, 76-80, dir pharmacol, 80-86, sr sci writer, 86-90, LECTR & CONSULT, SCI COMMUN INC, 91- *Concurrent Pos:* Mem, coun high blood pressure, Am Heart Asn; adj assoc prof, Univ Mo-Kansas City. *Mem:* Soc Toxicol; Am Soc Pharmacol & Exp Therapeut. *Mailing Add:* 12316 Wenonga Lane Leawood KS 66209

BROWNE, SHEILA EWING, b Danville, Ky, Feb 3, 49; m 74; c 2. PHYSICAL ORGANIC CHEMISTRY. *Educ:* Univ Tenn, Knoxville, BS, 71; Univ Calif, Berkeley, PhD(org chem), 74. *Prof Exp:* Teaching assoc chem, Worcester Polytech Inst, 74-75; fel phys org chem, grad dept biochem, Brandeis Univ, 75-76; ASSOC PROF CHEM, MT HOLYOKE COL, 76-, CHMN, CHEM DEPT, 90- *Mem:* Sigma Xi; Am Chem Soc; Am Indian Sci & Eng Soc. *Res:* Polymers-structural changes of polyelectrolytes; structure of carbanions using 13C nuclear magnetic resonance; mechanism of hydrolysis of amino acid esters. *Mailing Add:* Dept Chem Mt Holyoke Col South Hadley MA 01075

BROWNE, VANCE D'ARMOND, mechanical engineering, for more information see previous edition

BROWNE, W(ILLIAM) H(OOPER), b White Plains, NY, Aug 2, 15; m 39; c 3. MECHANICAL ENGINEERING. *Educ:* Rensselaer Polytech Inst, ME, 36. *Prof Exp:* Apprentice engr, Caterpillar Tractor Co, 36-38, res engr diesel engines, 38-42; res engr fuels, Battelle Mem Inst, 42-45, asst supvr

mech eng, 45-49, supvr, 49-51, supvr petrol eng, 51-53, mgr mech eng, 53-66, asst dir admin, Columbus Labs, 66-77; RETIRED. *Mem:* Am Soc Mech Engrs. *Res:* Fuels utilization; mechanical design, development and research. *Mailing Add:* PO Box 352 Jackson NH 03846

BROWNELL, ANNA GALE, b Leipzig, Ger, June 17, 42; US citizen. CELLULAR DIFFERENTIATION, PROTEIN BIOCHEMISTRY. *Educ:* Bowling Green State Univ, BA, 64; Northwestern Univ, PhD(biochem), 75. *Prof Exp:* Fel, Univ Southern Calif, Dent Sch, 75-77, res asst prof develop biol, 77-81; res biochemist, dept surg & orthop, Univ Calif, Los Angeles, 82-85; asst prof, 85-88, ASSOC PROF CELL BIOL & GENETICS, CHAPMAN COL, 88- *Concurrent Pos:* Prin investr, NIH grant, 86-90. *Mem:* Am Soc Cell Biol; Am Soc Bone & Mineral Res; AAAS; Sigma Xi. *Res:* Proteins of the extracellular matrix and their interactions with matrix synthesizing cells; specifically how bone matrix proteins modulate the differentiation and metabolism of bone-forming cells (osteoblasts) and bone-degrading cells (osteoclasts). *Mailing Add:* Dept Biol Chapman Col Orange CA 92666

BROWNELL, FRANK HERBERT, III, b New York, NY, Sept 20, 22; m 50; c 4. MATHEMATICS. *Educ:* Yale Univ, BA, 43, MS, 47; Princeton Univ, PhD(math), 49. *Prof Exp:* Fine instr, Princeton Univ, 49-50; from asst prof to prof math, Univ Wash, 50-90; RETIRED. *Concurrent Pos:* Ford fel, Inst Advan Study, 53-54. *Mem:* Am Math Soc. *Res:* Nonlinear delay differential equations; operations and measures on Hilbert space and applications to quantum mechanics. *Mailing Add:* 1525 Tor Jan Hill Rd Barnbridge Island WA 98110

BROWNELL, GEORGE H, b Minneapolis, Minn, Oct 23, 38; m 67; c 2. MICROBIOLOGY, BACTERIAL GENETICS. *Educ:* Univ Minn, BA, 61; Univ SDak, MA, 63, PhD(microbiol), 67. *Prof Exp:* From asst prof to assoc prof microbiol, 67-80, PROF IMMUNOL & MICROBIOL, MED COL GA, 80- *Concurrent Pos:* At Lab Molecular Virol, Nat Cancer Inst, 78-79; recipient, Bicentennial Order, Univ Los Andres, 90. *Mem:* AAAS; Am Soc Microbiol; Sigma Xi. *Res:* Nocardial genetics; microbial genetics. *Mailing Add:* Dept Immunol & Microbiol Med Col Ga Augusta GA 30912

BROWNELL, GEORGE L, b Hoosick Falls, NY, May 20, 23; m 56; c 3. ORGANIC CHEMISTRY. *Educ:* Rensselaer Polytech, BS, 47, MS, 48; Ohio State Univ, PhD(org chem), 53. *Prof Exp:* Res chemist, Aluminum Res Labs, 53-55; asst prof, Lehigh Univ, 55-58; res chemist, Miles Chem Co, 58-60; supvr indust chem res, Chem Div, Aristech Chem Corp, 60-67, assoc res consult, 75-83, SECT SUPVR RES LAB, ARISTECH CHEM CORP, 67-, SR STAFF SCIENTIST, 90- *Concurrent Pos:* Res consult, Aristech Chem Corp, 83- *Mem:* Am Chem Soc; Sigma Xi; Royal Soc Chem. *Res:* Electrolytic reduction, oxidation and substitution of organic compounds; pharmaceutical chemistry; chemistry of coal; lignites; esterifications; liquid phase oxidations; maleic anhydride chemistry; polymerization studies; hydroformylation studies; polyester resins; plasticizers; homogeneous catalysis; maleimido chemistry; engineering resins. *Mailing Add:* Aristech Chem Corp Res Lab 1000 Tech Ctr Dr Monroeville PA 15146-3051

BROWNELL, GORDON LEE, b Duncan, Okla, Apr 8, 22; m 44, 86; c 4. MEDICAL PHYSICS. *Educ:* Bucknell Univ, BS, 43; Mass Inst Technol, PhD(physics), 50. *Prof Exp:* Asst physics, 48-50, res assoc, 50-57, from asst prof to assoc prof, 57-70, PROF NUCLEAR ENG, MASS INST TECHNOL, 70-; LECTR, HARVARD MED SCH, 50- *Concurrent Pos:* From asst physicist to assoc physicist, Mass Gen Hosp, 49-61, physicist, 61-, head physics res lab; trustee, Neuro-Res Found; hon fac mem, Cuyo Univ, Arg, 54. *Honors & Awards:* Aebersold Award, Soc Nuclear Med; von Heuesy Medal, Europ Soc Nuclear Med; Collidge Award, Am Asn Physicists Med. *Mem:* AAAS; Am Asn Physicists in Med; Soc Nuclear Med; Am Phys Soc; Biophys Soc. *Res:* Imaging of positron emitting isotopes; computerized axiol tomography; reactor applications; analysis of tracer data; radiation effects and dosimetry. *Mailing Add:* Phys Res Lab Mass Gen Hosp Fruit St Boston MA 02114

BROWNELL, JAMES RICHARD, b Jeannette, Pa, Apr 20, 32; m 51; c 4. SOILS, IRRIGATION. *Educ:* Pa State Univ, BS, 54; Univ Minn, MS, 57; Univ Calif, Davis, PhD(geobiol), 70. *Prof Exp:* Jr voc instr soils & irrig, Fresno State Col, 58-59, asst prof chem, 59-60; agriculturist, Di Giorgio Fruit Corp, 60-64; technician soil fertil, Univ Calif, Davis, 64-69; PROF SOILS, CALIF STATE UNIV, FRESNO, 69-,. *Concurrent Pos:* Consult, USAID, Solar Energy Res Inst, Boyle Eng, J G Boswell, Western US & Int Egypt, Nigeria, Australia, Cent Am & Caribbean, Food & Fertilizer Tech Ctr, Taipei; Fulbright scholar, Egypt. *Mem:* Soil Sci Soc Am; Soil Conserv Soc Am. *Res:* Land use planning; waste disposal; environmental impact studies; reclamation of saline and sodic soils; remote sensing; water measurement; irrigation scheduling; soil selection for chemically stabilized brick fabrication; technology transfer; water application technology, crop and soil affects; agricultural energy use; integration of energy consumers with nitrogen and water use efficiency as major concerns; bio and solar energy inputs; solar greenhouse design; composting and methane generation (for co-generation of heat, electricity, and bio fertilizer). *Mailing Add:* 5541 N Sixth St Fresno CA 93710

BROWNELL, JOHN HOWARD, b Oneonta, NY, Oct 4, 42; m 65; c 2. INERTIAL CONFINEMENT FUSION, PLASMA INSTABILITY THEORY. *Educ:* Mass Inst Technol, BS, 64; Stanford Univ, MS, 70, PhD(physics), 71. *Prof Exp:* Vis scientist, Inst Atomic & Molecular Physics, Neth, 71-72; staff mem, asst group leader, assoc group leader, GROUP LEADER, LOS ALAMOS NAT LAB, 72- *Concurrent Pos:* Consult, Quantum Physics Corp, 69-72; prin investr, High Density Plasma Physics Proj, 78- *Mem:* Am Phys Soc; Sigma Xi. *Res:* Theoretical investigation of the physics of inertial confinement and hybrid fusion schemes, including the development and use of large computer simulation codes; high pulsed power and radiation source development; technical management. *Mailing Add:* Los Alamos Nat Lab MS-B259 Los Alamos NM 87545

BROWNELL, PHILIP HARRY, b Castro Valley, Calif, Aug 2, 47; m 69; c 2. CELLULAR NEUROBIOLOGY, SENSORY PHYSIOLOGY. *Educ:* Univ Calif, Berkeley, AB, 70; Univ Calif, Riverside, PhD(biol), 76. *Prof Exp:* Postdoctoral fel physiol & neurobiol, Univ Calif, San Francisco, 76-79; asst prof, 79-85, ASSOC PROF ZOOL, ORE STATE UNIV, 85- *Concurrent Pos:* Sloan Found fel, 76-77 & NIH, 77-79; prin investr, NIH & Am Heart Asn, 79-; Ger Acad Exchange Serv fel, Univ Regensburg, 90. *Mem:* Soc Neurosci; AAAS; Am Heart Asn; Sigma Xi. *Res:* Neuronal mechanisms of behavior; neural circuitry; neuroethology; neuropeptide actions in central nervous system; intracellular response mechanisms; arthropod sensory systems and physiology. *Mailing Add:* Dept Zool Ore State Univ Corvallis OR 97331

BROWNELL, R(ICHARD) M(ILLER), b Detroit, Mich, Jan 26, 30; m 53; c 5. ELECTRICAL ENGINEERING, MAGNETISM. *Educ:* Yale Univ, BS, 51, MEng, 56, DEng(magnetics), 61. *Prof Exp:* Mem tech staff, Bell Tel Labs Inc, 60-65, supvr, 65-89; RETIRED. *Concurrent Pos:* Lectr elec eng, Lehigh Univ, 62; mem tech prog comt, Int Nonlinear Magnetics Conf, 63-64. *Mem:* Inst Elec & Electronics Engrs. *Res:* Testing magnetic materials and devices; studies of magnetization processes by electrical techniques; sealed reed contacts; coaxial switches; ferreed and remreed switches. *Mailing Add:* 58 S Ashby Ave Livingston NJ 07039

BROWNELL, ROBERT LEO, JR, b San Pedro, Calif, July 8, 43. MARINE MAMMALS, CETACEANS. *Educ:* San Diego State Univ, BS, 67; Univ Tokyo, DAgr, 75. *Prof Exp:* Res assoc marine mammals, Cetacean Res Lab, Little Co Mary Hosp, Calif, 63-67; CHIEF MARINE MAMMAL SECT, DEPT INTERIOR, NAT ECOL RES CTR, US FISH & WILDLIFE SERV, 77- *Concurrent Pos:* Chmn, Subcomt Small Cetaceans, 75-78 & 84-86; sci adv, Comt Sci Adv Marine Mammals, Marine Mammal Comn, Wash, DC, 75-78 & 85-90; vchmn Sci Comt Int Whaling Comn, Eng, 86-88, chmn, 89-91. *Mem:* Am Soc Mammalogists; Soc Marine Mammal (pres elec, 86-88, pres 88-90); Am Asn Zool Nomenclature; Soc Conservation Biol; Xerces Soc; Soc Vertebrate Paleont. *Res:* Natural history of marine mammals especially cetaceans. *Mailing Add:* US Fish & Wildlife Serv PO Box 70 San Simeon CA 93452-0070

BROWNELL, WAYNE E(RNEST), b Hornell, NY, Feb 13, 18; m 41; c 3. CHEMISTRY, CERAMICS. *Educ:* Alfred Univ, BS, 40, MS, 49; Pa State Univ, PhD(ceramics), 53. *Prof Exp:* Asst ceramics, Alfred Univ, 40-42; mech engr, US Air Force, 42-43; from asst prof to prof ceramics, 46-71, chmn dept ceramic sci, 64-70, prof 70-84, EMER PROF, CERAMIC SCI, STATE UNIV NY COL CERAMICS, ALFRED UNIV, 84- *Mem:* Am Ceramic Soc; fel Am Inst Chem; Am Chem Soc; Am Soc Eng Educ; Brit Ceramic Soc. *Res:* Kinetics of solid state reaction; nonmetallic mineral resources; clay mineralogy and clay water systems; structural clay products. *Mailing Add:* RR PO Box 227 Almond NY 14804

BROWNELL, WILLIAM EDWARD, b Augusta, Ga, Oct 17, 42; m 68; c 3. CELL BIOLOGY OF HEARING. *Educ:* Univ Chicago, SB, 68, PhD(physiol), 73. *Prof Exp:* From asst prof to assoc prof neurosci & surg, Univ Fla, Gainesville, 74-85; DIR RES & ASSOC PROF OTOLARYNGOL & NEUROSCI, SCH MED, JOHNS HOPKINS UNIV, 85- *Concurrent Pos:* Prin investr, res grants NIH, NSF, Navy, 75-, Max Kade Found, 89-90, Hasselblad Found, 90-91; P & P Tech Comt, Acoust Soc Am, 81-84; Res Award, Roche Res Found, 82; invited prof physiol, Univ Geneva, Switz, 82-83; mem Rev Panel Commun Dis, NIH-Nat Inst Neurol & Commun Dis & Stroke, 82-85; res prof physiol, Univ Montpellier, France, 84; consult, Dept Army, 79, Off Naval Res, 84-89; mem Task Force, Nat Strategic Plan NIDCD, 89. *Honors & Awards:* Fulbright Award, French Fulbright Found, 84; Claude Pepper Award, Nat Inst Deafness & Other Commun Dis, 90. *Mem:* AAAS; Soc Neurosci; Centurions Deafness Res Found; Am Acad Otolaryngol; Asn Res Otolaryngol; Acoust Soc Am. *Res:* Electromechanical transduction in outer hair cells from the mammalian inner ear; fine structure of the intracochlear electric field. *Mailing Add:* Johns Hopkins Univ Sch Med 522 Traylor Bldg 720 Rutland Ave Baltimore MD 21205-2196

BROWNER, ROBERT HERMAN, b New York, NY, June 19, 43; m; c 2. NEUROBIOLOGY, COMPARATIVE NEUROLOGY. *Educ:* NY Univ, BA, 65, MA, 70, PhD(biol & neurobiol), 73. *Prof Exp:* Instr, 73-75, ASST PROF ANAT, NY MED COL, VALHALLA, NY, 75- *Mem:* Soc Neurosci; Am Zoologists; Sigma Xi. *Res:* Comparative neurobiology of the central auditory pathway in vertebrates; light microscopy, electron microscopy; regeneration, neurophysiology. *Mailing Add:* Five Fern Oval W Orangeburg NY 10962

BROWNIE, ALEXANDER C, b Bathgate, Scotland, Mar 6, 31. BIOCHEMISTRY. *Educ:* Univ Edinburgh, BSc, 52, PhD(biochem), 55, DSc, 81. *Prof Exp:* Asst biochem, Univ Edinburgh, 53-55; Scottish Hosps Endowment Res Trust fel, Royal Infirmary, Edinburgh, 55-56; Organon res fel pharmacol, Univ St Andrews, 56-59, med res fel, 59-62, hon lectr, 60-62; USPHS fel, Univ Utah, 62-63; asst prof path & biochem, 63-67, assoc prof biochem, 67-70, res assoc prof path, 67-76, PROF BIOCHEM, STATE UNIV NY, BUFFALO, 70-, CHMN DEPT, 76-, RES PROF PATH, 76-, RES PROF MED, 84- *Mem:* Endocrine Soc; Am Soc Biol Chemists. *Res:* Adrenal steroid biosynthesis; pituitary-adrenal function in disease; mechanism of action of ACTH; hormones and hypertension. *Mailing Add:* Dept Biochem 102 Cary Hall State Univ NY Buffalo NY 14214

BROWNING, CHARLES BENTON, b Houston, Tex, Sept 16, 31; m 56; c 6. ANIMAL NUTRITION. *Educ:* Tex Technol Col, BS, 55; Kans State Univ, MS, 56, PhD(animal nutrit), 58. *Prof Exp:* From asst prof to assoc prof dairy prod, Miss State Univ, 58-62, prof dairy sci, 62-66; chmn, Dept Dairy Sci, Univ Fla, 66-69, dean, Col Agr, 69-79; DEAN & DIR, COL AGR, AGR EXP STA, COOP EXTEN, OKLA STATE UNIV, 79- *Mem:* Am Soc Animal Sci; Am Dairy Sci Asn. *Res:* Dairy cattle nutrition and physiology. *Mailing Add:* Col Agr Okla State Univ Stillwater OK 74078-0500

BROWNING, DANIEL DWIGHT, b New Albany, Miss, Mar 24, 21; m 47; c 5. RADON. *Educ:* Miss Col, BA, 41. *Prof Exp:* Control chemist, E I du Pont de Nemours & Co, 41; fel, Mellon Inst, 42-44 & 46-50; res chemist, 50-58, mgr paint res, 58-60, mgr plastic flooring res, 60-62, gen mgr bldg prod res, 62-68, asst dir res, 68-76, vpres & dir, Bus Info Serv, Armstrong Cork, Co, 76-81, vpres & dir res, Armstrong World Ind, 81-86; PRES, SCI MGT SERV LTD, 86- *Concurrent Pos:* Mem, Bldg Res Inst. *Mem:* Am Chem Soc. *Res:* Building products. *Mailing Add:* 25 Eshelman Rd Lancaster PA 17601

BROWNING, DAVID GUNTER, b Wakefield, RI, May 3, 37. ACOUSTICS, PHYSICAL OCEANOGRAPHY. *Educ:* Univ RI, BS, 58; Mich State Univ, MS, 61. *Prof Exp:* Res physicist acoust, Naval Underwater Systs Ctr, 65-72; exchange scientist, Defense Sci Estab, NZ, 72-73; res physicist, Naval Underwater Systs Ctr, 73-79; vis scientist oceanog, Naval Postgrad Sch, 79-80; res physicist acoust, Naval Underwater Systs Ctr, 80-82; exchange scientist, Defense Res Estab-Pac Can, 83; head, Environ Acoust Br, 84-88, SR PHYSICIST, ENVIRON SYST DEPT, NAVAL UNDERWATER SYST CTR, 89- *Mem:* Am Phys Soc; AAAS; fel Acoust Soc Am; Australia-NZ Asn Advan Sci; Sigma Xi; US Naval Inst. *Res:* Attenuation of sound in the sea; acoustic characteristics of the oceans of the Southern Hemisphere; physical properties of ocean fronts and eddys. *Mailing Add:* 139 Old North Rd Kingston RI 02881

BROWNING, EDWARD T, b Cleveland, Ohio, July 15, 39; div; c 2. PHARMACOLOGY, NEUROCHEMISTRY. *Educ:* Purdue Univ, BS, 61; Univ Ill Med Ctr, Chicago, MS, 64, PhD(pharmacol), 66. *Prof Exp:* Johnson Res Found fel phys biochem, Univ Pa, 66-68; from asst prof to assoc prof pharmacol, Rutgers Med Sch, 74-84; PROF PHARMACOL, ROBERT WOOD JOHNSON MED SCH, UNIV MED & DENT NJ, 84- *Mem:* Fedn Am Socs Exp Biol; Soc Neurosci; Am Soc Neurochem. *Res:* Function of astrocytes; biochemistry and molecular biology of synaptic structures. *Mailing Add:* Dept Pharmacol Robert Wood Johnson Med Sch 675 Hoes Lane Piscataway NJ 08854

BROWNING, HORACE LAWRENCE, JR, b Overton, Tex, Oct 8, 32; m 53; c 2. POLYMER CHEMISTRY, PHYSICAL CHEMISTRY. *Educ:* Stephen F Austin State Col, BA, 54; La State Univ, MS, 56, PhD(phys chem), 60. *Prof Exp:* Asst prof chem, Northeast La State Col, 60-63; res chemist, 63-65, SR RES CHEMIST, TENN EASTMAN CO, 65- *Concurrent Pos:* Spec lectr chem, King Col, 66-71. *Mem:* Am Chem Soc. *Res:* Gel permeation chromatography; ultracentrifugation; characterization of polymers; solution thermodynamics. *Mailing Add:* 199 Painter Rd Fall Branch TN 37656

BROWNING, J(AMES) S(COTT), metallurgy, for more information see previous edition

BROWNING, JOE LEON, b Huntington, WVa, June 14, 25; div; c 1. APPLIED CHEMISTRY, CHEMICAL ENGINEERING. *Educ:* Marshall Univ, BS, 47. *Prof Exp:* Chemist, C&O Rwy Co, 47-49; chemist, Allied Chem & Dye Corp, 49-50; chemist, Int Nickel Co, 50-51; head combustion studies, Naval Powder Factory, 51-53, head prod dept labs, 53-56, polaris rocket motor develop engr, Polaris Prog, Navy Spec Projs Off, 56-58, dir res & develop dept, Naval Propellant Plant, 58-62, tech dir, Naval Ord Sta, Navy Dept, 62-75; CONSULT ENERGY & DEFENSE, 76-; CHMN, GEC TECHNOL, 80-; PRES, PAC FLEX INC, 81- *Concurrent Pos:* Mem, Md Govs Sci Adv Coun; mem, Navy Sr Scientists Coun; adj prof world bus, Thunderbird Grad Sch Int Mgt, 72-74. *Honors & Awards:* Pioneer in Space Award; Distinguished Scientists Award, USN, 75. *Mailing Add:* 7580 Bayside Lane Miami Beach FL 33141

BROWNING, JOHN ARTIE, b Kosse, Tex, Oct 3, 23; m 46; c 3. PLANT PATHOLOGY, GENETICS. *Educ:* Baylor Univ, BS, 47; Cornell Univ, PhD(plant path), 53. *Prof Exp:* Asst plant path, Cornell Univ, 49-53; from asst prof to prof Plant Path, Iowa State Univ, 53-81; prof & head plant sci, 81-85, PROF & HEAD PLANT PATH & MICROBIOL, TEX A&M UNIV, 85- *Concurrent Pos:* Mem staff, Rockefeller Found agr prog, Colombia, 63-64; vis prof, Tel-Aviv Univ, 69, 78, 82, Plant Breeding Inst, Cambridge Univ, 78; bd dirs, Am Inst Biol Sci, 83-86. *Mem:* Fel AAAS; fel Am Phytopath Soc (pres, 81-82); Am Soc Agron; Crop Sci Soc Am; Am Inst Biol Sci. *Res:* Epidemiology of cereal rusts; teaching epidemiology and plant pathology; conservation and utilization of germ plasm of indigenous crop species; how indigenous plants protect themselves from diseases and relevance of this to agroecosystems. *Mailing Add:* Dept Plant Path & Microbiol Tex A&M Univ College Station TX 77843

BROWNING, RONALD ANTHONY, b Washington, DC, Nov 7, 40; m 65; c 1. NEUROPHARMACOLOGY, NEUROBIOLOGY. *Educ:* George Washington Univ, BS, 63, MS, 68; Univ Ill, PhD(pharmacol), 71. *Prof Exp:* Fel pharmacol, Univ Chicago, 71-73; from asst prof to assoc prof, 73-90, PROF PHYSIOL & PHARMACOL, SCH MED, SOUTHERN ILL UNIV, 90- *Mem:* Am Soc Pharmacol & Exp Therapeut; Soc Neurosci; AAAS. *Res:* Understanding the neurochemistry of seizures and the mechanism of action of drugs which suppress tonic seizures; importance of subcortical pathways in seizure control and spread. *Mailing Add:* Dept Med Physiol & Pharmacol Sch Med Southern Ill Univ Carbondale IL 62901

BROWNLEE, DONALD E, b Las Vegas, Nev, Dec 21, 43; m 76; c 1. ASTRONOMY. *Educ:* Univ Calif, Berkeley, BSEE, 65; Univ Wash, PhD(astron), 70. *Prof Exp:* PROF ASTRON, UNIV WASH, 77- *Concurrent Pos:* Consult & prin investr, NASA, 76-; vis assoc geochem, Calif Inst Technol, 77- *Honors & Awards:* Medal Outstanding Sci Achievement, NASA. *Mem:* Int Astron Union; Meteoritical Soc; Am Astron Soc. *Res:* Collection and analysis of interplanetary dust to provide information on comets and the origin of the solar system. *Mailing Add:* Dept Astron Physics Hall Univ Wash Seattle WA 98195

BROWNLEE, PAULA PIMLOTT, b London, Eng, June 23, 34; m 61; c 3. ORGANIC CHEMISTRY. *Educ:* Oxford Univ, MA, 57, PhD(org chem), 59. *Hon Degrees:* LHD, Hollins Col. *Prof Exp:* Res fel org chem, Univ Rochester, 59-61; res chemist, Stamford Res Labs, Am Cyanamid Co, 61-62; from instr to assoc prof chem, Rutgers Univ, 70-76; dean fac & prof chem, Union Col, Schenectady, NY, 76-81; pres & prof chem, Hollins Col, Va, 81-90; PRES, ASN AM COLS, WASHINGTON, DC, 90- *Concurrent Pos:* Assoc dean, Douglass Col, 72-75, actg dean, 75-76, dir & vchair, Educ Testing Serv; dir, Nat Humanities Ctr. *Mem:* Soc for Values Higher Educ; Royal Soc Chem; Am Chem Soc; Am Asn Higher Educ; Sigma Xi. *Res:* Peptide synthesis; organic sulfur compounds; charge-transfer complexes incorporating polymers. *Mailing Add:* Asn Am Cols 1818 R St NW Washington DC 20009

BROWNLEE, ROBERT REX, b Zenith, Kans, Mar 4, 24; m 43; c 5. ASTROPHYSICS. *Educ:* Sterling Col, AB, 47; Univ Kans, MA, 51; Ind Univ, PhD(astron), 55. *Hon Degrees:* DSc, Sterling Col, 66. *Prof Exp:* Staff mem, Los Alamos Nat Lab, 55-68, nuclear tech dir, 68-70, group leader, 71-74, assoc div leader, 74-75, alt div leader, 75-77, div leader, 77-81, prog mgr, 81-90, ASSOC, LOS ALAMOS NAT LAB, 90- *Concurrent Pos:* Mem, Joint Hazard Eval Group, 64-75, chmn, 64-66 & 70-75; partic, Solar Eclipse Exped, 66, 70 & 80; mem test eval panel, US AEC, 66-81, sci adv, Nev Test Site, 67- & Div Mil Appln, 70-; sci dep comdr, Joint Task Force 8, 70-72. *Mem:* Am Astron Soc; Royal Astron Soc; Int Astron Union. *Res:* Stellar evolution; geothermal energy; underground nuclear testing; effects of nuclear explosions; hazard evaluation; mesospheric clouds. *Mailing Add:* 3007 Villa Los Alamos NM 87544

BROWNLOW, ARTHUR HUME, b Helena, Mont, July 25, 33; m 85; c 4. GEOCHEMISTRY OF NATURAL WATERS. *Educ:* Mass Inst Technol, SB, 55, PhD(geol), 60. *Prof Exp:* Asst prof geol, Univ Mo-Rolla, 60-65; asst prof, 65-67, assoc prof, 67-81, chmn dept, 75-79, PROF GEOL, BOSTON UNIV, 81- *Concurrent Pos:* NSF res grant, 70-72. *Mem:* AAAS; Geol Soc Am; Geochem Soc; Mineral Soc Am; Am Geophys Union. *Res:* Water pollution; gemstones. *Mailing Add:* Dept Geol Boston Univ 675 Commonwealth Ave Boston MA 02215

BROWNLOW, ROGER W, b Cleveland, Ohio, Dec 26, 22; m 46; c 2. AUTOMATIC CONTROLS, FLUID MECHANICS. *Educ:* Bucknell Univ, BS, 48; Cornell Univ, MS, 56; Ore State Univ, PhD(mech eng), 76. *Prof Exp:* Test engr, E I du Pont de Nemours, Inc, 48-53; instr decision geom, Cornell Univ, 53-56; proj engr flight control, Rockwell Int, 56-67; lectr thermo mech, Calif State Univ, Long Beach, 67-69; asst prof mech, Gonzaga Univ, 76-79; asst prof mech controls, Mich Technol Univ, 79-81; assoc prof, Western Mich Univ, 81-84; SR RES SCIENTIST, EASTMAN KODAK CO, 84- *Concurrent Pos:* Vis scholar, Stanford Univ, 80-81; Vis scholar, Stanford Univ, 80-81. *Mem:* Sigma Xi; Am Soc Mech Eng; Am Soc Eng Educ; Am Asn Univ Professors. *Res:* Wind tunnel simulation and field measurements of atmospheric boundary layer flow; computer modeling of turbulent flow; mathematical modeling and computer simulation of a flux summing redundant, servo valve control; modal analysis. *Mailing Add:* 29 Woodshire Lane Rochester NY 14606

BROWNSCHEIDLE, CAROL MARY, b Buffalo, NY, July 2, 46. HISTOCHEMISTRY, MATERNAL DIABETES. *Educ:* Niagara Univ, BS, 68; State Univ NY, Buffalo, PhD(anat), 74. *Prof Exp:* Asst prof anat, Col Med, Univ Cincinnati, 74-81; ASSOC RES SCIENTIST, MED FOUND BUFFALO, 81- *Mem:* NY Acad Sci; Sigma Xi; Am Asn Anatomists; Am Diabetes Asn; AAAS. *Res:* Placental morphology and function in maternal diabetes mellitus; teratology. *Mailing Add:* 6878 Omphalius Rd Colden NY 14033

BROWNSCOMBE, EUGENE RUSSELL, b National City, Calif, July 9, 06; m 32; c 1. PHYSICAL CHEMISTRY. *Educ:* Yale Univ, BS, 28, PhD(phys chem), 32. *Prof Exp:* Res chemist, Biol Lab, Cold Spring Harbor, 32-34; res scientist, Atlantic Richfield Co, Pa & Tex, 34-69; dir technol, Sonics Int, Inc, 69-77; tech dir, Diag Serv Inc, 78-86, RETIRED. *Concurrent Pos:* Instr calculus, Bishop Col, 69-71. *Honors & Awards:* Anthony F Lucas Gold Medal, Soc Petrol Engrs; Pioneer of Enhance Oil Recovery, Soc Petrol Engrs & Dept Energy. *Mem:* Am Soc Petrol Engrs; Am Chem Soc. *Res:* Band spectra; chemical effects of x-rays; solvent extraction; petroleum production and refining; flow of fluids through porous media; theoretical reservoir analysis; exploration methods. *Mailing Add:* 2822 Lewis Dr LaVerne CA 91750-4308

BROWNSON, ROBERT HENRY, b Evanston, Ill, Mar 14, 25; m 57; c 4. NEUROANATOMY, NEUROPATHOLOGY. *Educ:* John Carroll Univ, BS, 48; George Washington Univ, MS, 50, PhD, 53. *Prof Exp:* Res analyst, Nat Res Coun, 49-51; instr, Univ Southern Calif, 52-54; from asst prof to prof, Med Col Va, 54-68, chmn dept, 67-68; actg chmn dept, Sch Med, Univ Calif, Davis, 70-71, prof human anat, 68-78, vchmn dept, 71-78; chmn dept, 78-83, PROF ANAT, EASTERN VA MED SCH, NORFOLK, 78- *Concurrent Pos:* NIH grants, 56-64; Greenwell fel, NY Univ-Bellevue Med Ctr, 58; Am Cancer grant, 61-63; USAEC grant, 62-64; NIS spec fel, Univ C grant, 62-64; NIH spec fel, Univ Calif, Berkeley, 64-66; vis prof, Div Med Physics, Donner Lab, Univ Calif, Berkeley, 66-67; vis prof, Dept Anat, Univ of Helsinki, Finland, 75; Vet Admin grant, 81-83. *Mem:* Am Asn Neuropath; Radiation Res Soc; Am Asn Anat; Am Acad Neurol; Electron Micros Soc Am; Am Physiol Soc; World Fedn Neurol; Soc Marine Mammal; Soc Neurosci. *Res:* Exploration of the cytoarchitectonics of sea mammal brain in a variety of odontocetes with special attention to factors of symmetry, age and environment. *Mailing Add:* Dept Anat Eastern Va Med Sch PO Box 1980 Norfolk VA 23501

BROWNSTEIN, BARBARA L, b Philadelphia, Pa, Sept 8, 31. CELL BIOLOGY, MOLECULAR GENETICS. *Educ:* Univ Pa, BA, 57, PhD(microbiol), 61. *Prof Exp:* Fel virol, Wistar Inst, 61-62; USPHS fel genetics, Karolinska Inst, Sweden, 62-64; assoc, Wistar Inst, 64-68; assoc prof, 68-75, PROF BIOL, TEMPLE UNIV, 75-; UNIV PROVOST, 82- *Mem:* AAAS; Am Soc Microbiol; Am Soc Cell Biol. *Res:* Molecular biology of differentiation; mode of action of antibiotics; control of replication and motility in normal and malignant animal cells. *Mailing Add:* Provost/Dir Temple Univ Philadelphia PA 19122

BROWNSTEIN, KENNETH ROBERT, b New York, NY, May 6, 36. QUANTUM MECHANICS. *Educ:* Rensselaer Polytech Inst, BS, 57, PhD(physics), 66. *Prof Exp:* assoc prof, 65-73, PROF PHYSICS, UNIV MAINE, ORONO, 73- *Mem:* Am Phys Soc; Am Asn Physics Teachers; Am Soc Eng Educ; Sigma Xi. *Res:* Bounds for scattering phase shifts; approximation methods for bound and scattering states. *Mailing Add:* Dept Physics Univ Maine Orono ME 04469

BROWNSTEIN, MICHAEL JAY, b Washington, DC, Feb 27, 43; m 66; c 2. NEUROENDOCRINOLOGY, PHARMACOLOGY. *Educ:* Columbia Univ, AB, 64; Univ Chicago, MD & PhD(pharmacol), 71. *Prof Exp:* Intern pediat, Children's Hosp Med Ctr, 71-72; res assoc pharmacol, Pharmacol Res Assoc Training Prog, Nat Inst Gen Med Sci, NIH, Bethesda, 72-74; sr staff fel, 74-76; med officer res, Lab Clin Sci, Alcohol, Drug Abuse & Ment Health Admin, NIMH, Bethesda, 76-, chief, Neuroendocrinol Unit, 78-; AT LAB CELL BIOL, NIMH. *Mem:* Sigma Xi. *Res:* Neuroendocrine regulation; peptide biosynthesis; pineal gland; neuropharmacology; circadian rhythms. *Mailing Add:* 5307 Elsmers Ave Bethesda MD 20814

BROWNSTEIN, SYDNEY KENNETH, b Regina, Sask, Mar 5, 31; m 54; c 3. ORGANIC CHEMISTRY. *Educ:* Univ Sask, BA, 52; Univ Chicago, PhD(chem), 55. *Prof Exp:* Chemist, Argonne Cancer Res Hosp, 55; fel & hon asst, Univ Col, Univ London, 55-56; instr chem, Cornell Univ, 56-59; RES OFFICER, NAT RES COUN CAN, 59- *Concurrent Pos:* Weizmann fel, 65-66; vis scholar, Cambridge, 89-90. *Mem:* Chem Inst Can; Royal Soc Chem. *Res:* Application of nuclear magnetic resonance techniques to problems in chemistry. *Mailing Add:* Inst Environ Chem Nat Res Coun Ottawa ON K1A 0R6 Can

BROWNSTONE, YEHOSHUA SHIEKY, b Winnipeg, Man, May 12, 29; m 50; c 3. CLINICAL BIOCHEMISTRY. *Educ:* Univ Man, BScA, 50, MSc, 56; McGill Univ, PhD(biochem), 58. *Prof Exp:* Chemist, Dominion Linseed Oil Ltd, 51-52; asst dept physiol & med res, Univ Man, 54-56; demonstr & asst dept biochem, McGill Univ, 56-58, res assoc, 58-59, lectr, 60; biochemist, Dept Biochem & Radioisotopes, Queen Mary Vet Hosp, 59-60; from asst prof to prof path chem, 61-71, PROF, DIV CLIN BIOCHEM, DEPT BIOCHEM, UNIV WESTERN ONT, 71-, DIR, DEPT CLIN PATH, VICTORIA HOSP, 61- *Concurrent Pos:* Vis sci, Carlsberg Lab, Copenhagen, Denmark, 69-70. *Mem:* NY Acad Sci; Can Fedn Biol Soc; Can Biochem Soc; Can Soc Clin Chemists (pres, 73-74). *Res:* Clinical chemistry; enzymology of the normocyte; the pentose phosphate metabolic pathway; enzymes in neoplastic disease; blood preservation; effect of hormones on enzyme action; drugs; metabolism and detection; mass spectrometry. *Mailing Add:* Dept Biochem Univ Western Ont London ON N6A 5B8 Can

BROWZIN, BORIS S(ERGEEVICH), b St Petersburg, Russia, Oct 10, 12; m 34; c 1. CIVIL ENGINEERING, HYDROLOGY & WATER RESOURCES. *Educ:* Leningrad Polytech Inst, dipl, 38; Aachen Tech, DEng, 61; Univ Grenoble, DUniv, 63, DSc(d'etat), 64. *Prof Exp:* Engr, govt industs, USSR, 30-36, from engr to sr engr, 36-42; engr, Hoch & Tiefbau, Ger, 43-45; sci asst, Stuttgart Tech, 45; engr, Societe Nouvelle Froment Clavier, France, 45-46, Bur Veritas, 46-48 & Fargo Eng Co, Tex, 49-51; prof agrege, Laval Univ, 52-58; assoc prof civil eng & eng mech, Case & Ohio State Univs, 58-62; prof civil eng, Cath Univ Am, 62-71; consult civil engr, Ebasco Servs Inc, NY, 72-75; RES STRUCT ENGR & PROJS MGR, US NUCLEAR REGULATORY COMN, 75- *Concurrent Pos:* Sr engr, Harza Eng Co, Ill, 51-57; consult engr, Shawinigan Eng Co, Can, 55-56; consult ed, Soviet Hydrol J, 68-83; ed spec issue, Nuclear Eng & Design, 77, 79, 81, 83 & 85. *Mem:* Fel Am Soc Civil Engrs; Am Geophys Union; Soc Struct Mech in Reaction Technol; Sigma Xi; Int Asn Hydraul Res. *Res:* Water resources development; hydro-power plant design; structural mechanics for nuclear application; structures; soil mechanics and foundation engineering; hydrology. *Mailing Add:* 7905 Glendale Rd Chevy Chase MD 20815

BROXMEYER, HAL EDWARD, b Brooklyn, NY, Nov 27, 44; m 69; c 2. MEDICAL SCIENCE, BIOLOGY. *Educ:* Brooklyn Col, BS, 66; Long Island Univ, MS, 69; NY Univ, PhD(biol), 73. *Prof Exp:* Res assoc med biol, Queens Univ, Ont, 73-75; assoc researcher, 75-76, res assoc, 76-78, assoc leukemia res, Sloan Kettering Inst Cancer Res, Cornell Univ, 78-83; asst prof biol, Med Sch, 80-83, assoc prof med, 83-86, PROF MED, MICROBIOL-IMMUNOL, IND UNIV PURDUE, 86-; SCI DIR, WALTHER ONCOL CTR, IND UNIV SCH MED, 88- *Concurrent Pos:* Leukemia Soc Am spec fel, 76-78 & scholar award, 78-83; Am Cancer Soc grant, 78-80, Nat Cancer Inst grants, 78-81 & 81-86 & 86- *Honors & Awards:* Mellor Award, Sloan Kettering Inst, 76 & 77, Boyer Award, 83; Merit Award, Nat Cancer Inst, 87. *Mem:* AAAS; NY Acad Sci; Am Soc Hemat; Int Soc Exp Hemat; Am Asn Cancer Res; Am Asn Immunol. *Res:* Regulation of hematopoiesis; normal and leukemic bone marrow and blood cell proliferation; maturation in vitro and in vivo. *Mailing Add:* Med/Walther Oncol Ctr Ind Univ Sch Med 975 W Walnut St Indianapolis IN 46202-5121

BROXTON, DAVID EDWARD, b Winston-Salem, NC, Aug 19, 52; m 80; c 2. IGNEOUS PETROLOGY, MINERALOGY & GEOCHEMISTRY. *Educ:* Univ NC, Chapel Hill, BS, 74; Univ NMex, Albuquerque, MS, 76. *Prof Exp:* STAFF MEM GEOL & GEOCHEM, LOS ALAMOS NAT LAB, 77- *Concurrent Pos:* Prin investr, Nat Uranium Resource Eval Prog, 80-83 & Yucca Mountain Proj, 84- *Mem:* Am Geophys Union; Geol Soc Am. *Res:* Secondary minerals in tuffaceous rocks and their potential use as natural radionuclide migration barriers at potential nuclear waste repositories; chemical evolution of magmas associated large volcanic complexes. *Mailing Add:* Los Alamos Nat Lab Mail Stop 0462 Los Alamos NM 87545

BROYLES, ARTHUR AUGUSTUS, b Atlanta, Ga, May 16, 23; m 43; c 4. EFFECTS OF NUCLEAR EXPLOSIONS. *Educ:* Univ Fla, BS, 42; Yale Univ, PhD(physics), 49. *Prof Exp:* Asst prof physics, Univ Fla, 49-50; mem staff, Los Alamos Sci Lab, 50-53; res physicist, Rand Corp, 53-59; from assoc prof to prof, 59-86, EMER PROF PHYSICS, UNIV FLA, 86- *Concurrent Pos:* Vis prof, Univ Adelaide, Australia, 80; consult, Lawrence Livermore Nat Lab, 82-, physicist, 85-86. *Mem:* Am Asn Physics Teachers; fel Am Phys Soc. *Res:* Quantum electrodynamics; nonperturbative solution of the Schwinger-Dyson equations; global effects of nuclear war; effects of gamma rays on man. *Mailing Add:* Dept Physics Univ Fla Gainesville FL 32611

BROYLES, CARTER D, b Eckman, WVa, Nov 1, 24; m 55; c 2. PHYSICS, ENGINEERING SCIENCE. *Educ:* Univ Chattanooga, BS, 48; Vanderbilt Univ, PhD(physics), 52. *Prof Exp:* Res physicist, Sandia Lab, 52-56, supvr, Radiation Physics Div, 56-64, mgr, High Altitude Burst Physics Dept, 64-68, mgr test sci dept 9110, 68-72, dir effects exp orgn, 72-75, dir field eng, 75-89; RETIRED. *Concurrent Pos:* Mem staff, Vanderbilt Univ, 52. *Mem:* AAAS; fel Am Phys Soc; Am Asn Physics Teachers; Sigma Xi. *Res:* Magnetic spectrometer studies of radioactive isotopes; fluid dynamics; plasma physics; nuclear radiation measurements; radioactive waste disposal; in-situ fossil fuel technology. *Mailing Add:* 5310 Los Poblanos Lane NW Albuquerque NM 87107

BROYLES, ROBERT HERMAN, b Kingsport, Tenn, Feb 16, 43; m 66; c 2. DEVELOPMENTAL BIOLOGY, TOXICOLOGY. *Educ:* Wake Forest Univ, BS, 65, Bowman Gray Sch Med, PhD(biochem), 70. *Prof Exp:* NIH fel, & res assoc biochem, Fla State Univ, 70-72; asst prof zool, Univ Wis-Milwaukee, 72-77; assoc prof, 77-85, PROF BIOCHEM & MOLECULAR BIOL, COL MED, UNIV OKLA HEALTH SCI CTR, 85- *Concurrent Pos:* Cottrell grant, Res Corp, Inc, 73; res grant, US Sea Grant Prog, 75; NIH res grant, Nat Inst Arthritis, Metab & Digestive Dis, 76-88; sr scientist, NIH, NIDDK, 89-91. *Mem:* Am Soc Cell Biol; Sigma Xi; AAAS; Am Soc Zoologists; Soc Develop Biologists; Am Soc Biol Chem; Am Soc Hemat. *Res:* Developmental biology; gene regulation and DNA-binding proteins in red blood cell differentiation; effects of drugs and environmental contaminants on vertebrate embryos and on red blood cell differentiation. *Mailing Add:* Dept Biochem & Molecular Biol Univ Okla Health Sci Ctr PO Box 26901 Oklahoma City OK 73190

BRUALDI, RICHARD ANTHONY, b Derby, Conn, Sept 2, 39; m 63, 78; c 2. MATHEMATICS. *Educ:* Univ Conn, BA, 60; Syracuse Univ, MS, 62, PhD(math), 64. *Prof Exp:* Nat Acad Sci-Nat Res Coun fel, Nat Bur Stand, 64-65; from asst prof to assoc prof, 65-73, PROF MATH, UNIV WIS-MADISON, 73- *Concurrent Pos:* NATO fel, Univ Sheffield, 69-70. *Mem:* Am Math Soc; Math Asn Am. *Res:* Matrix theory and combinatorics. *Mailing Add:* Dept Math Univ Wis 480 Lincoln Dr Madison WI 53706

BRUBACHER, ELAINE SCHNITKER, PRE-OCLAMPSIA PREGNANCY. *Educ:* Univ Mich, PhD(physiol), 67. *Prof Exp:* ASSOC PROF BIOL, PHYSIOL & ANAT, DEPT BIOL, UNIV REDLANDS, 80- *Mailing Add:* University Redlands 1200 Colton Ave Redlands CA 92374

BRUBAKER, BURTON D(ALE), b Tiffin, Ohio, June 15, 35; m 59; c 2. CERAMICS ENGINEERING. *Educ:* Ohio State Univ, BCerE & MSc, 58, PhD(ceramic eng), 62. *Prof Exp:* Res assoc ceramics eng, Exp Sta, Ohio State Univ, 58-62; res engr, 62-68, sr res engr, 68-72, sr res specialist, 72-80, RES ASSOC, DOW CHEM CO, 80- *Mem:* Am Ceramic Soc; Sigma Xi; Nat Inst Ceramic Engrs; Soc Plastic Engrs. *Res:* Composite structures and properties; ceramic and glass foams; thermal insulations; polymer foams; ceramic armor; whisker fibers. *Mailing Add:* 1702 Sylvan Lane Midland MI 48640-2538

BRUBAKER, CARL H, JR, b Passaic, NJ, July 13, 25; m 49; c 2. INORGANIC CHEMISTRY, ORGANOMETALLIC CHEMISTRY. *Educ:* Franklin & Marshall Col, BS, 49; Mass Inst Technol, PhD(chem), 52. *Prof Exp:* From asst prof to assoc prof, 52-61, PROF CHEM, MICH STATE UNIV, 61- *Concurrent Pos:* Fulbright prof radiochem, Univ Chile, 58; assoc ed, J Am Chem Soc, 64-71, 74-88. *Mem:* Am Chem Soc; Royal Soc Chem; AAAS. *Res:* Oxidation-reduction reaction mechanisms; organometallic compounds of transition elements in lower oxidation states; organometallic compounds as catalysts for hydrogenation and nitrogen fixation. *Mailing Add:* PO Box 128 Okemos MI 48805-0128

BRUBAKER, GEORGE RANDELL, b New York, NY, July 17, 39; m 64; c 2. INORGANIC CHEMISTRY, BIOINORGANIC CHEMISTRY. *Educ:* Columbia Univ, BA, 60; Ohio State Univ, MSc, 63, PhD(chem), 65. *Prof Exp:* Res assoc inorg chem, Univ Pittsburgh, 65-66; asst prof, 66-72, ASSOC PROF CHEM, ILL INST TECHNOL, 72- *Mem:* Am Chem Soc; Royal Soc Chem. *Res:* Synthesis, characterization and reactions of transition metal complexes; biological significance of transition metal chemistry. *Mailing Add:* US Food & Drug Admin Ten W 35th St Chicago IL 60616-3703

BRUBAKER, INARA MENCIS, b Riga, Latvia, May 1, 38; US citizen; m 64; c 2. PHYSICAL CHEMISTRY. *Educ:* Ohio Northern Univ, BS, 59; Ohio State Univ, MS, 61, PhD(anal chem), 63. *Prof Exp:* Res chemist electro chem, Universal Oil Prod Co, 66-70; res assoc inorg chem, Ill Inst Technol, 76; res chemist, UOP Res Ctr, 78-82; res specialist, Allied Signal Eng Mat Res Ctr, 82-84, assoc res scientist, 84-86, group leader separations, 86-88, chemicals, 88-90, MGR PROCESS TECHNOL, ALLIED SIGNAL RES & TECHNOL, 90- *Concurrent Pos:* Chmn, 80-82, task force on occup safety & health, Am Chem Soc, 80-, chmn, 89- *Mem:* Fel Am Chem Soc. *Res:* Process technology; separations; hydromettalurgy; metals recovery and management; occupational safety and health. *Mailing Add:* 126 Ardmore Rd Des Plaines IL 60016-2117

BRUBAKER, KENTON KAYLOR, b Elizabethtown, Pa, Feb 17, 32; m 55; c 4. BIOLOGY, HORTICULTURE. *Educ:* Eastern Mennonite Col, BS, 54; Ohio State Univ, MSc, 57, PhD, 59. *Prof Exp:* Assoc prof biol, Eastern Mennonite Col, 59-62; horticulturist, Congo Polytech Inst, 62-64; chg de cours bot, Free Univ of the Congo, 64-65; chmn dept, 65-74, PROF BIOL, EASTERN MENNONITE COL, 65-, COORDR, INT AGR PROG, 85- *Mem:* Am Sci Affil. *Res:* Chemical weed control in vegetable crops; carbohydrate translocation in greenhouse tomatoes; tropical horticulture; protein content of tropical vegetables; alternative fertilizers. *Mailing Add:* Dept Biol Eastern Mennonite Col Harrisonburg VA 22801

BRUBAKER, LEONARD HATHAWAY, b Macon, Ga, July 14, 34; m 57; c 3. MEDICAL ONCOLOGY, HEMATOLOGY. *Educ:* Duke Univ, AB, 56; Emory Univ, MS & MD, 64. *Prof Exp:* Intern & resident internal med, Vet Admin Hosp & Emory Univ Hosp, 64-66; clin assoc leukemia, Nat Cancer Inst, 66-68; from asst prof to assoc prof med hemat & oncol, Sch Med, Univ Mo, 69-78; PROF HEMAT & ONCOL, MED COL GA, 78- *Concurrent Pos:* Fel hematol & oncol, Ohio State Univ Hosp, 68. *Mem:* Southern Soc Clin Investigation; fel Am Col Physicians; Am Soc Hemat; Am Soc Clin Oncol; Am Fedn Clin Res. *Res:* Neutrophil kinetics in neutropenic patients; paroxysmal nocturnal hemoglobinuria; marrow neutrophil reserves; neutrophil function in malnourished cancer patients; genetic metabolic characteristics in bladder cancer. *Mailing Add:* Dept Med Med Col Ga Augusta GA 30912

BRUBAKER, MERLIN L, b Live Oak, Calif, July 1, 22; m 43; c 6. INTERNAL MEDICINE, PUBLIC HEALTH. *Educ:* Calif Col Med, MD, 46; Univ London, dipl trop med & hyg, 52; Univ La Verne, BA, 56; Univ Calif, Los Angeles, MA, 62. *Prof Exp:* Priv pract med, Fullerton, Calif, 47-52; med supt, Garkida Leprosarium, Nigeria, 52-55; assoc prof med, Calif Col Med, 55-62, chmn curric & admis comt, 56-62, dir clin pract, 59-62; chief prog planning & eval med div, Peace Corps, 62-64; asst dir, USPHS Hosp, Staten Island, NY, 64-65 & dir, USPHS hosp, Carville, La, 65-68, dir career develop global community health, 68-70; regional adv leprosy, venereal dis & treponematosis, WHO-Pan Am Health Orgn, 70-77; USPHS rep to Commonwealth PR, 77-80; dir off refugee health affairs, Off Surgeon Gen, USPHS, 80-81, dep dir Nat Med Audiovisual Ctr, Nat Libr Med, 81-82; consult int health, 82-85; UNIV PHYSICIAN & PROF HEALTH CARE ADMIN, UNIV LA VERNE, 85- *Concurrent Pos:* Col physician & dir student health servs, La Verne Col, 56-60; asst med officer in chg, USPHS Hosp, Staten Island, NY, 64-65 & dir prof training & res, Carville, La, 65, dir, 65-68; clin assoc prof, Sch Med, La State Univ, 65-68; assoc clin prof community med & int health, Sch Med, Georgetown Univ, 68-; La State Univ fel trop med & parasitol to Mid Am. *Mem:* Am Soc Trop Med & Hyg; fel Am Pub Health Asn; fel Royal Soc Med; Int Leprosy Asn; fel Royal Soc Trop Med. *Res:* Preventive medicine; medical education. *Mailing Add:* PO Box 9342 Santurce PR 00908

BRUBAKER, PAUL EUGENE, II, b Indiana, Pa, Dec 31, 34; m 65; c 5. TOXICOLOGY. *Educ:* St Vincent Col, BA, 59; Cath Univ Am, MS, 65, PhD(biol), 68. *Prof Exp:* Jr staff fel cell biol, Nat Inst Environ Health Sci, 68-70, sr staff fel, 70-71; res biologist molecular biol, US Environ Protection Agency, 71-73, sr sci adv toxicol, 73-75; chief pathophysiol, Greenfield, Ahaway & Tyler, Consults, 75-76; sr toxicologist, Mobil Oil Corp, 76-78; mgr corp toxicol, Warner-Lambert Co, 78-79; REGULATORY AFFAIRS COORDR, TOXICOL, EXXON CORP, 79- *Concurrent Pos:* Adj prof biochem, Univ NC, 72- *Mem:* AAAS; Soc Toxicol; Air Pollution Coastal Asn; Am Petrol Inst; Sigma Xi. *Res:* Environmental health; toxicology; epidemiology; molecular biology; agriculture risk analysis and air pollution. *Mailing Add:* Three Halstead Rd Mendham NJ 07945

BRUBAKER, ROBERT ROBINSON, b Wilmington, Del, Jan 15, 33; wid; c 2. MICROBIOLOGY. *Educ:* Univ Del, BA, 57; George Washington Univ, MA, 61; Univ Chicago, PhD(microbiol), 66. *Prof Exp:* Microbiologist, US Army Biol Labs, 64-66; from asst to assoc prof, 66-79, PROF MICROBIOL, MICH STATE UNIV, 79- *Mem:* AAAS; Am Soc Microbiol; Sigma Xi. *Res:* Biochemical mechanisms of microbial virulence; genetic determinants of virulence. *Mailing Add:* 11042 W Scipio Hwy Vermontville MI 49096

BRUBAKER, THOMAS ALLEN, b Greeley, Colo, Jan 7, 35; m 62. ELECTRICAL ENGINEERING. *Educ:* Univ Wyo, BS, 57, MS, 58; Univ Ariz, PhD(elec eng), 63. *Prof Exp:* Instr elec eng, Univ Ariz, 61-63; engr, Burr-Brown Res Corp, 63; asst prof elec eng, Univ Wyo, 63-65 & Univ Mo-Columbia, 65-69; from asst prof to assoc prof, 69-77, PROF ELEC ENG, COLO STATE UNIV, 77- *Concurrent Pos:* Consult, Lawrence Radiation Lab, Calif, 64-; mem, Simulation Coun, Inc. *Res:* Hybrid computers; optimum control systems; random process theory; engineering education. *Mailing Add:* Dept Elec Eng Colo State Univ Ft Collins CO 80523

BRUCAT, PHILIP JOHN, b New York, NY, Aug 22, 56. CLUSTER ION SPECTROSCOPY, PHOTOCHEMISTRY. *Educ:* Mass Inst Technol, BS, 77; Stanford Univ, PhD(chem), 84. *Prof Exp:* PRF postdoctoral, Rice Univ, 84-86; ASST PROF, UNIV FLA, 86- *Concurrent Pos:* Am Soc Mass Spectrometry res award, 87. *Mem:* Am Chem Soc; Am Phys Soc; Sigma Xi. *Res:* High resolution optical spectroscopy of cold molecular ions in the gas phase. *Mailing Add:* Dept Chem Univ Fla Gainesville FL 32611-2046

BRUCE, ALAN KENNETH, b Nashua, NH, Aug 24, 27; m 48; c 4. RADIOBIOLOGY. *Educ:* Univ NH, BS, 51; Univ Rochester, MS, 54, PhD(radiation biol), 56. *Prof Exp:* Assoc Atomic Energy Proj, Univ Rochester, 53-56, jr scientist, 56; assoc biol div, Oak Ridge Nat Lab, 56-57; asst prof, 57-62, ASSOC PROF BIOL, STATE UNIV NY, BUFFALO, 62-, RADIATION SAFETY OFFICER, 57- *Mem:* AAAS; Am Soc Microbiol; Health Physics Soc; Radiation Res Soc; Sigma Xi. *Res:* Effects of radiation on cell permeability characteristics; applications of isotopic tracer techniques to biology; modification of radiation action on cells. *Mailing Add:* Dept Biol State Univ NY Buffalo NY 14214

BRUCE, CHARLES ROBERT, b Topeka, Kans, Jan 3, 25; m 49; c 5. PHYSICS. *Educ:* Wash Univ, AB, 50, PhD(physics), 56. *Prof Exp:* Res assoc, Wash Univ, 56-57; sr res scientist, Marathon Oil Co, 57-87; RETIRED. *Mem:* Am Phys Soc; Sigma Xi. *Res:* Nuclear magnetic resonance; geophysics and well logging; instrumentation; data acquisition; improving prospecting techniques. *Mailing Add:* 7155 S Grant Littleton CO 80122

BRUCE, CHARLES WIKOFF, b Syracuse, NY, Jan 13, 37; m 61; c 2. PHYSICS. *Educ:* Union Col, Schenectady, NY, BS, 59; NMex State Univ, Las Cruces, MS, 68, PhD(physics), 70. *Prof Exp:* Physicist plasmas & lasers, Air Force Weapons Lab, Albuquerque, NM, 62-65; phys scientist systs anal, Safeguard Syst Eval Agency, 71-74, RES PHYSICIST GAS/AEROSOL SPECTROS, ATMOSPHERIC SCI LAB, WHITE SANDS MISSILE RANGE, NMEX, 74-; PROF PHYSICS, NMEX STATE UNIV, 74- *Res:* Pulsed laser diagnostics; laser calorimetry; interferometry of laser induced plasmas; RF excited plasmoids; photo-acoustical spectroscopic techniques and in-situ spectroscopy of atmospheric aerosols. *Mailing Add:* Atmospheric Sci Lab White Sands Missile Range NM 88002-5501

BRUCE, DAVID LIONEL, b Champaign, Ill, Oct 27, 33; m 85; c 2. ANESTHESIOLOGY, MEDICINE. *Educ:* Univ Ill, MD, 60. *Prof Exp:* Resident anesthesia, Univ Pa, 61-62, USPHS fel anat, 62-63, resident anesthesia, 63-64; from instr to asst prof anesthesia, Univ Ky, 64-66; from asst prof to prof anesthesia, Med Sch, Northwestern Univ, Chicago, 66-77; chief anesthesiol, Vet Admin Hosp, Long Beach, 78-81; PROF ANESTHESIOL, UNIV CALIF, IRVINE, 77-; PROF & CHMN ANESTHESIOL, UNIV MISS, 84- *Concurrent Pos:* USPHS career develop award, Northwestern Univ, 67-72; consult, Food & Drug Admin, 74-77; prof anesthesiol, NY Univ Med Ctr, 81-84. *Mem:* Am Soc Anesthesiol; fel Royal Soc Med Eng; Asn Univ Anesthetists; Int Anesthesia Res Soc. *Res:* Effects of volatile anesthetics on cell structure and function; toxicity of occupational exposure to anesthetics. *Mailing Add:* Dept Anesthesiol Univ Miss Med Ctr 2500 N State St Jackson MS 39216

BRUCE, DAVID STEWART, b Amherst, Ohio, Sept 17, 39; m 61; c 2. ANIMAL PHYSIOLOGY, ENVIRONMENTAL PHYSIOLOGY. *Educ:* Taylor Univ, BA & BS, 62; Purdue Univ, MS, 65, PhD(physiol), 68. *Prof Exp:* Instr physiol, DePauw Univ, 67; instr environ biol, Purdue Univ, 67, vis asst prof, 68; asst prof, Seattle Pac Col, 68-73, assoc prof biol, 73-74; assoc prof, 74-78, acting chmn, Biol Dept, 89-90, PROF BIOL, WHEATON COL, 78- *Concurrent Pos:* Muscle researcher, Univ St Andrews, Scotland, 79 & 87; Wellcome res travel grant, Scotland, 80. *Mem:* AAAS; Am Physiol Soc; Am Sci Affil; Sigma Xi. *Res:* Phenomenon of hibernation, especially in the bat, Myotis lucifugus, ground squirrels, Citellus tridecemlineatus and Citellus lateralis; black bear, (Ursus americanus), and polar bear (Ursus maritimus). *Mailing Add:* Dept Biol Wheaton Col Wheaton IL 60187

BRUCE, E IVAN, JR, b Center, Tex, Aug 20, 17; m 46; c 3. PSYCHIATRY. *Educ:* Univ Tex, BA, 39, MD, 42; Am Bd Psychiat & Neurol, dipl. *Prof Exp:* Resident psychiat, 46-49, from instr to assoc prof, 49-65, asst dir med br, 49-52, asst adminr, 52-59, actg chmn dept, 74-76, chmn, Ad Interim Dept, 76-77, DIR, UNIV TEX MED BR GALVESTON, 59-, PROF PSYCHIAT, 65-,. *Mem:* Am Psychiat Asn; AMA; Am Col Psychiat. *Res:* Clinical and administrative psychiatry. *Mailing Add:* Dept Psychiat & Behav Sci Univ Tex Med Sch 3817 Brookhaven Circle Ft Worth TX 76109

BRUCE, JAMES DONALD, b Livingston, Tex, June 28, 36; m 59; c 3. ELECTRICAL ENGINEERING. *Educ:* Lamar State Col, BS(elec eng) & BS(math), 58; Mass Inst Technol, SM, 60, ScD(optimum quantization), 64. *Prof Exp:* Teaching asst, 58-60, from instr to assoc prof, 60-79, assoc dean, Sch Eng, 71-78, dir indust liaison, 79-82, actg dean, 77-78, dir Info Systs, 83-86, PROF ELEC ENG, 73-, VPRES INFO SYSTS, 86- *Concurrent Pos:* Ford Found fel eng, 64-65; consult, Farmer Elec Prod Co, 61 & Missile Systs Div, Raytheon Corp, 65-71; lectr, Bolt, Beranek & Newman's Prog Advan Study, 68-70, Cambridge Res Inst, 68-70 & Oak Ridge Nat Lab, 70; trustee, Harvard Coop Soc, 74-84, dir, Concurrent Comput Corp, NJ, 86, consult, gov't & indust, Consortium, Sci Comput, trustee, 84, vchmn, 86- *Mem:* AAAS; Inst Elec & Electronics Engrs; Am Soc Eng Educ; Soc Mgt Info Systs; Sigma Xi. *Res:* Digital signal processing; data management systems. *Mailing Add:* 12 Woodpark Circle Lexington MA 02173

BRUCE, JOHN GOODALL, JR, b Richmond, Va, Oct 19, 27; m 53; c 3. PHYSICAL OCEANOGRAPHY, OCEAN CIRCULATION. *Educ:* Va Polytech Inst, BS, 51, MS, 53. *Prof Exp:* Res assoc oceanog, Woods Hole Oceanog Inst, 54-83; OCEANOGR, STENNIS SPACE CTR, 83- *Concurrent Pos:* Fulbright scholar, Univ Liverpool, 56-57; consult, US Oceanog Off, 81- *Res:* Physical oceanography of the Arabian Sea with special interest in the dynamic response of the Somali current circulation to the monsoon winds; characteristics of the large mesoscale ocean eddies ocurring during the strong southwest monsoon; circulation equatorial Atlantic Ocean. *Mailing Add:* Code OPTA Naval Oceanog Off Stennis Space Center MS 39522

BRUCE, JOHN IRVIN, b Ellicott City, Md, Aug 1, 27; m 67; c 1. PARASITOLOGY, PHYSIOLOGY. *Educ:* Morgan State Col, BS, 53; Howard Univ, MS, 65, PhD(physiol, biochem), 68. *Prof Exp:* Jr chemist, Metrop Hosp, New York, NY, 53-54; res asst, Columbia Univ, 54-55; med res technician, Walter Reed Army Inst Res, 55-59, parasitologist, 59-68, chief drug screening unit, 406th Med Lab, Camp Zama, Japan, 68-71, chief, SRS, 71-73; prof biol sci, Lowell Technol Inst, 73-75, dean Col Pure & Appl Sci, 73-78; dep asst adminr, AID, 78-80; PROF BIOL SCI, 75-, DIR CTR TROP DIS, UNIV LOWELL, 80- *Concurrent Pos:* Adv comt, Gorgas Mem Inst Trop & Prev Med, 85-; mem, Microbiol & Infectious Dis Adv Comt, Nat Inst Allergy & Infectious Dis, NIH, 84- *Mem:* Am Soc Parasitologists; Am Soc Trop Med & Hyg; NY Acad Sci; AAAS; Am Pub Health Asn; Sigma Xi; fel Royal Soc Trop Med & Hyg; Wildlife Dis Asn; Am Inst Biol Sci; Japanese Soc Parasitol. *Res:* Chemotherapy and drug-resistance of parasitic diseases, especially schistosomiasis; physiological and biochemical studies of parasites and host parasite relationship; aspects of parasitic immunology pertaining to host parasite relationship, epidemiology of parasitic diseases and biology of parasites. *Mailing Add:* Univ Lowell 450 Aiken St Lowell MA 01854

BRUCE, KIM BARRY, b Dearborn, Mich, Oct 16, 48; m 81; c 4. SEMANTICS OF PROGRAMMING LANGUAGES, MODELS OF LAMBDA CALCULUS. *Educ:* Pomona Col, BA, 70; Univ Wis-Madison, PhD(math), 75. *Prof Exp:* Instr math, Princeton Univ, 75-77; PROF COMPUTER SCI, WILLIAMS COL, 77- *Concurrent Pos:* Lectr math, Minority Eng Prog, Univ Wis, 75, 78, 79; vis scholar, Stanford Univ, 77; vis scientist, Mass Inst Technol, 80-81; vis prof, Univ Pisa, Italy, 85 & Stanford Univ, 91; mem, Asn Comput Mach & Inst Elec & Electronics Engrs Computer Soc Joint Curric Task Force, 88-91. *Mem:* Asn Comput Mach; Inst Elec & Electronics Engrs Computer Soc; Asn Symbolic Logic. *Res:* Semantics of programming languages, especially polymorphic and object-oriented languages; models of extensions of typed lambda calculus; programming language design; mathematical logic-model theory; computer science education. *Mailing Add:* Dept Computer Sci Bronfman Sci Ctr Williams Col Williamstown MA 01267-2680

BRUCE, RICHARD CONRAD, b Cambridge, Mass, June 2, 36; m 58; c 3. ZOOLOGY. *Educ:* Tufts Univ, BS, 58; Duke Univ, MA, 61, PhD(zool), 68. *Prof Exp:* PROF BIOL, WESTERN CAROLINA UNIV, 63- *Concurrent Pos:* Exec dir, Highland Biol Sta, Univ NC, 72-; exec secy, Highland Biol Found, Inc, 77- *Mem:* Am Soc Ichthyologists & Herpetologists; Soc Study Amphibians & Reptiles; Herpetologists League; Sigma Xi. *Res:* Ecology, evolution and reproductive biology of plethodontid salamanders; natural diversity in Southern Appalachian ecosystems. *Mailing Add:* Dept Biol Western Carolina Univ Cullowhee NC 28723

BRUCE, ROBERT ARTHUR, b Somerville, Mass, Nov 20, 16; m 40; c 3. MEDICINE. *Educ:* Boston Univ, BS, 38; Univ Rochester, MS, 40, MD, 43. *Prof Exp:* Asst res physician, Strong Mem Hosp, Univ Rochester, 44-45, chief res physician, 45-46, instr med, 46-50; from asst prof to assoc prof med, 50-59, head div cardiol, 50-71, co-dir div cardiol, 72-82, prof, 59-87, EMER PROF MED, SCH MED, UNIV WASH, 87- *Concurrent Pos:* Buswell fel, Univ Rochester, 46-50; Commonwealth Fund fel, Sch Med, Univ Wash, 65-66; fel, Coun Clin Cardiol, Am Heart Asn; vis prof, Royal Infirmary, Edinburgh, Nottingham, Birmingham, Kuala Lumpor, Life Planning Inst, Tokyo, Sir Ch Gardner Hosp, Perth, WAustralia. *Honors & Awards:* Honor Award, Am Col Sports Med, 83. *Mem:* Asn Am Physicians; Asn Univ Cardiol (secy-treas, 64-67, vpres, 67-68, pres, 68-69); fel Am Col Cardiol; Aerospace Med Asn. *Res:* Cardiology; exercise physiology; electrocardiography; cardiac physiology and rehabilitation; cardiovascular epidemiology. *Mailing Add:* Dept Med RG-22 Univ Wash Sch Med Seattle WA 98195

BRUCE, ROBERT NOLAN, JR, b Melville, La, Oct 11, 30; m 58; c 2. CIVIL ENGINEERING. *Educ:* Tulane Univ, BS, 52, MS, 53; Univ Ill, PhD(civil eng), 62. *Prof Exp:* Engr, Raymond Int, Inc, 53-60; assoc prof civil eng, 62-68, PROF CIVIL ENG, TULANE UNIV, 68- *Concurrent Pos:* Spec lectr, George Washington Univ, 64 & 65, Univ Miami & Univ Maine, 66, Univ PR, 67, Rangoon Inst Tech, 79, Univ Dundee, 85; grants, US Dept Defense, 64-65, NSF, 64-66, La Dept Hwys, 65-87; consult, Off Civil Defense, USN; chmn Comt Prestressed Concrete, Am Concrete Inst, 85-89. *Honors & Awards:* PCI Korn Award, BOH Chair Civil Engr, 86. *Mem:* Am Soc Civil Engrs; Int Fedn Prestressing; Am Concrete Inst; Int Asn Bridge & Struct Engr. *Res:* Prestressed concrete in shear; foundations; structural response to nuclear weapons effects; fatigue in prestressed concrete; bridge durability. *Mailing Add:* Dept Civil Eng Tulane Univ New Orleans LA 70118

BRUCE, ROBERT RUSSELL, b Jasper, Ont, Mar 27, 26; m 51; c 1. SOIL PHYSICS. *Educ:* Ont Agr Col, BSA, 47; Cornell Univ, MS, 51; Univ Ill, PhD, 56. *Prof Exp:* Soil specialist, Ont Agr Col, 47-49; asst soil physics, Cornell Univ, 49-51; lectr, Ont Agr Col, 51-53; asst, Univ Ill, 53-55; from asst prof & asst agronomist to assoc prof & assoc agronomist, Miss State Univ, 55-65; RES SOIL SCIENTIST, SOUTHERN PIEDMONT CONSERV RES CTR, FED RES, SCI & EDUC ADMIN, USDA, 65- *Concurrent Pos:* Vis prof, Auburn Univ, 65-; chmn, Soil Sci Soc Am, 82. *Mem:* Fel AAAS; fel Soil Sci Soc Am; fel Am Soc Agron; Soil Conserv Soc Am; Can Soil Sci Soc. *Res:* Soil moisture in relation to plant growth; soil moisture, air and heat movement in saturated and unsaturated media; water and solute distribution in layered soils; soil erosion and crop production; modification of soil rooting environment by erosion. *Mailing Add:* Southern Piedmont Conserv Res Ctr PO Box 555 Watkinsville GA 30677

BRUCE, RUFUS ELBRIDGE, JR, b New Orleans, La, Mar 20, 26; m 51; c 4. PHYSICS. *Educ:* La State Univ, BS, 49; Okla State Univ, MS, 62, PhD(physics, math), 66. *Prof Exp:* Engr, Liberty Mutual Ins Co, 50-54, sales rep, 54-55, resident mgr, 55-58; instr math & physics, Northeastern La State Col, 59-60; from res asst physics to res assoc, Res Found, Okla State Univ, 60-65; head dept math & physics, Ark State Col, 65-66; PROF PHYSICS, UNIV TEX, EL PASO, 66- *Concurrent Pos:* Res physicist, Atmospheric Sci Off, Tex, 67-69. *Mem:* AAAS; Soc Photo-Optic Instrumentation Engrs. *Res:* Optical properties of materials; atmospheric effects of electromagnetic wave properties. *Mailing Add:* Dept Physics Univ Tex El Paso TX 79968

BRUCH, JOHN C(LARENCE), JR, b Kenosha, Wis, Oct 11, 40; m 67. NUMERICAL ANALYSIS, STRUCTURAL VIBRATION CONTROL. *Educ:* Univ Notre Dame, BS, 62; Stanford Univ, MS, 63, PhD(civil eng), 66. *Prof Exp:* From asst prof to prof, 66-78, PROF ENG MECH, UNIV CALIF, SANTA BARBARA, 78- *Concurrent Pos:* Res engr, Gen Motors Defense Res Lab, Calif, 67; consult, TEMPO, Gen Elec Co, Calif, 67-68, Oceanog Serv Inc, 69-70 & Pub Works Dept City Santa Barbara, 77-78. *Mem:* Am Soc Civil Engrs; Sigma Xi; US Asn Computational Mech; Am Sci Affil. *Res:* Ground-water flow; free boundary value problems; finite element analysis; numerical analysis; parallel computing, structural vibration control. *Mailing Add:* Dept Mech & Environ Eng Univ Calif Santa Barbara CA 93106

BRUCH, LUDWIG WALTER, b Rockford, Ill, Jan 23, 40; m 66; c 2. THEORETICAL PHYSICS. *Educ:* Univ Wis, BA, 59; Oxford Univ, BA, 61, MA, 65; Univ Calif, San Diego, PhD(physics), 64. *Prof Exp:* From asst prof to assoc prof, 66-75, PROF PHYSICS, UNIV WIS-MADISON, 75-

Concurrent Pos: Partic NSF visiting scientist prog, Dept Physics, Tokyo Univ Educ, Japan, 72-73; vis scientist, Univ Utrecht, Netherlands, 77-78; Sci & Eng Res Coun fel, Univ Sussex, 83; assoc prof, Univ Marseille II, France, 84; William C Foster Fel, US Arms Control & Disarmament Agency, 88; vis prof, Tech Univ, Denmark, 91. *Mem:* Am Phys Soc. *Res:* Chemical physics; theory of superfluid helium; statistical mechanics; physical adsorption. *Mailing Add:* Dept Physics Univ Wis Madison WI 53706

BRUCHOVSKY, NICHOLAS, b Toronto, Ont, Sept 21, 36; m 68; c 1. ENDOCRINOLOGY, ONCOLOGY. *Educ:* Univ Toronto, MD, 61, PhD(biophys), 66; FRCP(C), 75. *Prof Exp:* Intern med, Toronto Gen Hosp, 61-62; fel med, Univ Tex Southwestern Med Sch Dallas, 66-68; from asst prof to prof med, Univ Alta, 69-80; MEM STAFF, CANCER CTR AGENCY BC, 80- *Concurrent Pos:* Med Res Coun Can scholar, 70-75. *Mem:* Can Soc Clin Invest; Can Soc Endocrinol & Metab; Can Oncol Soc; Endocrine Soc. *Res:* Mechanism of action of androgens in prostate; control of cell proliferation in hormone-responsive and unresponsive tumors of the breast, endometrium and prostate. *Mailing Add:* Cancer Ctr Agency BC 600 W 10th Ave Vancouver BC V5Z 4E6 Can

BRUCK, DAVID LEWIS, b New York, NY, July 17, 33; m 69; c 1. GENETICS, ECOLOGY. *Educ:* Columbia Univ, BS, 55, AM, 57; NC State Univ, PhD(genetics), 65. *Prof Exp:* Res assoc genetics, NC State Univ, 65; res fel, Comt Math Biol, Univ Chicago, 65-66; asst prof biol, Queens Col, NY, 66-69; from asst prof to assoc prof, 69-88, PROF BIOL, UNIV PR, RIO PIEDRAS, 88- *Mem:* Sigma Xi; Soc Study Evolution; Int Soc Trop Ecol; Asn Trop Biol; Am Soc Naturalists. *Res:* Ecology and evolutionary genetics of Caribbean Drosophilia, especially the dunni subgroup, including mating behavior, dispersal and larval niche. *Mailing Add:* Dept Biol Univ PR Rio Piedras PR 00931

BRUCK, ERIKA, b Breslau, Ger, Apr 5, 08; nat US. PEDIATRICS. *Educ:* Friedrich-Wilhelms Univ, MD, 35. *Prof Exp:* Dir clin lab, 2nd Dept Internal Med, Istanbul Univ, 35-39; from instr to prof, 46-78, EMER PROF PEDIAT, SCH MED, STATE UNIV NY, BUFFALO, 78- *Concurrent Pos:* Attend physician, Dept Pediat, Children's Hosp, Buffalo, 46-82; consult, 82- *Mem:* Soc Pediat Res; Am Acad Pediat; Am Pediat Soc. *Res:* Renal and respiratory functions in children; electrolyte metabolism; diabetes mellitus. *Mailing Add:* Children's Hosp 219 Bryant Ave Buffalo NY 14222

BRUCK, GEORGE, b Budapest, Hungary, Oct 20, 04; nat US; m 29. APPLIED PHYSICS. *Educ:* Tech Univ, Vienna, PhD(applied physics), 27. *Prof Exp:* Asst chief engr, Compagnie des Lamps, 29-36; resident engr, Fabbr Ital Magn Marelli, 36-41; res engr, Crosley Corp, 41-43; dir special studies, Nat Union, 43-44; chief engr, Hudson Am Corp, 44; res engr, Raytheon Mfg, 44-46; chief photo engr, Specialities Inc, 46-53; pres, Bruck Indust, Inc & consult, Radiation, Inc, 53-54; staff sci adv, Avco Corp, Ohio, 54-67, chief scientist, Electronic Div, 67-69; sci adv, 69-76; RETIRED. *Mem:* Fel Inst Elec & Electronics Engrs; Am Phys Soc. *Res:* Electronic devices; solid state; military electronics. *Mailing Add:* 501 Portola Rd PO Box 8082 Portola Valley CA 94028-7603

BRUCK, RICHARD HUBERT, b Pembroke, Ont, Dec 26, 14; nat US; m 40. MATHEMATICS. *Educ:* Univ Toronto, BA, 37, MA, 38, PhD(abstract algebra), 40. *Prof Exp:* Instr math, Univ Ala, 40-42; from instr to assoc prof, 42-52, PROF MATH, UNIV WIS-MADISON, 52- *Concurrent Pos:* Asst ed, Bulletin, Am Math Soc, 45-57 & Proceedings, 55-57; Guggenheim fel & Univ res fel, Univ Wis, 46-47; vis prof, Univ NC, Chapel Hill, 63-64. *Honors & Awards:* Chauvenet Prize, Math Asn Am, 56. *Mem:* Am Math Soc (assoc secy, 45-48); Math Asn Am. *Res:* Representation theory; tensor algebra; linear non-associative algebra; theory of loops; projective planes; theory of groups. *Mailing Add:* 6342 Inner Dr Madison WI 53705

BRUCK, ROBERT IAN, b June 25, 52; US citizen; m 73; c 2. ENVIRONMENTAL EPIDEMIOLOGY, FOREST PATHOLOGY. *Educ:* State Univ NY, Buffalo, BA, 73; State Univ NY, Syracuse, PhD(plant path), 77; Syracuse Univ, PhD(forestry), 77. *Prof Exp:* Res assoc plant path, Cornell Univ, 77-79; from asst prof to assoc prof, 79-90, PROF PLANT PATH & FORESTRY, NC STATE UNIV, 90- *Mem:* Am Phytopath Soc; Can Phytopath Soc; Sigma Xi; AAAS; Fedn Am Scientists. *Res:* Environmental epidemiology of plant diseases; discovery of boreal montane (spruce-fir); forest decline in eastern North America. *Mailing Add:* Dept Plant Path NC State Univ Box 7616 Raleigh NC 27616

BRUCK, STEPHEN DESIDERIUS, b Budapest, Hungary, Feb 4, 27; nat US; m 54; c 3. BIOMATERIALS, BIOMEDICAL ENGINEERING. *Educ:* Boston Col, BS, 51; Johns Hopkins Univ, MA, 53, PhD(biochem), 55. *Hon Degrees:* DMC, Tokyo Med & Dent Univ, 83. *Prof Exp:* Jr instr, Johns Hopkins Univ, 51-55; res chemist, Dacron Res Lab, E I du Pont de Nemours & Co, NC, 55-56, res chemist, Carothers Res Lab, Textile Fibers Dept, Exp Sta, Del, 56-60; proj leader, Nat Bur Standards, Washington, DC, 61-62; actg sect head coatings mat, NASA, 62; mem sr sci staff chem, Appl Physics Lab, Johns Hopkins Univ, 62-66; actg mgr polymer chem, Watson Res Ctr, IBM Corp, 66-67; res prof chem eng, Cath Univ Am, 67-69; prog dir biomat, Nat Heart, Lung & Blood Inst, NIH, 69-78; SCI DIR BIOMAT & MED DEVICES CONSULT GROUP, STEPHEN D BRUCK ASSOCS, INC, 78- *Concurrent Pos:* Affil prof, Wash Univ, St Louis, 78-80; adj prof, Univ Tenn, 78- & vis prof, Univ Va, 79; ed-in-chief, J Long-Term Effects Med Implants, sect ed biomat, Int J Artificial Organs, ed, CRC Crit Rev Therapeut Drug Carrier Systs, NAm ed, Med Progress through Technol; former adv bd mem, Artificial Organs, Biomat & Med Devices; Nat Acad Sci exchange scholar, Czech, 83; scholar, Japan Soc for the Promotion of Sci, 83; grants, NIH, NSF & Dept Interior. *Mem:* Fel AAAS; sr mem Am Chem Soc; Biomat Soc; Am Soc Artificial Internal Organs; Japan Soc Prom Sci Scholar; Sigma Xi; NY Acad Sci. *Res:* Research and development administration; structure and property correlation of biomedical materials; biocompatibility and toxicity; technological risk assessment of health-care products; biopolymers; macromolecular pharmacology. *Mailing Add:* 3247 St Augustine Olney MD 20832

BRUCKENSTEIN, STANLEY, b Brooklyn, NY, Nov 1, 27; m 50; c 3. ANALYTICAL CHEMISTRY, ELECTROCHEMISTRY. *Educ:* Polytech Inst Brooklyn, BS, 50; Univ Minn, PhD(chem), 54. *Prof Exp:* From instr to assoc prof anal chem, Univ Minn, 54-62, prof anal chem & chief div, 62-68; prof chem, 68-73, chmn dept, 74-83, A CONGER GOODYEAR PROF, STATE UNIV NY, BUFFALO, 73- *Honors & Awards:* C N Reilley Award. *Mem:* AAAS; Am Chem Soc; Electrochem Soc. *Res:* Analytical chemistry; electroanalytical methods; phys electrochem. *Mailing Add:* Dept Chem State Univ NY Buffalo NY 14214

BRUCKER, EDWARD BYERLY, b Baltimore, Md, Mar 23, 31; m 63. OPTICS. *Educ:* Johns Hopkins Univ, BE, 52, PhD(physics), 59. *Prof Exp:* Res assoc physics, Duke Univ, 58-61, vis asst prof, 59-60; asst prof, Fla State Univ, 61-63; vis lectr, Johns Hopkins Univ, 63-64; asst prof, Fla State Univ, 64-69; sr res assoc, Rutgers Univ, 69-70; assoc prof, 71-74, asst dean res, 80-85, PROF PHYSICS, STEVENS INST TECHNOL, 75- *Mem:* Am Phys Soc; Sigma Xi; Int Soc Optical Eng. *Res:* High energy physics; interactions of fundamental particles; optical information processing. *Mailing Add:* Dept Physics Stevens Inst Technol Hoboken NJ 07030

BRUCKER, PAUL CHARLES, b Philadelphia, Pa, Nov 15, 31; m 57; c 3. FAMILY MEDICINE. *Educ:* Muhlenberg Col, BS, 53; Univ Pa, MD, 57. *Prof Exp:* Intern, Lankenau Hosp, Philadelphia, 57-58; resident, Hunterdon Med Ctr, Flemington, NJ, 58-59 & Lankenau Hosp, Philadelphia, 59-60; pvt family pract, Ambler, Pa, 60-73; ALUMNI PROF FAMILY MED & CHMN DEPT, JEFFERSON MED COL, 73-, CLIN ASSOC PROF MED, 76- *Concurrent Pos:* Mem staff, Jefferson Med Col, Temple Univ & Univ Pa, 62-72; mem, Patient Care Systs Coun. *Mem:* Am Acad Family Physicians; AMA; Soc Teachers Family Med; Am Heart Asn. *Mailing Add:* Dept Family Med Thomas Jefferson Univ Hosp 1015 Walnut St Philadelphia PA 19107

BRUCKNER, ADAM PETER, b Istanbul, Turkey, Mar 21, 43; m 72; c 1. SPACE SYSTEMS, PROPULSION. *Educ:* McGill Univ, BEngr, 66; Princeton Univ, MA, 68, PhD(aerospace & mech sci), 72. *Prof Exp:* Res assoc aerospace & energetics, 72-75, lectr asst prof, 75- 78, res assoc prof, 78-85, RES PROF AERONAUT & ASTRONAUT, UNIV WASH, 88- *Concurrent Pos:* Consult, Math Sci Northwest, 78-84, Spectra Technol, 84-86, United Technol Res Ctr, 84-86, Olin/Rocket Res Co, 88-, Eli Lilly Co, 89- *Honors & Awards:* Brit Asn Medal, 66; PNW Award, Am Inst Aeronaut & Astronaut, 73. *Mem:* Am Inst Aeronaut & Astronaut; Optical Soc Am; Sigma Xi. *Res:* Energy conversion for space applications; heat transfer, laser applications; advanced space propulsion techniques; hypervelocity accelerators, hypersonics. *Mailing Add:* FL-10 Aerospace & Energetics Res Prog Univ Wash Seattle WA 98195

BRUCKNER, ANDREW M, b Berlin, Ger, Dec 17, 32; US citizen; m 57; c 2. MATHEMATICS. *Educ:* Univ Calif, Los Angeles, BA, 55, PhD(math), 59. *Prof Exp:* From asst prof to assoc prof, 59-68, actg dean grad div, 66-69, PROF MATH, UNIV CALIF, SANTA BARBARA, 68- *Concurrent Pos:* NSF res grant, 62-74 & 78- *Mem:* Am Math Soc; Math Asn Am. *Res:* Theory of functions of a real variable. *Mailing Add:* Dept Math Univ Calif Santa Barbara CA 93018

BRUCKNER, DAVID ALAN, b Rhinelander, Wis, June 12, 41; m 65; c 2. MEDICAL PARASITOLOGY. *Educ:* Wis State Univ, Stevens Point, BS, 66; NDak State Univ, MS, 68; Johns Hopkins Univ, ScD(parasitol), 72. *Prof Exp:* Fel parasitol, 72-75, lectr parasitol, 75-76, ASSOC PROF, DEPT PATH & LAB MED, UNIV CALIF, LOS ANGELES, 78- *Mem:* Am Soc Trop Med & Hyg; Am Soc Parasitologists; Am Soc Microbiol. *Res:* Neurophysiological interactions of Schistosoma cercariae; in vitro cultivation of larval stages of cestodes and schistosomes; development test applications for clinical microbiology and immunology. *Mailing Add:* Dept Path & Lab Med 171315 Univ Calif 10833 Le Conte Ave Los Angeles CA 90024-1713

BRUCKNER, JAMES VICTOR, b Lubbock, Tex, Apr 4, 44; m 68; c 2. TOXICOLOGY, PHARMACOLOGY. *Educ:* Univ Tex, Austin, BS, 68, MS, 71; Univ Mich, Ann Arbor, PhD(toxicol), 74. *Prof Exp:* Asst prof pharm & toxicol, Sch Pharm, Univ Kans, 74-75; ASST PROF PHARM & TOXICOL, UNIV TEX MED SCH HOUSTON, 75- *Concurrent Pos:* Res contract, Nat Inst Drug Abuse, 75-78, consult, 76-77; res grants, Am Heart Asn, 77-78 & Nat Heart, Lung & Blood Int, 77-79; mem toxicol subcomt, Safe Drinking Water Comt, Nat Acad Sci, 78-79; consult, US Environ Protection Agency, 78-, res grants, 80- *Mem:* NY Acad Sci; AAAS; Soc Toxicol; Am Soc Pharmacol & Exp Therapeut. *Res:* Influence of environmental pollutants on therapeutic efficacy of drugs; mechanisms of gastrointestinal drug-absorption interactions; toxic interactions of hydrocarbon solvents and drugs; in vitro screening tests for detection of toxic chemicals. *Mailing Add:* 2351 College Station Rd Suite 568 Athens GA 30605

BRUCKNER, LAWRENCE ADAM, b Brooklyn, NY, Feb 16, 40; div; c 3. STATISTICAL ANALYSIS. *Educ:* Cath Univ, BA, 62, MA, 64, PhD(probability & statist), 68. *Prof Exp:* From instr to asst prof, Cath Univ, 66-68; mem tech staff, Sandia Corp, 68-71; assoc prof math, Col Santa Fe, 71-73; asst prof, Univ Maine, Portland, 73-74; MEM TECH STAFF, LOS ALAMOS NAT LAB, UNIV CALIF, 74- *Mem:* Am Statist Asn; Am Math Asn; Am Soc Qual Control; Am Soc Testing & Mat; Sigma Xi. *Res:* Statistical process control; statistical analysis. *Mailing Add:* Statist Group A-1/MS 600 Los Alamos Nat Lab Los Alamos NM 87545

BRUCKNER, ROBERT JOSEPH, b Jersey City, NJ, May 29, 21; m 54; c 3. DENTISTRY. *Educ:* Univ Md, DDS, 44; Western Reserve Univ, MS, 48; Am Bd Oral Path, dipl. *Prof Exp:* Teaching fel histol, embryol & anat, Sch Dent, Western Reserve Univ, 46-48, fel histol & embryol, Sch Med, 46-49, from instr to asst prof, 49-55, instr anat & oral diag, Sch Dent, 48-50, asst prof anat & path, 50-53, assoc prof path, 53-58, assoc prof path & periodont, 58-59, actg chmn dept oral path, 56-59; from assoc prof to prof, 59-86, assoc dean, 69-85, EMER PROF PATH, SCH DENT, UNIV ORE, 86- *Mem:* Fel Am Acad Oral Path; fel Am Col Dent. *Res:* Calcification of dentin; aplasia of cementum; hypophosphatasia; dental preceptorships. *Mailing Add:* 74 Greenridge Ct Lake Oswego OR 94703

BRUDER, JOSPEH ALBERT, b Brooklyn, NY, Oct 28, 38; m 65; c 4. RADIO FREQUENCY, WAVEGUIDE IMPLEMENTATION. *Educ:* Polytech Inst NY, BSEE, 62; State Univ NY, Buffalo, MSEE, 67. *Prof Exp:* Elect engr, Sylvania Electronic Syst, 62-65; res asst, State Univ NY, Buffalo, 65-66; prin engr, Calspan Corp, 66-79; SR RES ENG, ENG EXP STA, GA INST TECHNOL, 79- *Mem:* Sigma Xi; sr mem Inst Elec & Electronics Engrs. *Res:* Design and calibrate millimeter and monopulse instrumentation radar systems; conduct field measurements for backscatter data and monopulse signal characteristics; evaluate radar systems and recommend modification for system improvement. *Mailing Add:* 2404 Riverglenn Circle Dunwoody GA 30338

BRUDEVOLD, FINN, dentistry, for more information see previous edition

BRUDNER, HARVEY JEROME, b New York, NY, May 29, 31; m 63; c 3. BIOPHYSICS, ELECTRONICS. *Educ:* NY Univ, BS, 52, MS, 54, PhD(physics), 59. *Prof Exp:* Electronics engr, Pioneer Instrument Div, Bendix Corp, NJ, 52; instr atomic physics, NY Univ, 53-54; physicist, US Naval Ord Lab, 54; prin physicist, Emerson Res Labs, 54-57; res scientist, Courant Inst Math Sci, NY Univ, 57-62; prof math & physics, NY Inst Technol, 62-63, dean sci & technol, 63-64; sr res assoc, Princeton Lab, Am Can Co, 64-67; dir res & develop, Westinghouse Learning Corp, 67, from vpres to pres, 67-76; PRES, HJB ENTERPRISES, 76- *Concurrent Pos:* Consult, Emerson Radio & Phonograph Corp, 57-59 & Emertron, Inc, 57-62; mem adv comt, Middlesex County Col; mem, Coun Latin Am; adv, William Patterson Col, NJ, 75-; res & develop, NY Power Authority, 80-84; expert adv, comput in educ, US House Rep. *Mem:* AAAS; Am Phys Soc; fel Inst Elec & Electronics Engrs; Am Educ Res Asn; Soc Motion Picture & TV Engrs. *Res:* Computers and multi-media as applied to advanced educational and training systems; Thomas-Fermi techniques for determining wave functions of excited states; electromagnetic wave propagation; biological effects of microwave radiation; medical instrumentation; operations research; babylonian doubles and pythagorean triples; symmetry of powered whole numbers; fermats last theorem and N =2 triangles. *Mailing Add:* 812 Abbott St Highland Park NJ 08904-2909

BRUECK, STEVEN ROY JULIEN, b New York, NY, Aug 16, 44; m 68; c 3. QUANTUM ELECTRONICS. *Educ:* Columbia Univ, BS, 65; Mass Inst Technol, SM, 67, PhD(elec eng), 71. *Prof Exp:* Mem staff physics, Bell Tel Labs, 68; res asst elec eng, Lincoln Lab, Mass Inst Technol, 68-71, mem staff quantum electronics, 71-85; PROF ELEC ENG & PHYSICS, UNIV NMEX, 85-, DIR, CTR HIGH TECHNOL MAT, 86- *Concurrent Pos:* Assoc ed, Optics Let, 84-86 & Inst Elec & Electronics Engrs, J Quantum Electronics, 86-89, ed, 89; pub chmn, Lasers & Electrooptics Soc, Inst Elec & Electronics Engrs, 88. *Mem:* Am Phys Soc; sr mem Inst Elec & Electronics Engrs; fel Optical Soc Am; AAAS; Mat Res Soc. *Res:* Optoelectronics, semiconductors and lasers; metrology for semiconductor manufacturing; nonlinear optics. *Mailing Add:* Elec & Comput Eng Univ NMex EECE Bldg Albuquerque NM 87131

BRUECKNER, GUENTER ERICH, b Dresden, Ger, Dec 26, 34; US citizen; m 58; c 2. DESIGN OF OPTICAL INSTRUMENTATION FOR SPACEFLIGHT, DESIGN OF SPACECRAFTS. *Educ:* Univ Goettingen, Ger, Dr rer nat(astron & physics), 61, Habilitation, 67. *Hon Degrees:* PhD, Univ Oslo, Norway, 81. *Prof Exp:* Observer astron, Univ Goettingen, Ger, 65-67; asst prof physics & astrophys, Univ Md, 67-68; consult, 68-72, head, Solar Spectros Sect, 72-78, HEAD & CHIEF SCIENTIST, SOLAR PHYSICS PROG, NAVAL RES LAB, 78- *Concurrent Pos:* Prin investr, NASA, 74-; mem, Space Physics Subcomt, NASA, 88- *Mem:* Corresp mem Int Acad Astronaut; Am Astron Soc; Am Geophys Union; Optical Soc Am; fel AAAS; NY Acad Sci. *Res:* Physics of the solar atmosphere explored by means of optical instruments in the ultraviolet spectrum which are flown on sounding rockets and spacecraft; interaction between the sun and the upper earth atmosphere investigated by means of ultraviolet instrumentation flown on satellites. *Mailing Add:* Naval Res Lab Code 4160 4555 Overlook Ave SW Washington DC 20375-5000

BRUECKNER, HANNES KURT, b San Francisco, Calif, Apr 6, 40; div; c 2. GEOCHEMISTRY, STRUCTURAL GEOLOGY. *Educ:* Cornell Univ, BS, 62; Yale Univ, MS, 65, PhD(geol), 68. *Prof Exp:* Res assoc geosci, Univ Tex, Dallas, 67-70; asst prof, 70-75, assoc prof, 75-78, PROF, DEPT EARTH & ENVIRON SCI, QUEENS COL NY, 78- *Concurrent Pos:* Sr res assoc, Lamont-Doherty Geol Observ, 70-; NSF fels, 73-75, 75-78 & 79-81. *Mem:* Am Geophys Union; Sigma Xi. *Res:* Structural, petrological and geochemical studies of relationships between mantle and crust in orogenic zones; geochemistry of sediments, particularly chert. *Mailing Add:* Dept Earth & Environ Sci Queens Col Flushing NY 11367

BRUECKNER, KEITH ALLAN, b Minneapolis, Minn, Mar 19, 24; div; c 3. THEORETICAL PHYSICS. *Educ:* Univ Minn, BA, 45, MA, 47; Univ Calif, PhD(physics), 50. *Prof Exp:* From asst prof to prof physics, 51-59, chmn dept, 59-61, dean let & sci, 63-65, dean grad studies, Univ Pa, 65-; dir inst radiation physics & aerodyn, 65-67, dir inst pure & appl phys sci, 67-69, PROF PHYSICS, UNIV CALIF, SAN DIEGO, 59- *Concurrent Pos:* Consult, AEC, 53-70; physicist, Brookhaven Nat Lab, 55-56; vpres & dir res, Inst Defense Anal, Washington, DC, 61-62; vpres & tech dir, KMS Industs, 68-70, exec vpres, KMS Fusion, Inc, 71-74. *Honors & Awards:* Dannie Heinemann Prize, 63. *Mem:* Nat Acad Sci. *Res:* Theoretical nuclear physics; statistical mechanics, plasma physics; interactions of lasers with matter; magnetohydrodynamics and theory of metals. *Mailing Add:* Dept Physics B-019 Univ Calif San Diego La Jolla CA 92093

BRUEGGEMEIER, ROBERT WAYNE, b Toledo, Ohio, June 14, 50; m 72; c 2. BIOMEDICINAL CHEMISTRY. *Educ:* Mich State Univ, BA, 72; Univ Mich, MS, 75, PhD(med chem), 77. *Prof Exp:* Fel assoc biochem, Harvard Med Sch, 77-79; ASST PROF MED CHEM, OHIO STATE UNIV, 79- *Concurrent Pos:* Dir, Radiochem & Anal Labs, Ohio State Univ Cancer Ctr, 79- *Mem:* Am Chem Soc; AAAS; Sigma Xi. *Res:* Steroid chemistry and biochemistry; enqymology and enzyme inhibitors; hormones and cancer. *Mailing Add:* Col Pharm Ohio State Univ 500 W 12th Ave Columbus OH 43210

BRUEHL, GEORGE WILLIAM, b Mentone, Ind, Sept 10, 19; m 49; c 2. PLANT PATHOLOGY. *Educ:* Univ Ark, BSA, 41; Univ Wis, PhD(plant path), 48. *Prof Exp:* Assoc pathologist, Div Cereal Crops & Dis, USDA, SDak State Col Exp Sta, 48-52, pathologist, Div Sugar Plant Invests, 52-54; pathologist, dept plant path, Wash State Univ, 54-84; RETIRED. *Mem:* Am Phytopath Soc; Sigma Xi. *Res:* Root rots of cereal crops; general pathology of cereals. *Mailing Add:* Dept Plant Path Wash State Univ Pullman WA 99164

BRUENING, GEORGE, b Chicago, Ill, Aug 10, 38; m 60; c 3. PLANT VIROLOGY, BIOCHEMISTRY. *Educ:* Carroll Col, Wis, 60; Univ Wis, Madison, MS, 63, PhD(biochem), 65. *Prof Exp:* NSF fel, Virus Lab, Univ Calif, Berkeley, 65-66, from asst prof to prof biochem, 67-84, PROF PLANT PATH, UNIV CALIF, DAVIS, 85- *Concurrent Pos:* Vis scientist plant path, Cornell Univ, 74-75; Guggenheim Mem Found fel, 74-75; vis scientist biochem, Univ Adelaide, Australia, 81; vis scientist plant indust, CSIRO, Canberra, Australia, 89. *Mem:* AAAS; fel Am Phytopath Soc; Am Soc Biochem & Molecular Biol; Soc Microbiol UK. *Res:* Biochemistry and chemistry of plant viruses, satellite RNA, nucleic acids and proteins. *Mailing Add:* Dept Plant Path Univ Calif Davis CA 95616

BRUENING, JAMES THEODORE, b Cape Girardeau, Mo, Feb 27, 49; m 73; c 2. MATHEMATICS. *Educ:* Univ Mo-Rolla, BS, 71, MS, 72, PhD(math), 77. *Prof Exp:* Instr math, Westmar Col, 76-77; ASST PROF MATH, BAKER UNIV, 77- *Mem:* Am Math Soc. *Res:* Inverses of transfer function matrices; linear sequential circuits. *Mailing Add:* Southeast Mo State Univ Cape Girardeau MO 63701

BRUENNER, ROLF SYLVESTER, b Magdeburg, Ger, Dec 31, 21; US citizen; m 56; c 4. APPLIED CHEMISTRY. *Educ:* Munich Tech Univ, MS, 55, PhD(photochem), 57. *Prof Exp:* Sci asst photochem, Phys Chem Inst, Munich Tech Univ, 55-57; res chemist, US Army Signals Res & Develop Labs, Ft Monmouth, NJ, 57-59; from res chemist to sr res chemist, 59-75, SCIENTIST, AEROJET STRATEGIC PROPULSION CO, GEN TIRE & RUBBER CO, 75- *Honors & Awards:* Wyld Propulsion Award, Am Inst Aeronaut & Astronaut, 85. *Mem:* Am Chem Soc; Soc Ger Chem; Sigma Xi. *Res:* Mechanism of energy transfer from adsorbed dye molecules to crystal lattices; photopolymerization; photographic stabilization processing; catalysis of urethane formation; polymer chemistry; composite materials; propellants and explosives. *Mailing Add:* 7748 Excelsior Ave Orangevale CA 95662

BRUES, ALICE MOSSIE, b Boston, Mass, Oct 9, 13. PHYSICAL ANTHROPOLOGY. *Educ:* Bryn Mawr Col, AB, 33; Radcliffe Col, PhD(phys anthrop), 40. *Prof Exp:* Res assoc phys anthrop, Peabody Mus, Harvard Univ, 41-42; statistician, Anthropometric Unit, Wright Field, 42-44; anthropologist, Chem Warfare Serv, 45; from asst prof to prof anat, Sch Med, Univ Okla, 46-65; chmn dept anthrop, 68-71, PROF ANTHROP, UNIV COLO, BOULDER, 65- *Honors & Awards:* Phys Anthrop Award, Am Acad of Forensic Sci, 86. *Mem:* Am Asn Phys Anthrop (vpres, 66-68, pres, 71-); Soc Study Evolution; Am Soc Naturalists. *Res:* Regional variation in man; pigmentation; computer simulation of evolutionary change; forensic identification, cranial. *Mailing Add:* Box 233 Univ Colo Boulder CO 80309-0233

BRUES, AUSTIN, medicine, radiobiology; deceased, see previous edition for last biography

BRUESCH, SIMON RULIN, b Norman, Okla, July 7, 14. ANATOMY. *Educ:* La Verne Col, AB, 35; Northwestern Univ, MS, 39, MB, 40, MD, 41. *Hon Degrees:* DSc, La Verne Col, 67. *Prof Exp:* From instr to prof, 41-60, Goodman prof anat, 60-83, EMER PROF ANAT & HIST MED, UNIV TENN, MEMPHIS, 83- *Mem:* AAAS; Am Hist Sci Soc; Am Asn Hist Med; Am Asn Anat. *Res:* Visual system of vertebrates; afferent components of facial nerve; peripheral nerve endings; sweating by skin resistance methods. *Mailing Add:* 400 Madison Ave Memphis TN 38163

BRUESKE, CHARLES H, b New Ulm, Minn, Dec 18, 37; m 61; c 2. PLANT PHYSIOLOGY. *Educ:* Elmhurst Col, BS, 59; Univ Southern Ill, MS, 61; Ariz State Univ PhD(bot), 65. *Prof Exp:* From asst prof to assoc prof, 64-77, PROF BIOL, MT UNION COL, 77-, DEPT CHMN, 79- *Concurrent Pos:* Vis prof plant path, Univ Mo, 71, Ohio Agr Res & Develop Ctr, Wooster, 79; res nematologist, Agr Res Serv, USDA, Beltsville, 80. *Mem:* AAAS; Am Soc Plant Physiol; Am Soc Micros; Am Soc Adv Sci. *Res:* Pathogenicity of human fungal pathogens; physiology of plant diseases. *Mailing Add:* Dept Biol Mt Union Col Alliance OH 44601

BRUETMAN, MARTIN EDGARDO, b Buenos Aires, Arg, Aug 17, 32; US citizen; m 56; c 2. NEUROLOGY. *Educ:* Nat Col Buenos Aires, BS & BA, 49; Univ Buenos Aires, MD, 55. *Prof Exp:* Instr neurol, Baylor Col Med, 62-63, mem, Fac Comt Res Projs, 63-64; clin asst prof neurol & head sect, Inst Med Res, Univ Buenos Aires, 64-68; from assoc prof to prof, Chicago Med Sch, 68-74, chmn dept, 70-74; chmn, Dept Neurol & dir, Residency Training Prog, Mt Sinai Hosp Med Ctr, Chicago, 74-76; coordr, Ill Regional Med Prog, 69-73; PROF NEUROL, RUSH-PRESBY ST LUKE'S, 75-; CHIEF, DIV NEUROL, CHRIST HOSP, 75- *Concurrent Pos:* Fel neurol, Baylor Col Med, 61-62; Nat Heart Inst grant, 64-68; coordr coop study cerebrovascular insufficiency, NIH, 61-64; assoc investr, Neurol Res Ctr, Inst Torcuato di Tella, Arg, 64-68; adv to secy pub health training progs, Govt Arg, 66-68; mem, Coun Cerebrovascular Dis, Am Heart Asn, 68- *Mem:* Am Epilepsy Soc; Am Acad Neurol; NY Acad Sci; Arg Soc Neurol; Arg Col Neurol. *Res:* Medical education; natural history and incidence of cerebrovascular diseases; epidemiology; community resources in relation to cerebrovascular diseases. *Mailing Add:* Neurol Inst 11800 Southwest Hwy Palos Heights IL 60463-1018

BRUGAM, RICHARD BLAIR, b Philadelphia, Pa, Dec 23, 46; m 70; c 2. LIMNOLOGY, ECOLOGY. *Educ:* Lehigh Univ, BA, 68; Yale Univ, MPhil, 74, PhD(biol), 75. *Prof Exp:* Res assoc limnol, Limnol Res Ctr, Univ Minn, 75-78; from asst prof to assoc prof, 78-84, PROF BIOL, SOUTHERN ILL UNIV, EDWARDSVILLE, 85- *Concurrent Pos:* Vis scholar zool, Univ Wash, Seattle, 84-85. *Honors & Awards:* J Willard Gibbs Prize, 68. *Mem:* Ecol Soc Am; Am Soc Limnol & Oceanog; Am Quaternary Asn; NAm Lake Mgt Asn; Sigma Xi. *Res:* Paleolimnology, especially the long-term history of cultural eutrophication in lakes; fossil diatoms to reconstruct limnological conditions over post-glacial time; neutralization of acidic lakes on coal mine sites. *Mailing Add:* Dept Biol Sci Southern Ill Univ Edwardsville IL 62025

BRUGGE, JOHN F, b Brooklyn, NY, Apr 24, 37; m 60; c 3. PHYSIOLOGY. *Educ:* Luther Col, Iowa, BA, 59; Univ Ill, Urbana-Champaign, MS, 61, PhD(physiol), 63. *Prof Exp:* NIH fel, 63-66, asst prof, 66-71, assoc prof, 71-77, PROF NEUROPHYSIOL, SCH MED, UNIV WIS-MADISON, 77- *Mem:* Soc Neurosci; Am Physiol Soc; Int Brain Res Orgn; fel Acoust Soc Am. *Res:* Physiology and anatomy of auditory system. *Mailing Add:* Dept Neurophys Univ Wis Med Sch 1300 University Ave Madison WI 53706

BRUGGEMAN, GORDON ARTHUR, b Detroit, Mich, Sept 24, 33; m 56; c 2. PHYSICAL METALLURGY, MATERIALS PROCESSING. *Educ:* Mass Inst Technol, SB, 55, ScD(phys metall), 60. *Prof Exp:* Instr phys metall, Mass Inst Technol, 55-57; phys metallurgist, Manlabs, Inc, Mass, 60-61; phys metallurgist, Mat & Mech Res Ctr, US Army Mat Technol Lab, 62-81, chief process technol div, 82-86, dir emergency mat div, 87-89, ASSOC DIR, US ARMY MAT TECHNOL LAB, 90- *Concurrent Pos:* Lectr, Northeastern Univ, 63-74. *Mem:* Am Soc Metals; Am Inst Mining, Metall & Petrol Engrs. *Res:* Phase transformations and their relation to mechanical properties; failure analysis and fracture; materials processing; grain boundary structure. *Mailing Add:* US Army Mat Technol Lab Watertown MA 02172

BRUGGER, JOHN EDWARD, b Erie, Pa, Sept 3, 23; m 66. PHYSICAL CHEMISTRY. *Educ:* Gannon Col, BS, 45; Pa State Univ, MS, 46; Univ Chicago, PhD(phys chem), 54. *Prof Exp:* Asst prof chem, Gannon Col, 46-49; res & admin assoc, Res Insts, Univ Chicago, 52-57, chemist, Lab Appl Sci, 57-63; assoc chemist, Chem Eng Div, Argonne Nat Lab, 63-68; sr res chemist & mgr res, Microstatics Lab, Smith-Corona Marchant Div, SCM Corp, Ill, 68-69, res chemist & supvr phys systs group, Smith-Corona Marchant Res & Develop Lab, Palo Alto, 70-71; phys scientist & adminr, Hazardous Waste Eng Res Lab, 71-81, CHIEF SCIENTIST, RELEASES CONTROL BR, RISK REDUCTION ENG LAB, US ENVIRON PROTECTION AGENCY, EDISON, NJ, 81- *Concurrent Pos:* Mem, panel waste technol handling, US Off Sci & Technol (Exec Off President), 79-80; mem, Comt Environ Improv, Am Chem Soc, 81-83. *Mem:* Am Chem Soc; Am Phys Soc; Am Nuclear Soc; Soc Imaging Sci & Technol; Am Soc Testing & Mat; Am Inst Chem Engrs; Sigma Xi; Soc Info Display; Am Asn Univ Professors. *Res:* Environmental science; chemical-physical methods for control of toxic releases; electrophotography; photophysics; high temperature calorimetry; refractory nonmetallics; infrared detectors; fluorescence and phosphorescence; photosynthesis; isotope applications; reaction kinetics. *Mailing Add:* PO Box 15 Metuchen NJ 08840-0015

BRUGGER, ROBERT MELVIN, b Oklahoma City, Okla, Jan 13, 29; m 53; c 2. NEUTRON PHYSICS. *Educ:* Colo Col, BA, 51; Rice Inst, MA, 53, PhD(physics), 55. *Prof Exp:* Asst, Rice Inst, 51-54; res physicist, Nuclear Physics Br, Atomic Energy Div, Phillips Petrol Co, 55-61, head solid state physics sect, 61-66; head solid state physics sect, Idaho Nuclear Corp, 66-68; mgr nuclear technol div, Aerojet Nuclear Co, 68-74; DIR RES REACTOR FACIL, UNIV MO, 74- *Concurrent Pos:* Mem, UK Atomic Energy Res Estab, Harwell, 62-63; comr, Idaho Nuclear Energy Comn, 70-74; sabbatical leave, Los Alamos Nat Lab, 81-82. *Mem:* Fel Am Phys Soc; fel Am Nuclear Soc. *Res:* Neutron cross sections; neutron inelastic scattering; neutron diffraction; gamma detection; high pressure research; reactor conception; design and operation; metallurgy and materials testing. *Mailing Add:* Nuclear Eng Dept 333EE Bldg Univ Mo Columbia MO 65211

BRUGGER, THOMAS C, b Fond du Lac, Wis, Jan 19, 27; m 54; c 5. CHILD PSYCHIATRY. *Educ:* Univ Wis, BNS, 48, BS, 50, MD, 53. *Prof Exp:* Staff psychiatrist, Winebago State Hosp, 53; resident, Cincinnati Dept Psychiat, Cincinnati Gen Hosp, 56-58; trainee child psychiat, Dept Child Psychiat, Child Guid Home, Ohio, 58-60; staff psychiatrist, instr & asst dir children's clins, Sch Med, Univ Cincinnati, 60-61; asst prof child psychiat & dir child guid & eval clin, Wash Univ, 61-75; COORDR CHILDREN'S SERV & ASST PROF CHILD PSYCHIAT, CASE WESTERN RESERVE UNIV, 75-, ACTG DIR, DIV CHILD PSYCHIAT, 87- *Concurrent Pos:* Consult, Cincinnati Speech & Hearing Ctr, Ohio, 59-61, Diag Clin Ment Retarded Children, 59-61 & Miriam Sch, 63-75. *Mem:* AAAS; Am Psychiat Asn; Am Orthopsychiat Asn. *Res:* Child psychiatry. *Mailing Add:* Div Child Psychiat Case Western Reserve Univ-Hosp 2847 Coleridge Cleveland OH 44106

BRUGH, MAX, JR, b Roanoke, Va, Dec 16, 38. VETERINARY MEDICINE, MEDICAL MICROBIOLOGY. *Educ:* Univ Ga, DVM, 64, PhD(med microbiol), 68. *Prof Exp:* Res veterinarian foot & mouth dis, Plum Island Animal Dis Lab, USDA, 68-70; virologist, Pan Am Foot & Mouth Dis Ctr, Pan Am Health Orgn, 70-73; RES VETERINARIAN NEWCASTLE DIS, SOUTHEAST POULTRY RES LAB, USDA, 73- *Mem:* Am Vet Med Asn; Am Asn Avian Pathologists; Am Asn Vet Lab Diagnosticians; World Poultry Sci Asn; Sigma Xi. *Res:* Development of methods for control of viral diseases in food animals; development and improvement of vaccines, diagnostic methods and serologic tests for control of Newcastle disease and avian influenza. *Mailing Add:* USDA-Agr Res Serv-SEPRL PO Box 5657 Athens GA 30613

BRUHN, H(JALMAR) D(IEHL), b Spring Green, Wis, Aug 5, 07; m 38; c 1. MECHANICAL & AGRICULTURAL ENGINEERING. *Educ:* Univ Wis, BS, 31 & 33; Mass Inst Technol, MS, 37. *Prof Exp:* From instr to prof agr eng, 33-78, chmn dept agr eng, 62-66, EMER PROF AGR ENG, UNIV WIS, MADISON, 78- *Concurrent Pos:* Consult agr eng, 78- *Honors & Awards:* Engr of the Year, Wis Sect, Am Soc Agr Eng, 71. *Mem:* Am Soc Agr Eng; AAAS; Am Forage & Grassland Coun; Nat Soc Prof Engrs; NY Acad Sci; Soc Green Veg Res. *Res:* Farm power and machinery with emphasis on forage harvesting, crushing, pelleting and drying; irrigation equipment; seeding machinery; fruit harvesting equipment; harvesting and wet fractionation of terrestrial and aquatic vegetation; spontaneous ignition of agricultural crops. *Mailing Add:* 5418 Lake Mendota Dr Madison WI 53705

BRUHN, JOHN G, b Norfolk, Nebr, Apr 27, 34. MEDICAL SOCIOLOGY, PREVENTIVE MEDICINE. *Educ:* Univ Nebr, BA, 56, MA, 58; Yale Univ, PhD(med sociol), 61. *Prof Exp:* Res sociologist, Dept Psychol Med, Univ Edinburg, 61-62; instr, med sociol, 62-63, asst prof, Dept Psychiat & Behav Scis, 63-64, asst prof, 64-67, assoc prof med, prev med & pub health, 67-70, prof & chmn, Dept Human Ecol, Sch Health, Univ Okla Med Ctr, 69-72; assoc dean, community affairs, 72-81, DEAN, SCH ALLIED HEALTH SCIS & SPEC ASST TO THE PRES, COMMUNITY AFFAIRS, UNIV TEX MED BR, GALVESTON, 81-; PROF HUMAN ECOL, UNIV TEX SCH PUB HEALTH, HOUSTON, 75- & ACTG CHAIR, DEPT PREV MED & COMMUNITY HEALTH, 90- *Concurrent Pos:* Res, Conn Dept Ment Health, 58-59; instr social, Southern Coun Col, 60-61; res sociologist, Grace New Haven Community Hosp, 60-61; USPHS fel, 60-61, US Fulbright fel, 61-62, John E Fogarty Health Scientist Exchange fel, 89 & WHO fel, 91; consult, Am Soc Anesthesiologists, 66-71; assoc prof sociol, Univ Okla, Norman, 67-72; career develop award, Nat Heart Inst, 68-69; mem, Task Force on Funding, Am Pub Health Asn, 73-74; Danforth Found assoc, 73-86. *Mem:* Sigma Xi; fel Am Orthopsychiat Asn; Am Psychosom Soc; Am Asn Univ Prof; fel Royal Soc Health. *Res:* Study of the psychological and social factors that influence the clinical course of AIDS; published numerous articles in various journals. *Mailing Add:* Off Dean & Allied Health Sci Univ Tex Med Br 1100 Mechanic Galveston TX 77550

BRUHWEILER, FREDERICK CARLTON, JR, b Quincy, Ill, Dec 4, 46; m 77. ASTRONOMY, ASTROPHYSICS. *Educ:* Coe Col, BA, 68; Univ Tex, Austin, MA, 73, PhD(astron), 77. *Prof Exp:* MEM STAFF, LAB ASTRON & SOLAR PHYSICS, NASA, GODDARD SPACE FLIGHT CTR, 77- *Concurrent Pos:* Assoc, Nat Acad Sci-Nat Res Coun, 77- *Mem:* Am Astron Soc. *Res:* Ultraviolet astronomy, early type stars and the interstellar medium. *Mailing Add:* 10102 Gardiner Ave NASA Goddard Space Flight Ctr Silver Spring MD 20902

BRUICE, THOMAS CHARLES, b Los Angeles, Calif, Aug 25, 25; m; c 3. BIO-ORGANIC CHEMISTRY. *Educ:* Univ Southern Calif, BA, 50, PhD(biochem), 54. *Prof Exp:* Lilly fel, Univ Calif, Los Angeles, 54-55; from instr to asst prof, Yale Univ, 55-58; asst prof, Sch Med, Johns Hopkins Univ, 58-60; prof chem, Cornell Univ, 60-64; PROF CHEM, UNIV CALIF, SANTA BARBARA, 64- *Concurrent Pos:* USPHS sr fel, 57-58; career investr, NIH, 62- *Mem:* Nat Acad Sci; Am Chem Soc; Royal Soc Chem; Am Soc Biochem; Am Acad Arts & Sci. *Res:* Physical organics; bioorganics. *Mailing Add:* Dept Chem Univ Calif Santa Barbara CA 93106

BRUINS, PAUL F(ASTENAU), b Albert Lea, Minn, Dec 22, 05; m 29; c 5. CHEMICAL ENGINEERING. *Educ:* Cent Col, Iowa, BS, 26; Iowa State Univ, MS, 27, PhD(chem eng), 30. *Hon Degrees:* DSc, Cent Col, Iowa, 60 & Polytech Univ, NY, 78. *Prof Exp:* Instr, Iowa State Univ, 27-30; chem engr, A O Smith Corp, 30-32; chief chemist, Geuder, Paeschke & Frey Co, 32-34; asst supt, Fulton Co, 34-35; from res assoc to prof, 35-74, EMER PROF CHEM ENG, POLYTECH UNIV, NY, 74-; PLASTICS CONSULT, 35- *Concurrent Pos:* Chmn, dept chem, Defense Training Inst, Polytech Inst Brooklyn, 40-43. *Honors & Awards:* Outstanding Achievement Award, Soc Plastics Eng, Plastics Educ Award. *Mem:* Fel Am Inst Chem Engrs; Am Chem Soc; Soc Plastics Eng. *Res:* Plastics science and technology; coatings; adhesives; synthetic resins; microporous membranes and implants; electrochemistry; battery separators; conductive plastic film. *Mailing Add:* 708 Harris Ave Austin TX 78705

BRULEY, DUANE FREDERICK, b Chippewa Falls, Wis, Aug 3, 33; m 59; c 3. PROCESS CONTROL BIOCHEMICAL ENGINEERING, SYSTEMS PHYSIOLOGY. *Educ:* Univ Wis, BS, 56; Stanford Univ, MS, 59; Univ Tenn, PhD(chem eng), 62. *Prof Exp:* Assoc develop engr, Oak Ridge Nat Labs, Union Carbide Co, 56-59; from asst prof to prof chem eng, Clemson Univ, 62-72; prof chem eng & head dept, Tulane Univ, 72-76; vpres acad affairs & dean fac, Rose-Hulman Inst Technol, 76-81; prof & head biomed eng, La Tech Univ, 81-84; dean eng, 84-87, PROF ENG, CALIF POLYTECH STATE UNIV, 87- *Concurrent Pos:* NSF grant, 64-65; consult, WVa Pulp & Paper Co, 64-, Am Enka, Med Univ SC, E I DuPont de Nemours & Co, 73-, Exxon Corp, 78- & El Paso Polyolifens Co, 81-82; NIH res grant, 67-73; vis prof, Princeton Univ, 70- & Univ Yamagata, Japan, 75; adj prof, Dept Physiol & Dept Anat, Tulane Univ, 73-78; prog dir & sect head, bioeng & environ systs, NSF, 87-90. *Honors & Awards:* Outstanding Contribution to Res Award, Am Soc Eng Educ, 67; Chem Eng Educator Award, Chem Eng Educ Mag, 70; Charls M Kerr Award, 83. *Mem:* Am Inst Chem Engrs; Biomed Eng Soc; Int Soc Oxygen Transp Tissue (pres, 82); Nat Soc Prof Engrs; Am Soc Mech Engrs. *Res:* Process dynamics and control; biomedical and biochemical engineering. *Mailing Add:* Sch Eng Calif Polytech State Univ San Luis Obispo CA 93407

BRUMAGE, WILLIAM HARRY, b Slick, Okla, Nov 18, 23; m 47; c 1. SOLID STATE PHYSICS. *Educ:* Okla State Univ, BS, 48, MS, 49; Univ Okla, PhD(physics), 64. *Prof Exp:* Instr physics, Ark Agr & Mech Col, 49-52; from asst prof to assoc prof, 52-64, PROF PHYSICS, LA TECH UNIV, 64-, CHMN DEPT, 65- *Mem:* Am Phys Soc; Am Asn Physics Teachers; Sigma Xi; Optical Soc Am. *Res:* Paramagnetic susceptibilities of ions in host crystals; laser induced chemistry. *Mailing Add:* Dept Physics La Tech Univ Ruston LA 71270

BRUMBAUGH, DONALD VERWEY, b Rochester, NY, Sept 18, 52. PHYSICAL CHEMISTRY, MOLECULAR SPECTROSCOPY. *Educ:* Williams Col, AB, 74; State Univ NY, Binghamton, PhD(chem), 80. *Prof Exp:* Res assoc spectros, dept chem, James Franck Inst, Univ Chicago, 79-81; RES CHEMIST, PHOTOGRAPHIC RES LAB, EASTMAN KODAK CO, 81- *Mem:* Am Chem Soc; Sigma Xi. *Res:* Laser induced fluorescence spectra of molecules; spectral sensitization of photographic emulsions. *Mailing Add:* 38 Eastgate Dr Rochester NY 14617

BRUMBAUGH, JOE H, b Eldorado, Ohio, Sept 14, 30; m 61; c 3. MARINE BIOLOGY, MARINE & FRESHWATER INVERTEBRATES. *Educ:* Miami Univ, BSEd, 52; Purdue Univ, MS, 56; Stanford Univ, PhD(biol), 65. *Prof Exp:* High sch teacher, Ohio, 52-54; instr biol & bot, Wabash Col, 56-59; from asst prof to assoc prof biol, 64-71, chmn, Div Natural Sci, 71-74, PROF BIOL, SONOMA STATE COL, 71- *Mem:* Am Soc Zoologists; AAAS; Western Soc Naturalists; Sigma Xi. *Res:* Anatomy, diet and feeding mechanisms in marine invertebrates, particularly holothurian echinoderms; functional morphology of marine and freshwater invertebrates; natural history of marine and freshwater invertebrates. *Mailing Add:* Dept Biol Sonoma State Col Rohnert Park CA 94928

BRUMBAUGH, JOHN (ALBERT), b Buffalo, NY, May 3, 35; m 58; c 3. GENETIC CONTROL GENE EXPRESSION. *Educ:* Cedarville Col, BS, 58; Iowa State Univ, PhD(develop genetics), 63. *Prof Exp:* From instr to assoc prof biol, Cedarville Col, 59-64; from asst prof to assoc prof, 64-73, PROF GENETICS, SCH BIOL SCI, UNIV NEBR, LINCOLN, 73- *Concurrent Pos:* Attend, NSF Col Teachers Res Partic Prog, Purdue Univ, 64; Nat Inst Gen Med Sci res career develop award, 69-74; Lasby vis prof human & oral genetics, Univ Minn, 77-78; Prog dir eukaryotic genetics, NSF, 89-90. *Mem:* Am Genetic Asn; Am Soc Cell Biol; Genetics Soc Am; Int Pigment & Cell Soc; Poultry Sci Asn. *Res:* Melanocyte differentiation in the fowl; genetic regulation in higher organisms; somatic cell genetics of melanocytes in culture including melanoma; molecular biology of tyrosinase; automated DNA sequencing. *Mailing Add:* Sch Biol Sci Univ Nebr Lincoln NE 68588

BRUMBAUGH, PHILIP, b St Louis, Mo, Nov 14, 32. QUALITY CONTROL, INDUSTRIAL ENGINEERING. *Educ:* Wash Univ, AB, 54, MBA, 58, PhD(prod mgt), 63. *Prof Exp:* Asst indust rels, Granite City Steel Co, 57; instr indust mgt, Univ Kansas City, 58; analyst, Humble Oil & Refining Co, 61-63; asst prof comput sci, Wash Univ, 64-70, asst chmn, dept appl math & comput sci, 67-70; assoc prof prod mgt, Univ Mo-St Louis, 70-74; pres, Qual Assurance Inc, 74-83; CONSULT, 74- *Concurrent Pos:* Pres, Qualtech Systs, Inc, 77-85; dir, Palco Equipment, Inc, 78-85; Lighthouse for the Blind, 85-; adj prof eng, Wash Univ, 86- *Mem:* AAAS; Am Inst Indust Engrs; Am Statist Asn; Am Soc Qual Control. *Res:* Production and inventory theory; statistical quality control. *Mailing Add:* 1359 S Mason Rd St Louis MO 63131

BRUMBERGER, HARRY, b Vienna, Austria, Aug 28, 26; nat US; m 50; c 2. CATALYSIS, MATERIALS SCIENCE. *Educ:* Polytech Inst Brooklyn, BS, 49, MS, 52, PhD(chem), 55. *Prof Exp:* Res assoc chem, Cornell Univ, 54-57; from asst prof to assoc prof, 57-69, dir grad biophys prog, 77-83, PROF PHYS CHEM, SYRACUSE UNIV, 69-, DIR SOLID STATE SCI & TECHNOL PROG, 87- *Concurrent Pos:* Res leave, Graz Univ & Nat Bur Standards, 62-63, Weizmann Inst Sci, 74, Univ Cambridge, 83-84 & Swiss Fed Inst Technol, 90- 91; mem policy comt, Nat Ctr Small-Angle Scattering Res, Oak Ridge Nat Lab, 78-81. *Mem:* AAAS; Am Crystallog Asn; Mat Res Soc. *Res:* Supported-metal catalysts, structure-property correlations, sintering and redispersion phenomena, small-angle x-ray scattering, scattering models; application and theory of small-angle scattering for multicomponent systems. *Mailing Add:* Dept Chem Syracuse Univ Syracuse NY 13244-1200

BRUMER, MILTON, b Philadelphia, Pa, Jan 21, 02; m 26; c 2. CIVIL ENGINEERING. *Educ:* Rensselaer Polytech Inst, ME, 23; Univ Ala, MS, 24. *Hon Degrees:* DE, Rensselaer Polytech Inst, 70. *Prof Exp:* Asst engr, Port NY Auth, 25-38; chief engr tunnel design, Pa Turnpike Comn, 38-40; prin engr, O H Ammann & Ammann & Whitney, 40-50, pres, Ammann & Whitney, Inc & Ammann & Whitney Int, Ltd, 50-75, partner & prin engr, Ammann & Whitney, Inc, 75-79; RETIRED. *Concurrent Pos:* Lectr, numerous cols & univs. *Mem:* Nat Acad Eng; fel NY Acad Sci; Am Inst Consult Engrs; fel am Soc Civil Engrs. *Res:* Suspension bridge design; structural steel bridges; highways; turnpikes; hangars; vehicular tunnels. *Mailing Add:* 6750 Entrada Pl Boca Raton FL 33433

BRUMER, PAUL WILLIAM, b New York, NY, June 8, 45; m 68; c 4. THEORETICAL CHEMISTRY, CHEMICAL PHYSICS. *Educ:* Brooklyn Col, BS, 66; Harvard Univ, PhD(chem physics), 72. *Prof Exp:* Fel chem physics, Weizmann Inst Sci, 72-73 & molecular physics, Harvard Col Observ, 73-74, lectr astron, 74-75; from asst prof to assoc prof, 75-83, PROF CHEM, UNIV TORONTO, 83- *Concurrent Pos:* Alfred P Sloan Found fel, 77-81; Killam res fel, 81-83. *Honors & Awards:* Hudnall lectr Award, Univ Chicago, 83; Noranda Award, CIC, 85. *Mem:* Am Phys Soc; Sigma Xi. *Res:* Theories of intermolecular and intramolecular nonlinear mechanics and chemical dynamics; classical quantum correspondence; laser induced chemistry. *Mailing Add:* Dept Chem Univ Toronto Toronto ON M5S 1A1 Can

BRUMLEVE, STANLEY JOHN, b Teutopolis, Ill, July 3, 24; m 55; c 4. PHYSIOLOGY. *Educ:* St Louis Univ, BS, 50, MS, 54, PhD(physiol, biochem), 57. *Prof Exp:* Asst physiol, St Louis Univ, 54; instr biol, Webster Col, 55-57; asst prof physiol & pharmacol, 57-64, actg chmn dept, 64-65, assoc prof, 64-73, PROF PHYSIOL & PHARMACOL, MED SCH, UNIV NDAK, 73-, CHMN DEPT, 72- *Concurrent Pos:* Asst, Univ Alaska, 54. *Mem:* Am Physiol Soc. *Res:* Environmental physiology; temperature regulation; high pressure and other stresses. *Mailing Add:* Dept Physiol Univ NDak Med Sch Grand Forks ND 58201

BRUMM, DOUGLAS B(RUCE), b Nashville, Mich, Aug 4, 40; m 65; c 2. ELECTRICAL ENGINEERING, OPTICS. *Educ:* Univ Mich, MSE, 64, PhD(elec eng), 70. *Prof Exp:* Assoc eng, Wayland Lab, Raytheon Co, 62-63; teaching fel elec eng, Univ Mich, 63-64, res asst holography, Radar & Optics Lab, 64, res assoc, 66-70, res asst, Electro-Optical Sci Lab, 65-66; asst prof, 70-81, ASSOC PROF ELEC ENG, MICH TECHNOL UNIV, 81- *Concurrent Pos:* Res engr, USDA, 75- *Honors & Awards:* Teetor Award, Soc Automotive Engrs, 84. *Mem:* Inst Elec & Electronics Engrs. *Res:* Electronics; microprocessor applications; electronic applications to the wood products industry; laser applications; fiber optics; holography. *Mailing Add:* Dept Elec Eng Col Eng Mich Technol Univ Houghton MI 49931-1295

BRUMMER, JOHANNES J, b Graaff Reinet, SAfrica, Sept 2, 21; Can citizen; m 58; c 2. ECONOMIC GEOLOGY, EXPLORATION GEOLOGY. *Educ:* Univ Witwatersrand, c, 44 & 45, MSc, 51; McGill Univ, PhD(geol), 55. *Prof Exp:* Mine surveyor, E Geduld Mines Ltd, SAfrica, 45-47; mine geologist, Roan Antelope Copper Mines Ltd, Zambia, 47-49; chief geologist, Mufulira Copper Mines Ltd, 49-51 & Rhodesian Selection Trust Serv Ltd, 51-53; sr geologist, Kennco Explor Ltd, Que, 55-57, Vancouver, 57-58 & Toronto, 58-61; resident geologist, Cent Can, Falconbridge Nickel Mines Ltd, 61-68, explor mgr, Cent Div, 68-70; vpres explor, Occidental Minerals Corp Can, 70-72, explor mgr, Minerals Div, Can Occidental Petrol Ltd, Toronto, Ont, 72-83; CONSULT GEOLOGIST, 84- *Honors & Awards:* Duncan R Derry Medal, Geol Asn Can, 84. *Mem:* Soc Econ Geologists; Geol Soc Am; Can Inst Mining & Metall; Geol Asn Can; Geol Soc SAfrica; Can Soc Petrol Geologists. *Res:* Exploration for economic deposits of nickel in ultramafic rocks; sedimentary copper and porphyry copper deposits; volcanic sulfide and pyrometasomatic copper deposits; unconformity controlled (Saskatchewan-type) uranium deposits; gold deposits associated with Precambrian volcanic belts; platinum group elements in layered mafic complexes. *Mailing Add:* Six Wilgar Rd Toronto ON M8X 1J4 Can

BRUMMETT, ANNA RUTH, embryology, cytology; deceased, see previous edition for last biography

BRUMMETT, ROBERT E, b Concordia, Kans, Feb 11, 34; m 54; c 4. PHARMACOLOGY, OTOLARYNGOLOGY. *Educ:* Ore State Univ, BS, 59, MS, 60; Univ Ore, PhD, 64. *Prof Exp:* Asst prof pharmacog & pharmacol, Sch Pharm, Ore State Univ, 61-62; from asst to assoc prof otolaryngol, 64-80, assoc prof pharmacol, 64-80, PROF OTOLARYNGOL & PHARMACOL, SCH MED, UNIV ORE, 81- *Mem:* AAAS; Soc Exp Biol & Med; Am Acad Otolaryngol; Soc Toxicol; Soc Neurosci; Sigma Xi. *Res:* Effect of drugs on hearing, ototoxicity; cochlear function; clinical pharmacology of drugs used in otolaryngology. *Mailing Add:* 545 N Hayden Bay Dr Portland OR 97217

BRUMMOND, DEWEY OTTO, b Towner, NDak, July 28, 25; m 54; c 2. BIOCHEMISTRY. *Educ:* NDak State Univ, BS, 50; Univ Wis, MS, 52, PhD(biochem), 54. *Prof Exp:* Fel, Nat Found Infantile Paralysis, 54-56; prin scientist, Radioisotope Serv, Vet Admin Hosp, Cleveland, Ohio, 56-64; assoc prof biochem, NDak State Univ, 65-66; PROF CHEM, MOORHEAD STATE UNIV, 66- *Concurrent Pos:* Sr instr, Western Reserve Univ, 56-64. *Mem:* Am Soc Biol Chemists; Sigma Xi. *Res:* Enzymatic synthesis of cellulose. *Mailing Add:* Dept Chem Moorhead State Univ Moorhead MN 56560

BRUMUND, WILLIAM FRANK, b Joliet, Ill, June 12, 42; m 64; c 2. CIVIL ENGINEERING, SOIL MECHANICS. *Educ:* Purdue Univ, BSCE, 64, MSCE, 65, PhD(civil eng, soil mech), 69. *Prof Exp:* Asst prof civil eng, Ga Inst Technol, 69-74, assoc prof, 74-75; PRES, GOLDEN ASSOCS INC, 74- *Mem:* Am Soc Civil Engrs. *Res:* Soil dynamics of granular systems. *Mailing Add:* 2809 Woodland Park Dr NE Atlanta GA 30345

BRUN, MILIVOJ KONSTANTIN, b Novi Sad, Yugoslavia, Jan 8, 48; US citzen; m 80; c 2. CERAMICS ENGINEERING. *Educ:* Pa State Univ, BSc, 70, PhD(ceramic sci), 74. *Prof Exp:* Asst inorg chem, Tech Univ Norway, Trondheim, 76-77; res assoc ceramics, Mat Res Lab, Pa State Univ, 79; STAFF SCIENTIST CERAMICS, RES & DEVELOP, GEN ELEC CORP, 79- *Mem:* Am Ceramic Soc; AAAS. *Res:* Processing of ceramics; hot isostatic pressing; ceramic composites. *Mailing Add:* 4001 Jockey St Ballston Lake NY 12019

BRUN, WILLIAM ALEXANDER, b NJ, Sept 10, 25; m 50; c 3. CROP PHYSIOLOGY. *Educ:* Univ Miami, BS, 50; Univ Ill, MS, 51, PhD(bot), 54. *Prof Exp:* Asst bot & plant physiol, Univ Ill, 50-54; asst prof bot, NC State Col, 54-57; plant physiologist, Agr Res Serv, USDA, 57-59 & United Fruit Co, 59-65; from asst prof to assoc prof, Univ Minn, St Paul, 65-75, prof agron & plant genetics, 76-; RETIRED. *Mem:* Am Soc Plant Physiol; Scand Soc Plant Physiol. *Res:* Growth and reproduction; photosynthesis; water relations; nitrogen fixation. *Mailing Add:* 10117 SW 36th Pl Portland OR 97219

BRUNDA, MICHAEL J, b Passaic, NJ, Dec 16, 50; m 79; c 2. CYTOKINES, MACROPHAGES. *Educ:* Univ Rochester, AB, 71; Stanford Univ, PhD(med microbiol), 75. *Prof Exp:* Postdoctoral fel, Nat Jewish Hosp & Res Ctr, 75-78; immunologist, Nat Cancer Inst, 78-81; sr scientist, 82-86, res investr, 86-91, RES LEADER, DEPT ONCOL, HOFFMAN LA-ROCHE, INC, 91- *Mem:* Am Asn Immunologists; Am Asn Cancer Res; Soc Leukocyte Biol; AAAS. *Res:* Antitumor effects of interferon and other cytokines. *Mailing Add:* Dept Oncol Hoffmann-La Roche Inc Nutley NJ 07110

BRUNDAGE, ARTHUR LAIN, b Wallkill, NY, Dec 19, 27; m 51; c 4. DAIRY HUSBANDRY. *Educ:* Cornell Univ, BS, 50; Univ Minn, MS, 52, PhD, 55. *Prof Exp:* Prof animal sci, Univ Alaska, 68-85; RETIRED. *Mem:* AAAS; Am Dairy Sci Asn; Am Inst Biol Sci. *Mailing Add:* PO Box 616 Palmer AK 99645-0616

BRUNDAGE, DONALD KEITH, b Traverse City, Mich, Oct 24, 13; m 38; c 5. ORGANIC CHEMISTRY. *Educ:* Eastern Mich Univ, AB, 33; Univ Mich, PhD(phys org chem), 40. *Prof Exp:* Asst chem, Univ Mich, 35-38; chemist, Bendix Aviation Corp, Mich, 41; instr chem, St Cloud State Col, 41-46; from assoc prof to prof, 46-78, EMER PROF CHEM, UNIV TOLEDO, 78- *Concurrent Pos:* Res assoc, Univ Calif, Irvine, 67-68. *Mem:* Fel AAAS; Am Chem Soc; fel Am Inst Chemists. *Res:* Equilibrium; kinetics. *Mailing Add:* 2532 Glenwood Ave Toledo OH 43610-1327

BRUNDAGE, WILLIAM GREGORY, medical microbiology, immunochemistry, for more information see previous edition

BRUNDEN, KURT RUSSELL, b East Lansing, Mich, Mar 19, 58; m 88. MYELINATION, GLIAL RESEARCH. *Educ:* Western Mich Univ, BS, 80; Purdue Univ, PhD(biochem), 85. *Prof Exp:* Postdoctoral fel neurosci, Dept Neurol & Biochem, Mayo Clin, 85-87, res assoc & instr, 88; asst prof biochem, Univ Miss Med Ctr, 88-91; ASST PROF NEUROSCI, DEPT NEUROSCI, CASE WESTERN RESERVE UNIV, 91-; HEAD BIOMOLECULAR RES, GLIATECH, INC, 91- *Mem:* Am Soc Cell Biol; Am Soc Neurochem. *Res:* Biomolecular research; development of proprietary products based on glial cell properties; axonal regulation of myelin protein expression. *Mailing Add:* Gliatech Inc 23420 Commerce Park Rd Beachwood OH 44122

BRUNDEN, MARSHALL NILS, b Coloma, Mich, May 5, 34; m 57; c 3. BIO-STATISTICS. *Educ:* Mich State Univ, BS, 59; Iowa State Univ, MS, 62; Western Mich Univ, MBA, 71; Univ Mich, PhD(biostatist), 77. *Prof Exp:* Appl math group leader res & develop, Great Northern Paper Co, 63-67; SR BIOSTATISTICAL SCIENTIST, UPJOHN LABS, 67- *Mem:* Am Statist Asn; Biomet Soc; fel Royal Statist Soc. *Res:* Application of statistics to toxicity and teratology studies; distribution free statistical methods including rank sum statistics and randomization tests; biomathematical modeling. *Mailing Add:* Upjohn Co Henrietta St Kalamazoo MI 49001

BRUNDIDGE, KENNETH CLOUD, b St Louis, Mo, May 31, 27; m 47; c 2. METEOROLOGY. *Educ:* Univ Chicago, BA, 52, MS, 53; Tex A&M Univ, PhD(meteorol), 61. *Prof Exp:* Lab instr & res asst cloud physics, Chicago Midway Labs, 52, asst meteorologist, 54; meteorologist jet stream res, Col Geosci, Tex A&M Univ, 55, from instr to assoc prof, 55-68, asst dean student affairs, Col Geosci, 71-73, asst dean acad affairs, 73-75, head dept, 75-80, prof meteorol, 68-90; RETIRED. *Mem:* Am Meteorol Soc; Sigma Xi. *Res:* Dynamic and synoptic meteorology; mesoscale circulations; numerical simulation. *Mailing Add:* Dept Meteorol Tex A&M Univ College Station TX 77843

BRUNE, JAMES N, b Modesto, Calif, Nov 23, 34; m 57; c 4. SEISMOLOGY, GEOPHYSICS. *Educ:* Univ Nev, BS, 56; Columbia Univ, PhD(seismol), 61. *Prof Exp:* Res scientist, Lamont Geol Observ, Columbia Univ, 58-64; geophysicist, US Coast & Geod Surv, 64; assoc prof geophys, Calif Inst Technol, 65-69; assoc dir, Inst Geophys & Planetary Physics, 71-76, chmn, Geol Res Div, 74-76, PROF GEOPHYS, SCRIPPS INST OCEANOG, UNIV CALIF, SAN DIEGO, 69- *Concurrent Pos:* Adj assoc prof geol, Columbia Univ, 64. *Honors & Awards:* Macelwane Award, Am Geophys Union, 62, Karl Gilbert Award, 67. *Mem:* Fel Geol Soc Am; fel Am Geophys Union; Seismol Soc Am (pres, 70). *Res:* Earthquake hazard; seismology; earth structure; earth quake source mechanism; heat flow; geology. *Mailing Add:* Dept Geol Univ Nev Reno NV 89557

BRUNEAU, LESLIE HERBERT, b Cornwall, Ont, Nov 28, 28; nat US; m 53; c 2. GENETICS. *Educ:* McGill Univ, BSc, 50; Univ Tex, MA, 52, PhD(cytogenetics), 56. *Prof Exp:* From asst prof to assoc prof, 55-66, chmn biol sci, 60-74, PROF ZOOL, OKLA STATE UNIV, 66- *Mailing Add:* Dept Zool Okla State Univ Stillwater OK 74078

BRUNEL, PIERRE, b Montreal, Que, Mar 21, 31; m 55; c 4. MARINE ECOLOGY. *Educ:* Univ Montreal, BSc, 53; Univ Toronto, MA, 57; McGill Univ, PhD, 68. *Prof Exp:* Asst invertebrates, Royal Ont Mus Zool, 53-55; zoologist, Marine Biol Sta Grande-Riviere, 55-66; sr lectr, 66-68, from asst prof to assoc prof, 68-79, PROF BIOL SCI, UNIV MONTREAL, 79- *Concurrent Pos:* Vpres, Que Interuniv Oceanog Res Group, 70-88; pres, Mus Coord Comt, Dept Biol Sci, Univ Montreal, 83-; mem, In house & liaison comts, Joint MA Prog Museology, Univ Montreal & Univ Que, Montreal, 86- *Mem:* Am Inst Biol Sci; Can Soc Zool; Soc Syst Zool; Sigma Xi; Sci Res Soc; Soc Preserv Natural Hist Collections; Int Oceanog Found. *Res:* Ecology of marine bottom invertebrates and communities of northern seas; carcinology; taxonomy; zoogeography, ecology of Peracarid Crustacea, Amphipoda. *Mailing Add:* Dept Biol Sci Univ Montreal Box 6128 Montreal PQ H3C 3J7 Can

BRUNELL, GLORIA FLORETTE, b Meriden, Conn, Oct 6, 25. MATHEMATICS. *Educ:* Albertus Magnus Col, BA, 47; Fordham Univ, MA, 57; Yale Univ, PhD(math), 64. *Prof Exp:* Teacher, Holy Trinity High Sch, Ohio, 49-50, St Mary's Acad, 50-51, Holy Trinity High Sch, 51-55, Dominican Acad, NY, 55-57 & Watterson High Sch, Ohio, 57-58; from instr to assoc prof math, Albertus Magnus Col, 58-68; assoc prof, 68-71, prof math, 71-77, dean, Sch Arts & Sci, 77-81, PROF MATH, WESTERN CONN STATE COL, 81- *Concurrent Pos:* Vis lectr, Col St Mary of the Springs, 50-55. *Mem:* Am Math Soc; Math Asn Am. *Res:* Analysis; in service work with mathematics teachers. *Mailing Add:* Eight Howe Rd New Milford CT 06776

BRUNELL, KARL, b Cologne, Ger, Oct 7, 22; US citizen; m 47; c 2. MECHANICAL ENGINEERING. *Educ:* City Col NY, BME, 58. *Prof Exp:* Plant engr graphic arts, A A Watts Co, Inc, 51-54; sr proj engr, Mergenthaler Linotype Co, 54-59; asst to dir res, Bell & Howell Co, Ill, 59-61; proj consult optics, Mech Res Div, Am Mach & Foundry Co, 61-62; mgr res & develop mech eng, Universal Box Mach Corp, 62-64; sr res assoc eng sci, Princeton Lab, Am Can Co, 64-69; SR TECH ADV, JOHN DUSENBERY CO, CLIFTON, 69- *Mem:* Am Soc Mech Engrs. *Res:* Cams; optics; graphic arts; high speed machines; engineering management. *Mailing Add:* 25 Scarsdale Dr Livingston NJ 07039

BRUNELL, PHILIP ALFRED, b New York, NY, Feb 1, 31; m 52; c 3. PEDIATRICS, VIROLOGY. *Educ:* City Col New York, BS, 50; Univ Ill, MS, 52; Univ Buffalo, MD, 57. *Prof Exp:* Asst physiol, Univ Ill, 51-53; intern pediat, Meyer Mem Hosp, Buffalo, NY, 57-58; resident, Children's Hosp, Buffalo, 58-60; med off in chg virus reference unit, Nat Commun Dis Ctr, 63-64; from asst prof to assoc prof pediat, Sch Med, NY Univ, 69-75; prof pediat & chmn dept, Univ Tex Health Sci Ctr, San Antonio, 75-81; prof pediat & head pediat infectious dis, 81-87, DIR INFECTIOUS DIS & ASSOC DIR DEPT PEDIAT, CEDARS SINAI MED CTR, 87-; PROF PEDIAT, UNIV CALIF, LOS ANGELES, 87- *Concurrent Pos:* Mem adv comt, Human Resources Admin, Head Start, New York, 64-; res grants, NIH, 64-, Nat Commun Dis Ctr, 68- & WHO, 69-78 & 80-; consult, Am Acad Pediat, 66-, US Mil Acad, 69- & Northwick Pk Hosp, Harrow, Eng, 71-72; vis scientist, Clin Res Ctr, Harrow, 71-72; dir pediat, Bexar County Hosp, San Antonio, 75-81. *Mem:* Am Acad Pediat; Sigma Xi; Soc Pediat Res; Infectious Dis Soc Am; Am Pediat Soc; Am Soc Microbiol. *Res:* Clinical virology; infectious diseases of infants and children; varicellazoster virus; passive immunization. *Mailing Add:* Dept Pediat Cedars Sinai Med Ctr 8700 Beverly Blvd Los Angeles CA 90048

BRUNELLE, EUGENE JOHN, JR, b Montpelier, Vt, Mar 17, 32; m 55; c 3. AERONAUTICS. *Educ:* Univ Mich, BSE, 54, MSE, 55; Mass Inst Technol, ScD(aerothermoelasticity), 62. *Prof Exp:* Asst prof aeronaut, Princeton Univ, 60-64; ASSOC PROF AERONAUT, RENSSELAER POLYTECH INST, 64- *Concurrent Pos:* Vis prof mech, USAF Inst Technol, 83-85. *Mem:* Am Inst Aeronaut & Astronaut; Am Soc Mech Engrs; Am Acad Mech; Soc Indust & Appl Math; Mathematical Asn Am. *Res:* Solid mechanics; effects of transverse isotropy on beams and plates; non-linear beam and plate theory; theory of joined shells; higher order beam, plate, and shell theories; initially stressed and deformed solids; aeroelastic effects of composite material structures; mechanics of composite plates and shells; fluid flow in affine spaces; computational productivity. *Mailing Add:* Dept ME/AE/Mech Rensselaer Polytech Inst Troy NY 12180-3590

BRUNELLE, PAUL-EDOUARD, b Sherbrooke, Que, May 27, 36; m 55; c 2. HYDRAULICS. *Educ:* Univ Montreal, BASc, 58; Laval Univ, MASc, 63; Univ Toulouse, CES, 65, Dr Ing(hydraul), 68. *Prof Exp:* Lectr civil eng, 58-62, from asst prof to assoc prof, 62-76, head dept hydraul, 65-70, PROF CIVIL ENG & CHIEF HYDRAUL, SECT FLUID MECH, UNIV SHERBROOKE, 76-, CHMN DEPT CIVIL ENG, 68- *Concurrent Pos:* Fels, Nat Res Coun, 64, Asn Stages Tech France, 65 & Defence Res Bd, 67-69; consult, Rajasthan Power Proj, 65-66, Pickering Hydro, Ont, 66-67, Karachi Power Proj, 66-67, Hydro-Que, 67-70 & Sherbrooke Hydro Syst, 68-69; partic, Int Asn Hydraul Res Conf, Ft Collins, 67 & Kyoto, 69. *Mem:* Eng Inst Can; Int Asn Hydraul Res. *Res:* Stability of surge tanks and hydroelectric systems; optimal control and use of hydraulic resources; analog and numerical simulation of power systems. *Mailing Add:* Dept Civil Eng Univ Sherbrooke 2500 Univ Blvd Sherbrooke PQ J1K 2R1 Can

BRUNELLE, RICHARD LEON, b Littleton, Mass, May 9, 37; m 58; c 2. FORENSIC SCIENCE, INK DATING CHEMISTRY. *Educ:* Clark Univ, BA, 60; George Washington Univ, MS, 72. *Prof Exp:* Res chemist, Worcester Found Exp Biol, 60-61; anal chemist, US Food & Drug Admin, 61-63; forensic & regulatory chemist, Bur Alcohol, Tobacco & Firearms, US Dept Treas, 63-74, chief, Identification Lab, 74-78, Forensic Sci Br, 78-85, dep dir, Lab Serv, 85-87, dir, Forensic Sci Lab, 87-88; OWNER, BRUNELLE FORENSIC LAB, 88- *Concurrent Pos:* Mem fac, George Washington Univ, 74-82, Antioch Sch Law, 78-82 & Northeastern Univ; ed, J Forensic Sci. *Honors & Awards:* John A Dondero Mem Award, Int Asn Identification, 71. *Mem:* Int Asn Identification (pres, 81); fel Asn Off Anal Chemists; Fel Am Acad Forensic Sci. *Res:* Forensic science: methods for analysis and dating documents, ink and paper. *Mailing Add:* 3820 Acosta Rd Fairfax VA 22031

BRUNELLE, THOMAS E, b Crookston, Minn, Feb 12, 35; m 58; c 3. BIOCHEMISTRY, ORGANIC CHEMISTRY. *Educ:* Col St Thomas, BS, 57; Univ Minn, MS, 62, PhD(biochem), 68. *Prof Exp:* Chemist, 57-60, group leader org chem, 60-64, asst mgr org & biol res, 67-69, MGR CORP RES SECT, RES & DEVELOP DEPT, ECON LAB, INC, 69-, VPRES CORP SCI & TECHNOL, 75- *Mem:* AAAS; Am Chem Soc; Am Oil Chemists' Soc. *Res:* Acid polysaccharide chemistry; surface active agents. *Mailing Add:* 1252 Ohio St West St Paul MN 55118

BRUNENGRABER, HENRI, b Apr 2, 1940. METABOLISM. *Educ:* Univ Brussels, Belg, MD, 68, PhD(metab), 76. *Prof Exp:* Prof physiol chem, Univ Montreal, 84-90; PROF & CHMN DEPT NUTRIT, CASE WESTERN RES UNIV, CLEVELAND, OHIO, 90- *Mem:* Am Soc Biochem & Molecular Biol; Biochem Soc; Am Diabetes Asn. *Res:* Design and testing of artificial nutrients; cardiac metabolism; metabolic regulation; ketone body metabolism. *Mailing Add:* Div Nutrit Mt Sinai Med Ctr One Mt Sinai Dr Cleveland OH 44106-4198

BRUNER, BARBARA STEPHENSON, b Pittsburgh, Pa, Dec 10, 31; m 60; c 2. PEDIATRICS. *Educ:* Chatham Col, BS, 52; Emory Univ, MD, 56. *Prof Exp:* Instr, 59-61, assoc, 61-69, asst prof, 69-74, ASSOC PROF PEDIAT, SCH MED, EMORY UNIV, 74- *Concurrent Pos:* Asst dir pediat, Ambulatory Care, Grady Mem Hosp, 68-84, asst dir, Poison Control Ctr, 68-, dir, Pediat Emergency Clin, 85- *Mem:* Am Acad Pediat; AMA; Ambulatory Pediat Asn. *Mailing Add:* Dept Pediat Emory Univ Sch Med Woodruff MC Adm Bldg Atlanta GA 30322

BRUNER, DORSEY WILLIAM, b Windber, Pa, Dec 25, 06; m 40. MICROBIOLOGY. *Educ:* Albright Col, BS, 29; Cornell Univ, PhD(bact), 33, DVM, 37. *Prof Exp:* Teacher high sch, Pa, 29-30; instr res, Dept Path & Bact, State Univ NY Vet Col, Cornell Univ, 31-37; asst bacteriologist, Dept Animal Path, Exp Sta, Univ Ky, 41-42, virologist, 46-48, bacteriologist, 48-49; vet bacteriologist, dept path & bact, 49-65, vet microbiologist & chmn, dept vet microbiol, 65-72, EMER PROF VET MICROBIOL, STATE UNIV NY VET COL, CORNELL UNIV, 72- *Honors & Awards:* 12th Int Vet Cong Prize,

Am Vet Med Asn, 72. *Mem:* Soc Exp Biol & Med; Am Soc Microbiol; Am Vet Med Asn. *Res:* Genus Salmonella; equine virus abortion; neonatal isoerythrolysis; acidfast bacteria. *Mailing Add:* 40 Horizon Dr Ithaca NY 14850

BRUNER, HARRY DAVIS, b Jeffersonville, Ind, July 18, 11; m 31; c 2. RADIOBIOLOGY, RESEARCH ADMINISTRATION. *Educ:* Univ Louisville, SB, 30, MD, 34, SM, 36; Univ Chicago, PhD(physiol), 39. *Prof Exp:* Asst, Univ Louisville, 33-34, asst physiol & pharmacol, 34-36; instr physiol, Med Col SC, 36-38; asst prof, Univ NC, 39-42; vis fel, Harrison Dept Surg Res, Univ Pa, 42-43, res assoc, 43-45, asst prof, 45-47; prof pharmacol, Univ NC, 47-49; chief scientist, Med Div, Oak Ridge Inst Nuclear Studies, 49-52; prof physiol, Emory Univ, 52-56; chief med res br, Div Biol & Med, AEC, 56-60, asst dir health & med res, 60-69, asst dir, 69-72, spec asst to chmn, 72-75; spec asst to asst secy for Oceans, Int Environ & Sci Affairs, Dept of State, 75; med sci adv, US Energy Res & Develop Admin, Dept Energy, 75-78, Consult, Div Biomed & Environ Res, 78-; RETIRED. *Concurrent Pos:* Tech adv, US deleg, UN Sci Comt Effects Atomic Radiations. *Mem:* Am Soc Pharmacol & Exp Therapeut; Soc Exp Biol & Med; Radiation Res Soc; Health Physics Soc. *Res:* Hematology; pulmonary physiology; isotopes. *Mailing Add:* 27143 Flossmoor Dr SE Bonita Springs FL 33923

BRUNER, LEON JAMES, b Ponca City, Okla, Dec 28, 31; m 60; c 3. BIOPHYSICS. *Educ:* Univ Chicago, AB, 52, MS, 55, PhD(physics), 59. *Prof Exp:* Staff mem physics, Watson Sci Comput Lab, Int Bus Mach Corp, 59-62; from asst prof to assoc prof, 62-73, PROF PHYSICS, UNIV CALIF, RIVERSIDE, 73- *Mem:* AAAS; Biophys Soc; Am Phys Soc. *Res:* Bioelectric phenomena; structure and transport properties of membranes. *Mailing Add:* Dept Physics Univ Calif Riverside CA 92521

BRUNER, LEONARD BRETZ, JR, b Wichita Falls, Tex, Oct 12, 21; m 47; c 1. CHEMISTRY. *Educ:* Univ Tulsa, ChB, 43; Univ Mich, MS, 49, PhD(chem), 57. *Prof Exp:* Jr chemist, Phillips Petrol Co, 43-44; instr chem, Univ Tulsa, 46-47; res chemist, Dow Corning Corp, 55-59; sr scientist, Trionics Corp, 59-61; assoc prof, Eureka Col, 61-65; tech dir, Stauffer-Wacker Silicone Corp, 66-75, vpres & gen mgr, SWS Silicone Corp, 75-84; RETIRED. *Concurrent Pos:* Consult, Stauffer Chem Co, 63-65. *Mem:* Am Chem Soc. *Res:* Grignard reactions; organic azides; siloxane chemistry; polymerization; properties of siloxane polymers; electrical properties of ceramics. *Mailing Add:* 11233 Bemis Rd Manchester MI 48158

BRUNER, MARILYN E, b Oklahoma City, Okla, July 27, 34; div; c 3. SPACE PHYSICS. *Educ:* Univ Ariz, BS, 57, MS, 59; Univ Colo, PhD(physics), 64. *Prof Exp:* Phys sci aide, Naval Ord Test Sta, China Lake, Calif, 55; teaching asst, Univ Ariz, 57-58; physicist, US Naval Ord Test Sta, China Lake, 58-60; teaching asst, Lab Atmospheric & Space Physics, Univ Colo, 60-62, res asst, 62-64, res assoc, 64-76; MEM STAFF, RES LAB, LOCKHEED MISSILE & SPACE CO, 76- *Concurrent Pos:* Consult, Los Alamos Sci Lab, 75-76; prin investr, numerous NASA prog, 70- *Mem:* Am Astron Soc; Astron Soc Pac. *Res:* High resolution ultraviolet and x-ray spectroscopy of the sun; structure and dynamics of the solar chromosphere and transition zone; design of space instrumentation; absolute ultraviolet spectroradiometry. *Mailing Add:* Lockheed Orgn 91-30 Bldg 256 3251 Hanover St Palo Alto CA 94304

BRUNER, RALPH CLAYBURN, forensic science, process development, for more information see previous edition

BRUNETT, EMERY W, b Ovando, Mont, Dec 3, 27; m 60; c 4. MEDICINAL CHEMISTRY, PHARMACY. *Educ:* Univ Mont, BS, 53, MS, 56; Univ Wash, PhD(pharmaceut chem), 66. *Prof Exp:* Instr pharm, Univ Mont, 57-58, vis asst prof, 64-65; instr pharm, Univ Wash, 61-62; asst prof pharmaceut chem, Drake Univ, 66-69; assoc prof, 69-75, ASSOC PROF PHARM, UNIV WY, 75- *Mem:* AAAS; Am Asn Cols Pharm; fel Pharaceut Mfrs Asn. *Res:* Aminothiophene chemistry; vitamin A analogs; hay fever pollens; home remedies; microelements; tableting. *Mailing Add:* 315 S 11th St Laramie WY 82070

BRUNETTE, DONALD MAXWELL, b Hamilton, Ont, Aug 4, 44; m 71; c 1. CELL BIOLOGY. *Educ:* Univ Toronto, BSc, 66, MSc, 68, PhD(med biophys), 72. *Prof Exp:* Fel cell biol, York Univ, 72-74; asst prof, Fac Dent, 74-78, assoc prof cell biol, Med Res Coun Group, Univ Toronto, 78-79; assoc prof, 79-85, PROF & HEAD DEPT ORAL BIOL, FAC DENT, UNIV BC, 85- *Mem:* Int Asn Dent Res; Bromeliad Soc; Can Soc Cell Biol; Sigma Xi. *Res:* In vitro studies of cell function in epithelial and fibroblast-like cells derived from periodontal ligament; effects of surface topography on cell behavior in vitro & in vivo. *Mailing Add:* 4134 W Tenth Ave Vancouver BC V6R 2H3 Can

BRUNGRABER, ROBERT J, b Birmingham, Mich, Dec 20, 29; m 51; c 2. CIVIL ENGINEERING, STRUCTURAL ENGINEERING. *Educ:* Univ Mich, BSE, 51; Cornell Univ, MS, 56; Carnegie Inst Technol, PhD(civil eng), 63. *Prof Exp:* Field engr, Porter-Urquhart, Skidmore, Owings & Merrill, 51-53; instr civil eng, Cornell Univ, 53-56; res engr, Alcoa Res Labs, 56-60; res asst civil eng, Carnegie Inst Technol, 60-62; asst prof, Princeton Univ, 62-66; assoc prof, Union Col, NY, 66-68; prof & chmn dept, Bucknell Univ, 68-74; intergovt personnel act appointee, Nat Bur Standards, 74-76; PRESIDENTIAL PROF CIVIL ENG, BUCKNELL UNIV, 76-; PRES, SLIP-TEST, 76- *Concurrent Pos:* Dir & treas, Nat Inst Bldg Sci, 76-80. *Mem:* Am Soc Civil Engrs; Am Soc Test & Mat. *Res:* Welded aluminum structures; flow of low density detergent foam; shear strength of reinforced concrete structures; behavior of pile foundations; slip-resistance of walkway surfaces; reinforcement of steel and iron through truss bridges. *Mailing Add:* Dept Mech Eng Bucknell Univ Lewisburg PA 17837

BRUNGS, ROBERT ANTHONY, b Cincinnati, Ohio, July 7, 31. SOLID STATE PHYSICS. *Educ:* Bellarmine Col, NY, AB, 55; Loyola Sem, PhL, 56; St Louis Univ, PhD(physics), 62; Woodstock Col, Md, STL, 65. *Prof Exp:* Fel & res assoc physics, 66-70, asst prof physics, 70-75, asst prof dogmatic & syst theol, 72-75, assoc prof physics, 75-85, counr, Med Ctr, 70-73, ASSOC PROF THEOL STUDIES, ST LOUIS UNIV, 75-, ADJ PROF PHYSICS, 85- *Concurrent Pos:* Consult, US Cath Bishops Comt on Sci, Technol & Human Values, 85- *Mem:* Am Phys Soc; AAAS. *Res:* Experimental investigation of semiconductor properties of crystalline beta-rhombohedral boron. *Mailing Add:* St Louis Univ Dept Physics 221 N Grand Blvd St Louis MO 63103

BRUNGS, WILLIAM ALOYSIUS, b Covington, Ky, Aug 10, 32; m 62; c 2. POLLUTION BIOLOGY. *Educ:* Ohio State Univ, BSc, 58, MSc, 59, PhD(aquatic toxicol), 63. *Prof Exp:* Aquatic biologist, USPHS, 61-64; mem staff, Cincinnati Water Res lab, Fed Water Pollution Control Admin, Narragansett, 64-68, chief, Newtown Fish Toxicol lab, 68-71, asst dir water qual criteria, Environ Res Lab Duluth, 71-80, dep dir, 80-83, dir Environ Res lab, US Environ Protection Agency, 83-86; CONSULT, 87- *Concurrent Pos:* NSF fels, 59 & 61. *Mem:* Soc Environ Toxicol & Chem. *Res:* Distribution of radionuclides in fresh-water environments; determination of acute and chronic effects of water pollution on fish and the relationship of environmental variables on these effects. *Mailing Add:* 391 Yawgoo Valley Rd Slocum RI 02877

BRUNING, DONALD FRANCIS, b Boulder, Colo, Dec 18, 42; m 69; c 3. ORNITHOLOGY. *Educ:* Univ Colo, BA, 66, MA, 67, PhD(zool), 74. *Prof Exp:* Curatorial trainee ornith, 67-69, from asst cur to assoc cur, 69-74, CUR ORNITH, NEW YORK ZOOL SOC, 75-, CHMN, 87- *Concurrent Pos:* Res assoc, Ctr Field Biol & Conserv, New York Zool Soc, 73-; consult, Time/Life Bks, 73-; adj assoc prof, Fordham Univ, 74-; bd dirs, Am Asn Zool Parks & Aquaria, 82-85; tour leader, NY Zool Soc; chmn, Parrot Specialist Group, Int Comt Bird Preserv & Int Union Conserv Nature. *Mem:* AAAS; Am Ornithologists Union; Wilson Ornith Soc; Am Asn Zool Parks & Aquaria; Nat Audubon Soc; Nat Geog Soc. *Res:* Conducting captive and field research on birds of paradise and parrots while encouraging their conservation and establishing parks and reserves in Papua New Guinea and Indonesia. *Mailing Add:* New York Zool Soc Bronx NY 10460

BRUNINGS, KARL JOHN, organic chemistry, medicinal chemistry; deceased, see previous edition for last biography

BRUNJES, A(USTIN) S, b New York, NY, June 29, 06; m 35; c 2. CHEMICAL ENGINEERING. *Educ:* Polytech Inst Brooklyn, ChE, 29, MS, 31; Yale Univ, PhD(chem eng, distillation), 35. *Prof Exp:* Sr instr chem, Polytech Inst Brooklyn, 29-32, lectr chem micros, 31-32; jr engr, Lummus Co, NY, 35-37, sr process engr, 37-47, mgr, Tech Info Dept, 47-67; process consult, Stone & Webster Eng Corp, 67-78; RETIRED. *Concurrent Pos:* Adj prof chem eng, Polytech Inst Brooklyn, 38-58; consult, 78-82. *Mem:* Am Chem Soc; fel Am Inst Chem Engrs; Nat Soc Prof Eng. *Res:* Distillation; heat transfer; absorption; pilot plant development; solvent recovery; computer correlation of technical data. *Mailing Add:* Two Oakwood Lane Plandome NY 11030-1508

BRUNJES, PETER CRAWFORD, b Columbus, Ohio, June 19, 53; m 76; c 1. NEUROBIOLOGY. *Educ:* Mich State Univ, BS, 74; Ind Univ, PhD(psychol), 79. *Prof Exp:* Fel, Univ Ill, Champaign, 79-80; ASST PROF PSYCHOL, UNIV VA, 81- *Concurrent Pos:* Prin investr grants, NIH, 81- *Mem:* AAAS; Soc Neurosci; Int Soc Develop Psychobiol. *Res:* Investigations of the development of the brain and behavior; sensory development; neuroanatomical maturation; the role of experience during early life. *Mailing Add:* Dept Psychol 102 Gilmer Hall Univ Va Charlottesville VA 22903

BRUNK, CLIFFORD FRANKLIN, b Detroit, Mich, Feb 11, 40; m 62; c 3. BIOPHYSICS. *Educ:* Mich State Univ, BS, 61; Stanford Univ, MS, 62, PhD(biophys), 67. *Prof Exp:* Asst prof, 67-73, ASSOC PROF BIOL, UNIV CALIF, LOS ANGELES, 73- *Concurrent Pos:* USPHS fel, Carlsberg Biol Inst, Copenhagen, Denmark, 67-68; vis prof, Weismann Inst, 74-75; York Univ, Toronto, Can, 80-81; Univ Copenhagen, 86-87. *Mem:* Biophys Soc; Am Soc Cell Biol; Sigma Xi. *Res:* Genome reorganization in tetrahymena; DNA sequence elimination in somatic cell lines; molecular phylogenics of protists. *Mailing Add:* Dept Biol Univ Calif Los Angeles CA 90024-1606

BRUNK, HUGH DANIEL, b Manteca, Calif, Aug 22, 19; m 42; c 3. STATISTICS, MATHEMATICS. *Educ:* Univ Calif, AB, 40; Rice Inst, MA, 42, PhD(math), 44. *Prof Exp:* Asst math, Rice Inst, 40-44, from instr to asst prof, 46-51; mathematician, Sandia Corp, 51-52; from assoc prof to prof math, Univ Mo, 52-61; prof, Univ Calif, Riverside, 61-63; prof statist, Univ Mo, 63-69; PROF STATIST, ORE STATE UNIV, 69- *Concurrent Pos:* Fulbright lectr, Univ Copenhagen, 58-59; vis sr lectr, Univ Col Wales, 66-67; hon res assoc statist, Univ Col London, 71. *Mem:* Fel Am Statist Asn; Am Math Soc; Math Asn Am; fel Inst Math Statist; Int Statist Inst. *Res:* Analysis; mathematical statistics; probability. *Mailing Add:* 28087 Ewelty Way Corvallis OR 97330

BRUNK, WILLIAM EDWARD, b Cleveland, Ohio, Nov 24, 28; div; c 1. ASTRONOMY. *Educ:* Case Inst Technol, BS, 52, MS, 54, PhD(astron), 63. *Prof Exp:* Res scientist, Lewis Flight Propulsion Lab, Nat Adv Comt, Aeronaut, 54-58; aerospace res engr, Lewis Res Ctr, NASA, 58-64, staff scientist, 64-65, actg chief, 65, prog chief, Planetary Astron, 65-82, chief, Planetary Sci Br, NASA HQS, 82-85; PROG CHIEF, COMETARY SCI, UNIV SPACE RES ASN, 85- *Mem:* Am Astron Soc; Int Astron Union; fel AAAS. *Res:* Stellar and planetary astronomy; aerodynamics; heat transfer; trajectory analysis. *Mailing Add:* PO Box 3466 Annapolis MD 21403

BRUNKARD, KATHLEEN MARIE, b New Haven, Conn, June 21, 53; m 83; c 2. MEMBRANE BIOLOGY, ELECTROPHYSIOLOGY. *Educ:* Southern Conn State Col, BS, 77; Syracuse Univ, MS, 79; Univ Mass, PhD(plant & soil sci), 82. *Prof Exp:* Postdoctoral res assoc, Wash Univ, 82-84; ASSOC PROF

BIOL, EAST STROUDSBURG UNIV, 84- *Mem:* Am Soc Plant Physiologists; Bot Soc Am; Am Inst Biol Sci; Sigma Xi. *Res:* Effects of environmental factors on membranes and membrane transport systems; effects of environmental stress on plant growth and development. *Mailing Add:* Biol Sci Dept E Stroudsburg Univ East Stroudsburg PA 18301

BRUNKE, KAREN J, b Portland, Ore, Mar 17, 52; m 78; c 2. GENE REGULATION, PROMOTER RESEARCH. *Educ:* Univ Pa, PhD(microbiol), 80. *Prof Exp:* Postdoctoral fel molecular biol, Inst Cancer Res, Fox Chase, Pa, 80-83; sr scientist, Zeocon Corp, Palo Alto, CA, 83-85; sect leader, Sandoz Crop Protection Corp, Palo Alto, CA, 85-87, sr sect leader, 87-89, mgr, plant biotechnology, 89-91. *Concurrent Pos:* Lectr, molecular biol grad group, Univ Pa, 81-83. *Mem:* Plant Molecular Biol Asn; Am Soc Microbiol; AAAS; Am Soc Cell Biol. *Res:* Regulation of gene expression, including how promoters are involved in transcriptional regulation of gene expression in higher plants; expression of agronomically important genes in plants. *Mailing Add:* 975 California Ave Palo Alto CA 94304

BRUNNER, CARL ALAN, b Cincinnati, Ohio, Feb 21, 34; m 59; c 2. CHEMICAL ENGINEERING. *Educ:* Univ Cincinnati, ChemE, 57, MS, 59, PhD(chem eng), 63. *Prof Exp:* Instr chem eng, Univ Cincinnati, 58-59; chem eng, USPHS, 63-67; CHEM ENGR, US ENVIRON PROTECTION AGENCY, 67- *Mem:* Am Inst Chem Engrs; Water Pollution Control Fedn. *Res:* Wastewater treatment; wastewater reuse; storm and combined sewer discharges; hazardous wastes. *Mailing Add:* US Environ Protection Agency 26 W Martin Luther King Cincinnati OH 45268

BRUNNER, CHARLOTTE, b Helena, Mont, Oct 9, 48. MARINE MICROPALEONTOLOGY. *Educ:* Univ RI, BA, 70; Grad Sch Oceanog, PhD(oceanog), 78. *Prof Exp:* Res asst, Grad Sch Oceanog, Univ RI, 70-73; asst prof micropaleont, Univ Calif, Berkeley, 79-87; lectr, Calif State Univ, Hayward, 87-88; ASSOC PROF MARINE SCI, UNIV SOUTHERN MISS, 88- *Mem:* Geol Soc Am; Soc Econ Paleontologists & Mineralogists; AAAS. *Res:* Paleoceanography, taphonomy and marine stratigraphy of the neogene, based on quantitative analyses of microfossils, principally foraminifer. *Mailing Add:* Ctr Marine Sci Univ Southern Miss Stennis Space Center MS 39529

BRUNNER, EDWARD A, b Erie, Pa, July 18, 29; m 55; c 4. PHARMACOLOGY, ANESTHESIOLOGY. *Educ:* Villanova Univ, BS, 52; Hahnemann Med Col & Hosp, MD, 59, PhD(pharmacol), 62. *Prof Exp:* Instr pharmacol, Hahnemann Med Col & Hosp, 60-62, instr anesthesia, Sch Med, Univ Pa, 62-65, instr pharmacol, 64-65; from asst prof to assoc prof anesthesia, 66-71, PROF ANESTHESIA & CHMN DEPT, MED SCH, NORTHWESTERN UNIV, 71- *Mem:* Sigma Xi. *Res:* Carbohydrate metabolism in liver, brain and muscle; effects of anesthetic agents on metabolism; muscle relaxants; medical education. *Mailing Add:* Dept Anesthesia Med Sch Northwestern Univ 303 E Superior Ave Chicago IL 60611

BRUNNER, GORDON FRANCIS, b Des Plaines, Ill, Nov 6, 38; m 63; c 3. RESEARCH ADMINISTRATION. *Educ:* Univ Wis, BS, 61; Xavier Univ, MBA, 65. *Prof Exp:* Eng asst, Universal Oil Prods Co, 59-61; engr, Procter & Gamble Co, 61-71, assoc dir, 71-77, mgr prod coordr, Europe, 77-83, mgr res & develop, 83-85, vpres, 85-87, SR VPRES, PROCTER & GAMBLE CO, 87-, BD DIRS, 91- *Res:* Management of research and development of consumer goods and health care technologies; patentee in field. *Mailing Add:* Procter & Gamble Co One Procter & Gamble Plaza Cincinnati OH 45201

BRUNNER, JAY ROBERT, b Royersford, Pa, Sept 17, 18; m 47; c 3. FOOD SCIENCE, DAIRY CHEMISTRY. *Educ:* Pa State Univ, BS, 40; Univ Calif, MS, 42; Mich State Univ, PhD(dairy & phys chem), 52. *Prof Exp:* Asst, Univ Calif, 40-42; from instr to assoc prof, 46-59, PROF FOOD SCI & HUMAN NUTRIT, MICH STATE UNIV, 59- *Concurrent Pos:* Mem EBC study sect, USPHS; Sigma Xi sr res award, Mich State Univ, 79. *Honors & Awards:* Borden Award, Am Chem Soc, 64. *Mem:* Am Dairy Sci Asn; Inst Food Technol; Am Chem Soc; AAAS; Sigma Xi. *Res:* Physical and chemical properties of milkfat and milk proteins; dairy products processing; chemistry of milk. *Mailing Add:* Dept Food Sci & Human Nutrit 3940 Zimmer Rd Rte 2 Williamston MI 48895

BRUNNER, MATHIAS J, b Brooklyn, NY, May 28, 22; m 51; c 3. MECHANICAL, AEROSPACE & GAS TURBINE ENGINEERING. *Educ:* Pratt Inst, BME, 43; Univ Pa, MSME, 47. *Prof Exp:* Jr design engr, Steam Div, Westinghouse Elec Corp, Pa, 43-50, design engr, 50-54, test design engr, 54-56, fel engr, 56-57; specialist advan studies, Gen Elec Co, 57-68, consult engr, 68-79, sr aerothermodynamics engr, Re-entry Systs Div, 79-82; PRES & CONSULT ENGR, BRUNNER ASSOCS, INC, 82- *Concurrent Pos:* Lectr, exten, Pa State Univ, 48-50; mem, Nat Tech Comt Thermophysics, Am Inst Aeronaut & Astronaut. *Mem:* Am Soc Mech Engrs; assoc fel Am Inst Aeronaut & Astronaut; NY Acad Sci; Nat Soc Prof Engrs. *Res:* Fluid mechanics; internal aerodynamics; aerothermodynamics; heat transfer; aerothermophysics; space and re-entry vehicle research and development; axial flow compression research and development design for wind tunnel and gas turbine applications. *Mailing Add:* 1247 Berwyn Paoli Rd Berwyn PA 19312

BRUNNER, MICHAEL, b Brooklyn, NY, Apr 3, 43. MOLECULAR BIOLOGY. *Educ:* City Col New York, BS, 65; Pa State Univ, MS, 67, PhD(microbiol), 69. *Prof Exp:* Fel growth regulation, Sch Med, Washington Univ, 69-71; fel adenovirus, Inst Molecular Virol, 71-73; instr RNA tumor virus, Harvard Med Sch, 73-75; asst prof biol sci, St John's Univ, NY, 75-80; CONSULT, 80- *Concurrent Pos:* USPHS trainee grant, Sch Med, Washington Univ, 71-73; Leukemia Soc Am spec fel, 73. *Res:* Genetic interaction between leukemia-sarcoma viruses and the cells in which oncogenic changes are produced. *Mailing Add:* 84-29 153 Ave Howard Beach NY 11414

BRUNNER, ROBERT LEE, b Chicago, Ill, Sept 11, 45. NEUROPSYCHOLOGY, PHYSIOLOGICAL PSYCHOLOGY. *Educ:* Univ Calif, Los Angeles, BA, 67; Univ Cincinnati, MA, 69, PhD(psychol), 71. *Prof Exp:* Fel & res assoc, Dept Biol Sci, Purdue Univ, 71-74; res scholar, Inst Develop Res, 74-75; asst prof, 75-80, ASSOC PROF, COL MED, UNIV CINCINNATI & CHILDREN'S MED CTR, 80- *Concurrent Pos:* Prof consult, Cincinnati Ctr Develop Dis, 75- *Mem:* Am Psychol Asn; Int Neuropsychol Soc; Nat Acad Neuropsychologists. *Res:* Neuropsychological aspects of central nervous system related disorders in children, with emphasis on recovery over time; metabolic diseases; anorexia nervosa. *Mailing Add:* Dept Pediat Univ Cincinnati Col Med 231 Bethesda Ave Cincinnati OH 45267

BRUNNING, RICHARD DALE, b Grand Forks, NDak, Mar 5, 32. HEMATOLOGY, PATHOLOGY. *Educ:* Univ NDak, BSc, 57; McGill Univ, MD, 59. *Prof Exp:* Intern, Ancker Hosp, St Paul, Minn, 59-60; fel path, Univ Minn, Minneapolis, 60-62, fel lab med, 63-64; from instr to assoc prof lab med & assoc dir hemat labs, 65-74, PROF LAB MED & DIR HEMAT LABS, UNIV MINN, MINNEAPOLIS, 74- *Concurrent Pos:* Am Cancer Soc fel, 63-64. *Mem:* Am Soc Hemat; Am Asn Cancer Res. *Res:* Morphologic, ultrastructural and cytochemical characteristics of peripheral blood and bone marrow cells in inherited disorders; cytochemical and morphologic features of blast cells in leukemic disorders. *Mailing Add:* Dept Lab Med Univ Minn Health Sci Ctr Minneapolis MN 55455

BRUNO, CHARLES FRANK, b Westerly, RI, June 23, 36; m 58; c 6. BIOCHEMISTRY, BACTERIOLOGY. *Educ:* Roanoke Col, BS, 58; Va Polytech Inst, MS, 61, PhD(biochem), 63. *Prof Exp:* Sr res scientist microbiol & biochem, 63-71, process develop mgr fermentation, 71-75, dept head, 75-80, DIR FERMENTATION, SQUIBB INST MED RES, 80- *Mem:* Am Chem Soc; Am Soc Microbiol. *Res:* Fermentation, especially antibiotics from fungi and actinomyces; production and isolation of bacterial and fungal enzymes; production and purification of polysaccharide vaccine. *Mailing Add:* 11 Westwood Rd East Brunswick NJ 08816

BRUNO, DAVID JOSEPH, b Martins Ferry, Ohio, Sept 6, 51; m 88; c 2. PROCESS DEVELOPMENT, NEW TECHNOLOGIES FOR FOODS. *Educ:* Case Inst Technol, BS, 73; Univ Cincinnati, MS, 79. *Prof Exp:* Sect head shortening & oils, Foods Div, Procter & Gamble, 79-81, soy protein analog technol, 81-84, salted snacks, 84-85, new products Crisco, Crisco oil & Puritan, Edible Oils Div, 85-87, ASSOC DIR OLESTRA DIV, PROCTER & GAMBLE, 87- *Mem:* Am Inst Chem Engrs; Am Oil Chemist Soc; Inst Food Technologists. *Res:* Development of new technologies to satisfy consumer needs; soy analogs; shortening and oil products; savory snacks; fat substitutes such as olestra. *Mailing Add:* 6071 Center Hill Rd Cincinnati OH 45224

BRUNO, MERLE SANFORD, b Schenectady, NY, Jan 30, 39; m 82. SENSORY PHYSIOLOGY, SCIENCE EDUCATION. *Educ:* Syracuse Univ, BS, 60; Harvard Univ, MA, 63, PhD(biol), 71. *Prof Exp:* Staff developer, Educ Develop Ctr, 66-69; res assoc biol, Yale Univ, 71; asst prof, 71-77, dean natural sci, 85-89, ASSOC PROF BIOL, HAMPSHIRE COL, 77- *Concurrent Pos:* Consult, Workshop for Learning Things, 69-71; mem bd dirs, Yurt Found, 73-; NSF sch systs grant, 74-75; mem bd trustees, Hampshire Col, 80-82; mem adv bd, Mass Turning Points Proj, Proj Spark; proj dir, Partners Elem Sci, 87-90, Five Col Pub Sch Partnership Steering Comt, 87- *Mem:* Sigma Xi. *Res:* Comparative biochemistry of invertebrate visual systems; implementation of innovative science programs in elementary schools and middle schools; adult science education; recruiting women and minorities into science. *Mailing Add:* Sch Natural Sci Hampshire Col Amherst MA 01002

BRUNO, MICHAEL STEPHEN, b Newark, NJ, Apr 16, 58; m 88. COASTAL ENGINEERING, COASTAL WATER QUALITY. *Educ:* NJ Inst Technol, BS, 80; Univ Calif, Berkeley, MS, 81; Mass Inst Technol, PhD(civil & oceanog eng), 86. *Prof Exp:* Prin engr coastal eng, NJ Dept Environ Protection, 81-82; asst prof civil eng, NJ Inst Technol, 86-89; ASSOC PROF CIVIL & OCEANOG ENG & DIR, DAVIDSON LAB, STEVENS INST TECHNOL, 89- *Concurrent Pos:* Mem sci & tech adv comt, Del Estuary Prog, Environ Protection Agency, 88-89; young investr, Off Naval Res, 91; mem US deleg, NSF Workshop US-Arg Res Marine Sci, 91- *Mem:* Am Geophys Union; Am Soc Civil Eng; Am Soc Mech Eng; Oceanog Soc; Soc Naval Architects & Marine Eng. *Res:* Hydrodynamics, with emphasis on problems in coastal engineering, including the computer modeling of coastal circulation and pollution transport, wave transformation, sediment transport and beach erosion. *Mailing Add:* 254 Runnymeade Rd West Caldwell NJ 07006

BRUNO, RONALD C, b New York, NY, Nov 25, 46; m 77; c 2. PHYSICS, ENERGY TECHNOLOGY. *Educ:* City Col NY, BS, 67; Mass Inst Technol, PhD(physics), 72. *Prof Exp:* Asst prof physics, Southern Ill Univ, 72-77, assoc prof, 77-80; consult, 80-83; PROJ MGR, STANFORD TELECOM, 83- *Concurrent Pos:* Am Phys Soc Cong Sci fel, US Cong, 76-77. *Mem:* Am Phys Soc; AAAS; Inst Elec & Electronics Engrs. *Res:* Behavior of superconductors in high magnetic fields; applications of superconductivity; energy technology and policy; electrical and communications engineering. *Mailing Add:* 3203 N Fourth St Arlington VA 22201

BRUNO, STEPHEN FRANCIS, b New York, NY, Jan 1, 48; m 71; c 2. MARINE ECOLOGY, PLANT PHYSIOLOGY. *Educ:* St John's Univ, BS, 69; Fordham Univ, MS, 71, PhD(biol), 75. *Prof Exp:* Instr biol, Col Mt St Vincent, 73-75; asst res scientist marine sci, NY Ocean Sci Lab, Affil Col & Univ Inc, 75-80; CONSULT, 80- *Mem:* Am Soc Limnol & Oceanog; Phycol Soc Am; Sigma Xi. *Res:* Environmental regulation of phytoplankton productivity in estuaries and coastal waters; role of vitamins in marine phytoplankton ecology; nutritional physiology of marine phytoplankton in culture and in situ. *Mailing Add:* 36 Holland Lane Colts Neck NJ 07722-1632

BRUNO, THOMAS J, b New York, NY, Oct 12, 54; m 87. INSTRUMENT DESIGN, EQUATIONS OF STATE. *Educ:* Polytec Inst Brooklyn, BS, 76; Georgetown Univ, MS, 78, PhD (phys chem), 81. *Prof Exp:* Res assoc, 81-83, SCIENTIST, NAT BUR STANDARDS, 83-; ADJ PROF INSTRUMENTATION, COLO SCH MINES, 85- *Concurrent Pos:* Mem, Bd Dir, Mamie Doud Eisenhower Libr, 87- *Honors & Awards:* Eugene Holt Award, Eugene Holt Found, 76. *Mem:* Am Chem Soc; Am Inst Chem Engrs. *Res:* Experimental thermophysics of fluids, including phase equilibria and transport properties; methods in analytical chemistry, component design. *Mailing Add:* Thermophysics Div Nat Bur Standards Boulder CO 80303-3328

BRUNS, CHARLES ALAN, b Baltimore, Md, Nov 1, 30; m 53; c 3. PHYSICS. *Educ:* Tufts Univ, BS, 52; Johns Hopkins Univ, PhD (physics), 61. *Prof Exp:* From instr to asst prof physics, Univ Mich, 61-64; asst prof, 64-69, ASSOC PROF PHYSICS, FRANKLIN & MARSHALL COL, 69- *Mem:* Am Phys Soc; Optical Soc Am; Sigma Xi. *Res:* Holography; lasers. *Mailing Add:* Dept Physics Franklin & Marshall Col Lancaster PA 17604

BRUNS, DAVID EUGENE, b St Louis, Mo; c 2. CLINICAL CHEMISTRY. *Educ:* Wash Univ, St Louis, BS, 63 & AB, 65; St Louis Univ, MD, 73. *Prof Exp:* From asst prof to assoc prof, 77-90, PROF PATH, UNIV VA, 90- *Concurrent Pos:* Prin investr, Nat Inst Health, 78-81, Am Cancer Soc, 82-85 & Nat Dairy Coun, 88-90; chmn, lab utilization comt, Asn Am Clin Chem, 86-; vis prof, Wash Univ, 85-86; assoc dir clin chem, Univ Va Hosp, 77-; co-prin investr, NIH, 78-; ed, Clin Chem, 90. *Honors & Awards:* Roche Res Award, Am Asn Clin Chem, 87; Sunderman Award, Asn Clin Sci, 87. *Mem:* Am Asn Clin Chem; Asn Clin Scientists (pres, 80); Exec Coun, Acad Clin Lab Physicians & Scientists, 90- *Res:* Developmental aspects of mammalian calcium transport; new methods in clinical chemistry. *Mailing Add:* Dept Path Univ Va Med Ctr Charlottesville VA 22908

BRUNS, DONALD GENE, b Hebron, Nebr, Feb 9, 52; m 80. LASERS, ADAPTIVE OPTICS. *Educ:* Univ Colo, BA, 74; Univ Ill, MS, 75, PhD (physics), 78. *Prof Exp:* Sr staff physicist, Electro-Optical & Data Systs Group, Hughes Aircraft Co, 78-87; SR SCIENTIST, PHOTONICS DIV, KAMAN SCI CO, 87- *Res:* Optical systems design; lasers; FLIRs; adaptive optics; telescopes; high speed streak cameras; fiber optics; data communication systems; radiometry; photography. *Mailing Add:* 5195 Hearthstone Lane Colorado Springs CO 80919

BRUNS, HERBERT ARNOLD, b Richmond, Mo, Apr 6, 50. CROP PHYSIOLOGY, CROP ECOLOGY. *Educ:* Univ Mo, Columbia, BS, 72, MS, 75; Okla State Univ, PhD (crop sci), 81. *Prof Exp:* Agronomist, coop exten, Univ Mo, 75-78; res asst, crop physiol, Okla State Univ, 78-82; ASST PROF CROP PHYSIOL, UNIV MD, 82- *Mem:* Am Soc Agron; Crop Sci Soc Am. *Res:* Biochemistry and biophysics of crop growth and development; photosynthesis; respiration; carbon partitioning. *Mailing Add:* Dept Agron Univ Md College Park MD 20742

BRUNS, LESTER GEORGE, b Robertsville, Mo, Feb 11, 33; m 54; c 2. PHARMACY, BIOCHEMISTRY. *Educ:* St Louis Col Pharm, BS, 55, MS, 57; St Louis Univ, PhD, 72. *Prof Exp:* From instr to asst prof pharmaceut chem, St Louis Col Pharm, 57-62; grad fel, 62-65, from asst prof to assoc prof, 65-78, PROF BIOCHEM, ST LOUIS COL PHARM, 78- *Concurrent Pos:* Biopharmaceut chemist, Norcliff Thayer Co, St Louis, Mo, 77-89. *Mem:* Am Asn Cols Pharm; Nutrit Today Soc; Am Chem Soc. *Res:* Bile acid metabolism and nutritional biochemistry. *Mailing Add:* St Louis Col Pharm Dept Chem 4588 Parkview Pl St Louis MO 63110

BRUNS, PAUL DONALD, b Sioux City, Iowa, Oct 21, 14; m 43; c 7. OBSTETRICS & GYNECOLOGY. *Educ:* Trinity Col, Iowa, BS, 39; Univ Iowa, MS, 41. *Prof Exp:* Intern, Broadlawn Hosp, Des Moines, 41-42; resident, Univ Hosp, Johns Hopkins Univ, 45-49; from asst prof to prof, Med Ctr, Univ Colo, 49-67; prof obstet & gynec & chmn dept, Georgetown Univ Hosp, 67-80; PROF OBSTET & GYNEC, SCH MED, UNIV NDAK, 80- *Mem:* AMA; Am Col Obstet & Gynec; Am Fedn Clin Res. *Res:* Reproductive physiology and pathology; gynecological physiology and pathology. *Mailing Add:* Dept Obstet/Gynec Med Educ Ctr Univ NDak Med Sch 1919 N Elm Fargo ND 58102

BRUNS, PAUL ERIC, b New York, NY, July 30, 15; m 44; c 4. FORESTRY. *Educ:* NY Univ, AB, 37; Yale Univ, MF, 40; Univ Wash, PhD (forestry), 56. *Prof Exp:* Timber cruiser, Hamilton Veneer Co, SC, 41-42; instr aircraft engines & aerodyn, Southeastern Air Serv, 42-44, asst dir acads, 43 & dir, 44; assoc forester, Northeastern Forest Exp Sta, US Forest Serv, 44-45; woodlands mgr, N Troy Div, Blair Veneer Co, Vt, 45-47; assoc prof forestry, Univ Mont, 46-55; consult, 55-58; chmn dept, 59-68, prof, 59-80, EMER PROF FOREST RESOURCES, UNIV NH, 80- *Concurrent Pos:* Vis prof forest mgt, Univ Minn, 57-58; mem adv comt, Northeastern Forest Exp Sta, US Forest, 59-68, chmn 65-66; mem, Nat Coun Forestry Sch Exec, 58-68, chmn 64-65; consult, Glastenbury Timberlands, Bennington, Vt, 69-83; vis scientist remote sensing, Univ Ariz, 72. *Mem:* Soc Am Foresters; emer mem Am Soc Photogram. *Res:* Forest management; remote sensing. *Mailing Add:* 4930 Squires Dr Titusville FL 32796-1067

BRUNS, PETER JOHN, b Syracuse, NY, May 2, 42; m 67, 85; c 2. GENETICS, DEVELOPMENTAL BIOLOGY. *Educ:* Syracuse Univ, AB, 63; Univ Ill, PhD (cell biol), 69. *Prof Exp:* From asst prof to assoc prof genetics, chmn sect genetics & develop, 80-85, PROF GENETICS, CORNELL UNIV, 82- *Concurrent Pos:* Fel John Guggenheim Mem Found, 77-78; vis scientist, Biol Inst Carlsberg Found, Copenhagen, 77-78; mem, Biomed Sci Study Sect, NIH, 83-87; assoc dir, Cornell Biotech Prog, 84- & dir, Div Biol Sci, Cornell Univ, 87-; consult examr, Genetics & Biotech, Charter Oak Col, 86-88. *Mem:* AAAS; Genetics Soc Am; Soc Protozoologists. *Res:* Developmental genetics of tetrahymena thermophila; germinal and somatic genetic organization; genetic and molecular studies of conjugation. *Mailing Add:* Div Biol Sci Cornell Univ 169 Biotechnol Bldg Ithaca NY 14853

BRUNS, ROBERT FREDERICK, m; c 4. NEUROPHARMACOLOGY. *Educ:* Univ Calif, San Diego, PhD (neurosci), 78. *Prof Exp:* GROUP LEADER, WARNER-LAMBERT PARKE-DAVIS, 85- *Mem:* Am Soc Pharmacol Exp Therapeut; Soc Neuroscience. *Res:* Biochemical pharmacology. *Mailing Add:* Warner-Lambert Parke-Davis 2800 Plymouth Rd Ann Arbor MI 48105

BRUNS, ROY EDWARD, b Breese, Ill, Sept 10, 41; m 74; c 2. THEORETICAL CHEMISTRY. *Educ:* Southern Ill Univ, BA, 63; Okla State Univ, PhD (phys chem), 68. *Prof Exp:* Asst prof chem, Univ Fla, 68-71; PROF CHEM, UNIV CAMPINAS, BRAZIL, 71- *Mem:* Chemometrics Soc; Brazilian Chem Soc. *Res:* Classical and quantum mechanical investigations of the rotational vibrational spectral properties of molecules; pattern recognition and multivariate analysis of chemical data. *Mailing Add:* Chem Inst Univ Campinas Campinas SP 13100 Brazil

BRUNSCHWIG, BRUCE SAMUEL, b Philadelphia, Pa, July 29, 44; m 66. PHYSICAL INORGANIC CHEMISTRY. *Educ:* Univ Rochester, BA, 66; Polytech Inst NY, PhD (chem physics), 72. *Prof Exp:* Asst prof chem, Hofstra Univ, 72-79, chmn dept, 75-78; assoc chemist, 79-84, CHEMIST, BROOKHAVEN NAT LAB, 84- *Concurrent Pos:* Res collabr, Brookhaven Nat Labs, 74-77, vis chemist, 77-79. *Mem:* Am Chem Soc; AAAS; Sigma Xi. *Res:* Study of photochemical and thermal electron-transfer reactions of transition metals in inorganic complexes and metalloproteins; modelling of electron-transfer reactions; chemical kinetics. *Mailing Add:* Brookhaven Nat Lab Dept Chem Upton NY 11973

BRUNSER, OSCAR, b Buenos Aires, Arg, Nov 24, 35; Chilean citizen; m 64; c 3. PEDIATRIC GASTROENTEROLOGY. *Educ:* Univ Chile, Baccalaureate, 53, MD, 61; State Univ NY, Albany, MD, 75. *Prof Exp:* Instr Dept Pediat, Sch Med Univ Chile, 61-66, Internal Med, Sch Med, Univ Wash, Seattle, 67-69; asst prof pediat, Sch Med, 69-76, assoc prof, 76-78, PROF PEDIAT, INST NUTRIT FOOD TECHNOL, UNIV CHILE, 78- *Concurrent Pos:* John Simon Guggenheim Mem Found fel, 65 & 67; fel anat, Dept Biol Struct, Univ Wash, Seattle, 66-69; Williams Waterman Fund Res Corp fel, 68-69; asst prof pediat, Albert Einstein Sch Med, Bronx, 73-76. *Mem:* Am Inst Nutrit; Am Soc Clin Nutrit; Soc Res Int Nutrit; Am Acad Pediat. *Res:* Reaction of the intestine of children living in a less developed country to food deprivation and the microbiological contamination of the environment, developing chronic environmental enteropathy, malabsorption of nutrients and diarrhea. *Mailing Add:* Inst Nutrit & Food Technol Univ Chile Casilla 138-11 Santiago 11 Chile

BRUNSKI, JOHN BEYER, b Philadelphia, Pa, June 10, 49. BIOMATERIALS. *Educ:* Univ Pa, BS, 70, PhD (metall, mat sci), 77; Stanford Univ, MS, 72. *Prof Exp:* ASSOC PROF, CTR BIOMED ENG, RENSSELAER POLYTECH INST, 77- *Concurrent Pos:* Sr investr, Rehab Serv Admin, NY State Rehab Hosp, NY, 77-81; Nat Inst Dent Res-NIH grant, Rensselaer Polytech Inst, 79-86, Vet Admin grant, 84-89. *Mem:* Sigma Xi; Soc Biomat; Int Asn Dent Res; Am Soc Mech Engrs. *Res:* Dental implants, especially engineering aspects of their design and use; physiology and mechanical properties of bone. *Mailing Add:* Jonsson Eng Ctr Rensselaer Polytech Inst Rm 7040 Troy NY 12181

BRUNSKILL, GREGG JOHN, b Kansas City, Kans, Sept 10, 41; m 70; c 2. LIMNOLOGY, ENVIRONMENTAL SCIENCES. *Educ:* Augustana Col, SDak, 63; Cornell Univ, PhD (biogeochem), 68. *Prof Exp:* RES SCIENTIST, FRESHWATER INST, ENVIRON CAN, FISHERIES & OCEANS CAN, 67- *Concurrent Pos:* Adj prof earth sci, Univ Man, 71-74 & 80-86; vis scientist, Lamont-Doherty Geol Observ, Columbia Univ, 77. *Mem:* AAAS; Geochem Soc; Am Soc Limnol & Oceanog; Int Soc Limnol; Int Asn Gt Lakes Res; Arctic Inst N Am. *Res:* Aqueous and recent sediment geochemistry and radiochemistry of fresh and estuarine waters; lakes; aquatic ecology of arctic and subarctic ecosystems. *Mailing Add:* Freshwater Inst 501 University Crescent Winnipeg MB R3T 2N6 Can

BRUNSON, CLAYTON (CODY), b Colfax, La, Nov 3, 21; m 45; c 2. POULTRY HUSBANDRY. *Educ:* La State Univ, BS, 49; Okla Agr & Mech Col, MS, 51, PhD (animal breeding), 55. *Prof Exp:* Instr poultry husb, Okla Agr & Mech Col, 50-52; from asst prof to prof, La State Univ, 53-63; VPRES POULTRY RES, CAMPBELL INST AGR RES, CAMPBELL SOUP CO, 63- *Mem:* Poultry Sci Asn. *Res:* Poultry breeding. *Mailing Add:* 1732 Carolyn Dr Fayetteville AR 72701

BRUNSON, JOHN TAYLOR, b Kalamazoo, Mich, June 4, 40; m 71; c 2. PHYSIOLOGY. *Educ:* Hope Col, AB, 62; Univ Mich, Ann Arbor, MS, 64; Colo State Univ, PhD (zool), 73. *Prof Exp:* Instr biol & chem, State Univ NY Col Oneonta, 65-68; asst prof, 71-80, ASSOC PROF ZOOL, STATE UNIV NY COL OSWEGO, 71- *Mem:* Am Soc Zoologists; AAAS. *Res:* Physiological ecology, particularly respiratory physiology at low oxygen tension or high altitude; history of biology; insect physiology. *Mailing Add:* Dept Biol State Univ NY Col Oswego Oswego NY 13126

BRUNSON, ROYAL BRUCE, b DeKalb, Ill, Feb 16, 14; m 36; c 2. ZOOLOGY. *Educ:* Western Mich Col Educ, BS, 38; Univ Mich, MS, 45, PhD (zool), 47. *Prof Exp:* Teaching fel zool, Univ Mich, 42-46; from instr to prof zool, Univ Mont, 46-80; RETIRED. *Mem:* Am Malacol Union; Am Soc Ichthyologists & Herpetologists; Am Soc Limnol & Oceanog; Am Micros Soc; Soc Syst Zool. *Res:* Taxonomy and natural history of North American Gastrotricha; taxonomy and distribution of Western Montana invertebrates; limnology of Western Montana lakes. *Mailing Add:* 1522 34th St Missoula MT 59801

BRUNSTING, ELMER H(ENRY), b Hull, Iowa, Sept 29, 21; m 46; c 5. CHEMICAL ENGINEERING. *Educ:* Cent Col, Iowa, BS, 44; Iowa State Col, PhD (chem eng), 50. *Prof Exp:* Process engr, Vulcan Copper & Supply Co, 50-53; sr process engr, Mathieson Chem Corp, 53-54; chief process engr, John Deere Chem Co, 54-65; sr process engr, Eng Dept, Gulf Oil Chem Co,

65-67, dir process eng, 67-72, mgr, specialty chem, 72-83; pres, Graff Eng, 83-86; sr staff specialist, Stubbs Overbeck, 87-90; RETIRED. *Mem:* Am Chem Soc; Am Inst Chem Engrs. *Res:* Acidulation characteristics of various rock phosphates. *Mailing Add:* 13630 Pinerock Houston TX 77079

BRUNTON, GEORGE DELBERT, b McGill, Nev, Dec 20, 24. GEOLOGY, MINERALOGY. *Educ:* Univ Nev, BS, 50; Univ NMex, MS, 52; Ind Univ, PhD, 57. *Prof Exp:* Res asst, Pa State Univ, 52-53; geologist, Res Explor & Prod Res Lab, Shell Develop Co, 53-57, Pure Oil Res Ctr, 57-63 & US Gypsum Co, 63-64; geologist, Oak Ridge Nat Lab, 64-80; tech mgr, Battelle Mem Inst, 80-; AT DEPT GEOL & GEOL ENG, UNIV MISS. *Mem:* Mineral Soc Am; Soc Mining Engrs. *Res:* Clay mineralogy; carbonate mineralogy; crystal structures; phase equilibria. *Mailing Add:* Dept Geol & Geol Eng Univ Miss University MS 38677

BRUNTON, LAURENCE, b Charlottesville, Va, Mar 26, 47; c 1. CELLULAR PHYSIOLOGY, MOLECULAR PHARMACOLOGY. *Educ:* Harvard Col, BA, 69; Univ Va, PhD(pharmacol), 76. *Prof Exp:* Instr math & sci, Windsor Sch, 70-71; scholar, 76-78, asst res pharmacologist & lectr med, 78-80, ASST PROF MED, DIV PHARMACOL & CARDIOL, SCH MED, UNIV CALIF, SAN DIEGO, 81- *Concurrent Pos:* Nat Res Serv Award fel, NIH, 76-78, Res Career Develop Award, 81-86. *Mem:* Am Soc Pharmacol & Exp Therapeut; Am Fedn Clin Res. *Res:* Cyclic nucleotide metabolism and transport; mechanisms of specifically of cyclic nucleotide action; regulation of cell growth; interactions of drugs and hormones with the cell membrane. *Mailing Add:* Dept Med Div Pharmacol M-036 Univ Calif San Diego La Jolla CA 92093

BRUS, LOUIS EUGENE, b Cleveland, Ohio, Aug 10, 43; m 70; c 3. CHEMICAL PHYSICS & CLUSTERS. *Educ:* Rice Univ, BA, 65; Columbia Univ, PhD(chem physics), 69. *Prof Exp:* Sci staff officer physics & chem, US Naval Res Lab, 69-73; MEM TECH STAFF PHYS CHEM, BELL LABS, 73- *Mem:* Fel Am Phys Soc; Am Chem Soc. *Res:* Molecular radiationless transitions and condensed phase kinetics; photochemistry near surfaces and semiconductor clusters. *Mailing Add:* Bell Labs Murray Hill NJ 07974

BRUSCA, GARY J, b Bell, Calif, Oct 10, 39; m 62; c 5. ZOOLOGY, BIOLOGY. *Educ:* Calif State Polytech Col, BS, 60; Univ of the Pac, MA, 61; Univ Southern Calif, PhD(biol), 65. *Prof Exp:* Asst prof biol & asst dir Pac Marine Sta, Univ of the Pac, 64-67; from asst prof to assoc prof, 67-74, PROF BIOL, HUMBOLDT STATE UNIV, 74- *Concurrent Pos:* Dir, Humboldt State Marine Lab, Trinidad, Calif, 70- *Res:* crustacean systematics. *Mailing Add:* Dept Biol Humboldt State Univ Arcata CA 95521

BRUSCA, RICHARD CHARLES, b Los Angeles, Calif, Jan 25, 45; div; c 2. INVERTEBRATE ZOOLOGY, MARINE ECOLOGY. *Educ:* Calif State Polytech Univ, San Luis Obispo, BS, 67; Calif State Univ, Los Angeles, MS, 70; Univ Ariz, PhD(biol), 75. *Prof Exp:* Res biochemist, CalBioChem, Los Angeles, 69; teaching asst zool & cur insects & asst investr, Aquatic Insect Labs, Calif State Univ, Los Angeles, 69-70; resident dir, Univ Ariz-Univ Sonora Coop Marine Sta, Puerto Penasco, Mex, 70-72; teaching asst zool, Univ Ariz, 72-74; marine biologist, Biol Educ Exped, Vista, Calif, 74-75; from asst prof to assoc prof biol, Univ Southern Calif, Los Angeles, 75-84, cur crustacea, Allan Hancock Found, 75-84; cur & head invert zool sect, Los Angeles County Mus Natural Hist, 84-87; JOSHUA L BAILY CUR & CHMN, DEPT MARINE INVERT, SAN DIEGO NATURAL HIST MUS, 87- *Concurrent Pos:* Vpres, Panamic Environ Consult, Ariz, 73-75; sci ed, Allan Hancock Found Publ, 76-84; consult shallow water trop ecosystems, Univ Costa Rica, 79-81; dir acad progs, Catalina Marine Sci Ctr, Univ Southern Calif, 81-83; mem bd dirs, Orgn Trop Biol, 84-86; field reviewer, Inst Mus Serv, 85-; biol syst prog, NSF, 86-89, ad hoc mem adv panel, Numerous res grants; adj prof biol, Calif State Univ, Long Beach, 86-, San Diego, 87-; res assoc, Scripps Inst Oceanog, 90-; panel mem, Nat Acad Sci & Charles Lindberg Fund. *Honors & Awards:* Cong Medal, Serv Antarctica, 68. *Mem:* Am Soc Zool; AAAS; Sigma Xi; Soc Syst Zool; Willig Hennig Soc; Crustacean Soc; Western Soc Naturalists. *Res:* Evolutionary biology of invertebrates; crustacean biology; shallow water marine ecology; tropical marine ecosystems; marine symbiosis; parasites of marine fishes; biogeography. *Mailing Add:* San Diego Natural Hist Mus PO Box 1390 San Diego CA 92112

BRUSENBACK, ROBERT A(LLEN), b Chicago, Ill, Sept 24, 27; m 51; c 3. CHEMICAL ENGINEERING. *Educ:* Northwestern Univ, BS, 50, MS, 53, PhD(chem eng), 63. *Prof Exp:* Process develop engr, Houdry Process Corp, 53-56; instr chem eng, Northwestern Univ, 56-59; process develop engr, 59-63, ANALOG COMPUT SPECIALIST, ABBOTT LABS, 63- *Mem:* Am Inst Chem Engrs; Instrument Soc Am; Sigma Xi. *Res:* Microscopic analysis of fluidized bed heat transfer; application of analog computation to problems in the pharmaceutical industry. *Mailing Add:* Abbott Lab North Chicago IL 60064

BRUSEWITZ, GERALD HENRY, b Green Bay, Wis, June 1, 42; m 65; c 2. AGRICULTURAL ENGINEERING. *Educ:* Univ Wis, BS, 64 & 65, MS, 67; Mich State Univ, PhD(agr eng), 69. *Prof Exp:* From asst prof to assoc prof, 69-80, PROF AGR ENG, OKLA STATE UNIV, 80- *Concurrent Pos:* Eng consult, Solar Energy Dept, Kuwait Inst Sci Res, 80; vis prof, Univ Calif, Davis, 79, Cornell Univ, 88. *Mem:* Am Asn Cereal Chemists; Am Soc Eng Educ; fel Am Soc Agr Engrs; Inst Food Technol. *Res:* Physical and mechanical properties of biological materials; food process engineering. *Mailing Add:* Dept Agr Eng Okla State Univ Stillwater OK 74078

BRUSH, ALAN HOWARD, b Rochester, NY, Sept 29, 34; div; c 2. ZOOLOGY. *Educ:* Univ Southern Calif, BA, 56; Univ Calif, Los Angeles, MA, 57, PhD(zool), 64. *Prof Exp:* Res technician nuclear med & biophys, Univ Calif, Los Angeles, 56-61; postdoctoral fel zool, Cornell Univ, 64-65; asst prof zool, 65-69, assoc prof biol sci group, 69-76, PROF BIOL SCI GROUP, UNIV CONN, 76- *Concurrent Pos:* NIH spec fel, Univ Calif, Berkeley, 71-72; NSF int fel, Commonwealth Sci & Indust Res Orgn, Melbourne, Australia; ed, The Auk, 84- *Mem:* AAAS; Cooper Ornith Soc; Am Soc Zool; Am Physiol Soc; fel Am Ornith Union. *Res:* Comparative biochemistry and physiology; biochemical evolution; functional morphology. *Mailing Add:* Physiol & Neurobiol Univ Conn Storrs CT 06268

BRUSH, F(RANKLIN) ROBERT, b Phoenixville, Pa, Nov 24, 29; c 2. PSYCHOBIOLOGY, PSYCHONEUROENDOCRINOLOGY. *Educ:* Princeton Univ, BA, 51; Harvard Univ, MA, 53, PhD(social psychol), 56. *Prof Exp:* Asst prof psychol, Univ Md, 56-59 & Univ Pa, 59-65; assoc prof med psychol, Med Sch, Univ Ore, 65-67, prof, 67-71; prof psychol, Syracuse Univ, 71-80; head, dept psychol, 81-83, PROF, PURDUE UNIV, 81- *Concurrent Pos:* Vis asst prof & vis scientist, Dept Psychol & Regional Primate Res Ctr, Univ Wis, 64-65; res scientist develop award, NIMH, 67-71; vis prof, Sherrington Sch Physiol, St Thomas Hosp Med Sch, London, 76. *Mem:* Am Psychol Soc; Psychonomic Soc; Soc Neurosci; Behav Genetics Asn; Sigma Xi. *Res:* Psychobiology of stress and aversively motivated behaviors, focusing especially on behavioral effects of pituitary adrenocortical hormones and genetic determinants of individual differences in animal learning. *Mailing Add:* Dept Psychol Sci Purdue Univ West Lafayette IN 47907

BRUSH, GRACE SOMERS, b Antigonish, NS, Jan 18, 31; US citizen; m 53; c 3. FOREST ECOLOGY, PALYNOLOGY. *Educ:* St Francis Xavier Univ, BA, 49; Univ Ill, Urbana-Champaign, MS, 51; Radcliffe Col, PhD(biol), 56. *Prof Exp:* Technician, Geol Surv Can, 49-50, 51-52 & US Geol Surv, 56-57; lectr bot, George Washington Univ, 57-58; asst prof, Univ Iowa, 59-63; res assoc geol, Princeton Univ, 64-66, res staff mem, 66-70; res scientist, 70-75 & 78-81, PROF & PRIN RES SCIENTIST, DEPT GEOG & ENVIRON ENG, JOHNS HOPKINS UNIV, 81- *Concurrent Pos:* Asst prof, Rutgers Univ, 64-66. *Mem:* AAAS; Am Soc Limnol & Oceanog; Am Quaternary Asn; Ecol Soc Am; Am Asn Stratig Palynologists; Am Inst Biol Sci. *Res:* Relations between modern pollen distributions in water and surface sediments and vegetation; settling properties of pollen in water; mapping terrestrial vegetation; forest patterns; estuarine biostratigraphy. *Mailing Add:* Dept Environ Eng Johns Hopkins Univ 34th & Charles St Baltimore MD 21218

BRUSH, JAMES S, b Mankato, Minn, Mar 21, 29; m 58; c 4. BIOCHEMISTRY. *Educ:* Univ Chicago, PhD(biochem), 56. *Prof Exp:* Res assoc biochem, Univ Chicago, 56-57; clin chemist, Hurley Hosp, Flint, Mich, 58-60; fel biochem, Okla Med Res Inst, Oklahoma City, 60-63; biochemist, US Vet Admin, Wood, Wis, 63-69; from asst prof to assoc prof biochem, Col Basic Med Sci, Univ Tenn, Memphis, 69-74; assoc prof, Fac Med, Univ PR Med Sci Campus, San Juan, 75-83, prof, Dept Biochem & Nutrit, 83-91; vis scientist, NIH, Phoenix, 88-90; RETIRED. *Concurrent Pos:* Instr biochem, Sch Med, Marquette Univ, 64-69; prin investr Vet Admin grants, 69-72 & NIH, NSF grants, 74-77, 78-80; biochemist, US Vet Admin, Memphis, 69-74. *Mem:* AAAS; Am Soc Biochem & Molecular Biol; Endocrine Soc. *Res:* Protein sulfhydryl chemistry; enzyme kinetic mechanism of gulonolactone oxidase; insulin metabolism and mechanism of action; mechanisms of cyclic nucleotide action; cosmology of Swedenborg's Principia. *Mailing Add:* 7665 E Rancho Vista Dr Scottsdale AZ 85251

BRUSH, JOHN (BURKE), b New York, NY, Nov 25, 12; m 41; c 5. MECHANICAL ENGINEERING. *Educ:* Cornell Univ, ME, 34. *Prof Exp:* Trainee, Procter & Gamble, Inc, 34, engr packaging mach, 36; chief engr, Philippine Mfg Co, 41; assoc chief engr packing mach, Procter & Gamble Co, 49, assoc chief engr opers & planning, 51-56, mgr airplane opers, 53-57, head, mech dept, Overseas Eng Div, 55-68, Patent Div, 69-70, Eng Develop Div, 70-77; PRES, BRUSH AUTOBOAT CO, 58- *Mem:* Am Soc Mech Engrs; Soc Automotive Engrs. *Res:* Automatic packaging machinery; pneumatic conveying; materials handling; converting standard autos for amphibious operation; auto cooling; fuel systems cooling; increased reliability; reduced emissions and increased air conditioning capacity; vehicle fire prevention; 6 patents. *Mailing Add:* Two Beech Knoll Dr N College Hill Cincinnati OH 45224

BRUSH, LUCIEN M(UNSON), JR, b Pittsburgh, Pa, Dec 10, 29; m 53; c 3. CIVIL & GEOLOGICAL ENGINEERING. *Educ:* Princeton Univ, BSE, 52; Harvard Univ, PhD(geol), 56. *Prof Exp:* Geologist, US Geol Surv, 56-58; from asst prof to assoc prof & res engr, Iowa Inst Hydraul Res, 58-63; assoc prof civil & geol eng, Princeton Univ, 63-69; PROF HYDRAUL, JOHNS HOPKINS UNIV, 69- *Mem:* Am Soc Civil Engrs; Am Geophys Union; Int Asn Sci Hydrol; Int Asn Hydraul Res; Soc Econ Paleont & Mineral; Int Water Resources Asn. *Res:* Hydraulics; water pollution; sediment transportation; open channel flow; geomorphology. *Mailing Add:* Dept Geog & Environ Eng Johns Hopkins Univ Baltimore MD 21218

BRUSH, MIRIAM KELLY, b Boston, Mass, Nov 9, 15; m 42; c 4. NUTRITION. *Educ:* Mt Holyoke Col, AB, 37; Oberlin Col, AM, 39; Iowa State Col, PhD(nutrit), 46. *Prof Exp:* Assoc nutrit, Iowa State Col, 46-47; assoc cancer res, McArdle Mem Lab, Med Sch, Univ Wis, 47-50; assoc geront, Med Sch, Washington Univ, 51; lectr nutrit, Col Nursing, Rutgers Univ, 56-58; lectr, 57-69, PROF NUTRIT, RUTGERS UNIV, 69- *Mem:* AAAS; Am Inst Nutrit; Am Home Econ Asn; Am Dietetic Asn; Am Pub Health Asn; Sigma Xi. *Res:* Protein and amino acid metabolism; nutrition and the aging process; protein-vitamin B6 inter-relationships; vitamin C nutrition in school children; nutrition in infancy and early childhood. *Mailing Add:* Dept Home Econ Douglas Col Rutgers Univ New Brunswick NJ 08903

BRUSH, STEPHEN GEORGE, b Bangor, Maine; m 60; c 2. HISTORY OF SCIENCE. *Educ:* Harvard Univ, AB, 55; Oxford Univ, DPhil(phys sci), 58. *Prof Exp:* NSF fel, 58-59; physicist, Lawrence Radiation Lab, Univ Calif, 59-65; res assoc physics & ed, Harvard Proj Physics, 65-68, lectr physics & hist of sci, Harvard Univ, 66-68; assoc prof hist & res assoc prof, Inst Fluid Dynamics & Appl Math, 68-71, PROF HIST OF SCI, UNIV MD, COLLEGE PARK, 71- *Honors & Awards:* Pfizer Award, Hist Sci Soc, 77. *Mem:* Acad Int Hist Sci; Am Phys Soc; Hist of Sci Soc (vpres, 88, pres, 90-91). *Res:* History of physical science in 19th and 20th centuries, especially geophysics, astrophysics, kinetic theory and statistical mechanics; use of historical approach in teaching physics. *Mailing Add:* Inst Phys Sci & Technol Univ Md College Park MD 20742-2431

BRUSIE, JAMES POWERS, b North Egremont, Mass, July 3, 18; m 47; c 3. PHYSICAL CHEMISTRY. *Educ:* Lafayette Col, BS, 40; Yale Univ, PhD(phys chem), 43. *Prof Exp:* Res chemist, Manhattan Proj & Off Sci Res & Develop contract, Div War Res, Yale Univ, 42-43 & Columbia Univ, 43-44; sect supvr, Carbide & Carbon Chem Corp, Oak Ridge, 44-46; res chemist, Gen Aniline & Film Corp, 46-50, group leader, 50-52, res assoc, 52-55, prog mgr, 55-62, sr tech assoc, 62-64; tech dir, Girdler Catalysts Dept, Chemetron Corp, 64-68; prod mgr, Alrac Corp, 69-74; res assoc, Pullman Kellogg, 74-76, res supur, M W Kellogg, 76-80; RETIRED. *Mem:* Am Chem Soc. *Res:* Catalysis, pressure acetylene reactions; polymerization of alpha-halo-acrylic esters; polymerization of 2-pyrrolidone to nylon-4; catalysis and coal conversion; flue gas desulfurization. *Mailing Add:* 10546 Idlebrook Dr Houston TX 77070

BRUSILOW, SAUL W, b Brooklyn, NY, June 7, 27; wid; c 2. PEDIATRICS, PHYSIOLOGY. *Educ:* Princeton Univ, AB, 50; Yale Univ, MD, 54. *Prof Exp:* Intern, Grace-New Haven Hosp, Conn, 54-55, asst resident, 55-56; asst resident, Johns Hopkins Hosp, 56-57, Nat Found Infantile Paralysis fel, Johns Hopkins Sch Med, 57-59, from instr to assoc prof, 59-74, PROF PEDIAT, SCH MED, JOHNS HOPKINS UNIV, 74- *Mem:* Am Pediat Soc; Am Physiol Soc; Soc Pediat Res. *Res:* Inborn errors of metabolism; nephrology. *Mailing Add:* 4804 Keswick Rd Baltimore MD 21210

BRUSSARD, PETER FRANS, b Reno, Nev, June 20, 38; m 69; c 2. POPULATION BIOLOGY, CONSERVATION BIOLOGY. *Educ:* Stanford Univ, AB, 60, PhD(biol), 69; Univ Nev, Reno, MS, 66. *Prof Exp:* From asst prof to assoc prof ecol, Cornell Univ, 75-85; dir, Rocky Mountain Biol Lab, 78-82; prof & head biol dept, Mont State Univ, 85-89; PROF & CHAIR BIOL DEPT, UNIV NEV, RENO, 89- *Concurrent Pos:* Vis prof, Univ Tex, 77, Univ Dunedin, Otago NZ, 89; vis scholar, Univ Ariz, 84. *Mem:* Fel AAAS; Ecol Soc Am; Am Soc Naturalists (treas, 80-83); Soc Study Evolution; Soc Conservation Biol (secy-treas, 85-90, pres-elect, 90-). *Res:* Relations between genetic variation, distribution and ecology of natural populations; conservation biology. *Mailing Add:* Univ Nev Mon State Univ Reno NV 89557-0050

BRUSSEL, MORTON KREMEN, b New Haven, Conn, Mar 31, 29; m 57; c 2. PHYSICS. *Educ:* Yale Univ, BS, 51; Univ Minn, PhD(physics), 57. *Prof Exp:* Res assoc fel physics, Brookhaven Nat Lab, 57-59, asst physicist, 59-60; res asst prof, 60-64, assoc prof, 64-68, PROF PHYSICS, UNIV ILL, URBANA-CHAMPAIGN, 68- *Concurrent Pos:* Assoc prog dir nuclear physics, Physics Sect, NSF, 72-73; sabbatical leaves, Inst of Nuclear Physics, Orsay, France, 68-69, Div de Physique Nucleaire-High Energy, Centre D'Etudes Nucleaire, Saclay, France, 78-79, 87-88. *Mem:* Am Phys Soc; Fedn Am Scientists; Am Asn Univ Profs; Sigma Xi; Europ Phys Soc. *Res:* Experimental nuclear physics; nuclear structure studies; electron and photon reactions especially from light nuclei. *Mailing Add:* Dept Physics Univ Ill 1110 W Green St Urbana IL 61801

BRUST, DAVID PHILIP, b Albion, NY, Aug 13, 34; m 59; c 3. PHOTOGRAPHIC CHEMISTRY, MAGNETIC MEDIA CHEMISTRY. *Educ:* State Univ NY Buffalo, BA, 55; Univ Rochester, PhD(org chem), 65. *Prof Exp:* Res chemist polymer chem, 58-61, sr res chemist, 65-72, res assoc photog chem, 72-83, res assoc, magnetic formulations, 83-87, RES ASSOC COLOR HARDCOPY, 88- *Mem:* Am Chem Soc; AAAS; Soc Photog Scientists & Engrs. *Res:* Polymers in photographic systems. *Mailing Add:* 239 Southridge Dr Rochester NY 14626

BRUST, HARRY FRANCIS, b Milwaukee, Wis, Jan 2, 14; m 39; c 3. ORGANIC CHEMISTRY. *Educ:* Univ Wis, BS, 38; Univ Pa, MS, 40, PhD(chem), 43. *Prof Exp:* Asst instr chem, Univ Pa, 38-43; chemist & group leader, Dow Chem Co, 43-68, sr res chemist, 68-73, sr res specialist, 73-79; RETIRED. *Mem:* Am Chem Soc; Sigma Xi. *Res:* Chlorination; bromination of a variety of organic chemicals; agricultural chemicals, chiefly herbicides; new process development and reaction kinetics. *Mailing Add:* 601 Hillcrest St Midland MI 48640

BRUST, MANFRED, b Chemnitz, Ger, Oct 22, 23; nat US; m 57; c 3. PHYSIOLOGY. *Educ:* NY Univ, BA, 44, MSc, 46; Univ Ill, PhD(physiol), 54. *Prof Exp:* Asst physiol, Univ Chicago, 46-48; asst entom, Univ Ill, 50-53; res assoc zool, Syracuse Univ, 53-55; res assoc biol, Wash Sq Col, NY Univ, 55-59; res assoc, Div Physiol, Inst Muscle Dis, Inc, New York, 59-61, asst mem, 61-63; from asst prof to assoc prof rehab med, 63-70, ASSOC PROF PHYSIOL, STATE UNIV NY DOWNSTATE MED CTR, 70- *Concurrent Pos:* Asst, Babies Hosp, Columbia-Presby Med Ctr, 57. *Mem:* AAAS; Am Physiol Soc; NY Acad Sci; Biophys Soc. *Res:* Basic physiology of muscle contraction; excitation and muscle pharmacology in both normal and diseased conditions; muscular dystrophy; physiology of neuromuscular disease; fast and slow muscle; fatigue; ultrasound; atherosclerosis; cardiovascular control. *Mailing Add:* 450 Clarkson Ave Dept Physiol Box 31 State Univ NY Health Sci Ctr Brooklyn NY 11203

BRUST, REINHART A, b Sibbald, Alta, Feb 7, 34; m 59; c 2. ENTOMOLOGY, INSECT ECOLOGY. *Educ:* Univ Man, BSc, 59, MSc, 60; Univ Ill, Urbana, PhD(entom), 64. *Prof Exp:* From asst prof to assoc prof, 64-72, PROF ENTOM, UNIV MAN, 73- *Mem:* Entom Soc Am; fel, Entom Soc Can. *Res:* Ecology of North American species of mosquitoes, pathogen transmission, biology and systematics of mosquitoes. *Mailing Add:* Dept Entomol Univ Man Winnipeg MB R3T 2N2 Can

BRUST-CARMONA, HECTOR, physiology, neurophysiology, for more information see previous edition

BRUSVEN, MERLYN ARDEL, b Powers Lake, NDak, Mar 23, 37; m 59; c 3. ENTOMOLOGY. *Educ:* NDak State Univ, BS, 59, MS, 61; Kans State Univ, PhD(entom), 65. *Prof Exp:* Instr biol, Friends Univ, 61-63; PROF ENTOM, UNIV IDAHO, 65- *Mem:* Am Soc Limnol & Oceanog; Entom Soc Am. *Res:* Insect ecology, especially grasshopper taxonomy and aquatic entomology; pollution; population and community dynamics. *Mailing Add:* Dept Entom Univ Idaho Moscow ID 83843

BRUTCHER, FREDERICK VINCENT, JR, b Dorchester, Mass, Dec 5, 22. ORGANIC CHEMISTRY. *Educ:* Univ Mass, BS, 47; Yale Univ, MS, 49, PhD(org chem), 51. *Prof Exp:* Fel, Harvard Univ, 51-53; asst prof chem, 53-60, ASSOC PROF CHEM, UNIV PA, 60- *Mem:* Am Chem Soc. *Res:* Conformational analysis of substituted cyclopentanes; mathematics of organic chemistry; synthesis of indole and alicyclic compounds. *Mailing Add:* Harrison Lab Univ Pa Dept Chem Philadelphia PA 19104

BRUTLAG, DOUGLAS LEE, b Alexandria, Minn, Dec 19, 46; m 75; c 2. MOLECULAR BIOLOGY. *Educ:* Calif Inst Technol, BS, 68; Stanford Univ, PhD(biochem), 72. *Prof Exp:* Res scientist genetics, Commonwealth Sci & Indust Res Orgn, 72-74; asst prof, 74-81, ASSOC PROF BIOCHEM, STANFORD UNIV, 81-; dir, Intellicorp, 80-85, Intelligenerics, 85-90. *Concurrent Pos:* Genetics study sect, NIH, 82-86; mem Bd Sci Counr, Nat Libr Med, 89-93. *Mem:* Fedn Am Soc Exp Biol; Am Soc Biol Chemists; Am Asn Artificial Intel. *Res:* The application of computer science methods to molecular biology. *Mailing Add:* Dept Biochem Beckman Ctr B400 Stanford Med Sch Stanford CA 94305-5307

BRUTSAERT, WILFRIED, b Ghent, Belg, May 28, 34; nat US; m 71; c 4. HYDROLOGY. *Educ:* State Univ Ghent, BEng, 58; Univ Calif, MSc, 60, PhD(eng), 62. *Prof Exp:* From asst prof to assoc prof, 62-74, PROF HYDROL, CORNELL UNIV, 74- *Concurrent Pos:* Vis scholar, Tohoku Univ, Japan, 69-70 & 83-84; Am Soc Civil Engrs Freeman fel, 75; Fulbright-Hays vis prof, Wageningen, Netherlands, 76-77; pres-elect, Hydrol Sect, Am Geophys Union, 90-92, pres, 92-94. *Honors & Awards:* Horton Award, Am Geophys Union, 88. *Mem:* Fel Am Geophys Union; Am Soc Civil Engrs; Am Meteorol Soc; Sigma Xi. *Res:* Flow through porous media; permeability; infiltration and drainage; microclimatology; evaporation; surface water hydrology; hydrologic systems. *Mailing Add:* Sch Civil & Environ Eng Hollister Hall Cornell Univ Ithaca NY 14853

BRUTTEN, GENE J, b New York, NY, May 28, 28; m 55; c 2. COMMUNICATION DISORDERS, STUTTERING. *Educ:* Kent State Univ, BA, 51; Brooklyn Col, MA, 52; Univ Ill, PhD, 57. *Prof Exp:* Asst prof, New York State Univ, 54-56; asst prof, 57-61, assoc prof, 62-64, RES PROF COMMUN DIS, DEPT COMMUN DIS, SOUTHERN ILL UNIV, 67-, CHAIR, 87- *Concurrent Pos:* Coord, Speech & Hearing Serv, Southern Ill Univ, 61-65; vis prof, Univ Minn, 64, City Univ New York, Grad Ctr, 66-67; vis dir, Hunter Col, City Univ, New York, 66-67; Fulbright vis prof, Acad Med Sch, Holland, 71-72, res prof, 78-79. *Mem:* Am Speech & Hearing Assoc; Am Psychol Assoc; Am Assoc Behavior Therapy. *Res:* Molecular analysis of the fluency failures of stutterers in order to compare them with the normal disfluencies of nonstutterers so as to improve differential diagnosis and assessment and to enhance the strategy and tactics of therapy. *Mailing Add:* Dept Commun Disorders & Sci Southern Ill Univ Carbondale IL 62901

BRUTVAN, DONALD RICHARD, b Johnson City, NY, Oct 14, 24; m 50; c 4. CHEMICAL ENGINEERING. *Educ:* Rensselaer Polytech Inst, BChE, 50, MChE, 54, PhD(chem eng), 58. *Prof Exp:* Asst res engr, Behr Manning, Norton Abrasives, NY, 50-53; instr chem eng, Rensselaer Polytech Inst, 53-57; res engr, Metals Res Labs, Union Carbide Metals Co, 57-61; asst dean, Millard Fillmore Col, 65-69, actg univ dean, 68-69 & 75-78, assoc dean, Div Continuing Educ, 70-78, ASSOC PROF CHEM ENG, STATE UNIV NY BUFFALO, 61- *Honors & Awards:* Prof Achievement Award, Am Inst Chem Engrs, 67-68. *Mem:* Am Inst Chem Engrs; Am Soc Eng Educ; Sigma Xi. *Res:* Phase equilibria and staged operations; economics and process design. *Mailing Add:* 149 Somerton Ave Buffalo NY 14217

BRYAN, ASHLEY MONROE, b British West Indies, Apr 29, 17; m 48; c 4. PLANT BIOCHEMISTRY. *Educ:* Hampton Inst, BS, 48; Iowa State Univ, PhD(plant physiol), 53. *Prof Exp:* Prof biol, Talladega Col, 53-56; res assoc, Univ Pa, 56-59; assoc prof, Morgan State Col, 59-61; prof biochem, State Univ NY, Albany, 61-87; RETIRED. *Concurrent Pos:* Mem comn on higher educ, Mid States Asn Cols & Sec Schs. *Mem:* Am Chem Soc; Sigma Xi. *Res:* Thermal studies on nucleic acids by differential scanning calorimetry; interaction of protein-nucleic acids and the effects of environmental stressors on their structure and function. *Mailing Add:* 27 Norwood Ave Albany NY 12228

BRYAN, BILLY BIRD, b Forrest City, Ark, Aug 8, 21; m 42; c 2. AGRICULTURAL ENGINEERING. *Educ:* Univ Nebr, BS, 50, MS, 54. *Prof Exp:* Exten agr engr, Kans State Univ, 51-53; from asst prof to assoc prof agr eng, 53-61, actg head dept, 64-69, head dept, 69-85, PROF AGR ENG, UNIV ARK, FAYETTEVILLE, 61- *Mem:* Am Soc Agr Engrs; Am Soc Eng Educ; Nat Soc Prof Engrs. *Res:* Soil and water, especially hydraulic applications in agriculture and water resources; machinery and mechanization of cotton production. *Mailing Add:* Dept Agr Eng Univ Ark Fayetteville AR 72701

BRYAN, CARL EDDINGTON, b West Point, Miss, Jan 12, 17; m 42; c 2. ORGANIC CHEMISTRY, POLYMER CHEMISTRY. *Educ:* Univ Miss, BA, 37; Univ Minn, PhD(org chem), 42. *Prof Exp:* Instr chem, Delta Jr Col, 37-38; asst chem, Univ Minn, 38-42; asst, Univ Ill, 42-44; res chemist, E I du Pont de Nemours & Co, 44-46, res chemist, Southern Res Inst, 46-50, head org sect, 48-50; res chemist, Res Ctr, US Rubber Co, 50-59, res scientist, 59-61; sr res chemist, Chemstrand Res Ctr, Monsanto Co, 61-67; res chemist, Mallinckrodt Chem Works, Mo, 67-68; res assoc textiles chem, NC State Univ, 68-76, lectr & lab supvr org chem, 76-84; RETIRED. *Concurrent Pos:* With Off Sci Res & Develop; with Off Rubber Reserve; prin investr res projs, Water Resources Res Inst, US Environ Protection Agency & Am Dye Mfrs Inst, NC State Univ. *Mem:* AAAS; NY Acad Sci; Am Chem Soc; Am Inst Chem; Am Asn Textile Chemists & Colorists. *Res:* Organic synthesis and structure determinations; polymers and polymerization; textile chemistry and waste control. *Mailing Add:* 2631 St Mary's St Raleigh NC 27609-6632

BRYAN, CHARLES A, mathematics, for more information see previous edition

BRYAN, CHARLES F, b Louisville, Ky, Jan 17, 37; m 62. ICHTHYOLOGY, LIMNOLOGY. *Educ:* Bellarmine Col, Ky, BA, 60; Univ Louisville, PhD(zool), 64. *Prof Exp:* Asst cur fish, John G Shedd Aquarium, 64-66; fishery biologist, Marion Inserv Training Sch, US Bur Sport Fisheries, 66-67; asst leader, Calif Coop Fishery Unit, 67-70; LEADER, LA COOP FISH & WILDLIFE RES UNIT, 70- *Concurrent Pos:* Adj prof, La State Univ; sr scientist, Am Fisheries Soc. *Mem:* Am Soc Limnol & Oceanog; Am Soc Ichthyologists & Herpetologists; Am Fisheries Soc; Int Asn Theoret & Appl Limnol; fel Am Fisheries Res Biol; fel Am Inst Fisheries Res Biol. *Res:* Stream and swamp limnology; water quality; ecology of fishes. *Mailing Add:* 124 Forestry La State Univ Baton Rouge LA 70803

BRYAN, CHRISTOPHER F, b Augusta, Ga, Apr 10, 51. ORGAN TRANSPLANTATION, IMMUNOLOGY. *Educ:* Baylor Univ, BS, 73; Tex A&M Univ, MS, 75, PhD(immunogenetics), 78. *Prof Exp:* Supvr, Immunogenetics Lab, Med Sch, La State Univ, 80-84; lab dir, Transplantation-Immunol Lab, Baylor Univ Med Ctr, 84-87; dir histocompatibility, Iowa Methodist Med Ctr, 87-89; LAB DIR, MIDWEST ORGAN BANK, 89- *Mem:* Transplantation Soc; Am Soc Histocompatibility & Immunogenetics; Clin Immunol Soc. *Mailing Add:* Midwest Organ Bank 1900 W 47th Pl Suite 400 Westwood KS 66205

BRYAN, DAVID A, b Austin, Tex, July 29, 46; m 74; c 5. SOLID-STATE LASERS, OPTICAL WAVEGUIDES. *Educ:* Rice Univ, BA, 68; Univ Mo, Rolla, MS, 73, PhD(physics), 76. *Prof Exp:* Tech specialist, McDonnell Douglas Corp, 76-89; RES ENGR, LASER DIODE, INC, EARTH CITY, 89- *Concurrent Pos:* Reviewer, Appl Physics Lett, 85- *Mem:* Optical Soc Am; Int Laser Commun Soc. *Res:* Developed optical waveguide components for integrated optic wavelength demultiplexer; magnesium-doped lithium niobate for improved optical damage resistance; semiconductor waveguide components and coupling to fibers; analysis and experiment on laser beam combining with fibers; compact, high-energy Nd:YAG laser with unstable resonator; one United States patent. *Mailing Add:* Laser Diode Inc 4014 Wedgeway Ct Earth City MO 63045

BRYAN, EDWARD H, b Milwaukee, Wis, Oct 30, 24; m 49; c 3. WATER SUPPLY, WASTEWATER MANAGEMENT. *Educ:* Univ Wis, BS, 49, MS, 50, PhD(civil & sanit eng), 54. *Prof Exp:* Res assoc sanit eng, Univ Wis, 49-50, instr civil eng, 50-53; sanit engr, Dow Chem Co, 53-54, develop engr, plastics tech serv, 54-57, mkt res specialist, 57-58, develop engr tech serv & develop, 58-59, environ eng consult, Chem Dept, 59-60; assoc prof civil & environ eng, Duke Univ, 60-64, prof & dir undergrad studies civil eng, 64-70; mgr, Process Systs Sect, Rex Chainbelt Inc, 70, mgr environ projs group, Ecol Div, 70-71, mgr tech mkt & environ projs, 71-72; prog mgr, Environ Systs & Resources, NSF, 72-74, Systs Integration & Anal, 74-79 & Appropriate Technol, 79-81, prog dir, Water Resources & Environ Eng, 81-82, Environ & Water Qual Eng, 82-84, Environ Eng, 84-88, PROG DIR, ENVIRON & OCEAN SYSTS, NSF, 88- *Concurrent Pos:* Prof engr, Wis, 54-; diplomate, Am Acad Environ Eng, 75-; mem, Coun Pub Health Consults, Nat Sanit Found, 89-92. *Mem:* Am Water Works Asn; Water Pollution Control Fedn; Am Soc Civil Engrs; Int Asn Water Pollution Res & Control; Fed Water Qual Asn; Fed Con Environ Engr; Am Acad Environ Engr. *Res:* Water supply and treatment; pollution control; industrial and municipal waste control and treatment; water resources; radiological health; public health engineering; bacteriology of water and wastes; urban storm and sanitary drainage; spills of hazardous substances. *Mailing Add:* 4006 Thornapple St Chevy Chase MD 20815

BRYAN, FRANK LEON, b Indianapolis, Ind, Aug 29, 30; m 52; c 2. BACTERIOLOGY, FOOD MICROBIOLOGY. *Educ:* Ind Univ, BS, 53; Univ Mich, MPH, 56; Iowa State Univ, PhD(bact), 65. *Prof Exp:* Pub health aide milk & environ sanit, USPHS, Durham & Chapel Hill Local Health Dept, NC, 53; sanitarian, Ind State Bd Health, 55; training off pub health, New Eng Field Training Sta, USPHS, 56-58, environ health, Nat Commun Dis Ctr, 58-63, scientist dir & chief Foodborne Dis Training, Ctr for Dis Control, 65-85, DIR, FOOD SAFETY CONSULT & TRAINING, USPHS, 85- *Concurrent Pos:* Lectr, Univ Mass, 56-58; consult, WHO, UN Food & Agr Orgn, Pan Am Health Orgn; adj fac, Emory Univ, 85 & 87. *Honors & Awards:* Sherman Res Awards. *Mem:* Am Soc Microbiol; Inst Food Technol; Sigma Xi; Int Comn Microbiol Spec Foods; Am Pub Health Asn; Int Asn Milk, Food & Environ Sanitarians; World Asn Vet Food Hyg. *Res:* Salmonella in turkey products; time-temperature factors in thawing, cooking, chilling and reheating turkey products rice, oriental foods and beef; clostridium perfringens in beef; staphylococcal intoxication; foodborne infections and intoxications and miscellaneous foodborne diseases; hazard analysis food service and catering operations; hazard analyses of street vended foods and homes in developing countries. *Mailing Add:* 8233 Pleasant Hill Lithonia GA 30058

BRYAN, GEORGE TERRELL, b Antigo, Wis, July 29, 32; m 54; c 5. MEDICINE, PHARMACOLOGY. *Educ:* Univ Wis, BS, 54, MD, 57, PhD(oncol, biochem), 63. *Prof Exp:* Intern med, NC Baptist Hosp, Winston-Salem, 57-58; instr cancer res, 61-63, from asst prof to prof clin oncol & surg, 63-75, PROF HUMAN ONCOL, SCH MED, UNIV WIS-MADISON, 75- *Concurrent Pos:* Consult, Food & Drug Admin, Environ Protection Agency & Nat Cancer Inst. *Mem:* fel Am Col Physicians; Am Asn Cancer Res; Soc Surg Oncol; Am Soc Exp Path; Am Soc Biol Chemists. *Res:* Chemical carcinogenesis; clinical and experimental cancer chemotherapy; metabolic disorders and normal human metabolism; nutrition. *Mailing Add:* K4/528 Clin Sci Ctr 600 Highland Ave Madison WI 53792

BRYAN, GEORGE THOMAS, b Sewanee, Tenn, Nov 19, 30; m 52; c 2. MEDICAL EDUCATION. *Educ:* Univ Tenn, MD, 55; Am Bd Pediat, dipl, 69. *Prof Exp:* Intern, DC Gen Hosp, 55-56; resident pediat, State Univ Iowa Hosps, 56-58, fel pediat endocrinol, 58-59; clin assoc pediat, Nat Inst Allergy & Infectious Dis, 59-60, pediatrician, Clin Endocrinol Br, 61-63; from asst prof to assoc prof pediat, 63-73, asst dir clin study ctr & dir, Div Endocrinol, dept pediat, 63-70, assoc dean curric affairs, 74-77, PROF PEDIAT, UNIV TEX MED BR, GALVESTON, 73-, DEAN MED, 77-, VPRES ACAD AFFAIRS, 83- *Concurrent Pos:* Markle Found scholar acad med, 67-72. *Mem:* AMA; Am Pediat Soc; Soc Pediat Res; Endocrine Soc; Am Fedn Clin Res. *Res:* Adrenal steroid biosynthesis; carcadian periodicity; hypertension; hypopituitarism. *Mailing Add:* Off Dean Med Univ Tex Med Br Galveston TX 77550

BRYAN, GORDON HENRY, b Ashton, Idaho, Oct 22, 15; m 46; c 2. PHARMACOLOGY. *Educ:* Mont State Univ, BS, 42, MS, 47; Univ Md, PhD(pharmacol), 56. *Prof Exp:* Instr pharm, 47-49, from asst prof to assoc prof pharmacol, 50-56, prof, 56-80, EMER PROF PHARMACOL, UNIV MONT, 80- *Concurrent Pos:* NIH fel, 64-65; vis lectr med, Univ Utah, 57. *Mem:* AAAS; Am Pharmaceut Asn. *Res:* Cardiology. *Mailing Add:* 515 Dickinson Missoula MT 59802

BRYAN, HERBERT HARRIS, b Brunswick, Ga, Jan 25, 32; m 56; c 2. HORTICULTURE, PLANT PHYSIOLOGY. *Educ:* Univ Fla, BSA, 53; Cornell Univ, MS, 61, PhD(veg crops), 64. *Prof Exp:* Asst prof hort, NFla Exp Sta, Univ Fla, 64-66, asst prof, Agr Res & Educ Ctr, 67-74, assoc prof hort & assoc horticulturist, 74-80, prof hort, Agr Res & Educ Ctr, 80- *Mem:* Am Soc Hort Sci; Int Soc Hort Sci. *Res:* Nutrition and herbicides of vegetables and peaches; mechanical harvesting of tomatoes for fresh market; cultural practices; mulches; pole bean breeding; cover crops rotation with potatoes; bean and tomato growth regulators and plant populations. *Mailing Add:* 18905 SW 280th St Homestead FL 33031

BRYAN, HORACE ALDEN, b Bluff City, Tenn, Jan 1, 28. INORGANIC CHEMISTRY. *Educ:* King Col, BS, 50; Univ Tenn, PhD, 55. *Prof Exp:* Asst inorg chem, Univ Tenn, 51-55; from asst prof to assoc prof, 55-67, PROF CHEM, DAVIDSON COL, 67- *Concurrent Pos:* Vis prof chem, Univ Hawaii, 85. *Mem:* Am Chem Soc. *Res:* Polarography of azo compounds; extraction of metal-containing anions; cation-ligand equilibria. *Mailing Add:* PO Box 503 Davidson NC 28036

BRYAN, HUGH D, pharmaceutical chemistry, for more information see previous edition

BRYAN, JAMES BEVAN, b Alamada, Calif, May 6, 26; m 50; c 2. PRECISION ENGINEERING. *Educ:* Univ Calif, BS, 51. *Prof Exp:* Marine engr, US Maritime Serv, 44-49; mfg engr, Westinghouse Elec Corp, 51-55; CHIEF METROLOGIST, LAWRENCE LIVERMORE LAB, UNIV CALIF, 55- *Concurrent Pos:* Consult precision eng, 60-; mem, B-89 Comt Dimensional Metrol, B-46 Comt Surface Texture, Am Nat Standards Inst, 65- & B-89 Comt Coord Measuring Mach; mem eval panel, Nat Bur Standards-Nat Res Coun, 68-73. *Honors & Awards:* Soc Mfg Engrs Res Medal, 77. *Mem:* Soc Mfg Engrs; Int Inst Prod Eng Res. *Res:* Design of precision machine tools and measuring machines with emphasis on the problem of thermal stability. *Mailing Add:* 5196 Golden Rd Pleasanton CA 94566

BRYAN, JAMES CLARENCE, b Tenn, Jan 26, 23; m 48; c 2. CHEMICAL ENGINEERING. *Educ:* Univ Tenn, BS, 48, MS, 52, PhD(chem eng), 54. *Prof Exp:* Process engr, Phillips Petrol Co, 48-50; sr engr nylon, E I du Pont de Nemours & Co, Inc, 54-62, sr res engr, 62-65 & 70-84, process supvr, 65-70; RETIRED. *Mem:* Am Chem Soc; Am Inst Chem Engrs; Sigma Xi. *Res:* Polymerization technology; process control analysis; chemical process analysis; mathematical statistics; probability. *Mailing Add:* 1900 Brook Dr Camden SC 29020

BRYAN, JOHN HENRY DONALD, b London, Eng, Sept 18, 26; nat US; m 52; c 2. GENETIC DISEASES, MALE STERILITY & HYDROCEPHALUS. *Educ:* Univ Sheffield, BSc, 47; Columbia Univ, AM, 49, PhD(zool), 52. *Prof Exp:* Lectr zool, Columbia Univ, 49-50; instr biol, Mass Inst Technol, 51-54; asst prof genetics, Iowa State Univ, 54-60, from asst prof to assoc prof zool, 60-67, chmn comt cell biol, 61-67; PROF ZOOL, UNIV GA, 67- *Concurrent Pos:* Consult, NSF-AID Indian Educ Prog, 68. *Honors & Awards:* Paxton lectr, Int Asn Torch Clubs, 86. *Mem:* Fel AAAS; Am Soc Cell Biol; Am Soc Zool; Genetics Soc Am; Sigma Xi. *Res:* Cell structure and function, especially the nucleus; cytochemistry; chromosome ultrastructure; genetic control of cell differentiation, especially gametogenesis; use of mutations as probes of cell function, including male infertility and neurological development. *Mailing Add:* Dept Zool Univ Ga Athens GA 30602

BRYAN, JOHN KENT, b Indianapolis, Ind, Mar 20, 36; m 58. PLANT PHYSIOLOGY, BIOCHEMISTRY. *Educ:* Butler Univ, BS, 58; Univ Tex, PhD(cellular physiol), 62. *Prof Exp:* Res asst biochem, Inst Psychiat Res, Sch Med, Ind Univ, 58-59; res asst electron micros, Plant Res Inst, Univ Tex, 62; res assoc molecular biol, Wash Univ, 62-64; asst prof bot, 64-69, assoc prof biol, 70-76, PROF BIOL, SYRACUSE UNIV, 77- *Mem:* AAAS; Am Soc Plant Physiol; Am Soc Biol Chem & Molecualr Biol; Int Soc Plant Molecular Biol; Am Inst Biol Sci. *Res:* Biochemistry of plant growth and development; control mechanisms of amino acid biosynthesis. *Mailing Add:* Dept Biol Biol Res Labs Syracuse Univ 130 College Pl Syracuse NY 13244

BRYAN, JOSEPH, b Pittsburgh, Pa. BIOCHEMISTRY, CELL BIOLOGY. *Educ:* Rensselaer Polytech Inst, BS, 63, Univ Pa, PhD(anat), 67. *Prof Exp:* Instr med, Univ Pa, 63-64; fel cell biol, Univ Calif, Berkeley, 67-71, actg asst prof zool, 70-71; from asst prof to assoc prof biol, Univ Pa, 71-78; assoc prof, 78-80, PROF CELL BIOL, BAYLOR COL MED, 80- *Concurrent Pos:* Mem bd, NIH Ad Hoc Cell Biol Study Sect, 78-79, Molecular Biol Study Sect, 80-81 & 81-83 & Cell Biol Study Sect, 83-85. *Mem:* Am Soc Cell Biol; AAAS. *Res:* Biochemistry and cell biology of actin and actin associated proteins. *Mailing Add:* Dept Cell Biol Baylor Col Med One Baylor Plaza Houston TX 77030

BRYAN, KIRK, (JR), b Albuquerque, NMex, July 21, 29; m 56; c 2. METEOROLOGY. *Educ:* Yale Univ, BS, 51; Mass Inst Technol, PhD(meteorol), 57. *Prof Exp:* Res assoc meteorol, Woods Hole Oceanog Inst, 58-61; res meteorologist, Gen Circulation Res Lab, US Weather Bur, 61-68; OCEANOGR, GEOPHYS FLUID DYNAMICS LAB, NAT OCEANOG

& ATMOSPHERIC ADMIN, PRINCETON UNIV, 68- *Concurrent Pos:* Vis lectr, Princeton Univ, 68-78; mem Panel Climatic Variation, Global Atmos Res Prog, Nat Acad Sci, 72-74; chmn working group numerical models, Sci Comt Ocean Res, 75-77; pres, Oceanog Sect, Am Geophys Union, 76-78. *Honors & Awards:* Sverdrup Gold Medal, Am Meteorol Soc, 70. *Mem:* Fel Am Meteorol Soc; Am Geophys Union; fel Am Geophys Soc. *Res:* Dynamic meteorology; physical oceanography; general circulation of the atmosphere and the oceans. *Mailing Add:* 100 Gulich Rd Princeton NJ 08540

BRYAN, PHILIP STEVEN, b Lima, Ohio, Oct 5, 44; m 67; c 2. INORGANIC CHEMISTRY. *Educ:* Ohio State Univ, BS, 66; Univ Mich, PhD(chem), 70. *Prof Exp:* Asst prof chem, Macalester Col, 71-72; asst prof chem, Colgate Univ, 72-77; res chemist, 77-79, SR RES CHEMIST, EASTMAN KODAK CO, 80- *Mem:* Am Chem Soc. *Res:* Synthesis and characterization of transition metal complexes containing multidentate ligands; polymer-bound ligands and complexes and other new inorganic materials. *Mailing Add:* 1245 Stafford Crescent Webster NY 14580-9408

BRYAN, ROBERT FINLAY, b Rhu, Scotland, May 15, 33. CRYSTALLOGRAPHY. *Educ:* Glasgow Univ, BSc, 54, PhD(chem), 57. *Prof Exp:* Res assoc chem, Fed Inst Technol, Zurich, 57-59; res assoc biol, Mass Inst Technol, 59-61; asst prof biophys, Sch Med, Johns Hopkins Univ, 61-67; assoc prof, 67-82, PROF CHEM, UNIV VA, 82- *Concurrent Pos:* Fels, Battelle, 57-59, Sloane, 59-60 & NIH, 60-61. *Mem:* Am Crystallog Asn; Royal Soc Chem. *Res:* X-ray diffraction studies of molecular and crystal structure, especially of materials showing liquid-crystalline behavior. *Mailing Add:* Dept Chem Univ Va Charlottesville VA 22901

BRYAN, ROBERT H(OWELL), b Hendersonville, Tenn, May 19, 24; m 46; c 4. MATERIALS SCIENCE, NUCLEAR ENGINEERING. *Educ:* US Mil Acad, BS, 46; NC State Col, MS, 53, PhD(nuclear eng), 54. *Prof Exp:* Spec sci employee, Argonne Nat Lab, 54-56; tech analyst, US Atomic Energy Comn, 56-58, reactor engr, 58-61, chief, Res & Power Reactor Safety Br, 61-64, chief, Facilities Standards Br, 64-67; asst dir nuclear safety prog, Oak Ridge Nat Lab, 67-74, asst dir heavy-sect steel technol prog, 74-82, struct testing & anal group leader, 82-89; CONSULT, 89- *Mem:* Am Phys Soc. *Res:* Physics of fast nuclear reactors; neutron production by fission; reactor safety; fracture mechanics; large vessel testing. *Mailing Add:* 5424 Yosemite Trail Knoxville TN 37909

BRYAN, ROBERT NEFF, b Salt Lake City, Utah, Mar 18, 39; m 88; c 3. LINEAR ALGEBRA. *Educ:* Univ Utah, BA, 61, MA, 62, PhD(differential equations), 65. *Prof Exp:* Asst prof math, Ithaca Col, 65-69; asst prof, 69-70, ASSOC PROF MATH, UNIV WESTERN ONT, 70- *Concurrent Pos:* vis prof, Univ Ulm, WGer, 78-79; vis scholar, Univ Utah, 85-86. *Mem:* Math Asn Am; Can Math Soc. *Res:* Linear differential systems with general boundary conditions; unbounded linear operators. *Mailing Add:* Dept Math Univ Western Ont London ON N6A 5B7 Can

BRYAN, RONALD ARTHUR, b Portland, Ore, June 16, 32; m 53, 63; c 4. NUCLEAR & PARTICLE THEORY. *Educ:* Yale Univ, BS, 54; Univ Rochester, PhD(theoret physics), 61. *Prof Exp:* Res assoc theoret physics, Univ Calif, Los Angeles, 60-63; physicist, Lab Nuclear Physics, Univ Paris, 63-64; physicist, Lawrence Radiation Lab, 64-67; vis lectr, 68-69, assoc prof, 69-73, PROF PHYSICS, TEX A&M UNIV, 73- *Concurrent Pos:* NATO fel, France, 63-64; long term vis staff mem, Los Alamos Sci Lab, 73-74; Nordic Inst Theoret Atomic Physics fel, Univ Helsinki, Finland, 74-75; vis prof, Univ Nijmegen, Netherlands, 79-80. *Mem:* Fel Am Phys Soc; AAAS. *Res:* Scattering theory; two-nucleon interaction; elementary particles. *Mailing Add:* Dept Physics Tex A&M Univ College Station TX 77843

BRYAN, SARA E, b Yantley, Ala, Sept 22, 22. BIOCHEMISTRY. *Educ:* Auburn Univ, BS, 44; Baylor Univ, MS, 58, PhD(biochem), 64. *Prof Exp:* Med technologist, Hermann Hosp, Houston, 46-54; res assoc hemat, Southwestern Med Sch, Univ Tex, 54-56; instr chem, Univ New Orleans, 58-65, from asst prof to assoc prof biol sci, 66-77, prof biol sci, 77-; AT LA STATE UNIV. *Concurrent Pos:* Res assoc, Fla State Univ, 65-66; Brown-Hazen Fund Res Corp res grant, 67-; NIH res grant, 72- *Mem:* Am Chem Soc; Sigma Xi. *Res:* Role of metals in biological processes, especially interactions of divalent metals with nucleic acids and proteins. *Mailing Add:* Dept Biol Univ New Orleans New Orleans LA 70148

BRYAN, THOMAS T, b Yugoslavia, Sept 27, 25; US citizen; m 49; c 4. PHYSICAL CHEMISTRY, BIOMATERIALS. *Educ:* Univ Belgrade, MS, 52. *Prof Exp:* Res chemist photochem, Lindy Ain, 54-56; res chemist crystal growth, Carborundum, 56-58; res chemist photochem, Am Photog Equip Co, 58-61; RES SPECIALIST BIOMAT, 3M CO, 61- *Mem:* Am Chem Soc; Soc Biomat. *Res:* Biomaterials as artificial prosthetic devices; particularly hard tissue replacement and dental applications; surface and interface reactions to biomedical systems; dental materials. *Mailing Add:* Bldg 260-2A-10 3M Co St Paul MN 55144-1000

BRYAN, THORNTON EMBRY, b Frankfort, Ky, Mar 16, 27. FAMILY MEDICINE. *Educ:* Univ Ky, BS, 49; Univ Louisville, MD, 54. *Prof Exp:* Assoc prof family pract, Col Med, Univ Iowa, 71-74; PROF FAMILY MED & CHMN DEPT, CTR HEALTH SCI, UNIV TENN, 74- *Concurrent Pos:* AMA Physicians Recognition Award Continuing Med Educ, 70-73. *Mem:* AMA; Am Acad Family Physicians. *Res:* Clinical health care; ambulatory patient data, collection, storage and retrieval. *Mailing Add:* Univ Ala 201 Governors Dr SW Huntsville AL 35801

BRYAN, WILBUR LOWELL, b South River, NJ, Feb 20, 21; m 47. PHARMACEUTICAL CHEMISTRY. *Educ:* Lafayette Col, BA, 42. *Prof Exp:* Chemist explosives, Hercules Powder Co, 42-43; chemist, 45-61, res investr, 61-69, SR RES INVESTR PHARMACEUT, E R SQUIBB & SONS, 69- *Mem:* Am Chem Soc; NY Acad Sci. *Res:* Isolation techniques of natural products, particularly antibiotics. *Mailing Add:* 585 Canal Rd Somerset NJ 08873

BRYAN, WILFRED BOTTRILL, b Waterbury, Conn, Feb 18, 32; m 53; c 3. MARINE GEOLOGY, PLANETARY GEOLOGY. *Educ:* Dartmouth Col, BA, 54; Univ Wis-Madison, MA, 56, PhD(petrol), 59. *Prof Exp:* Geologist, M A Hanna Co, 59-61; sr lectr petrol, Univ Queensland, 61-67; sr fel, Carnegie Inst, Geophys Lab, 67-70; assoc scientist, 70-82, SR SCIENTIST, DEPT GEOL & GEOPHYS, WOODS HOLE OCEANOG INST, 82- *Concurrent Pos:* Field asst, US Geol Survey, 53-59; consult, Exoil Proprietary Ltd & Magellan Petrol Co, 63-65 & Tennant Mineral Develop Co Proprietary Ltd, 65-67; mem subcomt volcanology, Australian Acad Sci, 66-67; leader, Nat Geog-Smithsonian Exped, Tonga Islands, 69; chief scientist oceanog expeds, 71-88; mem team 5, Planetary Basaltic Volcanism Proj, NASA, 76-79; vis prof, Dept Oceanog, Univ Wash, 80, Dept Earth Planetary Sci, McDonnell Ctr, 83, Dept Geol, Univ Hawaii, 89. *Mem:* Geol Soc Australia; Geol Soc Am; Am Geophys Union; Int Asn Volcanology; Mineral Soc Gt Brit & Ireland; Sigma Xi; Math Asn Am. *Res:* Lunar and planetary volcanology; distribution and geological relationships of volcanic rocks in and around the ocean basins; compositional relationships and fractionation effects within genetically related rock suites; relations between geophysical phenomena and igneous activity; seafloor structure and morphology. *Mailing Add:* Dept Geol & Geophys Woods Hole Oceanog Inst Woods Hole MA 02543

BRYAN, WILLIAM L, b Gainesville, Fla, Sept 25, 28; m 57; c 3. BIOTECHNOLOGY. *Educ:* Univ Fla, BChE, 49; Yale Univ, DEng, 59. *Prof Exp:* Develop engr, Rohm & Haas Co, 54-59, res engr ballistics, Redstone Arsenal Res Div, 59; res engr, E I du Pont de Nemours & Co, Inc, 59-63, staff engr film dept, 63-65, sr res engr textile fibers dept, 65-71; res chem engr, US Citrus & Subtrop Prod Lab, USDA, 71-79, res chem engr, Southern Agr Energy Ctr, 80-85, chem engr, Northern Regional Res Ctr, 85-90, CHEM ENGR, NAT CTR AGR UTILIZATION RES, 91- *Mem:* Am Inst Chem Engrs; Am Chem Soc; AAAS. *Res:* Research and development for manufacture of agricultural chemicals, ion exchange resins, rocket propellants, industrial films and fibers, processing of citrus fruit, sweet sorghum processing, solid substrate fermentation, viscosity measurement, manufacture of acetate road deicers. *Mailing Add:* USDA-Nat Ctr Agr Utilization Res 1815 N University St Peoria IL 61604

BRYAN, WILLIAM PHELAN, b Chicago, Ill, June 4, 30; m 61. BIOCHEMISTRY, BIOPHYSICS. *Educ:* Univ Calif, Los Angeles, BS, 52, MS, 53; Univ Calif, Berkeley, PhD(chem), 57. *Prof Exp:* Grant, Calif Inst Technol, 57-59; fel, Carlsberg Lab, Copenhagen, Denmark, 59-60; chemist, Danish Atomic Energy Comn, Riso, 60-62; grant, Cornell Univ, 62-64; asst prof chem, Boston Univ, 64-69; ASSOC PROF BIOCHEM, MED CTR, IND UNIV, INDIANAPOLIS, 69- *Mem:* AAAS; Am Soc Biol Chemists; Am Chem Soc; Biophys Soc. *Res:* Physical chemistry of proteins, water-protein interactions; membrane structure and function. *Mailing Add:* Dept Biochem & Molecular Biol Ind Univ Med Ctr Indianapolis IN 46223

BRYANS, ALEXANDER (MCKELVEY), b Toronto, Ont, Sept 16, 21; m 54; c 3. MEDICINE. *Educ:* Univ Toronto, MD, 44; Mich State Univ, MA, 71; FRCP, 52. *Prof Exp:* McLaughlin traveling fel, 56-57; assoc prof, 59-60, dir, Health Sci Off Educ, 71-87, PROF PEDIAT, QUEEN'S UNIV, ONT, 60- *Mem:* Fel Am Acad Pediat; Can Med Asn; Can Pediat Soc; Can Physicians Prev War (pres, 90-91). *Res:* Medical education. *Mailing Add:* 31 Lakeland Pt Dr Kingston ON K7M 4E8 Can

BRYANS, CHARLES IVERSON, JR, b Augusta, Ga, May 11, 19; m 46; c 5. OBSTETRICS & GYNECOLOGY. *Educ:* Univ Ga, BS, 40, MD, 43; Am Bd Obstet & Gynec, dipl. *Prof Exp:* Fel obstet & gynec, 48, PROF OBSTET & GYNEC, MED COL GA, 62- *Concurrent Pos:* Consult, US Army Hosp, Ft Gordon, Ga, 53-, Milledgeville State Hosp, Macon Hosp, Univ Hosp, Augusta & St Joseph's Hosp, 62-; consult, Greenville City Hosp & Mem Med Ctr, Savannah. *Mem:* Am Col Obstet & Gynecol; AMA; Sigma Xi. *Res:* Obstetrical analgesia and anesthesia; trophoblastic and other placental abnormalities; toxemia of pregnancy. *Mailing Add:* Dept Obstet & Gynec Med Col Ga Augusta GA 30912

BRYANS, JOHN THOMAS, b Paterson, NJ, June 1, 24; m 59; c 2. ANIMAL VIROLOGY. *Educ:* Fla Southern Col, BS, 49; Univ Ky, MS, 51; Cornell Univ, PhD(vet bact & path), 54. *Hon Degrees:* DMV, Univ Bern, Switz, 78. *Prof Exp:* Virologist, Univ Ky, 54-60, chmn dept, 74-87, prof vet sci, Agr Exp Sta, 60-91, distinguished alumni prof, 87-91, EMER PROF VET SCI, UNIV KY, 91- *Mem:* AAAS; Am Soc Microbiol; Am Vet Med Asn. *Res:* Etiology, immunology and epizootiology of infectious diseases of domestic animals; pathogenesis and immunology of herpesviral infection. *Mailing Add:* Dept Vet Sci Univ Ky Lexington KY 40546-0099

BRYANS, TRABUE DALEY, b Atlanta, Ga, Mar 20, 52. MICROBIOLOGY, BIOLOGY. *Educ:* Univ Ga, BS, 75. *Prof Exp:* Chief microbiologist, MacMillan Res Ltd, 75-76; med technologist microbiol, Med Col Ga, 77-78; dir microbiol, 78-85, VPRES, MACMILLAN RES, LTD, 85- *Mem:* Am Soc Clin Pathologists; Am Soc Micro Biol. *Res:* Sterility testing of medical products, microbiology of foods and water. *Mailing Add:* Mac Millan Res 1221 Barcley Circle Marrietta GA 30060-2903

BRYANT, BARBARA EVERITT, b Ann Arbor, Mich, Apr 5, 26; m 48; c 3. SURVEY RESEARCH. *Educ:* Cornell Univ, BA, 47; Mich State Univ, MA, 67, PhD(commun), 70. *Prof Exp:* Art ed, Chem Eng Mag, McGraw-Hill Publ Co, NY, 47-48; ed res asst, Grad Col, Univ Ill, 48-49; freelance ed, 50-61; dir pub rels & sci course coordr, Div Continuing Educ, Oakland Univ, Rochester, Mich, 61-65; res asst, Mich State Univ, 65-70; sr res analyst & vpres, Mkt Opinion Res, Detroit, 70-89; SR VPRES & DIR, BUR CENSUS, US DEPT COM, WASHINGTON, DC, 89- *Concurrent Pos:* Mem, Job Develop Auth, State Mich, 80-85; mem adv comt, US Census Bur, 80-86; adj prof, Grad Sch Bus Admin, Univ Mich, 84-85; nat headliner, Women Commun, Inc, 80. *Mem:* Am Mkt Asn (vpres, 70-80 & 82-84); Women Commun; Am Statist Asn; Am Asn Pub Opinion Res; Pub Rels Soc Am. *Res:* Combining attitudes, interests and behaviors with traditional demographics to segment populations into life style clusters. *Mailing Add:* Bur Census Fed Ctr Washington DC 20233

BRYANT, BEN S, b Seattle, Wash, Mar 28, 23; m 47; c 3. WOOD TECHNOLOGY. *Educ:* Univ Wash, BS, 47, MS, 48; Yale Univ, DFor(wood technol), 51. *Prof Exp:* From instr to assoc prof, 49-69, dir inst forest prod, 60-64, PROF WOOD SCI & TECHNOL, COL FOREST RESOURCES, UNIV WASH, 69- *Concurrent Pos:* Consult, forest prod industs, asn, chem & mach co, US & foreign govts in forest prod develop, adhesive develop, forest prod industs develop & surv, 51- *Mem:* Soc Am Foresters; Forest Prod Res Soc; Tech Asn Pulp & Paper Indust; Soc Wood Sci & Technol. *Res:* Structural utilization of wood; adhesion and gluing process technology; product development methods; forest utilization. *Mailing Add:* 4102 51st Ave Seattle WA 98105

BRYANT, BILLY FINNEY, b McKenzie, Tenn, Nov 29, 22; m 46; c 3. MATHEMATICS. *Educ:* Univ SC, BS, 45; Peabody Col, BA, 48; Vanderbilt Univ, PhD(math), 54. *Prof Exp:* From instr to prof, 48-86, chmn dept, 70-76, EMER PROF MATH, VANDERBILT UNIV, 86- *Concurrent Pos:* Ford Found fac fel, Princeton Univ, 55-56; sci fac fel, Univ Calif, Berkeley, 67-68. *Mem:* Am Math Soc; Math Asn Am. *Res:* Point set topology. *Mailing Add:* 6020 Sherwood Dr Nashville TN 37215

BRYANT, BRUCE HAZELTON, b New York, NY, Sept 25, 30; m 55; c 4. REGIONAL GEOLOGY. *Educ:* Dartmouth Col, AB, 51; Univ Wash, PhD(geol), 55. *Prof Exp:* Geologist, 55-63, res geologist, 63-69, GEOLOGIST, US GEOL SURV, 69- *Mem:* Mineral Soc Am; Geol Soc Am; AAAS; Am Geophys Union. *Res:* Geologic mapping; structure; petrology of igneous and metamorphic rocks; boundary between the basin and range and Colorado plateau provinces in West Central Arizona. *Mailing Add:* US Geol Surv Denver Fed Ctr Denver CO 80225

BRYANT, DONALD G, b Hollywood, Calif, June 21, 27. MINING. *Educ:* Univ Ariz, BA, 54; Calif Tech Inst, MA, 55; Stanford Univ, PhD(geol), 64. *Prof Exp:* Chief explor geologist, Molybdenum Corp, 64-70; pres, Donald G Bryant Inc, 70-87; DESIGNING ENGR, MARTIN MARIETTA, 86- *Mem:* Fel Geol Soc Am; Soc Econ Geologists. *Mailing Add:* 714 S Filmore St Denver CO 80209

BRYANT, EDWARD CLARK, b Hat Creek, Wyo, June 28, 15; m 41; c 2. STATISTICS. *Educ:* Univ Wyo, BS, 38, MS, 40; Iowa State Univ, PhD(statist), 55. *Prof Exp:* Jr statistician & economist, Interstate Com Comn, 40-42; statistician, War Prod Bd, 42-43; chief statistician, Armed Forces, Mid Pac, 46-47; from asst prof to prof statist, Univ Wyo, 47-61; pres, 61-78, CHMN, STATIST & MGT, WESTAT INC, 78- *Concurrent Pos:* Vis prof math, Ariz State Univ, 61-63. *Mem:* Fel Am Statist Asn; fel AAAS; Int Asn Survey Statisticians; Prof Servs Coun; Am Soc Qual Control. *Res:* Efficiency of sample survey designs; statistical analysis; evaluation of government programs; information retrieval; statistical process control. *Mailing Add:* Westat Inc 1650 Research Blvd Rockville MD 20850

BRYANT, ERNEST ATHERTON, b Brewster, Mass, Oct 13, 31; m 53; c 2. NUCLEAR CHEMISTRY, GEOCHEMISTRY. *Educ:* Univ NMex, BS, 53; Wash Univ, PhD(radiochem), 56. *Prof Exp:* Instr chem, Wash Univ, 56-57; mem staff radiochem, 57-71, alt group leader radiochem, 71-80, group leader isotope geochem, 80-87, STAFF MEM ISOTOPE GEOCHEM, LOS ALAMOS NAT LAB, UNIV CALIF, 87- *Mem:* AAAS; Am Chem Soc. *Res:* High temperature chemistry; nuclear chemistry; radiochemical analysis; reactor technology; geologic behavior of radioactive waste; atmospheric chemistry; mass spectrometry; photo acoustic spectroscopy. *Mailing Add:* 125 La Senda Los Alamos NM 87544

BRYANT, GEORGE MACON, b Anniston, Ala, Aug 3, 26; m 50; c 2. POLYMER CHEMISTRY, TEXTILE & PAPER TECHNOLOGY. *Educ:* Auburn Univ, BS, 48; Inst Textile Technol, MS, 50; Princeton Univ, MA, 52, PhD(chem), 54. *Prof Exp:* Res chemist, Chem Div, Union Carbide Corp, 54-58, group leader, 58-63, res & develop mgr textile prod, 63-66, technol mgr, Fibers & Fabrics Div, 66-71, develop assoc res & develop, 71-75, corp res fel, 75-86; CONSULT, 86- *Concurrent Pos:* Lectr, Fiber Soc, 66 & 81; consult, fiber, textile & paper fields, 86- *Honors & Awards:* Award, Fiber Soc, 64; Henry Milson Award, Am Inst Textile Chemists & Colorists, 82; Sci Achievement Award, Am Chem Soc, 79. *Mem:* Am Chem Soc; Sigma Xi; Am Asn Textile Chemists & Colorists; Fiber Soc (pres, 68). *Res:* Physical chemistry of textile fibers; chemicals and polymers for oil recovery; foam finishing and dyeing of fabrics and paper; mechanical behavior of polymers; textile and paper chemicals. *Mailing Add:* 1204 Williamsburg Way Charleston WV 25314

BRYANT, GLENN D(ONALD), b Columbus, Ohio, Dec 10, 18; m 44; c 3. AERONAUTICS, MARINE ENGINEERING. *Educ:* Miss State Col, BS, 49, MS, 51. *Prof Exp:* Aeronaut engr, Aerodyn Res Aerophys Dept, Miss State Univ, 51-64; chief engr & vpres prod, Burns Aircraft Co, 64-67; asst prof, 69-77, ASSOC PROF MARINE ENG & MECH ENG, MISS STATE UNIV, 77- *Concurrent Pos:* Proprietor, Bryant Aircraft Co, 68- *Mem:* Am Inst Aeronaut & Astronaut. *Res:* Aerodynamics, specializing in secondary problems related to boundary layer control; marine vehicle design; causal modeling in physics and related experiments. *Mailing Add:* Rte 4 Box 329 Starkville MS 39759

BRYANT, HOWARD CARNES, b Fresno, Calif, July 9, 33; m 60; c 3. HIGH ENERGY ATOMIC PHYSICS. *Educ:* Univ Calif, Berkeley, BA, 55; Univ Mich, MS, 57, PhD(physics), 61. *Prof Exp:* From asst prof to assoc prof, 60-71, PROF PHYSICS, UNIV NMEX, 71- *Concurrent Pos:* Vis scientist, Stanford Linear Accelerator Ctr, 67-69; consult, Los Alamos Sci Lab, 61-; sr vis fel, Queen Mary Col, Univ London, 75-76; Fulbright guest prof, Univ Innsbruck, Austria, 83-84; fac res lectr, 85. *Mem:* Optical Soc Am; fel Am Phys Soc; Am Asn Physics Teachers. *Res:* Experimental particle and atomic physics; physical optics. *Mailing Add:* Dept Physics Univ NMex Albuquerque NM 87131

BRYANT, HOWARD SEWALL, b Kansas City, Mo, May 26, 28; m 53; c 2. CHEMICAL ENGINEERING. *Educ:* Ga Inst Technol, BChE, 50; Mass Inst Technol, SM, 52, ScD(chem eng), 56. *Prof Exp:* With Union Carbide Nuclear Co, 52-54, Courtaulds N Am, Inc, 56-61, Mobil Chem Co, 61-71 & Mobil Res & Develop Corp, 71-74; CORP VPRES ENG, WITCO CHEM CORP, 74- *Mem:* Am Inst Chem Engrs; Am Chem Soc. *Res:* Development, design, engineering and construction of processes and facilities for chemical and petroleum products. *Mailing Add:* 155 Tice Blvd Woodcliffe Lake NJ 07675

BRYANT, JAMES BERRY, JR, microbiology, for more information see previous edition

BRYANT, JOHN H(AROLD), b Baird, Tex, Apr 15, 20; m 48; c 3. PHYSICAL ELECTRONICS. *Educ:* Agr & Mech Col Tex, BS, 42; Univ Ill, MS, 47, PhD(elec eng), 49. *Prof Exp:* Head div microwave electron tubes, Bendix Corp, 49-55, supvr engr res labs div, 55-62; pres, Omni Spectra, Inc, Farmington, 62-80. *Mem:* Am Phys Soc; fel Inst Elec & Electronics Engrs. *Res:* Microwave components; radar and missile guidance; electron tubes. *Mailing Add:* 1505 Sheridan Dr Ann Arbor MI 48104

BRYANT, JOHN HARLAND, b Tucson, Ariz, Mar 8, 25; m 56; c 3. PUBLIC HEALTH. *Educ:* Univ Ariz, BA, 49; Columbia Univ, MD, 53. *Prof Exp:* Intern & asst resident, Presby Hosp, NY, 53-56; res fel biochem, Nat Found, NIH, Md, 56-57 & 58-59, Munich, 57-58; spec trainee hemat, Dept Med, Sch Med, Wash Univ, 59-60; asst prof med, Col Med, Univ Vt, 60-64, assoc prof & asst dean, 64-65; staff mem, Rockefeller Found & prof, Fac Med, Ramathibodi Hosp, Bangkok, Thailand, 65-71; dir, Sch Pub Health, Col Physicians & Surgeons, Columbia Univ, 71-78; dep asst secy int health & dir off int health, HEW, 78-82, spec asst to asst secy health, Dept Health & Human Serv, 82-85; PROF & CHMN, DEPT COMMUNITY HEALTH SERVS, AGA KHAN UNIV, KARACHI, PAKISTAN, 85- *Concurrent Pos:* Asst attend physician, Mary Fletcher Hosp, 60-; asst attend physician, Degoesbriand Mem Hosp, 60-63, assoc attend physician, 63-; spec staff mem, Study Med Educ Develop Nations, Rockefeller Found, 64-65; chmn, Christian Med Comn, World Coun Churches, Geneva, 68-; Joseph R DeLamar prof pub health, Columbia Univ, 73-; consult, WHO & CIOMS, 84-88. *Mem:* Inst Med-Nat Acad Sci. *Res:* National and international approaches to evaluation and design of health care systems; education of health personnel; national health insurance and health manpower policies. *Mailing Add:* c/o Aga Khan Univ PO Box 3500 Stadium Rd Karachi-5 Pakistan

BRYANT, JOHN LOGAN, b Corinth, Miss, Aug, 27, 40; m 58; c 2. MATHEMATICS. *Educ:* Univ Miss, BS & MS, 62; Univ Ga, PhD(math), 65. *Prof Exp:* Asst prof math, Univ Miss, 65-66; from asst prof to assoc prof, 66-74, chmn, Dept Math & Comput Sci, 80-84 PROF MATH, FLA STATE UNIV, 74- *Mem:* Am Math Soc; Math Asn Am. *Res:* Geometric topology; properties of topological embeddings of polyhedra. *Mailing Add:* Dept Math Fla State Univ Tallahassee FL 32306

BRYANT, LAWRENCE E, JR, b Nashville, Tenn, Mar 9, 36; m 71; c 2. MECHANICAL ENGINEERING, NUCLEAR ENGINEERING. *Educ:* Vanderbilt Univ, BS, 58, MS, 59. *Prof Exp:* Naval officer mech & nuclear eng, Naval Res Lab, Washington, DC, 59-61; SUPVR NONDESTRUCTIVE TESTING, LOS ALAMOS NAT LAB, 61- *Mem:* Fel Am Soc Nondestructive Testing. *Res:* Industrial radiography; ranging from low energy, microradiography to high energy radiography; flash (high speed) radiography and cine radiography, infrared imaging, high speed videography. *Mailing Add:* Los Alamos Nat Lab PO Box 1663 MS-C914 Los Alamos NM 87545

BRYANT, LESTER RICHARD, b Louisville, Ky, Sept 8, 30; m 51; c 2. MEDICINE, THORACIC SURGERY. *Educ:* Univ Ky, BS, 51; Univ Cincinnati, MD, 55, DSc(surg), 62; Am Bd Surg, dipl, 62; Am Bd Thoracic Surg, dipl, 63. *Prof Exp:* Fel physiol, Col Med, Baylor Univ, 61; instr, Univ Cincinnati, 61-62; from instr to prof surg, Med Ctr, Univ Ky, 62-73; prof surg & chief thoracic & cardiovasc surg, La State Univ Med Ctr, New Orleans, 73-77; prof surg & chmn dept, Col Med, E Tenn State Univ, 77-85; PROF SURG, DEAN & VPRES, SCH MED, MARSHALL UNIV, HUNTINGTON, WVA, 85- *Concurrent Pos:* Resident coordr, Surg Adjuvant Breast Proj, Nat Coop Study, 61-62; responsible investr coop study on coronary artery dis & coop study of cirrhosis & esophago-gastric varices, Vet Admin, 62-63, consult, 63-; consult, USPHS, 65-, Charity Hosp of La & Vet Admin Hosp, New Orleans, 73-77; mem assoc staff, Southern Baptist Hosp, New Orleans & Hotel Dieu Hosp, New Orleans, 74-77. *Mem:* Am Asn Thoracic Surg; Am Col Surg; AMA; Asn Acad Surg; Int Soc Surg. *Res:* Cardiothoracic surgery; development of primate smoking model; evaluation of prophylactic antibiotics in thoracic trauma; comparison of incisions for tracheostomy; study of bacterial colonization profile in patients with tracheostomy. *Mailing Add:* Univ Mo MA204 Med Sci Bldg Columbia OH 65212

BRYANT, MARVIN PIERCE, b Boise, Idaho, July 4, 25; m 46; c 5. MICROBIAL ECOLOGY. *Educ:* State Col Wash, BS, 49, MS, 50; Univ Md, PhD(bact), 55. *Prof Exp:* Res asst, State Col Wash, 49-51; bacteriologist, Agr Res Serv, USDA, 51-64; assoc prof bact, 64-66, PROF MICROBIOL, UNIV ILL, URBANA-CHAMPAIGN, 66- *Concurrent Pos:* Ed, Appl Environ Microbiol, 68-71, ed-in-chief, 71-; trustee, Bergey's Manual of Determinative Bact, 75-, vchmn trust, 76-; vis scientist, Inst Microbiol, Gottingen, WGer, 76-77. *Honors & Awards:* Superior Serv Award, USDA, 59; Borden Award, Am Dairy Sci Asn, 78. *Mem:* Nat Acad Sci; fel AAAS; Am Soc Microbiol; Am Dairy Sci Asn; Brit Soc Gen Microbiol. *Res:* Ruminal bacteriology; systematics; nutrition; physiology; ecology; nonsporeforming anaerobic bacteria and ciliate protozoa; methanogenic bacteria. *Mailing Add:* 1003 S Orchard St Urbana IL 61801

BRYANT, MICHAEL DAVID, b Danville, Ill, Feb 8, 51; m 78; c 5. SOLID MECHANICS & TRIBOLOGY. *Educ:* Univ Ill, BS, 72; Northwestern Univ, MS, 80, PhD(eng sci), 81. *Prof Exp:* Asst prof, 81-85, assoc prof mech & aerospace eng, NC State Univ, 85-88; ASSOC PROF MECH ENG, UNIV TEX, AUSTIN, 88- *Concurrent Pos:* Consult, 85-; presidential young investr award, NSF, 85. *Mem:* Am Soc Mech Engrs; Soc Mfg Engrs; Inst Elec & Electronics Engrs. *Res:* Determination of failure in bearings, wheels and sliders using fracture mechanics; wear of sliding contacts; active control of structures; precision positioning systems; precision machining systems; physics of electrical contacts; robotics and control theory. *Mailing Add:* Dept Mech Eng Univ Tex Austin TX 78712-1063

BRYANT, PATRICIA SAND, b Minot, NDak, July 18, 40; m 75; c 2. SCIENCE POLICY, PSYCHOLOGY. *Educ:* Univ Wash, BS, 61, MS, 63, PhD(psychol), 64. *Prof Exp:* Res asst psychol, Univ Wash, 61-64, Sch Med, 64-66, asst prof, Sch Nursing, 66-67, from asst prof to assoc prof, Dept Rehab Med, 68-75; HEALTH SCIENTIST ADMIN, CRANEOFACIAL ANOMALIES PAIN CONTROL & BEHAV RES BR, NAT INST DENT RES, NIH, 76- *Concurrent Pos:* Consult, Seattle Artificial Kidney Ctr, 65-67; SRS grant, 73-75; adj assoc prof, Uniformed Serv Univ Health Sci, Bethseda, Md, 81-84. *Honors & Awards:* Merit Award, NIH, 85. *Mem:* Am Psychol Asn; Am Asn Dent Res; Int Asn Dent Res; Am Cong Rehab Med; Behav Sci Dent Res (pres, 85-86); Int Asn Study Pain. *Res:* Behavioral research relating to oral health; psychological aspects of illness-health; rehabilitation; behavioral factors in chronic pain; behavioral factors in oral health and dental care. *Mailing Add:* Nat Inst Dent Res Extramural Prog NIH Westwood Bldg Rm 506 Bethesda MD 20892

BRYANT, PAUL JAMES, b Kansas City, Mo, May 11, 29; m 60. SOLID STATE PHYSICS. *Educ:* Rockhurst Col, BS, 51; St Louis Univ, MS, 53, PhD(physics), 57. *Prof Exp:* Prin physicist, Midwest Res Inst, 58-68; assoc prof, 68-74, PROF PHYSICS, UNIV MO, KANSAS CITY, 74- *Mem:* Am Phys Soc; Am Vacuum Soc. *Res:* Solid state structure studies; surface physics; field emission microscopy; friction of solids; ultra high vacuum technology. *Mailing Add:* Dept Physics Univ Mo 1110 E 48th St Kansas City MO 64110

BRYANT, PETER JAMES, Brit citizen. DROSOPHILA GENETICS, MOLECULAR GENETICS. *Educ:* Kings Col, London, BSc, 64; Univ Col, London, MSc, 65; Univ Sussex, Falmer, DPhil, 67. *Prof Exp:* Scholar, Case Western Univ & Univ Calif, Irvine, 67-70; from lectr to assoc prof, 70-77, PROF GENETICS, DEPT DEVELOP & CELL BIOL, UNIV CALIF, IRVINE, 77-, DIR, DEVELOP BIOL CTR, 79- *Concurrent Pos:* Prin investr, NIH grants, 74-; dir, NIH Training grant, 79-; consult, NSF, 83-85; ed-in-chief, Develop Biol, 85- *Mem:* Soc Develop Biol; Genetics Soc Am; AAAS; Int Soc Develop Biol; Soc Exp Biol; Am Cetacean Soc (pres, 82). *Res:* Analysis of pattern formation and growth control during organ development in animals, using molecular genetic approaches in the fruitfly Drosophila melanogaster. *Mailing Add:* Dept Cell & Molecular Biol Univ Calif Irvine CA 92717

BRYANT, RALPH CLEMENT, b New Haven, Conn, Sept 27, 13; m 37; c 2. FOREST MANAGEMENT. *Educ:* Yale Univ, BS, 35, MF, 36; Duke Univ, PhD, 51. *Prof Exp:* From foreman to dist forest ranger, US Forest Serv, 36-46; assoc prof forest mgt & utilization & actg head dept, Colo Agr & Mech Col, 47-52; prof, 52-79, EMER PROF FOREST MGT, NC STATE UNIV, 79- *Concurrent Pos:* Mem & vchmn, NC Forestry Coun, 75-78. *Mem:* Soc Am Foresters; Can Pulp & Paper Asn. *Res:* Prescribed burning. *Mailing Add:* 4213 Brockton Dr Raleigh NC 27604

BRYANT, REBECCA SMITH, b Checotah, Okla, May 23, 55; m 84; c 1. MICROBIAL ENHANCED OIL RECOVERY. *Educ:* Okla State Univ, BS, 76, MS, 79, PhD(microbiol), 82. *Prof Exp:* Vis asst prof microbiol, Univ Tulsa, 82-83; PROJ LEADER & SR BIOLOGIST, NAT INST PETROL & ENERGY RES, 83- *Concurrent Pos:* Adj instr cell biol, Univ Ctr Tulsa, 83 & adj prof, Univ Tulsa, 84-; consult, Biotechnol Res Workshop, 85-91. *Mem:* Am Soc Microbiol; Soc Indust Microbiol; Soc Petrol Engrs; Am Asn Univ Women; Sigma Xi; Am Soc Testing & Mat. *Res:* Use of micro-organisms for petroleum production and environmental pollutant degradation for government agencies and industry; directing microbial research and writing proposals relating to above topics. *Mailing Add:* Nat Inst Petrol & Energy Res PO Box 2128 Bartlesville OK 74005

BRYANT, RHYS, b Swansea, Wales, Nov 28, 36; m 60; c 3. ORGANIC CHEMISTRY. *Educ:* Univ Wales, BSc, 57, PhD(org chem), 60. *Prof Exp:* Fulbright res scholar chem, Yale Univ, 60-61; fel org chem, Mellon Inst, 61; res chemist, Unilever Res Lab, Eng, 61-63; asst lectr chem, Univ Manchester, 63-65; group leader instrumentation & methods develop, Res Ctr, Mead Johnson & Co, 65-67; sect leader, 67-68, dir pharmaceut qual control, 68-76; vpres qual control, Plough Div, Schering-Plough Corp, Memphis, 76-78; group dir qual assurance, Pharmaceut Prod Group, Searle Worldwide Pharmaceut, 78-79, vpres qual assurance, NAm Region, 79-80, group dir qual assurance, Searle Pharmaceut Inc, 81-86, SR DIR, TECH OPERS, G D SEARLE & CO, 87- *Mem:* Am Chem Soc; Sigma Xi; Chem Soc; Royal Inst Chem. *Res:* Synthesis of oxygen heterocyclic compounds; sesquiterpenoids; analysis of experimental drug products; author of Quality Control Pharmaceutical Handbook. *Mailing Add:* 15 Whitby Circle Lincolnshire IL 60069

BRYANT, ROBERT EMORY, b Charleston, SC, Feb 26, 42; m 68; c 1. CELL BIOLOGY, GENETICS. *Educ:* Univ Ala, BS, 64; Purdue Univ, MS, 66; Univ Ill, PhD (microbiol), 73. *Prof Exp:* Fel cell biol, Sch Med, Yale Univ, 73-76; ASST PROF BIOCHEM, UNIV TENN, 77- *Concurrent Pos:* Jane Coffin Childs Fund Med Res fel, 73-75. *Mem:* Am Soc Microbiol; AAAS. *Res:* Genetics of cultured animal cells; nucleotide metabolism. *Mailing Add:* 1200 Foxcrolf Dr Knoxville TN 37923

BRYANT, ROBERT GEORGE, b Mineola, NY, Sept 13, 43; m 65; c 3. BIOPHYSICAL CHEMISTRY. *Educ:* Colgate Univ, AB, 65; Stanford Univ, PhD(chem), 69. *Prof Exp:* From asst prof to assoc prof, Univ Minn, Minneapolis, 69-78, prof chem, 78-87; UNIV ROCHESTER MED CTR, DEPT BIOPHYS, 88- *Concurrent Pos:* Teacher-scholar grant, Dreyfus Found, 74. *Mem:* AAAS; Am Chem Soc; Biophys Soc; Sigma Xi. *Res:* Application of nuclear magnetic resonance spectroscopy to investigation of metal ion protein interactions, water macromolecule interactions, metalloenzymes, and macromolecule dynamics. *Mailing Add:* Dept Biophys Univ Rochester 601 Elmwood Ave Rochester NY 14642

BRYANT, ROBERT L, b New York, NY, Jan 3, 28; m 49; c 3. FORESTRY. *Educ:* La State Univ, BS, 53, MS, 54; Mich State Univ, PhD(forestry), 63. *Prof Exp:* Biologist wildlife invest, La Wildlife & Fisheries Comn, 52-53, res biologist refuge mgt, 53-54; asst prof forestry, McNeese State Col, 54-59; asst, Mich State Univ, 59-62; from asst prof to prof forestry, McNeese State Univ, 62-76; forest & biol consult, 76-91; Agt, Erwin Heirs Inc, Property Mgt, 78-91; RETIRED. *Concurrent Pos:* Collabr, Lake States Forest Exp Sta, 59-62. *Mem:* Soc Am Foresters; Am Forestry Asn; Ecol Soc Am. *Res:* Forest ecology; ecology of lowland hardwood forests, soils and ground-water; remote sensing of environment within forest structures. *Mailing Add:* 111 Orchard Dr Lake Charles LA 70605

BRYANT, ROBERT WESLEY, LEUKOTRIENE, PROSTAGLANDIN. *Educ:* Fla State Univ, PhD(biochem), 73. *Prof Exp:* SR PRIN SCIENTIST, SCHERING-PLOUGH CORP, 83- *Mailing Add:* Allergy & Inflammation Dept Schering-Plough Corp 60 Orange St Bloomfield NJ 07003

BRYANT, ROBERT WILLIAM, b Oxford, Ohio, June 8, 25; m 46; c 4. APPLIED MATHEMATICS. *Educ:* Miami Univ, AB, 47; Univ Ala, MA, 48. *Prof Exp:* Instr math, Miami Univ, 48-51; rating exam sci & tech personnel, Potomac River Naval Command, 51-52; mathematician, Underwater Sound Propagation, US Naval Res Lab, 52-56, supvry mathematician, 58-59; mathematician, Celestial Mech & Appl Orbit Anal, Goddard Space Flight Ctr, NASA, 59-66; mathematician, Systs Anal Staff, Anti-Submarine Warfare Spec Proj Off, HQS, Washington, DC, 66-72; systs analyst, Anti-Submarine Warfare Systs Proj Off, 72-80; RETIRED. *Mem:* Fel AAAS; Am Math Soc. *Res:* Celestial mechanics; density of upper atmosphere; solar radiation pressure effects on echo type satellites; density interference from drag; signal processing. *Mailing Add:* Rte 2 Box 690 Shannonda Harpers Ferry WV 25425

BRYANT, SHIRLEY HILLS, b Pittsfield, Mass, Dec 28, 24; m 46, 63, 87; c 3. PHARMACOLOGY, PHYSIOLOGY. *Educ:* Aurora Col, BS, 48; Univ Chicago, PhD(physiol), 54. *Prof Exp:* Asst physiol, Univ Chicago, 50-54; from instr to assoc prof pharmacol, 55-69, PROF PHARMACOL, COL MED, UNIV CINCINNATI, 69- *Concurrent Pos:* USPHS career develop award, Univ Cincinnati, 59-69; vis prof, Univ PR, 65-66 & Cambridge Univ, 72-73. *Mem:* AAAS; Soc Neurosci; Biophys Soc; Am Physiol Soc. *Res:* Physiology and pharmacology of excitation, conduction and synaptic transmission in excitable tissues; biophysics nerve and muscle; abnormal repetitive impulse production and excitation-contraction coupling mechanisms in skeletal muscle; myotonia in goats. *Mailing Add:* Dept Pharmacol & Cell Biophysics Univ Cincinnati Col Med Cincinnati OH 45267

BRYANT, STEPHEN G, b New York, NY, Nov 15, 51. POSTMARKETING DRUG SURVEILLANCE. *Educ:* St Johns Univ, BS, 75; Univ Tex, Austin, Pharm D, 79. *Prof Exp:* Fel psychopharmacol, Tex Dept Mental Health, 78-79; clin coordr psychopharmacol, NY Hosp-Cornell Med Ctr, 79-82; ASSOC PROF PSYCHIAT & PSYCHOPHARMACOL, UNIV TEX MED BR, GALVESTON, 82- *Concurrent Pos:* Prin investr, Food & Drug Admin Psychopharmacol Violence Study, 82-; mem credentials comt, Am Col Clin Pharm, 85-87; mem med adv comt, Tex Bd Mental Health, 85- *Mem:* Sigma Xi; AAAS; Am Col Clin Pharm; Am Col Clin Pharm; Am Soc Hosp Pharmacists. *Res:* Clinical psychopharmacology of violence and depression; postmarketing drug surveillance. *Mailing Add:* Dept Psychiat Univ Tex Med Br 1200 Graves Bldg Galveston TX 77550

BRYANT, SUSAN VICTORIA, b Sheffield, Eng, May 24, 43. DEVELOPMENTAL BIOLOGY, CELL BIOLOGY. *Educ:* Univ London, BSc, 64, PhD(zool), 67. *Prof Exp:* Res fel zool, Case Western Reserve Univ, 67-69; from lectr to assoc prof, 69-81, PROF BIOL, UNIV CALIF, IRVINE, 81- *Concurrent Pos:* Prog dir, develop biol, NSF, 81-82. *Mem:* Am Soc Zool; Soc Develop Biol; Asn Women Sci; Int Soc Develop Biologists. *Res:* Regeneration in vertebrates. *Mailing Add:* Dept Cell & Molecular Biol Univ Calif Irvine CA 92717

BRYANT, THOMAS EDWARD, b Bellamy, Ala, Jan 17, 36; m 61; c 2. PUBLIC HEALTH. *Educ:* Emory Univ, AB, 58, MD, 62, JD, 67. *Prof Exp:* Intern, Grady Mem Hosp, Atlanta, Ga, 62-63; dir health affairs, Off Econ Opportunity, 69-71; pres, Nat Drug Abuse Coun, 71-77; chmn & dir, President's Comn Ment Health, 77-79; chmn, Pub Comt Ment Health, 79-87; PRES, FRIENDS NAT LIBR MED, 86- *Concurrent Pos:* Vis lectr, Georgetown Univ. *Mem:* Inst Med Nat Acad Sci; Am Pub Health Asn. *Mailing Add:* 1527 Wisconsin Ave NW Washington DC 20007

BRYANT, VAUGHN MOTLEY, JR, b Dallas, Tex, Oct 5, 40; m 64; c 3. ANTHROPOLOGY, BOTANY. *Educ:* Univ Tex, Austin, BA, 64, MA, 66, PhD(bot), 69. *Prof Exp:* Asst prof anthrop, Wash State Univ, 69-71; from asst prof to prof, 71-80, HEAD, DEPT ANTHROP, TEX A&M UNIV, 75-, DIR PALYNOLOGY LAB, 90- *Concurrent Pos:* Managing ed, Am Asn Stratig Palynologists, 79-83, pres, 85, secy, 90- *Mem:* Am Asn Stratig Palynologists; Soc Am Archaeol; Am Quaternary Asn. *Res:* Pollen analytical studies of Quaternary paleoenvironments with special emphasis on areas in the American Southwest; prehistoric diets; pollen and macrofossil analysis of fossil human coprolites; pollen analytical and paleoethnobotanical studies of archaeological deposits; forensic palynology. *Mailing Add:* Dept Anthrop Tex A&M Univ College Station TX 77843

BRYANT, WILLIAM RICHARDS, b Chicago, Ill, Feb 12, 30; m 65; c 3. MARINE GEOLOGY, OCEANOGRAPHY. *Educ:* Univ Chicago, MS, 60, PhD(geol), 66. *Prof Exp:* Oceanogr, Off Naval Res, 60-62; res scientist, 62-64, from asst prof to assoc prof oceanog, 64-71, PROF OCEANOG, TEX A&M UNIV, 71- *Concurrent Pos:* Proj supvr, Off Naval Res grant, 66-; NSF grant, 69- *Mem:* AAAS; Sigma Xi. *Res:* Marine geotechnique; geology and geophysics of the Gulf of Mexico and Caribbean; acoustic characteristics of marine sediments; sediment transport. *Mailing Add:* 2811 Cherry Creek Bryan TX 77802-2927

BRYANT, WILLIAM STANLEY, b Frankfort, Ky, Nov 9, 43; m 69; c 2. FOREST ECOLOGY, GRASSLAND ECOLOGY. *Educ:* Tenn Technol Univ, BS, 66; Southern Ill Univ, MS, 69, PhD(bot), 73. *Prof Exp:* Evening instr ecol, Ky State Univ, 70-72; from instr to assoc prof biol, 71-80, PROF BIOL & CHMN DEPT, THOMAS MORE COL, 80- *Mem:* Ecol Soc Am; Sigma Xi. *Res:* Structure and dynamics of forest and grassland systems in the Midwest and Upper South. *Mailing Add:* Dept Biol Thomas More Col Crestview Hills KY 41017

BRYANT-GREENWOOD, GILLIAN DOREEN, b Oxfordshire, Eng, July 20, 42; m 76; c 2. ENDOCRINOLGOY, REPRODUCTIVE BIOLOGY. *Educ:* Brunel Univ, BSc, 65, PhD(biol, endocrinol), 68. *Prof Exp:* Res staff, Imperial Cancer Res Fund, London, 65-68; asst researcher, Dept Biochem & Biophys, 68-72, asst prof reproductive endocrinol, 72-75, assoc prof, 75-81, PROF, DEPT ANAT & REPROD BIOL, UNIV HAWAII, 81- *Concurrent Pos:* Res career develop award, NIH, 72-78, grants, 72 & 81-; sr Fogerty Int fel, 85-86. *Mem:* Endocrine Soc; Soc Endocrinol; Soc Study Reprod & Fertility; Asn Women Sci. *Res:* The hormone relaxin, its isolation from ovaries and uteri of different species, its chemical structure; the development of radioimmunoassays and elucidation of its biological and reproductive functions. *Mailing Add:* Dept Anat Univ Hawaii Burns Med Sch 1960 East-West Rd Honolulu HI 96822

BRYCE, GALE REX, b Safford, Ariz, Nov 18, 39; m 63; c 9. QUALITY IMPROVEMENT. *Educ:* Ariz State Univ, BS, 67; Brigham Young Univ, MS, 70; Univ Ky, PhD(exp statist), 74. *Prof Exp:* Math programmer, Semiconductor Prod Div, Motorola Inc, 63-68; asst prof, 72-75, assoc prof, 75-80, PROF STATIST, BRIGHAM YOUNG UNIV, 80- *Concurrent Pos:* Mgr qual technol, Intel Corp, 86-88. *Mem:* Am Statist Asn; Int Biomet Soc; Sigma Xi. *Res:* Linear models with special emphasis in the analysis of unbalanced designs where a mixed model is appropriate; quality management, quality improvement, industrial applications of statistics. *Mailing Add:* Dept Statist Brigham Young Univ Provo UT 84602

BRYCE, HUGH GLENDINNING, b Melville, Sask, Nov 30, 17; m 43; c 2. PHYSICAL CHEMISTRY. *Educ:* Univ Sask, BA, 38, MA, 40; Columbia Univ, PhD(chem), 43. *Prof Exp:* Res chemist, Off Sci Res & Develop contract, Columbia Univ, 42-43; res chemist, Los Alamos Lab, Univ Calif, 43-46; res chemist, Res Labs, Sharples Corp, Philadelphia, 46-49; head prod develop, Fluorochems Dept, 3M Co, 49-59, develop mgr, 59-61, asst tech dir, 61-66, tech dir, Chem Div, 66-73, exec dir, Cent Res Labs, 73-75, staff vpres, Cent Res Labs, 75-82; RETIRED. *Mem:* AAAS; Am Chem Soc; Indust Res Inst; Sigma Xi. *Res:* Ultracentrifuge; low temperatures; ion exchange resins; ultracentrifugal behavior of starch and its triacetate and trimethyl derivatives; chemistry of carbon fluorine compounds. *Mailing Add:* 461 Stage Line Rd Hudson WI 54016

BRYDEN, HARRY LEONARD, b Providence, RI, July 9, 46; m 69, 88; c 4. PHYSICAL OCEANOGRAPHY. *Educ:* Dartmouth Col, AB, 68; Mass Inst Technol, PhD(oceanog), 75. *Prof Exp:* Mathematician, US Naval Oceanog Off, 69-70 & US Naval Underwater Sound Lab, 70; res asst phys oceanog, Woods Hole Oceanog Inst, 70-75; res assoc phys oceanog, Ore State Univ, 75-77; asst scientist, 77-80, assoc scientist, 80-88, SR SCIENTIST, PHYS OCEANOG, WOODS HOLE OCEANOG INST, 88- *Concurrent Pos:* Vis fel, Wolfson Col, Oxford, 88-89. *Mem:* Am Geophys Union; Am Meteorol Soc. *Res:* Dynamics of mid-ocean low-frequency current and their effects on mean ocean circulation; role of ocean heat transport in the global heat budget; dynamics of the exchange through the Strait of Gilbraltor. *Mailing Add:* Dept Phys Oceanog Woods Hole Oceanog Inst Woods Hole MA 02543

BRYDEN, JOHN HEILNER, b Nampa, Idaho, Sept 9, 20; m 45; c 2. X-RAY CRYSTALLOGRAPHY. *Educ:* Col Idaho, BS, 42; Calif Inst Technol, MS, 45; Univ Calif, Los Angeles, PhD(chem), 51. *Prof Exp:* Chemist, Hercules Powder Co, 42-43; fel, Calif Inst Technol, 43-45; chemist, E I du Pont de Nemours & Co, 45-46; asst, Univ Calif, Los Angeles, 47-51; chemist, Res Dept, US Naval Ord Test Sta, 51-59; mem tech staff, Hughes Aircraft Co, 59-61; PROF CHEM, CALIF STATE UNIV, FULLERTON, 61- *Mem:* AAAS; Am Crystallog Asn; Am Chem Soc. *Res:* Crystal structures of organic and inorganic substances by x-ray diffraction. *Mailing Add:* 1416 Vista Del Mar Dr Fullerton CA 92631

BRYDEN, ROBERT RICHMOND, b Erie, Pa, July 19, 16; m 43; c 4. ANIMAL ECOLOGY. *Educ:* Mt Union Col, BS, 38; Ohio State Univ, MS, 41; Vanderbilt Univ, PhD(zool), 51. *Prof Exp:* Instr biol, Univ Akron, 40-41; asst sci, Goodyear Aircraft Corp, 41-43; assoc prof biol, Mid Tenn State Univ, 46-51; biologist, AEC, 51-54; head biol dept & chmn sci div, Union Col, 54-58; prof biol, High Point Col, 58-61; Dana prof biol & head dept, 61-83, EMER DANA PROF BIOL, GUILFORD COL, 83- *Concurrent Pos:* AAAS & Tenn Acad res grant, 49; Am Acad Arts & Sci res grant, 58; Sigma Xi res grant, 59. *Mem:* AAAS; Am Soc Zoologists. *Res:* Limnology; physiology; invertebrate zoology. *Mailing Add:* 1805 Evans St Morehead City NC 28557

BRYDEN, WAYNE A, b Chestertown, Md, Aug 18, 55; m 79; c 3. WIDE BANDGAP SEMICONDUCTOR PHYSICS, PHYSICAL VAPOR DEPOSITION. *Educ:* Frostburg State Univ, BS, 77; Johns Hopkins Univ, MS, 82, PhD(chem), 83. *Prof Exp:* Postdoctoral fel, 82-83, sr staff chemist, 83-90, PROG SUPVR, APPL PHYSICS LAB, JOHNS HOPKINS UNIV, 90- *Concurrent Pos:* Fac mem, Continuing Prof Progs, Eng Sch, 86-, lectr, Interdisciplinary Courses, Johns Hopkins Univ, 87- *Mem:* Am Phys Soc; Am Vacuum Soc; Mat Res Soc; Am Chem Soc. *Res:* Physical vapor deposition of wide bandgap semiconductors; electrical, optical and morphological characterization of semiconductor thin films; applications of magnetic resonance imaging technique to damage characterization of solids; electron paramagnetic resonance spectroscopy; cryogenic technology. *Mailing Add:* Appl Physics Lab Johns Hopkins Univ John Hopkins Rd Laurel MD 20723-6099

BRYDON, JAMES EMERSON, b Portage la Prairie, Man, June 28, 28; m 51; c 3. MINERALOGY, SOIL SCIENCE. *Educ:* Univ Man, BSc, 51; Univ Mo, MSc, 54, PhD(soils), 56. *Prof Exp:* Res officer soil mineral, Soil Res Inst, Can Dept Agr, 51-71; res mgr, Environ Protection Serv, 72-76, dir Contaminants Control Br, Environ Can, 76-84; HEAD CHEM DIV ENVIRON DIRECTORATE OECD, PARIS, FRANCE, 85- *Mem:* Agr Inst Can; Can Soc Soil Sci; Chem Inst Can. *Res:* Regulatory assessment and control of environmental contaminants; soil and clay mineralogy, as related to weathering reactions, plant nutrient uptake and pedogenesis. *Mailing Add:* Environ Directorate OECD Two rue Anadré-Pascal Paris France

BRYMAN, DOUGLAS ANDREW, b Brooklyn, NY, June 12, 45; m 68; c 2. PHYSICS. *Educ:* Syracuse Univ, BS, 66; Rutgers Univ, MS, 69; Va Polytech Inst & State Univ, PhD(physics), 72. *Prof Exp:* Res fel, Univ Victoria, BC, 72-75; SR RES SCIENTIST PHYSICS, TRI-UNIV, MESON FACIL, 75- *Concurrent Pos:* Prog dir intermediate energy physics, NSF, 77-78; adj prof, Univ Victoria, 80- *Mem:* Fel Am Phys Soc; Can Asn Physics; Inst Particle Physics. *Res:* Elementary particle physics. *Mailing Add:* TRIUMF 4004 Wesbrook Mall Vancouver BC V6T 2A3 Can

BRYNER, CHARLES LESLIE, b Dunbar, Pa, Oct 15, 14; m 47; c 2. BOTANY. *Educ:* Waynesburg Col, BS, 40; Univ WVa, MS, 48, PhD(bot), 57. *Hon Degrees:* DSc, Waynesburg Col, 82. *Prof Exp:* Teacher & supvry prin, Pub Schs, Pa, 40-42; from assoc prof to prof bot, Waynesburg Col, 46-81, dean men, 53-58, chmn, Div Sci, 58-71, asst acad vpres, 59-81; RETIRED. *Res:* Genus Helianthus in West Virginia; taxonomic botany. *Mailing Add:* RD 3 Box 42 Waynesburg PA 15370

BRYNER, JOHN C, b Salt Lake City, July 22, 31; m 53; c 4. APPLIED PHYSICS. *Educ:* Univ Utah, BA, 58, PhD(physics), 62. *Prof Exp:* Teaching asst physics, Univ Utah, 58-60; res specialist, NAm Rockwell, Inc, 62-70 & 72-76; res specialist, Hycon Mfg Co, 70-71; mem tech staff, Aerojet Electrosysts Co, Calif, 71-72; SR RES SCIENTIST, EYRING RES INST, 76- *Concurrent Pos:* Vis prof, Univ Southern Calif, 63-70. *Mem:* Am Asn Physics Teachers. *Res:* Plasma; lasers; infrared technology; electrostatics. *Mailing Add:* 631 James St Clearfield UT 84015

BRYNER, JOHN HENRY, b Washington, Pa, Oct 28, 24; m 45; c 4. MICROBIOLOGY. *Educ:* Eastern Nazarene Col, AB, 49; Univ Wis, PhD, 68. *Prof Exp:* Bacteriologist, Animal Dis Sta, Bur Animal Indust, USDA, 50-61, sr res microbiologist, Nat Animal Dis Lab, Agr Res Serv, 61-89; RETIRED. *Concurrent Pos:* US AID consult, Vet Sch, Porto Alegre, Brazil, 72; managing ed, J Wildlife Dis, Wildlife Dis Asn, 73-; consult, Off Int Coop & Develop, Novisad, Yugoslavia & Pećs, Hungary, 87-88; consult, Pan Am Health Org & vis prof, Vet Sch, Maracay, Venezuela, 76, Bogota, Colombia, 79-80, & Mexico City, Mexico, 81. *Mem:* Am Soc Microbiol; Wildlife Dis Asn; US Animal Health Asn; Am Asn Vet Lab Diagnosticians; Sigma Xi. *Res:* Infectious causes of sterility in livestock, specifically morphologic, serologic and biochemic characterization of the genus Vibrio, characterized and named Flexispira Rapini; intensive study of wild animals. *Mailing Add:* 929 Brookridge Ave Ames IA 50010

BRYNER, JOSEPH S(ANSON), b Pittsburgh, Pa, Dec 15, 20; m 49; c 3. METALLURGY. *Educ:* Washington & Jefferson Col, AB, 42; Carnegie Inst Technol, BS, 42; Pa State Univ, MS, 49, PhD(metall), 57. *Prof Exp:* Res metallurgist magnesium alloys, Dow Chem Co, 43-44; res metallurgist nickel alloys, Int Nickel Co, 46-47; asst, Pa State Univ, 48-50; assoc metallurgist liquid metals, Brookhaven Nat Lab, 51-59, metallurgist, 59-69; sr engr, Bettis Atomic Power Lab, Westinghouse Elec Corp, 70-86; RETIRED. *Mem:* Am Soc Metals. *Res:* Electron microscopy of magnesium alloys; development of inconel arc-welding electrodes; production and behavior of dispersions in liquid metals; electron microscopy of radiation damage, corrosion and stress corrosion. *Mailing Add:* 114 Telstar Dr Pleasant Hills PA 15236-4557

BRYNGDAHL, OLOF, b Stockholm, Sweden, Sept 26, 33; US citizen; m 59. DIFFRACTIVE OPTICS, IMAGE QUANTIZATION. *Educ:* Royal Inst Technol, Stockholm, BS, 56, MS, 60, PhD(physics), 62. *Prof Exp:* Scientist physics, Royal Inst Technol, Stockholm, 56-62, asst prof, 63-64; scientist optics, Xerox Res Ctr, Webster, 64-65; mgr optics, IBM, San Jose & Yorktown Heights, 66-70; prin scientist, Xerox Res Ctr, Palo Alto, 70-77; PROF PHYSICS, UNIV ESSEN, FED REPUB GER, 77- *Concurrent Pos:* Assoc prof optics, Optics Inst, Univ Paris, 75-76. *Mem:* Fel Optical Soc Am; Ger Soc Appl Optics. *Res:* Optical and digital handling of pictorial information; optical computing, image formation, processing, storage and transmission; diffractive optics, optical and digital holography; image coding and optical metrology. *Mailing Add:* Univ Essen 4300 Essen 1 Germany

BRYNILDSON, OSCAR MARIUS, b Wis, Mar 5, 16. ZOOLOGY. *Educ:* Univ Wis, BA, 47, MA, 50, PhD(zool), 58. *Prof Exp:* Asst fishery biol, Univ Wis, 48-52; fishery biologist, Wis Conserv Dept, 52-57, group leader cold water res, 57-69; LIMNOLOGIST, WIS DEPT NATURAL RESOURCES, 69- *Mem:* Am Soc Limnol & Oceanog; Am Fisheries Soc. *Res:* Ecology of trout and other cold water fishes in lakes and streams. *Mailing Add:* Rte 2 Box 25-A-5 Black River Falls WI 54615

BRYNJOLFSSON, ARI, b Akureyri, Iceland, Dec 7, 26; US citizen; m 50; c 5. NUCLEAR PHYSICS, GEOPHYSICS. *Educ:* Univ Iceland, BSc, 48; Univ Copenhagen, Cand Phil, 49, Cand Mag & Mag Sci, 54; Niels Bohr Inst, Dr Phil(theoret nuclear physics), 73; Harvard Univ, AMP, 71. *Prof Exp:* Res physicist, Danish AEC Res Estab, Riso, Denmark, 57-58, head radiation res group, 57-59, dir radiation res labs, 59-62; consult, US Army Natick Labs, 62-63; dir radiation res labs, Danish AEC, Riso, 63-65; chief, Radiation Sources Div, US Army, 65-72, actg dir food irradiation, 72-74, chief radiation preserve, Food Div & dir, Radiation Lab, 74-80, chief, Physics & Math Sci Advan Technol Lab, 80-83, res chemist, Natick Res & Develop Lab, 83-88; PROJ DIR, INT FACIL FOOD IRRADIATION TECHNOL, JOINT INT ATOMIC ENERGY AGENCY & FOOD & AGR ORGN DIV, VIENNA, AUSTRIA, 88- *Concurrent Pos:* Consult, Int Atomic Energy Agency Panels, 70- & Iranian Govt, 72-; vis lectr, Mass Inst Technol, Cambridge, 81- *Mem:* Am Phys Soc; Radiation Res Soc; Am Asn Physicists in Med; Inst Food Technologists; AAAS. *Res:* Theory of stopping of charged particles and degradation of radiation; basic radiation physics and effects of radiation on biological systems; design and operation of isotopes and electron accelerator irradiation facilities; irradiation of foods and medical products; research and industrial application; astrophysics; theory of general relativity; redshift. *Mailing Add:* Seven Bridle Path Wayland MA 01778

BRYSK, MIRIAM MASON, b Warsaw, Poland, Mar 10, 35; US citizen; m 55; c 2. BIOLOGICAL CHEMISTRY. *Educ:* NY Univ, BA, 55; Univ Mich, MS, 58; Columbia Univ, PhD(biol sci), 67. *Prof Exp:* Lectr biol, Queens Col, NY, 60-61; fel protein chem, Inst Muscle Dis, 67-69; res biologist, dept biol, Univ Calif, San Diego, 79-71; NIH fel cellular chem, Univ Mich, 74-77; res asst prof biochem & med & res scientist,Cancer Res & Treatment Ctr, Sch Med, Univ NMex, 77-80; from asst prof to assoc prof, 80-88, DIR, DERMAT RES LAB, 80-, PROF DERMAT, MICROBIOL HUMAN BIOL CHEM & GENETICS, UNIV TEX MED BR, GALVESTON, 88- *Mem:* Am Soc Microbiol; Soc Invest Dermat; affil mem Am Acad Dermat; Sigma Xi; Am Soc Biol Chemists; Am Soc Cell Biol. *Res:* Amino acid and cyanide metabolism in microorganisms; synthesis of cell walls in plants; differentiation and carcinogenesis of the epidermis; biochemistry of epidermis. *Mailing Add:* Dept Dermat Univ Tex Med Br Galveston TX 77550

BRYSON, ARTHUR E(ARL), JR, b Evanston, Ill, Oct 7, 25; m 46; c 4. CONTROL SYSTEMS. *Educ:* Iowa State Col, BS, 46; Calif Inst Technol, PhD(aeronaut), 51. *Prof Exp:* Aerodyn engr, Hughes Res & Develop Labs, 50-53; from asst prof to prof mech eng, Harvard Univ, 53-68; chmn, Dept Appl Mech, 69-71, chmn, Dept Aeronaut & Astronaut, 71-79, PROF APPL MECH, AERONAUT & ASTRONAUT, STANFORD UNIV, 68-, PAUL PIGOTT PROF ENG, 72- *Concurrent Pos:* Chmn, Aeronaut & Space Eng Bd, Nat Res Coun, 76-78. *Honors & Awards:* Pendray Award, Am Inst Aeronaut & Astronaut, 68; Westinghouse Award, Am Soc Eng Educ, 69; Mechanics & Control of Flight Award, Am Inst Aeronaut & Astronaut, 80; Rufus Oldenburger Medal, Am Soc Mech Engrs, 80; Control Systs Sci & Eng Award, Inst Elec & Electronics Engrs, 86; Control Heritage Award, Am Control Coun, 90. *Mem:* Nat Acad Sci; Nat Acad Eng; Am Soc Eng Educ; hon fel Am Inst Aeronaut & Astronaut; Am Acad Arts & Sci. *Res:* Automatic control; flight mechanics; systems engineering; fluid mechanics. *Mailing Add:* Dept Aeronautics & Astronautics Sch Eng Stanford Univ Stanford CA 94305

BRYSON, GEORGE GARDNER, b Santa Barbara, Calif, Dec 16, 35. PHYSIOLOGY, PSYCHOPHYSIOLOGY. *Educ:* Univ Calif, Santa Barbara, BA, 57, PhD, 81; San Francisco State Col, MA, 71. *Prof Exp:* Zoologist, 57-71, PHYSIOLOGIST, SANTA BARBARA COTTAGE HOSP RES INST, 71- *Mem:* Am Asn Cancer Res. *Res:* Psychophysiology of steroid hormones and catecholamines, pituitary-adrenal-gonadal function in aggression and reproductive behavior, carcinogenic mechanisms, solid state carcinogenesis; biologic transport of steroids. *Mailing Add:* 2412 Chapala St Santa Barbara CA 93105

BRYSON, MARION RITCHIE, b Centralia, Mo, Aug 26, 27; m 47; c 4. OPERATIONS RESEARCH. *Educ:* Univ Mo, BSEd, 49, MA, 50; Iowa State Univ, PhD(statist), 58. *Prof Exp:* Teacher math, Elgin High Sch, Ill, 50-52; instr, Univ Idaho, 52-53 & Drake Univ, 53-55; res assoc statist, Iowa State Univ, 55-58; dir spec res, Duke Univ, 58-68, from asst prof to assoc prof math & community health sci, 62-68; tech dir opers res, US Army Combat Develop Command Systs Anal Group, 68-72, sci adv, US Army Combat Develop Exp Command, 72-83, DIR, US ARMY TEXCOM EXPERIMENTATION CTR, 83-, CHIEF SCIENTIST, 90- *Concurrent Pos:* Vis prof, Okla State Univ, 65; statist ed, J Parapsychol, 66-69; lectr, Dept Path, Vet Admin Hosp, 67-68. *Honors & Awards:* Wanner Mem Award, Mil Opers Res Soc, 85; SS Wilks Award statist, US Army, 88. *Mem:* Inst Math Statist; Am Statist Asn; Opers Res Soc; fel Mil Opers Res Soc, 90 (pres, 75-76). *Res:* Military operations research; sampling theory; medical research. *Mailing Add:* US Army TEXCOM Experimentation Ctr Ft Ord CA 93941

BRYSON, MELVIN JOSEPH, b Providence, Utah, June 7, 16; m 42; c 2. CLINICAL BIOCHEMISTRY. *Educ:* Utah State Agr Univ, BS, 46; Univ Utah, MS, 48; Agr & Mech Tex State Univ, PhD(biochem & nutrit), 52. *Prof Exp:* Asst, Utah State Agr Col, 48-49; res biochemist & head drug enzym unit, Eaton Labs, 52-57; res assoc, Univ Utah, 57-64, asst res prof, 65-70, assoc res prof obstet & gynec, 71-83, adj assoc prof pathol, Col Med, 81-83; RETIRED. *Concurrent Pos:* Dir, Intermountain Perinatal Res & Serv Lab, 70-83; Interwest Endocrine Lab, 84-91. *Mem:* Endocrine Soc. *Res:* Synthesis and metabolism of steroid hormones and relation to intermediary metabolism; affect of drugs on enzymes systems; intestinal absorption; clinical endocrinology; clinical chemistry. *Mailing Add:* 5262 Woodcrest Dr Salt Lake City UT 84117

BRYSON, REID ALLEN, b Detroit, Mich, June 7, 20; m 42; c 4. CLIMATOLOGY, PALEOCLIMATOLOGY. *Educ:* Denison Univ, BA, 41, Univ Chicago, PhD(meteorol), 48. *Hon Degrees:* DSc, Denison Univ, 71. *Prof Exp:* Asst prof meteorol & geol, Univ Wis, 46-48, from asst prof to assoc prof meteorol, 48-56, chmn dept, 48-50 & 52-54; prof, Univ Ariz, 56-57; prof, 57-68, chmn dept, 57-61, dir Inst Environ Studies, 70-85, prof meteorol & geog, 68-86, SR SCIENTIST, CTR CLIM RES, UNIV WIS-MADISON, 86- *Concurrent Pos:* Trustee, Univ Corp Atmospheric Res, 59-66; mem environ studies bd, Nat Acad Sci-Nat Acad Eng, 70-73; sci adv comt, Arctic & Alpine Res Inst. *Honors & Awards:* Fel, AAAS, 87. *Mem:* Soc Am Archaeol; Am Meteorol Soc; Am Quaternary Assn (pres, 88-90). *Res:* Physical limnology; paleoclimatology; dynamic climatology; interdisciplinary environmental studies. *Mailing Add:* Ctr Climatic Res Univ Wis 1225 W Dayton St Madison WI 53706

BRYSON, THOMAS ALLAN, b Pittsburgh, Pa, July 24, 44; m 66; c 2. ORGANIC BIOCHEMISTRY. *Educ:* Washington & Jefferson Col, BA, 66; Univ Pittsburgh, PhD(org chem), 70. *Prof Exp:* NIH fel synthetic org chem, Stanford Univ, 70-71; asst prof, 71-75, ASSOC PROF ORG CHEM, UNIV SC, 75- *Concurrent Pos:* NSF trainee, 66-76. *Mem:* Am Chem Soc; Sigma Xi. *Res:* Synthetic organic chemistry; natural products chemistry; key biological compounds, particularly cortisone, juvenile hormone and nicotinamide-adenine dinucleotide. *Mailing Add:* Dept Chem Univ SC Columbia SC 29208

BRYTCZUK, WALTER L(UCAS), b Carteret, NJ, Oct 18, 07; m 40; c 3. EXTRACTIVE METALLURGY, ENVIRONMENTAL ENGINEERING. *Educ:* NY Univ, BS, 36; Polytech Inst Brooklyn, MS, 42; Environ Engrs Intersoc, dipl, 72. *Prof Exp:* Asst supt electrolytic ref, US Metals Ref Co, 29-36, supt, 36-47, ref metallurgist, 47-52, dir res, 52-56; assoc dir res, non-ferrous metals, Am Metal Climax, Inc, 56-62; plant metallurgist, US Metals Ref Co, 62-69, metall asst to plant mgr, 69-74; consult, 74-82; RETIRED. *Concurrent Pos:* Adj prof, Polytech Inst Brooklyn, 53-56 & grad sch, Newark Col Eng, 56-66; consult, Electrolytic Copper Ref Powder Metall; mem, LAN Assoc, 68-; vol, Int Exec Serv Corp, 77- *Mem:* Am Soc Metals; Am Inst Mining, Metall & Petrol Engrs; assoc mem Sigma Xi; Am Acad Environ Engrs. *Res:* Production of metals powders; germanium and semiconductors; extractive metallurgy of non-ferrous metals; electrolytic refining of copper; air and water pollution control. *Mailing Add:* 430 Galloping Hill Rd Roselle Park NJ 07204-2134

BRZENK, RONALD MICHAEL, b Jersey City, NJ, Jan 30, 49; m 71; c 2. NUMBER THEORY. *Educ:* St Peter's Col, NJ, BS, 69; Univ Notre Dame, PhD(math), 74. *Prof Exp:* asst prof, 74-80, ASSOC PROF MATH, HARTWICK COL, 80- *Mem:* Am Math Soc; Math Asn Am. *Res:* Axiomatic approaches to class field theory; computer applications in number theory and algebra. *Mailing Add:* Dept Math Hartwick Col Oneonta NY 13820

BRZOZOWSKI, JANUSZ ANTONI, b Warsaw, Poland, May 10, 35; Can citizen; m 59; c 2. HARDWARE SYSTEMS. *Educ:* Univ Toronto, BASc, 57, MASc, 59; Princeton Univ, MA & PhD(elec eng), 62. *Prof Exp:* Asst prof elec eng, Univ Ottawa, 62-65, assoc prof, 65-67; chmn dept, 78-83 & 87-89, PROF COMPUTER SCI, UNIV WATERLOO, 67- *Mem:* Inst Elec & Electronics Engrs; Asn Comput Mach. *Res:* Asynchronous circuits; testing. *Mailing Add:* 289 Shakespeare Dr Waterloo ON N2L 2T9 Can

BRZUSTOWSKI, THOMAS ANTHONY, b Warsaw, Poland, Apr 4, 37; Can citizen; m 64; c 3. MECHANICAL ENGINEERING. *Educ:* Univ Toronto, BASc, 58; Princeton Univ, AM, 60, PhD(aeronaut eng), 63. *Prof Exp:* From asst prof to assoc prof, 62-66, chmn dept, 67-70, assoc dean eng grad studies, 70-77, PROF MECH ENG, UNIV WATERLOO, 66-, VPRES ACADEMIC, 75- *Concurrent Pos:* Am Soc Eng Educ-Ford Found resident prof, environ control & safety dir, Esso Res & Eng Co, 70-71. *Mem:* Am Soc Mech Engrs; Am Inst Aeronaut & Astronaut; Am Soc Eng Educ; Sigma Xi; fel Inst Energy UK; fel Eng Inst Can; Can Soc Mech Eng. *Res:* Combustion of liquid fuels; flaring; applied thermodynamics-second law analysis. *Mailing Add:* Ministry Col & Univ 900 Bay St 3rd Floor Mowat Block Queen's Park Toronto ON M7A 1L2 Can

BUBAR, JOHN STEPHEN, b NB, Sept 13, 29; m 54; c 2. AGRONOMY. *Educ:* McGill Univ, BSc, 52, PhD(genetics), 57; Pa State Univ, MS, 54. *Prof Exp:* Asst agron, Pa State Univ, 52-53; asst, Macdonald Col, McGill Univ, 53-54, lectr, 54-59, asst prof, 59-67; assoc prof, 67-71, head dept, 67-83, PROF AGRON, NS AGR COL, 71- *Concurrent Pos:* Mem, Expert Comt Forage Crops Breeding, 74-77; AK Publications Comt, 85- *Mem:* Agr Inst Can; Can Soc Agron (pres, 71-72); hon mem Can Seed Growers Asn. *Res:* crops production and breeding, specifically red clover alfalfa and birdsfoot trefoil breeding and genetics; adaptation of cultivars to Atlantic provinces of Canada. *Mailing Add:* Dept Plant Sci NS Agr Col Truro NS B2N 5E3 Can

BUBE, RICHARD HOWARD, b Providence, RI, Aug 10, 27; m 48; c 4. SOLID STATE PHYSICS, MATERIALS SCIENCE. *Educ:* Brown Univ, ScB, 46; Princeton Univ, MA, 48, PhD(physics), 50. *Prof Exp:* Res physicist, Labs Div, Radio Corp Am, 48-62; assoc prof, 62-64, chmn dept, 75-86, ASSOC CHMN DEPT, 90-, PROF MAT SCI & ELEC ENG, STANFORD UNIV, 64- *Concurrent Pos:* Ed, J Am Sci Affil, 69-83; assoc ed, Ann Rev Mat Sci, 69-83,; consult ed, Mat Lett, 82-89. *Mem:* Fel AAAS; fel Am Phys Soc; fel Am Sci Affil; Am Soc Eng Educ; Sigma Xi; Am Div Int Solar Energy Soc. *Res:* Luminescence; photoconductivity; trapping in solids; phosphors; electronic crystal defect phenomena; semiconductors; photoelectronic properties of materials and devices; photovoltaics; amorphous materials. *Mailing Add:* Dept Mat Sci & Eng Stanford Univ Stanford CA 94305-2205

BUBECK, ROBERT CLAYTON, b Wilkes-Barre, Pa, Mar 20, 37; m 65; c 2. PHYSICAL & CHEMICAL LIMNOLOGY. *Educ:* Univ Rochester, BA, 60, PhD(geol sci), 72; Pa State Univ, MS, 65. *Prof Exp:* Asst prof chem geol, Chem Dept, US Naval Acad, 65-67; estuarine oceanographer chem, Fed Water Pollution Control Admin, US Dept Interior, 67-69; res asst, Dept Geol Sci, Univ Rochester, 69-73; hydrologist geochem & limnol, Water Resources Div, US Geol Surv, 73-76; chief inorg chem unit & staff hydrologist, Region III Lab, US Environ Protection Agency, 76-81; SUPVR HYDROLOGIST GEOCHEM, WATER RESOURCES DIV, US GEOL SURV, 81- *Concurrent Pos:* Res assoc sea grant, Dept Geol Sci, Univ Rochester, Nat Oceanic & Atmospheric Admin, 72-73; adj prof, Dept Biol & Ctr Marine Studies, Am Univ, 80- *Mem:* Am Soc Limnol & Oceanog; Int Soc Theoret & Appl Limnol; AAAS; Sigma Xi; Am Water Resources Asn. *Res:* Limnology of lakes and reservoirs; geochemistry of estuaries, rivers and streams; chemical characteristics of ground water. *Mailing Add:* 3940 Bibbits Dr Palo Alto CA 94303

BUBEL, HANS CURT, b Mannheim, Ger, Oct 3, 25; nat US; m 58; c 2. MEDICAL BACTERIOLOGY. *Educ:* Univ Utah, BS, 50, MS, 53, PhD(bact), 58. *Prof Exp:* Res assoc virol, Univ Utah, 53-58; from asst prof to assoc prof microbiol, 58-70, PROF MICROBIOL, UNIV CINCINNATI, 70- *Mem:* NY Acad Sci; Am Soc Microbiol. *Res:* Host-cell virus interactions; polio-virus and herpes simplex virus; vaccinia virus; host resistance to virus infection; biology of the macrophage; antiviral immunity. *Mailing Add:* Dept Microbiol Univ Cincinnati Col Med 231 Bethesda Ave Cincinnati OH 45267

BUBENZER, GARY DEAN, b Bicknell, Ind, Aug 21, 40; m 62; c 2. SOIL & WATER CONSERVATION, EROSION CONTROL. *Educ:* Purdue Univ, BS, 62, MS, 64; Univ Ill, PhD(agr eng), 70. *Prof Exp:* Instr conserv eng & surv, Univ Ill, 64-69; from asst prof to assoc prof, 69-78, chmn, 83-88, PROF AGR ENG, UNIV WIS-MADISON, 78- *Concurrent Pos:* Vis assoc prof, Univ Idaho, 77-78; vis lectr, Kyoto Univ, Japan, 81. *Mem:* Am Soc Agr Engrs; Soil Sci Soc Am; Am Soc Agron; Am Soc Eng Educ. *Res:* Rainfall simulation; soil erosion mechanics and small watershed hydrology. *Mailing Add:* 460 Henry Mall Madison WI 53706

BUBLITZ, CLARK, b Merrill, Wis, Dec 8, 27; m 58; c 5. BIOCHEMISTRY. *Educ:* Univ Chicago, PhB, 49, PhD(biochem), 55. *Prof Exp:* Am Cancer Soc fel, 55-58; sr instr pharmacol, Sch Med, St Louis Univ, 59-60; asst prof biochem, 60-71, ASSOC PROF BIOCHEM, UNIV COLO MED CTR, 71- *Mem:* AAAS; Am Chem Soc; Am Soc Biol Chemists. *Res:* Enzymology. *Mailing Add:* Biophys & Genetics Univ Colo Med Ctr 4200 E Ninth Ave Denver CO 80220

BUBLITZ, DONALD EDWARD, b Chicago, Ill, Dec 6, 35; m 57; c 2. ORGANIC CHEMISTRY. *Educ:* Univ Calif, Riverside, AB, 57; Univ Kans, PhD(chem), 61. *Prof Exp:* Res assoc chem, Univ Ill, 61-62; res chemist, Dow Chem USA, 62-70, sr res chemist, 70-72, res specialist, Western Div, 72-82. *Mem:* Am Chem Soc; NY Acad Sci. *Res:* Synthesis of agricultural and pharmaceutical chemicals; organometallic compounds. *Mailing Add:* 1592 Hoytt Dr Concord CA 94521

BUBLITZ, WALTER JOHN, JR, b Kansas City, Mo, Sept 26, 20; m 54; c 1. PULP CHEMISTRY, PAPER CHEMISTRY. *Educ:* Univ Ariz, BS, 41; Lawrence Col, PhD(paper chem), 49. *Prof Exp:* Res chemist, Lithographic Tech Found, 49-50; chemist, Munising Paper Co, 50-52; res chemist, Kimberly Clark Corp, 52-59; res chemist, Duplicating Prod Div, Minn Mining & Mfg Co, 59-60 & Paper Prod Div, 60-66; prof pulp & paper res, Ore State Univ, 66-83; RETIRED. *Concurrent Pos:* Guest lectr, SChina Inst Technol, Canton, China, 84; Tasman fel, Univ Canterbury, Christ Church, NZ, 85. *Mem:* Fel Tech Asn Pulp & Paper Indust; Can Pulp & Paper Asn. *Res:* Research and development of specialty paper products; technical problems of lithography; 3M Brand Action paper; pulping research. *Mailing Add:* 1430 NW 14th Place Corvallis OR 97330

BUCCAFUSCO, JERRY JOSEPH, b Jersey City, NJ, Aug 20, 49; m 73; c 2. NEUROPHARMACOLOGY. *Educ:* St Peter's Col, BS, 71; Canisius Col, MS, 73; Univ Med & Dent, NJ, PhD(pharmacol), 78. *Prof Exp:* Fel pharmacol, Roche Inst Molecular Biol, Roche Pharmaceut, 77-79; asst prof pharmacol & toxicol, 79-85, asst prof psychiat, 82-85, ASSOC PROF PHARMACOL, TOXICOL & PSYCHIAT, MED COL GA, 85- *Concurrent Pos:* Dir, neuropharmacol lab, Vet Admin Med Ctr, 82-; mem, Int Adv Bd, Giornale Italiano D I Pathologia Clinica; Pharmacol II Study Sect, Nat Inst Drug Abuse. *Mem:* Am Soc Pharmacol & Exp Therapeut; Soc Neurosci; Soc Exp Biol & Med; Sigma Xi; AAAS; Am Soc Hypertension. *Res:* Neuropharmacology and neurochemical studies on central cardiovascular regulation and the role of the brain in hypertensive disease; brain mechanisms mediating the symptoms of narcotic withdrawal syndrome; protection against the toxic effects of agricultural insecticides; learning and memory in non-human primates. *Mailing Add:* Dept Pharmacol & Toxicol Med Col Ga Augusta GA 30912

BUCCI, ENRICO, b Este, Italy, May 27, 32; US citizen; m 59; c 4. PHYSICAL BIOCHEMISTRY. *Educ:* Univ Rome, MD, 56, PhD(biochem), 62. *Prof Exp:* From vol asst to asst biochem, Univ Rome, 56-65; asst prof, Med Sch, Ind Univ, 65-69; assoc prof, 69-74, PROF BIOCHEM, SCH MED, UNIV MD, 74- *Mem:* Am Soc Biol Chemists; Am Chem Soc; NY Acad Sci; Biophys Soc; Am Soc Photobiol; Sigma Xi. *Res:* Structure function relationships and protein folding of hemoglobin and membrane proteins systems as seen in the physico-biochemical behavior of the entire molecules, their subunits and large peptides obtained from them. *Mailing Add:* 5210 Tilbury W Baltimore MD 21212

BUCCI, ROBERT JAMES, b Cambridge, Mass, Aug 18, 41; m 69; c 1. MECHANICAL ENGINEERING. *Educ:* Northeastern Univ, BS, 64; Brown Univ, MS, 67; Lehigh Univ, PhD(appl mech), 70. *Prof Exp:* Vpres res fracture mech, Del Res Corp, 69-73; staff engr, Alcoa Res Labs, 73-80, tech supvr fatigue & fracture, 80-85, tech specialist, 85-87, sr tech specialist, 87-90, TECH CONSULT, ALCOA RES LABS, 90- *Concurrent Pos:* Chmn task group fatigue crack growth testing, Am Soc Testing & Mat, 73-80, chmn subcomt on subcritical crack growth, 80- *Mem:* Am Soc Testing & Mat; Mineral, Metals & Mat Soc; Am Soc Metals Int; Soc Advan Mat & Process Eng; Sigma Xi. *Res:* Fatigue and fracture mechanics; test method development; aluminum and composite material characterization; material selection; design; failure analysis. *Mailing Add:* Alcoa Labs Aluminum Co Am Alcoa Center PA 15069

BUCCI, THOMAS JOSEPH, b Smithfield, RI, July 21, 34; m 57; c 4. COMPARATIVE PATHOLOGY. *Educ:* Univ Pa, VMD, 59; Univ Rochester, MS, 62; Univ Colo Med Ctr, PhD(path), 74. *Prof Exp:* Chief, Vet Div, 5th US Army Med Lab, 59-61, vet pathologist, Div Nuclear Med, Walter Reed Army Inst Res, 65-66, chief, Path Div, US Army Med Res & Nutrit Lab, 66-69, chief, Dept Comp Med, Letterman Army Inst Res, 74-78; head, Vet Med Dept, Med Res Unit, US Navy, Cairo, Egypt, 78-81; DIR PATH SERV, NAT CTR TOXICOL RES, DEPT HEALTH & HUMAN SERV, JEFFERSON, ARK, 81- *Mem:* Am Col Vet Pathologists; Int Acad Pathologists; Am Asn Pathologists; Am Vet Med Asn. *Res:* Radiation pathology; exercise physiology; comparative nutrition; vitamin metabolism; reproduction of nonhuman primates; infectious disease. *Mailing Add:* Path Serv Nat Ctr Toxicol Res M923 Jefferson AR 72079

BUCCINO, ALPHONSE, b New York, NY, Mar 14, 31; m 53; c 1. SCIENCE ADMINISTRATION, MATHEMATICS. *Educ:* Univ Chicago, BS, 58, MS, 59, PhD(math), 67. *Prof Exp:* Asst prof math, Roosevelt Univ, 61-63; from asst prof to assoc prof math, DePaul Univ, 63-70, chmn dept, 63-69; sci adminr, NSF, 70-81, actg dep asst dir sci & eng educ, 81-84; DEAN, COL EDUC, UNIV GA, 84- *Mem:* AAAS; Nat Coun Teachers Math; Am Math Soc; Math Asn Am. *Res:* Matrices and linear algebra; research administration. *Mailing Add:* Off Dean Aderhold Hall Univ Ga Athens GA 30602

BUCCINO, SALVATORE GEORGE, b New Haven, Conn, Dec 23, 33; m 69; c 2. NUCLEAR PHYSICS. *Educ:* Yale Univ, BS, 56; Duke Univ, PhD(nuclear physics), 63. *Prof Exp:* Res assoc nuclear physics, Duke Univ, 62-63; asst physicist, Argonne Nat Lab, 63-65; from asst prof to assoc prof, 65-70, PROF PHYSICS, TULANE UNIV, 70- *Concurrent Pos:* Consult, Radiation Lab, Aberdeen Proving Grounds, 73 & 74; vis scholar physics, Duke Univ, 74-75. *Mem:* Am Phys Soc; Sigma Xi. *Res:* Coulomb excitation; fast time-of-flight techniques; gamma ray spectroscopy, particularly high spin systems. *Mailing Add:* Dept Physics Tulane Univ New Orleans LA 70118

BUCCOLA, STEVEN THOMAS, b Pasadena, Calif, Dec 30, 44; m 73; c 2. AGRICULTURAL MARKETING. *Educ:* Saint Mary's Col Calif, BA, 66; Univ Calif, Davis, MS, 72, PhD(agr econ), 76. *Prof Exp:* Asst prof, Va Polytech Inst & State Univ, 76-80; from asst prof to assoc prof, 80-88, PROF AGR ECON, ORE STATE UNIV, 88- *Concurrent Pos:* Co-ed, Am J Agr Econ, 91-94. *Mem:* Am Agr Econ Asn; West Agr Econ Asn. *Res:* Application of decision theory to farmer and food processor decisions under risk; analysis of pricing mechanisms and contractual arrangements with a view to evaluating market performance. *Mailing Add:* Dept Agr & Resource Econ Ore State Univ Corvallis OR 97331-3601

BUCHAL, ROBERT NORMAN, b Passaic, NJ, Apr 14, 30; m 57; c 3. APPLIED MATHEMATICS. *Educ:* St Lawrence Univ, BS, 51; Univ Conn, MA, 52; NY Univ, PhD(math), 59. *Prof Exp:* Anal engr, Bendix Aviation Corp, 52-54; res asst, Inst Math Sci, NY Univ, 54-58; asst prof, Math Res Ctr, US Army, Univ Wis, 58-60; assoc mathematician, Argonne Nat Lab, 60-70; mathematician, Off Naval Res, 70-75; MATHEMATICIAN, AIR FORCE OFF SCI RES, 75- *Mem:* Soc Indust & Appl Math. *Res:* Asymptotic solutions and generalized solutions of differential equations; wave equations; mathematical biology. *Mailing Add:* 8707 Highgate Rd Alexandria VA 22308

BUCHAN, GEORGE COLIN, b Seattle, Wash, Aug 30, 27; m 62; c 2. MEDICINE, NEUROPATHOLOGY. *Educ:* McGill Univ, MD, CM, 58. *Prof Exp:* From asst prof to assoc prof, 67-68, PROF PATH & HEAD DIV NEUROPATH, MED SCH, UNIV ORE, 68- *Concurrent Pos:* Fel neuropath, Univ Wash, 63-65. *Mem:* Am Asn Neuropath. *Res:* Factors influencing myelination in the nervous system. *Mailing Add:* Div Neuropath Med Sch Univ Ore 3181 SW Sam Jackson Park Rd Portland OR 97201

BUCHANAN, BOB BRANCH, b Richmond, Va, Aug 7, 37; m 65; c 4. BIOCHEMISTRY, MICROBIOLOGY. *Educ:* Emory & Henry Col, AB, 58; Duke Univ, PhD(microbiol), 62. *Prof Exp:* NIH fel biochem, 62-63, chair div plant biol, 82-88, PROF PLANT BIOL, UNIV CALIF, BERKELEY, 63- *Concurrent Pos:* Guggenheim Found fel, 74, Off Comprehensive Employment fel, 84; vis prof, Univ Paris-Sud, Orsay, 84 & 91. *Honors & Awards:* Spec Creativity Award, NSF, 82, Sr Scientist Award, 84; Bessenyei Medal, Hungary, 87. *Mem:* AAAS; Am Soc Biol Chemists; Am Soc Microbiol; NY Acad Sci; Am Soc Plant Physiol. *Res:* Plant biochemistry; photosynthesis; enzymology; bacterial metabolism. *Mailing Add:* Dept Plant Biol Univ Calif Berkeley CA 94720

BUCHANAN, BRUCE G, b St Louis, Mo, July 7, 40; m 61; c 2. COMPUTER SCIENCE. *Educ:* Ohio Wesleyan Univ, AB, 61; Mich State Univ, MA & PhD(philos), 66. *Prof Exp:* Res assoc comput sci, Stanford Univ, 66-71, res comput scientist, 72-76, adj prof comput sci, 76-; DEPT COMPUTER SCI, UNIV PITTSBURGH. *Concurrent Pos:* NIH career develop award, 71-76; prog chmn, Int Joint Conf Artificial Intel, 79. *Mem:* Am Asn Artificial Intel; Philos Sci Asn; AAAS. *Res:* Application of artificial intelligence to scientific inference. *Mailing Add:* Dept Computer Sci Univ Pittsburgh Main Campus 206 Mineral Indust Bldg Pittsburgh PA 15260

BUCHANAN, CHRISTINE ELIZABETH, b St Louis, Mo, July 5, 46; m 69. MICROBIAL PHYSIOLOGY, MICROBIAL GENETICS. *Educ:* Drake Univ, BA, 68; Univ Chicago, PhD(microbiol), 73. *Prof Exp:* Postdoctoral biochem & molecular biol, Harvard Univ, 73-77; from asst prof to assoc prof, 77-89, dept chair, 88-90, PROF BIOL SCI, SOUTHERN METHODIST UNIV, 89- *Concurrent Pos:* Prin investr, funded res proj, NSF, 80-82, NIH, 83-; mem, Study Sect Microbial, Physiol & Genetics, NIH, 85-89; chair, Div K, Am Soc Microbiol, 88-89. *Mem:* Am Soc Microbiol; AAAS; Sigma Xi. *Res:* Role of penicillin-binding proteins in bacterial cell division, morphogenesis, and sporulation. *Mailing Add:* Dept Biol Sci Southern Methodist Univ Dallas TX 75275

BUCHANAN, DAVID HAMILTON, b Indiana, Pa, July 4, 42; m 67; c 2. PHYSICAL ORGANIC CHEMISTRY. *Educ:* Case Inst Technol, BS, 64; Univ Wis-Madison, PhD(org chem), 69. *Prof Exp:* NIH fel organometallic chem, Univ Calif, Berkeley, 68-71; from asst prof to assoc prof, 71-81, PROF CHEM, EASTERN ILL UNIV, 81-, CHMN CHEM, 89- *Concurrent Pos:* Vis scientist, SRI, Inc, 78-79; secy, Fuel Chem Div, Am Chem Soc, 90- *Mem:* Am Chem Soc; AAAS. *Res:* Organic chemistry of coal and coal liquefaction; mechanisms of SN2 reactions; chemistry of ethers; coal desulfurization. *Mailing Add:* Dept Chem Eastern Ill Univ Charleston IL 61920

BUCHANAN, DAVID ROYAL, b Mansfield, Ohio, Mar 28, 34; m 57; c 4. TEXTILE PHYSICS. *Educ:* Capital Univ, BSc, 56; Ohio State Univ, PhD(chem), 62. *Prof Exp:* Res chemist, Chemstrand Res Ctr, Inc, NC, 62-68; sr res chemist, Phillips Petrol Co, Okla, 68-70, proj mgr, Phillips Fibers Corp, SC, 70-72, mgr textile testing, 72-75; assoc prof design & environ anal, Cornell Univ, 75-78; prof textile mat & mgt & head dept, 78-83, PROF TEXTILE ENG & SCI, NC STATE UNIV, 83-, ASSOC DEAN, COL TEXTILES, 88- *Concurrent Pos:* Consult; gov coun, Fiber Soc, 80-83. *Mem:* Am Phys Soc; Am Chem Soc; Fiber Soc; Nat Coun Textile Educ; Sigma Xi; Soc Mfg Engr, Robotics Int. *Res:* Structure and properties of polymers, fibers and textiles; nonwoven processes and products; fiber extrusion processes and products; process automation in the textile and allied industries; applications of machine vision and robotics to process automation. *Mailing Add:* Dept Textile Eng & Sci NC State Univ Raleigh NC 27695-8301

BUCHANAN, DAVID SHANE, b Fargo, NDak, Mar 20, 53; m 75; c 3. ANIMAL BREEDING, BEEF CATTLE GENETICS. *Educ:* NDak State Univ, BS, 75; Univ Nebr-Lincoln, MS, 77, PhD(animal sci), 79. *Prof Exp:* From asst prof to assoc prof, 80-88, PROF ANIMAL SCI, OKLAHOMA STATE UNIV, 88- *Concurrent Pos:* Ed, Am Soc Am Sci, 84-86; res comt, Nat Pork Producer's Coun, 85-; dir Nat Swine Improvement Fedr, 87. *Mem:* Am Soc Animal Sci; Am Genetic Asn; Coun Agr Sci & Technol; Am Registry Prof Animal Scientists; Sigma Xi. *Res:* Animal breeding and genetics; beef cattle; swine and sheep; author of 34 articles in refereed journals and over 170 reports, abstracts and articles. *Mailing Add:* Div Agr Exp Sta Oklahoma State Univ 206 Animal Sci Stillwater OK 74078-0425

BUCHANAN, EDWARD BRACY, JR, b Detroit, Mich, Sept 12, 27. ANALYTICAL CHEMISTRY. *Educ:* Univ Detroit, BS, 50, MS, 54; Iowa State Univ, PhD, 59. *Prof Exp:* Instr chem, Iowa State Univ, 59-60; from asst prof to assoc prof, 60-88, PROF CHEM, UNIV IOWA, 67- *Mem:* Am Chem Soc; Soc Electroanal Chem. *Res:* Electrochemistry; design of instrumentation; hardware and software systems. *Mailing Add:* Dept Chem Univ Iowa Iowa City IA 52242

BUCHANAN, GEORGE DALE, b Wichita Falls, Tex, Oct 1, 28. SCIENCE EDUCATION, ANATOMY. *Educ:* Rice Inst, BA, 50, MA, 54, PhD(biol), 56. *Prof Exp:* From asst prof to assoc prof biol, Houston State Col, 56-58; res assoc, Rice Inst, 58-59; from instr to assoc prof anat, Med Units, Univ Tenn, 59-68; assoc prof anat, Med Col Ohio, 68-75; assoc prof nursing & anat, 75-84, prof nursing & anat, 84- 88, PROF BIOMED SCI, MCMASTER UNIV, 88- *Concurrent Pos:* Vis prof morphol & biol, Univ Valle, Colombia, 66-67. *Mem:* AAAS; Am Asn Anat; Am Soc Mammalogists; Brit Soc Study Fertil; Soc Study Reproduction; Sigma Xi. *Res:* Reproductive biology; blastocyst implantation; reproductive asymmetry; reproduction in bats. *Mailing Add:* 710-121 Hunter W Hamilton ON L8P 1R2 Can

BUCHANAN, GEORGE R(ICHARD), b Campbellsville, Ky, June 29, 39; m 61; c 3. ENGINEERING, ENGINEERING MECHANICS. *Educ:* Univ Ky, BS, 61, MS, 62; Va Polytech Inst, PhD(civil eng), 66. *Prof Exp:* From asst prof to assoc prof, 65-76, PROF ENG SCI, TENN TECHNOL UNIV, 76-, CHMN DEPT, 78- *Mem:* Assoc Am Soc Civil Engrs; Soc Eng Sci. *Res:* Structural mechanics; nonlinear analysis; wave propagation; finite elements. *Mailing Add:* Dept Civil Eng Tenn Technol Univ Cookeville TN 38505

BUCHANAN, GERALD WALLACE, b London, Ont, Mar 1, 43; m 67; c 1. ORGANIC CHEMISTRY. *Educ:* Univ Western Ont, BSc, 65, PhD(org chem), 69. *Prof Exp:* Nat Res Coun Can fel, 69-70; vis asst prof chem, Univ Windsor, 70-71; from asst prof to assoc prof chem, Carleton Univ, 71-79; assoc prof, Louis Pasteur Univ, Strasbourg, France, 79-80; PROF, CARLETON UNIV, 80- *Mem:* Am Chem Soc; Chem Inst Can. *Res:* Conformational analysis of organic molecules; multinuclear magnetic resonance; organic reaction mechanisms; nuclear magnetic resonance spectroscopy. *Mailing Add:* Dept Chem Carleton Univ Ottawa ON K1S 5B6 Can

BUCHANAN, HAYLE, b Teasdale, Utah, Apr 3, 25; m 53; c 4. PLANT ECOLOGY. *Educ:* Brigham Young Univ, BA, 51, MA, 53; Univ Utah, PhD(bot), 60. *Prof Exp:* Teacher & prin, Tabiona & Hurricane Latter-day Saints Sem, 53-56; teacher high sch, 58-60; asst prof bot, Church Col Hawaii, 60-62; instr life sci & bot, Am River Jr Col, 62-65; from asst prof to assoc prof, 65-71, PROF BOT, WEBER STATE COL, 71- *Concurrent Pos:* Partic range mgt res, US Forest Serv, 66- *Mem:* Ecol Soc Am; Nat Parks Asn; Sigma Xi. *Res:* Ecological study of an area for outdoor education; management of forests in national parks; ecological life history studies. *Mailing Add:* Dept Bot 3730 Harrison Blvd Weber State Col Ogden UT 84403

BUCHANAN, J ROBERT, b Newark, NJ, Mar 8, 28. MEDICINE. *Educ:* Amherst Col, AB, 50; Cornell Univ Med Sch, MD, 54; Nat Bd Med Examr & Am Bd Int Med, dipl. *Prof Exp:* Intern med, NY Hosp, 54-55, asst resident physician, 57-58, res fel endocrinol, 56-57; fes fel endocrinol, Cornell Univ Med Col, 60-61, from instr to asst prof med, 61-67, assoc dean, 65-69, assoc prof med, 69-71, dean, 69-76; prof med, Univ Chicago, 77-82; PROF MED, HARVARD MED SCH, 82-; GEN DIR, MASS GEN HOSP. *Concurrent Pos:* Fel, World Health Orgn, 63; asst to chmn, Dept Med, Cornell Univ Med Col, 64-65, clin assoc prof med, 67-69; assoc dean, Pritzker Sch Med, 78-82; physician, NY Hosp, 56-57 & 60-62, Michael Reese Hosp & Med Ctr, 77-82 & Mass Gen Hosp, 83-; asst chief & chief med, APO 20 Korea Patterson Army Hosp, 58-59; from asst attend physician to attend physician, NY Hosp, 58-59; pres, Michael Reese Hosp & Med Ctr, 77-82, Spaulding Rehab Hosp & Prof Serv Corp, Mass Gen Hosp, 84 & gen dir 82, assoc med sch NY & NJ, 72-76; consult endocrinol & metab dis, Riverview Hosp, 60-68; asst dir, Comprehensive Care & Teaching Prog, Cornell Univ Med Sch, 61-65; assoc dir, Welfare Med Care Proj, NY Hosp, 61-64, asst dir, Eugene F DuBois Clin Res Ctr, 63-65; mem, Comt Sci Policy, mem Sloan-Kettering Cancer Ctr, 69-76, Coun Deans, Asn Med Col, 69-76 & chmn, 75-76 & mgt advan prog steering comt, 72, bd dirs, Winifred Masterson Burke Relief Found, 72 & 80 & Pub Health Res Inst NY City, 69-76, psychiat comt, NY Hosp Bd Gov, 69, Cornell Med Ctr Affil Comt, Hosp Spec Surg, NY Hosp, 69-76, bd dirs, Marie & John Zimmerman Fund, 72 & vpres, 84-, Conf Jewish Hosp Dirs, 77, adv comt, Edwin L Crosby & W K Kellogg Found Fel, Am Hosp Asn, 79-80, comt review Inst Med, NAS, & pres search comt, Asn Am Med Col, 84-86, spec comt on mission & priorities, Am Hosp Asn, 86-, bd dirs, Charles River Labs, 87, Priv Ind Coun Boston, 88-; vchmn bd dirs, Asn Mgt Resources, 78-82, Ill Hosp Res & Educ Found, 78-79; chmn, Coun Teaching Hosp, Admin Bd, Asn Am Med Col, 88- *Mem:* Nat Acad Sci; fel Am Col Physicians; Sigma Xi. *Res:* Author of over 30 publications. *Mailing Add:* Off Gen Dir Mass Gen Hosp Fruit St Boston MA 02114

BUCHANAN, JAMES WESLEY, b Union Mills, NC, May 5, 37; m 59; c 2. PHYSICAL CHEMISTRY, INORGANIC CHEMISTRY. *Educ:* Univ NC, AB, 59; Univ Fla, MS, 62, PhD(radiation chem), 68. *Prof Exp:* Instr chem, Ga Col Milledgeville, 62-63; head dept, Gaston Jr Col, 64-66; from asst prof to assoc prof, Salem Col, NC, 70-75, chmn Dept Chem & Physics, 71-75; sr chemist, Environ Measurements Dept, Res Triangle Inst, NC, 75-77; chmn dept, 77-, PROF CHEM, APPALACHIAN STATE UNIV, 77- *Mem:* Am Chem Soc. *Res:* Radiation chemistry of both liquid and gas phase; radiolysis of ammonia-carbon tetrachloride solutions; radiolysis of phosphine and phosphine-containing gas mixtures. *Mailing Add:* 129 Howards Knob Rd Boone NC 28607

BUCHANAN, JOHN DONALD, b Mesa, Ariz, Oct 1, 27; m 55; c 4. HEALTH PHYSICS, RADIOCHEMISTRY. *Educ:* Univ Ariz, BS, 49; Am Bd Health Physics, cert, 70. *Prof Exp:* Radiochemist, Tracerlab, Inc, 50-56, sr chemist & sect head, 56-59; sr chemist, Gen Atomic, Inc, 59-62; sr radiochemist, Hazelton-Nuclear Sci Corp, 62-67, mgr nuclear measurements dept, 65-67, mgr nuclear prod, appln & measurements, Palo Alto Labs, Teledyne-Isotopes, Inc, 67-71; mgr appl res & radiation safety officer, Int Nutronics, Inc, 71-73; supvr, Radiol Monitoring Progs & Radiol Serv, NUS Corp, 73-75; RADIOCHEMIST & HEALTH PHYSICIST & SR HEALTH PHYSICIST, US NUCLEAR REGULATORY COMN, 75- *Honors & Awards:* Meritorious Serv, High Qual Serv & Spec Achievement Awards, US Nuclear Regulatory Comn. *Mem:* Am Chem Soc; Am Nuclear Soc; Health Physics Soc; fel Am Inst Chemists; Am Acad Health Physics; fel AAAS. *Res:* Radiochemical analysis; environmental radiation and radioactivity, radioassay, radiological monitoring, radiation protection standards, nuclear regulatory standards. *Mailing Add:* 7508 Dew Wood Dr Derwood MD 20855

BUCHANAN, JOHN MACHLIN, b Winamac, Ind, Sept 29, 17; m 48; c 4. BIOCHEMISTRY. *Educ:* DePauw Univ, BS, 38; Univ Mich, MS, 39; Harvard Univ, PhD(biochem), 43. *Hon Degrees:* DSc, Univ Mich, 61, DePauw Univ, 75. *Prof Exp:* Instr biochem, Univ Pa, 43-46; Nat Res Coun fel med, Nobel Inst, Stockholm, 46-48; from asst to prof biochem, Univ Pa, 48-53; prof biochem & head div, 53-67, Wilson prof, 67-88, WILSON EMER PROF BIOCHEM, MASS INST TECHNOL, 88- *Concurrent Pos:* Sabbatical leave, Salk Inst Biol Studies, 64-65; mem fel comt, Div Med Sci, Nat Res Coun, 54-63, mem subcomt nomenclature of biochem, 60-66; mem biochem study sect, NIH, 59-63, chmn, 61-63; mem nat comt, Int Union Biochemists, 57-60, 63-69, secy-treas, 63-65, vchmn, 65-66; mem sci adv bd, St Jude Res Hosp, 64-65; mem bd, Fedn Am Socs Exp Biol, 64-67; mem, Biochem Training Grant Comt, 65-69. *Honors & Awards:* Eli Lilly Award Biol Chem, Am Chem Soc, 51. *Mem:* Nat Acad Sci; Am Soc Biol Chemists (secy, 69-72); Am Chem Soc; Am Acad Arts & Sci; Sigma Xi. *Res:* Synthesis of glycogen; oxidation of fatty acids and ketone bodies; synthesis of purines; isolation and purification of enzymes; metabolism of bacteriophage infected bacteria; metabolic responses of eukaryotic cells to growth factors and oncogenic viruses. *Mailing Add:* 56 Meriam St Lexington MA 02173

BUCHANAN, RELVA CHESTER, b Port Antonio, Jamaica, WI, Apr 10, 36; m 57; c 1. CERAMIC SCIENCE. *Educ:* Alfred Univ, BS, 60; Mass Inst Technol, ScD(ceramic sci), 64. *Prof Exp:* Sr assoc engr, IBM Corp, 64-66, staff engr, 66-70, adv engr, 71-73; assoc prof ceramic eng, 74-84, PROF CERAMIC SCI ENG, DEPT MATS SCI & ENG, UNIV ILL, URBANA-CHAMPAIGN, 84- *Concurrent Pos:* Vis prof, Dept Ceramic Eng, Univ Ill, Urbana, 71-72; consult, Ctr Prof Advan, East Brunswick, NJ; adj prof Dept Mats Sci & Engr, Purdue Univ, 83; mem, Comt Appln Ferroelectrics, Inst Elec & Electronic Engrs. *Mem:* Fel Am Ceramic Soc; Nat Inst Ceramic Engrs; found mem Int Soc Hybrid Microelectronics; NAm Thermal Anal Soc; Electrochem Soc; Mats Res Sci; AAAS. *Res:* Electrical properties of glass and ceramic materials; ceramic substrate, glass-metal and electronic ceramic components; surface properties of ceramics and glasses; low temperature densification of refractory oxides; high TC superconductors; thin films; thick films; packaging; co-authored and edited two books; published over 60 technical works. *Mailing Add:* Dept Mat Sci & Eng Univ Ill Urbana IL 61801

BUCHANAN, ROBERT ALEXANDER, b Detroit, Mich, Sept 8, 32; m 62; c 3. PEDIATRICS, PHARMACOLOGY. *Educ:* Univ Mich, MD, 57; Am Bd Pediat, dipl, 64. *Prof Exp:* Intern, Philadelphia Gen Hosp, Pa, 57-58; resident pediat, Med Ctr, Univ Mich, 60-62; pvt pract, 62-66; investr, Parke, Davis & Co, 66-67, asst dir, 67-68, assoc dir, 69-74, dir, 74-83, vpres, Clin Res Dept, 83-88, VPRES, SCI LIAISON, PARKE, DAVIS & CO, 89- *Concurrent Pos:* Sr exam, Fed Aviation Agency, 63-; dir, Guthrie Clin Res Found, 68-; clin assoc prof pediat, Med Sch, Univ Mich, 68- *Mem:* AMA; fel Am Acad Pediat; Am Therapeut Soc; Am Epilepsy Soc. *Res:* Pediatric pharmacology, especially new drug development in anticonvulsants and antibiotics; medical electronic diagnostic instrument development. *Mailing Add:* Clin Res Dept Parke, Davis & Co Ann Arbor MI 48105

BUCHANAN, ROBERT LESTER, b Seattle, Wash, July 3, 46; m 74; c 3. MICROBIAL FOOD SAFETY, MYCOTOXICOLOGY. *Educ:* Rutgers Univ, BS, 69, MS, 71, MPhil, 72, PhD(food sci), 74. *Prof Exp:* Res assoc, Univ Ga, 74-75; from asst prof to assoc prof res, Drexel Univ, 75-82; RES LEADER, RES ADMIN, US DEPT AGR, 82- *Concurrent Pos:* Am Soc Microbiol; Inst Food Technologists. *Mem:* Fel Am Acad Microbiol. *Res:* Conducted research on the microbiology of foods and food safety including work with pathogenic bacteria and mycotoxigenic fungi. *Mailing Add:* Microbial Food Safety Res Unit 600 E Mermaid Lane Philadelphia PA 19118

BUCHANAN, ROBERT MARTIN, b Baltimore, Md, May 29, 51; m 75. ELECTROANALYTICAL, CATALYSIS. *Educ:* Western Md Col, BA, 73; WVa Univ, MS, 75; Univ Colo, PhD(chem), 80. *Prof Exp:* Res assoc, Solar Energy Res Inst, Univ Colo, 80-81, Mass Inst Technol, 81-82; ASST PROF INORG CHEM, UNIV LOUISVILLE, 82- *Mem:* Am Chem Soc; Sigma Xi; Electrochem Soc; AAAS; Am Asn Univ Prof. *Res:* Synthesis, characterization nuclear magnetic and electron spin resonance spectroscopy of organo-transition metal complexes; electrocatalytic reactions at chemically modified electrode surfaces; electroanalytical investigations of biologically important compounds; transition metal chemistry; bioinorganic; organometallic chemistry; electrocatalysis; synthesis of multifunctional ligands; activation of small molecules; magnetic exchange interactions and EPR; binuclear Cu, Fe and Mn complexes as models of protein active sites. *Mailing Add:* Dept Chem Univ Louisville Louisville KY 40292

BUCHANAN, RONALD JAMES, b Regina, Sask, Sept 15, 38; m 61; c 2. AQUATIC ECOLOGY, PHYCOLOGY. *Educ:* Univ BC, BS, 61, PhD(oceanog), 66. *Prof Exp:* Res assoc biol, Johns Hopkins Univ, 66-68; sr res assoc limnol, Univ Wash, 68-69; res officer aquatic ecol, Water Mgt Br, Ministry Environ, Govt BC, 69-74, sr biologist, 74-80, div chief, Environ Studies Div, Water Invests Br, 80-82, dir, Aquatic Studies Br, 82-90, MGR, RESOURCE QUAL SECT, WATER QUAL BR, 90-, DIR, WATER QUAL BR, MINISTRY ENVIRON, GOVT BC, 90- *Mem:* Am Soc Limnol & Oceanog; Ecol Soc Am; AAAS; NAm Lake Mgt Soc; Int Asn Theoret & Appl Limnol. *Res:* Taxonomy and ecology of marine and freshwater planktonic algae and protozoa; lake eutrophication; planktonic microbiota as water quality indicators; watershed management for water quality control. *Mailing Add:* Water Qual Br Ministry Environ Parliament Bldgs Victoria BC V8V 1X5 Can

BUCHANAN, RONALD LESLIE, b Kirkland Lake, Ont, May 15, 37; m 61; c 3. PHARMACEUTICAL LICENSING. *Educ:* Univ Western Ont, BSc, 59, PhD(org chem), 63. *Prof Exp:* Res fel, Univ BC, 63-64; sr res chemist, Bristol-Myers Co, 64-76, from asst dir to assoc dir res planning & licensing, 76-82, assoc dir licensing, 82-89; DIR LICENSING, BRISTOL-MYERS SQUIBB CO, 89- *Mem:* Am Chem Soc; NY Acad Sci; AAAS. *Res:* Synthesis of biologically interesting monfluorinated compounds; biogenetic-type synthesis of natural products; synthesis of general medicinal agents. *Mailing Add:* Bristol-Myers Squibb Co Five Research Pkwy PO Box 5100 Wallingford CT 06492-7660

BUCHANAN, RONNIE JOE, b Mulberry, Kans, July 14, 44; m 64; c 1. RADIATION PHYSICS, BIOPHYSICS. *Educ:* Univ Kans, Lawrence, BA, 73, PhD(radiation physics), 76. *Prof Exp:* Technician health physics, US Navy, Antarctica, 69-70; radiation physicist, Los Alamos Nat Lab, 77-81; mem staff, Nuclear Sci Ctr, Tex A&M Univ, 81-84; RES PHYSICIST, HALLIBURTON SERV, 85- *Concurrent Pos:* Radiation physicist, Univ Kans, Lawrence, 70-74, res asst, 74-77. *Mem:* Health Physics Soc. *Res:* Neutron detection and reactions; methods of determining neutron dose; developing new and better methods of determining neutron spectrums; determining pulsed neutron flux and dose; gamma-ray detection and radiation instrumentation. *Mailing Add:* Erd Dept Halliburton Serv PO Box 1431 Duncan OK 73536-0450

BUCHANAN, RUSSELL ALLEN, b Indianapolis, Ind, Aug 24, 28; m 50; c 10. POLYMER CHEMISTRY, NATURAL PRODUCTS. *Educ:* Ind Univ, BS, 50. *Prof Exp:* Res chemist food oils & fats, Standard Brands Inc, 50-56; chief chemist prod, Com Solvents Corp, 56-60; res chemist chemurgy, Northern Regional Res Ctr, USDA, 60-79; sr chem, Soil & Land Use Technol, Inc, 79-84; ENVIRON CHEMIST, MARION COUNTY HEALTH DEPT, 85- *Mem:* Am Chem Soc; Soc Econ Bot. *Res:* Food products; industrial uses for starch; hydrocarbon producing plants; processing of multi-use crops; botanochemicals and biomass conversions. *Mailing Add:* Rte 1 Box 510 Clayton IN 46118

BUCHANAN, THOMAS JOSEPH, b Albany, NY, Oct 26, 29; m 51; c 6. WATER RESOURCES ENGINEERING. *Educ:* Rensselaer Polytech Inst, BSCE, 51; Princeton Univ, MSCE, 67. *Prof Exp:* Jr engr, Water Resources Div, US Geol Surv, NY, 51-52, asst engr, 52-55, assoc engr, 55-56, off engr, 56-58, engr-in-charge, 58-61, staff engr, Washington DC, 62-64, asst dist engr, NJ, 64-66, asst dist chief, 67-68, subdist chief, Fla, 68-76, ASST CHIEF HYDROLOGIST OPERS, US GEOL SURV, 76- *Concurrent Pos:* Chmn, Interagency Task Force on Water Policy-Instream Flows, 78-79. *Mem:* Am Soc Civil Engrs; Am Water Works Asn; Am Water Resources Asn; Nat Soc Prof Engrs. *Res:* Updated techniques used in measurement of stream discharge; developed techniques for use in time of travel studies in streams using fluorescent tracers; longitudinal dispersion in streams. *Mailing Add:* 1834 St Boniface St Vienna VA 22182

BUCHANAN-DAVIDSON, DOROTHY JEAN, b Monmouth, Ill, Dec 22, 25; m 57; c 3. BIOCHEMISTRY. *Educ:* Wash State Col, BS, 47; Univ Cincinnati, MS, 49, PhD(biochem), 51. *Prof Exp:* Asst inorg chem, Univ Cincinnati, 47-48, res fel, Children's Hosp, 48-50; instr biochem, Sch Med, Vanderbilt Univ, 50-53; Nat Res Coun fel, Lister Inst, London, 53-55; USPHS fel immunochem, Pasteur Inst, Paris, 55-56; abstractor, Chem Abstr, 60-76; proj assoc biochem, Univ Wis, 56-60, prog coordr, Ctr Affective Disorders, 83-86, ed, Sch Nursing, 86-90; RETIRED. *Concurrent Pos:* Consult ed, Mother's Manual, 64-69; instr, Madison Area Tech Col, 74-76; proj specialist, Water Resources Ctr, Univ Wis, 75-76; sci writer, Wis Clin Cancer Ctr, 76-80; ed, Admin Med Prog, Univ Wis, 80-83 & Sch Nursing, 87- *Mem:* Am Med Writers Asn; Soc Tech Commun; Anciens Eleves Inst Pasteur; Sigma Xi. *Res:* Immunochemistry; chemistry of blood-group-specific substances and meconium; phosphorous poisoning; x-irradiation; immunochemical agar-gel diffusion techniques; immunochemistry and biological properties of synthetic polypeptides and modified proteins; immunochemistry of plant proteins; cancer; administrative medicine; writing on chemistry, biochemistry, cancer, and psychiatry. *Mailing Add:* 6278 Sun Valley Pkwy Oregon WI 53575

BUCHANAN-SMITH, JOCK GORDON, b Edinburgh, Scotland, Mar 9, 40; m 64; c 2. ANIMAL SCIENCE, ANIMAL NUTRITION. *Educ:* Aberdeen Univ, BSc, 62; Iowa State Univ, BS, 63; Tex Tech Col, MS, 65; Okla State Univ, PhD(animal sci), 69. *Prof Exp:* Res asst animal sci, Tex Tech Col, 64-65; res asst, Okla State Univ, 65-69; from asst prof to assoc prof, 69-84, PROF ANIMAL SCI, UNIV GUELPH, 85- *Concurrent Pos:* Nat Res Coun Can operating grants, 69-88; Agr Can operating grant, 75-78; ed, Can J Animal Sci, 85-87. *Mem:* AAAS; Am Soc Animal Sci; Am Dairy Sci Asn; Agr Inst Can; Sigma Xi. *Res:* Utilization of silages by ruminants; digesta passage in ruminants silage additives. *Mailing Add:* Pitcaple Farm RR 2 Hespeler Cambridge ON N3C 2V5 Can

BUCHBERG, HARRY, b Detroit, Mich, Oct 17, 17; m 43; c 3. ENGINEERING. *Educ:* Univ Calif, Berkeley, BS, 41; Univ Calif, Los Angeles, MS, 54. *Prof Exp:* Res & design engr, Lockheed Aircraft Corp, 41-47; res engr, 47-57, PROF ENG, UNIV CALIF, LOS ANGELES, 57- *Concurrent Pos:* Consult thermal & solar design & anal, 55-; Fulbright scholar, 62 & 72; vis prof, Technion, Israel, 62 & Bogagici Univ, Turkey 71-72. *Mem:* Am Soc Mech Engrs; Am Soc Eng Educ; Am Soc Heating, Refrig & Air-Conditioning Engrs; Solar Energy Soc. *Res:* Thermal and solar engineering; solar energy collectors and energy conversion; published numerous scientific paper and reports. *Mailing Add:* 9833 Vicar St Los Angeles CA 90034

BUCHEL, JOHANNES A, b Netherlands, July 31, 26; m 71. FLAVOR CHEMISTRY. *Educ:* Hogere Tech Sch, Netherlands, BSc, 46. *Prof Exp:* Tech dir, Naardeh, SAfrica, 56-64, vpres tech dir, USA, 64-71, tech dir, Eng, 72-73; VPRES INT RES & DEVELOP, PEPSI-COLA INC, 74- *Concurrent Pos:* Pres, Flavor & Extract Mfg Asn, 90. *Mem:* Am Chem Soc; Inst Food Technologists; Soc Flavor Chemists. *Res:* Management of soft drink research and development, including materials research, flavor research, and flavor creation. *Mailing Add:* Old Logging Rd W Box 312 RFD 1 Yorktown Heights NY 10598

BUCHELE, W(ESLEY) F(ISHER), b Cedar Vale, Kans, Mar 18, 20; m 45; c 4. AGRICULTURAL ENGINEERING. *Educ:* Kans State Univ, BS, 43; Univ Ark, MS, 51; Iowa State Univ, PhD(soil physics, agr eng), 54. *Prof Exp:* Student engr, John Deere Waterloo Tractor Works, 46-48, jr engr, 48; asst prof agr eng, Univ Ark, 48-51; asst prof & agr engr, Res Br, Agr Res Serv, USDA, Iowa State Univ, 54-56; assoc prof, Mich State Univ, 56-63; prof, 68-89, EMER PROF AGR ENG, IOWA STATE UNIV, 89- *Concurrent Pos:* Vis prof & head div, Univ Ghana, 68-69; vis scientist, CSIRO, Melbourne, Australia, 79; res engr, IITA, Ibadan, Nigeria, 79; vis prof, Beijing Inst Agr Mechanization, 83, Nat Agrarian Univ, La Molina, Peru, 86 & 88; exec engr, special purpose vehicles, Sydney, Australia, 87, JAC Tractor Co, Sydney, 88; deleg numerous int confs, 61-88; consult various cos, 53-90; mem numerous comts, Am Soc Agr Engrs, 57-; vis lectr, Manila, 63, WGer, 64, Moscow, 68, Ghana, 69, Los Banos, 79, NZ, 79, Australia, 79, 87 & 88, Africa, 79, Hungary, 80, Japan, 83, China, 83, Peru, 86. *Honors & Awards:* Cyrus Hall McCormick-Jerome Increase Case Medalist, 88. *Mem:* Am Soc Agron; fel Am Soc Agr Engrs; Soc Automotive Eng; Int Asn Mechanization Field Exp; Int Soc Terrain-Vehicle Systs; Osborne Asn; fel Nat Inst Agr Engrs; Int Platform Asn; AAAS; Sigma Xi. *Res:* Tillage; soil compaction; vehicular mechanics; torque meters; threshing cylinders and undisturbed soil samplers; agricultural safety; harvesting of grain and hay; energy conversion systems; biomass furnaces; producer gas generators; alcohol distillation; design of slot and punch planters for conservation production system of farming; vegetable-oil expelling and use; author or co-author of approximately 220 technical articles, 19 books and recipient of numerous patents. *Mailing Add:* 239 Parkridge Circle Ames IA 50010

BUCHENAU, GEORGE WILLIAM, b Amarillo, Tex, May 23, 32; m 55, 72; c 2. PLANT PATHOLOGY. *Educ:* NMex State Univ, BS, 54, MS, 55; Iowa State Univ, PhD(plant path), 60. *Prof Exp:* From asst prof to assoc prof plant sci, 59-80, PROF PLANT PATH, SDAK STATE UNIV, 80- *Mem:* Am Phytopath Soc. *Res:* Wheat disease; resistance, chemical control and biological control. *Mailing Add:* Dept Plant Sci SDak State Univ Brookings SD 57006

BUCHER, JOHN HENRY, b Newport, Ky, June 23, 39. STEEL RESEARCH & DEVELOPMENT, QUALITY CONTROL. *Educ:* Univ Cincinnati, BEng, 61; Ohio State Univ, MS, 62, PhD(metall eng), 64. *Prof Exp:* Res eng, Jones & Laughlin Steel Corp, 64-65, sr res eng, 65-66, res supvr, 66-72, chief metall, 72-73, mgr quality control, 73-74, staff engr, 74-76, supvr prod develop, Jones & Laughlin Steel Corp, 76-84; DIR, RES & QUAL CONTROL, LUKENS STEEL CO, 84- *Concurrent Pos:* Instr, mat eng, Carnegie-Mellon Univ, 67-70; pres, J H Bucher & Assoc Metall Consults Inc, 81- *Honors & Awards:* Sigma Xi. *Mem:* Am Soc Metals; Am Inst Mining; Metall & Petrol Eng; Am Iron & Steel Inst; Am Iron & Steel Engr. *Res:* Invented and developed steels with uniquely valuable properties such as improved motor lamination steels, high strength low alloy flat rolled and bar steels as well as new steelmaking practice. *Mailing Add:* 143 Barton Dr Spring City PA 19475

BUCHER, NANCY L R, b Baltimore, Md, May 4, 13. MEDICINE. *Educ:* Bryn Mawr Col, AB, 35; Johns Hopkins Univ, MD, 43. *Prof Exp:* Intern, Mass Mem Hosp, Boston, 43-44, clin fel, 44-45; res fel & assoc med, 45-59, clin assoc, 59-68, asst clin prof med, 68-72, assoc prof med, 72-80, ASSOC PROF SURG, HARVARD MED SCH, 80-, ASSOC BIOLOGIST, MASS GEN HOSP, 52- *Mem:* Am Asn Study Liver Dis; Am Soc Biol Chemists; Int Asn Study Liver; Tissue Culture Asn; Am Soc Cell Biol; Sigma Xi. *Res:* Cancer; growth regulation; cholesterol biosynthesis; liver regeneration. *Mailing Add:* One Longfellow Pl Boston MA 02114-2438

BUCHER, T(HOMAS) T(ALBOT) NELSON, b Philadelphia, Pa, June 29, 19; m 46; c 3. RADIO COMMUNICATIONS SYSTMS ENGINEERING. *Educ:* Drexel Inst Technol, BS, 41; Univ Pa, MS, 47, PhD(elec eng), 58. *Prof Exp:* Patent engr, Proctor Elec Co, 41-42; engr, Spec Apparatus Div, RCA Corp, 42-46, eng leader, surface Commun Div, 46-47, eng mgr, 57-63, staff engr, Commun Systs Div, 63-84; CONSULT ENG, 84- *Concurrent Pos:* Lectr, Grad Sch, Villanova Univ, 59-63. *Mem:* Inst Elec & Electronics Engrs; Nat Soc Prof Engrs. *Res:* Communications equipment and system design; communication theory. *Mailing Add:* 36 E Central Ave Moorestown NJ 08057

BUCHHEIM, H PAUL, b Vallejo, Calif, Apr 21, 47; m 72; c 3. PALEOENVIRONMENTS, TAPHONOMY. *Educ:* Pac Union Col, BA, 71, MA, 72; Univ Wyo, PhD(geol), 78. *Prof Exp:* From asst prof to assoc prof, 79-88, PROF, LOMA LINDA UNIV, 89- *Concurrent Pos:* Fel geol, Johns Hopkins Univ, 78-79. *Honors & Awards:* President's Award, Am Asn Petrol Geologists. *Mem:* Geol Soc Am; Soc Econ Paleontologists & Mineralogists; Am Asn Petrol Geologists. *Res:* Lacustrine deposits and their sedimentology, paleontology and geochemistry, both ancient and modern; taphonomy and sediment-organism relationships; sedimentary petrology of laminated rocks. *Mailing Add:* Dept Natural Sci Loma Linda Univ Loma Linda CA 92350

BUCHHEIT, RICHARD D(ALE), b Indiana, Pa, Aug 20, 24; m 49, 77; c 5. METALLURGY. *Educ:* Columbia Univ, BS, 45. *Prof Exp:* Metall observer, Midland Works, Crucible Steel Co Am, 46-48; res engr, Columbus Labs, Battelle Mem Inst, 48-52, asst div chief, 52-74, prin res scientist, 74-89; RETIRED. *Mem:* Am Soc Metals; Int Metallog Soc (treas, 71-91). *Res:* Metallography; physical metallurgy; failure analysis. *Mailing Add:* 3786 Quail Hollow Dr Columbus OH 43228

BUCHHOLZ, ALLAN C, b Manchester, NH, July 20, 40; m 63; c 3. PHYSICAL ORGANIC CHEMISTRY, POLYMER CHEMISTRY. *Educ:* Univ Mass, BS, 62; Univ Ill, MSc, 64, PhD, 67. *Prof Exp:* Sr res chemist, 67-73, supvr indust tape div, 73-77, mgr indust tape div, 3M Co, 77-; AT CONKLIN CO, INC. *Mem:* Sigma Xi; Am Chem Soc. *Res:* Preparation, nuclear magnetic resonance and mass spectral studies of carbon-13-labeled compounds; polymer synthesis and characterization, especially light-scattering of macromolecular systems; isocyanate and polyurethane chemistry; development and economic evaluation of new products and applications; encapsulation; pressure sensitive tapes and adhesives. *Mailing Add:* 3M Co 3M Ctr Bldg 230 St Paul MN 55101

BUCHHOLZ, JEFFREY CARL, b Sheboygan, Wis, June 10, 47; m 70. SURFACE PHYSICS. *Educ:* Univ Wis, Eau Claire, BS, 69; Univ Wis, Madison, MS, 71 & 73, PhD(mat sci), 74. *Prof Exp:* Res assoc chem, Univ Calif, Berkeley, 74-76; STAFF RES SCIENTIST, GEN MOTORS RES LABS, 76- *Mem:* Am Vacuum Soc; Am Phys Soc; Sigma Xi; Mat Res Soc; Soc Photoptical Instrumentation Engrs. *Res:* Study of the solid-gas and solid-liquid interface; modification of surface properties using ion beams and laser treatment; thin film materials. *Mailing Add:* Physics Dept Gen Motors Res Labs Warren MI 48090

BUCHHOLZ, R ALAN, CARDIOVASCULAR PHYSIOLOGY. *Educ:* Univ Tenn, Knoxville, PhD(physiol psychol), 79. *Prof Exp:* PROF & RES ASST PHARMACOL, HEALTH SCI CTR, UNIV TEX, 81- *Res:* Neutral-control circulation. *Mailing Add:* Dept Pharmacol Sterling-Winthrop Res Inst 81 Columbia Turnpike Rensselaer NY 12144

BUCHHOLZ, ROBERT HENRY, b Wakeeney, Kans, Aug 26, 24; m 47; c 3. PHYSIOLOGY. *Educ:* Ft Hays Kans State Col, BS, 49; Kans State Col, MS, 50; Univ Mo, PhD, 57. *Prof Exp:* Asst prof zool, 50-57, assoc prof biol, 57-63, PROF BIOL, MONMOUTH COL, ILL, 63- *Concurrent Pos:* Mem staff, Argonne Nat Lab, 62-63; NIH fel, Dept Physiol, Univ Edinburgh, 63-64. *Mem:* AAAS; Am Soc Zoologists; Am Soc Mammal; assoc mem Am Physiol Soc; Sigma Xi. *Res:* Gastrointestinal physiology; digestive enzymes in bat, rat and mole; pancreatic lipase in the mole and rat. *Mailing Add:* Dept Biol Monmouth Col Monmouth IL 61462

BUCHHOLZ SHAW, DONNA MARIE, b Chicago, Ill, May 27, 50; m; c 1. PHARMACEUTICAL RESEARCH & DEVELOPMENT, TECHNOLOGY TRANSFER. *Educ:* Quincy Col, BS, 72; Univ Ill, MS, 75, PhD(microbiol), 78. *Prof Exp:* Res assoc immunol, Argonne Nat Lab, 78-80; res info scientist, Abbott Labs, 80-82, sr proj mgr, 82-87, opers mgr, Thrombolytics Venture, 87-89, mgr, In-Licensing, 89-90, DIR, TECHNOL ASSESSMENT & LICENSING, ABBOTT LABS, 90- *Concurrent Pos:* Guest scientist, Argonne Nat Lab, 80-; part-time fac, NEastern Ill Univ, 85- *Honors & Awards:* Res Award, Sigma Xi, 78. *Mem:* Am Soc Microbiol; Am Acad Microbiol; Am Asn Immunologists; Sigma Xi; Licensing Execs Soc. *Res:* Discovery, development and clinical appraisal of new thrombolytic agents; immunological responses to cancer and immunological approaches to treatment of metastic disease. *Mailing Add:* Abbott Labs One Abbott Park Rd Abbott Park IL 60064

BUCHI, GEORGE, b Baden, Switz, Aug 1, 21. ORGANIC CHEMISTRY. *Educ:* Swiss Fed Inst Technol, DSc, 47. *Hon Degrees:* Dr, Univ Heidelberg, 84, Switz Fed Inst Technol, 87. *Prof Exp:* Firestone fel chem, Univ Chicago, 48-49, instr, 49-51; from asst prof to prof, 51-71, CAMILLE DREYFUS PROF CHEM, MASS INST TECHNOL, 71- *Honors & Awards:* Ruzicka Award, 57; Fritzche Award, 58; Am Chem Soc Award, 73; Hon Foreign Mem Pharmaceut Soc Japan, 84. *Mem:* Nat Acad Sci; fel Chem Soc London; Am Chem Soc; Swiss Chem Soc; German Chem Soc; Japanese Chem Soc. *Res:* Synthetic organic chemistry; natural products; free radical reactions. *Mailing Add:* Dept Chem Mass Inst Tech 77 Massachusetts Ave Cambridge MA 02139

BUCHIN, IRVING D, orthodontics; deceased, see previous edition for last biography

BUCHLER, EDWARD RAYMOND, b Chicago, Ill, Sept 6, 42; m 65; c 3. ETHOLOGY. *Educ:* Calif State Polytech Col, BS, 64; Univ Calif, Santa Barbara, MA, 66; Univ Mont, PhD(zool), 72. *Prof Exp:* Res zoologist, US Army, Ft Detrick, MD, 66-68; res assoc ethology, Rockefeller Univ, 72-75; asst prof ethology, univ Md, Col Park, 75-; AT ORI INC. *Concurrent Pos:*

NSF grant; consult bat conserv & control. *Honors & Awards:* Original Res Contribs Achievement Award, US Army, 68. *Mem:* AAAS; Animal Behav Soc; Am Soc Mammalogists; Am Asn Zool Parks & Aquaria; Sigma Xi. *Res:* Acoustic communication systems in animals; echolocation in small mammals; hunting strategies, prey selection and communication in insectivorous bats; environmental cues used by migrating and foraging bats. *Mailing Add:* 1303 Palmyra Lane Bowie MD 20716

BUCHLER, JEAN-ROBERT, b Luxembourg City, Luxembourg, Apr 4, 42; US citizen; m 73; c 3. STELLAR EVOLUTION & PULSATION. *Educ:* Univ Liege, Belg, Licence, 65; Univ Calif, San Diego, MS, 67, PhD(physics), 69. *Prof Exp:* Res fel physics, Calif Inst Technol, 69-71; asst prof physics, Belfer Grad Sch, Yeshiva Univ, 71-74; assoc prof, 74-80, PROF PHYSICS, UNIV FLA, 80- *Concurrent Pos:* Consult, Los Alamos Nat Lab, 77-; guest prof, Niels Bohr Inst, Copenhagen, 78-79 & Univ Paris VI, France; vis scientist, Observ Nice, France, 82, 84, 86 & Inst Astropys, Univ Liege, Belg. *Mem:* Int Astron Union. *Res:* Theoretical astrophysics; stellar evolution; variable stars; nonlinear dynamics; chaotic behavior; supernovae; equation of state of dense matter; radiative transfer; hydrodynamics. *Mailing Add:* Physics Dept Univ Fla Gainesville FL 32611

BUCHMAN, ELWOOD, b Iowa City, Iowa, June 10, 23; m 44; c 2. INTERNAL MEDICINE, GASTROENTEROLOGY. *Educ:* Univ Iowa, BA, 40, MD, 43. *Prof Exp:* Set chief gastroenterol, Vet Admin Hosp, Dearborn, Mich, 48-52; asst chief med, Vet Admin Hosp, Iowa City, 52-69; chief med serv, Vet Admin Hosp, Des Moines, 69-73; dir gastroenterol, Eaton Labs, Norwich, NY, 73-78; assoc group dir gastroenterol, Merrell Nat Labs, 78-80; MED DIR, CINTEST, 80- *Concurrent Pos:* Asst prof gastroenterol, Wayne State Univ, 48-52; assoc prof med, Univ Iowa, 73-78. *Mem:* Fel Am Col Physicians; Am Col Gastroenterol; Am Col Clin Pharmacol; Sigma Xi; Am Soc Clin Pharm & Therapeut. *Res:* Gastrointestinal drugs; clinical pharmacology. *Mailing Add:* 6080 Miami Rd Cincinnati OH 45243

BUCHMAN, RUSSELL, b New York, NY, Dec 19, 47; m 70; c 2. PROCESS DEVELOPMENT, PROCESS CHEMISTRY. *Educ:* State Univ NY, Buffalo, BA, 69; Purdue Univ, PhD(med chem), 73. *Prof Exp:* Res assoc med chem, Univ Kans, 73-75; res chemist agr chem & pharmaceut chem, Diamond Shamrock Corp, 75-78, sr res chemist agr chem, 78-79, res supvr process develop, 79-82, group leader, 82-85, GROUP LEADER ANALYTICAL SERV, SDS BIOTECH CORP, 85- *Mem:* Am Chem Soc. *Res:* Scale-up of intermediates for analog synthesis programs and final products for biological evaluations; process scouting activities evaluate different synthetic approaches to target compounds; process development activities define and optimize selected synthetic routes. *Mailing Add:* 2092 Kilbirnie Dr Germantown TN 38139

BUCHNER, MORGAN M(ALLORY), JR, b Baltimore, Md, Mar 5, 39; m 63; c 2. ELECTRICAL ENGINEERING. *Educ:* Johns Hopkins Univ, BES, 61, PhD(elec eng), 65. *Prof Exp:* Mem tech staff, 65-69, supvr traffic res, 69-72, head, Performance Measurement Dept, 72-75, dir, Traffic Network Planning, 75-77, DIR, OPERS SYSTS PLANNING, BELL TEL LABS, 77- *Mem:* Inst Elec & Electronics Engrs. *Mailing Add:* AT&T Network Systs 2f602 Holmdel NJ 07733

BUCHNER, STEPHEN PETER, b Cape Town, SAfrica, Sept 20, 45; m 76. SOLID STATE PHYSICS. *Educ:* Princeton Univ, BA, 68; Univ Pa, MS, 70, PhD(physics), 75. *Prof Exp:* Res assoc physics, Univ Md, 76-77; RES SCIENTIST PHYSICS, MARTIN MARIETTA LABS, 77- *Mem:* Am Phys Soc. *Res:* Semiconductor infrared detectors; metal oxide semiconductor devices for charge coupled device; surface physics; Raman scattering. *Mailing Add:* Martin Marietta Labs 1450 Rolling Rd Baltimore MD 21227

BUCHOLTZ, DENNIS LEE, b Greenville, Ohio, Jan 23, 51. AGRONOMY, PLANT PHYSIOLOGY. *Educ:* Manchester Col, BS, 73; Univ Nebr, Lincoln, MS, 75, PhD(agron), 78. *Prof Exp:* Fel, Agron Dept, 78-79, FEL, BOT & PLANT PATH DEPT, PURDUE UNIV, 80-, GRANTS ADMINR, MIDWEST PLANT BIOTECHNOL CONSORTIUM, 90- *Mem:* Weed Sci Soc Am. *Res:* Effects of herbicides on the physiology and biochemistry of plants; processes involved in the absorption of herbicides into plants; biosynthesis of lignin in normal and mutant sorghum. *Mailing Add:* 819 Vine West Lafayette IN 47906

BUCHSBAUM, DAVID ALVIN, b New York, NY, Nov 6, 29; m 49; c 3. MATHEMATICS. *Educ:* Columbia Col, NY, AB, 49; Columbia Univ, PhD, 54. *Prof Exp:* Instr math, Princeton, 53-54, NSF fel, 54-55; instr math, Univ Chicago, 55-56; asst prof, Brown Univ, 56-59; assoc prof, 59-60; assoc prof, 60-63, PROF MATH, BRANDEIS UNIV, 63- *Concurrent Pos:* NSF fel, 60-61; Guggenheim fel, 64-65. *Mem:* Am Math Soc. *Res:* Foundations and applications to commutative ring theory of homological algebra. *Mailing Add:* Three Victoria Circle Newton Center MA 02159

BUCHSBAUM, GERSHON, b Tel Aviv, Israel, July 24, 49; US citizen; m 76; c 3. VISION & COLOR, IMAGE ANALYSIS. *Educ:* Tel-Aviv Univ, BSc, 74, MSc, 75, PhD(eng sci), 79. *Hon Degrees:* MA, Univ Pa, 86. *Prof Exp:* Asst prof, 79-85, grad group chmn, 87-91, ASSOC PROF BIOENG, UNIV PA, 85- *Concurrent Pos:* Prin investr, NSF, 84-90, NIH, 86-89 & AFOSR, 90-; NSF presidential young investr, 84. *Mem:* Sr mem Inst Elec & Electronics Engrs; Optical Soc Am; Asn Res Vision Ophthal. *Res:* Image coding and analysis by the visual system; coloring of images in the visual system in space, time and color. *Mailing Add:* Dept Bioeng Univ Pa 220 S 33rd St Philadelphia PA 19104-6315

BUCHSBAUM, HERBERT JOSEPH, b Vienna, Austria, June 7, 34; US citizen; m 62; c 2. OBSTETRICS & GYNECOLOGY, ONCOLOGY. *Educ:* NY Univ, BA, 55; Univ Munich, MD, 61. *Prof Exp:* Fel gynec oncol, State Univ NY, Downstate Med Ctr, 66-68, asst prof, 68-70; assoc prof, Col Med, Univ Iowa, 71-75, prof, 75-78; prof obstet & gynec & dir, div gynec oncol, Univ Tex Southwestern Med Sch, Dallas, 78-84; prof obstet & gynec, Univ

Pittsburgh, 84-87; PROF & CHMN OBSTET & GYNEC, MED COL WIS, MILWAUKEE, 87- *Mem:* Am Col Obstetricians & Gynecologists; Am Col Surgeons; Soc Gynec Oncologists; Sigma Xi. *Res:* Gynecologic oncology; tumor immunology. *Mailing Add:* Dept Obstet & Gynec Med Col Wis 8701 Watertown Plank Rd Milwaukee WI 53226

BUCHSBAUM, MONTE STUART, b Chicago, Ill, Apr 15, 40; m 68; c 2. MEDICINE. *Educ:* Univ Pittsburgh, BS, 61; Univ Calif, San Francisco, MD, 65. *Prof Exp:* Res asst, Univ Pittsburgh, 61-62, Langley Porter Neuropsychiat Inst, San Francisco, 63-65; intern, Univ Calif Med Ctr, 65-66; sr asst surg, 66-68, med officer, 68-75, actg chief, Sect Perceptual & Cognitive Studies, 70-75, chief, 75-78, CHIEF, SECT CLIN PSYCHOPHYSIOL, USPHS, 78-; PROF PSYCHIAT, UNIV CALIF, IRVINE. *Concurrent Pos:* Co-ed-in-chief, Psychiat Res. *Honors & Awards:* Borden Award, 65. *Mem:* Soc Psychiat Res; Soc Psychophysiol Res. *Res:* Clinical studies of the psychophysiology of psychiatric patients and electrophysiological correlates of pharmacological treatment; positron emission tomography. *Mailing Add:* Dept Psychiat Univ Calif Irvine CA 92717

BUCHSBAUM, RALPH, b Chickasha, Okla, Jan 2, 07; m 33; c 2. INVERTEBRATE ZOOLOGY, CELL BIOLOGY. *Educ:* Univ Chicago, BS, 28, PhD(zool), 32. *Prof Exp:* Asst zool, Univ Wis, 28-29; asst, Univ Chicago, 29-30, from instr biol to asst prof zool, 31-42, asst prof, 45-47, res assoc, Inst Radiobiol & Biophys, 47-50; prof, 50-71, EMER PROF BIOL, UNIV PITTSBURGH, 71- *Concurrent Pos:* Instr, Gary Col, 33-35; cur, Mus Sci & Indust, Univ Chicago, 45-47; hon res assoc, Carnegie Mus, 55-; chief adv biol, Encycl Britannica Films, 59-; Fulbright award, Thailand, 59-60; dir, Ctr Studies Learning, 62; mem staff, UNESCO educ mission, India, 64, chief consult biol teaching, Africa, 65-66; US AID-Univ Pittsburgh Ecuador Educ Proj, 65-67; ed, Boxwood Press, 72- *Honors & Awards:* Univ Chicago Prize, 40; Condon lectr, 54. *Mem:* Am Soc Zoologists; Ecol Soc Am; NY Acad Sci; AAAS; Biol Photog Asn. *Res:* Invertebrate biology; ecology of invertebrates; effects of irradiating living cells with various radiations in tissue cultures; cellular physiology; perfusion techniques in tissue culture; invertebrate physiology; ecology; environmental sciences. *Mailing Add:* 183 Ocean View Blvd Pacific Grove CA 93950-3093

BUCHSBAUM, SOLOMON JAN, b Stryj, Poland, Dec 4, 29; nat US; m 55; c 3. PHYSICS. *Educ:* McGill Univ, BS, 52, MSc, 53; Mass Inst Technol, PhD(physics), 57. *Prof Exp:* Asst physics, Mass Inst Technol, 53-55, staff mem, Res Lab Electronics, 57-58; mem tech staff, Bell Tel Labs, 58-65, head dept, 61-65, dir electronics res lab, 65-68; vpres res, Sandia Labs, 68-71; exec dir res commun sci div, AT&T Bell Labs, 71-75, exec dir transmission systs div, 75-76, vpres network planning & customer systs, 76-79, SR VPRES TECHNOL SYSTS, AT&T BELL LABS, 79- *Concurrent Pos:* Assoc ed, Physics of Fluids, 63-64 & Rev of Mod Physics, 68-76; mem AEC standing comt controlled thermonuclear res, 65-72; chmn div plasma physics, Am Phys Soc, 68; mem, President's Sci Adv Comt, 70-73; consult; chmn, Defense Sci Bd, 72-77, mem, 72-; mem fusion power coord comt, Energy Res & Develop Admin, 72-76; mem exec comt, Assembly Eng, Nat Res Coun, 75-78; mem, Naval Res Adv Comt, 78-81; chmn, Energy Res Adv Bd, 78-81; cons, Off Sci & Technol Policy, 76-82; trustee, Argonne Univ Asn, 79-82, Rand Corp, 82-; mem, Coun Nat Acad Eng, 80-86; chmn, White House Sci Coun, 82; mem, Draper Lab Corp, 83-, dir, 86-; mem bd gov, Argonne Nat Lab, 85-88; mem, Stanford Univ Sch Eng Adv Coun, 86-; mem, President's Coun Sci & Tech Adv, 90- *Honors & Awards:* Ann Molson Medal, 52; Nat Medal of Sci, 86; Frederik Philips Award, Inst Elec & Electronic Engrs, 87; Arthur M Bueche Award, Nat Acad Eng, 90. *Mem:* Nat Acad Sci; Nat Acad Eng; fel Am Phys Soc; fel Inst Elec & Electronics Engrs; fel Am Acad Arts & Sci. *Res:* Plasma physics; gaseous electronics; plasmas in solids; controlled thermonuclear fusion; optical communications; communications systems. *Mailing Add:* AT&T Bell Labs Holmdel NJ 07733

BUCHTA, JOHN C(HARLES), b Minneapolis, Minn, Apr 13, 27; m 53; c 3. ELECTRICAL ENGINEERING. *Educ:* Univ Minn, BA, 49; Univ Ill, MS, 51 & 55, PhD(elec eng), 59. *Prof Exp:* Res assoc, Control Syst Lab, Univ Ill, 51-53; proj engr, Electronics Lab, 57-66, consult engr, 66-74, mgr, Eng Underseas Electronics Prod Dept, 74-76, mgr advan develop eng, Heavy Mil Electronics Dept, 76-80, mgr electronics technol, Turbine Bus Group, Gen Elec Co, Schenectady, NY, 80-86; mgr adv technol prog, Ocean Systs Div, Gen Elec, Syracuse, NY, 86-90; RETIRED. *Mem:* Asn Comput Mach; Inst Elec & Electronics Engrs; Sigma Xi; AAAS. *Res:* Computer science; signal processing; digital equipment; electrical power systems. *Mailing Add:* 113 Hiawatha Trail Liverpool NY 13088-4432

BUCHTA, RAYMOND CHARLES, b Philadelphia, Pa, Dec 17, 42; m 68; c 2. ANALYTICAL CHEMISTRY. *Educ:* Pa State Univ, BS, 65; Univ Wis-Madison, PhD(anal chem), 69. *Prof Exp:* Technician, Rohm and Haas Co, 62-63, chemist, 65; res chemist, 69-75, SR RES CHEMIST, EXP STA, E I DU PONT DE NEMOURS & CO, INC, 75- *Mem:* Am Chem Soc. *Res:* Electroanalytical chemistry; electrochemical detection in liquid chromategraphy; general trace analysis; gas chromatography; determination of polychlorinated biphenyl; syncrude analysis. *Mailing Add:* 2415 Allendale Rd Wilmington DE 19803

BUCHTEL, HENRY AUGUSTUS, IV, b Denver, Colo, Nov 29, 42; m 75; c 2. NEUROPSYCHOLOGY, PSYCHOLOGY. *Educ:* Dartmouth Col, BA, 64; McGill Univ, MA, 65, PhD(psychol), 69. *Prof Exp:* Res fel, Inst Physiol, Univ Pisa, Italy, 69-72; clin neuropsychologist & res assoc, Dept Psychol, Nat Hosp Nervous Diseases, London, 72-75; res assoc & vis prof psychol, Inst Human Physiol, Univ Parma, Italy, 75-78; clin neuropsychologist & res assoc, Psycol Dept, Montreal Neurol Inst & Hosp, 78-80; staff psychologist, 80-85, ACTG CHIEF, PSYCHOL SERV, VET ADMIN MED CTR, ANN ARBOR, 85-; ASSOC PROF PSYCHOL & NEUROPSYCHOL, UNIV MICH, ANN ARBOR, 80- *Concurrent Pos:* Vis prof, Scuola Normale Superiore & Univ Pisa, Italy, 71-72; Am Acad Sci fel, Prague, Czech, 73; vis prof, Sch Specialization Neurol & Psychiat, Univ Parma, Italy, 75-78. *Mem:* Am Psychol Asn; Neurosci Soc; Sigma Xi; Int Brain Res Orgn. *Res:* Biological basis of mind, especially the brain mechanisms of perception and attention; cerebral localization of functions and the nature of hemispheric dominance. *Mailing Add:* Dept Psychol Univ Mich Main Campus Ann Arbor MI 48109-1259

BUCHTHAL, DAVID C, b Chicago, Ill Oct 10, 43. MATHEMATICS. *Educ:* Loyola Univ, BS, 67; Purdue Univ, MS, 68, PhD(math), 71. *Prof Exp:* PROF MATH, UNIV AKRON, OHIO, 71-, ASST HEAD DEPT MATH SCI, 86- *Mem:* Am Math Soc; Math Asn Am. *Mailing Add:* Dept Math Univ Akron Akron OH 44325

BUCHWALD, CARYL EDWARD, b Medford, Mass, Oct 15, 37; m 59; c 3. ENVIRONMENTAL GEOLOGY. *Educ:* Union Col, BS, 60; Syracuse Univ, MS, 63; Univ Kans, PhD(geol), 66. *Prof Exp:* Teaching fel geol, McMaster Univ, 66-67; from asst prof to asssoc prof, 67-77, PROF GEOL, CARLETON COL, 77-, LLOYD MCBRIDE PROF ENVIRON STUDIES, 85- *Concurrent Pos:* Mem, Minn Environ Qual Bd, 79-89; dir & supv naturalist, Carleton Arboretum, 78-87. *Mem:* Fel Geol Soc Am; Sigma Xi. *Res:* River hydrology; land use planning; geological limnology. *Mailing Add:* Dept Geol Carleton Col One N College St Northfield MN 55057

BUCHWALD, HENRY, b Vienna, Austria, June 21, 32; US citizen; m 54; c 4. SURGERY. *Educ:* Columbia Univ, BA, 54, MD, 57; Univ Minn, Minneapolis, MS & PhD(surg), 66. *Prof Exp:* Intern, Columbia Presby Med Ctr, 57-58; from instr to assoc prof, 65-77, PROF SURG, UNIV MINN, MINNEAPOLIS, 77- *Concurrent Pos:* Helen Hay Whitney fel, 63-65; estab investr, Am Heart Asn, 65-70. *Honors & Awards:* Schering Award, 57; Sam D Gross Award, 67; Cine Clins Award, Am Col Surg, 69; Distinguished Serv Award, Asn Acad Surg, 76. *Mem:* Am Surg Asn; Am Col Surg; Soc Univ Surg; Am Heart Asn; Am Soc Artificial Internal Organs. *Res:* Atherosclerosis and the hyperlipidemias; development of surgical procedures for lipid reduction; cholesterol and bile acid metabolism; implantable infusion devices; hemodynamic rheology; morbid obesity. *Mailing Add:* Box 290 Univ Minn Hosp Minneapolis MN 55455

BUCHWALD, JENNIFER S, b Okmulgee, Okla, Oct 20, 30; m 52; c 3. NEUROPHYSIOLOGY, HUMAN-ELECTROPHYSIOLOGY. *Educ:* Lindenwood Col, AB, 51; Tulane Univ, PhD(neuroanat), 59. *Hon Degrees:* LLD, Lindenwood Col, 70. *Prof Exp:* From asst prof to assoc prof physiol, 65-73, assoc dir Brain Res Inst, 78-89, PROF PHYSIOL, SCH MED, UNIV CALIF, LOS ANGELES, 73- *Concurrent Pos:* Consult, NIH, NSF & NIMH; Ment Health Training Prog fel, 61; Parkinsonism Found sr fel, 63-66; USPHS res career develop grant, 64-69, spec fel award, 64-66; assoc dir, Brain Res Inst, Univ Calif, Los Angeles, 78-89, mem, Ment Retardation Res Ctr; vchair dept physiol, Univ Calif, Los Angeles, 85-87. *Mem:* Asn Women Sci; Neurosci Soc; Am Physiol Soc; Am Asn Anat. *Res:* Mechanisms of acoustic communication; auditory physiology; natural and synthesized vocalizations; animal and human subjects. *Mailing Add:* Dept Physiol Sch Med Univ Calif Los Angeles CA 90024

BUCHWALD, NATHANIEL AVROM, b Brooklyn, NY, July 19, 25. NEUROPHYSIOLOGY, NEUROANATOMY. *Educ:* Univ Minn, Minneapolis, PhD(neuroanat, neurophysiol), 53. *Prof Exp:* Instr anat, Sch Med, Tulane Univ, 53-57; from asst res anatomist to assoc res anatomist, 57-61, from assoc prof to prof anat, 61-69, assoc dir res, Ment Retardation Res Ctr, 71-74, PROF ANAT & PSYCHIAT, UNIV CALIF, LOS ANGELES, 69-, DIR, MENT RETARDATION RES CTR, 74- *Concurrent Pos:* Consult, Fed and state agencies, 57-; USPHS career develop award, 58-67. *Mem:* AAAS; Am Physiol Soc; Soc Neurosci; Am Asn Anat. *Res:* Brain function with relation to behavior and mental retardation. *Mailing Add:* Ment Retardation Res Ctr Univ Calif Los Angeles CA 90024

BUCHWEITZ, ELLEN, psychopharmacology, psychiatry, for more information see previous edition

BUCK, ALFRED A, b Hamburg, WGer, Mar 9, 21; m 62; c 2. EPIDEMIOLOGY, TROPICAL MEDICINE. *Educ:* Univ Hamburg, MD, 45; Johns Hopkins Univ, MPH, 59, PhD(epidemiol), 61. *Prof Exp:* Resident internal med, Hamburg City Hosp Altona, Ger, 46-52; physician in chg internal & trop med, Gen Hosp, Makassar-Celebes, Repub Indonesia, 52-55; chief, Dept Med, Ger Red Cross Hosp, Pusan, Korea, 56-58; asst prof epidemiol & pub health admin, Sch Hyg & Pub Health, Johns Hopkins Univ, 61-63, from assoc prof to prof epidemiol & int health, 63-71, res dir geog epidemiol, 64-71, dir, Div Bact & Mycol, 67-71; sr med officer & chief epidemiol method & clin path, Div Malaria & Other Parasitic Dis, WHO, Geneva, 71-74, chief med officer, Res Coord Epidemiol & Training, 74-78; trop med adv, Dept State, Agency for Int Develop, 78-85; ADJ PROF, INFECTIOUS DIS & IMMUNOL, JOHNS HOPKINS UNIV, 86- *Concurrent Pos:* Consult, US Agency Int Develop, Ethiopia, 62, Cent & WAfrica & WHO, Geneva, 71 & Sci & Technol Bur, 85; sr assoc immunol, infectious dis, epidemiol & int health, Johns Hopkins Univ, 80-; adj prof trop med, Tulane Univ, 81-; lectr trop pub health, Harvard Univ, 82-; mem sci working group & steering comt on epidemiol & appl field res on malaria, WHO, 82-; assoc ed, Trop Med & Parasitol, 82-; hon fel trop med, Liverpool Sch Trop Med & Hyg, 83; consult, Sci & Tech Bur, 85- *Honors & Awards:* Craig Lectr, Am Soc Trop Med & Hyg, 79; Bernhard Nocht Medal, 81. *Mem:* Fel Am Col Epidemiol; Am Soc Trop Med & Hyg; fel Am Pub Health Asn; Ger Soc Trop Med; Int Soc Epidemiol; fel Am Col Epidemiol; Am Soc Epidemiol Res. *Res:* International health; prevalence and interaction of diseases in Peru, Chad, Afghanistan and Cameroon by comprehensive epidemiologic investigations and laboratory studies; epidemiology and control of onchocerciasis and treponematoses; disease control and primary health care. *Mailing Add:* Sch Pub Health Rm 4013 Johns Hopkins Univ 615 N Wolfe St Baltimore MD 21205

BUCK, CARL JOHN, b Philadelphia, Pa, Mar 6, 29; m 53; c 3. POLYMER CHEMISTRY. *Educ:* Temple Univ, BA, 50, MA, 54; Cornell Univ, PhD(org chem), 58. *Prof Exp:* Instr biochem & dent mat, Sch Dent, Temple Univ, 50-51; anal chemist petrol, Sinclair Refining Co, 51-54; res chemist org synthesis, Pfizer Inc, 57-63, Shell Chem, 63-70; RES CHEMIST ORG & POLYMER SYNTHESIS, JOHNSON & JOHNSON RES CTR, 70- *Honors & Awards:* Distinguished Scientist Award, Johnson & Johnson, 77. *Mem:* Sigma Xi; Am Chem Soc. *Res:* Organic synthesis of antibiotics; medicinal agents; industrial chemicals; monomers and polymers for dental, orthopedic and wound care applications. *Mailing Add:* Johnson & Johnson Res Ctr US Rte 1 New Brunswick NJ 08903

BUCK, CAROL WHITLOW, b London, Ont, Apr 2, 25; m 46; c 2. PREVENTIVE MEDICINE. *Educ:* Univ Western Ont, MD, 47, PhD, 50; London Sch Trop Med, DPH, 51. *Prof Exp:* From asst prof to prof, Dept Psychiat & Prev Med, 52-67, chmn dept epidemiol & prev med, 67-77, PROF EPIDEMIOL, UNIV WESTERN ONT, 67- *Concurrent Pos:* Mem, Sci Coun Can, 70-73. *Mem:* Soc Epidemiol Res; Int Epidemiol Asn; Can Pub Health Asn; fel Am Col Epidemiol. *Res:* Epidemiology; medical statistics. *Mailing Add:* Dept Epidemiol Univ Western Ont Fac Med London ON N6A 5C1 Can

BUCK, CHARLES (CARPENTER), b Metamora, Ohio, June 28, 15; m 51; c 2. MATHEMATICS. *Educ:* Univ Mich, BS, 40, MS, 47, PhD, 54. *Prof Exp:* Instr math, Univ Nebr, 49-52 & Wayne Univ, 53; from asst prof to assoc prof, Univ Ala, 53-61; res assoc, Educ Res Coun, 61-67; ASSOC PROF MATH, CLEVELAND STATE UNIV, 67- *Concurrent Pos:* Consult, Educ Res Coun, 71- *Mem:* AAAS; Am Math Soc; Math Asn Am. *Res:* Foundations, non-Euclidean and differential geometry. *Mailing Add:* 1803 Wilton Rd Cleveland Heights OH 44118

BUCK, CHARLES ELON, b Linton, NDak, Jan 5, 19; m 48; c 3. BACTERIOLOGY. *Educ:* NDak State Col, BS, 42, MS, 47; Ohio State Univ, PhD(bact), 51. *Prof Exp:* From asst prof to assoc prof bact, 51-74, prof microbiol, Univ Maine, Orono, 74-, EMER PROF, UNIV MAINE, ORONO. *Mem:* Am Soc Microbiol. *Res:* Virology and tissue culture. *Mailing Add:* Dept Microbiol Univ Maine Orono ME 04473

BUCK, CHARLES FRANK, b Grayson, Ky, June 19, 20; m 46; c 2. ANIMAL HUSBANDRY. *Educ:* Univ Ky, BS, 42, MS, 51; Cornell Univ, PhD(animal husb), 53. *Prof Exp:* Teacher high schs, 42-50; asst animal husb, Univ Ky, 50-51; asst, Cornell Univ, 51-53; from asst prof to assoc prof, Univ Ky, 53-67, prof animal sci, 67-; RETIRED. *Mem:* Am Soc Animal Sci. *Res:* Pasture utilization; protein study for ruminants; pasture grazing. *Mailing Add:* 2075 Parkers Mill Rd Lexington KY 40513

BUCK, CLAYTON ARTHUR, b Lyons, Kans, Mar 2, 37; m 61; c 2. BIOCHEMISTRY, CANCER. *Educ:* Kans State Univ, BS, 59; Mont State Univ, PhD(bact), 64. *Prof Exp:* NIH fel, Univ Calif, Irvine, 64-67; asst prof therapeut res, Univ Pa, 67-70; assoc prof biol, Kans State Univ, 70-75; PROF, WISTAR INST, 75- *Mem:* Am Soc Biol Chemists; Am Soc Microbiol. *Res:* Virology, protein biosynthesis and animal cell membranes; tRNA in mitochondria and glycoproteins from the surface of virus-transformed and normal animal cells and their role in controlling cellular adhesion and morphology. *Mailing Add:* Wistar Inst 3601 Spruce St Philadelphia PA 19104-4268

BUCK, DAVID HOMER, b Clifton, Ariz, Dec 31, 20; m 53; c 4. FISHERIES. *Educ:* Agr & Mech Col, Tex, BS, 43; Okla State Univ, PhD(zool), 51. *Prof Exp:* Aquatic biologist, State Game & Fish Comn, Tex, 46-48; dist biologist, Corps Engrs, Ft Worth Dist, 51-52; fisheries investr, State Game & Fish Dept, Okla, 54-56; assoc aquatic biologist, 56-68, AQUATIC BIOLOGIST, ILL NATURAL HIST SURV, 68-; ASSOC PROF, DEPT ANIMAL SCI, UNIV ILL, 80- *Concurrent Pos:* Mem, Int Comt Prod Use Wastes Develop Countries & chmn, Sect Aquacult, Nat Acad Sci, 80-; mem task force, UN Food & Agr Org, China, 81-; consult, USAID/Aquacult, Rwanda, 88. *Mem:* Am Fisheries Soc; Am Soc Limnol & Oceanog; World Aquacult Soc. *Res:* Productivity and population dynamics of freshwater fisheries; effects of turbidity on fish production; species interrelationships of pond fishes; culture of channel catfish; use of specialized fishes for control of eutrophication; recycling of animal wastes for polyculture of fishes. *Mailing Add:* Ill Natural Hist Surv Rte 1 Box 174 Kinmundy IL 62854

BUCK, DOUGLAS L, b Frederic, Wis, May 9, 31; m 54; c 2. DENTISTRY. *Educ:* Univ Minn, BS, 54, DDS, 60, MSD, 62. *Prof Exp:* Asst orthod, Sch Dent, Univ Minn, 61-62; from asst prof to assoc prof, 62-71, PROF ORTHOD, DENT SCH, UNIV ORE, 71-, CHMN DEPT, 73- *Concurrent Pos:* Fulbright lectr, Ecuador, 68-69; vis prof, Sch Dent, Hokkaido Univ, 75. *Mem:* Am Asn Orthod; Int Asn Dent Res; Am Cleft Palate Asn; Sigma Xi. *Res:* Histology of tooth movement. *Mailing Add:* Dent Sch Univ Ore 611 SW Campus Dr Portland OR 97201

BUCK, ERNEST MAURO, b Hartford, Conn, Apr 21, 30; m 54; c 4. MEAT SCIENCE. *Educ:* Univ Conn, BS, 55; NC State Col, MS, 57; Univ Mass, Amherst, PhD(food sci & technol), 66. *Prof Exp:* Assoc dean, 69-76, PROF FOOD SCI & NUTRIT, UNIV MASS, AMHERST, 76- *Mem:* Am Meat Sci Asn; Inst Food Technol. *Res:* Meats; physico-chemical changes in muscle after death; seafood analogs. *Mailing Add:* Dept Nutrit Univ Mass Amherst Campus Amherst MA 01003

BUCK, F(RANK) A(LAN) MACKINNON, b Ottawa, Ont, June 26, 20; nat US; m 45; c 2. CHEMICAL ENGINEERING. *Educ:* Univ BC, BApSc, 43, MApSc, 44; Purdue Univ, PhD(phys chem), 48. *Prof Exp:* Develop engr, Powell River Co, Can, 44-45; sessional lectr chem, McGill Univ, 45-46; instr chem eng, Univ Wash, 48-49; chem engr exec, Shell Chem Co, 49-68; tech specialist, Off Oil & Gas, US Dept Interior, Washington, DC, 68-69; chem engr exec, Shell Chem Co, 69-76, bus rep, 76-79; managing partner, 79-80, MANAGING PRIN, KING, BUCK & ASSOCS, INC, 81- *Concurrent Pos:* Vchmn, San Diego County, Air Pollution hearing bd, 86- *Mem:* Am Chem Soc; Am Inst Chem Engrs. *Res:* Electron diffraction of gases; synthesis of fuels by Fischer-Tropsch process; petroleum research and plant operations. *Mailing Add:* 4058 Southview Dr San Diego CA 92117

BUCK, GEORGE SUMNER, JR, b Hartford, Conn, Sept 12, 14; m 42; c 4. CHEMICAL ENGINEERING. *Educ:* Johns Hopkins Univ, BS, 35. *Hon Degrees:* DTextileSci, Philadelphia Col Textiles Sci, 63. *Prof Exp:* Asst chemist, William Hooper & Sons, Md, 35-37, overseer, Finishing Plant, 37-41, supt, 41-46; tech supt tech serv dept, Nat Cotton Coun Am, 46-47, tech serv dir, 47-49, tech dir, 49-57, asst to exec vpres, 57-68, dir res, 68-70; PRES, RAMCON, INC, 70-, CHMN RAMCON ENVIRON CORP, 70-; PRES, FIBERLOK, INC, 70- *Concurrent Pos:* Res consult, Cotton Producers Inst, 62-70; pres, Mid-South Res Assoc, 68-70; exec vpres, Oscar Johnston Cotton Found & The Cotton Found, 68-73. *Mem:* Textile Res Inst; AAAS; Am Chem Soc; Am Soc Testing & Mat; Am Asn Textile Chemists & Colorists. *Res:* Textile fire-resistance and flammability; textile finishes; market research; consumer surveys; management research. *Mailing Add:* Ramcon Inc 223 Scott St Memphis TN 38112-3911

BUCK, GRIFFITH J, b Cincinnati, Iowa, Apr 19, 15; m 47; c 2. HORTICULTURE. *Educ:* Iowa State Col, PhD(hort, bot), 53. *Prof Exp:* From asst prof to assoc prof, 53-74, PROF HORT, IOWA STATE UNIV, 74- *Mem:* Am Soc Hort Sci; Sigma Xi. *Res:* Pelargonium and rose breeding; propagation and disease control. *Mailing Add:* 1108 Scott St Ames IA 50000

BUCK, JAMES R, b Big Rapids, Mich, Feb 22, 30; m 62; c 3. HUMAN FACTORS-ERGONOMICS, ENGINEER ECONOMICS. *Educ:* Mich Technol Univ, BS, 52, MS, 53; Univ Mich, PhD(indust eng), 64. *Prof Exp:* Struct engr, Austin Engrs Ltd, Detroit, 56-57; field engr, Cabinet Flexicore Corp, 57-58; instr math, Ferris State Col, 58-59; asst prof indust eng, Univ Mich, Ann Arbor & Dearborn, 60-66; from assoc prof to prof, Purdue Univ, 66-81; PROF INDUST ENG, UNIV IOWA, 81- *Concurrent Pos:* Consult, 76- *Honors & Awards:* Eugene Grant Award, Am Soc Eng Educ, 79 & 87. *Mem:* Fel Inst Indust Engrs; Am Soc Eng Educ; Inst Mgt Sci; Human Factors Soc; Ergonomics Soc. *Res:* Development of techniques to facilitate design or analysis such as sampling, simulation, visual abilities, inspection, economic analysis, risk measurement; human performance prediction. *Mailing Add:* 2353 Cae Dr Iowa City IA 52246

BUCK, JEAN COBERG, b Schenectady, NY, Apr 25, 35; m 68; c 1. POLYMER CHEMISTRY, ANALYTICAL CHEMISTRY. *Educ:* Philadelphia Col Pharm & Sci, BSc, 57; Univ Del, PhD(chem), 68. *Prof Exp:* Jr anal res chemist, Rohm & Haas Co, Philadelphia, 57-61; sr res chemist polymer chem, 67-71; teacher, Booth Sch, Rosemont, Pa, 71-74; sr anal res chemist, Betz Labs Inc, 74-85; MGR, MERCK, SHARP & DOHME, WEST POINT, PA, 85- *Mem:* Am Chem Soc; NAm Thermal Anal Soc. *Res:* Organic analysis of proprietary materials; characterization of aqueous polymer systems. *Mailing Add:* Qual Control Tech Serv Merck Sharp & Dohme West Point PA 19486

BUCK, JOHN BONNER, b Hartford, Conn, Sept 26, 12; m 39; c 4. EVOLUTION. *Educ:* Johns Hopkins Univ, AB, 33, PhD(zool), 36. *Prof Exp:* Asst zool, Johns Hopkins Univ, 33-36; Nat Res fel, Calif Inst Technol, 36-37; res assoc embryol, Carnegie Inst Wash, 37-39; from instr to asst prof zool, Univ Rochester, 39-45; cytologist, 45-47, sr biologist, 47-56, prin physiologist, 56-62, chief lab phys biol, 62-75, chief sect comp physiol, 75-85, EMER SCIENTIST, NIH 85- *Concurrent Pos:* Mem, Hopkins Exped, Jamaica, 36, 41 & 63; mem corp, Marine Biol Lab, Woods Hole, 37-, instr, 42-44 & 57-59, trustee, 63-80; vis prof, Univ Wash, 51, Calif Inst Technol, 53 & Cambridge Univ, 63; chief scientist, Alpha Helix Exped, New Guinea, 69. *Mem:* Am Soc Zool (vpres, 56); Soc Gen Physiol (secy-treas, 53-55, pres, 60). *Res:* Physiology and behavior of fireflies; chromosome and physiological cytology; biochemistry and physiology of insect blood; insect respiration; bioluminescence in invertebrates. *Mailing Add:* Nat Inst Diabetes Digestive & Kidney Dis Rm 112 Bldg 6 NIH Bethesda MD 20892

BUCK, JOHN DAVID, b Hartford, Conn, Oct 3, 35; m 60; c 3. MARINE MICROBIOLOGY. *Educ:* Univ Conn, BA, 57, MS, 60; Univ Miami, PhD(marine sci), 65. *Prof Exp:* Asst instr bact, Univ Conn, 60-61; res instr marine microbiol, Univ Miami, 61-62 & 64-65; asst prof bact, 65-71, assoc prof biol, 71-80, PROF MARINE SCI, UNIV CONN, 80- *Concurrent Pos:* Adj scientist, Mote Marine Lab, Sarasota, Fla. *Mem:* AAAS; Am Soc Microbiol. *Res:* Enumeration and sampling methods in marine microbiology; diseases of marine mammals. *Mailing Add:* Marine Res Lab Univ Conn Noank CT 06340

BUCK, JOHN HENRY, b London, Eng, Sept 22, 12; US citizen; m 40; c 2. NUCLEAR POWER DEVELOPMENT. *Educ:* Univ Sask, BSc, 33, MSc, 35; Univ Rochester, PhD(nuclear physics), 38. *Prof Exp:* Teaching, fel physics, Univ Rochester, 36-38; instr physics, Mass Inst Technol, 38-41; staff mem, Mass Inst Technol Radiation Lab, 41-46; mgr, Mobil Oil Co, 46-50; proj mgr, Oak Ridge Nat Lab, 50-53; vpres & gen mgr, Well Surveys Inc, 53-59; vpres & gen mgr, instruments div, Budd Co, 59-69; vchmn, Appeal Panel, AEC, 69-74; CONSULT NUCLEAR POWER, 84- *Concurrent Pos:* Mem, Sr Rev Team, Commanche Peak Nuclear Power Plants, 85- *Mem:* Fel Am Phys Soc; Am Nuclear Soc; Sigma Xi. *Res:* Early nuclear physics investigations with cyclotrons; molten salt reactor and material testing reactor. *Mailing Add:* 5606 Marengo Rd Bethesda MD 20816-3313

BUCK, KEITH TAYLOR, b Norwalk, Conn, Aug 1, 40; m 80. FLAVOR CHEMICALS, PROCESS DESIGN. *Educ:* Univ Mich, BS, 61; Ohio State Univ, PhD(organ chem), 66. *Prof Exp:* Fel chem, Pa State Univ, 65-66; asst prof org chem, Dickinson Col, 66-68; fel & asst prof, Univ Idaho, 68-70; fel chem, Univ Pa, 71-74; dir res, Frankincense Co, 75-77; dir aroma chem, Northville Labs, 77-79; MGR ORG SYNTHESIS, FRIES & FRIES DIV, MALLINCKRODT, INC, 79-, SR RES FEL, 85- *Mem:* Am Chem Soc. *Res:* Synthesis of flavor chemicals; isoquinoline alkaloids. *Mailing Add:* Fries & Fries Div Mallinckrodt Inc 110 E 70th St Cincinnati OH 45216

BUCK, MARION GILMOUR, b New Hope, Pa, May 24, 36; m 69; c 2. ENZYMES, CELL MEDIATORS. *Educ:* Wilmington Col, Ohio, BS, 58; Rutgers Univ, PhD(biochem), 65. *Prof Exp:* Fel biochem, Royal North Shore Hosp, Sydney, Australia, 65-66; fel, Univ Pa, 66-67; asst prof, Sch Med, Univ Louisville, 67-70; ASSOC RES SCIENTIST BIOCHEM, YALE UNIV, 70- *Concurrent Pos:* Muscular Dystrophy Asn Am fel, 65-67; vis sr res fac, Univ Utrecht, 79; prin investr, 80-86. *Mem:* Am Chem Soc; AAAS. *Res:* Mechanisms of cell to cell communication including cell mediators and effects of toxic inhalants; isolation, purification and identification of pharmacologically active plant compounds; effect of ascorbic acid on c-AMP phosphodiesterase; characterization of cytochrome oxidase and cytochrome b-5. *Mailing Add:* Old Town St Hadlyme CT 06439

BUCK, OTTO, b Ger, May 14, 33; m 61, 85; c 2. METAL PHYSICS. *Educ:* Univ Stuttgart, BS, 56, MS, 59, PhD(physics), 61. *Prof Exp:* Asst metal physics, Univ Stuttgart, 61-64; mem tech staff, sci ctr, NAm Aviation Inc, 64-66 & Siemens Co, Ger, 66-68; mem tech staff, Sci Ctr Rockwell Int, 68-75, group leader, 75-80; SR SCIENTIST & PROF, AMES LAB, IOWA STATE UNIV, 80-, PROG DIR, 89- *Mem:* Am Soc Testing & Mat; Am Inst Mining, Metall & Petrol Engrs; Mat Res Soc. *Res:* Plasticity of metals; dislocation theory; nonlinear theory of elasticity; internal friction; point-defect migration; fracture mechanics; surface properties; fatigue; nondestructive evaluation. *Mailing Add:* Ames Lab Iowa State Univ Ames IA 50011

BUCK, PAUL, b Highland Park, Mich, Sept 9, 27; m 50; c 2. PLANT ECOLOGY, AEROBIOLOGY. *Educ:* Univ Tulsa, BS, 58, MS, 59; Univ Okla, PhD(bot), 62. *Prof Exp:* From asst prof to assoc prof bot, Univ Tulsa, 62-88; RETIRED. *Concurrent Pos:* Summer instr & sr researcher, Rocky Mountain Biol Lab, 79-; instr, Tulsa Jr Col & Univ Ctr, Tulsa. *Mem:* AAAS; Ecol Soc Am. *Res:* Airborne allergenic pollen and mold spores in Oklahoma; sub-alpine habitats of Colorado; plant systematics with primary focus on woody vegetation of Oklahoma. *Mailing Add:* Dept Life Sci Univ Tulsa Tulsa OK 74104

BUCK, RAYMOND WILBUR, JR, b Monticello, Maine, Apr 20, 19. CYTOGENETICS. *Educ:* Univ Maine, BS, 41; Univ Md, MS, 50, PhD(bot), 52. *Prof Exp:* Instr bot, Univ Maine, 46-48; geneticist, bur plant indust, soils & agr eng, USDA, 52-53 & hort crops res br, Agr Res Serv, 53-68; from assoc prof to prof biol, 68-84, Montgomery Col, chmn dept, 70-76; RETIRED. *Mem:* AAAS; Bot Soc Am; Am Genetic Asn; Am Inst Biol Sci; Sigma Xi. *Res:* Cytogenetics of tuber-bearing Solanum species. *Mailing Add:* 4902 Laguna Rd College Park MD 20740

BUCK, RICHARD F, b Enterprise, Kans, Dec 8, 21; m 44; c 6. ATMOSPHERIC PHYSICS. *Educ:* Univ Kans, BS, 43; Okla State Univ, MS, 60. *Prof Exp:* Appl engr, radio tube dept, Tung Sol Lamp Works, Inc, 46-48; res physicist, Okla State Univ, 48-53, from asst proj dir to proj dir, Electronics Lab, 53-60, proj dir res found & instr physics, 60-67, dir, Electronics Lab, Res Found & asst prof physics, Sch Arts & Sci, 67-76, dir, Electronics Lab-CEAT & assoc prof eng res, 76-84, EMER ENG TECH, OKLA STATE UNIV, 84- *Concurrent Pos:* Consult electronic eng, 63- *Mem:* Inst Elec & Electronics Engrs; Nat Soc Prof Engrs; AAAS; Sigma Xi. *Res:* Electronic instrumentation of rockets and satellites for measurement of physical parameters of space. *Mailing Add:* 1301 Westwood Dr Stillwater OK 74074

BUCK, RICHARD PIERSON, b Los Angeles, Calif, July 29, 29; m 59; c 3. PHYSICAL CHEMISTRY, ANALYTICAL CHEMISTRY. *Educ:* Calif Inst Technol, BS, 50, MS, 51; Mass Inst Technol, PhD(chem), 54. *Prof Exp:* Res chemist polarography & electrochem, Calif Res Corp, 54-56, asst to gen mgr, 56-58, res chemist combustion & high temperature inorg chem, 58-60, electrochem fuel cells, 60-61; prin res chemist, electrochem, electroanal chem & instrumentation, Bell & Howell Res Ctr, 61-65; sr scientist, Beckman Instruments, Inc, 65-67; assoc prof, 67-75, PROF CHEM, UNIV NC, CHAPEL HILL, 75- *Concurrent Pos:* VChmn, Gordon Res Conf Electrochem, 64, chmn, 65; vis prof, Bristol Univ, Eng, 76-77, Imp Col, London, Eng, 87-88, Bundeswehr Univ, Munich, Ger, 89 & 91; adj prof biomed eng & math, Univ NC, Chapel Hill, 90-; organizer symp, Ion-Selective Electrodes Biol & Med, Dortmund, 80 & Erlangen, Ger, 83 Electrochem Soc, Hollywood Fla, 80, Neutral Carrier Mechan, NY, 85, Kinetics Liquid & Liquid Interfaces, Boston, Mass, 86, fundamental processes sensors, Hollywood, Fla, 89, invivo sensors, Montreal, 90; mem rev panels & site visit teams, NIH; exec bd, Int Soc Electrochem, 90-91; chmn sensors group, Electrochem Soc, 91-92; titular mem, Int Union Pure & Appl Chem, 90-95. *Honors & Awards:* Von Humboldt Preis, WGer, 89. *Mem:* Int Soc Electrochem; Am Chem Soc; Electrochem Soc; Soc Electroanal Chem. *Res:* Electrochemistry; electrode processes; electroanalysis; chemical instrumentation; bioelectrochemistry and biosensors; trace analysis; solid state transport; membrane electrochemistry. *Mailing Add:* Dept Chem Univ NC Chapel Hill NC 27599-3290

BUCK, ROBERT CRAWFORTH, b London, Ont, Sept 23, 23; m 46; c 2. ANATOMY, HISTOLOGY. *Educ:* Univ Western Ont, MD, 47, MSc, 50; Univ London, PhD(path), 52. *Prof Exp:* From asst prof to assoc prof, 53-62, chmn dept, 67-73, PROF ANAT, UNIV WESTERN ONT, 62- *Mem:* Am Asn Anat; Can Asn Anat; Anat Soc Gt Brit & Ireland; Path Soc Gt Brit. *Res:* Cytology; electron microscopy in fields of blood vascular pathology and tumor pathology. *Mailing Add:* 181 Elmwood E London ON N6C 1K1 Can

BUCK, ROBERT CREIGHTON, b Cincinnati, Ohio, Aug 30, 20; m 44; c 2. MATHEMATICAL ANALYSIS. *Educ:* Univ Cincinnati, BA, 41, MA, 42; Harvard Univ, PhD(math), 47. *Prof Exp:* Jr fel, Soc Fels, 42-43 & 45-47; asst prof math, Brown Univ, 47-49; from assoc prof to prof, 50-80, chmn dept, 64-66, dir, Math Res Ctr, 73-75, HILLDALE PROF MATH, UNIV WIS-MADISON, 80- *Concurrent Pos:* Ed, Proc, Am Math Soc, 53-55, 64-67; Guggenheim fel, 58-59; vis prof, Stanford Univ, 58-59; mem staff proj focus, Inst Defense Anal, 59-60; mem math div, Nat Res Coun, 61-64, travelling lectr, Math Asn Am, 62-63; mem, Nat Security Agency Adv Bd, 63-71; mem, US Comn Math Instruct, 63-67; mem math adv panel, NSF, 65-70; adv bd, Sch Math Study Group, 66-69; Nat Adv Coun Educ Prof Develop, 70-71, 73-76; trustee, Soc Indust & Appl Math, 73-74; mem panel to evaluate, Nat Bur Stand Appl Math Prog, 74-77. *Honors & Awards:* Piedmont lectr, 70; Ford Award, 81. *Mem:* Fel AAAS; Am Math Soc (vpres, 72-74); Math Asn Am (vpres, 75-77); Soc Indust & Appl Math. *Res:* Complex variable theory; algebraic analysis; number theory; approximation theory; mathematics education; history of mathematics. *Mailing Add:* Dept Math van Vleck Hall Univ Wis Madison WI 53706

BUCK, ROBERT EDWARD, b Altmar, NY, Oct 16, 12; m 44; c 3. FOOD SCIENCE. *Educ:* Cornell Univ, AB, 33; Univ Mass, MS, 34, PhD(chem), 36. *Prof Exp:* Lab instr, Univ Mass, 36-42; res chemist, eastern regional res lab, Bur Agr Chem & Eng, USDA, 42-47; technologist, Quartermaster Food & Container Inst, Chicago, 47-48; chemist & head food technol, H J Heinz Co, 48-73, sr mgr food res, 73-77; RETIRED. *Mem:* Am Chem Soc; Inst Food Technologists. *Res:* Process and product development in foods; pectin and pectic enzymes; technology and rheology ingredient materials as starch, flour, gums; flavoring materials. *Mailing Add:* 500 Bay Ave 605 S Bldg Ocean City NJ 08226

BUCK, STEPHEN HENDERSON, b Decatur, Ill, Mar 30, 47. NEUROPEPTIDES, NEUROPHARMACOLOGY. *Educ:* Bradley Univ, BS, 69; Univ Kans, MS, 71; Univ Del, MBA, 79; Univ Ariz, PhD(pharmacol), 82. *Prof Exp:* Res asst pharmacol, Univ Colo Sch Med, 71-73; res pharmacologist, William S Rorer, Inc, 73-76 & ICI Americas Inc, 76-79; postdoctoral assoc pharmacol, Univ Ariz Col Med, 83; PRAT staff fel, Nat Heart, Lung & Blood Inst, NIH, 83-85; SR SCIENTIST, MARION MERRELL DOW RES INST, 85- *Mem:* Soc Neurosci; Am Soc Pharmacol & Exp Therapeut; Am Pain Soc; Am Chem Soc. *Res:* Physiology, pharmacology, neurobiology, and biochemistry of neuropeptides, their receptors and the cells that secrete and-or respond to them. *Mailing Add:* Marion Merrell Dow Res Inst 2110 E Galbraith Rd Cincinnati OH 45215

BUCK, THOMAS M, b Elizabeth, Pa, June 29, 20; m 44; c 4. SURFACE PHYSICS & CHEMISTRY, ION SCATTERING. *Educ:* Muskingum Col, BS, 42; Univ Pittsburgh, MS, 48, PhD(chem), 50. *Prof Exp:* Asst, Univ Pittsburgh, 46-50; res chemist, Nat Lead Co, 50-52; mem tech staff, Bell Tel Labs, 52-63, supvr, Surface Studies Group, Semiconductor Device Technol Dept, 63-68, mem tech staff, Chem Electronics Res Dept, 68-71 & Radiation Physics Dept, Bell Labs, 71-89; ADJ PROF, DEPT MAT SCI & ENG, UNIV PA, PHILADELPHIA, 89- *Concurrent Pos:* Adj prof, Mat Sci Eng Dept, Univ Pa, 88- *Mem:* Am Phys Soc; Sigma Xi; Am Vacuum Soc. *Res:* Composition and atomic arrangement of single-crystal alloy surfaces, using low energy ion scattering, low energy electron diffraction and Auger electron spectroscopy. *Mailing Add:* 108 Old Farms Rd Basking Ridge NJ 07920

BUCK, WARREN HOWARD, b Neptune, NJ, June 22, 42; m 68; c 1. PHYSICAL CHEMISTRY, POLYMER CHEMISTRY. *Educ:* Lehigh Univ, BA, 64; Univ Del, PhD(phys chem), 70. *Prof Exp:* res chemist, polymer prod dept, 69-80, sr res chemist, 80-87, RES ASSOC, E I DU PONT DE NEMOURS & CO INC, 87- *Mem:* Am Chem Soc. *Res:* Physical chemistry of macromolecules; morphology of thermoplastic elastomers, polyurethane foams and elastomers, polymer blends, ethylene copolymers, fire retardance, fluoropolymer characterization; perfluorodioxole polymers. *Mailing Add:* Polymer Prod Dept E I du Pont de Nemours & Co Exp Sta PO Box 80353 Wilmington DE 19880-0353

BUCK, WARREN LOUIS, b Normal, Ill, Jan 20, 21; m 44; c 2. RADIATION PHYSICS. *Educ:* Ill State Norm Univ, BS, 46; Univ Ill, MS, 47. *Prof Exp:* Asst physics, Univ Ill, 46-50; assoc physicist, Argonne Nat Lab, 50-72, physicist, 72-82; RETIRED. *Mem:* Am Phys Soc; Sigma Xi. *Res:* Scintillation counters; luminescent materials; effects of radiation on materials; neutron flux monitors. *Mailing Add:* 453 Lakewood Blvd Park Forest IL 60466-1626

BUCK, WARREN W, III, b Washington, DC, Feb 16, 46; m 77; c 2. RELATAVISTIC INTERMEDIATE ENERGY & NUCLEAR PHYSICS, QUARK MODELS. *Educ:* Morgan State Univ, BS, 68; Col William & Mary, MS, 70, PhD(theoret physics), 76. *Prof Exp:* Post doc, State Univ NY, Stony Brook, 76-79; res assoc physics, Inst Nuclear Physics, Orsay, France, 79-80; vis assoc prof, Col William & Mary, 84, ASSOC PROF PHYSICS, HAMPTON UNIV, 84- *Concurrent Pos:* Vis staff mem, Los Alamos Nat Lab, 77-80; prin investr, NASA, 84-, NSF, 87-, US Dept Energy, 87-; consult, 86-, Old Dominion Univ, 88- *Honors & Awards:* Outstanding Serv Award, Continious Electron Beam Accelerator Facil Users Group. *Mem:* Life mem Am Phys Soc; AAAS; NY Acad Sci; Nat Soc Black Physicists. *Res:* Theoretical studies of few body systems including nucleon- nucleon and nucleon-antinucleon systems as well as multiquark interactions. *Mailing Add:* Dept Physics Hampton Univ Hampton VA 23668

BUCK, WILLIAM BOYD, b Mexico, Mo, May 17, 33; m 52; c 6. VETERINARY TOXICOLOGY. *Educ:* Univ Mo, BS & DVM, 56; Iowa State Univ, MS, 63; Am Bd Vet Toxicol, dipl. *Prof Exp:* Area vet, animal dis eradication div, Agr Res Serv, USDA, 56-58, from vet to res vet, 58-64; assoc prof, Iowa State Univ, 64-68, prof vet toxicol, 68-76; PROF VET TOXICOL, UNIV ILL, URBANA, 76- *Concurrent Pos:* Mem panel on copper, Comt Med & Biol Effects Environ Pollutants, Nat Acad Sci, 73-76, mem panel on arsenic, 73-76. *Mem:* Am Col Vet Toxicol; Soc Toxicol; Am Vet Med Asn; Conf Res Workers Animal Dis. *Res:* Toxicology of economic poisons and plants in animals; environmental toxicology; central nervous system physiopathology. *Mailing Add:* RR 2 Box 14 Tolono IL 61880-9507

BUCK, WILLIAM R, b Jacksonville, Fla, Dec 27, 50. BRYOLOGY. *Educ:* Univ Fla, BS, 72, MS, 74; Univ Mich, PhD, 79. *Prof Exp:* Assoc cur, 79-86, CUR BRYOPHYTES, NY BOT GARDEN, 86- *Honors & Awards:* Greenman Award, 81. *Mem:* Am Bryological & Lichenological Soc; Am Fern Soc; Am Soc Plant Taxonomists; Brit Bryological Soc; Brit Pteridological Soc; Sigma Xi. *Res:* Family concepts in the large moss order Hypnobryales and Isobryales and monographic studies on many of the genera traditionally placed in these orders. *Mailing Add:* NY Bot Garden Bronx NY 10458-5126

BUCKALEW, LOUIS WALTER, b Bloomsburg, Pa, Apr 21, 44; m 85. HUMAN ENGINEERING, RESEARCH DESIGN. *Educ:* Ga Southern Col, BA, 67; Univ Southern Miss, MS, 69; Howard Univ, PhD, 89. *Prof Exp:* Admin specialist, US Army, Vietnam, 69-70; instr psychol, SC State Col, 70-73; managing partner, Buckalew & Davis Assocs, 73-75; assoc prof psychol, Ala A&M Univ, 75-85; SR RES PSYCHOLOGIST, US ARMY RES INST, 85-; DIR RES, UNIV CENT TEX, 86- *Concurrent Pos:* Reviewer Psychological Reports-Perceptual & Motor Skills, 79-; J Alcohol & Drug Educ, 84-; asst prof, Univ Ala, Huntsville, 80-85; res assoc, Aerospace Med Res Lab-Sch, USAF, 81-84. *Mem:* Sigma Xi; Psychonomic Soc; Am Psychol Soc. *Res:* Psychopharmacology; medication, physical properties of drugs, patient compliance, placebo effects and methodology, alcohol effects, sugar effects and drug education; author of over 90 journal articles and books. *Mailing Add:* 2317 Tiffany Dr Copperas Cove TX 76522

BUCKALEW, VARDAMAN M, JR, b Mobile, Ala, Mar 19, 33. NEPHROLOGY, HYPERTENSION. *Educ:* Univ Pa, MD, 58. *Prof Exp:* PROF MED & PHYSIOL, BOWMAN-GRAY SCH MED, FOREST UNIV, 73-, CHIEF NEPHROLOGY SECT, 74- *Mem:* Royal Soc Med; Am Fedn Clin Res; Am Soc Nephrology; Am Soc Hypertension. *Mailing Add:* Dept Med Bowman-Gray Sch Med Wake Forest Univ 300 S Hawthorne Rd Winston-Salem NC 27103

BUCKARDT, HENRY LLOYD, b Leland, Ill, Oct 24, 04; m 37. AGRONOMY. *Educ:* Univ Ill, BS, 26, MS, 29, PhD(agron), 32. *Prof Exp:* Asst prin & instr high sch, Ill, 26-27; asst crop prod, Dept Agron, Univ Ill, 28-33, Standard Oil Co fel, 33, erosion exten specialist coordr, 35-36; agronomist, Soil Conserv Serv, USDA, 34, Iowa, 37, DC, 38-40; head admin off, US Civil Serv Comn, 40-46; civil serv adminr & adv, Civil Affairs Div, US War Dept, Korea, 46; personnel policy specialist, Off Secy Defense, Nat Mil Estab, 48-50, staff mem personnel policy bd, US Dept Defense, 50-52, spec asst to Asst Secy Defense, Manpower & Personnel, 52-55; agr attache, Seoul, Korea, 56-57, Far Eastern are off, Foreign Agr Serv, USDA, Washington, DC, 58, agr attache, Montevideo, Uruguay, 58-69; PRES, AM WORK HORSE MUS, 70-; PRES, PAEONIAN SPRINGS COUN, 86- *Concurrent Pos:* Hon mem, Asn Intensification Com Uruguay, Montevideo, 69. *Mem:* AAAS; Am Soc Agron; Assoc Soc Personnel Admin. *Res:* Crop productions; physiological botany; soil conservation; weed eradication; personnel administration and training; civil service systems. *Mailing Add:* PO Box 88 Paeonian Springs VA 22129

BUCKELEW, ALBERT RHOADES, JR, b Washington, DC, Oct 22, 42; m 65; c 2. MICROBIOLOGY. *Educ:* Fairleigh Dickinson Univ, BS, 64; Univ NH, PhD(microbiol), 68. *Prof Exp:* From asst prof to assoc prof, 69-84, PROF BIOL, BETHANY COL, WVA, 84- *Concurrent Pos:* Ed, Redstart; co-dir, WVa Breeding Bird Atlas. *Mem:* AAAS; Am Soc Microbiol; Wilson Ornith Soc; Am Ornith Union. *Res:* Interaction of staphylococcal alpha toxin with monolayer films; properties of pulmonary surfactants. *Mailing Add:* Dept Biol Bethany Col Bethany WV 26032

BUCKELEW, THOMAS PAUL, b Easton, Pa, Aug 2, 43; m 71; c 3. PARASITOLOGY, ELECTRON MICROSCOPY. *Educ:* Muhlenberg Col, BS, 65; Univ SC, MS, 67, PhD(biol), 71. *Prof Exp:* PROF BIOL, CALIF UNIV PA, 69- *Concurrent Pos:* Consult inst teaching electron micros, Asn Pa State Col & Univ Biologists, 78-80; consult inst training in-serv teachers molecular biol, NSF, 78-79; fel, Commonwealth Pa, 78. *Honors & Awards:* Darbaker Award, 81. *Res:* Electron microscopy of parasitic protozoa and helminths; fungal electron microscopy; histology and ultrastructure of hypothalamic nuclei. *Mailing Add:* Dept Biol Calif Univ Pa California PA 15419

BUCKER, HOMER PARK, JR, b Ponca City, Okla, July 23, 33; m 55; c 2. ACOUSTICS, SPECTROSCOPY. *Educ:* Univ Okla, BS, 55, PhD(physics), 62. *Prof Exp:* Engr, Univ Okla, 55-60; physicist, US Navy Electronics Lab, 62-67; res physicist, Tracor, Inc, Tex, 67-69; head sound propagation br, Naval Res Lab, Naval Undersea Ctr, 69-70, res physicist, 70-; AT NAVAL OCEAN SYSTS CTR. *Mem:* Acoust Soc Am. *Res:* Propagation of sound in the ocean; infrared and Raman molecular spectroscopy. *Mailing Add:* 808 Moana Dr San Diego CA 92106

BUCKEYE, DONALD ANDREW, b Lakewood, Ohio, Mar 12, 30; m 62; c 2. MATHEMATICS. *Educ:* Ashland Col, BS, 53; Ind Univ, MA, 61, EdD(math), 68. *Prof Exp:* Teacher, high sch, 53-54; instr math, Army Educ Ctr, Sendai, Japan, 54-56; teacher, high sch, 56-66; teaching assoc, Ind Univ, 66-68; PROF MATH, EASTERN MICH UNIV, 68- *Concurrent Pos:* Assoc instr, Ohio State Univ, Cleveland, 64-66. *Mem:* Nat Coun Teachers Math; Math Asn Am. *Res:* Ways of increasing the creative ability of students in mathematics; use of a laboratory approach to the teaching of mathematics and its effects on attitude, creativity and achievement. *Mailing Add:* 1823 Witmire Blvd Ypsilanti MI 48197

BUCKHAM, JAMES A(NDREW), b Minneapolis, Minn, Sept 15, 25; m 54; c 3. CHEMICAL ENGINEERING. *Educ:* Univ Wash, BS, 45, BS & MS, 48, PhD(chem eng), 53. *Prof Exp:* Design engr, Standard Oil Co, Calif, 48-50; instr chem eng, Univ Wash, 51-53; res engr, Calif Res & Develop Co, 50 & 53-54; sr res engr, Atomic Energy Div, Phillips Petrol Co, 54-55, group leader dissolution & extraction studies, 55-60, chief, Tech Proj Sect, 60-61, sr tech staff, 61-62, chief, Develop Eng Sect, 62-63, mgr, Chem & Process Develop Br, 63-66; mgr, Chem Technol Br, Idaho Nuclear Corp, 66-69, asst mgr nuclear & chem technol, 69-70, mgr, Chem Progs Div, 70-76; asst gen mgr, Idaho Chem Progs Opers Off, Allied Chem Corp, 76-77; exec vpres & chief operating officer, Allied Gen Nuclear Serv, 77-84; CONSULT, 84- *Concurrent Pos:* Prof educ prog, Nat Reactor Testing Sta, Univ Idaho, 55- *Honors & Awards:* Robert E Wilson Award, Am Inst Chem Engrs. *Mem:* Am Chem Soc; Am Nuclear Soc; Am Inst Chem Engrs. *Res:* Spray drying; heat transfer; nuclear fuel dissolution; fluidized bed calcination; transport phenomena. *Mailing Add:* 161 Griffin Ave Aiken SC 29803

BUCKHOLTZ, JAMES DONNELL, b Little Rock, Ark, Feb 25, 35; m 59; c 2. MATHEMATICAL ANALYSIS. *Educ:* Univ Tex, BA, 57, PhD(math), 60. *Prof Exp:* Instr math, Univ NC, 60-61, asst prof, 61-62; vis asst prof, Univ Wis, 62-63; asst prof math, Univ NC, 63-64; assoc prof, 64-69, PROF MATH, UNIV KY, 69- *Mem:* Am Math Soc. *Res:* Analytic functions; zeros of polynomials; polynomial expansions. *Mailing Add:* 685 Providence Rd Lexington KY 40502

BUCKHOUSE, JOHN CHAPPLE, b Billings, Mont, Feb 23, 44. RANGE HYDROLOGY. *Educ:* Univ Calif, Davis, BS, 66; Utah State Univ, MS, 68, PhD(watershed sci), 74. *Prof Exp:* ASST & ASSOC PROF RANGE WATERSHED MGT, ORE STATE UNIV, 74-, RES SCIENTIST RANGE HYDROL, 74- *Concurrent Pos:* Consult, Nat Acad Sci, 80-81 & Off Tech Assessment for Cong. *Mem:* Soc Range Mgt; Soil & Water Conserv Soc Am; Sigma Xi. *Res:* Natural resource and watershed valves associated with land uses and water quality; effects of livestock grazing in riparian zones. *Mailing Add:* Rangeland Resources Ore State Univ Corvallis OR 97331

BUCKINGHAM, ALFRED CARMICHAEL, b New York City, NY, June 10, 31; m 60. FLUID PHYSICS, THEORETICAL MECHANICS. *Educ:* Univ Okla, BS, 58, MS, 59; Univ Calif, PhD(aero sci), 63. *Prof Exp:* Teaching asst & res asst eng sci & thermodyn, Univ Okla, 58-59; aerodynamicist re-entry gasdynamics, Douglas Aircraft Co, Santa Monica, Calif, 59-60; res scientist fluid physics, Lockheed Aircraft Corp, Palo Alto, Calif, 61-69; sr physicist, Physics Int Co, San Leandro, Calif, 69-73; staff engr & scientist gasdynamics, Aerotherm/Acurex Corp, Mountain View, Calif, 73-74; PHYSICIST FLUID PHYSICS, TURBULENT FLOW EROSION & DRAG REDUCTION, LAWRENCE LIVERMORE NAT LAB, UNIV CALIF, 74-, PROG LEADER, 80- *Concurrent Pos:* Organizer, Ctr Compressible Turbulence, Lawrence Livermore Nat Lab, 88-; assoc ed, Int J Numerical Methods in Heat & Fluid Flow, 90- *Mem:* Fel Am Inst Aeronaut & Astronaut; Sigma Xi; Am Rocket Soc; Combustion Inst Am; Am Physical Soc. *Res:* Fluid physics; computational (numerical) fluid dynamics; turbulent flow; theoretical transport processes; three-dimensional viscous flow. *Mailing Add:* Lawrence Livermore Nat Lab MC/L-16 PO Box 808 Livermore CA 94550

BUCKINGHAM, WILLIAM THOMAS, b Lancaster, Ohio, Dec 21, 21; m 55; c 3. COMMUNICATIONS SCIENCE, PLANT BREEDING. *Educ:* Otterbein Col, BA, 45, BS, 46; Ohio Univ, MBA, 79. *Prof Exp:* Prin chemist, Battelle Mem Inst, 42-46; asst prof metall chem, Ohio State Univ, 47-54; prin chemist, Battelle Mem Inst, 54-56; lab mgr & spec proj off, Foseco, Inc, 56-68; PRES, BUCKINGHAM ELECTRONIC DIV, BUCKINGHAM PROD, PRES & GEN MGR, BUCKINGHAM ORCHARDS, 68- *Concurrent Pos:* Consult, Metall/Chem & Commun, 68-; consult & assoc sponsor, Dale Carnegie Courses, 68-85; asst prof, Emerson E Evans Sch Bus, Univ Rio Grande. *Mem:* Am Chem Soc. *Res:* Laboratory and general management; chemical treatment of metals; exothermics and explosives; development and growth of new fruit cultivars from origination to full scale commercial production. *Mailing Add:* Buckingham Orchards 8803 Cheshire Rd Sunbury OH 43074

BUCKLAND, ROGER BASIL, b Jemseg, NB, May 18, 42; m 65; c 2. POULTRY GENETICS, REPRODUCTIVE PHYSIOLOGY. *Educ:* McGill Univ, BSc, 63, MSc, 65; Univ Md, PhD(physiol genetics), 68. *Prof Exp:* Res scientist poultry physiol, Can Dept Agr, 67-71; from asst prof to assoc prof, McGill Univ, 71-81, chmn, 79-85, dean, fac agr, 85-90, PROF ANIMAL SCI, MACDONALD COL, MCGILL UNIV, 81-, VPRIN, 85-, DEAN, FAC AGR & ENVIRON SCI, 90- *Honors & Awards:* Poultry Sci Res Award, Poultry Sci Asn, 72. *Mem:* Can Soc Animal Sci; Poultry Sci Asn; World Poultry Sci Asn; Genetics Soc Can. *Res:* Genetics of reproduction, metabolism of chicken sperm; resistance to stress in poultry. *Mailing Add:* Macdonald Col McGill Univ 21111 Lakeshore Rd Ste Anne de Bellevue PQ H9X 1C0 Can

BUCKLER, ERNEST JACK, b Birmingham, Eng, June 3, 14; m 42. CHEMISTRY, POLYMER CHEMISTRY. *Educ:* Cambridge Univ, MA, 35, PhD(phys chem), 38. *Hon Degrees:* LLD, Queens Univ, Ont, 59. *Prof Exp:* Res chemist, Trinidad Leaseholds, BWI, 38-41 & Imp Oil, Sarnia, 41-42; prod controller, St Clair Processing Corp, 42-45, tech supt, 45-48; mgr res, Polymer Corp, 48-53, vpres res & develop, 53-79; CONSULT, POLYSAR LTD, 79- *Honors & Awards:* Tech Award, Int Inst Synthetic Rubber Producers, 80; R S Jane Mem Award, Chem Inst Can, 83. *Mem:* Fel Chem Inst Can; Am Chem Soc; Soc Chem Indust. *Res:* Chemical kinetics, gas phase; petroleum refining and petrochemicals; polymer synthesis and characterization; research management. *Mailing Add:* Polysar Ltd Sarnia ON N7T 7M2 Can

BUCKLER, SHELDON A, b New York, NY, May 18, 31; m 52, 78; c 3. THERMODYNAMICS & MATERIAL PROPERTIES, TECHNICAL MANAGEMENT. *Educ:* NY Univ, BA, 51; Columbia Univ, MA, 52, PhD(chem), 54. *Prof Exp:* Asst, Columbia Univ, 51-53; res fel, Univ Md, 55-56; res chemist & group leader basic res dept, Am Cyanamid Co, 56-62; mgr org res, Am Mach & Foundry Co, 62-64; mgr chem develop, 64-66, dir chem res & develop div, 66-69, asst vpres res, 69-72, vpres res, 72-75, group vpres, 75-77, sr vpres, 77-80, EXEC VPRES, POLAROID CORP, 80- *Concurrent Pos:* Dir, Advan Color technol, Lord Corp. *Mem:* Am Chem Soc; Soc Photog Sci & Eng. *Res:* Technical management; process research and development; synthetic and theoretical organophosphorus chemistry; photographic chemicals; carbohydrates; organometallics; organic peroxides. *Mailing Add:* Polaroid Corp 549 Technology Sq Cambridge MA 02139

BUCKLES, MARJORIE FOX, micro-chemistry; deceased, see previous edition for last biography

BUCKLES, ROBERT EDWIN, b Fallon, Nev, Aug 11, 17; m 44; c 2. PHYSICAL ORGANIC CHEMISTRY. *Educ:* Univ Calif, BS, 39, MS, 40; Univ Calif, Los Angeles, PhD(phys-org chem), 42. *Prof Exp:* Asst, Ill Inst Technol, 40-41 & Univ Calif, Los Angeles, 41-42; du Pont fel & instr org chem, Univ Minn, 43-45; from instr to assoc prof, 45-59, PROF ORG

CHEM, UNIV IOWA, 59- *Mem:* Am Chem Soc. *Res:* Molecular structure; addition reactions; replacement reactions; reaction mechanisms; halogen chemistry; chemistry education. *Mailing Add:* 3113 Cambridge Rd Cameron Pk Shingle Springs CA 95682

BUCKLEY, CHARLES EDWARD, b Charleston, WVa, Sept 2, 29; m 55; c 4. INTERNAL MEDICINE, ALLERGY. *Educ:* Va Polytech Inst, BS, 50; Duke Univ, MD, 54. *Prof Exp:* Assoc med & immunol, 58-67, assoc immunol, 67-70, asst prof med, 67-70, asst prof microbiol & immunol, 70-74, assoc prof med, 70-78, PROF MED, DUKE UNIV MED CTR, 78-, DIR ALLERGY & CLIN, IMMUNOL LAB, 70- *Concurrent Pos:* USPHS res career develop award, 61-69; mem adv coun, Nat Inst Allergy & Infectious Dis, 75-79. *Mem:* Am Fedn Clin Res; Am Thoracic Soc; Am Acad Allergy; Am Col Physicians; Am Asn Immunol; Sigma Xi. *Res:* Allergy, clinical immunology; immunochemistry; biochemical and immunochemical characterization of antigens important in human disease; genetics and epidemiology of immunity. *Mailing Add:* 3621 Westover Rd Durham NC 27707

BUCKLEY, DALE ELIOT, b Wolfville, NS, May 3, 36; m 64; c 3. GEOCHEMISTRY. *Educ:* Acadia Univ, BSc, 59; Univ Western Ont, MSc, 63; Univ Alaska, PhD(marine sci), 69. *Prof Exp:* Tech officer, Can Hydrographic Surv, Dept Mines & Tech Surv, 60-61 & Oceanog Res Div, 61-62; sci officer marine geol, Atlantic Oceanog Lab, 62-69, phys scientist environ marine geol, Atlantic Geosci Ctr, 70-79, RES SCIENTIST, BEDFORD INST OCEANOG, 79- *Concurrent Pos:* Lectr, Calif State Col, Los Angeles, 65-66. *Mem:* Prof Inst Pub Serv Can; fel Geol Asn Can; Can Meteorol & Oceanog Soc; Am Chem Soc. *Res:* Marine inorganic geochemistry, particularly processes of chemical exchange between natural sediments and ionic metals in seawater, research to be applied to problems of coastal pollution and waste disposal, including deep sea disposal of nuclear waste. *Mailing Add:* 21 Dumbarton Ave Dartmouth NS B2X 1Z7 Can

BUCKLEY, DONALD HENRY, tribology, for more information see previous edition

BUCKLEY, EDWARD HARLAND, b Brandon, Man, Jan 30, 31; m 56; c 3. DEVELOPMENTAL BIOLOGY, STRESS PHYSIOLOGY. *Educ:* Ont Agr Col, BSA, 54; Columbia Univ, MA, 56, PhD(plant physiol), 59. *Prof Exp:* Plant physiologist, Cent Res Labs, United Fruit Co, 58-60, plant biochemist, 61-65; plant biochemist, Boyce Thompson Inst Plant Res, Cornell Univ, 65-; RETIRED. *Concurrent Pos:* Prin investr, ecol zoning proj estuary Hudson River, 68-74. *Mem:* NY Acad Sci; Am Soc Plant Physiol; Can Soc Plant Physiol; Soc Wetland Scientist; Am Soc Cell Biol. *Res:* Host-parasite relationships in plants; biosynthesis of flavor and aroma compounds and of phenolic compounds in banana; biochemical aspects of plant gametogenesis; development of fungi; estuarine ecology; ecotoxicology. *Mailing Add:* 165 Iradell Rd Ithaca NY 14850

BUCKLEY, FRED THOMAS, b Newark, NJ, Feb 4, 51; m 75. MATHEMATICS, GRAPH THEORY. *Educ:* Pace Univ, BA, 73; City Univ New York, PhD(math), 78. *Prof Exp:* Asst prof math, St John's Univ, 78-82; from asst prof to assoc prof, 82-88, PROF MATH, BARUCH COL, NY, 88- *Concurrent Pos:* Vchmn math, NY Acad Sci, 90-91. *Mem:* Math Asn Am; NY Acad Sci. *Res:* Ramsey theory; graph operations; line graphs; distance in graphs; Ramsey-type matrix problems; centrality in graphs. *Mailing Add:* Dept Math Baruch Col 17 Lexington New York NY 10010

BUCKLEY, GLENN R, b Detroit, Mich, Apr 19, 43. PETROLEUM. *Educ:* Univ Ill, Champaign, PhD(geol), 73. *Prof Exp:* MGR, EXXON CORP, 73- *Mem:* Fel Geol Soc Am; Am Petrol Geologists. *Mailing Add:* 66 Hillcrest Ave Morris Township NJ 07960-5087

BUCKLEY, JAMES THOMAS, b Ottawa, Ont, Oct 3, 42; div; c 2. BIOCHEMISTRY. *Educ:* McGill Univ, BSc, 65, PhD(biochem), 69. *Prof Exp:* From asst prof to assoc prof, 72-82, PROF BIOCHEM, UNIV VICTORIA, 82- *Mem:* Can Biochem Soc. *Res:* Protein lipid interactions; mechanism of protein export by gram negative bacteria; mechanism action of hole-forming toxins. *Mailing Add:* Dept Biochem Univ Victoria Box 1700 Victoria BC V8W 2Y2 Can

BUCKLEY, JAY SELLECK, JR, b Ansonia, Conn, Feb 16, 24; m 48; c 4. INFORMATION SCIENCE. *Educ:* Williams Col, BS, 44; Univ Minn, PhD(org chem), 49. *Prof Exp:* Res chemist, Winthrop Chem Co, 45-46; res chemist, 49-51, res supvr, 51-54, res mgr, 54-68, mgr, Comput Based Info Serv, 68-74; DIR TECH INFO, PFIZER, INC, 74- *Mem:* Am Chem Soc. *Res:* Organic chemistry; mechanisms of organic reactions and medicinal products; systems for encoding and retrieval of chemical and biological information; methods for efficient and economical utilization of information resources available to research organizations. *Mailing Add:* 16 Birch Lane Groton CT 06340

BUCKLEY, JOHN DENNIS, b Saranac Lake, NY, Oct 28, 28; m 59; c 3. MATERIALS SCIENCE, SPACE SCIENCES. *Educ:* St Lawrence Univ, BS, 50; Clemson Univ, BS, 59, MS, 61; Iowa State Univ, PhD(eng), 68. *Prof Exp:* Aerospace technologist mat sci, 59-70, supvr aerospace eng & thermodyn, 70-71, supvr mat eng, 71-73, TECH ASST MAT PROCESSING, FABRICATION DIV, LANGLEY RES CTR, NASA, 73- *Concurrent Pos:* Assoc prof eng, George Washington Univ, 68-, assoc prof gen studies, 70-; fac mem, NASA- Langley-George Washington Univ Joint Inst Advan Flight Sci, 71-; mem, Va State Bd Health, 72-76; mem City Newport News Planning Comn; lectr eng, Thomas Nelson Community Col, 75-; lectr mat sci, Old Dominion Univ, 80-; lectr physics, Christopher Newport Col, 81- *Honors & Awards:* Exceptional Serv Medal, NASA, 77 & Technol Utilization Award, 78; IR-100 Award, 79. *Mem:* Am Inst Aeronaut & Astronaut; fel Am Ceramic Soc; Am Soc Metals; Soc Mfg Engrs; Nat Inst Ceramic Engrs. *Res:* Physical and thermal properties of materials, metals, ceramics, polymers; design and fabrication of composite materials. *Mailing Add:* Dept Eng Technol Thomas Nelson Community Col Hampton VA 23670

BUCKLEY, JOSEPH J, b Methuen, Mass, Sept 11, 22; m 45; c 4. ANESTHESIOLOGY. *Educ:* Dartmouth Col, AB, 44; NY Med Col, MD, 46; Univ Minn, MS, 58. *Prof Exp:* From instr to assoc prof, 54-61, PROF ANESTHESIOL, MED COL, UNIV MINN, MINNEAPOLIS, 61-, HEAD DEPT, 79- *Concurrent Pos:* Consult, US Vet Admin, 58-; vis prof, Col Physicians & Surgeons, Columbia Univ, 62, Univ Manitoba, 76, Univ Tex, San Antonio, 77, Wash Univ, 85. *Mem:* AMA; Am Soc Anesthesiol; Acad Anesthesiol. *Res:* Cardiac and pulmonary physiology; mass spectrometry. *Mailing Add:* Rm B 515 Univ Minn Hosp Minneapolis MN 55455

BUCKLEY, JOSEPH PAUL, b Bridgeport, Conn, Jan 12, 24; m 47. PHARMACOLOGY. *Educ:* Univ Conn, BS, 49; Purdue Univ, MS, 51, PhD(pharmacol), 52. *Prof Exp:* Instr, St Elizabeth Sch Nursing, 52; from asst prof to prof pharmacol & head dept, Sch Pharm, Univ Pittsburgh, 58-73, assoc dean, 69-73; dean, Col Pharm, 73-87, PROF PHARMACOL, UNIV HOUSTON, 73-, DIR, INST CARDIOVASC STUDIES, 76- *Concurrent Pos:* Hon prof, San Carlos Univ, Guatemala City. *Honors & Awards:* APHA Eli Lilly Award, 66. *Mem:* AAAS (secy, 62-66, vpres, 69); Am Pharmaceut Asn; Am Soc Pharmacol & Exp Therapeut; NY Acad Sci; Interam Soc Hypertension; US Pharmacopeal Conv. *Res:* Mechanisms of action of antihypertensive drugs; effects of stress on blood pressure and animal behavior; brain renin-anoiotensin system. *Mailing Add:* Col Pharm 1441 Moursund Houston TX 77030

BUCKLEY, JOSEPH THADDEUS, b Boston, Mass, Apr 13, 37; m 61; c 4. MATHEMATICS. *Educ:* Boston Col, BS, 58; Ind Univ, PhD(math), 64. *Prof Exp:* Young res instr math, Dartmouth Col, 64-67; asst prof, Univ Mass, 67-69; asst prof, 69-74, assoc prof, 74-80, PROF MATH, WESTERN MICH UNIV, 80- *Mem:* Am Math Soc; Math Asn Am. *Res:* Group theory; group rings; cohomology of groups; extensions; nilpotent groups; homological algebra. *Mailing Add:* Dept Math Western Mich Univ Kalamazoo MI 49008-5152

BUCKLEY, NANCY MARGARET, b Philadelphia, Pa, Dec 2, 24; m 84. PHYSIOLOGY. *Educ:* Univ Pa, AB, 45, MD, 50. *Prof Exp:* Asst med dir, Columbus Blood Prog, Am Red Cross, 51-52; res assoc med, Ohio State Univ, 52, asst prof & res assoc physiol, 52-55; from asst prof to prof physiol, Albert Einstein Col ed, 55- 89, prof biophys, 81-89; RETIRED. *Concurrent Pos:* Consult physiol, Div Pediat Cardiol, Long Island Jewish Med Ctr, 75-89. *Mem:* AAAS; Am Physiol Soc; NY Acad Sci; Am Heart Asn. *Res:* Cardiodynamics; cardiovascular development. *Mailing Add:* 372 Central Park W Apt 2-Y New York NY 10025

BUCKLEY, PAGE SCOTT, b Hampton, Va, June 23, 18; m 48; c 4. SYSTEMS DESIGN & SYSTEMS SCIENCE. *Educ:* Columbia Univ, BS, 40, BA, 40. *Hon Degrees:* DEng, Lehigh Univ, 75. *Prof Exp:* Asst to consult engr, Can, 40; develop engr chem process design, Monsanto Chem Co, Mo, 41-45, instrument engr, Tenn, 45-46, supvr engr, 45-47, develop engr, Tex, 47-49; instrument engr, Instrument Div, Sabine River Works, E I du Pont de Nemours & Co, 49-50, group leader, Tech Sect, 50-54, res proj engr, 54-56, res assoc automatic process control, Eng Res Lab, 56-62, sr engr, Design Div, 62-68, prin consult, 68-87; CONSULT, 87- *Mem:* Nat Acad Eng; Instrument Soc Am; Am Inst Chem Engrs. *Res:* Mathematical analysis and design of control systems for chemical process. *Mailing Add:* Nine N Kingston Rd Newark DE 19713-3707

BUCKLEY, R RUSS, b Jackson, Miss, May 7, 39; m 60; c 3. CORPORATE MANAGEMENT, TECHNICAL MANAGEMENT. *Educ:* Millsaps Col, BS, 61; Univ SC, PhD(inorg chem), 66. *Prof Exp:* tech supvr, Bell Tel Labs, 65-85; co-founder & vpres, Gain Electronics Corp, 85-88, corp secy, 86-88; VPRES & GEN MGR, HYBRID PROD DIV, MAT RES CORP, 89- *Mem:* Am Mgt Asn; Sigma Xi. *Res:* Management of research, development, and improvement of materials and processes of interest to the semiconductor and electronics industry; manufacture thin film alumina ceramic substrates; metallization and fabrication of high performance thin film hybrid integrated circuits. *Mailing Add:* 94 Tulip St Summit NJ 07901

BUCKLEY, REBECCA HATCHER, b Hamlet, NC, Apr 1, 33; m 55; c 4. PEDIATRICS, ALLERGY AND IMMUNOLOGY. *Educ:* Duke Univ, AB, 54; Univ NC, MD, 58. *Prof Exp:* Intern pediat, 58-59, from asst resident to resident, 59-61, from instr to prof pediat, 61-79, from asst prof to assoc prof immunol, 69-79, J BUREN SIDBURY PROF PEDIAT & PROF IMMUNOL, SCH MED, DUKE UNIV, 79-, CHIEF, DIV ALLERGY & IMMUNOL, 74- *Concurrent Pos:* Fel allergy, Sch Med, Duke Univ, 61-63, fel immunol, 63-65; dir Am Bd Allergy & Immunol, 71-73 & 82-87; mem, immunol sci study sect, NIH, 76-79, chmn, 79-80; Merit Res Award, NIH, 87- *Honors & Awards:* Allergic Dis Acad Award. *Mem:* Am Acad Allergy & Immunol (pres 79-80); Am Acad Pediat; Soc Pediat Res; Am Asn Immunologists; Sigma Xi (pres 80-81); Am Pediat Soc. *Res:* Cellular and molecular bases of human primary immunodeficiency; cell-mediated immune responsiveness in the atopic diseases; regulation of human synthesis; bone marrow transplantation; stem cell education in the human thymus. *Mailing Add:* Dept Pediat Box 2898 Duke Univ Sch Med Durham NC 27710

BUCKLIN, ROBERT VAN ZANDT, b Chicago, Ill, June 25, 16; m 67; c 5. LEGAL MEDICINE. *Educ:* Loyola Univ, Ill, BSM, 38, MD, 41; STex Col Law, JD, 69; Am Bd Path, dipl path anat, clin & forensic path. *Prof Exp:* Intern, St Josephs Hosp, Tacoma, Wash, 40-41; resident path, Tacoma Gen Hosp, 41-42; chief lab serv path, St Mary's Hosp & Saginaw Gen Hosp, Mich, 46-63; assoc med examr, Harris County, Tex, 64-68; prof path, Univ Tex Med Br, Galveston, 69-71; chief med examr, Galveston County, Tex, 69-71, dir med legal affairs, Health Commun, Inc, 72-74; dep med examr, Los Angeles County, 74-77; chief med examr, Travis County, Tex, 77-78; dep chief med examr, Harris County, Tex, 78-80; dep med examr, Los Angeles County, Calif, 80-83; pathologist, San Diego County Coroner, 83-87; Clin prof path, Univ Southern Calif, 80-86; CONSULT, LEGAL MED & FORENSIC PATH, 87- *Concurrent Pos:* Forensic path consult, 69- *Mem:* Am Col Physicians; Am Soc Clin Path; Am Col Legal Med; Am Acad Forensic Sci; Nat Asn Med Examrs. *Res:* Medical and legal aspects of the Crucifixion; aging and dating of traumatic lesions. *Mailing Add:* 7500 Estero Blvd Suite 1204 Ft Myers Beach FL 33931

BUCKMAN, ALVIN BRUCE, b Omaha, Nebr, Dec 7, 41; m 66. ELECTROOPTICS, OPTICAL PHYSICS. *Educ:* Mass Inst Technol, BS, 64; Univ Nebr, MS, 66, PhD(elec eng), 68. *Prof Exp:* Instr elec eng, Univ Nebr, Lincoln, 66-68, from asst prof to assoc prof, 68-74, assoc prof elec eng, 73-74; ASSOC PROF ELEC ENG, UNIV TEX, AUSTIN, 74- *Concurrent Pos:* Consult, 74- *Mem:* Am Phys Soc; Optical Soc Am. *Res:* Optical properties and electronic structure of solids; especially surface and thin film effects; integrated optical devices. *Mailing Add:* Elec Eng Dept ENS103 Univ Tex Austin TX 78712

BUCKMAN, WILLIAM GORDON, b Morganfield, Ky, Dec 1, 34; m 59; c 3. SOLID STATE PHYSICS. *Educ:* Western Ky Univ, BS, 60; Vanderbilt Univ, MS, 62; Univ NC, PhD(pub health, radiation physics), 67. *Prof Exp:* Instr radiation hyg, Univ NC, 62-64; chmn dept physics & math, Ky Wesleyan Col, 66-67; from asst prof to assoc prof, 67-73, PROF PHYSICS, WESTERN KY UNIV, 73- *Concurrent Pos:* Consult, environ radioactivity & dosimetry. *Mem:* Am Phys Soc; Health Physics Soc. *Res:* Luminescence; solid state radiation dosimetry. *Mailing Add:* Dept Physics Western Ky Univ Bowling Green KY 42101

BUCKMASTER, HARVEY ALLEN, b Calgary, Alta, Apr 8, 29; m 56, 68. MAGNETIC RESONANCE. *Educ:* Univ Alta, BSc, 50; Univ BC, MA, 52, PhD(physics), 56. *Prof Exp:* Nat Res Coun Can Overseas fel, Cavendish Lab, Cambridge Univ, 56-57; asst prof physics, Univ Alta, 57-60; from asst prof to assoc prof, 60-67, PROF PHYSICS, UNIV CALGARY, 67- *Concurrent Pos:* Leverhulme vis res fel, Univ Keele, 64-65; mem sci adv comt, Environ Conserv Authority, Prov Alta, 71-77; mem bd gov, Univ Calgary, 75-78; chmn sci adv comt, Environ Coun Alta, 77-79, mem, 77-88; trustee, Univ Alta Pension Plan, 72-78, Alta Univs Pension Plan, 78-86; mem Nat Sci & Eng Res Coun Can Mainline Phys Grant Selection Comt, 82-85. *Honors & Awards:* Queen Elizabeth II Can Silver Jubilee Medal 78. *Mem:* Sr mem Inst Elec & Electronics Engrs; Can Asn Physicists; fel Brit Inst Physics; Am Asn Physics Teachers; AAAS; Int Soc Magnetic Resonance; Int Electron Paramagnetic Resonance Soc; Sigma Xi. *Res:* Electronic paramagnetic resonance and endor lanthanide elements in hydrated, deuterated and anhydrous lattices; application determinations in biophysics; dielectric properties of biological tissue at millimeter wave lengths; electron paramagnetic resonance; energy consumption and climate change studies; energy scenarios; dielectric properties of high loss liquids (water, deuterium); electron paramagnetic resonance instrumentation development; microwave complex permittivity spectroscopy instrumentation development; fossil fuel studies using electron paramagnetic resonance; application of pulsed electron paramagnetic resonance to fossil fuel and biophysics studies; electron paramagnetic resonance spin-label studies of biological tissue. *Mailing Add:* Dept Physics & Astron Univ Calgary Calgary AB T2N 1N4 Can

BUCKMASTER, JOHN DAVID, b Belfast, Northern Ireland, Feb 2, 41; m 66; c 1. FLUID MECHANICS, APPLIED MATHEMATICS. *Educ:* Univ London, BSc, 62; Cornell Univ, PhD(appl math), 69. *Prof Exp:* Res engr fluids, Cornell Aeronaut Lab, Buffalo, NY, 62-65; lectr, Univ London, 68-69; asst prof, NY Univ, 69-72; asst prof eng & appl sci, Yale Univ, 72-74; assoc prof, 74-78, PROF MATH & MECH, UNIV ILL, URBANA, 78- *Mem:* Am Phys Soc; Soc Indust & Appl Math. *Mailing Add:* 101 Trans Bldg Univ Ill 104 S Mathews Ave Urbana IL 61801

BUCKMIRE, REGINALD EUGENE, b Oct 3, 38; Brit citizen; m 75; c 3. FOOD CHEMISTRY, TROPICAL AGRICULTURE. *Educ:* Univ WI, BS, 61; Univ Mass, Amherst, MS, 71, PhD(food sci & nutrit), 74. *Prof Exp:* Agr exten officer, USDA, 61-68; asst soils, Univ Mass, 68-74; group leader foods, Quaker Oats Co, Ill, 74-78; proj leader, Foremost Foods, Calif, 78; proj appraisal & food consult, Caribbean Develop Bank, 78-81, proj leader, 81-83, food consult, Technol Transfer & Tech Serv, 84-87; INDUST DEVELOP EXPERT, GRENADA DEVELOP BANK, 87- *Concurrent Pos:* Dir, Agro-Indust Prog, Grenada, 81-82; agro-indust consult, feasibility studies & food mfg, 81-83; chief exec officer, Superior Foods Ltd, 83-84; Caribbean tech consult & serv network coordr, technol networking, 83-84; vis lectr food & nutrit, Barbados Polytech, 84-; assoc consult, Redma Consult Ltd, Can, 84-, Indust Develop, 87- *Mem:* Inst Food Technologists. *Res:* Food product development; food program development for lesser developed countries. *Mailing Add:* Perris Pharm Grenville St Andrew's Grenada British West Indies

BUCKNAM, ROBERT CAMPBELL, b Lander, Wyo, Sept 29, 40; m 63; c 3. GEOLOGY. *Educ:* Colo Sch Mines, Geol Eng, 62; Univ Colo, PhD(geol), 69. *Prof Exp:* GEOLOGIST, US GEOL SURV, 68- *Mem:* Fel Geol Soc Am; Seismol Soc Am; Am Geophys Union. *Res:* Structural geology; geology of earthquakes. *Mailing Add:* US Geol Surv Stop 966 Box 25046 Fed Ctr Denver CO 80225

BUCKNELL, ROGER W(INSTON), chemical engineering, for more information see previous edition

BUCKNER, CARL KENNETH, AUTONOMIC-PHARMACOLOGY, IMMUNO-PHARMACOLOGY. *Educ:* Ohio State Univ, PhD(pharmacol), 70. *Prof Exp:* prof pharmacol, Univ Wis, Madison, 81-; AT ICI AMERICAS INC. *Mailing Add:* ICI Pharmaceut Group ICI Americas Inc Wilmington DE 19897

BUCKNER, CHARLES HENRY, b Toronto, Ont, Oct 8, 28; m 53; c 2. WILDLIFE ECOLOGY. *Educ:* Univ Toronto, BA, 52; Univ Man, MSc, 54; Univ Western Ont, PhD(zool), 59. *Prof Exp:* Res officer mammal, Can Dept Agr, 52-60 & Can Dept Fisheries & Forestry, 60-66; head vert biol, 66-70, dep dir, 75-76, dir, 76-78, HEAD ECOL IMPACT, CHEM CONTROL RES INST, 70- *Concurrent Pos:* Prog mgr, Int Spruce Budworms Agreement, 78- *Mem:* Am Soc Mammal; Entom Soc Can. *Res:* Impact of pesticides on non-target organisms; vertebrate damage to forests; populations feeding and behavior; ornithology; population dynamics; vertebrate predation on forest insects. *Mailing Add:* 54 Moorcroft Rd Ottawa ON K2G 0M7 Can

BUCKNER, DAVID LEE, b Boulder, Colo, Feb 4, 48; m 73; c 1. PLANT ECOLOGY, SOIL SCIENCE. *Educ:* Univ Colo, BA, 70, MA, 73, PhD(plant ecol), 77. *Prof Exp:* Consult plant ecol & soil sci, Genge Environ Consults, 75-77; CONSULT PLANT ECOL & SOIL SCI, ENVIRON SCI DIV, CAMP DRESSER & MCKEE, 77- *Mem:* Ecol Soc Am; Soc Range Mgt; Sigma Xi. *Res:* Recovery of alpine tundra following disturbance; recovery rates of semi-desert to subalpine ecosystems following pipeline construction; dynamics of high elevation forest-meadow ecosystems. *Mailing Add:* 1077 S Cherryvale Rd Boulder CO 80303

BUCKNER, JAMES STEWART, b Portland, Ore, June 26, 42; m 66; c 1. BIOCHEMISTRY. *Educ:* Univ Idaho, BS, 64; NDak State Univ, MS, 69, PhD(chem), 71. *Prof Exp:* Res scientist, Dept Agr Chem, Wash State Univ, 71-76; RES BIOCHEMIST, BIOSCI RES LAB, AGR RES SERV, FED RES, USDA, 76- *Mem:* Am Soc Biochem & Molecular Biol. *Res:* Elucidation and description of essential physiological processes in insects, such as nitrogen metabolism and surface lipid metabolism, which are controlled by the neuroendocrine system. *Mailing Add:* Biosci Res Lab USDA Agr Res Serv-Fed Res Fargo ND 58109

BUCKNER, RALPH GUPTON, b Casper, Wyo, May 12, 21; m 49; c 2. VETERINARY MEDICINE. *Educ:* Westminster Col, AB, 47; Kans State Univ, BS, DVM, 56; Univ Okla, MS, 66. *Prof Exp:* Inst vet med & surg, Okla State Univ, 56-58, from asst prof to assoc prof vet path, 58-67, assoc prof vet med & surg, 67-69, prof vet path, 69-73, prof vet med & chief small animal med, 74-86; RETIRED. *Concurrent Pos:* Consult & lobbyist, Okla Vet Med Asn, 86- *Mem:* Am Vet Med Asn; Am Soc Vet Clin Path; NY Acad Sci; Am Col Vet Int Med; Sigma Xi. *Res:* Clinical pathology, especially hematology and blood coagulation diseases; canine hemophilia; theriogenology. *Mailing Add:* 65 University Circle Stillwater OK 74074

BUCKNER, RICHARD LEE, b Owensboro, Ky, Jan 9, 47; m 73; c 1. PARASITOLOGY. *Educ:* Western Ky Univ, BS, 70, MS, 72; Univ Nebr-Lincoln, PhD(zool), 76. *Prof Exp:* Vis asst prof, Ind State Univ, , Evansville, 76-78; asst prof, 78-84, ASSOC PROF BIOL, LIVINGSTON UNIV, 84- *Mem:* Am Soc Parasitologists; Am Inst Biol Sci. *Res:* Taxonomy and host-parasite relationships of the acanthocephalans. *Mailing Add:* Div Natural Sci & Math Livingston Univ Livingston AL 35470

BUCKWALTER, GARY LEE, b Kittanning, Pa, June 29, 34; m 57; c 2. PHYSICS. *Educ:* Pa State Univ, BS, 56; Cath Univ Am, MS, 61, PhD(physics), 66. *Prof Exp:* From instr to asst prof, US Naval Acad, 59-65; fel, Cath Univ Am, 66; assoc prof, 66-69, chmn dept, 75-83, prof physics, Indiana Univ Pa, 69-87; INSTR, DAYTONA BEACH COMMUNITY COL, 87- *Concurrent Pos:* Res Corp Am res grant, 67- *Mem:* Am Asn Physics Teachers; Am Phys Soc. *Res:* High energy cosmic ray physics. *Mailing Add:* Dept Sci Daytona Beach Community Col Daytona Beach FL 32115-2811

BUCKWALTER, JOSEPH ADDISON, b Royersford, Pa, Jan 4, 20; m 46; c 4. SURGERY. *Educ:* Colgate Univ, AB, 41; Univ Pa, MD, 44; Am Bd Surg, dipl. *Prof Exp:* Intern, Col Med, Univ Iowa, 44-45, resident gen surg, 45-46 & 49-52, assoc surg, 52-54, from asst prof to prof, 54-70; PROF SURG, SCH MED, UNIV NC, CHAPEL HILL, 70- *Concurrent Pos:* Fel anat, Sch Med, Univ NC, 48, fel path, 48-49; sr registr, Postgrad Med Sch, Univ London, 53-54; dir surg, Dorothea Dix Hosp, Raleigh, NC. *Mem:* Soc Exp Biol & Med; AMA; Am Col Surg; Soc Univ Surg; Int Soc Surg. *Res:* Clinical and basic studies of thyroid disorders, gastroenterology, blood groups, human genetics, parathyroid physiology and disease, blood volume dynamics. *Mailing Add:* Dept Surg NC Mem Hosp Univ NC Chapel Hill NC 27514

BUCKWALTER, TRACY VERE, JR, geology; deceased, see previous edition for last biography

BUCKWOLD, SIDNEY JOSHUA, b Winnipeg, Can, Oct 30, 19; nat US; m 45; c 3. ANALYTICAL CHEMISTRY, SOIL CHEMISTRY. *Educ:* Univ Manitoba, BSA, 42, MSc, 44; Hebrew Univ, Israel, PhD(soil chem), 54. *Prof Exp:* Asst prof soils, NDak State Univ, 45-46; sr asst irrig & soil chem, Univ Calif, Davis, 46-47; soil chemist & chief soil & water sect, Ministry Agr, Govt Israel, 48-50; fac assoc soil chem, Hebrew Univ, Israel, 50-54; res assoc clay mineral, Pa State Univ, 55-56; asst prof chem, NMex Inst Mining & Technol, 56-58; assoc prof & head dept, Col St Joseph on the Rio Grande, 58-61; assoc prof, Am Int Col, 61-64, chmn dept, 63-69, prof, 65-69; ASSOC PROF CHEM, CENT CONN STATE COL, 69- *Mem:* Am Chem Soc; Sigma Xi; Am Asn Univ Professors. *Res:* Quantitative analysis; instrumental methods; ion exchange; physical chemistry of clays and soils; soil-water relations. *Mailing Add:* Mt Scopus PO Box 24358 Jerusalem Israel

BUCOVAZ, EDSEL TONY, b Eldorado, Ill, May 8, 28; m 53; c 2. BIOCHEMISTRY, ORGANIC CHEMISTRY. *Educ:* Southern Ill Univ, BA, 55, MA, 57; St Louis Univ, PhD(biochem), 62. *Prof Exp:* Control & res chemist, US Chem Co, 53-54; asst prof biochem, Ctr Health Sci, Univ Tenn, Memphis, 64-69; assoc prof, 69-75, prof biochem, 75, ASSOC DIR BASIC RES, MEMPHIS REGIONAL CANCER CTR, 74- *Concurrent Pos:* Res assoc fel biochem, Sch Med, Univ Tenn, 62-64. *Mem:* Sigma Xi; AAAS; Am Chem Soc; Am Asn Cancer Res; Am Soc Biol Chem. *Res:* Protein biosynthesis; cancer research; chemical carcinogenesis; human research related to fetal maturity and respiratory depression caused by analgysics; the comorosan effect on biological systems; developer of B-protein assay for the detection of cancer. *Mailing Add:* 4929 Mockingbird Lane Memphis TN 38117

BUCY, J FRED, b Tahoka, Tex, July 29, 28; m 47; c 3. SEMICONDUCTORS, GEOPHYSICS. *Educ:* Tex Tech Col, BA, 51; Univ Tex, Austin, MA, 53. *Prof Exp:* Res physicist, Defense Res Lab, Tex, 53; design engr, Ctr Res Librs, 53-54, proj engr, 54-58, prog mgr, 58-59, proj engr, VenSun Flow Sta in Maracaibo, 59-61, dept head, Petrol Div, 59-61, group mgr, Indust Prod Div, 61-63, vpres, Apparatus Div, 63-67, vpres, Semiconductor-Components Div, 67, group vpres, Components, 67-72, exec vpres, 72-76, mem bd dirs, 74, chief opers officer, Tex Instruments Inc, 75-84, pres, 76-84, chief exec officer, 84; RETIRED. *Concurrent Pos:* Mem Defense Sci Bd, Dept Defense, 70-; mem, Brazil/US Bus Coun; mem, Adv Coun, Nat Strategy Info Ctr; mem, Subcomt Export Admin, President's Export Coun; mem, Techol Assessment Adv Coun, Off Technol Assessment US Cong; mem, Adv Comt Assessment Technol & World Trade, Off Technol Assessment. *Honors & Awards:* Distinguished Engr Award, Tex Tech Univ, Lubbock. *Mem:* Nat Acad Eng; fel Inst Elec & Electronics Engrs; Soc Explor Geophysicists; Am Phys Soc. *Res:* Integrated circuit design and development; data processing systems--research, design, technology and engineering; author and co-author of technical reports and publications; holder of patents for data processing. *Mailing Add:* Investment & Consult PO Box 780929 Dallas TX 75378-0929

BUCY, PAUL CLANCY, b Hubbard, Iowa, Nov 13, 04; m 27; c 2. NEUROSURGERY, NEUROLOGY. *Educ:* Univ Iowa, BS, 25, MS & MD, 27. *Hon Degrees:* MD, Univ Thessaloniki, 70; Dr, Hon Causa, Univ Utrecht, 71. *Prof Exp:* Asst neuropath, Univ Iowa, 25-27; intern, Henry Ford Hosp, Detroit, 27-28; instr neurosurg, Univ Chicago, 28-33, from asst prof to assoc prof, 33-41, head div neurol & neurosurg, 39-41; from assoc prof to prof neurol & neurosurg, Univ Ill, 41-54; prof, Med Sch, Northwestern Univ, 54-73, emer prof surg, 73-88; clin prof neurol & neurosurg, Bowman Gray Med Sch, 74-88; RETIRED. *Concurrent Pos:* Gorgas lectr, Univ Ala, 44; Commonwealth vis prof, Univ Louisville, 50; mem, World Fedn Neurosurg Socs, pres, 57-61, hon pres, 61-; mem adv coun, Nat Inst Neurol Dis & Blindness, 61-64, prog proj comt, 65-69; vis lectr, Free Univ Berlin, 63 & Sch Med, Johns Hopkins Univ, 64; vis prof, Southwest Med Sch, Univ Tex, 63, Univ Rochester, Montreal Neurol Inst & Harvard Univ, 69, Mass Gen Hosp & State Univ Utrecht, 69; mem, Ill Psychiat Training & Res Authority; chmn, Nat Comt Res in Neurol Dis, 69-75; ed, Surg Neurol, 72-86. *Honors & Awards:* Gorgas lectr, Univ Ala, 44; John Black Johnson lectr, Univ Minn, 49; Max Minor Peet lectr, Univ Mich, 56. *Mem:* Soc Neurol Surg (pres, 59-60); Am Surg Asn; Am Physiol Soc; Am Asn Neurol Surg (pres, 51-52); Am Neurol Asn (vpres, 54-55, pres, 71-72). *Res:* Spinal cord injury; structure and function of the cerebral cortex; involuntary movements; intracranial neoplasms; clinical neurology and neurological surgery. *Mailing Add:* Dept Neurol & Neurosurg Bowman-Gray Med Sch 300 S Hawthorne Rd Winston-Salem NC 27103

BUCY, RICHARD SNOWDEN, b Washington, DC, July 20, 35; m 61; c 2. APPLIED MATHEMATICS. *Educ:* Mass Inst Technol, BS, 57; Univ Calif, Berkeley, PhD(probability), 63. *Prof Exp:* Assoc mathematician, Appl Physics Lab, Johns Hopkins Univ, 57-60; mathematician, Res Inst Advan Studies, 60-61 & 63-64; asst prof math, Univ Md, College Park, 64-65; assoc prof, Univ Colo, Boulder, 65-67; assoc prof aerospace & math, 67-70, PROF AEROSPACE ENG & MATH, UNIV SOUTHERN CALIF, 70-*Concurrent Pos:* Consult math dept, Rand Corp, 63-, Thompson Ramo Wooldridge Corp, Aerospace Corp & Electrac Inc; vis prof, Univ Paul Sabitier, Toulouse, France, 72-73; vis prof, French Govt, 73, 83 & 91, Tech Univ Berlin, 75-76, Univ Nice, France, 83-84 & 90-91; deleg, Inst Elec & Electronic Engrs Soviet Acad Info Theory Meeting, Moscow, 75; dir, advan study prog nonlinear stochastic probs, NATO, Algarve, Portugal, 83. *Honors & Awards:* Humboldt Prize, WGer Govt, 75. *Mem:* Am Math Soc; fel Inst Elec & Electronics Engrs. *Res:* Probability and control theory; optimum control and filtering theory; adaptive control; Markov processes; discrete potential theory; analysis and synthesis of optimal devices to separate signal from noise, with emphasis on direction finding. *Mailing Add:* Aerospace Eng & Math Dept Univ Southern Calif Los Angeles CA 90089-0192

BUDAI, JOHN DAVID, b Poughkeepsie, NY, Dec 2, 52; m 82; c 1. CONDENSED MATTER PHYSICS. *Educ:* Dartmouth Col, BA, 74; Cornell Univ, MS, 79, PhD(physics), 82. *Prof Exp:* Postdoctoral assoc, AT&T Bell Labs, 82-84; PHYSICIST, OAK RIDGE NAT LAB, 84- *Mem:* Am Phys Soc; Am Soc Metals. *Res:* X-ray diffraction studies of the structure of condensed matter systems; with emphasis on the areas of crystal interfaces, defects in quasicrystals, and superconducting epitaxial thin films. *Mailing Add:* Oak Ridge Nat Lab PO Box 2008 Bldg 3025 Oak Ridge TN 37831-6024

BUDAY, PAUL VINCENT, b Jersey City, NJ, Aug 19, 31; m 56; c 6. PHARMACOLOGY. *Educ:* Fordham Univ, BS, 53; Temple Univ, MS, 55; Purdue Univ, PhD(pharmacol), 58, Seton Hall Sch Law, JD, 77. *Prof Exp:* Asst chem, Temple Univ, 53-55; asst pharmacol, Purdue Univ, 55-56; asst prof biol sci, Fordham Univ, 58-59; asst prof pharmacol, col pharm, Univ RI, 59-61; pharmacol ed, Lederle Labs, NY, 61-63; head sci info, Vick Div Res & Develop, Richardson-Merrell Inc, 63-65; med res assoc, Warner-Chilcott Labs, NJ, 65-66, asst to dir med serv, Warner-Lambert Res Inst, 66-67; head drug regulatory affairs, Sandoz, 67-69; dir regulatory affairs, Ethicon, Inc, 69-77; dir extramural affairs, Block Drug Co, 77-78; corp dir regulatory affairs, G D Searle, 78-80; dir int regulatory affairs, Warner-Lambert Co, 80-84; mgr regulatory/clin affairs, Novametrix Med Syst, 84-86; regulatory affairs, QA Lukens Corp, 86-88; DIR, REGULATORY AFFAIRS, PORTON INT, 88- *Concurrent Pos:* Gustavus A Pfeiffer Mem res fel, Am Found Pharmaceut Ed, 58-59. *Mem:* Fel AAAS; Sigma Xi; Drug Info Asn; Regulatory Affairs Prof Soc. *Res:* Pharmacology of neurohormones; interrelationships between endocrine hormones and drug action; drug interactions and preclinical toxicity; medical writing; clinical testing of drugs; food and drug law. *Mailing Add:* Porton Int Inc 727 15th St NW Washington DC 20005

BUDD, GEOFFREY COLIN, b London, Eng, Dec 2, 35; m 59; c 4. CELL BIOLOGY, PHYSIOLOGY. *Educ:* Chelsea Col Sci & Technol, Univ London, BSc, 58; Birkbeck Col, Univ London, BSc, 61; Royal Free Hosp Med, Univ London, PhD(zool), 66. *Prof Exp:* Tech officer, Med Res Coun Biophys Res Unit, Kings Col, Univ London, 58-61; asst lectr biol & radiation biol, Royal Free Hosp Sch Med, Univ London, 61-62, lectr, 62-66; res assoc appl physics, neurobiol & behav, Cornell Univ, 66-68, vis asst prof, 68-69; from asst prof to assoc prof, 69-74, PROF PHYSIOL, MED COL OHIO, 74- *Mem:* Fel AAAS; Am Soc Cell Biol; Electron Micros Soc Am; Tissue Cult Asn; Histochem Soc; Sigma Xi. *Res:* Extra pancreatic insulin synthesis; development and application of quantitative electron microscope autoradiography. *Mailing Add:* Physiol CS 10008 Med Col Ohio Toledo OH 43689

BUDD, THOMAS WAYNE, b Pittsburgh, Pa, Oct 1, 46. PLANT PHYSIOLOGY, MOLECULAR BIOLOGY. *Educ:* NDak State Univ, BS, 68, PhD(bot), 72. *Prof Exp:* ASSOC PROF BIOL, ST LAWRENCE UNIV, 72- *Concurrent Pos:* Vis asst prof, Dept Med, Univ Tenn Ctr Health Sci, 81-82. *Mem:* AAAS; Am Soc Plant Physiologists; Sigma Xi; NY Acad Sci; Int Asn Aerobiol. *Res:* Cancer immunology - tumor escape mechanisms, cell tumor surveilance; allergy immunology - mold allergen identification; electron microscopist. *Mailing Add:* Dept Biol St Lawrence Univ Canton NY 13617

BUDDE, MARY LAURENCE, b Covington, Ky, May 17, 29. ZOOLOGY, MICROBIOLOGY. *Educ:* Villa Madonna Col, AB, 53; Cath Univ Am, MS, 55, PhD(zool), 58. *Prof Exp:* Instr, 58-61, from asst prof to assoc prof, 61-75, acad dean, 81-85, PROF BIOL, THOMAS MORE COL, KY, 75- *Concurrent Pos:* dir, Fresh-Water Biol Sta. *Mem:* AAAS; Sigma Xi. *Res:* Endocrinology: regulation of calcium and phosphorus metabolism in bony fish; aquatic biology: power plant operations and microorganisms and fish population effects; medical technology education. *Mailing Add:* Thomas More Col Crestview Hills KY 41017

BUDDE, PAUL BERNARD, b Covington, Ky, June 24, 26; m 59; c 6. AGRICULTURAL CHEMISTRY. *Educ:* Xavier Univ, Ohio, BS, 50, MS, 51. *Prof Exp:* Chemist, 59-62, res chemist, 62-71, SR RES CHEMIST, AGR PROD DEPT, DOW CHEM CO, 71- *Mem:* Am Chem Soc. *Res:* Agricultural formulations. *Mailing Add:* 1109 Mattes Dr Midland MI 48640-3773

BUDDE, WILLIAM L, b Cincinnati, Ohio, Dec 18, 34; m 59; c 6. ANALYTICAL CHEMISTRY. *Educ:* Xavier Univ, BS, 57, MS, 59; Univ Cincinnati, PhD(chem), 63. *Prof Exp:* Fel, Univ Calif, Riverside, 63-64; vis asst prof chem, Univ Kans, 64-65; assoc to sr chemist, Midwest Res Inst, 65-69; assoc dir, Mass Spectrometry Ctr, Purdue Univ, 69-71; sect chief advan instrumentation, Off Res & Develop, 71-88, DIR, CHEM RES DIV, US ENVIRON PROTECTION AGENCY, 88- *Mem:* Am Chem Soc; Am Soc Mass Spectrometry. *Res:* Mass spectrometry: trace organic analysis by computerized gas chromatography; laboratory automation with digital computers. *Mailing Add:* Nat Environ Res Ctr Environ Protection Agency Cincinnati OH 45268

BUDDENHAGEN, IVAN WILLIAM, b Ventura, Calif, Apr 26, 30; m 50; c 4. PLANT PATHOLOGY. *Educ:* Ore State Col, BS, 53, MS, 54, PhD, 57. *Prof Exp:* Plant pathologist, United Fruit Co, Honduras, 57-64; prof plant path, Univ Hawaii, 64-75; MEM STAFF CEREAL IMPROV PROG, INT INST TROP AGR, 75- *Mem:* Am Phytopath Soc. *Res:* Bacterial plant diseases; international plant path; host-parasite interaction; wilt diseases. *Mailing Add:* Dept Agron & Range Sci Univ Calif Davis CA 95616

BUDDING, ANTONIUS JACOB, b Amsterdam, Neth, Jan 30, 22; m 53; c 2. GEOLOGY. *Educ:* Univ Amsterdam, BSc, 42, MSc, 48, PhD(geol), 51. *Prof Exp:* Instr geol & petrol, Geol Inst, Amsterdam, 46-51; geologist, Geol Surv, Sask, 51-52, prin geologist, 52-56; from asst prof to assoc prof, 56-72, chmn, dept geosci, 81-84, PROF GEOL, NMEX INST MINING & TECHNOL, 72- *Concurrent Pos:* Consult, Los Alamos Sci Lab, 72-81. *Mem:* Fel Geol Soc Am; Ger Geol Soc; Am Geophys Union; fel AAAS; Sigma Xi. *Res:* Metamorphic and igneous petrology; structural geology. *Mailing Add:* Dept Geosci NMex Inst Mining & Technol Socorro NM 87801

BUDEN, ROSEMARY V, b New Haven, Conn, Mar 27, 31; m 51, 84; c 6. GEOCHEMISTRY, MINERALOGY-PETROLOGY. *Educ:* Oberlin Col, BA, 52; Univ Mich, MS, 54; Yale Univ, PhD(geol), 68. *Prof Exp:* From asst prof to assoc prof geol, State Univ NY, Binghamton, 68-75; res fel, Carnegie Inst, Washington, DC, 74-75; staff mem, Los Alamos Nat Lab, 75-77, assoc group leader nuclear chem, 78-80, assoc group leader Isotope Geochem, 80-85, consult, 85-88; prog dir exp & theoret geochem, NSF, 88-90; PRIN SCIENTIST, IDAHO NAT ENG LAB, 90- *Mem:* Fel AAAS; Geochem Soc; fel Am Mineral Soc; Sigma Xi; fel Geol Soc Am. *Res:* Experimental element transport and retention; geothermal systems. *Mailing Add:* 4770 E 65 S Idaho Falls ID 83406

BUDENSTEIN, PAUL PHILIP, b Philadelphia, Pa, June 27, 28; m 57; c 5. PHYSICS. *Educ:* Temple Univ, BA, 49; Lehigh Univ, MS, 51, PhD(physics), 57. *Prof Exp:* Instr physics, Lehigh Univ, 52-56; mem tech staff, Bell Tel Labs, 57-58; asst prof, 59-62, ASSOC PROF PHYSICS, AUBURN UNIV, 63- *Concurrent Pos:* Consult, US Army Missile Command, 69-77. *Mem:* Am Vacuum Soc; Electrochem Soc; Am Phys Soc; Am Asn Physics Teachers. *Res:* Dielectric breakdown in solids; current filamentation in solids. *Mailing Add:* 370 Bowden Dr Auburn AL 36830

BUDERER, MELVIN CHARLES, b Sacramento, Calif, May 12, 41; m 84; c 2. PHYSIOLOGY. *Educ:* Univ Calif, Berkeley, AB, 63, PhD(physiol), 70. *Prof Exp:* Nat res coun resident res assoc, L B Johnson Space Ctr, 70-72, res physiologist, Technol Inc, 72-79, mgr, sci support & mission opers, gen elect, 79-84, PROJ SCIENTIST, L B JOHNSON SPACE CTR, NASA, 84-, CHIEF, PAYLOAD ENG BR, 88- *Mem:* Aerospace Med Asn; Sigma Xi; AAAS. *Res:* Environmental physiology; space physiology. *Mailing Add:* 16523 Kentwood Houston TX 77058

BUDGE, WALLACE DON, b Pleasant View, Utah, Apr 25, 33; m 57; c 2. CIVIL ENGINEERING. *Educ:* Utah State Univ, BS, 59, MS, 61; Univ Colo, PhD(civil eng), 64. *Prof Exp:* Instr civil eng, Utah State Univ, 59-61; res assoc, Univ Colo, 63-64; assoc prof, 64-77, PROF CIVIL ENG, BRIGHAM YOUNG UNIV, 77- *Concurrent Pos:* Univ liaison rep, Hwy Res Bd, Nat Acad Sci-Nat Res Coun, 65- *Mem:* Am Soc Civil Engrs; Am Soc Eng Educ; Am Soc Testing & Mat. *Res:* Concrete; swelling soils. *Mailing Add:* Dept Civil Eng Brigham Young Univ Provo UT 84602

BUDGOR, AARON BERNARD, b Munich, Ger, July 6, 48; US citizen; m 77; c 2. STATISTICAL PHYSICS, LASER PHYSICS. *Educ:* Univ Calif, Los Angeles, BS, 69; Univ Rochester, MS, 71, PhD(phys chem), 74. *Prof Exp:* Statist physics, Univ Calif, San Diego, 74-76; staff laser physicist, Lawrence Livermore Lab, Univ Calif, 76-80; laser physicist & prog mgr, Rocketdyne Corp, 80-82; dept mgr, Allied Corp, 82-84; sect mgr, Northrop Electron Div, 84-87; prog mgr optical discrimination, Dept Navy, 87-90; CONSULT, 90- *Concurrent Pos:* Vis prof, Univ Mex, 73-76; res consult, Phys Dynamics Inc, 75-87; prof, Dept Appl Sci, Univ Calif, Davis, 78-79; appointee, Strategic Defense Initiative Orgn, interactive discrimination study, 86, laser radar study, 88, midcourse & terminal tier rev, 90. *Mem:* Laser Inst Am; Am Phys Soc. *Res:* Nonlinear deterministic and stochastic processes; laser systems design; electronic and vibrational structure of solids and amorphous materials; critical phenomena in liquids. *Mailing Add:* Dept Navy Naval Res Lab Washington DC 20375-5000

BUDIANSKY, BERNARD, b New York, NY, March 8, 25; m 52; c 2. SOLID MECHANICS, STRUCTURAL MECHANICS. *Educ:* City Col, BCE, 44; Brown Univ, PhD(appl math), 50. *Hon Degrees:* DSc, Northwestern Univ, 86. *Prof Exp:* Aeronaut res scientist, Nat Adv Comt Aeronaut, Va, 44-52, head, Struc Mech Br, 52-55; assoc prof struct mech, 55-61, PROF STRUCT MECH, HARVARD UNIV, 61- *Concurrent Pos:* Guggenheim fel, Tech Univ Denmark, 61; consult, Avco Corp, 58-70 & Gen Motors Res Lab, 63-78; mem, US Nat Comt Theoret & Appl Mech, 70-80, Mat Res Coun, Defense Advan Res Proj Agency, 68-, NASA Space Systs & Technol Adv Comt, 78-84 & Aeronaut & Space Eng Bd, Nat Res Coun, 85-90; vis prof, Technion, Haifa, 76; Sackler fel, Tel Aviv Univ, 83. *Honors & Awards:* Von Karman Medal, Am Soc Civil Engrs, 82; Eringen Medal, Soc Eng Sci, 85; Timoshenko Medal, Am Soc Mech Engrs, 89. *Mem:* Nat Acad Sci; Nat Acad Eng; fel Am Acad Arts & Sci; fel Am Soc Mech Engrs; fel Am Inst Aeronaut & Astronaut. *Res:* Elastic stability; plasticity; shell theory; fracture mechanics; composite materials. *Mailing Add:* Div Appl Sci Harvard Univ Cambridge MA 02138

BUDICK, BURTON, b Bronx, NY, May 22, 38; m 64; c 2. NUCLEAR PHYSICS, ATOMIC PHYSICS. *Educ:* Harvard Univ, AB, 59; Univ Calif, Berkeley, PhD(physics), 62. *Prof Exp:* Res assoc, Radiation Lab, Columbia Univ, 62-63, instr, 63-64; lectr, Hebrew Univ, Jerusalem, 64-67, sr lectr, 67-68; from asst prof to assoc prof, 68-81, PROF PHYSICS, NY UNIV, 81- *Concurrent Pos:* NSF fel Hebrew Univ, 65-67. *Mem:* Am Phys Soc. *Res:* Experimental nuclear physics; muonic and pionic atoms; atomic hyperfine structure; beta-decay. *Mailing Add:* Dept Physics NY Univ New York NY 10003

BUDINGER, THOMAS FRANCIS, b Evanston, Ill, Oct 25, 32; m 65; c 3. MEDICAL PHYSICS, NUCLEAR MEDICINE. *Educ:* Regis Col, Colo, BS, 54; Univ Wash, MS, 57; Univ Colo, MD, 64; Univ Calif, Berkeley, PhD, 71. *Prof Exp:* Asst chem, Regis Col, Colo, 53-54; anal chemist, Indust Labs, 54; sr oceanogr, Univ Wash, 61-66; physicist, Lawrence Livermore Lab, Univ Calif, 66-67, res physician, Donner Lab & Lawrence Berkeley Lab, 67-76, H MILLER PROF MED RES & GROUP LEADER RES MED, DONNER LAB, & PROF ELEC ENG & COMPUT SCI, UNIV CALIF, 76-, SR STAFF SCIENTIST, LAWRENCE BERKELEY LAB, 80- & PROF RADIOL, UNIV CALIF, SAN FRANCISCO, 84- *Concurrent Pos:* Peter Bent Brigham Hosp, Boston, 64; dir med serv, Lawrence Berkeley Lab, 68-76; chmn, study sect, NIH, 81-84. *Honors & Awards:* Special Award, Am Nuclear Soc, 84. *Mem:* AAAS; Am Geophys Union; NY Acad Sci; Soc Nuclear Med; Soc Magnetic Res Med (pres, 84-85). *Res:* Imaging body functions; electrical, magnetic, sound and photon radiation fields; electron microscopy; mathematics and computers in biology; polar oceanography; nuclear magnetic resonance; reconstruction tomography and instrument development; cardiology. *Mailing Add:* Lawrence Berkeley Lab Univ Calif Bldg 55 Berkeley CA 94720

BUDINSKI, KENNETH GERARD, b Rochester, NY, June 29, 39; m 62; c 3. TRIBOLOGY, MACHINE DESIGN. *Educ:* Gen Motors Inst, BS, 61; Mich Technol Univ, MS,63. *Prof Exp:* Develop engr, Rochester Prod Div, GMC, 60-62; metallurgist, 64-68, sr metallurgist, 68-85, TECH ASSOC, KODAK PARK DIV, EASTMAN KODAK CO, 85- *Concurrent Pos:* Consult, Mat Technol Inc, 65-70; instr, Rochester Inst Technol, 70-72; chmn, Rochester Chap, Am Soc Metals, 73; assoc prof, Monroe Community Col, 72-74; chmn, hardace ed comt, Welding Res Coun, 85, 62 comt wear & erosion, Am Soc Testing & Mat, 86-88; ed, Welding Inst, Surface Eng Soc. *Mem:* Fel Am Soc Metals; Am Soc Testing & Mat; Soc Automotive Engrs; Surface Eng Soc. *Res:* Tribological properties of engineering materials, particularly tool material and engineering plastics. *Mailing Add:* Mat Sci Div Eastman Kodak Co 5/23 Kodak Park Rochester NY 14652

BUDKE, CLIFFORD CHARLES, b Cincinnati, Ohio, Apr 3, 32; m 63; c 3. ANALYTICAL CHEMISTRY. *Educ:* Univ Cincinnati, BS, 54, MS, 67. *Prof Exp:* Anal chemist, 56-68, res supvr, 68-78, res assoc, 78-84, sr res assoc, US Indust Chem Co, Div Nat Distillers & Chem Corp, 84- 87; sr proj leader, 87-89, ASSOC SCIENTIST, USI DIV, QUANTUM CHEM CORP, 89- *Mem:* Am Chem Soc; hon fel Am Soc Testing & Mat. *Res:* Chemical and instrumental methods of analysis, especially trace analysis, functional group analysis, electrometric methods of analysis, and liquid chromatography. *Mailing Add:* Quantum Chem Corp USI Div 1275 Section Rd Cincinnati OH 45237

BUDNICK, JOSEPH IGNATIUS, b Jersey City, NJ, July 9, 29. PHYSICS. *Educ:* St Peters Col, BS, 51; Rutgers Univ, PhD(physics), 55. *Prof Exp:* Asst physics lab, Rutgers Univ, 51-52, asst atomic & nuclear lab, 52-53; proj physicist superconductivity, Res Ctr, IBM, NY, 55; from assoc prof to prof, Fordham Univ, 55-74; PROF PHYSICS, UNIV CONN, 74-, HEAD DEPT, 77- *Concurrent Pos:* Instr, IBM Gen Educ, 57-58. *Mem:* AAAS; Am Phys Soc. *Res:* Superconductivity; low temperature solid state physics of metals. *Mailing Add:* Dept Physics Rm 107 Univ Conn 2152 Hillside Rd Storrs CT 06269

BUDNITZ, ROBERT JAY, b Pittsfield, Mass, Oct 12, 40; m 61; c 3. ENVIRONMENTAL SYSTEMS AND TECHNOLOGY. *Educ:* Yale Univ, BA, 61; Harvard Univ, MA, 62, PhD(physics), 68. *Prof Exp:* Physicist, 67-71, coordr environ prog, 74-75, head , Energy & Environ Div, 74-75, assoc dir, 75-78, sr staff scientist environ sci, Lawrence Berkeley Lab, Univ Calif, 71-80; PRES, FUTURE RESOURCES ASSOC, INC, 81- *Concurrent Pos:* Mem, Reactor Safety Study Group, Am Phys Soc, 74-75, reactor hazards comt, Berkeley Res Reactor, Univ Calif, Berkeley, 74-78; ed-in-chief, CRC Forums Energy, CRC Press, West Palm Beach, Fla, 76-80; US rep, Comt on the Safety Nuclear Installations, Nuclear Energy Agency, Orgn Economic Coop & Develop, Paris, 79-80; tech coord, Spec Inquiry Group into the Three Mile Island, US Nuclear Regulatory Comn, 79-80; chmn, Peer Review Group, Dept Energy Off Crystalline Repository Develop, Battelle Prog Mgt Div, 83-86, Expert Panel Seismic Design Margins, Peer Review Expert Panel Plant Aging, Battelle Pac NW Lab, US Nuclear Regulatory Comn, 87-88; chmn, Peer Review Group, Off Crystalline Respository Develop, Batelle Proj Mgt Div, Dept Energy, 83-86; mem, explor comt future nuclear power generation, Nat Acad Sci, 84 & Energy Eng Bd, 83-86; mem, sr comt environ, safety & econ aspects fusion energy, US Dept Energy, 85-87; mem, Chernobyl Adv Panel, Elec Power Res Inst, 86-87, bd dirs, Pac Energy & Resources Ctr, 85-, Adv Group Treatment Hazards Probabilistic Safety Assessment, Int Atomic Energy Agency, Vienna, 86-; mem, US Army Chem Munitions Disposal Risk Assessment Rev Panel, GA Technol, Inc, 87-88 & Performance Assessment Peer Rev Panel, Dept Energy, Waste Isolation Pilot Plant, Sandia Nat Labs, 87- *Mem:* AAAS; Am Phys Soc; Am Nuclear Soc; Soc Risk Anal. *Res:* Instrumentation for measuring radiation in the environment; nuclear reactor environmental impact and safety analysis; radioactive waste disposal technology. *Mailing Add:* 734 Alameda Berkeley CA 94707

BUDNY, ROBERT VIERLING, b Honolulu, Hawaii, July 27, 42. PLASMA PHYSICS, CONTROLLED THERMONUCLEAR FUSION. *Educ:* Mass Inst Technol, BS, 63; Univ Paris, CES, 65; Univ Md, PhD(physics), 71. *Prof Exp:* Fel physics, Oxford Univ, 71-73 & Stanford Univ, 73-75; res assoc, Rockefeller Univ, 75-77; instr physics, 77-78, RES STAFF PLASMA PHYSICS, PRINCETON UNIV, 79- *Concurrent Pos:* Consult, US Dept Defense, 67; vis physicist, Aspen Inst Theoret Physics, 71, 76 & Conseil Europ Rescherche Nucleaire, 78; tutorial fel, Worcester Col, Oxford Univ, 72-73. *Mem:* Am Phys Soc; Am Vacuum Soc. *Res:* Limiter design observation and modeling of edge plasmas in Tokamaks unified theories of weak and electromagnetic interactions of elementary particles; electron-positron annihilations; neutrino scattering; observation and modeling of edge plasmas in Tokamaks. *Mailing Add:* Plasma Physics Lab Princeton Univ PO Box 451 Princeton NJ 08544

BUDOWSKI, GERARDO, b Berlin, Ger, June 10, 25; m 58; c 2. FORESTRY, PLANT ECOLOGY. *Educ:* Univ Venezuela, BS, 48; Inter-Am Inst Agr Sci, MS, 54; Yale Univ, PhD(forest ecol), 62. *Prof Exp:* Forester, Ministry Agr, Venezuela, 48-50, head dept forest res, 50-52, forester, Northern Zone, Tech Coop Prog, Orgn Am States, 53-55 & Forestry Prog, Inter-Am Inst Agr Sci, 56-57, head dept forestry, 57-67; prog specialist ecol & conserv, Natural Resources Res Div, UNESCO, Paris, 67-70; dir gen, Int Union Conserv Nature & Natural Resources, Switz, 70-76; head, Natural Renewable Resources Prog, Trop Agr Ctr Res & Training, 76-86; DIR, NATURAL RESOURCES, UNIV PEACE, 86- *Concurrent Pos:* Vis prof, Univ Calif, Berkeley, 67; chmn working group agroforestry, Int Union Forestry Res Orgn. *Mem:* Int Union Forestry Res Orgn; hon mem Soc Am Foresters; Int Soc Trop Foresters (hon vpres). *Res:* Tropical forest ecology and dendrology; silviculture; conservation; national park and forest education planning. *Mailing Add:* PO Box 199 Escazu 1250 Costa Rica

BUDRYS, RIMGAUDAS S, b Kaunas, Lithuania, Aug 2, 25; m 65; c 1. SURFACE CHEMISTRY. *Educ:* Univ Queensland, BSc, 59; Ill Inst Technol, PhD(phys chem), 64. *Prof Exp:* Chief chemist, Brisbane City Coun, Tennyson Power Sta, Queensland, Australia, 59-60; sr res chemist, res & develop ctr, Swift & Co, 63-65; sr res chemist, 65-67, ADV SCIENTIST, CONTINENTAL GROUP, INC, ILL, 68- *Mem:* AAAS; Am Chem Soc; Am Inst Physics; Sigma Xi. *Res:* Intermolecular bonding conformations and structural rearrangements in high polymers; characterization of metal surfaces; adhesion of polymers to metals; infrared spectroscopy of interfaces; physical chemistry of high polymers, coatings and surfaces. *Mailing Add:* 5833 S Sacramento Ave Chicago IL 60629

BUDZILOVICH, GLEB NICHOLAS, b Zlobin, Russia, Sept 7, 23; US citizen; m 50; c 1. PATHOLOGY, NEUROPATHOLOGY. *Educ:* Univ Munich, MD, 53. *Prof Exp:* Provisional asst pathologist, Med Ctr, Cornell Univ, 55-57; resident, Manhattan Vet Admin Hosp, 57-58, pathologist, 60-61; fel, NY Univ Med Ctr, 61-64; asst prof, 64-67, ASSOC PROF PATH & NEUROPATH, NY UNIV MED CTR, 67- *Concurrent Pos:* Am Cancer Soc fel, Francis Delafield Hosp, New York, 58-60, resident, 58-60; consult path, Manhattan Vet Admin Hosp, 63-; vis pathologist, Bellevue Hosp Med Ctr, New York, 65-; attend pathologist, Univ Hosp, NY Univ Med Ctr, 65-; neuropath consult, Off Med Examr, Suffolk County, NY, 67-; off chief med examr, NY. *Mem:* Am Asn Neuropathologists. *Res:* General pathology of peripheral sensory and sympathetic ganglia; pathology of sympathetic nervous system in diabetes mellitus; role of diabetic neuropathy in pathogenesis of peripheral vascular disease. *Mailing Add:* NY Univ Sch Med NY Univ Bellevue Hosp 234 Villard Ave Hastings on Hudson NY 10706

BUDZINSKI, WALTER VALERIAN, b Buffalo, NY, Aug 12, 37; m 65; c 3. EXPERIMENTAL SOLID STATE PHYSICS, SUPERCONDUCTIVITY. *Educ:* Canisius Col, BS, 59; Univ Pittsburgh, MS, 64, PhD(physics, 67. *Prof Exp:* Asst prof, 67-70, ASSOC PROF PHYSICS, ST BONAVENTURE UNIV, 70- *Concurrent Pos:* Vis assoc prof physics, Univ Pittsburgh, 74-75. *Mem:* Am Phys Soc; Sigma Xi; Am Asn Physics Teachers. *Res:* Tunneling into metals; microwave absorption by superconductors; thermal and optical properties of matter. *Mailing Add:* 110 E Oviatt St Olean NY 14760

BUECH, RICHARD REED, b Milwaukee, Wis, April 13, 40; m 68; c 3. BEHAVIOR, ECOLOGY & SOCIOBIOLOGY OF BEAVERS. *Educ:* Univ Mont, BS, 62; Yale Univ, MFS, 68; Univ Minn, PhD(wildlife), 88. *Prof Exp:* Res Forester, 64-69, res wildlife biologist, 69-81, proj leader, 81-84, PRIN WILDLIFE BIOLOGIST, USDA, FOREST SERV, 84- *Mem:* Am Soc Mammologists; Ecol Soc Am; Wildlife Soc. *Res:* Behavior, ecology & sociobiology of beavers; modeling wildlife-habitat relations; study of the spatial & temperal aspects of wildlife-habitat management; natural history of wood turtles. *Mailing Add:* 1499 Eighteenth Ave NW New Brighton MN 55112

BUECHE, FREDERICK JOSEPH, b Flushing, Mich, Aug 12, 23; m; c 2. SCIENCE EDUCATION. *Educ:* Univ Mich, BS, 44; Cornell Univ, PhD(physics), 48. *Prof Exp:* Asst eng physics, Cornell Univ, 44-48, res wildlife phys chem, 48-52; from asst prof to prof pysics, Univ Wyo, 52-59; physicist, Rohm and Haas Co, 53-54; prof physics, Univ Akron, 59-61; PROF PHYSICS, UNIV DAYTON, 61- *Concurrent Pos:* Peace Corps, 64-66. *Mem:* Am Phys Soc. *Res:* Physical properties of polymers; properties of liquids and gases; author of physics texts. *Mailing Add:* 45 Oakland Hills Pl Rotonda West FL 33947

BUECHLER, PETER ROBERT, chemistry, for more information see previous edition

BUECHNER, HOWARD ALBERT, internal medicine, for more information see previous edition

BUEGE, DENNIS RICHARD, b Juneau, Wis, Apr 2, 45; m 66; c 2. MEAT SCIENCE. *Educ:* Univ Wis-Madison, BS, 67, PhD(meat & animal sci), 75; Cornell Univ, MS, 69. *Prof Exp:* Sr food technologist, George A Hormel & Co, 75-77; EXTEN SPECIALIST MEATS, UNIV WIS, 77- *Concurrent Pos:* Military food inspector. 70-72. *Mem:* Inst Food Technologists; Am Meat Sci Asn. *Res:* Meat tenderness; basic causes and practical methods of improving; carry out extension education programs in meat technology aimed at meat processors, livestock producers and consumers; special interest in processed meats. *Mailing Add:* Univ Wis 1805 Linden Dr Madison WI 53706

BUEHLER, EDWIN VERNON, b Alliance, Ohio, Sept 23, 29; m 50; c 4. TOXICOLOGY. *Educ:* Ohio State Univ, BA, 51, MSc, 56, PhD(bact), 58; Am Bd Toxicol, dipl, 80; Acad Toxicol Sci, dipl, 84. *Prof Exp:* Bacteriologist, Res Div, Miami Valley Labs, Procter & Gamble Co, 58-61, res toxicologist, 61-63, group leader, 64-70, head toxicol, 70-78, head immunotoxicol, 78-84; dir toxicol, 84-87, VPRES SCI AFFAIRS, HILLTOP BIOLABS INC, 87- *Mem:* Soc Toxicol; Soc Cosmetic Chem. *Res:* Immunology; toxicology; immediate hypersensitivity; contact sensitization. *Mailing Add:* Hill Top Biolabs Inc PO Box 429501 Cincinnati OH 45242

BUEHLER, ROBERT JOSEPH, b Alma, Wis, May 1, 25; m 64; c 3. MATHEMATICAL STATISTICS. *Educ:* Univ Wis, BS, 48, MS, 49, PhD(math), 52. *Prof Exp:* Mem staff, Sandia Corp, NMex, 51-55; proj assoc chem & instr math, Univ Wis, 55-57; from asst prof to assoc prof, Iowa State Univ, 57-63; PROF STATIST, UNIV MINN, MINNEAPOLIS, 63- *Mem:* Math Asn Am; fel Am Statist Asn; fel Inst Math Statist; fel Royal Statist Soc; Int Statist Inst. *Res:* Statistical inference. *Mailing Add:* Sch Statist Vincent Hall Univ Minn Minneapolis MN 55455

BUEHRING, GERTRUDE CASE, b Chicago, Ill, May 28, 40; m 62; c 2. CANCER. *Educ:* Stanford Univ, BA, 62; Univ Calif, Berkeley, PhD(genetics), 72. *Prof Exp:* Fel breast cancer, Nat Cancer Inst, 72-73; asst prof, 73-80, ASSOC PROF MED MICROBIOL & TUMOR BIOL, SCH PUB HEALTH, UNIV CALIF, BERKELEY, 80- *Concurrent Pos:* Co-prin investr, Nat Cancer Inst Res Contract, 74-77; Systemside CRCC res grant, Univ Calif, 83-84; res grant, Elsa Pardee Found, 85-86. *Mem:* Sigma Xi; Am Asn Cancer Res; Tissue Cult Asn Am; Am Soc Microbiol; AAAS. *Res:* Human breast cancer - cell biology, virology, biochemistry, molecular biology, and genetics. *Mailing Add:* Sch Pub Health Warren Hall Univ Calif Berkeley CA 94720

BUELL, CARLETON EUGENE, mathematics, meteorology, for more information see previous edition

BUELL, DUNCAN ALAN, b Detroit, Mich, Oct 17, 50. NUMBER THEORY, COMPUTER SCIENCE. *Educ:* Univ Ariz, BS, 71; Univ Mich, MA, 72; Univ Ill, Chicago, PhD(math), 76. *Prof Exp:* Res assoc math, Carleton Univ, Ottawa, 76-77; asst prof comput sci, Bowling Green State Univ, 77-79; from asst prof to assoc prof, La State Univ, Baton Rouge, 79-85; res staff mem, 86-87, dir, algorithms res, 87-91, SR RES SCI, SUPERCOMPUT RES CTR, BOWIE, MD, 91- *Mem:* Asn Comput Mach; Am Math Soc; Inst Elec & Electronics Engrs; AAAS. *Res:* Computational number theory, parallel computations and computer retrieval systems. *Mailing Add:* Supercomput Res Ctr 17100 Sci Dr Bowie MD 20715

BUELL, GLEN R, b Lee's Summit, Mo, Jan 10, 31; m 61; c 2. ORGANIC CHEMISTRY, ANALYTICAL CHEMISTRY. *Educ:* Univ Mo, BS, 53, MA, 55; Univ Kans, PhD(chem), 61. *Prof Exp:* Res chemist, Pittsburgh Plate Glass Co, 61-62; vis res assoc organo-silicon, Ohio State Res Found, 62-63; res chemist, Aerospace Res Labs, 63-75, RES CHEMIST, AIR FORCE MAT LAB, WRIGHT-PATTERSON AFB, 75- *Mem:* Am Chem Soc; Sigma Xi. *Res:* Use of instrumental techniques for identification and characterization of organic and organo-metallic compounds with respect to structure and bonding. *Mailing Add:* 3548 Eastern Dr Dayton OH 45432

BUELL, KATHERINE MAYHEW, anatomy, morphology, for more information see previous edition

BUELOW, FREDERICK H(ENRY), b Minot, NDak, Mar 13, 29; m 54; c 4. AGRICULTURAL ENGINEERING. *Educ:* NDak Agr Col, BS, 51; Purdue Univ, MSE, 52; Mich State Univ, PhD(agr eng), 56. *Prof Exp:* From asst prof to prof agr eng, Mich State Univ, 56-66; chmn dept, 66-83, PROF AGR ENG, UNIV WIS-MADISON, 66- *Concurrent Pos:* Mem Eng Accreditation Comn, Accreditation Bd Eng & Technol, 80-85. *Mem:* Fel Am Soc Agr Engrs; Am Soc Eng Educ. *Res:* Agricultural building environment research; instrumentation for agricultural engineering research; agricultural materials handling. *Mailing Add:* Dept Agr Eng 460 Henry Mall Univ Wis Madison WI 53706

BUENING, GERALD MATTHEW, b Decatur Co, Ind, Sept 14, 40; m 64; c 2. VETERINARY IMMUNOLOGY, VETERINARY VIROLOGY. *Educ:* Iowa State Univ, MS, 66; Purdue Univ, DVM, 64, PhD(vet virol), 69; Am Col Vet Microbiologists, dipl, 70. *Prof Exp:* Instr vet microbiol, Purdue Univ, 66-69; asst prof, 69-72, assoc prof, 72-81, PROF VET MICROBIOL, UNIV MO-COLUMBIA, 81- *Honors & Awards:* Beecham Award for Res Excellence, Mo Vet Med, 87. *Mem:* Am Vet Med Asn; Conf Res Workers Animal Dis; Am Soc Microbiol; Asn Am Vet Cols. *Res:* Cell-mediated immunity of domestic animals; in vitro methods are utilized to follow the cell-mediated immunity response during disease processes and recovery; bovine hemotrophic diseases; biotechnology. *Mailing Add:* Dept Vet Microbiol Univ Mo Columbia MO 65211

BUESCHER, BRENT J, b Galesburg, Ill, Sept 15, 40; m 65; c 2. NUCLEAR REACTOR SAFETY RESEARCH. *Educ:* Univ NC, BS, 62; Univ Ariz, PhD(physics), 69. *Prof Exp:* Asst, Univ NC, 61-62 & Univ Ariz, 62-69; res assoc physics, Rensselaer Polytech Inst, 69-71; assoc, Argonne Nat Lab, 71-73; prin engr, Power Generation Group, Babcock & Wilcox Co, 73-79; SR ENG SPECIALIST, IDAHO NAT ENG LAB, EG&G IDAHO INC, 79- *Concurrent Pos:* US Nuclear Res Ctr deleg, Karlsruhe Nuclear Res Ctr; mem, ANS 5-4 subcomt on Plenum Gas, ANS 5-3 subcomt on Fission Prod Release. *Mem:* Am Phys Soc; Am Nuclear Soc. *Res:* Performance of nuclear reactor fuel under various conditions; normal (non-accident) reactor operation and behavior during accidents; radiation effects on ceramic and metal reactor fuel components; reactor safety and nuclear plant aging. *Mailing Add:* EG&G Idaho Inc PO Box 1625 Idaho Falls ID 83415-7121

BUESKING, CLARENCE W, b Ft Wayne, Ind, Apr 10, 19; m 43; c 4. MATHEMATICS. *Educ:* Ball State Univ, BA, 47; Purdue Univ, MS, 52. *Prof Exp:* Teacher math, Tri-State Col, 47, teacher math & sci, Burris Lab Sch, Ball State Univ, 47-49 & pub schs, Ind, 50-57; prof math & dir comput ctr, 57-78, head comput sci dept, 70-78, ASSOC PROF MATH, UNIV EVANSVILLE, 78- *Mem:* Math Asn Am; Asn Comput Mach. *Res:* Undergraduate mathematics; computer programming. *Mailing Add:* 4617 Taylor Ave Evansville IN 47714

BUETOW, DENNIS EDWARD, b Chicago, Ill, June 20, 32; m 60; c 4. CELL BIOLOGY. *Educ:* Univ Calif, Los Angeles, AB, 54, MA, 57, PhD(zool), 59. *Prof Exp:* Biochemist, Baltimore City Hosps & biologist, biophys lab, NIH, Md, 59-65; assoc prof, 65-70, head, Dept Physiol & Biophys, 83-88, PROF PHYSIOL, UNIV ILL, URBANA-CHAMPAIGN, 70- *Concurrent Pos:* NIH & NSF res grants, 73- *Mem:* Fel AAAS; Am Inst Biol Sci; Am Soc Cell Biol; Am Physiol Soc; fel Geront Soc; Sigma Xi. *Res:* Molecular biology of chloroplasts and mitochondria; protein biosynthesis in subcellular organelles; biology of aging. *Mailing Add:* Dept Physiol & Biophys 530 Burrill Hall Univ Ill 407 S Goodwin Urbana IL 61801

BUETTNER, GARRY RICHARD, b Vinton, Iowa, Oct 11, 45; c 4. PHYSICAL CHEMISTRY, PHOTOBIOLOGY. *Educ:* Univ Northern Iowa, BA, 67; Univ Iowa, MS, 69, PhD(chem), 76. *Prof Exp:* Fel radiation biol, Radiation Res Lab, Univ Iowa, 76-78; asst prof chem, Wabash Col, 78-83; Fulbright fel, Neuherberg, WGer, 85-87; EXPERT, NAT INST ENVIRON HEALTH SCI & NIH, RES TRIANGLE PARK, 87- *Concurrent Pos:* Vis prof, Med Col Wis, 81. *Mem:* Am Chem Soc; Sigma Xi; Am Soc Photobiol; Soc Free Radical Res; Oxygen Soc. *Res:* Superoxide and superoxide dismutase in human health. *Mailing Add:* EMRB 68/ESR Cir Univ Iowa Col Med Iowa City IA 52242

BUETTNER, MARK ROLAND, b Cottonwood, Idaho, Sept 9, 49; m 78; c 3. AGRONOMY, ANIMAL SCIENCE. *Educ:* Univ Idaho, BS, 72, MS, 75; Purdue Univ, PhD(agron), 78. *Prof Exp:* Asst prof agron, Mountain Meadow Res Ctr, Colo State Univ, 78-79; asst prof agron, Klamath Exp Sta, Ore State Univ, 79-; AGRONOMIST, BRANDT CONSOL. *Concurrent Pos:* From asst dir to dir limnol, Student Originated Studies, NSF, 71-72. *Mem:* Am Soc Agron; Crop Sci Soc Am. *Res:* Legume establishment and production; grass production; meadow and range species; forage physiology; digestion coefficients; total N content; cell wall; lignin; lignin-hemicellulose; infra-red absorbance; nuclear magnetic resonance. *Mailing Add:* Brandt Consol PO Box 277 Pleasant Plains IL 62677

BUFE, CHARLES GLENN, b Duluth, Minn, Jan 2, 38; m 67; c 3. SEISMOLOGY, TECTONICS. *Educ:* Mich Tech Univ, BS, 60, MS, 62; Univ Mich, PhD(geol), 69. *Prof Exp:* Res asst geophys, Willow Run Labs, Univ Mich, 64, res assoc, 67-69, assoc res geophysicist, 69; geophysicist, US Earthquake Mech Lab, Nat Oceanic & Atmospheric Admin, 69-73; GEOPHYSICIST, OFF EARTHQUAKES, VOLCANOES & ENG, US GEOL SURV, 73- *Concurrent Pos:* Vis assoc prof, Univ Wis, Milwaukee, 73; geothermal liaison, Dept Energy, Washington, DC, 80-83; tech adv to Kingdom Saudi Arabia, Riyadh, 85-86; prin investr seismicity, Yucca Mountain Proj, Nev, 88-89. *Mem:* Seismol Soc Am; Am Geophys Union; Soc Explor Geophys; Earthquake Eng & Res Inst. *Res:* Travel times and spectra of short-period body waves; seismicity of geothermal areas; micro-earthquake and fault-creep studies; earthquake recurrence, prediction and control; induced seismicity; plate tectonics. *Mailing Add:* US Geol Surv MS 967 Box 25046 DFC Denver CO 80225

BUFF, FRANK PAUL, b Munich, Ger, Feb 13, 24; nat US; m 56; c 2. PHYSICAL CHEMISTRY, STATISTICAL MECHANICS. *Educ:* Univ Calif, BA, 44; Calif Inst Technol, PhD(chem), 49. *Prof Exp:* from instr to assoc prof chem, 50-61, PROF CHEM, UNIV ROCHESTER, 61- *Concurrent Pos:* AEC fel, Calif Inst Technol, 49-50; NSF sr fel, Inst Theoret Physcis, Utrecht, Netherlands, 59-60; consult, Scony mobil Co, Inc, Mobil Oil Corp. *Mem:* AAAS; Am Chem Soc; fel Am Phys Soc; Sigma Xi. *Res:* Molecular theories of fluids; surface phenomena; solutions and chemical kinetics; nucleation processes; electrochemistry. *Mailing Add:* Dept Chem River Campus Sta Univ of Rochester Rochester NY 14627

BUFFALOE, NEAL DOLLISON, b Leachville, Ark, Nov 15, 24; m 47; c 5. BIOLOGY. *Educ:* David Lipscomb Col, BS, 49; Vanderbilt Univ, MS, 52, PhD(biol), 57. *Prof Exp:* Instr biol, David Lipscomb Col, 49-54 & Vanderbilt Univ, 54-56; prof, 57-87, EMER PROF BIOL, UNIV CENT ARK, 88- *Concurrent Pos:* Instr biol, George Peabody Col, 55. *Mem:* AAAS. *Res:* Cytology and cytogenetics. *Mailing Add:* Box 1721 Univ Cent Ark Conway AR 72032

BUFFETT, RITA FRANCES, b Beverly, Mass, Jan 18, 17. BIOLOGY, PHYSIOLOGY. *Educ:* Boston Univ, AB, 47, AM, 48, PhD(biol), 54. *Prof Exp:* Instr biol, Boston Univ, 50-54; res asst exp path, Children's Cancer Res Found, Mass, 54-59; res scientist exp path, 59-61, res scientist viral oncol, 61-69, asst & assoc res prof microbiol, Roswell Park Mem Inst, 69-80; RETIRED. *Mem:* AAAS; Am Soc Microbiol; Am Asn Cancer Res; NY Acad Sci; Sigma Xi. *Res:* Oncogenesis; radiation, hormonal and viral. *Mailing Add:* 1782 Sharon Dr Concord CA 94519

BUFFINGTON, ANDREW, b Fall River, Mass, Dec 25, 38; m 68; c 2. ASTROPHYSICS. *Educ:* Mass Inst Technol, BS, 61, PhD(physics), 66. *Prof Exp:* Res asst physics, Mass Inst Technol, 66-68; res physicist, Space Sci Lab & Lawrence Berkeley Lab, Univ Calif, Berkeley, 68-79; res assoc, Calif Inst Technol, Pasadena, 79-84; res physicist, Univ Calif, San Diego, 84-86. *Mem:* Sigma Xi; Am Phys Soc; Am Astronaut Soc. *Res:* Measurement of cosmic-ray, nuclear and isotopic abundances; interpretation of these to deduce cosmic-ray history; search for cosmic ray antimatter and measurement of antiproton flux; development of flexible telescope techniques to restore atmospherically perturbed images from astronomical telescopes; development of photoelectric astrometric telescopes. *Mailing Add:* 3166 Bremerton Pl La Jolla CA 92037-2211

BUFFINGTON, EDWIN CONGER, b Ontario, Calif, Aug 21, 20; m 43; c 4. MARINE GEOLOGY. *Educ:* Carleton Col, BA, 41; Calif Inst Technol, MSc, 47; Univ Southern Calif, PhD, 73. *Prof Exp:* Marine geologist, USN Electronics Lab, 48-69; head, Marine Geol Br, Navy Undersea Res & Develop Ctr, 69-74; assoc chief, Marine Geol Br, US Geol Surv, 74-80; RETIRED. *Concurrent Pos:* Dir, Gen Oceanog, Inc. *Mem:* Fel Geol Soc Am; Soc Econ Paleontologists & Mineralogists; Am Asn Petrol Geologists; Asn Eng Geol; Am Geophys Union; Sigma Xi; fel AAAS. *Res:* Bathymetry; sea floor geomorphology; plate tectonics. *Mailing Add:* 1015 Devonshire Dr San Diego CA 92107

BUFFINGTON, F(RANCIS) S(TEPHAN), materials science; deceased, see previous edition for last biography

BUFFINGTON, JOHN DOUGLAS, b Jersey City, NJ, Nov 26, 41; m 65; c 2. ECOLOGY, FISH & WILDLIFE SCIENCES. *Educ:* St Peter's Col, NJ, BS, 63; Univ Ill, Urbana-Champaign, MS, 65, PhD(zool), 67. *Prof Exp:* Asst prof, dept biol sci, Ill State Univ, 69-72; biologist, Argonne Nat Lab, 72-77, asst div dir, 76-77; sr staff, Coun Environ Qual, Washington, DC, 77-80; chief, Off Biol Serv, 80-82, dep assoc dir res & develop, 82-89, REGIONAL DIR RES & DEVELOP, US FISH & WILDLIFE SERV, 89- *Concurrent Pos:* Adj assoc prof, Northern Ill Univ, 75-77; dir, WAEPA, 83- *Mem:* AAAS; Ecol Soc Am. *Res:* Fish and wildlife management. *Mailing Add:* Res & Develop Dept Fish & Wildlife Serv Washington DC 20240

BUFFLER, CHARLES ROGERS, b Bryn Mawr, Pa, Apr 8, 34; m 56; c 3. MICROWAVE PROCESSING. *Educ:* Univ Tex, BS, 55; Harvard Univ, MS, 56, PhD(eng & appl physics), 60. *Prof Exp:* Res scientist, Varian Assocs, 59-66; mgr, Ferrite Eng, Microwave Assocs, 66-68; pres, Microwave Magnetics, 68-73; assoc dir, Advan Eng Resources, Leeds & Northrop, 73-75; dir res & develop, Litton Microwave Cooking, 75-85; PRES, ASSOC SCI RES FOUND, INC, 85- *Concurrent Pos:* Dir, Int Microwave Power Inst, 77-82; expert witness, Litton Microwave Cooking, 85-; consult, Assoc Sci Res Found, 85- *Mem:* Fel Int Microwave Power Inst; Bioelectromagnetics Soc; sr mem Inst Elec & Electronics Engrs. *Res:* Microwave processing, oven design and microwave interaction with foods. *Mailing Add:* Assoc Sci Res Found Inc 126 Water St Marlborough NH 03455-9701

BUFFLER, PATRICIA ANN HAPP, b Doylestown, Pa, Aug 1, 38; m 62; c 2. EPIDEMIOLOGY, PUBLIC HEALTH. *Educ:* Cath Univ Am, BSN, 60; Univ Calif, Berkeley, MPH, 65, PhD(epidemiol), 73. *Prof Exp:* Asst prof epidemiol, Sch Pub Health, Univ Tex, 70-72; dir, Ctr Health Sci & asst prof health sci, Alaska Methodist Univ, 73-74; dir, Epidemiol Res Unit & asst prof, Dept Prev Med & Community Health, Univ Tex Med Br, 74-78; ASSOC PROF EPIDEMIOL, SCH PUB HEALTH, UNIV TEX HEALTH SCI CTR, 79- *Concurrent Pos:* Consult to various pvt, pub & govt agencies, 73-79; mem, Med Adv Bd, Tex Air Qual Control Bd, 77-79. *Mem:* Soc Epidemiol Res; Am Pub Health Asn; Am Soc Prev Oncol; NY Acad Sci. *Res:* Chronic disease epidemiology; cancer; trauma; mental illness and occupational diseases. *Mailing Add:* Health Sci Ctr Univ Tex 1100 Holcombe Houston TX 77225

BUFFLER, RICHARD THURMAN, b Troy, NY, Nov 4, 37; m 62; c 2. MARINE GEOLOGY, MARINE GEOPHYSICS. *Educ:* Univ Tex, Austin, BS, 59; Univ Calif, Berkeley, PhD(geol), 67. *Prof Exp:* Geologist explor res, Shell Oil Co, 67-71; assoc prof geol, Univ Alaska, 71-74; RES SCIENTIST MARINE GEOL & GEOPHYS, UNIV TEX INST GEOPHYS, 75- *Mem:* Am Asn Petrol Geologists; Geol Soc Am. *Res:* Sedimentology; stratigraphy; depositional environments; petroleum geology; seismic (sequence) stratigraphy. *Mailing Add:* Inst Geophys Univ Tex 8701 Mopac Blvd Austin TX 78759-8345

BUFFUM, C(HARLES) EMERY, b Toulon, Ill, June 14, 10; wid; c 4. ELECTRICAL ENGINEERING. *Educ:* Calif Inst Technol, BS, 31, MS, 32. *Prof Exp:* Asst test engr, Consol Steel Corp, Calif, 33-34; computer, Western Geophys Co, 34-35, party chief in charge seismog field party, 35-38; seismog party chief, Pan Am Petrol Corp, 38-41, res engr, Amoco Prod Co, 41-44, tech group supvr, 44-46, supvr design, Construct & Tech Serv Sect, 46-49, res group supvr, 49-50, res sect supvr, 50-52, lab serv supt, 53-72; RETIRED. *Mem:* Soc Explor Geophys; Inst Elec & Electronics Engrs. *Res:* Design of seismograph amplifiers, oscillography, radio transmitters and auxiliary equipment. *Mailing Add:* 9524 E 71st St No 316 Tulsa OK 74133-5218

BUFFUM, DONALD C, b Narberth, Pa, Apr 13, 18; m 43; c 2. PHYSICAL METALLURGY. *Educ:* Tufts Univ, BS, 39. *Prof Exp:* Phys sci aide, Army Mat & Mech Res Ctr, 40-42, jr metall engr, 42, metall engr, 42-44, physicist, 46-51, phys metallurgist, 51-55, supvry phys metallurgist, 55-78; RETIRED. *Res:* Metals joining; alloy development of ferrous and titanium alloys; the temper brittleness of steel and welding of metals and alloys. *Mailing Add:* Seven Ivy Circle Arlington MA 02174

BUFKIN, BILLY GEORGE, b Columbia, Miss, Oct 8, 46; m 66; c 3. POLYMER SCIENCE & TECHNOLOGY. *Educ:* Univ Southern Miss, BS, 68, PhD(org chem), 72. *Prof Exp:* from asst prof to assoc prof polymer sci, Univ Southern Miss, 71-80, dept chmn, 76-80; SR VPRES SCI AFFAIRS, DAP, INC, 81-; VPRES RES, CONSUMER DIV, SHERWIN-WILLIAMS CO. *Concurrent Pos:* Consult, Tenn Eastman Co, 73-76. *Mem:* Am Chem Soc; Fedn Soc Coatings Technol; Soc Plastics Engrs. *Res:* Crosslinkable emulsions; emulsion polymerization; antifouling materials which function by controllable release mechanisms; utilization of renewable resources; organic chemistry. *Mailing Add:* Sherwin-Williams Co 601 Canal Rd Cleveland OH 44113

BUGENHAGEN, THOMAS GORDON, b Derby, NY, Dec, 16, 32; m 56; c 3. QUEUEING THEORY, TECHNICAL MANAGEMENT. *Educ:* Maryville Col, BA, 56; Univ Tenn, MA, 59. *Prof Exp:* Instr math, Maryville Col, 57-58, Univ Tenn, 58-59; mathematician opers anal, 59-81, GROUP SUPVR OPERS ANAL, APPL PHYSICS LAB, JOHNS HOPKINS UNIV, 81- *Concurrent Pos:* Prin staff mem, Appl Physics Lab, Johns Hopkins Univ, 80. *Honors & Awards:* David Rist Prize, Mil Opers Res Soc, 76. *Mem:* Fel AAAS; Opers Res Soc Am; Mil Opers Res Soc. *Res:* Application of operations research technique, queueing theory, to the solution of operational problems for anti-air warfare, Kalman filters; for ship tracking algorithms and simulations for determining requirements. *Mailing Add:* Appl Physics Lab Johns Hopkins Univ Johns Hopkins Rd Laurel MD 20723-6099

BUGG, CHARLES EDWARD, b Durham, NC, June 5, 41; m 62. PHYSICAL CHEMISTRY. *Educ:* Duke Univ, AB, 62; Rice Univ, PhD(phys chem), 65. *Prof Exp:* Res fel, Calif Inst Technol, 65-66; res chemist, Dacron Res Lab, E I du Pont de Nemours & Co, 66-67; res fel, Calif Inst Technol, 67-68; asst prof biochem, 68-74, assoc prof biochem, 74-77, PROF BIOCHEM, UNIV ALA MED CTR, 77-, ASSOC DIR, UNIV ALA COMPREHENSIVE CANCER CTR, 76- *Mem:* Am Crystallog Asn; Sigma Xi. *Res:* Crystal structures of compounds of biological interest. *Mailing Add:* Dept Biochem Box 79THT-UAB Sta Birmingham AL 35294

BUGG, STERLING L(OWE), b Harrodsburg, Ky, June 12, 20; m 47; c 3. CIVIL ENGINEERING. *Educ:* Univ Ky, BS, 44; Purdue Univ, MS, 48. *Prof Exp:* Eng aide, Ky Hwy Dept, 42; mat engr, Dept Hwys, 45-46; asst, Purdue Univ, 46-48; asst prof civil eng, Univ Fla, 48-52; struct engr, 52-55, supvr struct res engr, 55-63, head dept civil eng, 63-69, head dept ocean eng, 69-75, HEAD DEPT CIVIL ENG, US NAVAL CIVIL ENG LAB, 75- *Mem:* Am Soc Civil Engrs (pres, 58); Am Concrete Inst; Sigma Xi. *Res:* Engineering materials; highway design; research structure. *Mailing Add:* 600 Janetwood Dr Oxnard CA 93030

BUGG, WILLIAM MAURICE, b St Louis, Mo, Jan 23, 31; m 54; c 4. PHYSICS. *Educ:* Wash Univ, AB, 52; Univ Tenn, PhD(physics), 59. *Prof Exp:* From asst prof to assoc prof, 59-69, PROF PHYSICS & HEAD DEPT, UNIV TENN, KNOXVILLE, 69- *Concurrent Pos:* Consult, Oak Ridge Nat Lab, 59-79, Danforth Assocs, 69- *Mem:* Fel Am Phys Soc; Inst Elec & Electronics Engrs. *Res:* SSC detector research on silicon detectors; electron positron interactions at high energy. *Mailing Add:* Dept Physics Univ Tenn Knoxville TN 37996-1200

BUGGS, CHARLES WESLEY, b Brunswick, Ga, Aug 6, 06; m 27; c 1. BACTERIOLOGY. *Educ:* Morehouse Col, AB, 28; Univ Minn, MS, 32, PhD(bact), 34. *Prof Exp:* Instr biol, Dover State Col, 28-29; prof chem, Bishop Col, 34-35; prof biol & chmn, div sci, Dillard Univ, 35-43; from instr to assoc prof bact, Sch Med, Wayne Univ, 43-49; prof biol & chmn, div sci, Dillard Univ, 49-56; prof microbiol, Col Med, Howard Univ, 56-71, head dept, 58-70; proj dir, Fac Allied Health Sci, Charles R Drew Postgrad Med Sch, Univ Calif, Los Angeles, 69-72, dean, 72; prof microbiol, Calif State Univ, Long Beach, 73-83; RETIRED. *Concurrent Pos:* Rosenwald fel, Woods Hole, 43; Off Sci Res & Develop, Wayne Univ, 43-47; vis prof, Sch Med, Univ Calif, Los Angeles, 69-72 & Univ Southern Calif, 69-76. *Mem:* Am Soc Microbiol; fel Am Acad Microbiol. *Res:* Resistance of bacteria to antibiotics. *Mailing Add:* 5600 Verdun Ave Los Angeles CA 90043

BUGLIARELLO, GEORGE, b Trieste, Italy, May 20, 27; US citizen; c 2. SCIENCE ADMINISTRATION. *Educ:* Univ Padua, Dott Ing BS, 51; Univ Minn, MS, 54; Mass Inst Technol, ScD, 59. *Hon Degrees:* Dr, Univ Trieste, 89. *Prof Exp:* Fulbright scholar, Univ Minn, 52-54; asst to chair hydraul proj, Univ Padua, 54-55; res asst, Hydrodynamics Lab, Mass Inst Technol, 56-59, res asst, 59; from asst prof to assoc prof civil eng, Carnegie-Mellon Univ, 59-66, chmn, Biotech Prog, 64-69, prof biotechnol & civil eng, 66-69; dean eng & prof civil eng & biotechnol, dept systs eng, Univ Ill Chicago Circle, 69-73; PRES, POLYTECH UNIV, 73- *Concurrent Pos:* NATO sr postdoc fel, Tech Univ Berlin, 68; pub health serv spec fel, sci policy res, Harvard Univ, 68; chmn, comt fluid dynamics, Am Soc Civil Engrs, 66-68, task comt biol flows,

67-71, task comt problem-oriented lang, 67-72, task comt civil eng biomed & health care systs, 69-73, mem, res comt, 69-73, adv bd & Res Coun Comput Pract, 72-76; chair, ad hoc comt use comput & comput lang hydraul, Int Asn Hydraul Res, 69-75; mem, eng design comt, Am Soc Eng Educ, 72-73, conf theme comt, 73 & comt govt rel, 72-75; mem, Comn Educ, Nat Acad Eng, 70-73 & chmn, comt educ systs, 71-73, joint Nat Acad Eng & Nat Acad Sci comt role of US eng sch tech foreign assitance, 71-74, steering comt, Modern technol methods improving qual educ, 78, int affairs adv comt, 88-; mem, US Panel joint comt coop prog on sci & technol planning, Nat Res Coun, 82 & Comn Int Rel, 81-82; mem, bio med eng training comt, Nat Inst Gen Med Sci, 66-70 & proj comt, Gen Med Res Prog, 71-74; consult, Chemother Prog, Nat Cancer Inst, 72-74; mem, comt sci & pub policy, NY Acad Sci, 75; mem, adv comt sci educ, NSF, 79-80 & chair, 80-81; co-chair, prog planning & agenda subcomt, comt sci, eng & pub policy & conf prog & tours comt, AAAS, 84, review panel Philip Hauge Abelsm Prize, 85, chair, Panel Phys Sci & Eng, Carnegie & AAAS Proj 2061, 85- & comt sci, eng & pub policy, 86-; sci adv panel, Armed Forces Safety Explosives Bd, Dept Defence, 68-69; univ prog panel, Energy Res Adv Bd, Dept Energy, 82-83; chair, adv panel technol transfer to Mid E, Off Technol Assessment, 82-84; US rep, Prog Sci Stability, NAtlantic Treaty Orgn, 84-; eval comt, Nat Med Technol Nomination, Dept Com, 87-; secy, Coun Sci & Technol Develop, 77-82; consult, State Univ NY, Buffalo, 71; various positions different comt, nat & int orgn, 69-; mem Mayor's Coun Environ, 74-77, Comn Sci & Technol, 84-, chair, 87-, Task Force, Water Conserv, 85-88. *Honors & Awards:* Huber Res Prize, Am Soc Civil Engrs, 67; Alza lectr, Biomed Eng Soc, 76; Golden San Giusto Award, Trieste, Italy, 78; Centennial Award, Am Soc Mech Engrs, 80. *Mem:* Nat Acad Sci; Nat Acad Eng; fel Am Soc Eng Educ; Int Asn Hydraul Res; NY Acad Sci; fel AAAS; Am Arbitration Asn; char mem Cardiovasc Syst Dynamics Soc; Asn Independent Technol Univ (pres, 86-88); Acad Educ Develop; Int Asn Sci Parks (vpres, 85-); Int Soc Hemorheol (secy, 66-69); char mem, Int Water Resources Asn; Italian Hydrotech Asn; Sigma Xi (pres-elect, 90); Soc Natural Philos; Soc Rheol; Soc Col & Univ Planning. *Res:* Author of 10 books on engineering; over 200 publications on engineering and related subjects. *Mailing Add:* Polytech Univ 333 Jay St Brooklyn NY 11201

BUGNOLO, DIMITRI SPARTACO, b Atlantic City, NJ, Feb 3, 29; m 58; c 3. STOCHASTIC WAVE PROPAGATION. *Educ:* Univ Pa, BSEE, 52; Yale Univ, MEng, 55; Columbia Univ, ScD(elec eng), 60. *Prof Exp:* Computer components res, Burroughs Res Lab, 52-54; res asst, Yale Univ, 54-56; instr elec eng, Columbia Univ, 56-60; consult to lab dir & mem tech staff, Bell Tel Labs, 60-61, 63-66; asst prof elec eng, Columbia Univ, 61-63, sr res scientist, Hudson Labs, 66-67; prin engr adv res, Submarine Signal Div, Raytheon Corp, 67-69; consult engr & physicist, 69-80; assoc prof elec & comput eng, Fla Inst Technol, 80-83; ASSOC PROF ELEC & COMPUT ENG, UNIV ALA, HUNTSVILLE, 83- *Concurrent Pos:* Mem comn, Int Sci Radio Union, 64-68; consult & full prof, Nat Inst Sci Res, Univ Quebec, 71-73. *Mem:* Am Phys Soc; sr mem Inst Elec & Electronics Eng; Sigma Xi; Optical Soc Am. *Res:* Applied plasma physics; wave propagation in random media; plasma turbulence; propagation theory; electromagnetic theory and practice; antenna theory and practice; atmospheric turbulence; computer simulation. *Mailing Add:* Mitre Corp Burlington Rd MS N101 Bedford MA 02018

BUGOSH, JOHN, b Cleveland, Ohio, July 1, 24; m 52; c 3. PHYSICAL CHEMISTRY, COLLOID CHEMISTRY. *Educ:* Heidelberg Col & Adelbert Col, BS, 45; Western Reserve Univ, MS, 47, PhD(chem), 49; Escuela Interamericana de Verano, dipl, 48. *Prof Exp:* Res assoc ultrasonics, Western Reserve Univ, 48-49, lectr physics, 49-50; res supvr, Del, 50-68, planning consult, Pa, 68-72, DEVELOP MGR, E I DU PONT DE NEMOURS & CO, INC, 72-, CONSULT, 89- *Honors & Awards:* Award, Am Chem Soc, 62. *Mem:* AAAS; Am Chem Soc. *Res:* Ultrasonic transducers; electrolytic solutions, inorganic colloids; determination of masses of ions by ultrasonics; new form of film forming fibrous boehmite. *Mailing Add:* 1071 Squire Cheney Dr West Chester PA 19382-8046

BUHAC, IVO, b Dubrovnik, Yugoslavia, Sept 4, 26; m 62; c 1. GASTROENTEROLOGY, INTERNAL MEDICINE. *Educ:* Univ Zagreb, MD, 52, ScD, 64; Univ Erlangen, MD, 62. *Prof Exp:* Intern med, Zagreb, Yugoslavia, 52-53; resident, 57-60, staff physician med & gastroenterol, 62-68; instr med, Med Col Va, 69-70; from asst prof to assoc prof, 70-82, PROF MED, ALBANY MED COL, 82-; CHIEF GASTROENTEROL, ALBANY VET ADMIN HOSP, 70- *Concurrent Pos:* Alexander von Humboldt fel, Med Sch, Univ Hamburg, 64-65; fel gastroenterol, Vet Admin Hosp, Richmond, Va, 68-70. *Mem:* Am Gastroenterol Asn; Am Asn Study Liver Dis; fel Am Col Physicians; NY Acad Sci; Am Soc Gastrointestinal Endoscopy. *Res:* Peritoneal permeability; portal hypertension; albumin distribution. *Mailing Add:* Div Gastroenterol Albany Med Col Albany NY 12208

BUHKS, EPHRAIM, b Kishinev, USSR, Apr 30, 49; m 83; c 2. EDUCATIONAL ADMINISTRATION. *Educ:* Kishinev Univ, USSR, BS, 71; Tel-Aviv Univ, Israel, PhD(chem), 80. *Prof Exp:* Res fel photoelectrochem, Physics Dept, Univ Del, 80-81; proj leader photovoltaics, Solavolt Int, Shell Oil Co, 81-83; proj mgr opto-electronic mat, B F Goodrich Res & Develop Ctr, 83-87; tech dir IR imaging & detectors, Sunstone Inc, 87-89; ASST DIR COL ADMIN, ORT OPERS USA, ORT TECH INSTS, 90- *Concurrent Pos:* Consult biophysics, Johnson Res Found, Univ Pa, 80-83, IR detectors, Belor Technol Co, Inc, 87-89, fiber-optics, Kingston Technol, Inc, 89, SBIR, Energia Inc, 90. *Honors & Awards:* Von Humboldt Found Award, 80. *Mem:* Optical Soc Am; Phys Soc Am; Int Soc Optical Eng. *Res:* Electron transfer processes; photosynthesis; photoelectrochemistry; opto-electronic materials; solar cells; luminescent materials; IR imaging; IR detectors; fiber-optics; organic electronic materials; conducting polymers; photoresists; electrochromics; photochromics; optical data storage. *Mailing Add:* ORT Opers USA 200 Park Ave S New York NY 10003

BUHL, ALLEN EDWIN, b Evanston, Ill, Nov 10, 47. ZOOLOGY, PHYSIOLOGY. *Educ:* Carthage Col, BA, 69; Univ Ill, MS, 71, PhD(zool), 76. *Prof Exp:* Res assoc reprod physiol, Ore Regional Primate Res Ctr, 77-78, asst scientist, 78-80; scientist reprod physiol, 81-85, res scientist reprod physiol, 85-87, SR RES SCIENTIST CARDIOVASC RES, UPJOHN CO, 87-, SR RES SCIENTIST HAIR GROWTH RES, 87- *Mem:* Soc Investigative Dermat; Soc Study Reprod; AAAS. *Res:* Mensenchymal-epithelial interactions in organ development and function; control of growth and differentiation of cutaneous structures especially hair. *Mailing Add:* Hair Growth Res Upjohn Co Kalamazoo MI 49001

BUHL, DAVID, b Newark, NJ, Nov 20, 36; m 62; c 1. RADIO ASTRONOMY. *Educ:* Mass Inst Technol, BS & MS, 60; Univ Calif, Berkeley, PhD(elec eng), 67. *Prof Exp:* Electronics engr, Lawrence Radiation Lab, 61-64; from asst scientist to scientist, Nat Radio Astron Observ, 67-74; SPACE SCIENTIST, NASA GODDARD SPACE FLIGHT CTR, 74- *Mem:* Inst Elec & Electronics Engrs; Am Geophys Union; Am Astron Soc; Int Astron Union. *Res:* Radio astronomy of the sun, moon, and planets; observations of complex molecules in interstellar clouds; chemical evolution of the interstellar medium; infrared molecular astronomy; infrared receiver technology. *Mailing Add:* Code 693 Planetary Systs NASA Goddard Space Flight Ctr Greenbelt MD 20771

BUHL, ROBERT FRANK, b Utica, NY, Apr 20, 22; m 45; c 5. CORROSION, PHYSICS. *Educ:* Ohio State Univ, BS, 43. *Prof Exp:* Res asst, Ohio State Univ Res Found, Off Sci & Res Develop, Nat Defense Res Comt, 44-45; res assoc, Pure Oil Co, Ill, 45-61, group supvr, 62-66; res physicist, 66-67, sr res physicist, 67-75, res assoc, res ctr, 75-84, SR RES ASSOC, SCI & TECHNOL DIV, UNOCAL CORP, 84- *Mem:* Nat Asn Corrosion Eng; Inst Noise Control Eng. *Res:* Analytical instrument design and construction; spectro- metric analysis method development for lubes, fuels and petro-chemicals; infrared and ultraviolet analysis; corrosion control and coatings evaluation; industrial noise control; nondestructive testing. *Mailing Add:* 1191 N Richman Ave Fullerton CA 92635

BUHLE, EMMETT LOREN, b Moline, Ill, May 7, 18; m 47; c 4. INDUSTRIAL ORGANIC CHEMISTRY. *Educ:* Johns Hopkins Univ, AB, 41, MA, 42, PhD(chem), 48. *Prof Exp:* Res assoc, Mass Inst Technol, 48-50; res asst pharmacol, Johns Hopkins Univ, 50-54, instr, 52-54; chemist, Exp Sta, E I du Pont de Nemours & Co, 54-58; sr res chemist, 58-82, prod develop coordr, Wyeth Lab, 82-84; RETIRED. *Concurrent Pos:* Mem surv antimalarial drugs, Nat Res Coun, 42-46. *Mem:* Am Chem Soc; AAAS. *Res:* Antimalarial drugs; synthetic steroids; semisynthetic penicillins; prostaglandins; development of chemical processes. *Mailing Add:* 714 Hemlock Rd Media PA 19063

BUHLER, DONALD RAYMOND, b San Francisco, Calif, Oct 11, 25; m 76. BIOCHEMICAL PHARMACOLOGY. *Educ:* Ore State Col, BA & BS, 50, Ore State Univ, PhD(biochem), 56. *Prof Exp:* Asst, Ore State Col, 51-55; staff biochemist, div exp med, Med Sch, Univ Ore, 55-56, USPHS fel, 56-58; biochemist, western fish nutrit lab, US Fish & Wildlife Serv, 58-59, Upjohn Co, Mich, 59-64, western fish nutrit lab, US Fish & Wildlife Serv, 64-66 & Pac Northwest Labs, Battelle Mem Inst, 66-68; assoc prof, 68-74, PROF, DEPT AGR CHEM, ENVIRON HEALTH SCI CTR, ORE STATE UNIV, 74-, CHMN TOXICOL PROG, 83- *Concurrent Pos:* Mem, Toxicol Study Sect, NIH, 77-81. *Mem:* Am Chem Soc; Am Soc Biol Chemists; Soc Toxicol; Am Soc Pharm Exp Therapeut. *Res:* Biochemical mechanisms for toxicity; drug metabolism and pharmacokinetics; synthesis of radioactive compounds; fate of chemicals in the environment; heavy metal toxicity; biochemistry of fishes. *Mailing Add:* Dept Agr Chem Ore State Univ Corvallis OR 97331

BUHR, ROBERT K, METALLURGY. *Educ:* McGill Univ, BS, 51. *Prof Exp:* Head, Foundry Sect, Metals Technol Labs, Can Met, 53-89; RETIRED. *Mem:* Fel Am Soc Metals. *Mailing Add:* Seven Tiffany Crescent Kanata ON K2K 1W1 Can

BUHRMAN, ROBERT ALAN, b Waynesboro, Pa, Apr 24, 45; m 72; c 3. ELECTRON DEVICE, SUPERCONDUCTIVITY. *Educ:* Johns Hopkins Univ, BES, 67; Cornell Univ, MS, 70, PhD(appl physics), 73. *Prof Exp:* From asst prof to assoc prof, 73-83, assoc dir, Nat Res & Resource Facil, Submicron Struct, 80-83, PROF APPL & ENG PHYSICS, CORNELL UNIV, 83- *Mem:* Am Phys Soc. *Res:* Superconducting quantum devices; low temperature properties of metals; size effects in metals; optical properties of composite materials; submicron electron and x-ray lithography; high temperature superconductivity; defects states in semiconductors and insulators. *Mailing Add:* Sch Appl & Eng Physics Clark Hall Cornell Univ Ithaca NY 14853

BUHSE, HOWARD EDWARD, JR, b Chicago, Ill, July 22, 34; m 57; c 3. ZOOLOGY. *Educ:* Grinnell Col, AB, 57; Univ Iowa, MS, 60, PhD(zool), 63. *Prof Exp:* Res fel microbiol, State Univ NY Upstate Med Ctr, 63-65; asst prof biol sci, 65-68, assoc prof, 68-77, PROF BIOL SCI, UNIV ILL, CHICAGO CIRCLE, 77- *Concurrent Pos:* NIH fel, 63-65; NSF develop biol grant, 67-69. *Mem:* Soc Protozool; Sigma Xi. *Res:* Morphogenesis in the ciliated protozoa. *Mailing Add:* Dept Biol Sci Univ Ill Box 4348 Chicago IL 60680

BUHSMER, CHARLES P, b Wilkes-Barre, Pa, Sept 4, 37; m 65; c 2. CERAMICS, PHYSICAL CHEMISTRY. *Educ:* Kings Col, Pa, BS, 59; Pa State Univ, MS, 62; State Univ NY Col Ceramics, Alfred, PhD(ceramic sci), 68. *Prof Exp:* Res scientist, Airco-Speer Res Labs, Niagara Falls, 62-64, 67-70, mgr carbon & graphite res, 70-71; assoc scientist, Dexter Hysol, 71-73; new venture mgr, Horizons Res Inc, 73-75; DIR ABRASIVE TECH CTR, CARBORUNDUM CO, 76- *Honors & Awards:* Medal, Am Chem Soc, 59. *Mem:* Am Ceramic Soc; NY Acad Sci; Am Chem Soc. *Res:* Kinetics of solid state reactions; diffusion in the solid state; high temperature composite materials; thick film microelectronics; carbon and graphite; abrasive components and systems. *Mailing Add:* 6915 Lexington Ct E Amherst NY 14051

BUI, TIEN DAI, b Bac Ninh, Vietnam, Mar 22, 45; Can citizen; m 69; c 3. COMPUTER SCIENCE. *Educ:* Univ Saigon, BSc, 64; Univ Ottawa, BASc, 68; Carleton Univ, MEng, 68; York Univ, PhD(space sci), 71. *Prof Exp:* Res engr, Inst Aerospace Studies, Univ Toronto, 68-69; res assoc & asst prof lasers & comput, Dept Mech Eng, McGill Univ, 71-74; from asst prof to assoc prof, 74-85, PROF COMPUTER SCI, CONCORDIA UNIV, 85-, CHMN 86- *Concurrent Pos:* Prof Asn Res & Develop Fund assoc, McGill Univ, 71-74; vis prof, Inst Appln Calcolo, Romo, 78-79, Univ Calif Berkeley, 83-84. *Mem:* Brit Inst Physics; Soc Comput Simul; Asn Comput Mach; Soc Indust Appl Math; Am Math Soc. *Res:* Computational methods for analysis of complex systems; numerical modeling; optimization and simulation of various physical phenomena; numerical analysis. *Mailing Add:* Dept Comput Sci Concordia Univ 1455 De Maisonneuve St W Montreal PQ H3G 1M8 Can

BUI, TIEN RUNG, b Hanoi, Vietnam, Feb 18, 35; Can citizen; m; c 4. CONTROL SYSTEMS. *Educ:* Ecole Navale, Brest, France, Eng, 56; Naval Postgrad Sch, PhD, 64. *Prof Exp:* Prof eng, Naval Acad, Nhatrang, Vietnam, 64-67 & Nat Tech Ctr, Saigon, Vietnam, 67-70; rep, Econ Comn Asia & Far East, Bangkok, Thailand, 70-74; prof eng, Univ Que, 75-90; CHAIR PROF, NATURAL SCI & ENG-RES COUN CAN, ALCAN, 90- *Concurrent Pos:* Louis Beauchamp sci res award, Quebec, 90. *Mem:* Metall Soc; Sigma Xi; Inst Elec & Electronics Engrs; Can Inst Mining & Metall. *Res:* Mathematical modeling of industrial processes of a thermo-hydrodynamic nature; industrial process control. *Mailing Add:* Dept Appl Sci Univ Que Chicoutimi PQ G7H 2B1 Can

BUIE, BENNETT FRANK, b Patrick, SC, Jan 9, 10; m 38; c 4. ECONOMIC GEOLOGY. *Educ:* Univ SC, BS, 30; Lehigh Univ, MS, 32; Harvard Univ, MA, 34, PhD(petrog, struct geol), 39. *Prof Exp:* Asst geol, Harvard Univ, 32-37; geologist, Seaboard Oil Co, NY & Amiranian Oil Co, Iran, 37-38; geologist, Indian Oil Concessions, Ltd, Standard Oil Co Calif & affil co, 39-46; prof, Univ SC & geologist, State Develop Bd, SC, 46-56; prof, 56-80, head dept, 56-61, chmn, 61-64, EMER PROF GEOL, FLA STATE UNIV, 81- *Concurrent Pos:* Geologist in chg, Persian Gulf Command, US Army, 43-45; Fulbright res award, Iran, 51; chief geologist, Resources Develop Corp, Iran, 52; geologist, US Geol Surv, 53-57; consult geol, J M Huber Corp, 58-83 & 88- *Honors & Awards:* Order Red Star, USSR. *Mem:* Sigma Xi; fel Geol Soc Am; Am Asn Petrol Geol; Soc Mining Engrs, Am Inst Mining Metall & Petrol Eng; Soc Econ Geologists; fel Mineral Soc Am. *Res:* Economic geology; petrography; industrial minerals; economic geology of mineral deposits, especially industrial minerals; world resources of kaolin clays and phospate rock; optical mineralogy; tritium in ground water; dewatering of Florida phospate slimes. *Mailing Add:* RR 1 Box 2 Bakersville NC 28705

BUIKEMA, ARTHUR L, JR, b Evergreen Park, Ill, Feb 4, 41; m 62; c 11. AQUATIC TOXICOLOGY. *Educ:* Elmhurst Col, BS, 62; Univ Kans, MA, 65, PhD(zool), 70. *Prof Exp:* Asst prof biol, St Olaf Col, 67-71; from asst to assoc prof, 71-80, asst dir, Univ Ctr Environ Studies, 78-83, PROF ZOOL, VA POLYTECH INST & STATE UNIV, 80- *Concurrent Pos:* Sr ecologist, Ecol Soc Am. *Honors & Awards:* G Burke Johnson "Renaissance Man" Award. *Mem:* AAAS; Am Inst Biol Sci; Am Soc Testing Mat; Ecol Soc Am; NAm Benthol Soc. *Res:* Pollution biology; ecology and physiology of invertebrates; bioassay technique development; environmental physiology. *Mailing Add:* Dept Biol Va Polytech Inst & State Univ Blacksburg VA 24061

BUIKSTRA, JANE ELLEN, b Evansville, Ind, Nov 2, 45. BIOLOGICAL ANTHROPOLOGY, ARCHAEOLOGY. *Educ:* DePauw Univ, BA, 67; Univ Chicago, MA, 69, PhD(anthrop), 72. *Prof Exp:* From instr to prof anthrop, Northwestern Univ, Evanston, 70-86, assoc dean, Col Arts & Sci, 81-84; HAROLD H SWIFT DISTINGUISHED SERV PROF ANTHROP, UNIV CHICAGO, 86- *Concurrent Pos:* NIH bio-med sci res grant, Northwestern Univ, 71; NSF res grants, 74-76, 77-79, 83-85, 88-89 & 90-92; Wenner-Gren Found grant, 78-79; assoc ed, Am J Phys Anthrop, 78-81; resident scholar, Sch Am Res, 84-85; adj prof anthrop, Wash Univ, 86- *Mem:* Nat Acad Sci; Am Anthrop Asn; Am Asn Phys Anthrop (secy-treas, 81-85, pres, 85-87); fel Am Acad Forensic Sci; Soc Prof Archaeologists; fel AAAS; Sigma Xi. *Res:* Intensive regional approach to the study of prehistoric skeletal populations emphasizing micro-evolutionary change and biological response to environmental stress; author of numerous technical publications. *Mailing Add:* Dept Anthrop Univ Chicago 1126 E 59th St Chicago IL 60637

BUIS, OTTO J, b St Joseph, Mo, June 9, 31; wid; c 4. GEOLOGY, GEOPHYSICS. *Educ:* Southwestern La Inst, BS, 53; La State Univ, MS; 58. *Prof Exp:* Geophysicist, La Oil Explor Co, 53-54; dist geologist, Texaco Inc, Houston & Bogota, Colombia, 59-68; vpres & bd dir, Shenandoah Oil Corp, Ft Worth & Quito, Ecuador, 68-79; pres & bd dir, OKC Corp, 79-81; pres & bd dir, CBK Assoc, 81-89; PRES & CHIEF EXEC OFFICER, MSR EXPLOR, LTD, 89- *Mem:* Ecuadorian Geol & Geophys Soc (founder & pres, 69); Colombian Soc Geologists & Geophysicists (pres, 67). *Mailing Add:* 3800 Encanto Dr Ft Worth TX 76109

BUIS, PATRICIA FRANCES, b Jersey City, NJ, Dec 29, 53. COAL GEOCHEMISTRY. *Educ:* Rutgers Univ, BA, 76; Queens Col, MA, 83; Univ Pittsburgh, PhD(geol), 87. *Prof Exp:* Environ technician, Pennrun Corp, 88-89; geologist, Pa geol Surv, 89-91; PVT CONSULT GEOL & ENVIRON SCI, 91- *Concurrent Pos:* Adj lectr geol, Buhl Sci Mus, 85-88; environ intern, US Bur Mines, 87. *Mem:* Sigma Xi; Am Minerologist; Geol Soc Am. *Res:* X-ray diffraction analysis of coal ash for mineral components and chemistry, statistical analysis of S and ash, allowing the differentiation of different coal beds; four publications. *Mailing Add:* 285 E Main St Apt 5 Middletown PA 17057

BUIST, ALINE SONIA, b June 27, 40; m; c 3. PULMONARY EPIDEMIOLOGY. *Educ:* Univ St Andrews, Scotland, MD, 64. *Prof Exp:* PROF MED & HEAD, PULMONARY & CRIT CARE DIV, ORE HEALTH SCI UNIV, 66- *Mem:* Am Thoracic Soc (pres); Am Physiol Soc; Am Fedn Clin Res; Am Col Epidemiol. *Res:* Pulmonary epidemiology. *Mailing Add:* Dept Med Ore Health Sci Univ 3181 SW Sam Jackson Park Rd Portland OR 97201

BUIST, NEIL R M, b Karachi, India, July 11, 32; UK citizen; c 3. AMINO ACID DISORDERS, MUSCLE DISEASES. *Educ:* Univ St Andrews, Scotland, MB & ChB, 56; Inst Child Health, London, DCH, 60; Royal Col Physicians, Edinburgh, MRCPE, 61; FRCP(Edinburgh), 77. *Prof Exp:* House officer internal med, St Andrews Univ, 56-57, resident, 60-62, lectr child health, 62-64; fel metab, Med Sch, Univ Colo, 64-66; asst prof pediat, Med Sch, Univ Ore, 66-70; assoc prof pediat & med genetics, 70-76, PROF PEDIAT & MED GENETICS, ORE HEALTH SCI UNIV, 76- *Concurrent Pos:* House officer pediat, Royal Hosp Sick Children, Edinburgh, 57; vis scientist, Univ Montreal, 77-78; overseas adv, Soc Study Inborn Errors of Metab, 76-; ed, Kellys Pract Pediat, 80-; travelling pediatrician, Int relief work with Refugees, Kampuchea, 80, Ethiopia & Sudan, 85. *Mem:* Soc Inherited Metab Dis (treas, 78-); Soc Study Inborn Errors of Metab; Soc Pediat Res; Am Pediat Soc; Brit Med Asn. *Res:* Recognition, diagnosis, evaluation and treatment of inherited or acquired metabolic diseases; pediatrics; metabolism; inborn errors of metabolism. *Mailing Add:* Pediat Metab Lab L473 Ore Health Sci Univ 3181 SW San Jackson Pk Rd Portland OR 97201-3098

BUITING, FRANCIS P, b Belfeld, Neth, Jan 25, 24; nat US; m 52; c 3. ELECTRONICS ENGINEERING, INDUSTRIAL CONTROLS. *Educ:* Technol Univ Delft, Engr, 52. *Prof Exp:* Engr, Philips Telecommun Indust, Neth, 50-55, dept chief elec mech & instrumentation, 55-56; proj engr, AMP, Inc, Pa, 56-58; prin engr, Fenwal, Inc, Mass, 58-60, chief develop engr, 60-61, chief electronics engr, 61-64; mgr res & develop, Metals & Controls Inc, Div Tex Instruments, Inc, 64-69; vpres & gen mgr, Matrix Res & Develop Corp, 69-71; vpres & gen mgr, Europe for Transitron Electronic Corp, 72-75 & Transitron Electronics Corp, 75-77; vpres & gen mgr, Com Prod Sparton Electronics Div, Sparton Corp, 77-80; VPRES BUS PLANNING & RES & DEVELOP, THOMAS & BETTS CORP, 80-, VPRES EUROPE, THOMAS & BETTS INT. *Mem:* Inst Elec & Electronics Engrs; Neth Royal Inst Eng. *Res:* Development and design of automatic test equipment; development of temperature monitoring, indicating and control equipment; infrared sensing and flame and explosion protection equipment; aerospace power supplies; semiconductor manufacturing; electronic sensing systems; interconnect systems. *Mailing Add:* Thomas & Betts Corp 1001 Frontier Rd Bridgewater NJ 08807

BUJAKE, JOHN EDWARD, JR, b New York, NY, May 23, 33; m 64; c 4. PHYSICAL CHEMISTRY, FOOD SCIENCE. *Educ:* Manhattan Col, BS, 54; Col Holy Cross, MS, 55; Columbia Univ, PhD(phys chem), 59; NY Univ, MBA, 63. *Prof Exp:* Res assoc phys chem & new prod develop, Lever Brothers Co, NY, 59-68; develop assoc, Foods Div, Coca-Cola Co, 68-69, mgr tech serv, 69-71, dir, Res & Develop, 71-72; dir res & develop technol, Quaker Oats Co, 72-75, dir res & develop, Foods Div, 75-77; dir, Seven Up Co, 77-78, vpres res & develop, 78-87; VPRES RES & DEVELOP, BROWN-FORMAN BEVERAGE CO, 87- *Mem:* Am Chem Soc; Inst Food Technol; Indust Res Inst. *Res:* Alcoholic beverages, soft drinks, cereal, citrus, coffee, frozen foods, baked goods, snack protein food product development; processing and packaging; colloid chemistry; rheology, reaction kinetics. *Mailing Add:* Brown Forman Beverage Co PO Box 1080 Louisville KY 40201-1080

BUKACEK, RICHARD F, b Cedar Rapids, Iowa, Nov 3, 27; m 52; c 7. CHEMICAL ENGINEERING. *Educ:* Univ Iowa, BS, 49; Ill Inst Technol, MS & MGasTechnol, 51, PhD(chem eng), 60. *Prof Exp:* Engr, North Shore Gas Co, 51-52; res supvr, Inst Gas Technol, Ill Inst Technol, 53-57, instr chem eng, 57-60; asst prof, 60-62; sr engr, Air Prod & Chem, Inc, 62-63; chmn educ prog, Inst Gas Technol, Ill Inst Technol, 63-65, adj assoc prof gas technol & dir educ, 65-80. *Mem:* Am Inst Chem Engrs. *Res:* Fluid mechanics; thermodynamics; systems analysis. *Mailing Add:* 3424 S State St Chicago IL 60616

BUKANTZ, SAMUEL CHARLES, b New York, NY, Sept 12, 11; m 41; c 2. CLINICAL MEDICINE. *Educ:* NY Univ, BS, 30, MD, 34; Am Bd Int Med, dipl; Am Bd Allergy, dipl. *Prof Exp:* From instr to assoc prof med, Sch Med, Wash Univ, 47-54, assoc prof clin med, 54-58, asst dean, Sch Med, 48-54; med & res dir, Jewish Nat Home Asthmatic Children & Children's Asthma Res Inst & Hosp, Denver & assoc prof clin med, Med Ctr, Univ Colo, 58-63; assoc prof clin med, Sch Med, NY Univ, 64-72; chief, 72-84, EMER CHIEF SECT ALLERGY, MED SERV, VET ADMIN HOSP, TAMPA, 84-; PROF MED, UNIV SFLA, 72- *Concurrent Pos:* Baruch fel, Harlem Hosp, New York, 38-40; fel allergy, Internal Med Dept, Sch Med, Wash Univ, 46-47; sr attend physician, Jefferson Barracks Vet Hosp, 46-48; chmn res comt, Vet Admin Hosp, St Louis, 52-54, secy dean's comt, 51-54; consult, Beth Israel Hosp, Newark, NJ, 63- & Clara Maass Hosp, Belleville, 63-; assoc dir med res clin invest, Schering Corp, 63-65; assoc vis physician, Bellevue Med Ctr, NY, 64-72; dir dept clin res, Hoffmann-La Roche Inc, NJ, 65-67; ed, Hosp Pract, 69- *Mem:* Am Soc Exp Biol; AMA; fel Am Col Physicians; Am Col Chest Physicians; Am Asn Immunol; fel Am Acad Allergy & Immunol. *Res:* Clinical investigation with new drugs; experimental hyper-sensitive states; pathogenesis and therapy of bronchial asthma; immunochemistry. *Mailing Add:* 4940 San Rafael St Tampa FL 33629-5435

BUKER, ROBERT JOSEPH, b Vancouver, Wash, June 2, 30; m 52; c 5. PLANT BREEDING. *Educ:* Wash State Univ, BS, 53; Purdue Univ, MS, 59, PhD(plant breeding, genetics), 63. *Prof Exp:* Instr agron, Purdue Univ, 56-61; dir res, Farmers Forge Res Coop, FFR Coop, 61-73, exec vpres &gen mgr, 73-; res dir & prof, Univ Wyo, Baidoa Somalia, 85-87; PLANT BREEDER, INT PROG AGR, OHIO STATE UNIV, KAMPALA, UGANDA, 89- *Mem:* Fel Am Soc Agron; Am Forage & Grassland Coun (pres); Nat Coun Com Plant Breeders (pres); Sigma Xi. *Res:* Breeding forage crop varieties; directing forage, turf, soybean and corn breeding programs; improving field plot techniques and equipment. *Mailing Add:* Int Progs Agr Ohio State Univ 113 Agron Admin Bldg Columbus OH 43210-1099

BUKHARI, AHMAD IQBAL, b Punjab, India, Jan 5, 43. MOLECULAR BIOLOGY, GENETICS. *Educ:* D J Col, Karachi, BSc, 61; Univ Karachi, MSc, 63; Brown Univ, MS, 66; Univ Colo, PhD(microbiol), 70. *Prof Exp:* Asst lectr microbiol, Univ Karachi, 63-64; staff investr, Cold Spring Harbor Lab,

72-75, sr staff investr, 75-78; asst prof, 74-79, ASSOC PROF MICROBIOL, STATE UNIV NY, STONY BROOK, 79-; SR SCIENTIST, COLD SPRING HARBOR LAB, 78- Concurrent Pos: Fel, Cold Spring Harbor Lab, 70-71, Jan Coffin Childs fel, 71-72; NSF career develop award, 75- Mem: AAAS; Am Soc Microbiol; Genetics Soc Am. Res: DNA rearrangement; transposable elements and plasmids; genetic engineering; intracellular protein turnover. Mailing Add: Cold Spring Harbor Lab Cold Spring Harbor NY 11724

BUKOVAC, MARTIN JOHN, b Johnston City, Ill, Nov 12, 29; m 56; c 1. HORTICULTURE, PLANT PHYSIOLOGY. Educ: Mich State Univ, BS, 51, MS, 54, PhD, 57. Prof Exp: Res asst, 54-56, res assoc, 56-57, from asst prof to assoc prof, 57-63, PROF HORT, MICH STATE UNIV, 63- Concurrent Pos: Lectr, Japan Atomic Energy Res Inst, Tokyo, 58; adv, Int AEC, Vienna, 61; mem, subcomt effect of pesticides on physiol fruits & veg, Nat Acad Sci-Nat Res Coun, 64-66; NSF sr fel, Oxford Univ & Univ Bristol, 65-66; Nat Acad Sci exchange visitor, Coun Acad, Yugoslavia, 71; guest lectr, Polish Acad Sci, Warsaw & Serbian Sci Coun, Fruit Res Inst, Cacak, Yugoslavia, 79; mem, US Nat Comn, Int Union Biol Sci, 71-76, Agr Res Adv Comt, Eli Lilly, 71-88; distinguished vis prof, NMex State Univ, Las Cruces, 76; vis prof, Japan Soc Promotion Sci, Univ Osaka, 77 & Univ Guelph, Ont, Can, 82, Ohio State Univ, 90; US rep, Coun Int Soc Hort Sci, 77-78; guest res, Hort Res Inst, Budapest, Hungary, 83; vis lectr, Inst Hort, Univ Zagreb, Yugoslavia, 83; distinguished lectr, dept sci & technol, Min Agr, Animal Husbandry & Fisheries, People's Repub China, 84. Honors & Awards: Joseph Harvey Gourley Award, Am Soc Hort Sci, 69 & 76, M A Blake Award & Marion W Meadows Award, 75, Carrol R Miller Award, 80 & Outstanding Res Award, 88; Dennis R Hoagland Award, Am Soc Plant Physiologists, 88. Mem: Nat Acad Sci; fel AAAS; fel Am Soc Hort Sci (pres, 74-75); Am Hort Soc; Am Soc Plant Physiol; Bot Soc Am; Am Chem Soc. Res: Plant growth and development; plant growth substances; mechanisms of foliar penetration. Mailing Add: Dept Hort Mich State Univ East Lansing MI 48823

BUKOVSAN, LAURA A, b Norfolk, Va, Jan 2, 40; m 64; c 2. GENETICS, MICROBIOLOGY. Educ: Univ Richmond, BS, 61; Ind Univ, MA, 63, PhD(genetics), 69. Prof Exp: Teaching asst zool, Ind Univ, 62-63; lectr genetics, State Univ NY, Oneota, 74-78; lectr biol, Hartwick Col, 78-80; from asst prof to assoc prof, 80-90, PROF MICROBIOL, STATE UNIV NY, ONEONTA, 90- Mem: Genetics Soc Am; AAAS; Am Soc Microbiol; Soc Protozoologists. Res: Cytology of conjugation in Tokophrya lemnarum; genetics of mating type inheritance in the suctorian Tokophrya lemnarum; biology of aging in Tokophrya lemnarum. Mailing Add: Dept Biol State Univ NY Oneonta NY 13820

BUKOVSAN, WILLIAM, b Chicago, Ill, June 4, 29; m 64; c 2. ENDOCRINOLOGY. Educ: Univ Ill, BS, 54, MS, 58; Ind Univ, PhD(zool), 67. Prof Exp: PROF BIOL, STATE UNIV NY COL ONEONTA, 67- Mem: AAAS; Am Soc Zool; NY Acad Sci; Sigma Xi. Res: Adrenal phosphorus metabolism of the newly hatched chick. Mailing Add: Dept Biol State Univ NY Oneonta NY 13820

BUKOWSKI, RICHARD WILLIAM, b Chicago, Ill, July 4, 47; m 68; c 1. FIRE PROTECTION ENGINEERING. Educ: Ill Inst Technol, BS, 70. Prof Exp: Sr proj engr, Underwriters Labs, Inc, 70-75; res engr, Nat Bur Standards, 75-80, head hazard anal, 80-89, MGR TECHNOL TRANSFER, CTR FIRE RES, NAT INST STANDARDS & TECHNOL 89- Honors & Awards: Silver Medal, Dept of Com, 90. Mem: Soc Fire Protection Engrs; Int Asn Fire Safety Sci; Nat Fire Protection Asn. Res: Assessment of the hazards to building occupants from fires. Mailing Add: Ctr Fire Res Nat Inst Standards & Technol Gaithersburg MD 20899

BUKREY, RICHARD ROBERT, b St Paul, Minn, June 26, 41; m 64; c 2. SOLID STATE PHYSICS. Educ: St Mary's Col, Minn, BA, 63; Marquette Univ, MS, 65; Wayne State Univ, PhD(physics), 72. Prof Exp: Asst prof, 72-76, chmn dept 76-85, ASSOC PROF PHYSICS, LOYOLA UNIV, 76- Concurrent Pos: Undergrad res partic prog dir, NSF, 76-77, instrnl sci equip prog proj dir, 77-79. Mem: Am Phys Soc. Res: Structure and properties of amorphus magnetic materials, studies by Mossbauer effect; x-ray diffraction; magnetic susceptibility; electron spin resonance. Mailing Add: Dept Physics Loyola Univ 6525 N Sheridan Rd Chicago IL 60626

BUKRY, JOHN DAVID, b Baltimore, Md, May 17, 41. MICROPALEONTOLOGY. Educ: Johns Hopkins Univ, AB, 63; Princeton Univ, AM, 65, PhD(geol), 67. Prof Exp: Geologist, US Army CEngr, 63; res asst micropaleont, Socony-Mobil Oil Co, 65; geologist, US Geol Surv, La Jolla, 67-84, geologist, US Minerals Mgt Serv, 84-86, GEOLOGIST, US GEOL SURV, MENLO PARK, 86- Concurrent Pos: Biostratig consult, Deep Sea Drilling Proj, 68-87; res assoc, Scripps Inst Oceanog, 70-; ed, Marine Micropaleont, 76-83, Micropaleont, 85-90. Mem: Fel AAAS; fel Geol Soc Am; Am Asn Petrol Geologists; fel Explorers Club; Sigma Xi; Int Nannoplankton Asn. Res: Phytoplankton micropaleontology and ocean stratigraphy; evolutionary trends, paleoecology, and taxonomy of calcareous nannoplankton and silicoflagellates (300 new species described). Mailing Add: US Geol Surv MS-915 345 Middlefield Rd Menlo Park CA 94025

BULA, RAYMOND J, b Antigo, Wis, Aug 3, 27; m 52; c 8. PLANT PHYSIOLOGY, ECOLOGY. Educ: Univ Wis, BS, 49, MS, 50, PhD(agron, bot), 52. Prof Exp: Asst prof, seed analyst & agronomist, NY Agr Exp Sta, Geneva, 52-53; agronomist, Agr Res Serv, Agr Exp Sta, USDA, Alaska, 53-56 & Crops Res Div, Purdue Univ, 56-74; area dir, Agr Res Serv, USDA, 74-79; dir, US Dairy Forage Res Ctr, Agr Res Serv, Univ Wis-Madison, USDA, 79-84; dir res, Phytofarms Am, Dekalb, Ill, 84-86; SR SCIENTIST & EXEC DIR, WIS CTR SPACE AUTOMATION & ROBOTICS, UNIV WIS-MADISON, 86- Mem: Fel AAAS; fel Crop Sci Soc Am; fel Am Soc Agron; Am Soc Plant Physiol. Res: Forage crop production; weed control; physiology of cold resistance; forage crop physiology; influence of environment on genetic stability and plant growth and development; controlled environment growth of plants; hydroponics; space biology. Mailing Add: 7872 Deer Run Rd Cross Plains WI 53528

BULANI, WALTER, b Sask, July 7, 28; m 51; c 5. PHYSICAL CHEMISTRY, CHEMICAL ENGINEERING. Educ: Univ Sask, BScChEng, 49, MSc, 50; McGill Univ, PhD(phys chem), 55. Prof Exp: Res officer chem, Can Chem Co, 55-56 & E I du Pont de Nemours & Co, Inc, 56-59; asst prof eng sci, 59-64, assoc prof chem eng, 64-67, chmn dept, 64-70, PROF CHEM ENG, UNIV WESTERN ONT, 67- Concurrent Pos: Mem air pollution comt, Can Standards Asn, 70-; secy, treas & vchmn develop, Can Nat Comt, Int Asn Water Pollution Res & Control, 72-84. Mem: Am Inst Chem Engrs; Chem Inst Can; Can Soc Chem Engrs; Am Soc Eng Educ; NY Acad Sci; Can Asn Water Pollution Res & Control (pres, 85-); Int Asn Water Pollution Res & Control. Res: Wastewater treatment; fluidization; froth flotation; mineral recovery from oil-sand tailings. Mailing Add: 22 Regency Rd London ON N6H 4A8 Can

BULAS, ROMUALD, b Chabno, Russia, Feb 2, 22; US citizen; m 50; c 3. PHYSICAL CHEMISTRY. Educ: Univ Heidelburg, BS, 49; Polytech Inst Brooklyn, MS, 63. Prof Exp: RES ASSOC, CENT RES LABS, BASF CORP, CLIFTON, 51- Mem: Am Chem Soc; Soc Rheology. Res: Colloid chemistry; rheology; mass spectrometry. Mailing Add: 388 Beech Spring Rd South Orange NJ 07079

BULBENKO, GEORGE FEDIR, b Moshoryno, Ukraine, June 29, 27; nat US; m 53; c 2. ORGANIC CHEMISTRY. Educ: Hanover Col, AB, 52; Ind Univ, MA, 54, PhD(org chem), 58. Prof Exp: Sr res chemist, Org Sect, Thiokol Chem Corp, 58-59, supvr, Org & Polysulfides Sect, 60-68, head, Org Polysulfides & Anal Sect, 68-69; dir chem sci, Princeton Biomedix Inc, 69-72, vpres, 72-76, dir, prod develop, 76-78, gen mgr, Div Becton Dickinson & Co, 78-82. Concurrent Pos: Asst prof, Trenton Jr Col, 60-66; assoc prof, Mercer County Commun Col, 67-71. Mem: AAAS; Am Chem Soc; Am Inst Chem; Am Asn Clin Chem; Sigma Xi. Res: Synthetic organic chemistry; sulfur and clinical chemistry; polymers; research on clinical chemical assays. Mailing Add: 800 Trenton Rd Apt 246 Langhorne PA 19047

BULGER, ROGER JAMES, b New York, NY, May 18, 33; m 60; c 2. INFECTIOUS DISEASES, INTERNAL MEDICINE. Educ: Harvard Univ, AB, 55, MD, 60. Prof Exp: From asst prof to assoc prof med, Sch Med, Univ Wash, 66-70; prof community health sci & assoc dean, Sch Med, Duke Univ, 70-72; exec officer health policy, Inst Med, Nat Acad Sci, 72-76; chancellor-dean med, Univ Mass Med Ctr, 76-78; pres, Health Sci Ctr, Univ Tex, 78-88; PRES & CHIEF EXEC OFFICER, ASN ACAD HEALTH CTRS, 88- Concurrent Pos: Clin investr, US Vet Admin, 67-; mem adv comt, Allied Health Prof, NC State Bd Educ & Higher Educ, 70-71; mem consortium bd, NC Health Manpower Develop Comt, 70-71; consult, Inst Human Values Med, 72-74; mem adv panel, Nat Health Insurance, Subcomt Health, Comt & Means, US House Rep, 75-76. Mem: Inst Med-Nat Acad Sci; Am Soc Microbiol; Am Soc Nephrology; Infectious Dis Soc Am; fel Am Col Physicians. Res: Clinical pharmacology; antibiotics; bacterial resistance; hospital epidemiology and clinical infectious disease; health and science policy. Mailing Add: 1400 16th St NW Suite 410 Washington DC 20036

BULGER, RUTH ELLEN, b Kansas City, Mo, 36; m 60; c 2. ANATOMY, PATHOLOGY. Educ: Vassar Col, AB, 58; Radcliffe Col, AM, 59; Univ Wash, PhD(anat), 62. Prof Exp: Asst prof path, Med Sch, Univ Wash, 67-70; assoc prof anat, Univ NC, 70-72; from assoc prof to prof, Univ Md, Baltimore City, 72-76; prof path, Med Sch, Univ Mass, 76-77, prof anat, 77-78, chmn dept, 77-78; prof path & lab med, depts neurobiol & anat, Sch Med, Univ Tex Health Sci Ctr Houston, 78-88; DIR, DIV HEALTH SCI POLICY, INST MED-NAT ACAD SCI, 88- Concurrent Pos: Fel anat, Harvard Med Sch, 63-65; fel path, Univ Wash, 64-67. Honors & Awards: Minnie Stevens Piper Award, 85; John Freeman Award, 85. Mem: Am Soc Nephrology; Fedn Am Soc Exp Path; Electron Micros Soc Am; Int Soc Nephrology; Am Asn Anatomists; Am Soc Cell Biol. Res: Kidney morphology and function. Mailing Add: Inst Med-Nat Acad Sci 2101 Constitution Ave NW Washington DC 20418

BULGREN, WILLIAM GERALD, b Anamosa, Iowa, Aug 7, 37; m 60; c 3. COMPUTER SCIENCE, STATISTICS. Educ: Univ Iowa, BA, 59, MS, 61, PhD(statist), 65. Prof Exp: Asst prof statist, Univ Mo, Columbia, 65-68; assoc prof, 68-72, assoc chmn dept, 74-75, PROF COMPUT SCI, UNIV KANS, 72-, CHMN COMPUT SCI, 83- Concurrent Pos: Actg dir comput ctr, Univ Mo, Columbia, 67-68. Mem: Asn Comput Mach; Am Statist Asn; Sigma Xi. Res: System simulation; modelling; computer performance evaluation; computational statistics. Mailing Add: Dept Comput Sci Univ Kans Lawrence KS 66045

BULGRIN, VERNON CARL, b Cuyahoga Falls, Ohio, May 10, 23; m 54; c 2. PHYSICAL CHEMISTRY. Educ: Univ Akron, BS, 48; Iowa State Univ, PhD(chem), 53. Prof Exp: Asst, Iowa State Univ, 48-53; instr, 53-56, from asst prof to prof, 56-85, head dept, 62-67, EMER PROF CHEM, UNIV WYO, 85- Concurrent Pos: Res Corp grant, 56, 58; vis prof, Purdue Univ, 67-68. Mem: Fel AAAS; Am Chem Soc. Res: Kinetics of specific oxidation reactions; homogeneous equilibria. Mailing Add: 2311 Twin Lakes Dr Uniontown OH 44685-9717

BULICH, ANTHONY ANDREW, b San Pedro, Calif, Feb 16, 44; m 67; c 2. MICROBIOLOGY. Educ: Univ Southern Calif, BA, 66; Iowa State Univ, MS, 68, PhD(microbiol), 72. Prof Exp: Res microbiologist, A E Staley Mfg Co, 72-75, sr res microbiologist fermentation res, 75-76; sr develop microbiologist, Beckman Microbics, 76-77, mgr appln develop, 77-81, prin develop microbiologist, 77-86, VPRES SCI & TECHNOL, MICROBICS CORP, 86- Mem: AAAS; Am Soc Microbiol; Soc Indust Microbiol; Am Chem Soc. Res: Isolation and characterization of carbohydrase producing microorganism; strain improvement thru genetic manipulation and fermentation optimization; use of luminescent bacteria for environmental monitoring; characterization of thermophilic and acidophilic amylase producing microorganisms. Mailing Add: Microbics Corp 2232 Rutherford Rd Carlsbad CA 92008-8883

BULKIN, BERNARD JOSEPH, b Trenton, NJ, Mar 9, 42; m 66, 75; c 3. PHYSICAL CHEMISTRY, ANALYTICAL CHEMISTRY. *Educ:* Polytech Inst Brooklyn, BS, 62; Purdue Univ, PhD(phys chem), 66. *Prof Exp:* NSF fel, Swiss Fed Inst Technol, 66-67; from asst prof to prof chem, Hunter Col, 67-75, chmn dept, 73-75; prof & dean arts & sci, Polytech Inst NY, 75-82; vpres, Res & Grad Affairs & dir, Inst Imaging Sci, 82-85; dir, Anal Sci Lab, Standard Oil Co, Ohio, 85-88; dir, Downstream Oil Res & Develop, 88; mgr, Prod Div, 89-90, HEAD OIL RES, BRIT PETROL CO, 91- *Concurrent Pos:* Petrol Res Fund grant, 67-69; Am Cancer Soc res grant, 68-75; Army Res Off-Durham res grant, 69-75; dir, NSF High Sch Inst, 70-71 & NSF Women in Sci career facilitation grant, 76-78; mem, NSF Comt Equal Opportunity Sci & Technol, 83-86; bd dir, Optronics Int, 84-87, Spex Group Inc, 84-88; prin investr, NSF res grant, 84-87, Corn Refiners Asn res grant, 85-86. *Honors & Awards:* Coblentz Award, 75; Oscar Foster Award, 78. *Mem:* Soc Appl Spectros; Coblentz Soc (pres, 79-81); Am Chem Soc. *Res:* Oil products and processes. *Mailing Add:* BP Res Sunbury KT10 ODE England

BULKLEY, GEORGE, b Pa, Apr 16, 17; m 41; c 4. MEDICINE, UROLOGY. *Educ:* Northwestern Univ, BS, 38, MD, 42. *Prof Exp:* Instr urol, Sch Med, Yale Univ, 49-51; assoc, 51-55, from asst prof to assoc prof, 56-73, PROF UROL, MED SCH, NORTHWESTERN UNIV, 73- *Concurrent Pos:* From assoc attend to sr attend urologist, Chicago Wesley Mem Hosp, 53-; attend urologist, Vet Admin Res Hosp, 54-; consult urologist, Passavant Mem Hosp, 70- *Mem:* AMA; Am Urol Asn; Am Col Surg; Int Soc Urol. *Res:* Urologic clinical research; use of radioactive materials in treatment of carcinoma of the prostate; surgical and chemotherapy carcinoma of bladder. *Mailing Add:* 251 E Chicago Ave Chicago IL 60611

BULKLEY, JONATHAN WILLIAM, b Kansas City, Mo, May 17, 38; m 62; c 2. WATER RESOURCE POLICY, CIVIL ENGINEERING. *Educ:* Mass Inst Technol, SB(polit sci) & SB(civil eng), 61, SM, 63, PhD(polit sci), 66. *Prof Exp:* Res asst civil eng, Mass Inst Technol, 61-63, instr, 63-66; from asst prof to assoc prof nat resources, 68-77, from asst prof to assoc prof civil eng, 68-77, PROF, SCH NAT RESOURCES & COL ENG, UNIV MICH, ANN ARBOR, 77- *Concurrent Pos:* VChmn, Mich Environ Rev Bd, 76-79. *Mem:* Am Soc Civil Engrs; Inst Asn Great Lakes Res; Sigma Xi; Water Pollution Cent Fed. *Res:* Water policy; risk analysis; institutional arrangements; water planning and management. *Mailing Add:* 2506 B Dana Univ Mich Ann Arbor MI 48109

BULKLEY, ROSS VIVIAN, fisheries; deceased, see previous edition for last biography

BULKOWSKI, JOHN EDMUND, b Weymouth, Mass, Sept 7, 42; m 66; c 2. CATALYSIS, BIOINORGANIC. *Educ:* Brown Univ, ScB, 65; Carnegie-Mellon Univ, MS, 71, PhD(chem), 74. *Prof Exp:* Officer, USN Nuclear Prog, 65-67; engr, Westinghouse Elec Corp, 67-69; res fel, Harvard Univ, 73-75; ASST PROF INORG CHEM, CHEM DEPT, UNIV DEL, 75- *Mem:* Am Chem Soc; Sigma Xi. *Res:* Inorganic and organometallic chemistry: specifically the use of binuclear transition metal compounds as catalysts and biological models; and the use of photosensitive arrays for light assisted catalysis and solar energy conversion. *Mailing Add:* Dept Chem Univ Del Newark DE 19716

BULL, ALICE LOUISE, b White Plains, NY, May 16, 24. DEVELOPMENTAL BIOLOGY. *Educ:* Middlebury Col, AB, 46; Mt Holyoke Col, MA, 48; Yale Univ, PhD(zool), 53. *Prof Exp:* Asst zool, Yale Univ, 48-50; asst, Mt Holyoke Col, 46-48, instr, 53-55; from instr to asst prof, Wellesley Col, 55-64; assoc prof zool, 64-76, from assoc prof to prof biol, 76-90, EMER PROF BIOL, HOLLINS COL, 90- *Concurrent Pos:* Am Asn Univ Women fel, Univ Zurich, 52-53. *Mem:* AAAS; Genetics Soc Am; Am Soc Zoologists; Soc Develop Biol; Sigma Xi. *Res:* Oogenesis, fertility regulation and developmental genetics in Drosophila; embryogenesis in Mormoniella. *Mailing Add:* Dept Biol Hollins Col PO Box 9633 Hollins College VA 24020

BULL, BRIAN S, b London, Eng, Sept 14, 37; US citizen; m 63; c 2. PATHOLOGY. *Educ:* Walla Walla Col, BS, 57; Loma Linda Univ, MD, 61. *Prof Exp:* Intern, Grace New Haven Med Ctr, 61-62, resident, 62-63; resident, NIH, 63-65, staff hematologist, 66-67; from asst prof to assoc prof, 68-72, PROF PATH & CHMN DEPT, LOMA LINDA UNIV, 73- *Concurrent Pos:* Fel lab med, NIH, 65-66, spec fel hemat, Nat Inst Arthritis & Metab Dis, 67-68; vis prof, Inst Cellular Path, Paris, 72, Royal Postgrad Med Sch, London, 72, Univ Wis-Madison, 73, Univ Ohio, Columbus, 74, Univ Minn, Minneapolis, 79, Univ Hawaii, Honolulu, 81, St Thomas Hosp & Med Sch, Linden, 81, Mayo Clin & Med Sch, Rochester, Minn, 85; ed-in-chief, Blood Cells, 85- *Honors & Awards:* Merk Manual Award, 61; E B Cotlove Mem Lectr, 72. *Mem:* Am Soc Clin Path; AMA; Soc Acad Clin Lab Physicians & Scientists; Col Am Path; Am Soc Hemat. *Res:* Physiology of blood cells, particularly erythrocytes and platelets; clinical laboratory test development and automation; quality control of clinical laboratory tests. *Mailing Add:* Loma Linda Med Ctr Loma Linda Univ Rm 2516 Loma Linda CA 92350

BULL, COLIN BRUCE BRADLEY, b Birmingham, Eng, June 13, 28; m 56; c 3. GEOPHYSICS. *Educ:* Univ Birmingham, BS, 48, PhD(physics), 51. *Prof Exp:* Geophysicist, Cambridge Univ, 52-55; Imp Chem Industs res fel geophys, Univ Birmingham, 55-56; sr lectr physics, Victoria Univ, NZ, 56-61; vis asst prof, 61-62, assoc prof, 62-65, dir inst polar studies, 65-69, chmn dept geol, 69-72, dean col math & phys sci, 72-85, PROF GEOL, OHIO STATE UNIV, 65-, EDUC & SCI CONSULT, 85- *Concurrent Pos:* Geophysicist & chief scientist, Brit N Greenland Exped, 52-55; mem, Ross Dependency Res Comt, NZ Govt, 59-61; mem panels glaciol, geol & geophys & comt polar res, Nat Acad Sci; US rep, Sci Comt Antarctic Res, 78. *Honors & Awards:* Polar Medal, 55; Antarctic Serv Award, US Cong, 74. *Mem:* Am Geophys Union; Arctic Inst NAm; Glaciol Soc; Geol Soc Am; Royal Soc Arts. *Res:* Geophysical and glaciological investigations in polar regions, particularly Greenland and Antarctica. *Mailing Add:* Box 4675 Rolling Bay WA 98061-0675

BULL, DANIEL NEWELL, b Highland Park, Mich, July 5, 39; m 67. CHEMICAL ENGINEERING. *Educ:* Univ Mich, BSE, 62, MSE, 64, PhD(chem eng), 68. *Prof Exp:* Res chem engr, Continental Oil Co, Okla, 69-71; sr chem engr, Hoffmann-La Roche Inc, 71-77, res chem engr, 77-80; vpres & tech dir, New Brunswick Sci Co, 80-82; pres, Sartori Corp, NJ, 82-90; DIR ENG, HENDERSON INDUSTS, WEST CALDWELL, NJ, 90- *Res:* Mass transfer; fermentation kinetics; hydrocarbon dermentations; continuous fermentations; separation of biologicals; tissue culture. *Mailing Add:* Henderson Industs 26 Fairfield Pl West Caldwell NJ 07006

BULL, DON LEE, b Raymondville, Tex, Sept 3, 29; m 54; c 1. ENTOMOLOGY. *Educ:* Tex A&M Univ, BS, 53, MS, 60, PhD(entom), 62. *Prof Exp:* Analyst, Amoco Chem Corp, 56-57 & Shell Chem Co, 57-58; RES ENTOMOLOGIST TOXICOL, AGR RES SERV, USDA, 61- *Mem:* Entom Soc Am; Am Chem Soc. *Res:* Insect toxicology and physiology. *Mailing Add:* 2510 Towering Oaks Bryan TX 77802

BULL, EVERETT L, JR, b Stockton, Calif, Aug 12, 49. PROGRAMMING LANGUAGE SEMATICS & VERIFICATION. *Educ:* Pomona Col, BA, 71; Mass Inst Technol, PhD(math), 76. *Prof Exp:* Adj asst prof math, Univ Calif, Los Angeles, 76-78; vis asst prof math, 78-80, asst prof, 81-, ASSOC PROF MATH, POMONA COL. *Concurrent Pos:* Res fel, Stanford Univ, 80-81; vis assoc prof, Cornell Univ, 85-86. *Mem:* Am Math Soc; Asn Symbolic Logic; Asn Comput Mach; Inst Elec & Electronics Engrs Computer Soc. *Res:* Application of logic to computer science. *Mailing Add:* Dept Math Pomona Col Claremont CA 91711

BULL, LEONARD SETH, b Westfield, Mass, Jan 31, 41; m 71; c 3. ANIMAL NUTRITION, PHYSIOLOGY. *Educ:* Okla State Univ, BSc, 63, MSc, 64; Cornell Univ, PhD(nutrit), 69. *Prof Exp:* NASA fel physiol, Med Sch Univ Va, 68-70; from asst prof to assoc prof, Univ Md, 70-75; assoc prof nutrit, Univ Ky, 75-79; prof nutrit, Univ Maine, 79-81; PROF & HEAD, DEPT ANIMAL SCI, UNIV VT, 81- *Mem:* Am Dairy Sci Asn; Am Soc Animal Sci; Am Reg Prof Animal Sci. *Res:* Animal nutrition with special interest in energy metabolism, digestive physiology and trace element interactions on metabolism; body compositon and animal calorimetry plus regulation of energy balance. *Mailing Add:* Dept Animal Sci NC State Univ Box 7621 Raleigh NC 27695

BULL, RICHARD C, b Cedaredge, Colo, Nov 25, 34; m 59; c 2. ANIMAL NUTRITION, BIOCHEMISTRY. *Educ:* Colo State Univ, BS, 57, MS, 60; Ore State Univ, PhD(animal nutrit, biochem), 66. *Prof Exp:* Res assoc animal nutrit, Ore State Univ, 63-67; asst prof, Univ Idaho, 67-72, assoc prof animal sci & assoc animal scientist, 72-; AT TOXICOL & MICROBIOL DIV, ENVIRON PROTECTION AGENCY. *Mem:* Am Inst Biol Sci; Am Soc Animal Sci. *Res:* Nutritional biochemistry, especially vitamin E and selenium. *Mailing Add:* Animal Sci Univ Idaho Moscow ID 83843

BULL, STANLEY R(AYMOND), b Montezuma, Iowa, May 15, 41; m 63; c 3. RESEARCH ADMINISTRATION. *Educ:* Univ Mo, BS, 63; Stanford Univ, MS, 64, PhD(mech eng), 67. *Prof Exp:* From asst prof to prof eng, Univ Mo-Columbia, 67-80; sr eng, acad & univ prog, 80-81, sr sci adv, 81-82, mgr, planning & eval off, 82-84, dep dir, Solar Fuels Res Div, 84-86, DIV DIR, FUELS & CHEMICALS RES & ENG DIV, SOLAR ENERGY INST, GOLDEN, COLO, 86- *Concurrent Pos:* Vis scientist, Argonne Nat Lab, Ill, 69, consult, 69 & 71; consult, Ellis Fischel State Cancer Hosp, 72-80; sr Fulbright-Hays fel, Int Educ & Cult Exchange Prog, 73; vis scientist, Ctr d'Etudes Nucleaires, Grenoble, France, 73-74. *Mem:* Am Inst Chem Engrs; Am Soc Mech Engrs; Int Solar Energy Soc; Soc Automotive Engrs. *Res:* Energy systems and resources; renewable fuel options and processes; evaluation renewable energy technology; biotechnology, chemical conversion and photo conversion; research and development resource allocations. *Mailing Add:* Solar Energy Res Inst 1617 Cole Blvd Golden CO 80127

BULL, W(ILLIAM) REX, b Hull, Eng, Oct 27, 29; m 56; c 4. MINERAL PROCESSING. *Educ:* Univ Leeds, BSc, 51, Grad Dipl mineral processing, 52; Univ Queensland, PhD(mineral processing), 66. *Prof Exp:* Sci officer mineral processing, Can Dept Mines & Technol Surv, 52-56; from lectr to sr lectr, Univ Queensland, 56-67; ASSOC PROF METALL ENG, COLO SCH MINES, 67- *Concurrent Pos:* Vis lectr, McGill Univ, 62-63; consult to mining & metall co, 64- *Mem:* Am Inst Mining, Metall & Petrol Engrs; Australian Inst Mining & Metall. *Res:* Mathematical modelling; control of mineral treatment processes. *Mailing Add:* Dept Metall Eng Colo Sch Mines Golden CO 80401

BULL, WILLIAM BENHAM, b San Francisco, Calif, Apr 19, 30; m 53; c 2. GEOMORPHOLOGY. *Educ:* Univ Colo, BA, 53; Stanford Univ, MS, 57, PhD(geol), 60. *Prof Exp:* Geologist & res hydrologist, Water Resources Div, US Geol Surv, 56-76; PROF GEOL, UNIV ARIZ, 68- *Mem:* Am Geophys Union; Geol Soc Am; Nat Asn Geol Teachers; Sigma Xi. *Res:* Tectonic and climatic geomorphology; fluvial geomorphology of arid regions; land subsidence due to ground water withdrawal and due to collapse of soils upon wetting. *Mailing Add:* Dept Geosci Univ Ariz Tucson AZ 85721

BULL, WILLIAM EARNEST, b Franklin Co, Mo, Jan 17, 33; m 55; c 3. INORGANIC CHEMISTRY. *Educ:* Southern Ill Univ, AB, 54; Univ Ill, AM, 55, PhD(chem), 57. *Prof Exp:* From asst prof to assoc prof, 57-70, PROF CHEM, UNIV TENN, KNOXVILLE, 70-, DIR GEN CHEM, 74-, ASSOC HEAD, 79- *Concurrent Pos:* Fulbright-Hayes award, Univ Col, Dublin, 70-71. *Mem:* AAAS; Am Chem Soc; Sigma Xi. *Res:* Nonaqueous solvents; coordination compounds; photoelectron spectroscopy. *Mailing Add:* Dept Chem Univ Tenn Knoxville TN 37996-1600

BULLA, LEE AUSTIN, JR, b Oklahoma City, Okla, May 1, 41; m 62; c 3. MICROBIOLOGY, BIOCHEMISTRY. *Educ:* Midwestern Univ, BS, 65; Ore State Univ, PhD(microbiol), 68. *Prof Exp:* Bacteriologist & chemist water purification, City of Wichita Falls, Tex, 65; microbiologist, Northern Regional Res Lab, USDA, 68-73, microbiologist, US Grain Mkt Res Ctr, 73-81; prof, div biol, Kans State Univ, 77-81; prof dept basteriology & biochem & dir, Inst Molecular & Agr Genetic Eng, Univ Idaho, Moscow, 81-84; PROF DEPT MICROBIOL & BIOCHEM & DEAN, COL AGR, UNIV WYO, 84- *Concurrent Pos:* Assoc dean Col Agr & dir Idaho Agr Exp Sta, 81-84. *Mem:* AAAS; Am Soc Microbiol; Soc Invertebrate Path; NY Acad Sci. *Res:* Molecular biology of bacterial sporulation; biochemical and biophysical characterization of baculoviruses. *Mailing Add:* Dept Molecular Biol Univ Wyo 3944 University Sta Laramie WY 82071

BULLARD, CLARK W, b Springfield, Ill, Mar 22, 44; m 70; c 3. TECHNOLOGY POLICY ANALYSIS. *Educ:* Univ Ill, Urbana-Champaign, BS, 66, MS, 67, PhD(aeronaut & astronaut eng), 71. *Prof Exp:* Assoc engr-scientist, Flight Kinetics Dept, Missile & Space Systs Div, Douglas Aircraft Co, Santa Monica, 66; engr-scientist specialist, Advan Aerothermodyn Dept, McDonnell Douglas Astronaut Co, Huntington Beach, 67-69; teaching & res asst aeronaut & astronaut eng, Univ Ill, Urbana-Champaign, 69-71, res assoc prof, Ctr Advan Comput, 71-77; dir, Off Conserv & Advan Energy Systs Policy, US Dept Energy, Washington, DC, 77-79; sr analyst, Off Technol Assessment, US Cong, Washington, DC, 79-80; prof mech eng & dir, Off Energy Res, 80-90, AFFIL NUCLEAR ENG, UNIV ILL, URBANA-CHAMPAIGN, 84-, DIR, AIR CONDITIONING & REFRIG CTR, 90- *Concurrent Pos:* Mem, comt nuclear & alternative energy systs, Demand-Conserv Panel, Nat Acad Sci, 76-80; chmn adv panel, Workshop Solar Power Satellite & Alternatives, Off Technol Assessment, US Cong, 80, Workshop Nuclear Power, 82; consult, Off Technol Assessment, US Cong, SAIC, Radian Corp, ETA, Inc, GRI, 80-; deleg, Midwest Univs Energy Consortium, 80-; mem, Orgn Econ Coop & Develop- Int Energy Agency Res Adv Bd, 82-83 & Gov's Task Force Utility Reform, State Ill, 82-84; chmn, Cent Midwest Compact Comn Low Level Radioactive Waste Mgt, 84-; mem, Am Coun Energy Efficient Econ; Fulbright scholar, Sussex Univ, 86. *Honors & Awards:* Chevron Nat Conserv Award, 90. *Mem:* Am Soc Mech Engrs; Am Soc Heating, Refrig & Air Conditioning Engrs; Fedn Am Scientist; AAAS. *Res:* Optimization of cogeneration system design for heating, ventilation and air conditioning applications; design and siting of radioactive waste disposal technologies; acid emission reduction strategies for electric utilities. *Mailing Add:* Dept Mech Eng Univ Ill 1206 W Green St Urbana IL 61801

BULLARD, EDWIN ROSCOE, JR, b El Paso, Tex, Dec 15, 21; m 45; c 1. GEOPHYSICS, GEOLOGY. *Educ:* Univ Tex, El Paso, BS, 49 & 63. *Prof Exp:* Subsurface geologist, Kerr-McGee Oil Indust, Inc, 49-51, Tex Gulf Prod Co, 51-56, Ambassador Oil Corp, 56-60 & Ralph Lowe of Midland, Tex, 60-62; proj physicist, Schellenger Res Lab, Univ Tex, El Paso, 63-64; res geophysicist, Water Resources Div, US Geol Surv, 64-66; res geophysicist, Globe Explor Co, 66-69, dir, Globe Universal Sci, Advan Proj Div, Shell Oil Co, 69-70, Atlantic Richfield, 70-83 & Tex Eastern, 83-85; RETIRED. *Concurrent Pos:* Geophys petrol grant, Gus Mfg. *Mem:* Am Asn Petrol Geologists; Am Inst Prof Geologists. *Res:* Bringing the state of the art to seismic and borehole geophysics for oil exploration. *Mailing Add:* 7405 Capulin Rd NE Albuquerque NM 87109-4907

BULLARD, ERVIN TROWBRIDGE, b New York, NY, May 25, 20; m 48; c 3. HORTICULTURE. *Educ:* NC State Col, BS, 43; Cornell Univ, MS, 46; Purdue Univ, PhD(hort), 50. *Prof Exp:* Asst prof veg crops, Univ RI, 46-47; asst horticulturist, Purdue Univ, 50. Prof Exp: Asst prof veg crops, Univ RI, 46-47; asst horticulturist, Purdue Univ, 48-50; assoc prof veg res, Univ Idaho, 50-54; agr res adv, USAID, 54-64, agr prod adv, Brazil, 64-68, dep rural develop off, Repub Panama, 69-72, chief, Agr Div, India, 72-74, chief agr prod, Pakistan, 74-77, dep rural develop off, Honduras, 77-79, Chief of Party, Burma, 84-86; AT OHIO STATE UNIV. *Concurrent Pos:* Fulbright res scholar, Univ Cairo, 51-52; vis prof, Dept Hort, Ohio State Univ, 68-69. *Mem:* Am Soc Hort Sci; Am Genetic Asn; Sigma Xi. *Res:* Cacao; coffee; vegetable crops. *Mailing Add:* 129 Cimmaron Dr Palm Coast FL 32137

BULLARD, FRED MASON, b McLoud, Okla, July 20, 01; m; c 2. GEOLOGY, VOLCANOLOGY. *Educ:* Univ Okla, BS, 21, MS, 22; Univ Mich, PhD(geol), 28. *Prof Exp:* Field geologist, Okla Geol Surv, Norman, 21-23; consult geologist, 23-24; from instr to prof, 24-71, chmn dept, 29-37, EMER PROF GEOL SCI, UNIV TEX, AUSTIN, 71- *Concurrent Pos:* Vis prof, Vassar Col, Poughkeepsie, NY, 49, Univ Mich, Ann Arbor, 33 & 37, Nat Univ Mex, 43-46, Columbia Univ, NY, 49-51, Northern Ariz Univ Flagstaff, 67, 69-70; Fulbright res scholar, Italy, 53; Fulbright lectr, Peru, 59; lectr on Paricutin Volcano, Am Asn Petrol Geologists, 43-45 & a volcanic cycle, 54; vis prof & chief party, US Tech Assistance Prog, Univ Baghdad, 62-64. *Mem:* Fel Geol Soc Am; Sigma Xi; Am Asn Petrol Geologists. *Res:* Igneous geology; meteorites; Paricutin Volcano, Mexico; volcanoes and geology of Latin America; active volcanoes of the world; Holocene volcanic activity of western United States. *Mailing Add:* Dept Geol Sci Univ Tex Austin TX 78712

BULLARD, TRUMAN ROBERT, b Miami, Fla, Jan 15, 39; m 60; c 2. PHYSIOLOGY, SCIENCE EDUCATION. *Educ:* Asbury Col, AB, 61; WVa Univ, MS, 63, PhD(physiol), 66. *Prof Exp:* Res assoc dept theoret & appl mech, WVa Univ, 64-65; asst prof, Christian Med Col, Ludhiana, Punjab, India, 68-69; asst prof, Rutgers Univ, 69-71; reader physiol, Christian Med Col, Vellore, India, 71-84; EDUC CONSULT MIDAS REX INST, FT WORTH, TEX, 85- *Concurrent Pos:* Fel physiol, Sch Med, Univ Va, 65-68; vis assoc prof pharmacol, Sch Med, Tex Tech Univ, 80-81 & 84-85; mem bd, Ludhiana Christian Med Col, 89- *Mem:* Am Chem Soc; Am Soc Bone & Mineral Res; Am Hosp Asn. *Res:* Effects of exercise on body composition and energy metabolism; interaction of endocrines and atrophy of disuse in bone; vitamin D metabolism. *Mailing Add:* 1408 Navaho St Arlington TX 76012-4341

BULLAS, LEONARD RAYMOND, b Lismore, New South Wales, Dec 8, 29; wid; c 2. MICROBIAL GENETICS. *Educ:* Univ Adelaide, BSc, 53, MSc, 57; Mont State Col, PhD(genetics), 63. *Prof Exp:* Instr bact, Univ Adelaide, 53-58; asst genetics, Mont State Col, 59-62; from instr to assoc prof, 62-80, PROF MICROBIOL, SCH MED, LOMA LINDA UNIV, 80- *Concurrent Pos:* Vis prof, Univ Louvain, Belg, 73-74 & Europ Molecular Biol Lab, Ger, 81-82; invited prof, Univ Louvain, Belg, 89. *Mem:* Am Soc Microbiol; Genetics Soc Am; Sigma Xi. *Res:* Genetics and molecular biology of the genes for modification and restriction of DNA in salmonella. *Mailing Add:* Sch Med Loma Linda Univ Loma Linda CA 92350

BULLEN, ALLAN GRAHAM ROBERT, b Wellington, NZ, Dec 8, 36. TRANSPORTATION ENGINEERING. *Educ:* Univ NZ, BSc, 60; Univ Wellington, MSc, 62; Northwestern Univ, MS Eng, 65, PhD(transp), 69. *Prof Exp:* Traffic eng cadet, NZ Govt Transp Dept, 55-60, traffic engr, 60-66; res engr traffic flow, Northwestern Univ, 67-69; asst prof civil eng & res asst prof, Coord Sci Lab, Univ Ill, Urbana-Champaign, 69-70; civil engr traffic control, Ill Div Hwys, 70; ASSOC PROF CIVIL ENG & ENVIRON SYST ENG, UNIV PITTSBURGH, 70- *Concurrent Pos:* NZ Govt rep, World Traffic Eng Conf, London, 64; mem, Hwy Res Bd, Nat Acad Sci-Nat Res Coun. *Honors & Awards:* O'Farrill Hwy Award, Int Road Fedn, 64. *Mem:* Inst Transp Eng; Am Soc Civil Engrs; Transp Res Forum; Sigma Xi. *Res:* Multi-lane models of traffic flow, traffic flow theory; traffic control, expressway control systems; impact of new urban transportation facilities; travel structure. *Mailing Add:* 4626 Hidden Pond Dr Allison PA 15101

BULLEN, PETER SOUTHCOTT, b Portsmouth, Eng, Jan 19, 28; m 52; c 4. MATHEMATICS. *Educ:* Univ Natal, MSc, 49; Univ Cambridge, PhD(math), 55. *Prof Exp:* Lectr, Dept Pure & Appl Math, Univ Natal, 47-50, 53-56; from instr to assoc prof math, 56-70, PROF MATH, UNIV BC, 70- *Res:* Non-absolute integration. *Mailing Add:* Dept Math Univ BC 121 1984 Math Rd Vancouver BC V6T 1Y4 Can

BULLER, CLARENCE S, b McPherson, Kans, June 21, 32; m 59; c 2. MICROBIOLOGY. *Educ:* Univ Kans, BA, 58, MA, 60, PhD(bact), 63. *Prof Exp:* Res fel microbiol, Western Reserve Univ, 63-66; res fel, 66-73, from asst prof to assoc prof, 66-77, PROF MICROBIOL, UNIV KANS, 77- *Mem:* Am Soc Microbiol. *Res:* Physiology of bacteriophage infected coliforms; function of phospholipases; virus induced changes in cell surfaces. *Mailing Add:* Dept Microbiol Univ Kans Lawrence KS 66045

BULLERMAN, LLOYD BERNARD, b Adrian, Minn, June 20, 39; m 60; c 4. FOOD MICROBIOLOGY, MYCOLOGY. *Educ:* SDak State Univ, BS, 61, MS, 65; Iowa State Univ, PhD(bact, food technol), 68. *Prof Exp:* Lab technician qual control, Dairy Prod, Inc, SDak, 61-63; food scientist prod develop, Green Giant Co, Minn, 68-70; from asst prof to assoc prof, 70-79 acting dept head, 80-81, PROF, FOOD SCI & TECHNOL, UNIV NEBR, LINCOLN, 79- *Mem:* Inst Food Technol; Am Soc Microbiol; Int Asn Milk, Food & Environ Sanit; Am Asn Cereal Chemists; Sigma Xi. *Res:* Food microbiology, food safety and toxicology, mycology and mycotoxins, particularly food-borne microorganisms that may be toxic or pathogenic to humans; food product development; waste disposal; biotechnology of solid state fungal fermentations. *Mailing Add:* Dept Food Sci & Technol Univ Nebr 349 FIB East Campus Lincoln NE 68583-0919

BULLIS, GEORGE LEROY, b Eau Claire, Wis, Apr 26, 22; m 45; c 4. MATHEMATICS. *Educ:* Wis State Univ, Eau Claire, BS, 41; Univ Wis, MA, 42. *Prof Exp:* Instr math, Univ Wis, 43-45; assoc prof, Univ Wis, Platteville, 45-68, from actg dean to dean, sch arts & sci, 68-77, prof math, 77- *Mem:* Math Asn Am. *Mailing Add:* 405 A Feltham Trail Sun City Center FL 33573

BULLIS, HARVEY RAYMOND, JR, b Milwaukee, Wis, June 14, 24; m 44; c 4. MARINE BIOLOGY. *Educ:* Wis State Col, BS, 49; Univ Miami, MS, 51. *Prof Exp:* Asst zool, Univ Miami, 49-50; marine biologist, US Fish & Wildlife Serv, 50-70; assoc dir marine fisheries serv, Nat Oceanic & Atmospheric Admin, Dept Com, 70-71, dir, 71-78, sr scientist, southeast fisheries ctr, nat marine fisheries serv, US Dept Com, 78-79; bromeliad hort develop prog, collecting in Costa Rica, Panama, Brazil, Trinidad, 79-88; fishery surv, Red Sea, Egyptian Acad Sci, 81-83; fishery develop prog, Quintana Roo, Mex, 83-85; RETIRED. *Concurrent Pos:* Dir, Bur Com Fisheries Explor Fishing Base, 55-56; consult, USAID, surv WAfrican fisheries, 61; mem, Nat Acad Sci comt technol & sci base Puerto Rican econ, 66-67, Sino-Am colloquium ocean resources, 71; US deleg, WCent Atlantic Fisheries Comn, 75; US asst nat coordr fisheries, Coop Invests Caribbean & Adjacent Regions, 71-76, asst int coordr, 75-76; consult trop marine fisheries, 79-91. *Honors & Awards:* Award, US Dept Interior, 54. *Mem:* Fel Am Inst Fishery Res Biologists. *Res:* Zoogeography of tropical west Atlantic; fishery exploration and resource analysis. *Mailing Add:* 12420 SW 248th St Princeton FL 33032

BULLIS, WILLIAM MURRAY, b Cincinnati, Ohio, Aug 29, 30; m 53; c 3. SEMICONDUCTOR ELECTRONICS. *Educ:* Miami Univ, AB, 51; Mass Inst Technol, PhD(physics), 56. *Prof Exp:* Asst physics, Mass Inst Technol, 52-54; staff mem solid state physics, Lincoln Lab, 54 & Los Alamos Sci Lab, 56-57; physicist, Int Tel & Tel Labs, 57-59; mem tech staff, Tex Instruments Inc, Dallas, 59-65; res physicist, Nat Bur Standards, 65-68, chief, Semiconductor Characterization Sect, 68-75, prog mgr semiconductor technol, Electronic Technol Div, 75-78, chief Electron Devices Div, 78-81; mgr, semiconductor mat, Res & Develop Lab, Fairchild Adv, 81-83; dir technol, 83-88, vpres res & develop, 88-91, PRIN MAT & METROL, 91- *Concurrent Pos:* Lectr elec eng, Univ Md, 66-70; chmn, Electronics Div, Electrochem Soc, 87-89; hon mem, Am Soc Testing Mat Comt F-1, 87. *Honors & Awards:* Standards Honor Award, Semiconductor Equip & Mat Inst, 87; Thurber Award, Am Soc Testing Mat Comt F-1, 84, Award of Merit, 79; Rosa Award, Nat Bur Standards, 80; Silver Medal, US Dept Com, 79. *Mem:* Am Phys Soc; Electrochem Soc; fel Am Soc Testing & Mat; sr mem Inst Elec & Electronics Engrs; Mat Res Soc. *Res:* Transport and defect phenomena in silicon and other high-punity crystalline semiconducting materials; transport phenomena in solids and fluids; semiconductor measurement technology. *Mailing Add:* Mat & Metrol 1477 Enderby Way Sunnyvale CA 94087-4015

BULLOCK, FRANCIS JEREMIAH, b Brookline, Mass, Jan 14, 37; m 61; c 2. PHARMACY. *Educ:* Mass Col Pharm, BS, 58; Harvard Univ, AM, 61, PhD(chem), 63. *Prof Exp:* Res assoc, dept pharmacol, Harvard Med Sch, 63-64 & chem biodynamics, Univ Calif, Berkeley, 64-65; proj staff, Arthur D Little, Inc, 65-72; mgr med chem, Abbott Labs, 72-79; vpres new drug discovery 79-81, SR VPRES RES OPERS, SCHERING-PLOUGH, 81- *Concurrent Pos:* Instr dept food & nutrit sci, Mass Inst Technol & Sch Pharm, Northeastern Univ, 71-72; vpres, DNAX Res Inst Molecular Biol & Immunol, 82- *Mem:* Am Chem Soc; Am Soc Microbiol; Am Soc Pharmacol & Exp Therapeut; Royal Soc Chem. *Res:* New drug discovery research; drug safety evaluation. *Mailing Add:* 60 Orange St Bloomfield NJ 07003

BULLOCK, GRAHAM LAMBERT, b Martinsburg, WVa, Mar 6, 35; m 55; c 5. BACTERIOLOGY, FISH PATHOLOGY. *Educ:* Shepherd Col, WVa, BS, 57; Univ Wis-Madison, MS, 59; Fordham Univ, PhD(biol sci), 70. *Prof Exp:* Bacteriologist qual control, Thomas J Lipton Inc, NJ, 59-60; RESEARCHER BACT FISH DIS, EASTERN FISH DIS LAB, FISH & WILDLIFE SERV, DEPT INTERIOR, 60- *Concurrent Pos:* Vis prof, Ore State Univ, 73-74. *Mem:* Am Soc Microbiol; Am Fisheries Soc; Sigma Xi. *Res:* Nature and control of bacterial infections of fish including the role of environmental factors, chemotherapy and serological methods for detection and identification of bacterial pathogens. *Mailing Add:* Box 45B Rte 1 Kearneysville WV 25430

BULLOCK, J BRUCE, b Lindsay, Okla, June 23, 40; m 59; c 3. AGRICULTURAL FINANCE. *Educ:* Okla State Univ, BS, 62, MS, 64; Univ Calif, Berkeley, PhD(agr econ), 68. *Prof Exp:* From asst prof to assoc prof agr econ, NC State Univ, 69-76; res dir, Farmbank Serv, 76-80; assoc prof, Okla State Univ, 80-82; dept chair, Dept Agr Econ, 82-88, ASSOC DEAN RES, COL AGR, UNIV MO, 88- *Mem:* Am Agr Econ Asn. *Res:* Economics of information; agricultural finance; agricultural commodity markets. *Mailing Add:* 5113 Brock Rogers Columbia MO 65201

BULLOCK, JOHN, b Orange, NJ, Apr 24, 32; m 58; c 3. ENDOCRINOLOGY. *Educ:* Seton Hall Univ, BS, 57; Rutgers Univ, MS, 62; Colo State Univ, PhD(physiol), 65. *Prof Exp:* Res technician, Merck Inst Therapeut Res, 56-57; res asst, 57-61, staff mem & res assoc, 61-62, asst prof, 65-75, ASSOC PROF PHYSIOL, COL MED & DENT NJ, NEWARK, 75- *Mem:* Microcirculatory Soc; Am Physiol Soc; Sigma Xi. *Res:* Androgenic control of connective tissue; biochemistry of tumors; mechanism of action of androgenic esters; connective tissue physiology; biochemistry of cancer; physiology of aging; bile acid metabolism. *Mailing Add:* Univ Med & Dent NJ 185 Bergen St Newark NJ 07103-2757

BULLOCK, JONATHAN S, IV, b Mobile, Ala, Dec 8, 42; m 69; c 2. ELECTRO CHEMISTRY, INORGANIC CHEMISTRY. *Educ:* Ala Col, BS, 64; Tulane Univ, PhD(phys chem), 69. *Prof Exp:* Develop chemist, Union Carbide Corp, 68-76, mem develop staff, Nuclear Div, 76-84; MEM DEVELOP STAFF, MARTIN MARIETTA ENERGY SYSTS, 84- *Mem:* Am Chem Soc; Sigma Xi; AAAS. *Res:* Solvent effect in kinetics; electrochemistry of fluid-electrode and solid-electrode systems; gas-solid interface reactions; corrosion; electrochemical energy systems; electrochemical machining; high-temperature thermochemistry; electrodeposition; fused-salt processes. *Mailing Add:* 120 Euclid Circle Oak Ridge TN 37830

BULLOCK, KATHRYN RICE, b Bartlesville, Okla, Sept 24, 45; m 67; c 2. ELECTROCHEMISTRY, BATTERY TECHNOLOGY. *Educ:* Colo Univ, BA, 67; Northwestern Univ, MS, 69, PhD(chem), 73. *Prof Exp:* Res tech, electrochem, Gates Rubber Co, Denver, 67; res asst, chem, Northwestern Univ, 67-68; res, spec, electrochem, Gates Rubber Co, 72-76; sr electrochemist, electrochem, 77-80, MGR CHEM, CHEM RES, JOHNSON CONTROLS INC, 80- *Concurrent Pos:* Adjunct asst prof, Chem Dept, Univ Wis, Milwaukee, 79. *Honors & Awards:* Electrochem Soc Award, 80. *Mem:* Electrochem Soc; fel Royal Soc Chem, London; fel Am Inst Chemists; Int Union Pure & Applied Chem; Am Chem Soc; Int Soc Electrochem. *Res:* Kinetics of electrochemical reactions, electro chemical cell design and fabrication, corrosion, secondary batteries, electrochemical analysis, solubilities of salts in liquids; publications in various journals. *Mailing Add:* N 32 W 22180 Shady Lane Pewaukee WI 53072

BULLOCK, KENNETH C, b Pleasant Grove, Utah, Sept 8, 18; m 38; c 4. GEOLOGY. *Educ:* Brigham Young Univ, BS, 40, MA, 42; Univ Wis, PhD(geol), 49. *Prof Exp:* From instr to assoc prof, 43-57, chmn dept, 56-62, PROF GEOL, BRIGHAM YOUNG UNIV, 57- *Concurrent Pos:* Geol engr, US Steel Corp, 53-54; geologist, Utah Geol & Mineral Surv, 67, 74; res geologist, Columbia Iron Mining Co, 60 & US Bur Mines, 75. *Mem:* Fel Geol Soc Am; Am Inst Mining, Metall & Petrol Engrs; Nat Asn Geol Teachers. Mineral Soc Am. *Res:* Economic geology; mineralogy; petrology; iron and fluorite deposits of Utah. *Mailing Add:* 1035 N 900 East Provo UT 84604

BULLOCK, LESLIE PATRICIA, b Los Angeles, Calif, May 20, 36. ENDOCRINOLOGY. *Educ:* Pomona Col, BA, 57; Univ Calif, Davis, DVM, 63. *Prof Exp:* Res fel endocrinol, Nat Cancer Inst, 68-70 & Med Col, Cornell Univ, 67-68; intern & mem staff, Angell Mem Hosp, Boston, 63-67; from asst prof to assoc prof comp med, Milton S Hershey Med Ctr, Pa State Univ, 71-89, res assoc endocrinol, 71-76, sr res assoc, 76-89; CAMPUS VET, BOYDEN LAB, UNIV CALIF, RIVERSIDE, 89- *Mem:* AAAS; Endocrine Soc; Soc Study Reproductive Biol; Am Fedn Clin Res; Am Vet Med Asn. *Res:* Interaction of androgens and other steroid and protein hormones at the biologic and molecular level; molecular mechanism of androgen action. *Mailing Add:* Boyden Lab Univ Calif Riverside CA 92521

BULLOCK, R MORRIS, b Greensboro, NC, Apr 18, 57; m 83. MECHANISTIC ORGANOMETALLIC CHEMISTRY, HOMOGENEOUS CATALYSIS. *Educ:* Univ NC, Chapel Hill, BS, 79; Univ Wis-Madison, PhD(chem), 84. *Prof Exp:* Res assoc, Colo State Univ, 84-85; from asst chemist to assoc chemist, 85-89, CHEMIST, CHEM DEPT, BROOKHAVEN NAT LAB, 89- *Mem:* Am Chem Soc. *Res:* Mechanistic and synthetic transition metal organometallic chemistry; reactions of metal hydrides with unsaturated substrates; mechanistic studies of the hydroformylation reaction; hydrogen atom transfer reactions of metal hydrides. *Mailing Add:* Chem Dept Brookhaven Nat Lab Upton NY 11973

BULLOCK, RICHARD MELVIN, b Glasco, Kans, July 8, 18; m 47; c 2. HORTICULTURE, RESEARCH ADMINISTRATION. *Educ:* Kans State Univ, BS, 40; Wash State Univ, MS, 42, PhD, 50. *Prof Exp:* Asst hort, Wash State Univ, 40-42, asst & assoc horticulturist, Tree Fruit Exp Sta, 46-52; prof & head dept hort, Utah State Agr Col, 52-53; supt & horticulturist, Southwestern Wash Exp Sta, 53-58, North Willamette Exp Sta, Ore State Univ, 58-69; horticulturist, 69-72, asst dir, Hawaii Agr Exp Sta, 69-75, agronomist, 72-75, prof hort, 72-81, prof agron, 74-75, horticulturist, 75-81, EMER PROF HORT, UNIV HAWAII, MANOA, 81-; CONSULT HORTICULTURIST, 81- *Mem:* Fel AAAS; Am Soc Plant Physiol; fel Am Soc Hort Sci; Sigma Xi; Int Plant Propagators Soc. *Res:* Nutrition, production, culture and management of horticultural crops, flower and fruit cycling; plant physiology; plant growth regulation; herbicides; research administration. *Mailing Add:* 1151 Upper Devon Lane Lake Oswego OR 97034

BULLOCK, ROBERT CROSSLEY, b New York, NY, Oct 16, 24; m 52; c 2. ENTOMOLOGY. *Educ:* St Lawrence Univ, BS, 48; Univ Conn, MS, 50, PhD(entom), 54. *Prof Exp:* Entomologist, Trop Res Dept, Tela RR Co, Repub Honduras, 54-61; ASSOC ENTOMOLOGIST, AGR RES CTR, UNIV FLA, 61- *Mem:* Entom Soc Am; Am Registry Prof Entomologists. *Res:* Biology and control of insect pests of citrus. *Mailing Add:* Agr Res & Educ Ctr PO Box 248 Ft Pierce FL 34954

BULLOCK, ROBERT M, III, b Cincinnati, Ohio, June 30, 37; m 63; c 2. AUDIO APPLICATIONS. *Educ:* Univ Cincinnati, BS, 61, MA, 62, PhD(math), 66. *Prof Exp:* Teacher pvt sch, Ohio, 62-63; instr math, Univ Cincinnati, 63-66; asst prof, 66-82, ASSOC PROF MATH & STATIST, MIAMI UNIV, 82- *Mem:* Inst Elec & Electronics Engrs; Soc Indust & Appl Math; Audio Eng Soc. *Res:* Mathematical models of loudspeaker and loudspeaker-crossover systems; design and construction methods based on the models. *Mailing Add:* Dept Math Miami Univ Oxford OH 45056

BULLOCK, RONALD ELVIN, b Camden, Ark, June 17, 34; m 64; c 1. NUCLEAR ENGINEERING, MATERIALS SCIENCE. *Educ:* La Polytech Inst, BS, 56; Tex Christian Univ, MA, 63, MS, 69. *Prof Exp:* From nuclear engr to sr nuclear engr, Gen Dynamics Corp, 56-72, sr design engr mat, 72-73; staff nuclear scientist, Gen Atomic Co, 73-86; SR ENG SPECIALIST, CONVAIR DIV, GEN DYNAMICS, INC, 86- *Mem:* Am Ceramics Soc; Am Carbon Soc. *Res:* Irradiation effects on materials; nuclear fuel performance; fracture mechanics; brittle materials design; advanced composite structures; high-temperature materials. *Mailing Add:* Gen Dynamics Inc Mail Zone 41-6850 PO Box 85357 San Diego CA 92186

BULLOCK, THEODORE HOLMES, b Nanking, China, May 16, 15; m 37; c 2. NEUROBIOLOGY. *Educ:* Univ Calif, AB, 36, PhD(zool), 40. *Hon Degrees:* Doctorate, Univ Frankfurt, Ger, 88. *Prof Exp:* Sterling fel, Yale Univ, 40-41, Rockefeller fel, 41-42, res asst pharmacol & neuroanat, 42-43, instr neuroanat, 43-44; asst prof anat, Sch Med, Univ Mo, 44-46; from asst prof to prof zool, Univ Calif, Los Angeles, 46-66; prof, 66-83, EMER PROF NEUROSCI, UNIV CALIF, SAN DIEGO, 83- *Concurrent Pos:* Mem corp, Marine Biol Lab, Woods Hole, 44-, trustee, 56-58; fel, Ctr Advan Study Behav Sci, 59-60; mem exec comt, Assembly of Life Sci, 73-76. *Honors & Awards:* Gerard Prize, Soc Neurosci, 84. *Mem:* Nat Acad Sci; Am Soc Zool (pres, 65); Am Philos Soc; Soc Neurosci (pres, 75); Int Soc Neuroethology (pres, 84-86); Am Acad Arts & Sci; Am Physiol Soc; Int Brain Res Orgn. *Res:* Comparative neurophysiology. *Mailing Add:* Dept Neurosci Univ Calif San Diego La Jolla CA 92093

BULLOCK, THOMAS EDWARD, JR, b Abilene, Tex, July 21, 38; m 80; c 5. ELECTRICAL ENGINEERING. *Educ:* Rice Univ, BA, 60, BSEE, 61; Stanford Univ, MSEE, 62, PhD(elec eng), 66. *Prof Exp:* Res assoc elec eng, Stanford Univ, 62-63; from asst prof to assoc prof, 66-74, PROF ELEC ENG, UNIV FLA, 74- *Mem:* Inst Elec & Electronics Engrs; Soc Indust & Appl Math. *Res:* Mathematical systems theory; digital control; applications of microprocessors in control. *Mailing Add:* Dept Elec Eng 911 CSE Univ Fla Gainesville FL 32601

BULLOCK, WARD ERVIN, JR, b Cincinnati, Ohio, June 21, 31. EXPERIMENTAL BIOLOGY. *Educ:* State Univ NY, Buffalo, BA, 54; Temple Univ, MS & MD, 59; Am Bd Internal Med, dipl, 66. *Prof Exp:* Med intern, Univ Minn Hosp, 59-60, fel internal med, 60-62; sr resident internal med, Yale Univ Med Ctr, 62-63, advan postdoctoral res fel immunol, 63-64; dir, Clin Res Ward, US Naval Med Res Unit No 2, Taipei, Taiwan, Repub China, 64-66, vis lectr, Infectious Dis & Clin Immunol, Nat Defense Med Ctr Hosp, Taiwan, 64-66; sr instr internal med, Univ Rochester Col Med, 66-67, asst prof microbiol, 66-68; from assoc prof to prof med & dir, Div Infectious Dis, Univ Ky Col Med, 70-80; assoc chmn res, Dept Med, Col Med, 88-89, sr assoc dean, Col Med, 89-91, DIR, INFECTIOUS DIS DIV & ARTHUR RUSSELL MORGAN PROF MED, UNIV CINCINNATI COL MED, 80- *Concurrent Pos:* Grants, USPHS & WHO, 67-95; mem, US Leprosy Panel, US-Japan Coop Med Sci Prog, NIH, 72-77 & 83-85, Bacterial & Mycotic Dis Study Sect, US Army Res & Develop Command, 75-79, Bact & Mycol Study Sect, Nat Inst Allergy & Infectious Dis, NIH, 76-80, sci adv bd, Am Leprosy Found, 81-84, Microbiol & Infectious Dis Res Comt, Nat Inst Allergy & Infectious Dis, NIH, 88-, chmn, 90-; adj prof, Dept Molecular Genetics, Biochem & Microbiol, Univ Cincinnati, Col Med, 80-; consult staff, Cincinnati Children's Hosp Med Ctr, 80- *Mem:* Sigma Xi; Am Fedn Clin Res; Am Soc Clin Invest; Asn Am Physicians; Am Asn Immunologists; fel Am Col Physicians; fel Infectious Dis Soc Am; AAAS; Am Soc Microbiol; Int Leprosy Asn. *Res:* Microbiology; immunology; molecular genetics; infectious diseases; leprosy. *Mailing Add:* Div Infectious Dis Dept Internal Med Col Med Univ Cincinnati 231 Bethesda Ave ML 560 Cincinnati OH 45267-0560

BULLOCK, WILBUR LEWIS, b New York, NY, Mar 8, 22; m 44; c 4. PARASITOLOGY. *Educ:* Queen's Col, NY, BS, 42; Univ Ill, MS, 47, PhD(zool), 48. *Prof Exp:* Univ Va fel, Mountain Lake Biol Sta, 48; from instr to assoc prof, Univ NH, 48-61, actg chmn dept, 58-59, prof zool, 61-87, EMER PROF ZOOL, UNIV NH, 87- *Concurrent Pos:* Asst instr, US Army Univ, France, 45; res fel biol, Rice Inst, 55-56; vis res prof, Fla Presby Col, 62-63; res affil, Harold W Manter Lab, Univ Nebr State Mus, 72- *Mem:* Am Soc Parasitol. *Res:* Morphology and taxonomy of Acanthocephala; fish intestinal histology and histopathology; blood and intestinal protozoa of marine fish. *Mailing Add:* Dept Zool Univ NH Durham NH 03824

BULLOFF, JACK JOHN, b New York, NY, Dec 9, 14; m 42, 52; c 4. TECHNOLOGY ASSESSMENT, TECHNOLOGICAL FORECASTING. *Educ:* City Col New York, BS, 39; Rensselaer Polytech Inst, PhD, 53. *Prof Exp:* Jr chemist, Los Alamos Nat Lab, 46; asst prof chem, assoc Col Upper NY State, 46-50; asst, Rensselaer Polytech Inst, 50-52; proj supvr, Commonwealth Eng Co, 53-56; prin chemist, Battelle Mem Inst, 56-68; prof hist sci & dir, Sci & Technol Studies, State Univ NY, 68-76; res assoc, Fla State Univ, Tallahassee, 77-78; safety & wastes consult, JT Baker Chem Co, US & Can, 79-85; CHIEF CONSULT SCIENTIST, NY STATE LEGIS COMN SCI & TECHNOL, ALBANY, 85- *Concurrent Pos:* Consult, 68- *Mem:* AAAS; Am Chem Soc; Asn Consult Chemists & Chem Engrs; NY Acad Sci; Am Inst Chemists; Sigma Xi. *Res:* Actinides; metallic soaps; polyoxyanions; particle technology; graphic arts; photopolymerization; technological forecasting; science and technology assessment; research and development planning; oil and chemical spill amelioration; environmental chemistry; science policy; science education; history of chemistry; chemical safety; hazardous wastes management; science and technology legislation; technology transfer; superconductor synthesis; photoultramicrominiaturization and diazophotosensitization. *Mailing Add:* 399 Ridge Hill Rd Schenectady NY 12303-5716

BULLOUGH, VERN L, b Salt Lake City, Utah, July 24, 28; m 47; c 5. SEX & HISTORY. *Educ:* Univ Utah, BA, 51; Univ Chicago, MA, 51, PhD(hist & hist sci), 54; Calif State Univ, Long Beach, BSN, 81. *Prof Exp:* Assoc prof hist & social sci, Youngstown Univ, 54-59; from asst prof to prof, Calif State Univ, Northridge, 59-80; prof, 80-87, dean col, 80-89, DISTINGUISHED PROF HIST & HIST SCI, STATE UNIV COL NY, BUFFALO, 87-,. *Concurrent Pos:* Co-prin investr, Proj Michaelson grant, USN, 61-64; adj prof, Calif Col Med, 61-68; Fulbright prof, Ain Shams Univ, Cairo, Egypt, 66-67; prin investr, US Off Educ grant, 66-69, Erickson Found grant, 69-76; vis prof, Univ Southern Calif, 69; lectr, Sch Pub Health, Univ Calif, Los Angeles, 70-79; comnr Bldg & Safety, City Los Angeles, 74-77; adj prof nursing, State Univ NY, Buffalo, 80- *Honors & Awards:* Garrison lectr, Am Asn Hist Med, 88; Distinguished Sci Achievement Award, Soc Sci Study Sex, 90. *Mem:* Soc Sci Study Sex (pres, 81-83); Am Asn Hist Med; Hist Sci Soc; Am Asn Hist Nursing; Am Hist Asn; AAAS; fel Am Acad Nursing. *Res:* Human sexuality (history & sociology); history of intellectual and creative, achievement, sex research and medicine and nursing; influences of biological factors upon history. *Mailing Add:* Hist & Soc Studies State Univ NY 1300 Elmwood Ave Buffalo NY 14222

BULMER, GLENN STUART, b Windsor, Ont, May 3, 31; US citizen; m 53; c 4. MEDICAL MYCOLOGY, MICROBIOLOGY. *Educ:* Mich State Univ, BS, 53, PhD(mycol), 60. *Prof Exp:* From instr to prof microbiol, Med Sch, Univ Okla, 60-; RETIRED. *Concurrent Pos:* NIH fel, Med Sch, Univ Okla, 60-62 & career develop award, 65-72; consult, Southeast Asian Ministers of Educ Orgn, 71-; instr, Sch Med, Univ Geneva, 72-73; assoc prof, Sch Med, Univ Saigon, 73- *Mem:* Am Soc Microbiol; Med Mycol Soc of the Americas; Trop Med & Microbiol Soc SVietnam; Mycol Soc Am. *Res:* Fungus diseases; pathogenesis of cryptococcosis. *Mailing Add:* 25 Atlas Dr Filinbest II Quezon City Manila Philippines

BULOW, FRANK JOSEPH, b Oaklawn, Ill, July 11, 41; m 63; c 4. FISH BIOLOGY. *Educ:* Southern Ill Univ, BA, 64, MA, 66; Iowa State Univ, PhD(fisheries biol), 69. *Prof Exp:* From asst prof to assoc prof, 69-80, PROF BIOL, TENN TECHNOL UNIV, 80- *Concurrent Pos:* Res grants, Sport Fishing Inst, 68-69, Tenn Technol Univ, 69-80, Bur Sport Fisheries & Wildlife Dingell-Johnson grant, 69-75 & US Army CEngr, 81-85 & 85-87; pres, Tenn chap, Am Fisheries Soc, 79-80; cert fisheries scientist. *Honors & Awards:* Sigma Xi Res Award, 79. *Mem:* Am Fisheries Soc; Sigma Xi. *Res:* Propagation and rearing of channel catfish; relationship between nucleic acid concentrations and growth rate of fishes; water pollution biology; factors limiting fish production in reservoirs, streams and ponds. *Mailing Add:* Dept Biol Tenn Technol Univ Cookeville TN 38501

BULTMAN, JOHN D, b Chicago, Ill, Apr 11, 24; m 52; c 1. CHEMISTRY, MARINE BIOLOGY. *Educ:* George Washington Univ, BS, 50, MS, 54; Georgetown Univ, PhD(chem), 62. *Prof Exp:* Mem staff, E I du Pont de Nemours & Co, 42-43; chemist, Bur Mines, US Dept Interior, 50, RES CHEMIST, NAVAL RES LAB, 50- *Concurrent Pos:* Assoc Ed, Biotropica. *Mem:* Am Chem Soc; Int Res Group Wood Preserv; Am Wood Preserv Asn; Asn Trop Biol; Am Soc Naval Engrs; Pan-Am Biodeterioration Soc. *Res:* Biological deterioration of materials in the terrestrial and marine environments. *Mailing Add:* US Naval Res Lab Code 6127 Washington DC 20390

BULUSU, SURYANARAYANA, b Ellore, India, Aug 22, 27; US citizen. ORGANIC CHEMISTRY, RADIATION CHEMISTRY. *Educ:* Andhra Univ, India, BSc, 48; Univ Bombay, BSc, 50, PhD(org chem), 54. *Prof Exp:* J N Tata Endowment Higher Educ Indians overseas fel chem, Yale Univ, 54, univ fel, 54-56; res assoc, Brookhaven Nat Lab, 56-59 & Univ Wis, Madison, 59-60; RES CHEMIST, EXPLOSIVES LAB, PICATINNY ARSENAL, US ARMY, 60- *Mem:* AAAS; Am Chem Soc; Sigma Xi; Am Soc Mass Spectrometry. *Res:* Organic mass spectrometry; applications to study of structural chemistry of explosives; nuclear magnetic resonance spectroscopy; decomposition chemistry of energetic materials. *Mailing Add:* 22 Skyview Terr Morris Plains NJ 07950

BUMBY, RICHARD THOMAS, b Brooklyn, NY. NUMBER THEORY. *Educ:* Mass Inst Technol, SB, 57; Princeton Univ, MA, 59, PhD(math), 62. *Prof Exp:* From instr to asst prof, 60-68, assoc prof, 68-77; PROF MATH, RUTGERS UNIV, 77- *Mem:* Am Math Soc; Am Asn Univ Professors; Math Asn Am. *Res:* Number theory, especially diophantine problems; analogs of Littlewood's approximation problem; the Markoff spectrum; algebraic and combinatorial problems in number theory. *Mailing Add:* Dept Math Rutgers Univ Brunswick NJ 08903

BUMCROT, ROBERT J, b Kansas City, Mo, Nov 24, 36; m 60, 75; c 2. MATHEMATICS. *Educ:* Univ Chicago, BS, 59, MS, 60; Univ Mo, PhD(math), 62. *Prof Exp:* Asst prof math, Ohio State Univ, 62-66; lectr, Univ Sussex, 66-67; asst prof, Ohio State Univ, 67-68; assoc prof, 68-74, chmn dept, 69-75, PROF MATH, HOFSTRA UNIV, 74- *Mem:* Am Math Soc; Math Asn Am. *Res:* Geometry; abstract structures; lattice theory. *Mailing Add:* Dept Math Hofstra Univ Hempstead NY 11550

BUMGARDNER, CARL LEE, b Belmont, NC, Jan 8, 25; m 56; c 5. ORGANIC CHEMISTRY. *Educ:* Univ Toronto, BASc, 52; Mass Inst Technol, PhD(org chem), 56. *Prof Exp:* Res assoc, Mass Inst Technol, 56; res chemist, Redstone Res Div, Rohm & Haas Co, 56-64; assoc prof chem, 64-67, PROF CHEM, NC STATE UNIV, 67-, HEAD DEPT, 76- *Mem:* Am Chem Soc. *Res:* Elimination reactions; photochemistry; nitrenes. *Mailing Add:* 4113 Glen Laurel Dr Raleigh NC 27612

BUMP, CHARLES KILBOURNE, b Pittsfield, Mass, June 1, 07; m 34; c 2. CHEMISTRY. *Educ:* Amherst Col, AB, 29; Cornell Univ, PhD(phys chem), 33. *Prof Exp:* Fel chem, Amherst Col, 34-35; teacher sci, Dorland-Bell Sch, NC, 35-36; inspector, Food & Drug Admin, USDA, 36-37; res chemist, plastics div, Monsanto Chem Co, 37-38, group leader, 38-46, asst dir res, 46-57, res personnel mgr, Monsanto Co, 58-70, res specialist, plastics prod & resins div, 70-72; RETIRED. *Mem:* Am Chem Soc. *Res:* Industrial problems dealing with plastics; resins for surface coatings. *Mailing Add:* 78 North Rd Hampden MA 01036

BUMPUS, FRANCIS MERLIN, b Rome, Ky, Dec 6, 22; m 47; c 4. ORGANIC CHEMISTRY, BIOCHEMISTRY. *Educ:* Purdue Univ, BS, 44; Univ Wis, MS, 47, PhD(biochem), 49. *Prof Exp:* Chemist, Allied Chem & Dye Co, 44-45; chemist, Cleveland State Univ, 49-61, asst dir res, 61-66, sci dir cardiovasc res, 66-67, chmn, Res Div, Cleveland Clin, 67-85, adj prof biol, 70-85; CONSULT, CLEVELAND CLIN FOUND, 85- *Concurrent Pos:* Chmn, Coun High Blood Pressure Res, Am Heart Asn, mem, Coun Basic Sci & Study Sect. *Honors & Awards:* Purdue Frederich Award, 67; Stouffer Prize, 68; Sci Achievement Award, Am Heart Asn, 89. *Mem:* AAAS; Am Chem Soc; Am Soc Biol Chemists; Am Heart Asn; Soc Exp Biol & Med; Am Pharm Soc; Coun High Blood Pressure Res; Int Soc Hypertension; Inter Am Soc Hypertension. *Res:* Synthesis of vaccenic acid and peptides; structure of antimycin A; purification and structure determination of angiotonin; vapor phase oxidation, isolation and identification of vasopressor substances; mechanisms of action of angiotensin; alodsterone biosynthesis and release phenomena; peptide chemistry and pharmacology; etiology of cardiovascular diseases; peptide synthesis, brain hormones. *Mailing Add:* Cleveland Clin Found 9500 Euclid Ave Cleveland OH 44195

BUMPUS, JOHN ARTHUR, b Syracuse, NY, Apr 25, 48; m 79; c 1. HEMEPROTEINS, MYCOLOGY. *Educ:* State Univ NY, Oswego, BS, 71, MS, 73, Binghamton, MA, 77; St Louis Univ, PhD(biochem), 80. *Prof Exp:* Fel biochem, St Louis Univ, 80-81; asst prof chem, Lake Superior State Col, 81-84; at Mich State Univ, 84-87; res assoc prof, Utah State Univ, 87-90; ASSOC FAC FEL, NOTRE DAME, 91- *Mem:* AAAS; Am Chem Soc; Am Soc Microbiol; Brit Mycol Soc; AAAS; Soc Toxicol; Soc Indust Microbiol. *Res:* Structure and function of hemeproteins, especially peroxidages and cytochrome P-450 containing monooxygenases; biodegradation of environmental pollutants by micro-organisms. *Mailing Add:* Dept Chem & Biochem Notre Dame IN 46556

BUNAG, RUBEN DAVID, b Manila, Philippines, June 3, 31; m 56; c 2. CARDIOVASCULAR PHYSIOLOGY, PHARMACOLOGY. *Educ:* Univ Philippines, MD, 55; Univ Kans, MA, 62. *Prof Exp:* Instr physiol, Univ Philippines, 55-57; from asst prof to assoc prof pharmacol, Med Ctr, Univ of the East, Manila, 57-60; USPHS int res fel, 60-62; Life Ins Med Res Fund res fel, Med Sch, Case Western Reserve Univ, 62-63; res fel, Cleveland Clin Res Div, 63-69, assoc res staff, 69-70; assoc prof, 70-75, PROF PHARMACOL, MED CTR, UNIV KANS, 75- *Concurrent Pos:* Res pharmacologist, Vet Admin Hosp, Kansas City, Kans, 70-72. *Mem:* Sigma Xi; Soc Exp Biol & Med; Neurosci Soc; Am Physiol Soc; Coun High Blood Pressure Res. *Res:* Cardiovascular pharmacology; experimental hypertension; central neural mechanisms for cardiovascular regulation; autonomic pharmacology. *Mailing Add:* Dept Pharmacol Univ Kans Med Ctr 39th & Rainbow Kansas City KS 66103

BUNBURY, DAVID LESLIE, b Georgetown, Brit Guiana, Feb 12, 33; m 57; c 4. PHYSICAL ORGANIC CHEMISTRY. *Educ:* Berea Col, BA, 52; Univ Notre Dame, PhD(phys chem), 56. *Prof Exp:* Fel chem, Univ Colo, 56-58; asst prof, 58-65, ASSOC PROF CHEM, ST FRANCIS XAVIER UNIV, 65- *Concurrent Pos:* Nat Res Coun Can grant, 59-71. *Mem:* Am Chem Soc; Chem Inst Can; Royal Soc Chem. *Res:* The photochemical decomposition of some ketones. *Mailing Add:* Dept Chem St Francis Xavier Univ Antigonish NS B2G 1C0 Can

BUNCE, ELIZABETH THOMPSON, b Mineola, NY, Apr 25, 15. GEOPHYSICS. *Educ:* Smith Col, AB, 37, MA, 49. *Hon Degrees:* ScD, Smith Col, 71. *Prof Exp:* Instr physics, Smith Col, 49-51; res asst, Woods Hole Oceanog Inst, 51-52, assoc scientist, 52-75, sr scientist, 75-80, EMER SCIENTIST, WOODS HOLE OCEANOG INST, 80- *Mem:* Fel Geol Soc Am; Soc Explor Geophys; Am Geophys Union; Am Asn Petrol Geologists. *Res:* Marine seismology; underwater acoustics. *Mailing Add:* PO Box 623 West Falmouth MA 02574

BUNCE, GEORGE EDWIN, b Nashville, Tenn, Dec 13, 32; m 55; c 3. NUTRITION, BIOCHEMISTRY. *Educ:* Va Polytech Inst, BS, 54, MS, 56; Univ Wis, PhD(biochem), 61. *Prof Exp:* Biochemist, US Army Med Res & Nutrit Lab, Fitzsimons Gen Hosp, Denver, Colo, 61-63; chief clin chem, Tripler Gen Hosps, Honolulu, Hawaii, 63-65; asst prof, 65-70, assoc prof, 70-78, PROF BIOCHEM, VA POLYTECH INST & STATE UNIV, 78- *Concurrent Pos:* Vis scientist, chem physiol lab, Cath Univ Louvain, 70-71, Rowett Inst, Aberdeen, 81-82. *Mem:* Am Inst Nutrit. *Res:* Mineral nutrition, particularly magnesium and its relation to soft tissue calcinosis and urolithiasis; nutritional effects on cataract; zinc and steroids. *Mailing Add:* Dept Biochem & Nutrit Va Polytech Inst & State Univ Blacksburg VA 24061

BUNCE, GERRY MICHAEL, b Pontiac, Mich; m; c 2. HIGH ENERGY PHYSICS. *Educ:* Mass Inst Technol, BS, 67; Univ Mich, PhD(physics), 71. *Prof Exp:* Res assoc physics, Univ Mich, 71; vis scientist, Centre d'Etudes l'Energie Nucleaire, Saclay, France, 71-73; res assoc, Univ Wis-Madison, 73-77; assoc physicist, 77-80, PHYSICIST, BROOKHAVEN NAT LAB, 80- *Concurrent Pos:* Vis physicist, Lab d' Annecy-Le-Vieux, France, 85-86, Univ Marseille, France, 86. *Mem:* Am Phys Soc. *Res:* Weak interactions; strong interaction dynamics. *Mailing Add:* Accelerator Dept Bldg 911B Brookhaven Nat Lab Upton NY 11973

BUNCE, JAMES ARTHUR, b Troy, NY, June 19, 49; m 75; c 2. PHYSIOLOGICAL ECOLOGY. *Educ:* Bates Col, BS, 71; Cornell Univ, PhD(phys ecol), 75. *Prof Exp:* res assoc crop physiol, Duke Univ, 75-77; PLANT PHYSIOLOGIST, USDA, 77- *Mem:* Sigma Xi; Ecol Soc Am; Am Soc Plant Physiologists. *Res:* Analysis of plant adaptations and responses to environmental factors, especially water, and consequences of these in determining distribution patterns and productivity. *Mailing Add:* Climate Stress Lab Beltsville Agr Res Ctr-West Beltsville MD 20705

BUNCE, NIGEL JAMES, b Sutton Coldfield, Eng, June 10, 43; m 67; c 2. ORGANIC CHEMISTRY, ENVIRONMENTAL CHEMISTRY & TOXICOLOGY. *Educ:* Oxford Univ, BA, 64, DPhil(chem), 69. *Prof Exp:* From asst prof to assoc prof, 69-80, PROF CHEM, UNIV GUELPH, 80- *Concurrent Pos:* Killam Mem fel, Univ Alta, 67-69. *Mem:* Chem Inst Can; Air Pollution Control Asn. *Res:* Mechanisms of free radical and photochemical reactions; photochemistry of aromatic compounds; environmental chemistry and toxicology of chlorinated aromatic compounds. *Mailing Add:* Dept Chem & Biochem Univ Guelph Guelph ON N1G 2W1 Can

BUNCE, PAUL LESLIE, b Fargo, NDak, Sept 24, 16; m 45; c 2. MEDICINE. *Educ:* Oberlin Col, AB, 38; Univ Chicago, MD, 42. *Prof Exp:* Asst instr pharmacol, Univ Pa, 46-48; house officer & resident urol, Johns Hopkins Univ, 48-51, instr, 50-52; from asst prof to prof surg, Univ NC, Chapel Hill, 52-81; RETIRED. *Mem:* AMA; Am Urol Asn. *Res:* Urology. *Mailing Add:* 1340 Old Lystra Rd Chapel Hill NC 27514

BUNCE, STANLEY CHALMERS, b Bayonne, NJ, Aug 21, 17; m 43; c 3. ORGANIC CHEMISTRY, SCIENCE EDUCATION. *Educ:* Lehigh Univ, BS, 38, MA, 42; Rensselaer Polytech Inst, PhD(chem), 51. *Prof Exp:* Teacher, high sch, Pa, 39-41 & NJ, 41-43; res chemist, Johns-Manville Corp, 43-46; from instr to prof, 46-84, EMER PROF CHEM, RENSSELAER POLYTECH INST, 84- *Concurrent Pos:* Vis prof, Univ Oslo, 74. *Mem:* Fel AAAS; Am Chem Soc. *Res:* Organic reaction mechanisms; synthetic organic chemistry; medicinal chemistry. *Mailing Add:* Dept Chem Rennsselaer Polytech Inst Troy NY 12181

BUNCEL, ERWIN, b Presov, Czech, May 31, 31, Can citizen; m 56; c 2. PHYSICAL ORGANIC CHEMISTRY. *Educ:* Univ London, BSc, 54, PhD(chem), 57. *Hon Degrees:* DSc, Univ London, 70. *Prof Exp:* Res assoc chem & fel, Univ NC, 57-58; Nat Res Coun Can fel, McMaster Univ, 58-61; res chemist, Am Cyanamid Co, 61-62; from asst prof to assoc prof, 62-66, PROF CHEM, QUEEN'S UNIV, ONT, 70- *Concurrent Pos:* Ed, Can J Chem, Isotopes Org Chem, Comprehensive Carbanion Chem, Isotopes in the Phys & Biomed Sci. *Honors & Awards:* Syntex Award Phys Org Chem, Chem Inst Can, 85. *Mem:* Am Chem Soc; Chem Inst Can; Royal Soc Chem. *Res:* Isotope effects in organic reaction mechanisms; aromatic substitution; acid base catalysis; carbanion mechanisms; dynamics of nitroaromatic-base interactions; enzyme and protein structural and mechanistic investigations; metal ion biomolecule interactions; organometallic chemistry; crown ether/cryptand chemistry author of over 200 publications; reviews chapters, books and edited monographs. *Mailing Add:* Dept Chem Queen's Univ Kingston ON K7L 3N6 Can

BUNCH, DAVID WILLIAM, b Eldon, Mo, Jan 20, 36; m 56; c 2. CHEMICAL ENGINEERING. *Educ:* Mo Sch Mines, BS, 57, MS, 60, PhD(chem eng), 64. *Prof Exp:* Instr chem eng, Mo Sch Mines, 57-64; process design engr, Ethyl Corp, 64-66, Sr Chem engr, 66-69, supvr, 69-79, prod mgr, 79-82, mkt mgr, 82-87, MKT DEVELOP DIR, ETHYL CORP, 87- *Mem:* Am Inst Chem Engrs. *Res:* Mass transfer; vaporliquid equilibria; process dynamics; mathematical simulation. *Mailing Add:* Ethyl Corp 451 Florida Blvd Baton Rouge LA 70801

BUNCH, HARRY DEAN, b Yukon, Okla, Oct 3, 15; m 41; c 3. AGRONOMY. *Educ:* Okla State Univ, BS, 41; Univ Tenn, MS, 42, PhD(agron, seed technol), 59. *Prof Exp:* Instr agron, Univ Ky, 42-43; asst agronomist, Agr Exp Sta, Miss State Univ, 46-55, assoc agronomist & supvr seed technol lab, 55-64, prof agron, 64-85, dir, int progs agr & forestry, 67-85; RETIRED. *Mem:* Am Soc Agron. *Res:* Seed storage; mechanical injury of seeds; electrostatic separation of seeds; magnetic seed cleaning. *Mailing Add:* 101 E Wood St Starkville MS 39759

BUNCH, JAMES R, b Globe, Ariz, Sept 23, 40. NUMERICAL ANALYSIS. *Educ:* Univ Ariz, BS, 62; Univ Calif, Berkeley, MA, 65, PhD(appl math), 69. *Prof Exp:* Asst mathematician, Argonne Nat Lab, 69-70; instr math, Univ Chicago, 70-71; asst prof comput sci, Cornell Univ, 71-74; assoc prof, 74-79, PROF MATH, UNIV CALIF, SAN DIEGO, 79- *Concurrent Pos:* Consult, Argonne Nat Lab, 74-, Los Alamos Nat Lab, 80- *Mem:* Am Math Soc; Soc Indust & Appl Math; Asn Comput Mach. *Res:* Numerical linear algebra; linear equations, inertia, sparse matrics, Toeplitz matrices, eigenproblems, signal processing. *Mailing Add:* Dept Math C-012 Univ Calif San Diego La Jolla CA 92093-0112

BUNCH, PHILLIP CARTER, b Maryville, Tenn, Sept 14, 46; m 67. MEDICAL PHYSICS. *Educ:* Univ Chicago, AB, 69, MS, 71, PhD(med physics), 75. *Prof Exp:* res physicist, 75-80, SR RES PHYSICIST, RES LABS, EASTMAN KODAK CO, 80- *Mem:* Optical Soc Am; Am Asn Physicists in Med; Soc Photo-Optical Instrument Engrs; Soc Photog Scientists & Engrs. *Res:* Radiologic image analysis; Monte Carlo simulation of the optical characteristics of radiographic screen-film systems; objective measurement of optical and noise properties of radiologic imaging systems; application of visual psychophysics to radiography. *Mailing Add:* Eastman Kodak Co Health Sci Res Div Bldg 81 R1/BSMT Rochester NY 14650

BUNCH, ROBERT MAXWELL, b Natchez, Miss, Oct 29, 53. SOLID STATE PHYSICS. *Educ:* La State Univ, BS, 75; Kans Univ, PhD(physics), 81. *Prof Exp:* Res assoc, Kans Univ, 81; asst res prof & lectr, San Diego State Univ, 81-; ASSOC PROF PHYS, ROSE-HULMAN INST, 83- *Mem:* Am Phys Soc; Soc Photo Optical Instrumentation Engrs; Optical Soc Am; Am Asn Physics Teachers. *Res:* Light scattering in solids to study diffusion, aggregation and precipitation of impurities; fiber optics. *Mailing Add:* Physics Dept Rose-Hulman Inst 5500 Wabash Ave Terre Haute IN 47803

BUNCH, THEODORE EUGENE, b Huntsville, Ohio, Mar 1, 36; m 63; c 2. ASTROGEOLOGY, GEOLOGY. *Educ:* Miami Univ, BA, 59, MS, 62; Univ Pittsburgh, PhD(geol), 66. *Prof Exp:* Res asst geosci, Mellon Inst Sci, 61-64; res asst impact craters & meteorites, Univ Pittsburgh, 64-66; NSF-Nat Res Coun fel, 66-69, RES SCIENTIST PLANETARY SCI & METEORITES, NASA-AMES RES CTR, 69- *Concurrent Pos:* Vis prof, Univ Cologne, 71 & Sonoma State Col, 74-; res assoc, Scripps Inst Oceanog, Univ Calif, San Diego, 74- *Honors & Awards:* H J Allen Award, NASA, 81. *Mem:* Fel Mineral Soc Am; fel Meteoritical Soc (secy, 70-75). *Res:* Geological research on the moon, earth, Mars, asteroids, meteorites, interplanetary and interstellar dust. *Mailing Add:* NASA-Ames Res Ctr 245-3 Moffett Field CA 94035

BUNCH, WILBUR LYLE, b Pine Bluffs, Wyo, Apr 24, 25; m 46; c 5. PHYSICS. *Educ:* Univ Wyo, BS, 49, MS, 51. *Prof Exp:* Engr, Gen Elec Co, 51-62, sr physicist, 62-64; res assoc reactor shielding, Battelle Mem Inst, 65-66, mgr, 67-70; mgr radiation & shielding anal, Wadco, 70-72; mgr radiation & shield anal, Westinghouse Hanford Co, 72-88; RETIRED. *Mem:* Am Nuclear Soc. *Res:* Nuclear reactor shielding; nuclear instrumentation; reactor design. *Mailing Add:* 2403 Pullen St Richland WA 99352

BUNCH, WILTON HERBERT, b Walla Walla, Wash, Jan 12, 35; m 56; c 2. ORTHOPEDIC SURGERY. *Educ:* Walla Walla Col, BS, 56; Loma Linda Univ, MD, 60; Univ Minn, PhD(physiol), 67; Univ Chicago, MBA, 82. *Prof Exp:* Mem attend staff orthop, Gillette State Hosp, 68-69; instr, Univ Minn, 68-69; assoc prof orthop, 69-74, prof orthop & pediat, Univ Va, 74-75; Scholl prof orthop & pediat, Sch Med, Loyola Univ, 75-85; DEAN MED AFFAIRS, UNIV CHICAGO, 85- *Concurrent Pos:* Mem fac prosthetics & orthotics, Northwestern Univ, 60- *Honors & Awards:* Richards Award, Richards Co, Tenn, 68; Nicholas Andry Award, Asn Bone & Joint Surg, 70. *Mem:* Am Physiol Soc; Orthop Res Soc; Opers Res Soc Am; Scoliosis Res Soc. *Res:* Cartilage growth and maturation. *Mailing Add:* 12901 Bruce B Downs Blvd MDC Box 21 Tampa FL 33612

BUNCHER, CHARLES RALPH, b Dover, NJ, Jan 9, 38. BIOSTATISTICS, EPIDEMIOLOGY. *Educ:* Mass Inst Technol, BS, 60; Harvard Univ, MS, 64, ScD, 67. *Prof Exp:* Statistician, Atomic Bomb Casualty Comn, Nat Acad Sci, 67-70; chief biostatistician, Merrell-Nat Labs, 70-73; asst prof statist, 70-73, PROF & DIR DIV EPIDEMIOL & BIOSTATIST, MED COL, UNIV CINCINNATI, 73- *Mem:* AAAS; Am Pub Health Asn; Am Statist Asn; Biomet Soc; Soc Epidemiol Res. *Res:* Cancer epidemiology; screening, diagnosis, and treatment, as well as occupational and environmental epidemiology; statistical research; clinical trials; design of experiments; pharmaceutical research; biostatistical analysis, pharmaceutical statistics; ALS epidemiology; risk analysis. *Mailing Add:* Univ Cincinnati Med Col Mail Location 183 Cincinnati OH 45267-0183

BUNDE, CARL ALBERT, b Ashland Co, Wis, Apr 22, 07; m 30. CLINICAL PHARMACOLOGY. *Educ:* Univ Wis, AB, 33, AM, 34, PhD(zool), 37; Southwestern Med Col, MD, 48. *Prof Exp:* Asst zool, Univ Wis, 34-37; instr, Sch Med, Univ Okla, 38-42; asst prof physiol & pharm, Col Med, Baylor Univ, 42-43; assoc prof, Southwestern Med Col, 43-44, assoc prof physiol, 44-49; vpres res, Pitman-Moore Co, 49-60; dir med res, William S Merrell Co, 60-72; RES CONSULT, 72- *Concurrent Pos:* Fel, Sch Med, Univ Okla, 37-38. *Mem:* Endocrine Soc; Am Physiol Soc; Soc Exp Biol & Med; AMA; Am Fedn Clin Res; Sigma Xi. *Res:* Endocrinology; circulation; metabolism; clinical pharmacology. *Mailing Add:* 3738 Donegal Dr Cincinnati OH 45236

BUNDE, DARYL E, b Sioux Falls, SDak, Oct 29, 37; m 59; c 3. ZOOLOGY, PHYSIOLOGY. *Educ:* Augustana Col, SDak, BA, 59; Univ Tex, MA, 62; Mont State Univ, PhD(zool), 65. *Prof Exp:* Assoc Rocky Mountain Univs fac orientation grant, Los Alamos Sci Lab, 65; asst prof zool, 65-74, ASSOC PROF ZOOL, IDAHO STATE UNIV, 74- *Mem:* Am Soc Zoologists. *Res:* Amino acid metabolism of an insect; metabolism of radioactive elements in mammals. *Mailing Add:* Dept Biol Idaho State Univ Pocatello ID 83209

BUNDSCHUH, JAMES EDWARD, b St Louis, Mo, Nov 13, 41; m 65; c 4. PHYSICAL CHEMISTRY. *Educ:* St Louis Univ, BS, 63; Duquesne Univ, PhD(phys chem), 67. *Prof Exp:* Fel, Univ Ill, Chicago Circle, 67-68; asst prof chem, Western Ill Univ, 68-75, assoc prof & chmn dept, 75-80, prof, 80; mem staff, Dept Chem & Dean, Sch Sci & Humanities, Ind Univ-Purdue Univ, Ft Wayne, 80-86; DEAN COL ARTS & SCI, ST LOUIS UNIV, 86- *Concurrent Pos:* Guest lectr, Univ Stuttgart, 73-74. *Mem:* Am Chem Soc; Sigma Xi. *Res:* Surface chemistry; atmospheric chemistry; nuclear magnetic resonance. *Mailing Add:* St Louis Univ 221 N Grand Blvd St Louis MO 63103

BUNDY, BONITA MARIE, b Washington, DC, Jan 21, 48. CELLULAR IMMUNOLOGY, CLINICAL IMMUNOLOGY. *Educ:* George Washington Univ, BS, 69, PhD(microbiol, immunol), 82; Univ Md, College Park, MS, 72. *Prof Exp:* Instr, Dept Microbiol, Univ Md, 69-72; MICROBIOLOGIST, RES IMMUNOL, NAT CANCER INST, NIH, 73- *Concurrent Pos:* Nat Defense Educ Act fel, Dept Microbiol, Univ Md, 69-72; med technologist clin microbiol, George Washington Univ Med Ctr, 69-76. *Mem:* Am Soc Microbiol; Am Asn Immunologists; Sigma Xi. *Res:* Experimental and clinical research in cellular immunology with emphasis on human immune deficiency diseases; studies of non-immune and immune virus-mediated cytotoxic and other immune functions; major histocompatibility complex-directed interactions in humans. *Mailing Add:* PO Box 91635 Washington DC 20090

BUNDY, FRANCIS P, b Columbus, Ohio, Sept 1, 10; m 36; c 4. EARTH SCIENCES. *Educ:* Otterbein Col, BS, 31; Ohio State Univ, MS, 33, PhD(physics), 37. *Hon Degrees:* Otterbein Col, DSci(physics), 59. *Prof Exp:* Instr & asst prof physics, Ohio Univ, Athens, 37-42; res assoc physics, Harvard Underwater Sound Lab, 42-45, Gen Elec Res & Develop Ctr, 46-86; CONSULT, 86- *Honors & Awards:* Roozeboom Gold Medal, Neth Acad Arts & Sci, 69; Bridgman Gold Medal, Int Asn Adv High Pressure Sci & Technol, 87. *Mem:* Am Phys Soc; AAAS; Sigma Xi. *Res:* Fields of spectroscopy, electro-mechanical transducers, vacuum thermal insulation, high pressure science and technology; diamond synthesis by high pressure and temperature; behavior of matter at ultra-high pressure. *Mailing Add:* 250 Alplaus Ave PO Box 29 Alplaus NY 12008

BUNDY, GORDON LEONARD, b Akron, Ohio, Nov 27, 42; m 66; c 3. ORGANIC CHEMISTRY. *Educ:* Col Wooster, BA, 64; Northwestern Univ, PhD(org chem), 68. *Prof Exp:* RES CHEMIST, UPJOHN CO, 68- *Mem:* Am Chem Soc; The Chem Soc. *Res:* Chemistry and synthesis of naturally occurring materials, especially sesquiterpenes, steroids and prostaglandins; new synthetic methods; arachidonic acid cascade modulation; quinolone antibacterials; renin inhibitors. *Mailing Add:* Dept Med Chem Upjohn Co 7246-209-6 Kalamazoo MI 49001

BUNDY, HALLIE FLOWERS, b Los Angeles, Calif, Apr 2, 25. BIOCHEMISTRY. *Educ:* Mt St Mary's Col, BA, 47; Univ Southern Calif, MS, 55, PhD, 58. *Prof Exp:* Instr, Univ Southern Calif, 58-60; from asst prof to assoc prof, 60-67, PROF BIOCHEM, MT ST MARY'S COL, CALIF, 67- *Concurrent Pos:* Sci Fac Fel, Nat Sci Foundation, 69; Grant-in-Aid, Grad Women in Sci, 75. *Mem:* AAAS; Am Chem Soc; Sigma Xi. *Res:* Chemistry of proteins; proteolytic enzymes; zymogens and inhibitors; carbonic anhydrase; comparative biochemistry. *Mailing Add:* PO Box 4338 Sunriver OR 97707

BUNDY, KIRK JON, b Highland Park, Mich, May 21, 47. BIOMATERIALS, CORROSION SCIENCE & ENGINEERING. *Educ:* Mich State Univ, BS, 68; Stanford Univ, MS, 70, PhD(mat sci & eng), 75. *Prof Exp:* Res asst, dept aeronaut & astron, Stanford Univ, 71; sci assoc, Biomed Eng Inst, Swiss Fed Inst Technol, 71-75; teaching fel biomat & metall, Metall Prog, Ga Inst Technol, 75-78; asst prof biomat, mat sci & eng, biomed eng dept & mat sci & eng dept, Johns Hopkins Univ, 78-83; ASSOC PROF BIOMAT, BIOMED ENG DEPT, TULANE UNIV, 83- *Concurrent Pos:* Consult, Metall Div, Nat Bur Standards, 81-83; sabbatical, Lab Exp Surg, Davos, Switz, 90; Fogarty sr int fel, 90. *Mem:* Soc Biomat; Biomed Eng Soc; Acad Dent Mat; Am Soc Testing & Mat; Am Inst Mining, Metall & Petrol Engrs; Nat Asn Corrosion Eng. *Res:* Interaction of applied stresses and corrosion processes as related to surgical implant materials; application of composite theory to bone. *Mailing Add:* Biomed Eng Dept Tulane Univ New Orleans LA 70118

BUNDY, LARRY GENE, b Red Bud, Ill, Dec 21, 43; m 66; c 2. CORN NITROGEN MANAGEMENT. *Educ:* Univ Ill, Urbana, BS, 66, MS, 67; Iowa State Univ, PhD(soil chem), 73. *Prof Exp:* Soil specialist, Libby, McNeil & Libby, 73-77, Nestle Enterprises, Inc, 77-82; asst prof, 82-88, ASSOC PROF SOIL FERTILITY, UNIV WIS-MADISON, 88- *Mem:* Am Soc Agron; Soil Sci Soc Am. *Res:* Improvement of nitrogen management in corn production; evaluation of diagnostic tests to predict corn nitrogen needs and development of nitrogen management techniques for improved agronomic efficiency and reduced environmental risks. *Mailing Add:* Dept Soil Sci 1525 Observatory Dr Madison WI 53706-1298

BUNDY, ROBERT W(ENDEL), b Passaic, NY, Jan 25, 24; m 52; c 3. ENGINEERING MECHANICS. *Educ:* Va Polytech Inst, BS, 47, MS, 48; Pa State Univ, PhD(eng mech), 54. *Prof Exp:* Asst, Pa State Univ, 48-52, instr, 52-53; engr, Am Flexible Coupling Co, 51-52; res physicist, E I du Pont de Nemours & Co, 54-61, res supvr, 61-71; PRES, SECURITY ELECTRONICS, INC, 71- *Concurrent Pos:* Consult eng & security; chmn bd, PSA Security Network, 77- *Mem:* Soc Exp Stress Anal; Am Soc Indust Security. *Res:* Design and installation of closed circuit television and access control systems. *Mailing Add:* 3603 Woodmont Blvd Nashville TN 37215

BUNDY, ROY ELTON, b Zanesville, Ohio, July 8, 24; c 2. PHYSIOLOGY. *Educ:* Ohio State Univ, BS, 46, MS, 48; Univ Wis, PhD(zool), 55. *Prof Exp:* Teacher pub sch & instr physics, Ohio State Univ, 46-47; instr biol, Transylvania Col, 48-51; asst, Univ Wis, 51-55; asst prof biol, Wagner Col, 55-57; asst prof biol & physiol, Sch Dent, Fairleigh Dickinson Univ, 57-61, assoc prof physiol & chmn dept, 61-68; PROF & CHMN DEPT BIOL, OKALOOSA-WALTON COMMUNITY COL, 68- *Mem:* AAAS. *Mailing Add:* Dept Continuing Educ Okaloosa-Walton Jr Col Niceville FL 32578

BUNDY, WAYNE MILEY, b Anderson, Ind, Jan 10, 24; m 45; c 3. CLAY MINERALOGY. *Educ:* Ind Univ, AB, 50, MA, 54, PhD, 57. *Prof Exp:* Geologist, NMex State Bur Mines, 51-53; petrogr, Ind Geol Surv, 53-57; dir labs, Ga Kaolin Co, 57-67, dir res, 67-74, vpres res, 74-87, vpres technol, 87-91; RETIRED. *Mem:* Clays Minerals Soc; Am Chem Soc; Tech Asn Pulp & Paper Indust. *Res:* Geochemistry; petrology; surface chemistry. *Mailing Add:* 45 Bissell Rd Lebanon NJ 08833

BUNGAY, HENRY ROBERT, III, b Cleveland, Ohio, Jan 22, 28; m 52; c 3. MICROBIAL BIOENGINEERING. *Educ:* Cornell Univ, BChE, 49; Syracuse Univ, PhD(biochem), 54. *Prof Exp:* Biochemist, Eli Lilly & Co, Ind, 54-62; prof sanit eng, Va Polytech Inst, 62-67, prof bioeng, Clemson Univ, 67-73; vpres, Worthington Biochem Corp, 73-76; PROF CHEM & ENVIRON ENG, RENSSELAER POLYTECH INST, 76- *Honors & Awards:* James Van Lanen Award, Am Chem Soc. *Mem:* Am Chem Soc; fel Am Inst Chem Engrs; Am Soc Microbiol; Sigma Xi. *Res:* Bioengineering; fermentation processes; biological waste treatment. *Mailing Add:* Rensselaer Polytech Inst Troy NY 12181

BUNGAY, PETER M, b Brooklyn, NY, June 17, 41; m 67. SUSPENSION HYDRODYNAMICS. *Educ:* Cooper Union, BChE, 63; Carnegie-Mellon Univ, PhD(chem eng & biotechnol), 71. *Prof Exp:* Res fel, Montreal Gen Hosp, Univ Med Clin, McGill Univ, 71-73; grants assoc, Div Res Grants, 73-74, spec asst, Off Dir, 74-75, CHEM ENGR, NAT CTR RES RESOURCES, NIH, 75- *Mem:* Am Inst Chem Engrs; AAAS; NAm Membrane Soc; Europ Soc Membrane Sci & Technol. *Res:* Physiological pharmacokinetics of drugs and environmental agents; low-Reynolds-number hydrodynamics of suspensions; biomedical research applications of synthetic membranes; transport phenomena in the microcirculation. *Mailing Add:* Biomed Eng & Instrumentation Prog NIH Bldg 13 Rm 3W-13 Bethesda MD 20892

BUNGE, CARLOS FEDERICO, b Buenos Aires, Arg, Mar 27, 41; m 62; c 4. ATOMIC SPECTROSCOPY, ELECTRONIC STRUCTURE CALCULATIONS. *Educ:* Nat Univ Buenos Aires, BSc, 62; Univ Fla, PhD(chem), 66. *Prof Exp:* Res assoc chem, Ind Univ, 67-68; asst prof, Cent Univ Venezuela, 68-70; prof, Univ Sao Paulo, Sao Carlos, 71-76; PROF PHYSICS, NAT UNIV AUTONOMA, MEX, 76- *Res:* Atomic andmolecular electronic structure calculations; algorithms; calculations strategies; software development; atomic spectroscopy; search for new species; atom-atom interactions; bond formation; small molecules. *Mailing Add:* Instituto de Física Apartado 20-364 Mexico DF 01000 Mexico

BUNGE, MARIO AUGUSTO, b Buenos Aires, Arg, Sept 21, 19; Arg & Can citizen; m 40, 59; c 4. THEORETICAL PHYSICS, PHILOSOPHY. *Educ:* Univ La Plata, PhD(physics), 52;. *Hon Degrees:* LLD, Simon Fraser Univ, 81, Univ Rosario, 85, Univ La Plata, 87. *Prof Exp:* Prof physics, Univ Buenos Aires, 56-58, prof philos, 57-63; prof physics, Univ La Plata, 56-59; PROF PHILOS, MCGILL UNIV, 66- *Concurrent Pos:* Ed, Minerva, 44-45, Arg Physics Asn, 53-63, Episteme, D Reidel Publ Co, 74- & Found & Philos Sci & Technol, Pergamon Press, 79-; vis prof philos, Univ Pa, 60-61, physics & philos, Temple Univ, 63-64, Univ Del, 64-65 & Univ Freiburg, 65-66 & ETH Zürich, 73; res prof philos, Univ Mex, 75-76; nat lectr, Sigma Xi, 80-82 & Univ Geneva, 86-87. *Honors & Awards:* Prince of Asturia's Prize, 82; Humanist Laureate, Acad Humanism, 85. *Mem:* Int Acad Philos Sci; Int Inst Philos; fel AAAS; Brit Soc Philos Sci; Philos Sci Asn; Can Philos Asn; Can Soc Hist & Philos Sci. *Res:* Theoretical physics; foundations and philosophy of physics, psychology and social science; semantics, ontology, epistemology, sytems theory, value theory and ethics. *Mailing Add:* 29 Bellevue Ave Montreal PQ H3Y 1G4 Can

BUNGE, MARY BARTLETT, b New Haven, Conn, Apr 3, 31; m 56; c 2. CYTOLOGY. *Educ:* Simmons Col, BS, 53; Univ Wis, MS, 55, PhD(zool, cytol), 60. *Prof Exp:* Asst med, Univ Wis, 53-56, asst zool, 56-60; res assoc anat, Col Physicians & Surgeons, Columbia Univ, 62-70; from res asst prof to res assoc prof, Sch Med, Wash Univ, 70-74, assoc prof anat, 74-78, prof anat neurobiol, 78-89; PROF CELL BIOL ANAT & NEUROSURG SURG, UNIV MIAMI, 89- *Concurrent Pos:* Nat Inst Neurol Dis & Blindness fel, Col Physicians & Surgeons, Columbia Univ, 60-62; res assoc, Harvard Med Sch, 68-69; mem, Neurol Dis Prog Proj Rev Comt, Nat Inst Neurol & Commun Dis & Stroke, NIH, 83-87; Nat Adv Neurol & Commun Dis & Stroke Coun, NIH, 87- *Mem:* Soc Neurosci; Am Soc Cell Biol; Am Asn Anat; Electron Micros Soc Am. *Res:* Growth, differentiation, injury and repair of nervous tissue; normal and pathological nervous tissue; tissue culture and cell fine structure; cell to cell and cell to extra cellular matrix interaction in developing and regenerating nervous tissue. *Mailing Add:* Miami Proj Univ Miami Sch Med 1600 NW Tenth Ave R-48 Miami FL 33136

BUNGE, RICHARD PAUL, b Madison, SDak, Apr 15, 32; m 56; c 2. ANATOMY, CELL BIOLOGY. *Educ:* Univ Wis, BA, 54, MS, 56, MD, 60. *Prof Exp:* Asst anat, Univ Wis, 54-57, instr, 57-58; from asst prof to assoc prof anat, Col Physicians & Surgeons, Columbia Univ, 62-70; prof anat, Sch Med, Wash Univ, 70-89; SCI DIR, MIAMI PROJ/PARALYSIS DIV & PROF NEUROSURG CELL BIOL ANAT, MIAMI UNIV, 89- *Concurrent Pos:* Vis Auburn Univ prof, Harvard Med Sch, 68-69; Nat Multiple Sclerosis Soc fel surg, Col Physicians & Surgeons, Columbia Univ, 60-62; Lederle med fac award, 64-67. *Mem:* Am Asn Anat; Am Soc Cell Biol; Tissue Cult Asn; Am Asn Neuropath; Soc Neurosci. *Res:* Biology of cells of the nervous system in vivo and in vitro. *Mailing Add:* Miami Proj Univ Miami Med Sch 1600 NW Tenth Ave R-48 Miami FL 33136

BUNGER, JAMES WALTER, b Orange City, Iowa, Apr 7, 45; m; c 2. FUELS ENGINEERING, ANALYTICAL CHEMISTRY. *Educ:* Univ Wyo, BSc, 68; Univ Utah, PhD(fuels eng), 79. *Prof Exp:* Lab technician chem, Laramie Energy Technol Ctr, Dept Energy, 67-68, chemist, 68-75; res assoc fuels eng, 75-79, RES ASST PROF FUELS ENG & SCI ADV, UNIV UTAH, 79- *Mem:* Am Chem Soc; Am Inst Chem Engrs. *Res:* Characterization, processing and utilization research and development of tar sand, oil shale and coal resources. *Mailing Add:* Box 520037 Salt Lake City UT 84152

BÜNGER, ROLF, b Hamburg, Germany, Oct 19, 41; m 73; c 2. METABOLIC HEART & CORONARY RESEARCH, BIOENERGETICS OF THE HEART. *Educ:* Univ Heidelberg, Dr med, 70; Univ Hamburg, MD, 71; Univ Munich, Dr med habil(physiol), 79. *Prof Exp:* Intern clin med, Heidberg Hosp, Hamburg, 70; sci asst physiol, Tech Univ Aachen, Germany, 70-74 & Univ Munich, 74-79; asst prof, 79-82, ASSOC PROF PHYSIOL, UNIFORMED SERV UNIV HEALTH SCI, BETHESDA, MD, 82-

Concurrent Pos: Reviewer sci manuscripts Am & Europ journals physiol & related fields, 78-; vis prof, Uniformed Serv Univ Health Sci, Bethesda, Md, 78, prin investr, 79-; travel awards, German Res Asn, 78 & NSF, 83; NIH res grant, 82-90. *Mem:* German Physiol Soc; Int Study Group Heart Res; Am Physiol Soc; Am Heart Asn; AAAS. *Res:* Metabolic regulation of coronary flow in relation to cellular compartments of cardiac adenine nucleotides; energy and redox metabolism in normal and oxygen-deficient hearts; regulation of pyruvate metabolism. *Mailing Add:* Dept Physiol Uniformed Serv Univ 4301 Jones Bridge Rd Bethesda MD 20814-4799

BUNGO, MICHAEL WILLIAM, b Passaic, NJ, July 18, 50. CARDIOVASCULAR MEDICINE, AEROSPACE MEDICINE. *Educ:* Rensselaer Polytech Inst, BS, 71; NJ Med Sch, MD, 75. *Prof Exp:* Resident med, New Eng Deaconess Hosp, Boston, Mass, 75-78, fel cardiol, 78-80; head cardiovasc res, 80-86, dir, Space Biomed Res Inst, 86-90, CHIEF SCIENTIST, MED SCI DIV, JOHNSON SPACE CTR, NASA, HOUSTON, 90- *Concurrent Pos:* Clin asst prof med, Health Sci Ctr, Univ Tex, Houston, 84, Med Br, Galveston, 89- *Honors & Awards:* Louis H Bauer Founders Award, Aerospace Med Asn, 87. *Mem:* Fel Am Col Cardiol; Am Heart Asn; Aerospace Med Asn. *Res:* Aerospace medicine, particularly the cardiovascular changes occurring in humans exposed to space flight. *Mailing Add:* 18100 Hospital Blvd No 370 Houston TX 77058

BUNICK, GERARD JOHN, b Boston, Mass, May 27, 47; m 73; c 2. PROTEIN CRYSTALLOGRAPHY, NEUTRON SCATTERING. *Educ:* Univ Mass, BS, 69; Univ Pa, PhD(biol chem), 75. *Prof Exp:* Qual control chemist, Monsanto Co, 68; asst scientist, Anal Div, Silver Lab, Polaroid Corp, 69; res assoc, Dept Chem, Univ Pa, 74-77; res chemist, Oak Ridge Nat Lab, 77-81; RES ASST PROF, OAK RIDGE GRAD SCH BIOMED SCI, UNIV TENN, 82- *Concurrent Pos:* Lectr, Oak Ridge Grad Sch Biomed Sci, Univ Tenn, 79- *Mem:* Am Chem Soc; Am Crystallog Asn; AAAS; Am Inst Physics. *Res:* Application of biophysical techniques, especially x-ray diffraction and small angle neutron scattering, to elucidate the structure of biologically important macromolecules. *Mailing Add:* 117 Claymore Lane Oak Ridge TN 37830-7675

BUNKER, BRUCE ALAN, b Pasadena, Calif, Apr 20, 52. XRAY SPECTROSCOPY. *Educ:* Univ Wash, BSc, 74, PhD(physics), 80. *Prof Exp:* Res assoc, Univ Wash, 80 & NC State Univ, 81; IBM fel physics, Univ Ill, 81-83; asst prof, 83-87, ASSOC PROF PHYSICS, UNIV NOTRE DAME, 87- *Mem:* Am Phys Soc. *Res:* X-ray and ultraviolet spectroscopy using synchrotron radiation sources. *Mailing Add:* Dept Phys Univ Notre Dame Col Sci Notre Dame IN 46556

BUNKER, JAMES EDWARD, b Huntington Park, Calif, Aug 29, 36; m 63; c 1. MEDICINAL CHEMISTRY. *Educ:* Univ Calif, Los Angeles, BS, 58. *Prof Exp:* Chemist, Riker Labs, 58-73; sr res chemist med chem, 74-79, SR PROD DEVELOP CHEMIST, 3M CO, 80- *Mem:* Am Chem Soc. *Res:* Synthesis of central nervous system agents; chemotherapeutic nucleotides; biodegradable polymers; dental adhesive products. *Mailing Add:* 3M Ctr 230-25-05 St Paul MN 55144-1000

BUNKER, JOHN PHILIP, b Boston, Mass, Feb 13, 20; m 44; c 4. MEDICINE. *Educ:* Harvard Univ, BA, 42, MD, 45; Am Bd Anesthesiol, dipl. *Prof Exp:* Instr anesthesia, Harvard Med Sch, 50-52, assoc, 52-55, asst clin prof, 55-60; PROF ANESTHESIA, SCH MED, STANFORD UNIV, 60-, PROF FAMILY, COMMUNITY & PREV MED, 75-, DIR HEALTH SERV RES, 76- *Concurrent Pos:* Anesthetist, Mass Gen Hosp, Boston, 50-60. *Mem:* Am Soc Anesthesiol; Am Soc Pharmacol & Exp Therapeut. *Res:* Health services research; technology assessment. *Mailing Add:* Anesthesia & Health Res & Policy 13 The Green Twickenham Middlesex TW2 5TU England

BUNKER, MERLE E, b Kansas City, Mo, Feb 8, 23; m 43; c 2. NUCLEAR PHYSICS. *Educ:* Purdue Univ, BS, 46; Ind Univ, PhD(physics), 50. *Prof Exp:* Mem staff, Physics Div, Los Alamos Nat Lab, Univ Calif, 50-65, alt group leader, Group P-2, 65-74, group leader, Group P-2, 74-81, group leader, Group Inc-5, 81-90; RETIRED. *Concurrent Pos:* NSF sr fel, 64-65; supvr res reactor, 54-90; adv, Nuclear Data Group, Oak Ridge Nat Lab, 67-78; lab assoc, Los Alamos Nat Lab, Univ Calif, 90- *Mem:* Fel Am Phys Soc. *Res:* Gamma-ray spectroscopy; nuclear structure; neutron activation analysis. *Mailing Add:* Reactor Group INC-5 MS-G776 Los Alamos Nat Lab PO Box 1663 Los Alamos NM 87545

BUNN, CLIVE LEIGHTON, b Melbourne, Australia, May 25, 45; m 76. CELL BIOLOGY, BIOTECHNOLOGY. *Educ:* Monash Univ, BSc Hons, 69, PhD(biochem), 75. *Prof Exp:* Fel assoc res, Dept Human Genetics, Yale Univ, 72-75, res assoc, 76; vis scientist, Genetics Div, Nat Inst Med Res, UK, 76-79; asst prof, Dept Biol, Univ SC, 80-87; SR RES SCIENTIST, BIOTECH AUSTRALIA, 87- *Concurrent Pos:* Beit fel, 76-79. *Mem:* Am Soc Cell Biol; Tissue Culture Asn; Australian Biotechnol Asn; Australian NZ Soc Cell Biol. *Res:* Biology of animal cells, especially the use of cultured cells for the production of recombinant proteins with therapeutic applications; aging in cultured cells. *Mailing Add:* Biotech Australia 28 Barcoo St PO Box 20 Roseville NSW 2069 Australia

BUNN, HOWARD FRANKLIN, b Morristown, NJ, July 7, 35; m 62; c 3. HEMATOLOGY, BIOCHEMISTRY. *Educ:* Harvard Univ, AB, 57; Univ Pa, MD, 61. *Prof Exp:* Asst prof med, Albert Einstein Col Med, 68-69; from asst prof to assoc prof, 73-79, PROF MED, SCH MED, HARVARD UNIV, 79-; DIR HEMAT RES, PETER BENT BRIGHAM HOSP, 75- *Concurrent Pos:* Res career develop award, Nat Heart & Lung Inst, NIH, 69-74; scholar in residence, Fogarty Int Ctr, NIH, 82-84. *Honors & Awards:* Merit Award, NIH, 90. *Mem:* Am Soc Biol Chemists; Am Soc Clin Invest; Asn Am Physicians. *Res:* Erythropoiesis; hemoglobinopathies; sickle cell anemia; red cell metabolism; hemoglobin structure and function. *Mailing Add:* Div Hemat 721 Huntington Ave Boston MA 02115

BUNN, JOE M(ILLARD), b Goldsboro, NC, Jan 20, 32; m 55; c 2. AGRICULTURAL ENGINEERING. *Educ:* NC State Col, BS, 55, MS, 57; Iowa State Univ, PhD(agr eng, math), 60. *Prof Exp:* From asst prof to assoc prof agr eng, Univ Ky, 60-78; PROF AGR ENG, CLEMSON UNIV, 78- *Concurrent Pos:* Engr, Univ Ky-Agency Int Develop contract team, Thailand, 68-69. *Mem:* Am Soc Agr Engrs. *Res:* Engineering aspects of agricultural and food processing and materials handling; corn and soybean drying and storage. *Mailing Add:* Agr Eng Dept Clemson Univ Clemson SC 29634-0357

BUNN, PAUL A, JR, b New York, NY, Mar 16, 45; m 68; c 3. MEDICAL ONCOLOGY, HEMATOLOGY. *Educ:* Amherst Col, BA, 67; Cornell Univ, MD, 71. *Prof Exp:* Intern med, H C Moffitt Hosp, Univ Calif, San Francisco, 71-72, resident, 72-73; fel, Med Br, 73-76, sr investr, Med Oncol Br, 76-81, CHIEF MED ONCOL, CELLULAR KINETIC SECT, NAVY MED ONCOL BR, NAT CANCER INST, 81- *Concurrent Pos:* Asst prof med, Med Col, Georgetown Univ, 77-81; assoc prof med, Uniformed Serv Univ Health Sci, 81- *Mem:* Am Soc Hemat; Am Soc Clin Oncol; Am Asn Cancer Res; Cell Kinetic Soc; Am Fedn Clin Res. *Res:* Growth and characterization of malignant tritium cell disorders; cell biology of malignant tritium cell disorders; clinical studies of malignant lymplomas and small cell lung cancer. *Mailing Add:* Univ Colo Univ Hosp 4200 E Ninth Ave Denver CO 80220

BUNN, WILLIAM BERNICE, III, b Raleigh, NC, June 28, 52; m 83; c 1. OCCUPATIONAL CARCINOGENS. *Educ:* Duke Univ, AB, 74, MD & JD, 79; Univ NC, MPH, 83. *Prof Exp:* Intern & resident internal med, Duke Univ Med Ctr, 81-83, from asst prof to assoc prof, Dept Community & Family Med, Div Occup Med, 84-85, dir res, 85; asst clin prof, Dept Epidemiol, Sch Med, Yale Univ, 86; dir occup health & environ affairs, Pharmaceut Res & Develop Div, Bristol-Myers Co, 86, sr dir, 87; VPRES & CORP MED DIR, MANVILLE SALES CORP, 88-, VPRES & SR DIR HEALTH SAFETY & ENVIRON, 89- *Concurrent Pos:* OSHA fel, 73, Mercury Res fel, 78-80, NIH Cancer Res fel, 82-83, NIOSH training fel, 83-84. *Mem:* Am Col Physicians; Am Acad Occup Med; Am Occup Med Asn; AMA; Am Col Occup Med. *Res:* Cancer epidemiology; nasal carcinogens; toxic effects of mercury; legal research; author of over 40 technical publication; fiber toxicity and carcinogenicity. *Mailing Add:* 72 N Ranch Rd Littleton CO 80217

BUNNELL, FREDERICK LINDSLEY, b Vancouver, BC, May 8, 42; m 78; c 1. WILDLIFE ECOLOGY, SYSTEMS ANALYSIS. *Educ:* Univ BC, BSF, 65; Univ Calif, Berkeley, PhD(wildland resource sci), 73. *Prof Exp:* Res assoc forestry, Univ Calif, Berkeley, 66-69, from instr II to assoc prof wildlife, 71-79; PROF WILDLIFE, UNIV BC, 80- *Concurrent Pos:* Consult, BC Fish & Wildlife Br, 75-; hon consult, Int Union Conserv Nature & Natural Resources, 76-80; Nuffield Found fel, 78-79; NSERC sr indust fel, 85-86. *Honors & Awards:* Gold Medal, Can Inst Forestry, Can Forestry Sci Award, 89. *Mem:* Can Wildlife Soc (vpres, 81-83); Can Inst Forestry; Int Union Forest Res Orgn; Wildlife Soc; Ecol Soc Am. *Res:* Wildlife habitat relationships, particularly impacts of forestry practices upon wildlife; simulation models of wildlife energetics, behaviour and range relations. *Mailing Add:* Fac Forestry Univ BC Vancouver BC V6T 1W5 Can

BUNNER, ALAN NEWTON, b St Catharines, Ont, Jan 11, 38; US citizen; m 73; c 2. HIGH ENERGY ASTROPHYSICS, X-RAY ASTRONOMY. *Educ:* Univ Toronto, BA, 60; Cornell Univ, MS, 64, PhD(physics), 67. *Prof Exp:* Instr physics, Cornell Univ, 66-67; assoc scientist space physics, Univ Wis, 67-79; sr scientist, Perkin-Elmer Corp, 79-85; BR CHIEF ASTROPHYSICS, NASA HQ, 86- *Honors & Awards:* Skylab Achievement Award, NASA, 73. *Mem:* Am Astron Soc; Int Astron Union; Soc Photo-Optical Instrumentation Engrs; Am Inst Aeronaut & Astronaut. *Mailing Add:* 7200 Burtonwood Dr Alexandria VA 22307

BUNNETT, JOSEPH FREDERICK, b Portland, Ore, Nov 26, 21; m 42; c 3. ORGANIC CHEMISTRY. *Educ:* Reed Col, BA, 42; Univ Rochester, PhD(org chem), 45. *Prof Exp:* Res chemist, Western Pine Asn, Portland, 45-46; instr org chem,Reed Col, 46-48, asst prof chem, 48-52; from asst prof to assoc prof chem, Univ NC, 52-58; from assoc prof to prof, Brown Univ, 58-66, chmn dept, 61-64; PROF CHEM, UNIV CALIF, SANTA CRUZ, 66- *Concurrent Pos:* Fulbright fel, Univ London, 49-50; res grants, Res Corp, NSF Petrol Res Fund, Am Chem Soc & NIH; Fulbright & Guggenheim fels, Univ Munich, 60-61; ed, Accounts Chem Res, 66-86, trustee, Reed Col; chmn, Comn on Phys Org Chem, Int Union Pure & Appl Chem, 78-83, pres, Org Chem Div, 85-87. *Mem:* Am Acad Arts & Sci; Am Chem Soc; Royal Soc Chem (UK); hon mem Chem Soc Italy; hon mem Pharmaceut Soc Japan; fel Japan Soc Prom Sci; hon Accademia Gioenia di Catania (Italy). *Res:* Mechanisms of reactions of aromatic compounds with basic or nucleophilic reagents, of reactions of diazonium salts, of olefin-forming elimination reactions, of reactions involving electron transfer steps. *Mailing Add:* Thimann Labs Univ Calif Santa Cruz CA 95064

BUNNEY, BENJAMIN STEPHENSON, b Lansing, Mich, Sept 27, 38; m 60; c 3. PHARMACOLOGY, NEUROSCIENCE. *Educ:* NY Univ, BA, 60, MD, 64;. *Prof Exp:* Intern internal med, Bellevue Hosp, 64-65, resident, 65-66; with US Air Force, 66-68; fel psychiat, 68-71, from asst prof to assoc prof psychiat, 71-84, from asst prof to assoc prof pharmacol, 74-84, PROF PSYCHIAT & PHARMACOL, SCH MED, YALE UNIV, 84- *Concurrent Pos:* Mem, basic psychopharmacol Res Review Subcomt, 77-81, Res Scientists Develop Rev Comt, 82-86, Bd Sci Counselors, NIMH, 89-; actg chmn, 87-88, chmn, Dept Psychiat, NIMH, 88- *Honors & Awards:* Daniel H Efron Res Award, Am Col Neuropsychopharmacol; Leiber Prize, Nat Alliance Res Schizophrenia & Depression. *Mem:* NY Acad Sci; Psychiat Res Soc; Am Col Neuropsychopharmacol; Soc Neurosci; Am Psychiat Asn. *Res:* Electrophysiological and anatomical techniques to study the central site and mechanism of action of psychoactive drugs and central nervous system circuitry and function. *Mailing Add:* Neuropsychopharmacol Pharmacol Res Unit Yale Univ Sch Med PO Box 3333 New Haven CT 06510

BUNNEY, WILLIAM E, b Boston, Mass, June 27, 30. PSYCHIATRY, CLINICAL PSYCHOBIOLOGY. *Educ:* Oberlin Col, BA, 52; Univ Pa Med Sch, MD, 56. *Prof Exp:* Rotating intern, Henry Ford Hosp, 56-57; jr asst resident, VA Hosp, 57-58; asst resident psychiat, Outpatient Dept, Grace-New Haven Community Hosp, 58-59; resident psychiat, Student Mental Hyg Dept, Yale Univ, 59-60; clin assoc, Sect Psychosomatic Med, Adult Psychiat Br, Nat Inst Mental Health, 60-62, proj chief, Studies Biochem & Behav Factors Depressive Reactions, 62-66, chief, 66-68, chief, Sect Psychiat, 68-71, dir, Div Narcotic Addiction & Drug Abuse, 71-73, chief, Biol Psychiat Br, 73-82; DISTINGUISHED PROF & CHMN, DEPT PSYCHIAT & PROF PHARMACOL, COL MED & PROF PSYCHOBIOL, UNIV CALIF, IRVINE, 82- *Concurrent Pos:* Dep clin dir, Div Clin & Behav Res, Nat Inst Ment Health, 77-81, actg dir, Intramural Res Prog, 81-82; mem comt, Biol Psychiat Sect, World Psychiat Asn, 76-89, comt pharmacopsychiat, 78-, Ment Health Adv Coun, WHO, 84-; sci adv bd, Max Planck Inst Psychiat, 86; chmn & orgnr, Nat Conf Pain, Discomfort & Humanitarian Care, 79; assoc ed, Am J Psychiat; consult, Pres Comn Drug Abuse, Rio de Janeiro & Mexico City; HEW, Bellagio, Italy, First Nat Conf Drug Abuse, Bogota, Colombia; mem, US del comt challenges modern soc, NATO, Brussels, Belgium; orgnr & chair, Int Neuropsychopharmacol Col, Copenhagen, Denmark; invited lectr, Dept Psychol Med, Royal Victoria Infirmary & Univ, Newcastle-on-Tyne, Eng, Birmingham Univ, Korolinska Inst, Stockholm, Sweden, 3rd Peking Hosp, Shanghai Psychiat Hosp, China, Bombay, Lucknow & New Delhi, India & various univs in US; US rep, Planning Conf Collab Int Res Biol Psychiat, WHO, Moscow, USSR, Copenhagen, Denmark, Munich, W Ger, Sopporo, Japan & Basel, Switz; mem sci adv bd, Dept Ment Health, Calif, 85-; Lieber award, Nat Alliance Res Schizophrenia & Depression, 87-; prin investr, Neurosci & Schizophrenia Ctr grant, NIMH, 89- *Honors & Awards:* Rush Gold Medal Award, Am Psychiat Asn, 70; Int Anna-Monika Award Psychiat Res, 71; Pauline Goldman Lectr, Metrop Hosp, NY; DiMascio Mem Lectr, Tufts Univ; Janssen Award, Int Col Neuropsychopharmacol, 89. *Mem:* Inst Med-Nat Acad Sci; Psychiat Res Soc (pres, 74-75); Am Col Neuropsychopharmacol (vpres, 80 & pres, 83-84); Col Int Neuropsychopharmacol (pres, 86-88). *Res:* Clinical psychobiological studies of manic-depressive illness, schizophrenia and childhood mental illness, which includes behavioral studies, neuroendocrine, electrolyte and amine metabolism and studies of efficacy and mode of action of pharmacological agents which have been developed for treatment of these illnesses; author of over 300 publications. *Mailing Add:* Dept Psychiat Univ Calif Irvine CA 92717

BUNSHAH, ROINTAN F(RAMROZE), b Bombay, India, Dec 18, 27; US citizen. METALLURGY. *Educ:* Benares Hindu Univ, BSc, 48; Carnegie Inst Technol, MS, 51, DSc(metall). 52. *Prof Exp:* Res metallurgist & instr metall, Metals Res Lab, Carnegie Inst Technol, 52-54; res scientist & adj prof, Res Div, Col Eng, NY Univ, 54-60; metallurgist, Lawrence Radiation Lab, Univ Calif, Livermore, 60-69; PROF ENG, UNIV CALIF, LOS ANGELES, 69- *Honors & Awards:* Hadfield Medal, Geol, Mining & Metall Inst India, 48; Gaede-Langmuir Prize, Am Vac Soc, 86. *Mem:* Fel Am Soc Metals; Am Vacuum Soc (pres, 70-71); Indian Vacuum Soc. *Res:* Vacuum metallurgy; physical metallurgy of titanium and beryllium purification; dispersion strengthening of metals; phase transformations; ultra-high vacuum techniques; high rate physical vapor deposition processes; space processing of materials; materials synthesis, plasma-assisted vapor deposition techniques; electronic and opto-electronic materials, super hard coatings for wear resistance; biomaterials. *Mailing Add:* Dept Mat Sci Univ Calif 405 Hilgard Ave Los Angeles CA 90024

BUNTINAS, MARTIN GEORGE, b Klaipeda, Lithuania, Sept 1, 41; US citizen; m 66; c 3. TOPOLOGICAL SEQUENCE SPACES WITH APPLICATIONS TO FOURIER ANALYSIS. *Educ:* Univ Chicago, AB, 64; Ill Inst Technol, MS, 67, PhD(math), 70. *Prof Exp:* Instr math, Ill Inst Technol, 67-70; from asst prof math to assoc prof, 70-80, PROF MATH, LOYOLA UNIV, CHICAGO, 80- *Concurrent Pos:* Res fel, Alexander von Humboldt Found, 71-72 & 77-78, Fulbright Prog, 77-78. *Mem:* Am Math Soc; Math Asn Am. *Res:* Functional analysis and topological sequence spaces. *Mailing Add:* Dept Math Sci Loyola Univ 6525 N Sheridan Rd Chicago IL 60626

BUNTING, BRIAN TALBOT, b Sheffield, Eng, Oct 15, 32; m 58; c 2. ENVIRONMENTAL SCIENCES. *Educ:* Univ Sheffield, BA, 53, MA, 57; Univ London, PhD, 70. *Prof Exp:* Demonstr geog, Univ Keele, Eng, 54-55; lectr, Birkbeck Col, London, 57-68; assoc prof, 68-74, mem, Arctic Res Group, 69-76, PROF GEOG & PEDOLOGY, MCMASTER UNIV, 75- *Concurrent Pos:* Kemsley fel, Copenhagen Univ, 53-54; Brit Coun Interchange, Sorbonne, Paris, 65; instr, Field Studies Coun, UK, 64-68; vis lectr, Univ Ibadan, 65-66; guest prof, Univ Montana, 69, Aarhus Univ, 74-75 & Lakehead Univ, 78. *Mem:* Soil Sci Soc Am; Can Soc Soil Sci; Int Soil Sci; Asn Scand Studies in Can. *Res:* Studies of soil genesis and soil amelioration; geography of soils, especially factorial influences on microfabrics of soils; impact of forest fire on soils; soil pollution from landfill sites, land zoning analysis. *Mailing Add:* Dept Geog BS 313 McMaster Univ Hamilton ON L8S 4K1 Can

BUNTING, BRUCE GORDON, b Detroit, Mich, Feb, 22, 48; m 73. INTERNAL COMBUSTION ENGINES, TRIBOLOGY. *Educ:* Mich State Univ, BS, 70; Mich Technol Univ, MS, 75; Rensselaer Polytech Inst, PhD(mech eng), 87. *Prof Exp:* Res engr, Mich Technol Univ, 75-78; tech specialist, Cummins Engine Co, 78-81; sr engr, United Technologies Diesel Systs, 81-85; sr scientist, MRC Bearings, SKF USA Inc, 87-89; STAFF ENGR, AMOCO OIL CO, 89- *Mem:* Soc Automotive Engrs; Soc Tribologists & Lubrication Engrs; Am Soc Mech Engrs. *Res:* Hydrocarbon derived fuels and lubricants in relation to engine deposits, combustion, and lubrication. *Mailing Add:* 722 E Willow Ave Wheaton IL 60187

BUNTING, CHRISTOPHER DAVID, b Lytham, Gt Brit, Oct 2, 44; m 88. ELECTRICAL ENGINEERING. *Educ:* Cambridge Univ, BA, 66, MA, 70, PhD(elec eng), 71. *Prof Exp:* Sr electronics engr elec eng, Eurotherm Ltd, UK, 73-75, vpres, 77-84, DIR RES & DEVELOP ELEC ENG, EUROTHERM CORP, 75-; DIR ENG, HUNTER ASSOCS LAB INC, 84- *Concurrent Pos:* Fel, Trinity Col, Cambridge Univ, 70-74. *Res:* Application of advanced electronics to the control of processes in industry and research. *Mailing Add:* 9924 Brownsmill Rd Vienna VA 22182

BUNTING, DEWEY LEE, II, b Louisville, Ky, Sept 1, 32; m 57; c 3. RESEARCH ADMINISTRATION. *Educ:* Univ Louisville, BA, 57, MS, 59; Okla State Univ, PhD(zool), 63. *Prof Exp:* From asst prof to assoc prof, 68-74, PROF ZOOL, UNIV TENN, 74-, DIR, ECOL PROG, 81- *Concurrent Pos:* USPHS res grant, 66-69; Fed Water Pollution Control Admin res grants, 67-69; res grants, Off Water Resources Res, 69-74, Ecol Prog Subcontract Admin, 81- *Mem:* Am Soc Limnol & Oceanog; Int Asn Theoret & Appl Limnol; Ecol Soc Am. *Res:* Ecological effects of water pollution; systematics of fresh-water invertebrates. *Mailing Add:* Ecol Prog Univ Tenn Knoxville TN 37996

BUNTING, GEORGE SYDNEY, JR, plant taxonomy, for more information see previous edition

BUNTING, JACKIE ONDRA, b Poquoson, Va, Jan 27, 38; m 61; c 2. AERONAUTICAL & ASTRONAUTICAL ENGINEERING. *Educ:* Univ Va, BS, 60, MS, 62; Stanford Univ, PhD(aeronaut, astronaut), 67. *Prof Exp:* Asst, Stanford Univ, 62-67; asst prof aerodyn, Univ Va, 67-69; res scientist, 69-76, sr group engr, 76-80, DIR ENG OPERS, DENVER DIV, MARTIN MARIETTA CORP, 81- *Mem:* Am Inst Aeronaut & Astronaut. *Res:* Experimental measurements of transport properties of high temperature gases; measurement of velocity distributions for aerodynamic molecular beams. *Mailing Add:* Martin Marietta Astronaut PO Box 179 Denver CO 80201

BUNTING, JOHN WILLIAM, b Newcastle, Australia, Mar 30, 43; m 68; c 2. PHYSICAL ORGANIC CHEMISTRY. *Educ:* Univ NSW, BSc, 63; Australian Nat Univ, PhD(med chem), 67. *Prof Exp:* NIH fel chem, Northwestern Univ, 67-68; from asst prof to assoc prof, 68-80, PROF CHEM, UNIV TORONTO, 80- *Mem:* Am Chem Soc; Royal Australian Chem Inst; Chem Inst Can. *Res:* Physical organic chemistry; heterocyclic chemistry; enzymology. *Mailing Add:* Dept Chem Univ Toronto Toronto ON M5S 1A1 Can

BUNTING, ROGER KENT, b Creston, Ill, Nov 27, 35; m 58; c 4. INORGANIC CHEMISTRY. *Educ:* Univ Ill, BS, 58, MS, 61; Pa State Univ, PhD(chem), 65. *Prof Exp:* Vis res assoc chem, Ohio State Univ, 65-66; from asst prof to assoc prof, 66-85, chmn dept, 80-81, PROF CHEM, ILL STATE UNIV, 85- *Concurrent Pos:* Hon res fel, Birkbeck Col, Univ London, 74-75; Air Sci Eng fel, USAF Academy, 78; consult, Advan Sci Div, Eureka Co, 76-; res fel, Univ Florence, Italy, 82; vis prof, Ohio State Univ, 83; distinguished vis prof, US Air Force Acad, 87-88. *Mem:* Am Chem Soc; Royal Soc Chem; Sigma Xi; Royal Photographic Soc. *Res:* Phosphorus-nitrogen and boron-nitrogen chemistry; inorganic heterocycles; molten salt electrolytes; coordination chemistry; photographic chemistry. *Mailing Add:* Dept Chem Ill State Univ Normal IL 61761

BUNTLEY, GEORGE JULE, b Sheldon, Iowa, Feb 20, 24; m 47; c 2. SOIL MORPHOLOGY. *Educ:* SDak State Univ, BS, 49, MS, 50, PhD(soils), 62. *Prof Exp:* From asst prof to assoc prof agron, SDak State Univ, 50-68; assoc prof agron, 68-81, PROF PLANT & SOIL SCI, UNIV TENN, KNOXVILLE, 81- *Concurrent Pos:* Vis prof agron, Univ Tenn, 66-67. *Mem:* Am Soc Agron; Soil Sci Soc Am; Int Soil Sci Soc; Soil Conserv Soc Am; Coun Agr Sci & Technol. *Res:* Relationships between the physical and chemical properties of soils and specific land use alternatives. *Mailing Add:* 516 Gila Trail Knoxville TN 37919

BUNTON, CLIFFORD A, b Chesterfield, Eng, Jan 4, 20; US citizen; m 45; c 2. ORGANIC CHEMISTRY. *Educ:* Univ London, BSc, 42, PhD(org chem), 45. *Hon Degrees:* DSc, Univ Perugia, 86. *Prof Exp:* Lectr, Univ London, 45-58, reader, 58-63; prof chem, 63-90, chmn dept, 67-72, EMER PROF CHEM, UNIV CALIF, SANTA BARBARA, 90- *Concurrent Pos:* Commonwealth Fund fel, Columbia Univ, 48-49; Brit Coun vis lectr, Chile & Argentina, 60; vis lectr, Univ Calif, Los Angeles, 61 & Univ Toronto, 62; mem policy comt, Univ Chile-Univ Calif Coop Prog; vis prof, Univ Lausanne, Switz, 76-79. *Honors & Awards:* Tolman Medal, 84. *Mem:* Fel Am Chem Soc; Brit Chem Soc; fel AAAS; Chilean Acad Sci. *Res:* Mechanisms of organic and inorganic reactions; colloid chemistry. *Mailing Add:* Dept Chem Univ Calif Santa Barbara CA 93106

BUNTROCK, ROBERT EDWARD, b Minneapolis, Minn, Nov 19, 40; m 61; c 2. INFORMATION SCIENCE, ORGANIC CHEMISTRY. *Educ:* Univ Minn, Minneapolis, BChem, 62; Princeton Univ, MA, 64, PhD(org chem), 67. *Prof Exp:* Asst org chem, Princeton Univ, 62-63; res chemist pesticide chem, Air Prod & Chem, Inc, 67-70; proj chemist, Am Oil Co, 70-71; res info scientist, Standard Oil Co, 71-81, sr res info scientist, 81-85; RES ASSOC, AMOCO CORP, 85- *Concurrent Pos:* Chmn, Div Chem Info, Am Chem Soc, 81. *Mem:* AAAS; Am Chem Soc; Am Soc Info Sci; Sigma Xi. *Res:* Information retrieval; evaluation of information systems; selective dissemination of information; end-user training; amide hemiaminals; 1,2,4-oxadiazetidines; cyclization reactions of 2,2-disubstituted biphenyls; syntheses of pesticides; heterocyclic chemistry. *Mailing Add:* Amoco Corp Amoco Res Ctr PO Box 3011 Naperville IL 60566

BUNYAN, ELLEN LACKEY SPOTZ, b Clarks Mills, Pa, Aug 14, 21; m 44, 77; c 3. THEORETICAL CHEMISTRY, SCIENCE EDUCATION. *Educ:* Univ Pittsburgh, BS, 42; Univ Wis-Madison, PhD(chem), 50. *Prof Exp:* Chemist, Eastman Kodak Co, 42-44; instr chem, Univ Wis-Milwaukee, 46-47, res asst, Univ Wis-Madison, 47-50, proj assoc, 50-52; instr physics, St Agnes

Acad, Houston, Tex, 65; Welch fel chem, Rice Univ, 68-69; lectr, Montgomery Col, 70-72; asst prof, 72-80, ASSOC PROF CHEM, UNIV DC, 80- Concurrent Pos: Guest worker, Nat Bur Stand, 76-76. Mem: Am Chem Soc; Sigma Xi. Res: Molecular theory of gases; transport properties gases; intermolecular forces; theory of mass spectra-translational energy of fragments of ion decomposition of hydrocarbons; curriculum development reentering students. Mailing Add: 8215 Roanoke Ave Takoma Park MD 20912-6226

BUOL, STANLEY WALTER, b Madison, Wis, June 14, 34; m 60; c 2. SOIL SCIENCE. Educ: Univ Wis, BS, 56, MS, 58, PhD(soils), 60. Prof Exp: Asst prof agr chem & soils, Univ Ariz, 60-64, assoc prof, 64-66; assoc prof soil sci, 66-69, PROF SOIL SCI, NC STATE UNIV, 69- Concurrent Pos: Consult, Sensory Systs Lab, 63-65, West Va Co, 80 & Int Paper, 81. Honors & Awards: Int Soil Sci Award, Am Soc Agron. Mem: Fel Am Soc Agron; fel Soil Sci Soc Am. Res: Micromorphology of soil profiles with related work in soil genesis and classification; studies of soil formation and classification with related land use in intertropical areas of America, Asia and Africa. Mailing Add: Dept Soil Sci NC State Univ Raleigh NC 27695-7619

BUONAMICI, RINO, b Italy, Jan 13, 29; US citizen; m 54; c 2. MECHANICAL ENGINEERING. Educ: Univ Nev, BSME, 55. Prof Exp: Engr, Douglas Aircraft Co, 55-56; engr, Martin Co, 56-57; engr specialist & mgr, Aerojet Rocket Co, 57-63; fluidic engr, Naval Ammunition Depot, 73-74; PRIN ENGR, WESTINGHOUSE HANFORD CO, 74- Concurrent Pos: Prin investr, Aerojet Liquid Rocket Co & Hanford Engr Develop Lab; consult, Williams & Lane Co. Mem: Am Soc Mech Engrs. Res: Thermodynamic, fluid and heat transfer fields for materials in a cryogenic and high temperature environments; sophisticated rotating machinery for aerospace and nuclear applications; analysis and design, structural and dynamic testing of complex structures, deployment and actuation systems for engine/space vehicles; design and development of complex nuclear power plant systems; in the fluidic engineering field designed sophisticated hydraulic, pneumatic and mechanical systems including remote control operation; extensive analysis and design of major components: steam generators, boiler systems for steam power. Mailing Add: 507 S Taft St Kennewick WA 99336

BUONGIORNO, JOSEPH, b Golfech, France, Jan 15, 44; m 72; c 2. FOREST ECONOMICS. Educ: Advan Sch Forestry, Paris, Ingenieur, 67; State Univ NY Syracuse, MS, 69; Univ Calif, Berkeley, PhD(forest econ), 71. Prof Exp: Forestry officer, Food & Agr Orgn, UN, 71-75; from asst prof to assoc prof forest econ, 75-82, PROF FORESTRY, UNIV WIS-MADISON, 82- Concurrent Pos: Consult, Food & Agr Orgn, UN, World Bank, Int Trop Timber Orgn. Honors & Awards: Carl Schenk Award, Soc Am Foresters, 88. Mem: Am Econ Asn; Soc Am Forestry; Am Agr Econ Asn; Inst Mgt Sci. Res: Forestry and forest products economics; forestry sector planning; international trade of forest products; computer models. Mailing Add: Dept Forestry Univ Wis Madison WI 53706

BUONI, FREDERICK BUELL, b Brooklyn, NY, July 15, 34; m 58, 80; c 2. DECISION ANALYSIS, EXPERT SYSTEMS. Educ: Rutgers Univ, AB, 57; Ohio State Univ, MS, 67, PhD(nuclear eng), 71. Prof Exp: Chief, reactor eng br, Air Force Inst Technol, 65-68; chief, atmospheric br, Air Force Tech Appln Ctr, Patrick AFB, Fla, 71-76; dep dir advan technol, 76-78; assoc prof, 78-79, PROF OPERS RES & COMPUTER SCI & CHMN OPERS RES PROG, FLA INST TECHNOL, 79- Concurrent Pos: Adj prof, Fla Inst Technol, 72-79; pres, Fla Acad Scis, 90-91. Mem: Opers Res Soc Am; Inst Mgt Sci; Inst Elec & Electronics Engrs; Inst Indust Engrs; Am Asn Artificial Intel; Asn Comput Mach. Res: Development and use of computer-based decision aids for decision analysis and decision making; artificial intelligence; expert systems; decision support systems; multiobjective decision theory. Mailing Add: Opers Res Prog Fla Inst Technol 150 W University Blvd Melbourne FL 32901-6988

BUONI, JOHN J, b Philadelphia, Pa, Apr 25, 43; m 70; c 2. COMPUTATIONAL MATHEMATICS. Educ: St Joseph's Col, Pa, BS, 65; Univ Pittsburgh, MS, 68, PhD(math), 70. Prof Exp: From asst prof to assoc prof, 70-80, PROF MATH, YOUNGSTOWN STATE UNIV, 80- Concurrent Pos: Vis prof, Kent State Univ, 78-79. Mem: Am Math Soc; Soc Indust & Applied Math; Sigma Xi. Res: Numerical linear algebra; microcomputer software; operators on Banach spaces. Mailing Add: 79 Green Bay Dr Boardman OH 44512

BUONO, FREDERICK J, microbiology, biochemistry, for more information see previous edition

BUONO, JOHN ARTHUR, b Cambridge, Mass, Apr 29, 47. SURFACE SCIENCE, FAILURE ANALYSIS. Educ: Merrimack Col, BS, 69; Univ RI, PhD(anal chem), 75. Prof Exp: Lab mgr, Fisher Anal Inst Div, 73-77; DIR, PHOTOMETRICS INC, 77-, VPRES TECH DIV, 81- Mem: Am Chem Soc; Int Soc Hybrid Microelectronics; Soc Appl Spectros; Am Soc Qual Control. Res: Application of surface analysis techniques to research and development problems in microelectronics and materials. Mailing Add: 371 Old Beaver Brook Nagog Woods MA 01718

BUONOMO, FRANCES CATHERINE, b Brooklyn, NY, June 29, 55; m 79; c 1. ENDOCRINOLOGY, ANIMAL PHYSIOLOGY. Educ: Cook Col Agr & Environ Sci, BS, 77; Rutgers Univ, PhD(physiol), 81. Prof Exp: Res biologist, 81-89, ASSOC RES FEL, ANIMAL SCI DIV, MONSANTO CO, 89- Mem: Am Soc Animal Sci; Poultry Sci Asn; Endocrine Soc; Am Soc Zoologists. Res: Endocrine physiology of growth and metabolism in food-producing animal species. Mailing Add: Monsanto Co-BB3G 700 Chesterfield Village Pkwy Chesterfield MO 63198

BUONOPANE, RALPH A(NTHONY), b Boston, Mass, Dec 5, 38; m 65; c 2. CHEMICAL ENGINEERING, HEAT TRANSFER. Educ: Northeastern Univ, BS, 61, MS, 63, PhD(chem eng), 67. Prof Exp: Teaching asst, 61-63, instr chem eng, 63-64, asst prof, 66-71, actg dept chmn, 81-82, ASSOC PROF CHEM ENG, NORTHEASTERN UNIV, 71-, DEPT CHMN, 86- Concurrent Pos: Consult, Jet-Vac Corp, Mass, 63 & 66, Polaroid Corp, 69-81, Kendall Corp, 81, YTL Eng, Inc, 82-84; Off Energy Related Inventions & US Nat Bur Standards, 82-84; lectr, Lincoln Col, Northeastern Univ Eve Div, 64-77. Mem: fel Am Inst Chem Engrs; Am Soc Eng Educ; Sigma Xi. Res: Heat transfer equipment research; plate and spiral heat exchangers; evaporators. Mailing Add: Dept Chem Eng 360 Huntington Ave Boston MA 02115

BUPP, LAMAR PAUL, physical chemistry; deceased, see previous edition for last biography

BUR, ANTHONY J, b Philedelphia, Pa, Dec 4, 35; m 63; c 5. OPTICS. Educ: St Joseph's Col, Pa, BS, 57; Pa State Univ, PhD(physics), 62. Prof Exp: Fel mech, Johns Hopkins Univ, 62-63; physicist polymer dielectrics, 63-71, physicist dent res, 72-77, physicist polymer res, 77-86, PHYSICIST POLYMER PROCESSING, NAT BUR STANDARDS, 87- Mem: Am Chem Soc; Am Phys Soc; Soc Plastics Engr. Res: Electrical properties of insulators; polymer dielectrics and physics; solution properties of polymers; piezoelectric effect; thermodynamics; physical properties of teeth, bone and skin; fluorescence properties of polymers. Mailing Add: Nat Bur Standards Polymer Bldg B-320 Gaithersburg MD 20899

BURACK, WALTER RICHARD, INTERNAL MEDICINE, EPIDEMIOLOGY. Educ: Wake Forest Univ, MD, 51. Prof Exp: dir med serv, Allied Signal, Inc, 77-89; RETIRED. Res: Occupational medicine. Mailing Add: PO Box 5 Jackson NH 03846

BURAS, EDMUND MAURICE, b New Orleans, La, Oct 24, 21; m 43; c 3. ORGANIC CHEMISTRY. Educ: Tulane Univ, BS, 41, MS, 47. Prof Exp: Chemist, southern regional res lab, Bur Agr & Indust Chem, USDA, 42-57; group leader, Harris Res Labs, 57-69; PRIN SCIENTIST, GILLETTE RES INST, 69- Mem: Am Chem Soc; Fiber Soc; Am Inst Chemists; Am Asn Textile Chemists & Colorists; Brit Textile Inst. Res: Physics and chemistry of cutting and shaving; absorbency; chemical modification of natural and synthetic fibers; highway marking materials; instrumentation. Mailing Add: 824 Burnt Mills Ave Silver Springs MD 20901

BURAS, NATHAN, b Barlad, Romania, Aug 23, 21; US citizen; m 51; c 1. DESIGN OF WATER RESOURCES SYSTEMS, OPERATION OF REGIONAL WATER RESOURCES SYSTEMS. Educ: Univ Calif, Berkeley, BS, 49; Technion, Israel Inst Technol, MS, 58; UCLA, PhD(eng), 62. Prof Exp: Soil survr soil mapping, Israel Ministry Agr, 51-52; dist engr drainage, Water Planning Corp Israel, 54-57; jr res engr water resources, UCLA, 59-62; prof agr eng, Technion, Israel Inst Technol, 62-80; vis prof opers res, Stanford, 80-81; head, 81-89, PROF HYDROL & WATER RES, UNIV ARIZ, 81- Concurrent Pos: Consult, Water Planning Corp Israel, 63-79, World Bank, Mex Nat Water Plan, 72-76, World Bank, Narmada River Proj, India, 80-82; dean agr eng, Technion, Israel Inst Technol, 66-68; vis prof, Dept Opers Res, Stanford, 76-80. Honors & Awards: A Blanc Medal, Sixth Int Congress of Agr Eng, 64; Cert of Appreciation for Res Achievements, USDA, 70. Mem: Am Geophys Union; fel Am Soc Civil Engrs; Am Water Resources Asn; fel Int Water Resources Asn; AAAS. Res: Conjunctive management of surface and ground water resources with multiple decision makers; environmental effects of irrigated agriculture and its long term sustainability; global climatic change and its effect on regional water resources. Mailing Add: Dept Hydrol & Water Resources Univ Ariz Tucson AZ 85721

BURATTI, BONNIE J, b Bethlehem, Pa; c 3. PLANETARY ASTRONOMY. Educ: Mass Inst Technol, MS, 77; Cornell Univ, MS, 80, PhD(astron & space sci), 83. Prof Exp: Teaching & res asst astron, Cornell Univ, 77-83; MEM TECH STAFF PLANETARY ASTRON, JET PROPULSION LAB, CALIF INST TECNOL, 83- Mem: Am Astron Soc; Asn Women Sci. Res: Telescopic and spacecraft observation of planetary surfaces; modeling and interpretation of results; laboratory measurements of planetary surface analogues. Mailing Add: Jet Propulsion Lab 4800 Oak Grove Dr MS 183-301 Pasadena CA 91109

BURBA, JOHN VYTAUTAS, b Lithuania, Oct 1, 26; Can citizen; m 55; c 2. RADIOBIOLOGY, PHARMACOLOGY. Educ: Loyola Col Montreal, BSc, 51; Univ Ottawa, PhD(biochem), 63. Prof Exp: Control chemist, Merck & Co, Montreal, 51-52; tech sales rep, Can Lab Supplies, 52-54; tech sales rep, Mallinckrodt Chem Works, 54-57; med detail man, Frank W Horner, Ltd, 57-58; asst prof pharmacol, Univ Ottawa, 62-65; res scientist, 65-69, sci adv, Bur Drugs, 69-73, head radiopharmacol sect, 73-84, RADIOPHARMACOL ADV, RADIATION PROTECTION BUR, 84- Concurrent Pos: Fel, Dept Pharmacol, Yale Univ, 66-67. Mem: Soc Toxicol Can; Pharmacol Soc Can. Res: Toxicity of drugs and their metabolites; drug metabolizing activity of human placenta; stereospecificity in enzymic reactions; effect of drugs on hepatic azo reductase activity; effect of stannous ion on hepatic azo reductase and aromatic hydroxylase activity; effect of molybdenum-99 on sulfite and xanthine oxidase. Mailing Add: 1357 Fontenay Cresent Ottawa ON K1V 7K5 Can

BURBAGE, JOSEPH JAMES, b Rural, Ohio, June 11, 14; m 41; c 2. INORGANIC CHEMISTRY. Educ: Miami Univ, BS, 35; Ohio State Univ, MS, 42, PhD(chem), 47. Prof Exp: Asst chem, Ohio State Univ, 39-40, asst physics, 43; res chemist, Monsanto Indust Chem Co, 43-44; group leader, 44-45, opers mgr, 45-46, asst lab dir, 46-50, dir Mound Lab, 50-55, dir develop, Inorg Div, 55-62, dir prod planning, 62-68, dir financial control, 68-91; RETIRED. Mem: AAAS; Am Chem Soc. Res: Phase rule; liquid ammonia; radioactivity; alpha-emitters; heavy chemicals; market research. Mailing Add: 405 N Belleair Towers 1100 Ponce De Leon Blvd Clearwater FL 34616

BURBANCK, MADELINE PALMER, b Moorestown, NJ, Oct 27, 14; m 40; c 2. BOTANY, ECOLOGY. *Educ:* Wellesley Col, AB, 35, AM, 38; Univ Chicago, PhD, 41. *Prof Exp:* Asst bot, Wellesley Col, 36-39 & Univ Chicago, 40; guest, Columbia Univ, 41-42; special instr, Drury Col, 42-46, asst prof, 46-50; herbarium asst, 56-65; res assoc biol, Emory Univ, 50-83; RETIRED. *Concurrent Pos:* Mem corp, Marine Biol Lab, Woods Hole, Mass. *Mem:* Ecol Soc Am; Estuarine Res Fedn. *Res:* Cytology and morphology of Cyathura polita; plant succession on granite outcrops. *Mailing Add:* 1164 Clifton Rd NE Atlanta GA 30307

BURBANCK, WILLIAM DUDLEY, b Indianapolis, Ind, Aug 20, 13; m 40; c 2. PROTOZOOLOGY, ANIMAL ECOLOGY. *Educ:* Earlham Col, AB, 35; Haverford Col, MS, 36; Univ Chicago, PhD(zool), 41. *Prof Exp:* Instr biol, Earlham Col, 36-38, City Col NY, 41-42; guest, Dept Zool, Columbia Univ, 41-42; from asst prof to assoc prof, Drury Col, 42-45, prof, 45-50, chmn dept, 42-50; vis prof, 49-50, chmn dept, 52-57, prof, 50-80, EMER PROF BIOL, EMORY UNIV, 80- *Concurrent Pos:* Assoc ed, Estuaries, 77-83; mem corp, Marine Biol Lab, Woods Hole, Mass. *Mem:* Soc Protozool (vpres, 56-57); Int Asn Theoret & Appl Limnol; Ecol Soc Am; Marine Biol Asn UK; Estuarine Res Fedn. *Res:* Competition among protozoological populations; estuarine ecology, physiology and zoogeography of species of the isopod Cyathura. *Mailing Add:* 1164 Clifton Rd NE Atlanta GA 30307

BURBANK, NATHAN C, JR, b Wilton, Maine, July 20, 16; m 44. CIVIL & SANITARY ENGINEERING. *Educ:* Harvard Univ, AB, 38, SM, 40; Okla Agr & Mech Col, BSCE, 50; Mass Inst Technol, ScD(sanit eng), 55. *Prof Exp:* Pub health engr, USPHS, 40-46; instr chem, Okla Agr & Mech Col, 46-50, assoc prof civil eng, 50-52, 55; res assoc sanit eng, Mass Inst Technol, 52-55; sr sanit engr, Infilco, Inc, Ariz, 55-58; prof sanit eng, Wash Univ, 58-65, head dept civil eng, 58-64; prof environ health & sanit eng, Sch Pub Health, Univ Hawaii, 65-77; PRIN CIVIL ENGR, DEPT WASTE WATER MGT, PIMA CO, ARIZ, 77- *Concurrent Pos:* Sanit engr, USPHS, 43-46; consult ed, Water & Sewage Works Mag, 62; ed, Indust Water & Wastes Mag, 62-64; vis prof, Univ Hawaii, 63-64. *Honors & Awards:* Resources Div Award, Am Water Works Asn, 62. *Mem:* Am Chem Soc; Am Soc Civil Engrs; Am Water Works Asn; Water Pollution Control Fedn; Sigma Xi. *Res:* Industrial waste treatment; anaerobic digestion of waste; coagulation of water; solids waste disposal. *Mailing Add:* 5160 E Calle dos Cabezas Tucson AZ 85718

BURBECK, CHRISTINA ANDERSON, b Los Angeles, Calif, Sept 2, 48; m 79; c 1. VISUAL PSYCHOPHYSICS. *Educ:* Univ Calif, San Diego, BA, 70, MA, 74; Univ Calif, Irvine, PhD(psychol), 80. *Prof Exp:* res psychologist, 79-86, VISUAL SCI PROG MGR, SRI INT, 86- *Mem:* Optical Soc Am; Asn Res Vision & Ophthal; AAAS; Sigma Xi. *Res:* Human visual processing: psychophysical techniques with emphasis on spatial and temporal characteristics of the response to luminance patterns, retinal processing and methodological issues. *Mailing Add:* Dept Psychol Univ NC Chapel Hill CB No 3270 Chapel Hill NC 27599-3270

BURBIDGE, ELEANOR MARGARET, b Davenport, Eng, Aug 12, 19; US citizen; m 48; c 1. ASTRONOMY, ASTROPHYSICS. *Educ:* Univ London, BSc, 39, PhD(astrophys), 43. *Hon Degrees:* DSc, Smith Col, 63, Univ Sussex, 70, Univ Bristol, 72, Univ Leicester, 72, City Univ, London, 74, Univ Mich, 78, Univ Mass, 78, Williams Col, 79, SUNY, Stony Brook, 85, Univ Notre Dame, 86, Rensselaer Poly Inst, 86. *Prof Exp:* Actg dir, Univ London Observ, 43-51; res assoc astron, Yerkes Observ, 51-53; res fel astrophys, Calif Inst Technol, 55-57; Shirley Farr fel astron, Yerkes Observ, 57-59; assoc prof, Univ Chicago, 59-62; res assoc astrophys, 62-64, PROF ASTRON, UNIV CALIF, 64-, PROF PHYSICS & DIR, CTR ASTROPHYS & SPACE SCI, 79- *Concurrent Pos:* Abby Mauze Rockefeller prof, Mass Inst Technol, 68; mem, Space Sci Bd, Nat Acad Sci-Nat Res Coun, 71-74 & Astron Comt, 73-75, NSF Astron Adv Panel, 72-74, Steering Group for Large Space Telescope, 73-77 & Assoc Univs for Res Astron Bd, 74-79; dir, Royal Greenwich Observ, 72-73; Virginia Gildersleeve prof, Barnard Col, NY, 74; chmn bd, AAAS, 83; vis prof, Univ Calif, 84- *Honors & Awards:* Warner Prize, Am Astron Soc, 59, Russell Prize, 84; Bruce Medal, Astron Soc Pacific, 82; Nat Medal Sci, 85; David Elder lectr, Univ Strathclyde, 72; V Gildersleeve lectr, Barhard Col, 74; Jansky lectr, Nat Radio Astron Observ, 77; Henry Norris Russel lectr, 84; Brode lectr, Whitman Col, 86. *Mem:* Nat Acad Sci; Am Astron Soc (vpres, 72-74, pres, 76-78); Royal Astron Soc; Am Acad Arts & Sci; Int Astron Union; fel AAAS (pres, 82); Am Philos Soc; fel Royal Soc. *Res:* Astronomical spectroscopy; quasi-stellar objects; radio galaxies; masses and evolution of normal galaxies; nucleosynthesis; chemical evolution of the universe; chemical compo. *Mailing Add:* Dept Physics Univ Calif San Diego C-011 La Jolla CA 92093-0111

BURBIDGE, GEOFFREY, b Chipping Norton, Eng, Sept 24, 25; m 48; c 1. ASTRONOMY. *Educ:* Univ Bristol, BSc Hons, 46; Univ Col, London, PhD(theoret physics), 51. *Prof Exp:* Lectr astron, Univ London, 50-51; Agassiz fel astron, Harvard Univ, 51-52; res fel astron, Univ Chicago, 52-53, from asst prof to assoc prof astron, 57-62; from assoc prof to prof physics, 62-83, emer prof physics, 84-88, PROF PHYSICS, UNIV CALIF SAN DIEGO, 88- *Concurrent Pos:* Res fel physics, Cavendish Lab, Cambridge Univ, 53-55; Carnegie fel, Mt Wilson & Palomar Observ, 55-57; sr res fel, Calif Inst Technol, 58-62; Phillips vis prof, Harvard Univ, 68; mem coun, Am Astron Soc, 68-71; fel, Royal Soc London, 68, Univ Col London, 70; bd dirs, Astron Soc Pac, 69-75, pres, 74-76; dir, Kitt Peak Nat Observ, 78-84. *Honors & Awards:* Warner Prize, Am Astron Soc, 58. *Mem:* Am Astron Soc; Astron Soc Pac; fel Am Phys Soc; Royal Astron Soc; Int Astron Union; fel Am Acad Arts & Sci. *Res:* Research in extragalactic astronomy and high energy astrophysics. *Mailing Add:* Cass Mail Code 0001 Univ Calif San Diego La Jolla CA 92093

BURBUTIS, PAUL PHILIP, b Haverhill, Mass, Nov 1, 27; m 52; c 2. ENTOMOLOGY. *Educ:* Univ Mass, BS, 50; Rutgers Univ, MS, 52, PhD, 54. *Prof Exp:* Asst entom, Rutgers Univ, 50-54, asst res specialist, 54-58; from asst prof to prof, entom, 58-87, chmn, Entom & Appl Ecol Dept, 81-82 & 84-87, EMER PROF ENTOM, UNIV DEL, 87- *Mem:* Entom Soc Am; Int Orgn Biol Control; Am Entom Soc; Am Asn Univ Prof. *Res:* Economic entomology; biological insect control. *Mailing Add:* PO Box 111 Nobleboro ME 04555-0111

BURCH, BENJAMIN CLAY, b Wilson, NC, Oct 23, 48. MATHEMATICS. *Educ:* NC State Univ, BS, 70; Tulane Univ, MS, 73, PhD(math), 75. *Prof Exp:* Asst prof math, Northwestern Univ, Evanston, 75-76; from asst prof math to assoc prof, Univ Tex, El Paso, 76-87; PRES, RTR SOFTWARE, RALEIGH, NC, 87- *Mem:* Am Math Soc. *Res:* Initial value, boundary value and free boundary problems for nonlinear partial differential equations. *Mailing Add:* 513 Watauga St Raleigh NC 27604

BURCH, CLARK WAYNE, b Harrisonville, Mo, Sept 8, 07; m 35. VETERINARY MEDICINE. *Educ:* Kans State Univ, BS, 32, DVM, 37. *Prof Exp:* With livestock sanit, State Dept Agr, Wis, 37-38; practicing vet, 38-42, 46-52; prof, 52-73, EMER PROF VET SCI, COL AGR, UNIV WISMADISON, 73- *Mailing Add:* 10419 High Dr Leawood KS 66206

BURCH, DAVID STEWART, b Sapulpa, Okla, Oct 9, 26; m 49. PHYSICS. *Educ:* Univ Wash, BS, 50, MS, 54, PhD(physics), 56. *Prof Exp:* Physicist, atomic physics lab, Nat Bur Standards, 56-58; from asst prof to assoc prof, 58-67, PROF PHYSICS, ORE STATE UNIV, 68- *Res:* Electronic, ionic, atomic and molecular collisions; gaseous electronics; upper atmosphere physics; atomic and molecular structure; spectroscopy. *Mailing Add:* Dept Physics Ore State Univ Corvallis OR 97331

BURCH, DEREK GEORGE, b Caerphilly, UK, June 26, 33; m 61; c 2. TROPICAL ORNAMENTAL PLANTS, TAXONOMY. *Educ:* Univ Col Wales, BSc, 54, MSc, 57; Univ Fla, PhD(plant taxon), 65. *Prof Exp:* Asst prof bot, Washington Univ, 65-69; assoc prof biol, Univ SFla, 69-74; state specialist & assoc prof ornamental hort, 79-86, VIS PROF, UNIV FLA, 86- *Concurrent Pos:* Chief horticulturist, Mo Bot Garden, 65-69; dir, Bot Gardens, Univ SFla, 69-74; Bot & Hort consult, 86. *Mem:* Int Plant Propagators Soc; Asn Trop Biol; Int Asn Plant Taxon; Am Soc Hort Sci. *Res:* Taxonomy and biology of new world Euphorbieae; taxonomy of horticultural plants (tropical); propagation and cultural techniques for tropical ornamentals. *Mailing Add:* 3205 SW 70th Ave Ft Lauderdale FL 33314-7799

BURCH, GEORGE NELSON BLAIR, b Charlottetown, PEI, Sept 22, 21; nat US; m 44. POLYMER CHEMISTRY. *Educ:* Mt Allison, BSc, 43; McGill Univ, MSc, 46; Ohio State Univ, PhD(chem), 49. *Prof Exp:* Asst anal chem, Ohio State Univ, 46-48; res chemist, Hercules Powder Co, 49-56, res supvr, 56-57, lab supvr, 57-62, sr res chemist, 62-68, mgr auxiliary group, 68-75, mgr upholstery develop, 75-79, mgr, Tech Serv Labs, 79-82, MEM FIBER FUNDAMENTAL GROUP, HERCULES, INC, 82- *Mem:* Am Chem Soc. *Res:* Tall oil; organic fluorides; polyhydric alcohols; polymers; fibers, fiber design, processing and application in home furnishings. *Mailing Add:* Hercules Inc PO Box 8 Oxford GA 30267

BURCH, HELEN BULBROOK, biochemistry; deceased, see previous edition for last biography

BURCH, JAMES LEO, b San Antonio, Tex, Nov 28, 42; m 65; c 3. MAGNETOSPHERIC PHYSICS. *Educ:* St Mary's Univ, Tex, BS, 64; Rice Univ, PhD(space sci), 68; George Wash Univ, MSA, 73. *Prof Exp:* Physicist, Redstone Arsenal, Ala, 68-69; physicist, Sci Res Lab, US Mil Acad, 70-71; physicist, NASA-Goddard Space Flight Ctr, 71-74; physicist, NASA-Marshall Space Flight Ctr, 74-77; mgr, Space Physics Sect, 77-80, dir, Space Sci Dept, 80-85, VPRES, INSTRUMENTATION & SPACE RES DIV, SOUTHWEST RES INST, 85- *Concurrent Pos:* Assoc ed, J of Geophys Res & Geophys Res Lett, ed-in-chief; mem, Space Sci & Appln Adv Comt, NASA. *Mem:* Am Geophys Union; Sigma Xi. *Res:* Experimental study of the geophysics and plasma physics of the earth's magnetosphere, particularly those processes that couple the magnetosphere, ionosphere and upper atmosphere. *Mailing Add:* Southwest Res Inst PO Drawer 28510 San Antonio TX 78228-0510

BURCH, JOHN BAYARD, b Charlottesville, Va, Aug 12, 29; m 51; c 3. ZOOLOGY. *Educ:* Randolph-Macon Col, BS, 52; Univ Richmond, MS, 54; Univ Mich, PhD, 59. *Prof Exp:* Res assoc, 58-62, from asst prof to prof zool, 62-70, PROF BIOL, UNIV MICH, 70-, PROF NATURAL RESOURCES, 80-, CUR, MOLLUSKS, MUS, 62- *Concurrent Pos:* Res awards, Va Acad, 53 & NSF, 54; USPHS res career develop award, 64-69; res career develop award, USPHS, 64-69; regents fel, Smithsonian Inst, 83-84. *Mem:* Inst Malacol (exec secy-treas, 61-62, treas, 63-67, pres, 76-78); Soc Exp & Descriptive Malacol (pres, 68-80, exec secy-treas, 81-); French Soc Malacol; Malacol Soc Philippines; Unitas Malacologica; Int Soc Med & Appl Malacol (pres, 85); Am Malacol Union (vpres, 75). *Res:* Cytology, comparative biochemistry and systematics of mollusks; medical malacology; Protozoa of mollusks. *Mailing Add:* PO Box 3037 Ann Arbor MI 48106

BURCH, MARY KAPPEL, b St Paul, Minn, Mar 27, 57; m 83; c 1. MATERIAL SCIENCE COMPOSITES. *Educ:* Col St Catherine, BA, 79; Univ Calif, Berkeley, PhD(chem), 83. *Prof Exp:* Teaching asst chem lab, Col St Catherine, 78-79; student fel chem, Univ Minn, 78, 3M Central Labs, 78-79; res assoc gen chem & inorg, Univ Calif, Berkeley, 79-83; NIH postdoctorate biochem, La State Med Sch, New Orleans, 83-85; asst prof gen chem & thermo, Loyola Col, Baltimore, MD, 85-86; res chemist, 86-89, RES SECT MGR, ROHM & HAAS CO, 89- *Mem:* Am Chem Soc; Am Ceramic Soc. *Res:* Inorganic-organic solid state composite materials-physical characterization including polymer interactions with cementitious materials as well as interactions with refractory materials; inorganic solution thermodynamics and bioinorganics chemistry. *Mailing Add:* Rohm & Haas Co 727 Norristown Rd Bldg 3B Spring House PA 19477

BURCH, ROBERT EMMETT, b St Louis, Mo, Oct 9, 33; m 56; c 2. INTERNAL MEDICINE, NUTRITION. *Educ:* St Louis Univ, BS, 55, MD, 59. *Prof Exp:* Intern med & surg, Jewish Hosp, St Louis, 59-60; resident med, Firmin Desloge Hosp, St Louis Univ, 60-62; asst vis physician, Goldwater Mem Hosp, Columbia Univ, 65-68, assoc med, Univ, 65-69, asst prof, Col Physicians & Surgeons, 69-71; from assoc prof to prof med, Sch Med, Creighton Univ, 71-77; prof med & assoc chmn dept, Sch Med, Marshall

Univ, 77-80; PRICE-GOLDSMITH PROF CLIN NUTRIT, MED SCH, TULANE UNIV, 80- *Concurrent Pos:* Fel biochem, Western Reserve Univ, 62-65; asst vis physician, Francis Delafield Hosp, 68-70, assoc vis physician, 70-71; asst physician, Presby Hosp, 70; staff physician, Vet Admin Hosp, Omaha, 71-77; chief, Med Serv, Huntington, WVa, 77-; mem coun arteriosclerosis, Am Heart Asn; assoc chmn dept med, Sch Med, Marshall Univ, 77-80; chief sect clin nutrit, Tulane Univ Med Sch, 80-; assoc chief staff, New Orleans Vet Admin Hosp, 80- *Mem:* Am Physiol Soc; Am Inst Nutrit; Am Soc Clin Nutrit; Cent Soc Clin Res; Am Fedn Clin Res. *Res:* Trace element metabolism; zinc deficiency; role of zinc deficiency in ethanol metabolism. *Mailing Add:* Sch Med Tulane Univ 1430 Tulane Ave New Orleans LA 70112

BURCH, ROBERT RAY, b Edgard, La, May 28, 24; m 56; c 2. MEDICINE. *Educ:* Tulane Univ, BS, 48, MD, 51; Am Bd Internal Med, dipl, 58. *Prof Exp:* Intern med, Philadelphia Gen Hosp, 51-52; asst resident, Duke Univ Hosp, 52-53; from instr to assoc prof clin & prev med, 54-73, clin prof med, 73-91, EMER CLIN PROF MED, SCH MED, TULANE UNIV, 91- *Concurrent Pos:* Fel internal med, Sch Med, Tulane Univ, 53-54; Nat Heart Inst trainee, 54-55, fel internal med, 55-56; pvt pract, 56- *Mem:* AMA; Am Heart Asn; Am Geriat Soc; fel Am Col Physicians; NY Acad Sci. *Res:* Cardio-renal diseases. *Mailing Add:* 4427 S Robertson St New Orleans LA 70115

BURCH, ROBERT RAY, JR, b New Orleans, La, Oct 18, 56; m 83; c 1. INORGANIC & ORGANOMETALLIC CHEMISTRY, POLYMER CHEMISTRY. *Educ:* Tulane Univ, BS, 78; Univ Calif, Berkeley, PhD(chem), 82. *Prof Exp:* Vis scientist, 84-85, RES CHEMIST, E I DU PONT DE NEMOURS & CO, 85- *Mem:* Am Chem Soc. *Res:* Inorganic and organometallic chemistry; catalysis; polymer synthesis; fine chemical synthesis. *Mailing Add:* Cent Res & Develop Dept E I du Pont de Nemours & Co PO Box 80328 Wilmington DE 19880-0328

BURCH, RONALD MARTIN, b Annapolis, Md, Jan 16, 55; m 80; c 2. PHARMACOLOGY, PATHOLOGY. *Educ:* Col Charleston, BS(marine biol) & BS(chem), 77; Med Univ SC, PhD(pharmacol), 81, MD, 85. *Prof Exp:* Med staff fel, NIH, 85-87; staff scientist, 87-88, group leader, 88-90, DIR, NOVA PHARMACEUT CORP, 91- *Concurrent Pos:* Guest researcher, Lab Cell Biol, NIMH, 87- *Mem:* Am Soc Pharmacol & Exp Therapeut; Am Fedn Clin Res; Inflammation Res Asn. *Res:* Pharmacology of inflammatory mediators; cell biology of fibroblasts and neutrophils. *Mailing Add:* Nova Pharmaceut Corp 6200 Freeport Center Baltimore MD 21224

BURCH, THADDEUS JOSEPH, b Baltimore, Md, June 4, 30. SOLID STATE PHYSICS. *Educ:* Bellarmine Col, NY, AB, 54, PhL, 55; Fordham Univ, MA, 56, MS, 65, PhD(physics), 67; Woodstock Col, Md, STB, 60, STL, 62. *Prof Exp:* Asst prof physics, St Joseph's Col, Pa, 69-72; asst prof, Fordham Univ, New York, 72-74; assoc prof, Univ Conn, 74-76; assoc prof, 76-80, chmn dept, 76-86, actg dean grad sch, 85-86, dean grad sch, 86, PROF PHYSICS, MARQUETTE UNIV, 80- *Concurrent Pos:* Res collab, Brookhaven Nat Lab, 68-75. *Mem:* Am Phys Soc; Am Asn Physics Teachers. *Res:* Study of magnetic and crystallographic phase transitions in ferromagnetic alloys and magnetic semiconductors by nuclear magnetic resonance, Mossbauer effect, specific heat and resistivity measurements. *Mailing Add:* Grad Sch Marquette Univ Milwaukee WI 53233

BURCH, WILLIAM PAUL, dentistry, for more information see previous edition

BURCHALL, JAMES J, b Brooklyn, NY, Feb 25, 32; m 60; c 3. MOLECULAR BIOLOGY. *Educ:* St John's Univ, BS, 54; Brooklyn Col, MA, 59; Univ Ill, PhD(bact), 63. *Prof Exp:* Sr res microbiologist, Wellcome Res Labs, 62-68, HEAD, MICROBIOL DEPT, BURROUGHS WELLCOME & CO, 68-, DIR DIV MOLECULAR GENETICS & MICROBIOL, 89-; PRES, NC BR, AM SOC MICROBIOL, 86- *Concurrent Pos:* Adj prof, Duke Univ & Univ NC, 74-; mem, Bd of Dirs, NC Biotechnol, 86- *Mem:* Am Soc Biol Chemists; AAAS; Am Soc Microbiol. *Res:* Molecular genetics; chemotherapy; recombinant DNA. *Mailing Add:* Microbiol Dept Wellcome Res Lab Res Triangle Park NC 27709

BURCHAM, DONALD PRESTON, physics; deceased, see previous edition for last biography

BURCHAM, PAUL BAKER, b Fayette, Mo, Feb 22, 16; m 41; c 2. MATHEMATICS. *Educ:* Cent Col, Mo, BS, 35; Northwestern Univ, MA & PhD(math), 41. *Prof Exp:* Asst math, Northwestern Univ, 39-41; asst instr math, Cent Col, Mo, 41-42; asst prof, 46-54, PROF MATH, UNIV MO-COLUMBIA, 54- *Mem:* Am Math Soc; Math Asn Am. *Res:* Some inclusion relations in the domain of Hausdorff matrices; summability of series. *Mailing Add:* 401 Westmount Ave Columbia MO 65203

BURCHARD, HERMANN GEORG, b Würzburg, Ger, Dec 9, 34; m 61; c 4. MATHEMATICS. *Educ:* Univ Hamburg, Dipl-math, 63; Purdue Univ, PhD(comput sci), 68. *Prof Exp:* Res mathematician, Gen Motors Res Labs, Mich, 63-66; vis asst prof, Math Res Ctr, US Army, Univ Wis, Madison, 68-69; asst prof, Ind Univ, Bloomington, 69-72; assoc prof, 72-77, PROF MATH, OKLA STATE UNIV, 77- *Concurrent Pos:* Consult, Phillips Petroleum Co, 75- *Mem:* Am Math Soc; Soc Indust & Appl Math; Math Asn Am. *Res:* Approximation theory; numerical analysis; applied mathematics. *Mailing Add:* Dept Math Okla State Univ Stillwater OK 74074

BURCHARD, JEANETTE, b Wanesville, Mo, July 20, 17. MICROBIOLOGY, SEROLOGY. *Educ:* Southwest Mo State Col, BS, 39; Menorah Hosp Sch Med Technol, MT, 40; Univ Mich, MPH, 48. *Prof Exp:* Biologist, Southwest Br Lab, Mo Div Health, 41-42, biologist in charge, 42-45, sr biologist in charge, 45-58, dir, 48-63; microbiologist, Venereal Disease Res Lab, Commun Disease Ctr, USPHS, 63-65; chief microbiologist, Mo Div Health, 65-67, asst dir, Sect Lab Serv, 67-80; RETIRED. *Concurrent Pos:* Guest lectr, AID course, Sch Pub Health, Univ WI. *Mem:* Fel Am Pub Health Asn; Am Soc Microbiol; Am Soc Med Technol; Asn State & Territorial Pub Health Lab Dirs. *Res:* Culture and serology of brucellosis; serology of poliomyelitis; microbiology of foods; syphilis serology. *Mailing Add:* 2124 Wornall Place Lakewood Village Springfield MO 65804

BURCHARD, JOHN KENNETH, b St Louis, Mo, May 12, 36; m 58; c 2. TECHNICAL MANAGEMENT. *Educ:* Carnegie-Mellon Univ, BS, 57, MS, 59, PhD(chem eng), 62. *Prof Exp:* Staff scientist, United Technol Ctr, 61-68; chief scientist, Combustion Power Co, 68-70; lab dir, Environ Protection Agency, 70-80; div dir, Res Triangle Inst, 80-83; pres, Search Assocs, 83-85; dir res & sponsored progs, Univ Cent Ark, 85-87; asst dir, Res Admin, Ariz State Univ, 87-90; ENVIRON ENGR SPECIALIST, DEPT ENVIRON QUAL, STATE ARIZ, 90- *Mem:* Am Inst Chem Engrs; Sigma Xi. *Res:* Pollution control technology; aerosol physics; synthetic fuels technology; catalysis; chemical kinetics; process automatic control; combustion; environmental assessments. *Mailing Add:* 2101 E Redmon Dr Tempe AR 85283-2250

BURCHARD, ROBERT P, b New York, NY, Nov 7, 38; m 67; c 2. BACTERIOLOGY. *Educ:* Brown Univ, BA, 60, MSc, 62; Univ Minn, Minneapolis, PhD(microbiol), 65. *Prof Exp:* Lectr microbiol, Univ Ife, Nigeria, 65-66; from asst prof to assoc prof, 67-77, PROF BIOL SCI, UNIV MD, BALTIMORE COUNTY, 77- *Concurrent Pos:* NSF res grants, 68-70, 71-73, 75-77, 78-79 & 80-82; Off Naval Res, grant, 88-90 UK Sci Res Coun vis fel, dept of biochem, Univ Leeds, 73-74; vis prof, Univ Md Sch Med. *Mem:* Am Soc Microbiol; AAAS. *Res:* Behavior and mechanisms of motility and adhesion of gliding bacteria. *Mailing Add:* Dept Biol Univ Md Baltimore County Baltimore MD 21228-5398

BURCHELL, HOWARD BERTRAM, b Athens, Ont, Nov 28, 07; nat US; m 42; c 4. MEDICINE. *Educ:* Univ Toronto, MD, 32; Univ Minn, PhD(med), 40. *Prof Exp:* Instr med, Univ Pittsburgh, 36; from instr to prof, Mayo Found, 41-67, chief, Sect Cardiol, 67-74, prof med, Univ Hosp, Univ Minn, Minneapolis, 67-74; sr cardiologist, Northwestern Hosp, Minneapolis, 74-78; EMER PROF MED, UNIV MINN, MINNEAPOLIS, 74- *Concurrent Pos:* Ed, Circulation, 66-70; vis prof, Stanford Univ, 75-80. *Mem:* Am Physiol Soc; Am Fedn Clin; Asn Am Physicians; Am Heart Asn. *Res:* Clinical investigation; physiology of circulation. *Mailing Add:* 260 Woodlawn Ave St Paul MN 55105

BURCHENAL, JOSEPH HOLLAND, b Milford, Del, Dec 21, 12; m 48; c 7. MEDICINE. *Educ:* Princeton Educ: Univ Pa, MD, 37. *Prof Exp:* Intern, Union Mem Hosp, Baltimore, 37-38; resident dept pediat, NY Hosp, Cornell Univ, 38-39; asst resident, Boston City Hosp, 40-42; assoc, Sloan-Kettering Inst Cancer Res, 48-52, vpres, 64-72; asst prof clin med, Med Col, Cornell Univ, 49-50, asst prof med, 50-51, assoc prof, Sloan-Kettering Div, 51-52, prof, 52-55, PROF MED, MED COL, CORNELL UNIV, 55-; mem, 52-, Head Appl Ther Lab & Field Coordr Human Cancer, 73-, EMER MEM, SLOAN-KETTERING INST CANCER RES. *Concurrent Pos:* Res fel med, Harvard Med Sch, 40-42; spec fel, Mem Hosp, NY, 46-49; intern, NY Hosp, 38-39; asst attend physician, Med Serv, Mem Hosp, NY, 49-52, attend physician, 52-, chief chemother serv, 52-64, assoc dir clin invest, 64-66, dir clin invest, 66-; consult, USPHS, Am Cancer Soc & Div Med Sci, Nat Res Coun, 54-56; mem, Nat Panel Consult Conquest of Cancer, US Senate Comt Labor & Pub Welfare, 70; chmn, Chemother Adv Comt, Nat Cancer Inst, 70-71; vpres med & sci affairs & chmn, Med & Sci Adv Comt, Leukemia Soc Am, 70-75. *Honors & Awards:* Alfred P Sloan Award, 63; Albert Lasker Award Clin Cancer Chemother, 72; Prix Leopold Griffuel, 70; David A Kamoksky Mem Award, Am Soc Clin Oncol, 74; John Phillips Award, 74; James Ewing Soc Award, 75. *Mem:* Am Soc Cli; Soc Exp Biol & Med; Asn Cancer Res (vpres, 64-65, pres, 65-66); Am Soc Trop Med & Hyg; Am Fedn Clin Res. *Res:* Chemotherapy of cancer and leukemia. *Mailing Add:* Juniper Rd Norton CT 06820

BURCHETT, O NEILL J, b Seiling, Okla, June 11, 35; m 61; c 2. MECHANICAL ENGINEERING. *Educ:* Okla State Univ, BS, 58, MS, 60, PhD(mech eng), 66. *Prof Exp:* From instr to asst prof mech eng, Okla State Univ, 59-66; assoc prof, Ariz State Univ, 67-68; staff mem tech, 68-69, DIV SUPVR, SANDIA LABS, 69- *Mem:* Am Soc Mech Engrs. *Res:* Nondestructive testing; materials characterization; nonlinear vibrations. *Mailing Add:* Sandia Labs Orgn 7551 Bldg 860 Rm 201 Albuquerque NM 87185

BURCHFIEL, BURRELL CLARK, b Stockton, Calif, Mar 21, 34; m 83; c 3. GEOLOGY. *Educ:* Stanford Univ, BS, 57, MS, 58; Yale Univ, PhD(geol), 61. *Prof Exp:* Geologist, US Geol Surv, 61; from asst prof to assoc prof geol, Rice Univ, 61-70, prof, 70-75, Carey Croneis prof, 74-76; prof, 77-84, SCHLUMBERGER PROF GEOL, MASS INST TECHNOL, 84- *Concurrent Pos:* Exchange prof, Geol Inst Belgrade, Yugoslavia, 68 & Geol Inst Bucharest, Romania, 70; vis prof, Australian Nat Univ, 76; Guggenheim fel, 85-86. *Mem:* Nat Acad Sci; fel Geol Soc Am; Am Asn Petrol Geologists; Geol Soc Australia; fel Am Geophys Union; Am Acad Arts & Sci; hon foreign fel Europ Union Geologists. *Res:* Tectonics of the western United States; regional tectonics; orogenesis and its relation to plate boundary activity; structural geology; regional geology. *Mailing Add:* Dept Earth & Planetary Sci Mass Inst Technol Cambridge MA 02139

BURCHFIEL, JAMES LEE, b Los Angeles, Calif, Mar 16, 41; m 63; c 2. NEUROPHYSIOLOGY. *Educ:* Stanford Univ, BS, 63, PhD(pharmacol), 69. *Prof Exp:* From res assoc to prin res assoc neurol, Sch Med, Harvard Univ, 70-89; CO-DIR, COMPREHENSIVE EPILEPSY PROG, STRONG MEM HOSP, UNIV ROCHESTER, 89- *Concurrent Pos:* Mem, Comt Long-Term Monitoring, Am Electroencephalographic Soc, 83-85. *Mem:* Soc Neurosci; Eastern Asn Electroencephalographers (secy/treas, 83-); Am Epilepsy Soc. *Res:* Neuronal mechanisms of epilepsy; role of inhibitory mechanisms in central nervous system development; development and functional organization of the visual system; application of computer techniques to electrophysiological data. *Mailing Add:* Dept Neurol Strong Mem Hosp Univ Rochester 601 Elmswood Ave Box 673 Rochester NY 14642-8673

BURCHFIELD, HARRY P, b Pittsburgh, Pa, Dec 22, 15; m 42, 63; c 5. BIOCHEMISTRY. *Educ:* Columbia Univ, AB & MA, 38, PhD, 56. *Prof Exp:* Analyst, Nat Oil Prod Co, NJ, 38-40; res chemist, 40-41, group leader, 41-50, dir plantations res dept, Naugatuck Chem Div, US Rubber Co, 50-51; assoc dir, Boyce Thompson Inst Plant Res, 51-61; inst scientist & mgr anal & biochem, Southwest Res Inst, 61-65; officer-in-charge, pesticides res lab, USPHS, 65-67; sci dir & dir biol sci, Gulf South Res Inst, 67-76; res prof & head, Div Molecular Biol, Med Res Inst, Fla Inst Technol, 77-82; PRIN SCIENTIST, RES ASSOC, 76- *Concurrent Pos:* Adj prof, Univ Southwestern La, 67-; consult, Environ Protection Agency, 76- *Honors & Awards:* Prize, Chicago Rubber Group, 46. *Mem:* Am Inst Biol Sci; Soc Toxicol; Am Chem Soc. *Res:* Biochemical applications of gas chromatography; metabolism and analysis of drugs, pesticides and natural products; mechanism of action of biocides; chemical carcinogenesis and mutagenesis; armadillo biochemistry. *Mailing Add:* Res Assoc 72 Riverview Terr Indiatlantic FL 32903

BURCHFIELD, THOMAS ELWOOD, b Marshfield, Mo, Aug 6, 51; m 73; c 1. SOLUTION THERMODYNAMICS. *Educ:* Univ Mo, Rolla, BSc, 73, MSc, 75, PhD(phys chem), 77. *Prof Exp:* Res assoc, dept chem, Univ Lethbridge, Can, 77-79; res chemist, Bartlesville Energy Technol Ctr, US Dept Energy, 79-83; sr chemist, 83-84, mgr, recovery processes res, 84-89, DIR, ENERGY PROD RES, NAT INST PETROL & ENERGY RES, 89- *Concurrent Pos:* Mem, Calorimetry Conf; mem adv bd, Found Chem Res. *Honors & Awards:* Sumner Award, 86. *Mem:* Am Chem Soc; Sigma Xi; Soc Petrol Engrs; Calorimetry Conf; Int Union Pure & Appl Chem. *Res:* Thermodynamics properties of surfactant systems; experimental thermodynamics and solution calorimetry; understanding the thermodynamics of micellization, solubilization, and microemulsion formation by measuring the thermodynamics properties of surfactant systems; enhanced oil recovery; colloid chemistry. *Mailing Add:* 2705 SE Kensington Way Bartlesville OK 74006

BURCHILL, BROWER RENE, b El Dorado, Kans, Dec 29, 38; m 60; c 2. CELL BIOLOGY. *Educ:* Phillips Univ, BA, 60; Fla State Univ, MS, 63; Western Reserve Univ, PhD(biol), 66. *Prof Exp:* NIH trainee, Fla State Univ, 63; postdoctoral appointee, Los Alamos Sci Lab, Univ Calif, 66-68; asst prof zool, Univ Kans, 68-69, from asst to assoc prof physiol & cell biol, 69-76, chmn div biol sci, 73-79, ASSOC VCHMN ACAD AFFAIRS, UNIV KANSAS, 76-, PROF PHYSIOL & CELL BIOL, 84- *Mem:* AAAS; Am Soc Cell Biol; Soc Protozool (treas, 78-). *Res:* Nucleocytoplasmic interactions essential to oral differentiation in the ciliate protozoan Stentor coeruleus. *Mailing Add:* Dept Physiol & Cell Biol Univ Kans Haworth Hall Lawrence KS 66045

BURCHILL, CHARLES EUGENE, b Makwa, Sask, Dec 12, 32; m 57; c 4. PHYSICAL CHEMISTRY. *Educ:* Univ Sask, BA, 55, MA, 61; Univ Leeds, PhD(phys chem), 68. *Prof Exp:* Instr chem, Victoria Col, 56-63, demonstr phys chem, Univ Leeds, 63-66; asst prof chem, 66-68, ASSOC PROF CHEM, UNIV MAN, 68- *Res:* Kinetics and mechanisms of free-radical reactions in solution; radiation chemistry and photochemistry of aqueous solutions. *Mailing Add:* Dept Chem Univ Man Winnipeg MB R3T 2N2 Can

BURCK, LARRY HAROLD, b Detroit, Mich, Sept 25, 45; m 67; c 3. FAILURE ANALYSIS, FRACTURE MECHANICS. *Educ:* Mich State Univ, BS, 67; Rensselaer Polytech Inst, MS, 71; Northwestern Univ, PhD(mat sci), 75. *Prof Exp:* Res engr, Advan Mat & Develop Lab, Pratt & Whitney Aircraft, 67-72; from asst prof to assoc prof mat sci, Univ Wis-Milwaukee, 75-84; CONSULT & VPRES, WEISS & BURCK CONSULT ENGRS, 84- *Mem:* Am Soc Metals; Am Inst Mining Metall & Petrol Engrs; Am Soc Mech Engrs. *Res:* Fracture, fatigue and mechanical behavior of materials; analytical and experimental fracture mechanics; failure analysis; machine guarding. *Mailing Add:* Weiss & Burck, Ltd 744 W Fourth St Suite 580 Milwaukee WI 53203

BURCK, PHILIP JOHN, b Milwaukee, Wis, Sept 21, 36; m 68; c 1. BIOCHEMISTRY. *Educ:* Lawrence Univ, AB, 58; Univ Ill, MS, 60, PhD(biochem), 62. *Prof Exp:* RES SCIENTIST BIOCHEM, LILLY RES LABS, ELI LILLY & CO, 62- *Mem:* AAAS; Am Chem Soc. *Res:* Enzyme and enzyme inhibitor purification and characterization; physiological role of proteases and protease inhibitors. *Mailing Add:* Lilly Res Labs Biotech Res Div Lily Corp Res Ctr Indianapolis IN 46285

BURCKEL, ROBERT BRUCE, b Louisville, Ky, Dec 15, 39; m 67. MATHEMATICS. *Educ:* Univ Notre Dame, BS, 61; Yale Univ, MA, 63, PhD(math), 68. *Prof Exp:* From instr to asst prof, Univ Ore, 66-70; actg chmn dept, 74-75, assoc prof, 71-80, PROF MATH, KANS STATE UNIV, 81- *Concurrent Pos:* asst prof, Univ Saarland, 77-78; guest prof, Univ Erlangen-Nuremburg, 84-85. *Mem:* Am Math Soc; Deutscher Math Verein. *Res:* Harmonic analysis, classical and abstract; function algebras; hardy spaces, classical and abstract; Banach algebra; classical complex analysis. *Mailing Add:* Dept Math Kans State Univ Manhattan KS 66506

BURCKHARDT, CHRISTOPH B, b Zurich, Switz, Apr 25, 35; nat US; m 64; c 2. ELECTRICAL ENGINEERING. *Educ:* Swiss Fed Inst Technol, MS, 59, PhD(elec eng), 63. *Prof Exp:* Asst, Swiss Fed Inst Technol, 60-63; mem tech staff, Bell Tel Labs, 63-70; MEM TECH STAFF, HOFFMANN-LA ROCHE, INC, 70- *Mem:* Fel Inst Elec & Electronics Engrs. *Res:* Hall-effect, varactor diode applications; optical and acoustical holography; ultrasonic imaging for medical diagnosis. *Mailing Add:* Hoffmann-La Roche ZFE Basel Switzerland

BURD, JOHN FREDERICK, biochemistry, for more information see previous edition

BURD, LAURENCE IRA, b New York, NY. OBSTETRICS & GYNECOLOGY. *Educ:* Univ Rochester, AB, 62; Chicago Med Sch, MD, 66. *Prof Exp:* Res fel, Div Perinatal Med, Univ Colo, 73-75; dir obstet, Michael Reese Hosp, 76-83; co-dir, Perinatal Ctr & assoc prof obstet & gynec,

76-83, CLIN ASSOC PROF, UNIV CHICAGO, 83- *Concurrent Pos:* NIH res fel, 74-75; Basil O'Connor grant, Nat Found March of Dimes, 76-; attending physician obstet & gynec, Michael Reese Hosp, 76- & attending physician pediat, 78-83. *Mem:* Soc Gynec Invest; Perinatal Res Soc. *Res:* Maternal and fetal physiologic changes at the time of parturition; fetal and maternal pathophysiology. *Mailing Add:* Dept Obstet & Gynec Michael Reese Hosp 31st & Lake Shore Dr Chicago IL 60616

BURDASH, NICHOLAS MICHAEL, b Hazleton, Pa, Sept 18, 41; m 65; c 2. IMMUNOLOGY, BACTERIOLOGY. *Educ:* Univ Scranton, BS, 63; Duquesne Univ, MS, 65; Ohio State Univ, PhD(microbiol), 69. *Prof Exp:* asst prof microbiol & immunol, Med Univ SC, 69-80, assoc prof lab med & med technol, Med Lab Technol, 80-; DEPT MICROBIOL PUB HEALTH, PHILADELPHIA COL OSTEOP MED. *Mem:* AAAS; Am Soc Microbiol; Electron Micros Soc Am; Am Inst Biol Sci. *Res:* Humoral and cellular aspects of autoimmune diseases, especially experimental allergic encephalomyelitis and experimental allergic neuritis. *Mailing Add:* Dept Microbiol-Pub Health Philadelphia Col Osteop Med 4150 City Ave Philadelphia PA 19131

BURDEN, HARVEY WORTH, b Glendale, Calif, Nov 20, 33; m 61; c 2. AERONAUTICS, MECHANICAL ENGINEERING. *Educ:* US Naval Acad, BS, 55; US Naval Postgrad Sch, BS, 63; Calif Inst Technol, AeE, 64; Univ Pa, PhD(mech eng), 69. *Prof Exp:* Sr scientist, Teledyne McCormick Seiph, 75-78; prin staff mem, BDM Corp, 78-81; ADJ PROF AERONAUT, US NAVAL POSTGRAD SCH, MONTEREY, CALIF, 81- *Mem:* Sigma Xi; Am Inst Aeronaut & Astronaut; Am Soc Mech Engrs. *Res:* Heat and mass transfer in fluid flows; boundary layer flows. *Mailing Add:* 1013 Sombrero Rd Pebble Beach CA 93953

BURDEN, HUBERT WHITE, b Elizabeth City, NC, Sept 12, 43; m 67; c 2. REPRODUCTIVE ENDOCRINOLOGY. *Educ:* Atlantic Christian Col, AB, 65; ECarolina Univ, MA, 67; Tulane Univ, PhD(anat), 71. *Prof Exp:* Instr biol, ECarolina Univ, 67-68; teaching asst anat, Med Sch, Tulane Univ, 68-71; from asst prof to assoc prof, 71-79, PROF ANAT, EAST CAROLINA UNIV, 80- *Concurrent Pos:* USPHS res grants, 75 & 79. *Mem:* Soc Study Reproduction; Soc Develop Biol; Am Asn Anatomists; Am Soc Zoologists; Pan Am Asn Anatomists; Sigma Xi. *Res:* Research on the role of the peripheral autonomic nervous system in reproductive function. *Mailing Add:* 109 Dellwood Dr Greenville NC 27834

BURDEN, RICHARD L, MATHEMATICS. *Educ:* Albion Col, BA, 66; Case Inst Technol, MS, 68; Case Western Reserve Univ, PhD(math), 71; Univ Pittsburgh, MS, 81. *Prof Exp:* PROF MATH, YOUNGSTOWN STATE UNIV, 70- *Mem:* Soc Indust & Appl Math; Asn Comput Mach; Math Asn Am. *Res:* Numerical solution of partial differential equations, numerical algebra & programming languages. *Mailing Add:* Dept Math Youngstown State Univ Youngstown OH 44503

BURDEN, STANLEY LEE, JR, b Aurora, Ill, Mar 9, 39; m 62; c 2. CHEMICAL INSTRUMENTATION, ANALYTICAL CHEMISTRY. *Educ:* Taylor Univ, BS, 61; Ind Univ, PhD(anal chem) 66. *Prof Exp:* Teacher, high sch, Ind, 61-62; from instr to assoc prof, 66-74, chmn sci div, 85-89, PROF CHEM, TAYLOR UNIV, 75-, CHMN DEPT, 78-, ASSOC DEAN, DIV NATURAL SCI, 90- *Concurrent Pos:* NASA res fel, Manned Spacecraft Ctr, 69. *Mem:* Am Chem Soc; Am Sci Affiliation. *Res:* Instrumentation of analytical methods of analysis; environmental analysis and on-line computer applications in chemical instrumentation. *Mailing Add:* Sci Ctr Taylor Univ Upland IN 46989

BURDETT, JEREMY KEITH, b London, Eng, July 1, 47; m 72; c 2. PHYSICAL INORGANIC CHEMISTRY. *Educ:* Cambridge Univ, BA, 68, MS & PhD(chem), 72; Univ Mich, Ann Arbor, MS, 70; Univ Cambridge, ScD, 90. *Prof Exp:* Bye fel chem, Magdalene Col, Eng, 71-72; sr res officer chem, Univ Newcastle, Eng, 72-78; from asst prof to assoc prof, 78-86, PROF CHEM, UNIV CHICAGO, 86-, ASSOC DEAN PHYS SCI DIV, 87- *Concurrent Pos:* Fel Alfred P Sloan Found, 79-84, teacher scholar, Camille & Henry Dreyfus Found, 79-84; Wilsmore fel, Univ Melbourne, Australia, 85; Guggenheim fel, 90. *Honors & Awards:* Meldola Medal & Prize, Soc Maccabeans & Royal Inst Chem, 77. *Mem:* Am Chem Soc; Royal Inst Chem. *Res:* Theoretical studies on structural, electronic and kinetic properties of molecules and solids. *Mailing Add:* Searle Chem Lab Univ Chicago Chicago IL 60637

BURDETT, LORENZO WORTH, b Pocatello, Idaho, Aug 9, 16; m 45; c 7. ANALYTICAL CHEMISTRY. *Educ:* Univ Idaho, BS, 43; Univ Ill, PhD(anal chem), 49. *Prof Exp:* Jr chemist, Shell Develop Co, 43-46; res chemist, Standard Oil Co Ind, 49-52; res chemist, Union Oil Co Calif, 52-55, sect leader chem, 55-58, sr sect leader, 58-65, supvr res dept, 65-83; RETIRED. *Mem:* Am Chem Soc; Am Soc Testing Mat. *Res:* Method development and analytical research in electrochemical and chemical area. *Mailing Add:* 1607 N Mountain Oakes Dr Orem UT 84057-2309

BURDETTE, WALTER JAMES, b Hillsboro, Tex, Feb 5, 15; m 47; c 2. GENERAL & THORACIC SURGERY, GENETICS. *Educ:* Baylor Univ, AB, 35; Univ Tex, AM, 36, PhD(zool), 38; Yale Univ, MD, 42; Am Bd Surg, dipl, 50; Am Bd Thoracic Surg, dipl, 53. *Prof Exp:* Intern, Johns Hopkins Hosp, 42-43; Cushing fel surg, Yale Univ, 43-44; resident, New Haven Hosp, Conn, 44-46; from instr to assoc prof, Sch Med, La State Univ, 46-55, coordr cancer res & teaching, 48-55; prof surg & chmn dept, Sch Med, Univ Mo, 55-56; prof clin surg, Sch Med, St Louis Univ, 56-57; prof surg & head dept, Col Med, Univ Utah, 57-65; prof surg & assoc dir, Univ Tex M D Anderson Hosp & Tumor Inst, 65-75; PVT PRACT GEN & THORACIC SURG, 76- *Concurrent Pos:* Asst vis surgeon, Charity Hosp of La, New Orleans, 46-47, vis surgeon, 47-55; vis surgeon, Southern Baptist Hosp, 53-55; vis investr, Chester Beatty Inst Cancer Res, London, Eng, 53 & Univ Tubingen, W Germany, 57; surgeon-in-chief, Univ Hosp, Univ Mo, 55-56; dir, St Louis Univ Surg Serv, Vet Admin Hosp, 56-57; dir, Lab Clin Biol, 57-; surgeon-in-chief, Salt Lake Gen Hosp, 57-65; Gibson lectr advan surg, Oxford Univ, 66;

vis prof, Off Univ Congo, 68; lectr, Hosp Martinez, PR, 69, Univ Sendai, Japan, 70, Univ Melbourne, 70 & Univ Freiburg, 70. Consult, Oak Ridge Inst Nuclear Studies, 51 & Touro Infirmary, New Orleans, 53-55; chief surg consult, Vet Admin Hosp, Salt Lake City, 57-65, Mem morphol & genetics study sect, NIH, 55-58, chmn genetics study sect, 57-61, mem nat adv cancer coun, 61-65, mem nat adv heart coun, 65, dir working cadre on carcinoma of large intestine, Nat Cancer Inst; chmn res adv coun, Am Cancer Soc, 57-; mem adv comt smoking & health, Surgeon Gen US, 62-64; chmn Nat Acad Sci comt, Int Union Against Cancer, 62-; mem transplantation comt, Nat Acad Sci. *Mem:* Am Asn Cancer Res; Soc Exp Biol & Med; Am Col Surgeons; Soc Clin Surgeons (treas, 63-64); Am Surg Asn; Sigma Xi. *Res:* Genetics and cancer; metabolism of cardiac muscle; cardiovascular surgery; invertebrate hormones. *Mailing Add:* Park Plaza Hosp 1200 Binz St Suite 740 Houston TX 77004

BURDG, DONALD EUGENE, b Milford, Nebr, Aug 12, 29; m 52; c 2. MATHEMATICS. *Educ:* Colo State Univ, BS, 51; Univ Northern Colo, MA, 52; Ore State Univ, MS, 66. *Prof Exp:* Asst prof math, Cent Ore Community Col, 56-67; PROF MATH, SOUTHWESTERN ORE COMMUNITY COL, 67- *Mem:* Nat Coun Teachers Math; Math Asn Am; Am Math Asn Two Yr Cols. *Mailing Add:* Southwestern Ore Community Col Empire Lakes Coos Bay OR 97420

BURDGE, DAVID NEWMAN, b Sullivan, Ind, May 1, 31; m 53; c 2. FUEL TECHNOLOGY & PETROLEUM ENGINEERING. *Educ:* Purdue Univ, BS, 52, MS, 56, PhD(org chem), 59. *Prof Exp:* Res chemist, 59-67, mgr, Org Chem Dept, 67-70, mgr, Ref Sci & Eng Dept, 70-72, mgr, Petrol Chem Dept, 72-77, sr staff engr, 77-80, mgr, Tech Projs, Offshore Technol Div, 80-86, ASSOC DIR, TECHNOL SERV, MARATHON OIL CO, 86- *Concurrent Pos:* Res chemist, Eastman Kodak Co, 61-62. *Mem:* Am Chem Soc; Soc Petroleum Engrs; Sigma Xi. *Res:* Organic sulfur chemistry; oxidation of alkyl aromatics; petroleum and refining chemistry; petroleum production chemistry; enhanced recovery chemicals and techniques. *Mailing Add:* 2526 E Pine Bluff Highlands Ranch CO 80126

BURDGE, GEOFFREY LYNN, b Munich, Ger, Sept 20, 47; US citizen; m 71. NON-LINEAR OPTICS, QUANTUM ELECTRONICS. *Educ:* Univ Md, BS, 70, MS, 74, PhD(electronics eng), 81. *Prof Exp:* Mem tech staff, US Dept Defense, 70-80; MEM TECH STAFF, LAB PHYS SCI, 80- *Mem:* Optical Soc Am; Inst Elec & Electronics Engrs. *Res:* Nonlinear optics; phase conjugation and its application to optical processing; modelocked semiconductor lasers; high intensity, near resonant excitation of sodium vapor. *Mailing Add:* Lab Phys Sci 4928 College Ave College Park MD 20740

BURDI, ALPHONSE R, b Chicago, Ill, Aug 28, 35; m 69; c 2. DENTAL RESEARCH, CHILD GROWTH. *Educ:* Northern Ill Univ, BSEd, 57; Univ Ill, MS, 59; Univ Mich, MS, 61, PhD(anat), 63. *Prof Exp:* USPHS res trainee, 60-61, from instr to assoc prof, 62-74, PROF ANAT, UNIV MICH, 74-, DIR INTEGRATED PREMED-PROG, 82- *Concurrent Pos:* Fel Inst Advan Educ Dent Res, 64. *Mem:* AAAS; Am Asn Anat; Am Asn Phys Anthrop; Am Cleft Palate Asn; Tissue Cult Asn; Int Asn Dent Res. *Res:* Embryology and prenatal craniofacial growth; biomechanics of the facial skeleton; developmental craniofacial biology; birth defects; mechanisms of seven orofacial tissue interactions; cephalometrics; human teratology. *Mailing Add:* Dept Anat Univ Mich Ann Arbor MI 48104

BURDICK, ALLAN BERNARD, b Cincinnati, Ohio, Aug 16, 20; m 43; c 4. GENETICS, CYTOGENETICS. *Educ:* Iowa State Col, BS, 45, MS, 47; Univ Calif, Berkeley, PhD, 49. *Prof Exp:* Asst prof genetics & plant breeding, Univ Ark, 49-52; from asst prof to prof genetics, Purdue Univ, 52-63; prof & assoc dean sci, Am Univ Beirut, 63-66; prof biol & chmn dept, Adelphi Univ, 66-69, actg dir, Adelphi Inst Marine Sci, 67-68; chmn dept genetics, 69-70, group leader cytol & genetical sci, 70-75, prof, 69-86, EMER PROF GENETICS, UNIV MO-COLUMBIA, 86- *Concurrent Pos:* Res collabr, Brookhaven Nat Lab, 55, 57; Fulbright res prof, Kyoto Univ, 59-60; Guggenheim fel, 59-60; vis prof, Sch Med, Ind Univ, 82, Peking Univ, 86. *Mem:* Fel AAAS; Genetics Soc Am; Soc Craniofacial Genetics; Am Soc Human Genetics; Am Soc Cell Biol; Am Bd Med Genetics. *Res:* Quantitative inheritance; human genetics; Drosophila gene structure; tomato genetics; mutation; mammalian meiotic cytogenetics; clinical genetics. *Mailing Add:* Div Biol Sci 409 Tucker Hall Univ Mo-Columbia Columbia MO 65211

BURDICK, CHARLES LALOR, b Denver, Colo, Apr 14, 92; m 38; c 2. CHEMISTRY. *Educ:* Drake Univ, BS, 11; Mass Inst Technol, SB, 13, MS, 14; Univ Basel, PhD(chem), 15. *Hon Degrees:* DSc, Univ Del, 55; LLD, Drake Univ, 70; DEng, Widener Col, 76. *Prof Exp:* Res assoc phys chem, Mass Inst Technol & Calif Inst Technol, 16-17; metall engr, Chile Copper Co & Guggenheim Bros, 19-24; vpres & consult engr, Anglo-Chilean Consol Nitrate Corp, 24-28; asst chem dir, ammonia dept, E I Du Pont de Nemours & Co, 28-35 & develop dept, 36-39, asst to pres, 39-45, chmn bd, Du Pont, SA & Cia Mexicana de Explosivos, Mexico City, 45-46, secy, high polymer comt, Wilmington, 46-50 & polyfibers comt, 50-57. *Concurrent Pos:* Trustee & exec dir, Lalor Found, 37-; mem vis comt biol & Bussey Inst, Harvard Univ, 49-64; dir, Planned Parenthood-World Pop, 61-67; mem exec comt, Int Planned Parenthood Fedn, 62-68,; pres, Christiana Found, 60-73; emer trustee, Univ Del Res Found; mem corp, Marine Biol Lab, Woods Hole, 58-68. *Honors & Awards:* Marshall Medal, Soc Study of Fertil, 84. *Mem:* Fel AAAS; Am Chem Soc; NY Acad Sci; Am Inst Chem Engrs; Soc Study Fertil. *Res:* Administration of awards for research in mammalian reproductive physiology. *Mailing Add:* 900 Barley Dr Barley Mill Ct Wilmington DE 19807

BURDICK, DANIEL, b Syracuse, NY, Oct 21, 15; m 49; c 5. SURGERY. *Educ:* Syracuse Univ, AB, 37, MD, 50; Am Bd Surg, Dipl. *Prof Exp:* CLIN DIR TUMOR CLIN & ATTEND SURGEON, UNIV HOSP, STATE UNIV NY UPSTATE MED CTR, 54- *Concurrent Pos:* Attend, Syracuse Crouse-Irving Mem Hosp, 54-; surgeon, Community Gen Hosp. *Mem:* Am Col Surgeons; Soc Surg Oncol; Am Soc Clin Oncol. *Res:* Cancer research. *Mailing Add:* 475 Irving Ave Syracuse NY 13210

BURDICK, DONALD, biochemistry, plant chemistry, for more information see previous edition

BURDICK, DONALD SMILEY, b Newark, NJ, Feb 8, 37; m 58; c 3. MATHEMATICAL STATISTICS. *Educ:* Duke Univ, BS, 58; Princeton Univ, MA, 60, PhD(math statist), 61. *Prof Exp:* Instr, Princeton Univ, 61-62; from asst prof to assoc prof math, 62-77, assoc prof, 77-88, ASSOC PROF STATIST & MATH & BIOMED ENG, DUKE UNIV, 88- *Concurrent Pos:* Consult, Army Res Off, 62-66, Res Triangle Inst, 85- & Metametrics Corp, 86-; vis asst prof statist, Univ Wis, 65-66; statist ed, J Parapsychol, 81; vis assoc prof, Va Polytech Inst & State Univ, 85-86. *Mem:* Am Statist Asn; Royal Statist Soc. *Res:* Linear statistical models; multivariate statistics; application of statistical methods in chemometrics and psychometrics, including measurement of reading comprehension. *Mailing Add:* Inst Statist & Decision Sci Duke Univ Durham NC 27706

BURDICK, GLENN ARTHUR, b Pavillion, Wyo, Sept 9, 32; m 51; c 2. DEVICE PHYSICS, ACCIDENT RECONSTRUCTION. *Educ:* Ga Inst Technol, BS, 58, MS, 59; Mass Inst Technol, PhD(physics), 61. *Prof Exp:* Mem staff & spec tools design, Ga Technol Inst, 54-57, instr physics, 57-59; researcher, Div Sponsored Res, Mass Inst Technol, 59-61; sr mem res staff, Sperry Microwave, 61-65; assoc prof, 65-67, dean, 79-86, PROF ELEC ENG, UNIV SFLA, 67- *Concurrent Pos:* Prof, Univ Fla, 62-65; accident reconstruction expert, 65-; consult, Sperry Rand, 65-69, Fla Power Co, 67-, Los Alamos Sci Labs, 71, Seaboard Coastline Railroad, 76-; mem bd dirs, ABA Industs, 81-84. *Mem:* Int Soc Hybrid Microelectrons (pres, 74); Inst Elec & Electronics Engrs; Am Soc Eng Educ; NY Acad Sci; Acad Ambulatory Foot Surg. *Res:* Solid state physics; electromagnetic device physics; microelectronics; ferroelectrics; pyroelectrics; theoretical determination of energy bands and Fermi surface of copper. *Mailing Add:* PO Box 668 Tarpon Springs FL 34688

BURDINE, HOWARD WILLIAM, b Big Stone Gap, Va, July 1, 09; m 34. PLANT NUTRITION, PHYSIOLOGY. *Educ:* Berea Col, BS, 35; Univ Ky, MS, 51; Cornell Univ, PhD, 56. *Prof Exp:* County agr agent, Ky, 37-44; soil conservationist, Soil Conserv Serv, US Dept Agr, 44-47, teacher vocational agr, 47-51, asst horticulturist, 56-59, asst & assoc soils chemist, 59-68, plant physiologist, 68-70, prof, 70-77, EMER PROF PLANT PHYSIOL, EVERGLADES RES & EDUC CTR, UNIV FLA, 77- *Honors & Awards:* Fla Fruit & Veg Asn Res Award, 78. *Mem:* Am Soc Plant Physiol; Am Soc Hort Sci; Soil Sci Soc Am; Am Soc Agron. *Res:* Nutrition and physiology of vegetable crops grown on organic soils of the everglades area of Florida. *Mailing Add:* 233 Via Del Aqua Clewiston FL 33440

BURDINE, JOHN ALTON, b Austin, Tex, Feb 7, 36; m 59; c 3. NUCLEAR MEDICINE. *Educ:* Univ Tex, Austin, BA, 59; Univ Tex Med Br Galveston, MD, 61. *Prof Exp:* Intern med, Med Ctr, Ind Univ, Indianapolis, 61-62; resident internal med, Univ Tex Med Br Galveston, 62-65; instr nuclear med, Dept Radiol, 65-66, from asst prof to assoc prof radiol, 66-74, actg chmn dept, 68-71, PROF RADIOL, BAYLOR COL MED, 74-, CHIEF NUCLEAR MED SECT, 65- *Concurrent Pos:* Consult, US Army, 66; chief nuclear med serv, St Luke's Episcopal-Tex Children's Hosps, 69-85, pres & chief exec officer, 84-90, vchmn & chief exec officer, 85. *Mem:* Soc Nuclear Med. *Res:* Development and evaluation of new radiopharmaceuticals; design and implementation of a computer system for processing scintillation camera data, including development of radionuclide techniques for quantification of regional pulmonary function, bone densitometric studies. *Mailing Add:* Exec Offices St Luke's Episcopal Hosp 6270 Bertner Houston TX 77030-0269

BURDITT, ARTHUR KENDALL, JR, b Elizabeth, NJ, Feb 12, 28; m 52; c 4. ENTOMOLOGY. *Educ:* Rutgers Univ, BS, 50; Univ Minn, MS, 53, PhD(entom), 55. *Prof Exp:* Asst prof, Univ Mo, 55-57; res entomolgist, fruit fly lab, 57-60, invest leader, citrus insect invest, 60-64, asst to chief, fruit fly res br, 64-68, asst to dir, entom res div, 68-71, staff asst, plant sci & entom, 71-72, res leader, subtrop hort res, 72-80, dir, Yakima Agr Res Lab, 80-83, RES LEADER, QUARANTINE TREATMENT RES UNIT, AGR RES SERV, US DEPT ABR, YAKIMA, WASH, 83- *Concurrent Pos:* Affil prof, Univ Hawaii, 60; adj prof, Univ Fla, 72-81; courtesy prof, Univ Miami, Fla, 78-80; courtesy entomologist, Wash State Univ, 81-; consult, Agency Int Develop, 84-85, Int Atomic Energy Agency, 81- *Mem:* Fel AAAS; Entom Soc Am. *Res:* Treatments for pests infesting commodities of quarantine treatments, including fumigation, refrigeration and gamma radiation; techniques for detection of pests of quarantine importance. *Mailing Add:* Yakima Agr Res Lab 3706 W Nob Hill Blvd Yakima WA 98902

BURDSALL, HAROLD HUGH, JR, b Hamilton, Ohio, Nov 18, 40; div; c 2. MYCOLOGY. *Educ:* Miami Univ, BA, 62; Cornell Univ, PhD(mycol), 67. *Prof Exp:* Botanist, Forest Dis Lab, Laurel, Md, 67-71, botanist, 71-82, SUPVR BOTANIST & PROJ LEADER, FOREST PROD LAB, CTR FOREST MYCOL RES, FOREST SERV, USDA, 82- *Concurrent Pos:* Lectr, Dept Forest Prod, Univ Wis-Madison, 71-74, adj asst prof, Dept Plant Path, 74-80, adj assoc prof, 80-88, adj prof, 88- *Mem:* Mycol Soc Am (secy, 83-86); Brit Mycol Soc; Am Inst Biol Sci; Int Asn Plant Taxon. *Res:* Systematics of Armillaria, and other Basidiomycetous root rot and decay fungi, including culture studies, genetics, morphology and ecology. *Mailing Add:* Ctr Forest Mycol Res One Gifford Pinchot Dr Madison WI 53705-2398

BUREK, ANTHONY JOHN, b Rockville Ctr, NY, June 14, 46. SOLAR PHYSICS, X-RAY ASTRONOMY. *Educ:* Princeton Univ, AB, 68; Univ Chicago, MS, 69, PhD(physics), 75. *Prof Exp:* Fel physics, Los Alamos Sci Lab, 75-77; MEM STAFF, NAT BUR STANDARDS, 77- *Mem:* Am Phys Soc; Sigma Xi. *Res:* Solar x-ray spectroscopy; theory and evaluation of x-ray diffracting properties of crystals. *Mailing Add:* 2714 Parkard Rd Apt D Ann Arbor MI 48108

BUREK, JOE DALE, pathology, toxicology, for more information see previous edition

BURES, DONALD JOHN (CHARLES), b Winnipeg, Man, Jan 23, 38; m 76. MATHEMATICS. *Educ:* Queen's Univ, Ont, BA, 58; Princeton Univ, PhD(math), 61. *Prof Exp:* Asst prof math, Queen's Univ, Ont, 61-62; from asst prof to assoc prof, 62-71, head dept, 73-78, PROF MATH, UNIV BC, 71- *Mem:* Am Math Soc; Can Math Cong; Royal Soc Can. *Res:* Abstract analysis; von Neumann algebras. *Mailing Add:* Dept Math Univ BC 2075 Westbrook Pl Vancouver BC V6T 1W5 Can

BURES, MILAN F, b Prague, Czech, May 6, 32; m 59. EPIDEMIOLOGY, PUBLIC HEALTH. *Educ:* Charles Univ, Prague, MD, 58; Inst Postgrad Training Physicians, Prague, Czech, 64, MPH, 64. *Prof Exp:* Physician, County Hosp & Munic Med Ctrs, Czech, 58-61; res worker, Inst Radiation Hyg, Prague, 61-65; asst prof epidemiol & health, McGill Univ, 68-70; asst med dir, Northwestern Mutual Life Ins Co, 70-75; MED DIR, STATE MUTUAL LIFE ASSURANCE CO AM, 75- *Concurrent Pos:* Res fel radiation biol, Med Ctr, Univ Ala, 66-68. *Res:* Life insurance medicine; epidemiology of cardiovascular diseases; vital statistics; radiological health. *Mailing Add:* 440 Lincoln St Worcester MA 01605

BURFENING, PETER J, b Reno, Nev, Nov 8, 42; m 64; c 1. REPRODUCTIVE PHYSIOLOGY, GENETICS. *Educ:* Colo State Univ, BS, 64; NC State Univ, MS, 66, PhD(animal physiol), 68. *Prof Exp:* PROF ANIMAL PHYSIOL, MONT STATE UNIV, 60- *Mem:* Am Soc Animal Sci. *Res:* Reproductive physiology of domestic animals with special emphasis on the relationship between genotype and the environment as it relates to reproductive events. *Mailing Add:* Dept Animal Sci Mont State Univ Bozeman MT 59717

BURFORD, ARTHUR EDGAR, b Olean, NY, Mar 5, 28; m 53; c 7. PETROLEUM GEOLOGY, ECONOMIC GEOLOGY. *Educ:* Cornell Univ, AB, 52; Univ Tulsa, MS, 54; Univ Mich, PhD(geol), 60. *Prof Exp:* Geologist, Pan Am Petrol Corp, Utah & Colo, 58-60; from asst prof to assoc prof, WVa Univ, 60-68; actg head dept, Univ Akron, 70-71, head dept, 71-82, prof geol, 68-89, EMER PROF GEOL, UNIV AKRON, 90-; PRES, GEOCONCEPTS, INC, 90- *Concurrent Pos:* Geologist, US Geol Surv, 52, 53, 54, 79 & 80; fel, Nat Sci Found, 55-58; coop geologist, WVa Geol & Econ Surv, 62-69; chmn, N Cent Sect, Geol Soc Am, 87-88. *Mem:* Geol Soc Am; Am Asn Petrol Geol; Am Soc Photogram & Remote Sensing; Am Geophys Union; German Geol Union; Am Inst Prof Geologists; chmn N Cent Sect Geol Soc Am. *Res:* Geology of Rocky Mountains and Appalachians; geologic interpretation of remote sensing imagery; structural analysis of joints; geologic exploration for petroleum; cementation history of well-cemented sandstone. *Mailing Add:* Dept Geol Univ Akron Akron OH 44325

BURFORD, HUGH JONATHAN, b Memphis, Tenn, Aug 5, 31; m 57; c 2. MEDICOLEGAL ALCOHOL IMPAIRMENT, MEASUREMENT. *Educ:* Millsaps Col, BS, 54; Univ Miss, MS, 56; Univ Kans, PhD(pharmacol), 62. *Prof Exp:* Asst chem, Univ Miss, 54-56; asst pharmacol, Univ Kans, 57-60; asst prof, Bowman Gray Sch Med, 63-68; assoc prof pharmacol & dir MDL Labs, Med Sch, Northwestern Univ, 68-71; ASSOC PROF PHARMACOL & TEACHING ASSOC MED SCI, UNIV NC, CHAPEL HILL, 71- & ASSOC PROF PHARM, PHARM SCH, 73- *Concurrent Pos:* USPHS fel, Tulane Univ, 62-63; Am Heart Asn, USPHS & Nat Fund Med Educ grants, 64-; fac fel, Kellogg Ctr Teaching Professions, 70-71; mem, Asn Multidisc Educ in H H Sci, 69, health prof, Educ Spec Interest Group, Am Educ Res Asn, 70; consult, Am Soc Pharmacol & Exp Therapeut & Nat Lib Med, 80-86; Nat Comt Blood Alcohol Adv Comm, Div H H Servs, 84-; chair, teaching & evaluation materials, Educ Affairs Comm, Am Soc Pharm Exp Therapeut; mem, Responsible Bevarage Serv Coun. *Mem:* AAAS; Am Col Clin Pharmacol; Am Soc Pharmacol & Exp Therapeut; Am Educ Res Asn. *Res:* Optimizing education in pharmacology; production and evaluation of pharmacology teaching materials for health profession students including textual, visual and electronic media formatted for individualized and group learning; research on alcohol and health policy which focuses on server intervention and highway safety. *Mailing Add:* 1026 Flob Bldg B7365 Univ NC Sch Med Chapel Hill NC 27599

BURFORD, M GILBERT, b Brooklyn, NY, Oct 27, 10; m 34. PHYSICAL CHEMISTRY. *Educ:* Wesleyan Univ, AB, 32; Princeton Univ, AM, 33, PhD(chem), 35. *Prof Exp:* Asst chem, Princeton Univ, 33-34; instr, Cornell Univ, 35-36; from instr to assoc prof, 36-47, E B Nye Prof Chem, 47-77, assoc provost, 69-76, chmn, Col Sci Soc, 78-79, EMER PROF CHEM, WESLEYAN UNIV, 77-, ADMIN CONSULT, 77- *Concurrent Pos:* Consult & supvr, Conn State Dept Environ Protection, 42-80 & New Eng Interstate Water Pollution Control Comn, 49-80; Fund Advan of Educ fac fel, 55-56; chmn bd dirs, Univ Res Inst Conn, 74-77. *Mem:* Fel AAAS; Am Chem Soc; fel Am Inst Chemists. *Res:* Chemical reactions of hydriodic acid and ammonium iodide; analytical chemistry of difficulty soluble compounds; spectrophotometric methods of analysis; industrial trade waste disposal; chemical kinetics. *Mailing Add:* Hall Lab Wesleyan Univ Middletown CT 06457

BURFORD, NEIL, b Liverpool, Eng, Apr 29, 58; m 80; c 2. NON-METAL SYNTHETIC CHEMISTRY, CHEMISTRY OF PHOSPHORUS & ARSENIC. *Educ:* Univ Wales Col, Cardiff, BSc, 79; Univ Calgary, Alta, PhD(inorg chem), 83. *Prof Exp:* Postdoctoral fel inorg chem, Univ Alta, Edmonton, 83-84; res assoc inorg chem, Univ NB, Fredericton, 84-86; asst prof, 87-91, ASSOC PROF INORG CHEM, DALHOUSIE UNIV, HALIFAX, NS, 91- *Concurrent Pos:* Jour referee, Can J Chem, J Chem Soc, Dalton Trans, J Am Chem Soc. *Mem:* Am Chem Soc; Chem Inst Can. *Mailing Add:* Dept Chem Dalhousie Univ Halifax B3H 4J3 Can

BURFORD, ROGER LEWIS, b Independence, Miss, Jan 19, 30; m 48; c 2. STATISTICS, ECONOMICS. *Educ:* Univ Miss, BBA, 56, MA, 57; Ind Univ, PhD(econ, statist), 61. *Prof Exp:* Lectr econ & statist, Ind Univ, 59-60; asst prof, Ga State Univ, 60-63; assoc prof, 63-67, PROF STATIST, LA STATE UNIV, 67- *Concurrent Pos:* Ford fac fel, Carnegie Inst Technol, 62; Fulbright prof, Nat Taiwan Univ, 67-68; dir bus res econ, La State Univ, 69-74; exec vpres statist & econ, Econ & Indust Res Inc, 69-; chmn, La Gov's Coun Econ Advan, 73-75, mem, 73-79. *Mem:* Am Statist Asn; Econometric Soc; Regional Sci Asn; Am Inst Decision Sci. *Res:* Random variate generation; input-output multipliers; mathematical methods and models in regional economic and demographic research. *Mailing Add:* 590 Castle Kirk Ave Baton Rouge LA 70808

BURG, ANTON BEHME, b Dallas City, Ill, Oct 18, 04. INORGANIC CHEMISTRY. *Educ:* Univ Chicago, BS, 27, MS, 28, PhD(chem), 31. *Prof Exp:* Instr inorg chem, Univ Chicago, 31-39; from asst prof to prof, 39-74, head dept, 40-50, EMER PROF CHEM, UNIV SOUTHERN CALIF, 74- *Concurrent Pos:* Consult var govt agencies, 47- *Honors & Awards:* Tolman Medal, Am Chem Soc, 61, Award Distinguished Serv Advan Inorg Chem, 69. *Mem:* Fel AAAS; Am Chem Soc; Sigma Xi. *Res:* Boron and silicon hydrides; fluorine compounds; vacuum technique; non-aqueous solvents and addition compounds; organo-phosphorus and inorganic polymers; metal-carbonyl analogues; fluorocarbon phosphines; study of new syntheses and compound-types in both fields. *Mailing Add:* Dept Chem Univ Southern Calif Los Angeles CA 90089-0744

BURG, JOHN PARKER, b Great Bend, Kans, Dec 17, 31; m 76; c 5. DEVELOPMENT OF ENTROPIC CONCEPTS, DIGITAL CONCEPTS. *Educ:* Univ Tex, Austin, BA, 53, BS, 53; Mass Inst Technol, MS, 60; Stanford Univ,PhD(geophys), 75. *Prof Exp:* Eng, Tex Instruments Inc, 56-60, geophysicist, 60-62, sr res geophysicist, 62-73; pres, Time & Space Processing, Inc, 73-78, chmn, 78-83; CHMN, ENTROPIC PROCESSING INC, 83-; CHMN, ENTROPIC SPEECH INC, 86- *Concurrent Pos:* Chmn, Entropic Geophysical Inc, 86- *Honors & Awards:* Best Presentation Award, Soc Explor Geophysicists, 67; Naval Res Lab Res Publ Award, 84. *Mem:* Fel Inst Elec & Electronics Engrs; Soc Explor Geophysicists. *Res:* Development of new signal processing methods and entropic processing concepts with applications to speech compression to low bit rates (2400 bps), to geophysical seismic data processing as well as to other fields such as anti submarine warfare. *Mailing Add:* Entropic Processing Inc 10011 N Foothill Blvd Cupertino CA 95014

BURG, MARION, b Bridgeport, Conn, May 25, 21. ORGANIC CHEMISTRY. *Educ:* Queens Col, BS, 42; Cornell Univ, PhD(org chem), 47. *Prof Exp:* Asst instr chem, Queens 42-44; res assoc org chem, Mass Inst Technol, 47-51; with E I du Pont de Nemours Co, Inc, 51-86; RETIRED. *Mem:* Am Chem Soc. *Res:* Synthetic organic chemistry; photochemistry. *Mailing Add:* 2300 Riddle Ave Apt 6 Wilmington DE 19806

BURG, MAURICE B, b Boston, Mass, Apr 9, 31; m 66; c 4. NEPHROLOGY. *Educ:* Harvard Univ, BA, 52, MD, 55. *Prof Exp:* Res assoc, NIH, 57-75, CHIEF LAB KIDNEY & ELECTROLYTE METAB, NIH, 75- *Mem:* Nat Acad Sci; Am Soc Clin Invest; Am Fedn Clin Res; Biophys Soc; Soc Gen Physiol; Am Physiol Soc. *Res:* Renal and electrolyte physiology. *Mailing Add:* Lab Kidney & Electrolyte Metab Nat Heart, Lung & Blood Inst NIH Bethesda MD 20205

BURG, RICHARD WILLIAM, b Ft Wayne, Ind, June 22, 32; m 61; c 2. MICROBIAL BIOCHEMISTRY. *Educ:* Wabash Col, AB, 54; Univ Ill, PhD(biochem), 58. *Prof Exp:* NSF res fel, Univ Calif, Berkeley, 58-60; sr res microbiol, 60-69, from res fel to sr res fel, 70-81, SR INVESTR THERAPEUT RES, MERCK INST THERAPEUT RES, 82- *Mem:* AAAS; Am Chem Soc; Am Soc Microbiol; Soc Indust Microbiol; Am Soc Pharmacog. *Res:* Biochemistry of viruses; antibiotics and physiologically active products of microorganisms. *Mailing Add:* 100 Walnut St Murray Hill NJ 07974

BURG, WILLIAM ROBERT, b Venango, Nebr, Aug 18, 29; m 61; c 2. INDUSTRIAL HYGIENE, ANALYTICAL CHEMISTRY. *Educ:* Nebr State Teachers Col, BS, 59; Univ Nebr, MS, 61; Kent State Univ, PhD(chem), 64. *Prof Exp:* Asst prof chem, Eastern NMex Univ, 64-67 & Nichols State Col, 67-69; ASST PROF ENVIRON HEALTH, UNIV CINCINNATI, 69- *Mem:* Am Chem Soc; Am Indust Hyg Asn. *Res:* Development of analytical methodology for the evaluation of the industrial environment. *Mailing Add:* Kettering Lab Univ Cincinnati 3223 Eden Ave Cincinnati OH 45267

BURGAUER, PAUL DAVID, b St Gallen, Switz, May 21, 26; nat US; m 51; c 2. ORGANIC CHEMISTRY, PHARMACEUTICAL CHEMISTRY. *Educ:* Swiss Fed Inst Technol, dipl ing, PhD(org & pharmaceut chem), 52. *Prof Exp:* Res chemist, Hilton-Davis Chem Co, 53-54; patent liaison, E I du Pont de Nemours & Co, 54-59; mem staff, Patent Dept, Abbott Labs, 59-72, int patent mgr, 72-81; RETIRED. *Concurrent Pos:* Patent agent, 61- *Res:* Polymer chemistry; natural and synthetic drugs and polymers. *Mailing Add:* 813 Hayes Libertyville IL 60048

BURGE, DENNIS KNIGHT, b Ogden, Utah, Dec 8, 35; div; c 2. PHYSICAL OPTICS. *Educ:* Univ Nev, Reno, BS, 56, MS, 61. *Prof Exp:* PHYSICIST, NAVAL WEAPONS CTR, 57- *Mem:* Optical Soc Am; Sigma Xi. *Res:* Optical properties of solids; thin films; ellipsometry. *Mailing Add:* Code 3810 Michelson Lab Naval Weapons Ctr China Lake CA 93555-6001

BURGE, JEAN C, b Detroit, MI, Mar 10, 47; m; c 2. RENAL DISEASE. *Educ:* Mich State Univ, PhD(nutrit), 79. *Prof Exp:* ASST PROF NUTRIT, UNIV NC, 80- *Honors & Awards:* Mary P Huddleson Award, Am Dietetic Found. *Mem:* Am Inst Nutrit; Am Pediat Asn; Am Pub Health Asn; Am Dietetic Asn. *Mailing Add:* Dept Nutrit CB 7405 Univ NC 315 Pittsboro St Chapel Hill NC 27514

BURGE, ROBERT ERNEST, JR, b Kansas City, Mo, Aug 21, 25; m 50. ORGANIC CHEMISTRY. *Educ:* Univ Ill, BS, 48; Cornell Univ, PhD(chem), 52; Bernard M Baruch Col, MBA, 69. *Prof Exp:* Asst org chem, Cornell Univ, 48-51; res chemist, Shell Chem Corp, 51-54, group leader, 54-58, sr chemist, 57-59, sr technologist, 59-60, sr chemist, 60-63, supvr, 63-68; from asst prof to prof chem, Suffolk County Community Col, 68-90; RETIRED. *Mem:* Soc Invest Recurring Events. *Res:* Organic chemical synthesis; macrocyclic carbon compounds; epoxy resins. *Mailing Add:* 43 Huntington Rd Garden City NY 11530

BURGE, WYLIE D, b Denver, Colo, June 2, 25; m 58; c 2. MICROBIOLOGY, BIOCHEMISTRY. *Educ:* Colo State Univ, BS, 50, MS, 56; Univ Calif, Davis, PhD(soil microbiol), 58. *Prof Exp:* Jr soil scientist soil microbiol, Univ Calif, Berkeley, 58-60; assoc specialist soils res & sta supt, Antelope Valley Field Sta, Univ Calif, Riverside, 60-66; res soil scientist, Agr Res Serv, USDA, 66-73, microbiologist, Agr Environ Qual Inst Sci & Educ Admin, 73-86; RETIRED. *Mem:* Soil Sci Soc Am; Am Soc Microbiol; Sigma Xi. *Res:* Ammonium fixation by soil organic matter; biochemistry of nitrification; chemistry of ammonium potassium fixation by soils; microbiology of pesticide degradation in soils; fate of pathogens during land application and composting of sewage sludges; fate in the soil mutagens derived from hazardous wastes in farmed land; molecular processes in the infection of plants by microorganisms. *Mailing Add:* 1106 Edgevale Rd Silver Spring MD 20910-1638

BURGENER, FRANCIS ANDRE, b Visp, Switz, May 21, 42; m 70; c 2. RADIOLOGY. *Educ:* Univ Bern, MD, 67, Diss, 69; Am Bd Radiol, dipl & cert diag radiol, 72. *Prof Exp:* Resident radiol, Univ Bern, 67-68; fel exp med, Univ Zurich, 68-69; resident radiol, Univ Bern, 69-70; instr, Univ Mich, 70-71; from asst prof to assoc prof, 71-81, PROF RADIOL, UNIV ROCHESTER, 82- *Concurrent Pos:* Panelist, US Pharmacopeia Adv Panel, 75-80; mem bd investr radiol, 77-90; ed, J Med Imaging, 87-90. *Mem:* AAAS; NY Acad Sci. *Res:* Radiological and MRI contrast media; liver imaging and portal hypertension. *Mailing Add:* Dept Radiol Univ Rochester Med Ctr Rochester NY 14642

BURGER, ALFRED, b Vienna, Austria, Sept 6, 05; nat US; m 36; c 1. MEDICINAL CHEMISTRY. *Educ:* Univ Vienna, PhD(chem), 28. *Hon Degrees:* DSc, Col Pharm & Sci, Philadelphia, 71. *Prof Exp:* Chemist, Hoffmann-La Roche Co, 28-29; res assoc, Drug Addiction Lab, Nat Res Coun, 29-38, actg asst prof, 38-39, from asst prof to assoc prof, 39-52, prof, 52-70, chmn dept, 62-63, EMER PROF CHEM, UNIV VA, 70- *Concurrent Pos:* Chmn, Med Chem Div, Am Chem Soc, 54; USPHS, 56-59; mem chem panel, Cancer Chemother Nat Serv Ctr, 56-59, med chem, 60-64; vchmn, Gordon Res Conf Med Chem, 58, chmn, 59; mem study sect on exp ther & pharmacol; vis lectr, Univ Calif, 63; NIH spec fel biochem, Univ Hawaii, 65; consult, Smith Kline & French Labs & Philip Morris Res Ctr, Dept Health, Mex; ed, J Med Chem; mem psychopharmacol study comt, Nat Inst Ment Health, 67-71 & 76-80. *Honors & Awards:* Pasteur Medal, 53; Smissman Award, Am Chem Soc, 77; Am Pharm Soc Found Award, 71. *Mem:* Am Chem Soc; Am Pharmacol Soc; AAAS. *Res:* General organic chemistry; chemistry of opium alkaloids; syntheses of morphine substitutes; chemotherapy; antimalarials; antituberculous drugs; organic phosphorus compounds; antimetabolites; psychopharmacological drugs; synthesis and design of drugs. *Mailing Add:* 510 Wiley Dr Charlottesville VA 22901-3245

BURGER, AMBROSE WILLIAM, b Jasper, Ind, Nov 27, 23; m 46, 67; c 6. AGRONOMY, SCIENCE EDUCATION. *Educ:* Purdue Univ, BSA, 47; Univ Wis, MSA, 48, PhD(agron & plant physiol), 50. *Prof Exp:* Asst prof agron, Univ Md, 50-53; prof agron, 53-86, EMER PROF AGRON, UNIV ILL, URBANA-CHAMPAIGN, 86- *Honors & Awards:* Agron Educ Award, Am Soc Agron, 64. *Mem:* Crop Sci Soc Am; AAAS; fel Nat Asn Col Teachers Agr; fel Am Soc Agron; Sigma Xi. *Res:* Forage crop production; pasture investigations; field crop science. *Mailing Add:* Dept Agron AE-106 Turner Hall Univ Ill 1102 S Goodwin Ave Urbana IL 61801

BURGER, CAROL J, b Flushing, NY, Jan 3, 44; m; c 3. EXPERIMENTAL BIOLOGY. *Educ:* Rosary Col, BA, 65; Va Polytech Inst & State Univ, PhD(immunol), 83. *Prof Exp:* Med technologist, St Francis Hosp, Peoria, Ill, 65-66, pediat hemato/oncol, Riley Hosp Children, Indianapolis, Ind, 71-73; res tech, Animal Sci Dept, Purdue Univ, 74; med technologist, Patterson-Coleman Labs, Gainesville, Fla, 76-77, pediat hemat/oncol, Shands Teaching Hosp, Gainesville, 77-79; grad res asst, NIH, Biol Dept, Va Tech, 81-83, res assoc, Div Vet Biol, Col Vet Med, 84-85, Dept Vet Biosci, 86-87, vis asst prof, Dept Biol, 87-88, ASST PROF MICROBIOL & IMMUNOL, DEPT BIOL, VA TECH & DIR, WOMEN'S RES INST, 89- *Concurrent Pos:* Grants, var orgns & corps, 80-92. *Mem:* Sigma Xi; Am Soc Microbiol; Am Asn Immunologists; Soc Leukocyte Biol. *Res:* Biology; immunology; pediatric hematology/oncology. *Mailing Add:* Women's Res Inst VPI & State Univ Tenn Sandy Hall Blacksburg VA 24061-0338

BURGER, CHRISTIAN P, b George, SAfrica, Dec 30, 29; US citizen; m 56; c 3. EXPERIMENTAL MECHANICS & PHOTOMECHANICS, ELECTRO OPTICS. *Educ:* Univ Stellenbosch, BSc, 52; Univ Cape Town, PhD(eng), 67. *Prof Exp:* Engr, Vacuum Oil Co, SAfrica, 55-62; sr lectr eng, Univ Cape Town, 62-71; prof eng mech, Iowa State Univ, 71-86; PROF MECH ENG, TEX A&M UNIV, COLLEGE STATION, 86- *Honors & Awards:* Peterson Award, Soc Exp Stress Anal, 74 & 75, Hetenyi Award, 84, Frocht Award, 84. *Mem:* Fel Soc Exp Stress Anal; Am Soc Nondestructive Testing; Am Soc Mech Eng; Soc Photo-Optical Instrumentation Engrs; Am Soc Testing Mat; Brit Soc Strain Measurement; fel AAAS. *Res:* Thermal stress analysis; experimental mechanics; nondestructive evaluation; fracture mechanics; fiber optic sensors; high temperature and severe environment testing. *Mailing Add:* Dept Mech Eng Tex A&M Univ College Station TX 77843

BURGER, DIONYS, b Ambarawa, Indonesia, May 29, 23; Can citizen; m 52; c 2. FOREST ECOLOGY, SOILS. *Educ:* State Agr Univ Wageningen, MF, 52; Univ Toronto, PhD(bot), 65. *Prof Exp:* Forester, Ont Ministry Natural Resources, 53-61, res scientist forest soil, Res Br, 61-77, head, Ont Forest Res Ctr, 77-81, prin sci, Ecol & Silvicult, Ont Tree Improvement & Forest Biomass Inst, 81-88; RETIRED. *Concurrent Pos:* Leader, Working Group Site Classification, Int Union Forest Res Orgn, 70-80; mem, Ont Joint Forest Res Comn, 75-; mem, Can Nat Res Coun, Assoc Comn Univ Res, 78-80; adv, forest site classification & evaluation, Forestry Fac, Agr Univ Malaysia, Serdang, 80; adj prof land classification, Forestry Fac, Univ Toronto, 81-87; forest soil specialist, F A O prof, Forest Res Inst, Yezin, Burma, 82; consult forest soil & ecosystems, Can Int Develop Agency, Lanxiang, Heilongjang, People's Repub China; consult forest ecologist, 88- *Mem:* Can Inst Forestry; Int Soc Trop Foresters; Int Soc Soil Sci; Ger Soc Forest Site Sci & Tree Breeding. *Res:* Physiographic site classification; ecological regions; mapping on aerial photographs; surface geology; forest nutrition and humus. *Mailing Add:* Forest Ecologist RR 1 Gilford ON L0L 1R0 Can

BURGER, EDWARD JAMES, JR, b Cleveland, Ohio, Feb 23, 33; m 60; c 2. MEDICINE, PHYSIOLOGY. *Educ:* McGill Univ, BSc, 54, MDCM, 58; Harvard Univ, MIH, 60, ScD(physiol), 66. *Prof Exp:* Intern med, Royal Victoria Hosp, Montreal, Que, 58-59; res assoc physiol, Sch Pub Health, Harvard Univ, 66-69; staff asst, Off Sci & Technol, Exec Off of the President, 69-76; clin asst prof, 76-81, assoc prof prev med, 81-88, DIR, INST HEALTH POLICY ANALYSIS, 81-, PROF PREV MED, GEORGETOWN UNIV, 88- *Concurrent Pos:* Consult, President's Adv Comn, 67; assoc, J F Kennedy Sch Pub Admin, 67-68; mem, Nat Cancer Adv Bd & Nat Heart, Lung & Blood Inst Adv Coun, NIH, 73-76; deleg, Orgn Econ Coop & Develop, Paris, 73-76; sr sci adv, Econ Comn Europe, UN, 74. *Mem:* Am Col Prev Med; AAAS. *Res:* Oxygen toxicity and its effects on the lung; effects of environmental contaminants on human health; public policy issues concerning health. *Mailing Add:* Inst Health Policy Analysis 2115 Wisconsin Ave NW Suite 600 Washington DC 20007-2292

BURGER, GEORGE VANDERKARR, b Woodstock, Ill, Jan 22, 27; m 49; c 3. WILDLIFE CONSERVATION. *Educ:* Beloit Col, BS, 50; Univ Calif, MA, 52; Univ Wis, PhD(wildlife mgt), 59. *Prof Exp:* Asst zool & wildlife mgt, Univ Calif, 50-52; instr zool, bot & conserv, Contra Costa Jr Col, 52-54; asst wildlife mgt, Univ Wis, 54-58; field rep, Sportsmen's Serv Bur, NY, 58-62; mgr wildlife mgt, Remington Arms Co, 62-66; gen mgr, 66-88, DIR RES, MCGRAW WILDLIFE FOUND, 88- *Concurrent Pos:* Ed, Wildlife Soc Bull, Wildlife Soc, 72-75. *Honors & Awards:* Leopold Award, Green Tree Club, 58; Nature Conservancy Nat Award, 54. *Mem:* Hon mem Wildlife Soc; Am Fisheries Soc; Outdoor Writers Asn Am. *Res:* Waterfowl and upland game bird management and ecology; shooting preserve and game farm management; wildlife and general conservation education. *Mailing Add:* PO Box 9 Dundee IL 60118

BURGER, HENRY ROBERT, III, b Pittsburgh, Pa, Aug 2, 40; m 63; c 2. STRUCTURAL GEOLOGY, GEOPHYSICS. *Educ:* Yale Univ, BS, 62; Ind Univ, AM, 64, PhD(geol), 66. *Prof Exp:* From asst prof to assoc prof, 66-75, PROF GEOL, SMITH COL, 75- *Mem:* AAAS; Am Geophys Union; Geol Soc Am; Nat Asn Geol Teachers (pres, 85-86); Sigma Xi. *Res:* Gravity and magnetics of rifts; fracture analysis; petrofabrics; computer assisted instruction. *Mailing Add:* Dept Geol Smith Col Northampton MA 01063

BURGER, JOANNA, b Schenectady, NY, Jan 18, 41; m; c 2. ETHOLOGY, ECOLOGY. *Educ:* State Univ NY Albany, BS, 63; Cornell Univ, MS, 64; Univ Minn, PhD(ecol & behav), 72. *Prof Exp:* Instr biol, State Univ NY Buffalo, 64-68; asst, Univ Minn, 68-72; res assoc ethology, Rutgers Inst Animal Behav, 72-73; asst prof, 73-76, assoc prof, 76-81, PROF BIOL, RUTGERS UNIV, 81- *Concurrent Pos:* Mem, BEST, Nat Res Coun. *Mem:* Fel Am Ornith Union; Animal Behav Soc; Ecol Soc; AAAS; Soc Toxicol. *Res:* Examination of the relationship between an animal's environment and its behavior, particularly in the ways animals partition the environment, such as food resources and nesting habitat; effects of heavy metal on behavioral development. *Mailing Add:* Dept Biol Rutgers Univ Piscataway NJ 08855-1059

BURGER, JOHN ALLAN, geology, for more information see previous edition

BURGER, LELAND LEONARD, b Buffalo, Wyo, Nov 5, 17; m 42; c 3. PHYSICAL INORGANIC CHEMISTRY. *Educ:* Univ Wyo, BA, 39; Univ Wash, PhD(chem), 48. *Prof Exp:* Res chemist, Div War Res, Columbia Univ, 42-45 & Hanford Labs, Gen Elec Co, 48-64; res assoc chem, 65-77, STAFF SCIENTIST, PAC NORTHWEST LAB, BATTELLE MEM INST, 77- *Concurrent Pos:* Adj prof chem, Wash State Univ, 73- *Mem:* Fel AAAS; Am Chem Soc; Am Inst Physics; Am Nuclear Soc. *Res:* Solvent extraction mechanisms, absorption spectroscopy; chemistry and control volatile fission products; nuclear fuel reprocessing; fluorine chemistry; radiation chemistry. *Mailing Add:* 1925 Howell Ave Richland WA 99352

BURGER, RICHARD MELTON, b New York, NY, Mar 23, 41; m 77; c 2. MOLECULAR PHARMACOLOGY. *Educ:* Adelphi Col, BA, 62; Princeton Univ, PhD(biol), 69. *Prof Exp:* Spec trainee microbiol, Harvard Med Sch, 64-65, spec trainee biochem, Brandeis Univ, 65-68; Nat Cancer Inst trainee, Univ Calif, Berkeley, 68-71; asst prof biol, Mid E Tech Univ, Turkey, 71-72; assoc, Sloan-Kettering Inst Caner Res, 72-77; assoc molecular pharmacol, Albert Einstein Col Med, 77-82, prin assoc, 82-86; ASSOC, PUB HEALTH RES INST, 86- *Concurrent Pos:* Res assoc, Haskins Labs, New York, 59-; asst prof biochem, Sloan-Kettering Div, Cornell Univ, 73-77; assoc biol sci, Columbia Univ, 76-77; adj assoc prof biol sci, Lehman Col, City Univ NY, 80-81; adj assoc prof pharmacol, NY Univ Sch Med, 86-; Stohlman mem scholar, Leukemia Soc Am, 87. *Mem:* Soc Biol Chem; Harvey Soc; Am Chem Soc. *Res:* Drug-induced DNA damage; metal-oxygen interactions. *Mailing Add:* Pub Health Res Inst 455 First Ave New York NY 10016

BURGER, ROBERT M, b Frederick, Md, Feb 14, 27; m 49; c 3. SOLID STATE PHYSICS, ELECTRONICS. *Educ:* Col William & Mary, BS, 49; Brown Univ, ScM, 52, PhD(physics), 55. *Prof Exp:* Physicist, US Dept Defense, 55-59; fel eng, Solid State Lab, Air Arm Div, Westinghouse Elec Corp, 59-62; dir solid state lab, Res Triangle Inst, 62-67, dir eng & environ sci div, 67-71, chief scientist, 71-82; VPRES, SEMICONDUCTOR RES CORP, 82- *Concurrent Pos:* Res affiliate, Univ Md, 56-61; adj assoc prof elec eng, Duke Univ, 62-69; chmn scientific review comt, Dental Res Ctr, Univ NC, 74-81; consult mem, Adv Group Electron Devices, Dept Dent, 83- *Mem:* AAAS; Am Phys Soc; fel Inst Elec & ElectronicS Engrs; Electrochem Soc. *Res:* Solid state electronics and physics; electronic systems; materials; public policy; technology management; instrumentation; research utilization and analysis; sensors. *Mailing Add:* PO Box 12053 Semiconductor Res Corp Res Triangle Park NC 27709

BURGER, WARREN CLARK, b Ripon, Wis, Feb 11, 23; m 44; c 2. PLANT SCIENCE. *Educ:* Univ Wis, BS, 48, MS, 50, PhD(biochem), 52. *Prof Exp:* Res chemist, 52-73, DIR, BARLEY & MALT LAB, AGR RES SERV, USDA, 73-; PROF AGRON, UNIV WIS-MADISON, 79- *Concurrent Pos:* Adj assoc prof agron, Univ Wis-Madison, 74-79. *Mem:* Am Soc Plant Physiol; Am Chem Soc; Am Soc Brewing Chemists. *Res:* Plant proteolytic enzymes; malting and brewing biochemistry; barley nutritional quality. *Mailing Add:* Rt 1 1928 Lewis Rd Mt Horeb WI 53572

BURGERT, BILL E, b Horton, Kans, Aug 9, 29; m 72. ORGANIC CHEMISTRY. *Educ:* Kans State Teachers Col, AB, 51; Kans State Col, MS, 52; Northwestern Univ, PhD(chem), 55. *Prof Exp:* Res chemist, Dow Chem Co, 55-58, proj leader, 58-60, group leader, 60-64, div leader, 64-68, dir, Phys Res Lab, 68-79, tech dir, saran & converted prod res, 79-84, proj dir, Mich Div Res, 84-86; RETIRED. *Mem:* Am Chem Soc; Sigma Xi. *Res:* Polymer chemistry. *Mailing Add:* 1920 Sylvan Lane Midland MI 48640

BURGES, STEPHEN JOHN, b Newcastle, Australia, Aug 26, 44; m 70. SURFACE WATER HYDROLOGY. *Educ:* Univ Newcastle, Australia, BSc & BE Hons, 67; Stanford Univ, MS, 68, PhD(civil eng), 70. *Prof Exp:* Res asst civil eng, Stanford Univ, 67-70; from asst prof to assoc prof, 70-79, PROF CIVIL ENG, UNIV WASH, 79- *Concurrent Pos:* Co-ed, Water Resources Res,Am Geophys Union, 81-84, guest ed, 85-86, ed, Water Resources Monogr, 85-87; mem, Water Sci & Technol Bd, US Nat Res Coun, 85-89; chmn, WSTB Steering Comt Climate Change & Water Resource Mgt, Nat Res Coun, 89- *Mem:* Fel Am Soc Civil Engrs; fel Am Geophys Union; AAAS; Am Water Resources Asn; Sigma Xi; Int Asn Hydraul Res. *Res:* Surface water hydrology; hydrologic process representation; rainfall-runoff modeling; stochastic hydrology; water resource system analysis, design and operation; urban hydrology; integrated learning systems. *Mailing Add:* Dept Civil Eng 160 Wilcox Hall FX-10 Univ Wash Seattle WA 98195

BURGESON, ROBERT EUGENE, b Newhall, Calif, Aug 5, 45. BIOCHEMISTRY, CELL BIOLOGY. *Educ:* Univ Calif, Irvine, BS, 68; Univ Calif, Los Angeles, PhD(molecular biol), 74. *Prof Exp:* Res biochemist molecular biol, Sch Dent, Univ Calif, Los Angeles, 73-74, asst res biochemist med genetics, assoc prof pediat, Harbor Med Ctr, 76-87; assoc res dir, Shriners Hosp, 86-88; PROF, BIOCHEM & MOLECULAR BIOL, CELL BIOL & ANAT & DERMAT, ORE HEALTH SCI UNIV, 88- *Concurrent Pos:* NIH fel, 74-76. *Mem:* AAAS; Am Soc Cell Biol. *Res:* Investigations of the structure and function of connective tissue macromolecules with regard to human growth, development and inheritable disease. *Mailing Add:* Res Dept Shriners Hosp Crippled Children 3101 SW Sam Jackson Pk Rd Portland OR 97201

BURGESS, ANN BAKER, b Madison, Wis, Sept 20, 42; m 67; c 2. CELLULAR BIOLOGY. *Educ:* Univ Wis-Madison, BS, 64; Harvard Univ, PhD(molecular biol), 69. *Prof Exp:* Fel, Inst Molecular Biol, Univ Geneva, Switz, 69-71; fel, McArdle Lab, Univ Wis-Madison, 71-72, asst scientist, 72-73, lectr biol core curric, 73-87, assoc chmn, 80-90, SR LECTR, UNIV WIS-MADISON, 87-, DIR, 90- *Concurrent Pos:* Adv, Fac Adv Serv, Univ Wis-Madison, 78-; vis scientist, Dept Genetics, Univ Wash-Seattle, 83-84. *Mem:* Asn Biol Lab Educ; Am Asn Higher Educ. *Mailing Add:* Biol Core Curric Univ Wis 361 Noland Hall Madison WI 53706

BURGESS, CECIL EDMUND, b Happy, Tex, Jan 21, 20; m 48; c 2. MATHEMATICS. *Educ:* WTex State Univ, BS, 41; Univ Tex, PhD(math), 51. *Prof Exp:* Instr math, Univ Tex, 41-42; with Naval Ord Lab, 42-43; instr math, Univ Tex, 46-51; from instr to assoc prof, 51-61, chmn dept, 67-77, PROF MATH, UNIV UTAH, 61- *Concurrent Pos:* Vis lectr, Univ Wis, 56-57; vis mem, Inst Advan Study, 62-63; vis prof, Univ Tex, Austin, 77-78; vis mem, Math Inst, Univ Warwick, Eng, 78, comt undergrad prog math, 65-67. *Honors & Awards:* Cert of Meritorious Serv, Math Asn Am, 85. *Mem:* Am Math Soc; Math Asn Am. *Res:* Geometric topology; embeddings of surfaces in Euclidean 3-dimensional space. *Mailing Add:* 2236 Logan Ave Salt Lake City UT 84108

BURGESS, DAVID RAY, b Hobbs, NMex, Nov 21, 47; c 4. CELL BIOLOGY. *Educ:* Calif Polytech State Univ, BS, 69, MS, 71; Univ Calif, Davis, PhD(zool), 74. *Prof Exp:* Res assoc zool, Univ Wash, 74-76; asst prof, Dartmouth Col, 76-82; from assoc prof to prof, Sch Med, Univ Miami, 82-90; PROF & CHMN, DEPT BIOL SCI, UNIV PITTSBURGH, 90- *Concurrent Pos:* NIH fel, Univ Wash, 74-76; NIH career develop award, 81-86; mem, Cell Biol Study Sect, NIH, 82-86; instr physiol, Marine Biol Lab, Woods Hole, Mass, 87-88. *Mem:* AAAS; Am Soc Cell Biol; Soc Develop Biol. *Res:* Cellular morphogenesis; biochemistry of actin-based motility in single cells and in organ development. *Mailing Add:* Dept Biol Sci Univ Pittsburgh Pittsburgh PA 15260

BURGESS, DONALD WAYNE, b Okmulgee, Okla, July 11, 47; m 72; c 2. ATMOSPHERIC SCIENCE, RADAR METEOROLOGY. *Educ:* Univ Okla, BS, 71, MS, 74. *Prof Exp:* RES METEOROLOGIST RADAR METEOROL, NAT SEVERE STORMS LAB, ENVIRON RES LAB, NAT OCEANIC & ATMOSPHERIC ADMIN, DEPT OF COM, 70- *Honors & Awards:* Special Achievement Award, Nat Oceanic & Atmospheric Admin, 76; Silver Medal Award, Dept Com, 79; Outstanding Proj, Nat Soc Prof Engrs, 80. *Mem:* Am Meteorol Soc; Nat Weather Asn. *Res:* Radar, particularly Doppler, applied to detection of and further understanding of severe local storms and tornados. *Mailing Add:* Nat Severe Storms Lab NOAA 1313 Halley Circle Norman OK 73069

BURGESS, EDWARD MEREDITH, b Birmingham, Ala, June 8, 34; m 57; c 2. ORGANIC CHEMISTRY. *Educ:* Auburn Univ, BS, 56; Mass Inst Technol, PhD(org chem), 62. *Prof Exp:* Instr chem, Yale Univ, 62-64; from asst prof to assoc prof, Ga Inst Technol, 64-74, PROF CHEM, GA INST TECHNOL, 74- *Mem:* Am Chem Soc; The Chem Soc. *Res:* Organic photochemistry; small ring heterocycle synthesis; new functional groups. *Mailing Add:* Dept Chem Ga Inst Technol Atlanta GA 30332

BURGESS, FRED J, b La Grande, Ore, June 7, 26; m 49; c 3. CIVIL & SANITARY ENGINEERING. *Educ:* Ore State Univ, BS, 50; Harvard Univ, MS, 55. *Prof Exp:* From instr to assoc prof, Ore State Univ, 53-62, head dept, 66-74, prof civil eng, 62-, dean eng, 70-; RETIRED. *Concurrent Pos:* Consult, USPHS, 63-; mem, Am Environ Eng Intersoc Bd, Gov Capitol Planning Comn, Univ-Gov Environ Sci Adv Comt & Univ Marine Sci Coun; chmn nat environ sci training comt, USPHS, 62-63, mem training & demonstration grants comt, 62- *Honors & Awards:* Clemens Award, Harvard Univ, 55. *Mem:* Am Soc Civil Engrs; Am Water Works Asn; Nat Soc Prof Eng; Water Pollution Control Fedn; Sigma Xi. *Res:* Water supply and waste disposal; water resources engineering. *Mailing Add:* 8535 SW Curry Dr No B Wilsonville OR 97070

BURGESS, HOVEY MANN, b Stoneham, Mass, Oct 19, 16; m 39; c 1. FOOD SCIENCE, NUTRITION. *Educ:* Bowdoin Col, BS, 38; Columbia Univ, AM, 40. *Prof Exp:* From jr chemist to sect head, Cent Labs, Gen Foods Corp, 40-52, mgr res & develop, Gaines Div, 52-57, lab mgr, Post div, 57-61, corp res, 61-74, mgr tech eval, 61-65, group res mgr technol, 65-69, dir tech appln, 69-70, fel basic sci, 70-74; CONSULT FOOD SCI, 74- *Res:* Animal nutrition; dehydration; soluble coffee; dessert products; process for dehydrated potato product; cereals. *Mailing Add:* 555 Port Side Dr Naples FL 33940

BURGESS, JACK D, b Moline, Ill, July 16, 24; m 52; c 2. GEOLOGY, BOTANY. *Educ:* Univ Ill, BS, 49; Univ Mo, MA, 55. *Prof Exp:* Inspector qual control, Deere & Co, 50-52; mining geologist, Northern Pac RR, 55-56; geologist, sub-surface geol, Gulf Res & Develop Co, 56, field geol, 56-59, paleontologist, 59-72, sr res geologist, 72-80, staff geologist, 80-85; GEOCHEM COORDR, CHEVRON USA, SOUTHERN REGION, 85- *Mem:* Am Asn Petrol Geol; Brit Paleont Asn; Am Asn Stratig Palynologists (pres, 78-79); Int Comn Coal Petrol; Soc Org Petrog (pres-elect, 87-88). *Res:* Fossil acid-insoluble palynomorphs from the entire geologic column within the Rocky Mountain Region; integrating microscopic kerogen description and thermal maturation with other hydrocarbon source rock studies in geochemistry of Southern United States. *Mailing Add:* Chevron USA Southern Region PO Box 1635 Houston TX 77251-1635

BURGESS, JAMES HARLAND, b Portland, Ore, May 11, 29; m 51; c 3. MAGNETIC RESONANCE. *Educ:* State Col Wash, BS, 49, MS, 51; Wash Univ, PhD(physics), 55. *Prof Exp:* Sr engr, Sylvania Elec Prod, Inc, 55-56; res assoc physics, Stanford Univ, 56-67; from instr to assoc prof, 57-73, PROF PHYSICS, WASH UNIV, 73- *Mem:* Am Phys Soc; Am Asn Physics Teachers; Sigma Xi. *Res:* Nuclear magnetic resonance; paramagnetic resonance; magnetic cooperative phenomena; molecular dynamics. *Mailing Add:* Dept Physics Wash Univ St Louis MO 63130

BURGESS, JOHN C(ARGILL), b Providence, RI, Oct 29, 23; m 53; c 3. ACOUSTICS. *Educ:* Brown Univ, BS, 44; Stanford Univ, MS, 49, PhD(eng mech), 55. *Prof Exp:* Instr eng drawing & descriptive geometry, Brown Univ, 46-48; asst, Stanford Univ, 48-49 & 50-51, res asst, 51-53; instr eng mech, Univ PR, 49-50; mech engr, Stanford Res Inst, 53-61; head, Appl Mech Dept, United Tech Ctr, 61-62, asst mgr, Eng Sci Br, 62-64, mgr, 64-66; chmn dept, 66-68, PROF MECH ENG, UNIV HAWAII, 66- *Concurrent Pos:* US Int Educ Exchange Serv lectr, Cordoba Nat Univ, 56; pres, Univ Hawaii Col Eng faculty senate, 69-71; chmn, Adv Comt Noise Control, Hawaii State Dept Health, 70-72; inv researcher, Nat Res Coun Can, 73, 87 & 90-91; resident visitor, Bell Labs, 79-80. *Honors & Awards:* Distinguished Serv Award, Acoust Soc Japan, 78 & Medal of Special Merit, 88. *Mem:* Am Soc Eng Educ; Am Soc Mech Engrs; Acoust Soc Am; Inst Elec & Electronics Engrs. *Res:* Dynamics of rigid and flexible bodies; vibration, stress waves, acoustics; structural dynamics; automatic control; signal processing; computer modelling of dynamic systems; adaptive sound control. *Mailing Add:* Univ Hawaii Dept Mech Eng 2540 Dole St Honolulu HI 96822

BURGESS, JOHN HERBERT, b Montreal, Que, May 24, 33; m 58; c 4. PULMONARY CIRCULATION, CARDIAC FAILURE. *Educ:* McGill Univ, BSc, 54, MD, 58. *Prof Exp:* Med resident, McGill Univ, 58-60; res fel, Univ Birmingham, Eng, 60-62; med resident, McGill Univ, 62-64; res fel, Cardiovasc Res Inst, Univ Calif, San Francisco, 64-66; from asst prof to assoc prof, 66-75, PROF MED, MCGILL UNIV, 75 -; DIR, CARDIOL, MONTREAL GEN HOSP, 73 - *Concurrent Pos:* Nuffield Found fel, 60-62; R S McLaughlin Found fel, 64-66; Med Res Coun Scholar, 66-71; gov, Am Col Physicians, 79-83; Pres, Royal Col of Physicians & Surgeons Can, 90-92. *Honors & Awards:* Gold Medal, McGill Univ, 58; Order of Can, 87. *Mem:* Am Col Physicians; Royal Col Can; Am Physiol Soc; AAAS; Am Heart Asn; Can Heart Found. *Res:* Pulmonary circulation; pulmonary diffusive capacity; pulmonary capillary blood volume; nor-epinephrine extraction in normal and failing heart. *Mailing Add:* Montreal Gen Hosp 1650 Cedar Ave Montreal PQ H3G 1A4 Can

BURGESS, JOHN S(TANLEY), b Milwaukee, Wis, May 1, 18; m 48; c 3. PHYSICS, ELECTRONICS. *Educ:* St Lawrence Univ, BS, 40; Univ Notre Dame, MS, 42; Ohio State Univ, PhD(physics), 49. *Prof Exp:* Asst microwaves, Res Lab, Gen Elec Co, 42-46; asst prof physics, St Lawrence Univ, 49-51; proj engr, Rome Air Develop Ctr, 51-53, sect chief radar lab, 53-56, chief, 56-57, tech dir control & guid, 57-60, chief scientist ground electronics, 60-71, dep dir, Shape Tech Ctr, 71-76, chief scientist, 76-80; RETIRED. *Concurrent Pos:* Mem adv comt electron tubes, US Dept Defense, 52-58; mem tech mgt coun, US Air Force Systs Command, 60-65 & US Air Force Coun Scientists, 61-64. *Mem:* Fel Inst Elec & Electronics Engrs; Am Phys Soc. *Res:* Thermionic and secondary electron emission; infrared spectrum of methane and deutero-ammonia; microwave tubes; radar technology. *Mailing Add:* 36 Julie Lane Newark DE 19711

BURGESS, PAUL RICHARDS, b Logan, Utah, May 2, 34; m 63. NEUROBIOLOGY. *Educ:* Reed Col, BA, 56; Oxford Univ, BA, 59; Rockefeller Inst, PhD(physiol), 65. *Prof Exp:* From instr to assoc prof, Univ Utah, 67-72; assoc prof, 72-75, PROF PHYSIOL, MED SCH, UNIV UTAH, 75- *Concurrent Pos:* Fel physiol, Univ Utah, 65-67. *Mem:* Soc Neurosci; Am Physiol Soc. *Res:* Somatic sensation; nerve growth. *Mailing Add:* Dept Physiol Univ Utah Sch Med 50 N Medical Dr Salt Lake City UT 84132

BURGESS, RICHARD RAY, b Mt Vernon, Wash, Sept 8, 42; m 67; c 2. ONCOLOGY, SCIENCE EDUCATION. *Educ:* Calif Inst Technol, BS, 64; Harvard Univ, PhD(molecular biol & biochem), 69. *Prof Exp:* From asst prof to assoc prof, 71-82, PROF ONCOL, MCARDLE LAB CANCER RES, UNIV WIS-MADISON, 82-, DIR, BIOTECHNOLOGY CTR, 84- *Concurrent Pos:* Helen Hay Whitney Found fel, Inst Molecular Biol, Geneva, Switz, 69-71; mem, Molecular Biol Study Sect, NSF, 81-84; Guggenheim fel, 83-84. *Honors & Awards:* Pfizer Award for Enzyme Chem, 82. *Mem:* Am Soc Biochem & Molecular Biol; Am Chem Soc; Am Asn Cancer Res; Protein Soc. *Res:* RNA polymerase and the regulation of gene expression; purification, subunit structure, physical and enzymatic properties, and regulation of the DNA-dependent RNA polymerases from prokaryotic and eukaryotic cells; protein biotechnology, immunoaffinity purification. *Mailing Add:* McArdle Lab Cancer Res Univ Wis Madison WI 53706

BURGESS, ROBERT LEWIS, b Kalamazoo, Mich, Sept 12, 31; m 55; c 5. PLANT ECOLOGY, BOTANY. *Educ:* Univ Wis-Milwaukee, BS, 57; Univ Wis-Madison, MS, 59, PhD(plant ecol), 61. *Prof Exp:* Asst prof bot, Ariz State Univ, 60-63, dir desert inst, 63; from asst prof to assoc prof bot, NDak State Univ, 63-71; dep dir, Eastern Deciduous Forest Biome, Int Biol Prog, Oak Ridge Nat Lab, 71-77, prog dir, sect head & sr res ecologist, Environ Sci Div, 72-81; PROF & CHMN, DEPT ENVIRON & FOREST BIOL, STATE UNIV NY COL ENVIRON SCI & FORESTRY, 81- *Concurrent Pos:* res collab, Nat Park Serv, Dept Interior, 64; vis prof bot, Pahlavi Univ, Iran, 65-66; consult, NSF, 74-75, EPA, 85-87; mem bd govs, Am Inst Biol Sci, 81-84, bd dir, 91- *Honors & Awards:* Distinguished Serv Citation, Ecol Soc Am, 88. *Mem:* Ecol Soc Am; Wilderness Soc; Sigma Xi; fel AAAS; Am Inst Biol Sci; Nature Conservancy; Brit Ecol Soc; Int Asn Trop Ecol. *Res:* Regional vegetation studies on composition, structure and dynamics of native plant communities in Wisconsin, Arizona, New York and North Dakota; application of statistical methods to community analysis and interpretation; vegetation mapping; landscape dynamics; history of ecology. *Mailing Add:* Dept Environ & Forest Biol State Univ NY Col Environ Sci & Forestry Syracuse NY 13210

BURGESS, TERESA LYNN, b San Francisco, Calif, Apr 12, 57; m 83; c 1. CELL BIOLOGY, MEMBRANE TRAFFICKING. *Educ:* Univ Calif, Berkeley, BA, 80; Univ Calif, San Francisco, PhD(biochem), 87. *Prof Exp:* Fel, Dept Cell Biol, Yale Univ, 87; fel, Neurosci Res Inst, 88-90, RESEARCHER, DEPT BIOL SCI, UNIV CALIF, SANTA BARBARA, 90- *Mem:* Am Soc Cell Biol; AAAS. *Res:* Interaction of the microtubular cytoskeleton with membrane bound organelles; post-translational modifications of tubulin on the structure/organization of the Golgi apparatus. *Mailing Add:* Dept Biol Sci Univ Calif Santa Barbara Santa Barbara CA 93106

BURGESS, THOMAS EDWARD, b Pontiac, RI, Nov 19, 23; m 49. INORGANIC CHEMISTRY. *Educ:* RI State Col, BS, 49; Syracuse Univ, MS, 52; Univ Conn, PhD(chem), 59. *Prof Exp:* Chemist, Am Optical Co, 52-54; chemist, Sprague Elec Co, North Adams, 59-71; TECH DIR, BURGESS ANALYTICAL LAB, 71- *Mem:* Am Soc Test & Mat; Am Ind Hygiene Asn; Am Ceramic Soc. *Res:* Analytical chemistry of semiconductor thin films; radiotracer studies of diffusion in solids; atomic absorption spectroscopy; optical emission spectroscopy. *Mailing Add:* 47 Grandview Dr Williamstown MA 01267

BURGESS, WILLIAM HOWARD, b Boston, Mass, Mar 13, 24; m 49; c 2. APPLIED CHEMISTRY. *Educ:* Cornell Univ, BChE, 49, MFS, 50, PhD(dairy chem), 54. *Prof Exp:* Asst food sci, Cornell Univ, 49-50, dairy chem, 50-53, nutrit & biochem, 53-54l from asst prof to assoc prof, 54-68, PROF CHEM ENG, UNIV TORONTO, 68- *Mem:* Am Chem Soc. *Res:* Reaction kinetics. *Mailing Add:* Dept Chem Eng 200 College St Rm 217 Toronto ON M5S 1A4 Can

BURGESS, WILSON HALES, b Alexandria, Va, Sept 29, 54. PROTEIN FUNCTION. *Educ:* Univ Va, BA, 76, PhD(biol), 81. *Prof Exp:* Teaching asst biol, Univ Va, 76-77; Governor's fel, 77-78, res asst, 78-81; res assoc, Dept Pharmacol, Sch Med, Vanderbilt Univ, Nashville, Tenn & assoc, Howard Hughes Med Inst, 81-84; sr scientist, Div Cell Biol, Revlon Biotechnol Res Ctr, Rockville, Md, 84-86; sect mgr, Protein Biochem Div, Rorer Biotechnol Inc, Rockville, Md, 86-87; SR SCIENTIST, MOLECULAR BIOL LAB, AM RED CROSS, JEROME HOLLAND LAB BIOMED SCI, ROCKVILLE, MD, 87- *Concurrent Pos:* Lectr cell biol, Univ Va, 80; mem ad hoc rev, NSF, peer rev subcomt, Am Cancer Soc & Am Heart Asn. *Mem:* Am Soc Cell Biol; AAAS. *Res:* Molecular biology; author of numerous publications. *Mailing Add:* Dept Molecular Biol Am Red Cross 15601 Crabbs Branch Way Rockville MD 20855

BURGGRAF, ODUS R, b Ft Wayne, Ind, Feb 27, 29; m 50; c 3. AERONAUTICAL ENGINEERING, COMPUTATIONAL FLUID DYNAMICS. *Educ:* Ohio State Univ, BAeroE & MSc, 52; Calif Inst Technol, PhD(aeronaut, physics), 55. *Prof Exp:* Aerodynamicist, Douglas Aircraft Corp, 52; asst prof mech eng, US Air Force Inst Technol, 54-56; eng specialist, Curtiss-Wright Corp, 56-59; staff scientist, Lockheed Missiles & Space Co, 60-64; from assoc prof to prof, 64-88, EMER PROF AERONAUT & ASTRONAUT ENG, OHIO STATE UNIV, 89- *Concurrent Pos:* Consult, Astro Res Corp, Calif, 59-69 & Lockheed Missiles & Space Co, 64-65; hon res fel, Math Dept, Univ Col London, 72, spec lectr, 78; sr postdoctoral fel, Nat Center Atmospheric Res, 77; mem, Nat Tech Comt on Fluid Dynamics, Am Inst Aeronaut & Astronaut, 78-80. *Mem:* Assoc fel Am Inst Aeronaut & Astronaut. *Res:* Boundary layer stability theory, separated flows; turbulent boundary layers; vortex flows; ducted-propeller theory; stall propagation in axial compressors, optimum filamentary structures; trajectory statistical analysis. *Mailing Add:* Dept Aeronaut & Astronaut Eng Ohio State Univ 2036 Neil Ave Columbus OH 43221

BURGGREN, WARREN WILLIAM, b Edmonton, Alta, Aug 14, 51; m 85; c 3. COMPARATIVE PHYSIOLOGY, EVOLUTIONARY & DEVELOPMENTAL BIOLOGY. *Educ:* Univ Calgary, BSc, 73; Univ East Anglia, Eng, PhD(physiol), 76. *Prof Exp:* Asst prof, 78-82, assoc prof, 82-87, PROF ZOOL, UNIV MASS, 87- *Concurrent Pos:* Prin investr, NSF, 79-; Ed Physiol Zoo, 88- *Mem:* Can Soc Zoologists; Soc Exp Biol; Am Soc Zoologists. *Res:* Comparative animal physiology; evolution of respiratory and cardiovascular physiology; growth of regulatory processes during development from embryo to adult. *Mailing Add:* Dept Biol Univ Mass Amherst MA 01003-0027

BURGHARDT, GORDON MARTIN, b Milwaukee, Wis, Oct 11, 41; m 83; c 3. CHEMORECEPTION, BIOPSYCHOLOGY. *Educ:* Univ Chicago, BS, 63, PhD(biopsychol), 66. *Prof Exp:* Instr biol, Univ Chicago, 66-67; from asst prof to assoc prof, 68-74, PROF PSYCHOL, UNIV TENN, KNOXVILLE, 74-, PROF ECOL, 77-, PROF ZOOL, 80-, DIR, GRAD PROG ETHOLOGY (LIFE SCI), 81- *Concurrent Pos:* NIMH res grant, 67-75; NSF res grant, 75-; J S Guggenheim fel, 76-77; ed, Ethology, 81-87. *Mem:* Am Soc Ichthyol & Herpet; Animal Behav Soc (2nd pres-elect, 84-85, 1st pres-elect, 85-86, pres, 86-87); Am Psychol Asn; Psychonomic Soc; Soc Study Amphibians & Reptiles; Int Soc Develop Psychobiol; Am Phychol Soc. *Res:* Chemical and visual perception; theoretical and historical issues; ontogeny and evolution of behavior; behavior of reptiles; precocial behavior, growth and play. *Mailing Add:* Dept Psychol Univ Tenn Knoxville TN 37996-0900

BURGHARDT, ROBERT CASEY, b Detroit, Mich, Apr 15, 47; m 69; c 3. CELL BIOLOGY, REPRODUCTIVE BIOLOGY. *Educ:* Univ Mich, BS, 69; Wayne State Univ, MS, 73, PhD(biol), 76. *Prof Exp:* Fel anat, Lab Human Reproduction & Reproductive Biol, Harvard Med Sch, 76-78; asst prof biol, 78-84, ASSOC PROF VET ANAT, TEX A&M UNIV, 84-, ASSOC PROF MED PHYSIOL, 90- *Mem:* Sigma Xi; Am Soc Cell Biol; Soc Study Reproduction. *Res:* Reproductive physiology; gamete surfaces; cell biology. *Mailing Add:* Dept Vet Anat Tex A&M Univ Coll Station TX 77843

BURGHART, JAMES H(ENRY), b Erie, Pa, July 18, 38; m 61; c 2. CONTROL SYSTEMS, ELECTRICAL ENGINEERING. *Educ:* Case Western Reserve Univ, BS, 60, MS, 62, PhD(eng), 65. *Prof Exp:* Asst prof elec eng, US Air Force Inst Technol, 65-68; from asst prof to assoc prof, State Univ NY, Buffalo, 68-75; dir grad studies, 70-75; PROF ELEC ENG, CLEVELAND STATE UNIV, 75-, CHMN DEPT, 75-85 & 89- *Concurrent Pos:* Chmn, Student Activ Comt, Indust Appln Soc, 78-88; secy, Region 2, Inst Elec & Electronics Engrs, 88- *Mem:* Inst Elec & Electronic Engrs; Am Soc Eng Educ. *Res:* Control systems; optimization; hybrid vehicle design and control; artifical intelligence. *Mailing Add:* Dept Elec Eng Cleveland State Univ Cleveland OH 44115-2440

BURGHOFF, HENRY L, b Yalesville, Conn, June 14, 07. PHYSICAL METALLURGY & PHYSICAL METALLURGICAL ENGINEERING. *Educ:* Yale Univ, BS, 28, MS, 30, DSc(eng), 39. *Prof Exp:* Dir res & develop, Chase Brass Copper Co, 60-68; RETIRED. *Mem:* Fel Am Soc Metals. *Res:* Copper and copper alloys composition and processing treatment and properties; processing treatment for non-copper based materials such as rhenium, titanium and zirconium. *Mailing Add:* 12-C Heritage Circle Southbury CT 06488

BURGI, ERNEST, JR, b Salt Lake City, Utah, Mar 22, 24; c 2. AUDIOLOGY, SPEECH PATHOLOGY. *Educ:* Ariz State Univ, BA, 50; Univ Denver, MA, 51; Univ Pittsburgh, PhD(audiol), 57. *Prof Exp:* Instr speech & speech path, Univ Nebr, 51-55; res assoc speech & hearing dis, Univ Pittsburgh, 55-57; asst prof audiol, Bowling Green State Univ, 57-58; assoc prof audiol & speech path, Univ Nebr, 59-62; assoc prof, 62-69, PROF SPEECH & THEATRE ARTS, UNIV PITTSBURGH, 69- *Concurrent Pos:* Consult, Cleft Palate Proj, Univ Pittsburgh, 58. *Mem:* Am Speech & Hearing Asn; Speech Asn Am. *Mailing Add:* RD 2 Box 291A Cheswick PA 15024

BURGIN, GEORGE HANS, b Switz, Feb 13, 30; US citizen; m 60; c 3. APPLIED MATHEMATICS, ELECTRICAL ENGINEERING. *Educ:* Swiss Fed Inst Technol, Dipl EEng, 54, PhD(elec eng), 60. *Prof Exp:* Mgr comput prog, Omni Ray, Switz, 60-62; design specialist, Gen Dynamics/Convair, 62-66; sr scientist, Decision Sci Inc, 66-83; CHIEF SCIENTIST, TITAN SYSTS, INC, 83- *Concurrent Pos:* Consult, Narmco Res & Develop Div, Whittaker Corp, 64-66; instr, Exten Div, Univ Calif, San Diego, 66-88; lectr, Dept Math, San Diego State Univ, 79- *Mem:* Asn Comput Mach; Simulation Coun; Inst Elec & Electronics Engrs; Int Soc Neural Networks. *Res:* Application of modern control theory to the design of aircraft stability; augmentation system; systems engineering; computer simulation; neural networks. *Mailing Add:* 6284 Avenida Cresta La Jolla CA 92037-6505

BURGINYON, GARY ALFRED, b Spokane, Wash, June 29, 35; m 58; c 2. NUCLEAR PHYSICS. *Educ:* Wash State Univ, BS, 58; Yale Univ, MS, 61, PhD(physics), 66. *Prof Exp:* Physicist, Naval Res Lab, 61-62; res staff physicist, Yale Univ, 66-68; PHYSICIST, LAWRENCE LIVERMORE LAB, 68- *Mem:* Am Phys Soc; Am Astron Soc. *Res:* Collective nuclear structure studies among rare earth nuclei; x-ray astronomy; x-ray plasma diagnostics. *Mailing Add:* Lawrence Livermore Lab PO Box 808 Livermore CA 94550

BURGISON, RAYMOND MERRITT, pharmacology; deceased, see previous edition for last biography

BURGMAIER, GEORGE JOHN, b Utica, NY, Jan 28, 44; m 68; c 2. ORGANIC CHEMISTRY, PHOTOGRAPHY. *Educ:* Univ Toronto, BSc, 66; Yale Univ, PhD(org chem), 70. *Prof Exp:* SR RES CHEMIST, EASTMAN KODAK CO, 70- *Mem:* Am Chem Soc; Sigma Xi. *Res:* Chemistry of highly strained small ring hydrocarbons; preparation of chemicals with photographic applications. *Mailing Add:* 74 Stuyvesant Rd Pittsford NY 14534

BURGNER, ROBERT LOUIS, b Yakima, Wash, Jan 16, 19; m 42; c 2. FISH BIOLOGY. *Educ:* Univ Wash, BS, 42, PhD(fisheries), 58. *Prof Exp:* Fishery biologist, 46-54, asst dir, 55, res assoc prof, 57-64, assoc prof, 64-69, asst dir, 57-67, PROF FISHERIES, FISHERIES RES INST, UNIV WASH, 67- DIR INST, 67- *Mem:* Am Fisheries Soc; Am Inst Fisheries Res Biol; Int Asn Theoret & Appl Limnol. *Res:* Life history and survival of salmonid; lake ecology and limnology. *Mailing Add:* 14370 Edgewater Lane NE Seattle WA 98125

BURGOA, BENALI, b Panama, Aug 24, 56; m. WATER MOVEMENT & CHEMICAL TRANSPORT, SOIL-SOLUTE INTERACTION. *Educ:* Univ Panama, BS, 79; Univ Fla, MS, 84, PhD(soil sci), 89. *Prof Exp:* Res asst, Univ Panama, 79-81; res asst, Univ Fla, 81-89, postdoctoral, 89-90; RES CHEMIST, AGR RES SERV, USDA, 90- *Concurrent Pos:* Soil scientist, Southeast Soil Serv, Inc, 88- *Mem:* Am Soc Agron; Am Geophys Union; Soil & Water Conserv Soc. *Res:* Environmental fate and transport of pollutants from agricultural & industrial activities; hydrologic effects on the transport of chemicals in urban and agricultural watersheds; kinetic evaluation of ion retention in soils; model development and application for environmental pollution remediation. *Mailing Add:* PO Box 748 Tifton GA 31793

BURGOYNE, EDWARD EYNON, b Montpelier, Idaho, Sept 26, 18; m 50; c 4. ORGANIC CHEMISTRY. *Educ:* Utah State Univ, BS, 41; Univ Chicago, cert, 43; Univ Wis, MS, 47, PhD(chem), 49. *Prof Exp:* Alumni Res Found asst org chem, Univ Wis, 46-48; res chemist, Phillips Petrol Co, 49-51; from asst prof to prof, 51-83, EMER PROF CHEM, ARIZ STATE UNIV, 84- *Mem:* Fel AAAS; Am Chem Soc. *Mailing Add:* 223 E 15th St Tempe AZ 85281

BURGOYNE, GEORGE HARVEY, b Flint, Mich, June 5, 39; m 83; c 8. MICROBIOLOGY. *Educ:* Mich State Univ, BS, 63, MS, 66, PhD(virol), 72. *Prof Exp:* Res microbiologist virol, Regional Poultry Res Lab, USDA, 63-72; chief, Tuberculosis Unit, Diag Microbiol Sect, 72-76, CHIEF VIROL VACCINES SECT, MICH DEPT PUB HEALTH, BUR LAB SERV & DIS CONTROL, 76- *Mem:* Am Soc Microbiol. *Res:* Development for human use of an inactivated rabies virus vaccine of tissue culture origin and of a live attenuated vaccine for influenza using temperature sensitive mutants. *Mailing Add:* Mich Dept Pub Health 3500 N Logan Lansing MI 48906

BURGOYNE, PETER NICHOLAS, b Bracknell, Eng, Oct 8, 32; Can citizen; m 62. MATHEMATICAL PHYSICS. *Educ:* McGill Univ, BSc, 55, MSc, 56; Princeton Univ, PhD(math physics), 61. *Prof Exp:* Instr math, Princeton Univ, 59-61; asst prof, Univ Calif, Berkeley, 61-66; assoc prof, Univ Ill, Chicago, 66-68; assoc prof, 68-74, PROF MATH, UNIV CALIF, SANTA CRUZ, 74- *Concurrent Pos:* Sloan Found fel, 63-65. *Res:* Theoretical physics, particularly problems in the theory of elementary particles; pure mathematics, particulary finite group theory and number theory. *Mailing Add:* Dept Math Univ Calif Santa Cruz CA 95064

BURGSTAHLER, ALBERT WILLIAM, b Grand Rapids, Mich, July 10, 28; m 57; c 5. ORGANIC CHEMISTRY. *Educ:* Univ Notre Dame, BS, 49; Harvard Univ, MA, 50, PhD(chem), 53. *Prof Exp:* Instr org chem, Univ Notre Dame, 53-54; proj assoc, Univ Wis, 55-56; instr & res assoc, 56-57, from asst prof to assoc prof chem, 57-65, PROF CHEM, UNIV KANS, 65- *Concurrent Pos:* Sloan fel, 61-64. *Honors & Awards:* Notre Dame Centennial Sci Award, 65. *Mem:* Am Chem Soc; Int Soc Fluoride Res (pres, 70-73). *Res:* Environmental chemistry; reactions and synthesis of natural products; newer synthetic methods and applications; stereochemistry and circular dichroism. *Mailing Add:* Dept Chem Malott Hall Univ Kans Lawrence KS 66045

BURGSTAHLER, SYLVAN, b Corvuso, Minn, Nov 7, 28; m 61; c 3. MATHEMATICS. *Educ:* Univ Minn, BS, 51, MS, 53, PhD(math), 63. *Prof Exp:* Asst prof, 61-70, head dept, 64-72, assoc prof, 74-77, PROF MATH, UNIV MINN, DULUTH, 77- *Mem:* Math Asn Am; Am Math Soc. *Res:* Flow of heat between materials having different physical constants; solution of polynomial equations. *Mailing Add:* 27 W Kent Rd Duluth MN 55812

BURGUS, ROGER CECIL, b Osceola, Iowa, Sept 10, 34; m 55; c 3. BIOCHEMISTRY, NEUROENDOCRINOLOGY. *Educ:* Iowa State Univ, BS, 57, MS, 60, PhD(biochem), 62. *Prof Exp:* Asst chem, Iowa State Univ, 57-60, asst biochem, 60-62; res assoc physiol, Wayne State Univ, 62-65; from asst prof to assoc prof physiol & biochem, Col Med, Baylor Univ, 65-70; sr res assoc, Salk Inst, 70-73, assoc res prof, 73-78; prof biochem, Sch Med, Oral Roberts Univ, 78-; AT DEPT PATH, CITY OF FAITH MED & RES CTR. *Mem:* Am Soc Biol Chem; Int Soc Neuroendocrinol; AAAS; Am Chem Soc; Endocrine Soc; Sigma Xi. *Res:* Isolation and characterization of natural products; metabolism of pyrroles; vitamin B-twelve and related compounds in bacteria; chemistry of hormones originating in the hypothalmus and other areas of the brain. *Mailing Add:* 6246 E 116th St Tulsa OK 74137-9990

BURGUS, WARREN HAROLD, b Burlington, Iowa, Oct 28, 19; m 51; c 2. RADIOCHEMISTRY. *Educ:* Univ Iowa, BS, 41; Wash Univ, PhD(chem), 49. *Prof Exp:* Asst metall lab, Univ Chicago, 42-43; res assoc, Oak Ridge Nat Lab, 43-46; asst, Wash Univ, 46-47; mem staff, Los Alamos Nat Lab, 48-54; head mat testing reactor chem sect, Atomic Energy Div, Phillips Petrol Co, 54-68, sr tech consult, Water Reactor Safety Prog Off, Atomic Energy Div, 68-69; sr scientist, Aerojet Nuclear Co, 71-76; sr scientist, Earth & Life Sci Div, EG&G, Idaho, Inc, 76-82; RETIRED. *Mem:* Fel AAAS; Am Chem Soc. *Res:* Radiochemical procedures; hot atom chemistry; radioactive decay schemes; fission products; chemical processing of reactor fuels; cross sections. *Mailing Add:* 13027 Ballad Dr Sun City West AZ 85375

BURHANS, RALPH W(ELLMAN), b Cleveland, Ohio, May 30, 22; m 54; c 2. ELECTRICAL ENGINEERING. *Educ:* Oberlin Col, AB, 47. *Prof Exp:* Res assoc, Standard Oil Co, Ohio, 47-65; res assoc, Ohio Univ, 65-66; sta mgr, Nat Radio Astron Observ, Kitt Peak, Ariz, 66-67; res engr, Dept Elec Eng, Ohio Univ, 67-85, lectr, 70-85; CONSULT, 85- *Mem:* Inst Elec & Electronics Engrs; Audio Eng Soc; Am Inst Navigation; Int OMEGA Asn; Sigma Xi. *Res:* Instrumentation; avionics; time and phase measurements; electronic music; VLF and Loran-C navigation. *Mailing Add:* 161 Grosvenor St Athens OH 45701

BURHOLT, DENNIS ROBERT, b New York, NY, Sept 22, 42; m 64; c 2. CELL BIOLOGY, CANCER. *Educ:* Adelphi Col, BA, 63; Univ Fla, PhD(bot, zool), 68. *Prof Exp:* Asst prof biol, Southampton Col, 68-69; NIH fel, Brookhaven Nat Lab, 69-71; res asst, Inst Med Radiation Sci, Würzburg, WGer, 71-73; res asst cancer res, Allegheny Gen Hosp, 73-75, res assoc cancer res, 75-82, sr res scientist, Cancer Res Ctr, 82-89, DIR REPROD SCI RES, DEPT OBSTET & GYNEC, ALLEGHENY GEN HOSP, 89- *Mem:* Radiation Res Soc; Am Asn Cancer Res. *Res:* Influence of radiation and chemotherapeutic agents on gastrointestinal epithelial cell proliferation. *Mailing Add:* 502 Kerwood Rd Pittsburgh PA 15215

BURHOP, KENNETH EUGENE, b Port Washington, Wis, May 7, 53; m 77; c 3. CARDIOPULMONARY PHYSIOLOGY, ARTIFICIAL BLOOD RESEARCH. *Educ:* Univ Wis-Milwaukee, BA, 75; Univ Wis-Madison, MS, 79, PhD(vet sci & physiol), 84. *Prof Exp:* Res asst physiol, Dept Vet Sci, Univ Wis-Madison, 76-79, res specialist, Dept Prev Med, 79-80, res asst, Dept Vet Sci, 80-84; NIH pulmonary res postdoctoral fel physiol, Dept Physiol, Albany Med Col, 84-86; postdoctoral physiol & pharm, Dept Appl Sci, 86-87, mgr res sci, 87-90, MGR RES SCI, DEPT BLOOD SUBSTITUTES, BAXTER HEALTHCARE CORP, 90- *Concurrent Pos:* Co-investr, Lung Prog Proj, Dept Physiol, Albany Med Col, 84-86. *Mem:* Am Physiol Soc; NY Acad Sci; Am Fedn Clin Res; Shock Soc; Am Soc Artificial Organs. *Res:* Biocompatibility assessment of materials and compounds, especially as it relates to acute lung injury; shock and the development and evaluation of modified hemoglobin solutions and other blood substitute solutions. *Mailing Add:* Dept Blood Substitutes Baxter Healthcare Corp Baxter Technol Park WG2-1S Round Lake IL 60073-0490

BURIAN, RICHARD M, b Hanover, NH, Sept 14, 41; m 63, 89; c 3. RELATIONS BETWEEN GENETICS & EVOLUTIONARY BIOLOGY, DEVELOPMENTAL BIOLOGY & GENETICS. *Educ:* Reed Col, Portland, Ore, BA, 63; Univ Pittsburgh, PhD(philos), 71. *Prof Exp:* From instr to assoc dean, Col Philos, Brandeis Univ, 67-76; res assoc, Mus Comp Zool, Harvard Univ, 76-77; assoc prof philos, Drexel Univ, 77-83; PROF & DEPT HEAD PHILOS, VA POLYTECH INST, 83- *Concurrent Pos:* Vis asst prof, Fla A&M Univ, 68-69; vis assoc prof, Univ Pittsburg, 78, Univ Calif, Davis, 82; adj prof sci studies, Va Polytech Inst & State Univ, 83-; NSF grants, 84-85, 85-87 & 89; res fel, Nat Humanities Ctr, Research Triangle Park, 91-92. *Mem:* AAAS; Am Philos Asn; Am Soc Zool; Philos Sci Asn; Hist Sci Soc; Int Soc Hist, Philos & Soc Studies Biol. *Res:* History and philosophy of biology, especially developmental and evolutionary biology plus genetics; relations of embryology, evolution, genetics; French treatments of heredity; interdisciplinary relations, epistemological evaluation of experiments, and of evidence for biological theories. *Mailing Add:* Dept Philos Va Polytech Inst & State Univ Blacksburg VA 24061

BURICK, RICHARD JOSEPH, b Pueblo, Colo, Aug 10, 39; m 61; c 3. CHEMICAL ENGINEERING, NUCLEAR ENGINEERING. *Educ:* Colo State Univ, BS, 61; Purdue Univ, MS, 64, PhD(mech eng), 67. *Prof Exp:* Res asst, Purdue Univ, 62-67; res scientist, Martin-Marietta, 67-69; staff tech mgt, Rockwell Int, 69-75; staff & group leader, 75-80, prog mgr, 80-84, DEP ASSOC DIR, LOS ALAMOS NAT LAB, 84- *Mem:* AAAS; Laser Inst Am. *Res:* High energy lasers; laser isotope separation; neutral particle beams; combustion; fluid dynamics. *Mailing Add:* 1600 Camino Redwood Los Alamos NM 87544

BURIOK, GERALD MICHAEL, b Heilwood, Pa, Nov 6, 43; m 66; c 2. FINITE MATHEMATICS, APPLIED PROGRAM LANGUAGE. *Educ:* Ind State Col, BS, 65; Pa State Univ, MA, 67, DEd(math), 71. *Prof Exp:* Asst prof math, State Univ Col, Oneonta, NY, 67-70; prof math, 71-77, dept chmn & prof comput sci, 79-84, actg dean, Col Natural Sci & Math, 84-85, PROF MATH, IND UNIV PA, 85- *Mem:* Math Asn Am; Nat Coun Teachers Math. *Res:* use of computer in teaching math; developing advance program language tools for mathematics educ. *Mailing Add:* 986 Barclay Rd Indiana PA 15701

BURISH, THOMAS GERARD, b Menominee, Mich, May 4, 50; m 76; c 2. BEHAVIORAL MEDICINE, CLINICAL PSYCHOLOGY. *Educ:* Univ Notre Dame, AB, 72; Univ Kans, MA, 75, PhD, 76. *Prof Exp:* Asst instruct, Univ Kans, 75-76; from asst prof to assoc prof, 76-86, chmn, 84-86, PROF, 86-, ASSOC PROVOST, VANDERBILT UNIV, 86- *Concurrent Pos:* consult, 83- *Honors & Awards:* David Shulman Award for Clinical Psychol. *Mem:* Am Psychol Asn. *Res:* Stress and coping; etiology and chemotherapy. *Mailing Add:* 221 Kirkland Hall Vanderbilt Univ Nashville TN 37240

BURK, CARL JOHN, b Troy, Ohio, Dec 30, 35; m 66; c 2. PLANT TAXONOMY, PLANT ECOLOGY. *Educ:* Miami Univ, AB, 57; Univ NC, MA, 59, PhD(bot), 61. *Prof Exp:* Asst bot, Univ NC, 58-59, part-time instr, 60-61; from instr to assoc prof bot, 61-73, actg chmn dept biol sci, 70-71, chmn, 72-75 & 83-86, PROF BIOL SCI, SMITH COL, 73-, GATES PROF, 82- *Honors & Awards:* Katherine Asher Engel lectr, 87. *Mem:* Int Asn Plant Taxon; New Eng Bot Club; Am Soc Plant Taxonomists; Sigma Xi; Ecol Soc Am. *Res:* Biosystematics of Quercus, Hepatica, Heterotheca; ecology, biogeography and floristics of coastal and freshwater ecosystems. *Mailing Add:* Clark Sci Ctr Smith Col Northampton MA 01063

BURK, CORNELIUS FRANKLIN, JR, b Sarnia, Ont, Jan 8, 33; m 59; c 2. INFORMATION MANAGEMENT, AUDITING. *Educ:* Univ Western Ont, BSc, 56; Northwestern Univ, PhD(geol), 59. *Prof Exp:* Explor geologist, Texaco, Inc, 59-60; geologist, Geol Surv Can, Dept Energy, Mines & Resources, 60-70, res scientist, 67-68, nat coordr, Secretariat Geosci Data, 68-70, dir, Can Ctr Geosci Data, 70-81, adv, Info Resources, 81-82, dir, Off Auditor Gen, 82-88; ASSOC DIR, DMR GROUP INC, 88- *Concurrent Pos:* Mem, Ad Hoc Comt Storage & Retrieval Geol Data, Nat Adv Comt Res Geol Sci, Can, 65-68, mem, Subcomt Comput Appln, 68-73, mem, Comt Storage, Automatic Processing & Retrieval Geol Data, Comt Geol Data for Sci & Technol, secy, 73-76, ed, 75- 80; ed, Bull Can Petrol Geol & Cogeodata News; chmn, Can Nat Comt for Comt Data for Sci & Technol, 77-82, chmn, Cogeodata Task Group, Geol Data Sources, 78- *Mem:* Asn Fed Info

Resources Mgt; Info Res Mgt Asn Can. *Res:* Regional stratigraphy, especially Silurian of eastern Canada and Upper Cretaceous of western Canada; computer-based storage and retrieval of geoscience data and national information systems for science; information management; development and management of science information services; information management. *Mailing Add:* DMR Group Inc 360 Albert St Ottawa ON K1R 7X7 Can

BURK, CREIGHTON, b Laramie, Wyo, Feb 1, 29; m 49; c 3. GEOLOGY. *Educ:* Univ Wyo, BSc, 52, MA, 53; Princeton Univ, PhD(geol), 64. *Prof Exp:* Field geologist, Stanolind Oil & Gas Co, 52-53; explor geologist, Richfield Oil Corp, 53-60; chief scientist, Am Miscellaneous Soc, Mohole Proj, Nat Acad Sci, 62-64; corp explor adv, Socony Mobil Oil Co, Inc, 64-69, chief geologist, & mgr regional geol, Mobil Oil Corp, 69-75; PROF GEOL SCI, CHMN DEPT MARINE STUDIES & DIR, MARINE SCI INST, UNIV TEX, AUSTIN, 75- *Concurrent Pos:* Instr, Princeton Univ, 60-61, vis prof geol & geophys sci, 67-75; NSF fel, 61-62; mem, US Geodynamics Comt; Pres Comn Marine Sci, Eng & Resources, Nat Oceanic Adv Comt, 68-, Circum-Pac Int Map Prog, 73- & Int Geol Correlation Proj, 73-; US del, World Petrol Cong, 75; mem & del, US-USSR Protocol on Oceanog, 74- *Honors & Awards:* Frank A Morgan Award, 56. *Mem:* Geol Soc Am; Am Geophys Union; Am Asn Petrol Geol; Marine Technol Soc (vpres, 74-); Geol Soc London. *Res:* Regional, structural and historical geology throughout the world; marine geology and geophysics; geology of continental margins. *Mailing Add:* Northwest Meditlex 5301 Duval Rd Austin TX 78727

BURK, DAVID LAWRENCE, b Minn, July 20, 29; m 53; c 3. QUALITY ASSURANCE, INDUSTRIAL & MANUFACTURING ENGINEERING. *Educ:* Carnegie Inst Technol, BS, 51, MS, 55, PhD(physics), 57. *Prof Exp:* Asst, Carnegie Inst Technol, 51-54; res specialist physics, Allegheny Ludlum Steel Corp, 57-67, mgr, Appl Physics Dept, 67-70, mgr qual assurance, 70-78, dir mfg methods, 78-84, dir mfg eng, 84-89, DIR ADV MFG TECH, ALLEGHENY LUDLUM STEEL CORP, 89- *Mem:* Am Phys Soc; Am Soc Metals; Am Iron & Steel Engrs. *Res:* Physics of metals. *Mailing Add:* 105 Yorkshire Dr Pittsburgh PA 15238

BURK, LAWRENCE G, b New Orleans, La, June 1, 20; m 43. GENETICS. *Educ:* Univ Ga, BSA, 48, MS, 49. *Prof Exp:* Geneticist, Rubber Res Sta, USDA, 51-53, geneticist tobacco invest, Field Crops Res Br, Md, 53-67, res geneticist, Tobacco Res Lab, Southern Region, Agr Res Serv, 67-81; from assoc prof to prof genetics, NC State Univ, 67-; RETIRED. *Concurrent Pos:* Collaborator, USDA, 81- *Mem:* Am Genetic Asn (treas, 56-62); Am Soc Agron; Bot Soc Am. *Res:* Crop improvement through interspecific hybridization; development of bridge-cross methods for overcoming sterility barriers; elucidation of tri-partite nature of tunica in Nicotiana tabacum; development of methods for the induction of haploid plantlets from aseptically cultured anthers of tobacco and converting haploids to diploids. *Mailing Add:* PO Box 215 Pine Mt GA 31822

BURK, RAYMOND FRANKLIN, JR, b Kosciusko, Miss, Dec 9, 42; m 67; c 2. LIVER DISEASES. *Educ:* Univ Miss, BA, 63; Vanderbilt Univ, MD, 68. *Prof Exp:* Res internist nutrit, US Army Med Res & Nutrit Lab, 70-73; teaching fel liver dis, Southwestern Med Sch, Dallas, Tex, 73-74, asst prof med, 75-78, asst prof biochem, 77-78; assoc prof med & biochem, Sch Med, La State Univ, Shreveport, 78-80; assoc prof med, Health Sci Ctr, Univ Tex, San Antonio, 80-82, prof med, 82-87; PROF MED & DIR GASTROENTEROL, VANDERBILT UNIV SCH MED, NASHVILLE, 87- *Mem:* Am Inst Nutrit; Am Soc Biol Chemists; Am Soc Clin Invest. *Res:* Selenium metabolism; role of lipid peroxidation in cell injury. *Mailing Add:* 217 Harpeth Wood Dr Nashville TN 37221

BURKA, MARIA KARPATI, b Budapest, Hungary, June 24, 48; m 68; c 3. ENGINEERING. *Educ:* Mass Inst Technol, BS, 69 & MS, 70; Princeton Univ, MA, 73 & PhD(chem eng), 78. *Prof Exp:* Process design engr, Sci Design Co, 70-71; asst prof chem eng, Univ Md, 78-81; environ scientist, Environ Protection Agency, 81-82; PROG DIR, NAT SCI FOUND, 84- *Mem:* Sigma Xi; Am Inst Chem Eng (secy & treas, 88-); Am Asn Univ Women; AAAS. *Res:* Process design; process control; reaction engineering; polymerization. *Mailing Add:* 5056 Macomb St NW Washington DC 20016

BURKART, BURKE, b Feb 23, 33; US citizen; m 66; c 2. GEOLOGY, TECTONICS GEOCHEMISTRY. *Educ:* Univ Tex, BS, 54, MA, 60; Rice Univ, PhD(geol), 65. *Prof Exp:* Fel geol, Univ Tex, 65; asst prof, Temple Univ, 65-70; from asst prof to assoc prof, 70-82, PROF GEOL, UNIV TEX, ARLINGTON, 82- *Concurrent Pos:* Consult geologist, 81- *Mem:* Sigma Xi; fel Geol Soc Am; Am Geophys Union; Am Asn Petrol Geol. *Res:* Regional tectonics of Central America primarily the major transform faults; neogene tectonics of Guatemala and southern Mexico; geomorphological aspects of strike-slip faults. *Mailing Add:* Dept Geol Univ Tex Arlington TX 76019

BURKART, LEONARD F, b Mosier, Ore, Oct 8, 20; m 49; c 3. WOOD CHEMISTRY, WOOD TECHNOLOGY. *Educ:* Univ Wash, BSF, 49, MF, 50; Univ Minn, MS, 61, PhD(wood chem), 63. *Prof Exp:* Technologist, Wash Veneer Corp, 50-52; staff missionary, Student Missionary Coun, Ore, 52-54; asst prof wood technol, Stephen F Austin State Col, 63-66, from assoc prof to prof forestry, 66-87; RETIRED. *Mem:* Tech Asn Pulp & Paper Indust; Soc Am Foresters; Forest Prod Res Soc; Sigma Xi. *Res:* Utilization of wood residues as sources of organic chemicals, especially work with lignin from waste waters of the pulp and paper industry and lignin and carbohydrates from sawdust. *Mailing Add:* 106 Summerfield Rd Longview TX 75601

BURKART, MILTON W(ALTER), b Chicago, Ill, Sept 1, 22; m 48. PHYSICAL METALLURGY. *Educ:* Ill Inst Technol, BS, 44, MS, 48; Columbia Univ, PhD(metall), 53. *Prof Exp:* Lectr, City Col New York, 49; res assoc, Columbia Univ, 50-53; sr scientist, Westinghouse Elec Corp, 53-55, fel, 55-58, supvr irradiations, 58-63, mgr naval reactor mat irradiation 63-67, mgr mat irradiation, 67-72, mgr fuel element develop labs, Bettis Atomic Power Lab, 72-82, ADV SCIENTIST, WESTINGHOUSE ELEC CORP, 82- *Concurrent Pos:* Lectr, Carnegie Inst Technol, 56. *Mem:* Am Soc Metals; Am Inst Mining, Metall & Petrol Engrs; NY Acad Sci; Sigma Xi. *Res:* Diffusionless transformations in metals; corrosion; radiation damage. *Mailing Add:* 7444 Ben Hur St Pittsburgh PA 15208

BURKE, BERNARD FLOOD, b Boston, Mass, June 7, 28; m 53; c 4. PHYSICS. *Educ:* Mass Inst Technol, SB, 50, PhD, 53. *Prof Exp:* Asst physics, 50-53, chmn, prof, 65-81, chmn, Astrophysic Dir, 70-83, WILLIAM BURDEN PROF PHYSICS, MASS INST TECHNOL, 81- *Concurrent Pos:* Jr res assoc, Brookhaven Nat Lab, 52-53; mem, Nat Radio Astron Observ Adv Comt, 58-62; astron adv comt, NSF, 58-63, 73-76, consult, Found, 63-, NASA, 68-; assoc ed, Astron, J, 62-64; vis prof, Univ Leiden, 71-72; trustee, Assoc Univs Inc, 72-90; trustee & chmn, Northeast Radio Observ Corp, 73-, chmn, 82-; Sherman Fairchild Scholar, Calif Inst Technol, 84-85; Smithsonian Regents Fel, 85. *Honors & Awards:* Warner Prize, Am Astron Soc, 63; Rumford Prize, Am Acad Arts & Sci, 71. *Mem:* Nat Acad Sci; fel AAAS; Am Acad Arts & Sci; Am Phys Soc; Am Astron Soc (pres, 86); Royal Astron Soc; Int Scientific Radio Union; Nat Sci Bd. *Res:* Radio astronomy; galactic structure; microwave interferometry; microwave spectroscopy; antenna arrays; low-noise electronic circuitry. *Mailing Add:* Room 26-335 Mass Inst Technol Cambridge MA 02139

BURKE, CARROLL N, b New Haven, Conn, Aug, 16, 29; m 52; c 3. VIROLOGY. *Educ:* Univ Conn, BS, 55, MS, 59, PhD(microbiol), 65. *Prof Exp:* From asst prof to assoc prof, 66-76,dir, Inst Water Rescources, 80-87, PROF PATHOBIOL, UNIV CONN, 76- *Honors & Awards:* Outstanding Res Award, Am Asn Avian Pathologists, 67. *Mem:* Electron Micros Soc Am; NY Acad Sci. *Res:* Virus diseases of importance to humans, using animal models. *Mailing Add:* Dept Pathobiol U-89 Univ Conn 61 N Eagleville Storrs CT 06268

BURKE, DAVID H, b Philadelphia, Pa, Feb 16, 39; m 64; c 2. PHARMACOLOGY, TOXICOLOGY. *Educ:* Temple Univ, BS, 68, MS, 70, PhD(pharmacol), 72. *Prof Exp:* Asst prof pharmacol, Sch Pharm, Univ Mont, 72-74 & Kansas City Osteop Col Med, 74-75; asst prof, 75-80, ASSOC PROF & HEAD, DEPT PHARMACOL, SCH DENT, MARQUETTE UNIV, 80- *Mem:* Am Pharmaceut Asn. *Res:* Thermoregulatory and fetal pharmacology. *Mailing Add:* 604 N 16th St Milwaukee WI 53233

BURKE, DENNIS GARTH, b Bracken, Sask, June 4, 35; m 60; c 3. PHYSICS. *Educ:* Univ Sask, BE, 57, MS, 58; McMaster Univ, PhD(physics), 63. *Prof Exp:* Sci officer upper atmosphere physics, Defence Res Bd, 58-60; fel, Niels Bohr Inst, Copenhagen, 63-65; from asst prof to assoc prof physics, 65-74, PROF PHYSICS, McMASTER UNIV, 74- *Concurrent Pos:* Alfred P Sloan Found fel, 66-70. *Mem:* Can Asn Physicists. *Res:* Instrumentation of rocket nose cones for upper atmosphere studies; nuclear spectroscopy using decay scheme studies; nuclear structure studies using nuclear reactions. *Mailing Add:* Dept Physics McMaster Univ Hamilton ON L8S 4L8 Can

BURKE, DENNIS KEITH, b Omaha, Nebr, June 29, 43; m 63; c 3. SET-THEORETIC TOPOLOGY. *Educ:* Univ Wyo, BA, 64, MS, 66; Wash State Univ, PhD(math), 69. *Prof Exp:* From asst prof to assoc prof, 69-79, PROF MATH, MIAMI UNIV, 79- *Concurrent Pos:* Ed, Proceedings of the Am Math Soc, 84- *Mem:* Math Asn Am; Am Math Soc; Sigma Xi. *Res:* Study of covering properties; metrization theorems, continuous mappings, counter examples and base axioms for topological spaces. *Mailing Add:* 7783 Fairfield Rd Oxford OH 45056

BURKE, DEREK CLISSOLD, molecular virology, for more information see previous edition

BURKE, EDMUND C, b Fargo, NDak, Nov 23, 19; m 45; c 9. PEDIATRICS. *Educ:* St Thomas Col, BS, 41; Univ Minn, MB, 44, MD, 45, MS, 51. *Prof Exp:* Consult, Mayo Clin, 52, from instr to asst prof, Mayo Found, 53-64, assoc prof clin pediat, Mayo Grad Sch Med, Univ Minn, 64-72, prof pediat, 72-86, EMER PROF PEDIAT, MAYO MED SCH, 86- *Concurrent Pos:* Fel pediat, Mayo Found, Univ Minn, 48-51; prof pediat & adolescent med, Sch Med, Univ SDak, 86-88. *Mem:* Sigma Xi. *Res:* Renal disease; salt and electrolyte disturbance; migraine in children. *Mailing Add:* Mayo Clin Rochester MN 55905

BURKE, EDMUND C(HARLES), b Bridgeport, Conn, June 21, 21; m 47. PHYSICAL METALLURGY. *Educ:* Mo Sch Mines, BS, 43; Case Western Reserve Univ, MS, 48; Yale Univ, PhD(phys metall), 51. *Prof Exp:* Res metallographer, Aluminum Co Am, Ohio, 43-48; res metallurgist, Dow Chem Co, 51-56; mgr res lab, Hunter Eng Co, 56-58; res scientist, Lockheed Aircraft Corp, 58-67, dir mat sci lab, Lockheed Palo Alto Res Lab, 67-77, dir mat & struct, Res & Develop Dept, Lockheed Missiles & Space Co, 77-84; RETIRED. *Concurrent Pos:* Mem light aluminium alloys, Mat Adv Bd, Nat Acad Sci, 59-68 & high temp mat comt, 65-66. *Mem:* Am Inst Mining, Metall & Petrol Engrs; fel Am Soc Metalls. *Res:* Plastic deformation of metals; physical metallurgy of aluminum, magnesium, and beryllium alloys; high temperature materials; composite materials; materials science. *Mailing Add:* 241 Ferne Ave Palo Alto CA 94306-4602

BURKE, EDMUND R, b New York, NY, Aug, 23, 49. ATHLETIC PERFORMANCE, EXERCISE PHYSIOLOGY. *Educ:* Ball State Univ, MA, 76; Ohio State Univ, PhD(exercise physiol), 79. *Prof Exp:* Fel, Univ Iowa Hosp, 79-81, res scientist, 81-82; DIR SPORTS SCI & TECHNOL, US CYCLING FEDN, 82- *Mem:* Fel Am Col Sports Med; Am Alliance Health Phys Educ. *Mailing Add:* 3508 Queen Anne Way Colorado Springs CO 80917

BURKE, EDWARD ALOYSIUS, b White Plains, NY, Oct 21, 29; m 55; c 7. THEORETICAL PHYSICS. *Educ:* NY Univ, BA, 54; Fordham Univ, MS, 55, PhD(physics), 59. *Prof Exp:* Electronics engr, Raytheon Mfg Co, 54; instr physics, US Maritime Acad, 55; asst, Fordham Univ, 55-58; asst prof phys sci, Montclair State Col, 58-59; from asst prof to assoc prof physics, St John's Univ, NY, 59-66, chmn dept, 62-65; assoc prof, 66-69, dir space related sci, 66-70, PROF PHYSICS, ADELPHI UNIV, 69- *Concurrent Pos:* Vis prof, Univ Del, 80-81. *Mem:* Am Asn Physics Teachers; Am Phys Soc; Am Asn Univ Professors; Sigma Xi. *Res:* Theoretical atomic physics. *Mailing Add:* 11 Indian Hill Rd Wobern MA 01801

BURKE, EDWARD WALTER, JR, b Macon, Ga, Sept 16, 24; m 46; c 2. PHYSICS. *Educ:* Presby Col, BS, 47; Univ Wis, MS, 49, PhD(physics), 54. *Prof Exp:* Vpres acad affairs, 77-79, MARY REYNOLDS BABCOCK PROF PHYSICS, KING COL, 49-, CHMN DIV NATURAL SCI & MATH, 61- *Concurrent Pos:* Fulbright lectr, Univ Chile, 59. *Honors & Awards:* Pegram Award, Am Phys Soc. *Mem:* Sigma Xi; Am Astron Soc; Am Asn Physics Teachers; Astron Soc Pac. *Res:* Spectroscopy; isotope shift in the atomic spectra of boron; astrophysics; photoelectric study of variable stars. *Mailing Add:* King Col Bristol TN 37620

BURKE, HANNA SUSS, b Karlsruhe, Ger, Oct 5, 26; nat US; m 65. ORGANIC CHEMISTRY. *Educ:* Goucher Col, BA, 48; Univ Pa, MS, 52, PhD(org chem), 56. *Prof Exp:* Chemist, Johns Hopkins Univ, 48-51 & Rohm & Haas Co, 55-63; tech info specialist, Fibers & Plastics Co, Allied Chem Corp, 63-90; RETIRED. *Mem:* Am Chem Soc; Am Inst Chemists. *Res:* Separation of amino acids; isolation and synthesis of natural products; water soluble polymers; literature and patent searching; surveillance and indexing of literature and patents in fields of fiber technology and related subjects; registered patent agent. *Mailing Add:* 4802 Cutshaw Ave Richmond VA 23230

BURKE, J ANTHONY, b New York, NY, Jan 29, 37. ASTRONOMY, ASTROPHYSICS. *Educ:* Harvard Univ, AB, 58, AM, 59, PhD(astron), 65. *Prof Exp:* Res assoc astrophys, Brandeis Univ, 64-67; lectr & res assoc, Dept Astron, Boston Univ, 67-68; ASSOC PROF ASTRON, UNIV VICTORIA, 68- *Mem:* AAAS; Am Astron Soc; Can Astron Soc; Royal Astron Soc; Int Astron Union. *Res:* Cosmogony; interstellar medium. *Mailing Add:* Dept Physics & Astron Univ Victoria Victoria BC V8W 3P6 Can

BURKE, J(OSEPH) E(LDRID), b Berkeley, Calif, Sept 1, 14; m 39; c 2. MICROSTRUCTURE & PROPERTIES OF MATERIALS. *Educ:* McMaster Univ, BA, 35; Cornell Univ, PhD(chem), 40. *Prof Exp:* Chemist, Int Nickel Co, NJ, 40-41; chemist, Norton Co, Mass, 41-43, metallurgist & group leader, Manhattan dist, Los Alamos, 43-46; assoc prof metall inst study metals, Univ Chicago, 46-49; res assoc, Knolls Atomic Power Lab, 49-52, mgr metall, 52-54, mgr, Ceramic Br, Gen Elec Co Res & Develop Ctr, 54-72, mgr spec projs, 72-79; CONSULT MAT SCI & ENG, 79 - *Concurrent Pos:* Adj prof mat sci, Renssalaer Polytech Inst, 79-84. *Honors & Awards:* Jeppson Medal, Am Ceramic Soc, 81; Frenkel Prize, Int Inst Sintering, 81. *Mem:* Nat Acad Eng; fel Am Soc Metals; fel Am Ceramic Soc (pres, 74-75); fel Am Nuclear Soc. *Res:* Grain growth and recrystallization; metallography; metallurgy of uranium; materials for nuclear reactors; sintering; structure and properties of ceramics. *Mailing Add:* 33 Forest Rd Burnt Hills NY 12027

BURKE, JAMES DAVID, b Darby, Pa, June 30, 37; m 68; c 4. ORGANIC CHEMISTRY. *Educ:* Spring Hill Col, BS, 61; Univ Calif, Berkeley, PhD(org chem), 65; Univ Pa, dipl, 77. *Prof Exp:* NIH fel, Columbia Univ, 65-66; sr res scientist, 66-72, placement supvr, 72-79, sr scientist, 79-81, mgr res security, 81-83, RECRUITMENT SUPVR, ROHM & HAAS CO, 84- *Mem:* Am Chem Soc. *Res:* Surface active agents for aqueous and non-aqueous systems; acrylate polymers; technology evaluation; dispersants; technology security. *Mailing Add:* ROHM/HAAS Co Independence Mall W Philadelphia PA 19105

BURKE, JAMES EDWARD, b Los Angeles, Calif, Oct 8, 31; m 54; c 3. APPLIED MATHEMATICS, MATHEMATICAL PHYSICS. *Educ:* San Jose State Col, BA, 53; Stanford Univ, MS, 55, PhD(math), 58. *Prof Exp:* Mathematician, Edwards AFB, 53-54; aeronaut res scientist syst anal, Ames Lab, NASA, 54-55; asst, Stanford Univ, 54-57; scattering specialist, Sylvania Electronics Defense Lab, Gen Tel & Electronics Co, 58-69; LAB MGR, ELECTROMAGNETIC SYST LABS, INC, 69- *Mem:* Sigma Xi; Am Math Soc; Soc Indust & Appl Math; Optical Soc Am; fel Acoust Soc Am. *Res:* Scattering of electromagnetic and acoustic waves by single and many objects. *Mailing Add:* 19960 Angus Ct Saratoga CA 95070

BURKE, JAMES JOSEPH, b Northampton, Mass, June 11, 37; m 65. POLYMER PHYSICS. *Educ:* Univ Mass, BS, 58; Mass Inst Technol, PhD(phys chem), 62. *Prof Exp:* Res assoc phys chem, Mass Inst Technol, 62-63; res chemist, Chemstrand Res Ctr, Inc, 63-66, sr res chemist, 66-67, res specialist, 67-68, res group leader, 68-72, sr res group leader, 72-74; sr res group leader, Monsanto Triangle Park 74-77, Monsanto fel 77-89, MONSANTO SR FEL, MONSANTO CO, 89- *Mem:* AAAS; Am Chem Soc; Am Soc Testing & Mat; NY Acad Sci; Fiber Soc; NAm Membrane Soc. *Res:* Polymers in biological applications; hollow fiber membranes; bioseparations; controlled delivery gas and liquid separations; material characterization and product development. *Mailing Add:* Monsanto Co 800 N Lindbergh Blvd St Louis MO 63167-0001

BURKE, JAMES JOSEPH, JR, b Chicago, Ill, July 26, 31; m 63; c 1. OPTICS. *Educ:* Univ Chicago, MS, 59; Univ Ariz, PhD(optics), 72. *Prof Exp:* Asst physicist optics, Ill Inst Technol Res Inst, 59-61; from res physicist to sr physicist, Optics Technol, Inc, 61-67; res assoc, 67-74, staff scientist & lectr, 74-82, PROF, OPTICAL SCI CTR, UNIV ARIZ, 82-, DIR, OPTICAL DATA STORAGE CTR, 87- *Mem:* Optical Soc Am; Soc Photo Optical Instrumentation Engrs. *Res:* Optical waveguides and integrated optics, image analysis and processing; optical recording. *Mailing Add:* Optical Sci Ctr Univ Ariz Tucson AZ 85721

BURKE, JAMES L(EE), b Jamestown, NY, Sept 25, 18; m 42; c 4. SYSTEMS INTERACTION. *Educ:* Univ Ariz, BS, 72, MA, 78. *Prof Exp:* Regular serv positions, US Army, 41-50, comdr, Res & Develop Procedurement Off, Ft Monmouth, NJ, 50-54, signal off, Allied Land Forces, Denmark, 55-58, chief, Mutual Security Div, Off of Chief Signal Officer, 58-61, chief, Instrumentation & Range Develop & Combat Surveillance Dept, Electronic Proving Ground, 62-65, chief, Logistics-Commun Div, Army Concept Team, Vietnam, 65, asst to commanding gen for aviation & aviation electronics, Army Electronics Command, 66-69, dir, Prof Study Groups, 64-69, asst commandant, Electronic Warfare Sch, 69-70; VPRES, SYSTS ENG INC, ARIZ, 70-; EXEC DIR, COCHISE HEALTH SYSTS INC, 77- *Concurrent Pos:* Dir bus finance, Univ Hosp, Univ Ariz, 74; mil com pilot; consult & dir systs planning, Mentic Corp, Calif, 70- *Mem:* Sr mem Inst Elec & Electronics Engrs; assoc fel Am Inst Aeronaut & Astronaut; Army Aviation Asn Am. *Res:* Research and development, test and evaluation in electronic, surveillance, avionics and optics; development and evaluation in electromagnetic compatibility and radio intelligibility; human engineering and resources research and application; 05380790xtation development; specialization in systems integration. *Mailing Add:* Box BB Bisbee AZ 85603

BURKE, JAMES OTEY, b Richmond, Va, May 20, 12; m 42; c 4. MEDICINE. *Educ:* Va Mil Inst, BS, 33; Med Col Va, MD, 37. *Prof Exp:* Assoc prof med, 54-59, assoc prof clin med, 59-78, ASST PROF MED, MED COL VA, 78- *Concurrent Pos:* Consult, Surgeon Gen, US Dept Army, 50-54 & Vet Admin Hosp, 59-60; med dir, A H Robins Co, Inc, 58-60, dir prev med, 60-76. *Mem:* Fel Am Col Physicians; AMA; Am Fedn Clin Res; sr mem Am Gastroenterol Asn; fel Am Occup Med Asn. *Mailing Add:* St Mary's Hosp Richmond VA 23226

BURKE, JAMES PATRICK, BIOCHEMISTRY, ZINC DEFICIENCY. *Educ:* Mt Sinai Sch Med, PhD(biochem), 74. *Prof Exp:* ASSOC PROF PHYSIOL SCI, PA COL PODIATRIC MED, 75- *Mailing Add:* Dept Physiol Sci Pa Col Podiatric Med Eighth St at Race Philadelphia PA 19107

BURKE, JANICE MARIE, RETINAL CELL BIOLOGY, OCULAR PATHOLOGY. *Educ:* Univ Mass, PhD(cell & develop biol), 72. *Prof Exp:* ASSOC PROF OPHTHAL, ANAT & CELL BIOL, MED COL WIS, 82- *Res:* Cell proliferation. *Mailing Add:* Dept Ophthal Med Col Wis 8700 W Wisconsin Ave Milwaukee WI 53226

BURKE, JOHN A, b Eastland, Tex, Dec 14, 36; m 62; c 3. INORGANIC CHEMISTRY. *Educ:* Tex Tech Col, BS, 59; Ohio State Univ, MSc, 61, PhD(inorg chem), 63. *Prof Exp:* From asst prof to assoc prof chem, Trinity Univ, 63-74, chmn dept, 69-76, dean, Div Sci Math & Eng, 76-86, chmn dept, 88-89, PROF CHEM, TRINITY UNIV, 74- *Mem:* Sigma Xi; fel Am Inst Chemists; Am Chem Soc. *Res:* Inorganic chemistry, especially coordination compounds. *Mailing Add:* Dept Chem Trinity Univ 715 Stadium Dr San Antonio TX 78284

BURKE, JOHN FRANCIS, b Chicago, Ill, July 22, 22; m 50; c 4. SURGERY. *Educ:* Univ Ill, BS, 47; Harvard Med Sch, MD, 51. *Prof Exp:* Intern surg, Mass Gen Hosp, Boston, 51-52, resident, 52-56, resident surgeon, 57; instr surg, Havard Med Sch, 58-60, tudor med sci, 60-65, clin assoc surg, 60-66, asst clin prof, 66-69, assoc prof surg, 69-75; chief staff, 69-81, chief surg, Shriners Burns Inst, Boston Unit, 68-69; PROF SURG, HARVARD MED SCH, 75-, HELEN ANDRUS BENEDICT PROF, 76- *Concurrent Pos:* Asst surg, Mass Gen Hosp, 58-60, asst surgeon, 61-63, assoc vis surgeon, 64-68, vis surgeon, 68-, chief, Trauma Sevices, 81. *Mem:* Am Thoracic Soc; Soc Univ Surg; Infectious Dis Soc Am; Am Surg Asn; Am Asn Surg Trauma. *Res:* Host defense against infection and the treatment of traumatic injury and thermal injury. *Mailing Add:* Dept Surg Harvard Med Sch Mass Gen Hosp Boston MA 02114

BURKE, JOHN MICHAEL, b Takoma Park, Md, Apr 27, 46; m 75; c 2. PHYSICAL CHEMISTRY, LASER APPLICATIONS. *Educ:* Thomas More Col, AB, 66; Case Western Reserve Univ, PhD(chem), 71. *Prof Exp:* Res assoc chem, Case Western Reserve Univ, 71-72; presidential intern, Nat Bur Standards, 72-73; res assoc, Princeton Univ, 73-77; RES & DEVELOP STAFF MEM CHEM, PROCTER & GAMBLE CO, 77- *Concurrent Pos:* NIH fel, Princeton Univ, 75-76; adv bd, Ohio LAser Tech Consortium, 84-85; adv comt, Cincinnati Tech Col, 84- *Mem:* Optical Soc Am. *Res:* Spectroscopy; laser chemistry; raman spectroscopy; catalysis; bioinorganic chemistry; biophysical chemistry; laser applications in manufacturing. *Mailing Add:* 15 Baldwin St North Easton MA 02356-1501

BURKE, JOHN T, b Missouri Valley, Iowa, Mar 26, 29; m 57; c 2. INTERNAL MEDICINE. *Educ:* Univ Iowa, MD, 55. *Prof Exp:* Asst resident internal med, Vet Admin Hosp, San Francisco, 59-61; resident, Presby Med Ctr, 61-62; clin assoc med res, Merck Sharp & Dohme, 62-65, dir clin res, 65-73; dir clin res, Merrell Int Res Ctr, Strasbourg, 73-77, MED DIR, MERRELL FRANCE, 77-; ASSOC DIR, DOW CHEM, EUROPE HUMAN HEALTH, PARIS, 81- *Mem:* AAAS; NY Acad Sci; Europ Soc Study Drug Toxicity; Royal Soc Health; AMA. *Res:* Clinical pharmacology. *Mailing Add:* Merrell Dow Res Inst 2110 E Galbraith Rd Cincinnati OH 45215

BURKE, KENNETH B S, b Willenhall, Eng, Dec 28, 35; m 61; c 2. GEOPHYSICS. *Educ:* Univ Leeds, BSc, 58, dipl, 59, PhD(appl geophys), 61. *Prof Exp:* Mining geophysicist, Henry Krumb Sch Mines, Columbia Univ, 61-62; asst prof appl geophys, Univ Sask, 62-68; vis prof, Univ Man, 68-69; assoc prof geol, 69-76, PROF GEOL, UNIV NB, 76- *Mem:* Fel Geol Asn Can; Soc Explor Geophys; Asn Prof Eng; Seismol Soc Am. *Res:* Earthquake seismology; application of geophysics to regional geological mapping; historical seismicity studies; intraplate seismicity; environmental geophysics. *Mailing Add:* Dept Geol Univ NB Fredericton NB E3B 5A3 Can

BURKE, KEVIN CHARLES, b London, Eng, Nov 13, 29; m 61; c 3. GEOLOGY. *Educ:* Univ London, BSc, 51, PhD(geol), 53. *Prof Exp:* Lectr geol, Univ Ghana, 53-56; geologist nuclear raw mat, Geol Surv, Gt Brit, 56-60; adv, Govt Korea, Int Atomic Energy Agency, 60-61; sr lectr geol, Univ West Indies, 61-65; prof, Univ Ibadan, 65-71; vis prof, Univ Toronto, 71-72; prof geol, State Univ NY Albany, 72-87; Lunar & Planetary Inst, Houston, TX, 83-88; DEPT GEOSCI, UNIV HOUSTON. *Concurrent Pos:* Vis prof, Calif Inst Technol, 76, Univ Minn, 78 & Univ Calgary, 79; consult, Oil Indust & NASA; Nat Res Coun, Wash. *Mem:* Geol Soc Am; Am Geophys Union; Nigerian Mining, Geol & Metall Soc. *Res:* Application of the findings of plate tectonics to interpretation of the geological history of the earth. *Mailing Add:* 912 Green St Alexandria VA 22314

BURKE, LEONARDA, mathematics, data processing, for more information see previous edition

BURKE, MICHAEL FRANCIS, b Gallup, NMex, Jan 29, 39; m 60; c 4. ANALYTICAL CHEMISTRY. *Educ:* Regis Col, BS, 60; Va Polytech Inst, PhD, 65. *Prof Exp:* Teacher, New Orleans, La, 60-61; fel, Purdue Univ, 65-66, asst prof, 66-67; asst prof, 67-74, ASSOC PROF CHEM, UNIV ARIZ, 74- *Mem:* Am Chem Soc. *Res:* Digital data handling and control of analytical instrumentation by means of small high speed computers; chromatographic separations; pyrolysis-gas chromatography; thermodynamics of gas-solid chromatography; high pressure chromatography. *Mailing Add:* Dept Chem Univ Ariz Tucson AZ 85721

BURKE, MICHAEL JOHN, b Westchester, Ill, July 7, 42; m 64; c 2. BIOPHYSICS. *Educ:* Blackburn Col, BA, 64; Iowa State Univ, PhD(biophys), 69. *Prof Exp:* Res asst biophys, Iowa State Univ, 64-69; fel chem, Univ Minn, 69-71, fel bot, 71-72, from res assoc to asst prof hort, Univ Minn, 72-76; assoc prof hort, Colo State Univ, 76-79; prof & chmn, dept fruit crops, Univ Fla, 79-84; ASSOC DEAN, COL AGR SCI, ORE STATE UNIV, 84- *Honors & Awards:* Darrow Award, Am Soc Hort Sci. *Mem:* Am Soc Hort Sci; AAAS; Am Chem Soc; Am Soc Plant Physiologists; Cryobiol Soc. *Res:* Plant stress physiology and physiological ecology; emphasis on biophysical properties of plants which allow them to survive low temperature and drought. *Mailing Add:* Col Agr Sci Ore State Univ Corvallis OR 97331-2201

BURKE, MORRIS, b Hong Kong, Oct 25, 38; m 64; c 2. PROTEIN CHEMISTRY, ORGANIC CHEMISTRY. *Educ:* Univ Sydney, BSc, 60; Univ New South Wales, MSc, 63, PhD(wool chem), 66. *Prof Exp:* Res assoc radiation biochem, Mich State Univ, 66-67; res assoc phys biochem, Johns Hopkins Univ, 67-74, fel biol, 68-74; asst prof physiol, Sch Dent, Univ Md, 74-78; asst prof, 78-82, ASSOC PROF BIOL, CASE WESTERN RESERVE UNIV, 82- *Mem:* Am Chem Soc; Biophys Soc; Protein Soc. *Res:* Molecular basis of muscle contraction; mechanism of myosin and actomyosin Mg ATPase; selective chemical modification of proteins; peptide fractionation; associating macromolecules. *Mailing Add:* Dept Biol Case Western Reserve Univ Cleveland OH 44106

BURKE, PAUL J, b New York, NY, Apr 21, 20. OPERATIONS RESEARCH. *Educ:* City Col New York, BS, 40; Harvard Univ, EdM, 50; Columbia Univ, PhD(math statist), 66. *Prof Exp:* MEM TECH STAFF PROBABILITY & STATIST, BELL LABS, 53- *Mem:* Fel AAAS; Opers Res Soc Am. *Res:* Telephone traffic theory; queuing theory. *Mailing Add:* 430 W 24th St Apt 1F New York NY 10011

BURKE, RICHARD LERDA, b San Francisco, Calif, Aug 30, 25; m 50; c 4. ORGANIC CHEMISTRY. *Educ:* Univ San Francisco, BS, 44, MS, 48; Mich State Univ, PhD(org chem), 52. *Prof Exp:* Asst chem, Univ San Francisco, 47-48 & Mich State Univ, 48-52; chemist, Calif Res Corp, 52-57; tech rep prod develop, Oronite Chem Co, 57-60, prod specialist indust chem, Oronite Div, Calif Chem Co, 60-63; sr res chemist, Res & Develop Dept, Colgate Palmolive Co, 64-69; tech dir, Pac Soap Co, 69-88; RETIRED. *Mem:* Am Chem Soc. *Res:* Detergents and bleaches. *Mailing Add:* 5322 Soledad Rancho Ct San Diego CA 92109

BURKE, ROBERT D, b Swift Current, Sask, July, 31, 51; m 81; c 2. MORPHOGENESIS. *Educ:* Univ Alberta, BSc, 74, PhD(embryol), 78. *Prof Exp:* Nat Oceanic & Atmospheric Admin Sea Grant fel, dept zool, Univ Md, 78-79; Smithsonian Inst fel, Washington, DC, 79-80; NSERC univ res fel, 80-85, asst prof 85-87, ASSOC PROF EMBRYOL, DEPT BIOL, UNIV VICTORIA, CAN, 87- *Concurrent Pos:* Vis asst prof, Bamfield Marine Sta, BC, 83 & 84 & Friday Harbor Labs, Univ Wash, 84. *Mem:* AAAS; Am Soc Zool; Soc Develop Biol. *Res:* Developmental and cell biology; morphogenetic mechanisms involved in embryonic and larval development of sea urchins; application of immunological techniques to problems of differentiation of cell surface molecules, metamorphosis and evolution. *Mailing Add:* Dept Biol Univ Victoria Victoria BC V8W 2Y2 Can

BURKE, ROBERT EMMETT, b New York, NY, July 26, 34; m 60; c 4. NEUROPHYSIOLOGY. *Educ:* St Bonaventure Univ, BS, 56; Univ Rochester, MD, 61. *Prof Exp:* Intern & asst resident internal med, Mass Gen Hosp, 61-63, asst resident neurol, 63-64; res assoc spinal cord sect, Lab Neurophysiol, Nat Inst Neurol & Commun Disorders & Stroke, NIH, 64-67; vis scientist, Dept Physiol, Univ Goteborg, Sweden, 67-68; staff scientist, 68-75, LAB CHIEF, LAB NEURAL CONTROL, NIH, 75- *Concurrent Pos:* Mem sci adv bd, Amyotrophic Lateral Sclerosis Soc Am, 76- *Mem:* Am Physiol Soc; Soc for Neurosci; Int Brain Res Orgn. *Res:* Structure and function of spinal cord; neural control of movement; synaptic transmission; cellular biology of neurons; neurobiology of motor units. *Mailing Add:* Lab Neural Control Bldg 36 Rm 5A29 NINCDS NIH Bethesda MD 20892

BURKE, ROBERT F, b Joliet, Ill, June 13, 25; m 57; c 2. CHEMICAL ENGINEERING, PROCESS DESIGN. *Educ:* Carnegie Inst Technol, BS & MS, 49, DSc(chem eng), 52. *Prof Exp:* Design engr, Foster Wheeler Corp, NY, 51-55; res engr, Esso Res & Eng Co, NJ, 55-59; sr res staff mem, Raytheon Co, Mass, 59-60; sr res engr, T J Lipton Inc, NJ, 60-64; sr process design engr, Lummus Co, 64-67; sr chem engr, 67-75, ENG FEL, HOFFMANN-LA ROCHE, INC, 75- *Mem:* Am Chem Soc; Sigma Xi. *Res:* Food dehydration; freeze drying; high temperature decomposition and deposition; chemical process design and research; fluidization; reaction kinetics; vitamin plant design and operation. *Mailing Add:* 530 Brook Ave Hillsdale NJ 07042

BURKE, ROBERT WAYNE, b Ridgecrest, Calif, Feb 2, 58; m 91; c 2. NP COMPLETENESS, ALGORITHM DEVELOPMENT. *Educ:* Univ Calif, Santa Barbara, BS & BA, 82; Calif State Univ, Chico, MS, 91. *Prof Exp:* Electronics engr, Naval Weapons Ctr, China Lake, 82-85; HEAD INFO SYSTS, NAVAL RES LAB, STENNIS SPACE CENTER, 91- *Concurrent Pos:* Mem, Subcomt Parallel & Heterogeneous Distrib Comput, Defense Acquisition Bd, 90-91. *Mem:* Inst Elec & Electronics Engrs Computer Soc; Soc Computer Simulation. *Res:* Parallel and heterogeneous distributed computing; theoretical computer science, particularly "The Knapsack Problem". *Mailing Add:* PO Box 3363 Bay St Louis MS 39521-3363

BURKE, ROGER E, industrial organic chemistry, for more information see previous edition

BURKE, SHAWN EDMUND, b Waterville, Maine, Oct 28, 59; m 85. DISTRIBUTED PARAMETER CONTROL SYSTEMS, SONAR TRANSDUCTION. *Educ:* Princeton Univ, BSE, 81; Mass Inst Technol, MS, 83, PhD(mech eng), 89. *Prof Exp:* Instr mech eng, 84-85, fel, C S Draper Lab, 85-89, MEM TECH STAFF, C S DRAPER LAB, 89- *Concurrent Pos:* Sr engr, Chase Inc, 84-86; consult scientist, Technol Integration & Develop Group, 85-86. *Mem:* Sigma Xi; Int Soc Optical Eng. *Res:* Application of distributed transducer technology for structural vibration and shape control; sonar; hydrodynamics; hydroacoustics. *Mailing Add:* C S Draper Lab 555 Technology Sq Cambridge MA 02139

BURKE, STEVEN DOUGLAS, b Eau Claire, Wis, June 29, 51. SYNTHETIC ORGANIC & NATURAL PRODUCTS CHEMISTRY. *Educ:* Univ Wis, BS, 73; Univ Pittsburgh, PhD(chem), 78. *Prof Exp:* From asst prof to Carolina res prof chem, Univ SC, 78-87; PROF CHEM, UNIV WIS, MADISON, 87- *Concurrent Pos:* Vis assoc prof chem, Univ Wis, Madison, 84; consult, Wyeth-Ayerst Pharmaceut, 85-; reviewer, Biomed Sci Study Sect, NIH, 80-86, Biorg & Natural Prod Study Sect, 87; Alfred P Sloan res fel, 84. *Mem:* Am Chem Soc; Royal Chem Soc. *Res:* Total synthesis of biologically active natural products; organometallics; asymmetric synthesis; structure-activity relationships; molecular modeling; ion recognition; transport by synthetic macrocydic ligands; pharmaceutical chemistry. *Mailing Add:* Dept Chem Univ Wis Madison WI 53706

BURKE, THOMAS JOSEPH, b Baltimore, Md, Mar 15, 38; m 65; c 2. PHYSIOLOGY. *Educ:* Niagara Univ, BS, 61; Adelphi Univ, MS, 66; Univ Houston, PhD(physiol), 70. *Prof Exp:* ASST PROF PHYSIOL, MED SCH, UNIV COLO MED CTR, DENVER, 73- *Concurrent Pos:* Nat Heat & Lung Inst fel, Med Sch, Duke Univ, 69-73. *Mem:* Am Soc Nephrol; Int Soc Nephrol; Am Fedn Clin Res. *Res:* Renal autoregulation of blood flow and filtration rate; micropuncture techniques; nephron function; sodium balance; renin-angiotension system. *Mailing Add:* Med Sch Dept Physiol C281 Univ Colo Med Ctr 4200 E 9th Ave Denver CO 80262

BURKE, WILLIAM HENRY, b Dallas, Tex, Sept 1, 24; m 53; c 3. GEOPHYSICS. *Educ:* Rice Inst, PhD(physics), 51. *Prof Exp:* Sr res physicist, Mobil Res & Develop Corp, 51-56, res assoc, Field Res Labs, 65-85; RETIRED. *Concurrent Pos:* Consult, Grad Res Ctr, 64-65. *Mem:* Am Phys Soc; Am Soc Mass Spectrometry. *Res:* Isotope geology; nuclear reactions induced by proton, deuteron and neutron bombardment; extremely low level radioactivity measurements; mass spectrometry; geochronometry; terrestrial heat flow; geochemistry. *Mailing Add:* 3246 Tower Trail Dallas TX 75229

BURKE, WILLIAM J, b Newton, Mass, June 9, 35; m 73; c 3. AURORAL PHYSICS, WAVE PLASMA INTERACTIONS. *Educ:* Boston Col, BS, 60; Mass Inst Technol, PhD(physics), 71. *Prof Exp:* Postdoctoral fel, dept physics & astron, Rice Univ, 71-73; fel, Earth Resources Div, Nat Res Coun, Johnson Space Ctr, Houston, 73-75; res physicist, Regis Col, 75-78; res physicist, physics dept, Boston Col, 78-80; res physicist, 80-84, PHYSICIST, AIR FORCE GEOPHYSICS LAB, SPACE PHYSICS DIV, HANSCOM AFB, MA, 84- *Mem:* Am Geophys Union; AAAS; Am Phys Soc. *Res:* Electrodynamics of coupling between the ionosphere and magnetosphere at equatorial and high magnetic latitudes. *Mailing Add:* Air Force Geophys Lab Hanscom AFB MA 01731

BURKE, WILLIAM JAMES, b Lowellville, Ohio, May 24, 12; m 40; c 4. POLYMER CHEMISTRY. *Educ:* Ohio Univ, AB, 34; Ohio State Univ, PhD(chem), 37. *Prof Exp:* Res chemist, Cent Chem Dept, E I du Pont de Nemours & Co, 37-46; assoc prof chem, Ohio Univ, 46-47; assoc prof, Univ Utah, 47-50, head dept, 49-62, prof, 50-62; vpres, 62-76, prof chem, 62-83 dean grad col, 63-76, EMER PROF CHEM, ARIZ STATE UNIV, 83- *Concurrent Pos:* Chmn, Midwest Conf Grad Study & Res, 69-70; generalist consult, Nat Coun Archit Registr Bd, 69-72; pres, Western Asn Grad Schs, 71-72; mem, Nat Archit Accrediting Bd, 72-77. *Mem:* Fel AAAS; Am Chem Soc; Am Rock Art Res Asn; Australian Rock Art Res Asn; Sigma Xi. *Res:* Heterocyclic nitrogen compounds; organic sulfur compounds; carbohydrates; phenol-formaldehyde polymers; preservation of rock art; synthesis and characterization of polymers. *Mailing Add:* 501 E Bishop Dr Tempe AZ 85282

BURKE, WILLIAM JOSEPH, b Milwaukee, Wis, July, 24, 40; m 77; c 4. NEUROLOGY. *Educ:* Marquette Univ, BS, 62; St Louis Univ, MD & PhD(biochem), 72. *Prof Exp:* Intern med, 72-73, resident neurol, 73-76, asst prof, 76-83, ASSOC PROF NEUROL, MED SCH, ST LOUIS UNIV, 83- *Concurrent Pos:* Prin investr, Vet Admin Med Ctr, St Louis, 77- *Mem:* Am Soc Biol Chemists. *Res:* Aging; biochemical mechanisms regulating catecholamine synthesizing enzymes; pathogenic mechanisms in Alzheimer's disease. *Mailing Add:* Vet Admin Med Ctr Jefferson Barracks 128 JB St Louis MO 63125

BURKE, WILLIAM L, b Bennington, Vt, July 5, 41; m. THEORETICAL PHYSICS. *Educ:* Calif Inst Technol, BS, 63, PhD(physics), 69. *Prof Exp:* Lectr astrophys, 70-71, ASST PROF PHYSICS & ASTROPHYS, LICK OBSERV, UNIV CALIF, SANTA CRUZ, 71- *Res:* General relativity; gravitational waves; singular perturbation theory; general astrophysics. *Mailing Add:* Dept Astron Univ Calif Santa Cruz CA 95064

BURKE, WILLIAM THOMAS, JR, b Rochester, NY, July 30, 24; m 47; c 7. BIOLOGY, BIOCHEMISTRY. *Educ:* Univ Rochester, BA, 50, PhD(biol), 53. *Prof Exp:* Nat Cancer Inst fel, Atomic Energy Proj, Univ Rochester, 53-56; from instr to asst prof biol chem, Col Med & Dent, Seton Hall Univ, 56-60; assoc prof biochem, Sch Med, WVa Univ, 60-64; assoc prof exp path & biochem, NY Med Col, 64-67; dir div natural sci, Southampton Col, 67-73, dean Col, 73- 80, PROF BIOL, SOUTHAMPTON COL, LONG ISLAND UNIV, 67- *Mem:* AAAS; Am Soc Cell Biol; Am Asn Cancer; NY Acad Sci.

Res: Protein and amino acid metabolism; metabolism of tumors; liver pathology; metabolic diseases; environmental action; social responsibility of scientists. *Mailing Add:* Div Nat Sci Long Island Univ Southampton NY 11968

BURKEL, WILLIAM E, b Mankato, Minn, Oct 4, 38; m 66; c 7. ANATOMY, CELL BIOLOGY. *Educ:* St John's Univ, Minn, BA, 60; Univ NDak, MS, 62, PhD(anat), 64. *Prof Exp:* NIH fel anat, Lab Electron Micros, Univ NDak, 64-66; from asst prof to assoc prof, 66-82, PROF ANAT & CELL BIOL, UNIV MICH, 82- *Mem:* AAAS; Electron Micros Soc Am; Am Soc Cell Biol; Am Asn Anat; Am Soc Artificial Internal Organs. *Res:* Electron microscopy; development of artificial blood vessels. *Mailing Add:* Dept Anat & Cell Biol 4643 Med Sci 2 Univ Mich 1301 Catherine Rd Ann Arbor MI 48109-0616

BURKET, GEORGE EDWARD, b Kingman, Kans, Dec 10, 12; m 38; c 3. FAMILY MEDICINE. *Educ:* Univ Kans, MD, 37. *Prof Exp:* Pvt pract med, 40-73; prof family med, Univ Kans, 73-78, clin prof, 78-88, RETIRED. *Mem:* Inst Med-Nat Acad Sci; Am Acad Family Physicians (pres, 68); Am Bd Family Pract (pres, 75-77); Asn Am Med Cols. *Res:* Health care delivery. *Mailing Add:* Spring Lake Rte 1 Kingman KS 67068

BURKETT, HOWARD (BENTON), b Putnam Co, Ind, Feb 26, 16; m 36; c 4. ORGANIC CHEMISTRY. *Educ:* DePauw Univ, BA, 38; Univ Wis, PhD(org chem), 42. *Prof Exp:* Asst org chem, DePauw Univ, 37-38; asst, Univ Wis, 38-40; res chemist, Eli Lilly & Co, Ind, 42-45; from asst prof to prof chem, 54-81, rotating head dept, 64-81, EMER PROF CHEM, DEPAUW UNIV, 81- *Concurrent Pos:* Fel, Univ Wash, 53-54, Petrol Res Fund grant, 62-63; sr eng assoc, Naka Works, Hitachi, Ltd, Japan, 70-71; org chem sect, Nat Inst Arthritis & Metab & Digestive Dis, NIH, Bethesda, Md, 77-78. *Mem:* Am Chem Soc. *Res:* Organic synthesis; barbituric acid derivatives; aliphatic nitro-compounds, basic condensations, physical-organic; acid catalysis. *Mailing Add:* 700 Shadowlawn Ave Greencastle IN 46135

BURKEY, BRUCE CURTISS, b Ravenna, Ohio, Nov 29, 38. TECHNICAL MANAGEMENT. *Educ:* Hiram Col, BA, 61; Mich State Univ, MS, 63, PhD(physics), 67. *Prof Exp:* PHYSICIST, EASTMAN KODAK CO, 67- *Mem:* Am Phys Soc; Sigma Xi; Inst Elec & Electronics Engrs. *Res:* Device physics and solid state image sensor research and development. *Mailing Add:* 37 Berkshire Dr Rochester NY 14626

BURKEY, RONALD STEVEN, b Columbis, Ohio, March 24, 57. ENGINEERING. *Educ:* Ohio State Univ, BA, 78; Univ Tex, Dallas, MS, 83. *Prof Exp:* SR ENGR, HEADS-UP TECHNOL, 86- *Mem:* Am Phys Soc; Inst Elec & Electronics Engrs; Sigma Xi. *Mailing Add:* 8208 Spring Valley Rd Apt 124 Dallas TX 75240

BURKHALTER, ALAN, b Bloomington, Ind, Aug 11, 32; m 54; c 4. PHARMACOLOGY. *Educ:* DePauw Univ, BA, 54; Univ Iowa, MS, 56, PhD(pharmacol), 57. *Prof Exp:* Sr asst scientist, Lab Chem Pharmacol, Nat Heart Inst, 57-59; PROF PHARMACOL, MED CTR, UNIV CALIF, SAN FRANCISCO, 59- *Concurrent Pos:* USPHS grants, 60-63, 64-67; NIH spec fel, 63-64; consult, State of Calif, 60-62. *Mem:* AAAS; Am Soc Pharmacol & Exp Therapeut; Tissue Cult Asn. *Res:* Biochemical pharmacology; effect of drugs on enzyme regulation; histamine metabolism; effect of drugs during fetal and neonatal development; use of cell and organ culture in pharmacologic research. *Mailing Add:* Dept Pharmacol Univ Calif Med Ctr San Francisco CA 94143

BURKHALTER, BARTON R, b Toledo, Ohio, July 22, 38; m 63, 78; c 5. OPERATIONS RESEARCH & EVALUATION, HEALTH INFORMATION SYSTEMS. *Educ:* Univ Mich, BS(eng mech) & BS(eng math), 61, MS, 62, PhD(indust eng), 64. *Prof Exp:* Res asst regional develop, Inst Sci & Technol, Univ Mich, 61-62, res asst systs eng, 63-64, teaching fel indust eng univ, 63, instr, 64; lectr mgt sci, Wayne State Univ Exten, 62-63; exec dir, Community Systs Found, 64-69, pres, 69-73, chmn & sr scientist, CSF Ltd, 73-76, assoc & sr scientist, 75-88; prin, Performance Inst, 81-85; vpres, Zana Int Inc, 83-84, WV Int Indust Inc, 84-85; pres, Winch Inst Am Inc, 84-87; assoc dir, Wellstart, 86-88; RES PROF, UNIV ARIZ, 80-, DIR, CTR INT HEALTH INFO, 88- *Concurrent Pos:* Adj prof, PhD Prog Urban & Regional Planning, Univ Mich, 70-77; ed, Nutrit Planning, 77-80; consult, Papago Tribe of Ariz, 78-79, consult, Hopi Tribe, 81-82. *Mem:* Hosp Mgt Systs Soc (pres, 72-73); Opers Res Soc Am; Am Pub Health Asn; AAAS. *Res:* Planning; public health; science and American Indian communities; measurement of system performance. *Mailing Add:* 1601 N Kent St Suite 1001 Arlington VA 22209

BURKHALTER, PHILIP GARY, b Ft Wayne, Ind, July 20, 37; m 59; c 4. PHYSICS. *Educ:* Purdue Univ, BS, 59; Drexel Inst Technol, MS, 65. *Prof Exp:* Assoc engr x-ray diffraction, Martin-Marietta Corp, 59-61; scientist x-ray analysis, US Bur Mines, 61-70; RES PHYSICIST ATOMIC PHYSICS, NAVAL RES LAB, 70- *Concurrent Pos:* Group leader, extreme ultraviolet spectros, Naval Res Lab, prog mgr, Systs/Prog Off, Strategic Defense Initiative, 87. *Mem:* Optical Soc Am; Am Phys Soc. *Res:* X-ray diagnostics of high-temperature plasmas; atomic spectroscopy of highly-ionized atoms; x-ray detection devices. *Mailing Add:* Naval Res Lab Code 4681 Washington DC 20375-5000

BURKHARD, DONALD GEORGE, b NJ, Feb 10, 18; m 48; c 3. PHYSICS. *Educ:* Univ Calif, AB, 41; Univ Mich, MS, 47, PhD(physics), 50. *Prof Exp:* Eng aide, State Div Hwy, Calif, 36-39; asst eng aide, US Dept Eng, Los Angeles, 40-41; physicist, Radiation Lab, Univ Calif, 44; res physicist, Carter Oil Co, Okla, 44-45; res assoc, Ohio State Univ, 49-50; from asst prof to prof physics, Univ Colo, 50-64; PROF PHYSICS & ASTRON, UNIV GA, 65-; RES DIR, PHYSICS, ENG & CHEM RES ASSOCS, INC, 56- *Mem:* AAAS; Am Phys Soc; Optical Soc Am; Am Acoustical Soc; Am Asn Physics Teachers; Sigma Xi. *Res:* Theoretical physics; theoretical atomic and molecular structure; infrared and microwave spectroscopy; psychometrics; solid state physics; physical and geometrical optics. *Mailing Add:* Dept Physics & Astron Univ Ga Athens GA 30601

BURKHARD, MAHLON DANIEL, b Seward, Nebr, Jan 14, 23; m 45; c 4. ACOUSTICS. *Educ:* Nebr Wesleyan Univ, AB, 46; Pa State Col, MS, 50. *Prof Exp:* Teacher high sch, Nebr, 46-47; asst acoustics, Dept Physics, Pa State Univ, 47-50; physicist, Nat Bur Standards, 50-57; supvr, Acoustics Sect, IIT Res Inst, 57-60; mgr acoust res, Indust Res Prod, Inc, 60-75, mgr res, 75-87, vpres & gen mgr, Prof Sound Prod, 87-89; PRES, SONIC PERCEPTIONS, INC, 90- *Concurrent Pos:* Cert Superior Accomplishment, US Dept Com, 54; chmn standard working group, Am Nat Standards Inst, 58-; exec coun, Acoustical Soc Am, 83-86. *Mem:* Sr mem Inst Elec & Electronics Engr; fel Acoust Soc Am; fel Audio Eng Soc; Nat Soc Prof Engrs; Sigma Xi. *Res:* Atmospheric sound propagation; edge tone and whistle sound sources; calibration and standardization of audiometric instruments and microphones; acoustic impedance measurements; ultrasonic phenomena in gases, liquids, and solids; electro mechanical transducers; sound and noise control; hearing and hearing aids; auditorium acoustics; Sound reproduction; digital processing of sound and speech. *Mailing Add:* 48 Burchard Lane Norwalk CT 06853

BURKHARD, RAYMOND KENNETH, b Tempe, Ariz, Aug 6, 24; m 48; c 4. BIOCHEMISTRY. *Educ:* Ariz State Univ, AB, 47; Northwestern Univ, PhD(chem), 50. *Prof Exp:* From instr to prof, 50-89, EMER PROF BIOCHEM, KANS STATE UNIV, 89- *Concurrent Pos:* Res fel, Inst Enzyme Res, Wis, 59-60; vis prof, Yale Univ, 77. *Mem:* Am Soc Biol Chem; Am Chem Soc. *Res:* Protein interactions; microcalorimetry. *Mailing Add:* Dept Biochem Kans State Univ Manhattan KS 66506-3702

BURKHARDT, ALAN ELMER, b Buffalo, NY, Jan 19, 47; m 71. BIOCHEMISTRY, CHEMISTRY. *Educ:* Clarkson Col Technol, BS, 68; Cornell Univ, MS, 73, PhD(chem), 75. *Prof Exp:* Fel biochem, NIH, 75-77; res scientist, 77-80, supvr, 80-83, mgr, 83-85, DIR, DRY REAGENT CHEM LAB, AMES DIV, MILES LABS, 85- *Mem:* Am Chem Soc; AAAS; Am Asn Clin Chem; fel Nat Acad Clin Biochem. *Res:* Clinical chemistry, biochemistry; rapid diagnostic tests. *Mailing Add:* CIBA Corning Diag Corp 333 Coney St East Walpole MA 02032-1516

BURKHARDT, CHRISTIAN CARL, b Palmer, Nebr, Dec 26, 24; m 46; c 3. ENTOMOLOGY. *Educ:* Kans State Col, BS, 50, MS, 51; Univ Mo, PhD(entom), 67. *Prof Exp:* Field aide, European corn borer surv, Bur Entom & Plant Quarantine, USDA, 51; asst entomologist, Agr Exp Sta, Kans State Univ, 51-64, asst prof entom, 55-64; res entomologist, Univ Mo, 64-67; prof entom, Univ Wyo, 67-; RETIRED. *Mem:* AAAS; Entom Soc Am; Am Soc Sugar Beet Technol; Sigma Xi. *Res:* Life history, biology and control of insects attacking cereal and forage crops; soil insects; biological control; ecological studies. *Mailing Add:* 1214 Shields Laramie WY 82070

BURKHARDT, KENNETH J, b Elizabeth, NJ, Mar 30, 45; m 81; c 3. MICROCOMPUTER & WORKSTATION DESIGN, INTEGRATION OF VOICE & COMPUTERS. *Educ:* Cornell Univ, AB, 67; Rutgers Univ, MS, 70; Univ Wash, PhC, 72, PhD(comp sci), 75. *Prof Exp:* Systs programmer, Am Cyanamid Co, 67-70; res assoc comp sci, Univ Wash, 71-76; vpres systs, Intec Inc, 75-78; asst prof elec eng, Rutgers Univ, 76-82; pres, KJ Burkhardt & Co, Inc, 78-90; syst architect, Burroughs Corp, 81-87; EXEC VPRES, DIALOGIC CORP, 87- *Concurrent Pos:* Res award, Am Cyanamid Co, 71-72; res assoc, NIH, 73-74. *Mem:* Inst Elec & Electronics Engrs; Asn Comput Mach. *Res:* Computer systems design; microcomputer system design; computer architecture. *Mailing Add:* Box 420 Quakertown NJ 08868

BURKHARDT, WALTER H, b Stuttgart, Ger; US citizen; m 77; c 2. COMPUTER SCIENCE, HARDWARE & SOFTWARE SYSTEMS. *Educ:* Univ Stuttgart, Vordiplom, 51, Dipl Phys, 54, Dr rer nat(physics), 59. *Prof Exp:* Sr asst programmer, World Trade Co & Data Syst Div, IBM Corp, 61-64; prin systs programmer, Univac Div, Sperry Rand Co, 64-66; sr staff res & develop, Comput Control Div, Honeywell, Inc, 65-66; staff prog mgr advan systs, Info Systs Div, RCA Corp, 66-69; assoc prof comput sci, Univ Pittsburgh, 69-74; CHMN & PROF HARDWARE, DIR, INST INFO, UNIV STUTTGART, 74- *Concurrent Pos:* Reviewer, Comput Rev, 66; dir, DPM Comput Leasing Co, 69; mem exec comt, Comput Ctr, Univ Pittsburgh, 72-74. *Mem:* Inst Mgt Sci; Asn Comput Mach; Inst Elec & Electronics Engrs; Am Mgt Asn. *Res:* Microcomputer systems; micro-programming and micro-implementation; multi-processor systems; design automation. *Mailing Add:* 12 Azenbergst Stuttgart D 7000 Germany

BURKHART, HAROLD EUGENE, b Wellington, Kans, Feb 29, 44; m 71; c 1. FOREST BIOMETRY. *Educ:* Okla State Univ, BS, 65; Univ Ga, MS, 67, PhD(forest biomet), 69. *Prof Exp:* From asst prof to prof, 69-81, THOMAS M BROOKS PROF FOREST BIOMET, VA POLYTECH INST & STATE UNIV, 81- *Concurrent Pos:* Sr res fel, Forest Res Inst, Rotorua, New Zealand, 76-77. *Honors & Awards:* J Shelton Horsley Res Award; Sci Achievement Award, Int Union Forestry Res Orgn. *Mem:* Fel AAAS; fel Soc Am Foresters; Biomet Soc. *Res:* Forest growth and yield. *Mailing Add:* Dept Forestry Va Polytech Inst & State Univ Blacksburg VA 24061-0324

BURKHART, LAWRENCE E, chemical engineering, materials science; deceased, see previous edition for last biography

BURKHART, RICHARD DELMAR, b Kersey, Colo, June 26, 34; m 58; c 3. PHYSICAL CHEMISTRY. *Educ:* Dartmouth Col, AB, 56; Univ Colo, PhD(phys chem), 60. *Prof Exp:* Res chemist, Chem Div, Union Carbide Corp, 60-63; res assoc chem kinetics, Univ Ore, 63-65; from asst prof to assoc prof, 65-71, PROF PHYS CHEM, UNIV NEV, RENO, 71- *Mem:* Am Chem Soc; Sigma Xi. *Res:* Energy migration in amorphous solids; diffusion of excited species; photochemistry. *Mailing Add:* Dept Chem Univ Nev Reno NV 89507

BURKHART, RICHARD HENRY, b Tacoma, Wash, Dec 17, 46; m 71; c 3. APPLIED OPTIMIZATION, STOCHASTIC PROCESSES. *Educ:* Reed Col, BA, 69; Dartmouth Col, AM, 74, PhD(math), 76. *Prof Exp:* asst prof math, Univ NC, Wilmington, 76-80; APPL MATHEMATICIAN, BOEING

COMPUT SERV, 81- *Mem:* Am Math Soc; Math Asn Am; Soc Indust & Appl Math; Asn Comput Mach. *Res:* Numerical analysis; linear algebra; a symptotic analysis and optimization methods applied to the numerical solution of partial differential equations especially in aerodynamics and acoustics. *Mailing Add:* Boeing Comput Serv PO Box 24346 MS-7L-21 Seattle WA 98124-0346

BURKHEAD, MARTIN SAMUEL, b Ogden, Utah, May 23, 33; m 56; c 2. ASTROPHYSICS. *Educ:* Tex A&M Univ, BS, 55; Univ Calif, Los Angeles, MS, 57, PhD(astron), 64. *Prof Exp:* Sta chief, Smithsonian Satellite Tracking Prog, 57-60; res assoc astron, Univ Wis, 61-64; asst prof, 64-71, ASSOC PROF ASTRON, IND UNIV, BLOOMINGTON, 71-, ASSOC DIR, GOETHE LINK OBSERV, 74- *Concurrent Pos:* Vis prof USAID, Seoul Nat Univ, Korea, 77-78. *Res:* Photoelectric photometry. *Mailing Add:* Dept Astron Indiana Univ Bloomington IN 47401

BURKHOLDER, DAVID FREDERICK, b Grand Rapids, Mich, Mar 1, 31; m 54; c 1. PHARMACY. *Educ:* Ferris State Col, BS, 53; Univ Mich, MS, 61, PharD, 62. *Prof Exp:* Dir drug info ctr, asst dir pharm & res assoc, Med Ctr, Univ Ky, 62-67; assoc prof pharm & dir ctr pharmaceut pract, Sch Pharm, State Univ NY Buffalo, 67-70; from assoc prof to prof pharm, Sch Pharm, Univ Mo-Kansas City, 70-87, asst dean, 74-87; RETIRED. *Res:* Hospital pharmacy; rational drug therapy; statistical studies on drug selection and usage; drug literature organization and evaluation. *Mailing Add:* 71901 CR-388 South Haven MI 49090

BURKHOLDER, DONALD LYMAN, b Octavia, Nebr, Jan 19, 27; m 50; c 3. MATHEMATICS. *Educ:* Earlham Col, BA, 50; Univ Wis, MS, 53; Univ NC, PhD(math statist), 55. *Prof Exp:* From asst prof to assoc prof, 55-64, PROF MATH, UNIV ILL, URBANA, 64- *Concurrent Pos:* Prof, Ctr Advan Study, Univ Ill, 78- *Mem:* Am Math Soc; fel Inst Math Statist (pres, 75-76); AAAS. *Res:* Probability and its applications to analysis. *Mailing Add:* Dept Math Univ Ill Urbana IL 61801

BURKHOLDER, JOHN HENRY, b Octavia, Nebr, July 11, 25; m 50; c 4. ZOOLOGY. *Educ:* McPherson Col, AB, 49; Univ Chicago, PhD(zool), 54. *Prof Exp:* Head dept, 54-70, PROF BIOL, McPHERSON COL, 53-, CHMN DIV NATURAL SCI, 67- *Honors & Awards:* Burkholder Res Award. *Mem:* AAAS; Genetics Soc Am; Am Inst Biol Sci; Sigma Xi. *Res:* Genetics; position effects in Drosophila melanogaster. *Mailing Add:* 127 N Charles St McPherson KS 67460

BURKHOLDER, PETER M, b Cambridge, Mass, May 7, 33; m 56; c 2. PATHOLOGY, HEALTH CARE MANAGEMENT. *Educ:* Yale Univ, BS, 55; Cornell Univ, MD, 59; Am Bd Path, dipl, 64. *Prof Exp:* Intern path, New York Hosp-Cornell Med Ctr, 59-60; asst, Med Col, Cornell Univ, 60-62, from instr to asst prof, 62-65; asst prof, Med Ctr, Duke Univ, 65-70; assoc prof, Sch Med, Univ Wis-Madison, 70-72, chmn dept, 72-74, prof path, 72-79; dir, Kidney Disease Inst, 79-81; actg dir, NY State Dept Health, Lab Med Inst, 80, dep dir, Div Lab, 81-82; chief of staff, 82-84, STATE PATHOLOGIST, VET ADMIN MED CTR, 84-; asst dean, 82-84, PROF MED SCH, UNIV MICH, 84- *Concurrent Pos:* USPHS trainee, 60-63; asst pathologist, New York Hosp, 60-63; asst attend pathologist, 63-65; consult pathologist, Nat Nephrotic Sydrome Study & Adult Glomerula Dis Study. *Mem:* Am Asn Pathologists; Am Asn Immunol; Am Soc Nephrology; Int Soc Path; Int Soc Nephrology; Sigma Xi. *Res:* Experimental pathology; immunochemistry of serum complement; immunohistochemistry and electron microscopy of renal disease. *Mailing Add:* Dept Pathology Univ Arizona Sch Medicine, Med 2601 E Roosevelt, Box 5099 Phoenix AZ 85010

BURKHOLDER, TIMOTHY JAY, b Orrville, Ohio, Jan 15, 41; m 64; c 3. VERTEBRATE PHYSIOLOGY. *Educ:* Taylor Univ, BA, 63; Ohio State Univ, MS, 65, PhD(biol sci), 70. *Prof Exp:* Instr biol, Wooster High Sch, 65-67; PROF BIOL, TAYLOR UNIV, 70- *Mem:* Sigma Xi; Am Sci Affiliation; Nat Asn Biol Teachers; Creation Res Soc. *Res:* Relation of serum calcium levels to the reproductive cycle and egg production of the red-winged blackbird, Agelaius Phoeniceus. *Mailing Add:* Dept Biol Taylor Univ Upland IN 46989

BURKHOLDER, WENDELL EUGENE, b Octavia, Nebr, June 24, 28; m 51; c 4. ENTOMOLOGY. *Educ:* McPherson Col, AB, 50; Univ Nebr, MSc, 56; Univ Wis, PhD(entom), 67. *Prof Exp:* From asst prof to assoc prof, 67-70, PROF ENTOM, UNIV WIS-MADISON, 75-; ENTOMOLOGIST, USDA, 56- *Honors & Awards:* C V Riley Achievement Award in Entomol, 88. *Mem:* Entom Soc Am; Soc Invert Path; AAAS; Sigma Xi; Int Soc Chem Ecol. *Res:* Insect pheromones; reproductive biology; sex attractants; insect behavior; dermestid beetles, stored grain insects and other stored product insects; insect management technology. *Mailing Add:* Dept Entom Univ Wis Madison WI 53706

BURKI, HENRY JOHN, cell biology, radiation biology; deceased, see previous edition for last biography

BURKI, NAUSHERWAN KHAN, INTERNAL MEDICINE, PHYSIOLOGY. *Educ:* Univ Pakistan, MD; London Univ, Eng, PhD(med). *Prof Exp:* PROF INTERNAL MED & CHIEF PULMONARY DIV, KY MED CTR. *Res:* High altitude studies. *Mailing Add:* Pulmonary Div Ky Med Ctr 800 Rose St Lexington KY 40536

BURKLE, JOSEPH S, b Philadelphia, Pa, July 28, 19; m 44; c 3. NUCLEAR MEDICINE. *Educ:* Univ Pa, AB, 40, MD, 43; Am Bd Internal Med, dipl, 55 & 77; Am Bd Nuclear Med, dipl, 72. *Prof Exp:* Intern, US Naval Hosp, Pa, US Navy, 44, resident med, Pa Hosp, 47-48, resident med, US Naval Hosp, Pa, 50-51, clin & lab instr, Radioisotope Lab, US Naval Hosp, St Albans, NY, 56-60, mem staff, Naval Hosp, Bethesda, Md, 60-63 & US Naval Sta Hosp, 63-66, cmndg officer, Armed Forces Radiobiol Res Inst, Defense Atomic Support Agency, 66-67; chmn, Dept Nuclear Med, York Hosp, 67-; asst clin prof med, Univ Md, 70-; RETIRED. *Concurrent Pos:* Adj prof, Millersville

State Col & Harrisburg Community Col; staff, Dept Nuclear Med, Harrisburg Hosp; pres, Am Col Nuclear Med, 87-88, pres, Col Nuclear Med, site visitor, Joint Rev Comt; radiation mgt consult, 87- *Mem:* AMA; fel Am Col Physicians; Soc Nuclear Med; Am Col Nuclear Med. *Res:* Internal medicine; nuclear medicine; ultrasound. *Mailing Add:* 75 Bridlewood Way A32 York PA 17402

BURKLEY, RICHARD M, b Pittsburgh, Pa, Sept 5, 41; m 65; c 5. RESEARCH ADMINISTRATION. *Educ:* Georgetown Univ, BS, 61; Ohio State Univ, MS, 63. *Prof Exp:* Physicist, David Taylor Model Basin, US Navy, 60-63; programmer, IBM, 64-68, mgr, 68-81; dir, 81-86, VPRES, AUTO-TROL, 86- *Concurrent Pos:* Lectr, Univ Calif, Los Angeles, 69-70. *Mem:* Asn Comput Mach; Nat Comput Graphic Asn; Am Inst Maintenance. *Mailing Add:* 1333 Marble Dr Boulder CO 80303

BURKMAN, ALLAN MAURICE, b Waterbury, Conn, Apr 23, 32; m 65; c 2. PHARMACOLOGY. *Educ:* Univ Conn, BS, 54; Ohio State Univ, MSc, 55, PhD(pharmacol), 58. *Prof Exp:* Teaching asst pharmacol, Ohio State Univ, 54-57; asst prof, Univ Ill, 58-63; assoc prof, Butler Univ, 63-66; assoc prof, 66-71, chmn div, 78-87, PROF PHARMACOL, OHIO STATE UNIV, 71- *Concurrent Pos:* Fel, Am Found Pharmaceut Educ, 57-83; vis prof, Univ Utah, 81-82. *Honors & Awards:* Mead Johnson Award for Undergrad Res Direction, 65. *Mem:* Am Pharmaceut Asn; Soc Exp Biol & Med; Am Soc Pharmacol & Exp Therapeut; Am Asn Univ. *Res:* Bioassay development; drug screening; emetic and antiemetic mechanism; drug-receptor theory. *Mailing Add:* Div Pharmacol Col Pharm Ohio State Univ Columbus OH 43210

BURKMAN, ERNEST, b Detroit, Mich, Oct 4, 29; m 54; c 4. SCHOOL IMPROVEMENT, TEXTBOOK DESIGN. *Educ:* Eastern Mich Univ, BS, 52; Univ Mich, MS, 56, MA, 59, EdD(ed), 62. *Prof Exp:* Teacher high sch, Mich, 55-60; from asst prof to assoc prof, 60-67, head dept, 65-67, PROF SCI EDUC & INST SYST, FLA STATE UNIV, 66- *Concurrent Pos:* Consult sci curriculum develop, Various Sch Systs, 60-90; co-dir, Turkish Nat High Sch Sci Proj, Ford, 62-65; regional consult biol sci curriculum study, 62-63; dir, NSF Intermediate Sci Currie Study, 66-71; dir, Educ Res Inst, Fla State Univ, 70-72 & Div Instrnl design, 72-74; dir, NSF Individualized Sci Instrnl Syst Proj, 72-85; consult, ministries of educ in Turkey, 62-65, Philippines, 67-68, Uruguay, 75, Saudi Arabia, 78, Botswana, 87-89, US Dept Defense, 69, 71, 77, Arthur Andersen & Co, 85. *Honors & Awards:* Fulbright Lectr, Uruguay, 75. *Mem:* Fel AAAS; Nat Asn Biol Teachers; Nat Sci Teachers Asn; Nat Asn Res Sci Teaching. *Res:* Science curriculum development; science education; school improvement; textbook design. *Mailing Add:* 705 N Ride Fla State Univ Tallahassee FL 32303

BURKS, CHRISTIAN, b Wash, DC, Mar 23, 54; m 76; c 2. BIOINFORMATICS, COMPUTATIONAL BIOLOGY. *Educ:* St Johns Col, Santa Fe, BA, 76; Yale Univ, PhD(molecular biol), 82. *Prof Exp:* Postdoctoral fel, 82-84, staff mem, 84-90, GROUP LEADER THEORET BIOL & BIOPHYSICS, LOS ALAMOS NAT LAB, 90- *Mem:* AAAS; Biophys Soc. *Res:* Pattern recognition in nucleotide sequences; development of databases pertinent to molecular biology. *Mailing Add:* Los Alamos Nat Lab T-10 MS K710 Los Alamos NM 87545

BURKS, G EDWIN, b Apr 10, 01; US citizen. ENGINEERING. *Hon Degrees:* DSc, Bradley Univ, 72. *Prof Exp:* Asst chief engr, Caterpillar Tractor Co, 38-42, chief engr, 42-54, dir eng & res, 54-55, vpres eng & res, 55-67; RETIRED. *Mem:* Nat Acad Eng; Am Soc Metals; Am Soc Mech Engrs; Soc Automotive Engrs (pres, 66). *Mailing Add:* 3320 N Bigelow St Peoria IL 61604

BURKS, JAMES KENNETH, b Great Falls, Mont, Apr 3, 45; m 68. ENDOCRINOLOGY, BIOCHEMISTRY. *Educ:* Tex Tech Univ, BS, 67; Univ Tex Southwestern Med Sch, MD, 71. *Prof Exp:* Intern & resident med, Vanderbilt Univ Hosp, 71-73; staff assoc med & biochem, Nat Inst Arthritis, Metab & Digestive Dis, Nat Inst Dent Res, NIH, 73-75; fel endocrinol, Strong Mem Hosp, Univ Rochester, 75-76; fel, 76-78, INSTR & ASSOC MED, JEWISH HOSP, SCH MED, WASH UNIV, 78- *Concurrent Pos:* Consult artificial kidney-chronic uremic prog, Nat Inst Arthritis, Metab & Digestive Dis, NIH, 76-78, co-investr grant, 78-81. *Mem:* Nat Asn Advan Sci; Am Fedn Clin Res. *Res:* Factors that affect growth and differentiation of bone cells; purification of growth factors; development of a serum-free tissue culture medium. *Mailing Add:* Eight Medical Pkwy Ste E-307 Farmers Branch TX 75234

BURKS, STERLING LEON, b Reydon, Okla, Mar 3, 38. ECOLOGY, LIMNOLOGY. *Educ:* Southwestern State Col, Okla, BS, 63; Okla State Univ, MS, 65, PhD(zool), 69. *Prof Exp:* Res assoc water pollution, 69-70, from asst prof to assoc prof, 71-83, PROF ZOOL, OKLA STATE UNIV, 83-, DIR ANALYTICAL LAB RESERVOIR RES CTR, 70-, DIR, OKLA STATE UNIV WATER QUAL RES LAB, 80- *Concurrent Pos:* Fel, Fed Water Pollution Control Admin, US Dept Interior, 68-70, proj dir, Off Water Resources Res, 70-72. *Mem:* Water Pollution Control Fedn; Am Chem Soc; Soc Environ Toxicol & Anal Chem. *Res:* Identification of fish toxicants in surface waters with gas chromatography, mass spectrometry and atomic absorption spectrophotometry; determination of effects of chemical contaminants upon aquatic organisms. *Mailing Add:* Water Qual Res Lab Okla State Univ Stillwater OK 74074

BURKS, THOMAS F, b Houston, Tex, Apr 3, 38; m 62; c 2. PHARMACOLOGY. *Educ:* Univ Tex, BS, 62, MS, 64; Univ Iowa, PhD(pharmacol), 67. *Prof Exp:* From instr to assoc prof pharmacol, Sch Med, Univ NMex, 67-71; from assoc prof to prof pharmacol, Univ Tex Med Sch, Houston, 74-77; prof & head, Pharmacol, 77-84, ASSOC DEAN RES, COL MED, UNIV ARIZ, 84- *Concurrent Pos:* USPHS fel, Nat Inst Med Res, Eng, 67-68; USPHS res grant, 69- *Mem:* Am Soc Pharmacol & Exp Therapeut; Am Fedn Clin Res; Soc Exp Biol & Med; AAAS; Am Heart Asn. *Res:* Pharmacology of peripheral and central neuro transmission; mode of action of narcotic analgesic agents. *Mailing Add:* Dept Pharmacol Col Med Univ Ariz 1501 N Campbell Ave Tucson AZ 85724

BURKSTRAND, JAMES MICHAEL, b Philadelphia, Pa, Jan 29, 46; m 67; c 2. SURFACE PHYSICS. *Educ:* Rensselaer Polytech Inst, BS, 67; Univ Ill, Urbana, MS, 69, PhD(physics), 72. *Prof Exp:* Staff res physicist, Gen Motors Res Labs, 72-81; staff scientist, 81-82, DIR MKT, PHYS ELECTRONICS DIV, PERKIN-ELMER CORP, 82- *Mem:* Am Vacuum Soc; Am Phys Soc; Sigma Xi. *Res:* Experimental investigation of chemical and electronic properties of solid surfaces and interfaces and the application of this information to the solution of technological problems. *Mailing Add:* 18215 23rd Ave N Plymouth MN 55447

BURKWALL, MORRIS PATON, JR, b Kansas City, Mo, May 16, 39; m 63; c 3. BIOCHEMISTRY, FOOD CHEMISTRY. *Educ:* Fla State Univ, BS, 61; Univ Minn, MS, 63, PhD(biochem), 66. *Prof Exp:* Group leader food chem, Quaker Oats Co, 66-71; sr group leader, 70-71, sect mgr semi-moist pet foods, 71-72, sect mgr dog food res, 72-75, mgr, 75-77, sr mgr, explor pet foods, 77-90, SR MGR, TECHNOL DEVELOP, JOHN STUART RES LAB, QUAKER OATS CO, 90- *Honors & Awards:* Sherwood Award, Am Asn Cereal Chem, 64. *Mem:* Am Chem Soc; Inst Food Technol; Am Asn Cereal Chem; AAAS. *Res:* Biochemical modification of meat and cereal byproducts; theory of flavor and palatability; microstability of semi-moist foods and pet foods; toxicology of food ingredients; enzymatic modification of proteins and fats; 11 patents. *Mailing Add:* Quaker Oats Res Div 617 W Main St Barrington IL 60010

BURKY, ALBERT JOHN, b Utica, NY, June 30, 42; m 70; c 4. PHYSIOLOGICAL ECOLOGY. *Educ:* Hartwick Col, BA, 64; Syracuse Univ, PhD(zool), 69. *Prof Exp:* Vis asst prof zool, Syracuse Univ, 69; fel, Cent Univ Venezuela, 69-70; instr biol, Case Western Reserve Univ, 71-73; from asst prof to assoc prof, 73-90, PROF BIOL, UNIV DAYTON, 91- *Mem:* AAAS; Am Soc Zoologists; Ecol Soc Am; Am Malacol Union; Malacol Soc London. *Res:* Physiological ecology of freshwater clams, snails and terrestrial snails; energy budgets of natural populations, especially growth, reproduction and metabolism, and the physiological metabolic adaptations to environmental conditions. *Mailing Add:* Dept Biol Univ Dayton Dayton OH 45469-2320

BURLAGA, LEONARD F, b Superior, Wis, Oct 1, 38; c 2. PHYSICS. *Educ:* Univ Chicago, BS, 60; Univ Minn, MS, 62, PhD(physics), 66. *Prof Exp:* From teaching asst to teaching assoc physics, 60-63 res assoc, Univ Minn, 63-66; Nat Acad Sci-Nat Res Coun res assoc, 66-68, ASTROPHYSICIST, GODDARD SPACE FLIGHT CTR, NASA, 68- *Concurrent Pos:* Goddard sr fel, 88. *Honors & Awards:* Except Sci Achievement Medal, NASA, 79; Lindsay Award, 88. *Mem:* Am Geophys Union; Int Astron Union. *Res:* Interplanetary medium; propagation of cosmic rays. *Mailing Add:* Code 692 NASA Goddard Space Flight Ctr Greenbelt MD 20771

BURLAND, DONALD MAXWELL, b La Jolla, Calif, Sept 14, 43; m 68; c 1. MOLECULAR CRYSTAL PHYSICS. *Educ:* Dartmouth Col, AB, 65; Calif Inst Technol, PhD(chem & physics), 70. *Prof Exp:* Asst physics, Univ Leiden, Neth, 70-71; res staff mem chem, 71-80, mgr optical mat physics, 80-82, mgr rec media, 82-83, mgr electrophotography, IBM Res Lab, 83-89, MGR NONLINEAR OPTICAL MAT, IBM ALMADEN RES CTR, SAN JOSE, CALIF, 89- *Mem:* Fel Am Phys Soc; Am Chem Soc. *Res:* Physics of energy and charge transport in molecular crystals and amorphous solids; holographic techniques to investigate solid state photochemistry; materials for electrophotography; organic nonlinear optical materials. *Mailing Add:* IBM Almaden Res Ctr Dept K95-801 650 Harry Rd San Jose CA 95120-6099

BURLANT, WILLIAM JACK, b Chicago, Ill, Oct 20, 28; m 55; c 2. ORGANIC CHEMISTRY. *Educ:* City Col New York, BS, 49; Brooklyn Col, MA, 51; Polytech Inst Brooklyn, PhD(chem), 55. *Prof Exp:* USPHS asst, Brooklyn Col, 50-52, instr org & gen chem, 53-54; Res Corp asst, Polytech Inst Brooklyn, 52-53; instr org & gen chem, Cooper Union, 54-55; chemist, Sci Lab, Ford Motor Co, 55, mgr res staff, Mat Applications Dept, 55-69, asst dir chem sci, 69-79, exec engr, 72-79; mgr technol opers, Gen Elec Co, 79-; tech dir, Kendall Co, 81-; AT GAF CORP. *Mem:* Am Chem Soc. *Res:* Radiation chemistry of organic systems; structure and properties of polymers; plastic applications; composites. *Mailing Add:* GAF Corp 1361 Alps Rd Wayne NJ 07470

BURLEIGH, BRUCE DANIEL, JR, b Augusta, Ga, June 23, 42; div; c 1. BIOCHEMISTRY, PROTEIN CHEMISTRY. *Educ:* Carnegie-Mellon Univ, BS, 64; Univ Mich, MS, 67, PhD(biochem), 70. *Prof Exp:* Vis scholar molecular biol, Med Res Coun Lab, Cambridge, Eng, 70-72, res staff, 72-73; asst prof biochem, M D Anderson Hosp & Tumor Inst, Univ Tex Systs Cancer Ctr & Grad Sch Biomed Sci, 73-79, assoc biochemist, 80-81; res scientist, Int Minerals & Chem Corp, 81-83, sr res scientist, 83-87;; PRIN RES SCIENTIST, PITMAN-MOORE INC, 88- *Concurrent Pos:* Am Cancer Soc fel, 70-72. *Mem:* AAAS; Am Chem Soc; Am Soc Biol Chem & Molecular Biol; Endocrine Soc; NY Acad Sci; Sigma Xi. *Res:* Structure and function of peptide hormones; hormonal control of cell states; animal growth regulation. *Mailing Add:* Pitman-Moore Inc 1331 S First St PO Box 207 Terre Haute IN 47808

BURLEIGH, JAMES REYNOLDS, b Fresno, Calif, Sept 6, 36; m 59; c 4. PLANT PATHOLOGY, GENETICS. *Educ:* Fresno State Col, BS, 58; Wash State Univ, MS, 62, PhD(plant path), 65. *Prof Exp:* Res asst agron, Wash State Univ, 59-61, res asst plant path, 61-64; RES PLANT PATHOLOGIST, SCI & EDUC ADMIN-AGR RES, USDA, 64-; ASSOC PROF PLANT SCI, CALIF STATE UNIV, CHICO, 71- *Concurrent Pos:* Asst prof plant physiol, Kans State Univ, 64-71. *Mem:* Am Phytopath Soc; Am Mycol Soc. *Res:* Epidemiology of cereal rust diseases. *Mailing Add:* Sch Agr Calif State Univ Chico CA 95929-0310

BURLEIGH, JOSEPH GAYNOR, b Crowley, La, Jan 20, 42; m 69; c 2. INSECT ECOLOGY, PSYCHOLOGY. *Educ:* Univ Southwestern La, BS, 64; La State Univ, MS, 66, PhD(entom), 70; Univ Cent Ark, MS, 79. *Prof Exp:* Mem planning & recreation dept, La State Parks Comn, 66-68; res assoc entom, Okla State Univ, 70-72; PROF ENTOM, UNIV ARK, PINE BLUFF, 72- *Concurrent Pos:* Psychotherapist, 80-91. *Mem:* Entom Soc Am. *Res:* Ecology and population dynamics of the bollworm and tobacco budworm; effects of cultural practices on soybean insects, sunflower insects; aquatic insect populations in minnow production ponds. *Mailing Add:* Dept Agr Univ Ark Pine Bluff AR 71601

BURLESON, GEORGE ROBERT, b Baton Rouge, La, Oct 12, 33; m 60; c 2. EXPERIMENTAL NUCLEAR PHYSICS. *Educ:* La State Univ, BS, 55; Stanford Univ, MS, 57, PhD(physics), 60. *Prof Exp:* Res assoc high energy physics, Argonne Nat Lab, 60-62, asst scientist, 62-64; asst prof physics, Northwestern Univ, 64-71; assoc prof, 71-78, PROF PHYSICS, NMEX STATE UNIV, 78- *Concurrent Pos:* Vis staff mem, Los Alamos Sci Lab, 72, 73 & 74. *Mem:* Am Phys Soc; Sigma Xi. *Res:* Electron scattering at medium energies; hyperon polarization; mesonic x-rays; polarization in hadron-nucleon scattering; pion-nucleus scattering at medium energies, nucleon-nucleon interactions. *Mailing Add:* Dept Physics Box 3D NMex State Univ Las Cruces NM 88003

BURLETT, DONALD JAMES, b East Orange, NJ, Dec 20, 49; m 73. POLYMER CHEMISTRY. *Educ:* State Univ NY Col Oswego, BS, 71; Univ Cincinnati, PhD(org chem), 76. *Prof Exp:* Res asst org chem, Mass Inst Technol, 76-77; assoc scientist, 77-89, SR RES CHEMIST ORG CHEM, GOODYEAR TIRE & RUBBER CO, 89- *Mem:* Am Chem Soc, Rubber & Polymer Div; North Am Thermal Anal Soc; Tire Soc. *Res:* Synthetic organic chemistry; polymer bound reagents; thermal analysis; heterocyclic chemistry; vulcanization chemistry; oxidation and antioxidants for polymers. *Mailing Add:* Goodyear Tire & Rubber Co 142 Goodyear Blvd Akron OK 44305

BURLEY, CARLTON EDWIN, b Auburn, NY. PHYSICS, NONDESTRUCTIVE EVALUATION. *Educ:* Syracuse Univ, AB, 41, MS, 42, MS, 49. *Prof Exp:* Instr physics, Utica Col, Syracuse Univ, 49-53; res scientist, 53-56, supvr physics & elec, 56-65, DIR PHYSICS & INSTRUMENTATION, METALL RES DIV, REYNOLDS METALS CO, 65- *Concurrent Pos:* Adj fac math, Va Commonwealth Univ, 59- *Mem:* Am Phys Soc; Am Soc Metals; Am Soc Testing & Mat; fel Am Soc Non Destructive Testing; Inst Elec & Electronics Engrs. *Res:* Materials physical properties; applied mechanics; electrical and thermal properties; vacuum techniques; non destructive test development and operator training. *Mailing Add:* 8207 Metcalf Dr Richmond VA 23227

BURLEY, DAVID RICHARD, b Crooksville, Ohio, Nov 1, 42; m 66; c 2. PHYSICAL CHEMISTRY, COMPUTER SCIENCE. *Educ:* Ohio State Univ, BS, 64; Univ Calif, PhD(phys chem), 69; Rider Col, MBA, 78. *Prof Exp:* Fel photochem, Univ Col Swansea, Wales, 69-71; head chem res, Tile Coun Am, 71-75; sr chemist, Am Cyanamid Co, 75-76; group leader, 76-78, bus planner, 78-79, mkt mgr, 79-81, bus mgr, 81-89; PRES, TECHLINE, INC, 89- *Mem:* Am Chem Soc; Int Soc Pharmaceut Engrs. *Res:* Photochemistry; surface chemistry; chemical technology; management. *Mailing Add:* 652 Beechwood Lane Kinnelon NJ 08512

BURLEY, GORDON, b Giessen, Ger, Feb 15, 25; nat US; m 57, 72. ENVIRONMENTAL CHEMISTRY, SCIENCE ADMINISTRATION. *Educ:* Temple Univ, AB, 48; Univ Md, MS, 52; Georgetown Univ, PhD, 62. *Prof Exp:* Res assoc, Geophys Lab, Carnegie Inst Wash, 50-52; phys chemist, Nat Bur Standards, 52-67; sr scientist, Div Reactor Licensing, US AEC, 67-72; br chief, 72-79, dep div dir, 79-81, actg dir, 81-82, SCI ADV, OFF RADIATION PROGS, US ENVIRON PROTECTION AGENCY, 82- *Mem:* Am Chem Soc; AAAS. *Res:* Solid state chemistry; environmental radiation standards; mathematical models; transuranium elements; science policy. *Mailing Add:* Off Radiation Progs ANR-458 US Environ Protection Agency Washington DC 20460

BURLEY, J WILLIAM ATKINSON, b Moundsville, WVa, Mar 31, 28; m 49; c 4. PLANT PHYSIOLOGY, AGRICULTURAL RESEARCH. *Educ:* WVa Univ, BA & MS, 54; Ohio State Univ, PhD(bot), 60. *Prof Exp:* From instr to assoc prof bot, Ohio State Univ, 57-68; prof, Bowling Green State Univ, 68-72; prof biol sci & head dept, Drexel Univ, 72-84, dean, grad sch, 84-85, assoc dean, 85-87, dean, Col Sci, 87-90, SR ASSOC DEAN, COL ARTS & SCIS, DREXEL UNIV, 90- *Concurrent Pos:* Consult radioisotope assayist, 61-; NSF res grant, 63-; consult, Agr Res. *Mem:* Sigma Xi; Am Soc Plant Physiol. *Res:* Synthesis and translocation of C14-labeled compounds in vascular plants; uptake and translocation of radioactive labeled ions in corn and pea roots; analysis of light intensity and quality in natural environments. *Mailing Add:* Col Arts & Scis Drexel Univ Philadelphia PA 19104

BURLEY, NANCY, b Syracuse, NY, July 5, 49. EVOLUTION OF BEHAVIOR, SEXUAL SELECTION. *Educ:* Syracuse Univ, BS, 71; Univ Cincinnati, MS, 73; Univ Tex, Austin, PhD(zool), 77. *Prof Exp:* Asst prof biol, McGill Univ, 77-79; asst prof, 79-84, ASSOC PROF ECOL, ETHOLOGY & EVOLUTION, UNIV ILL, URBANA-CHAMPAIGN, 84- *Concurrent Pos:* Assoc prof scientist, Ill Nat Hist Surv, 85- *Mem:* AAAS; Animal Behav Soc; Soc Study Evolution; Am Soc Naturalists; Am Ornithologists Union. *Res:* Evolution of behavior; sexual selection theory; experimental investigation of male choice in monogamous species of birds; sex-ratio manipulation. *Mailing Add:* Dept Ecol Ethology & Evolution Univ Ill 505 S Goodwin Ave Urbana IL 61801

BURLING, JAMES P, b Baltimore, Md, May 29, 30; m 61; c 2. MATHEMATICS. *Educ:* Grinnell Col, BA, 52; State Univ NY, Albany, MA, 57; Univ Colo, PhD(math), 65. *Prof Exp:* High sch teacher, NY, 55-56; asst prof math, State Univ NY, Col Oneonta, 57-60; PROF MATH, STATE UNIV NY, COL OSWEGO, 65- *Mem:* AAAS; Math Asn Am; Am Math Soc. *Res:* Coloring problems of families of convex bodies. *Mailing Add:* Dept Math State Univ NY Col Oswego NY 13126

BURLING, RONALD WILLIAM, b Masterton, NZ, Aug 4, 20; m 48; c 1. OCEANOGRAPHY. *Educ:* Univ NZ, BSc, 49, MSc, 50; Univ London, PhD(meteorol), 55. *Prof Exp:* Scientist, Oceanog Observ, Dept Sci & Indust Res, NZ, 49-51; mem staff, Nat Inst Oceanog, Eng, 53-55; scientist, Oceanog Observ, Dept Sci & Indust Res, NZ, 55-60; from asst prof to assoc prof phys & dynamical oceanog, Univ BC, 60-67, prof oceanog, Dept Oceanog, 67-85; RETIRED. *Res:* Physical and dynamical oceanography; air sea interaction; transfer of momentum, energy, heat and water vapor across the sea surface; inlet studies; carbon dioxide in the ocean. *Mailing Add:* 1881 Allison Rd Vancouver BC V6T 1T1 Can

BURLINGAME, ALMA L, b Cranston, RI, Apr 29, 37; div; c 2. PHYSICAL CHEMISTRY, BIOCHEMISTRY. *Educ:* Univ RI, BS, 59; Mass Inst Technol, PhD(chem), 62. *Prof Exp:* res chemist, Space Sci Lab, 63-84, DIR, BIOMED & ENVIRON MASS SPECTROMETRY RESOURCE, UNIV CALIF, BERKELEY, 73- *Concurrent Pos:* Mem, Lunar Sample Anal Planning Team & Lunar Sample Preliminary Exam Team, NASA Johnson Space Ctr, 69-73; Guggenheim fel, 70-71; prof pharmaceut chem & chem, Univ Calif, San Francisco, 78- *Mem:* Fel AAAS; Am Chem Soc; Am Soc Mass Spectrometry; Am Soc Biochem & Molecular Biol. *Res:* Development of mass spectrometry; elucidation of organic molecular structures and biological macromolecules; applications to biomedical and environmental problems; molecular basis of xenobiotic substances toxicity; sequencing of proteins and study of post-translational modifications; structures of glycoconjugates. *Mailing Add:* Dept Pharmaceut Chem Univ Calif San Francisco CA 94143-0446

BURLINGTON, HAROLD, b Yonkers, NY, June 26, 25. PHYSIOLOGY. *Educ:* Franklin & Marshall Col, BS, 48; Syracuse Univ, MS, 49, PhD(physiol), 51. *Prof Exp:* Asst zool, Syracuse Univ, 48-51; instr physiol, NY State Col Med, Syracuse, 51-55; from asst prof to prof, Col Med, Univ Cincinnati, 55-68; PROF PHYSIOL & BIOPHYSICS, MT SINAI SCH MED, 68- *Mem:* AAAS; Am Physiol Soc; NY Acad Sci. *Res:* Cellular metabolism; regulation of growth and function. *Mailing Add:* Dept Physiol Mt Sinai Sch Med Fifth Ave & 100th St New York NY 10029

BURLINGTON, ROY FREDERICK, b South Bend, Ind, Oct 2, 36; m 59; c 3. ENVIRONMENTAL PHYSIOLOGY. *Educ:* Purdue Univ, BS, 59, MS, 61, PhD(environ physiol), 64. *Prof Exp:* Res physiologist, US Army Res Inst Environ Med, 66-70; PROF BIOL, CENT MICH UNIV, 70- *Mem:* Am Soc Zoologists; Am Physiol Soc. *Res:* Effects of pollution on aquatic invertebrate populations; hypoxic tolerance in and cellular physiology of hibernating mammals; effects of starvation on intermediary metabolism; myocardial metabolism and nutrition at high altitude. *Mailing Add:* Dept Biol Cent Mich Univ Mt Pleasant MI 48859

BURLITCH, JAMES MICHAEL, b Wheeling, WVa. MATERIALS CHEMISTRY, CERAMIC PRECURSORS. *Educ:* Wheeling Col, BS, 60; Mass Inst Technol, PhD(inorg chem), 65. *Prof Exp:* Postdoctoral assoc, Mass Inst Technol, 64-65; asst prof, 65-70, ASSOC PROF INORG CHEM, CORNELL UNIV, 70- *Concurrent Pos:* Alfred P Sloan found fel, 70-72. *Mem:* Am Chem Soc; Am Ceramic Soc. *Res:* Inorganic chemistry of materials; syntheses of ceramics with geological value and technological interest; metal-ceramic adhesion. *Mailing Add:* Dept Chem Baker Lab Cornell Univ Ithaca NY 14853-1301

BURMAN, KENNETH DALE, b St Louis, Mo, Aug 9, 44; m 72; c 5. MEDICINE. *Educ:* Wash Univ, AB, 66; Univ Mo, MD, 70. *Prof Exp:* House officer, Internal Med, Barnes Hosp, Wash Univ Med Ctr, St Louis, Mo, 70-72; endocrine fel, 72-74, staff endocrine, 74-91, CHIEF, ENDOCRINE, WALTER REED ARMY MED CTR, 91- *Concurrent Pos:* Consult nutrit, Surgeon Gen US Army, 75-; prof, Uniformed Serv Univ Health Sci, Bethesda, Md, 77-; chmn, Future Meetings Comt, Endocrine Soc, 88- *Honors & Awards:* Van Meter Award, Am Thyroid Asn, 75; Young Investr Award, Am Fedn Clin Res, 75. *Mem:* Endocrine Soc; Am Thyroid Asn; Am Soc Clin Invest. *Res:* Clinical immunology and molecular biology processes involved in the genesis and propogation of thyroid diseases. *Mailing Add:* Walter Reed Army Med Ctr Washington DC 20307-5001

BURMAN, ROBERT L, b Chicago, Ill, June 13, 33; m 56; c 3. NUCLEAR PHYSICS. *Educ:* Mass Inst Technol, BS, 55; Univ Ill, MS, 57, PhD(physics), 61. *Prof Exp:* NSF fel, Bohr Inst Theoret Physics, Copenhagen, Denmark, 61-62, Ford Found fel, 62-63; asst prof physics, Univ Rochester, 63-68; staff mem, 68-81, FEL, LOS ALAMOS NAT LAB, 81- *Mem:* Fel Am Phys Soc. *Res:* Beta decay; nucleon-nucleon forces; pion-nucleus interactions; neutrino physics; gamma ray astrophysics. *Mailing Add:* Los Alamos Nat Lab MP-Div MS 846 Los Alamos NM 87545

BURMAN, SUDHIR, b Farrukhabad, India, July 17, 55; m 82; c 2. HIGH PERFORMANCE LIQUID CHROMATOGRAPHY. *Educ:* Univ Delhi, BS, 74; Indian Inst Technol, Delhi, MS, 76; Indian Inst Sci, Bangalore, PhD(chem), 81. *Prof Exp:* Postdoctoral chem, Drexel Univ, Philadelphia, 80-82; res assoc, Univ Va, Charlottesville, 82-85 & Cornell Univ Med Col, NY, 85-88; ASSOC SR INVESTR, SMITH KLINE BEECHAM PHARMACEUTICALS, 88- *Concurrent Pos:* Instr, Drexel Univ, 80-82. *Mem:* Am Soc Biochem & Molecular Biol; Am Chem Soc. *Res:* Development of high performance liquid chromatography based analytical methods for the characterization of biopharmaceuticals from recombinant sources; solution properties-aggregation and degradation of proteins; elucidation of primary structure of proteins. *Mailing Add:* Smith Kline Beecham Pharmaceut PO Box 1539 King of Prussia PA 19406

BURMEISTER, HARLAND RENO, b Seymour, Wis, Jan 10, 29; m 68; c 2. MICROBIOLOGY. *Educ:* Univ Wis, BS, 57; Iowa State Univ, PhD(bact), 64. *Prof Exp:* Jr chemist, A O Smith Corp, 58-60; RES MICROBIOLOGIST, NORTHERN UTILIZATION RES & DEVELOP DIV, AGR RES SERV, USDA, 64- *Mem:* Am Soc Microbiol; AAAS. *Res:* Mycotoxin investigations: Fusarium antibiotics and toxins. *Mailing Add:* 3935 N Brookside Dr Peoria IL 61615

BURMEISTER, JOHN LUTHER, b Fountain Springs, Pa, Feb 20, 38; m 60; c 2. INORGANIC CHEMISTRY. *Educ:* Franklin & Marshall Col, BS, 59; Northwestern Univ, PhD(coord chem), 64. *Prof Exp:* Instr inorg chem, Univ Ill, 63-64; from asst prof to assoc prof, 64-69, assoc dir, Ctr Catalytic Sci & Technol, 77-79, PROF CHEM, UNIV DEL, 73-, ASSOC CHMN DEPT, 74- *Concurrent Pos:* Mem ed bd, Inorganica Chimica Acta & Synthesis & Reactivity in Inorg & Metalorg Chem; exec secy, Intercollegiate Student Chemists, 68-71; consult, Sun Oil Co, 69-74, AMP, Inc, 71-73 & Control Data Corp, 81-85. *Honors & Awards:* Catalyst Award, Chem Mfrs Asn, 81. *Mem:* Am Chem Soc (secy-treas, Inorg Chem Div, 75-77); Sigma Xi. *Res:* Coordination chemistry of ambidentate ligands; inorganic linkage isomerism; generation and stabilization of unusual oxidation states; synthesis of metal-ylide complexes. *Mailing Add:* Dept Chem & Biochem Univ Del Newark DE 19716

BURMEISTER, LOUIS C, b Great Bend, Kans, July 20, 35; m 59; c 1. MECHANICAL ENGINEERING, HEAT TRANSFER. *Educ:* Kans State Univ, BS, 57, MS, 59; Purdue Univ, PhD(mech eng), 66. *Prof Exp:* Mech engr, Sandia Corp, NMex, 59-62; from asst prof to assoc prof, 66-76, PROF MECH ENG, UNIV KANS, 76- *Mem:* Am Soc Mech Engrs; Sigma Xi. *Res:* Effect of pressure fluctuations on film boiling; spray production through use of strong, nonuniform electric fields; transient pressure effects on convective heat transfer in a cylinder. *Mailing Add:* 2604 Stratford Rd Lawrence KS 66049

BURMEISTER, ROBERT ALFRED, b Milwaukee, Wis, Dec 28, 39; m 66. MATERIALS SCIENCE, SOLID STATE PHYSICS. *Educ:* Univ Wis, BS, 61; Stanford Univ, MS, 62, PhD(mat sci), 65. *Prof Exp:* Asst mat res, Stanford Univ, 61-65; HEAD MAT RES DEPT, HEWLETT-PACKARD LABS, 65- *Concurrent Pos:* Mem tech staff, Bell Tel Labs, NJ, 63; lectr, Stanford Univ & Univ Calif; mem, Solid State Sci Adv Panel, Nat Res Coun. *Mem:* Am Phys Soc; Am Inst Mining, Metall & Petrol Engrs; Inst Elec & Electronics Engrs; Electrochem Soc. *Res:* Electronic materials of elemental and compound semiconductors, optical and magnetic materials; measurement of optical and electrical transport properties; physical property measurements; solid state devices. *Mailing Add:* Saratoga Technol Assocs 12520 Saratoga Creek Dr Saratoga CA 95070

BURN, IAN, b Dewsbury, Eng, Nov 25, 37; US citizen; m 61; c 2. CERAMICS. *Educ:* Durham Univ, BSc, 60; Leeds Univ, PhD(ceramics), 66. *Prof Exp:* Physicist, Brit Oxygen Co Ltd, 60-62; res asst ceramics, Univ Leeds, 62-66; sr scientist, Pilkington Brothers Ltd, 66-67; sr tech staff ceramics, Sprague Elec Co, 67-83; SR RES ASSOC, DUPONT CO, 83- *Concurrent Pos:* Assoc prof, North Adams State Col, 68-80. *Mem:* Brit Inst Physics; fel Am Ceramic Soc; Int Soc Hybrid Microelectronics. *Res:* Ceramic dielectrics, including interactions with base metals and glass-ceramic systems. *Mailing Add:* Dupont Co Exp Sta PO Box 80334 Wilmington DE 19880-0334

BURNELL, EDWIN ELLIOTT, b St John's, Nfld, Dec 4, 43; m 71; c 3. CHEMICAL PHYSICS, NUCLEAR MAGNETIC RESONANCE. *Educ:* Mem Univ Nfld, BSc, 65, MSc, 68; Bristol Univ, PhD(chem), 70. *Prof Exp:* Fel physics, Univ BC, 69-71 & Univ Basel, 71-72; from asst prof to assoc prof, 78-84, PROF CHEM, UNIV BC, 84- *Concurrent Pos:* Sabbatical leave, Univ Utrecht, 78-79. *Res:* Nuclear magnetic resonance studies of liquid crystals and of molecules partially oriented in liquid crystal solvents; understanding the intermolecular forces leading to orientational order in liquids. *Mailing Add:* Dept Chem 2036 Main Mall Univ BC Vancouver BC V6T 1Y6 Can

BURNELL, JAMES MCINDOE, b Manila, Philippines, July 17, 21; US citizen; m 49; c 5. MEDICINE, PHYSIOLOGY. *Educ:* Stanford Univ, BA, 45, MD, 49. *Prof Exp:* Instr path, Columbia Univ, 49-50; asst med, 50-54, res assoc prof, 60-71, RES PROF MED, UNIV WASH, 71- *Concurrent Pos:* Res fel, Univ Wash & Pfizer Corp, 49-50. *Res:* Physiology acid-base metabolism; bone mineral. *Mailing Add:* 515 Minor Seattle WA 98104

BURNELL, LOUIS A, b Belg, Sept 15, 28; m 55; c 2. THEORETICAL CHEMISTRY. *Educ:* Univ Liege, BSc, 50, PhD(chem), 55. *Prof Exp:* Researcher molecular physics, Univ Liege, 55-60, lectr, 60-63; res physicist, Res Inst Advan Study, 63-66; ASSOC PROF CHEM, NY UNIV, 66- *Concurrent Pos:* Belg govt travel grant, 55; Brit Coun scholar, Oxford Univ, 56-57; Nat Acad Sci grant, Ind Univ & Mass Inst Technol, 59-60; expert, UNESCO, Colombia, 74- *Honors & Awards:* Stas-Spring Prize, Royal Acad Belg, 55. *Mem:* Am Chem Soc. *Res:* Quantum chemistry; interpretation of the electronic spectra of polyatomic molecules; theoretical problems concerned with the nature of the chemical bonds and molecular reactivity; history of science. *Mailing Add:* Cassilla Guayaquil 8551 Equador

BURNELL, ROBERT H, b Tondu, Wales, Nov 23, 29; m 50; c 3. ORGANIC CHEMISTRY. *Educ:* Sir George Williams Univ, BSc, 52; Univ NB, PhD(org chem), 55. *Prof Exp:* Lectr org chem, Univ West Indies, 55-62; investr nat prod, Venezuelan Inst Sci Res, 62-64; PROF CHEM, LAVAL UNIV, 64- *Mem:* Chem Inst Can; The Chem Soc; Am Soc Pharmacog. *Res:* Extraction and isolation of alkaloids from plants; structure elucidation of alkaloids; synthesis of natural products, particularly the group of polyhydroxylated diterpenes. *Mailing Add:* Dept Chem Laval Univ Quebec PQ G1K 7P4 Can

BURNELL, S JOCELYN BELL, b UK, July 15, 43; m 68; c 1. NEUTRON STARS, INFRARED ASTRONOMY. *Educ:* Univ Glasow, BSc, 65; Univ Cambridge, PhD(radio astron), 69. *Prof Exp:* Fel, Univ Southhampton, Eng, 68-70, teaching fel, physics, 70-73; grad programmer, 74-76, assoc res fel, Mullard Space Sci Lab, Univ Col, London, 76-82; sr res fel, 82-86, sr sci officer, 86-89, GRADE 7, ROYAL OBSERV, EDINBURGH, SCOTLAND, 89- *Concurrent Pos:* Tutor, consult, examr, lectr, The Open Univ, UK, 73-87; ed, The Observ, 73-76; mem, Brit Nat Space Ctr, rolling grant assessment panel 86- *Honors & Awards:* Michelson Medal, Franklin Inst, Philadelphia, 73; J Robert Oppenheimer Mem Prize, Ctr for Theort Studies, Miami, 78; Beatrice M Tinsley Prize, Am Astron Soc, 87; Herschel Medal, Royal Astron Soc, London, 89. *Mem:* Royal Astron Soc; Int Astron Union. *Res:* Research in astronomy at many wavelengths - radio to gamma rays; neutron stars. *Mailing Add:* Royal Observ Blackford Hill Edinburgh Scotland

BURNER, ALPHEUS WILSON, JR, b Jacksonville, Fla, Dec 4, 47; m 68; c 2. FLOW VISUALIZATION, CAMERA CALIBRATION. *Educ:* Univ SC, BS, 69; Univ Rochester, MS, 76. *Prof Exp:* PHYSICIST, LANGLEY RES CTR, NASA, 69- *Mem:* Optical Soc Am. *Res:* Holographic flow visualization; video photogrammetry; video camera calibration. *Mailing Add:* NASA Langley Res Ctr MS 236 Hampton VA 23665

BURNESS, ALFRED THOMAS HENRY, b Birmingham, Eng, Feb 10, 34; m 59; c 2. VIROLOGY, MOLECULAR BIOLOGY. *Educ:* Univ Liverpool, BSc, 55, PhD(biochem), 59. *Prof Exp:* Mem staff, Virus Res Unit, Brit Med Res Coun, 58-68; assoc, Sloan-Kettering Inst Cancer Res, 68-73, assoc mem, 74-76; assoc prof molecular biol, 76-79, PROF MOLECULAR VIROL, MEM UNIV NFLD, 79- *Concurrent Pos:* USPHS int res fel, Virus Lab, Univ Calif, Berkeley, 62-63; asst prof biol, Grad Sch Med Sci, Cornell Univ, 71-72, assoc prof, Sloan-Kettering, Div, 74-76; vis fel, biochem dept, Australian Nat Univ, Canberra, 83-84. *Mem:* Brit Soc Gen Microbiol; Brit Biochem Soc; Can Biochem Soc; Am Soc Biol Chem; Am Soc Microbiol; Am Soc Virol. *Res:* Relationship between structure and function in viruses; nature of cell receptors for viruses; structure of eukaryotic cell membranes. *Mailing Add:* Fac Med Mem Univ Nfld St John's NF A1B 3V6 Can

BURNESS, JAMES HUBERT, b Philadelphia, Pa, Nov 20, 49; m 71; c 2. BIOINORGANIC CHEMISTRY. *Educ:* Rutgers Univ, BA, 71; Va Polytech Inst & State Univ, PhD(inorg chem), 75. *Prof Exp:* Fel biophys, Mich State Univ, 75-76; ASST PROF CHEM, PA STATE UNIV, YORK CAMPUS, 76- *Concurrent Pos:* Vis prof, Univ MD, Munich Campus, 83-85; guest researcher, Tech Univ Munich, 83-85. *Honors & Awards:* J Shelton Horsley Res Award, Va Acad Sci; Sigma Xi Res Award, 75. *Mem:* Am Chem Soc; Sigma Xi. *Res:* Model systems for vitamin B-12 and hemoglobin; binding of oxygen to metals; x-ray photoelectron spectroscopy of transition metal complexes; platinum complexes in cancer chemotherapy; inverse gas chromatography. *Mailing Add:* Pa State Univ 1031 Edgecomb Ave York PA 17403

BURNET, GEORGE, b Ft Dodge, Iowa, Jan 30, 24; m 44; c 6. CHEMICAL ENGINEERING. *Educ:* Iowa State Univ, BS, 48, MS, 49, PhD(chem eng), 51. *Prof Exp:* Instr chem eng, Iowa State Univ, 50-51; process engr, Commercial Sovents Corp, Ind, 52-56; prof chem eng, Iowa State Univ, 56-61, chmn dept, 61-78; assoc eng, Ames Lab, Dept Energy, 56-61, chief chem eng div, 61-78, sr engr, 78-90; chmn, Nuclear Eng Dept, 78-83, coordr, Eng Educ Projs Off, 78-90, ASSOC DEAN ENG, IOWA STATE UNIV, 90- *Concurrent Pos:* Consult, Frye Mfg Co, Iowa, 57-60; Int Minerals & Chem Corp, Ill, 63-65 & Agency Int Develop higher educ prog, India, 67; Phillips lectr, Okla State Univ, 70; mem educ & accreditation comt, Engrs Coun Prof Develop, 70-77, chmn, 76-77; mem bd dirs, Engrs Joint Coun, 76-77; chmn, Educ Affairs Coun, Am Asn Eng Soc, 80-82; bd dirs, Accreditation Bd Eng & Technol, 77-83, exec comt, 81-83; US Rep Comt Educ & Training, World Fedn Eng Socs, 84-88. *Honors & Awards:* Iowa Citizen Chemical Engr Award, Iowa Sect, Am Inst Chem Engrs, 70; Founders Award, Am Inst Chem Engrs, 81; Mikol Schmitz Award & Lamme Medal, Am Soc Eng Educ, 82; Grinter Award, Accreditation Bd Eng & Technol, 84. *Mem:* Am Chem Soc; fel Am Inst Chem Engrs; hon mem Am Soc Eng Educ (pres, 76-77); Nat Soc Prof Eng; AAAS. *Res:* Fertilizer technology; coal waste utilization; separations chemistry. *Mailing Add:* 104 Marston Hall Iowa State Univ Ames IA 50011

BURNETT, BRUCE BURTON, b Norristown, Pa, July 26, 27; m 50; c 2. ANALYTICAL CHEMISTRY, POLYMER PHYSICS. *Educ:* Lehigh Univ, BS, 50; Univ Ill, PhD(anal chem), 53. *Prof Exp:* Res chemist, Textile Fibers Dept, E I du Pont de Nemours & Co, 53-60, sr res chemist, 60-62; group leader anal chem, 62-69, sect mgr anal & phys measurements, 69-74, SECT MGR PULP & PAPER, UNION-CAMP CO, 74- *Mem:* AAAS; Am Chem Soc; Tech Asn Pulp & Paper Indust. *Res:* Instrumental methods of analysis; spectroscopy; chromatography; x-ray physics of polymers and fibers; pulping and bleaching chemistry; paper chemistry and physics; crystallization kinetics. *Mailing Add:* Union Camp PO Box 3301 Princeton NJ 08543-3301

BURNETT, BRYAN REEDER, b Bethlehem, Pa, Aug 10, 45; m 75; c 2. MARINE ECOLOGY, CRUSTACEAN CIRCULATORY MORPHOLOGY. *Educ:* San Diego State Univ, BS, 68, MS, 71; Scripps Inst Oceanog, Univ Calif, San Diego, MS, 75. *Prof Exp:* Sr res analyst, Univ Calif, San Diego, 75-82; FORENSIC CONSULT, 82- *Mem:* Am Acad Forensic Sci; Am Micros Soc. *Res:* Analysis of inorganic particulates in lung tissue; analysis of gunshot residue; forensic applications of scanning electron microscopy; energy dispersive x-ray analysis. *Mailing Add:* Meixa Tech PO Box 844 Cardiff CA 92007

BURNETT, CLYDE RAY, b Nora Springs, Iowa, Dec 23, 23; m 47; c 5. PHYSICS. *Educ:* Univ Upper Iowa, BS, 46; Univ Wis, MS, 48, PhD(physics), 51. *Prof Exp:* Asst prof physics, SDak State Col, 50-53 & Pa State Univ, 53-57; physicist, Proj Matterhorn, Princeton Univ, 57-58; assoc prof physics, Pa State Univ, 58-63; res assoc & lectr, Univ Wis, 63-64; assoc prof & chmn dept, 64-70, PROF PHYSICS, FLA ATLANTIC UNIV, 70- *Mem:* Am Phys Soc; Optical Soc Am; Am Meteorol Soc; Am Asn Physics Teachers. *Res:* Atomic spectroscopy; aeronomy. *Mailing Add:* PO Box 59 Rollinsville CO 80474

BURNETT, DONALD STACY, b Dayton, Ohio, June 25, 37; m 59; c 2. GEOCHEMISTRY. *Educ:* Univ Chicago, BS, 59; Univ Calif, Berkeley, PhD(chem), 63. *Prof Exp:* NSF res fel physics, 63-65, asst prof nuclear geochem, 65-68, assoc prof, 68-77, PROF NUCLEAR GEOCHEM, CALIF INST TECHNOL, 77- *Res:* Origin and abundances of the elements; nuclear chemistry. *Mailing Add:* Geol & Planetary Sci Calif Inst of Technol Pasadena CA 91125

BURNETT, EARL, soil science, for more information see previous edition

BURNETT, GEORGE WESLEY, microbiology, dentistry, for more information see previous edition

BURNETT, JAMES R, b Eldorado, Ill, Nov 27, 25; m 47; c 4. ENGINEERING. *Educ:* Purdue Univ, BS, 46, MS, 47, PhD, 49. *Hon Degrees:* DEng, Purdue Univ, 69. *Prof Exp:* Mem fac, Purdue Univ, 52-56; mem staff, TRW Inc, 56-75, vpres & asst gen mgr, defense & space systs group, 75-82, vpres & gen mgr, defense systs group, 82-85, EXEC VPRES & DEP GEN MGR, SPACE & DEFENSE SECTOR, TRW INC, 85- *Concurrent Pos:* Mem selection bd, NSF; consult, Argonne Nat Lab, Dept Defense & other govt agencies. *Honors & Awards:* Charles A Coffin Award. *Mem:* Nat Acad Eng; fel Am Inst Aeronaut & Astronaut; Sigma Xi. *Mailing Add:* TRW Space & Defense Sector One Space Park Redondo Beach CA 90278

BURNETT, JEAN BULLARD, b Flint, Mich, Feb 19, 24; m 47. BIOLOGICAL CHEMISTRY, HISTOPATHOLOGY & ANATOMY. *Educ:* Mich State Univ, 44, MS, 45, PhD(chem & math), 52. *Prof Exp:* Instr math, Mich State Univ, 46-49, 50-52, natural sci, 52- 54, res assoc zool, 54-59, asst prof res, Biol Chem, 59-62; vis prof dermat, Harvard Univ, 62-64, assoc biol chem & dermat, 64-70, prin assoc biol chem & dermat, 70-73, mem fac med, 64-73; assoc biochemist, Mass Gen Hosp, 70-73; assoc prof & asst to chmn dept biochem, Col Osteop Med, 73-82, prof dept anat, 82-84, PROF DEPT ZOOL, COL NATURAL SCI, MICH STATE UNIV, 84- *Concurrent Pos:* Vis prof, Dept Biol, Univ Ariz, Tucson, 79-80. *Mem:* AAAS; Am Chem Soc; Genetics Soc Am; Am Inst Biol Sci; Soc Invest Dermat; Sigma Xi. *Res:* Biochemical genetics; genetic control of protein synthesis; enzymes; biology of melanin; detection, diagnosis, prognosis and treatment of malignant melanoma; identification, physiological role, and mode of action of neurotrophic factors. *Mailing Add:* Dept Zool Natural Sci Bldg Mich State Univ East Lansing MI 48824

BURNETT, JERROLD J, b Mt Pleasant, Tex, May 31, 31; m 53; c 2. RADIATION DETECTION, SCIENCE ADMINISTRATION. *Educ:* Tex A&M Univ, BA, 53; Tex A&I Univ, MS, 59; Univ Okla, PhD(eng sci), 66. *Prof Exp:* Equip engr, Southwest Bell Tel Co, Tex, 55-57; instr physics & math, Univ Dallas, 58-59; res physicist, NMex Inst Mining & Tech, 59-61; assoc prof physics & head dept, Northwestern State Col, Okla, 61-64; from asst prof to assoc prof, 66-75, prof physics, 75-86, EMER PROF, COLO SCH MINES, 86-; MGR, MEASUREMENT STANDARDS LAB, KING FAHD UNIV PETROL & MINERALS, DHAHRAN, SAUDI ARABIA, 89- *Concurrent Pos:* Prog mgr, Grad & Postdoctoral Progs, NSF, 77-78; lab mgr, Optics & Radiation, Measurement Standards, Univ Petrol & Minerals, Dhahran, Saudi Arabia, 82-84; sr engr, Los Alamos Tech Assoc, Hanford, WA, 87-88. *Mem:* AAAS; Am Asn Physics Teachers; Nat Sci Supvrs Asn; Nat Sci Teachers Asn; Sigma Xi. *Res:* Neutron radiography; activation analysis in bore-holes; radiological health; uranium and tritium in water analysis; low energy radiation detection; equipment and system design; metrology; quality assurance. *Mailing Add:* 13590 W Colfax Ave Golden CO 80401

BURNETT, JOHN L, b Wichita, Kans, Aug 28, 32; div; c 2. EXPLORATION & DEVELOPMENT OF ORE DEPOSITS. *Educ:* Univ Calif, Berkeley, AB, 57, MS, 60. *Prof Exp:* GEOLOGIST, CALIF DIV MINES & GEOL, 58- *Concurrent Pos:* Lectr, Univ Calif Exten, 67-73; instr, Cosumnes River Col, 81- *Mem:* Fel Geol Soc Am; Asn Eng Geologists; Soc Mining Metall & Explor. *Res:* Exploration and development of ore deposits using geochemical and other techniques; development of industrial mineral deposits; geological research in northern and eastern California. *Mailing Add:* 8227 Union House Way Sacramento CA 95823

BURNETT, JOHN LAMBE, b Bismarck, NDak, Dec 1, 34; m 57; c 3. INORGANIC CHEMISTRY, PHYSICAL CHEMISTRY. *Educ:* NDak State Univ, BS, 56; Univ Calif, PhD(chem), 64. *Prof Exp:* Chemist, Isotopes Div, Oak Ridge Nat Lab, 65-66 & Chem Div, 66-69; chemist, Div Phys Res, US AEC, 69-75 & Div Phys Res, US Energy Res & Develop Admin, 75-77; nuclear & heavy element chemist, Div Nuclear Sci/BES, 77-79 CHEMIST, DIV CHEMICAL SCI/BES, DEPT ENERGY, 79- *Mem:* AAAS. *Res:* Lanthanide and actinide thermodynamics and inorganic chemistry. *Mailing Add:* Div Chemical Sci/BES Dept Energy ER-142 Washington DC 20545

BURNETT, JOHN NICHOLAS, b Atlanta, Ga, Aug 19, 39. ANALYTICAL CHEMISTRY. *Educ:* Emory Univ, BA, 61, MS, 63, PhD(chem), 65. *Prof Exp:* Res chemist, Org Chem Dept, E I du Pont de Nemours & Co, Del, 65-66; res assoc chem, Univ NC, Chapel Hill, 66-68; from asst prof to prof chem, 68-80, chmn dept, 72-85, assoc dean fac res & develop, 80-85, dir, Lib Arts Prog Technol, 82-85, MAXWELL CHAMBERS PROF CHEM, DAVIDSON COL, 81- *Concurrent Pos:* Vis fel, civil eng dept, Princeton Univ, 85-86. *Honors & Awards:* Thomas Jefferson Award, 80- *Mem:* AAAS; Am Chem Soc; Chem Soc; Sigma Xi; Am Asn Univ Prof; Am Inst Chem Engrs. *Res:* Oxidation reduction processes in biologically important compounds, organometallic complexes and electro-spectrochemical techniques; history of chemical technology. *Mailing Add:* Dept Chem Davidson Col PO Box 238 Davidson NC 28036-0238

BURNETT, JOSEPH W, b Oil City, Pa, Mar 21, 33; m 60; c 3. DERMATOLOGY. *Educ:* Yale Univ, AB, 54; Harvard Med Sch, MD, 58. *Prof Exp:* Intern & asst resident, Johns Hopkins Univ, 58-61; resident dermat, Harvard Med Sch, 61-63; asst prof, 65-69; assoc prof med, 69-76, PROF INTERNAL MED, SCH MED, UNIV MD, BALTIMORE CITY, 76- *Concurrent Pos:* Fel med, Johns Hopkins Univ, 58; fel dermat, Mass Gen Hosp, Boston, 61-63; fel trop pub health, Harvard Med Sch, 63-65. *Mem:* Am Soc Exp Path; Soc Invest Dermat; Am Fedn Clin Res; AMA; Am Acad Dermat. *Res:* Toxins and pathogens of skin. *Mailing Add:* Univ Md Hosp 4401 Roland Ave Baltimore MD 21210

BURNETT, LOWELL JAY, b Portland, Ore, June 15, 41; m 61; c 2. INSTRUMENTATION DEVELOPMENT, SYNTHETIC MEMBRANES. *Educ:* Portland State Univ, BS, 64; Univ Wyo, MS, 67, PhD(physics), 70. *Prof Exp:* Presidential fel chem, Los Alamos Nat Lab, 70-72; from asst prof to assoc prof, 72-78, chmn physics dept, 79- 87, PROF PHYSICS, SAN DIEGO

STATE UNIV, 78-; PRES, QUANTUM MAGNETICS, 87- *Concurrent Pos:* Prin investr numerous grants and contracts, 73-; fac fel, Assoc Western Univs, 73 & 74; mem, int adv panel, Electronics, McGraw-Hill, 74; consult, UOP Inc, 74-85, Gillette Co, 75-76, Los Alamos Nat Lab, 78-85, IRT Corp, 82-83, Sci Applns Inc, 82- & Nat Inst Petrol & Energy Res, 85-86, Quantum Design, 87- *Mem:* AAAS; Am Chem Soc; Am Phys Soc; Sigma Xi. *Res:* Design of nuclear magnetic resonance and microwave instrumentation for remote and local detection applications; development of synthetic polymer membranes for the separation of mixed gas streams. *Mailing Add:* 10937 Quail Canyon Rd El Cajon CA 92021

BURNETT, ROBERT WALTER, b Lima, Ohio, Aug 5, 44. CLINICAL CHEMISTRY, PATHOLOGY INFORMATICS. *Educ:* Purdue Univ, BS, 66; Emory Univ, PhD(chem), 69; Rensselaer Polytech Inst, MS, 85. *Prof Exp:* ASSOC DIR CLIN CHEM LAB, HARTFORD HOSP, 69-, DIR LAB INFO SYSTS, 80-; ASST PROF LAB MED, SCH MED, UNIV CONN, 74- *Concurrent Pos:* Mem, Expert Panel on Blood ph & gases, Int Fedn Clin Chemists, 82- *Mem:* Am Asn Clin Chem; Sigma Xi. *Res:* Reference methods and materials for clinical analyses; high accuracy spectrophotometry; blood pH and gas analysis; applied mathematics; computer applications in laboratory medicine. *Mailing Add:* Dept Path Hartford Hosp Hartford CT 06115

BURNETT, ROGER MACDONALD, b London, Eng, Jan 10, 41; m. PROTEIN CRYSTALLOGRAPHY, VIRUS STRUCTURE. *Educ:* Univ London, BSc, 64; Purdue Univ, PhD(protein crystallog), 70; Univ Basel, Habilitation, 75. *Prof Exp:* Assoc res biophysicist, Biophysics Res Div, Univ Mich, 70-73; vollassistent Biozentrum, Univ Basel, Switzerland, 73-75, privatdozent, 75-80, projektleiter, 77-80; assoc prof biochem, Col Physicians & Surgeons, Columbia Univ, 80-88; PROF, WISTAR INST, 88-; PROF CHEM, UNIV PA, 88- *Concurrent Pos:* Prin investr, Nat Inst Allergy & Infectious Dis, 80-, NSF, 85-; mem, adv panel, Biophysics Prog, Div Molecular Biosci, NSF, 86-90. *Mem:* Am Crystallog Soc; Biophys Soc; NY Acad Sci; Am Asn Advan Sci; Harvey Soc. *Res:* X-ray crystallography to determine the molecular structure of proteins; definition of the architecture, the principles of construction, and the manner of assembly of viruses, in particular adenovirus. *Mailing Add:* Wistar Inst 3601 Spruce St Philadelphia PA 19104-4268

BURNETT, THOMPSON HUMPHREY, b Bethlehem, Pa, July 25, 41; m 64; c 3. EXPERIMENTAL HIGH ENERGY PHYSICS. *Educ:* Univ Calif, Berkeley, AB, 63; Univ Calif, San Diego, PhD(physics), 68. *Prof Exp:* NSF fel, 69; res assoc physics, Princeton Univ, 68-70 & Univ Calif, San Diego, 70-76; from asst prof to assoc prof, 76-85, PROF PHYSICS, UNIV WASH, 85- *Concurrent Pos:* Alfred P Sloan Found fel, 77-79. *Mem:* Am Phys Soc. *Mailing Add:* Dept Physics FM 15 Univ Wash Seattle WA 98195

BURNETT, WILLIAM CRAIG, b Lynn, Mass, Sept 20, 45; m 75; c 1. MARINE GEOCHEMISTRY, CHEMICAL OCEANOGRAPHY. *Educ:* Upsala Col, BS, 68; Univ Hawaii, MS, 71, PhD(geol, geophys), 74. *Prof Exp:* Fel geochem, Dept Earth & Space Sci, State Univ NY, Stony Brook, 74-76; vis scientist, Inst Fisica, Univ Fed Bahia, Salvador, Brazil, 76-77; asst prof, 77-81, ASSOC PROF CHEM OCEANOG, FLA STATE UNIV, 81- *Concurrent Pos:* Co-proj leader, UNESCO Phosphate Proj, Int Geol Correlation Proj 156, 84-; res fel, E-W Ctr Resources Systs Inst, 79; various res grants from fed & state agencies. *Mem:* Geochem Soc; Am Geophys Union; Geol Soc Am; AAAS. *Res:* Uranium-series isotopes in authigenic mineral deposits of the sea floor and natural waters; origin and geochemistry of phosphate deposits; water quality studies in marine and fresh water systems. *Mailing Add:* Dept Oceanog Fla State Univ Tallahassee FL 32306

BURNETTE, MAHLON ADMIRE, III, b Lynchburg, Va, May 18, 46; m 84; c 2. FOOD SCIENCE, NUTRITION. *Educ:* Va Polytech Inst, BS, 68; Rutgers Univ, MPhil, 73, PhD(nutrit), 74. *Prof Exp:* Dir sci affairs, Grocery Mfrs Am, Inc, 74-80; exec officer, Nat Nutrit Consortium, 80-81, League Int Food Educ, 82-83; PRES, ARNON FOOD POLICY, 83- *Mem:* Inst Food Technologists; AAAS; Asn Food & Drug Officials; Sigma Xi. *Res:* Population nutrition; food safety; federal food and drug regulations. *Mailing Add:* 631 Walker Rd Great Falls VA 22066

BURNEY, CURTIS MICHAEL, b Beatrice, Nebr, Sept 26, 47; m 72; c 3. CHEMICAL ECOLOGY, MICROBIAL ECOLOGY. *Educ:* Nebr Wesleyan Univ, BS, 69; Univ RI, MS, 76, PhD(oceanog), 80. *Prof Exp:* Res asst, Grad Sch Oceanog, Univ RI, 74-80; res assoc, 80-81, asst prof, 81-90, ASSOC PROF, OCEANOG CTR, NOVA UNIV, 90- *Concurrent Pos:* Prog co-dir, Inst Marine & Coastal Studies, Nova Univ. *Mem:* Am Soc Microbiol; Am Soc Limnol & Oceanog; Sigma Xi. *Res:* Dissolved organic matter in natural waters, its nature, dynamics and interactions with the biota. *Mailing Add:* Oceanog Ctr Nova Univ 8000 N Ocean Dr Dania FL 33004

BURNEY, DONALD EUGENE, b Hartington, Nebr, Oct 17, 15; m 42; c 4. INDUSTRIAL ORGANIC CHEMISTRY. *Educ:* Univ SDak, BA, 37; Univ Ill, PhD(org chem), 41. *Prof Exp:* Res chemist, Amoco Chem Corp, 41-47, group leader, 47-57, sect leader, 57-65, div dir, 65-69, mgr org chem res & develop, 69-71, exec dir, Amoco Found, Inc, 71-80; RETIRED. *Mem:* Am Chem Soc. *Res:* Chemicals from petroleum; propylene polymerization to make synthetic lubricating oils; synthetic detergents; reactions of 2, 8-dihydoxynaphtaldehyde; oxidation of hydrocarbons-production of aromatic polybasic acids; polymerization of olefins to thermoplastics. *Mailing Add:* 1005 Anne Rd Naperville IL 60540

BURNHAM, ALAN KENT, b Decorah, Iowa, Oct 10, 50; m 70; c 3. PHYSICAL CHEMISTRY. *Educ:* Iowa State Univ, BS, 72; Univ Ill, Urbana, PhD(phys chem), 77. *Prof Exp:* Jr chemist, Ames Lab, 72; CHEMIST, LAWRENCE LIVERMORE NAT LAB, UNIV CALIF, 77- *Concurrent Pos:* Ed Bd, Energy & Fuels, 87-; sci ed, Energy & Technol Rev, 89-90. *Mem:* Am Phys Soc; Am Chem Soc, Geochem Div (secy, 91-). *Res:* Petroleum formation; chemistry of oil-shale retorting and fabrication of laser fusion targets; electro-optics; dielectric and structural properties of liquids; analysis of trace organics in water; kinetics of carbon gasification and organic pyrolysis reactions. *Mailing Add:* Lawrence Livermore Nat Lab Univ Calif Livermore CA 94551-0808

BURNHAM, BRUCE FRANKLIN, b New York, NY, Nov 12, 31; m 55; c 2. BIOCHEMISTRY. *Educ:* Univ Utah, BS, 53, MS, 54; Univ Calif, Berkeley, PhD(biochem), 60. *Prof Exp:* Nat Found fel, Nobel Med Inst, Sweden, 60-61; Jane Coffin Childs Mem Fund fel med res, Oxford Univ, 61-62, sr fel, C F Kettering Res Lab, Yellow Springs, Ohio, 62-63; vis asst prof chem, Cornell Univ, 63-64, asst prof, 64-66; asst prof med & biochem, Univ Minn, 66-68; from assoc prof to prof chem, Utah State Univ, 68-74, chmn div biochem, 69-74; PRES, PORPHYRIN PRODS, 74- *Concurrent Pos:* Vis prof chem, Univ Calif, Los Angeles, 72-73. *Mem:* Biochem Soc; Am Soc Biol Chem; Am Chem Soc. *Res:* Enzymology; microbiology; metal ion incorporation into poryphyrins and corrins; control mechanisms of tetraphrrole biosynthesis. *Mailing Add:* PO Box 31 Logan UT 84321-0031

BURNHAM, CHARLES WILSON, b Detroit, Mich, Apr 6, 33; m 58; c 2. MINERALOGY. *Educ:* Mass Inst Technol, SB, 54, PhD(mineral, petrol), 61. *Hon Degrees:* AM, Harvard Univ, 66. *Prof Exp:* Fel, Geophys Lab, Carnegie Inst, 61-63, petrologist, 63-66; assoc prof, 66-69, PROF MINERAL, HARVARD UNIV, 69- *Concurrent Pos:* Assoc ed, Am Mineralogist, 74-76; vis scientist, Los Alamos Nat Lab, 81, 88, 90; vis fel, Clare Hall, Univ Cambridge, 87-88. *Mem:* AAAS; fel Mineral Soc Am (vpres, 88, pres, 89); Am Crystallog Asn; Am Geophys Union. *Res:* Determination and refinement of the crystal structures of minerals; theoretical and experimental aspects of relationships between crystal structures, crystal chemistry and phase relations of minerals in natural systems. *Mailing Add:* Harvard Univ Hoffman Lab 20 Oxford St Cambridge MA 02138

BURNHAM, DONALD C, b Athol, Mass, Jan 28, 15; m 37; c 5. MECHANICAL, INDUSTRIAL & MANUFACTURING ENGINEERING. *Educ:* Purdue Univ, BS, 36. *Hon Degrees:* DEng, Purdue Univ, 59, Ind Inst Technol, 52, 63, Drexel Inst Technol, 64 & Polytech Inst Brooklyn, 67. *Prof Exp:* With Gen Motors Corp, 36-54, asst chief engr, Oldsmobile Div, 53-54; vpres mfg, Westinghouse Elec Corp, 54-62, head, Div Indust Prod, 62-63, pres, 63-69, chmn, 69-75, dir officer, 76-80; RETIRED. *Concurrent Pos:* Dir, Mellon Nat Bank & Trust Co & Logistics Mgt Inst; vchmn, Nat Comt Productivity & Work Qual, 74-75; dir, Nat Ctr Productivity & Qual of Work Life, 75-76 & Am Productivity Work Qual, 74-75. *Honors & Awards:* Richards Mem Award, Am Soc Mech Engrs, 58; Outstanding Achievement in Mgt Award, Am Inst Indust Engrs, 64; Hoover Medal, 78; Nat Engr Award, Am Asn Engrs Soc, 81. *Mem:* Nat Acad Eng; Am Soc Mech Engrs; Soc Automotive Eng; Inst Elec & Electronics Engrs; Am Soc Metals; Inst Indust Engrs. *Res:* Industrial production; productivity improvement. *Mailing Add:* 615 Osage Rd Pittsburgh PA 15243

BURNHAM, DONALD LOVE, b Lebanon, NH, Dec 6, 22; m 46; c 3. PSYCHIATRY. *Educ:* Dartmouth Col, BA, 43; Cornell Univ, MD, 46. *Prof Exp:* From resident psychiat to dir res, Chestnut Lodge, Md, 50-63, Ford Found grant, 57-63; res psychiatrist, Div Clin & Behav Res, HIMH, 63-82; RETIRED. *Concurrent Pos:* From instr to pres, Wash Psychoanal Inst, 56-; trustee, William Alanson White Psychiat Found, 62-; ed, J Psychiat, 62-85. *Mem:* Am Psychoanal Asn; fel Am Psychiat Asn. *Res:* Personality development and mental illness, particularly schizophrenia, thought and language, and psychotherapy; psycho-biographic study of August Strindberg. *Mailing Add:* 5003 Edgemoor Lane Bethesda MD 20814

BURNHAM, DWIGHT COMBER, b Macomb, Ill, Mar 17, 22; m 47; c 3. ELECTROGRAPHIC TECHNOLOGY, INVESTMENT MANAGEMENT. *Educ:* Iowa State Univ, BS, 43; US Mil Acad, BS, 46; Univ Ill, MS, 56, PhD(physics), 60. *Prof Exp:* Teaching & res asst physics, Univ Ill, 54-59; physicist, Mil Photog Dept, Kodak Res Labs, Eastman Kodak Co, NY, 59-61 & Solid State Phys Dept, 61-67, prod planning specialist, Bus Systs Mkt Div, 67-73, prod planning coordr, Bus Systs Mkt Div, 73-81, res assoc develop, Electrographic Technol Div, 81-86; RETIRED. *Concurrent Pos:* Investment Mgt, Burney Co, 86-88. *Mem:* Am Phys Soc. *Res:* Ionic crystals; emphasis on magnetic resonance techniques and problems; computer-microfilm information systems; electrographic technology. *Mailing Add:* 42 Creek Ridge Pittsford NY 14534-4404

BURNHAM, JEFFREY C, b Beverly, Mass, Aug 6, 42; m 64; c 2. MICROBIOLOGY. *Educ:* Dartmouth Col, AB, 64; Univ NH, PhD(microbiol), 67. *Prof Exp:* USPHS fel, Univ Ky, 67-69; from asst prof to assoc prof, 69-84, actg chmn dept, 77-78, PROF MICROBIOL, MED COL OHIO, 85- *Concurrent Pos:* Adj prof biol, Bowling Green State Univ, 72- *Honors & Awards:* Pres Award, Ohio Pub Health Asn. *Mem:* AAAS; Am Soc Microbiol; Electron Micros Soc Am; Phycol Soc Am; Can Soc Microbiol. *Res:* Relationship between microbial cell function and structure; bacterial control of algae populations; microbiol adherence to epithelial cells; water pollution microbiology; electron cytochemistry; infectious diseases. *Mailing Add:* Dept Microbiol Med Col Ohio CS 10008 Toledo OH 43699

BURNHAM, KENNETH DONALD, b Chicago, Ill, Aug 10, 22; m 81; c 1. ZOOLOGY. *Educ:* Roosevelt Univ, BS, 48; Univ Iowa, MS, 51, PhD(zool), 57. *Prof Exp:* Teacher & asst prin pub schs, 48-49; asst zool, Univ Iowa, 50-53, zool & biol, 55-57; instr, Calif State Polytech Col, 53-55; from asst prof to assoc prof zool, Southeast Mo State Col, 57-63, inter-Am fel trop med & parasitol, Caribbean, 63; asst prof biol, Ball State Univ, 63-66; UNESCO prof biol, Brazil, 73; assoc prof zool, Univ Tenn, Knoxville, 67-72, 74-; RETIRED. *Concurrent Pos:* NSF res grant, Inst Marine Biol, Univ Ore, 60; consult biol, Rensselaer Polytech Inst, 63; AAAS Chautaugua courses biol & human affairs, Clark Col, Atlanta, Ga, 71, genetics & societal probs, 74; instituted Univ Tenn trop & marine biol course, Jamaica; instituted Bioethics, 74-; instr, Int Student Orientation Camp, 81-; pre-health adv, Lib Arts Adv Ctr, 75- *Mem:* Am Inst Biol Sci; Nat Asn Biol Teachers; Inst Soc, Ethics & Life Sci; Sigma Xi. *Res:* Bioethics; improvement of teaching biology; environmental ethics. *Mailing Add:* 4816 Shady Dell Trail Knoxville TN 37914

BURNHAM, MARVIN WILLIAM, b Wichita, Kans, May 15, 25; m 54; c 2. ENGINEERING MECHANICS, MANUFACTURING PRODUCTIVITY. *Educ:* Univ Kans, BS, 50; Univ Ark, MS, 57; Univ Colo, PhD(mech eng), 66. *Prof Exp:* Instr mech eng, Univ Ark, 54-57; asst prof, Univ Denver & res engr, Denver Res Inst, 57-62; sr res engr, Falcon Res & Develop Co, 62-67; sr res engr, Dow Chem Co, 67-75; sr res engr, Rockwell Int, 75-80, assoc scientist, 80-87; CONSULT, 87- *Concurrent Pos:* Chmn, Soc Mfg Engrs Nat Mat Removal Coun, 77-80; chmn, Rockwell Corp Task Group, 84-87. *Honors & Awards:* Citation Prof Achievement, Soc Mfg Engrs, 77, Award of Merit, 85; Award of Merit, Calif Coun Bus Prof Asn, 87. *Mem:* Soc Exp Stress Anal; NY Acad Sci; fel Soc Mfg Engrs; Sigma Xi. *Res:* Solid mechanics; elasticity; shell theory, energy conversion; experimental mechanics; mechanics of metal cutting; high accuracy manufacturing and metrology; manufacturing automation; product design innovation and improvement; remote manufacturing. *Mailing Add:* 4705 Ricara Dr Boulder CO 80303

BURNHAM, ROBERT DANNER, b Havre de Grace, Md, Mar 21, 44; m 65; c 3. SEMICONDUCTORS. *Educ:* Univ Ill, BS, 66, MS, 68, PhD(elec eng), 71. *Prof Exp:* Teaching asst, Univ Ill, 69-70, res asst, 66-71; res staff, Palo Alto Res Ctr, Xerox, Calif, 71-77, mat res specialist, 77-80, prin scientist, 80-83, res fel, 83-86; sr res assoc, Amoco Res Ctr, 86-89, TECH DIR, AMOCO TECHNOL CO, NAPERVILLE, ILL, 89- *Concurrent Pos:* Edward J James Scholar, 62-66, NDEA fel, 66-69, Gen Tel & Electronics, 69-71. *Honors & Awards:* Jack A Morton Award, Inst Elec & Electronics Engrs, 85. *Mem:* Nat Acad Eng; fel Inst Elec & Electronics Engrs; Metall Soc; fel Optical Soc Am. *Res:* Growth and fabrication of gallium aluminum arsenide; quantum well heterostructures with emphasis on lasers for optoelectronic integration, using growth technique of metalorganic chemical vapor deposition; author or co-author of over 355 publications and holder of 75 patents; high power cw diode laser. *Mailing Add:* Amoco Technol Co PO Box 3011 MS F-4 Warrenville Rd & Mill St Naperville IL 60566

BURNHAM, THOMAS K, b Berlin, Ger, June 6, 27; m 52; c 4. DERMATOLOGY, IMMUNOLOGY. *Educ:* Univ London, MB, BS, 52; Am Bd Dermat, dipl, 63. *Prof Exp:* Staff physician dermat, Henry Ford Hosp, 62-86, dir dermat res lab, 68-86; CONSULT, 86- *Honors & Awards:* Gold Medal, Am Acad Dermat, 66. *Mem:* Am Acad Dermat; AMA; Am Dermat Asn. *Res:* Auto-immunity; antinuclear factors; fluorescent antibody; evaluation of the diagnostic and prognostic aspects of antinuclear antibodies detected by indirect immunofluorescence, being analyzed now immunologically by other immunologic techniques to determine responsible nuclear antigens; discovered Lupus "Band". *Mailing Add:* 962 Santa Helena Park Court Solana Beach CA 92075

BURNISON, BRYAN KENT, b Great Falls, Mont, Feb 26, 43; m 71; c 3. MICROBIAL ECOLOGY. *Educ:* Mont State Col, BS, 65; Ore State Univ, MS, 68, PhD(microbiol), 71. *Prof Exp:* head, Nutrient Pathways Sect, 79-86, RES SCIENTIST MICROBIAL ECOL, CAN CTR INLAND WATERS, 73-, RES SCIENTIST, 87- *Concurrent Pos:* Fel, Univ BC, 71-73. *Mem:* Am Soc Limnol & Oceanog. *Res:* Bacterial decomposition of organic contaminants in the lake environment; improving techniques for the measurement of microbial and productivity, and in situ metabolic activities; organic colloid chemistry in lake waters - isolation, purification and chemical composition. *Mailing Add:* Lakes Res Br/NWRI Can Ctr Inland Waters Burlington ON L7R 4A6 Can

BURNISTON, ERNEST EDMUND, b Sheffield, Eng, Oct 26, 37; m 59; c 3. APPLIED MATHEMATICS. *Educ:* Univ London, BSc, 60, PhD(math), 62. *Prof Exp:* Lectr math, Univ London, 62-65; from asst prof to assoc prof, 65-72, PROF MATH, NC STATE UNIV, 72-, HEAD DEPT, 80- *Mem:* Soc Indust & Appl Math; Math Asn Am. *Res:* Fracture mechanics; transport theory. *Mailing Add:* Dept Math NC State Univ Raleigh NC 27695-8205

BURNS, ALLAN FIELDING, b Washington, DC, Nov 22, 36; m 58; c 4. SILICAS, SILICATES. *Educ:* Cornell Univ, BS, 58; Purdue Univ, MS, 60, PhD(soil chem), 62. *Prof Exp:* Res chemist silicates, Johns-Manville Res & Eng Ctr, Johns-Manville Corp, 62-66, sr res chemist, 66-69, sect chief basic chem, 69-71; chemist pollution abatement, US Army Picatinny Arsenal, 72-73; tech mgr silica & silicates, Philadelphia Quartz Co, 73-77, tech mgr specialty chemicals, 77-78, mgr corp planning, 78-82, mgr mkt develop, 82-85, mkt mgr, 85-89, ASSOC DIR RES & DEVELOP, PQ CORP, 89- *Mem:* Am Chem Soc; Indust Res Inst. *Res:* Clay chemistry; structural and surface modification of clays; chemistry of soluble alkali and organic ammonium silicates; amorphous synthetic and natural silicas; synthetic insoluble silicates; zeolites. *Mailing Add:* PQ Corp Res & Develop Ctr 280 Cedar Rd Conshohocken PA 19428-2240

BURNS, BRUCE PETER, b Mt Vernon, NY, Feb 11, 42; m 68; c 2. COMPOSITE MATERIALS & APPLICATIONS, TRANSIENT STRUCTURAL ANALYSIS. *Educ:* Drexel Inst Technol, BS, 64, MS, 66; Drexel Univ, PhD(appl mech), 69. *Prof Exp:* Postdoctoral, Drexel Univ, 69-70; officer, US Army Ballistic Res Lab, 70-72, mech engr, 72-77, asst to dir, 77-78, mech engr & team leader, 78-82, SUPV GEN ENGR, US ARMY BALLISTIC RES LAB, 82- *Concurrent Pos:* Adj prof mech engr & short course coordr, Drexel Univ, 90- *Honors & Awards:* R H Kent Award. *Res:* Basic and applied research pertaining to the structural performance of projectiles and guns subjected to launch forces; structural behavior of complicated projectiles; author of more than 50 publications; awarded nine patents. *Mailing Add:* 309 Windsor Ct Churchville MD 21028

BURNS, C(HARLES) M(ICHAEL), b Detroit, Mich, June 7, 38; Can citizen; m 63; c 2. CHEMICAL ENGINEERING, POLYMER SCIENCE. *Educ:* Univ Toronto, BASc, 61, MASc, 62; Polytech Inst Brooklyn, PhD(chem), 67. *Prof Exp:* From asst prof to assoc prof, 67-84, assoc chmn undergrad studies chem eng, 84-87, PROF CHEM ENG, UNIV WATERLOO, 84- *Concurrent Pos:* Consult, Steel Co Can, 69-75 & Dominion Chain Co, 80-82. *Mem:* Am Chem Soc; Chem Inst Can; Can Soc Chem Eng. *Res:* Correlation of polymer structure and properties; synthesis and characterization of new monomers and polymers; characterization of copolymers and mixed polymer systems; polymer-polymer compatability; adhesion of polymers; gas permeation; polyblends. *Mailing Add:* Dept Chem Eng Univ Waterloo Waterloo ON N2L 3G1 Can

BURNS, CHESTER RAY, b Nashville, Tenn, Dec 5, 37; m 62; c 2. HISTORY OF MEDICINE. *Educ:* Vanderbilt Univ, BA, 59, MD, 63; Johns Hopkins Univ, PhD(hist med), 69. *Prof Exp:* Asst prof hist, 69-71, from James Wade Rockwell asst prof to assoc prof, 71-79, asst prof to assoc prof, Dept Prev Med & Community Health, 74-79, JAMES WADE ROCKWELL PROF HIST MED & PROF, DEPT PREV MED & COMMUNITY HEALTH, UNIV TEX MED BR, GALVESTON, 79-, MEM INST MED HUMANITIES, 74- *Concurrent Pos:* Dir, Hist Med Div, Univ Tex Med Br, Galveston, 69-74, assoc dir, Inst Med Humanities, 74-80, assoc, Grad Sch Biomed Sci, 74-77, mem, 79- *Mem:* Am Asn Hist Med; Soc Health & Human Values (pres, 75-76); Int Soc Hist Med; Hist Sci Soc; Am Hist Asn. *Res:* History of medicine, especially medical education, medical science, medical ethics and medical jurisprudence. *Mailing Add:* Dept Prev Med-Pub Health Univ Tex Med Sch 301 University Blvd Galveston TX 77550

BURNS, DANIEL ROBERT, b Boston, Mass, Sept 25, 55; m 78; c 2. SEISMIC WAVE PROPAGATION, BOREHOLE GEOPHYSICS. *Educ:* Bridgewater State Col, BA, 77; Colo Sch Mines, MS, 79; Mass Inst Technol, PhD, 86. *Prof Exp:* Geophysicist, Union Oil Co Calif, 79-83; sr res geophysicist, Unocal Sci & Technol Div, 87-88; ASST SCIENTIST, WOODS HOLE OCEANOG INST, 88-; EXEC VPRES, NEW ENG RES, INC, 89- *Concurrent Pos:* Staff scientist, Arthur D Little, Inc, 80-81; vis scientist, Mass Inst Technol, 88-; prin investr, NSF, 89-91, Off Naval Res & Off Naval Technol, 89-, Gas Res Inst, 90-; mem assoc panel, Nat Res Coun, 91- *Mem:* Soc Explor Geophysicists; Am Geophys Union. *Res:* Numerical modeling of elastic wave propagation; borehole acoustics; property estimation from geophysical data; in-situ permeability estimation; subsurface characterization for energy, environmental and engineering applications. *Mailing Add:* New Eng Res Inc-Geosci 1100 Crown Colony Dr Quincy MA 02169

BURNS, DAVID JEROME, b Hobart, La, July 13, 22; m 46; c 2. AGRICULTURAL ECONOMICS, RESOURCE ECONOMICS. *Educ:* Univ Md, BS, 48, MS, 49, PhD(agr econ), 54. *Prof Exp:* From instr to asst prof agr econ, Univ Md, 51-56; assoc prof, 56-60, PROF AGR ECON, RUTGERS UNIV, NEW BRUNSWICK, 60-, ACAD ADMINR, 80- *Mem:* Am Agr Econ Asn. *Res:* Marketing of fruits and vegetables and land use planning. *Mailing Add:* 21 Hale St New Brunswick NJ 08901

BURNS, DENVER P, b Bryan, Ohio, Oct 27, 40; m 65; c 1. RESEARCH ADMINISTRATION. *Educ:* Ohio State Univ, BSc, 62, MSc, 64, PhD(entom), 67; Harvard Univ, MPA, 81. *Prof Exp:* From asst entomologist, 62-68, res entomologist, 68-72, asst dir, Southern Forest Exp Sta, 72-74, staff asst to dep chief for res, US Forest Serv, 74-76, dep dir, NCent Exp Sta, 76-81, DIR NORTHEASTERN FOREST EXP STA, 81- *Mem:* AAAS; Soc Am Foresters. *Res:* Research administration; insect-host tree interactions; biology and bionomics of insects; biology, ecology and behavior of borers attacking living hardwoods. *Mailing Add:* Five Radnor Corp Ctr 100 Matsonford Rd Suite 200 Radnor PA 19087

BURNS, DONAL JOSEPH, b Belfast, Ireland, Mar 5, 41; m 72; c 2. ATOMIC PHYSICS. *Educ:* Queen's Univ, Belfast, BSc, 62, PhD(physics), 65. *Prof Exp:* Lectr physics, Queen's Univ, Belfast, 65-68; res assoc, Univ Nebr, Lincoln, 68, Univ Res Coun jr fac fel, 69, from asst prof to assoc prof, 68-76, assoc dean arts & sci, 77-82, dir, Univ Studies Prog, 80-86, asst exec vpres & provost, 86-89, PROF PHYSICS, UNIV NEBR, LINCOLN, 76-, ASSOC EXEC VPRES & PROVOST, 89- *Concurrent Pos:* Vis scientist, Meudon Observ, France, 64. *Mem:* Am Phys Soc; fel Inst Physics & Phys Soc. *Res:* Observations on forbidden transitions; measurement of atomic and molecular excitation cross sections; lifetimes of excited states; coherent beam foil spectroscopy; atomic alignment and orientation in coincidence experiments. *Mailing Add:* 111 Varner Hall Univ Nebr 3835 Holdrege St Lincoln NE 68583-0743

BURNS, EDWARD EUGENE, b Ft Wayne, Ind, Apr 13, 26; m 48; c 4. FOOD SCIENCE. *Educ:* Purdue Univ, BS, 50, MS, 52, PhD(food technol), 56. *Prof Exp:* Asst, Purdue Univ, 47-50, instr, 52-56; PROF HORT, TEX A&M UNIV, 56- *Honors & Awards:* Cruess Award, Inst Food Technol, 72. *Mem:* Am Hort Soc; Sigma Xi; Inst Food Technol; Am Soc Hort Sci. *Res:* Quality control techniques and methods; chemical and physiological indicators of quality; human foods of plant origin. *Mailing Add:* 1203 Park Pl College Station TX 77840

BURNS, EDWARD ROBERT, b Catskill, NY, Nov 6, 39; m 60; c 2. MICROSCOPIC ANATOMY, EXPERIMENTAL EMBRYOLOGY. *Educ:* Hartwick Col, BA, 61; Univ Maine, MS, 63; Tulane Univ, PhD(anat), 67. *Prof Exp:* From instr to assoc prof, 68-79, PROF ANAT, MED COL, UNIV ARK, LITTLE ROCK, 79- *Concurrent Pos:* NIH fel exp path, George Washington Univ, 67-68; NIH res career develop award, 74-79. *Mem:* AAAS; Am Asn Anatomists; Am Soc Zoologists; Int Soc Chronobiol; Am Asn Cancer Res; Sigma Xi. *Res:* Experimental oncology; control of growth and differentiation in normal and neoplastic tissues; chronochemotherapy; chronobiology of neoplasia. *Mailing Add:* 48 Kingspark Rd Little Rock AR 72207

BURNS, ELIZABETH MARY, NEUROSCIENCE, AGING. *Educ:* Univ Colo, PhD(physiol), 72. *Prof Exp:* PROF BIOPHYSIOL, COL NURSING, UNIV IOWA, 82- *Res:* Effects of alcohol on the developing brain. *Mailing Add:* Dept Life Span Process Ohio State Univ 1585 Neil Ave Columbus OH 43210

BURNS, ERSKINE JOHN THOMAS, b Calgary, Alta, Oct 23, 44; m 82. NUCLEAR PHYSICS. *Educ:* Occidental Col, BA, 66; Calif State Univ, Los Angeles, MS, 68; Univ Calif, Davis, PhD(physics), 71. *Prof Exp:* Res physicist, 71-75, chief, X-ray Diag Group, Air Force Weapons Lab, 75-76; MEM TECH STAFF, SANDIA LABS, 76- *Mem:* Am Phys Soc. *Res:* Ion sources for nuclear fusion. *Mailing Add:* Orgn 2564 Sandia Labs Albuquerque NM 87185

BURNS, FRANK BERNARD, b Oak Park, Ill, July 11, 28. ANALYTICAL CHEMISTRY, PHYSICAL ORGANIC CHEMISTRY. *Educ:* Calif State Univ, Long Beach, BS, 72; Brown Univ, PhD(phys org chem), 76. *Prof Exp:* STAFF MEM ANALYTICAL CHEM, SANDIA LABS, 76- *Mem:* Am Chem Soc; Am Inst Physics; Am Soc Spectronomy. *Res:* Develop quantitative parallel axis gaschromatograph. *Mailing Add:* Sandia Labs Orgn 7343 Bldg 892 Rm 1040 Albuquerque NM 87185

BURNS, FRED PAUL, b New York, NY, Dec 27, 22; m 47; c 2. PHYSICS. *Educ:* City Col New York, BME, 47; Columbia Univ, PhD(physics), 54. *Prof Exp:* Asst prof mech eng, City Col New York, 47-54; mem tech staff, Bell Tel Labs, 54-57; mgr silicon rectifier develop, Tung Sol Elec, 57-59; mgr indust transistor design, Radio Corp Am, 59-60; mgr semiconductor devices, Solid State Radiations, Inc, 60-62; mgr opers, Korad Corp, 62-68; pres, Apollo Lasers Inc, 68-86; CONSULT, 86- *Mem:* Am Phys Soc; Inst Elec & Electronics Engrs. *Res:* Semiconductor device development; germanium silicon transistors and rectifiers; solid state physics; lasers systems development and interaction of high intensity light with solids. *Mailing Add:* 101 California Ave Apt 904 Santa Monica CA 90403

BURNS, FREDRIC JAY, b Wilmington, Del, Apr 20, 37; m 62; c 2. TOXICOLOGY. *Educ:* Harvard Col, AB, 59; Columbia Univ, MA, 61; NY Univ, PhD(biol), 67. *Prof Exp:* Asst prof, 69-74, assoc prof, 75-80, PROF ENVIRON MED, MED CTR, NY UNIV, 80- *Concurrent Pos:* NIH fel, Inst Cancer Res, Sutton, Eng, 67-69. *Mem:* Radiation Res Soc; Am Asn Cancer Res. *Res:* Cell population kinetics; radiation and chemical carcinogenesis in skin and liver; control of cell division; cell cycle models. *Mailing Add:* Inst Environ Med NY Univ Med Ctr 550 First Ave New York NY 10016

BURNS, GEORGE, b Russia, Apr 6, 25; US citizen; c 2. CHEMISTRY. *Educ:* Columbia Univ, BS, 51; Princeton Univ, PhD(chem), 61. *Prof Exp:* Chemist, Gen Elec Co, 52-56; Nat Acad Sci-Nat Res Coun fel phys chem, Cambridge Univ, 61-62; from asst prof to prof, 62-71, assoc chmn dept, 74-75, PROF CHEM, UNIV TORONTO, 71- *Mem:* Am Phys Soc; Am Chem Soc. *Res:* theoretical and experimental gas phase chemical kinetics; recombination-dissociation reactions, theory and experiment; environmental archaeological chemistry. *Mailing Add:* Dept Chem Lash Miller Chem Lab Univ Toronto Toronto ON M5S 1A1 Can

BURNS, GEORGE ROBERT, b Lineville, Ala, Nov 12, 31; m 54; c 3. SOIL FERTILITY. *Educ:* Auburn Univ, BS, 54; NC State Univ, MS, 56; Iowa State Univ, PhD(soil fertil), 61. *Prof Exp:* Res soil scientist, US Soil Lab, 62-64, asst to dir agr res, Sulpher Inst, 64-67, br chief, Soil & Water Conserv Res Div, 67-72, AREA DIR, SOUTHERN REGION, USDA, 72- *Mem:* Am Soc Agron; Soil Sci Soc Am; Soil Conserv Soc Am; Sigma Xi. *Res:* Chemistry of soil and plant interaction; chemistry of nitrogen and phosphorous in soil. *Mailing Add:* PO Box 5677 Athens GA 30613

BURNS, GEORGE W, b Cincinnati, Ohio, Nov 20, 13; m 42; c 3. GENETICS. *Educ:* Univ Cincinnati, AB, 37; Univ Minn, PhD(bot), 41. *Prof Exp:* Teaching fel bot, Univ Minn, 37-41, instr, 46; from asst prof to prof, Ohio Wesleyan Univ, 46-79, chmn dept, 54-70, actg vpres & dean, 57-60, actg pres, 58-59, vpres & dean, 60-61, EMER PROF BOT, OHIO WESLEYAN UNIV, 79- *Concurrent Pos:* Head insts sect, NSF, 61-62. *Mem:* Fel AAAS; Am Genetic Asn; Am Soc Human Genetics; Bot Soc Am; Sigma Xi. *Res:* Pleistocene flora. *Mailing Add:* 354 Troy Rd Delaware OH 43015-1010

BURNS, GERALD, b New York, NY, Oct 5, 32; m 54; c 2. PHYSICS. *Educ:* Rensselaer Polytech Inst, BS, 54; Columbia Univ, AM, 57, PhD, 62. *Prof Exp:* Res engr, Cornell Aerodyn Labs, 54; asst physics, Watson Labs, 54-57; RES PHYSICIST, RES LABS, IBM CORP, 57- *Mem:* Am Phys Soc; Inst Elec & Electronics Engrs. *Res:* Solid state physics. *Mailing Add:* IBM T J Watson Res Ctr Box 218 Yorktown Heights NY 10598

BURNS, GROVER PRESTON, b Putnam Co, WVa, Apr 25, 18; m 41; c 2. APPLIED MECHANICS, THEORETICAL PHYSICS. *Educ:* Marshall Col, AB, 37; WVa Univ, MS, 41. *Hon Degrees:* DSc, Colo State Christian Col, 73. *Prof Exp:* Teacher high sch, WVa, 37-40; instr physics, Univ Conn, 41-42; asst prof, Miss State Col, 42-44, actg head dept, 44-45; asst prof, Tex Tech Col, 46; assoc prof math, Marshall Col, 46-47; res physicist, Naval Res Lab, 47-48; from asst prof to assoc prof physics, Mary Washington Col, Univ Va, 48-69, chmn dept, 48-69; supvr statist anal sect, Am Viscose Div, FMC Corp, 50-67; mathematician, Naval Surface Weapons Ctr, 67-81; staff mathematican, Sperry Corp, 82-87; RETIRED. *Concurrent Pos:* Pres, Burns Enterprises, Inc, 58-; consult, FMC Corp, 84-86. *Mem:* Am Phys Soc; Am Asn Physics Teachers; Am Defense Preparedness Asn; Am Asn Univ Prof. *Res:* Superconductivity; electricity; mathematics; numerical integration; exterior ballistics; mathematical models for gunfire control systems; effects of Coriolis acceleration on projectile motion. *Mailing Add:* 600 Virginia Ave Fredericksburg VA 22401

BURNS, H DONALD, b Scranton, Pa, Apr 17, 46; m 69; c 2. ORGANIC CHEMISTRY, NUCLEAR MEDICINE. *Educ:* Univ Scranton, BS, 68; Lehigh Univ, MS, 72, PhD(org chem), 74. *Prof Exp:* NIH fel, Johns Hopkins Med Inst, 74-75, asst prof nuclear med, Div Nuclear Med, 75-76; from asst prof to assoc prof radiol, environ health sci & radiol, 76-87; sr res fel, dept res imaging, 87-90, HEAD DEPT RADIOPHARMACOL, 90- *Concurrent Pos:* Assoc prof, Dept Radiation Therapy & Nuclear Med, Hahnemann Univ, 78-79; assoc dir res, Div Radiation Health Sci, Johns Hopkins Univ, 82-87. *Honors & Awards:* Nat Res Serv Award, Nat Heart & Lung Inst, 75. *Mem:* Soc Nuclear Med; Am Chem Soc; Sigma Xi; AAAS; Am Asn Pharmaceut Scientists. *Res:* Application of nuclear imaging to drug discovery, development and approval. *Mailing Add:* Merck Sharp & Dohme Res Labs West Point PA 19486

BURNS, JACK O'NEAL, b Ayer, Mass, Jan 2, 53; m 80; c 2. RADIO ASTRONOMY. *Educ:* Univ Mass, BS, 74; Ind Univ, MA, 76, PhD(astron), 78. *Prof Exp:* res asst radio astron, Nat Radio Astron Observ, 77-78, res assoc, 78-80; from asst prof to assoc prof, 80-90, PROF ASTRON & DEPT HEAD, NMEX STATE UNIV, 89- *Concurrent Pos:* Consult, Sandia Nat Labs, 82-89, Los Alamos Nat Labs, 86- *Mem:* Royal Astron Soc; Am Astron Soc; Int Astron Union; Sigma Xi; AAAS. *Res:* Radio interferometry; extended extragalactic radio sources; study of radio galaxies and clusters of galaxies at radio, optical and x-ray frequencies. *Mailing Add:* Dept Astron NMex State Univ Las Cruces NM 88003

BURNS, JAY, III, b Lake Wales, Fla, Mar 22, 24; m 48; c 3. PHYSICS. *Educ:* Northwestern Univ, BS, 47; Univ Chicago, MS, 51, PhD, 59. *Prof Exp:* Physicist, Chicago Midway Labs, 51-57; sr physicist, Labs Appl Sci, Univ Chicago, 57-63, dir, 63-65; assoc prof astron, Northwestern Univ, 65-67; assoc dir res, Rauland Div, Zenith Radio Corp, 67-68; assoc prof astron & head lab exp astrophys, Northwestern Univ, 68-71, dir, Astro-Sci Workshops; prof physics & head dept physics & space sci, 76-87, PROF PHYSICS & SPACE SCI, FLA INST TECHNOL, 87- *Concurrent Pos:* Vis prof physics, Univ Ill, Chicago Circle & sr res consult, Zenith Radio Corp, 71-76. *Mem:* AAAS; Am Phys Soc; Am Astron Soc; Am Phys Soc; Acoust Soc Am. *Res:* Secondary electron emission; photoelectric emission; solid state; surface physics; astronomical image tubes; atomic and molecular transition probabilities; electron emission from solids; solid surface physics; polymer physics; biophysics. *Mailing Add:* Dept Physics & Space Sci Fla Inst Technol Melbourne FL 32901

BURNS, JOHN ALLEN, b Little Rock, Ark, Aug 15, 45; m 65; c 1. APPLIED MATHEMATICS. *Educ:* Ark State Univ, BSE, 67, MSE, 68; Univ Okla, MA, 70, PhD(math), 73. *Prof Exp:* Spec instr math, Univ Okla, 72-73; asst prof, Lefschetz Ctr Dynamical Syst, Brown Univ, 73-74; ASSOC PROF MATH, VA POLYTECH INST & STATE UNIV, 74- *Mem:* Soc Indust & Appl Math. *Res:* Control theory; ordinary differential equations; functional differential equations. *Mailing Add:* Dept of Math Va Polytech Inst & State Univ Blacksburg VA 24061

BURNS, JOHN FRANCIS, b Minneapolis, Minn, Jan 10, 01; m 29; c 2. EXPERIMENTAL ATOMIC PHYSICS. *Educ:* Loras Col, BA, 22; Univ Wis, MA, 27; Univ Tenn, PhD(physics), 54. *Prof Exp:* Instr physics, Univ Wis, 25-27; instr physics & math, Gen Beadle State Teachers Col, 29-40; instr physics, Va Polytech Inst, 40-41; physicist, Radford Proving Ground, Va, 41-42, ballistics engr, Radford Proving Ground, Hercules Powder Co, 41-45; physicist, Oak Ridge Gaseous Diffusion Plant, 45-56; sr res physicist, Oak Ridge Nat Lab, Union Carbide Corp, Tenn, 56-66; assoc prof physics, Univ Tenn, Knoxville, 66-71; RETIRED. *Concurrent Pos:* Vis lectr, Univ Col, Dublin, 66; consult, Oak Ridge Nat Lab, 66-68. *Mem:* Am Phys Soc. *Res:* Gaseous ionization under electron impact; atomic and molecular structure; mass spectrometry; nondestructive testing; interior and external ballistics. *Mailing Add:* 131 Georgia Ave Oak Ridge TN 37830

BURNS, JOHN HOWARD, b Kilgore, Tex, Oct 8, 30; m 52, 76; c 2. XRAY DIFFRACTION. *Educ:* Rice Inst, BA, 51, MA, 53, PhD(chem), 55. *Prof Exp:* Res assoc chem, Oak Ridge Nat Lab, 55-56; asst prof, Univ Ky, 56-57; asst chemist, Argonne Nat Lab, 57-59, assoc chemist, 59-60; chemist, 60-78, SR RES STAFF, OAK RIDGE NAT LAB, 78- *Mem:* Am Chem Soc; Am Crystallog Asn. *Res:* X-ray and neutron diffraction; chemistry of actinide elements. *Mailing Add:* Oak Ridge Nat Lab Chem Div PO Box 2008 Oak Ridge TN 37831-6119

BURNS, JOHN J, b Flushing, NY, Oct 8, 20. BIOCHEMISTRY. *Educ:* Queens Col, BS, 42; Columbia Univ, AM, 48, PhD(chem), 50. *Hon Degrees:* DSc, Queens Col, CUNY. *Prof Exp:* Deputy chief, Lab Chem Pharmacol, NIH, 58-60; dir res pharmacodynamics div, Wellcome Res Labs, 60-66; vpres res, Hoffmann-La Roche Inc, 67-84, ADJ MEM, ROCHE INST MOLECULAR BIOL, 84- *Concurrent Pos:* Mem pharmacol & exp therapeut study sect, NIH, 58-62 & drug res bd, Nat Acad Sci-Nat Res Coun, 64-; vis prof, Albert Einstein Sch Med, 60-68 & Cornell Univ Med Col, 68-; adj prof, Rockefeller Univ, 84- *Mem:* Nat Acad Sci; Inst Med-Nat Acad Sci; Am Soc Pharmacol & Exp Therapeut; Am Soc Biol Chem; NY Acad Sci (vpres, 64); AAAS; Am Chem Soc. *Res:* Metabolism of vitamin C, pentoses, uronic acids; antirheumatic drugs; muscular relaxants; barbiturates; anticoagulants; local anesthetics; mechanism of action of adrenergic blocking drugs. *Mailing Add:* Roche Inst Molecular Biol Nutley NJ 07110

BURNS, JOHN JOSEPH, JR, b Chicago, Ill, Dec 16, 25; m 50; c 4. SOLID MECHANICS, ENGINEERING. *Educ:* Ill Inst Technol, BS, 45; Drexel Inst Technol, MS, 53 & 55; Univ Colo, Boulder, PhD(appl mech), 67. *Prof Exp:* Engr, Bell Aircraft Corp, 45-46; sr engr, Fairchild Engine & Airplane Corp, 47-51; chief struct engr, Aerospace Div, Kaiser Metal Prod Co, 51-55; tech specialist, Missile Syst Div, Repub Aviation Corp, 55-59; sect head, Denver Div, Martin-Marietta Corp, 59-66; staff engr, Lockheed-Ga Corp, 66-67; assoc prof eng sci & mech, Univ Fla, 67-72; prof mech & engr, Univ Ill, Champaign, 72-76; RETIRED. *Concurrent Pos:* Indust mem, Mil Spec 25 Comt on Composite Mat, 50-67; consult, Gen Elec Co, 67-, Martin-Marietta Corp, 68- & US Navy, 68-; head struct res, US Nuclear Regulatory Comn, 73-; assoc ed, J Mech Design, Am Soc Mech Engrs. *Mem:* AAAS; Am Inst Aeronaut & Astronaut; Am Soc Eng Educ; Sigma Xi; Am Soc Mech Engrs; Am Acad Mech; Am Asn Eng Educ. *Res:* Effects of load, dynamics, thermal environment on civil, mechanical and aerospace structures; shell stability and deformation process, mechanical and structural reliability; fracture mechanics; fatigue and shock load failures of materials and components; high speed aerodynamics. *Mailing Add:* 12304 Pueblo Rd Gaithersburg MD 20878

BURNS, JOHN MCLAUREN, b Rochester, NY, June 6, 32; m 54; c 3. SYSTEMATICS, EVOLUTION. *Educ:* Johns Hopkins Univ, AB, 54; Univ Calif, Berkeley, MA, 57, PhD(zool), 61. *Prof Exp:* Asst zool, Univ Calif, Berkeley, 54-57, asst entom, 57-58; asst prof biol, Wesleyan Univ, 61-69; assoc cur lepidoptera, Mus Comp Zool, Harvard Univ, 69-75, assoc prof biol, 72-75; assoc cur, 75-77, CUR ENTOM, NAT MUS NATURAL HIST, SMITHSONIAN INST, 78- *Mem:* Fel AAAS; Soc Study Evolution; Am Soc Naturalists; Soc Syst Zool; Genetics Soc Am; Lepidopterists Soc. *Res:* Population differentiation and speciation in sexually reproducing animals; genetics and ecology of polymorphism, including electrophoretically detectable protein polymorphisms; systematics and behavior of Lepidoptera, primarily skipper butterflies (Hesperiidae); biological poetry. *Mailing Add:* Dept Entom Smithsonian Inst Washington DC 20560

BURNS, JOHN MITCHELL, b Hobbs, NMex, Dec 18, 40; m 63; c 2. PHYSIOLOGY, ENDOCRINOLOGY. *Educ:* NMex State Univ, BS, 63, MS, 66; Ind Univ, PhD(zool), 69. *Prof Exp:* Sci aide & comput programmer, Phys Sci Lab, NMex State Univ, 63-65, res asst microbial physiol, 65-66; teaching assoc zool, Ind Univ, 66-69; from asst prof to assoc prof biol sci, 69-80, PROF BIOL SCI, TEX TECH UNIV, 80- *Mem:* AAAS; Am Soc Zoologists; Am Chem Soc; Endocrine Soc; Soc Develop Biol. *Res:* Steroid mechanism of action. *Mailing Add:* Dept Biol Tex Tech Univ Lubbock TX 74909

BURNS, JOHN THOMAS, b Richmond, Ind, Feb 22, 43; m 75; c 2. CHRONOBIOLOGY. *Educ:* Wabash Col, Crawfordsville, Ind, BA, 65; La State Univ, MS, 68, PhD(zool), 77. *Prof Exp:* Asst prof biol, Baker Univ, 68-77; asst prof anat, WVa Sch Osteop Med, 77-79; exec dir, Found Study Cycles, 79-85; ASST PROF BIOL, BETHANY COL, 85- *Mem:* AAAS; NY Acad Sci; Found Study Cycles (secy, 85-); Int Soc Chronobiol; Soc Biol Rhythm Res. *Res:* Biological rhythms in the physiology of vertebrates; hormonal and neurochemical regulation of seasonal physiological and behavioral changes in migrating birds and hibernating mammals. *Mailing Add:* Dept Biol Bethany Col Bethany WV 26032

BURNS, JOSEPH A, b New York, NY, Mar 22, 41; m 67; c 2. PLANETARY SCIENCE, CELESTIAL MECHANICS. *Educ:* Webb Inst Naval Archit, BS, 62; Cornell Univ, PhD(space mechanics), 66. *Prof Exp:* asst prof mech, 66-75, assoc prof, 75-81, PROF MECH & ASTRON, CORNELL UNIV, 81-, CHMN DEPT, 87- *Concurrent Pos:* Nat Acad Sci-Nat Res Coun res assoc theoret div, Goddard Space Flight Ctr, NASA, 67-68; Nat Acad Sci exchange fel, Schmidt Inst, Moscow & Astron Inst, Prague, 73; sr scientist, Space Sci Div, NASA Ames Res Ctr, 75-76, 82-83; ed, Icarus, 79-; prof astron, Paris Observ, 80, Univ Calif, Berkeley, 82-83, Univ Ariz, 89-90; prin investr, NY Coun on Arts, 72, NY Sci Found, 74, NASA, 74-, NSF, 83-86; adv comt, Space & Earth Sci, NASA, 83-87; mem, Space Studies Bd, Nat Acad Sci, 89- *Mem:* AAAS; Am Geophys Union; Am Astron Soc; Int Astron Union; Sigma Xi; Comt Space Res. *Res:* Mechanics of solar system; celestial rotation; asteroid collisions; origin of solar system; orbital evolution; planetary satellites, dust dynamics; Saturn's rings celestial mechanics; teaching methods; art and science. *Mailing Add:* Dept Theoret & Appl Mech 209 Kimball Hall Cornell Univ Ithaca NY 14853

BURNS, JOSEPH CHARLES, b Iowa City, Iowa, Feb 16, 37; m 56; c 6. FORAGE QUALITY, SECONDARY COMPOUNDS. *Educ:* Iowa State Univ, BS, 60, MS, 63; Purdue Univ, PhD(plant physiol, ecol), 66. *Prof Exp:* Dist sales mgr, Boeke Feed Co, Iowa, 61; res asst forage crop prod, Iowa State Univ, 61-63; res asst plant physiol & ruminant nutrit, Purdue Univ, 63-66; prof forage physiol, Tex A&M Univ, 66; from asst prof to assoc prof, 67-76, PROF CROP SCI, NC STATE UNIV & RES PLANT PHYSIOLOGIST, AGR RES SERV, USDA, 76- *Mem:* Fel Am Soc Agron; Am Forage & Grassland Coun; Fel Crop Sci Soc Am. *Res:* Initiation and accumulation of secondary plant products from primary plant metabolite induced through management, environment and genetic control and relationship of these constituents to animal acceptance, intake and conversion and utilization of forages by grazing ruminants. *Mailing Add:* Dept Crop Sci 1119 Williams Hall NC State Univ Raleigh NC 27695-7620

BURNS, KENNETH FRANKLIN, b Lebanon, Ind, June 6, 16; m 39. VIROLOGY. *Educ:* Ont Vet Col, DVM, 40; Univ Toronto, DVSc, 50; Univ Tokyo, PhD(microbiol), 52; Am Bd Vet Pub Health, dipl, 55; Am Col Lab Animal Med, dipl, 61. *Prof Exp:* Chief virol & vet br, Army Med Lab, Ft McPherson, Ga, 41-43, chief virol & vet br, 4th Army Med Lab, Ft Sam Houston, Tex, 46-47, chief dept virus & rickettsial dis, 406th Med Gen Lab, Japan, 47-51, dep chief vet microbiol div, Chem Corps Biol Warfare Labs, Ft Detrick, Md, 51-53, chief vet & virol br, 4th Army Area Med Lab, Brooke Army Med Ctr, Tex, 53-58, chief animal colonies div, Directorate Med Res, US Army Chem Warfare Labs, Army Chem Ctr, Md, 59-61, chief pub health officer, Civil Affairs Group & Lab & consult to High Comnr, Ryukyu Islands, 61-62; prof comp med & chmn dept vivarial sci & res, 62-76, PROF & CHMN COMP MED, SCH MED, TULANE MED CTR, TULANE UNIV, 76- *Concurrent Pos:* Consult, US Ord Command, 59-61. *Mem:* Am Vet Med Asn; Animal Care Panel. *Res:* Laboratory animal medicine. *Mailing Add:* 2344 Killdear St New Orleans LA 70122

BURNS, LAWRENCE ANTHONY, b Washington, DC, Aug 12, 40; m 75; c 3. SYSTEMS ECOLOGY, ECOTOXICOLOGY. *Educ:* NY Univ, BA, 68; Univ NC, Chapel Hill, PhD(ecol), 78. *Prof Exp:* Lab technician microbiol, NY Univ, 66-68; res asst ecol, Univ NC, Chapel Hill, 68-71; res aquatic biologist, Region IV Surveillance & Anal Div, US Environ Protection Agency, 71-73; res asst, Ctr Wetlands, Phelps Lab, Univ Fla, 73-76; RES ECOLOGIST, ATHENS ENVIRON RES LAB, US ENVIRON PROTECTION AGENCY, 77- *Honors & Awards:* Civil Serv Silver Medal, US Environ Protection Agency, 73, Sci Achievement Award, 85. *Mem:* Ecol Soc Am; Am Soc Limnol & Oceanog; Int Soc Ecol Model; AAAS; Am Inst Biol Sci; Soc Environ Toxicol Chem. *Res:* Mathematical and computer modeling of ecosystem dynamics; transport, fate, and effects of nutrients and toxic chemicals in aquatic systems; ecological risk assessment of synthetic organic chemicals. *Mailing Add:* Athens Environ Res Lab College Station Rd Athens GA 30613-7799

BURNS, MICHAEL J, b Hollywood, Calif, Oct 6, 56. SURFACE SCIENCE, ELECTRONIC TRANSPORT. *Educ:* Univ Calif, Los Angeles, BS, 78, MS, 80, PhD(physics), 84. *Prof Exp:* MTS engr, Hughes Aircraft Co, 78-80; res physicist, Univ Pa, 83-85; postdoctoral fel, Harvard Univ, 85-87; PROF PHYSICS, UNIV FLA, 87- *Mem:* Am Physical Soc; Mat Res Soc. *Res:* Magnetic field induced metal-insulator transitions in semiconductors; magnetic flux quantization in aperiodic superconducting structures; electronic and magnetic properties of monolayer films and organic molecule formation in deep space. *Mailing Add:* Phys Dept Univ Fla 215 Williamson Hall Gainesville FL 32611

BURNS, MOORE J, b Wedowee, Ala, May 31, 17; m 39; c 3. BIOCHEMISTRY, PHYSIOLOGY. *Educ:* Auburn Univ, BS, 40, MS, 46; Purdue Univ, PhD(biochem), 50. *Prof Exp:* Instr animal sci, Auburn Univ, 46-47; res asst biochem, Purdue Univ, 47-50; assoc prof animal nutrit, 50-55, prof physiol, Auburn Univ, 56-82; RETIRED. *Mem:* NY Acad Sci; Am Inst Nutrit. *Res:* Vitamin A stability and utilization; nutrition in relation to cancer; parenteral nutrition; factors affecting cholesterol metabolism and atherosclerosis. *Mailing Add:* Sch Vet Med Auburn Univ Auburn AL 36830

BURNS, NANCY A, b Ft Worth, Tex, Aug 21, 36; m 56; c 2. RURAL HEALTH, HOSPICES. *Educ:* Tex Christian Univ, BS, 57; Tex Women's Univ, MS, 74, PhD(nursing), 81. *Prof Exp:* Pub health nurse, Ft Worth Health Dept, 57-59; nurse, All Saints Hosp, Ft Worth, 61-68; instr, All Saints Voc Nursing Sch, 68-72, dir, 72-74; instr, Syst Sch Nursing, Univ Tex, Ft Worth, 74-76; dir, Rural Hosp Outreach Prog, Univ Tex, Arlington, 75-81, dir continuing educ, 75-81, from asst prof to assoc prof, 78-86, PROF NURSING, UNIV TEX, ARLINGTON, 86-, DIR, CTR NURSING RES, 87- *Concurrent Pos:* Consult, var hosps, 74-; area chmn, Cancer Prev Study, Am Cancer Soc, 82-88. *Honors & Awards:* Sword of Hope Award, Am Cancer Soc, 83. *Mem:* Am Nurses Asn; Nat League Nursing; Oncol Nursing Soc; Nat Rural Health Asn. *Res:* Rural health; health service; hospice care; oncology; smoking among nurses; author of two books and numerous publications. *Mailing Add:* Sch Nursing Univ Tex Box 19407 Arlington TX 76019-0407

BURNS, NED HAMILTON, b Magnolia, Ark, Nov 25, 32; m 55; c 3. CIVIL ENGINEERING. *Educ:* Univ Tex, BS, 54, MS, 58; Univ Ill, PhD(civil eng), 62. *Prof Exp:* Engr, Mosher Steel Co, Tex, 54-55; instr civil eng, Univ Tex, 57-59, res engr struct mech, Balcones Res Ctr, 59; asst civil eng, Univ Ill, 59-62; from asst prof to prof civil eng, 62-80, grad adv, 63-77, 80-86, PROF, UNIV TEX, 80- *Concurrent Pos:* Res reinforced concrete, USAF Spec Weapons Ctr, 63-; NSF res grant, 64-65; Tex Hwy Dept Res, 67-; consult, struct eng prods on reinforced & prestressed concrete. *Mem:* Am Concrete Inst; Am Soc Civil Engrs; Nat Soc Prof Engrs; Sigma Xi. *Res:* Prestressed concrete flat slabs with unbonded tendons; long-span prestressed concrete bridges with segmental construction and other research in prestressed concrete. *Mailing Add:* Dept Civil Eng Univ Tex Austin TX 78712

BURNS, PAUL YODER, b Tulsa, Okla, July 4, 20; m 42; c 3. FORESTRY. *Educ:* Univ Tulsa, BS, 41; Yale Univ, MF, 46, PhD, 49. *Prof Exp:* From asst prof to assoc prof forestry, Univ Mo, 48-55; prof, 55-86, EMER PROF FORESTRY, LA STATE UNIV, BATON ROUGE, 86- *Mem:* Fel Soc Am Foresters; Ecol Soc Am; Am Soc Photogram. *Res:* silvics; silviculture; mensuration. *Mailing Add:* Sch Forestry La State Univ Baton Rouge LA 70803-6200

BURNS, RICHARD CHARLES, b Chicago, Ill, Oct 8, 30; m 55; c 4. BIOCHEMISTRY. *Educ:* Univ Wis, BS, 52, MS, 61, PhD(biochem), 63. *Prof Exp:* Fel biochem, C F Kettering Res Lab, 63-64, staff scientist, 64-67; biochemist, Exp Sta, 67-73, proj supvr, Photo Prod Dept, 73-80, RES SUPVR, CENT RES & DEVELOP DEPT, E I DU PONT DE NEMOURS & CO, INC, 80- *Mem:* Fedn Am Soc Exp Biol. *Res:* Biological nitrogen fixation; hydrogen metabolism; intermediary metabolism of microorganisms; enzymology. *Mailing Add:* IPD Glasgow Site 300 E I du Pont de Nemours & Co Inc Wilmington DE 19898

BURNS, RICHARD PRICE, b Bartlesville, Okla, Sept 19, 32; m 59; c 2. PHYSICAL CHEMISTRY, INORGANIC CHEMISTRY. *Educ:* Okla Baptist Univ, AB, 54; Univ Chicago, PhD(chem), 65. *Prof Exp:* Jr chemist, Lawrence Radiation Lab, Univ Calif, 55-58; res asst physics, Univ Chicago, 59-65; asst prof, 65-70, ASSOC PROF CHEM, UNIV ILL, CHICAGO CIRCLE, 70- *Mem:* Am Chem Soc; Am Phys Soc. *Res:* Applications of the mass spectrometer and thermal imaging techniques to high temperature chemistry, thermodynamics, kinetics and materials. *Mailing Add:* Dept Chem M/C 111 Univ Ill at Chicago PO Box 4348 Chicago IL 60680

BURNS, ROBERT ALEXANDER, INFANT NUTRITION, PROTEIN NUTRITION. *Educ:* Queen's Univ, Belfast, Northern Ireland, PhD(nutrit), 76. *Prof Exp:* SR RES SCIENTIST, MEAD JOHNSON NUTRIT GROUP, 82- *Res:* Mineral availability. *Mailing Add:* Dept Nutrit Sci Mead Johnson Nutrit Group 2400 W Lloyd Expwy Evansville IN 47721

BURNS, ROBERT DAVID, b Detroit, Mich, Feb 28, 29; m 60; c 2. ZOOLOGY. *Educ:* Mich State Univ, BS, 51, MS, 54, PhD(zool), 58. *Prof Exp:* Asst prof zool, Univ Okla, 58-63; from asst prof to prof biol, Kenyon Col, 63-92, chmn dept, 69-71 & 79-84; RETIRED. *Concurrent Pos:* Vis prof, Biol Sta, Mich State Univ, 61; vpres, Ohio Acad Sci, 74-75. *Mem:* AAAS; Ecol Soc Am; Am Soc Mammal; Am Ornith Union; Wilson Ornith Soc. *Res:* Feeding and behavior in kangaroo rats; reproductive behavior in birds and mammals; physiology of the vestibular portion of the inner ear in kangaroo rats. *Mailing Add:* PO Box 204 Gambier OH 43022

BURNS, ROBERT EARLE, b New York, NY, May 1, 25; m 51; c 2. OCEANOGRAPHY. *Educ:* Col Wooster, AB, 47; Lehigh Univ, MS, 50; Univ Wash, PhD(oceanog), 62. *Prof Exp:* Instr geol, Bucknell Univ, 49-51; oceanogr, div oceanog, US Navy Hydrographic Off, 51, head tides & currents unit, 51-52, dep head regional sect, 52-53, tech asst, 53-54, head geol oceanog unit, 54-56, supv oceanogr, 56-57; assoc oceanog, Univ Wash, 57-62; res oceanogr, Off Res & Develop, US Coast & Geod Surv, 62-65; chief environ sci serv admin complement, Joint Oceanog Res Group, Univ Wash, 65-69; res oceanogr, Pac Oceanog Labs, Nat Oceanic & Atmospheric Admin, 70-73, actg dir, Pac Marine Environ Lab, 73-76, proj mgr, deep ocean mining environ studies, 76-79, mgr, long range effects studies, Off Marine Pollution Assessment, 80-83, mgr, marine resources res div, Pac Marine Environ Lab, 84-85; RETIRED. *Concurrent Pos:* Chmn, Fed Interagency Subcomt Sedimentation, 63-64; vis geoscientist prog, Am Geol Inst, 64-65 & Am Geophys Union, 66-67, 70; chmn, Joint Oceanog Insts Deep Earth Sampling Pac Adv Panel, 68-73. *Mem:* Geophys Union; Marine Technol Soc. *Res:* Assessment of potential impact of man's activities on the marine environment. *Mailing Add:* 3602 47th Ave NE Seattle WA 98105

BURNS, ROBERT EMMETT, b Oxford, Iowa, May 9, 18; m 52; c 4. PLANT PHYSIOLOGY. *Educ:* Univ Iowa, BA, 40, MS, 47, PhD(plant physiol), 49. *Prof Exp:* Instr, Univ Calif, Santa Barbara, 49-50; plant physiologist, 50-64, assoc plant physiologist, 64-84, EMER ASSOC PROF, EXP STA, AGR RES SERV, USDA, UNIV GA, 84- *Mem:* Am Soc Plant Physiol. *Res:* Seed physiology; turf physiology. *Mailing Add:* 1010 E McIntosh Rd Griffin GA 30223

BURNS, ROBERT WARD, b Fresno, Calif, Nov 30, 41; m 67; c 2. MICROELECTRONIC PACKAGING & ASSEMBLY PROCESSES, FIBER OPTIC SENSING MANUFACTURING EQUIPMENT. *Educ:* Univ Calif, Berkeley, BS, 64, MS, 66; Univ Wash, Seattle, PhD(ceramic eng), 72. *Prof Exp:* Sr engr, Tektronix Inc, 74-80; opers mgr, PM Industs, 80-81; develop mgr, Brush Wellman, 81-82; eng sect mgr, John Fluke Mfg, 82-86; prin engr, Boeing Electronics Co, 86-87; MFG MGR, METRICOR, 87- *Mem:* Fel Int Soc Hybrid Microelectronics; Am Ceramic Soc. *Res:* Processing of advanced materials for electronic applications; thick and thin film hybrids; ceramic substrates; surface mount technology; integrated circuit packaging; polymer thick film. *Mailing Add:* 18232 Butternut Rd Lynnwood WA 98037

BURNS, ROGER GEORGE, b Wellington, NZ, Dec 28, 37; m 63; c 2. GEOCHEMISTRY, MINERALOGY. *Educ:* Victoria Univ, NZ, BSc, 59, MSc, 61; Univ Calif, Berkeley, PhD(geochem), 65. *Hon Degrees:* MA, Wadham Col, Oxford Univ, 68 & DSc, 85. *Prof Exp:* Demonstr chem, Victoria Univ, NZ, 58-60, sci officer, Dept Sci & Indust Res, 60-61; res assoc geochem, Dept Mineral Technol, Univ Calif, Berkeley, 64-65; sr res visitor, Cambridge Univ, 65-66; sr lectr, Victoria Univ, NZ, 66-67; lectr, Oxford Univ, 68-70; PROF GEOCHEM, DEPT EARTH & PLANETARY SCI, MASS INST TECHNOL, 70- *Concurrent Pos:* Vis lectr, Oxford & Cambridge Univs, 65-66; res visitor, Brit Coun, 65-66; sr res visitor, Natural Environ Coun, Eng, 65-66; ed, Chem Geol, 68-82; UNESCO vis prof, Jadavpur Univ, India, 81; vis prof, Scripps Inst Oceanog, 76; assoc ed, Geochemica et Cosmochimica Acta, 78-, Can Mineralogist, 88-; consult, Battery Prod Div, Eveready Battery, 75-, Nat Oceanic & Atmospheric Admin Marine Minerals Project, 76, NASA Lunar Res Proposal Review, 78-80, Spacelab Proposal Review, 79-81, Planetary Instrument Definition Prog, 81; mem, Comt Mineral Physics, Am Geophys Union, 83-86; Guggenheim Prof, Manchester Univ, Eng, 91. *Honors & Awards:* Min Soc Am Award, 75; Hallimond lectr, Mineral Soc, Eng, 87. *Mem:* Fel Geochem Soc; fel Mineral Soc Am (vpres-elect, 88-); fel Mineral Soc Gt Brit & Ireland; fel The Chem Soc; fel Am Geophys Union; Electrochem Soc. *Res:* Transition metal geochemistry and spectroscopic studies of minerals; bonding, distribution and properties of transition elements in silicate and ore minerals; Mossbauer, infrared and electronic absorption spectroscopy of minerals; geochemistry of deep-sea ferromanganese nodules; metallogenesis and plate tectonics; minerals as battery anodes; disposal of nuclear waste. *Mailing Add:* 54-816 Dept Earth & Planetary Sci Mass Inst Technol Cambridge MA 02139

BURNS, RUSSELL MACBAIN, b New York, NY, Aug 25, 26; m 48; c 3. SILVICULTURE, PLANT PHYSIOLOGY. *Educ:* Mich State Univ, BS, 50; Univ Miss, MS, 59; Univ Fla, PhD, 71. *Prof Exp:* Res forester, 51-67, silviculturist, 67-77, PRIN CONIFER SILVICULTURIST, US FOREST SERV, 77- *Concurrent Pos:* Res consult, Arg govt; US rep to USSR, US Man & Biosphere Prog, UNESCO; USDA Forest Serv rep to India, US-INDO Agr Coop; chmn, Nat Capital Soc Am Foresters, 91-92. *Mem:* Soc Am Foresters; Am Soc Plant Physiol; Soil Conserv Soc Am. *Res:* Silviculture problems; tree physiology and soil chemistry. *Mailing Add:* US Forest Serv PO Box 96090 Washington DC 20090-6090

BURNS, STEPHEN JAMES, b New York, NY, Jan 31, 39; m 64; c 3. MATERIALS SCIENCE, MECHANICS. *Educ:* Pratt Inst, BS, 61; Cornell Univ, MS, 65, PhD(mat sci), 67. *Prof Exp:* Res assoc eng, Brown Univ, 66-67, res & teaching fel, 67-68, asst prof, 68-72; assoc prof, 72-79, PROF, DEPT MECH ENG, UNIV ROCHESTER, 79- *Concurrent Pos:* NSF res initiation grant, 68-69. *Mem:* Am Inst Mining, Metall & Petrol Engrs; Am Soc Metals; Am Soc Mech Engrs. *Res:* Cleavage fractures in single crystals of zinc and lithium fluoride; dislocation motion in non-uniform time dependent stress fields; x-ray diffraction topography; mechanical properties of solids; fracture mechanics. *Mailing Add:* Dept Mech Eng Univ Rochester Rochester NY 14627

BURNS, THOMAS WADE, b Dayton, Ohio, July 29, 24; m 52; c 4. INTERNAL MEDICINE, ENDOCRINOLOGY. *Educ:* Univ Utah, BA, 45, MD, 47, MS, 48; Am Bd Internal Med, dipl, 55. *Prof Exp:* Intern, Boston City Hosp, 48-49; asst resident, Harvard Med Sch, 49-50; clin investr, US Naval Hosp, Calif, 51-52; clin investr, US Naval Med Res Unit, Egypt, 52-54; res physician, Med Ctr , Univ Calif, 55; asst prof, 55-57, assoc prof, 57-65, PROF MED, SCH MED, UNIV MO,COLUMBIA, 65-, PHYSICIAN, MED CTR, 55-, DIR DIV ENDOCRINOL & METAB, 69- *Concurrent Pos:* Fel, Harvard Med Sch, 49-50; fel med, Sch Med, Duke Univ, 50-51. *Mem:* AAAS; Endocrine Soc; Am Diabetes Asn; AMA; Am Col Physicians; Am Fedn Clin Res; Am Soc Internal Med. *Res:* Endocrine and metabolic disorders. *Mailing Add:* Univ Mo Columbia Med Ctr D110 A Columbia MO 65212

BURNS, TIMOTHY JOHN, b Ames, Iowa, June 28, 49; m 78; c 1. ORDINARY & PARTIAL DIFFERENTIAL EQUATIONS, ASYMPTOTIC METHODS. *Educ:* Univ NMex, BS, 70, MA, 72, PhD(math), 77. *Prof Exp:* Mem tech staff appl math, Solid Dynamics Res Dept, Sandia Nat Labs, 77-86; MATHEMATICIAN APPL MATH, COMPUT & APPL MATH LAB, NAT INST STANDARDS & TECHNOL, 86- *Mem:* Soc Indust & Appl Math; Am Soc Mech Engrs. *Res:* Asymptotic, numerical methods and mathematical modeling for problems in applied and classical mechanics. *Mailing Add:* Nat Inst Standards & Technol Admin A231 Gaithersburg MD 20899

BURNS, VICTOR WILL, b Los Angeles, Calif, Nov 16, 25; m 50; c 3. BIOPHYSICS, CELL PHYSIOLOGY. *Educ:* Univ Calif, PhD(biophys), 55. *Prof Exp:* Assoc med physics, Univ Calif, 53-55; dir high energy proton biophys, Royal Univ Uppsala, 55-56; from res assoc to lectr, Stanford Univ, 56-64; from assoc prof to prof, 64-90, EMER PROF BIOPHYS, UNIV CALIF, DAVIS, 90- *Mem:* Radiation Res Soc; Biophys Soc. *Res:* Cell division and intracellular coordination; biological and physical effects of ionizing radiation; fluorescence analysis of cells and nucleic acids. *Mailing Add:* Dept Physiol Sci Univ Calif Davis CA 95616

BURNS, WILLIAM, JR, b New Castle, Pa, Dec 18, 38; m 66; c 2. TECHNICAL MANAGEMENT, HARDWARE SYSTEMS. *Educ:* Carnegie Mellon Univ, BS, 61, MS, 62, PhD(elec eng), 65. *Prof Exp:* Res geophysicist, Gulf Res & Develop Ctr, 65-66; sr researcher astrophys, Observ Paris, 66-67; asst scientist radio astron, 67-68, assoc scientist radio astron, 68-69, HEAD COMPUT DIV, NAT RADIO ASTRON OBSERV, 69- *Concurrent Pos:* Fel, Nat Ctr Sci Res, France, 66; mem, Site Vis Comt, NSF, 85, bd dirs, R-Tech Inc, 85-91; consult, var co, 83-91. *Mem:* Inst Elec & Electronics Engrs; Int Union Radio Sci. *Res:* Signal processing; investigations of computer hardware systems; communications networking; computer management both from applied and research consideration. *Mailing Add:* 2320 Westover Dr Charlottesville VA 22901

BURNS, WILLIAM CHANDLER, b Fargo, NDak, Jan 5, 26; m 56; c 4. PARASITOLOGY. *Educ:* Ore State Col, BS, 50, MA, 52; Univ Wis-Madison, PhD(zool), 58. *Prof Exp:* Teaching asst zool, Ore State Col, 50-52; teaching asst, 52-57, actg instr, 57-58, from instr to assoc prof, 58-68, PROF ZOOL, UNIV WIS-MADISON, 68- *Concurrent Pos:* NSF fel, Ore State Col, 59-61. *Mem:* Am Soc Parasitol. *Res:* Ecology of helminths; coccidiosis; host-parasite relationships. *Mailing Add:* Dept Zool Univ Wis 412 Birge Hall Madison WI 53706

BURNSIDE, EDWARD BLAIR, b Madoc, Ont, Mar 16, 37; m 59; c 3. POPULATION GENETICS, ANIMAL BREEDING. *Educ:* Univ Toronto, BSA, 59, MSA, 60; NC State Univ, PhD(animal breeding, genetics, statist), 65. *Prof Exp:* Lectr animal breeding, Ont Agr Col, Univ Guelph, 60-61; res asst, NC State Univ, 61-64; asst prof, 64-66, assoc prof, 66-77, PROF ANIMAL BREEDING, UNIV GUELPH, 77- *Mem:* Am Dairy Sci Asn; Can Soc Animal Prod; Sigma Xi. *Res:* Estimation of genetic trend in populations; effect of environmental factors on milk yield; genotype by environment interaction; pedigree indexing of dairy cattle; sire evaluation methods. *Mailing Add:* Dept Animal & Poultry Sci Univ Guelph Guelph ON N1G 2W1 Can

BURNSIDE, MARY BETH, b San Antonio, Tex, Apr 23, 43. CELL BIOLOGY, NEUROBIOLOGY. *Educ:* Univ Tex, Austin, BA, 65, MA, 67, PhD(zool), 68. *Prof Exp:* Instr anat, Harvard Med Sch, 71-72; asst prof anat, Univ Pa, 72-75; from asst prof to assoc prof, 75-82, dean biol sci, 83-90, PROF CELL & DEVELOP BIOL, UNIV CALIF, BERKELEY, 82- *Concurrent Pos:* Am Asn Univ Women fel, Hubrecht Lab, Utrecht, Holland, 68-69; NIH fels, Harvard Univ Biolabs, Cambridge, 69-70 & Harvard Med Sch, Boston, 70-71; vis scholar, Harvard Univ, 80-81. *Honors & Awards:* Merit Award, Nat Eye Inst, NIH. *Mem:* AAAS; Am Soc Cell Biol; Am Asn Anat; Am Soc Zool; Asn Res Vision & Ophthal. *Res:* Roles of actin and microtubules in cell shape determination; roles of dopamine and cyclic nucleotides in diurnal and circadiah regulation of photoreceptor motility in teleost retinas; techniques are electron microscopy, biochemistry and immunocytochemistry. *Mailing Add:* Dept Molecular & Cell Biol 335 Life Sci Annex Univ Calif Berkeley CA 94720

BURNSIDE, ORVIN C, b Hawley, Minn, June 9, 32; m 54; c 2. WEED SCIENCE, AGRONOMY. *Educ:* NDak State Univ, BS, 54; Univ Minn, MS, 58, PhD(weed sci), 59. *Prof Exp:* From asst prof to prof agron, Univ Nebr, Lincoln, 59-85; head, dept agron & plant genetics, 85-90, PROF WEED SCI, UNIV MINN, 91- *Honors & Awards:* Res Award, Weed Sci Soc Am, 79. *Mem:* Weed Sci Soc Am; Int Weed Sci Soc; Am Soc Agron; Crop Sci Soc Am; Coun Agr Sci Technol. *Res:* Weed control in agronomic crops; phenology and life history of weeds; integrated weed management. *Mailing Add:* Dept Agron & Plant Genetics Univ Minn 1991 Buford Circle 308 Agron St Paul MN 55108

BURNSIDE, PHILLIPS BROOKS, b Columbus, Ohio, July 20, 27; m 54; c 3. PHYSICS. *Educ:* Ohio State Univ, BSc, 51, MSc, 54, PhD(physics), 58. *Prof Exp:* Instr physics, Ohio State Univ, 58-59; from asst prof to assoc prof, 59-69, PROF PHYSICS, OHIO WESLEYAN UNIV, 69- *Mem:* AAAS; Optical Soc Am; Am Asn Physics Teachers; Hist Sci Soc. *Res:* Fourier methods; fluid physics; optical physics; philosophy of science. *Mailing Add:* Dept Physics Ohio Wesleyan Univ Delaware OH 43015

BURNSTEIN, RAY A, b Harrisburg, Pa, Oct 5, 30; m 63. ELEMENTARY PARTICLE PHYSICS. *Educ:* Univ Chicago, BS, 52; Univ Wash, MS, 56; Univ Mich, PhD(physics), 60. *Prof Exp:* Asst prof physics, Univ Md, 60-66, vis assoc prof, 72-73; assoc prof, 66-73, PROF PHYSICS, ILL INST TECHNOL, 73- *Concurrent Pos:* Guest scientist, Physics Dept, Fermilab, 80-81, 86-87. *Mem:* Fel Am Phys Soc. *Res:* Experiments using visual detectors and counter techniques at high energy; neutrino interactions. *Mailing Add:* Dept Physics Ill Inst Technol Chicago IL 60616

BURNSTEIN, THEODORE, b Denver, Colo, Mar 18, 25; m 53, 71; c 2. VIROLOGY. *Educ:* Colo State Univ, DVM, 49; Cornell Univ, MS, 51, PhD(microbiol), 53. *Prof Exp:* Res fel biol, Johns Hopkins Univ, 53-55; res assoc, Merck, 55-59; res assoc prof microbiol, Med Sch, Univ Miami, 59-61; assoc prof, 61-65, PROF VIROL, SCH VET MED, PURDUE UNIV, 65- *Mem:* Am Asn Immunol; Am Soc Microbiol; Soc Leucocyte Biol. *Res:* Pathogenesis of viral pneumonia; monoclonal antibody therapy for influenza; immunobiology of pneumocystosis. *Mailing Add:* Dept Vet Pathobiol Purdue Univ Sch Vet Med West Lafayette IN 47906

BUROW, DUANE FRUEH, b San Antonio, Tex, June 12, 40; m 69. SULFUR DIOXIDE CHEMISTRY, FOSSIL FUEL PROCESSING. *Educ:* Univ Tex, Austin, BA, 61, PhD(chem), 66. *Prof Exp:* Instr chem, St Edward's Univ, 64-65; res phys chemist, Eng Physics Lab, E I du Pont de Nemours & Co, Inc, Del, 65-67; asst prof chem, Mich State Univ, 67-69; asst prof, 69-73, assoc prof, 73-78, PROF CHEM, UNIV TOLEDO, 78- *Concurrent Pos:* Petrol Res Fund grant, 68-70; Res Corp res grant, 70; NSF res grant, 74 & 77; Dept Energy res grant, 78; Ohio Coal Res Lab Grant, 80. *Mem:* Am Chem Soc; Optical Soc Am; The Chem Soc; Sigma Xi; Soc Appl Spectroscopy. *Res:* Chemistry of sulfur dioxide; utilization of sulfur dioxide for fossil fuel and minerals processing; infrared and raman spectros; matrix isolation chemistry; raman optical activity; molecular optical properties; instrument development. *Mailing Add:* 2620 Crissey Rd Sylvania OH 43560

BUROW, KENNETH WAYNE, JR, b Humboldt, Nebr, Oct 24, 46; m 66; c 3. SYNTHETIC ORGANIC CHEMISTRY. *Educ:* Univ Nebr, BS, 68; Iowa State Univ, PhD(org chem), 73. *Prof Exp:* Instr org chem, Univ Nebr, 72-73; sr org chemist, 73-79, res scientist, 79-80, mgr, org chem, 80-83, mgr, org chem & biochem, 83-85, DIR, ORG CHEM & BIOCHEM, ELI LILLY & CO, 85- *Mem:* Am Chem Soc. *Res:* Modification of ionophoric compounds, including synthesis of crown ethers; ruminant nutrition; synthesis of sesquiterpanes; fungicidal and entomological research; generation of agricultural products. *Mailing Add:* 6467 Johnson Rd Indianapolis IN 46220

BURR, ALEXANDER FULLER, b Cambridge, Mass, July 18, 31; m 62; c 3. X-RAY PHYSICS, PHYSICS EDUCATION. *Educ:* Jamestown Col, BS, 53; Univ Edinburgh, MS, 58; Johns Hopkins Univ, PhD(physics), 66. *Prof Exp:* Res asst, Univ NDak, 50 & 53; physicist, Ballistics Res Lab, Aberdeen Proving Ground, Md, 54; res physicist, US Naval Res Lab, 65-66; asst prof, 66-69, assoc prof, 69-77, PROF PHYSICS, NMEX STATE UNIV, 77- *Concurrent Pos:* Nat Acad Sci-Nat Res Coun res assoc, 65-66; sr vis fel, Univ Strathclyde, Scotland, 73; chief scientist, Duntech Indust, 79-80; consult, Int Atomic Energy Agency, 81, 83, 85 & 88-91. *Mem:* Fel AAAS; Am Asn Physics Teachers; Am Phys Soc. *Res:* X-ray wavelengths and the x-ray wavelength scale; determination of electron binding energies; physics of soft x-rays; application of computers to physics education. *Mailing Add:* Dept Physics NMex State Univ Las Cruces NM 88003-0001

BURR, ARTHUR ALBERT, b Manor, Can, Aug 23, 13; nat US; m 41; c 2. PHYSICAL METALLURGY. *Educ:* Univ Sask, BS, 38, MS, 40; Pa State Col, PhD(physics), 43. *Prof Exp:* Instr physics, Univ Sask, 38-40 & Pa State Col, 40-43; res physicist, Armstrong Cork Co, 43-46; head dept, Rensselaer Polytechnic Inst, 55-62, dean, Sch Eng, 62-74, chair Rensselaer prof, 74-78, from asst prof to prof metall eng, 46-78, emer prof & emer dean, 78-; RETIRED. *Honors & Awards:* Outstanding Teacher Award, Am Soc Metals, 52. *Mem:* Am Soc Metals; Am Soc Eng Educ; Am Inst Mining, Metall & Petrol Engrs. *Res:* Properties of thin films; physical metallurgy of alloys; thermodynamics and kinetics of metallurgical reactions; nondestructive testing; applications of x-rays; instrumentation. *Mailing Add:* Sch Eng JEC 3103 Rensselaer Polytech Inst Troy NY 12180-3590

BURR, BROOKS MILO, b Toledo, Ohio, Aug 15, 49; div; c 1. ICHTHYOLOGY, SYSTEMATICS. *Educ:* Greenville Col, BA, 71; Univ Ill, Urbana-Champaign, MS, 74, PhD(zool), 77. *Prof Exp:* Lab instr biol, Greenville Col, 71-72; res asst ichthyol, Ill Natural Hist Surv, 72-77; from asst prof to assoc prof, 77-87, PROF ZOOL, SOUTHERN ILL UNIV, CARBONDALE, 87- *Concurrent Pos:* Res assoc & vis prof ichthyol, Inst Marine Sci, Univ NC, Chapel Hill, 83-84; dir grad studies zool, S Ill Univ, Carbondale, 84-88; res assoc, Ill Nat Hist Survey, 88-; vis prof & res assoc, Dept Biol, Univ NMex, Albuquerque, 90-91. *Honors & Awards:* Leo Kaplan Mem Res Award, Sigma Xi, 90. *Mem:* Am Soc Ichthyologists & Herpetologists (secy, 90-); Soc Syst Zool; Am Fish Soc; AAAS; Asn Syst Collectors. *Res:* Systematics and ecology of freshwater fishes; fishes of Kentucky; field guide to North American freshwater fishes. *Mailing Add:* Dept Zool Southern Ill Univ Carbondale IL 62901-6501

BURR, GEORGE OSWALD, b Conway, Ark, Oct 6, 96. BIOCHEMISTRY, PLANT PHYSIOLOGY. *Educ:* Hendrix Col, AB, 16, LLD, 36; Univ Ark, AM, 20; Univ Minn, PhD(biol chem), 24. *Prof Exp:* Prin high sch, Ark, 16-17; prof chem & physics, Ky Wesleyan Col, 17-18; asst agr biochem, Univ Minn, 20-22; Nat Res Fel chem, Univ Calif, 22-24, res assoc, 22-27; from assoc prof to prof plant physiol, Univ Minn, 27-40, head div physiol chem, 40-46; head dept physiol & biochem, Exp Sta, Hawaiian Sugar Planters Asn, 46-62; consult, 62-66; res adv, Taiwan Sugar Corp, 66-72; RETIRED. *Concurrent Pos:* Guggenheim Mem Found fel, 34; mem indust adv comt, Sugar Res Found; deleg, First Int Conf on Peaceful Uses of Atomic Energy, Geneva, 55. *Mem:* AAAS; fel Am Inst Nutrit; Am Soc Biol Chem; Am Chem Soc. *Res:* Protein chemistry; nutrition; chemistry of reproduction of animals; chemistry and physiology of fats; photosynthesis. *Mailing Add:* 112 Niuiki Circle Honolulu HI 96821

BURR, HELEN GUNDERSON, b Iowa City, Iowa, Dec 30, 18; m 54; c 1. SPEECH PATHOLOGY, AUDIOLOGY. *Educ:* Stanford Univ, BA, 37; Univ Southern Calif, MA, 40; Columbia Univ, PhD(speech path & audiol), 49. *Prof Exp:* Speech pathologist, NY, 44-50; asst prof speech path & audiol & dir speech clin, State Univ NY, 50-53; from asst prof to prof speech path & audiol, 53-84, chmn dept, 62-78, dir, Speech & Hearing Ctr, 61-78, EMER PROF AUDIOL, UNIV VA, 84- *Concurrent Pos:* Rehab Serv Admin training grant, 61-80; US Off Educ training grant, 64-80, res grant, 67-69; USPHS training grant, 67-70. *Mem:* AAAS; fel Am Speech & Hearing Asn; Ling Soc Am; NY Acad Sci. *Res:* Linguistics; verbal behavior. *Mailing Add:* Carrsgove Stribling Ave Univ Va Charlottesville VA 22903

BURR, JOHN GREEN, b Ft Sill, Okla, Mar 12, 18; m 43; c 7. PHOTOCHEMISTRY, PHOTOBIOLOGY. *Educ:* Mass Inst Technol, BS & MS, 40; Northwestern Univ, PhD(org chem), 48. *Prof Exp:* Chemist, Jackson Lab, E I du Pont de Nemours & Co, Inc, 40-41, 42; asst, Manhattan Dist, Chicago, 43-44; asst prof chem, Miami Univ, 47-48; sr chemist, Oak Ridge Nat Lab, 48-57; supvr, Atomics Int, 57-62; group leader, NAm Sci Ctr, 62-69; prof, 69-86, EMER PROF CHEM & RADIOL SCI, UNIV OKLA, 86- *Concurrent Pos:* USPHS spec fel, Univ London, Eng, 53-54; Guggenheim fel, Cambridge Univ, 65-66; Rosetta Brigel Bartin lectr, Univ Okla, 68; USPHS serv fel, 75-76; vis prof, Hebrew Univ & Max Planck Inst Radiation Chem, 75; NIH Serv fel, Nat Cancer Inst, 75-76. *Mem:* AAAS; Am Chem Soc; Radiation Res Soc; fel Am Inst Chem; Am Photobiol Asn. *Res:* Isotope tracers; organic radiation chemistry; organic reaction mechanisms; nucleic acid photochemistry; aqueous photochemistry; energy transfer; coal chemistry and properties; chemiluminescence. *Mailing Add:* Dept Chem Univ Okla 620 Parrington Oval Norman OK 73019

BURR, WILLIAM WESLEY, JR, b Lincoln, Nebr, Mar 12, 23; m 50; c 6. BIOCHEMISTRY, MEDICINE. *Educ:* Univ Nebr, AB, 47; Univ Ill, MS, 48, PhD(chem), 51; Univ Tex, MD, 60. *Prof Exp:* Asst prof biochem, Southwestern Med Sch, Tex, 51-53, assoc prof, 53-60, prof, 60-63; chief med res br, US AEC, 61-69, asst dir med & health res, Div Biol & Med, dep dir, 70-72, dep dir, Div Biomed & Environ Res, 72-75; dep dir, Div Biomed & Environ Res, US Energy Res & Develop Admin, 75-77; dep dir, 77-78, actg dir, div biomed & environ res, US Dept Energy, 78- biomed & environ res, US Energy Res & Develop Admin, 70-; AT OAK RIDGE ASSOC UNIV. *Concurrent Pos:* Prof lectr, George Washington Univ, 63- *Mem:* Fel AAAS; Am Inst Nutrit; Am Chem Soc; Am Soc Exp Biol & Med; Am Soc Biol Chem. *Res:* Protein and lipid metabolism. *Mailing Add:* PO Box 1068 Oak Ridge TN 37831

BURRAGE, LAWRENCE MINOTT, b Cleveland, Ohio, Nov 18, 25; m 50; c 4. ELECTRICAL ENGINEERING, PHYSICS. *Educ:* Rensselaer Polytech Inst, BEE, 51; Marquette Univ, MS, 68. *Prof Exp:* Jr engr, Line Mat Industs, 51-52, asst engr test, 52-55, engr, 55-59, engr design, 59-61; sr engr res & develop, Cooper Power Systs, 61-69, chief engr, 69-78, prog mgr, 78-80, SR STAFF ENGR, COOPER POWER SYSTS, 80- *Honors & Awards:* Centennial Medal, Inst Elec & Electronics Engrs, 84. *Mem:* Sr mem & fel Inst Elec & Electronics Engrs; Am Vacuum Soc; Gaseous Electronics Conf; Am Phys Soc; Nuclear & Plasma Sci Soc (secy, 80). *Res:* Electric arc; electrical insulation; arc interruption phenomena. *Mailing Add:* Cooper Power Systs Box 100 Franksville WI 53126

BURRELL, DAVID COLIN, b Eng. CHEMICAL OCEANOGRAPHY. *Educ:* Univ Nottingham, BS, 61, PhD(geochem), 64. *Prof Exp:* From asst prof to assoc prof, Univ Alaska, 65-76, prof marine sci, 76-; RETIRED. *Mem:* Geochem Soc; Am Geophys Union; Arctic Inst NAm. *Res:* Marine trace metal chemistry. *Mailing Add:* Inst Marine Sci Univ Alaska Fairbanks AK 99701

BURRELL, ELLIOTT JOSEPH, JR, b Phoenix, Ariz, Feb 3, 29; m 50; c 4. PHYSICAL CHEMISTRY. *Educ:* Univ Notre Dame, BS, 50, MS, 51; Pa State Univ, PhD(phys chem), 54. *Prof Exp:* Res chemist, Fabrics & Finishes Div, E I du Pont de Nemours & Co, Inc, 54-58, from res chemist to sr res chemist, Radiation Physics Lab, 58-63; from asst dean to assoc dean sci, Col Arts & Sci, 70-74, asst prof, 63-65, ASSOC PROF CHEM, LOYOLA UNIV, CHICAGO, 65- *Concurrent Pos:* Lectr, Temple Univ, 55-57, 58-63 & Univ Pa, 57-58. *Mem:* Am Chem Soc. *Res:* Nuclear and radiation chemistry; pulse radiolysis; photochemistry; kinetics; physical-organic chemistry; magnetic resonance spectroscopy; flash photolysis. *Mailing Add:* Dept Chem Loyola Univ 6525 N Sheridan Rd Chicago IL 60626

BURRELL, ROBERT, b Springfield, Ohio, Aug 26, 33; m 55; c 2. IMMUNOLOGY, MICROBIOLOGY. *Educ:* Ohio State Univ, BS, 55, MS, 56, PhD(bact), 58. *Prof Exp:* Asst prof biol sci, Carnegie Inst Technol, 58-61; asst prof, 61-69, PROF MICROBIOL, SCH MED, W VA UNIV, 69- *Mem:* Am Asn Immunol; fel Am Acad Microbiol; Am Acad Allergy & Clin Immunol. *Res:* Immune injury in pulmonary diseases; tissue antigens; responses to inhaled microbial antigens. *Mailing Add:* Dept of Microbiol Med Ctr W Va Univ Morgantown WV 26506

BURRELL, VICTOR GREGORY, JR, b Wilmington, NC, Sept 12, 25; m 56; c 4. MARINE SCIENCES. *Educ:* Col Charleston, BS, 49; Col William & Mary, MA, 68, PhD(marine sci), 72. *Prof Exp:* Res assoc, Va Inst Marine Sci, 66-68, from asst marine scientist to assoc marine scientist, 68-72; assoc marine scientist, 72-73, asst dir, 73-74, DIR, MARINE RESOURCES RES INST, SC, 74- *Concurrent Pos:* Adj prof biol, Col Charleston, 74-; mem, Outer Continental Shelf Tech Adv Comt Fisheries & adj assoc prof marine biol med, Univ SC, 81- *Mem:* Am Inst Fishery Res Biologists; Estuarine Res Fedn (secy, 77-79); Nat Shellfish Asn (pres, 82-83); Gulf & Caribbean Fisheries Inst; World Mariculture Soc. *Res:* Estuarine and marine zooplankton dynamics; copepod taxonomy; marine fisheries science; shellfish biology. *Mailing Add:* Marine Resources Res Inst PO Box 12559 Charleston SC 29412

BURRESON, EUGENE M, b Seattle, Wash, July 25, 44; m 66; c 2. PROTOZOOLOGY, FISH DISEASES. *Educ:* Eastern Ore State Col, BS, 66; Ore State Univ, MS, 73, PhD(zool), 75. *Prof Exp:* Proj dir, Normandeau Assoc, Inc, 75-77; SR MARINE SCIENTIST PARASITOL & ASST PROF, VA INST MARINE SCI, COL WILLIAM & MARY, 77- *Mem:* Am Soc Parasitologists; Soc Protozoologists; Soc Syst Zool. *Mailing Add:* Dept Marine Sci William & Mary Sch Marine Sci Gloucester Point VA 23062

BURRI, BETTY JANE, b San Francisco, Calif, Jan 23, 55; m 84. NUTRITION RESEARCH. *Educ:* San Francisco State Univ, BA, 76; Calif State Univ, Long Beach, MS, 78; Univ Calif, San Diego, PhD(chem), 82. *Prof Exp:* Res assoc molecular immunol, Res Inst Scripps Clin, 82-85; res chemist GS, 85-91, RES CHEMIST GM NUTRIT, WESTERN HUMAN NUTRIT RES CTR, 91- *Mem:* Am Inst Nutrit; Am Soc Clin Nutrit; NY Acad Sci; Asn Women Sci. *Res:* Establishing the intakes, requirements and statutes for fat soluble vitamins and antioxidants; methods research, especially in human nutrition; determining groups at risk for fat soluble vitamin and antioxidant nutrition. *Mailing Add:* Western Human Nutrit Ctr USDA Agr Res Serv PWA Bldg 1110 Presidio San Francisco CA 94129

BURRIDGE, MICHAEL JOHN, b St Albans, Eng, Apr 27, 42; m; c 1. TROPICAL ANIMAL HEALTH, BIOTECHNOLOGY. *Educ:* Univ Edinburgh, BVM&S, 66; Univ Calif, Davis, MPVM, 74, PhD(epidemiol), 76. *Prof Exp:* Res asst, EAfrican Trypanosomiasis Res Org, Uganda, 66; vet, Grant & Arnold, Woking, UK, 67-68; animal health officer, Food & Agr Org UN, Kenya, 68-73; grad res asst, Univ Calif, Davis, 73-76; assoc prof, 76-82, PROF EPIDEMIOL, UNIV FLA, 82-, CHMN, DEPT INFECTIOUS DIS, 84- *Concurrent Pos:* mem, Nat Acad Sci Comt Animal Health, 80-83; dir, Ctr Trop Animal Health, 82- *Mem:* Fel Am Col Epidemiol; Royal Col Vet Surgeons; Am Vet Med Asn. *Res:* Epidemiology of Zoonotic and livestock diseases; development using biotechnology of improved vaccines for tropical animal diseases that form important constraints to food production in developing countries. *Mailing Add:* Col Vet Med Univ Fla Bldg 471 Mowry Rd Gainesville FL 32611-0633

BURRIDGE, ROBERT, b Westcliff-on-Sea, Essex, UK, Dec 6, 37; m 87; c 3. ASYMPTOTIC ANALYSIS, WAVE MOTION. *Educ:* Univ Cambridge, BA, 59, PhD(math), 63, ScD, 80. *Prof Exp:* Res fel seismol, Calif Inst Technol, 63-64; res geophysicist, Univ Calif, Los Angeles, 64-65; asst lectr math, Univ Cambridge, 65-67; res fel math, UK Atomic Energy Authority, 67-71; from assoc prof to prof math, Courant Inst, NY Univ, 71-86; SCI ADV, SCHLUMBERGER-DOLL RES, 86- *Concurrent Pos:* Prin investr, NSF, 71-86; consult, Gulf Oil, Pittsburgh, Pa, 78-80 & Exxon, Clinton, NJ, 80-86. *Mem:* Am Math Soc; Soc Indust & Appl Math; Soc Explor Geophysicists. *Res:* Wave propagation; high frequency asymptotics; caustics; waves in finely layered and heterogeneous media; seismology; earthquake mechanisms; fracture mechanics; mechanics of fluid-filled porous media and other types of media with microstructure. *Mailing Add:* 43 Rockwell Rd Ridgefield CT 06877

BURRIER, ROBERT E, b York, Pa, Oct 2, 57; m. BIOCHEMISTRY. *Educ:* Univ Mass, BS, 79; Boston Univ, PhD(biochem), 83. *Prof Exp:* Postdoctoral fel, Lipoprotein Res Lab, Wake Forest Univ, Winston-Salem, NC, 83-85; postdoctoral fel, Cardiovasc Res Inst, Boston Univ Sch Med, 85-86, asst res prof biochem, 86-87; sr scientist, 87-88, prin scientist, 88-89, SR PRIN SCIENTIST, SCHERING-PLOUGH, BLOOMFIELD, NJ, 89- *Concurrent Pos:* NIH training grants, 80-83, 83-85 & 85-86. *Mem:* Am Heart Asn; assoc mem Am Soc Biol Chemists. *Res:* Effect of cholestyramine and a ACAT inhibitor on cholesterol metabolism in the hamster; circulation; author of several articles. *Mailing Add:* Schering-Plough Bloomfield NJ 07003

BURRILL, CLAUDE WESLEY, b Akron, Iowa, Feb 19, 25; m 55; c 2. DATA PROCESSING MANAGEMENT, MATHEMATICS. *Educ:* Univ Iowa, BS, 48, MS, 50, PhD(math), 52. *Hon Degrees:* LHD, William Paterson Col NJ, 79. *Prof Exp:* Instr math, Univ Iowa, 48-52; asst prof, NY Univ, 53-56; mathematician, Int Bus Mach Corp, 56-57; assoc prof math, NY Univ, 57-67; consult, IBM Systs Res Syst, 57-67, sr staff mem, 67-72, IBM Systs Sci Inst, 72-81, SR STAFF MEM, IBM SYSTS RES INST, 81- *Concurrent Pos:* Fulbright Scholar, 52-53; chmn Bd Trustees, William Paterson Col NJ, 74-76 , actg pres, 76-77. *Mem:* Am Math Soc; Math Asn Am; Opers Res Soc Am. *Res:* Mathematical analysis; financial model building; project methodology; data processing quality. *Mailing Add:* 26 Birchwood Pl Tenafly NJ 07670

BURRILL, MELINDA JANE, b Washington, DC, Mar 31, 47. GENETICS, ANIMAL BREEDING. *Educ:* Univ Ariz, BS, 69; Ore State Univ, PhD(genetics), 74. *Prof Exp:* Lab technologist toxicity, Indust Bio-Test Inc, 69; teaching asst animal sci, Ore State Univ, 70-74; from asst prof animal sci to assoc prof, 76-85, dir, small animal res lab, 76-87, PROF & GRAD STUDIES COORDR, CALIF STATE POLYTECH UNIV, 85- *Concurrent Pos:* Res fel, Univ Minn, 75-76; CIES, res fel, INRA, Jouy-en-Josas, France, 80-; Women in Develop fel, The Gambia, 84. *Mem:* Am Soc Animal Sci; Brit Soc Animal Prod; Can Soc Animal Sci; Am Genetics Asn; Sigma Xi. *Res:* Large animal breeding; selection theory and breed evaluation; sheep and goat semen evaluation and storage; histology of cattle hot brand scars. *Mailing Add:* Dept Animal Sci Calif State Polytech Univ Pomona CA 91768

BURRILL, ROBERT MEREDITH, b Oklahoma City, Okla, Oct 23, 33; m 73. BIOGEOGRAPHY, PHYSICAL GEOGRAPHY. *Educ:* Wesleyan Univ, BA, 56; Univ Chicago, MS, 61; Univ Kans, PhD(geog), 70. *Prof Exp:* Instr geog, Ohio Univ, 61-63; asst prof geog, Univ Ga, 66-79; vis assoc prof, Univ W Fla, 80-82; ADJ PROF, TROY STATE UNIV, FLA REGION, 84- *Honors & Awards:* Outstanding Hon Prof, Univ Ga, 76. *Mem:* Am Geog Soc; Asn Am Geogrs; Ecol Soc Am. *Res:* Investigation of weather effects upon honeybee activity using an invention which makes measure of honeybee activity possible; geographic analysis of animal behavior; environmental resource management (coastal zone). *Mailing Add:* 541 Woodbine Dr Pensacola FL 32503

BURRIS, CONRAD TIMOTHY, b Edmonton, Alta, May 17, 24; US citizen. CHEMICAL ENGINEERING, PHYSICAL CHEMISTRY. *Educ:* Univ Alta, BChE, 46, MChE, 48; Cath Univ Am, PhD(phys chem), 55. *Prof Exp:* Instr chem, Manhattan Col, 55-58, dir chem eng, 58-61, head dept, 61-71, dean eng, 71-80, prof chem eng, 80-89, head dept, 83-89, PROF LECTR, MANHATTAN COL, 89- *Mem:* Fel Am Inst Chem Engrs; Am Chem Soc; Am Soc Eng Educ; Sigma Xi. *Res:* Pressure effects on liquid phase reactions; dissolved oxygen analysis by polarography; factors influencing oxygen uptake rate in aqueous systems. *Mailing Add:* Sch Eng Manhattan Col Bronx NY 10471

BURRIS, JAMES F, b Mauston, Wis, Apr 15, 47; m 71; c 1. HYPERTENSION, PREVENTIVE CARDIOLOGY. *Educ:* Brown Univ, BA & BSc, 70; Col Physicians & Surgeons, Columbia Univ, MD, 74; Am Bd Internal Med, dipl, 79, dipl geriat med, 88. *Prof Exp:* Intern, Roosevelt Hosp, 74-75; gen med off, USPHS, 75-77; resident, Georgetown Univ Med Ctr, 77-79; fel, hypertension, Vet Admin Med Ctr, 79-81; from asst prof to assoc prof internal med, Georgetown Univ Med Sch, 81-91, asst dean, sponsored res, 87-90, ASSOC PROF MED & PHARMACOL, GEORGETOWN UNIV MED CTR, 86-, ATTEND PHYSICIAN MED, 87-, ASSOC DEAN RES OPERS, 90- & PROF INTERNAL MED, 91- *Concurrent Pos:* Res assoc, Hypertension Res Unit, Vet Admin Med Ctr, Washington, DC, 81-; attending physician, Georgetown Univ Med Ctr, 81-85; vis investr, div nephrol & hypertension, Univ Vaudois Med Ctr, Lausanne, Switzerland, 82-83; consult hypertension, Cardiovasc Ctr, Northern, Va, 84-; staff physician, geriat unit, Vet Admin Med Ctr, Washington, DC, 85-87; mem, bd dirs, Nations Capital Affil, Am Heart Asn, Washington, DC, 87- & Metropolitan Wash Chap, Am Geriat Soc, 89-; bd regents, Am Col Clin Pharmacol, 90- *Honors & Awards:* Physicians Recognition Award, Am Med Asn, 82, 85 & 88. *Mem:* Fel Am Col Cardiol; fel Am Col Physicians; Am Fed Clin Res; Am Soc Clin Pharmacol & Therapeut; Am Soc Hypertension; Nat Coun Univ Res Adminrs. *Res:* Administer medical center research programs; investigator for clinical trials specifically hypertension, hyperlipidemia, angina and basic research on glucocorticoid-induced hypertension in rats and sodium-potassium pump in human hypertension. *Mailing Add:* Off Sponsored Res NE 120 Med-Dent Georgetown Univ Med Sch 3900 Reservoir Rd NW Washington DC 20007

BURRIS, JOHN EDWARD, b Madison, Wis, Feb 1, 49; m 74; c 3. PLANT PHYSIOLOGY, MARINE BIOLOGY. *Educ:* Harvard Univ, AB, 71; Univ Calif, San Diego, PhD(marine biol), 76. *Prof Exp:* From asst prof to assoc prof biol, Pa State Univ, 76-85; dir, bd biol, 84-89, EXEC DIR, COMN LIFE SCI, NAT RES COUN, 88- *Concurrent Pos:* Adj assoc prof biol, Pa State Univ, 85- *Mem:* Am Soc Plant Physiologists; AAAS. *Res:* Science policy and administration; plant ecology and physiology; photosynthesis and photorespiration in aquatic and terrestrial plants; coral reef biology. *Mailing Add:* Nat Res Coun 2101 Constitution Ave NW Washington DC 20418

BURRIS, JOSEPH STEPHEN, b Cleveland, Ohio, Apr 18, 42; m 63; c 4. PLANT PHYSIOLOGY. *Educ:* Iowa State Univ, BSc, 64; Va Polytech Inst, MSc, 66, PhD(agron), 68. *Prof Exp:* Instr agron, Va Polytech Inst, 67; SEED PHYSIOLOGIST, SEED SCI CTR, IOWA STATE UNIV, 68- *Mem:* Am Soc Agron; Crop Sci Soc Am; Asn Official Seed Analysts; Int Seed Testing Asn. *Res:* Seed quality, production and technology with emphasis on the drying management of seed corn and the production of soybean seed and subsequent quality assurance programs. *Mailing Add:* Seed Sci Ctr Iowa State Univ Ames IA 50011

BURRIS, LESLIE, b Carmi, Ill, Sept 1, 22; m 44; c 3. NUCLEAR ENGINEERING. *Educ:* Univ Colo, BS, 43; Ill Inst Technol, MS, 56. *Prof Exp:* Chem engr, Monsanto Chem Co, 43-48; spec assignment, Tracerlab, 48; assoc chem eng, Argonne Nat Lab, 48-62, sr chem engr, 62-89, assoc div dir, 69-73, div dir, 73-84, prog mgr, 84-89; RETIRED. *Honors & Awards:* Robert E Wilson Award, nuclear chem eng, 84. *Mem:* Am Nuclear Soc; fel Am Inst Chem Engrs; Sigma Xi. *Res:* Atomic energy. *Mailing Add:* 206 Douglas Naperville IL 60540

BURRIS, MARTIN JOE, b Hebron, Nebr, Mar 30, 27; wid; c 4. GENETICS, ANIMAL HUSBANDRY. *Educ:* Univ Nebr, BS, 49, MS, 51; Ore State Col, PhD(genetics), 53. *Prof Exp:* Asst animal breeding, Univ Nebr, 49-51; asst prof, Univ Ark, 53-54; assoc prof, Va Agr Exp Sta, 54-57; animal geneticist, Coop State Res Serv, USDA, 57-58, prin animal geneticist, 58-66, asst dir, Mont Agr Exp Sta, Mont State Univ, 66-67, assoc dir, 67-80, prof, Animal & Range Sci Dept, 80-; RETIRED. *Concurrent Pos:* Vis prof animal sci, Purdue Univ, 64-65. *Mem:* Am Soc Animal Sci; Genetics Soc Am. *Res:* Beef cattle breeding; hormone physiology; meat processing; growth in domestic animals; livestock breeding and genetics; research administration. *Mailing Add:* 2503 Spring Creek Dr Bozeman MT 59717

BURRIS, ROBERT HARZA, b Brookings, SDak, Apr 13, 14; m 45; c 3. BIOCHEMISTRY, MICROBIOLOGY. *Educ:* SDak State Univ, BS, 36; Univ Wis, MS, 38, PhD(agr bact), 40. *Hon Degrees:* DSc, SDak State Univ, 66. *Prof Exp:* Asst agr bact, 36-40, from instr to prof, 41-84, chmn dept, 58-70, EMER PROF BIOCHEM, UNIV WIS-MADISON, 84- *Concurrent Pos:* Nat Res Coun fel, Columbia Univ, 40-41; consult, NSF, 53-57 & NIH, 61-71; Guggenheim fel, 54; chmn bot sect, Nat Acad Sci, 71; Indo-US Sci Technol Initiative, NASA-CELSS, Bd Sci Tech Int Develop. *Honors & Awards:* Stephen Hales Award & Charles Reid Barnes Award, Am Soc Plant Physiol; Thom Award, Soc Indust Microbiol; Browning Award, Am Soc Agron; Nat Medal Sci & Carty Award, Nat Acad Sci; Wolf Award; Spencer Award, Am Chem soc. *Mem:* Nat Acad Sci; fel Am Soc Biochem & Molecular Biol; Am Chem Soc; Soc Plant Physiol (pres, 60); Am Acad Arts & Sci; Am Philos Soc; foreign fel Indian Nat Sci Acad. *Res:* Biological nitrogen fixation; respiration of plants; nitrogen metabolism of plants; photosynthesis; biological oxidations; cytochromes; hydrobiology. *Mailing Add:* Dept Biochem 420 Henry Mall Univ Wis Madison WI 53706-1569

BURRIS, WILLIAM EDMON, b Okla, Nov 14, 24; m 47; c 2. ENVIRONMENTAL SCIENCES. *Educ:* Okla State Univ, BS, 48, MS, 49, PhD(zool), 56. *Prof Exp:* Instr zool, Okla State Univ, 52; biologist, US Army Eng Dist Fort Worth, 52-59; from asst prof to prof biol, San Antonio Col, 59-67, from actg chmn to chmn dept, 60-67; biologist, Southwestern Div, CEngrs, 67-70, recreation resource specialist, 70-73, environ resources planner, 73-76; environ planner, Bd Engrs for Rivers & Harbors, 76-79; SR ENVIRON ADV, OFF CHIEF ENGRS, US ARMY CORP ENGRS, 79- *Concurrent Pos:* Environ planner, Off Asst Secy Army Civil Works, 79. *Mem:* Am Fisheries Soc; Sigma Xi. *Res:* Limnology; fisheries biology; environmental planning; planning, development and operation of multiple purpose reservoirs and other water resource projects. *Mailing Add:* HQ USACE (CECW-PO) Washington DC 20314-1000

BURROS, RAYMOND HERBERT, b Scranton, Pa, Sept 24, 22; m 50; c 2. BAYESIAN STATISTICS. *Educ:* Trinity Col, BS, 45; Yale Univ, MA, 46, PhD(psychol), 49. *Prof Exp:* Asst prof psychol, Univ Ark, 49-51 & Univ Ill, 51-54; vis assoc prof, Univ Houston, 54-55; dir statist controls, Inst Motivational Res Inc, 55; sr opers res analyst, Tech Opers Inc, 55-61; indust design specialist human factors, Gen Elec Co, 61-62; opers res scientist sr, Syst Develop Corp, 62-66; sr opers res analyst, Sanders Assoc Inc, 67-68; sr engr, Sperry Rand Corp, 68-69; res engr, Port Authority NY & NJ, 69-79; staff engr, Lockheed Elec Co Inc, 80-81, eng sect head & staff engr, Harris Corp, 81-84, sr engr, Eaton Corp, 84-90; RETIRED. *Mem:* Am Statist Asn; Sigma Xi. *Res:* Application theory of stochastic processes to structural engineering, including spectral analysis and fire protection; logical foundations of statistical decision theory; application of Bayesian statistics to electronic warfare. *Mailing Add:* 55 Fleets Cove Rd Huntington NY 11743

BURROUGHS, RICHARD, b New Haven, Conn, July 5, 46; m 84; c 2. OCEANOGRAPHY. *Educ:* Princeton Univ, AB, 69; Mass Inst Technol, PhD(oceanog & marine geol), 75. *Prof Exp:* Staff off, Nat Res Coun, Nat Acad Sci, 74-77; sr fel, Marine Biol Lab, 77-79, 81, sci advr to dir, Bur Land Mgt, US Dept Interior, 79-81; res scientist, John E Gray Inst, Lamar Univ, 82-83; ASST/ASSOC PROF, GRAD PROG MARINE AFFAIRS, UNIV RI, KINGSTON, 83- *Concurrent Pos:* vis lectr, Sch Forestry & Environ Studies, Yale Univ, 78-79; vis prof, Col Atlantic, 88-90; vis assoc prof, Sch Forestry & Environ Studies, Yale Univ, 90-91. *Mem:* AAAS; Am Geophys Union; Geol Soc Am; Sigma Xi. *Res:* Management of marine resources. *Mailing Add:* Grad Prog Marine Affairs Univ RI Kingston RI 02881

BURROUGHS, RICHARD LEE, b Detroit, Mich, Aug 30, 32; m 61; c 2. PETROLEUM GEOLOGY. *Educ:* Okla State Univ, BS, 55; Univ Ariz, MS, 60; Univ NMex, PhD(geol), 72. *Prof Exp:* Explor geologist, Pure Oil Co, 56-58 & Monsanto Chem Co, 60-63; prof geol & head dept, 63-85, EMER PROF GEOL, ADAMS STATE COL, 88- *Concurrent Pos:* Consult geology, 73-; sr consult geologist, Binglidesh Oil, Gas & Mineral Resources, Dhaka, 85-86. *Mem:* Geol Soc Am; Am Asn Petrol Geologists. *Res:* Subsurface geology of the San Luis Basin and Rio Grande rift; geothermal hydrocarbons. *Mailing Add:* 1008 Douglas Dr Alamosa CO 81101

BURROUS, MERWYN LEE, organic chemistry, for more information see previous edition

BURROUS, STANLEY EMERSON, b Elkhart, Ind, Mar 14, 28; div; c 3. BIOCHEMISTRY. *Educ:* Manchester Col, BA, 50; Ind Univ, MA, 53; Univ Ill, PhD(microbiol), 62. *Prof Exp:* Res asst virol, Upjohn Co, 58-60; asst prof bioeng, Okla State Univ, 62-63; sr res scientist, E R Squibb Div, Olin Mathieson Chem Corp, 63-65; sr biochemist, Smith Kline & French Labs, Pa, 65-67; unit leader, Eaton Labs Div, Norwich Pharmacal Co, 67-75; pres, Burrous Enterprises, 75-77; EXEC SECY, REV BR, DIV RES GRANTS, NIH, 77- *Mem:* AAAS; Am Soc Microbiol; Am Chem Soc. *Mailing Add:* Rte 3 Box 780 Boone NC 28607

BURROW, GERARD N, b Boston, Mass, Jan 9, 33; m 56; c 3. INTERNAL MEDICINE. *Educ:* Brown Univ, AB, 54; Yale Univ, MD, 58; FRCPS(C); Am Bd Internal Med, cert, 65 & 74. *Prof Exp:* Intern internal med, Yale New Haven Hosp, 58-59, resident med, 61-63; instr, 65-66, from asst prof to assoc prof internal med, Sch Med, Yale Univ, 70-75; prof med, Univ Toronto, 75-81, Sir John & Lady Eaton prof & chmn, Dept Med, 81-87; VCHANCELLOR HEALTH SCI & DEAN, SCH MED, UNIV CALIF, SAN DIEGO, 88-, PROF, DEPT MED, 88- *Concurrent Pos:* Fel metab, Yale New Haven Hosp, 63-65; res career develop award, 68-73; chief med res, Yale New Haven Hosp, 65-66; vis scientist, Fac Med, Univ Marseille, 72-73; dir, Endocrinol Sect & sr physician, Toronto Gen Hosp, 76-81, physician-in-chief, 81-87; mem comt, Cancer Ctrs Prog, Inst Med, 89; mem bd dirs, La Jolla Inst Allergy & Immunol, 89-90; mem, Pub Issues Comt, Am Cancer Soc, 90-; mem develop comt, Arnold & Mabel Beckman Ctr, Inst Med-Nat Acad Sci, 90-; mem planning comt, Asn Acad Health Ctrs, 91; chmn, Rev Panel Int Res Scholars Prog, Howard Hughes Med Inst, 91. *Mem:* Inst Med-Nat Acad Sci; Can Asn Professors Med (vpres, 85, pres, 86); Asn Am Physicians; Can Soc Clin Invest; Endocrine Soc; fel Am Col Physicians; Am Thyroid Asn (pres-elect, 85, pres, 86); Am Fedn Clin Res; AMA; Am Soc Cell Biol. *Res:* Protein synthesis and hormone action; thyroid gland. *Mailing Add:* Off Dean M-002 Sch Med Univ Calif San Diego 9500 Gilman Dr La Jolla CA 92093-0602

BURROW, PAUL DAVID, b Okla City, Okla, Aug 3, 38; m 65; c 2. ELECTRON SCATTERING. *Educ:* Mass Inst Technol, BS, 60; Univ Calif, PhD(physics), 66. *Prof Exp:* From asst prof to assoc prof eng & appl sci, Yale Univ, 67-76; assoc prof 76-78, PROF PHYSICS, UNIV NEBR, 78- *Concurrent Pos:* Exchange prof, Univ Paris-Sud, Orsay, France, 82. *Mem:* Fel Am Phys Soc. *Res:* Low electron scattering studies particularly temporary negative ion formation in complex molecules. *Mailing Add:* Behlen Lab Univ Nebr Lincoln NE 68588

BURROWS, ADAM SETH, b Salt Lake City, Utah. HYDRODYNAMICS, RADIATIVE TRANSFER. *Educ:* Princeton Univ, AB, 75; Mass Inst Technol, PhD(physics), 79. *Prof Exp:* Lectr, physics, Univ Mich, Ann Arbor, 79-80; fel, State Univ NY, Stony Brook, 80-83, asst prof physics, 83-86; ASSOC PROF PHYSICS & ASTRON, UNIV ARIZ, 86- *Honors & Awards:* Alfred Sloan Found award, 87. *Mem:* Am Astron Soc; Int Astron Union; Am Phys Soc; AAAS; NY Acad Sci; Sigma Xi. *Res:* Supernova theory, neutron stars, neutrino astronomy, brown dwarf theory, nuclear astrophysics; published articles in various journals. *Mailing Add:* Dept Physics Univ Ariz Tucson AZ 85721

BURROWS, BENJAMIN, b New York, NY, Dec 16, 27; m 49; c 4. INTERNAL MEDICINE. *Educ:* Johns Hopkins Univ, MD, 49. *Prof Exp:* Instr med, Univ Chicago Clins, 55-56, asst prof, 56-61, assoc prof, 61-68; PROF MED & DIR DIV RESPIRATORY SCI, UNIV ARIZ, 68- *Concurrent Pos:* Consult, Tucson Vet Admin Hosp. *Mem:* Am Physiol Soc; Am Fedn Clin Res; Am Thoracic Soc; fel Am Col Chest Physicians; fel Am Col Physicians; Asn Am Physicians; emer mem Am Soc Clin Invest. *Res:* Pulmonary disease; pulmonary physiology; diffusion; epidemiology. *Mailing Add:* Div Respiratory Sci Ariz Health Sci Ctr Tucson AZ 85724

BURROWS, CYNTHIA JANE, b St Paul, Minn, Sept 23, 53. PHYSICAL ORGANIC CHEMISTRY, BIO-ORGANIC CHEMISTRY. *Educ:* Univ Colo, Boulder, BA, 75; Cornell Univ, MS, 78, PhD(chem), 82. *Prof Exp:* Res assoc org chem, Univ Louis Pasteur, Strasbourg, France, 81-83; ASSOC PROF CHEM, STATE UNIV NY, STONY BROOK, 83- *Mem:* Am Chem Soc; Sigma Xi; AAAS. *Res:* Synthetic macrocyclic compounds for formation of molecular complexes; enzyme mimetic chemistry; reaction mechanisms. *Mailing Add:* Dept Chem State Univ NY Stony Brook NY 11794-3400

BURROWS, ELIZABETH PARKER, b Pittsburgh, Pa, Nov 5, 30; m 58. MASS SPECTROMETRY, TRACE ORGANIC ANALYSIS. *Educ:* Middlebury Col, AB, 52; Stanford Univ, MS, 54, PhD(chem), 57. *Prof Exp:* Res assoc pharmacol, Stanford Univ, 56-57; fel chem, Wayne State Univ, 57-58; fel, Mass Inst Technol, 58-61, res assoc, 62-66; res assoc, Children's Cancer Res Found, Mass, 66-67; staff scientist, Worcester Found Exp Biol, 67-70; res assoc chem, Oakland Univ, 70-71; res assoc, 71-75, res asst prof, Vanderbilt Univ, 75-76; res assoc, Johns Hopkins Univ, 76-79; lectr, Univ Maryland, 79-80; RES CHEMIST, BIOMED RES & DEVELOP LAB, US ARMY, 80- *Mem:* Am Chem Soc; Am Inst Chemists; Sigma Xi. *Res:* Mechanistic and analytical problems associated with trace organics in the environment; organic structural determination by spectroscopic methods. *Mailing Add:* US Army Biomed Res & Develop Lab Ft Detrick Frederick MD 21702-5010

BURROWS, GEORGE EDWARD, b Seattle, Wash, Feb 17, 35; m 55; c 6. VETERINARY PHARMACOLOGY, VETERINARY TOXICOLOGY. *Educ:* Univ Calif, Davis, BS, 64, DVM, 66; Wash State Univ, MS, 69, PhD(pharmacol, toxicol), 72. *Prof Exp:* From instr to asst prof vet med, Col Vet Med, Wash State Univ, 66-72; vis lectr, Univ Nairobi & Colo State Univ, 72-74; asst prof, Col Vet Med, Wash State Univ, 74-75; assoc prof vet pharmacol & toxicol, Univ Idaho & Northwest Col Vet Med, 75-78; PROF CLIN PHARMACOL & TOXICOL, OKLA STATE UNIV, 78- *Concurrent Pos:* Vis prof, Univ Zimbabwe, 86-87. *Mem:* Am Vet Med Asn; Am Col Vet Toxicologists. *Res:* Cyanide intoxication and its therapy and prevention; plants toxic to domestic animals; diet and its effects on xenobiotic disposition in animals; clinical pharmacology of antibiotics in domestic animals. *Mailing Add:* Dept Physiol Sci Okla State Univ Stillwater OK 74074

BURROWS, KERILYN CHRISTINE, b Catasauqua, Pa, Jan 15, 51; m 79. POLAROGRAPHY, VOLTAMETRY. *Educ:* Muhlenberg Col, BS, 72; Lehigh Univ, PhD(chem), 79. *Prof Exp:* Instr chem, Muhlenberg Col, 76-78; ASST PROF CHEM, MADONNA COL, 78- *Mem:* Am Chem Soc; Nat Wildlife Fedn. *Res:* Linear sweep differential pulse voltametry; analysis of metals in environmental systems. *Mailing Add:* 3481 Dogwood Lane Box 363 Bath OH 44210-0363

BURROWS, MICHAEL L(EONARD), b London, Eng, July 30, 34; m 60; c 3. ELECTRICAL ENGINEERING. *Educ:* Univ London, BSc, 56; Univ Mich, MS, 59, PhD(elec eng), 64. *Prof Exp:* Apprentice elec eng, Brit Insulated Callenders Cables, 56-58, sci off electronics, 59-60; STAFF MEM, LINCOLN LAB, MASS INST TECHNOL, 64- *Mem:* Sr mem Inst Elec & Electronics Engrs; Brit Inst Elec Engrs. *Res:* Antennas, electromagnetic scattering and propagation; eddy-current testing; communications. *Mailing Add:* Lincoln Lab Mass Inst Technol PO Box 73 Lexington MA 02173

BURROWS, ROBERT BECK, b Columbia, SC, Oct 29, 07; wid; c 3. PARASITOLOGY. *Educ:* Emory Univ, AB, 29, MS, 30; Yale Univ, PhD(zool), 36. *Prof Exp:* Lab asst, Emory Univ, 29-30 & Yale Univ, 30-33; lab technician, State Bd Health, Ga, 33-34; biologist, US Bur Fisheries, WVa, 34-35, jr biologist, Conn, 35-36; assoc prof biol, Elon Col, 36-37; adj prof, Am Univ Beirut, 37-40, head dept, 39-40; parasitologist, State Hosp, SC, 40-43 & 46-49; actg chief dept parasitol, Army Med Ctr, 49-50; cmndg officer, Trop Res Lab, US Dept Army, PR, 50-51; chief parasitol br, Sixth Army Area Med Lab, 52-55; head, Parasitol Sect, Wellcome Res Labs, 55-72; RETIRED. *Concurrent Pos:* Med consult, 1st US Army Area, 63-68. *Honors & Awards:* Jefferson Award, SC Acad, 47. *Mem:* AAAS; Am Soc Parasitol; Am Soc Trop Med & Hyg. *Res:* Parathyroidism and bone; diagnosis, incidences, treatment and pathology of parasitic infections; morphology of amoebae; experimental chemotherapy of helminth infections. *Mailing Add:* Five Maple Tree Dr Mt Holly NJ 08060

BURROWS, THOMAS WESLEY, b Janesville, Wis, May 17, 43; m 68; c 2. NUCLEAR PHYSICS, COMPUTER PROGRAMMING. *Educ:* Univ Wis-Madison, BS, 65, PhD(physics), 72. *Prof Exp:* Nuclear Info Res assoc, Univ Ky, 72-74; asst physicist, Nat Neutron Cross Sect Ctr, 74-76, assoc physicist nuclear physics, 76-78, PHYSICIST, NAT NUCLEAR DATA CTR, BROOKHAVEN NAT LAB, 78- *Concurrent Pos:* Secy, Panel Ref Nuclear Data, 76-81; file mgr, Evaluated Nuclear Struct Data File, 81-82; secy & ed, NEANDC Specialists Meeting on Yields & Decay Data Fission Prod Nuclides, 83-84. *Mem:* Am Phys Soc; Inst Physics; affil mem Am Chem Soc; AAAS. *Res:* Nuclear structure; neutron physics; evaluating nuclear structure and decay data for publication in nuclear data sheets; supervise other evaluators; develop, modify and write analysis programs for these activities. *Mailing Add:* Nat Nuclear Data Ctr Bldg 197D Brookhaven Nat Lab Upton NY 11973

BURROWS, VERNON DOUGLAS, b Winnipeg, Man, Jan 9, 30; m 54; c 2. PLANT BREEDING. *Educ:* Univ Man, BSA, 51, MSc, 53; Calif Inst Technol, PhD, 58. *Prof Exp:* Plant physiologist, 58-69, HEAD CEREAL SECT, RES STA, AGR CAN, 69- *Honors & Awards:* Grindley Medal, Agr Inst Can, 75. *Res:* Plant breeding in general and oats breeding in particular. *Mailing Add:* 32 Scholar Ct Nepean ON K2E 7S2 Can

BURROWS, WALTER HERBERT, b Columbia, SC, Oct 23, 11; m 35; c 2. INDUSTRIAL CHEMISTRY. *Educ:* Emory Univ, AB, 33, MS, 38. *Prof Exp:* Prof sci & math, Spartanburg Jr Col, SC, 38-41; from instr to asst prof chem, Ga Inst Technol, 41-56, res assoc prof & head, Indust Prod Br, Eng Exp Sta, 56-67, prin res chemist & head, Spec Projs Br, 67-73, head, Indust Chem

Lab, Eng Exp Sta, 73-76; RETIRED. *Concurrent Pos:* Pres, W H Burrows, Consult, 72- *Mem:* AAAS; Am Chem Soc; Am Inst Chemists; Am Inst Chem Engrs; Nat Fire Protection Asn; Am Soc Testing & Mat. *Res:* Surface chemistry, lubricant, coatings, adhesives sealants, detergents, corrosion; fire and combustion technology; nomography, graphical and mechanical computation; bioconversion of energy, anaerobic fermentation, waste utilization. *Mailing Add:* 3007 Northbrook Dr Atlanta GA 30340

BURROWS, WILLIAM DICKINSON, b Saginaw, Mich, Dec 31, 30; m 58. ENVIRONMENTAL CHEMISTRY. *Educ:* Cornell Univ, BA, 53; Stanford Univ, PhD(chem), 56; Vanderbilt Univ, MS, 73. *Prof Exp:* Res assoc chem, Mass Inst Technol, 58-61; asst prof, Clarkson Col Technol, 61-62; res assoc, Mass Inst Technol, 62-66; sr vis scientist, US Army Natick Labs, 66-68; sr chemist, Garrett Res & Develop Co, Occidental Petrol Corp, 68-69; sr engr, Assoc Water & Air Resources Engrs Inc, 73-75; ENVIRON ENGR, US ARMY MED & BIOMED RES & DEVELOP LAB, 75- *Concurrent Pos:* Adj prof. George Washington Univ, 82-87; vpres, Schaub Burrows Assoc, Ltd, 85- *Mem:* Am Chem Soc; Water Pollution Control Fedn; Sigma Xi; Am Water Works Asn. *Res:* Aquatic chemistry and treatment of munitions production wastes; military water supply and sanitation. *Mailing Add:* US Army Biomed Res & Develop Lab Ft Detrick MD 21701

BURRUS, CHARLES ANDREW, JR, b Shelby, NC, July 16, 27; m 57; c 3. OPTICS. *Educ:* Davidson Col, BS, 50; Emory Univ, MS, 51; Duke Univ, PhD, 55. *Prof Exp:* Res assoc physics, Duke Univ, 54-55; MEM TECH STAFF, AT&T BELL LABS, 55- *Concurrent Pos:* Distinguished mem tech staff, AT&T Bell Labs, 82. *Honors & Awards:* David Richardson Medal, Optical Soc Am, 82. *Mem:* Fel AAAS; fel Inst Elec & Electronics Engrs; fel Am Phys Soc; fel Optical Soc Am. *Res:* Microwave spectroscopy in the shorter millimeter and sub-millimeter wave region; millimeter-wave diodes; semiconductor lasers, electroluminescent diodes and high speed photodetectors; optical communications; optical-fiber devices; single-crystal fibers. *Mailing Add:* Crawford Hill Lab AT&T Bell Labs Holmdel NJ 07733

BURRUS, CHARLES SIDNEY, b Abilene, Tex, Oct 9, 34; m 58; c 2. ELECTRICAL ENGINEERING. *Educ:* Rice Univ, BA & BS, 58, MS, 60; Stanford Univ, PhD(elec eng), 65. *Prof Exp:* Res asst elec eng, Stanford Univ, 62-65; from asst prof to assoc prof, 65-74, PROF ELEC ENG, RICE UNIV, 74- *Concurrent Pos:* Grants, Ford Found, 62-66, NSF, 66-68, 70-85; sr award, Alexander von Humboldt Found, 75; vis prof, Univ Erlangen, Ger, 75-76 & 79-80; Fulbright fel, 79; vis fel, Trinity Col, Cambridge, 84; chmn, ECE dept, Rice Univ, 84-; vis prof, Mass Inst Technol, 89-90. *Honors & Awards:* Tech Achievement Award, Inst Elec & Electronics Engrs, 85. *Mem:* Fel Inst Elec & Electronics Engrs. *Res:* System theory, especially signal theory; optimization theory; digital signal processing. *Mailing Add:* Dept Elec Eng Rice Univ Houston TX 77251-1892

BURRUS, ROBERT TILDEN, b High Point, NC, July 15, 35; m 64; c 2. INORGANIC & PHYSICAL CHEMISTRY, FIBER & POLYMER PROCESSING. *Educ:* Univ NC, BS, 57; Univ Tenn, PhD(Chem), 62. *Prof Exp:* Trainee, Westinghouse Elec Corp, 57-58; asst chem, Univ Tenn, 58-62; res chemist, Fiber Surface Res Sect, 62-69, SR RES CHEMIST DACRON STAPLE & YARN, E I DU PONT DE NEMOURS & CO, INC, 69- *Mem:* Am Chem Soc. *Res:* Physical inorganic chemistry; electrochemistry in nonaqueous media; friction and surface chemistry of polymers; finish development for synthetic fibers; fiber and polymer chemistry; physics of fiber formation and quenching; fiber spinning and processing; texturing of continous filament yarns. *Mailing Add:* Dacron Yarn Res & Develop Lab Box 800 E I du Pont de Nemours & Co Inc Kinston NC 28501

BURRY, JOHN HENRY WILLIAM, b Nfld, June 10, 38; m 65; c 2. MATHEMATICS. *Educ:* Mem Univ Nfld, BAEd, 58; Dalhousie Univ, MSc, 61; Queen's Univ, Ont, PhD(math), 71. *Prof Exp:* Lectr, 61-63, from asst prof to assoc prof, 63-75, head dept math statist & comput sci, 76-85, PROF MATH & STATIST, MEM UNIV NFLD, 86- *Concurrent Pos:* Chmn, Nat Can Olympiad Comt, 78-80. *Mem:* Can Math Soc; Can Asn Univ Teachers; Am Math Soc; Nat Coun Teachers Math; Math Asn Am. *Res:* Radon measures in Euclidian spaces; measure theory; information theory. *Mailing Add:* Dept Math Mem Univ Nfld St John's NF A1C 5S7 Can

BURRY, KENNETH A, b Monterey Park, Calif, Oct 2, 42; m 82; c 2. REPRODUCTIVE ENDOCRINOLOGY, INFERTILITY. *Educ:* Whittier Col, BA, 64; Univ Calif, Irvine, MD, 68. *Prof Exp:* Fel reprod endocrinol, Univ Wash, 74-76; from asst prof to assoc prof, obstet & gynec, Ore Health Sci Univ, 76-89, co-dir infertil serv, 85-89, dir, Ore Reproduction Res & Fertil Prog, 82-89, PROF & ASST CHMN OBSTET & GYNEC, ORE HEALTH SCI UNIV, 89- *Concurrent Pos:* Dir, fel training reprod endocrinol & infertil, Ore Health Sci Univ, 84- *Honors & Awards:* Lange Award, Univ Calif, 66. *Mem:* Fel Am Col Obstet & Gynec; Endocrine Soc; Am Fertil Soc; Soc Reprod Endocrinologists; Soc Reprod Surgeons; Am Fedn Clin Res; Soc Assisted Reprod Technol. *Res:* Evaluation of endocrine and physical causes of reproductive failure; emotional sequence of infertility specifically ovulation induction, in vitro fertilization and endometriosis. *Mailing Add:* 3181 SW Sam Jackson Park Rd 3181 SW Jackson Park Road Portland OR 97201

BURSCHKA, MARTIN A, b Troisdors, WGer, May 11, 52. STATISTICAL PHYSICS. *Educ:* Univ Aachen, MD, 81, MS, 78; Rikks Univ, Utrecht, PhD(math), 85. *Prof Exp:* RES ASSOC, ROCKEFELLER UNIV, 85- *Mem:* AAAS; Am Phys Soc; NY Acad Sci. *Mailing Add:* Biophys Lab 262 Rockefeller Univ 1230 York Ave New York NY 10021

BURSE, RICHARD LUCK, b Boston, Mass, Feb 6, 36; c 1. ENVIRONMENTAL PHYSIOLOGY, ENGINEERING PSYCHOLOGY. *Educ:* Mass Inst Technol, BS, 58; Harvard Univ, MAT, 62; Univ Pittsburgh, MSc, 70. *Hon Degrees:* ScD, Univ Pittsburgh, 72. *Prof Exp:* Sci teacher biol & chem, Newton High Sch, Mass, 62-63; eng psychol human factors, US Army Natick Labs, Mass, 63-69; asst prof physiol, Sch Med, St Louis Univ, 73; PHYSIOLOGIST ENVIRON PHYSIOL, US ARMY RES INST ENVIRON MED, 74- *Concurrent Pos:* Mallinkrodt vis prof, Sch Med, St Louis Univ, 73; guest lectr, Sch Allied Health Prof, Boston Univ, 75-80, adj asst prof, 81-; res & develop coordr, Hq US Army Mat Command, Alexandria, Va, 79- *Mem:* Am Physiol Soc; Human Factors Soc; Sigma Xi. *Res:* Environmental and work physiology; human thermoregulatory and altitude adaptation; control of breathing; human strength capabilities. *Mailing Add:* 149 Jackson Rd Newton MA 02158

BURSEY, CHARLES ROBERT, b Paris, Tenn, Feb 13, 40; m 64; c 2. INVERTEBRATE PHYSIOLOGY. *Educ:* Kalamazoo Col, BA, 62; Mich State Univ, MS, 65, PhD(zool), 69. *Prof Exp:* NIH fel, Inst Marine Sci, Univ Miami, 69-70; from asst prof to assoc prof, 70-83, PROF BIOL, PA STATE UNIV, 83- *Concurrent Pos:* Partic, Univ Miami Deep Sea Exped to PR Trench, R/V Pillsbury, 70; chief scientist student training cruises R/V Annandale, Marine Sci Consortium, Millersville, 72-75; instr NSF Sci Student Training Prog Coastal Biol, Marine Sci Consortium, Wallops Island, Va, 77. *Mem:* Am Soc Zoologists; Crustacean Soc. *Res:* Microanatomy of the eyes, optic ganglia, and supraesophageal ganglion of Munida iris; feeding behavior of the sea anemone Condylactis; osmoregulation, salinity and temperature tolerance of the mole crab, Emerita. *Mailing Add:* Dept Biol Shenango Valley Campus Sharon PA 16146

BURSEY, JOAN TESAREK, b Omaha, Nebr, Mar 25, 43; m 70; c 2. MASS SPECTROMETRY, ENVIRONMENTAL ANALYSIS. *Educ:* Creighton Univ, BSChem, 65; Univ Calif, Berkeley, PhD(chem), 69. *Prof Exp:* Res assoc chem, Univ NC, Chapel Hill, 69-71; jr chemist, Res Triangle Inst, 71-74, chemist, 74-78, sr chemist, 78-81, mgr, Mass Spectrometry Lab, 81-84; sr scientist, 84-87, SR STAFF SCIENTIST, RADIAN CORP, 87- *Mem:* Am Soc Mass Spectrometry; Am Chem Soc; Royal Soc Chem. *Res:* Gas chromatography-mass spectrometry; environmental and biomedical analysis; organic mass spectrometry. *Mailing Add:* 101 Longwood Pl Chapel Hill NC 27514-9584

BURSEY, MAURICE MOYER, b Baltimore, Md, July 27, 39; m 70; c 2. ANALYTICAL CHEMISTRY. *Educ:* Johns Hopkins Univ, BA, 59, MA, 60, PhD(chem), 63. *Prof Exp:* Lectr chem, Johns Hopkins Univ, 63-64; asst prof, Purdue Univ, 64-66; from asst prof to assoc prof, 66-74, dir undergrad studies chem, 85-88, PROF CHEM, UNIV NC, CHAPEL HILL, 74- *Concurrent Pos:* Sloan fel, 69-71. *Mem:* Am Chem Soc; Am Soc Mass Spectrometry; The Royal Soc Chem; Japanese Soc Mass Spectros. *Res:* Mass spectrometry; chemistry of gaseous ions by mass spectrometry; chemistry of formation of gaseous ions from condensed phases. *Mailing Add:* CB 3290 Venable Hall Univ NC Chapel Hill NC 27599-3290

BURSH, TALMAGE POUTAU, b Leesville, La, Dec 24, 32; m 54; c 3. PHYSICAL CHEMISTRY. *Educ:* Southern Univ, BS, 56; Alfred Univ, PhD(chem), 63. *Prof Exp:* Asst instr phys sci, Southern Univ, 57, from assoc prof to prof chem, 68-80; AT SOUTHERN UNIV. *Res:* Interaction of water with silica, alumina and silica alumina co-oxide surfaces and the relationship of these interactions to catalytic activities of these oxides. *Mailing Add:* Dept Chem Southern Univ A&M Col Baton Rouge LA 70813

BURSIAN, STEVEN JOHN, b Petoskey, Mich, Sept 7, 47; div; c 2. TOXICOLOGY. *Educ:* Univ Mich, BS, 69; Univ Minn, MS, 71; NC State Univ, PhD(physiol), 78. *Prof Exp:* Res biologist, US Environ Protection Agency, 71-79; from asst prof to assoc prof, 79-90, PROF, DEPT ANIMAL SCI, MICH STATE UNIV, 90- *Mem:* AAAS; Sigma Xi; Soc Toxicol; Soc Neurosci; Soc Exp Biol & Med. *Res:* Examine the neurotoxicological effects of environmental pollutants in various species of animals. *Mailing Add:* Dept Animal Sci 132 Anthony Hall Mich State Univ East Lansing MI 48824

BURSNALL, JOHN TREHARNE, b Neath, UK, May 9, 40; m 66, 85; c 2. ARCHEAN GEOLOGY. *Educ:* Univ London, BSc, 68; Univ Cambridge, PhD(geol), 75. *Prof Exp:* Demonstr-tutor geol, Univ Cambridge, 68-72; tutor-organizer natural sci, Workers' Educ Asn, UK, 72-75; asst prof struct geol, Syracuse Univ, 75-82; asst prof struct geol, St Lawrence Univ, 82-83; prof struct geol, Southampton Col, 83-86; prof struct geol, Univ Mass, 86-87 & Univ Idaho, 87-88; ASST PROF MINERAL & PETROL, ST LAWRENCE UNIV, 89- *Concurrent Pos:* Vis lectr geol, Cambridgeshire Col Arts & Technol, 68-72; res fel, Geophys, Univ Toronto, 89. *Mem:* Geol Soc Can; Sigma Xi. *Res:* Geological structure of parts of the northern Appalachians, central Maine, northwest Newfoundland; ophiolite emplacement and deformation history within orogenic belts, shear zones in foliated rocks; mid to lower Archean crust, Kapukasing structual zone, Ontario. *Mailing Add:* Dept Geol Saint Lawrence Univ Canton NY 13617

BURSON, BYRON LYNN, b Hobart, Okla, Feb 24, 40; m 61; c 2. PLANT CYTOGENETICS, PLANT GENETICS. *Educ:* Okla State Univ, BS, 62; Tex A&M Univ, MS, 65, PhD(plant breeding, cytogenetics), 67. *Prof Exp:* Asst prof grass cytogenetics & asst agronomist, Miss State Univ, 67-71, assoc prof grass cytogenetics & assoc agronomist, 71-75; RES GENETICIST, GRASSLAND, SOIL & WATER RES LAB, AGR RES SERV, USDA, 75- *Mem:* Am Soc Agron; Crop Sci Soc Am; Am Genetic Asn; Genetics Soc Can. *Res:* Cytogenetics of grasses; reproductive systems, especially apomixis; species relationships within genus paspalum; cell and tissue culture. interspecific hybridization. *Mailing Add:* Grassland Soil & Water Res Lab 808 E Blackland Rd Box 6112 Temple TX 76502

BURSON, SHERMAN LEROY, JR, b Pittsburgh, Pa, Dec 24, 23; m 44; c 4. ORGANIC CHEMISTRY. *Educ:* Univ Pittsburgh, BS, 47, PhD(org chem), 53. *Prof Exp:* Asst chem, Univ Pittsburgh, 47-48, 52; chemist, Lederle Lab Div, Am Cyanamid Co, 52-57; assoc prof chem, Pfeiffer Col, 57-60, prof & chmn dept, 60-63; prof chem, Univ NC, Charlotte, 63-67, chmn dept, 63-75, Charles H Stone prof chem, 67-85, dean, Col Sci & Math, 75-80, dean, Col Arts & Sci, 80-85; RETIRED. *Mem:* Am Chem Soc. *Res:* Biotin analogues; polysaccharides; natural products. *Mailing Add:* 52 Dusty Miller Lane South Chatham MA 02659-1340

BURST, JOHN FREDERICK, b St Louis, Mo, Oct 16, 23; m 56; c 4. CLAY MINERALOGY. *Educ:* Univ Mo-Rolla, MS, 47; Univ Mo-Columbia, PhD(geol), 50. *Prof Exp:* Mem staff geol, Shell Develop Co, Tex, 50-65; dir res & develop, Gen Refractories Co, 65, dir res, 65-70, tech dir, 70-73; tech adv, Certain-Teed Prods Co, 73-75; tech dir, Dresser Minerals, 75-80, dir planning, Harbison-Walker Group, 80-82, vpres resource develop, Dresser Minerals Div, Dresser Indust, 82-85; exec dir, Mo Inc Found, 86-88; PRIN, IMMI CONSULT GROUP, 85-; PRES, TRIANGLE ENVIRON SCI & ENG, 90- *Mem:* Fel Geol Soc Am; fel Mineral Soc Am; Clay Minerals Soc (pres, 71); Soc Petrol Geol; Am Ceramic Soc; distinguished mem, Soc Mining Engrs; Mineral Soc London. *Res:* Chemistry minerals and ceramics. *Mailing Add:* 1605 Lincoln Lane Rolla MO 65401-2613

BURSTEIN, DAVID, b Englewood, NJ, May 19, 47; m 71; c 2. COSMOLOGY, STRUCTURE OF GALAXIES. *Educ:* Wesleyan Univ, BA, 69; Univ Calif, Santa Cruz, PhD(astron), 78. *Prof Exp:* Res fel astron, Dept Terrestrial Magnetism, Carnegie Inst Washington, 77-79; res assoc, Nat Radio Astron Observ, 79-82; asst prof, 82-88, ASSOC PROF PHYSICS DEPT, ARIZ STATE UNIV, 88- *Mem:* Am Astron Soc; Sigma Xi; Int Astron Union. *Res:* Stellar content, structure and dynamics of galaxies; galactic structure; large scale structure of universe cosmology. *Mailing Add:* Dept Physics & Astron Ariz State Univ Tempe AZ 85287-1504

BURSTEIN, ELIAS, b New York, NY, Sept 30, 17; m 43; c 3. SOLID STATE PHYSICS. *Educ:* Brooklyn Col, AB, 38; Univ Kans, AM, 41. *Hon Degrees:* DTech, Chalmers Univ Technol, Gothenburg, Sweden, 82; DSc, Brooklyn Col, 85. *Prof Exp:* Assoc, Nat Defense Res Comt Proj, Mass Inst Technol, 42-44; res & develop proj engr, White Res Assocs Mass, 44-45; physicist, Crystal Br, Naval Res Lab, 45-57, head, Semiconductor Br, 58; prof, 58-82, Mary Amanda Wood prof, 82-88, EMER MARY AMANDA WOOD PROF PHYSICS, UNIV PA, 88- *Concurrent Pos:* Chmn solid state sci comt, Nat Res Coun-Nat Acad Sci, 77-78; ed-in-chief, Solid State Commun; consult, Thomas J Watson Res Ctr, IBM, 71-; co-ed, Comments on Solid State Physics, 72-; vis Jubilee prof physics, Chalmers Univ Technol, Sweden, 81; Guggenheim fel, 80-81; Alexander von Humboldt sr US scientist award, 88-89. *Honors & Awards:* Res Soc Am Pure Sci Award, 58; John Price Wetherill Medal, Franklin Inst, 79; Isakson Prize, Am Phys Soc, 86. *Mem:* Nat Acad Sci; fel Optical Soc Am; Am Asn Physics Teachers; fel Am Phys Soc. *Res:* Optical and acoustical spectroscopy of solids; dielectrics; semiconductors; photoconductors; crystal physics; electron tunneling; lattice dynamics; surface elastic and electromagnetic waves; electromagnetic phenomena at metal surfaces; surface enhanced Raman scattering; Raman scattering spectroscopy of adsorbates on metal surfaces; nonlinear optical properties of metal and semiconductor surfaces and interfaces. *Mailing Add:* Dept Physics Univ Pa Philadelphia PA 19104

BURSTEIN, SAMUEL Z, b Brooklyn, NY, Oct 4, 35; m 59; c 2. APPLIED MATHEMATICS. *Educ:* Polytech Inst Brooklyn, BME, 57, MME, 58, PhD(mech eng), 62. *Prof Exp:* Instr mech eng, Polytech Inst Brooklyn, 57-62; ASSOC PROF APPL MATH, COURANT INST MATH SCI, NY UNIV, 62- *Concurrent Pos:* Assoc ed, J Comput Physics, 68-74. *Mem:* Soc Indust & Appl Math. *Res:* Combustion instability in rocket motors; numerical methods in hydrodynamics; computation of shock waves and hypersonic flow; transonic flow; biomechanics; reactive gas dynamics; deformation of nonlinear elastic-plastic materials. *Mailing Add:* Eight Bay Ave Larchmont NY 10538

BURSTEIN, SOL, b Chelsea, Mass, Dec 25, 22. NUCLEAR & MECHANICAL ENGINEERING. *Educ:* Northeastern Univ, BS, 44. *Prof Exp:* Chief power plants, Wis Elec Power Co, 65-66, vpres, 66-69, sr vpres, 69-73, dir & exec vpres, 83-84, vchmn bd, 84-87; CONSULT, 87- *Mem:* Nat Acad Eng; fel Am Soc Mech Eng; Am Soc Metals; Am Soc Testing & Mat. *Mailing Add:* 7475 N Crossway Rd Milwaukee WI 53219

BURSTEIN, SUMNER, b Boston, Mass, Feb 22, 32. MECHANISM DRUG ACTION. *Educ:* Wayne State Univ, PhD(chem), 59. *Prof Exp:* ASSOC PROF BIOCHEM, SCH MED, UNIV MASS, 76- *Mem:* AAAS; Am Chem Soc; Am Soc Pharmacol & Exp Therapeut. *Mailing Add:* Dept Biochem Sch Med Univ Mass 55 Lake Ave N Worcester MA 01605

BURSTEN, BRUCE EDWARD, b Chicago, Ill, Mar 8, 54; m 77; c 1. INORGANIC CHEMISTRY. *Educ:* Univ Chicago, SB, 74; Univ Wis, PhD(chem), 78. *Prof Exp:* NSF fel chem, Tex A&M Univ, 78-80; from asst prof to assoc prof, 80-90, PROF CHEM, OHIO STATE UNIV, 90- *Concurrent Pos:* Sloan Found fel; teacher-scholar, Dreyfus Found. *Mem:* Am Chem Soc. *Res:* Bonding and electronic structural studies of transition metal complexes, particularly organometallic, organoactinide, and metal cluster systems; molecular orbital theory; low-temperature crystal spectroscopy. *Mailing Add:* Dept Chem Ohio State Univ 120 W 18th Ave Columbus OH 43210

BURSTONE, CHARLES JUSTIN, b Kansas City, Mo, Apr 4, 28. ORTHODONTICS. *Educ:* Wash Univ, DDS, 50; Ind Univ, MS, 55. *Hon Degrees:* PhD, Royal Deutol Col. *Prof Exp:* From asst prof to assoc prof orthod, Sch Dent, Ind Univ, 55-70; HEAD DEPT ORTHOD, SCH DENT MED, UNIV CONN, 70- *Mem:* AAAS; Am Asn Orthod; Am Dent Asn. *Res:* Application of biophysics to orthodontics; soft tissue morphology of the face; segmented arch therapy; growth and development; biomaterials. *Mailing Add:* Dept Orthod Sch Dent Med Univ Conn Farmington CT 06032

BURSTYN, HAROLD LEWIS, b Boston, Mass, Feb 26, 30; m 58; c 3. HISTORY OF SCIENCE, HISTORY OF TECHNOLOGY. *Educ:* Harvard Col, AB, 51; Univ Calif, MS, 57; Harvard Univ, PhD(hist sci), 64. *Prof Exp:* Instr hist sci, Brandeis Univ, 62-65; asst prof, Carnegie-Mellon Univ, 66-69, assoc prof hist sci & technol, 69-73; prof math & natural sci, William Paterson Col, NJ, 73-77; HISTORIAN, US GEOL SURV, 76- *Concurrent Pos:* NSF fel, Imp Col, Univ London, 65-66; dean grad & res progs, William Paterson Col, NJ, 73-75; adv ed, Isis, 76-80; assoc ed natural sci, Environ Rev, 77-; adj

fac mem, Physics Dept, Rutgers Univ, New Brunswick, NJ, 77- *Honors & Awards:* Henry Schuman Prize, Hist Sci Soc, 60. *Mem:* Hist Sci Soc; AAAS; Am Geophys Union; Soc Hist Technol; Int Ctr Hist Oceanog. *Res:* History of earth sciences and technologies, especially in the United States; history of marine science and technology, especially physical oceanography and its applications; history of marine exploration, especially the Challenger Expedition; environmental history, especially the fertilizer industry. *Mailing Add:* 950 Nat Ctr US Geol Surv Reston VA 22092

BURSUKER, ISIA, b USSR, Sept 15, 41; Israel citizen; m 67; c 1. TUMOR IMMUNOLOGY, IMMUNOREGULATION. *Educ:* Czernowitz State Univ, USSR, MSc, 68; Weizmann Inst Sci, Israel, PhD(biol), 82. *Prof Exp:* Engr, membrane res, Weizmann Inst Sci, 73-78; res assoc tumor immunol, Trudeau Inst, Saranac Lake, NY, 82-84; sr res scientist, 84-90, SR RES INVESTR II, TUMOR IMMUNOL, BRISTOL-MYERS SQUIBB CO, 90- *Concurrent Pos:* Aharon Katzir Travel Award, Aharon Katzir Ctr, Israel, 81; adj assoc prof, Univ Conn, Storrs, 88- *Mem:* Am Asn Immunologists; Soc Leukocyte Biol; AAAS. *Res:* Research in tumor immunology; down-regulation of antitumor immunity by suppressor cells; immunomodulation; immunotherapy of cancer; modulation of hemopoiesis. *Mailing Add:* Dept Immunol Bristol-Myers Squibb Co Five Res Pkwy Wallingford CT 06492

BURSZTAJN, SHERRY, b Poland, Feb 19, 46; US citizen; m 84. NEUROBIOLOGY, TRANSPORT OF SYNAPTIC PROTEINS. *Educ:* Fairleigh Dickinson Univ, BS, 69; Syracuse Univ, PhD(develop biol), 74. *Prof Exp:* Teaching fel neurobiol, Med Col, Cornell Univ, 74-76; teaching fel immunol, Med Sch, Yale Univ, 76-77; res assoc neurobiol, Med Sch, Harvard Univ, 77-80; ASST PROF NEUROBIOL, BAYLOR COL MED, 80- *Concurrent Pos:* Corp mem, Marine Biol Lab, Woods Hole, Mass, 81-; adv, Mult Sclerosis Soc, 85-; vis scholar, Harvard Univ, 87-88. *Honors & Awards:* Res Career Develop Award. *Mem:* Am Soc Cell Biol; Soc Neurosci; AAAS. *Res:* Mechanisms involved in the organization of the postsynaptic membrane during early stages of synapse formation. *Mailing Add:* Dept Neurol Baylor Col Med 6501 Fannin Houston TX 77030

BURT, ALVIN MILLER, III, b Bridgeport, Conn, Aug 14, 35; div; c 2. NEUROCHEMISTRY, NEUROANATOMY. *Educ:* Amherst Col, BA, 57; Univ Kans, PhD(anat), 62. *Prof Exp:* Asst prof anat, Med Col Va, 62-63; instr, Sch Med, Yale Univ, 63-66; from asst prof to prof anat, 66-85, PROF CELL BIOL, VANDERBILT UNIV, MED SCH, 85- *Concurrent Pos:* USPHS res career develop award, Vanderbilt Univ, 68-73; vis scientist neurochem, Agr Res Coun, Inst Animal Physiol, Babraham, Cambridge, Eng, 72-73. *Mem:* AAAS; Am Soc Neurochem; Am Asn Anat; Soc Neurosci; Int Soc Neurochem. *Res:* Developmental neurochemistry with particular reference to neurotransmitter systems and intermediary metabolism; writer of neurobiology textbooks. *Mailing Add:* Dept Cell Biol Vanderbilt Univ Nashville TN 37232

BURT, BRIAN AUBREY, b Melbourne, Australia, Jan 14, 39; m 65; c 2. DENTAL EPIDEMIOLOGY. *Educ:* Univ Western Australia, BDSc, 60; Univ Mich, Ann Arbor, MPH, 66; Univ London, PhD(dent epidemiol), 73. *Prof Exp:* Sr clin dentist, Perth Dent Hosp, Western Australia, 60-65; pub health adv dent, Div Dent Health, USPHS, 66-67; lectr, London Hosp Dent Sch, 68-74; assoc prof, 74-81, PROF, DENT PUB HEALTH, SCH PUB HEALTH & DENT, UNIV MICH, ANN ARBOR, 81- *Concurrent Pos:* Chmn, Dept Community Health Progs, Sch Pub Health, Univ Mich. *Mem:* Am Pub Health Asn; Int Asn Dent Res; Am Asn Pub Health Dentists. *Res:* Epidemiology of dental disease; prevention of dental disease; delivery of dental care. *Mailing Add:* Prog in Dent Pub Health Univ Mich Sch Pub Health Ann Arbor MI 48109-2029

BURT, CHARLES TYLER, b Tampa, Fla, Dec 20, 42; m 68. BIOPHYSICAL CHEMISTRY, PHYSIOLOGY. *Educ:* Carnegie Inst Technol, BS, 64; Univ Chicago, MS, 67; Fla State Univ, PhD(phys chem), 72. *Prof Exp:* Fel biochem, Inst Muscle Dis, 72-74; fel, dept biol chem, Med Ctr Univ Ill, 74-75, res assoc, 75-76, asst prof 76-79; asst prof, Reed Col, 79-80; asst prof, dept radiol, Harvard Med Sch, Mass Gen Hosp, 81-83; vis scientist, Mass Inst Technol, 83-85; nuclear magnetic resonance expert, Nat Inst Environ Health Sci, 85-89; AT DEPT RADIOL, COL MED, UNIV ILL, 90- *Mem:* Am Chem Soc; Biophys Soc; AAAS; Soc Magnetic Resonance Med. *Res:* Mechanisms of contraction and diseases that effect them; nuclear magnetic resonance and its application to biological processes from the molecular level to intact organs; nuclear magnetic resonance imaging. *Mailing Add:* Magnetic Resonance Ctr Univ Ill Med Ctr M/C 711 Chicago IL 60680

BURT, DAVID REED, b East Orange, NJ, Oct 28, 43; div. MOLECULAR PHARMACOLOGY, NEUROSCIENCE. *Educ:* Amherst Col, AB, 65; Johns Hopkins Univ, PhD(biophys), 72. *Prof Exp:* Assoc res scientist biophys, Johns Hopkins Univ, 72-73, fel pharmacol, Sch Med, 73-76; asst prof, 76-82, ASSOC PROF PHARMACOL, SCH MED, UNIV MD, 82- *Concurrent Pos:* Sr Fogarty int fel, MRC molecular neurobiol unit, Univ Cambridge Med Sch, Eng, 85-86. *Mem:* Am Soc Neurochem; Am Soc Pharmacol & Exp Therapeut; Soc Neurosci; Endocrine Soc; Biochem Soc. *Res:* Neuropharmacology; neuroendocrinology; peptide neurotransmitters; molecular pharmacology and molecular biology of neurotransmitter receptors. *Mailing Add:* Dept Pharmacol & Exp Therapeut Univ Md Sch Med Baltimore MD 21201

BURT, DONALD MCLAIN, b East Orange, NJ, Oct 27, 43; m 72; c 2. MINERALOGY, ORE DEPOSITS. *Educ:* Princeton Univ, AB, 65; Harvard Univ, AM, 68, PhD(geol), 72. *Prof Exp:* Lectr geochem, State Univ Utrecht, 72-73; Gibbs instr, Yale Univ, 73-75; from asst prof to assoc prof, 75-83, PROF, ARIZ STATE UNIV, 83- *Concurrent Pos:* Assoc ed, Am Mineralogist, 77-80; distinguished lectr, Can Inst Mining & Metall, 81; vis scientist, Lunar & Planetary Inst, Houston, 87-88. *Mem:* Mineral Soc Am; Mineral Asn Can; Geol Soc Am; Am Geophys Union; Soc Econ Geologists; Geochem Soc; Am Ceramic Soc. *Res:* Mineralogy and petrology of skarn, greisen, pegmatite and volcanogenic ore deposits; phase equilibria; geochemistry of acid-base processes; crystal chemistry; new mineral named "Burtite". *Mailing Add:* Dept Geol Ariz State Univ Tempe AZ 85287-1404

BURT, GERALD DENNIS, b Dovray, Minn, Jan 3, 36. ORGANIC CHEMISTRY. *Educ:* Luther Col, BA, 57; Univ Iowa, MS, 60, PhD(org chem), 61. *Prof Exp:* Asst proj chemist, Am Oil Co, Ind, 61-64; res fel chem, Univ Tex, Austin, 64-65; sr res chemist, 66-76, RES ASSOC, HARSHAW CHEM CO DIV, GULF OIL CO, 76- *Mem:* AAAS; Am Chem Soc; Sigma Xi; Coblentz Soc. *Res:* Synthetic organic chemistry; flame retardants; organometallic chemistry; surface modified resin fillers; rheology; pryolysis-IR; plastics additives. *Mailing Add:* 38375 Miles Rd Chagrin Falls OH 44022

BURT, JAMES KAY, b Des Moines, Mar 5, 34; m 54; c 5. VETERINARY RADIOLOGY. *Educ:* Iowa State Univ, DVM, 62, MS, 67; Am Col Vet Radiol, dipl. *Prof Exp:* Instr obstet & radiol, Iowa State Univ, 62-65, vet clin sci, 65-66, asst prof, 66-67; asst prof vet surg & radiol, 67-70, from assoc prof vet clin sci to prof, 70-87, EMER PROF VET CLIN SCI, OHIO STATE UNIV, 87- *Mem:* Am Vet Med Asn; Am Col Vet Radiol. *Res:* Teratologic effects of drugs on the embryo-fetus; dynamics of bone metabolism in eosinophilic panosteitis. *Mailing Add:* 6633 Guyer St Worthington OH 43085

BURT, JANIS MAE, EXCITATION CONTRACTION COUPLING. *Educ:* Univ Calif, Irvine, PhD(cell-physiol), 80. *Prof Exp:* ASST PROF CARDIAC-PHYSIOL, UNIV ARIZ, 82- *Res:* Cardiac physiology; intercelluar communication. *Mailing Add:* Dept Physiol Col Med Univ Ariz 1501 N Campbell Tucson AZ 85724

BURT, MICHAEL DAVID BRUNSKILL, b Colombo, Ceylon, Jan 19, 38; m 60; c 4. BIOLOGY, PARASITOLOGY. *Educ:* Univ St Andrews, BSc, 61, PhD(parasitol), 67. *Prof Exp:* Brit Coun scholar, Inst Zool, Univ Neuchatel, 63; from asst prof to assoc prof, 61-75, chmn dept, 74-85, PROF BIOL, UNIV NB, 85- *Concurrent Pos:* N Atlantic Treaty Orgn Advan Study Inst Sr Scientist Fel; Nova Scotia Fel Royal Col; Royal Soc exchange fel. *Honors & Awards:* W A Wardle Award in Parasitol, Can Soc Zool. *Mem:* Am Soc Parasitol; Can Soc Zool; Brit Soc Parasitol; World Fedn Parasitologists; Sci Coun Can. *Res:* Cestode and Platyhelminth biology with reference to functional morphology and developmental changes using light and electron microscopy, life cycles, seasonal variation, and geographical distribution and systematics, host specificity, and evolution. *Mailing Add:* Dept Biol Univ NB Fredericton NB E3B 5A3 Can

BURT, PHILIP BARNES, b Memphis, Tenn, July 1, 34; m 55; c 4. THEORETICAL PHYSICS. *Educ:* Univ Tenn, AB, 56, MS, 58, PhD(physics), 61. *Prof Exp:* Sr scientist, Jet Propulsion Lab, Pasadena, Calif, 61-65; from asst prof to assoc prof, 65-73, head, Dept Physics & Astron, 82-85, PROF PHYSICS, CLEMSON UNIV, 73- *Concurrent Pos:* Vis asst prof, Univ Southern Calif, 62; consult, Oak Ridge Nat Lab, assoc physicist, 60, res partic, 66; vis, Stanford Linear Accelerator Ctr, 75-76. *Honors & Awards:* Sigma Xi. *Mem:* Am Math Soc; Am Phys Soc. *Res:* Plasma quantum and field theories; quantum mechanics and nonlinear waves; scattering; theoretical high energy physics. *Mailing Add:* Dept Physics Clemson Univ Clemson SC 29634-1911

BURT, WAYNE VINCENT, b South Shore, SDak, May 10, 17; m 41; c 4. OCEANOGRAPHY. *Educ:* Pac Col, BS, 39; Univ Calif, MS, 48, PhD(phys oceanog), 52; cert aerol, US Naval Acad Post Grad Sch, cert aerol, 44. *Hon Degrees:* ScD, George Fox Col, 63. *Prof Exp:* Mat engr, Kaiser Co, Inc, Wash, 42; instr math, Univ Ore, 46; asst oceanogr, Scripps Inst, Univ Calif, 46-48, assoc oceanogr, 48-49; asst prof oceanog & res oceanogr, Chesapeake Bay Inst, Johns Hopkins Univ, 49-53, asst dir, Inst, 53; res oceanogr, Univ Wash, 53-54; from assoc prof to prof, 54-82, chmn dept oceanog, 59-68, dir marine sci ctr, 64-72, assoc dean res, 68-76, assoc dean oceanog, 76-82, EMER PROF OCEANOG, ORE STATE, 82- *Concurrent Pos:* Mem, Nat Acad Sci Comt on Oceanog, 65-70; distinguished prof, Ore State Univ, 68; mem, Nat Comt on Oceans & the Atmosphere, 71-75; liaison scientist, US Off Naval Res, London Eng, 79-80; sci attache Am Embassy New Delhi, India, 86-87. *Honors & Awards:* Centennial Award, 68; Gov Scientist Award, Ore Mus Sci & Indust, 69; Ocean Sci Award, Am Geophys Union, 84. *Mem:* Fel Am Meteorol Soc; Am Geophys Union; Nat Sea Grant Rev Panel (90-). *Res:* Interaction between ocean and atmosphere; estuarine and inshore physical oceanography; light transmission in turbid water; physical limnology; reservoir temperature; air-sea interaction; historical climatology. *Mailing Add:* Sch Oceanog Ore State Univ Corvallis OR 97331

BURTE, HARRIS M(ERL), b New York, NY, Nov 14, 27; m 55; c 3. MATERIALS SCIENCE, CHEMICAL ENGINEERING. *Educ:* NY Univ, BS, 46; Mass Inst Technol, MS, 47; Princeton Univ, PhD(chem eng), 50. *Prof Exp:* Sr res engr, Textile Res Inst, 47-53; proj engr res & develop, USAF, 53-55, consult, Mat Laab, Wright Air Develop Ctr, 55-56, chief prog br, Mat Lab, 56-59, adv metall studies br, 59-60, chief engr mat, Dept Systs Eng, Aeronaut Systs Div, 60-61, chief, Metals & Ceramic Div, 61-77, chief scientist, Air Force Mat Lab, 77-78, dir, Metals & Ceramics Div, 78-81, CHIEF SCIENTIST, AERONAUT SYSTS DIV, AIR FORCE MAT LAB, WRIGHT AIR DEVELOP CTR, 81- *Concurrent Pos:* Mem, Nat Acad Sci/Nat Acad Eng/Nat Res Coun panels adv Nat Bur Standards, 67-73, NASA subcomt mat processing in space, 76-78 & NATO adv group aerospace res & develop struct & mat panel, 79-84; adj prof, Univ Cincinnati, 69-78 & Wright State Univ, 80-; chmn, comt eng mgt, 71-72, comt govt, energy & mat, Metall Soc, 77-80 & dir Metall Soc, Am Inst Mining, Metall & Petrol Engrs, 81-84. *Mem:* Am Chem Soc; Fiber Soc; Am Inst Aeronaut & Astronaut; Am Inst Mining, Metall & Petrol Engrs; fel Am Soc Metals. *Res:* Properties of materials; textiles; plastics; metals; ceramics; nondestructive evaluation; technology and societal issues. *Mailing Add:* 3006 Hedge Run Ct Dayton OH 45415-2808

BURTIS, CARL A, JR, b Flagstaff, Ariz, July 3, 37; m 59; c 3. BIOCHEMISTRY, ANALYTICAL CHEMISTRY. *Educ:* Colo State Univ, BS, 59; Purdue Univ, MS, 64, PhD(biochem), 67. *Prof Exp:* Chemist, Ind State Control Labs, 63-66; fel bioanal, Oak Ridge Nat Lab, 66-67, res assoc molecular anat, 67-69; sr chemist, Varian Aerograph, 69-70; group leader gemsaec fast analyzer proj, Molecular Anat Prog, Oak Ridge Nat Lab, 70-76, coordr biotechnol prog, 73-77; chief, Anal Biochem Br, Ctr Dis Control, 76-79; CHIEF CLIN CHEM, OAK RIDGE NAT LAB, 79- *Mem:* Am Asn Clin Chemists; Am Soc Biol Chemists; AAAS; Am Chem Soc. *Res:* Chemical inducement of ovine ketosis; separation and quantitation of body fluid constituents by liquid chromatography; separation of nucleic acid constituents by high efficiency liquid chromatography; rapid clinical analysis; reference methods and materials. *Mailing Add:* Oak Ridge Nat Lab Oak Ridge TN 37830

BURTIS, THEODORE A, b Jamaica, NY, May 17, 22; m; c 3. ENGINEERING. *Educ:* Carnegie Inst Tech, BS; Texas A&M, MS. *Prof Exp:* chmn & chief exec officer, Sun Co, Inc; RETIRED. *Mem:* Nat Acad Eng; Am Inst Chem Engrs; Am Petrol Inst. *Mailing Add:* Sun Co Inc 100 Matsonford Rd Radnor PA 19087

BURTNER, DALE CHARLES, b Portland, Ore, Oct 20, 26; m 50; c 2. ANALYTICAL CHEMISTRY. *Educ:* Reed Col, BA, 48; Univ Wash, Seattle, MS, 51, PhD(chem), 54. *Prof Exp:* Chemist, Shell Develop Co, 54-58; from asst prof to assoc prof, 58-67, chmn dept, 65-67, dean sch arts & sci, 67-69, PROF CHEM, CALIF STATE UNIV, 67- *Mem:* Am Chem Soc. *Res:* Trace metal analysis; coordination compounds; spectrophotometry; flame photometry; microchemistry. *Mailing Add:* Dept Chem Calif State Univ Fresno CA 93740

BURTNER, ROGER LEE, b Hershey, Pa, Mar 31, 36; m 65; c 1. SEDIMENTARY PETROLOGY & CLAY MINEROLOGY, INORGANIC GEOCHEMISTRY. *Educ:* Franklin & Marshall Col, BS, 58; Stanford Univ, MS, 59; Harvard Univ, PhD(geol), 65. *Prof Exp:* From assoc res geologist to res geologist, Calif Res Corp Div, 63; explor geologist, Standard Oil Co Tex, 68-69, from res geologist to sr res geologist, 69-77, SR RES ASSOC, CHEVRON OIL FIELD RES CO, STANDARD OIL CO CALIF, 77- *Concurrent Pos:* Chmn, res proj adv comt, Am Petrol Inst, 72-75; petrol group proj leader, Chevron Oil Field Res Co, 75-80, supvr, Electron Micros Lab, 77-82; counr, Clay Minerals Soc, 81-84. *Mem:* Geol Soc Am; Geochemistry Soc; Soc Econ Paleont & Mineral; Am Asn Petrol Geol; Clay Minerals Soc; Int Asn Geochem & Cosmochem; Sigma Xi. *Res:* Environmental studies of ancient sediments; sedimentary structures; origin and evolution of formation waters; sandstone petrography; apatite fission track thermochronology of thrust belts; basin evaluation; diagenesis of sandstones and shales; isotope geochemistry of authigenic minerals in sandstone; geochemistry of 129 iodine & 127 iodine. *Mailing Add:* Chevron Oil Field Res Co PO Box 446 La Habra CA 90633

BURTNESS, ROGER WILLIAM, b Chicago, Ill, May 13, 25; m 47; c 3. ELECTRICAL ENGINEERING. *Educ:* Univ Ill, BS, 46, MS, 51, PhD(elec eng), 53; St Olaf Col, BA, 47. *Prof Exp:* Test engr, Gen Elec Co, 47-48, field engr, 48-49; asst elec eng, Univ Ill, 49-52, instr, 52-53; sect engr, Electro-Motive Div, Gen Motors Corp, 53-56; proj engr, radar & digital control, Stewart Warner Electronics, 56-58, mgr eng & res mil electronics, 58-59; assoc prof, 59-87, EMER ASSOC PROF ELEC ENG, UNIV ILL, URBANA, 87- *Concurrent Pos:* Consult, Trionics, Wis, 59-62, Control Systs Inc, Ill, 59-62, Borg Warner Res Ctr, 62-65 & Electro-Motive Div, Gen Motors Corp, 65-70. *Mem:* Sr mem Inst Elec & Electronics Engrs. *Res:* Linear integrated circuits; solid state power conversion. *Mailing Add:* 339 Elec Eng Bldg Univ Ill 1406 W Green St Urbana IL 61801

BURTON, ALBERT FREDERICK, b Brandon, Man, Feb 7, 29; m 53; c 2. BIOCHEMISTRY, CANCER. *Educ:* Brandon Col, BSc, 53; Univ Western Ont, MSc, 56; Univ Sask, PhD, 58. *Prof Exp:* Asst prof, 62-67, ASSOC PROF BIOCHEM, UNIV BC, 67- *Concurrent Pos:* Nat Cancer Inst Can res fel biochem, Univ BC, 59-62. *Mem:* Soc Exp Biol & Med; Can Biochem Soc; Can Investrs Reproduction; Sigma Xi. *Res:* Metabolism of corticosteroids and their biological action, especially in fetal tissues and in relation to cancer. *Mailing Add:* Dept Biochem Univ BC Vancouver BC V6T 1W5 Can

BURTON, ALEXIS LUCIEN, b Paris, France, Feb 4, 22; US citizen; m 52; c 3. HISTOLOGY, CYTOLOGY. *Educ:* Univ Geneva, BMSc, 43; Univ Strasbourg, MD, 46. *Prof Exp:* Surgeon, Dept Pub Health, Morocco, 51-56; asst prof microanat, Fac Med, Univ Montreal, 57-63; from asst prof to prof anat, Med Sch, Univ Tex, San Antonio, 64-; RETIRED. *Concurrent Pos:* Fel anat, Univ Tex Med Sch San Antonio, 63-64; vis prof anat, Nat Yang Ming Med Col, Taipei, Repub China. *Mem:* Am Asn Anat; Tissue Cult Asn; Soc Motion Picture & TV Eng. *Res:* Scientific cinematography; cytology of the connective tissue; tissue culture and cinemicrography. *Mailing Add:* 100 W Elbaado San Antonio TX 78212

BURTON, ALICE JEAN, b Peiping, China, May 19, 34; US citizen. BIOCHEMISTRY, MICROBIOLOGY. *Educ:* Univ Mich, BS, 57; Univ Ill, PhD(biol). *Prof Exp:* NIH res fel, Calif Inst Technol, 61-63, res fel biol, 63-64; from asst biochemist to biochemist, Brookhaven Nat Lab, 64-70; ASSOC PROF BIOL, ST OLAF COL, 70- *Concurrent Pos:* Mem US nat comt, Int Union Pure & Appl Biophys, 75-78. *Mem:* Biophys Soc; Am Chem Soc; AAAS; Sigma Xi. *Res:* Biochemistry of nucleic acids, particularly the replication of bacteriophages and host modification of the replicative process. *Mailing Add:* Dept Biol St Olaf Col Northfield MN 55057

BURTON, BENJAMIN THEODORE, b Wiesbaden, Ger, Aug 29, 19; nat US; m 52; c 2. NUTRITION, NEPHROLOGY. *Educ:* Univ Calif, BS, 41, MS, 43, PhD(microbiol, biochem), 47. *Prof Exp:* Res chemist, Mills Orchards Corp, 41-42; outside supvr & tech consult, Rosenberg Bros & Co, 42-48; vpres & tech dir, Pac States Labs, Inc, 52-55; staff consult nutrit, H J Heinz Co, 55-60; from nutritionist & spec asst to dir, 60-67; assoc dir & chief artificial kidney prog, Nat Inst Diabetes, Digestive & Kidney Dis, 67-83, ASSOC DIR DIS PREV & TECHNOL TRANSFER, NAT INST DIABETES, DIGESTIVE & KIDNEY DIS, 83- *Honors & Awards:* Superior Serv Award, US Dept Health, Educ & Welfare, 70; NIH Dir Award, 79. *Mem:* Am Inst Nutrit; Europ Dialysis & Transplant Asn; Int Soc Nephrol; Am Soc Artificial

Internal Organs; Am Soc Nephrology; hon mem Am Soc Extracorporeal Technol. *Res:* Human nutrition; intermediate metabolism; nutritional and metabolic diseases; malnutrition; chronic kidney failure and uremia; artificial kidney development; dialysis. *Mailing Add:* NIDDK NIH Rm 9A03 Bldg 31 Bethesda MD 20892

BURTON, CHARLES JEWELL, physics; deceased, see previous edition for last biography

BURTON, CLYDE LEAON, b Battle Creek, Mich. PLANT PATHOLOGY, HORTICULTURE. *Educ:* Mich State Univ, BS, 50, MS, 52, PhD(plant path), 60. *Prof Exp:* Mgr tech servs, Agr Chem Div, FMC Corp, 60-66; RES PLANT PATHOLOGIST, AGR RES SERV, USDA, 66- *Mem:* Am Phytopath Soc; Am Soc Hort Sci; Am Soc Agr Engrs; Sigma Xi. *Res:* Postharvest quality of fruits and vegetables harvested, handled and stored. *Mailing Add:* Dept Bot & Plant Path Mich State Univ East Lansing MI 48824

BURTON, DANIEL FREDERICK, b Chicago, Ill, Oct 3, 15; m 50. BOTANY. *Educ:* Univ Chicago, MS, 40, PhD(bot), 47. *Prof Exp:* Instr bot, Miss State Col, 46-48; from asst prof to prof bot, Mankato State Col, 48-75; admin asst, vchancellor for mgt, Minn Comn Col, 75-77; dir, Adult Voc Educ Funding Study, 77-79; RETIRED. *Concurrent Pos:* Mem, Minn State Bd Educ, 71-77. *Mem:* AAAS; Am Fern Soc; Sigma Xi. *Res:* Leaf morphogenesis; formative effects of certain substituted phenoxy compounds on bean leaves. *Mailing Add:* 512 Hickory St Mankato MN 56001

BURTON, DAVID LEE, horticulture, soils, for more information see previous edition

BURTON, DAVID NORMAN, b Birkenhead, Eng, Feb 3, 41; m 65; c 3. BIOCHEMISTRY, MICROBIOLOGY. *Educ:* Univ Liverpool, BSc, 62, PhD(biochem), 65. *Prof Exp:* Fel biochem, Univ Calif, Davis, 65-66; chemist, Lipid Metab Lab, Vet Admin Hosp, Madison, Wis, 66-68; fel, 68-69, from asst prof to assoc prof, 69-77, PROF MICROBIOL, 77-, UNIV MAN, ASSOC DEAN SCI, 87- *Concurrent Pos:* Nat Res Coun Can res operating grant, 69- *Mem:* Can Biochem Soc. *Res:* Mechanisms and control of lipid metabolism in microorganisms and animals; membrane and surface properties of cultured mammalian cells. *Mailing Add:* 95 Thatcher Winnipeg MB R3T 2L6 Can

BURTON, DENNIS THORPE, b Richmond, Va, Apr 16, 41; m 69; c 3. AQUATIC ECOLOGY, PHYSIOLOGICAL ECOLOGY. *Educ:* Va Commonwealth Univ, BS, 65; Va Polytech Inst & State Univ, PhD(zool), 70. *Prof Exp:* Res assoc aquatic ecol, Va Polytech Inst & State Univ, 70-71; asst cur, Acad Natural Sci, Philadelphia, 71-77, assoc cur limnol & ecol, 77-80; sr prof staff biologist, Johns Hopkins Univ, 80-85, prin prof staff biologist, appl physics lab, 85-91; SR RES SCIENTIST, WYE RES & EDUC CTR, UNIV MD SYST, 91- *Concurrent Pos:* Fel, Va Polytech Inst & State Univ, 70-71; consult, utility & indust corp, 71-; mem, Environ Res Guidance Comt, State Md Power Plant Siting Prog, 73-78 & 84-87, chmn, 78-80, mem, Power Plant Siting Adv Comt, 78-; assoc ed, Chesapeake Sci, 74-78; adj res assoc, Ctr Environ & Estuarine Studies, Univ Md, 76-77, adj assoc prof, Dept Path, Sch Med, 79-85; adj res scientist, Chesapeake Bay Inst, Johns Hopkins Univ, 84-; mem Environ Biol Rev Panel, US Environ Protection Agency, 86-; vchmn, exec comt sci adv bd, Md Dept Nat Res, 87-89. *Honors & Awards:* Prof Fisheries Scientist Cert, Am Fishery Soc, 71. *Mem:* AAAS; Am Fisheries Soc; Am Soc Zoologists; Sigma Xi; fel Am Inst Fishery Res Bd; Soc Environ Toxicol Chem; Estuarine Res Fedn. *Res:* Physiological responses of aquatic organisms exposed to temperature changes; low dissolved oxygen conditions; biocides and organic compounds; heavy metals; as well as multifactorial combinations of these variables; effects of low oxygen on anaerobic metabolism of fishes; biofouling control at industrial facilities; bioluminescence in marine organisms; toxicity evaluation of industrial and municipal discharges; toxicity evaluation of groundwater; toxicity evaluation of nitrated organic compounds. *Mailing Add:* Wye Res & Educ Ctr Univ Md Systs PO Box 67 Queenstown MD 21658

BURTON, DONALD EUGENE, b Kansas City, Mo, Apr 24, 41; m 70; c 2. NUMERICAL HYDRODYNAMICS, NUMERICAL ANALYSIS. *Educ:* Univ Mo, Rolla, BS, 62; Kans State Univ, PhD(theoret nuclear physics), 69. *Prof Exp:* Res & develop coordr, radiol safety, Nuclear Cratering Group, 68-69, nuclear cratering, Eng Explosive Excavation Res Lab, US Army CEngrs, 70-73; physicist, 73-77, proj leader, Earth Sci Div, 80-85, group leader, Geomechanics Code Develop, Lawrence Livermore Nat Lab, Univ Calif, 77-82; dep leader, Geotech Group, Earth Sci Div, 82-85, LEADER, GROK DESIGN CODE GROUP, A-DIV, DEFENSE SCI DEPT, LAWRENCE LIVERMORE NAT LAB, UNIV CALIF, 85- *Honors & Awards:* US Army Commendation Medal Res, 73. *Mem:* Am Geophys Union; Int Soc Rock Mech. *Res:* Development of large multidimensional computer code systems used principally for geoscience applications; develop constitutive model for earth materials; energy coupling from nuclear explosions; development of nuclear weapons design codes. *Mailing Add:* 2802 Waverly Way Livermore CA 94550

BURTON, DONALD JOSEPH, b Baltimore, Md, July 16, 34; m 58; c 5. ORGANIC CHEMISTRY. *Educ:* Loyola Col, Md, BS, 56; Cornell Univ, PhD(org chem), 61. *Prof Exp:* Fel, Purdue Univ, 61-62; from asst prof to assoc prof, 62-70, PROF CHEM, UNIV IOWA, 70- *Concurrent Pos:* Carver-Shriner prof, 89. *Honors & Awards:* Am Chem Soc Award, Fluorine Div, 84. *Mem:* Am Chem Soc; The Chem Soc. *Res:* Synthesis and chemistry of polyfluorinated compounds; chemistry of polyhalogenated organoboranes; metal halide catalysis of halogenated olefins; preparation and reactions of halogenated ylides, polyfluorinated organ metallics. *Mailing Add:* Dept Chem Univ Iowa Iowa City IA 52242

BURTON, FREDERICK GLENN, b Greensburg, Pa, Nov 30, 39; m 68. ANALYTICAL CHEMISTRY, ORGANOPHOSPHORUS CHEMISTRY. *Educ:* Col Wooster, BA, 62; Wesleyan Univ, MA, 66; Univ Rochester, PhD(biophysics), 71. *Prof Exp:* Instr dairy sci, Ohio Agr Exp Sta, 62-64; fel

origins of life, Salk Inst Biol Studies, 71-73; res scientist controlled release, Battelle Mem Inst, 74-76, sr res scientist bioanal chem, Pac Northwest Div, 76-85, proj mgr, Surety Agent Res, 85-89, LAB DIR, BATTELLE TOOELE OPERS, BATTELLE MEM INST, 90- *Concurrent Pos:* Consult, Immunodiagnostics Inc, 73-74. *Honors & Awards:* Spec Award for Excellence in Technol Transfer, Fed Lab Consortium, 86. *Mem:* AAAS; Sigma Xi; Soc Appl Spectros; Int Soc Study Origins Life; Controlled Release Soc; Am Chem Soc; Asn Off Anal Chemists; Am Defense Preparedness Asn. *Res:* Sustained release devices; bioanalytical applications of toxicology; surface chemistry; chemical surety agents. *Mailing Add:* PO Box 1179 Tooele UT 84074-1179

BURTON, GEORGE JOSEPH, b New York, NY, Mar 20, 19. CANCER ENVIRONMENTAL EPIDEMIOLOGY, MEDICAL ENTOMOLOGY. *Educ:* City Col New York, BS, 39, MS, 40; NY Univ, PhD(entom), 46, Northeastern Univ, electron micros cert, 70. *Prof Exp:* Teacher biol, Theodore Roosevelt High Sch, New York, 39-40; biostatistician agr statist, Census Bur, Dept of Com, Washington, DC, 40-41; biostatistician med statist, War Dept, Surgeon-General's Off, Washington, DC, 41-42; sr topogr cartogr, Hydrographic Off, Dept of Navy, Suitland, Md, 42-43; malariologist, US Navy, 43-45; agr entomologist, Bur Entom & Plant Quarantine, USDA, New York, 45-47; assoc prof biol, State Teachers Col, East Stroudsburg, Pa, 47-50; epidemiologist & med entomologist cmndg offices, US Army, German, Italy, 50-52, res entomologist insect vector control, Eng Res & Develop Labs, US Army, Ft Belvoir, Va, 52-53; epidemiologist & malariologist malaria, Am Embassy, Monrovia, Liberia, US AID, 53-55 & Am Embassy, Kathmandu, Nepal, 55-58; filariologist, Am Embassy, New Delhi & Ernakulam, India, US Tech Coop, 58-61; epidemiologist & vectorborne dis specialist filariasis, Am Consulate, Georgetown, Brit Guiana, US AID, 61-63; epidemiologist & vectorborne dis specialist onchocerciasis & malaria, NIH (US), NIH (Ghana), Joint Res Prog, Nat Cancer Inst, Accra, Ghana, 63-65; scientist dir cancer, virol & epidemiol, Spec Animal Leukemia Ecol Segment, Virus Cancer Prog, 65-70, coordr epidemiol studies, immunol-epidemiol segment, 71-75, SCIENTIST DIR CANCER EPIDEMIOL, ENVIRON EPIDEMIOL BR & EPIDEMIOL BIOSTATIST PROG, NAT CANCER INST, NIH, 75- *Concurrent Pos:* Exec secy & vchmn spec animal leukemia ecol segment, Virus Cancer Prog, Nat Cancer Inst, NIH, 65-70, exec secy in-house source eval group, 75- & cancer contracts coordr, NCI/EPA Collab Prog, 77-; mem adv panel insect viruses in insect control, Food & Drug Admin, Washington, DC, 73-74; sci adv panel onchocerciasis control prog, WHO, Geneva, Switz, 74-; consult onchocerciasis, Dept of State, US AID, Washington, DC, 76- *Mem:* Int Asn Comp Res on Leukemia & Rel Dis; Am Soc Trop Med & Hyg; Am Asn Cancer Res; Royal Soc Trop Med & Hyg; Am Soc Parasitologists. *Res:* Cancer epidemiology and etiology; viral oncology; epidemiology of vectorborne diseases, especially malaria, filariasis, onchocerciasis, yellow fever, cancer viruses; medical parasitology; photomicrography; bionomics of arthropods and parasites of medical importance. *Mailing Add:* Nat Cancer Inst Landow Bldg Rm C318 NIH Bethesda MD 20892

BURTON, GILBERT W, b Covington, Va, Mar 2, 36; m 64; c 1. ORGANIC CHEMISTRY, POLYMER CHEMISTRY. *Educ:* Univ Calif, BS, 58; Univ Ill, PhD(org chem), 64. *Prof Exp:* Res Prof Exp: From res chemist to sr res chemist, Esso Res & Eng Co, 64-75; staff chemist, 75-84, SR STAFF CHEMIST, EXXON CHEM CO, 84- *Mem:* AAAS; Am Chem Soc. *Res:* Organic photochemistry; chemistry of ozonization of organic compounds; anionic coordination polymerization; elastomer chemistry; environmental affairs. *Mailing Add:* 1179 Puddingstone Rd Mountainside NJ 07092

BURTON, GLENN WILLARD, b Clatonia, Nebr, May 5, 10; m 34; c 5. AGRONOMY. *Educ:* Univ Nebr, BA, 32; Rutgers Univ, MS, 33, PhD(agron), 36. *Hon Degrees:* ScD, Rutgers Univ, 55; ScD, Univ Nebr, 62. *Prof Exp:* Asst agron, Rutgers Univ, 32-36; from agent to sr geneticist, Div Forage Crops & Dis, Bur Plant Indust, 36-53, RES GENETICIST & LEADER, AGR RES SERV, US DEPT AGR, 53- *Concurrent Pos:* Chmn agron div, Univ Ga, 50-64. *Honors & Awards:* Nat Medal Sci Award, 83; Stevenson Award & South Seedsman Asn Award, 49; Sears Roebuck Award, 53, 61; John Scott Award, 57; Superior Serv Award, USDA, 55; Edward W Browning Award, Am Soc Agron, 75; Alexander von Humboldt Found Award, 88. *Mem:* Nat Acad Sci; fel Am Soc Agron (pres, 62); Am Genetic Asn; Range Soc Am; fel Crop Sci Soc Am; hon mem Grassland Soc SAfr; Sigma Xi; Am Forage & Grassland Soc; Coun Agr Sci & Technol. *Res:* Grass breeding and genetics; grass cytology; grass seed production; grass fertilization and management; revegation of native ranges. *Mailing Add:* Ga Coastal Plain Exp Sta Tifton GA 31794

BURTON, HAROLD, b New York, NY, Jan 19, 43; m 63; c 1. NEUROPHYSIOLOGY. *Educ:* Univ Mich, BA, 64; Univ Wis, PhD(physiol), 68. *Prof Exp:* Nat Inst Neurol Dis & Blindness trainee neurophysiol, Med Sch, Univ Wis, 68-70; asst prof neurobiol & physiol, 70-76, assoc prof anat neurobiol & physiol, 76-83, PROF ANAT & NEUROBIOL, MED SCH, WASHINGTON UNIV, 83- *Mem:* Soc Neurosci. *Res:* Somato sensory system physiology and anatomy. *Mailing Add:* Dept Anat Washington Univ Sch Med 660 S Euclid Ave St Louis MO 63110

BURTON, JOHN HESLOP, b Ottawa, Can, Aug 27, 38; m 69; c 2. ANIMAL NUTRITION. *Educ:* Univ Toronto, BSc, 62; Cornell Univ, MSc, 67, PhD(nutrit), 70. *Prof Exp:* Res assoc, 69-70, asst prof, 71-79, ASSOC PROF ANIMAL NUTRIT, UNIV GUELPH, 79- *Mem:* Can Nutrit Soc; Agr Inst Can; Am Dairy Sci Asn; Am Soc Animal Sci; NY Acad Sci; Sigma Xi. *Res:* Role of metabolic hormones in controlling milk synthesis and production in dairy cattle and growth in the bovine; growth and energy metabolism in the equine. *Mailing Add:* Dept Animal-Poultry Sci Univ Guelph Guelph ON N1G 1K9 Can

BURTON, JOHN WILLIAMS, b Atlanta, Ga, Apr 15, 37; m 59; c 4. COMPUTER SCIENCES, EXPERIMENTAL NUCLEAR PHYSICS. *Educ:* Carson-Newman Col, BS, 59; Univ Ill, MS, 61, PhD(physics), 65; Univ Tenn, MS, 77. *Prof Exp:* Asst physics, Univ Ill, 61-64; assoc prof & head dept,

64-81, dir, comput ctr, 81-89, PROF PHYSICS, CARSON-NEWMAN COL, 66- *Concurrent Pos:* Consult, Oak Ridge Nat Lab, 65-75. *Mem:* Am Phys Soc; Am Asn Physics Teachers. *Res:* Mossbauer effect on the surface of tungsten, involving ultra high vacuum techniques; Mossbauer spectroscopy; computer programming for data analysis; computer assisted instruction. *Mailing Add:* Dept Physics Carson-Newman Col Jefferson City TN 37760

BURTON, KAREN POLINER, b Albuquerque, NMex, Feb 18, 52; m 74; c 2. CARDIOLOGY, ULTRASTRUCTURE. *Educ:* Southern Methodist Univ, BS, 74; Univ Tex Health Sci Ctr, PhD(physiol), 78. *Prof Exp:* Fel physiol, Univ Tex Health Sci Ctr, Dallas, 78-79, fac assoc, 79, instr, 79-82; asst prof pharmacol & physiol, Col Med, Univ S Ala, 81-82; asst prof physiol, 83-88, asst prof radiol, 89-90, ASSOC PROF RADIOL, SOUTHWESTERN MED CTR, UNIV TEX, DALLAS, 90- *Concurrent Pos:* Investr, NIH Res Award, 82-85, & res career develop award, 86-91; prin investr, NIH Grant, 85-88, 88-93. *Honors & Awards:* New Invest Res Award, NIH, 82-85. *Mem:* Am Physiol Soc; Am Soc Heart Res; AAAS; Electron Micros Soc Am; Sigma Xi; Am Heart Asn; Oxygen Soc. *Res:* Cardiovascular pathophysiology, role of calcium and electrolyte alterations in the development of ischemic myocardial damage; role of superoxide free radicals in heart damage; analytical electron microscopy; fluorescent indicators for calcium measurements. *Mailing Add:* Dept Radiol Univ Tex Southwestern Med Ctr 5323 Harry Hines Dallas TX 75235-9071

BURTON, LARRY C(LARK), b Neola, Utah, Nov 2, 39; m 62; c 2. PHYSICAL ELECTRONICS, SEMICONDUCTOR & SURFACE PHYSICS. *Educ:* Temple Univ, BA, 62, MS, 66; Pa State Univ, PhD(physics), 70. *Prof Exp:* Engr advan develop group, Philco Corp, Pa, 63-65; scientist res & develop div, Leeds & Northrup Co, 65-66; asst prof elec eng, Tex Tech Univ, 70-73; mem staff, Inst Energy Conversion, Univ Del, 73-77; ASSOC PROF ELEC ENG, VA POLY INST & STATE UNIV, 77- *Mem:* AAAS; Inst Elec & Electronics Engrs; Am Phys Soc; Am Soc Eng Educ. *Res:* Thermionic, photoelectric and field emission of electrons from semiconductor surfaces; semiconductor surface preparation and evaluation; electronic transport in solids; analyses of ultra-high frequency semiconductor devices and switching circuits. *Mailing Add:* Dept Elec Eng Va Poly Inst & State Univ Blacksburg VA 24061

BURTON, LEONARD PATTILLO, b Jasper, Ala, June 8, 18; m 42; c 3. MATHEMATICS. *Educ:* Univ Ala, AB, 39, MA, 40; Univ NC, PhD(math), 51. *Prof Exp:* Instr math, Univ Ala, 46-48; from instr to asst prof, Univ Calif, 51-54; from asst prof to prof, 54-65, head prof, 65-77, PROF MATH, AUBURN UNIV, 77- *Mem:* Am Math Soc; Math Asn Am; Sigma Xi. *Res:* Ordinary differential equations. *Mailing Add:* 3802 Heritage Pl Opelika AL 36801

BURTON, LESTER PERCY JOSEPH, b Sudbury, Ont, Mar 5, 53; m 78; c 1. POLYMER STABILIZATION. *Educ:* Univ Western Ont, BSc, 76; Ore State Univ, PhD(org chem), 81. *Prof Exp:* SR RES SCIENTIST, ETHYL CORP, 81- *Res:* Design and synthesis of stabilizers for polymers. *Mailing Add:* Himont Res & Develop Ctr 800 Green Bank Rd Wilmington DE 19808-5905

BURTON, LLOYD EDWARD, b Phoenix, Ariz, May 13, 22; m 68; c 3. PHARMACY, PUBLIC HEALTH. *Educ:* Univ Ariz, BS, 54, MS, 56, PhD(pharmacol), 64. *Prof Exp:* Instr pharm, 54-59, res assoc pharmacol, 59-60, instr pharm, 60-64, from asst prof to prof pharmacol, Col Pharm, 64-78, PROF COMMUNITY HEALTH, SCH HEALTH RELATED PROF, UNIV ARIZ, 78- *Mem:* Am Pharmaceut Asn; Am Pub Health Asn; Sigma Xi. *Res:* Brain neurohormone studies; development of effective antidote for oleander poisoning; investigation of toxicity of plant drugs; development of health care services. *Mailing Add:* 2802 E First St Tucson AZ 85716

BURTON, LOUIS, b Ft Riley, Kans, June 15, 49; c 2. BIOCHEMISTRY. *Educ:* Mich State Univ, PhD(biochem), 76. *Prof Exp:* Scientist, protein & biochem, 82-85, SR SCIENTIST, PROCESS RES & DEVELOP, GENENTECH INC, 85- *Mem:* Am Soc Biochem & Molecular Biol. *Res:* Biotechnology; recovery of recombinant proteins. *Mailing Add:* Dept Process Res & Develop Genentech Inc 460 Pt San Bruno S San Francisco CA 94080

BURTON, PAUL RAY, b Burnsville, NC, Dec 7, 31; m 59; c 2. ZOOLOGY, CELL BIOLOGY. *Educ:* Western Carolina Col, BS, 54; Univ Miami, MS, 56; Univ NC, PhD(zool), 60. *Prof Exp:* Asst prof biol, St Olaf Col, 60-63; NIH fel anat, Sch Med, Univ Wis, 63-64; from asst prof to assoc prof zool, 64-69, chmn dept physiol & cell biol, 73-76, PROF PHYSIOL & CELL BIOL, UNIV KANS, 69- *Concurrent Pos:* USPHS Career Develop Award, 68-73. *Mem:* Am Micros Soc; Electron Micros Soc Am; Am Soc Parasitol; AAAS; Am Soc Cell Biol. *Res:* Electron microscopy; studies on the structure and function of microtubular elements, particularly in neuronal systems. *Mailing Add:* Dept Physiol & Cell Biol Univ Kans Haworth Hall Lawrence KS 66045

BURTON, RALPH A(SHBY), b Shreveport, La, Oct 31, 25; m 48; c 1. MECHANICAL ENGINEERING. *Educ:* Univ Ark, BS, 47; Univ Tex, MS, 51, PhD(mech eng), 52. *Prof Exp:* Instr, Univ Ark, 47-49; asst prof mech eng, Mass Inst Technol, 52-54; assoc prof, Univ Mo, 54-58; sr res engr, Southwest Res Inst, 58-61, staff scientist & mgr lubrication res sect, 61-67; physicist on leave, Off Naval Res London, 67, liaison scientist, 67-69; prof mech eng & astronaut sci, Northwestern Univ, 69-80; prof mech & aerospace eng & head dept, NC State Univ, 80-85; PVT RESEARCHER, BURTON TECHNOL, INC, 85- *Concurrent Pos:* Consult, Westinghouse Labs, 77-78; on leave, US Off Naval Res, Arlington, 78-79. *Mem:* AAAS; fel Am Soc Mech Engrs; Am Soc Metals. *Res:* Surface mechanics; wear mechanisms; thermoelastic effects; applied mechanics; liquid metal current collectors. *Mailing Add:* Burton Technol Inc 554 Pylon Dr Raleigh NC 27606

BURTON, RALPH GAINES, b Columbia, Mo, Jan 3, 55; m 78. CARBON-METAL COMPOSITES, TRIBOLOGY. *Educ:* Northwestern Univ, BS, 78. *Prof Exp:* Sr engr, Newport News Shipbuilding, 79-80, sr shift test engr, 80-84; VPRES, BURTON TECHNOLOGIES INC, 84- *Concurrent Pos:* Co-prin investr, Burton Technologies Inc, 88-91. *Mem:* Am Soc Metals; AAAS. *Res:* Liquid-metal electrical contacts; active, feed forward and feedback, control of gas bearings and metrology systems; development of a family of vitreous carbon based composite materials for seals, bearings, electrical, and structural applications. *Mailing Add:* Burton Technologies Inc PO Box 33809 Raleigh NC 27636-0809

BURTON, RALPH L, b Afton, Wyo, Mar 20, 36. ENTOMOLOGY, PESTICIDE REGISTRATION. *Educ:* Va Polytech Inst, DEntom. *Prof Exp:* MGR, ISK BIOTECH CORP, 68- *Mem:* Sigma Xi; Am Phytopath Soc; Weed Sci Soc Am; Agr Res Inst. *Res:* Commercial development. *Mailing Add:* ISK Biotech Corp 5966 Heisley Rd PO Box 8000 Mentor OH 44061

BURTON, ROBERT CLYDE, b Borger, Tex, Feb 27, 29; m 51; c 4. GEOLOGY. *Educ:* Tex Tech Col, BA, 57, MSc, 59; Univ NMex, PhD(geol), 65. *Prof Exp:* Instr geol, Tex Tech Col, 58-59; from instr to asst prof, West Tex State Univ, 59-63; vis lectr, Univ NMex, 65; assoc prof, 65-66, head dept, 66-76, PROF GEOL, W TEX STATE UNIV, 59- *Concurrent Pos:* Petrol Res Fund grant, Am Chem Soc, 69-71; owner, Geomag Surv. *Mem:* Nat Asn Geol Teachers (secy, 68, vpres, 69, pres, 70). *Res:* Biostratigraphy; conodont biostratigraphy in New Mexico; exploration magnetics. *Mailing Add:* 711 Taylor Lane Canyon TX 79015

BURTON, ROBERT LEE, b Bluefield, Va, May 29, 43; m 65; c 1. PROPULSION SYSTEMS, GAS DYNAMICS. *Educ:* Va Polytechnic Inst & State Univ, BS, 66; George Washington Univ, MS, 73. *Prof Exp:* Res engr Saturn V/Apollo, Boeing Aerospace, NASA Hq Consult, 66-72; lead engr jet eng controls, Boeing Com Airplane, 747, 72-79; eng supvr gas centrifuge, Boeing Eng Co SE, 79-85; technol mgr gas centrifuge, Boeing Tenn, Inc, 85-89; MGR ENG, BOEING AEROSPACE & ELECTRONICS, OAK RIDGE, 89- *Concurrent Pos:* Recorder, NASA Apollo Exp, Photog Team, 70-72; Boeing rep, NASA Apollo Mission Readiness Rev, 70-72, Indust/FAA/EPA Jet Engine Emissions Controls, 77-79, Dept Energy Gas Centrifuge Gas Dynamics Working Group, 81-85; moderator, NASA, Indust Post-Apollo Mission Study Team, 72; prin investr, Nat Aerospace Plane Hypersonic Wind Tunnel, 85-88; consult, Dept Energy, Indust Centrifuge Chem Separation, 87-88; bd dirs, WATTec, Inc, 90-91. *Mem:* Nat Space Soc. *Res:* High speed gas dynamics analysis and design associated with aircraft, jet engines, rockets, spacecraft and gas centrifuges; avenger missile launcher engineering; thermal design for laser isotope separation systems; gas dynamic lasers; composite materials; granted one patent. *Mailing Add:* Boeing Aerospace & Electronics 767 Boeing Rd Oak Ridge TN 37830

BURTON, ROBERT MAIN, b Oklahoma City, Okla, Mar 5, 27; m 47; c 5. BIOCHEMISTRY, PHARMACOLOGY. *Educ:* Univ Md, BS, 50; Georgetown Univ, MS, 52; Johns Hopkins Univ, PhD(biol, biochem), 55. *Prof Exp:* Chemist, USDA, Md, 50; chemist, Nat Heart Inst, 50-52; chemist, Nat Inst Neurol Dis & Blindness, 55-57; from asst prof to assoc prof pharmacol, Sch Med, Wash Univ, 57-78; prof & chmn, Dept Pharmacol, Sch Med, Oral Roberts Univ, 78-79; CONSULT, BURTON INT, 79- *Concurrent Pos:* Lectr, Georgetown Univ, 56-57; Am Chem Soc & Am Soc Biol Chemists travel award, Int Biochem Cong, Vienna, 58, Moscow, 61 & Tokyo, 67; NSF sr fel, Inst Phys Chem, Univ Cologne, 63; mem child health & human develop prog comt, NIH, 68-69, mem pop res & training comt, 69-70, chmn, 70-72; sci adv, Ctr Biochem Studies, Portugal, 65-; assoc ed, Lipids, 65-; mem med adv bd, Nat Tay-Sachs & Allied Dis, Inc, 75-; mem genetic dis rev and adv comt, Health Serv Admin, USPHS, 81-; Fulbright res scholar, Univ Porto, Port, 81- *Mem:* Am Soc Pharmacol & Exp Therapeut; Am Chem Soc; Am Soc Biol Chemists; Am Soc Neurochem; fel Am Inst Chemists; Sigma Xi. *Res:* Biochemistry of nervous system; lipid and glycolipid metabolism; biochemical basis for neuropharmacological activity of drugs; hyperlipoproteinemias in children. *Mailing Add:* Burton Int Box 13135 St Louis MO 63119

BURTON, RUSSELL ROHAN, b Chico, Calif, Jan 15, 32; m; c 4. PHYSIOLOGY, MANAGEMENT. *Educ:* Univ Calif, Davis, BS, 54, DVM, 56, MS, 65, PhD, 70. *Prof Exp:* Pvt pract, 56-62; from assoc res specialist to res specialist, Univ Calif, Davis, 62-71; res physiologist, 71-80, br chief, 80-88, chief scientist, USAF Sch Aerospace Med, 88-90; CHIEF SCIENTIST, ARMSTRONG LAB, BROOKS AFB, 90- *Concurrent Pos:* Prof, Univ Tex, San Antonio, 84-; non panel expert, Adv Group Aerospace Res & Develop. *Honors & Awards:* Paul Bert Award, 75, Arnold D Tuttle Award, 76 & Environ Sci Award, 80, Aerospace Med Asn; Eric Liljencrantz Award, 88. *Mem:* AAAS; Sigma Xi; fel Aerospace Med Asn; Am Physiol Soc. *Res:* Pathophysiologic effects of high-levels of acceleration resulting in methods to protect aircrew. *Mailing Add:* Chief Scientist DET 4 AL/CA Armstrong Lab Brooks AFB TX 78235-5301

BURTON, SHERIL DALE, b Malad, Idaho, May 10, 35; m 59; c 3. MICROBIOLOGY, BIOCHEMISTRY. *Educ:* Brigham Young Univ, BS, 59, MS, 61; Ore State Univ, PhD(microbiol), 64. *Prof Exp:* Res asst bact, Brigham Young Univ, 59-61 & microbiol, Univ Nebr, 61-62; res asst microbiol, Ore State Univ, 62-64, asst prof & res grantee, 64-65; asst prof, Inst Marine Sci, Univ Alaska, 65-67; assoc prof, 67-78, PROF MICROBIOL, BRIGHAM YOUNG UNIV, 78- *Mem:* Am Soc Microbiol. *Res:* Ultrastructure, metabolism and ecology of aquatic microbes, particularly of Beggiatoa; bioconversions; genetic engineering; drilling fluids (oil fields); biopolymers; sugars. *Mailing Add:* Dept Microbiol Brigham Young Univ Provo UT 84602

BURTON, THEODORE ALLEN, b Longton, Kans, Sept 7, 35; m 61. MATHEMATICS. *Educ:* Wash State Univ, BS, 59, MA, 62, PhD(math), 64. *Prof Exp:* Asst prof math, Univ Alta, 64-66; assoc prof, 66-71, PROF MATH, SOUTHERN ILL UNIV, CARBONDALE, 71- *Concurrent Pos:* Fulbright Scholar. *Mem:* Am Math Soc; Math Asn Am; Soc Indust Appl Math. *Res:* Stability theory of ordinary differential equations and integral equations. *Mailing Add:* Dept Math Southern Ill Univ Carbondale IL 62901

BURTON, THOMAS MAXIE, b West Carroll Parish, La, Nov 24, 41; m; c 1. AQUATIC ECOLOGY. *Educ:* Northeast La Univ, BS, 63, MS, 65; Cornell Univ, PhD(ecol), 73. *Prof Exp:* Teacher sci, Winnsboro High Sch, 65-66; instr biol, Univ NC, Charlotte, 66-67; res technician, US Army Res Inst Environ Med, Mass, 67-69; ecol consult, Eng-Sci, Cincinnati, 73; res assoc aquatic ecol, Fla State Univ, 74-75; asst prof, Inst Water Res, 75-79, assoc prof, Fisheries & Wildlife, 79-84, PROF AQUATIC ECOL, DEPT ZOOL, FISHERIES & WILDLIFE, MICH STATE UNIV, 84- *Concurrent Pos:* Indo-Am fel, 88-89. *Mem:* Ecol Soc Am; Am Soc Limnol & Oceanog; Sigma Xi; AAAS; NAm Benthological Soc; Int Soc Limnol; Wetlands Soc. *Res:* Studies of stream and marsh biota at the community and ecosystem level; acidification effects on stream biota; logging effects on stream biota; intrasystem nutrient cycles in streams and marshes. *Mailing Add:* Dept Zool Mich State Univ East Lansing MI 48824

BURTON, VERONA DEVINE, b Reading, Pa, Nov 23, 22; m 50; c 1. BOTANY. *Educ:* Hunter Col, AB, 44; State Univ Iowa, MS, 46, PhD(plant anat), 48. *Prof Exp:* Asst, State Univ Iowa, 44-48; from asst prof to assoc prof, 48-70, prof, 70-86, EMER PROF BIOL SCI, MANKATO STATE UNIV, 86- *Mem:* AAAS; Bot Soc Am; Am Fern Soc; Int Soc Plant Morphol. *Res:* Floral abscission and plant embryology. *Mailing Add:* 512 Hickory St Mankato MN 56001

BURTON, WILLARD WHITE, b Richmond, Va, Feb 24, 22; m 48; c 3. ORGANIC CHEMISTRY. *Educ:* Univ Richmond, BS, 43. *Prof Exp:* From jr res assoc to res assoc, Am Tobacco Co, 46-68, supvr org chem sect, 68-70, supvr chem develop, Process Develop Lab, 70-71, supvr res, 71-81, asst res mgr, Dept Res & Develop, 81-87; RETIRED. *Mem:* Am Chem Soc; fel Am Inst Chem. *Res:* Tobacco chemical composition; browning reaction pigments; phenols; amino acids; carbohydrates and organic acids; analytical methods development. *Mailing Add:* 6808 Greenvale Dr Richmond VA 23225

BURTON, WILLIAM BUTLER, b Richmond, Va, July 13, 40; m 72; c 3. ASTRONOMY. *Educ:* Swarthmore Col, BA, 62; State Univ Leiden, Drs, 65, PhD(astron), 70. *Prof Exp:* Asst astron, State Univ Leiden, 64-65, sci officer, 66-70; from res assoc to scientist, Nat Radio Observ, 71-78; prof & chmn, Dept Astron, Univ Minn, 78-81; PROF, UNIV LEIDEN, 81- *Mem:* Am Astron Soc; Int Astron Union; Netherlands Astron Soc. *Res:* Galactic structure; interstellar medium. *Mailing Add:* Sterrewacht PO Box 9513 2300 RA Leiden Netherlands

BURTS, EVERETT C, b Mae, Wash, Sept 25, 31; m 54; c 2. ENTOMOLOGY. *Educ:* State Col Wash, BS, 54; Ore State Col, MS, 57, PhD(entom), 59. *Prof Exp:* Jr entomologist, Ore State Col & agent, Div Entom, Agr Res Serv, USDA, 54-58; ENTOMOLOGIST, WASH STATE UNIV, 58- *Mem:* AAAS; Entom Soc Am. *Res:* Biology and control of insects and mites of tree fruits; insect vectors of plant viruses; deciduous fruit insect pest management. *Mailing Add:* Tree Fruit Res Ctr 1100 N Western Ave Wenatchee WA 98801

BURTT, BENJAMIN PICKERING, b Newburyport, Mass, June 7, 21; m 45; c 3. PHYSICAL CHEMISTRY, RADIATION CHEMISTRY. *Educ:* Ohio State Univ, BA, 42, PhD(chem), 46. *Prof Exp:* Asst chem & instr physics, Ohio State Univ, 42-45, instr chem, 45-46; instr, 46-47, from instr to assoc prof, 49-59, PROF CHEM, SYRACUSE UNIV, 59 - *Concurrent Pos:* Assoc scientist, Brookhaven Nat Lab, 48-49. *Mem:* Am Chem Soc. *Res:* Mass spectrometric studies of electron impact phenomena; ion-molecule reactions. *Mailing Add:* Dept Chem Syracuse Univ Syracuse NY 13244

BURTT, EDWARD HOWLAND, JR, b Waltham, Mass, Apr 22, 48; m 72; c 2. BIOMECHANICS, ELECTRON MICROSCOPY. *Educ:* Bowdoin Col, AB, 70; Univ Wis-Madison, MS, 73, PhD(zool), 77. *Prof Exp:* Vis instr psychol, Univ Tenn, 76-77; asst prof, 77-83, assoc prof, 83-87, PROF ZOOL, OHIO WESLEYAN UNIV, 87- *Concurrent Pos:* Rev ed, J Field Ornith, 78-85; res fel zool, Ohio State Univ, 81-82; first elect parliamentarian, Animal Behav Soc, 83-89; ed, J Field Ornithol, 85-90; exec coun, Wilson Ornith Soc, 89-91. *Mem:* Fel Am Ornithologists Union; Brit Ornithologists Union; Wilson Ornith Soc; Animal Behav Soc; Soc Am Naturalists; Cooper Orinth Soc; Asn Field Ornithologists (vpres, 89-91, pres, 91-). *Res:* Evolution and function of color and pattern in animals; avian reproductive ecology; maintenance behavior. *Mailing Add:* Dept Zool Ohio Wesleyan Univ Delaware OH 43015

BURWASH, RONALD ALLAN, b Edmonton, Alta, Sept 20, 25; m 52; c 3. GEOLOGY. *Educ:* Univ Alta, BSc, 45, BEd, 47, MSc, 51; Univ Minn, PhD(geol), 55. *Prof Exp:* Geologist, Sullivan Mines, BC, 51-52; petrogr, Shell Oil Co, Alta, 55-56; from asst prof to assoc prof, 56-65, PROF GEOL, UNIV ALTA, 65- *Mem:* Geochem Soc; Geol Asn Can; Can Soc Petrol Geologists. *Res:* Igneous and metamorphic petrology; Precambrian geology; geochemistry of Continental crust. *Mailing Add:* Dept Geol Univ Alta Edmonton AB T6G 2E3 Can

BURWASSER, HERMAN, b New York, NY, June 21, 27; m 54; c 3. PHYSICAL CHEMISTRY. *Educ:* Rutgers Univ, BS, 50; NY Univ, PhD(phys chem), 54. *Prof Exp:* Asst combustion chem, Princeton Univ, 54-56; res chemist, Atlantic Refining Co, 56-59; res scientist, AeroChem Res Labs, Inc, 59-60; proj chemist, Thiokol Chem Corp, 60-68; tech assoc, GAF Corp, 66-82; tech assoc, Transcopy Inc, 82-85; CONSULT, 85- *Mem:* Am Chem Soc. *Res:* Photo and radiation chemistry of organic materials, especially reversible photochemical reactions; organic photoconductors; electrophotography. *Mailing Add:* 21 Dogwood Lane Rd 1 Boonton NJ 07005

BURWELL, CALVIN C, b Lafayette, Ind, Sept 20, 30; m 51, 76; c 5. CHEMICAL & NUCLEAR ENGINEERING. *Educ:* Purdue Univ, BS, 52; Univ NMex, MS, 60. *Prof Exp:* Tech engr, Procter & Gamble Mfg Co, 52-55; prod mgr, 55-57; staff mem mat studies, Los Alamos Sci Lab, Calif, 57-63, eng design, 63-65; asst dir nuclear desalination, Oak Ridge Nat Lab, 65-69, tech dir Mid East studies, 69-72, lab prog planning & anal, 72-78, res engr, Oak Ridge Assoc Univs, 78-88; PVT CONSULT, 89- *Concurrent Pos:* Traveling lectr, Oak Ridge Nat Lab, 66-67. *Res:* Nuclear desalination; liquid metal containment; remote equipment design; nuclear siting policy; national energy policy; solar biomass energy; electricity end-use efficiency. *Mailing Add:* Rte 1 Box 183A1 Clinton TN 37716

BURWELL, E LANGDON, INTERNAL MEDICINE. *Educ:* Harvard Univ, MD, 44. *Prof Exp:* Asst path, internal med, Univ Zurich, Switz, 48-49; res fel, internal med, Univ Wash, 49-50; sr res med, internal med, King Co Hosp, Seattle, 50-51; attending staff, Internal Med, Falmouth Hosp, Mass, 51-88; RETIRED. *Concurrent Pos:* Clin instr med, Harvard Univ; sr asst surg, USPHS. *Mem:* Inst Med-Nat Acad Sci; Am Med Asn; Am Col Physicians. *Mailing Add:* PO Box 14 Woods Hole MA 02543

BURWELL, JAMES ROBERT, b Anderson, Ind, Mar 28, 29; m 51; c 3. ELEMENTARY PARTICLE PHYSICS, HIGH ENERGY PHYSICS. *Educ:* Ind Univ, BS, 52, MS, 54, PhD(physics), 57. *Prof Exp:* Res assoc high energy physics, Ind Univ, 57-58 & nuclear physics, 58-59; from asst prof to prof physics, Univ Okla, 59-85, chmn dept, 64-65, from asst dean to dean, Col Arts & Sci, 66-84, Regents prof, 85-88, EMER REGENTS PROF PHYSICS, UNIV OKLA, 88- *Mem:* Am Phys Soc; Sigma Xi; AAAS. *Res:* High energy elementary particle physics. *Mailing Add:* 708 Cedarbrook Dr Norman OK 73072

BURWELL, ROBERT LEMMON, JR, b Baltimore, Md, May 6, 12; m 39; c 2. CATALYSIS, SURFACE CHEMISTRY. *Educ:* St John's Col, Md, AB, 32; Princeton Univ, AM, 34, PhD(phys chem), 36. *Prof Exp:* Instr chem, Trinity Col, Conn, 36-39; from instr to prof, 39-70, chmn dept, 52-57, Ipatieff prof, 70-80, EMER IPATIEFF PROF CHEM, NORTHWESTERN UNIV, 80- *Concurrent Pos:* Chemist, Naval Res Lab, 43-45; mem bd dirs, Int Cong Catalysis, 56-64, mem coun, 64-, secy, 68-72, vpres, 72-76, pres-elect, 76-80, pres, 80-84; chmn, Gordon Res Conf Catalysis, 57; mem subcomt heterogeneous catalysis, Int Union Pure & Appl Chem, 62-67, assoc mem comn colloid & surface chem, 67-69, mem, 69-77; mem chem adv panel, NSF, 68-71; Alexander von Humboldt Found sr US scientist award, 81; vis prof, Univ Pierre & Marie Curie, Paris, 82. *Honors & Awards:* Kendall Award, Am Chem Soc, 73, Lubrizol Award Petrol Chem, 83; Robert Burwell Award, Catalysis Soc, 83. *Mem:* Am Chem Soc; Catalysis Soc (vpres, 68-73, pres, 73-77); Royal Soc Chem. *Res:* Heterogeneous catalysis and surface chemistry. *Mailing Add:* Dept Chem Northwestern Univ Evanston IL 60208-3113

BURWELL, WAYNE GREGORY, b New Haven, Conn, Feb 21, 33; m 55; c 4. COMBUSTION SCIENCE, LASERS. *Educ:* Yale Univ, BEng, 55, MEng, 56; Yale Univ, DEng, 62; Harvard Univ, AMP, 78. *Prof Exp:* Res engr, Gen Motors Corp, 56-59; sr res scientist, chief kinetics & thermal sci, 67-72, mgr kinetics & gas lasers, 72-74, mgr energy res, 74-77, asst dir, 77-81, dep dir res, 81-82, DIR RES, UNITED TECHNOLOGIES RES CTR, 82- *Concurrent Pos:* Lectr, Yale Univ, 60-61 & Univ Conn, 63-68; consult, Lycoming Div, Avco Corp, 60-61. *Mem:* Fel Am Inst Aeronaut & Astronaut; Combustion Inst; Indust Res Inst; Asn Res Dir; Indust Res Inst (vpres); Software Productivity Consortium; Res & Develop Coun Am Mgt Asn (vpres). *Res:* Propulsion technology; combustion processes; gas lasers, energy conversion systems; systems analysis. *Mailing Add:* 27 Robbinswood Dr Wethersfield CT 06109

BURWEN, SUSAN JO, b New York, NY, Nov 9, 46; m 68. CELL BIOLOGY, PHYSIOLOGY. *Educ:* Brandeis Univ, AB, 68; Harvard Univ, PhD(physiol), 72. *Prof Exp:* Fel molecular biol, Syntex Res, 72-73, fel cell biol, Univ Calif, Berkeley, 73-80; MEM STAFF, VET ADMIN MED CTR, 80- *Mem:* Am Soc Cell Biol; Women Cell Biol; AAAS; Sigma Xi. *Res:* Cell physiology: hepatocyte protein transport function, including endocytosis, intracellular processing and biliary secretion. *Mailing Add:* Vet Admin Med Ctr 151E 4150 Clement St San Francisco CA 94121

BURZYNSKI, NORBERT J, b South Bend, Ind, July 7, 29; m 61; c 3. EXPERIMENTAL PATHOLOGY, CLINICAL GENETICS. *Educ:* Ind Univ, AB, 52; St Louis Univ, DDS, 60, MS, 63. *Prof Exp:* USPHS fel oral path, St Louis Univ, 60-62; from asst prof to assoc prof, 65-73, PROF ORAL DIAG & ORAL MED & CHMN DEPT, SCH DENT, UNIV LOUISVILLE, 73-, PROF & CHMN DIAG SCI, 83- *Concurrent Pos:* Fel med genetics, Ind Univ, 70-72; consult, Nat Cancer Inst, 78-80. *Mem:* Fel AAAS; Am Soc Human Genetics; Int Asn Dent; fel Am Col Dentists; Am Dent Asn. *Res:* Chemistry of mucous membrane; genetics and neoplasia; salivary gland metabolism. *Mailing Add:* Sch Dent E-4 Univ Louisville Louisville KY 40292

BURZYNSKI, STANISLAW RAJMUND, b Lublin, Poland, Jan 23, 43; m; c 3. ONCOLOGY, PROTEIN CHEMISTRY. *Educ:* Med Acad, Lublin, Poland, MD, 67, PhD(biochem), 68. *Prof Exp:* Teaching asst chem, Med Acad, Lublin, Poland, 62-67, from intern to resident internal med, 67-70; res assoc biochem, 70-72, asst prof, Baylor Col Med, 72-77; PRES, BURZYNSKI RES INST, 77- *Concurrent Pos:* Nat Cancer Inst grant, 74; West Found grant, 75. *Mem:* AAAS; AMA; World Med Asn; Soc Neurosci; Sigma Xi; Am Chem Soc. *Res:* Discovery of antineoplastons-components of biochemical defense system against cancer; elucidation of the structure of Ameletin-first substance known to be responsible for remembering sound in animal's brain. *Mailing Add:* 6221 Corporate Dr Houston TX 77036

BUS, JAMES STANLEY, b Kalamazoo, Mich, June 27, 49; m 74; c 3. TOXICOLOGY, PHARMACOLOGY. *Educ:* Univ Mich, Ann Arbor, BS, 71; Mich State Univ, PhD(pharmacol), 75. *Prof Exp:* Asst prof environ health toxicol, Univ Cincinnati, 75-76; SCIENTIST BIOCHEM TOXICOL, CHEM INDUST INST TOXICOL, 77- *Mem:* Sigma Xi; AAAS; Teratology Soc; Soc Toxicol; Am Thoracic Soc; Am Soc Pharmacol & Exp Therapeut. *Res:* Biochemical and developmental toxicology; disposition of industrial chemicals after parenteral or inhalation dosing. *Mailing Add:* Chem Indust Inst of Toxicol PO Box 12137 Research Triangle Park NC 27709

BUSBY, EDWARD OLIVER, b Macomb, Ill, June 22, 26; m 50; c 3. CIVIL ENGINEERING. *Educ:* Univ Wis, BS, 50, MS, 62 & PhD, 71. *Prof Exp:* Engr, State Hwy Comn, Wis, 50-51; asst engr, City La Crosse, 51-53; sales engr, Wis Culvert Co, 53-59; lectr civil eng, Univ Wis, 59-66; assoc prof, Col Eng Univ Wis-Platteville, 66-68, dean, 66-84, prof civil eng, 66-88, EMER DEAN, COL ENG UNIV WIS-PLATTEVILLE, 88-; PRES, BUSBY & ASSOC, 88- *Concurrent Pos:* Engrs sect, Wis Exam Bd Architects & Prof Engrs; vis prof, Univ Tenn, 84-85. *Mem:* Nat Soc Prof Engrs; fel Am Soc Civil Engrs; Am Soc Eng Educ. *Res:* Hardening of asphaltic concrete; flexural properties of bituminous paving mixture, including modulus of rupture and modulus of elasticity in flexure at low temperatures. *Mailing Add:* Wexford Crossing 7486 Old Saule Rd Madison WI 53717

BUSBY, HUBBARD TAYLOR, JR, b Birmingham, Ala, May 20, 41; m 63; c 3. ORGANIC CHEMISTRY, INDUSTRIAL ORGANIC CHEMISTRY. *Educ:* Miss Col, BS, 63; Univ NC, PhD(org chem), 68. *Prof Exp:* Res chemist, Southern Dyestuff Co, 67-73; group leader, Martin Marietta Chem Co, 73-76, area prod mgr, Disperse Dyes & Chem Intermediates, 76-81, gen mgr, 81-84; GEN MGR, SANDOZ CHEM CORP, 85- *Mem:* Am Chem Soc; Am Asn Textile Chem & Colorists; Com Develop Asn. *Res:* Interannular substituent effects in ferrocene; synthesis and development of marketable dye stuff; organic chemical process development. *Mailing Add:* 6736 Commerce Ave Port Richey FL 34668-6814

BUSBY, ROBERT CLARK, b Darby, Pa, July 1, 40; m 64; c 2. MATHEMATICS. *Educ:* Drexel Inst, BS, 63; Univ Pa, AM, 64, PhD(math), 66. *Prof Exp:* Instr math, Drexel Inst, 65-67; asst prof, Oakland Univ, 67-69; from asst prof to assoc prof, 69-81, PROF MATH, DREXEL UNIV, 82- *Concurrent Pos:* Consult math, Off Emergency Preparedness, Exec Off Pres, 68-71. *Mem:* Am Math Soc; Soc Indust & Appl Math. *Res:* Modern analysis and applications; algebras of operators on a Hilbert space and group representations; mathematical demography; integral operators. *Mailing Add:* Dept Math Drexel Univ Philadelphia PA 19104

BUSBY, WILLIAM FISHER, JR, b Pawtucket, RI, Oct 29, 39; m 62; c 1. CARCINOGENESIS. *Educ:* Univ RI, BS, 61; Univ Calif, San Diego, PhD(marine biol), 66. *Prof Exp:* Res assoc biochem, Worcester Found Exp Biol, 66-71; res assoc toxicol, 71-81, RES SCIENTIST TOXICOL, MASS INST TECHNOL, 81-; SR STAFF SCIENTIST, HEALTH EFFECTS INST, 90- *Mem:* Am Asn Cancer Res; Soc Toxicol. *Res:* Development of short-term animal bioassays for carcinogenicity testing; toxicokinetics of xenobiotics; molecular epidemiology; mycotoxins; limited-term animal bioassays for testing polynuclear aromatic hydrocarbons and their derivatives; carcinogenic interactions of components of combustion-derived complex mixtures. *Mailing Add:* Health Effects Inst 141 Portland St Suite 7300 Cambridge MA 02139

BUSCEMI, PHILIP AUGUSTUS, b Mt Pleasant, Iowa, Mar 1, 26; div; c 3. LIMNOLOGY, ENVIRONMENTAL MANAGEMENT. *Educ:* Univ Colo, BS, 50, MA, 52, PhD(biol & zool), 57. *Prof Exp:* Asst gen biol & invert zool, Univ Colo, 50-52, instr, 52-53, gen biol, Denver Exten Cent, 55-56; from instr to asst prof zool, Univ Idaho, 56-65; asst prof, Okla State Univ, 65-66; from assoc prof to prof, Eastern NMew Univ, 66-85, chmn, Dept Biol Sci, 66-75, pres, Interdisciplinary Environ Inst, 72-76, EMER PROF BIOL, EASTERN NMEX UNIV, 85- *Concurrent Pos:* NSF fel, Ore Inst Marine Biol, 58; consult, Arctic Health Res Ctr, Anchorage, Alaska, 59; mem, Bio-Med Inst, Lawrence Radiation Lab, 70- *Mem:* Int Asn Theoret & Appl Limnol; Sigma Xi; Exp Aircraft Asn. *Res:* Hydrobiology; animal ecology of mountain lakes; macroscopic bottom fauna; organic production in fresh water; zooplankton population dynamics; biochemical energy pathways in aquatic ecosystems. *Mailing Add:* PO Box 1759 Cave Junction OR 97523

BUSCH, ARTHUR WINSTON, b Houston, Tex, Oct 9, 26; m 48; c 2. ENGINEERING. *Educ:* Tex Tech Col, BS, 50; Mass Inst Technol, SM, 52. *Prof Exp:* Asst sanit eng, Mass Inst Technol, 50-52; asst to dir res & develop, Infilco, Inc, Ariz, 52-55; asst prof civil eng, Rice Univ, 55-61, from assoc prof to prof environ eng, 61-75, chmn, Dept Environ Sci & Eng, 67-70; regional adminr, US Environ Protection Agency, Region VI, 72-75; vpres, SW Res Inst, 75-76; RETIRED. *Concurrent Pos:* Mem, President's Air Qual Adv Bd, 71-; adj prof, Univ Tex, San Antonio, 75-76; consult, 76-; adj sr res scientist, NTex State Univ, 78-80; vis prof, Univ Tex, Dallas, 80-81. *Honors & Awards:* Harrison Prescott Eddy Medal, Water Pollution Control Fedn, 61; Environ Award, Am Inst Chem Eng, 75. *Mem:* Am Chem Soc; Water Pollution Control Fedn; Am Inst Chem Engrs. *Res:* Environmental science; kinetics and mechanisms of biological oxidation processes in measurement and control of organic pollution of water; systems response; biodegradability. *Mailing Add:* PO Box 823 Wimberley TX 78676

BUSCH, DANIEL ADOLPH, b St Paul, Minn, May 31, 12; m 39; c 2. GEOLOGY. *Educ:* Capital Univ, BS, 34; Ohio State Univ, AM, 36, PhD(geol), 39. *Hon Degrees:* DSc, Capital Univ, 60. *Prof Exp:* Asst geol, State Geol Surv, Ohio, 35-36; instr, Univ Pittsburgh, 38-43; petrol geologist, State Topog & Geol Surv, Pa, 43-44; consult geologist, Huntley & Huntley, Pittsburgh, 44-46; sr res geologist, Carter Res Lab, 46-49; staff geologist, Carter Oil Co, 49-51; explor mgr, Zephyr Petrol Co, 51-54; CONSULT GEOLOGIST, 54- *Concurrent Pos:* Carnegie Mus field exped, 41; vis prof, Ohio State Univ, 63, Univ Tulsa, 63-64 & Univ Okla, 64-74; world-wide lectr, Oil & Gas Consults Int, Inc, 69- *Honors & Awards:* Orton Award, Ohio State Univ, 59, 60; Matson Award, Am Asn Petrol Geol, 59, Levorsen Award, 71; President's Award, Am Asn Petrol Geol, 76, Sidney Powers Mem Medal, 82. *Mem:* Hon mem Am Asn Petrol Geol (vpres, 66-67, pres, 73-74); fel Geol Soc Am; Sigma Xi. *Res:* Silurian and Devonian stratigraphy of Appalachian Basin; Pennsylvania stratigraphy and sedimentology of eastern and north-central; subsurface stratigraphy of mid-continent; stratigraphy and structure of Mexican Gulf Coast. *Mailing Add:* 3757 S Wheeling Ave Tulsa OK 74105

BUSCH, DARYLE HADLEY, b Carterville, Ill, Mar 30, 28; m 51; c 5. INORGANIC CHEMISTRY. *Educ:* Univ Southern Ill, BA, 51; Univ Ill, MS, 52, PhD, 54. *Prof Exp:* From asst prof to assoc prof, 54-63, PROF CHEM, OHIO STATE UNIV, 63- *Concurrent Pos:* Consult, E I du Pont de Nemours & Co, 56-, Div Res Grants, NIH, 61-65, Div Res Grants, NSF, 65-68, Beaunit Fibers, 66-68 & Chem Abstr Serv, 67-; consult ed, Allyn & Bacon, Inc, 63-; Guggenheim fel, 81-82. *Honors & Awards:* Inorg Chem Award, Am Chem Soc, 63, Morley Medal, 75 & Distinguished Serv in Advan of Inorg Chem, 76; Dwyer Medal, Australia, 78; Bailar Medal. *Mem:* Am Chem Soc; fel AAAS; Sigma Xi. *Res:* Coordination chemistry; stereo chemistry mechanisms of substitution reactions; magnetochemistry of transition metal ions; complexes of macrocyclic ligands; reactions of coordinated ligands; bioinorganic chemistry. *Mailing Add:* Dept Chem Univ Kans Malott Hall Lawrence KS 66045

BUSCH, GEORGE JACOB, IMMUNO-PATHOLOGY, TRANSPLANTATION PATHOLOGY. *Educ:* Johns Hopkins Univ, MD, 62. *Prof Exp:* ASSOC PROF PATH, MED SCH, HARVARD UNIV, 65-, CHIEF PATH, VET ADMIN MED CTR, 81- *Mailing Add:* Vet Admin Med Ctr 1400 VFW Pkwy West Roxbury MA 02132

BUSCH, HARRIS, b Chicago, Ill, May 23, 23; m 45; c 4. PHARMACOLOGY, BIOCHEMISTRY. *Educ:* Univ Ill, MD, 46; Univ Wis, PhD(biochem), 52. *Prof Exp:* Asst & sr asst surgeon, USPHS, 47-49; asst prof biochem, Yale Univ, 52-54, asst prof med & biochem, 54-55; assoc prof pharmacol, Univ Ill, 55-59, prof, 59-60; prof biochem & chmn dept, 60-62, PROF PHARMACOL & CHMN DEPT, BAYLOR COL MED, 60- *Concurrent Pos:* Baldwin scholar oncol, Sch Med, Yale Univ, 51-55, scholar cancer res, 54-55; vis prof, Univ Chicago, 68, Northwestern Univ, Ore State Univ, Ind Univ, Vet Hosp & Methodist Hosp, Houston; dir, Cancer Res Ctr; mem consult bd, Eli Lilly Co; mem pathogenesis panel, Am Cancer Soc; mem cancer chemother study sect, USPHS; consult, Uniformed Servs Univ, Bethesda, Md, 75; mem bd sci counrs, Div Cancer Treatment, Nat Cancer Inst, 75; vis prof, Univ Toronto, 75. *Mem:* AAAS; Biochem Soc; Am Chem Soc; Am Asn Cancer Res; Am Soc Biol Chemists. *Res:* Metabolism of nuclear proteins in tumor nuclei; isolation and metabolism of tumor nucleoli. *Mailing Add:* Dept Pharmacol Baylor Col Med Houston TX 77030

BUSCH, JOSEPH SHERMAN, b Chicago, Ill, Feb 20, 27; m 51; c 2. NUCLEAR & CHEMICAL ENGINEERING. *Educ:* Northwestern Univ, BS, 46 & 49; Johns Hopkins Univ, MS, 53; Carnegie Mellon Univ, PhD, 60. *Prof Exp:* Test engr munitions, US Army Chem Corps, Ft Detrick, 49-52; group leader nuclear naval vessels, Westinghouse Corp, Bettis, 53-60; group leader indust nuclear power, Atomic Power Develop Assocs, 60-62; PROJ MGR NUCLEAR ENG, KAISER ENGRS, 62- *Concurrent Pos:* Consult, Italian Atomic Energy Comn, 68, Argonne Nat Lab. *Mem:* Defense Preparedness Asn; Am Chem Asn; Am Inst Chem Engrs; Am Nuclear Soc. *Res:* Heat transfer; fluid flow and air pollution abatement. *Mailing Add:* Kaiser Engrs 1800 Harrison St PO Box 23210 Oakland CA 94623

BUSCH, KENNETH LOUIS, b New York, NY, Oct 26, 53. ANALYTICAL CHEMISTRY, BIOENGINEERING & BIOMEDICAL ENGINEERING. *Educ:* Univ Md, BS, 75; Univ NC, PhD(chem), 79. *Prof Exp:* Asst res scientist, Purdue Univ, W Lafayette, Ind, 81-83; ASST PROF, IND UNIV, BLOOMINGTON, 83- *Concurrent Pos:* Secy anal chem div, Am Chem Soc, 88- *Mem:* Am Chem Soc; Am Soc Mass Spectrometry; AAAS; Sigma Xi. *Res:* Analytical mass spectrometry; fast atom bombardment; the coupling of chromatography with secondary ion mass spectrometry. *Mailing Add:* Sch Chem & Biochem Ga Inst Tech Atlanta GA 30332

BUSCH, KENNETH WALTER, b Mt Vernon, NY, Mar 29, 44; m 68. INFRARED SPECTROSCOPY, COLLOID CHEMISTRY. *Educ:* Fla Atlantic Univ, BS, 66; Fla State Univ, PhD(anal chem), 71. *Prof Exp:* Teaching fel chem, Fla State Univ, 71-72; res assoc, Cornell Univ, 72-74; from asst prof to assoc prof, 74-88, PROF CHEM, BAYLOR UNIV, 88- *Concurrent Pos:* Vis assoc, Calif Inst Technol, 83; vis scientist, Cornell Univ, 85, 86, 87 & 88. *Mem:* Am Chem Soc; Soc Appl Spectros; Int Soc Optical Eng; fel Am Inst Chemists; Am Water Works Asn; AAAS. *Res:* Design of spectroscopic instrumentation; infrared systems design; infrared physics; applications of infrared emission. *Mailing Add:* Dept Chem Baylor Univ PO Box 97348 Waco TX 76798-7348

BUSCH, LLOYD VICTOR, b London, Ont, Nov 15, 18; m 43; c 4. PHYTOPATHOLOGY. *Educ:* Ont Agr Col, BSA, 42; Univ Toronto, MSA, 48; Univ Wis, PhD(plant path), 55. *Prof Exp:* From lectr to assoc prof, Univ Guelph, 46-71, prof environ biol, Ont Agr Col, 72-83; RETIRED. *Mem:* Can Phytopath Soc (pres, 74-75); Am Phytopath Soc. *Res:* Physiology of broad leafed plants and alfalfa wilt; physiology and histology of wilting; fusarium stock and cob rot of corn mycotoxins; diseases of greenhouse ornamentals. *Mailing Add:* 55 Mary St Guelph ON N1G 2A9 Can

BUSCH, MARIANNA ANDERSON, b Pittsburgh, Pa, Oct 28, 43; m 68. INFRARED SPECTROSCOPY, COLLOID CHEMISTRY. *Educ:* Randolph-Macon Woman's Col, BA, 65; Fla State Univ, PhD(inorg chem), 72. *Prof Exp:* Lectr chem, Cornell Univ, 72, res assoc, 73-74; lectr, 75 & 77, from asst prof to assoc prof, 77-91, PROF CHEM, BAYLOR UNIV, 91- *Concurrent Pos:* Robert A Welch fel, 74-75 & 77, NSF energy-related fel, 76-77; vis assoc, Dept Environ & Eng Sci, Calif Inst Technol, 83, vis scientist, Chem Dept, Cornell Univ, 85-88; consult scientist, Nat Bur Standards, 84-; Fulbright fel, Victoria Univ, Wellington, NZ, 66; vis scientist, DuPont, Wilminton, Del, 86. *Honors & Awards:* James Lewis Howe Award, Am Chem Soc, 65. *Mem:* Am Chem Soc; Sigma Xi; Am Water Works Asn; fel Am Inst Chemists; Soc Appl Spectros; Int Soc Optical Eng. *Res:* Design and application of spectroscopic instrumentation; infrared emission spectrometry; colloid chemistry. *Mailing Add:* Dept Chem Baylor Univ PO Box 97348 Waco TX 76798-7348

BUSCH, ROBERT EDWARD, b Independence, Mo, May 13, 24; m 54; c 2. ANIMAL GENETICS. *Educ:* Univ Mo, BS, 45, MEd, 55; Ore State Univ, PhD(genetics), 63. *Prof Exp:* Instr high schs, Mo, 45-59; res geneticist, US Fish & Wildlife Serv, 62-63; assoc prof animal sci, 63-77, PROF ANIMAL SCI, CALIF STATE UNIV, CHICO, 77- *Mem:* Am Soc Animal Sci; Int Embryo Transfer Soc. *Res:* Sheep genetics; genetics of fish; swine genetics and nutrition. *Mailing Add:* Dept Animal Sci Calif State Univ Chico CA 95929

BUSCH, ROBERT HENRY, b Jefferson, Iowa, Oct 22, 37; m 58; c 2. GENETICS, STATISTICS. *Educ:* Iowa State Univ, BS, 59, MS, 63; Purdue Univ, PhD(genetics, plant breeding), 67. *Prof Exp:* Technician, Iowa State Univ, 59-61, res assoc agron, 61-63; instr, Purdue Univ, 66-67; from asst prof to assoc prof agron, NDak State Univ, 67-77, prof, 77-; AT DEPT AGRON, UNIV MINN. *Concurrent Pos:* Res geneticist agron & plant genetics, Sci & Educ Admin-Agr Res, USDA, Univ Minn, 78. *Mem:* Fel Am Soc Agron; fel Crop Sci Soc Am. *Res:* Quantitative inheritance of economic characteristics of spring wheat and more effective methods of effecting genetic control and selection; premier seedsman. *Mailing Add:* Dept Agron Univ Minn St Paul MN 55108

BUSCHBACH, THOMAS CHARLES, b Cicero, Ill, May 12, 23; m 47; c 3. TECTONICS. *Educ:* Univ Ill, BS, 50, MS, 51, PhD, 59. *Prof Exp:* From asst geologist to geologist, Ill State Geol Surv, 51-78; res prof geol, St Louis Univ, 78-85; RETIRED. *Concurrent Pos:* Consult, US Nuclear Regulatory Comn, 75-85; geol consult, 78- *Mem:* fel Geol Soc Am; Am Asn Petrol Geol. *Res:* Cambrian and Ordovician stratigraphy; underground gas storage; siting of nuclear facilities; tectonics of midcontinent-US. *Mailing Add:* PO Box 1620 Champaign IL 61820

BUSCHE, ROBERT M(ARION), b St Louis, Mo, June 14, 26; m 50, 80; c 4. BIOTECHNOLOGY, RESEARCH PLANNING. *Educ:* Washington Univ, St Louis, BS, 48, MS, 49, DSc(chem eng), 52. *Prof Exp:* Chem engr, Coal to Oil Demonstr Plant, Bur Mines, 50-53; asst tech supt, Belle Works, E I Du Pont de Nemours & Co, Inc, 53-56 & Sabine River Works, 56-59, res & develop supvr, Exp Sta Labs,62-64, prod develop mgr, Newport Labs, 64-66, tech mgr, Heat Transfer Prod Div, 66-67, staff consult, Mgt Serv Div, 67-70, res planning consult life sci, Cent Res & Develop Dept, 70-85; PRES, BIO EN-GENE-ER ASSOCS, INC, 85- *Concurrent Pos:* Res Found fel; adj prof chem eng, Univ Pa; consult, biotechnol. *Mem:* Am Chem Soc; Am Inst Chem Engrs; Bio-Energy Coun. *Res:* Chemicals from renewable resources via biotechnology; high-temperature reaction systems; coal gasification; spray drying; nylon intermediates; hydrocarbon polymers; plastics fabrication; business analysis; new venture development; single-cell protein; renewable resource utilization. *Mailing Add:* Bio En-Gene-Er Assoc Inc 533 Rothbury Rd Wilmington DE 19803-2439

BUSCHERT, ROBERT CECIL, b Preston, Ont, Nov 28, 24; m 48; c 3. SOLID STATE PHYSICS. *Educ:* Goshen Col, BA, 48; Purdue Univ, MS, 52, PhD(physics), 57. *Prof Exp:* Instr physics & math, Goshen Col, 48-50; mem tech staff, Bell Tel Labs, 56-58; asst prof physics, Purdue Univ, 58-65; PROF PHYSICS, GOSHEN COL, 65- *Mem:* Am Phys Soc. *Res:* Structure of semiconductors; germanium and silicon. *Mailing Add:* Dept Physics Goshen Col Goshen IN 46526

BUSCHKE, HERMAN, b Berlin, Ger, Oct 15, 32; US citizen; div; c 2. NEUROLOGY, PSYCHOLOGY. *Educ:* Reed Col, BA, 54; Western Reserve Univ, MD, 58; Am Bd Psychiat & Neurol, dipl, 66. *Prof Exp:* Intern med, Bronx Munic Hosp Ctr, NY, 58-59, resident neurol, 59-62; from instr to asst prof, Sch Med, Stanford Univ, 62-69, res assoc, Inst Math Studies in Soc Sci, 68-69; assoc prof, 69-74, PROF NEUROL & NEUROSCI, ALBERT EINSTEIN COL MED, 74-, GLUCK DISTINGUISHED SCHOLAR NEUROL, 73- *Concurrent Pos:* USPHS fel, Albert Einstein Col Med, 59-62, res scientist develop award, 64-69; consult, Vet Admin Hosp, Palo Alto, Calif, 63-69; prin investr, USPHS Res Grant, 64-; sr investr, Rose F Kennedy Ctr Res Ment Retardation & Human Develop, 69- *Mem:* Psychonomic Soc; Acad Aphasia; Int Neuropsychol Soc; Am Acad Neurol; Am Psychol Asn; Soc Neurosci. *Res:* Human memory; learning; cognition; linguistic and cognitive dysfunction; neuropsychology. *Mailing Add:* Dept Neurol 1300 Morris Park Ave Albert Einstein Col of Med Bronx NY 10461

BUSCHMANN, MARYBETH TANK, b Evanston, Ill, Oct 10, 42; m 70. GERONTOLOGY, NEUROBIOLOGY. *Educ:* Augustana Col, BSN, 65; Univ Ill, MS, 68, PhD(anat), 75. *Prof Exp:* Staff regist nurse med surg, McKinley Univ Hosp & Mercy Hosp, 66-67; res asst zool, Univ Ill, Urbana, 68, teaching asst, 68-69; res dir, Renal Lab, Univ Wis, 69-70; teaching asst anat, 71-74, from asst prof to prof gen nursing, 75-85, ASST PROF ANAT, UNIV ILL MED CTR, 75-, PROF MED-SURG NURSING, 85- *Concurrent Pos:* NIH fel, 75-76; prin investr, Univ Ill, 75-79 & Am Nurses' Found, 76-78; Am Nurses Found scholar, 79. *Mem:* AAAS; Nat Geront Nursing Asn; Soc Neurosci; Am Soc on Aging; Geront Soc; Sigma Xi. *Res:* Gerontological nursing; neurobiology of aging and depression in the elderly employing the intervention of touch. *Mailing Add:* Univ Ill M/C 802 845 S Damen Chicago IL 60612

BUSCHMANN, ROBERT J, b Chicago, Ill, July 30, 42; m 70. ULTRASTRUCTURAL PATHOLOGY, MORPHOMETRY. *Educ:* Loyola Univ, Ill, BS, 64; Univ Ill, Urbana, MS, 66, PhD(physiol), 69. *Prof Exp:* Physiologist, 69-72, DIR ELECTRON MICROS PROG, VET ADMIN WEST SIDE MED CTR, 72- *Mem:* AAAS; Int Soc Stereology; Am Soc Cell Biol; Electron Micros Soc Am; Am Asn Pathol; AAAS. *Res:* Liver cirrhosis. *Mailing Add:* Dept Path MP113 Vet Affairs West Side Med Ctr 820 S Damen Ave Chicago IL 60612

BUSCOMBE, WILLIAM, b Hamilton, Ont, Feb 12, 18; m 42; c 8. ASTROPHYSICS. *Educ:* Univ Toronto, BA, 40, MA, 48; Princeton Univ, PhD(astron), 50. *Prof Exp:* Meteorologist, Dept Transp, Govt Can, 41-45; instr math & astron, Univ Sask, 45-49; res fel, Mt Wilson & Palomar Observ, 50-52; astronr, Mt Stromlo Observ, Australian Nat Univ, 52-68; PROF

ASTRON, NORTHWESTERN UNIV, EVANSTON, 68- *Concurrent Pos:* Vis prof, Univ Pa, 64-65, Northern Ill Univ, 70-73; nat lectr, Am Astron Soc, 84- *Mem:* Am Astron Soc; Royal Astron Soc Can; Int Astron Union. *Res:* Stellar spectra; radial velocities; binary orbits; galactic kinematics; interstellar gas; variable stars; compiler catalogs of stellar data. *Mailing Add:* Dearborn Observ Northwestern Univ Evanston IL 60208

BUSE, JOHN FREDERICK, b Charleston, SC, Oct 17, 21; m 56; c 1. INTERNAL MEDICINE. *Educ:* Univ SC, BS, 44, MD, 50. *Prof Exp:* Intern, Roper Hosp, 50-51, asst resident med, 51-52; asst resident, Univ Va, 52-53; assoc & assoc prof, 56-70, PROF MED, MED UNIV SC, 70. *Concurrent Pos:* Fel, Med Col SC, 53-54 & Cox Inst, Univ Pa, 54-56. *Mem:* Am Fedn Clin Res; Endocrine Soc; Am Diabetes Asn. *Res:* Diabetes mellitus; influence of muscular activity on insulin requirement. *Mailing Add:* 80 Barre St Charleston SC 29401

BUSE, MARIA F GORDON, b Budapest, Hungary, July 17, 27; m 56; c 3. MEDICINE. *Educ:* Univ Buenos Aires, MD, 54. *Prof Exp:* Intern, Rivadavia Hosp, Arg, 53-54; res fel physiol, Inst Exp Biol & Med, Univ Buenos Aires, 54-55; fel med, Univ Pa, 55-56; res assoc med, 56-59, instr biochem, 59-61, assoc, 61-62, from asst prof to assoc prof res med, 62-72, PROF MED, MED UNIV SC, 72-, PROF BIOCHEM, 74- *Concurrent Pos:* NIH res career develop award, 63. *Mem:* AMA; Am Fedn Clin Res; Soc Nuclear Med; Am Phys Soc; Am Soc Biol Chemists; Am Diabetes Asn; Endocrine Soc. *Res:* Diabetes; action of insulin; intermediary metabolism. *Mailing Add:* Med Univ SC Dept Med 171 Ashley Ave Charleston SC 29425

BUSE, REUBEN CHARLES, b Fergus Falls, Minn, July 8, 32; div; c 2. AGRICULTURAL ECONOMICS. *Educ:* Univ Minn, BS, 54, MS, 56; Pa State Univ, PhD(agr econ), 59. *Prof Exp:* Res asst agr econ, Univ Minn, 54-56; instr, Pa State Univ, 56-59; from asst prof to assoc prof, 59-69, PROF AGR ECON, UNIV WIS, 69- *Concurrent Pos:* Consult, Brazilian & Peruvian Ministry of Planning, USDA, AID & several private firms. *Honors & Awards:* Qual of Commun Award, Am Agr Econ Asn. *Mem:* Am Agr Econ Asn; Am Econ Asn; Int Asn Agr Economists. *Res:* Quantitative methods in agricultural economics; consumer expenditure behavior. *Mailing Add:* 3436 Valley Creek Circle Middleton WI 53562

BUSECK, PETER R, b Sept 30, 35; US citizen; m 60; c 4. GEOCHEMISTRY, MINERALOGY. *Educ:* Antioch Col, AB, 57; Columbia Univ, MA, 59, PhD(geol), 62. *Prof Exp:* Fel, Geophys Lab, Carnegie Inst Washington, 61-63; REGENTS' PROF GEOL & CHEM, ARIZ STATE UNIV, 63- *Concurrent Pos:* Vis prof, Oxford Univ, 70-71, Stanford Univ, 79-80, Univ Paris, 86-87; distinguished res prof, Ariz State Univ, 78-79. *Honors & Awards:* Geol Award, Microbeam Anal Soc, 81. *Mem:* Geochem Soc (secy, 76-79); fel Meteoritical Soc; Am Geophys Union; fel Geol Soc Am; fel AAAS; fel Mineral Soc Am. *Res:* Electron microscopy and diffraction of minerals; electron microprobe analysis of air pollutants and minerals; scanning electron microscopy; meteoritics. *Mailing Add:* Depts Geol & Chem Ariz State Univ Tempe AZ 85287

BUSEMANN, ADOLF, aeronautics; deceased, see previous edition for last biography

BUSENBERG, EURYBIADES, b Jerusalem, Israel, Nov 13, 39; m 74. GEOCHEMISTRY. *Educ:* NY Univ, BA, 61, MS, 67; State Univ NY Buffalo, PhD(geochem), 75. *Prof Exp:* Chemist, Am Petrol Inst, 65-67; petrol geologist, Texaco Inc, 69-71; vis asst prof, 74-75, asst prof geochem, State Univ NY Buffalo, 75-78; HYDROLOGIST, US GEOL SURV, 78- *Mem:* Sigma Xi; Clay Minerals Soc. *Res:* The dissolutions kinetics of aluminosilicate minerals; natural weathering reactions of aluminosilicates and their products; the aquatic chemistry of natural waters. *Mailing Add:* 11628 Quail Ridge Ct Reston VA 22094

BUSENBERG, STAVROS NICHOLAS, b Jerusalem, Palestine, Oct 16, 41; nat US; m 69; c 2. DIFFERENTIAL EQUATIONS, MATHEMATICAL MODELLING. *Educ:* Cooper Union, BME, 62; Ill Inst Technol, MS, 64 & 65, PhD(math), 67. *Prof Exp:* Instr math, Loyola Univ, Ill, 66-67; fel, Sci Ctr, NAm Rockwell Corp, Calif, 67-68; from asst prof to assoc prof, 68-79, PROF MATH, HARVEY MUDD COL, 79- *Concurrent Pos:* Vis scholar, Stanford Univ, 75-76; consult, Oak-Ridge Nat Lab, 79-82; vis prof, Univ Trenton, 81-82, Univ Victoria, 89; prin investr, NSF res grants, 83-; vis assoc, Calif Inst Technol, 85; sr vis, Oxford Univ, 86; assoc ed, J Math Anal & Appln, 87-; Fulbright res prof, Massey Univ, NZ, 89. *Mem:* Soc Indust & Appl Math; Am Math Soc; Math Asn Am; Soc Math Biol; Ital Math Soc. *Res:* Differential equations; applied mathematics; industrial mathematics. *Mailing Add:* Dept Math Harvey Mudd Col Claremont CA 91711

BUSER, KENNETH RENE, b Bloomington, Ind, Apr 13, 25; m 46; c 3. ORGANIC CHEMISTRY. *Educ:* Wabash Col, BA, 50; Purdue Univ, MS, 52, PhD(chem), 54. *Prof Exp:* Asst, Purdue Univ, 50-54; res chemist, Exp Sta, 54-69, staff chemist, Fabrics & Finishes Dept, 69-85, RES ASSOC, E I DU PONT DE NEMOURS & CO, INC, 85- *Mem:* Am Chem Soc. *Res:* Organic chemistry of sulfur, polymer chemistry, finishes. *Mailing Add:* 2813 Ambler Ct Skyline Crest Wilmington DE 19808

BUSER, MARY PAUL, b Wichita, Kans, Sept 7, 28. MATHEMATICS, COMPUTER SCIENCE. *Educ:* Marymount Col, Kans, AB, 57; St Louis Univ, AM, 59, PhD(math), 61; Ohio State Univ, MS, 74. *Prof Exp:* From instr to assoc prof math, Marymount Col, Kans, 61-75, pres, 75-81; VIS ASSOC PROF MATH, WICHITA STATE UNIV, 81- *Concurrent Pos:* NSF, Inst Comput Sci, Univ Mo, Rolla, 64; partic, Asn Comput Mach, Inst Comput Sci, Purdue Univ, 71. *Mem:* Math Asn Am; Am Math Soc. *Res:* Approximating polynomial for the number of partitions of a positive integer into unit and prime summands; Waring's problem for congruences. *Mailing Add:* 2356 Marigold St Wichita KS 67204

BUSEY, RICHARD HOOVER, b Cairo, Ill, Apr 6, 19; m 42; c 4. PHYSICAL CHEMISTRY, THERMODYNAMICS. *Educ:* Southern Methodist Univ, BS, 41; Univ Calif, PhD(chem), 50. *Prof Exp:* Res anal chemist, Magnolia Petrol Co, 41-45; assoc, Univ Calif, 50-52; res chemist, Oak Ridge Nat Lab, 52-84; RETIRED. *Mem:* AAAS; Am Chem Soc. *Res:* Low temperature calorimetry; spectroscopy; chemistry of Technetium and Rhenium; high temperature aqueous inorganic species equilibria; high temperature and pressure calorimetry on aqueous solutions. *Mailing Add:* 106 S Tampa Oak Ridge TN 37830

BUSH, ALFRED LERNER, b Rochester, NY, Dec 21, 19; m 42, 65; c 8. ECONOMIC GEOLOGY, ENVIRONMENTAL GEOLOGY. *Educ:* Univ Rochester, AB, 41, MS, 46. *Prof Exp:* Geologist, 46-87, EMER GEOLOGIST, US GEOL SURV, 87- *Concurrent Pos:* Sci secy for mineral resources, US deleg, UN Conf Sci & Technol, Switz, 63. *Mem:* Fel Geol Soc Am; Soc Econ Geol; Int Asn Genesis Ore Deposits. *Res:* Uranium deposits of the Colorado Plateau; geology of the western San Juan Mountains, Colorado; economic geology of lightweight aggregates, United States and worldwide. *Mailing Add:* US Geol Surv MS 941 Denver Fed Ctr Box 25046 Denver CO 80225

BUSH, C ALLEN, b Rochester, NY, Aug 2, 38. BIOPHYSICAL CHEMISTRY. *Educ:* Cornell Univ, BA, 61; Univ Calif, Berkeley, PhD(chem), 65. *Prof Exp:* Res assoc chem, Cornell Univ, 66-68; from asst prof to assoc prof, 68-81, PROF CHEM, ILL INST TECHNOL, 81- *Mem:* Am Soc Biol Chemists; Am Chem Soc; Biophys Soc. *Res:* Optical rotatory dispersion and circular dichroism of biopolymers; oligosaccharide structure and conformation glycoproteins; nuclear magnetic resonance. *Mailing Add:* Dept Chem Univ Md 5401 Wilkens Ave Baltimore MD 21228-5329

BUSH, CHARLES EDWARD, b Miami, Fla, Aug 21, 38; m 66; c 2. FUSION ENERGY RESEARCH, BOLOMETRIC & INFRARED DIAGNOSTICS. *Educ:* Knoxville & Lafayette Col, BS, 62; Univ Wis, Madison, MS, PhD (nuclear eng), 72. *Prof Exp:* Proj engr, US Army, Edgewood Arsenal, Md, 62-64; RES STAFF FUSION, OAK RIDGE NAT LAB, 72- *Concurrent Pos:* Vis fac fusion res, Princeton Plasma Physics Lab, Princeton Univ, 83-; chmn comt minorities in physics, Am Phys Soc, 78-82. *Mem:* Am Phys Soc; Am Nuclear Soc; Sigma Xi; NY Acad Sci; Am Asn Advan Sci. *Res:* Energy and power balance and confinement of high temperature plasma in tokamaks; bolometric and infrared diagnostics in fusion. *Mailing Add:* 192 Loomis Ct Princeton Univ C-Site LOB B233 PO Box 451 Princeton NJ 08540-3439

BUSH, DAVID CLAIR, b Malad City, Idaho, Nov 6, 22; m 44; c 4. ORGANIC CHEMISTRY. *Educ:* Univ Idaho, BS, 48; Ore State Col, MS, 50, PhD(org chem), 53. *Prof Exp:* Res chemist, Pittsburgh Plate Glass Co, 52-54; VPRES, IDAHO CHEM INDUSTS, 54- *Mem:* Am Chem Soc. *Res:* Metal pyridine salts; Mannich reaction; synthetic polymers; polyesters and epoxy resins. *Mailing Add:* 6820 McMullen Boise ID 83709

BUSH, DAVID E, b Washington, DC, Feb 24, 52. CARDIOLOGY, HOMODYNAMICS. *Educ:* Univ Calif, San Francisco, MD, 77. *Prof Exp:* Intern med, 77-78, resident med, 78-80, cardiol fel, 80-83, asst med, 83-85, instr med, 85-87, ASST PROF MED, JOHNS HOPKINS UNIV SCH MED, 87- *Mem:* Am Heart Asn. *Res:* Effects of drugs and aging on cardiovascular (ventricular-vascular) couplings; cardiac magnetic resonances imaging; human atherosclerosis, pathogenesis and drug effects. *Mailing Add:* Francis Scott Key Med Ctr 4940 Eastern Ave Baltimore MD 21224

BUSH, DAVID GRAVES, b Westfield, Mass, Apr 21, 22; m 45; c 3. ENVIRONMENTAL ANALYSIS, INDUSTRIAL HYGIENE. *Educ:* Univ Mass, BS, 47; Univ Minn, MS, 50. *Prof Exp:* Asst synthetic rubber res, Univ Minn, 52; from res chemist to sr res chemist, Eastern Kodak Co, 52-64, res assoc anal res & develop, 64-75, lab head, Res Labs, 75-78, anal consult, Health & Environ Lab, 79-84; OCCUP HEALTH CONSULT, DAVID G BUSH ASSOCS, 84- *Honors & Awards:* Rochester Sect Award, Am Chem Soc, 87. *Mem:* Am Chem Soc; AAAS; Am Indust Hyg Asn; Am Soc Safety Eng. *Res:* Gas chromatography; functional group analysis; liquid chromatography; polymer identification and analysis; light microscopy; atomic absorption spectroscopy. *Mailing Add:* 147 Landing Park Rochester NY 14625-1717

BUSH, FRANCIS M, b Bloomfield, Ky, Sept 5, 33; m 59. ZOOLOGY. *Educ:* Univ Ky, BS, 55, MS, 57, DMD, 75; Univ Ga, PhD(zool), 62. *Prof Exp:* From asst prof to assoc prof biol, Stanford Univ, 60-64; asst prof, 64-69, ASSOC PROF GEN DENT, MED COL VA, VA COMMONWEALTH UNIV, 69- *Concurrent Pos:* Mary Glide Goethe travel awards, 59-60; Sigma Xi res grant, 63; Am Philos Soc grant, 64; USPHS grant, 65-68. *Mem:* Am Dent Soc; Int Asn Dent Res. *Res:* Epidemiology of mandibular dysfunction; facial pain; effect of local anesthetics on skeletal muscle. *Mailing Add:* Med Col Va Sch Dent 520 N 12th St Box 566 Richmond VA 23298

BUSH, GARY GRAHAM, b Atlanta, Ga, Dec 29, 50; m 82; c 2. MAGNETICS, INSTRUMENTATION DESIGN. *Educ:* Ga Inst Technol, BS, 72, MS, 73, MS(physics) & MS(elec eng), 76, PhD(elec eng), 83. *Prof Exp:* Sr engr, Intel Magnetics, 84-85; res specialist, 85-89, STAFF SCIENTIST, LOCKHEED PALO ALTO RES LABS, 89-, PRIN SCIENTIST, MICROMAGNETICS LAB, 89- *Concurrent Pos:* Chief engr & owner, DosTek, Computer Systs Integrator, 87-; consult appl math models, Becket & Co, 90- *Mem:* Sr mem Inst Elec & Electronics Engrs; assoc mem Sigma Xi. *Res:* Artificial magnetodielectric materials and their modeling and measurement; electromagnetic applications of materials and design and analysis of thin film materials and bulk materials. *Mailing Add:* 5505 Castle Manor Dr San Jose CA 95129

BUSH, GEORGE CLARK, b St Catharines, Ont, Oct 18, 30; m 60; c 2. ALGEBRA, DATA STRUCTURES. *Educ:* McMaster Univ, BA, 54; Mass Inst Technol, SM, 56; Queen's Univ, Ont, PhD(math), 61; Inst Retraining Computer Sci, cert, 85. *Prof Exp:* Asst prof math, Queen's Univ, Ont, 61-66;

assoc prof, Beirut Univ Col, 67-71; spec lectr, Queen's Univ, Ont, 71-73; assoc prof, Col Arts & Sci, Univ Shiraz, 73-76; assoc prof, Acadia Univ, 76-78; from asst prof to assoc prof math, Univ Col Cape Breton, 78-88; ASSOC PROF, MID E TECH UNIV, TURKEY, 88- *Mem:* Math Asn Am; Am Math Soc; Can Math Soc. *Res:* Algebra; numerical analysis; data structures. *Mailing Add:* Dept Math Mid E Tech Univ Ankara 06531 Turkey

BUSH, GEORGE EDWARD, b Jan 9, 37; US citizen; m 62. COMPUTER SECURITY SYSTEM DESIGN. *Educ:* Purdue Univ, BS, 60; San Jose State Univ, MS, 75. *Prof Exp:* Physicist, Midwestern Univ Res Asn, 60-66; physicist, 66-82, SECURITY ADMIN, LAWRENCE NAT LAB, 82- *Concurrent Pos:* Consult, Solar Systs Inc, 76- *Mem:* Int Solar Energy Soc; AAAS. *Res:* Development of secure network, development of tempest-approved switch to allow a microcomputer system to operate in both a classified and an unclassified mode; data base management development. *Mailing Add:* Lawrence Livermore Nat Lab PO Box 808L-303 Livermore CA 94550

BUSH, GEORGE F(RANKLIN), b Philadelphia, Pa, July 1, 09; m 39; c 3. BIOMEDICAL & MECHANICAL ENGINEERING. *Educ:* Lafayette Col, BS, 32; Univ Pa, MS, 47. *Prof Exp:* From instr to assoc prof mech & asst dean, George Washington Univ, 39-42; asst prof eng, Princeton Univ, 42-45; FOUNDER, OWNER & DIR, G F BUSH ASSOCS, 47- *Concurrent Pos:* Mem, Bioinstrumentation Adv Coun, US Navy. *Mem:* AAAS; Sigma Xi. *Res:* New scientific instrument creation and design; mechanical, electromechanical, electrochemical and biomedical devices. *Mailing Add:* Box 905 Stonington ME 04681

BUSH, GLENN W, b Bellwood, Pa, Sept 9, 33; m 56; c 3. METALLURGY. *Educ:* Pa State Univ, BS, 55, MS, 58, PhD(metall), 60. *Prof Exp:* Metallurgist, Allegheny Ludlum Steel Corp, 55-56; res metallurgist, Nat Steel Corp, 60-63, sr res metallurgist, 63-65, supvr coated prod res, 65-69, asst div chief, 69-78, div chief, 78-84, sr res consult, 85, mgr coated prod/process, 85-87, mgr, finished prod/processes, 87-88; PRES, BUSH & ASSOCS, 88- *Mem:* Am Soc Metals; Am Inst Mining, Metall & Petrol Engrs; Am Vacuum Soc; Metall Soc; Sigma Xi. *Res:* Physical metallurgy; effect of a residual element on properties of steel; measurement of residual stress in electrodeposited metals; relationship of micro-structure of properties of metals. *Mailing Add:* 213 Timberyoke Rd Coraopolis PA 15108

BUSH, GUY L, b Greenfield, Iowa, July 9, 29; m 59; c 3. EVOLUTIONARY BIOLOGY. *Educ:* Iowa State Col, BS, 53; Va Polytech Inst, MS, 60; Harvard Univ, PhD(biol), 64. *Prof Exp:* Entomologist, USDA, 55-57; NIH fel, Univ Melbourne, 64-66; asst prof, 66-73, assoc prof, 73-76, prof zool, Univ Tex, Austin, 76-82; HANNAH PROF, DISTINGUISHED PROF, EVOLUTIONARY BIOL, MICH STATE UNIV, 82- *Concurrent Pos:* Guggenheim fel, Univ Calif, Berkeley, 77; vis res scientist, Swiss Fed Res Sta Arboracult, Hort & Viticulture, W-deuswil, 71; vis prof, Dept Biol Sci, Univ Sao Paulo, 77; vis fel, Oxford Univ, 90-91. *Honors & Awards:* Founders Mem Award, Entom Soc Am, 87. *Mem:* Fel AAAS; Soc Study Evolution (vpres, 80, pres, 81); Genetics Soc Am; Entom Soc Am; Soc Syst Zool; Sigma Xi; Am Soc Naturalists (vpres, 80). *Res:* Modes of speciation; systematics of higher Diptera; evolutionary cytogenetics; genetics of host-parasite interactions. *Mailing Add:* Dept Zool Mich State Univ East Lansing MI 48824-1115

BUSH, KAREN JEAN, b Evansville, Ind, Oct 5, 43; m 73; c 2. MICROBIOLOGY. *Educ:* Monmouth Col, Ill, BA, 65; Ind Univ, Bloomington, PhD(biochem), 70. *Prof Exp:* Fel biol, Univ Calif, Santa Barbara, 70-71; instr biochem, Sch Med, Univ NC, Chapel Hill, 71-72; asst prof chem, Univ Del, 72-73; res investr, Squibb Inst Med Res, 73-81, sr res investr, 81-88, res fel, 88-89, RES LEADER, BRISTOL-MYERS SQUIBB PHARMACEUT RES INST, 89- *Mem:* Am Chem Soc; AAAS; Am Soc Microbiol. *Res:* B-lactamases: characterization, mechanism of action, inhibition; mechanism of action, inhibition; monobactams; penicillin-binding proteins; bacterial resistance to B-lactam antibiotics. *Mailing Add:* PO Box 4000 Princeton NJ 08543-4000

BUSH, LEON F, b Brooksville, Ky, Jan 28, 24; m 49; c 4. ANIMAL NUTRITION. *Educ:* Univ Ky, BS, 50, MS, 51; Cornell Univ, PhD(nutrit & physiol), 54. *Prof Exp:* ASSOC PROF ANIMAL SCI, S DAK STATE UNIV, 54- *Mem:* Am Soc Animal Sci. *Res:* Sheep management and nutrition; physiology of reproduction studies with sheep. *Mailing Add:* RR 3 Box 25 Brookings SD 57006

BUSH, LINVILLE JOHN, b Winchester, Ky, May 7, 28; m 52; c 4. DAIRY NUTRITION. *Educ:* Univ Ky, BSc, 48; Ohio State Univ, MSc, 49; Iowa State Univ, PhD(dairy nutrit), 58. *Prof Exp:* Field agent, Dairying, Ky, 53-55; assoc prof, 58-77, prof dairy sci, 77-88, EMER PROF, OKLA STATE UNIV, 88- *Concurrent Pos:* Consult, US Feed Grains Coun, Mexico & Venezuela. *Honors & Awards:* Tyler Prof Distinction, 78. *Mem:* Am Dairy Sci Asn; Prof Animal Scientists; Coun Agricult Sci & Technol. *Res:* Nutrition of young dairy calf; dietary requirements of lactating dairy cows; factors affecting rumen function. *Mailing Add:* Dept Animal Sci Okla State Univ Stillwater OK 74078

BUSH, LOWELL PALMER, b Tyler, Minn, Mar 31, 39; m 62; c 3. AGRONOMY. *Educ:* Macalester Col, BA, 61; Iowa State Univ, MS, 63, PhD(plant physiol), 64. *Prof Exp:* Fel plant path, Univ Minn, 64-66; assoc prof, 66-75, PROF AGRON, UNIV KY, 75- *Concurrent Pos:* Fulbright res scholar, Agr Inst, Repub Ireland, 87; Philip Morris Prof, 87- *Honors & Awards:* Philip Morris Distinguished Sci Award. *Mem:* Am Soc Plant Physiol; Phytochem Soc; fel Am Soc Agron; fel Crop Sci Soc Am. *Res:* Alkaloid metabolism in plants. *Mailing Add:* Dept Agron Univ Ky Lexington KY 40546-0091

BUSH, MAURICE E, b Chicago, Ill, Dec 9, 36. CELLULAR IMMUNOLOGY. *Educ:* Univ Chicago, BS, 60; Howard Univ, MS, 65; Brandeis Univ, PhD(biochem), 70. *Prof Exp:* Fel immunol, Univ Calif, San Francisco, 71-73; res assoc endocrinol, Montifiore Med Ctr, 74-76; res assoc rheumatology, 76-78, instr, 78-80, DIR CLIN RES CORE LAB, DOWNSTATE MED CTR, NY, 80- *Mem:* AAAS. *Res:* Examination of the products of arachidonic acid metabolites from platelets, monocytes and polymorphonuclears and the determination of their various roles in the inflammation observed in patients with rheumatic diseases. *Mailing Add:* Dept Med Downstate Med Ctr State Univ NY 450 Clarkson Ave Brooklyn NY 11203

BUSH, NORMAN, b New York, NY, Dec 10, 29; m 52; c 3. STATISTICS. *Educ:* City Col New York, BBA, 51, MBA, 52; NC State Col, PhD, 62. *Prof Exp:* Statistician, Army Chem Ctr, Md, 52-56, Patrick AFB, Fla, 56-61 & Instrument Corp, Fla, 61-63; res dir statist, D Brown Assocs, Fla, 63-64; prin engr, Pan Am World Airways, 64-72; PRES, ENSCO, INC, VA, 72- *Concurrent Pos:* Statistician, Comput Ctr, Univ NC, 59-62; adj asst prof, Grad Eng Educ Syst, Univ Fla, Cape Kennedy, 67-70. *Mem:* Am Statist Asn. *Res:* Technometrics; regression analysis; experimental design; computers. *Mailing Add:* ENSCO Inc 5400 Port Royal Rd Springfield VA 22151

BUSH, RAYMOND SYDNEY, oncology, radiation biology; deceased, see previous edition for last biography

BUSH, RICHARD WAYNE, b Cleveland, Ohio, Nov 5, 34; m 69; c 2. POLYMER CHEMISTRY. *Educ:* Mass Inst Technol, SB, 56; Univ Ill, PhD(org chem), 60; Johns Hopkins Univ, MAS, 78. *Prof Exp:* Res chemist, 60-64, sr res chemist, 64-69, res supvr, 69-72, RES ASSOC, W R GRACE & CO, 72- *Mem:* Am Chem Soc; Soc Plastics Engrs. *Res:* Photopolymers; adhesives. *Mailing Add:* W R Grace & Co 7379 Rte 32 Columbia MD 21044

BUSH, ROY SIDNEY, b Aug 31, 46; Can citizen; m 73; c 1. ANIMAL SCIENCE & NUTRITION, BIOCHEMISTRY. *Educ:* Univ Alta, BSA, 68, MSc, 70; Univ Man, PhD(animal biochem), 75. *Prof Exp:* RES SCIENTIST, AGR CAN, 77- *Concurrent Pos:* Fel, Animal Res Inst, Agr Can, 74-77; sabbatical, Lab du Jeune Ruminants, INRA, Rennes, 88-89. *Mem:* Am Dairy Sci Asn; Am Soc Animal Sci; Can Soc Animal Sci; Agr Inst Can. *Res:* Development of the preruminant calf and lamb into a functional ruminant; feed additives for the young ruminant; digestive immune reactions. *Mailing Add:* PO Box 20280 Agr Can Fredericton NB E3B 4Z7 Can

BUSH, S(PENCER) H(ARRISON), b Flint, Mich, Apr 4, 20; m 48; c 2. CODES & STANDARDS, SAFETY. *Educ:* Univ Mich, BS(chem eng), 48, BS, 48, MS, 50, PhD(metall eng), 53. *Prof Exp:* Assoc, Eng Res Inst, Univ Mich, 47-53, res asst, Off Naval Res, 50-53, instr dent mat, 51-53; metallurgist, Hanford Atomic Prods Oper, Gen Elec Co, 53-54; metallurgist, Hanford Atomic Prods Oper, Gen Elec Co, 53-54, supvr phys metall, 54-57, fuels fabrication develop, 57-60, metall specialist, 60-63, consult metallurgist, 63-65; consult to dir, Pac Northwest Lab, Battelle Mem Inst, 65-70, sr staff consult, 70-81; PRES, REV & SYNTHESIS ASSOCS, 82- *Concurrent Pos:* Adj prof metall, Joint Ctr Grad Study, Richland, Wash; mem, US Atomic Energy Comn adv comt on reactor safeguards, 66-77, v chmn, 70, chmn, 71; regents prof metall eng, Univ Calif, Berkeley, 73-74. *Honors & Awards:* Gillett lectr, Am Soc Testing & Mat, 74; Mehl lectr, Am Soc Nondestructive Testing, 81; Bernard Langer Nuclear Codes & Standards Award, Am Soc Mech Eng, 83; TJ Thompson Award, Am Nuclear Soc, 87. *Mem:* Nat Acad Eng; fel Am Soc Mech Engrs; fel Am Soc Metals; Am Inst Mining, Metall & Petrol Engrs; fel Am Nuclear Soc. *Res:* Temper embrittlement, stress corrosion of steels; metallurgy of reactor metals; safety of nuclear reactor systems. *Mailing Add:* 630 Cedar Ave Richland WA 99352

BUSH, STEWART FOWLER, b Charlotte, NC, Jan 8, 41; m 65; c 2. PHYSICAL CHEMISTRY. *Educ:* Erskine Col, AB, 63; Univ SC, PhD(phys chem), 67. *Prof Exp:* Asst prof chem, Erskine Col, 67-69; asst prof, 69-74, ASSOC PROF CHEM, UNIV NC, CHARLOTTE, 74- *Concurrent Pos:* NSF prof develop award, 78; vis scientist, NIH, Washington, DC, 78-79. *Honors & Awards:* NSF Prof Develop Award, 78; Intergovt Personal Act Award, NIH, 79. *Mem:* Am Chem Soc; Coblentz Soc. *Res:* Infrared, far-infrared and Raman spectroscopy as applied to vibrational analysis and molecular structure determination; mechanisms of interaction between cell membrane componets. *Mailing Add:* Dept Chem Univ NC University Sta Charlotte NC 28223

BUSH, WARREN VAN NESS, b Montclair, NJ, June 10, 31; m 57; c 3. PETROLEUM CHEMISTRY, ENVIRONMENTAL SCIENCE. *Educ:* Princeton Univ, BS, 53; Calif Inst Technol, PhD, 58. *Prof Exp:* Chemist, Shell Develop Co, Calif, 57-65, engr, Shell Oil Co, NY, 65-67, chemist, Shell Develop Co, 67-68, supvr, 68, group leader, Shell Oil, Martinez Refinery, 68-72, sr res engr, Shell Develop Co, 72-74, supvr staff, 74-76, staff res engr, 76-77, sr staff res engr, 77-89, RES ASSOC, SHELL DEVELOP CO, 89- *Mem:* AAAS; Am Chem Soc; Am Inst Chem Engrs; Sigma Xi. *Res:* Organic chemistry and chemical engineering, principally as applied to petroleum refining, petrochemical manufacture, gas treating, sulfur manufacture, and air pollution control; environmental engineering. *Mailing Add:* Shell Develop Co PO Box 1380 Houston TX 77251-1380

BUSHAW, DONALD (WAYNE), b Anacortes, Wash, May 5, 26; m 46; c 5. MATHEMATICS. *Educ:* State Col Wash, BA, 49; Princeton Univ, PhD(math), 52. *Prof Exp:* Instr math, State Col Wash, 48-49; asst, Princeton Univ, 49-51; math consult, Stevens Inst Technol, 51-52; from instr to assoc prof, Wash State Univ, 52-62, actg chmn dept, 66-68, actg dir libr, 84-85 & 89-90, PROF MATH, WASH STATE UNIV, 62-, VPROVOST INSTRUCTION, 86- *Concurrent Pos:* Vis scientist, Res Inst Advan Study, 62-63; NAS-PAN exchangee, Jagiellonian Univ, Cracow, Poland, 72-73. *Mem:* Math Asn Am; Polish Math Soc. *Res:* Qualitative theory of ordinary differential equations; general topology; mathematical economics; history of mathematics; postsecondary mathematics education. *Mailing Add:* PO Box 124 Pullman WA 99163-0124

BUSHEY, ALBERT HENRY, b Mansfield, Ohio, Mar 18, 11; m 40; c 2. PHYSICAL CHEMISTRY, ANALYTICAL CHEMISTRY. *Educ:* Wittenberg Col, AB, 32; Univ Minn, PhD(phys chem), 40. *Prof Exp:* Asst chem, Univ Minn, 36-40; chemist, Res Lab, Aluminum Co Am, 40-44, asst chief anal div, 44-48; head anal res, Hanford Atomic Prod Oper, Gen Elec Co, 48-53, head chem res, 53-54, instr, Sch Nuclear Eng, 49-54, consult chemist, Gen Eng Lab, 54-62; head finishing res, Aluminum Res Div, Kaiser Aluminum & Chem Corp, 62-67, tech supvr, 67-76; RETIRED. *Concurrent Pos:* Instr, Pa State Univ, 42-43. *Mem:* Am Chem Soc. *Res:* Coprecipitation and aging of precipitates; chemistry and analyses of aluminum, uranium, transuranium and fission products; instrumental analyses; industrial separations processes for uranium, transurancis and fission products; electrochemistry; finishing of aluminum. *Mailing Add:* 48 Baja Loma Ct Danville CA 94526

BUSHEY, DEAN FRANKLIN, b Gettysburg, Pa, Oct 27, 50; m 73; c 2. SYNTHETIC ORGANIC CHEMISTRY. *Educ:* Gettysburg Col, BA, 72; Univ SC, PhD(org chem), 76. *Prof Exp:* NIH fel, Mass Inst Technol, 76-77; res chemist, Union Carbide Corp, 77-84; biochemist, 86-90, MGR BIOCHEM & BIOTECHNOL, RHÔNE POULENC AGR CO, 90- *Concurrent Pos:* Vis scientist, Cornell Biotechnol Inst, 84-86. *Mem:* Sigma Xi; Am Chem Soc. *Res:* Synthesis of natural products and small ring heterocycles; pesticide metabolism. *Mailing Add:* 5336 Cherry Crest Ct Raleigh NC 27609

BUSHEY, GORDON LAKE, b Mission, Tex, Jan 13, 22; m 50; c 3. PHYSICAL CHEMISTRY. *Educ:* Rice Univ, BS, 43, MA, 44, PhD(chem), 48. *Prof Exp:* Instr chem, Univ Ill, 48-51; sci asst, Off Chief Chem Officer, US Army Mat Develop & Readiness Command, 51-62, asst sci affairs, US Defense Off NAtlantic & Medit Areas, Paris, 62-64, asst chief scientist, US Army Mat Command, 64-70, sci asst, Dept Foreign Labs, 70-75, phys sci adminr, Off Asst Dep Sci & Technol, 75-85; RETIRED. *Concurrent Pos:* Exec officer, Div Chem & Chem Technol, Nat Res Coun, 51-53 & Nat War Col, 59-60. *Mem:* Am Inst Chem; AAAS; Am Chem Soc. *Res:* Absorption by hydrous oxides; physical chemistry of soap crystals; x-ray diffraction studies of colloids and mixed crystals; arms control; international science organizations; technical information; physical sciences of military interest. *Mailing Add:* T-3 3364 Woodburn Rd Annandale VA 22003

BUSHEY, MICHELLE MARIE, b Boston, Mass, Apr 26, 60; m 83. BIOTECHNOLOGY, BIOANALYTICAL CHEMISTRY. *Educ:* Oberlin Col, BA, 82; Univ NC, PhD(anal chem), 90. *Prof Exp:* ASST PROF ANALYTICAL CHEM, TRINITY UNIV, 90- *Mem:* Am Chem Soc; Coun Undergrad Res; AAAS. *Res:* Protein structure and properties by capillary electrophoresis; optimal separation strategies in micellar electrokinetic capillary chromatography; application of capillary electrophoresis to biologically significant problems. *Mailing Add:* Dept Chem Trinity Univ 715 Stadium Dr San Antonio TX 78212

BUSHINSKY, DAVID ALLEN, b Elizabeth, NJ, March 16, 49; m 78; c 2. MEDICINE, PHYSIOLOGY. *Educ:* Lehigh Univ, BS, 71; Tufts Univ, MD, 75. *Prof Exp:* Intern med, Tufts-New Eng Med Ctr, 75-76, resident med, 76-77, clin fel nephrology, 77-78, res fel nephrology, 78-79; instr med, Sch Med, Tufts Univ, 79-80; asst prof, 80-87, ASSOC PROF MED, UNIV CHICAGO PRITZKER SCH MED, 87- *Concurrent Pos:* Attend physician, Michael Reese Hosp & Med Ctr, 80- & Univ Chicago Hosp, 82- *Mem:* Am Soc Nephrology; Int Soc Nephrology; Am Fed Clin Res; Am Soc Bone & Mineral Res; Central Soc Clin Res; Am Physiol Soc. *Res:* The effects of hydrogen ions on cultured bone and the regulation of serum 1.25 dihydroxy vitamin D3 by ionic calcium. *Mailing Add:* Unit Chief Nephrol Univ Rochester 601 Elmwood Ave Box MED Rochester NY 14642

BUSHKIN, YURI, b Riga, USSR, Oct 22, 49. MOLECULAR IMMUNOLOGY. *Educ:* Latvian State Univ & Inst Cytol, USSR, MSc, 71; Weizmann Inst Sci, Israel, PhD(genetics), 78. *Prof Exp:* Vis res fel immunogenetics, Lab Cell Surface Immunogenetics, Sloan-Kettering Inst Cancer Res, 77-81, res assoc immunol, Lab Molecular Immunol, 81-86; assoc, Dept Retroviral Biol, 86-88, ASST MEM & HEAD, LAB MOLECULAR IMMUNOL, PUB HEALTH RES INST, 88-; RES ASSOC PROF PATH, DEPT PATH, NY UNIV MED CTR, 89- *Concurrent Pos:* Cancer res award, Milberg Women's Cancer Res Fund, 85; prin investr grants, Biochem Studies on Human T Cell Receptor Molecules, NIH, New Investr Award AI21579, 84-87, Biochem & Functional Studies on Fc Receptor Molecule in Human Nonlymphoid Malignancy, Nat Cancer Cytology Ctr, 86-87, A Complex of MHC Class I Heavy Chain & T8 on the membrane of T Lymphoid Cells, Am Chem Soc Grant IM463, 87 & T Cell Receptor-Like/T8/HLA Interactions in Recognition, NIH Grant CA46965, 87-91. *Res:* Elucidation of structures and specific molecular interactions of the T cell surface receptors and medium soluble factors and role of these interactions in functional activities of T cells. *Mailing Add:* Pub Health Res Inst 455 First Ave New York NY 10016

BUSHMAN, JESS RICHARD, b American Fork, Utah, May 12, 21; c 5. GEOLOGY. *Educ:* Brigham Young Univ, BA, 49; Princeton Univ, PhD, 58. *Prof Exp:* From asst prof to assoc prof, 55-70, PROF GEOL, BRIGHAM YOUNG UNIV, 70- *Concurrent Pos:* Consult-geologist, Ministry Mines & Hydrocarbons, Venezuela, 53-55, 58 & 59-60; vis lectr sponsored by Ministry Coal Ind, Peoples Republic of China. *Mem:* Geol Soc Am; Am Asn Petrol Geol; Asn Mining & Petrol Geol Venezuela; Nat Asn Geol Teachers; Am Asn Stratig Palynologists. *Res:* Sedimentation; palynology; stratigraphy; history of geology; science education. *Mailing Add:* Dept Geol Brigham Young Univ Provo UT 84602

BUSHMAN, JOHN BRANSON, b Farmington, NMex, Aug 26, 26; m 51; c 5. ECOLOGY, VERTEBRATE ZOOLOGY. *Educ:* Univ Utah, BS, 52, MS, 55, PhD, 83. *Prof Exp:* Asst cur birds, Univ Utah, 46 & 49-52; from asst ecologist to ecologist, Univ Utah, 54-62; zoologist, US Army Desert Test Ctr, Salt Lake City, Utah, 62-72, ECOLOGIST, US ARMY CORPS ENGRS, WASHINGTON, DC, 72- *Mem:* Wilson Ornith Soc; Cooper Ornith Soc; Am Ornith Union; Am Soc Mammal; Wildlife Soc. *Res:* Endangered species; avian and mammalian ecology; life histories; distribution and population dynamics. *Mailing Add:* US Army Corps of Engrs Civil Works Washington DC 20314-1000

BUSHNELL, DAVID L, b Platteville, Wis, Apr 14, 29; m; c 3. PHYSICS. *Educ:* Univ Wis, BS, 51, MS, 53; Va Polytech Inst, PhD(physics), 61. *Prof Exp:* Instr physics, Mankato State Col, 53-55; from instr to asst prof, Va Polytech Inst, 56-61; assoc prof, 61-75, PROF PHYSICS, NORTHERN ILL UNIV, 75- *Mem:* Am Phys Soc; Am Asn Physics Teachers; Int Solar Energy Soc. *Res:* Neutron capture gamma-ray studies; gamma-ray spectroscopy and nuclear level schemes; solar energy applications; high temperature solar oven with heat storage. *Mailing Add:* Dept Physics Northern Ill Univ DeKalb IL 60115

BUSHNELL, GORDON WILLIAM, b Oxford, Eng, June 17, 36; m 59; c 2. STRUCTURAL CHEMISTRY, CRYSTALLOGRAPHY. *Educ:* Oxford Univ, MA, 59, BSc, 63; Univ WI, PhD(chem), 66. *Prof Exp:* Sci officer, UK Atomic Energy Auth, 59-63; lectr chem, Univ WI, 63-67; assoc prof, 67-87, PROF CHEM, UNIV VICTORIA, 87- *Mem:* Chem Inst Can; Am Crystallog Asn. *Res:* X-ray crystallography of coordination compounds; structure determination by single crystal x-ray diffraction of proteins. *Mailing Add:* Dept Chem Elliot Bldg Univ Victoria PO Box 3055 Victoria BC V8W 3P6 Can

BUSHNELL, JAMES JUDSON, b Madison, Wis, Nov 13, 34; m 55; c 3. AIR CONDITIONING SYSTEM DESIGN, VEHICLE DESIGN. *Educ:* Univ Wis-Madison, BSME, 57. *Prof Exp:* Designer, Hamilton Standard Div, United Technol, 57-59; sr engr, Lockheed Aircraft, 69-70 & Rohr Industs, 70-77; supvr, RMI Inc, 77-82; sr engr, 59-69, ENG SPECIALIST, CONVAIR DIV, GEN DYNAMICS, 82- *Concurrent Pos:* Mem, Tech Comts, Am Soc Heating Refrig & Air Conditioning Engrs, 83-, chmn, Transp Air Conditioning Comt, 85-87 & Transp Refrig Comt, 89- *Mem:* Am Soc Heating Refrig & Air Conditioning Engrs; Soc Automotive Engrs. *Res:* Human comfort related to transportation environments. *Mailing Add:* 215 Carmelita Pl Solana Beach CA 92075

BUSHNELL, JOHN HORACE, b Grand Rapids, Mich, Mar 17, 25; m 51; c 4. INVERTEBRATE ZOOLOGY. *Educ:* Vanderbilt Univ, BA, 48; Mich State Univ, MS, 56, PhD(zool), 61. *Prof Exp:* US State Dept spec lectr biol & English, Habibia Col, Kabul, Afghanistan, 48-50; asst secy & sales mgr, Wagemaker Boat Co & US Molded Shapes, Inc, 50-55; instr zool, Mich State Univ, 59-60; asst prof biol, Washington & Jefferson Col, 60-64; from asst prof to assoc prof, 64-71, PROF ENVIRON, POP & ORGANISMIC BIOL, UNIV COLO, BOULDER, 71- *Concurrent Pos:* AEC summer study & continuing isotopes res grant, 63-; assoc ed, Am Midland Naturalist, 67-74; consult ed, Barnes & Noble Publ House, 70-; NSF alpine flowage system grant. *Mem:* Sigma Xi; Am Soc Zool; Am Micros Soc (vpres & prog chmn, 73-74, pres, 76-77); Int Bryozool Asn; NY Acad Sci. *Res:* Ecology, zoogeography and systematics of Ectoprocta and other aquatic published studies on invertebrates include crustacea, oligochaetes, sponges and hydrozoans; phenotypic plasticity, adaptation and physiological ecology of aquatic invertebrates; nutritional and pollution studies and sexual cycle studies on Ectoprocta and other invertebrates; Ectoproct population studies. *Mailing Add:* Dept Environ Pop & Organic Biol Univ Colo Boulder CO 80302

BUSHNELL, KENT O, b Tolland, Conn, Dec 12, 29; m 55. ENVIRONMENTAL GEOLOGY, GEOPHYSICS. *Educ:* Univ Conn, BA, 51; Yale Univ, MS, 52, PhD(geol), 55. *Prof Exp:* Geologist, Stand Oil Co Calif, 55-68; assoc prof, 68-74, PROF GEOL, SLIPPERY ROCK STATE COL, 74-, CHAIRPERSON DEPT, 76- *Concurrent Pos:* Dir, NSF seismic equip grant, 70-72. *Mem:* Am Asn Petrol Geol; Geol Soc Am; Soc Explor Geophys; Sigma Xi. *Res:* Gravity surveys for buried valleys; Pleistocene history of western Pennsylvania. *Mailing Add:* Dept Geol Slippery Rock State Col Slippery Rock PA 16057

BUSHNELL, ROBERT HEMPSTEAD, b Wooster, Ohio, May 11, 24; m 65; c 2. ATMOSPHERIC PHYSICS. *Educ:* Ohio State Univ, BSc, 47, MSc, 48; Univ Wis, PhD(meteorol), 62. *Prof Exp:* Physicist, Hoover Co, 48-50, Goodyear Aircraft Co, 50-56, Radio Corp Am, 57-58, Univ Wis, 58-62 & Nat Ctr Atmospheric Res, Boulder, Colo, 62-74; consult, 74-85; CONSULT, NAT OCEANIC ATMOSPHERIC ADMIN, US DEPT COM, 86- *Concurrent Pos:* Univ Corp Atmospheric Res fel, 61. *Mem:* AAAS; Sigma Xi; Am Meteorol Soc; Am Soc Testing & Mat; Int Solar Energy Soc. *Res:* Climatology and statistics of sunshine and temperature related to solar heating; thunderstorm structure; weather radar. *Mailing Add:* 502 Ord Dr Boulder CO 80303-4732

BUSHNELL, WILLIAM RODGERS, b Wooster, Ohio, Aug 19, 31; m 52; c 3. PHYSIOLOGY OF PLANT DISEASE. *Educ:* Univ Chicago, AB, 51; Ohio State Univ, BS, 53, MS, 55; Univ Wis, PhD(bot), 60. *Prof Exp:* PLANT PHYSIOLOGIST, USDA, 60- *Concurrent Pos:* From asst prof to prof, Dept Plant Path, Univ Minn, 66-73. *Honors & Awards:* Alexander Von Humboldt Sr US Scientist Award, 84. *Mem:* Am Soc Plant Physiol; fel Am Phytopath Soc; Am Asn Advan Sci. *Res:* Physiology of parasitism, especially rust and mildew diseases of cereals. *Mailing Add:* Cereal Rust Lab Univ Minn St Paul MN 55108

BUSHONG, JEROLD WARD, b Dayton, Ohio, Sept 22, 35; m 55; c 2. PLANT PATHOLOGY. *Educ:* Miami Univ, BA, 57; Univ Ill, MS, 59, PhD(plant path), 61. *Prof Exp:* Sr plant pathologist, Fungicide Res, Niagara Chem Div, FMC Corp, 61-66; asst dir pesticide res, W R Grace & Co, 66-67; sr agr chem scientist, 67-69, supvr agr eval, 69-70, mgr agr chem, 70-73, mgr biol & field develop, 73-75, mgr int mkt & prod develop, Minn Mining & Mfg Co, 75-80, mgr int mkt & prod develop, 3M DO Brazil, 80-83, prod mgr, hearing health prods, 84-86, MKT MGR, MICROBIOL PROD, 3M EUROPE, 87- *Concurrent Pos:* Ed staff, Plant Dis Reporter, 74-76. *Mem:* Am Chem Soc; Int Weed Sci Soc; Am Phytopath Soc; Weed Sci Soc Am. *Res:* Fungicide chemistry; legume and fruit pathology; mycology; entomology. *Mailing Add:* Minn Mining & Mfg Co 3M Ctr Bldg 225-5S-01 St Paul MN 55144-1000

BUSHONG, STEWART CARLYLE, b Washington, DC, Nov 25, 36; m 58; c 3. RADIOLOGICAL HEALTH. *Educ:* Univ Md, BS, 59; Univ Pittsburgh, MS, 63, ScD(radiol health), 67. *Prof Exp:* Health serv officer, USPHS, Washington, DC, 59-61; health physicist, Univ Pittsburgh, 62-64; asst prof to assoc prof 66-76, PROF RADIOL SCI, BAYLOR COL MED, 76- *Concurrent Pos:* Assoc prof radiol, Univ Tex Dent Br, 70-; assoc prof radiol sci, Houston Community Col, 71- *Mem:* Am Asn Physicists Med; Health Physics Soc; Soc Nuclear Med; Am Indust Hyg Asn; Radiation Res Soc. *Res:* Investigations of radiation dose and dose distribution in patients receiving radiologic examination for medical diagnosis. *Mailing Add:* Baylor Col Med Houston TX 77030

BUSHUK, WALTER, b Pruzana, Poland, Jan 2, 29; Can citizen; m 55; c 2. CEREAL CHEMISTRY. *Educ:* Univ Man, BSc, 52, MSc, 53; McGill Univ, PhD(phys chem), 56. *Hon Degrees:* Dr, Agr Acad, Poznan, Poland. *Prof Exp:* Chemist, Grain Res Lab, Man, Can, 53-61, sect head, 61-62; dir, Ogilvie Flour Mills Co, Ltd, 62-64; res chemist, Grain Res Lab, 64, head wheat sect, 64-66; provost, 80-84, PROF PLANT SCI, UNIV MAN, 66-, PROF FOOD SCI, 84- *Concurrent Pos:* Nat Res Coun Can overseas fel, Macromolecule Res Ctr, France, 57-58. *Honors & Awards:* Osborne Medal; Neumann Medal. *Mem:* Am Asn Cereal Chem; fel Chem Inst Can; Can Inst Food Technol; fel Royal Soc Can; Sigma Xi; fel Agr Inst Can. *Res:* Enzyme kinetics; surface properties of flour; physicochemical properties of polymers; mechanism of flour quality improvement; wheat proteins. *Mailing Add:* Dept Food Sci Univ Man Winnipeg MB R3T 2N2 Can

BUSHWELLER, CHARLES HACKETT, b Port Jervis, NY, June 21, 39; m 61; c 3. PHYSICAL ORGANIC CHEMISTRY, STRUCTURAL CHEMISTRY. *Educ:* Hamilton Col, AB, 61; Middlebury Col, MS, 63; Univ Calif, Berkeley, PhD(org chem), 66. *Prof Exp:* Sr chemist, Mobil Chem Co, 66-68; asst prof, 68-72, assoc prof chem, Worcester Polytech Inst, 72-76; prof chem, State Univ NY Albany, 76-78; PROF CHEM & CHMN DEPT, UNIV VT, 78- Prof Exp: Alfred P Sloan res fel, 71-73; Camille & Henry Dreyfus teacher scholar, 72-75. *Mem:* Am Chem Soc. *Res:* Nuclear magnetic resonance spectroscopy; stereodynamics of organic and inorganic systems. *Mailing Add:* Chem Dept Univ Vt Cook Bldg Burlington VT 05405-0125

BUSING, WILLIAM RICHARD, b Brooklyn, NY, June 21, 23; m 51; c 3. PHYSICAL CHEMISTRY. *Educ:* Swarthmore Col, BA, 43; Princeton Univ, MA, 48, PhD(chem), 49. *Prof Exp:* Res assoc chem, Brown Univ, 49-51; instr, Yale Univ, 51-54; chemist, 54-90, CONSULT, OAK RIDGE NAT LABS, 90- *Concurrent Pos:* Mem, US Nat Comt Crystallog, Nat Res Coun, 70-76 & 79-81, secy-treas, 74-76; US deleg, Int Cong Crystallog, Kyoto, 72 & Amsterdam, 75. *Honors & Awards:* Martin J Buerger Award, Am Crystallog Asn, 85; Mat Sci Award, Dept Energy, 88. *Mem:* Am Crystallog Asn (pres, 71). *Res:* Molecular structure and spectroscopy; neutron diffraction; crystallographic computation; interionic and intermolecular forces in crystals; fiber crystallography. *Mailing Add:* Chem Div Oak Ridge Nat Lab PO Box 2008 Oak Ridge TN 37831-6197

BUSINGER, JOOST ALOIS, b Haarlem, Neth, Mar 29, 24; m 49, 87; c 3. METEOROLOGY. *Educ:* State Univ Utrecht, BS, 47, MSc, 50, PhD(meteorol), 54. *Prof Exp:* Technician, Inst Heating Eng, Neth, 48-51 & Inst Hort Eng, 51-56; assoc & proj dir, Dept Meteorol, Univ Wis, 56-58; from asst prof to prof meteorol, Univ Wash, 58-83, chmn geophys exec comt, 63-65, chmn fac sen, 76-77; sr scientist, Nat Ctr Atmospheric Res, 86-89; EMER PROF METEROL, UNIV WASH, 83- *Concurrent Pos:* Corresp mem, Royal Neth Acad Sci, 80; vis scientist, Nat Ctr Atmospheric Res, 83-86. *Honors & Awards:* Second Half Century Award, Am Meteorol Soc, 78. *Mem:* Fel Am Meteorol Soc; Am Geophys Union; fel AAAS. *Res:* Energy transfer in atmosphere; turbulence; radiation; micro and physical meteorology; air-sea interaction. *Mailing Add:* 422 Guemes Island Rd Amacortes WA 98221

BUSK, GRANT CURTIS, JR, b American Fork, Utah, May 28, 49; m 80; c 2. PRODUCT & PROCESS DEVELOPMENT, FOOD RESEARCH MANAGEMENT. *Educ:* Ore State Univ, BS, 71; Univ Minn, PhD(food sci & technol), 78. *Prof Exp:* Food technologist, Hunt Wesson Foods Inc, 71-73; teaching assoc food chem, Univ Minn, 75-77, lab asst, 77-78; group leader prod develop, Uncle Bens Inc, 78-81; mgr, Snack-master Inc, 81-85; mgr prod develop, Kal Kan Inc, 85-86; DIR PROD DEVELOP, NABISCO BISCUIT CO, 86- *Concurrent Pos:* Consult, Nat Food & Nutrit Consults, 76-78. *Mem:* Inst Food Technol; Am Asn Cereal Chemists; Am Oil Chemists Soc. *Res:* New and improved food products ranging from dry mix to extruded to baked; human and pet-animal foods. *Mailing Add:* RR 1 Box D11 Chester NJ 07930

BUSKE, NORMAN L, b Milwaukee, Wis, Oct 11, 43; m 80; c 4. ALTERNATIVE METHODOLOGIES. *Educ:* Univ Conn, BA, 64, MS, 65; Johns Hopkins Univ, MA, 67. *Prof Exp:* Oceanogr, Ocean Sci & Eng, Inc, 68-71; oceanogr, BK Dynamics, Inc, 71-72; prin res & design, Sea-Test Co, 72-76; sr scientist, Van Gulick & Assocs, Inc, 76-77; dir res, Pac Eng Corp, 77-78; PRIN RES & DESIGN, SEARCH TECH SERVS, 78- *Concurrent Pos:* Researcher, US Naval Ship Eltanin cruises, NSF, 67 & 69; consult, Physicians for Social Responsibility, 80-81; sci adv, Greenpeace, NW & USA, 83-; sr scientist, Hanford Reach Proj, 85- *Mem:* Am Phys Soc; Am Soc Mech Engrs; Am Soc Testing and Mat; Inst Elec & Electronics Engrs; Nat Soc Prof Engrs; NY Acad Sci. *Res:* Development of temporal mechanical description of physical reality; methodological alternatives in science; pathways analysis for environmental assessments; design of 25,000 RPM diesel engine; development of probabilistic scenarios as a research tool. *Mailing Add:* HCR 11 Box 17 Davenport WA 99122-9404

BUSKIRK, ELSWORTH ROBERT, b Beloit, Wis, Aug 11, 25; m 48; c 2. HUMAN APPLIED PHYSIOLOGY, ENVIRONMENTAL PHYSIOLOGY. *Educ:* St Olaf Col, BA, 50; Univ Minn, MA, 51, PhD(physiol hyg), 54. *Prof Exp:* Res assoc physiol, Lab Physiol Hyg, Univ Minn, 54; physiologist, Environ Protection Res Div, Q Res & Develop Ctr, Mass, 54-56, chief environ physiol sect, 56-57; physiologist, Metab Dis Br, Inst Arthritis & Metab Dis, NIH, 57-63; PROF APPL PHYSIOL & DIR,

LAB HUMAN PERFORMANCE, CHMN GRAD PHYSIOL PROG, PA STATE UNIV, 63-, MARIE A NOLL PROF HUMAN PERFORMANCE RES. *Concurrent Pos:* Mem, coun, Sci Adv Comt, President's Coun Phys Fitness, 59- 61, thermal factors subcomt, Nat Acad Sci, 63-64, human adaptability subcomt, Int Biol Prog, 68-72, comt interaction eng with biol & med, Nat Acad Eng-Nat Acad Sci, 68-75, Educ Mat Rev Comt, Asn Am Med Cols, 76-84 & epidemiol sect, Am Heart Asn; chmn, study sect appl physiol & bioeng, NIH, 64-68 & appl physiol & orthop, 76-80; mem, subcomt calories, food & nutrit bd, Nat Acad Sci-Nat Res Coun, 68-70, comt personnel protection, 71-74 & comt mil nutrit res, 79-; sect ed, J Appl Physiol, Am Physiol Soc, 73-84, assoc ed, 76-84; NATO Sr Fel Sci, NSF, 77; ed-in-chief, Med & Sci in Sports & Exercise, 85-89. *Honors & Awards:* Res Citation, Am Col Sports Med, 73, Honor Award, 84. *Mem:* AAAS; Am Physiol Soc; Aerospace Med Soc; Am Inst Nutrit; Am Soc Heating, Refrig & Air-Conditioning Engrs; Am Heart Asn; Am Col Sports Med (pres, 64-65); Sigma Xi; Am Soc Clin Nutrit. *Res:* Human applied physiology with an emphasis on exercise and environmental physiology; the effects of age, gender, physical condition and fitness, body composition and life style; metabolism; growth, development and aging; epidemiology of coronary heart disease and obesity. *Mailing Add:* 216 Hunter Ave State College PA 16801

BUSKIRK, FRED RAMON, b Indianapolis, Ind, Dec 29, 28; m 58. THEORETICAL PHYSICS, EXPERIMENTAL PHYSICS. *Educ:* Western Reserve Univ, BS, 51, MS, 55, PhD(physics), 58. *Prof Exp:* Proj assoc physics, Univ Wis, 58-59; instr, Case Western Reserve Univ, 59-60; from asst prof to assoc prof, 60-74, PROF PHYSICS, US NAVAL POSTGRAD SCH, 74- *Res:* Nuclear and field theory; nuclear and accelerator experiments. *Mailing Add:* Code 61 Bs Physics Dept Naval Postgrad Sch Monterey CA 93943

BUSKIRK, RUTH ELIZABETH, b Indianapolis, Ind, Mar 9, 44; m 75; c 3. ANIMAL BEHAVIOR, ANIMAL ECOLOGY. *Educ:* Earlham Col, AB, 65; Harvard Univ, MAT, 66; Univ Calif, Davis, PhD(zool), 72. *Prof Exp:* Instr biol, Southern Univ, 66-68; fel, Univ Calif, Davis, 72-74; asst prof behav, Cornell Univ, 74-78; lectr, 78-90, RES SCIENTIST, UNIV TEX, 78-, SR LECTR, 90- *Concurrent Pos:* Mem, Orgn Trop Studies, 75- *Mem:* Animal Behav Soc; Ecol Soc Am; Am Geophys Union; Am Soc Arachnologists; Asn Trop Biol. *Res:* Animal social behavior; behavioral ecology of predators; animal response to geophysical phenomena. *Mailing Add:* Dept Microbiol Univ Tex Austin TX 78712-1095

BUSLIG, BELA STEPHEN, b Budapest, Hungary, Aug 27, 38; m 64; c 2. BIOCHEMISTRY, PLANT PHYSIOLOGY. *Educ:* Queen's Univ, BA, 62; Fla State Univ, MS, 67; Univ Fla, PhD(plant biochem), 70. *Prof Exp:* Chemist, 67-69, res biochemist, 69-77, RES SCIENTIST BIOCHEM CITRUS, FLA DEPT CITRUS, 77- *Concurrent Pos:* Res assoc, Univ Fla, 67-71, adj asst prof, 71-84, adj assoc prof, 84- *Mem:* Am Chem Soc; Can Soc Plant Physiologists; Am Soc Plant Physiologists; Genetics Soc Am; AAAS; fel Am Inst Chemists. *Res:* Metabolism of citrus plants; organic acids and energetics of respiration; enzymic regulation of metabolites; constituents of processed citrus products; process control instruments; color measurements. *Mailing Add:* Fla Dept Citrus c/o USDA PO Box 1909 Winter Haven FL 33883-1909

BUSLIK, ARTHUR J, b Philadelphia, Pa, Mar 7, 33. REACTOR PHYSICS. *Educ:* Univ Pa, BA, 54; Columbia Univ, MA, 56; Univ Pittsburgh, PhD(physics), 62. *Prof Exp:* From assoc scientist to sr scientist, Bettis Atomic Power Lab, Westinghouse Elec Corp, 58-72; adv nuclear engr, Gulf United Nuclear Fuels Corp, 72-74; assoc physicist, 79-80, PHYSICIST, BROOKHAVEN NAT LAB, 80- *Concurrent Pos:* Vis prof nuclear eng, Ben Gurion Univ Negev, Israel, 78-79. *Mem:* Am Phys Soc; Am Nuclear Soc. *Res:* Theoretical solid state physics; magnetic resonance line width; kinetics and stability; neutron resonance absorption; variational principles in reactor physics. *Mailing Add:* 4701 Willard Ave Apt 226 Chevy Chase MD 20815

BUSNAINA, AHMED A, b Sept 2, 53; US citizen; m 77; c 3. MICRO CONTAMINATION CONTROL, COMPUTATIONAL FLUID DYNAMICS. *Educ:* Univ Tripoli, BSc hons, 76; Okla State Univ, MS, 79, PhD(mech eng), 83. *Prof Exp:* Engr, Gen Elec Co, 76, Esso Oil Co, 77; instr eng, Univ Garyounis, 77; teaching & res asst eng, Okla State Univ, 78-79, teaching & res assoc, 80-83; vis asst prof eng, San Diego State Univ, 83-84; asst prof, 84-88, ASSOC PROF ENG, CLARKSON UNIV, 88- *Concurrent Pos:* Consult, IBM, 85-, Du Pont, 88-89 & Motorola, 89-90; pres, Advan Comput Sci, 86-; dir, Microcontamination Lab, 87-, Ctr Particulate Control, 90-; mem exec comt, Fine Particle Soc, 88-; mem, Computer Eng Comt, Am Soc Mech Engrs, 88-; mem adv bd, Inst Environ Sci, 89- *Honors & Awards:* John W Graham Jr Award, Clarkson Univ, 90. *Mem:* Am Inst Aeronaut & Astronaut; Am Soc Mech Engrs; Am Soc Eng Educ; Inst Environ Sci; Fine Particle Soc; Am Phys Soc. *Res:* Numerical computation of fluid flow phenomena with turbulence, diffusion and swirl, with applications to combustion; microcontamination control; microcontaminant particle adhesion, removal and deposition. *Mailing Add:* Mech Eng Dept Clarkson Univ Potsdam NY 13699-5725

BUSS, DARYL DEAN, b Rock Rapids, Iowa, Sept 20, 45; m 68; c 1. CARDIOVASCULAR PHYSIOLOGY, CORONARY PHYSIOLOGY. *Educ:* Univ Minn, St Paul, BS, 66, DVM, 68; Univ Wis-Madison, MS, 74, PhD(vet sci), 75. *Prof Exp:* Intern vet med, Animal Med Ctr, NY, 68-69; assoc, Hillcrest Animal Hosp, Whitebear Lake, Minn, 69-70; instr, Col Vet Med, Univ Minn, 70-71; instr, Sch Vet Med, Univ Pa, 71-72; NIH fel physiol, Univ Wis, 72-75; assoc prof, 76-80, PROF & CHMN, DEPT PHYSIOL SCI, COL VET MED, UNIV FLA, 80- *Concurrent Pos:* Fel physiol, Max Planck Inst Physiol & Clin Res, 75-76; prin investr, Am Heart Asn-NIH, 76-; mem, sci adv bd, Morris Animal Found, 84-87; mem, coun basic sci, Am Heart Asn; mem, Res Peer Rev Comt, Am Heart Asn, Fla Affil, 87-88. *Mem:* Am Vet Med Asn; Am Heart Asn; Am Physiol Soc. *Res:* Regulation of coronary blood flow to the heart, transitional physiological changes in the coronary circulation in the newborn, and tolerance of the newborn heart to global ischemia. *Mailing Add:* Dept Physiol Sci Univ Fla Box J-144 JHMHC Gainesville FL 32610

BUSS, DAVID R(ICHARD), b Chicago, Ill, May 6, 39; m 62; c 3. CHEMICAL ENGINEERING, CHEMISTRY. *Educ:* Harvey Mudd Col, BS, 62; Univ Calif, Berkeley, PhD(chem eng), 66. *Prof Exp:* Chem eng Res & Develop, Upjohn Co, 66-71, mkt develop, 71-73, res scientist, 73-77, prod mgt, 77-78, Res & Develop mgt, 78-80, dir mat planning, 80-89, DIR ADMIN SERV, UPJOHN CO, 89- *Honors & Awards:* UpJohn Award, 84. *Mem:* Am Chem Soc; Am Inst Chem Engrs. *Res:* Resolution of optical isomers; chemical process research and development; fermentation research and development. *Mailing Add:* Upjohn Co Kalamazoo MI 49001-0199

BUSS, DENNIS DARCY, b Gainesville, Fla, July 11, 42; m 64; c 2. SOLID STATE ELECTRONICS, SEMICONDUCTORS. *Educ:* Mass Inst Technol, SB, 63, SM, 64, PhD(solid state physics), 68. *Prof Exp:* Asst prof elec eng, Mass Inst Technol, 68-69; res staff theoret physics, Tex Instruments Inc, 69-74; vis assoc prof elec eng, Mass Inst Technol, 75; BR MGR SOLID STATE RES, TEX INSTRUMENTS INC, 75-, VPRES SEMICONDUCTOR PROCESS & DESIGN CTR. *Concurrent Pos:* Mem prog comts, Int Electron Device Meeting & Int Solid State Circuits Conf, 75-77; assoc ed, Inst Elec & Electronics Engrs Trans Electron Devices, 76 & Inst Elec & Electronics Engrs J Solid State Circuits, 77-; Tex Instruments fel, 78. *Mem:* Inst Elec & Electronics Engrs. *Res:* Development of advanced, charge-coupled solid state structures for digital memory and analog signal processing. *Mailing Add:* 22 Railroad Ave Andover MA 01810

BUSS, EDWARD GEORGE, b Concordia, Kans, Aug 28, 21; m 49; c 2. POULTRY GENETICS. *Educ:* Kans State Col, BS, 43; Purdue Univ, MS, 49, PhD(poultry genetics), 56. *Prof Exp:* Asst prof poultry husb, Colo State Univ, 49-55, actg head dept, 50-55; instr, Purdue Univ, 55-56; from assoc prof to prof poultry sci, 56-83, chmn grad sch interdept prog genetics, 69-74, prof agr, 83-86, EMER PROF AGR, PA STATE UNIV, 87- *Concurrent Pos:* Prin investr, NIH Res Grant, 61-80 & Small Bus Inovation Res; mem, Comn Undergrad Educ Biol Sci panel, 65-67; mem exec comt, Am Inst Biol Sci, 73-75; rep gov bd, Am Inst Biol Sci, 65-73 & 77-78, rep, Coun Agr & Sci Technol, 78-80; coun mem, Am Genetic Asn, 74-76; assoc ed, Breeding & Genetics, 81-83; consult, poultry breeding, Indonesia & India, 86-; Fulbright lectr, Sierra Leone, W Africa, 88. *Honors & Awards:* Corrispondente Award, Ital Soc Zootechnol, 72. *Mem:* Am Soc Zool; fel Poultry Sci Asn; Am Genetic Asn; Genetics Soc Am; Am Inst Biol Sci; Soc Study Reproduction; AAAS. *Res:* Poultry genetics and physiology of reproduction; parthenogenesis in turkeys; biochemical nature of gene action (riboflavin, chondrodystrophy, diabetes insipidus), infertility, obesity and eggshell quality. *Mailing Add:* 1420 S Garner St State College PA 16801

BUSS, GLENN RICHARD, b Easton, Pa, Apr 12, 40; m 61; c 2. SOYBEAN BREEDING, PLANT GENETICS. *Educ:* Pa State Univ, BS, 62, MS, 64, PhD(genetics), 67. *Prof Exp:* From asst prof to assoc prof, 67-90, PROF AGRON, VA POLYTECH INST & STATE UNIV, 90- *Mem:* Am Soc Agron; Crop Sci Soc Am; Am Genetic Asn; Coun Agr Sci & Tech. *Res:* Qualitative genetics and cultivar development in soybeans; genetics of host-pathogen interactions among soybean genotypes and soybean mosaic virus strains. *Mailing Add:* Crop & Soil Environ Sci Dept Va Polytech Inst & State Univ Blacksburg VA 24061

BUSS, JACK THEODORE, b Benton Harbor, Mich, Oct 2, 43; m 65; c 2. VERTEBRATE PHYSIOLOGY. *Educ:* Bethel Col, BA, 65; Wayne State Univ, MS, 68; Univ Minn, PhD(zool), 71. *Prof Exp:* Instr biol, Bethel Col, 66-68; assoc prof biol, William Jewell Col, 71-75; from asst prof to assoc prof, 75-85, PROF BIOL, FERRIS STATE UNIV, 85- *Concurrent Pos:* Dir res, Antigen Labs, Inc, 72-75. *Mem:* AAAS; Asn Res Vision Ophthal. *Res:* Ocular development; corneal neovascularization. *Mailing Add:* Dept Biol Sci Ferris State Univ Big Rapids MI 49307

BUSS, LEO WILLIAM, b Alexandria, Va, Sept 27, 53; m; c 2. EVOLUTIONARY BIOLOGY. *Educ:* Johns Hopkins Univ, BA, 75, MA, 77, PhD(geol), 79. *Hon Degrees:* MA, Yale Univ, 91. *Prof Exp:* From asst prof to assoc prof biol, 79-90, assoc prof geol, 89-90, PROF BIOL & GEOL, YALE UNIV, 90-; CUR, PEABODY MUS NATURAL HIST, 79- *Concurrent Pos:* Fel, John Simon Guggenheim Found, 84; prize fel, John T & Catherine B MacArthur Found, 89; dir, Yale Inst Biospheric Studies, 91- *Mem:* AAAS. *Res:* Structure and confirmation of evolutionary theory, with particular reference to expanding current theory to accommodate multiple units of selection. *Mailing Add:* Dept Biol Yale Univ New Haven CT 06511

BUSS, WILLIAM CHARLES, b Bremerton, Wash, Aug 13, 38; m 65, 79; c 1. ANTIBIOTICS, TRANSCRIPTION. *Educ:* Portland State Univ, BSc, 65; Univ Alta, MSc, 67; Univ Ore, PhD(pharmacol), 71. *Prof Exp:* Fel biochem, Univ Calif, San Francisco, 72; from asst prof to assoc prof, 73-89, PROF PHARMACOL, SCH MED, 89-, ACTG CHMN DEPT, UNIV NMEX, 89- *Concurrent Pos:* Vis assoc prof molecular biol, Sch Med, Univ Calif, San Diego, 83. *Mem:* Am Soc Pharmacol & Exp Therapeut; AAAS. *Res:* Effects of chemical agents on eukaryotic transcription and translation; mechanisms of action and toxicity of antibiotics. *Mailing Add:* Dept Pharmacol BMSB Rm 145 Sch Med Univ NMex Albuquerque NM 87131

BUSSE, EWALD WILLIAM, b St Louis, Mo, Aug 18, 17; m 41; c 4. PSYCHIATRY, GERIATRICS. *Educ:* Westminster Col, AB, 38; Wash Univ, MD, 42; Am Bd Psychiat & Neurol, dipl. *Hon Degrees:* ScD, Westminster Col, 60. *Prof Exp:* From instr to prof psychiat, Med Ctr, Univ Colo, 46-53; prof psychiat & chmn dept, Sch Med, 53-74, assoc provost & dean, 74-82, emer dean med & allied health educ, 82-87, EMER DEAN & EMER PROF PSYCHIAT, DUKE UNIV MED CTR, 87-; PRES & CHIEF EXEC OFFICER, NC INST MED, 87- *Concurrent Pos:* Dir, EEG Lab, Colo Psychopath Hosp, 46-53; lectr, Sch Grad Educ, Univ Denver, 47-53; head, Div Psychosom Med, Colo Gen Hosp, 50-53; actg head ment hyg & child guid clin, Univ Colo, 52-53; dir, Ctr Study Aging, Duke Univ, 57-70; dir, Am Bd Psychiat & Neurol, 61-69; chmn sect psychogeriat, WHO, 69-72; consult, US Navy, US Army, Vet Admin & NIH; mem, President's Biomed Res Panel, 75-76; spec med adv group, Vet Admin, 78-87, chmn comt geriat & geront,

81-85; recipient res awards, numerous insts & socs. *Honors & Awards:* Strecker Award, Pa Hosp Inst, 67; Robert W Kleemeier Award, Geront Soc, 68,; William C Menninger Award, 71; Mod Med Award, 72; Seltzer Award, 75; Malford W Thewlis Award, Am Geriat Soc, 79; Sandoz Prize Gerontol, 83; Brookdale Found Award, Geront Soc Am, 82; Jack Weinberg Mem Award, Am Psychiat Asn, 83, Distinguished Serv Award, 88; Awards named in honor, Busse Res Award & Ewald Busse Award, 90. *Mem:* Inst Med-Nat Acad Sci; life fel Am Psychiat Asn (pres, 71-72); fel Am Col Physicians; fel Am Geriat Soc (pres, 75-76); fel Geront Soc (pres, 67-68); Int Asn Geront (pres, 83-89); Sigma Xi. *Res:* Electroencephalography; gerontology; author of numerous articles and publications. *Mailing Add:* NC Inst Med 905 W Main St Box 25 Durham NC 27701

BUSSE, FRIEDRICH HERMANN, b Berlin, Germany, Sept 30, 36. GEOPHYSICS, FLUID DYNAMICS. *Educ:* Univ Munich, Dr rer nat, 62. *Prof Exp:* Sci asst physics, Univ Munich, 61-65; res fel math, Mass Inst Technol, 65-66; assoc res geophysicist, Univ Calif, Los Angeles, 66-67; sr scientist astrophysics, Max Planck Inst Physics, 67-70; PROF GEOPHYSICS, UNIV CALIF, LOS ANGELES, 70- *Concurrent Pos:* Guggenheim Mem Found fel, 72. *Mem:* Fel Am Phys Soc; fel Am Geophys Union. *Res:* Theory of rotating fluids; bounds on properties of turbulent flows; origin of planetary magnetism and dynamo theory; theory and experiments of thermal convection. *Mailing Add:* Inst Geophys & Planetary Physics Univ Calif Los Angeles CA 90024

BUSSE, ROBERT FRANKLYN, b Plainfield, NJ, Apr 30, 37; m 59; c 2. BIOCHEMISTRY, ORGANIC CHEMISTRY. *Educ:* Univ Notre Dame, BS, 59; Rutgers Univ, PhD(org chem), 63. *Prof Exp:* Res chemist, Celanese Corp Am, 63-67; mgr tech progs, Celanese Fibers Mkt Co, 67-69; tech develop mgr, 69-72; chmn sci dept, Providence Day Sch, Charlotte, NC, 73-78; CHMN SCI DEPT, BENJAMIN SCH, NORTH PALM BEACH, FLA, 78- *Mem:* Am Chem Soc. *Res:* Inhibition of enzymes and metabolism; separation scheme for the isolation of plant sterols; new product and process development and evaluation. *Mailing Add:* 11690 Ficus St Palm Beach Gardens FL 33410

BUSSELL, BERTRAM, b New York, NY, Sept 30, 23; m 52, 70; c 3. COMPUTER SCIENCE. *Educ:* Univ Calif, Los Angeles, BS, 50, MS, 52, PhD(eng), 62. *Prof Exp:* Jr engr, 50-55, assoc res engr, 55-63, from asst prof to assoc prof eng, 63-70, assoc prof comput sci & vchmn dept, 70-76, PROF ENG & APPL SCI, UNIV CALIF, LOS ANGELES, 76- *Mem:* Inst Elec & Electronics Engrs; Asn Comput Mach. *Res:* Optimized design of digital computer systems for the optimal solution of important classes of scientific and nonscientific problems. *Mailing Add:* West Coast Univ 440 Shatto Pl Los Angeles CA 90020

BUSSELL, WILLIAM HARRISON, b Rochelle, Ga, Jan 29, 23; m 50; c 4. MECHANICAL ENGINEERING. *Educ:* Univ Fla, BME, 48, MS, 53; Mich State Univ, PhD(mech eng), 64. *Prof Exp:* Instr & asst res, Univ Fla, 48-52, asst prof, 53-57; mech engr, Southeastern Forest Exp Sta, US Forest Serv, 52-53; assoc prof mech eng, La Polytech Inst, 57-65; prof mech eng, Auburn Univ, 65-89; RETIRED. *Mem:* Am Soc Mech Engrs; Sigma Xi. *Res:* Design; mechanisms; levered vehicles; lubrication. *Mailing Add:* 230 Singleton St Auburn AL 36830

BUSSERT, JACK FRANCIS, b Chicago, Ill, Dec 13, 22; m 54; c 2. ORGANIC CHEMISTRY. *Educ:* DePaul Univ, BSc, 47; Purdue Univ, MSc, 50; Ohio State Univ, PhD(org chem), 55; Univ Chicago, MBA, 65. *Prof Exp:* Res chemist, Universal Oil Prod Co, 48 & Nat Aluminate Corp, 50-51; RES CHEMIST, AMOCO CHEM CORP, STANDARD OIL CO, IND, 55- *Mem:* Am Chem Soc. *Res:* Hydrocarbon synthesis; anti-oxidants-petroleum products; agricultural pesticides; hydrocarbon oxidation; chemicals marketing research; chemicals commercial development. *Mailing Add:* 319 Montclair Glen Ellyn IL 60137

BUSSEY, ARTHUR HOWARD, Can citizen. BIOLOGY. *Educ:* Univ Bristol, BSc, 63, PhD(microbiol), 66. *Prof Exp:* Fulbright fel microbial regulation & travel scholar, Dept Biol Sci, Purdue Univ, Lafayette, 66-69; asst prof, 69-73, ASSOC PROF BIOL, MCGILL UNIV, 73- *Mem:* Can Genetics Soc; Am Soc Microbiol; AAAS. *Res:* Eukaryotic cell surface; energy coupling; yeast killer factor; cell-cell recognition; regulatory switching. *Mailing Add:* Dept Biol McGill Univ 853 Sherbrooke St W Montreal PQ H3G 2T6 Can

BUSSEY, HOWARD EMERSON, b Yankton, SDak, Sept 14, 17; m 46; c 4. ELECTROMAGNETISM. *Educ:* George Washington Univ, BA, 43, MS, 51; Univ Colo, PhD(physics), 64. *Prof Exp:* Sci aide physics, Nat Bur Standards, 41-42, meteorologist, synoptic & radar, 43-46, phys scientist, 46-49, physicist, 50-80; RETIRED. *Concurrent Pos:* Guest worker, Nat Bur Standards, 80- *Mem:* Am Phys Soc; Inst Elec & Electronics Eng; Sigma Xi; Int Sci Radio Union. *Res:* Microwave measurements; dielectric and ferrimagnetic measurements; antenna theory; electromagnetic wave scattering theory and experiment. *Mailing Add:* 1860 Bluebell Ave Boulder CO 80302

BUSSGANG, J(ULIAN) J(AKOB), b Lwow, Poland, Mar 26, 25; nat US; m 60; c 3. ELECTRICAL ENGINEERING. *Educ:* Univ London, BSc, 49; Mass Inst Technol, MS, 51; Harvard Univ, PhD(appl physics), 55. *Prof Exp:* Asst res lab electronics, Mass Inst Technol, 50-52; mem staff, Lincoln Lab, 52-53; asst, Gordon McKay Lab, Harvard Univ, 54-55; mgr radar develop & appl res, RCA, 55-62; pres, Signatron, Inc, 62-87; CONSULT, 87- *Concurrent Pos:* Consult, Lincoln Lab, 53-55, Mitre Corp, 62-65 & Rand Corp, 62-69; lectr, Northeastern Univ, 61-65 & Harvard Univ, 64; mem, Comn C, Int Sci Radio Union; assoc ed, Radio Sci, 76-78. *Mem:* Fel Inst Elec & Electronics Engrs. *Res:* Electronic systems; radar; statistical theory of communications; random processes; sequential analysis. *Mailing Add:* Two Forest St Lexington MA 02173

BUSSIAN, ALFRED ERICH, b Milwaukee, Wis, Sept 9, 33; div; c 2. GEOPHYSICS, PETROLEUM ENGINEERING. *Educ:* Ripon Col, BA, 55; Univ Colo, Boulder, PhD(physics), 64. *Prof Exp:* Engr, A C Spark Plug Div, GMC, Wis, 57-58; res asst physics, Univ Colo, Boulder, 58-64; res assoc, Max Planck Inst Physics & Astrophys, 64-66; vis scientist, Nat Ctr Atmospheric Res, Colo, 66-67; res assoc physics, Univ Mich, 67-73; prof physics, Community Col Denver, Red Rocks Campus, 73-79; RES PHYSICIST, TEXACO INC, 79- *Mem:* Am Phys Soc. *Res:* Geophysical research and development of instrumentation as it applies to the understanding and search for hydrocarbons. *Mailing Add:* PO Box 425 Bellaire TX 77402

BUSSMAN, JOHN W, b Mankato, Minn, 24; m; c 6. PEDIATRICS. *Educ:* Univ Minn, MD, 47; Am Bd Pediat, dipl, 53; Am Bd Pediat Cardiol, dipl, 62. *Prof Exp:* Intern, Sioux Valley Hosp, 48; assoc staff, St Vincent Hosp, 51-53; from asst prof to prof pediat, Univ Ore, 63-91; MED DIR, ORE MED PROF REV ORGN, 91- *Concurrent Pos:* Mem bd dirs, Am Asn Foundations Med Care, 73-86; mem, Nat Health Care Technol Coun, 79-80; consult, Rand Corp; mem, Conf Health Care Cost Containment, 83. *Mem:* Inst Med-Nat Acad Sci; Am Acad Pediat Dent; Am Col Clin Pharm; fel Am Acad Pediat; fel Am Col Chest Physicians; fel Am Col Cardiol. *Mailing Add:* Ore Med Prof Rev Orgn 1220 SW Morrison Suite 200 Portland OR 97205

BUSTA, FRANCIS FREDRICK, b Faribault, Minn, Sept 25, 35; m 57; c 2. FOOD MICROBIOLOGY. *Educ:* Univ Minn, Minneapolis, BA, 57; Univ Minn, St Paul, MS, 61; Univ Ill, PhD(food sci), 63. *Prof Exp:* From asst prof to assoc prof food sci, NC State Univ, 63-67; prof & head dept food sci & nutrit, Univ Fla, 84-87; from assoc prof to prof 67-84, PROF & HEAD DEPT FOOD SCI & NUTRIT, UNIV MINN, 87- *Concurrent Pos:* Vis scientist, Commonwealth Sci & Indust Res Orgn, Sydney, Australia, 74-75; educ res grant, Campbell Soup Co Res Inst, 74; mgr, Int Mkt Res Capsule Labs, St Paul, Minn, 81-82. *Honors & Awards:* Educator Award, Int Asn Milk Food & Environ Sanit. *Mem:* Am Meat Sci Asn; Int Asn Milk, Food & Environ Sanit; Am Soc Microbiol; fel Inst Food Technologists; Brit Soc Appl Bact; fel Am Acad Microbiol; Am Asn Cereal Chemists. *Res:* Microbiological aspects of food processing; environmental stress on microorganisms; microbiological quality of food; food substances inhibitory or stimulative to microorganisms; bacterial spore physiology. *Mailing Add:* Dept Food Sci & Nutrit Univ Minn St Paul MN 55108

BUSTAD, LEO KENNETH, b Stanwood, Wash, Jan 10, 20; m 42; c 3. ANIMAL ASSISTED THERAPY. *Educ:* Wash State Univ, BS, 41, MS, 48, DVM, 49; Univ Wash, PhD, 60. *Prof Exp:* Mgr exp farm, Biol Labs, Hanford Labs, Gen Elec Co, 49-64; prof radiation biol & dir radiobiol & comp oncol labs, Univ Calif, Davis, 65-73; prof physiol & dean, Col Vet Med, 73-83, EMER PROF & EMER DEAN, COL VET MED, WASH STATE UNIV, 84- *Concurrent Pos:* Consult, USAF-AEC, 58-63; guest fac, Univ Wash, 60-65; mem subcomt, Adv Comt Civil Defense, Nat Res Coun-Nat Acad Sci, 60-65, comt prof educ, Inst Lab Animal Resources, 66-68, subcomt standards large animals, 67-70, adv coun, 69-74, subcomt radioactivity in food, 67-70, food protection comt, 67-75; mem adv comt interdisciplinary conf prog, Am Inst Biol Sci, 63-70; consult scientist, Biol Dept, Battelle Mem Inst, 65; mem nat adv comt, Regional Primate Res Ctr, Univ Wash, 65-; assoc ed, Lab Animal Care, 67-77; mem, Nat Coun Radiation Protection & Measurements, 69-75; mem panel consult, Nev Opers Off, AEC, 69-78; mem bd regents, Calif Lutheran Col, 70-73; mem bd gov, Found Human Ecol, 70-76; mem, Adv Res Resources Coun, NIH, 75-78; chmn comt vet med sci, Nat Res Coun-Nat Acad Sci, 75-79; nat consult vet med, Surgeon Gen, US Air Force, 75-, consult, Lovelace Biomed & Environ Res Inst, 77-89; exec dean, Wash, Ore, Idaho, Regional Prog Vet Med; vis prof, Murdoch Univ, 83, Sch Med, Univ Wash, 83; distinguished vis bicentennial prof, Univ Ga, 84; vis lectr, Univ Tenn, Knoxville, 86 & 88, Pac Lutheran Univ, 86 & La State Univ, 88; Miller vis prof, Univ Ill, 87. *Honors & Awards:* John G Rutherford Mem lectr, 79; Wesley Spink lectr, 79; LaCroix Mem lectr, 80; Prof Larry Smith & John O'Donaghue Mem lectr, 85; Prof Mark Allam lectr, 86; Award of Merit, Am Animal Hosp Asn, 84; Int Award for Serv to Prof, World Small Animal Vet Asn, 91. *Mem:* Sr mem Inst Med-Nat Acad Sci; Am Asn Lab Animal Sci; Am Vet Med Asn; Am Physiol Soc; Am Inst Biol Sci; Soc Health Human Values; Delta Soc (pres, 81-88). *Res:* Thyroid physiology; metabolism and toxicity of radionuclides; physiological response to irradiation; laboratory animal biology and medicine; veterinary medical education; cooperative regional curriculum; carcinogenesis and aging; health sciences education; human companion animal bond; animal assisted therapy; ethics and values in animal-oriented professions. *Mailing Add:* Col Vet Med 110 Bustad Hall Wash State Univ Pullman WA 99164-7010

BUSTARD, THOMAS STRATTON, b Baltimore, Md, Feb 18, 34; m 57; c 2. NUCLEAR ENGINEERING. *Educ:* Johns Hopkins Univ, BES, 55; Drexel Inst Technol, MSME, 61; Univ Md, PhD(nuclear eng), 65. *Prof Exp:* Technologist, E I du Pont de Nemours & Co, 55-56; process engr, Davison Chem Co, 56-57; staff engr, Martin-Marietta Corp, 57-65; vpres, Hittman Assocs, Inc, 65-78; dir, Div Technol Develop, US Dept Energy, 78-80; PRES, ENERGETICS, INC, 80- *Mem:* Am Nuclear Soc; Am Soc Mech Engrs; Sigma Xi; Health Physics Soc; Asn Advan Med Instrumentation. *Res:* Systems for nuclear auxiliary power devices; production of electricity from radioisotopes by advanced energy conversion methods; radioactive waste disposal; industrial energy conservation; cogeneration; fuel switching. *Mailing Add:* 10364 Old Frederick Rd Woodstock MD 21163

BUSTEAD, RONALD LORIMA, JR, b Woburn, Mass, Mar 29, 30; m 61. FOOD TECHNOLOGY, FOOD CHEMISTRY. *Educ:* Mass Inst Technol, BS, 52; Northeastern Univ, MS, 66. *Prof Exp:* Dairy technologist, H P Hood & Sons, 56-57; chemist, Nat Lead Co, Mass, 57-58; food technologist, Schroeder Industs, Inc, 58-59, vending machine engr, 59-62; food technologist, H A Johnson Co, 62; qual control specialist, Armed Forces Food & Container Inst, 62-64, food technologist, 64-70, opers res analyst, US Army Natick Labs, 70-85; RETIRED. *Mem:* Inst Food Technol; Am Chem Soc. *Res:* Feeding systems; military rations; space foods; new convenience foods; quality assurance for foods; application of operations research, systems analysis and computer sciences to feeding situations. *Mailing Add:* 124 Winter St Framingham MA 01701

BUSTEED, ROBERT CHARLES, biology; deceased, see previous edition for last biography

BUSTIN, MICHAEL, b Bucharest, Rumania, Apr 19, 37; US citizen; m 67; c 2. MOLECULAR BIOLOGY, BIOCHEMISTRY. *Educ:* Univ Denver, BS, 64; Univ Calif, Berkeley, PhD(biochem), 68. *Prof Exp:* Res assoc biochem, Rockefeller Univ, 68-69; res assoc, Weizmann Inst, 69-73, jr scientist immunochem, 74-76, assoc prof, 76-78; vis scientist biochem, 75-85, SUPVRY CHEMIST BIOCHEM, NIH, 85-, SECT HEAD BIOCHEM, 85- *Concurrent Pos:* Adj prof, Sch Dent, Georgetown Univ, 84- *Honors & Awards:* Lubell Prize, Weizmann Inst, 75. *Mem:* Am Soc Biol Chemists; Am Soc Cell Biol. *Res:* Gene structure and function; chromatin; chromosomal proteins and the manner in which they regulate gene expression; use of immunochemical techniques for studies on chromatin and chromosomes. *Mailing Add:* Nat Cancer Inst NIH Bldg 37 Rm 3D 12 Bethesda MD 20892

BUSTIN, ROBERT MARC, b Sask, Can, July 29, 52. COAL GEOLOGY. *Educ:* Univ Calgary, BSc, 75, MSc, 77; Univ BC, PhD(geol), 80. *Prof Exp:* Geologist explor, Mobil Oil Can, 76, Gulf Oil Can, 77-78; ASSOC PROF GEOL, UNIV BC, 85- *Concurrent Pos:* Consult, Coal & Oil Explor, 79- *Mem:* Can Soc Petrol Geologists; Soc Econ Paleontologists & Mineralogists; Geol Asn Can; Int Coal Petrol Asn; Am Asn Petrol Geologists. *Res:* Sedimentology, structure and petrology of coal and coal measures. *Mailing Add:* Dept Geol Sci Univ BC 2075 Westbrook Pl Vancouver BC V6T 1W5 Can

BUSTOS-VALDES, SERGIO ENRIQUE, b Constitucion, Chile, July 6, 32; m 63; c 3. BIOCHEMISTRY. *Educ:* Univ Concepcion, Chile, DDS, 58; Univ Rochester, PhD(biochem), 68. *Prof Exp:* From instr to asst prof physiol sch med, Univ Concepcion, Chile, 58-67, from assoc prof to prof, Inst Biomed Sci, 67-71; assoc prof, 71-74, actg coordr biochem, 78-79, PROF BIOCHEM, DEPT ORAL BIOL, MED COL GA, 74-, ASST PROF CELL & MOLECULAR BIOL, 71-, COORDR BIOCHEM, 79- *Concurrent Pos:* Fulbright scholar & US State Dept grant, Univ Rochester, 62-63; consult, World Health Orgn, 79; vis prof, Univ Concepcion, Chile, 75, 83. *Mem:* Sigma Xi; Int Asn Dent Res; Am Soc Cell Biol. *Res:* Biochemical analysis of proteins and proteoglucans of the extra cellular matrix. *Mailing Add:* Dept Oral Biol Med Col Ga 1120 15th St Augusta GA 30912

BUSZA, WIT, b Ploesti, Roumania, Jan 14, 40; m 64; c 2. EDUCATION, ELEMENTARY PARTICLE PHYSICS. *Educ:* Univ London, BSc, 60, PhD(nuclear physics), 64. *Prof Exp:* Res assoc physics, Univ Col, Univ London, 63-66 & Stanford Linear Accelerator Ctr, 66-69; from asst prof to assoc prof, 69-79, PROF PHYSICS, MASS INST TECHNOL, 79- *Mem:* Am Inst Physics. *Res:* Study of elementary particles by observing their interactions with nuclei at high energies; study of nucleon structure and fundamental interactions using muon as a probe. *Mailing Add:* Rm 24-518 Mass Inst Technol Cambridge MA 02139

BUTCHBAKER, ALLEN F, b Three Rivers, Mich, May 22, 35; m 84; c 4. AGRICULTURAL ENGINEERING. *Educ:* Mich State Univ, BS, 57, MS, 60; Univ Mo, PhD(agr eng), 64. *Prof Exp:* Asst agr eng, Mich State Univ, 58-60; instr, Univ Mo, 60-63; from asst prof to assoc prof, NDak State Univ, 63-70; assoc prof, Okla State Univ, 70-75; pres, B&C Irrig, Inc, 75-80; CONSULT, 80- *Concurrent Pos:* Consult engr. *Mem:* Am Soc Agr Engrs; Am Soc Heating, Refrig & Air-Conditioning Engrs. *Res:* Livestock facilities with emphasis on environment and animal waste management; irrigation system design; electrical applications. *Mailing Add:* 57865 Kaylor Rd Jones MI 49061

BUTCHER, BRIAN T, b Guildford, Eng, Dec 28, 40; m 62; c 1. IMMUNOLOGY, ALLERGY. *Educ:* Inst Med Lab Technol, London, FIMLT, 64; Univ NC, Chapel Hill, MS, 68; La State Univ, New Orleans, PhD(microbiol), 73. *Prof Exp:* Lab technician clin lab, King George V Hosp, 57-63; lab supvr clin lab, St Thomas Hosp, 63-64; lab supvr epidemiol, Sch Pub Health, Univ NC, 64-68; instr epidemiol, Med Ctr, Tulane Univ, 68-71; res assoc med, Med Ctr, La State Univ, 73-75; asst prof, 75-79, ASSOC PROF MED, MED CTR, TULANE UNIV, 79- *Mem:* Am Acad Allergy; Am Thoracic Soc; Am Soc Microbiol; AAAS; Soc Exp Biol & Med. *Res:* Pathogenesis of occupational asthma; immunopharmacology. *Mailing Add:* Arthritis Found 1314 Spring St NW Atlanta GA 30309

BUTCHER, FRED RAY, b Rochester, Pa, Aug 11, 43; m 65; c 2. BIOCHEMISTRY. *Educ:* Ohio State Univ, BSc, 65, PhD(biochem), 69. *Prof Exp:* NIH fel oncol, Univ Wis, 69-71; from asst prof to assoc prof biochem, Brown Univ, 71-78; PROF BIOCHEM, WVA UNIV, 78-, CHMN, 81- *Mem:* Am Soc Biol Chemists. *Res:* Role of cyclic nucleotides and calcium in the regulation of target tissue physiology and biochemistry by regulatory agents; the rat parotid gland, primary culture of liver cells, and tissue culture cell lines HL-60 and friend cells are used as models. *Mailing Add:* MBR Cancer Ctr WVa Univ 103 Mary Babb Randolph Cancer Ctr Morgantown WV 26506

BUTCHER, HARVEY RAYMOND, JR, surgery, for more information see previous edition

BUTCHER, HARVEY RAYMOND, III, b Salem, Mass, Aug 3, 47; m 71; c 2. ASTRONOMY. *Educ:* Calif Inst Technol, BS, 69; Australian Nat Univ, PhD(astron), 74. *Prof Exp:* Bart Bok fel, Steward Observ, Univ Ariz, 74-76; asst astronr, Kitt Peak Nat Observ, Asn Univs Res Astron, 76-79, assoc astronr, 79-81, astronr, 81-83; prof astron & dir, Kapteyn Observ, Univ Groningen, Neth, 83-91; DIR GEN, NETH FOUND RES ASTRON, 91- *Mem:* Am Astron Soc; Int Astron Soc; Royal Astron Soc. *Res:* Elucidation of the evolution of galaxies over cosmological time. *Mailing Add:* Floralaan 8 Roden 9301 KE Netherlands

BUTCHER, HENRY CLAY, IV, b Newton, Mass, July 31, 33; m 58; c 3. PLANT PHYSIOLOGY, BIOCHEMISTRY. *Educ:* Tufts Univ, BSc, 55; Ohio State Univ, MS, 61, PhD(plant physiol), 64. *Prof Exp:* Fel, Res Inst Advan Studies, Md, 64-65; from asst prof to assoc prof, 69-77, chmn dept, 71-75, 78-83, PROF BIOL, LOYOLA COL, MD, 77- *Concurrent Pos:* Guest plant physiologist, Brookhaven Nat Lab, 75-76; vis sr scientist, Johns Hopkins Univ, 83-84. *Mem:* AAAS; Am Soc Plant Physiol; Am Inst Biol Sci. *Res:* Biochemistry of plant cell vacuoles; translocation of organic solutes in plants; photosynthesis; tissue culture; differentiation; molecular biology of aging. *Mailing Add:* 6621 Queens Ferry Rd Baltimore MD 21212

BUTCHER, JOHN EDWARD, b Belle Fourche, SDak, Aug 4, 23; m 51; c 3. ANIMAL PRODUCTION. *Educ:* Mont State Col, BS, 50, MS, 52; Utah State Univ, PhD(animal prod), 56. *Prof Exp:* Instr animal indust, Mont State Col, 49-50, range mgt, 52-53; from asst prof to prof animal sci, Utah State Univ, 55-86; CONSULT, 86- *Concurrent Pos:* Mem subcomt sheep nutrit, Nat Res Coun, 75; mem, Nat Pub Land Adv Coun, Bur Land Mgt, US Dept Interior, 82-84. *Mem:* Fel AAAS; fel Am Soc Animal Sci; Soc Range Mgt; Am Soc Farm Mgr & Rural Appraisers. *Res:* Ruminant nutrition and environment. *Mailing Add:* 1703 E 1030 N Logan UT 84321

BUTCHER, LARRY L, b Richmond, Ind, Feb 21, 40. BRAIN HISTOCHEMISTRY. *Educ:* Univ Mich, PhD(psychol), 66. *Prof Exp:* PROF PSYCHOL, UNIV CALIF, LOS ANGELES, 80- *Mem:* Soc Neurosci; Western Pharmacol Soc; Am Soc Pharmacol & Exp Therapeut. *Mailing Add:* Dept Psychol Univ Calif Los Angeles CA 90024

BUTCHER, RAYMOND JOHN, b Greymouth, NZ, Oct 11, 45; US citizen; m 79; c 2. X-RAY CRYSTALLOGRAPHY, BIOINORGANIC CHEMISTRY. *Educ:* Univ Canterbury, NZ, BSc Hons, 68, PhD(chem), 74; Christchurch Teachers Col, NZ, dipl physics & math, 70. *Prof Exp:* Instr chem, Univ Va, Charlottesville, 74-76; postdoctoral fel inorg chem, Georgetown Univ, Wash, DC, 76-77; asst prof, 77-83, ASSOC PROF CHEM, HOWARD UNIV, WASH, DC, 83- *Concurrent Pos:* Vis assoc prof, Univ Va, Charlottesville, 85-86; fac fel, Goddard Space Flight Ctr, 87 & 88; adj prof, Hampton Univ, 87-90; sr scientist, Naval Res Lab, Wash, DC, 89 & 90; int Fulbright fel, Univ Poona, India, 89. *Mem:* Am Chem Soc; Sigma Xi; NY Acad Sci; AAAS; Am Crystallog Asn. *Res:* Magneto-structural correlations in exchange-coupled systems; bioinorganic chemistry of manganese, iron, copper and molybdenum; promotion of electrophilic and nucleophilic substitution reactions by use of coordinated exchange inert metal complexes. *Mailing Add:* Dept Chem Howard Univ Washington DC 20059

BUTCHER, REGINALD WILLIAM, b Bayshore, NY, May 4, 30; m 53; c 3. BIOCHEMISTRY, PHYSIOLOGY. *Educ:* US Naval Acad, BS, 53; Western Reserve Univ, PhD(pharmacol), 63. *Prof Exp:* From instr to assoc prof physiol, Vanderbilt Univ, 63-69, investr, Howard Hughes Med Inst, 66-69; prof biochem & chmn dept, Med Sch, Univ Mass, 69-79, assoc dean sci affairs, 76-79; DEAN, GRAD SCH BIOMED SCI, UNIV TEX HEALTH DCI CTR, HOUSTON, 79-, PROF BIOCHEM & MOLECULAR BIOL, MED SCH, 79- DIR, INST TECHNOL DEVELOP & ASSESSMENT, 84- *Concurrent Pos:* Vis prof, Univ Libre de Bruxelles, Belgium, 77-; Fulbright Sr Res Scholar, 77-78; dir, Speech & Hearing Inst, Univ Tex Health Sci Ctr, Houston, 82-, actg dir, Position Diag & Res Ctr,87- *Mem:* Endocrine Soc; NY Acad Sci; Am Physiol Soc; Am Soc Biol Chemists; Sigma Xi. *Res:* Mechanism of hormone action at the molecular level; control of cyclic adenosinemonophosphate levels by adenyl cyclase and phosphodiesterase activities and effects of cyclic AMP on cellular processes. *Mailing Add:* GSBS Univ Tex Health Sci Ctr PO Box 20334 Houston TX 77225

BUTCHER, ROY LOVELL, b Reedy, WVa, Dec 15, 30; m 55; c 3. REPRODUCTIVE PHYSIOLOGY. *Educ:* WVa Univ, BS, 53, MS, 59; Iowa State Univ, PhD(animal reproduction), 62. *Prof Exp:* From instr to assoc prof, 62-78, PROF OBSTET & GYNEC, WVA UNIV, 78- *Mem:* Endocrine Soc; Soc Study Reproduction; Am Soc Animal Sci; Soc Gynec Invest; Brit Soc Study Fertil. *Res:* Control of luteal function; effects of delayed ovulation on congenital anomalies. *Mailing Add:* Dept Obstet & Gynec WVa Univ HSN Morgantown WV 26506-6302

BUTCHER, SAMUEL SHIPP, b Gaylord, Mich, Nov 12, 36; m 61; c 2. PHYSICAL CHEMISTRY. *Educ:* Albion Col, AB, 58; Harvard Univ, AM, 61, PhD(chem), 63. *Prof Exp:* Fel spectros, Nat Res Coun Can, 62-64; from asst prof to assoc prof, 64-74, PROF CHEM, BOWDOIN COL, 74- *Mem:* AAAS; Am Chem Soc; Am Meteorol Soc. *Res:* Atmospheric chemistry. *Mailing Add:* Dept Chem Bowdoin Col Brunswick ME 04011

BUTEAU, L(EON) J, b Teaneck, NJ, Nov 19, 32. MATERIALS SCIENCE, THERMODYNAMICS. *Educ:* Newark Col Eng, BS, 58; Stanford Univ, MS, 59; Univ Fla, PhD(mat eng), 63. *Prof Exp:* Mem tech staff systs eng, Bell Tel Labs, 63-65; PROF MAT SCI, NJ INST TECHNOL, 65- *Mem:* Sigma Xi. *Res:* Topology of powder metallurgy, the neuron networks in the brain and the law of supply and demand; twinning systems in zirconium; cycle theories in economics. *Mailing Add:* Dept Physics NJ Inst Technol 323 High St Newark NJ 07102

BUTEL, JANET SUSAN, b Overbrook, Kans, May 24, 41; m 67; c 2. VIROLOGY, CELL BIOLOGY. *Educ:* Kans State Univ, BS, 63; Baylor Univ, PhD(virol), 66. *Prof Exp:* Fel, 66-68, from asst prof to assoc prof, 68-76, PROF VIROL, BAYLOR COL MED, 76-, HEAD, DIV MOLECULAR VIROL, 89- *Concurrent Pos:* Exp virol study sect, NIH, 80-84; bd scientific counr, Div Cancer Etiol, Nat Cancer Inst, 85-89; co-ed, Med Microbiol, Lange Med Pub, 89- *Honors & Awards:* Am Cancer Soc Fac Res Award, 72; Joseph L Melnick Prof Virol, 86. *Mem:* Fel AAAS; Am Soc Microbiol; Am Soc Virol; Sigma Xi; Am Asn Cell Biol; Int Asn Breast Cancer Res. *Res:* Tumor viruses; oncogenes; transformation of cells. *Mailing Add:* 4051 Mischire Houston TX 77025

BUTENSKY, IRWIN, b New York, NY, Jan 27, 36; m 69. COSMETIC CHEMISTRY, PHARMACEUTICAL CHEMISTRY. *Educ:* Columbia Univ, BS, 56; Univ Mich, MS, 59, PhD(pharmaceut chem), 61. *Prof Exp:* Develop chemist, Lederle Labs, Am Cyanamid Co, NY, 61-66, group leader process improve lab, 66-67; develop chemist pharmaceut res & develop, Vicks Div Res & Develop, Richardson-Merrell, Inc. 67-69, sect head, Vicks Div Res, 69-71, asst dir pharmaceut develop, 71-72, dir skin care & toiletries, 72-79; vpres res & develop, Int Playtex Inc, 79-90, SR VPRES, PLAYTEX FAMILY PRODS CORP, 90- *Mem:* AAAS; Am Pharmaceut Asn; Am Chem Soc; Soc Cosmetic Chem. *Res:* Tablet coating technology; spraying techniques; fine particle technology; suspensions of natural product; proprietary drug development; aerosols, creams and all other dosage forms; plastics; development of new cosmetic and dermatological products. *Mailing Add:* 533 Winthrop Rd Teaneck NJ 07666-2917

BUTENSKY, MARTIN SAMUEL, b Brooklyn, NY, Nov 28, 37; m 60; c 3. CHEMICAL ENGINEERING. *Educ:* Columbia Col, BA, 57; Columbia Sch Eng, BS, 58; Mass Inst Technol, ScD(chem eng), 63. *Prof Exp:* Res chem engr, 62-69, sr res chem engr, 69-75, res group leader process develop, 75-86, PROJS MGR PROCESS TECHNOL, CHEM RES DIV, AM CYANAMID CO, 86- *Mem:* Am Inst Chem Engrs. *Mailing Add:* 90 Nutmeg Lane Stamford CT 06905

BUTERA, RICHARD ANTHONY, b Carnegie, Pa, Nov 30, 34; m; c 2. SOLID STATE CHEMISTRY, PHYSICAL CHEMISTRY. *Educ:* Univ Pittsburgh, BS, 60; Univ Calif, Berkeley, PhD(phys chem), 63. *Prof Exp:* Res assoc phys chem, Univ Pittsburgh, 57-60; res asst low temperature chem, Univ Calif, Berkeley, 60-63; asst prof to assoc prof, 63-79, PROF PHYS CHEM, UNIV PITTSBURGH, 79- *Mem:* Am Phys Soc; Am Chem Soc; AAAS; NY Acad Sci; fel Am Inst Chem. *Res:* Magneto-thermo-dynamic studies of magnetic phase transitions; heat capacity studies of intermetallic compounds and high temp superconductors; computer controlled experimentation; metal/semiconductor interfaces; high temperature superconductors. *Mailing Add:* Dept Chem Univ Pittsburgh Pittsburgh PA 15260

BUTH, CARL EUGENE, b Gatesville, Tex, June 17, 40; m 61; c 3. CIVIL ENGINEERING. *Educ:* Tex A&M Univ, BS, 63, ME, 64, PhD(civil eng), 72. *Prof Exp:* Res asst, Tex Transp Inst, 63-66, engr res assoc, 66-68, asst res engr, 68-74, head fabrication & testing sect, Tech Support Servs, 69-70, head eng & construct, Proving Grounds Res Prog, 70-73, head, 73-78, from instr to asst prof civil eng, Tex A&M Univ, assoc res engr, Tex Transp Inst, 74-79, ASSOC PROF CIVIL ENG, TEX A&M UNIV, 77-, HEAD, SAFETY DIV, TEX TRANSP INST, 78-, RES ENGR, 79- *Mem:* Nat Soc Prof Engrs; Am Soc Civil Engrs; Am Soc Eng Educ; Sigma Xi. *Res:* Structural analysis and design; highway safety. *Mailing Add:* Tex Transp Inst Tex A&M Univ College Station TX 77843

BUTH, DONALD GEORGE, b Chicago, Ill, Feb 23, 49; m 89; c 1. PHYLOGENETIC SYSTEMATICS, ICHTHYOLOGY. *Educ:* Univ Ill, BS, 71, AB, 72, MS, 74, PhD(ecol, ethology & evolution), 78. *Prof Exp:* Res fel systematics, 78-79, lectr pop genetics, 80, asst prof, 80-86, ASSOC PROF BIOL, UNIV CALIF, LOS ANGELES, 86- *Concurrent Pos:* Consult, endangered species recovery team, 79-; sectional ed, Copeia, 85-; ed, Isozyme Bulletin, 90- *Mem:* Am Soc Ichthyologists & Herpetologists; Soc Systematic Zool; Am Fisheries Soc; Soc Study Evolution; fel AAAS. *Res:* Biochemical systematics of reptiles, gasterosteiform, cypriniform and perciform fishes; polyploid gene expression in fishes. *Mailing Add:* Dept Biol Univ Calif Los Angeles CA 90024-1606

BUTKIEWICZ, EDWARD THOMAS, b Philadelphia, Pa, Mar 15, 35; m 58; c 3. SYSTEMS ENGINEERING, PHYSICS. *Educ:* La Salle Col, BA, 56. *Prof Exp:* Physicist, Exp Div, Naval Air Eng Facil, 56-59 & Off Chief Scientist, Naval Air Mat Ctr, 59-61; physicist, Launcher Div, Naval Air Eng Ctr, 61-69 & Aircraft/Weapons/Ship Div, 69-70, prog mgr Carrier Aircraft Support Off & dir, Tech Info Syst for Carrier Aviation, 70-75, staff physicist, Ship Installations Eng Dept, 75-81; PRES, CARRIER AVIATION INFO SERV, 75-; PROG MGR, OCCUP SAFETY & HEALTH OFF, PHILADELPHIA NAVAL SHIPYARD, PA, 81- *Mem:* AAAS; Marine Tech Soc; Am Soc Oceanog; NY Acad Sci. *Res:* Computer information retrieval systems; management of information systems and decision making computer systems; aircraft catapult launching performance; airflow visualization; carrier aircraft compatibility; carrier aircraft support systems; designed and developed the hazardous materials/hazardous waste program and the hazard deficiency abatement program; software development of navy occupational health information monitoring system version 2.0; structured injured workers rehabilitation program and asbestos related materials records system. *Mailing Add:* 4248 Barnett St Philadelphia PA 19135

BUTKOV, EUGENE, b Pancevo, Yugoslavia, Sept 16, 28; US citizen. THEORETICAL PHYSICS. *Educ:* Univ BC, BASc, 54, MA, 56; McGill Univ, PhD(theoret physics), 60. *Prof Exp:* Asst prof physics, St John's Univ, NY, 59-62 & Hunter Col, 62-65; asst prof, 65-68, chmn dept, 74-80, ASSOC PROF PHYSICS, ST JOHN'S UNIV, NY, 68- *Mem:* Am Phys Soc; Math Asn Am; Soc Ind Appl Math. *Res:* Theory of non-linear differential equations; quantum field theory. *Mailing Add:* Dept Physics St John's Univ Grand Central & Utopia Jamaica NY 11439

BUTKUS, ANTANAS, b Lithuania, June 20, 18; US citizen; m 43; c 2. ANALYTICAL CHEMISTRY, CARDIOVASCULAR DISEASES. *Educ:* Univ Halle, BAgrSci, 43; Univ Bonn, DrAgrSci, 47. *Prof Exp:* Chief chemist, Processing Lab. Southland Frozen Meat, NZ, 54-62; chemist, Water Control Comn, Md, 62; asst staff mem lipids, Cleveland Clin, 62-67, assoc staff mem, 67-69, STAFF MEM, CLEVELAND CLIN FOUND, 69- *Mem:* AAAS; Am Chem Soc; Am Oil Chem Soc; Am Soc Exp Path. *Res:* Disorders of lipid metabolism and arteriosclerosis. *Mailing Add:* 19406 Van Aken Blvd Cleveland OH 44122

BUTLER, ANN BENEDICT, b Wilmington, Del, Dec 2, 45; m 68. NEUROANATOMY. *Educ:* Oberlin Col, BA, 67; Case Western Reserve Univ, PhD(anat), 71. *Prof Exp:* NIH fel neuroanat, Brown Univ, 71-72 & Univ Va, 72-73; asst prof anat, George Washington Univ, 73-75; from asst prof to assoc prof anat, Georgetown Univ, 74-84; INDEPENDENT NEURO & ANAT RES SCIENTIST, 86- *Mem:* Soc Neurosci. *Res:* Comparative neuroanatomy of the visual system; evolution of sensory system organization in the thalamus and telencephalon; evolution of neocortex and of limbic system. *Mailing Add:* ITNI 4433 N 33rd St Arlington VA 22207

BUTLER, ARTHUR P, b Morristown, NJ, June 23, 08. RESOURCES. *Educ:* Harvard Univ, BA, 30, MA, 37, PhD(geol), 46. *Prof Exp:* Geologist, US Geol Surv, 41; RETIRED. *Mem:* Soc Econ Geologist; Sigma Xi. *Mailing Add:* 9625 W 36th Ave Wheat Ridge CO 80033

BUTLER, BLAINE R(AYMOND), JR, b Johnstown, NY, Aug 11, 25; m 48; c 3. ASTRONAUTICAL & AERONAUTICAL ENGINEERING. *Educ:* US Mil Acad, BS, 48; Purdue Univ, MS, 61, PhD(aeronaut eng), 65. *Prof Exp:* USAF, 48-72, from instr to asst prof aeronaut, USAF Acad, 63-66, prof & head dept, 66-69, dep comdr, Space Defense Ctr, NAm Air Defense Command, 70-71, analyst, Joint Strategic Planning Staff & Joint Chiefs of Staff, Washington, DC, 71-72; prof eng, Purdue Univ, 72-83; PROF & DEPT HEAD, AEROSPACE ENG, EMBRY-RIDDLE AERONAUT UNIV, 83- *Mem:* Am Inst Aeronaut & Astronaut; Am Soc Eng Educ; assoc fel Can Aeronaut & Space Inst; fel Brit Interplanetary Soc. *Mailing Add:* 2242 Loma Rica Circle Prescott AZ 86301

BUTLER, BRUCE DAVID, b Houston, Tx, Jan 27, 53. PULMONARY PHYSIOLOGY, SPACE MEDICINE. *Educ:* Univ Tex, BA, 75, PhD(physiol & biophys), 80. *Prof Exp:* Fel, Dept Anesthesiol, 81-82, from instr to asst prff, 82-89, ASSOC PROF ANESTHESIOL, UNIV TEX MED SCH, 89- *Concurrent Pos:* Mem, Sch Biomed Sci, 84-, adj asst prof, Univ Tex Health Sci Ctr, 87- *Mem:* Undersea Med Soc; Am Physiol Soc; Aerospace Med Asn; Sigma Xi; Int Union Physiol Scientists. *Res:* Pulmonary pathophysiology & decompression biology; studies in physiology, biochemistry, & biophysical indicies in decompression sickness experience by commercial deep sea divers or astronauts; pulmonary pathophysiological studies related to surfactant molecules and their role in lungs. *Mailing Add:* Dept Anesthesiol Univ Tex Sch Med 6431 Fannin 5-020 MSMB Houston TX 77030

BUTLER, BYRON C, b Carroll, Iowa, Aug 10, 18; m 58, 75; c 5. GEMOLOGY, GYNECOLOGY. *Educ:* Columbia Univ, MD, 43, ScD, 51; Gemological Inst Am, GG, 87. *Prof Exp:* Instr path, Col Physicians & Surgeons, Columbia Univ, 41-42, instr obstet & gynec, 50-53; dir biophys, Butler Res Found, 63-87. *Concurrent Pos:* Am Cancer Soc grant, 51 & NIH grant, 52-53. *Mem:* Fel AAAS; AMA; Am Soc Fertil; Am Gemological Trade Asn; Int Colored Gemstone Asn; Laser Med & Surg Asn. *Res:* Hypnosis in relation to relief of pain in cancer; fibrinolysin enzyme system in relation to blood clotting defects in pregnancy; bacterial vaginal flora; magnetic field inhibition of cancer cells in tissue culture; infertility; diagnosis and treatment of cancer of cervix and early detection; diagnosis of cancer and viral diseases in gynecology; laser surgery in treatment of cancer and viral disease in gynecology; gemologic research in color, inclusions and identification; computer science. *Mailing Add:* 6302 N 38th St Paradise Valley AZ 85253

BUTLER, CALVIN CHARLES, b Fowler, Colo, Nov 8, 37; m 59; c 4. MATHEMATICAL STATISTICS. *Educ:* Colo State Univ, BS, 62, MS, 63, PhD(math statist), 66. *Prof Exp:* Asst prof math, Univ Colo, Boulder, 66-71; PROF MATH, COL SOUTHERN IDAHO, 71- *Mem:* Am Statist Asn. *Res:* Applied probability theory; stochastic processes. *Mailing Add:* Dept Math Col Southern Idaho Twin Falls ID 83301

BUTLER, CHARLES MORGAN, b Columbia, Tenn, Dec 16, 29; m 58; c 3. STATISTICS, OPERATIONS RESEARCH. *Educ:* US Mil Acad, BS, 53; Mass Inst Technol, SM, 62; Univ Ala, MA, 66, PhD(math), 70. *Prof Exp:* Asst prof indust eng, Univ Ala, 68-70; assoc prof, 70-76, PROF STATIST, MISS STATE UNIV, 76- *Res:* Application of statistical and other quantitative techniques for the solution of business and industrial problems; air pollution data processing. *Mailing Add:* Drawer DB-Bus Info Miss State Univ Mississippi State MS 39762

BUTLER, CHARLES THOMAS, b Muskogee, Okla, Nov 30, 32; m 81; c 3. NEURAL NETWORKS. *Educ:* Iowa State Univ, BS, 54; Tex A&M Univ, MS, 57; Okla State Univ, PhD(physics), 72. *Prof Exp:* Sr scientist, Jet Propulsion Lab, Calif Inst Technol, 56-60; physicist, Oak Ridge Nat Lab, 60-70; mem fac, dept physics, Okla State Univ, 70-78; mem fac, Dept Physics, Va Commonwealth Univ, 78-83; PRIN RES PHYSICIST, PHYS SCI, INC, 83- *Mem:* Am Phys Soc; AAAS; Sigma Xi; Inst Elec & Electronics Engrs; Int Neurol Network Soc. *Res:* Metal hydrides; radiation damage in insulators; crystal growth; neural network applications. *Mailing Add:* Phys Sci Inc 635 Slaters Lane Alexandria VA 22314

BUTLER, DON, b Camden, NJ, July 2, 60. EXPLOSION BONDING. *Educ:* Tulane Univ, BS, 82. *Prof Exp:* SR PROJ ENGR, NORTHWEST TECH INDUSTS, 84- *Mem:* Am Soc Mech Engrs. *Res:* explosion bonding and forming materials which have low ductility. *Mailing Add:* Northwest Tech Ind Inc 547 Diamond Pt Sequim WA 98382

BUTLER, DONALD EUGENE, b Detroit, Mich, Dec 1, 33; m 62; c 3. ORGANIC CHEMISTRY, PHARMACEUTICAL CHEMISTRY. *Educ:* Wayne State Univ, BS, 55; Univ Fla, PhD(chem), 58. *Prof Exp:* From assoc res chemist to sr res chemist, 58-77, res assoc, 77-81, SR RES ASSOC & SR GROUP MGR, WARNER LAMBERT-PARKE DAVIS PHARMACEUT RES DIV, PARKE, DAVIS & CO, 77- *Mem:* Am Chem Soc; AAAS. *Res:* Piperazine derivatives; synthetic organic medicinals; gastrointestinal and central nervous system pharmacology; heterocyclic chemistry, pyrrollidines and pyrazoles; learning and memory; new routes to known compounds and large scale syntheses. *Mailing Add:* Warner Lambert-Parke Davis 188 Howard Holland MI 49424-6596

BUTLER, DONALD J(OSEPH), b New York, NY, May 16, 25; m 61; c 1. CIVIL ENGINEERING. *Educ:* Columbia Univ, BS, 49, MS, 53, PhD(eng mech), 58. *Prof Exp:* From instr to assoc prof civil eng, Columbia Univ, 49-66; PROF CIVIL ENG, RUTGERS UNIV, 66- *Concurrent Pos:* Gulbenkian fel, Nat Lab Civil Eng, Lisbon, 65-66; vis scientist, Building Res Estab, Garston, Eng, 75-76. *Mem:* Am Soc Civil Engrs; Soc Exp Stress Anal; Am Soc Eng Educ. *Res:* Stability, realiability and dynamics of structural systems; experimental mechanics; structural materials. *Mailing Add:* PO Box 405 Cuttingsville VT 05738

BUTLER, DOUGLAS NEVE, b Melbourne, Victoria, Australia, Mar 30, 36; m 73; c 1. STRUCTURAL CHEMISTRY, ORGANIC CHEMISTRY. *Educ:* Univ NSW, BSc, 58; Univ Manchester, Victoria, PhD(chem), 63. *Prof Exp:* Asst prof, 68-71, ASSOC PROF, CHEM DEPT, YORK UNIV, 71- *Concurrent Pos:* Vis prof, Chem Dept, Univ Capetown, SAfrica, 71 & Bond Univ, Queensland, Australia, 89; vis fel, Chem Dept, Australia Nat Univ, 74-75 & 81-82. *Res:* Design and synthesis of rigid polycyclic organic chemical frameworks bearing fixed or flexibly attached functional groups as specific binding agents on catalysts. *Mailing Add:* Dept Chem York Univ North York ON M3J 1P3 Can

BUTLER, DWAIN KENT, b Stamford, Tex, Jan 6, 46; m 68; c 2. GEOPHYSICS, GEOSCIENCE ENGINEERING. *Educ:* Tex Tech Univ, BS, 68; Univ Md, MS, 71; Texas A&M Univ, PhD, 83. *Prof Exp:* Res physicist, Naval Ord Lab. 68-71, res physicist & geophys, Mo River Div Lab, 71-73, RES GEOPHYSICIST, WATERWAYS EXP STA, CORPS ENGRS, 73- *Concurrent Pos:* Instr physics, St Aloysius High Sch, Vicksburg, 75-76; advan study sabbatical, Waterways Exp Sta, Tex A&M Univ, 77-78; adj prof geosci, Univ Southern Miss & Miss State Univ, 80-88; prof, Geophys, Texas A & M Univ, 89- *Honors & Awards:* Herbert D Vogel Award, Outstanding Scientist, Waterways Exp Sta, 85, Outstanding Tech Presentation Award, 85. *Mem:* Am Phys Soc; Soc Explor Geophysicists; Am Inst Prof Geologists; Int Soc Rock Mech; Am Geophys Union; Soc Am Military Engrs. *Res:* Engineering geophysics; microgravimetry for site investigations; water resources; global plate tectonics, global gravity field. *Mailing Add:* Waterways Exp Sta PO Box 631 Vicksburg MS 39180

BUTLER, EDWARD EUGENE, b Wilmington, Del, Dec 8, 19; m 47; c 5. PHYTOPATHOLOGY, MYCOLOGY. *Educ:* Univ Del, BS, 43; Mich State Univ, MS, 48; Univ Minn, PhD(plant path), 54. *Prof Exp:* Asst bot, Mich State Univ, 46-49; instr plant path, Univ Minn, 51-54; jr plant pathologist, Univ Calif, Davis, 55-56, lectr plant path & asst plant pathologist, 57-61, from assoc prof to prof, 61-90, EMER PROF PLANT PATH, UNIV CALIF, DAVIS, 90- *Concurrent Pos:* Vis prof, Univ PR, 66-67; assoc ed, Phytopath, 73-76; vis mycologist, Rancho Santa Ana Bot Garden, 83-84. *Honors & Awards:* WH Weston Award, Mycol Soc Am. *Mem:* Fel AAAS; Am Phytopath Soc; Mycol Soc Am; Brit Mycol Soc. *Res:* Ecology and taxonomy of plant pathogenic fungi especially soil borne species causing root diseases; mycoparasitisan. *Mailing Add:* Dept Plant Path Univ Calif Davis CA 95616

BUTLER, ELIOT ANDREW, b Snowflake, Ariz, Feb 13, 26; m 49; c 4. ANALYTICAL CHEMISTRY. *Educ:* Calif Inst Technol, BS, 52, PhD, 55. *Prof Exp:* From asst prof to assoc prof chem, 55-65, dean, Col Phys & Math Sci, 77-80, PROF CHEM, BRIGHAM YOUNG UNIV, 65-, ASSOC ACAD VPRES, 80- *Mem:* Am Chem Soc; Sigma Xi. *Res:* Precipitation from homogeneous solution; equilibria of ionic and molecular species in aqueous solutions; electrode reactions. *Mailing Add:* 220 ESC Brigham Young Univ D380-ASB Provo UT 84602

BUTLER, G(ERARD), b Austria, Sept 12, 27; nat US; m 53; c 2. CHEMICAL ENGINEERING. *Educ:* NY Univ, BS, 49, MS, 63. *Prof Exp:* From res engr to group leader, Am Cyanamid Co, 51-66, asst dir res & develop, 66-71, mgr tech serv, 71-77, sr engr, 77-80, tech mgr, 80-88; TECH MGR, CRITERION CATALYST CO, 88- *Mem:* Am Chem Soc; Am Inst Chem Engrs. *Res:* Pigment technology; engineering fundamentals; high temperature reactions; radiation technology; heterogeneous catalysts. *Mailing Add:* Criterion Catalyst Co 1001 N Todd Ave Azusa CA 91702-1602

BUTLER, GEORGE BERGEN, b Liberty, Miss, Apr 15, 16; m 44; c 2. ORGANIC CHEMISTRY. *Educ:* Miss Col, BA, 38; Univ NC, PhD(org chem), 42. *Prof Exp:* Res chemist, Rohm & Haas Co, Philadelphia, 42-46; from instr to assoc prof, 46-57, dir ctr macromolecular sci, 70, RES PROF CHEM, UNIV FLA, 57- *Concurrent Pos:* Past vpres, Peninsular Chem Res, Inc; consult, TRW, Inc, IMC, Inc, & Allied Chem Co; co-ed, J Macromolecular Sci Rev & Rev Macromolecular Chem. *Honors & Awards:* Polymer Award, Am Chem Soc, 80. *Mem:* AAAS; Am Chem Soc; Sigma Xi. *Res:* Biologically active polymers; pharmaceutical chemicals; water-soluble polymers, polymer modification; polyelectrolytes; ion exchange resins; reaction mechanisms; cyclopolymerization, triazolinedione polymers, quaternary ammonium compounds and polymers; polymerization mechanisms. *Mailing Add:* 1906 NW 26th St Gainesville FL 32605

BUTLER, GEORGE DANIEL, JR, b Newark, NJ, Apr 13, 23; m 49; c 3. ENTOMOLOGY. *Educ:* Univ Mass, BS, 48; Cornell Univ, PhD(entom), 51. *Prof Exp:* Asst prof entom, Univ Ariz, 51-59, assoc prof, 59-66; res entomologist, Cotton Insects Br, 66-72, res entomologist Western Cotton Res Lab, 72-86, COLLABR, WESTERN COTTON RES LAB, USDA, 87- *Mem:* Am Entom Soc; Ecol Soc Am. *Res:* Biological control of insects; biology and control of emisia tabaci. *Mailing Add:* USDA Cotton Res Lab 4135 E Broadway Phoenix AZ 85040

BUTLER, GILBERT W, b Ft Collins, Colo, Feb 22, 41; m 78. NUCLEAR CHEMISTRY. *Educ:* Ore State Univ, BS, 63; Univ Calif, Berkeley, PhD(nuclear chem), 67. *Prof Exp:* Res nuclear chemist, Lawrence Radiation Lab, 67-70 & Argonne Nat Lab, 70-72; STAFF MEM & CHEMIST, LOS ALAMOS SCI LAB, 72- *Mem:* Am Phys Soc; Am Chem Soc. *Res:* Study of high energy nuclear reactions using solid state detectors; high resolution gamma ray spectroscopy. *Mailing Add:* Los Alamos Sci Lab Group INC-11 MS-J514 Los Alamos NM 87545

BUTLER, GORDON CECIL, b Ingersoll, Ont, Sept 4, 13; m 37; c 4. BIOCHEMISTRY. *Educ:* Univ Toronto, BA, 35, PhD(biochem), 38. *Prof Exp:* 1851 Exhib scholar, Univ London, 38-40; res chemist, Chas E Frosst & Co, 40-42; mem, Nat Res Coun Can, 45-47; from assoc prof to prof biochem, Univ Toronto, 47-57; dir biol & health physics div, Atomic Energy of Can Ltd, 57-65; dir div radiation biol, 65-68, dir div biol sci, 68-78, CONSULT, DIV BIOL SCI, NAT RES COUN CAN, 78- *Concurrent Pos:* Int Found Sci, 75-81, pres, 82-87. *Mem:* Health Phys Soc; Can Physiol Soc; Am Soc Biol Chem; Royal Soc Can; Can Biochem Soc. *Res:* Steroids; immunochemistry; radiation chemistry; nucleic acids and nucleoproteins; scientific criteria for environmental quality; radiation protection. *Mailing Add:* 4694 W 13th Ave Vancouver BC V6R 2V7 Can

BUTLER, HAROLD S, b Ardmore, Okla, Sept 14, 31; m 54; c 1. COMPUTER CONTROL SYSTEMS. *Educ:* Phillips Univ, BA, 53; Kans State Univ, MS, 56; Stanford Univ, PhD(physics), 61; Univ NMex, MS, 80. *Prof Exp:* Scientist, Lockheed Missile & Space Co, 56-61; staff mem, Stanford Linear Accelerator Ctr, 61-63; STAFF MEM, LOS ALAMOS NAT LAB, 63- *Concurrent Pos:* Vis scientist, Europ Orgn Nuclear Res, 74; adj prof, Univ NMex, 81. *Mem:* Asn Comput Mach. *Res:* Computer control of accelerators; beam dynamics and magnet system design; application of computers to solution of problems in physics. *Mailing Add:* Ten Escondido Los Alamos NM 87544

BUTLER, HARRY, b Birmingham, Eng, Dec 29, 16; m 42; c 1. EMBRYOLOGY. *Educ:* Univ Cambridge, BA, 38, MB, BCh, 41, MA, 46, MD, 51, PhD, 76. *Prof Exp:* Demonstr anat, Univ Cambridge, 46-50, lectr, 50-51; sr lectr, St Bartholomew's Hosp Med Col, London, 51; reader, Univ London, 51-55; prof, Univ Khartoum, 55-64; from assoc prof to prof anat, 64-84, EMER PROF, COL MED, UNIV SASK, 84- *Mem:* Can Asn Anat. *Res:* Primatology; early development; implantation; placentation and reproductive cycle of the Prosimii. *Mailing Add:* Dept Anat Univ Sask Saskatoon SK S7N 0W0 Can

BUTLER, HERBERT I, b New York, NY, July 24, 14; m 40; c 2. PHYSICS, ELECTRICAL ENGINEERING. *Educ:* Monmouth Col, BS, 59, MAT, 76. *Prof Exp:* Chief engr, Bruno Labs, Inc, NY, 38-41; proj & equip develop engr, US Army Signal Res & Develop Labs, 41-54, chief data processing sect, 54-57, mem spec task force, Proj Cyclops, 57-58, Signal Corps tech rep, US Army Satellite Prog, 58-60, Tiros Proj mgr, 60, chief aoc chief projs, Aeronomy & Meteorol Div, Goddard Space Flight Ctr, NASA, 61-65, chief oper satellites off, 66-69, staff asst meteorol planning, 69-72; consult, Herbert Butler-Aerospace, 72-; RETIRED. *Concurrent Pos:* Chmn subcomt meteorol satellites, Intdept Comt Appl Meteorol Res, 63- *Mem:* AAAS; Am Inst Aeronaut & Astronaut; sr mem Inst Elec & Electronics Engrs; Planetary Soc. *Res:* Development of meteorological satellite systems and radio direction finding equipment. *Mailing Add:* Meadow Lakes No 501-L Hightstown NJ 08520

BUTLER, HOWARD W(ALLACE), b Cranston, RI, Oct 14, 16; m 41; c 4. MECHANICAL ENGINEERING. *Educ:* Univ RI, BS, 40; Yale Univ, MEng, 52, DEng, 58. *Prof Exp:* Instr mech eng, Univ Conn, 41-44; asst dir res, New Britain Mach Co, 44-47; from asst prof to assoc prof, Univ Conn, 47-57; prof, Rensselaer Polytech Inst, 57-65; prof & chem dept, WVa Univ, 65-78; prof mech eng, Brown Univ, 77; RETIRED. *Concurrent Pos:* Consult, New Britain Mach Co, 47-50, Am Thermos Bottle Co & Raymond Eng Labs, 49, Whiton Mach Co, 51-53, Eastern Indust, Inc, 52-53, Terry Stream Turbine Co, 55 & 58-59, Liberty Mutual Life Ins Co, 61, 66, 69 & 71, Sikorsky Aircraft Div, 63, Kaman Aircraft, 64 & FMC Corp, 68. *Mem:* AAAS; Am Soc Mech Engrs; Am Soc Eng Educ; Sigma Xi. *Res:* Irreversible thermodynamics; turbulence; gas dynamics; underground coal gasification. *Mailing Add:* 2175 Boston Neck Rd Saunderstown RI 02874

BUTLER, HUGH C, b Helena, Mont, Jan 7, 25; m 48; c 3. SURGERY, PHYSIOLOGY. *Educ:* Mont State Col, BS, 50; Wash State Univ, DVM, 54, MS, 68; Am Col Vet Surgeons, dipl. *Prof Exp:* Asst state veterinarian, Livestock Sanit Bd, Mont, 54-55; from instr to assoc prof clin med & surg, Wash State Univ, 55-64, assoc prof physiol & pharmacol, 64-66; head dept surg, Animal Med Ctr, New York, 66-68; prof surg, Kans State Univ, 68-86; RETIRED. *Concurrent Pos:* Vis scientist exp surg, Mayo Clin, 63. *Honors & Awards:* McLean Mem Award. *Mem:* Am Vet Med Asn; Am Col Vet Surgeons. *Res:* Experimental surgery with emphasis on cardiovascular orthopedic and tissue transplantation. *Mailing Add:* 290 N Miny Lakes Talspell MT 59905

BUTLER, IAN SYDNEY, b Newhaven, Eng, Aug 22, 39; m 66; c 4. INORGANIC CHEMISTRY. *Educ:* Bristol Univ, BSc, 61, PhD(inorg chem), 65. *Prof Exp:* Res assoc, Ind Univ, 64-65 & Northwestern Univ, 65-66; from asst prof to assoc prof, 66-75, PROF CHEM, McGILL UNIV, 75-, DIR GRAD STUDIES, 85- *Concurrent Pos:* Co-ed, Can J Spectros, 72-86, sr ed, 86-; vis prof, CNRS Lab Chem Coord, 84 & Univ Montreal, 77. *Mem:* Am Chem Soc; Fel Chem Inst Can; Spectros Soc Can; The Chem Soc; Sigma Xi. *Res:* Synthesis and studies of the physical properties of transition metal organometallic compounds, particularly metal carbonyls; infrared, nuclear magnetic resonance and laser Raman spectroscopy. *Mailing Add:* Dept Chem McGill Univ 801 Sherbrooke St W Montreal PQ H3A 2K6 Can

BUTLER, JACK F, b El Centro, Calif, July 18, 33; m 59; c 4. SOLID STATE PHYSICS, ELECTRICAL ENGINEERING. *Educ:* Univ Calif, BS, 59, MS, 60, PhD(elec eng), 63. *Prof Exp:* Assoc elec eng, Univ Calif, 61-62; staff mem solid state physics, Lincoln Lab, Mass Inst Technol, 62-68; staff scientist, Gen Dynamics Pomona Div, 68-71; sr scientist, Arthur D Little, Inc, 71-74; pres, Laser Analytics, Inc, 74-80; pres, Butler Res & Eng, Inc, 80-85; PRES, SAN DIEGO SEMICONDUCTORS, INC, 85- *Mem:* Am Phys Soc; Inst Elec & Electronics Engrs; AAAS. *Res:* Applied research in the optical and electronic properties of semiconductors; lead-tin-chalcogenide tunable infrared lasers; photodetectors; nuclear detectors. *Mailing Add:* PO Box 1333 Rancho Santa Fe CA 92067

BUTLER, JACKIE DEAN, b Raleigh, Ill, Mar 1, 31; m 57; c 2. HORTICULTURE. *Educ:* Univ Ill, Urbana, BS, 57, MS, 59, PhD, 66. *Prof Exp:* Asst county farm adv agr, Univ Ill, Urbana, 57-58, from res asst to res assoc hort, 58-63, from instr to assoc prof, 63-71; assoc prof, 71-76, PROF HORT, COLO STATE UNIV, 76- *Mem:* Am Soc Hort Sci; Am Soc Agron. *Res:* Principles and practices associated with turfgrass establishment and maintenance. *Mailing Add:* PO Box 416 Bay Center WA 98527

BUTLER, JAMES EHRICH, b Tenafly, NJ, Nov 29, 44; m 69. CHEMICAL PHYSICS. *Educ:* Mass Inst Technol, BS, 66; Univ Chicago, PhD(chem physics), 72. *Prof Exp:* Fel chem physics, NIH, Univ Chicago, 72-74; res assoc, James Franck Inst, Univ Chicago, 74-75; RES CHEMIST, CHEM DIV, US NAVAL RES LAB, 75- *Mem:* Am Phys Soc; Optical Soc Am; Japan Chem Soc. *Res:* Photochemistry; chemical kinetics; chemical vapor reposition; surface vibrational spectroscopy; diamond film growth. *Mailing Add:* Chem Div Code 6174 Naval Res Lab Washington DC 20375

BUTLER, JAMES HANSEL, b Canton, Ohio, Dec 7, 36; m 61; c 3. DENTISTRY. *Educ:* Denison Univ, BA, 58; Ohio State Univ, DDS, 62; Univ Rochester, MS, 67; Eastman Dent Ctr, Rochester, NY, cert periodont, 67. *Prof Exp:* Assoc prof periodont, assoc mem grad fac & mem hosp staff, Univ Minn, 67-74; assoc prof occlusion & chmn dept, 74-81, prof gen dent & chmn dept, 81-88, PROF PERIODONT, SCH DENT, VA COMMONWEALTH UNIV-MED COL VA, 88- *Concurrent Pos:* Consult, US Army, Ft Meade, Md, 75-82, Vet Admin Hosp, Richmond, 75-, J Am Dent Asn, 76-, Asn Am Med Cols, 76-80, J Prosthetics Dent, 78- & J Dent Educ, 80- *Mem:* Am Dent Asn; Am Acad Periodont; Int Asn Dent Res; Am Asn Dent Schs; fel Int Col Dentists; Sigma Xi. *Res:* Occlusion; mandibular dysfunction; biofeedback; relaxation therapy. *Mailing Add:* Dept Periodont Va Commonwealth Univ Sch Dent 520 N 12th St Richmond VA 23298

BUTLER, JAMES JOHNSON, b Jackson, Tenn, May 5, 26; m 52; c 2. PATHOLOGY. *Educ:* Univ Mich, MD, 52. *Prof Exp:* Intern, Cincinnati Gen Hosp, Ohio, 52-53; resident path, Univ Iowa & Univ Cincinnati, 53-57; from jr pathologist to assoc pathologist, Armed Forces Inst Path, 57-61; asst pathologist, 59-60 & 62-64, assoc pathologist, 64-74, PATHOLOGIST, UNIV TEX M D ANDERSON HOSP & TUMOR INST HOUSTON, 74-, PROF PATH, 75- *Concurrent Pos:* Mem path panel, Lymphoma Clin Trials. *Mem:* AAAS; AMA; Am Soc Hemat; Int Acad Path. *Res:* Path-physiology of the reticuloendothelial system; cancer research. *Mailing Add:* Univ Tex M D Anderson Hosp Houston TX 77030

BUTLER, JAMES KEITH, b Temple, Tex, Sept 26, 26; m 49; c 1. HOLOCRINE SECRETION, HISTOTECHNIQUE. *Educ:* Univ Tex, BS, 50, MA, 52, PhD(zool), 61. *Prof Exp:* Asst anat, Med Br, Univ Tex, 52-53, res assoc orthop surg, 53-55; ASSOC PROF CELL BIOL, UNIV TEX, ARLINGTON, 60- *Concurrent Pos:* Genetics Found fel, Univ Tex, Austin, 65; vis assoc prof anat, Univ Tex Southwestern Med Sch, Dallas, 67; vis scientist, Harvard Sch Pub Health, 74. *Mem:* AAAS. *Res:* Holocrine secretory mechanisms in invertebrate endocrine and reproductive systems; invention and development of an instrument system for preparing specimens for high resolution light microscopical histochemistry. *Mailing Add:* Dept Biol Univ Tex Box 19498 Arlington TX 76019

BUTLER, JAMES LEE, b Sevierville, Tenn, Jan 8, 27; m 48; c 3. AGRICULTURAL ENGINEERING. *Educ:* Univ Tenn, BS, 50, MS, 51; Mich State Univ, PhD(agr eng), 58. *Prof Exp:* Asst agr engr, Ga Exp Sta, 51-56, 58-59; asst, Mich State Univ, 56-58; assoc agr engr, Ga Exp Sta, 59-60; proj leader, Harvesting & Farm Processing Res Br, Sci & Educ Admin-Agr Res, USDA, 60-62, invests leder, Southern Region, Ga-SC Area, Agr Eng Res Div, 62-72, res leader & tech adv, 71-79, MGR, S&E SOUTHERN AGR ENERGY CTR, 79- *Concurrent Pos:* Consult to Minister Agr, El Salvador, Cent Am, 64; consult to Secy Agr, Repub Philippines & USAID, 70; consult, Ministry of Agr, Govt Pakistan, 71. *Honors & Awards:* Bailey Award, Am Peanut Res & Educ Asn, 78. *Mem:* Am Soc Agr Engrs; Am Peanut Res & Educ Asn (pres, 81-82); Int Solar Energy Soc; Sigma Xi. *Res:* Engineering fundamentals related to harvesting and processing forage; effect of plant properties, environmental conditions, harvesting mechanisms and procedures on peanut quality and mycotoxin development. *Mailing Add:* Ga Coastal Planin Exp Sta USDA PO Box 748 Tifton GA 31793

BUTLER, JAMES NEWTON, b Cleveland, Ohio, Mar 27, 34; m 57, 66; c 3. PHYSICAL CHEMISTRY, ENVIRONMENTAL & MARINE SCIENCES. *Educ:* Rensselaer Polytech Inst, BSc, 55; Harvard Univ, PhD(chem physics), 59. *Prof Exp:* Fel chem, Harvard Univ, 59; from instr to asst prof, Univ BC, 59-63; sr scientist chem physics, Tyco Labs, Inc, 63-66, head phys chem dept, 66-71; GORDON McKAY PROF APPL CHEM, HARVARD UNIV, 71-, MEM FAC EARTH & PLANETARY SCI, 72- *Concurrent Pos:* Lectr, Harvard Univ, 70-71; mem comt oceanog, 72-; trustee, Bermuda Biol Sta, 72-, vpres, 85-86, pres, 86-89; consult various companies & agencies, 68-; mem comt, Petrol Marine Environ, Nat Res Coun-Nat Acad Sci, 73-75, Environ Decision Making, 75-76 & Petrol Marine Environ, 80-82; NSF fac sci fel, 77; chmn, Comt on Effectiveness of Oil spill Dispersants, Nat Res Coun-Nat Acad Sci, 85-88. *Mem:* Am Soc Limnol & Oceanog; AAAS; Am Chem Soc. *Res:* chemical oceanography, oil pollution; environmental quality. *Mailing Add:* Div of Appl Sci Harvard Univ Pierce Hall Cambridge MA 02138

BUTLER, JAMES PRESTON, b Boston, Mass, Aug 19, 45; m 73; c 2. BIOMECHANICS, STEREOLOGY. *Educ:* Pomona Col, BA, 67; Harvard Univ, MA, 68, PhD(physics), 74. *Prof Exp:* From asst prof to assoc prof, 78-89, ADJ ASSOC PROF BIOMATH, HARVARD UNIV SCH PUB HEALTH, BOSTON, 89-; SR SCIENTIST, BIOMECH INST, BOSTON, 89- *Concurrent Pos:* Henry Luce Found scholar, 74. *Mem:* Math Asn Am. *Res:* Biomechanics of airflow in mammalian and avian lungs; light scattering; stereology; intracellular magnetometry; applied mathematics; endocrinology. *Mailing Add:* 176 Coolidge St Brookline MA 02146

BUTLER, JAMES ROBERT, b Macon, Ga, Apr 17, 30; m 89; c 3. GEOLOGY, PETROLOGY. *Educ:* Univ Ga, BS, 52; Univ Colo, MS, 55; Columbia Univ, PhD(geol), 62. *Prof Exp:* Lectr geol, Columbia Univ, 59-60; from asst prof to assoc prof, 60-72, PROF GEOL, UNIV NC, CHAPEL HILL, 72- *Mem:* Fel Geol Soc Am; Mineral Soc Am; Am Asn Geol Teachers; Am Geophys Union. *Res:* Igneous, structural and metamorphic evolution of the crystalline southern Appalachians; petrology of Precambrian metamorphic rocks, Blue Ridge, North Carolina and Tennessee; differentiation and textural development of igneous rocks. *Mailing Add:* Dept Geol Box 3315 Univ NC Chapel Hill NC 27599-3315

BUTLER, JERRY FRANK, b Lingle, Wyo, Apr 17, 38; m 61; c 2. VETERINARY ENTOMOLOGY, MEDICAL ENTOMOLOGY. *Educ:* Univ Wyo, BS, 62, MS, 64; Cornell Univ, PhD(entom), 68. *Prof Exp:* Res asst entom, Univ Wyo, 62-64; res asst, Cornell Univ, 64-68, res technician, 68; asst prof, 68-74, assoc prof, 74-79, PROF ENTOM, UNIV FLA, 79- *Mem:* Entom Soc Am; Acarology Soc Am. *Res:* Bionomics and control of arthropods of veterinary importance; special interest in population dynamics of ectoparasites. *Mailing Add:* Dept Entom & Nematol Univ Fla One Melab Bldg Gainesville FL 32601

BUTLER, JOHN, b Grantham, Eng, Nov 28, 23; nat US; m 60, 80; c 5. MEDICINE. *Educ:* Birmingham Univ, MB, ChB, 46, MD, 57; FRCP(E), 69. *Prof Exp:* Lectr med, Birmingham Univ, 56-58; sr lectr & consult physician, Manchester Univ, 58-60; assoc clin prof & lectr, Cardiovasc Res Inst, San Francisco, 60-65; PROF & HEAD DIV RESPIRATORY DIS, UNIV WASH, 65- *Mem:* Am Physiol Soc; Am Fedn Clin Res; Brit Med Res Soc. *Res:* Mechanical forces in the lungs; pulmonary circulation. *Mailing Add:* Dept Med Div Respiratory Dis Univ Wash MS Rm-12 Seattle WA 98195

BUTLER, JOHN BEN, JR, b New York, NY, Aug 26, 23; m 46; c 2. MATHEMATICS. *Educ:* Swarthmore Col, BS, 45; NY Univ, MS, 47; Univ Calif, PhD, 54. *Prof Exp:* Instr math, Univ Calif, 55-56; asst prof, Univ Wash, 57-59 & Univ Ariz, 59-60; assoc prof, 61-65, PROF MATH, PORTLAND STATE UNIV, 66- *Concurrent Pos:* AID prof, Middle East Tech Univ, Ankara, 69-70. *Honors & Awards:* Fulbright lectr, Robert Col, Istanbul, 65. *Mem:* Soc Indust & Appl Math; Am Math Soc. *Res:* Abstract analysis; differential equations. *Mailing Add:* 6815 SW 14th Ave Portland OR 97219

BUTLER, JOHN C, b Port Clinton, Ohio, Oct 31, 41; m 65; c 2. GEOLOGY, MINERALOGY. *Educ:* Miami Univ, BA, 63, MS, 65; Ohio State Univ, PhD(mineral), 68. *Prof Exp:* Instr geol, Miami Univ, 66-68; asst prof, 68-71, assoc prof, 71-80, PROF GEOL, UNIV HOUSTON, 80-, CHMN DEPT. *Mem:* Mineral Soc Am. *Res:* Structural states of alkali feldspars; mineralogy of contact metamorphosed marbles; size analysis of lunar regolith. *Mailing Add:* Dept Geosci Univ Houston Houston TX 77204-5503

BUTLER, JOHN E, b Rice Lake, Wis, Jan 10, 38; c 1. IMMUNOCHEMISTRY, MOLECULAR IMMUNOLOGY. *Educ:* Univ Wis River Falls, BS, 61; Univ Kans, PhD(zool & biochem), 66. *Prof Exp:* Park ranger, Nat Park Serv, 61-63; res biologist, USDA, 67-71; from asst prof to assoc prof, 71-80, PROF MICROBIOL, UNIV IOWA, 80- *Concurrent Pos:* Res scientist, Anton Bruun Oceanog Exped, 67; vis prof, Max Planck Inst, WGer, 73-74; Fogarty Int sr fel, Inst Animal Sci, Ger, 82-83. *Mem:* Am Asn Immunologists; NY Acad Sci. *Res:* Immunochemistry of solid-phase immunoassay; swine immunoglobulins and immunoglobulin genes; bovine immunoglobulins; maternal-neonatal immune regulation. *Mailing Add:* 3-450 Bowen Sci Bldg Univ Iowa Iowa City IA 52242

BUTLER, JOHN EARL, b Stockton, Kans, Sept 7, 18; m 75; c 2. BOTANY. *Educ:* Ft Hayes Kans State Col, AB, 40; Univ Wis, MS, 49, PhD(bot), 54. *Prof Exp:* With AEC, Univ Calif, 48-51; teacher bot & agron, Wis State Col, 54-55; asst prof, Fresno State Col, 55-58; assoc prof bot, 58-71, PROF BIOL & SCI EDUC, HUMBOLDT STATE UNIV, 71- *Mem:* Ecol Soc Am; Phycol Soc Am; Nat Asn Biol Teachers. *Res:* Ecology of life histories; secondary science education; conservation survival education. *Mailing Add:* Dept Biol Humboldt State Univ Arcata CA 95521

BUTLER, JOHN EDWARD, b Rice Lake, Wis, Jan 10, 38; div; c 1. IMMUNOLOGY. *Educ:* Univ Wis-River Falls, BS, 61; Univ Kans, PhD(zool), 66. *Prof Exp:* Actg asst prof zool, Univ Kans, 65-66, res assoc immunobiol, 66-67; res biologist, Agr Res Serv, USDA, 67-71; from asst prof to assoc prof, 71-80, PROF MICROBIOL, UNIV IOWA, 80- *Concurrent Pos:* USPHS fel & NSF fel, 64-66; vis prof, Max Planck Inst, WGer, 73-74; Fogarty Int Sr fel, WGer, 82-83; dir, Iowa Biotechnology Training Prog, 84- *Mem:* Am Asn Immunologists; NY Acad Sci. *Res:* Bovine and swine immunology; secretory immunity; enzyme-linked immunosorbent assay; immunochemistry; maternal-neonatal immune regulation. *Mailing Add:* Dept Microbiol Univ Iowa Sch Med Iowa City IA 52241

BUTLER, JOHN F(RANCIS), b McKeesport, Pa, June 26, 32; m 61; c 4. PHYSICAL METALLURGY. *Educ:* Carnegie Inst Technol, BS, 54, MS, 57, PhD(metall eng), 58. *Prof Exp:* Sr res engr, Jones & Laughlin Steel Corp, 58-63, res assoc phys metall & vacuum coating, 63-69, res supvr vacuum vapor coating, 69-72, res supvr, alloy design, 72-73, dir, Prod Metall, Graham Res Lab, 73-84; DIR, PROD DEVELOP, LTV STEEL CORP, 84- *Concurrent Pos:* Vis scientist, Fr Iron & Steel Res Inst, 62-63; instr, Carnegie Inst Technol, 63-64. *Mem:* Am Soc Metals; Am Inst Mining, Metall & Petrol Engrs; Soc Automotive Engrs. *Res:* Ferrous physical metallurgical research in the areas of yielding, strain aging and quench aging; development work in vacuum metallic, inorganic and organic coating of continuous sheet material; alloy design of high strength and corrosion resistant steels; management of steel product research and development programs. *Mailing Add:* Jones/Laughlin Steel Corp 900 Agnew Rd Pittsburgh PA 15263

BUTLER, JOHN JOSEPH, b Rochester, NY, Oct 18, 20; m 48; c 5. HEMATOLOGY, MEDICAL ONCOLOGY. *Educ:* Univ Toronto, BA, 42; Univ Rochester, MD, 44; Am Bd Internal Med, dipl, 51, dipl hemat, 76, dipl med oncol, 81. *Prof Exp:* From instr to asst prof med, Georgetown Univ, 52-59; assoc prof, Seton Hall Col Med & Dent, 59-66; CHIEF HEMAT, CATH MED CTR BROOKLYN & QUEENS, 66-; ASSOC CLIN PROF MED, NYU MED SCH, 74- *Mem:* Am Col Physicians; Am Fedn Clin Res; Am Soc Hemat; Int Soc Hemat; Am Soc Clin Oncol. *Res:* Red and white cell metabolism. *Mailing Add:* 152-11 89th Ave Jamaica NY 11432

BUTLER, JOHN LOUIS, b Brockton, Mass, Aug 23, 34; m 59; c 5. PHYSICS. *Educ:* Northeastern Univ, BS, 57, PhD(acoust), 67; Brown Univ, ScM, 62. *Prof Exp:* Res engr, Melpar Inc, 61-62; sr engr, Harris ASW Div, Gen Instrument Corp, 62-66; res assoc physics & acoust, Parke Math Labs, Inc, 66-70; develop engr acoust, Massa Div Dynamics Corp of Am, 70-72; sr engr, Raytheon Co, Submarine Signal Div, 72-75; PRES, IMAGE ACOUST, INC, 75- *Concurrent Pos:* Lectr, Northeastern Univ, 68, 70, 73 & 75. *Mem:* Fel Acoust Soc Am; Audio Eng Soc. *Res:* Acoustical radiation from underwater sound sources and arrays; acoustic transducers and arrays. *Mailing Add:* Image Acoust Inc 97 Elm St Cohasset MA 02025

BUTLER, JOHN MANN, b Richmond, Va, Mar 23, 17; m 42; c 2. ORGANIC CHEMISTRY, POLYMER CHEMISTRY. *Educ:* Richmond Univ, BS, 37; Ohio State Univ, PhD(org chem), 40. *Prof Exp:* Res chemist, Bakelite Corp, NJ, 40-41; res chemist, Monsanto Co, 41-45, group leader, 45-53, sect leader, 53-54, asst dir res, 54-59, res specialist, 59-61, mgr org & polymer res, Monsanto Res Corp, 61-78; SR POLYMER SPECIALIST, RES INST, UNIV DAYTON, 78- *Mem:* Am Chem Soc; Soc Plastics Engrs; Sigma Xi. *Res:* Polymer synthesis and application; organic synthesis. *Mailing Add:* 7820 Normandy Lane Dayton OH 45459

BUTLER, JON TERRY, b Baltimore, Md, Dec 26, 43; m 68; c 1. DIGITAL LOGIC DESIGN, SYSTEM DIAGNOSIS. *Educ:* Rensselaer Polytech Inst, BEE, 66, MEng, 67; Ohio State Univ, PhD(elec eng), 73. *Prof Exp:* USAF capt, Computer Eng, Air Force Avionics Lab, Wright-Patterson AFB, Ohio, 67-70, Nat Res Coun postdoctoral, 73-74; assoc prof, Northwestern Univ, Evanston, 74-87; PROF COMPUTER ENG, NAVAL POSTGRAD SCH, 87- *Concurrent Pos:* Chmn, Mult-Valued Logic Tech Comt, Inst Elec & Electronics Engrs, Computer Soc, 80-81, mem, distinguished vis, 82-85, vchair, Tech Activ Bd, 87-88, chmn, Mag Adv Comt, 90-91; ed, Inst Elec & Electronic Engrs Trans Computers, 82-86, ed-in-chief, Computer, 91-; Navalex chair prof, Naval Postgrad Sch, 85-87; prin investr, NSF, USN, USAF, NATO & Sigma Xi. *Honors & Awards:* Meritorious Serv, Inst Elec & Electronics Engrs, Computer Soc, 88, TAB Pioneer Award, 89. *Mem:* Fel Inst Elec & Electronics Engrs; Inst Elec & Electronics Engrs, Computer Soc. *Res:* Computer hardware; multiple-valued logic and reliable computing; development of the first multiple-valued programmable logic array, the first computer-aided design tool for multiple-valued logic, and discovery of a flaw in almost all commercially available programmable logic arrays; reliable multiprocessing systems; binary logic design; cellular automata; analysis of computer programs. *Mailing Add:* Dept Elec & Computer Eng Naval Postgrad Sch Code EC/BU Monterey CA 93943-5004

BUTLER, KARL DOUGLAS, SR, b Douglas, Ariz, Feb 4, 10; m; c 5. AGRICULTURE. *Educ:* Univ Ariz, BS, 31, MS, 33; Cornell Univ, PhD(plant sci), 40. *Prof Exp:* Asst plant path, Iowa State Col, 33-34; jr forest pathologist, Div Forest Path, USDA, 34; res asst & instr plant path, Univ Ariz, 34-36, asst plant pathologist, 37-40; agent rubber dis invests, Bur Plant Indust, USDA, 40-43, dir res agr chem, Coop Grange League Fedn Exchange, Inc, 43-45, dir educ & res, 45-47; pres, Am Inst Coop, Washington, DC, 47-50; farm counr, 50-67; BUS & AGR CONSULT, 67- *Concurrent Pos:* Mem, Am Rubber Surv, Bolivia & Brazil, 40-41; columnist & consult, Syracuse Post Standard, 51-; pres, Comn Increased Industs Use Agr Prods, 56-57; USDA farm mech exchange deleg to USSR, 58; agr specialist, Peace Corps, Latin Am Countries, 63-64, US AID, Taiwan & India, 64 & WAfrican Studies for Afro-Am Labor Ctr, New York, 66; chmn exec comt & mem res comt, Chemurgic Coun; mem farmers exchange progs, Italy, India, Ireland & Latin Am; consult to Chief of Staff, USAF. *Honors & Awards:* Am Meat Inst Award, 63. *Mem:* Fel AAAS; Am Inst Animal Agr (secy); Am Soc Agr Engrs. *Res:* Farmer cooperatives; diseases of cotton, dates, lettuce, trees, cereal crops and watermelons. *Mailing Add:* 1583 E Shore Dr Ithaca NY 14850

BUTLER, KATHARINE GORRELL, b Chicago Heights, Ill, March 15, 25; m 44; c 3. SPEECH & LANGUAGE PATHOLOGY. *Educ:* Western Mich Univ, BA, 50, MA, 53; Mich State Univ, PhD(speech & hearing sci), 67. *Prof Exp:* Asst prof psychol, Western Mich Univ, 61-64; asst prof speech path & audiol, San Jose State Univ, 64-68, assoc prof, 68-69, chmn, Dept Spec Educ, 69-74, prof & dir, Speech & Hearing Ctr, 69-77, assoc dean grad studies & res, 75-77, actg dean, 77-79; dir div spec educ & rehab, 79-83, DIR CTR RES, SYRACUSE UNIV, 83- *Concurrent Pos:* Ed, Topics Language Dis, 79-; chair, Joint Nat Comt Learning Disabilities, 81; ed, Topics Language Dis, 79- *Mem:* Fel Am Speech-Language-Hearing Asn (pres, 77-78); Am Psychol Asn; Int Asn Logopedics & Phoniatrics (pres, 83-86, 86-89); Nat Asn Hearing & Speech Action (pres, 79-87). *Res:* Language processing of children and adults suffering from language disorders, including children 3 to 6 who exhibit semantic processing disorders. *Mailing Add:* Syracuse Univ 805 S Crouse Ave Syracuse NY 13244-2280

BUTLER, KEITH WINSTON, b St Boniface, Man, Apr 12, 41; m 66; c 2. BIOPHYSICS. *Educ:* Univ Toronto, BSA, 63; Duke Univ, PhD(physiol), 68. *Prof Exp:* Univ fel, Aarhus Univ, 68-69; Nat Res Coun Can fel, 69-71, vis scientist, 71-72, asst res officer biophys, 72-77, asst res officer, 77-86, SR RES OFFICER, NAT RES COUN CAN, 86- *Res:* Electron spin resonance and nuclear magnetic resonance studies of metabolism. *Mailing Add:* Biol Sci Div Nat Res Coun Can Ottawa ON K1A 0R6 Can

BUTLER, LARRY G, b Elkhart, Kans, Dec 14, 33; m 53; c 4. INTERNATIONAL AGRICULTURE, CHEMICAL ECOLOGY. *Educ:* Okla State Univ, BS, 60; Univ Calif, Los Angeles, PhD(biochem), 64. *Prof Exp:* Asst prof natural sci & chmn dept, Los Angeles Baptist Col, 64-65; res assoc biochem, Univ Ariz, 65-66; PROF BIOCHEM, PURDUE UNIV, WEST LAFAYETTE, 66- *Concurrent Pos:* NIH res career develop award, 70-75; vis prof chem, Univ Ore, 74-75. *Mem:* Am Chem Soc; Am Soc Biol Chem; Int Soc Chem Ecol; Int Asn Plant Tissue Cult; Crop Sci Soc Am. *Res:* Chemistry and biochemistry of tannins and other plant phenolics and their nutritional effects; chemical ecology. *Mailing Add:* Dept Biochem Purdue Univ West Lafayette IN 47907

BUTLER, LEWIS CLARK, b Hornell, NY, July 11, 23; m 48; c 4. MATHEMATICS. *Educ:* Alfred Univ, BA, 44; Rutgers Univ, MS, 48; Univ Ill, PhD(math), 57. *Prof Exp:* Instr math, Alfred Univ, 47-49; asst, Univ Ill, 49-54; instr, Pa State Univ, 54-57; from asst prof to assoc prof, State Univ NY Col Ceramics, Alfred Univ, 57-62; DEAN GRAD SCH, ALFRED UNIV, 63- *Concurrent Pos:* Dir cent inst math, Concepcion Univ, 61-63. *Mem:* Am Math Soc. *Res:* Algebraic topology; fiber spaces. *Mailing Add:* Eight Terrace St Alfred NY 14802

BUTLER, LILLIAN CATHERINE, ENDOCRINOLOGY, HORMONES. *Educ:* Univ Calif, Berkeley, PhD(nutrit), 53. *Prof Exp:* Fel, NIH, 50-55, endocrinol res, 55-64; diabetes res, Vet Admin Hosp, Birmingham, 64-66; assoc prof nutrit & biochem, Trace mineral res, Univ Md, 67-78; RETIRED. *Mailing Add:* 9125 E Indian Hills Tucson AZ 85749

BUTLER, LILLIAN IDA, b Saginaw, Mich, Jan 31, 10. ANALYTICAL CHEMISTRY, BIOLOGICAL CHEMISTRY. *Educ:* Univ Mich, BS, 29, MS, 30. *Prof Exp:* Asst chem, Mich State Col, 30-42; assoc human nutrit & home econ, USDA, 42-43, chemist, 43-47, Div Bee Cult, 47-51, pesticides chem res, 51-53, asst head chemist, 53-58, actg in charge, 57-68, res chemist & asst head, 58-81; RETIRED. *Concurrent Pos:* Mem, Gov Status of Women Comn, Wash, 63-65; collabr, USDA, Agr Res Serv. *Mem:* Fel AAAS; Am Chem Soc; fel Am Inst Chem; Entom Soc Am. *Res:* Analytical methods, primarily in foods and crops; inorganic minerals; vitamins; insecticides; antibiotics and virus; development of methods; use of attractants of insects in monitoring of flights in field for insect control; isolation of sex attractants from insects. *Mailing Add:* 2804 W Arlington Yakima WA 98902

BUTLER, LINDA, b Martin, Tenn, Nov 11, 43. ENTOMOLOGY. *Educ:* Univ Ga, BS, 65, MS, 66, PhD(entom), 68. *Prof Exp:* From asst prof to assoc prof, 68-75, PROF ENTOM, WVA UNIV, 75- *Honors & Awards:* J Everett Bussart Mem Award, Entom Soc Am, 74. *Mem:* AAAS; Entom Soc Am; Lepidopterists Soc. *Res:* Insect pest management; lepidoptera; forest entomology. *Mailing Add:* Dept Plant Sci Box 6108 WVa Univ Morgantown WV 26506

BUTLER, LOUIS PETER, b Baltimore, Md, Oct 9, 53; m 77; c 1. ELECTRONIC WARFARE SOFTWARE SYSTEMS, ENGINEERING MANAGEMENT. *Educ:* Univ Md, College Park, BS, 75; George Washington Univ, MS, 81. *Prof Exp:* Apprentice engr, ILSD, 74, engr, 75-88, SUPVRY ENGR, ESG-BWI, WESTINGHOUSE ELEC, 88- *Concurrent Pos:* Consult, Onal's Agency, 81-88. *Mem:* Am Inst Aeronaut & Astronaut; Asn Old Crows. *Res:* Electronic warfare software systems with a transition of these software systems into commercial venues for application and production utilization in national and international programs. *Mailing Add:* PO Box 746 MS 432B Baltimore MD 21203-0746

BUTLER, MARGARET K, b Evansville, Ind, Mar 7, 24; m 51; c 1. MATHEMATICS. *Educ:* Ind Univ, AB, 44. *Prof Exp:* Statistician, US Bur Labor Statist, 45-46 & US Air Forces in Europe, 46-48; mathematician, Argonne Nat Lab, 48-49; statistician, US Bur Labor Statist, 49-51; mathematician, 51-80, DIR, NAT ENERGY SOFTWARE CTR, 60-, SR COMPUTER SCIENTIST, 80- *Concurrent Pos:* Adv comt comput info systs, Col DuPage, 87-; chair, Prog Comt Nat Comput Conf, 87; indust adv bd comput sci, Bradley Univ. *Mem:* Fel Am Nuclear Soc; Asn Comput Mach; AAAS; Inst Elec & Electronics Engrs Comput Soc; Asn Women Sci; Asn Women Comput. *Res:* Information systems; computer program interchange; computer system performance; documentation standards; scientific and engineering applications. *Mailing Add:* Nat Energy Software Ctr Argonne Nat Lab 9700 S Cass Ave Argonne IL 60439

BUTLER, MICHAEL ALFRED, b Chesterfield, Eng, Nov 24, 43; m 66; c 2. SOLID STATE PHYSICS. *Educ:* Rensselaer Polytech Inst, BS, 64; Univ Calif, Santa Barbara, MA, 66, PhD(physics), 69. *Prof Exp:* Res fel, Univ Calif, Santa Barbara, 69-70; mem tech staff, Bell Tel Labs, Inc, NJ, 70-75; MEM TECH STAFF, SANDIA LABS, 75- *Mem:* Am Phys Soc; Electrochem Soc. *Res:* Development of new sensor technology based on the interaction of light with matter. *Mailing Add:* Div 1163 Sandia Nat Labs Albuquerque NM 87185

BUTLER, OGBOURNE DUKE, JR, b Orange, Tex, Sept 29, 18; m 43; c 3. MEATS RESEARCH, DIET-HEALTH ISSUE. *Educ:* Agr & Mech Col Tex, BS, 39, MS, 47; Mich State Univ, PhD(animal husb), 53. *Prof Exp:* From instr to assoc prof, Tex A&M Univ, 47-57, prof animal sci & head dept, 57-78, assoc vpres agr & renewable resources, 78-80, assoc dep chancellor agr, 80-86; RETIRED. *Concurrent Pos:* Tech adv, Am Nat Cattlemen's Asn, 61; mem, Coun Agr Sci & Technol, pres, 80. *Mem:* Fel Am Soc Animal Sci (pres, 71); Am Inst Food Technologists; Am Meat Sci Asn; Am Dairy Sci Asn. *Res:* Meats; microbiology; biochemistry; statistics; nutrition and its effects on the diet-health issue. *Mailing Add:* Four Buttonbush Ct Spring TX 77380

BUTLER, PHILIP ALAN, b Upper Montclair, NJ, May 15, 14; m 41; c 2. MARINE BIOLOGY. *Educ:* Northwestern Univ, BS, 35, PhD(zool), 40. *Prof Exp:* Fishery res biologist, US Fish & Wildlife Serv, 46-58, dir biol lab, 58-68, res consult, 68-70; res consult, Environ Protection Agency, 71-79; RETIRED. *Mem:* AAAS; Am Soc Limnol & Oceanog; Nat Shellfisheries Asn (secy-treas, 57-59, vpres, 59-61, pres, 61-63); Am Fisheries Soc. *Res:* Biology and economics of shellfish; effects of pesticides in estuaries. *Mailing Add:* 106 Matamoros Dr Gulf Breeze FL 32561

BUTLER, R(OGER) M(OORE), b London, Eng, Oct 14, 27; m 54; c 3. CHEMICAL ENGINEERING. *Educ:* Univ London, BSc, 48, PhD(eng), 51, DIC, 51. *Prof Exp:* Asst prof chem eng, Queen's Univ, Ont, 51-55; res chemist, Imp Oil Enterprises, Ltd, 55-64, sr res chemist, New Proj, 64-68, res adv, 68-69, asst to refinery mgr, 69-70, mgr bus & tech serv, Sarnia Refinery, 70-71; planning adv, Exxon Corp, 71-73; new technol mgr, Logistics Dept, Imp Oil Enterprises, Ltd, 73-74; scientific adv, New Energy Resources Dept, Imp Oil Ltd, 74-75; mgr, heavy crude recovery res, Esso Resources Can, 75-82 & dir tech progs, Alta Oil Sands Technol Res Authority, 82-83; ENDOWED CHMN, PETROL ENG, UNIV CALGARY, 83- *Honors & Awards:* R S Jane Award, Can Soc Chem Eng, 87; Distinguished lectr, Soc Petrol Engrs, 87-88. *Mem:* Fel Inst Chem Engrs UK; fel Chem Inst Can; fel Am Inst Chem Engrs; Sigma Xi; Soc Petrol Engrs; Can Inst Mining & Metall. *Res:* In-situ recovery of heavy crudes. *Mailing Add:* Seven Bayview Dr SW Calgary AB T2N 1N4 Can

BUTLER, RICHARD GORDON, b Toronto, Can, Mar 17, 43. ANIMAL PHYSIOLOGY. *Educ:* Univ Toronto, BSc, 65, MSc, 68; Australian Nat Univ, PhD(neurobiol), 72. *Prof Exp:* Fel physiol, Oxford Univ, 72-74; asst prof, 74-80, ASSOC PROF ANAT, MCMASTER UNIV, 80- *Mem:* Can Asn Anatomists; Am Asn Anatomists; Can Physiol Soc; Soc Neurosci. *Res:* Competitive reinnervation of mammalian muscle spindles following injury, by appropriate and inappropriate fusimotor neurones; development of muscle spindles in fetal and neonatal animals. *Mailing Add:* Dept Anat McMaster Univ 1200 Main St W Hamilton ON L8N 3Z5 Can

BUTLER, ROBERT ALLAN, b Pittsfield, Mass, Mar 29, 23; m 52; c 4. PSYCHOACOUSTICS. *Educ:* Univ Fla, BA, 47; Univ Chicago, PhD(psychol), 51. *Prof Exp:* Instr psychol, Univ Wis, 51-53; res psychologist, Walter Reed Army Hosp, 53-57; res assoc biopsychol, 57-65, assoc prof surg & psychol, 65-72, PROF SURG & BEHAV SCI, UNIV CHICAGO, 72-, CHMN, DEPT BEHAV SCI, 79- *Mem:* Am Otol Soc; fel Acoustical Soc Am; Asn Res Otolaryngol (pres, 75-76). *Res:* Psychophysics of hearing and the electrophysiology of the auditory system. *Mailing Add:* Dept Psychol Univ Chicago 5848 S University Ave Chicago IL 60637

BUTLER, ROBERT FRANKLIN, b Eugene, Ore, July 6, 46; m 67; c 1. PALEOMAGNETISM. *Educ:* Ore State Univ, BS, 68; Stanford Univ, MS, 70, PhD(geophys), 72. *Prof Exp:* Fel geophys, Univ Minn, Minneapolis, 72-74; from asst prof to assoc prof, 74-83, PROF GEOPHYS, UNIV ARIZ, 83- *Concurrent Pos:* Assoc ed, J Geophys Res, 78-81, Geol Soc Am Bull, 90-92. *Mem:* Am Geophys Union; AAAS; Geol Soc Am. *Res:* Paleomagnetic polarity stratigraphy of Cretaceous-lower Tertiary deposits; Mesozoic and Cenozoic apparent polar wander, plate tectonics and microplate tectonics. *Mailing Add:* Dept Geosci Bldg 77 Univ Ariz Tucson AZ 85721

BUTLER, ROBERT NEIL, b New York, NY, Jan 21, 27; m 50; c 3. GERONTOLOGY, PSYCHIATRY. *Educ:* Columbia Univ, BA, 49, MD, 53. *Hon Degrees:* DSc, Univ Southern Calif, 81. *Prof Exp:* Intern med, St Luke's Hosp, New York, 53-54; resident psychiat, Univ Calif, 54-55; res psychiatrist, NIMH, 55-62; clin prof psychiat, Sch Med, George Wash Univ 62-82; dir, Nat Inst Aging, NIH, 76-82; BROOKDALE PROF GERIAT & ADULT DEVELOP & CHMN DEPT, MT SINAI MED CTR, NEW YORK, 82- *Concurrent Pos:* Resident psychiat, NIMH, Chestnut Lodge, Md, 57-58, clin adminr psychother aged, 58-59, consult, 59-68; res psychiatrist, Wash Sch Psychiat, 62-; lectr, Wash Psychoanal Inst, 62-76; consult, St Elizabeth's Hosp, Washington, DC, 62-76; US Senate Comt Aging, 71-76 & NIMH; chmn, Washington, DC Adv Comt on Aging; mem bd, Nat Coun Aging; chmn comt aging, Group Advan Psychiat. *Honors & Awards:* Pulitzer Prize, 76; Leo Laks Award, 76. *Mem:* Inst Med-Nat Acad Sci; fel Am Psychiat Asn; fel Am Geriat Soc; Geront Soc. *Res:* Psychotherapy theory and technique; nature of health and disorder in the aged; aging; methods in psychiatry; creativity. *Mailing Add:* Dept Geriat & Adult Develop Mt Sinai Med Ctr One Gustave L Levy Pl New York NY 10029

BUTLER, RONALD GEORGE, b Oswego, NY, Feb 21, 51. AVIAN ECOLOGY, VERTEBRATE BEHAVIORAL ECOLOGY. *Educ:* State Univ NY, Oswego, BA, 73; Syracuse Univ & State Univ NY, Syracuse, PhD(zool), 78. *Prof Exp:* Res assoc behav ecol, State Univ NY Res Found, 77-78; assoc res scientist ecol & toxicol, Mt Desert Island Biol Lab, Maine, 77-81; asst prof, 81-85, assoc prof animal ecol, Dept Biol, Duquesne Univ, 85-86; ASSOC PROF ANIMAL ECOL, DEPT BIOL, UNIV MAINE AT FARMINGTON, 86- *Concurrent Pos:* Vis scientist, Polish Acad Sci, Antarctica, 80-81; res fel, Mem Univ, St John's, Nfld, 83-84. *Mem:* Am Ornithologists Union; Animal Behav Soc; Cooper Ornith Soc; Sigma Xi; Wilson Ornith Soc. *Res:* Ecology of marine birds; evolution and ecology of vertebrate behavior; effects of environmental pollutants on animal populations. *Mailing Add:* Dept Math & Sci Univ Maine Farmington ME 04938

BUTLER, STANLEY S, hydrology, hydraulic engineering; deceased, see previous edition for last biography

BUTLER, THOMAS ARTHUR, b Farmington, Ark, May 14, 19; m 45, 53; c 3. CHEMISTRY. *Educ:* Southwest Mo State Col, AB, 41; Iowa State Col, MS, 51. *Prof Exp:* Asst chem, Iowa State Col, 41-42, jr res chemist, 42-51; SUPT, OAK RIDGE NAT LAB, 51- *Mem:* Am Chem Soc. *Res:* Preparation of uranium compounds; separation of rare earth elements; radioisotope research and development. *Mailing Add:* 119 Dana Dr Oak Ridge TN 37830

BUTLER, THOMAS DANIEL, b Oklahoma City, Okla, Apr 19, 38; m 58; c 3. FLUID DYNAMICS. *Educ:* NMex Inst Mining & Technol, BS, 60; Univ NMex, MS, 64. *Prof Exp:* Staff mem group T-3, Los Alamos Nat Lab, 60-72, assoc group leader T-3, 72-73, alt group leader T-3, 73-79, group leader T-3 Fluid Dynamics, 80-84, DEP GROUP LEADER T-3, 85- *Concurrent Pos:* Consult, Battelle Columbus Labs, Durham Opers Off, 75-80. *Res:* Development of numerical methods to solve a wide variety of multidimensional, transient fluid dynamics problems. *Mailing Add:* Theoret Div Los Alamos Nat Lab Mail Stop B 216 Group T-3 Los Alamos NM 87545

BUTLER, THOMAS MICHAEL, b Pittsburgh, Pa, May 17, 48. PHYSIOLOGY, BIOCHEMISTRY. *Educ:* La Salle Col, BA, 70; Univ Pa, PhD(molecular biol), 74. *Prof Exp:* Fel, Univ Pa, 74-76; asst prof, 76-81, ASSOC PROF PHYSIOL, THOMAS JEFFERSON UNIV, 81- *Mem:* Am Physiol Soc; Biophys Soc. *Res:* Muscle energetics. *Mailing Add:* Dept Physiol Thomas Jefferson Univ 1020 Locust St Philadelphia PA 19107

BUTLER, THOMAS W(ESLEY), b Baltimore, Md, Oct 22, 38; m 61; c 2. APPLIED MECHANICS, MATERIALS SCIENCE. *Educ:* Johns Hopkins Univ, BES, 61; George Washington Univ, MSE, 65; Brown Univ, PhD(mech, mat), 69. *Prof Exp:* Mech engr, eng mech sect, Nat Bur Standards, 61-65; from asst prof to assoc prof, 68-77, dept chmn, 80-82, PROF MECH ENG, US NAVAL ACAD, 77- *Concurrent Pos:* Lectr, Johns Hopkins Univ, 69-; consult mat failures, 70- *Mem:* Am Soc Testing Mat; Am Soc Metals; Sigma Xi. *Res:* Analysis of materials failures; experimental stress analysis and continuum plasticity; fracture mechanics. *Mailing Add:* Dept Mech Eng US Naval Acad Annapolis MD 21402

BUTLER, VINCENT PAUL, JR, b Jersey City, NJ, Feb 16, 29. INTERNAL MEDICINE, IMMUNOCHEMISTRY. *Educ:* St Peter's Col, AB, 49; Columbia Univ, MD, 54. *Prof Exp:* Intern med, Presby Hosp, New York, 54-55, asst resident, 55-56 & 58-59; asst med & microbiol, Sch Med & Dent, Univ Rochester, 59-60, instr med, 60-61; vis fel, Dept Microbiol, 61-63, from asst prof to assoc prof, 63-74, PROF MED, COL PHYSICIANS & SURGEONS, COLUMBIA UNIV, 74-; ATTEND PHYSICIAN, PRESBY HOSP, NEW YORK, 74- *Concurrent Pos:* Helen Hay Whitney Found fel, 60-63; vis res fel microbiol, Col Physicians & Surgeons, Columbia Univ, 61-63; asst attend physician, First Med Div, Bellevue Hosp, 63-68; sr investr, Arthritis Found, 63-68; from asst to assoc attend physician, Presby Hosp, New York, 68-74; asst vis physician, Harlem Hosp, 68-88; res career develop award, Nat Heart, Lung & Blood Inst, 68-73; career scientist, Irma T Hirschl Found, 73-78; vis fel, Tumour Immunol Unit, Univ Col London, 79-80; mem, Immunol Sci Study Sect, NIH, 79-83, chmn, 80-83; mem, Res Comt, Arthritis Found, 86-91, Chmn, 89-91; mem bd trustees, 88-90. *Honors & Awards:* Joseph Mather Smith Prize, Col Physicians & Surgeons, Columbia Univ, 73. *Mem:* Asn Am Physicians; Am Asn Immunol; Am Heart Asn; Am Soc Clin Invest; Am Soc Pharmacol & Exp Therapeut; fel AAAS. *Res:* Immunochemical studies of rheumatoid factor; purine specific antibodies and their cross reactions with DNA; digoxin-specific antibodies; immunochemical studies of fibrinogen. *Mailing Add:* Dept Med Columbia Univ Col Phys & Surg New York NY 10032

BUTLER, WALTER CASSIUS, b Rifle, Colo, Feb 10, 10; m 34; c 2. MATHEMATICS. *Educ:* Colo Agr & Mech Col, BS, 32; Colo State Col, MA, 42. *Prof Exp:* Teacher high sch, Colo, 32-33 & 34-35; miner, 33-34; prin high sch, Ill, 35-38; supt sch, Wyo, 38-46; prof math, Ft Lewis Agr & Mech Col, 46-50; from asst prof to assoc prof, 50-74, EMER ASSOC PROF MATH, COLO STATE UNIV, 74- *Concurrent Pos:* Dir, Jr Engr Scientists' Summer Inst, 60- *Mem:* Am Soc Mech Eng; Math Asn Am. *Res:* Training of junior and senior high school mathematics teachers. *Mailing Add:* 612 W Magnolia St Ft Collins CO 80521

BUTLER, WALTER JOHN, b Cavan, Ireland, Oct 11, 36; US citizen; m 66; c 4. ELECTRICAL ENGINEERING, SOLID STATE PHYSICS. *Educ:* Dublin Univ, BA & BAI, 66; McMaster Univ, PhD(elec eng), 70. *Prof Exp:* Elec engr, 70-75, MGR ELECTRONIC CONTROLS PROG, GEN ELEC CO, 75- *Mem:* Sr mem Inst Elec & Electronics Engrs; Brit Inst Elec Engrs. *Res:* Solid state electronics; microcomputer applications; digital control systems. *Mailing Add:* Electronics Lab Electronics Park PO Box 4840 Syracuse NY 13221

BUTLER, WILLIAM ALBERT, b Independence, Kans, Jan 4, 22; m 44; c 3. PHYSICS. *Educ:* Univ Kansas City, BA, 42; Univ Ill, MS, 44, PhD(physics), 52. *Prof Exp:* Asst physics, Univ Ill, 46-52 & Carleton Col, 52-70; prof physics & head dept, Eastern Ill Univ, 70-87; RETIRED. *Concurrent Pos:* Vis assoc prof, Univ Calif, Berkeley, 60-61; vis scientist, Univ Ill, 67-68; res assoc, Alliance Save Energy, Washington, DC, 78. *Mem:* Am Phys Soc; Am Asn Physics Teachers; Sigma Xi. *Res:* Nuclear and solid state physics. *Mailing Add:* Dept Physics Eastern Ill Univ Charleston IL 61920

BUTLER, WILLIAM BARKLEY, b Ithaca, NY, Feb 10, 43; m 64, 86; c 2. BIOCHEMISTRY, CELL BIOLOGY. *Educ:* Miami Univ, BA, 64; Univ Wis-Madison, MS, 68, PhD(biochem), 69. *Prof Exp:* Fel oncol, McArdle Lab Cancer Res, Med Sch, Univ Wis-Madison, 69-74; res scientist biochem, Mich Cancer Found, 74-80; asst prof, Dept Biochem, Sch Med, Wayne State Univ, 80-88; ASSOC PROF, BIOL DEPT, INDIANA UNIV PA, 88- *Concurrent Pos:* Asst mem, Mich Cancer Found, 80-88. *Mem:* Am Soc Cell Biol; Am Asn Cancer Res. *Res:* Mechanism of steroid hormone action; breast cancer. *Mailing Add:* Dept Biol Indiana Univ Pa 114 Weyandt Hall Indiana PA 15705

BUTLER, WILLIAM H, b Tallassee, Ala, Aug 13, 43; m 67; c 2. SOLID STATE PHYSICS, THEORY OF MATERIALS. *Educ:* Univ Calif, San Diego, PhD(physics), 69. *Prof Exp:* Asst prof physics, Auburn Univ, 69-72; mgr comput planning, 84-85, MEM RES STAFF, OAK RIDGE NAT LAB, 72-, GROUP LEADER, 85- *Honors & Awards:* Award for Outstanding Res, Dept Energy. *Mem:* AAAS; fel Am Phys Soc; Mat Res Soc. *Res:* Theory of electronic states in metals and alloys; theory of disordered alloys; electron phonon interaction in transition metals; transport and superconductivity in metals. *Mailing Add:* Metals & Ceramic Div Oak Ridge Nat Lab Oak Ridge TN 37831-6114

BUTLER, WILLIAM T, b Boston, Mass, Aug 10, 32. MEDICINE. *Prof Exp:* PRES, BAYLOR COL MED, HOUSTON, TEX, 79- *Mem:* Inst Med-Nat Acad Sci. *Mailing Add:* Baylor Col Med One Baylor Plaza Houston TX 77030

BUTLER, WILLIAM THOMAS, b Ranger, Tex, Dec 10, 35; m 59; c 2. BIOCHEMISTRY. *Educ:* Baylor Univ, BS, 58; Vanderbilt Univ, PhD(biochem), 66. *Prof Exp:* Pub sch teacher, Tex, 58-60; res asst biochem, Vanderbilt Univ, 61-63; staff fel, Nat Inst Dent Res, 66-67, asst prof, 67-72, assoc prof biochem & asst prof physiol & biophys, 72-75, PROF BIOCHEM, MED CTR, UNIV ALA, BIRMINGHAM, 75-, SR SCIENTIST, INST DENT RES, 75- *Concurrent Pos:* Vis scientist, Bone Res Lab, Univ Calif, Los Angeles, 75. *Mem:* Am Soc Biol Chemists; Int Asn Dent Res; Am Soc Bone & Mineral Res. *Res:* Primary structure of collagen; proteins and proteoglycans of bone and dentin; proteins of periodontal tissues; mechanism of formation of calcified tissues. *Mailing Add:* 2806 15th Ave Northport AL 35476

BUTMAN, BRYAN TIMOTHY, b Massilon, Ohio, May 25, 52; m 74; c 3. BIOTECHNOLOGY, IMMUNOCHEMISTRY. *Educ:* Univ Detroit, BS, 74; Wayne State Univ, MS, 77, PhD(biol sci), 81. *Prof Exp:* Scientist, Warner-Lambert Co, 83-85; SR SCIENTIST, BIOTECHNOL RES INST, 85- *Mem:* Am Asn Clin Chem; Am Soc Microbiol; Inst Food Technologists; Am Soc Cell Biol; AAAS. *Res:* Hybridoma; immunoassay development; infectious disease; oncology; development and characterization of monoclonal antibodies with diagnostic significance; antigens as diverse as infectious agents (listeria, chlamydia, human immunodeficiency virus), cancer markers (carcinoembryonic antigen, alpha-fetoprotein), coagulation factors, cardiovascular disease markers and toxins; several monoclonal antibodies have been used for diagnostic immunoassays. *Mailing Add:* Diag Prod Res Biotechnol Res Inst 1330A Piccard Dr Rockville MD 20850

BUTOW, RONALD A, b Aug 1, 36; US citizen; m 58; c 1. BIOCHEMISTRY. *Educ:* Hobart Col, BS, 58; Cornell Univ, MNS, 60, PhD(biochem), 63. *Prof Exp:* NSF res fel biochem, Pub Health Res Inst New York, 63-65; asst prof, Princeton Univ, 65-71; PROF BIOCHEM, UNIV TEX HEALTH SCI CTR DALLAS, 71- *Concurrent Pos:* USPHS res grant, 66. *Mem:* AAAS; Am Soc Biol Chemists. *Res:* Biological electron transport systems; oxidative phosphorylation bioenergetic reactions; structure; function and biogenesis of membranes. *Mailing Add:* Dept Biochem Univ Tex Health Sci Ctr 5323 Harry Hines Blvd Dallas TX 72535

BUTRUM, RITVA RAUANHEIMO, b Helsinki, Finland, Oct 30, 27; US citizen; m 55; c 4. FOOD SCIENCE. *Educ:* Col Home Econ, Univ Helsinki, BS, 51; Univ Calif, Los Angeles, MS, 56; Univ Md, PhD(food sci), 77. *Prof Exp:* Nutritionist, NIH, 66-69 & Fedn Am Soc Exp Biol, 70-73; nutrit analyst, Consumer & Food Econ Inst, USDA, 73-77, res nutritionist, Nutrient Compos Lab, 77-78, assoc prog mgr, Competitive Res Grants Off, 78-83; ACTG CHIEF, DIET & CANCER BR, DIV CANCER PREV & CONTROL, NAT CANCER INST, NIH, 83- *Concurrent Pos:* Fulbright scholarship, Univ Calif, Los Angeles, 54-56; Fulbright lectr grant, Univ Helsinki, Finland, 82. *Mem:* Am Inst Nutrit; Inst Food Technologists; Am Chem Soc; Soc Nutrit Educ; Asn Consumer Res. *Res:* Protein quality; zinc and copper nutrition; folacin methodology and content in foods; food sampling designs for nutrient analysis; data management systems for nutrients and contaminants; nomenclature of foods, parts and processes; diet and cancer; fat and breast cancer; fiber and colon cancer. *Mailing Add:* Nat Cancer Inst Div Cancer Prev & Control EPN Bldg Rm 632 Bethesda MD 20892-4000

BUTSCH, ROBERT STEARNS, b Owatonna, Minn, July 10, 14; m 41; c 1. MAMMALOGY. *Educ:* Univ Iowa, BA, 36, MA, 41; Univ Mich, PhD(zool), 54. *Prof Exp:* Cur, Arrowhead Mus, 36-37; chief preparator marine zool, Barbados Mus & Hist Soc, Barbados, BWI, 38-39; asst to dir, Univ Mich, 51-55, assoc cur, 56-64, cur exhibits, 64-74, actg dir, 74-75, dir exhibit mus, 75-85; RETIRED. *Mem:* Am Soc Mammal; Am Asn Mus. *Res:* Ecology of vertebrates; design and construction of natural history exhibits; teaching of museum methods. *Mailing Add:* 1219 S Forest Ann Arbor MI 48109

BUTSON, ALTON THOMAS, b Pa, Feb 18, 26; wid; c 4. MATHEMATICS. *Educ:* Franklin & Marshall Col, BS, 50; Mich State Univ, MS, 51, PhD(math), 55. *Prof Exp:* Asst math, Mich State Univ, 50-54, instr, 54-55; from asst prof to assoc prof, Univ Fla, 55-59; res specialist, Boeing Co, Wash, 59-61; PROF MATH, UNIV MIAMI, 61- *Mem:* Am Math Soc; Math Asn Am. *Res:* Lattice, group and number theory; combinatorial analysis. *Mailing Add:* Dept Math Univ Miami Box 249085 Coral Gables FL 33124

BUTT, BILLY ARTHUR, b Brazil, Ind, June 9, 31; div; c 2. ENTOMOLOGY, AGRICULTURE. *Educ:* Purdue Univ, BS, 53. *Prof Exp:* Entomologist med, Agr Res Serv, USDA, Beltsville, Md, 55, entomologist plants, Brownsville, Tex, 56-63, res entomologist fruit insects, Yakima, Wash, 63-72; first officer entom, Int Atomic Energy Agency, Vienna, 72-74; res entomologist veg insects, USDA, 74-79, dir, Appalachian Fruit Res Sta, Agr Res Serv, 79-; RETIRED. *Concurrent Pos:* Affil fac, Univ Idaho, 77-; courtesy entomologist, Wash State Univ, 75-; adj prof, WVa Univ, 79- *Mem:* Entom Soc Am. *Res:* Agricultural entomology, including pest management and sex pheromones. *Mailing Add:* Rte 1 Box 540 Shepherdstown WA 25443

BUTT, HUGH ROLAND, GASTROENTEROLOGY, NUTRITION. *Educ:* Univ Va, MD, 33. *Prof Exp:* Chmn sci adv comt, Lugwid Inst Cancer Res, 79; RETIRED. *Res:* Liver disease. *Mailing Add:* 1014 Seventh St SW Rochester MN 55902

BUTT, JOHN B(AECHER), b Norfolk, Va, Sept 10, 35; m 63; c 1. CHEMICAL ENGINEERING. *Educ:* Clemson Univ, BChE, 56; Yale Univ, MEng, 58, DEng(chem eng), 60. *Prof Exp:* Instr chem eng, Yale Univ, 59-60, from asst prof to assoc prof eng & appl sci, 60-70; prof chem eng, 70-81, WALTER P MURPHY PROF CHEM ENG, TECHNOL INST, NORTHWESTERN UNIV, 81- *Concurrent Pos:* Asst master, Calhoun Col, 60-64; vis prof chem eng, Univ Brussels, 71; assoc ed, Indust & Eng Chem Process Design & Develop, 78-87. *Honors & Awards:* Alan P Colburn Award, 68 & Prof Prog Award, 78, Am Inst Chem Engrs; Alexander von Humboldt Sr US Scientist Award, 86. *Mem:* AAAS; Am Inst Chem Engrs; Am Chem Soc; Catalysis Soc; Sigma Xi. *Res:* Catalysis and kinetics of chemical reactions; simultaneous mass and heat transport in reacting systems. *Mailing Add:* Dept Chem Eng Northwestern Univ Evanston IL 60201

BUTTAR, HARPAL SINGH, b Nanonkee, India, Nov 2, 39; Can citizen; m 62; c 3. PHARMACOLOGY, TOXICOLOGY. *Educ:* Col Vet Med, India, BVSc, 61; Univ Alta, MSc, 66, PhD(pharmacol), 71. *Prof Exp:* Lectr pharmacol, Col Vet Med, India, 61-64; teaching asst, Univ Alta, 64-70; res fel med, Sch Med, Wayne State Univ, 70-71; RES SCIENTIST & HEAD, REPRODUCTIVE TOXICOL SECT, BUR DRUG RES, HEALTH PROTECTION BR, HEALTH & WELFARE CAN, 71- *Mem:* NY Acad Sci; Chem Inst Can; Soc Toxicol Can; Soc Toxicol; Teratology Soc; Can Col Neuropsychopharmacol. *Res:* Toxicological assessment of drugs, drug metabolites and cosmetics with reference to their absorption, distribution, biotransformation and excretion as well as drug interactions and embryotoxicity/teratogenicity; author & co-author of 50 scientific publications and three book chapters. *Mailing Add:* Bur Drug Res Health Protection Br Health & Welfare Can Ottawa ON K1A 0L2 Can

BUTTEMER, WILLIAM ASHLEY, b San Diego, Calif, Jan 9, 47. PHYSIOLOGICAL ECOLOGY, BIOPHYSICAL ECOLOGY. *Educ:* San Diego State Univ, AB, 71; Univ Mich, PhD(biol), 81. *Prof Exp:* Postdoctor researcher avian sci, Univ Calif, Davis, 81-83; asst prof comp physiol, Dept Biol, Rutgers Univ, 83-85; lectr avian energetics, Sch Biol Sci, Univ NSW, 85-86; res assoc avian endocrinol, Dept Zool, Univ Wash, 88-91; lectr, 86-88, LECTR HUMAN & ANIMAL PHYSIOL, DEPT PHYSIOL, UNIV TASMANIA, 91- *Concurrent Pos:* Consult, Orca Seafarms, Inc, 85-; lectr, Dept Zool, Univ Wash, 90-91; vis scholar, Ocean Sci Ctr, Mem Univ Nfld, 91- *Mem:* Sigma Xi; Am Soc Zool; Am Ornith Soc. *Res:* Renal physiology; endocrinology; thermoregulation. *Mailing Add:* Dept Physiol Univ Tasmania Hobart 7001 Australia

BUTTER, STEPHEN ALLAN, b New York, NY, May 15, 37; m 63; c 1. INORGANIC CHEMISTRY. *Educ:* Brooklyn Col, BS, 59; Univ Del, PhD(inorg chem), 65. *Prof Exp:* Teaching asst, Univ Md, 59-60; chemist, Thiokol Chem Corp, 63; res chemist, Indust & Biochem Dept, E I du Pont de Nemours & Co, 64-66; res fel, Univ Sussex, 66-67; sr res chemist, Mobil Chem Co, 67-76, mem staff, Res Dept, Paulsburo Lab, Mobil Res & Develop Corp, 76-80; prog mgr, US Dept Energy, 80-81; mgr chem res, Corp Res Dept, Air Prod & Chemicals, Inc, 81-86; PROG MGR, DIV CHEM SCI, OFF BASIC ENERGY SCI, US DEPT ENERGY, 86- *Honors & Awards:* Excellence Catalysis Award, Catalysis Soc, 84. *Mem:* AAAS; Catalysis Soc; NY Acad Sci; Am Chem Soc. *Res:* Basic research in chemical sciences: fossil chemistry, synthesis-gas chemistry, homogeneous and heterogeneous catalysis involving transitional metals and zeolites; energy and synthetic fuels; boron chemistry; coordination complexes; zeolites; homogeneous and heterogeneous catalysis. *Mailing Add:* Off Energy Res Div Chem Sci US Dept Energy Washington DC 20585

BUTTERFIELD, DAVID ALLAN, b Milo, Maine, Jan 14, 46; m 68; c 1. BIOPHYSICAL CHEMISTRY. *Educ:* Univ Maine, BA, 68; Duke Univ, PhD(phys chem), 74. *Prof Exp:* Teacher chem high sch, United Methodist Church, Rhodesia, 68-71; NIH fel neurosci, Med Sch, Duke Univ, 74-75; from asst prof to assoc prof, 75-83, PROF CHEM, UNIV KY, 83-, DIR, CTR MEMBRANE SCI, 86- *Concurrent Pos:* Vis scientist, St Jude Children's Res Hosp. *Mem:* Am Chem Soc; Sigma Xi; Biophys Soc; Am Soc Biochem & Molecular Biol. *Res:* Biological applications of electron spin resonance, with particular emphasis on membrane systems; effects of neurotoxins, environmental pollutants and pharmacological agents on membrane structure and function; muscular dystrophy; Huntington's disease; Alzheimer's disease; relationships between skeletal membrane proteins and cell surface carbohydrates; transmembrane signaling. *Mailing Add:* Dept Chem & Ctr Memb Sci Univ Ky Lexington KY 40506-0055

BUTTERFIELD, EARLE JAMES, b Gilroy, Calif, Feb 11, 49; m 69; c 3. PLANT PATHOLOGY. *Educ:* Univ Calif, Davis, BS, 71, PhD(plant path), 75. *Prof Exp:* Fel plant path, Univ Calif, Davis, 75-76; fel, 76-78, ASST PLANT PATHOLOGIST, BOYCE THOMPSON INST, 78- *Mem:* Am Phytopath Soc; Sigma Xi. *Res:* Soil microbiology; epidemiology; chemical control of pathogens. *Mailing Add:* 4507 Highgate Dr Durham NC 27713

BUTTERFIELD, VELOY HANSEN, JR, b Fillmore, Utah, Sept 24, 42; m 66; c 4. COMPUTER SCIENCES. *Educ:* Univ Utah, BSEE, 67, MEA, 73. *Prof Exp:* Electronic engr, Design Div, Mare Island Naval Shipyard, 68-70; assoc scientist, Res Ctr, Kennecott Minerals Co, 70-80, sect head process instrumentation, control & technol, 80-83; prog mgr, Leeds & Northrup Systs, 83-85. *Honors & Awards:* Res Award, Kennecott Copper Corp, 74. *Mem:* Inst Elec & Electronics Engrs; Instrument Soc Am. *Res:* Computer applications; cybernetics; microelectronics applications. *Mailing Add:* Evans & Sutherland Computer Co PO Box 58700 Salt Lake City UT 84158

BUTTERWORTH, BERNARD BERT, b Woodbine, Iowa, May 2, 23. ANATOMY. *Educ:* Univ Mich, BS, 52, MS, 54; Univ Southern Calif, PhD(zool), 60. *Prof Exp:* Asst biol, Univ Southern Calif, 55-58; lectr, Los Angeles State Col, 58-59; asst prof, Univ Wichita, 60-62, Sch Dent, Univ Southern Calif, 62-64 & Univ Ariz, 64; from assoc prof to prof anat, Sch Dent, Univ Mo, Kansas City, 64-86, dir admis, 73-80, chmn dept, 75-86, EMER PROF ANAT, SCH DENT, UNIV MO, KANSAS CITY, 90- *Concurrent Pos:* Res partic, Inst Trop Biol, Costa Rica, 62. *Mem:* Soc Study Evolution; Ecol Soc Am; Soc Syst Zool; Am Soc Mammal; Am Asn Anatomists. *Res:* Mammalian behavior and reproduction; Heteromyidae; growth and development human, head and neck. *Mailing Add:* 6501 E 55th Terr Kansas City MO 64129

BUTTERWORTH, BYRON EDWIN, b Salt Lake City, Utah, Dec 9, 41; m 66; c 4. GENETIC TOXICOLOGY, VIROLOGY. *Educ:* Brigham Young Univ, BS, 68; Univ Wis-Madison, PhD(biochem), 72. *Prof Exp:* Mem res staff virol, E I du Pont de Nemours & Co, 72-75; chief molecular biol, Haskell Lab Toxicol & Indust Med, 75-77; CHIEF GENETIC TOXICOL, CHEM INDUST INST TOXICOL, 77- *Concurrent Pos:* Adj assoc prof path, Univ NC, Chapel Hill, 78; adj prof toxicol, Duke Univ, NC. *Mem:* Environ Mutagen Soc; Am Asn Cancer Res; Soc Toxicol. *Res:* Mechanism of action of antiviral agents; short-term tests for predicting the mutagenic-carciogenic potential of chemicals; nongenotoxic mechanisms in carcinogenesis. *Mailing Add:* Chem Indust Inst Toxicol PO Box 12137 Research Triangle Park NC 27709

BUTTERWORTH, CHARLES E, JR, b Lynchburg, Va, Mar 11, 23; m 46; c 3. INTERNAL MEDICINE. *Educ:* Univ Va, BA, 44, MD, 48. *Prof Exp:* Intern, Med Col Ala, 48-49, resident hemat, 49-50, resident internal med, 51-53; mem staff, Trop Res Med Lab, San Juan, PR, 55-57; mem staff, Walter Reed Army Inst Res, Washington, DC, 57-58; from instr to assoc prof internal med, 58-66, prof & chmn, Dept Nutrit Sci, 76, PROF INTERNAL MED, SCH MED, UNIV ALA, 66- *Concurrent Pos:* Consult, NIH, 69- & WHO, 71; mem, Food & Nutrit Bd, Nat Res Coun, Nat Acad Sci, 77-80. *Honors & Awards:* Goldberger Award, AMA, 76; McCollum Award, Am Soc Clin Nutrit, 77; Lederle Award, Am Inst Nutrit, 83. *Mem:* Am Fedn Clin Res; Am Clin & Climat Asn; hon mem Am Dietetic Asn; Am Col Physicians; Am Soc Clin Nutrit (pres, 74-75). *Res:* Hematology; nutrition support of hospitalized patients; studies of nutritional anemia; studies on nutrition and the chemoprevention of cancer. *Mailing Add:* Dept Nutrit Sci Univ Ala Sch Med Birmingham AL 35294

BUTTERWORTH, FRANCIS M, b Philadelphia, Pa, Mar 29, 35; m 61; c 2. CELL BIOLOGY, GENOTOXICITY. *Educ:* Columbia Univ, BA, 57; Northwestern Univ, PhD(genetics), 65. *Prof Exp:* USPHS fel, Univ Va, 65-66; from asst prof to assoc prof, 66-74, PROF GENETICS, OAKLAND UNIV, 74- *Concurrent Pos:* Guest prof, Univ Nijmegen, Holland, 73-74; adj prof, Mich State Univ, 79-, Wayne State Univ, 82-88; vis prof, Wayne State Univ, 85; grants from NSF, NIH, Res Corp, Environ Protection Agency. *Mem:* AAAS; Environ Mutagen Soc; Genetics Soc Am; Sigma Xi; Am Soc Cell Biol. *Res:* Genetic, cellular, and molecular control of development, aging, and protein folding; genotoxicity of toxic waste metabolites. *Mailing Add:* Dept of Biol Oakland Univ Rochester MI 48309

BUTTERWORTH, GEORGE A M, b Bramhall, Eng, Sept 3, 35; m 60; c 2. PHYSICS, MECHANICAL ENGINEERING. *Educ:* Univ Manchester, BSc, 58; Mass Inst Technol, SM, 61. *Prof Exp:* Res assoc, Fabric Res Labs, Inc, Dedham, 60-61; sr res assoc, 61-64, asst dir, 64-69, vpres, 69-72; mgr res & develop, Johnson & Johnson, 72-78; DIR PROD DEVELOP, AM HOSP SUPPLY, 79- *Honors & Awards:* McKay Mem Award, 57. *Mem:* AAAS; Am Soc Mech Engrs; Fiber Soc (pres, 80-81); fel Brit Textile Inst; Plastics & Rubber Inst. *Res:* Materials technology and engineering, especially structural and interactive behavior of fibrous, organic and related materials; physical and mechanical properties of high polymers; biomedical materials; tire cord reinforcements and non-woven structures; disposable medical products. *Mailing Add:* Tambrands Inc PO Box 271 Bridge & Springfield St Palmer MA 01069

BUTTERWORTH, THOMAS AUSTIN, b Los Angeles, Calif, Apr 30, 44; m 68; c 2. FOOD ENGINEERING, CHEMICAL ENGINEERING. *Educ:* Univ Calif, Berkeley, BS, 67; Mass Inst Technol, PhD(biochem eng), 73. *Prof Exp:* Res engr chem eng, Western Regional Res Lab, USDA, 68; assoc dir & res develop, Beatrice/Hunt-Wesson Inc, 73-88; AUSTIN FOOD TECH INC. *Mem:* Inst Food Technologists; Am Inst Chem Engrs; AAAS. *Res:* Food engineering topics related to industrial application. *Mailing Add:* Austin Food Tech Inc 114 Pearl Ave Newport Beach CA 92662

BUTTIKER, MARKUS, b Wolfwil, Switz, July 18, 50; m 88. TRANSPORT IN MESOSCOPIC CONDUCTORS. *Educ:* Swiss Fed Inst Technol, dipl theoret physics, 75; Univ Basel, Switz, PhD(theoret physics), 78. *Prof Exp:* Res asst theoret physics, Univ Basel, Switz, 78-79; MEM RES STAFF, IBM RES DIV, T J WATSON RES CTR, 79- *Mem:* Fel Am Phys Soc. *Res:* Transport in mesoscopic conductors; persistent currents, Aharonov-Bohm effects, noise properties. *Mailing Add:* IBM Res Div T J Watson Res Ctr PO Box 218 Yorktown Heights NY 10598

BUTTKE, THOMAS MARTIN, b Bay Shore, NY, Dec 5, 52; m 77; c 2. LIPID METABOLISM, MEMBRANE STRUCTURE FUNCTION. *Educ:* Univ NC, Wilmington, BA, 74; Univ Fla, MS, 75, PhD(microbiol), 78. *Prof Exp:* Fel chem, Harvard Univ, 78-80; from asst prof to assoc prof microbiol, Med Ctr, Univ Miss, 85-86; ASSOC PROF MICROBIOL, E CAROLINA UNIV SCH MED, 86- *Mem:* Am Asn Immunologists; Am Soc Microbiol; Am Chem Soc; AAAS. *Res:* Lipid metabolism and function, with particular emphasis on the role of lipids in cells associated with the immune system; regulation of membrane lipid composition in yeast. *Mailing Add:* Dept Microbiol & Immunol E Carolina Univ Sch Med Greenville NC 27858-4354

BUTTLAIRE, DANIEL HOWARD, b New York, NY, Sept 5, 41; m 63; c 2. BIOCHEMISTRY, BIOPHYSICS. *Educ:* Univ Denver, BA, 63; Univ Kans, PhD(biochem), 70. *Prof Exp:* Res assoc biochem, Univ Kans, 70-71; res fel biophysics, Johnson Res Found, Sch Med, Univ Pa, 71-74; from asst prof to prof, 75-83 CHMN DEPT CHEM, SAN FRANCISCO STATE UNIV, 80- *Concurrent Pos:* USPHS fel, 71-73; Cottrell res grant, Res Corp, 77-79; NASA-Univ Consortium res grant, Ames Res Ctr, 78-79. *Mem:* Am Chem Soc; Sigma Xi; Int Soc Magnetic Resonance; Am Soc Biol Chem. *Res:* Structure-function relationships in macromolecular systems; mechanisms of enzyme action; roles of metal ions in biological systems. *Mailing Add:* Dept Chem & Biochem San Francisco State Univ San Francisco CA 94132

BUTTLAR, RUDOLPH O, b Chicago, Ill, Dec 31, 34; m 55; c 3. INORGANIC CHEMISTRY. *Educ:* Wheaton Col, BS, 56; Ind Univ, PhD(inorg chem), 62. *Prof Exp:* Asst prof chem, 62-71, asst dean col arts & sci, 67-70, assoc dean, 70-74, actg dean, 74-75, ASSOC PROF CHEM, KENT STATE UNIV, 71-, DEAN COL ARTS & SCI, 75- *Mem:* Am Chem Soc. *Res:* Boron hydride and boron-nitrogen chemistry, especially cyclic boron-nitrogen compounds. *Mailing Add:* Col Arts & Sci Kent State Univ Kent OH 44240

BUTTNER, F(REDERICK) H(OWARD), b Cleveland, Ohio, Oct 11, 20. PHYSICAL METALLURGY. *Educ:* Mich State Univ, BS, 44; Mass Inst Technol, MS, 49, DSc, 51. *Prof Exp:* Asst prof chem & math, Muskegon Community Col, 46-47; instr & res assoc, Mass Inst Technol, 47-51; staff metallurgist, Nat Acad Sci-Nat Res Coun, 51-53; staff metallurgist tech serv & develop, Union Carbide Metals, 53-57, mgr new prod eng, 57-62; res assoc, 62-66, FEL, BATTELLE MEM INST, 66- *Concurrent Pos:* Consult, Saudi Arabian Standards Orgn. *Mem:* AAAS; Am Chem Soc; Am Soc Metals; Am Inst Mining, Metall & Petrol Engrs. *Res:* Advanced construction and manufacturing practices; corporate investigations and profitability analyses of new venture options, leading to long term implementation plans and monitoring of them; metal commodity analyses to forecast supply/demand/price relationships; technology assessment; technology forecasting. *Mailing Add:* 4153 Rowanne Columbus OH 43214

BUTTON, DON K, b Kenosha, Wis, Oct 2, 33; m 61; c 5. BIOCHEMISTRY, MARINE MICROBIOLOGY. *Educ:* Wis State Col, Superior, BS, 55; Univ Wis, MS, 61, PhD(biochem), 64. *Prof Exp:* From asst prof marine biochem to assoc prof marine sci, 64-74, PROF MARINE SCI, INST MARINE SCI, UNIV ALASKA, 74- *Concurrent Pos:* Res assoc, Dept Biophys, Univ Colo Med Ctr, 69-70; consult, continuous cult oper & design, oil slick biodegradation. *Mem:* Am Soc Microbiol; Am Soc Limnol & Oceanog; Am Chem Soc; Sigma Xi. *Res:* Microbial biochemistry; metabolism kinetics of marine heterotrophs. *Mailing Add:* Inst Marine Sci Univ Alaska Fairbanks AK 99701

BUTTON, KENNETH J, physics, for more information see previous edition

BUTTON-SHAFER, JANICE, b Cincinnati, Ohio, Sept 13, 31; m 62; c 3. EXPERIMENTAL HIGH ENERGY PHYSICS, NUCLEAR STRUCTURE. *Educ:* Cornell Univ, BEngPhys, 54; Univ Calif, Berkeley, PhD(physics), 59. *Prof Exp:* Physicist, Lawrence Radiation Lab, Univ Calif, 59-66, lectr physics, Univ Calif, Berkeley, 62-64; assoc prof, 66-70, PROF PHYSICS, UNIV MASS, AMHERST, 70- *Concurrent Pos:* Fulbright fel, Germany, 54-55; G E Coffin fel, Berkeley, 56-57; prin investr, High Energy Physics, Dept Energy, Univ Mass, 66-85; Dept Energy fel, Assoc Western Univs, 90. *Mem:* Fel Am Phys Soc; AAAS; Sigma Xi. *Res:* High-energy and nuclear experimental physics. *Mailing Add:* Dept Physics Univ Mass Amherst MA 01003

BUTTREY, BENTON WILSON, b Craigmont, Idaho, Mar 25, 19; m 61. PROTOZOOLOGY. *Educ:* Univ Idaho, BS, 47, MS, 49; Univ Pa, PhD(zool), 53. *Prof Exp:* Asst zool, Univ Idaho, 47-49; asst instr, Univ Pa, 49-52; from asst prof to prof, Univ SDak, 53-61; prof entom, Iowa State Univ, 61-76, prof zool, 61-88; RETIRED. *Mem:* AAAS; Soc Protozool; Am Soc Zool; Am Soc Parasitol. *Res:* Morphology; taxonomy and host-parasite relationships of the protozoa from amphibia and swine; morphological variations in trichomonad protozoa. *Mailing Add:* 2020 Country Club Blvd Ames IA 50010

BUTTRILL, SIDNEY EUGENE, JR, b Corpus Christi, Tex, July 21, 44; m 65; c 3. ANALYTICAL CHEMISTRY. *Educ:* Mass Inst Technol, BS, 66; Stanford Univ, PhD(chem), 70. *Prof Exp:* Asst prof chem, Univ Minn, Minneapolis, 69-76; sr scientist mass spectros, SRI Int, 76-85, mgr, Mass Spectros Develop, 78-85; prof & chmn dept, 74-85, Aderhold distinguished prof sci educ, Univ Ga, 85-; AT CHARLES EVANS & ASSOC. *Honors & Awards:* Robert H Carleson Award, 79. *Mem:* AAAS; Am Chem Soc; Am Soc Mass Spectrometry. *Res:* Field ionization mass spectrometry; chemical ionization mass spectrometry; ion-molecule chemistry; fossil fuels analysis by field ionization mass spectrometry; drug analysis by gas chromatography and field ionization mass spectrometry. *Mailing Add:* 1417 Parkinson Palo Alto CA 94301-3455

BUTTS, DAVID, b Rochester, NY, May 9, 32; m 58; c 2. BIOLOGY, SCIENCE EDUCATION. *Educ:* Butler Univ, BS, 54; Univ Ill, MS, 60, PhD(bot, sci educ), 62. *Prof Exp:* Asst prof sci educ, Olivet Nazarene Col, 61-62; from asst prof to assoc prof, Univ Tex, Austin, 62-71, prof curric & instr, 71-74; prof & chmn dept, 74-85, ADERHOLD DISTINGUISHED PROF SCI EDUC, UNIV GA, 85- *Concurrent Pos:* Consult, Robert E Lee Sci Curric Proj, 62-63; consult, Eval Ctr & mem, Stanford Writing Conf, Am Asn Advan Sci Elem Sci Proj, 63 & 64; dir, Sci In-Serv Proj, 65-66, TAB Sci Test Proj, 65-69 & Elem Process-Oriented Curric Proj; proj dir, Personalized Teacher Educ Prog, 66-74. *Honors & Awards:* Robert Carleton Award, Nat Sci Teachers Asn, 79. *Mem:* AAAS; Nat Asn Res Sci Teaching. *Res:* Curriculum materials in sciences for elementary and junior high students. *Mailing Add:* Dept Sci Educ Univ Ga Athens GA 30602

BUTTS, HUBERT S, b Burkburnett, Tex, Nov 7, 23; m 47; c 3. MATHEMATICS. *Educ:* NTex State Univ, BS, 47, MS, 48; Ohio State Univ, PhD(math), 53. *Prof Exp:* Instr math, NTex State Univ, 47-48; from asst prof to assoc prof, La State Univ, Baton Rouge, 53-62, prof math, 62-; RETIRED. *Mem:* Am Math Soc. *Res:* Algebraic number theory; ideal theory in commutative rings. *Mailing Add:* 610 Lowray Dr Baton Rouge LA 70808

BUTTS, JEFFREY A, b Lansing, Mich, May 25, 47; m 69; c 3. PARASITOLOGY. *Educ:* Albion Col, AB, 69; Bowling Green State Univ, MS, 71, PhD(biol), 73. *Prof Exp:* From asst prof to assoc prof biol, Univ NC, Charlotte, 73-81; assoc prof, 81-84, PROF BIOL, APPALACHIAN STATE UNIV, 84-, CHAIR DEPT, 81- *Concurrent Pos:* Pres, NC conf, Am Asn Univ Prof, 86-87. *Mem:* AAAS; Am Soc Parasitologists; Am Asn Univ Profs; Sigma Xi. *Res:* Filariasis specifically heartworm (Dirofilaria immitis) and toxoplasmosis; coccidia of amphibians. *Mailing Add:* Biol Dept Applachian State Univ Boone NC 28608

BUTTS, WILLIAM CUNNINGHAM, b Evanston, Ill, Jan 13, 42; m 63; c 2. LABORATORY MEDICINE. *Educ:* Purdue Univ, BS, 63; Iowa State Univ, PhD(anal chem), 68. *Prof Exp:* Res chemist, Oak Ridge Nat Lab, 68-71; fel clin chem, Univ Wash, 71-73; sr clin chemist, Insts Med Sci, San Francisco, 73-75; DIR CLIN CHEM, GROUP HEALTH OF PUGET SOUND, 75-

Concurrent Pos: Clin asst prof, Univ Wash, 76- *Mem:* Am Asn Clin Chem; Am Chem Soc; AAAS. *Res:* Identification of constituents isolated from physiologic fluids; development of separation techniques for biochemical systems; gas chromatography of inorganic species; individual biochemical profiles. *Mailing Add:* Chem Lab Group Health 801 SW 16th St Renton WA 98057

BUTTS, WILLIAM LESTER, b Reynoldsburg, Ohio, Dec 7, 31; m 54; c 3. ENTOMOLOGY, ORNITHOLOGY. *Educ:* Wilmington Col, Ohio, BSc, 53; Ohio State Univ, MSc, 54, PhD(entom), 64. *Prof Exp:* From instr to assoc prof entom, Purdue Univ, 57-66; assoc prof, 66-69, PROF BIOL, STATE UNIV NY COL ONEONTA, 69- *Mem:* Am Mosquito Control Asn; Sigma Xi; NY Acad Sci. *Res:* Medical and public health entomology. *Mailing Add:* Dept Biol State Univ NY Col Oneonta NY 13820

BUTZ, ANDREW, b Perth Amboy, NJ, Feb 15, 31. INSECT PHYSIOLOGY. *Educ:* St Peter's Col, BS, 52; Fordham Univ, MS, 54, PhD, 56. *Prof Exp:* Asst physiol, Fordham Univ, 52-56; asst prof, 58-64, ASSOC PROF ZOOL, UNIV CINCINNATI, 64- *Mem:* AAAS; Am Soc Zoologists; Entom Soc Am. *Res:* Circadian rhythm; RNA studies; protein studies; pheromones; metabolism; insect ecology; soil arthropods; insecticidal relationships; insect immunology and toxicology; plant/extracts and their development and or metamorphosis; factors which influence vitellogensis. *Mailing Add:* Dept Biol Sci Univ Cincinnati Cincinnati OH 45221

BUTZ, DONALD JOSEF, b Marion, Ohio, Oct 3, 58; m 82. ORDNANCE RESEARCH & DEVELOPMENT, HIGH ENERGY DENSITY PHENOMENA. *Educ:* Ohio State Univ, BS, 81. *Prof Exp:* Student res assoc aeronaut eng, Aeronaut & Astronaut Res Lab, Ohio State Univ, 79-81; researcher ord eng, Battelle Mem Inst, 81-83, from res scientist to prin res scientist ord eng, 83-88, SR RES SCIENTIST ORD ENG, BATTELLE MEM INST, 88- *Mem:* Sigma Xi; Am Inst Aeronaut & Astronaut; Am Defense Preparedness Asn; Soc Explosives Engrs; Int Asn Bomb Technicians & Investigators. *Mailing Add:* Battelle Mem Inst 505 King Ave Columbus OH 43201

BUTZ, ROBERT FREDERICK, b Valparaiso, Ind, May 7, 49; m 88; c 2. CLINICAL RESEARCH & MEDICAL TECHNOLOGY. *Educ:* Kenyon Col, BS, 71; Duke Univ, PhD(pharmacol), 80. *Prof Exp:* Res scientist pharmacol, Burroughs Wellcome Co, 74-82, sr clin scientist med, 82-85; vpres, sci affairs clin, DAR, Inc, 85-87; vpres clin, 87-90, VPRES BUS DEVELOP, QUINTILES TRANSNAT CORP, 90- *Concurrent Pos:* Lectr, Duke Univ Grad Sch, 85-87. *Mem:* Am Soc Clin Pharmacol & Therapeut; Am Soc Pharmacol & Exp Therapeut; Drug Info Asn. *Res:* Ways to direct antibody-fragment-linked antineoplastics specifically against solid tumors; patents in radioimmunoassay & discovery of acrivastine; computerized approaches to collection and management of clinical information. *Mailing Add:* 812 Queensbury Circle Durham NC 27713

BUTZER, KARL W(ILHELM), b Ger. ARCHAEOLOGY, HUMAN GEOGRAPHY. *Educ:* McGill Univ, BSc, 54, MSc, 55; Univ Bonn, Dr rer nat(geog), 57. *Prof Exp:* Ger Acad grant, Univ Bonn, 57-59; from asst prof to assoc prof geog, Univ Wis, 59-66; prof anthrop & geog, Univ Chicago, 66-80, Henry Schultz prof environ archeol, 80-84; DICKSON CENTENNIAL PROF LIB ARTS, UNIV TEX, AUSTIN, 84- *Concurrent Pos:* NSF grants, 64-; Wenner-Gren grants, 65-; chair prof human geog, Swiss Fed Inst Technol, Zurich, 81-82. *Honors & Awards:* Meritorious Contrib Award, Asn Am Geogr, 68; Busk Medal, Royal Geog Soc, 79; Fryxell Medal, Soc Am Archeol, 81; Stopes Medal, Geologists Asn London, 82. *Mem:* Am Quaternary Asn; Asn Am Geogr; fel Am Acad Arts & Sci; Asn Field Archeol. *Res:* Environmental archaeology; cultural ecology; environmental geomorphology. *Mailing Add:* Dept Geog Univ Tex Austin TX 78712

BUTZOW, JAMES J, b Chicago, Ill, Oct 17, 35; m 80. BIOCHEMISTRY, MOLECULAR BIOLOGY. *Educ:* St Bonaventure Univ, BS, 57; Stanford Univ, PhD(biochem), 63. *Prof Exp:* Res chemist, Geront Br, Nat Heart Inst, 62-65, res chemist, Geront Res Ctr, Nat Inst Child Health & Human Develop, 67-75, RES CHEMIST, GERONT RES CTR, NAT INST AGING, 75- *Concurrent Pos:* Fel, Dept Biochem, Univ Göteborg, Sweden, 65-67; mem, Carroll House Found (res in med ethics), Baltimore, 73- *Mem:* Am Chem Soc. *Res:* Physical chemistry of nucleic acids and proteins; interaction of metals with nucleic acids; RNA polymerase. *Mailing Add:* Geront Res Ctr Nat Inst on Aging NIH Francis Scott Key Med Ctr Baltimore MD 21224

BUURMAN, CLARENCE HAROLD, b Orange City, Iowa, July 31, 15; m 43; c 4. INDUSTRIAL ORGANIC CHEMISTRY. *Educ:* Univ Iowa, AB, 36, MS, 38, PhD(org chem), 41. *Prof Exp:* Semi works chemist, Iowa State Dept Health, 39-41; chief chemist, GAF Corp, 41-52, chief chemist surfactants sulfur color & chlorine area, 52-56, prod mgr vats-intermediates & org chem dept, 56-65, plant mgr, 67-70; vpres & gen mgr, Mauldin Oper, Emery Indust, Inc, 70-84; RETIRED. *Mem:* Am Chem Soc; Am Inst Chemists. *Res:* Organic research chemistry; photographic products; dye and pigment chemistry; surface active agents; agricultural chemistry; textile chemistry. *Mailing Add:* 19 Red Fox Trail Greenville SC 29615-3737

BUXBAUM, JOEL N, b New York, NY, Feb 22, 38; m 61; c 2. IMMUNOGLOBULINS & IMMUNOGLOBULIN GENES. *Educ:* Union Col, BS, 58; Tufts Med Sch, MD, 62. *Prof Exp:* Instr cell biol, Albert Einstein Sch Med, 69-71; from asst prof to assoc prof, 71-81, PROF MED, NY UNIV MED SCH, 81- *Concurrent Pos:* Chief, Cell Biol Res Lab, NY Vet Admin Med Ctr, 71- *Mem:* Am Soc Clin Invest; AAAS; Am Asn Immunologists; Am Rheumatism Asn. *Res:* Molecular mechanisms responsible for diseases in which abnormal proteins plan a role in the pathogenesis; human immunoglobulin heavy chain disease; light chain amyloidosis; familial amyloidotic poly-neuropathy; carcinoma of the colon. *Mailing Add:* Res Serv N Y Vet Admin Med Ctr 408 First Ave New York NY 10010

BUXTON, DWAYNE REVERE, b Tremonton, Utah, Apr 14, 39; m 64; c 5. CROP PHYSIOLOGY & PRODUCTION. *Educ:* Utah State Univ, BS, 64, MS, 66; Iowa State Univ, PhD(agron), 69. *Prof Exp:* From asst prof to prof agron & plant genetics, Univ Ariz, 69-77; PROF AGRON & SUPT MALHEUR EXP STA, ORE STATE UNIV, 77- *Mem:* AAAS; Soil Sci Soc Am; Am Soc Agron; Crop Sci Soc Am. *Res:* Irrigated crop production and physiology. *Mailing Add:* Dept Agron Iowa State Univ Ames IA 50011

BUXTON, IAIN LAURIE OFFORD, b Haulton, Eng, Oct 1, 50; nat US; m 81; c 1. CYCLIC NUCLEOTIDES, RECEPTOR-EFFECTOR COUPLING. *Educ:* Univ Calif, San Diego, BA, 73; Univ Pac, DPh(pharmaceut sci), 78. *Prof Exp:* Res assoc biochem, Univ NC, Raleigh, 73-74 & res assoc cell biol, Salk Inst, La Jolla, Calif, 74-75; clin instr, clin pharm, Univ Southern Calif, Los Angeles, 78-80; resident clin pharm & therapeut, Vet Admin Med Ctr, San Diego, 78-81; teaching fel pharmacol, Sch Med, Univ Calif, San Diego, 81-84, asst res pharmacologist, 84-85; ASST PROF PHARMACOL, SCH MED, UNIV NEV, 85- *Concurrent Pos:* Lectr, cardiovasc drug ther, Clin Clerkship Prog, Univ Pac, 78-80 & drug abuse pharmacol, Nat Univ, San Diego, 80-81; teaching fel, NIH, 81-84; prin investr, NIH, 84-87. *Honors & Awards:* New Investr Res Award, NIH, 84. *Mem:* Am Heart Asn; AAAS; Am Col Clin Pharm; Am Soc Pharmacol & Exp Therapeut. *Res:* Investigation of the mechanisms of action of adrenergic receptors in mammalian cells; regulation of cyclic adrenosine monophosphate compartmentation in cardiac tissue; hormonal control of adrenylate cyclase and inositol-phospholipid metabolism in mammalian cells. *Mailing Add:* Dept Pharmacol Sch Med Univ Nev Honard Med Sci Bldg Reno NV 89557-0046

BUXTON, JAY A, b Carrara, Italy, May 29, 19; m 46; c 3. ENTOMOLOGY. *Educ:* Southwest Tex State Teachers Col, BS, 48; Univ Tex, MA, 50; Ohio State Univ, PhD(entom), 57. *Prof Exp:* Tech res asst entom, Ohio State Res Found, 50-52; entomologist, Battelle Mem Inst, 52-57; assoc prof biol, Tex Col Art & Indust, 57-61; asst prof entom, Clemson Univ & asst entomologist, Exp Sta, 61-67; prof & chmn dept, Catawba Col, 67-87, emer prof, 87-; RETIRED. *Mem:* AAAS; Entom Soc Am; Am Inst Biol Sci; Sigma Xi. *Res:* Taxonomy of immature Orthoptera and genus Melanoplus; fruit insects. *Mailing Add:* 502 Wellington Hills Salisbury NC 28144

BUYERS, WILLIAM JAMES LESLIE, b Aboyne, Scotland, Apr 10, 37; Can citizen; m 66; c 3. SOLID STATE PHYSICS. *Educ:* Aberdeen Univ, BSc, 59, PhD(physics), 63. *Prof Exp:* From asst lectr to lectr physics, Aberdeen Univ, 62-65; fel, Atomic Energy Can, Ltd, 65-66, assoc res officer, 66- 75, sr res officer physics, 75-85, MGR, NEUTRON & SOLID STATE PHYICS, ACEL RES, 85- *Concurrent Pos:* Sr res fel, Oxford Univ, 71-72; vis scientist, Oak Ridge Nat Lab, 72, Riso Nat Lab, Denmark, 75, Inst Theoret Physics, Santa Barbara, 80-81, Riso Nat Lab, Denmark, 84, Univ Lausanne, 85; chmn, Can Nat Comt, Int Union Crystallog, 82-85; dir, NATO Advan Studies Inst, Vancouver Island, 83; mem, Can Nat Comt, Int Union Pure & Appl Physics; secy, Comn Magnetism, Int Union Pure & Appl Physics, 84- *Honors & Awards:* Rutherford Gold Medal, Royal Soc Can, 86. *Mem:* Can Asn Physicists; fel Am Phys Soc; fel Inst Physics; fel Royal Soc Can. *Res:* Neutron scattering from spin waves, phonons in solids and liquids; study of crystal fields, exchange, soft modes, phase transitions, defects and random alloys, guentum spins, one- dimensional magnetism, actinide magnetism, heavy-fermion systems. *Mailing Add:* AECL Res Chalk River ON K0J 1J0 Can

BUYNISKI, JOSEPH P, b Worcester, Mass, July 18, 41; m 73; c 2. PHARMACOLOGY. *Educ:* Univ Cincinnati, BS, 63, PhD(pharmacol), 67. *Prof Exp:* NIH cardiovasc res fel, Bowman Gray Sch Med, 67-69, instr, 68-69; head cardiovasc res, 69-72, asst dir, 72-74, DIR, DEPT PHARMACOL, BRISTOL LABS, 75-, PROJ DIR CARDIOVASC DRUGS, 79- *Concurrent Pos:* NC Heart Asn grants, 68 & 69. *Mem:* Am Soc Pharmacol & Exp Therapeut; assoc Am Physiol Soc; Am Heart Asn; NY Acad Sci. *Res:* Antidysrhythmic, antithrombotic and antihypertensive drug research; myocardial infarction and lethal arrhythmias; tissue catecholamine levels and cardiovascular hemodynamics; endotoxemia and platelet function; control mechanisms affecting cerebral blood flow. *Mailing Add:* Pharmacol Dept Bristol-Myers Co PO Box 5100 Wallingford CT 06492-7660

BUYSKE, DONALD ALBERT, biochemistry, organic chemistry; deceased, see previous edition for last biography

BUZACOTT, J(OHN) A(LAN), b Burwood, Australia, May 21, 37; m 63; c 2. OPERATIONS RESEARCH, INDUSTRIAL ENGINEERING. *Educ:* Univ Sydney, BSc, 57, BE, 59; Univ Birmingham, MSc, 62, PhD(eng prod), 67. *Prof Exp:* Engr, Assoc Elec Indust Ltd, 59-61 & Opers Res Off, Hotpoint Ltd, 63-64; from asst prof to prof indust eng, Univ Toronto, 67-83; prof mgt sci, Univ Waterloo, 84-91; PROF ADMIN STUDIES, YORK UNIV, 91. *Concurrent Pos:* Assoc ed, Int J Prod Res, 70-81, Naval Res Logistics Quart, 84-88, Transactions, Inst Indust Engrs, 84-88, Queueing Syst Theory & Appln, 86-; res scientist, IIASA, Laxenburg, Austria, 79-80; mem, NATO Adv Comt, Adv Study Inst & Adv Res Workshops, 82-85. *Mem:* Opers Res Soc Am; Can Opers Res Soc (pres, 84-85). *Res:* Production and inventory control; reliability; manufacturing system modeling; technological innovation and productivity. *Mailing Add:* Fac Admin Studies York Univ North York ON M3J 1P3 Can

BUZARD, JAMES ALBERT, b Warren, Ohio, Nov 2, 27; m 51; c 2. BIOCHEMISTRY. *Educ:* Kent State Univ, BS, 49; Univ Buffalo, MA, 51, PhD(biochem), 54. *Prof Exp:* Sr res biochemist, Sect Biol, Eaton Labs, Norwich Pharmacal Co, 54-55, unit head, 55-58, group leader, 58-59, chief biochem sect, 59-62, from asst dir to dir biol res, 62-65, dir res, 65-68; dir develop, 68-69, dir res, 69-70, vpres res & develop, 70-72, pres, Searle Labs, 72-75, vpres, 72-75, exec vpres opers, G D Searle & Co, 75-79; EXEC VPRES, MERRELL DOW PHARMACEUT, INC, 79- *Mem:* Am Chem Soc; Am Soc Biol Chem. *Res:* Intermediary metabolism; drug metabolism and mode of action; chemotherapy; analytical biochemistry; research management; commercial management worldwide. *Mailing Add:* Merrell Dow Pharmaceut Inc 10123 Alliance Rd Cincinnati OH 45242

BUZAS, MARTIN A, b Bridgeport, Conn, Jan 30, 34; m 58; c 3. PALEOECOLOGY. *Educ:* Univ Conn, BA, 58; Brown Univ, MSc, 60; Yale Univ, PhD(geol), 63. *Prof Exp:* CUR INVERT PALEONT, SMITHSONIAN INST, 63- *Concurrent Pos:* Mem, Cushman Found. *Mem:* AAAS; Cushman Found; Paleont Soc. *Res:* Quantitative ecology, paleoecology; Benthonic Forminifera. *Mailing Add:* Dept Paleobiol Nat Mus Natural Hist Smithsonian Inst Washington DC 20560

BUZBEE, BILLY LEWIS, b Gorman, Tex, Sept 12, 36; m 62; c 3. NUMERICAL ANALYSIS, MATHEMATICS. *Educ:* Univ Tex, Austin, BA, 61, MA, 62; Univ NMex, PhD(math), 72. *Prof Exp:* Staff mem, Los Alamos Sci Lab, 62-67 & 68-73; analyst, Esso Prod Res Co, 67-68; vis prof comput sci, Chalmers Inst Technol, 73-74; group leader, 74-80, ASST COMPUTER DIV LEADER, LOS ALAMOS NAT LAB, 80- *Concurrent Pos:* Dir, Sci Comput Div, Nat Ctr Atmospheric Res. *Mem:* Asn Comput Mach; Sigma Xi. *Res:* Numerical solution of elliptic and parabolic difference approximations. *Mailing Add:* Sci Computer Div Nat Ctr Atmospheric Res PO Box 3000 Boulder CO 80307

BUZEN, JEFFREY PETER, b Brooklyn, NY, May 28, 43. COMPUTER SCIENCE, QUEUEING THEORY. *Educ:* Brown Univ, ScB, 65; Harvard Univ, ScM, 66, PhD(appl math), 71. *Prof Exp:* Tech staff, Honeywell, 71-75; VPRES, BGS SYSTS INC, 75- *Concurrent Pos:* Adj fac, Harvard Univ, Brown Univ. *Honors & Awards:* A A Michelson Award, 79. *Mem:* Sigmetrics; Asn Comput Mach; Comput Measurement Group. *Res:* Mathematical models of computer system performance; stochastic and operational analysis; queueing network models; capacity planning. *Mailing Add:* BGS Systs Inc 128 Technol Ctr Waltham MA 02254

BUZIN, CAROLYN HATTOX, BIOCHEMISTRY. *Educ:* Univ Tex, Austin, PhD(biochem), 67. *Prof Exp:* ASST RES SCIENTIST, CITY OF HOPE MED CTR, DUARTE, CALIF, 75- *Res:* Differentiation of Drosophila embryonic cells; heat shock protein. *Mailing Add:* 2219 Canyon Rd Arcadia CA 91006

BUZINA, RATKO, b Samobor, Croatia, Yugoslavia, Nov 27, 20. ENDEMIC GOITRE, CORONARY HEART DISEASE. *Educ:* Univ Zagreb, MD, 44, Dipl, 53, ScD, 62. *Prof Exp:* Res asst biochem, Inst Biochem, Univ Basel, 50-51; chief physiol nutrit, Inst Pub Health, Zagreb, 52-54; res assoc physiol nutrit, Lab Physiol Hygiene, Univ Minn, 55-56; assoc prof nutrit, Sch Pub Health, Univ Zagreb, 56-63; head, Dept Nutrit, 72-85, RES CONSULT NUTRIT, INST PUB HEALTH, ZAGREB, 86-; PROF NUTRIT, SCH PUB HEALTH, ZAGREB, 63- *Concurrent Pos:* vis researcher, Dept Nutrit, Vanderbilt Univ, Nashville, 56; prin investr, Epidemiol Coronary Heart Dis, Yugoslavia, 59-66. *Mem:* Am Soc Clin Nutrit; Am Inst Nutrit; Am Pub Health Asn; Int Union Nutrit Sci; Int Soc & Fedn Cardiol. *Res:* Pathophysiology of endemic goitre including the role of goitrogenic factors; epidemiology of coronary heart disease (seven country study) and effects of subclinical micronutrient deficiencies on functional status of the body, primarily physical activity, immunity and cognitive functions. *Mailing Add:* Inst Pub Health Mirogojska 16 Zagreb 41000 Yugoslavia

BUZYNA, GEORGE, b Kovel, Poland, Apr 25, 38; m 65; c 2. FLUID DYNAMICS, HEAT TRANSFER. *Educ:* Ill Inst Technol, BS, 60, MS, 62; Yale Univ, PhD(eng & appl sci), 67. *Prof Exp:* Res staff, Yale Univ, 67-69; asst prof meteorol, 74-75, RES ASSOC, GEOPHYS FLUID DYNAMICS, FLA STATE UNIV, 69-, ASSOC PROF & ASSOC CHMN, DEPT MECH ENG, FAMU/FSU COL ENG, 89- *Mem:* Sigma Xi; Am Soc Mech Engrs; Am Phys Soc; Am Geophys Union. *Res:* Experimental gas dynamics; geophysical fluid dynamics; rotating differentially heated fluid convection; large-scale atmospheric circulation. *Mailing Add:* Dept Mech Engr Fla A&M Univ/Fla State Univ Col Eng Tallahassee FL 32316-2175

BUZZELL, JOHN GIBSON, b Delavan, Wis, Apr 29, 22; m 50; c 3. POLYMER CHEMISTRY. *Educ:* Univ Wis, BS, 48; Univ Iowa, PhD(chem), 55. *Prof Exp:* Chemist, Sherwin-Williams, 48-51; res chemist, 55-67, SR RES CHEMIST, E I DU PONT DE NEMOURS & CO, INC, 67- *Mem:* Am Chem Soc; Sigma Xi. *Res:* Photo polymer process research. *Mailing Add:* 1708 Woodledge Circle State College PA 16803

BUZZELL, RICHARD IRVING, b Lincolnville, Maine, May 1, 29; m 56; c 4. PLANT BREEDING, PLANT GENETICS. *Educ:* Univ Maine, BS, 58; Iowa State Univ, MS, 60, PhD(plant breeding), 62. *Prof Exp:* RES SCIENTIST BREEDING & GENETICS, CAN DEPT AGR, 62-, HEAD, CROP SCI SECT, HARROW RES STA, 76- *Concurrent Pos:* Mem assoc grad fac, crop sci dept, Univ Guelph; ed, Can J Plant Sci, 89-91. *Mem:* Can Soc Agron; hon mem Can Seed Growers Asn. *Res:* Genetics of soybeans and corn; variety development. *Mailing Add:* Box 373 Harrow ON N0R 1G0 Can

BUZZELLI, DONALD EDWARD, b Detroit, Mich, July 24, 36. MISCONDUCT IN SCIENCE-ENGINEERING RESEARCH & EDUCATION, SCIENCE TECHNOLOGY STATISTICS. *Educ:* Cornell Univ, BChE, 59, Mass Inst Technol, ScD(chem eng), 64; Fordham Univ, MA, 71, PhD(philos), 74. *Prof Exp:* From res engr to sr res engr chem eng, Chevron Res Co, 64-69; prog analyst, 75-89, STAFF ASSOC FOR OVERSIGHT, OFF INSPECTOR GEN, NSF, 89- *Mem:* AAAS. *Mailing Add:* 2400 Pennsylvania Ave NW Washington DC 20037

BUZZELLI, EDWARD S, b Cleveland, Ohio, July 31, 39; m 60; c 4. PHYSICS, ELECTROCHEMISTRY. *Educ:* John Carroll Univ, BS, 62, MS, 65. *Prof Exp:* From jr physicist to physicist, Res & Develop Lab, Standard Oil Co Ohio, 62-65, sr physicist & proj leader, 65-67, sr res physicist, 67-70; sr engr, Westinghouse Elec Corp, 70-73, proj mgr, 73-80, prog mgr, 80, spec projs venture, 84-90, MGR ELECTROCHEM TECHNOL, RES & DEVELOP CTR, WESTINGHOUSE ELEC CORP, 80-, MGR, ADVAN BATTERIES, 90- *Mem:* Electrochem Soc; Marine Tech Soc; Am Defense Preparedness Asn. *Res:* Fused salt electrochemistry; secondary batteries; reactions in fused salt media; solid state reactions and devices; air electrode and metal-air battery development; alkaline and lithium batteries; fuel cells; hydrogen production; specialty batteries; electrolysis. *Mailing Add:* 39 Morris St Export PA 15632

BYALL, ELLIOTT BRUCE, b Fall River, Mass, May 31, 40; m 62; c 4. FORENSIC SCIENCE. *Educ:* Sterling Col, BS, 62; Ore State Univ, PhD(org chem), 67. *Prof Exp:* Res chemist, E I Du Pont de Nemours & Co, Inc, 67-71; forensic chemist, 71-80, FORENSIC BR CHIEF, BUR ALCOHOL, TOBACCO & FIREARMS, SAN FRANCISCO LAB CTR, 81- *Mem:* Int Asn Identification. *Res:* New methods for forensic analysis; explosive and gunshot residue examinations. *Mailing Add:* San Francisco Lab Ctr Bur Alcohol Tobacco & Firearms 355 N Wiget Lane Suite 8 Walnut Creek CA 94598

BYAR, DAVID PEERY, b Lockland, Ohio, Feb 23, 38. BIOMETRY RESEARCH. *Educ:* Emory Univ, AB, 60; Harvard Med Sch, MD, 64. *Prof Exp:* Am Urol Asn fel, Armed Forces Inst Path, 65-66, Nat Cancer Inst fel, 66-68; med officer res, Biomet Br, DCCP, 68-72, head, Clin & Diag Trails Sect, 72-85, CHIEF, BIOMET BR, DIV CANCER PREV & CONTROL, NAT CANCER INST, 85- *Mem:* Am Statist Asn; Soc Epidemiol Res; Soc Clin Trials; Int Statist Inst. *Res:* Biostatistics as applied to medicine, special interest etiology, behavior, prevention and treatment of cancer. *Mailing Add:* Nat Cancer Inst Biomet Br NIH Exec Plaza N Rm 344 6130 Executive Blvd Bethesda MD 20892

BYARD, JAMES LEONARD, b Hartwick, NY, Dec 1, 41; c 2. TOXICOLOGY, BIOCHEMISTRY. *Educ:* Cornell Univ, BS, 64; Univ Wis, MS, 66, PhD(biochem), 68; Am Bd Toxicol, dipl. *Prof Exp:* Arthritis fel, Harvard Med Sch, 68-70; asst prof toxicol & biochem, Albany Med Col, 70-74; from asst prof to assoc prof environ toxicol, Univ Calif, Davis, 74-84; CONSULT TOXICOLOGIST, 84- *Mem:* Soc Toxicol; Soc Environ Toxicol & Chem; AAAS. *Res:* Metabolism and molecular mechanism of toxic action of environmental chemicals; primary hepatocyte culture; chemical carcinogenesis. *Mailing Add:* PO Box 2556 El Macero CA 95618

BYARD, PAMELA JOY, b Rochester, NY, Sept 16, 51. POPULATION GENETICS, HUMAN GROWTH. *Educ:* State Univ NY, Albany, BA, 75, MA, 77; Univ Kans, PhD(biol anthrop), 81. *Prof Exp:* Res analyst, Shook Hardy & Bacon, 80-81; res assoc growth & genetics, Fels Res Inst, Sch Med, Wright State Univ, 81-; DEPT ANTHROP, RAINBOW BABY'S & CHILDREN HOSP, CASE WESTERN RESERVE UNIV. *Mem:* Am Anthrop Asn; Am Asn Phys Anthrop; Human Biol Coun. *Res:* Demography and population history of St Lawrence Island; human growth, fertility, population genetics and quantitative genetics. *Mailing Add:* Pulmonary Pediat Rainbow Baby's & Children's Hosp Cleveland OH 44106

BYARS, EDWARD F(ORD), b Lincolnton, NC, Mar 22, 25; m 50; c 4. MECHANICAL ENGINEERING & ENGINEERING MECHANICS. *Educ:* Clemson Univ, BME, 46, MCE, 53; Univ Ill, PhD(appl mech), 57. *Prof Exp:* From asst prof to assoc prof eng mech, Clemson Univ, 47-60; prof mech eng & chmn dept, WVa Univ, 60-80; exec asst to pres, Clemson Univ, 80-85; RETIRED. *Concurrent Pos:* Consult, biomech & exp mech. *Mem:* Am Soc Eng Educ; Soc Exp Stress Anal; Am; Sigma Xi. *Res:* Author engineering mechanics textbooks. *Mailing Add:* Four River Point Clemson SC 29631

BYATT, PAMELA HILDA, b Cape Town, SAfrica, Nov 28, 16; nat US. VIROLOGY. *Educ:* Univ Calif, Los Angeles, BA, 43, MA, 47, PhD(microbiol), 50. *Prof Exp:* Asst microbiol, Univ Wash, 50-52; virologist, Dept Med Microbiol & Immunol & Clin Labs, Univ Calif, Los Angeles, 53-73, chief Virol Lab, Microbiol Sect, Clin, Ctr Health Sci, 73-80; RETIRED. *Mem:* AAAS; Am Soc Microbiol; Tissue Cult Asn; NY Acad Sci. *Res:* Staphylococci; antibiotics; viruses; microbiology. *Mailing Add:* 1223 19th St Santa Monica CA 90404

BYCK, ROBERT, b Newark, NJ, Apr 26, 33; m 76; c 4. PHARMACOLOGY, PSYCHIATRY. *Educ:* Univ Pa, AB, 54, MD, 59. *Hon Degrees:* MA, Yale Univ, 77. *Prof Exp:* Res assoc, NIH, 60-62; sr fel pharm, Yeshiva Univ, 62-64, asst prof rehab med & pharmacol, 64-69; fel psychiat & lectr pharmacol, 69-72, assoc prof, 72-77, PROF PSYCHIAT & PHARMACOL, YALE UNIV, 77- *Concurrent Pos:* NIMH sr fel & res career develop award, Dept Pharmacol, Albert Einstein Col Med, 67-69; Burroughs Wellcome Fund scholar clin pharmacol, 72-; dir psychiat consult & assoc physician, Yale New Haven Hosp, 72-; mem adv bd, Med Letter on Drugs & Therapeut, 75-90; resident fel, Pierson Col, 76-80. *Mem:* Am Soc Pharmacol & Exp Therapeut; AAAS; Am Soc Clin Pharmacol & Therapeut; Am Col Neuropsychopharmacol. *Res:* Neuropharmacology; clinical pharmacology; psychopharmacology; magnetic resonance imaging; biological basis of positive mood using pharmacological probes and magnetic resonance imaging. *Mailing Add:* Dept Pharmacol Yale Univ Sch Med New Haven CT 06510-8066

BYDAL, BRUCE A, b Chicago, Ill, Mar 2, 37; m 64; c 3. CHEMICAL ENGINEERING, APPLIED MATHEMATICS. *Educ:* Univ Ill, BS, 59; Univ Minn, PhD(chem eng), 62. *Prof Exp:* Res engr, Eng Res Lab, Del, 62-63; res engr, Textile Fibers Dept, Va, 63-67, sr res engr, 67-73, sr engr, Appl Math Group, 73-76, CONSULT, APPL MATH GROUP, E I DU PONT DE NEMOURS & CO, INC, 76- *Res:* Two phase flow; mixing; initiating explosives. *Mailing Add:* 2912 Jaffe Rd Wilmington DE 19808-1918

BYDALEK, THOMAS JOSEPH, b Grand Rapids, Mich, Apr 22, 35; m 57; c 4. ANALYTICAL CHEMISTRY, INORGANIC CHEMISTRY. *Educ:* Aquinas Col, BS, 57; Purdue Univ, PhD(anal chem), 61. *Prof Exp:* From instr to asst prof chem, Univ Wis, 61-65; assoc prof, 65-68, PROF CHEM, UNIV MINN, DULUTH, 68- *Mem:* Am Chem Soc. *Res:* Electrochemical methods development for analysis of metal ions at trace levels in natural waters; computer interfacing of analytical instrumentation; kinetics of multidentate Ligand complexes of the transition metals and their application in analysis. *Mailing Add:* Dept Chem Univ Minn Duluth MN 55812

BYER, NORMAN ELLIS, b Brooklyn, NY, July 17, 40; m 64; c 3. SOLID STATE PHYSICS. *Educ:* Cooper Union, BEE, 62; Princeton Univ, MSE, 63; Cornell Univ, PhD(appl physics), 67. *Prof Exp:* Res assoc appl physics, Cornell Univ, 66-67; mem sci staff injection laser studies, RCA Labs, 67-69; res scientist infrared quantum counter detector res, Res Inst Advan Studies, 69-74, sr res scientist infrared detector develop, 74-79, MGR, SEMICONDUCTOR PHYSIC DEPT, MARTIN MARIETTA LABS, 79- *Honors & Awards:* Indust Res 100 Award, 78. *Mem:* Am Phys Soc; Inst Elec & Electronics Engrs. *Res:* Research and development of infrared and microwave detection including narrow bandgap semiconductor detectors; pyroelectric detectors and millimeter wave devices such as impatt's and mixers; optically active defects in infrared detection materials; studies of pyroelectric and other infrared detectors. *Mailing Add:* Martin Marietta Labs 1450 S Rolling Rd Baltimore MD 21227

BYER, PHILIP HOWARD, b Washington, DC, Jan 16, 48; m 70; c 2. CIVIL ENGINEERING, PROJECT DECISION MAKING. *Educ:* Mass Inst Technol, SB, 72, SM, 72, PhD(civil eng), 75. *Prof Exp:* Asst prof, 75-80, ASSOC PROF CIVIL ENG, UNIV TORONTO, 80- *Concurrent Pos:* Chmn, Ont Environ Assessment Adv Comt, 86- *Mem:* Am Soc Civil Eng; Soc Risk Anal. *Res:* Development and application of systems analysis, risk assessment and evaluation techniques for public projects; solid waste management, transportation and water resources planning; environmental impact assessment. *Mailing Add:* Dept Civil Eng Univ Toronto Toronto ON M5S 1A4 Can

BYER, ROBERT L, b Glendale, Calif, May 9, 42; m; c 4. RESEARCH ADMINISTRATION. *Educ:* Univ Calif, Berkeley, BA, 64; Stanford Univ, MS, 67, PhD(appl physics), 69. *Prof Exp:* Mem, Spectra Physics, Mountain View, Calif, 64-65; from asst prof to assoc prof appl physics, Stanford Univ, 69-79, dept chair, 80-83, assoc dean humanities & sci, 84-86, PROF APPL PHYSICS, STANFORD UNIV, 79-, DEAN RES, 87- *Concurrent Pos:* Sloan fel, 74-77; mem, Educ Comt, Optical Soc Am, 83-86, bd dirs, 86-89; guest prof, Tshinghua Univ, Beijing, China, 86; mem, Calif Coun Sci & Technol, 89-; mem, Eng Adv Comn, NSF, 90-; consult lasers & nonlinear optics. *Honors & Awards:* Adolph Lomb Medal, Optical Soc Am, 72; R V Pole Mem Lectr, Conf on Lasers & Electrooptics, 87. *Mem:* Nat Acad Eng; AAAS; Am Phys Soc; fel Inst Elec & Electronics Engrs; Inst Elec & Electronics Engrs Lasers & Electrooptics Soc (pres, 84-85); fel Optical Soc Am. *Res:* Diode pumped solid state lasers; nonlinear optics; nonlinear materials; optical parametric oscillators; coherent antistokes raman spectroscopy; laser remote sensing; optical tomography; slab lasers; soft x-ray lithography and microscopy; single crystal fiber growth and applications; laser diode pumped solid state laser sources. *Mailing Add:* Stanford Univ Cypress Hall Rm C-14 Stanford CA 94305-4147

BYERLEY, LAURI OLSON, b Peoria, Ill, Mar 25, 57; m 81. TUMOR METABOLISM, SEPSIS CYTOKINES. *Educ:* Iowa State Univ, BS, 79; Purdue Univ, MS, 81; Univ Calif, Los Angeles, PhD(pub health & nutrit sci), 87. *Prof Exp:* Clin dietitian specialist, Nutrit Dept, Stanford Univ Hosp, Calif, 88-89; res dietitian, Clin Study Ctr, 81-83, res assoc, Res & Educ Inst, 83-87, POSTDOCTORAL FEL, DEPT ENDOCRINOL, NUTRIT & METAB, HARBOR-UNIV CALIF LOS ANGELES MED CTR, TORRANCE, 89-, INSTR, DEPT PEDIAT, 90- *Concurrent Pos:* Chief dietitian, Div Clin Nutrit, Sch Med, Univ Calif, Los Angeles, 84-88; postdoctoral fel, Div Gen Surg, Stanford Med Ctr, Calif, 88-89; lectr nutrit, Dept Home Econ, Calif State Univ, Long Beach, 90-91; prin investr, individual NRSA, Nat Cancer Inst, NIH, 91- *Mem:* Am Soc Clin Nutrit; Am Inst Nutrit; Am Dietetic Asn; Sigma Xi. *Res:* Stable isotopes to trace cellular metabolism; tumor cell metabolism; cytokines effects on cellular metabolism. *Mailing Add:* Div Endocrinol Nutrit & Metab Harbor-Univ Calif Los Angeles Med Ctr 1124 W Carson St RBI Torrance CA 90502

BYERLY, DON WAYNE, b Wilmington, Del, Mar 8, 33; m 55; c 4. GEOLOGY. *Educ:* Col Wooster, AB, 55; Univ Tenn, MS, 57, PhD(geol), 66. *Prof Exp:* Instr geol, Univ Tenn, 56-66; asst prof, Murray State Univ, 66-67; asst prof, 67-81, ASSOC PROF GEOL, UNIV TENN, KNOXVILLE, 82- *Concurrent Pos:* Consult, Tenn Div Geol, 63-65 & Oak Ridge Nat Lab, 73-82. *Mem:* Sigma Xi; Geol Soc Am; Nat Asn Geol Teachers; Am Assoc Prof Geologists; Assoc Ground Water Scientists & Engrs; Asn Eng Geol. *Res:* Areal and environmental geological mapping; high and low nuclear waste disposal in geological media; abandoned mine sealing; strip mine reclamation; acid drainage related to mining and highway construction. *Mailing Add:* Dept Geol Univ Tenn Knoxville TN 37916

BYERLY, GARY RAY, b Paducah, Ky, Apr 13, 48. PETROLOGY. *Educ:* Mich State Univ, BS, 70, MS, 72, PhD(geol), 74. *Prof Exp:* Fel geol, Smithsonian Inst, 74-75; res fel, US Geol Surv, 75-76; asst prof, 77-80, ASSOC PROF, LA STATE UNIV, BATON ROUGE, 80- *Concurrent Pos:* Res assoc, Smithsonian Inst, 75-; consult prof, US Geol Surv, 76- *Mem:* Geol Soc Am; Am Geophys Union. *Res:* Chemical and mineralogical variation of basalts and formation of ocean floor; the textures of igneous and metamorphic rocks; early evolution of the earth's crust, particularly the Archean of South Africa and Australia. *Mailing Add:* 2611 Lydia Ave Baton Rouge LA 70808

BYERLY, PAUL ROBERTSON, JR, b Lancaster, Pa, Sept 4, 22; m 47; c 5. PHYSICS. *Educ:* Washington & Jefferson Col, AB, 43; Univ Pa, PhD, 51. *Prof Exp:* Instr physics, Washington & Jefferson Col, 43-44; jr physicist, Tenn Eastman Corp, 44-45; instr physics, Univ Pa, 45-50; physicist, Radiation Lab, Univ Calif, 50-58; educ adv physics, Govt Philippines, 58-63; ASSOC PROF PHYSICS, UNIV NEBR, LINCOLN, 63- *Concurrent Pos:* Regional counr, physics, Nebr, 65-71. *Mem:* Am Phys Soc; Am Asn Physics Teachers. *Res:* Mass spectroscopy; nuclear physics; Mossbauer effect; computer graphics; computer assisted instruction. *Mailing Add:* Dept Physics Univ Nebr Lincoln NE 68508

BYERLY, PERRY EDWARD, b Berkeley, Calif, Feb 2, 26; m 59; c 1. GEOPHYSICS. *Educ:* Univ Calif, AB, 49; Harvard Univ, AM, 51, PhD(geophys), 54. *Prof Exp:* From geophysicist to sr res geophysicist, US Geol Surv, 52-63; geophys rep, Calif Res Corp, Houston, 64-66; sr geophysicist, Chevron Explor Co, 67-69, sr res geophysicist, Chevron Oilfield Res Co, 69-70; SR RES SCIENTIST, CONTINENTAL OIL CO, 70- *Concurrent Pos:* Asst prof, Johns Hopkins Univ, 54-55. *Mem:* Am Geophys Union; Seismol Soc Am; Soc Explor Geophys. *Res:* Explosion seismology; gravity and magnetics; geothermal problems; general geophysics. *Mailing Add:* 4310 E 66th St No C Tulsa OK 74136

BYERRUM, RICHARD UGLOW, b Aurora, Ill, Sept 22, 20; m 45; c 4. BIOCHEMISTRY. *Educ:* Wabash Col, AB, 42; Univ Ill, Urbana, PhD(biochem, org chem), 47. *Hon Degrees:* DSc, Wabash Col, 67. *Prof Exp:* From instr to prof chem, 47-59, asst provost, 59-62, actg dir biol & med, 61-62, dean, Col Natural Sci, 62-86,PROF BIOCHEM, MICH STATE UNIV, 57- *Concurrent Pos:* Sabbatical leave, Calif Polytech Inst, 56 & Scripts Inst Oceanog, 73; travel awards, Int Cong Biochem, Vienna, 58 & Montreal, 59; consult, Am Chem Soc, 58-62; vis scientist, 62-64; mem bd, Mich Health Coun, 61-; consult, NCent Asn Sec Schs & Cols, 62- & Nat Acad Sci, 72-74. *Mem:* AAAS; Am Chem Soc; Am Soc Plant Physiol; Soc Exp Biol & Med; Am Soc Biochem Molecular Biol. *Res:* Metabolism in plants with special emphasis on the production of glycerol; glycerol phosphate during photosynthesis. *Mailing Add:* 305 Biochem Bldg Mich State Univ East Lansing MI 48824

BYERS, BENJAMIN ROWE, b Austin, Tex, Nov 7, 36; m 57; c 2. MICROBIOLOGY. *Educ:* Univ Tex, Austin, BA, 58, MA, 60, PhD(microbiol), 65. *Prof Exp:* Bacteriologist, Standard Brands, Inc, 60-62; NIH fel, Univ Tex, Austin, 65-66, asst prof microbiol, 66; from asst prof to assoc prof, 67-73, PROF MICROBIOL, MED CTR, UNIV MISS, 73- *Concurrent Pos:* NIH res career develop award, Med Ctr, Univ Miss, 71-75, NIH res grant, 71-, NIH contract, 74-78; Cooley's Anemia Found res grant, 78-82; Am Heart Asn res grant, 83-85. *Mem:* AAAS; Am Soc Microbiol. *Res:* Iron and trace metal metabolism; mechanisms of iron and trace metal transport; iron chelating agents in treatment of iron storage disease; iron and infection. *Mailing Add:* Dept Microbiol Univ Miss Med Ctr Jackson MS 39216-4505

BYERS, BRECK EDWARD, b St Louis, Mo, July 4, 39; m 64; c 2. GENETICS OF YEAST CELL CYCLE & MEOISIS. *Educ:* Univ Colo, BS, 61; Harvard Univ, PhD(biol), 67. *Prof Exp:* PROF GENETICS & ADJ PROF BIOCHEM, UNIV WASH, 71-; CHAIR, 91- *Mem:* Am Soc Cell Biol; Am Soc Microbiol; Genetics Soc Am. *Res:* Employ molecular genetics and cytological methods in yeast to analyze mechanisms that control microtubule; distribution and cell cycle progression in mitosis and that regulate chromosomal organization and recombination in meiosis. *Mailing Add:* Dept Genetics Univ Wash SK50 Seattle WA 98195

BYERS, CHARLES HARRY, b Buckingham, Que, Feb 29, 40; m 64; c 3. CHEMICAL ENGINEERING. *Educ:* McGill Univ, BEng, 62; Univ Calif, Berkeley, PhD(chem eng), 66. *Prof Exp:* Asst prof chem eng, Univ Rochester, 66-70; res chem engr, Merichem Co, 70-77; prin engr, Polaroid Corp, 77-80; GROUP LEADER, OAK RIDGE NAT LAB, 80-; PROF CHEM ENG, UNIV TENN, 84- *Mem:* Am Inst Chem Engrs; Am Chem Soc. *Res:* Heat and mass transport; interfacial phenomena; laser light scattering. *Mailing Add:* 129 Valley Ct Oak Ridge TN 37830

BYERS, CHARLES WESLEY, b Philadelphia, Pa, Aug 23, 46; m 82; c 2. GEOLOGY. *Educ:* Marietta Col, BS, 68; Yale Univ, MPh, 71, PhD(geol), 73. *Prof Exp:* PROF GEOL, UNIV WIS-MADISON, 73- *Mem:* Geol Soc Am; Paleont Soc; Soc Econ Paleontologists & Mineralogists; Sigma Xi; AAAS. *Res:* Stratigraphy and paleoecology of cretaceous rocks; deposition of shales; cambrian paleoenvironments; trace fossils. *Mailing Add:* 1215 W Dayton St Madison WI 53706

BYERS, FLOYD MICHAEL, b Rush City, Minn, Feb 8, 47; m 67; c 3. ANIMAL NUTRITION. *Educ:* Univ Minn, BS, 69; SDak State Univ, MS, 72; Colo State Univ, PhD(ruminant nutrit), 74. *Prof Exp:* Lab technician plant res, Univ Minn, 65-67, nutrit res, 67-69; feedlot mgr, SDak State Univ, 70-71, dir, Ruminant Nutrit Lab, 71-72; res asst energy metab, Colo State Univ, 72-73, NDEA res fel, 73-74; asst prof beef cattle nutrit, Ohio Agr Res & Develop Ctr, 75-80; assoc prof, 80-86, PROF BEEF CATTLE NUTRIT & GROWTH, TEX A&M UNIV, 86- *Concurrent Pos:* Consult, Food Safety Inspection Serv, USDA, 87. *Mem:* Am Soc Animal Sci; Am Inst Nutrit; Animal Nutrit Res Coun; Sigma Xi. *Res:* Investigations concerning protein and energy requirements and efficiency of protein and energy utilization for maintenance and production of cattle, varying in mature size, fed rations varying in available energy. *Mailing Add:* Animal Sci Dept Texas A&M Univ College Station TX 77843

BYERS, FRANK MILTON, JR, b Moline, Ill, Mar 5, 16; m 45; c 4. PETROGRAPHY OF IGNEOUS ROCKS, GEOCHEMISTRY OF IGNEOUS ROCKS. *Educ:* Augustana Col, AB, 38; Univ Chicago, PhD(geol), 55. *Prof Exp:* Geologist, US Geol Surv, 41-81; petrologist, 82-88, LAB ASSOC, LOS ALAMOS NAT LAB, 88- *Concurrent Pos:* Guest assoc ed, J Geophys Res, 87-88. *Mem:* Fel Geol Soc Am; Am Geophys Union. *Res:* Mineralogy, petrology and geochemistry of silicic volcanic rocks. *Mailing Add:* 125 Everett St Lakewood CO 80226

BYERS, GEORGE WILLIAM, b Washington, DC, May 16, 23; m 55. ENTOMOLOGY. *Educ:* Purdue Univ, BS, 47; Univ Mich, MS, 48, PhD, 52. *Prof Exp:* Rackham fel zool, Univ Mich, 52-53; res med entomologist, US Army, Korea & Japan, 53-56; from asst prof to prof entom, Univ Kans, 56-88, asst cur, Snow Entom Mus, 56-73, chmn dept entom, 69-72, 84-87, cur, Snow Entom Mus, 74-88, dir, 83-88; RETIRED. *Concurrent Pos:* Ed, Syst Zool, Soc Syst Zool, 63-66; vis prof, Biol Sta, Univ Va, 61-90 & Univ Minn, 70. *Mem:* Soc Study Evolution; Soc Syst Zool; Entom Soc Am; Am Soc Naturalists; Entom Soc Can. *Res:* Biology and classification of Tipulidae and Mecoptera. *Mailing Add:* Dept Entom Univ Kans Lawrence KS 66045-2119

BYERS, HORACE ROBERT, b Seattle, Wash, Mar 12, 06; m 27; c 1. METEOROLOGY, CLOUD PHYSICS. *Educ:* Univ Calif, AB, 29; Mass Inst Technol, MS, 32, ScD(meteorol), 35. *Prof Exp:* Asst, Mass Inst Technol, 30-32 & Scripps Inst, Univ Calif, 32-33; instr meteorol, Transcontinental & Western Air, 33-35; meteorologist, US Weather Bur, 35-40; from assoc prof to prof meteorol, Univ Chicago, 40-65, chmn dept, 48-60; dean col geosci, 65-68, acad vpres, 68-71, distinguished prof, 65-74, EMER PROF METEOROL, TEX A&M UNIV, 74- *Concurrent Pos:* Dir, US Interdept Thunderstorm Proj, 46-50; vpres, Int Asn Meteorol & Atmospheric Physics, 54-60, pres, 60-63; chmn bd, Univ Corp Atmospheric Res, 63-65; chmn bd, Gulf Univ Res Corp, 66-69; vis prof, Univ Clermont-Ferrand, France, 75. *Honors & Awards:* Losey Award, Am Inst Aeronaut & Astronaut, 41; Brooks Award, Am Meteorol Soc, 60, Cleveland Abbe Award, 78. *Mem:* Nat Acad Sci; Am Meteorol Soc (pres, 52-53). *Res:* Thunderstorms; physical and dynamic meteorology. *Mailing Add:* 300 Hot Springs Rd No 178 Santa Barbara CA 93108

BYERS, JOHN ROBERT, b Stoughton, Sask, July 27, 37; m 63; c 2. ENTOMOLOGY. *Educ:* Univ Sask, BSA, 62, MSc, 63, PhD(entom, physiol), 66. *Prof Exp:* Nat Res Coun Can overseas fel, 66-67; asst prof biol, Univ Sask, 67-68; RES SCIENTIST, CAN DEPT AGR, 68- *Mem:* Can Soc Zoologists; Entom Soc Can; Entom Soc Am. *Res:* Behavior and ecology of noctuid moths; agricultural entomology. *Mailing Add:* Res Sta Agr Can Lethbridge AB T1J 4B1 Can

BYERS, LARRY DOUGLAS, b Los Angeles, Calif, Feb 18, 47. BIOCHEMISTRY. *Educ:* Univ Calif, Los Angeles, BS, 68; Princeton Univ, MS, 70, PhD(biochem), 72. *Prof Exp:* Fel enzym, Univ Calif, Berkeley, 72-75; from asst prof to assoc prof chem, 75-86, PROF CHEM, TULANE UNIV, 86- *Concurrent Pos:* NIH fel, 73-75. *Mem:* Am Chem Soc; Am Soc Biochemists & Molecular Biol. *Res:* Mechanism of action of enzymes; design of enzyme inhibitors. *Mailing Add:* Dept Chem Tulane Univ New Orleans LA 70118

BYERS, LAWRENCE WALLACE, b Pulaski, Pa, Nov 12, 16; m 56; c 2. BIOCHEMISTRY, NUTRITION. *Educ:* Westminster Col, Pa, BS, 38; Oberlin Col, MA, 40; Univ Ill, PhD(biochem), 48. *Prof Exp:* Teaching asst chem, Oberlin Col, 38-40; res asst, Exp Sta, Mich State Univ, 40-42; asst org res, Dow Chem Co, Mich, 42-45; teaching asst, Univ Ill, 45-46; res biochemist, Bristol Labs, 48-52; res biochemist, Ft Detrick, Md, 52-56; res biochemist, Dept Psychiat & Neurol, Sch Med, Tulane Univ, 56-66, asst prof biochem, 60-66; from asst prof to assoc prof, 66-78, RES PROF BIOCHEM, CTR HEALTH SCI, UNIV TENN, MEMPHIS, 78- *Concurrent Pos:* Biochem consult, Vet Hosp, Gulfport, Miss, 64-66 & USPHS grant; res biochemist, dept path, Baptist Mem Hosp, Memphis, 66-85, Univ Tenn, Memphis, 85- *Honors & Awards:* Sigma Xi. *Mem:* AAAS; NY Acad Sci; Am Asn Clin Chemists. *Res:* Carotenoids; fatty acids; silicones; phospholipids; antibiotics; nutrition; intermediary metabolism, biochemical aspects of mental illness; antihypertensive factors of kidney medulla. *Mailing Add:* 3138 Dumbarton Rd Memphis TN 38128

BYERS, NINA, b Los Angeles, Calif, Jan 19, 30; wid. PARTICLE PHYSICS, THEORETICAL PHYSICS. *Educ:* Univ Calif, BA, 50; Univ Chicago, MS, 53, PhD(physics), 56. *Hon Degrees:* MA, Oxford Univ, 67. *Prof Exp:* Res fel math physics, Univ Birmingham, 56-58; res assoc, Stanford Univ, 58-59, asst prof, 59-61; from asst prof to assoc prof, 61-67, PROF PHYSICS, UNIV CALIF, LOS ANGELES, 67- *Concurrent Pos:* Guggenheim fel & vis mem, Inst Advan Study, 63-64; fel, Somerville Col, 67-68, Janet Watson vis fel, 68-74; fac lectr, Oxford Univ, 67-68, vis scientist, 69-74 & 77-78; sci res council fel, Oxford Univ, 78; councillor-at-large, Am Phys Soc, 77-81, comt mem on opportunities in physics 80-83, panel on public affairs, 80-82, vice chmn forum on physics, 81, chmn forum on physics, 82; comt mem, Sect Physics, AAAS, 83-87. *Mem:* Fel Am Phys Soc; Fedn Am Scientists; fel AAAS. *Res:* Superconductivity, theoritical particle physics. *Mailing Add:* Dept Physics Univ Calif Los Angeles CA 90024

BYERS, R LEE, b Hagerstown, Md, Dec 20, 35; m 57; c 4. CHEMICAL ENGINEERING, AIR POLLUTION. *Educ:* Juniata Col, BA, 58; Carnegie-Mellon Univ, BS, 58; Univ Rochester, MS, 60; Pa State Univ, PhD(chem eng), 67. *Prof Exp:* From instr to asst prof physics & pre-eng, Elizabethtown Col, 59-63; asst prof mech eng, Pa State Univ, 68-70; assoc prof chem eng, Univ NH, 70-72; mem eng staff, Environ Control & Safety Div, Exxon Res & Eng Co, 72-80. *Mem:* Sigma Xi; Am Inst Chem Engrs; Air Pollution Control Asn. *Res:* Gas purification; small particle dynamics; ambient air quality evaluation. *Mailing Add:* 131 Boxfield Rd Pittsburgh PA 15241

BYERS, ROBERT ALLAN, b Latrobe, Pa, Dec 6, 36; m 60; c 2. ENTOMOLOGY. *Educ:* Pa State Univ, BSc, 60; Ohio State Univ, MSc, 61; Purdue Univ, PhD(entom), 71. *Prof Exp:* Res entomologist, Entom Res Div, Tifton, Ga, 61-66; Lafayette, Ind, 66-70, RES ENTOMOLOGIST, US REGIONAL PASTURE RES LAB, AGR RES SERV, USDA, UNIVERSITY PARK, PA, 70- *Mem:* Entom Soc Am. *Res:* Biology and control of the two-lined spittlebug on Coastal bermuda grass; effect of a complex of insects on the yield of Coastal bermuda grass; ability of the Hessian fly to stunt winter wheat; host plant resistance to alfalfa blotch leafminer, and clover root curculio in alfalfa; invertebrate control on legumes in conservation-tillage systems. *Mailing Add:* US Regional Pasture Res Lab University Park PA 16802

BYERS, ROLAND O, b Boston, Ohio, Sept 9, 19; m 44; c 2. INDUSTRIAL TECHNOLOGY, ENGINEERING GRAPHICS. *Educ:* Ohio Univ, BS, 46, MS, 49. *Prof Exp:* Assoc prof eng graphics, Univ Wichita, 46-51, head dept, 53-54; prof gen eng & chmn dept, Univ Idaho, 54-81; RETIRED. *Mem:* Am Soc Eng Educ. *Res:* Engineering descriptive geometry. *Mailing Add:* 821 Nez Perce St Moscow ID 83843

BYERS, RONALD ELNER, b Everett, Wash, Nov 8, 36; m 60; c 2. THEORETICAL PHYSICS. *Educ:* Wash State Univ, BS, 66, PhD(physics), 74. *Prof Exp:* Mathematician, Strategy & Tactics-Anal Group, US Army, 60-62; technologist, Battelle Northwest Lab, 62-65; teacher, 70-77, ASSOC PROF PHYSICS, CENT COL, 78- *Mem:* Am Asn Physics Teachers. *Res:* Quantum hidden variables theory unifying essentials of quantum mechanics and general relativity; evolution processes in scientific theory making; philosophical grounding of scientific theories. *Mailing Add:* Dept Nat Sci Cent Col Pella IA 50219

BYERS, SANFORD OSCAR, b New York, NY, May 1, 18; m 42; c 5. BIOCHEMISTRY. *Educ:* Rensselaer Polytech Inst, BSc, 39; Univ Cincinnati, PhD(biochem), 44. *Prof Exp:* Instr biochem & physiol, Washington Univ, 44-45; res biochemist, Nat Drug Co, Philadelphia, 45-46; res biochemist, Harold Brunn Inst, Mt Zion Hosp, 46-78; RETIRED. *Concurrent Pos:* Fel coun arteriosclerosis, Am Heart Asn. *Mem:* AAAS; Reticuloendothelial Soc; Am Chem Soc; Am Soc Exp Biol & Med; Am Physiol Soc. *Res:* Atherosclerosis; metabolism of digitalis, cholesterol, and lipids; catecholamines. *Mailing Add:* 40 New Hall Dr San Rafael CA 94901-1319

BYERS, STANLEY A, b Ashland, Ohio, Jan 9, 31; m 66; c 2. CERAMICS. *Educ:* Antioch Col, BA, 55; Mich State Univ, BS, 57; Case Western Reserve Univ, PhD(mat sci), 74. *Prof Exp:* Metall & ceramics engr, Ohio Brass Co, 58-63; ceramic engr, Am Standard, Inc, 63-69 & Carborundum Co, 69-70; res scientist mat, Res Exp Sta, Ga Inst Technol, 72-75; mem staff, 75-80, MGR MAT RES, BALL CORP, 80-, ASSOC PROF INDUST & TECHNOL. *Mem:* Am Ceramic Soc; Am Soc Metals. *Res:* Glass processing and formulation; glass coatings. *Mailing Add:* Dept Indust & Technol Ball State Univ Muncie IN 47306

BYERS, THOMAS JONES, b Philadelphia, Pa, Oct 12, 35; m 60; c 2. GENETICS. *Educ:* Cornell Univ, AB, 58; Univ Pa, PhD(zool), 62. *Prof Exp:* USPHS fel biophys, Carnegie Inst Dept Terrestrial Magnetism, 62-64; asst prof zool & entom, Ohio State Univ, 64-68, dir develop biol prog, 72-75, assoc prof microbiol & cell biol, 68-77, prof microbiol, 77-87, PROF MOLECULAR GENETICS, OHIO STATE UNIV, 87-, ASSOC DEAN BIOL SCI, 90- *Mem:* Am Soc Cell Biol; Soc Protozool. *Res:* Developmental biology of protoza; molecular biology of protozoa. *Mailing Add:* Dept Molecular Genetics Ohio State Univ Columbus OH 43210

BYERS, VERA STEINBERGER, b Houston, Tex, Nov 18, 42; m 73. IMMUNOBIOLOGY. *Educ:* Univ Calif, Los Angeles, BA, 65, MS, 67, PhD(immunobiol), 69; Univ Calif, San Francisco, MD, 81. *Prof Exp:* Bench chemist, Res Div, Abbott Labs, 69-71; res immunologist, Dept Med, Univ Calif, San Francisco, 71-74, asst prof dermat, 75-; DIR CORP DEVELOP IMMUNOL, VILMA XOMA CORP. *Concurrent Pos:* Nat Inst Allergy & Infectious Dis fels, 68 & 72; Arthritis Found fel, 72. *Res:* Human cellular immunology; lymphocyte regulation and maturation; cancer immunobiology. *Mailing Add:* 500 Sutter St No 511 San Francisco CA 94102

BYERS, WALTER HAYDEN, b Johnstown, Pa, July 6, 14; m 45; c 3. PHYSICS. *Educ:* Univ Fla, BS, 36, MS, 38; Pa State Col, PhD(physics), 42. *Prof Exp:* Physicist, USAF, 43-46; asst prof physics, Univ Okla, 46-47; asst prof elec eng, Univ Ill, 46-51; physicist, USAF, 51-54; weapon syst analyst, Sandia Corp, 54-56; physicist, USAF, 56-71; asst prof eng, Clark Tech Col, 71-73, assoc prof eng & natural sci, 73-84; RETIRED. *Mem:* Sigma Xi. *Res:* Ultrasonics; astronautics; space system engineering. *Mailing Add:* 3155 Meadow Dr Springfield OH 45505

BYFIELD, JOHN ERIC, b Toronto, Ont, Nov 26, 36; US citizen; m 73; c 4. RADIOTHERAPY, ONCOLOGY. *Educ:* Univ Calif, Los Angeles, BA, 60, MD, 65, PhD(physiol), 70. *Prof Exp:* Intern path, 66-67, radiation therapist, 67-70, instr radiol, 69-70, asst prof & asst res physician, 70-74, assoc prof radiol & assoc res physician, Lab Nuclear Med & Radiation Biol, Sch Med, Univ Calif, Los Angeles, 74-75; assoc prof radiol & chief radiation ther, Univ San Diego, 76-82, prof radiol & med, 76-83; DIR RADIATION THER ASN, BAKERSFIELD, CALIF, 86- *Concurrent Pos:* Chief radiation ther, Harbor Gen Hosp, Torrance, Calif, 70-75. *Mem:* Am Soc Clin Oncol; fel Am Col Radiol; Am Soc Therapeut Radiol; Am Soc Cell Biologists; Am Asn Cancer Res. *Res:* Biochemistry of anticancer drugs; clinical chemotherapy; radiation biology; clinical radiation therapy. *Mailing Add:* 3550 Q St Bakersfield CA 93301

BYINGTON, KEITH H, b Plymouth, Iowa, Mar 14, 35; m 55; c 4. BIOCHEMICAL PHARMACOLOGY, PHARMACOLOGY. *Educ:* Univ Iowa, BS, 58; Univ SDak, PhD(pharmacol), 64. *Prof Exp:* Org chemist, 58-60; instr pharmacol, Univ SDak, 63-64; fel, Univ Fla, 64-65; fel biochem, Inst Enzyme Res, Univ Wis, 65-68; asst prof, 68-74, ASSOC PROF PHARMACOL, SCH MED, UNIV MO, COLUMBIA, 74- *Mem:* Am Soc Pharmacol & Exp Therapeut; Am Chem Soc. *Res:* Drug metabolism; drug enzyme interaction; drug-membrane interactions; toxicology. *Mailing Add:* Dept Pharmacol Univ Mo Sch Med Columbia MO 65212

BYLER, DAVID MICHAEL, b Mishawaka, Ind, Dec 23, 45; div; c 2. INFRARED & RAMAN SPECTROSCOPY. *Educ:* Univ NC, Chapel Hill, AB, 68; Northwestern Univ, Evanston, MS, 69, PhD(inorg chem), 74. *Prof Exp:* Instr chem, Dept Phys Sci, Wilbur Wright Col, Kennedy-King Col & Chicago City Cols, 70 & 73; instr, dept chem, Drexel Univ, 73-75; vis asst prof, dept chem, Temple Univ, 75-78; Nat Res Coun res fel, 78-79, res chemist, Eastern Regional Res Ctr, USDA, 79-90; ASSOC PROF, DEPT CHEM & PHYS SCI, PHILADELPHIA COL TEXTILES & SCI, 90- *Honors & Awards:* Sigma Xi, 74. *Mem:* Soc Appl Spectros; Can Spectros Soc; Am Chem Soc; The Chem Soc; Sigma Xi; Coblentz Soc. *Res:* Estimation of protein conformation by resolution-enhanced infrared spectroscopy; vibrational spectroscopy of amino acids, polypeptides, and other biomolecules; vibrational spectroscopic characterization of clays and their intercalates with small molecules and ions. *Mailing Add:* Dept Chem & Phys Sci Philadelphia Col Textile & Sci Philadelphia PA 19144-5497

BYLES, PETER HENRY, b Exeter, Eng, 1931; m 58; c 2. ANESTHESIOLOGY. *Educ:* Univ London, MB & BS, 55; Am Bd Anesthesiol, dipl, 65. *Prof Exp:* Intern, King's Col Hosp, London, 55-56, sr house officer anesthesiol, 57-58; sr house officer, Plymouth Hosps Group, 56-57; res asst, Western Hosp, London, 58-61; resident, State Univ Hosp, Syracuse, NY, 62-63; from asst prof to assoc prof, 63-72, asst dir, Sch Allied Health Prof, 68-72, PROF ANESTHESIOL, STATE UNIV NY HEALTH SCI CTR, 72- *Concurrent Pos:* Asst prof, State Univ Hosp, Syracuse, 63-67, assoc prof, 67-72; attend anesthesiologist, Vet Admin Hosp, Syracuse, 64- *Mem:* AMA; Am Soc Anesthesiol; Am Asn Advan Med Instrumentation; fel Am Col Anesthesiol; Int Anesthesia Res Soc. *Res:* Medical instrumentation and electronics; new anesthetic drugs. *Mailing Add:* Dept Anesthesiol State Univ NY Health Sci Ctr Syracuse NY 13210

BYLUND, DAVID BRUCE, b Spanish Fork, UT, Apr 16, 46; m 70; c 7. NEUROPHARMACOLOGY, BIOCHEMICAL PHARMACOLOGY. *Educ:* Univ Calif, PhD(biochem), 74. *Prof Exp:* PROF PHARMACOL, SCH MED, UNIV MO, 77- *Mem:* Soc Neurosci; Am Soc Pharmacol & Exp Therapeut. *Res:* Regulation classification and mechanism of action of adrenergic receptors. *Mailing Add:* Dept Pharmacol Univ NE Med Ctr 600 42nd St Omaha NE 68198-6260

BYNUM, BARBARA S, b Washington, DC, June 13, 36. SCIENCE ADMINISTRATION. *Prof Exp:* SCI ADMINR & DIR, EXTRAMURAL ACTIV, NAT CANCER INST, NIH, 81- *Mailing Add:* Nat Cancer Inst NIH Bldg 31 Rm 10A03 Bethesda MD 20205

BYNUM, T E, b Chickasha, Okla, Sept 13, 39. CLINICAL GASTROENTEROLOGY. *Educ:* Univ Okla, MD. *Prof Exp:* CHIEF CLIN GASTROENTEROL, BRIGHAM & WOMEN'S HOSP, HARVARD MED SCH, 79- *Mailing Add:* Dept Clin Gastroenterol Brigham & Women's Hosp Harvard Med Sch Boston MA 02115

BYNUM, WILLIAM LEE, b Carlsbad, NMex, June 28, 36; m 65. MATHEMATICS. *Educ:* Tex Technol Col, BS, 57; Univ NC, MA, 64, PhD(math), 66. *Prof Exp:* Engr, Bell Helicopter Corp, Tex, 57-61; instr math, Univ NC, 65-66; asst prof, La State Univ, 66-69; asst prof, 69-74, assoc prof math, 74-83, PROF COMPUT SCI, COL WILLIAM & MARY, 83- *Concurrent Pos:* Mem, Comput Soc, Inst Elec & Electronics Engrs. *Mem:* Asn Comput Mach; Inst Elec & Electronics Engrs. *Mailing Add:* Dept Comput Sci Col William & Mary Williamsburg VA 23185

BYRAM, GEORGE WAYNE, b Asheville, NC, Nov 25, 38. INSTRUMENTATION, ELECTRICAL ENGINEERING. *Educ:* Ga Inst Technol, BEE, 61, MSEE, 62; Mass Inst Technol, SM, 64, ScD(instrumentation), 67. *Prof Exp:* Electronic engr, Naval Undersea Res & Develop Ctr, Pasadena, 61-68; res engr, Naval Res Lab, 68-70; ELECTRONIC ENGR, NAVAL OCEAN SYSTS CTR, SAN DIEGO, 70- *Res:* Signal processing algorithms and implementations. *Mailing Add:* 3519 Emerson St San Diego CA 92106-2547

BYRD, BENJAMIN FRANKLIN, JR, b Nashville, Tenn, May 18, 18; m 50; c 6. SURGERY, ONCOLOGY. *Educ:* Vanderbilt Univ, AB, 37, MD, 41. *Prof Exp:* CLIN PROF SURG, MEHARRY MED COL, 69- & SCH MED, VANDERBILT UNIV, 71- *Concurrent Pos:* Surg, Pvt Pract, 48-; chmn breast cancer task force, Am Cancer Soc, 71-; chmn cancer adv comt, Joint Comn Accreditation of Hosp, 72-76; cancer control community activ rev comt, Div Cancer Control & Rehab, Nat Cancer Inst, 76-77. *Mem:* Am Cancer Soc (pres, 75-76); Sigma Xi; Soc Surg Oncol; AAAS; Am Col Surgeons; Am Surg Asn. *Res:* Diagnosis and management of breast cancer; long-term effects of estrogen administration. *Mailing Add:* 2611 W End Ave No 201 Nashville TN 37203-1424

BYRD, DANIEL MADISON, III, b Detroit, Mich, Dec 30, 40; m 63, 85; c 4. SAFETY EVALUATION, RISK ASSESSMENT. *Educ:* Yale Univ, BA, 63, PhD(pharmacol), 71; Am Bd Toxicol, dipl, 82. *Prof Exp:* Res assoc cell biol, Sch Med, Univ Md, 70; vis scientist pharmacol, Nat Cancer Inst, 71; res assoc oncol, Med Sch, Johns Hopkins Univ, 71-72; from asst cancer res scientist to cancer res scientist, Roswell Park Mem Inst, 72-75; asst prof, pharmacol, Col Med, Univ Okla, 75-79; sci adv premanufacture rev, Off Toxic Substances, US Environ Protection Agency, pharmacologist hazard eval, Off Pesticides & toxologist, carcinogen assessment group, Off Res Develop, 80-81 & exec secy adv bd, 84-87; sci adv, Am Petrol Inst, 82-83; dir, sci affairs, Cammer Assoc, 87-88; CONSULT TOXICOLOGIST, 88- *Concurrent Pos:* Liaison, Soc Risk Anal & Am Col Toxicol; assoc ed, Toxicol Indust Health; prog comt, Soc Toxicol; dir, Med Adv Comt, Distilled Spirits Coun; mem, sci comt & Delivered Dose Workshop, Am Indust Health Coun; bd dirs, Sci Legis Regulatory Prof Inc. *Mem:* Soc Risk Anal; Soc Toxicol; Am Col Toxicol; Am Col Epidemiol; Soc Epidemiol Res; Int Soc Regulatory Toxicol Pharmacol. *Res:* Toxicology; risk assessment models, uncertainty anal, chronic disease (cancer, develop, immunol, mutation, neurol & reproductive), pharmacogenetics, chemotherapy. *Mailing Add:* 6322 Cavalier Corridor Falls Church VA 22044-1301

BYRD, DAVID LAMAR, b Houston, Tex, June 3, 22; m 47; c 1. DENTISTRY. *Educ:* Univ Tex, DDS, 46; Northwestern Univ, MSD, 49. *Prof Exp:* Resident oral surg, Jackson Mem Hosp, Fla, 49-50; intern, Charity Hosp La, 50-51; prof oral surg & chmn dept, Baylor Col Dent, 51-84, chief dent, Baylor Med Ctr, 64-84, dir ann oral surg seminar, Col Dent, 77-84; RETIRED. *Concurrent Pos:* Consult, Attend Staff, Baylor Med Ctr, Parkland Mem Hosp; ed, Current Ther Dent, Vols I-III; past mem Am Bd Oral Surg. *Mem:* Am Soc Oral Surg; Sigma Xi; fel Am Col Dent. *Res:* Clinical oral pathology; diseases of soft tissue of bone of head and neck; research in facial and oral pain. *Mailing Add:* 3409 Worth St Suite 610 Sammons Tower Dallas TX 75246

BYRD, DAVID SHELTON, b Stephens, Ark, May 28, 30; m 53; c 2. ORGANIC CHEMISTRY, PHYSICAL CHEMISTRY. *Educ:* Southern State Col, BS, 52; Univ Ky, MS, 55; Univ Louisville, PhD(org chem), 60. *Prof Exp:* Instr chem, Univ Louisville, 57-60; from asst prof to assoc prof, 60-69, PROF CHEM, NORTHEAST LA UNIV, 69- *Mem:* Am Chem Soc; Sigma Xi. *Res:* Trace metal analysis of soil and water. *Mailing Add:* Dept Chem Northeast La Univ Monroe LA 71209

BYRD, EARL WILLIAM, JR, b Pomona, Calif, Mar 17, 46; m; c 1. DEVELOPMENTAL BIOLOGY. *Educ:* San Francisco State Univ, BA, 68, MA, 70; Univ BC, PhD(zool), 73. *Prof Exp:* Fel develop biol, Scripps Inst Oceanog, 73-75; asst prof embryol, La State Univ, Baton Rouge, 76-82; res scientist, Univ Tex-Houston Sci Ctr, 82-84; asst prof, 84-90, ASSOC PROF, DEPT OBSTET & GYNEC, SOUTHWESTERN MED SCH, 90- *Concurrent Pos:* Chair, Reproductive Biol SIG, Am Fertil Soc, 90-91. *Mem:* Soc Develop Biol; Am Soc Cell Biol; Sigma Xi; Am Soc Zoologists; AAAS; Soc Study Reprod; Am Fertility Soc. *Res:* Human sperm maturation and capacitation in vitro. *Mailing Add:* Dept Obstet & Gynec Southwestern Med Sch 5323 Harry Hines Blvd Dallas TX 75235

BYRD, ISAAC BURLIN, b Canoe, Ala, Mar 14, 25; m 49; c 3. FISHERIES MANAGEMENT, MARINE BIOLOGY. *Educ:* Auburn Univ, BS, 48, MS, 50. *Prof Exp:* Fisheries res asst, Auburn Univ, 49-51; chief fisheries sect, Ala Dept Conserv, 51-65; chief div fed aid, Fisheries Res & Develop, Nat Marine Fisheries Serv, 65-74, chief, Fisheries Mgt Div, 74-76, chief, Grants Admin Br, 76-84, asst red dir, 84-91; RETIRED. *Concurrent Pos:* Chmn, Gulf of Mex, State/Fed Fisheries Mgt Bd, 85-86; mem, S Atlantic State/Fed Fisheries mgt Bd. *Honors & Awards:* Ala Gov Conserv Award, 64; Ala Fisheries Asn Freshwater Div Award, 65. *Mem:* Am Fisheries Soc (pres, 65); Gulf & Caribbean Fisheries Inst; World Maricult Soc; fel Am Inst Fishery Res Biol; Int Asn Fish & Wildlife Agencies. *Res:* Marine and freshwater fisheries; estuarine waters; fish population dynamics; reservoir management; aquaculture; aquatic botany and biology; ecology; water pollution; oceanography; public fishing lakes; hydrology. *Mailing Add:* 11105 Seventh St E Treasure Island FL 33706

BYRD, J ROGERS, b Henderson, NC, Apr 30, 31; m 63. CYTOGENETICS, GENETICS. *Educ:* Wake Forest Col, BS; Univ Mich, MS, 57, PhD(zool), 60. *Prof Exp:* Asst prof biol, Col William & Mary, 59-63; NIH res fel endocrinol, 63-65, asst prof, 65-69, assoc prof endocrinol & pediat, 69-81, PROF PHYSIOL & ENDOCRINOL, MED COL GA, 81- *Mem:* Am Soc Human Genetics. *Res:* Human chromosome studies related to the inheritance of chromosomal aberrations as a causative factor in habitual spontaneous abortion and sexual dysfunction. *Mailing Add:* 3105 Shelley Ct Augusta GA 30909

BYRD, JAMES DOTSON, b Jackson, Miss, July 9, 32; m 54; c 3. ORGANIC POLYMER CHEMISTRY, MANAGEMENT. *Educ:* Miss Col, BS, 54; Univ Ala, MS, 74. *Prof Exp:* Res asst chem, Purdue Univ, 54-55; chemist, Ethyl Corp, 55-61; sr chemist, Geigy Chem Corp, 61-63; unit chief polymer chem, George C Marshall Space Flight Ctr, NASA, 63-67; sr res chemist, 67-74, MAT GROUP SUPVR, HUNTSVILLE DIV, THIOKOL CORP, 74- *Honors & Awards:* NASA Awards, 54-67. *Mem:* Am Chem Soc; fel Am Inst Chem; Soc Aerospace Mat & Process Eng; Am Mgt Asn. *Res:* New adhesive and insulation materials used in a rocket motor environment, such as high and low temperature, vacuum and radiation, including the development of new formulations. *Mailing Add:* 9032 Craigmont Rd SW Huntsville AL 35802

BYRD, JAMES WILLIAM, b Mt Olive, NC, Dec 4, 36; m 59; c 3. ENERGY. *Educ:* NC State Univ, BS, 59, MS, 61; Pa State Univ, PhD(physics), 63. *Prof Exp:* From assoc prof to prof physics, ECarolina Univ, 62-84, chmn dept, 65-84; dean, Col Arts & Sci, 84-90, PROF PHYSICS, APPALACHIAN STATE UNIV, 84- *Mem:* Am Phys Soc; Am Asn Physics Teachers; Sigma Xi; Am Asn Higher Educ. *Res:* Plasma and fluid physics; mathematical methods; energy. *Mailing Add:* Dept Physics & Astron Appalachian State Univ Boone NC 28608

BYRD, KENNETH ALFRED, b Erwin, NC, Sept 17, 40; m 64; c 2. ALGEBRA. *Educ:* Duke Univ, BS, 62; NC State Univ, PhD(math), 69. *Prof Exp:* ASST PROF MATH, UNIV NC, GREENSBORO, 69- *Mem:* Am Math Soc; Math Asn Am. *Res:* Non-commutative ring theory; quotient rings; homological algebra as it applies to the module category of a ring. *Mailing Add:* Dept Math Univ NC Greensboro NC 27412

BYRD, LARRY DONALD, b Salisbury, NC, July 14, 36; m 61; c 4. BEHAVIORAL PHARMACOLOGY, PSYCHOBIOLOGY. *Educ:* ECarolina Univ, AB, 62, MA, 64; Univ NC, Chapel Hill, PhD(exp psychol), 68. *Prof Exp:* Instr psychol, E Carolina Univ, 63-64; res asst, Univ NC, Chapel Hill, 65-67; USPHS res fel pharmacol, Med Sch, Harvard Univ, 67-70, instr psychiat, 70-73; res asst pharmacol psychiat, 73-74; psychobiologist & chmn, Div Primate Behav, Yerkes Ctr, 74-80, assoc prof, & chief, 80-82, RES PROF & CHIEF, DIV BEHAV BIOL, YERKES REGIONAL PRIMATE RES CTR, EMORY UNIV, 82-, ASSOC PROF PHARMACOL, 81- *Concurrent Pos:* Assoc scientist, NEng Regional Primate Res Ctr, 69-74, prin assoc psychobiology, 74; Assoc ed, J Exp Anal Behav, 70-76; prin investr, Nat Inst Drug Abuse res grant, 75-; consult, Univ Chicago, 73, Mass Inst Technol press, 75, US Food & Drug Admin, 76-77, Southwest Found Res & Educ, 77, US Navy, 77, Nat Inst Drug Abuse, 79-, Neurobehav Toxicity Test Standards Comn, Am Psychol Asn, 80-84, US Vet Admin, 83, NIH, 83-84, NSF, 84; Fed Am Socs Exp Biol, 84 & Int Study Group Investigating Drugs and Reinforcers; adj prof psychol, Ga Inst Technol, 75- & Emory Univ, 81-; mem, Spec Rev Comt, Nat Inst Drug Abuse, 79-81 & 84, Spec Study Sect, NIH, 84; consult ed, Am J Primatol, 80-83. *Mem:* Am Soc Pharmacol & Exp Therapeut; Behav Pharmacol Soc (pres, 84-86); fel Am Psychol Asn (pres, 82-83); Soc Exp Anal Behav (vpres, 75-76); Am Soc Primatologists. *Res:* Behavioral effects of drugs; cardiovascular effects of psychoactive drugs; behavioral modulation of cardiovascular activity; drug effects on the central nervous system; conditioned behavior; experimental analysis of behavior; behavior of primates. *Mailing Add:* Yerkes Regional Primate Res Ctr Emory Univ Atlanta GA 30322

BYRD, LLOYD G, SCIENCE ADMINISTRATION. *Educ:* Ohio State Univ, BS, 50. *Prof Exp:* Engr, Geomet Design Sect, Ohio Dept Hwys, 49-52; field engr, Ohio Turnpike Comn, 52-56, maintenance engr, 55-60; assoc ed, Pub Works Publ, Ridgewood, NJ, 60-63; partner, Byrd, Tallamy, MacDonald & Lewis, consult engrs, 63-72; sr vpres & dir, Wilbur Smith & Assocs, 72-84; interim dir, Strategic Hwy Res Prog, Am Asn State Hwy & Transp Officials, 84-86; CONSULT ENGR, 86- *Concurrent Pos:* Mgr, Byrd, Tallamy, MacDonald & Lewis, 72-84; chmn, Nat Transp Policy Comt, Am Soc Civil Engrs, mem, Nat Comt Govt Affairs, Hwy Div Res Comt. *Honors & Awards:* Wilbur S Smith Award, Am Soc Civil Engrs, 85; Roy W Crum Distinguished Serv Award, Transp Res Bd, 87. *Mem:* Nat Acad Eng; Am Soc Civil Engrs; Transp Res Bd. *Mailing Add:* 3904 Rive Dr Alexandria VA 22309

BYRD, MITCHELL AGEE, b Franklin, Va, Aug 16, 28; m 54. ZOOLOGY. *Educ:* Va Polytech Inst, BS, 49, MS, 51, PhD(biol), 54. *Prof Exp:* From asst prof to assoc prof, 56-63, PROF BIOL, COL WILLIAM & MARY, 63-, HEAD DEPT, 62- *Mem:* Am Soc Parasitol; Am Soc Mammal; Am Ornith Union. *Res:* Mammalogy; ecology; taxonomy; wildlife diseases and parasites; avian ecology. *Mailing Add:* Dept Biol Col William & Mary Williamsburg VA 23185

BYRD, NORMAN ROBERT, b New York, NY, Mar 10, 21; m 46; c 2. ORGANIC POLYMER CHEMISTRY. *Educ:* Polytech Inst Brooklyn, BS, 49, PhD(org chem), 55. *Prof Exp:* Res chemist, Nat Starch Prod, NY, 48-51; asst, Polytech Inst Brooklyn, 51-52 & US Navy Proj, 52-54; res chemist, E I Du Pont de Nemours & Co, Inc, 54-55; sr res chemist, Goodyear Tire & Rubber Co, 55-58; head fundamental sect, Rayonier, Inc, 58-60; res scientist, Aeronutronic, Calif, 61-62; head org-polymer sect, Astropower Labs, Douglas Aircraft Co, 62-69, STAFF MGR-MCDONNEL DOUGLAS CORP FEL, CHEM RES, DOUGLAS AIRCRAFT DIV, MCDONNEL DOUGLAS CORP, 69- *Mem:* AAAS; Am Chem Soc; Am Inst Chem; The Chem Soc; NY Acad Sci. *Res:* Starch and cellulose chemistry; synthetic resins; elastomers; graft and block polymers; organic reactions macromolecules; electrical properties of organic polymers; semiconductor; adhesion studies; fire-resistant materials; icephobics. *Mailing Add:* 17991 Athens Ave Villa Park CA 92667

BYRD, RICHARD DOWELL, b Newport, Ark, Mar 14, 33; m 64; c 1. MATHEMATICS. *Educ:* Hendrix Col, BA, 58; Univ Ark, MS, 59; Tulane Univ, PhD(math). 66. *Prof Exp:* Instr math, La State Univ, New Orleans, 60-64; asst prof, Lehigh Univ, 66-67; from asst prof to assoc prof, 67-72, PROF MATH, UNIV HOUSTON, 72- *Mem:* Am Math Soc; Math Asn Am. *Res:* Lattice-ordered groups. *Mailing Add:* Univ Houston-Univ Park Houston TX 77204-3476

BYRD, WILBERT PRESTON, b Burlington, NC, July 7, 26; m 47; c 1. STATISTICS, GENETICS. *Educ:* NC State Univ, BS, 49, MS, 52; Iowa State Univ, PhD(crop breeding), 55. *Prof Exp:* Res assoc, NC State Univ, 49-52; asst, Iowa State Univ, 52-55; asst prof, Ohio State Univ, 55-56; assoc prof & statistician, 56-66, PROF EXP STATIST & CHMN DEPT, CLEMSON UNIV, 66- *Concurrent Pos:* Agent, USDA, 52-54. *Mem:* Am Soc Agron. *Res:* Experimental design; data processing; statistical genetics. *Mailing Add:* Dept Exp Statist F148 Peas Bldg Clemson Univ Clemson SC 29631

BYRD, WILLIS EDWARD, physical chemistry, for more information see previous edition

BYRKIT, DONALD RAYMOND, b Indianapolis, Ind, March 19, 33; m 58, 75; c 2. MATHEMATICS, APPLIED MATHEMATICS. *Educ:* Ill Inst Technol, BS, 55; Ill State Univ, MS, 58; Fla State Univ, PhD(math educ), 68. *Prof Exp:* From asst prof to assoc prof, 67-75, PROF MATH & STATIST, UNIV WFLA, 75- *Concurrent Pos:* Vis prof, Kaohsiung Teachers Col, Taiwan, 78-79; chmn, Univ Fla, 72-76 & 87- *Mem:* Am Statist Asn; Math Asn Am; Nat Coun Teachers Math & Sch Sci. *Res:* Writer on topics of statistics, business calculus, number theory. *Mailing Add:* 4995 Woodcliff Dr Pensacola FL 32504

BYRN, ERNEST EDWARD, b Frederick, Okla, Apr 4, 24; m 52; c 2. PHYSICAL CHEMISTRY, ANALYTICAL CHEMISTRY. *Educ:* Univ Tenn, BS, 50, PhD, 54. *Prof Exp:* Res chemist, Buckeye Cellulose Corp, 54-55; asst prof chem, Univ Okla, 55-61; asst res dir, MacDermid, Inc, 61-62, res dir, 62-64; prof, George Peabody Col, 64-66; chmn dept, 66-70, PROF CHEM, EASTERN KY UNIV, 66- *Mem:* AAAS; Am Chem Soc; Sigma Xi. *Res:* Organic reagents for inorganic analysis; titrations in nonaqueous solvents; liquid-liquid extractions; surface activity and detergency; chemical education. *Mailing Add:* 125 Buckwood Dr Eastern Ky Univ Richmond KY 40475

BYRN, STEPHEN ROBERT, b New Albany, Ind, Oct 7, 44; m 69; c 6. BIOPHYSICAL CHEMISTRY, SOLID STATE CHEMISTRY. *Educ:* DePauw Univ, BA, 66; Univ Ill, Urbana, PhD(chem), 70. *Prof Exp:* Scholar, Univ Calif, Los Angeles, 71-72; from asst prof to assoc prof, 72-81, PROF MED CHEM, PURDUE UNIV, 81-, ASSOC DEPT HEAD, 79-, DEPT HEAD, 88- *Concurrent Pos:* From asst dean to assoc dean, Grad Sch, Purdue Univ, 85-88. *Mem:* Am Chem Soc; Am Crystallog Asn; Am Asn Pharm Sci. *Res:* Solid state reactions of organic crystals, particularly drugs; interaction of drugs with DNA; design and synthesis of anti-AIDS agents; solid state NMR spectroscopy. *Mailing Add:* Dept Med Chem & Pharmacog Purdue Univ Sch Pharm West Lafayette IN 47907

BYRNE, BARBARA JEAN MCMANAMY, b Baraboo, Wis, Aug 9, 41; m 68; c 2. SURFACE PROTEINS OF PARAMECIUM, PARAMECIUM BEHAVIOR. *Educ:* Blackburn Col, BA, 62; Ind Univ, Bloomington, MA, 63, PhD(genetics), 69. *Prof Exp:* Lectr biol, Ind Univ, 71-72; res specialist genetics, Cornell Univ, 72-74; from asst prof to assoc prof, Wells Col, 74-87, assoc dean & registrar, 84-87, prof biol, 87-90; DEAN NATURAL SCI & MATH, STOCKTON STATE COL, 90- *Concurrent Pos:* Vis scholar, Ind Univ, 81-82; NIH Area Grant, PI, 87-89; NIH Extramural Assoc, 89. *Mem:*

Sigma Xi; Genetics Soc Am; AAAS. *Res:* Behavioral genetics of Paramecium aurelia; ciliary antigens of paramecium; molecular investigations of the structure of surface protein genes and their control regions; emphasis on comparison of stocks with different frequency of expression of selected genes. *Mailing Add:* Dean of Natural Sci & Math Stockton State Col Pomona NJ 08240-9988

BYRNE, BRUCE CAMPBELL, b Hammond, Ind, May 13, 45; m 68; c 2. MOLECULAR BIOLOGY, MEDICINE. *Educ:* Ind Univ, Bloomington, AB, 67, PhD(genetics), 72. *Prof Exp:* USPHS trainee genetics, Cornell Univ, 72-74; from asst prof to prof biol, Wells Col, 74-89; ADJ PROJ MED, ROBERT WOOD JOHNSON MED SCH, CAMDEN, 89- *Concurrent Pos:* Vis scientist, State Univ NY, HSC at Syracuse, 87- *Mem:* AAAS; Genetics Soc; Am Soc Microbiol. *Res:* Development of new molecular probes for human immunodeficiency virus. *Mailing Add:* Dept Med UMDNJ RW Johnson Med Sch Three Cooper Plaza Suite 220 Camden NJ 08103

BYRNE, FRANCIS PATRICK, b Kansas City, Mo, July 25, 13; m 42; c 6. ANALYTICAL CHEMISTRY. *Educ:* Rockhurst Col, BS, 35; Creighton Univ, MS, 38; Univ Tenn, PhD(chem), 49. *Prof Exp:* Instr chem, St Joseph Col, 40-42; res chemist, Westvaco Chlorine Prods, 42-46; instr chem, Univ Tenn, 46-49; mgr, Anal Chem Dept, Westinghouse Res & Develop Labs, 49-78; ADJ PROF, DEPT CHEM, WESTERN KY UNIV, 78- *Concurrent Pos:* Pres, Pittsburgh Conf Anal Chem & Appl Spectros, 64; mem, NSF Adv Panel to Anal Div, Nat Inst Standards & Technol, 72-75; adj prof, Western Ky Univ, 78- *Honors & Awards:* Lundel-Bright Award, Am Soc Testing & Mat, 74. *Mem:* Am Chem Soc; Am Soc Testing & Mat. *Res:* Detection and determination of pertinent materials in environmental situations. *Mailing Add:* Dept Chem Western Ky Univ Bowling Green KY 42101

BYRNE, GEORGE D, b Earlham, Iowa, June 15, 33; m 60; c 8. NUMERICAL ANALYSIS. *Educ:* Creighton Univ, BS, 55; Iowa State Univ, MS, 61, PhD(appl math), 63. *Prof Exp:* Mathematician, White Sands Proving Grounds, NMex, 55-56; programmer, Sandia Corp, NMex, 56-58; asst prof math & comput sci, Univ Pittsburgh, 63-67, assoc prof math, 67-80; SR STAFF MATHEMATICIAN, COMPUT TECHNOL & SERV, EXXON RES & ENG CO, LINDEN, NJ, 80- *Concurrent Pos:* Adj assoc prof, Dept Chem & Petrol Eng, Univ Pittsburgh, 67-80; consult, Lawrence Livermore Lab, Univ Calif, 73-80; vis scientist, Appl Math Div, Argonne Nat Lab, 74-75; consult, Argonne Nat Lab, 75-80 & Sandia Labs, 76-78; assoc ed, Asn Comput Mach Trans on Math Software, 76-78. *Mem:* Am Math Soc; Soc Indust & Appl Math; Asn Comput Mach. *Res:* Numerical solution of ordinary differential equations and related software; software; solution of nonlinear systems; numerical solution of models. *Mailing Add:* 820 Nancy Way Westfield NJ 07090

BYRNE, J(OSEPH) GERALD, b New York, NY, July 15, 30; m 57; c 4. METALLURGY. *Educ:* Stevens Inst Technol, ME, 53, MS, 57; Northwestern Univ, PhD(metall), 60. *Prof Exp:* Mech engr, Heat-X-Changer Corp, 53-54; metallurgist, Crucible Steel Co, 54-55 & Am Brake Shoe Co, 55-57; res asst metall, Northwestern Univ, 57-60; res engr, Dow Chem Co, 60-62; from asst prof to assoc prof, Stevens Inst Technol, 62-66; assoc prof metall, Univ Utah, 66-69, prof mat sci & eng, 69-85, prof metall, 85-87, IVOR D THOMAS PROF PHYS METALL, UNIV UTAH, 87- *Concurrent Pos:* Mem, Nat Mat Adv Bd, Nat Acad Sci, 72; vis prof, Cent Univ Venezuela, 72-73; mem exec comt, Int Confs Mat Technol, 77-; assoc ed, Mat Sci & Eng. *Mem:* Am Inst Mining, Metall & Petrol Engrs; Am Soc Testing & Mat; Am Soc Metals; Acad Metals & Mats; Am Powder Metall Inst; fel Am Soc Metals Int. *Res:* Strengthening of solids; phase transformations; radiation damage; recrystallization; dislocation theory; position annihilation applied to damage in solids. *Mailing Add:* 1159 First Ave Salt Lake City UT 84103

BYRNE, JEFFREY EDWARD, b Minneapolis, Minn, July 15, 39; m 78; c 3. PHARMACOLOGY, CARDIOVASCULAR RESEARCH. *Educ:* Univ NDak, BA, 62; Univ SDak, MA, 64, PhD(physiol & pharmacol), 66. *Prof Exp:* Fel & lectr pharmacol & exp therapeut, Univ Man, 66-69; sr scientist cardiovasc pharmacol, 69-73, sr investr, 73-81, prin res scientist, Mead Johnson & Co, 81-87; SR RES SCIENTIST II, BRISTOL-MYERS PROD, 87- *Concurrent Pos:* Adj fac mem, Sch Nursing, Evansville Univ & Sch Med, Ind Univ, 72-87. *Mem:* AAAS; NY Acad Sci; Sigma Xi; Am Heart Asn; Am Soc Pharmacol & Exp Therapeut. *Res:* Cardiovascular pharmacology; cardiac arrhythmia, antiarrhythmic drugs; myocardial ischemia; electrophysiology. *Mailing Add:* Dept Pharmacol-Bristol Myers Squibb Co PO Box 4000 Princeton NJ 08540

BYRNE, JOHN HOWARD, NEUROBIOLOGY. *Educ:* Polytech Inst Brooklyn, PhD(bioeng), 73. *Prof Exp:* PROF PHYSIOL & CELL BIOL, UNIV TEX HEALTH SCI CTR, HOUSTON, 82- *Res:* Learning and memory; synaptic transmission. *Mailing Add:* 5754 Valkeith Houston TX 77096

BYRNE, JOHN JOSEPH, b Morristown, NJ, Aug 30, 16. SURGERY. *Educ:* Princeton Univ, AB, 37; Harvard Univ, MD, 41. *Prof Exp:* Mem staff, Sch Med, Boston Univ, 47-57, prof surg, 57-73, prof sociol-med sci, 73; CHIEF SURG, FRAMINGHAM UNION HOSP, 73- *Concurrent Pos:* Asst chief, Neurosurg Sect, May Gen Hosp, Ill; dir hand serv & vis surgeon, Boston City Hosp; chief surg, Framingham Union Hosp; mem staff, Boston Univ Hosp. *Mem:* AAAS; Am Surg Asn; Am Soc Surg of Hand; Am Asn Surg Trauma; Am Col Surgeons; Sigma Xi. *Res:* Physiology of pulmonary embolism; serum amylase activity in intestinal obstruction; diseases of the hand; phlebitis; medical history; shock; pancreatitis. *Mailing Add:* Framingham Union Hosp 115 Lincoln St Framingham MA 01701

BYRNE, JOHN MAXWELL, b Gassaway, WVa, May 7, 33; m 60; c 1. PLANT ANATOMY. *Educ:* Glenville State Col, BA, 60; Univ Miami, MA, 64, PhD(bot), 69. *Prof Exp:* Instr biol, Univ Miami, 64-66; asst prof, Va Polytech Inst & State Univ, 69-75; asst prof, 75-, ASSOC PROF BIOL, KENT STATE UNIV. *Mem:* AAAS; Bot Soc Am; Am Inst Biol Sci. *Res:*

BYRNES, WILLIAM RICHARD, b Barnesboro, Pa, Oct 12, 24; m 47; c 5. FORESTRY. *Educ:* Pa State Univ, BS, 50, MF, 51, PhD(agron, soils), 61. *Prof Exp:* Asst forestry, Pa State Univ, 50-51; soil scientist, Soil Conserv Serv, USDA, 51-52; instr forestry, Pa State Univ, 52-61, assoc prof, 61-62; from assoc prof to prof, 62-75, asst head & dir res, 75-85, HEAD DEPT FORESTRY, PURDUE UNIV, 85- *Mem:* Soc Am Foresters; Am Soc Agron; Weed Sci Soc Am; Soil Sci Soc Am; Nat Walnut Coun. *Res:* Forest soils; watershed management; ecology; silviculture; physiology. *Mailing Add:* Dept Forestry & Natural Resources Purdue Univ West Lafayette IN 47907

BYRNE, JOHN RICHARD, b Portland, Ore, Mar 23, 26; m 48; c 2. MATHEMATICS. *Educ:* Reed Col, BA, 47; Univ Wash, MSc, 51, PhD, 53. *Prof Exp:* Asst prof math, San Jose State Col, 53-54; instr, Portland State Univ, 54-56; asst prof, San Jose State Col, 56-57; from asst prof to assoc prof, 57-63, head dept, 67-77, PROF MATH, PORTLAND STATE UNIV, 63- *Concurrent Pos:* Asst consult curric study, Portland High Sch, 59; dir, NSF Insts Math Teachers, 61-66 & Ore Math Contest. *Mem:* Math Asn Am. *Res:* Modern algebra. *Mailing Add:* Dept Math Portland State Univ Portland OR 97207

BYRNE, JOHN VINCENT, b Hempstead, NY, May 9, 28; m 54; c 4. OCEANOGRAPHY. *Educ:* Hamilton Col, AB, 51; Columbia Univ, MA, 53; Univ Southern Calif, PhD(geol), 57. *Prof Exp:* Field asst, Newell Reef Study, Am Mus, 51 & Raroia Exped, Pac Sci Bd, 52; lab assoc, Univ Southern Calif, 53-55; geologist, Res Sect, Humble Oil & Refining Co, Tex, 57-60; from assoc prof oceanog to prof, Ore State Univ, 60-81, chmn dept, 68-81, dean, Sch Oceanog, 72-77, dean res, 77-80, vpres res & grad studies, 80-81; admin, Nat Oceanic & Atmospheric Admin, US Dept Com, 81-84; PRES, OREGON STATE UNIV, 84- *Concurrent Pos:* Dir oceanog prog, NSF, 66-67. *Mem:* AAAS; Am Geophys Union; Geol Soc Am; Am Asn Petrol Geologists; Soc Econ Paleontologists & Mineralogists. *Res:* Marine geology. *Mailing Add:* Oregon State Univ 3520 NW Hayes St Corvallis OR 97330

BYRNE, KEVIN J, b Philadelphia, Pa, Mar 23, 44; m 81; c 3. CHEMISTRY. *Educ:* La Salle Col, BA, 66; Villanova Univ, MS, 69, PhD(chem), 73. *Prof Exp:* Res assoc chem, Col Environ Sci & Forestry, State Univ NY, 73-74 & Monell Chem Senses Ctr, Univ Pa, 74-76; res scientist, 76-83, SR RES SCIENTIST, RES & DEVELOP, JOSEPH E SEAGRAM & SONS, INC, 83- *Mem:* Am Chem Soc. *Res:* Identification and organoleptic evaluation of flavor compounds in alcoholic beverages; development of analysis for regulated compounds in alcoholic beverages. *Mailing Add:* Box 196 Gaylordsville CT 06755

BYRNE, KEVIN M, b Aug 6, 48; m 76; c 2. FERMENTATION MICROBIOL, BIOCHEMISTRY. *Educ:* Fairfield Univ, BS, 71; Rutgers Univ, PhD(biochem), 76. *Prof Exp:* Res assoc, Sch Chem Sci, Univ Ill, Urbana, 76-78; scientist I, microbiol sect, fermentation prog, Frederick Cancer Res Facil, Nat Cancer Inst, 78-80; scientist II, 80-85; RES FEL FERMENTATION MICROBIOL, MERCK SHARP & DOHME RES LABS, MERCK & CO, INC, 85- *Mem:* Am Soc Microbiol; Am Chem Soc; Am Soc Biochem Molecular Biol. *Res:* Discovery, biosynthesis, mechanism of action and regulation of natural products. *Mailing Add:* Merck Sharp & Dohme Res Labs PO Box 2000 Rahway NJ 07065

BYRNE, NELSON, b Pittsfield, Mass, Sept 23, 37; m; c 2. PLASMA PHYSICS, COMPUTATIONAL PHYSICS. *Educ:* Calif Inst Technol, BS, 59; Stanford Univ, PhD(physics), 65. *Prof Exp:* Group leader transport & diffusion, Lawrence Livermore Labs, 65-71; mem staff transport & diffusion, 71-74, ASST DIR PLASMA PHYSICS LABS, SCI APPLN INC, 74-, SR SCIENTIST. *Mem:* Am Phys Soc. *Res:* Applications of numerical models to analysis of physical systems. *Mailing Add:* MS No 12 SALC 10210 Campus Pt Dr San Diego CA 92121

BYRNE, PETER M, b Dublin, Ireland, Mar 3, 36; Can citizen; m 62; c 2. GEOTECHNICAL ENGINEERING. *Educ:* Nat Univ Ireland, BE, 59; Univ BC, MASc, 66, PhD(soil mech), 69. *Prof Exp:* Design engr, George Wimpey, Eng, 59-60 & CBA Eng, BC, 60-63; ASST PROF CIVIL ENG, UNIV BC, 67- *Concurrent Pos:* Nat Res Coun Can study grant earth sci; partner, Peacs Eng, Comput & Eng Serv. *Mem:* Eng Inst Can. *Res:* Application of the finite element method of analysis to the solution of both static and dynamic problems in porous media such as soil, concrete and rock. *Mailing Add:* Dept Civil Eng Univ BC 2075 Westbrook Pl Vancouver BC V6T 1W5 Can

BYRNE, ROBERT HOWARD, b Omaha, Nebr, Apr 15, 41; m 67; c 1. OCEANOGRAPHY. *Educ:* Univ Chicago, BS, 64; DePaul Univ, MS, 67; Boston Univ, MA, 70; Univ RI, PhD(oceanog), 74. *Prof Exp:* Res assoc oceanog, Univ RI, 74-; PROF CHEM, UNIV S FLA, ST PETERSBURG. *Res:* Investigation of the physical chemistry of seawater; and trace metal speciation in seawater with special emphasis on the speciation of ferric ions. *Mailing Add:* Dept Marine Sci Univ SFla MSL 119 St Petersburg FL 33701

BYRNE, ROBERT JOHN, b Chicago, Ill, Sept 17, 32; m 56; c 2. OCEANOGRAPHY. *Educ:* Univ Chicago, MS, 61, PhD(geophys sci), 64. *Prof Exp:* Ford Found fel, Woods Hole Oceanog Inst, 64-65, asst scientist, Dept Geophys & Geol, 65-67; res oceanogr, Land & Sea Interaction Lab, Environ Sci Serv Admin, 67-69; assoc prof marine sci, 69-72, SR MARINE SCIENTIST, VA INST MARINE SCI, 72-, PROF MARINE SCI & ASSOC DIR RES. *Concurrent Pos:* Assoc prof, Sch Marine Sci, Col William & Mary, 72- *Mem:* Am Geophys Union. *Res:* Geological oceanography; coastal engineering; mechanics of sediment transport; coastal fluid and sediment processes; coastal zone planning. *Mailing Add:* Sch Marine Sci Va Inst Marine Sci Gloucester Point VA 23062

BYRNES, EUGENE WILLIAM, b Roselle, NJ, July 3, 33; m 62; c 2. ORGANIC CHEMISTRY. *Educ:* Rensselaer Polytech Inst, BS, 56; Univ NH, PhD(org chem), 64. *Prof Exp:* Prod chemist, Schering Corp, 56; instr chem, Ohio Northern Univ, 62-63; NIH fel biochem, Mich State Univ, 63-65; asst prof chem, Liberty State Col, 65-66 & Alliance Col, 66-68; from asst prof to assoc prof, 68-85, chem div natural sci & math, 72-77, PROF CHEM, ASSUMPTION COL, MASS, 85- *Concurrent Pos:* consult, Astra Pharmaceut Prod, Inc; vis scientist, AB Hassle, Molndal, Sweden, 76. *Honors & Awards:* C P Snow Lectr, Ithaca Col, 64. *Mem:* Sigma Xi; Am Chem Soc. *Res:* Synthesis of antiarrhythmic and local anesthetic drugs; antiarrhythmic drugs. *Mailing Add:* 191 Nola Dr Holden MA 01520

BYRNES, WILLIAM RICHARD, b Barnesboro, Pa, Oct 12, 24; m 47; c 5. FORESTRY. *Educ:* Pa State Univ, BS, 50, MF, 51, PhD(agron, soils), 61. *Prof Exp:* Asst forestry, Pa State Univ, 50-51; soil scientist, Soil Conserv Serv, USDA, 51-52; instr forestry, Pa State Univ, 52-61, assoc prof, 61-62; from assoc prof to prof, 62-75, asst head & dir res, 75-85, HEAD DEPT FORESTRY, PURDUE UNIV, 85- *Mem:* Soc Am Foresters; Am Soc Agron; Weed Sci Soc Am; Soil Sci Soc Am; Nat Walnut Coun. *Res:* Forest soils; watershed management; ecology; silviculture; physiology. *Mailing Add:* Dept Forestry & Natural Resources Purdue Univ West Lafayette IN 47907

BYRON, JOSEPH WINSTON, b New York, NY, Apr 23, 30; m 61; c 1. PHARMACOLOGY. *Educ:* Fordham Univ, BSc, 52; Philadelphia Col Pharm, MSc, 55; Univ Buffalo, PhD(pharmacol), 59. *Prof Exp:* Asst pharmacol, Philadelphia Col Pharm, 53-55; assoc, Univ Buffalo, 56-59; NSF fel, Oxford Univ, 59-60, Am Cancer Soc Brit-Am Exchange fel, 61-62; from sr res scientist to prin res scientist, Christie Hosp & Holt Radium Inst, Eng, 62-73; assoc prof pharmacol, Sch Med, Univ Md, Baltimore City, 73-76, prof, 76-82; PROF & CHMN, DEPT PHARMACOL, SCH MED, MEHARRY MED COL, NASHVILLE, 82- *Concurrent Pos:* Hon lectr, Univ Manchester, 68-73; consult radiation hemat, WHO, 69; mem bd sci counselors, div cancer treat, Nat Cancer Inst, 80-82; develop therapeut contracts rev comt, Div Cancer Treat, Nat Cancer Inst, NIH, 84-88, fel panelist, Nat Res Coun, 86- *Mem:* Int Soc Exp Hematol; Am Soc Pharmacol & Exp Therapeut; Cell Kinetics Soc; Nat Asn Minority Med Educators. *Res:* Hematology; radiation biology; bone marrow toxicology; oncology; AIDS. *Mailing Add:* Dept Pharmacol Sch Med Meharry Med Col 1005 18th Ave N Nashville TN 37208

BYSTROFF, ROMAN IVAN, b San Francisco, Calif, Nov 6, 31; m 56; c 4. ANALYTICAL CHEMISTRY. *Educ:* Univ Calif, Berkeley, BS, 53; Iowa State Univ, PhD(chem), 59. *Prof Exp:* SR CHEMIST, LAWRENCE LIVERMORE LAB, 58- *Mem:* Am Chem Soc. *Res:* Stability of coordination compounds in solution; inorganic analysis by solution chemistry; ultraviolet and visible spectrophotometry; flame spectroscopy, both emission and atomic absorption. *Mailing Add:* 330 Scott St Livermore CA 94550

BYSTRYN, JEAN-CLAUDE, b Paris, France, May 8, 38. DERMATOLOGY, IMMUNODERMATOPATHOLOGY. *Prof Exp:* PROF DERMATOL, NY UNIV, 70-, DIR, IMMUNOFLUORESCENCE LAB, 72- *Mem:* Am Dermatol Asn; Am Asn Immunologist; Am Asn Cancer Res; Soc Investigative Dermatol; Am Soc Cell Biol. *Mailing Add:* NY Univ Med Ctr 560 First Ave New York NY 10016

BYVIK, CHARLES EDWARD, b Ladd, Ill, Mar 26, 40; m 64; c 3. SOLID STATE PHYSICS. *Educ:* Ill Inst Technol, BS, 63; Univ Mo, Rolla, MS, 64; Va Polytech Inst & State Univ, PhD(physics), 72. *Prof Exp:* PHYSICIST, NASA, HAMPTON, VA, 64- *Mem:* Sigma Xi; AAAS. *Res:* Identification and characterization of tunable solid state laser materials; spectroscopy of laser ions in solid state hosts and modeling of ion-host interactions. *Mailing Add:* 17 Garrett Dr Hampton VA 23669

BYWATER, ANTHONY COLIN, b Malacca, Malayia, Sept 24, 48; UK citizen; m 76; c 2. LIVESTOCK ECONOMICS. *Educ:* Univ Reading, Eng, BSc, 71; Univ Nottingham, Eng, PhD(farm mgt), 77. *Prof Exp:* Fel farm mgt & econ, Dept Farm Mgt, Lincoln Col, New Zealand, 76-77, res assoc, Agr Econ Res Unit, 77; ASST PROF LIVESTOCK MGT, DEPT ANIMAL SCI, UNIV CALIF, DAVIS, 78- *Mem:* Am Soc Animal Sci; Am Agr Econ Asn; Brit Soc Animal Prod. *Res:* Development of management systems and decision aids for beef, dairy and sheep production; economics of livestock feeding and production; analysis of the role of livestock in food production; systems analysis and simulation. *Mailing Add:* Farm Mgt Lincoln Col Canterbury New Zealand

BZOCH, KENNETH R, b Chicago, Ill, Nov 6, 27; m 50; c 2. SPEECH PATHOLOGY, AUDIOLOGY. *Educ:* DePaul Univ, BA, 50; Northwestern Univ, MA & PhD(speech path), 56. *Prof Exp:* Instr, DePaul Acad, Ill, 50; lectr speech, Univ Col, Northwestern Univ, 51; asst prof, Grad Sch, Loyola Univ Ill, 53-57; prof grad fac & coordr, Cleft Lip & Palate Inst, Northwestern Univ, Chicago, 57-59 & 60-64; assoc prof, 64-80, PROF COMMUN DIS, COL HEALTH RELATED PROFESSIONS, UNIV FLA, 80-, CHMN DEPT, 64- *Concurrent Pos:* Consult, St Francis Hosp, Evanston, Ill, 53-58, Vet Res Hosp, Chicago, 57-59, Nat Inst Dent Res, 64, 66-70 & 80-84, HEW Div, Hosp & Med Facil, 64 & Vet Hops, Gainesville, 70-; HEW dir, Hosp & Med Fac, Nat Inst Dent Res, 80-84. *Mem:* Fel Am Speech & Hearing Asn; Speech Asn Am; Am Cleft Palate Asn (pres, 76). *Res:* Cineflurographic studies of co-articulation in normal and abnormal speech; efficacy of cleft palate habilitation procedures; normal and abnormal language development in infancy; physiological phonetics. *Mailing Add:* Col Health Related Prof Univ Fla Gainesville FL 32601

BZOCH, RONALD CHARLES, b Chicago, Ill, Mar 16, 30; m 51; c 2. MATHEMATICS. *Educ:* DePaul Univ, AB, 53, MS, 54; Ill Inst Technol, PhD(math), 57. *Prof Exp:* Asst math, Ill Inst Technol, 54-57; asst prof math, Univ Minn, 57-60 & Univ Utah, 60-61; assoc prof math, La State Univ, 61-66, assoc chmn dept, 64-66; PROF MATH & CHMN DEPT, UNIV NDAK, 66- *Mem:* Math Asn Am. *Res:* Real variables; integration theory. *Mailing Add:* Univ NDak Grand Forks ND 58202